eleventh edition

Prescott's Microbiology

Joanne M. Willey
HOFSTRA UNIVERSITY

Kathleen M. Sandman

Dorothy H. Wood
DURHAM TECHNICAL COMMUNITY COLLEGE

Mc Graw Hill Education

PRESCOTT'S MICROBIOLOGY, ELEVENTH EDITION

This book is printed on acid-free paper.

1 2 3 4 5 6 7 8 9 LWI 22 21 20 19

ISBN 978-1-260-21188-7
MHID 1-260-21188-6

Portfolio Manager: *Marija Magner*
Product Developer: *Darlene M. Schueller*
Marketing Manager: *Valerie L. Kramer*
Content Project Managers: *Laura Bies, Tammy Juran & Sandra Schnee*
Buyer: *Sandy Ludovissy*
Design: *Matt Backhaus*
Content Licensing Specialists: *Shawntel Schmitt & Beth Cray*
Cover Image: *Source: Photo by De Wood, digital colorization by Chris Pooley, USDA-ARS*
Compositor: *MPS Limited*

Library of Congress Cataloging-in-Publication Data

Willey, Joanne M., author.
Prescott's microbiology / Joanne M. Willey, Hofstra University,
Kathleen M. Sandman, Dorothy H. Wood, Durham Technical Community College.
Microbiology
Eleventh edition. | New York, NY : McGraw-Hill Education, [2020] |
Includes bibliographical references and index.
LCCN 2018026319 | ISBN 9781260211887 (student edition)
LCSH: Microbiology—Textbooks.
LCC QR41.2 .P74 2020 | DDC 579—dc23
LC record available at https://lccn.loc.gov/2018026319

mheducation.com/highered

Brief Contents

About the Authors

Courtesy of Joanne Willey

Joanne M. Willey has been a professor at Hofstra University on Long Island, New York, since 1993, where she is the Leo A. Guthart Professor of Biomedical Science and Chair of the Department of Science Education at the Donald and Barbara Zucker School of Medicine at Hofstra/Northwell. Dr. Willey received her B.A. in Biology from the University of Pennsylvania, where her interest in microbiology began with work on cyanobacterial growth in eutrophic streams. She earned her Ph.D. in biological oceanography (specializing in marine microbiology) from the Massachusetts Institute of Technology–Woods Hole Oceanographic Institution Joint Program in 1987. She then went to Harvard University, where she spent her postdoctoral fellowship studying the filamentous soil bacterium *Streptomyces coelicolor*. Dr. Willey has coauthored a number of publications that focus on its complex developmental cycle. She is an active member of the American Society for Microbiology (ASM), and served on the editorial board of the journal *Applied and Environmental Microbiology* for nine years and as Chair of the Division of General Microbiology. Dr. Willey taught microbiology to biology majors for 20 years and now teaches microbiology and infectious disease to medical students. She has taught courses in cell biology, marine microbiology, and laboratory techniques in molecular genetics. Dr. Willey lives on the north shore of Long Island and has two grown sons. She is an avid runner and enjoys skiing, hiking, sailing, and reading. She can be reached at joanne.m.willey@hofstra.edu.

Courtesy of Adele Anderson

Kathleen M. Sandman received her B.A. in Biology from La Salle University and her Ph.D. in Cellular and Developmental Biology from Harvard University. She was inspired to a career in science by her older brother's experience as an organic chemist and by the developing technology in recombinant DNA in the 1970s. Her graduate work used a transposable element as a mutagen in *Bacillus subtilis* to study gene expression during endospore formation. She continued in the genetics of Gram-positive bacteria with a postdoctoral year studying *Bacillus thuringiensis* at the University of Cambridge in the United Kingdom. Another postdoctoral opportunity at The Ohio State University provided an introduction to the emerging field of archaeal molecular biology, where Dr. Sandman discovered archaeal histones and continued research in the structural biology of archaeal chromatin for about 20 years. She served the National Science Foundation as a research grant reviewer and panelist for the Life in Extreme Environments program, and has organized conference sessions on archaeal molecular biology and proteins from extremophiles. Dr. Sandman has taught microbiology to hundreds of students, at both the introductory level and in an advanced molecular microbiology laboratory. Dr. Sandman has worked as a consultant in a variety of industries, including industrial microbiology, environmental geomicrobiology, and technical publishing. She lives with her husband in Columbus, Ohio, and has two grown daughters. She enjoys biking, fabric arts, reading, and genealogy, and can be reached at kathleenmsandman@gmail.com.

Courtesy of Dorothy Wood

Dorothy H. Wood has taught microbiology and general biology at Durham Technical Community College in North Carolina since 2004. Dr. Wood received her B.A. in Biology from Rhode Island College where her love of microbes began, nurtured by Dr. Charles Owens. She earned her Ph.D. in Cell and Molecular Pathology from the University of North Carolina at Chapel Hill, focusing on pancreatic damage caused by antimicrobial drugs, and investigated alternative therapies based on receptor binding by novel compounds. After three years as Assistant Professor at NC Central University, Dr. Wood made the move to the NC Community College System to focus her attention on her primary interest of teaching. Throughout her career she has developed several courses, including graduate bacteriology, pathophysiology, and biotechnology. She serves as a visiting scholar at Duke University where she is a mentor for the Preparing Future Faculty program. Dr. Wood is a member of the American Society for Microbiology and the Association of College and University Biology Educators, as well as several local organizations that foster pedagogy. She is a digital faculty consultant for McGraw-Hill and has worked on several textbooks in a variety of disciplines, developing and editing digital content to accompany the texts. Outside of the classroom, Dr. Wood is a fitness professional, leads health and wellness seminars, and has been the treasurer of a nonprofit organization for the past 10 years. She enjoys life in North Carolina with her husband and two grown children and can be reached at woodd@durhamtech.edu.

Digital Tools for Your Success

McGraw-Hill Education connect®

Save time with auto-graded assessments. Gather powerful performance data.

McGraw-Hill Connect for Prescott's Microbiology provides online presentation, assignment, and assessment solutions, connecting your students with the tools and resources they'll need to achieve success.

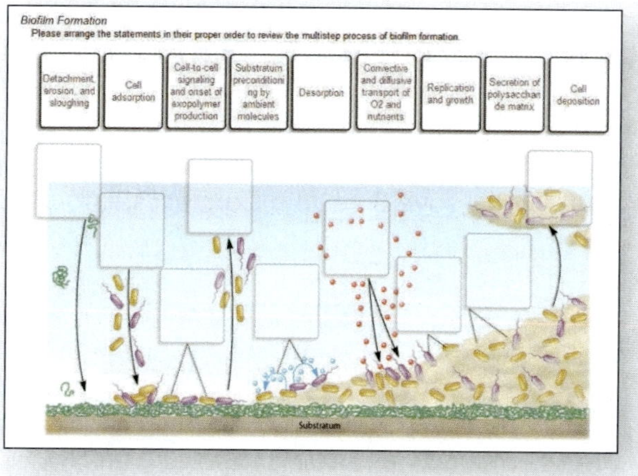

Homework and Assessment

With **Connect for Prescott's Microbiology,** you can deliver auto-graded assignments, quizzes, and tests online. Choose from a robust set of interactive questions and activities using high-quality art from the textbook and animations. Assignable content is available for every Learning Outcome in the book and is categorized according to the **ASM Curriculum Guidelines.** As an instructor, you can edit existing questions and author entirely new ones.

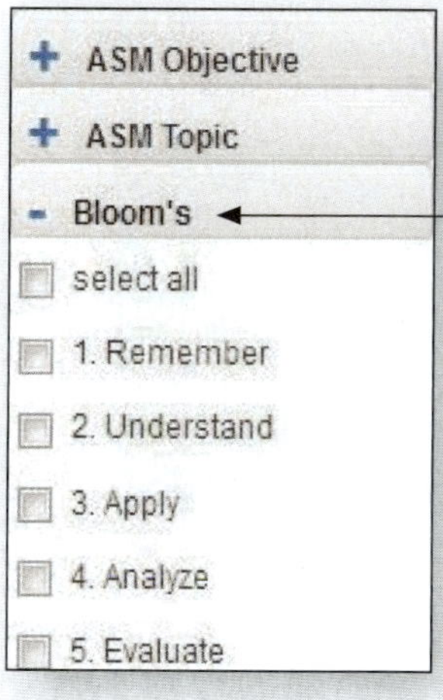

Faculty demand for assignments at higher Bloom's levels led us to examine the quality and consistency of our Connect content, to develop a scientific approach to systemically increase critical-thinking levels, and develop balanced digital assessments that promote student learning. The increased challenge at higher Bloom's levels will help the student grow intellectually and be better prepared to contribute to society.

Detailed Reports

Track individual student performance—by question, by assignment, or in relation to the class overall—with detailed grade reports. Integrate grade reports easily with your Learning Management Systems (LMS).

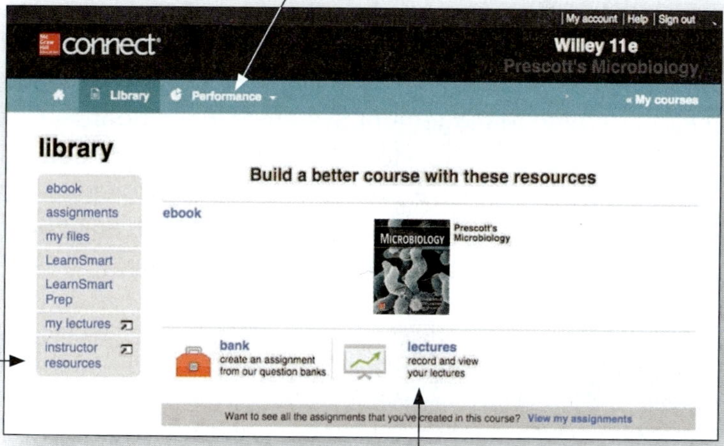

Instructor Resources

Customize your lecture with tools such as PowerPoint® presentations, animations, and editable art from the textbook. An instructor's manual for the text saves you time in developing your course.

Lecture Capture

McGraw-Hill Tegrity® records and distributes your class lecture with just a click of a button. Students can view anytime, anywhere via computer or mobile device. Indexed as you record, students can use keywords to find exactly what they want to study.

Learn more at connect.mheducation.com.

For Students

©Shutterstock/wavebreakmedia

Effective, efficient studying.

Connect helps you be more productive with your study time and get better grades using tools like SmartBook, which highlights key concepts and creates a personalized study plan. Connect sets you up for success, so you walk into class with confidence and walk out with better grades.

"I really liked this app—it made it easy to study when you don't have your textbook in front of you."

- Jordan Cunningham,
Eastern Washington University

Study anytime, anywhere.

Download the free ReadAnywhere app and access your online eBook when it's convenient, even if you're offline. And since the app automatically syncs with your eBook in Connect, all of your notes are available every time you open it. Find out more at **www.mheducation.com/readanywhere**

No surprises.

The Connect Calendar and Reports tools keep you on track with the work you need to get done and your assignment scores. Life gets busy; Connect tools help you keep learning through it all.

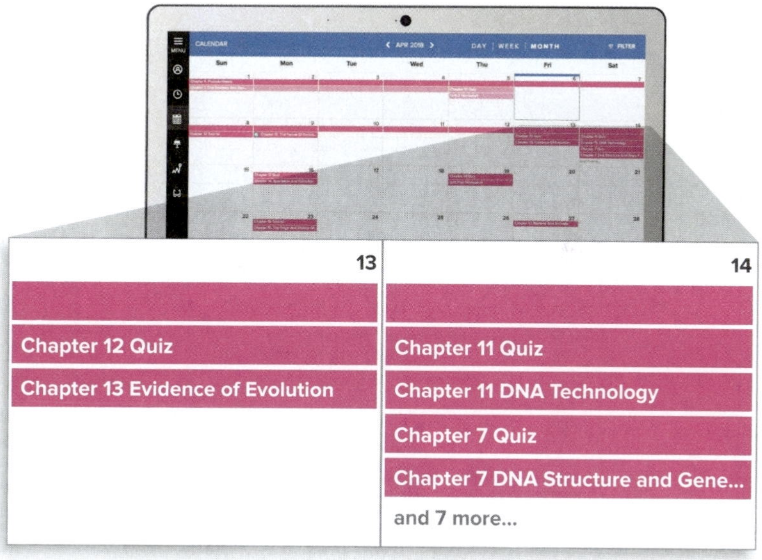

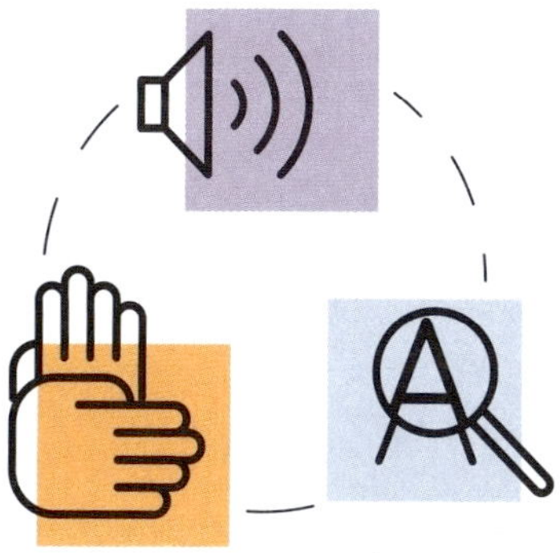

Learning for everyone.

McGraw-Hill works directly with Accessibility Services Departments and faculty to meet the learning needs of all students. Please contact your Accessibility Services office and ask them to email accessibility@mheducation.com, or visit **www.mheducation.com/about/accessibility.html** for more information.

A Modern Approach to Microbiology

Evolution as a Framework

Introduced immediately in chapter 1 and used as an overarching theme throughout, evolution helps unite microbiological concepts and provides a framework upon which students can build their knowledge.

An Introduction to the Entire Microbial World

Covered in chapters 3–6, separate chapters on the structure and function of bacteria and archaea are followed by the discussion of eukaryotic cells and viruses.

Broad Coverage of Microbial Ecology

The importance and multidisciplinary nature of microbial ecology are demonstrated by content that ranges from global climate change to the human microbiome.

Brief Contents

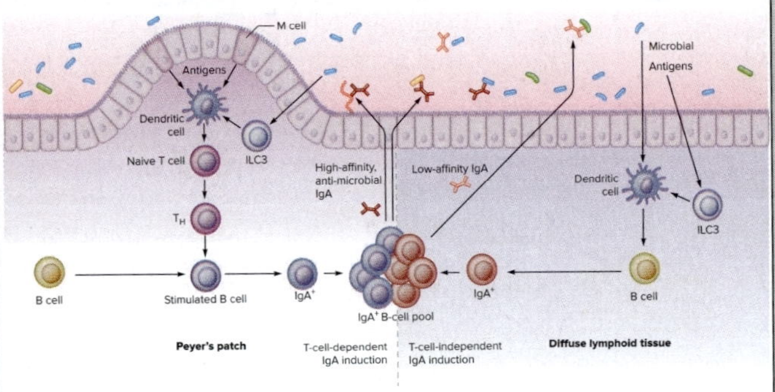

Figure 34.7 **Production of SIgA Requires ILC3 and Microbial Stimulation.** B-cell stimulation can be T-cell dependent (left, Peyer's patch) or T-cell independent (right, isolated lymphoid follicle). In both cases, innate lymphoid cells-3 (ILC3s) optimization of dendritic cell activity is needed. ILC3 function requires interaction with host microbes or their products.

Molecular Microbiology and Immunology

The eleventh edition includes updates on genetics, biotechnology, genomics and metagenomics, immunology, and the human microbiome. A streamlined discussion of immunity, with enhanced detail between innate and adaptive linkages, helps students grasp the complexity and specificity of immune responses. A new chapter, The Microbe-Human Ecosystem, introduces students to the development and impact of the human microbiome.

A Modern Approach to Microbiology

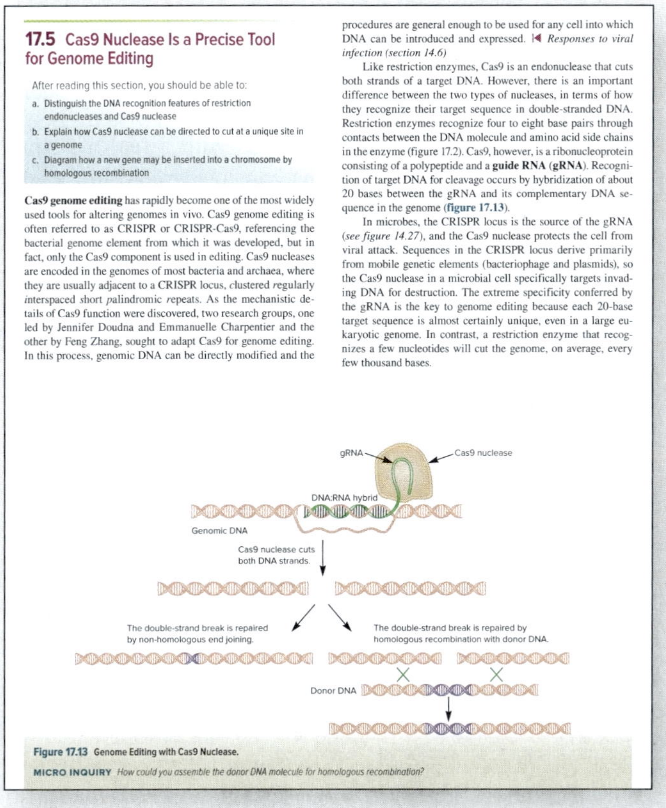
17.5 Cas9 Nuclease Is a Precise Tool for Genome Editing

After reading this section, you should be able to:

a. Distinguish the DNA recognition features of restriction endonucleases and Cas9 nuclease
b. Explain how Cas9 nuclease can be directed to cut at a unique site in a genome
c. Diagram how a new gene may be inserted into a chromosome by homologous recombination

Cas9 genome editing has rapidly become one of the most widely used tools for altering genomes in vivo. Cas9 genome editing is often referred to as CRISPR or CRISPR-Cas9, referencing the bacterial genome element from which it was developed, but in fact, only the Cas9 component is used in editing. Cas9 nucleases are encoded in the genomes of most bacteria and archaea, where they are usually adjacent to a CRISPR locus, clustered regularly interspaced short palindromic repeats. As the mechanistic details of Cas9 function were discovered, two research groups, one led by Jennifer Doudna and Emmanuelle Charpentier and the other by Feng Zhang, sought to adapt Cas9 for genome editing. In this process, genomic DNA can be directly modified and the procedures are general enough to be used for any cell into which DNA can be introduced and expressed. ◀ *Responses to viral infection (section 14.6)*

Like restriction enzymes, Cas9 is an endonuclease that cuts both strands of a target DNA. However, there is an important difference between the two types of nucleases, in terms of how they recognize their target sequence in double-stranded DNA. Restriction enzymes recognize four to eight base pairs through contacts between the DNA molecule and amino acid side chains in the enzyme (figure 17.2). Cas9, however, is a ribonucleoprotein consisting of a polypeptide and a **guide RNA (gRNA)**. Recognition of target DNA for cleavage occurs by hybridization of about 20 bases between the gRNA and its complementary DNA sequence in the genome (**figure 17.13**).

In microbes, the CRISPR locus is the source of the gRNA (*see figure 14.27*), and the Cas9 nuclease protects the cell from viral attack. Sequences in the CRISPR locus derive primarily from mobile genetic elements (bacteriophage and plasmids), so the Cas9 nuclease in a microbial cell specifically targets invading DNA for destruction. The extreme specificity conferred by the gRNA is the key to genome editing because each 20-base target sequence is almost certainly unique, even in a large eukaryotic genome. In contrast, a restriction enzyme that recognizes a few nucleotides will cut the genome, on average, every few thousand bases.

Figure 17.13 Genome Editing with Cas9 Nuclease.

MICRO INQUIRY *How could you assemble the donor DNA molecule for homologous recombination?*

21st-Century Microbiology

Prescott's Microbiology leads the way with text devoted to global climate change, biofuels, and microbial fuel cells. For more, see chapters 28, 30, 42, and 43.

Metagenomics and the Human Microbiome

The importance of metagenomics in understanding the role of microbes in all environments and in exploring symbionts of invertebrates is threaded throughout the text. A new chapter, The Microbe-Human Ecosystem, explores the human microbiome.

Laboratory Safety

Reflecting recommendations from the Centers for Disease Control and Prevention, along with the American Society for Microbiology, chapter 37 provides specific guidance for laboratory best practices to help instructors provide safe conditions during the teaching of laboratory exercises.

Special Interest Essays

Organized into four themes—Microbial Diversity & Ecology, Techniques & Applications, Historical Highlights, and Disease—these focused and interesting essays provide additional insight into relevant topics.

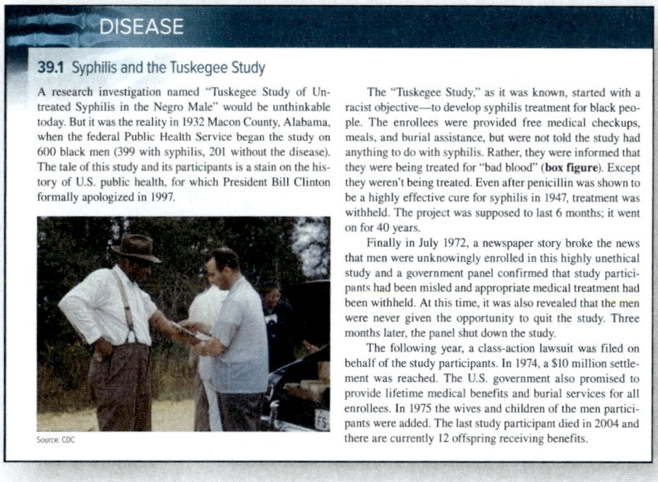
DISEASE

39.1 Syphilis and the Tuskegee Study

A research investigation named "Tuskegee Study of Untreated Syphilis in the Negro Male" would be unthinkable today. But it was the reality in 1932 Macon County, Alabama, when the federal Public Health Service began the study on 600 black men (399 with syphilis, 201 without the disease). The tale of this study and its participants is a stain on the history of U.S. public health, for which President Bill Clinton formally apologized in 1997.

Source: CDC

The "Tuskegee Study," as it was known, started with a racist objective—to develop syphilis treatment for black people. The enrollees were provided free medical checkups, meals, and burial assistance, but were not told the study had anything to do with syphilis. Rather, they were informed that they were being treated for "bad blood" (**box figure**). Except they weren't being treated. Even after penicillin was shown to be a highly effective cure for syphilis in 1947, treatment was withheld. The project was supposed to last 6 months; it went on for 40 years.

Finally in July 1972, a newspaper story broke the news that men were unknowingly enrolled in this highly unethical study and a government panel confirmed that study participants had been misled and appropriate medical treatment had been withheld. At this time, it was also revealed that the men were never given the opportunity to quit the study. Three months later, the panel shut down the study.

The following year, a class-action lawsuit was filed on behalf of the study participants. In 1974, a $10 million settlement was reached. The U.S. government also promised to provide lifetime medical benefits and burial services for all enrollees. In 1975 the wives and children of the men participants were added. The last study participant died in 2004 and there are currently 12 offspring receiving benefits.

MICROBIAL DIVERSITY & ECOLOGY

27.1 *Wolbachia pipientis*: The World's Most Infectious Microbe?

Most people have never heard of the bacterium *Wolbachia pipientis*, but this rickettsia infects more organisms than any other microbe. It is known to infect a broad range of crustaceans, spiders, mosquitoes, millipedes, and nematodes, and may infect more than 2 million insect species worldwide. To what does *W. pipientis* owe its extraordinary success? Quite simply, this endosymbiont is a master at manipulating its hosts' reproductive biology.

W. pipientis inhabits the cytoplasm of its host's cells and is transferred from one generation to the next through the eggs of infected females. To survive, *Wolbachia* must ensure the fertilization and viability of infected eggs while decreasing the likelihood that uninfected eggs survive. The mechanism by which this is accomplished depends on the host. In wasps and mosquitoes, *W. pipientis* causes cytoplasmic incompatibility, which means that embryonic development will be abnormal if only the male is infected. For instance, when infected sperm of the wasp *Nasonia vitripennis* fertilizes uninfected eggs, chromosomes from the *W. pipientis*–laden sperm prematurely try to align with the egg's chromosomes. These eggs then divide as if never fertilized. However, chromosomes behave normally when an infected female mates with an uninfected male. This yields a normal sex distribution, and all progeny are infected with the rickettsia.

In other infected insects, *W. pipientis* may simply kill all the male offspring and induce parthenogenesis in infected females; that is, the females simply clone themselves. This limits genetic diversity but allows 100% transmission of rickettsias to the next generation. In still other hosts, the microbe modifies male hormones so that the males become feminized and produce eggs.

Another effect of *Wolbachia* infection is interference with viral replication. Viruses normally spread by insects may not be transmissible if the insect is infected with *Wolbachia*. This has led to the notion that this infection could be used as a means of biological control. *Aedes* mosquitoes transmit numerous viruses, like Zika, dengue, chikungunya, and West Nile. In humans, these viral infections have no cure and no treatment beyond supportive care, so insect control is the best prevention.

However, the mosquito vectors of these viruses, which belong the genus *Aedes* (**box figure**), are not natural hosts to *Wolbachia*. Nonetheless, infection can be established by transinfection, the process of transferring the microbes from another insect species. Many *Wolbachia* isolates have been screened for those that have the most severe effects on *Aedes*, and several stable insect strains have been established with *Wolbachia* from *Drosophila*. Cytoplasmic interference is extensive in these strains of *Wolbachia*. To establish a stable population of *Aedes* outside the laboratory, it is important that the infection not impose a dramatic burden on the insect, as it must compete with and integrate into wild populations.

In a promising development, *Wolbachia*-infected *Aedes* mosquitoes are effective at blocking the transmission of both Zika and dengue viruses. Experiments in Australia have confirmed the ability of these insects to persist and spread in local mosquito populations, validating this approach to insect vector control.

Female *Aedes aegypti* mosquito. Infection with *Wolbachia* may render these insects incapable of spreading arboviruses to humans.
Source: CDC/James Gathany

Student-Friendly Organization

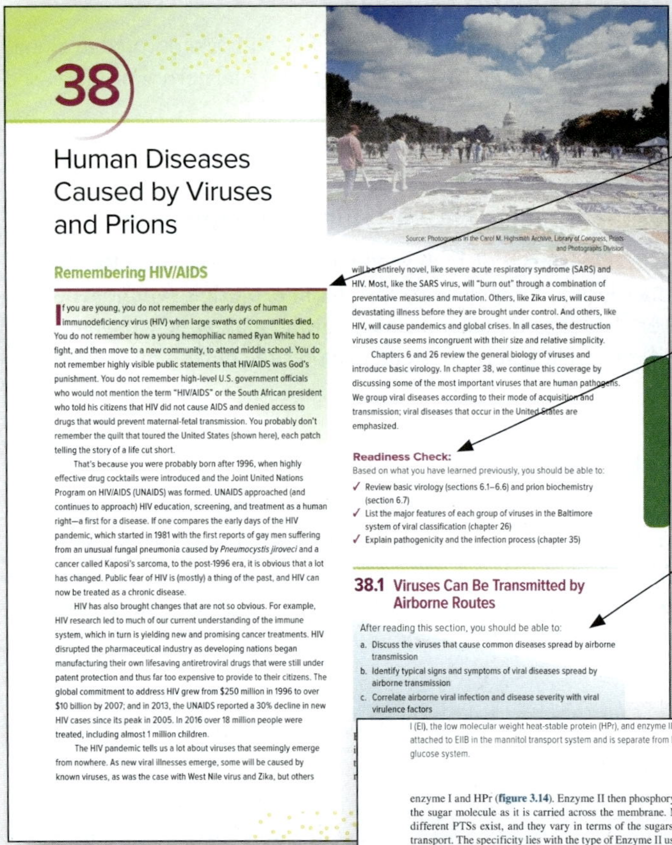

38

Human Diseases Caused by Viruses and Prions

Source: Photographs in the Carol M. Highsmith Archive, Library of Congress, Prints and Photographs Division

Remembering HIV/AIDS

If you are young, you do not remember the early days of human immunodeficiency virus (HIV) when large swaths of communities died. You do not remember how a young hemophiliac named Ryan White had to fight, and then move to a new community, to attend middle school. You do not remember highly visible public statements that HIV/AIDS was God's punishment. You do not remember high-level U.S. government officials who would not mention the term "HIV/AIDS" or the South African president who told his citizens that HIV did not cause AIDS and denied access to drugs that would prevent maternal-fetal transmission. You probably don't remember the quilt that toured the United States (shown here), each patch telling the story of a life cut short.

That's because you were probably born after 1996, when highly effective drug cocktails were introduced and the Joint United Nations Program on HIV/AIDS (UNAIDS) was formed. UNAIDS approached (and continues to approach) HIV education, screening, and treatment as a human right—a first for a disease. If one compares the early days of the HIV pandemic, which started in 1981 with the first reports of gay men suffering from an unusual fungal pneumonia caused by *Pneumocystis jiroveci* and a cancer called Kaposi's sarcoma, to the post-1996 era, it is obvious that a lot has changed. Public fear of HIV is (mostly) a thing of the past, and HIV can now be treated as a chronic disease.

HIV has also brought changes that are not so obvious. For example, HIV research led to much of our current understanding of the immune system, which in turn is yielding new and promising cancer treatments. HIV disrupted the pharmaceutical industry as developing nations began manufacturing their own lifesaving antiretroviral drugs that were still under patent protection and thus far too expensive to provide to their citizens. The global commitment to address HIV grew from $250 million in 1996 to over $10 billion by 2007; and in 2013, the UNAIDS reported a 30% decline in new HIV cases since its peak in 2005. In 2016 over 18 million people were treated, including almost 1 million children.

The HIV pandemic tells us a lot about viruses that seemingly emerge from nowhere. As new viral illnesses emerge, some will be caused by known viruses, as was the case with West Nile virus and Zika, but others will be entirely novel, like severe acute respiratory syndrome (SARS) and HIV. Most, like the SARS virus, will "burn out" through a combination of preventative measures and mutation. Others, like Zika virus, will cause devastating illness before they are brought under control. And others, like HIV, will cause pandemics and global crises. In all cases, the destruction viruses cause seems incongruent with their size and relative simplicity.

Chapters 6 and 26 review the general biology of viruses and introduce basic virology. In chapter 38, we continue this coverage by discussing some of the most important viruses that are human pathogens. We group viral diseases according to their mode of acquisition and transmission; viral diseases that occur in the United States are emphasized.

Readiness Check:

Based on what you have learned previously, you should be able to:

- ✓ Review basic virology (sections 6.1–6.6) and prion biochemistry (section 6.7)
- ✓ List the major features of each group of viruses in the Baltimore system of viral classification (chapter 26)
- ✓ Explain pathogenicity and the infection process (chapter 35)

38.1 Viruses Can Be Transmitted by Airborne Routes

After reading this section, you should be able to:

a. Discuss the viruses that cause common diseases spread by airborne transmission
b. Identify typical signs and symptoms of viral diseases spread by airborne transmission
c. Correlate airborne viral infection and disease severity with viral virulence factors

Micro Focus—Each chapter begins with a real-life story illustrating the relevance of the content covered in the upcoming text.

Readiness Check—The introduction to each chapter includes a skills checklist that defines the prior knowledge students need to understand the material that follows.

Learning Outcomes—Every section in each chapter begins with a list of content-based activities students should be able to perform after reading.

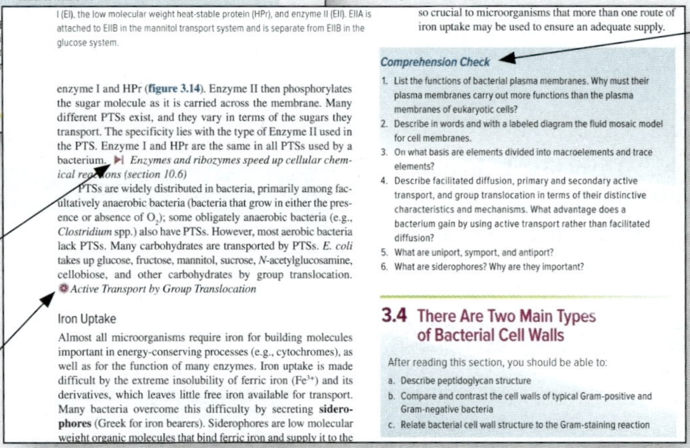

I (EI), the low molecular weight heat-stable protein (HPr), and enzyme II (EII). EIIA is attached to EIIB in the mannitol transport system and is separate from EIIB in the glucose system.

enzyme I and HPr (**figure 3.14**). Enzyme II then phosphorylates the sugar molecule as it is carried across the membrane. Many different PTSs exist, and they vary in terms of the sugars they transport. The specificity lies with the type of Enzyme II used in the PTS. Enzyme I and HPr are the same in all PTSs used by a bacterium. ▶| *Enzymes and ribozymes speed up cellular chemical reactions (section 10.6)*

PTSs are widely distributed in bacteria, primarily among facultatively anaerobic bacteria (bacteria that grow in either the presence or absence of O_2); some obligately anaerobic bacteria (e.g., *Clostridium* spp.) also have PTSs. However, most aerobic bacteria lack PTSs. Many carbohydrates are transported by PTSs. *E. coli* takes up glucose, fructose, mannitol, sucrose, *N*-acetylglucosamine, cellobiose, and other carbohydrates by group translocation. *Active Transport by Group Translocation*

Iron Uptake

Almost all microorganisms require iron for building molecules important in energy-conserving processes (e.g., cytochromes), as well as for the function of many enzymes. Iron uptake is made difficult by the extreme insolubility of ferric iron (Fe^{3+}) and its derivatives, which leaves little free iron available for transport. Many bacteria overcome this difficulty by secreting **siderophores** (Greek for iron bearers). Siderophores are low molecular weight organic molecules that bind ferric iron and supply it to the

so crucial to microorganisms that more than one route of iron uptake may be used to ensure an adequate supply.

Comprehension Check

1. List the functions of bacterial plasma membranes. Why must their plasma membranes carry out more functions than the plasma membranes of eukaryotic cells?
2. Describe in words and with a labeled diagram the fluid mosaic model for cell membranes.
3. On what basis are elements divided into macroelements and trace elements?
4. Describe facilitated diffusion, primary and secondary active transport, and group translocation in terms of their distinctive characteristics and mechanisms. What advantage does a bacterium gain by using active transport rather than facilitated diffusion?
5. What are uniport, symport, and antiport?
6. What are siderophores? Why are they important?

3.4 There Are Two Main Types of Bacterial Cell Walls

After reading this section, you should be able to:

a. Describe peptidoglycan structure
b. Compare and contrast the cell walls of typical Gram-positive and Gram-negative bacteria
c. Relate bacterial cell wall structure to the Gram-staining reaction

Comprehension Check—Questions within the narrative of each chapter help students master section concepts before moving on to other topics.

Cross-Referenced Notes—In-text references refer students to other parts of the book to review.

Animation Icon—This symbol indicates that material presented in the text is accompanied by an animation within Instructor Resources in Connect. Create a file attachment assignment in Connect to have your students view the animation, or post it to your Learning Management System for students.

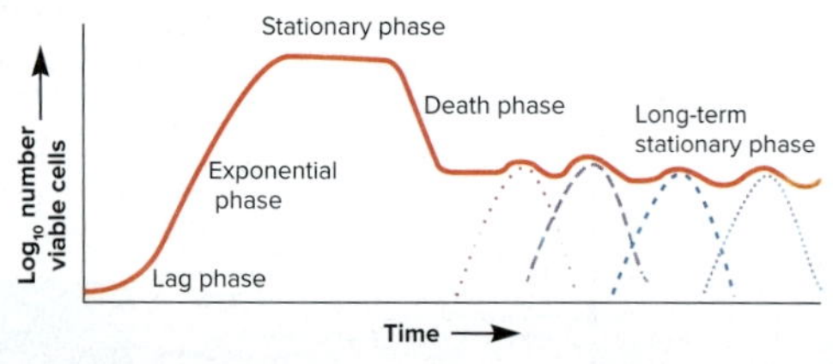

Figure 7.10 **Microbial Growth Curve in a Closed System.** The five phases of the growth curve are identified. The dotted lines shown during the long-term stationary phase represent successive waves of genetic variants that evolve during this phase of the growth curve.

MICRO INQUIRY *Identify the regions of the growth curve in which (1) nutrients are rapidly declining and (2) wastes accumulate.*

Micro Inquiry—Selected figures in every chapter contain probing questions, adding another assessment opportunity for the student.

Student-Friendly Organization

Vivid Instructional Art—Three-dimensional renditions and bright, attractive colors enhance learning.

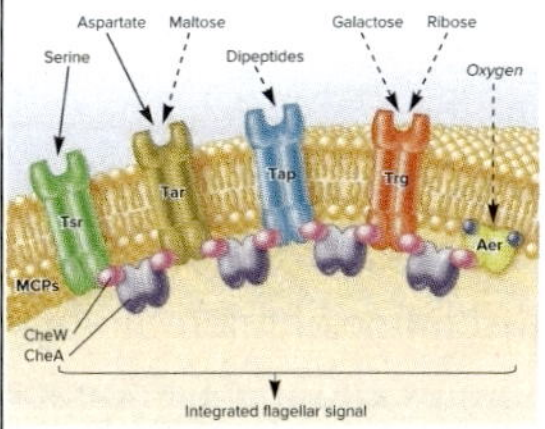

Figure 14.23 *E. coli* **Methyl-Accepting Chemotaxis Proteins.** The attractants sensed by each methyl-accepting chemotaxis protein (MCP) are shown. Some are sensed directly, when the attractant binds the MCP (solid lines). Others are sensed indirectly (dashed lines). Maltose, dipeptides, galactose, and ribose are detected by their interaction with periplasmic binding proteins. Oxygen is detected indirectly by the Aer chemoreceptor, which lacks a periplasmic sensing domain. Instead, the cytoplasmic domain has a binding site for FAD. FAD is an electron carrier found in many electron transport systems. The redox state of the MCP-bound FAD molecule is used to monitor the functioning of the electron transport system. This in turn mediates a tactic response to oxygen.

MICRO INQUIRY *Why doesn't the Aer receptor need a periplasmic domain?*

density. Quorum sensing is also used to control genes whose products are needed for maintenance of the symbiotic relationship between *V. fisheri* and its host. As a result, the squid/*V. fisheri* symbiosis has become an important model for understanding animal-bacterial associations. Our focus is on the regulation of a single operon, that involved with bioluminescence. However, it should be kept in mind that **quorum sensing** regulates multiple genes and operons. *Cell-cell communication within microbial populations (section 7.6)*

Quorum sensing in *V. fischeri* and many other Gram-negative bacteria uses an ***N*-acylhomoserine lactone** (**AHL**) signal (**figure 14.24**). Synthesis of this small molecule is catalyzed by an enzyme called AHL synthase, the product of the *luxI* gene. The *luxI* gene is subject to positive autoregulation; that is, transcription of *luxI* increases as AHL accumulates in the cell. This is accomplished through the transcriptional activator LuxR, which is active only when AHL binds to it. Thus a simple feedback loop is created. Without AHL-activated LuxR, the *luxI* gene is transcribed only at basal levels. When *V. fisheri* cell density within the squid light organ is low, the small amounts of AHL produced by the bacterial cells freely diffuse out of each cell and accumulate in the environment. As cell density increases, the concentration of AHL outside each cell eventually exceeds that inside the cell, and the concentration gradient is reversed. As the intracellular AHL concentration increases, it binds and activates LuxR. LuxR then increases transcription of *luxI* and the genes whose products are needed for bioluminescence (*luxCDABEG*). Quorum sensing is often called **autoinduction,** and the AHL signal is termed the **autoinducer** (**AI**) to reflect the autoregulatory nature of this system. *Quorum Sensing*

Annotated Figures—All key metabolic pathways and molecular processes are annotated, so each step is clearly illustrated and explained.

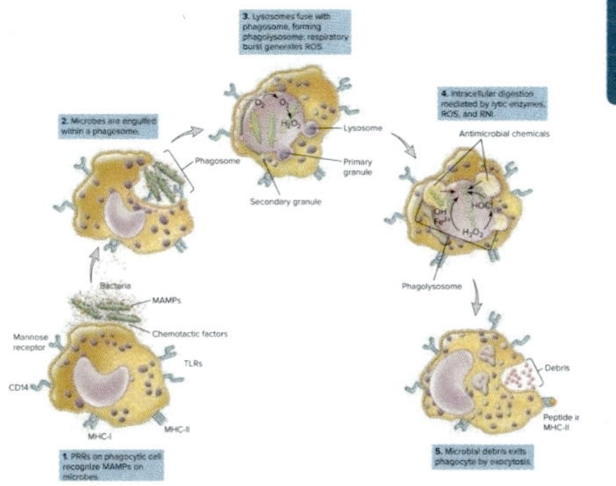

Figure 32.17 **Phagocytosis.** CD14 is a lipopolysaccharide receptor that associates with TLR-4; MAMPs: microbe-associated molecular patterns; MHC-I: class I major histocompatibility protein; MHC-II: class II major histocompatibility protein (MHC proteins are discussed in chapter 33); RNI: reactive nitrogen intermediates; ROS: radical oxygen species; TLRs: toll-like receptors.

Key Concepts

2.1 Lenses Create Images by Bending Light

- A light ray moving from air to glass or vice versa is bent in a process known as refraction (**figure 2.1**).
- Lenses focus light rays at a focal point and magnify images (**figure 2.2**).

2.2 There Are Several Types of Light Microscopes

- In a compound microscope such as the bright-field microscope, the primary image is an enlarged image formed by the objective lens. The primary image is further enlarged by the ocular lens to yield the final image (**figure 2.3**).
- Microscope resolution increases as the wavelength of radiation used to illuminate the specimen decreases and as the numerical aperture lens increases. The maximum resolution of a light microscope is about 0.2 μm (**figure 2.4**).
- The dark-field microscope uses only refracted light to form an image, and objects appear light against a black background (**figure 2.6**).
- The phase-contrast microscope converts variations in the refractive index into changes in light intensity and thus makes colorless, unstained, live cells visible (**figures 2.8–2.10**).
- The differential interference contrast microscope uses two beams of light to create high-contrast images of live specimens (**figure 2.11**).
- The fluorescence microscope illuminates a fluorochrome-labeled specimen and forms an image from its fluorescence (**figures 2.12–2.14**).
- The confocal microscope is used to study thick, complex specimens. It creates an image by using only the light emanating from the plane of focus, while blocking out light from above and below the plane of focus (**figures 2.15** and **2.16**).

2.3 Staining Specimens Helps to Visualize and Identify Microbes

- Specimens are often fixed and stained before viewing them in the bright-field microscope. There are two fixation methods: heat fixation and chemical fixation.
- Most dyes are either positively charged basic dyes or negatively charged acidic dyes that bind to ionized parts of cells.
- In simple staining, a single dye is used to stain microorganisms (**figure 2.17**).
- Differential staining procedures such as Gram and acid-fast staining distinguish between microbial groups by staining them differently (**figures 2.18** and **2.19*a,b***). Other differential staining techniques are specific for particular structures such as bacterial capsules and flagella (**figure 2.19*c,d***).

2.4 Electron Microscopes Use Beams of Electrons to Create Highly Magnified Images

- The transmission electron microscope (TEM) uses magnetic lenses to form an image from electrons that have passed through a very thin section of a specimen (**figure 2.22**). Resolution is high because the wavelength of a beam of electrons is very short.
- Specimens for TEM are usually prepared by methods that increase contrast. Specimens can be stained by treatment with solutions of heavy metals such as osmium tetroxide, uranium, and lead. They can also be prepared for TEM by negative staining, shadowing with metal, or freeze-etching (**figures 2.24** and **2.25**).
- The scanning electron microscope is used to study external surface features of microorganisms (**figures 2.26** and **2.27**).
- Electron cryotomography freezes specimens rapidly, keeps them frozen while being examined, and creates images from a series of directions that are combined and processed to form a three-dimensional reconstruction of the object (**figure 2.28**).

2.5 Scanning Probe Microscopy Can Visualize Molecules and Atoms

- Scanning probe microscopes reach very high magnifications that allow scientists to observe biological molecules (**figures 2.29** and **2.31**).
- Scanning tunneling microscopy enables the visualization of molecular surfaces using electron interaction between the probe and the specimen, whereas atomic force microscopy can scan the surface of molecules that do not conduct electricity well (**figure 2.30**).

Key Concepts—At the end of each chapter, organized by numbered headings, this feature distills the content to its essential components with cross-references to figures and tables.

Active Learning

1. You have prepared a specimen for light microscopy, stained it using the Gram staining procedure, but failed to see anything when you looked through your light microscope. Discuss the things that you may have done incorrectly.
2. In a journal article, find an example of a light micrograph, a scanning or transmission electron micrograph, or a confocal image. Discuss why the figure was included in the article and why that particular type of microscopy was the method of choice for the research. What other figures would you like to see used in this study? Outline the steps that the investigators would take to obtain such photographs or figures.

Active Learning—Includes questions taken from current literature; designed to stimulate analytical problem-solving skills.

List of Content Changes

Each chapter has been thoroughly reviewed.

Part One

Chapter 1—Evolution is the driving force of all biological systems; this is made clear by introducing essential concepts of microbial evolution first. Advances in the discipline of microbiology and the increasing contributions of genomics and metagenomics are discussed.

Chapter 2—Microscopy was and is critical to the study of microorganisms and this chapter considers the most commonly used methods, including expanded coverage of phase-contrast microscopy.

Chapter 3—Coverage of bacterial cellular structure and function. New material includes a discussion of membrane microdomains, and the effect of macromolecular crowding in the cytoplasm.

Chapter 4—Discussion of archaea has been updated to include recent discoveries, including expanded taxonomy, polyploidy, and the role of nucleoid-associated proteins. Comparisons to bacteria are made throughout the chapter.

Chapter 5—An introduction to eukaryotic cell structure and function, with emphasis on eukaryotic microbes. More detailed information on protist and fungal cells is presented in chapters 24 (Protists) and 25 (*Fungi*), which also focus on the diversity of these microbes. The current understanding of the evolution of mitochondria and mitochondria-like organelles is considered. Comparisons between bacteria, archaea, and eukaryotes are included throughout the chapter.

Chapter 6—This chapter surveys essential morphological, physiological, and genetic elements of viruses as well as viroids, satellites, and prions. Images and descriptions of archaeal viruses have been incorporated. This chapter completes our four-chapter introduction to microbial life.

Part Two

Chapter 7—Discussion of the growth of microbes has been updated to include new information about chromosome partitioning and the archaeal cell cycle.

Chapter 8—A new chapter-opening story and updated tables reflect the challenges associated with controlling prions. A new Microbial Diversity & Ecology box describes the conditions required in NASA spacecraft assembly facilities.

Chapter 9—Content focuses on the mechanism of action of each class of antimicrobial agents and introduces mechanisms of drug resistance.

Part Three

Chapter 10—This introduction to metabolism includes a section outlining the nature of biochemical pathways. The concept of metabolic flux is presented by discussing the interconnected biochemical pathways used by cells.

Chapter 11—An introduction to metabolic diversity and nutritional types is followed by an exploration of the energy-conserving process of each nutritional type. An introduction to flavin-based electron bifurcation has been added.

Chapter 12—New comparison of pathways used to synthesize lipids in bacteria and archaea.

Part Four

Chapter 13—A revised section now covers posttranslational modifications, protein folding, and secretion systems. Membrane vesicles are introduced.

Chapter 14—The regulation of bacterial cellular processes, with updated coverage of regulation by messengers like c-di-GMP. A new section on responses to viral infection includes a discussion of restriction-modification and CRISPR.

Chapter 15—Recent developments in archaeal replication, gene regulation, and protein secretion have been included.

Chapter 16—Covers mutation, repair, and recombination in the context of processes that introduce genetic variation into populations. Updated coverage of integrative conjugative elements and mobilizable genomic islands.

Chapter 17—This chapter has been completely reorganized to update the content on gene cloning and heterologous gene expression. Cas9 genome engineering methodologies are described.

Chapter 18—Next-generation nucleotide sequencing and single-cell genome sequencing are covered in the context of metagenomics as it relates to the microbial ecology of natural systems, including the human microbiome.

Part Five

Chapter 19—This overview of microbial evolution has been updated to include whole genome comparison and related computational techniques in determining relatedness.

Chapter 20—The discussion of archaeal taxonomy has been revised and updated to reflect the new diversity uncovered by metagenomics. The methanogenesis discussion has been updated to include the mechanism of flavin-based electron bifurcation.

Chapter 21—In addition to the ecology and physiology of photosynthetic bacteria, the recently described *Planctomycetes,*

Verrucomicrobia, Chlamydia (PVC) superphylum is introduced with an updated review of each of these genera. New information about the *Deinococcus* radiation response is included.

Chapter 22—This chapter's coverage includes a discussion of the proteobacterial origin of mitochondria.

Chapter 23—This overview of Gram-positive bacteria includes firmicutes and actinobacteria. The discussion of the evolutionary aspects of diderm firmicutes is expanded.

Chapter 24—This chapter introduces protist morphology and diversity, with an emphasis on physiological adaptation and ecology.

Chapter 25—Fungal diversity is presented within a phylogenetic framework. Morphology, ecology, and reproductive strategies are stressed.

Chapter 26—Updated discussion of the molecular mechanisms in the bacteriophage T4 life cycle.

Chapter 27—Important model systems for the exploration of microbial symbioses are presented. Updated discussion of *Wolbachia*-infected insects.

Part Six

Chapter 28—The description of each nutrient cycle is accompanied by a "student-friendly" figure that distinguishes between reductive and oxidative reactions. Updated coverage of the role of biogeochemical cycling in global climate change.

Chapter 29—This chapter continues to emphasize culture-based techniques as the "gold standard" and reviews culture-independent approaches such as mass spectrometry in the identification of microbial taxa as well as metatranscriptomics and metaproteomics in the study of community activity.

Chapter 30—Updated discussion of the role of marine microbes in the global carbon budget as well as an update on subsurface microbes.

Chapter 31—New coverage of the microbial ecology of the phyllosphere, rhizoplane, and rhizosphere. Expanded discussion of fungal plant pathogens.

Part Seven

Chapter 32—Streamlined and updated, this chapter on innate host resistance provides in-depth coverage of physical and chemical components of the nonspecific host response, followed by an overview of cells, tissues, and organs of the immune system. The chapter concludes with an overview of the molecular mechanisms that drive phagocytosis and inflammation.

Chapter 33—Updated to enhance linkages between innate and adaptive immune activities. Discussions integrate concepts of cell biology, physiology, and genetics to present the immune system as a unified response having various components. Implications of dysfunctional immune actions are also discussed.

Chapter 34—This new chapter introduces the establishment of a human microbiome as a developmental process from infancy through adulthood. The importance of the microbiome to host homeostasis is emphasized by discussion of its role in metabolism, immune function, and the gut-brain axis as well as an introduction to the consequences of dysbiosis.

Chapter 35—This chapter has been reorganized to delineate the development of disease from microbial transmission to host cell damage. Emphasis is placed on the overlap between microbial molecules that facilitate survival and those that act as virulence factors. This chapter is placed after the immunology chapters to stress that the host-parasite relationship is dynamic, with adaptations and responses offered by both host and parasite.

Part Eight

Chapter 36—This chapter presents the development of modern epidemiology as an investigative science, emphasizing its role in preventative medicine. The latest epidemiological data from the Centers for Disease Control and Prevention are reported.

Chapter 37—This chapter has been updated to reflect the technological advances in the modern clinical laboratory. Emphasis is on modern diagnostic testing to identify infectious disease.

Chapter 38—Updated and expanded coverage includes viral pathogenesis, common viral infections, and prion-mediated diseases.

Chapter 39—Updated coverage of bacterial organisms and the ways in which they commonly lead to human disease.

Chapter 40—Updated and expanded coverage of fungal and protozoal diseases.

Part Nine

Chapter 41—The essentials of both food safety and microbial processes involved in food production have been updated.

Chapter 42—Includes updated coverage of biofuel production (first introduced in chapter 21) and an introduction to synthetic biology.

Chapter 43—This chapter complements our 21st-century approach to microbiology by emphasizing the importance of clean water and the power of microbial environmental remediation.

Lab Tools for Your Success

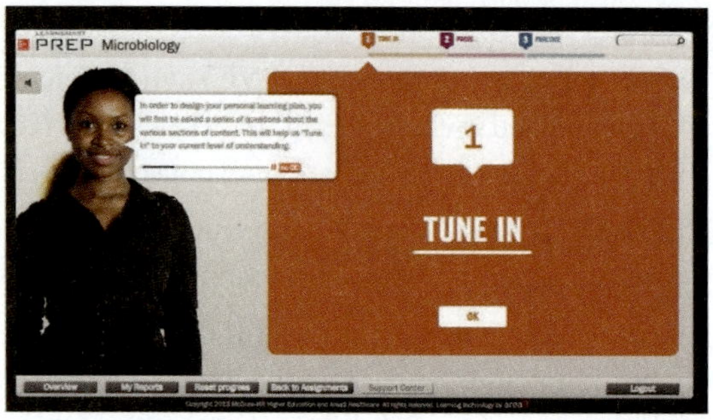

LearnSmart® Prep is an adaptive learning tool that prepares students for college-level work in Microbiology. LearnSmart Prep individually identifies concepts the student does not fully understand and provides learning resources to teach essential concepts so he or she enters the classroom prepared. Data-driven reports highlight areas where students are struggling, helping to accurately identify weak areas.

Acknowledgements

In the preparation of each edition, we are guided by the collective wisdom of reviewers who are expert microbiologists and excellent teachers. They represent experience in community colleges, liberal arts colleges, comprehensive institutions, and research universities. We have followed their recommendations, while remaining true to our overriding goal of writing readable, student-centered content. Each feature incorporated into this edition has been carefully considered in terms of how it may be used to support student learning in both the traditional and the flipped learning environment.

Also in this edition, we are very excited to incorporate real student data points and input, derived from thousands of our LearnSmart users, to help guide our revision. With this information, we were able to hone both book and digital content.

The authors wish to extend their gratitude to our team at McGraw-Hill Education, including Marija Magner, Darlene Schueller, Valerie Kramer, Laura Bies, Matt Backhaus, Tammy Juran, and Beth Cray. Finally, we thank our spouses and children, who provided support and tolerated our absences (mental, if not physical) while we completed this demanding project.

Contents

Contents

Contents

Contents

Contents

Contents

Contents

Contents

Contents

eleventh edition

Prescott's Microbiology

1

The Evolution of Microorganisms and Microbiology

©alex-mit/iStock/Getty Images

The Microbial Universe

If you have ever gazed at the night sky on a cloudless evening in a region far from light pollution, you have probably been amazed and perhaps a little humbled by the vast number of stars. It's hard to estimate just how many stars are out there; best estimates start with our own Milky Way Galaxy, which has roughly 100 billion (1 x 10^{11}) stars. Astronomers figure there is something like 10 trillion (1 x 10^{13}) galaxies in the universe. Assuming all galaxies are roughly the size of the Milky Way, you wind up with a number around 1 x 10^{24} stars in the universe—not the most accurate number, but nonetheless daunting.

If you turn your gaze back to Earth, however, you will find even more awesome abundance. For example, if you stacked all of the 1 x 10^{31} viruses on Earth one on top of the other, they would stretch about 100 million light-years. That's about 43 times further away than the Andromeda Galaxy. And next time you take a dip in the sea, consider that there are at least a hundred thousand (1 x 10^{5}) more bacteria in the ocean (about 1.3 x 10^{29}) than stars in the universe. Or perhaps you prefer staying on land where a teaspoon of soil has about a billion (1 x 10^{9}) microorganisms.

What are all these microorganisms doing? The short answer is, making life for the rest of us possible. How? Starting about 2.4 billion years ago, bacteria called cyanobacteria were the first organisms to release oxygen in abundance into the atmosphere. This has been dubbed the "great oxidation event," and it set the stage for oxygen-consuming organisms (like us) to evolve. Microorganisms have another starring role in the evolution of life because only bacteria can fix nitrogen—that is, take gaseous nitrogen and convert it to organic nitrogen used by plants, animals, and other microbes. Finally, can you imagine what life on Earth would look like if dead organic material were not degraded? Probably best not to. Much better to think of foods like beer, wine, chocolate, cheese, and yogurt—all microbial products.

But of course not all microorganisms make life possible, or even easier. Each year about 16 million (1.6 x 10^{7}) people die from infectious disease; and many of these deaths are preventable by either vaccination or antibiotic treatments. Ironically most vaccines and antibiotics are also microbial products. Although we know the most about disease-causing microorganisms, because less than 1% of all microorganisms cause disease, there is a lot left to learn about microbes. In fact, like the number of stars, the number of microbial species (including bacteria, viruses, fungi, and protists) is actively debated. What is certain is that microbes are important for the life of the planet and its plants and animals.

Our goal in this chapter is to introduce you to this amazing world of microorganisms and to outline the history of their evolution and discovery. Microbiology is a biological science, and as such, much of what you will learn in this text is similar to what you have learned in high school and college biology classes that focus on large organisms. But microbes have unique properties, so microbiology has unique approaches to understanding them. These too will be introduced. But before you delve into this chapter, check to see if you have the background needed to get the most from it.

Readiness Check:

Based on what you have learned previously, you should be able to:

✓ List the features of eukaryotic cells that distinguish them from other cell types

✓ Understand the basic structure of the macromolecules, nucleic acids, proteins, carbohydrates, and lipids (*see appendix I*)

1.1 Members of the Microbial World

After reading this section, you should be able to:

a. Define the term *microbiology*
b. Explain Carl Woese's contributions in establishing the three-domain system for classifying cellular life
c. Determine the type of microbe (bacterium, fungus, etc.) when given a description of a newly discovered one
d. Provide an example of the importance to humans of each of the major types of microbes

Microorganisms are defined as those organisms too small to be seen clearly by the unaided eye (**figure 1.1**). They are generally 1 millimeter or less in diameter. Although small size is an important characteristic of microbes, it is not sufficient to define them.

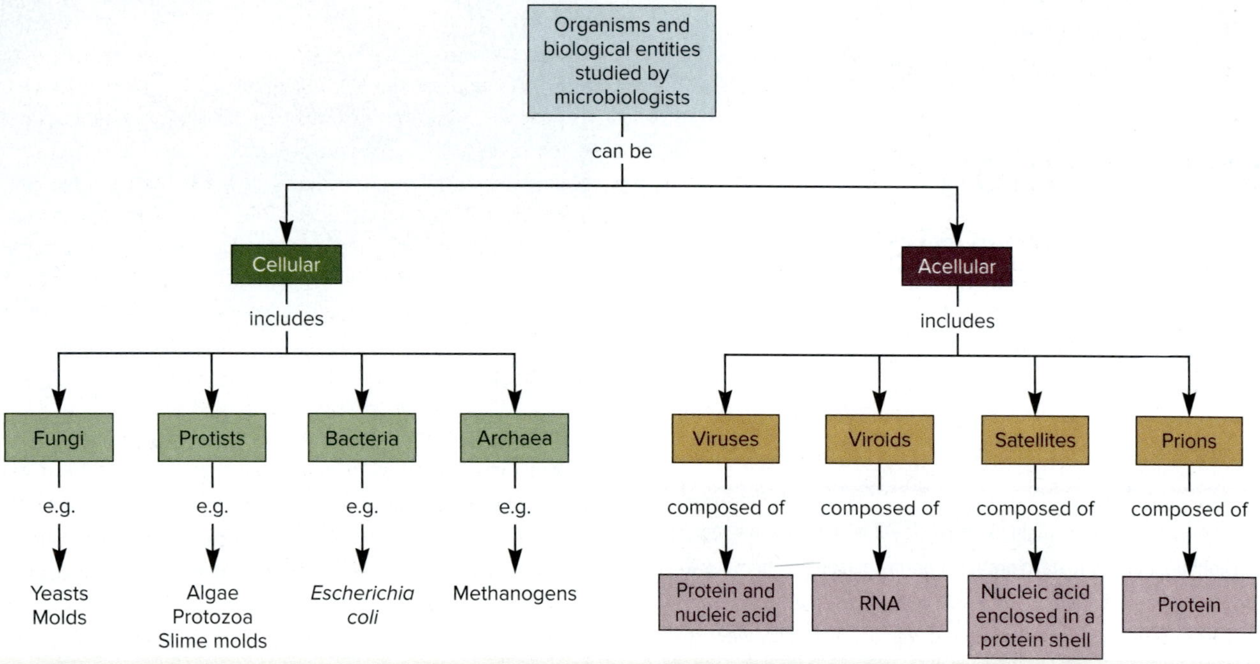

Figure 1.1 Concept Map Showing the Types of Biological Entities Studied by Microbiologists.

MICRO INQUIRY *How would you alter this concept map so that it also distinguishes cellular organisms from each other?*

Some microbes, such as bread molds and filamentous photosynthetic microbes (e.g., "pond scum"), are actually visible without microscopes. These macroscopic microbes often consist of small aggregations of cells. Some macroscopic microorganisms are multicellular. They are distinguished from other multicellular life forms such as plants and animals by their lack of highly differentiated tissues. Most unicellular microbes are microscopic. In summary, cellular microbes are usually smaller than 1 millimeter in diameter, often unicellular and, if multicellular, lack differentiated tissues.

In addition to microorganisms, microbiologists study a variety of acellular biological entities (figure 1.1). These include viruses and subviral agents. The terms "microorganism" and "microbe" are sometimes applied to these acellular agents as well.

The diversity of microorganisms has always presented a challenge to microbial taxonomists. Early descriptions of cellular microbes as either plants or animals were too simple. For instance, some microbes are motile like animals but also have cell walls and are photosynthetic like plants. An important breakthrough in microbial taxonomy arose from studies of their cellular architecture, when it was discovered that cells exhibited one of two possible "floor plans." Cells that came to be called **prokaryotic cells** (Greek *pro,* before; *karyon,* nut or kernel) have an open floor plan. That is, their contents are not divided into compartments ("rooms") by membranes. Only **eukaryotic cells** (Greek *eu,* true) have a nucleus and other membrane-bound organelles (e.g., mitochondria, chloroplasts) that separate some cellular materials and processes from others.

These observations eventually led to the development of a classification scheme that divided organisms into five kingdoms: *Monera, Protista, Fungi, Animalia,* and *Plantae.* Microorganisms (except for viruses and other acellular infectious agents, which have their own classification system) were placed in the first three kingdoms. In this scheme, all organisms with prokaryotic cell structure were placed in *Monera.* The five-kingdom system was an important development in microbial taxonomy, but it is no longer accepted by microbiologists. This is because "prokaryotes" are too diverse to be grouped together in a single kingdom. Furthermore, it is currently argued that the term *prokaryote* is not meaningful and should be abandoned. ▶| *Use of the term "prokaryote" is controversial (section 3.1)*

Great progress has been made in three areas that profoundly affect microbial classification. First, much has been learned about the detailed structure of microbial cells from the use of electron microscopy. Second, microbiologists have determined the biochemical and physiological characteristics of many different microorganisms. Third, the sequences of nucleic acids and proteins from a wide variety of organisms have been compared. The comparison of ribosomal RNA (rRNA) nucleic acid sequences, begun by Carl Woese (1928–2012) in the 1970s, was instrumental in demonstrating that there are two very different groups of organisms with prokaryotic cell architecture: *Bacteria* and *Archaea.* Later studies based on rRNA comparisons showed that *Protista* is not a cohesive taxonomic unit (i.e., taxon) and that it should be divided into three or more kingdoms. These studies and others led many taxonomists to reject the five-kingdom system in favor of one that divides cellular organisms

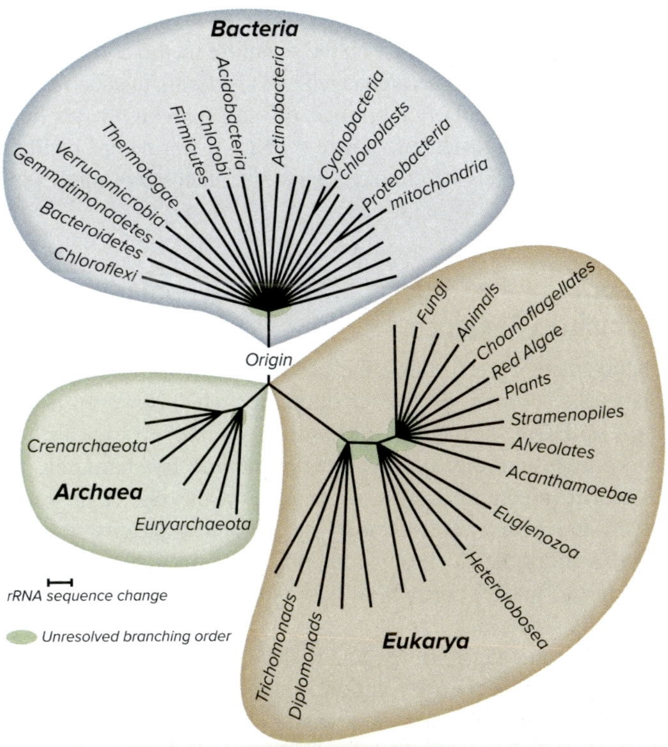

Figure 1.2 Universal Phylogenetic Tree. These evolutionary relationships are based on rRNA sequence comparisons. Only representative lineages have been identified.

MICRO INQUIRY *How many of the taxa listed in the figure include microbes?*

into three domains: *Bacteria*, *Archaea*, and *Eukarya* (all eukaryotic organisms) (**figure 1.2**). ▶| *Nucleic acids (appendix I); Proteins (appendix I)*

Members of domain ***Bacteria*** are usually single-celled organisms.[1] Most have cell walls that contain the structural molecule peptidoglycan. Bacteria are abundant in soil, water, and air, including sites that have extreme temperatures, pH, or salinity. Bacteria are also major inhabitants of our bodies, forming the human **microbiome.** Indeed, more microbial cells are found in and on the human body than there are human cells. These microbes begin to colonize humans shortly after birth. As the microbes establish themselves, they contribute to the development of the body's immune system. Those microbes that inhabit the large intestine help the body digest food and produce vitamins. In these and many other ways, the human microbiome helps maintain our health and well-being. ▶| *Overview of bacterial cell wall structure (section 3.4); The microbe-human ecosystem (chapter 34)*

Unfortunately some bacteria cause disease, and some of these diseases have had a huge impact on human history. In 1347 the plague (Black Death), a disease carried by bacteria living in fleas, struck Europe with brutal force, killing one-third of the population within 4 years. Over the next 80 years, the disease struck repeatedly, eventually wiping out roughly half of the European population. The plague's effect was so great that most historians believe it changed European culture and prepared the way for the Renaissance.

Members of domain ***Archaea*** are distinguished from bacteria by many features, most notably their distinctive rRNA sequences, lack of peptidoglycan in their cell walls, and unique membrane lipids. Some have unusual metabolic characteristics, such as the ability to generate methane (natural) gas. Many archaea are found in extreme environments, including those with high temperatures (thermophiles) and high concentrations of salt (extreme halophiles). Although some archaea are members of a community of microbes involved in gum disease in humans, their role in causing disease has not been clearly established.

Domain ***Eukarya*** includes plants, animals, and microorganisms classified as protists or fungi. **Protists** are generally unicellular but larger than most bacteria and archaea. They have traditionally been divided into protozoa and algae. However, these terms lack taxonomic value because protists, algae, and protozoa do not form three groups, each with a single evolutionary history. Nonetheless, for convenience, we use these terms here.

The major types of protists are algae, protozoa, slime molds, and water molds. **Algae** are photosynthetic. They, together with cyanobacteria, produce about 50% of the planet's oxygen and are the foundation of aquatic food chains. **Protozoa** are usually motile and many free-living protozoa function as the principal hunters and grazers of the microbial world. They obtain nutrients by ingesting organic matter and other microbes. They can be found in many different environments, and some are normal inhabitants of the intestinal tracts of animals, where they aid in digestion of complex materials such as cellulose. A few cause disease in humans and other animals. **Slime molds** are protists that behave like protozoa in one stage of their life cycle but like fungi in another. In the protozoan phase, they hunt for and engulf food particles, consuming decaying vegetation and other microbes. **Water molds** are protists that grow on the surface of freshwater and moist soil. They feed on decaying plant material. Some water molds have produced devastating plant infections, including the Great Potato Famine of 1846–1847 in Ireland, which led to the mass exodus of Irish to the United States and other countries. Although slime and water molds are protists, they were once thought to be fungi, thus the confusing nomenclature. ▶| *Protists (chapter 24)*

Fungi are a diverse group of microorganisms that range from unicellular forms (yeasts) to molds and mushrooms. Molds and mushrooms are multicellular fungi that form thin, threadlike structures called hyphae. They absorb all their nutrients from their environment. Because of their metabolic capabilities, many fungi play beneficial roles, including making bread dough rise, producing antibiotics, and decomposing dead organisms. Some fungi associate with plant roots to form mycorrhizae. Mycorrhizal fungi transfer nutrients to the roots, improving growth of the plants, especially in poor soils. Other fungi cause plant diseases (e.g., rusts, powdery mildews, and smuts) and diseases in humans and other animals. ▶| Fungi *(chapter 25)*

1 In this text, the term *bacteria* (s., *bacterium*) is used to refer to those microbes belonging to domain *Bacteria*, and the term *archaea* (s., *archaeon*) is used to refer to those that belong to domain *Archaea*. In some publications, the term *bacteria* is used to refer to all cells having prokaryotic cell structure. That is not the case in this text.

The microbial world also includes numerous acellular infectious agents. **Viruses** are acellular entities that must invade a host cell to multiply. The simplest virus particles (also called virions) are composed only of proteins and a nucleic acid, and can be extremely small (the smallest is 10,000 times smaller than a typical bacterium). However, their small size belies their power. They cause many animal and plant diseases and have caused epidemics that have shaped human history. Viral diseases include smallpox, rabies, influenza, AIDS, the common cold, and some cancers. Viruses also play important roles in aquatic environments, where they play a critical role in shaping aquatic microbial communities. **Viroids** are infectious agents composed only of ribonucleic acid (RNA). They cause numerous plant diseases. **Satellites** are composed of a nucleic acid enclosed in a protein shell. They must coinfect a host cell with virus, called a helper virus, to complete their life cycle. Viroids, on the other hand, cause only plant diseases while satellites and their helper viruses cause both plant and animal diseases. Finally, **prions,** infectious agents composed only of protein, are responsible for causing a variety of spongiform encephalopathies such as scrapie and "mad cow disease." ▶| *Viruses and other acellular infectious agents (chapter 6)*

Comprehension Check

1. How did the methods used to classify microbes change, particularly in the last half of the twentieth century? What was the result of these technological advances?
2. Identify one characteristic for each of these types of microbes that distinguishes it from the other types: bacteria, archaea, protists, fungi, viruses, viroids, satellites, and prions.

1.2 Microbes Have Evolved and Diversified for Billions of Years

After reading this section, you should be able to:

a. Propose a timeline of the origin and history of microbial life and integrate supporting evidence into it
b. Design a set of experiments that could be used to place a newly discovered cellular microbe on a phylogenetic tree based on small subunit (SSU) rRNA sequences
c. Compare and contrast the definitions of plant and animal species, microbial species, and microbial strains

A review of figure 1.2 reminds us that in terms of the number of taxa, microbes are the dominant organisms on Earth. How has microbial life been able to radiate to such an astonishing level of diversity? To answer this question, we must consider microbial evolution. The field of microbial evolution, like any other scientific endeavor, is based on the formulation of hypotheses, the gathering and analysis of data, and the reformation of hypotheses based on newly acquired evidence. That is to say, the study of microbial evolution is based on the scientific method. To be sure, it is difficult to amass evidence when considering events that occurred millions, and often billions, of years ago, but the application of molecular methods has revealed a living record of life's ancient history. This section describes the outcome of this scientific research.

Theories of the Origin of Life Depend Primarily on Indirect Evidence

Dating meteorites through the use of radioisotopes places our planet at an estimated 4.5 to 4.6 billion years old. However, conditions on Earth for the first 100 million years or so were far too harsh to sustain any type of life. Eventually bombardment by meteorites decreased, water appeared on the planet in liquid form, and gases were released by geological activity to form Earth's atmosphere. These conditions were amenable to the origin of the first life forms. But how did this occur, and what did these life forms look like?

To find evidence of life and to develop hypotheses about its origin and subsequent evolution, scientists must be able to define life. Although even very young children can examine an object and correctly determine whether it is living or not, defining life succinctly has proven elusive for scientists. Thus most definitions of life consist of a set of attributes. The attributes of particular importance to paleobiologists are an orderly structure, the ability to obtain and use energy (i.e., metabolism), and the ability to reproduce. Just as NASA scientists are using the characteristics of microbes on Earth today to search for life elsewhere, so too are scientists examining **extant organisms,** those organisms present today, to explore the origin of life. Some extant organisms have structures and molecules that represent "relics" of ancient life forms. These can provide scientists with ideas about the type of evidence to seek when testing hypotheses.

The best direct evidence for the nature of primitive life would be a fossil record. There have been reports of microbial fossil discoveries since 1977 (**figure 1.3**). These have always met with skepticism because finding them involves preparing thin slices of ancient rocks and examining the slices for objects that look like cells. Unfortunately some things that look like cells can be formed by geological forces that occurred as the rock was formed. The result is that the fossil record for microbes is sparse and always open to reinterpretation. Despite these problems most scientists agree that life was present on Earth about 3.5–3.7 billion years ago (**figure 1.4**). To reach this conclusion, biologists rely primarily on indirect evidence. Among the indirect evidence used are molecular fossils. These are chemicals found in rock or sediment that are chemically related to molecules found in cells. For instance, the presence in a rock of molecules called hopanes is an indication that when the rock was formed, bacteria were present. This conclusion is reached because hopanes are formed from hopanoids, which are found in the plasma membranes of extant bacteria. As you can see, no single piece of evidence can

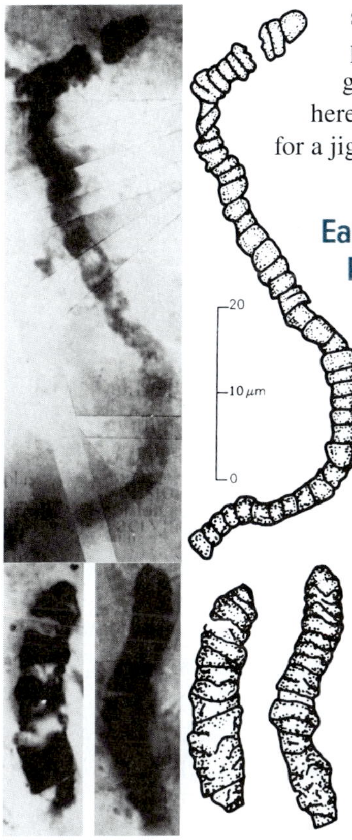

Figure 1.3 **Possible Microfossils Found in the Archaeon Apex Chert of Australia.** Chert is a type of granular sedimentary rock rich in silica. These structures were discovered in 1977. Because of their similarity to filamentous cyanobacteria they were proposed to be microfossils. In 2011 scientists reported that similar structures from the same chert were not biological in origin. They used spectrometry and microscopy techniques not available in 1977 to show that the structures were fractures in the rock filled with quartz and hematite. Scientists are still debating whether or not these truly are microfossils.
©J. William Schopf

stand alone. Instead many pieces of evidence are put together in an attempt to get a coherent picture to emerge, much as for a jigsaw puzzle.

Early Life Was Probably RNA Based

The origin of life rests on a single question: How did early cells arise? At a minimum, modern cells consist of a plasma membrane enclosing water in which numerous chemicals are dissolved and subcellular structures float. It seems likely that the first self-replicating entity was much simpler than even the most primitive modern living cells. Before there was life, most evidence suggests that Earth was a very different place: hot and anoxic, with an atmosphere rich in water vapor, carbon dioxide, and nitrogen. In the oceans, hydrogen, methane, and carboxylic acids were formed by geological and chemical processes. Areas near hydrothermal vents or in shallow pools may have provided the conditions that allowed chemicals to react with one another, randomly "testing" the usefulness of the reaction and the stability of its products. Some reactions released energy and would eventually become the basis of modern cellular metabolism. Other reactions generated molecules that functioned as catalysts, some aggregated with other molecules to form the predecessors of modern cell structures, and others were able to replicate and act as units of hereditary information.

In modern cells, three different molecules fulfill the roles of catalysts, structural molecules, and hereditary molecules (**figure 1.5**). Proteins have two major roles in modern cells: catalytic and structural. Catalytic proteins are **enzymes** and structural proteins serve a myriad of functions such as transport, attachment, and motility. DNA stores hereditary information that is replicated and passed on to the next generation. RNA is involved in converting the information stored in DNA into protein. Any hypothesis about the origin of life must account for the evolution of these molecules, but the very nature of their relationships to each other in modern cells complicates attempts to imagine how they evolved. As demonstrated in figure 1.5, proteins can do cellular work, but their synthesis involves other proteins and RNA, and uses information stored in DNA. DNA can't do cellular work and proteins are needed for its replication. RNA synthesis requires both DNA as the template and proteins as the catalysts for the reaction.

Based on these considerations, it is hypothesized that at some time in the evolution of life, there must have been a single molecule that could do both cellular work and replicate. This idea was supported in 1981 when Thomas Cech discovered a catalytic RNA molecule in a protist (*Tetrahymena* sp.) that could cut out an internal section of itself and splice the remaining sections back together. Since then, other catalytic RNA molecules have been discovered, including an RNA found in ribosomes that is responsible for forming peptide bonds—the bonds that hold together amino acids, the building blocks of proteins. Catalytic RNA molecules are now called **ribozymes.**

The discovery of ribozymes suggested that RNA at some time was capable of storing, copying, and expressing genetic information, as well as catalyzing other chemical reactions. In 1986 Nobel laureate Walter Gilbert coined the term **RNA world** to describe this precellular stage in the evolution of life. However, for this precellular RNA-based stage to proceed to the evolution of cellular life forms, a lipid membrane must have formed around the RNA (**figure 1.6**). This important evolutionary step is easier to imagine than other events in the origin of cellular life forms because lipids, major structural components of the membranes of modern organisms, spontaneously form liposomes—vesicles bounded by a lipid bilayer. ▶| *Lipids (appendix I)*

Jack Szostak, also a Noble laureate, is a leader in exploring how RNA-containing cells, so-called protocells, may have formed. When his group created liposomes using simpler fatty acids than those found in membranes today, the liposomes were leaky. These leaky liposomes allowed single RNA nucleotides to move into the liposome, but prevented large RNA chains from moving out. Furthermore, researchers could prod the liposomes into growing and dividing. Dr. Szostak's group has also been able to create conditions in which an RNA molecule could serve as a template for synthesis of a complementary RNA strand. Thus their experiments may have recapitulated the early steps of the evolution of cells. As seen in figure 1.6, several other processes would need to occur to reach the level of complexity found in extant cells.

Apart from its ability to perform catalytic activities, the function of RNA suggests its ancient origin. Consider that much

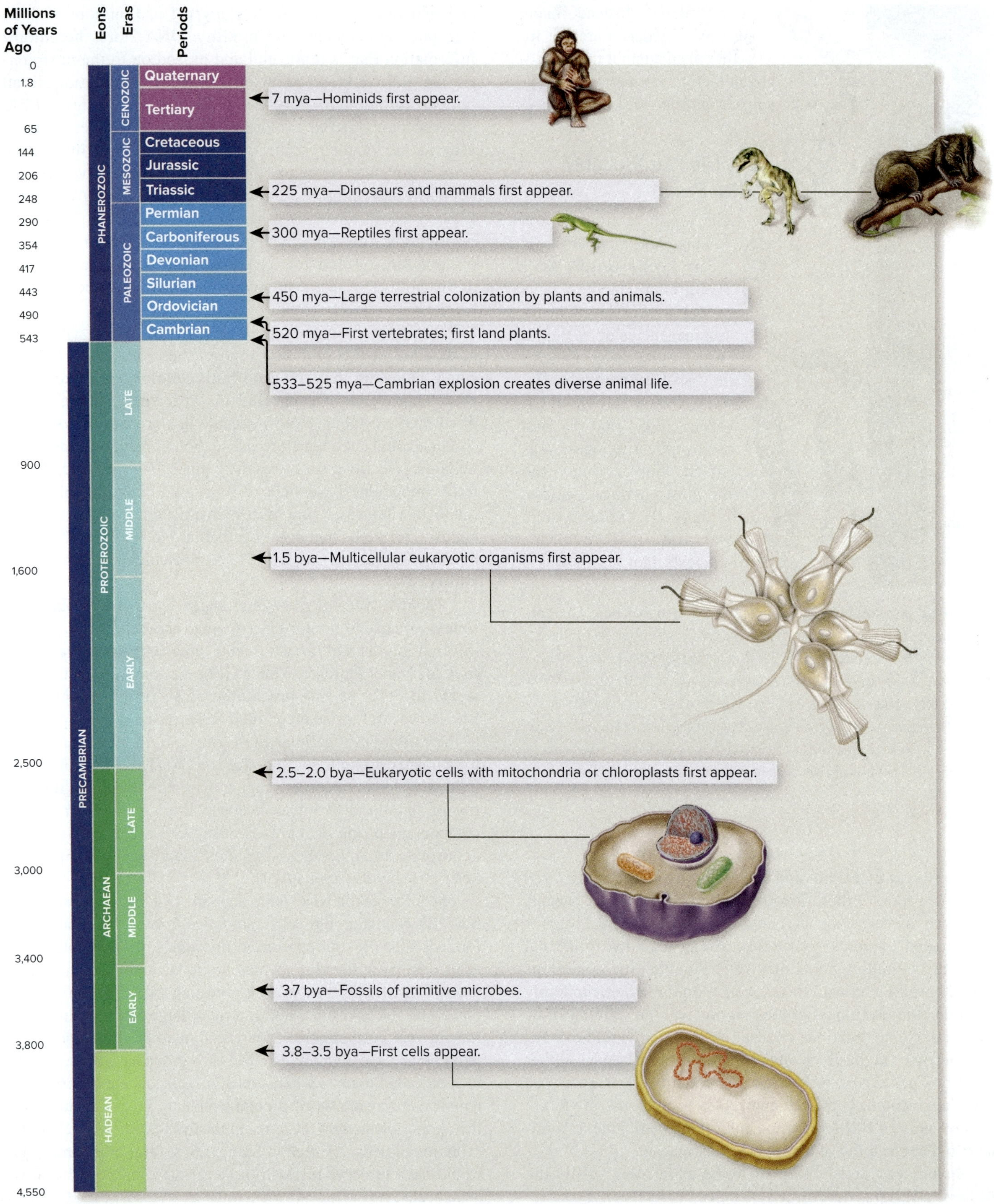

Figure 1.4 An Overview of the History of Life on Earth. mya = million years ago; bya = billion years ago.

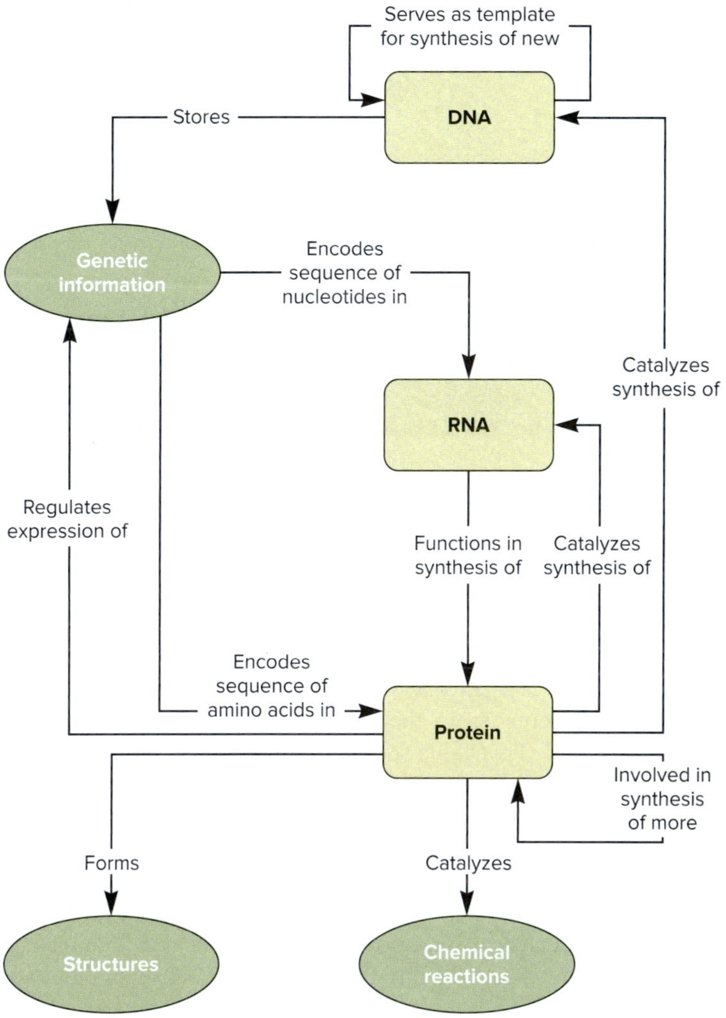

Figure 1.5 **Functions of DNA, RNA, and Protein, and Their Relationships to Each Other in Extant Cells.**

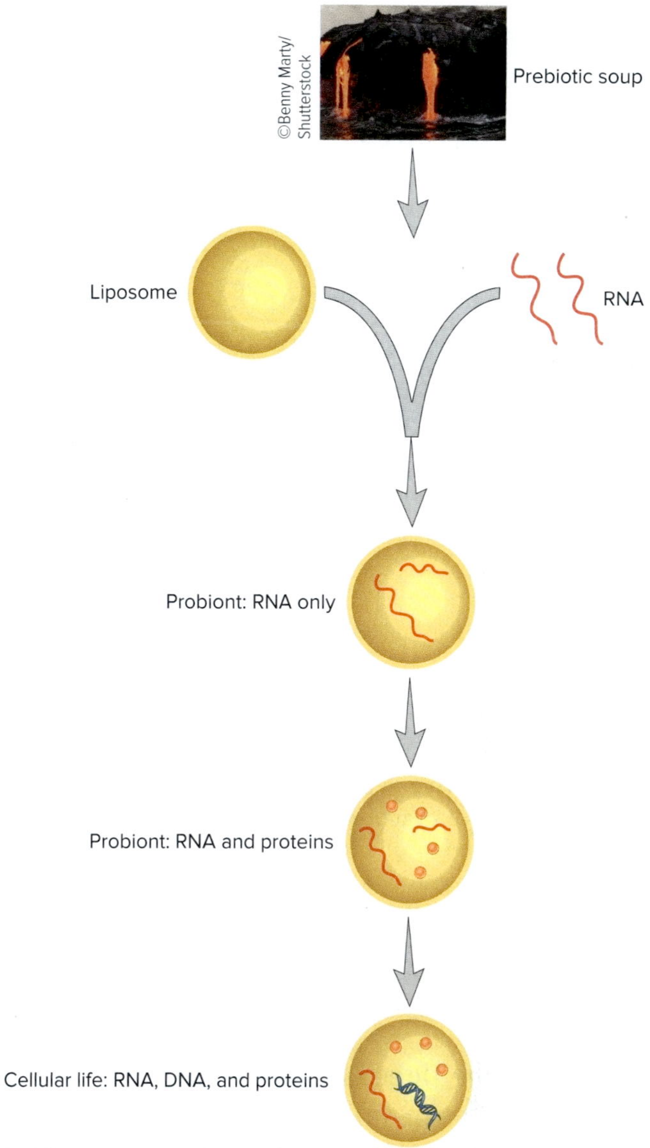

Figure 1.6 **The RNA World Hypothesis for the Origin of Life.**

MICRO INQUIRY *Why are the probionts pictured above not considered cellular life?*

of the cellular pool of RNA in modern cells exists in the ribosome, a structure that consists largely of rRNA and uses messenger RNA (mRNA) and transfer RNA (tRNA) to construct proteins. Also rRNA itself catalyzes peptide bond formation during protein synthesis. Thus RNA seems to be well poised for its importance in the development of proteins. Because RNA and DNA are structurally similar, RNA could have given rise to double-stranded DNA. It is suggested that once DNA evolved, it became the storage facility for genetic information because it provided a more chemically stable structure. Two other pieces of evidence support the RNA world hypothesis: the fact that the energy currency of cells, ATP, is a ribonucleotide and the discovery that RNA can regulate gene expression. So it would seem that proteins, DNA, and cellular energy can be traced back to RNA. ▶| *ATP: the major energy currency of cells (section 10.2); Riboswitches (section 14.3); Translational riboswitches (section 14.4)*

Despite evidence supporting the RNA world hypothesis, it is not without problems, and many argue against it. Another area of research also fraught with considerable debate is the evolution of metabolism, in particular the evolution of energy-conserving metabolic processes. The early Earth was a hot environment that lacked oxygen. Thus the cells that arose there must have been able to use the available energy sources under these harsh conditions. Today there are heat-loving archaea capable of using inorganic molecules such as FeS as a source of energy. Some suggest that this interesting metabolic capability is a remnant of the first form of energy metabolism. Another metabolic strategy, oxygen-releasing photosynthesis (oxygenic photosynthesis), appears to have evolved perhaps as early as 2.7 billion years ago. Fossils of cyanobacteria-like cells found in rocks dating to that time

Figure 1.7 **Stromatolites.** (a) Section of a fossilized stromatolite. Evolutionary biologists think the layers of material were formed when mats of cyanobacteria, layered one on top of the other, became mineralized. (b) Modern stromatolites from Western Australia. Each stromatolite is a rocklike structure, typically 1 m in diameter, containing layers of cyanobacteria.

support this hypothesis, as does the discovery of ancient stromatolites (**figure 1.7*a***). Stromatolites are layered rocks formed by the incorporation of mineral sediments into layers of microorganisms growing in thick mats on surfaces (figure 1.7*b*). The appearance of cyanobacteria-like cells was an important step in the evolution of life on Earth. The oxygen they released is thought to have altered Earth's atmosphere to its current oxygen-rich state, allowing the evolution of additional energy-capturing strategies such as aerobic respiration, the oxygen-consuming metabolic process that is used by many microbes and animals.

Evolution of the Three Domains of Life

As noted in section 1.1, rRNA comparisons were an important breakthrough in the classification of microbes; this analysis also provides insights into the evolutionary history of all life. What began with the examination of rRNA from relatively few organisms has been expanded by the work of many others, including Norman Pace. Dr. Pace has developed a **universal phylogenetic tree** (figure 1.2) based on comparisons of small subunit rRNA molecules (SSU rRNA), the rRNA found in the small subunit of the ribosome. Here we examine how these comparisons are made and what the universal phylogenetic tree tells us. ▶| *Bacterial ribosomes (section 3.6); Microbial taxonomy and phylogeny are largely based on molecular characterization (section 19.3)*

Comparing SSU rRNA Molecules

The details of phylogenetic tree construction are discussed in chapter 19. However, the general concept is not difficult to understand. In one approach, the sequences of nucleotides in the genes that encode SSU rRNAs from diverse organisms are aligned, and pair-wise comparisons of the sequences are made. For each pair of SSU rRNA gene sequences, the number of differences in the nucleotide sequences is counted (**figure 1.8**). This value serves as a measure of the evolutionary distance between the organisms; the more differences counted, the greater the evolutionary distance. The evolutionary distances from many comparisons are used by sophisticated computer programs to construct the tree. The tip of each branch in the tree represents one of the organisms used in the comparison. The distance from the tip of one branch to the tip of another is the evolutionary distance between the two organisms.

Two things should be kept in mind when examining phylogenetic trees developed in this way. The first is that they are molecular trees, not organismal trees. In other words, they represent, as accurately as possible, the evolutionary history of a molecule and the gene that encodes it. Second, the distance between branch tips is a measure of relatedness, not of time. If the distance along the lines is very long, then the two organisms are more evolutionarily diverged (i.e., less related). However, we do not know when they diverged from each other. This concept is analogous to a printed map that accurately shows the distance between two cities but because of many factors (traffic, road conditions, etc.) cannot show the time needed to travel that distance.

LUCA

What does the universal phylogenetic tree tell us about the evolution of life? At the center of the tree is a line labeled "Origin" (figure 1.2). This is where data indicate the *l*ast *u*niversal *c*ommon *a*ncestor (LUCA) to all three domains should be placed. LUCA is on the bacterial branch, which means that *Archaea* and *Eukarya* evolved independently, separate from *Bacteria*. Thus the universal phylogenetic tree presents a picture in which all

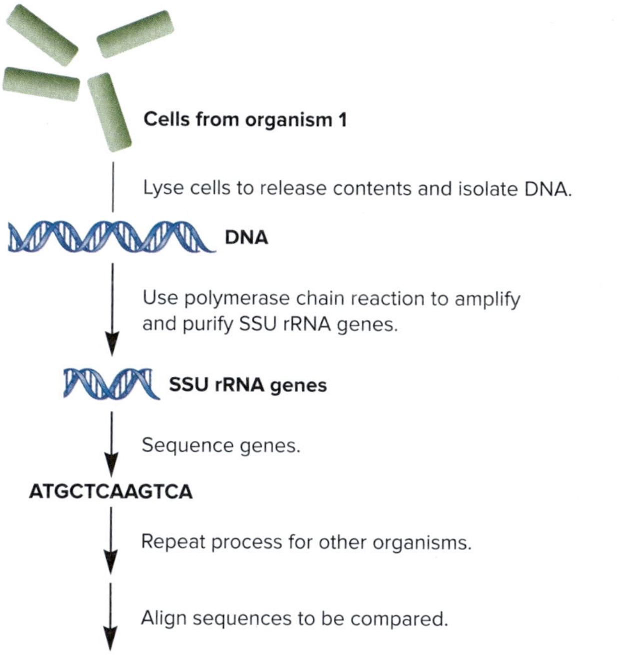

Organism	SSU rRNA sequence
1	ATGCTCAAGTCA
2	TAGCTCGTGTAA
3	AAGCTCTAGTTA
4	AACCTCATGTTA

Count the number of nucleotide differences between each pair of sequences and calculate evolutionary distance (E_D).

Pair compared	E_D	Corrected E_D
1 → 2	0.42	0.61
1 → 3	0.25	0.30
1 → 4	0.33	0.44
2 → 3	0.33	0.44
2 → 4	0.33	0.44
3 → 4	0.25	0.30

For organisms 1 and 2, 5 of the 12 nucleotides are different: $E_D = 5/12 = 0.42$.

The initial ED calculated is corrected using a statistical method that considers for each site the probability of a mutation back to the original nucleotide or of additional forward mutations.

Computer analysis is used to construct phylogenetic tree.

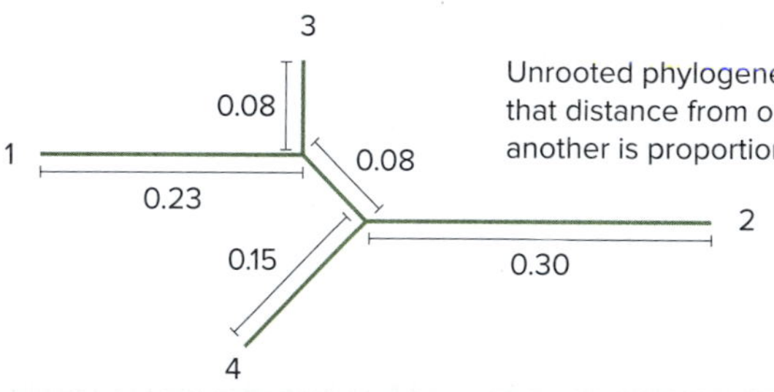

Unrooted phylogenetic tree. Note that distance from one tip to another is proportional to the E_D.

Figure 1.8 The Construction of Phylogenetic Trees Using a Distance Method. The polymerase chain reaction is described in chapter 17.

MICRO INQUIRY *Why does the branch length indicate amount of evolutionary change but not the time it took for that change to occur?*

life, regardless of eventual domain, arose from a single common ancestor. One can envision the universal tree of life as a real tree that grows from a single seed.

The evolutionary relationship of *Archaea* and *Eukarya* is still the matter of considerable debate. According to the universal phylogenetic tree we show here, *Archaea* and *Eukarya* shared common ancestry but diverged and became separate domains. Other versions suggest that *Eukarya* evolved out of *Archaea*. The close evolutionary relationship of these two forms of life is still evident in the manner in which they process genetic information. For instance, certain protein subunits of archaeal and eukaryotic RNA polymerases, the enzymes that catalyze RNA synthesis, resemble each other to the exclusion of those of bacteria. However, archaea have other features that are most similar to their counterparts in bacteria (e.g., mechanisms for conserving energy). This has further complicated and fueled the debate. The evolution of the nucleus and endoplasmic reticulum is also at the center of many controversies. However, hypotheses regarding the evolution of other membrane-bound organelles are more widely accepted and are considered next.

Mitochondria, Mitochondria-Like Organelles, and Chloroplasts Evolved from Endosymbionts

The **endosymbiotic hypothesis** is generally accepted as the origin of several eukaryotic organelles, including mitochondria, chloroplasts, and hydrogenosomes. **Endosymbiosis** is an interaction between two organisms in which one organism lives inside the other. The original endosymbiotic hypothesis proposed that over time a bacterial endosymbiont of an ancestral cell in the eukaryotic lineage lost its ability to live independently, becoming either a mitochondrion, if the intracellular bacterium used aerobic respiration, or a chloroplast, if the endosymbiont was a photosynthetic bacterium (*see figure 19.7*).

Although the mechanism by which the endosymbiotic relationship was established is unknown, there is considerable evidence to support the hypothesis. Mitochondria and chloroplasts contain DNA and ribosomes; both are similar to bacterial DNA and ribosomes. Peptidoglycan, the unique bacterial cell wall molecule, has even been found between the two membranes that enclose the chloroplasts of some algae. Indeed, inspection of figure 1.2 shows that both organelles belong to the bacterial lineage based on SSU rRNA analysis. More specifically, mitochondria are most closely related to bacteria called proteobacteria. The chloroplasts of plants and green algae are thought to have descended from an ancestor of the cyanobacterial genus *Prochloron*, which contains species that live within marine invertebrates. ▶| *Phylum* Cyanobacteria: *oxygenic photosynthetic bacteria (section 21.4); The proteobacterial origin of mitochondria (section 22.1)*

Recently the endosymbiotic hypothesis for mitochondria has been modified by the **hydrogen hypothesis.** This asserts that the endosymbiont was an anaerobic bacterium that produced H_2 and CO_2 as end products of its metabolism. Over time, the host became dependent on the H_2 produced by the endosymbiont. Ultimately the endosymbiont evolved into one of several organelles (*see figure 5.13*). If the endosymbiont developed the capacity to perform aerobic respiration, it evolved into a mitochondrion. Other endosymbionts evolved into other organelles such as a hydrogenosome—an organelle that produces ATP by a process called fermentation found in some extant protists *(see figure 5.15)*.

Evolution of Cellular Microbes

Although the history of early cellular life forms may never be known, we know that once they arose, they were subjected to the same evolutionary processes as modern organisms. The ancestral bacteria, archaea, and eukaryotes possessed genetic information that could be duplicated, lost, or mutated in other ways. These mutations could have many outcomes. Some led to the death of the mutant microbe, but others allowed new functions and characteristics to evolve. Those mutations that allowed the organism to increase its reproductive ability were selected and passed on to subsequent generations. In addition to selective forces, geographic isolation of populations allowed some groups to evolve separately from others. Thus selection and isolation led to the eventual development of new collections of genes (i.e., genotypes) and many new species.

In addition to mutation, other mechanisms exist for reconfiguring the genotypes of a species and therefore creating genetic diversity. Most eukaryotic species increase their genetic diversity by reproducing sexually. Thus each offspring of the two parents has a mixture of parental genes and a unique genotype. Bacteria and archaea do not reproduce sexually. They increase their genetic diversity by mutation and horizontal (or lateral) gene transfer (HGT). During HGT, genetic information from a donor organism is transferred to a recipient, creating a new genotype. Thus genetic information is passed between individuals of the same generation and even between species found in different domains of life. Genome sequencing has revealed that HGT has played an important role in the evolution of all microbial species. Importantly, HGT still occurs in bacteria and archaea leading to the rapid evolution of microorganisms with antibiotic resistance, new virulence properties, and novel metabolic capabilities. The outcome of HGT is that most microbes have mosaic genomes composed of bits and pieces of the genomes of other organisms. ▶| *Horizontal gene transfer: creating genetic variation the asexual way (section 16.4)*

Microbial Species

All students of biology are introduced early in their careers to the concept of a species. But the term has different meanings for sexual and asexual organisms. Taxonomists working with plants and animals define a **species** as a group of interbreeding or potentially interbreeding natural populations that is reproductively isolated from other groups. This definition also is appropriate for the many eukaryotic microbes that reproduce sexually. However, bacterial and archaeal species cannot be defined by this criterion, since they do not reproduce sexually. Therefore, comparisons of genome sequences are often used to distinguish one species from another. An appropriate definition is currently the topic of considerable discussion. A common definition is that bacterial and archaeal species are a collection of strains that share many stable properties and differ significantly from other groups of strains. A **strain** consists of the descendants of a single, pure microbial culture. Strains within a species may be described in a number of different ways. Biovars are variant strains characterized by biochemical or physiological differences, morphovars differ morphologically, serovars have distinctive properties that can be detected by antibodies, and pathovars are pathogenic strains distinguished by the plants in which they cause disease. ▶| *What is a microbial species? (section 19.5)*

Microbiologists name microbes using the binomial system of the eighteenth-century biologist and physician Carl Linnaeus. The Latinized, italicized name consists of two parts. The first part, which is capitalized, is the generic name (i.e., the name of the genus to which the microbe belongs), and the second is the uncapitalized species epithet. For example, the bacterium that causes plague is called *Yersinia pestis*. Often the name of an organism will be shortened by abbreviating the genus name with a single capital letter (e.g., *Y. pestis*).

Comprehension Check

1. Describe two reasons RNA is thought to be the first self-replicating biomolecule.
2. Explain the endosymbiotic hypothesis of the origin of mitochondria, hydrogenosomes, and chloroplasts. List two pieces of evidence that support this hypothesis.
3. What is the difference between mutation and horizontal gene transfer?
4. What is the correct way to write this microbe's name: *bacillus subtilis*, Bacillus subtilis, *Bacillus Subtilis*, or *Bacillus subtilis*? Identify the genus name and the species epithet.

1.3 Microbiology Advanced as New Tools for Studying Microbes Were Developed

After reading this section, you should be able to:

a. Evaluate the importance of the contributions to microbiology made by Hooke, Leeuwenhoek, Pasteur, Lister, Koch, Beijerinck, von Behring, Kitasato, Metchnikoff, and Winogradsky
b. Outline a set of experiments that might be used to decide if a particular microbe is the causative agent of a disease
c. Predict the difficulties that might arise when using Koch's postulates to determine if a microbe causes a disease unique to humans

Even before microorganisms were seen, some investigators suspected their existence and role in disease. Among others, the Roman philosopher Lucretius (about 98–55 BCE) and the physician Girolamo Fracastoro (1478–1553) suggested that disease was caused by invisible living creatures. However, until microbes could actually be seen and studied in some other way, their existence remained a matter of conjecture. Therefore **microbiology** is defined not only by the organisms it studies but also by the tools used to study them. The development of microscopes was the critical first step in the evolution of the discipline. However, microscopy alone is unable to answer the many questions microbiologists ask about microbes. A distinct feature of microbiology is that microbiologists often remove microorganisms from their normal habitats and grow them isolated from other microbes. This is called a **pure** or **axenic culture.** The development of techniques for isolating microbes in pure culture was another critical step in microbiology's history. However, it is now recognized as having limitations. Microbes in pure culture are in some ways like animals in a zoo; just as a zoologist cannot fully understand the ecology of animals by studying them in zoos, microbiologists cannot fully understand microbes by studying them in pure culture. Today molecular genetic techniques and genomic analyses are providing new insights into the lives of microbes. ▶| *Methods in microbial ecology (chapter 29); Microbial genomics (chapter 18)*

Here we describe how the tools used by microbiologists have influenced the development of the field. As microbiology evolved as a science, it contributed greatly to the well-being of humans. This is exemplified by the number of microbiologists who have won the Nobel Prize. The historical context of some of the important discoveries in microbiology is shown in **figure 1.9**.

Microscopy Led to the Discovery of Microorganisms

The earliest microscopic observations of organisms appear to have been made between 1625 and 1630 on bees and weevils by the Italian Francesco Stelluti (1577–1652), using a microscope probably supplied by Galileo (1564–1642). Robert Hooke (1635–1703) is credited with publishing the first drawings of microorganisms in the scientific literature. In 1665 he published a highly detailed drawing of the fungus *Mucor* in his book *Micrographia*. *Micrographia* is important not only for its exquisite drawings but also for the information it provided on building microscopes. One design discussed in *Micrographia* was probably a prototype for the microscopes built and used by the amateur microscopist Antony van Leeuwenhoek (1632–1723) of Delft, the Netherlands (**figure 1.10*a***). Leeuwenhoek earned his living as a draper and haberdasher (a dealer in men's clothing and accessories) but spent much of his spare time constructing simple microscopes composed of double convex glass lenses held between two silver plates (figure 1.10*b*). His microscopes could magnify about 50 to 300 times, and he may have illuminated his liquid specimens by placing them between two pieces of glass and shining light on them at a 45-degree angle to the specimen plane. This would have provided a form of dark-field illumination whereby organisms appeared as bright objects against a dark background. Beginning in 1673, Leeuwenhoek sent detailed letters describing his discoveries to the Royal Society of London. It is clear from his descriptions that he saw both bacteria and protists (figure 1.10*c*). ▶| *Dark-field microscope: bright object, dark background (section 2.2)*

Culture-Based Methods for Studying Microorganisms Were a Major Development

As important as Leeuwenhoek's observations were, the development of microbiology essentially languished for the next 200 years until techniques for isolating and culturing microbes in the laboratory were formulated. Many of these techniques were developed as scientists grappled with the conflict over the theory of spontaneous generation. This conflict and the subsequent studies on the role played by microorganisms in causing disease ultimately led to what is now called the golden age of microbiology.

Spontaneous Generation

From earliest times, people had believed in **spontaneous generation**—that living organisms could develop from nonliving matter. This view finally was challenged by the Italian physician Francesco Redi (1626–1697), who carried out a series of experiments on decaying meat, which was thought to produce maggots spontaneously. Using covered and uncovered containers of meat, Redi clearly demonstrated that maggots on decaying meat resulted from the presence of fly eggs, and meat did not spontaneously generate maggots. Other experiments helped discredit the theory for larger organisms. However, Leeuwenhoek's communications on microorganisms renewed the controversy. Some proposed that microbes arose by spontaneous generation but larger organisms did not. They pointed out that boiled extracts of hay or meat gave rise to microorganisms after sitting for a while. In 1748 the English priest John Needham (1713–1781) suggested that the organic matter in these extracts contained a "vital force" that could confer the properties of life on nonliving matter.

A few years after Needham's experiments, the Italian priest and naturalist Lazzaro Spallanzani (1729–1799) sealed glass flasks that contained water and seeds and then placed the flasks in boiling water for about 45 minutes. He found that no growth took place as long as the flasks remained sealed. He proposed that air carried germs to the culture medium but also commented that external air might be required for growth of animals already in the medium. The supporters of spontaneous generation responded that heating the air in sealed flasks destroyed its ability to support life, and therefore did not discredit the theory of spontaneous generation.

In the mid-1800s, several investigators attempted to counter such arguments. These experiments involved allowing air to enter a flask containing a nutrient solution after boiling. The air was either also very hot or it was filtered through sterile cotton wool. In all cases, no microbial growth occurred in the medium. Despite

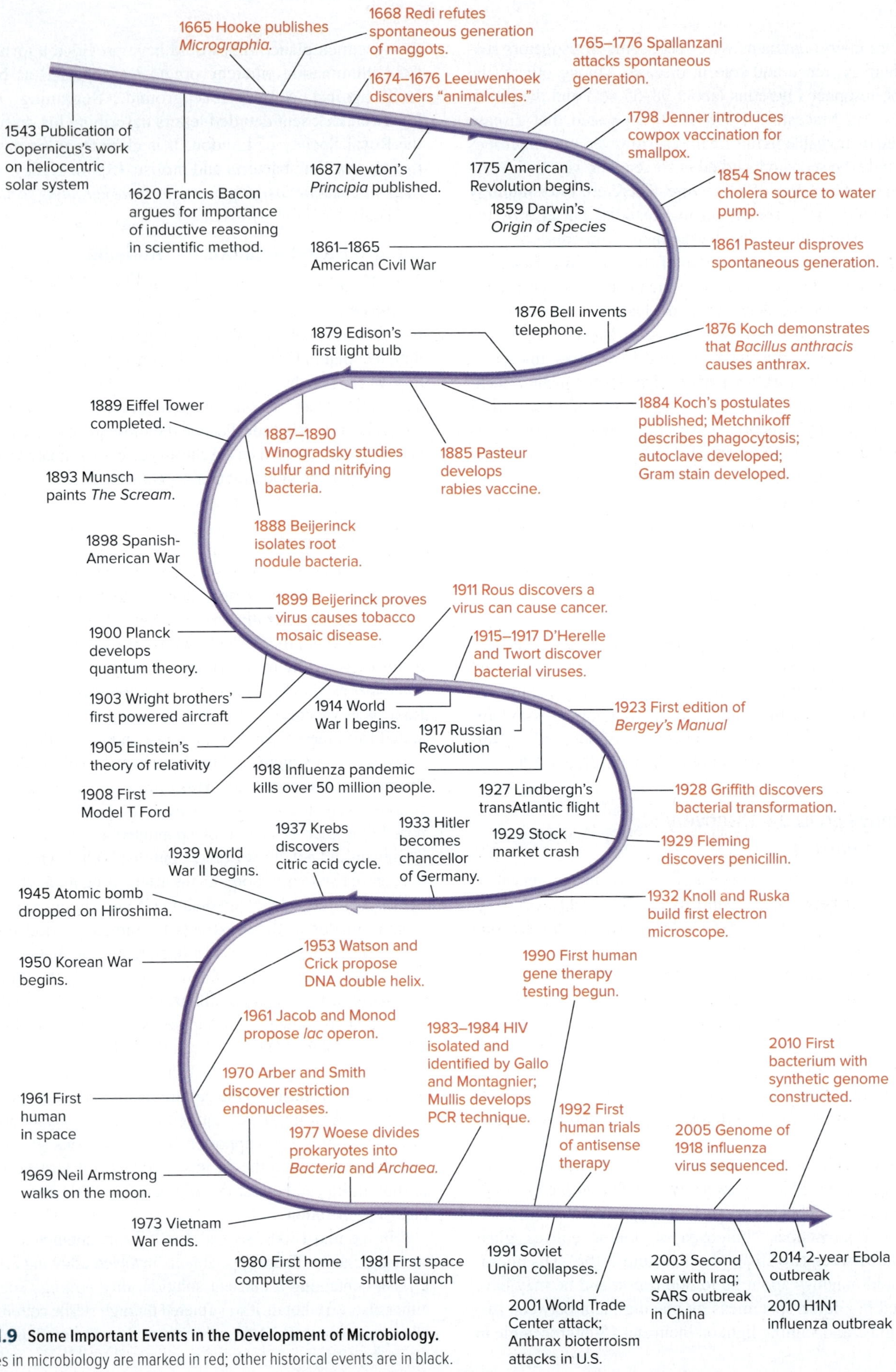

Figure 1.9 **Some Important Events in the Development of Microbiology.** Milestones in microbiology are marked in red; other historical events are in black.

(a)

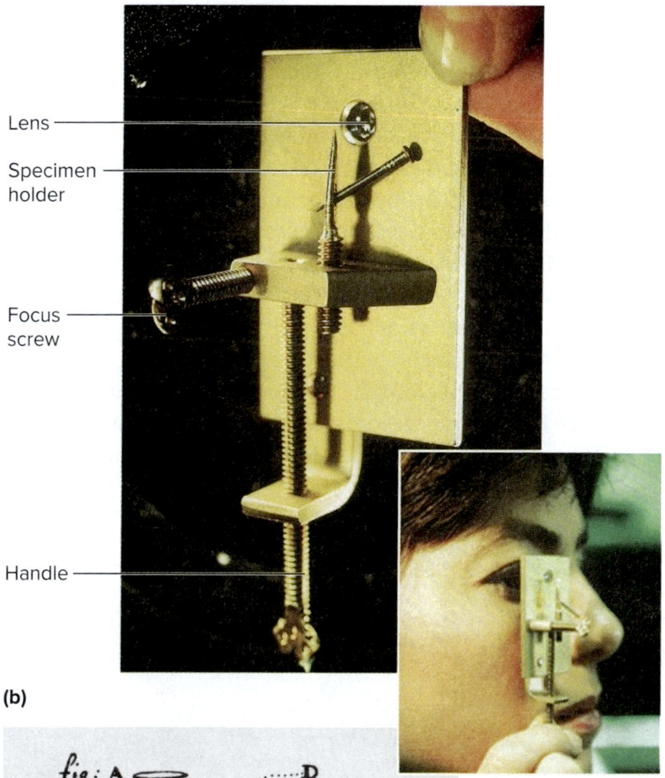

(b)

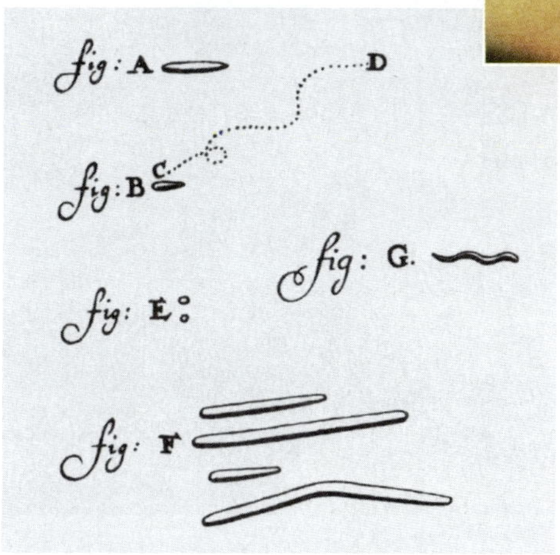

(c)

Figure 1.10 Antony van Leeuwenhoek. (a) An oil painting of Leeuwenhoek. (b) A brass replica of the Leeuwenhoek microscope. Inset photo shows how it is held. (c) Leeuwenhoek's drawings of bacteria from the human mouth.

(a) ©Bettmann/Getty Images; (b) ©Kathy Park Talaro/Pasadena City College; (c) ©Dr. Jeremy Byrgess/SPL/Getty Images

these experiments, the French naturalist Felix Pouchet (1800–1872) claimed in 1859 to have carried out experiments conclusively proving that microbial growth could occur without contact with air.

Pouchet's claim provoked Louis Pasteur (1822–1895) to settle the matter of spontaneous generation. Pasteur (**figure 1.11**) first filtered air through cotton and found that objects resembling plant spores had been trapped. If a piece of the cotton was placed in sterile medium after air had been filtered through it, microbial growth occurred. Next he placed nutrient solutions in flasks, heated their necks in a flame, and pulled them into a variety of curves. The swan-neck flasks that he produced in this way had necks open to the atmosphere (**figure 1.12**). Pasteur then boiled the solutions for a few minutes and allowed them to cool. No growth took place even though the contents of the flasks were exposed to the air (figure 1.12). Pasteur inferred that growth did not occur because dust and germs had been trapped on the walls of the curved necks. If the necks were broken, growth commenced immediately. Pasteur had not only resolved the controversy by 1861 but also had shown how to keep solutions sterile.

Figure 1.11 Louis Pasteur.

©Pixtal/age fotostock

The English physicist John Tyndall (1820–1893) and the German botanist Ferdinand Cohn (1828–1898) dealt a final blow to spontaneous generation. In 1877 Tyndall demonstrated that dust

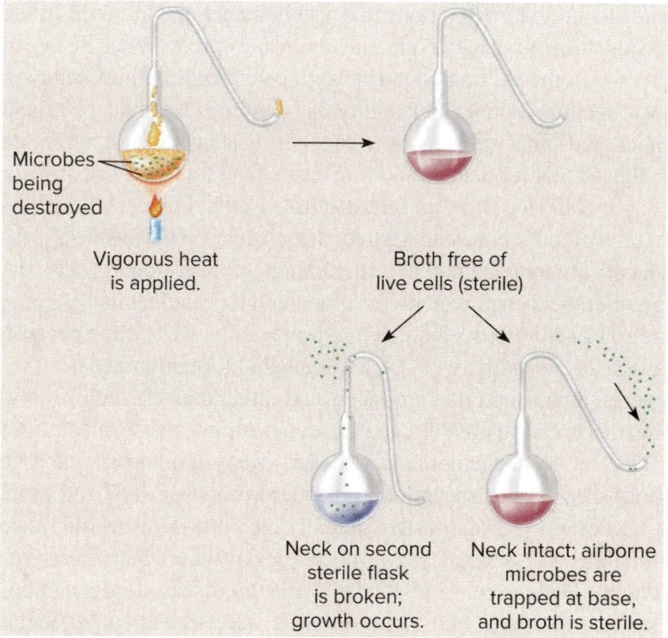

Figure 1.12 Pasteur's Experiments with Swan-Neck Flasks.

did indeed carry germs and that if dust was absent, broth remained sterile even if directly exposed to air. During the course of his studies, Tyndall provided evidence for the existence of exceptionally heat-resistant forms of bacteria. Working independently, Cohn discovered that the heat-resistant bacteria recognized by Tyndall were species capable of producing bacterial endospores. Cohn later played an instrumental role in establishing a classification system for bacteria based on their morphology and physiology.
▶| *Bacterial endospores are a survival strategy (section 3.9)*

These early microbiologists not only disproved spontaneous generation but also contributed to the rebirth of microbiology. They developed liquid media and the methods for sterilizing it so that microbes could be cultured. These techniques were next applied to understanding the role of microorganisms in disease.

Microorganisms and Disease

For hundreds of years, most people believed that disease was caused by supernatural forces, poisonous vapors, and imbalances among the four humors thought to be present in the body. The role of the four humors (blood, phlegm, yellow bile [choler], and black bile [melancholy]) in disease had been widely accepted since the time of the Greek physician Galen (129–199). Support for the idea that microorganisms cause disease—that is, the germ theory of disease—began to accumulate in the early nineteenth century from diverse fields. Agostino Bassi (1773–1856) demonstrated in 1835 that a silkworm disease was due to a fungal infection. In 1845 M. J. Berkeley (1803–1889) proved that the great potato blight of Ireland was caused by a water mold (then thought to be a fungus), and in 1853 Heinrich de Bary (1831–1888) showed that fungi caused cereal crop diseases.

Pasteur also contributed to this area of research in several ways. His contributions began in what may seem an unlikely way. Pasteur was trained as a chemist and spent many years studying the alcoholic fermentations that yield ethanol and are used in the production of wine and other alcoholic beverages. When he began his work, the leading chemists were convinced that fermentation was due to a chemical instability that degraded the sugars in grape juice and other substances to alcohol. Pasteur did not agree; he believed that fermentations were carried out by living organisms.

In 1856 M. Bigo, an industrialist in Lille, France, where Pasteur worked, requested Pasteur's assistance. His business produced ethanol from the fermentation of beet sugars, and the alcohol yields had recently declined and the product had become sour. Pasteur discovered that the fermentation was failing because the yeast normally responsible for alcohol formation had been replaced by bacteria that produced acid rather than ethanol. In solving this practical problem, Pasteur demonstrated that fermentations were due to the activities of specific yeasts and bacteria, and he published several papers on fermentation between 1857 and 1860.

Pasteur was also called upon by the wine industry in France for help. For several years, poor-quality wines had been produced. Pasteur referred to the wines as diseased and demonstrated that particular wine diseases were linked to particular microbes contaminating the wine. He eventually suggested a method for heating the wines to destroy the undesirable microbes. The process is now called pasteurization.

Indirect evidence for the germ theory of disease came from the work of the English surgeon Joseph Lister (1827–1912) on the prevention of wound infections. Lister, impressed with Pasteur's studies on fermentation, developed a system of antiseptic surgery designed to prevent microorganisms from entering wounds. Instruments were heat sterilized, and phenol was used on surgical dressings and at times sprayed over the surgical area. The approach was remarkably successful in reducing the rate of surgical infection. It also provided strong indirect evidence for the role of microorganisms in disease.

Koch's Postulates

The first direct demonstration that bacteria cause disease came from the study of anthrax by the German physician Robert Koch (1843–1910). Koch (**figure 1.13**) used the criteria proposed by his former teacher Jacob Henle (1809–1885) and others to establish the relationship between *Bacillus anthracis* and anthrax. In these studies, published in 1876, Koch used mice as his model or test organism. Following the steps outlined in **figure 1.14**, he next used guinea pigs to show that *Mycobacterium tuberculosis* causes tuberculosis (TB). The importance of this discovery cannot be understated as TB was a leading cause of death in Europe at the time. In 1905 Koch was awarded the Nobel Prize in Physiology or Medicine and his criteria for proving the causal relationship between a microorganism and a specific disease are known as **Koch's postulates.** Koch's postulates have since been used to discover the causative microorganisms for many infectious diseases.

While Koch's postulates are still widely used, their application is at times not feasible. For instance, viruses and organisms that must

Figure 1.13 Robert Koch. Koch examining a specimen in his laboratory

Postulate	Experimentation
1. The microorganism must be present in every case of the disease but absent from healthy organisms.	Koch developed a staining technique to examine human tissue. *Mycobacterium tuberculosis* could be identified in diseased tissue.
2. The suspected microorganisms must be isolated and grown in a pure culture.	Koch grew *M. tuberculosis* in pure culture on coagulated blood serum.
3. The same disease must result when the isolated microorganism is inoculated into a healthy host.	Koch injected cells from the pure culture of *M. tuberculosis* into guinea pigs. The guinea pigs subsequently died of tuberculosis.
4. The same microorganisms must be isolated again from the diseased host.	Koch isolated *M. tuberculosis* in pure culture on coagulated blood serum from the dead guinea pigs.

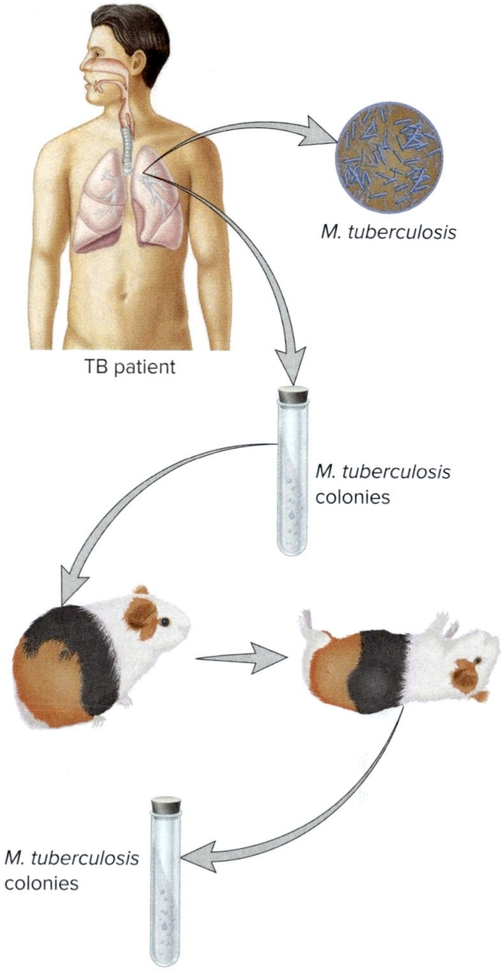

Figure 1.14 **Koch's Postulates Applied to Tuberculosis.**

MICRO INQUIRY *Why is the fourth postulate necessary?*

live within host cells such as *Mycobacterium leprae,* the causative agent of leprosy, cannot be isolated in pure culture. Some human diseases lack an appropriate animal model so the postulates cannot be fully met. To avoid some of these difficulties, microbiologists sometimes use molecular and genetic evidence. For instance, molecular methods might be used to detect the nucleic acid of a microorganism in body tissues, rather than isolating it, or the genes thought to be associated with the virulence of a disease-causing microbe (pathogen) might be mutated. In this case, the mutant organism should have decreased ability to cause disease. Introduction of the normal gene back into the mutant should restore the pathogen's virulence.

Pure Culture Methods

During Koch's studies on bacterial diseases, it became necessary to isolate suspected bacterial pathogens in pure culture. At first Koch cultured bacteria on the sterile surfaces of cut, boiled potatoes, but the bacteria did not always grow well. Eventually he developed culture media using meat extracts and protein digests, reasoning these were similar to body fluids. Initially he tried to solidify the media by adding gelatin. Separate bacterial colonies developed after the surface of the solidified medium had been streaked with a bacterial sample. The sample could also be mixed with liquefied gelatin medium. When the medium hardened, individual bacteria produced separate colonies.

Despite its advantages, gelatin was not an ideal solidifying agent because it can be digested by many microbes and melts at temperatures above 28°C. A better alternative was provided by Fanny Eilshemius Hesse (1850–1934), the wife of Walther Hesse (1846–1911), one of Koch's students and a physician. She often assisted her husband with his experiments and suggested using agar, which she used to make jellies, as a solidifying agent. Agar was not broken down by most bacteria. Furthermore, it did not melt until reaching a temperature of 100°C and, once melted, did not solidify until reaching a temperature of 50°C; this eliminated the need to handle boiling liquid. Some of the media developed by Koch and his associates, such as nutrient broth and nutrient agar, are still widely used. Another important tool developed in Koch's laboratory was a container for holding solidified media—the Petri dish (plate), named after Richard Petri (1852–1921), who devised it. These developments directly stimulated progress in all areas of microbiology. ▶| *Culture media (section 7.7); Enrichment and isolation of pure cultures (section 7.7)*

Our focus thus far has been on the development of methods for culturing bacteria. But viral pathogens were also being studied during this time, and methods for culturing them were also being developed. The discovery of viruses and their role in disease was made possible when Charles Chamberland (1851–1908), one of Pasteur's associates, constructed a porcelain filter to remove bacteria. Dimitri Ivanowski (1864–1920) and Martinus Beijerinck (pronounced "by-a-rink"; 1851–1931) used the filter to study tobacco mosaic disease. They found that plant extracts and sap from diseased plants were infectious, even after being filtered with Chamberland's filter. Because the infectious agent passed through a filter that was designed to trap bacterial cells, they reasoned that the agent must be something smaller than a bacterium. Beijerinck proposed that the agent was a "filterable virus." Eventually viruses were shown to be tiny, acellular infectious agents.

Immunology

The ability to culture microbes also played an important role in early immunological studies. During studies on the bacterium that causes chicken cholera, Pasteur and Pierre Roux (1853–1933) discovered that bacteria incubated in cultures for long periods of time lost their ability to cause disease. These bacteria were said to be attenuated or weakened. When chickens were injected with attenuated bacteria, they remained healthy and were surprisingly able to resist the disease when exposed to virulent bacteria. Pasteur called the attenuated bacteria a vaccine (Latin *vacca,* cow) in honor of Edward Jenner (1749–1823) because, many years earlier, Jenner had used material from cowpox lesions to protect people against smallpox (*see Historical Highlights 36.5*). Shortly after this, Pasteur and Chamberland developed an attenuated anthrax vaccine. ▶| *Vaccines immunize susceptible populations (section 36.6)*

Pasteur also prepared a vaccine using an attenuated strain of rabies virus. During the course of these studies, a nine-year-old boy named Joseph Meister was bitten by a rabid dog and was brought to Pasteur. Since Joseph's death was certain in the absence of treatment, Pasteur agreed to try vaccination. Joseph was injected 13 times over the next 10 days with increasingly virulent preparations of the attenuated virus. His survival marked a huge advance in the use of vaccines. To thank Pasteur, people from around the world contributed to the construction of the Pasteur Institute in Paris, France. One of the initial tasks of the institute was vaccine production.

These early advances in immunology were made without any concrete knowledge about how the immune system works. Immunologists now know that white blood cells and the chemicals they produce play a central role in immunity. Among the chemicals are soluble proteins called antibodies, found in blood, lymph, and other body fluids. The role of soluble substances in preventing disease was recognized by Emil von Behring (1854–1917) and Shibasaburo Kitasato (1852–1931). After the discovery that diphtheria was caused by a toxin produced by bacteria, they injected inactivated diphtheria toxin into rabbits. The inactivated toxin induced rabbits to produce an antitoxin, which protected against the disease. Antitoxins are now known to be antibodies that specifically bind and neutralize toxins. The first immune system cells were discovered when Elie Metchnikoff (1845–1916) found that some white blood cells could engulf disease-causing bacteria. He called these cells phagocytes and the process phagocytosis (Greek *phagein,* eating).

Microbial Ecology

Culture-based techniques were also applied to the study of microbes in soil and aquatic habitats. Early microbial ecologists studied microbial involvement in the carbon, nitrogen, and sulfur cycles. The Russian microbiologist Sergei Winogradsky (1856–1953) made many contributions to soil microbiology, including the discovery that soil bacteria could oxidize iron, sulfur, and ammonia to obtain energy and that many of these bacteria could incorporate CO_2 into organic matter much as photosynthetic organisms do. Winogradsky also isolated anaerobic nitrogen-fixing soil bacteria and studied the decomposition of cellulose. Martinus Beijerinck made fundamental contributions to microbial ecology as well as virology. He isolated several kinds of nitrogen-fixing bacteria and sulfate-reducing bacteria. Beijerinck and Winogradsky also developed enrichment culture techniques and selective media, which have been of great importance in microbiology. ▶| *Biogeochemical cycling sustains life on Earth (section 28.1); Culture media (section 7.7); Enrichment cultures (section 7.7)*

Comprehension Check

1. What does the theory of spontaneous generation propose? What did Pasteur prove when he showed that a cotton plug that had filtered air would trigger microbial growth when transferred to a sterile medium? What argument made previously was he addressing?
2. Discuss the contributions of Lister, Pasteur, and Koch to the germ theory of disease and the treatment or prevention of diseases.
3. Describe Koch's postulates. What is a pure culture? Why are pure cultures important to Koch's postulates?
4. How did Jenner, Pasteur, von Behring, Kitasato, and Metchnikoff contribute to the development of immunology? How was the ability to culture microbes important to their studies?
5. How did Winogradsky and Beijerinck contribute to the study of microbial ecology? What new culturing techniques did they develop in their studies?

1.4 Microbiology Encompasses Many Subdisciplines

After reading this section, you should be able to:

a. Construct a concept map, table, or drawing that illustrates the diverse nature of microbiology and how it has improved human conditions
b. Discuss the belief held by many microbiologists that microbiology is experiencing its second golden age

Microbiology today is as diverse as the organisms it studies. It has both basic and applied aspects. The basic aspects are concerned with the biology of microorganisms themselves. The applied aspects are concerned with practical problems such as disease, water and wastewater treatment, food spoilage and food production, and industrial uses of microbes. Despite this apparent dichotomy, the basic and applied aspects of microbiology are intertwined. Basic research is often conducted in applied fields, and applications often arise out of basic research.

An important recent development in microbiology is the increasing use of molecular and genomic methods to study microbes and their interactions with other organisms. These methods have led to a time of rapid advancement that rivals the golden age of microbiology. Indeed, many feel that microbiology is in its second golden age. ▶| *Microbial DNA technologies (chapter 17); Microbial genomics (chapter 18)*

Major Fields in Microbiology

Microbiology is commonly divided into subdisciplines based on the type of microbe studied. Thus microbiology encompasses bacteriology and virology as well as other microbe-specific fields. Microbiology can also be divided based on the activities of microbes—for instance, environmental microbiology and agricultural microbiology. Finally, microbiologists may study only one aspect of the biology of microbes, leading to subdisciplines such as microbial genetics and microbial physiology.

One of the most active and important fields in microbiology is medical microbiology, which deals with diseases of humans and animals. Medical microbiologists investigate agents causing infectious diseases and measures for their control and elimination. They are involved in tracking down new, unidentified pathogens such as those causing hantavirus pulmonary syndrome, West Nile encephalitis, and Middle East respiratory syndrome. These microbiologists also study how microorganisms cause disease. Clinical laboratory scientists, the microbiologists who work in hospital and other clinical laboratories, use culture and molecular techniques to provide information needed by physicians to diagnose and treat infectious disease.

Major epidemics have regularly affected human history. For example, human immunodeficiency virus/acquired immune deficiency syndrome (HIV/AIDS) has killed over 35 million people since the beginning of the HIV epidemic in 1981. Public health microbiology is concerned with the control and spread of such communicable diseases. Public health microbiologists and epidemiologists monitor the amount of disease in populations. Based on their observations, they can detect outbreaks and epidemics as they begin, and implement appropriate control measures. They also conduct surveillance for new diseases as well as bioterrorism events. Public health microbiologists working for local governments monitor community food establishments and water supplies to ensure they are safe and free from pathogens. ▶| *Epidemiology and public health microbiology (chapter 36)*

To understand, treat, and control infectious disease, it is important to understand how the immune system protects the body from pathogens; this question is the concern of immunology. Immunology is one of the fastest growing areas in science. Many advances have been in response to the discovery of HIV, which specifically targets cells of the immune system. Immunology also deals with the nature and treatment of allergies and autoimmune diseases such as rheumatoid arthritis. ▶| *Innate host resistance (chapter 32); Adaptive immunity (chapter 33)*

Microbial ecology is another important field in microbiology. Microbial ecologists employ a variety of culture and molecular approaches to describe the vast diversity of microbes in terms of their morphology, physiology, and relationships with organisms and the components of their habitats. The importance of microbes in local and global cycling of carbon, nitrogen, and sulfur has been long studied; however, these studies have acquired new urgency as our climate changes. Of particular interest is the role of microbes in both the production and removal of greenhouse gases such as carbon dioxide and methane. Microbial ecologists also are employing microorganisms in bioremediation to reduce pollution. An exciting frontier in microbial ecology is the study of the microbes normally associated with the human body—so-called human microbiota. ▶| *Global climate change (section 28.2); Methods in microbial ecology (chapter 29); The microbe-human ecosystem (chapter 34); Biodegradation and bioremediation harness microbes to clean the environment (section 43.4)*

Agricultural microbiology is a field related to both medical microbiology and microbial ecology. It is concerned with the impact of microorganisms on food production, such as nitrogen-fixing bacteria, which affect soil fertility. Other microbes live in the digestive tracts of ruminants such as cattle and break down the plant materials these animals ingest. There are also plant and animal pathogens that have significant economic impact if not controlled. Furthermore, some pathogens of domestic animals also cause human disease. Agricultural microbiologists work on methods to increase soil fertility and crop yields, study rumen microorganisms to increase meat and milk production, and try to combat plant and animal diseases. Currently many agricultural microbiologists are studying the use of bacterial and viral insect pathogens as substitutes for chemical pesticides. ▶| *Microorganisms in terrestrial ecosystems (chapter 31)*

In addition to agricultural microbiology, food microbiology has contributed to the ready supply of high-quality foods. Food microbiologists study the microbes used to make food and beverages (e.g., yogurt, cheese, beer) as well as the microbes that cause food spoilage or are pathogens that are spread through food. For example, periodic outbreaks of certain *Escherichia coli* strains cause disease outbreaks that in some cases can lead to renal failure and death. To protect the public, the specific strain must be traced back to the contaminated food source. Food microbiologists also work to prevent microbial spoilage of food and conduct research on the use of microorganisms as nutrient sources for livestock and humans. ▶| *Microbiology of food (chapter 41)*

Industrial microbiology involves the use of microbes to make products helpful to humans. An important advance occurred in 1929 when Alexander Fleming rediscovered that the fungus *Penicillium* sp. produced what he called penicillin, the first antibiotic that could successfully control bacterial infections. Although it took World War II for scientists to learn how to mass-produce penicillin, scientists soon found other microorganisms capable of producing additional antibiotics. Today industrial microbiologists also use microorganisms to make products such as vaccines, steroids, alcohols and other solvents, vitamins, amino acids, enzymes, and biofuels. These alternative fuels are renewable and may help decrease pollution associated with burning fossil fuels. ▶| *Biofuel production is a dynamic field (section 42.2); Microbial fuel cells: batteries powered by microbes (section 43.3)*

The advances in medical microbiology, agricultural microbiology, food and dairy microbiology, and industrial microbiology are outgrowths of the labor of many microbiologists doing basic research in areas such as microbial physiology, microbial genetics, molecular biology, and bioinformatics. Microbial physiologists study many aspects of the biology of microorganisms, including their diverse metabolic capabilities. They also study the synthesis of antibiotics and toxins, the ways in which microorganisms survive harsh environmental conditions, and the effects of chemical and physical agents on microbial growth and survival. Microbial geneticists, molecular biologists, and bioinformaticists study the nature of genetic information and how it regulates the development and function of cells and organisms. The bacteria *E. coli* and *Bacillus subtilis,* the yeast *Saccharomyces cerevisiae* (baker's yeast), and bacterial viruses such as T4 and lambda continue to be important model organisms used to understand biological phenomena.

Clearly, the future of microbiology is bright. Genomics in particular has revolutionized microbiology, as scientists are now beginning to understand organisms holistically, rather than in a reductionist, piecemeal manner. How the genomes of microbes evolve, the nature of host-pathogen interactions, the minimum set of genes required for an organism to survive, and many more topics are aggressively being examined by molecular and genomic analyses. This is an exciting time to be a microbiologist. Enjoy the journey.

Comprehension Check

1. Briefly describe the major subdisciplines in microbiology. Which do you consider to be applied fields? Which are basic?
2. Log all the microbial products you use in a week. Be sure to consider all foods and medications (including vitamins). Consider how your life might be different without these items.
3. List all the activities or businesses you can think of in your community that directly depend on microbiology.

Key Concepts

1.1 Members of the Microbial World

- Microbiology studies microscopic cellular organisms that are often unicellular or, if multicellular, do not have highly differentiated tissues. Microbiology also focuses on biological entities that are acellular (**figure 1.1**).
- Microbiologists divide cellular organisms into three domains: *Bacteria, Archaea,* and *Eukarya* (**figure 1.2**).
- Domains *Bacteria* and *Archaea* have a prokaryotic cellular architecture. Eukaryotic microbes (protists and fungi) are placed in *Eukarya.* Viruses, viroids, satellites, and prions are acellular entities not placed in any domain but are classified by a separate system.

1.2 Microbes Have Evolved and Diversified for Billions of Years

- Earth is approximately 4.5 billion years old. Within the first 1 billion years of its existence, life arose (**figure 1.4**).
- The RNA world hypothesis posits that the earliest self-replicating entity on the planet had RNA enclosed in a lipid bilayer. RNA stores genetic information and conducts cellular processes (**figure 1.6**).
- Comparisons of small subunit (SSU) rRNA genes have been useful in creating universal phylogenetic trees. These trees provide information about the origin and evolution of life (**figure 1.8**).
- The last universal common ancestor (LUCA) is placed on the bacterial branch of the universal phylogenetic tree (figure 1.2). Thus, *Bacteria* are thought to have diverged first, and *Archaea* and *Eukarya* arose later.
- Mitochondria, chloroplasts, and hydrogenosomes are thought to have evolved from bacterial endosymbionts of ancestral cells in the eukaryotic lineage.
- The concept of a bacterial or archaeal species is difficult to define because these microbes do not reproduce sexually. Species are named using the binomial system of Linnaeus.

1.3 Microbiology Advanced as New Tools for Studying Microbes Were Developed

- Microbiology is defined not only by the organisms it studies but also by the tools it uses. Microscopy and culture-based techniques play important roles in the evolution of the discipline.

- Antony van Leeuwenhoek used simple microscopes and was the first person to extensively describe microorganisms (**figure 1.10**).
- Culture-based techniques for studying microbes began to develop as scientists debated the theory of spontaneous generation. Experiments by Redi and others disproved the theory of spontaneous generation of larger organisms. The spontaneous generation of microorganisms was disproved by Pasteur and others (**figure 1.12**).
- The availability of culture-based techniques played an important role in the study of microbes as the causative agents of disease. Support for the germ theory of disease came from the work of Pasteur, Koch, Lister, and others.
- Koch's postulates are used to prove a direct relationship between a suspected pathogen and a disease. Koch and his coworkers developed techniques to grow bacteria on solid media and isolate pure cultures of pathogens (**figure 1.14**).
- Viruses were discovered following the invention of a bacterial filter by Chamberland. Dimitri Ivanowski and Martinus Beijerinck were important contributors to the field of virology.
- Microbiologists founded the field of immunology; they created vaccines and discovered antibodies and phagocytic cells. Pasteur, Mechnikof, and others made important contributions.
- Microbial ecology grew out of the work of Winogradsky and Beijerinck. They studied the role of microorganisms in carbon, nitrogen, and sulfur cycles and developed enrichment culture techniques and selective media.

1.4 Microbiology Encompasses Many Subdisciplines

- There are many fields in microbiology. These include medical, public health, industrial, and food and dairy microbiology. Microbial ecology, physiology, and genetics are important subdisciplines of microbiology.

Active Learning

1. Why aren't viruses, viroids, satellites, and prions included in the three-domain system?
2. Would microbiology have developed more slowly if Fanny Hesse had not suggested the use of agar? Give your reasoning.
3. Some individuals can be infected by a pathogen yet not develop disease. In fact, some become chronic carriers of the pathogen. How does this observation affect Koch's postulates? How might the postulates be modified to account for the existence of chronic carriers?
4. Support this statement: "Vaccinations against various childhood diseases have contributed to the entry of women, particularly mothers, into the full-time workplace."
5. For many years, the oldest stromatolites with compelling evidence of ancient life on Earth were found in Western Australia. The microbial communities that built these sedimentary structures were dated at roughly 3.5 billion years old. However, in 2016 stromatolites in Greenland were reported to be 3.7 billion years old. Interestingly, the complexity and morphology of the Greenland stromatolites is similar to that of those in Australia. Why is it of interest to document the oldest evidence of life on Earth? What might this suggest about the evolution of life during the intervening 200 million years? What is the significance, if any, of the geographic distribution of these stromatolites?

   Read the original paper: Nutman, A.P., et al. 2016. Rapid emergence of life shown by discovery of 3,700-million-year-old microbial structures. *Nature*. 537:535–539. doi:10.1038/nature19355.

2

Microscopy

©Kenneth Lambert/AP Images

Anthrax Bioterrorism Attack

While on a trip to North Carolina in late September 2001, a sixty-three-year-old photojournalist and outdoor enthusiast from Florida began feeling ill. His muscles ached, he felt nauseous, and he had a fever. His symptoms were annoying and varied in severity during his trip. After returning home, his symptoms worsened considerably. On October 2, he awoke with vomiting. Even more alarming was his mental confusion. When his wife took him to a local emergency room, doctors found that he was unable to answer simple questions not only about his illness but also about where he was and what time it was. His doctors performed a spinal tap and collected cerebral spinal fluid (CSF). The fluid was sent to the hospital laboratory where it was stained using the Gram-staining technique. Much to the surprise of the clinical lab scientists, the CSF contained long, Gram-positive rods that formed chains, a morphology unlike that of typical meningitis-causing bacteria. Based on the Gram-stain results, doctors made an initial diagnosis of inhalation anthrax. What made this diagnosis so surprising was that inhalation anthrax is extremely rare in the United States. Indeed, the last diagnosed case was in 1976. Doctors immediately began treatment with antibiotics, but the photojournalist's condition worsened, and he died October 5.

Thus began the first anthrax bioterrorism attack in the United States. Over the next several weeks, 16 other people developed inhalation anthrax; four of these individuals died. Much was learned about the disease from this attack. Much was also learned about the nation's readiness to deal with bioterrorism. This event is also a reminder of the continuing importance of microscopy in microbiology and in diagnosing disease.

In this chapter, we introduce some of the most commonly used types of microscopy. We begin with light microscopes and then describe other common types of microscopes, as well as how specimens are prepared for examination by microscopes.

Readiness Check:

Based on what you have learned previously, you should be able to:

- ✓ Explain why microscopy is a major tool used by microbiologists (sections 1.1 and 1.3)
- ✓ Define magnification
- ✓ Express the size of organisms using the metric system (**table 2.1**)

Table 2.1 Common Units of Measurement

Unit	Abbreviation	Value
1 centimeter	cm	10^{-2} meter or 0.394 inches
1 millimeter	mm	10^{-3} meter
1 micrometer	μm	10^{-6} meter
1 nanometer	nm	10^{-9} meter
1 Angstrom	Å	10^{-10} meter

2.1 Lenses Create Images by Bending Light

After reading this section, you should be able to:

a. Relate the refractive indices of glass and air to the path light takes when it passes through a prism or convex lens
b. Correlate lens strength and focal length

Light microscopes were the first microscopes invented and they continue to be the most commonly used type. To understand light microscopy, we must consider the way lenses bend and focus light to form images. When a ray of light passes from one medium to another, refraction occurs; that is, the ray is bent at the interface. The **refractive index** is a measure of how much a substance slows the velocity of light; the direction and magnitude of bending are determined by the refractive indices of the two media forming the interface. For example, when light passes from air into glass, which has a greater refractive index, it is slowed and bent toward the normal, a line perpendicular to the surface (**figure 2.1**). As light leaves glass and returns to air, a medium with a lower refractive index, it accelerates and is bent away from the normal. Thus a prism bends light because glass has a different refractive index from air and the light strikes its surface at an angle.

Lenses act like a collection of prisms operating as a unit. When the light source is distant so parallel rays of light

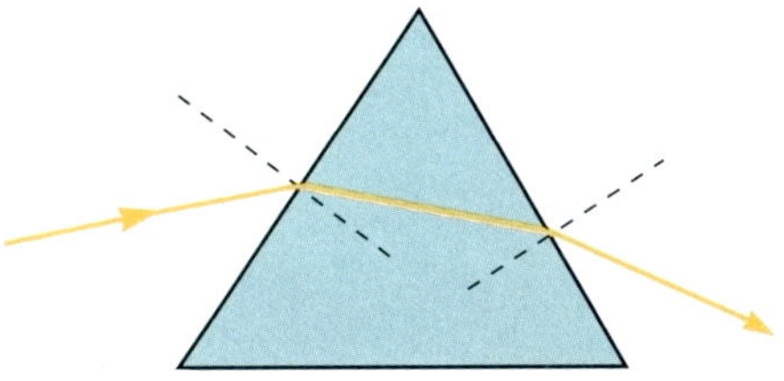

Figure 2.1 The Bending of Light by a Prism. Lines perpendicular to the surface of the prism are called normal; they are indicated by dashed lines. As light enters the glass, it is bent toward the first normal. When light leaves the glass and returns to air, it is bent away from the second normal.

strike the lens, a convex lens focuses the rays at a specific point, the focal point (*F* in **figure 2.2**). The distance between the center of the lens and the focal point is called the focal length (*f* in figure 2.2).

Our eyes cannot focus on objects nearer than about 25 cm (i.e., about 10 inches). This limitation may be overcome by using a convex lens as a simple magnifier (or microscope) and holding it close to an object. A magnifying glass provides a clear image at much closer range, and the object appears larger. Lens strength is related to focal length; a lens with a short focal length magnifies an object more than a lens having a longer focal length.

Comprehension Check

1. How do refraction, refractive index, focal point, and focal length differ?
2. Draw the path of a light ray through a thin and thick lens.
3. If the lenses in the above question were in corrective eyeglasses, which would be used for reading and which would improve distance vision?

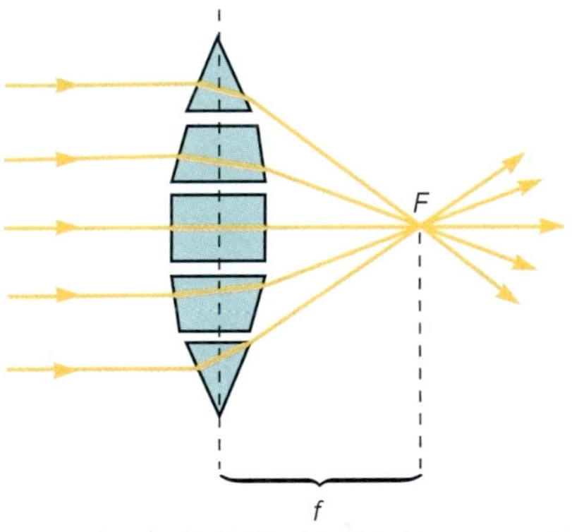

Figure 2.2 Lens Function. A lens functions somewhat like a collection of prisms. Light rays from a distant source are focused at the focal point *F*. The focal point lies a distance *f*, the focal length, from the lens center.

MICRO INQUIRY *How would the focal length change if the lens shown here were thicker?*

2.2 There Are Several Types of Light Microscopes

After reading this section, you should be able to:

a. Evaluate the parts of a light microscope in terms of their contributions to image production and use of the microscope
b. Predict the relative degree of resolution based on light wavelength and numerical aperture of the lens used to examine a specimen
c. Create a table that compares and contrasts the various types of light microscopes in terms of their uses, how images are created, and the quality of images produced

Microbiologists currently employ a variety of light microscopes in their work, each designed for specific applications. Modern microscopes are compound microscopes; that is, they have two sets of lenses. The **objective lens** is the lens closest to the specimen. It forms a magnified image that is further enlarged by one or more additional lenses.

Bright-Field Microscope: Dark Object, Bright Background

The **bright-field microscope** is routinely used to examine both stained and unstained specimens. It forms a dark image against a lighter background, thus it has a "bright field." It consists of a metal stand composed of a base and an arm to which the remaining parts are attached (**figure 2.3**). A light source, either a mirror or an electric illuminator, is located in the base. Two focusing knobs, the fine and coarse adjustment knobs, are located on the arm and move either the stage or the nosepiece vertically to focus the image.

The stage is positioned about halfway up the arm. Microscope slides are clipped to the stage, which can be moved during viewing by rotating control knobs. The substage **condenser** lens (or simply, condenser) is within or beneath the stage and focuses a cone of light on the slide. Its position may be fixed in simpler microscopes but can be adjusted vertically in more advanced models.

The curved upper part of the arm holds the body assembly, to which a nosepiece and one or more **ocular lenses** (also called **eyepieces**) are attached. Binocular microscopes have eyepieces for both eyes. The body assembly contains a series of mirrors and prisms so the barrel holding the eyepiece may be tilted for ease in viewing. The nosepiece holds three to five objective lenses of differing magnifying power and can be rotated to change magnification. Ideally a microscope should be **parfocal;** that is, the image should remain in focus when objective lenses are changed.

The image seen when viewing a specimen with a compound microscope is created by the objective and ocular lenses working together. Light from the illuminated specimen is focused by the objective lens, creating an enlarged image within the

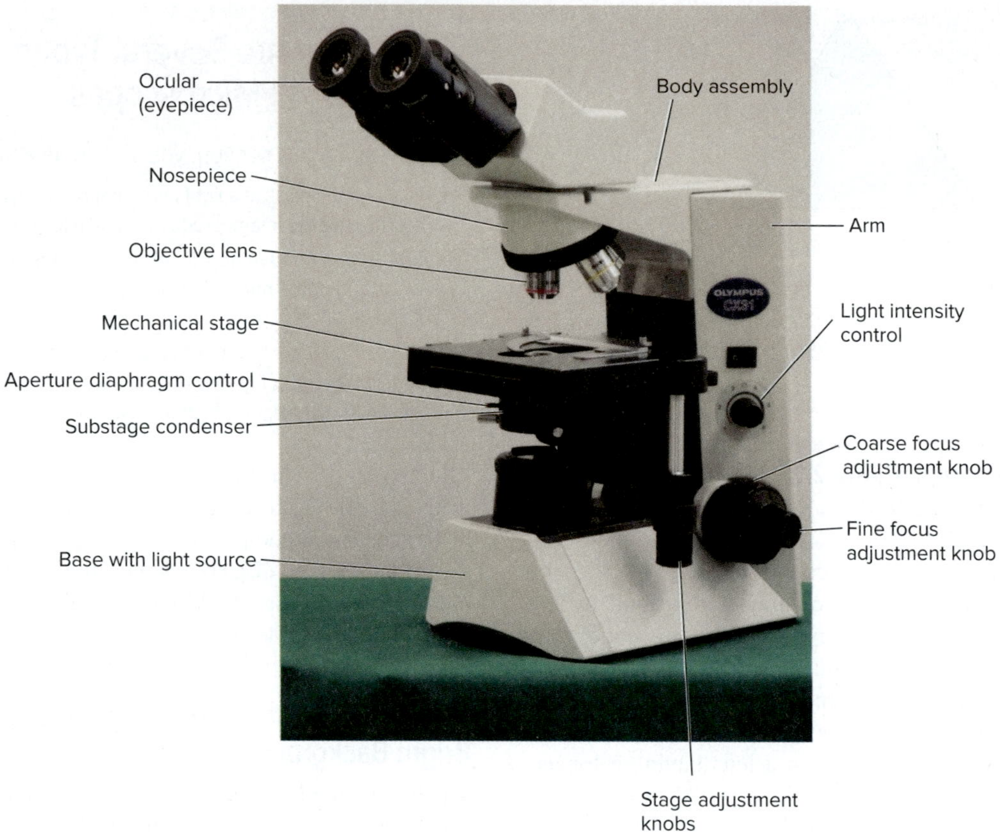

Figure 2.3 **A Bright-Field Microscope.**
©McGraw-Hill Education/James Redfearn, photographer

microscope. The ocular lens further magnifies this primary image. The total magnification is calculated by multiplying the objective and eyepiece magnifications together. For example, if a 45× objective lens is used with a 10× eyepiece, the overall magnification of the specimen is 450×.

Better Microscope Resolution Means a Clearer Image

The most important part of the microscope is the objective lens, which must produce a clear image, not just a magnified one. **Resolution** is the ability of a lens to separate or distinguish between small objects that are close together. At best, the resolution of a bright-field microscope is 0.2 μm, which is about the size of a very small bacterium. Why is this the case? Resolution is in part dependent on the **numerical aperture** (n sin θ) of a lens. Numerical aperture is defined by two components: n is the refractive index of the medium in which the lens works (e.g., air = 1) and θ is 1/2 the angle of the cone of light entering an objective (**figure 2.4**). A cone with a narrow angle does not adequately separate the rays of light emanating from closely packed objects, and the images are not resolved. A cone of light with a very wide angle is able to separate the rays, and the closely packed objects appear widely separated and resolved. Recall that the refractive indices of all materials through which light waves pass determine the direction of the light rays emanating from the specimen. Some objective lenses work in air; since sin θ cannot be greater than 1 (the maximum θ is 90° and sin 90° is 1.00), no lens working in air can have a numerical aperture greater than 1.00. The only practical way to raise the numerical aperture above 1.00, and therefore achieve higher resolution, is

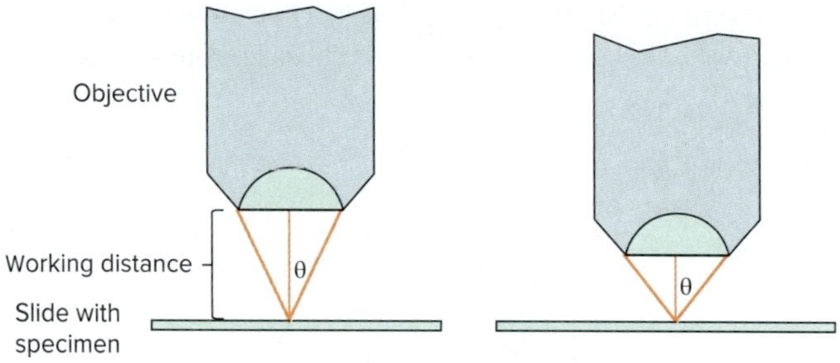

Figure 2.4 **Numerical Aperture in Microscopy.** The numerical aperture of a lens is related to a value called the angular aperture (symbolized by θ), which is 1/2 the angle of the cone of light that enters a lens from a specimen. The equation for numerical aperture is n sin θ. In the right-hand illustration, the lens has larger angular and numerical apertures; its resolution is greater and its working distance smaller.

Table 2.2 Properties of Objective Lenses

	OBJECTIVE			
Property	**Scanning**	**Low Power**	**High Power**	**Oil Immersion**
Magnification	4×	10×	40–45×	90–100×
Numerical aperture	0.10	0.25	0.55–0.65	1.25–1.4
Approximate focal length (*f*)	40 mm	16 mm	4 mm	1.8–2.0 mm
Working distance	17–20 mm	4–8 mm	0.5–0.7 mm	0.1 mm
Approximate resolving power with light of 450 nm (blue light)	2.3 µm	0.9 µm	0.35 µm	0.18 µm

to increase the refractive index with immersion oil, a colorless liquid with the same refractive index as glass (**table 2.2**). If air is replaced with immersion oil, many light rays that would otherwise not enter the objective due to reflection and refraction at the surfaces of the objective lens and slide will now do so (**figure 2.5**). This results in an increase in numerical aperture and resolution.

Resolution is described mathematically by an equation developed in the 1870s by Ernst Abbé (1840–1905), a German physicist responsible for much of the optical theory underlying microscope design. The Abbé equation states that the minimal distance (*d*) between two objects that reveals them as separate entities depends on the wavelength of light (λ) used to illuminate the specimen and on the numerical aperture of the lens ($n \sin \theta$), which is the ability of the lens to gather light.

$$d = \frac{0.5\,\lambda}{n \sin \theta}$$

The smaller *d* is, the better the resolution, and finer detail can be discerned in a specimen; *d* becomes smaller as the wavelength of light used decreases and as the numerical aperture increases. Thus the greatest resolution is obtained using a lens with the largest possible numerical aperture and light of the shortest wavelength, at the blue end of the visible spectrum (in the range of 450 to 500 nm).

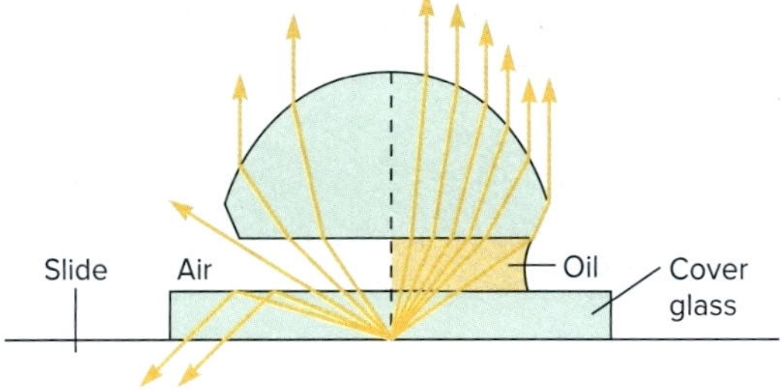

Figure 2.5 The Oil Immersion Objective. An oil immersion objective operating in air and with immersion oil.

Examination of figure 2.4 shows that one objective is much closer to the specimen than the other. Specifically, the objective on the right has a higher numerical aperture and a shorter working distance. The working distance of an objective is the distance between the surface of the lens and the surface of the cover glass (if one is used) or the specimen when it is in sharp focus. As illustrated here, objectives with large numerical apertures and great resolving power have short working distances (table 2.2).

Numerical aperture is also an important feature of a microscope's condenser. The condenser is a large, light-gathering lens used to project a wide cone of light through the slide and into the objective lens. Although most condensers have a numerical aperture between 1.2 and 1.4, the numerical aperture will not exceed about 0.9 unless the top of the condenser is oiled to the bottom of the slide. During routine microscope operation, the condenser usually is not oiled, and this limits the overall resolution, even with an oil immersion objective.

The most accurate calculation of a microscope's resolving power considers both the numerical aperture of the objective lens and that of the condenser, as is evident from the following equation, where NA is the numerical aperture.

$$d_{\text{microscope}} = \frac{\lambda}{(\text{NA}_{\text{objective}} + \text{NA}_{\text{condenser}})}$$

However, in most cases the limit of resolution of a light microscope is calculated using the Abbé equation, which considers the objective lens only. We now see why the maximum theoretical resolving power of a microscope when viewing a specimen using an oil immersion objective (numerical aperture of 1.25) and blue-green light is approximately 0.2 µm.

$$d = \frac{(0.5)(530 \text{ nm})}{1.25} = 212 \text{ nm or } 0.2\ \mu\text{m}$$

Given the limit of resolution of a light microscope, the largest useful magnification—the level of magnification needed to increase the size of the smallest resolvable object to be visible with the light microscope—can be determined. Our eye can just detect a speck 0.2 mm in diameter. When the acuity of the eye and the resolution of the microscope are considered together, it is calculated that the useful limit of magnification is about 1,000 times the numerical aperture of the objective lens. Most standard microscopes have 10× eyepieces and have an upper limit of about 1,000× with oil immersion. A 15× eyepiece may be used with good objective lenses to achieve a useful magnification of 1,500×. Any further magnification does not enable a person to see more detail. Indeed, a light microscope can be built to yield a final magnification of 10,000×, but it would simply be magnifying a blur. Only the electron microscope provides sufficient resolution to make higher magnifications useful.

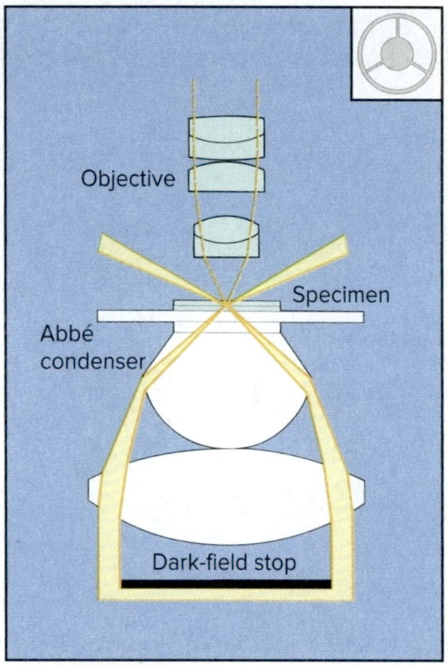

Figure 2.6 Dark-Field Microscopy. In dark-field microscopy, a dark-field stop (inset) is placed underneath the condenser lens system. The condenser then produces a hollow cone of light so that the only light entering the objective is reflected or refracted by the specimen.

Visualizing Living, Unstained Microbes

Bright-field microscopes are probably the most common microscope found in teaching, research, and clinical laboratories. However, many microbes are unpigmented so are not clearly visible because there is little difference in contrast between the cells, subcellular structures, and water. As we discuss in section 2.3, one solution to this problem is to stain cells before observation. Unfortunately staining procedures usually kill cells. But what if an investigator wishes to view living cells? Three types of light microscopes create detailed, clear images of living specimens: dark-field microscopes, phase-contrast microscopes, and differential interference contrast microscopes.

Dark-Field Microscope: Bright Object, Dark Background

The **dark-field microscope** produces detailed images of living, unstained cells and organisms by simply changing the way in which they are illuminated. A hollow cone of light is focused on the specimen in such a way that unreflected and unrefracted rays do not enter the objective. Only light that has been reflected or refracted by the specimen forms an image (**figure 2.6**). The field surrounding a specimen appears black, while the object itself is brightly illuminated (**figure 2.7**). The dark-field microscope can reveal considerable internal structure in larger eukaryotic microorganisms (figure 2.7*b*). It also can be used to identify certain bacteria such as the thin and distinctively shaped *Treponema pallidum*, the causative agent of syphilis (figure 2.7*a*).

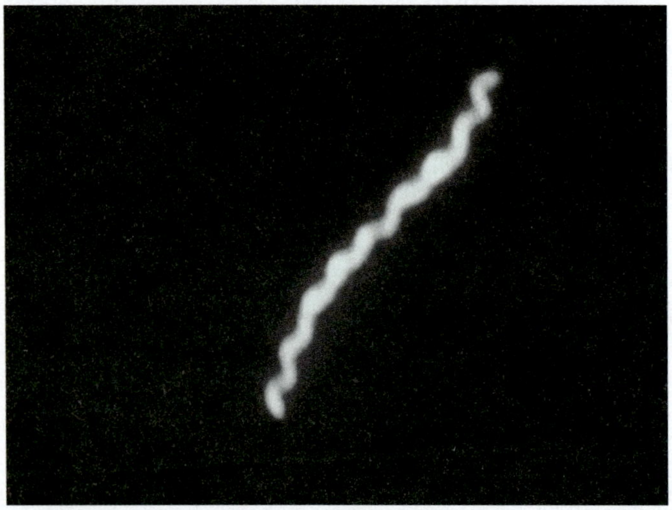

(a) *T. pallidum*

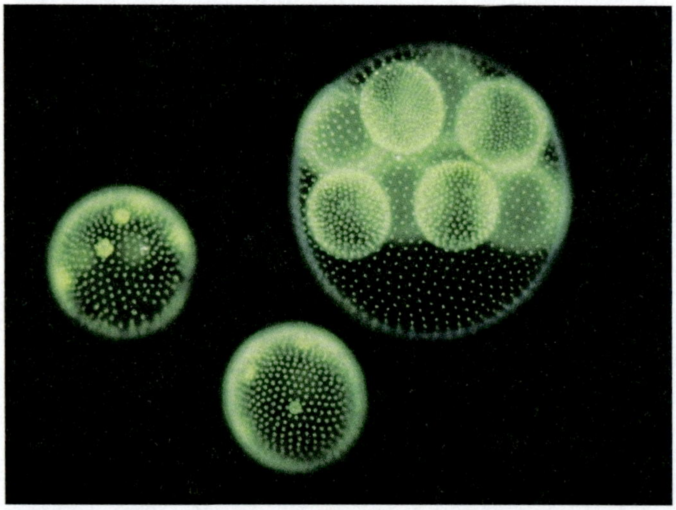

(b) *Volvox*

Figure 2.7 Examples of Dark-Field Microscopy. (a) *Treponema pallidum*, the spirochete that causes syphilis (×400). (b) The protist *Volvox*. Note daughter colonies within the mature *Volvox* colony.

(a) Source: CDC/Schwartz; (b) ©McGraw-Hill Education/Stephen Durr, photographer

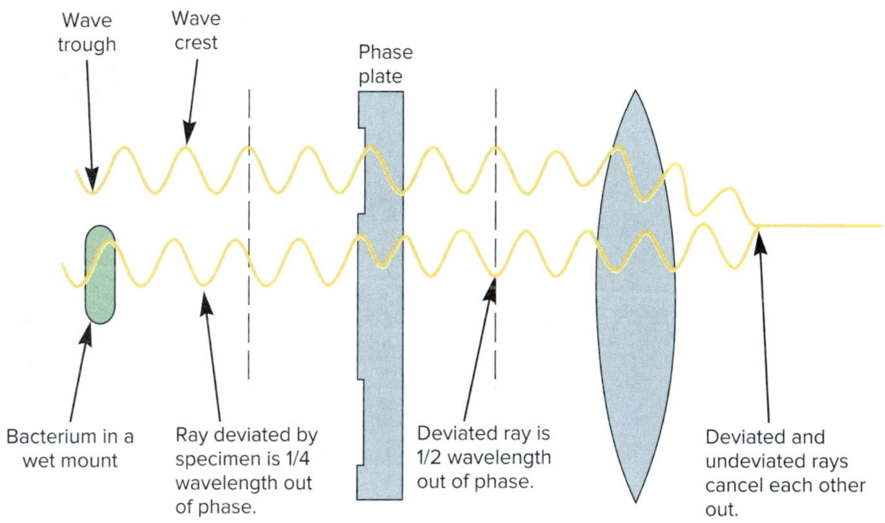

Figure 2.8 The Production of Contrast in Phase-Contrast Microscopy. The behavior of deviated and undeviated (i.e., undiffracted) light rays in the positive-phase-contrast microscope. Because the light rays tend to cancel each other out, the image of the specimen will be dark against a brighter background.

Phase-Contrast Microscope

To understand phase-contrast microscopy, consider a bacterium in a drop of water on a microscope slide (i.e., a wet mount) as illustrated in **figure 2.8**. The refractive indices of bacterial cell structures are greater than that of water. Therefore, light waves passing through a cell structure will be diffracted and slowed more than light waves passing through the water inside and outside the cell. Thus both deviated light waves that interact with bacterial cell structures and undeviated light waves that pass around and through the cell are produced. Because the deviated light waves are slowed relative to the undeviated light waves, they are said to be out of phase. That is, the crests and troughs of the deviated and undeviated waves do not align. Typically the deviated light waves are slowed by about ¼ wavelength compared to the undeviated light (figure 2.8).

Phase-contrast microscopes take advantage of this phenomenon to create differences in light intensity that provide contrast to allow the viewer to see a clearer, more detailed image of the specimen (**figure 2.9**). They do so by separating the two types of light so that the undeviated light (primarily from the surroundings) can be manipulated and then recombined with the deviated light (from the bacterium) to form an image. Two components allow this to occur: a condenser annulus and a phase plate (**figure 2.10**). The condenser annulus is an opaque disk with a thin transparent ring. A ring of light is directed by the condenser annulus to the condenser, which focuses the light on the specimen as shown in figure 2.10. Deviated and undeviated light then pass through the objective toward the phase plate. The phase plate has a thin ring through which the undeviated light (i.e., from the surroundings) is focused (figure 2.8). In a common type of phase-contrast microscopy (positive phase contrast), the ring is coated with a substance that advances the phase of the undeviated light by ¼ wavelength. The deviated light is focused on the rest of the phase plate, which lets the deviated light pass through unchanged.

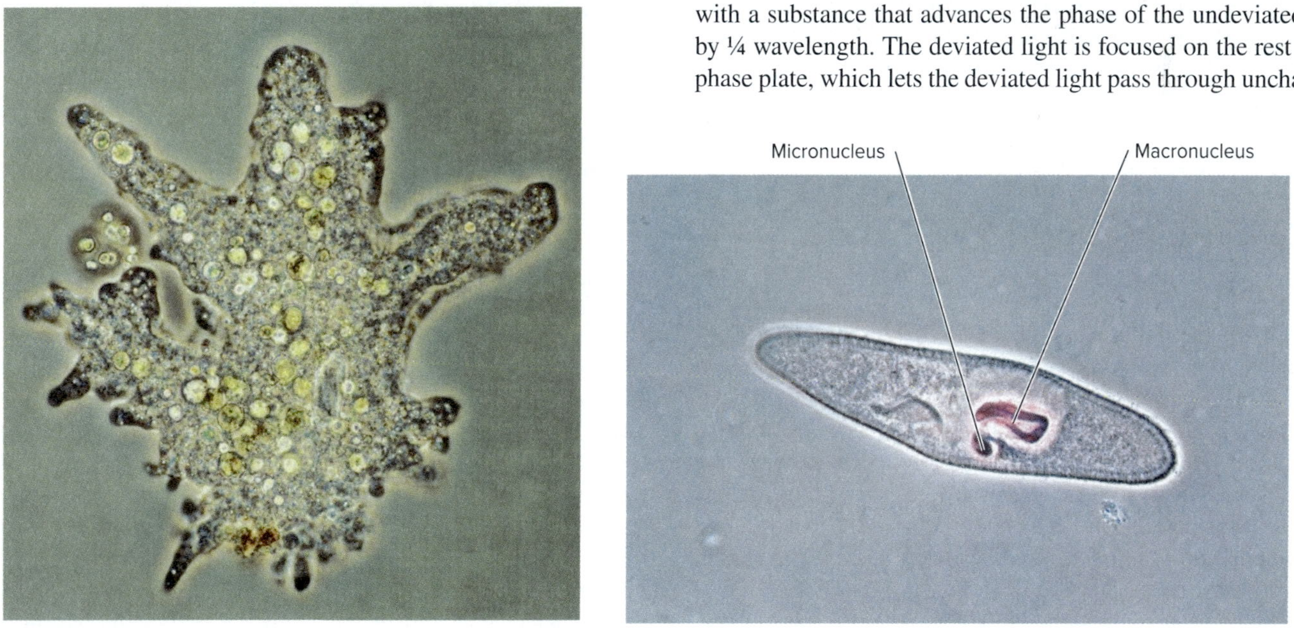

(a) An amoeba **(b)** *Paramecium* sp.

Figure 2.9 Examples of Phase-Contrast Microscopy. (a) An amoeba, a eukaryotic microbe that moves by means of pseudopodia, which extend out from the main part of the cell body. (b) *Paramecium* sp. stained to show a large central macronucleus with a small spherical micronucleus at its side (×400).

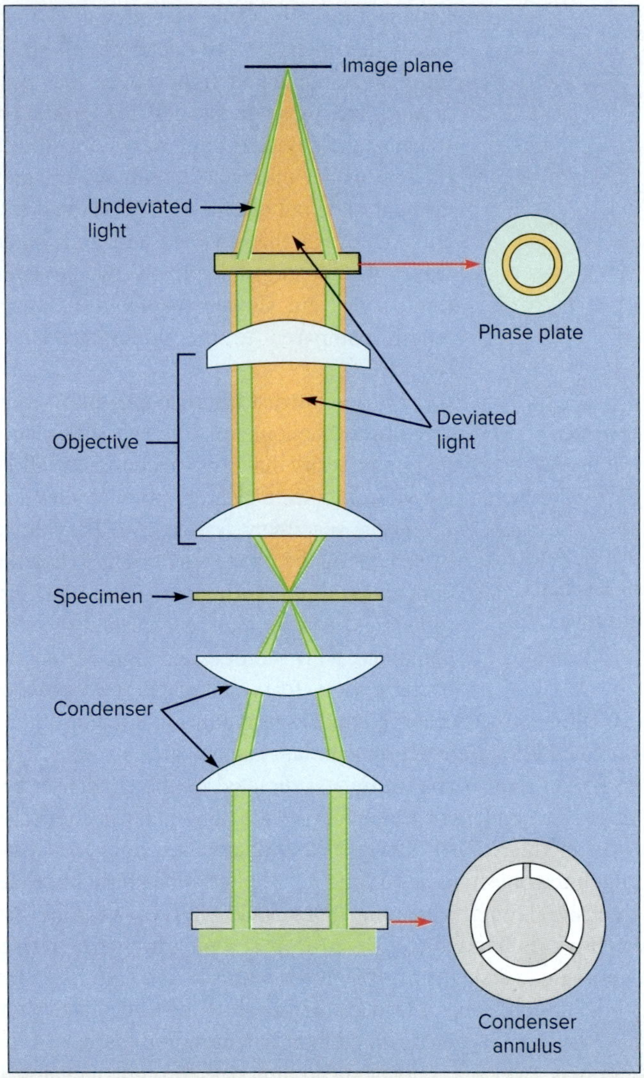

Figure 2.10 Phase-Contrast Microscopy. The optics of a positive-phase-contrast microscope.

MICRO INQUIRY *What is the purpose of the annular stop in a phase-contrast microscope? Is it found in any other kinds of light microscopes?*

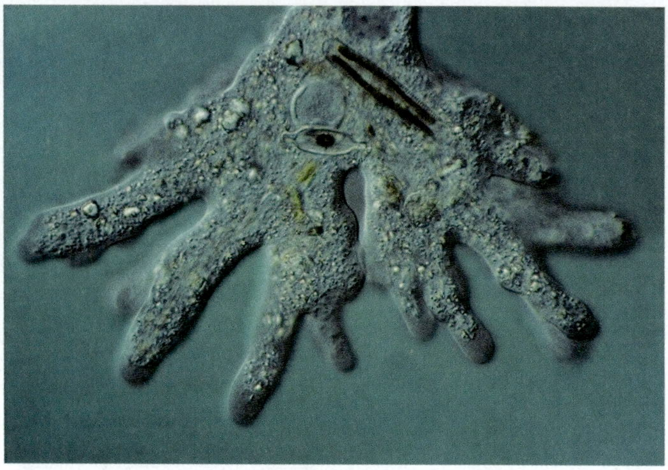

Figure 2.11 Differential Interference Contrast Microscopy. This image of the protozoan *Amoeba proteus* appears three-dimensional and contains considerable detail.

After leaving the phase plate, the deviated and undeviated light are now out of phase by ½ wavelength (figure 2.8). When the two rays of light recombine to form an image, they cancel each other out, a phenomenon called destructive interference. Destructive interference is also seen if the crest of a wave of water meets the trough of another wave—the two cancel each other out and the surface of the water remains calm at the point where they meet. In our example, the resulting image consists of a darker bacterium against a lighter background.

Phase-contrast microscopy is especially useful for studying microbial motility, determining the shape of living cells, and detecting bacterial structures such as endospores and inclusions. These are clearly visible because they have refractive indices markedly different from that of water. Phase-contrast microscopes also are widely used to study eukaryotic cells. ▶| *Bacterial endospores are a survival strategy (section 3.9); Inclusions (section 3.6)*

Differential Interference Contrast Microscope

The **differential interference contrast (DIC) microscope** is similar to the phase-contrast microscope in that it creates an image by detecting differences in refractive indices and thickness. Two beams of plane-polarized light at right angles to each other are generated by prisms. In one design, the object beam passes through the specimen, while the reference beam passes through a clear area of the slide. After passing through the specimen, the two beams combine and interfere with each other to form an image. A live, unstained specimen appears brightly colored and seems to pop out from the background, giving the viewer the sense that a three-dimensional image is being viewed (**figure 2.11**). Structures such as cell walls, endospores, granules, vacuoles, and nuclei are clearly visible.

Fluorescence Microscopes Use Emitted Light to Create Images

The light microscopes thus far considered produce an image from light that passes through a specimen. An object also can be seen because it emits light. This is the basis of fluorescence microscopy. When some molecules absorb radiant energy, they become excited and release much of their trapped energy as light. Any light emitted by an excited molecule has a longer wavelength (i.e., has lower energy) than the radiation originally absorbed. **Fluorescent light** is emitted very quickly by the excited molecule as it gives up its trapped energy and returns to a more stable state.

The **fluorescence microscope** excites a specimen with a specific wavelength of light that triggers the emission of fluorescent light by the object, which forms the image. The most

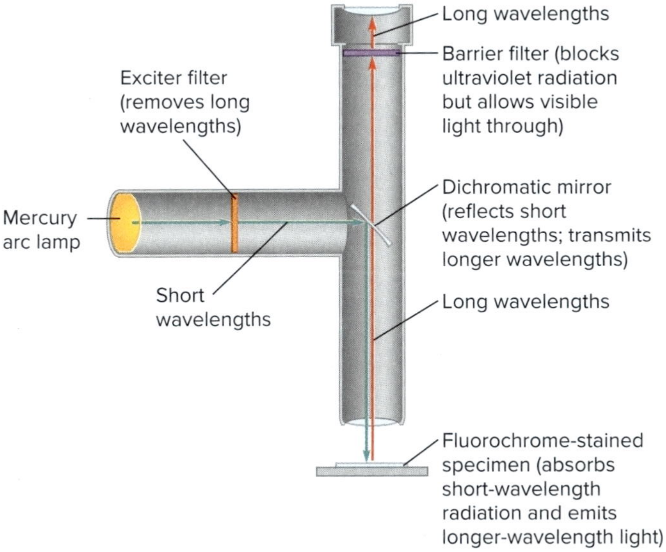

Figure 2.12 Epifluorescence Microscopy. The principles of operation of an epifluorescence microscope.

commonly used fluorescence microscopy is epifluorescence microscopy, also called incident light or reflected light fluorescence microscopy. Epifluorescence (Greek *epi*, upon) microscopes illuminate specimens from above. The objective lens also acts as a condenser (**figure 2.12**). A mercury vapor arc lamp or other source produces an intense beam of light that passes through an exciter filter. The exciter filter transmits only the desired wavelength of light. The excitation light is directed down the microscope by the dichromatic mirror. This mirror reflects light of shorter wavelengths (i.e., the excitation light) but allows light of longer wavelengths to pass through. The excitation light continues down, passing through the objective lens to the specimen, which is usually stained with molecules called **fluorochromes** (**table 2.3**). The fluorochrome absorbs light energy from the excitation light and emits fluorescent light that travels up through the objective lens into the microscope. Because the emitted fluorescent light has a longer wavelength, it passes through the dichromatic mirror to a barrier filter, which blocks out any residual excitation light. Finally, the emitted light passes through the barrier filter to the eyepieces.

The fluorescence microscope has become an essential tool in microbiology. Bacterial pathogens can be identified after staining with fluorochromes or specifically tagging them with fluorescently labeled antibodies using immunofluorescence procedures. In ecological studies, fluorescence microscopy is used to observe microorganisms stained with fluorochrome-labeled probes or fluorochromes that bind specific cell constituents (table 2.3). In addition, microbial ecologists use epifluorescence microscopy to visualize photosynthetic microbes, as their pigments naturally fluoresce when excited by light of specific wavelengths. It is even possible to distinguish live bacteria from dead bacteria by the color they fluoresce after treatment with a specific mixture of stains (**figure 2.13*a***). ▶| *Genetic methods are used to assess microbial diversity (section 29.2); Identification of microorganisms from specimens (section 37.2)*

Another important use of fluorescence microscopy is the localization of specific proteins within cells. One approach is to use genetic engineering techniques that fuse the gene for the protein of interest to a gene isolated from jellyfish belonging

Table 2.3 Commonly Used Fluorochromes

Fluorochrome	Uses
Acridine orange	Stains DNA
Diamidino-2-phenyl indole (DAPI)	Stains DNA
Fluorescein isothiocyanate (FITC)	Often attached to DNA probes or to antibodies that bind specific cellular components
Tetramethyl rhodamine isothiocyanate (TRITC or rhodamine)	Often attached to antibodies that bind specific cellular components

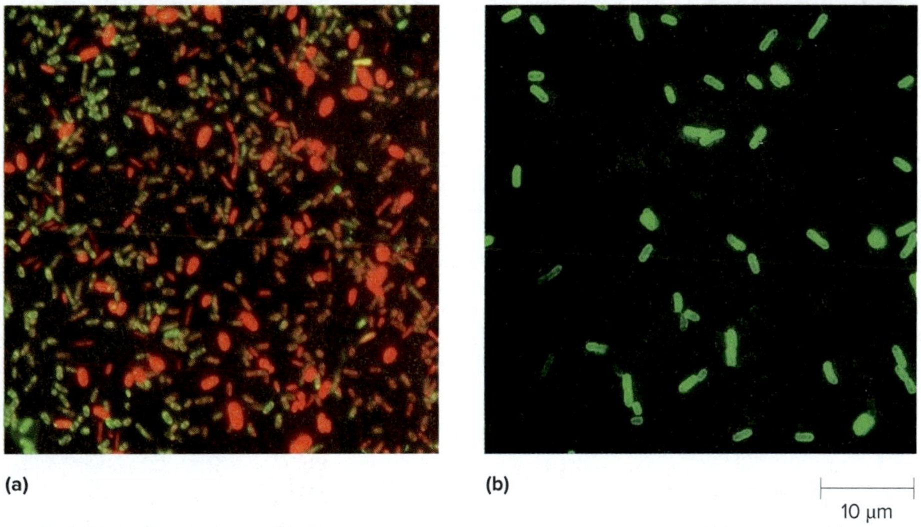

Figure 2.13 Fluorescent Dyes and Tags. (a) Dyes that cause live cells to fluoresce green and dead ones red. (b) Fluorescent antibodies tag specific molecules. In this case, the antibody binds to a molecule that is unique to *Yersinia pestis,* the bacterium that causes plague.

(a) ©Dr. Rita B. Moyes; (b) Source: CDC/Courtesy of Larry Stauffer, Oregon State Public Health Laboratory

MICRO INQUIRY *How might the fluorescently labeled antibody used in figure 2.13b be used to diagnose strep throat?*

Figure 2.14 Green Fluorescent Protein. Visualization of Mbl, a cytoskeletal protein of *Bacillus subtilis.* The Mbl protein has been fused with green fluorescent protein and therefore fluoresces green.

to the genus *Aequorea.* This jellyfish gene encodes a protein that naturally fluoresces green when exposed to light of a particular wavelength and is called green fluorescent protein (GFP). Thus when the protein is made by the cell, it is fluorescent. GFP has been used extensively in studies on bacterial cell division and related phenomena (**figure 2.14**). In fact, the 2008 Nobel Prize in Chemistry was awarded to Osamu Shimomura of Japan and Americans Martin Chalfie and Roger Tsien for their development of this important tool. ▶| *Fluorescence labeling (section 17.4)*

Confocal Microscopy

Like the large and small beads illustrated in **figure 2.15*a***, biological specimens are three-dimensional. When three-dimensional objects are viewed with traditional light microscopes, light from all areas of the object, not just the plane of focus, enters the microscope and is used to create an image. The resulting image is murky and fuzzy (figure 2.15*b*). This problem has been solved by the confocal scanning laser microscope

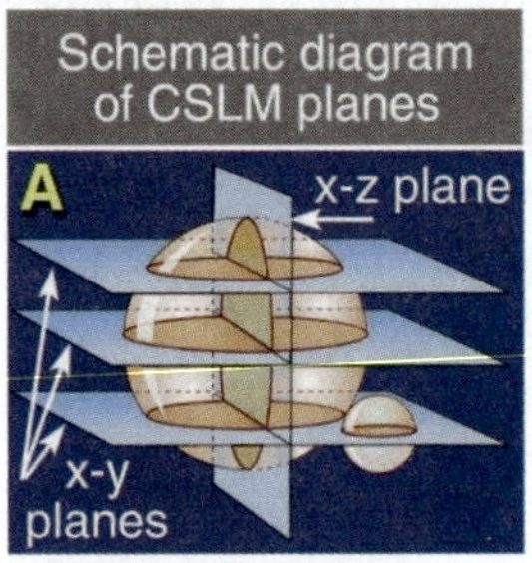

(a) All objects are defined by three axes: x, y, and z. The confocal scanning laser microscope (CSLM) creates images of planes formed by the x and y axes (x-y planes) and planes formed by the x and z axes (x-z planes).

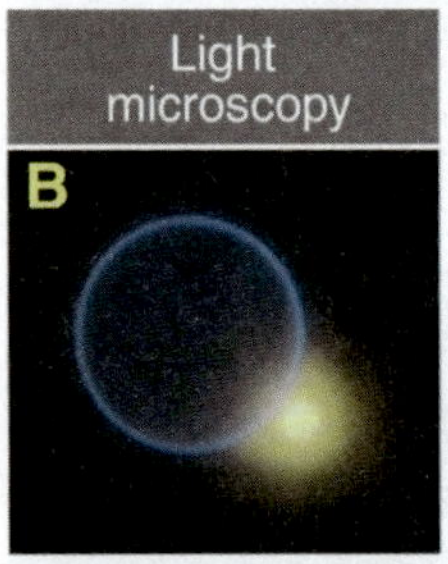

(b) The light microscope image of the two beads shown in **(a)**. Note that neither bead is clear, because the image is generated from light emanating from multiple planes of focus.

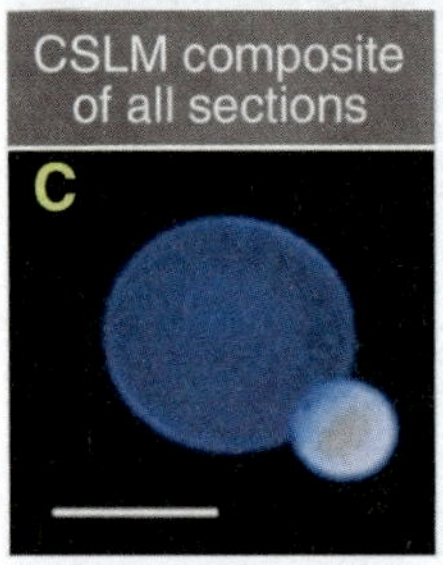

(c) A CSLM uses light from a single plane of focus to generate an image. A composite image of the two beads is generated using digitized information collected from multiple planes within the beads.

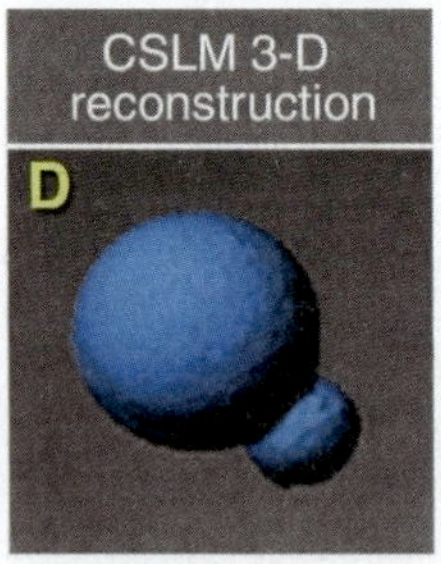

(d) Digitized information is collected from multiple planes within the beads to generate a three-dimensional reconstruction.

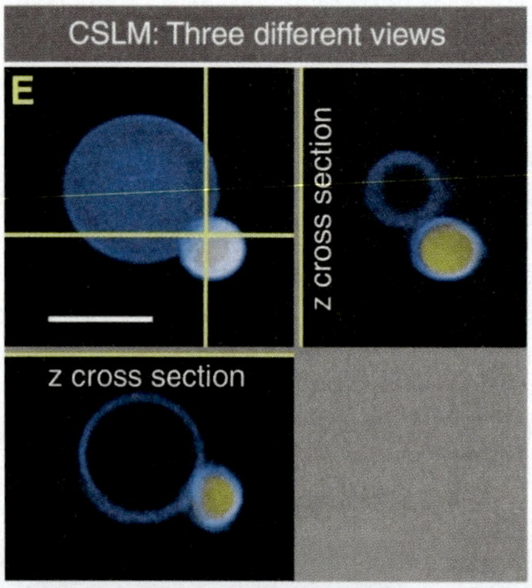

(e) Views are also generated using different planes. The top left panel is the image of a single x-y plane (i.e., looking down from the top of the beads). The two lines represent the two x-z planes imaged in the other two panels. The vertical line indicates the x-z plane shown in the top right panel (i.e., a view from the right side of the beads) and the horizontal line indicates the x-z plane shown in the bottom panel (i.e., a view from the front face of the beads).

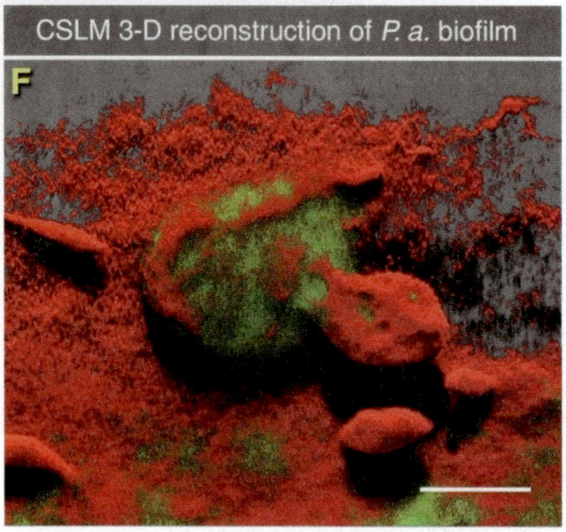

(f) A three-dimensional reconstruction of a *Pseudomonas aeruginosa* biofilm. The biofilm was exposed to an antibacterial agent and then stained with dyes that distinguish living (green) from dead (red) cells. The cells on the surface of the biofilm have been killed, but those in the lower layers are still alive. This image clearly demonstrates the difficulty of killing all the cells in a biofilm.

Figure 2.15 Light and Confocal Microscopy. (a–e) Two beads examined by light and confocal microscopy. (f) Three-dimensional reconstruction of a biofilm.

(CSLM), or simply, **confocal microscope.** The confocal microscope uses a laser beam to illuminate a specimen that has been fluorescently stained. A major component of the confocal microscope is an opening (that is, an aperture) placed above the objective lens. The aperture eliminates stray light from parts of the specimen that lie above and below the plane of focus (**figure 2.16**). Thus the only light used to create the image is from the plane of focus, and a much sharper image is formed.

To generate a confocal image, a computer interfaced with the confocal microscope receives digitized information from each plane in the specimen. This information can be used to create a composite image that is very clear and detailed (figure 2.15*c*) or a three-dimensional reconstruction of the specimen (figure 2.15*d*). Images of x-z plane cross sections of the specimen can also be generated, giving the observer views of the specimen from three perspectives (figure 2.15*e*). Confocal microscopy has numerous applications. One is the study of biofilms, which can form on many different types of surfaces, including indwelling medical devices such as hip joint replacements. As shown in figure 2.15*f*, it is difficult to kill all cells in a biofilm. This makes biofilms a particular concern because their formation on medical devices can result in infections that are difficult to treat. ▶| *Biofilms are common in nature (section 7.6)*

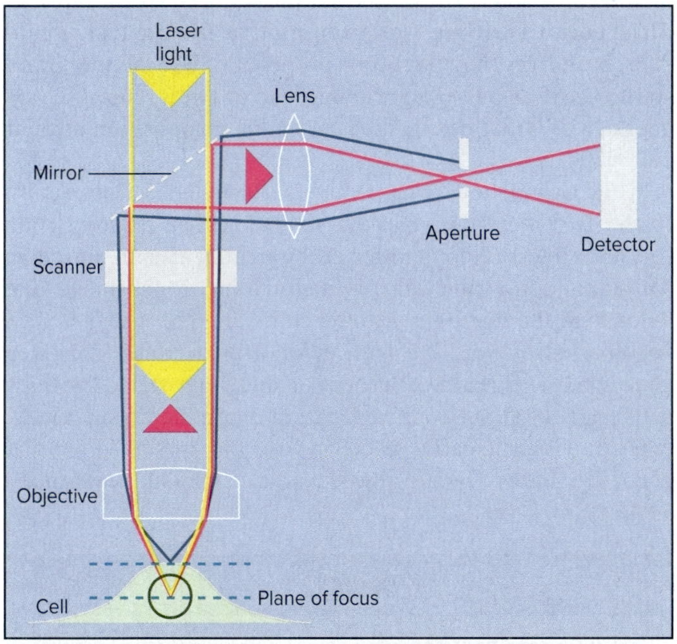

Figure 2.16 Ray Diagram of a Confocal Microscope. The yellow lines represent laser light used for illumination. Red lines symbolize the light arising from the plane of focus, and the blue lines stand for light from parts of the specimen above and below the focal plane.

MICRO INQUIRY *How does the light source differ between a confocal light microscope and other light microscopes?*

Comprehension Check

1. Compare and contrast the function of the condenser, objective, and eyepiece lenses.
2. What happens to resolution if each of the following increase: wavelength of light, refractive index, and numerical aperture? How are resolution and magnification related?
3. What might a microscopist see through an oil immersion objective if no immersion oil is present?
4. Why don't most light microscopes use 30× ocular lenses for greater magnification?
5. Compare and contrast how dark-field, phase-contrast, differential interference contrast, epifluorescence, and confocal microscopes work and the kind of image or images provided by each. Give a specific use for each type.
6. Which type of microscope(s) would be used to examine the following specimens? Photosynthetic bacteria, swimming bacteria, GFP-labeled microorganisms, unstained microbes, and pond scum; explain each of your answers.

2.3 Staining Specimens Helps to Visualize and Identify Microbes

After reading this section, you should be able to:

a. Recommend a fixation process to use when the microbe is a bacterium or archaeon and when the microbe is a protist
b. Plan a series of appropriate staining procedures to describe an unknown bacterium as fully as possible
c. Compare what happens to Gram-positive and Gram-negative bacterial cells during each step of the Gram-staining procedure

As noted in section 2.2, specimens examined by bright-field microscopy are often fixed and stained before being examined. Such preparation serves to increase the visibility of the microorganisms, accentuate specific morphological features, and preserve them for future study. Importantly, some staining procedures help microbiologists identify the organism being examined.

Fixation

Stained cells seen in a microscope should resemble living cells as closely as possible. **Fixation** is the process by which the internal and external structures of specimens are preserved and fixed in position. It inactivates enzymes that might disrupt cell morphology and toughens cell structures so that they do not change during staining and observation. A microorganism usually is killed and attached firmly to the microscope slide during fixation.

There are two fundamentally different types of fixation: heat fixation and chemical fixation. **Heat fixation** is routinely used to observe bacteria and archaea. Typically, a film of cells (a smear) is gently heated. Heat fixation preserves overall morphology and inactivates enzymes. However, it also destroys proteins in subcellular structures, which may distort their appearance. **Chemical fixation** is used to protect fine cellular substructure as well as morphology. It is used when examining microorganisms by many electron microscopy techniques. Chemical fixatives penetrate cells and react with cellular components, usually proteins and lipids, to render them inactive, insoluble, and immobile. Common fixative mixtures contain ethanol, acetic acid, mercuric chloride, formaldehyde, and glutaraldehyde.

Dyes and Simple Staining

The many types of dyes used to stain microorganisms have two features in common. First, they have **chromophore groups**—chemical moieties with conjugated double bonds that give the dye its color. Second, they bind cells by ionic, covalent, or hydrophobic bonds.

Dyes that bind cells by ionic interactions are probably the most commonly used dyes. These ionizable dyes may be divided into two general classes based on the nature of their charged group: basic dyes and acidic dyes (**table 2.4**). The staining effectiveness of ionizable dyes may be altered by pH, since the nature and number of the charged moieties on cell components change with pH. Thus acidic dyes stain best under acidic conditions when proteins and many other molecules carry a positive charge; basic dyes are most effective at higher pH values.

Dyes that bind through covalent bonds or that have certain solubility characteristics are also useful. For instance, DNA can be stained by the Feulgen procedure in which the staining compound (Schiff's reagent) is covalently attached to the deoxyribose sugars of DNA. Sudan III (Sudan Black) selectively stains lipids because it is lipid soluble but does not dissolve in aqueous portions of the cell.

Microorganisms can be stained by **simple staining,** in which a single dye is used (**figure 2.17**). Simple staining's value lies in its ease of use. The fixed smear is covered with stain for a short time, excess stain is washed off with water, and the slide is blotted dry. Basic dyes such as crystal violet, methylene blue, and carbolfuchsin are frequently used in simple staining to determine the size, shape, and arrangement of bacterial and archaeal cells.

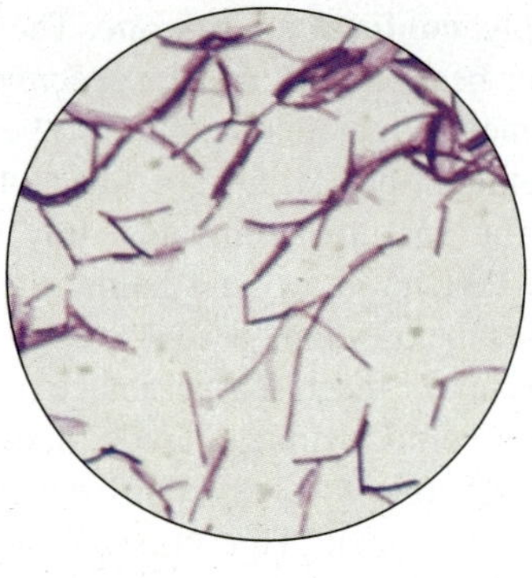

Crystal violet stain
of slender, rod shaped bacteria.

Figure 2.17 **Simple Staining Illustrates Cell Size and Morphology.**
©Dr. Rita B. Moyes

While most dyes directly stain the cell or object of interest, some dyes (e.g., India ink and nigrosin) are used in **negative staining.** In negative staining, the background is stained, not the cell; instead, the unstained cells appear as bright objects against a dark background.

Differential Staining

The **Gram stain,** developed in 1884 by the Danish physician Christian Gram, is the most widely employed staining method in bacteriology. The Gram stain is an example of **differential staining**—procedures that distinguish organisms based on their staining properties. Use of the Gram stain classifies most bacteria into one of two groups—Gram negative or Gram positive, based on the composition of their cell walls.

The Gram-staining procedure is illustrated in **figure 2.18.** In the first step, the smear is stained with a primary stain (crystal violet). This is followed by treatment with an iodine solution. Iodine functions as a **mordant,** a substance that helps bind the dye tightly to the cell wall. The smear is then decolorized by washing with alcohol or acetone. This step generates the differential aspect of the Gram stain; the thick cells walls of Gram-positive bacteria retain the crystal violet, whereas Gram-negative bacteria lose the crystal violet and become colorless. Finally, the smear is counterstained, usually

Table 2.4 Ionizable Dyes

Type of Dye	Examples	Chromophore Characteristics
Basic dyes	Methylene blue, basic fuchsin, crystal violet, safranin, malachite green	Have positively charged groups; bind to negatively charged molecules such as nucleic acids, many proteins, and the surfaces of bacterial and archaeal cells
Acidic dyes	Eosin, rose bengal, and acid fuchsin—possess groups such as carboxyls (—COOH) and phenolic hydroxyls (—OH).	Have negatively charged groups; bind to positively charged cell structures

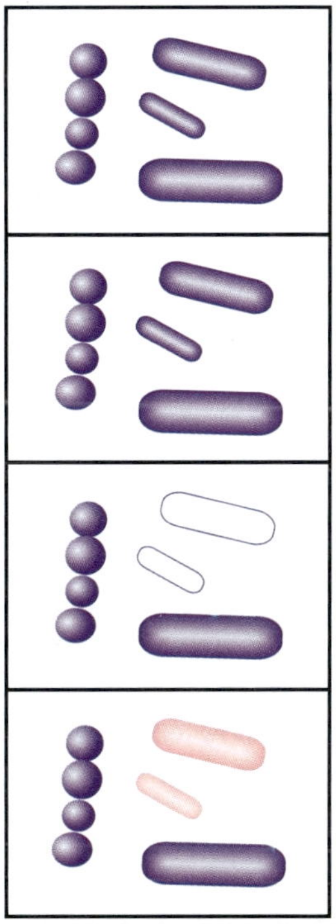

Steps in Staining	State of Bacteria
Step 1: Crystal violet (primary stain) for 1 minute. Water rinse.	Cells stain purple.
Step 2: Iodine (mordant) for 1 minute. Water rinse.	Cells remain purple.
Step 3: Alcohol (decolorizer) for 10–30 seconds. Water rinse.	Gram-positive cells remain purple. Gram-negative cells become colorless.
Step 4: Safranin (counterstain) for 30–60 seconds. Water rinse. Blot dry.	Gram-positive cells remain purple. Gram-negative cells appear red.

Figure 2.18 **Gram Stain.** Steps in the Gram-staining procedure.

MICRO INQUIRY *Why is the decolorization step considered the most critical in the Gram-staining procedure?*

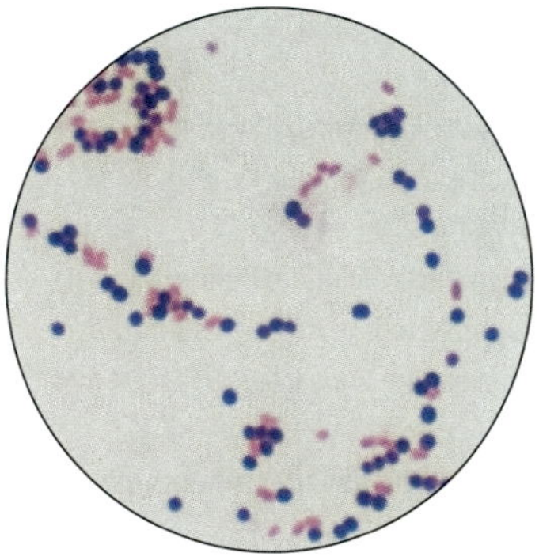

(a) Gram stain
Purple cells are Gram positive.
Red cells are Gram negative.

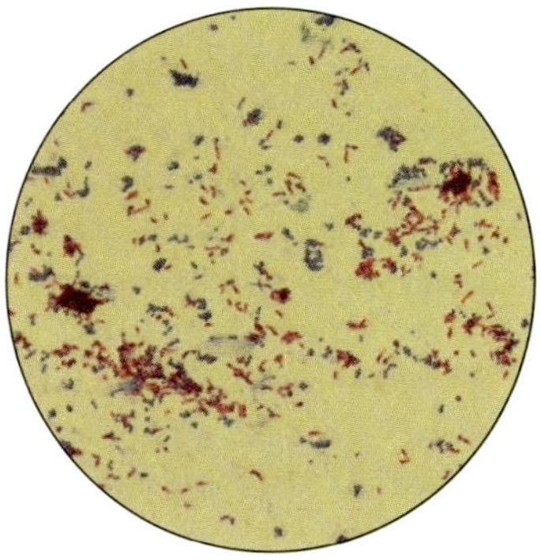

(b) Acid-fast stain
Red cells are acid-fast.
Blue cells are non-acid-fast.

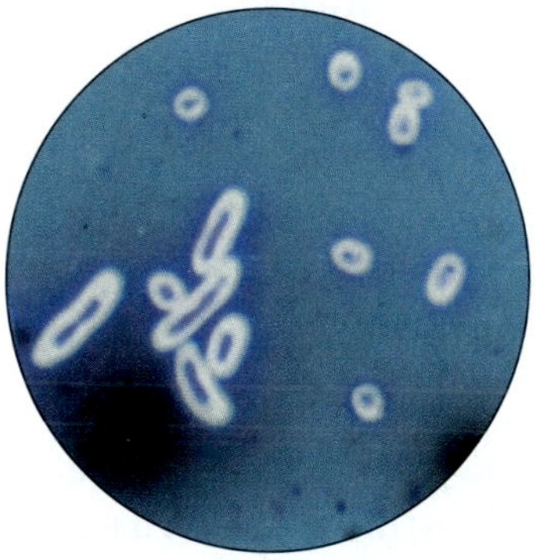

(c) Capsule stain of *Klebsiella pneumoniae*

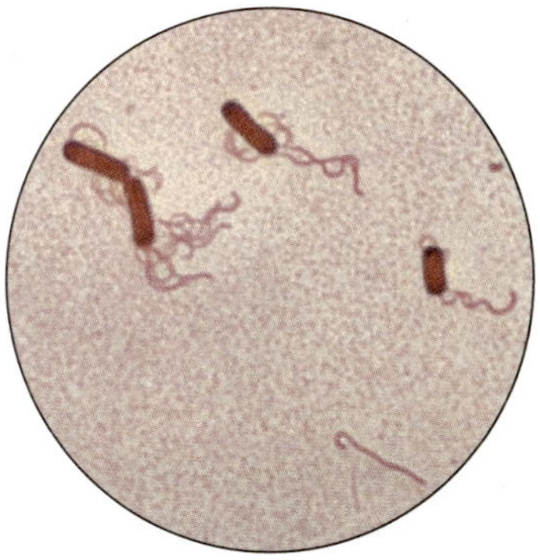

(d) Flagellar stain of *Bacillus brevis*

Figure 2.19 **Differential Stains.**

(a, c) ©McGraw-Hill Education/Lisa Burgess, photographer; (b) ©McGraw-Hill Education/James Redfearn, photographer; (d) Source: CDC/Dr. William A. Clark

with safranin, which colors Gram-negative bacteria pink to red and leaves Gram-positive bacteria dark purple (**figure 2.19*a***). ▶| *There are two main types of bacterial cell walls (section 3.4)*

Acid-fast staining is another important differential staining procedure. It can be used to identify *Mycobacterium tuberculosis* and *M. leprae* (figure 2.19*b*), the pathogens responsible for tuberculosis and leprosy, respectively. These bacteria, as well as other mycobacteria, have cell walls containing lipids constructed from mycolic acids, a group of branched-chain hydroxy fatty acids, which prevent dyes from readily binding to the cells (*see figure 23.7*). A commonly used staining procedure, the cold Ziehl-Neelsen method, uses high concentrations of phenol and carbol fuchsin, as well as a wetting agent, to drive the stain carbol fuchsin into mycobacterial cells. Once this dye has penetrated, the cells are not easily decolorized by acidified alcohol (acid-alcohol) and thus are said to be acid-fast. Non-acid-fast bacteria are easily decolorized by acid-alcohol and thus are stained another color by a second dye called a counterstain.

One of the simplest staining procedures is **capsule staining** (figure 2.19*c*), a technique that reveals the presence of capsules, a network usually made of polysaccharides that surrounds many bacteria and some fungi. Cells are mixed with India ink or nigrosin dye and spread out in a thin film on a slide. After air-drying, cells appear as bodies surrounded by a halo of capsule in the midst of a blue-black background because ink and dye particles cannot penetrate either the cell or its capsule. Thus capsule staining is an example of negative staining. There is little distortion of cell shape, and the cell can be counterstained for even greater visibility. ▶| *Capsules and slime layers (section 3.5)*

Flagella staining provides taxonomically valuable information about the presence and distribution pattern of flagella on bacterial and archaeal cells (figure 2.19*d*; *see also figure 3.37*). Their flagella are fine, threadlike organelles of locomotion that are so slender (about 10 to 30 nm in diameter) they can only be seen directly using an electron microscope (although bundles of flagella can be visualized by dark-field microscopy). To observe bacterial flagella with a light microscope, their thickness is increased by coating them with mordants such as tannic acid and potassium alum and then staining with a dye such as basic fuchsin. ▶| *Bacterial flagella (section 3.7)*

Comprehension Check

1. Describe the two general types of fixation. Which would you use when Gram staining a bacterium? Which would you use before observing the organelles of a protist?
2. Why would basic dyes be more effective under alkaline conditions?
3. What procedural error(s) might be responsible if you Gram stained a mixture of bacteria known to be Gram-positive and Gram-negative and see only red cells?
4. Are capsular and flagellar staining differential staining procedures? Explain your answer.

2.4 Electron Microscopes Use Beams of Electrons to Create Highly Magnified Images

After reading this section, you should be able to:

a. Create a concept map, illustration, or table that compares transmission electron microscopes (TEMs) to light microscopes
b. Decide when it would be best to examine a microbe by TEM, scanning electron microscopy (SEM), and electron cryotomography

For centuries the light microscope has been the most important instrument for studying microorganisms. However, even the best light microscopes have a resolution limit of about 0.2 μm, which greatly compromises their usefulness for detailed studies of many microorganisms (**figure 2.20**). Viruses, for example, are too small to be seen with light microscopes (with the exception of some recently discovered giant viruses). Bacteria and archaea can be observed, but because they are usually only 1 to 2 μm in diameter, only their general shape and major morphological features are visible. The detailed internal structure of microorganisms therefore cannot be effectively studied by light microscopy.

Recall that the resolution of a light microscope increases as the wavelength of the light it uses for illumination decreases. In electron microscopes, electrons replace light as the illuminating beam. The electron beam can be focused, much as light is in a light microscope, but its wavelength is about 100,000 times shorter than that of visible light. Therefore electron microscopes have a practical resolution roughly 1,000 times better than the light microscope; with many electron microscopes, points closer than 0.5 nm can be distinguished, and the useful magnification is well over 100,000× (figure 2.20). The value of the electron microscope is evident on comparison of the photographs in **figure 2.21**. Microbial morphology can now be studied in great detail. We briefly review the most common types of electron microscopy.

Transmission Electron Microscope

A **transmission electron microscope** (**TEM**) uses a heated tungsten filament in the electron gun to generate a beam of

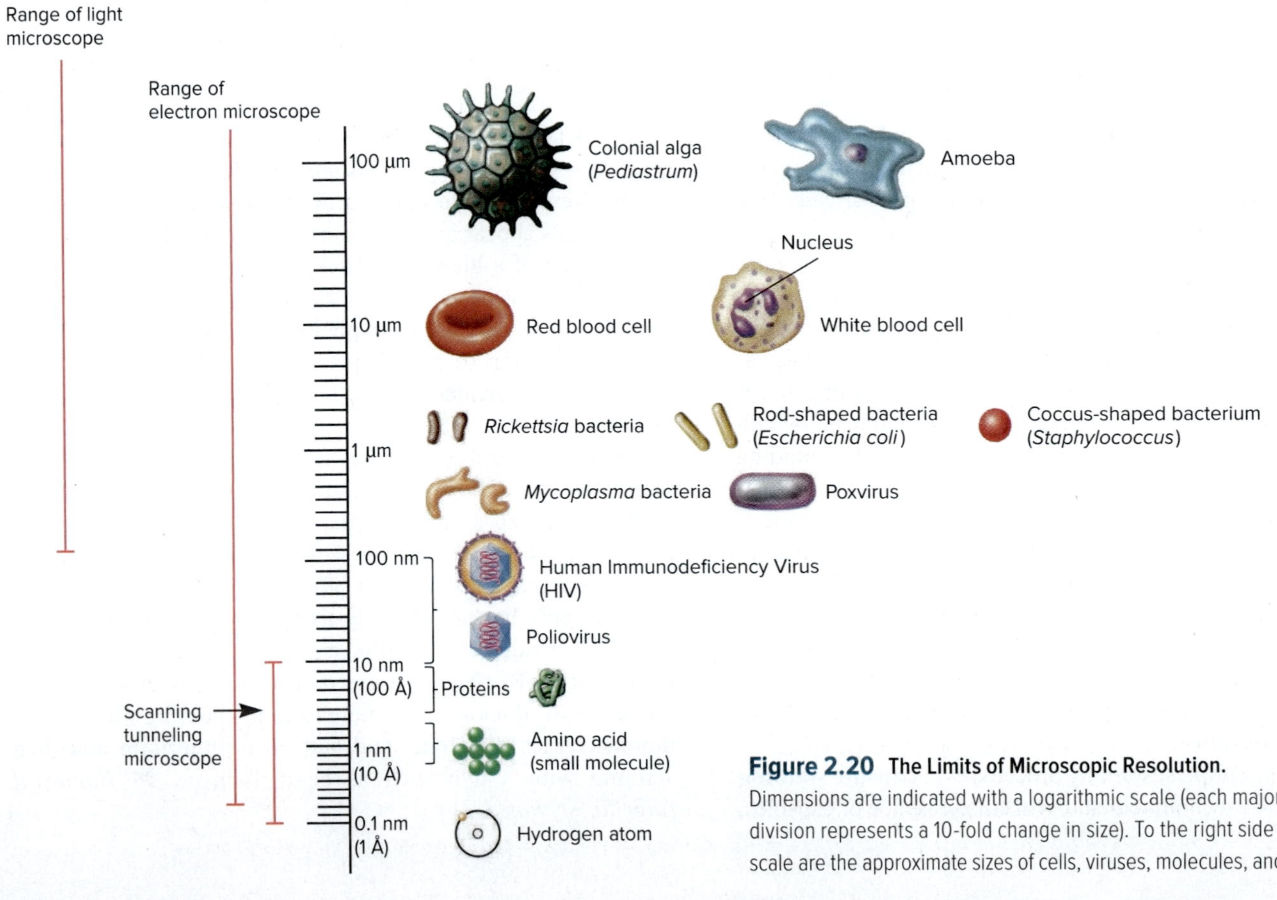

Figure 2.20 The Limits of Microscopic Resolution. Dimensions are indicated with a logarithmic scale (each major division represents a 10-fold change in size). To the right side of the scale are the approximate sizes of cells, viruses, molecules, and atoms.

cut such a thin slice, specimens must be embedded in a supportive plastic matrix. To prepare specimens, they are first fixed with chemicals such as glutaraldehyde and osmium tetroxide to stabilize cell structure. The specimen is then dehydrated with organic solvents (e.g., acetone or ethanol). Next the specimen is soaked in unpolymerized, liquid epoxy plastic until it is completely permeated, and then the plastic is hardened to form a solid block. Thin sections are skillfully cut from the block with a glass or diamond knife using a device called an ultramicrotome.

As with bright-field light microscopy, cells are usually stained so they can be seen clearly with a TEM. The probability of electron scattering is determined by the density (atomic number) of atoms in the specimen. Biological molecules are composed primarily of atoms with low atomic numbers (H, C, N, and O), and electron scattering is fairly constant throughout an unstained cell or virus. Therefore specimens are further

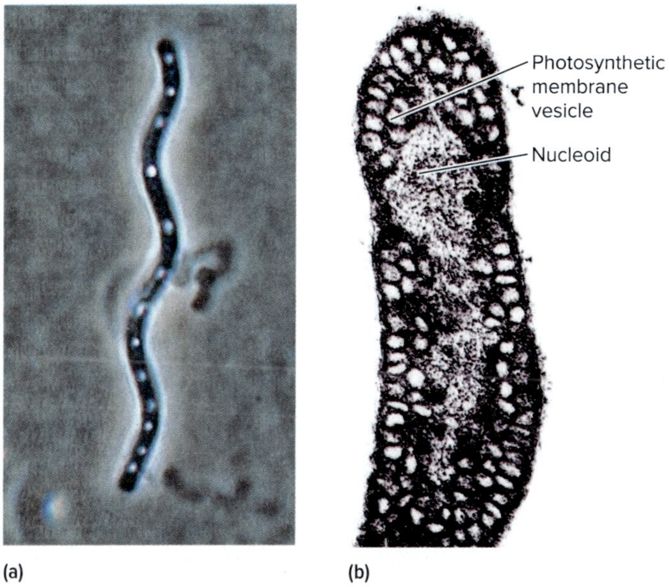

Figure 2.21 Light and Electron Microscopy. A comparison of light and electron microscopic resolution. (a) The proteobacterium *Spirillum volutans* in phase-contrast light microscope (×1,000). (b) A thin section of another spiral-shaped proteobacterium *Rhodospirillum rubrum* in transmission electron microscope (×100,000).
(a) ©McGraw-Hill Education/James Redfearn, photographer; (b) ©Biology Media/Science Source

electrons that is focused on the specimen by the condenser (**figure 2.22**). Since electrons cannot pass through a glass lens, doughnut-shaped electromagnets called magnetic lenses focus the beam. The column containing the lenses and specimen must be under vacuum to obtain a clear image because electrons are deflected by collisions with air molecules. The specimen scatters some electrons, but those that pass through are used to form an enlarged image of the specimen on a fluorescent screen that interfaces with a computer monitor (**figure 2.23**). Denser regions in the specimen scatter more electrons and therefore appear darker because fewer electrons strike that area of the screen; these regions are said to be "electron dense." In contrast, electron-transparent regions are brighter.

Table 2.5 compares some of the important features of light and transmission electron microscopes. The TEM has distinctive features that place harsh restrictions on the nature of samples that can be viewed and the means by which those samples must be prepared. Specimens must be viewed in a vacuum and only extremely thin slices (20 to 100 nm) of a specimen can be viewed because electron beams are easily absorbed and scattered by solid matter. To

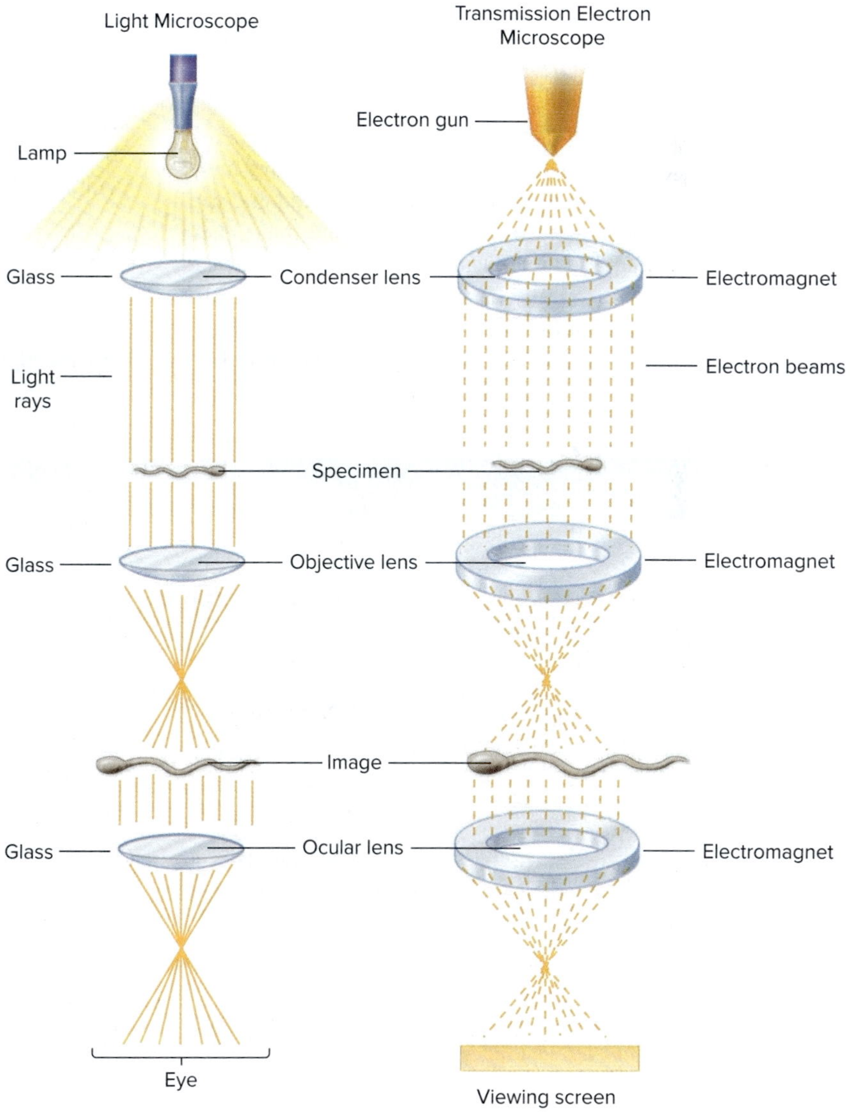

Figure 2.22 A Comparison of Light and Transmission Electron Microscopes.

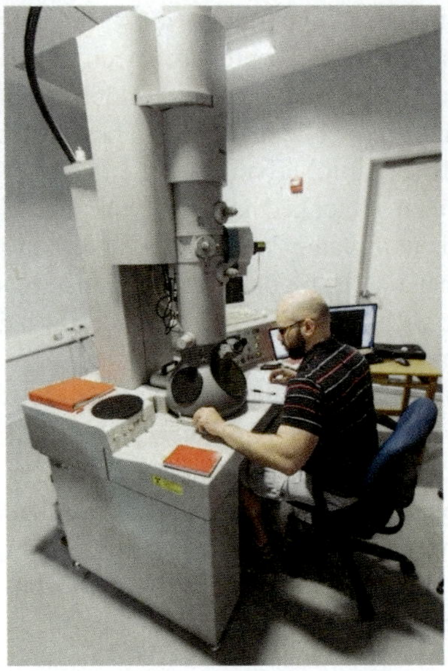

Figure 2.23 A Transmission Electron Microscope. The electron gun is at the top of the central column, and the magnetic lenses are within the column. The image on the fluorescent screen is also viewed on a computer monitor.

prepared by soaking thin sections with solutions of heavy metal salts such as lead citrate and uranyl acetate. The lead and uranium ions bind to structures in the specimen and make them more electron opaque, thus increasing contrast in the material. Heavy osmium atoms from the osmium tetroxide fixative also stain specimens and increase their contrast. The stained thin sections are then mounted on tiny copper grids and viewed.

Two other important techniques for preparing specimens for TEM include negative staining and shadowing. In negative staining, the specimen is spread out in a thin film with either phosphotungstic acid or uranyl acetate. Just as in negative staining for light microscopy, the specimen appears bright against a dark background, in this case because the heavy metals do not penetrate biological material. Negative staining enables visualization of viruses and cellular microbes, but unlike thin sections, internal structures cannot be discerned (**figure 2.24*a***). In shadowing, a specimen is coated with a thin film of platinum or other heavy metal by evaporation at an angle of about 45 degrees from horizontal so that the metal strikes the microorganism on only one side. In one commonly used imaging method, the area coated with metal appears dark in photographs, whereas the uncoated side and the shadow region created by the object are light (figure 2.24*b*). This technique is particularly useful in studying virus particle morphology, bacterial and archaeal flagella, and DNA.

The process of chemical fixation and dehydration can introduce artifacts that can alter cellular morphology. This can be minimized or avoided by using a freeze-etching procedure. When cells are rapidly frozen in liquid nitrogen, they become very brittle and can be broken along lines of greatest weakness, usually down the middle of internal membranes (**figure 2.25**). The exposed surfaces are then shadowed and coated with layers of platinum and carbon to form a replica of the surface. After the specimen has been removed chemically, this replica is studied in the TEM, providing a detailed view of intracellular structure.

Scanning Electron Microscope

Transmission electron microscopes form an image from radiation that has passed through a specimen. The **scanning electron microscope** (**SEM**) produces an image from electrons released

Table 2.5 Characteristics of Light and Transmission Electron Microscopes

Feature	Light Microscope	Transmission Electron Microscope
Highest practical magnification	About 1,000–1,500	Over 100,000
Best resolution[1]	0.2 µm	0.2 nm
Radiation source	Visible light	Electron beam
Medium of travel	Air	High vacuum
Type of lens	Glass	Electromagnet
Source of contrast	Differential light absorption	Scattering of electrons
Focusing mechanism	Adjust lens position mechanically	Adjust current to the magnetic lens
Method of changing magnification	Switch the objective lens or eyepiece	Adjust current to the magnetic lens
Specimen mount	Glass slide	Metal grid (usually copper)

1 The resolution limit of a human eye is about 0.2 mm.

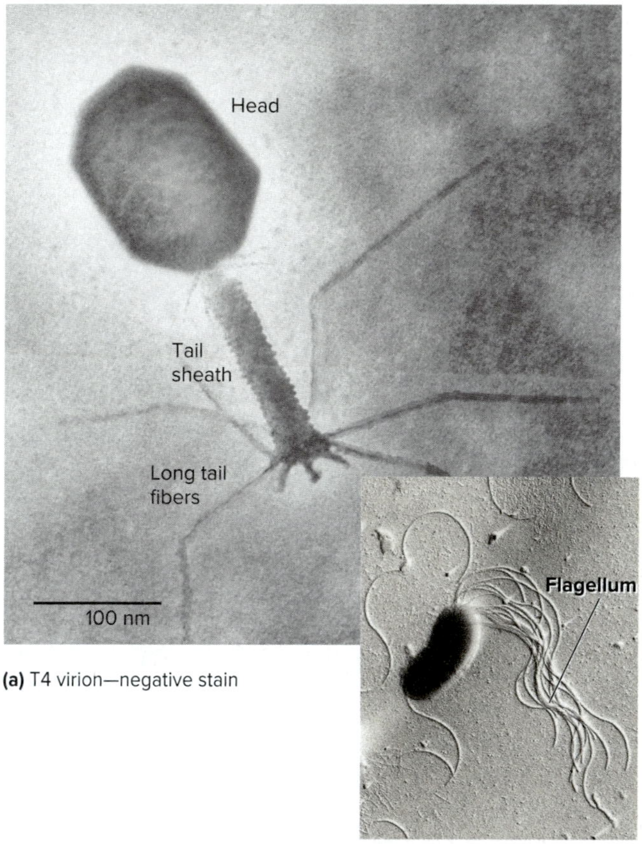

(a) T4 virion—negative stain

(b) *P. fluorescens*—after shadowing

Figure 2.24 Stainied Microorganisms Visualized by TEM. (a) T4 is a virus that infects *Escherichia coli*. (b) *Pseudomonas fluorescens* with its polar flagella.
(a) ©AMI Images/Science Source; (b) ©Dr. Tony Brain/SPL/Science Source

MICRO INQUIRY *Why are all electron micrographs black and white (although they are sometimes artificially colorized after printing)?*

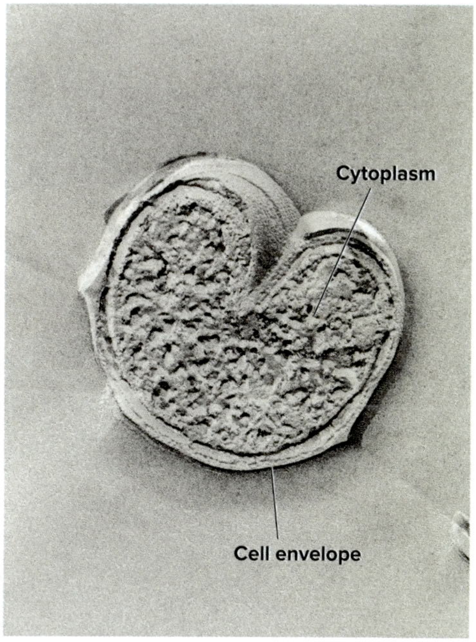

Figure 2.25 Example of Freeze-Etching. A freeze-etched preparation of the bacterium *Nitrospira* sp.
©Dr. Kari Lounatmaa/SPL/Science Source

from an object's surface. The surfaces of microorganisms are visualized in great detail; most SEMs have a resolution of about 10 nm.

Specimen preparation for SEM is relatively easy, and in some cases, air-dried material can be examined directly. However, microorganisms usually must first be fixed, dehydrated, and dried to preserve surface structure and prevent collapse of cells when they are exposed to the SEM's vacuum. Before viewing, dried samples are mounted and coated with a thin layer of metal to prevent the buildup of an electrical charge on the surface and to give a better image.

To create an image, the SEM scans a narrow, tapered electron beam back and forth over the specimen (**figure 2.26**). When the beam strikes a particular area, surface atoms discharge a tiny shower of electrons called secondary electrons, and these are trapped by a detector. Secondary electrons strike a material in the detector that emits light when struck by electrons (the material is called a scintillator). The flashes of light are converted to an electrical current and amplified by a photomultiplier. The signal is digitized and sent to a computer, where it can be viewed.

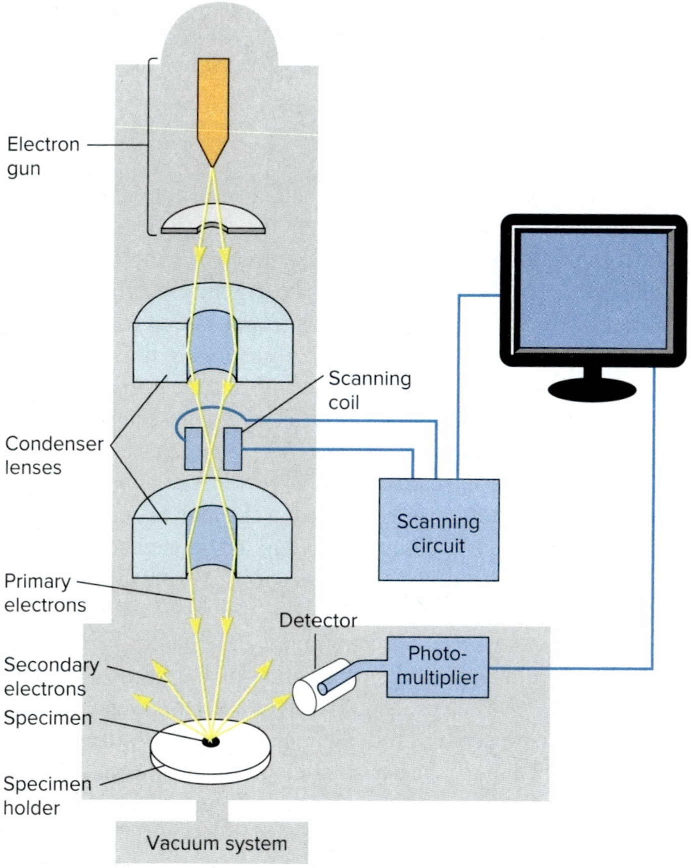

Figure 2.26 The Scanning Electron Microscope.

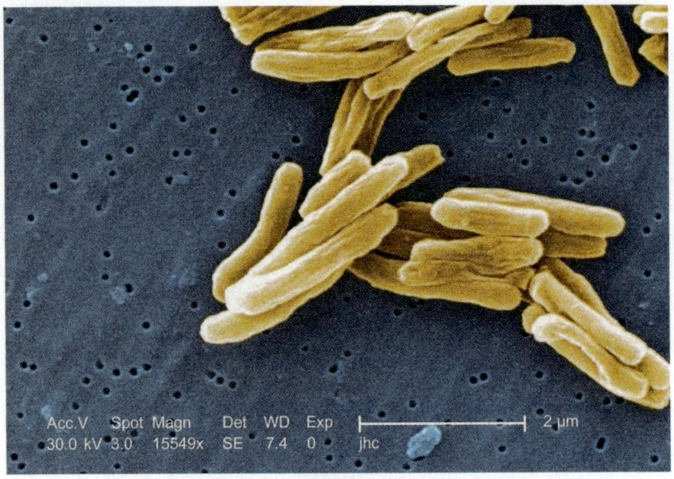

Figure 2.27 **Scanning Electron Micrograph of *Mycobacterium tuberculosis.*** Colorized image (×15,549).
Source: CDC/Janice Haney Carr

The number of secondary electrons reaching the detector depends on the nature of the specimen's surface. When the electron beam strikes a raised area, a large number of secondary electrons enter the detector; in contrast, fewer electrons escape a depression in the surface and reach the detector. Thus raised areas appear lighter on the screen and depressions are darker. A realistic three-dimensional image of the microorganism's surface results (**figure 2.27**). A single instrument can house both transmission and scanning electron microscopes (S/TEM).

Electron Cryotomography

Beginning approximately 40 years ago, a series of advances in electron microscopy paved the way for the development of **electron cryotomography,** a technique that since the 1990s has been providing exciting insights into the structure and function of cells and viruses. *Cryo-* refers to sample preparation and visualization. Samples are prepared by rapidly plunging the specimen into an extremely cold liquid (e.g., ethane) and the sample is kept frozen while being examined. Rapid freezing of the sample forms vitreous ice rather than ice crystals. Vitreous ice is a glasslike solid that preserves the native state of structures and immobilizes the specimen so that it can be viewed in the high vacuum of the electron microscope. *Tomography* refers to the method used to create images. The object is viewed from many directions, referred to as a tilt series. The individual images are recorded and processed by computer programs, and finally merged to form a three-dimensional reconstruction of the object. Three-dimensional views, slices, and other types of representations of the object can be derived from the reconstruction (**figure 2.28**). The ultrastructure of bacterial and archaeal cells has been the focus of numerous studies using electron cryotomography. Some of these studies have revealed new cytoskeletal elements, such as those associated with magnetosomes—the inclusions used by some bacteria to orient themselves in magnetic fields *(see figure 22.14)*.

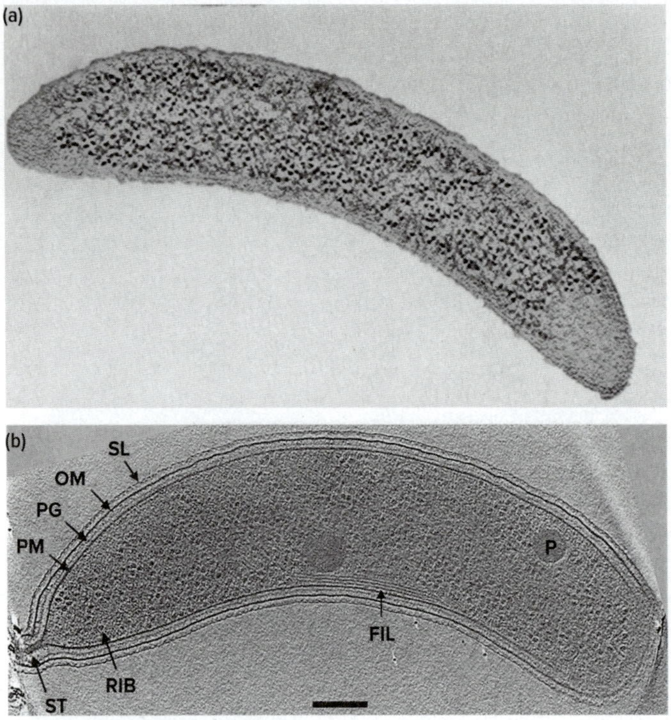

Figure 2.28 **TEM and Electron Cryotomography.** A comparison of (a) a thin section of a *Caulobacter crescentus* cell prepared using conventional TEM procedures and (b) a central slice from a three-dimensional reconstruction of an intact *C. crescentus* cell. FIL, a bundle of filaments; OM, outer membrane; P, a presumed inclusion; PG, peptidoglycan; PM, plasma membrane; RIB, ribosome; SL, S-layer; ST, stalk.
(a) ©AMI Images/SPL/Science Source; (b) ©Grant J. Jensen

Comprehension Check

1. Why does an electron microscope have much greater resolution than a light microscope?
2. Where is the electron gun located relative to the sample? Why must a TEM use high vacuum and very thin sections?
3. Under what circumstances would it be desirable to prepare specimens for a TEM by thin section? Negative staining? Shadowing? Freeze-etching?
4. How does a scanning electron microscope differ from a TEM? If you were shown a micrograph, how would you know if it were a TEM, SEM, or electron cryotomography?

2.5 Scanning Probe Microscopy Can Visualize Molecules and Atoms

After reading this section, you should be able to:

a. Distinguish scanning tunneling from atomic force microscopes in terms of how they create images and their uses
b. Compare and contrast light microscopy, electron microscopy, and scanning probe microscopy in terms of their uses, resolution, and the quality of the images created

Among the most powerful microscopes are **scanning probe microscopes (SPMs).** These microscopes measure surface features of an object by moving a sharp probe over the object's surface. One type of

SPM is the **scanning tunneling microscope.** It can achieve magnifications of 100 million times, and it allows scientists to view atoms on the surface of a solid. The scanning tunneling microscope has a needlelike probe with a point so sharp that often there is only one atom at its tip. The probe is lowered toward the specimen surface until its electron cloud just touches that of the surface atoms. When a small voltage is applied between the tip and specimen, electrons flow through a narrow channel in the electron clouds. This tunneling current, as it is called, is extraordinarily sensitive to distance and will decrease about a thousandfold if the probe is moved away from the surface by a distance equivalent to the diameter of an atom.

The arrangement of atoms on the specimen surface is determined by scanning the probe tip back and forth over the surface while keeping the probe at a constant height above the specimen. As the tip follows the surface contours by moving up and down, its motion is recorded and analyzed by a computer to create an accurate three-dimensional image of the surface atoms. The surface map can be displayed on a computer screen or plotted on paper. The resolution is so great that individual atoms are observed easily. Even more exciting is that the microscope can examine objects when they are immersed in water. Therefore it can be used to study biological molecules such as DNA (**figure 2.29**). The microscope's inventors, Gerd Binnig and Heinrich Rohrer, shared the 1986 Nobel Prize in Physics for their work, together with Ernst Ruska, the designer of the first transmission electron microscope.

A second type of SPM is the **atomic force microscope,** which moves a sharp probe over the specimen surface while keeping the distance between the probe tip and the surface constant. It does this by exerting a very small amount of force on the tip, just enough to maintain a constant distance but not enough force to damage the surface. The vertical motion of the tip usually is followed by measuring the deflection of a laser beam that strikes the lever holding the probe (**figure 2.30**). Unlike the scanning tunneling microscope, the atomic force microscope can be used to study surfaces that do not conduct electricity well. The atomic force microscope has been used to study the interactions of proteins, to follow the behavior of living bacteria and other cells, and to visualize membrane proteins such as aquaporins (**figure 2.31**).

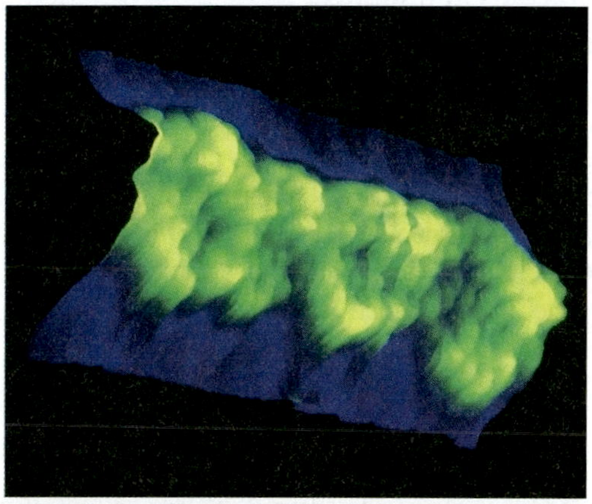

Figure 2.29 Scanning Tunneling Microscopy of DNA. The DNA double helix with approximately three turns shown (false color; ×2,000,000).
©Driscoll, Youngquist & Baldeschwieler, Cal-tech/SPL/Science Source

MICRO INQUIRY *Compare the magnification of this image with that shown in figure 2.27 (SEM). How many times greater is the magnification shown here?*

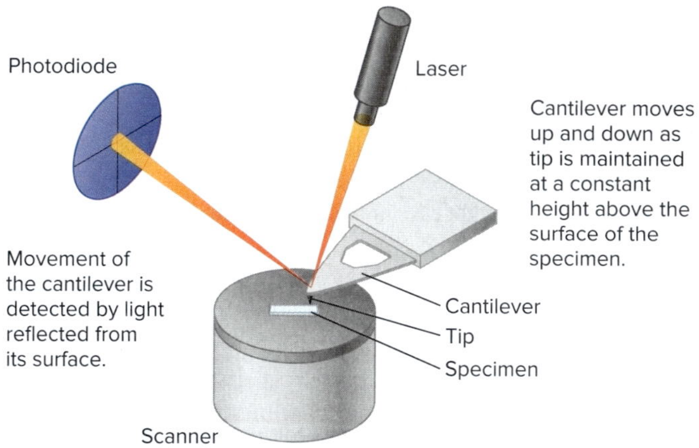

Figure 2.30 The Basic Elements of an Atomic Force Microscope. The tip used to probe the specimen is attached to a cantilever. As the probe passes over the "hills and valleys" of the specimen's surface, the cantilever is deflected vertically. A laser beam directed at the cantilever is used to monitor these vertical movements. Light reflected from the cantilever is detected by the photodiode and used to generate an image of the specimen.

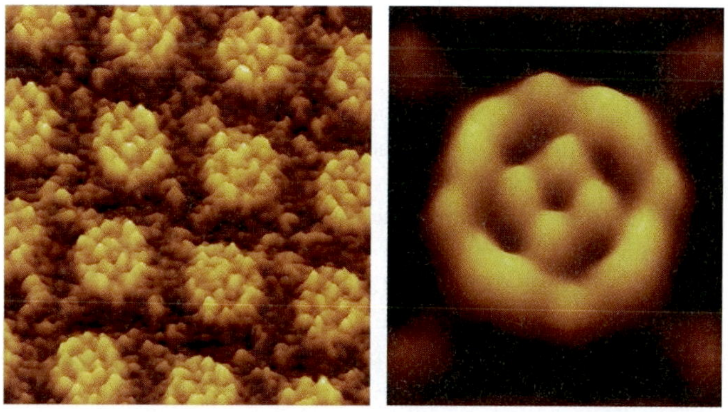

(a) (b)

Figure 2.31 The Membrane Protein Aquaporin Visualized by Atomic Force Microscopy. Aquaporin is a membrane-spanning protein that allows water to move across the membrane. (a) Surface view of aquaporin proteins. (b) A single aquaporin protein observed in more detail and at higher magnification.
(a–b) ©Simon Scheuring

Comprehension Check

1. Briefly describe each of the most popular versions of the scanning probe microscope. Discuss a structure that might be visualized.
2. Scanning probe microscopy can be applied to samples that have not been dehydrated, but electron microscopy cannot. Why is this an important advantage?

Key Concepts

2.1 Lenses Create Images by Bending Light

- A light ray moving from air to glass or vice versa is bent in a process known as refraction (**figure 2.1**).
- Lenses focus light rays at a focal point and magnify images (**figure 2.2**).

2.2 There Are Several Types of Light Microscopes

- In a compound microscope such as the bright-field microscope, the primary image is an enlarged image formed by the objective lens. The primary image is further enlarged by the ocular lens to yield the final image (**figure 2.3**).
- Microscope resolution increases as the wavelength of radiation used to illuminate the specimen decreases and as the numerical aperture increases. The maximum resolution of a light microscope is about 0.2 μm (**figure 2.4**).
- The dark-field microscope uses only refracted light to form an image, and objects appear light against a black background (**figure 2.6**).
- The phase-contrast microscope converts variations in the refractive index into changes in light intensity and thus makes colorless, unstained, live cells visible (**figures 2.8–2.10**).
- The differential interference contrast microscope uses two beams of light to create high-contrast images of live specimens (**figure 2.11**).
- The fluorescence microscope illuminates a fluorochrome-labeled specimen and forms an image from its fluorescence (**figures 2.12–2.14**).
- The confocal microscope is used to study thick, complex specimens. It creates an image by using only the light emanating from the plane of focus, while blocking out light from above and below the plane of focus (**figures 2.15** and **2.16**).

2.3 Staining Specimens Helps to Visualize and Identify Microbes

- Specimens are often fixed and stained before viewing them in the bright-field microscope. There are two fixation methods: heat fixation and chemical fixation.
- Most dyes are either positively charged basic dyes or negatively charged acidic dyes that bind to ionized parts of cells.
- In simple staining, a single dye is used to stain microorganisms (**figure 2.17**).
- Differential staining procedures such as Gram and acid-fast staining distinguish between microbial groups by staining them differently (**figures 2.18** and **2.19*a,b***). Other differential staining techniques are specific for particular structures such as bacterial capsules and flagella (**figure 2.19*c,d***).

2.4 Electron Microscopes Use Beams of Electrons to Create Highly Magnified Images

- The transmission electron microscope (TEM) uses magnetic lenses to form an image from electrons that have passed through a very thin section of a specimen (**figure 2.22**). Resolution is high because the wavelength of a beam of electrons is very short.
- Specimens for TEM are usually prepared by methods that increase contrast. Specimens can be stained by treatment with solutions of heavy metals such as osmium tetroxide, uranium, and lead. They can also be prepared for TEM by negative staining, shadowing with metal, or freeze-etching (**figures 2.24** and **2.25**).
- The scanning electron microscope is used to study external surface features of microorganisms (**figures 2.26** and **2.27**).
- Electron cryotomography freezes specimens rapidly, keeps them frozen while being examined, and creates images from a series of directions that are combined and processed to form a three-dimensional reconstruction of the object (**figure 2.28**).

2.5 Scanning Probe Microscopy Can Visualize Molecules and Atoms

- Scanning probe microscopes reach very high magnifications that allow scientists to observe biological molecules (**figures 2.29** and **2.31**).
- Scanning tunneling microscopy enables the visualization of molecular surfaces using electron interaction between the probe and the specimen, whereas atomic force microscopy can scan the surface of molecules that do not conduct electricity well (**figure 2.30**).

Active Learning

1. You have prepared a specimen for light microscopy, stained it using the Gram staining procedure, but failed to see anything when you looked through your light microscope. Discuss the things that you may have done incorrectly.
2. In a journal article, find an example of a light micrograph, a scanning or transmission electron micrograph, or a confocal image. Discuss why the figure was included in the article and why that particular type of microscopy was the method of choice for the research. What other figures would you like to see used in this study? Outline the steps that the investigators would take to obtain such photographs or figures.

3. Stimulated emission depletion (STED) microscopes were first developed about 10 years ago. They are modifications of confocal microscopes that use two highly focused beams of light rather than one as in typical confocal microscopy. The first beam uses a wavelength that excites the fluorophore with which a specimen is labeled. This creates a fluorescent spot where the fluorophore is located. The second beam is a hollow cone. It strikes the outermost edges of the fluorescent spot illuminated by the first beam. The wavelength used by the second beam returns the fluorophore to its ground state, thus only the center of the fluorescent spot is visible. The result is that the resolution of STED microscopes is greater and objects smaller than 200 nm can be resolved. If you could use STED microscopy, what questions in biological science would you want to ask and why would you ask these questions? You may want to refer to figures 2.15 and 2.20 to help formulate your answer.

4. Bacterial viruses with double-stranded DNA genomes package their genetic material in a hollow protein shell called a capsid. The proteins that make up the capsid are assembled without the help of enzymes, so no covalent bonds are formed between individual proteins. Nonetheless, the storage of DNA within the capsid subjects the proteins to pressures 50 times greater than atmospheric pressure. A group of scientists from Kansas and Indiana recently used electron cryotomography to resolve the capsid to a resolution of 2.9 Å. First, refer to figure 2.20 and convert this to nanometers. Next, discuss why you believe these investigators visualized the bacteriophage capsid using electron cryotomography. What specific questions do you think the researchers were asking?

   Read the original paper: Zhao, H., et al. 2017. Structure of a headful DNA-packaging bacterial virus at 2.9Å resolution by electron cryo-microscopy. *Proc. Natl. Adad. Sci. USA*. 114:3601–3606. doi:10.1073/pnas.1615025114.

3

Bacterial Cell Structure

©Design Pics/Hammond HSN

Hooking Up

Each year over 100 million people around the world become infected with *Neisseria gonorrhoeae*, the bacterium that causes gonorrhea. This troubling statistic is made even more disturbing by the increasing resistance of the bacterium to the antibiotics used to treat the disease. In men, infection is usually readily detected, but for women, infection is often asymptomatic and can lead to serious consequences such as pelvic inflammatory disease (PID) and sterility. These concerns have led scientists to consider methods for preventing infection. One method is to block transmission. Unfortunately, relatively little is known about the transmission process except that it occurs during sexual intercourse and that numerous hairlike structures (called pili) covering the surface of the bacterium play a role in establishing infection. The bacterium uses pili for a type of movement called twitching motility and to adhere to surfaces such as the sperm and epithelial cells of its human host. It has long been thought that by attaching to sperm cells the bacterium could hitch a ride during sexual intercourse. This explained transmission between partners. However, it did not clarify how transmission from women to men occurs.

It turns out that exposure of *N. gonorrhoeae* to seminal fluid increases its twitching motility and enhances formation of small clumps of bacteria. In addition, seminal fluid proteins appear to alter the morphology and function of pili. In particular these proteins cause bundles of pili to separate into single filaments, enhancing the interaction of bacterial cells with each other and with host surfaces. These changes evolved to promote infection of host epithelial cells and increase the likelihood of transmission during sexual intercourse from females to their male partner.

As this story illustrates, even small, seemingly simple organisms such as bacteria can exhibit complex behaviors. To understand these amazing microbes, we must first examine their cell structure and begin to relate it to the functions they carry out. As we consider bacterial cell structure, it is important to remember that only about 1% of bacterial species have been cultured. Of the cultivated species, only a few have been studied in great detail. From this small sample, many generalizations are made, and it is presumed that most other bacteria are like the well-studied model organisms. However, part of the wonder and fun of science is that nature is full of surprises. As the biology of more and more bacteria is explored, our understanding of them may change in interesting and exciting ways.

Readiness Check:

Based on what you have learned previously, you should be able to:

- ✓ Describe the application of small subunit (SSU) rRNA analysis to the establishment of the three domain classification system proposed by Carl Woese (section 1.2)
- ✓ Identify the following structures or regions of a plant or animal cell and describe their functions: cell wall, plasma membrane, cytoplasm, mitochondria, chloroplasts, and ribosomes
- ✓ Define and give examples of essential nutrients; describe how they are used by cells

3.1 Use of the Term "Prokaryote" Is Controversial

After reading this section, you should be able to:

a. List the characteristics originally used to describe prokaryotic cells
b. Debate the "prokaryote" controversy using current evidence about bacterial cells

Bacteria and archaea have long been lumped together and referred to as prokaryotes. Although the term was first introduced early in the twentieth century, the concept of a prokaryote was not fully outlined until 1962, when R. Stanier and C. B. van Niel described prokaryotes in terms of what they lacked in comparison to eukaryotic cells. For instance, Stanier and van Niel pointed out that prokaryotes lack a membrane-bound nucleus, a cytoskeleton, membrane-bound organelles, and internal membranous structures such as the endoplasmic reticulum and Golgi apparatus. Since the 1960s, biochemical, genetic, and genomic analyses have shown that *Bacteria* and *Archaea* are distinct taxa. Because of these discoveries, Norman Pace proposed in 2006 that the term *prokaryote* should be abandoned and most microbiologists are in agreement.

This controversy illustrates that microbiology is an exciting, dynamic, and rapidly changing field of study. Throughout this text, we avoid the term *prokaryote* and are as explicit as possible about which characteristics are associated with members of *Bacteria*, which with members of *Archaea*, and which with members of both taxa.

3.2 Bacteria Are Diverse but Share Some Common Features

After reading this section, you should be able to:

a. Distinguish a typical bacterial cell from a typical plant or animal cell in terms of cell shapes and arrangements, size, and cell structures
b. Discuss the factors that determine the size and shape of a bacterial cell

Much of this chapter is devoted to a discussion of individual cell components. Therefore a preliminary overview of the features common to many bacterial cells is in order. We begin by considering overall cell morphology and then move to cell structures.

Shape, Arrangement, and Size

It might be expected that bacterial cells, being small and relatively simple, would be uniform in shape and size. This is not the case, as the microbial world offers considerable variety in terms of morphology. However, the two most common shapes are cocci and rods (**figure 3.1**). **Cocci** (s., **coccus**) are roughly spherical cells. They can exist singly or can be associated in characteristic arrangements that can be useful in their identification. **Diplococci** (s., **diplococcus**) arise when cocci divide and remain together to form pairs. Long chains of cocci result when cells adhere after repeated divisions in one plane; this pattern is seen in the genera *Streptococcus, Enterococcus,* and *Lactococcus* (figure 3.1*a*). Members of the genus *Staphylococcus* divide in random planes to generate irregular, grapelike clusters (figure 3.1*b*). Divisions in two or three planes can produce symmetrical groupings of cocci. Bacteria in the genus *Micrococcus* often divide in two planes to form square groups of four cells called tetrads. In the genus *Sarcina,* cocci divide in three planes, producing cubical packets of eight cells.

Legionella pneumophila is an example of a bacterium with a **rod** shape (figure 3.1*c*). Rods, sometimes called **bacilli** (s., **bacillus**), differ considerably in their length-to-width ratio, the coccobacilli being so short and wide that they resemble cocci. The shape of the rod's end often varies between species and may be flat, rounded, football-shaped, or bifurcated. Although many rods occur singly, some remain together after division to form pairs or chains (e.g., *Bacillus megaterium* is found in long chains).

There are several less common cell shapes and arrangements. **Vibrios** are comma-shaped (**figure 3.2*a***). **Spirilla** are rigid, spiral-shaped cells (figure 3.2*b*). Many have tufts of flagella at one or both ends. **Spirochetes** are flexible, spiral-shaped bacteria that have a unique, internal flagellar arrangement (figure 3.2*c*). These bacteria are distinctive in other ways, and all belong to a single phylum, *Spirochaetes.* Some bacteria form stalks (e.g., *Caulobacter crescentus*) (figure 3.2*d*). Other bacteria are **pleomorphic,** being variable in shape and lacking a single, characteristic form. ▶| *Phylum* Spirochaetes *(section 21.6);* Caulobacteraceae *and* Hyphomicrobiaceae *bacteria reproduce in unusual ways (section 22.1); Order* Vibrionales *includes aquatic bioluminescent bacteria and pathogens (section 22.3)*

Some bacteria can be thought of as multicellular. Many actinobacteria form long filaments called hyphae. The hyphae form a network called a **mycelium** (figure 3.2*e*), and in this sense, they are similar to eukaryotic filamentous fungi. Many cyanobacteria, a group of photosynthetic bacteria, are also filamentous. Being filamentous allows some degree of

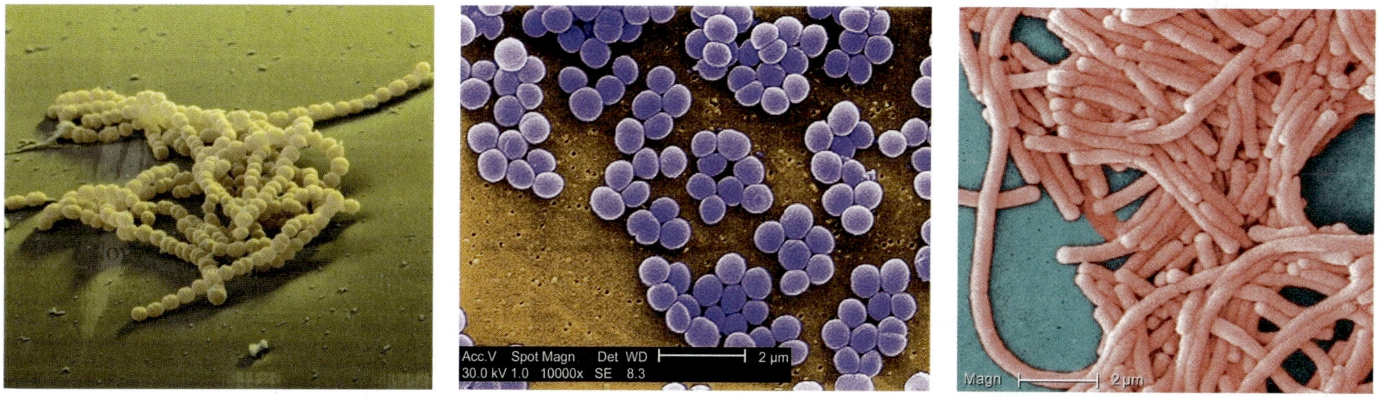

(a) *S. agalactiae*—cocci in chains **(b)** *S. aureus*—cocci in clusters **(c)** *L. pneumophila*—rods in chains

Figure 3.1 Cocci and Rods Are the Most Common Bacterial Shapes. These images are color-enhanced scanning electron micrographs. (a) *Streptococcus agalactiae,* the cause of group B streptococcal infections (×4,800). (b) *Staphylococcus aureus.* (c) *Legionella pneumophila,* the cause of Legionnaires' disease.

(a) ©Science Source; *(b, c)* Source: CDC/Janice Haney Carr,

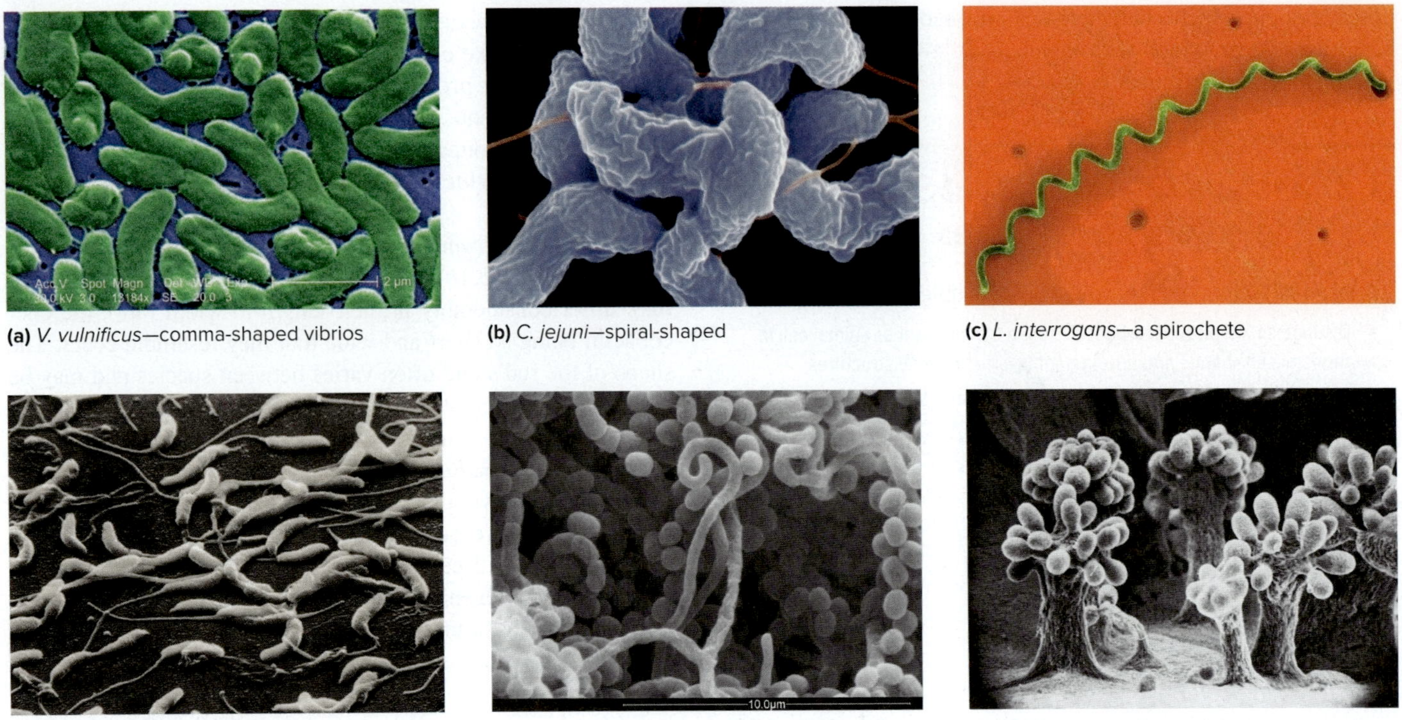

(a) *V. vulnificus*—comma-shaped vibrios

(b) *C. jejuni*—spiral-shaped

(c) *L. interrogans*—a spirochete

(d) *C. crescentus*—a stalked bacterium

(e) *Streptomyces*—a filamentous bacterium

(f) *C. crocatus* fruiting body

Figure 3.2 Other Cell Shapes and Aggregations. (a) *Vibrio vulnificus,* scanning electron micrograph (SEM, X13,184). (b) *Campylobacter jejuni,* SEM. (c) *Leptospira interrogans,* the spirochete that causes the disease leptospirosis. (d) *Caulobacter crescentus,* SEM. (e) *Streptomyces* sp., SEM. (f) Fruiting body of the myxobacterium *Chondromyces crocatus.* The fruiting body is composed of thousands of cells.

(*a*) ©Media for Medical/Getty Images; (*b*) Source: Photo by DeWood, digital colorization by Stephen Ausmus/USDA-ARS; (*c*) ©Sebastian Kaulitzki/Getty Images; (*d*) ©Biology Pics/Science Source; (*e*) ©Dr. Amy Gehring; (*f*) ©Yoav Levy/DIOMEDIA

differentiation among cells in the filament. For instance, some filamentous cyanobacteria form specialized cells within the filament, heterocysts, that carry out nitrogen fixation *(see figure 21.10c)*. Myxobacteria are of particular note. These bacteria sometimes aggregate to form complex structures called fruiting bodies (figure 3.2*f*). ▶| *Order Streptomycetales: an important source of antibiotics (section 23.1); Phylum* Cyanobacteria: *oxygenic photosynthetic bacteria (section 21.4); Order* Myxococcales: *bacteria with morphological complexity and multicellularity (section 22.4)*

Escherichia coli is an excellent representative of an average-sized bacterium. This rod-shaped bacterium is 1.1 to 1.5 µm wide by 2.0 to 6.0 µm long. However, the size range of bacterial cells extends far beyond this average (**figure 3.3**). Near the small end of the size continuum are members of the genus *Mycoplasma* (0.3 µm in diameter). At the other end of the continuum are bacteria such as some spirochetes, which can reach 500 µm in length, and the cyanobacterium *Oscillatoria,* which is about 7 µm in diameter (the same diameter as a red blood cell). Some bacteria are huge by "bacterial standards." For instance, *Epulopiscium fishelsoni* grows as large as 600 by 80 µm, a little smaller than a printed hyphen and clearly larger than the well-known eukaryote *Paramecium* (**figure 3.4**). An even larger bacterium, *Thiomargarita*

Specimen	Approximate diameter or width × length in µm
Red blood cell	7
E. coli	1.3 × 4.0
Streptococcus	0.8–1.0
Poxvirus	0.23 × 0.32
Influenza virus	0.085
T2 *E. coli* bacteriophage	0.065 × 0.095
Tobacco mosaic virus	0.015 × 0.300
Poliovirus	0.027

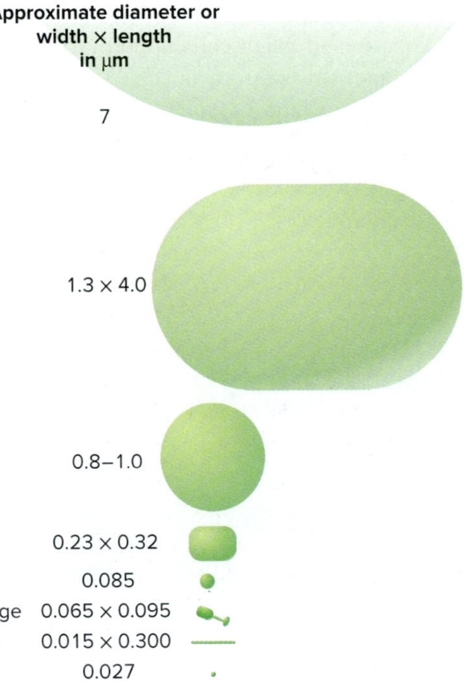

Figure 3.3 Sizes of Bacteria Relative to a Red Blood Cell and Viruses. The larger viruses are comparable in size to the smaller bacteria.

Figure 3.4 A Giant Bacterium. This phase-contrast micrograph shows *Epulopiscium fishelsoni* dwarfing the paramecia, which are protozoa. *E. fishelsoni* cells are about 530 µm long.

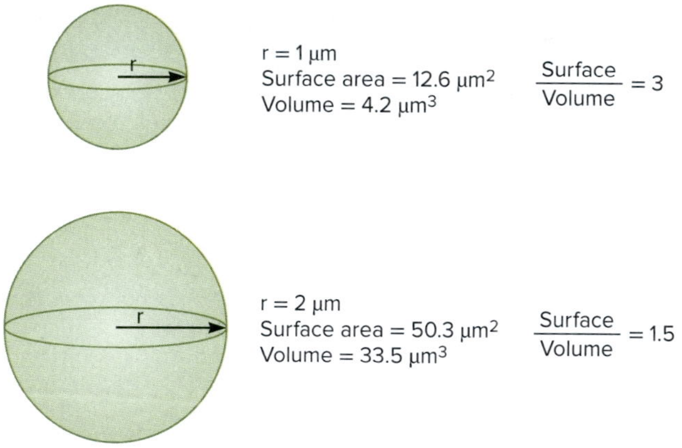

Figure 3.5 The Surface-to-Volume Ratio Is an Important Determinant of Cell Size. Surface area is calculated by the formula $4\pi r^2$. Volume is calculated by the formula $4/3\pi r^3$. Shape also affects the S/V ratio; rods with the same volume as a coccus have a greater S/V ratio.

namibiensis, lives in ocean sediment (*see figure 22.20*). Thus a few bacteria are much larger than the average eukaryotic cell (typical plant and animal cells are around 10 to 50 µm in diameter).

The variety of sizes and shapes exhibited by bacteria raises a fundamental question: What causes a bacterial species to have a particular size and shape? Although far from being answered, recent discoveries have fueled a renewed interest in this question, and it is clear that size and shape determination are related and have been selected for during the evolutionary history of each bacterial species. For many years it was thought that microbes had to be small to increase the surface area-to-volume ratio (S/V ratio; **figure 3.5**). As this ratio increases, the uptake of nutrients and the diffusion of these and other molecules within the cell become more efficient, which in turn facilitates a rapid growth rate. Shape affects the S/V ratio. A rod with the same volume as a coccus has a higher S/V ratio than does the coccus. This means that a rod can have greater nutrient flux across its plasma membrane. However, the discovery of *E. fishelsoni* demonstrates that bacteria can be very large. For bacteria to be large, they must have other characteristics that maximize their S/V ratio, or their size must be beneficial in some way. For instance, *E. fishelsoni* has a highly convoluted plasma membrane, which increases its S/V ratio. In addition, large cells are less likely to be eaten by predatory protists. Cells that are filamentous, have stalks, or are oddly shaped are also less susceptible to predation.

Cell Organization

Structures often observed in bacterial cells are summarized and illustrated in **table 3.1** and **figure 3.6**. Note that no single bacterium possesses all of these structures at all times. Some are found only in certain cells in certain conditions or in certain phases of the life cycle.

There are several common features of bacterial cell structure. Bacterial cells are surrounded by several layers, which are collectively called the cell envelope. The most common cell envelope layers are the plasma membrane, cell wall, and capsule or slime layer. The innermost layer of the cell envelope is the plasma membrane, which surrounds the cytoplasm. Most bacteria have a chemically complex cell wall, which covers the plasma membrane. Many bacteria surround the cell wall with a capsule or slime layer. Because most bacteria do not contain internal, membrane-bound organelles, their interior appears morphologically simple. The genetic material is localized in a discrete region called the nucleoid and is not separated from the surrounding cytoplasm by membranes. Ribosomes and larger masses called inclusions are scattered about the cytoplasm. Finally, many bacteria use flagella for locomotion. In the remaining sections of this chapter, we describe these major structures observed in bacterial cells in more detail.

Comprehension Check

1. Why is the term *prokaryote* considered an inadequate descriptor by some microbiologists?
2. What characteristic shapes can bacteria assume? Describe the ways in which bacterial cells cluster together.
3. What advantages might a bacterial species that forms multicellular arrangements (e.g., clusters or chains) have that are not afforded unicellular bacteria?
4. What is the relevance of the surface area-to-volume ratio?

Table 3.1	Common Bacterial Structures and Their Functions
Plasma membrane	Selectively permeable barrier, mechanical boundary of cell, nutrient and waste transport, location of many metabolic processes (respiration, photosynthesis), detection of environmental cues for chemotaxis
Gas vacuole	An inclusion that provides buoyancy for floating in aquatic environments
Ribosomes	Protein synthesis
Inclusions	Storage of carbon, phosphate, and other substances; site of chemical reactions (microcompartments); movement
Nucleoid	Localization of genetic material (DNA)
Periplasmic space	In typical Gram-negative bacteria, contains hydrolytic enzymes and binding proteins for nutrient processing and uptake; in typical Gram-positive bacteria, may be smaller or absent
Cell wall	Protection from osmotic stress, helps maintain cell shape
Capsules and slime layers	Resistance to phagocytosis, adherence to surfaces
Fimbriae and pili	Attachment to surfaces, bacterial conjugation and transformation, twitching
Flagella	Swimming and swarming motility
Endospore	Survival under harsh environmental conditions

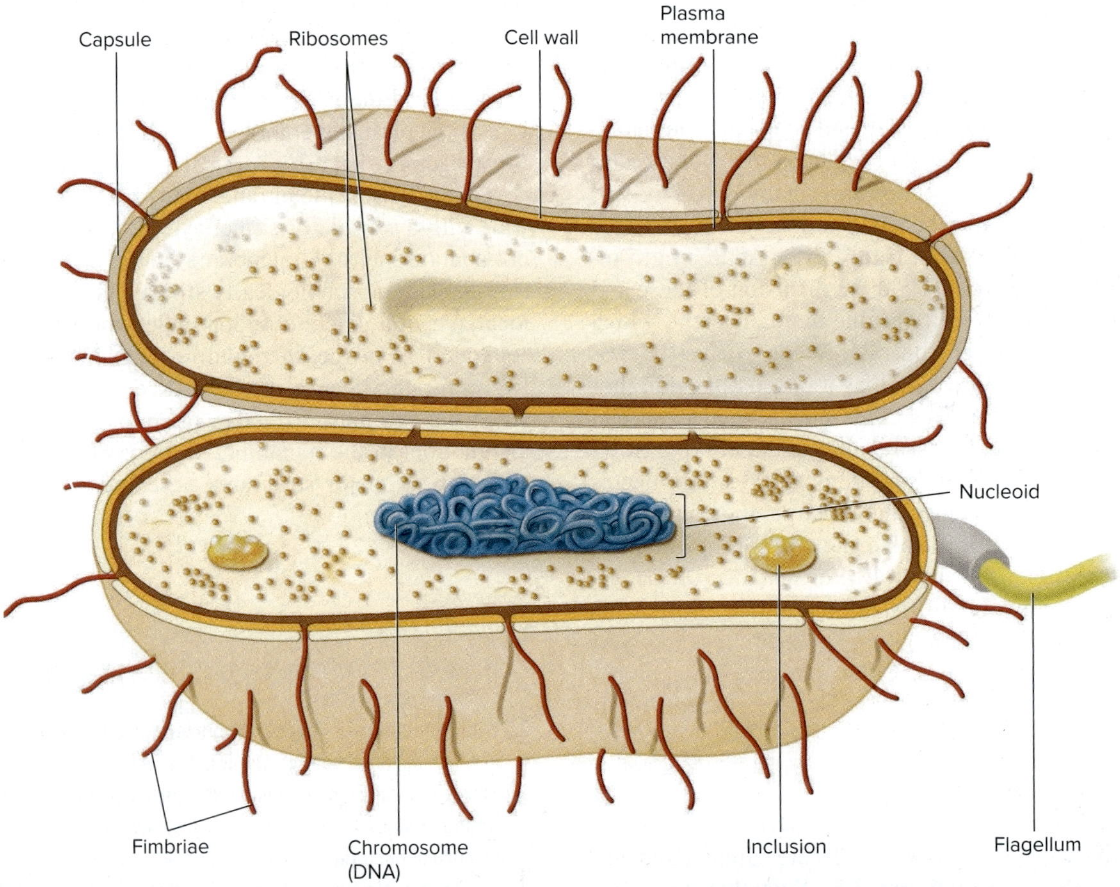

Figure 3.6 Structure of a Bacterial Cell.

3.3 Bacterial Plasma Membranes Control What Enters and Leaves the Cell

After reading this section, you should be able to:

a. Describe the fluid mosaic model of membrane structure and identify the types of lipids typically found in bacterial membranes
b. Distinguish macroelements (macronutrients) from trace elements (micronutrients) and provide examples of each
c. Provide examples of growth factors needed by some microorganisms
d. Compare and contrast passive diffusion, facilitated diffusion, active transport, and group translocation, and provide examples of each
e. Discuss the challenge of iron uptake and describe how bacteria overcome this difficulty

The **cell envelope** is defined as the plasma membrane and all the surrounding layers external to it. The cell envelopes of many bacteria consist of the plasma membrane, cell wall, and at least one additional layer (e.g., capsule or slime layer). Of all these layers, the **plasma membrane** is the most important because it encompasses the cytoplasm and defines the cell. If it is removed or compromised, the cell's contents spill into the environment and the cell dies. Furthermore, despite being the innermost layer of the cell envelope, the plasma membrane is responsible for much of the cell's relationship with the outside world. Thus we begin our consideration of bacterial cell structure by describing the plasma membrane.

First, let's consider what cells do to survive. Cells must interact in a selective fashion with their environment, acquire nutrients, and eliminate waste. They also have to maintain their interior in a constant, highly organized state in the face of external changes. Plasma membranes are an absolute requirement for all living organisms because they are involved in carrying out these cellular tasks.

A primary role of all plasma membranes is that they are selectively permeable barriers: They allow particular ions and molecules to pass either into or out of the cell, while preventing the movement of others. Thus the plasma membrane prevents the loss of essential components through leakage while allowing the movement of other molecules. Bacterial plasma membranes play additional critical roles. They are the location of several crucial metabolic processes: respiration, photosynthesis, and the synthesis of lipids and cell wall constituents.

In addition to the plasma membrane, some bacteria have extensive intracytoplasmic membrane systems. These internal membranes and the plasma membrane share a basic design. However, they can differ significantly in the lipids and proteins they contain. To understand these chemical differences and the many functions of the plasma membrane and other membranes, it is necessary to become familiar with membrane structure.

Fluid Mosaic Model of Membrane Structure

The most widely accepted model for membrane structure is the **fluid mosaic model** of Singer and Nicholson, which proposes that membranes are lipid bilayers within which proteins float (**figure 3.7**). Bacterial membranes have roughly equal amounts of lipids and proteins. Cell membranes are very thin structures, about 2 to 3 nm thick, that look like two dark lines on either side of a light interior when imaged by TEM. This characteristic appearance is evidence that the membrane is composed of two sheets of lipid molecules arranged end-to-end (figure 3.7). Cleavage of membranes by freeze-etching, a technique that reveals fine detail, exposes the proteins lying within the membrane lipid bilayer. ◀ *Electron microscopes use beams of electrons to create highly magnified images (section 2.4); Scanning probe microscopy can visualize molecules and atoms (section 2.5)*

The chemical nature of membrane lipids is critical to their ability to form bilayers. Most membrane-associated lipids (e.g., the phospholipids shown in figure 3.7) are **amphipathic:** They are structurally asymmetric, with polar and nonpolar ends

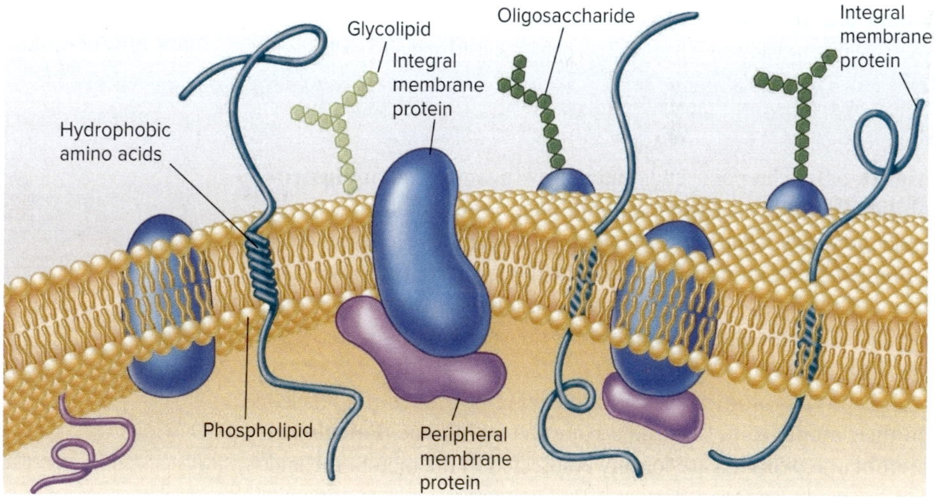

Figure 3.7 The Fluid Mosaic Model of Bacterial Membrane Structure. This diagram shows the integral membrane proteins (blue) floating in a lipid bilayer. Peripheral membrane proteins (purple) are associated loosely with the inner membrane surface and/or integral membrane proteins. Small tan spheres represent the hydrophilic ends of membrane phospholipids, and wiggly tails are the hydrophobic fatty acid chains. Phospholipids are drawn much larger than their actual size. Oligosaccharides (chains of carbohydrates) protrude into the environment and may be attached to proteins or to membrane phospholipids (glycolipids).

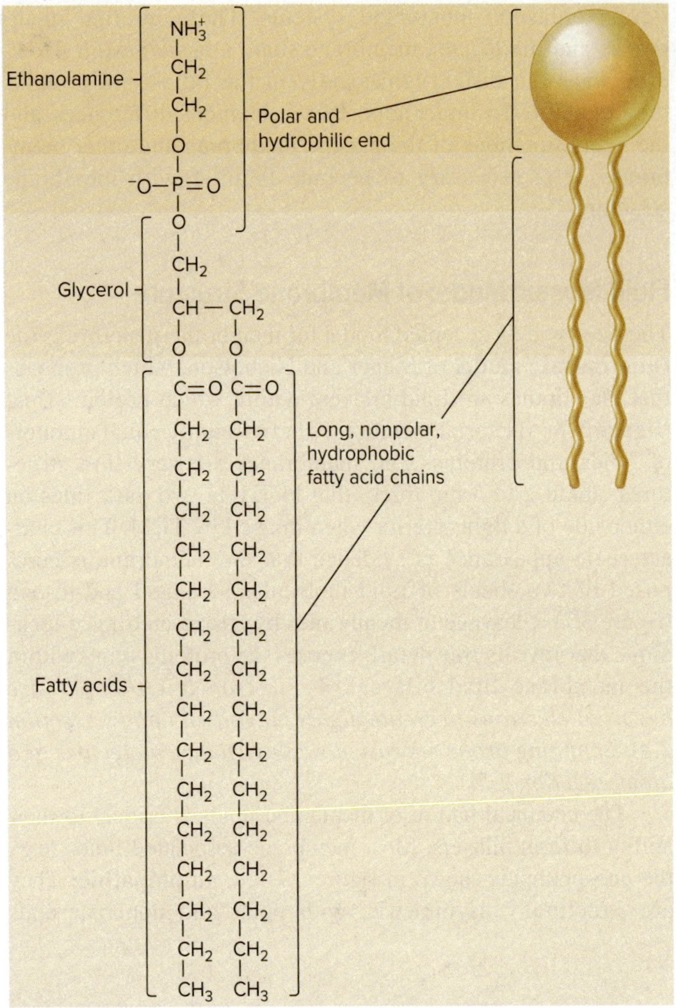

Figure 3.8 **The Structure of a Phospholipid.**
Phosphatidylethanolamine, a phospholipid often found in bacterial membranes.

(**figure 3.8**). The polar ends interact with water and are **hydrophilic;** the nonpolar **hydrophobic** ends are insoluble in water and tend to associate with one another. In aqueous environments, amphipathic lipids can interact to form a bilayer. The outer surfaces of the bilayer are hydrophilic, whereas hydrophobic ends are buried in the interior away from the surrounding water (figure 3.7). ▶| *Lipids (appendix I)*

Two types of membrane proteins have been identified based on their ability to be separated from the membrane. **Peripheral membrane proteins** are loosely connected to the membrane and can be easily removed (figure 3.7). They are soluble in aqueous solutions and make up about 20 to 30% of total membrane protein. The remaining proteins are **integral membrane proteins.** These are not easily extracted from membranes and are insoluble in aqueous solutions when freed of lipids. Integral membrane proteins, like membrane lipids, are amphipathic; their hydrophobic regions are buried in membrane lipids while the hydrophilic portions project from the membrane surface (figure 3.7).

Integral membrane proteins carry out some of the most important functions of the membrane. Many are transport proteins used to move materials either into or out of the cell. Others are involved in energy-conserving processes, such as the proteins found in electron transport chains. Those integral membrane proteins with regions exposed to the outside of the cell enable the cell to interact with its environment. ▶| *Proteins (appendix I)*

Bacterial Plasma Membranes Are Dynamic

Bacterial membranes are lipid bilayers and many of their amphipathic lipids are phospholipids (figure 3.8). The plasma membrane is dynamic: The lipid composition varies with environmental temperature in such a way that the membrane remains fluid during growth. For example, bacteria growing at lower temperatures have more unsaturated fatty acids in their membrane phospholipids; that is, there are one or more double covalent bonds in the long hydrocarbon chains. At higher temperatures, their phospholipids have more saturated fatty acids—those in which the carbon atoms are connected only with single covalent bonds. ▶| *Environmental factors affect microbial growth (section 7.5)*

Although most aspects of the fluid mosaic model are well supported by experimentation, the suggestion that membrane lipids and integral proteins are homogeneously distributed has been challenged by the discovery of **microdomains.** These regions of the membrane are formed by phospholipids in conjunction with other lipids like farnesol, hopanoids, and carotenoids (*see figure 11.31*). **Hopanoids** are similar in structure to cholesterol found in eukaryotic membranes (**figure 3.9**), and their rigid planar structure makes them more hydrophobic than phospholipids. The specific lipids found in microdomains vary among organisms, but they all regulate membrane rigidity and mark microdomain boundaries.

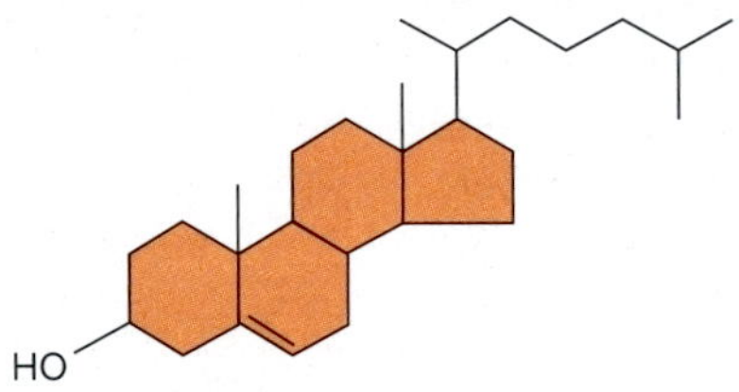

(a) Cholesterol (a steroid) is found in the membranes of eukaryotes.

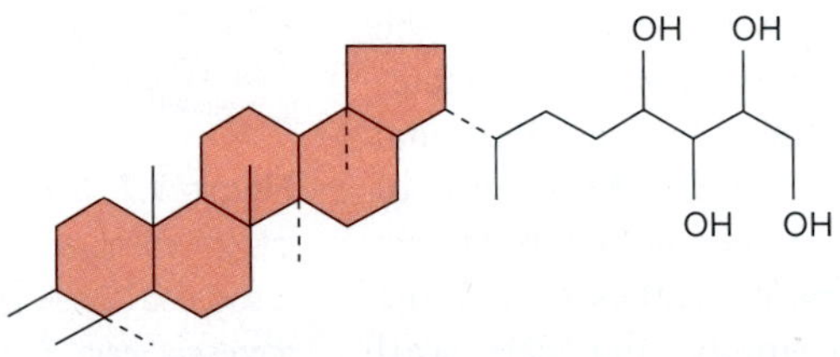

(b) Bacteriohopanetetrol (a hopanoid) is found in many bacterial membranes.

Figure 3.9 **Membrane Steroids and Hopanoids.**

The integral membrane proteins in microdomains differ from those in the bulk membrane, and are organized by proteins called **flotillins.** Flotillins are themselves integral membrane proteins, and they function to assemble large protein complexes like secretion systems for transporting molecules out of the cell and complexes that transmit signals from the environment to molecules in the cytoplasm.

Bacteria Use Many Mechanisms to Bring Nutrients into the Cell

All plasma membranes function as barriers. Yet they must also allow movement of nutrients into the cell. If a microbe does not obtain nutrients from its environment, it will quickly exhaust its supply of amino acids, nucleotides, and other molecules needed to survive. In addition, if a microbe is to thrive and reproduce, it must have a source of energy. The energy source is used to generate the cell's major energy currency: the high-energy molecule ATP. Clearly, obtaining energy and nutrient sources is one of the most important jobs an organism has, and it is primarily a function of the bacterial plasma membrane. Here we discuss nutrient uptake, but first let's define some terms used to describe the nutrients needed by cells.

Microbiologists refer to carbon, oxygen, hydrogen, nitrogen, sulfur, and phosphorus as **macroelements** or macronutrients because they are required in relatively large amounts. They are found in organic molecules such as proteins, lipids, nucleic acids, and carbohydrates. Other macroelements are potassium, calcium, magnesium, and iron. They exist as cations and generally are associated with and contribute to the activity and stability of molecules and cell structures such as enzymes and ribosomes. Thus they are important in many cellular processes, including protein synthesis and energy conservation. ▶| *Enzymes and ribozymes speed up cellular chemical reactions (section 10.6); Translation in bacteria (section 13.7); Electron transport chains: sets of sequential redox reactions (section 10.4)*

Other elements are required in small amounts—amounts so small that in the lab they are often obtained as contaminants in water, glassware, and growth media. Likewise in nature, they are ubiquitous and usually present in adequate amounts to support the growth of microbes. Microbiologists call these elements **micronutrients** or **trace elements.** The micronutrients—manganese, zinc, cobalt, molybdenum, nickel, and copper—are needed by most cells. Micronutrients are part of certain enzymes, and they aid in catalysis of reactions and maintenance of protein structure.

Some microbes are able to synthesize all the organic molecules they need from macroelements. However, some microbes are unable to synthesize certain molecules needed for survival. These molecules are called **growth factors,** and they must be obtained from the environment. There are three types of growth factors: amino acids, purines and pyrimidines, and vitamins.

What are the common features of nutrient uptake by bacteria? Bacteria can only take in dissolved molecules. Uptake mechanisms are specific; that is, the necessary substances, and not others, are acquired. It does a cell no good to take in a substance that it cannot use. Bacteria are able to transport nutrients into the cell even when the concentration of a nutrient inside the cell is higher than the concentration outside. Thus they are able to move nutrients up a concentration gradient. This is important because bacteria often live in nutrient-poor habitats. In view of the enormous variety of nutrients and the complexity of the task, it is not surprising that bacteria use several different transport mechanisms: passive diffusion, facilitated diffusion, primary and secondary active transport, and group translocation.

Passive Diffusion

Passive diffusion, often called diffusion or simple diffusion, is the process by which molecules move from a region of higher concentration to one of lower concentration; that is, the molecules move down the concentration gradient. The rate of passive diffusion depends on the size of the concentration gradient between a cell's exterior and its interior (**figure 3.10**). A large concentration gradient is required for adequate nutrient uptake by passive diffusion (i.e., the external nutrient concentration must be high while the internal concentration is low). The rate of diffusion decreases as more nutrient accumulates in the cell. This occurs if the nutrient is not used immediately upon entry.

Most substances cannot freely diffuse into a cell. However, water and some gases, including O_2 and CO_2, easily cross the

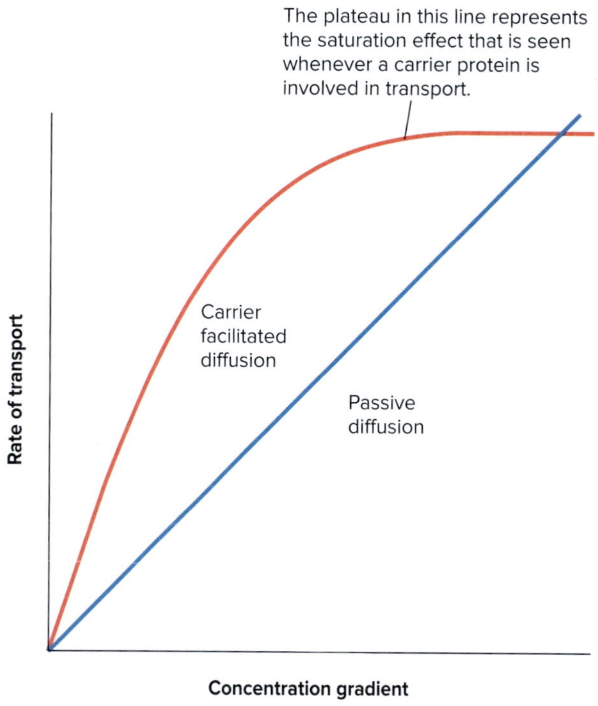

Figure 3.10 Passive and Facilitated Diffusion. The rate of diffusion depends on the size of the solute's concentration gradient (the ratio of the extracellular concentration to the intracellular concentration). This example of facilitated diffusion involves a carrier protein that can be saturated. Sometimes facilitated diffusion is mediated by a channel. Channels often do not exhibit a saturation effect.

plasma membrane by passive diffusion. H_2O also moves across membranes by passive diffusion. Larger molecules, ions, and polar substances must enter the cell by other mechanisms, all of which involve specialized proteins that are referred to as transport proteins. *How Diffusion Works*

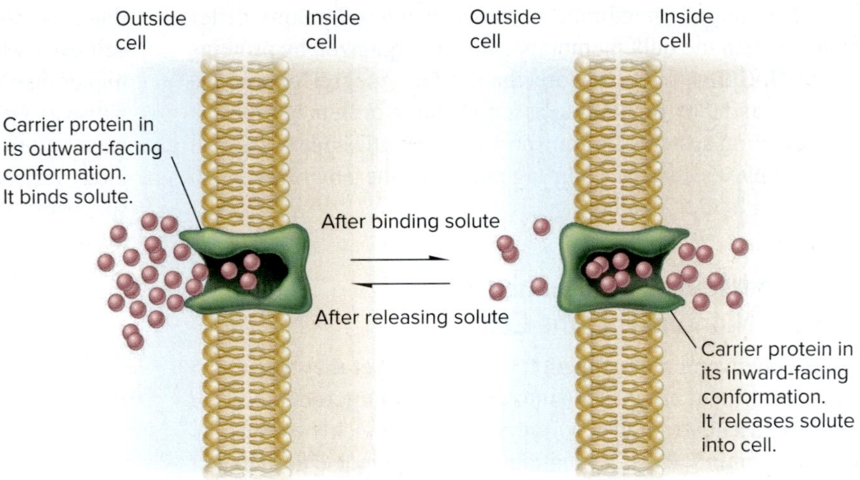

Figure 3.11 A Model of Facilitated Diffusion. Because there is no energy input, molecules continue to enter only as long as their concentration is greater on the outside.

Facilitated Diffusion

During **facilitated diffusion,** substances move across the plasma membrane with the assistance of transport proteins that are either channels or carriers. Channels, as their name indicates, are proteins that form pores in membranes through which substances can pass; they are often involved in facilitated diffusion. Channels show some specificity for the substances that pass through them, but this is considerably less than that shown by carriers, which are far more substrate specific. The rate of facilitated diffusion increases with the concentration gradient much more rapidly and at lower concentrations of the diffusing molecule than that of passive diffusion (figure 3.10). When the transporter is a carrier, the diffusion rate reaches a plateau above a specific gradient value because the carrier protein is saturated; that is, it is transporting as many solute molecules as possible. The resulting curve resembles an enzyme-substrate curve *(see figure 10.16)* and is different from the linear response seen with passive diffusion. An example of channel-mediated facilitated diffusion is that involving aquaporins (*see figure 2.31*), which transport water. Aquaporins are members of the major intrinsic protein (MIP) family of proteins. MIPs facilitate diffusion of small polar molecules, and they are observed in all organisms.

Although facilitated diffusion relies on transport proteins, it is truly diffusion because a concentration gradient spanning the membrane drives the movement of molecules, and no energy is used. If the concentration gradient disappears, net inward movement ceases. The gradient can be maintained by converting the transported nutrient to another compound, as occurs when a nutrient is metabolized.

Considerable work has been done on the mechanism of carrier-mediated facilitated diffusion. When the solute molecule binds to the outside of the carrier, it changes conformation and releases the molecule on the cell interior (**figure 3.11**). The carrier subsequently changes back to its original shape and is ready to pick up another molecule. The net effect is that a hydrophilic molecule can enter the cell in response to its concentration gradient.

Facilitated diffusion has been documented in some bacteria but it does not seem to be the major uptake mechanism for these microbes. Recall that many bacteria live in environments where nutrient concentrations are low, and facilitated diffusion cannot concentrate nutrients inside cells. Therefore energy-dependent transport mechanisms capable of concentrating nutrients are significantly more important uptake mechanisms for bacterial cells.

Primary and Secondary Active Transport

Active transport is the transport of solute molecules to higher concentrations (i.e., against a concentration gradient) with the input of metabolic energy. Three types of active transport are observed in bacteria: primary active transport, secondary active transport, and group translocation. They differ in terms of the energy used to drive transport and whether or not the transported molecule is modified as it enters.

Active transport resembles facilitated diffusion in that it involves carrier proteins. Recall that carrier proteins bind particular solutes with great specificity. Active transport is also characterized by the carrier saturation effect at high solute concentrations (figure 3.10). Nevertheless, active transport differs from facilitated diffusion because it uses metabolic energy and can concentrate substances within the cell.

Primary active transport is mediated by carriers called primary active transporters. They use energy provided by ATP hydrolysis to move substances against a concentration gradient without modifying them. Primary active transporters are **uniporters;** that is, they move a single molecule across the membrane (**figure 3.12**). ***ATP-binding cassette transporters*** (**ABC transporters**) are important primary active transporters. Our focus here is on those ABC transporters that are used for import of substances. Other ABC transporters are used for export of substances, in particular proteins; these exporters are described in chapter 13. ▶| *Protein maturation and secretion (section 13.8)*

Most ABC transporters consist of two hydrophobic membrane-spanning regions (domains) with two ATP-binding

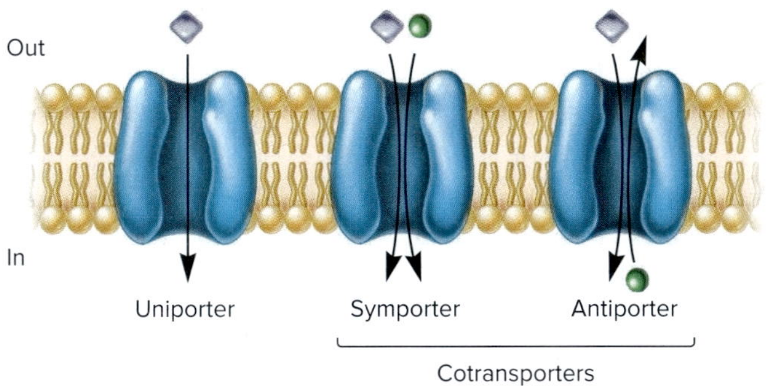

Figure 3.12 Carrier Proteins Can Be Uniporters or Cotransporters. Uniporters move a single substance into the cell. Cotransporters simultaneously move two substances across the membrane. When both substances move in the same direction, the carrier is a symporter. When the two substances move in opposite directions, the carrier is an antiporter.

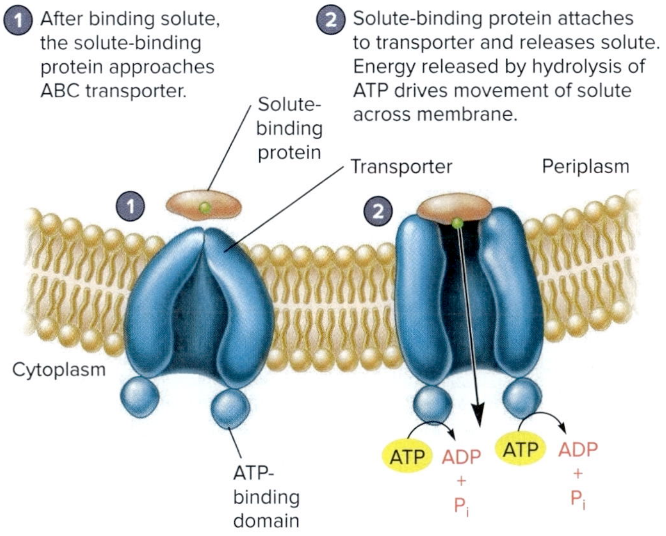

Figure 3.13 ABC Transporter Function. Shown here is a transporter that works with a solute-binding protein free in the periplasm. Other solute-binding proteins are associated with the plasma membrane, always associated with the transporter, or even fused to the transporter.

domains facing the cytoplasm (**figure 3.13**). The membrane-spanning domains form a pore in the membrane, and the ATP-binding domains bind and hydrolyze ATP to drive uptake. Most ABC transporters employ solute-binding proteins to deliver the molecule to be transported to the transporter.

Secondary active transport couples the potential energy of ion gradients to transport of substances without modifying them. Secondary active transporters are cotransporters (figure 3.12). They move two substances simultaneously: the ion whose gradient powers transport and the substance being moved across the membrane. When the ion and other substance both move in the same direction, it is called **symport.** When they move in opposite directions, it is called **antiport.** *Cotransport (Symport and Antiport)*

The ion gradients used by secondary active transporters arise primarily in three ways. The first results from bacterial metabolic activity. During energy-conserving processes, electron transport generates a proton gradient in which protons are at a higher concentration outside the cell than inside. The proton gradient is used to do cellular work, including secondary active transport. Some bacteria use the second method, in which an enzyme called a V-type ATPase hydrolyzes ATP and uses the energy released to create either a proton gradient or a sodium gradient across the plasma membrane. Finally, a proton gradient can be used to create another ion gradient such as a sodium gradient. This is accomplished by an antiporter that brings protons in as sodium ions are moved out of the cell. The sodium gradient can then be used to drive uptake of nutrients by a symport mechanism. ▶| *Electron transport and oxidative phosphorylation (step 3) generate the most ATP (section 11.6)*

The lactose permease of *E. coli* is a well-studied symport secondary active transporter. It is a single protein that transports a lactose molecule inward as a proton simultaneously enters the cell. The proton is moving down a proton gradient, and the energy released drives solute transport. X-ray diffraction studies show that the carrier protein exists in outward- and inward-facing conformations. When lactose and a proton bind to separate sites on the outward-facing conformation, the protein changes to its inward-facing conformation, and the sugar and proton are released into the cytoplasm.

Bacteria often have more than one transport system for a nutrient, as can be seen with *E. coli.* This bacterium has at least five transport systems for the sugar galactose, three systems each for the amino acids glutamate and leucine, and two potassium transport complexes. When several transport systems exist for the same substance, the systems differ in such properties as their energy source, their affinity for the solute transported, and the nature of their regulation. This diversity gives the bacterium an added competitive advantage in a variable environment.

Group Translocation

The distinguishing characteristic of **group translocation** is that a molecule is chemically modified as it is brought into the cell. The best-known group translocation system is the **phosphoenolpyruvate: sugar phosphotransferase system (PTS),** which is observed in many bacteria. The PTS transports a variety of sugars while phosphorylating them, using phosphoenolpyruvate (PEP) as the phosphate donor. PEP is a high-energy molecule that can be used to synthesize ATP, the cell's energy currency. However, when it is used in PTS reactions, the energy present in PEP is used to energize sugar uptake rather than ATP synthesis. ▶| *ATP: The major energy currency of cells (section 10.2)*

The transfer of phosphate from PEP to the incoming molecule involves several proteins and is an example of a **phosphorelay system.** In *E. coli* and *Salmonella,* the PTS consists of two enzymes and a low molecular weight heat-stable protein (HPr). A phosphate is transferred from PEP to enzyme II with the aid of

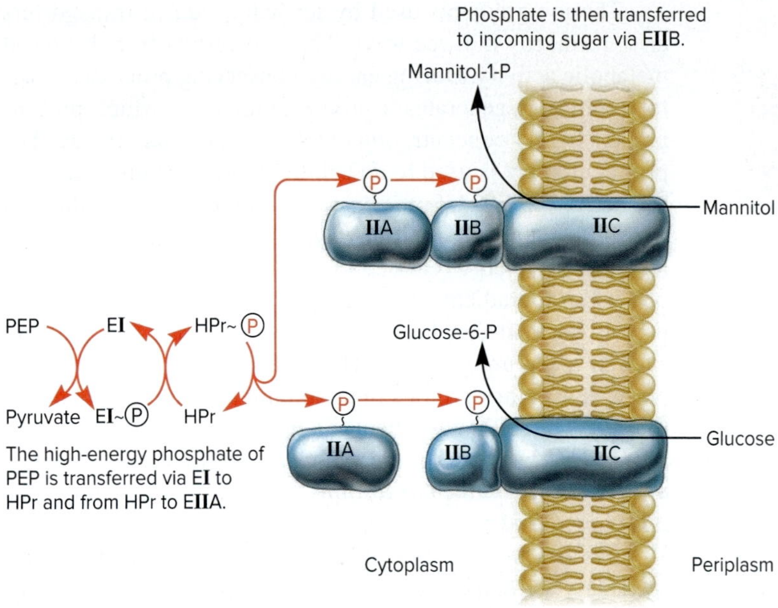

Figure 3.14 Group Translocation: Bacterial PTS Transport. Two examples of the phosphoenolpyruvate: sugar phosphotransferase system (PTS) are illustrated. The following components are involved in the system: phosphoenolpyruvate (PEP), enzyme I (EI), the low molecular weight heat-stable protein (HPr), and enzyme II (EII). EIIA is attached to EIIB in the mannitol transport system and is separate from EIIB in the glucose system.

enzyme I and HPr (**figure 3.14**). Enzyme II then phosphorylates the sugar molecule as it is carried across the membrane. Many different PTSs exist, and they vary in terms of the sugars they transport. The specificity lies with the type of Enzyme II used in the PTS. Enzyme I and HPr are the same in all PTSs used by a bacterium. ▶| *Enzymes and ribozymes speed up cellular chemical reactions (section 10.6)*

PTSs are widely distributed in bacteria, primarily among facultatively anaerobic bacteria (bacteria that grow in either the presence or absence of O_2); some obligately anaerobic bacteria (e.g., *Clostridium* spp.) also have PTSs. However, most aerobic bacteria lack PTSs. Many carbohydrates are transported by PTSs. *E. coli* takes up glucose, fructose, mannitol, sucrose, *N*-acetylglucosamine, cellobiose, and other carbohydrates by group translocation. *Active Transport by Group Translocation*

Iron Uptake

Almost all microorganisms require iron for building molecules important in energy-conserving processes (e.g., cytochromes), as well as for the function of many enzymes. Iron uptake is made difficult by the extreme insolubility of ferric iron (Fe^{3+}) and its derivatives, which leaves little free iron available for transport. Many bacteria overcome this difficulty by secreting **siderophores** (Greek for iron bearers). Siderophores are low molecular weight organic molecules that bind ferric iron and supply it to the cell (**figure 3.15**). ▶| *Electron transport chains: sets of sequential redox reactions (section 10.4)*

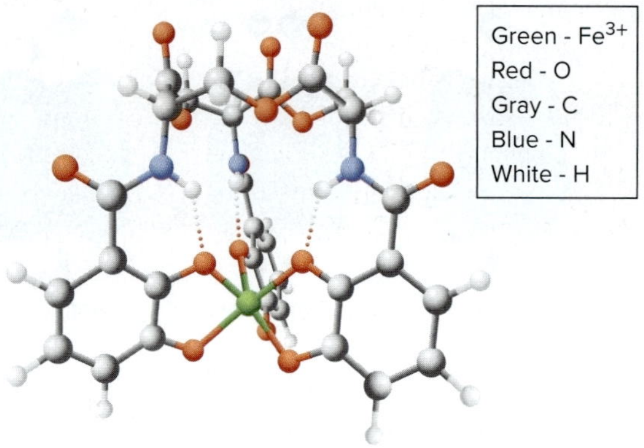

Figure 3.15 Enterobactin: A Siderophore Produced by *E. coli*. Ball-and-stick model of enterobactin complexed with Fe^{3+}.

Microorganisms secrete siderophores when iron is scarce in the medium. Once the iron-siderophore complex has reached the cell surface, it binds to a siderophore-receptor protein. Then either the iron is released to enter the cell directly or the whole iron-siderophore complex is transported inside by an ABC transporter. Iron is so crucial to microorganisms that more than one route of iron uptake may be used to ensure an adequate supply.

Comprehension Check

1. List the functions of bacterial plasma membranes. Why must their plasma membranes carry out more functions than the plasma membranes of eukaryotic cells?
2. Describe in words and with a labeled diagram the fluid mosaic model for cell membranes.
3. On what basis are elements divided into macroelements and trace elements?
4. Describe facilitated diffusion, primary and secondary active transport, and group translocation in terms of their distinctive characteristics and mechanisms. What advantage does a bacterium gain by using active transport rather than facilitated diffusion?
5. What are uniport, symport, and antiport?
6. What are siderophores? Why are they important?

3.4 There Are Two Main Types of Bacterial Cell Walls

After reading this section, you should be able to:

a. Describe peptidoglycan structure
b. Compare and contrast the cell walls of typical Gram-positive and Gram-negative bacteria
c. Relate bacterial cell wall structure to the Gram-staining reaction

The **cell wall** is the layer that lies just outside the plasma membrane. It is one of the most important structures for several reasons:

It helps maintain cell shape and protect the cell from osmotic lysis; it can protect the cell from toxic substances; and in pathogens, it can contribute to pathogenicity. Cell walls are so important that most bacteria have them. Those that lack them have other features that fulfill cell wall function. Bacterial cell wall synthesis is an important target for many antibiotics. ▶| *Antibacterial drugs (section 9.4)*

Overview of Bacterial Cell Wall Structure

After Christian Gram developed the Gram stain in 1884, it soon became evident that most bacteria could be divided into two major groups based on their response to the Gram-staining procedure. Gram-positive bacteria stained purple, whereas Gram-negative bacteria were pink or red. The true structural difference between these two groups did not become clear until the advent of the transmission electron microscope. Here we describe the long-held models of Gram-positive and Gram-negative cell walls developed from these studies. More recent studies of diverse groups of bacteria have shown that these models do not hold true for all bacteria (**Microbial Diversity & Ecology 3.1**). Because of ongoing discussions related to these new studies, we will refer to bacteria that fit the models as being typical Gram-positive or typical Gram-negative bacteria. |◀ *Differential staining (section 2.3)*

The cell walls of *Bacillus subtilis* and many other typical Gram-positive bacteria consist of a single, 20- to 80-nm-thick homogeneous layer of **peptidoglycan (murein)** lying outside the plasma membrane (**figure 3.16**). In contrast, the cell walls of *E. coli* and many other typical Gram-negative bacteria have two distinct layers: a 2- to 7-nm-thick peptidoglycan layer covered by a 7- to 8-nm-thick **outer membrane.**

One important feature seen in typical Gram-negative bacteria is a space between the plasma membrane and the outer membrane. It also is sometimes observed between the plasma membrane and cell wall in typical Gram-positive bacteria. This space is called the **periplasmic space.** The substance that occupies the periplasmic space is the **periplasm.**

Peptidoglycan Structure

The feature common to nearly all bacterial cell walls is the presence of peptidoglycan, which forms an enormous meshlike structure often referred to as the peptidoglycan sacculus. Peptidoglycan is composed of many identical subunits. Each subunit within the sacculus contains two sugar derivatives, *N*-acetylglucosamine (NAG) and *N*-acetylmuramic acid (NAM), and several different amino acids. The amino acids form a short peptide, sometimes called the stem peptide, consisting of four alternating D- and L-amino acids; the peptide is connected to the carboxyl group of NAM (**figure 3.17**). Three of the amino acids are not found in proteins: D-glutamic acid, D-alanine, and *meso*-diaminopimelic acid. The presence of D-amino acids in the stem peptide protects against degradation by most peptidases, which recognize only the L-isomers of amino acid residues. The peptidoglycan subunit of many bacteria is shown in figure 3.17. ▶| *Carbohydrates (appendix I); Proteins (appendix I); Proteins are polymers of amino acids (section 13.2)*

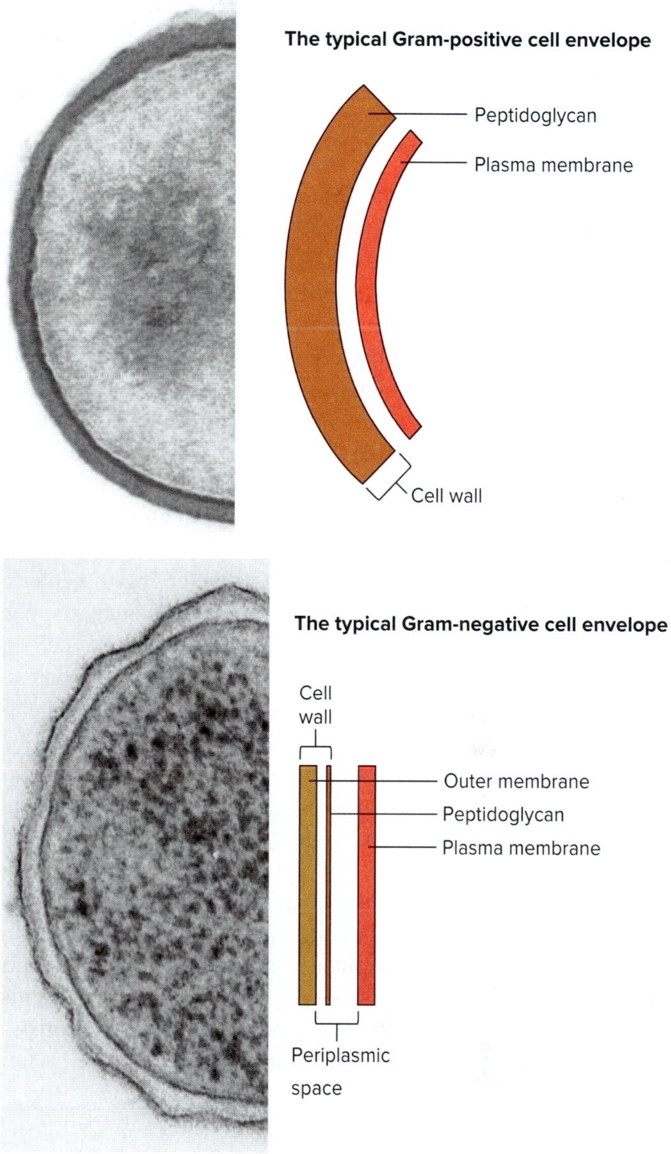

©Egbert Hoiczyk

Figure 3.16 Cell Envelopes of Typical Gram-Positive and Gram-Negative Bacteria. Cell envelopes consist of the plasma membrane and any layers (e.g., cell wall) exterior to it. For simplicity, we show only the plasma membrane and cell wall. *Staphylococcus aureus* (top) has a typical Gram-positive cell wall that consists primarily of peptidoglycan. *Myxococcus xanthus* (bottom) has a typical Gram-negative cell wall consisting of a thin layer of peptidoglycan, an outer membrane, and the periplasmic space.

The peptidoglycan sacculus is formed by linking the sugars of the peptidoglycan subunits together to form a strand; the strands are then cross-linked to each other by covalent bonds formed between the stem peptides extending from each strand. As seen in **figure 3.18**, the backbone of each strand is composed of alternating NAG and NAM residues. The strand is helical, and the stem peptides extend out from the backbone in different directions. There are two types of cross-links: direct and indirect

MICROBIAL DIVERSITY & ECOLOGY

3.1 Gram Positive and Gram Negative or Monoderms and Diderms?

The importance of the Gram stain in the history of microbiology cannot be overstated. The Gram stain reaction was for many years one of the critical pieces of information used by bacterial taxonomists to construct taxa, and it is still useful in identifying bacteria in clinical settings. The initial studies done to differentiate bacteria that stained Gram positive from those that stain Gram negative were done using model organisms such as *Bacillus subtilis* (Gram positive) and *Escherichia coli* (Gram negative). At the time, it was thought that all other bacteria would have similar cell wall structures. However, the long-held models of Gram-positive and Gram-negative cell walls do not hold true for all bacteria. Iain Sutcliffe has proposed that microbiologists stop referring to bacteria as either Gram positive or Gram negative. He suggests that instead we should more precisely describe bacterial cell envelope architectures by focusing on the observation that some bacteria have envelopes with a single membrane—the plasma membrane as seen in typical Gram-positive bacteria—while others have envelopes with two membranes—the plasma membrane and an outer membrane as seen in typical Gram-negative bacteria. He proposed calling the former monoderms and the latter diderms.

But why make this change? Sutcliffe begins by pointing out that some bacteria staining Gram positive are actually diderms and some staining Gram negative are actually monoderms. By referring to Gram-positive-staining diderms as Gram-positive bacteria, it is too easy to mislead scientists and many a budding microbiologist into thinking that the bacterium has a typical Gram-positive envelope. He also argues that by relating cell envelope architecture to the phylogenies of various bacterial taxa, we may gain insight into the evolution of these architectures. He notes that the phyla *Firmicutes* and *Actinobacteria* are composed almost completely of monoderm bacteria, whereas almost all other bacterial phyla consist of diderms.

There are interesting exceptions to the relationship of phylogeny and cell envelope structure. For instance, members of the genus *Mycobacterium* (e.g., *M. tuberculosis*) belong to the predominantly monoderm phylum *Actinobacteria*. Mycobacteria have cell walls that consist of peptidoglycan and an outer membrane. The outer membrane is composed of mycolic acids rather than the phospholipids and lipopolysaccharides (LPSs) found in the typical Gram-negative cells' outer membrane. ▶| *Order* Corynebacteriales *includes important human pathogens (section 23.1)*

Members of the genus *Deinococcus* are another interesting exception. These bacteria stain Gram positive but are diderms. Their cell envelopes consist of the plasma membrane, what appears to be a typical Gram-negative cell wall, and an S-layer. Their outer membrane is distinctive because it lacks LPS. Deinococci are not unique in this respect, however. It is now known that members of several taxa have outer membranes that lack LPS.

Source: Sutcliffe, I. C. 2010. A phylum level perspective on bacterial cell envelope architecture. Trends Microbiol. *18(10):464–70.*

via a peptide interbridge. A direct cross-link is characterized by connecting the carboxyl group of an amino acid in one stem peptide to the amino group of an amino acid in another stem peptide. For instance, many bacteria cross-link the strands by connecting the carboxyl group of the D-alanine at position 4 of the stem peptide directly to the amino group of diaminopimelic acid (position 3) of the other peptidoglycan strand's stem peptide (the position 5 D-alanine is removed as the cross-link is formed). Bacteria that have indirect linkage use a **peptide interbridge** (also called an interpeptide bridge), a short chain of amino acids that links the stem peptide of one peptidoglycan strand to that of another (**figure 3.19**). Cross-linking results in one dense, interconnected network of peptidoglycan strands ▶| *Synthesis of peptidoglycan occurs in the cytoplasm, at the plasma membrane, and in the periplasmic space (section 12.4)*

The peptidoglycan sacculus is strong but elastic. It is able to stretch and contract in response to osmotic pressure. This is due to the rigidity of the backbone coupled with the flexibility of the cross-links. Peptidoglycan sacculi are also rather porous, allowing globular proteins having a molecular weight as large as 50,000 to pass through, depending on whether the sacculus is relaxed or stretched; thus only extremely large proteins are unable to pass through peptidoglycan.

Variants of peptidoglycan are found, particularly among typical Gram-positive bacteria. For example, some substitute the diamino acid lysine for *meso*-diaminopimelic acid (**figure 3.20**) and cross-link chains via interpeptide bridges. These interpeptide bridges can vary considerably (**figure 3.21**). Peptidoglycan can also vary in terms of the length of the peptidoglycan strands and the amount of cross-linking. Bacteria that stain Gram positive tend to have much more cross-linking, whereas those that stain Gram negative have considerably less.

Typical Gram-Positive Cell Walls Consist Primarily of Peptidoglycan

Most cultured bacteria that stain Gram positive belong to the phyla *Firmicutes* and *Actinobacteria*, and most of these bacteria have

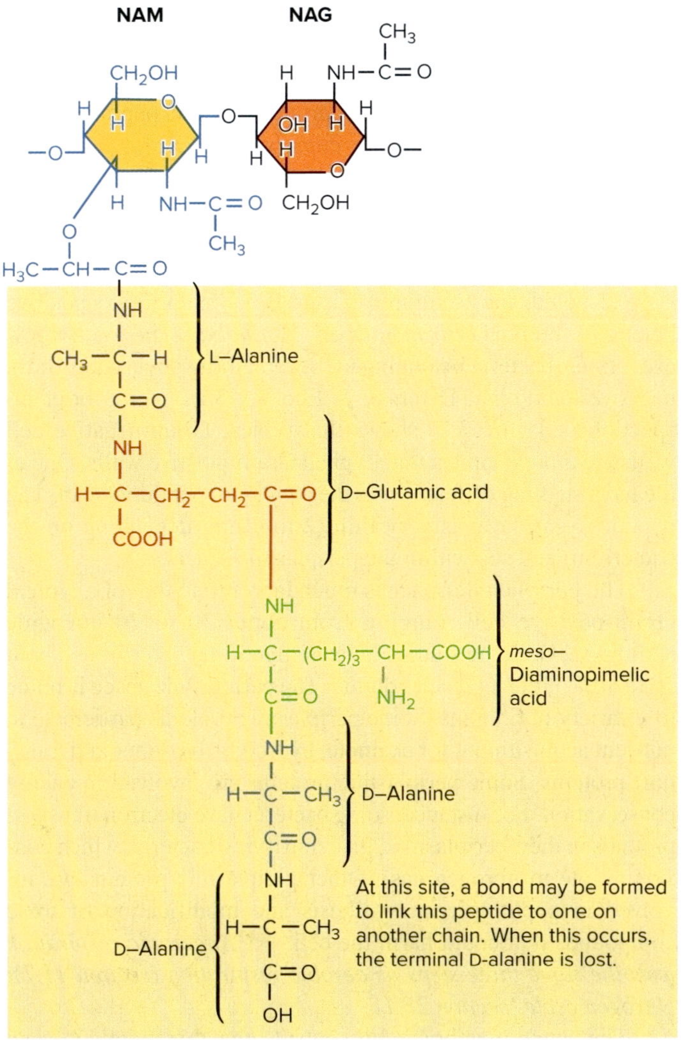

Figure 3.17 Peptidoglycan Subunit Composition. Shown is the peptidoglycan subunit of *E. coli*, many other typical Gram-negative bacteria, and many typical Gram-positive bacteria. This illustration shows the subunit before it has been inserted into the existing peptidoglycan polymer. NAG is *N*-acetylglucosamine. NAM is *N*-acetylmuramic acid. The stem peptide is composed of alternating D- and L-amino acids; it terminates with two D-alanines. The amino acids are shown in different colors for clarity.

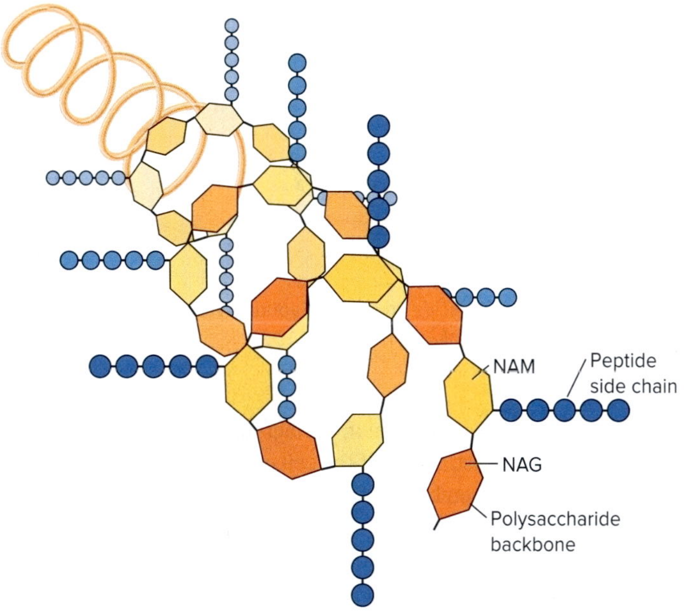

Figure 3.18 A Helical Peptidoglycan Strand. Because of the strand's helical nature, the stem peptides project out in different directions from the NAM-NAG backbone. Here the stem peptides are shown projecting out at 90-degree angles. Some studies suggest that the angle is actually 120 degrees.

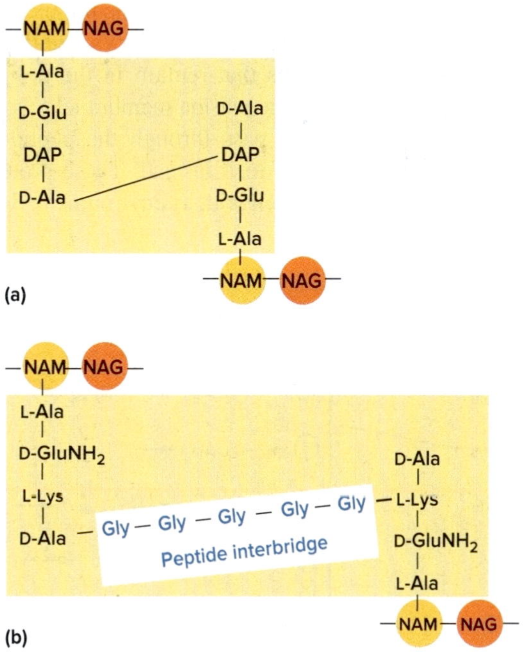

Figure 3.19 Peptidoglycan Cross-Links Can Be Direct or Indirect via a Peptide Interbridge. (a) *E. coli* peptidoglycan with direct cross-linking, typical of many Gram-negative bacteria. (b) *Staphylococcus aureus* peptidoglycan with an interbridge. *S. aureus* stains Gram positive. Gly is glycine. D-GluNH$_2$ is D-glutamic acid with an NH$_2$ group attached to the α carbon (the carbon next to the carboxyl group).

thick cell walls composed of peptidoglycan and large amounts of other polymers such as teichoic acids (**figure 3.22**). **Teichoic acids** are polymers of glycerol or ribitol joined by phosphate groups (**figure 3.23**). Some teichoic acids are covalently linked to peptidoglycan and are referred to as wall teichoic acids. Others are covalently connected to the plasma membrane; they are called lipoteichoic acids. Wall teichoic acids extend beyond the surface of the peptidoglycan. They are negatively charged and help give the cell wall its negative charge. Teichoic acids are not present in other bacteria.

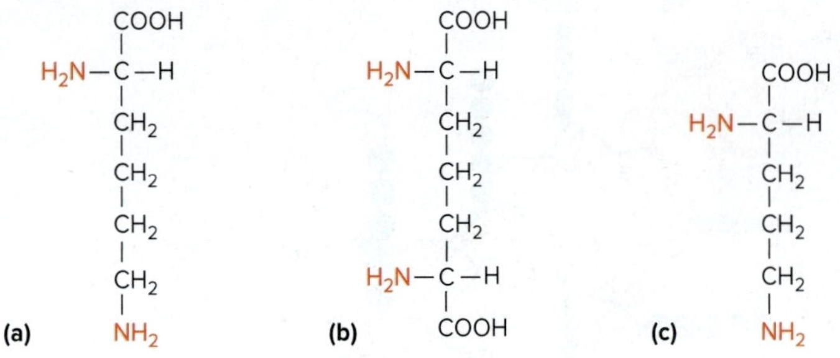

Figure 3.20 Diamino Acids Present in Peptidoglycan. (a) L-lysine, (b) *meso*-diaminopimelic acid, (c) D-ornithine.

Teichoic acids have several important functions. They help create and maintain the structure of the cell envelope by anchoring the wall to the plasma membrane. They are important during cell division, and they protect the cell from harmful substances in the environment (e.g., antibiotics and host defense molecules). In addition, they function in ion uptake and are involved in binding pathogenic species to host tissues, thus initiating the infectious disease process.

The periplasmic space lies between the plasma membrane and the cell wall and is so narrow that it is often not visible by electron microscopy. The periplasm has relatively few proteins; this is probably because the peptidoglycan sacculus is so porous that many proteins translocated across the plasma membrane pass through the sacculus. Some secreted proteins are enzymes called **exoenzymes.** Exoenzymes often serve to degrade polymers such as proteins and polysaccharides that would otherwise be too large for transport across the plasma membrane; the degradation products, the monomer building blocks, are then taken up by the cell. Those proteins that remain in the periplasmic space are usually attached to the plasma membrane.

Not all secreted proteins pass through the peptidoglycan sacculus; some become bound to it instead. These proteins are involved in interactions of the cell with its environment. Some are noncovalently bound to teichoic acids or other cell wall polymers, while others are covalently attached to the peptidoglycan. Membrane-bound enzymes called sortases catalyze the formation of covalent bonds that join these proteins to the peptidoglycan. ▶| *Protein maturation and secretion (section 13.8)*

Typical Gram-Negative Cell Walls Include Additional Layers Besides Peptidoglycan

As just noted, most cultured bacteria that stain Gram positive belong to the phyla *Firmicutes* and *Actinobacteria.* With a few exceptions, bacteria belonging to the remaining phyla stain Gram negative (Microbial Diversity & Ecology 3.1). Even a brief inspection of figure 3.16 shows that typical Gram-negative cell walls are more complex than typical Gram-positive walls. One of the most striking differences is the paucity of peptidoglycan. The peptidoglycan layer is very thin (2 to 7 nm, depending on the bacterium) and sits within the periplasmic space.

The periplasmic space is much larger than that of a typical Gram-positive cell, ranging from about 30 to 70 nm wide (**figure 3.24**). Some studies indicate that it may constitute about 20 to 40% of the total cell volume. The periplasmic space is home to a variety of proteins. Some periplasmic proteins participate in nutrient acquisition—for example, hydrolytic enzymes and transport proteins. Some periplasmic proteins are involved in energy conservation. For instance, some bacteria have electron transport proteins in their periplasm (e.g., denitrifying bacteria, which convert nitrate to nitrogen gas). Other periplasmic proteins are involved in peptidoglycan synthesis and modification of toxic compounds that could harm the cell. ▶| *Anaerobic respiration uses the same three steps as aerobic respiration (section 11.7); Nitrogen cycle (section 28.1)*

The outer membrane lies outside the thin peptidoglycan layer. It is linked to the cell by Braun's lipoprotein, the most abundant protein in the outer membrane (figure 3.24). This small

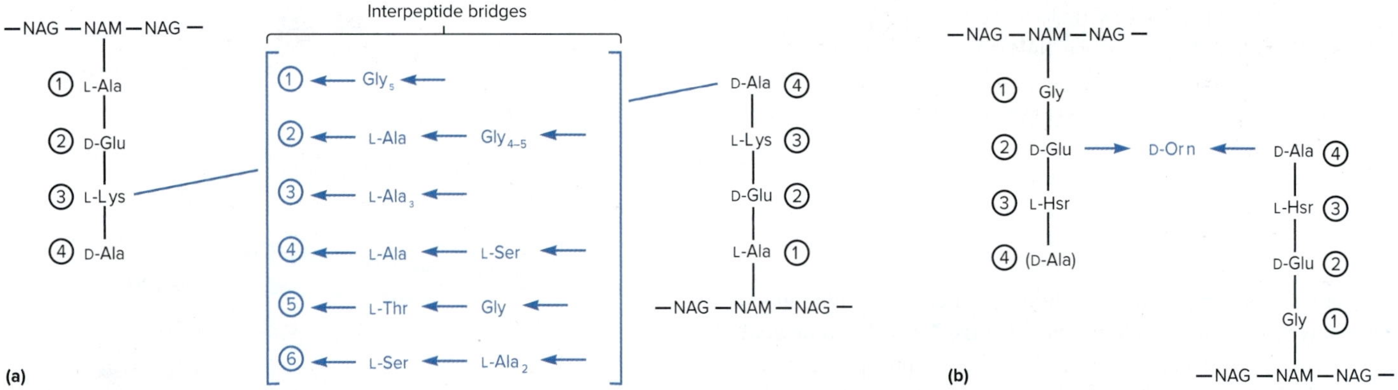

Figure 3.21 Examples of Peptidoglycan Cross-Links. Most variation in peptidoglycan structure occurs in the composition of the stem peptide and the method by which peptidoglycan strands are cross-linked. (a) Examples of interpeptide bridges that link D-alanine in position 4 with L-lysine in position 3. The bracket contains six typical bridges: (1) *Staphylococcus aureus,* (2) *S. epidermidis,* (3) *Micrococcus roseus* and *Streptococcus thermophilus,* (4) *Lactobacillus viridescens,* (5) *Streptococcus salivarius,* and (6) *Leuconostoc cremoris*. The arrows indicate the polarity of peptide bonds running in the C to N direction. (b) An interpeptide bridge observed in *Corynebacterium poinsettiae*. The bridge extends between positions 2 and 4 and consists of the D-diamino acid ornithine (figure 3.20c). Note that L-homoserine (L-Hsr) is in position 3 rather than *meso*-diaminopimelic acid or L-lysine.

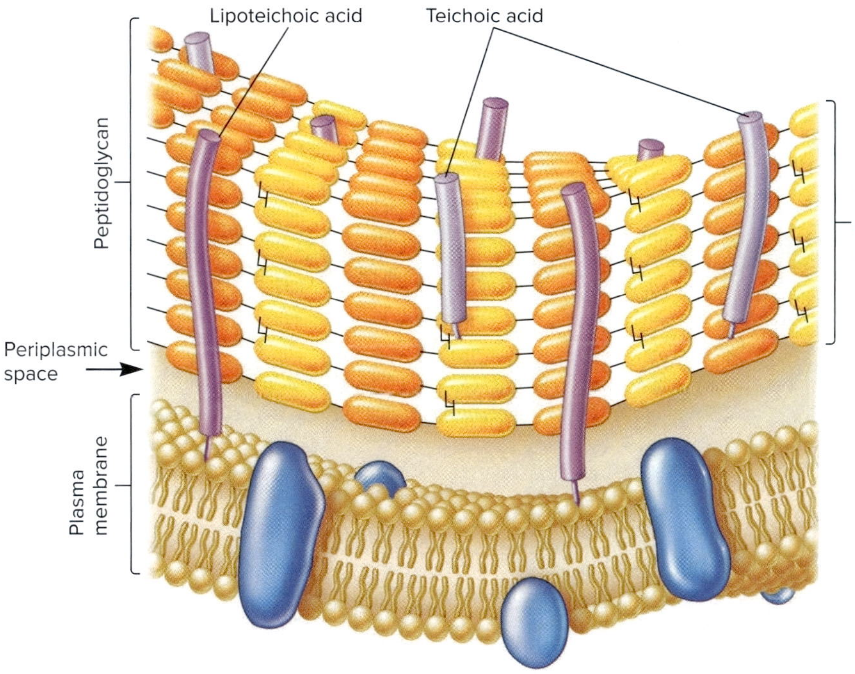

Figure 3.22 **Typical Gram-Positive Cell Wall.** This component of the cell envelope lies just outside the plasma membrane.

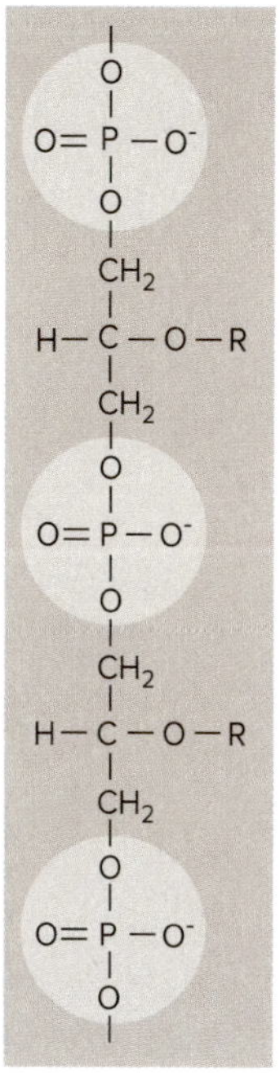

Figure 3.23 **Teichoic Acid Structure.** The segment of a teichoic acid made of phosphate, glycerol, and a side chain, R. R may represent D-alanine, glucose, or other molecules.

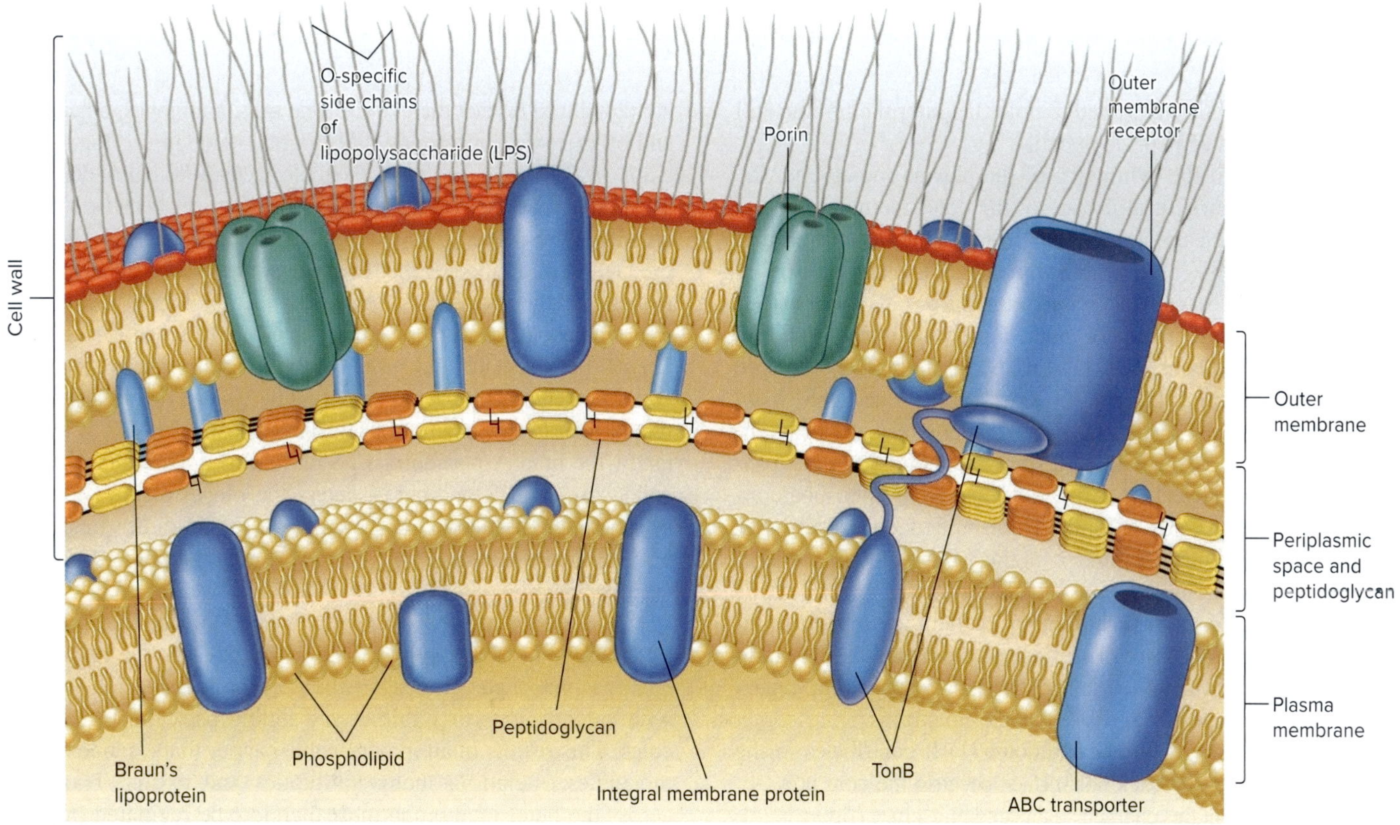

Figure 3.24 **Typical Gram-Negative Cell Wall.** Notice that these bacteria are bounded by two membranes, the plasma membrane and the outer membrane of the cell wall.

MICRO INQUIRY *How does the outer membrane of the cell wall differ from the plasma membrane?*

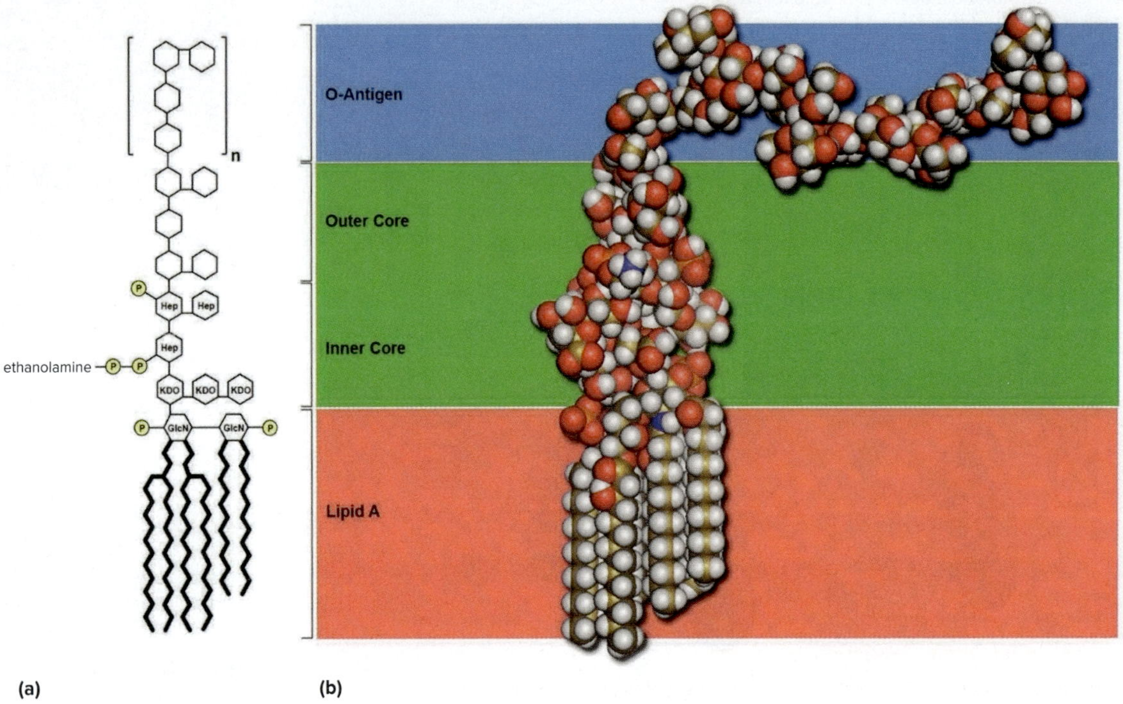

Figure 3.25 Lipopolysaccharide Structure. (a) A simplified diagram of LPS. Abbreviations: GlcN, glucosamine; Hep, heptulose; KDO, 2-keto-3-deoxyoctonate; P, phosphate. Lipid A is buried in the outer membrane. (b) Molecular model of a lipopolysaccharide. The lipid A and core polysaccharide are straight; the O side chain is bent at an angle in this model.

lipoprotein is covalently joined to both the outer membrane and the peptide chain of peptidoglycan.

In contrast to the symmetry of the plasma membrane, the most distinctive structural feature of the outer membrane is the difference between the inner leaflet (facing the periplasm) and the outer leaflet (facing the environment). This external layer of the Gram-negative cell is comprised of **lipopolysaccharides (LPSs)**. These large, complex molecules contain both lipid and carbohydrate, and consist of three parts: (1) lipid A, (2) the core polysaccharide, and (3) the O side chain. The LPS from *Salmonella* spp. has been studied most, and its general structure is described here (**figure 3.25**). **Lipid A** contains two glucosamine sugar derivatives, each with fatty acids and phosphate attached. The fatty acids of lipid A are embedded in the outer membrane, while the remainder of the LPS molecule projects from the surface. The **core polysaccharide** is joined to lipid A and is constructed of 10 sugars, many of them unusual in structure. The **O side chain** or **O antigen** is a polysaccharide chain extending outward from the core. It has several peculiar sugars and varies in composition between bacterial strains.

LPS has many important functions. (1) It contributes to the negative charge on the bacterial surface because the core polysaccharide usually contains charged sugars and phosphate (figure 3.25). (2) It helps stabilize outer membrane structure because lipid A is a major constituent of the exterior leaflet of the outer membrane. (3) It helps create a permeability barrier. The geometry of LPS (figure 3.25*b*) and interactions between neighboring LPS molecules are thought to restrict the entry of bile salts, antibiotics, detergents, and other toxic substances that might kill or injure the bacterium. (4) LPS helps protect pathogenic bacteria from host defenses. The O side chain of LPS is also called the O antigen because it elicits an immune response by an infected host. This response involves the production of antibodies that bind the strain-specific form of LPS that elicited the response. For example, microbiologists refer to specific strains of Gram-negative bacteria using the O antigen, such as *E. coli* O157; here the O side chain is the antigenic type, number 157. Unfortunately, many bacteria can rapidly change the antigenic nature of their O side chains, thus thwarting host defenses. (5) Importantly, the lipid A portion of LPS can act as a toxin and is called endotoxin. If LPS or lipid A enters the bloodstream, a form of septic shock develops for which there is no direct treatment. ▶| *Endotoxins (section 35.4); Antibodies bind specific 3-D antigens (section 33.7)*

As the outermost layer of the cell envelope, the LPS presents an effective barrier for the cell. Within the LPS, the negatively charged phosphate groups on the core polysaccharide interact with calcium ions to stabilize and tightly pack the LPS molecules. This results in an impermeable barrier that excludes small molecules, including many antibiotics and toxins. Transport across the outer membrane must therefore be regulated by integral membrane proteins, including one class of proteins termed **porins.** Most porin proteins cluster together to form a trimer in the outer membrane (figure 3.24 and **figure 3.26**). Each porin protein spans the outer membrane and is more or less

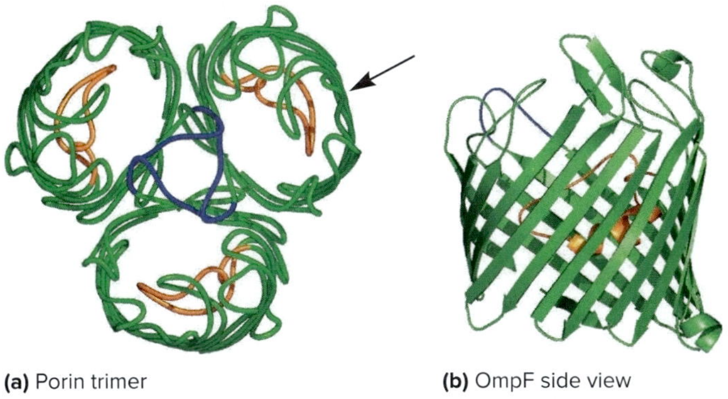

Figure 3.26 Porin Proteins. Two views of the OmpF porin of *E. coli.* (a) Structure of a trimeric porin observed when looking down at the outer surface of the outer membrane (i.e., top view). The center of each porin monomer is a water-filled channel. Each OmpF monomer can be divided into three parts: the green barrel forms the channel, the blue segment interacts with other porin proteins to help form the trimer, and the orange loop narrows the channel and dictates the chemical selectivity of the porin. The arrow indicates the area of a porin molecule viewed from the side in panel (b). Side view of a porin monomer showing the β-barrel structure characteristic of porin proteins.

MICRO INQUIRY *Are these transporter proteins categorized as channels or carriers?*

tube-shaped; its narrow, water-filled channel allows passage of nutrients and hydrophilic molecules smaller than about 600 daltons. Larger molecules such as vitamin B_{12}, carbohydrates, and iron complexes are too large for porins and cross the outer membrane through the action of specific outer membrane receptors (figure 3.24). The receptor-solute complex in the outer membrane interacts with the TonB complex of proteins spanning the periplasm. TonB transfers the energy for the ligand transport from the plasma membrane across the periplasm to the receptor. Upon entering the periplasm, a periplasmic transport protein shuttles the ligand to an ABC-transporter in the plasma membrane for transfer to the cytoplasm.

Mechanism of Gram Staining

The difference between typical Gram-positive bacteria and typical Gram-negative bacteria is due to the physical nature of their cell walls. If the cell wall is removed, typical Gram-positive bacteria stain Gram negative. Furthermore, bacteria that never make cell walls, such as mycoplasmas, also stain Gram negative. ◀ *Differential staining (section 2.3)*

During the Gram-staining procedure, bacteria are first stained with crystal violet, a dye with a positive charge that is attracted to the bacterial cell's net negative charge. They are next treated with iodine, a mordant that interacts with the crystal violet, forming an insoluble complex and thus promoting dye retention. When bacteria are then treated with ethanol in the decolorization step, the pores of the thick peptidoglycan found in the cell walls of typical Gram-positive bacteria appear to shrink, causing the peptidoglycan to act as a permeability barrier that prevents loss of crystal violet. Thus the dye-iodine complex is retained during the decolorization step and the bacteria remain purple, even after the addition of a second dye. In contrast, the peptidoglycan in typical Gram-negative cell walls is very thin, not as highly cross-linked, and has larger pores. Alcohol treatment also may extract enough lipid from the outer membrane to increase the cell wall's porosity further. For these reasons, alcohol more readily removes the crystal violet–iodine complex, decolorizing the cells. The counterstain safranin, also a dye with a net negative charge, easily stains the decolorized cells so that they appear red or pink.

Cell Walls and Osmotic Protection

Microbes have several mechanisms for responding to changes in osmotic pressure. Osmotic stress arises when the concentration of solutes inside the cell differs from that outside, and the responses work to equalize the solute concentrations. However, in certain situations, osmotic pressure can exceed the cell's ability to acclimate. In these cases, additional protection is provided by the cell wall. When cells are in hypotonic solutions—in which the solute concentration is less than that in the cytoplasm—water diffuses into the cell to dilute the solutes, causing it to swell. Without the peptidoglycan layer of the cell wall, the pressure on the plasma membrane would become so great that the membrane would be disrupted and the cell would burst—a process called **lysis.** Conversely, in hypertonic solutions, water flows out and the cytoplasm shrivels up—a process called **plasmolysis.**

The protective nature of peptidoglycan is most clearly demonstrated when bacterial cells are treated with lysozyme or penicillin. The enzyme **lysozyme** attacks peptidoglycan by hydrolyzing the bond that connects *N*-acetylmuramic acid with *N*-acetylglucosamine (figure 3.17; *see also figure 32.4*). Penicillin works by a different mechanism. It inhibits the enzyme transpeptidase, which is responsible for making the cross-links between peptidoglycan chains. If bacteria are treated with either of these substances while in a hypotonic solution, they lyse. However, if they are in an isotonic solution, they can survive and grow normally. Treatment of typical Gram-positive bacteria with lysozyme or penicillin results in the complete loss of the cell wall, and the cell becomes a protoplast. When typical Gram-negative bacteria are exposed to lysozyme or penicillin, the peptidoglycan sacculus is destroyed, but the outer membrane remains. These cells are called **spheroplasts.** Both protoplasts and spheroplasts are osmotically sensitive. If they are transferred to a hypotonic solution, they lyse due to uncontrolled water influx (**figure 3.27**). ▶ *Antibacterial drugs (section 9.4)*

Bacteria That Lack Cell Walls

Most bacteria are defined by their cell wall—they are either Gram positive or Gram negative. A few bacteria, most notably a group called the mycoplasmas, are defined by their lack of a cell wall. Without a cell wall, they are pleomorphic and osmotically sensitive

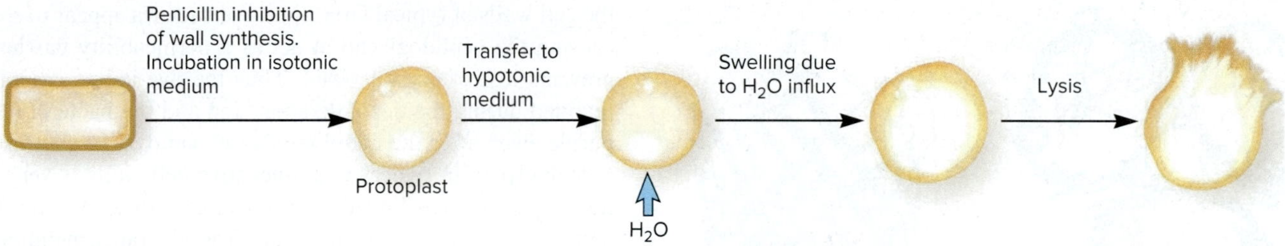

Figure 3.27 Protoplast Formation and Lysis. Protoplast formation induced by incubation with penicillin in an isotonic medium. Transfer to a hypotonic medium will result in lysis.

(see figure 21.3). Despite this, they often grow in dilute media or terrestrial environments because their plasma membranes are more resistant to osmotic pressure than those of walled bacteria. The precise reason for this is not clear, although the presence of sterols in the membranes of many species may provide added strength.

Comprehension Check

1. Describe in detail the composition and structure of peptidoglycan. Why does peptidoglycan contain the unusual D-isomers of alanine and glutamic acid rather than the L-isomers observed in proteins?
2. Draw a Venn diagram to classify the major molecules that make up typical Gram-positive and Gram-negative cell walls and those common to both types of cell wall. How do these molecules contribute to the functions of the cell wall?
3. When protoplasts and spheroplasts are made, the shape of the cell becomes spherical regardless of the original cell shape. Why does this occur?
4. The cell walls of most members of the phyla *Firmicutes* and *Actinobacteria* lack porins. Why is this the case?
5. What two mechanisms allow the passage of nutrients across the outer membrane of typical Gram-negative bacteria?

3.5 The Cell Envelope Often Includes Layers Outside the Cell Wall

After reading this section, you should be able to:

a. List the structures found in all the layers of bacterial cell envelopes
b. Identify the functions and major component molecules in cell envelope structures

Figures 3.22 and 3.24 show the bacterial envelope as consisting solely of the plasma membrane and cell wall. However, many bacteria have another layer in their cell envelopes that lies outside the cell wall. This layer is given different names depending on its makeup and how it is organized.

Capsules and Slime Layers

Capsules are layers that are well organized and not easily washed off (**figure 3.28**). They are most often composed of polysaccharides, but some are constructed of other materials. For example,

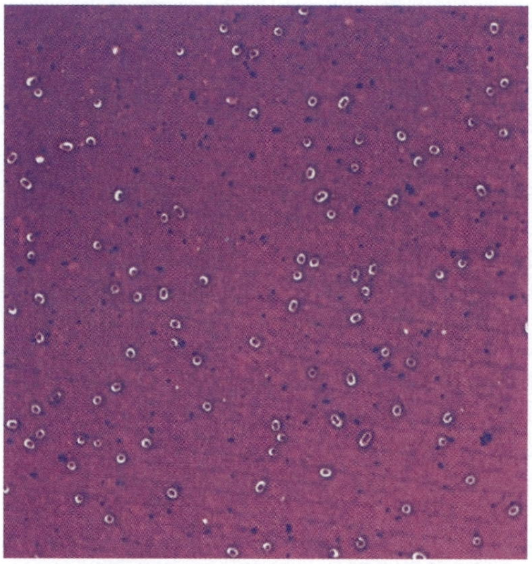

K. pneumoniae

Figure 3.28 Bacterial Capsules. Capsule stain (a negative stain) of *Klebsiella pneumoniae*, bright-field light microscopy. (X1,000)

Bacillus anthracis (the anthrax bacterium) has a proteinaceous capsule composed of poly-D-glutamic acid. Capsules are clearly visible in the light microscope when negative stains or specific capsule stains are employed; they also can be studied with the electron microscope.

Capsules are not required for growth and reproduction in laboratory cultures. However, they confer several advantages when bacteria grow in their normal habitats. They help pathogenic bacteria resist phagocytosis by host phagocytes. *Streptococcus pneumoniae*, which causes ear infections, pneumonia, and other diseases, provides a dramatic example. When it lacks a capsule, it is phagocytosed easily and does not cause disease. On the other hand, the capsulated variant commonly causes disease. Capsules can also protect against desiccation because they contain a great deal of water. They exclude viruses and most hydrophobic toxic materials such as detergents.

A **slime layer** is a zone of diffuse, unorganized material that is removed easily. It is usually composed of polysaccharides but is not as easily observed by light microscopy. Gliding bacteria often produce slime, which in some cases has been shown to facilitate motility (section 3.8).

A bacterial **glycocalyx** is a layer consisting of a network of polysaccharides extending from the surface of the cell. The term can encompass both capsules and slime layers because they usually are composed of polysaccharides. The glycocalyx aids in attachment to solid surfaces, including tissue surfaces in plant and animal hosts. ▶| *Pathogenicity islands encode virulence factors (section 35.4)*

S-Layers

Many bacteria have an ordered covering called an **S-layer** on their surface. The S-layer has a pattern something like floor tiles and is composed of protein or glycoprotein (**figure 3.29**). In typical Gram-negative bacteria, the S-layer adheres noncovalently to the outer membrane; it is associated with the peptidoglycan surface of typical Gram-positive cell walls.

S-layers are of considerable interest not only for their biological roles but also in the growing field of nanotechnology. Their biological roles include protecting the cell against ion and pH fluctuations, osmotic stress, enzymes, or predatory bacteria. The S-layer also helps maintain the shape and envelope rigidity of some cells, and it can promote cell adhesion to surfaces. Finally, the S-layer seems to protect some bacterial pathogens against host defenses, thus contributing to their virulence. The potential use of S-layers in nanotechnology is due to the ability of S-layer proteins to self-assemble; that is, S-layer proteins contain the information required to spontaneously associate and form the S-layer without the aid of any additional enzymes or other factors. Thus S-layer proteins could be used as building blocks for the creation of technologies such as drug-delivery systems and novel detection systems for toxic chemicals or bioterrorism agents.

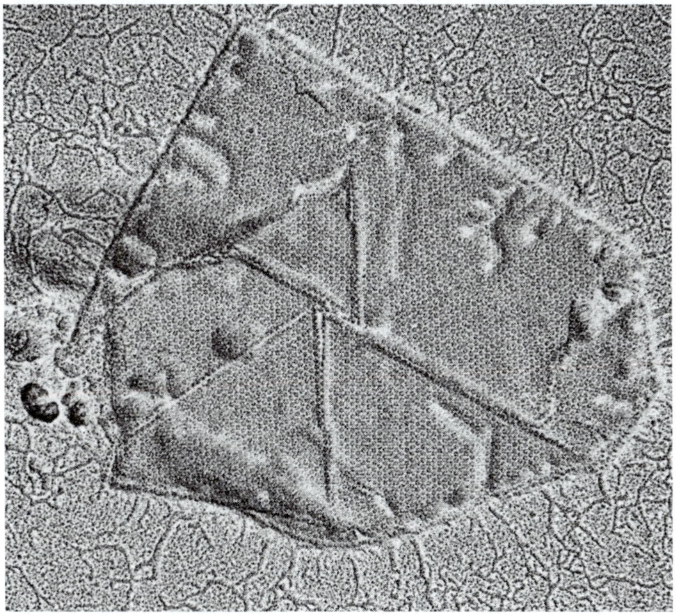

Figure 3.29 The S-Layer. An electron micrograph of the S-layer of the bacterium *Deinococcus radiodurans* after shadowing.

Comprehension Check

1. What is the difference between a capsule and a slime layer? Why does the term *glycocalyx* usually encompass both?
2. S-layers and some capsules are composed of proteins. How does an S-layer differ from a proteinaceous capsule?

3.6 The Bacterial Cytoplasm Is More Complex than Once Thought

After reading this section, you should be able to:

a. Describe the function of three types of bacterial cytoskeletal proteins and compare their structure with those of eukaryotes
b. Compare and contrast storage inclusions and microcompartments, citing specific examples
c. List the composition of bacterial ribosomes and their spatial organization within the cell
d. Differentiate the structure and function of bacterial chromosomes and plasmids

The plasma membrane and everything within is called the **protoplast.** The **cytoplasm** is the material bounded by the plasma membrane; thus the cytoplasm is a major part of the protoplast. The cytoplasm is further divided into the liquid component, called the **cytosol;** those structures such as inclusions, ribosomes, and plasmids that float in the cytosol; and the many molecules dissolved in the cytosol. The high concentration of these macromolecules and their precursors and metabolites creates a phenomenon known as macromolecular crowding. Abundant solutes result in less available space around each molecule, a familiar experience to anyone who travels around a city at rush hour. The bacterial cytoplasm is estimated to be about 10 times more viscous than water, and physical and chemical processes like diffusion are affected.

For many years, bacterial cells were thought of as bags of biochemicals, but the exciting discovery of cytoskeletal proteins in bacterial cells has forever changed that view. Although the bacterial cytoskeleton is less complex than that of eukaryotes, it clearly helps organize the cytoplasm. In this section, we consider the bacterial cytoskeleton and then a variety of other structures in the cytoplasm.

Bacterial Cytoskeleton

Eukaryotes possess three major cytoskeletal elements: actin filaments, microtubules, and intermediate filaments. Actin filaments are made from actin, and microtubules are made from tubulin. Intermediate filaments are composed of a mixture of one or more members of different classes of proteins. Homologues of all three types of eukaryotic proteins have been identified in bacteria (**table 3.2**). Bacterial cytoskeletal proteins are structurally similar to their eukaryotic counterparts and carry out similar functions: They participate in cell division, localize proteins to certain sites in the cell, and determine cell shape. In addition, some bacterial cytoskeletal proteins appear to be unique. Thus it is likely that the evolution of the cytoskeleton was an early event in the history of

Table 3.2 Some Bacterial Cytoskeletal Proteins

Type	Function	Comments
Tubulin Homologues		
FtsZ	Cell division	Widely observed in bacteria and archaea
TubZ	Bacteriophage replication	Observed in multiple bacteriophages
Actin Homologues		
FtsA	Cell division	Attaches the FtsZ ring to the cell envelope
MamK	Positioning magnetosomes	Observed only in magnetotactic species
MreB/Mbl	Helps determine cell shape; may be involved in chromosome segregation, localizing proteins, motility, and establishing cell polarity	Observed in most rod-shaped bacteria
ParM	Plasmid segregation	Plasmid encoded
Intermediate Filament Homologues		
CreS (crescentin)	Induces curvature in curved rods	*Caulobacter crescentus*
Unique Bacterial Cytoskeletal Proteins		
MinD	Prevents polymerization of FtsZ at cell poles	Observed in many rod-shaped bacteria
ParA	Segregates chromosomes and plasmids, helps localize chemotaxis proteins and type IV pili to one pole of certain rod-shaped bacteria	Observed in many species, including *Vibrio cholerae, C. crescentus,* and *Thermus thermophilus*
Bactofilin	Scaffold for protein and chromosome positioning	Widely observed in bacteria

life on Earth. ▶| *The eukaryotic cytoplasm contains a cytoskeleton and organelles (section 5.3); Bacterial cell cycles can be divided into three phases (section 7.2)*

The **cytoskeletons** of *Escherichia coli, Bacillus subtilis,* and *Caulobacter crescentus* are the most important bacterial model systems and are the focus of our discussion. *E. coli* is a Gram-negative rod that has been extensively studied and can be easily manipulated. *B. subtilis* is a Gram-positive rod found in soil. It is an endospore-forming bacterium, making it a good model for cellular differentiation (section 3.9). Both were models for cell wall structure as described in section 3.4. *C. crescentus* is a curved rod found in aquatic habitats. It is interesting in part because it exhibits a complex life cycle that includes two different stages: a motile swarmer cell and a sessile, stalked cell that attaches to surfaces by a holdfast (*see figure 22.8*). ▶| Caulobacteraceae *and* Hyphomicrobiaceae *reproduce in unusual ways (section 22.1)*

The best studied bacterial cytoskeletal proteins are FtsZ, MreB, and CreS (also known as crescentin). FtsZ was one of the first bacterial cytoskeletal proteins identified and has since been found in most bacteria as well as mitochondria and chloroplasts. FtsZ is a homologue of the eukaryotic protein tubulin. It forms a ring at the center of a dividing cell and is required for the formation of the septum that will separate the daughter cells (**figure 3.30*a***). MreB and its relative Mbl are actin homologues. Their major function is to determine cell shape in rod-shaped cells so it is not found in cocci. MreB and Mbl determine cell shape by properly positioning the machinery needed for peptidoglycan synthesis (figure 3.30*b,c*). CreS (crescentin) was discovered in *C. crescentus* and is responsible for its curved shape. CreS is a homologue of lamin and keratin, two intermediate filament proteins. ▶| *Cellular growth and determination of cell shape (section 7.2)*

The movement of newly replicated DNA to daughter cells occurs in eukaryotes by DNA attachment to the mitotic spindle. Although it is not a general phenomenon in bacteria, there are a few examples of cytoskeletal filaments contributing to chromosome segregation (*see figure 7.4*). Certain plasmids, in particular, rely on partition (Par) proteins to ensure equal distribution in daughter cells following cell division. ▶| *Chromosome replication and partitioning (section 7.2)*

Intracytoplasmic Membranes

Although members of *Bacteria* do not contain complex membranous organelles like mitochondria or chloroplasts, internal membranous structures are observed in some bacteria (**figure 3.31**). These can be extensive and complex in photosynthetic bacteria and in bacteria with very high respiratory activity, such as nitrifying bacteria. The internal membranes of the photosynthetic cyanobacteria are called thylakoids and are analogous to the thylakoids of chloroplasts. ▶| *Photosynthetic bacteria are diverse (section 21.4); Nitrifying*

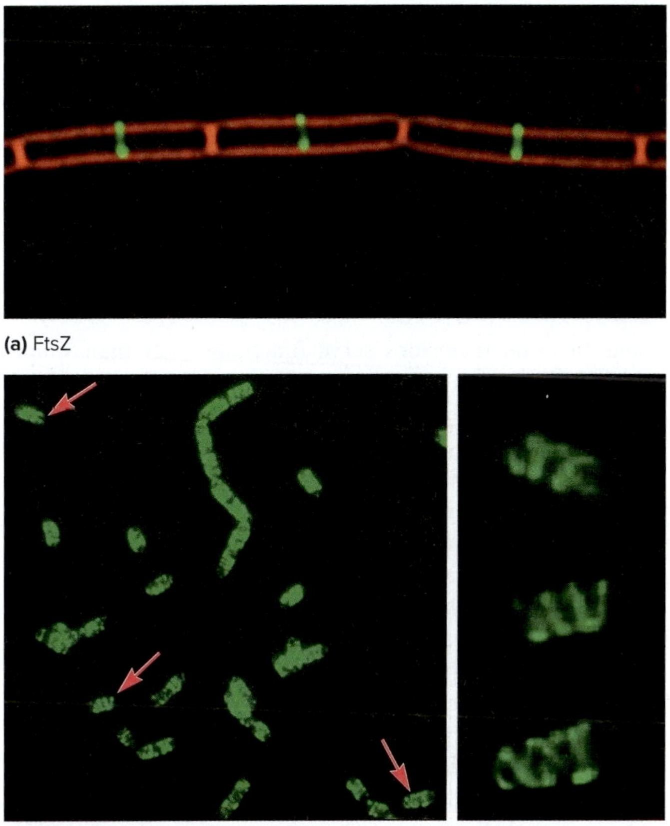

(a) FtsZ

(b) Mbl (c) Mbl

Figure 3.30 The Bacterial Cytoskeleton. (a) FtsZ protein in a chain of *Bacillus subtilis* cells; FtsZ-green fluorescent (GFP) fusion protein viewed by fluorescence microscopy. (b) The MreB-like cytoskeletal protein (Mbl) of *B. subtilis;* Mbl-GFP in live cells was examined by fluorescence microscopy. With this method, Mbl seems to form helices (arrows). More sensitive methods show that Mbl forms patches that move perpendicular to the long axis of the cell. (c) Three of the cells from (b) are shown at a higher magnification.

bacteria oxidize ammonium or nitrite to gain energy and electrons (section 22.1); Mitochondria, related organelles, and chloroplasts are involved in energy conservation (section 5.6)

The internal membranous structures observed in bacteria may be aggregates of spherical vesicles, flattened vesicles, or tubular membranes (*see figure 22.3*). They are often connected to the plasma membrane and arise from it by invagination. However, these internal membranes differ from the plasma membrane by being enriched for proteins and other molecules involved in energy conservation. For instance, the thylakoids of cyanobacteria contain the chlorophyll and photosynthetic reaction centers responsible for converting light energy into ATP. Thus the function of internal membranes may be to provide a larger membrane surface for greater metabolic activity.

Inclusions

Inclusions are common in all cells. They are formed by the aggregation of substances that may be either organic or inorganic. Inclusions

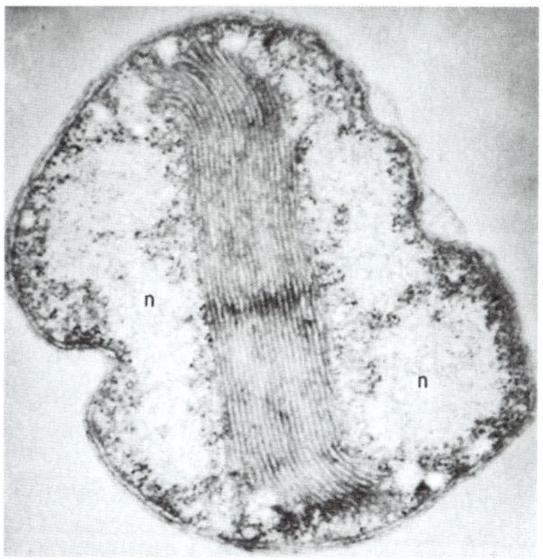

Figure 3.31 Internal Bacterial Membranes. The nitrifying bacterium *Nitrosococcus oceani* has parallel membranes traversing the whole cell. Note the nucleoid (n) with fibrillar structure.

can take the form of granules, crystals, or globules; some are amorphous. Some inclusions lie free in the cytoplasm. Other inclusions are enclosed by a shell that is single-layered and may consist of proteins or of both proteins and phospholipids. Some inclusions are surrounded by invaginations of the plasma membrane. Many inclusions are used for storage (e.g., of carbon compounds, inorganic substances, and energy) or to reduce osmotic pressure by tying up molecules in particulate form. The quantity of inclusions used for storage varies with the nutritional status of the cell. Some inclusions are so distinctive that they are increasingly being referred to as microcompartments. A brief description of several important inclusions follows.

Storage Inclusions

Many storage inclusions form when one nutrient is in ready supply but another nutrient is not. Others store end products of metabolic processes. In some cases, these end products are used by the microbe when it is in different environmental conditions. The most common storage inclusions are glycogen inclusions, polyhydroxyalkonate granules, sulfur globules, and polyphosphate granules. Some storage inclusions are observed only in certain organisms, such as the cyanophycin granules in cyanobacteria. ▶| *Phylum* Cyanobacteria: *oxygenic photosynthetic bacteria (section 21.4)*

Carbon is often stored as polyhydroxyalkonate (PHA) granules, also termed carbonosomes. Several types of PHA granules have been identified, but the most common contain **poly-β-hydroxybutyrate (PHB)** (**figure 3.32**). The structure of PHB inclusions has been well studied, and PHB granules are surrounded by a single-layered shell composed of proteins. Much of the interest in PHB and other PHA granules is due to their industrial use in making biodegradeable plastics. ▶| *Biopolymers (section 42.1)*

Polyphosphate granules and sulfur globules are inorganic inclusions observed in many organisms. **Polyphosphate granules** store the phosphate needed for synthesis of important cell constituents such as nucleic acids. In some cells, they act as an energy reserve, and polyphosphate also can serve as an energy source in some reactions, when the bond linking the final phosphate in the polyphosphate chain is hydrolyzed. Sulfur globules

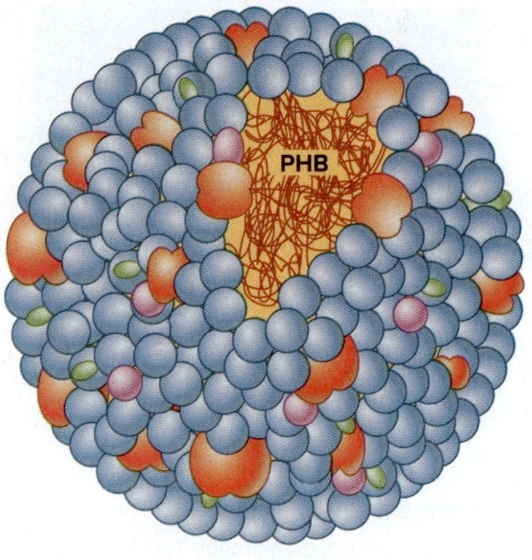

Figure 3.32 PHB Inclusions in Bacteria. PHB is a polymer of β-hydroxybutyrate molecules joined by ester bonds between the carboxyl and hydroxyl groups of adjacent molecules. PHB inclusions are around 0.2 to 0.7 µm in diameter. PHB (red lines) is enclosed by a shell composed of several different proteins, including the PHB-synthesizing enzyme (red) and the PHB-degrading enzyme (green).

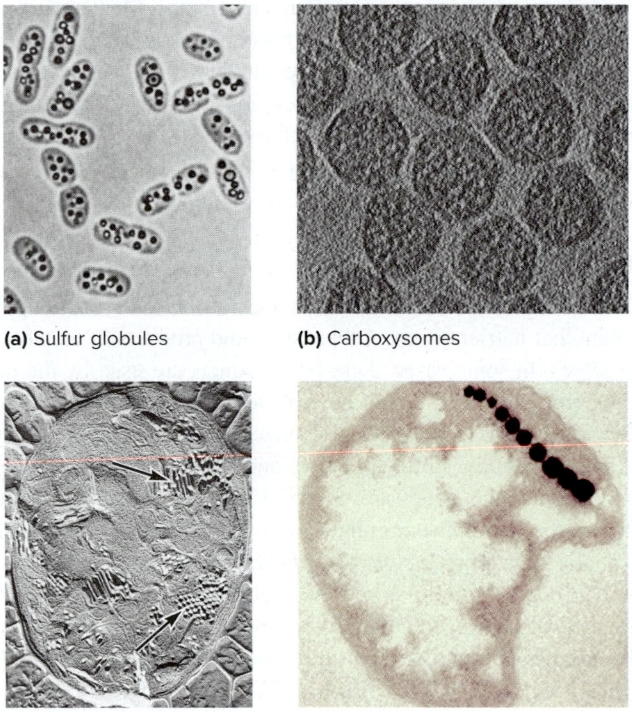

(a) Sulfur globules **(b)** Carboxysomes **(c)** Gas vacuoles **(d)** Magnetosomes

Figure 3.33 Bacterial Cell Inclusions. (a) *Chromatium vinosum,* a purple sulfur bacterium, with intracellular sulfur globules. (b) Carboxysomes in the bacterium *Halothiobacillus neapolitanus.* Each carboxysome is approximately 100 nm in diameter. (c) A freeze-fracture preparation of *Anabaena flosaquae* showing gas vesicles and gas vacuoles. Both longitudinal and cross-section views of gas vesicles are indicated by arrows. (d) *Magnetococcus marinum* cells are roughly spherical with a single chain of magnetite crystals.

are formed by bacteria that use reduced sulfur-containing compounds as a source of electrons during their energy-conserving metabolic processes (**figure 3.33*a***). For example, some photosynthetic bacteria use hydrogen sulfide (rather than water) as an electron donor and accumulate the resulting sulfur either externally or internally. ▶| *Light reactions in anoxygenic photosynthesis (section 11.12); Purple sulfur bacteria perform anoxygenic photosynthesis (section 22.3)*

Microcompartments

Some bacterial inclusions serve functions other than simply storing substances for later use. These inclusions, called microcompartments, share several characteristics. They are relatively large polyhedrons formed by one or more different proteins. Enclosed within the protein shell are one or more enzymes. Microcompartments include the ethanolamine utilization (Eut) microcompartment, the propandiol utilization (Pdu) microcompartment, and carboxysomes. We focus here on carboxysomes as they are the best studied.

Carboxysomes are present in many cyanobacteria and other CO_2-fixing bacteria (figure 3.33*b*). Their polyhedral coat is composed of three different proteins and is about 100 nm in diameter. Associated with the shell is the enzyme carbonic anhydrase that converts carbonic acid and bicarbonate into CO_2. Recall that biological membranes allow the free diffusion of CO_2. However, the carboxysome shell prevents CO_2 from escaping so it can accumulate. Enclosed within the polyhedron is the enzyme ribulose-1, 5-bisphosphate carboxylase/oxygenase (RubisCO). RubisCO is the critical enzyme for CO_2 fixation, the process of converting CO_2 into sugar. Thus the carboxysome serves as a site for CO_2 fixation. As such, it is critical that carboxysomes be distributed to both daughter cells during cell division. ParA, a cytoskeletal protein (table 3.2), helps ensure appropriate segregation of carboxysomes. ▶| *CO_2 fixation: reduction and assimilation of CO_2 carbon (section 12.3)*

Other Inclusions

Inclusions can be used for functions other than storage or as microcompartments. Two of the most remarkable inclusions are gas vacuoles and magnetosomes. Both are involved in bacterial movement.

The **gas vacuole** provides buoyancy to some aquatic bacteria, many of which are photosynthetic. Gas vacuoles are aggregates of enormous numbers of small, hollow, cylindrical structures called **gas vesicles** (figure 3.33*c*). Gas vesicle walls are composed of many copies of a single small protein. These protein subunits assemble to form a rigid cylinder that is impermeable to water but freely permeable to atmospheric gases. Cells with gas vacuoles can regulate their buoyancy to float at the depth necessary for proper light intensity, oxygen concentration, and nutrient levels. They descend by simply collapsing vesicles and float upward when new ones are constructed.

Aquatic magnetotactic bacteria use **magnetosomes** to orient themselves in Earth's magnetic field. Northern Hemisphere bacteria use their magnetosome chain to determine northward and downward directions, and swim down to nutrient-rich sediments

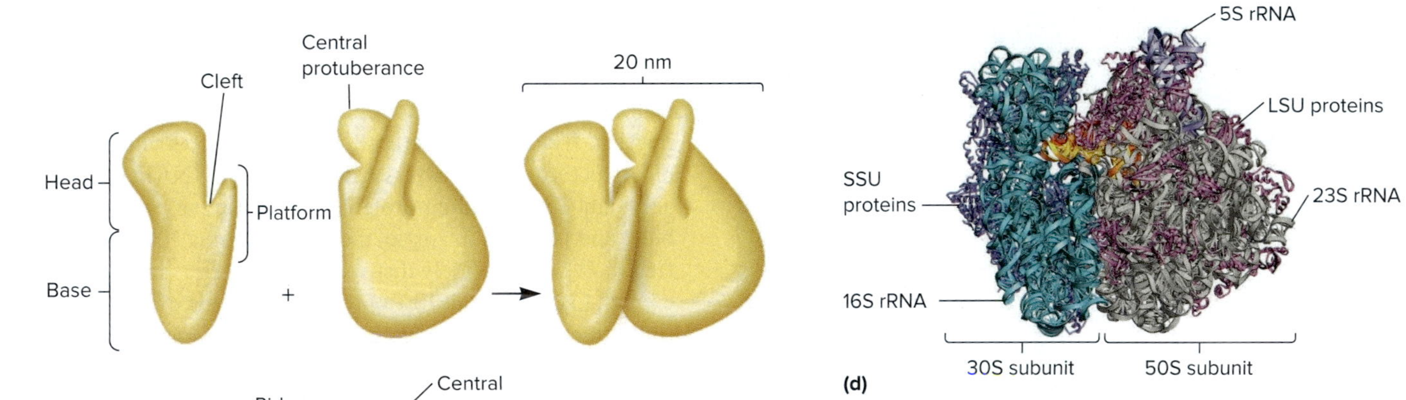

Figure 3.34 Bacterial Ribosomes. (a–c) Schematic representation of the two subunits and the complete 70S ribosome of *E. coli.* (d) Molecular structure of the 70S ribosome of *Thermus thermophilus.* The 50S subunit (LSU) includes 23S rRNA (gray) and 5S rRNA (lavender), while 16S rRNA (turquoise) is found in the 30S subunit (SSU). A molecule of tRNA (gold) is shown in the A site. To generate this structural model, crystals of purified bacterial ribosomes were prepared and exposed to X rays, and the resulting diffraction pattern analyzed.

or locate the optimum depth in freshwater and marine habitats. Magnetotactic bacteria in the Southern Hemisphere generally orient southward and downward, with the same result. Magnetosomes are intracellular chains of magnetite (Fe_3O_4) or greigite (Fe_3S_4) particles (figure 3.33*d*). They are around 35 to 125 nm in diameter and enclosed within invaginations of the plasma membrane. The invaginations contain distinctive proteins that are not found elsewhere in the plasma membrane. For the cell to move properly within a magnetic field, magnetosomes must be arranged in a chain. A cytoskeletal protein called MamK is responsible for establishing a framework upon which the chain can form (table 3.2; *see figure 22.14*).

Bacterial Ribosomes

Ribosomes are the site of protein synthesis, and large numbers (10,000 to 20,000) are found in nearly all cells. The cytoplasm of rapidly growing bacterial cells is often packed with ribosomes, and additional ribosomes may be loosely attached to the plasma membrane. The cytoplasmic ribosomes synthesize proteins destined to remain within the cell, whereas plasma membrane–associated ribosomes make proteins that will reside in the cell envelope or be transported to the outside.

Translation, the process of protein synthesis, is amazingly complex and is discussed in detail in chapter 13. This complexity is evidenced in part by the structure of ribosomes, which are made of dozens of proteins and several ribonucleic acid (RNA) molecules. The ribosomes of all three domains of life share similarities. However, there are important differences. Here we focus strictly on bacterial ribosomes. We compare bacterial and archaeal ribosomes in chapter 4 and compare all three types of ribosomes in chapter 5.

Bacterial ribosomes are called 70S ribosomes and are constructed of a 50S and a 30S subunit (**figure 3.34**). The S in these values stands for **Svedberg unit.** This is the unit of the sedimentation coefficient, a measure of sedimentation velocity in a centrifuge; the faster a particle travels when centrifuged, the greater its Svedberg value. The sedimentation coefficient is a function of a particle's molecular weight, volume, and shape. Heavier and more compact particles normally have larger Svedberg numbers.

Bacterial ribosomes are composed primarily of ribosomal RNA (rRNA) molecules. The small subunit contains 16S rRNA, whereas the large subunit consists of 23S and 5S rRNA molecules. Approximately 55 proteins make up the rest of the mass of the ribosome: 21 in the small subunit, and 34 in the large subunit.

Nucleoid

The **nucleoid** is an ellipsoidal region that contains the cell's chromosome and numerous proteins (**figure 3.35**). Although the nucleoid is not separated from the cytoplasm by a membrane, it is a distinct region of the cell, as observed by microscopy. Ribosomes are excluded from the nucleoid. The chromosome itself has an orientation, and particular regions occupy discrete locations within the nucleoid.

The chromosomes of most bacteria are a circle of double-stranded **deoxyribonucleic acid (DNA)**, but some bacteria have a linear chromosome. Most bacteria have a single chromosome, but some bacteria, such as *Vibrio cholerae* and *Borrelia burgdorferi* (causative agents of cholera and Lyme disease, respectively), have more than one chromosome. Many bacteria carry a single copy of their chromosome (monoploid), but others are

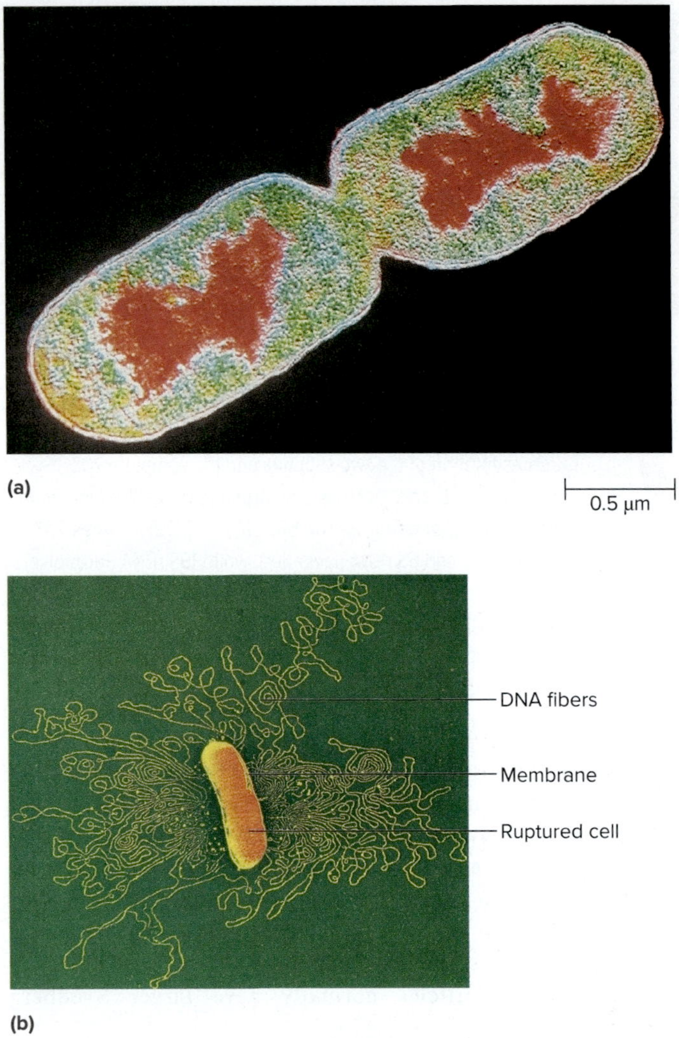

Figure 3.35 *E. coli* **Nucleoids and Chromosomes.** Bacterial chromosomes are located in the nucleoid, an area in the cytoplasm. (a) A color-enhanced transmission electron micrograph of a thin section of a dividing *E. coli* cell. The red areas are the nucleoids present in the two daughter cells. (b) Chromosome released from a gently lysed *E. coli* cell. Note how tightly packaged the DNA must be to fit inside the cell.

(a) ©CNRI/SPL/Science Source; *(b)* ©Dr. Gopal Murti/SPL/Science Source

known to be polyploid, often with more than 10 copies per cell. In *Deinococcus radiodurans,* this is believed to be necessary to allow the repair of damaged DNA under adverse environmental conditions, and the giant bacterium *Epulopiscium fishelsoni* (figure 3.4) appears to require tens of thousands of chromosome copies.

Bacterial chromosomes are longer than the length of the cell. Therefore, an important and unanswered question is how these microbes manage to fit their chromosomes into the relatively small space occupied by the nucleoid. For instance, *E. coli*'s circular chromosome measures approximately 1,400 μm, or about 230–700 times longer than the cell (figure 3.35*b*). Thus the chromosome must be organized and packaged in a manner that decreases its overall size yet retains easy accessibility to the genetic information. Physical factors like macromolecular crowding and supercoiling contribute to nucleoid organization. Supercoiling produces a dense, central core of DNA with loops of DNA extending out from the core. Besides physical factors, architectural proteins also contribute to nucleoid structure. Nucleoid-associated proteins (NAPs) are small, abundant DNA binding proteins that cause the chromosome to bend and fold. For example, the HU protein, a NAP first identified in *E. coli,* is widely distributed in bacteria, and is abundant enough to cover about 10% of the chromosome. NAPs have multiple functions: They form bridges between one section of the chromosome and another and they participate in the many processes that occur on the chromosome. Cells have several different NAPs, but HU is the only one found in almost all bacteria. NAPs are particularly important during cell division, when they further compact the chromosomes. This extra level of packing is important for proper segregation of daughter chromosomes during cell division. ▶| *DNA is a polymer of deoxyribonucleotides (section 13.2)*

Plasmids

In addition to the genetic material present in the nucleoid, many bacteria contain extrachromosomal DNA molecules called plasmids. The genome of an organism includes all DNA, including plasmids. Indeed, most of the bacterial genomes sequenced thus far include plasmids. In some cases, numerous different plasmids within a single species have been identified. For instance, *B. burgdorferi* carries 12 linear and 9 circular plasmids. Plasmids play many important roles in the lives of the organisms that have them. They also have proved invaluable to microbiologists and molecular geneticists in constructing and transferring new genetic combinations and in cloning genes, as described in chapter 17.

Plasmids are double-stranded DNA molecules that can exist independently of the chromosome. Both circular and linear plasmids have been documented, but most known plasmids are circular. Plasmids have relatively few genes, generally less than 30. Their genetic information is not essential to the bacterium, and cells that lack them usually function normally. However, many plasmids carry genes that confer a selective advantage to the bacterium in certain environments.

Plasmids use the cell's DNA-synthesizing machinery to replicate, but their replication is not linked to any particular stage of the cell cycle. Thus regulation of plasmid and chromosomal replication are independent. However, some plasmids are able to integrate into the chromosome. Such plasmids are called **episomes** and when integrated are replicated as part of the chromosome. Plasmids are inherited stably during cell division, but they are not always equally apportioned into daughter cells and sometimes are lost. The loss of a plasmid is called **curing.** It can occur spontaneously or be induced by treatments that inhibit plasmid replication but not host cell reproduction. Some commonly used curing treatments are acridine mutagens, ultraviolet and ionizing

Table 3.3 Major Types of Bacterial Plasmids

Type	Function	Example	Size (kbp)	Hosts	Phenotypic Features[1]
Conjugative Plasmids[2]	Transfer of DNA from one cell to another	F factor	95–100	*E. coli, Salmonella, Citrobacter*	Sex pilus, conjugation
R Plasmids	Carry antibiotic-resistance genes	RP4	54	*Pseudomonas* and many other Gram-negative bacteria	Sex pilus, conjugation, resistance to Amp, Km, Nm, Tet
Col Plasmids	Produce bacteriocins, substances that destroy closely related species	ColE1	9	*E. coli*	Colicin E1 production
Virulence Plasmids	Carry virulence genes	Ti	200	*Agrobacterium tumefaciens*	Tumor induction in plants

1 Abbreviations used for antibiotics: Amp, ampicillin; Km, kanamycin; Nm, neomycin; Tet, tetracycline.
2 Many R plasmids and others are also conjugative.

radiation, thymine starvation, antibiotics, and growth above optimal temperatures.

Plasmids may be classified in terms of their mode of existence, spread, and function. A brief summary of the types of bacterial plasmids and their properties is given in **table 3.3**. These various types of plasmids can also differ in terms of the number of copies found within the cell. Single-copy plasmids produce only one copy per host cell. Multicopy plasmids may be present at concentrations of 100 or more per cell.

Comprehension Check

1. Briefly describe the nature and function of the cytoplasm, and the regions and structures within it. How is the cytosol different from the cytoplasm?
2. List the most common kinds of inclusions. How are they similar to eukaryotic organelles such as mitochondria and chloroplasts? How do they differ?
3. How do plasmids differ from chromosomes? What is an episome?
4. Explain the importance of each of the following plasmids: conjugative plasmid, R plasmid, Col plasmid, and virulence plasmid.

3.7 Many Bacteria Have External Structures Used for Attachment and Motility

After reading this section, you should be able to:

a. Distinguish pili (fimbriae) and flagella
b. Illustrate the various patterns of flagella distribution

Bacteria must constantly respond to their changing environments. Sometimes it is advantageous to attach to a surface. Other times it is better to move toward or away from something in the environment. Many bacteria have structures that extend beyond the cell envelope and are involved in either attachment to surfaces or motility. In addition, these external structures can function in protection and horizontal gene transfer. Several are discussed in this section.

Bacterial Pili and Fimbriae

Many bacteria have fine, hairlike appendages that are thinner and typically shorter than flagella. These are called **fimbriae** (s., **fimbria**) or **pili** (s., **pilus**). The terms are synonymous, although certain structures are historically called pilus (e.g., sex pilus), while others are called fimbriae. We will use the terms interchangeably, except in those instances. A cell may be covered with up to 1,000 fimbriae, but they are only visible in an electron microscope due to their small size (**figure 3.36**). They are slender tubes composed of helically arranged protein subunits and are about 3 to 10 nm in diameter and up to several

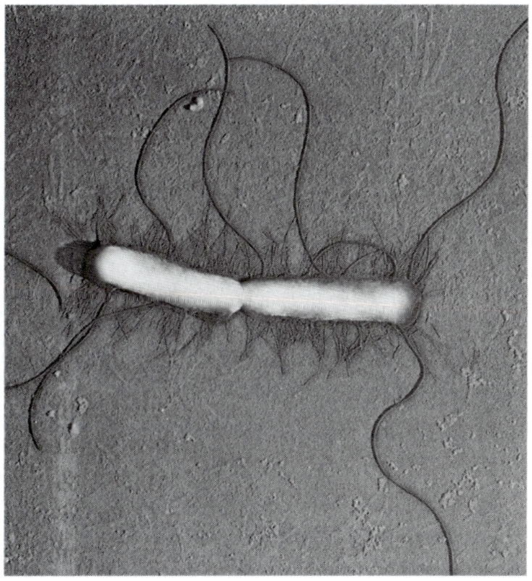

Figure 3.36 Flagella and Fimbriae. The long flagella and numerous shorter fimbriae are evident in this SEM of the bacterium *Proteus vulgaris*.
©Thomas Deerinck, NCMIR/Science Source

micrometers long. Pili grow by adding protein subunits to their base. Several different types have been identified and most function to attach cells to solid surfaces such as rocks in streams and host tissues. Type IV pili are involved in motility (section 3.8) and two gene transfer mechanisms: bacterial transformation and bacterial conjugation. ▶| *Bacterial conjugation requires cell–cell contact (section 16.6); Bacterial transformation is the uptake of free DNA from the environment (section 16.7)*

Many bacteria have up to 10 **sex pili** (s., **sex pilus**) per cell. These hairlike structures differ from other pili in the following ways. Sex pili often are larger than other pili (around 9 to 10 nm in diameter). They are genetically determined by conjugative plasmids and are required for conjugation. Some bacterial viruses attach specifically to sex pili at the start of their infection cycle.

Bacterial Flagella

Many motile bacteria move by means of **flagella** (s., **flagellum**), threadlike locomotor appendages extending outward from the plasma membrane and cell wall. Although the main function of flagella is motility, they can have other roles. They can be involved in attachment to surfaces, and in some bacteria, they are virulence factors, that is, they contribute to the ability of the bacterium to cause disease.

Bacterial flagella are slender, rigid structures about 20 nm across and up to 20 μm long. Flagella are so thin they cannot be observed directly with a light microscope but must be stained with techniques designed to increase their thickness. The detailed structure of a flagellum can only be seen in the electron microscope.

Bacterial species often differ in their patterns of flagella distribution, and these patterns are useful in identifying bacteria. **Monotrichous** bacteria (Greek *trikhos,* hair) have one flagellum; if it is located at an end, it is said to be a **polar flagellum** (**figure 3.37*a***). **Amphitrichous** bacteria (Greek *amphi*, on both sides) have a single flagellum at each pole. In contrast, **lophotrichous** bacteria (Greek *lopho*, crest or tuft) have a cluster of flagella at one or both ends (figure 3.37*b*). Flagella are spread evenly over the whole surface of **peritrichous** (Greek *peri*, around) bacteria (figure 3.37*c*).

Transmission electron microscope studies have shown that the bacterial flagellum is composed of three parts (**figure 3.38**). (1) The **filament** is the longest and most obvious portion. It extends from the cell surface to the tip. (2) The **basal body** is embedded in the cell envelope; and (3) a short, curved segment, the **hook,** links the filament to its basal body and acts as a flexible coupling.

The filament is a hollow, rigid cylinder constructed of subunits of the protein **flagellin,** which ranges in molecular mass from 30,000 to 60,000 daltons, depending on the bacterial species. The filament ends with a capping protein. Some bacteria have sheaths surrounding their flagella. For example, *Vibrio cholerae* flagella have lipopolysaccharide sheaths.

The hook and basal body are quite different from the filament (figure 3.38). Slightly wider than the filament, the hook is made of different protein subunits. The basal body is the most complex part of a flagellum. The basal bodies of *E. coli* and most other typical Gram-negative bacteria have four rings: L, P, MS, and C, which are connected to a central rod (figure 3.38*a*). The L, P, and MS rings are embedded in the cell envelope, and the C ring is on the cytoplasmic side of the MS ring. Typical Gram-positive bacteria have only two rings: an inner ring connected to the plasma membrane and an outer one probably attached to the peptidoglycan (figure 3.38*b*).

The synthesis of bacterial flagella is complex and involves at least 20 to 30 genes. Besides the gene for flagellin, 10 or more genes code for hook and basal body proteins; other genes are concerned with control of flagellar construction or function. Many components of the flagellum lie outside the cell envelope and must be transported out of the cell for assembly. Interestingly, the basal body is a specialized version of the type III protein secretion system observed in typical Gram-negative bacteria. Type III secretion systems have a needlelike structure through which proteins are secreted. In the flagellar type III secretion system, the filament replaces

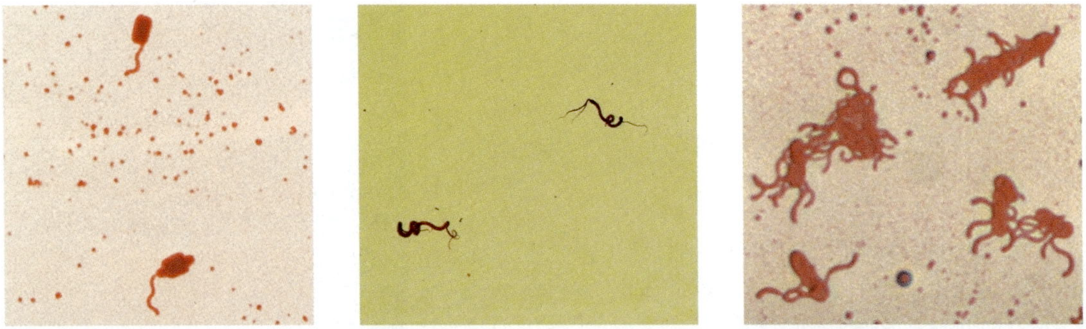

(a) *P. mendocina*—monotrichous polar flagellation

(b) *S. volutans*—lophotrichous flagellation

(c) *B. alvei*—peritrichous flagellation

Figure 3.37 Flagellar Distribution. Examples of various patterns of flagellation as seen in the light microscope. (a) Flagella stain of *Pseudomonas mendocina*. (b) *Spirillum volutans* (×1,000). Only the tuft of flagella at one end is clearly visible. (c) *Bacillus alvei*.

(*a, c*) Source: CDC/Dr. William A. Clark; (*b*) ©McGraw-Hill Education/James Redfearn, photographer,

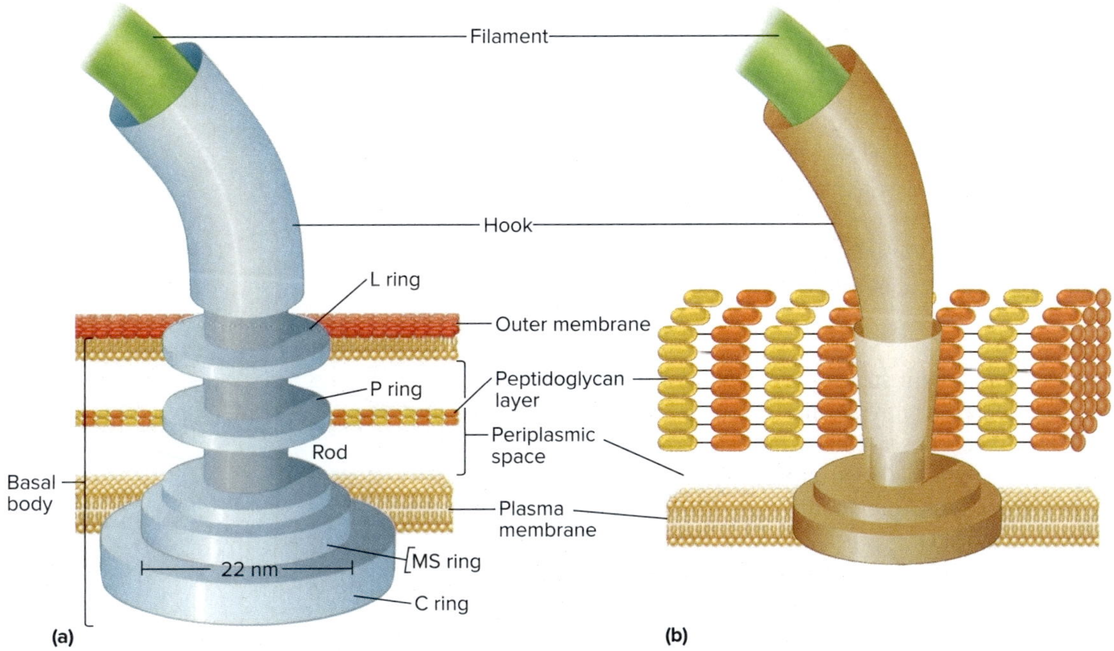

Figure 3.38 The Ultrastructure of Bacterial Flagella. Flagellar basal bodies and hooks in (a) typical Gram-negative and (b) typical Gram-positive bacteria.

the needle. Individual flagellin subunits are transported through the hollow filament. When the subunits reach the tip, they spontaneously aggregate under the direction of a protein called the filament cap; thus the filament grows at its tip rather than at the base (**figure 3.39**). Filament synthesis, like S-layer formation, is an example of **self-assembly.** ▶| *Protein maturation and secretion (section 13.8)*

Comprehension Check

1. What are the functions of fimbriae (pili) and sex pili?
2. What terms are used to describe the different flagella distribution patterns observed among bacteria?
3. What is self-assembly? Why does it make sense that the filament of a flagellum is assembled in this way?

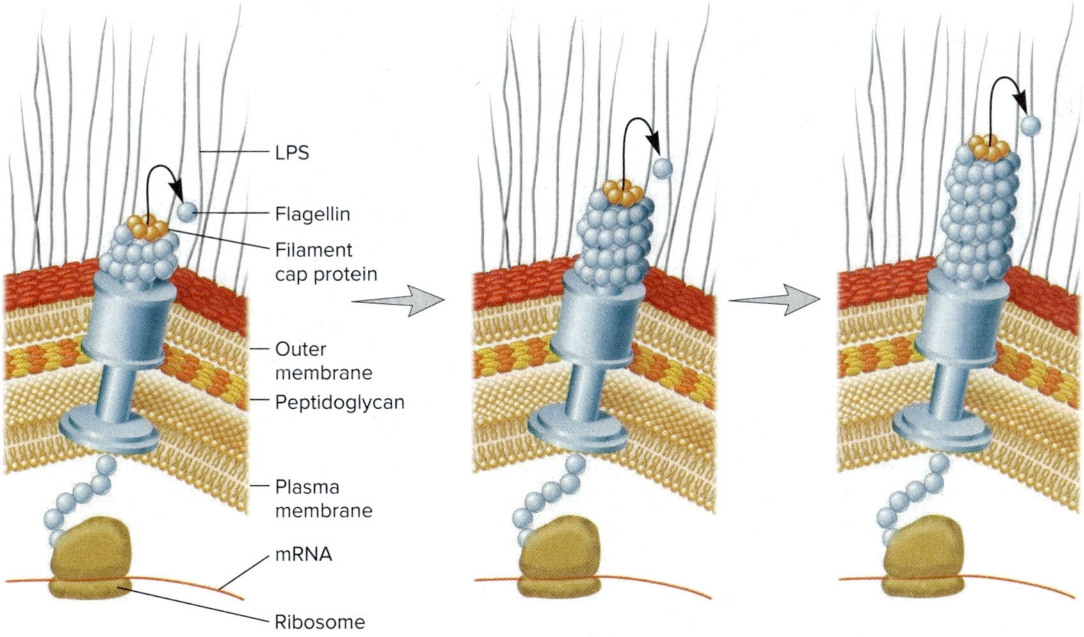

Figure 3.39 Flagellar Filaments Grow at the Tip. Flagellin subunits travel through the flagellar core and attach to the growing tip. Their attachment is directed by the filament cap protein.

MICRO INQUIRY *How does flagellum growth compare to pilus growth?*

3.8 Bacteria Move in Response to Environmental Conditions

After reading this section, you should be able to:

a. Compare and contrast flagellar swimming motility, swarming, spirochete flagellar motility, and twitching and gliding motility
b. State the source of energy that powers flagellar motility
c. Explain why bacterial chemotaxis is referred to as a "biased random walk"

As we note in section 3.7, several structures extending beyond bacterial cell envelopes contribute to motility. Five major methods of movement have been observed: swimming movement conferred by flagella, flagella-mediated swarming, corkscrew movement of spirochetes, twitching motility associated with type IV pili, and gliding motility.

Motile bacteria do not move aimlessly. Rather, motility is used to move toward nutrients such as sugars and amino acids and away from many harmful substances and bacterial waste products. Movement toward chemical attractants and away from repellents is known as chemotaxis. Motile bacteria also can move in response to environmental cues such as temperature (thermotaxis), light (phototaxis), oxygen (aerotaxis), osmotic pressure (osmotaxis), and gravity.

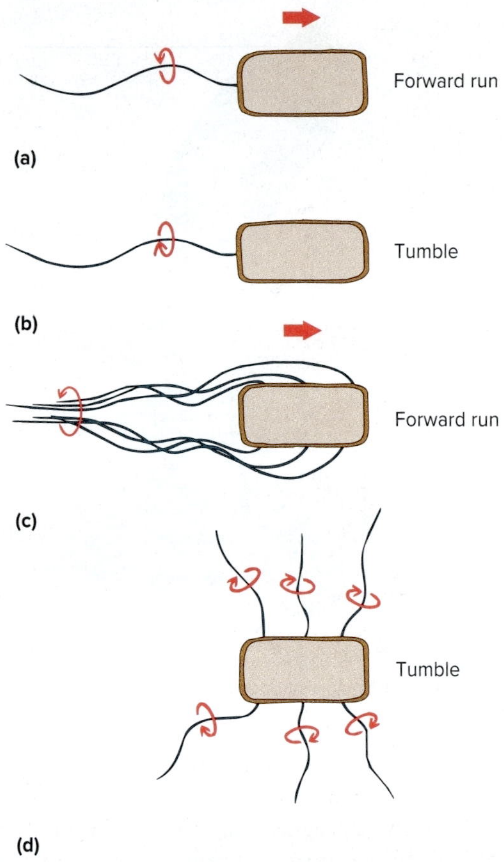

Figure 3.40 The Direction of Flagellar Rotation Often Determines the Way a Bacterium Moves. Parts (a) and (b) describe the motion of bacteria with monotrichous, polar flagella. Parts (c) and (d) illustrate the movements of bacteria with peritrichous flagella.

Flagellar Movement

Swimming

The filament of a bacterial flagellum is in the shape of a rigid helix, and the cell moves when this helix rotates like a propeller on a boat. The flagellar motor can rotate very rapidly. The *E. coli* motor rotates 270 revolutions per second (rps); *Vibrio alginolyticus* averages 1,100 rps. For many bacteria in an aquatic environment, flagellar rotation results in two types of movement: a smooth swimming movement often called a **run,** which actually moves the cell from one spot to another, and a **tumble,** which serves to reorient the cell. As we shall see in our discussion of chemotaxis, alternating between runs and tumbles is important for responding to environmental conditions. Often, the direction of flagellar rotation determines whether a run or a tumble occurs. For example, many bacteria with monotrichous, polar flagella use a counterclockwise rotation for a run (**figure 3.40**). When rotation is reversed, the cell tumbles. Many peritrichously flagellated bacteria operate in a somewhat similar way. To move forward in a run, the flagella rotate counterclockwise. As they do so, the flagella bend at their hooks to form a rotating bundle that propels the cell forward. Clockwise rotation of the flagella disrupts the bundle and the cell tumbles.

Not all bacteria use runs and tumbles for swimming motility. For instance, *Rhodobacter sphaeroides* cells alternate between rotating their single flagellum in one direction (run) and no rotation, a so-called run-stop motility. When the cells are stopped, molecules in the environment bombard the cells and cause them to make small changes in orientation. When they resume a run, they move in a new direction. Another type of swimming motility is seen with the monotrichous bacterium *Vibrio alginolyticus*. It uses a run-reverse-flick pattern. It swims forward (run) when the flagellum rotates in one direction. When rotation reverses, the cell moves backward (reverse). Just as the rotation switches again for a run, the flagellum flicks, causing the cell to change its orientation and move in a new direction.

The motor that drives flagellar rotation is located at the base of the flagellum. Torque generated by the motor is transmitted to the hook and filament. The motor is composed of two components: the rotor (the moving parts) and the stator (the stationary components). It is thought to function like an electrical motor, where the rotor turns in the center of a ring of electromagnets, the stator. In typical Gram-negative bacteria, the rotor is composed of the central rod and all four rings (**figure 3.41**). The C ring protein, FliG, is particularly important because it interacts with the stator. The stator is composed of the proteins MotA and MotB, which form a channel through the plasma membrane. MotB also anchors MotA to cell wall peptidoglycan.

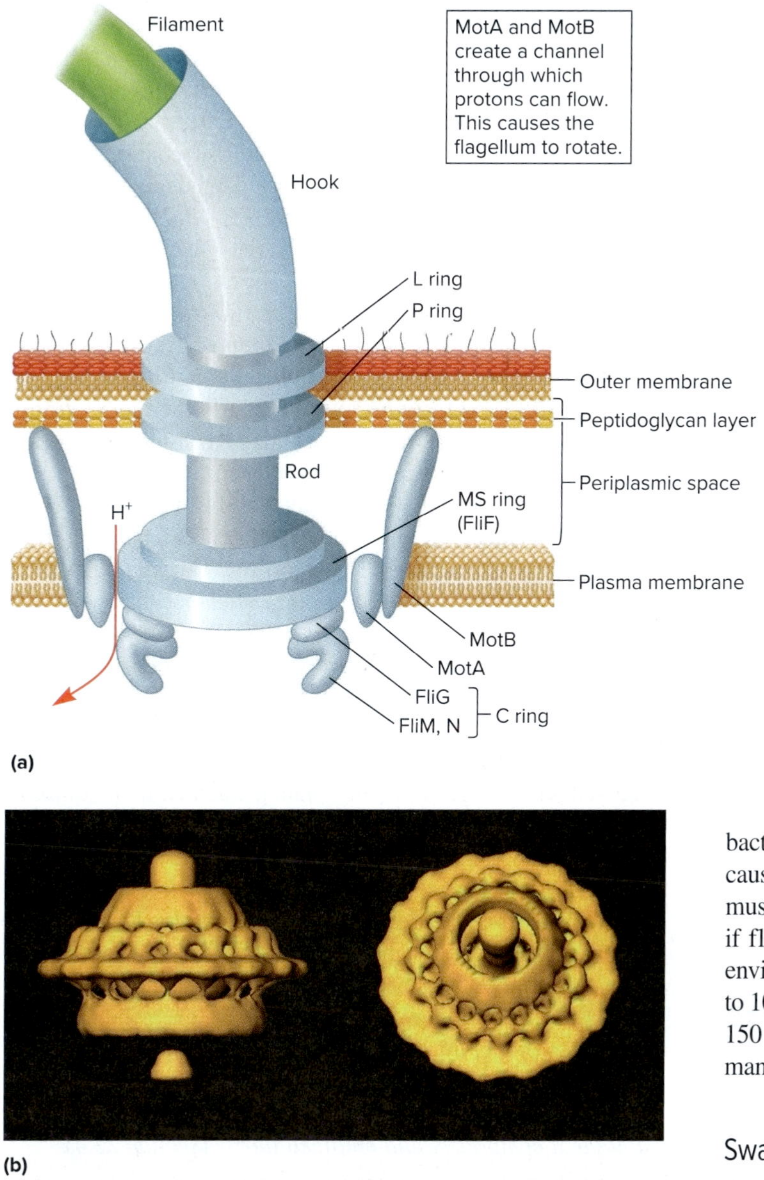

Figure 3.41 Mechanism of Flagellar Movement. (a) This diagram of a flagellum shows some of the more important components and the flow of protons that drives rotation. Some of the many flagellar proteins are labeled. (b) A three-dimensional electron cryotomographic reconstruction of the flagellar motor.

(b)©Gavin Murphy/Nature/Science Source

MICRO INQUIRY *Would this flagellum be found in a typical Gram-negative or Gram-positive bacterium?*

As with all motors, the flagellar motor must have a power source that allows it to generate torque and cause flagellar rotation. The power used by most flagellar motors is a difference in charge and pH across the plasma membrane. This difference is called the proton motive force (PMF). PMF is largely created by the metabolic activities of organisms, as described in chapter 11. One important metabolic process is the transfer of electrons from an electron donor to a terminal electron acceptor via a chain of electron carriers called the electron transport chain (ETC). In bacterial cells, most components of the ETC are located in the plasma membrane. As electrons are transported down the ETC, protons are transported from the cytoplasm to the outside of the cell. Because there are more protons outside the cell than inside, the outside has more positively charged ions (the protons) and a lower pH. PMF is a type of potential energy that can be used to do work: mechanical work, as in the case of flagellar rotation; transport work, the movement of materials into or out of the cell (section 3.3); or chemical work such as the synthesis of ATP, the cell's major energy currency.

How can PMF be used to power the flagellar motor? The channels created by the MotA and MotB proteins allow protons to move across the plasma membrane from the outside to the inside (figure 3.41). Thus the protons move down the charge and pH gradient. This movement releases energy that is used to rotate the flagellum. In essence, the entry of a proton into the channel is like the entry of a person into a revolving door. The "power" of the proton generates torque, rather like a person pushing the revolving door. Indeed, the speed of flagellar rotation is proportional to the magnitude of the PMF.

The flagellum is a very effective swimming device. From the bacterium's point of view, swimming is quite a difficult task because the surrounding water seems as viscous as molasses. The cell must bore through the water with its corkscrew-shaped flagella, and if flagellar rotation ceases, it stops almost instantly. Despite such environmental resistance to movement, bacteria can swim from 10 to 100 μm per second. This is equivalent to traveling from 2 to over 150 cell lengths per second. In contrast, an exceptionally fast human might be able to run around 5 to 6 body lengths per second.

Swarming

An increasing number of bacterial species has been found to exhibit an interesting type of motility called swarming. This motility occurs on moist surfaces and is a type of group behavior in which cells move in unison across the surface. Most bacteria that swarm have peritrichous flagella. Many also produce and secrete molecules that help them move across the substrate. When bacteria that swarm are cultured in the laboratory on appropriate solid media, they produce characteristic colony morphologies (**figure 3.42**).

Spirochete Motility

Spirochetes have flagella that work in a distinctive manner. In many spirochetes, multiple flagella arise from each end of the cell and wind around the cell (**figure 3.43**). The flagella do not extend outside the cell wall but rather remain in the periplasmic space and are covered by the outer membrane. They are called periplasmic flagella and rotate like the external flagella of other

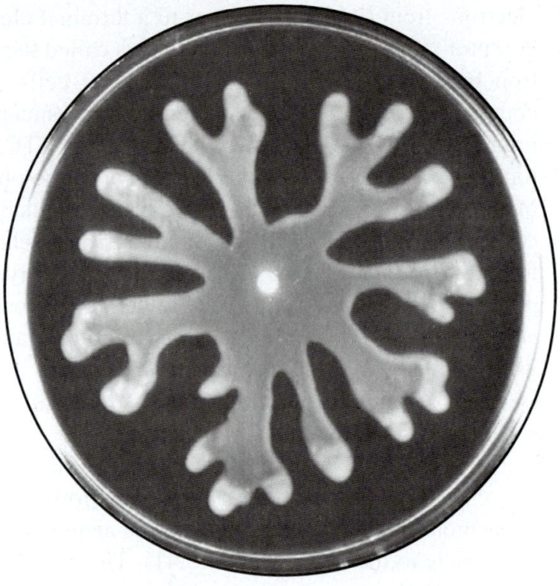

Figure 3.42 Swarming Bacteria Often Produce Distinctive Patterns on a Solid Growth Medium. These bacterial cells swarmed from the center of the plate and produced a branching pattern called dendrites.

bacteria, causing the corkscrew-shaped outer membrane to rotate and move the cell through the surrounding medium, even very viscous liquids. Flagellar rotation may also flex or bend the cell and account for the creeping or crawling movement observed when spirochetes are in contact with a solid surface. ▶| *Phylum* Spirochaetes *(section 21.6)*

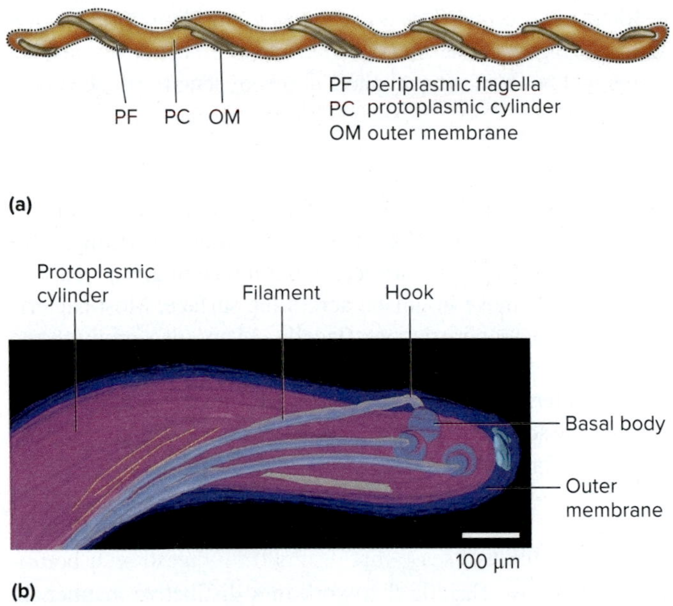

Figure 3.43 Spirochete Flagella. (a) Numerous flagella arise from each end of the spirochete. These intertwine and wind around the cell, usually overlapping in the middle. (b) Electron cryotomographic image of the spirochete *Treponema denticola* showing three flagella arising from the tip of the cell.

Twitching and Gliding Motility

Twitching and gliding motility occur when cells are on a solid surface. However, unlike swarming, neither involves flagella. Gliding motility varies greatly in rate (from 2 to over 600 μm per minute) and in the nature of the motion. Although first observed over 100 years ago, the mechanism by which many bacteria glide remains a mystery. We describe the gliding motility of *Flavobacterium* spp. and *Mycoplasma* spp. in detail in chapter 21. ▶| *Phylum* Bacteroidetes *includes important gut microbiota (section 21.7); Class* Mollicutes, *Phylum* Tenericutes: *bacteria that lack cell walls (section 21.3)*

Here we focus on *Myxococcus xanthus* because it exhibits both twitching and gliding motility. Its twitching motility is called social (S) motility because it occurs when large groups of cells move together in a coordinated fashion. **Twitching motility** is characterized by short, intermittent, jerky motions of up to several micrometers in length and is normally seen on very moist surfaces. Type IV pili alternately extend and retract to move cells during twitching motility. The extended pilus contacts the surface at a point some distance from the cell body. When the pilus retracts, the cell is pulled forward. ATP hydrolysis powers the extension/retraction process.

In contrast to the jerky movement of twitching motility, **gliding motility** is smooth. The gliding motility exhibited by *Myxococcus xanthus* is called adventurous (A) motility; it is observed when single cells move independently. In *M. xanthus,* proteins similar to MotA and MotB of the flagellar motor function as the motors for gliding motility and these are also powered by PMF. They are located in the plasma membrane (like flagellar motors). Additional motility proteins are located in the periplasm and the outer membrane where they move independently. It is only when the full set of protein complexes aligns across the entire cell envelope that they can exert the mechanical force to propel the cell. Adhesin proteins on the cell surface are compressed and slide on the substrate, pushing the body of the cell forward. ▶| *Order* Myxococcales: *bacteria with morphological complexity and multicellularity (section 22.4)*

Chemotaxis

Imagine a motile bacterium in an ocean. Depending on where it is in the water column, light, oxygen, and nutrient levels will vary. These also vary over time due to the activities of all the organisms present. To position itself in the most beneficial location in the water column, this bacterium needs to sense changes in the environment and move accordingly. The movement toward or away from attractants or repellents is called taxis. As noted earlier, bacteria exhibit taxes to a variety of stimuli, including light and oxygen. The movement of cells toward chemical attractants or away from chemical repellents (**chemotaxis**) is the best-studied type of taxis. Chemotaxis is readily observed in Petri dish cultures. If bacteria are placed in the center of a dish of semisolid agar containing an attractant, the bacteria will exhaust the local supply of the nutrient and swim outward following the attractant gradient they have created. The result is an expanding ring of bacteria. When a disk of repellent is placed in a Petri dish of semisolid agar and bacteria, the bacteria will swim away from the repellent, creating a clear zone around the disk.

Attractants and repellents are detected by **chemoreceptors,** proteins that bind chemicals and transmit signals to other components of the chemosensing system. Chemosensing systems, located in the plasma membrane, are sensitive and allow the cell to respond to low levels of attractants (about 10^{-8} M for some sugars). Some receptors also participate in the initial stages of sugar transport into the cell.

The chemotactic behavior of *E. coli* has been extensively studied and is our focus here. Its movements can be followed using a tracking microscope, a microscope with a moving stage that automatically keeps an individual bacterium in view. In the absence of a chemical gradient, *E. coli* cells move randomly, switching back and forth between runs and tumbles. During a run, the bacterium swims in a straight or slightly curved line. After a few seconds, the bacterium stops and tumbles. The tumble randomly reorients the cell so that it is facing in a different direction. Therefore when it begins the next run, it usually goes in a different direction (**figure 3.44*a***). In contrast, when *E. coli* is exposed to an attractant, it tumbles less frequently (or has longer runs) when traveling toward the attractant. Although the tumbles can still orient the bacterium away from the attractant, over time the cell gets closer and closer to the attractant (figure 3.44*b*). The opposite response occurs with a repellent. Tumbling frequency decreases (the run time lengthens) when the bacterium moves away from the repellent.

E. coli must have some mechanism for sensing that it is getting closer to the attractant (or moving away from the repellent).

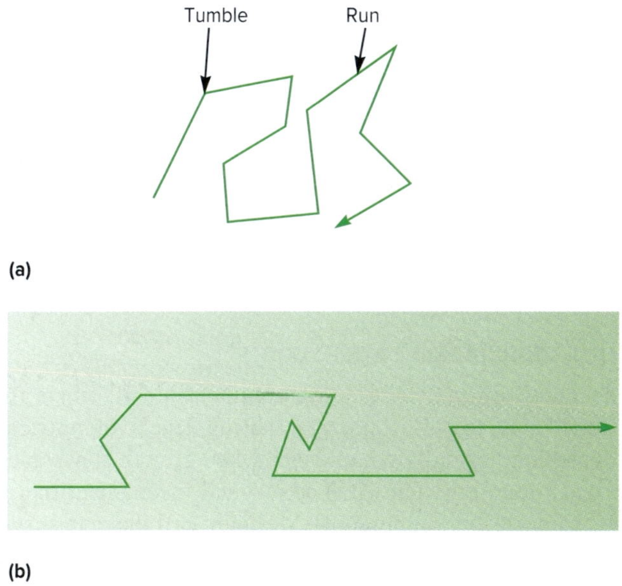

Figure 3.44 Tumbles and Runs Are Used to Direct the Movement of Many Bacteria Toward a Chemical Attractant. (a) Random movement of a bacterium in the absence of a concentration gradient. Tumbling frequency is fairly constant. (b) Movement in an attractant gradient. Tumbling frequency is reduced when the bacterium is moving up the gradient. Therefore runs in the direction of increasing attractant are longer.

The behavior of the bacterium is shaped by temporal changes in chemical concentration. The cell is able to compare the current concentration with the concentration a few seconds earlier. If the concentration of the attractant is increasing, tumbling is suppressed. Likewise, *E. coli* moves away from a repellent because it senses that the concentration of the repellent is decreasing. The bacterium's chemoreceptors play a critical role in this process. The molecular events that enable *E. coli* cells to sense a chemical gradient and respond appropriately are presented in chapter 14.

Comprehension Check

1. Describe the way many flagella operate to move a bacterium.
2. How does swimming differ from swarming?
3. Explain in a general way how bacteria move toward substances such as nutrients and away from toxic materials.
4. Suggest why chemotaxis is sometimes called a "biased random walk."

3.9 Bacterial Endospores Are a Survival Strategy

After reading this section, you should be able to:

a. Describe the structure of a bacterial endospore
b. Explain why bacterial endospores are of particular concern to the food industry and why endospore-forming bacteria are important model organisms
c. Describe in general terms the process of sporulation
d. Describe those properties of endospores that are thought to contribute to its resistance to environmental stresses
e. Describe the three stages that transform an endospore into an active vegetative cell

Endospores, dormant cells formed within a so-called mother cell, are fascinating bacterial structures only produced by certain members of the genera *Bacillus* and *Clostridium* (rods), and *Sporosarcina* (cocci) within the phylum *Firmicutes.* It is important to distinguish bacterial endospores from spores formed by fungi and plants. Endospore formation is not a reproductive strategy as only a single endospore is produced by a bacterial cell. There are several reasons why endospores have long held the interest of microbiologists. The primary reason is that endospores are extraordinarily resistant to environmental stresses such as heat, ultraviolet radiation, gamma radiation, chemical disinfectants, and desiccation. In fact, some endospores have remained viable for around 100,000 years. Another important reason is that several species of endospore-forming bacteria are dangerous pathogens. For example, *Clostridium botulinum* causes botulism, a food-borne disease that results from ingestion of botulinum toxin, the deadliest toxin known. The extreme heat resistance of *C. botulinum*'s endospores is a major concern of the food industry. Other medically important endospore-forming bacteria include *B. anthracis* (causes anthrax), *C. tetani* (causes tetanus), and *C. perfringens* (causes gas gangrene and food poisoning). Endospores also are of considerable interest to scientists because of their complex structures and the equally complex process that generates them. Over the decades, microbiologists have asked many questions about endospores and endospore formation. (1) How are endospores structurally different from vegetative cells? (2) What makes them so resistant to harsh environmental conditions? (3) What triggers endospore formation? (4) What are the steps in endospore formation and how is the process regulated? (5) How do endospores convert back to vegetative cells? In this section, we address these questions. We save the consideration of regulating endospore formation for chapter 14. ▶| *Sporulation in* B. subtilis *(section 14.6)*

Endospore Structure and Resistance

Electron microscopy has been an important tool for dissecting endospore structure. Electron micrographs show that the endospore consists of a core surrounded by several layers that vary significantly in composition. The core has normal cell structures such as ribosomes and a nucleoid but has very low water content (**figure 3.45**). The core is surrounded by an inner membrane, which is in turn covered by the core wall (also called the germ cell wall, as it contains the peptidoglycan that will form the wall of the vegetative cell that grows out of the endospore following germination). Next is the cortex, which may occupy as much as half the endospore volume. It is made of peptidoglycan that is less cross-linked than that in vegetative cells. The cortex is surrounded by a phospholipid bilayer called the outer membrane. Outside of the outer membrane is the coat, a complex structure composed of several layers. It is composed of more than 70 different proteins, which are highly cross-linked to each other. Finally, many endospore-forming bacteria produce endospores enclosed by a thin covering called the exosporium, which is made up of glycoproteins.

The ability of the endospore to survive heat, radiation, and damaging chemicals requires that its enzymes and DNA be protected. The various layers of the endospore contribute to this resistance in several ways. The coat protects the endospore from chemicals and various lytic enzymes such as lysozyme. The inner membrane is extremely impermeable to various chemicals, including those that cause DNA damage. The core has very low water content, high amounts of dipicolinic acid complexed with calcium ions (Ca-DPA), and a slightly lower pH, all of which contribute to the endospore's resistance to harsh conditions. The low water content seems to be especially important for endospore resistance. Evidence exists that the water content of the core is low enough to prevent rotation of enzymes and other proteins present. However, it is not low enough to prevent protein denaturation. It has been suggested that the immobilization prevents them from interacting with each other and becoming entangled. Thus, even though they may become denatured, they are able to refold to their proper active structure as the endospore germinates. The endospore's DNA is protected by two main mechanisms. Ca-DPA complexes are inserted between the nitrogenous bases of DNA, which helps stabilize it. In addition, small, acid-soluble DNA-binding proteins (SASPs) saturate endospore DNA. SASPs alter the three-dimensional structure of DNA in the endospore core, converting it from the common B form to A form. The A form DNA is less susceptible to damage by ultraviolet (UV) light, which accounts for the ability of endospores to tolerate UV exposure. ▶| *DNA is a polymer of deoxyribonucleotides (section 13.2)*

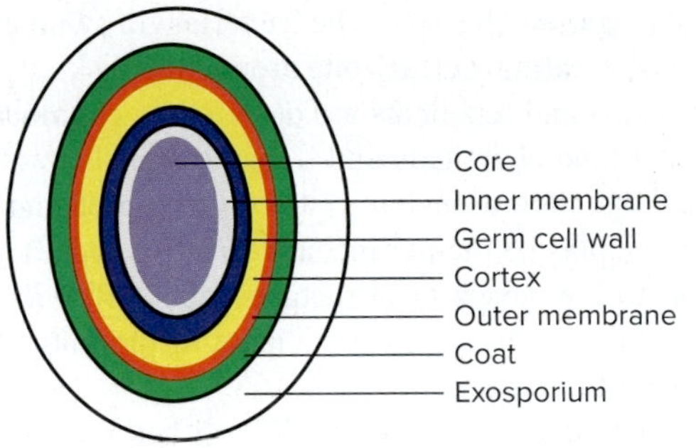

(a)

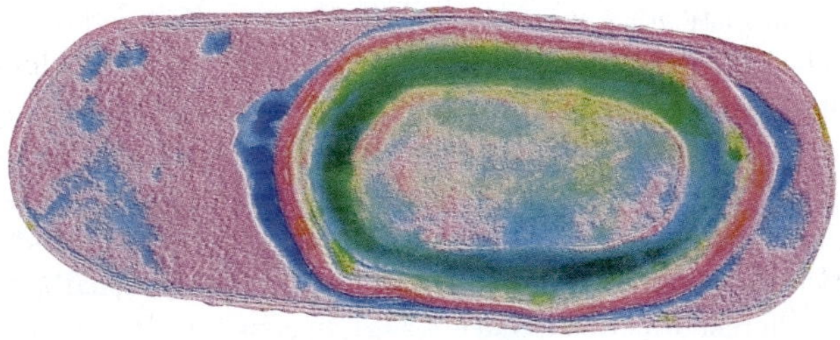

(b)

Figure 3.45 Bacterial Endospores. (a) A schematic showing the various layers of an endospore. They are not drawn to scale. (b) A colorized cross section of a *Bacillus subtilis* cell undergoing sporulation. The oval in the center is an endospore that is almost mature; when it reaches maturity, the mother cell will lyse to release it.

(b) ©CNRI/Science Source

Sporulation: Making Endospores

Endospore-forming bacteria are common in soil, where they must be able to withstand fluctuating levels of nutrients. **Sporulation** normally commences when growth slows due to nutrient limitation. Thus it is a survival mechanism that allows the bacterium to produce a dormant cell that can survive until nutrients are again available and vegetative growth can resume. These bacteria cycle between two states: vegetative growth and survival as an endospore. Vegetative growth is the normal, continuous cycle of growth and division. By contrast, sporulation is a complex process that occurs in a highly organized fashion over several hours. The mature endospore occupies a characteristic location in the mother cell (referred to as the sporangium), depending on the species of bacteria.

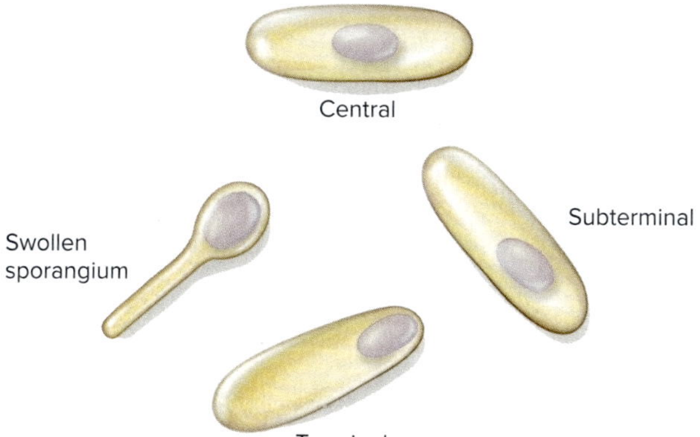

Figure 3.46 Examples of Endospore Location and Size.

Endospores may be centrally located, close to one end (subterminal), or terminal (**figure 3.46**). Sometimes an endospore is so large that it swells the sporangium. One of the best-studied endospore formers is *Bacillus subtilis*, which is an important model organism.

Sporulation may be divided into seven stages (**figure 3.47**). The cell's DNA is replicated (stage I), followed by an inward folding of the cell membrane to enclose part of the DNA and produce the forespore septum (stage II). The mother cell membrane continues to grow and engulfs the immature endospore in a second membrane (stage III). Next, cortex is laid down in the space between the two membranes, and both calcium and dipicolinic acid are accumulated (stage IV). Protein coats are formed around the cortex (stage V), and maturation of the endospore occurs (stage VI). Finally, lytic enzymes destroy the sporangium, releasing the endospore (stage VII). Sporulation requires about 8 to 10 hours. *Bacterial Endospore Formation*

Endospore to Vegetative Cell

The transformation of dormant endospores into active vegetative cells is almost as complex as sporulation. It occurs in three stages: (1) activation, (2) germination, and (3) outgrowth. Activation is a process that prepares endospores for germination and can result from treatments such as heating. This is followed by **germination,** the breaking of the endospore's dormant state. It begins when proteins called germinant receptors, located in the inner membrane and the cortex, detect small molecules such as sugars and amino acids. Upon detection of these molecules by the germinant receptors, a series of events occur. These activated receptors trigger the release of the Ca-DPA complexes, breakdown of the peptidoglycan in the cortex, and water uptake. Eventually

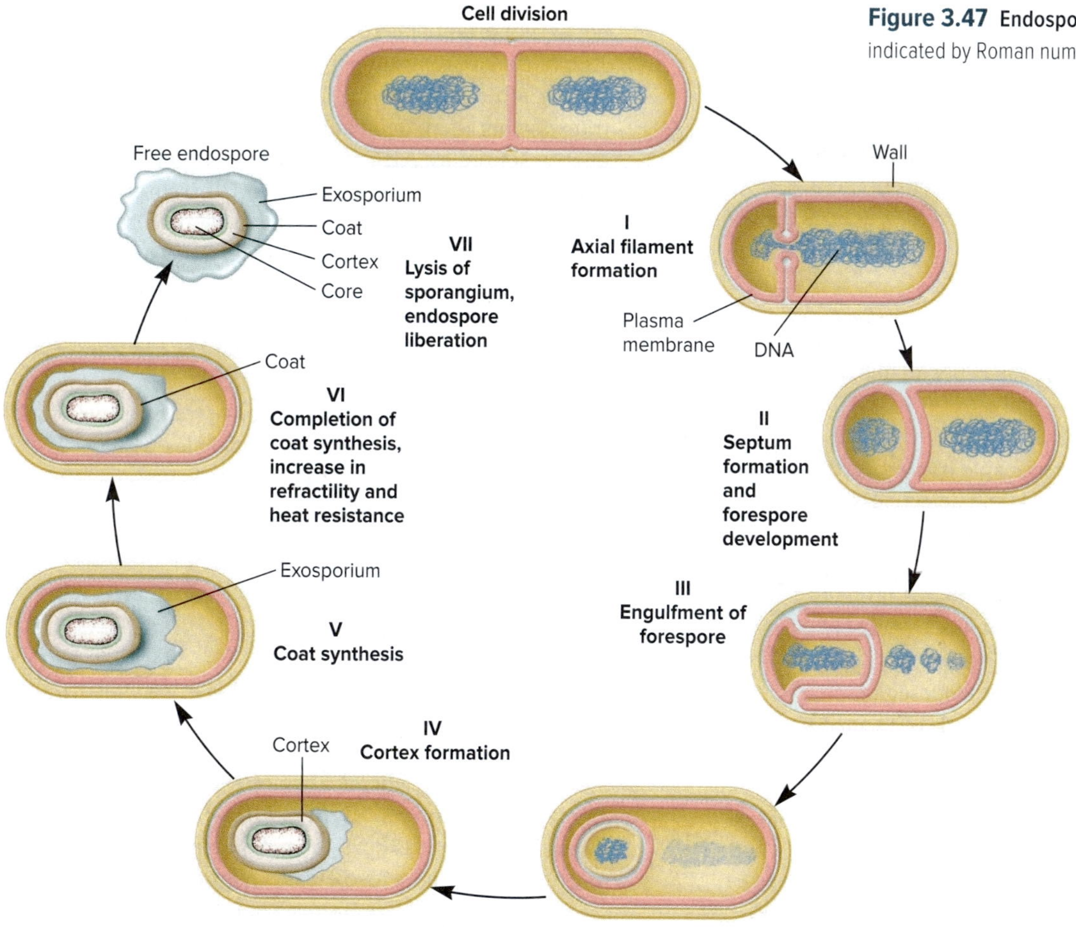

Figure 3.47 Endospore Formation. The stages are indicated by Roman numerals.

water levels inside the germinating endospore reach those characteristic of vegetative cells and enzymes in the core become active. This allows the endospore to begin synthesizing various molecules needed to initiate endospore outgrowth and return to a vegetative state. Because germination of some species' endospores results in disease or toxin production, inhibiting germination by blocking germinant receptors is an active area of research.

Comprehension Check

1. Describe the structure of the bacterial endospore using a labeled diagram.
2. Briefly describe endospore formation and germination. What is the importance of the endospore?
3. What features of the endospore contribute to its resistance to harsh conditions?

Key Concepts

3.1 Use of the Term "Prokaryote" Is Controversial

- Bacteria and archaea are related and share a common cell architecture. However, they have distinct characteristics and evolutionary lineages.

3.2 Bacteria Are Diverse but Share Some Common Features

- Rods (bacilli) and cocci (spheres) are common bacterial shapes. In addition, bacteria may be comma-shaped (vibrio), spiral (spirillum and spirochetes), or filamentous; they may form buds and stalks; or they may have no characteristic shape (pleomorphic) (**figures 3.1 and 3.2**).
- Some cells remain together after division to form pairs, chains, and clusters of various sizes and shapes.
- Frequently observed bacterial cell structures include a cell wall, plasma membrane, cytoplasm, nucleoid, fimbriae, inclusions, capsule, ribosomes, and flagella (**figure 3.6**). **Table 3.1** summarizes the major functions of these structures.

3.3 Bacterial Plasma Membranes Control What Enters and Leaves the Cell

- The cell envelope consists of the plasma membrane and all external coverings, including the cell wall and other layers (e.g., capsules).
- The bacterial plasma membrane fulfills many roles, including acting as a semipermeable barrier, carrying out respiration and photosynthesis, and detecting and responding to chemicals in the environment.
- The fluid mosaic model proposes that cell membranes are lipid bilayers in which integral membrane proteins are buried. Peripheral membrane proteins are loosely associated with the membrane (**figure 3.7**).
- Bacterial membranes are bilayers composed of phospholipids constructed of fatty acids connected to glycerol by ester linkages (**figure 3.8**). Bacterial membranes include microdomains characterized by an altered lipid composition, in which large protein complexes are assembled by flotillins.
- Microorganisms require nutrients, materials that are used in energy conservation and biosynthesis. Macronutrients are needed in relatively large quantities. Micronutrients (trace elements) are used in very small amounts.
- In passive diffusion, a substance moves through the membrane on its own. Movement is down a concentration gradient and does not require an input of energy. Only a few substances enter bacteria by passive diffusion.
- Transport proteins are divided into two major types: channels and carriers. Channels create a pore through which a substance moves. Carriers are so called because they carry the molecule across the membrane.
- In facilitated diffusion, a transport protein (either a channel or a carrier) helps move substances in the direction of decreasing concentration; no metabolic energy is required (**figure 3.11**). Facilitated diffusion is not an important uptake mechanism for bacterial cells.
- Active transport systems use metabolic energy and carrier proteins to concentrate substances by transporting them against a gradient. ATP is used as an energy source by ABC transporters, a type of primary active transporter (**figure 3.13**). ABC transporters are uniporters (one substance is transported). Gradients of ions drive solute uptake in secondary active transport systems. Secondary active transporters are either symporters (two substances are transported in the same direction) or antiporters (two substances are transported in opposite directions) (**figure 3.12**).
- Group translocation is a type of active transport in which bacterial cells transport organic molecules while modifying them (**figure 3.14**).
- The secretion of siderophores, which bind ferric iron, enables the accumulation of iron within bacterial cells (**figure 3.15**).

3.4 There Are Two Main Types of Bacterial Cell Walls

- The vast majority of bacteria have a cell wall outside the plasma membrane to help maintain their shape and protect them from osmotic stress.
- Bacterial walls are chemically complex and usually contain peptidoglycan (**figures 3.16–3.21**). Typical Gram-positive walls have thick, homogeneous layers of peptidoglycan and teichoic acids (**figures 3.22** and **3.23**). Typical Gram-negative bacteria have a thin peptidoglycan

layer surrounded by a complex outer membrane containing lipopolysaccharides (LPSs) and other components (**figures 3.24** and **3.25**). The lipid A portion of LPS is also called endotoxin. Its release into the human body can lead to septic shock.

- The mechanism of the Gram stain is thought to depend on the thickness of peptidoglycan. The thick peptidoglycan of typical Gram-positive bacteria binds crystal violet tightly, preventing its loss during the ethanol wash.

3.5 The Cell Envelope Often Includes Layers Outside the Cell Wall

- Capsules, slime layers, and glycocalyces are layers of material lying outside the cell wall. They can protect cells from certain environmental conditions, allow cells to attach to surfaces, and protect pathogenic bacteria from host defenses (**figure 3.28**).
- S-layers are the external-most layer in some bacteria. They are composed of proteins or glycoprotein and have a characteristic geometric shape (**figure 3.29**).

3.6 The Bacterial Cytoplasm Is More Complex than Once Thought

- The bacterial cytoplasm is a highly concentrated and crowded solution of biochemicals.
- The bacterial cytoplasm contains proteins that are similar in structure and function to the cytoskeletal proteins observed in eukaryotes (**figure 3.30** and **table 3.2**).
- Some bacteria have simple internal membrane systems containing photosynthetic and respiratory machinery (**figure 3.31**).
- Inclusions are observed in all cells (**figure 3.33**). Most are used for storage (e.g., PHB inclusions and polyphosphate granules), but some are used for other purposes (e.g., magnetosomes and gas vacuoles). Microcompartments such as carboxysomes contain enzymes that catalyze important reactions (e.g., CO_2 fixation).
- Bacterial ribosomes are 70S in size. They are composed of numerous proteins and several rRNA molecules (**figure 3.34**).
- The genetic material of bacterial cells is located in an area within the cytoplasm called the nucleoid. The nucleoid is not usually enclosed by a membrane (**figure 3.35**). The chromosome usually consists of a double-stranded, covalently closed, circular DNA molecule.
- Plasmids are extrachromosomal DNA molecules found in many bacteria. Some are episomes—plasmids that exist freely in the cytoplasm or can be integrated into the chromosome. Although plasmids are not required for survival in most conditions, they can encode traits that confer selective advantage in some environments. Many types of plasmids have been identified (**table 3.3**).

3.7 Many Bacteria Have External Structures Used for Attachment and Motility

- Many bacteria have hairlike appendages called fimbriae or pili. Fimbriae function primarily in attachment to surfaces, but type IV pili are involved in twitching motility. Sex pili participate in the transfer of DNA from one bacterium to another (**figure 3.36**).
- Many bacteria are motile, often by means of threadlike, locomotory organelles called flagella. Bacterial species differ in the number and distribution of their flagella (**figure 3.37**). Each bacterial flagellum is composed of a filament, hook, and basal body (**figure 3.38**).

3.8 Bacteria Move in Response to Environmental Conditions

- Several types of bacterial motility have been observed: swimming by flagella, swarming by flagella, spirochete motility, twitching motility, and gliding motility.
- The bacterial flagellar filament is a rigid helix that rotates like a propeller to push the bacterium through water (**figures 3.40** and **3.41**). When many bacteria swim, they alternate between two types of movement: runs and tumbles.
- Some bacterial species move as a group in a behavior called swarming. Swarming is mediated by flagella and occurs on moist surfaces.
- Spirochete motility is brought about by flagella that are wound around the cell and remain within the periplasmic space. When they rotate, the outer membrane of the spirochete is thought to rotate, thus moving the cell (**figure 3.43**).
- *Myxococcus* spp. exhibit both twitching and gliding motility. Twitching motility is a jerky movement brought about by type IV pili, whereas gliding motility is smooth.
- Motile cells can respond to gradients of attractants and repellents, a phenomenon known as chemotaxis. *E. coli* and many other peritrichously flagellated bacteria accomplish movement toward an attractant by increasing the length of time spent moving toward the attractant and shortening the time spent tumbling (**figure 3.44**). Similarly, they increase their run time when they move away from a repellent.

3.9 Bacterial Endospores Are a Survival Strategy

- Some bacteria survive adverse environmental conditions by forming endospores, dormant structures resistant to heat, desiccation, and many chemicals (**figure 3.45**).
- Both endospore formation and germination are complex processes made in response to certain environmental signals (**figure 3.47**).

Active Learning

1. Propose a model for the assembly of a flagellum in a typical Gram-positive cell envelope.
2. The peptidoglycan of bacteria has been compared with the chain mail worn beneath a medieval knight's suit of armor. It provides both protection and flexibility. Describe other structures in biology that have an analogous function. How are they replaced or modified to accommodate the growth of the inhabitant?
3. Why might a microbe have more than one uptake system for certain substrates?
4. Design a demonstration to illustrate the cell wall's role in protecting against lysis.
5. What would you expect to observe if you were able to "transplant" CreS into a rod-shaped bacterium such as *Bacillus subtilis*?
6. Develop a hypothesis to explain why gas vacuoles are bounded by proteins rather than a lipid bilayer membrane.
7. Support or refute the statement that the periplasm is an organelle because it is a membrane-bound compartment of the bacterial cell.
8. Calculate the surface area-to-volume ratio of an ultrasmall bacterium, a coccus of 0.2 μm diameter, and compare it to the ratios calculated in figure 3.5.
9. *Kingella kingae* colonizes the pharynx in young children. It has three distinct surface components that may be responsible for attachment to host cells: a type IV pilus (T4P) that confers twitching motility, a protein adhesin, and a polysaccharide capsule on its outer membrane. To study attachment, four variants were created: one lacking the T4P, one with a T4P unable to twitch, one lacking the adhesin, and one lacking the capsule. Why do you think the researchers created the strain with the disabled T4P?

   Dimensions of these components were measured: The T4P was several microns in length, the adhesin extended about 110 nm from the surface, and the capsule was 700 nm deep. Micrographs of *Kingella* cells adhered to host cells show that the capsule was compressed significantly when the bacteria were in contact with a host cell. Sketch these structures to help visualize the cell surface. The most important determinants in adherence were a functional, twitching T4P and the adhesin. Considering the geometry of the cell surface, how might these two components collaborate in attachment of the bacterium?

Read the original paper: Kern, B. K., et al. 2017. Defining the mechanical determinants of *Kingella kingae* adherence to host cells. *J. Bacteriol.* 199(23):e00314-17.

4

Archaeal Cell Structure

©Ingram Publishing

Cows and Buffaloes and Sheep, Oh My!

The next time you drive or walk through the countryside, count the number of domesticated ruminants you see—dairy cows, beef cattle, sheep, and goats, among others. It is estimated that there are about 3 billion head worldwide. These animals are important because they are a substantial protein source. But they do something else that is noteworthy: Each year they exhale about 200 million metric tons of methane, a potent greenhouse gas (GHG). The manure these and other domesticated animals produce is thought to release another 18 million metric tons of methane. As detailed in chapter 28, carbon dioxide is an important GHG and is considered to be a significant contributor to global warming. Methane and nitrous oxide are also GHGs. Carbon dioxide is the dominant GHG in the atmosphere. However, methane has about 30 times the heat-retaining capacity of carbon dioxide. Thus, although the contribution of methane by domesticated ruminants and their manure represents only about 5% of the total GHG emissions worldwide each year, the impact of these emissions is of concern. But don't blame the animals; they are not directly responsible for generating methane. Rather, it is a unique group of archaea—the methanogens—that make methane from hydrogen gas, carbon dioxide, and acetate. Furthermore, these substrates arise in the animals as their food is digested with the help of a complex community of microbes living in their rumen. For these animals, methane is a waste product; for farmers and ranchers, methane represents a loss in production. There is considerable interest in figuring out how to decrease methane production so that GHG emissions are lowered and farmers and ranchers can earn more from each animal.

Methanogens are only one physiological group in domain *Archaea*. Other archaea have attracted attention because of their ability to live in extreme environments. All are of interest because of their chimeric natures: They have some features that are similar to bacteria and some that are similar to eukaryotes. Their cell structure is like that of bacteria; that is, they look like the canonical prokaryotic cell. However, the molecules used to construct their cell structures are often unique or are similar to those found in eukaryotes. As we describe in chapters 15 and 20, their physiologies exhibit this dichotomy. In general, the processes and molecules used to conserve energy are like those of bacteria, whereas the processes and molecules used to replicate and express their genomes are like those of eukaryotes. In this chapter, we describe some of the molecular and structural aspects of archaeal cells that set them apart.

Readiness Check:

Based on what you have learned previously, you should be able to:

- ✓ Summarize the structures observed in bacterial cells and their molecular makeup (chapter 3)
- ✓ Describe the mechanisms used by bacteria to obtain nutrients (section 3.3)
- ✓ Describe the mechanisms used by bacteria to move in response to stimuli (section 3.8)

4.1 Archaea Are Diverse but Share Some Common Features

After reading this section, you should be able to:

a. Describe a typical archaeal cell
b. Discuss key differences between bacteria and archaea

Recall that for many years, **archaea** and bacteria had been lumped together and referred to as prokaryotes. Today *Archaea* and *Bacteria* are recognized as distinct taxa. The separation of bacteria and archaea into distinct taxa correlates with the observation that they each have unique and distinguishing characteristics. Some of these are summarized in **table 4.1**.

Keep in mind that only a small subset of the archaea can be grown in the laboratory. Many archaea inhabit niches that constitute extreme environments from a human perspective. Their requirements for high temperature, low pH, or high salt present challenges for culturing them. Historically, environments dominated by archaea were not believed to support life. Once microbiologists broadened their exploration of earth's microbial habitats, archaea have been found to be ubiquitous, but in small numbers. Surveys of nonextreme environments have established that archaea reside in most habitats—again, in small numbers. Most of the studies described in this chapter were performed on archaea that are readily grown in the lab. Methanogens (microbes that generate methane gas) are killed by exposure to oxygen, and must be handled in sealed containers. The names of these organisms begin with *Methano-*. Halophiles (salt-lovers) require high salt conditions and are named *Halo-*. Another commonly studied archaeon is *Sulfolobus,* which requires both high temperature and low pH for growth.

Many archaea are known only in the sense that their 16S rRNA sequence or a partial genome sequence has been determined. Archaeal taxonomy is currently in a state of flux because of the vast

Table 4.1 Comparison of Bacterial and Archaeal Cells

Property	Bacteria	Archaea
Plasma membrane lipids	Ester-linked phospholipids form a lipid bilayer	Glycerol diethers form lipid bilayers; glycerol tetraethers form lipid monolayers
Cell wall constituents	Peptidoglycan is present in nearly all; some lack cell walls	Very diverse but peptidoglycan is always absent: some consist of S-layer only, others combine S-layer with polysaccharides or proteins or both; some lack cell walls
Inclusions present	Yes, including gas vacuoles	Yes, including gas vacuoles
Ribosome size	70S	70S
Chromosome structure	Most are circular, double-stranded (ds) DNA	All known are circular, dsDNA
Plasmids present	Yes; circular and linear dsDNA	Yes; circular dsDNA
External structures	Flagella, fimbriae (pili) common	Archaella, pili common
Capsules or slime layers	Common	Rare

amount of genomic data that has identified new archaea. There is controversy over the taxonomic designations used for these new organisms, but three phyla are well established: *Crenarchaeota, Euryarchaeota,* and *Thaumarchaeota*. We begin our discussion of archaeal cells by considering overall cell morphology and organization and then specific cell structures.

Shape, Arrangement, and Size

Archaeal cells, like bacterial cells, exhibit a variety of shapes. **Cocci** and **rods** are common (**figure 4.1*a***). Both usually exist singly, but some cocci form clusters and some rods form chains. Curved rods, spiral shapes, and pleomorphic (many shaped) archaea have also been observed. To date, no spirochete-like and mycelial archaea have been discovered. However, some archaea exhibit unique shapes, such as the branched form of *Thermoproteus tenax* (figure 4.1*b*) and the flat, postage-stamp-shaped *Haloquadratum walsbyi,* an archaeon that lives in salt ponds and measures about 2 μm by 2 to 4 μm and only 0.25 μm thick. This shape has the advantage of greatly increasing the surface area-to-volume (S/V) ratio (*see figure 3.5*), which in turn increases efficiency of nutrient uptake, diffusion of molecules, and growth rate. ◀ *Bacteria are diverse but share some common features (section 3.2)*

Archaeal cells also vary in size as much as in shape. Typical rods are 1 to 2 μm wide by 1 to 5 μm long; cocci are typically 1 to 3 μm in diameter. However, extremely small and extremely large archaea have been identified. Several free-living, acid-loving (acidophilic), mine-dwelling microbes measure a mere 0.2 to 0.4 μm in diameter. Also at the small end of the size continuum is the parasitic archaeon *Nanoarchaeum equitans* (0.4 μm in diameter). At the other extreme are two recently observed giant archaea that form long

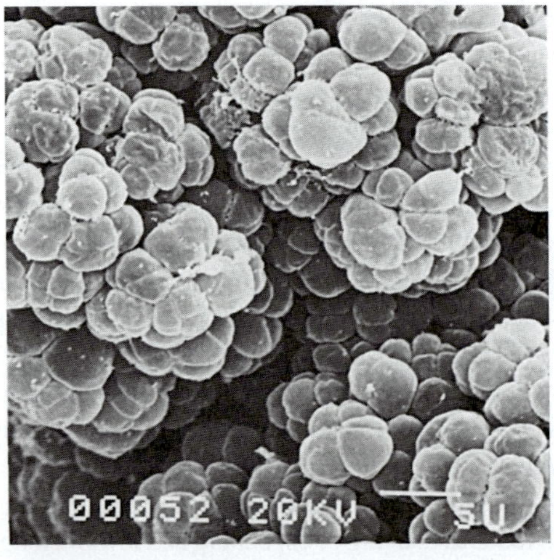

(a) *Methanosarcina mazei*—a coccus that forms clusters

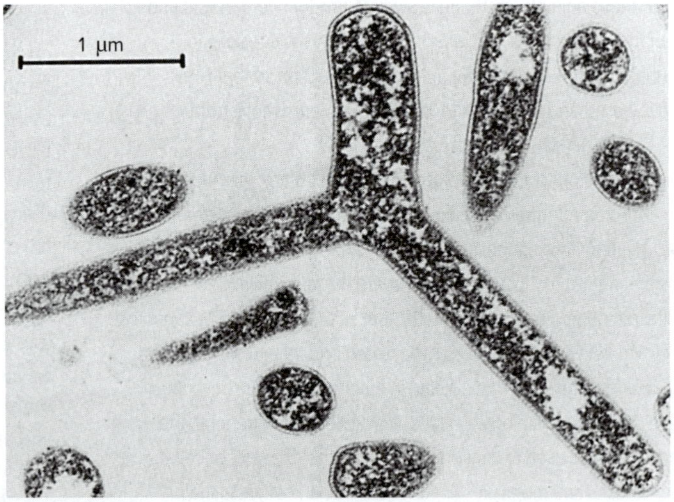

(b) *Thermoproteus tenax*—a branched archaeal cell

Figure 4.1 Archaeal Cell Morphology. (a) Scanning electron micrograph. Bar = 5 μm. (b) Transmission electron micrograph.
From J.T. Staley, M.P. Bryant, N. Pfenning and J.G. Holt (Eds), Bergey's Manual of Systematic Bacteriology, Vol. 3. ©1989 Williams and Wilkins Co., Baltimore

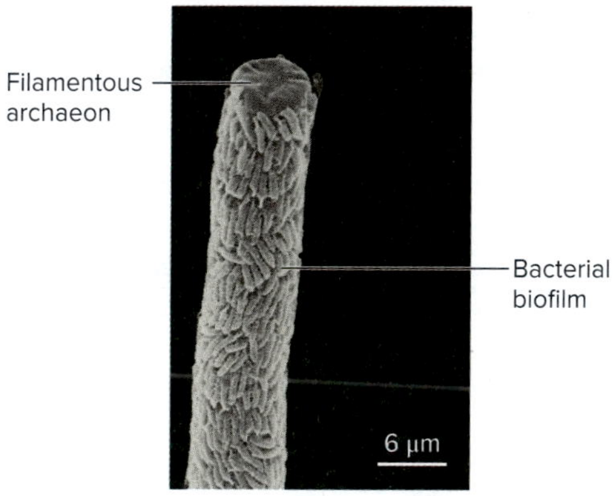

Figure 4.2 A Giant Archaeon. *Candidatus* Giganthauma karukerense forms long filaments that are covered with a bacterial biofilm. The bacteria may be symbionts of the host archaeon.

filaments up to 30 mm in length. For one archaeon, the filament is composed of numerous cells, each measuring 8 to 10 µm wide by 20 to 24 µm long. An interesting characteristic of this archaeon is that its filaments are coated with a biofilm formed by bacteria of a single species (**figure 4.2**). The nature of this interaction is not known, but it is thought that the archaeon is the host and the bacteria are symbionts.

Cell Organization

Structures observed in members of both *Bacteria* and *Archaea* are summarized and their differences noted in table 4.1. The archaeal plasma membrane is composed of strikingly different lipids than those found in bacterial membranes; in fact, the unusual lipids were one of the first pieces of evidence to suggest that these microbes are phylogenetically distinct from bacteria. Most archaea have a cell wall, but their walls are considerably more diverse than bacterial walls. Notably, archaeal cell walls lack peptidoglycan. Capsules are not widespread among archaea examined thus far. Within the archaeal cytoplasm, a nucleoid, ribosomes, and inclusions can be found. Finally, many archaea use flagella for locomotion and attachment, but these structures differ sufficiently from bacterial flagella so they have been named archaella (s., archaellum).

In the remaining sections of this chapter, we describe the major structures observed in archaea in more detail, noting the aspects that distinguish them from the analogous bacterial structures. Understanding archaeal structures at the molecular level is an area of intense research because many archaea are found in extreme habitats. We note those structural adaptations that contribute to this lifestyle. However, the topic of survival in extreme habitats is considered more thoroughly in chapter 7. ▶| *Environmental factors affect microbial growth (section 7.5)*

Comprehension Check

1. Which cell shapes are observed in members of both *Bacteria* and *Archaea*? Which are unique to bacteria? Which to archaea?
2. *Archaea* was first defined as a distinct taxon by comparisons of ribosomal RNA sequences. Identify two other molecules that could be used to determine if a microbe having a typical prokaryotic architecture is a bacterium or an archaeon.

4.2 Archaeal Cell Envelopes Are Structurally Diverse

After reading this section, you should be able to:

a. Draw an archaeal cell envelope and identify the component layers
b. Compare and contrast archaeal and bacterial cell envelopes in terms of their structure, molecular makeup, and functions
c. Compare and contrast nutrient uptake mechanisms observed in bacteria and archaea

We define the **cell envelope** as the plasma membrane and any layers external to it. For bacteria, these extra layers include cell walls, S-layers, capsules, and slime layers. One of the most distinctive features of archaeal cells is the nature of their cell envelopes. Their uniqueness begins with their plasma membranes. Recall from chapter 3 that the fluid mosaic model describes membranes as lipid bilayers within which proteins float (*see figure 3.7*). The model is based on studies of eukaryotic and bacterial membranes and is well established. Imagine the surprise of microbiologists when they found that some archaea have monolayer membranes that function like bilayers. This structural difference is due to the presence of unique lipids in archaeal plasma membranes. Archaeal cell envelopes also differ in terms of organization. For many archaea, an S-layer is the major, and sometimes only, component of the cell wall. Slime layers have been observed in some archaea where they appear to mediate cell-cell interactions, but little is known about their composition and regulation. Finally, capsules are relatively rare among archaea and thus are not discussed.

Archaeal Plasma Membranes Are Composed of Unique Lipids but Function Like Bacterial Membranes

Archaeal membranes are composed primarily of lipids that differ from bacterial and eukaryotic lipids in two ways. First, they contain hydrocarbons derived from isoprene units—five-carbon, branched molecules (**figure 4.3**). Thus the hydrocarbons are branched as shown in **figure 4.4**. This affects the way the lipids pack together, which in turn affects the fluidity and permeability of the membrane. This is especially important for extremophilic archaea for which membrane fluidity and permeability could be compromised by extreme conditions. Second, the hydrocarbons are attached to glycerol by ether links rather than ester links (figure 4.4*a*). Ether linkages are more resistant to chemical attack and heat than are ester links.

Two major types of archaeal lipids have been identified: glycerol diethers and diglycerol tetraethers. **Glycerol diether lipids** are

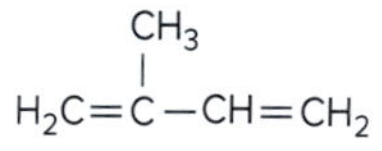

Isoprene

Figure 4.3 Isoprene. This five-carbon, branched molecule is the building block of archaeal lipids.

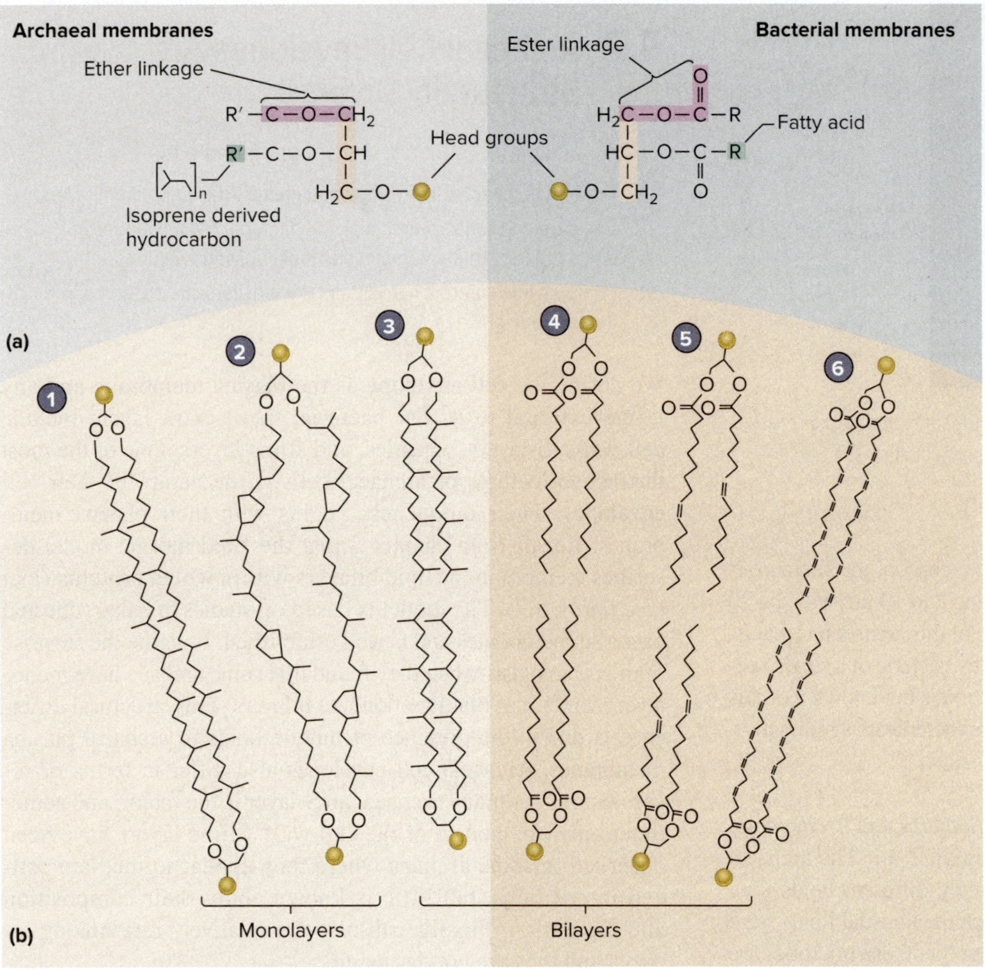

Figure 4.4 Comparison of Archaeal and Bacterial Membranes. (a) Archaeal membrane lipids are attached to glycerol by ether linkages instead of ester linkages, as found in bacteria and eukaryotes. The stereochemistry also differs. In archaeal lipids, the stereoisomer of glycerol is *sn*-glycerol-1-phosphate; in bacterial lipids, the stereoisomer is *sn*-glycerol-3-phosphate. Thus in archaeal lipids, the side chains are attached to carbons 2 and 3 of glycerol, and in bacterial lipids, the side chains are attached to carbons 1 and 2. (b) Examples of archaeal lipids are lipids 1, 2, and 3. Lipids 4, 5, and 6 are bacterial lipids. Note that some archaeal lipids form monolayers (figure 4.5), whereas all bacterial lipids form bilayers.

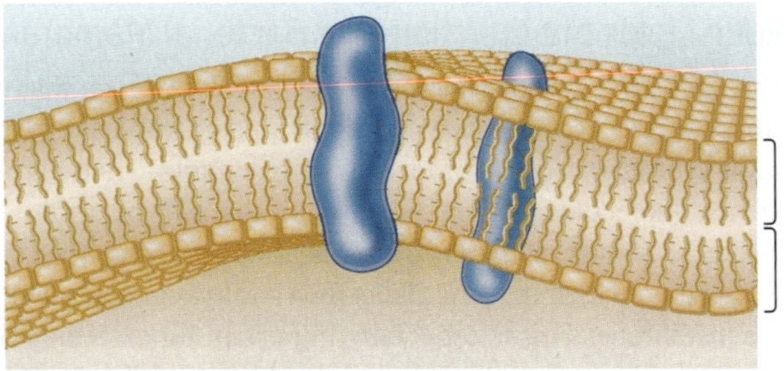

(a) Bilayer of C_{20} diethers

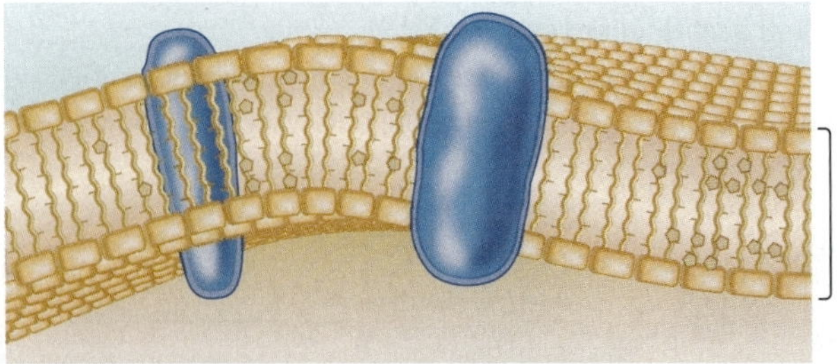

(b) Monolayer of C_{40} tetraethers

Figure 4.5 Examples of Archaeal Plasma Membranes.

formed when two hydrocarbons are attached to glycerol (figure 4.4*b*, lipid 3). Usually, the hydrocarbon chains in glycerol diethers are 20 carbons in length. **Diglycerol tetraether lipids** are formed when two glycerol residues are linked by two long hydrocarbons that are 40 carbons in length (figure 4.4*b*, lipids 1 and 2). Tetraethers are more rigid lipids than diethers. Cells can adjust the overall length of the tetraethers by cyclizing the chains to form pentacyclic rings (figure 4.4*b*, lipid 2). Phosphorus-, sulfur-, amino-, and sugar-containing groups can be attached to the glycerol moieties in the diethers and tetraethers, just like the phospholipids observed in bacterial and eukaryotic membranes.

Despite the significant differences in membrane lipids, the basic design of archaeal membranes is similar to that of bacterial and eukaryotic membranes: There are two hydrophilic surfaces and a hydrophobic core. When C_{20} diethers are used, a typical bilayer membrane is formed (**figure 4.5*a***). When the membrane is constructed of C_{40} tetraethers, a monolayer membrane with much more rigidity is formed (figure 4.5*b*). The addition of pentacyclic rings further increases this rigidity. As might be expected from their need for stability, the membranes of extreme thermophiles such as *Sulfolobus* spp., which grow best at temperatures over 85°C, are almost completely tetraether monolayers. Archaea that live in moderately hot environments such as

Thermoplasma spp., have membranes containing some regions with monolayers and some with bilayers.

Archaeal membranes are believed to include microdomains, as described in chapter 3 for bacterial membranes. Flotillin-like proteins have been identified in *Pyrococcus* membranes where they cluster into complexes as observed in both bacterial and eukaryotic membranes.

As we discuss in chapter 3, organisms have several options for obtaining the nutrients they need from their environment. Importantly, microbes often find themselves in nutrient-poor environments and therefore must be able to accumulate nutrients in their cytoplasm at concentrations higher than the external milieu. Thus, although passive and facilitated diffusion have been observed, archaea primarily use active transport for nutrient uptake. Both primary (e.g., ABC transport) and secondary active transport systems (e.g., symport and antiport) have been identified in archaeal cells, often by searching an archaeon's genome for the genes encoding components of these systems (i.e., genomic analysis). These systems tend to be similar to those seen in bacteria. The group translocation system phosphoenolpyruvate:sugar phosphotransferase system (PTS) also functions in some archaea. |◀ *Bacteria use many mechanisms to bring nutrients into the cell (section 3.3)* ▶| *Microbial genomics (chapter 18)*

There Are Many Different Types of Archaeal Cell Walls

Before they were distinguished as a unique domain of life, archaeal species were characterized as being either Gram positive or Gram negative based on their response to Gram staining. Thus, like all cells (even eukaryotic cells), they will stain either purple or pink when Gram stained. However, their staining reaction does not correlate reliably with a particular cell wall structure as it does for bacteria. Archaeal cell walls exhibit considerable variety in terms of their chemical makeup. Furthermore, their cell walls lack peptidoglycan.

The most common type of archaeal cell wall is an **S-layer** composed of either glycoprotein or protein (**figure 4.6*a***). The S-layer may be as thick as 70 nm. Some methanogens (*Methanolobus* and *Methanococcus* spp.), salt-loving archaea (*Halobacterium* spp.), and extreme thermophiles (*Sulfolobus, Thermoproteus,* and *Pyrodictium* spp.) have S-layer cell walls.

Other archaea have additional layers of material outside the S-layer. For instance, *Methanospirillum* spp. have a protein sheath external to the S-layer (figure 4.6*b*). Other methanogens (*Methanosarcina* spp.) have a polysaccharide layer covering the S-layer (figure 4.6*c*). This material, called methanochondroitin, is similar to the chondroitin sulfate of animal connective tissue.

In some archaea, the S-layer is the outermost layer and is separated from the plasma membrane by a peptidoglycan-like molecule called pseudomurein (figure 4.6*d*). Pseudomurein differs from peptidoglycan in that it has L-amino acids instead of D-amino acids in its cross-links, *N*-acetyltalosaminuronic acid instead of *N*-acetylmuramic acid, and β(1→3) glycosidic bonds instead of β(1→4) glycosidic bonds (**figure 4.7**). These differences mean that lysozyme, penicillin, and other chemicals that

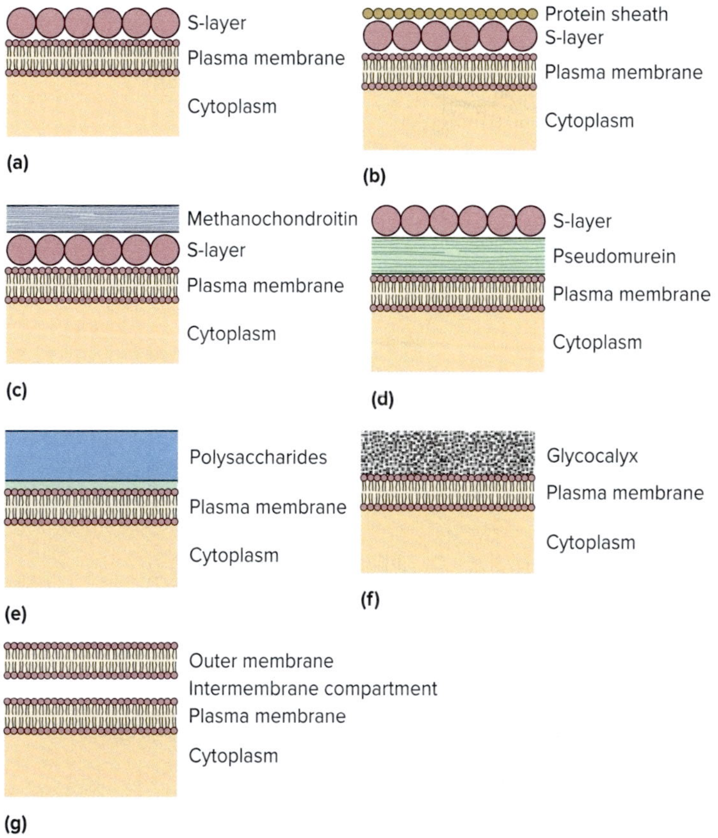

Figure 4.6 Archaeal Cell Envelopes. (a) *Methanococcus, Halobacterium, Pyrodictium, Sulfolobus,* and *Thermoproteus* spp. cell envelopes. (b) *Methanospirillum* spp. cell envelope. (c) *Methanosarcina* spp. cell envelope. (d) *Methanothermus* and *Methanopyrus* spp. cell envelopes. (e) *Methanothermobacter, Methanosphaera, Methanobrevibacter, Halococcus,* and *Natronococcus* spp. cell envelopes. For *Methanothermobacter* and *Methanosphaera* spp., the polysaccharide layer is composed of pseudomurein. (f) *Thermoplasma* cell envelope. (g) *Ignicoccus hospitalis* cell envelope.

affect peptidoglycan structure and synthesis in bacterial cell walls have no effect on pseudomurein-containing archaeal cell walls. |◀ *Cell walls and osmotic protection (section 3.4)*

Another type of archaeal cell wall consists of a single, thick homogeneous layer resembling that in Gram-positive bacteria (figure 4.6*e*). These archaea lack an S-layer and often stain Gram positive. Their wall chemistry varies from species to species but usually consists of complex polysaccharides such as pseudomurein.

Some archaea lack any layer resembling a cell wall. For instance, members of the acidophilic genera *Ferroplasma* and *Thermoplasma* have envelopes consisting only of a plasma membrane covered by a layer of slime. The slime, which is referred to as a glycocalyx, may provide some of the protection needed for these archaea to survive in their acidic habitats (figure 4.6*f*). The most unique wall-less archaeon is *Ignicoccus hospitalis.* Its envelope consists only of the plasma membrane and an outer membrane, with an intermembrane compartment between them (figure 4.6*g* and **figure 4.8**). The outer membrane contains protein complexes that form pores, much like bacterial porin proteins create pores in the outer membrane of typical Gram-negative bacteria.

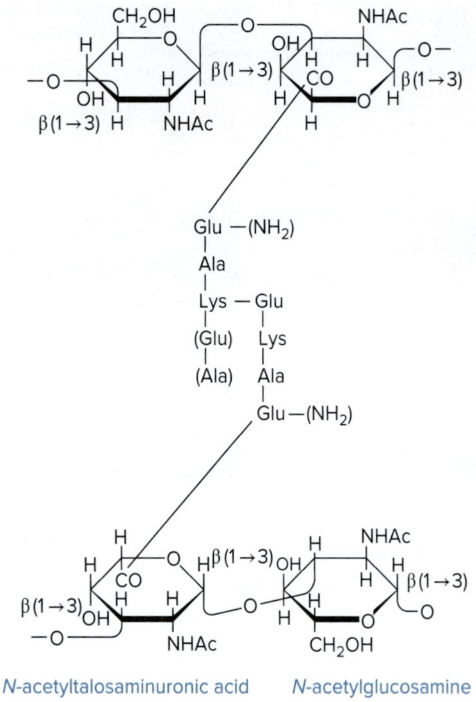

Figure 4.7 Pseudomurein. The amino acids and amino groups in parentheses are not always present. Ac, acetyl group.

MICRO INQUIRY *How is pseudomurein similar to peptidoglycan? How does it differ?*

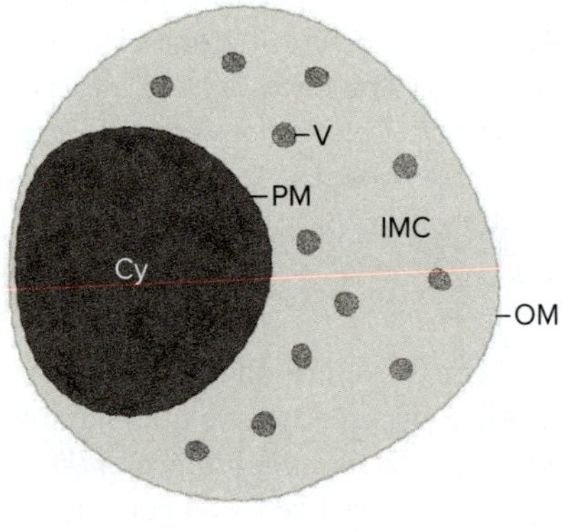

Figure 4.8 Cell Envelope of the Wall-less Archaeon *Ignicoccus hospitalis*. Cy, cytoplasm; IMC, intermembrane compartment; OM, outer membrane; PM, plasma membrane; V, periplasmic vesicles.

Many archaeal energy conservation reactions occur in the space between the plasma membrane and the S-layer. This region of the cell is sometimes referred to as the pseudo-periplasm, by analogy to the structure in Gram-negative bacteria.

Comprehension Check

1. Identify three features that distinguish archaeal plasma membranes from those of bacteria.
2. Both bacteria and archaea can have S-layers. How does their use within the cells differ?

4.3 Archaeal Cytoplasm Is Similar to Bacterial Cytoplasm

After reading this section, you should be able to:

a. Compare and contrast the cytoplasm of bacterial and archaeal cells
b. Discuss the organization of archaeal DNA and similarities to both bacteria and eukaryotes

Overall, the cytoplasm of archaeal cells is similar to that of bacteria. Within it can be found inclusions—polyhydroxyalkonates, polyphosphate granules, glycogen granules, and gas vacuoles; ribosomes; a nucleoid; and plasmids. Proteins that might form a **cytoskeleton** have also been identified, including FtsZ (tubulin homologue), MreB (actin homologue), and crenactin, an actin homologue unique to certain members of the *Crenarchaeota*. In those archaea having FtsZ, the protein participates in cell division, as it does in bacteria. Evidence suggests that the actin homologues MreB and crenactin are involved in conferring a rod shape. Interestingly, a tubulin homologue called CetZ is observed in some members of the *Euryarchaeota*. CetZ confers a rod shape to cells rather than functioning in cytokinesis as FtsZ does. Some structures found in the cytoplasm of archaeal cells have distinctive molecular makeup. In this section, we focus on those structures. ▶| *Phylum* Crenarchaeota *(section 20.2); Phylum Euryarchaeota: methanogens, haloarchaea, and others (section 20.3)*

Archaeal Ribosomes Are the Same Size as Bacterial Ribosomes but Are Composed of Different Molecules

Like bacterial cells, the **ribosomes** of archaeal cells are 70S in size and are constructed of a 50S and a 30S subunit. However, their shape is somewhat different and their component molecules are not the same. Both have **ribosomal RNA (rRNA)** molecules of similar size: 16S in the small subunit, and 23S and 5S in the large subunit. However, as we discuss in chapter 1, the differences in nucleotide sequences of these molecules were the initial basis for establishing the taxon *Archaea*. Furthermore, at least one archaeon has an additional rRNA, a 5.8S rRNA, in the large subunit. This is of interest because the large subunit of eukaryotic ribosomes contains both 5S and 5.8S rRNA molecules.

The protein composition of bacterial and archaeal ribosomes also differs. Archaeal ribosomes have more proteins: about 68 rather than about 55. Archaeal ribosomal proteins can be sorted into three groups based on their similarity to proteins found in bacterial and eukaryotic ribosomes: (1) those observed in all three domains of life, (2) those unique to archaea, (3) those observed in both archaea and eukaryotes. There are only a few ribosomal proteins unique to archaea; the rest are about evenly divided between the other two groups. Importantly, all the ribosomal proteins present in both archaeal and bacterial ribosomes are also seen in eukaryotic ribosomes. Thus there are no ribosomal proteins that are present only in the ribosomes of archaea and bacteria. The difference in bacterial and archaeal ribosomes correlates with the fact that archaeal ribosomes are unaffected by antibiotics that bind and inhibit bacterial ribosomes ▶| *Protein synthesis inhibitors (section 9.4)*

Nucleoid

The **nucleoid** provides another example of the difference between archaea and bacteria. This region in the cytoplasm contains the cell's **chromosome** and numerous proteins. The chromosomes of all known archaea are circular, double-stranded **deoxyribonucleic acid** (**DNA**). Members of the *Euryarchaeota* are polyploid; that is, they have multiple copies of their chromosomes throughout their life cycles. In contrast, members of the *Crenarchaeota* are monoploid, with a single copy of the chromosome. It is believed that this difference reflects the mechanism for chromosome partitioning to daughter cells during cell division. *Crenarchaeota* have a highly regulated chromosome segregation mechanism, but the process appears to be random in the *Euryarchaeota*. Maintaining multiple copies of the genetic information is good insurance for an intact copy to end up in each daughter cell. ▶| *Chromosome replication and partitioning (section 7.2)*

The problem of compacting a genome to fit into a cell is common to archaea and bacteria, and the solution is similar. Both supercoiling and the presence of nucleoid-associated proteins (NAPs) on the genomic DNA contribute to chromosome organization. As in bacteria, there is a variety of NAPs, but they bear little resemblance to bacterial NAPs. The common features of these proteins are small size and positive charge. Each archaeon contains at least two different NAPs that may act at different scales of chromosome structure. On short segments of DNA, certain NAPs bend or stiffen the chromosome; on a global level, other NAPs bridge nonadjacent regions of the chromosome.

Many members of the *Euryarchaeota* have histones associated with their chromosomes. There is a strong correlation between the presence of histones and multiple chromosome copies, although it is not clear how the two features are related. These **histones** form **nucleosomes** that are related to the nucleosomes observed in eukaryotes (*see figure 5.10*). In *Haloferax volcanii,* the nucleosomes differ from eukaryotic nucleosomes in that they consist of four histones rather than eight and they organize a shorter length of DNA. In *Thermococcus kodakarensis,* however, histones assemble into various-sized complexes containing from two histones to more than a dozen. It is hypothesized that in thermophilic archaea, histones help prevent denaturation of the chromosomes. |◀ *Nucleoid (section 3.6);* ▶| *Nucleus (section 5.5)*

Comprehension Check

1. Thus far, homologues of intermediate filaments have not been identified in archaea. Is it likely that they will be identified eventually? Explain your answer.
2. Archaea are often described as chimeric organisms having some features that are similar to bacteria and some that are similar to eukaryotes. Based on what you have just read in this section, provide two examples that illustrate the similarity of archaea to bacteria; list two examples of their similarity to eukaryotes.

4.4 Many Archaea Have External Structures Used for Attachment and Motility

After reading this section, you should be able to:

a. Compare and contrast bacterial and archaeal pili
b. Compare and contrast bacterial and archaeal flagella in terms of their structure and function

Like bacteria, many archaea have structures that extend beyond the cell envelope, namely pili and flagella. These external structures and their functions are discussed in this section.

Pili

Many archaea have **pili**, and these appendages are found in a range of lengths and diameters, some solid and others hollow. Pili are composed of many copies of single protein subunits termed **pilins**, but pilins are not related to one another, thus explaining the diversity in the structures observed. *Methanococcus maripaludis* makes a major pilin that comprises the bulk of its pilus, and several other less abundant pilins. Its genome encodes at least six other pilins, and changing the relative proportions of pilins may be a means of responding to environmental stimuli.

Archaeal pili are **type IV pili**, which describes their mechanism of assembly within the cell envelope. Pilins are synthesized on ribosomes in the cytoplasm and then anchored to a protein complex in the plasma membrane. Pilin proteins are secreted through the complex to assemble the mature pilus structure.

Pili perform a range of functions. *Sulfolobus solfataricus* can display at least two distinct types of pili. The archaeal adhesive pilus (Aap) is responsible for cell adhesion to surfaces under certain growth conditions. The ultraviolet (UV)–inducible pilus (Ups) is only produced following exposure of the cells to UV light. As UV light can damage DNA, the Ups mediates the aggregation of cells and promotes the transfer of DNA between cells. |◀ *Bacterial pili and fimbriae (section 3.7)*

Two particular pili are noteworthy: cannulae and hami. **Cannulae** are hollow, tubelike structures observed on the surface of thermophilic archaea belonging to the genus *Pyrodictium* (**figure 4.9**). The function of cannulae is unknown, but it is known that daughter cells arising from a single round of cell division remain connected to each other by cannulae. After many rounds of cell division, a network of cells is formed. **Hami** are of particular interest because of their shape (**figure 4.10**). They look like tiny grappling hooks, which suggests they might function to attach cells to surfaces. Indeed, archaeal cells that produce hami are members of biofilm communities generally consisting of a hami-producing archaeon and a bacterium.

Archaella and Motility

Archaeal flagella, or archaella, have been studied in detail in only a few model archaea. They are superficially similar to

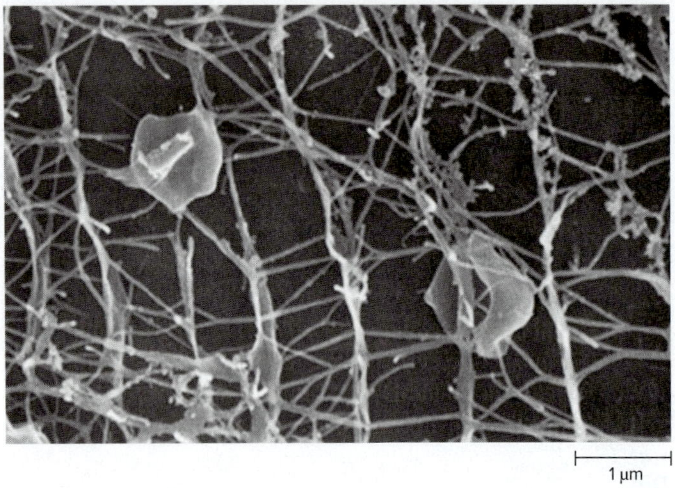

Figure 4.9 Cannulae. Cannulae are tubular structures about 25 nm in diameter. They have only been observed on *Pyrodictium* spp. They connect daughter cells, ultimately forming a dense network of cells. SEM, bar = 1 µm. ©Karl O. Stetter

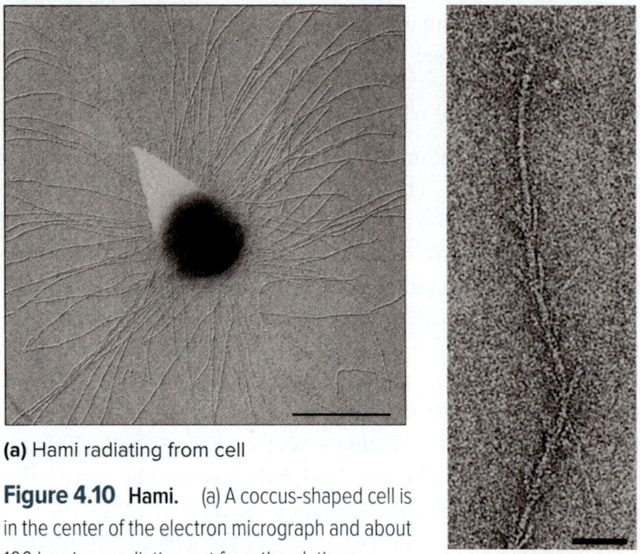

(a) Hami radiating from cell

(b) "Grappling hooks" at distal ends of hami

Figure 4.10 Hami. (a) A coccus-shaped cell is in the center of the electron micrograph and about 100 hami are radiating out from the platinum-shadowed cell. TEM, bar = 500 nm. (b) At the ends of the hami are grappling hook–like structures that are thought to allow cells to adhere to surfaces, including other cells. TEM, bar = 50 nm. ©Dr. Christine Moissl-Eichinger

their bacterial counterparts, but important differences have been identified. Archaella are thinner than bacterial flagella (10 to 14 nm rather than 18 to 22 nm) and some are composed of more than one type of protein subunit (**figure 4.11**). The filament is not hollow. Assembly of the archaellum also differs from bacterial flagellum assembly. In a manner similar to the assembly of bacterial type IV pili, the archaellin proteins are exported through the motor complex (figure 4.11) and added to the base of the filament. Additional proteins anchor the archaellum to the S-layer. Cryotomography has revealed bundles of archaella at one pole of the cell, organized by a proteinaceous structure in the membrane. ◀ *Bacterial flagella (section 3.7)*

Archaella from the two major phyla, *Euryarchaeota* and *Crenarchaeota*, have characteristic structural variations. Crenarchaeal filaments are composed of a single archaellin subunit. In contrast, euryarchaeal filaments have multiple archaellins, one of which is the primary subunit with others in lower abundance. Halophiles have been demonstrated to synthesize archaellins in different proportions under different culture conditions. It is not known how the filament properties vary with subunit composition. The motor complex also differs between these phyla, and it is believed that this variation is related to the presence of a bacterial-like chemotaxis system in *Euryarchaeota*.

Archaella work in a manner similar to bacterial flagella: Rotation propels the cell. Again, there are some important differences. First, rotation is powered by ATP hydrolysis rather than by proton motive force. Second, when the direction of rotation switches, it causes the cell to move in either the forward or reverse direction, just as is seen for some bacteria having polar flagella. Thus far, alternation between runs and tumbles has not been observed.

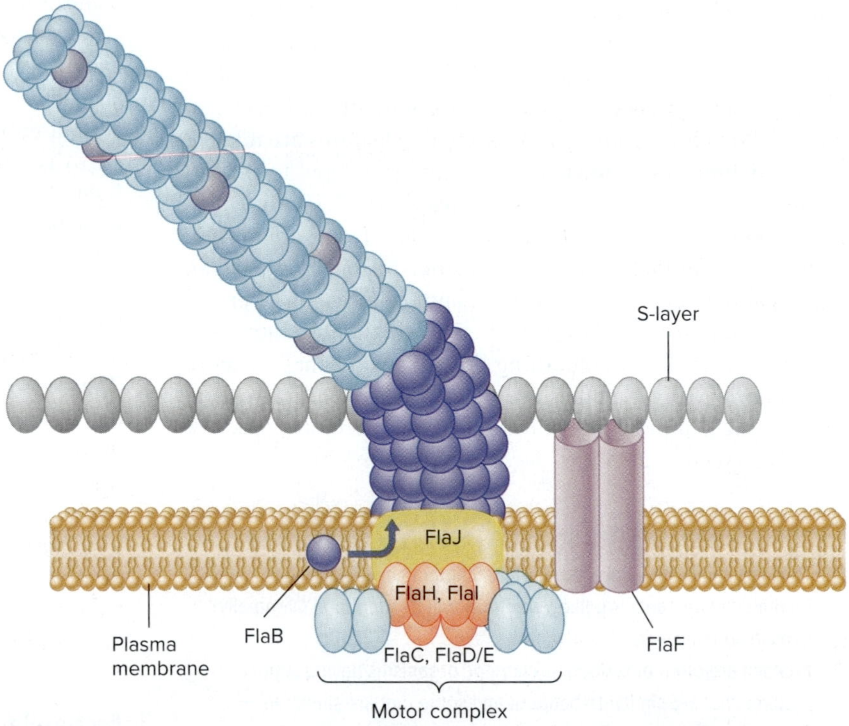

Figure 4.11 Structure of the Archaellum, Based on *Pyrococcus*. The archaellin protein FlaB comprises the filament, and monomers are added through FlaJ. The motor complex is composed of multiple Fla proteins, and the FlaF protein anchors the rotating archaellum in the cell envelope.

Halobacterium salinarum flagellar movement is the best studied. In this archaeon, clockwise rotation of the flagella pushes the cell forward, and counterclockwise rotation pulls the cell (i.e., the flagella are in front of the cell as it moves). *H. salinarum* exhibits both chemotaxis and phototaxis. Phototaxis is used to position the archaeon properly to absorb light when carrying out rhodopsin-based phototrophy. Surprisingly, considering the difference in flagellar architecture, much of the machinery controlling chemotaxis and phototaxis in *H. salinarum* is homologous to that found in bacterial systems. The archaellum in *Sulfolobus* also switches rotational direction, but the switch appears to be a random event. ▶| *Rhodopsin-based phototrophy (section 11.12)*

Archaea have been observed to engage in two types of swimming behaviors, termed a relocate-and-seek strategy. Fast speeds in a straight line (up to ~500 μm/sec) result in a rapid change of location. A slower zigzag motion (~50 to 100 μm/sec) constitutes the seek, a scan for favorable conditions. The fastest known swimmers, *Methanocaldococcus villosus* and *M. jannaschii,* can travel at about 500 body lengths per second (bps), compared to 20 bps for *E. coli* and 11 bps for humans. This extreme swimming behavior is an advantage in their deep sea hydrothermal vent environment. Temperature gradients force these microbes to inhabit a narrow space between the 400°C vent fluid and the surrounding 2°C seawater. When fast swimming identifies a suitable niche (50° to 90°C), the microbes use the archaella for attachment.

In addition to their role in motility, archaella are involved in adhesion to substrates, the formation of bioflms, and cell-cell interactions.

Comprehension Check

1. What observations about cannulae and hami suggest that they allow archaeal cells to adhere to surfaces, including other cells?
2. List three aspects of archaella and their motility that are similar to bacterial flagella and their motility.
3. Defend the proposal that the archaellum is a type of pilus.

Key Concepts

4.1 Archaea Are Diverse but Share Some Common Features

- Members of *Bacteria* and *Archaea* share a common cell architecture. However, they have characteristics that define them as distinct taxa. **Table 4.1** summarizes some of the differences.
- Many archaea have been identified only through 16S rRNA or partial genome sequences.
- Rods and cocci are the most common archaeal shapes. Archaea also can be curved rods, spirals, branched, square, and pleomorphic (**figure 4.1**).
- Most archaea are similar in size to bacteria. Archaea that are extremely small or are extremely large have also been identified (**figure 4.2**).

4.2 Archaeal Cell Envelopes Are Structurally Diverse

- The cell envelope consists of the plasma membrane and all external coverings, including cell walls and other layers. Archaeal cell envelopes usually consist of only the plasma membrane and the cell wall.
- Archaeal membranes are composed of glycerol diether and diglycerol tetraether lipids (**figure 4.4**). Membranes composed of glycerol diether are lipid bilayers. Membranes composed of diglycerol tetraethers are lipid monolayers. The overall structure of a monolayer membrane is similar to that of the bilayer membrane in that the membrane has a hydrophobic core and its surfaces are hydrophilic (**figure 4.5**).
- Archaea typically use active transport systems to obtain nutrients from their environment.
- Archaeal cell walls do not contain peptidoglycan, and they exhibit great diversity in their makeup. The most common type of cell wall is one consisting of an S-layer only (**figure 4.6**).

4.3 Archaeal Cytoplasm Is Similar to Bacterial Cytoplasm

- Cytoskeletal proteins have been identified in archaeal cells. They include FtsZ (tubulin homologue), MreB (actin homologue), and crenactin (a unique archaeal actin homologue).
- Numerous inclusions are observed in archaeal cells, including gas vesicles.
- Bacterial and archaeal ribosomes are 70S in size but differ slightly in their morphology. They also differ in terms of their protein content, with many archaeal ribosomal proteins being more similar to those in eukaryotic ribosomes than to those in bacterial ribosomes.
- The genetic material of archaeal cells is located in the nucleoid, which is not enclosed by a membrane. All known archaeal chromosomes consist of a double-stranded, covalently closed, circular DNA molecule. In many archaea, the nucleoid contains a single chromosome. However, some archaea are polyploid, having more than one copy of their chromosome. Like bacteria, archaea use nucleoid-associated proteins (NAPs) to organize their chromosomes.

4.4 Many Archaea Have External Structures Used for Attachment and Motility

- Many archaea have pili, similar to bacterial type IV pili.
- Many archaea are motile by means of flagella called archaella, which are structurally related to bacterial type IV pili (**figure 4.11**). They are rigid helices that rotate, and the direction of rotation determines if the cell moves forward or backward. Rotation is powered by ATP hydrolysis.
- Motile archaea exhibit chemotaxis. Some are also phototactic. The archaeal taxis machinery is similar to that of bacteria.
- Archaea engage in relocate-and-seek swimming behaviors that may aid in positioning cells optimally in a temperature gradient.

Active Learning

1. Archaea with cell walls consisting of a thick, homogeneous layer of complex polysaccharides often retain crystal violet dye when stained using the Gram-staining procedure. Suggest a reason for this staining reaction.
2. Isoprene serves as a building block not only for the hydrocarbons observed in archaeal membranes but also for sterols, carotenoids, retinal, and quinones. Use any resources necessary to identify the function of these other isoprene-based molecules and to determine their distribution in nature. What does the use of isoprene to make this diverse array of molecules suggest about the nature of the last universal common ancestor (LUCA)?
3. As we note in section 4.3, archaeal ribosomal proteins can be divided into three classes: (1) those with homologues in all three domains of life, (2) those unique to archaea, and (3) those with homologues in eukaryotic ribosomes. Predict a class of archaeal ribosomal proteins that would refute the argument made by some microbiologists that the term *prokaryote* should be abandoned.
4. Consider the cell shape of *Haloquadratum walsbyi*, a rectangular prism with dimensions of 2 μm by 2 μm by 0.25 μm. Calculate the surface area and volume of a single cell with these dimensions and calculate the ratio of these values. Compare your results to the values given for a coccus in figure 3.5.
5. Identify some challenges to working with archaea in the laboratory, given their habitats in locations like salt ponds, acid mine drainage sites, and deep sea hydrothermal vents.
6. Scientists studying the human archaeome, the subset of the human microbiome that comprises archaea, found that some typical laboratory procedures bias the types of organisms detected. As you might expect, using techniques devised for working with bacteria resulted in those organisms being detected more frequently. The scientists developed strategies to survey the types of archaea present in several regions of the human body and documented archaeal populations among the resident microbes. Why might breaking open a mixture of microbial cells to extract genomic DNA be biased against archaeal cells? Why might procedures to determine the 16S rRNA sequences be biased against archaeal 16S rRNA sequences?

   Read the original paper: Koskinen, K., et al. 2017. First insights into the diverse human archaeome. *mBio* 8(6): e00824-17.

5

Eukaryotic Cell Structure

Source: USDA

Red Means Dead

Anyone driving down a highway in the Rocky Mountains of Canada and the United States will observe an astonishing sight: large stands of rust-colored, dead pine trees in once verdant mountain forests. In British Columbia alone, over half the commercial lodgepole pine trees have been killed, while in northern Colorado and southern Wyoming, these trees have been completely eliminated.

Two culprits working together have caused this destruction: the mountain pine beetle and its symbiont, the blue stain fungus *Grosmannia clavigera*. How they kill is an interesting example of a mutualistic relationship—one in which both benefit. The fungus is helped by the beetle, which carries the fungus on its mouth parts and transports it from one tree to another. The beetle is helped by the fungus, which protects the beetle from the tree's defense mechanisms and nourishes the beetle's young.

The killing process begins when the beetle chews through the bark of a pine tree to reach underlying tissues. The tree fights back by producing a resin toxic to the beetle. However, because the fungus degrades the toxic resin and uses it as a carbon and energy source, the beetle is protected. Thus, the fungus enables the beetle to continue its attack, ultimately laying its eggs in the tree and carrying the fungus deeper into the tree. The fungus, now deep inside the tree, produces masses of filaments that help feed the beetle's progeny. These filaments are lethal as they literally clog the tree's circulatory system. Nutrients and water can no longer flow through the tree, so it dries up and starves to death.

Much of this text is devoted to discussions of bacteria and archaea. But as *G. clavigera* shows us, eukaryotic microbes also deserve our attention. They are prominent members of ecosystems, important model organisms, and exceptionally useful in industrial microbiology. A number are also major human pathogens.

Eukaryotic microbes can be divided into two major groups: protists and fungi. Protists are very diverse. Some use organic molecules as energy sources just as animals do; others are photosynthetic like plants. The many lineages of protists cannot be grouped together in a single monophyletic taxon—one that includes all descendants of the most recent common ancestor. Thus the term *protist* is a common name for these microbes rather than a valid taxon. Fungi, on the other hand, all share a common evolutionary history, so they are placed in one taxon (*Fungi*).

The division of eukaryotes into appropriate taxa is currently being debated, as is their evolution. The taxonomy and evolution of eukaryotes are discussed more completely in chapters 19, 24, and 25. In this chapter, we introduce eukaryotic cell structures and their functions. As we do so, comparisons are made to plants and animals. Thus it is important that you review what you already know about them.

Readiness Check:

Based on what you have learned previously, you should be able to:

✓ List the three domains of life (section 1.1)
✓ Describe the application of small subunit (SSU) rRNA analysis to the establishment of the three-domain classification system proposed by Carl Woese (section 1.2)
✓ Identify the following structures or regions of a plant or animal cell and describe their functions: cell wall, plasma membrane, cytoplasm, mitochondria, chloroplasts, and ribosomes
✓ Identify the major cell structures observed in bacterial and archaeal cells, and describe the function and molecular makeup of each (chapters 3 and 4)

5.1 Eukaryotic Cells Are Diverse but Share Some Common Features

After reading this section, you should be able to:

a. Compare and contrast eukaryotic, bacterial, and archaeal cells in terms of their use of membranes, size, morphological diversity, and organelles
b. Explain why compartmentation of the cell interior is advantageous to eukaryotic cells

Before we begin our discussion of the numerous structures and organelles observed in eukaryotic cells, it is wise to begin with an overview. Here we make broad generalizations about eukaryotic cells. As we proceed in this chapter, we will give examples of deviations from the common features.

Shape and Size

One hallmark of eukaryotic cells is their morphological diversity. This is especially true for eukaryotic microbes, which are found in numerous habitats and have evolved morphological adaptations accordingly. Some of the diverse shapes of eukaryotic microbes are shown in **figure 5.1**.

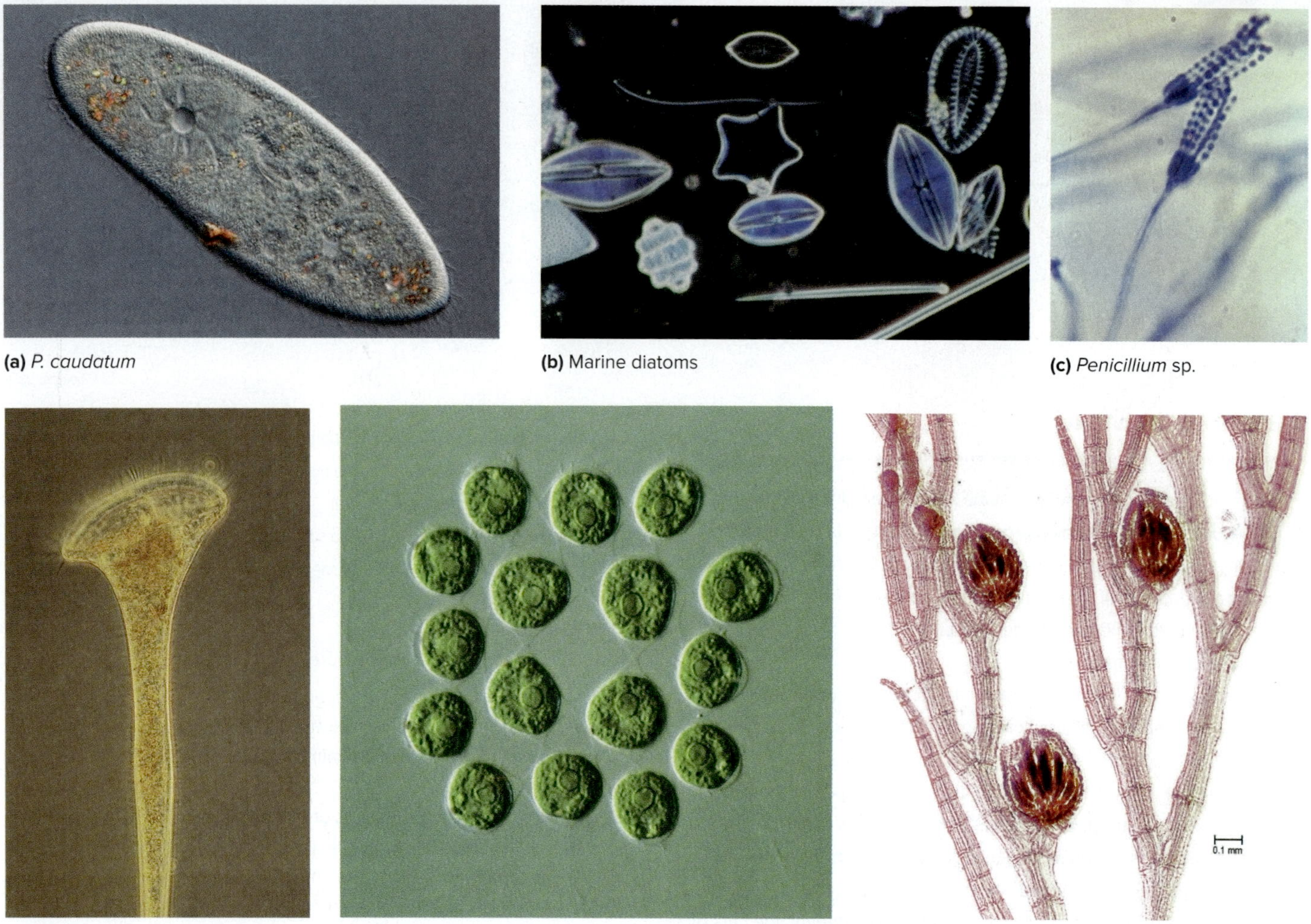

(a) *P. caudatum* **(b)** Marine diatoms **(c)** *Penicillium* sp.

(d) *Stentor* sp. **(e)** *Gonium* sp. **(f)** *Polysiphonia* sp.

Figure 5.1 Some Eukaryotic Microorganisms. (a) *Paramecium caudatum,* a protist; interference-contrast microscopy. (b) Mixed diatom frustules. (c) *Penicillium* sp., the mold's hyphae and conidia. (d) *Stentor* sp., a ciliated protist in its extended, actively feeding form; phase-contrast microscopy (×100). (e) *Gonium* sp., a freshwater, colonial alga (×40). (f) *Polysiphonia* sp., a filamentous, red alga (photosynthetic protist).

In general, eukaryotic microbes are larger than bacterial and archaeal cells (*see figure 2.20*). However, just as some bacteria and archaea are very large, even larger than some of the largest protists (*see figure 3.4*), so too are there some protists smaller than many bacteria and archaea. These small eukaryotic microbes include *Ostreococcus tauri,* which at about 0.8 μm in diameter, and is the smallest free-living alga known. In fact, these and other picoeukaryotes (Spanish *pico,* little bit) in marine environments can reach densities as high as 10^3 to 10^4 per milliliter.

Cell Organization

Eukaryotic cells are distinctive because of their use of membranes (**figures 5.2** and **5.3**; *compare these figures with figure 3.6*). They have membrane-delimited nuclei, and membranes play a prominent part in the structure of many other organelles. **Organelles** are intracellular structures that perform specific functions in cells analogous to the functions of organs in the body of a multicellular organism. The partitioning of the eukaryotic cell interior into organelles makes possible the placement of different cellular functions in separate compartments so they can more easily take place simultaneously under independent control and proper coordination. Large membrane surfaces make possible greater respiratory and photosynthetic activity because these processes are located exclusively in membranes. The intracytoplasmic membrane complex also serves as a transport system to move materials between different cell locations. Abundant membrane systems probably are necessary in eukaryotic cells because of their large volume and the need for adequate regulation, metabolic activity, and transport.

Figures 5.2 and 5.3 illustrate most of the organelles discussed in this chapter. Our discussion of eukaryotic cell structure begins with eukaryotic cell envelopes. We then proceed to the cytoplasm, organelles within the cytoplasm, and finally to external structures.

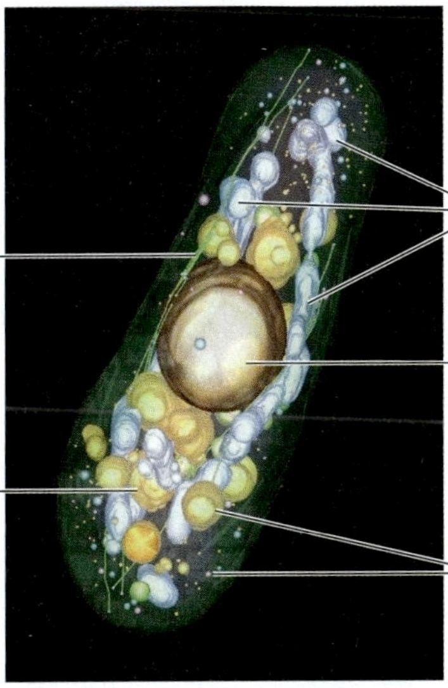

Figure 5.2 Eukaryotic Cell Ultrastructure. An electron cryotomogram of the yeast *Schizosaccharomyces pombe*. This image is the first high-resolution, three-dimensional reconstruction of a complete eukaryotic cell. ©Johanna Höög/European Molecular Biology Laboratory/Science Source

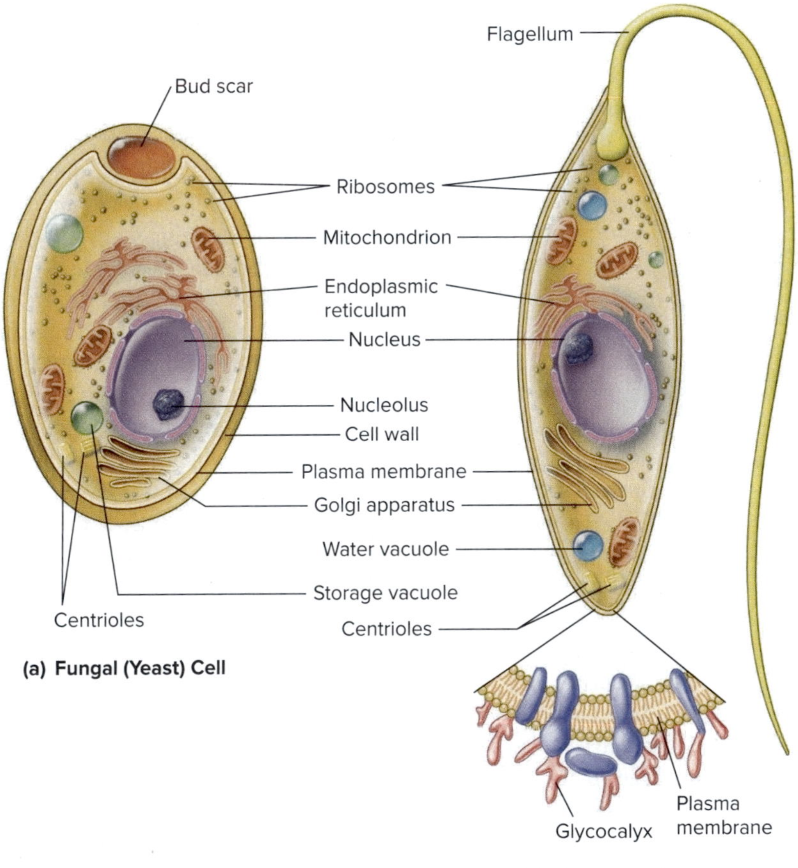

Figure 5.3 The Structure of Two Representative Eukaryotic Microbes. Illustrations of (a) a yeast cell (fungus) and (b) the flagellated protist *Peranema*.

MICRO INQUIRY *In addition to separating each organelle from the rest of the cytoplasm, what other functions do internal membranes serve?*

5.2 Eukaryotic Cell Envelopes

After reading this section, you should be able to:

a. Compare and contrast the lipids found in bacteria and eukaryotes
b. Identify the types of eukaryotic microbes that have cell walls and describe their composition

The cell envelope consists of the plasma membrane and all coverings outside it. Many eukaryotic microbes have only a plasma membrane without a cell wall, so they have very simple cell envelopes. However, when cell walls are present, they are quite different from those of bacteria and archaea.

The plasma membrane of eukaryotes is a lipid bilayer composed of a high proportion of sphingolipids and sterols (e.g., cholesterol and ergosterol) in addition to the phospholipids observed in bacterial membranes (**figure 5.4**). In phospholipids and sphingolipids, the fatty acids are attached to glycerol with ester links. Sphingolipids and sterols can pack together very tightly, so they affect the fluidity and permeability of the plasma membrane. Eukaryotic cells therefore alter the quantity of sterols and sphingolipids in response to stresses such as high or low temperatures.

The distribution of lipids in the plasma membrane is asymmetric. This means that the lipids in the outer leaflet differ from those of the inner leaflet. Although most lipids in each leaflet mix freely with each other, microdomains that differ in lipid and protein composition are found. These microdomains are sometimes referred to as lipid rafts. They have been implicated in a variety of cellular processes (e.g., signal transduction, assembly and release of virus particles, and endocytosis). As shown in figure 5.3*b*, many eukaryotic cells have integral membrane proteins and lipids decorated with carbohydrates. This often forms a carbohydrate-rich layer on the surface of the cell called the **glycocalyx.**

Like bacterial and archaeal cells, eukaryotic cells must be able to move materials across their plasma membranes. Facilitated diffusion is much more common in eukaryotic cells than in bacterial and archaeal cells (*see figure 3.10*). Eukaryotes also use primary and secondary active transport and passive diffusion. In addition, eukaryotes have another option for bringing materials into cells: endocytosis. We focus on this unique process in section 5.4.

Eukaryotic microbes that have cell walls include the photosynthetic protists (commonly called algae), cysts made by many types of protists, and most fungi. Not surprisingly given this vast diversity, the chemical composition of the cell walls of eukaryotic microbes varies considerably. Algal

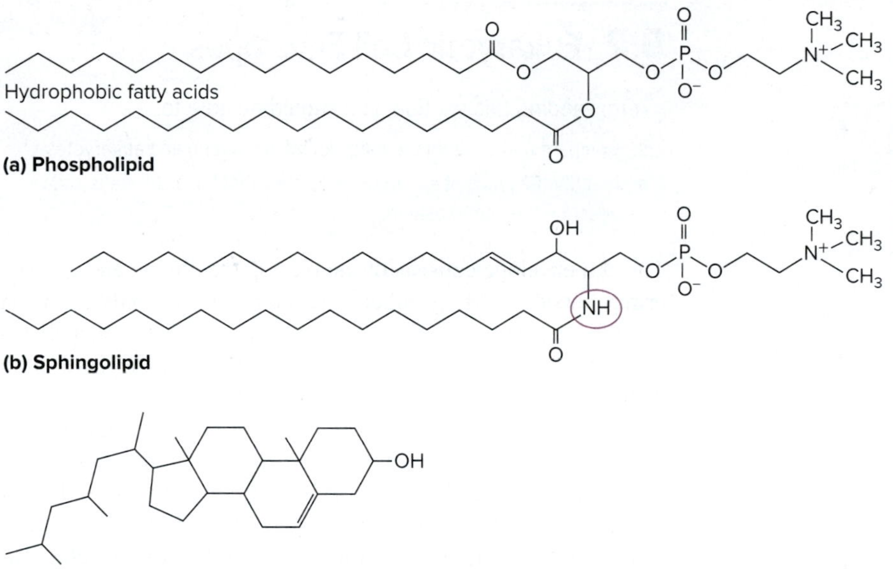

(a) Phospholipid

(b) Sphingolipid

(c) Sterol

Figure 5.4 **Examples of Eukaryotic Membrane Lipids.** (a) Phosphatidylcholine, a phospholipid. (b) Sphingomyelin, a sphingolipid. The amine group is circled as this differentiates a sphingolipid from a phospholipid. (c) Cholesterol, a sterol.

cell walls usually have a layered appearance and contain large quantities of the polysaccharides cellulose and pectin. In addition, inorganic substances such as silica or calcium carbonate may be present. The dormant cysts formed by many protists consist of multiple layers of chitin. Fungal cell walls normally are rigid. Their exact composition varies with the organism; usually cellulose, chitin, and glucan (a glucose polymer different from cellulose) are present. These materials are chemically simpler than bacterial peptidoglycan. ◀ *There are two main types of bacterial cell walls (section 3.4)*

5.3 The Eukaryotic Cytoplasm Contains a Cytoskeleton and Organelles

After reading this section, you should be able to:

a. Describe the functions of the cytoplasm
b. Identify the three filaments that make up the cytoskeleton of eukaryotic cells and describe their functions

The **cytoplasm** is one of the most important and complex parts of a eukaryotic cell. It consists of a liquid component, the **cytosol,** in which many organelles are located. It is the location of many important biochemical processes, and several physical changes (e.g., viscosity changes and cytoplasmic streaming) are due to cytoplasmic activity. Because so many different kinds of organelles are observed in the eukaryotic cytoplasm, we address them separately in sections 5.4–5.6. Here we focus on the cytoskeleton, which helps organize the contents of the cytoplasm.

The **cytoskeleton** is a vast network composed of three types of interconnected filaments: actin filaments (also called microfilaments), intermediate filaments, and microtubules (**figure 5.5**). The cytoskeletal network is like a highway that guides the movement of cellular structures. Three motor proteins (myosin, kinesin, and dynein) associated with actin filaments and microtubules track along the cytoskeletal filaments, helping to move cell structures from one location to another.

Actin filaments are minute protein filaments, 4 to 7 nm in diameter, organized to form a variety of structures depending on the cell type (**figure 5.6**). They are composed of an actin protein that is similar to the actin contractile protein of muscle tissue. In addition to moving cell structures within the cell, actin filaments help cells change shape; thus they are involved in amoeboid movement, endocytosis, and cytokinesis.

Intermediate filaments (IFs) are flexible yet very strong. They are not associated with motor proteins. They therefore play a variety of structural roles. They are about 10 nm in diameter and are assembled from a group of proteins that can be divided into several classes (e.g., keratin and vimentin). IFs having different functions are assembled from one or more of these classes of proteins. Fungi and plants lack intermediate filaments. In animals and certain protozoa intermediate filaments form the nuclear lamina, a structure that provides support for the nuclear envelope. Additional IFs in animals help position organelles within the cell and help link cells together to form tissues.

Microtubules are shaped like thin cylinders about 25 nm in diameter. They are constructed of two spherical protein subunits—α-tubulin and β-tubulin (figure 5.5). The two proteins are the same molecular weight and differ only slightly in terms of their amino acid sequence and tertiary structure. Each tubulin is approximately 4 to 5 nm in diameter. These subunits are assembled in a helical arrangement to form a cylinder with an average of 13 subunits in one turn.

Microtubules have several important functions. They form the spindle apparatus that separates chromosomes during mitosis and meiosis. They form tracks along which numerous types of organelles and vesicles are moved about the cell, and they are found in cilia and flagella, two organelles that confer motility. Microtubules also are found in cell structures requiring support such as long, slender, rigid pseudopodia (called axopodia) observed in some protists (*see figure 24.7*).

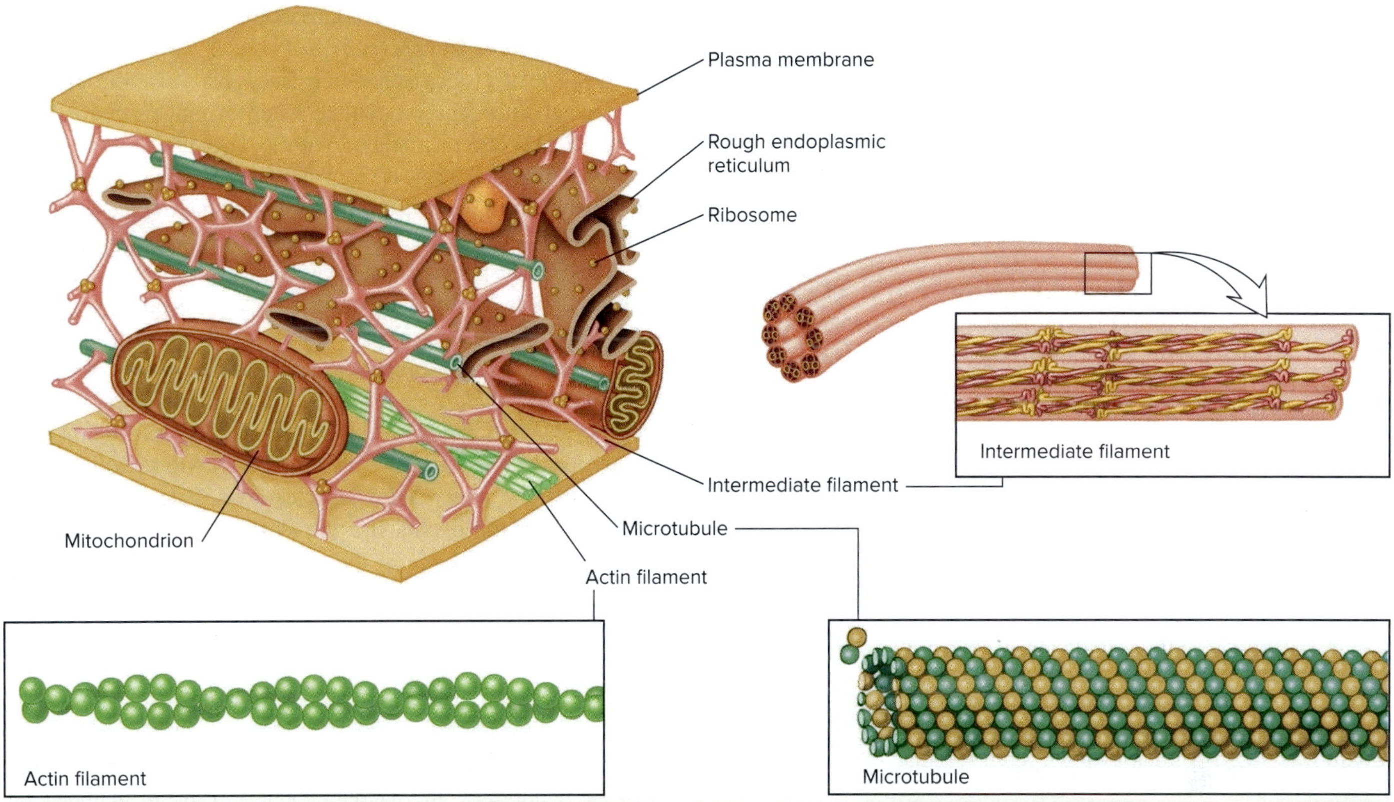

Figure 5.5 The Eukaryotic Cytoplasm and Cytoskeleton. The cytoplasm of eukaryotic cells contains many organelles. The cytoskeleton helps form a framework within which the organelles lie. The cytoskeleton is composed of three elements: actin filaments, microtubules, and intermediate filaments.

MICRO INQUIRY *Which cytoskeletal filament is made of the proteins α- and β-tubulin?*

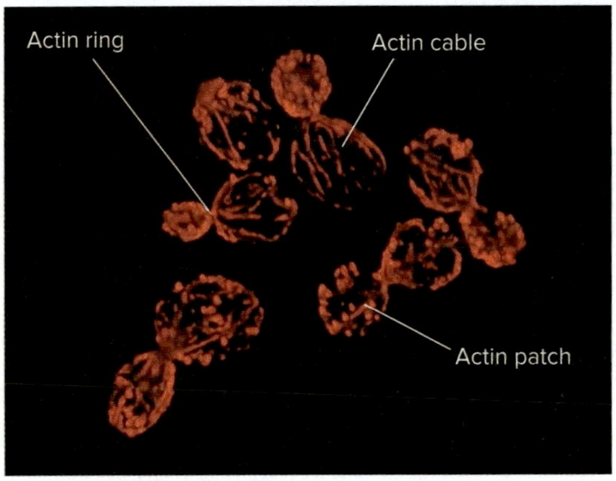

Figure 5.6 The Actin Cytoskeleton of the Yeast *Saccharomyces cerevisiae*. Three kinds of structures made from actin filaments can be observed: actin patches, actin rings, and actin cables. Actin patches are sites where clathrin-dependent endocytosis occurs. Actin rings participate in cytokinesis and actin cables function in transport of material through the cytoplasm. The actin structures appear red in this micrograph.

Comprehension Check

1. What are the major lipids in eukaryotic membranes? Which other domain of life possesses similar lipids?
2. Which types of eukaryotic microbes have cell walls? How do their walls differ from those of bacterial and archaeal cells?
3. What are the three cytoskeletal elements observed in eukaryotic cells? What are their functions? Do all eukaryotic microbes have each of the three elements?

5.4 Several Organelles Function in the Secretory and Endocytic Pathways

After reading this section, you should be able to:

a. Differentiate the two types of endoplasmic reticulum in terms of structure and function
b. Outline the pathway of molecules through transport and secretory pathways, noting the structures involved and their roles
c. List the endocytic pathways observed in mammalian cells, noting the structures involved and their role in the process, and noting those pathways observed in eukaryotic microbes

The cytoplasm of eukaryotic cells is permeated with an intricate complex of membranous organelles and vesicles that move materials into the cell from the outside (endocytic pathway) and from the inside of the cell out, as well as from location to location within the cell (secretory pathway). In this section, some of these organelles are described.

Endoplasmic Reticulum

The **endoplasmic reticulum** (**ER**) (figure 5.3) is an irregular network of branching and fusing membranous tubules, around 40 to 70 nm in diameter, and many flattened sacs called cisternae (s., cisterna). The nature of the ER varies with the functional and physiological status of the cell. In cells synthesizing a great deal of protein to be secreted, a large part of the ER is studded on its outer surface with ribosomes and is called **rough endoplasmic reticulum** (**RER**). Other cells, such as those producing large quantities of lipids, have ER without ribosomes. This is **smooth endoplasmic reticulum** (**SER**).

The endoplasmic reticulum is an important hub of molecular synthesis and transport. Lipids and proteins are synthesized by ER-associated enzymes and ribosomes. Polypeptide chains synthesized on RER-bound ribosomes may be inserted either into the ER membrane or into its lumen for transport elsewhere. The ER is also a major site of cell membrane synthesis.

The Golgi Apparatus Prepares Materials for Secretion

The **Golgi apparatus** is composed of flattened, saclike cisternae (**figure 5.7**). A complex network of tubules and vesicles is located at the edges of the cisternae. In many eukaryotes, the cisternae are stacked on each other, forming a structure called a dictyosome. There are usually around four to eight cisternae in a stack. In many cells, multiple stacks are linked to form a Golgi ribbon. Most eukaryotic microbes have just a single stack of cisternae, and in some, the cisternae are not in stacks (e.g., the yeast *Saccharomyces cerevisiae*).

A stack of cisternae has two faces that are quite different from one another. The sacs on the cis or forming face are closest to the ER and often associated with it. The sacs on the trans or maturing face are farthest from the ER. The two faces of the Golgi differ in thickness, enzyme content, and degree of vesicle formation.

The Golgi apparatus packages materials and prepares them for secretion; the exact nature of its role depends on the organism. For instance, the surface scales of some flagellated photosynthetic and radiolarian protists are constructed within the Golgi apparatus and then transported to the surface in vesicles. The Golgi often participates in the development of cell membranes and the packaging of cell products. The growth of some fungal hyphae occurs when Golgi vesicles contribute their contents to the wall at the hyphal tip. ▶| *Protists (chapter 24);* Fungi *(chapter 25)*

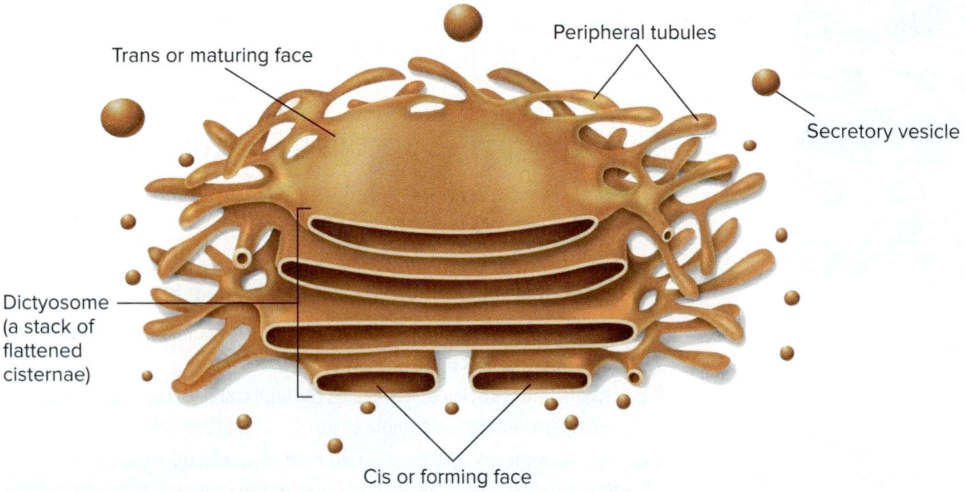

Figure 5.7 Golgi Apparatus. Each cisterna is 15 to 20 nm thick and separated from other cisternae by 20 to 30 nm.

Lysosomes Are Degradative Organelles That Function in Endocytic Pathways

Lysosomes are found in animal cells because they are involved in digesting nutrients. They are roughly spherical, are enclosed in a single membrane, and average about 500 nm in diameter but range from 50 nm to several micrometers in size. They contain the enzymes needed to digest all types of macromolecules. These enzymes, called hydrolases, catalyze the hydrolysis of molecules and function best under slightly acidic conditions (usually around pH 3.5 to 5.0). Lysosomes maintain an acidic environment by pumping protons into their interior. *Lysosomes*

Organelles with the same function as lysosomes are found in fungal and protist cells, where they instead may be called vacuoles, phagocytic vacuoles, or food vacuoles. In addition to functioning in intracellular digestion, they may have other functions, including storage of calcium ions, phosphate, and amino acids. They also are components of the endocytic pathways observed in protists and fungi.

Transport and Secretory Pathways Move Materials to Other Locations, Including Outside the Cell

Various pathways are used to move materials within a cell. The early steps in these pathways are the same as those used to move materials into the plasma membrane or outside the cell. Thus they are often described together and referred to as the **secretory pathway.** The process of

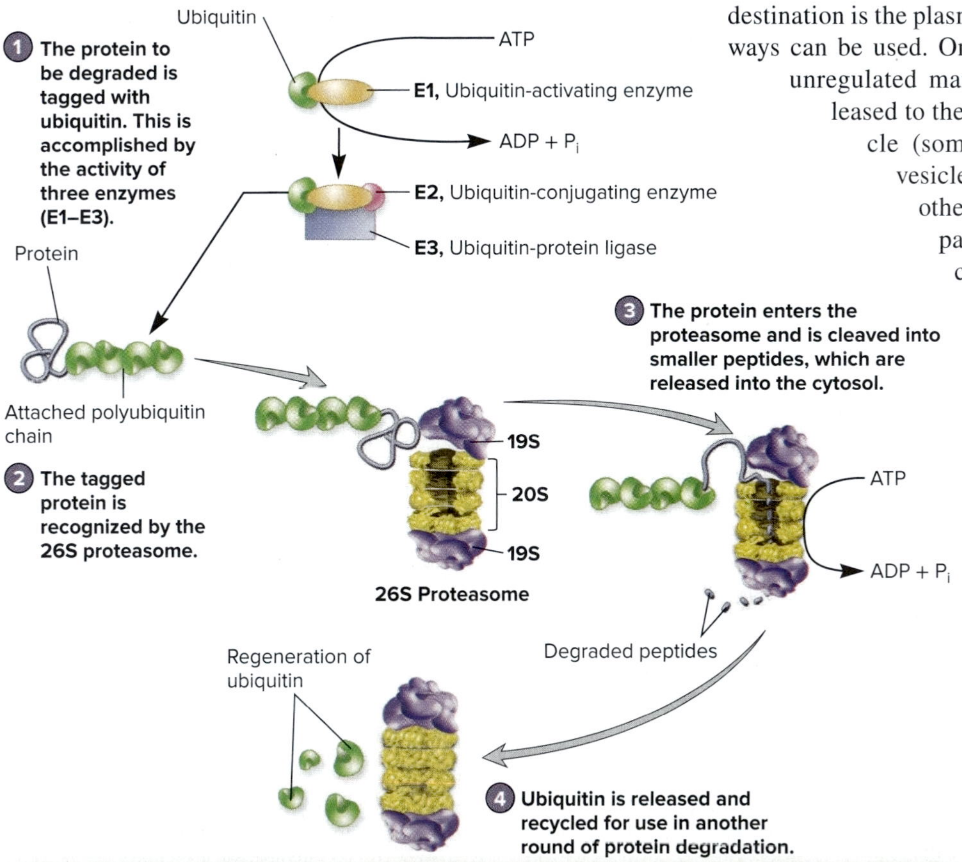

Figure 5.8 Proteasome Degradation of Proteins.

MICRO INQUIRY *Why must proteins be unfolded when they enter the proteasome? What do you think happens to the peptides after exiting the proteasome?*

moving materials such as proteins and lipids within or out of the cell is complex and not fully understood. However, it involves numerous molecules that help target the protein or lipid to its proper location. The movement of proteins is of particular importance and is our focus.

Proteins destined for a cellular target or secretion follow a similar path to their final location. Proteins synthesized by ribosomes attached to the rough endoplasmic reticulum (RER) have sequences of amino acids that target them to the lumen of the RER. These proteins move through the RER lumen until they are released in small vesicles that bud from the ER. On their journey through the ER, sugars are often added to the proteins—a process known as glycosylation. ▶| *Protein localization and secretion in eukaryotes (section 15.4)*

Vesicles released from the ER travel to the cis face of the Golgi apparatus (the side closest to the ER; figure 5.7) to which they may fuse. The proteins then proceed from the cis to the trans side of the Golgi (the side farthest from the ER) and are further modified. Some of these modifications target the proteins for their final location.

Transport vesicles released from the trans face of the Golgi deliver their contents to various locations in the cell. If the destination is the plasma membrane, one of two secretory pathways can be used. One constitutively delivers proteins in an unregulated manner. In this pathway, proteins are released to the outside of the cell as the transport vesicle (sometimes called a constitutive secretory vesicle) fuses with the plasma membrane. The other pathway, called the regulated secretory pathway, requires a signal before its vesicles (regulated secretory vesicles) fuse with the plasma membrane and release their contents.

Secretory pathways efficiently move proteins, but what if the protein acquired from the RER was made incorrectly? The cell has a built-in quality assurance mechanism to make sure proteins that fail to fold or are misfolded are not transported to their final destination. Instead they are released into the cytosol, where they are destroyed. However, because there are many normal proteins in the cytosol, proteins with mistakes must be labeled so only they are eliminated. This is accomplished by attaching several small polypeptides, called ubiquitin, to the protein (**figure 5.8**). The addition of ubiquitin marks a protein for degradation by a huge, cylindrical complex called a 26S **proteasome.** The ubiquitin-tagged protein is fed into the 26S proteasome much like a log into a wood chipper. It is broken down to smaller peptides using energy supplied by adenosine triphosphate (ATP) hydrolysis. As the protein is broken down, ubiquitins are released and reused. Proteasomes and proteasome-like structures have also been identified in some bacteria and archaea. However, they are simpler in structure and the proteins to be degraded are targeted to the degradation machinery by mechanisms other than ubiquitin.

There Are Several Endocytic Pathways

Endocytosis is used by all eukaryotic cells to bring materials into the cell from the outside. During endocytosis, a cell takes up solutes (pinocytosis) or particles (phagocytosis) by enclosing them in vesicles pinched off from the plasma membrane. Endocytosis occurs regularly in all eukaryotic cells as a mechanism for recycling molecules in the membrane. In addition, some eukaryotic cells have specialized **endocytic pathways** that allow them to concentrate materials outside the cell before bringing them in. Others use endocytic pathways as a feeding mechanism. Endocytosis is not found in either bacteria or archaea; on the contrary,

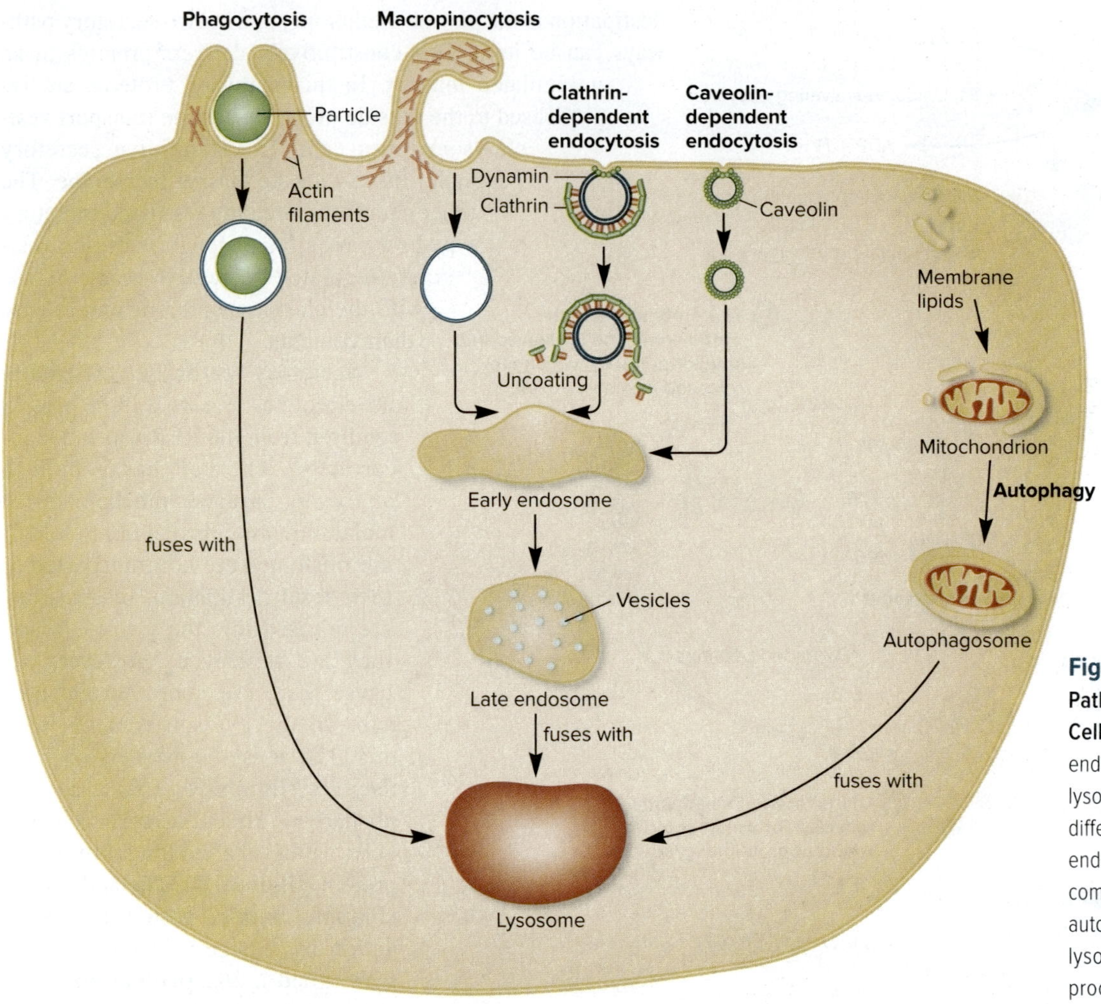

Figure 5.9 **Some Endocytic Pathways Observed in Mammalian Cells.** Materials ingested by endocytic processes are delivered to lysosomes. The pathway to lysosomes differs, depending on the type of endocytosis. In addition, cell components are recycled when autophagosomes deliver them to lysosomes for digestion. This process is called autophagy.

eukaryotic cells use endocytosis to engulf many viruses and other intracellular pathogens.

Mammalian cells have several types of endocytic pathways (**figure 5.9**). All generate an endocytic vesicle of some type, and the substances in the vesicle are eventually delivered to a lysosome for degradation. **Phagocytosis** involves the use of protrusions from the cell surface to surround and engulf particulates. The endocytic vesicles formed by phagocytosis are called **phagosomes. Clathrin-dependent endocytosis** begins with specialized membrane regions coated on the cytoplasmic side with the protein clathrin; these are called coated pits. The endocytic vesicles formed when coated pits invaginate are then called coated vesicles. Clathrin-dependent endocytosis is used to internalize hormones, growth factors, iron, and cholesterol. **Caveolin-dependent endocytosis** involves caveolae ("little caves"), tiny, cup-shaped invaginations of the plasma membrane (about 50 to 80 nm in diameter) enriched in cholesterol and the membrane protein caveolin. The vesicles formed when caveolae bud into the cytoplasm are called caveolin-coated vesicles. Caveolin-dependent endocytosis has been implicated in signal transduction, transport of small molecules such as folic acid, as well as transport of macromolecules. Caveolae also are used by many viruses, bacteria, and protozoa to enter host cells.

When does a cell "know" when to undergo endocytosis? Endocytosis is often triggered by the binding of a molecule outside the cell, called a **ligand** to a **receptor** located in the cell membrane. This is referred to as **receptor-mediated endocytosis.** The nature of ligands varies (e.g., hormones or molecules on the surface of a virus). Sometimes receptors are concentrated in certain areas of the membrane, and receptor-mediated endocytosis serves to concentrate the ligand before it is endocytosed.

How materials are delivered to lysosomes depends on the endocytic pathway. Clathrin-coated vesicles and caveolin-coated vesicles first deliver their contents to small organelles containing hydrolytic enzymes. These organelles are called early **endosomes** (figure 5.9). Early endosomes mature into late endosomes and then fuse with lysosomes, which deliver more degradative compounds. Phagosomes take a different route to lysosomes. They skip the early endosome stage and simply fuse directly with lysosomes (figure 5.9) to form a phagolysosome.

Endocytosis in eukaryotic microbes has not been studied as extensively as it has in mammals. It is known that some eukaryotic microbes carry out phagocytosis and probably macropinocytosis (figure 5.9). Clathrin-dependent endocytosis has been observed in fungi and is best characterized in the yeast *Saccharomyces cerevisiae*.

Lysosomes are involved in another important process called autophagy (Greek *auto*, self; *phagia*, to eat). Several types of autophagy have been identified. Some selectively sequester and destroy bacteria, viruses, aggregated proteins, and mitochondria. However, these selective processes are less well studied than **macroautophagy,** which nonselectively digests and recycles cytoplasmic components (including organelles such as mitochondria). This is a normal process that helps maintain cellular homeostasis.

During macroautophagy, materials to be digested are surrounded by a double membrane to form a structure called an **autophagosome.** Autophagosome formation begins with creation of an isolation membrane. Numerous autophagy-related proteins and lipids are involved in the formation of the isolation membrane and its subsequent maturation into the autophagosome. The mature autophagosome ultimately fuses with a lysosome.

No matter the route taken, digestion occurs once the targeted materials have been delivered to a lysosome. As the contents of the lysosome are digested, small products of digestion leave the lysosome and are used as nutrients or for other purposes. The resulting lysosome containing undigested material is sometimes called a **residual body.** In some cases, the residual body can release its contents to the cell exterior.

Comprehension Check

1. How do RER and SER differ from one another in terms of structure and function? List the processes in which the ER is involved.
2. What is a proteasome? Why is it important to the proper functioning of the ER?
3. Describe the structure of a Golgi apparatus. How do the cis and trans faces of the Golgi apparatus differ? List the major functions of the Golgi.
4. What are lysosomes? How do they participate in intracellular digestion? What might happen if lysosomes released their enzymatic contents into the cytoplasm?
5. Describe the secretory pathway. To what destinations does this pathway deliver proteins and other materials?
6. Define endocytosis. Describe the routes that deliver materials to lysosomes for digestion.

5.5 The Nucleus and Ribosomes Are Involved in Genetic Control of the Cell

After reading this section, you should be able to:

a. Describe the structure and function of the nucleus, chromosomes, nucleolus, and ribosomes
b. Compare and contrast the chromosomes and ribosomes of bacterial, archaeal, and eukaryotic cells

Perhaps the most profound way in which eukaryotic cells differ from bacterial and archaeal cells is the eukaryotic nucleus and the way DNA is stored and used. In this section, the organelles involved with these important cellular functions are introduced.

Nucleus

The nucleus is by far the most visually prominent organelle in eukaryotic cells. It was discovered early in the study of cell structure and was shown by Robert Brown (1773–1858) in 1831 to be a constant feature of eukaryotic cells. The **nucleus** is home to the cell's DNA; RNA is made in the nucleus but is then exported.

A cell's nucleus is a membane-bound spherical body about 5 to 7 μm in diameter (figures 5.2 and 5.3). The exact number of chromosomes within a given nucleus depends on the organism, cell type, and stage in the life cycle. Each eukaryotic **chromosome** is composed of chromatin. **Chromatin** is a complex of DNA and proteins, including histones. Histones are small basic proteins rich in the amino acids lysine, arginine, or both. There are five types of histones in most eukaryotic cells: H1, H2A, H2B, H3, and H4. As basic proteins, histones are positively charged; recall that DNA is acidic and negatively charged. This allows histones to associate closely with DNA; some histones serve a spool-like function, with DNA wrapped around the proteins. Eight histone molecules plus its associated DNA form a "beads-on-a-string" appearance (**figure 5.10**). Each bead is called a **nucleosome.** ◀ *Nucleoid (sections 3.6 and 4.3)*

The nucleus is surrounded by the **nuclear envelope** (figure 5.3), a complex structure consisting of two lipid bilayer membranes, the inner and outer membranes, separated by a perinuclear space. The nuclear envelope is continuous with the ER at several points, and its outer membrane is covered with ribosomes. Many nuclear pores penetrate the envelope, and each pore is formed by about 30 proteins; each pore plus the associated proteins is called a **nuclear pore complex** (**figure 5.11**). Pores are

Figure 5.10 Nucleosomes. DNA is wrapped around an ellipsoid composed of eight histone proteins (two each of H2A, H2B, H3, and H4). In most eukaryotes, the histone H1 associates with the nucleosome.

(a) Nucleus with nuclear pores

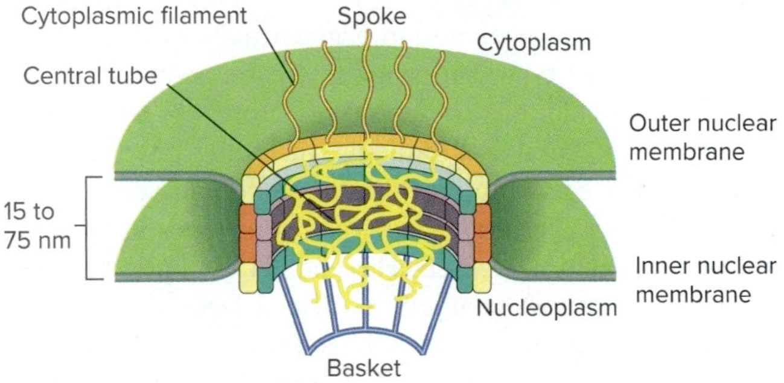

(b) Nuclear pore complex

Figure 5.11 The Nucleus and Nuclear Pore Complex. (a) A transmission electron micrograph (TEM) of a freeze-etch preparation of a cell, showing the large, convex nuclear surface with pores scattered over it. (b) Schematic of the nuclear pore complex. Note that the nuclear envelope consists of two lipid bilayer membranes: outer nuclear membrane and inner nuclear membrane.

(a) ©Biophoto Associates/Science Source

about 70 nm in diameter and collectively occupy about 10 to 25% of the nuclear surface. The nuclear pore complexes serve as transport routes between the nucleus and surrounding cytoplasm. Small molecules move through the nuclear pore complex unaided. Large molecules are transported with the help of nuclear pore complex proteins.

Often the most noticeable structure within the nucleus is the **nucleolus**—the site of ribosomal RNA (rRNA) synthesis (figure 5.3). Recall that rRNA is needed to build ribosomes, the site of protein synthesis. Therefore, depending on how many ribosomes the cell needs, a single nucleus may contain from one to many nucleoli. Although the nucleolus is not membrane-enclosed, it is considered a complex organelle with separate granular and fibrillar regions.

The DNA that encodes a cell's rRNA is located on a highly condensed chromosomal region called the nucleolar organizer. This RNA is synthesized in a single long piece that is cut to form the final rRNA molecules. The processed rRNAs combine with ribosomal proteins (which have been synthesized in the cytoplasm) to form partially completed ribosomal subunits. The granules seen in the nucleolus are probably these subunits. Immature ribosomal subunits then leave the nucleus, presumably by way of the nuclear pore complexes, and mature in thc cytoplasm.

Eukaryotic Ribosomes Are Larger and Contain Different Molecules than Bacterial and Archaeal Ribosomes

Eukaryotic ribosomes are larger than bacterial and archaeal 70S ribosomes with the exception of those found in mitochondria and chloroplasts. Each ribosome is a dimer of a 60S and a 40S subunit. The 60S subunit is composed of three rRNA molecules (5S, 28S, and 5.8S rRNAs) and about 50 proteins. The 40S subunit is composed of an 18S rRNA and about 30 proteins. Each ribosome is about 22 nm in diameter and has a sedimentation coefficient of 80S and a molecular weight of 4 million (**figure 5.12**). Eukaryotic ribosomes are either associated with the endoplasmic reticulum or free in the cytoplasm. When bound to the endoplasmic reticulum to form rough ER, they are attached through their 60S subunits.

Both free and ER-bound ribosomes synthesize proteins. Recall that proteins made on the ribosomes of the RER are often secreted or are inserted into the ER membrane as integral membrane proteins. Free ribosomes are the sites of synthesis for nonsecreted and nonmembrane proteins. Some proteins synthesized by free

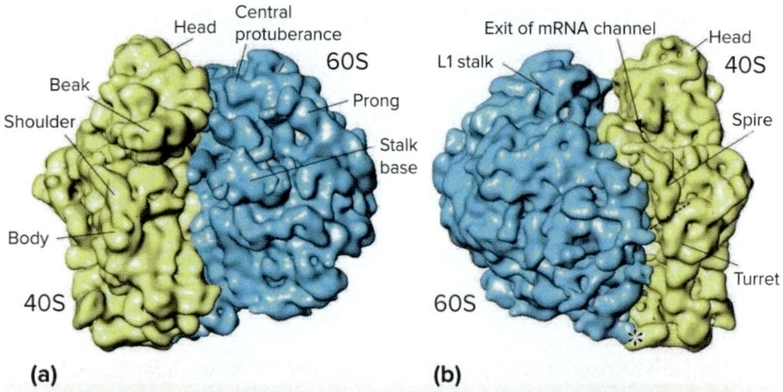

Figure 5.12 Eukaryotic Ribosome. This image showing two views of the 80S ribosome of *Trypanosoma cruzi* was generated using electron cryomicroscopy. The image is a density map with a resolution to 12 Å. Landmarks of the ribosome are shown; compare these to bacterial 70S ribosomes shown in figure 3.34.

MICRO INQUIRY *Which subunit attaches to the rough ER?*

ribosomes are inserted into organelles such as the nucleus, mitochondrion, and chloroplast. As we discuss in chapters 13 and 15, proteins called molecular chaperones aid the proper folding of proteins after synthesis. They also assist the transport of proteins into eukaryotic organelles such as mitochondria.

Comprehension Check

1. What proteins contribute to the folding of eukaryotic chromosomes?
2. Why is the nucleolus needed for protein synthesis to occur?
3. Describe the structure of the eukaryotic 80S ribosome and contrast it with bacterial and archaeal ribosomes.
4. How do free ribosomes and those bound to the ER differ in function?

5.6 Mitochondria, Related Organelles, and Chloroplasts Are Involved in Energy Conservation

After reading this section, you should be able to:

a. Draw a mitochondrion and identify its component parts
b. Compare and contrast mitochondria and hydrogenosomes in terms of their structure and the chemical processes they carry out
c. Draw a chloroplast and identify its component parts

Three important energy-conserving organelles are the focus of this section: mitochondria, hydrogenosomes, and chloroplasts. They are of scientific interest not only for this role but also because of their evolutionary history. All are thought to be derived from bacterial cells that invaded or were ingested by early ancestors of eukaryotic cells (*see figure 19.7*). ▶| *Evolution of the three domains of life (section 19.5); The proteobacterial origin of mitochondria (section 22.1)*

The Aerobic Mitochondrion Belongs to the Family of Mitochondrial-Related Organelles

In the last decade or so it has become apparent that eukaryotic microbes contain numerous organelles related to mitochondria. It is thought that they arose from a single endosymbiotic event, as we describe in chapter 1, but then different evolutionary paths were taken to yield at least five different mitochondrial-related organelles, or MROs (**figure 5.13**). In this section we describe two MROs: the aerobic mitochondrion and the hydrogenosome.

The aerobic mitochondrion is the organelle in this family most familiar to biology students. For our discussion, we will simply refer to it as the mitochondrion. Found in most eukaryotic cells, these **mitochondria** (s., **mitochondrion**) frequently are called the "powerhouses" of the cell (**figure 5.14**). Metabolic processes such as the tricarboxylic acid cycle and generation of ATP, the major energy currency of all life forms, take place here, but only when oxygen is available. Many eukaryotes have cylindrical mitochondria that measure approximately 0.3 to 1.0 μm by 5 to 10 μm. In other words, they are about the same size as bacterial cells. Some cells possess 1,000 or more mitochondria; some yeasts, unicellular algae, and trypanosome protists have a single, giant, tubular mitochondrion twisted into a continuous network permeating the cytoplasm. ▶| *Pyruvate to carbon dioxide (step 2) is accomplished by the tricarboxylic acid cycle (section 11.5); Electron transport and oxidative phosphorylation (step 3) generate the most ATP (section 11.6)*

Like Gram-negative bacteria, mitochondria are bounded by two membranes: an outer mitochondrial membrane separated from an inner mitochondrial membrane by a 6- to 8-nm intermembrane space (figure 5.14). The outer mitochondrial membrane contains porins also like Gram-negative bacteria. The inner membrane has infoldings called **cristae** (s., **crista**), which greatly increase its surface area. The shape of cristae differs in mitochondria from various species. Platelike (laminar) cristae, cristae shaped like disks, tubular cristae, and cristae in the shape of vesicles have all been observed. The inner membrane encloses the mitochondrial matrix, a dense material containing ribosomes, DNA, and often large calcium phosphate granules. In many organisms, mitochondrial DNA is a closed circle, like most bacterial DNA. However, in some protists, mitochondrial DNA is linear.

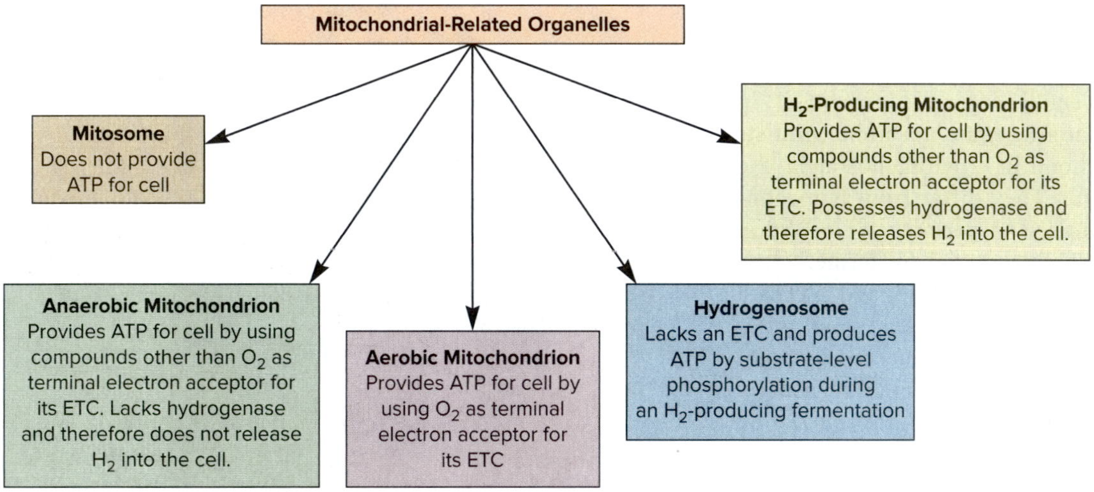

Figure 5.13 Mitochondrial-Related Organelles. It is thought that the ancestral endosymbiont evolved in different cell lines to give rise to these five different organelles. Descriptions of the metabolic processes noted are described in chapter 11. ETC is electron transport chain.

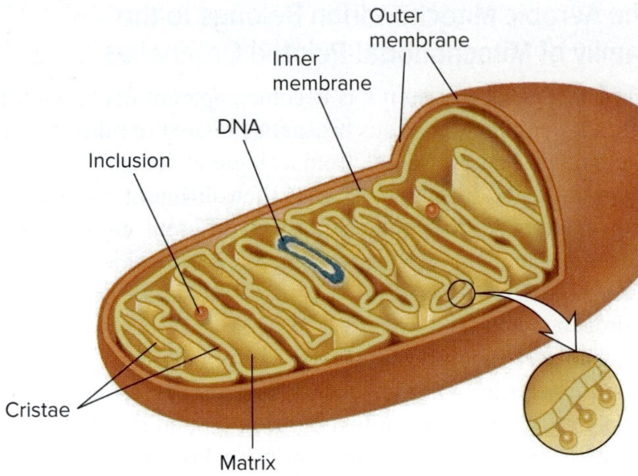

(a)

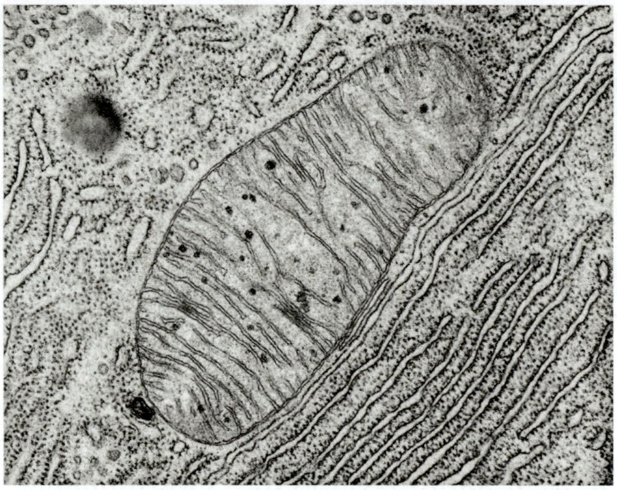

(b)

Figure 5.14 Mitochondrial Structure. (a) A diagram of mitochondrial structure. The insert shows the ATP-synthesizing enzyme ATP synthase lining the inner surface of the cristae. (b) Transmission electron micrograph (×85,000) of a mitochondrion showing the cristae.

(b) ©Keith R. Porter/Science Source

Each mitochondrial compartment has a characteristic chemical and enzymatic composition. For example, the inner mitochondrial membranes possess enzymes and electron carriers involved in electron transport and oxidative phosphorylation (the formation of ATP during respiration). Enzymes of the tricarboxylic acid cycle *(see figure 11.8)* and those involved with the catabolism (breaking down) of fatty acids are located in the matrix. ▶| *Lipid catabolism (section 11.9)*

Mitochondria are unlike other organelles in possessing their own DNA and ribosomes. This reflects their bacterial ancestry; however, mitochondria synthesize only some of their own proteins. Most mitochondrial proteins are manufactured under the direction of the nucleus and must be transported into the mitochondrion. In another similarity to bacteria, mitochondria reproduce by binary fission. ▶| *Bacterial cell cycles can be divided into three phases (section 7.2); Protein localization and secretion in eukaryotes (section 15.4)*

Hydrogenosomes are small organelles involved in energy capture or conservation in some anaerobic protists (**figure 5.15**). Like mitochondria, hydrogenosomes are bound by a double membrane. They differ from mitochondria in terms of the method used to generate ATP. Within hydrogenosomes, pyruvate is catabolized by fermentation rather than respiration, and CO_2, H_2, acetate, and ATP are formed. Because ATP is made during fermentation rather than by oxidative phosphorylation, hydrogenosomes do not require extensive internal membrane networks so usually lack cristae. In some hydrogenosome-bearing protists, the metabolic products are consumed by symbiotic bacteria and archaea living within the protist. ▶| *Fermentation does not involve an electron transport chain (section 11.8); Mutualism and cooperation are two-way interactions (section 27.2)*

Chloroplasts

Plastids are cytoplasmic organelles found in algae (photosynthetic protists) and plants. They often possess pigments such as chlorophylls and carotenoids, and are the sites of synthesis and storage of food reserves. The most important type of plastid is the chloroplast. **Chloroplasts** contain chlorophyll and use light energy to convert CO_2 and water to carbohydrates and O_2; that is, they are the site of photosynthesis. Two major types of chloroplasts have been identified: those that evolved from a primary endosymbiotic event and those that evolved from a secondary (or tertiary) event (**Microbial Diversity & Ecology 5.1**). Plant and

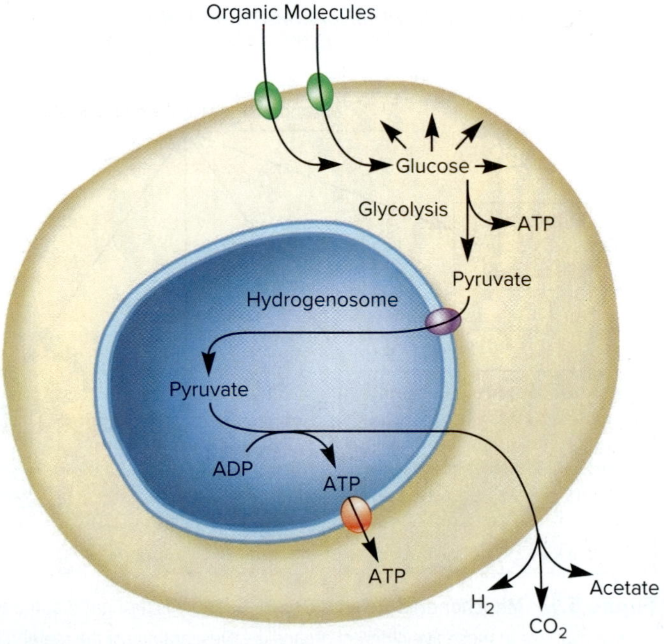

Figure 5.15 The Metabolic Activities of Hydrogenosomes.

MICROBIAL DIVERSITY & ECOLOGY

5.1 There Was an Old Woman Who Swallowed a Fly

The children's song "There Was an Old Woman Who Swallowed a Fly" describes the unusual eating habits of an old woman. Her story began when she ate a fly but continued as she ate a series of other animals, all in an attempt to get rid of the fly. The evolutionary history of chloroplasts also appears to involve a series of meals. Molecular studies of the chloroplasts of a wide variety of photosynthetic protists have revealed that some chloroplasts arose when a phagocytic ancestor of a eukaryotic cell engulfed the ancestor of a cyanobacterium (**box figure**). This evolutionary event eventually gave rise to plants and the photosynthetic protists commonly called green algae and red algae. These organisms are now classified together in the supergroup *Archaeplastida* (*see chapter 24*). The chloroplasts that evolved from this endosymbiosis are the typical chloroplasts that biology students are familiar with. These primary plastids have two membranes. It is thought that these primary plastids gave rise to at least two different types of plastids when two phagocytic plastid-free eukaryotes ingested an archaeplastid; one ingested a green alga, the other, a red alga. The ingested archaeplastid evolved into a chloroplast surrounded by three membranes. These chloroplasts are known as secondary plastids and are found in photosynthetic protists called stramenopiles, which include diatoms and brown algae. The final meal in the series was eaten by some dinoflagellates. These interesting protists can be chemoorganotrophic, phototrophic, or both (mixotrophic). It appears that some photosynthetic dinoflagellates ingested a protist containing secondary plastids. The ingested protist evolved into a tertiary plastid surrounded by four membranes.

Source: Reyes-Prieto, A., et al. 2007. The origin and establishment of the plastid in algae and plants. Annu. Rev. Genet. *41:147; Dorrell, R. G., and Smith, A. G. 2011. Do red and green make brown?: Perspectives on plastid acquisitions within chromalveolates.* Euk. Cell *10:856.*

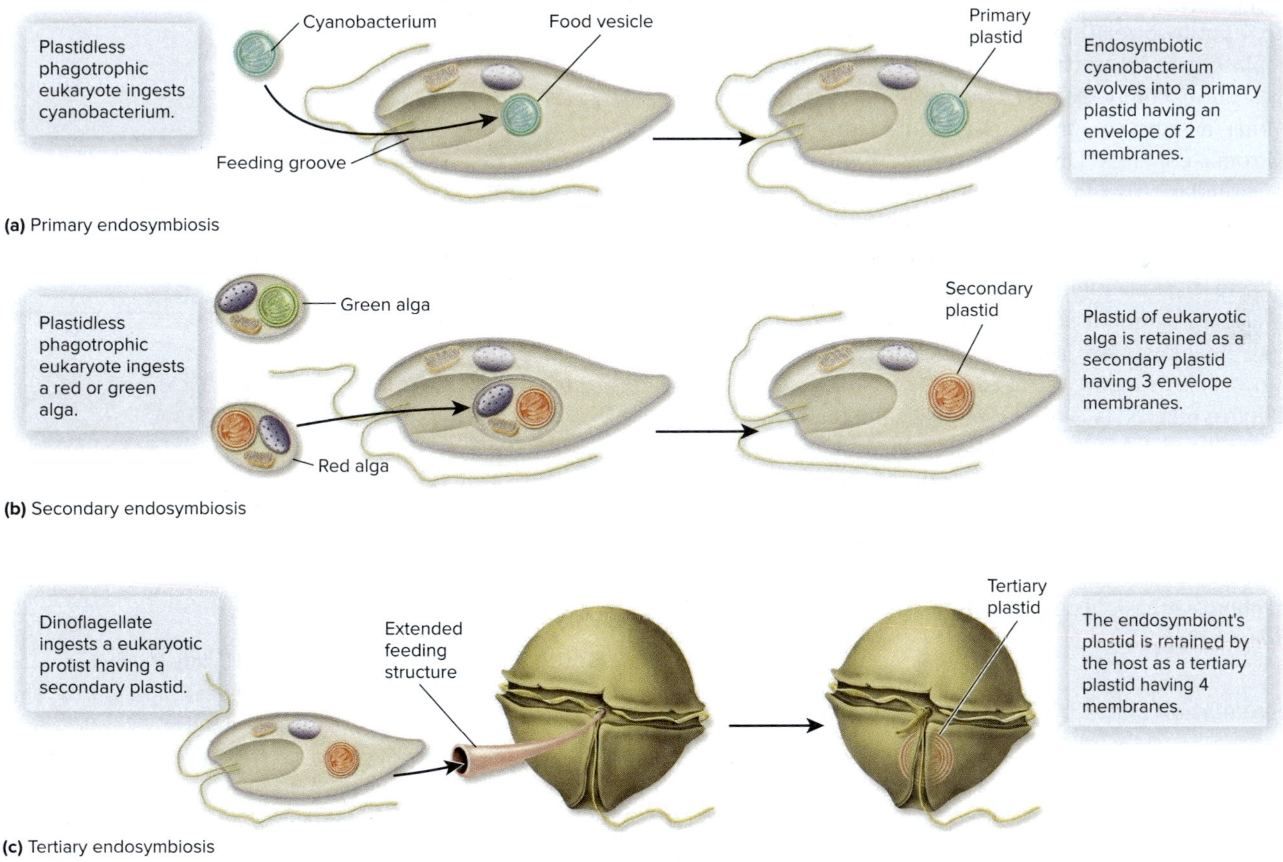

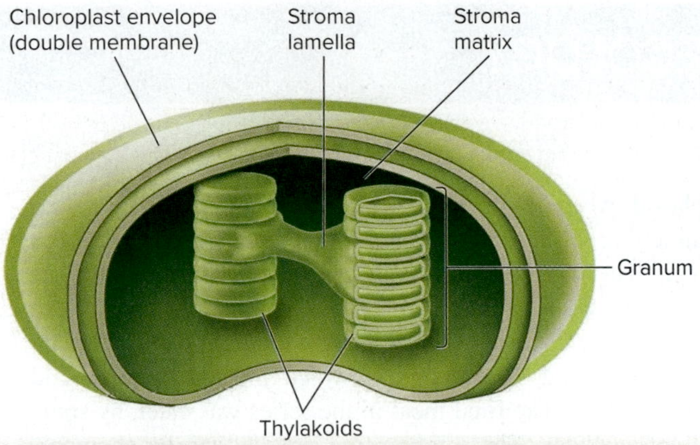

Figure 5.16 Diagram of Chloroplast Structure.

MICRO INQUIRY *In what regions of the chloroplast do light reactions occur? Dark reactions?*

some algal chloroplasts are primary plastids and are the focus of this discussion.

Chloroplasts are quite variable in size and shape, but they share many structural features. Most are oval with dimensions of 2 to 4 μm by 5 to 10 μm, but some algae possess one huge chloroplast that fills much of the cell. Like mitochondria, chloroplasts are encompassed by two membranes (**figure 5.16**). The inner membrane surrounds a matrix called the stroma. The stroma contains DNA, ribosomes, lipid droplets, and starch granules. Also located within the stroma is a complex internal membrane system that features flattened, membrane delimited sacs called **thylakoids.** Clusters of two or more thylakoids are dispersed within the stroma of most algal chloroplasts (figure 5.16). In some photosynthetic protists, several disklike thylakoids are stacked on each other like coins to form grana (s., granum).

Photosynthesis is defined as the combination of light-driven ATP and NADPH production, called the light reaction, and the use of that ATP and NADPH to convert CO_2 and water to carbohydrate, termed the dark reaction. The light reaction takes place in the thylakoid membranes, where chlorophyll and electron transport components are also found. The dark reaction of photosynthesis takes place in the stroma. ▶| *Phototrophy (section 11.12)*

Comprehension Check

1. Differentiate between aerobic mitochondria, hydrogenosomes, and chloroplasts. Where are the different components of mitochondria and chloroplast energy-trapping systems located?
2. What processes are used by aerobic mitochondria, hydrogenosomes, and chloroplasts to conserve energy?
3. What is the role of mitochondrial DNA?

5.7 Many Eukaryotic Microbes Have External Structures Used for Motility

After reading this section, you should be able to:

a. Differentiate between eukaryotic flagella and cilia
b. Compare and contrast bacterial flagella, archaella, and eukaryotic flagella
c. Discuss the types of motility observed in eukaryotic microbes

Cilia (s., **cilium**) and **flagella** (s., **flagellum**) are the most prominent external structures observed on eukaryotic cells; they are associated with motility. Although both serve to move the microorganism, they differ from one another in two ways. First, cilia are short hairlike structures, typically only 5 to 20 μm in length, whereas flagella are long whiplike filaments, 100 to 200 μm long. Second, their patterns of movement are usually distinctive (**figure 5.17**). The beating of flagella causes them to undulate and generate planar or helical waves from the base to the tip, or vice versa. If the wave moves from base to tip, the cell is pushed along; a beat traveling from the tip toward the base pulls the cell through the water. Some flagella have lateral hairs

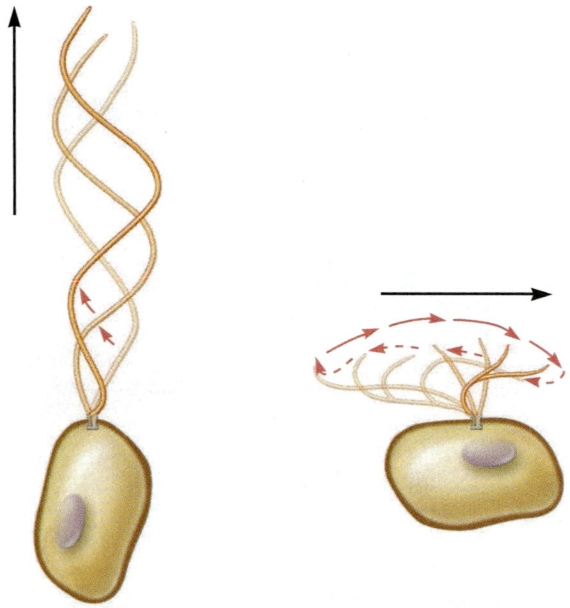

Figure 5.17 Patterns of Flagellar and Ciliary Movement. Flagellar and ciliary movement takes the form of waves. Flagella (left illustration) move either from the base of the flagellum to its tip or in the opposite direction. The motion of these waves propels the organism along. The beat of a cilium (right illustration) may be divided into two phases. In the effective stroke, the cilium remains fairly stiff as it swings through the water. This is followed by a recovery stroke in which the cilium bends and returns to its initial position. The black arrows indicate the direction of water movement in these examples.

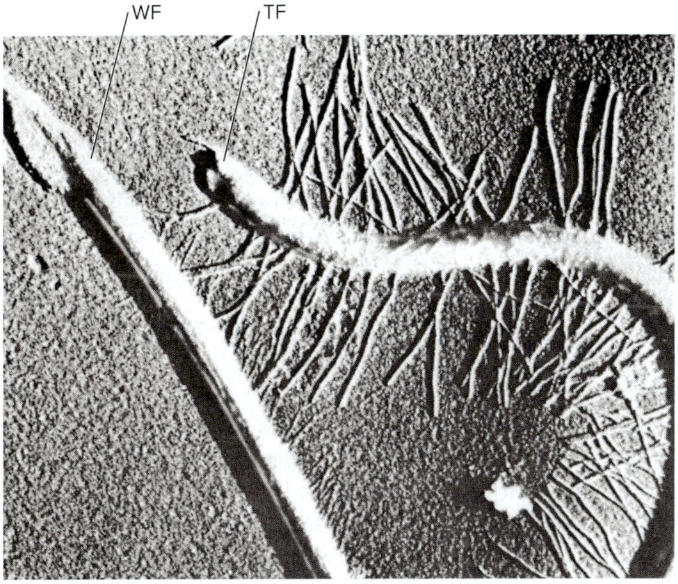

Figure 5.18 Whiplash and Tinsel Flagella. Transmission electron micrograph of a shadowed whiplash flagellum, WF, and a tinsel flagellum, TF, with mastigonemes.

From "Electron microscopic observations of the flagellar hairs of Phytophthora palmivora zoospores" by P. R. Desjardins, G. A. Zentmyer, D. A. Reynolds. Canadian Journal of Botany, 1969, 47(7): 1077-1079, 10.1139/b69-153

called flimmer filaments (thicker, stiffer hairs are called mastigonemes). These filaments change flagellar action so that a wave moving down the filament toward the tip pulls the cell along instead of pushing it. Such flagella often are called tinsel flagella, whereas naked flagella are referred to as whiplash flagella (**figure 5.18**). Cilia, on the other hand, normally have a beat with two distinctive phases. In the effective stroke, the cilium moves through the surrounding fluid like an oar, thereby propelling the organism. The cilium next bends along its length while it is pulled forward during the recovery stroke in preparation for another effective stroke (figure 5.17). A ciliated microorganism actually coordinates the beats so that some of its cilia are in the recovery phase while others are carrying out their effective stroke. This coordination allows the organism to move smoothly through the water.

Cilia and flagella are structurally the same. They were named long before their molecular structure was known. It was assumed that given their difference in size and distribution on cells, they were structurally unique. However, all cilia and flagella are membrane-bound cylinders about 0.2 μm in diameter. Each possesses an axoneme within the matrix. An axoneme usually consists of nine microtubule doublets arranged in a circle around two central microtubules (**figure 5.19**). This is called the 9 + 2 pattern of microtubules. Each doublet has pairs of arms projecting from subtubule A toward a neighboring doublet. A radial spoke extends from subtubule A toward the internal pair of microtubules with their central sheath. These microtubules are similar to those found in the cytoplasm.

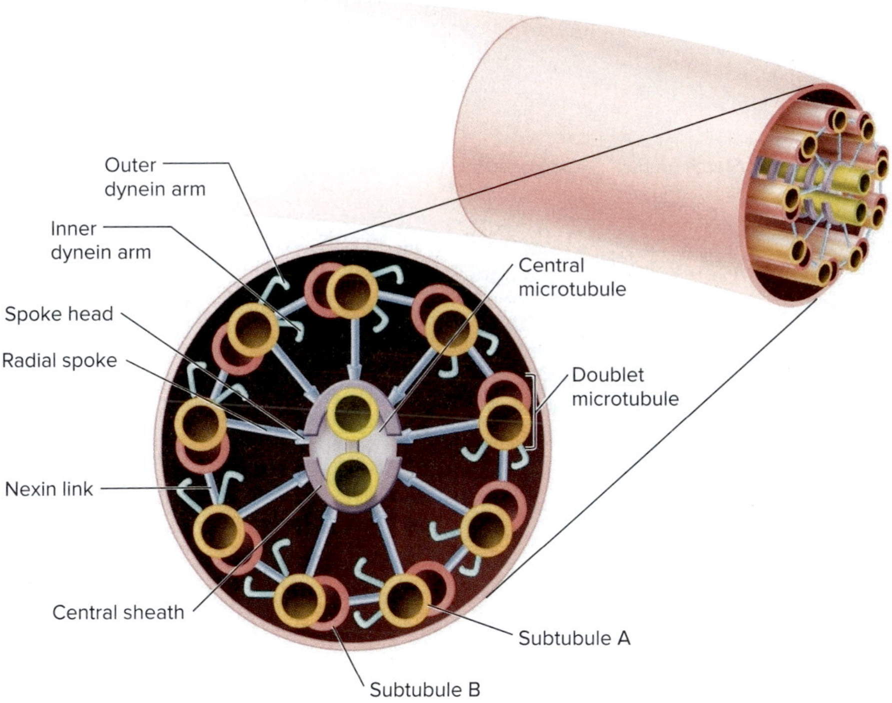

Figure 5.19 Diagram of Cilia and Flagella Structure.

A basal body lies in the cytoplasm at the base of each cilium or flagellum. It is a short cylinder with nine microtubule triplets around its periphery (a 9 + 0 pattern) and is separated from the rest of the organelle by a basal plate. The basal body directs the construction of these organelles. Cilia and flagella appear to grow through the addition of preformed microtubule subunits at their tips.

Cilia and flagella bend because adjacent microtubule doublets slide along one another while maintaining their individual lengths. The doublet arms, about 15 nm long, are made of the protein dynein (figure 5.19). Dynein arms interact with the B subtubules of adjacent doublets to cause the sliding. The radial spokes also participate in this sliding motion. ATP powers the movement of cilia and flagella.

Cilia and flagella beat at a rate of about 10 to 40 strokes or waves per second and propel microorganisms rapidly. The flagellated protist *Monas stigmatica* swims at a rate of 260 μm/second (approximately

40 cell lengths per second); the ciliated protist *Paramecium caudatum* swims about 2,700 μm/second (12 lengths per second). Such speeds are equivalent to or much faster than those seen in higher animals but not as fast as those in bacteria or archaea. ◀ *Flagellar movement (section 3.8); Archaella and motility (section 4.4)*

Comprehension Check

1. Draw a labeled diagram showing the detailed structure of a cilium or flagellum.
2. How do cilia and flagella move, and what is dynein's role in the process? Contrast the ways in which flagella and cilia propel eukaryotic microorganisms through water.
3. Calculate your running speed relative to that of *M. stigmatica*: 40 body lengths per second. Is this running speed humanly possible?

5.8 Comparison of Bacterial, Archaeal, and Eukaryotic Cells

After reading this section, you should be able to:

a. Create a Venn diagram or concept map that clearly distinguishes bacterial, archaeal, and eukaryotic cells in terms of their genome organization, organelles, cell envelopes, ribosome size and component molecules, and cytoskeleton
b. Determine the type of microbe when given a description of a newly discovered microbe

A comparison of archaeal, bacterial, and eukaryotic cells in **table 5.1** and **figure 5.20** demonstrates that there are many structural differences between these cells. In addition to structure, bacterial and archaeal cells are functionally simpler in several fundamental ways. They lack mitosis and meiosis, and have a simpler genetic organization. Many complex eukaryotic processes are absent in bacteria and archaea: Endocytosis, intracellular digestion, directed cytoplasmic streaming, and ameboid movement are just a few.

Despite the many significant differences, all cells are remarkably similar on the biochemical level, as we discuss in succeeding chapters. With a few exceptions, the genetic code is the same in all, as are the overall processes of replication, transcription, and translation. The principles underlying metabolic processes and many important metabolic pathways are identical. Thus beneath the profound structural and functional differences between bacterial, archaeal, and eukaryotic cells, there is a molecular unity basic to all known life processes.

Comprehension Check

1. Outline the major differences between bacterial, archaeal, and eukaryotic cells. How are they similar?
2. Examine table 5.2. What characteristics make members of *Archaea* more like eukaryotes? What features make them more like bacteria?

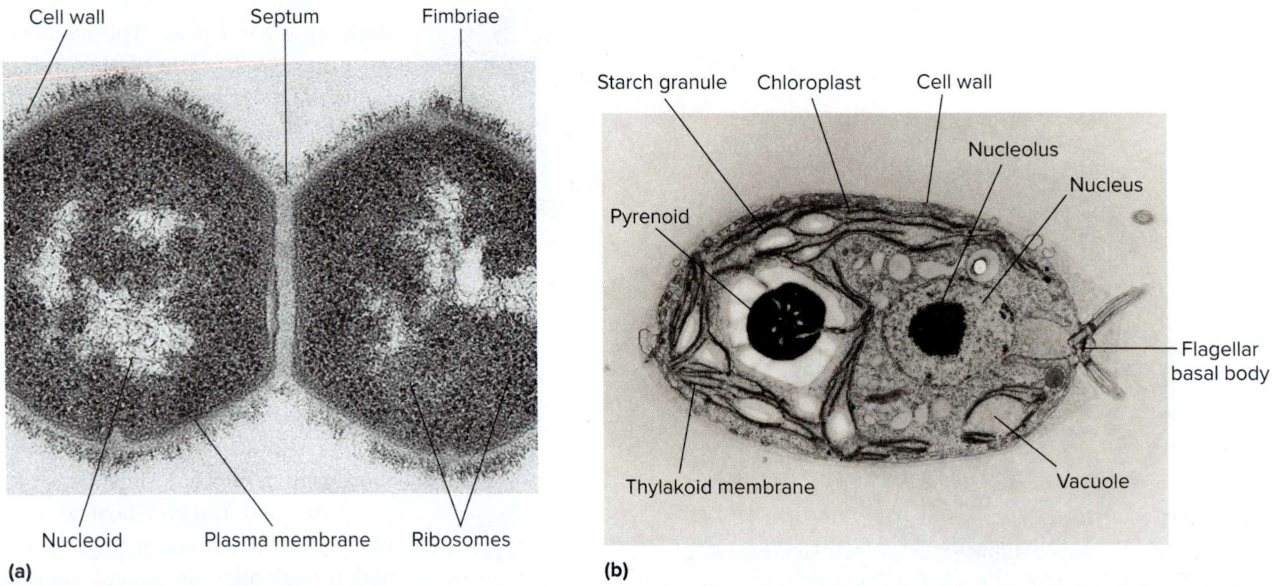

Figure 5.20 Comparison of Bacterial and Eukaryotic Cell Structure. (a) TEM of the bacterium *Streptococcus pyogenes* undergoing cell division. The septum separates the two daughter cells (×20,000). (b) TEM of the eukaryotic alga *Chlamydomonas reinhardtii*, which measures about 10 μm in diameter.

Table 5.1 Comparison of Bacterial, Archaeal, and Eukaryotic Cells

Property	Bacteria	Archaea	Eukaryotes
Organization of Genetic Material			
True membrane-bound nucleus	No	No	Yes
DNA complexed with histones	No	Some	Yes
Chromosomes	Usually one circular chromosome; chromosomes have single origin of replication; some are polyploid	One circular chromosome; some have chromosomes with multiple origins of replication; some are polyploid	Multiple, linear chromosomes with multiple origins of replication
Plasmids	Very common	Very common	Rare
Introns in genes	Rare	Rare	Yes
Nucleolus	No	No	Yes
Mitochondria, Chloroplasts, Endoplasmic Reticulum, Golgi, and Lysosomes Observed	No	No	Yes
Plasma Membrane Lipids	Ester-linked phospholipids; some have sterols	Glycerol diethers and diglycerol tetraethers	Ester-linked phospholipids and sterols
Flagella	Submicroscopic in size; filament composed of single type of flagellin protein	Submicroscopic in size; some filaments composed of more than one type of archaellin protein	Microscopic in size; membrane bound; usually 20 microtubules in 9 + 2 pattern
Peptidoglycan in Cell Walls	Yes	No	No
Ribosome Size and Structure	70S; 3 rRNAs; ~55 ribosomal proteins	70S; most have 3 rRNAs; ~68 ribosomal proteins	80S; 4 rRNAs and ~80 ribosomal proteins
Cytoskeleton	Rudimentary	Rudimentary	Yes
Gas Vesicles	Yes	Yes	No

Key Concepts

5.1 Eukaryotic Cells Are Diverse but Share Some Common Features

- Eukaryotic cells are generally larger than bacterial and archaeal cells, and they exhibit considerably more morphological diversity.
- The eukaryotic cell has a membrane-bound nucleus and many membranous organelles (**figures 5.2** and **5.3**).
- The membranous organelles form compartments in the cytoplasm of the cell. This allows the cell to carry out a variety of biochemical reactions simultaneously. It also provides more surface area for membrane-associated activities such as respiration.

5.2 Eukaryotic Cell Envelopes

- Eukaryotic membranes are similar in structure and function to bacterial membranes. The two differ in terms of their lipid composition (**figure 5.4**).
- Eukaryotic membranes contain microdomains that are enriched for certain lipids and proteins, and participate in a variety of cellular processes.
- When a cell wall is present, it is constructed from polysaccharides (e.g., cellulose) that are chemically simpler than peptidoglycan, the molecule found in bacterial cell walls.

5.3 The Eukaryotic Cytoplasm Contains a Cytoskeleton and Organelles

- The cytoplasm consists of a cytosol in which are found the organelles and cytoskeleton.
- The cytoskeleton is organized from actin filaments, intermediate filaments, and microtubules. These cytoskeletal elements are partly responsible for cell structure, endocytosis, movement of cell structures, cell division, and motility (**figure 5.5**).
- Actin filaments and microtubules are found in eukaryotic microbes. Actin filaments are composed of actin proteins; microtubules are composed of α-tubulin and β-tubulin.
- Intermediate filaments are assembled from a heterogeneous family of proteins. They are found in animal cells and some protozoa but have not been identified in fungi or plant cells.
- Uniquely eukaryotic organelles are numerous and include the endoplasmic reticulum, Golgi, mitochondria, and chloroplasts.

5.4 Several Organelles Function in the Secretory and Endocytic Pathways

- Specific organelles and vesicles are involved in the synthesis and secretion of materials (secretory pathways). Some are involved in the uptake of materials from the extracellular milieu (endocytic pathways).
- The endoplasmic reticulum (ER) is an irregular network of tubules and flattened sacs (cisternae). Rough ER has ribosomes attached to it and is active in protein synthesis. Smooth ER lacks ribosomes.
- The ER donates some materials to the Golgi apparatus, an organelle usually composed of one or more stacks of cisternae (**figure 5.7**). The Golgi prepares and packages cell products for secretion.
- The Golgi apparatus buds off vesicles that deliver hydrolytic enzymes and other proteins to lysosomes and lysosome-like organelles. Lysosomes contain digestive enzymes that aid in intracellular digestion of materials delivered to them by endocytosis and macroautophagy.
- Eukaryotic cells use several kinds of endocytosis (**figure 5.9**). These include phagocytosis, clathrin-dependent endocytosis, and caveolin-dependent endocytosis. Some macromolecules are bound to receptors prior to endocytosis in a process called receptor-mediated endocytosis.

5.5 The Nucleus and Ribosomes Are Involved in Genetic Control of the Cell

- The nucleus is a large organelle containing the cell's chromosomes. It is surrounded by a double-membrane envelope perforated by pores for the nuclear import and export of cellular materials (**figure 5.11**).
- The nucleolus lies within the nucleus and participates in the synthesis of ribosomal RNA and ribosomal subunits.
- Eukaryotic ribosomes are found either free in the cytoplasm or bound to the ER. They are 80S in size and also differ from bacterial and archaeal ribosomes in their size and molecular make up (**figure 5.12**).

5.6 Mitochondria, Related Organelles, and Chloroplasts Are Involved in Energy Conservation

- The family of mitochondrial-related organelles consists of at least five members, including aerobic mitochondria and hydrogenosomes (**figure 5.13**).
- Aerobic mitochondria are organelles bounded by two membranes, with the inner membrane folded into cristae (**figure 5.14**). They are the site of energy conservation by cell respiration, which involves electron transport and oxidative phosphorylation.
- Hydrogenosomes have a double membrane and carry out metabolic processes (fermentation) that generate hydrogen gas (**figure 5.15**). They are found in certain protozoa.
- Chloroplasts are pigment-containing organelles that serve as the site of photosynthesis (**figure 5.16**). The trapping of light energy takes place in the thylakoid membranes of the chloroplast, whereas CO_2 fixation occurs in the stroma.

5.7 Many Eukaryotic Microbes Have External Structures Used for Motility

- Cilia and flagella are membrane-enclosed organelles with nine microtubule doublets surrounding two central microtubules that provide motility for eukaryotic cells (**figure 5.19**).
- Cilia and flagella are structurally identical, although cilia are shorter and more numerous. Both exhibit whiplike motion. This occurs when the microtubule doublets slide along each other, causing the cilium or flagellum to bend.

5.8 Comparison of Bacterial, Archaeal, and Eukaryotic Cells

- Despite the fact that eukaryotic, bacterial, and archaeal cells differ structurally in many ways (**table 5.1** and **figure 5.20**), they are quite similar biochemically.

Active Learning

1. Discuss the statement: "The most obvious difference between eukaryotic, bacterial, and archaeal cells is in their use of membranes." Specifically, compare the roles of bacterial and archaeal plasma membranes with those of the organelle membranes in eukaryotes.
2. Bacterial and archaeal cell size is limited by the rate at which nutrients diffuse from one place in the cell to another. Are eukaryotic microorganisms similarly limited? Why or why not?
3. Compare the mechanisms by which most eukaryotic cells acquire nutrients with those used by bacteria and archaea. What are the implications in terms of survival in different habitats and competition between eukaryotic microbes such as protists and bacteria and archaea?
4. In an effort to learn about the origin of hydrogenosomes, an international group of scientists performed detailed analysis of all the mRNA produced by several protozoa species that possess these mitochondrial-related organelles (MROs). They discovered that metabolic changes did not occur in a smooth evolutionary progression. Rather, there were complex gains and losses of biochemical reactions within ancient hydrogenosomes. Furthermore, their evidence also suggests the existence of a previously undiscovered MRO that produces hydrogen but does generate ATP. How would figure 5.13 need to be modified if these results are confirmed? Discuss how you might group MROs based on metabolic similarity and how this might contribute to the development of hypotheses about how they evolved.

Read the original paper: Leger, M. M., et al. 2017. Organelles that illuminate the origins of *Trichomonas hydrogenosomes* and *Giardia* mitosomes. *Nature Ecol. Evol.* 1(4). doi:10.1038/s41559-017-0092.

6

Viruses and Other Acellular Infectious Agents

©Ingram Publishing/age fotostock

Mustard, Ketchup, and Viruses?

Twenty million hot dogs are sold annually to fans attending games at major league baseball parks in the United States. Hot dogs and lunch meats are popular at baseball games and in lunches carried to work or school. Yet each year in the United States, approximately 1,600 people are sickened by a bacterium that can contaminate the meat and, worse, survive and grow even when the meat is properly refrigerated.

The disease culprit is *Listeria monocytogenes*, a Gram-positive rod found in soil and many other environmental sites. It is not only cold tolerant but salt and acid tolerant as well. Although it is in the minor leagues when compared to some of the big hitters of food-borne disease (e.g., *Salmonella enterica*), it is of concern for two reasons: who it kills and how many it kills. *L. monocytogenes* targets the young and old, pregnant women, and immunocompromised individuals; about 15% of those infected die.

Its effect on pregnant women is particularly heartbreaking. The woman usually only suffers mild, flulike symptoms; however, these innocuous symptoms belie the fact that the child she carries is in serious danger. Her pregnancy often ends in miscarriage or stillbirth. Newborns infected with the bacterium are likely to develop meningitis. Many will die as a result. Those who survive often have neurological disorders.

Currently, pregnant women are counseled against eating ready-to-eat foods unless they have been cooked prior to consumption. However, *L. monocytogenes* is known to contaminate many foods other than hot dogs and these can't always be heated. In 2006 the U.S. Food and Drug Administration (FDA) approved a new approach to prevent listeriosis: spraying viruses that attack and destroy the bacterium on ready-to-eat cold cuts and luncheon meats. In other words, the viruses are a food additive! The method is safe because the viruses only attack *L. monocytogenes,* not human cells.

Since approval, the use of viruses to control the transmission of listeriosis by other foods has been studied. Unfortunately, those studies did not include fresh fruit. In 2011 *L. monocytogenes*–contaminated cantaloupe caused an outbreak of listeriosis in 20 U.S. states, which infected over 100 and killed over 20.

Viruses as agents of good will come as a surprise to many. Typically we think of them as major causes of disease. However, viruses are significant for other reasons. They are vital members of aquatic ecosystems. There they interact with cellular microbes and contribute to the movement of organic matter from particulate forms to dissolved forms. Bacterial viruses are being used in some European countries to treat bacterial infections, and a number of animal viruses are being used to target and destroy cancer cells. Finally, they are important model organisms. In this chapter, we introduce viruses and other acellular infectious agents. ▶| *Marine viruses: mortality at sea (section 30.2); Microorganisms can be controlled by biological methods (section 8.7)*

Readiness Check:

Based on what you have learned previously, you should be able to:

- ✓ Define the term *acellular*
- ✓ Compare and contrast in general terms viruses, viroids, satellites, and prions (section 1.1)

6.1 Viruses Are Acellular

After reading this section, you should be able to:

a. Define the terms *virology, bacteriophages,* and *phages*
b. List organisms that are hosts to viruses

The discipline of **virology** studies **viruses,** a group of infectious agents unique in their simple, acellular organization and pattern of multiplication. Despite this simplicity, viruses are major causes of disease. For instance, many human diseases are caused by viruses, and more are discovered every year, as demonstrated by the appearance of the H1N1 (swine) influenza virus in 2009, the H7N9 avian influenza virus in 2013, and the Middle East respiratory syndrome coronavirus (MERS-CoV) also in 2013. However, their simplicity also has made them attractive model organisms. They served as models for understanding DNA replication, RNA synthesis, and protein synthesis. Therefore the study of viruses has contributed significantly to the discipline of molecular biology. In fact, the field of genetic engineering is based in large part on the use of viruses and viral enzymes such as the enzyme reverse transcriptase. ▶| *Microbial DNA technologies (chapter 17)*

Viruses can exist either extracellularly or intracellularly. When extracellular, they are inactive because they possess few, if any, enzymes and cannot reproduce outside of living cells. When intracellular, viruses exist primarily as nucleic acids that can, at some point in the viral life cycle, commandeer host cells and use them to synthesize viral components from which progeny virions are assembled and eventually released.

Viruses can infect all cell types. Numerous viruses infect bacteria. They are called **bacteriophages,** or **phages** for short. Fewer archaeal viruses have been identified. Most known viruses infect eukaryotic organisms, including plants, animals, protists, and fungi. Viruses have been classified into numerous families based primarily on genome structure, life cycle, morphology, and genetic relatedness. ▶| *Virus phylogeny is difficult to establish (section 26.1)*

In this chapter, we introduce the structure and multiplication strategies of viruses. Their taxonomic diversity is presented in chapter 26, their ecological relevance is discussed in chapter 30, and their role in human disease is the topic of chapter 38.

6.2 Virion Structure Is Defined by Capsid Symmetry and Presence or Absence of an Envelope

After reading this section, you should be able to:

a. State the size range of virions
b. Identify the parts of a virion and describe their function
c. Distinguish enveloped viruses from nonenveloped viruses
d. Describe the types of capsid symmetry

A complete virus particle is called a **virion.** Virion morphology has been intensely studied because of the importance of viruses and the realization that virion structure is simple enough to be understood in detail. Progress has come from the use of several different techniques: electron microscopy, X-ray diffraction, biochemical analysis, and immunology. Although our knowledge is incomplete due to the vast diversity of viruses, we can discuss the general nature of virion structure.

General Structural Properties

Virions can be extraordinarily tiny (about 20 nm in diameter) to about the same size as a rod-shaped bacterial cell (1.5 × 0.5 μm) (**figure 6.1**). The smallest are a little larger than ribosomes, whereas mimiviruses, among the largest viruses known, can be seen in the light microscope. However, most virus particles must be viewed with electron microscopes. |◀ *Electron microscopes use beams of electrons to create highly magnified images (section 2.4)*

The simplest virions consist only of a **nucleocapsid,** which is composed of a nucleic acid, either DNA or RNA, and a protein coat called a **capsid** (**figure 6.2**). The capsid surrounds the viral nucleic acid, protects the viral genome, and often aids in its transfer between host cells. Some virions are covered by a lipid membrane, and these are termed enveloped viruses, whereas those lacking a membrane are called nonenveloped or naked viruses. Notice what is missing from viruses: ribosomes for protein synthesis and a mechanism for generating ATP. Cytoplasm is absent, and while a few enzymes may be found, there are not enough to sustain cellular processes.

Nonenveloped viruses construct a capsid from many copies of one protein and a few minor proteins. Each subunit is termed a protomer, and thousands of protomers self-assemble to form the capsid (**figure 6.3**). In contrast, enveloped viruses require both nucleocapsid proteins and additional proteins to anchor the membrane. Some viruses use noncapsid proteins as scaffolding upon which the capsids are assembled. Probably the most important advantage of this design strategy is that the viral genome is used with maximum efficiency. For example, the *t*obacco *m*osaic *v*irus (TMV) capsid is constructed using a single type of protomer. Recall that the building blocks of proteins are amino acids and that each amino acid is encoded by three nucleotides. The TMV protomer is 158 amino acids in length. Therefore only about 474 nucleotides are required to code for the coat protein. The entire TMV genome consists of only 6,400 nucleotides. Thus only a small fraction of the genome is used to code for the capsid. Suppose, however, that the TMV capsid were composed of six different protomers, all about 150 amino acids in length. If this were the case, about 2,700 nucleotides in the TMV genome would be required just for capsid construction, and much less genetic material would be available for other purposes.

Helical Capsids

Helical capsids are shaped like hollow tubes with protein walls. Tobacco mosaic virus is a well-studied example of helical capsid structure (figure 6.3). The self-assembly of TMV protomers into a helical arrangement produces a rigid tube. The capsid encloses an RNA genome, which is wound in a spiral and lies within a groove formed by the protein subunits. Not all helical capsids are as rigid as the TMV capsid. The nucleocapsids of influenza viruses are thin and flexible and are enclosed within an envelope (**figure 6.4**). ▶| *Tobacco mosaic virus (section 26.5)*

The size of a helical capsid is influenced by both its protomers and the viral genome. The diameter of the capsid is a function of the size, shape, and interactions of the protomers. The length of the capsid appears to be determined by the nucleic acid because a helical capsid does not extend much beyond the end of the viral genome.

Icosahedral Capsids

An icosahedron is a regular polyhedron with 20 equilateral triangular faces and 12 vertices (figure 6.1*b,g,h*). **Icosahedral capsids** are the most efficient way to enclose a space. They are constructed from ring- or knob-shaped assemblages of five or six protomers; the assemblages are called **capsomers** (**figure 6.5**). Capsomers composed of five protomers are called pentamers (pentons); hexamers (hexons) are capsomers that possess six protomers. Pentamers are usually at the vertices of the icosahedron, whereas hexamers generally form its edges and triangular faces. The virions of some RNA viruses have pentamers and hexamers constructed with only one type of subunit. Other virions have pentamers and hexamers composed of different proteins.

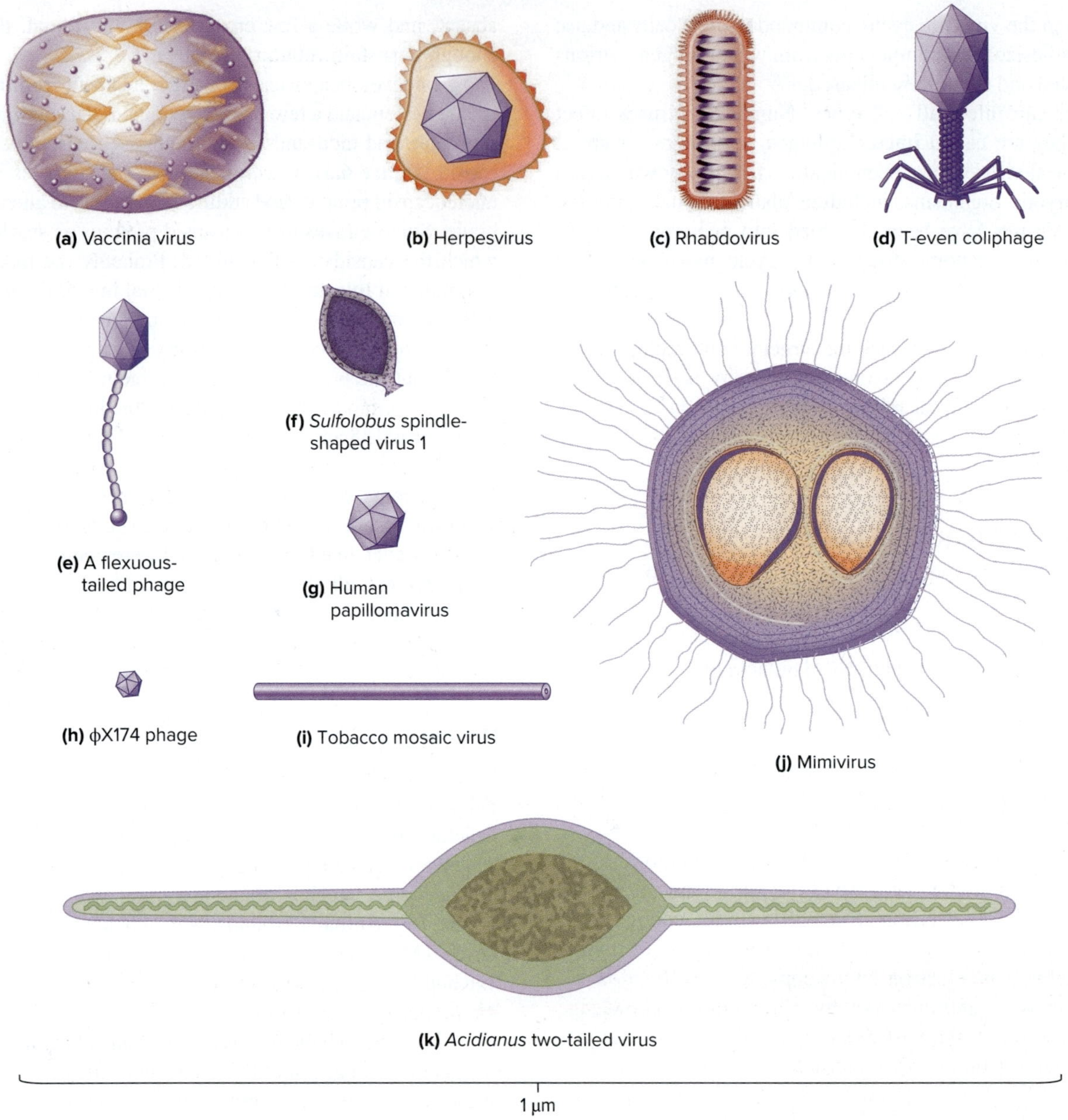

Figure 6.1 The Size and Virion Morphology of Selected Viruses. The virions are drawn to scale.

MICRO INQUIRY *Which capsids are icosahedral? Which are helical? Which have complex symmetry?*

Although many icosahedral capsids contain both pentamers and hexamers, some have only pentamers.

Capsids of Complex Symmetry

Most viruses have either icosahedral or helical capsids, but some viruses do not fit into either category. Poxviruses and large bacteriophages are two important examples.

Poxvirus virions are among the largest of the animal viruses (about 400 by 240 by 200 nm in size) and can be seen with a light microscope. They possess an exceptionally complex internal structure with an ovoid- to brick-shaped exterior. **Figure 6.6** shows the virion morphology of the vaccinia virus. Its double-stranded DNA genome is associated with proteins and contained in the core, a central structure shaped like a biconcave disk and surrounded by a membrane. Two lateral bodies lie between the core and the virion's outer envelope and contain viral enzymes.

Some large bacteriophages have virions that are even more elaborate than those of poxviruses. The virions of T2, T4, and T6 phages (T-even phages) that infect *Escherichia coli* are said to have **binal symmetry** because they have a head that resembles an icosahedron and a tail that is helical. The icosahedral head is

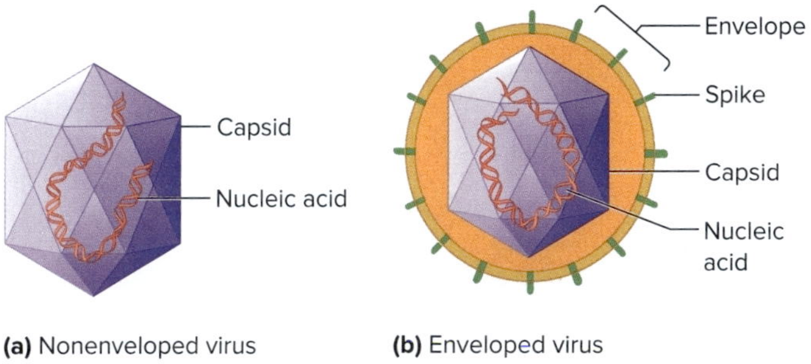

Figure 6.2 Generalized Structure of Virions. (a) The simplest virion is that of a nonenveloped virus, consisting of a capsid assembled around its nucleic acid (nucleocapsid). (b) Virions of enveloped viruses are composed of a nucleocapsid surrounded by a membrane called an envelope. The envelope usually has viral proteins called spikes inserted into it.

elongated by one or two rows of hexamers in the middle and contains the DNA genome (**figure 6.7**). The tail is composed of a collar joining it to the head, a central hollow tube, a sheath surrounding the tube, and a complex baseplate. In T-even phages, the baseplate is hexagonal and has a pin and a jointed long tail fiber and a short tail fiber at each corner. ▶| *Bacteriophage T4: a virulent bacteriophage (section 26.2)*

Viral Envelopes and Enzymes

The nucleocapsids of many animal viruses, some plant viruses, and at least one bacterial virus are surrounded by an outer membranous layer called an **envelope** (**figure 6.8**). Animal virus envelopes usually arise from the plasma or nuclear membranes of the host cell. Envelope lipids and carbohydrates are therefore acquired from the host. In contrast, envelope proteins are coded for by viral genes and may even project from the envelope surface as **spikes,** which are also called **peplomers** (figure 6.8). In many cases, spikes are involved in virion attachment to the host cell surface. Because spikes differ among viruses, they also can be used to identify some viruses. Many enveloped viruses have virions with a somewhat variable shape and are called pleomorphic. However, the envelopes of viruses such as the bullet-shaped rabies viruses are firmly attached to the underlying nucleocapsid and endow the virion with a constant, characteristic shape (figure 6.8*a*).

Influenza virus (figure 6.4) is a well-studied enveloped virus with two types of spikes. Some spikes consist of the enzyme **neuraminidase,** which functions in the release of mature virions from the host cell. Other spikes are **hemagglutinin** proteins, so named because they bind virions to red blood cells and cause the cells to clump together—a process called hemagglutination. Influenza virus's hemagglutinins participate in virion attachment to host cells. Most of its envelope proteins are glycoproteins—proteins that have carbohydrate attached to them. A nonglycosylated protein, the M1 (matrix) protein, is found on the inner surface of the envelope and stabilizes both the lipid envelope and the display of both types of spike.

In addition to enzymes associated with the envelope or capsid (e.g., influenza neuraminidase), some viruses have enzymes within their capsids. Such enzymes are usually involved in nucleic acid replication. For example, influenza virus virions have an RNA genome and carry an enzyme that synthesizes RNA using

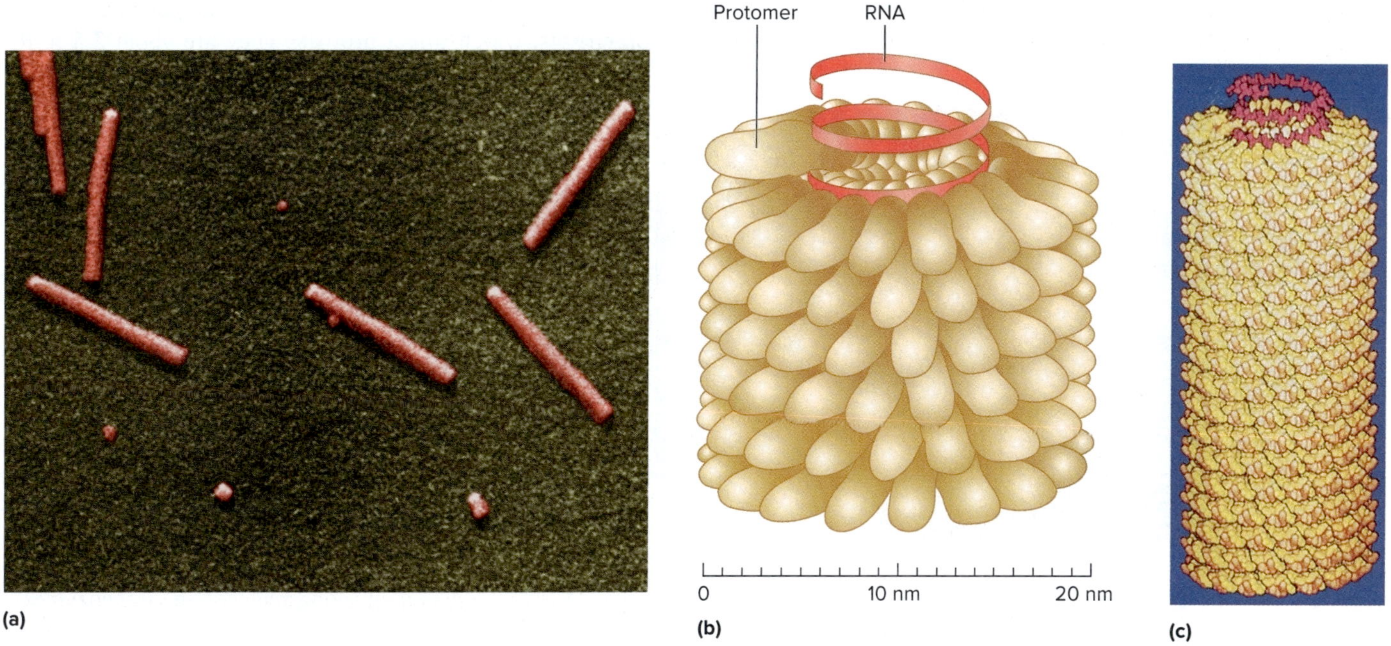

Figure 6.3 Tobacco Mosaic Virus (TMV) Virions. (a) An electron micrograph of negatively stained helical capsids. The virions are 15 to 18 nm in diameter and 300 nm long. (b) Illustration of a TMV nucleocapsid. It is composed of a helical array of protomers with the RNA spiraling on the inside. (c) A model of a TMV virion.

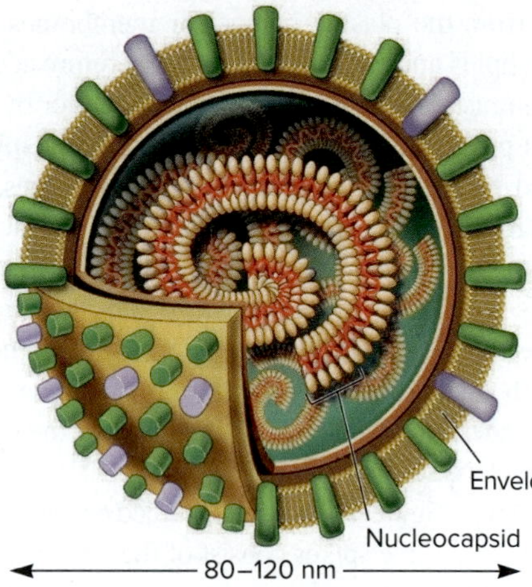

(a)

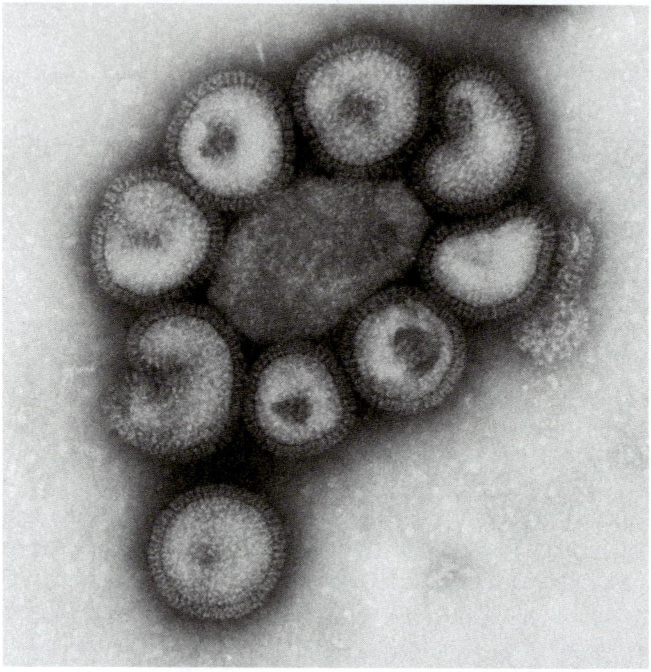

(b)

Figure 6.4 Influenza Virus Virions: An Enveloped Virus with Helical Nucleocapsids. (a) Schematic view. Influenza viruses have segmented genomes consisting of seven to eight different RNA molecules. Each is coated by capsid proteins to create seven to eight flexible nucleocapsids. The envelope has two types of spikes that project 10 nm from the surface at 7- to 8-nm intervals. (b) Because the nucleocapsids are flexible, the virions are pleomorphic. Electron micrograph.
(b) Source: CDC/Dr. F.A. Murphy

an RNA template (*see figure 26.24*). Thus although viruses lack true metabolism and cannot reproduce independently of living cells, their virions may carry one or more enzymes essential to the completion of their life cycles.

Viral Genomes Are Structurally Diverse

One clear distinction between cellular organisms and viruses is the nature of their genomes. Cellular genomes are always double-stranded (ds) DNA. Viruses, on the other hand, employ all four possible nucleic acid types: dsDNA, single-stranded (ss) DNA, ssRNA, and dsRNA. All four types are used by animal viruses. Most plant viruses have ssRNA genomes, and most bacterial and archaeal viruses have dsDNA. The size of viral genomes also varies greatly. Very small genomes are around 4,000 nucleotides—just large enough to code for three or four proteins. Some viruses save additional space by using overlapping genes. At the other extreme are the genomes of pandoraviruses, which infect protists. They are about 2.5×10^6

(a)

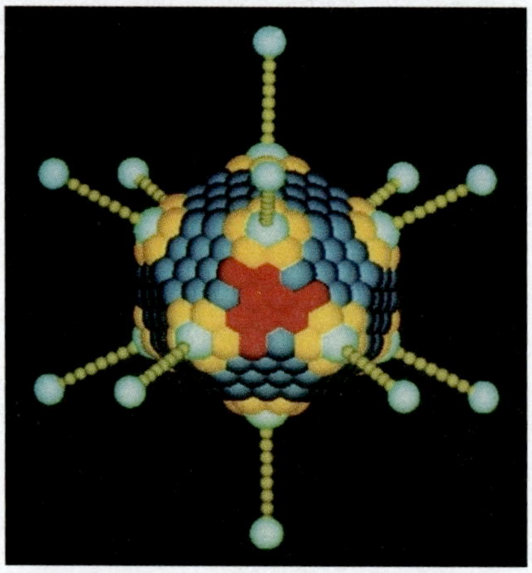

(b)

Figure 6.5 The Icosahedral Capsid of an Adenovirus. (a) Electron micrograph of adenovirus virions, 252 capsomers. (b) Computer-simulated model of an adenovirus virion.
(a) ©Biophoto Associates/Science Source; (b) ©Division of Computer Research & Technology, NIH/Science Source

MICRO INQUIRY *What is the difference between a pentamer and a hexamer? Are the capsomers at the vertices of an adenovirus virion pentamers or hexamers?*

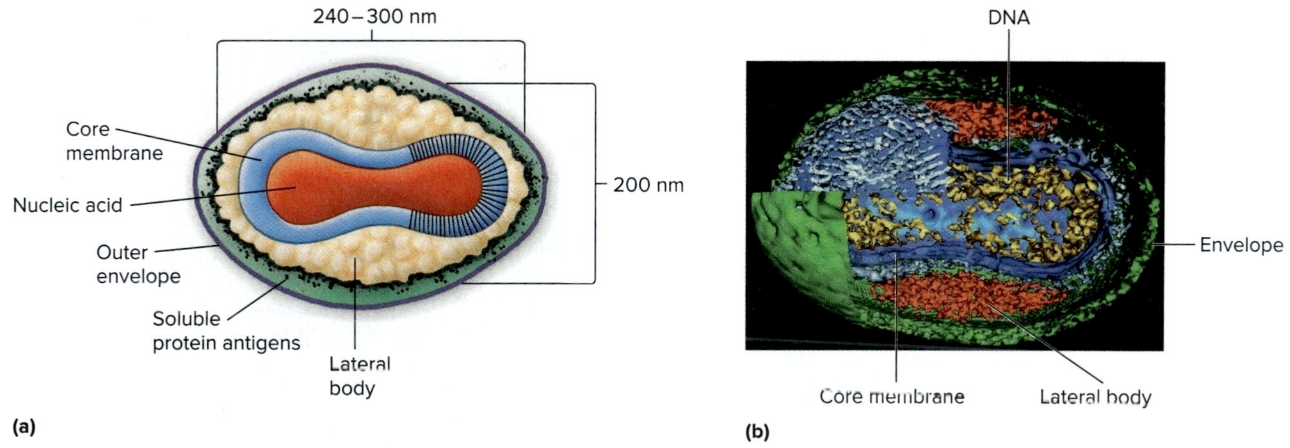

Figure 6.6 Morphology of Vaccinia Virus Virions. (a) Diagram of virion structure. (b) Electron cryotomogram of the virion.
(b) ©Marek Cyrklaff

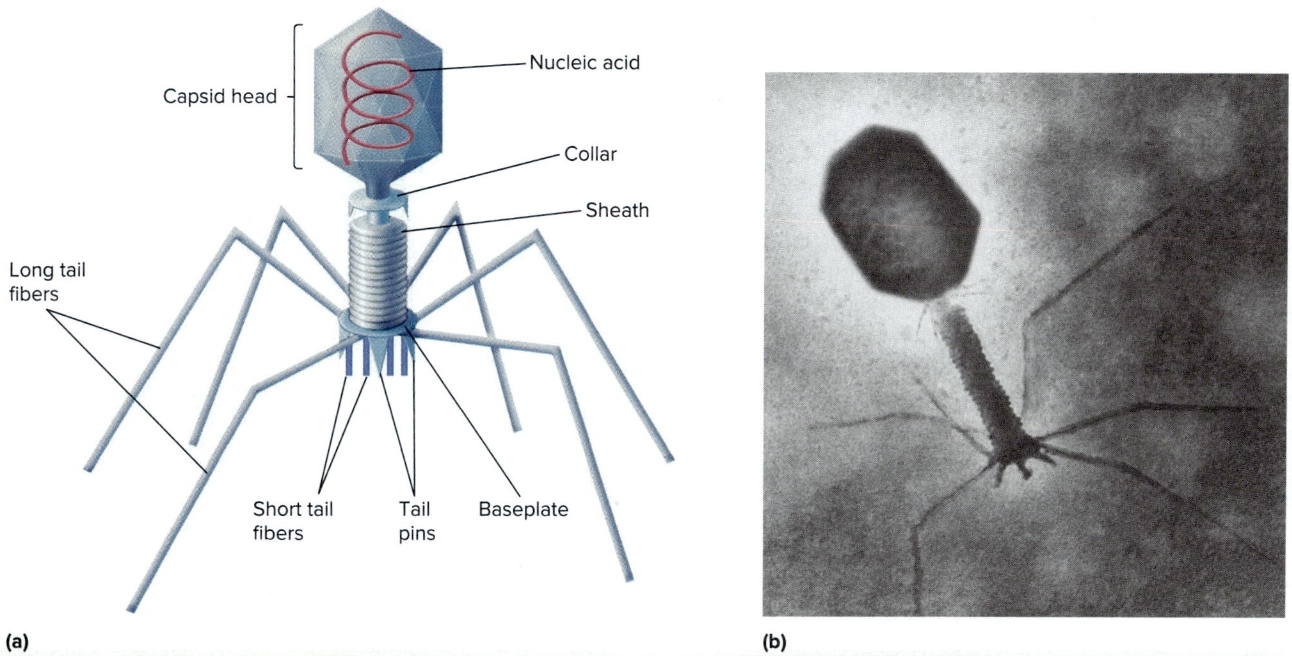

Figure 6.7 T4 Phages of *E. coli*. (a) The structure of T4 bacteriophage virion. (b) Electron micrograph of the virion.
(b) ©Ami Images/Science Source

MICRO INQUIRY *Why is T4 said to have binal symmetry?*

nucleotides long, exceeding some bacteria and archaea in coding capacity.

Most DNA viruses use dsDNA as their genetic material. However, some have ssDNA genomes (e.g., ϕX174 and M13). In both cases, the genomes may be either linear or circular. The ends of linear viral genomes may be covalently closed, attached to a protein, or otherwise masked. Some DNA genomes can switch from one form to the other. For instance, the *E. coli* phages lambda has a linear genome in its capsid, but it becomes circular once it enters the host cell. ▶| *Bacteriophages ϕX174 and Pf1 (section 26.3); Bacteriophage lambda: a temperate bacteriophage (section 26.2)*

Relatively few RNA viruses have dsRNA genomes. More common are viruses with ssRNA genomes. Polio, tobacco mosaic, rabies, influenza, and human immunodeficiency (HIV) are all ssRNA viruses.

Some RNA viruses have **segmented genomes**—genomes that consist of multiple pieces (segments) of RNA. In many cases, each segment codes for one protein and there may be as many as 10 to 12 segments. Usually all segments are enclosed in the same capsid (figure 6.4). However, the genome of brome mosaic virus, a virus that infects certain grasses, is composed of three segments distributed among three different virions.

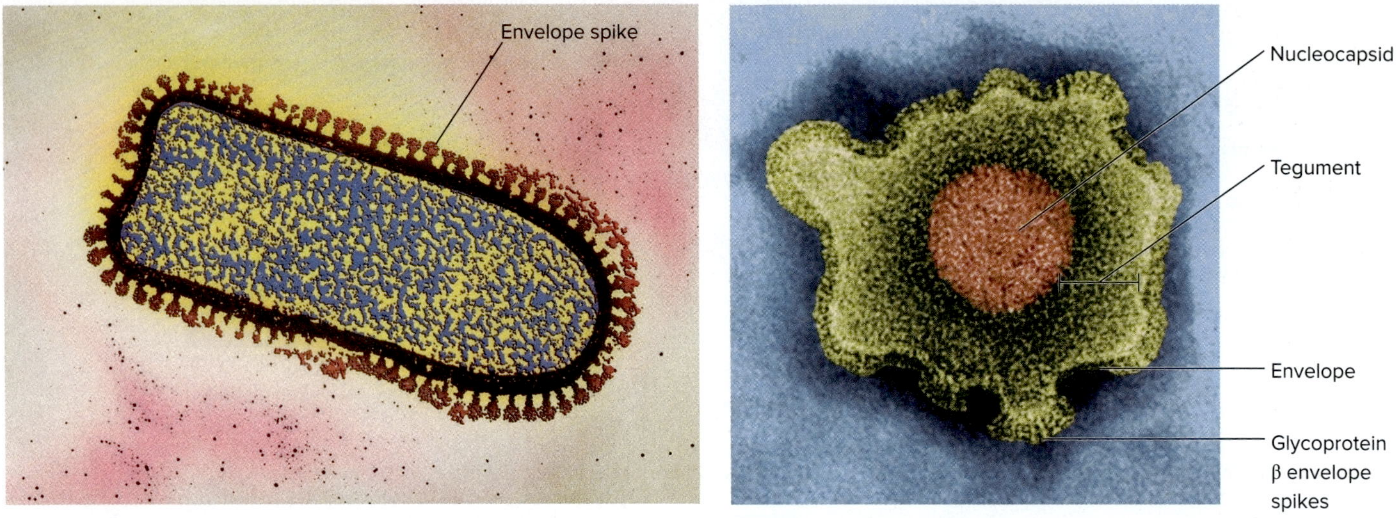

(a) Rabies virus

(b) Herpesvirus

Figure 6.8 Examples of Enveloped Viruses. (a) Negatively stained virion of a rabies virus. (b) Herpesvirus virion. Images are artificially colorized.
(a) ©Chris Bjornberg/Science Source; (b) ©Dr. Linda Stannard, UCT/Science Source

Comprehension Check

1. How are viruses similar to cellular organisms? In what fundamental way do they differ?
2. What is the difference between a nucleocapsid and a capsid?
3. Compare the structure of an icosahedral capsid with that of a helical capsid. How do pentamers and hexamers associate to form a complete icosahedron? Which virus would have a longer helical capsid: a virus with a 7,200 base pair DNA genome or one with an 11,000 base ssRNA genome?
4. What is an envelope? What are spikes (peplomers)? Why do some enveloped viruses have pleomorphic virions?
5. All four nucleic acid forms can serve as viral genomes. Describe each. What is a segmented RNA genome?

6.3 Viral Life Cycles Have Five Steps

After reading this section, you should be able to:

a. Describe the five steps common to the life cycles of viruses
b. Discuss the role of receptors, capsid proteins, and envelope proteins in the life cycles of viruses
c. Describe the two most common methods for virion release from a host cell

In the early years of virology, one of the fundamental questions was how progeny viruses are made. The one-step growth experiment devised in 1939 by Max Delbrück (1906–1981) and Emory Ellis (1906–2003) served as an experimental approach to answering this question. Delbrück and Ellis worked with bacteriophage T4. They knew that T4 lysed its host, *Escherichia coli*, and released progeny phages. In their experiment, *E. coli* cells were mixed with T4. After a short interval, the mixture was greatly diluted so that any virions released upon host cell lysis would not encounter and infect other cells. The diluted culture was then incubated, and over time samples were removed to determine the number of infectious phages particles in the culture. This was determined using a plaque assay as we describe in section 6.5. A plot of the number of phages particles versus time shows several distinct periods in the resulting growth curve (**figure 6.9**, red line). The latent period occurs immediately following addition of the phages. During this period, no virions are released. The rise period follows and is characterized by the rapid release of infective phages. Finally, a plateau is reached and no more virions are produced.

The one-step growth experiment was important for several reasons. It employed procedures that are still used today to culture and enumerate viruses, and it ushered in the modern era of phages biology. It also led to another fundamental question: What is occurring during the latent period? In a subsequent set of experiments, infected cells were artificially lysed during the latent period. This revealed that intracellular virions could not be detected early in the latent period. In essence, the phages disappeared once inside the cell. This period is called the eclipse period because the infecting virions were concealed or eclipsed within the host cell. These experiments also demonstrated that the number of completed, infective phages within the host cell increases after the end of the eclipse period. Still other experiments eventually showed that a carefully orchestrated series of events occurs during the latent period. These events are the focus of this section.

If a virus is to multiply and give rise to new progeny viruses, it must find and use (and in many cases abuse) a host cell. To accomplish this, a virus must use guile and subterfuge to access an appropriate host, enter the host, and avoid any defenses the host might employ to rid itself of the virus or prevent its multiplication. Once inside a host cell, a virus uses a repertoire of clever tricks to take control of cellular functions, thereby ensuring that viral genomes, mRNAs, and proteins are synthesized using ATP generated by the host. There are a plethora of distinctive viral life cycles (also called viral replicative cycles). Viral strategies are often related to virion structure, in particular the nature of its genome. Thus viruses with a similar type of genome (e.g., dsDNA, ssRNA) often employ similar tactics, as we explore in chapter 26.

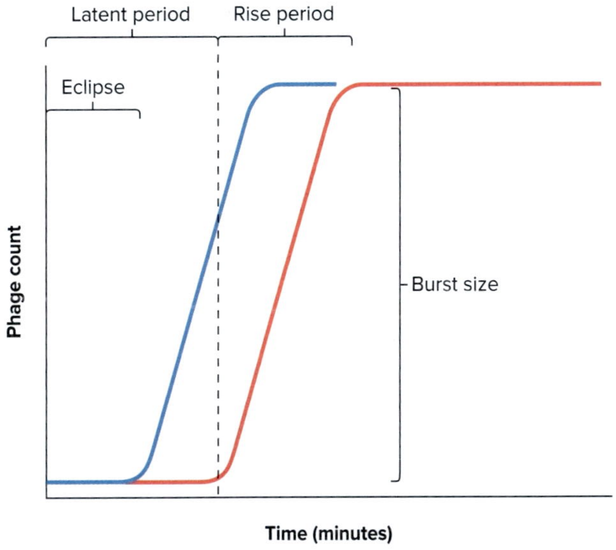

Figure 6.9 The One-Step Growth Curve. The eclipse period is the initial portion of the latent period in which host cells do not contain any complete, infective virions. During the remainder of the latent period, an increasing number of infective virions are present, but none are released. The latent period ends with host cell lysis and rapid release of virions (rise period). In this figure the blue line represents the total number of complete virions, both intracellular and free in the medium. The red line represents the number of free viruses (virions that were free in the culture and had not infected a host cell plus those virions that were released from infected cells). The burst size is the number of progeny viruses produced per infected cell.

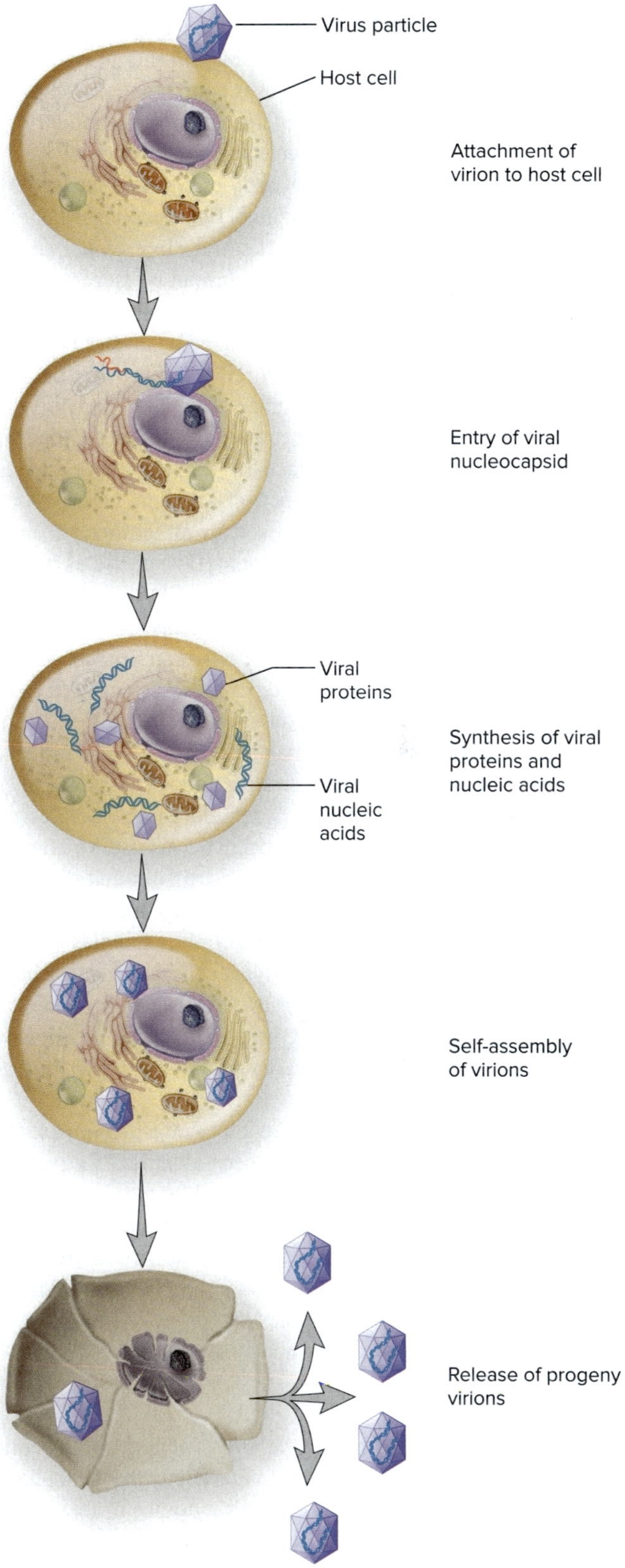

Figure 6.10 Generalized Illustration of Virus Multiplication. The life cycle of a virus that infects a eukaryotic cell is shown. Figure 6.16 illustrates the generalized life cycle of many bacterial and archaeal viruses. Virions are not drawn to scale.

Despite the diversity of viral life cycles, a general pattern of viral replicative cycles can be discerned; it can be divided into five steps. Because viruses need a host cell in which to multiply, the first step is usually attachment, often called adsorption (Latin *sorbere,* to suck in), to a host (**figure 6.10**). This is followed by entry of either the entire nucleocapsid or just the viral nucleic acid into the host. If the nucleocapsid enters, uncoating of the genome usually occurs before the life cycle continues. Once inside the host cell, the synthesis stage begins. During this stage, viral genes are transcribed and translated. This allows the virus to take control of the host cell, forcing it to manufacture viral genomes and viral proteins. Next is the assembly stage, during which new nucleocapsids are constructed by self-assembly of coat proteins and nucleic acids are packaged within. Finally, during the release step, mature virions escape the host.

Attachment (Adsorption)

All viruses, with the exception of plant viruses, must attach to a potential host cell long enough to gain entry into the cell. Attachment to the host is accomplished by specific interactions between molecules on the surface of the virion (ligands) and molecules on the surface of the host cell called **receptors.** In many cases, the viral surface ligand is a protein, but some animal viruses use lipids derived from the cell membrane as ligands. A recurring theme among archaeal viruses is the presence of virion appendages for host cell attachment. These include beardlike fiber clusters and structures resembling claws. Bacteriophages attach to cell wall lipopolysaccharides and proteins as receptors, while others use teichoic acids, flagella, or pili. Binding of an animal virus particle to its receptor often causes conformational changes in virion proteins that facilitate interaction with secondary receptors, entry into the host, and nucleocapsid uncoating.

Receptor specificity is at least partly responsible for the preferences viruses have for a particular host. Bacteriophages not only infect a particular bacterial species but often infect only certain strains within a given species. Likewise, animal viruses infect specific animals and, in some cases, only particular tissues within that host. However, if the receptor recognized by a virus is present in numerous animals, then the virus will infect more than one animal species. Such is the case with rabies viruses.

Viruses have evolved such that they use host receptors that are always present on the surface of the host cell and are important for normal host cell function. In some cases, two or more host cell receptors are involved in attachment. For instance, *h*uman *i*mmunodeficiency *v*irus (HIV) particles bind to two different proteins on human cells (e.g., CD4 and CCR5).

In complex animals like humans, the presence of receptor molecules on a cell surface varies with the organ and tissue. Viruses have a tropism; that is, particular cell types are infected based on the distribution of surface receptors in tissues. For example, poliovirus receptors are found only in the human nasopharynx, gut, and anterior horn cells of the spinal cord. Therefore polioviruses infect these tissues, causing disease that ranges from milder forms such as gastrointestinal disease to more serious paralytic disease. In contrast, measles virus receptors are present in most tissues and disease is disseminated throughout the body. Some viruses, including Ebola, mimic damaged cells by exposing certain phospholipids on their surface. This display is a hallmark of dying cells targeted for phagocytosis. By camouflaging themselves as damaged cells, viruses can gain entry to a wide variety of cell types. ⏮ *Eukaryotic cell envelopes (section 5.2)* ⏭ *Human diseases caused by viruses and prions (chapter 38)*

Plant viruses are a notable exception to attachment based on receptor binding as no receptors have been identified for plant viruses. Rather, damage of host cells is required for the virus particles to enter the host. Infection is often achieved by plant-eating insects that carry virions from one plant to another. The virions are deposited in plant tissues as the insect devours the plant. Interestingly, some viruses may alter their plant hosts to promote activity of the insects and thereby foster transmission to new plants.

Entry into the Host

After attachment to the host cell, the virus's genome or the entire nucleocapsid enters the cytoplasm. For many bacteriophages only their nucleic acid enters the host's cytoplasm, leaving the capsid outside and attached to the cell. In contrast to phages, many eukaryotic viral nucleocapsids enter the cytoplasm with the genome still enclosed. Once inside the cytoplasm, some shed their capsid proteins in a process called uncoating, whereas others remain encapsidated. Because penetration and uncoating are often coupled, we consider them together.

The mechanisms of penetration and uncoating vary with the type of virus, and for many animal viruses, detailed mechanisms of penetration are unclear. However, it appears that one of three different modes of entry is usually employed by animal viruses: fusion of the viral envelope with the host cell's plasma membrane, entry by endocytosis, and release of viral nucleic acid into the cytoplasm of the host cell (**figure 6.11**). *Entry of Animal Viruses into Host Cells*

Fusion of viral envelopes with the host cell's plasma membrane often involves viral envelope glycoproteins or phospholipids that interact with proteins in the plasma membrane of the host cell (figure 6.11*a*). This interaction sets into motion events that allow the nucleocapsid to enter. For example, after attachment of paramyxovirus virions (e.g., the measles virus), membrane lipids rearrange, the adjacent halves of the contacting membranes merge, and a proteinaceous fusion pore forms. The nucleocapsid then enters the host cell cytoplasm, where a viral enzyme carried within the nucleocapsid begins synthesizing viral mRNA while it is still within the capsid.

Virions of nonenveloped viruses and some enveloped viruses enter cells by one of the endocytic pathways, including clathrin-dependent endocytosis and macropinocytosis (figure 6.11*b*). The resulting endocytic vesicle contains the virion and fuses with an endosome; depending on the virus, escape of the nucleocapsid or its genome from the endocytic vesicle may occur either before or after fusion with an endosome. Endosomal enzymes can aid in virion uncoating, and low pH often triggers the uncoating process. For some enveloped viruses, the viral envelope fuses with the endosomal membrane, and the nucleocapsid is released into the cytosol (the capsid proteins may have been partially removed by endosomal enzymes). Once in the cytosol, the viral nucleic acid may be released from the capsid upon completion of uncoating or may function while still attached to capsid components. Nonenveloped animal viruses cannot employ the membrane fusion mechanism for release from the endosome (figure 6.11*c*). In this case, it is thought that the low pH of the endosome causes a conformational change in the capsid. The altered capsid contacts the endosome membrane and either releases the viral nucleic acid into the cytosol or ruptures the membrane to release the intact nucleocapsid. ⏮ *There are several endocytic pathways (section 5.4)*

Synthesis Stage

This stage of the viral life cycle differs dramatically among viruses because the genome of a virus dictates the events that occur. For dsDNA viruses, the synthesis stage can be very similar to the typical flow of information in cells. That is, the genetic information is stored in DNA and replicated by enzymes called DNA polymerases, recoded as mRNA (transcription), and decoded during protein synthesis (translation). Because of this similarity, some dsDNA viruses have the luxury of depending solely on their host cells' biosynthetic machinery to replicate their genomes and synthesize their proteins.

The same is not true for RNA viruses. Cellular organisms (except for plants) lack the enzymes needed to replicate RNA or to synthesize mRNA from an RNA genome. Therefore RNA viruses

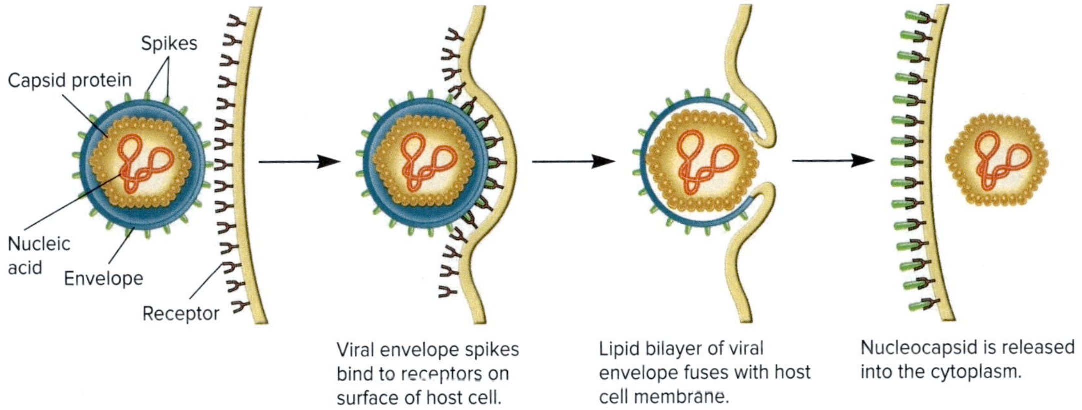

(a) Entry of enveloped virus by fusing with plasma membrane

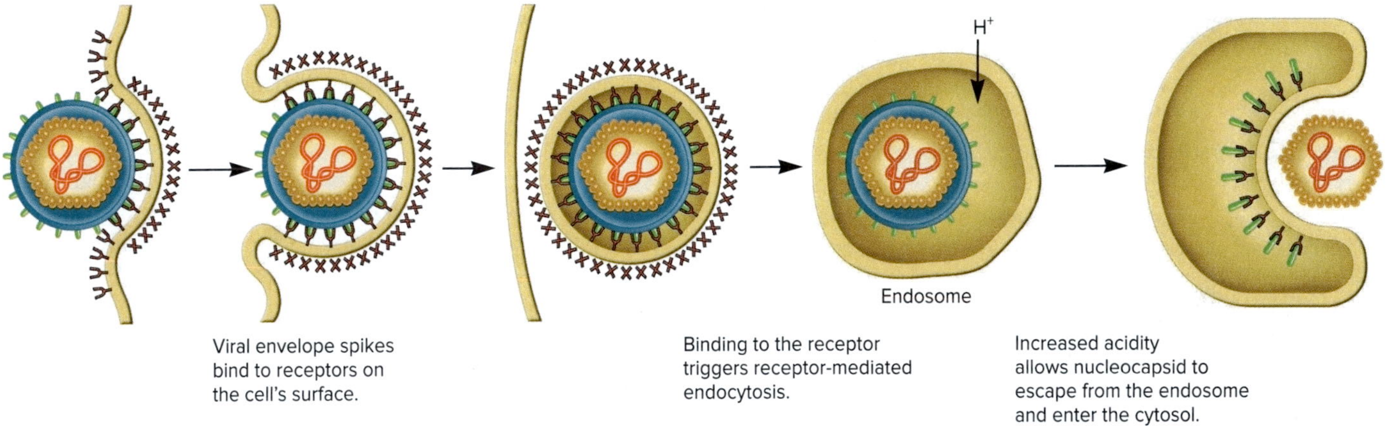

(b) Entry of enveloped virus by endocytosis

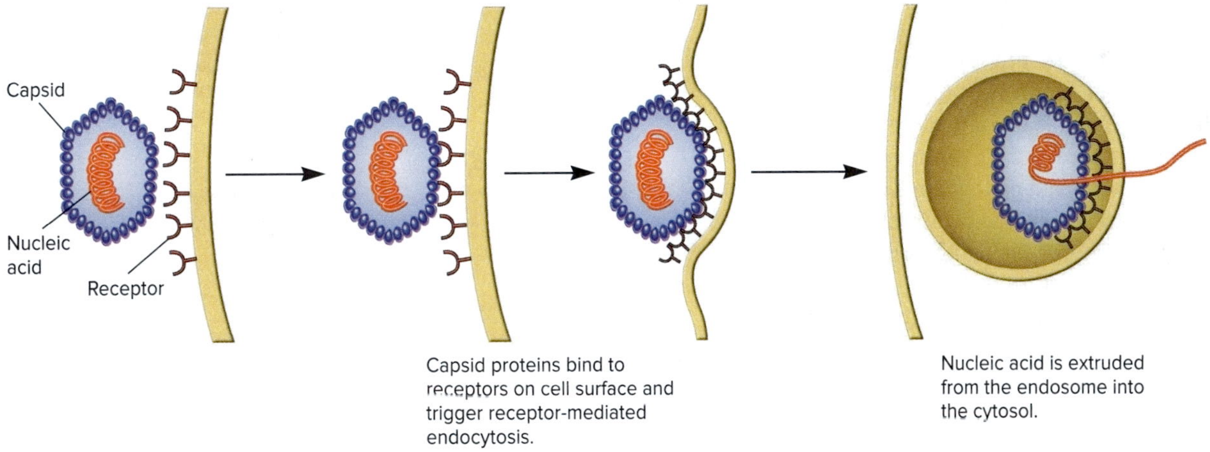

(c) Entry of nonenveloped virus by endocytosis

Figure 6.11 Animal Virus Entry. Examples of animal virus attachment and entry into host cells. Enveloped viruses can (a) enter after fusion of the envelope with the plasma membrane or (b) escape from an endosome after endocytosis. (c) Nonenveloped viruses may be taken up by endocytosis and then insert their nucleic acid into the cytosol through the vesicle membrane. It also is possible that they insert the nucleic acid directly through the plasma membrane.

MICRO INQUIRY *Which of these mechanisms involves the production of a vesicle within the host cell? Which involves the interaction between viral proteins and host cell membrane proteins?*

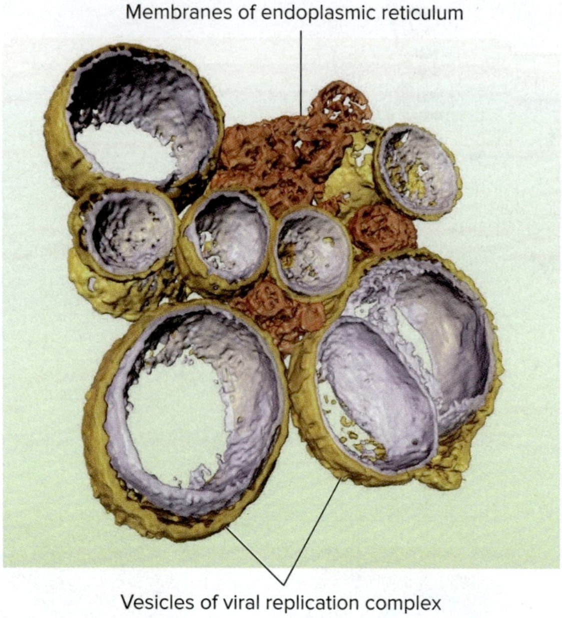

Figure 6.12 A Viral Replication Complex. An electron tomogram of cells infected with severe acute respiratory syndrome coronavirus (SARS-CoV). The vesicles of the viral replication complex have a double membrane derived from endoplasmic reticulum (ER) membranes. The membranes of vesicles are connected to the ER.

From "SARS-Coronavirus Replication Is Supported by a Reticulovesicular Network of Modified Endoplasmic Reticulum" Knoops K, Kikkert M, Worm SH, Zevenhoven-Dobbe JC, van der Meer Y, Koster AJ, Mommaas AM, Snijder EJ - (2008)

must carry in their nucleocapsids the enzymes needed to complete the synthesis stage, or the enzymes must be synthesized during the infection process. This is discussed in more detail in chapter 26.

Some animal and plant viruses carry out the synthesis stage and subsequent assembly step within the host's cytoplasm. To protect these processes from host defenses, some viruses bring about the reorganization of host cell membranes (e.g., membranes of the endoplasmic reticulum, Golgi apparatus, and lysosomes) to form membranous structures that enclose the machineries needed for genome replication, transcription, and protein synthesis (**figure 6.12**). The structures are called viral replication complexes, and they appear as vesicles, tubular structures, and other forms in electron micrographs of infected cells. Other viruses carry out synthesis and assembly in defined areas within the cytoplasm that are not enclosed by membranes. These areas of concentrated viral genomes, mRNAs, and proteins are called viroplasms, and they are also visible in electron micrographs of infected cells. Both viral replication complexes and viroplasms are sometimes referred to as virus factories.

One important feature of the synthesis stage is the tight regulation of gene expression and protein synthesis. Genes are often referred to as early, middle, or late genes based on when they are expressed. The proteins they encode are likewise referred to as early, middle, or late proteins. Many early proteins are involved in taking over the host cell. Middle proteins often participate in replication of the viral genome or activation of expression of late genes. Late proteins usually include capsid proteins and other proteins involved in self-assembly and release.

Assembly

Several kinds of late proteins are involved in the assembly of mature virions. Some are nucleocapsid proteins, some are not incorporated into the nucleocapsid but participate in its assembly, and still other late proteins are involved in virion release. In addition, proteins and other factors synthesized by the host may be involved in assembling mature virions.

The assembly process can be quite complex with multiple subassembly lines functioning independently and converging in later steps to complete nucleocapsid construction. As shown in **figure 6.13**, the baseplate, tail fibers, and head components of bacteriophage T4 are assembled separately. Once the baseplate is finished, the tail tube is built on it and the sheath is assembled around the tube. The phages prohead (procapsid) is constructed with the aid of scaffolding proteins that are degraded or removed after assembly is completed. DNA is incorporated into the prohead by a complex of proteins sometimes called the "packasome." The packasome consists of a protein called the portal protein, which is located at the base of the prohead, and an enzyme called terminase, which moves DNA into the prohead. The movement of DNA consumes energy in the form of ATP, which is supplied by the metabolic activity of the host bacterium. After the head is completed, it spontaneously combines with the tail assembly.

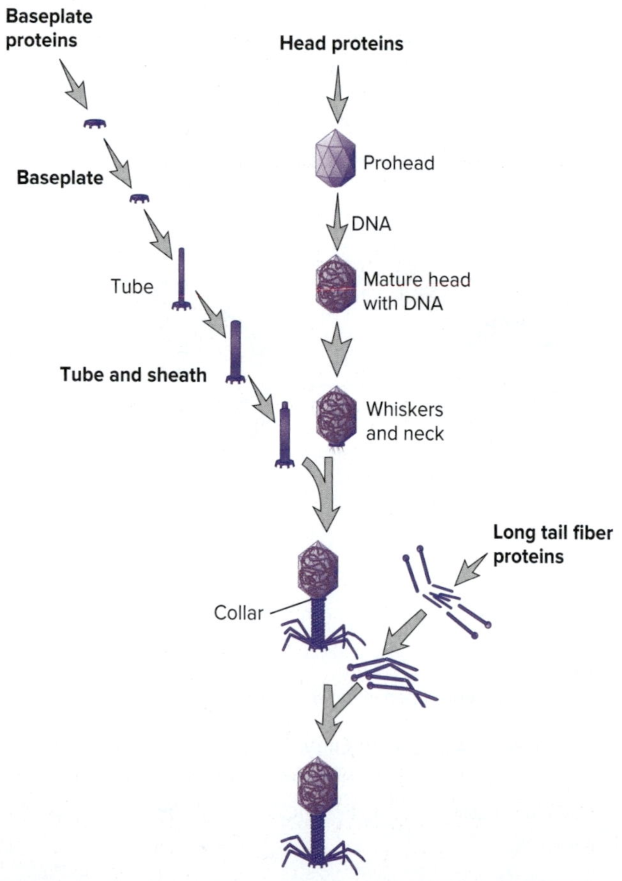

Figure 6.13 The Assembly of T4 Virions. Note the subassembly lines for the baseplate, tail tube and sheath, long tail fibers, and head.

Virion Release

Several release mechanisms have been identified. The two most common are release by lysing the host cell and release by budding. Release by lysis is especially common for bacterial viruses and some nonenveloped animal viruses. This process involves the activity of viral proteins. For instance, lysis of *E. coli* by T4 requires two specific proteins (**figure 6.14*a***). One is lysozyme, an enzyme that attacks peptidoglycan in the host's cell wall. The other, called holin, creates holes in *E. coli*'s plasma membrane, enabling T4 lysozyme to move from the cytoplasm to the peptidoglycan.

A novel release mechanism has been elucidated for some archaeal viruses infecting *Sulfolobus* spp. The virions are assembled in the cell cytoplasm, and a seven-sided pyramid structure forms near the cell envelope. These virus-associated pyramids puncture the envelope and open up much like the petals of a flower (figure 6.14*b*). This allows escape of the progeny virions, leaving the empty cell ghosts behind.

Budding is frequently observed for enveloped viruses; in fact, envelope formation and virion release are usually concurrent processes. When virions are released by budding, the host cell may survive and continue releasing virions for some time. All envelopes of animal viruses are derived from host cell membranes by a multistep process. First, virus-encoded proteins are incorporated into the membrane. Then the nucleocapsid is simultaneously released and the envelope formed by membrane budding (**figure 6.15**). In several virus families, a matrix (M) protein attaches to the plasma membrane and aids in budding. Most envelopes arise from the plasma membrane. The endoplasmic reticulum, Golgi apparatus, and other internal membranes also can be used to form envelopes. *Mechanism for Releasing Enveloped Virions*

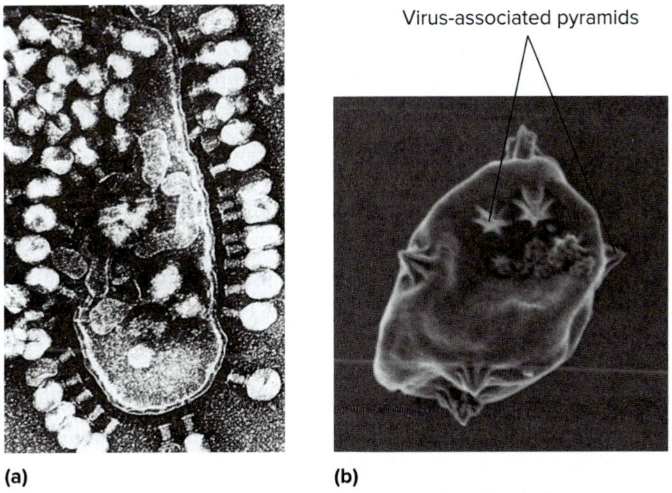

Figure 6.14 Viral Release Mechanisms. (a) Release of T4 virions by lysis of the host cell. The host cell has been lysed (upper left portion of the cell) and virions have been released. Progeny virions also can be seen in the cytoplasm. Empty capsids coat the outside of the cell. (b) *Sulfolobus solfataricus* infected with *Sulfolobus* turreted icosahedral virus (STIV). Virions are released through pyramid structures that form on the cell surface.
(*a*) ©Lee D. Simon/Science Source; (*b*) Courtesy of Susan Brumfield and Mark Young

MICRO INQUIRY *Why do the empty capsids remain attached to the cell after the viral genome enters the host cell?*

Budding has been observed as the archaeal SSV1 enveloped virus exits its host, *Sulfolobus shibatae*. Recall the structure of an archaeal cell envelope: a monolayer membrane with ether-linked lipids and a proteinaceous S-layer. Mature SSV1 particles

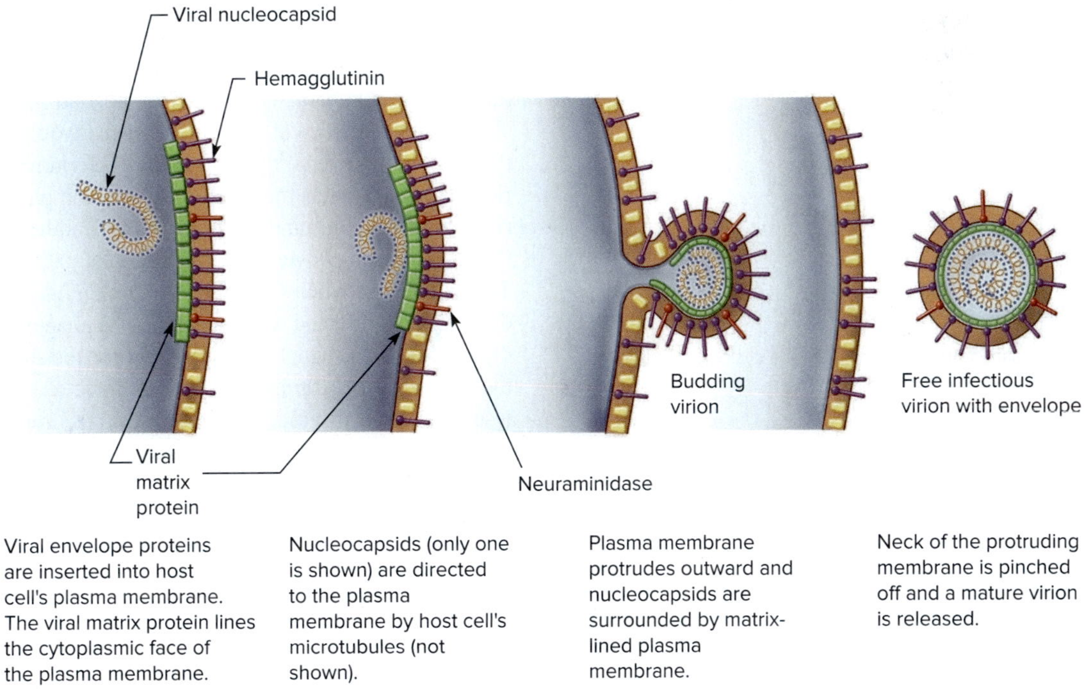

Figure 6.15 Release of Influenza Virus Virions by Budding. For simplicity, only one of the seven to eight possible nucleocapsids is shown.

covered in a host-derived membrane are initially trapped inside the S-layer but ultimately freed by structural alterations in the S-layer. Studies on the viral path through the cell envelope may reveal novel mechanisms for viral budding. ◀ *There are many different types of archaeal cell walls (section 4.2)*

Interestingly, some viruses are not released from their host cell into the surrounding environment. Rather, their virions move from one host cell directly to another host cell. Most fungal viruses lack an extracellular phase in their replicative cycles. Instead they are transmitted by cell division, spore formation, or during mating. Vaccinia viruses elicit the formation of long actin tails that propel nucleocapsids through the plasma membrane, directly into an adjacent cell. In this way, the virus avoids detection by the host immune system. The genomes or nucleocapsids of many plant viruses also move directly from cell to cell through small connections called plasmodesmata that link adjacent cells. This spread of the virus typically involves virus-encoded movement proteins. ◀ *The eukaryotic cytoplasm contains a cytoskeleton and organelles (section 5.3)*

Comprehension Check

1. Explain why the receptors that viruses have evolved to use are host surface proteins that serve very important, and sometimes essential, functions for the host cell.
2. What probably plays the most important role in determining the tissue and host specificity of viruses? Give some specific examples.
3. How do you think the complexity of the viral assembly process correlates with viral genome size?
4. In general, DNA viruses can be much more dependent on their host cells than can RNA viruses. Why is this so?
5. Consider the origin of viral envelopes and suggest why enveloped viruses that infect plants and bacteria are rare.
6. Why are the proteins involved in virion assembly synthesized late in the viral life cycle?

6.4 There Are Several Types of Viral Infections

After reading this section, you should be able to:

a. Compare and contrast the major steps of the life cycles of virulent phages and temperate phages
b. List examples of lysogenic conversion
c. Differentiate among the types of viral infections of eukaryotic cells
d. Summarize the current understanding of how oncoviruses cause cancer

So far, our discussion has focused on viral structures and modes of multiplication with little mention of the cost to host cells. The dependence of viruses on their host cells has many consequences—as anyone who has ever had a cold or the flu well understands. Next we discuss the interaction between virus and host cell.

Lytic and Lysogenic Infections Are Common for Bacterial and Archaeal Cells

Most bacteriophages are either virulent or temperate. A **virulent phages is** one that has only one option: to begin multiplying immediately upon entering its bacterial host, followed by release from the host by lysis. T4 is an example of a virulent phages. **Temperate phages** have two options: upon entry into the host, they can multiply like virulent phages and lyse the host cell, or they can remain within the host without destroying it (**figure 6.16**). Bacteriophage lambda is an example of this type of phages. The life cycles of both T4 and lambda are described in more detail in chapter 26. *Steps in the Replication of T4 Phages in* E. coli; *Lambda Phages Replication Cycle*

The relationship between a temperate phages and its host is called **lysogeny.** The form of the virus that remains within its host is called a **prophage.** A prophage is simply the viral genome either integrated into the bacterial chromosome or free in the cytoplasm. The infected bacteria are called **lysogens** or **lysogenic bacteria.** Lysogenic bacteria reproduce and in most other ways appear to be perfectly normal. However, they have two distinct characteristics. The first is that they cannot be reinfected by the same virus; that is, they have immunity to superinfection. The second is that as they reproduce, the prophage is replicated and inherited by progeny cells. This can continue for many generations until conditions arise that cause the prophage to initiate synthesis of phages proteins and to assemble new virions, a process called **induction.** Induction is commonly caused by changes in growth conditions or ultraviolet irradiation of the host cell. As a result of induction, the **lysogenic cycle** ends and the **lytic cycle** commences; the host cell lyses and progeny phages particles are released.

Another important outcome of lysogeny is **lysogenic conversion.** This occurs when a temperate phages changes the phenotype of its host. Lysogenic conversion often involves alteration in surface characteristics of the host. For example, when a member of the genus *Salmonella* is infected by epsilon phages, the phages changes the activities of several enzymes involved in construction of the carbohydrate component of the bacterium's lipopolysaccharide. This eliminates the receptor for epsilon phage, so the bacterium becomes immune to infection by another epsilon phages. Other lysogenic conversions give the host pathogenic properties. This is the case when *Corynebacterium diphtheriae,* the cause of diphtheria, is infected with phages β. The phages genome encodes diphtheria toxin, which is responsible for the disease. Thus only those strains of *C. diphtheriae* that are infected by the phages (i.e., lysogens) cause disease. ▶ *Diphtheria (section 39.1)*

Clearly the infection of a bacterium by a temperate phages has significant impact on the host, but why would viruses evolve this alternate cycle? Two advantages of lysogeny have been recognized. The first is that lysogeny allows the viral nucleic acid to be maintained within its host. Bacteria often become dormant due to nutrient deprivation, and while in this state, they do not synthesize nucleic acids or proteins. In such situations, a prophage would survive but most virulent bacteriophages would not be replicated, as they require active cellular biosynthetic machinery.

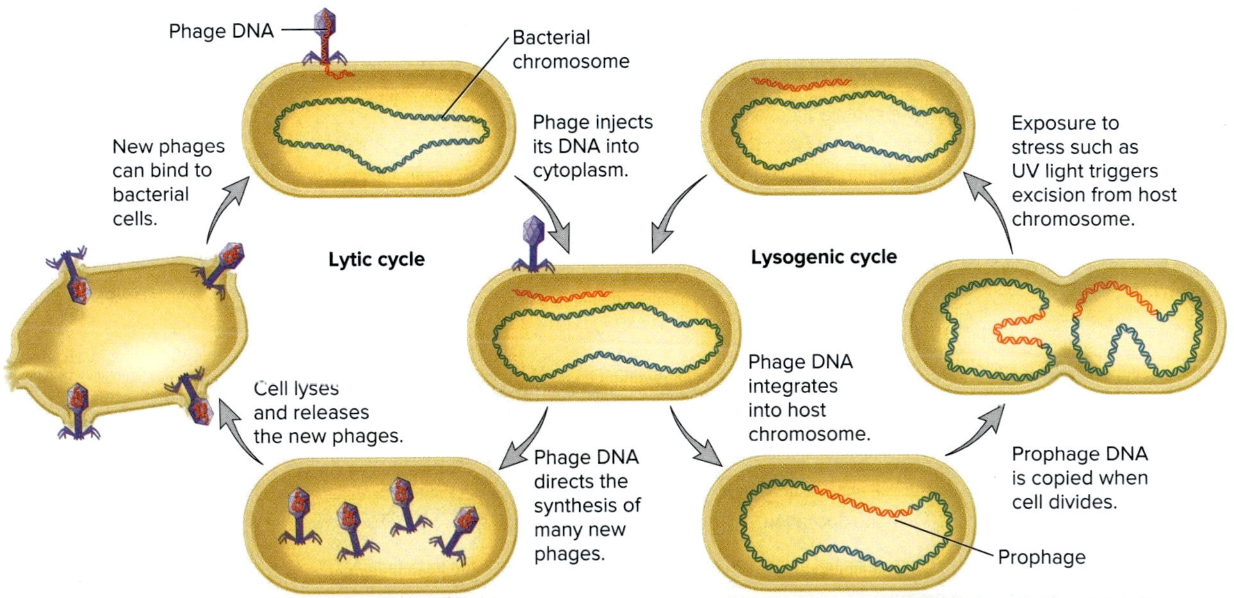

Figure 6.16 Lytic and Lysogenic Cycles of Temperate Phages. Temperate phages have two phases to their life cycles. The lysogenic cycle allows the genome of the virus to be replicated passively as the host cell's genome is replicated. Certain environmental factors such as UV light can cause a switch from the lysogenic cycle to the lytic cycle. In the lytic cycle, new virus particles are made and released when the host cell lyses. Virulent phages are limited to just the lytic cycle.

MICRO INQUIRY *Why is a lysogen considered a new or different strain of a given bacterial species?*

Furthermore, their genome would be degraded as the host cell entered dormancy. The second advantage arises when there are many more phages in an environment than there are host cells, a situation virologists refer to as a high multiplicity of infection (MOI). In these conditions, lysogeny enables the survival of infected host cells within a population that has few uninfected cells. When MOI is high, a virulent phages would rapidly destroy the available host cells in its environment. However, a prophage will be replicated as the host cell reproduces.

Archaeal viruses can also be virulent or temperate. In addition, many archaeal viruses establish chronic infections. Unfortunately, little is known about the mechanisms they use to regulate their replicative cycles.

Comprehension Check

1. Define the terms *lysogeny, temperate phages, lysogen, prophage, immunity,* and *induction.*
2. What advantages might a phages gain by being capable of lysogeny?
3. Describe lysogenic conversion and its significance.

Infections of Eukaryotic Cells

Viruses can harm their eukaryotic host cells in many ways. An infection that results in cell death is a cytocidal infection. As with bacterial and archaeal viruses, this can occur by lysis of the host (**figure 6.17*a***). Infection does not always result in lysis of host cells. Some viruses (e.g., herpesviruses) can establish persistent infections lasting many years (figure 6.17*b,c*). Eukaryotic viruses can cause microscopic or macroscopic degenerative changes or abnormalities in host cells and in tissues that are distinct from lysis. These are called **cytopathic effects** (**CPEs**). Viruses use a variety of mechanisms to cause cytopathic and cytocidal effects. Some important examples are noted in chapter 38. One mechanism of particular note is that some viruses cause the host cell to be transformed into a malignant cell (figure 6.17*d*). This is discussed next.

Viruses and Cancer

Cancer is one of the most serious medical problems in developed nations, and it is the focus of an immense amount of research. A tumor is a growth or lump of tissue resulting from **neoplasia**—unregulated abnormal new cell growth and reproduction. Tumor cells have aberrant shapes and altered plasma membranes that may contain distinctive molecules (tumor antigens). These changes result from the tumor cells becoming less differentiated. Their unregulated proliferation and loss of differentiation result in invasive growth that forms unorganized cell masses.

Two major types of tumor growth patterns exist. If the tumor cells remain in place to form a compact mass, the tumor is benign. In contrast, cells from malignant or cancerous tumors actively spread throughout the body in a process known as metastasis. Some cancers are not solid but cell suspensions. For example, leukemias are composed of undifferentiated malignant white blood cells that circulate throughout the body. Indeed, dozens of kinds of cancers arise from a variety of cell types and afflict all kinds of organisms.

Some viruses have been shown to cause cancer in animals, including humans; it is estimated that about 10 to 20% of human

Figure 6.17 Types of Viral Infections and Their Effects on Animal Cells. (a) Lytic infections of host cells can lead to disease states in animal hosts called acute infections. There are two types of persistent infections: (b) latent infections and (c) chronic infections. (d) Some infections of animal cells cause the cell to be transformed into a malignant cell, which can cause cancer in the animal host.

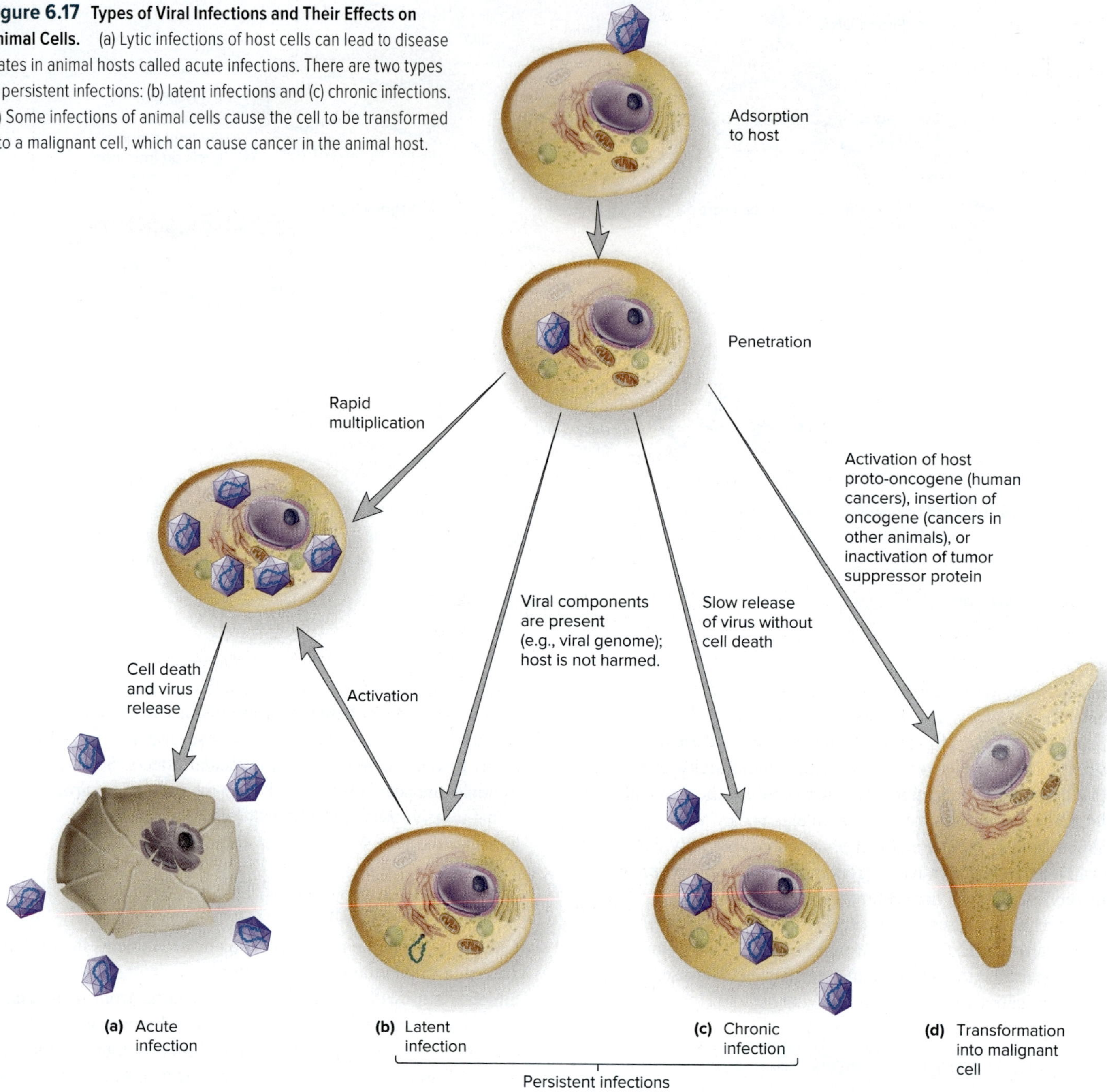

cancers have a viral etiology. To understand the role viruses play in cancer, we must begin by considering carcinogenesis when viruses are not involved. Carcinogenesis is a complex, multistep process caused by mutations in multiple genes. Some mutations lead to the unregulated proliferation that is a major characteristic of a cancer cell. Other mutations prevent cells from repairing DNA damage, thereby making the cells more inclined to mutate further. This leads to additional genotypic changes. For example, mutations that promote metastasis may occur, allowing the tumor to invade other tissues. ▶| *Mutations: heritable changes in a genome (section 16.1)*

Considerable research into the causes of cancer has focused on the mutations that allow cancerous cells to grow uncontrollably. Those studies show that there are two types of genes that must be mutated: proto-oncogenes and tumor suppressor genes. Their names suggest that these genes are somehow abnormal. But this is not the case. Proto-oncogenes and tumor suppressor genes are normal cellular genes. **Proto-oncogenes** must be expressed if cell division is to occur. However, they normally are expressed only if the cell receives an appropriate signal, such as the binding of a growth factor to a receptor on the cell surface. When such a signal is received, the cell initiates the cell cycle, but it cannot continue past a checkpoint controlled by the activity of proteins encoded by **tumor suppressor genes.** These proteins are called **tumor suppressor proteins.** When tumor suppressor proteins are active, they prevent progression through the cell cycle. Thus, for cell division to occur, proto-oncogene proteins (sometimes called proto-oncoproteins) must be active, and tumor suppressor proteins must be inactive.

During carcinogenesis, mutations arise that disrupt the normal activity of these two types of genes. The first type of mutation allows proto-onocogenes or their protein products to function without an appropriate signal. Thus the proto-oncogene or its protein is active even when it should not be. The second type of mutation prevents expression of tumor suppressor genes, yields nonfunctional tumor suppressor proteins, or in some other way leads to inactivation of a tumor suppressor protein. Thus the cell can continue cell division even when it shouldn't, such as in the presence of severe DNA damage. Numerous proto-oncogenes and tumor suppressor genes have been identified. An important proto-oncogene encodes a protein called Myc. Myc controls transcription of many genes, including some involved in DNA replication and other cell-cycle-related functions. Two of the most important tumor suppressor proteins are Rb and p53. The protein p53 is often referred to as "the guardian of the genome" because not only does it cause cell cycle arrest, but it also initiates programmed cell death in response to DNA damage. Programmed cell death, also called apoptosis, is a process that causes cells to kill themselves. Thus, when p53 is inactivated, the cell reproduces in an unregulated fashion, avoiding programmed cell death even when it should destroy itself. This is a catastrophic event for the cell because it can lead to the accumulation of additional mutations that contribute to carcinogenesis. *Tumor suppressor genes*

We now turn our attention to viruses and their role in causing cancer. These viruses are called **oncoviruses.** As you can see in **table 6.1**, most human oncoviruses have dsDNA genomes, and most of these viruses cause cancer in a similar way—by interacting with and inactivating either Rb or p53. Oncoviruses that are retroviruses (e.g., HTLV-1) exert their oncogenic powers in a different manner. Some carry genes called **oncogenes** that stimulate the activity of cellular proto-oncogenes. For example, HTLV-1 infects immune system cells called T cells. During infection HTLV-1 produces a regulatory protein that activates expression of numerous cellular proto-oncogenes. Other retroviruses transform cells into cancerous cells when the viral genome integrates into the host chromosome such that strong, viral regulatory elements are near a cellular proto-oncogene. These elements cause the nearby proto-oncogene to be transcribed at a high level, causing the cellular gene to be considered an oncogene.

Comprehension Check

1. How does a latent infection differ from a chronic infection?
2. What is a cytocidal infection? What is a cytopathic effect?
3. Define the following terms: *tumor, neoplasia, anaplasia, metastasis, proto-oncogene, oncogene,* and *tumor suppressor gene.*
4. Distinguish the mechanism by which dsDNA viruses cause cancer from that of retroviruses. Why do you think there is an environmental component to many kinds of cancer?

6.5 Cultivation and Enumeration of Viruses

After reading this section, you should be able to:

a. List the approaches used to cultivate viruses, noting which types of viruses are cultivated by each method
b. Describe three direct counting methods and two indirect counting methods used to enumerate viruses
c. Outline the events that lead to formation of a plaque in a lawn of bacterial cells
d. Distinguish lethal dose from infectious dose

Because they are unable to reproduce outside of living cells, viruses cannot be cultured in the same way as cellular microorganisms. It is relatively simple to culture lytic bacterial and archaeal viruses as long as the host cell is easily grown in culture. Cells and viruses are simply mixed together in a broth culture. Over time, more and more cells are infected and lysed, releasing virions into the broth. Culturing temperate viruses requires an additional step of inducing the lytic phase of their life cycles. When cultured with cells on a solid medium, lysis of the host cells is observed as **plaques** on the bacterial lawn (**figure 6.18**)

Table 6.1 Some Viruses Associated with Human Cancers

Virus	Genome Type	Cancer
Human herpesvirus 8 (HHV8)	Double-stranded (ds) DNA	Several, including Kaposi's sarcoma
Epstein-Barr virus (EBV)	dsDNA	Several, including Burkitt's lymphoma and nasopharyngeal carcinoma
Hepatitis B virus	dsDNA	Liver cancer
Hepatitis C virus	Single-stranded (ss) RNA	Liver cancer
Human papillomaviruses (HPV) strains 6, 11, 16, and 18	dsDNA	Cervical cancer and oral pharyngeal cancer
Human T-cell lymphotropic virus 1 (HTLV-1)	ssRNA (retrovirus)	T-cell leukemia

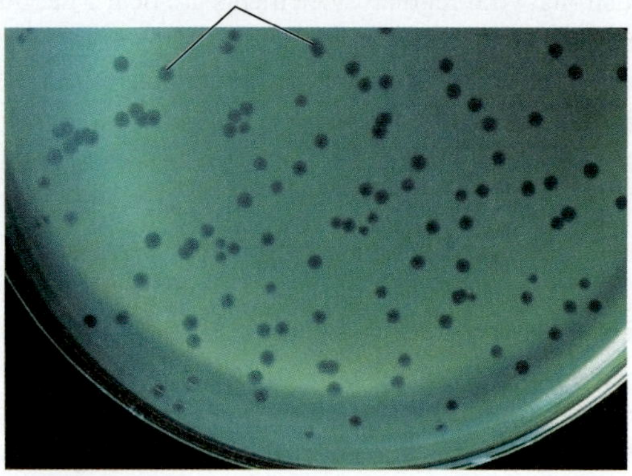

Figure 6.18 Viral Plaques. The plaques are formed in the bacterial cell lawn by the process illustrated in figure 6.20.
©Carolina Biological Supply Company/DIOMEDIA

Animal viruses are often cultivated by inoculating suitable host animals or embryonated eggs—fertilized chicken eggs incubated about 6 to 8 days after laying. Animal viruses also are grown in tissue (cell) culture on monolayers of animal cells. If an animal virus causes its host cell to lyse, plaques often are formed in the monolayer of cells. Viruses that cause cytopathic effects are also grown and detected in tissue culture.

Plant viruses are cultivated in a variety of ways. Plant tissue cultures, cultures of separated cells, or cultures of protoplasts (cells lacking cell walls) may be used. Viruses also can be grown in whole plants, usually after the leaves are mechanically inoculated by rubbing them with a mixture of viruses and an abrasive. When the cell walls are broken by the abrasive, the virions directly contact the plasma membrane and infect the exposed host cells. Some plant viruses can be transmitted only if a diseased part is grafted onto a healthy plant. A localized necrotic lesion often develops in infected plants due to the rapid death of cells in the infected area. Even when lesions do not occur, the infected plant may show symptoms such as changes in pigmentation or leaf shape (**figure 6.19**).

Figure 6.19 Necrotic Lesions on Plant Leaves. A tobacco plant leaf showing leaf color changes caused by TMV.
©Norm Thomas/SPL/Science Source

It is often necessary to determine the number of virions in a preparation of viruses or an environmental sample. One direct approach is to count virions using an electron microscope. However, this is a time-consuming process that requires a high concentration of virus particles. Another direct counting method uses epifluorescence microscopy. This approach is often used to enumerate viruses in aquatic habitats. The procedure stains the virions with fluorescent dyes prior to microscopic examination. Finally, sometimes rather than counting virions, the nucleic acid of a virus can be quantified by quantitative-polymerase chain reaction (qPCR). This is possible because PCR amplifies specific nucleic acids in a mixture of nucleic acids. ▶| *Marine viruses: mortality at sea (section 30.2); Polymerase chain reaction amplifies targeted DNA (section 17.2)*

An indirect method of counting animal viruses is the **hemagglutination assay.** Many animal virus particles bind to the surface of red blood cells (*see figure 37.8*). If the ratio of virions to cells is large enough, virions will join the red blood cells together; that is, they agglutinate, forming a network that keeps the red blood cells in suspension. In practice, red blood cells are mixed with diluted samples of the virus particles, and each mixture is examined. The hemagglutination titer is the highest dilution of the virus preparation (or the reciprocal of the dilution) that still causes hemagglutination. ▶| *Agglutination (section 37.3)*

Other indirect assays determine virion numbers based on their infectivity, and many of these are based on the same techniques used for virus cultivation. For example, in the **plaque assay,** several dilutions of a sample containing virus particles are plated with appropriate host cells. Theoretically, when the multiplicity of infection is very low, then each plaque in a layer of host cells should have arisen from the multiplication of a single virion (**figure 6.20**). Therefore a count of the plaques produced at a particular dilution can be used to estimate the number of virions in the original sample. The resulting value is expressed as **plaque-forming units (PFU)** rather than as virions, for several reasons. First, not all virions may be infective. Furthermore, even though there are far fewer viruses than host cells, it is still possible for more than one virion to infect the same cell (figure 6.14). Finally, it is possible for two infected cells to be plated to the same area, which would then give rise to a single plaque. Despite these caveats, the number of PFU is proportional to the number of viruses: a preparation with twice as many virions will have twice the plaque-forming units.

The same approach employed in the plaque assay may be used with embryos and plants. Chicken embryos and plant leaves can be inoculated with a diluted preparation of viruses. The number of pocks on embryonic membranes or necrotic lesions on leaves is used to calculate the concentration of infectious units.

When biological effects are not readily quantified by plaque and hemagglutination assays, the amount of virions required to cause disease or death can be determined. Organisms or cell cultures are inoculated with serial dilutions of a preparation of virus particles. The **lethal dose** (**LD_{50}**) is the dilution that contains a concentration (dose) of virions large enough to destroy 50% of the host cells or organisms. In a similar sense, the **infectious dose** (**ID_{50}**) is the dose that causes 50% of the host organisms to become infected (**figure 6.21**).

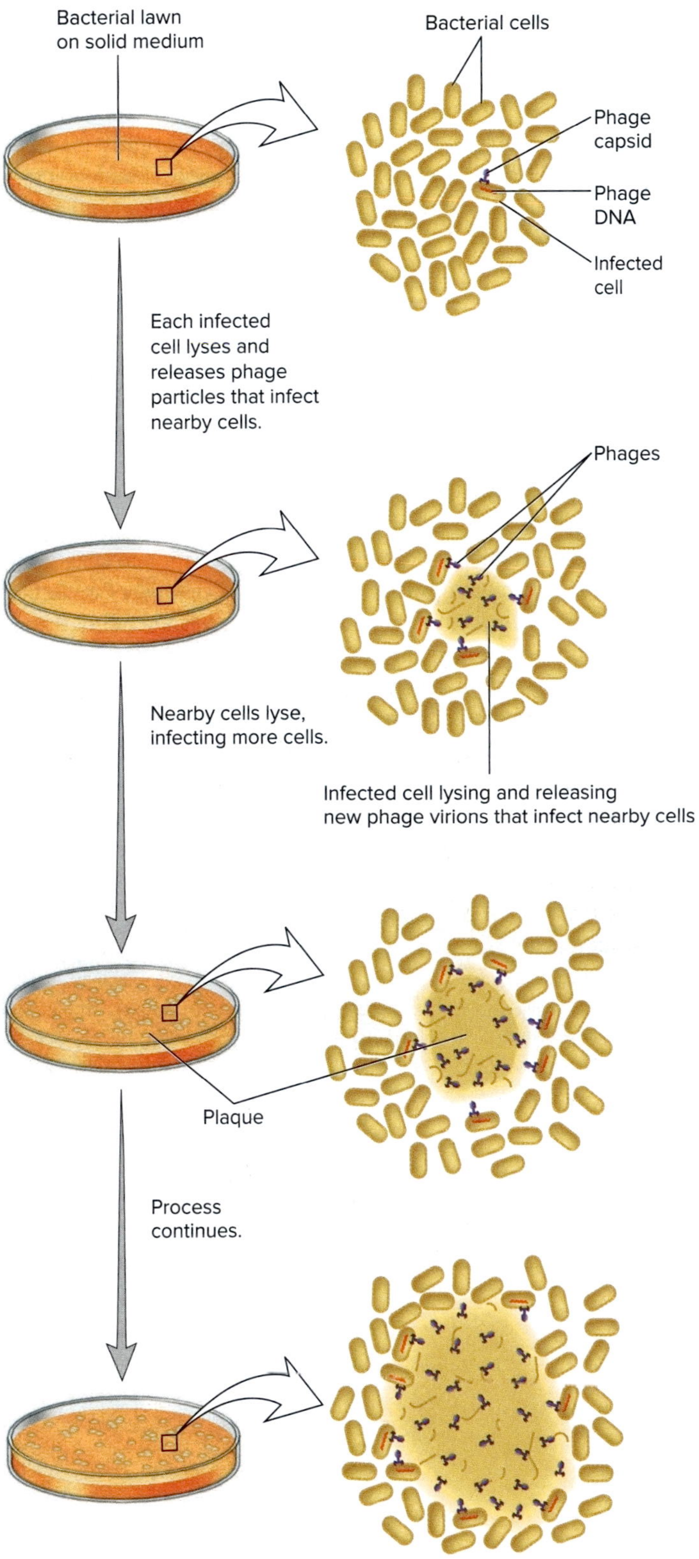

Figure 6.20 Formation of Phages Plaques. When phages and host bacterial cells are mixed at an appropriate ratio, only a portion of the cells are initially infected. When this mixture is plated, the infected cells will be separated from each other. The infected cells eventually lyse, releasing progeny phages. They infect nearby cells, which eventually lyse, releasing more phages. This continues and ultimately gives rise to a clear area within a lawn of bacteria. The clear area is a plaque.

MICRO INQUIRY *What would happen to a plate with individual plaques if it were returned to an incubator so that the viral infection could continue?*

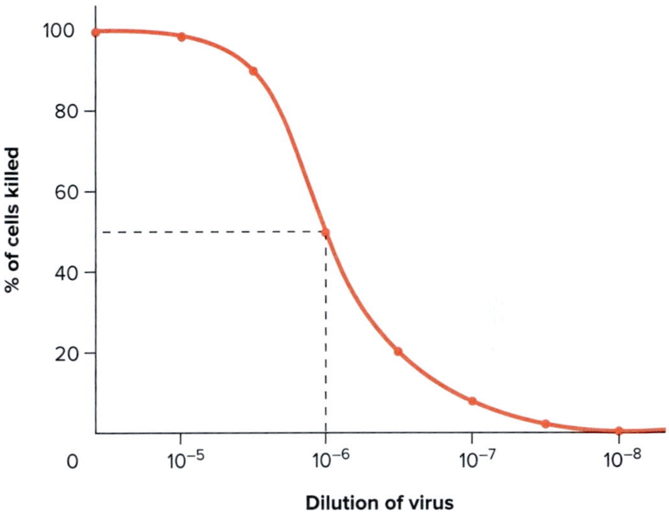

Figure 6.21 Graphical Determination of LD_{50}. The dashed line drawn from the y-axis at 50% to the curve identifies the LD_{50} of the virus.

Comprehension Check

1. Discuss the ways that viruses can be cultivated. Define the terms *plaque, cytopathic effect,* and *necrotic lesion.*
2. Given that viruses must be cultivated to make vaccines against viral diseases, discuss the advantages and disadvantages that might occur with each approach (embryonated egg versus cell culture) to growing animal viruses.

6.6 Viroids and Satellites: Nucleic Acid-Based Subviral Agents

After reading this section, you should be able to:

a. Describe the structure of a viroid and discuss the practical importance of viroids
b. Distinguish satellite viruses from satellite nucleic acids

Although viruses are exceedingly simple, even simpler infectious agents exist. **Viroids** are infectious agents that consist

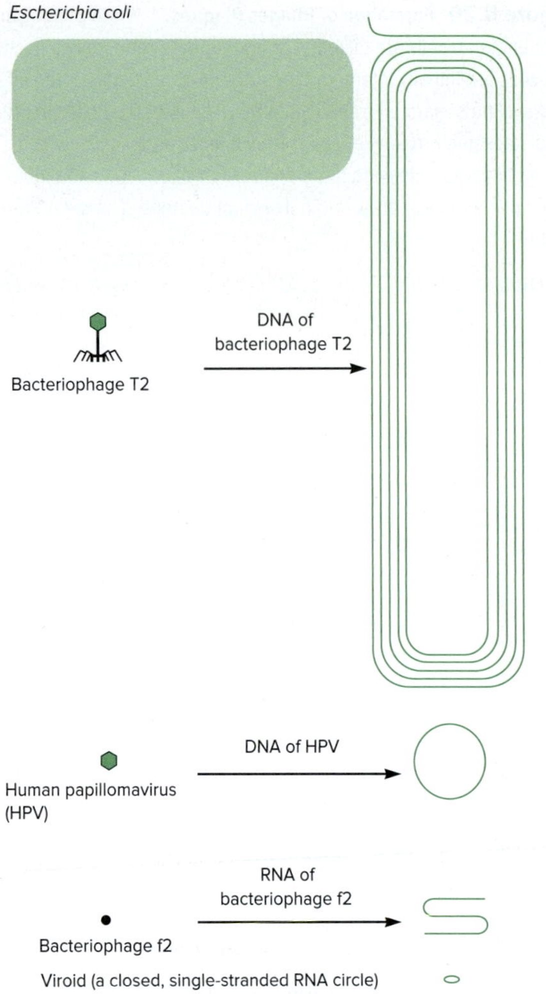

Figure 6.22 Viroids, Viruses, and Bacteria. A comparison of *E. coli*, several viruses, and the potato spindle-tuber viroid with respect to size and the amount of nucleic acid possessed. (All dimensions are enlarged approximately ×40,000.)

only of RNA. They cause over 20 different plant diseases, including potato spindle-tuber disease, exocortis disease of citrus trees, and chrysanthemum stunt disease. Viroids are covalently closed, circular ssRNAs, about 250 to 400 nucleotides long and they do not encode any proteins (**figure 6.22**). Viroids are currently divided into two families. Viroids in family *Pospiviroidae* have circular RNA that exists as a rodlike shape due to intrastrand base pairing, which forms double-stranded regions with single-stranded loops (**figure 6.23*a***). The circular RNA of viroids in the family *Avsunviroidae* is shaped like a rod with a highly branched structure at one end, rather like a tree trunk with its roots. Each branch is formed by intramolecular base pairing of the RNA that creates a stem-loop structure. The two types of viroids replicate in different locations within the infected plant cell. Pospiviroids replicate in the nucleus, whereas avsunviroids replicate in plant plastids such as chloroplasts. However, they replicate in similar ways. Because the RNA of viroids does not encode any proteins, they cannot replicate themselves. Rather, they use a host cell enzyme called a DNA-dependent RNA polymerase. This enzyme normally functions in the host to synthesize RNA using DNA as the template during transcription. However, when infected by a viroid, the host polymerase is able to use the viroid RNA as a template for RNA synthesis. The host polymerase synthesizes a complementary RNA molecule, which then serves as the template for synthesis of new viroid RNAs.

A plant may be infected with a viroid without showing symptoms; that is, it may have a latent infection. However, the same viroid in another host species may cause severe disease. The pathogenicity of viroids is not well understood, but it is known that particular regions of the RNA are required; studies have shown that removing these regions blocks the development of disease (figure 6.23*a*). Some data suggest that viroids cause disease by triggering a eukaryotic response called **RNA silencing,** which normally functions to protect against infection by RNA viruses. During RNA silencing, the cell detects the presence of dsRNA and cuts it into small fragments. These are used by the RNA silencing machinery to destroy target mRNA molecules or prevent their translation. Viroids may usurp this response by hybridizing to specific host mRNA molecules to which they have a complementary nucleotide sequence. Formation of the viroid-host hybrid dsRNA molecule is thought to elicit RNA silencing. This results in destruction of host mRNA and therefore silencing of the host genes. Failure to express required host genes leads to disease in the host plant.

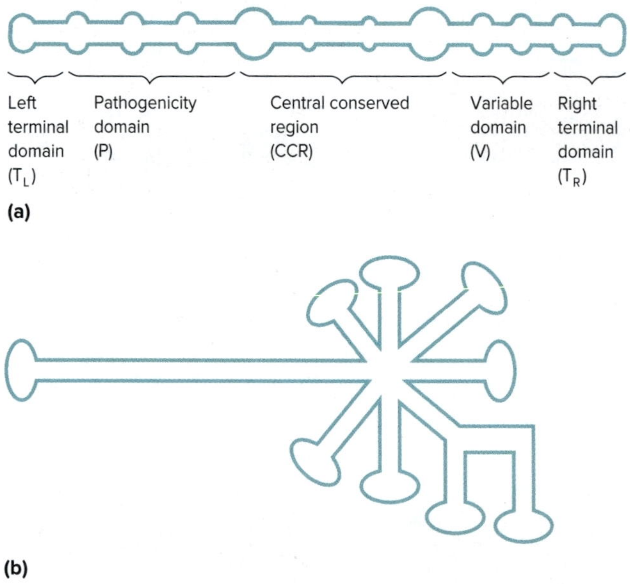

Figure 6.23 Viroid Structure. (a) This schematic diagram shows the general organization of a viroid belonging to the family *Pospiviroidae*. The closed single-stranded RNA circle has extensive intrastrand base pairing and interspersed unpaired loops. Also shown are the five domains identified in the molecule. Most changes in viroid pathogenicity arise from variations in the P and T_L domains. (b) Schematic diagram of a viroid belonging to *Avsunviroidae*. These viroids lack the central conserved region observed in pospiviroids.

Satellites are similar to viroids in that they also consist of a nucleic acid (either DNA or RNA). They differ from viroids in that their nucleic acid gets enclosed in a capsid and they need a helper virus to replicate. There is little or no homology between the nucleic acid of the satellite and that of its helper virus (i.e., a satellite is not a defective version of its helper virus). Satellites are further divided into three types: satellite viruses, satellite RNAs, and satellite DNAs. Satellite viruses encode their own capsid proteins, whereas satellite RNAs and DNAs do not. Most satellites use plant viruses as their helper viruses.

6.7 Prions Are Composed Only of Protein

After reading this section, you should be able to:

a. Describe prion structure and how prions are thought to replicate
b. List characteristics common to all animal diseases caused by prions
c. Name at least two human diseases caused by prions
d. Describe the mechanisms by which a prion protein might first appear in a brain cell

Prions (for *pro*teinaceous *in*fectious particles) are dramatically simpler than viruses, comprising only a single protein. They cause a variety of neurodegenerative diseases in humans and other animals, including scrapie in sheep, bovine spongiform encephalopathy (BSE or "mad cow disease"), and the human diseases kuru, fatal familial insomnia, Creutzfeldt-Jakob disease (CJD), and Gerstmann-Strässler-Scheinker syndrome (GSS). All result in progressive degeneration of the brain and eventual death. At present, no effective treatment exists. ▶| *Prion proteins transmit disease (section 38.6)*

The best-studied prion is the scrapie prion. Scrapie is caused by an abnormal form of a cellular protein. The normal cellular protein is called PrP^{C}. Although its exact function is unknown, it has been shown to reside on the cell surface of neurons and other cell types, and play a role in brain development. The abnormal form is called PrP^{Sc} (for *sc*rapie-associated *pr*ion *p*rotein). Evidence supports a model in which entry of PrP^{Sc} into the brain of an animal causes the PrP^{C} protein to change from its normal conformation to an abnormal form that aggregates (**figure 6.24**). The newly produced PrP^{Sc} molecules then convert more PrP^{C} molecules into the abnormal PrP^{Sc} form. How the PrP^{Sc} causes this conformational change is unclear. However, the best-supported model is that the PrP^{Sc} directly interacts with PrP^{C}, causing the change. It is noteworthy that mice lacking the PrP gene cannot be infected with PrP^{Sc}.

An important characteristic of PrP^{Sc} is its ability to oligomerize—that is, to form short chains of PrP^{Sc} proteins. This is important because oligomers may play a central role in continued conversion of PrP^{C} to PrP^{Sc}, as well as in the spread of PrP^{Sc} from cell to cell and from one individual to another. Furthermore, evidence suggests that the toxic form of PrP^{Sc} is a trimer. However, how the PrP^{Sc} trimer causes disease is not understood.

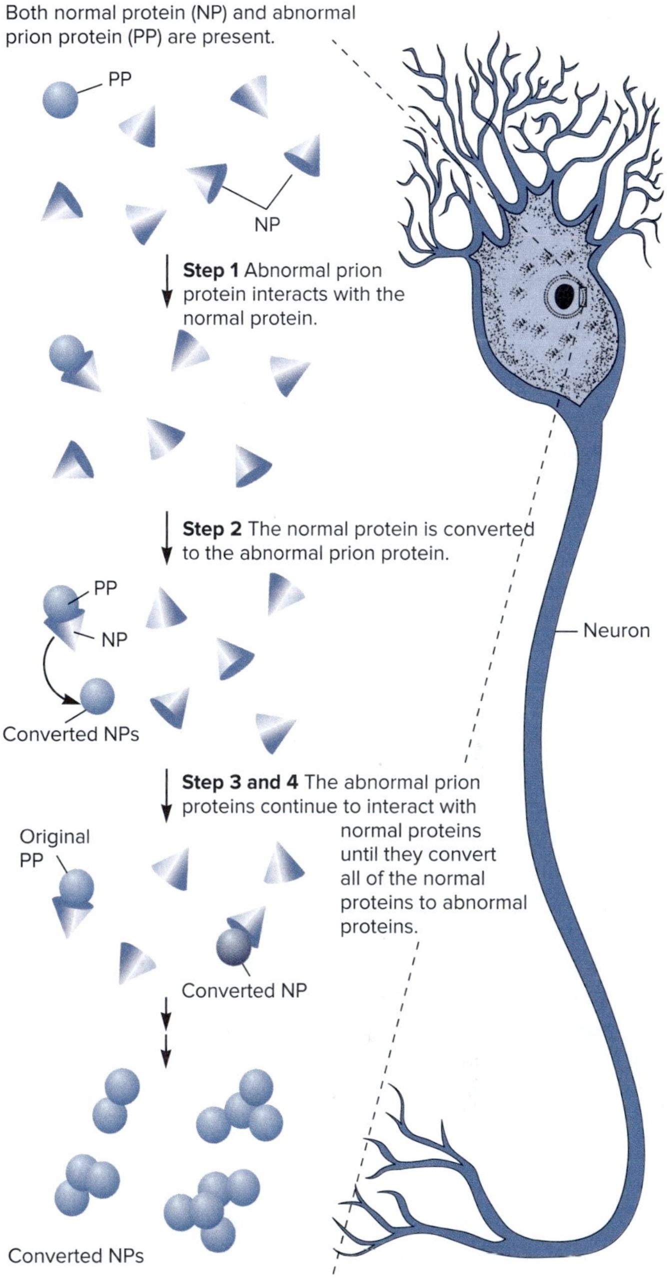

Figure 6.24 One Proposed Model for Prion Replication. The normal and prion proteins differ in their tertiary structures. Recent studies suggest that cofactors may function in prion replication. One proposed cofactor is RNA, which is thought to act as a catalyst for the conversion of PrP^{C} to PrP^{Sc}.

Comprehension Check

1. What are viroids and why are they of great interest?
2. How does a viroid differ from a virus? From a satellite?
3. What is a prion? In what way does a prion differ fundamentally from viruses and viroids?
4. Prions are difficult to detect in host tissues. Why do you think this is so? Why do you think we have not been able to develop effective treatments for prion-caused diseases?

Key Concepts

6.1 Viruses Are Acellular

- Virology is the study of viruses and other acellular infectious agents.
- Viruses cannot reproduce independently of living cells.

6.2 Virion Structure Is Defined by Capsid Symmetry and Presence or Absence of an Envelope

- All virions have a nucleocapsid composed of a nucleic acid, either DNA or RNA, held within a protein capsid made of one or more types of protein subunits called protomers (**figure 6.2**).
- Helical capsids resemble hollow protein tubes and may be either rigid or flexible (**figures 6.3** and **6.4**).
- Icosahedral capsids are usually constructed from two types of capsomers: pentamers (pentons) at the vertices and hexamers (hexons) on the edges and faces of the icosahedron (**figure 6.5**).
- Complex virions (e.g., poxviruses and large phages) have complicated morphology not characterized by icosahedral and helical capsid symmetry (**figure 6.6**). Virions of large phages often have binal symmetry—that is, their heads are icosahedral and their tails are helical (**figure 6.7**).
- Some viruses have a membranous envelope surrounding their nucleocapsid. The envelope lipids usually come from the host cell; in contrast, many envelope proteins are viral and may project from the envelope surface as spikes (**figure 6.8**).
- Viral nucleic acids can be either single stranded (ss) or double stranded (ds), DNA or RNA. Most DNA viruses have dsDNA genomes that may be linear or closed circles. RNA viruses usually have ssRNA. Some RNA genomes are segmented.
- Although viruses lack true metabolism, some contain a few enzymes necessary for their multiplication.

6.3 Viral Life Cycles Have Five Steps

- The one-step growth experiment devised by Delbruck and Ellis was an important first step in understanding how viruses multiply (**figure 6.9**). This type of experiment showed that there is an initial latent period, during which no virions are released, followed by a rise period where virions are released. Early in the latent period is an eclipse period, in which no virus particles are present in infected cells. Viral life cycles can be divided into five steps: (1) attachment to host; (2) entry into host; (3) synthesis of viral nucleic acids and proteins; (4) self-assembly of nucleocapsids; and (5) release from host (**figures 6.10–6.15**).
- The synthesis stage of a virus's life cycle depends on the nature of its genome. DNA viruses may use enzymes that are similar to host enzymes. In some cases, they rely solely on their hosts for synthesis of their nucleic acids and proteins.
- RNA viruses must either encode the enzymes they need to make mRNA and replicate their genomes or carry these enzymes in their capsids.

6.4 There Are Several Types of Viral Infections

- Virulent bacteriophages and archaeal viruses lyse their host. In addition to host cell lysis, temperate bacterial and archaeal viruses can enter the lysogenic cycle in which they remain dormant in the host cell. This is often accomplished by integrating the viral genome into that of the host (**figure 6.16**).
- Viruses that infect eukaryotic cells can cause either host cell lysis or a more insidious cellular death; such viruses cause cytocidal infection. One relatively rare outcome of animal virus infection is the transformation of normal host cells into malignant or cancerous cells (**figure 6.17**).

6.5 Cultivation and Enumeration of Viruses

- Viruses are cultivated using tissue cultures, embryonated eggs, bacterial cultures, and other living hosts.
- Phages produce plaques in bacterial lawns. Sites of animal viral infection may be characterized by cytopathic effects such as pocks and plaques. Plant viruses can cause localized necrotic lesions in plant tissues (**figures 6.18–6.20**).
- Virions can be counted directly by transmission electron microscopy or indirectly by hemagglutination and plaque assays.
- Infectivity assays can be used to estimate virion numbers in terms of plaque-forming units, lethal dose (LD_{50}), or infectious dose (ID_{50}) (**figure 6.21**).

6.6 Viroids and Satellites: Nucleic Acid-Based Subviral Agents

- Viroids are infectious agents of plants that consist only of circular ssRNA molecules (**figures 6.22** and **6.23**). They do not encode any proteins.
- Satellites are subviral infectious agents that may encode some proteins but require a helper virus for replication. Satellite viruses encode their own capsid proteins, whereas satellite RNAs and satellite DNAs do not.

6.7 Prions Are Composed Only of Proteins

- Prions are small proteinaceous agents associated with at least six degenerative nervous system disorders: scrapie, bovine spongiform encephalopathy, kuru, fatal familial insomnia, Gerstmann-Strässler-Scheinker syndrome, and Creutzfeldt-Jakob disease.
- Most evidence supports the hypothesis that prion proteins exist in two forms: the infectious, abnormally folded form and a normal cellular form. The interaction between the abnormal form and the cellular form converts the cellular form into the abnormal form (**figure 6.24**).

Active Learning

1. Many classification schemes are used to identify bacteria. These start with Gram staining, progress to morphology and arrangement characteristics, and include a battery of metabolic tests. Build an analogous scheme that could be used to identify viruses. You might start by considering the host, or you might start with viruses found in a particular environment, such as a marine filtrate.
2. The origin and evolution of viruses is controversial. Discuss whether you think viruses evolved before the first cell or whether they have coevolved and are perhaps still coevolving with their hosts.
3. Consider the separate stages of an animal virus life cycle. Assemble a short list of structures and processes that are unique to the virus and would make good drug targets for an antiviral agent. Explain your rationale for each choice.
4. *Klebsiella* are Gram-negative bacteria that can infect many human tissues. A thick polysaccharide capsule allows the bacteria to avoid host defenses, and it also presents a barrier to bacteriophage. Dozens of different capsule types have been described. These vary in the carbohydrate monomers and the linkages between monomers.

   Some *Klebsiella* bacteriophages are morphologically similar to T4 (figure 6.7). On their tail fibers, these phages carry capsule depolymerase, an enzyme that hydrolyzes glycosidic bonds in the capsule. Why would a capsule depolymerase be located on a tail fiber?

   Following the dissolution of the capsule, these bacteriophages contact their secondary receptor on the cell surface. On what part of the cell envelope would the secondary receptor be located on *Klebsiella*?

   Because of the variation in capsules, there is a corresponding set of depolymerases. A bacteriophage can only infect cells whose capsules are the substrate for its enzyme. A recently identified virus, ϕK64-1, is termed a multihost virus because it can infect nine different *Klebsiella* strains. This virus's genome was found to carry an array of depolymerase genes. Can you design an experiment to determine if each bacteriophage virion carries nine depolymerase enzymes on its tail fibers, or if bacteriophage ϕK64-1 is a mixture of nine viral types, each with a single enzyme on its tail fibers? (Hint: Examine figure 6.20.)

   Read the original article: Pan, Y. J., et al. 2017. *Klebsiella* phages ϕK64-1 encodes multiple depolymerases for multiple host capsular types. *J. Virol.* 91:e02457-16.
5. Of the known archaeal viruses, few are specific to methanogens. In a search for methanogen-specific viruses, scientists identified a new virus, which they termed *Methanosarcina* spherical virus (MetSV). As its name implies, MetSV is a lytic virus with spherical virions. In what type of environment might you find viruses specific for methanogens? (*See section 20.3 for a discussion of methanogens.*)

   Several species of *Methanosarcina*, and different strains of each species, were tested to see which are hosts for MetSV. Some *Methanosarcina* strains grow as single cells; others grow as aggregates (*see figure 4.1*). Most of the single-cell growth strains were lysed by MetSV; those that aggregate were unaffected. The scientists realized that there is a connection between these observations. What feature of the cell affects both cellular arrangement and susceptibility to viral infection?

   Read the original paper: Weidenbach, K., et al. 2017. *Methanosarcina* spherical virus, a novel archaeal lytic virus targeting *Methanosarcina* strains. *J. Virol.* 91:e00955-17.

7

Bacterial and Archaeal Growth

©Arthur Dorety/Stocktrek Images/Getty Images

How Low Can You Go?

Twenty million years ago, in a forest in what is now the western Pacific Ocean, organic debris accumulated to form peat. It supported a diverse microbial ecosystem that thrived on degrading plant matter. Over the millennia, peat was compressed into sedimentary rock and ultimately lignite coal, and the site receded into the ocean. The remains of the forest is now about 2 km below the sea floor. And that microbial ecosystem? It's still there.

We know the microbial ecosystem still exists because in the summer of 2012, scientists in the International Ocean Discovery Program drilled into this Miocene era coal bed. Core samples were crushed and set up in culture vessels incubated under conditions that mimic the original site: no oxygen, warm temperature (45°C), and simple carbon and nitrogen compounds. The scientific team patiently incubated these cultures for 2.5 years before breaking them open to examine their contents.

They then made two notable discoveries. First, 16S rRNA analysis revealed what looked like a contemporary peat microbial community, not a marine sediment community. And while incorporation of isotopes like $^{2}H_2O$ and $^{15}NH_4^+$ indicated that life processes were occurring, scientists calculated that their generation times ranged from several months to over 100 years. In other words, in the time it takes for an average human to be born, have children, grow very old, and die, many of these bacteria didn't divide a single time.

Only microbes can live at such a slow pace. Amazingly, these cells seemingly counteract chemical degradation and maintain cellular integrity. They remain intact and capable of reviving. Although we are only beginning to study microbes with such long generation times, these observations confirm that growth in nature can be challenging, not to mention difficult for microbiologists to observe and quantify.

Defining "growth" and "alive" confounds microbiologists. From our anthropomorphic ideas about what constitutes a habitable environment to our expectation of microbial cell division in a brief time frame, only in the last century have we realized that life exists in extreme environments, and it is almost exclusively microbial. In this chapter, we present bacterial and archaeal growth patterns, how they are measured, and the environmental conditions that affect growth.

Readiness Check:

Based on what you have learned previously, you should be able to:

- ✓ Describe the functions of eukaryotic cytoskeletal proteins (section 5.3)
- ✓ Define and list examples of essential nutrients, and describe how they are used by cells
- ✓ Distinguish macroelements (macronutrients) from trace elements (micronutrients), list examples of each, and describe how they are used (section 3.3)
- ✓ Provide examples of growth factors needed by some microorganisms (section 3.3)
- ✓ Describe the eukaryotic cell cycle
- ✓ Describe the major events of mitosis and meiosis

7.1 Most Bacteria and Archaea Reproduce by Binary Fission

After reading this section, you should be able to:

a. Describe binary fission as observed in bacteria and archaea
b. Compare binary fission with other bacterial reproductive strategies

Most bacterial and archaeal cells reproduce by **binary fission** (**figure 7.1**). Binary fission is a relatively simple type of cell division. The cell elongates as new cell envelope material is synthesized, and subcellular structures like ribosomes and inclusions are abundant enough to be evenly distributed in the cytoplasm. Only the nucleoid is present as a single entity, and its replication and partitioning into each half of the elongated cell is a critical

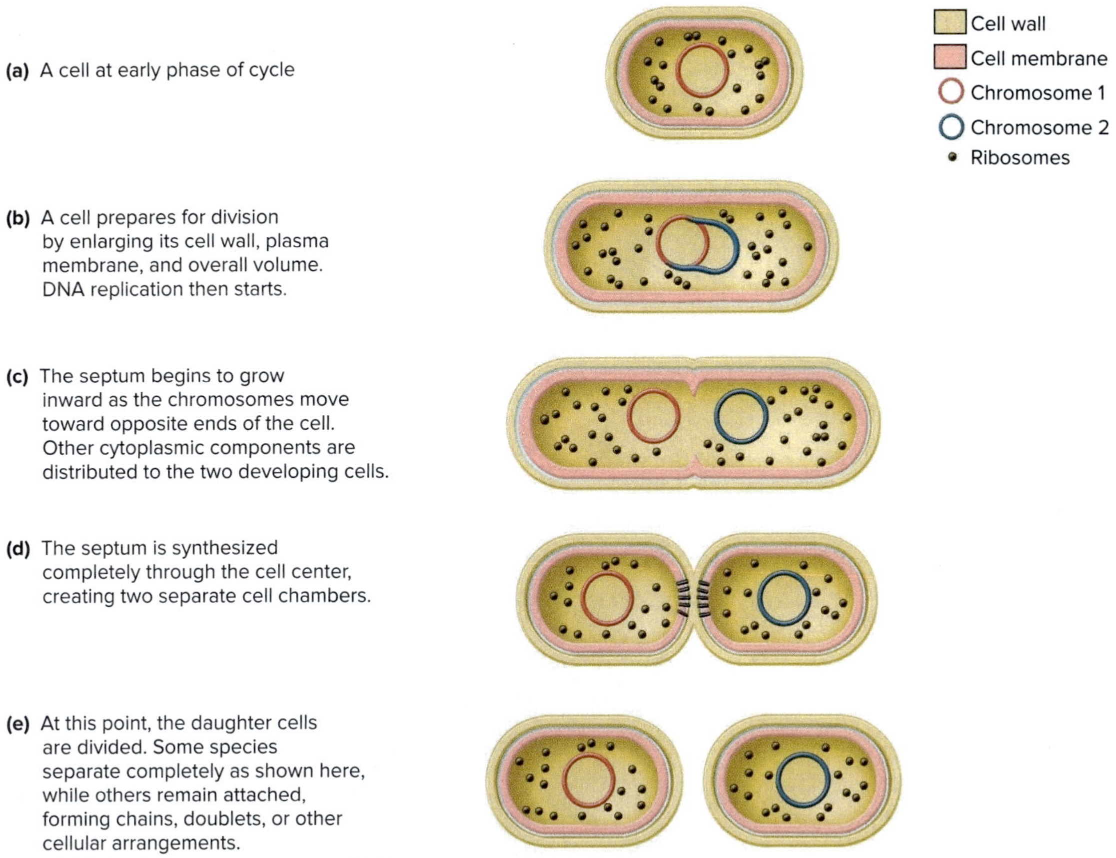

Figure 7.1 Binary Fission.

MICRO INQUIRY *In addition to chromosomes, what other cytoplasmic contents must be equally distributed between daughter cells?*

event in binary fission. Finally, a septum (cross wall) is formed at midcell, dividing the parent cell into two progeny cells, each having its own nucleoid and a complement of other cellular constituents.

Several other reproductive strategies have been identified in bacteria (**figure 7.2**). Some bacteria reproduce by forming a bud. Certain cyanobacteria undergo multiple fission. The progeny cells, called baeocytes, are held within the cell wall of the parent cell until they mature. Other bacteria, such as members of the genus *Streptomyces,* form multinucleoid filaments that eventually divide to form uninucleoid spores. These spores are readily dispersed, much like the dispersal spores formed by filamentous fungi. ▶| *Order* Streptomycetales: *an important source of antibiotics (section 23.1)*

Despite the diversity of bacterial reproductive strategies, certain features are shared. In all cases, the genome of the cell must be replicated and segregated to form distinct nucleoids. At some point during reproduction, each nucleoid and its surrounding cytoplasm becomes enclosed within its own plasma membrane. These processes are the major steps of the cell cycle. In section 7.2, we examine the bacterial cell cycle in more detail.

7.2 Bacterial Cell Cycles Can Be Divided into Three Phases

After reading this section, you should be able to:

a. Summarize the three phases in a typical bacterial cell cycle
b. Summarize current models for chromosome partitioning
c. State the functions of cytoskeletal proteins during cytokinesis and in determining cell shape

The **cell cycle** is the complete sequence of events extending from formation of a new cell through the next division. It is of intrinsic interest as a fundamental biological process. However, understanding the cell cycle has practical importance as well. For instance, synthesis of peptidoglycan during the cell cycle is the target of numerous antibiotics used to treat bacterial infections. ▶| *Inhibitors of cell wall synthesis (section 9.4)*

The cell cycles of several bacteria—*Escherichia coli, Bacillus subtilis,* and the aquatic bacterium *Caulobacter crescentus*—have been examined extensively, and our understanding of the bacterial cell cycle is based largely on

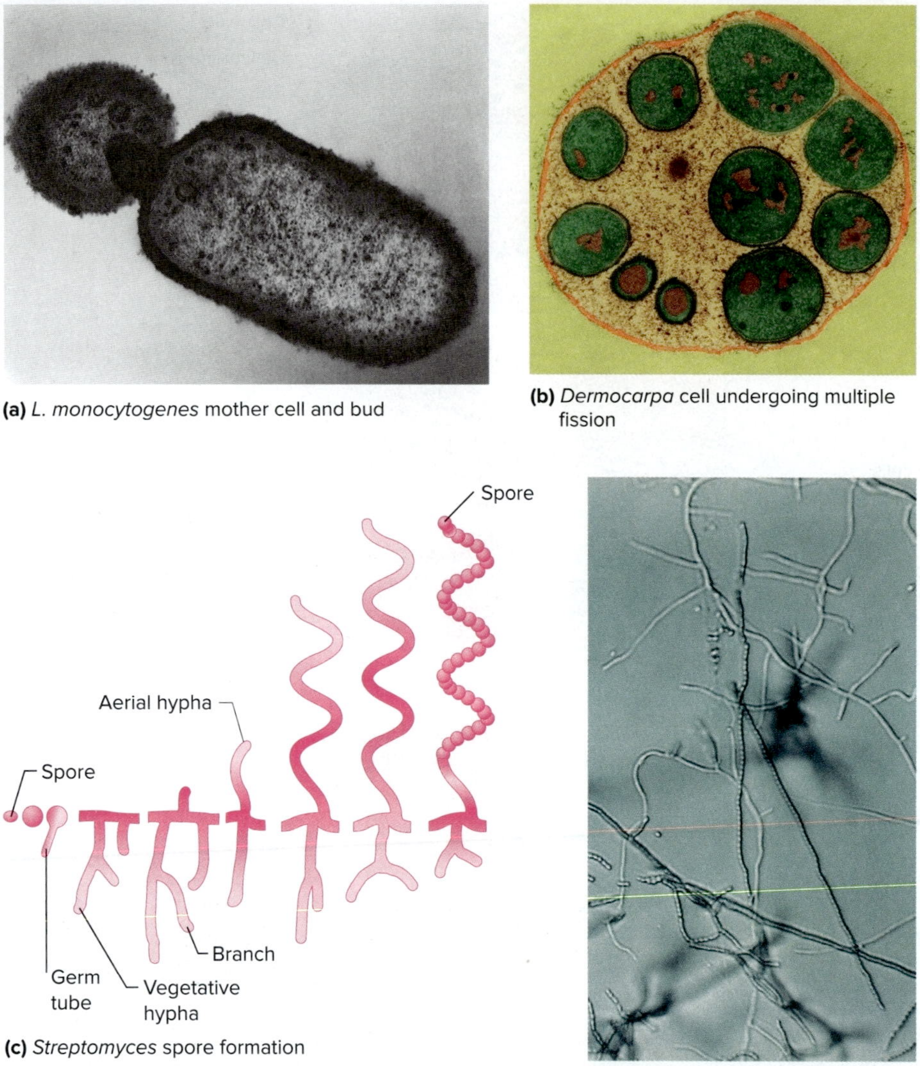

(a) *L. monocytogenes* mother cell and bud

(b) *Dermocarpa* cell undergoing multiple fission

(c) *Streptomyces* spore formation

Figure 7.2 Some Bacteria Reproduce by Methods Other than Binary Fission. (a) Transmission electron micrograph (TEM) of a budding *Listeria monocytogenes* cell. (b) Baeocytes produced by the cyanobacterium *Dermocarpa*. (c) Spore formation by *Streptomyces* spp. Branched filaments, aerial hyphae, and chains of spores are visible.

(a) ©Dr. Kari Lounatmaa/Science Source; (b) ©Dennis Kunkel Microscopy/Science Source (c) ©Smith Collection/Gado/Getty Images

these studies. The bacterial cell cycle consists of three phases: (1) a period of growth after the cell is born, which is similar to the G1 phase of the eukaryotic cell cycle; (2) chromosome replication and partitioning period, which functionally corresponds to the S and mitosis events of the M phase of the eukaryotic cycle; and (3) cytokinesis, during which a septum and daughter cells are formed (**figure 7.3**). Recall that in the eukaryotic cell cycle, the S phase is separated from the M phase by another period called G2. In G2, chromosome replication is completed and some time passes before chromosome segregation occurs. This is not the case for bacteria. As you will see in the discussion that follows, chromosome replication and partitioning occur concurrently. Furthermore, the initial events of cytokinesis actually occur before chromosome replication and partitioning are complete. Finally, some bacteria are able to initiate new rounds of replication before the first round of replication and cytokinesis is finished. Although chromosome replication and partitioning overlaps with cytokinesis, we consider them separately.

Chromosome Replication and Partitioning

Most bacteria have a single circular chromosome. Each circular chromosome has a single site at which replication starts called the **origin of replication,** or simply the origin (figure 7.3). Replication is completed at the terminus, located directly opposite the origin. In a newly formed *E. coli* cell, the chromosome is compacted and oriented so that the origin and terminus are in opposite halves of the cell. Early in the cell cycle, the origin and terminus move to midcell, and proteins needed for chromosome replication assemble at the origin. This DNA synthesizing machinery is called the **replisome,** and DNA replication proceeds in both directions from the origin. As the progeny chromosome is synthesized, the two

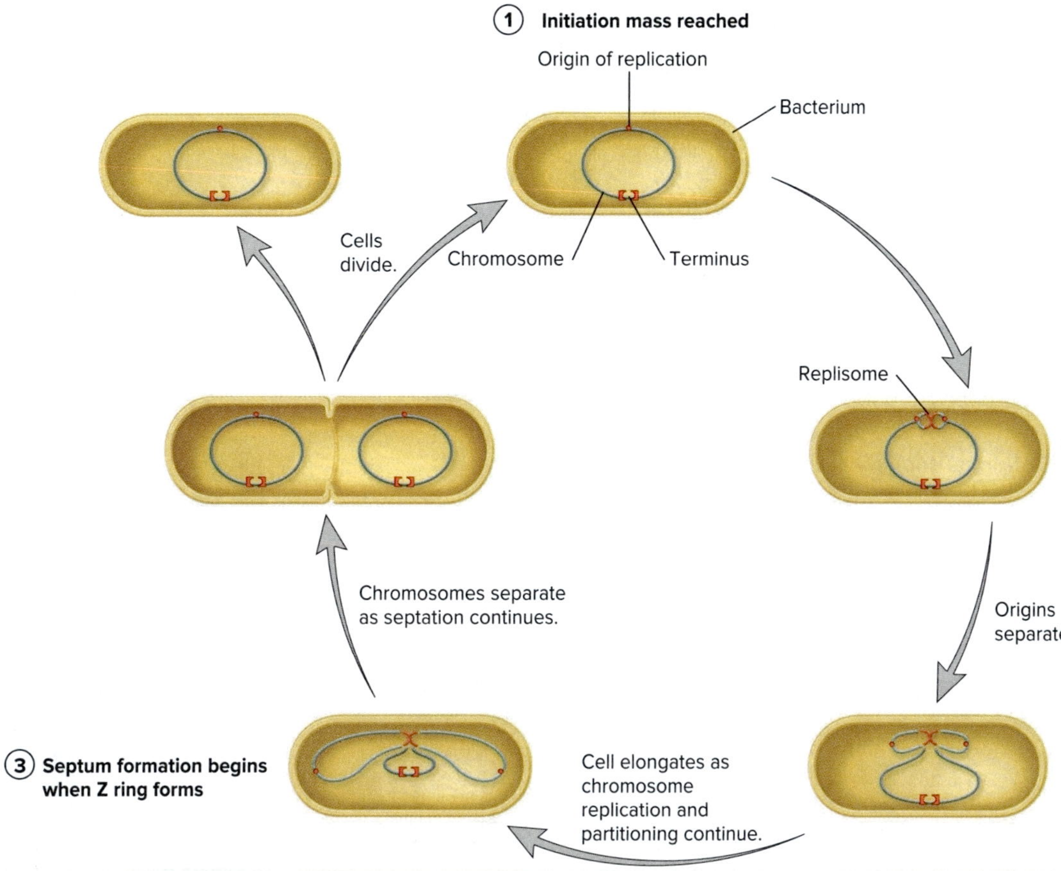

Figure 7.3 Cell Cycle of *E. coli*. An increase in cell mass results in cell cycle initiation and accumulation of the DnaA protein, which initiates DNA replication. As the cell readies for DNA replication, the origin of replication migrates to the center of the cell and proteins that make up the replisome assemble. In this illustration, one round of DNA replication is completed before the cell divides. In rapidly growing cultures (generation times of about 20 minutes), second and third rounds of DNA replication are initiated before division of the original cell is completed. Thus the daughter cells inherit partially replicated DNA.

MICRO INQUIRY *Why is it important that the origin of replication migrate to the center of the cell prior to replication?*

origins move toward opposite ends of the cell, and the rest of each chromosome follows in an orderly fashion.

Although the process of DNA synthesis and movement seems rather straightforward, the mechanism by which chromosomes are partitioned to each daughter cell varies somewhat among bacterial species. Studies of *C. crescentus* have provided the most complete understanding of the partitioning process. This bacterium has an interesting life cycle in which a sessile stalked cell divides to give rise to a slightly smaller, flagellated swarmer cell; the flagellum forms at the pole opposite the stalk (*see figure 22.8*).

The partitioning system for bacterial chromosomes has three components: ParA and ParB proteins and a *parS* region on the chromosome (**figure 7.4**). The *C. crescentus* chromosome has two *parS* sites close together and near the origin of replication, and this region functions like a eukaryotic centromere. Following replication initiation, ParB binds to the *parS* sites on each chromosome.

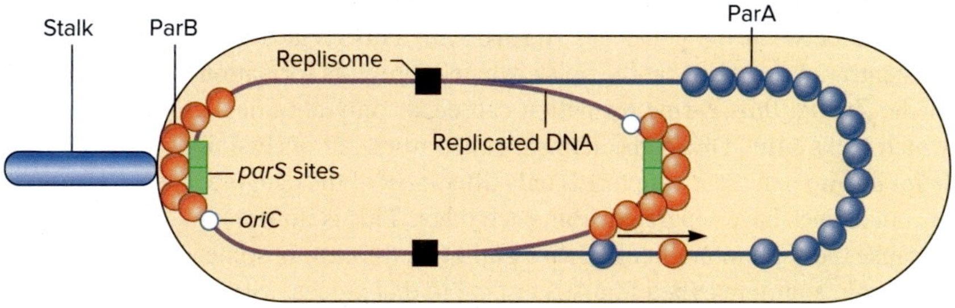

Figure 7.4 The ParAB/*parS* Partitioning System of *Caulobacter crescentus*. ParB binds each *parS* site on the two daughter chromosomes. A gradient of ParA proteins directs the partitioning complex of ParB/*parS* to the opposite pole of the cell in a relay mechanism.

Additional molecules of ParB bind nearby and ultimately cover a large chromosomal region. This protein-DNA assembly is termed the partitioning complex. One partitioning complex remains at the stalk pole of the dividing cell, while the other is guided by ParA to the opposite pole. The ParA-mediated movement of the second partitioning complex is described as a relay model. In a relay race, the baton is passed from one runner to the next; by the end of the race, the baton has traveled the full distance, but each runner has traveled only part of the distance. Similarly, movement of the newly replicated chromosome occurs by passing the partitioning complex to a series of ParA proteins. The ParA protein is distributed in a gradient in the cell and is most abundant at the opposite pole. ParA molecules in complex with ATP are bound loosely and nonspecifically to the chromosome. When a partitioning complex contacts ParA-ATP, ATP is hydrolyzed and the partitioning complex is moved in the direction of the opposite pole. Therefore, the abundance of ParA-ATP at the opposite pole directs chromosome movement from one pole to the other. It is still unclear how the ParA gradient forms and is maintained in the cell. ▶| Caulobacteraceae *and* Hyphomicrobiaceae *bacteria reproduce in unusual ways (section 22.1)* *Binary Fission; Bidirectional DNA Replication*

Cytokinesis

Septation is the process of forming a cross wall between two daughter cells. **Cytokinesis,** a term that was once used to describe the formation of two eukaryotic daughter cells, is now used to describe this process in all cells. In bacteria and archaea, septation is divided into several steps: (1) selection of the site where the septum will be formed; (2) assembly of the Z ring, which is composed of the cytoskeletal protein FtsZ; (3) assembly of the cell wall–synthesizing machinery (i.e., for synthesis of peptidoglycan and other cell wall constituents); and (4) constriction of the cell and septum formation. ▶| *Synthesis of peptidoglycan occurs in the cytoplasm, at the plasma membrane, and in the periplasmic space (section 12.4)*

Correct septation requires that the Z ring form at the proper place at the proper time. In *E. coli,* the MinCDE system limits Z-ring formation to the center of the cell. Three proteins compose the system (MinC, MinD, and MinE). These proteins oscillate between the cell poles (**figure 7.5**). This creates high concentrations of MinC at the poles, where it prevents formation of the Z ring; thus Z-ring formation can occur only at midcell, which lacks MinCDE. A mechanism called **nucleoid occlusion** helps ensure that the Z ring forms only after most of the daughter chromosomes have separated from each other. This is important because the septum might otherwise guillotine the chromosome. In *E. coli,* a protein called SlmA is critical to this process. SlmA binds to specific sites distributed around the chromosome except near the terminus of replication. DNA-bound Slm inhibits FtsZ polymerization, thereby preventing the FtsZ ring from forming while the chromosome is at midcell. Once enough of the

Figure 7.5 MinCDE Proteins Help Establish the Site of Septum Formation. MinC (not shown) blocks septum formation. It oscillates with MinD. MinD's oscillations are controlled by MinE, which also follows MinD movements. As shown, MinD-ATP interacts with the plasma membrane and forms filaments. MinE binding to MinD-ATP causes hydrolysis of ATP, yielding MinD-ADP. MinD-ADP is released from the filament into the cytosol. There MinD-ADP is converted back to MinD-ATP, which then forms filaments at the opposite end of the cell. MinC and MinE follow and the process begins again. As a result of these oscillations, MinC concentrations are highest at the poles forcing septum formation at the center of the cell.

MICRO INQUIRY *What would be the outcome if FtsZ formed a Z ring that was not in the center of the cell? Consider both the morphology of the daughter cells and the partitioning of chromosomes.*

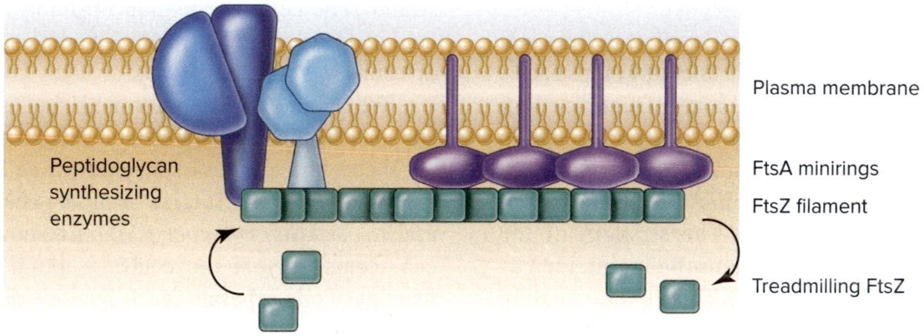

Figure 7.6 The *E. coli* Divisome. The cell division apparatus is composed of numerous proteins. The first step in divisome formation is FtsZ polymerization to form the Z ring. FtsA anchors the Z ring to the plasma membrane through insertion of a helix. Cytoplasmic FtsZ polymerizes in a treadmilling process, where it forms the platform for septal peptidoglycan synthesis.

daughter chromosomes and SlmA molecules have moved away from midcell, the Z ring can form.

Assembly of the Z ring is the critical early step in septation, as it forms the scaffold for the complete cell division machinery. The FtsZ protein is a homologue of eukaryotic tubulin, and like tubulin, it polymerizes, and its filaments create the meshwork that constitutes the Z ring. The Z ring is dynamic, with portions being exchanged constantly with newly formed, short FtsZ polymers from the cytosol. Although the detailed architecture of the Z ring has not yet been defined, FtsA and ZipA proteins are known to promote its attachment to the plasma membrane.

Once the Z ring forms, the late division proteins assemble into the **divisome,** as illustrated in **figure 7.6**. The cell wall–synthesizing machinery is assembled using the Z ring as a scaffold (**table 7.1**). Short polymers of FtsZ, termed protofilaments, move around the inner circumference of the cell near the Z ring in a motion called treadmilling. This motion occurs by the removal of individual FtsZ molecules from one end of the protofilament and the repolymerization at the other end. These treadmilling protofilaments associate with enzyme complexes that synthesize peptidoglycan to build the septum between the daughter cells. The Z ring constricts—that is, it is reduced in circumference as the septum grows inward. The plasma membrane invaginates, and the cytoplasms of the daughter cells are separate before the septal wall is complete. A final step involves amidases, enzymes that function to split the newly synthesized wall so that the daughter cells can separate.

So far, our discussion of the cell cycle describes what occurs in slowly growing *E. coli* cells. In these cells, the cell cycle takes approximately 60 minutes to complete. However, *E. coli* can reproduce at a much more rapid rate, completing the entire cell cycle in about 20 minutes, despite the fact that DNA replication always requires at least 40 minutes. *E. coli* accomplishes this by beginning a second round of DNA replication (and sometimes even a third or fourth round) before the first round of replication is completed. Thus the progeny cells receive a chromosome with two or more replication forks, and replication is continuous because the cells are always copying their DNA. *Bacterial Cell Cycle*

Table 7.1 Some *E. coli* Divisome Proteins and Their Functions

Divisome Protein	Function
FtsZ	Forms Z ring; positions peptidoglycan synthesis enzymes for septum growth
FtsA, ZipA	Anchor Z ring to plasma membrane
FtsK	Chromosome segregation and separation of chromosome dimers
FtsEX	Coordinate hydrolases for septum growth and cell separation
FtsI[1], FtsW	Peptidoglycan synthesis in the septum
FtsN, FtsQLB	Coordinate ring constriction with septum growth

1 FtsI is also known as penicillin-binding protein 3 (PBP3).

Cellular Growth and Determination of Cell Shape

As we have seen, bacterial and archaeal cells have species-specific defined shapes. These shapes are neither accidental nor random, as demonstrated by the faithful propagation of shape from one generation to the next. In addition, some microbes change their shape under certain circumstances. For instance, *Sinorhizobium meliloti* switches from rods to Y-shaped cells when living symbiotically with plants. Likewise, *Helicobacter pylori,* the causative agent of gastric ulcers and stomach cancer, changes from its characteristic helical shape to a sphere in stomach infections and in prolonged culture.

To consider the shape of a cell, we must first review cell wall function. The cell wall constrains the turgor pressure exerted by the cytoplasm, preventing the cell from swelling and bursting. Turgor pressure describes the force pushing against the cell wall because of the osmolarity of the cytoplasmic contents. Peptidoglycan is a rigid but elastic structure that protects the cell from lysis. The strength of the existing peptidoglycan in the cell wall must be maintained as new peptidoglycan subunits are added. Thus understanding peptidoglycan synthesis is critical to understanding the determination of cell shape. *Peptidoglycan structure (section 3.4)*

Peptidoglycan synthesis involves many proteins, including a group of enzymes called **penicillin-binding proteins (PBPs)** so named because they were first noted for their capacity to bind penicillin. While this property is important, their function is to link strands of peptidoglycan together or hydrolyze bonds in existing strands so that new units can be inserted during cell growth. The PBP enzymes that hydrolyze bonds of peptidoglycan are also called **autolysins;** other autolysins besides PBPs also help sculpt the peptidoglycan sacculus during cell growth and cell division. **Figure 7.7** illustrates some of the components of the peptidoglycan-synthesizing machinery and outlines a general scheme of peptidoglycan synthesis. Synthesis of the NAG-NAM-pentapeptide building block is completed while it is attached to a lipid carrier, bactoprenol, located in the plasma membrane. The carrier-bound building block is then flipped across the membrane by MurJ. Upon release of the NAG-NAM-pentapeptide into the periplasmic space, it is inserted into a peptidoglycan strand. The cellular location of autolysin activity and peptidoglycan export is not random and plays an important role in determining cell shape.

We begin our discussion with the simplest shape, the coccus. Although it has been stated that this is the "default" cellular shape, the growth of a spherical cell is more complicated than once thought. Studies of model cocci (e.g., *Enterococcus faecalis* and *Staphylococcus aureus*) show that new peptidoglycan forms only at the central septum (**figure 7.8*a***). When daughter cells separate, each has one new and one old hemisphere. As in most bacteria, proper placement of the septum depends on FtsZ localization. In mutant cells without this tubulin homologue, peptidoglycan synthesis occurs in a random pattern around the cell, leading to bloated cells that lyse. Thus FtsZ placement determines the site of cell wall growth by recruiting enzymes needed for peptidoglycan synthesis to the divisome.

The peptidoglycan of rod-shaped cells is synthesized by two molecular machines that share many proteins but differ in terms of placement and time of function. The first is responsible for elongation of the cell that occurs prior to septum formation. It is sometimes called the elongasome. The second molecular machine is the divisome, which synthesizes peptidoglycan during cytokinesis.

During elongation, proteins in the actin homologue MreB family play an essential role. In a manner similar to that of FtsZ and the Z ring in cytokinesis, MreB proteins polymerize, creating patches of filaments along the cytoplasmic face of the plasma membrane (figure 7.8*b*). Cell wall growth occurs in numerous bands around the circumference of the cell. The

1. Peptidoglycan synthesis starts in the cytoplasm with the attachment of uridine diphosphate (UDP) to the sugar *N*-acetylglucosamine (NAG). Some of the UDP-NAG molecules are converted to UDP-NAM. Amino acid addition to NAM is not shown for simplicity.

2. NAM is transferred from UDP to bactoprenol, a carrier embedded in the plasma membrane. NAG is then attached to bactoprenol-NAM generating bactoprenol-NAM-NAG, called lipid II. The divisome protein MurJ (not shown) "flips" lipid II across the plasma membrane so that the NAM-NAG units are available for insertion into the sacculus.

3. Autolysins (blue balls labeled "A") located at the divisome degrade bonds in the existing peptidoglycan sacculus. This permits the insertion of new NAM-NAG units into the sacculus.

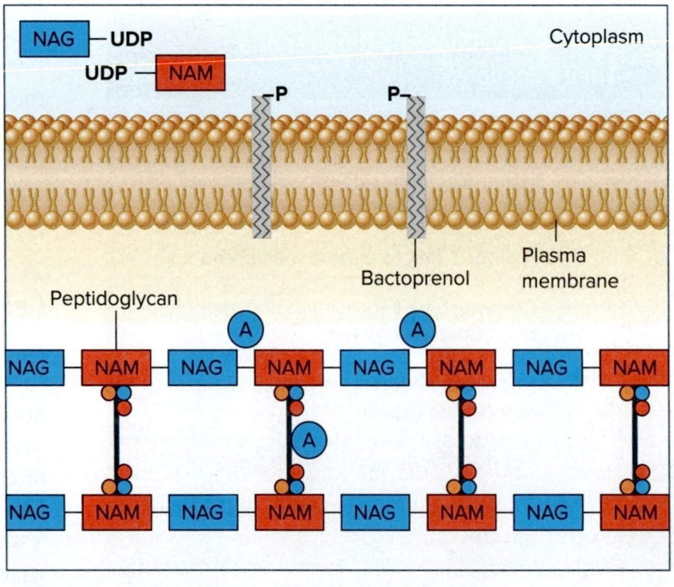

Figure 7.7 Components of the Peptidoglycan Synthesizing Machinery. See figures 9.6 and 12.9 for detailed diagrams of peptidoglycan synthesis.

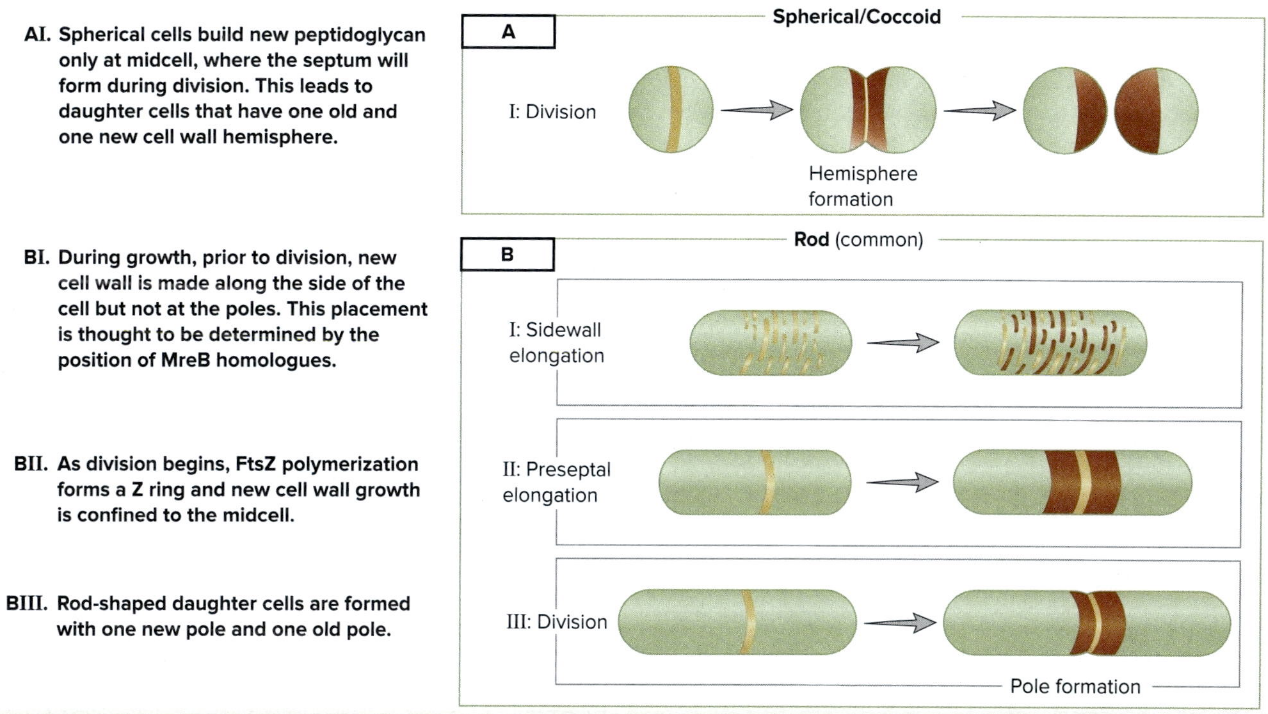

Figure 7.8 Cell Wall Biosynthesis and Determination of Cell Shape in Spherical and Rod-Shaped Cells.

MICRO INQUIRY *Which step in the development of rod-shaped cells is essential in determining cell morphology?*

bands are positioned along the length of the cell but not at the poles *(see figure 3.30b)*. MreB functions as a scaffold upon which the cell-wall synthesizing machinery assembles. As with FtsZ, MreB filaments also undergo treadmilling, but in this case, the activity of the peptidoglycan synthesizing enzymes is required for filament movement. As the FtsZ ring forms at the midcell in cytokinesis, MreB proteins and other elongasome proteins redeploy from the sidewalls to the midcell, where they contribute to cell wall synthesis during cytokinesis. Thus, cell wall growth switches from the side wall to the septum at this time. The importance of MreB proteins in determining cell shape is demonstrated by two observations. First, rod-shaped cells in which MreB has been depleted assume a spherical shape. In addition, while almost all rod-shaped bacteria and archaea synthesize at least one MreB homologue, coccoid-shaped cells lack proteins in the MreB family.

The last cell shape we consider is that of comma-shaped cells, as studied in the aquatic bacterium *Caulobacter crescentus*. In addition to the actin homologue MreB and the tubulin-like protein FtsZ, these cells (and other vibroid-shaped cells) produce a cytoskeletal protein called crescentin, a homologue of eukaryotic intermediate filaments. This protein localizes to one side of the cell, where it slows the insertion of new peptidoglycan units into the peptidoglycan sacculus (**figure 7.9**). The resulting asymmetric cell wall growth gives rise to the inner curvature that characterizes the comma shape.

It is clear from these examples that common microbial cell shapes require cytoskeletal proteins that bear startling similarity with those of eukaryotes. However, some bacterial shapes appear to be determined by other mechanisms. For

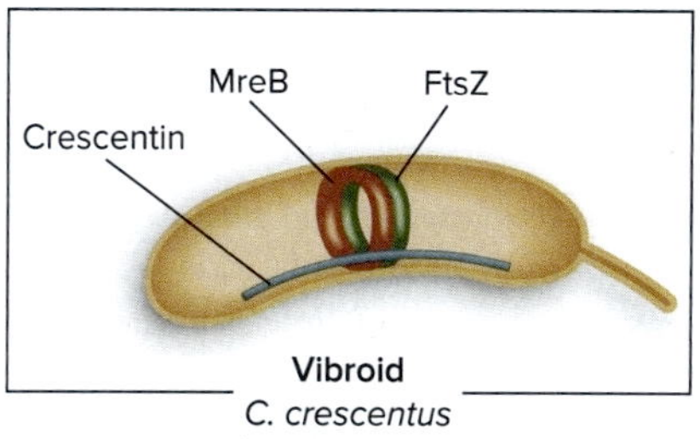

Figure 7.9 Vibroid Shape Is Determined by Crescentin in *Caulobacter crescentus.* The intermediate filament homologue crescentin polymerizes along the length of the inner curvature of the cell. Cells depleted of crescentin are rod-shaped. Note that *C. crescentus* has all three types of cytoskeleton protein homologues.

example, the spirochete *Borrelia burgdorferi,* which causes Lyme disease, relies on its periplasmic flagella to confer its spiral shape. As another example, *Spiroplasma,* a citrus pathogen, has no cell wall or flagella. It relies on contractile cytoplasmic fibrils to give its spiral shape. ▶| *Phylum* Spirochaetes *(section 21.6); Class* Mollicutes, *Phylum* Tenericutes: *bacteria that lack cell walls (section 21.3)*

The study of bacterial cytoskeletal elements and cell shape has important practical aspects. Bacterial cytoskeletal homologues provide a tractable model for the study of more complicated eukaryotic cytoskeletal proteins and their assembly. In addition, these proteins may prove to be valuable targets for drug development. Indeed, molecules that block FtsZ function inhibit growth of the pathogen *Staphylococcus aureus* and might be effective treatments for staphylococcal diseases. ▶| *Antimicrobial chemotherapy (chapter 9)*

Comprehension Check

1. Describe the three phases of a bacterial cell cycle. The overlapping of cytokinesis and chromosome partitioning could potentially create problems for a cell during the cell cycle. What mechanisms does the cell use to prevent any problems?
2. How does the bacterial cell cycle compare with the eukaryotic cell cycle? List two ways they are similar and two ways they differ.
3. Do you think MinCDE functions in coccoid-shaped cells? Explain your answer.
4. Do you think *Spiroplasma* produces FtsZ? What about MreB? Explain your reasoning.

7.3 Archaeal Cell Cycles Are Unique

After reading this section, you should be able to:

a. Compare and contrast the *Sulfolobus* spp. cell cycle and the typical eukaryotic cell cycle
b. Compare and contrast the *Sulfolobus* spp. cell cycle and a bacterial cell cycle

As is true with many other aspects of archaeal biology, understanding of archaeal cell cycles lags behind that of bacterial cell cycles. However, studies of model archaea such as *Sulfolobus* spp. have yielded intriguing results. *Sulfolobus* spp. are members of the phylum *Crenarchaeota* and grow best in environments that are hot (80°C) and acidic (pH 3). Their cell cycle is reminiscent of a mitotic cell cycle. After a growth phase (G1), their DNA is replicated (S) using replicative machinery similar to that of eukaryotes. Of note is the fact that unlike bacteria, the circular chromosomes of *Sulfolobus* spp. have three origins of replication. Furthermore, once replicated, the daughter chromosomes remain unsegregated for some time (G2); the *Sulfolobus* G2 phase consumes more than 50% of the cell cycle. G2 is followed by segregation of the chromosome and then cytokinesis occurs. ▶| *Archaea use homologues of eukaryotic replisome proteins (section 15.2)*

Advances have been made in elucidating how daughter chromosomes are segregated. Two proteins have been identified: SegA and SegB. SegA is similar to ParA proteins found in bacteria, and SegB is unique to archaea, but is thought to be the functional equivalent of bacterial ParB. The genes encoding the two proteins are located near one of the origins of replication, just as the genes for ParA and ParB are located near the bacterial origin. SegA polymerizes and is thought to exert the force, either pulling or pushing, that segregates the chromosomes. The role of SegB is unclear, although it is known to bind DNA and interact with SegA, enhancing SegA polymerization. However, *parS*-like sequences have not been clearly identified in *Sulfolobus* spp.

Unlike chromosome segregation, which appears to be bacteria-like, cytokinesis in *Sulfolobus* spp. is most similar to eukaryotic processes. These archaea undergo cytokinesis using proteins related to the ESCRT proteins of eukaryotes. ESCRT proteins are so named because of their roles in endocytosis: *e*ndosomal *s*orting *c*omplex *r*equired for *t*ransport. ESCRT proteins form four distinct complexes that are involved in formation of multivesicular bodies (MVB), a type of endosome (*see figure 5.9*). During MVB formation, some ESCRT proteins cause the endosome membrane to invaginate. Then a complex of ESCRT proteins called ESCRT-III circles the neck formed by the invaginating membrane and brings about membrane scission, releasing a small vesicle into the developing MVB. Three division proteins have been identified in *Sulfolobus* spp. CdvA binds the plasma membrane and forms a ring at midcell. ESCRT-III and Vps4 are recruited to this CdvA belt and have been proposed to form an hourglass structure that constricts the membranes until they separate.

Archaea belonging to the phylum *Euryarchaeota* have cell cycles that differ in several substantive ways from the *Sulfolobus* model just described. The euryarchaeal cell cycle is comparable to the bacterial cell cycle, featuring growth, DNA replication, and cytokinesis. FtsZ is present in most euryarchaea, often in two related forms. Recall that archaea lack peptidoglycan, so unlike the bacterial model where the Z ring is physically associated with the peptidoglycan synthesizing enzymes, the archaeal Z ring associates with new S-layer proteins and possibly other cell wall components. How this is accomplished awaits discovery.

As we note in chapter 4, euryarchaea are polyploid, often with a dozen or more chromosome copies. It is possible that a genome segregation system does not exist, and the large number of chromosomes may compensate for this deficiency. Only *Halobacterium salinarum,* a halophile, has been demonstrated to actively segregate daughter chromosomes. |◀ *Archaeal cytoplasm is similar to bacterial cytoplasm (section 4.3)*

Comprehension Check

1. What elements of the *Sulfolobus* spp. cell cycle are similar to that of *Caulobacter crescentus*? What elements are similar to the eukaryotic cell cycle?
2. Many archaea have genes encoding an FtsZ homologue. Describe how FtsZ might function in an archaeal cell cycle.

7.4 Growth Curves Consist of Five Phases

After reading this section, you should be able to:

a. Describe the five phases of a microbial growth curve observed when microbes are grown in a batch culture
b. Describe three hypotheses proposed to account for the decline in cell numbers during the death phase of a growth curve
c. Correlate changes in nutrient concentrations in natural environments with the five phases of a microbial growth curve
d. Relate growth rate constant to generation (doubling) time and suggest how these values might be used by microbiologists doing basic research or working in industrial settings

In section 7.2, we commented on the changes in bacterial cell size that accompany its preparation for division. This is one type of growth of concern to microbiologists. However, microbiologists are more frequently concerned with the increase in population size that follows cell division. Therefore the term *growth* is more commonly used to refer to growth in the size of a population.

Population growth is often studied by analyzing the growth of microbes in liquid (broth) culture. When microorganisms are cultivated in broth, they usually are grown in a **batch culture;** that is, they are incubated in a closed culture vessel like a test tube or a flask with a single batch of medium. Fresh medium is not provided during incubation, so as nutrients are consumed, their concentrations decline, and wastes accumulate. Population growth of microbes reproducing by binary fission in a batch culture can be plotted as the logarithm of the number of viable cells versus the incubation time. The resulting curve has five distinct phases (**figure 7.10**), which we examine in this section. Although this is "life in the lab," microbes do encounter conditions in their natural environments that mimic what occurs in a batch culture. Furthermore, humans routinely create artificial environments for microbes (e.g., the fermentation vessel in a pharmaceutical plant) that are batch cultures. Therefore understanding the growth curve is of paramount importance.

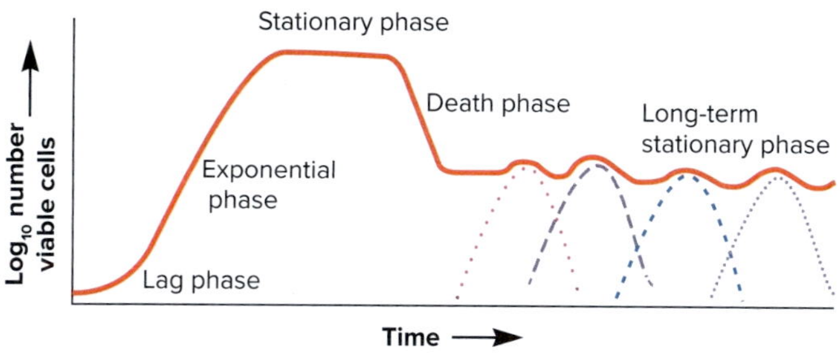

Figure 7.10 Microbial Growth Curve in a Closed System. The five phases of the growth curve are identified. The dotted lines shown during the long-term stationary phase represent successive waves of genetic variants that evolve during this phase of the growth curve.

MICRO INQUIRY *Identify the regions of the growth curve in which (1) nutrients are rapidly declining and (2) wastes accumulate.*

Lag Phase

When microorganisms are introduced into fresh culture medium, usually no immediate increase in cell number occurs. This period is called the **lag phase.** It is not a time of inactivity; rather cells are synthesizing new components. This can be necessary for a variety of reasons. The cells may be old and depleted of ATP, essential cofactors, and ribosomes; these must be synthesized before growth can begin. The medium may be different from the one the microorganism was growing in previously. In this case, new enzymes are needed to use different nutrients. Possibly the microorganisms have been injured and require time to recover. Eventually, however, the cells begin to replicate their DNA, increase in mass, and divide. As a result, the number of cells in the population begins to increase.

Exponential Phase

During the **exponential phase,** microorganisms grow and divide at the maximal rate possible given their genetic potential, the nature of the medium, and the environmental conditions. Their rate of growth is constant during the exponential phase; that is, they are completing the cell cycle and doubling in number at regular intervals (figure 7.10). The population is most uniform in terms of chemical and physiological properties during this phase; therefore exponential phase cultures are usually used in biochemical and physiological studies. The growth rate during exponential phase depends on several factors, including nutrient availability. When microbial growth is limited by the low concentration of a required nutrient, the final net growth or yield of cells increases with the initial amount of the limiting nutrient present (**figure 7.11*a***). The rate of growth also increases with nutrient concentration (figure 7.11*b*) but it saturates, much like what is seen with many enzymes (*see figure 10.16*). The shape of the curve is thought to reflect the rate of nutrient uptake by microbial transport proteins. At sufficiently high nutrient levels, the transport systems are saturated, and the growth rate does not rise

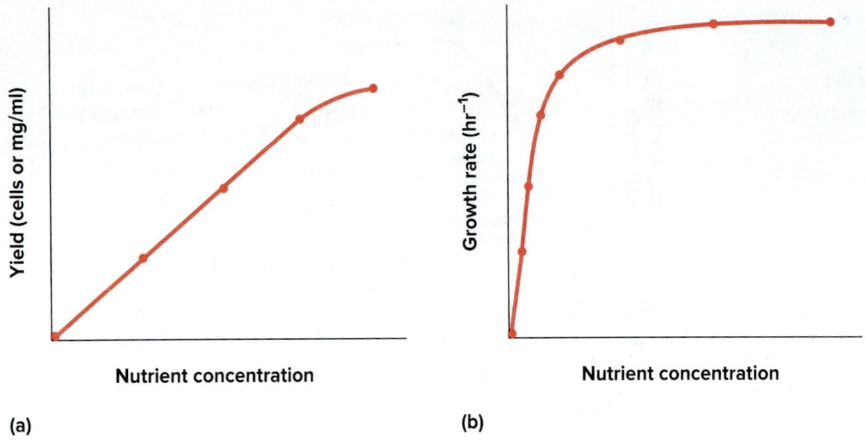

Figure 7.11 Nutrient Concentration and Growth. (a) The effect of changes in limiting nutrient concentration on total microbial yield. At sufficiently high concentrations, total growth will plateau. (b) The effect on growth rate.

further with increasing nutrient concentration. ◀ *Bacteria use many mechanisms to bring nutrients into the cell (section 3.3)*

Stationary Phase

In a closed system such as a batch culture, population growth eventually ceases and the growth curve becomes horizontal (figure 7.10). This **stationary phase** is attained by some bacteria at a population level of around 10^9 cells per milliliter. Protist cultures often have maximum concentrations of about 10^6 cells per milliliter. Final population size depends on nutrient availability and other factors, as well as the type of microorganism. In stationary phase, the total number of viable microorganisms remains constant. This may result from a balance between cell division and cell death, or the population may simply cease to divide but remain metabolically active.

Microbes enter the stationary phase for many reasons. One important reason is nutrient limitation; if an essential nutrient is severely depleted, population growth will slow and eventually stop. Aerobic organisms often are limited by O_2 availability. Oxygen is not very soluble and may be depleted so quickly that only the surface of a culture will have an O_2 concentration adequate for growth. Population growth also may cease due to the accumulation of toxic waste products. This seems to limit the growth of many cultures growing in the absence of O_2. For example, streptococci can produce so much acid from sugar fermentation that growth is inhibited *(see figure 11.20)*. Finally, some evidence exists that growth may cease when a critical population level is reached. Thus entrance into the stationary phase may result from several factors operating in concert.

Information about the nutrient level that affects the growth of the population must be transmitted to the molecules that affect the growth of individual cells. DnaA, the protein that binds to the chromosome's origin to initiate replication, becomes less active in stationary phase. Ongoing replication is completed, but no further initiation occurs. This is one of many ways the cell conserves energy by eliminating processes that are not essential to survival.

Death Phase

Cells growing in batch culture cannot remain in stationary phase indefinitely. Eventually they enter a phase known as the death phase (figure 7.10). During this phase, the number of viable cells declines exponentially, with cells dying at a constant rate. Detrimental environmental changes such as nutrient deprivation and the buildup of toxic wastes cause irreparable harm to the cells.

Long-Term Stationary Phase

Long-term growth experiments reveal that after a period of exponential death some microbes have a long period where the population size remains more or less constant. This **long-term stationary phase** (also called extended stationary phase) can last months to years (figure 7.10). During this time, the bacterial population continually evolves so that actively reproducing cells are those best able to use the nutrients released by their dying brethren and best able to tolerate the accumulated toxins. This dynamic process is marked by successive waves of genetically distinct variants. Thus natural selection can be witnessed within a single culture vessel.

Mathematics of Growth

Knowledge of microbial growth rates during the exponential phase is indispensable to microbiologists. Growth rate studies contribute to basic physiological and ecological research, and are applied in industry. The quantitative aspects of exponential phase growth discussed here apply to microorganisms that divide by binary fission.

During the exponential phase, each microorganism is dividing at constant intervals. Thus the population doubles in number during a specific length of time called the **generation (doubling) time** (g). This can be illustrated with a simple example. Suppose that a culture tube is inoculated with one cell that divides every 20 minutes (**table 7.2**). The population will be 2 cells after 20 minutes, 4 cells after 40 minutes, and so forth. Because the population is doubling every generation, the increase in population is always 2^n where n is the number of

Table 7.2 An Example of Exponential Growth[1]

Time[2]	Generation Number	2^n	Number of cells[3] ($N_0 \times 2^n$)	$\log_{10} N_t$
0	0	$2^0 = 1$	1	0.000
20	1	$2^1 = 2$	2	0.301
40	2	$2^2 = 4$	4	0.602
60	3	$2^3 = 8$	8	0.903
80	4	$2^4 = 16$	16	1.204

1 Definitions of n, N_0, N_t are as in figure 7.13.
2 The hypothetical culture begins with one cell having a 20-minute generation time.
3 Number of cells in the culture.

generations. The resulting population increase is exponential (**figure 7.12**).

The mathematics of growth during the exponential phase is illustrated in **figure 7.13**, which shows the calculation of two important values. The **growth rate constant** (*k*) is the number of generations per unit time and is often expressed as generations per hour (hr^{-1}). It can be used to calculate the generation time. As can be seen in figure 7.13, the generation time is simply the reciprocal of the growth rate constant. The generation time can also be determined directly from a semilogarithmic plot of growth curve data (**figure 7.14**). Once this is done, it can be used to calculate the growth rate constant.

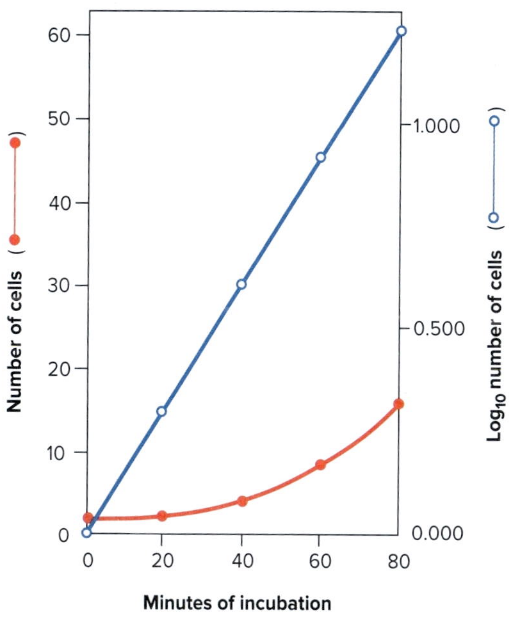

Figure 7.12 Exponential Microbial Growth. Four generations of growth are plotted directly (—) and in the logarithmic form (○—○). The growth curve is exponential, as shown by the linearity of the log plot.

Generation times vary markedly with the microbial species and environmental conditions. They range from less than 10 minutes (0.17 hours) to several days (**table 7.3**). Generation times in nature are usually much longer than in laboratory culture.

Calculation of the growth rate constant

Let N_0 = the initial population number

N_t = the population at time t

n = the number of generations in time t

For populations reproducing by binary fission

$$N_t = N_0 \times 2^n$$

Solving for n, the number of generations, where all logarithms are to the base 10,

$$\log N_t = \log N_0 + n \cdot \log 2, \text{ and}$$

$$n = \frac{\log N_t - \log N_0}{\log 2} = \frac{\log N_t - \log N_0}{0.301}$$

The growth rate constant (k) is the number of generations per unit time $\left(\frac{n}{t}\right)$. Thus

$$k = \frac{n}{t} = \frac{\log N_t - \log N_0}{0.301t}$$

Calculation of generation (doubling) time

If a population doubles, then

$$N_t = 2N_0$$

Substitute $2N_0$ into the growth rate constant equation and solve for

$$k = \frac{\log (2N_0) - \log N_0}{0.301g} = \frac{\log 2 + \log N_0 - \log N_0}{0.301g}$$

$$k = \frac{1}{g}$$

The generation time is the reciprocal of the growth rate constant.

$$g = \frac{1}{k}$$

Figure 7.13 Calculation of the Growth Rate Constant and Generation Time. The calculations are only valid for the exponential phase of growth, when the growth rate is constant.

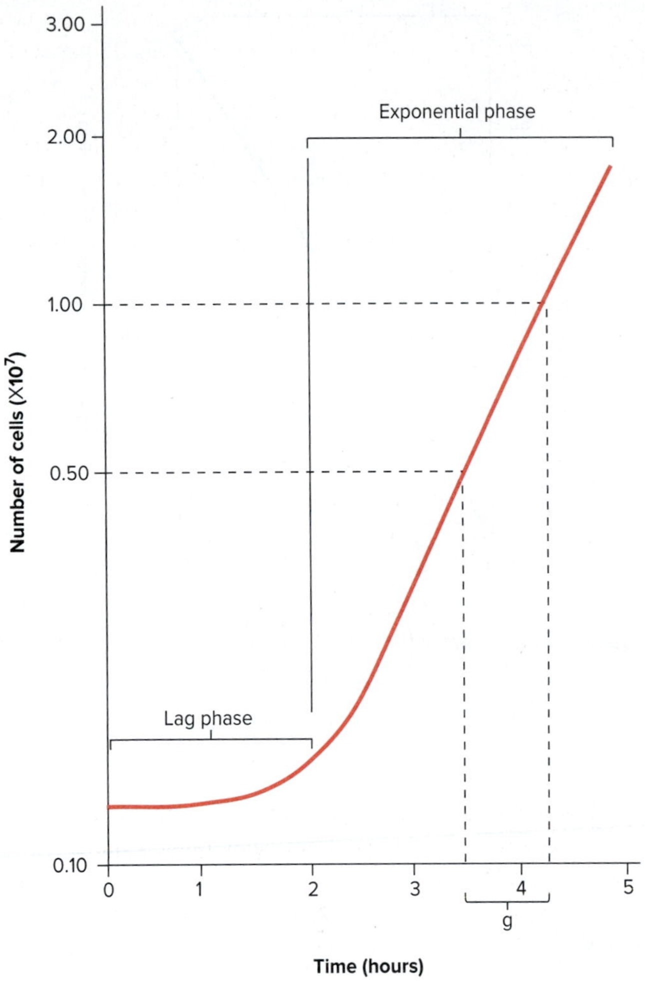

Figure 7.14 Generation Time Determination. The generation time can be determined from a microbial growth curve. The population data are plotted with the logarithmic axis used for the number of cells. The time to double the population number is then read directly from the plot. The log of the population number can also be plotted against time on regular axes.

Table 7.3 Examples of Generation Times[1]

Microorganism	Incubation Temperature (°C)	Generation Time (Hours)
Bacteria		
Escherichia coli	40	0.35
Staphylococcus aureus	37	0.47
Clostridium botulinum	37	0.58
Mycobacterium tuberculosis	37	~12
Treponema pallidum	37	33
Archaea		
Ignicoccus hospitalis	90	1
Pyrococcus abyssi	90	0.67
Sulfolobus tokodaii	75	6
Methanosphaera sedula	75	8
Nitrososphaera viennensis	37	45
Protists		
Tetrahymena geleii	24	2.2–4.2
Chlorella pyrenoidosa	25	7.75
Euglena gracilis	25	10.9
Ceratium tripos	20	82.8
Fungi		
Saccharomyces cerevisiae	30	2
Monilinia fructicola	25	30

1 Generation times differ depending on the growth medium and environmental conditions used.

Comprehension Check

1. Define microbial growth.
2. Describe the phases of the growth curve and discuss the causes of each.
3. Why would cells that are vigorously growing have a shorter lag phase than those that have been stored in a refrigerator when inoculated into fresh culture medium?
4. Calculate the growth rate constant and generation time of a culture that increases in the exponential phase from 5×10^2 to 1×10^8 in 12 hours.
5. Suppose the generation time of a bacterium is 90 minutes and the number of cells in a culture is 10^3 cells at the start of the exponential phase. How many bacteria will there be after 8 hours of exponential growth?

7.5 Environmental Factors Affect Microbial Growth

After reading this section, you should be able to:

a. Use terms that describe a microbe's growth range or requirement for each of the factors that influence microbial growth
b. Summarize the adaptations of extremophiles to their habitats
c. Summarize the strategies used by nonextremophiles to acclimate to changes in their environment
d. Describe enzymes observed in microbes that protect them against toxic O_2 products

Almost every year at Yellowstone National Park, a visitor is seriously burned or killed after falling into one of the park's hot springs. Yet, a rich microbial community lives in these same springs, as well as in the hot pots, fumaroles, and other thermal features of the park. Clearly, the adaptations of some microorganisms to what humans perceive as inhospitable, **extreme environments** are truly remarkable. Indeed, microbes are thought to be present nearly everywhere on Earth. Bacteria such as *Bacillus infernus* are able to live over 2.4 km below Earth's surface, without oxygen and at temperatures above 60°C. Other microbes live at great ocean depths or in lakes such as the Great Salt Lake in Utah (USA) that have high sodium chloride concentrations. Microorganisms that grow in such harsh conditions are called **extremophiles.** Whether an extremophile or not, all microbes must respond to changes in their environment. However, if the conditions exceed their ability to respond, they will not grow and eventually they may die. Thus for each environmental parameter, all microbes have a characteristic range at which growth occurs defined by high and low values beyond which the microbe cannot survive. Within the range is an optimal value at which growth is best. ▶| *Microorganisms in marine ecosystems (section 30.2); The subsurface biosphere is vast (section 31.4)*

To study the ecological distribution of microbes, it is important to understand the strategies they use to survive. An understanding of the environmental influences on microbes and their activity also aids in the control of microbial growth. In this section, we briefly review the effects of the most important environmental factors on microbial growth. Major emphasis is given to solutes and water activity, pH, temperature, oxygen level, pressure, and radiation. The adaptations that microbes have evolved to live in their environments also are discussed. **Table 7.4** summarizes how microorganisms are categorized in terms of their response to these factors.

Solutes Affect Osmosis and Water Activity

Water is critical to the survival of all organisms, but water can also be destructive. Solutes in an aqueous solution alter the behavior of water. One way this occurs is the phenomenon of osmosis, which is observed when two solutions are separated by a semipermeable membrane that allows movement of water but not solutes. If the solute concentration of one solution is higher than the other, water moves to equalize the concentrations. In other words, water moves from solutions with lower solute concentrations to those with higher solute concentrations. Because a selectively permeable plasma membrane separates a cell's cytoplasm from its environment, microbes can be affected by changes in the solute concentration of their surroundings. If a microorganism is placed in a hypotonic solution (one with a lower solute concentration; solute concentration is also referred to as osmotic concentration or osmolarity), water will enter the cell and cause it to burst unless something prevents the influx of water or inhibits plasma membrane expansion. Conversely, if the microbe is placed in a hypertonic solution (one with a higher osmotic concentration), water will flow out of the cell. In microbes that have cell walls, the membrane shrinks away from the cell wall. Dehydration of the cell in hypertonic environments may damage the plasma membrane and cause the cell to become metabolically inactive.

Because of the potential damaging effects of uncontrolled osmosis, it is important that microbes be able to respond to changes in the solute concentrations of their environment. Microbes in hypotonic environments are protected in part by their cell wall, which prevents overexpansion of the plasma membrane. However, not all microbes have cell walls. Wall-less microbes can be protected by reducing the osmotic concentration of their cytoplasm; this protective measure is also used by many walled microbes to provide protection in addition to their cell walls. Microbes use several mechanisms to lower the solute concentration of their cytoplasm. For example, some bacteria have mechanosensitive (MS) channels in their plasma membrane. In a hypotonic environment, the membrane stretches due to an increase in hydrostatic pressure and cellular swelling. MS channels then open and allow solutes to leave. Thus MS channels act as escape valves to protect cells from bursting. Many protists use contractile vacuoles to expel excess water.

Some microbes are adapted to extreme hypertonic environments, and can be called osmophiles. By definition, osmophiles include **halophiles,** which require the presence of NaCl at a concentration above about 0.2 M (**figure 7.15**). However, usually **osmophile** refers to organisms that require high concentrations of sugars. Here we focus on halophiles. Extreme halophiles have adapted so completely to hypertonic, saline conditions that they require NaCl concentrations between about 3 M and saturation (about 6.2 M). Archaeal halophiles can be isolated from the Dead Sea (a salt lake between Israel and Jordan), the Great Salt Lake in Utah, and other aquatic habitats with salt concentrations approaching saturation.

Halophiles generally are able to live in their high-salt habitats because they synthesize or obtain from their environment molecules called **compatible solutes.** Compatible solutes can be kept at high intracellular concentrations without interfering with metabolism and growth. Some compatible solutes are inorganic molecules such as potassium chloride (KCl). Others are organic molecules such as choline, betaines (neutral molecules having both negatively charged and positively charged functional groups), and amino acids such as proline and glutamic acid. The use of compatible solutes extends beyond halophiles. This is also a strategy utilized by many osmophiles. Furthermore, many microorganisms, whether in hypotonic or hypertonic environments, use compatible solutes to keep the osmotic concentration of their cytoplasm somewhat above that of the habitat so that the plasma membrane is always pressed firmly against the cell wall. For instance, fungi and photosynthetic protists employ the compatible solutes sucrose and polyols (e.g., arabitol, glycerol, and mannitol) for this purpose.

Some of the best-studied halophiles belong to the archaeal order *Halobacteriales.* These archaea accumulate potassium and chloride ions to remain hypertonic to their environment; the internal potassium concentration may reach 4 to 7 M. Their

Table 7.4 Microbial Responses to Environmental Factors

Descriptive Term	Definition	Representative Genera and Species
Solute Concentration and Water Activity		
Halophile	Requires high levels of sodium chloride, usually above about 0.2 M, to grow	*Halobacterium, Dunaliella, Ectothiorhodospira*
Osmotolerant	Able to grow over wide ranges of water activity or osmotic concentration	*Staphylococcus aureus, Zygosaccharomyces rouxii*
Xerophile	Organisms that grow best at low water activity, typically with optima at 0.85 or below	*Xeromyces bisporus*
pH		
Acidophile	Growth optimum between pH 0 and 5.5	*Sulfolobus, Picrophilus, Ferroplasma, Acontium*
Neutrophile	Growth optimum between pH 5.5 and 8.0	*Escherichia, Euglena, Paramecium*
Alkaliphile	Growth optimum between pH 8.0 and 11.5	*Bacillus alcalophilus, Natronobacterium*
Temperature		
Psychrophile	Grows at 0°C and has an optimum growth temperature of 15°C or lower	*Bacillus psychrophilus, Chlamydomonas nivalis*
Psychrotolerant	Can grow at 0–7°C; has an optimum between 20 and 30°C and a maximum around 35°C	*Listeria monocytogenes, Pseudomonas fluorescens*
Mesophile	Has growth optimum between 20 and 45°C	*Escherichia coli, Trichomonas vaginalis*
Thermophile	Can grow at 55°C or higher; optimum often between 55 and 65°C	*Geobacillus stearothermophilus, Thermus aquaticus, Cyanidium caldarium, Chaetomium thermophilum*
Hyperthermophile	Has an optimum between 85 and about 113°C	*Sulfolobus, Pyrococcus, Pyrodictium*
Oxygen Concentration		
Obligate aerobe	Completely dependent on atmospheric O_2 for growth	*Micrococcus luteus,* most protists and fungi
Facultative anaerobe	Does not require O_2 for growth but grows better in its presence	*Escherichia, Enterococcus, Saccharomyces cerevisiae*
Aerotolerant anaerobe	Grows equally well in presence or absence of O_2	*Streptococcus pyogenes*
Obligate anaerobe	Does not tolerate O_2 and dies in its presence	*Clostridium, Bacteroides, Methanothermobacter*
Microaerophile	Requires O_2 levels between 2 and 10% for growth and is damaged by atmospheric O_2 levels (20%)	*Campylobacter, Spirillum volutans, Treponema pallidum*
Pressure		
Piezophile (barophile)	Growth more rapid at high hydrostatic pressures	*Photobacterium profundum, Shewanella benthica*
Radiation		
Radiation resistant	Survives high doses of gamma radiation	*Deinococcus radiodurans, Thermococcus gammatolerans*

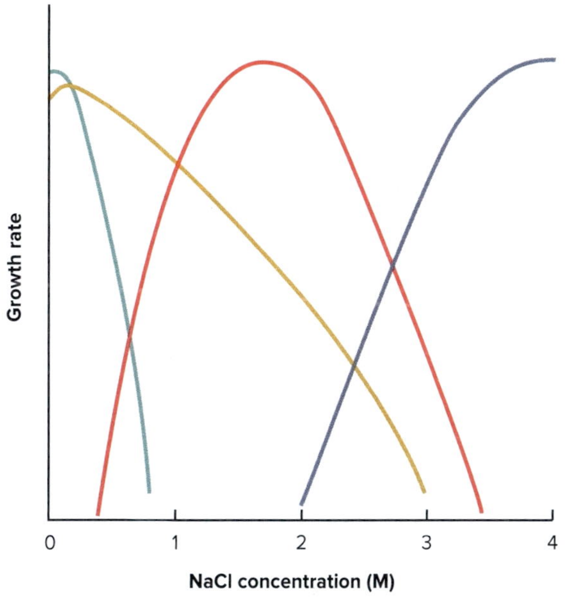

Figure 7.15 The Effects of Sodium Chloride on Microbial Growth. Four different patterns of microbial dependence on NaCl concentration are depicted. The curves are only illustrative and are not meant to provide precise shapes or salt concentrations required for growth.

MICRO INQUIRY *What is the difference between halophilic and halotolerant?*

enzymes, ribosomes, and transport proteins are unusual because they require high potassium levels for stability and activity. In addition, their plasma membranes and cell walls are stabilized by high concentrations of sodium ion. If the sodium concentration decreases too much, the wall and plasma membrane disintegrate. Although extreme halophiles have successfully adapted to environmental conditions that would destroy most organisms, they have become so specialized that they lack ecological flexibility. ▶| *Haloarchaea (section 20.3)*

Another way solutes change the behavior of water is by decreasing the availability of water to microbes. When solutes such as salts and sugars are present, the water is "tied up" by its interaction with the solutes. Microbiologists express quantitatively the degree of water availability by determining **water activity (a_w)**. The water activity of a solution is 1/100 the relative humidity of the solution (when expressed as a percent). It is also equivalent to the ratio of the solution's vapor pressure (P_{soln}) to that of pure water (P_{water}). Distilled water has an a_w of 1, milk has an a_w of 0.97, a saturated salt solution has an a_w of 0.75, and the a_w of dried fruits is only about 0.5.

To survive in a habitat with a low a_w value, microorganisms must maintain a high internal solute concentration to retain water. These **osmotolerant** microorganisms grow over wide ranges of water activity but optimally at higher levels. Osmotolerant organisms can be found in all domains of life. For example, *Staphylococcus aureus* is halotolerant, can be cultured in media containing up to about 3 M sodium chloride, and is well adapted for growth on the skin. The yeast *Zygosaccharomyces rouxii* grows in sugar solutions with a_w values as low as about 0.65. The photosynthetic protist *Dunaliella viridis* tolerates sodium chloride concentrations from 1.7 M to a saturated solution. In contrast, xerotolerant microbes can withstand either high solute concentrations or the effects of desiccation. These microbes exist in desert regions, household dust, and in preserved foods. However, most microorganisms only grow well at water activities around 0.98 (the approximate a_w for seawater) or higher. This is why drying food or adding large quantities of salt and sugar effectively prevents food spoilage. ▶| *Various methods are used to control food spoilage (section 41.2)*

Comprehension Check

1. How do microorganisms adapt to hypotonic and hypertonic environments?
2. Define water activity and briefly describe how it can be determined. Why is it difficult for microorganisms to grow at low a_w values?
3. What are halophiles and why do they require sodium and potassium ions?

pH

pH is a measure of the relative acidity of a solution and is defined as the negative logarithm of the hydrogen ion concentration (expressed in terms of molarity).

$$pH = -\log [H^+] = \log(1/[H^+])$$

The pH scale extends from pH 0.0 (1.0 M H^+) to pH 14.0 (1.0×10^{-14} M H^+), and each pH unit represents a 10-fold change in hydrogen ion concentration. **Figure 7.16** shows that microbial habitats vary widely in pH—from pH 0 to 2 at the acidic end to alkaline lakes and soil with pH values between 9 and 10.

Each species has a definite pH growth range and pH growth optimum. **Acidophiles** have their growth optimum between pH 0 and 5.5; **neutrophiles,** between pH 5.5 and 8.0; and **alkaliphiles (alkalophiles)**, between pH 8.0 and 11.5. In general, different microbial groups have characteristic pH preferences. Most known bacteria and protists are neutrophiles. Most fungi prefer more acidic surroundings, about pH 4 to 6; photosynthetic protists also seem to favor slight acidity. Many archaea are acidophiles. For example, the archaeon *Sulfolobus acidocaldarius* is a common inhabitant of acidic hot springs; it grows well from pH 1 to 3 and at high temperatures. The archaea *Ferroplasma acidarmanus* and *Picrophilus oshimae* can actually grow very close to pH 0. Alkaliphiles are distributed among all three domains of life. They include bacteria belonging to the genera *Bacillus, Micrococcus, Pseudomonas,* and *Streptomyces;* yeasts

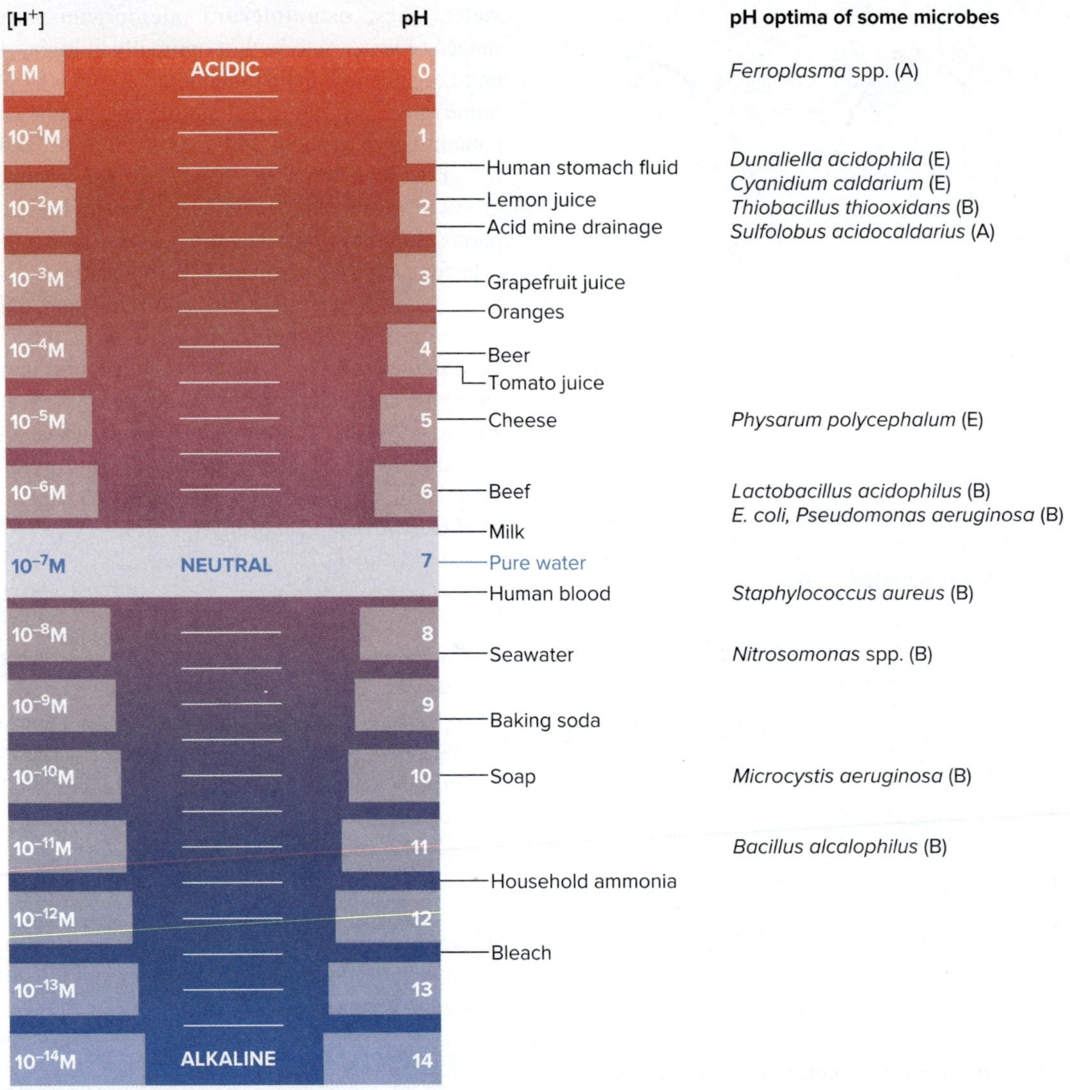

Figure 7.16 The pH Scale. The pH scale and examples of substances with different pH values. Several microorganisms are placed at their growth optima. A, archaeon; E, eukaryote; B, bacterium.

and filamentous fungi; and numerous archaea. Because seawater has a pH of about 8.3, marine microorganisms are alkaliphilic.

Although microorganisms often grow over wide ranges of pH and far from their optima, there are limits to their tolerance. When the external pH is low, the concentration of H^+ is much greater outside than inside, and H^+ will move into the cytoplasm and lower the cytoplasmic pH. Drastic variations in cytoplasmic pH can harm microorganisms by disrupting the plasma membrane or inhibiting the activity of enzymes and membrane transport proteins. Most microbes die if the internal pH drops much below 5.0 to 5.5. Changes in the external pH also can alter the ionization of nutrient molecules and thus reduce their availability to the organism.

Microorganisms respond to external pH changes using mechanisms that maintain a neutral cytoplasmic pH. Several responses to small changes in external pH have been identified. Neutrophiles appear to exchange potassium for protons using an antiport transport system. Internal buffering also contributes to pH homeostasis. However, if the external pH becomes too acidic, other mechanisms come into play. When the pH drops below about 5.5, *Salmonella enterica* serovar Typhimurium and *E. coli* synthesize an array of new proteins as part of what has been called their acid tolerance response. An ATPase enzyme contributes to this protective response by pumping protons out of the cell, at the expense of ATP. If the external pH decreases to 4.5 or lower, acid shock proteins and heat shock proteins are synthesized. These prevent the denaturation of proteins and aid in refolding denatured proteins in acidic conditions. ◀ *Primary and secondary active transport (section 3.3)* ▶ *Molecular chaperones: proteins that help proteins fold (section 13.8)*

What about microbes that live at pH extremes? Extreme alkaliphiles such as *Bacillus alcalophilus* maintain their internal pH close to neutrality by exchanging internal sodium ions for external protons. Acidophiles use a variety of measures to maintain a neutral internal pH. These include the transport of cations (e.g., potassium ions) into the cell, thus decreasing the movement of H^+ into the cell; proton transporters that pump H^+ out if they get in; and highly impermeable cell membranes.

Comprehension Check

1. Define pH, acidophile, neutrophile, and alkaliphile.
2. Classify each of the following organisms as an alkaliphile, a neutrophile, or an acidophile: *Staphylococcus aureus, Microcystis aeruginosa, Sulfolobus acidocaldarius,* and *Pseudomonas aeruginosa* (see figure 7.16). Which might be pathogens? Explain your choices.
3. Describe the mechanisms microbes use to maintain an internal neutral pH. Explain how extreme pH values might harm microbes.

Temperature

Microorganisms are particularly susceptible to external temperatures because they cannot regulate their internal temperature. An important factor influencing the effect of temperature on growth is the temperature sensitivity of enzyme-catalyzed reactions. Each enzyme has a temperature at which it functions optimally. At some temperature below the optimum, it ceases to be catalytic. As the temperature rises from this low point, the rate of catalysis increases to that observed for the optimal temperature. The velocity of the reaction roughly doubles for every 10°C rise in temperature. When all enzymes in a microbe are considered together, as the rate of each reaction increases, metabolism as a whole becomes more active and the microorganism grows faster. However, beyond a certain point, further temperature increases actually slow growth, and sufficiently high temperatures are lethal. High temperatures denature enzymes, as well as transport and other proteins. Temperature also has a significant effect on microbial membranes. At very low temperatures, membranes solidify. At high temperatures, the lipid bilayer simply melts and disintegrates. Thus when organisms are above or below their optimum temperature, both cell function and structure are affected.

Because of these opposing temperature influences, microbial growth has a characteristic temperature dependence with distinct **cardinal temperatures**—minimum, optimum, and maximum growth temperatures (**figure 7.17**). The peak of the graph, the optimum temperature, is where the generation time (*g*) is shortest. Although the shape of temperature dependence curves varies, the temperature optimum is always closer to the maximum than to the minimum. The cardinal temperatures are not rigidly fixed. Instead, they depend to some extent on other environmental factors such as pH and available nutrients. For example, *Crithidia fasciculata,* a flagellated protist living in the gut of mosquitoes, grows in a simple medium at 22° to 27°C. However, if its medium is supplemented with extra metals, amino acids, vitamins, and lipids, it can grow at 33° to 34°C.

The cardinal temperatures vary greatly among microorganisms (**table 7.5**). Optima usually range from 0° to 75°C, whereas microbial growth occurs at temperatures extending from less

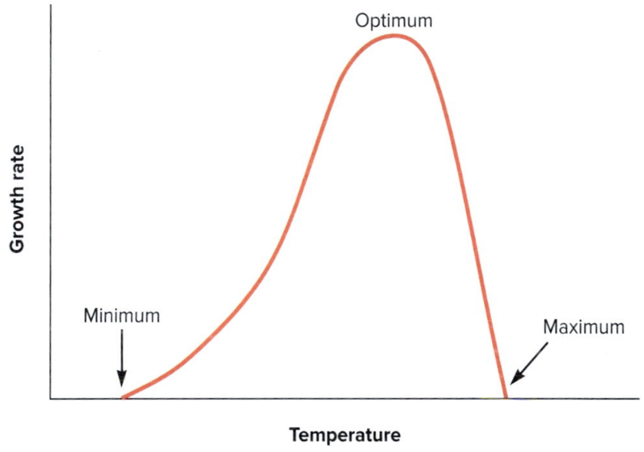

Figure 7.17 Temperature and Growth. The effect of temperature on growth rate. The y-axis is the growth rate constant, k.

MICRO INQUIRY *How would you generate a curve of this type with growth curve data, measured as in figure 7.14?*

Table 7.5 Temperature Ranges for Microbial Growth

	CARDINAL TEMPERATURES (°C)		
Microorganism	**Minimum**	**Optimum**	**Maximum**
Bacteria and Archaea			
Planococcus halocryophilus	–15	25	37
Bacillus psychrophilus	0–3	25	30
Pseudomonas fluorescens	4	25–30	40
Escherichia coli	10	37	45
Neisseria gonorrhoeae	30	35–36	38
Thermus aquaticus	40	70–72	79
Pyrolobus fumarii	90	106	113
Protists			
Chlamydomonas nivalis	–36	0	4
Amoeba proteus	4–6	22	35
Trichomonas vaginalis	25	32–39	42
Cyclidium citrullus	18	43	47
Fungi			
Candida scotti	0	4–15	15
Saccharomyces cerevisiae	1–3	28	40
Mucor pusillus	21–23	45–50	50–58

than –20°C to over 120°C. Some archaea even grow at 121°C (250°F), the temperature normally used in autoclaves. A major factor determining growth range is the availability of water. Even at the most extreme temperatures, microorganisms need liquid water to grow. The growth temperature range for a particular microorganism usually spans about 30°. Some species (e.g., *Neisseria gonorrhoeae*) have a small range; others grow over a wide range of temperatures. The major microbial groups differ from one another regarding their maximum growth temperatures. The upper limit for protists is around 50°C. Some fungi grow at temperatures as high as 55° to 60°C. Some bacteria and archaea can grow at much higher temperatures than eukaryotes. It has been suggested that eukaryotes are not able to manufacture stable and functional organellar membranes at temperatures above 60°C.

Several terms are used to describe microbes based on their temperature ranges for growth (**figure 7.18**). Microbes that grow in cold environments are either psychrotolerants or psychrophiles. **Psychrotolerants** (sometimes called **psychrotrophs**) grow at 0°C or higher and typically have maxima at about 35°C. **Psychrophiles** (sometimes called **cryophiles**) grow well at 0°C and have an optimum growth temperature of 15°C; the maximum is around 20°C. They are readily isolated from Arctic and Antarctic habitats. Oceans constitute an enormous habitat for psychrophiles because 90% of ocean water is 5°C or colder. The psychrophilic protist *Chlamydomonas nivalis* can actually turn a snowfield or glacier pink with its bright red spores (a phenomenon called "watermelon snow"). Psychrophiles are widespread among bacterial taxa and are found in such genera as *Pseudomonas, Vibrio, Alcaligenes, Bacillus, Photobacterium,* and *Shewanella.* Psychrophilic microorganisms have adapted to their environment in several ways. Their enzymes, transport systems, and protein synthetic machinery function well at low temperatures. The membranes of psychrophilic microorganisms have high levels of unsaturated fatty acids and remain semifluid when cold. Indeed, many psychrophiles begin to leak cellular constituents at temperatures higher than 20°C because of membrane disruption. Many psychrophiles accumulate compatible solutes. Rather than protecting against osmotic stress, in this case the compatible solutes decrease the freezing point of the cytosol. Still other psychrophiles use antifreeze proteins to decrease the freezing point of the cytosol. Psychrophilic bacteria and fungi are major causes of refrigerated food spoilage.

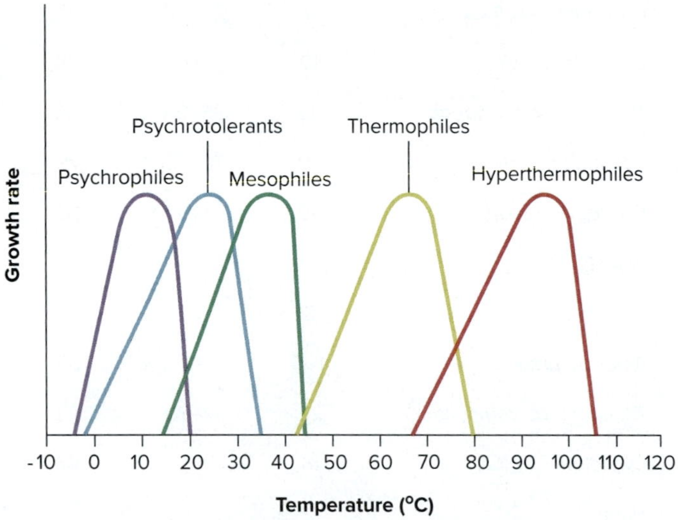

Figure 7.18 Temperature Ranges for Microbial Growth. Microorganisms are placed in different classes based on their temperature ranges for growth. They are ranked in order of increasing growth temperature range as psychrophiles, psychrotolerants, mesophiles, thermophiles, and hyperthermophiles. Representative ranges and optima for these five types are illustrated.

Mesophiles are microorganisms that grow in moderate temperatures. They have growth optima around 20° to 45°C and often have a temperature minimum of 15° to 20°C and a maximum of about 45°C. Almost all human pathogens are mesophiles, as might be expected because the human body is a fairly constant 37°C.

Microbes that grow best at high temperatures are thermophiles and hyperthermophiles. **Thermophiles** grow at temperatures between 45° and 85°C, and they often have optima between 55° and 65°C. The vast majority are members of *Bacteria* or *Archaea,* although a few photosynthetic protists and fungi are thermophilic. Thermophiles flourish in many habitats including composts, self-heating hay stacks, hot water lines, and hot springs. **Hyperthermophiles** have growth optima between 85° and 113°C. They usually do not grow below 55°C. *Pyrococcus abyssi* and *Pyrodictium occultum* are examples of marine hyperthermophiles found in hot areas of the seafloor.

Thermophiles and hyperthermophiles differ from mesophiles in many ways. They have heat-stable enzymes and protein synthesis systems that function at high temperatures. Heat-stable proteins have highly organized hydrophobic interiors and many hydrogen and other noncovalent bonds to stabilize their structure. Large quantities of amino acids such as proline make polypeptide chains less flexible and more heat stable. In addition, the proteins are stabilized and aided in folding by proteins called chaperones. Nucleoid-associated proteins appear to stabilize the DNA of thermophilic bacteria. Hyperthermophiles have an enzyme called reverse DNA gyrase that changes the topology of their DNA and enhances its stability. The membrane lipids of thermophiles and hyperthermophiles are also quite temperature stable. They tend to be more saturated, more branched, and of higher molecular weight. This increases the melting points of membrane lipids. Recall that archaea have membrane lipids with ether linkages. Such lipids are resistant to hydrolysis at high temperatures. Furthermore, the diglycerol tetraethers observed in the membranes of some archaeal thermophiles span the membrane to form a rigid, stable monolayer. ◀ *Archaeal cell envelopes are structurally diverse (section 4.2)* ▶ *Proteins (appendix I)*

Comprehension Check

1. What are cardinal temperatures?
2. Why does the growth rate rise with increasing temperature and then fall again at higher temperatures?
3. Identify environments where you could find microbes considered psychrophilic, psychrotolerant, mesophilic, thermophilic, and hyperthermophilic.
4. Compare and contrast the metabolic and structural adaptations for extreme temperatures found in psychrophiles and thermophiles.

Oxygen Concentration

The importance of oxygen to the growth of an organism correlates with the processes it uses to conserve energy. Almost all energy-conserving metabolic processes involve the movement of electrons through a series of membrane-associated electron carriers called an electron transport chain (ETC). For organisms that use chemical energy sources, an externally supplied terminal electron acceptor is critical to the functioning of the ETC. In many cases, the terminal electron acceptor is oxygen. ▶| *Electron transport chains: sets of sequential redox reactions (section 10.4)*

Traditionally, five types of relationships to oxygen have been described. Almost all multicellular organisms are completely dependent on atmospheric O_2 for growth; that is, they are **obligate aerobes** (table 7.4). Oxygen serves as the terminal electron acceptor for the ETC in the metabolic process called aerobic respiration. **Microaerophiles** are damaged by the atmospheric level of O_2 (20%) and require O_2 levels in the range of 2 to 10% for growth. **Facultative anaerobes** do not require O_2 for growth but grow better in its presence. In the presence of oxygen, they use O_2 as the terminal electron acceptor during aerobic respiration. **Aerotolerant anaerobes** grow equally well whether O_2 is present or not; they can tolerate O_2, but they do not make use of it. The ability to grow in both oxic and anoxic environments provides considerable flexibility and is an ecological advantage for facultative and aerotolerant anaerobes. In contrast, for strict or **obligate anaerobes,** O_2 is toxic, and they are usually killed by prolonged exposure to O_2. Strict anaerobes cannot generate energy through aerobic respiration and employ other metabolic strategies that don't require O_2. Although obligate anaerobes are killed by O_2, they may be recovered from habitats that appear to be oxic. In such cases, they associate with facultative anaerobes that use up the available O_2 and thus make the growth of strict anaerobes possible. For example, the strict anaerobe *Porphyromonas gingivalis* lives in the mouth, where it grows in the anoxic crevices around the teeth. ▶| *Aerobic respiration can be divided into three steps (section 11.3); Anaerobic respiration uses the same three steps as aerobic respiration (section 11.7); Fermentation does not involve an electron transport chain (section 11.8)*

The nature of bacterial O_2 responses can be readily determined by growing the microbe in a medium such as thioglycollate broth, which contains a reducing agent to lower O_2 levels (**figure 7.19**). All five responses are found among bacteria, archaea, and protists. Most fungi are aerobic, but a number of species—particularly among yeasts—are facultative anaerobes. Photosynthetic protists are usually obligate aerobes.

The different relationships with O_2 are due to several factors, including the inactivation of proteins and the effect of toxic O_2 derivatives. Enzymes can be inactivated when sensitive groups such as sulfhydryls are oxidized. However, even molecules that have evolved to function aerobically can be damaged by O_2. This is because the unpaired electrons in the outer shell of oxygen make it inherently unstable (*see figure AI.2*). Toxic O_2 derivatives are formed when cellular proteins such as flavoproteins transfer electrons to O_2. These toxic O_2 derivatives are called ***reactive oxygen species*** (**ROS**), and they can damage proteins, lipids, and nucleic acids. ROS include the superoxide radical, hydrogen peroxide, and the most dangerous hydroxyl radical.

$$O_2 + e^- \rightarrow O_2^{\bar{\bullet}} \text{ (superoxide radical)}$$

$$O_2^{\bar{\bullet}} + e^- + 2H^+ \rightarrow H_2O_2 \text{ (hydrogen peroxide)}$$

$$H_2O_2 + e^- + H^+ \rightarrow H_2O + OH\bullet \text{ (hydroxyl radical)}$$

A microorganism must be able to protect itself against ROS or it will be killed. Indeed, neutrophils and macrophages, two important

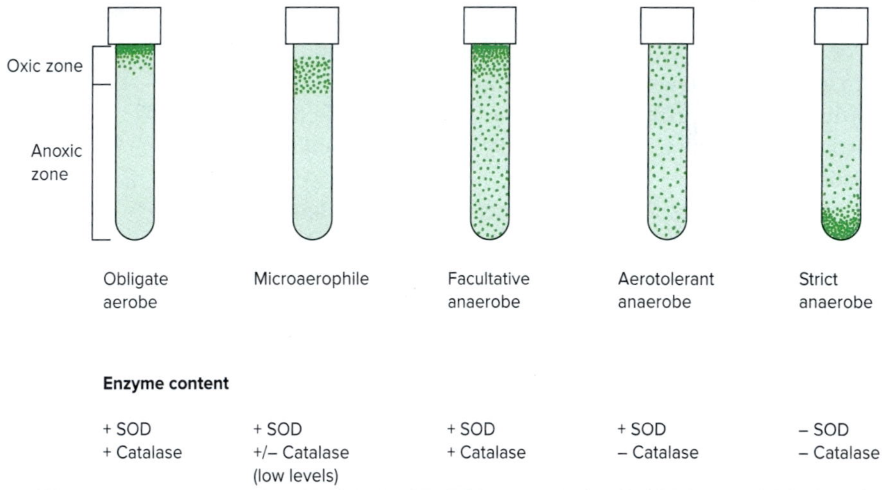

Figure 7.19 Oxygen and Bacterial Growth. Each dot represents an individual bacterial colony within the broth or on its surface. The surface, which is directly exposed to atmospheric oxygen, is oxic. The oxygen content of the medium decreases with depth until the medium becomes anoxic toward the bottom of the tube. The presence and absence of the enzymes superoxide dismutase (SOD) and catalase for each type are shown.

MICRO INQUIRY *Why do facultative anaerobes grow best at the surface of the tube while aerotolerant anaerobes demonstrate a uniform growth pattern throughout the tube?*

immune system cells, use ROS to destroy invading pathogens. ▶| *Redox reactions: reactions of central importance In metabolism (section 10.3); Phagocytosis destroys invaders (section 32.6)*

Many microorganisms possess enzymes that protect against toxic O_2 products (figure 7.19). Obligate aerobes and facultative anaerobes usually contain the enzymes **superoxide dismutase (SOD)** and **catalase,** which catalyze the destruction of superoxide radicals and hydrogen peroxide, respectively. Peroxidase also can be used to destroy hydrogen peroxide.

$$2O_2^{\bar{\bullet}} + 2H^+ \xrightarrow{\text{superoxide dismutase}} O_2 + H_2O_2$$

$$2H_2O_2 \xrightarrow{\text{catalase}} 2H_2O + O_2$$

$$H_2O_2 + NADH + H^+ \xrightarrow{\text{peroxidase}} 2H_2O + NAD^+$$

Strict anaerobes lack these enzymes or have them in very low concentrations and therefore cannot tolerate O_2.

Comprehension Check

1. Describe the five types of O_2 relationships seen in microorganisms.
2. What are the toxic effects of O_2? How do aerobes and other oxygen-tolerant microbes protect themselves from these effects?

Pressure

Organisms that spend their lives on land or the surface of water are always subjected to a pressure of 1 atmosphere (atm; 1 atm is ~0.1 megapascal, or MPa for short) and are never affected significantly by pressure. Other organisms, including many bacteria and archaea, live in the deep sea (ocean depths of 1,000 m or more), where the hydrostatic pressure can reach 600 to 1,100 atm and the temperature is about 2° to 3°C. These high hydrostatic pressures affect membrane fluidity and membrane-associated function. How do microbes survive under such conditions?

Many microbes found at great ocean depths are **barotolerant**—that is, increased pressure adversely affects them but not as much as it does nontolerant microbes. Some are truly **piezophilic (barophilic)**. A piezophile is defined as an organism that has a maximal growth rate at pressures greater than 1 atm. For instance, a piezophile recovered from the Mariana trench near the Philippines (depth about 10,500 m) grows only at pressures between about 400 to 500 atm when incubated at 2°C.

An important adaptation observed in piezophiles is that they change their membrane lipids in response to increasing pressure. For instance, bacterial piezophiles increase the amount of unsaturated fatty acids in their membrane lipids as pressure increases. They may also shorten the length of their fatty acids. Piezophiles are thought to play important roles in nutrient cycling in the deep sea. Thus far, they have been found among several bacterial genera (e.g., *Photobacterium, Shewanella, Colwellia*). ▶| *Microbial communities in benthic marine sediments are enormous and mysterious (section 30.2)*

Radiation

Our world is bombarded with electromagnetic radiation of various types (**figure 7.20**). Radiation behaves as if it were composed of waves like those traveling on the surface of water. The distance between two wave crests or troughs is the wavelength. As the wavelength of electromagnetic radiation decreases, the energy of the radiation increases; gamma rays and X rays are much more energetic than visible light or infrared waves. Electromagnetic radiation also acts like a stream of energy packets called photons, each photon having a quantum of energy whose value depends on the wavelength of the radiation.

Sunlight is the major source of radiation on Earth. It includes visible light, ultraviolet (UV) radiation, infrared rays, and radio waves. Visible light is a most conspicuous and important aspect of our environment because most life on Earth depends on the ability of photosynthetic organisms to trap the energy it holds. The visible spectrum consists of the ROYGBV wavelengths most of us memorized for early science classes. Sunlight is evenly distributed across these wavelengths and therefore appears white. Despite the importance of visible light, it is not the predominant component of sunlight. Infrared rays hold that honor, making up almost 60% of the sun's radiation. Infrared radiation is the major source of Earth's heat. Only about 3% of the light reaching Earth's surface is UV radiation, and this includes only certain types of UV light. There are three major types of UV radiation: UVA, UVB, and UVC, which range from longest (UVA) to shortest (UVC) wavelengths. At sea level, there is very little UV radiation at wavelengths below about 290 nm

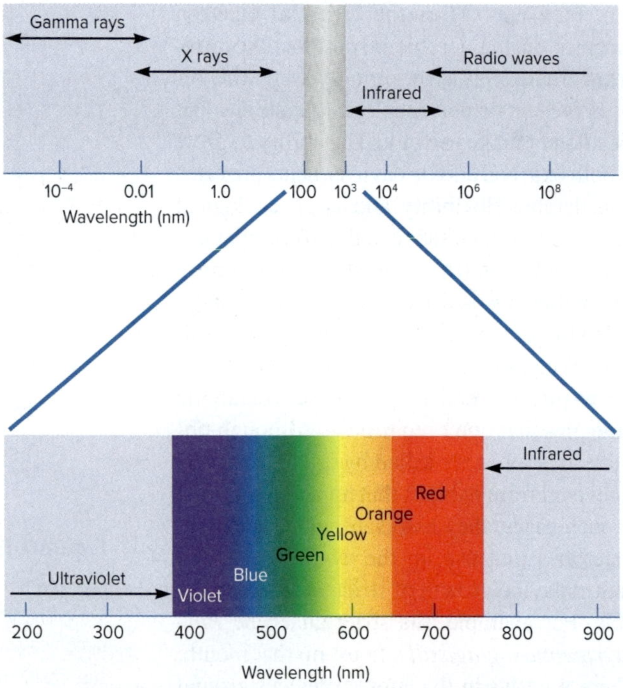

Figure 7.20 The Electromagnetic Spectrum. A portion of the spectrum showing the visible light range and the nearby ultraviolet and infrared wavelengths is expanded at the bottom of the figure.

(UVC and some UVB). This is because wavelengths shorter than this are absorbed by O_2 in Earth's atmosphere; this process forms a layer of ozone (O_3) between 40 and 48 km above Earth's surface. The ozone layer absorbs somewhat longer UV rays and reforms O_2. Thus the major form of UV radiation reaching Earth's surface is UVA radiation. ▶| *Phototrophy (section 11.12)*

Many forms of electromagnetic radiation are very harmful to microorganisms. One of the most damaging is **ionizing radiation,** radiation of very short wavelength and high energy, which causes atoms to lose electrons (ionize). Two major forms of ionizing radiation are X rays, which are artificially produced, and gamma rays, which are emitted during natural radioisotope decay. Low levels of ionizing radiation may produce mutations that indirectly result in death, whereas higher levels are directly lethal. Ionizing radiation causes a variety of changes in cells. It breaks hydrogen bonds, destroys ring structures, and polymerizes some molecules but the most severe effect is protein oxidation.

Although microorganisms are more resistant to ionizing radiation than larger organisms, they are still destroyed by a sufficiently large dose. Indeed, ionizing radiation can be used to sterilize items. However, bacterial endospores and bacteria such as *Deinococcus radiodurans* are extremely resistant to large doses of ionizing radiation. *D. radiodurans* is of particular note. This amazing microbe is able to piece together its genome after it is blasted apart by massive doses of radiation. Mn^{+2} ions exert a protective role, when they are complexed with small metabolites. ▶| *Radiation (section 8.4);* Deinococcus-Thermus *includes radiation-resistant bacteria (section 21.2)*

Ultraviolet (UV) radiation is another very damaging form of radiation. It can kill microorganisms due to its short wavelength (approximately from 10 to 400 nm) and high energy. The most lethal UV radiation has a wavelength of 260 nm, the wavelength most effectively absorbed by and damaging to DNA. The damage caused by UV light can be repaired by several DNA repair mechanisms, as we discuss in chapter 16. However, excessive exposure to UV light outstrips the organism's ability to repair the damage and death results. Longer wavelengths of UV light (near-UV radiation; 325 to 400 nm) can also harm microorganisms because they induce the breakdown of the amino acid tryptophan to toxic photoproducts. These toxic photoproducts plus the near-UV radiation itself produce breaks in DNA strands. ▶| *Mutations: heritable changes in a genome (section 16.1)*

Even visible light, when present in sufficient intensity, can damage or kill microbial cells. Usually pigments called photosensitizers and O_2 are involved. Photosensitizers include pigments such as chlorophyll, bacteriochlorophyll, cytochromes, and flavins, which can absorb light energy and become excited or activated. The excited photosensitizer (P) transfers its energy to O_2, generating singlet oxygen ($^1O_2$).

$$P \xrightarrow{\text{light}} P(\text{activated})$$

$$P(\text{activated}) + O_2 \longrightarrow P + {}^1O_2$$

Singlet oxygen is a very reactive, powerful oxidizing agent that quickly destroys a cell.

Many microorganisms that are airborne or live on exposed surfaces use carotenoid pigments for protection against photooxidation. Carotenoids effectively quench singlet oxygen; that is, they absorb energy from singlet oxygen and convert it back into the unexcited ground state. Both phototrophic and nonphototrophic microorganisms employ pigments in this way.

Comprehension Check

1. Where would you expect to find barotolerant and piezophilic bacteria? Explain your answer.
2. List the types of electromagnetic radiation in the order of decreasing energy or increasing wavelength.
3. What is the importance of ozone formation?
4. How do ionizing radiation, ultraviolet radiation, and visible light harm microorganisms? How do microorganisms protect themselves against damage from UV and visible light?

7.6 Microbial Growth in Natural Environments

After reading this section, you should be able to:

a. Discuss the mechanisms used by microbes to survive starvation
b. Predict how the presence of viable but nonculturable cells in food or water systems might impact public health
c. Distinguish sessile and planktonic microbial life styles
d. Describe the formation of biofilms and summarize their importance in natural environments, industrial settings, and medicine
e. Define quorum sensing and provide examples of cellular processes regulated by quorum sensing
f. Discuss in general terms the communication that occurs between rhizobia and their plant hosts

Natural microbial environments—that is, those not created by humans—are complex and constantly changing. They expose microbes to many overlapping gradients of nutrients and other environmental factors. These include the environmental parameters we described in section 7.5, as well as inhibitory substances that limit microbial growth. Furthermore, microbes do not exist alone. Microbial habitats contain both micro- and macroorganisms. Microbes often associate with each other and they frequently form relationships with plants and animals. In this section, we explore three aspects of life in nature: stress-induced growth arrest, biofilms, and the mechanisms used by microbes to communicate among and between species.

Most Microbes Live in Growth-Arrested States

Relatively few microbes are lucky enough to live in nutrient-rich **eutrophic** environments. Rather, most inhabit **oligotrophic** environments—ones in which nutrient levels are low—and some live in "feast or famine" environments. For microbes in oligotrophic or nutrient-labile environments, the ability to survive starvation conditions is paramount. Fortunately, microbes have evolved numerous responses to starvation and environmental stress.

Some bacteria respond to nutrient limitation with obvious morphological changes such as endospore formation. Recall that endospores are not only highly resistant to a variety of stressors (e.g., high temperatures and radiation) but are also metabolically dormant. Other microbes become dormant without such a drastic reorganization of the cell. Many eukaryotic microbes and some bacteria form cysts (*see figures 22.5 and 24.2*).

Most commonly, nutrient and environmental conditions leave microbes in a state termed growth arrest. They are not actively dividing, nor are they dead. They are stressed by lack of nutrients or by deviation from optimal conditions. Inspection of figure 7.10 illustrates this phenomenon in the growth curve. Entry into stationary phase is triggered by nutrient limitation or waste accumulation, and cells stop growing as they adjust to new conditions. As time goes on, cells may die, but the long-term stationary phase illustrates that many cells remain viable even though the population is not increasing. Natural populations, too, are comprised of a diverse group of microbes most of which are in some form of growth arrest.

Cells must maintain their integrity during growth arrest to be ready to respond when conditions improve. Energy-intensive processes like cell division are halted to conserve ATP and redirect it to cellular maintenance. Resources like storage inclusions are tapped first to provide nutrients. Prolonged starvation induces the use of cellular components as nutrients. For example, many growth-arrested cells decrease in size. This results from the loss of membrane lipids and ribosomes as their components are recycled within the cell. A further benefit of cell size decrease is an increase in the surface area-to-volume ratio, which in turn promotes more efficient nutrient transport.

In many bacteria, two molecules that regulate the transition from rapid growth to growth arrest are RpoS and ppGpp. RpoS is a component of RNA polymerase, the enzyme responsible for binding DNA and initiating RNA synthesis (transcription). RpoS directs the enzyme to genes encoding proteins that help the bacterium survive starvation, as well as proteins involved in a general stress response. ppGpp is a signaling molecule related to the guanine nucleotides in DNA and RNA that functions in a regulatory network called the stringent response, which we describe in more detail in chapter 14. ppGpp is often referred to as an alarmone because it is synthesized in response to many stressors and regulates the activity of numerous genes and proteins so that the cell can survive. ▶| *Transcription in bacteria (section 13.5) Stringent response in* E. coli *(section 14.5)*

The proteins made in response to starvation help ensure survival in several ways. Some increase peptidoglycan cross-linking and cell wall strength. The starvation protein Dps (*D*NA-binding *p*rotein from *s*tarved cells) protects DNA. Proteins called chaperones prevent protein denaturation and renature damaged proteins. Because of these and many other mechanisms, starved cells become harder to kill and more resistant to starvation, damaging temperature changes, oxidative and osmotic damage, and toxic chemicals such as chlorine. These changes are so effective that some members of the bacterial population survive growth arrest for years. There is even evidence that *Salmonella enterica* serovar Typhimurium (*S.* Typhimurium) and some other bacterial pathogens become more virulent when starved. Clearly, these considerations are of great practical importance in medical and industrial microbiology.

After an extended time in growth arrest, cells may enter a state termed viable but not culturable (VBNC). Their ability to grow is critically dependent on the conditions used to revive them. Once the appropriate conditions are available (e.g., a change in temperature or passage through an animal), VBNC microbes resume growth. VBNC microorganisms could pose a public health threat, as many assays that test food and drinking water are culture based.

An interesting phenomenon is often observed in a population of bacterial or archaeal cells growing in favorable conditions: the formation of persisters. **Persisters** are growth-arrested cells identified by their ability to survive exposure to an antibiotic (often multiple antibiotics) even though they do not harbor antibiotic-resistance genes. In these nongrowing cells, the target of an antibiotic is unavailable or inactive and would be unaffected by the antibiotic. Persisters are of considerable interest for several reasons, including their potential to cause recurrent infections after antibiotic treatment.

Two major questions about persister formation are when and how it occurs. Evidence suggests that in some cases, small subsets of cells in a population of growing cells spontaneously become persister cells, even when nutrients are readily available. These persisters can be thought of as the population's attempt to prepare for future possible starvation conditions. There is also increasing evidence that persisters arise in response to various triggers, with starvation probably being the most important. Whether spontaneous or induced, toxin-antitoxin (TA) modules and ppGpp are currently thought to play a role in persister formation. TA modules consist of an internal, nonsecreted toxin and a cognate antitoxin that protects the toxin-producing cell. The toxin, when functioning, disrupts normal cellular functions, causing growth arrest and the persister phenotype. The production of ppGpp is thought to set into motion destruction of the antitoxin of a TA module, thus releasing the toxin to attack its cellular target, cause growth to cease, and convert the cell into a persister.

Despite the attractiveness of this model of persistence, it may not hold true for all microbes. For instance, at least one bacterium produces persisters that actively divide. Their antibiotic tolerance is mediated by efflux pumps that pump out the antibiotic as soon as it enters the cell. Thus the story of persisters is still incomplete.

Biofilms Are Common in Nature

Many introductory microbiology students have scraped their teeth and examined the microbes in the scrapings, thus replicating the way van Leeuwenhoek (*see figure 1.10*) often entertained

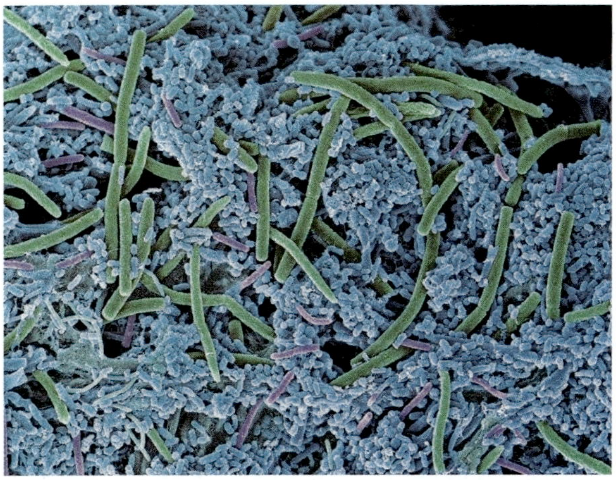

(a)

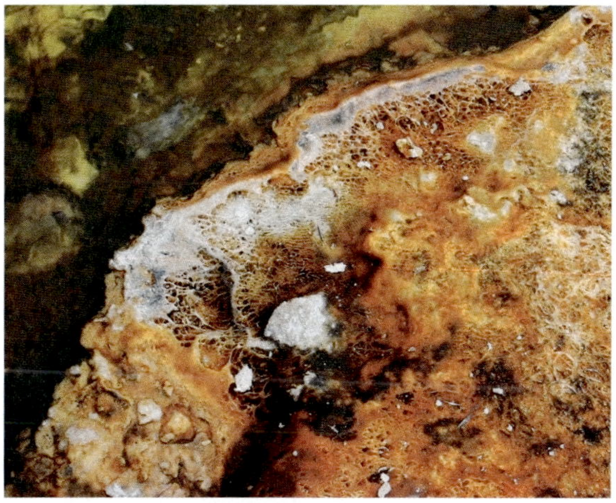

(b)

Figure 7.21 Examples of Biofilms. Biofilms form on almost any surface exposed to microorganisms. (a) Dental plaque biofilm. (b) Biofilm growing in a hot spring, Yellowstone National Park.

(a) ©Steve Gschmeissner/SPL/Getty Images; (b) ©toos/iStock/Getty Images

dinner guests. What was sampled by all was dental plaque, an important example of a biofilm (**figure 7.21*a***). Ecologists observed as early as the 1940s that more microbes in aquatic environments were found in biofilms on surfaces (sessile) than were free-floating (planktonic). However, only relatively recently has this fact gained the attention of microbiologists. **Biofilms** are complex, slime-encased communities of microbes. They are ubiquitous in nature, where they are often seen as layers of slime on rocks or other objects in water or at water-air interfaces (figure 7.21*b*). When they form on the hulls of boats and ships, they cause corrosion. Of major concern is the formation of biofilms on medical devices such as hip and knee implants and indwelling catheters. These biofilms often cause serious illness and failure of the medical device. Biofilms can also form on wound scabs and delay wound healing. Biofilm formation is apparently an ancient ability among microbes, as evidence for biofilms is found in the fossil record from about 3.4 billion years ago.

Biofilms can form on virtually any surface, once it has been conditioned by proteins and other molecules present in the environment (**figure 7.22**). Initially microbes attach to the

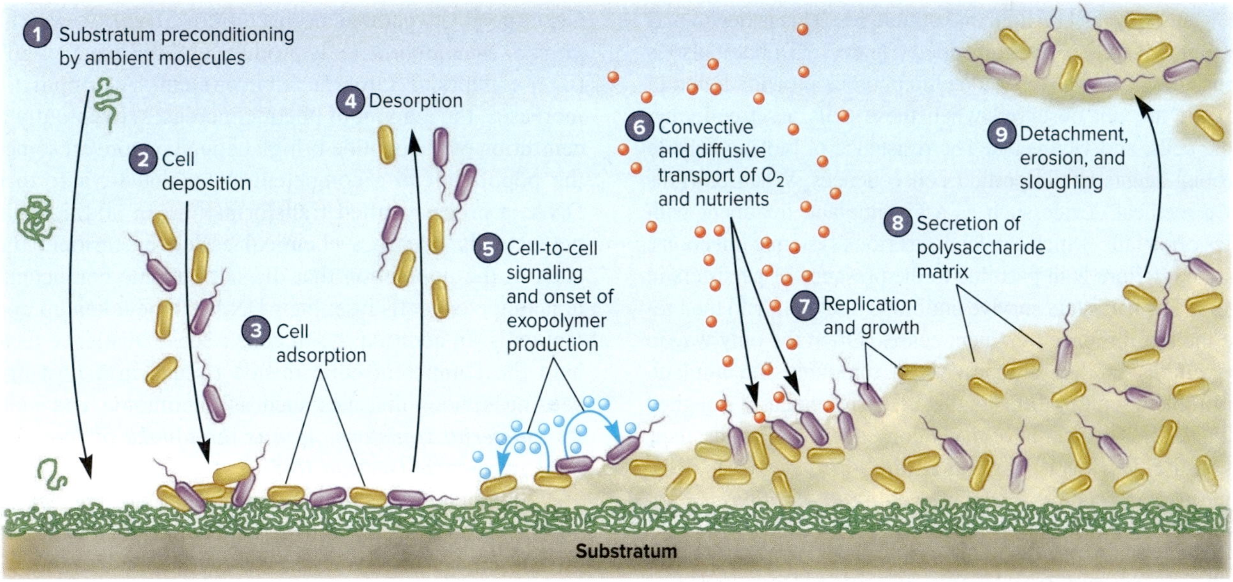

Figure 7.22 Biofilm Formation.

MICRO INQUIRY *What biomolecules make up the extracellular polymeric matrix, and what functions does the matrix serve?*

conditioned surface but can readily detach. Eventually they form a slimy matrix made up of various polymers, depending on the microbes in the biofilm. The polymers are collectively called extracellular polymeric substances (EPS) or extracellular matrix (ECM), and they include polysaccharides, proteins, glycoproteins, glycolipids, and DNA. The EPS allows the microbes to stick more stably to the surface. As the biofilm thickens and matures, the microbes reproduce and secrete additional polymers. Eventually, conditions at locations in the biofilm can become detrimental to the cells and it becomes beneficial for cells to detach and escape the biofilm. This is of considerable importance for medical device-associated biofilms because the escaping cells can seed sites of infection elsewhere in the body.

A mature biofilm is a complex, dynamic community of microorganisms. It exhibits considerable heterogeneity due to differences in the metabolic activity of microbes at various locations within the biofilm; some are persister cells (**figure 7.23**). Biofilm microbes interact in a variety of ways. For instance, the waste products of one microbe may be the energy source for another. The cells also use molecules to communicate with each other, as we describe next. Finally, DNA present in the EPS can be taken up by members of the biofilm community. Thus genes can be transferred from one cell (or species) to another.

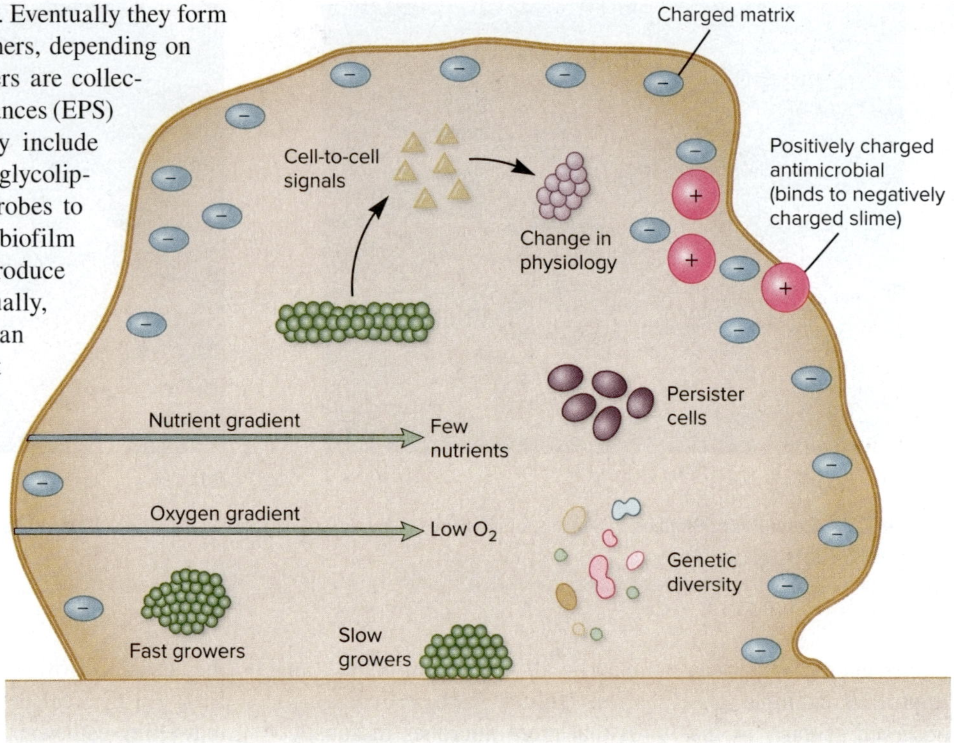

Figure 7.23 **Biofilm Heterogeneity.**

While in the biofilm, microbes are protected from numerous harmful agents such as UV light and antibiotics. This is due in part to the EPS in which they are embedded (figure 7.23), but it also is due to physiological changes. Indeed, numerous proteins found in biofilm cells are not observed when these cells are free-living, planktonic cells, and vice versa. The resistance of biofilm cells to antimicrobial agents has important consequences. When biofilms form on a medical device such as a hip implant, treatment with antibiotics often fails, which can lead to serious systemic infections. The treatment failure is in part due to the presence of persisters in the biofilm. The persisters survive antibiotic treatment and then repopulate the biofilm once treatment ceases. Often the only way to manage patients in this situation is by removing the implant. Another problem with biofilms is that cells are regularly sloughed off (figure 7.22). This can have many consequences. For instance, biofilms in a city's water distribution pipes can clog or foul the pipes and serve as a source of contamination. *Biofilms*

Cell-Cell Communication Within Microbial Populations

For decades, microbiologists tended to think of bacterial populations as collections of individual cells growing and behaving independently. But about 40 years ago, two examples of bacterial cells using molecular signals to communicate with each other in a density-dependent manner were discovered. This is now referred to as **quorum sensing;** a quorum usually refers to the minimum number of members in an organization, such as a legislative body, needed to conduct business.

The first example of quorum sensing was observed in the Gram-positive, pathogenic bacterium *Streptococcus pneumoniae. S. pneumoniae* cells produce and release a small protein (i.e., a peptide) into the environment. As population size increases, the amount of peptide increases. Eventually the concentration of the peptide is high enough to convert some cells in the population to a competent state, that is, able to take up DNA, a process called transformation. In addition, the competent cells release a chemical called bacteriocin that lyses cells in the population that did not become competent. When noncompetent cells lyse, their DNA can be taken up by competent cells. In addition, lysed cells release virulence factors that help the competent cells invade tissues in a host organism, causing serious diseases such as pneumonia and meningitis. ▶| *Bacterial transformation is the uptake of free DNA from the environment (section 16.7)*

Although the quorum-sensing system of *S. pneumoniae* was the first discovered, the term was not coined until years later when the quorum-sensing system of a very different bacterium was discovered. The marine luminescent bacterium *Vibrio fischeri* lives within the light organ of certain fish and squid. *V. fischeri* regulates its luminescence by producing a small, diffusible molecule called autoinducer. The autoinducer molecule is an ***N*-acylhomoserine**

Signal and Structure	Representative Organism	Function Regulated
N-acylhomoserine lactone (AHL) 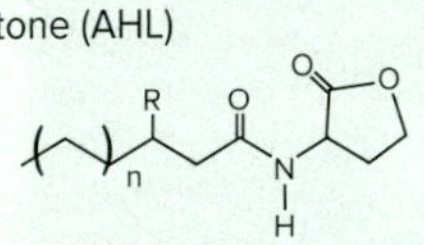	*Vibrio fischeri* *Agrobacterium tumefaciens* *Erwinia carotovora* *Pseudomonas aeruginosa* *Burkholderia cepacia*	Bioluminescence Plasmid transfer Virulence and antibiotic production Virulence and biofilm formation Virulence
Furanosylborate (AI-2) 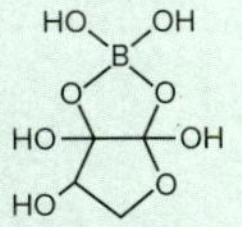	*Vibrio harveyi*[a]	Bioluminescence
Cyclic thiolactone (AIP-II) 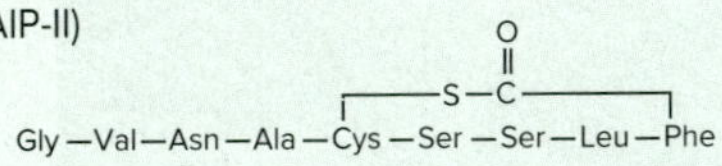	*Staphylococcus aureus*	Virulence
Methyl dodecenoic acid 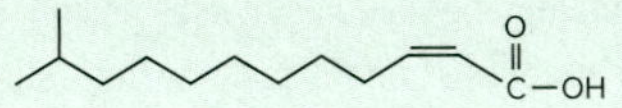	*Xanthomonas campestris*	Virulence
Farnesoic acid 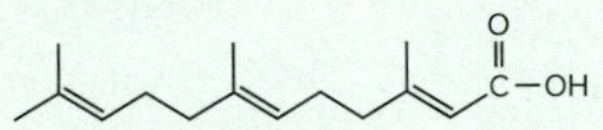	*Candida albicans*	Dimorphic transition and virulence
3-hydroxytridecan-4-one 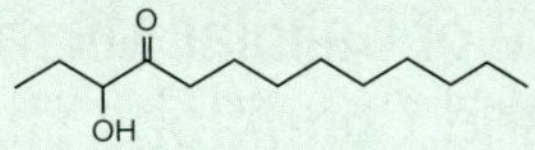	*Vibrio cholerae*	Virulence
3,5 dimethyl-pyrazin-2-ol 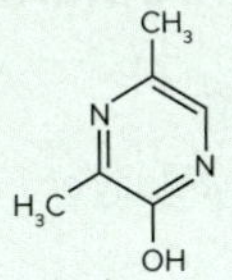	*Vibrio cholerae*	Biofilm formation

[a] Other bacteria make a form of AI-2 that lacks boron.

Figure 7.24 Representative Cell-Cell Communication Molecules.

lactone (**AHL**). It is now known that many Gram-negative bacteria make AHL molecular signals that vary in length and substitution at the third position of the acyl side chain (**figure 7.24**). In many of these species, AHL is freely diffusible across the plasma membrane. Thus at a low cell density, the diffusion gradient favors movement of AHL from the cytoplasm to the outside of the cell. However, when the cell population increases, the diffusion gradient is reversed so that movement of AHL into the cell is favored. Because the influx of AHL is cell density dependent, it enables individual cells to assess population density. When AHL reaches a threshold level inside the cell, it induces the expression of genes that regulate a number of functions, depending on the microbe. These functions are most effective only if a large number of microbes are present. For instance, the light produced by one *V. fischeri* cell would be invisible and energetically costly, so bioluminescence is inhibited. However, cell densities within the light organ of marine fish and squid reach 10^{10} cells per milliliter. At such densities, light is clearly visible. Thus quorum sensing works to promote bioluminescence at high densities. This provides fish with a flashlight effect. In squid, bioluminescence provides countershading, which protects the squid from predation as it hunts at night. In return, the microbes have a safe and nutrient-enriched habitat (**figure 7.25**). ▶| *Quorum sensing by* Vibrio *spp. (section 14.6)*

Scientists have learned that many of the processes regulated by quorum sensing involve host-microbe interactions, including pathogenicity. For instance, the Gram-negative, opportunistic bacterial pathogens *Burkholderia cepacia* and *Pseudomonas aeruginosa* use AHLs to regulate biofilm formation and the expression of virulence factors (figure 7.24). These bacteria cause debilitating pneumonia in people who are immunocompromised and are important pathogens in cystic fibrosis patients.

Like *S. pneumoniae,* other Gram-positive bacteria communicate using short peptides called oligopeptides. Examples include *Enterococcus faecalis,* whose oligopeptide signal is used to determine the best time to conjugate (transfer genes). Oligopeptide communication by *Staphylococcus aureus* and *B. subtilis* is also used to trigger the uptake of DNA from the environment.

The discovery of additional molecular signals made by a variety of microbes underscores the importance of cell-cell communication in regulating cellular processes. For instance, while only Gram-negative bacteria are known to make AHLs, many bacteria make autoinducer-2 (AI-2), a universal quorum-sensing signal. The soil microbe *Streptomyces griseus* produces a γ-butyrolactone known as A-factor. This small molecule regulates both morphological differentiation and the production of the antibiotic streptomycin. Eukaryotic microbes also rely on cell-cell communication to coordinate key activities within a population. For example, the pathogenic fungus *Candida albicans* secretes farnesoic acid to govern morphology and virulence (figure 7.24).

These examples of cell-cell communication demonstrate what might be called multicellular behavior in that many individual cells communicate and coordinate their activities to act as a unit. Other examples of such complex behavior are pattern formation in colonies and fruiting body formation in the myxobacteria. ▶| *Order* Myxococcales*: bacteria with morphological complexity and multicellularity (section 22.4)*

Interdomain Communication

The quorum-sensing phenomenon allows members of the same species to communicate. As we describe in chapter 14, this system also allows different *Vibrio* species to communicate. Thus the *Vibrio* system is an example of interspecies communication. Even more dramatic examples of interspecies communication exist, including bacteria and plants "talking" to each other.

(a) The bobtail squid, *E. scolopes*

(b) Light organ

Figure 7.25 Squid—*Vibrio* Symbiosis. (a) The bobtail squid remains buried in sand during the day and feeds at night. (b) When feeding, it uses its light organ (boxed, located on its ventral surface) to provide camouflage by projecting light downward. Thus the outline of the squid appears as bright as the water's surface to potential predators looking up through the water column. The light organ is colonized by a large number of *Vibrio fischeri* cells, so autoinducer accumulates to a threshold concentration, triggering light production. In addition to being a model for quorum sensing, this relationship is an important model for the most common type of animal-bacteria association, colonization of epithelial surfaces.

(a) ©Chris Frazee/UW-Madison; (b) ©Dr. Margaret Jean McFall-Ngai

One of the best-studied examples of interdomain communication is observed when members of the genus *Rhizobium* interact with a leguminous plant. A successful interaction ends with the invasion of a plant root by the rhizobial cells. Once inside, the rhizobia differentiate, becoming nitrogen-fixing forms called bacteroids. Nitrogen-fixation requires considerable energy, and the plant assists the bacterium by helping it meet its energy needs. In return, the bacterium supplies the plant with nitrogen in the form of amino acids.

A successful interaction is initiated when the bacterium and plant "talk" to each other using chemicals they secrete. The plant releases molecules called flavonoids, which are taken up by the bacterium and bind to a bacterial protein called NodD. This activates the production of chemicals by the bacterium called Nod factors. Some Nod factors affect the outer cells of plant root hairs. Ultimately this causes the root hairs to curl, entrapping the bacteria. Other Nod factors trigger formation of an infection tube, through which the bacteria enter the plant. Thus by "speaking" to each other, the plant and bacterium form a relationship that is beneficial to both. ▶| *Rhizobia are symbionts of leguminous plants (section 31.3)*

Comprehension Check

1. As a cell adapts to starvation conditions during growth arrest, which cell division molecules are no longer needed?
2. What is a biofilm? List two ways life in a biofilm is advantageous for microbes.
3. What medical challenges do biofilms present?
4. What is quorum sensing? Describe how it occurs, and briefly discuss its importance to microorganisms.
5. How is the communication that occurs between a rhizobium and a leguminous plant similar to that occurring between *Vibrio* species? How does it differ?

7.7 Laboratory Culture of Cellular Microbes Requires Media and Conditions That Mimic the Normal Habitat of a Microbe

After reading this section, you should be able to:

a. Describe the importance of culturing microbes to the study of microorganisms
b. Distinguish defined (synthetic) media from complex media and the uses of liquid from solid growth media
c. List the characteristics of agar that make it a particularly useful solidifying agent
d. Compare and contrast supportive (general purpose), enriched, selective, and differential media, listing examples of each and describing how each is used
e. Discuss the use of enrichment cultures in isolating microbes
f. Differentiate the streak-plate, spread-plate, and pour-plate methods for isolating pure cultures
g. Use the terms commonly used by microbiologists to describe colony morphology

For decades studying microbes in their natural habitats has proven to be a significant hurdle for microbiologists. For that reason, microbiology research has depended largely on the ability to grow and maintain microorganisms in the laboratory. Unfortunately, it is estimated that only 1 to 5% of all microbes are currently culturable. Recently nonculture-based alternatives have become available. Yet even with these techniques, an important goal is to eventually isolate microbes from their normal habitat. Thus culturing microbes continues to be an important tool used by microbiologists. ▶| *Microbial biology relies on cultures (section 29.1)*

Culture Media

Culturing microbes is possible only if suitable culture media are available. A **culture medium** is a solid or liquid preparation used to grow, transport, and store microorganisms. To be effective, the medium must contain all the nutrients the microorganism requires for growth. Specialized media are essential in the isolation and identification of microorganisms, the testing of antibiotic sensitivities, water and food analysis, industrial microbiology, and other activities. Although all microorganisms need sources of energy and macro- and micronutrients, the precise composition of a satisfactory medium depends on the species being cultivated due to the great variety of nutritional requirements. Knowledge of a microorganism's normal habitat often is useful in selecting an appropriate medium because its nutrient requirements reflect its natural surroundings. Frequently a medium is used to select and grow specific microorganisms or to help identify a particular species. Media can also be specifically designed to facilitate the growth of one type of microbe present in a sample from nature. |◀ *Bacteria use many mechanisms to bring nutrients into the cell (section 3.3)*

Culture media can be classified based on several parameters: the chemical constituents from which they are made, their physical nature, and their function (**table 7.6**). The types of media defined by these parameters are described here.

Table 7.6 Types of Media

Basis for Classification	Types
Chemical composition	Defined (synthetic), complex
Physical nature	Liquid, semisolid, solid
Function	Supportive (general purpose), enriched, selective, differential

Chemical and Physical Types of Culture Media

A **defined** or **synthetic medium** is one in which each ingredient can be defined with a chemical formula. It can be in a liquid form (broth) or solidified by an agent such as agar. Defined media are often used to culture photosynthetic microbes that use CO_2 as a carbon source and light as an energy source. Thus they can be grown on media containing sodium carbonate or bicarbonate (sources of CO_2), nitrate or ammonia as a nitrogen source, sulfate, phosphate, and other minerals (**table 7.7**). Many organisms that use reduced organic molecules as carbon and energy sources also can be grown in defined media. Thus a medium with glucose as a carbon source and an ammonium salt as a nitrogen source will support their growth. Not all defined media are as simple as the examples in table 7.7; some are constructed from dozens of components. Defined media are used widely in research, as it is often desirable to know exactly what the microorganism is metabolizing.

Media that contain some ingredients of unknown chemical composition are **complex media.** They are very useful because a single complex medium may be able to meet all the nutritional requirements of many different microorganisms. In addition,

Table 7.7 Examples of Defined Media

BG–11 Medium for Cyanobacteria	Amount (g/liter)	Medium for *Escherichia coli*	Amount (g/liter)
$NaNO_3$	1.5	Glucose	1.0
$K_2HPO_4 \cdot 3H_2O$	0.04	Na_2HPO_4	16.4
$MgSO_4 \cdot 7H_2O$	0.075	KH_2PO_4	1.5
$CaCl_2 \cdot 2H_2O$	0.036	$(NH_4)_2SO_4$	2.0
Citric acid	0.006	$MgSO_4 \cdot 7H_2O$	0.2
Ferric ammonium citrate	0.006	$CaCl_2$	0.01
EDTA (Na_2Mg salt)	0.001	$FeSO_4 \cdot 7H_2O$	0.0005
Na_2CO_3	0.02	Final pH 6.8–7.0	
Trace metal solution[1]	1.0 ml/liter		
Final pH 7.4			

Sources: Data from Rippka, R., et al. 1979. Journal of General Microbiology, *111:1–61; and Cohen, S. S., and Arbogast, R. 1950.* Journal of Experimental Medicine, *91:619.*

1 The trace metal solution contains H_3BO_3, $MnCl_2 \cdot 4H_2O$, $ZnSO_4 \cdot 7H_2O$, $Na_2Mo_4 \cdot 2H_2O$, $CuSO_4 \cdot 5H_2O$, and $Co(NO_3)_2 \cdot 6H_2O$.

Table 7.8 Some Common Complex Media

Nutrient Broth	Amount (g/liter)	Tryptic Soy Broth	Amount (g/liter)	MacConkey Agar	Amount (g/liter)
Peptone (gelatin hydrolysate)	5	Tryptone (pancreatic digest of casein)	17	Pancreatic digest of gelatin	17.0
Beef extract	3	Peptone (soybean digest)	3	Pancreatic digest of casein	1.5
Final pH 6.8		Glucose	2.5	Peptic digest of animal tissue	1.5
		NaCl	5	Lactose	10.0
		K_2HPO_4	2.5	Bile salts	1.5
		Final pH 7.3		NaCl	5.0
				Neutral red	0.03
				Crystal violet	0.001
				Agar	13.5
				Final pH 7.4	

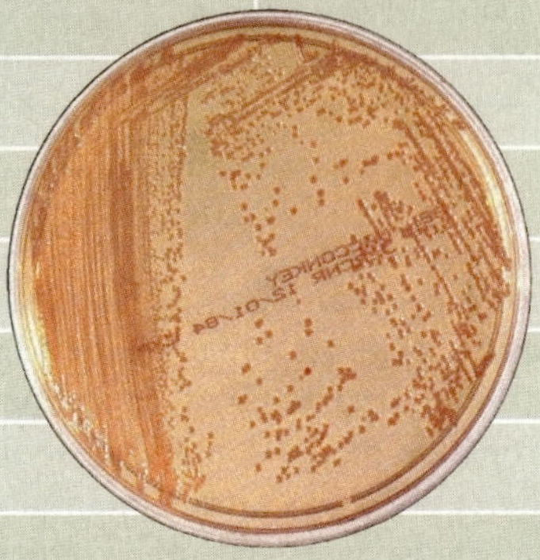

MacConkey Agar
Source: CDC

complex media often are needed because the nutritional requirements of a particular microorganism are unknown, and thus a defined medium cannot be formulated. Complex media are also used to culture fastidious microbes, those with complicated nutritional or cultural requirements.

Most complex media contain undefined components such as peptones, meat extract, and yeast extract. Peptones are protein hydrolysates prepared by partial proteolytic digestion of meat, casein, soya meal, gelatin, and other protein sources. They serve as sources of carbon, energy, and nitrogen. Beef extract and yeast extract are aqueous extracts of lean beef and brewer's yeast, respectively. Beef extract contains amino acids, peptides, nucleotides, organic acids, vitamins, and minerals. Yeast extract is an excellent source of B vitamins as well as nitrogen and carbon compounds. Three commonly used complex media are nutrient broth, tryptic soy broth, and MacConkey agar (**table 7.8**).

Both liquid and solidified media are routinely used in laboratories. However, solidified media are particularly important because they can be used to isolate different microbes from each other to establish pure cultures. As we discuss in chapter 1, this is a critical step when using Koch's postulates to demonstrate the relationship between a microbe and a disease. **Agar** is the most commonly used solidifying agent. It is a sulfated polymer composed mainly of D-galactose, 3,6-anhydro-L-galactose, and D-glucuronic acid. It usually is extracted from red algae. Agar is well suited as a solidifying agent for several reasons. One is that it melts at about 90°C but, once melted, does not harden until it reaches about 45°C. Thus after being melted in boiling water, it can be cooled to a temperature that is tolerated by human hands as well as microbes. Furthermore, microbes growing on agar medium can be incubated at a wide range of temperatures. Finally, agar is an excellent hardening agent because most microorganisms cannot degrade it.

Functional Types of Media

Media such as tryptic soy broth and tryptic soy agar are called general purpose or **supportive media** because they sustain the growth of many types of microorganisms. Blood and other nutrients may be added to supportive media to encourage the growth of fastidious microbes. These fortified media (e.g., blood agar) are called **enriched media** (**figure 7.26**).

Selective media allow the growth of particular microorganisms, while inhibiting the growth of others (**table 7.9**). For instance, many Gram-negative bacteria will grow on media containing bile salts or dyes such as basic fuchsin and crystal violet; however, the growth of Gram-positive bacteria is inhibited. Eosin methylene blue agar and MacConkey agar (table 7.9) are widely used for the detection of *E. coli* and related bacteria in water supplies and elsewhere. These media suppress the growth of Gram-positive bacteria.

Differential media are media that distinguish among different groups of microbes and even permit tentative identification of microorganisms based on their biological characteristics. Blood agar is both a differential medium and an enriched one. It distinguishes between hemolytic and nonhemolytic bacteria. Some hemolytic bacteria (e.g., many streptococci and staphylococci isolated from throats) produce clear zones around their colonies because of red blood cell destruction (figure 7.26*b*). Blood agar is an enriched growth medium in that blood (usually sheep blood) provides protein, carbohydrate, lipid, iron, and a

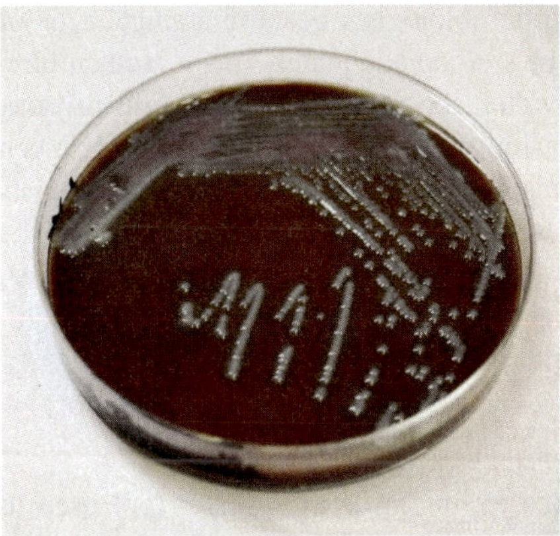

(a)

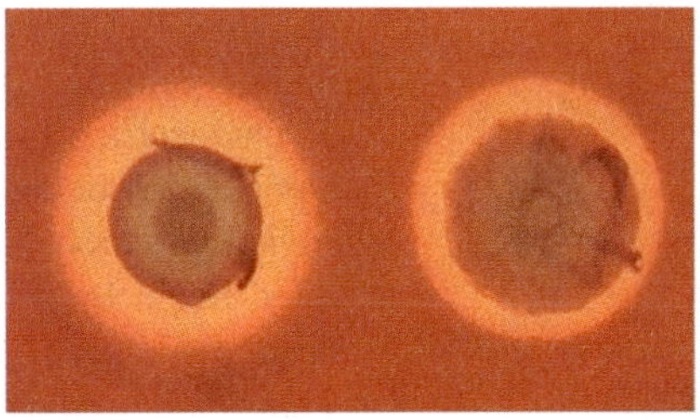

(b)

Figure 7.26 **Enriched Media.** (a) Chocolate agar is used to grow fastidious organisms such as *Franciscella tularensis*. It is called chocolate agar because of its color, the result of heated and lysed red blood cells added to the medium. (b) Blood agar culture of bacteria that have secreted a hemolysin. The blood cells around each colony have lysed to produce a yellow halo in the agar.

(a) Source: CDC/Megan Mathias and J. Todd Parker; (b) Source: CDC/Dr. Negut

number of growth factors and vitamins necessary for the cultivation of fastidious organisms. MacConkey agar is both differential and selective. Because it contains lactose and neutral red dye, bacteria that catabolize lactose by fermenting it release acidic waste products that make colonies appear pink to red in color. These are easily distinguished from colonies of bacteria that do not ferment lactose.

Cultivation of Aerobes and Anaerobes

Because aerobes need O_2 and anaerobes are killed by it, radically different approaches must be used when they are cultivated. When large volumes of aerobic microorganisms are cultured, they must be shaken or sterile air must be pumped through the culture vessel to aerate the culture medium. Without aeration, the low solubility of O_2 in liquid prevents them from obtaining an adequate supply.

Precisely the opposite problem arises with anaerobes, in that all O_2 must be excluded. This is accomplished in several ways: (1) Anaerobic media containing reducing agents such as thioglycollate or cysteine may be used. The medium is boiled during preparation to dissolve its components and drive off oxygen. The reducing agents eliminate any residual dissolved O_2 in the medium so that anaerobes can grow beneath its

Table 7.9 **Mechanisms of Action of Example Selective and Differential Media**

Medium	Functional Type	Mechanism of Action
Blood agar	Enriched and differential	Blood agar supports the growth of many fastidious bacteria. These can be differentiated based on their ability to produce hemolysins—proteins that lyse red blood cells. Hemolysis appears as a clear zone (β-hemolysis) or greenish halo around the colony (α-hemolysis) (e.g., *Streptococcus pyogenes,* a β-hemolytic streptococcus).
Eosin methylene blue (EMB) agar	Selective and differential	Two dyes, eosin Y and methylene blue, inhibit the growth of Gram-positive bacteria. They also react with acidic products released by certain Gram-negative bacteria when they use lactose or sucrose as carbon and energy sources. Colonies of Gram-negative bacteria that produce large amounts of acidic products have a green, metallic sheen (e.g., fecal bacteria such as *E. coli*).
MacConkey (MAC) agar	Selective and differential	The selective components in MAC are bile salts and crystal violet, which inhibit the growth of Gram-positive bacteria. The presence of lactose and neutral red, a pH indicator, allows the differentiation of Gram-negative bacteria based on the products released when they use lactose as a carbon and energy source. The colonies of those that release acidic products are red (e.g., *E. coli*).
Mannitol salt agar	Selective and differential	A concentration of 7.5% NaCl selects for the growth of staphylococci. Pathogenic staphylococci can be differentiated based on the release of acidic products when they use mannitol as a carbon and energy source. The acidic products cause a pH indicator (phenol red) in the medium to turn yellow (e.g., *Staphylococcus aureus*).

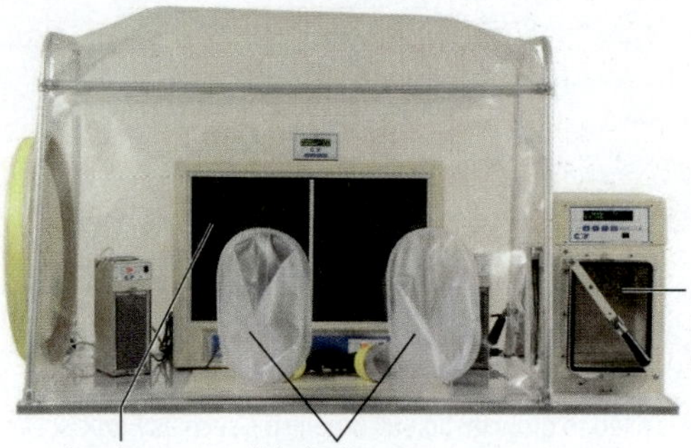

Figure 7.27 An Anaerobic Workstation and Incubator. This system contains an oxygen-free work area and an incubator. The airlock compartment allows materials to be transferred without exposing the interior to oxygen.

surface. (2) Oxygen also may be eliminated from an enclosed work area, often called an anaerobic chamber or anaerobic workstation. Most of the air is removed with a vacuum pump followed by purges with nitrogen gas (**figure 7.27**). A gas mix containing hydrogen is then introduced into the workstation. In the presence of a palladium catalyst, the hydrogen and last remaining molecules of O_2 react to form water, creating an anoxic environment. Often CO_2 is added to the chamber because many anaerobes require a small amount of CO_2 for best growth. (3) A method for culturing small numbers of anaerobes is the GasPak system (**figure 7.28**). (4) A similar approach uses plastic bags or pouches containing calcium carbonate and a catalyst, which produce an anoxic, carbon dioxide–rich atmosphere.

Enrichment and Isolation of Pure Cultures

In natural habitats, microorganisms often grow in complex, microbial communities containing many different species. This presents a problem for microbiologists because a single type of microorganism cannot be studied adequately in a mixed culture. To characterize an individual species, a **pure** or **axenic culture,** a population of cells arising from a single cell is needed. When isolating microbes from soil or aquatic habitats, microbiologists often begin by setting up conditions that favor growth of the desired microbes; that is, they enrich for the microbes of interest. This is followed by using other methods to obtain an axenic culture arising from a single cell.

Enrichment Cultures

The **enrichment culture** technique is a powerful tool used to encourage the growth of microbes having particular characteristics, while at the same time inhibiting the growth of other microbes. It has been used by microbiologists since it was first developed by Sergei Winogradsky and Martinus Beijerinck in the late 1800s. They used enrichment cultures to isolate bacteria with metabolic capabilities such as nitrogen fixation and the use of inorganic molecules (e.g., ammonia) as energy sources. To enrich for these organisms, they considered three factors: (1) a suitable source of the microbes, (2) nutrients that should and should not be included in the culture medium, and (3) environmental conditions provided during the incubation period.

Consider the following: You want to isolate an oil-degrading bacterium and study it in pure culture. A logical source of the bacterium (i.e., an inoculum) would be soil contaminated with oil. The assumption is that such a source would have exposed all bacteria in the soil to the oil but the population of those able to use oil as a sole source of energy and carbon will expand, or become enriched. To select for the oil-degrading bacteria, a medium would be formulated that contained inorganic nitrogen, sulfur, and phosphorus sources but no source of carbon and energy other than the types of molecules typically found in oil. Finally, a temperature, pH, and oxygen level similar to that of the soil environment would be used during incubation.

Despite its utility, the enrichment culture technique is not without problems. It is sometimes difficult to create the proper

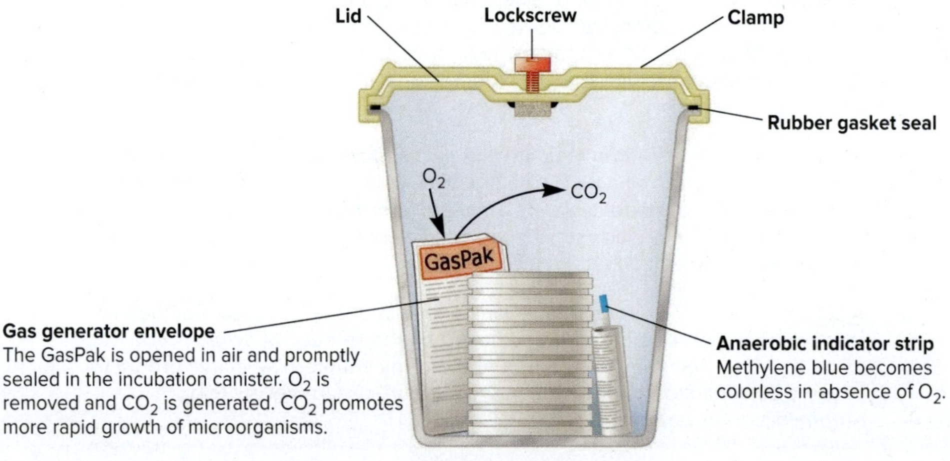

Figure 7.28 The GasPak Anaerobic System. Various types of GasPak envelopes can generate anaerobic or microaerophilic atmospheres inside the sealed container.

selective conditions to successfully isolate the desired organisms. Furthermore, even if microbes are successfully cultivated, they may not adequately represent the population as a whole at a sampling site. Generally, enrichment cultures yield the organisms that grow fastest, and these may not be the predominant organisms in the habitat. Finally, enrichment cultures are not pure cultures. They usually contain more than one species having similar characteristics. To obtain a pure culture, one of the methods described next must be used.

Streak Plate

Methods for isolating pure cultures were developed by the German bacteriologist Robert Koch. He used these techniques in a set of steps that now bear his name: Koch's postulates. As we describe in chapter 1, Koch's postulates are used to establish that a microbe is the causative agent of a particular disease. The use of these methods transformed microbiology, and within 20 years of their development, most pathogens responsible for major bacterial diseases had been isolated and identified. Koch reasoned that if cells from a mixture of microbes could be spatially isolated from each other, each cell would give rise to a completely separate **colony**—a macroscopically visible cluster of microorganisms in or on a solid medium. Because each colony arises from a single cell, each colony represents a pure culture.

In the **streak plate** technique, cells are transferred to the edge of an agar plate with an inoculating loop or swab and then streaked across the surface in one of several patterns (**figure 7.29**). After the first sector is streaked, the inoculating loop is sterilized and an inoculum for the second sector is obtained from the first sector. A similar process is followed for streaking the third sector, except that the inoculum is from the second sector. Thus the number of organisms is diluted as each sector is streaked. Eventually very few cells will be on the loop, and single cells will drop from it as it moves across the agar surface. These develop into separate colonies.

Spread Plate and Pour Plate

Spread-plate and pour-plate techniques are similar in that they both dilute a sample of cells before separating them spatially. They differ in that the spread plate spreads the cells on the surface of the agar, whereas the pour plate embeds the cells within the agar. Both methods can be used to determine the number of microorganisms in a sample.

For the **spread plate,** a small volume of a diluted mixture containing around 25 to 250 cells is transferred to the center of an agar plate and spread evenly over the surface with a sterile bent rod (**figure 7.30**). The dispersed cells develop into isolated colonies. The dilutions for a spread plate are usually made in the same way as for a pour plate—by preparing serial dilutions (**figure 7.31**).

The **pour plate** is extensively used with bacteria, archaea, and fungi. It is particularly useful when sampling a heterogeneous population of microbes, some of which might produce colonies that overgrow an agar surface if isolated by the streak-plate or spread-plate methods. In the pour-plate method, the original sample is serially diluted to reduce the microbial population sufficiently to obtain separate colonies when plating (figure 7.31). Then small volumes of several diluted samples are mixed with liquid agar that has been cooled to about 45°C, and the mixtures are poured immediately into sterile culture dishes. Most microbes survive a brief exposure to the warm agar. Each cell becomes fixed in place to form an individual colony after the agar hardens and the plate is incubated.

Although the preparation of serial dilutions is the bane of many microbiology students, it has many applications other than the spread-plate and pour-plate methods. The numbers of cells in a solution can be diluted to the point where no cells are present in the dilution tube. If replicate dilutions are prepared and the number of tubes yielding growth determined, then the number of viable cells in the sample can be estimated by the most probable number method (*see figure 29.3*). Similarly, the number of viruses in a solution or the number of antibodies in a blood sample can be determined by first diluting the sample and then testing for the presence of viruses or antibodies.

⏮ *Cultivation and enumeration of viruses (section 6.5)*
⏭ *Antibodies bind specific 3-D molecules (section 33.7)*

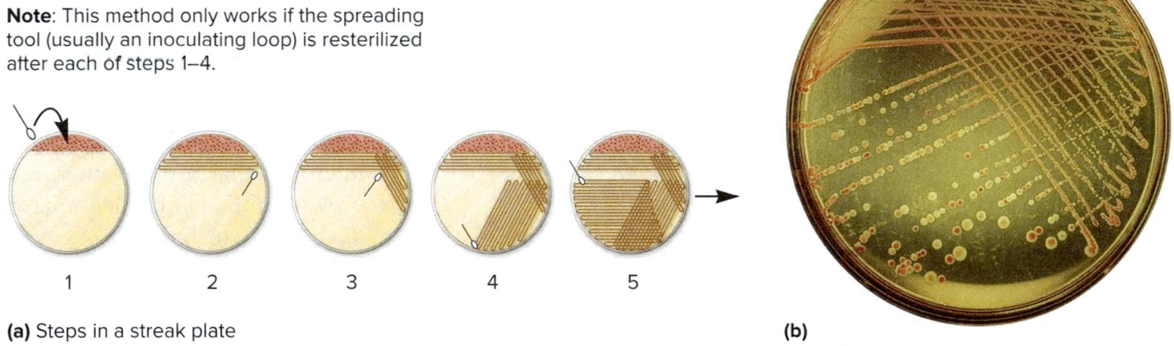

Figure 7.29 Streak-Plate Technique. A typical streaking pattern is shown (a), as well as an example of a streak plate (b).

(b) ©Kathy Park Talaro/Pasadena City College

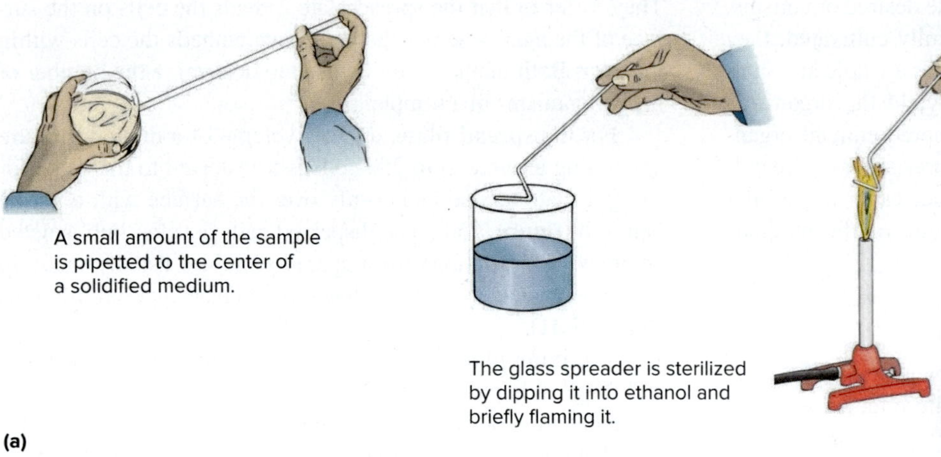

(a)

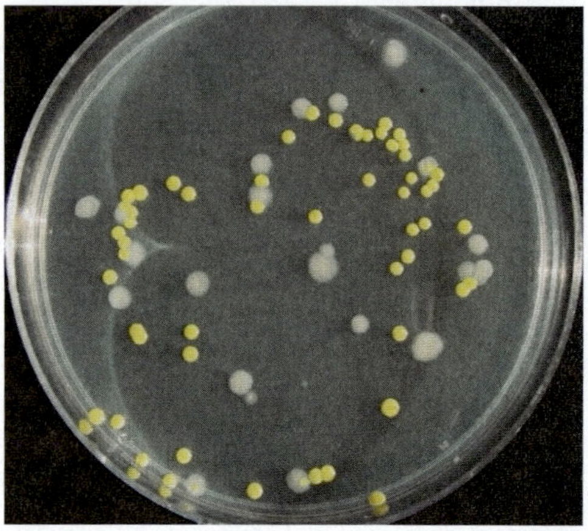

(b)

Figure 7.30 Spread-Plate Technique. (a) The preparation of a spread plate. (b) Typical result of spread-plate technique.

(b) ©Kathy Park Talaro/Pasadena City College

Microbial Growth on Solid Media

Colony development on agar surfaces aids microbiologists in identifying microorganisms because individual species often form colonies of characteristic appearance (**figure 7.32**; also figure 7.30*b*). When a mixed population has been plated properly, it sometimes is possible to identify the desired colony based on its overall appearance and use it to obtain a pure culture.

It is obvious from the colonies pictured in figure 7.32 that bacteria growing on solid surfaces such as agar can form quite complex colony shapes. Colony size and pattern depend on many factors, including nutrient diffusion and availability, bacterial chemotaxis, the presence of liquid on the surface, and hardness of the agar. Cell-cell communication is important as well. Much research is currently focused on understanding the formation of bacterial colonies.

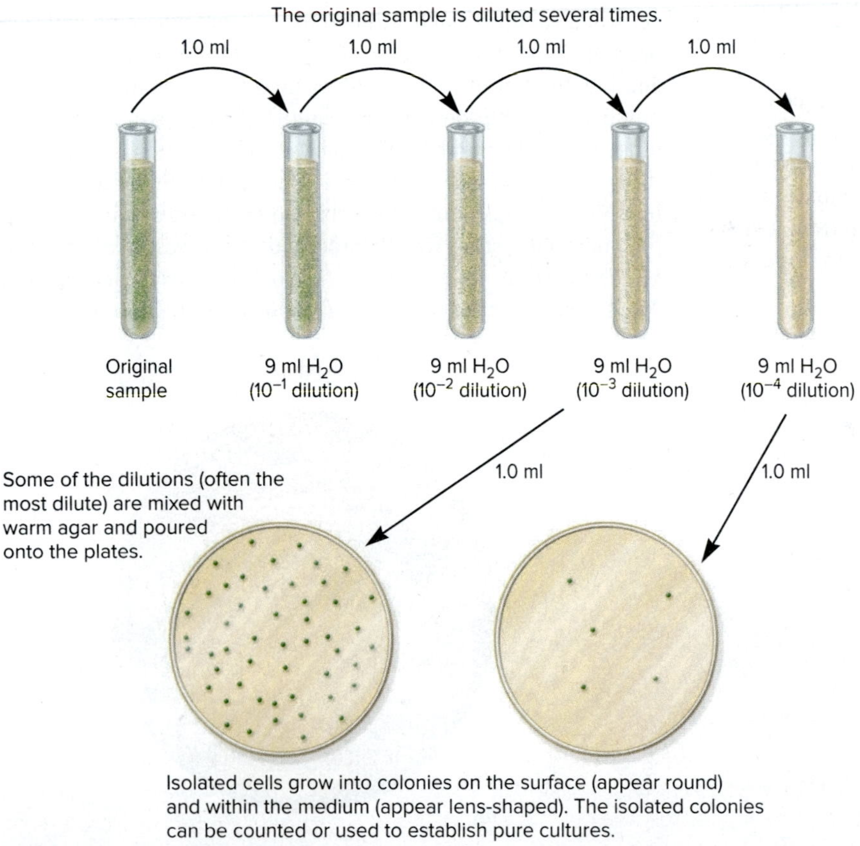

Figure 7.31 Pour-Plate Technique.

New Approaches to Culturing Microbes

The importance of culturing even those microbes that are the most recalcitrant to growth in the lab has led microbiologists to devise new techniques. As we have noted, traditional methods for culturing microbes have depended on being able to construct artificial habitats in a flask or Petri dish that mimic the microbes' natural environments. However, it is often difficult to know all the

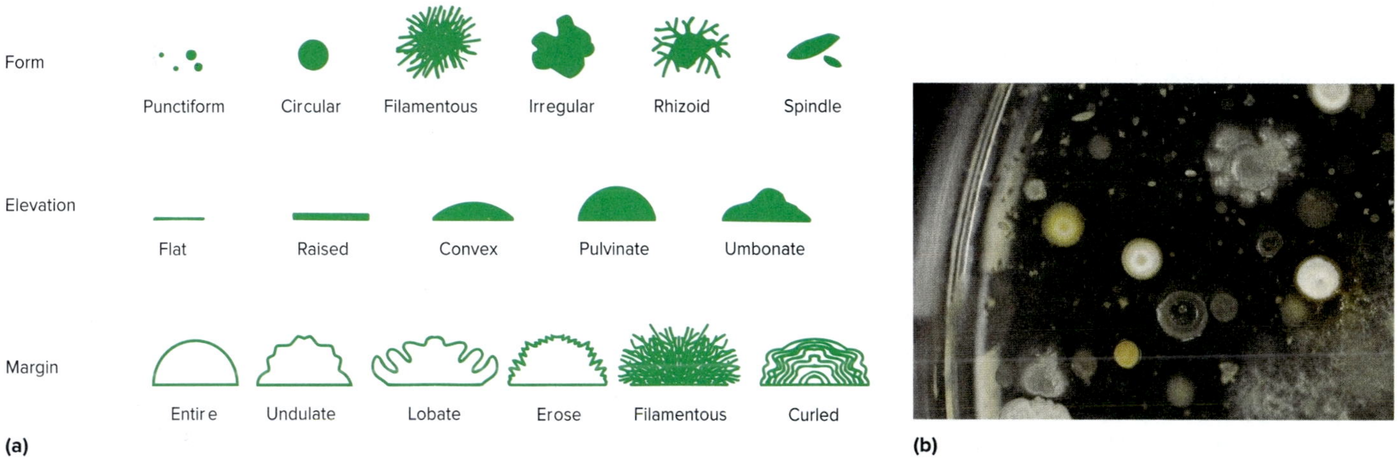

Figure 7.32 Bacterial Colony Morphology. (a) Variations in bacterial colony morphology seen with the naked eye. The general form of the colony and the shape of the edge or margin can be determined by looking down at the top of the colony. The nature of colony elevation is apparent when viewed from the side as the plate is held at eye level. (b) Portion of a Petri dish showing some commonly observed colony morphologies.

(b) ©McGraw-Hill Education/Lisa Burgess photographer

environmental conditions required for growth. One way to solve this problem is to bring part of the natural environment into the lab. For instance, an aquarium containing seawater, beach sand, marine flora, and macroscopic organisms collected from a natural habitat can be used to culture marine microbes. This works very well for maintaining a mixed population in the lab, but it still leaves the problem of isolating particular microbes from the mixture. One solution is to create an enclosure within the man-made "natural" habitat that allows free diffusion of nutrients and other factors from the surroundings but keeps the microbes in the enclosure. For our aquarium example, a very dilute suspension of the seawater theoretically containing a single cell would be used to inoculate the enclosure. Thus an axenic culture would form in the enclosure.

Another approach is useful for microbes that require the presence of a different species to survive. In this case the microbe of interest is co-cultured with the required organism. Sometimes the required organism is an animal, animal tissue, a plant, or plant tissue. The microbe must be maintained growing in or on that organism. Another example of co-culturing occurs in a Petri dish on which a dilute suspension of the desired microbe has been spread and then the helper microbe spotted onto the medium at one location. The factors produced by the helper microbe diffuse into the medium and support the growth of the microbe of interest. This method is particularly useful for long-term maintenance of cultures once the microbe has been isolated in axenic culture.

Microbiologists are exploring even more methods for culturing the unculturable. In the future it is expected that with these new techniques microbiologists will gain a better understanding of the diversity of microbes and their roles in many environments. ▶| *Microbial biology relies on cultures (section 29.1)*

Comprehension Check

1. Describe the following kinds of media and their uses: defined media, complex media, supportive (general purpose) media, enriched media, selective media, and differential media. Give an example of each.
2. What are peptones, yeast extract, beef extract, thioglycollate, and agar? Why are they used in media?
3. Describe four ways in which anaerobes may be cultured.
4. What are pure cultures and why are they important? How are spread plates, streak plates, and pour plates prepared?
5. It is known that growth of individual microbes varies within a colony. What factors might cause these variations?
6. How might an enrichment culture be used to isolate bacteria capable of growing photoautotrophically from a mixed microbial assemblage?

7.8 Microbial Population Size Can Be Measured Directly or Indirectly

After reading this section, you should be able to:

a. Evaluate direct cell counts, viable counting methods, and cell mass measurements for determining population size
b. Explain why plate count results are expressed in terms of colony forming units (CFU)
c. Design appropriate approaches for measuring the population size of different types of samples

There are many ways to measure microbial growth to determine growth rate constants and generation times. Either population number or mass can be followed because growth leads to increases in both. Here the most commonly employed techniques for determining population size are examined and the advantages and disadvantages of each noted. No single technique is always best; the most appropriate approach depends on the experimental situation.

Direct Measurement of Cell Numbers

The most obvious way to determine microbial numbers is by **direct counts** using a counting chamber (e.g., Petroff-Hausser counting chamber). This approach is easy, inexpensive, and relatively quick. It also gives information about the size and morphology of microorganisms. Counting chambers consist of specially designed slides and coverslips; the space between the slide and coverslip creates a chamber of known depth. On the bottom of the chamber is an etched grid that facilitates counting the cells (**figure 7.33**). The number of microorganisms in a sample can be calculated from the chamber's volume and any dilutions made of the sample before counting. One disadvantage of using counting chambers is that to determine population size accurately, the microbial population must be relatively large and evenly dispersed because only a small volume of the population is sampled.

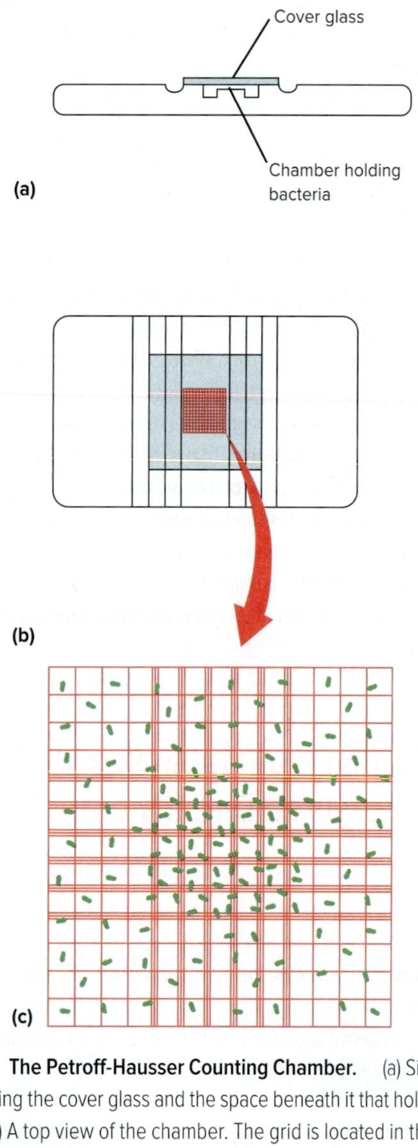

Figure 7.33 The Petroff-Hausser Counting Chamber. (a) Side view of the chamber showing the cover glass and the space beneath it that holds a bacterial suspension. (b) A top view of the chamber. The grid is located in the center of the slide. (c) An enlarged view of the grid. The bacteria in several of the central squares are counted, usually at ×400 to ×500 magnification. The average number of bacteria in these squares is used to calculate the concentration of cells in the original sample.

The number of bacteria in aquatic samples is frequently determined from direct counts after the bacteria have been trapped on membrane filters. In the membrane filter technique, the sample is first filtered through a black polycarbonate membrane filter. Then the bacteria are stained with nucleic acid fluorescent stains such as acridine orange or DAPI and observed microscopically (*see figure 29.5*). Alternatively, fluorescently labeled dyes that are specific for members of a given taxon may be used. The stained cells are easily observed against the black background of the membrane filter and can be counted when viewed with an epifluorescence microscope. Direct cell counts of environmental samples almost invariably result in higher cell densities than do methods that rely on culturing. This is because only a small percentage (about 1%) of cells growing in nature can be cultivated in the laboratory. ◀ *Fluorescence microscopes use emitted light to create images (section 2.2)*

Flow cytometry is increasingly being used to directly count microbes and to gain detailed information about them. A flow cytometer creates a stream of cells so narrow that one cell at a time passes through a beam of laser light. As each cell passes through the beam, the light is scattered. Scattered light is detected by the flow cytometer. Because cells are separated in space, each light-scattering event is detected independently. Thus the number of light-scattering events represents the number of cells in the sample (*see figure 29.4*). Cells of differing size, internal complexity, and other characteristics within a population can also be counted. This usually involves the use of fluorescent dyes or fluorescently labeled antibodies. These more sophisticated uses of flow cytometry can provide valuable information about characteristics of the population of cells.

Microorganisms also can be directly counted with electronic counters such as the Coulter counter. In the Coulter counter, a microbial suspension is forced through a small hole. Electrical current flows through the hole, and electrodes placed on both sides of the hole measure electrical resistance. Every time a microbial cell passes through the hole, electrical resistance increases (i.e., the conductivity drops), and the cell is counted.

Traditional methods for directly counting microbes in a sample usually yield cell densities that are much higher than the plating methods described next in part because direct counting procedures do not distinguish dead cells from culturable cells. Newer methods for direct counts help alleviate this problem. Fluorescent dyes can differentiate between live and dead cells, making it possible to count the number of live and dead microorganisms in a sample (*see figures 2.13a and 29.2*). Unfortunately, the method is not completely accurate. Just as scientists have difficulty defining life, so too do they have difficulty deciding when a cell is truly dead (**figure 7.34**).

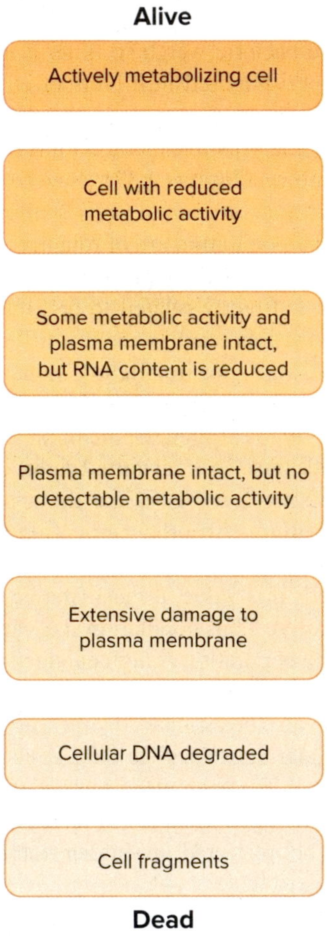

Figure 7.34 Alive or Dead? Microbiologists have traditionally defined microbes as dead when they could not be cultured. However, cells may be inactive or damaged and therefore unable to reproduce temporarily. With time and appropriate conditions, the cells may recover and begin to reproduce. Various criteria have been used to establish these states; however, microbiologists are still unable to define the point at which a microbe is truly dead and cannot be resuscitated.

Viable Counting Methods

Several plating methods can be used to determine the number of viable microbes in a sample. These are referred to as either **viable counting methods** or **standard plate counts** because they count only those cells able to reproduce when cultured. Two commonly used procedures are the spread-plate and the pour-plate techniques, which are described in section 7.7. When using these methods to determine population size, the samples should yield between 25 and 250 colonies per plate for accurate counting. Once the number of colonies is known, the original number of viable microorganisms in the sample can be calculated from that number and the sample dilution. For example, if 1.0 milliliter of a solution diluted by a factor of 1×10^6 yielded 150 colonies (i.e., 1.5×10^2 colonies), then the original sample contained around 1.5×10^8 cells per milliliter. However, because it is not possible to be certain that each colony arose from an individual cell, the results are often expressed in terms of **colony forming units** (**CFU**), rather than the number of microorganisms.

Another commonly used plating method first traps bacteria in aquatic samples on a membrane filter. The filter is then placed on an agar medium or on a pad soaked with liquid media (**figures 7.35** and **7.36**) and incubated until each cell forms a separate colony. A colony count gives the number of microorganisms in the filtered sample, and selective media can be used to select for specific microorganisms. This technique is especially useful in analyzing water purity. ▶| *Sanitary analysis of waters (section 43.1)*

Plating techniques are simple, sensitive, and widely used for viable counts of bacteria and other microorganisms in samples of food, water, and soil. However, several problems can lead to inaccurate counts. The population will be underestimated if clumps of cells are not broken up and microorganisms well dispersed. The hot agar used in the pour-plate technique may injure or kill sensitive cells; thus spread plates sometimes give higher counts than pour plates. A major problem is that plate counts will be

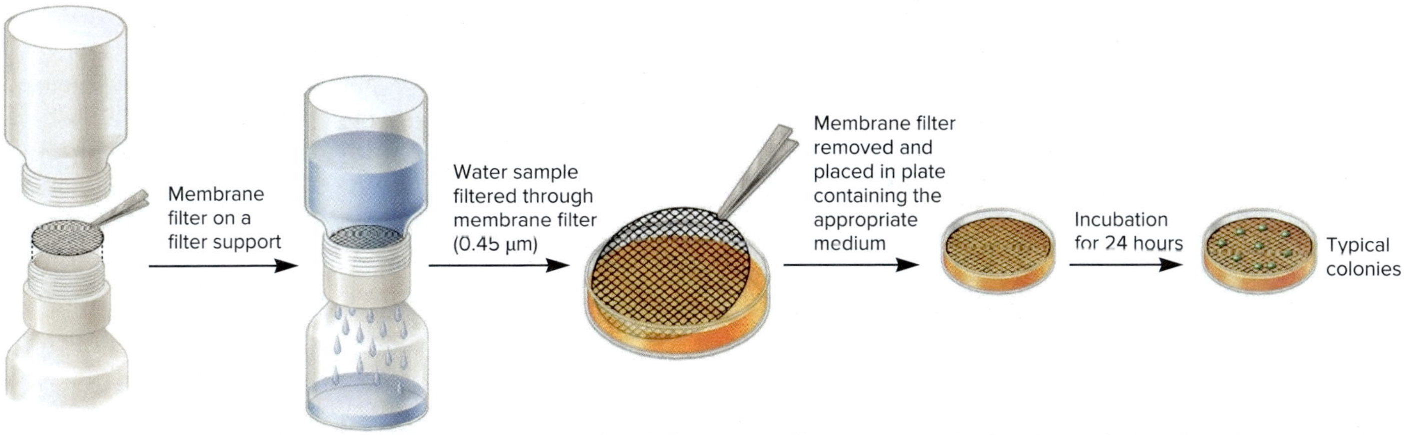

Figure 7.35 The Membrane Filtration Procedure. Membranes with different pore sizes are used to trap different microorganisms. Incubation times for membranes also vary with the medium and microorganism.

MICRO INQUIRY *Why is it important to have no more than about 250 colonies on the plate when counting colonies to determine population size?*

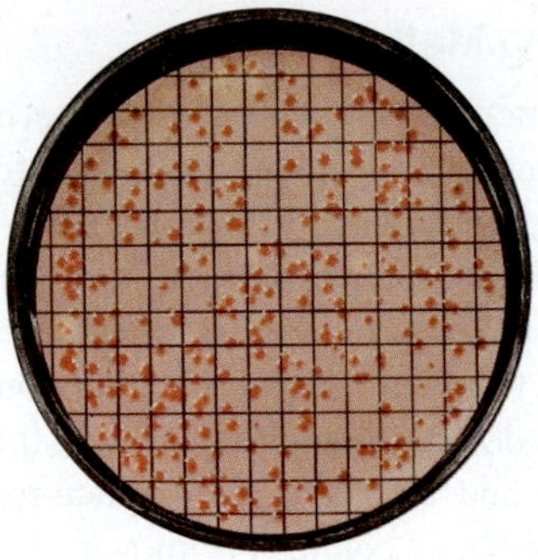

Figure 7.36 Colonies on a Membrane Filter. A sample was filtered using a membrane filter that retained bacterial cells. The filter was placed on a culture medium and incubated. The resulting colonies are counted using the grid and that value is used to determine colony-forming units in the sample.

artificially low if the medium employed cannot support growth of all the viable microorganisms present. This is a common problem encountered by microbial ecologists trying to understand the makeup of microbial communities. In such studies, the number of cells observed by microscopy is often much higher than the population size determined by plate counts. This discrepancy is sometimes referred to as "the great plate count anomaly" and is discussed in detail in chapter 29. ▶| *Genetic methods are used to assess microbial diversity (section 29.2)*

Sometimes plate counts cannot be used to measure population size. For instance, plate counts are not helpful if the microbe cannot be cultured on solid media or if large colonies overgrow the surface of the plate, making it impossible to get an accurate count. In these cases, another approach is used: **most probable number (MPN)** determination. In this method, numerous replicates of several dilutions of a culture are made and added to tubes containing a suitable liquid growth medium. After incubation, each tube is examined to determine if growth occurred. It is assumed that the last tube in the dilution series that demonstrates growth was inoculated with between 1 and 10 cells, while the next tube had between 11 and 100 cells, and so on. If no growth is observed, then the tube is assumed not to have received any cells. In this way, the number of cells in the original sample is estimated (*for an example, see figure 29.3*). MPN values are most commonly used when a selective medium can be employed that supports the growth of a specific type of microbe. It is often used in assessing microbial density in water samples.

Measurement of Cell Mass

Techniques for measuring changes in cell mass also can be used to determine population size. One approach is the determination of microbial dry weight. Cells growing in liquid medium are collected by centrifugation, washed, dried in an oven, and weighed. This is an especially useful technique for measuring the growth of filamentous fungi. However, it is time consuming and not very sensitive. Because bacteria weigh so little, it may be necessary to centrifuge several hundred milliliters of culture to collect a sufficient quantity.

A more rapid and sensitive method for measuring cell mass is spectrophotometry (**figure 7.37**). Spectrophotometry depends on the fact that microbial cells scatter light. Because microbial cells in a population are of roughly constant size, the amount of scattering is directly proportional to the biomass of cells present and indirectly related to cell number. When the concentration of bacteria reaches about a million (10^6) cells per milliliter, the medium appears slightly cloudy or turbid. Further increases in concentration result in greater turbidity, and less light is transmitted through the medium. The extent of light scattering (i.e., decrease in transmitted light) can be measured by a spectrophotometer and is called the absorbance (optical density) of the medium. Absorbance is almost linearly related to cell concentration at absorbance levels less than about 0.5. If the sample exceeds this value, it must first be diluted and then absorbance measured. Thus population size can be easily measured as long as the population is high enough to give detectable turbidity.

Cell mass can also be estimated by measuring the concentration of some cellular substance, as long as its concentration is constant in each cell. For example, a sample of cells can be analyzed for total protein or nitrogen. An increase in the microbial population will be reflected in higher total protein levels.

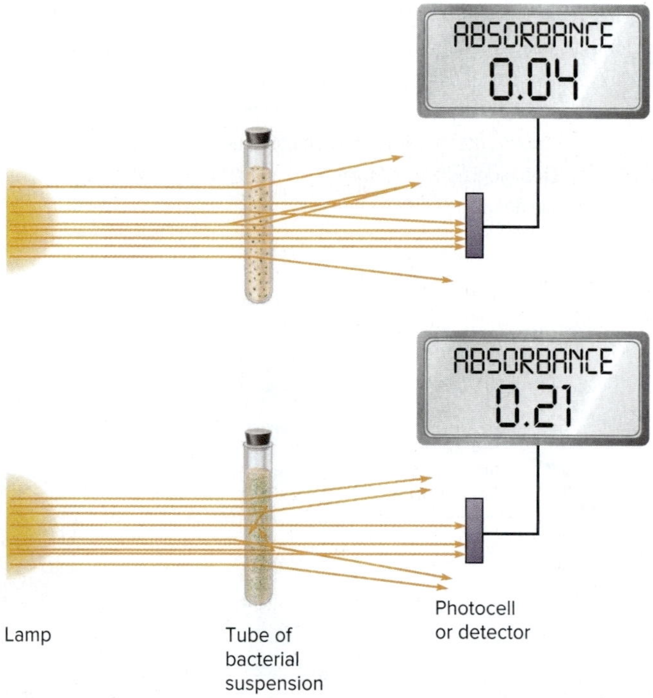

Figure 7.37 Turbidity and Microbial Mass Measurement. Determination of microbial mass by measurement of light absorption. As the population and turbidity increase, more light is scattered and the absorbance reading given by the spectrophotometer increases.

Similarly, chlorophyll determinations can be used to measure phototrophic protist and cyanobacterial populations, and the quantity of ATP can be used to estimate the amount of living microbial mass.

Comprehension Check

1. Briefly describe each technique by which microbial population numbers may be determined and give its advantages and disadvantages.
2. When using direct cell counts to follow the growth of a culture, it may be difficult to tell when the culture enters the death phase. Why?
3. Why are plate count results expressed as colony forming units?
4. For each of the following, which enumeration technique would you use? Explain your choice. (a) A pure culture of *Staphylococcus aureus;* (b) a water sample that needs to be checked for *E. coli* contamination; (c) a sample of yogurt.

7.9 Chemostats and Turbidostats Are Used for Continuous Culture of Microorganisms

After reading this section, you should be able to:

a. Distinguish batch culture and continuous culture
b. Differentiate chemostats and turbidostats
c. Discuss the relationship between the dilution rate of a chemostat and population size and growth rate

In section 7.4, our focus is on closed systems called batch cultures in which nutrients are not renewed nor wastes removed. Exponential growth lasts for a few generations and soon stationary phase is reached. However, it is possible to grow microorganisms in a system with constant environmental conditions maintained through continual supply of nutrients and removal of wastes. Such a system is called a **continuous culture system.** These systems maintain a microbial population in exponential growth, growing at a known rate and at a constant biomass concentration for extended periods. Continuous culture systems make possible the study of microbial growth at very low nutrient levels, concentrations close to those present in natural environments. These systems are essential for research in many areas, particularly microbial ecology. For example, interactions between microbial species in environmental conditions resembling those in a freshwater lake can be modeled. Continuous culture systems also are used in food and industrial microbiology. Two major types of continuous culture systems commonly are used: chemostats and turbidostats. ▶| *Microbiology of food (chapter 41); Biotechnology and industrial microbiology (chapter 42)*

Chemostats

A **chemostat** is constructed so that the rate at which a sterile medium is fed into a culture vessel is the same as the rate at which the medium containing microorganisms is removed (**figure 7.38**). The culture medium for a chemostat has a limited quantity of an essential nutrient (e.g., a vitamin). Because one nutrient is limiting, growth rate is determined by the rate at which sterile medium is fed into the growth chamber; the final cell density depends on the concentration of the limiting nutrient. The rate of nutrient exchange is expressed as the dilution rate (D), the rate at which medium flows through the culture vessel relative to the vessel volume, where f is the flow rate (milliliter/hr) and V is the vessel volume (milliliter).

$$D = f/V$$

For example, if f is 30 milliliters/hr and V is 100 milliliters, the dilution rate is 0.30 hr^{-1}.

Both population size and generation time are related to the dilution rate (**figure 7.39**). When dilution rates are very low, only a limited supply of nutrient is available and the microbes can conserve only a limited amount of energy. Much of that energy must be used for cell maintenance, not for growth and reproduction. Slightly higher dilution rates make more nutrients available to the microbes. When the dilution rate provides enough nutrients for both maintenance and reproduction, the

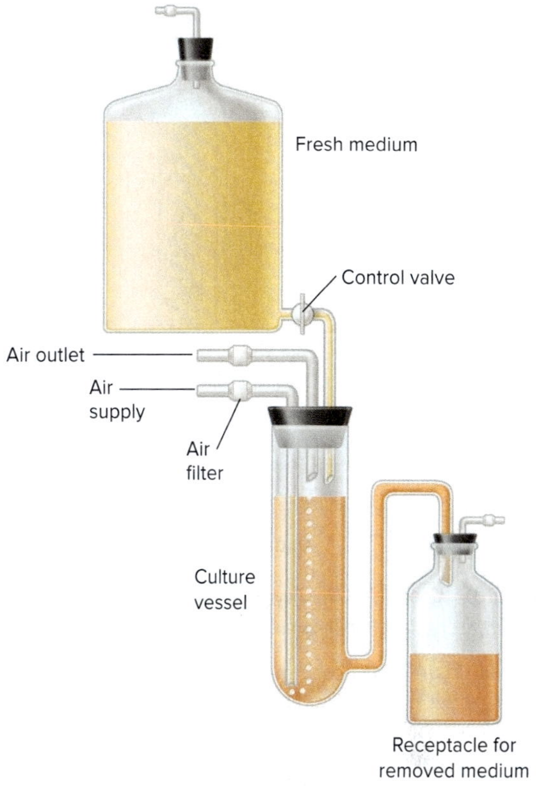

Figure 7.38 A Chemostat. Schematic diagram of the system. The fresh medium contains a limiting amount of an essential nutrient. Although not shown, usually the incoming medium is mechanically mixed with that already in the culture vessel. Growth rate is determined by the flow rate of medium through the culture vessel.

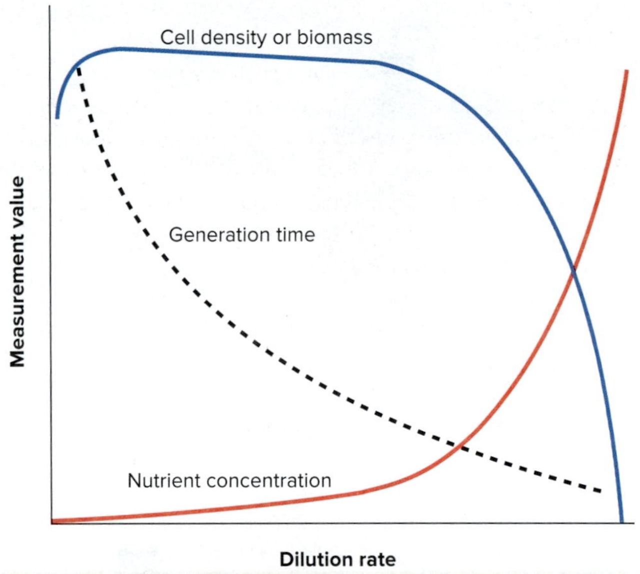

Figure 7.39 **Chemostat Dilution Rate and Microbial Growth.** The effects of changing the dilution rate in a chemostat.

MICRO INQUIRY *Why does the cell density stay relatively constant over such a large range of dilution rates?*

cell density will begin to rise. In other words, the growth rate can increase (and the generation time can decrease) when the total available energy provided by the nutrient supply exceeds the **maintenance energy.** Notice in figure 7.39 that for a wide range of dilution rates, the cell density remains stable. This is because the limiting nutrient is almost completely depleted by the rapid reproductive rate of the cells (as seen by the decreasing generation times). However, if the dilution rate is too high, microorganisms can actually be washed out of the culture vessel before reproducing because the dilution rate is greater than the maximum growth rate.

Turbidostats

The second type of continuous culture system, the **turbidostat,** has a photocell that measures the turbidity (defined as the amount of light scattered) of the culture in the growth vessel. The flow rate of media through the vessel is automatically regulated to maintain a predetermined turbidity. Because turbidity is related to cell density, the turbidostat maintains a desired cell density. Turbidostats differ from chemostats in several ways. The dilution rate in a turbidostat varies, rather than remaining constant, and a turbidostat's culture medium contains all nutrients in excess. That is, none of the nutrients is limiting. A turbidostat operates best at high dilution rates; a chemostat is most stable and effective at lower dilution rates.

Comprehension Check

1. How does a continuous culture system differ from a closed culture system (i.e., a batch culture)?
2. Describe how chemostats and turbidostats operate. How do they differ?
3. What is the dilution rate? What is maintenance energy? How are they related?

Key Concepts

7.1 Most Bacteria and Archaea Reproduce by Binary Fission

- The most common type of cell division observed among bacteria and archaea is binary fission. During binary fission, the cell elongates, the chromosome is replicated, and then it segregates to opposite poles of the cell prior to completing the septum, which divides the cell into two progeny cells (**figure 7.1**).
- Other bacterial reproductive strategies include budding, and baeocyte and spore formation (**figure 7.2**).

7.2 Bacterial Cell Cycles Can Be Divided into Three Phases

- Bacterial cell cycles consist of an initial growth phase followed by chromosome replication and segregation, and cytokinesis. The latter two phases overlap (**figure 7.3**).
- In bacterial cells, chromosome replication begins at a single site called the origin of replication. Many models for partitioning of the progeny chromosomes have been developed. The best-studied partitioning system is the ParAB/*parS* system of *Caulobacter crescentus* (**figure 7.4**).
- Septum formation in most bacteria is accomplished by Z ring formation and assembly of cell wall synthesizing machinery at the center of the cell once the chromosome has been partially segregated. This is brought about by Min proteins and nucleoid exclusion (**figures 7.5** and **7.6**).
- In rapidly dividing cells, initiation of DNA synthesis may occur before the previous round of synthesis is completed. This shortens the time needed for completing the cell cycle.
- The shape of bacterial cells is maintained by the cell wall. In rod-shaped bacteria, the cytoskeletal

proteins FtsZ and MreB are involved in determining where peptidoglycan will be synthesized. When rod-shaped cells are lengthening prior to cell division, MreB localizes along the length of the cell, and peptidoglycan synthesis occurs there as a result. During cell division, FtsZ localization confines peptidoglycan synthesis to the midcell (**figure 7.8**). In cocci, FtsZ and peptidoglycan synthesis localize to the midcell.

7.3 Archaeal Cell Cycles Are Unique

- Members of the order *Sulfolobales* have a cell cycle that consists of G1-, S-, and G2-like phases followed by partitioning and cytokinesis. The partitioning process is similar to the ParAB/*parS* system of *C. crescentus*. However, these archaea use homologues of eukaryotic ESCRT proteins for cytokinesis.
- The cell cycles of other archaea are more similar to the bacterial model. Many have homologues of FtsZ, which are thought to function in cytokinesis.

7.4 Growth Curves Consist of Five Phases

- Growth is an increase in cellular constituents and results in an increase in cell size, cell number, or both.
- When microorganisms are grown in a batch culture, the resulting growth curve usually has five phases: lag, exponential (log), stationary, death, and long-term stationary phase (**figure 7.10**).
- In the exponential phase, the population number of cells undergoing binary fission doubles at a constant interval called the generation or doubling time (**figure 7.12**). The growth rate constant (*k*) is the reciprocal of the generation time (**figure 7.13**).

7.5 Environmental Factors Affect Microbial Growth

- Solute concentrations can adversely affect microbes by causing lysis or shrinkage of the cytoplasm. Most bacteria, photosynthetic protists, and fungi have rigid cell walls that help protect them from osmotic stress. By synthesizing cytoplasmic compatible solutes, they remain hypertonic relative to their habitats.
- Solutes also affect microbes because they interact with water, making it less available for microbial activities. The amount of water available to microorganisms is expressed as the water activity (a_w). Most microorganisms do not grow well at water activities below 0.98; osmotolerant organisms survive and even flourish at low a_w values. Halophiles require high sodium chloride concentrations for growth (**figure 7.15** and **table 7.4**).
- Each microbial species has an optimum pH for growth and can be classified as an acidophile, neutrophile, or alkalophile (**figure 7.16** and **table 7.4**).
- Microorganisms have distinct temperature ranges for growth with minima, maxima, and optima—the cardinal temperatures. These ranges are determined by the effects of temperature on the rates of catalysis, protein denaturation, and membrane disruption. There are five major classes of microorganisms with respect to temperature preferences: (1) psychrophiles, (2) psychrotolerants, (3) mesophiles, (4) thermophiles, and (5) hyperthermophiles (**figures 7.17** and **7.18** and **tables 7.4** and **7.5**).
- Microorganisms can be placed into at least five different categories based on their O_2 requirements: obligate aerobes, microaerophiles, facultative anaerobes, aerotolerant anaerobes, and strict or obligate anaerobes (**figure 7.19** and **table 7.4**). Oxygen can become toxic because of the production of hydrogen peroxide, superoxide radical, and hydroxyl radical. These are destroyed by the enzymes superoxide dismutase, catalase, and peroxidase.
- Most deep-sea microorganisms are barotolerant, but some are piezophilic and require high pressure for optimal growth.
- High-energy or short-wavelength radiation harms organisms in several ways. Ionizing radiation (X rays and gamma rays) ionizes molecules and destroys DNA and other cell components. Ultraviolet (UV) radiation induces DNA damage. Visible light can provide energy for the formation of reactive singlet oxygen, which will destroy cells.

7.6 Microbial Growth in Natural Environments

- Microbial growth in natural environments is profoundly affected by nutrient limitations and other adverse factors. Microbes are often found in oligotrophic environments. They have evolved strategies to survive in these habitats, including spore formation and other types of dormancy in growth-arrested states.
- Many microbes form biofilms—aggregations of microbes growing on surfaces and held together by extracellular polymeric substances (EPS) (**figures 7.22** and **7.23**). Life in a biofilm has several advantages, including protection from harmful agents.
- Bacteria often use chemicals to communicate with one another in a density-dependent way and carry out a particular activity only when a certain population density is reached. This phenomenon is called quorum sensing (**figure 7.24**).
- Some bacteria are able to communicate via chemicals with large organisms such as plants. A well-studied example is the rhizobium-plant interaction.

7.7 Laboratory Culture of Cellular Microbes Requires Media and Conditions That Mimic the Normal Habitat of a Microbe

- Culture media can be constructed completely from chemically defined components (defined media or synthetic media) or constituents, such as peptones and yeast extract, whose precise composition is unknown (complex media).

Culture media can be solidified by the addition of agar, a complex polysaccharide from red algae.

- Culture media are classified based on function. Supportive media are used to culture a wide variety of microbes. Enriched media are supportive media that contain additional nutrients needed by fastidious microbes. Selective media contain components that select for the growth of some microbes. Differential media contain components that allow microbes to be differentiated from each other, usually based on some metabolic capability.
- Aerobes are cultivated by supplying oxygen. Anaerobes are cultivated using media and methods that exclude oxygen (**figures 7.27** and **7.28**).
- Enrichment cultures are usually the first step in isolating pure cultures. They use selective conditions (media, environmental conditions) to promote the growth of certain microbes.
- Pure cultures usually are obtained by isolating individual cells with any of three plating techniques: the streak-plate, spread-plate, and pour-plate methods. The streak-plate technique uses an inoculating loop to spread cells across an agar surface. The spread-plate and pour-plate methods involve diluting a culture or sample and then plating the dilutions. In the spread-plate technique, cells are spread on an agar surface using a bent rod; in the pour-plate technique, the cells are first mixed with cooled agar-containing media before being poured into a Petri dish (**figures 7.29–7.31**).
- Microorganisms growing on solid surfaces often form colonies with characteristic morphology (**figure 7.32**).
- New approaches are being developed to culture microbes that in the past have not been culturable. These include creating natural habitats within the lab setting and co-culturing microbes with organisms that help them survive.

7.8 Microbial Population Size Can Be Measured Directly or Indirectly

- Microbial populations can be counted directly with counting chambers, flow cytometers, electronic counters, or fluorescence microscopy. However, it is sometimes difficult to discern if cells are living or dead (**figures 7.33** and **7.34**).
- Indirect measurements include viable counting techniques (standard plate counts) such as the spread-plate, the pour-plate, or the membrane filter method (**figure 7.35**).
- Population size also can be determined indirectly by measuring microbial mass through the measurement of dry weight, turbidity, or the amount of a cell component (**figure 7.37**).

7.9 Chemostats and Turbidostats Are Used for Continuous Culture of Microorganisms

- Microorganisms can be grown in an open system in which nutrients are constantly provided and wastes removed.
- A continuous culture system can maintain a microbial population in log phase. There are two types of these systems: chemostats and turbidostats (**figure 7.38**).

Active Learning

1. As an alternative to diffusible signals, discuss another mechanism by which bacteria might quorum sense.
2. If you wished to obtain a pure culture of bacteria that could degrade benzene and use it as a carbon and energy source, how would you proceed?
3. How would you determine if a slow-growing microbial culture is in lag phase or exponential phase?
4. Suggest one specific mechanism underlying the observation that the cardinal temperatures of some microbes change depending on other environmental conditions (e.g., pH).
5. Consider cell-cell communication: Bacteria that "subvert" and "cheat" have been described. Discuss a situation in which it would be advantageous for one species to subvert another, that is, degrade an intercellular signal made by another species. Also, describe a scenario whereby bacterial cheaters—defined as bacteria that do not make a molecular signal but profit by the uptake and processing of signal made by another microbe—might have a growth advantage.
6. Suppose you discovered a new bacterial strain from the permafrost of Alaska. You are able to culture the bacterium, but it grows very slowly. What characteristics of the bacterium would you determine to see if you could optimize its growth rate? Explain your choices.
7. What changes in equipment and techniques would be necessary to work with an organism that is both hyperthermophilic and anaerobic? Or, an organism whose growth substrate is a gas, like CO_2?
8. The bacterium *Variovorax* can use *N*-acylhomoserine lactones as a carbon and energy source for growth. How would the presence of *Variovorax* in an ecosystem affect the microbial community?

9. When attempting to use a standard agar medium to grow microbes from environmental sources like a body of freshwater, microbiologists find that only a small fraction will grow into colonies. Scientists from Japan established that a chemical reaction between the agar used to solidify the medium and the phosphate buffer in the medium resulted in the production of hydrogen peroxide (H_2O_2). They hypothesized that H_2O_2 would inhibit microbial growth. In an effort to cultivate greater microbial diversity, they included a biological molecule in their medium to eliminate the H_2O_2. What additive might reduce the H_2O_2 concentration in bacterial media?

Read the original paper: Kawasaki, K., and Kamagata, Y. 2017. Phosphate-catalyzed hydrogen peroxide formation from agar, gellan, and κ-carrageenan. *Appl. Env. Microbiol.* 83(21):e01366-17.

8

Control of Microorganisms in the Environment

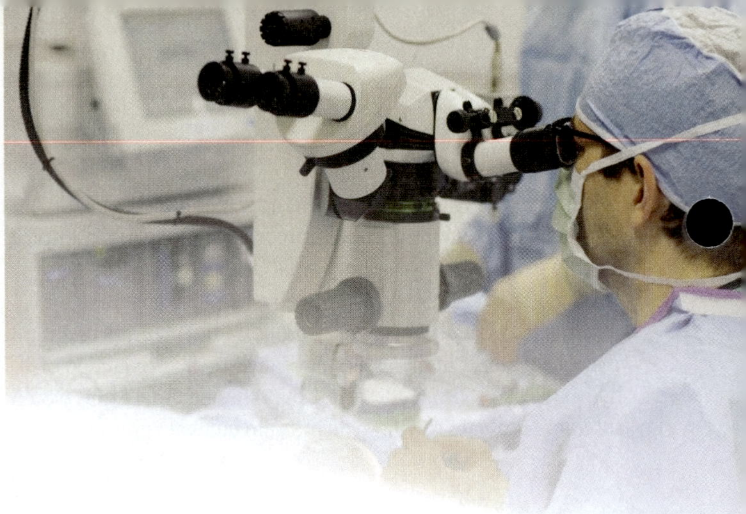

©Erproductions Ltd/Blend Images LLC

Keeping Infection at Bay

Imagine going to the hospital for neurosurgery, as thousands of people do every day. Now imagine finding out that as a consequence of this surgery you were infected with an agent that has a 100% mortality rate. Surely this couldn't happen in the twenty-first century, right? Sadly, the answer is yes it has happened, but we must stress it is exceedingly rare.

Since the days of Joseph Lister, disinfection in health care facilities has been the focus of intense scrutiny. However, in the 1980s, microbiologists were challenged when it became clear that a previously unknown infectious agent, the prion, was not susceptible to any of the disinfection methods in practice at that time. Composed entirely of protein, prions resist decontamination efforts that involve heat, radiation, and chemicals that attack cellular components. ⏮ *Prions are composed only of protein (section 6.7)*

The need to protect against prion contamination was made frighteningly clear when it was determined that infected individuals exhibit high levels of prions in brain and neural tissue and suffer neurological symptoms that always result in death. In cases of iatrogenic (Greek *iatros,* healer; *genic,* producing) prion disease, it was discovered that prion-containing corneas were transplanted or contaminated neurosurgical instruments were used. This has given rise to the use of disposable products that are later incinerated to over 1,000°C, decontamination of durable goods by ozone exposure, and tighter screening of donated corneas. In addition, U.S. blood banks screen donors who resided in Britain for at least 3 months between 1980 and 1996. During those years, the domestic beef supply was tainted with spinal cord tissue from practices in slaughterhouses, and meat from "mad cows" entered the food supply. Because low levels of prions can be detected in the blood of infected individuals, as a cautionary practice, those meeting these criteria are rejected for blood donation in several countries. ⏭ *Prion proteins transmit disease (section 38.6)*

Having introduced the breadth of microbial forms in chapters 3 through 6 and how they grow in chapter 7, we turn now to how to control microbial growth. This chapter addresses the means to prevent infection; chapter 9 addresses the means to treat an infection.

Microorganism control continues to be a hot topic as microorganisms evolve to resist current strategies. Control efforts have a substantial role in public health to prevent disease as well as in therapeutic use to treat disease. Thus this chapter focuses on the control of microorganisms by physical, chemical, and biological agents (such as engineered bacteria). In general, any chemical, physical, or biological product that controls microorganisms is referred to as an **antimicrobial agent.** Chemotherapeutic agents that are used inside the body to treat human disease are discussed in chapter 9.

Readiness Check:

Based on what you have learned previously, you should be able to:

✓ Identify the structures and their functions of bacteria, protists, fungi, viruses, and prions, including their replication processes and energy requirements (sections 3.3–3.9, 5.1–5.7, 6.1–6.7, 7.1–7.7)

8.1 Microbial Growth and Replication: Targets for Control

After reading this section, you should be able to:

a. Compare and contrast actions of disinfection, antisepsis, chemotherapy, and sterilization
b. Distinguish between cidal (killing) and static (inhibitory) agents

The principles of microbial control are rooted in microbial nutrition, growth, and development, which are discussed in chapter 7. If you can starve or poison microbes, or inhibit or prevent growth or replication, you can control microorganisms. Of course, it is not as simple as this sounds, and a complex vocabulary is applied to describe differences in population density, degree of killing, and even what is used to remove microorganisms. Terminology is especially important when the control of microorganisms is discussed because words such as *disinfectant* and *antiseptic* often are used loosely, yet they have different meanings. The situation is even more confusing because a particular treatment can either inhibit growth, inactivate replication, or kill, depending on the conditions. The types of control agents and their uses are

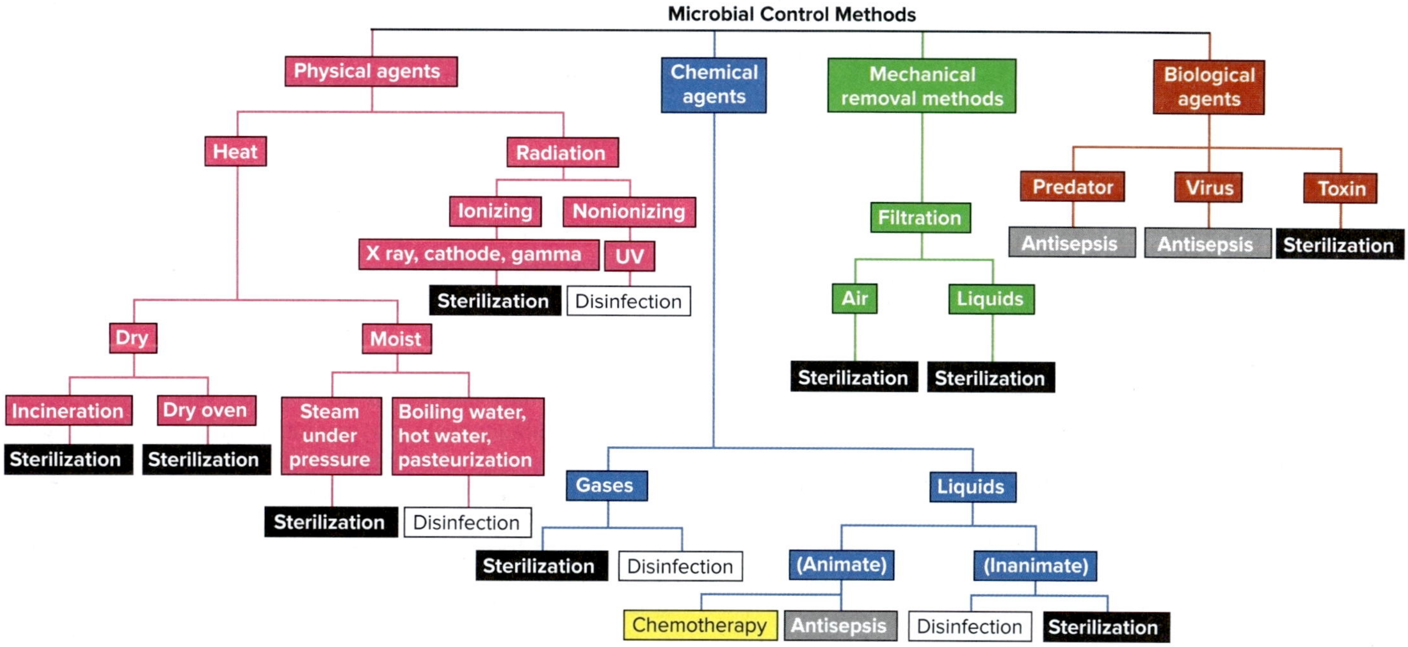

Figure 8.1 Microbial Control Methods.

MICRO INQUIRY *Which types of agents can be used for sterilization? Which can be used for antisepsis? What is the difference?*

outlined in **figure 8.1**. Note that not all agents are chemical. Rather, antimicrobial agents can be physical, chemical, mechanical, and biological. To simplify the terminology, we use the term **biocide** to mean all antimicrobial agents that can be used to control microorganisms. In general, to control microorganisms a biocide must be evaluated to determine the specific parameters under which it will be effective.

The range of microbial cell types presents a broad target for growth control. The outer layer of a microbe presents a permeability barrier that must be overcome for a treatment to be effective. Many bacterial cells are readily destroyed, with Gram-negatives somewhat less susceptible as a consequence of their outer membrane. Exterior surfaces on *Mycobacterium* spp., bacterial endospores, and protozoan cysts render these cell types especially recalcitrant. Microbial control methods are typically applied to populations of microbes, where they will affect individuals differently. It is critical to consider the diversity of microbes in a given situation when evaluating control options.

Sterilization (Latin *sterilis,* unable to produce offspring), in our context, is the process by which all living cells, spores, and acellular entities (e.g., viruses, viroids, and prions) are either destroyed or removed from an object or habitat. A sterile object is totally free of viable microorganisms, spores, and other infectious agents. When sterilization is achieved by a chemical agent, the chemical is called a sterilant. In contrast, **disinfection** is the killing, inhibition, or removal of microorganisms that may cause disease. Disinfection therefore results in the substantial reduction of the total microbial population and the destruction of potential pathogens. **Disinfectants** are agents, usually chemical, used to carry out disinfection and normally used only on inanimate objects. A disinfectant does not necessarily sterilize an object because viable spores and a few microorganisms may remain. In **sanitization**, the microbial population is reduced to levels that are considered safe by public health standards. The inanimate object is usually cleaned as well as partially disinfected. For example, sanitizers are used to clean eating utensils in restaurants. ◀ *Viroids and satellites: nucleic acid-based subviral agents (section 6.6); Prions are composed only of protein (section 6.7)*

It also is frequently necessary to control microorganisms on or in living tissue. **Antisepsis** (Greek *anti,* against; *sepsis,* putrefaction) is the destruction or inhibition of microorganisms on living tissue; it is the prevention of infection or sepsis. **Antiseptics** are chemical agents applied to tissue to prevent infection by killing or inhibiting pathogen growth; they also reduce the total microbial population. Because they must not cause too much harm to the host, antiseptics are generally not as toxic as disinfectants. The exposure of

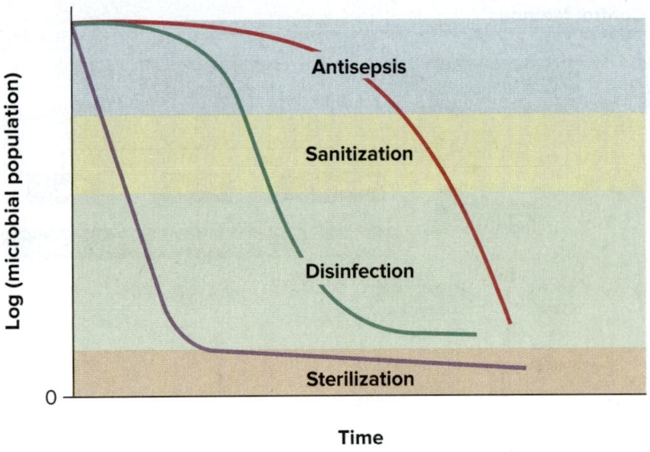

Figure 8.2 Impact of Biocide Exposure. Three exponential plots of survivors versus time of biocide exposure, indicating the potential kinetics of biocide action. Note how the general terms referencing microbial control are reflected by decreasing numbers of microbes. For example, sterilization refers to the absence of viable organisms regardless of biocide kinetics.

microorganisms to increasing biocide concentrations decreases the number of viable organisms. **Figure 8.2** shows three possible population reduction curves resulting from three different biocides. The shape of the curve reflects conditions that influence biocide effectiveness. Note that in each case, the eventual decline in viable microorganisms can occur as a staged interval of viability from antisepsis to sterilization. **Chemotherapy** is the use of chemical agents to kill or inhibit the growth of microorganisms within host tissue and is the topic of chapter 9.

The type of antimicrobial agent can be discerned from its suffix. Substances that kill organisms often have the suffix *-cide* (Latin *cida,* to kill); a cidal agent kills microbes but not necessarily endospores. A disinfectant or antiseptic can be particularly effective against a specific group, in which case it may be called a **bactericide, fungicide,** or **viricide.** Other chemicals do not kill but rather prevent growth. If these agents are removed, growth will resume. Their names end in *-static* (Greek *statikos,* causing to stand or stopping)—for example, **bacteriostatic** and **fungistatic.**

It is worth noting here that resistance to antimicrobial biocides has been increasing nearly as rapidly as resistance to antibiotics. Similar to antibiotic resistance mechanisms (discussed in chapter 9), biocide resistance mechanisms include the induction of efflux pumps, modified membrane permeability, and target modification. In fact, a number of studies have linked the acquisition of biocide resistance mechanisms to resistance to antibiotics. In many cases they are shared on plasmids and upregulated in response to ubiquitously low environmental biocide concentrations. It is clear that microorganisms adapt well to new situations, resulting in the sharing of successful resistance strategies.

Comprehension Check

1. Define the following terms: *sterilization, sterilant, disinfection, disinfectant, sanitization, antisepsis, antiseptic, chemotherapy, biocide*.
2. What is the difference between bactericidal and bacteriostatic? To which category do you think most household cleaners belong? Why?

8.2 The Pattern of Microbial Death Mirrors the Pattern of Microbial Growth

After reading this section, you should be able to:

a. Calculate the decimal reduction time (D value)
b. Correlate antisepsis, sanitization, disinfection, and sterilization with agent effectiveness

A microbial population is not killed instantly when exposed to a lethal agent. Population death is generally exponential; that is, the population is reduced by the same fraction at constant intervals (**table 8.1**). If the logarithm of the population number

Table 8.1 A Theoretical Microbial Heat-Killing Experiment

Minute	Microbial Number at Start of Minute	Microorganisms Killed in 1 Minute (90% of Total)[1]	Microorganisms at End of 1 Minute	Log_{10} of Survivors
1	10^6	9×10^5	10^5	5
2	10^5	9×10^4	10^4	4
3	10^4	9×10^3	10^3	3
4	10^3	9×10^2	10^2	2
5	10^2	9×10^1	10	1
6	10^1	9	1	0
7	1	0.9	0.1	–1

1 Assume that the initial sample contains 10^6 vegetative microorganisms per milliliter and that 90% of the organisms are killed during each minute of exposure. The temperature is 121°C.

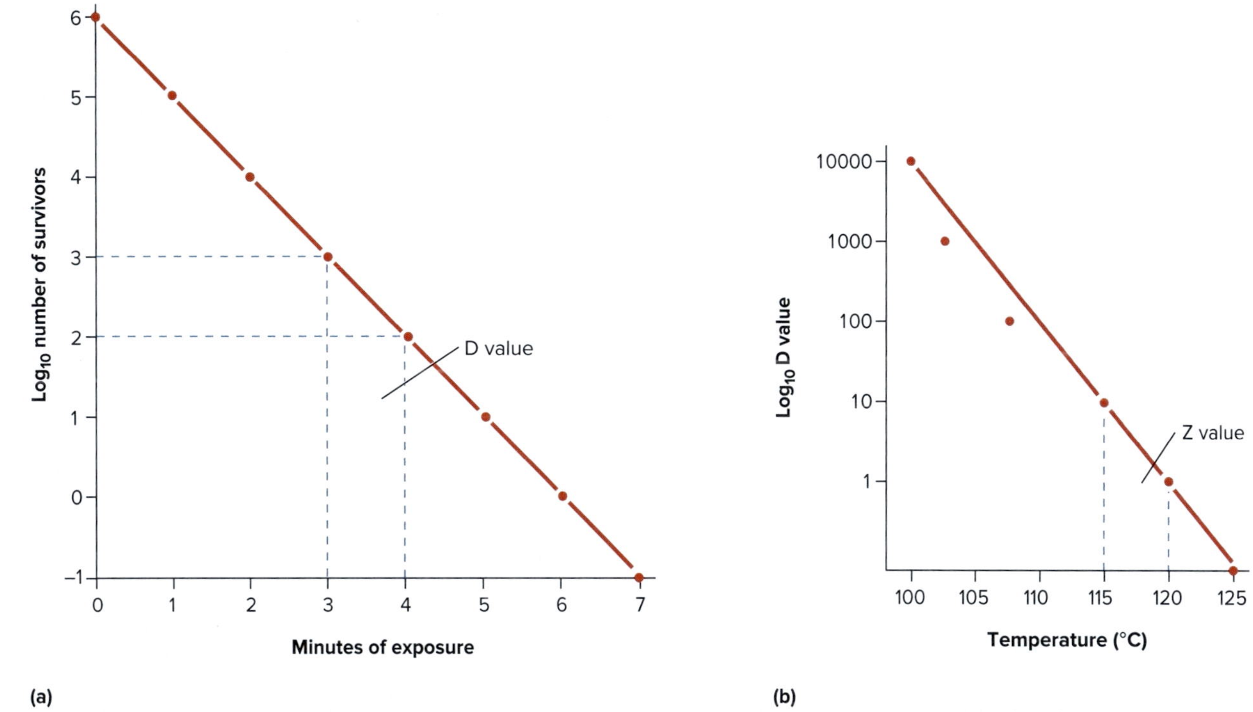

Figure 8.3 The Pattern of Microbial Death. (a) An exponential plot of the survivors versus the minutes of exposure to heating at 121°C. In this example, the D value is 1 minute. The data are from table 8.1. (b) An exponential plot of D values versus temperature. The temperature change at a given D value that reduces the population by one log unit is known as the Z value. In this example, the Z value is 5°C.

MICRO INQUIRY *Examine graph (a). How long would it take to kill one-half of the original population of microorganisms? Keep in mind that the y axis is exponential.*

remaining is plotted against the time of exposure of the microorganism to the agent, a straight-line plot will result (**figure 8.3**). When the population has been greatly reduced, the rate of killing may slow due to the survival of a more resistant strain of the microorganism.

One measure of an agent's killing efficiency is the **decimal reduction time** (D) or **D value.** The decimal reduction time is the time required to kill 90% of the microorganisms or endospores in a sample under specified conditions. For example, in a semilogarithmic plot of the population remaining versus the time of heating, the D value is the time required for the line to drop by one log cycle or 10-fold (figure 8.3*a*). It is also possible to determine the temperature change at a given D value that decreases the microbial population by one log cycle (90%). This temperature change is referred to as the Z value and is predicted from a semilogarithmic plot of D values versus temperature (figure 8.3*b*).

To study the effectiveness of a lethal agent, one must be able to decide when microorganisms are dead, which may present some challenges. A microbial cell is often defined as dead if it does not grow when inoculated into culture medium that would normally support its growth. In like manner, an inactive virus cannot infect a suitable host. This definition has flaws, however. It has been demonstrated that when bacteria are exposed to certain conditions, they can remain alive but are temporarily unable to reproduce (*see figure 7.34*). In conventional tests to demonstrate killing by an antimicrobial agent, these viable but nonculturable bacteria would be thought to be dead. This is a serious problem because the bacteria may regain their ability to reproduce and cause infection after a period of recovery. ◀ *Most microbes live in growth-arrested states (section 7.6)*

8.3 Mechanical Removal Methods Rely on Barriers

After reading this section, you should be able to:

a. Explain the mechanism by which filtration removes microorganisms
b. Propose filtration methods to selectively remove one type of microbe from a mixed population

Filtration is an excellent way to reduce the microbial population in solutions of heat-sensitive material and can be used to sterilize liquids and gases (including air). Rather than directly

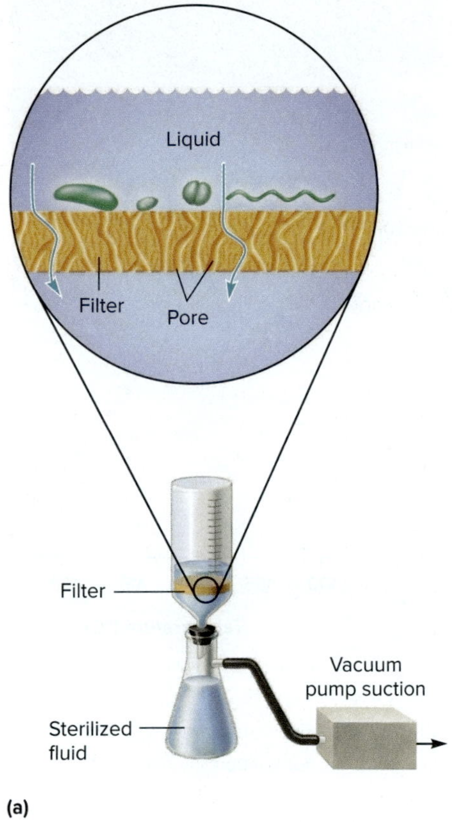

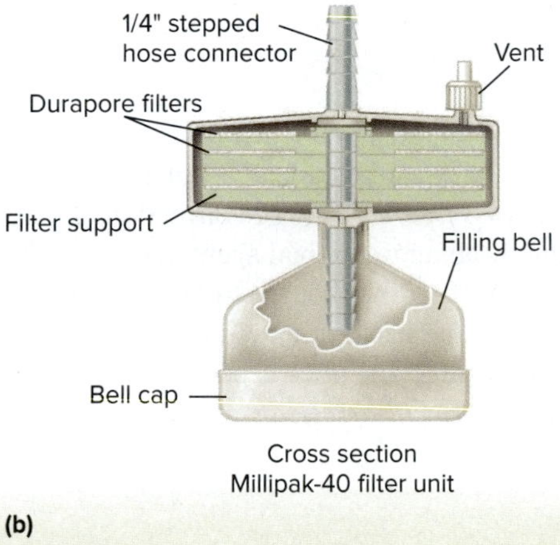

Figure 8.4 Membrane Filter Sterilization. The liquid to be sterilized is pumped through a membrane filter and into a sterile container. (a) Schematic representation of a membrane filtration setup that uses a vacuum pump to force liquid through the filter. The inset shows a cross section of the filter and its pores, which are too small for microbes to pass through. (b) Cross section of a membrane filtration unit. Several membranes are used to increase its capacity.

MICRO INQUIRY *How might one verify that filtration removed all microorganisms?*

destroying contaminating microorganisms, the filter simply acts as a barrier to retain them. There are two types of filters. **Depth filters** consist of fibrous or granular materials that have been bonded into a thick layer filled with narrow, twisting channels. The solution containing microorganisms is sucked through this layer under vacuum, and microbial cells are removed by entrapment and by adsorption to the surface of the filter material.

Membrane filters have replaced depth filters for many purposes. These filters are porous membranes, a little over 0.1 mm thick, made of cellulose acetate, cellulose nitrate, polycarbonate, polyvinylidene fluoride, or other synthetic materials. Although a wide variety of pore sizes are available, membranes with pores about 0.2 μm in diameter are used to remove most vegetative cells, but not viruses, from liquids. The membranes are held in special holders (**figure 8.4**), and their use is often preceded by depth filters to remove larger particles that might clog the membrane filter. The liquid is forced through the filter and collected in previously sterilized containers. Membrane filters remove microorganisms by screening them out much as a sieve separates large sand particles from small ones (**figure 8.5**). These filters are used to sterilize pharmaceuticals, ophthalmic solutions, culture media, oils, antibiotics, and other heat-sensitive solutions.

Air also can be filtered to remove microorganisms. Two common examples are N95 disposable masks used in hospitals and labs, and **high-efficiency particulate air (HEPA) filters** that let air move freely but restrict microorganisms. N95 masks exclude 95% of particles that are larger than 0.3 μm. HEPA filters (a type of depth filter made from fiberglass) remove 99.97% of particles 0.3 μm or larger by both physical retention and electrostatic interactions. HEPA filters sterilize air by removing viruses that are 0.1 μm and

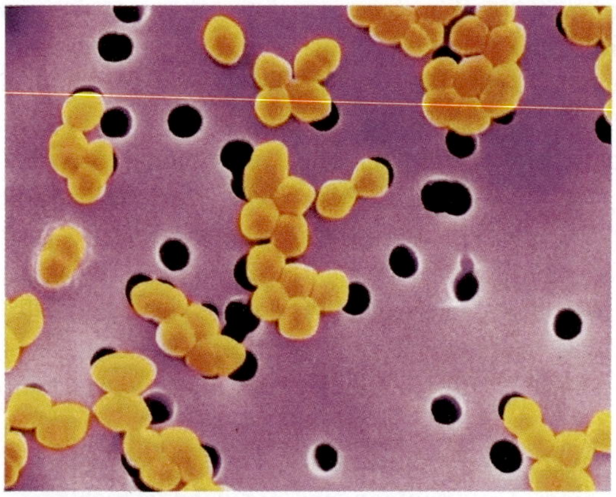

Figure 8.5 Membrane Filter. *Enterococcus faecalis* resting on a polycarbonate membrane filter with 0.4-μm pores.

MICROBIAL DIVERSITY & ECOLOGY

8.1 The Cleanest Place on Earth?

Many environments must be as devoid of microbes as possible, for example, operating rooms or pharmaceutical manufacturing facilities. Microbial contamination in these places can result in severe infection or fatality. Spacecraft assembly sites, too, must be free of microbes, but the consequences of contamination could affect an entire planet.

The National Air and Space Administration (NASA) is planning two missions to look for signs of past life or current habitable conditions elsewhere in the solar system. Mars 2020 and the Europa Clipper mission to Jupiter's moon take the search for extraterrestrial life directly to those locations.

Two concerns face astrobiologists: first, avoiding false-positive signals for life by detecting stowaway microbes; and second, avoiding contamination of extraterrestrial environments. Although complete spacecraft sterility is the ideal goal, it is currently unattainable, as sensitive instruments cannot withstand heat treatment. Extremophiles and spores are the primary concern because they have the greatest probability for survival in space. The major contamination source, the engineers who assemble the spacecraft, work in special suits to confine their natural microbial shedding. These clean rooms operate with HEPA-filtered air and surfaces are regularly treated with ultraviolet light.

Surprisingly, the lack of dust and organic matter in NASA clean rooms has created an extreme environment that enriches for the presence of certain microbes. *Tersicoccus phoenicis* has been isolated from only two places on Earth: spacecraft assembly clean rooms on two continents.

(a) (b)

NASA Clean Room Conditions. (a) Staff in NASA clean rooms dress in protective garb when working on spacecraft to minimize contamination. (b) *Tersicoccus phoenicis,* whose only known natural environment is clean rooms where spacecraft are assembled.

(a) Source: NASA/Leif Heimbold; (b) Source: NASA/JPL-Caltech

smaller. HEPA-filtered air is critical in clean room environments, as discussed in **Microbial Diversity & Ecology 8.1**. Laminar flow biological safety cabinets or hoods force air through HEPA filters, then project a vertical curtain of sterile air across the cabinet opening. This protects a worker from microorganisms being handled within the cabinet and prevents contamination of the room (**figure 8.6**). Dangerous agents such as *Mycobacterium tuberculosis,* pathogenic fungi, or tumor viruses must be confined to biological safety cabinets.

Comprehension Check

1. What are depth filters and membrane filters, and how are they used to sterilize liquids?
2. Describe the operation of a biological safety cabinet.

(a)

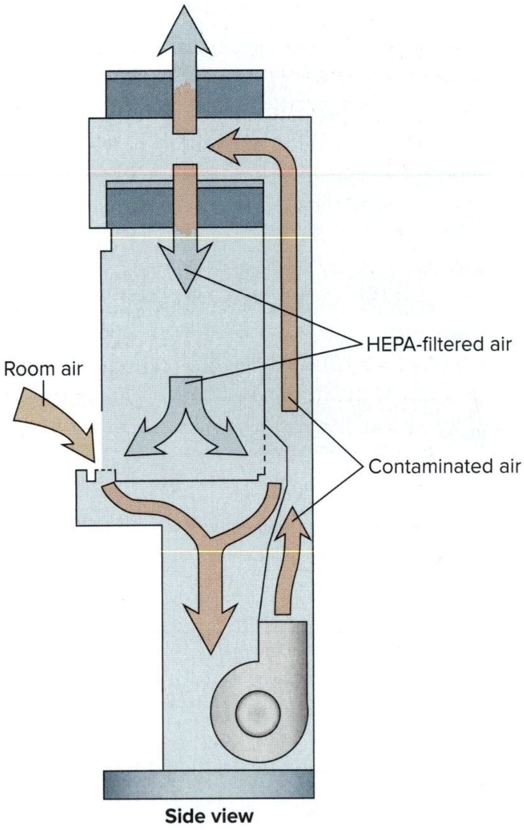

(b)

Figure 8.6 A Biological Safety Cabinet. (a) A technician pipetting potentially hazardous material in a safety cabinet. (b) A schematic diagram showing the airflow pattern within a class II safety cabinet.
(a) ©miR156/Alamy Stock Photo

8.4 Physical Control Methods Alter Microorganisms to Make Them Nonviable

After reading this section, you should be able to:

a. Describe the application of heat and radiation to control microorganisms
b. Explain the mechanisms by which heat and radiation kill microbes
c. Design novel antimicrobial control applications using heat and radiation

Heat and other physical agents are normally used to control microbial growth and sterilize objects, as can be seen from the operation of the autoclave. The most frequently employed physical agents are heat and radiation.

Heat

Most microorganisms require specific temperatures for normal growth and replication; temperatures that exceed those damage structures and alter chemical reactions. Moist and dry heat readily destroy viruses, bacteria, and fungi in this way (**table 8.2**).

Moist heat destroys cells and viruses by degrading nucleic acids, denaturing proteins, and disrupting cell membranes. Exposure to boiling water for 10 minutes is sufficient to destroy vegetative cells and eukaryotic spores. Unfortunately, the temperature of boiling water (100°C at sea level) is not sufficient to destroy bacterial endospores, which may survive hours of boiling. Therefore boiling can be used for disinfecting drinking water and objects not harmed by water, but boiling does not sterilize.

To destroy bacterial endospores, moist heat sterilization must be carried out at temperatures above 100°C, and this requires the use of saturated steam under pressure. Steam sterilization is carried out with an **autoclave** (**figure 8.7**), a device somewhat like a fancy pressure cooker. The development of the autoclave by Charles Chamberland in 1884 tremendously stimulated the growth of microbiology as a science. Steam is released

Table 8.2 Approximate Conditions for Moist Heat Inactivation

Entity	Vegetative Cells	Spores
Yeasts	5 minutes at 50–60°C	5 minutes at 70–80°C
Molds	30 minutes at 62°C	30 minutes at 80°C
Bacteria[1]	10 minutes at 60–70°C	2 to over 800 minutes at 100°C 0.5–12 minutes at 121°C
Viruses	30 minutes at 60°C	
Prions	90 minutes at 134°C	

1 Conditions for mesophilic bacteria.

(a)

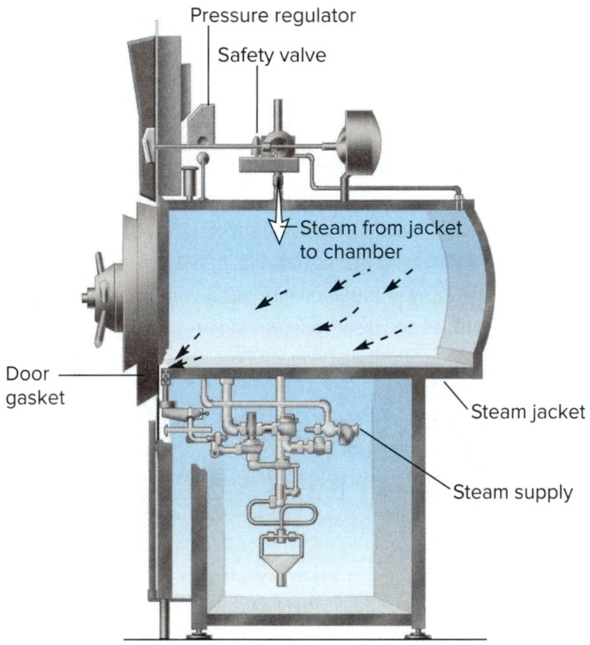

(b)

Figure 8.7 The Autoclave. (a) A modern, automatically controlled autoclave or sterilizer. (b) Longitudinal cross section of a typical autoclave showing some of its parts and the pathway of steam.

(a) ©BSIP SA/Alamy Stock Photo

into the autoclave chamber (figure 8.7*b*). The air initially present in the chamber is forced out until the chamber is filled with saturated steam and the outlets are closed. Hot, saturated steam continues to enter until the chamber reaches the desired temperature and pressure, usually 121°C and 15 pounds per square inch (psi) of pressure.

Autoclaving must be carried out properly so that the saturated steam can destroy all vegetative cells and endospores within the processed materials. If all air has not been flushed out of the chamber, it will not reach 121°C, even though it may reach a pressure of 15 psi. The chamber should not be packed too tightly because the steam needs to circulate freely and contact everything in the autoclave. Bacterial endospores will be killed only if they are kept at 121°C for 10 to 12 minutes. When a large volume of liquid must be sterilized, an extended sterilization time is needed because it takes longer for the center of the liquid to reach 121°C; 5 liters of liquid may require about 70 minutes. In view of these potential difficulties, a biological indicator is often included in the autoclave. This indicator commonly consists of a culture tube containing a sterile ampule of medium and a paper strip covered with endospores of *Geobacillus stearothermophilus.* After autoclaving, the ampule is aseptically broken and the culture incubated. If the test bacterium does not grow, the sterilization run has been successful. Sometimes indicator tape or paper that changes color upon heating is autoclaved with a batch of material. These indicate heating has occurred and are convenient but are not as reliable as techniques that directly demonstrate the inactivation of bacterial endospores.

Many heat-sensitive substances, such as milk, are treated with controlled heating at temperatures well below boiling, a process known as **pasteurization** in honor of its developer, Louis Pasteur (1822–1895). In the 1860s the French wine industry was plagued by wine spoilage, which made wine storage and shipping difficult. Pasteur examined spoiled wine under the microscope and detected microorganisms that looked like the bacteria responsible for lactic acid and acetic acid fermentations; these bacteria do not form endospores. He discovered that a brief heating at 55° to 60°C would destroy these microorganisms and preserve wine for long periods. In 1886 German chemists V. H. Soxhlet and F. Soxhlet adapted the technique for preserving milk. Milk pasteurization was introduced in the United States in 1889. Milk, beer, and many other beverages are now pasteurized. Pasteurization does not sterilize, but it does kill pathogens and drastically slows spoilage by reducing the level of nonpathogenic spoilage microorganisms (*see table 41.2*).

Some materials cannot withstand the high temperature of the autoclave, and endospore contamination precludes the use of other methods to sterilize them. For these materials, a process of intermittent sterilization, also known as **tyndallization** (for John Tyndall [1820–1893], the British physicist who used the technique to destroy heat-resistant microorganisms in dust) is used. The process also uses steam (30–60 minutes) to destroy vegetative bacteria. However, steam exposure is repeated for a total of three times with 23- to 24-hour incubations between exposures. The incubations permit remaining endospores to germinate into heat-sensitive vegetative cells that are destroyed upon subsequent steam exposures.

Many objects are best sterilized by dry heat. For instance, inoculating loops, which are used routinely in the laboratory, can be sterilized in a small, bench-top incinerator (**figure 8.8**). Other items are sterilized in an oven at 160° to 170°C for 2 to 3 hours. Microbial death results from the oxidation of cell constituents and protein denaturation. Dry air heat is less effective than moist heat. The endospores of *Clostridium botulinum,* the cause of botulism, are killed in 5 minutes at 121°C by moist heat but only after 2 hours at 160°C by dry heat. However, dry heat has some advantages. It does not

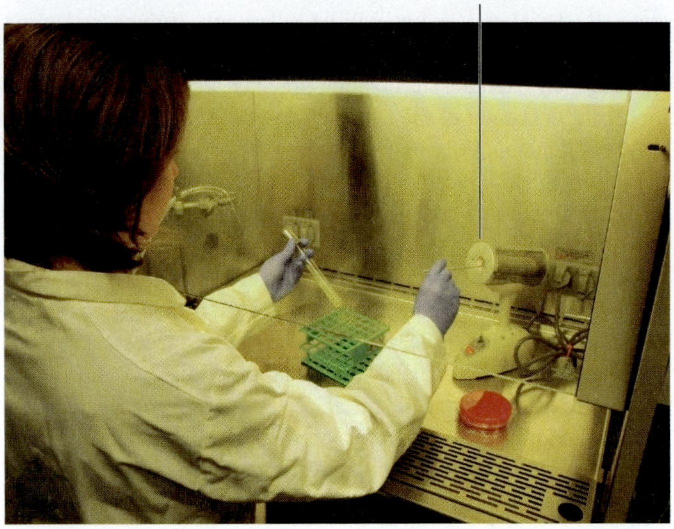

Figure 8.8 Dry Heat Incineration. Bench-top incinerators are routinely used to sterilize inoculating loops used in microbiology laboratories.

corrode glassware and metal instruments as moist heat does, and it can be used to sterilize powders, oils, and similar items. Despite these advantages, dry heat sterilization is slow and not suitable for heat-sensitive materials such as plastic and rubber items.

Radiation

Ultraviolet (UV) radiation around 260 nm (*see figure 7.20*) is quite lethal. It causes thymine-thymine dimerization of DNA, preventing replication and transcription (*see figure 16.5*). However, UV radiation does not penetrate glass, dirt films, water, and other substances effectively. Because of this disadvantage, UV radiation is used as a sterilizing agent only in a few situations. UV lamps are sometimes placed on lab ceilings or in biological safety cabinets to sterilize the air and any exposed surfaces. Because UV radiation burns the skin and damages eyes, the UV lamps are off when the areas are in use. Commercial UV units are available for water treatment. Microbes are destroyed when a thin layer of water is passed under the lamps. ▶| *Purification and sanitary analysis ensure safe drinking water (section 43.1)*

Ionizing radiation is an excellent sterilizing agent that penetrates deep into objects. Ionizing radiation has sufficient energy to dislodge electrons from atoms or molecules, producing chemically reactive free radicals. The free radicals react with nearby matter to weaken or destroy it. Ionizing radiation destroys bacterial endospores and all microbial cells; however, it is not always effective against viruses. Gamma radiation (from a cobalt 60 source) and beta radiation (accelerated electrons from high-voltage electricity) are used in the cold sterilization of antibiotics, hormones, sutures, and plastic disposable supplies such as syringes. Gamma radiation and electron beams have also been used to sterilize and "pasteurize" meat and other foods (**figure 8.9**). Irradiation can eliminate the threat of such pathogens as *E. coli* O157:H7, which causes a life-threatening intestinal disease; *Staphylococcus aureus,* which causes skin and blood infections, and readily colonizes medical devices used on patients; and *Campylobacter jejuni,* which contaminates poultry, causing intestinal disease when undercooked meat is eaten. Both the U.S. Food and Drug Administration and the World Health Organization have approved irradiated food and declared it safe for human consumption. Currently irradiation is used to treat meat, fruits, vegetables, and spices. ▶| *Human diseases caused by bacteria (chapter 39); Various methods are used to control food spoilage (section 41.2)*

Comprehension Check

1. Describe how an autoclave works. What conditions are required for sterilization by moist heat? What three things must one do when operating an autoclave to help ensure success?
2. In the past, spoiled milk was responsible for a significant proportion of infant deaths. Why is untreated milk easily spoiled?
3. List the advantages and disadvantages of ultraviolet light and ionizing radiation as sterilizing agents. Provide a few examples of how each is used for this purpose.
4. What is the correlation between radiation "energy" and the mechanisms of sterilization?

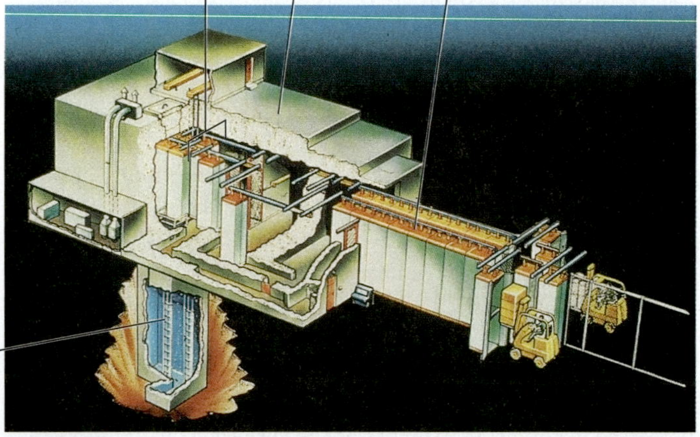

Figure 8.9 Sterilization with Ionizing Radiation. An irradiation facility that uses radioactive cobalt 60 as a gamma radiation source to sterilize fruits, vegetables, meats, fish, and spices.

8.5 Microorganisms Are Controlled with Chemical Agents

After reading this section, you should be able to:

a. Describe the use of and mechanism of action for phenolics, alcohols, halogens, heavy metals, quaternary ammonium compounds, aldehydes, and oxides to control microorganisms
b. Design novel antimicrobial control applications using phenolics, alcohols, halogens, heavy metals, quaternary ammonium compounds, aldehydes, and oxides

Chemicals can be employed for sterilization, disinfection, and antisepsis. The proper use of chemical agents is essential for personal safety. Chemicals also are employed to prevent microbial growth in food, and certain chemicals are used to treat infectious disease. The use of chemical agents for chemotherapy in humans is covered in chapter 9 and their use in food is discussed in chapter 41. Here we discuss chemicals used outside the body.

Many different chemicals have been specifically formulated as disinfectants, each with its own advantages and disadvantages. Ideally the biocide is effective against a wide variety of infectious agents (bacteria, bacterial endospores, fungi, viruses, and prions) at low concentrations and in the presence of organic matter. Although the chemical must be toxic for infectious agents, it should not be toxic to people or corrosive for common materials. In practice, this balance between effectiveness and low toxicity is hard to achieve. Some chemicals are used despite their low effectiveness because they are relatively nontoxic. The ideal disinfectant should be stable upon storage, odorless or with a pleasant odor, and soluble in water and lipids for penetration into microorganisms; have a low surface tension so that it can enter cracks in surfaces; and be relatively inexpensive.

Other chemical biocides are used as antiseptics. Recall that antiseptics are less toxic to humans than disinfectants and as such may be less effective at killing all the microorganisms that disinfectants can kill. In general, antiseptics should reduce the number of pathogens on human tissue to prevent infection. Examples of antiseptics include hand sanitizers, silver threads woven into clothing, and dilute iodine solutions that can be sprayed onto wounds. One potentially serious problem is the overuse of antiseptics. For instance, resistance to the antibacterial agent triclosan (found in products such as deodorants, mouthwashes, soaps, cutting boards, and baby toys) has become a problem. Because of its overuse, triclosan was banned from sale in the United States in 2017, but it persists in many products in consumers' homes. ▶| *There are several mechanisms of drug resistance (section 9.8)*

The properties and uses of common disinfectants and antiseptics are surveyed next. Many of the characteristics of disinfectants and antiseptics are summarized in **table 8.3**. Structures of some common agents are shown in **figure 8.10**.

Phenolics

Phenol was the first widely used antiseptic and disinfectant. In 1867 Joseph Lister employed it to reduce the risk of infection during surgery. Today phenol and phenol derivatives (phenolics) are used as disinfectants in laboratories and hospitals. The commercial disinfectant Lysol is a mixture of phenolics. Phenolics denature proteins and disrupt cell membranes. They have some important advantages as disinfectants: Phenolics are tuberculocidal, effective in the presence of organic material, and remain active on surfaces long after application. However, they have a disagreeable odor and can cause skin irritation.

Alcohols

Alcohols are among the most widely used disinfectants, antiseptics, and sanitizers. They are bactericidal and fungicidal but not sporicidal; some enveloped viruses are also destroyed. The two most popular alcohol germicides are ethanol and isopropanol, usually used in about 60 to 80% concentration. They act by denaturing proteins and possibly by dissolving membrane lipids. A 10- to 15-minute soaking is sufficient to disinfect small instruments, while rubbing hands with specially formulated alcohol products sanitizes them by killing many pathogens.

Halogens

The halogens iodine and chlorine are important antimicrobial agents. Iodine is used as a skin antiseptic and kills by oxidizing cell constituents and iodinating proteins. At higher concentrations, it may even kill some endospores. Iodine often is applied as tincture of iodine, 2% or more iodine in a water-ethanol solution of potassium iodide. Although it is an effective antiseptic, the skin may be damaged, a stain remains, and iodine allergies can result. Iodine can be complexed with an organic carrier to form an **iodophor.** Iodophors are water soluble, stable, and nonstaining, and release iodine slowly to minimize skin irritation. They are used in hospitals for cleansing preoperative skin and in hospitals and laboratories for disinfecting. Some popular brands are Wescodyne for skin and laboratory disinfection, and Betadine for wounds.

Chlorine is the usual disinfectant for municipal water supplies and swimming pools, and is also employed in the dairy and food industries. It may be applied as chlorine gas (Cl_2), sodium hypochlorite (bleach, NaOCl), or calcium hypochlorite [$Ca(OCl)_2$], all of which yield hypochlorous acid (HOCl):

$$Cl_2 + H_2O \rightarrow HCl + HOCl$$
$$NaOCl + H_2O \rightarrow NaOH + HOCl$$
$$Ca(OCl)_2 + 2H_2O \rightarrow Ca(OH)_2 + 2HOCl$$

The result is oxidation of cellular materials and destruction of vegetative bacteria and fungi. Death of almost all microorganisms usually occurs within 30 minutes. Two important eukaryotic pathogens are not killed by chlorine, *Cryptosporidium* and *Giardia*. Both are transmitted via water. ▶| *Food and water are vehicles for fungal and protozoal diseases (section 40.5)*

Chlorine is also an excellent disinfectant for individual use because it is effective, inexpensive, and easy to employ. Small quantities of drinking water can be disinfected with halazone

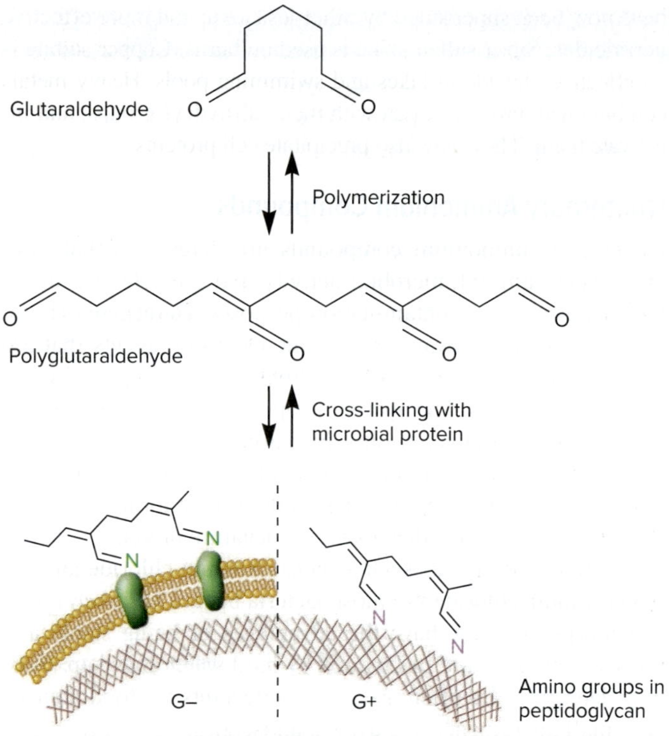

Figure 8.11 Effects of Glutaraldehyde. Glutaraldehyde polymerizes and then interacts with amino acids in proteins (left) or in peptidoglycan (right). As a result, the proteins are alkylated and cross-linked to other proteins, which inactivates them. The amino groups in peptidoglycan are also alkylated and cross-linked, which prevents them from participating in other chemical reactions such as those involved in peptidoglycan synthesis.

MICRO INQUIRY *Why are cross-linking agents such as glutaraldehyde often called "fixatives" or are said to "fix the cells"?*

Chlorine dioxide (ClO_2) gas is also used as a disinfectant. Of note is the fact that the chemistry of chlorine dioxide is very different from that of chlorine gas. ClO_2 is typically aerosolized in a humidified environment, 1 mg/liter of air in 60% relative humidity, for at least 4 hours. At this concentration, ClO_2 provides a greater than 6 log reduction of endospores and vegetative bacteria. It has been used to sterilize hospital operating and patient rooms. ClO_2 fumigation is used in the food industry to sanitize fruits and vegetables of contaminating yeasts and molds. As a disinfecting gas, it is best known for its use in sterilizing the U.S. Senate and Postal facilities after the 2001 anthrax attacks, and to kill molds that contaminated Gulf Coast homes in the aftermath of hurricane Harvey in 2017. ClO_2 has a broad killing spectrum, controlling bacteria, endospores, fungi, and protozoa. ClO_2 appears to have several mechanisms of action: It reacts readily with amino acids cysteine, tryptophan, and tyrosine, denaturing proteins; with free fatty acids, lysing membranes; and with nucleic acids, inhibiting replication. To date, no resistance to ClO_2 has been identified. ClO_2 is highly water soluble, especially in cold water, where it does not hydrolyze but remains as a dissolved gas in solution. Thus a number of municipal water treatment facilities use it for water disinfection.

Vaporized hydrogen peroxide (VHP) can also be used to decontaminate biological safety cabinets, operating rooms, and other large facilities. VHP is produced from a solution of hydrogen peroxide in water that is passed over a vaporizer to achieve a vapor concentration between 140 and 1,400 parts per million (ppm), depending on the agent to be destroyed. VHP is then introduced as a sterilizing vapor into the enclosure for some time, depending on the size of the enclosure and the materials within. Hydrogen peroxide and its oxy-radical by-products are toxic (75 ppm are dangerous to human health) and kill a wide variety of microorganisms. During the course of the decontamination process, VHP breaks down to water and oxygen, both of which are harmless.

Comprehension Check

1. Why are most antimicrobial chemical agents disinfectants rather than sterilants? What general characteristics should one look for in a disinfectant?
2. Construct a table that compares the chemical nature, mechanism of action, mode of application, common uses and effectiveness, and advantages and disadvantages between phenolics, alcohols, halogens, heavy metals, quaternary ammonium compounds, aldehydes, and ethylene oxide.

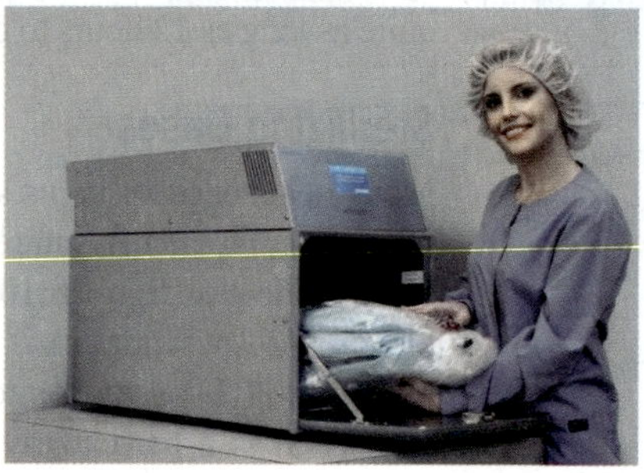

(a)

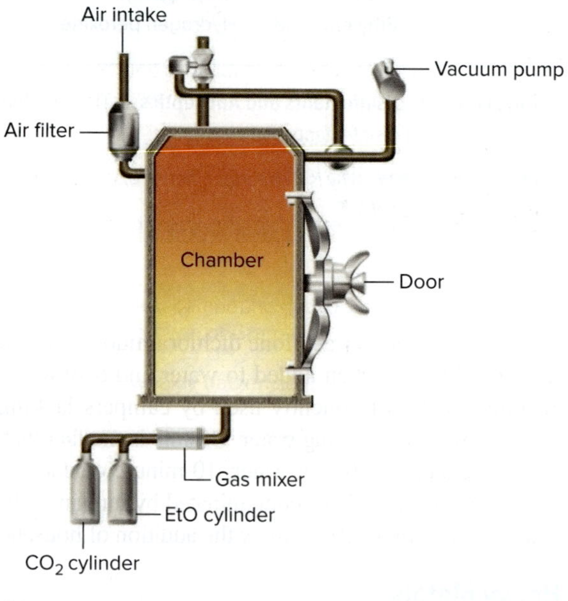

(b)

Figure 8.12 An Ethylene Oxide Sterilizer. (a) An automatic ethylene oxide (EtO) sterilizer. (b) Schematic of an EtO sterilizer. Items to be sterilized are placed in the chamber, and EtO and carbon dioxide are introduced. After the sterilization procedure is completed, the EtO and carbon dioxide are pumped out of the chamber and air enters.

3. Which disinfectants or antiseptics would be used to treat the following: laboratory bench top, drinking water, patch of skin before surgery, small medical instruments (probes, forceps, etc.)? Explain your choices.
4. How do phenolic agents differ from the other chemical control agents described in this chapter?
5. Which physical or chemical agent would be the best choice for sterilizing the following items: glass pipettes, tryptic soy broth tubes, nutrient agar, antibiotic solution, interior of a biological safety cabinet, wrapped package of plastic Petri plates? Explain your choices.

8.6 Antimicrobial Agents Must Be Evaluated for Effectiveness

After reading this section, you should be able to:

a. Predict the effects of (1) microbial population size and composition, (2) temperature, (3) exposure time, and (4) local environmental conditions on antimicrobial agent effectiveness
b. Describe the processes used to measure microbial killing rates, dilution testing, and in-use testing of antimicrobial agents

The assessment of antimicrobial agent effectiveness is a complex process regulated by two different federal agencies. The Environmental Protection Agency regulates disinfectants, whereas agents used on humans and animals are under the control of the Food and Drug Administration. They establish the guidelines under which these agents are used and agent effectiveness is measured. Importantly, there are a number of variables that must also be considered when evaluating antimicrobial agent effectiveness.

Destruction of microorganisms and inhibition of microbial growth are not simple matters because the effectiveness of an antimicrobial agent is affected by at least six factors.

1. **Population size.** Because an equal fraction of a microbial population is killed during each interval, a larger population requires a longer time to die than does a smaller one (table 8.1 and figure 8.3).
2. **Population composition.** The effectiveness of an agent varies greatly with the nature of the organisms being treated because microorganisms differ markedly in susceptibility. Bacterial endospores are much more resistant to most antimicrobial agents than vegetative forms, and younger cells are usually more readily destroyed than mature organisms. Some species are able to withstand adverse conditions better than others. For instance, *M. tuberculosis,* which causes tuberculosis, is much more resistant to antimicrobial agents than most other bacteria.
3. **Concentration or intensity of an antimicrobial agent.** Often, but not always, the more concentrated a chemical agent or intense a physical agent, the more rapidly microorganisms are destroyed. However, agent effectiveness usually is not directly related to concentration or intensity. Over a short range, a small increase in concentration leads to an exponential rise in effectiveness; beyond a certain point, increases may not raise the killing rate much at all. Sometimes an agent is more effective at lower concentrations. For example, 70% ethanol is more bactericidal than 95% ethanol because the activity of ethanol is enhanced by the presence of water.
4. **Contact time.** The longer a population is exposed to a microbicidal agent, the more organisms are killed (figures 8.2 and 8.3). To achieve sterilization, contact time should be long enough to reduce the probability of survival by at least 6 logs.
5. **Temperature.** An increase in the temperature at which a chemical acts often enhances its activity. Frequently a lower concentration of disinfectant or sterilizing agent can be used at a higher temperature.
6. **Local environment.** The population to be controlled is not isolated but surrounded by environmental factors that may either offer protection or aid in its destruction. For example, because heat kills more readily at an acidic pH, acidic foods and beverages such as fruits and tomatoes are easier to pasteurize than more alkaline foods such as milk. A second important environmental factor is organic matter, which can protect microorganisms against physical and chemical disinfecting agents. Biofilms are a good example. The organic matter in a biofilm protects the biofilm's microorganisms. Furthermore, bacteria in biofilms are altered physiologically, and this makes them less susceptible to many antimicrobial agents. Because of the impact of organic matter, it may be necessary to clean objects, especially medical and dental equipment, before they are disinfected or sterilized. ▶| *Biofilms are common in nature (section 7.6)*

The actual testing of antimicrobial agents often begins with an initial screening to see if they are effective and at what concentrations. This may be followed by more realistic in-use testing. The best-known disinfectant screening test is the **phenol coefficient test** in which the potency of a disinfectant is compared with that of phenol. A series of dilutions of phenol and the disinfectant being tested are prepared. Standard amounts of *Salmonella enterica* serovar Typhi and *Staphylococcus aureus* are added to each dilution; the dilutions are then placed in a 20° or 37°C water bath. At 5-minute intervals, samples are withdrawn from each dilution and used to inoculate growth medium, which is incubated and examined for growth. Growth indicates that the dilution at that particular time of sampling did not kill the bacteria. The highest dilution (i.e., the lowest concentration) that kills the bacteria after a 10-minute exposure but not after 5 minutes is used to calculate the phenol coefficient. The higher the phenol coefficient value, the more effective the disinfectant under these test conditions. A value greater than 1 means that the disinfectant is more effective than phenol.

The phenol coefficient test is a useful initial screening procedure, but the phenol coefficient can be misleading if taken as a direct indication of disinfectant potency during actual use. This is because the phenol coefficient is determined under carefully controlled conditions with pure bacterial cultures, whereas disinfectants are normally used on complex populations in the presence of organic matter and with significant variations in environmental factors such as pH, temperature, and presence of salts.

To more realistically estimate disinfectant effectiveness, other tests are often used. The rates at which selected bacteria are destroyed with various chemical agents may be experimentally determined and compared. A **use dilution test** can also be carried out. Stainless steel carriers are contaminated with one of three specific bacterial species under carefully controlled conditions. The carriers are dried briefly, immersed in the test disinfectants for 10 minutes, transferred to culture media, and incubated for 2 days. The disinfectant concentration that kills the bacteria on at least 59 of 60 carriers (a 95% level of confidence) is determined. Disinfectants also can be tested under conditions designed to simulate normal in-use situations. In-use testing techniques allow a more accurate determination of the proper disinfectant concentration for a particular situation.

Comprehension Check

1. Briefly explain how the effectiveness of antimicrobial agents varies with population size, population composition, concentration or intensity of the agent, contact time, temperature, and local environmental conditions.
2. How does being in a biofilm affect an organism's susceptibility to antimicrobial agents?
3. Suppose hospital custodians have been assigned the task of cleaning all showerheads in patient rooms to prevent the spread of infectious disease. What two factors would have the greatest impact on the effectiveness of the disinfectant the custodians use? Explain what that impact would be.
4. Briefly describe the phenol coefficient test.
5. Why might it be necessary to employ procedures such as the use dilution and in-use tests?

8.7 Microorganisms Can Be Controlled by Biological Methods

After reading this section, you should be able to:

a. Propose predation, competition, and other methods for biological control of microorganisms
b. Suggest alternative decontamination and medical therapies using viruses of bacteria, fungi, and protozoa

The emerging field of biological control of microorganisms holds great promise. Scientists are learning to exploit natural control processes such as predation of one microorganism on another, viral-mediated lysis, and toxin-mediated killing. While these mechanisms occur in nature, their use by humans is relatively new. Studies evaluating control of the human intestinal pathogens *Salmonella* spp., *Shigella* spp., and *E. coli* by Gram-negative predators such as *Bdellovibrio* spp. suggest that poultry farms may be sprayed with a predatory bacterium to reduce potential contamination. Another biological control method had its start in the early 1900s at the Pasteur Institute in France. Felix d'Herelle isolated bacteriophage from patients recovering from bacillary dysentery. After numerous tests in vitro, d'Herelle concluded that the bacteriophage participated in the destruction of dysentery-causing bacteria. Bacteriophage therapies were under development when penicillin ushered in the age of antibiotics; they are currently used only in Russia, Poland, and Georgia. However, control of pathogens using bacteriophage is regaining wide support and appears to be effective in the eradication of a number of bacterial species by lysing the pathogenic host. In fact, the U.S. FDA has now approved the use of a bacteriophage spray to eradicate *Listeria, Salmonella,* and *E. coli* in foods. Yet to be approved are several designs for treating human infectious diseases by similar methods. This seems intuitive, knowing that the virus lyses its specific bacterial host, yet unnerving when one thinks about swallowing, injecting, or applying a virus (albeit a bacteriophage) to the human body. In essence, microbial control relies on the lytic effect of bacteriophage on cells. The distaste of using bacteriophage as a therapeutic can be overcome by the use of enzybiotics. These are proteins purified from bacteriophage that cause host cell lysis. They are endolysins whose substrate is peptidoglycan, and the exposed Gram-positive cell wall renders these cells especially susceptible. Other potentially useful bacterial proteins are depolymerases that hydrolyze a biofilm matrix and microbial toxins (such as bacteriocins). ◀ *Viruses and other acellular infectious agents (chapter 6)* ▶ *Class* Deltaproteobacteria *includes chemoheterotrophic anaerobes and predators (section 22.4); Bacteriocins (section 32.3); Various methods are used to control food spoilage (section 41.2)*

Comprehension Check

1. How would you explain to a patient that a virus can be used to eliminate a bone infection caused by bacteria that do not respond to antibiotics?
2. Propose the use of specific bacterial, viral, or fungal products that might be used to kill other, more virulent bacteria, viruses, or fungi.

Key Concepts

8.1 Microbial Growth and Replication: Targets for Control

- Sterilization is the process by which all living cells, viable spores, viruses, viroids, and prions are either destroyed or removed from an object or habitat. Disinfection is the killing, inhibition, or removal of microorganisms (but not necessarily endospores) that can cause disease.
- The main goal of disinfection and antisepsis is the removal, inhibition, or killing of pathogenic microbes. Both processes also reduce the total number of microbes. Disinfectants are chemicals used to disinfect

inanimate objects; antiseptics are used on living tissue.

- Antimicrobial agents that kill organisms often have the suffix *-cide,* whereas agents that prevent growth and reproduction have the suffix *-static*.

8.2 The Pattern of Microbial Death Mirrors the Pattern of Microbial Growth

- Microbial death is usually exponential (**figure 8.3**).
- The decimal reduction time measures an agent's killing efficiency. It represents the time needed to kill 90% of the microbes under specified conditions.

8.3 Mechanical Removal Methods Rely on Barriers

- Microorganisms can be efficiently removed by filtration with either depth filters or membrane filters (**figure 8.4**).
- Biological safety cabinets with high-efficiency particulate filters sterilize air by filtration (**figure 8.6**).

8.4 Physical Control Methods Alter Microorganisms to Make Them Nonviable

- Moist heat kills by degrading nucleic acids, denaturing proteins, and disrupting cell membranes.
- Although treatment with boiling water for 10 minutes kills vegetative forms, an autoclave must be used to destroy endospores by heating at 121°C and 15 pounds of pressure (**figure 8.7**).
- Glassware and other heat-stable items may be sterilized by dry heat at 160° to 170°C for 2 to 3 hours.
- Radiation of short-wavelength or high-energy ultraviolet and ionizing radiation can be used to sterilize objects (**figure 8.9**).

8.5 Microorganisms Are Controlled with Chemical Agents

- Chemical agents usually act as disinfectants or antiseptics because they cannot readily destroy bacterial spores. Disinfectant effectiveness depends on concentration, treatment duration, temperature, and presence of organic material (**table 8.3**).
- Phenolics and alcohols are popular disinfectants and antiseptics that act by denaturing proteins and disrupting cell membranes (**figure 8.10**).
- Halogens (iodine and chlorine) kill by oxidizing cellular constituents; cell proteins may also be iodinated. Iodine is applied as a tincture or iodophor. Chlorine may be added to water as a gas, hypochlorite, or an organic chlorine derivative.
- Heavy metals tend to be bacteriostatic agents. They are employed in specialized situations such as the use of copper sulfate in lakes and pools.
- Cationic detergents (e.g., quaternary ammonium compounds) are often used as disinfectants and antiseptics; they disrupt membranes and denature proteins.
- Aldehydes such as formaldehyde and glutaraldehyde can sterilize as well as disinfect because they kill spores.
- Ethylene oxide gas penetrates plastic wrapping material and destroys all life forms by reacting with proteins. It is used to sterilize packaged, heat-sensitive materials.
- Chlorine dioxide is aerosolized at low concentrations to kill bacteria, spores, fungi, and protozoa that contaminate living spaces and agriculture. It has multiple killing mechanisms.
- Vaporized hydrogen peroxide is used to decontaminate enclosed spaces (e.g., safety cabinets and small rooms). The vaporized hydrogen peroxide is a vapor that can be circulated throughout the space. The peroxide and its oxy-radical by-products are toxic to most microorganisms.

8.6 Antimicrobial Agents Must Be Evaluated for Effectiveness

- The effectiveness of a disinfectant or sterilizing agent is influenced by population size, population composition, concentration or intensity of the agent, exposure duration, temperature, and nature of the local environment.
- The presence of a biofilm can dramatically alter the effectiveness of an antimicrobial agent.
- The phenol coefficient test is frequently used to evaluate the effectiveness of antimicrobial agents. However, it does so using conditions that do not replicate real-life use.
- Other procedures used to determine the effectiveness of disinfectants include measurement of killing rates with germicides, use dilution testing, and in-use testing.

8.7 Microorganisms Can Be Controlled by Biological Methods

- Control of microorganisms by natural means such as through predation, viral lysis, and toxins is emerging as a promising field.

Active Learning

1. Throughout history, spices have been used as preservatives and to cover up the smell or taste of food that is slightly spoiled. The success of some spices led to a magical, ritualized use, and possession of spices was often limited to priests or other powerful members of the community.
   a. Choose a spice and trace its use geographically and historically. What is its common use today?
   b. Spices grow and tend to be used predominantly in warmer climates. Explain.

2. Design an experiment to determine whether an antimicrobial agent is acting as a cidal or static agent. How would you determine whether an agent is suitable for use as an antiseptic rather than as a disinfectant?
3. Suppose you tested the effectiveness of disinfectants with the phenol coefficient test and obtained the results shown in the "Bacterial Growth After Treatment" table. What disinfectant can you safely say is the most effective? Can you determine its phenol coefficient from these results?

Bacterial Growth After Treatment

Dilution	Disinfectant A	Disinfectant B	Disinfectant C
1/20	–	–	–
1/40	+	–	–
1/80	+	–	+
1/160	+	+	+
1/320	+	–	+

4. *Pseudomonas aeruginosa* is a common pathogen infecting the sinuses of patients with chronic rhinosinusitis. In an experiment to test the feasibility of using virulent bacteriophage as a treatment for *Pseudomonas* infection, researchers collected 47 strains of *P. aeruginosa* from patients in hospitals on three continents. The bacterial strains were determined to be distinct from one another, and a cocktail (mixture) of four different bacteriophage was tested on each clinical isolate. When the bacterial strains were evaluated by looking for plaques on bacterial lawns *(see figure 6.18)*, 89% were found to be susceptible to the bacteriophage cocktail. In an alternate procedure, the same bacterial strains were grown as biofilms, and under these conditions 80% showed a reduction in biomass following application of the bacteriophage cocktail. Why was it important to use a mixture of four bacteriophage? Why was the bacteriophage cocktail less effective at killing *P. aeruginosa* in a biofilm?

Read the original paper: Fong, S. A., et al. 2017. Activity of bacteriophages in removing biofilms of *Pseudomonas aeruginosa* isolates from chronic rhinosinusitis patients. *Front. Cell. Infect. Microbiol.* doi:10.3389/fcimb.2017.00418.

9

Antimicrobial Chemotherapy

Source: The National Library of Medicine

The Story of Syphilis Treatment

As we detail later in the text, the sexually transmitted infection syphilis is caused by a spirochete called *Treponema pallidum*. This dry fact belies the fascinating history of the disease. Starting with the origin of syphilis, the story gets pretty twisted. Some believe Christopher Columbus brought *T. pallidum* back from the Americas; others believe it was always present in the Old World but was diagnosed as leprosy until the 1500s. The confusion with leprosy stems from the fact that syphilis trots through the unlucky host in three stages. First, the infected individual develops a chancre—an open ulcer usually on the genitals. These sores are teeming with spirochetes, but because they are not painful they can be overlooked or at least ignored. The next stage, secondary syphilis, is where the leprosy confusion arises. The most common sign of this syphilis stage is a rash that can be variable in nature, but can involve any part of the body including inside the mouth. In fact, the medical specialty of dermatology has its origins in the diagnosis and treatment of syphilis. The last or tertiary stage of syphilis can be manifested as neurosyphilis, once called "general paresis of the insane" because the patient may become blind, lose mental agility, and develop motor paralysis. These ailments result from inflammation of the central nervous system triggered by persistent infection with *T. pallidum*.

Apart from the stigma that might be attached to contracting syphilis, there were compelling reasons to develop a cure. After the first major outbreak of syphilis in the 1500s, two treatments were advocated: a potion made from the guaiac tree and mercury ointments. Both were thought to cleanse the patient of syphilis "toxins," the potion through severe sweating and diarrhea and the ointments through sweating and increased urination. Apart from dehydration, guaiac treatment was relatively benign, but mercury treatment caused some to die. Nonetheless, various forms of mercury were used to treat syphilis through the beginning of the twentieth century, when it was replaced with arsenic-based compounds. In fact, in 1908 the "father of antibiotics," Paul Ehrlich, won a Nobel Prize for his arsenic-based medication, Salvarsan. Later Salvarsan was used to treat many soldiers in World War I, for their wounds of love rather than war. But Salvarsan was difficult to administer and, not surprisingly, had a number of toxic side effects. Antibiotic development was at the stage where arsenic was considered an improvement over mercury.

Salvarsan, as imperfect as it was, was only (sometimes) effective against primary and secondary syphilis. But in the early twentieth century, 5 to 10% of psychiatric inpatients had neurosyphilis. Serendipity led to the development of pyrotherapy; here the idea was that very high fevers could "cook to death" the syphilis spirochete. As WW I was raging in 1917, Austrian physician Julius Wagner-Jauregg encountered a wounded soldier who had malaria, which he could treat with quinine. However, the soldier had also been diagnosed with the early stages of neurosyphilis. After allowing the soldier to endure rounds of high fever that characterize malaria, Wagner-Jauregg noted that the symptoms of neurosyphilis abated. In experiments that could never happen today for many reasons, Wagner-Jauregg took the soldier's malaria-laden blood and infected nine other neurosyphilis patients; six were cured. Thousands of neurosyphilis patients would eventually be treated with malaria and Wagner-Jauregg won the Nobel Prize in 1927.

As this story illustrates, the search for cures to infectious disease is sometimes barely believable. Now very few people living in the industrialized West reach adulthood without having taken an antibiotic. But what does the term *antibiotic* really mean? **Antibiotic** (Greek *anti*, against; *bios*, life) refers to microbial products and their derivatives that kill susceptible microorganisms or inhibit their growth. The rise in antibiotic resistance among microbes has forced modern medicine to develop new drugs and rethink antimicrobial strategies (discussed in chapter 42). This chapter introduces the principles of **antimicrobial** chemotherapy and resistance.

Readiness Check:

Based on what you have learned previously, you should be able to:

✓ Identify the major structures and their functions of bacteria, protists, fungi, and viruses (sections 3.3–3.8, 5.1–5.7, 6.2–6.3, and 7.1–7.7)

✓ Compare the various growth and replication processes of viruses and bacteria (sections 6.3, 6.5, 7.1–7.2, and 7.4–7.6)

9.1 Antimicrobial Chemotherapy Evolved from Antisepsis Efforts

After reading this section, you should be able to:

a. Trace the general history of antimicrobial chemotherapy
b. Propose natural sources of new antimicrobial agents

The modern era of chemotherapy began with the work of the German physician Paul Ehrlich (1854–1915). Ehrlich was fascinated with dyes that specifically bind to microbial cells. He reasoned that one of the dyes could be a chemical that would selectively destroy pathogens without harming human cells—a "magic bullet." As mentioned in the chapter opener, Ehrlich and a young Japanese scientist named Sahachiro Hata (1873–1938) developed a treatment for syphilis by testing a variety of arsenic-based chemicals on infected rabbits. They found that arsphenamine was active against the syphilis spirochete. Arsphenamine was made available in 1910 under the trade name Salvarsan.

Ehrlich and Hata's success led to the screening of a vast number of chemicals for other magic bullets. In 1927 Gerhard Domagk (1895–1964) discovered that Prontosil red, a new dye for staining leather, protected mice completely against pathogenic streptococci and staphylococci without apparent toxicity. Jacques (1897–1977) and Therese (1892–1978) Trefouel later showed that the body metabolized the dye to sulfanilamide. Domagk received the 1939 Nobel Prize in Physiology or Medicine for his discovery of sulfonamides, or sulfa drugs.

Penicillin, the first true antibiotic because it is a natural microbial product, was first discovered in 1896 by a twenty-one-year-old French medical student named Ernest Duchesne (1874–1912). His work was forgotten until Alexander Fleming (1881–1955) accidentally rediscovered penicillin in 1928. After returning from a weekend vacation, Fleming noticed that a Petri plate of staphylococci also had mold growing on it and there were no bacterial colonies surrounding it (**figure 9.1**). Although the precise events are still unclear, it has been suggested that a *Penicillium notatum* spore had contaminated the Petri dish before it had been inoculated with the staphylococci. The mold apparently grew before the bacteria and produced penicillin. The bacteria nearest the fungus were lysed. Fleming correctly deduced that the mold produced a diffusible substance, which he called penicillin. However, Fleming could not demonstrate that penicillin remained active in vivo long enough to destroy pathogens and he dropped the research.

Figure 9.1 Bactericidal Action of Penicillin. The *Penicillium* mold colony secretes penicillin that kills the *Staphylococcus aureus* streaked nearby.

Fleming's discovery languished until about a decade later when Howard Florey (1898–1968) and Ernst Chain (1906–1979), professors at Oxford University, obtained the *Penicillium* culture from Fleming and set about purifying the antibiotic. Norman Heatley (1911–2004), a biochemist, was enlisted to help. He devised the original assay, culture, and purification techniques needed to produce crude penicillin. When purified penicillin was injected into mice infected with streptococci or staphylococci, almost all the mice survived. This success was reported in 1940, and subsequent human trials were equally successful. Fleming, Florey, and Chain received the Nobel Prize in 1945 for the discovery and production of penicillin.

The discovery of penicillin stimulated the search for other antibiotics. Selman Waksman (1888–1973), while at Rutgers University, announced in 1944 that he and his associates had found a new antibiotic, streptomycin, produced by *Streptomyces griseus.* This discovery arose from the careful screening of about 10,000 strains of soil bacteria and fungi. The importance of streptomycin cannot be overstated, as it was the first drug to successfully treat tuberculosis. Waksman received the Nobel Prize in 1952, and his success led to a worldwide search for other antibiotic-producing soil microorganisms. Chloramphenicol, neomycin, oxytetracycline, and tetracycline were isolated from other *Streptomyces* spp. by 1953 (*see table 23.3*).

This golden era of antibiotics saw the rapid development of new antimicrobial agents that changed not only how infectious disease was treated, but also how medicine was and continues to be practiced. Prior to the advent of antibiotics, elective surgery (e.g., joint replacement and cosmetic surgery) was unthinkable. The use of immune-suppressing chemotherapeutic drugs to treat cancer would likewise be out of the question. Thus many of the tremendous gains in medicine are due in large part to the dawn of antibiotics and the decline of infectious disease.

Comprehension Check

1. Why did Ehrlich and others use the term "magic bullet" to describe agents with antimicrobial activity?
2. Louis Pasteur is often credited with saying, "Chance favors the prepared mind." How does this apply to Fleming?

9.2 Antimicrobial Drugs Have Selective Toxicity

After reading this section, you should be able to:

a. Explain the difference between a narrow and broad-spectrum drug
b. Correlate drug action with cidal and static effects

As Ehrlich so clearly saw, a successful chemotherapeutic agent has **selective toxicity,** meaning it kills or inhibits the microbial pathogen while damaging the host as little as possible. The degree of selective toxicity may be expressed in terms of (1) the therapeutic dose—the drug level required for treatment of a particular infection, and (2) the toxic dose—the drug level at which the agent becomes too toxic for the host. The **therapeutic index** is the ratio of the toxic dose to the therapeutic dose. The larger the therapeutic index, the better the chemotherapeutic agent in general.

A drug that disrupts a microbial structure or function not found in host cells often has greater selective toxicity, fewer side effects, and a higher therapeutic index. For example, penicillin inhibits bacterial peptidoglycan synthesis but has little effect on host cells because they lack cell walls; therefore penicillin's therapeutic index is high. A drug may have a low therapeutic index because it inhibits the same process in host cells or damages the host in other ways. This can lead to a diverse range of side effects that may involve almost any organ system. Because side effects can be severe, chemotherapeutic agents must be administered with great care.

Bacteria and fungi naturally produce many of the commonly employed antibiotics (**table 9.1**). In contrast, several important chemotherapeutic agents, such as sulfonamides, trimethoprim,

Table 9.1 Properties of Some Common Antibacterial Drugs

Antibiotic Group	Primary Effect	Mechanism of Action	Example Members	Spectrum	Common Side Effects
Cell Wall Synthesis Inhibition					
Penicillins	Cidal	Inhibit transpeptidation enzymes involved in cross-linking the polysaccharide chains of peptidoglycan. Activate cell wall lytic enzymes	Penicillin G, penicillin V, nafcillin, docloxacillin Ampicillin, carbenicillin	Narrow (Gram positive) Broad (Gram positive, some Gram negative)	Allergic reactions (diarrhea, anemia, hives, nausea, renal toxicity)
Cephalosporins	Cidal	Same as above	Cephalothin, cefoxitin, cefaperazone, ceftriaxone	Broad (Gram positive, some Gram negative)	Allergic reactions, thrombophlebitis, renal injury
Vancomycin	Cidal	Prevents transpeptidation of peptidoglycan subunits by binding to D-Ala-D-Ala amino acids at the end of peptide side chains. Thus it has a different binding site than that of the penicillins.	Vancomycin	Narrow (Gram positive)	Ototoxic (tinnitus and deafness), nephrotoxic, allergic reactions
Protein Synthesis Inhibition					
Aminoglycosides	Cidal	Bind to small ribosomal subunit (30S) and interfere with protein synthesis by directly causing misreading of mRNA	Neomycin, gentamicin, amikacin	Broad (Gram negative, mycobacteria)	Ototoxic, renal damage, loss of balance, nausea, allergic reactions
Tetracyclines	Static	Same as aminoglycosides	Oxytetracycline, doxycycline	Broad (including rickettsia and chlamydia)	Gastrointestinal upset, teeth discoloration, renal and hepatic injury

(Continued)

Table 9.1 Properties of Some Common Antibacterial Drugs (*continued*)

Macrolides	Static	Bind 23S rRNA of large ribosomal subunit (50S) to inhibit peptide chain elongation during protein synthesis	Erythromycin, azithromycin	Broad (aerobic and anaerobic Gram positive, some Gram negative)	Gastrointestinal upset, hepatic injury, anemia, allergic reactions
Lincosamides	Static	Act on the 50S ribosomal subunit, preventing transpeptidation by inhibiting peptidyl transferase activity	Clindamycin	Broad (most Gram positive except enterococci and most anaerobes)	Gastrointestinal upset, diarrhea, hepatic injury, overgrowth of *C. difficile*
Oxazolidinones	Static	Bind 50S ribosome and prevent formation of 70S initiation complex	Linezolid	Narrow (Gram positive)	Depressed bone marrow function, allergic reactions
Nucleic Acid Synthesis Inhibition					
Fluoroquinolones	Cidal	Inhibit DNA gyrase and topoisomerase II, thereby blocking DNA replication	Levofloxacin, ciprofloxacin	Broad spectrum	Tendonitis, headache, light-headedness, convulsions, allergic reactions
Rifamycins	Cidal	Inhibit bacterial DNA-dependent RNA polymerase	Rifamycin, rifampin	*Mycobacterium* infections and prophylaxis for bacterial meningitis exposure	Nausea, vomiting, diarrhea, fatigue, anemia, drowsiness, headache, mouth ulceration, liver damage
Cell Membrane Disruption					
Polymyxin B	Cidal	Binds to plasma membrane and disrupts its structure and permeability properties	Colistin, polymyxin topical ointment	Narrow (Gram-negatives)	Can cause severe kidney damage, drowsiness, dizziness
Daptomycin	Cidal	Forms membrane channels thereby depolarizing bacterial cell	Daptomycin	Narrow (Gram positives only)	Muscle aches and toxicity, GI upset
Antimetabolites					
Sulfonamides	Static	Inhibit folic acid synthesis by competing with *p*-aminobenzoic acid (PABA)	Silver sulfadiazine, sulfamethoxazole, sulfanilamide, sulfasalazine	Broad spectrum	Nausea, vomiting, and diarrhea; hypersensitivity reactions such as rashes, photosensitivity
Trimethoprim	Static	Blocks folic acid synthesis by inhibiting the enzyme tetrahydrofolate reductase	Trimethoprim (in combination with a sulfamethoxazole)	Broad spectrum	Same as sulfonamides but less frequent
Dapsone	Static	Thought to interfere with folic acid synthesis via same mechanism as sulfonamides	Dapsone	Narrow—mycobacterial infections, principally leprosy	Back, leg, or stomach pains; discolored fingernails, lips, or skin; breathing difficulties, fever, loss of appetite, skin rash, fatigue
Isoniazid	Cidal if bacteria are actively growing, static if bacteria are dormant	Exact mechanism is unclear but thought to inhibit lipid synthesis (especially mycolic acid); putative enoyl-reductase inhibitor	Isoniazid	Narrow—mycobacterial infections, principally tuberculosis	Nausea, vomiting, liver damage, seizures, "pins and needles" in extremities (peripheral neuropathy)

ciprofloxacin, isoniazid, and dapsone, are synthetic—manufactured by chemical processes. Some antibiotics are semisynthetic—natural antibiotics that have been structurally modified by the addition of chemical groups to make them less susceptible to stomach acids and inactivation by pathogens (e.g., ampicillin and amoxicillin).

Antimicrobial agents are often classified as either **narrow-spectrum drugs**—that is, they are effective only against a limited variety of pathogens—or **broad-spectrum drugs** that attack many different kinds of bacteria (table 9.1). Drugs may also be classified based on the general microbial group they act against: antibacterial, antifungal, antiprotozoan, and antiviral. A few agents can be used against more than one group; for example, sulfonamides are active against bacteria and some protozoa.

Finally, chemotherapeutic agents can be either **cidal** or **static** (table 9.1). Static agents reversibly inhibit growth; if the agent is removed, the microorganisms will recover and grow again. Although a cidal agent kills the target pathogen, it may be static at low levels. The effect of an agent also varies with the target species; that is, an agent may be cidal for one species and static for another. Because static agents do not directly destroy the pathogen, elimination of the infection depends on the host's own immunity mechanisms. A static agent may not be effective if the host is immunosuppressed. An estimate of the effectiveness of a chemotherapeutic agent against a pathogen can be obtained from the **minimal inhibitory concentration** (**MIC;** referred to as "mick"). The MIC is the lowest concentration of a drug that prevents growth of a particular pathogen. The **minimal lethal concentration** (**MLC**) is the lowest drug concentration that kills the pathogen. A cidal drug generally kills pathogens at levels only two to four times more than the MIC, whereas a static agent kills at much higher concentrations, if at all.

Comprehension Check

1. Define the following: *selective toxicity, therapeutic index, side effect, narrow-spectrum drug, broad-spectrum drug, synthetic and semisynthetic antibiotics, cidal and static agents, minimal inhibitory concentration,* and *minimal lethal concentration.*
2. Which drug would have more side effects: a drug with a high or low therapeutic index? How is this related to Ehrlich's notion of a "magic bullet"?
3. How can the MIC and MLC concepts be applied to distinguish between cidal and static agents?

9.3 Antimicrobial Activity Can Be Measured by Specific Tests

After reading this section, you should be able to:

a. Explain how to determine the level of antibacterial drug activity using the dilution susceptibility test, the disk diffusion test, and the Etest®
b. Predict antimicrobial drug levels in vivo from in vitro data

Determination of antimicrobial drug effectiveness against specific pathogens is essential for proper therapy. Testing can show which agents are most effective against a pathogen and give an estimate of the proper therapeutic dose.

Dilution Susceptibility Tests

Dilution susceptibility tests can be used to determine MIC and MLC values. Antibiotic dilution tests can be done in both agar and broth. Media called Mueller-Hinton broth or agar is used because all conditions must be standardized. In the broth dilution test, a series of tubes containing broth with antibiotic concentrations in the range of 0.1 to 128 mg per milliliter (twofold dilutions) is inoculated with a standard density of the test organism. The lowest concentration of the antibiotic resulting in no growth after 16 to 20 hours of incubation is the MIC. The MLC can be ascertained if the tubes showing no growth are then cultured into fresh medium lacking antibiotic. The lowest antibiotic concentration from which the microorganisms do not subsequently grow is the MLC. The agar dilution test is very similar to the broth dilution test. Plates containing Mueller-Hinton agar and various amounts of antibiotic are inoculated and examined for growth.

A number of automated systems for susceptibility and MIC determination have been commercialized. These products, along with specialized instruments to interpret the tests, miniaturize the broth dilution method to add convenience, decrease time to results, and reduce technical errors. Robotic instruments inoculate the microdilution plates, incubate the cultures, collect data, and interpret the results.

Disk Diffusion Tests

In many situations, such as in analyzing rapidly growing bacteria that cause urinary tract infections, a disk diffusion technique is used. This assay is fairly simple. Mueller-Hinton agar is inoculated with the bacterium, isolated from the clinical sample (urine in this case). Small paper disks, each impregnated with a different antibiotic, are placed on the inoculated agar. When the antibiotic diffuses radially outward through the agar, it produces a concentration gradient. The antibiotic is present at high concentrations near the disk and affects even minimally susceptible microorganisms. On the other hand, resistant organisms will grow close to the disk. As the distance from the disk increases, the antibiotic concentration decreases and only more susceptible pathogens are harmed. A clear zone or ring forms around an antibiotic disk after incubation if the agent inhibits bacterial growth. The wider the zone surrounding a disk, the more susceptible the pathogen is. Zone width also is a function of the antibiotic's initial concentration, its solubility, and its diffusion rate through agar. Thus zone width cannot be used to compare directly the effectiveness of different antibiotics.

The **Kirby-Bauer method** is the disk diffusion test most often used. It was developed in the early 1960s by William Kirby, A. W. Bauer, and their colleagues. (**figure 9.2**). Kirby-Bauer test results are interpreted using a table that relates zone diameter to the degree of microbial resistance (**table 9.2**). The values are derived by preparing

(a)

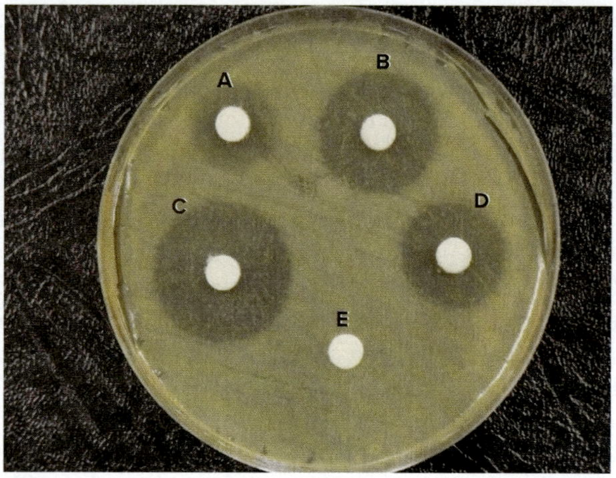

(b)

Figure 9.2 The Kirby-Bauer Method. (a) A multiple antibiotic disk dispenser and (b) disk diffusion test results.
(a-b) ©McGraw-Hill Education/James Redfearn, photographer

MICRO INQUIRY *To which antibiotic (A, B, C, D, or E) is the plated bacterium resistant?*

multiple graphs for drugs tested against specific organisms that represent the MIC as a function of the zone of inhibition diameter (**figure 9.3**). As the drug diffuses away from the disk into the agar, the concentration diminishes; once it drops below the efficacy level, the organism will grow. Thus, the border of the zone defines the MIC. These data are compiled for combinations of bacteria and drugs, and the resulting table is used as a first step to guide antimicrobial therapy. When a clinical sample is tested, the diameters of inhibition for the pathogen exposed to various drugs are measured and compared to the diameters listed in the table. The designation of resistance or susceptibility to each drug can be established because the data in the table approximately correlate with the expected level of drug in a patient's body when given a therapeutic dose.

The Etest®

The Etest® from bioM´erieux S.A. is often used in sensitivity testing. Unlike the Kirby-Bauer test, the MIC is also determined by the Etest®. In general, each bacterial isolate to be tested for antimicrobial sensitivities is inoculated on the surface of an agar medium and then Etest® strips are placed on the surface (**figure 9.4**). Each strip contains a gradient of an antibiotic and is labeled with a scale of MIC values. After 24 to 48 hours of incubation, an elliptical zone of inhibition appears. As shown in figure 9.4, MICs are determined from the point of intersection between the inhibition zone and the strip's scale of MIC values.

Table 9.2 Inhibition Zone Diameter of Selected Chemotherapeutic Drugs

		ZONE DIAMETER (NEAREST MM)		
Chemotherapeutic Drug	**Disk Content**	**Resistant**	**Intermediate**	**Susceptible**
Carbenicillin (with *Proteus* spp. and *E. coli*)	100 µg	≤17	18–22	≥23
Carbenicillin (with *Pseudomonas aeruginosa*)	100 µg	≤13	14–16	≥17
Erythromycin	15 µg	≤13	14–17	≥18
Penicillin G (with staphylococci)	10 U[1]	≤20	21–28	≥29
Penicillin G (with other bacteria)	10 U	≤11	12–21	≥22
Streptomycin	10 µg	≤11	12–14	≥15
Sulfonamides	250 or 300 µg	≤12	13–16	≥17

1 One milligram of penicillin G sodium = 1,600 units (U).

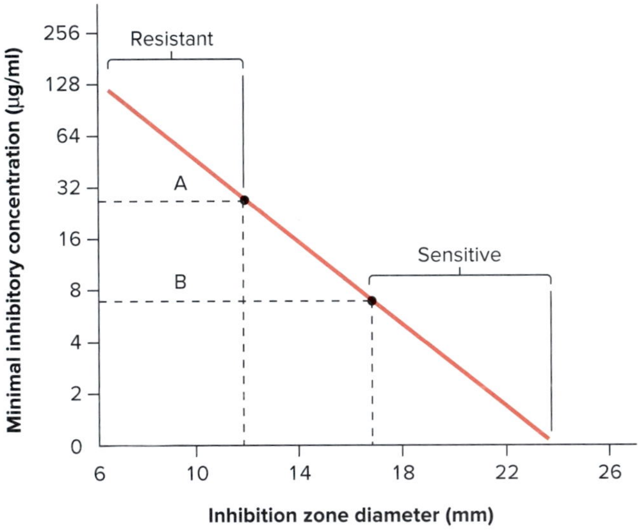

Figure 9.3 Interpretation of Kirby-Bauer Test Results. The relationship between MICs of a hypothetical drug and the size of the zone around a disk in which microbial growth is inhibited is plotted to predict drug effectiveness. As the sensitivity of microorganisms to the drug increases, the MIC value decreases and the inhibition zone grows larger. Suppose that this drug varies from 7 to 28 µg/mL in the body during treatment. Dashed line A shows that any pathogen with a zone of inhibition less than 12 mm in diameter will have a MIC value greater than 28 µg/mL and will be resistant to drug treatment. A pathogen with a zone diameter greater than 17 mm will have a MIC less than 7 µg/mL and will be sensitive to the drug (see line B). Zone diameters between 12 and 17 mm indicate intermediate sensitivity and usually signify evolving resistance.

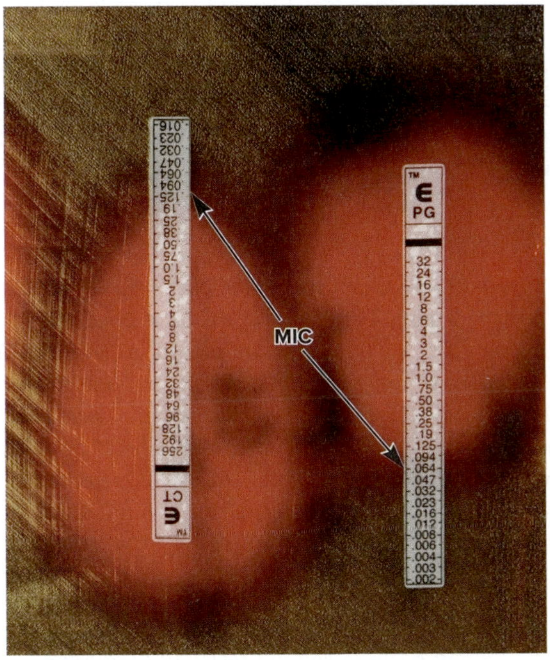

Figure 9.4 Etest®. An example of a bacterial culture plate with Etest® strips. The MIC value is read at the point where the inhibition ellipse intersects the scale on the strip, as shown by the arrows in this example. Etest® is a registered trademark of bioMérieux S.A. or one of its subsidiaries.

Comprehension Check

1. Describe the relationship between the diameter of growth inhibition and antibiotic susceptibility in the Kirby-Bauer test.
2. What would you surmise if you examined a Kirby-Bauer assay and found individual colonies of the plated microbe growing within the zone of inhibition?
3. How is the Etest® carried out? When might it be used instead of the Kirby-Bauer test?

9.4 Antibacterial Drugs

After reading this section, you should be able to:

a. Compare antibacterial drug mechanisms of action
b. Correlate lack of microbial growth with selective toxicity
c. Relate side effects of antibacterial drugs to mechanism of action
d. Explain the relative effectiveness of various antibacterial agents based on drug target

Although there are many different antibiotics, most fall into a limited number of classes. Part of the challenge in developing new antibiotics is to find bacterial structures or processes not already targeted. These novel targets have the best chance of beating antimicrobial resistance, at least for a while. Antimicrobial drug development is an active research area discussed in chapter 42. Here we present some of the more common antibiotics according to functional/structural class.

Inhibitors of Cell Wall Synthesis

The most selective antibiotics are those that interfere with bacterial cell wall synthesis. Drugs such as penicillins, cephalosporins, and vancomycin have a high therapeutic index because they target structures and functions not found in eukaryotic cells. ⏮ *There are two main types of bacterial cell walls (section 3.4)*

Penicillins

Most penicillins (e.g., penicillin G or benzylpenicillin) are derivatives of 6-aminopenicillanic acid and differ from one another with respect to the side chain attached to the amino group (**figure 9.5**). The most crucial feature of the molecule is the **β-lactam ring,** which is essential for bioactivity. Many penicillin-resistant bacteria produce **penicillinases** (also called **β-lactamases**), enzymes that inactivate the antibiotic by hydrolyzing a bond in the β-lactam ring.

Recall that when new peptidoglycan units are added to a growing cell wall, the peptide side chain on one peptidoglycan subunit must be linked to the peptide side chain on another subunit in a process known as transpeptidation (**figure 9.6**; *also see figure 3.19*). The structure of the penicillins resembles the two amino acids at the end of the side chain, called the terminal

Penicillin G

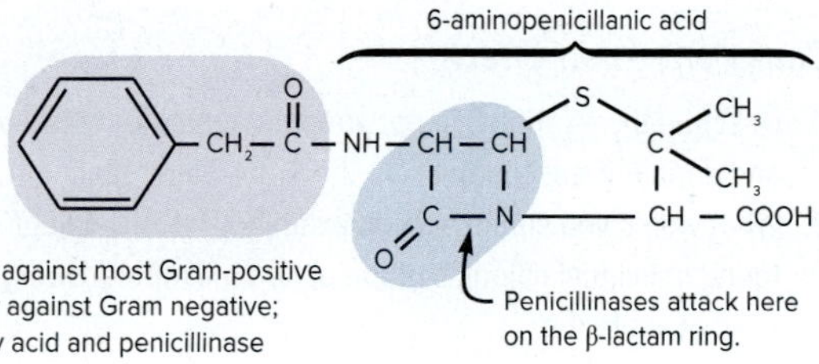

High activity against most Gram-positive bacteria, low against Gram negative; destroyed by acid and penicillinase

Penicillin V

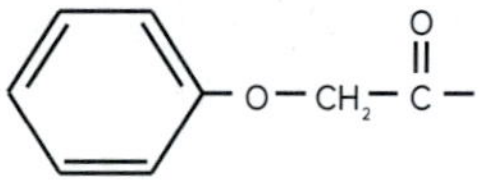

Same spectrum but more acid-resistant than penicillin G

Nafcillin

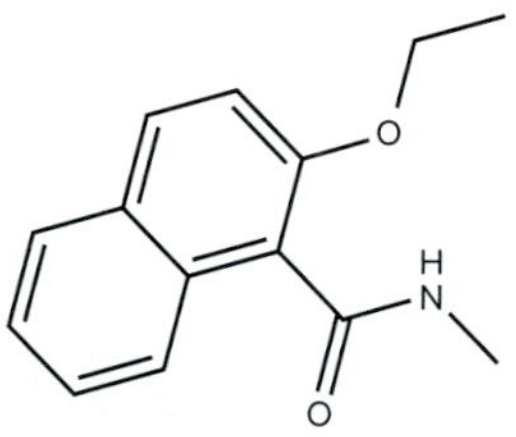

Active against staphylococci that produce β-lactamase enzymes

Ampicillin

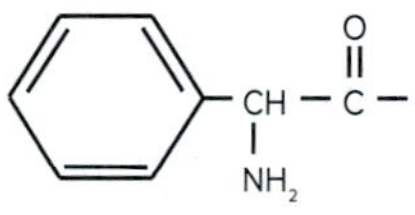

Active against Gram-positive and Gram-negative bacteria; acid stable

Carbenicillin

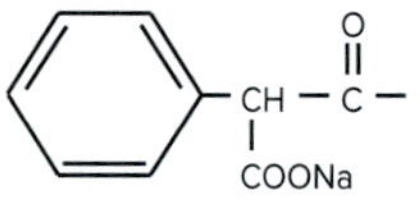

Active against Gram-negative bacteria such as *Pseudomonas* and *Proteus;* acid stable; not well absorbed by small intestine

Piperacillin

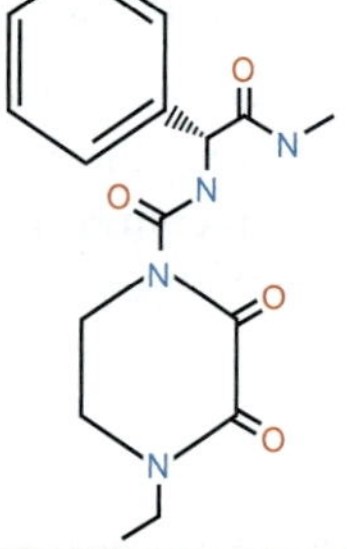

Extended spectrum; active against Gram-negative bacteria including *E. coli, Pseudomonas, Enterobacter, Proteus,* and *Klebsiella* spp.

Figure 9.5 Penicillins. The structures and characteristics of representative penicillins. All are derivatives of 6-aminopenicillanic acid; in each case, the shaded portion of penicillin G (purple) is replaced by the side chain indicated. The β-lactam ring is also shaded (blue).

MICRO INQUIRY *What is the difference between penicillin G and penicillin V? How do the semisynthetic penicillins differ from their parent compounds?*

D-alanyl-D-alanine. This structural similarity blocks the transpeptidation reaction that forms the peptidoglycan cross-links because the transpeptidation enzyme binds penicillin instead of its true substrate. In fact, these transpeptidases are commonly called penicillin-binding proteins (PBPs). Because cell wall formation is blocked, osmotic lysis occurs. Penicillins also destroy bacteria by activating their own autolytic enzymes. Thus penicillins act only on growing bacteria actively synthesizing new peptidoglycan.

Penicillins differ from each other in several ways. The two naturally occurring penicillins, penicillin G and penicillin V, are narrow-spectrum drugs (figure 9.5). Penicillin G is effective against many Gram-positive pathogens. Penicillin G must be administered by injection (parenterally) because it is destroyed by stomach acid. Penicillin V is similar to penicillin G in spectrum of activity but can be given orally because it is more resistant to stomach acid. Almost immediately after penicillin was introduced, bacterial resistance was discovered. To fight this and broaden the spectrum of activity, semisynthetic penicillins were developed. Among the first semisynthetic penicillins were the antistaphylococcal drugs nafcillin, oxacillin, dicloxacillin, and methicillin. These compounds have bulkier side chains than natural penicillin, making them more difficult for β-lactamase enzymes to degrade. Next came the aminopenicillins, which are more hydrophilic allowing passage through porins in Gram-negative outer membranes. While the antistaphylococcal penicillins retain the narrow Gram-positive spectrum of the parent compound, aminopenicillins like ampicillin and amoxicillin have broader coverage that includes many Gram negatives.

As an increasing number of bacteria became resistant to these semisynthetic penicillins, newer semisynthetic drugs were made. For instance, carbenicillin was developed; it specifically targets Gram-negative bacteria. Later, extended-spectrum penicillins, such as piperacillin (figure 9.5) and the penem—penicillin (penam) and cephalosporin (cephem) hybrid—class of drugs, were synthesized to specifically resist β-lactamase-producing Gram-negative and Gram-positive bacteria. Unfortunately, but not surprising, these newer drugs are not immune from bacterial resistance.

Cephalosporins

Cephalosporins are a family of antibiotics originally isolated in 1948 from the fungus *Cephalosporium.* They contain a β-lactam structure very similar to that of the penicillins (**figure 9.7**). As might be expected from their structural similarities to penicillins, cephalosporins also inhibit the transpeptidation reaction during peptidoglycan synthesis. The earlier generation cephalosporins have better Gram-positive coverage than later generations, which are more effective against a wider range of Gram negatives.

Carbapenems and Monobactam

With the recognition that β-lactamase enzymes continue to evolve and many bacteria acquire resistance to multiple classes of antibiotics, two new classes of β-lactam drugs, the **carbapenems** and **monobactams,** were developed by the mid-1980s. There are four members of the carbapenem family: imipenem, ertapenem, meropenem, and doripenem; but only one

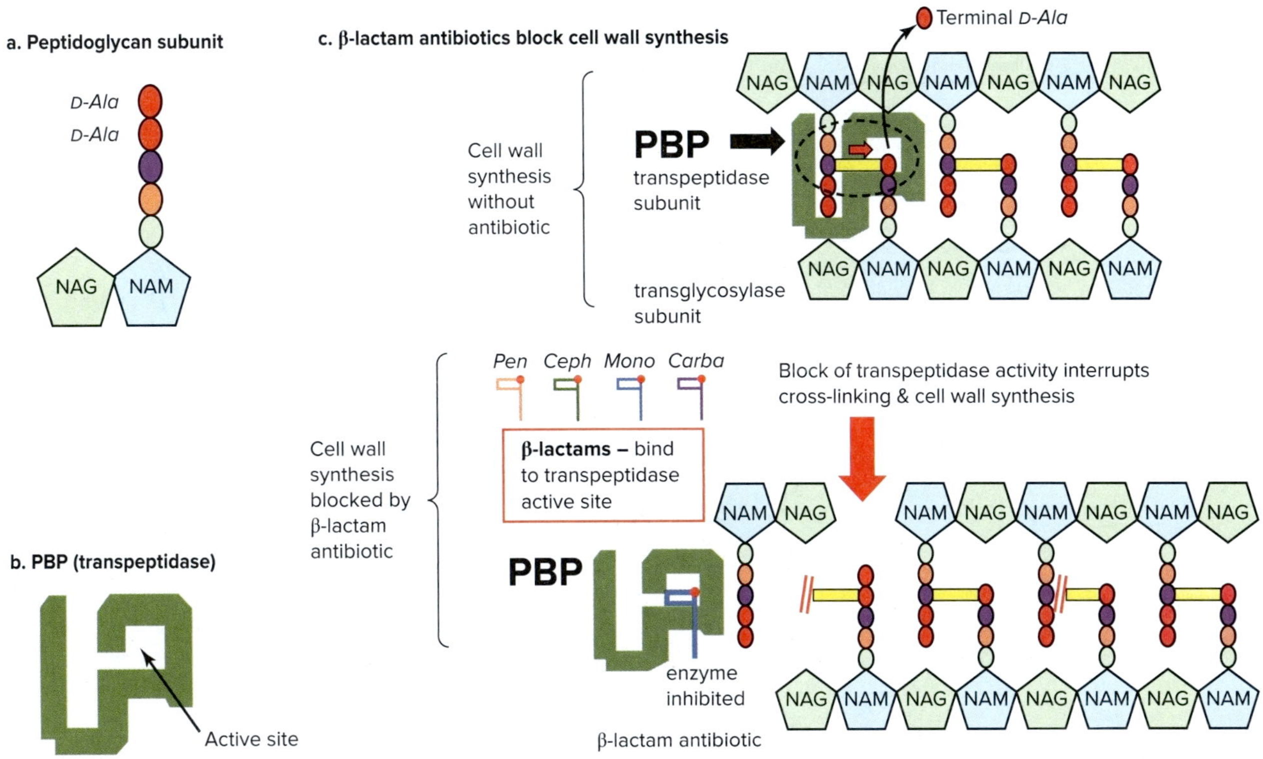

Figure 9.6 β-lactam Antibiotics Block Transpeptidation of Growing Peptidoglycan Chains. Pen, peniclIins; Ceph, cephalosporins; Mono, monobactams; Carba, carbapenems.

monobactam, aztreonam, is used clinically. Whereas the carbapenems are effective against a broad spectrum of bacteria including many Gram-positive, Gram-negative, and anaerobic bacteria, meropenem is active only against Gram-negative bacteria. These drugs cannot be taken orally and are used to treat complex infections involving antibiotic-resistant bacteria. Indeed, for many years, these drugs were considered impervious to the many different β-lactamase enzymes produced by a wide range of bacteria. However, more recently, Gram-negative bacteria producing β-lactamase enzymes now called carbapenemases have been discovered.

Vancomycin

Vancomycin is a glycopeptide antibiotic produced by the bacterium *Streptomyces orientalis*. It is a cup-shaped molecule composed of a peptide linked to a disaccharide. It binds specifically to the D-alanyl-D-alanine terminal sequence on the pentapeptide portion of peptidoglycan (**figure 9.8**). Thus like β-lactam antibiotics, vancomycin blocks transpeptidation, but it binds to the enzyme's substrate rather than the enzyme itself.

Vancomycin is bactericidal only for Gram-positive bacteria. It targets members of the genus *Staphylococcus* and some species in the genera *Bacillus* (food poisoning), *Streptococcus* ("strep" throat), *Enterococcus* (urinary tract infections), and *Clostridium* (gangrene and debilitating diarrhea). It is usually given intravenously, but is taken orally for *C. difficile* infections. It is an important drug in the treatment of antibiotic-resistant staphylococcal and enterococcal infections. However, vancomycin-resistant strains of *Enterococcus* have emerged, and a few cases of resistant *Staphylococcus aureus* have been reported. Vancomycin resistance poses a serious public health threat; vancomycin is one of the "drugs of last resort" in cases of antibiotic-resistant *S. aureus*.

Protein Synthesis Inhibitors

Many antibiotics inhibit protein synthesis by binding bacterial ribosomal proteins or rRNA. Because these drugs discriminate between bacterial and eukaryotic ribosomes, their therapeutic index is fairly high but not as high as that of cell wall synthesis inhibitors. Several different steps in protein synthesis can be affected by drugs in this category. ▶| *Translation in bacteria (section 13.7)*

Aminoglycosides

Although considerable variation in structure occurs among **aminoglycoside antibiotics,** all contain a cyclohexane ring and amino sugars (**figure 9.9**). Streptomycin**,** kanamycin, neomycin, and tobramycin are synthesized by different species of the

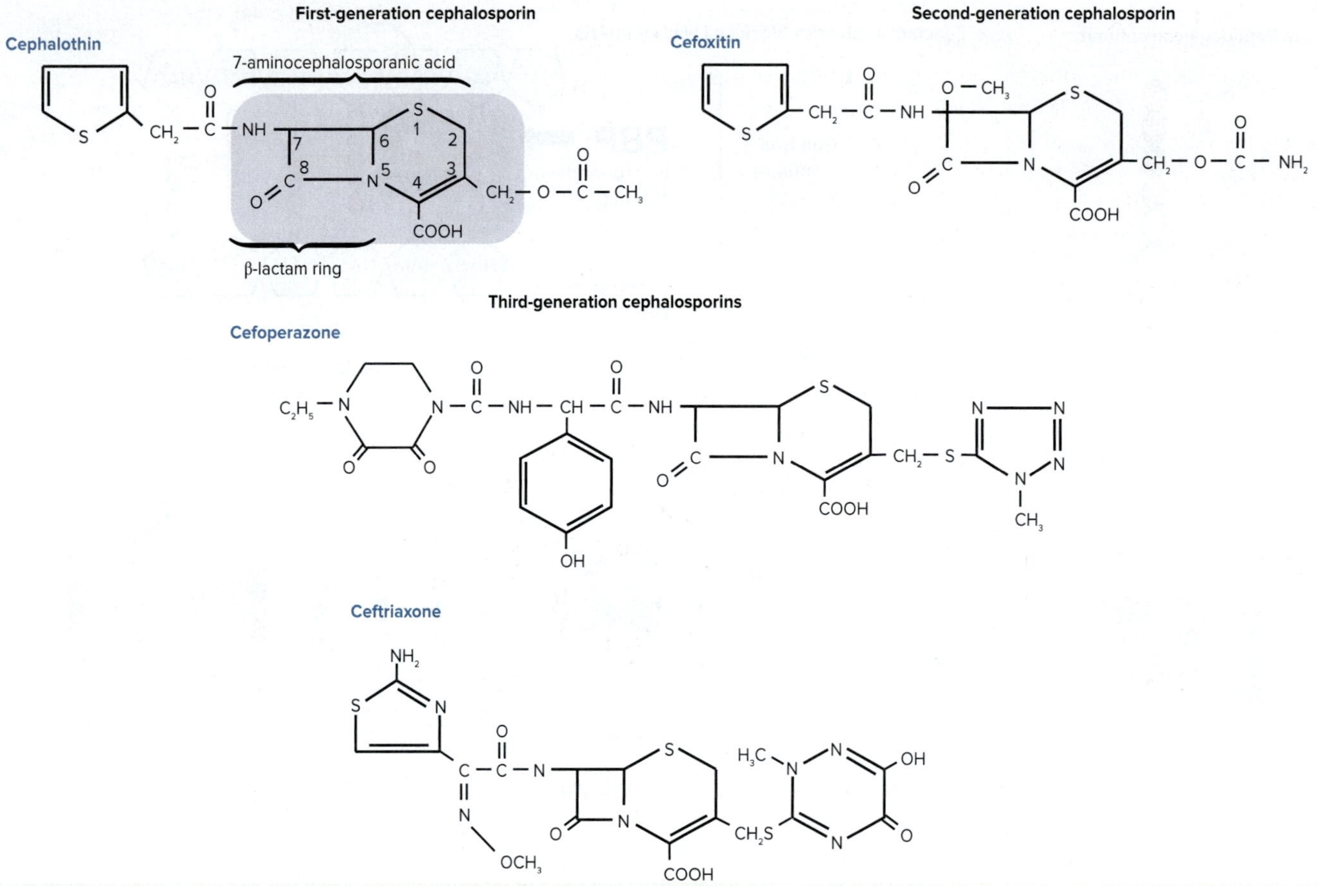

Figure 9.7 Cephalosporin Antibiotics. These drugs are derivatives of 7-aminocephalosporanic acid and contain a β-lactam ring. As newer or next-generation antibiotics are developed, they are numbered first generation, second generation, etc.

MICRO INQUIRY *Do you think cephalosporin antibiotics are susceptible to degradation by β-lactamase enzymes? Explain.*

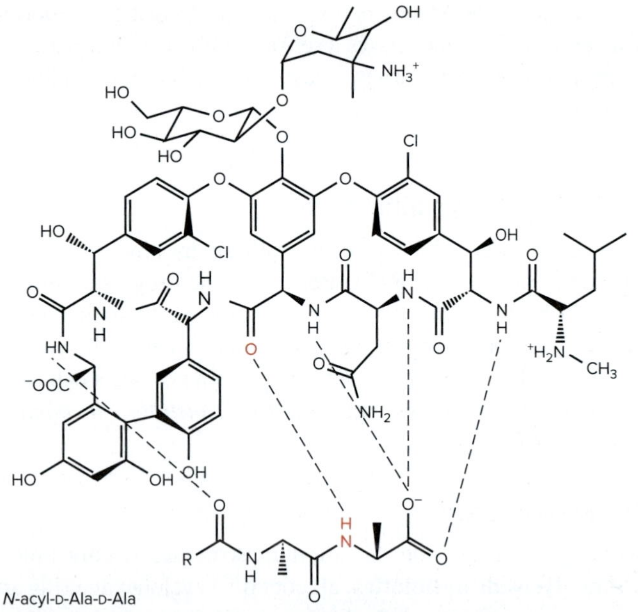

Figure 9.8 Vancomycin. The cup-shaped vancomycin molecule binds to the D-alanyl-D-alanine terminal sequence of peptidoglycan as shown by the dotted lines.

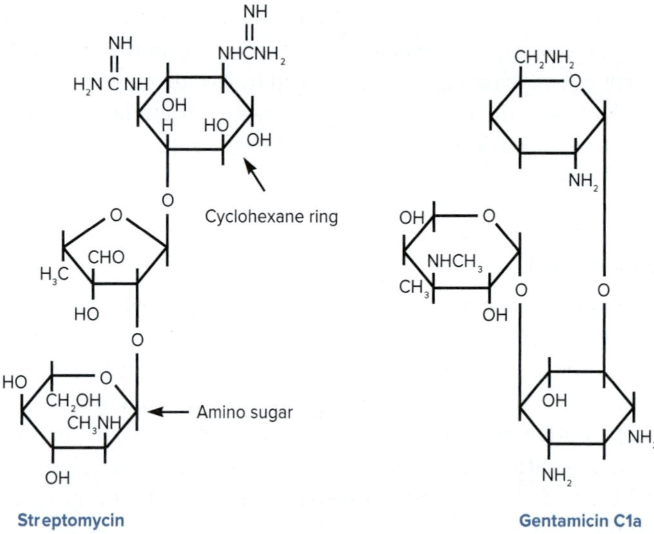

Figure 9.9 Representative Aminoglycoside Antibiotics. The cyclohexane ring and amino sugar are identified.

MICRO INQUIRY *How do these drugs inhibit protein synthesis?*

bacterial genus *Streptomyces,* whereas gentamicin comes from a related bacterium, *Micromonospora purpurea.* Streptomycin was the first drug to cure tuberculosis, but its usefulness has decreased greatly due to widespread drug resistance. Aminoglycosides are bactericidal and are generally used to treat certain infections caused by Gram-negative bacteria. However, because they can be quite toxic, causing hearing and renal damage, loss of balance, nausea, and allergic reactions, they are used sparingly.

All aminoglycoside antibiotics disrupt peptide elongation during translation. This occurs as aminoglycosides bind to ribosomal RNA of the bacterial 30S ribosomal subunit, interfering with mRNA reading and/or causing early termination of peptide synthesis. Aminoglycoside side effects are thought to be due to their ability to bind to host mitochondrial ribosomes, which share the same binding site as their bacterial ancestors.

Tetracyclines

The **tetracyclines** are a family of antibiotics with a common four-ring structure to which a variety of side chains are attached (**figure 9.10**). Oxytetracycline and chlortetracycline are produced naturally by *Streptomyces* spp., whereas other tetracyclines are semisynthetic. These antibiotics are similar to the aminoglycosides in that they target the 30S subunit of the ribosome, inhibiting protein synthesis. Their action is only bacteriostatic, though. Tetracyclines are broad-spectrum antibiotics that are active against most bacteria, including the intracellular pathogens rickettsias, chlamydiae, and mycoplasmas.

Macrolides

The **macrolide antibiotics** contain a ring structure consisting of 12 to 22 carbons called a lactone ring. The lactone ring is linked to one or more sugars (**figure 9.11**). Erythromycin binds to the 50S ribosomal subunit to inhibit bacterial protein elongation. Erythromycin is a relatively broad-spectrum antibiotic effective against Gram-positive bacteria, mycoplasmas, and some Gram-negative bacteria, but it is usually only bacteriostatic. It is used with patients who are allergic to penicillins and in the treatment of whooping cough, diphtheria, diarrhea caused by *Campylobacter,* and pneumonia from *Legionella* or *Mycoplasma* infections. Azithromycin has surpassed erythromycin in use; it is particularly effective against many bacteria, including the sexually transmitted *Chlamydia trachomatis.* Ketolides are semisynthetic macrolide derivatives with a similar mechanism of action. They benefit from better acid stability and oral bioavailability, which is the amount of a drug that reaches the circulation. Telithromycin is a ketolide used to treat community-acquired pneumonia.

Tetracycline (chlortetracycline, doxycycline)

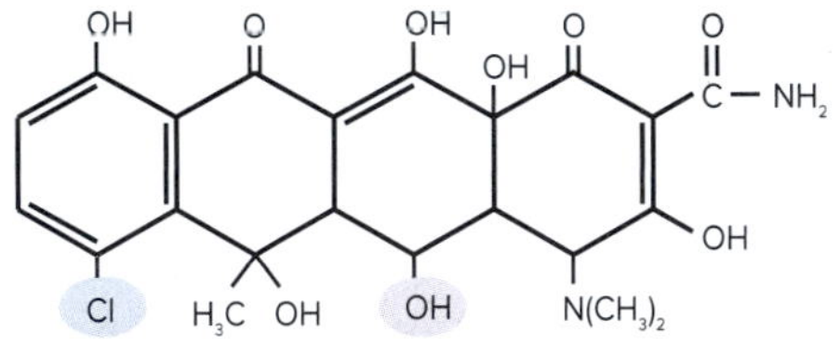

Figure 9.10 Tetracyclines. Three members of the tetracycline family. Tetracycline lacks both of the groups that are shaded. Chlortetracycline (aureomycin) differs from tetracycline in having a chlorine atom (blue); doxycycline consists of tetracycline with an extra hydroxyl (purple).

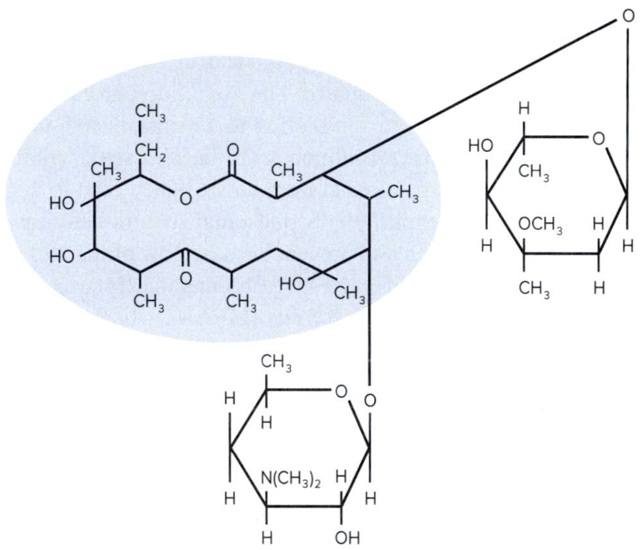

Figure 9.11 Erythromycin, a Macrolide Antibiotic. Erythromycin is composed of a 14-member lactone ring (shaded), so named because it contains a lactone or cyclic ester group that is connected to two sugars.

MICRO INQUIRY *How is the mechanism by which macrolides block protein synthesis similar to that of the tetracyclines? How is it different?*

Lincosamides

Lincosamide antibiotics are produced by *Streptomyces* bacteria. They have a broad spectrum of activity against anaerobic bacteria and many Gram-positive cocci are inhibited. Lincosamides are used sparingly, as they can support the overgrowth of *C. difficile,* the cause of a diarrheal infection that can be life threatening. Clindamycin is used to treat dental infections and some staphylococcal and streptococcal infections.

Chloramphenicol

Chloramphenicol was first produced from cultures of *Streptomyces venezuelae* but is now synthesized chemically. Like erythromycin, this antibiotic binds the 50S ribosomal subunit to inhibit bacterial protein synthesis. It has a very broad spectrum of activity but, unfortunately, is quite toxic. Consequently this antibiotic is used only in life-threatening situations when no other drug is adequate.

Oxazolidinones

The **oxazolidinones** are unique synthetic drugs with a heterocyclic five-membered ring structure. The oxazolidinone called linezolid is one of the rare antibiotics to be introduced in the twenty-first century. Oxazolidinones are bacteriostatic, inhibiting protein synthesis by a novel mechanism. They bind to a site on the 23S rRNA within the 50S ribosomal subunit before protein synthesis has begun and prevent the assembly of the 70S initiation complex. Linezolid is active against methicillin susceptible and resistant *Staphylococcus aureus* (MRSA), which is important because MRSA cannot be treated with any β-lactam. It can also be used to treat vancomycin-resistant enterococci, penicillin-resistant *Streptococcus pneumoniae*, and some Gram-positive anaerobes. It can be given by mouth or by injection and has few side effects, but is used primarily in hospitals in an attempt to prevent overuse and the development of resistance.

Metabolic Antagonists

Several valuable drugs act as **antimetabolites,** as they antagonize, or block, the functioning of metabolic pathways. Antimetabolites are structurally similar to substrates of key enzymes and compete with the metabolites for the binding site of these enzymes. However, once bound to the enzyme, the antimetabolites are different enough to block enzyme activity and further progression of the pathway. By preventing metabolism, they are broad spectrum but bacteriostatic; their removal reestablishes the metabolic activity. ▶| *Enzymes and ribozymes speed up cellular chemical reactions (section 10.6)*

Sulfonamides or Sulfa Drugs

Sulfonamides, or sulfa drugs, are structural analogues of *p*-aminobenzoic acid, or PABA (**figure 9.12**). PABA is needed for folic acid (folate) synthesis. Folic acid is required for the synthesis of purines. Thus by depleting the pool of folate in the cell, sulfonamides prevent synthesis of DNA, RNA, proteins, and other important cell constituents (e.g., ATP). When a sulfa drug enters a bacterial cell, it competes with PABA for the active site of an enzyme (dihydropteroate synthase) involved in one of the first steps in folic acid synthesis (*see figure 10.17*). Sulfonamides are selectively toxic for many bacteria and protozoa because these microbes manufacture their own folate and cannot effectively take up this cofactor, whereas humans do not synthesize folate; instead, we must obtain it in our diet. Sulfonamides thus have a high therapeutic index.

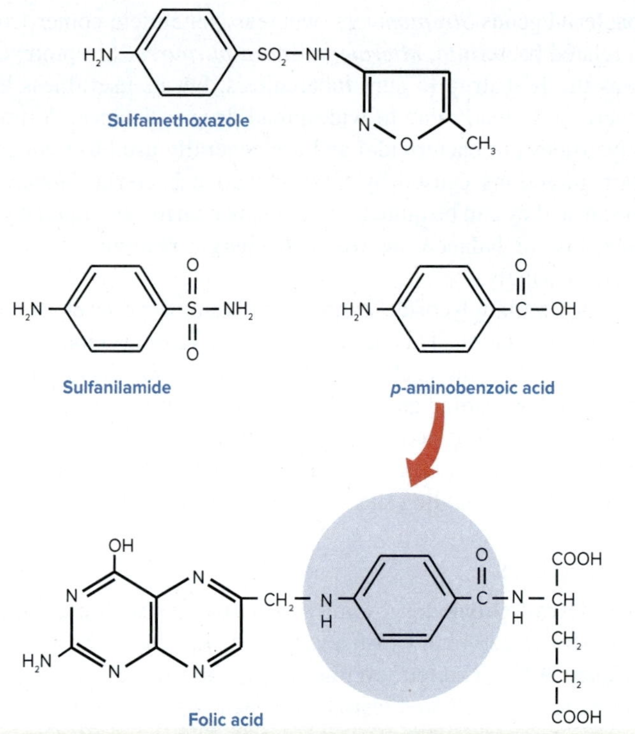

Figure 9.12 Sulfa Drugs. Both sulfanilamide and sulfamethoxazole compete with *p*-aminobenzoic acid to block folic acid synthesis.

MICRO INQUIRY *Why do sulfa drugs have a high therapeutic index?*

Trimethoprim

Trimethoprim is a synthetic antibiotic that also interferes with the production of folic acid but at a later step in the folate biosynthetic pathway. Trimethoprim binds to dihydrofolate reductase (DHFR), the enzyme responsible for converting dihydrofolic acid to tetrahydrofolic acid, competing with the dihydrofolic acid substrate (**figure 9.13**). Trimethoprim is a broad-spectrum antibiotic often used to treat urinary tract infections and traveler's diarrhea. It is usually combined with sulfa drugs to increase efficacy of treatment by blocking two key steps in the folic acid pathway. The inhibition of two successive steps in a single biochemical pathway means that less of each drug is needed in combination than when used alone. This is termed a synergistic (Greek *syn*, together; *ergos*, work) drug interaction.

Nucleic Acid Synthesis Inhibition

The most commonly used antibacterial drugs that inhibit nucleic acid synthesis function by inhibiting (1) DNA polymerase and topoisomerases (fluoroquinolones) or (2) RNA polymerase (rifamycins). Neither are as selectively toxic as other antibiotics because bacteria and eukaryotes do not differ greatly with respect to nucleic acid synthesis. The most commonly used drugs in this category are the fluoroquinolones.

Fluoroquinolones

Fluoroquinolones are synthetic drugs that feature a fluorinated four-quinolone ring (**figure 9.14**). They act by binding to the bacterial topoisomerases DNA gyrase and topoisomerase II. DNA gyrase introduces negative twist in DNA and helps separate its strands. Inhibition of DNA gyrase disrupts

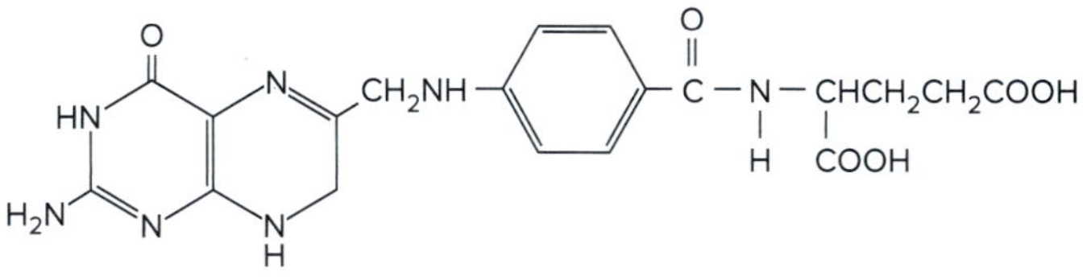

(a) Dihydrofolic acid (DFA)

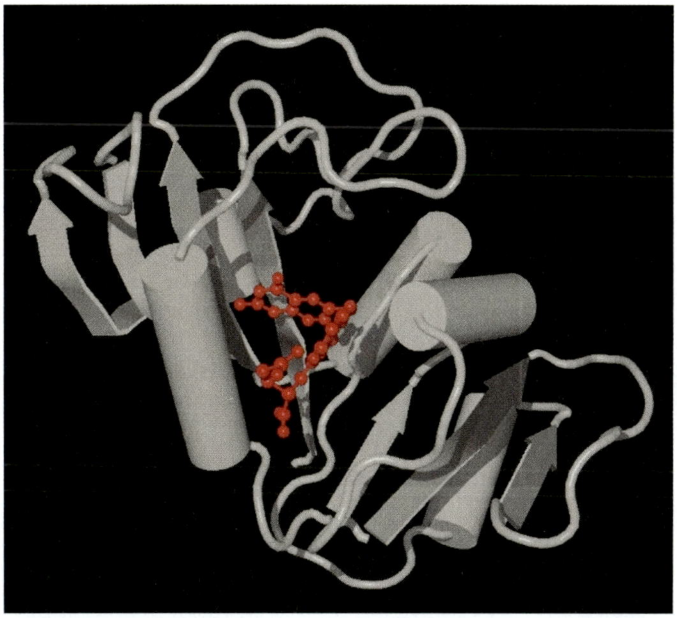

(b) Dihydrofolate reductase

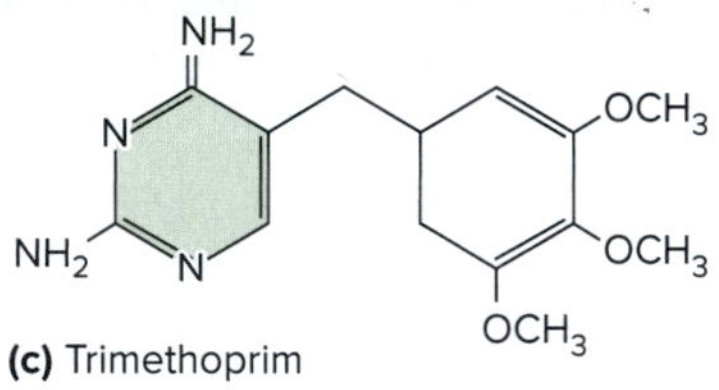

(c) Trimethoprim

Figure 9.13 Competitive Inhibition of Dihydrofolate Reductase (DHFR) by Trimethoprim. (a) Dihydrofolic acid (DFA) is the natural substrate for the DHFR enzyme of the folic acid pathway. (b) DHFR structure and its interaction with DFA (red), its natural substrate. Note the chemical structure and how it fits into the active site of the enzyme. (c) Trimethoprim mimics the structural orientation of the DFA and thus competes for the active site of the enzyme. This causes delayed or absent folic acid synthesis because the DFA cannot be converted to tetrahydrofolic acid when trimethoprim occupies the DHFR active site.

(b) ©Alex MacKerell, Ph.D.

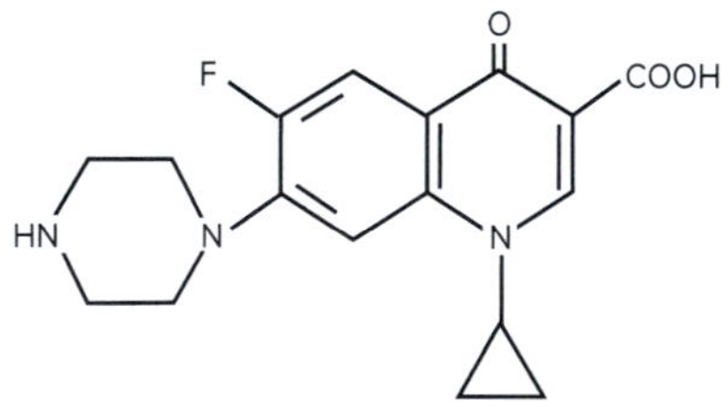

Ciprofloxacin

Figure 9.14 A Fluoroquinolone Antimicrobial Agent. Note the fluorine atom attached to the four-quinolone ring.

DNA replication and repair, bacterial chromosome separation during division, and other processes involving DNA. Fluoroquinolones also inhibit topoisomerase II, another enzyme that untangles DNA during replication. It is not surprising that quinolones are bactericidal. ▶| *DNA replication in bacteria (section 13.3)*

The fluoroquinolones are broad-spectrum antibiotics. They have been used to treat a wide variety of infections. This success has led to their overuse and fluoroquinolone resistance has become problematic. In addition, because they can cause disabling damage to tendons and nerves, the U.S. Food and Drug Administration has deemed them too dangerous to use for all but very severe infections.

Rifamycins

The most commonly used member of the **rifamycin** class is the semisynthetic derivative rifampin. These agents block bacterial transcription by binding to the β-subunit of RNA polymerase. Rifampin is an important member of the multidrug regimen used to treat tuberculosis and other mycobacterial infections. It can also be used to protect those who have come into contact with an individual diagnosed with meningitis caused by *Neisseria meningitidis* or *Haemophilus influenzae*. In addition to the side effects listed in table 9.1, rifamycins typically cause urine and sweat to appear red. This is perfectly harmless but can be upsetting to an uninformed patient.

Comprehension Check

1. Explain five mechanisms by which chemotherapeutic agents kill or damage bacterial pathogens.
2. Why do penicillins and cephalosporins have a higher therapeutic index than most other antibiotics?
3. Why is linezolid not more commonly prescribed?
4. What are antimetabolites? Why are these effective against protozoan pathogens but the other drugs presented here generally are not?

9.5 Antiviral Drugs

After reading this section, you should be able to:

a. Compare antiviral drug mechanisms of action
b. Provide a rationale for combination drug therapy in the treatment of HIV
c. Explain why there are far fewer antiviral agents than there are antibacterial agents

The discovery of small molecules that inhibit virus-specific enzymes and replication cycle processes has led to the development of antiviral drugs. However, antiviral drugs have had mixed success and the vast majority of viral infections cannot be cured. Some antiviral drugs simply limit the duration of the illness (e.g., flu) or its severity (e.g., herpes, HIV). Some important antiviral agents (recall these are not called antibiotics) are shown in **figure 9.15**. |◀ *Viral life cycles have five steps (section 6.3)*

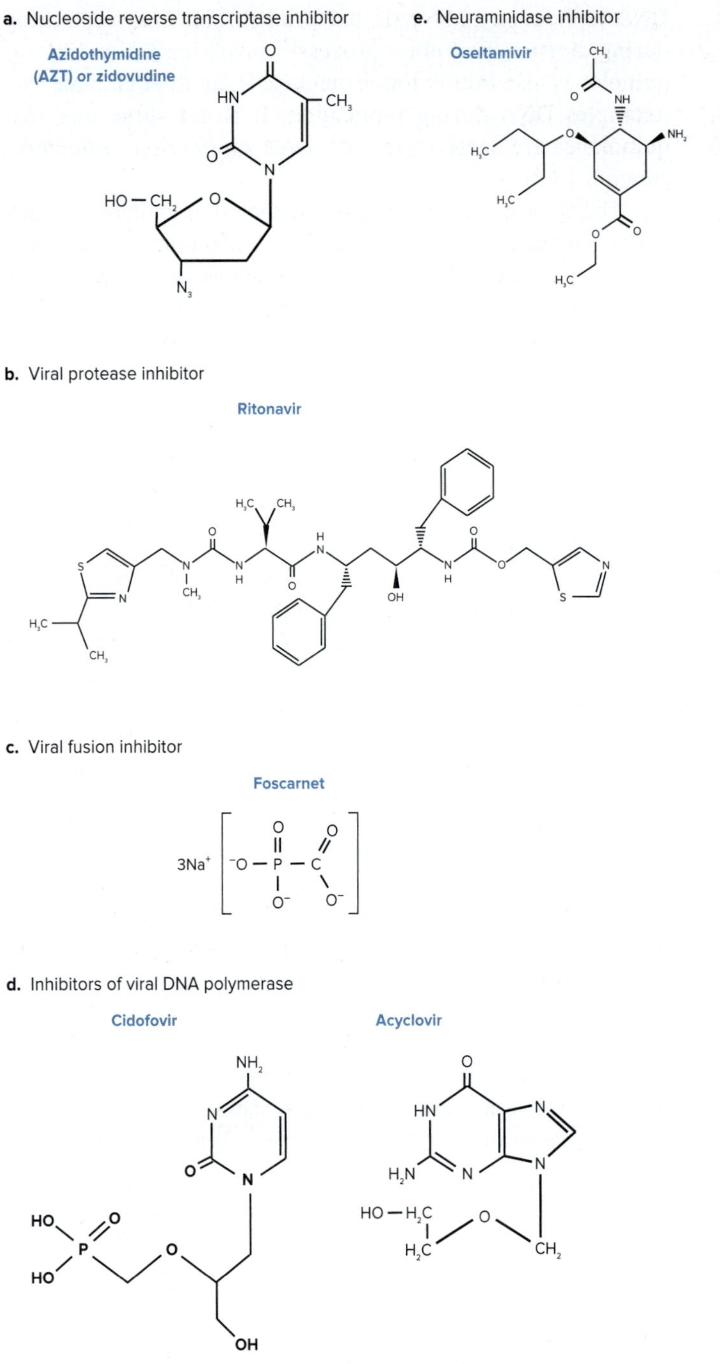

Figure 9.15 Antiviral Drugs Target the Viral Replication Cycle.

Most antiviral drugs disrupt critical stages in a virus's multiplication cycle. One of the most publicized antiviral agents is oseltamivir (Tamiflu). Oseltamivir (figure 9.15*e*) inhibits the viral enzyme neuraminidase, which is essential for release of newly synthesized influenza A and B virus particles from host cells. However, its widespread use has generated concerns about growing resistance in influenza virus, driving the 2014 CDC recommendation for limiting oseltamivir use. While it is not a cure, patients who take it within 48 hours of influenza infection are relieved of flu symptoms about a day and a half sooner than patients who do not take oseltamivir. Two other neuraminidase inhibitors, peramivi (for intravenous use) and inhaled zanamivir, can be used to treat influenza infection in adults. ▶| *Influenza (flu) (section 38.1)*

Several drugs are commonly used to treat illnesses caused by viruses with DNA genomes. These drugs rely on the fact that the viruses that cause herpes, chickenpox and shingles, and cytomegalovirus use their own (rather than the host's) enzyme to phosphorylate nucleosides to nucleotides. Thus these drugs are nucleoside or mononucleotide analogues that, once phosphorylated, are incorporated into the growing viral genome. Because they are not natural nucleotides, their addition blocks further DNA synthesis, a process called chain termination. Examples of these antiviral agents include acyclovir (figure 9.15*d*) and the ointment vidarabine, which are used to treat herpes infections and shingles. Ganciclovir is used to treat systemic cytomegalovirus illness, and cidofovir, a mononucleotide analog (figure 9.15*d*), is used to treat eye infections (retinitis) caused by this virus. In cases of acyclovir or ganciclovir resistance, foscarnet (figure 9.15*c*) may be used. It is a pyrophosphate analogue that does not require modification by viral enzymes, so can be used to treat illnesses caused by both herpes simplex viruses and cytomegalovirus. ▶| *Direct contact diseases can be caused by viruses (section 38.3)*

Since the early days of human immunodeficiency virus (HIV) treatment, much effort has been focused on developing new drugs. These drugs target and interfere with critical steps in the viral replicative processes (**figure 9.16**). There are now five categories of drugs used to manage HIV infection: (1) nucleoside reverse transcriptase inhibitors (NRTIs), which are nucleoside analogues that cause chain termination in much the same way as acyclovir and cidofovir (e.g., azidothymidine or AZT; figure 9.15*a*); (2) nonnucleoside reverse transcriptase inhibitors (NNRTIs), which prevent HIV DNA synthesis by selectively binding to and inhibiting the viral reverse transcriptase enzyme; (3) protease inhibitors (PIs) that block the activity of the HIV protease needed for the production of all viral proteins, which are translated as a single polypeptide (figure 9.15*b*); (4) integrase inhibitors that prevent the incorporation of the HIV genome into the host's chromosomes; and (5) fusion inhibitors (FIs), which prevent HIV entry into cells. ▶| *Retroviruses: plus-strand viruses that use reverse transcriptase in their life cycles (section 26.7)*

The most successful treatment approach to date in combating HIV/AIDS is to use drug combinations. Most effective is the simultaneous use of three agents to prevent the development of drug resistance. For example, the combination of two NRTIs and an integrase inhibitor reduces HIV concentrations in plasma to undetectable levels. However, the treatment does not eliminate proviral HIV DNA that still resides in certain cells of the immune system (e.g., memory T cells) and possibly other cells. Thus with proper treatment, the

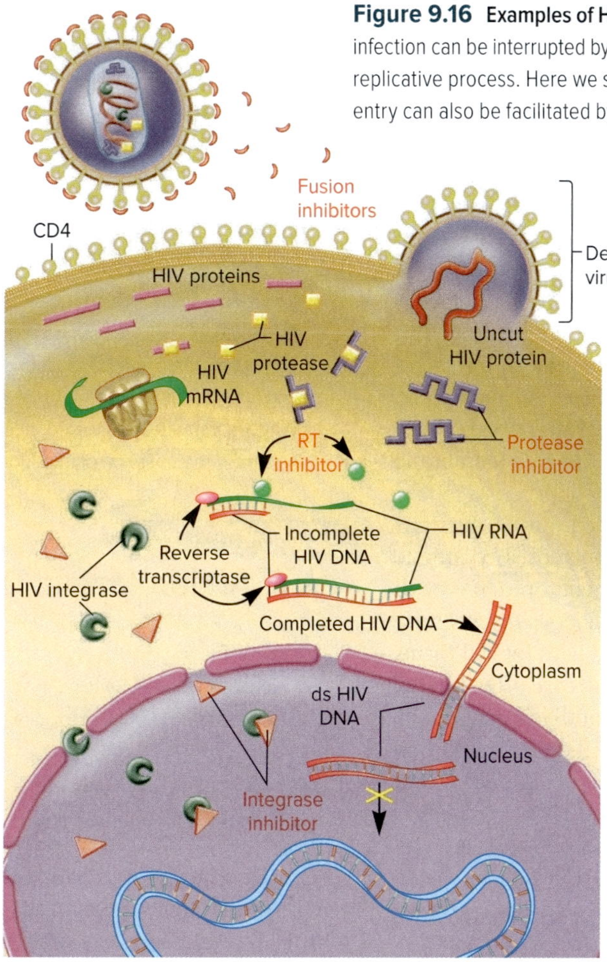

Figure 9.16 Examples of How Anti-HIV Agents Block HIV Replication. Viral infection can be interrupted by drugs that target specific components of the HIV replicative process. Here we show virus entry by membrane fusion. However, virus entry can also be facilitated by receptor-mediated endocytosis.

Infection begins with HIV fusion. Fusion inhibitors block this step.

Once inside a host cell, HIV uncoats and its reverse transcriptase (RT) makes DNA from the viral RNA genome. RT inhibitors block this step.

Viral DNA is transcribed and translated into polyproteins that are cut to release viral proteins. Protease inhibitors block this step.

Viral DNA is added to the host DNA by the action of a viral integrase. Integrase inhibitors block this step.

virus often disappears from a patient's blood and drug-resistant strains do not seem to arise. But because HIV remains dormant host cells, it will reactivate if drugs are stopped. Thus patients are never cured and require drug therapy for life. Globally, the vast majority of HIV-positive individuals do not have access to effective combination therapy. Recent studies have led the CDC to recommend a pre-HIV exposure prophylaxis (prevention) strategy to men and women who have sex with HIV-infected partners. The daily oral dose of two NRTIs is strongly correlated with reduced HIV transmission among uninfected individuals exposed to the virus through sexual contact. This is now offered to high-risk individuals as preexposure prophylaxis (PrEP). ▶| *Acquired immunodeficiency syndrome (AIDS) (section 38.3)*

The ability to cure hepatitis C virus (HCV) infection is a recent and exciting development. The CDC estimates that there are currently about 3.5 million people with hepatitis C infection in the United States, which is the leading cause of liver transplant. For many years, attempts to limit the damage caused by this RNA virus centered on treatment with the human immune system mediator (cytokine) interferon-α and the antiviral medication ribavirin. This treatment was not particularly effective and patients suffered severe side effects. In 2014 drugs called direct-acting antiviral agents were introduced and changed the prognosis for thousands of patients. These drugs are given in combination. Sofosbuvir is a nucleotide analogue that blocks the activity of the HCV RNA polymerase, and ledipasvir and velpatasvir target a viral enzyme needed for replication. The specific regimen prescribed depends on the genotype of HCV infecting the individual. About 70% of HCV patients in the United States are infected with genotype 1, so are candidates for ledipasvir-sofosbuvir or sofosbuvir-velpatasvir. Almost all others are infected with genotypes 2 and 3 and can be treated with sofosbuvir-velpatasvir. The cost of treatment has been a contentious factor that has limited the availability of this life-saving treatment.

Comprehension Check

1. Why do drugs used to treat herpes target the virus and spare the host cell?
2. Compare and contrast the activity of a nucleoside reverse transcriptase inhibitor and a nonnucleoside reverse transcriptase inhibitor.
3. Why must individuals with HIV infections take antiretroviral drugs for the rest of their lives?
4. Compare the older and newer drugs used to treat hepatitis C virus. Explain why the newer compounds are called direct-acting antiviral agents.

9.6 Antifungal Drugs

After reading this section, you should be able to:

a. Compare antifungal drug mechanisms of action
b. Explain why there are far fewer antifungal agents than there are antibacterial agents

Treatment of fungal infections generally has been less successful than that of bacterial infections largely because as eukaryotes, fungal cells are much more similar to human cells than are bacterial cells. Many drugs that inhibit or kill fungi are therefore quite toxic for humans and thus have a low therapeutic index. Most antifungals are fungistatic only as long as repeated application maintains high levels of drug. Nonetheless, a few drugs are useful in treating many major fungal diseases.

There are two major classes of antifungal drugs: polyenes and azoles; both block fungal cell membrane synthesis. This is a good drug target because fungal membranes require the sterol ergosterol, not found in human membranes. Polyenes bind directly to ergosterol, and azoles block the last step in ergosterol biosynthesis. Thus, both drugs disrupt fungal membrane formation.

Fungal infections are often subdivided into infections called superficial mycoses, subcutaneous mycoses, systemic mycoses, and opportunistic mycoses. Treatment for each type of mycosis differs. Because superficial and cutaneous mycoses are limited to the skin and hair, these are treated with topical forms of azole drugs. Superficial mycoses of the nails are often treated with the agent griseofulvin, which is taken orally (**figure 9.17**). This agent disrupts fungal mitotic spindles thereby preventing cell division. Nystatin (figure 9.17*a*), a polyene antibiotic from *Streptomyces noursei,* is used to control opportunistic *Candida* infections of the skin, vagina, or mouth and esophagus. It is too toxic to be taken systemically; when used as a "swish and swallow" agent for thrush, it is not absorbed from the gastrointestinal tract. ▶| *Human diseases caused by fungi and protists (chapter 40)*

Systemic fungal infections are very difficult to control and can be fatal. Drugs commonly used against systemic mycoses include amphotericin B, 5-flucytosine, and azoles such as fluconazole, itraconazole, and voriconazole. Amphotericin B from *Streptomyces* spp. is quite toxic to humans earning it the nickname "ampho-terrible." It is used only for serious, life-threatening infections. The synthetic oral agent 5-flucytosine (5-fluorocytosine) is effective against many systemic fungi, although drug resistance develops rapidly so it is typically used in combination with amphotericin B. The drug is converted to 5-fluorouracil (5FU) only by fungi. 5FU is an analogue of uracil, one of the building blocks of RNA. During transcription, it is incorporated into RNA in place of uracil and disrupts RNA function. It is most commonly used in combination with amphotericin B to treat cryptococcal meningitis. Fluconazole is used in the treatment of candidiasis and coccidioidomycosis (valley fever). Because adverse effects of fluconazole are relatively uncommon, it is used prophylactically to prevent life-threatening fungal infections in individuals who are severely immunosuppressed.

One such life-threatening infection of immunosuppressed individuals is caused by *Pneumocystis jiroveci* (formerly *P. carinii*). Because *P. jiroveci* membranes possess cholesterol rather than ergosterol, polyenes and azoles are ineffective. Rather, infections are treated with the combination of antimetabolite drugs trimethoprim and sulfisoxazole. This is effective because *P. jiroveci* synthesizes folic acid using enzymes that are very similar to those used by bacteria.

Comprehension Check

1. Compare and contrast the mechanism by which polyenes, azoles, and 5-flucytosine inhibit fungal growth.
2. Go online or visit a drugstore and compare the agents found in triple antibiotic cream with those found in athlete's foot medicine. What can you infer about athlete's foot from this?
3. Why are immunosuppressed individuals sometimes given antifungal agents in the absence of infection?

9.7 Antiprotozoan Drugs

After reading this section, you should be able to:

a. Compare the mechanisms of action of antiprotozoan drugs
b. Relate side effects and toxicity of antiprotozoan drugs to their mechanism of action
c. Explain why there are far fewer antiprotozoan agents than there are antibacterial agents

As with other antimicrobial therapies, antiprotozoan drug effectiveness starts by identifying a unique target to which a drug can bind and thus prevent some vital function. However, because protozoa are eukaryotes, the potential for drug action on host cells and tissues is greater than it is when targeting bacteria. Most of the drugs used to treat protozoan infections have significant side effects; nonetheless, the side effects are usually acceptable when weighed against the parasitic burden.

The number of antiprotozoan drugs is relatively small, and the mechanism of action for most is not completely understood. The drugs described here have potent antiprotozoan action and appear to act on protozoan nucleic acid or some metabolic process.

Malaria, caused by any of five species in the genus *Plasmodium*, kills about half a million people annually—most of them children. Drugs to treat and prevent malaria include everything from ancient Chinese remedies to quinine water. In fact, gin and tonic is said to have been the most popular drink among imperial British in South Asia due to its antimalarial activity. We now use more purified forms of **quinine**, including chloroquine and quinine sulfate to treat malaria (**figure 9.18**). These drugs suppress *Plasmodium* reproduction and are effective in eradicating asexual stages of the protozoan's life cycle that occur in red blood cells. An essential part of the *Plasmodium* spp. life cycle is degradation of hemoglobin. During this process, the microbe polymerizes a toxic heme metabolite into a nontoxic

a. Polyenes bind to sterols, resulting in membrane damage

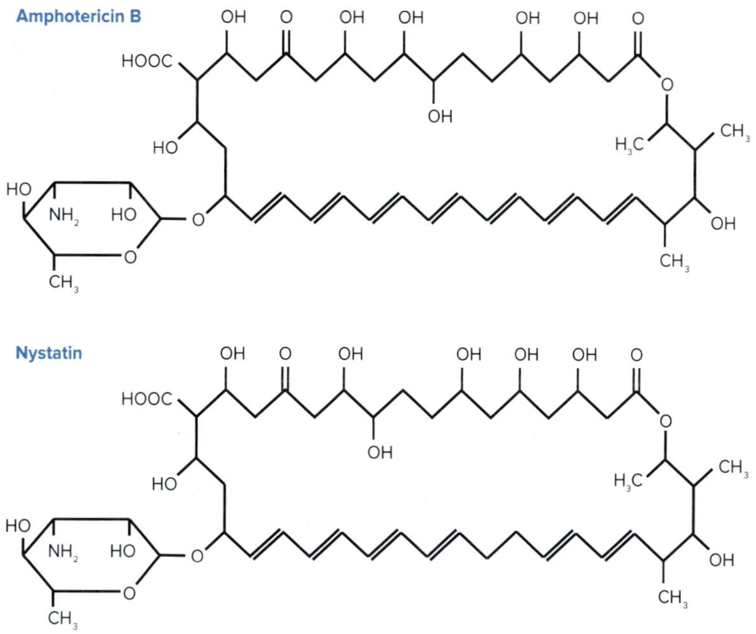

b. Azoles inhibit sterol synthesis, resulting in altered membrane permeability

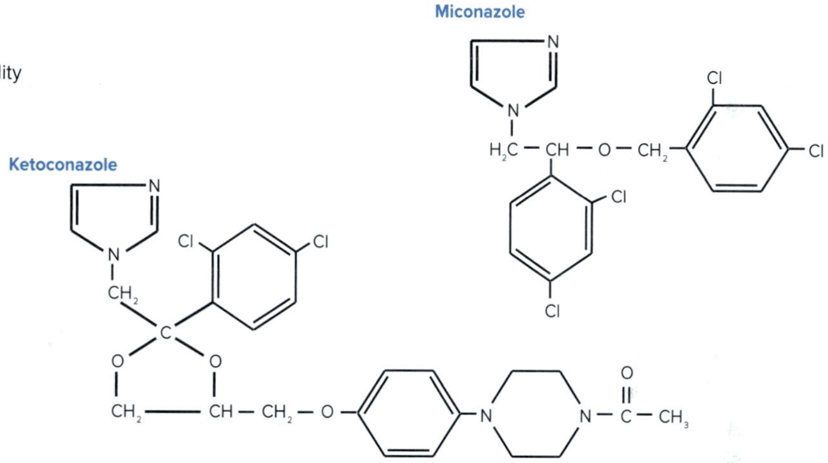

c. Drug that inhibits nucleic acid synthesis, protein synthesis, or cell division

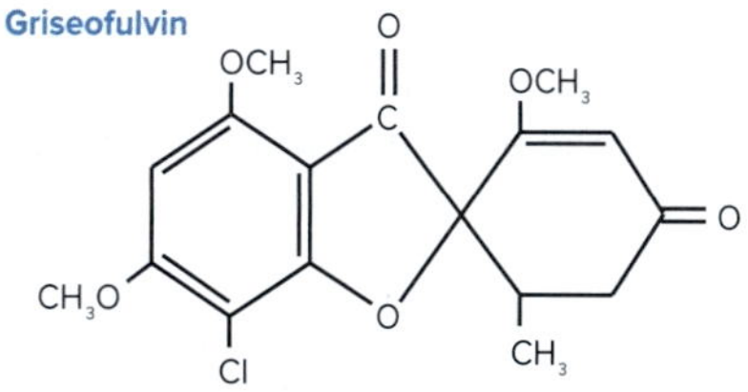

d. Drug that disrupts RNA function

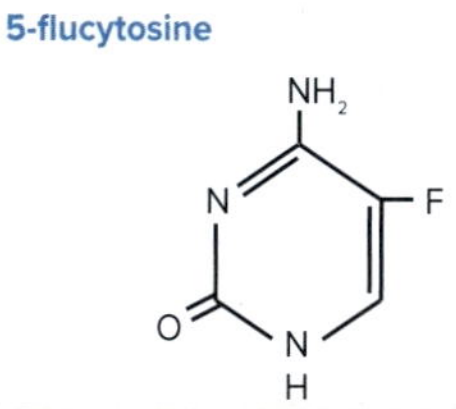

Figure 9.17 Antifungal Drugs.

MICRO INQUIRY *What is the mechanism by which nystatin inhibits growth? How does this compare to that of 5-flucytosine?*

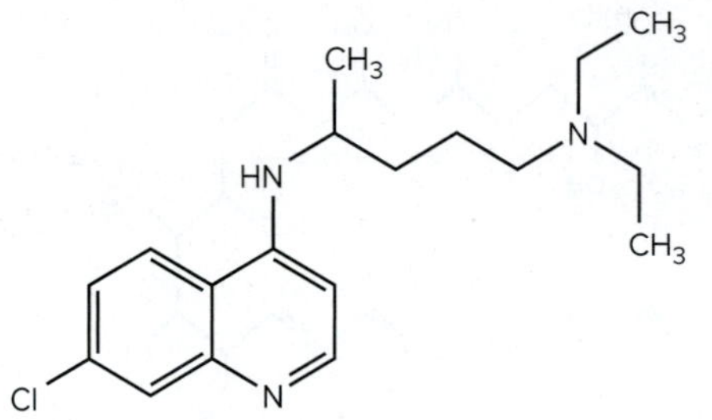

Figure 9.18 Chloroquine, a Quinine Drug. Quinine drugs have a long history of use as antimalarial agents.

form, called hemazoin. Quinine drugs block this process and the toxic metabolite accumulates. Another drug, mefloquine, has been found to swell *P. falciparum* food vacuoles, where it may act by forming toxic complexes that damage membranes and other plasmodial components. Primaquine is active against a dormant form of the protists (hypnozoites) found in the liver so it prevents relapses of malaria caused by *P. vivax* and *P. ovale*, the two species that produce hypnozoites. Individuals who travel to areas where malaria is endemic should adhere to chemoprophylactic treatment with antimalarial agents. ▶| *Malaria (section 40.3)*

In many places where malaria is endemic, the World Health Organization recommends a combination of a quinine drug and a derivative of **artemisinin**. Having its origins as a treatment in traditional Chinese medicine, artemisinin is a chemical found in the sweet wormwood plant (*Artemisia annua*). A semisynthetic artemisinin derivative used in combination therapies is an effective and inexpensive malaria treatment. The mechanism of artemisinin action is not well understood; it appears to form reactive oxygen intermediates inside *Plasmodium*-infected red blood cells, leading to altered hemoglobin catabolism and damage to the protist's electron transport chain. ▶| *Electron transport chains (section 11.6)*

There are many other diseases caused by protozoa, as discussed in chapter 40. Dysentery caused by the amoeba *Entamoeba histolytica* and vaginal infections of the protozoan *Trichomonas vaginalis* can be treated with **metronidazole**. These anaerobic organisms readily reduce metronidazole to a reactive nitrogen species within the cytoplasm. Reactive nitrogen species cause oxidative damage to a variety of cellular macromolecules. The drug nitazoxanide has a very similar mechanism of action and is used to treat the waterborne diarrheal diseases caused by the protozoa *Giardia intestinalis* and *Cryptosporidium parvum*. A number of antibiotics that inhibit bacterial protein synthesis (e.g., the aminoglycosides clindamycin and paromomycin) are also used to treat protozoan infection. These agents are used to prevent or treat infections caused by protozoa known as apicomplexans. These microbes possess an organelle derived from plastids, the apicoplast, that is essential for viability. These protein synthesis inhibitors bind to the ribosomes in these organelles. ▶| *Supergroup* Excavata: *primitive eukaryotes (section 24.2); Supergroup* Amoebozoa *includes protists with pseudopodia (section 24.3); Supergroup SAR: protists of global importance (section 24.4); Food and water are vehicles for fungal and protozoal diseases (section 40.5)*

Toxoplasmosis is a life-threatening infection in immunocompromised individuals and can cause severe birth defects in human fetuses. In the United States it is thought to be a leading cause of death associated with foodborne illness. The CDC estimates that 11% of the population over the age of 6 years has been infected with *Toxoplasma gondii*, but the immune system usually keeps the protist from causing symptoms or illness. Toxoplasmosis is typically treated with combinations of pyrimethamine, dapsone, and sulfadiazine, plus folinic acid. Pyrimethamine and dapsone are thought to act in the same way as trimethoprim—interfering with folic acid synthesis by inhibition of dihydrofolate reductase. ▶| *Toxoplasmosis (section 40.5)*

Comprehension Check

1. Why are drugs that inhibit bacterial protein synthesis also effective against some protists?
2. Why is malaria, like tuberculosis, now treated with several drugs simultaneously?
3. Synthesis of what metabolite is blocked by the drugs used to treat toxoplasmosis?

9.8 Antimicrobial Drug Resistance Is a Public Health Threat

After reading this section, you should be able to:

a. Report the common reasons for increasing drug resistance
b. Describe common mechanisms by which antimicrobial drug resistance occurs
c. Suggest strategies to overcome drug resistance

The spread of antibiotic-resistant bacteria has become so widespread that **methicillin-resistant *S. aureus*** (**MRSA**) has become well known. Fortunately most MRSA strains can still be treated, but the more recent emergence of **vancomycin-resistant enterococci** (**VRE**) and **carbapenem-resistant *Enterobacteriaceae*** (**CRE**) bacteria has caused great concern among public health officials worldwide. The spread of resistance from one pathogen to another coupled with the slow pace of the development of new antibiotics kills about 700,000 people per year. Some fear it may grow to as many as 10 million per year globally by 2050 if the tide of resistance is not stemmed and antibiotic research is not supported. While both these issues are important, we save our discussion of antibiotic development for chapter 42; here we discuss the mechanisms of antibiotic resistance.

There are two types of resistance: intrinsic and acquired. An example of intrinsic resistance is that of mycoplasma resistance to β-lactam antibiotics and other cell wall inhibitors simply because these bacteria lack a cell wall. Acquired resistance occurs when there is a change in the genome of a bacterium that converts it from one that is sensitive to an antibiotic to one that is resistant. Another distinction is drug tolerance as opposed to resistance. Drug-tolerant bacteria lack the mechanisms for antibiotic resistance we describe next. Instead these bacteria "ignore" the presence of antibiotics, usually because they are embedded in biofilms that antibiotics cannot effectively penetrate or are growing too slowly to be inhibited. These bacteria are called persisters and are discussed in chapter 7. When moved from the environment that supports such slow growth, persister cells demonstrate drug sensitivity. By contrast, antibiotic-resistant bacteria have acquired genome changes, either by mutation or horizontal gene transfer, that code for proteins that confer true resistance, regardless of growth rate. The mechanism by which these genetic changes occur is discussed in chapter 16.

There Are Several Mechanisms of Drug Resistance

Bacteria have evolved several resistance mechanisms (**figure 9.19**). One is to modify the target of the antibiotic. For example, MRSA is resistant to all β-lactam antibiotics because it has replaced its transpeptidase (or, as mentioned, penicillin-binding protein [PBP]) with another version of this enzyme (figure 9.6*c*). This altered PBP still functions as a transpeptidase, but has no binding site for β-lactam antibiotics. This explains why MRSA is resistant to all members of this class of antibiotics but remains sensitive to vancomycin. Recall that vancomycin binds a different target—the terminal D-ala of the pentapeptide on the peptidoglycan subunit (figure 9.6*a*). Resistance to this drug among some enterococci is another example of changing the target. Vancomycin resistance occurs when bacteria gain the capacity to change the last D-ala to a D-lactate or D-serine residue to which the drug cannot bind.

A second resistance strategy is drug inactivation. The best-known example is hydrolysis of the β-lactam ring of penicillins by penicillinase and other β-lactamase enzymes (**figure 9.20**). Drugs also are inactivated by the addition of chemical groups. For example, aminoglycosides can be modified and inactivated by acetylation of amino groups and phosphorylation or adenylylation of hydroxyl groups present on the aminoglycoside. Because these resistance strategies involve gaining enzymes that catalyze the inactivating reaction, the new function is most often acquired when bacteria share genes with each other, a process known as horizontal gene transfer.

The third resistance strategy minimizes the concentration of the antibiotic in the cell. This can be accomplished by altering membrane structure, particularly the outer membrane for Gram-negative bacteria, so that less antibiotic enters the cell. Another approach is to pump the drug out of the cell after it has entered, using translocases, called **efflux pumps**, that expel drugs. Efflux pumps are relatively nonspecific and pump many different drugs; therefore, they often confer multidrug resistance. Many efflux pumps are drug/proton antiporters; that is, protons enter the cell as the drug leaves.

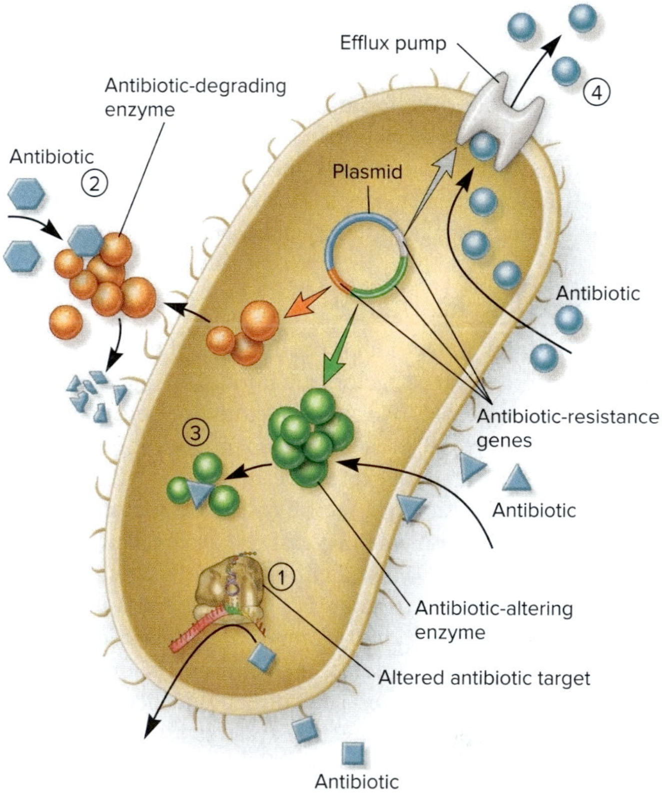

Figure 9.19 Antibiotic-Resistance Mechanisms. Bacteria can resist the action of antibiotics by (1) preventing access to (or altering) the target of the antibiotic, (2) degrading the antibiotic, (3) altering the antibiotic, or (4) rapidly extruding the antibiotic.

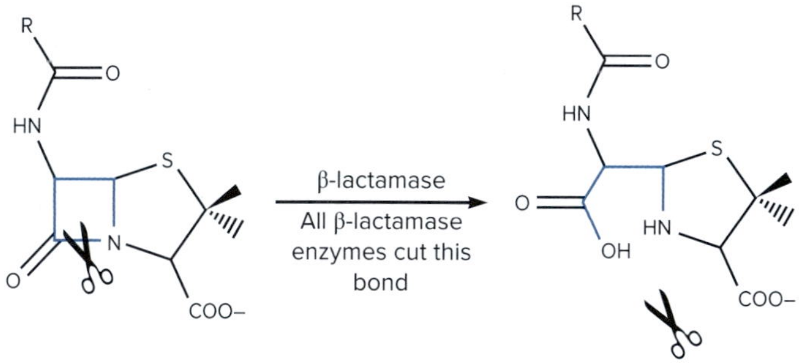

Figure 9.20 β-lactamase enzymes cleave the β-lactam ring. Some β-lactamases degrade only a few different antibiotics, while other, more dangerous enzymes have a broader β-lactam specificity.

Finally, resistant bacteria may either use an alternate pathway to bypass the biochemical reaction inhibited by the agent or increase the production of the target metabolite. For example, some bacteria are resistant to sulfonamides simply because they use preformed folic acid from their surroundings, rather than synthesize it themselves. Other strains increase their rate of folic acid production and thus counteract sulfonamide inhibition. Again, these changes are most often mediated by horizontal gene transfer.

Detecting Drug Resistance

It is important to know if a clinically isolated microorganism is expressing resistance factors and/or has the genetic potential to do so. Because resistance factors are genetically encoded, it is possible to test for the gene product or for the gene itself. Commercial gene expression systems are designed to identify the production of specific resistance factors, such as a target-modifying enzyme. Several test systems measure the color change induced when a chromophore is acted upon by either a beta-lactamase (figure 9.20) or an antibiotic-modifying enzyme. The resulting color changes are measured spectrophotometrically and protein concentration is extrapolated from a standardized curve. Thus direct evidence of induced resistance can be made. Detection systems are also commercially available to identify genes encoding drug resistance factors using polymerase chain reactions (PCRs). ▶| *Polymerase chain reaction amplifies targeted DNA (section 17.2)*

Overcoming Drug Resistance

The World Health Organization cites antibiotic resistance as "one of the biggest threats to global health, food security, and development today." The CDC has classified certain antibiotic-resistant bacteria as "urgent, serious, or concerning threats." These alarms are coupled with calls to action to ensure the effectiveness of antimicrobial agents continues. It has long been understood that the problem has resulted from antibiotic overuse, misuse, and abuse. Put simply, the more drugs are used, the more likely it is that bacteria will become resistant to them. That raises the question: Is there a way to overcome drug resistance?

Several strategies can be employed to limit the emergence of drug resistance. Tighter controls over indiscriminate drug use and use of antibiotics in animal feed must be enacted. Drugs should be given in a high enough concentration to destroy susceptible microbes and most spontaneous mutants that might arise during treatment. Patients must be educated to take their antimicrobial agents as prescribed and to finish the course of therapy. If a patient fails to complete treatment, resistant mutants survive and flourish because of their competitive advantage over nonresistant bacteria. Sometimes two or even three different drugs can be administered simultaneously with the hope that each drug will prevent the emergence of resistance to the other. This approach is used in treating tuberculosis, HIV, and malaria, for example.

Other strategies used to prevent drug resistance include the strict control on use of antimicrobial drugs, especially newly developed drugs to which bacteria have not become widely resistant. The use of broad-spectrum drugs must also be curtailed. If possible, the pathogen should be identified, drug sensitivity tests performed, and the proper narrow-spectrum drug employed. These ideas and others are now incorporated into antibiotic stewardship programs at most health-care facilities.

Despite efforts to control the emergence and spread of drug resistance, the situation continues to worsen. There is an urgent need for new antibiotics that microorganisms have never encountered. Pharmaceutical and biotechnology companies collect and analyze samples from around the world in a search for completely new antimicrobial agents. Both culture-based and metagenomics approaches are used. Structure-based or rational drug design is another option. If the three-dimensional structure of a susceptible target molecule such as an enzyme essential to microbial function is known, drugs can be designed that precisely fit the target molecule and disrupt its function. Information derived from the sequencing and analysis of pathogen genomes is also useful in identifying new targets for antimicrobial drugs. A most interesting response to the current crisis is the renewed interest in an idea first proposed early in the twentieth century by Felix d'Herelle, one of the discoverers of bacterial viruses (bacteriophages). d'Herelle proposed that bacteriophages could be used to treat bacterial diseases. Although most microbiologists did not pursue his proposal due to technical difficulties and the advent of antibiotics, Russian scientists developed the medical use of bacteriophages. The idea has reemerged and trials using bacteriophages to treat certain bacterial infections are underway. |◀ *Viruses and other acellular infectious agents (chapter 6)* ▶| *Microbial genomics (chapter 18); Biotechnology and industrial microbiology (chapter 42)*

Comprehsion Check

1. Compare and contrast the five mechanisms bacteria use to block antibiotic activity.
2. Which of the resistance mechanisms described here could be used by viruses? Explain your answer.
3. Why does the use of narrow-spectrum antibiotics help prevent the spread of drug resistance to a greater extent than broad-spectrum antibiotics?
4. What are the primary medical practices that result in antimicrobial drug resistance?

Key Concepts

9.1 Antimicrobial Chemotherapy Evolved from Antisepsis Efforts

- The modern era of chemotherapy began with Paul Ehrlich's work on drugs against African sleeping sickness and syphilis.
- Other early pioneers were Gerhard Domagk, Alexander Fleming, Howard Florey, Ernst Chain, Norman Heatley, and Selman Waksman.

9.2 Antimicrobial Drugs Have Selective Toxicity

- An effective chemotherapeutic agent must have selective toxicity. A drug with great selective toxicity has a high therapeutic index and usually disrupts a structure or process unique to the pathogen. It has fewer side effects.
- Antibiotics can be classified in terms of the range of target microorganisms (narrow spectrum versus broad spectrum); their source (natural, semisynthetic, or synthetic); and their general effect (static versus cidal) (**table 9.1**).

9.3 Antimicrobial Activity Can be Measured by Specific Tests

- Antibiotic effectiveness can be estimated through the determination of the minimal inhibitory concentration and the minimal lethal concentration with dilution susceptibility tests.
- Diffusion tests such as the Kirby-Bauer assay and the Etest® are often used to estimate a pathogen's susceptibility to drugs quickly (**figures 9.2–9.4**).
- Antibiotic resistance can also be determined using protein and gene detection strategies.

9.4 Antibacterial Drugs

- Members of the penicillin family have a β-lactam ring and disrupt bacterial cell wall synthesis, resulting in cell lysis (**figure 9.5**). Natural penicillins are effective against Gram-positive bacteria only. Semisynthetic penicillins are broader spectrum and are more resistant to penicillinases.
- Cephalosporins are similar to penicillins but are typically broader spectrum and more resistant to β-lactamase attack (**figure 9.7**).
- Vancomycin is a glycopeptide antibiotic that inhibits transpeptidation during peptidoglycan synthesis (**figure 9.8**). It has a narrow spectrum, targeting only Gram positives.
- Aminoglycoside antibiotics such as streptomycin and gentamicin bind to the small ribosomal subunit, inhibit protein synthesis, and are bactericidal (**figure 9.9**).
- Tetracyclines are broad-spectrum antibiotics having a four-ring nucleus with attached groups (**figure 9.10**). They bind to the small ribosomal subunit and inhibit protein synthesis.
- Erythromycin and azithromycin are bacteriostatic macrolide antibiotics that bind to the large ribosomal subunit and inhibit protein synthesis (**figure 9.11**).
- Chloramphenicol is a broad-spectrum, bacteriostatic antibiotic that inhibits protein synthesis. It is quite toxic and rarely used clinically.
- Linezolid blocks protein synthesis by binding to the large ribosomal subunit and preventing the formation of the 70S initiation complex. Linezolid is used to treat serious infections caused by drug-resistant Gram positives.
- Sulfonamides or sulfa drugs resemble *p*-aminobenzoic acid and competitively inhibit folic acid synthesis (**figure 9.12**).
- Trimethoprim is a synthetic antibiotic that inhibits dihydrofolate reductase, which is required by organisms in the manufacture of folic acid (**figure 9.13**).
- Fluoroquinolones are a family of bactericidal synthetic drugs that inhibit DNA gyrase and thus inhibit DNA replication (**figure 9.14**).
- Rifamycins are synthetic antibiotics that block transcription by binding to bacterial RNA polymerase.

9.5 Antiviral Drugs

- Antiviral drugs interfere with critical stages in the virus life cycle or inhibit the synthesis of virus-specific nucleic acids (**figure 9.15**).
- Drug combinations (cocktails) are more effective than monotherapies when treating HIV and hepatitis C.
- There are five classes of antiretroviral agents used to treat HIV. Drug cocktails often include two nucleoside reverse transcriptase inhibitors and an integrase inhibitor (**figure 9.16**).

9.6 Antifungal Drugs

- The two major classes of antifungal agents include the polyenes and the azoles. Both disrupt the fungal plasma membranes. Polyenes, such as amphotericin B, have a low therapeutic index.
- Superficial mycoses can be treated with miconazole, ketoconazole, clotrimazole, tolnaftate, nystatin, and griseofulvin (**figure 9.17**). Amphotericin B, 5-flucytosine, and azole drugs are used for systemic mycoses.

9.7 Antiprotozoan Drugs

- Antiprotozoan drugs interfere with critical steps in nucleic acid synthesis, protein synthesis, electron transport, or folic acid synthesis.

9.8 Antimicrobial Drug Resistance Is a Public Health Threat

- Mechanisms used by microorganisms to resist the toxic effects of antimicrobial agents include changing the drug target so the agent no longer binds, modifying the drug, degrading the drug, or preventing uptake of the drug.
- Drug resistance can often limit drug effectiveness. However, it can be overcome through the judicious use of antimicrobial agents (decreasing selective pressure), use of drug cocktails (attacking more than one microbial structure or function), and use of specific, narrow-spectrum drugs to target infecting microbes.

Active Learning

1. Some advocate stockpiling the drug oseltamivir in the event of an influenza pandemic. Others point out that wealthy, Western nations would have an unfair advantage because developing nations (where the pandemic is most likely to start) would not have access to this expensive antiviral. Furthermore, some point out that indiscriminate use of the drug promotes the evolution of resistant flu strains. Given these caveats, discuss the advantages and disadvantages of stockpiling oseltamivir in the event of an influenza pandemic.
2. You are a pediatrician treating a child with an upper respiratory infection that is clearly caused by a virus. The child's parents insist that you prescribe antibiotics. They aren't leaving without them! How do you convince the child's parents that antibiotics will do more harm than good?
3. Antibiotics that target molecules not previously exploited are desperately needed. One such target is the protein FtsZ (*see chapters 3 and 7*). The small molecule 3-methoxybenzamide (3-MBA) is known to inhibit FtsZ in *Bacillus subtilis* but is not bactericidal. Using 3-MBA as a good starting point for the synthesis of a molecule that might be a potential drug candidate, how would you go about finding similar molecules that are bactericidal to bacterial pathogens?
4. Newly developed peptide antibiotics are an active area of research. Some of these small proteins target two molecules on bacterial pathogens. For example, a peptide might bind to a cell wall structure and then insert a hole in the underlying plasma membrane thereby disrupting both structures. What might be the pros and cons of a drug that targets two structures or processes? If you were to design a drug, what two bacterial targets would you pursue?
5. *Pseudomonas aeruginosa* is a Gram-negative bacterium that is particularly dangerous for cystic fibrosis patients and patients receiving mechanical ventilation. This pathogen is intrinsically resistant to a variety of antibiotics and has acquired additional antibiotic resistance genes through horizontal gene transfer. Thus there is an urgent need for new drugs that target novel *P. aeruginosa* molecules. A European group of scientists reports one such promising compound that selectively inhibits the transport of *P. aeruginosa* lipopolysaccharide (LPS) to the outer membrane. A murine (mouse) model of *P. aeruginosa* pneumonia was used in an effort to demonstrate the drug's efficacy before moving to clinical (human) trials. Discuss the questions these researchers needed to answer in their animal experiments to make a strong case for clinical trials. Think about what control experiments they would need as well as how they could quantitatively measure efficacy.

   Read the original paper: Cigana, C., et al. 2016. Efficacy of the novel antibiotic POL7001 in preclinical models of *Pseudomonas aeruginosa* pneumonia. *Antimicrob Agents Chemother.* 60:4991–5000. doi:10.1128/AAC.00390-16.

10

Introduction to Metabolism

©McGraw-Hill Education/Roger Lowenberg, photographer

Flushed Away

For one day, count the number of times the toilet in your home is flushed. Is your toilet a low-volume toilet that sends no more than 1.6 gallons out of your house with each flush? Or is it an older, water-greedy toilet? And where do the wastes you flush go?

If you live in a city, your wastes travel through a labyrinth of pipes to a wastewater treatment plant (as shown in the accompanying figure). Now do some math: Multiply the number of gallons your house flushed by the number of households in your town. That's how much wastewater is sent each day to your town's wastewater treatment plant. If you live in a large city, the amount of wastewater generated is astronomical. In New York City, for example, it's about 1.4 billion gallons per day. The drought-prone state of California produces a whopping 4 billion gallons per day!

The cost of treating wastewater so that it is safe to release into the environment is also astronomical. New York City spends an estimated $400 million each year. Thus wastewater treatment is a big item in any city's budget. But can these wastes be used to decrease the cost of treatment or, better yet, generate funds for municipalities? The answer to that question is yes, thanks to the metabolic diversity of microbes.

As we describe in chapter 43, wastewater treatment plants are home to a complex microbial community. Working together, these microbes decrease the organic matter (sewage sludge) in wastewater and generate several by-products using their diverse metabolic abilities. One of the most important is methane, which is made when organic matter is degraded under anoxic conditions. For many years, wastewater treatment facilities simply burned the methane. Increasingly, wastewater treatment plants are using the methane to heat their buildings, and some municipalities, such as New York City, are working with natural gas providers to convert sewage sludge to pipeline quality natural gas to heat homes.

Other cities and towns are selling sludge to farmers and ranchers to use as a soil amendment. Some are exploring the use of sludge for paving roads and as building materials. Converting wastewater to drinking water is already a reality in many parts of the world. In fact, it may soon be your reality.

The treatment of wastewater is only one example of how humans exploit the metabolic prowess of microbes. As we describe in chapter 41, ethanol is both a biofuel and is used in the production of beer, ale, and wine. Propionic acid flavors Swiss cheese. Carbon dioxide leavens bread, and antibiotics save the lives of humans every day.

The impact of microbial metabolites goes beyond their use in industry and food production. It is now recognized that the microbes living in and on the human body contribute compounds needed for human homeostasis. Indeed, the metabolic capabilities of both the microbiota and their human hosts are highly integrated as the result of coevolution. ▶| *The microbe-human ecosystem (chapter 34)*

Clearly, microbial metabolism is extremely important to the well-being of humans, and much of the study of microbes has focused on this aspect of their biology. Chapters 10 through 12 focus on the metabolic processes that capture (conserve) energy supplied by an organism's energy source and how it is used to synthesize the building blocks from which an organism is constructed. In this chapter, we lay the foundation for that discussion.

Readiness Check:

Based on what you have learned previously, you should be able to:

✓ Distinguish metabolism, catabolism, and anabolism
✓ Describe the structure of bacterial and archaeal cell envelopes (sections 3.3–3.5 and 4.2)
✓ Describe mitochondrial structure and function (section 5.6)

10.1 Metabolism: Important Principles and Concepts

After reading this section, you should be able to:

a. List the features common to all types of metabolism
b. Describe the three types of work carried out by cells
c. State the relationship between cellular work and energy
d. Paraphrase the first and second laws of thermodynamics
e. Use the equation for free energy change and the equation that links standard free energy change to a reaction's equilibrium constant and explain their importance
f. Infer from a standard free energy change whether a reaction is exergonic or endergonic

It would be nice to think that microbes existed for the benefit of humans, but of course they don't. Rather, they use their vast repertoire of chemical reactions to survive and reproduce, just as all organisms do. Indeed, **metabolism** is central to all life. The forces of evolution have shaped the metabolic process used by organisms for billions of years. Despite the diversity of chemical reactions that have evolved, there are several aspects of metabolism that are common to all organisms. Here we list the common features, noting where they are discussed in this chapter.

- **Life obeys the laws of thermodynamics** (this section). Organisms must obtain energy from their environment; they cannot "make" energy, it must be captured or conserved.
- **The energy cells obtain from their environment is most often conserved as ATP** (section 10.2). ATP supplies the energy needed for certain chemical reactions to proceed.
- **Oxidation-reduction (redox) reactions play a critical role in energy conservation** (section 10.3). Many of these redox reactions occur in electron transport chains (section 10.4).
- **The chemical reactions that occur in cells are organized into pathways.** The product of one reaction is the substrate for the next reaction in the pathway (section 10.5).
- **Each reaction of a pathway is catalyzed by an enzyme or a ribozyme.** Enzymes and ribozymes are critical to all life as they speed up the reactions that enable life to exist (section 10.6).
- **The functioning of biochemical pathways is regulated** so that their products are made at the correct time (section 10.7).

Cellular Work and Energy Transfers

Examine the common features of metabolism just described. It should be clear that cells must do work to survive and reproduce. Cells carry out three major types of work. **Chemical work** involves the synthesis of complex biological molecules from much simpler precursors (i.e., anabolism); energy is needed to increase the molecular complexity of a cell. **Transport work** requires energy to take up nutrients, eliminate wastes, and maintain ion balances. Energy input is needed because molecules and ions often must be transported across cell membranes against concentration and electrochemical gradients. The third type of work is **mechanical work.** Energy is required for cell motility and the movement of structures within cells, such as partitioning chromosomes during cell division.

As just indicated, cells need energy to do work. Indeed, **energy** may be defined most simply as the capacity to do work. This is because all physical and chemical processes are the result of the application or movement of energy. Organisms obtain the energy they need from an energy source present in their environment. They convert the energy it provides into a useful form.

Cellular energy is most commonly stored in the nucleotide triphosphate ATP. To understand how energy is conserved in ATP and how ATP is used to do cellular work, some knowledge of the basic principles of thermodynamics is required. The science of **thermodynamics** analyzes energy changes in a collection of matter (e.g., a cell or a plant) called a system. All other matter in the universe is called the surroundings. Thermodynamics focuses on the energy differences between the initial state and the final state of a system. It is not concerned with the rate of the process. For instance, if a pan of water is heated to boiling, only the condition of the water at the start and at boiling is important in thermodynamics, not how fast it is heated.

Laws of Thermodynamics

Two important laws of thermodynamics are critical to understanding energy transfers. The **first law of thermodynamics** says that energy can be neither created nor destroyed. The total energy in the universe remains constant, although it can be redistributed, as it is during the many energy exchanges that occur during chemical reactions. For example, heat is given off by exothermic reactions and absorbed during endothermic reactions. However, the first law alone cannot explain why heat is released by one chemical reaction and absorbed by another. Explanations for this require the **second law of thermodynamics** and a condition of matter called entropy. **Entropy** is a measure of the randomness or disorder of a system. The greater the disorder of a system, the greater its entropy. The second law states that physical and chemical processes proceed in such a way that the randomness or disorder of the universe (the system and its surroundings) increases. However, even though the entropy of the universe increases, the entropy of any given system within the universe can increase, decrease, or remain unchanged.

Two types of energy units are employed to specify the amount of energy used in or evolving from a particular process. A **calorie** (cal) is the amount of heat energy needed to raise 1 gram of water from 14.5 to 15.5°C. The amount of energy also may be expressed in terms of **joules** (J), the units of work capable of being done. One cal of heat is equivalent to 4.1840 J of work. One thousand calories, or a kilocalorie (kcal), is enough energy to boil 1.9 milliliters of water. A kilojoule is enough energy to boil about 0.44 milliliters of water or enable a person weighing 70 kilograms to climb 35 steps.

Free Energy Change Predicts the Nature of a Chemical Reaction

The first and second laws of thermodynamics can be combined in a useful equation, relating the changes in energy that can occur in chemical reactions and other processes.

$$\Delta G = \Delta H - T\,\Delta S$$

ΔG is the change in free energy, ΔH is the change in enthalpy, T is the temperature in Kelvin (°C + 273), and ΔS is the change in entropy occurring during the reaction. The change in **enthalpy** is the change in heat content. Cellular reactions occur under conditions of constant pressure and volume. Thus the change in enthalpy is about the same as the change in total energy during the reaction. The **free energy change** is the amount of energy in a system (or cell) available to do useful work at constant temperature and pressure. Therefore the change in entropy (ΔS) is a measure of the proportion of the total energy change that the system cannot use in performing work. Free energy and entropy changes do not depend on how the system gets from start to finish. A reaction will occur spontaneously—that is, without any external cause—if the free energy of the system decreases during the reaction or, in other words, if ΔG is negative. It follows from the equation that a reaction with a large positive change in entropy will normally tend to have a negative ΔG value and therefore occur spontaneously. A decrease in entropy will tend to make ΔG more positive and the reaction less favorable.

It is helpful to think of the relationship between entropy (ΔS) and change in free energy (ΔG) in terms that are more concrete. Consider the Greek myth of Sisyphus, king of Corinth. For his assorted crimes against the gods, he was condemned to roll a large boulder to the top of a steep hill for all eternity. This represents a very negative change in entropy—a boulder poised at the top of a hill is neither random nor disordered—and this activity (reaction) has a very positive ΔG. That is to say, Sisyphus had to put a lot of energy into the system. Unfortunately for Sisyphus, as soon as the boulder was at the top of the hill, it spontaneously rolled back down. This represents a positive change in entropy and a negative ΔG. Sisyphus did not need to put energy into the system. He could simply stand at the top of the hill and watch the reaction proceed.

The change in free energy has a definite, concrete relationship to the direction of chemical reactions. Consider this simple reaction.

$$A + B \rightleftharpoons C + D$$

If molecules A and B are mixed, they will combine to form the products C and D. Eventually C and D will become concentrated enough to combine and produce A and B at the same rate as C and D are formed. The reaction is now at **equilibrium:** The rates in both directions are equal and no further net change occurs in the concentrations of reactants and products. This situation is described by the **equilibrium constant (K_{eq})**, relating the equilibrium concentrations of products and substrates to one another.

$$K_{eq} = \frac{[C]\,[D]}{[A]\,[B]}$$

If the equilibrium constant is greater than one, the products are in greater concentration than the reactants at equilibrium; that is, the reaction tends to go to completion as written.

Exergonic reactions	**Endergonic reactions**
$A + B \rightleftharpoons C + D$	$A + B \rightleftharpoons C + D$
$K_{eq} = \frac{[C]\,[D]}{[A]\,[B]} > 1.0$	$K_{eq} = \frac{[C]\,[D]}{[A]\,[B]} < 1.0$
$\Delta G^{\circ\prime}$ is negative.	**$\Delta G^{\circ\prime}$ is positive.**

Figure 10.1 The Relationship of $\Delta G^{\circ\prime}$ to the Equilibrium of Reactions. Note the differences between exergonic and endergonic reactions.

MICRO INQUIRY *Which reaction would release heat? Why?*

The equilibrium constant of a reaction is directly related to its change in free energy. When the free energy change for a process is determined at carefully defined standard conditions of concentration, pressure, pH, and temperature, it is called the **standard free energy change** (ΔG°). If the pH is set at 7.0 (which is close to the pH of living cells), the standard free energy change is indicated by the symbol $\Delta G^{\circ\prime}$. The change in standard free energy may be thought of as the maximum amount of energy available from the system for useful work under standard conditions. Using $\Delta G^{\circ\prime}$ values allows comparisons of reactions without considering variations in ΔG due to differences in environmental conditions. The relationship between $\Delta G^{\circ\prime}$ and K_{eq} is given by this equation.

$$\Delta G^{\circ\prime} + -2.303RT \cdot \log K_{eq}$$

R is the gas constant (1.9872 cal/mole-degree or 8.3145 J/mole-degree), and T is the absolute temperature. Inspection of this equation shows that when $\Delta G^{\circ\prime}$ is negative, the equilibrium constant is greater than one and the reaction goes to completion as written. It is said to be an **exergonic reaction** (**figure 10.1**). In an **endergonic reaction,** $\Delta G^{\circ\prime}$ is positive and the equilibrium constant is less than one. That is, the reaction is not favorable, and little product will be formed at equilibrium under standard conditions. Keep in mind that the $\Delta G^{\circ\prime}$ value shows only where the reaction lies at equilibrium, not how fast the reaction reaches equilibrium.

Comprehension Check

1. Suppose a bacterium was doing the following: synthesizing peptidoglycan, rotating its flagellum and swimming, and secreting siderophores. What type of work is the bacterium doing in each case?
2. In your own words, summarize the first and second laws of thermodynamics.
3. Define entropy, enthalpy, and free energy. How do exergonic and endergonic reactions differ?
4. Is a chemical reaction with a large negative $\Delta G^{\circ\prime}$ value endergonic or exergonic? What would this indicate about its equilibrium constant? If displaced from equilibrium, would the reaction proceed rapidly to completion? Would much or little free energy be made available?

10.2 ATP: The Major Energy Currency of Cells

After reading this section, you should be able to:

a. Draw a simple schematic that illustrates the structure of ATP
b. Describe ATP's role as a coupling agent that links exergonic and endergonic reactions
c. Compare ATP's phosphate transfer potential to that of glucose 6-phosphate and phosphoenolpyruvate (PEP) and relate this to ATP's function in cells
d. Describe the energy cycle observed in all organisms
e. Describe the metabolic functions of three nucleoside triphosphates (other than ATP)

Energy is released from a cell's energy source in exergonic reactions (i.e., those reactions with a negative ΔG). Rather than wasting this energy, much of it is trapped in a practical form that allows it to be used to do work. Cells carry out endergonic reactions (e.g., anabolism), and the energy captured by the cell is used to drive these reactions to completion. In living organisms, the most commonly used practical form of energy is the nucleotide **adenosine 5′-triphosphate (ATP; figure 10.2)**. In a sense, cells carry out certain processes so that they can "earn" ATP, which they "spend" in carrying out other processes. Thus ATP is often referred to as the cell's energy currency. In the cell's economy, ATP serves as the link between exergonic reactions and endergonic reactions (**figure 10.3**).

What makes ATP suited for its role as energy currency? ATP is a high-energy molecule. That is, it is hydrolyzed almost completely to the products **adenosine diphosphate (ADP)** and orthophosphate (P_i), with a strongly exergonic $\Delta G^{\circ\prime}$ of −7.3 kcal/mole (−30.5 kJ/mole).

$$ATP + H_2O \rightleftharpoons ADP + P_i + H^+$$

The energy released is used to power endergonic reactions (figure 10.3).

The very negative $\Delta G^{\circ\prime}$ of ATP hydrolysis is related to another important characteristic of ATP: its ability to transfer a phosphoryl group to another molecule. ATP is said to have a **high phosphate transfer potential** because it readily donates a phosphoryl group to other molecules. Note in **table 10.1** that there are other phosphorylated compounds found in cells. These molecules are generated during catabolism of organic molecules such as glucose. Some of these molecules have even higher phosphate transfer potentials than ATP. The fact that ATP does not have the highest phosphate transfer potential means that it can easily be made by cells from ADP, using molecules such as phosphoenolpyruvate (PEP) as the source of the phosphoryl group. This mechanism for making ATP is called **substrate-level phosphorylation.** ▶| *Glucose to pyruvate: the first step (section 11.4)*

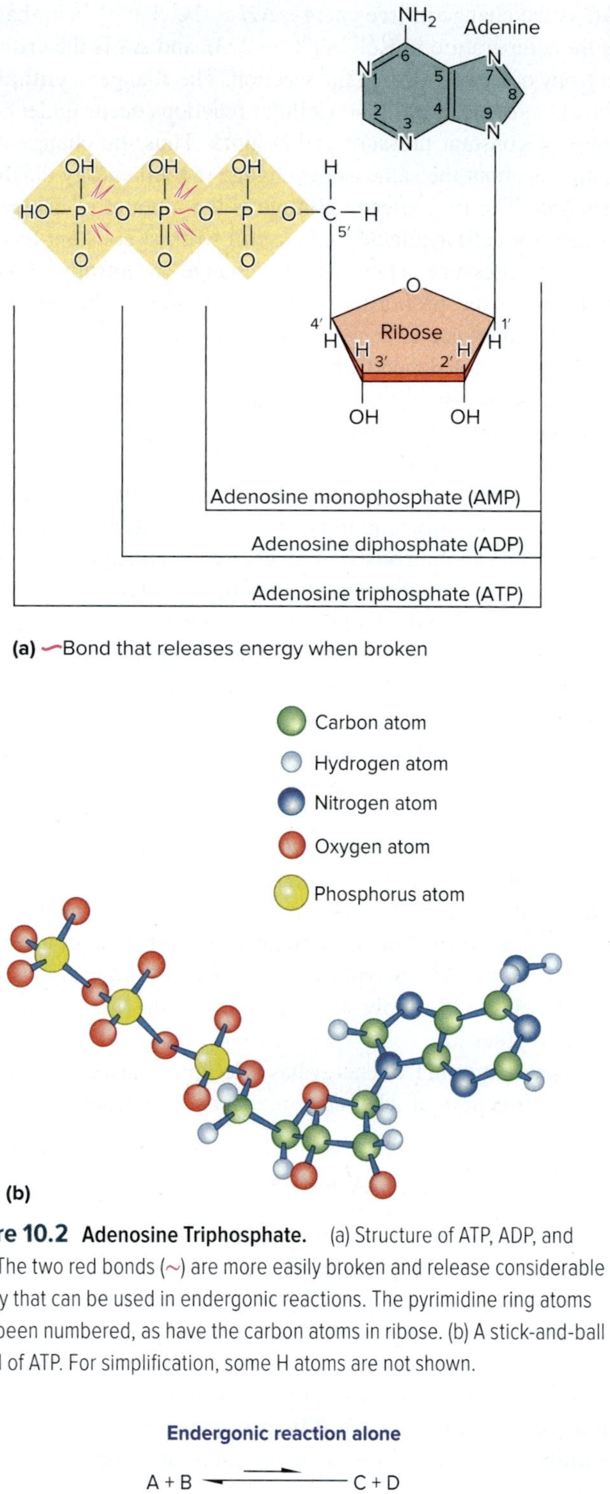

Figure 10.2 Adenosine Triphosphate. (a) Structure of ATP, ADP, and AMP. The two red bonds (~) are more easily broken and release considerable energy that can be used in endergonic reactions. The pyrimidine ring atoms have been numbered, as have the carbon atoms in ribose. (b) A stick-and-ball model of ATP. For simplification, some H atoms are not shown.

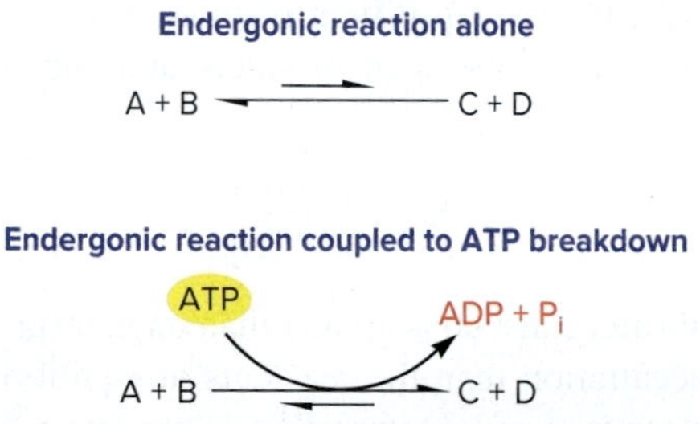

Figure 10.3 ATP as a Coupling Agent. ATP is formed by exergonic reactions and then used to drive endergonic reactions.

Table 10.1 Phosphate Transfer Potential of Common Phosphorylated Compounds[1]

Phosphorylated Molecule	$\Delta G^{\circ\prime}$ of Hydrolytic Removal of Phosphate (KJ/mol)	Phosphate Transfer Potential
High-Energy Phosphorylated Compounds		
Phosphoenolpyruvate[2]	−61.9	61.9
1, 3-bisphosphoglycerate[2]	−49.3	49.3
ATP (hydrolysis to AMP)	−45.6	45.6
ATP (hydrolysis to ADP)	−30.5	30.5
Low-Energy Phosphorylated Compounds		
Glucose 6-phosphate	−13.8	13.8
Glycerol 1-phosphate	−9.2	9.2

1 Phosphate transfer potential is defined as the negative of the standard free energy change ($\Delta G^{\circ\prime}$) for the hydrolytic removal of phosphate from the phosphorylated molecule.
2 Phosphoenolpyruvate and 1,3-bisphosphoglycerate are generated during the catabolism of organic molecules such as glucose.

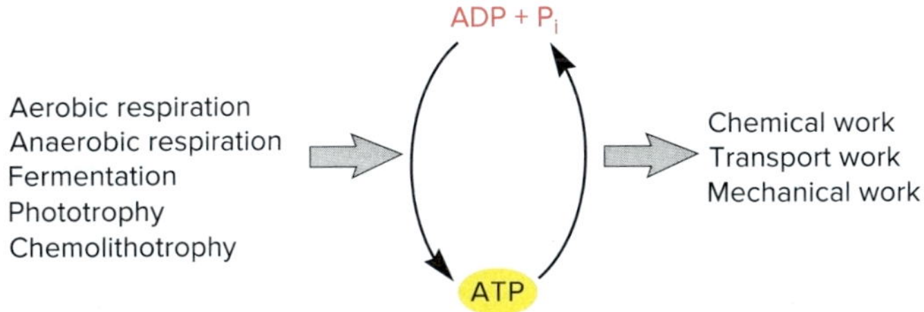

Figure 10.4 The Cell's Energy Cycle. ATP is formed from energy made available during aerobic respiration, anaerobic respiration, fermentation, chemolithotrophy, and phototrophy. Its breakdown to ADP and phosphate (P_i) makes chemical, transport, and mechanical work possible.

ATP, ADP, and P_i form an energy cycle. As seen in **figure 10.4**, the energy released from an energy source is used to synthesize ATP from ADP and P_i. When ATP is hydrolyzed, the energy released drives endergonic processes such as anabolism, transport, and mechanical work. The mechanisms for synthesizing ATP are described in more detail in chapter 11.

ATP is the major energy currency for cells, but it is not the only energy currency. Other nucleoside triphosphates (NTPs) have major roles in metabolism. Guanosine 5′-triphosphate (GTP) supplies some of the energy used during protein synthesis. Cytidine 5′-triphosphate (CTP) is used during lipid synthesis, and uridine 5′-triphosphate (UTP) is used for the synthesis of peptidoglycan and other polysaccharides. ▶| *Translation in bacteria (section 13.7); Lipid synthesis (section 12.7); Synthesis of carbohydrates (section 12.4)*

Comprehension Check

1. Why is ATP called a high-energy molecule? How is this related to its phosphate transfer potential?
2. Describe the energy cycle and ATP's role in it. What characteristics of ATP make it suitable for this role?

10.3 Redox Reactions: Reactions of Central Importance in Metabolism

After reading this section, you should be able to:

a. Describe a redox reaction, noting the role of the two half reactions and identifying the electron donor, electron acceptor, and conjugate redox pairs of the reaction
b. Relate the standard reduction potential of a conjugate redox pair to its tendency to act as an electron-donating half reaction
c. State the equation that links standard free energy change to the difference in standard reduction potentials of the two half reactions of a redox reaction
d. Predict which molecule will act as an electron donor, which molecule will act as an electron acceptor, and the relative amount of energy released by a redox reaction, using the standard reduction potentials of the reaction's conjugate redox pairs

Free energy changes are related to the equilibria of all chemical reactions, including the equilibria of oxidation-reduction reactions. The release of energy from an energy source normally involves oxidation-reduction reactions. **Oxidation-reduction (redox) reactions** are those in which electrons move from an **electron donor** to an **electron acceptor.**[1] As electrons transfer from donor to acceptor, the donor becomes less energy rich and the acceptor becomes more energy rich. Thus electrons can be thought of as packets of energy. The more electrons a molecule has and is able to donate in a redox reaction, the more energy rich the molecule is. This explains why molecules such as glucose, which can donate up to 24 electrons in redox reactions, are such excellent sources of energy.

1 In redox reactions, the electron donor is often called the reducing agent or reductant because it is donating electrons to the acceptor and thus reducing it. The electron acceptor is called the oxidizing agent or oxidant because it is removing electrons from the donor and oxidizing it.

Each redox reaction consists of two half reactions. One half reaction functions as the electron-donating half (i.e., an oxidation reaction), and the other functions as the electron-accepting half (i.e., the reduction). By convention, half reactions are written as reductions. As shown in **table 10.2,** the left side of each half reaction consists of a molecule that can accept electrons and the number (n) of electrons (e^-) it accepts, while the right side includes the molecule it becomes after accepting the electrons. This molecule is referred to as a donor, because it has electrons it can give up. The acceptor and donor of a half reaction are referred to as a **conjugate redox pair.**

$$\text{Acceptor} + ne^- \rightleftharpoons \text{donor}$$

The equilibrium constant for a redox half reaction is called the **standard reduction potential** (E_0) and is a measure of the tendency of the donor of a half reaction to lose electrons. By convention, the standard reduction potentials for half reactions, such as those in table 10.2, are determined at pH 7 and are represented by E'_0. Standard reduction potentials are measured in volts, a unit of electrical potential or electromotive force. Therefore conjugate redox pairs are a potential source of energy.

The reduction potential has a concrete meaning. Conjugate redox pairs with more negative reduction potentials will spontaneously donate electrons to pairs with more positive potentials and greater affinity for electrons. Thus electrons tend to move from donors at the top of the list in table 10.2 to acceptors at the bottom because the latter have more positive potentials. This may be expressed visually in the form of an electron tower in which the most negative reduction potentials are at the top (**figure 10.5**). Electrons "fall down" the tower from donors higher in the tower (i.e., those having more negative potentials) to acceptors lower in the tower (i.e., those having more positive potentials).

Consider the case of the electron acceptor **nicotinamide adenine dinucleotide (NAD^+)**. The NAD^+/NADH conjugate redox pair has a very negative E'_0, and NADH can therefore give electrons to many acceptors, including O_2.

$$NAD^+ + 2H^+ + 2e^- \rightleftharpoons NADH + H^+ \qquad E'_0 = -0.32 \text{ volts}$$

$$\tfrac{1}{2}O_2 + 2H^+ + 2e^- \rightleftharpoons H_2O \qquad E'_0 = +0.82 \text{ volts}$$

Because the reduction potential of NAD^+/NADH is more negative than that of $\frac{1}{2}O_2/H_2O$, electrons flow without added energy from NADH (the donor) to O_2 (the acceptor), as shown in figure 10.5. The redox reaction is shown here.

$$NADH + H^+ + \tfrac{1}{2}O_2 \rightarrow H_2O + NAD^+$$

The relatively negative E'_0 of the NAD^+/NADH pair also means that the pair stores more potential energy than redox pairs with less negative (or more positive) E'_0 values. It follows that when electrons move from a donor to an acceptor with a more positive redox potential, free energy is released. The $\Delta G^{\circ\prime}$ of the reaction is directly related to the magnitude of the difference between the reduction potentials of the two couples ($\Delta E'_0$). The larger the $\Delta E'_0$, the greater the amount of free energy made available, as is evident from the equation

$$\Delta G^{\circ\prime} = -nF \cdot \Delta E'_0$$

in which n is the number of electrons transferred, F is the Faraday constant (23,062 cal/mole-volt; 96,480 J/mole-volt), and $\Delta E'_0$ is the E'_0 of the acceptor minus the E'_0 of the donor. For every 0.1-volt change in $\Delta E'_0$, there is a corresponding 4.6-kcal (19.3-kJ) change in $\Delta G^{\circ\prime}$ when a two-electron transfer takes place. This is similar to the relationship of $\Delta G^{\circ\prime}$ and K_{eq} in other chemical reactions: The larger

Table 10.2 Selected Biologically Important Half Reactions

Half Reaction	E'_0 (Volts)[1]
$2H^+ + 2e^- \rightarrow H_2$	−0.42
Ferredoxin (Fe^{3+}) + e^- → ferredoxin (Fe^{2+})	−0.42
$NAD(P)^+ + H^+ + 2e^- \rightarrow NAD(P)H$	−0.32
$S + 2H^+ + 2e^- \rightarrow H_2S$	−0.27
Acetaldehyde + $2H^+ + 2e^-$ → ethanol	−0.20
$Pyruvate^-$ + $2H^+ + 2e^-$ → $lactate^{2-}$	−0.19
$FAD + 2H^+ + 2e^- \rightarrow FADH_2$	−0.18[2]
$Oxaloacetate^{2-}$ + $2H^+ + 2e^-$ → $malate^{2-}$	−0.17
$Fumarate^{2-}$ + $2H^+ + 2e^-$ → $succinate^{2-}$	0.03
Cytochrome *b* (Fe^{3+}) + e^- → cytochrome *b* (Fe^{2+})	0.08
Ubiquinone + $2H^+ + 2e^-$ → ubiquinone H_2	0.10
Cytochrome *c* (Fe^{3+}) + e^- → cytochrome *c* (Fe^{2+})	0.25
Cytochrome *a* (Fe^{3+}) + e^- → cytochrome *a* (Fe^{2+})	0.29
Cytochrome a_3 (Fe^{3+}) + e^- → cytochrome a_3 (Fe^{2+})	0.35
$NO_3^- + 2H^+ + 2e^- \rightarrow NO_2^- + H_2O$	0.42
$NO_2^- + 8H^+ + 6e^- \rightarrow NH_4^+ + 2H_2O$	0.44
$Fe^{3+} + e^- \rightarrow Fe^{2+}$	0.77[3]
$\frac{1}{2}O_2 + 2H^+ + 2e^- \rightarrow H_2O$	0.82

1 E'_0 is the standard reduction potential at pH 7.0.
2 The value for FAD/$FADH_2$ applies to the free cofactor because it can vary considerably when bound to an apoenzyme.
3 The value for free Fe, not Fe complexed with proteins (e.g., cytochromes).

MICRO INQUIRY *For the $Fe^{3+} + e^- \rightarrow Fe^{2+}$ half reaction, which form of iron is the acceptor and which is the donor?*

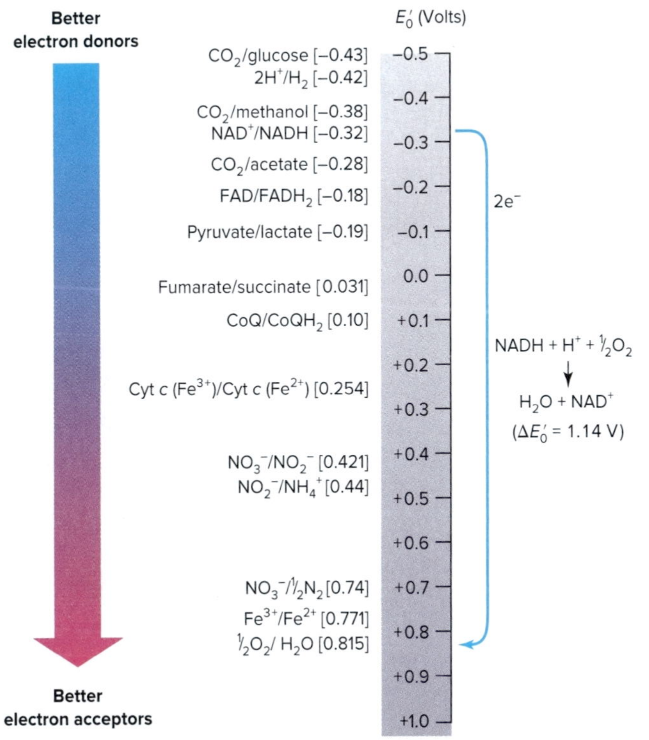

Figure 10.5 Electron Movement and Reduction Potentials. Electrons spontaneously move from donors higher on the tower (more negative potentials) to acceptors lower on the tower (more positive potentials). That is, the donor is higher on the tower than the acceptor. For example, NADH will donate electrons to oxygen and form water in the process. Some typical conjugate redox pairs are shown on the left, and their reduction potentials are given in brackets.

MICRO INQUIRY *Why would energy be required to move electrons "up" the tower, from water to nitrate, for example?*

the equilibrium constant, the greater the $\Delta G^{\circ\prime}$. The difference in reduction potentials between NAD^+/NADH and $\frac{1}{2}O_2/H_2O$ is 1.14 volts, a large $\Delta E_0'$ value. When electrons move from NADH to O_2, a large amount of free energy is made available and can be used to synthesize ATP and do other work. *How NAD^+ Works*

10.4 Electron Transport Chains: Sets of Sequential Redox Reactions

After reading this section, you should be able to:

a. List the molecules that are commonly found in electron transport chains (ETCs) and indicate if they transfer electrons and protons or just electrons
b. Arrange electron carriers in an ETC using their standard reduction potentials
c. Indicate the location of ETCs in bacterial, archaeal, and eukaryotic cells

We have focused our attention on the reduction of O_2 by NADH because NADH plays a central role in the metabolism of many organisms. For instance, glucose is a common organic energy source. As glucose is catabolized, it is oxidized. Many of the electrons released from glucose are accepted by NAD^+, reducing it to NADH, which transfers the electrons to O_2. However, it does not do so directly. Instead, the electrons are transferred to O_2 via a series of additional electron carriers. These electron carriers are membrane-bound and organized into a system called an **electron transport chain** (**ETC**). An ETC is similar to a bucket brigade. Each carrier represents a person receiving pails of water (electrons) that are destined for the fire (terminal electron acceptor). Each pail of water is passed down the line, just as electrons are passed from carrier to carrier. As soon as one person hands off a pail of water to the next person in line, a new bucket is received. Likewise, as electrons flow through the ETC, each carrier is sequentially reduced (given the pail full of water) and then reoxidized (passes the pail on to the next person in the brigade), and is ready to accept more electrons as catabolism continues.

The first electron carrier in an ETC has the most negative E_0', and each successive carrier is slightly less negative (**figure 10.6**). Thus electrons are transferred spontaneously from one carrier to the next. The carriers direct the electrons to the terminal electron acceptor (in this case, O_2). The E_0' of electron carriers must be relatively close to prevent the cell from losing energy by random reductions of other molecules in the cell. It may be helpful to think of electron carriers in an ETC as stepping stones. If the distance between each stone is too far, it is very hard to move from stone to stone in a controlled fashion.

Electron transport chains are within or associated with plasma membranes and intracytoplasmic membranes of bacterial and archaeal cells (figure 10.6). In eukaryotes they are localized to the internal membranes of mitochondria and chloroplasts. They are important to almost all types of energy-conserving processes, as we discuss in chapter 11.

The electron carriers associated with ETCs differ in terms of their chemical nature and the way they carry electrons. NADH, which donates electrons to the ETC, and its chemical relative **nicotinamide adenine dinucleotide phosphate** (**NADPH**), which donates electrons to other molecules during biosynthetic reactions, contain a nicotinamide ring (**figure 10.7**). This ring accepts two electrons and one proton from a donor (e.g., an intermediate formed during the catabolism of glucose), and a second proton is released. **Flavin adenine dinucleotide** (**FAD**) and **flavin mononucleotide** (**FMN**) bear two electrons and two protons on the complex ring system shown in **figure 10.8**. Proteins bearing FAD and FMN are often called flavoproteins. **Coenzyme Q** (**CoQ**) or **ubiquinone** is a quinone that transports two electrons and two protons (**figure 10.9**). **Cytochromes** and several other carriers use iron atoms to transport one electron at a time. In cytochromes, the iron atoms are part of a heme group or other similar iron-porphyrin rings (**figure 10.10**). There are several different cytochromes,

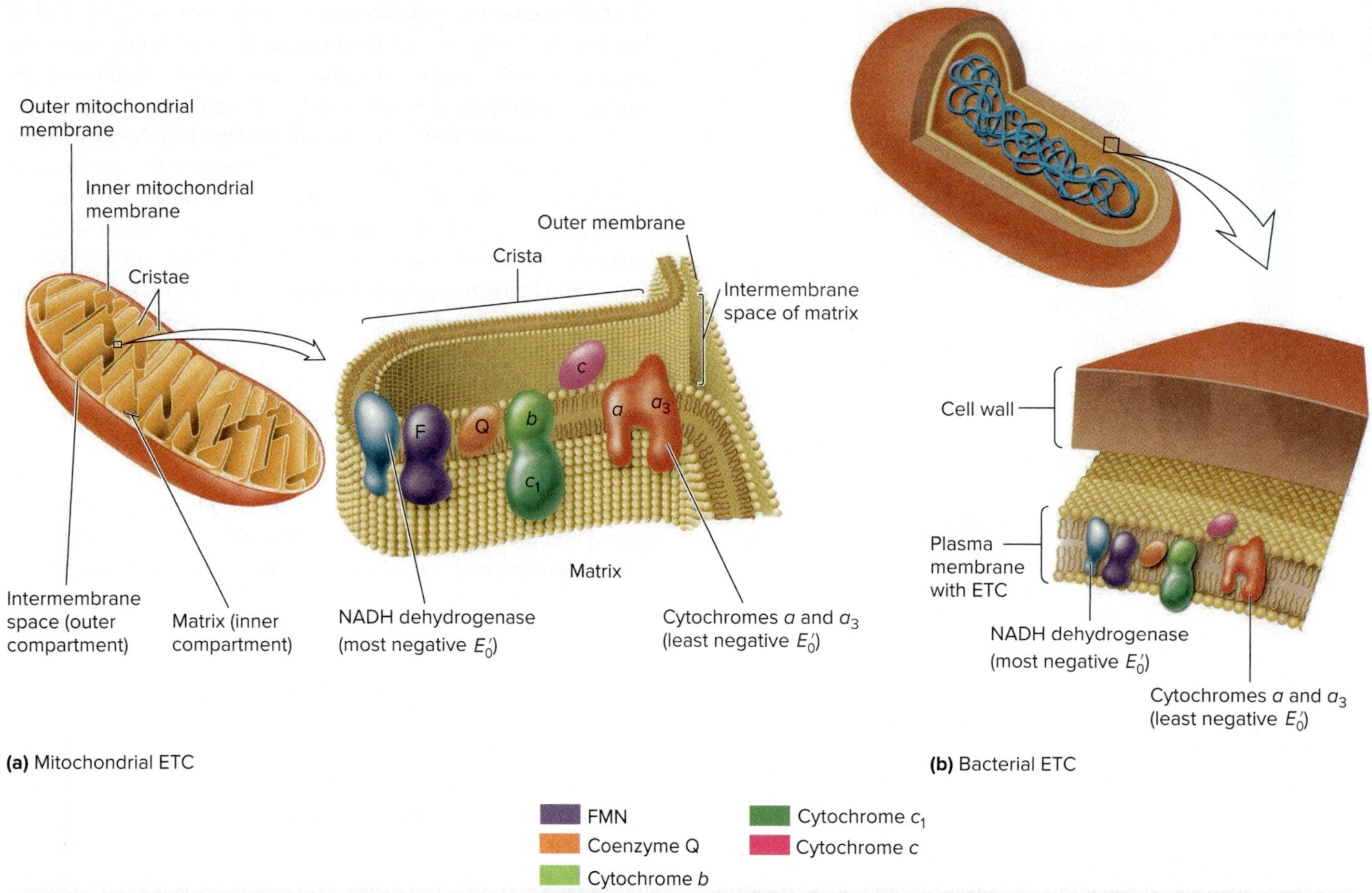

Figure 10.6 Electron Transport Chains. Electron transport chains (ETCs) are associated with membranes. Electrons flow from the electron carrier having the most negative reduction potential to the carrier having the most positive reduction potential. During respiratory processes (e.g., aerobic respiration; *see figure 11.2*), the source of electrons is a reduced molecule such as glucose and an exogenous molecule such as oxygen serves as the terminal electron acceptor (*see figures 11.11–11.13*). (a) The mitochondrial ETC. (b) A typical bacterial ETC.

MICRO INQUIRY *Refer to figure 10.5 and determine the E_0' for NAD^+/NADH and coenzyme Q/$CoQH_2$. Suggest a plausible E_0' value for FMN.*

each consisting of a protein and an iron-porphyrin ring. Some iron-containing electron-carrying proteins lack a heme group and are called **nonheme iron proteins.** They are often referred to as **iron-sulfur** (**Fe-S**) **proteins** because the iron is associated with sulfur atoms (**figure 10.11**). **Ferredoxin** is an Fe-S protein active in photosynthesis-related electron transport. Like cytochromes, Fe-S proteins carry only one electron at a time. The differences in the number of electrons and protons transported by electron carriers are of great importance to the operation of ETCs, and we discuss them further in chapter 11. *Proton Pump*

Comprehension Check

1. How is the direction of electron flow between conjugate redox pairs related to the standard reduction potential and the release of free energy?
2. When electrons flow from the NAD^+/NADH conjugate redox pair to the $\frac{1}{2}O_2/H_2O$ redox pair, does the reaction begin with NAD^+ or with NADH? What is produced—O_2 or H_2O?
3. Which among the following would be the best electron donor? Which would be the worst? Ubiquinone/ubiquinone H_2, NAD^+/NADH, FAD/$FADH_2$, NO_3^-/NO_2^-. Explain your answers.
4. In general terms, how is $\Delta G^{\circ\prime}$ related to $\Delta E'_0$? What is the $\Delta E'_0$ when electrons flow from the NAD^+/NADH redox pair to the Fe^{3+}/Fe^{2+} redox pair? How does this compare to the $\Delta E'_0$ when electrons flow from the Fe^{3+}/Fe^{2+} conjugate redox pair to the $\frac{1}{2}O_2/H_2O$ pair? Which will yield the largest amount of free energy to the cell?
5. Name and briefly describe the major electron carriers found in cells. Why is NADH a good electron donor? Why is ferredoxin an even better electron donor?

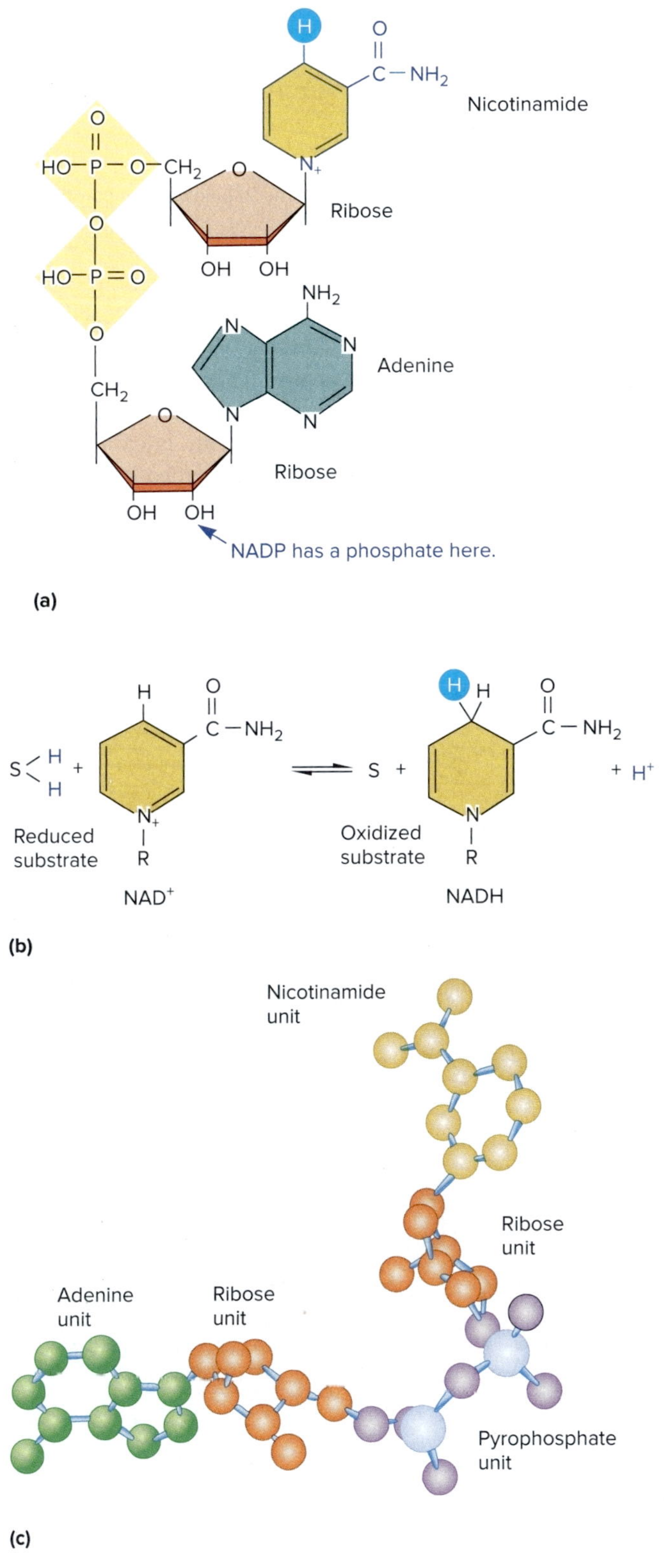

Figure 10.7 The Structure and Function of NAD. (a) The structure of NAD^+ and $NADP^+$. $NADP^+$ differs from NAD^+ in having an extra phosphate on one of its ribose sugar units. (b) NAD^+ can accept two electrons and one proton from a reduced substrate (S). In this example, the electrons and proton are supplied by the hydrogen atoms of the reduced substrate; recall that each hydrogen atom consists of one proton and one electron. (c) Stick-and-ball model of NAD^+.

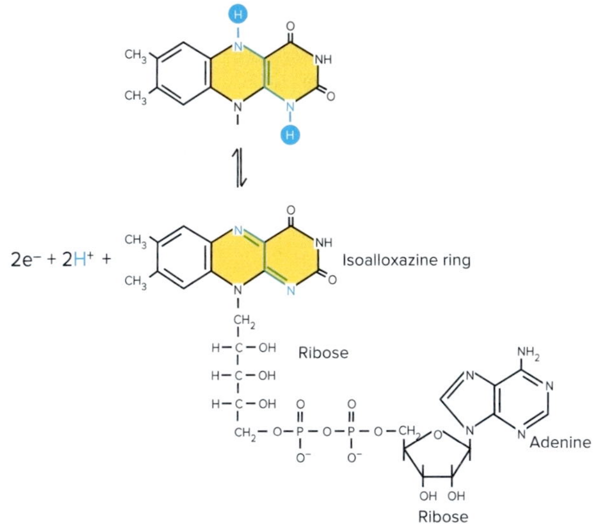

Figure 10.8 The Structure and Function of FAD. The vitamin riboflavin is composed of the isoalloxazine ring and its attached ribose sugar. FMN is riboflavin phosphate. The portion of the ring directly involved in oxidation-reduction reactions is in color.

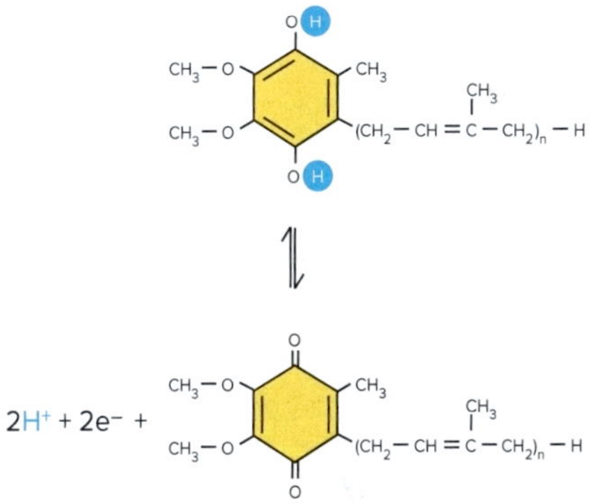

Figure 10.9 The Structure and Function of Coenzyme Q (Ubiquinone). The length of the side chain varies among organisms from n = 6 to n = 10.

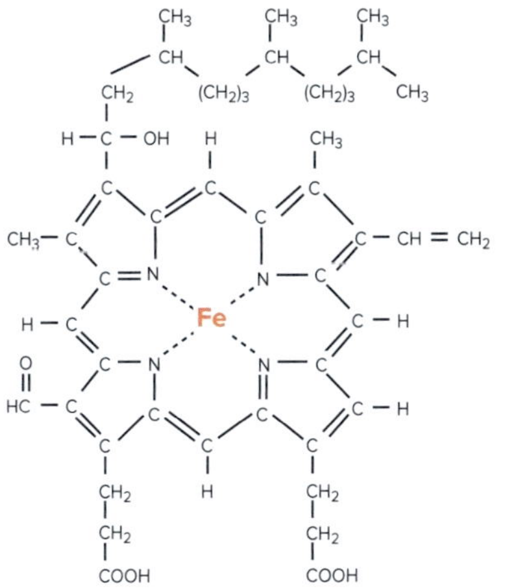

Figure 10.10 The Structure of Heme. Heme is composed of a porphyrin ring and an attached iron atom. It is the nonprotein component of many cytochromes. The iron atom alternatively accepts and releases an electron.

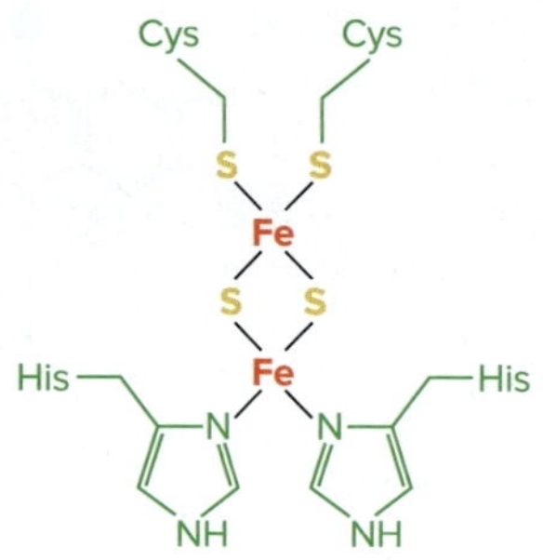

Figure 10.11 The Iron-Sulfur Cluster of an Iron-Sulfur Protein. This Fe-S protein consists of four S atoms and two Fe atoms. The cluster is bound to the polypeptide portion of the protein (green) by bonds between one Fe and the two S atoms provided by cysteine (Cys) residues and by bonds between the other Fe and two N atoms provided by histidine (His) residues. The iron atoms are responsible for transferring one electron at a time to the next carrier in an ETC.

10.5 Biochemical Pathways: Sets of Linked Chemical Reactions

After reading this section, you should be able to:

a. Describe the components of a biochemical pathway and how they are organized

Organisms carry out a myriad of chemical reactions, and the products of these reactions are called **metabolites.** Reactions are organized into **biochemical pathways,** which can take various forms. Some are linear, in which the first molecule in the pathway is often termed the starting molecule or the substrate of the pathway (**figure 10.12*a***). The last molecule is called the end product. Those metabolites in between are called intermediates. As seen in figure 10.12*b*, some linear pathways are branched and can yield more than one product. Biochemical pathways can also

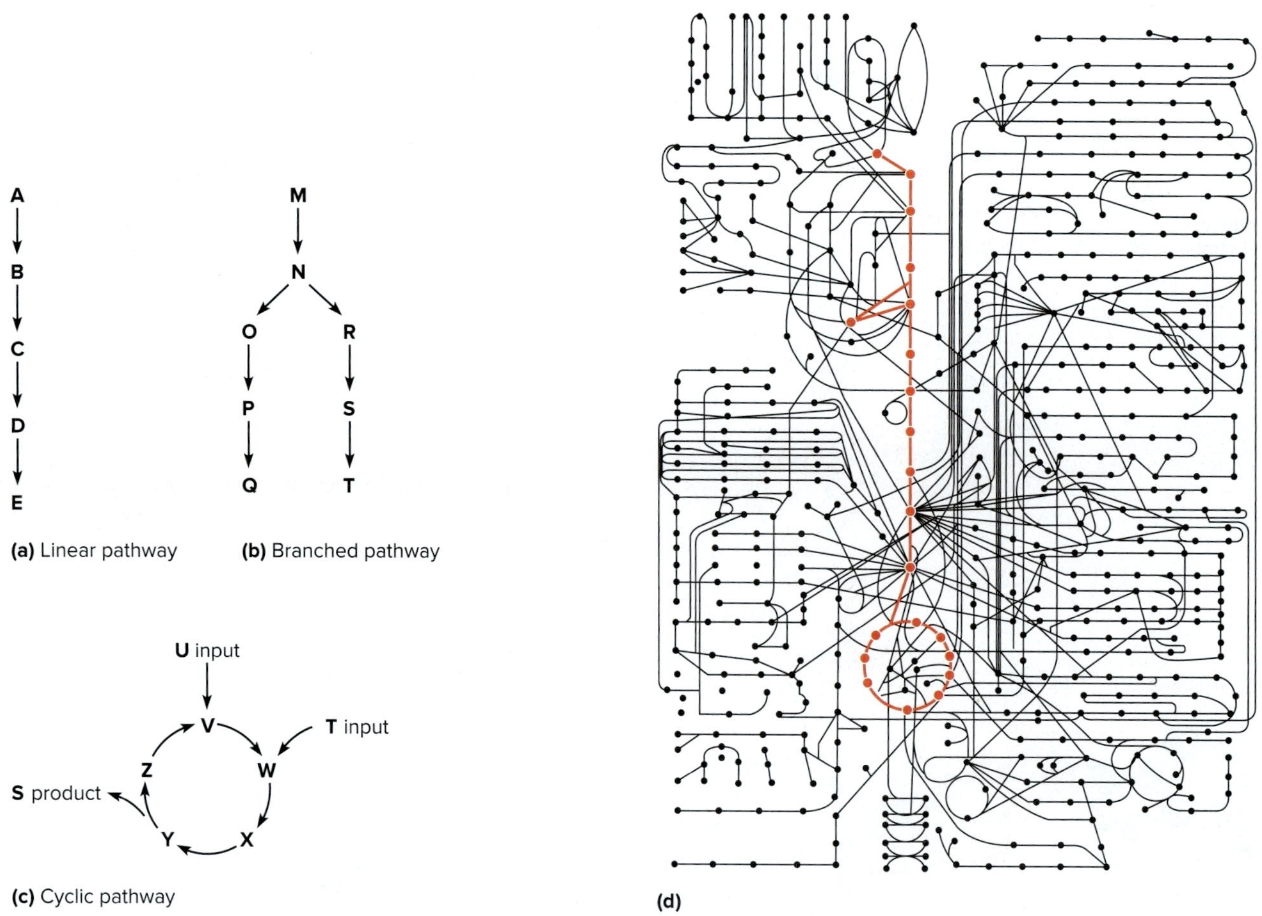

Figure 10.12 Chemical Reactions Carried Out by Cells Are Organized into Biochemical Pathways. Biochemical pathways take two main forms: linear and cyclic. (a) Molecules in the linear forms are distinguished based on their placement in the pathway. The first molecule (A) is the starting molecule (substrate) of the pathway. Those in the middle of the pathway are called intermediates (molecules B, C, and D). The last molecule of the pathway (E) is the product of the pathway. (b) Some linear biochemical pathways are branched and thus have more than one end product (molecules Q and T). (c) The molecules present in biochemical cycles are intermediates of the pathway (molecules V, W, X, Y, and Z). Biochemical cycles also generate products (molecule S); however, to keep the cycle running, there must be inputs into the cycle (molecules U and T). (d) Metabolic pathways are interconnected and form a complex network. In this representation of a metabolic network, each ball represents a metabolite formed by reactions in the network. Each line represents the enzyme-catalyzed reaction that converts one metabolite into another.

be cycles (figure 10.12*c*). In this case, all molecules in the pathway can be thought of as intermediates. For the cycle to continue running, there must be inputs into it. In both types of pathways, each reaction is represented by an arrow and is catalyzed by an enzyme or ribozyme, as discussed in section 10.6.

Biochemical pathways are often illustrated as though they exist in isolation of each other. Unfortunately, this is very misleading (especially to students). In reality, biochemical pathways are dynamic and form a complex network (figure 10.12*d*). Thus, an intermediate of a pathway may be diverted from one pathway to another pathway. As long as the starting molecules or inputs of the pathway are available and some end product is needed, these molecules will flow into and out of the many pathways that function in cells. Microbiologists are often concerned with metabolite flux. **Metabolite flux** is the rate of turnover of a metabolite—that is, the rate at which a metabolite is formed and then used. Metabolite flux can be used as a measure of pathway activity and to understand metabolic networks.

10.6 Enzymes and Ribozymes Speed Up Cellular Chemical Reactions

After reading this section, you should be able to:

a. Describe the function and chemical makeup of enzymes
b. Distinguish apoenzyme from holoenzyme and prosthetic group from coenzyme
c. Draw a diagram that shows the effect of an enzyme on the activation energy of a chemical reaction
d. Describe the effects of substrate concentration, pH, and temperature on enzyme activity
e. Differentiate competitive and noncompetitive inhibitors of enzymes
f. Compare and contrast ribozymes and enzymes
g. List two examples of ribozymes and the reactions they catalyze

As we discussed in section 10.1, an exergonic reaction is one with a negative $\Delta G^{\circ\prime}$ and an equilibrium constant greater than one. An exergonic reaction proceeds to completion in the direction written (i.e., toward the right of the equation). Nevertheless, reactants for an exergonic reaction often can be combined with no obvious result. For instance, if a polysaccharide such as starch is mixed in water, the hydrolysis of the starch into its component monosaccharides (glucose) is exergonic and will occur spontaneously, given enough time. However, the time needed is very long. Even if an organic chemist carried out this reaction in 6 moles/liter (M) HCl and at 100°C, it would still take several hours to go to completion. A cell, on the other hand, can accomplish the same reaction at neutral pH, at a much lower temperature, and in just fractions of a second. Cells can do this because they manufacture biological catalysts that speed up chemical reactions. Most of these catalysts are proteins called enzymes. Other catalysts are RNA molecules termed ribozymes. Enzymes and ribozymes are critically important to cells, since most biological reactions occur very slowly without them. Indeed, enzymes and ribozymes make life possible.

Enzyme Structure

Enzymes are protein catalysts that have great specificity for the reaction catalyzed, the molecules acted on, and the products they yield. A **catalyst** is a substance that increases the rate of a chemical reaction without being permanently altered itself. The reacting molecules are called **substrates,** and the substances formed are the **products.** Enzymes may be placed in one of six general classes and usually are named in terms of the substrates they act on and the type of reaction catalyzed (**table 10.3**). ▶| *Proteins (appendix I)*

Many enzymes are composed only of proteins. However, some are composed of two parts: a protein component called the **apoenzyme** and a nonprotein component called a **cofactor.** Cofactors include metal ions and a variety of organic molecules. The complete enzyme consisting of the apoenzyme and its

Table 10.3 Enzyme Classification

Type of Enzyme	Reaction Catalyzed by Enzyme	Example of Reaction
Oxidoreductase	Oxidation-reduction reactions	Lactate dehydrogenase: Pyruvate + NADH + H^+ ⇌ lactate + NAD^+
Transferase	Reactions involving the transfer of chemical groups between molecules	Aspartate carbamoyltransferase: Aspartate + carbamoylphosphate ⇌ carbamoylaspartate + phosphate
Hydrolase	Hydrolysis of molecules	Glucose 6-phosphatase: Glucose 6-phosphate + H_2O → glucose + P_i[1]
Lyase	Breaking of C—C, C—O, C—N and other bonds by a means other than hydrolysis	Fumarase: L-malate ⇌ fumarate + H_2O
Isomerase	Reactions involving isomerizations	Alanine racemase: L-alanine ⇌ D-alanine
Ligase	Joining of two molecules using ATP (or the energy of other nucleoside triphosphates)	Glutamine synthetase: Glutamate + NH_3 + ATP → glutamine + ADP + P_i

1 P_i is inorganic phosphate.

cofactor is called the **holoenzyme.** If the cofactor is firmly attached to the apoenzyme, it is a **prosthetic group.** If the cofactor is loosely attached and can dissociate from the apoenzyme after products have been formed, it is called a **coenzyme.** Many vitamins required by humans serve as coenzymes or as their precursors (e.g., riboflavin is incorporated into FAD). Coenzymes often carry one of the products of a chemical reaction to another enzyme or transfer chemical groups from one substrate to another (**figure 10.13**).

How Enzymes Speed Up Reactions

It is important to keep in mind that enzymes increase the rates of reactions but do not alter their equilibrium constants. If a reaction is endergonic, the presence of an enzyme will not shift its equilibrium so that more products are formed. Enzymes simply speed up the rate at which a reaction proceeds toward its final equilibrium. *How Enzymes Work*

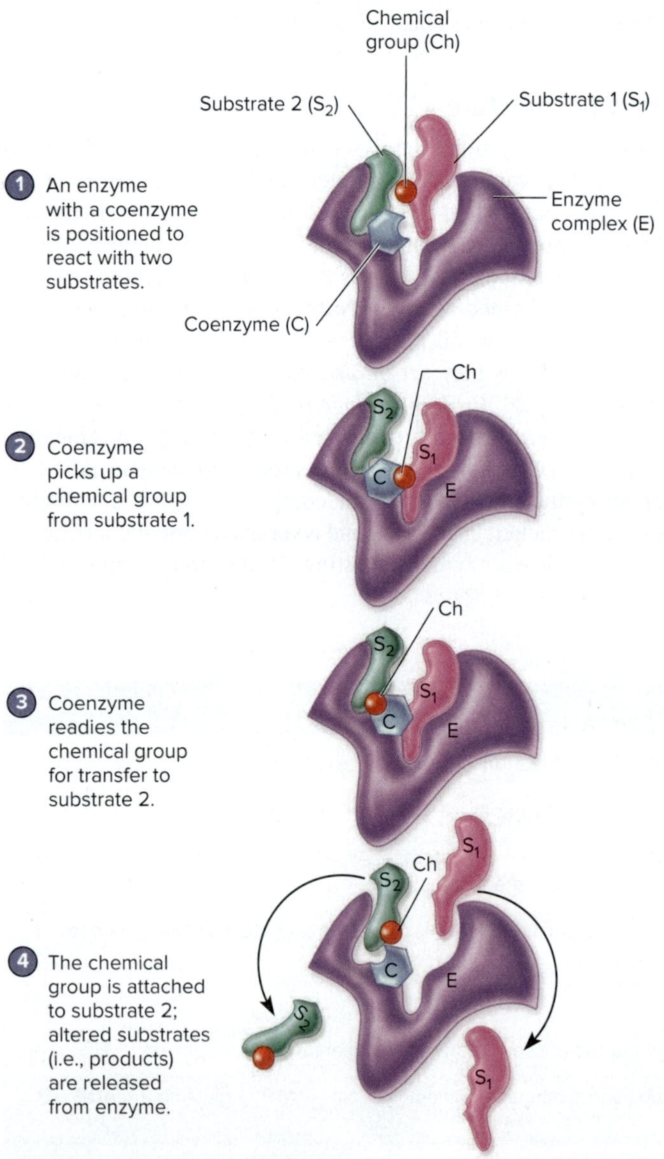

Figure 10.13 Coenzymes as Carriers.

How do enzymes catalyze reactions? Some understanding of the mechanism can be gained by considering the course of a simple exergonic chemical reaction.

$$A + B \rightleftharpoons C + D$$

When molecules A and B approach each other to react, they form a transition-state complex, which resembles both the substrates and the products (**figure 10.14**). **Activation energy** is required to bring the reacting molecules together in the correct way to reach the transition state. The transition-state complex can then resolve to yield the products C and D. The difference in free energy level between reactants and products is $\Delta G^{\circ\prime}$. Thus the equilibrium in our example lies toward the products because $\Delta G^{\circ\prime}$ is negative (i.e., the products are at a lower energy level than the substrates).

As seen in figure 10.14, A and B will not be converted to C and D without energy; the amount of energy is equivalent to the activation energy. Enzymes accelerate reactions by lowering the activation energy; therefore more substrate molecules will have sufficient energy to come together and form products. Even though the equilibrium constant (or $\Delta G^{\circ\prime}$) is unchanged, equilibrium is reached more rapidly in the presence of an enzyme because of this decrease in activation energy.

Researchers have worked hard to discover how enzymes lower the activation energy of reactions. Enzymes bring substrates together at a specific location in the enzyme called the

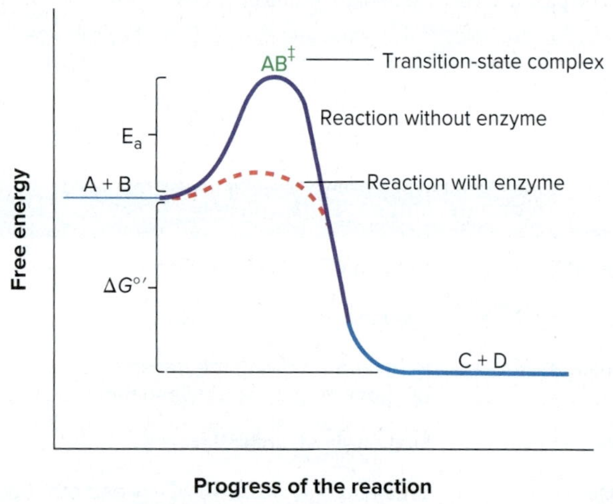

Figure 10.14 Enzymes Lower the Energy of Activation. This figure traces the course of a chemical reaction in which A and B are converted to C and D. The transition-state complex is represented by $AB^{\ddagger}$, and the activation energy required to reach it, by E_a. The red line represents the course of the reaction in the presence of an enzyme. Note that the activation energy is much lower in the enzyme-catalyzed reaction.

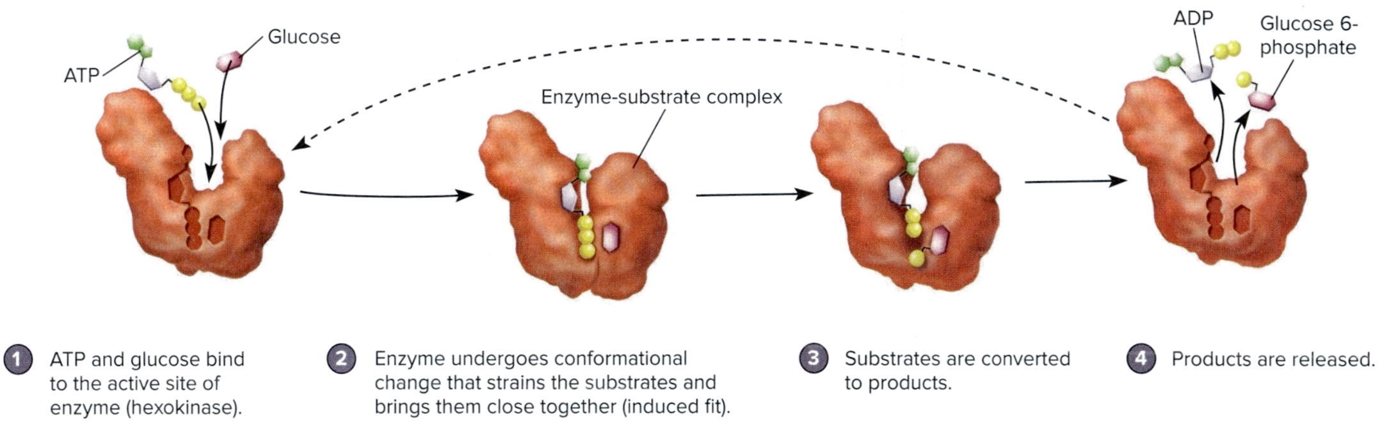

Figure 10.15 The Induced Fit Model of Enzyme Function. Shown is the reaction between glucose and ATP catalyzed by the enzyme hexokinase.

active or **catalytic site** to form an enzyme-substrate complex. How the enzyme and substrate interact still has not been fully elucidated. However, the induced fit model describes the interaction for many enzymes. In the induced fit model, the enzyme changes shape when it binds the substrate so that the active site surrounds and precisely fits the substrate. This mechanism is used by hexokinase, as illustrated in **figure 10.15**. The formation of an enzyme-substrate complex can lower the activation energy in many ways. For example, by bringing the substrates together at the active site, the enzyme is, in effect, concentrating them so the reaction can occur more quickly. An enzyme does not simply concentrate its substrates, however. It also binds them so they are correctly oriented with respect to each other. Such an orientation lowers the amount of energy required by the substrates to reach the transition state. These and other catalytic site activities speed up a reaction by hundreds of thousands of times.

Substrate Concentration Affects Enzyme Activity

Enzyme activity varies in response to substrate concentration. Examination of **figure 10.16** shows that the curve describing the change in enzyme activity (usually expressed as the rate of product formation) in response to substrate concentration is a hyperbola. Why is this so? At very low substrate concentrations, an enzyme makes product slowly because it seldom contacts a substrate molecule. If more substrate molecules are present, an enzyme binds substrate more often, and the velocity of the reaction is greater than at a lower substrate concentration. Eventually further increases in substrate concentration do not result in a greater reaction velocity because the available enzyme molecules are binding substrate and converting it to product as rapidly as possible. That is, the enzyme is saturated with substrate and operating at maximal velocity (V_{max}).

The substrate concentrations in cells are often low. Therefore it is useful to know the substrate concentration an enzyme needs to function adequately. By convention, the substrate concentration required for the enzyme to achieve half-maximal velocity is used as a measure of the apparent affinity of an enzyme for its substrate. This value is called the **Michaelis constant** (K_m).

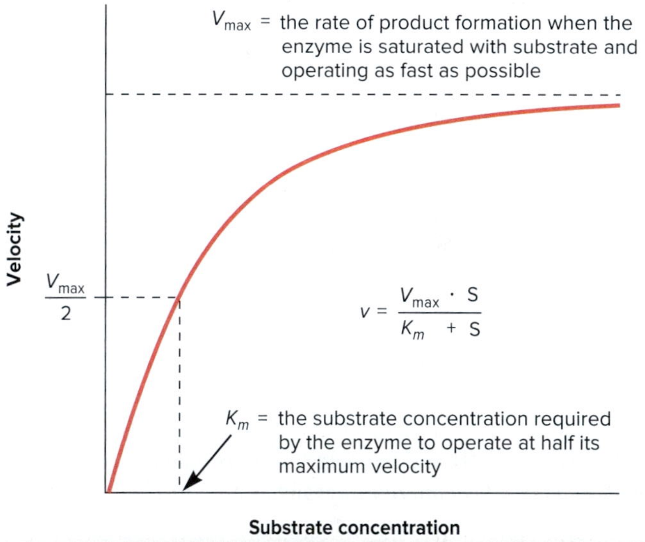

Figure 10.16 The Effect of Substrate Concentration on Enzyme Activity. Enzyme activity increases as substrate concentration increases until a maximum velocity is reached (V_{max}). The resulting curve is described by the Michaelis-Menten equation, which was derived by biochemists Leonor Michaelis (1875–1949) and Maud Menten (1879–1960). It relates reaction velocity (v) to substrate concentration (S) using V_{max} and the Michaelis constant (K_m).

MICRO INQUIRY *Will an enzyme with a relatively high K_m have a high or low affinity for its substrate? Explain.*

The lower the K_m value, the lower the substrate concentration at which an enzyme catalyzes its reaction. Enzymes with a low K_m value are said to have a high affinity for their substrates.

Enzyme Denaturation Destroys Enzyme Activity

Enzyme activity is changed not only by substrate concentration but also by alterations in pH and temperature. Each enzyme functions most rapidly at its specific pH optimum. When the pH deviates too greatly from its optimum, enzyme activity slows and

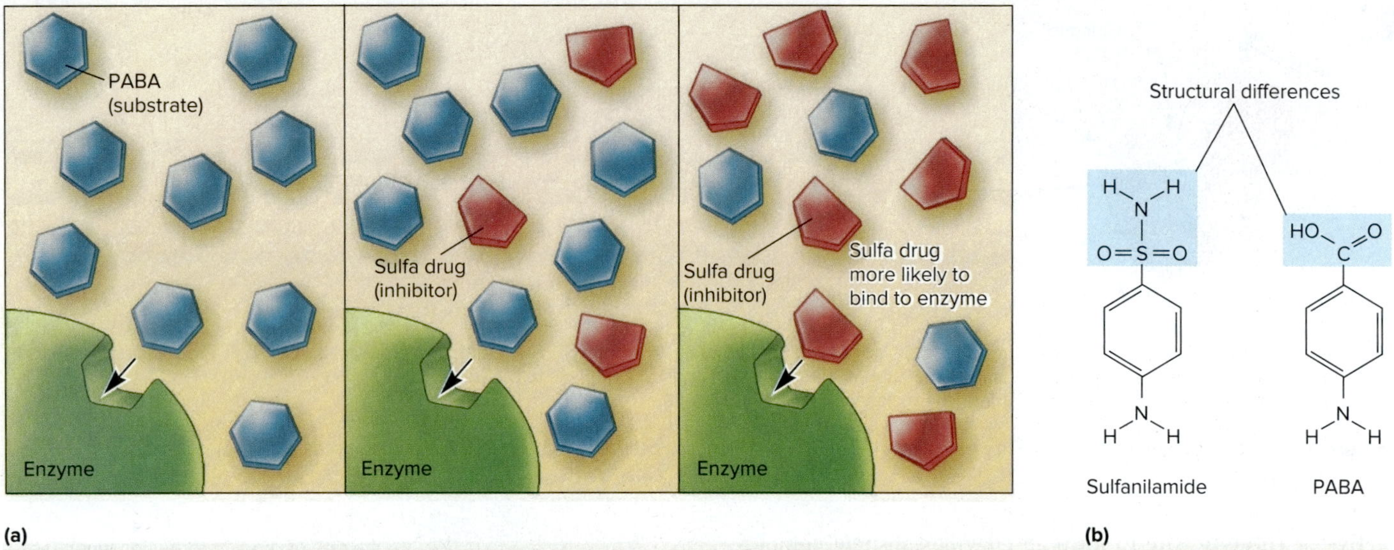

Figure 10.17 Competitive Inhibition of Enzyme Activity. (a) A competitive inhibitor is usually similar in shape to the normal substrate of the enzyme and therefore can bind the active site of the enzyme. This prevents the substrate from binding, and the reaction is blocked. (b) Structure of sulfanilamide, a structural analogue of PABA. PABA is the substrate of an enzyme involved in folic acid biosynthesis. When sulfanilamide binds the enzyme, activity of the enzyme is inhibited and synthesis of folic acid is stopped.

MICRO INQUIRY *Based on this figure, explain the notion of a minimum inhibitory concentration of sulfanilamide.*

the enzyme may be damaged. Enzymes likewise have temperature optima for maximum activity. If the temperature rises too much above the optimum, an enzyme's structure will be disrupted and its activity lost. This phenomenon is known as **denaturation.** The pH and temperature optima of a microorganism's enzymes often reflect the pH and temperature of its habitat. ◀ *Environmental factors affect microbial growth (section 7.5)*

Enzyme Inhibition

Microorganisms can be poisoned by a variety of chemicals (e.g., cyanide), and many of the most potent poisons are enzyme inhibitors. A **competitive inhibitor** directly competes with the substrate at an enzyme's catalytic site and prevents the enzyme from forming product (**figure 10.17**). Competitive inhibitors usually resemble normal substrates, but they cannot be converted to products.

Competitive inhibitors are important in the treatment of many microbial diseases. Sulfa drugs such as sulfanilamide (figure 10.17*b*) resemble *p*-aminobenzoate (PABA), a molecule used in the formation of the coenzyme folic acid. Sulfa drugs compete with PABA for the catalytic site of an enzyme involved in folic acid synthesis. This blocks the production of folic acid and inhibits the growth of organisms that require its synthesis. Humans are not harmed because they do not synthesize folic acid but rather obtain it in their diet. ◀ *Metabolic antagonists (section 9.4)*

Noncompetitive inhibitors affect enzyme activity by binding to the enzyme at a location that is not the active site. This alters the enzyme's shape, rendering it inactive or less active. These inhibitors are called noncompetitive because they do not directly compete with the substrate. Heavy metals such as mercury frequently are noncompetitive inhibitors of enzymes.

Ribozymes: Catalytic RNA Molecules

Biologists once thought that all cellular reactions were catalyzed by proteins. However, in the early 1980s Thomas Cech and Sidney Altman discovered that some RNA molecules also catalyze reactions. Catalytic RNA molecules are called **ribozymes.** One important ribozyme is located in ribosomes and is responsible for catalyzing peptide bond formation between amino acids during protein synthesis. Other ribozymes catalyze self-splicing, in which they cut themselves and then join segments of themselves back together (**table 10.4**). Just as with enzymes, the shape of a ribozyme is essential to catalytic efficiency. Ribozymes even have Michaelis-Menten kinetics (figure 10.16). ▶ *Translation in bacteria (section 13.7)*

Table 10.4 Examples of Self-Splicing Ribozymes

Function	Where Observed
Splicing of pre-rRNA	*Tetrahymena* spp.
Splicing of mitochondrial rRNA and mRNA	Numerous fungi
Splicing of chloroplast tRNA, rRNA, and mRNA	Plants and algae
Splicing of viral mRNA	Viruses (e.g., T4, hepatitis delta virus)

Comprehension Check

1. What is the difference between an apoenzyme and a holoenzyme? What are the two types of cofactors? What roles do cofactors play in enzyme function?
2. Draw a diagram showing the effect enzymes have on the activation energy of the reactions they catalyze and explain why enzymes do not change the equilibria of the reactions they catalyze.
3. How does enzyme activity change with substrate concentration? How does the Michaelis constant (K_m) relate to enzyme function?
4. What special properties might an enzyme isolated from a psychrophilic bacterium have? Will enzymes need to lower the activation energy more or less in thermophiles or psychrophiles?
5. What are competitive and noncompetitive inhibitors, and how do they inhibit enzymes?
6. How are enzymes and ribozymes similar? How do they differ?

10.7 Metabolism Must Be Regulated to Maintain Homeostasis

After reading this section, you should be able to:

a. List the three general approaches cells use to regulate metabolism
b. Describe metabolic channeling and one example of how it is accomplished
c. Distinguish allosteric regulation and covalent modification
d. Describe the structure of an allosteric enzyme
e. Create a concept map that illustrates how positive allosteric and negative allosteric effectors regulate the activity of an enzyme
f. List the three chemical groups commonly used to covalently modify an enzyme and its activity
g. Explain how feedback inhibition is used to control the functioning of biosynthetic pathways
h. Predict which enzymes of a biochemical pathway are likely to be regulated by either allosteric control or covalent modification
i. Develop a model illustrating how feedback inhibition can be used to regulate a multiply branched biosynthetic pathway

Microorganisms must regulate their metabolism to conserve raw materials and energy, and to maintain balance among various cell components. Microbes often live in environments where the nutrients, energy sources, and physical conditions change rapidly. Therefore microbes must continuously monitor internal and external conditions, and respond accordingly. This involves activating or inactivating metabolic pathways as needed. Metabolic pathways can be regulated in three major ways: (1) metabolic channeling, (2) regulation of the synthesis of a particular enzyme (often referred to as regulation of gene expression), and (3) directly controlling the activity of enzymes, referred to as posttranslational regulation. Although we focus here on individual pathways, figure 10.12*d* reminds us that pathways are interconnected. Thus regulating one pathway affects another. Cells often use multiple regulatory approaches to coordinate their complex metabolic activities.

Metabolic Channeling

Metabolic channeling influences pathway activity by localizing metabolites and enzymes to specific parts of a cell. One of the most common metabolic channeling mechanisms is **compartmentation,** the distribution of enzymes and metabolites among separate cell structures or organelles. Compartmentation is particularly important in eukaryotic microorganisms with their many membrane-bound organelles. For example, fatty acid catabolism is located within the mitochondrion, whereas fatty acid synthesis occurs in the cytosol. Microcompartments and the periplasm in Gram-negative bacteria are also examples of compartmentation. Compartmentation makes possible the simultaneous but separate operation and regulation of similar pathways. Furthermore, pathway activities can be coordinated through regulation of the transport of metabolites and coenzymes between cell compartments. ◀ *Microcompartments (section 3.6)*

Regulation of Gene Expression

The regulation of the synthesis of a particular enzyme involves changing the rate of transcription or translation to control the amount of an enzyme present in the cell. When the enzyme is not needed, very little if any of the enzyme is synthesized. Conversely, when the enzyme is needed, more enzyme molecules are synthesized. Regulation of gene expression is a relatively slow response to changes in the cell's state, but it saves the cell considerable energy and raw materials. Cells have evolved numerous mechanisms for regulating gene expression. We save the discussion of these mechanisms for chapters 14 and 15.

Posttranslational Regulation of Enzyme Activity

In contrast to regulation of gene expression, the direct stimulation or inhibition of the activity of critical enzymes rapidly alters pathway activity. It is often called **posttranslational regulation** because it occurs after the enzyme has been synthesized. A number of posttranslational regulatory mechanisms are known. Some are irreversible; for instance, cleavage of a protein can either increase or decrease its activity. Others, such as allosteric regulation and covalent modification, are reversible.

Allosteric Regulation

Most regulatory enzymes are **allosteric enzymes.** The activity of an allosteric enzyme is altered by a small molecule known as an **allosteric effector.** The effector binds reversibly by noncovalent forces to a **regulatory site** separate from the catalytic site and causes a change in the shape (conformation) of the enzyme (**figure 10.18**). The activity of the catalytic site is altered as a result. A positive effector increases enzyme activity, whereas a negative effector decreases activity (i.e., inhibits the enzyme).

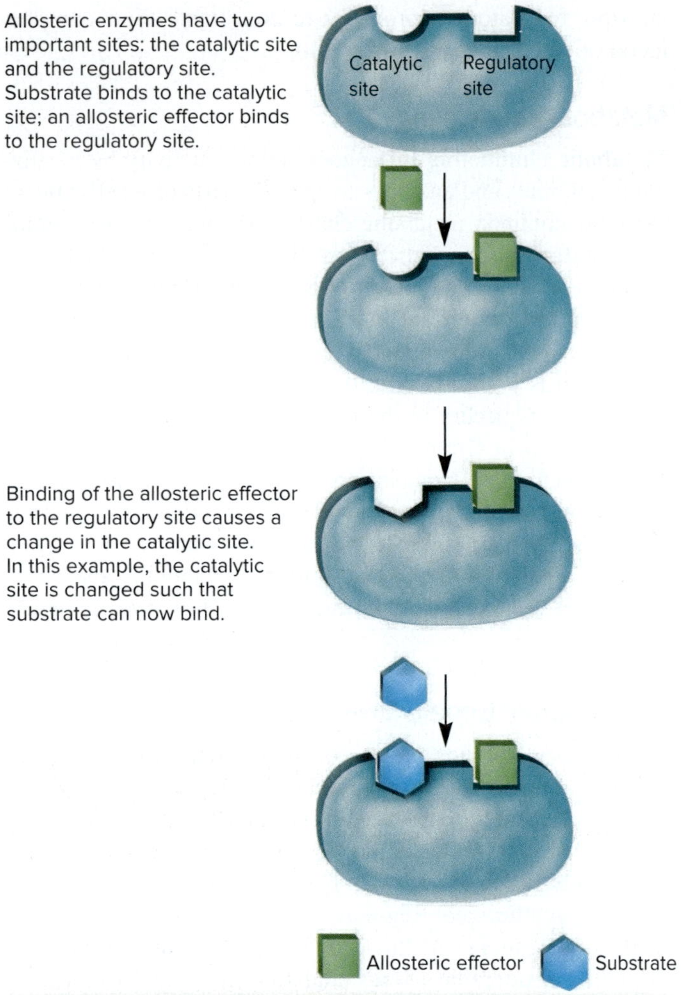

Figure 10.18 Allosteric Regulation. The structure and function of an allosteric enzyme.

MICRO INQUIRY *Is this allosteric effector a positive or negative effector? Is this true of all allosteric effectors?*

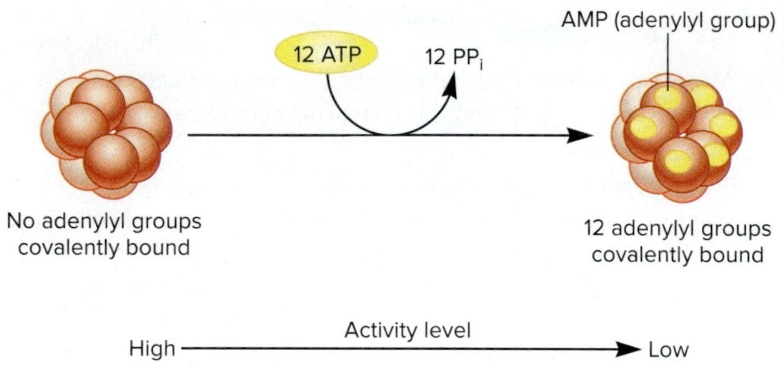

Figure 10.19 Regulation of Glutamine Synthetase Activity by Covalent Modification. Glutamine synthetase consists of 12 subunits. Six subunits form a ring, and one ring (darker color) is above the other (lighter color). Not all of the lower ring subunits are shown. Each subunit can be adenylylated. Only the adenylyl groups on the upper ring are shown. As the number of adenylyl groups increases, the activity of the enzyme decreases.

These changes in activity often result from alterations in the apparent affinity of the enzyme for its substrate, but changes in maximum velocity also can occur.

Covalent Modification of Enzymes

Regulatory enzymes also can be switched on and off by **reversible covalent modification.** Usually this occurs through the addition and removal of a particular chemical group, typically a phosphoryl, methyl, or adenylyl group.

One of the most intensively studied regulatory enzymes is *Escherichia coli*'s glutamine synthetase, an enzyme involved in nitrogen assimilation. It is a large, complex enzyme consisting of 12 subunits, each of which can be covalently modified by an adenylic acid residue (**figure 10.19**). When an adenylic acid residue is attached to all of its 12 subunits, glutamine synthetase is not very active. Removal of AMP groups produces more active deadenylylated glutamine synthetase, and glutamine is formed. ▶| *Inorganic nitrogen assimilation (section 12.5)*

Using covalent modification for the regulation of enzyme activity has some advantages. These interconvertible enzymes often are also allosteric. For instance, glutamine synthetase also is regulated allosterically. Because each form can respond differently to allosteric effectors, systems of covalently modified enzymes are able to respond to more stimuli in varied and sophisticated ways. Regulation can also be exerted on the enzymes that catalyze the covalent modifications, which adds a second level of regulation to the system.

Feedback Inhibition

The rate of many metabolic pathways is adjusted through control of a pacemaker enzyme, one that catalyzes the slowest or rate-limiting reaction in the pathway. Because other reactions proceed more rapidly than the pacemaker reaction, changes in the activity of this enzyme directly alter the speed with which the entire pathway operates. Usually the first reaction in a pathway is catalyzed by a pacemaker enzyme. The end product of the pathway often inhibits this regulatory enzyme, a process known as **feedback inhibition** or **end product inhibition.** Feedback inhibition ensures balanced production of a pathway end product. If the end product becomes too concentrated, it inhibits the regulatory enzyme and slows its own synthesis. As the concentration of the end product decreases, pathway activity again increases and more product is formed. In this way, feedback inhibition matches end product supply with demand. *A Biochemical Pathway, Feedback Inhibition of Biochemical Pathways*

Frequently a biosynthetic pathway branches to form more than one end product. In such a situation, the synthesis of all end products must be coordinated precisely. It would not do to have one end product present in excess while another is lacking. Branching biosynthetic pathways usually achieve a balance among end products by using regulatory enzymes at

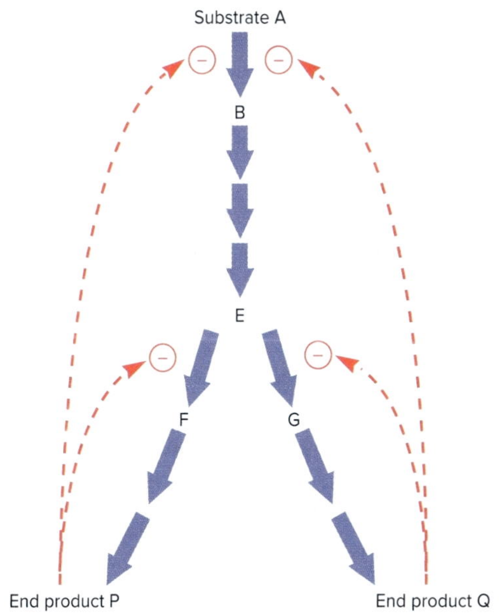

Figure 10.20 Feedback Inhibition. Feedback inhibition in a branching pathway with two end products. The branch-point enzymes, those catalyzing the conversion of intermediate E to F and G, are regulated by feedback inhibition. Products P and Q also inhibit the initial reaction in the pathway. A colored line with a minus sign at one end indicates that an end product, P or Q, inhibits the enzyme catalyzing the step next to the minus.

MICRO INQUIRY *Why does inhibition of the pacemaker enzyme by end product P also shut off production of end product Q?*

branch points (**figure 10.20**). If an end product is present in excess, it often inhibits the first enzyme on the branch leading to its formation, thereby regulating its own formation without affecting the synthesis of other products. In figure 10.20, notice that both products also inhibit the initial enzyme in the pathway. An excess of one product slows the flow of carbon (substrate A in figure 10.20) into the whole pathway and also inhibits the appropriate branch-point enzyme. Because less carbon is required when only one branch is functioning, feedback inhibition of the initial pacemaker enzyme helps match the supply with the demand in branching pathways. The regulation of branched pathways is often made even more sophisticated by the presence of **isoenzymes (isozymes)**, different forms of an enzyme that catalyze the same reaction. In the pathway shown in figure 10.20, end product P would control one isoenzyme and end product Q would control a different isoenzyme of the first enzyme in the pathway. If Q were present in excess, it would affect only its isoenzyme. The P isoenzyme would still function, and the pathway would still operate, but at a decreased level. Thus an initial pacemaker step may be catalyzed by several isoenzymes, each under separate control. Therefore, an excess of a single end product reduces pathway activity but does not completely block pathway function because some isoenzymes are still active.

Comprehension Check

1. Define the terms *metabolic channeling* and *compartmentation*. How are they involved in the regulation of metabolism?
2. Define *allosteric enzyme* and *allosteric effector*.
3. How can regulatory enzymes be influenced by reversible covalent modification?
4. What is a pacemaker enzyme? Feedback inhibition? How does feedback inhibition automatically adjust the concentration of a pathway end product?
5. Why are regulatory enzymes often located at pathway branch points? What are isoenzymes, and what is their role in pathway regulation?

Key Concepts

10.1 Metabolism: Important Principles and Concepts

- Despite the tremendous metabolic diversity observed in the microbial world, there are common features to all types of metabolisms. These are (1) life obeys the laws of thermodynamics, (2) ATP is the major form of energy currency for cells, (3) redox reactions are important in energy-conserving processes, (4) cellular chemical reactions are organized into biochemical pathways, (5) cellular chemical reactions are catalyzed by enzymes or ribozymes, and (6) the functioning of biochemical pathways is regulated.
- Living cells carry out three major kinds of work: chemical work of biosynthesis, transport work, and mechanical work. Energy is required to do cellular work and can be defined as the capacity to do work.
- Thermodynamics is the field that analyzes energy changes in systems such as cells. The first law of thermodynamics states that energy is neither created nor destroyed. The second law of thermodynamics states that changes occur in such a way that the randomness or disorder of the universe increases to the maximum possible. That is, entropy always increases during spontaneous processes.
- The first and second laws of thermodynamics can be combined to determine the amount of energy made available for useful work.

$$\Delta G = \Delta H - T \cdot \Delta S$$

  In this equation the change in free energy (ΔG) is the energy made available for useful work, the change in enthalpy (ΔH) is the change in heat content, and the change in entropy is ΔS.
- The standard free energy change ($\Delta G^{\circ\prime}$) for a chemical reaction is directly related to the equilibrium constant.

- In exergonic reactions, $\Delta G^{\circ\prime}$ is negative and the equilibrium constant is greater than one; the reaction goes to completion as written. Endergonic reactions have a positive $\Delta G^{\circ\prime}$ and an equilibrium constant less than one (**figure 10.1**).

10.2 ATP: The Major Energy Currency of Cells

- ATP is a high-energy molecule that carries energy in a useful form from one reaction or location in a cell to another (**figure 10.2**).
- ATP is readily synthesized from ADP and P_i using energy released from exergonic reactions; when hydrolyzed back to ADP and P_i, it releases the energy, which is used to drive endergonic reactions. This cycling of ATP with ADP and P_i is called the cell's energy cycle (**figure 10.4**).

10.3 Redox Reactions: Reactions of Central Importance in Metabolism

- In oxidation-reduction (redox) reactions, electrons move from an electron donor to an electron acceptor. Each redox reaction consists of two half reactions: One is the electron-donating half reaction; the other is the electron-accepting half reaction.
- Each half reaction of a redox reaction consists of a molecule that can accept electrons and one that can donate electrons. These two molecules are called a conjugate redox pair.
- The standard reduction potential measures the tendency of the donor of a conjugate redox pair to give up electrons.
- Conjugate redox pairs with more negative reduction potentials donate electrons to those with more positive potentials, and energy is made available during the transfer (**figure 10.5** and **table 10.2**).

10.4 Electron Transport Chains: Sets of Sequential Redox Reactions

- Many metabolic processes use a series of electron carriers to transport electrons from the primary electron donor (e.g., an organism's energy source) to a final electron acceptor (e.g., oxygen). A series of electron carriers is called an electron transport chain (ETC). ETCs are associated with membranes (**figure 10.6**).
- Some of the most important electron carriers in cells are NAD, NADP, FAD, FMN, coenzyme Q, cytochromes, and nonheme iron (Fe-S) proteins (**figures 10.7–10.11**). They differ in several ways, including how many electrons and protons they transfer.

10.5 Biochemical Pathways: Sets of Linked Chemical Reactions

- The chemical reactions carried out by cells are organized to form biochemical pathways. The intermediates and products of pathways are called metabolites (**figure 10.12**).
- Metabolic pathways are interconnected such that the intermediates and end product of one pathway may be the starting molecules of others (**figure 10.12*d***).

Metabolite flux refers to the turnover of metabolites as they are used in various biochemical pathways.

10.6 Enzymes and Ribozymes Speed Up Cellular Chemical Reactions

- Enzymes are proteins that catalyze specific reactions.
- Many enzymes consist of a protein component, the apoenzyme, and a cofactor. Cofactors may be metals or organic molecules. Some cofactors are prosthetic groups, which are tightly bound to the apoenzyme. Other cofactors are loosely bound and are called coenzymes.
- Enzymes and ribozymes speed chemical reactions by binding substrates at their active sites and lowering the activation energy (**figures 10.14** and **10.15**).
- The rate of an enzyme- or ribozyme-catalyzed reaction increases with substrate concentration at low substrate levels and reaches a plateau (the maximum velocity) at saturating substrate concentrations. The Michaelis constant is the substrate concentration that the enzyme or ribozyme requires to achieve half maximal velocity (**figure 10.16**).
- Enzymes have pH and temperature optima for activity. Their activity can be slowed by competitive and noncompetitive inhibitors (**figure 10.17**).
- Some ribozymes alter either their own structure or that of other RNAs. An important ribozyme is located in the ribosome, where it links amino acids together during protein synthesis.

10.7 Metabolism Must Be Regulated to Maintain Homeostasis

- The regulation of metabolism keeps cell components in proper balance and conserves metabolic energy and material. There are three major approaches to regulating metabolism: metabolic channeling, regulation of gene expression, and posttranslational regulation of enzyme activity.
- Metabolic channeling localizes metabolites and enzymes in different parts of the cell and influences pathway activity. A common channeling mechanism is compartmentation.
- Regulation of gene expression is brought about by controlling the rate of transcription and translation. This is discussed in chapters 14 and 15.
- Posttranslational regulation is the direct control of enzyme activity. There are two major types of reversible posttranslational regulation: allosteric control and covalent modification. In allosteric control, an allosteric effector binds noncovalently to an enzyme at a regulatory site separate from the catalytic site. This causes a conformational change in the enzyme and alters its activity (**figure 10.18**). In covalent modification, enzyme activity is regulated by reversible attachment of a chemical group, such as a phosphoryl, methyl, or adenylyl group (**figure 10.19**).
- The first enzyme in a pathway and enzymes at branch points often are subject to feedback inhibition by one or more end products. Excess end product slows its own synthesis (**figure 10.20**).

Active Learning

1. Examine the structures of macromolecules in appendix I. Which type has the most electrons to donate? Suggest why carbohydrates are usually the primary source of energy for bacteria that use organic molecules as energy sources.
2. Examine the branched pathway shown here for the synthesis of the amino acids aspartate, methionine, lysine, threonine, and isoleucine.

   Scenario 1: The microbe is cultured in a medium containing aspartate and lysine but lacking methionine, threonine, and isoleucine.

   Scenario 2: The microbe is cultured in a medium containing a rich supply of all five amino acids.

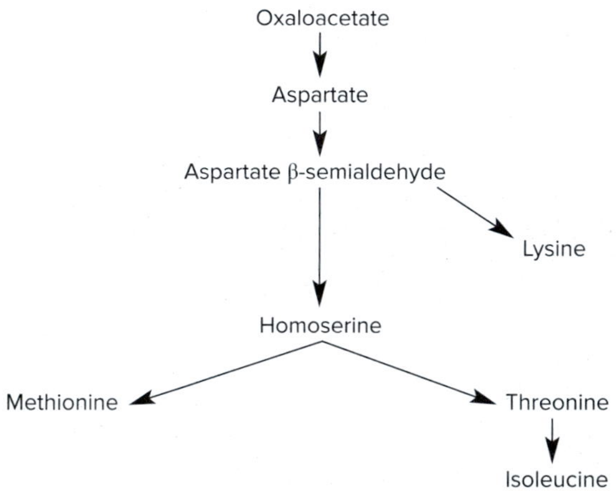

   For each of these two scenarios, answer the following questions:

   a. Which portion(s) of the pathway would need to be shut down in this situation?
   b. How might allosteric control be used to accomplish this?

3. Chlamydia are pathogenic bacteria that must be grown within a eukaryotic host cell. They rely on the host cell for many nutrients, including nucleotides, lipids, and amino acids. It is thus no surprise that when grown in coculture with a host depleted of any of these metabolites, chlamydial growth declines. However, high levels of certain amino acids also inhibit chlamydial growth. Specifically, the amino acids leucine, isoleucine, methionine, and phenylalanine retard growth by inhibiting the amino acid transporter called BrnQ. On the other hand, if valine is present in the medium these amino acids have no effect on BrnQ. Discuss how this might be possible.
4. *Listeria monocytogenes* is a food-borne human pathogen that lives inside host cells during infection. It can also live freely in water and soil when not causing infection. A group of Israeli scientists wanted to learn how *L. monocytogenes* "knows" when it is in the host and when it is not. This is because the bacterium only produces molecules needed for disease, called virulence factors, when in host cells. They found that the amino acid L-glutamine, which is present at high concentration inside host cells, serves as an off/on switch for virulence factors. Given that L-glutamine must be taken up into the bacterial cell, discuss ways in which it could regulate the production of virulence factors, some of which are proteins but many of which are assembled by enzymes.

   Read the original paper: Haber, A., et al. 2017. L-glutamine induces expression of *L. monocytogenes* virulence genes. *PLOS Pathog.* 13(1):e1006161. doi:10.1371/journal.ppat.1006161.

11

Catabolism: Energy Release and Conservation

The Richest Hill on Earth

The history of Butte, Montana, is filled with interesting characters, great tragedies, and the amazing resilience of its people. The city has gone through boom-and-bust cycles related to its location—it sits on a mountain rich in copper and other metals. From the 1860s until 1982, mining shaped life in Butte. It began with underground mining, which led to one of the worst mining disasters in U.S. history. In 1917 the Granite Mountain fire took the lives of 168 miners.

The advent of open-pit mining of the mountain in 1955 raised hopes that earlier mining tragedies would be averted. Over the next 20 years, the strip mine devoured the side of the mountain and parts of the city, creating the Berkeley Pit, a deep hole 1.5 miles across (see accompanying figure). In the process, over 1 billion tons of copper, gold, lead, and other metals were extracted, fueling a boon in Butte's economy.

As the mine operated, groundwater and runoff entered the pit, requiring water to be continuously pumped out. When the mine closed in 1982, the pumps were shut down and water accumulated. Now a new kind of hazard has emerged. The north edge of Butte borders on the Berkeley Pit Lake, a lake containing acidic water (pH about 2.5) contaminated with toxic metals. The danger of the water was tragically demonstrated in 1995 when 342 migrating snow geese died after landing in the lake.

But the Berkeley Pit Lake is not devoid of life. A team led by Montana Tech University scientist Grant Mitman discovered algae, fungi, and bacteria living in the lake leading to the possibility that microbes might be used to clean up the water. Later studies by Don and Andrea Stierle of the University of Montana-Missoula led to the discovery of microbes that produce interesting chemicals that may be useful for treating cancer, migraines, and other illnesses. Thus the Berkeley Pit Lake may be able to reclaim its title, "the richest hill on Earth."

The story of the Berkeley Pit clearly demonstrates that microbes are the most successful organisms on Earth as measured by their growth under almost every conceivable condition. In large part, their success results from the diversity of their fueling reactions—the focus of this chapter.

Source: NASA

Readiness Check:

Based on what you have learned previously, you should be able to:

- ✓ Distinguish metabolism, catabolism, and anabolism
- ✓ Discuss the function of ATP and how that function is related to its phosphate transfer potential (section 10.2)
- ✓ Relate the standard reduction potential of a conjugate redox pair to its tendency to act as an electron-donating half reaction (section 10.3)
- ✓ Explain the organization of the molecules commonly found in electron transport chains (ETCs) (section 10.4)
- ✓ Describe the organization of cellular chemical reactions into interconnected biochemical pathways (section 10.5)
- ✓ State the function of enzymes and ribozymes in cellular chemical reactions (section 10.6)

11.1 Metabolic Diversity and Nutritional Types

After reading this section, you should be able to:

a. State the carbon, energy, and electron sources of photolithoautotrophs, photoorganoheterotrophs, chemolithoautotrophs, chemolithoheterotrophs, and chemoorganoheterotrophs
b. Describe the products of the fueling reactions
c. Discuss the metabolic flexibility of microorganisms

As we discuss in chapter 10, all organisms require an energy and electron source to maintain order and complexity. Because life on Earth is carbon-based, all organisms must obtain carbon from their environments to synthesize the organic molecules they need to survive. The central importance of energy, electrons, and carbon has led biologists to describe organisms in terms of how these requirements are met. The terms can be combined to characterize an organism's nutritional type. The energy, electrons, and carbon that an organism obtains from its environment are used in chemical reactions referred to as fueling reactions. The nutritional types and fueling reactions are the focus of this section.

Nutritional Types Are Defined by the Source of an Organism's Energy, Electrons, and Carbon

For each requirement—energy, electrons, and carbon—a dichotomy exists for how an organism fulfills that need. Only two sources of energy are available to organisms: light and certain chemical molecules. **Phototrophs** use light as their energy source (**table 11.1**); **chemotrophs** obtain energy from the oxidation of chemical compounds (either organic or inorganic). Likewise, organisms have only two sources for electrons. **Lithotrophs** (i.e., "rock-eaters") use reduced inorganic substances as their electron source, whereas **organotrophs** extract electrons from reduced organic compounds. Finally, organisms may be **heterotrophs**—organisms that use reduced, preformed organic molecules as their carbon source for growth, or they may be **autotrophs**—organisms that use carbon dioxide (CO_2) as their sole or principal source of carbon.

By combining the roots of these terms, most organisms can be placed in one of five nutritional types based on their primary sources of carbon, energy, and electrons (**table 11.2**). For instance, plants are photo (energy from light), litho (electrons from the inorganic molecule, water), auto (carbon from CO_2), trophs (Greek *trophe,* nourishment); animals are quite the opposite: chemoorganoheterotrophs. The majority of microorganisms thus far studied are either photolithoautotrophic, chemolithoautotrophic, or chemoorganoheterotrophic. **Photolithoautotrophs,** often called simply **photoautotrophs,** include photosynthetic protists (algae) and cyanobacteria. These microorganisms are plantlike in that they employ water as the electron donor and release oxygen. Other photolithoautotrophs, such as the purple sulfur bacteria and the green sulfur bacteria, cannot oxidize water but extract electrons from inorganic donors such as hydrogen, hydrogen sulfide, and elemental sulfur. **Chemolithoautotrophs** oxidize reduced inorganic compounds such as iron-, nitrogen-, or sulfur-containing molecules to derive both energy and electrons for biosynthesis, using carbon dioxide as the carbon source. Chemolithoautotrophs contribute greatly to the chemical transformations of elements (e.g., the conversion of ammonia to nitrate or sulfur to sulfate) that continually occur in ecosystems. Photoautotrophs and chemolithoautotrophs also are important primary producers in ecosystems. That is, they fix CO_2, making reduced organic molecules that sustain the chemoorganoheterotrophs that share their habitats.

Table 11.1 Sources of Carbon, Energy, and Electrons

Carbon Sources	
Autotrophs	CO_2 sole or principal biosynthetic carbon source
Heterotrophs	Reduced, preformed, organic molecules from other organisms
Energy Sources	
Phototrophs	Light
Chemotrophs	Oxidation of organic or inorganic compounds
Electron Sources	
Lithotrophs	Reduced inorganic molecules
Organotrophs	Organic molecules

Table 11.2 Major Nutritional Types of Microorganisms

Nutritional Type	*Carbon Source*	*Energy Source*	*Electron Source*	*Representative Microorganisms*
Photolithoautotroph	CO_2	Light	Inorganic e^- donor	Purple and green sulfur bacteria, cyanobacteria, diatoms
Photoorganoheterotroph	Organic carbon	Light	Organic e^- donor	Purple nonsulfur bacteria, green nonsulfur bacteria
Chemolithoautotroph	CO_2	Inorganic chemicals	Inorganic e^- donor	Sulfur-oxidizing bacteria, hydrogen-oxidizing bacteria, methanogens, nitrifying bacteria, iron-oxidizing bacteria
Chemolithoheterotroph	Organic carbon	Inorganic chemicals	Inorganic e^- donor	Some sulfur-oxidizing bacteria (e.g., *Beggiatoa* spp.)
Chemoorganoheterotroph	Organic carbon	Organic chemicals, often same as C source	Organic e^- donor, often same as C source	Most nonphotosynthetic microbes, including most pathogens, fungi, and many protists

Chemoorganoheterotrophs (sometimes called **chemoheterotrophs** or chemoorganotrophs) use reduced organic compounds as sources of energy, electrons, and carbon. Frequently the same organic nutrient will satisfy all these requirements. Many chemoorganotrophs are used industrially to make foods (e.g., yogurt, pickles, cheese), medical products (e.g., antibiotics), and beverages (e.g., beer and wine). Nearly all pathogenic microorganisms are chemoorganoheterotrophs.

The other nutritional types have fewer known microorganisms but are very important ecologically. Some phototrophic bacteria (e.g., purple nonsulfur and green bacteria) use organic matter as their electron donor and carbon source. These **photoorganoheterotrophs** are common inhabitants of polluted lakes and streams. **Chemolithoheterotrophs** use reduced inorganic molecules as their energy and electron source but derive their carbon from organic sources. Like chemolithoautotrophs, they contribute to numerous biogeochemical cycles. ▶| *Purple nonsulfur bacteria perform anoxygenic photosynthesis (section 22.1)*

Although we have sorted microbes into particular nutritional types, microbes are not always so easily categorized. Some show great metabolic flexibility and alter their metabolism in response to environmental changes. For example, many purple nonsulfur bacteria act as photoorganoheterotrophs in the absence of oxygen but oxidize organic molecules and function chemoorganotrophically at normal oxygen levels (*see figure 22.4*). When oxygen is low, phototrophic and chemoorganotrophic metabolism may function simultaneously. This allows the bacterium to gain energy from both light and organic molecules, and still supply the carbon it needs for biosynthesis. Some of these bacteria also can grow as photolithoautotrophs with molecular hydrogen as an electron donor. This sort of flexibility gives these microbes a distinct advantage if environmental conditions frequently change.

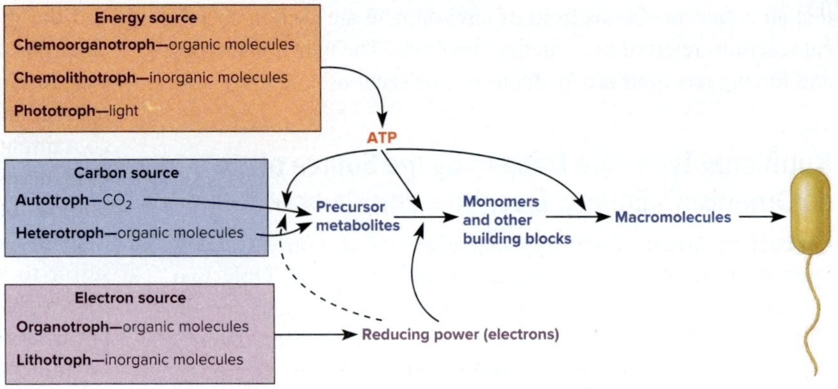

Figure 11.1 The Fueling Reactions Convert an Organism's Carbon, Energy, and Electron Sources into Precursor Metabolites, ATP, and Reducing Power. In autotrophs, the precursor metabolites arise from CO_2-fixation and related pathways, which are discussed in chapter 12. In heterotrophs, the precursor metabolites arise from reactions of the central metabolic pathways (*see figure 12.2*). ATP and reducing power are used in anabolic reactions that convert precursor metabolites into monomers and other building blocks. Monomers are used to generate macromolecules such as proteins, which are assembled into various cell structures.

Fueling Reactions Convert Energy, Electron, and Carbon Sources into ATP, Reducing Power, and Precursor Metabolites

Despite the diversity of energy, electron, and carbon sources used by organisms, all are used to generate three main products: **ATP,** the primary molecule used to conserve the energy supplied by an energy source; **reducing power,** molecules that serve as a ready supply of electrons for a variety of chemical reactions; and **precursor metabolites,** small organic molecules that provide the carbon skeletons needed for biosynthesis of important chemical building blocks (monomers) such as amino acids (**figure 11.1**). The processes that supply ATP, reducing power, and precursor metabolites are called **fueling reactions.** With the exception of CO_2-fixation, the fueling reactions are part of catabolism. As seen in figure 11.1, the generation of monomers using the products of the fueling reactions paves the way for anabolic reactions that synthesize macromolecules (e.g., polymers such as proteins). The macromolecules are used to construct cellular structures such as ribosomes and flagella.

Comprehension Check

1. Identify the energy, carbon, and electron source for each of the following types of microbes: photolithoautotroph, photolithoheterotroph, chemoorganoheterotroph, and chemolithoheterotroph.
2. Compare photolithoautotrophy with chemolithoautotrophy. Do you think it is possible for an ecosystem to exist solely on the organic carbon generated by chemolithotrophy (i.e., without any contribution by photosynthetic organisms)? Explain your reasoning.
3. What are the three major products generated by the fueling reactions? Summarize how they are used in anabolic processes.

11.2 There Are Three Chemoorganotrophic Fueling Processes

After reading this section, you should be able to:

a. List the three types of chemoorganotrophic metabolisms
b. List the pathways of major importance to chemoorganotrophs and explain their importance
c. Propose an explanation that accounts for the existence of amphibolic pathways

As illustrated in figure 11.1, chemoorganotrophs use reduced organic compounds for catabolic and anabolic processes. When an organic compound is oxidized to release energy (catabolism), it also provides the carbon and electrons needed for anabolism.

There are two general approaches a chemoorganotroph may use to catabolize its energy source: respiration and fermentation. When the organic energy source is oxidized, the electrons released must be accepted by electron carriers such as NAD^+ and FAD. When these reduced electron carriers (e.g., NADH, $FADH_2$) in turn donate the electrons to an electron transport chain, the metabolic process is called **respiration** and may be divided into two different types (**figure 11.2**). In aerobic respiration, the final electron acceptor is oxygen, whereas the terminal acceptor in anaerobic respiration is a different oxidized molecule such as NO_3^-, SO_4^{2-}, CO_2, Fe^{3+}, or SeO_4^{2-}. Organic acceptors such as fumarate and humic acids also may be used. During respiration, electrons pass through the electron transport chain to the final electron acceptor generating a type of potential energy called the proton motive force (PMF) The PMF is used to synthesize ATP from ADP and phosphate (P_i). In contrast, **fermentation** (Latin *fermentare,* to cause to rise) uses an electron acceptor that is endogenous (from within the cell) and does not involve an electron transport chain. The endogenous electron acceptor is usually an intermediate (e.g., pyruvate) of the catabolic pathway used to degrade and oxidize the organic energy source. During fermentation, ATP is synthesized almost exclusively by substrate-level phosphorylation, a process in which a phosphate is transferred to ADP from a high-energy molecule (e.g., phosphoenolpyruvate) generated by catabolism of the energy source.

By convention, aerobic respiration, anaerobic respiration, and fermentation are usually described with glucose as the energy source. This is done for several reasons. One is that glucose is used by many chemoorganotrophs as an energy source. But perhaps more important is the way catabolic pathways are organized. Most chemoorganotrophs use a wide variety of organic molecules as energy sources (**figure 11.3**). They are degraded by pathways that either generate glucose or intermediates of the pathways used in glucose catabolism. Thus nutrient molecules are funneled into ever fewer metabolic intermediates. Indeed, a common pathway often degrades many similar molecules (e.g., several different sugars). The existence of a few metabolic pathways that each break down many nutrients greatly increases metabolic efficiency by avoiding the need for a large number of less metabolically flexible pathways.

The diversity of organic molecules used as energy sources by chemoorganotrophic microbes contributes to their ecology. They are found in habitats rich in organic molecules. For instance, lactose (milk sugar) catabolizers are common in the intestinal tracts of mammals, where they may have a ready supply of this sugar. Clostridia and other soil bacteria that use amino acids as energy sources are important decomposers of proteins in protein-rich habitats. Other soil-dwelling bacteria such as the pseudomonads degrade unusual and often complex organic molecules, including pesticides, polystyrene, polychlorinated biphenyls (PCBs), and even the phenolic resins used as glue in the production of plywood and fiberboard. These microbes are useful in bioremediation. ▶| *Biodegradation and bioremediation harness microbes to clean the environment (section 43.4)*

The catabolic pathways of greatest importance to chemoorganotrophs are the glycolytic pathways (section 11.4) and the tricarboxylic acid (TCA) cycle (section 11.5). During catabolism organic substrates are oxidized thereby serving as a source of electrons and energy. Conversely, during anabolism organic substrates supply material needed for biosynthesis, including precursor metabolites and reducing power. ▶| *Precursor metabolites: starting molecules for biosynthesis (section 12.2)*

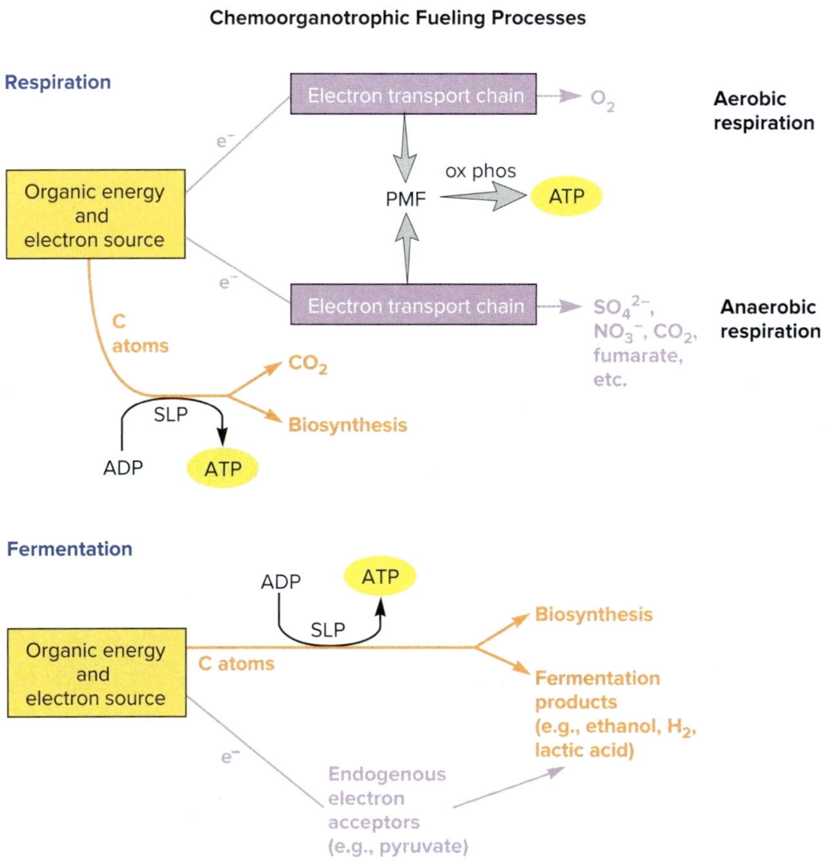

Figure 11.2 Chemoorganotrophic Fueling Processes. Organic molecules serve as energy and electron sources for all three fueling processes used by chemoorganotrophs. In aerobic respiration and anaerobic respiration, the electrons pass through an electron transport chain. This generates a proton motive force (PMF), which is used to synthesize most of the cellular ATP by a mechanism called oxidative phosphorylation (ox phos); a small amount of ATP is made by a process called substrate-level phosphorylation (SLP). In aerobic respiration, O_2 is the terminal electron acceptor, whereas in anaerobic respiration, exogenous molecules other than O_2 serve as electron acceptors. During fermentation, endogenous organic molecules act as electron acceptors, electron transport chains do not function, and most organisms synthesize ATP only by substrate-level phosphorylation.

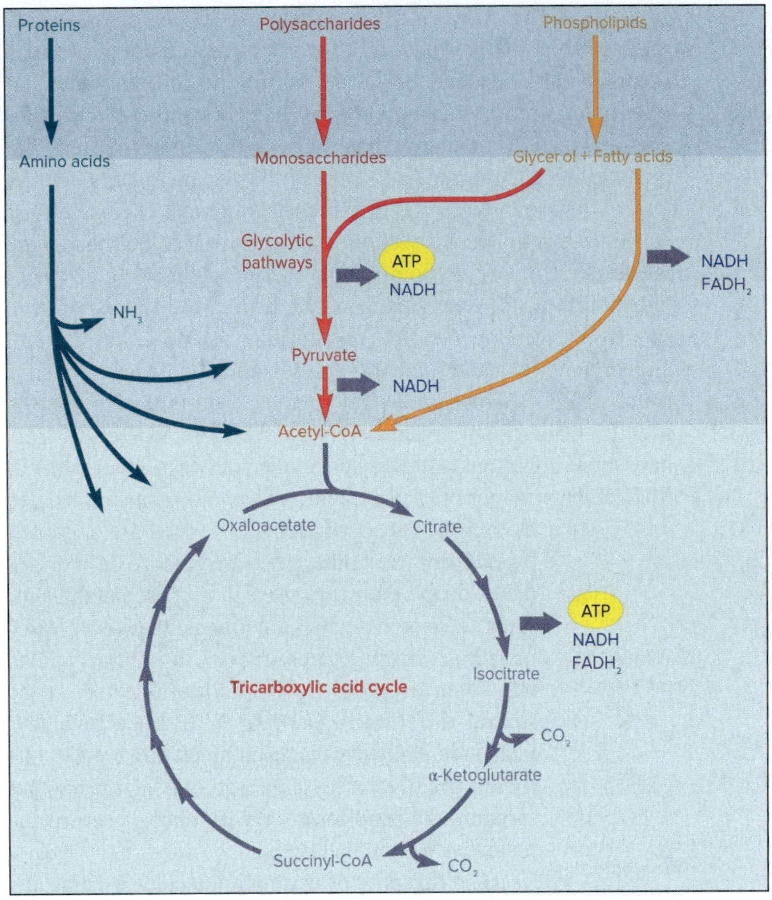

Figure 11.3 Pathways Used by Chemoorganotrophs to Catabolize Organic Energy Sources. Notice that these pathways funnel metabolites into the glycolytic pathways and the tricarboxylic acid cycle, thus increasing metabolic efficiency and flexibility.

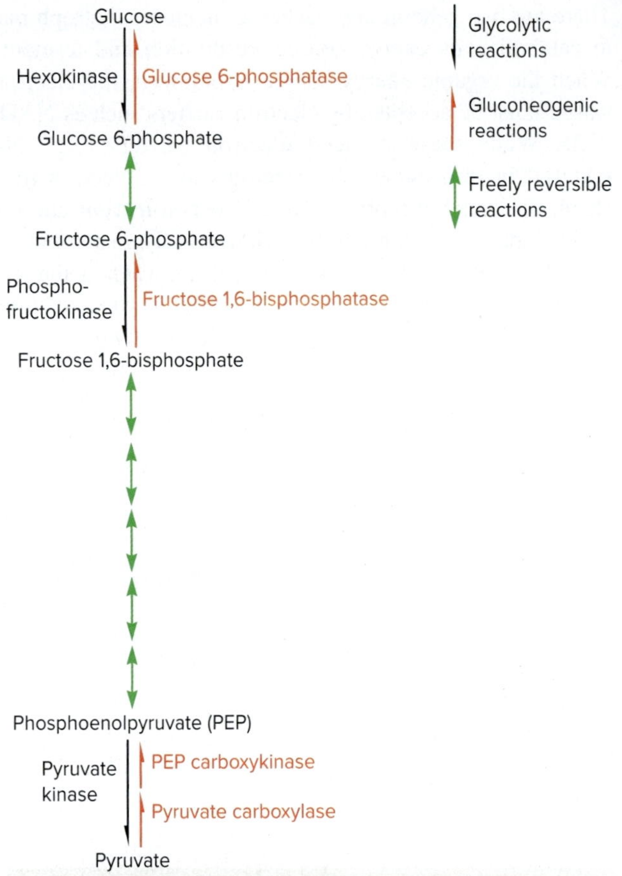

Figure 11.4 The Embden-Meyerhof Pathway Is an Example of an Amphibolic Pathway. Many reactions of this pathway are catalyzed by enzymes that function in glycolysis and in an anabolic pathway called gluconeogenesis. Gluconeogenesis reverses the glycolytic process and allows cells to synthesize glucose from smaller molecules such as pyruvate. Note that some glycolytic reactions are catalyzed by enzymes unique to that pathway. Likewise, some gluconeogenic reactions are catalyzed by enzymes unique to gluconeogenesis. Therefore, although they share some enzymes, the two pathways are distinct.

MICRO INQUIRY *Is NAD^+ reduced to NADH in the catabolic or anabolic direction of this pathway?*

Because a carbon source can function as an energy source or for anabolic purposes, some pathways include enzymes that function both catabolically and anabolically. These are called **amphibolic pathways** (Greek *amphi,* on both sides). For instance, many of the enzymes of the Embden-Meyerhof pathway (a glycolytic pathway) are freely reversible. They function catabolically during glycolysis but anabolically during gluconeogenesis, a pathway that generates glucose from pyruvate and other small molecules (**figure 11.4**). Other reactions in glycolysis are not reversible. During glycolysis, an enzyme catalyzes the reaction in a catabolic direction, and during gluconeogenesis, a different enzyme catalyzes the reverse, anabolic reaction. Thus, glycolysis and gluconeogenesis are two distinct pathways.

Our discussion of chemoorganotrophy begins with aerobic respiration and introduces the glycolytic pathways, TCA cycle, and other important processes. Many of these also occur during anaerobic respiration. Fermentation is quite distinct from respiration. It involves only a subset of the reactions that function during respiration and only partially catabolizes the energy source. However, it is widely used by microbes and has important practical applications, including the production of many foods. ▶| *Microbiology of fermented foods: beer, cheese, and much more (section 41.5)*

Comprehension Check

1. Give examples of the types of electron acceptors used by fermentation and respiration. What is the difference between aerobic respiration and anaerobic respiration?
2. Why is it to a cell's advantage to catabolize diverse organic energy sources by funneling them into a few common pathways?
3. What are amphibolic pathways? Why are they important?

11.3 Aerobic Respiration Can Be Divided into Three Steps

After reading this section, you should be able to:

a. Describe in general terms what happens to a molecule of glucose during aerobic respiration
b. List the end products made during aerobic respiration
c. Identify the process that generates the most ATP during aerobic respiration

Aerobic respiration is a process that can completely catabolize a reduced organic energy source to CO_2 using the glycolytic pathways and TCA cycle with O_2 as the terminal electron acceptor for an electron transport chain (figure 11.2). The catabolism of glucose can be divided into three steps. It begins with the formation of pyruvate, using one or more pathways described in section 11.4. These pathways also produce NADH, $FADH_2$, or both. Next pyruvate is fed into the TCA cycle and oxidized completely to CO_2 with the production of some GTP or ATP, NADH, and $FADH_2$ (section 11.5). Finally, the NADH and $FADH_2$ formed by glycolysis and the TCA cycle are oxidized by an electron transport chain, using O_2 as the terminal electron acceptor (section 11.6). It is the activity of the electron transport chain that conserves most of the energy used to make ATP during aerobic respiration.

11.4 Glucose to Pyruvate: The First Step

After reading this section, you should be able to:

a. List the three major pathways that catabolize glucose to pyruvate
b. Describe substrate-level phosphorylation
c. Diagram the major changes made to glucose as it is catabolized by the Embden-Meyerhof, Entner-Doudoroff, and pentose phosphate pathways
d. Identify those reactions of the Embden-Meyerhof, Entner-Doudoroff, and pentose phosphate pathways that consume ATP, produce ATP and NAD(P)H, generate precursor metabolites, or are redox reactions
e. Calculate the yields of ATP and NAD(P)H by the Embden-Meyerhof, Entner-Doudoroff, and pentose phosphate pathways
f. Summarize the function of the Embden-Meyerhof, Entner-Doudoroff, and pentose phosphate pathways
g. Draw a simple diagram that shows the connection between the Entner-Doudoroff pathway and the Embden-Meyerhof pathway and the connection between the pentose phosphate pathway and the Embden-Meyerhof pathway
h. Create a table that shows which types of organisms use each of the glycolytic pathways

Microorganisms employ several metabolic pathways to catabolize glucose to pyruvate in the cytosol, including (1) the Embden-Meyerhof pathway, (2) the Entner-Doudoroff pathway, and (3) the pentose phosphate pathway. We refer to these pathways collectively as **glycolytic pathways** or as **glycolysis** (Greek *glyco*, sweet; *lysis*, a loosening). However, in some texts, the term *glycolysis* refers only to the Embden-Meyerhof pathway. For the sake of simplicity, the detailed structures of some metabolic intermediates are not used in pathway diagrams. However, these can be found in appendix II.

Embden-Meyerhof Pathway: The Most Common Route to Pyruvate

The **Embden-Meyerhof pathway** is undoubtedly the most common pathway for glucose degradation to pyruvate. It is found in all major groups of microorganisms, as well as in plants and animals, and functions in the presence or absence of O_2. It provides several precursor metabolites, NADH, and ATP for the cell.

The pathway is divided into two parts: a 6-carbon phase and a 3-carbon phase (**figure 11.5** and *appendix II*). In the initial 6-carbon phase, ATP is used in two phosphorylation reactions, yielding fructose 1,6-bisphosphate. This preliminary phase "primes the pump" by adding phosphates to each end of the sugar. In essence, the organism invests some of its ATP so that more can be made later in the pathway.

The 3-carbon, energy-conserving phase begins when fructose 1,6-bisphosphate is cleaved into two 3-C molecules: dihydroxyacetone phosphate and glyceraldehyde 3-phosphate. Dihydroxyacetone phosphate is immediately converted to glyceraldehyde 3-phosphate. Thus two molecules of glyceraldehyde 3-phosphate are formed from a single glucose by the cleavage reaction. These are then catabolized to pyruvate in a five-step process.

In contrast to the 6-carbon phase when ATP is consumed, in the 3-carbon phase of the pathway NADH and ATP are produced. NADH is formed when glyceraldehyde 3-phosphate is oxidized with NAD^+ as the electron acceptor, and a phosphate is simultaneously incorporated to give a high-energy molecule called 1,3-bisphosphoglycerate. This reaction sets the stage for ATP production. The phosphate on the first carbon of 1,3-bisphosphoglycerate is donated to ADP to produce ATP. This is an example of **substrate-level phosphorylation** because ADP phosphorylation is coupled with the exergonic hydrolysis of a high-energy molecule having a higher phosphate transfer potential than ATP (*see table 10.1*). A second ATP is made by substrate-level phosphorylation when the phosphate on phosphoenolpyruvate (the last intermediate of the pathway) is donated to ADP. This reaction also yields pyruvate, the final product of the pathway.
◀ *ATP: the major energy currency of cells (section 10.2)*
How NAD^+ Works

The yields of ATP and NADH by the Embden-Meyerhof pathway may be calculated. In the 6-carbon phase, two ATP are used to form fructose 1,6-bisphosphate. For each glyceraldehyde 3-phosphate transformed into pyruvate, one NADH and two ATP are formed. Because two glyceraldehyde 3-phosphates arise from a single glucose, the 3-carbon phase generates four ATP and two NADH per glucose. Subtraction of the ATP used in the 6-carbon phase from ATP produced by substrate-level phosphorylation in the 3-carbon phase gives a net yield of two ATP per glucose.

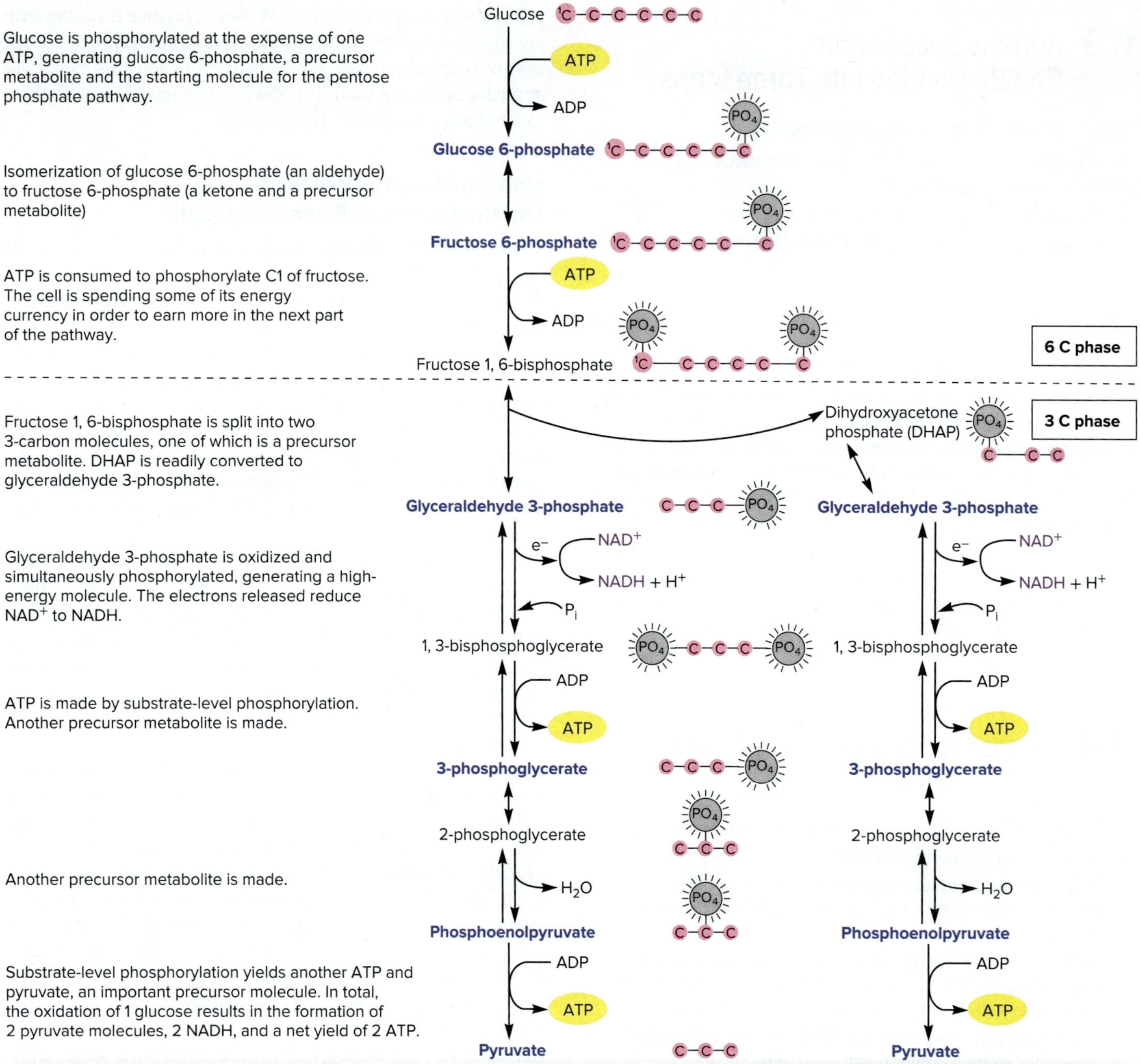

Figure 11.5 Embden-Meyerhof Pathway. This is one of three pathways used to catabolize glucose to pyruvate, and it can function during aerobic respiration, anaerobic respiration, and fermentation. When used during respiration, the electrons accepted by NAD^+ are transferred to an electron transport chain and are ultimately accepted by an exogenous electron acceptor. When used during fermentation (section 11.8), the electrons accepted by NAD^+ are donated to an endogenous electron acceptor (e.g., pyruvate). The Embden-Meyerhof pathway also generates several precursor metabolites (shown in blue).

MICRO INQUIRY *Which reactions are examples of substrate-level phosphorylation?*

Thus the catabolism of glucose to pyruvate via the Embden-Meyerhof pathway can be represented by the following equation.

$$\text{Glucose} + 2\text{ADP} + 2\text{P}_i + 2\text{NAD}^+ \rightarrow 2\text{ pyruvate} + 2\text{ATP} + 2\text{NADH} + 2\text{H}^+$$

How Glycolysis Works

Entner-Doudoroff Pathway

The **Entner-Doudoroff pathway** is used by some Gram-negative bacteria, especially those found in soil. To date, very few Gram-positive bacteria have been found to use this pathway, with the intestinal bacterium *Enterococcus faecalis* being a rare exception. It is not used by eukaryotes.

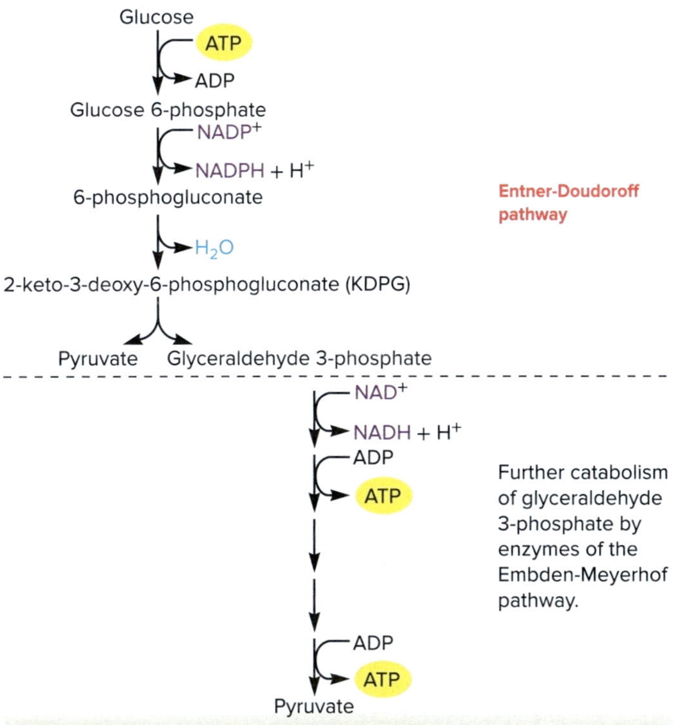

Figure 11.6 The Entner-Doudoroff Pathway.

MICRO INQUIRY *For what kinds of reactions is NADPH used?*

The Entner-Doudoroff pathway replaces the 6-carbon phase of the Embden-Meyerhof pathway to yield pyruvate and glyceraldehyde 3-phosphate instead of two molecules of glyceraldehyde 3-phosphate. A key intermediate of the Entner-Doudoroff pathway has the long but descriptive name 2-keto-3-deoxy-6-phosphogluconate (KDPG), which is formed from glucose by three reactions that consume one ATP and produce one NADPH (**figure 11.6** and *appendix II*). KDPG is then cleaved to pyruvate and glyceraldehyde 3-phosphate. Bacteria that use this pathway also have the enzymes that function in the 3-carbon phase of the Embden-Meyerhof pathway. These enzymes may be used to catabolize glyceraldehyde 3-phosphate to form a second pyruvate molecule. If this occurs, two ATP and one NADH are formed. Thus the catabolism of one glucose molecule to two pyruvates by way of the Entner-Doudoroff pathway coupled with the second half of the Embden-Meyerhof pathway has a net yield of one ATP, one NADH, and one NADPH. During aerobic respiration, the NADH is used to transport electrons to an ETC; the NADPH provides electrons for anabolic reactions.

Pentose Phosphate Pathway: A Major Producer of Reducing Power for Anabolic Reactions

The **pentose phosphate pathway** (also called the **hexose monophosphate pathway**) can be used aerobically or anaerobically. It may be used at the same time as either the Embden-Meyerhof or the Entner-Doudoroff pathways because it is important in biosynthesis as well as catabolism. Both eukaryotes and bacteria use it to provide reducing power (as NADPH) and important precursor metabolites for biosynthetic reactions. However, it has not yet been found in archaea.

The pentose phosphate pathway begins with a two-step oxidative phase when glucose 6-phosphate is oxidized to 6-phosphogluconate, which in turn is oxidized and decarboxylated to the 5-carbon sugar (i.e., a pentose sugar) ribulose 5-phosphate and CO_2 (**figure 11.7** and *appendix II*). Unlike the Embden-Meyerhof pathway where NAD^+ accepts electrons, thereby generating NADH, here $NADP^+$ is the electron acceptor and NADPH is produced. This leads to another important difference between the two pathways because NADH donates its electrons to the ETC so that energy can be conserved, while NADPH donates its electrons to biosynthetic reactions, which often consume energy.

Following the oxidative phase, ribulose 5-phosphate is converted to a mixture of 3- to 7-carbon sugar phosphates. Two enzymes play a central role in these transformations: (1) transketolase catalyzes the transfer of 2-carbon groups, and (2) transaldolase transfers a 3-carbon group from sedoheptulose 7-phosphate (a 7-carbon molecule) to glyceraldehyde 3-phosphate (a 3-carbon molecule). The overall result is that three glucose 6-phosphates are converted to two fructose 6-phosphates, glyceraldehyde 3-phosphate, and three CO_2 molecules, as shown in the following equation.

$$\text{3 glucose 6-phosphate} + 6NADP^+ + 3H_2O \rightarrow \text{2 fructose 6-phosphate} + \text{glyceraldehyde 3-phosphate} + 3CO_2 + 6NADPH + 6H^+$$

These intermediates are used in two ways. Fructose 6-phosphate can be changed back to glucose 6-phosphate while glyceraldehyde 3-phosphate is converted to pyruvate by enzymes of the Embden-Meyerhof pathway. Alternatively two glyceraldehyde 3-phosphates may combine to form fructose 1,6-bisphosphate, which is eventually converted back into glucose 6-phosphate. This results in the complete degradation of glucose 6-phosphate to CO_2 and the production of a great deal of NADPH.

$$\text{Glucose 6-phosphate} + 12NADP^+ + 7H_2O \rightarrow 6CO_2 + 12NADPH + 12H^+ + P_i$$

The pentose phosphate pathway is an important amphibolic pathway because: (1) NADPH serves as a source of electrons for the reduction of molecules during biosynthesis. Indeed, the pentose phosphate pathway is the major source of reducing power for cells, yielding two NADPH molecules for each glucose metabolized to pyruvate in this way. (2) The pathway produces two important precursor metabolites: erythrose 4-phosphate and ribose 5-phosphate. Erythrose 4-phosphate is used to synthesize aromatic amino acids and vitamin B_6; ribose 5-phosphate is a major component of nucleic acids. Furthermore, when a microorganism is growing on a 5-carbon sugar, the pathway can function biosynthetically to supply 6-carbon sugars (hexose sugars, such as the

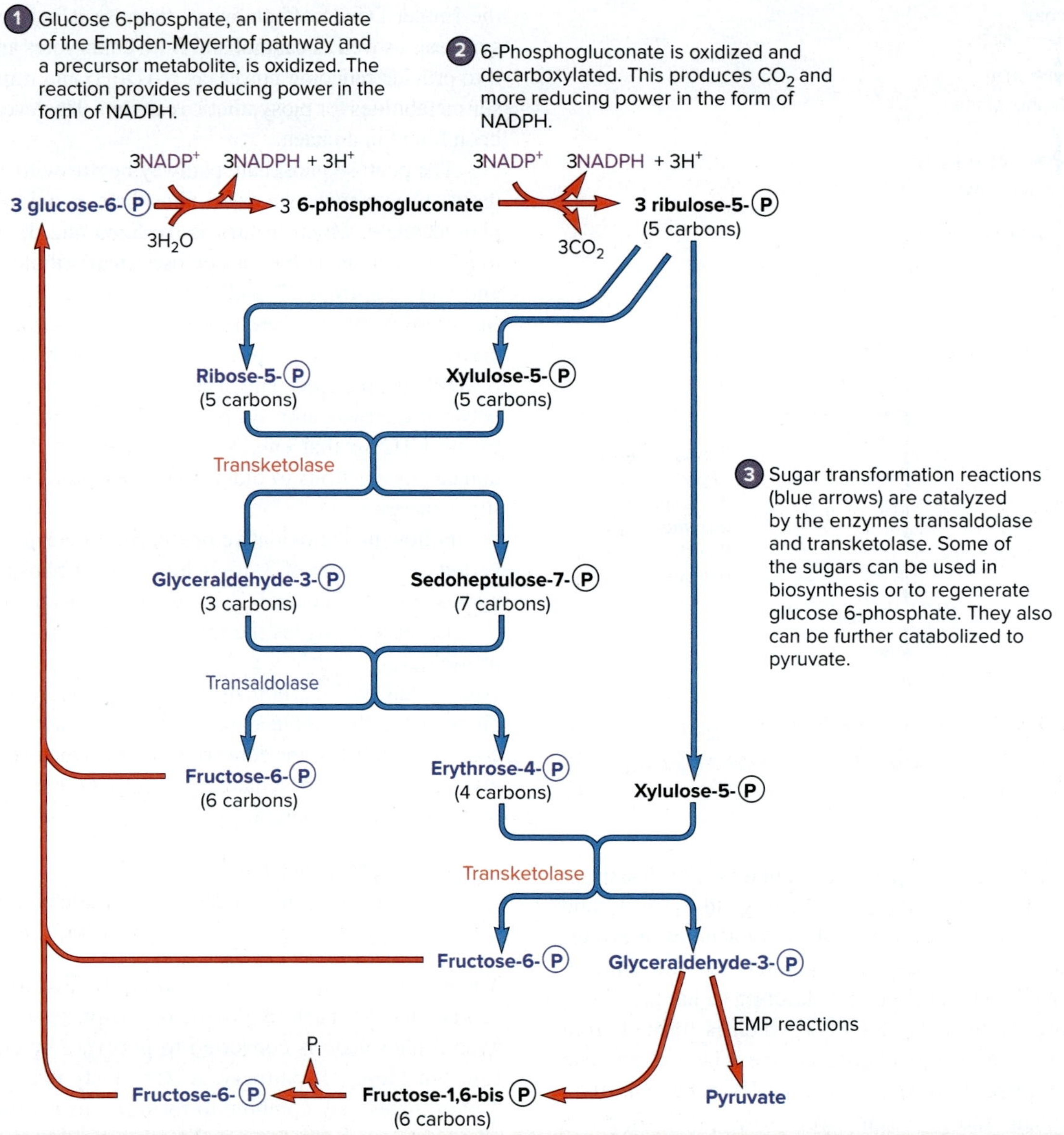

Figure 11.7 The Pentose Phosphate Pathway. The catabolism of three glucose 6-phosphate molecules to two fructose 6-phosphates, a glyceraldehyde 3-phosphate, and three CO_2 molecules is traced. Note that the pentose phosphate pathway generates several intermediates that are also intermediates of the Embden-Meyerhof pathway (EMP). These intermediates can be fed into the EMP with two results: (1) continued degradation to pyruvate or (2) regeneration of glucose 6-phosphate by gluconeogenesis. The pentose phosphate pathway also plays a major role in producing reducing power (NADPH) and several precursor metabolites (shown in blue) for biosynthesis. The sugar transformations are indicated with blue arrows.

MICRO INQUIRY *For what macromolecule is ribose 5-phosphate a precursor?*

glucose needed for peptidoglycan synthesis). (3) Intermediates in the pathway may be used to produce ATP. For instance, glyceraldehyde 3-phosphate can enter the 3-carbon phase of the Embden-Meyerhof pathway. As it is degraded to pyruvate, two ATP are formed by substrate-level phosphorylation. ▶| *Synthesis of amino acids consumes many precursor metabolites (section 12.5); Synthesis of purines, pyrimidines, and nucleotides (section 12.6)*

Comprehension Check

1. Summarize the major features of the Embden-Meyerhof, Entner-Doudoroff, and pentose phosphate pathways. Include the starting points, the products of the pathways, the ATP yields, and the metabolic roles each pathway has.
2. What is substrate-level phosphorylation?

11.5 Pyruvate to Carbon Dioxide (Step 2) Is Accomplished by the Tricarboxylic Acid Cycle

After reading this section, you should be able to:

a. State the alternate names for the tricarboxylic acid (TCA) cycle
b. Diagram the changes made to pyruvate as it is catabolized by the TCA cycle
c. Identify those reactions of the TCA cycle that produce GTP and NADH, generate precursor metabolites, or are redox reactions
d. Calculate the yields of GTP, NADH, and $FADH_2$ by the TCA cycle
e. Summarize the function of the TCA cycle
f. Diagram the connections between the various glycolytic pathways and the TCA cycle
g. Locate the TCA cycle enzymes in bacterial, archaeal, and eukaryotic cells

In the glycolytic pathways, glucose is oxidized to pyruvate. During aerobic respiration, the catabolic process continues by oxidizing pyruvate to three CO_2. The first step of this process employs a multienzyme system called the **pyruvate dehydrogenase complex (PDH)**. PDH oxidizes and cleaves pyruvate to form one NADH, one CO_2, and the 2-carbon molecule **acetyl-coenzyme A (acetyl-CoA)** (**figure 11.8**). The oxidization of a molecule to simultaneously release carbon as CO_2 and NADH (or other reduced electron carrier) is called oxidative decarboxylation. Acetyl-CoA is an energy-rich molecule because the bond that links acetic acid to coenzyme A (a thioester bond) has a large negative change in free energy when hydrolyzed.

Acetyl-CoA then enters the **tricarboxylic acid (TCA) cycle,** also called the **citric acid cycle** or the **Krebs cycle** (figure 11.8 and *appendix II*). In the first reaction, acetyl-CoA is condensed with (i.e., added to) the 4-carbon intermediate oxaloacetate to form citrate, a molecule with six carbons. This reaction is fueled by the cleavage of the high-energy thioester bond in acetyl-CoA. Citrate is rearranged to give isocitrate, a more readily oxidized alcohol. Isocitrate undergoes oxidative decarboxylation twice to yield α-ketoglutarate (five carbons) and then succinyl-CoA (four carbons), another high-energy molecule containing a thioester bond. Note that each decarboxylation is accompanied by the production of an NADH, so at this point two NADH and two CO_2 molecules have been produced in the cycle. Succinyl-CoA is then converted to succinate; hydrolysis of the high-energy thioester bond in succinyl-CoA drives the synthesis of one GTP by substrate-level phosphorylation. GTP is a high-energy molecule functionally equivalent to ATP. It is used in protein synthesis and to make other nucleoside triphosphates, including ATP. Two additional oxidation steps follow, yielding one $FADH_2$ and one NADH. The last oxidation step regenerates oxaloacetate, and as long as there is a supply of acetyl-CoA, the cycle can repeat itself. Inspection of figure 11.8 shows that the TCA cycle generates two CO_2 molecules, three NADH molecules, one $FADH_2$, and one GTP for each acetyl-CoA molecule oxidized. *How the Krebs Cycle Works*

TCA cycle enzymes are widely distributed among microorganisms. In bacteria and archaea, they are located in the cytosol. In eukaryotes, they are found in the mitochondrial matrix. The complete cycle appears to be functional in many aerobic bacteria, archaea, free-living protists, and fungi. This is not surprising because the cycle plays an important role in energy conservation by producing NADH and $FADH_2$ (section 11.6). Even microorganisms that lack the complete TCA cycle usually have most of its enzymes, because the TCA cycle is also a key source of precursor metabolites for use in biosynthesis. ▶| *Anaplerotic reactions replace the precursor metabolites used for amino acid biosynthesis (section 12.5)*

Comprehension Check

1. Identify the substrate and products of the TCA cycle. Describe its organization in general terms. What are its major functions?
2. What chemical intermediate links pyruvate to the TCA cycle?
3. How many times must the TCA cycle be performed to oxidize one molecule of glucose completely to six molecules of CO_2? Why?
4. In what eukaryotic organelle is the TCA cycle found? Where is the cycle located in bacterial and archaeal cells?
5. Why is it desirable for a microbe with the Embden-Meyerhof pathway and the TCA cycle also to have the pentose phosphate pathway?

11.6 Electron Transport and Oxidative Phosphorylation (Step 3) Generate the Most ATP

After reading this section, you should be able to:

a. Compare and contrast the mitochondrial electron transport chain (ETC) and bacterial ETCs
b. Describe the chemiosmotic hypothesis
c. Correlate length of an ETC and the carriers in it with the magnitude of the proton motive force (PMF) it generates
d. Explain how ATP synthase uses PMF to generate ATP
e. Draw a simple diagram that shows the connections between the glycolytic pathways, TCA cycle, ETC, and ATP synthesis
f. List the ways the PMF is used by bacterial cells in addition to ATP synthesis
g. Calculate the maximum possible ATP yields when glucose is completely catabolized to six molecules of CO_2 during aerobic respiration

During the oxidation of glucose to six CO_2 molecules by glycolysis and the TCA cycle, as many as four ATP molecules are generated by substrate-level phosphorylation. Thus at this point, the work done by the cell has yielded relatively little ATP. However, in oxidizing glucose, the cell has also generated lots of NADH and $FADH_2$. Both of these molecules have a relatively negative E'_0 and can be used to conserve energy. In fact, most of the ATP generated during respiration comes from the energy conserved when these electron carriers are oxidized by an ETC. We examine the mitochondrial electron transport chain first because it has been well studied and because it functions in fungi and many protists. We then turn to bacterial chains and a

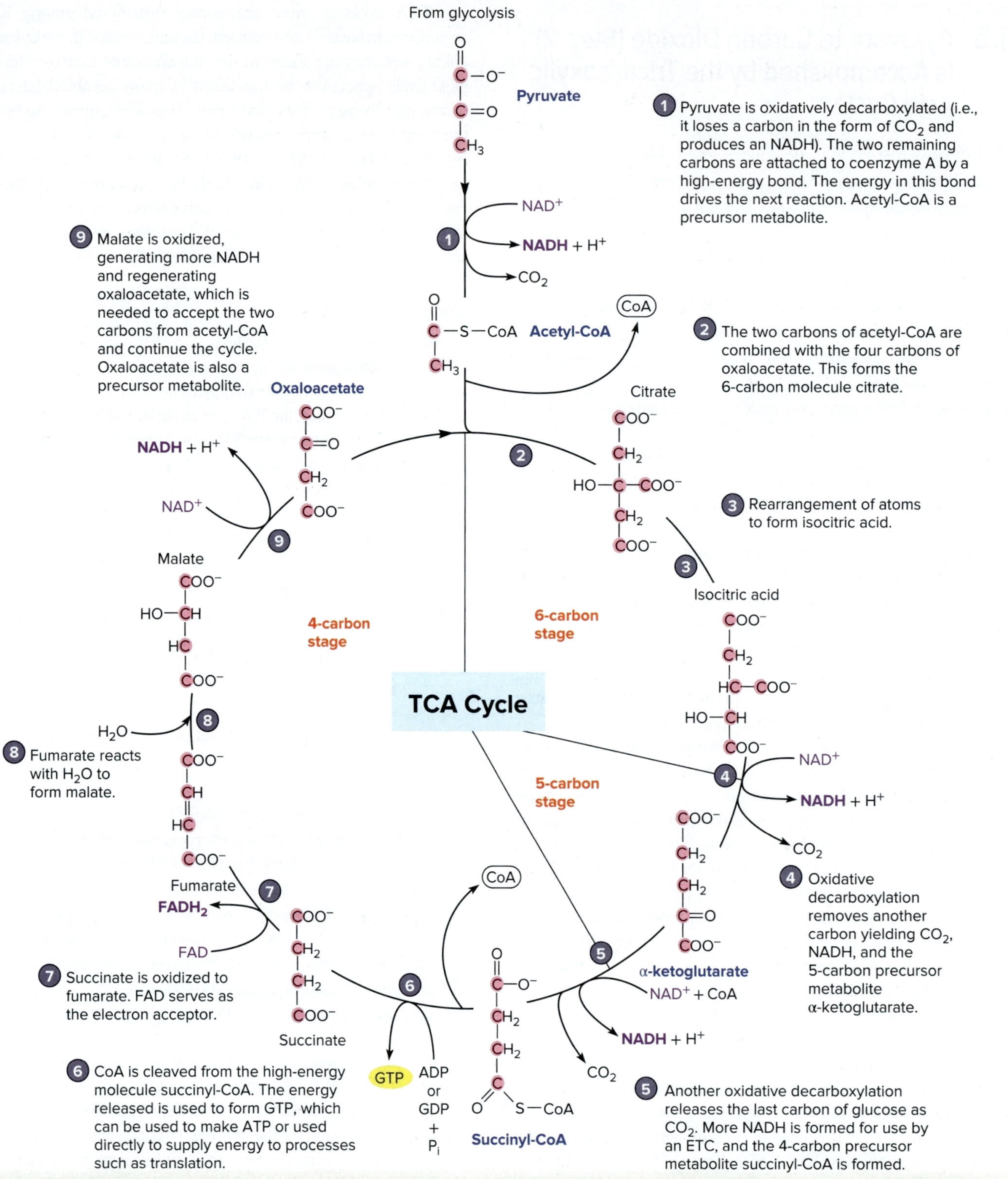

Figure 11.8 The Tricarboxylic Acid Cycle. The TCA cycle is linked to glycolysis by a connecting reaction catalyzed by the pyruvate dehydrogenase complex. The reaction decarboxylates pyruvate and generates acetyl-CoA. The TCA cycle may be divided into three stages based on the size of its intermediates. The three stages are separated from one another by two oxidative decarboxylation reactions. Precursor metabolites are shown in blue. NADH and $FADH_2$ are shown in purple; they can transfer electrons to an electron transport chain (ETC).

MICRO INQUIRY *What reaction provides the energy to fuel the condensation (joining) of the 4-carbon oxaloacetate with the 2-carbon acetyl-CoA to form citrate?*

discussion of ATP synthesis. ▶| *Electron transport chains: sets of sequential redox reactions (section 10.4)*

Electron Transport Chains

The mitochondrial **electron transport chain** (**ETC**) is composed of a series of electron carriers that operate together to transfer electrons from donors, such as NADH and $FADH_2$, to O_2 (**figure 11.9**). The electrons flow from carriers with more negative reduction potentials to those with more positive potentials and eventually combine with O_2 and H^+ to form water. Each carrier is reduced when it accepts electrons and then reoxidized when the electrons are transferred to the next carrier in the chain (**figure 11.10**). The electrons move down this potential gradient much like water flowing down a series of rapids. The difference in reduction potentials between O_2 and NADH is large, about 1.14 volts, which makes possible the release of a great deal of energy. However, the ETC breaks up the overall energy release (electron transfer) into small steps, preventing electrons from reducing cellular components outside the ETC. The energy released is sufficient for ATP production. Thus much as the energy from waterfalls can be harnessed by waterwheels and used to generate electricity, so too is the flow of electrons through an ETC harnessed to make ATP.

In eukaryotes, ETC carriers reside within the inner membrane of the mitochondrion. The ETC consists of four large, multiprotein complexes and two additional carriers, coenzyme Q and the small protein cytochrome *c*, that connect complex I to complex III and complex III to IV, respectively. The final complex (complex IV) is referred to as the terminal oxidase as it is responsible for the final step transferring electrons to O_2 (**figure 11.11**). As electrons move through the ETC, protons are moved across the inner membrane, forming a proton gradient.

Like mitochondrial ETCs, bacterial and archaeal ETCs are designed to generate a proton gradient. However, because bacteria and archaea do not have mitochondria, their ETCs are located primarily within the plasma membrane. Furthermore, some Gram-negative bacteria have ETC carriers in the periplasmic space and even the outer membrane (*see figure 22.27*). Also unlike mitochondrial ETCs, bacterial and archaeal ETCs can be composed of different electron carriers, employ different terminal oxidases, and be branched. That is, electrons may enter the chain at several points and leave through several terminal oxidases not just that which reduces O_2. Bacterial and archaeal ETCs also may be shorter, resulting in the transport of fewer protons across the membrane, and therefore less energy. Although microbial ETCs differ in details of construction, they operate using the same fundamental principles. *Electron Transport System and ATP Synthesis*

The ETCs of *Paracoccus denitrificans* and *Escherichia coli* will serve as examples of bacterial chains. *P. denitrificans* is a Gram-negative, soil bacterium that is a facultative anaerobe that can use several metabolic strategies. Under oxic conditions, it carries out aerobic respiration using the ETC shown in **figure 11.12**. This

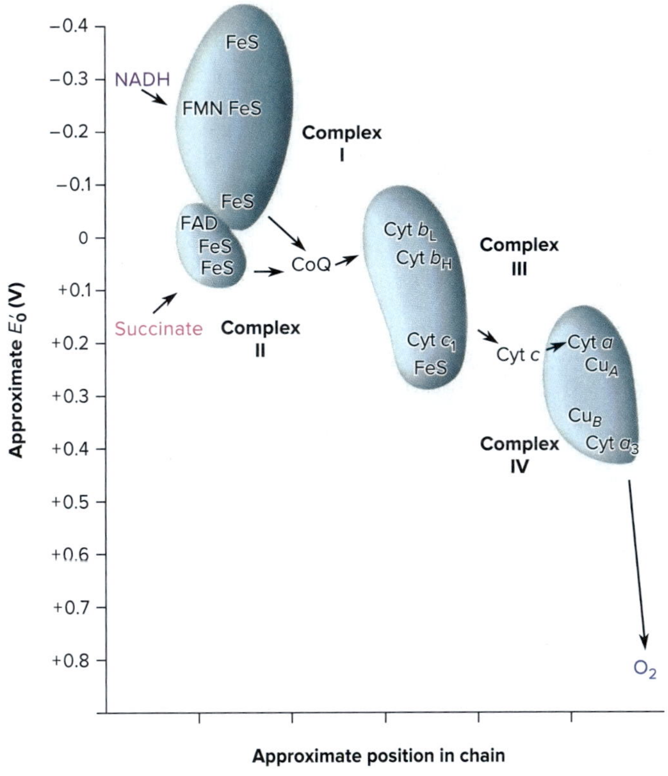

Figure 11.9 The Reduction Potentials of Carriers in the Mitochondrial Electron Transport Chain. In mitochondria, the electron carriers are organized into four complexes that are linked by the electron carriers coenzyme Q (CoQ) and cytochrome *c* (Cyt *c*). Electrons flow from NADH and succinate down the reduction potential gradient to oxygen.

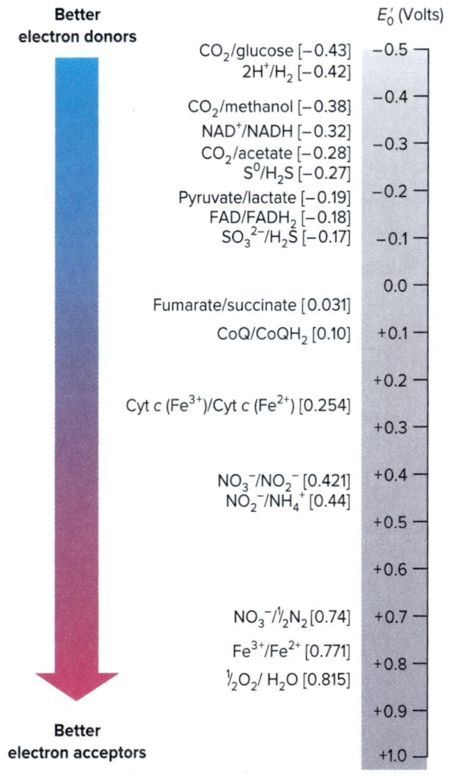

Figure 11.10 The Electron Tower. Electrons move spontaneously from donors with more negative reduction potentials to acceptors with more positive reduction potentials.

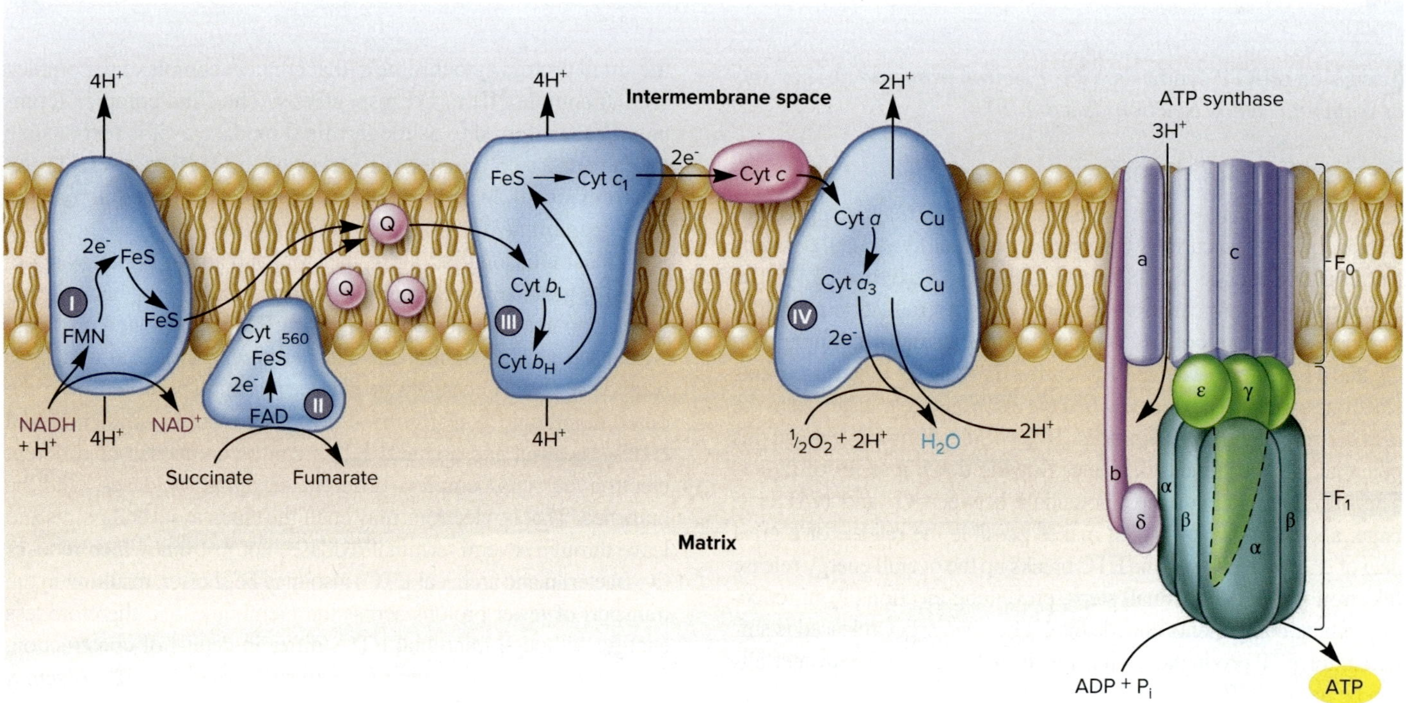

Figure 11.11 Creation of a Proton Gradient by the Mitochondrial Electron Transport Chain and Its Use in Synthesis of ATP. As electrons move along the ETC, protons are transported across the membrane creating a proton gradient. Proton release occurs when electrons are transferred from carriers, such as FMN and coenzyme Q (Q), that carry both electrons and protons to components such as nonheme iron proteins (FeS) and cytochromes (Cyt) that transport only electrons. Complex IV is the terminal oxidase that transfers electrons to oxygen. The number of protons moved across the membrane at each site per pair of electrons transported is still somewhat uncertain and may vary among species; the current consensus is that 10 protons move outward during NADH oxidation. One molecule of ATP is synthesized and released from the enzyme ATP synthase for every 2.5 protons that move back into the matrix by passing through the enzyme.

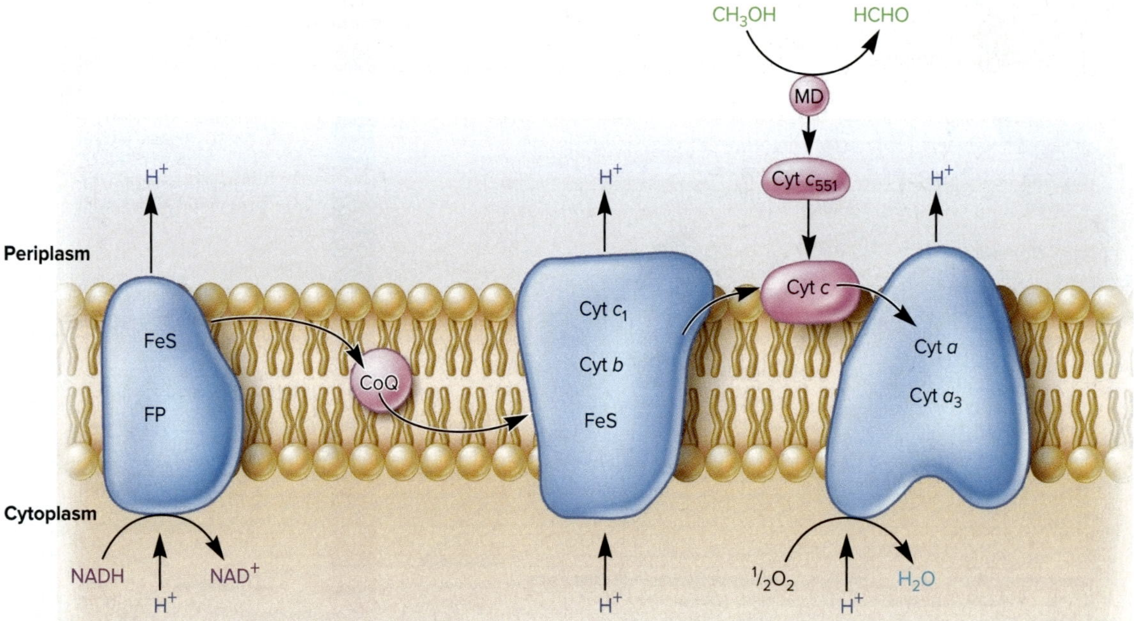

Figure 11.12 ***Paracoccus denitrificans*** **Electron Transport Chain Used During Aerobic Respiration.** *P. denitrificans* can respire both aerobically and anaerobically (figure 11.18). The aerobic ETC resembles a mitochondrial ETC and uses oxygen as the terminal electron acceptor. This metabolically versatile bacterium can also use methanol and methylamine as electron donors. They donate electrons to the chain at the level of cytochrome *c*. Locations of proton movement are shown, but not the number of protons involved. However, 8 to 10 protons are thought to be translocated. FP, Flavoprotein; MD, methanol dehydrogenase.

MICRO INQUIRY *How does the number of H^+ moved across the membrane as a result of NADH oxidation compare with the number of H^+ moved following methanol oxidation?*

chain is very similar to the mitochondrial chain. NADH generated by oxidation of organic substrates (during glycolysis and the TCA cycle) is oxidized to NAD^+ by the first component in the ETC, membrane-bound NADH dehydrogenase. The electrons from NADH are transferred to carriers with progressively more positive reduction potentials. As electrons move through the carriers, protons are moved across the plasma membrane to the periplasmic space, rather than to an intermembrane space, as seen in the mitochondrion (compare figures 11.11 and 11.12). The metabolic diversity of the bacterium is also demonstrated in figure 11.12. When the microbe is growing on an energy source such as glucose, electrons are transferred to the ETC by NADH, as just described. However, *P. denitrificans* also can use 1-carbon molecules such as methanol as a source of energy and electrons. In this case, NADH is not produced because methanol is oxidized by the enzyme methanol dehydrogenase, which in turn transfers the electrons directly to cytochrome *c*.

A simplified view of the *E. coli* chain is shown in **figure 11.13**. This chain differs from the mitochondrial chain in part because it contains a different group of cytochromes. Furthermore, the *E. coli* ETC has two branches that operate under different oxygen levels. When oxygen is readily available, the cytochrome *bo* branch is used (lower half of figure 11.13). When oxygen is less plentiful, the cytochrome *bd* branch is used because it has a higher affinity for oxygen (upper half of figure 11.13). However, it is less efficient than the *bo* branch because the *bd* branch moves fewer protons into the periplasmic space. *Electron Transport Systems and Formation of ATP*

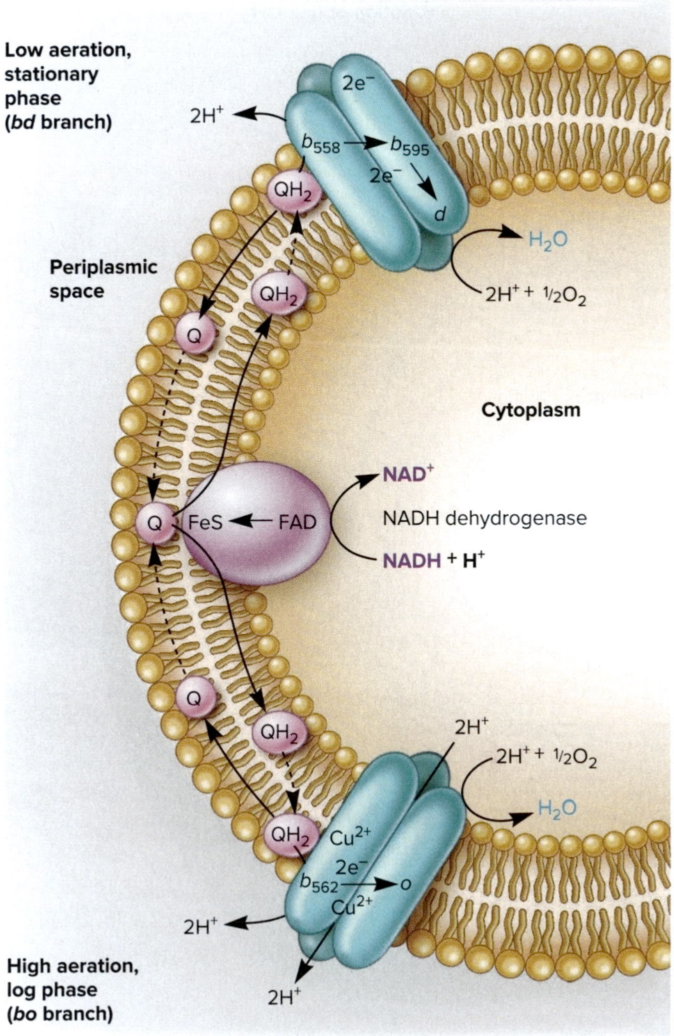

Figure 11.13 The Electron Transport Chain of *E. coli*. NADH transfers electrons from an organic energy and electron source to the ETC. Ubiquinone-8 (Q) connects the NADH dehydrogenase with two terminal oxidase systems. The upper (*bd*) branch operates when the bacterium is in stationary phase and there is little oxygen. It is called the *bd* branch because it uses the cytochromes b_{558}, b_{595}, and *d*. The lower *bo* branch functions when *E. coli* is growing rapidly with good aeration. It is so named because it uses two cytochromes, b_{562} and *o*.

Oxidative Phosphorylation

Oxidative phosphorylation is the process by which ATP is synthesized using the energy from electron transport, which in turn is driven by the oxidation of a chemical energy source. The mechanism by which oxidative phosphorylation takes place is best explained by the **chemiosmotic hypothesis,** formulated by British biochemist Peter Mitchell (1920–1992). According to the chemiosmotic hypothesis, mitochondrial ETCs are organized so that as electrons are transported down the chain, protons move across the inner membrane from the mitochondrial matrix to the intermembrane space (figure 11.11). In bacteria and archaea, the protons usually move across the plasma membrane from the cytoplasm to the periplasmic space (figures 11.12 and 11.13).

The movement of protons across the membrane is not completely understood. Some protons are actively pumped across the membrane (e.g., by complex IV of the mitochondrial chain; figure 11.11). In other cases, translocation of protons results from the juxtaposition of carriers that accept both electrons and protons with carriers that accept only electrons. For instance, coenzyme Q, a lipid-soluble molecule that moves freely in the membrane, accepts two electrons and two protons but delivers only two electrons to complex III of the mitochondrial chain. This transfer of electrons is complicated by the fact that cytochrome *b* and the FeS protein of complex III accept only one electron at a time and do not accept protons. This difference in protons and electrons carried sets the stage for a phenomenon called the **Q cycle**; the Q cycle ultimately moves four protons across the membrane (**figure 11.14**). The Q cycle involves the sequential oxidation of two reduced coenzyme Q molecules (QH_2), the transfer of two of the four electrons released by these oxidations to one molecule of the oxidized form of coenzyme Q (located nearby in the membrane) via the cytochrome *b* located in complex III, and the transfer of the remaining two electrons to the FeS protein of complex III.

The result of proton expulsion during electron transport is the formation of a proton concentration gradient (Δ pH; chemical potential energy) and a charge gradient (Δψ; electrical potential

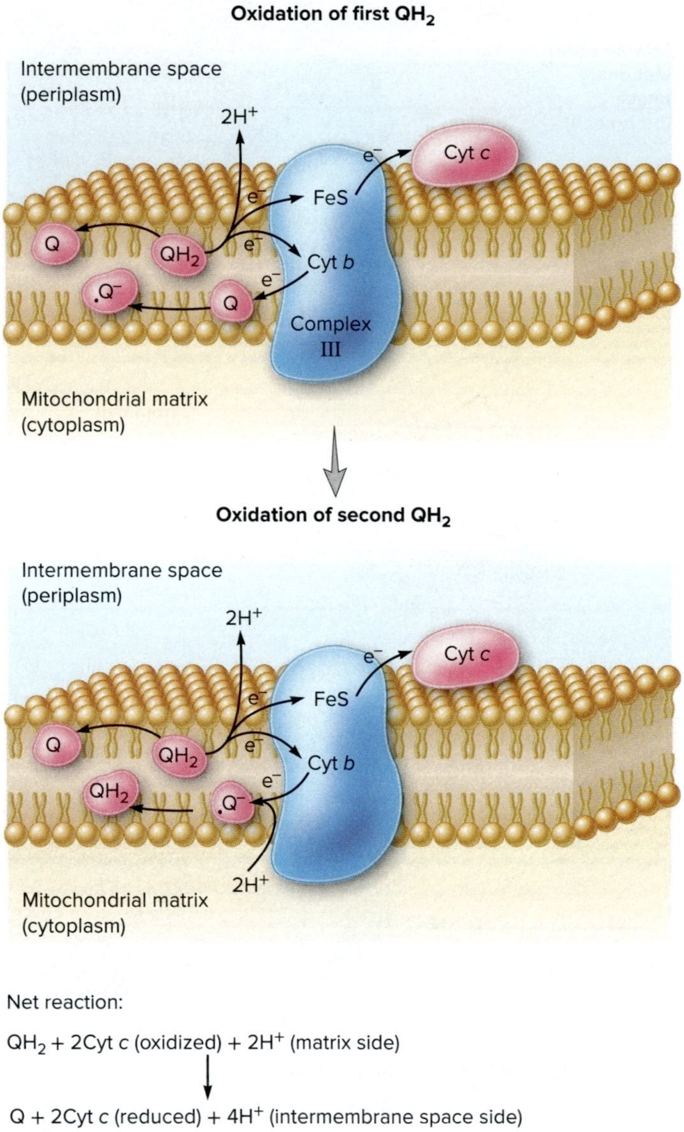

Figure 11.14 The Q Cycle. This simplified illustration shows the major events in the Q cycle.

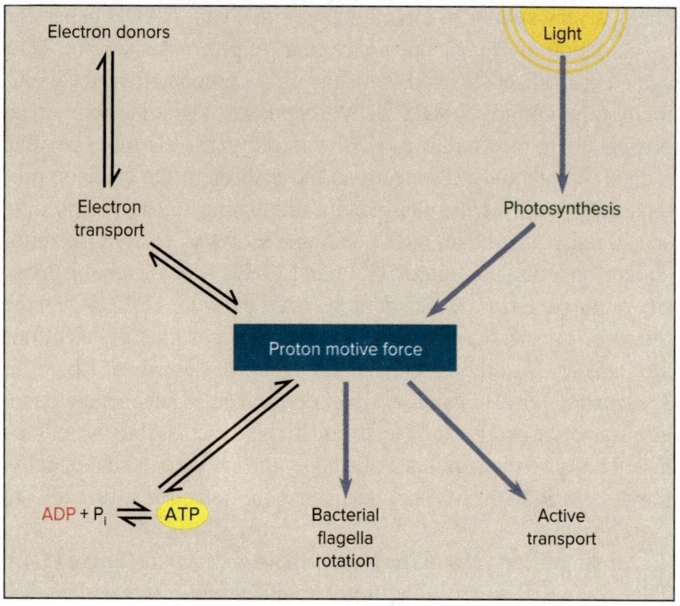

Figure 11.15 The Central Role of Proton Motive Force. Active transport is not always driven by PMF.

energy). Thus as long as the membrane is intact, the mitochondrial matrix has fewer protons so it is more alkaline and more negative than the intermembrane space. Likewise with bacterial and archaeal cells, the cytoplasm is more alkaline and more negative than the periplasmic space. The combined chemical and electrical potential differences make up the **proton motive force** (**PMF**). The PMF is used to perform work when protons flow back across the membrane, down the concentration and charge gradients, and into the mitochondrial matrix or the cytoplasm of bacterial and archaeal cells. This flow is exergonic and is often used to phosphorylate ADP to ATP. PMF is also used by many secondary active transport systems to directly move nutrients into the cell and to rotate the bacterial flagellar motor without the need for ATP hydrolysis (**figure 11.15**). ◀ *Primary and secondary active transport (section 3.3); Bacterial flagella (section 3.7)*

The use of the PMF for ATP synthesis is catalyzed by **ATP synthase.** The best-studied ATP synthases are the F_1F_0 ATP synthases found in mitochondria, chloroplasts, and bacteria. They are also known as F_1F_0 ATPases because they can catalyze ATP hydrolysis. The mitochondrial F_1 component appears as a spherical structure attached to the mitochondrial inner membrane surface by a stalk (**figure 11.16**). The F_0 component is embedded in the membrane. ATP synthase is on the cytoplasmic surface of the plasma membrane in bacterial cells. F_0 participates in proton movement across the membrane. F_1 is a large complex in which three α subunits alternate with three β subunits. The catalytic sites for ATP synthesis are located on the β subunits. At the center of F_1 is the γ subunit. The γ subunit extends through F_1 and interacts with F_0.

The ATP synthase functions like a rotary engine, much like the rotary motor of bacterial flagella. The flow of protons down the proton gradient through the F_0 subunit causes F_0 and the γ subunit to rotate. As the γ subunit rotates rapidly within the F_1 (much like blowing into a pinwheel), conformation changes occur in the β subunits (figure 11.16*b*). One conformation change (β_E to β_{HC}) allows entry of ADP and P_i into the catalytic site. Another conformation change (β_{HC} to β_{DP}) loosely binds ADP and P_i in the catalytic site. ATP is synthesized when the β_{DP} conformation is changed to the β_{TP} conformation, and ATP is released when β_{TP} changes to the β_E conformation, to start the synthesis cycle anew.

ATP Yield During Aerobic Respiration

It is possible to estimate the number of ATP molecules synthesized per pair of electrons donated by an NADH or $FADH_2$ to

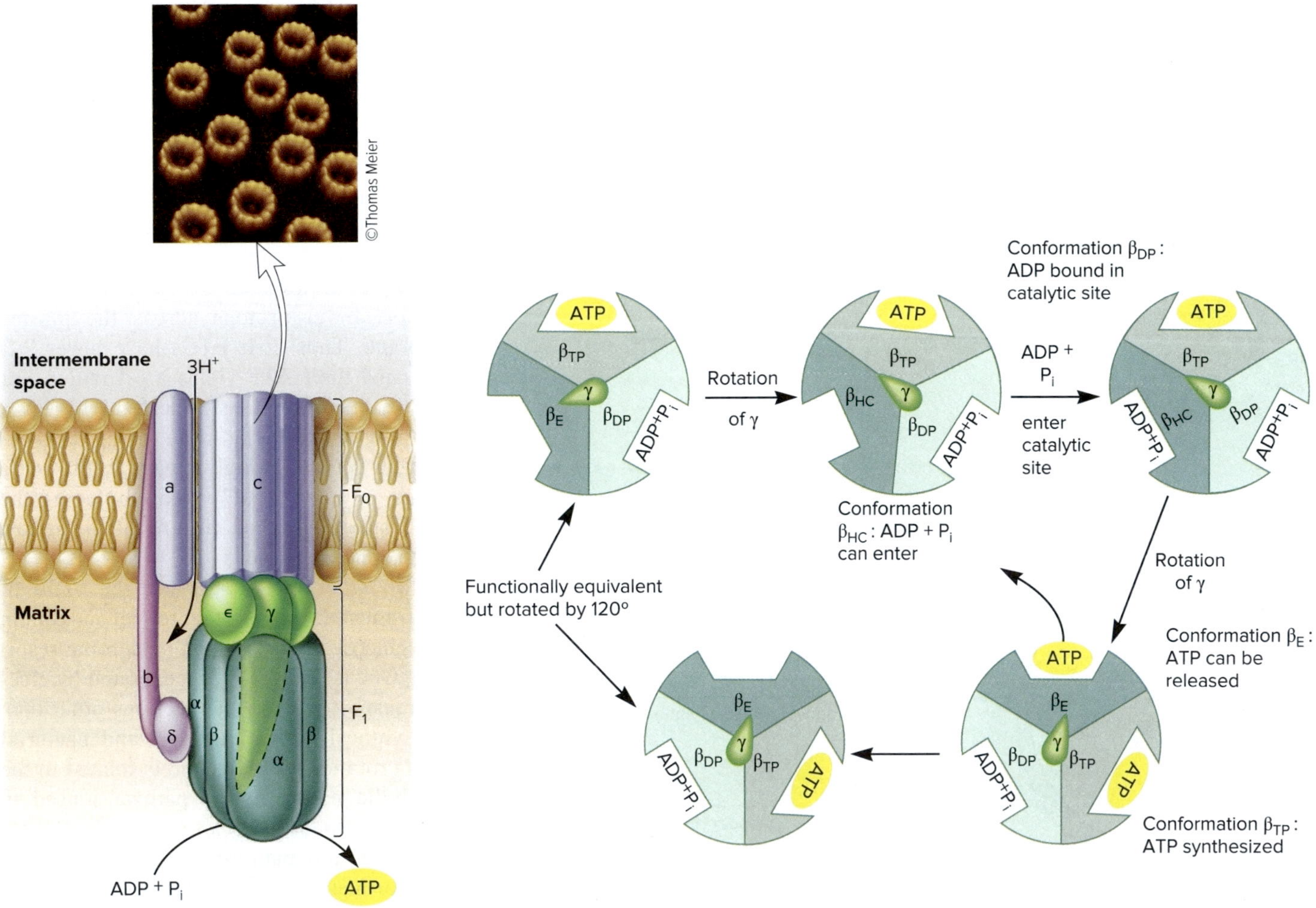

(a) (b)

Figure 11.16 ATP Synthase Structure and Function. (a) The major structural features of ATP synthase. F_1 is composed largely of alternating α and β subunits; the three active sites are on the β subunits. The γ subunit extends through the center and can rotate. The stalk (γ and ε subunits) connects the sphere to F_0, the membrane embedded complex that serves as a proton channel. F_0 contains one a subunit, two b subunits, and nine to twelve c subunits. The stator arm is composed of subunit a, two b subunits, and the δ subunit; it is embedded in the membrane and attached to F_1. A ring of c subunits in F_0 is connected to the stalk and may act as a rotor and move past the a subunit of the stator. As the c subunit ring turns, it rotates the shaft (γε subunits). The inset is an atomic force micrograph of the F_0 rotor. (b) The binding change mechanism is a widely accepted model of ATP synthesis. This simplified drawing of the model shows the three catalytic β subunits and the γ subunit, which is located at the center of the F_1 complex. As the γ subunit rotates, it causes conformational changes in each subunit. The β_E (empty) conformation is an open conformation, which does not bind nucleotides. When the γ subunit rotates, β_E is converted to the β_{HC} (half closed) conformation. P_i and ADP can enter the catalytic site when it is in this conformation. The next rotation by the γ subunit is critical because it brings about three significant conformational changes: (1) β_{HC} to β_{DP} (ADP bound), (2) β_{DP} to β_{TP} (ATP bound), and (3) β_{TP} to β_E. Change from β_{DP} to β_{TP} is accompanied by the formation of ATP; change from β_{TP} to β_E allows for release of ATP from ATP synthase.

the ETC and ultimately used to reduce an atom of oxygen to H_2O. For historical reasons, this is called the phosphorus to oxygen (P/O) ratio. Prior to the acceptance of the chemiosmotic hypothesis, the concept of the P/O ratio was used as a measure of the number of ATP molecules (phosphorus) generated per oxygen (O) reduced as NADH and $FADH_2$ were oxidized. With the acceptance of the chemiosmotic hypothesis, it was recognized that the important measurement was actually the number of protons transported across the membrane per NADH oxidation and the number of protons needed to drive the synthesis of a single ATP. Nonetheless, the amount of ATP synthesized as the result of the oxidation of one NADH or $FADH_2$ is still often referred to as a P/O ratio. The best estimates of the number of protons moved across the membrane due to NADH oxidation is 10; six electrons are thought to be transferred when $FADH_2$ oxidation initiates electron flow to oxygen. It is currently thought that four protons are consumed during ATP synthesis; three are

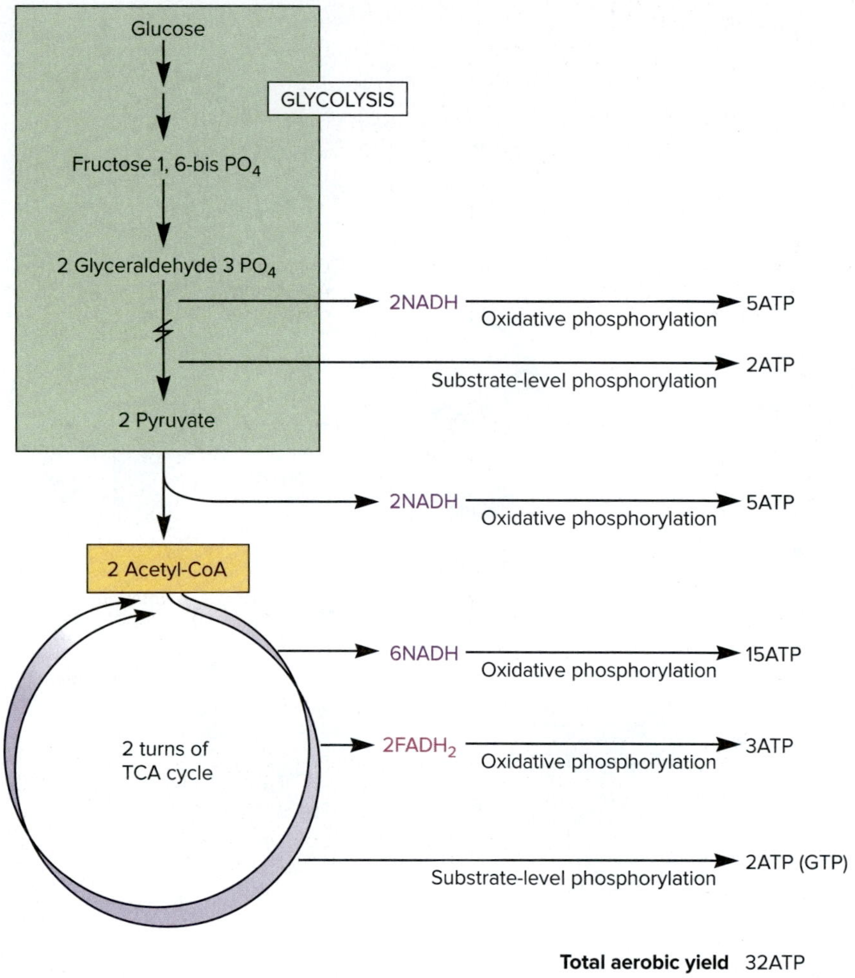

Figure 11.17 Maximum Theoretic ATP Yield from Aerobic Respiration. To calculate the theoretic maximum yield of ATP, P/O ratios of 2.5 for the oxidation of NADH and 1.5 for $FADH_2$ are assumed.

used by ATP synthase and one is used for transport of ATP, ADP, and P_i. Thus the P/O ratio is estimated to be 2.5 for NADH and 1.5 for $FADH_2$.

Given this information, the maximum ATP yield of aerobic respiration in a eukaryote can be calculated (**figure 11.17**). Substrate-level phosphorylation yields at most two ATP molecules per glucose converted to pyruvate (figure 11.5). Two GTP (ATP equivalent) are generated by substrate-level phosphorylation during the two turns of the TCA cycle needed to oxidize two acetyl-CoA molecules (figure 11.8). Thus during aerobic respiration of glucose, at most four ATP molecules are made by substrate-level phosphorylation. Most of the ATP made during aerobic respiration is generated by oxidative phosphorylation. Up to 10 NADH (2 from glycolysis, 2 from pyruvate conversion to acetyl-CoA, and 6 from the TCA cycle) and 2 $FADH_2$ (from the TCA cycle) are generated when glucose is oxidized completely to 6 CO_2. Assuming a P/O ratio of 2.5 for NADH oxidation and 1.5 for $FADH_2$ oxidation, the 10 NADH could theoretically drive the synthesis of 25 ATP, while oxidation of the 2 $FADH_2$ molecules would add another 3 ATP for a maximum of 28 ATP generated via oxidative phosphorylation. Thus oxidative phosphorylation accounts for seven times more ATP than does substrate-level phosphorylation.

The maximum total yield of ATP during aerobic respiration by eukaryotes is 32 ATP. Bacterial ETCs are often shorter and therefore transport fewer protons across the plasma membrane. Thus their ETCs have lower P/O ratios, and their ATP yields are smaller. For example, *E. coli,* with its truncated ETC, has a P/O ratio around 1.3 when using the cytochrome *bo* path at high oxygen levels and a ratio of only about 0.67 when employing the cytochrome *bd* branch (figure 11.13) at low oxygen concentrations.

Two other factors prevent the yield of ATP from reaching its theoretical maximum when glucose is catabolized by aerobic respiration. One is that the PMF generated by electron transport is used for functions other than ATP synthesis (e.g., transport and bacterial flagella rotation). The second is related to the amphibolic nature of the pathways used to catabolize glucose. Recall that an important function of these pathways is the generation of precursor metabolites for anabolism. For each molecule of glucose degraded, numerous precursor metabolites are made, and as each is made, a microbe must "decide" if that metabolite is needed for anabolism or if it can continue the catabolic process. If the precursor metabolite is used for biosynthesis, fewer NADH and $FADH_2$ molecules are made and fewer ATP molecules generated. Cells must carefully monitor and regulate the flow of the carbons from glucose such that ATP production is appropriately balanced with biosynthesis. The mechanisms for this regulation are described in chapters 10 and 14.

Comprehension Check

1. How do the location and structure of mitochondrial ETCs differ from bacterial and archaeal ETCs?
2. Describe the current model of oxidative phosphorylation. Briefly describe the structure of ATP synthase and explain how it is thought to function.
3. Compare and contrast substrate-level phosphorylation and oxidative phosphorylation.
4. Suppose a bacterium used the Entner-Doudoroff pathway to degrade glucose to pyruvate and then completed the catabolism of glucose via the TCA cycle. What is the total maximum ATP yield? Explain your reasoning.

11.7 Anaerobic Respiration Uses the Same Three Steps as Aerobic Respiration

After reading this section, you should be able to:

a. Compare and contrast aerobic respiration and anaerobic respiration using glucose as carbon source
b. List examples of terminal electron acceptors used during anaerobic respiration
c. Explain why the use of nitrate (NO_3^-) as a terminal electron acceptor is dissimilatory nitrate reduction.
d. Explain why less energy is conserved during anaerobic respiration, as compared to aerobic respiration
e. Discuss three examples of the importance of anaerobic respiration

As we have seen, during aerobic respiration, sugars and other organic molecules are oxidized and their electrons transferred to NAD^+ and FAD to generate NADH and $FADH_2$, respectively. These carriers then donate the electrons to an ETC that uses O_2 as the terminal electron acceptor, thereby generating a PMF. However, it is also possible for other terminal electron acceptors to be used for electron transport. **Anaerobic respiration** is the chemoorganotrophic process whereby a terminal electron acceptor other than O_2 is used for electron transport. It is carried out by many bacteria and archaea, and some eukaryotic microbes. The most common terminal electron acceptors used during anaerobic respiration are nitrate, sulfate, and CO_2, but metals and a few organic molecules can also be reduced (**table 11.3**).

In both aerobic and anaerobic respiration processes, NADH and $FADH_2$ are used as the electron donors to the ETC. However, for each NADH or $FADH_2$ oxidized, anaerobic respiration yields less ATP. The lower ATP yield is due to the fact that all terminal electron acceptors used during anaerobic respiration have less positive reduction potentials than O_2 (figure 11.10). Thus, the difference in standard reduction potentials between NADH and these electron acceptors is smaller than the difference between NADH and O_2. This results in a shorter ETC with fewer protons transported to the periplasm per oxidation of NADH or $FADH_2$ as compared to aerobic respiration. The smaller PMF and ATP yield generated by anaerobic respiration explains why facultative anaerobes capable of both aerobic and anaerobic respiration will use aerobic respiration whenever possible.

An example of a facultative anaerobe that can carry out both anaerobic respiration and aerobic respiration is *Paracoccus denitrificans*. Figure 11.12 shows the ETC used by this microbe during aerobic respiration. Under anoxic conditions, *P. denitrificans* uses NO_3^- as its electron acceptor, reducing it to gaseous dinitrogen (N_2). As shown in **figure 11.18**, the anaerobic ETC is more complex than the aerobic chain. The anaerobic ETC is branched and uses different electron carriers, some of which are located in the periplasmic space. Not as many protons are pumped across the membrane during anaerobic growth, but nonetheless, a PMF is established.

Table 11.3 Some Electron Acceptors Used in Respiration

	Electron Acceptor	Reduced Products	Examples of Microorganisms
Aerobic	O_2	H_2O	All aerobic bacteria, fungi, and protists
Anaerobic	NO_3^-	NO_2^-	*E. coli* and other enteric bacteria
	NO_3^-	NO_2^-, N_2O, N_2	*Pseudomonas*, *Bacillus*, and *Paracoccus* spp.
	SO_4^{2-}	H_2S	*Desulfovibrio* and *Desulfotomaculum*
	CO_2	CH_4	Methanogens
	CO_2	Acetate	Acetogens
	S^0	H_2S	*Desulfuromonas* and *Thermoproteus* spp.
	Fe^{3+}	Fe^{2+}	*Pseudomonas*, *Bacillus*, and *Geobacter* spp.
	$HAsO_4^{2-}$	$HAsO_2$	*Bacillus*, *Desulfotomaculum*, *Sulfurospirillum* spp.
	SeO_4^{2-}	Se, $HSeO_3^-$	*Aeromonas*, *Bacillus*, *Thauera* spp.
	Fumarate	Succinate	*E. coli*, *Bacteroides* spp.

The anaerobic reduction of NO_3^- makes this nitrogen unavailable for assimilation into the cell. That is, it can't be taken up by the cell to construct N-containing molecules such as amino acids and nucleotides. Therefore this process is called **dissimilatory nitrate reduction.** When dissimilatory nitrate reduction leads to the production of gaseous compounds such as N_2 that are released into the atmosphere, the process is also called **denitrification.** As can be seen in figure 11.18, denitrification carried out by *P. denitrificans* is a multistep process with four enzymes participating. These enzymes function sequentially to reduce NO_3^- to nitrite (NO_2^-), then nitrite to nitric oxide (NO), followed by NO to nitrous oxide (N_2O), and finally N_2O to N_2.

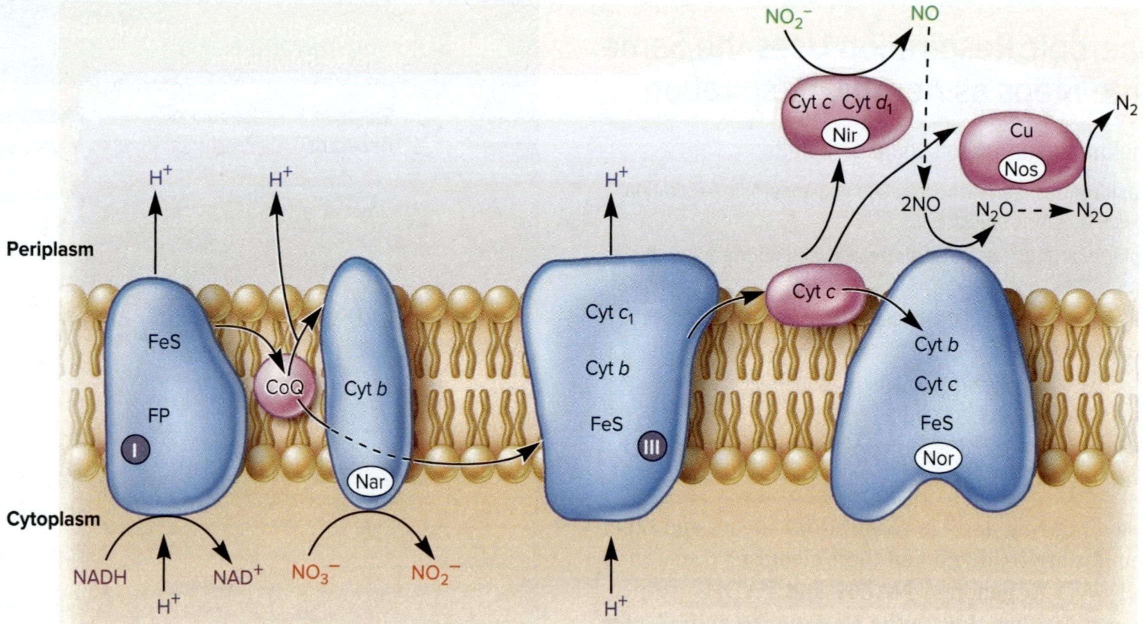

Figure 11.18 ***Paracoccus denitrificans*** **Electron Transport Chain Used During Anaerobic Respiration.** This branched ETC is made of both membrane and periplasmic proteins. Nitrate is reduced to diatomic nitrogen (N_2) by the collective action of four different reductases that receive electrons from CoQ and cytochrome *c*. Locations of proton movement are indicated. Four protons are pumped into the periplasm by complex I, two by nitrate reductase (Nar), and two by complex III. However, two protons are used by nitric oxide reductase (Nor) to reduce nitric oxide to nitrous oxide. Thus six protons, net, are used to create a PMF. FP, Flavoprotein; Nir, nitrite reductase; Nos, nitrous oxide reductase, which has an unusual copper (Cu) - sulfur center.

The use of alternate electron acceptors has important ecological and practical consequences. For instance, *P. denitrificans* and many other soil microbes use nitrate (NO_3^-) as an electron acceptor in **anoxic** soils. When this occurs in agricultural soils, it depletes the soil nitrogen, which in turn decreases crop yields. It also forces farmers to use nitrogen-containing fertilizers that can have detrimental consequences such as the contamination of nearby wells, groundwater, and rivers. On the other hand, if bacteria in a sewage treatment plant perform denitrification, NO_3^- levels in the plant effluent are decreased, decreasing the likelihood that the effluent will cause pollution when discharged. Finally, the use of NO_3^- and other electron acceptors by chemoorganotrophs links the carbon cycle to other biogeochemical cycles such as the nitrogen cycle and the sulfur cycle. This is the basis for many of the interactions that occur between microbes in their habitats. ▶| *Biogeochemical cycling sustains life on Earth (section 28.1)*

Comprehension Check

1. Why does anaerobic respiration yield less ATP than aerobic respiration?
2. What is denitrification? Why do farmers dislike this process?
3. *E. coli* can use O_2, fumarate, or nitrate as a terminal electron acceptor under different conditions. Refer to figure 11.10 to determine the order of energy yield from highest to lowest for these electron acceptors.

11.8 Fermentation Does Not Involve an Electron Transport Chain

After reading this section, you should be able to:

a. Compare and contrast the importance of oxidative phosphorylation and substrate level phosphorylation in aerobic respiration, anaerobic respiration, and fermentation
b. List the pathways that may function during fermentation if glucose is the organism's carbon and energy source
c. Create a table that lists some of the common fermentation pathways and their products and examples of their importance
d. Compare the use of ATP synthase during respiration and fermentation

Despite the tremendous ATP yield obtained by oxidative phosphorylation, some chemoorganotrophic microbes never respire (i.e., use an ETC to generate a PMF) and others may be unable to respire under certain conditions. Microbes completely incapable of respiration lack ETCs. Those that can carry out aerobic respiration may repress the synthesis of ETC components under anoxic conditions, making anaerobic respiration impossible. Finally, some facultative anaerobes can respire both aerobically and anaerobically but may find themselves in an environment that lacks both oxygen and the terminal electron acceptor they use for anaerobic respiration (e.g., nitrate). For each of these microbes, NADH produced by the Embden-Meyerhof pathway reactions during glycolysis (figure 11.5) must still be oxidized back to NAD^+ even though there is no ETC to accept electrons from NADH. If NAD^+ is not regenerated, the oxidation of glyceraldehyde 3-phosphate will cease and glycolysis will stop. Many microorganisms solve this problem by slowing or stopping pyruvate dehydrogenase, the enzyme complex that feeds pyruvate into the TCA cycle as acetyl-CoA. Instead, they use pyruvate or one of its derivatives as an electron acceptor for the reoxidation of NADH in a fermentation process (**figure 11.19**). There are many kinds of fermentations, and they often are characteristic of particular microbial groups (**figure 11.20**). A few of the more common fermentations are introduced here.

Four unifying themes should be kept in mind when microbial fermentations are examined: (1) O_2 is not needed; (2) the electron acceptor is usually either pyruvate or a pyruvate derivative; (3) NADH must be oxidized to NAD^+ without an ETC; and (4) the absence of an ETC reduces the ATP yield per glucose significantly. Thus in fermentation, the substrate (e.g., glucose) is only partially catabolized (i.e., not oxidized entirely to CO_2) and ATP is formed in most organisms only by substrate-level phosphorylation. However, not having a functioning ETC creates a problem for fermenting microbes: They still need a PMF to do work, in particular to drive secondary transport (i.e., antiport, symport, uniport). To solve this dilemma, they use their ATP synthase in the reverse direction. That is, instead of making ATP from ADP + P_i, the ATP synthase pumps protons out of the cell. The energy for this comes from the hydrolysis of ATP to ADP and P_i.

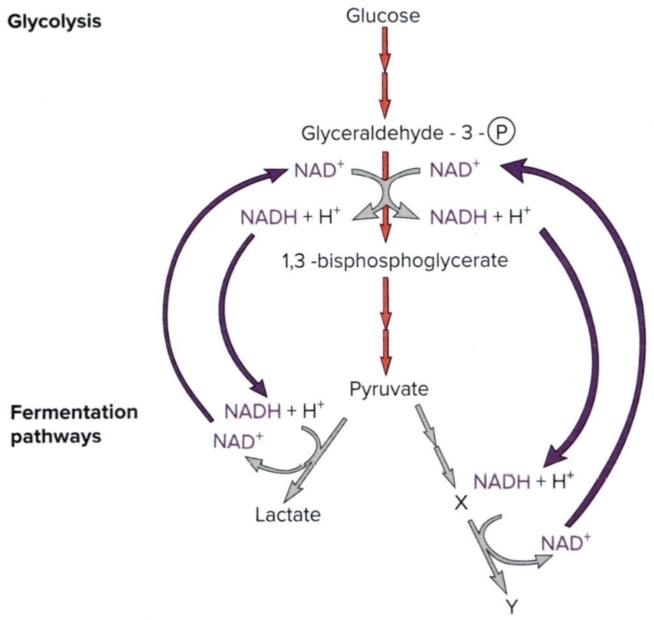

Figure 11.19 Reoxidation of NADH During Fermentation. NADH from glycolysis is reoxidized by being used to reduce pyruvate or a pyruvate derivative (X). Either lactate or reduced product Y result.

MICRO INQUIRY *How many NADH must be reoxidized to NAD^+ for each glucose catabolized?*

Fermentation pathways are named after the major acid or alcohol produced. The most common fermentation is lactic acid (lactate) fermentation, the reduction of pyruvate to lactate (figure 11.20, number 1). It occurs in bacteria (lactic acid bacteria, *Bacillus* spp.), protists (*Chlorella* spp. and some water molds), and animal skeletal muscle. Lactic acid fermenters can be separated into two groups. **Homolactic fermenters** use the Embden-Meyerhof pathway and directly reduce almost all their pyruvate to lactate with the enzyme lactate dehydrogenase. **Heterolactic fermenters** use the pentose phosphate pathway to form products other than lactate; many also produce ethanol and CO_2. Lactic acid bacteria are important to the food industry, where they are used to make a variety of fermented foods (e.g., yogurt, cheese). ▶| *Microbiology of fermented foods (section 41.5)*

Some fungi, protists, and bacteria ferment sugars to ethanol and CO_2 in a process called **alcoholic fermentation.** Pyruvate is decarboxylated to acetaldehyde, which is then reduced to ethanol by alcohol dehydrogenase with NADH as the electron donor (figure 11.20, number 2). This fermentation produces the ethanol desired by brewers making beer and ale and by vintners making wine.

Many bacteria, especially members of the family *Enterobacteriaceae,* metabolize pyruvate to numerous products using several pathways simultaneously. One such complex fermentation is the **mixed acid fermentation,** so called because it results in the excretion of a mixture of acids, particularly acetic, lactic, succinic, and formic acids as well as ethanol (**table 11.4**). Members of the genera *Escherichia, Salmonella,* and *Proteus* carry out mixed acid fermentation. They use pathways numbered 1, 5, 8, and 9 in figure 11.20 to make all the fermentation products, except succinate. If the enzyme formate hydrogenlyase is present (figure 11.20, number 6), formate will be degraded to H_2 and CO_2 (*see figure 22.27*). Another complex fermentation is the

Table 11.4 Mixed Acid Fermentation Products of *Escherichia coli*

	FERMENTATION BALANCE (μM PRODUCT/100 μM GLUCOSE)	
	Acid Growth (pH 6.0)	**Alkaline Growth (pH 8.0)**
Ethanol	50	50
Formic acid	2	86
Acetic acid	36	39
Lactic acid	80	70
Succinic acid	11	15
Carbon dioxide	88	2
Hydrogen gas	75	0.5

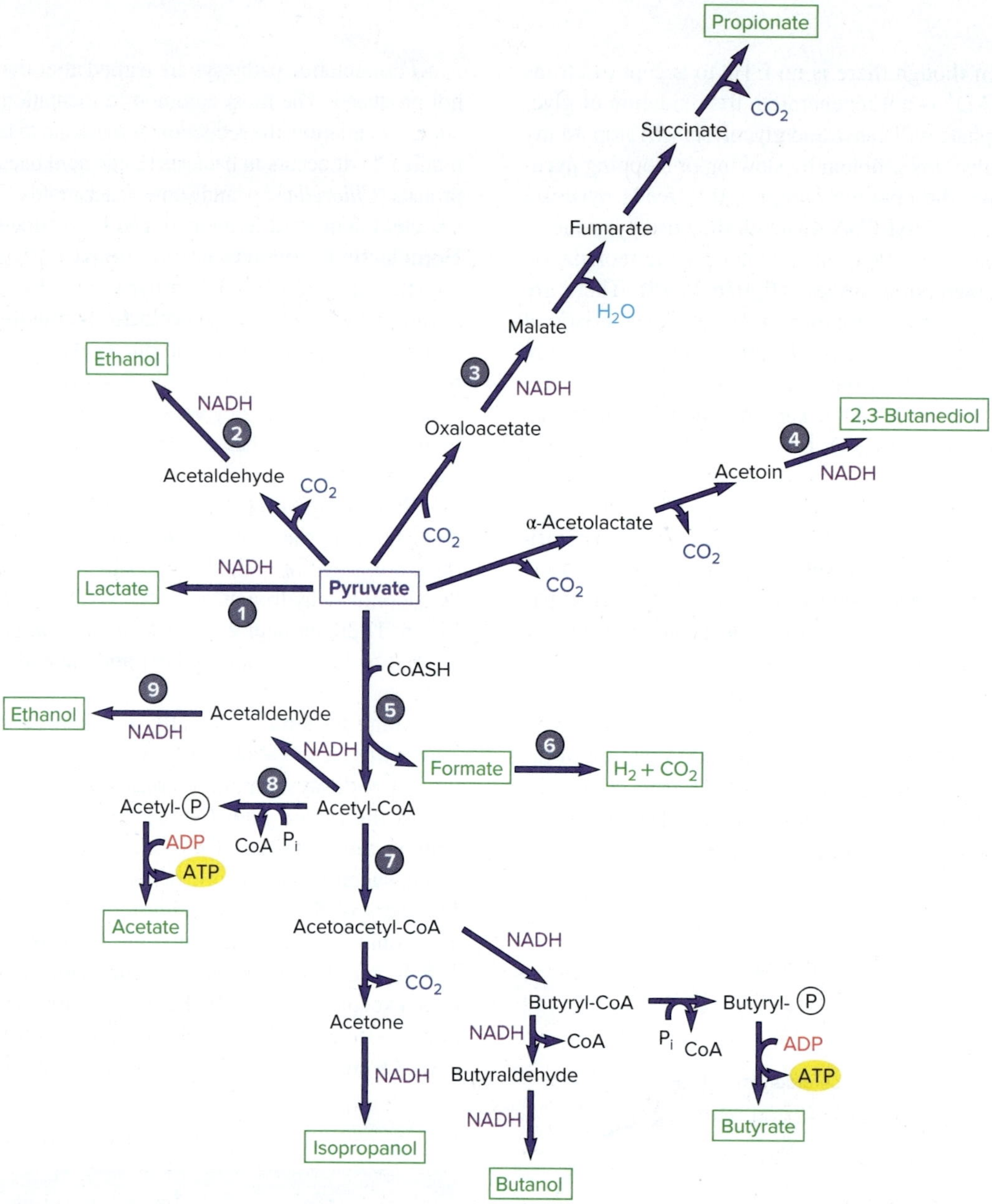

1. Lactic acid bacteria (*Streptococcus, Lactobacillus* species), *Bacillus* spp., enteric bacteria
2. Yeast, *Zymomonas* spp.
3. Propionic acid bacteria (*Propionibacterium* spp.)
4. *Enterobacter, Serratia, Bacillus* species
5. Enteric bacteria
6. Enteric bacteria
7. *Clostridium* spp.
8. Enteric bacteria
9. Enteric bacteria

Figure 11.20 Some Common Microbial Fermentations. Only pyruvate fermentations are shown for the sake of simplicity; many other organic molecules can be fermented. Most of these pathways have been simplified by deletion of one or more steps and intermediates. Pyruvate and major end products are shown in color. Enteric bacteria include members of the genera *Escherichia, Enterobacter, Salmonella,* and *Proteus*.

butanediol fermentation, which is characteristic of members of the genera *Enterobacter, Serratia, Erwinia,* and some species of *Bacillus*. As the name implies, butanediol is the major end product of this fermentation (figure 11.20, number 4). However, large amounts of ethanol also are produced (figure 11.20, number 9), as are smaller amounts of lactic acid (figure 11.20, number 1) and formate (figure 11.20, number 5); in some bacteria, formate is further catabolized to H_2 and CO_2 (figure 11.20, number 6). The use of either mixed acid fermentation or butanediol fermentation is important in differentiating members of the family *Enterobacteriaceae,* many of which are clinically important. ▶| *Order* Enterobacteriales *includes pathogens and mutualists (section 22.3); Class* Bacilli*: aerobic endospore-forming bacteria (section 23.2); Identification of microorganisms from specimens (section 37.2)*

Comprehension Check

1. What are fermentations and why are they useful to many microorganisms?
2. How do the electron acceptors used in fermentation differ from the terminal electron acceptors used during aerobic respiration and anaerobic respiration?
3. How do homolactic fermenters and heterolactic fermenters differ? How do mixed acid fermenters and butanediol fermenters differ?
4. What is the net yield of ATP during homolactic, acetate, and butyrate fermentations? How do these yields compare to aerobic respiration in terms of both quantity and mechanism of phosphorylation?
5. When bacteria carry out fermentation, only a few reactions of the TCA cycle operate. What purpose do you think these reactions might serve? Why are some parts of the cycle shut down?

11.9 Catabolism of Organic Molecules Other Than Glucose

After reading this section, you should be able to:

a. Discuss the role of ATP and UTP in the catabolism of monosaccharides other than glucose
b. Differentiate the catabolism of disaccharides and polysaccharides by hydrolysis from their catabolism by phosphorolysis
c. Discuss the fate of the fatty acid and glycerol components of triglycerides when triglycerides are catabolized
d. State the name of the enzymes responsible for hydrolyzing proteins into amino acids
e. Distinguish deamination from transamination and explain how the two are related
f. Draw a simple diagram that illustrates how the pathways used to catabolize reduced organic molecules other than glucose connect to the glycolytic pathways and the TCA cycle

So far our main focus has been on the catabolism of glucose. However, microorganisms catabolize many other organic molecules, including a variety of other carbohydrates, lipids, and amino acids. Many of these catabolic processes have been put to use by humans in the production of food and the removal of toxic compounds from the environment. The use of these energy sources is discussed next. ▶| *Microbiology of fermented foods (section 41.5); Biodegradation and bioremediation harness microbes to clean the environment (section 43.4)*

Carbohydrates

Carbohydrates catabolized by microbes may come either from outside the cell or from internal sources generated during normal metabolism. Often the initial steps in catabolism of external carbohydrate polymers differ from those employed with internal reserves. However, the goal is the same: to convert the carbohydrate into a monomer that can be fed into one of the glycolytic pathways previously discussed. **Figure 11.21** outlines some catabolic pathways for the monosaccharides (single sugars) glucose, fructose, mannose, and galactose. The first three are phosphorylated using ATP and easily enter the Embden-Meyerhof pathway. In contrast, galactose must be converted to uridine diphosphate galactose (UDP-gal, *see figure 12.7*) after initial phosphorylation, then converted to glucose 6-phosphate in a three-step process.

The common disaccharides are cleaved to monosaccharides by at least two mechanisms: hydrolysis and phosphorolysis (figure 11.21). Maltose, sucrose, and lactose can be directly hydrolyzed to their constituent sugars. Many disaccharides (e.g., maltose, cellobiose, and sucrose) are split by phosphorolysis, in which a phosphate attacks the bond joining the two sugars. This yields the two constituent monosaccharides, one of which is phosphorylated.

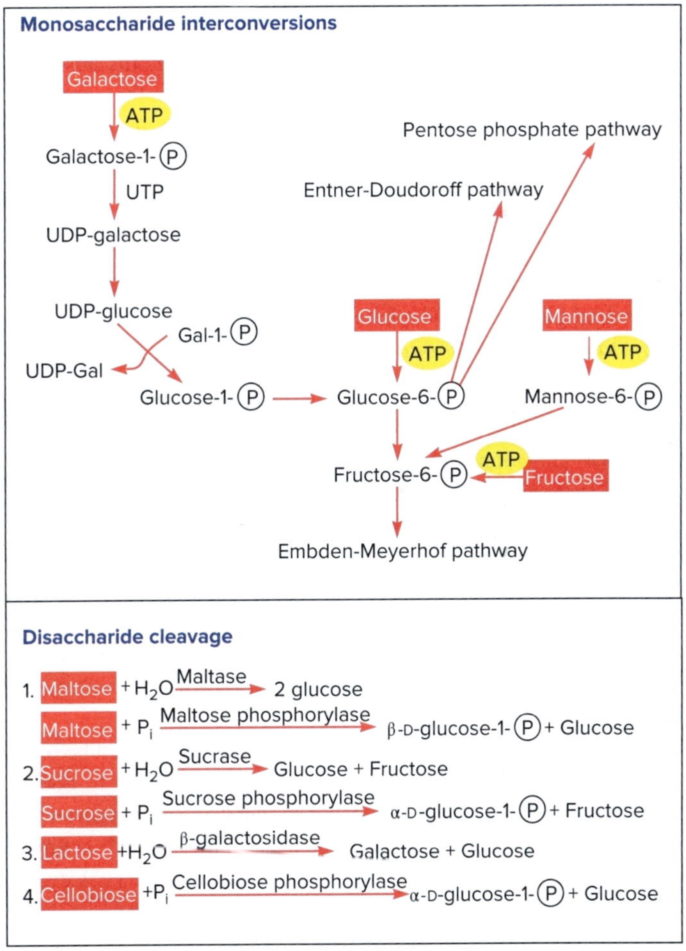

Figure 11.21 Carbohydrate Catabolism. Examples of enzymes and pathways used in disaccharide and monosaccharide catabolism. UDP, uridine diphosphate. See figure AI.9 to learn how monosaccharides are named.

MICRO INQUIRY *What is the difference between a hydrolase and phosphorylase?*

Polysaccharides, like disaccharides, are cleaved by both hydrolysis and phosphorolysis depending on whether they are extracellular or intracellular. Because of their large size, polysaccharides in the environment cannot be transported into the cell. Instead, bacteria, archaea, and fungi degrade external polysaccharides by secreting hydrolytic enzymes. These exoenzymes cleave polysaccharides into smaller molecules that can then be taken up and catabolized. Microbes can catabolize numerous polysaccharides in this way. For instance, starch and glycogen are hydrolyzed by amylases to glucose, maltose, and other products. Cellulose, which is more difficult to digest, is hydrolyzed by the cellulases produced by many fungi and a few bacteria, yielding cellobiose and glucose. Some bacteria excrete an agarase that degrades agar (e.g., *Cytophaga* spp.). Many soil bacteria and bacterial plant pathogens degrade pectin, a polymer of galacturonic acid (a galactose derivative) that is an important constituent of plant cell walls and tissues. Recall that in times of plenty, microorganisms store excess carbon by synthesizing intracellular reserves such as glycogen and starch. They use these to survive for long periods in the absence of exogenous nutrients. Under such circumstances, they catabolize their intracellular stores. Glycogen and starch are degraded by phosphorylases that shorten the polysaccharide chain by one glucose, yielding glucose 1-phosphate.

$$(\text{Glucose})_n + P_i \rightarrow (\text{glucose})_{n-1} + \text{glucose-1-P}$$

Glucose 1-phosphate can enter glycolytic pathways after conversion to glucose 6-phosphate (figure 11.21). ▶| *Soils are an important microbial habitat (section 31.1)*

Lipid Catabolism

Chemoorganotrophic microorganisms frequently use lipids as energy sources. Triglycerides (also called triacylglycerols) are common energy sources. They are composed of a glycerol core attached to three fatty acids (**figure 11.22**). Triglycerides can be hydrolyzed to glycerol and free fatty acids by enzymes called lipases. The glycerol is then phosphorylated and oxidized to dihydroxyacetone phosphate, an intermediate of the Embden-Meyerhof pathway (EMP), which allows for its continued degradation.

Fatty acids from triacylglycerols and other lipids can be oxidized in the **β-oxidation pathway** after being linked to coenzyme A (**figure 11.23**). In this pathway, fatty acids are shortened by two carbons with each turn of the cycle. The two carbon units are released as acetyl-CoA, which can be fed into the TCA cycle or used in biosynthesis. One turn of the cycle produces acetyl-CoA, NADH, and $FADH_2$; NADH and $FADH_2$ can be oxidized by an ETC to provide more ATP. Thus, fatty acids are a rich source of energy for microbes that perform respiration.

Protein and Amino Acid Catabolism

Some bacteria and fungi—particularly pathogenic and food spoilage microorganisms—use proteins as their source of carbon and energy. Like polysaccharides, proteins are too large to transport across the plasma membrane, so these microbes secrete enzymes called **proteases** that hydrolyze proteins to amino acids. These are then transported into the cell and catabolized.

The first step in amino acid catabolism is **deamination,** the removal of the amino group from an amino acid. This is often accomplished by **transamination.** The amino group from an amino acid is transferred to an α-keto acid acceptor (**figure 11.24**). The deaminated amino acid is now an organic acid that can be converted to pyruvate, acetyl-CoA, or a TCA cycle intermediate. The resulting organic acid may be used in several ways. Depending on the microbe, the organic acid may be fermented to release energy. For example, some members of the genus *Clostridium* ferment mixtures of amino acids using the Stickland reaction in which one amino acid is oxidized and a second amino acid acts as the electron acceptor (*see figure 23.26*). Most microbes feed the organic acid into the TCA cycle where it is oxidized, or use it as a source of carbon for the synthesis of cell constituents. Excess nitrogen from deamination may be excreted as ammonium ion. ▶| *Class* Clostridia *(section 23.2)*

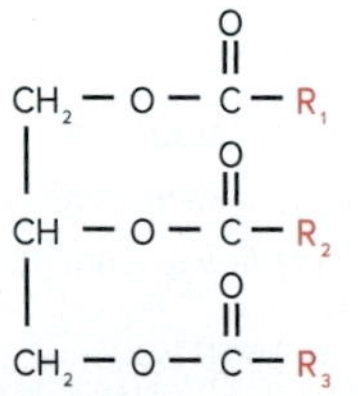

Figure 11.22 A Triacylglycerol (Triglyceride). The R groups represent the fatty acid side chains.

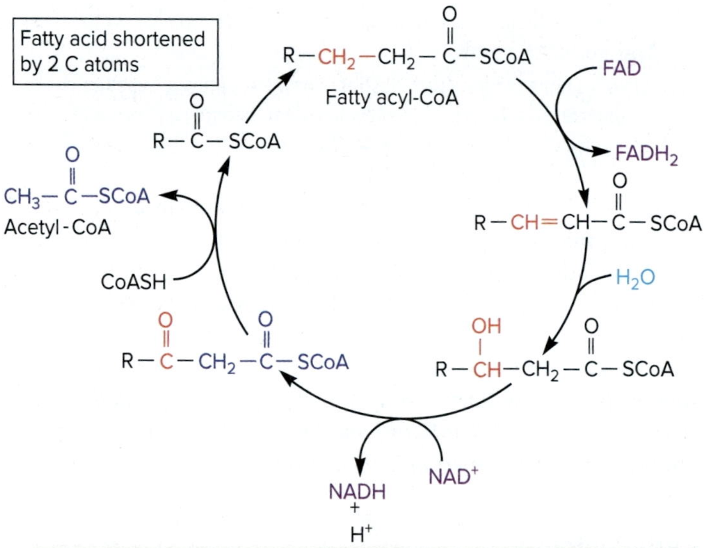

Figure 11.23 Fatty Acid β-Oxidation. The portions of the fatty acid being modified are shown in red.

MICRO INQUIRY *Why are fatty acids a rich source of energy even though no ATP is generated when they are degraded by the β-oxidation pathway?*

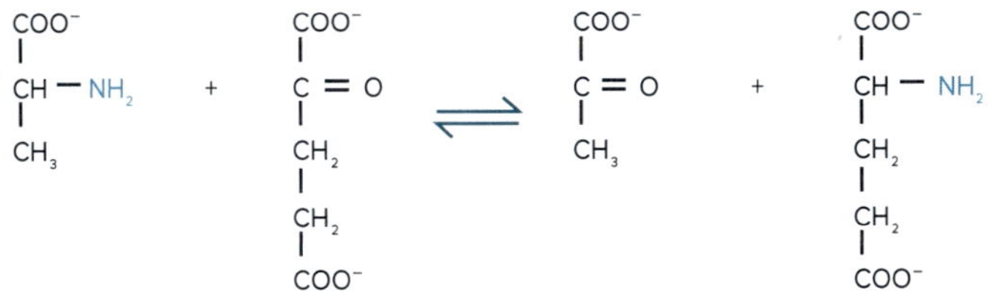

Figure 11.24 An Example of Transamination. The α-amino group (blue) of alanine is transferred to the acceptor α-ketoglutarate, forming pyruvate and glutamate. The pyruvate can be fermented, catabolized in the tricarboxylic acid cycle, or used in biosynthesis.

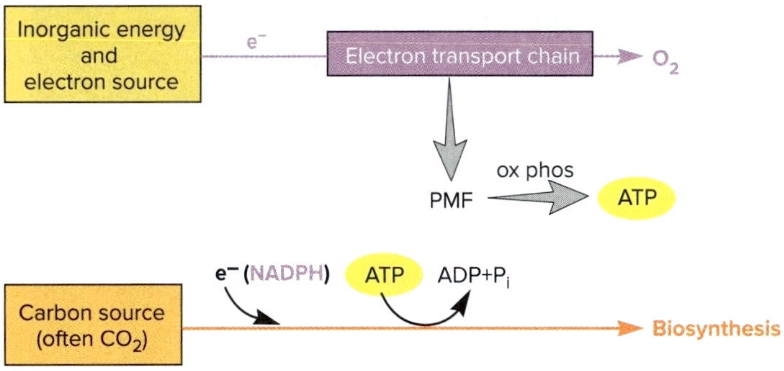

Figure 11.25 Chemolithotrophic Fueling Processes. Chemolithotrophic bacteria and archaea oxidize inorganic molecules (e.g., H_2S and NH_3), which serve as energy and electron sources. The electrons released pass through an electron transport chain, generating a proton motive force (PMF). ATP is synthesized by oxidative phosphorylation (ox phos). Most chemolithotrophs use O_2 as the terminal electron acceptor. However, some can use other exogenous molecules as terminal electron acceptors. Note that a molecule other than the energy source provides carbon for biosynthesis. Many chemolithotrophs are autotrophs.

Comprehension Check

1. Briefly discuss the ways in which microorganisms degrade and use common monosaccharides, disaccharides, and polysaccharides from both external and internal sources.
2. Why might the oxidation of triglycerides pose a problem for microbes capable only of fermentation?
3. Describe how a microorganism might derive carbon and energy from the lipids and proteins in its diet. What is β-oxidation? Deamination? Transamination?

11.10 Chemolithotrophy: "Eating Rocks"

After reading this section, you should be able to:

a. Describe in general terms the fueling reactions of chemolithotrophs
b. List the molecules commonly used as energy sources and electron donors by chemolithotrophs
c. Discuss the use of electron transport chains and oxidative phosphorylation by chemolithotrophs
d. Predict the relative amount of energy released for each of the commonly used energy sources of chemolithotrophs, as compared to energy released during aerobic and anaerobic respiration of glucose
e. Differentiate nitrification from denitrification
f. List three examples of important chemolithotrophic processes

So far, we have considered microbes that synthesize ATP with the energy liberated by the oxidation of organic substrates such as carbohydrates, lipids, and proteins. Certain bacteria and archaea are **chemolithotrophs.** These microbes donate electrons to their ETCs by oxidizing inorganic molecules rather than organic nutrients (**figure 11.25**). The most common energy sources (electron donors) are hydrogen, reduced nitrogen compounds, reduced sulfur compounds, and ferrous iron (Fe^{2+}). The acceptor is usually O_2, but sulfate and nitrate may also be used (**table 11.5**).

With the exception of H_2, much less energy is available from the oxidation of inorganic molecules than from the complete oxidation of glucose to CO_2 (**table 11.6**). This is because the reduction potentials of most inorganic substrates used by chemolithotrophs are much more positive than the reduction potentials of organic molecules such as glucose and the NADH and $FADH_2$ electron carriers (figure 11.10). Thus to have enough ATP to grow and reproduce, chemolithotrophs must oxidize a large quantity of inorganic material. This is particularly true of chemolithoautotrophs, those chemolithotrophs that use CO_2 as their source of carbon. CO_2-fixation

Table 11.5 Representative Chemolithotrophs and Their Energy Sources

Bacteria	Electron Donor	Electron Acceptor	Products
Alcaligenes, Hydrogenophaga, and *Pseudomonas* spp.	H_2	O_2	H_2O
Nitrobacter spp.	NO_2^-	O_2	NO_3^-, H_2O
Nitrosomonas spp.	NH_4^+	O_2	NO_2^-, H_2O
Thiobacillus denitrificans	S^0, H_2S	NO_3^-	SO_4^{2-}, N_2
Acidithiobacillus ferrooxidans	Fe^{2+}, S^0, H_2S	O_2	Fe^{3+}, H_2O, H_2SO_4

Table 11.6 Energy Yields from Oxidations Used by Chemolithotrophs

Reaction	$\Delta G^{\circ\prime}$ (kcal/mole)[1]
$H_2 + \frac{1}{2}O_2 \rightarrow H_2O$	−56.6
$NO_2^- + \frac{1}{2}O_2 \rightarrow NO_3^-$	−17.4
$NH_4^+ + 1\frac{1}{2}O_2 \rightarrow NO_2^- + H_2O + 2H^+$	−65.0
$S^0 + 1\frac{1}{2}O_2 + H_2O \rightarrow H_2SO_4$	−118.5
$S_2O_3^{2-} + 2O_2 + H_2O \rightarrow 2SO_4^{2-} + 2H^+$	−223.7
$2Fe^{2+} + 2H^+ + \frac{1}{2}O_2 \rightarrow 2Fe^{3+} + H_2O$	−11.2

1 The $\Delta G^{\circ\prime}$ for complete oxidation of glucose to CO_2 is −686 kcal/mole. A kcal is equivalent to 4.184 kJ.

pathways consume considerable amounts of ATP and reducing power (often NADPH). Because they must consume a large amount of inorganic material, chemolithotrophs have significant ecological impact. They make important contributions to several biogeochemical cycles, including the nitrogen, sulfur, and iron cycles. ▶| *CO_2 fixation: reduction and assimilation of CO_2 carbon (section 12.3); Biogeochemical cycling sustains life on Earth (section 28.1)*

Members of several bacterial and archaeal genera can oxidize hydrogen gas using a hydrogenase enzyme (table 11.5).

$$H_2 \rightarrow 2H^+ + 2e^-$$

The $2H^+/H_2$ conjugate redox pair has a very negative standard reduction potential (figure 11.10), so its electrons can be donated to either an ETC or NAD^+, depending on the hydrogenase. If NADH is produced, it can be used in ATP synthesis by electron transport and oxidative phosphorylation, with O_2, Fe^{3+}, S^0, and carbon monoxide (CO) as the terminal electron acceptors. Often these hydrogen-oxidizing microorganisms use organic compounds as energy sources when available.

Some bacteria and archaea use the oxidation of nitrogenous compounds as a source of electrons. Among these chemolithotrophs, the **nitrifying bacteria** are best understood. They are soil and aquatic bacteria of considerable ecological significance that carry out **nitrification**—the oxidation of ammonia to nitrate (*see figure 22.13*). Nitrification is a two-step process that until 2015 was thought to always depend on the activity of at least two different microbes. In the first step, ammonia is oxidized to nitrite. In the second step, the nitrite is oxidized to nitrate. When two bacteria are involved one performs the ammonia oxidation (e.g., *Nitrosomonas* spp.) while another oxidizes nitrite to yield nitrate (e.g., *Nitrobacter* spp.). Some members of the genus *Nitrospira* can perform both steps, a process called **comammox,** for *com*plete *amm*onia *ox*idation.

Care should be taken not to confuse nitrification and denitrification. Nitrification is a chemolithotrophic process in which ammonia is oxidized in a two-step process to yield nitrate. Electrons from ammonia and nitrite are donated to an ETC that uses oxygen as the terminal electron acceptor. By contrast, denitrification is a form of anaerobic respiration (figure 11.18).

A unique metabolic process that combines aspects of both nitrification and denitrification is carried out by anammox bacteria (phylum *Planctomycetes*). These bacteria carry out *an*aerobic *amm*onia *ox*idation, or **anammox,** using a membrane-bound organelle called the anammoxosome. In the anammoxosome, ammonia oxidation is coupled with the reduction of nitrite, generating N_2 and H_2O (*see figure 21.13*). The anammox reaction differs from comammox, which produces nitrate as the only end product.

Sulfur-oxidizing microbes are the third major group of chemolithotrophs. The metabolism of *Thiobacillus* and *Acidithiobacillus* spp. has been best studied. These bacteria use sulfur (S^0), hydrogen sulfide (H_2S), thiosulfate ($S_2O_3^{2-}$), and other reduced sulfur compounds as electron donors to their ETCs. Interestingly, they generate ATP by both oxidative phosphorylation and substrate-level phosphorylation (*see figure 22.17*). Substrate-level phosphorylation involves adenosine 5′-phosphosulfate (APS), a high-energy molecule formed from sulfite and adenosine monophosphate (**figure 11.26**).

Some sulfur-oxidizing microbes are extraordinarily flexible metabolically. For example, some bacteria and archaea can grow by oxidizing sulfur with oxygen as the electron acceptor (*see figure 20.9*). However, in the absence of O_2, they switch from chemolithotrophy to anaerobic respiration and oxidize organic material with sulfur as the electron acceptor. Furthermore, many sulfur-oxidizing chemolithotrophs use CO_2 as their carbon source but will grow heterotrophically if they are

(a) Direct oxidation of sulfite

$$SO_3^{2-} \xrightarrow{\text{sulfite oxidase}} SO_4^{2-} + 2e^-$$

(b) Formation of adenosine 5′-phosphosulfate

$$2SO_3^{2-} + 2AMP \rightarrow 2APS + 4e^-$$
$$2APS + 2P_i \rightarrow 2ADP + 2SO_4^{2-}$$
$$2ADP \rightarrow AMP + ATP$$

$$2SO_3^{2-} + AMP + 2P_i \rightarrow 2SO_4^{2-} + ATP + 4e^-$$

Figure 11.26 Energy Generation by Sulfur Oxidation. (a) Sulfite can be directly oxidized to provide electrons for electron transport and oxidative phosphorylation. (b) Sulfite can also be oxidized and converted to adenosine 5′-phosphosulfate (APS; *see figure 22.17*). This route produces electrons for use in electron transport and ATP by substrate-level phosphorylation.

supplied with reduced organic carbon sources such as glucose or amino acids.

Electrons released by the oxidation of ammonia, nitrite, and sulfur-containing compounds are donated to an ETC, thereby generating a PMF that can be used to make ATP by oxidative phosphorylation. As we have already noted, the ATP yields for these chemolithotrophs are low; this reflects their short ETCs. Complicating this already difficult lifestyle is the fact that many of these microbes are also autotrophs, and autotrophs need NAD(P)H (reducing power) as well as ATP to reduce CO_2 and other molecules (figure 11.25). Molecules such as ammonia, nitrite, and H_2S have more positive reduction potentials than the $NAD(P)^+$/NAD(P)H conjugate redox pair (figure 11.10). Therefore, they cannot directly donate their electrons to $NAD(P)^+$. Recall that electrons spontaneously move only from donors with more negative reduction potentials to acceptors with more positive potentials. Thus these chemolithotrophs face a major dilemma: how to make the NAD(P)H they need. They solve this problem by using a process called **reverse electron flow.** During reverse electron flow, electrons derived from oxidation of inorganic substrates (reduced nitrogen or sulfur compounds) are moved up their ETCs to reduce $NAD(P)^+$ to NAD(P)H (**figure 11.27**). Of course, this is not thermodynamically favorable, so energy in the form of the PMF must be diverted from performing other cellular work (e.g., ATP synthesis, transport, motility) to "push" the electrons from molecules of relatively positive reduction potentials to those that are more negative.

Overall, the use of inorganic molecules by microorganisms can be confusing because some of these molecules can be used as electron donors in chemolithotrophy or terminal electron acceptors in anaerobic respiration. There are two key questions to ask when considering the microbial use of inorganic chemicals for energy conservation. First, "What is the oxidation state of the inorganic molecule?" If it is fully reduced (e.g., H_2S, NH_3), then the molecule can only serve as an electron donor because it simply cannot accept any more electrons. Therefore, fully reduced molecules are used by chemolithotrophs to donate electrons to the ETC. Conversely, if the molecule is fully oxidized (e.g., SO_4^{2-}, NO_3^-), it can accept electrons but it has none to donate, so serves as a terminal electron acceptor during anaerobic respiration. The tricky molecules are those that are neither

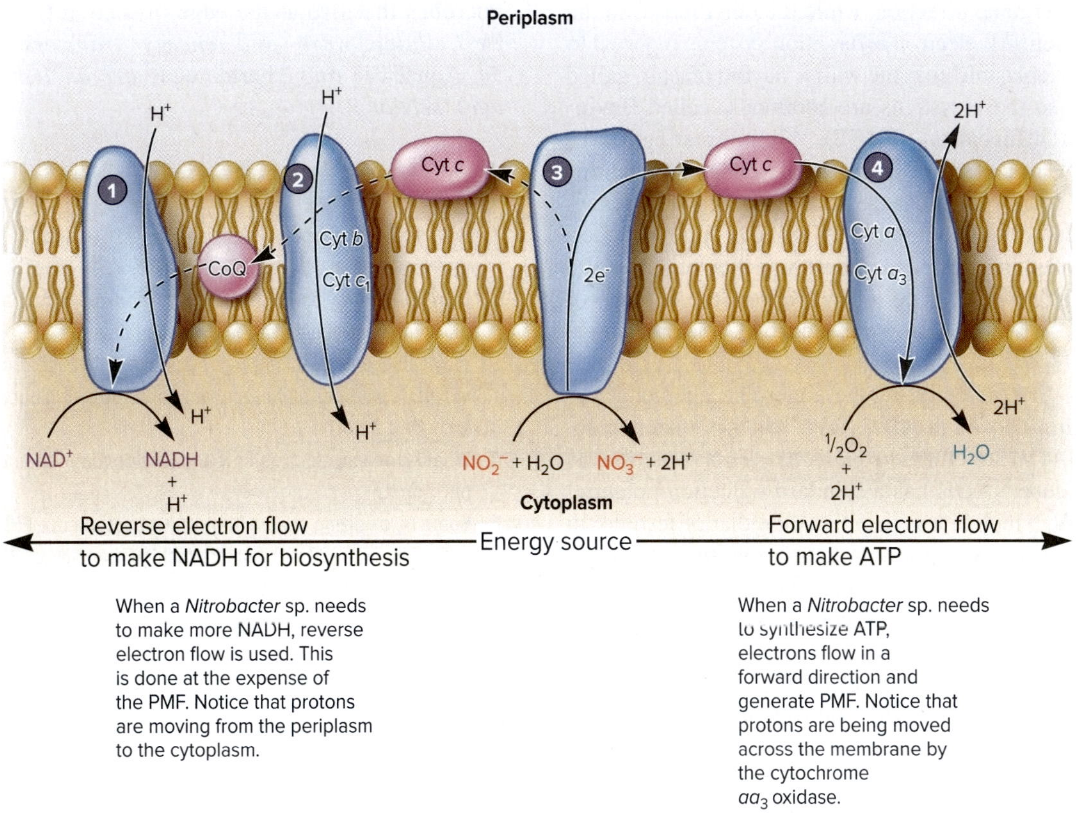

Figure 11.27 Electron Flow in a *Nitrobacter* spp. Electron Transport Chain. *Nitrobacter* spp. oxidize nitrite and carry out normal electron transport to generate proton motive force for ATP synthesis. This is the right-hand branch of the diagram. Some of the proton motive force also is used to force electrons to flow up the reduction potential gradient from nitrite to NAD^+ (left-hand branch). Cytochrome *c* (Cyt c), Coenzyme Q (CoQ), and four protein complexes make up this ETC. The four complexes are NADH-ubiquinone oxidoreductase (1), ubiquinol-cytochrome *c* oxidoreductase (2), nitrite oxidase (3), and cytochrome *aa*$_3$ oxidase (4).

fully oxidized nor reduced (e.g., S^0, NO_2^-), as these can either donate electrons to chemolithotrophic ETCs or serve as terminal electron acceptors in anaerobic respiration. In this case, one must know something about the physiology of the particular microbe.

11.11 Flavin-Based Electron Bifurcation

After reading this section, you should be able to:

a. Explain why electron bifurcation is a fundamental mechanism of energy conservation that differs from ATP hydrolysis or proton motive force
b. Discuss the concept of linking an endergonic reaction with an exergonic one to accomplish an otherwise thermodynamically unfavorable redox reaction

It was recently discovered that some anaerobic archaea and bacteria bypass the thermodynamic barrier of unfavorable, endergonic redox reactions by tightly coupling such reactions with favorable, exergonic ones. This process is known as electron bifurcation (Latin, *bifurcate,* divided into two forks) because the flow of electrons is branched: One electron (or electron pair) moves in the endergonic direction, while the other moves in the exergonic direction. All electron bifurcation systems reported to date feature a cytoplasmic enzyme with a flavin cofactor, called a flavoenzyme; so these systems are commonly called **flavin-based electron bifurcation** (**FBEB**). All bifurcating flavoenzymes oxidize one electron donor, while donating electrons simultaneously to two different electron acceptors. The reduction of one acceptor has to be exergonic so it can "pull" the thermodynamically less favorable endergonic reduction of the second acceptor. In this way FBEB performs work in the absence of ATP hydrolysis or by harnessing the energy of a proton motive force.

FBEB was first detected in the anaerobic chemoheterotrophic bacterium *Clostridium kluyveri,* which transfers electrons from NADH to ferredoxin (Fd). This reaction is endergonic because NADH has a standard reduction potential (E_0') of –320 mV, which is less negative than that of ferredoxin (E_0' about –500 mV). It is possible for NADH to "push electrons uphill" to reduce Fd only because this is accompanied by an exergonic reaction: The simultaneous flow of electrons from NADH to crotonyl-CoA, which is reduced to butyryl-CoA (E_0' = –30 mV). In this reaction electrons are flowing from more negative to less negative E_0'. The role of the cytoplasmic bifurcating flavoenzyme is much like a traffic cop directing the flow of electrons (**figure 11.28**). Since the discovery of FBEB in *C. kluyveri,* other FBEB systems have been discovered. For example, *Thermotoga maritima* is a chemolithoautotrophic bacterium that uses H_2 as a source of electrons and energy. It uses FBEB to transfer electrons from NADH to $2H^+$ (the $2H^+/H_2$ redox pair has an E_0' of about –420 mV) to generate H_2 by simultaneously reducing $2H^+$ with ferredoxin. FBEB appears to be an important form of energy conservation for microbes that live at the edge of their thermodynamic limit. ◀ *Cellular work and energy transfers (section 10.1)* ▶ Aquificae *and* Thermotogae *are ancient bacterial lineages (section 21.1)*

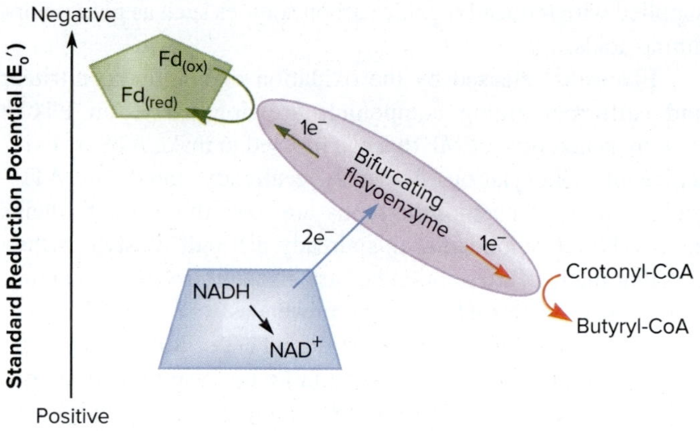

Figure 11.28 Flavin-Based Electron Bifurcation. The bifurcating flavoenzyme accepts 2 electrons from NADH and fuels the endergonic reduction of ferredoxin (Fd) by simultaneously reducing crotonyl-CoA, an exergonic reaction. To reduce both Fd and crotonyl-CoA, 2 NADH must be oxidized. NAD^+ returns to central metabolism where it contributes to energy conservation during glycolysis.

Comprehension Check

1. How do chemolithotrophs obtain their ATP and NAD(P)H? What is their most common source of carbon?
2. Describe energy production by hydrogen-oxidizing bacteria, nitrifying bacteria, and sulfur-oxidizing bacteria.
3. Why can hydrogen-oxidizing bacteria and archaea donate electrons to NAD^+, whereas sulfur- and ammonia-oxidizing bacteria and archaea cannot?
4. What is reverse electron flow and why do many chemolithotrophs perform it?
5. Some H_2 oxidizing chemolithotrophic bacteria use FBEB to oxidize H_2 to $2H^+$ (E_0' = –420mV) while reducing Fd and NAD^+ to NADH. Sketch a diagram showing the flow of electrons relative to the E_0' of these redox pairs.

11.12 Phototrophy

After reading this section, you should be able to:

a. Describe in general terms the fueling reactions of phototrophs
b. Differentiate phototrophy from photosynthesis
c. Describe the light and dark reactions that occur during photosynthesis
d. Summarize the structure and function of the light-absorbing pigments used by oxygenic and anoxygenic phototrophs

e. Explain why oxidative phosphorylation and photophosphorylation by chlorophyll-based phototrophs differ primarily in the energy source driving the process
f. Distinguish cyclic photophosphorylation from noncyclic photophosphorylation
g. Compare and contrast oxygenic photosynthesis, anoxygenic phototrophy, and rhodopsin-based phototrophy
h. List two examples of the importance of chlorophyll-based phototrophy

Many microbes capture the energy in light and use it to synthesize ATP and reducing power (e.g., NADPH). When the ATP and reducing power are used to reduce and incorporate CO_2 (CO_2 fixation), the process is called **photosynthesis.** Photosynthesis is one of the most significant metabolic processes on Earth because almost all our energy is ultimately derived from solar energy. Photosynthetic organisms serve as the base of most food webs in the biosphere. One type of photosynthesis is also responsible for replenishing the supply of O_2 in Earth's atmosphere. Although most people associate photosynthesis with plants, at least half the photosynthesis on Earth is carried out by microorganisms (**table 11.7**). Indeed, *Prochlorococcus*, a cyanobacterium, is thought to be the most abundant photosynthetic organism on Earth and thus a major contributor to the functioning of the biosphere.

Photosynthesis is divided into two parts. In the **light reactions,** light energy is trapped and converted to chemical energy (ATP) and reducing power (e.g., NADPH). ATP synthesis is accomplished by formation of a PMF, initiated by the absorption of light—a process called **photophosphorylation.** The **dark reaction** uses this ATP and reducing power to fix CO_2 and synthesize cell constituents. The term **phototrophy** refers to the use of light energy to fuel a variety of cellular activities but not necessarily CO_2 fixation. In this sense, photoheterotrophs, which use light to drive ATP synthesis but not carbon fixation, are considered phototrophs but are not photosynthetic. In this section, three types of phototrophy are discussed: oxygenic photosynthesis, anoxygenic photosynthesis, and rhodopsin-based phototrophy (**figure 11.29**). The dark reactions of photosynthesis are reviewed in chapter 12. ▶| *CO_2 fixation: reduction and assimilation of CO_2 carbon (section 12.3)*

Light Reactions in Oxygenic Photosynthesis

Photosynthetic eukaryotes and cyanobacteria carry out **oxygenic photosynthesis,** so named because oxygen is generated and released into the environment when light energy is converted to chemical energy. Central to this process, and to all other phototrophic processes, are light-absorbing pigments (**table 11.8**). In oxygenic phototrophs, the most important pigments are **chlorophylls.** Chlorophylls are large planar molecules composed of four pyrrole rings (tetrapyrrole) with a central magnesium atom coordinated to the four nitrogen atoms of the pyrrole rings (**figure 11.30**). The pyrrole rings are decorated with specific chemical groups that distinguish one type of

Table 11.7	Diversity of Phototrophic Microorganisms
Eukaryotes	Multicellular green, brown, and red algae; unicellular protists (e.g., euglenoids, dinoflagellates, diatoms)
Bacteria	Cyanobacteria, green sulfur bacteria, green nonsulfur bacteria, purple sulfur bacteria, purple nonsulfur bacteria, heliobacteria, acidobacteria
Archaea	Halophiles

chlorophyll from another. A long hydrophobic tail (the R side chain at the bottom left of figure 11.30) aids in chlorophyll's attachment to membranes, the site of the light reactions. Several chlorophylls are found in eukaryotes; the two most important are chlorophyll *a* and chlorophyll *b*. These two molecules

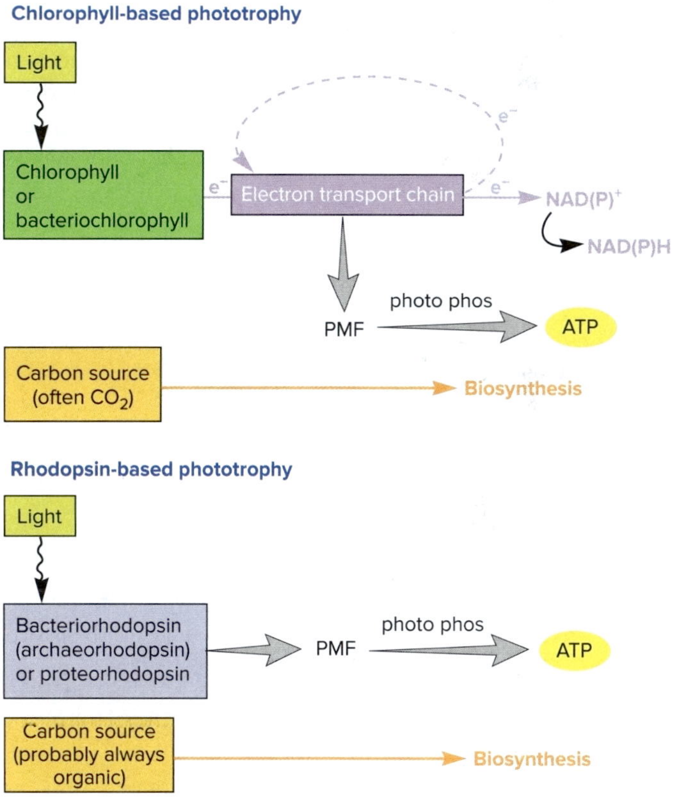

Figure 11.29 Phototrophic Fueling Reactions. Phototrophs use light to generate a proton motive force (PMF), which is then used to synthesize ATP by a process called photophosphorylation (photo phos). The process requires light-absorbing pigments. When the pigments are chlorophyll or bacteriochlorophyll, the absorption of light triggers electron flow through an electron transport chain, accompanied by the pumping of protons across a membrane. The electron flow can be either cyclic (dashed line) or noncyclic (solid line), depending on the organism and its needs. Rhodopsin-based phototrophy differs in that the PMF is formed directly by the light-absorbing pigment, which is a light-driven proton pump. Many phototrophs are autotrophs and must use much of the ATP and reducing power they make to fix CO_2.

Table 11.8 Properties of Chlorophyll-Based Photosynthetic Systems

Property	Eukaryotes	Cyanobacteria	Green Bacteria, Purple Bacteria, Heliobacteria, and Acidobacteria
Photosynthetic pigment	Chlorophyll *a*	Chlorophyll *a*[1]	Bacteriochlorophyll
Number of photosystems	2	2[2]	1
Photosynthetic electron donors	H_2O	H_2O	H_2, H_2S, S, organic matter
O_2 production pattern	Oxygenic	Oxygenic[3]	Anoxygenic
Primary products of energy conversion	ATP + NADPH	ATP + NADPH	ATP
Carbon source	CO_2	CO_2	Organic or CO_2

1 Members of the cyanobacterial genus *Prochlorococcus* have divinyl chlorophyll *a* and *b*.
2 A recently discovered cyanobacterium lacks photosystem II.
3 Some cyanobacteria can function anoxygenically under certain conditions. For example, *Oscillatoria* can use H_2S as an electron donor instead of H_2O.

differ slightly in their structure and spectral properties. They both absorb light in the red range of the visible light spectrum (*see figure 7.20*), but chlorophyll *a* has a light absorption peak at a slightly longer wavelength than chlorophyll *b* (665 nm versus 645 nm). In addition to absorbing red light, chlorophylls also absorb blue light strongly. Because chlorophylls absorb primarily in the red and blue ranges, green light is transmitted, and these organisms appear green.

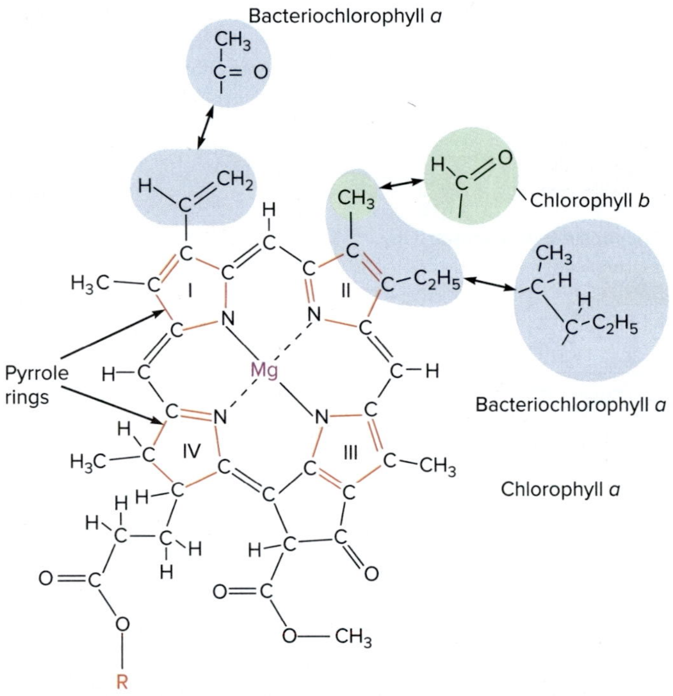

Figure 11.30 Chlorophyll Structure. The structures of chlorophyll *a*, chlorophyll *b*, and bacteriochlorophyll *a*. The complete structure of chlorophyll *a* is given. Only one group is altered to produce chlorophyll *b*, and two modifications in the ring system are found in bacteriochlorophyll *a*. The side chain (R) of bacteriochlorophyll *a* may be either phytyl (a 20-carbon chain also found in chlorophylls *a* and *b*) or geranylgeranyl (a 20-carbon side chain similar to phytyl but with three more double bonds). Pyrroles are numbered and shown in red.

Because chlorophylls absorb a narrow range of light wavelengths, other photosynthetic pigments assist in trapping light energy. The most widespread of these are carotenoids, long molecules, usually yellowish in color, that possess an extensive conjugated double bond system (**figure 11.31**). β-carotene is present in cyanobacteria belonging to the genus *Prochloron* and most photosynthetic protists; fucoxanthin is found in protists such as diatoms and dinoflagellates. Red algae and cyanobacteria have photosynthetic pigments called **phycobiliproteins,** consisting of a protein with a linear tetrapyrrole attached. These pigments give these microbes their characteristic colors.

Carotenoids and phycobiliproteins are often called **accessory pigments** because they help chlorophylls harvest light energy but they cannot harvest energy directly. They are important because they absorb light in the range not absorbed by chlorophylls (the blue-green through yellow range; about 470 to 630 nm; *see figure 21.6*). This light energy is transferred to chlorophyll. In this way, accessory pigments make photosynthesis possible over a broader range of wavelengths. In addition, this allows organisms to use light not used by other phototrophs in their habitat. Accessory pigments also serve as microbial sunscreen, protecting microorganisms from intense sunlight, which could oxidize and damage the photosynthetic apparatus.

About 300 chlorophyll molecules and the accessory pigments are assembled in highly organized arrays called antennas whose purpose is to create a large surface area to trap as many photons as possible. Light energy captured in an antenna is transferred from accessory pigment to chlorophyll and from chlorophyll to chlorophyll until it reaches a **reaction-center chlorophyll pair** that is directly involved in photosynthetic electron transport (*see figure 21.9*). In oxygenic phototrophs, there are two kinds of antennas associated with two different photosystems (**figure 11.32**).

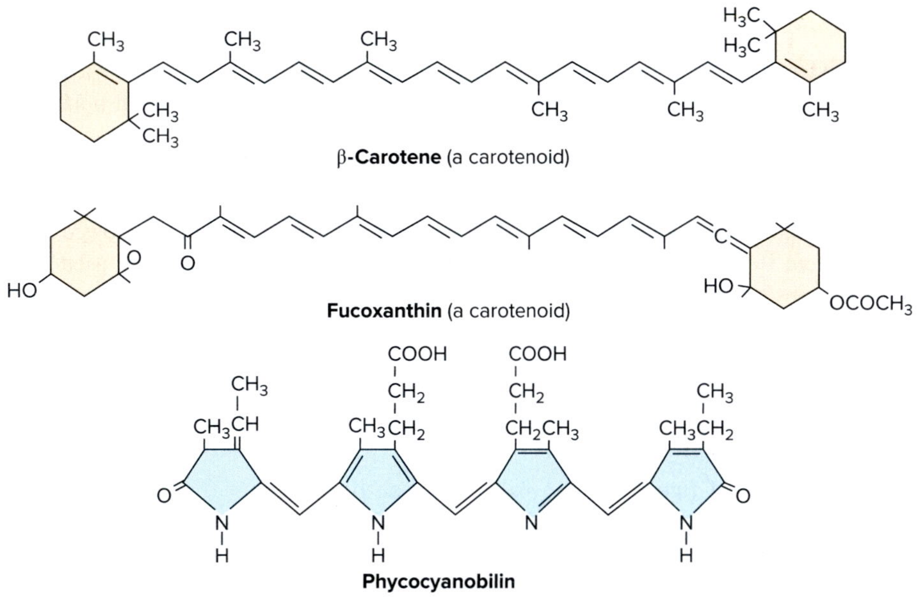

Figure 11.31 Representative Accessory Pigments. Beta-carotene is a carotenoid found in photosynthetic protists and plants. Note that it has a long chain of alternating double and single bonds called conjugated double bonds. Fucoxanthin is a carotenoid accessory pigment in several types of algae. Phycocyanobilin is a linear molecule consisting of four pyrrole rings (tetrapyrrole; see figure 11.30); it is attached to a protein to form a phycobiliprotein.

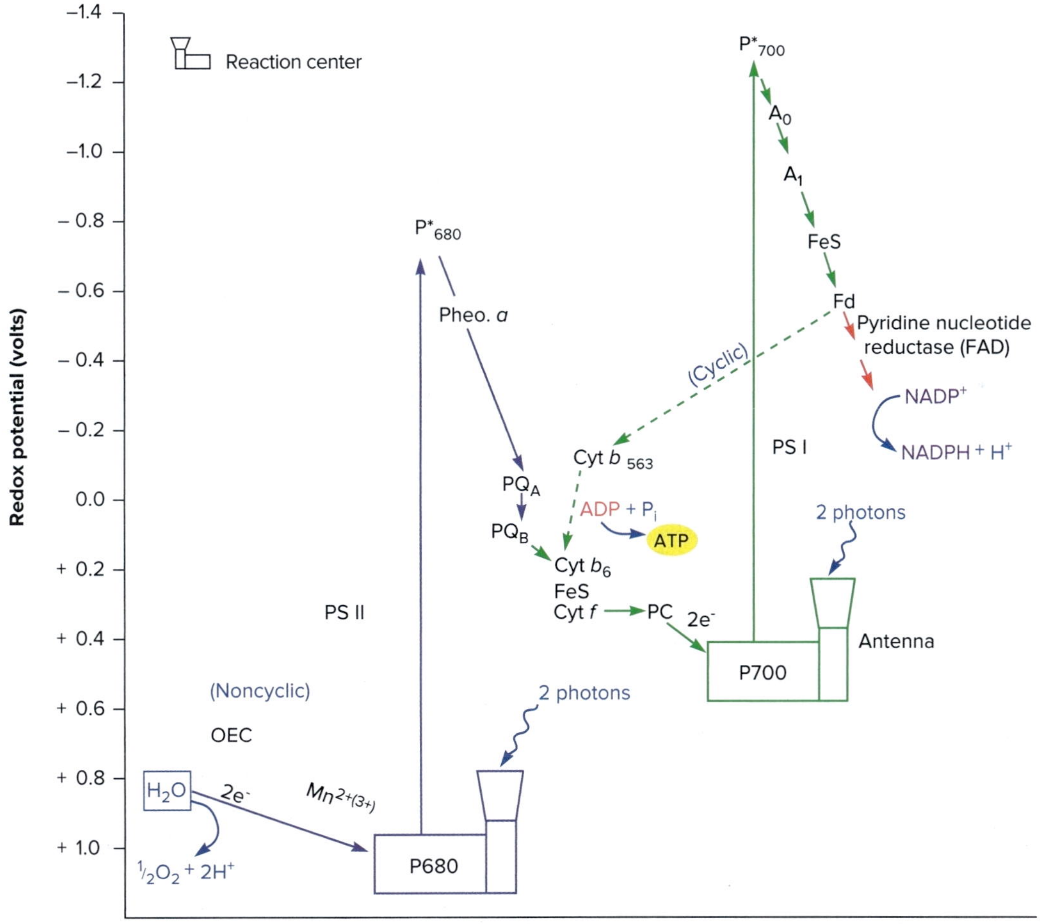

Figure 11.32 Oxygenic Photosynthesis. Cyanobacteria and eukaryotic algae are similar in having two photosystems, although they may differ in some details. The carriers involved in electron transport are ferredoxin (Fd) and other FeS proteins; cytochromes b_6, b_{563}, and *f*; plastiquinones A and B (PQ_A and PQ_B); copper-containing plastocyanin (PC); pheophytin *a* (Pheo. *a*); possibly a specialized form of chlorophyll (A_0); and phylloquinone (A_1). Both photosystem I (PS I) and photosystem II (PS II) are involved in noncyclic photophosphorylation; only PS I participates in cyclic photophosphorylation. The oxygen-evolving complex (OEC) that extracts electrons from water contains manganese ions that enable electron transfer to the PS II reaction center.

MICRO INQUIRY *When electrons from P700 are used to reduce $NADP^+$, what compound supplies electrons for the re-reduction of P700?*

Photosystem I absorbs longer wavelength light and funnels the energy to a reaction center chlorophyll *a* pair called P700. The term P700 signifies that this molecule most effectively absorbs light at a wavelength of 700 nm. **Photosystem II** traps light at shorter wavelengths and transfers its energy to the reaction center chlorophyll pair P680. Depending on the needs of the organism, photosystem I can function independently, while at other times photosystems I and II are used together. We begin our discussion with the two photosystems working together.

Two Photosystems Working Together Generate ATP and NADPH; Oxygen Is Released

Close inspection of figure 11.32 reveals that when light is absorbed by a reaction-center chlorophyll pair in a photosystem (i.e., P680 and P700), the chlorophyll molecules change from a very positive standard reduction potential to a very negative standard reduction potential. That is, the absorption of light causes the reaction-center chlorophylls to become excellent electron donors. With that in mind, let's first examine photosystem II. When it absorbs light, the energized P680 (P^*_{680} in figure 11.32) donates electrons to a series of electron carriers that ultimately donate the electrons to P700 of photosystem I. The flow of electrons through the P680-associated ETC generates a PMF, which can be used by ATP synthase to make ATP from ADP and P_i. In the meantime, photosystem I has also absorbed light and its energized reaction-center chlorophylls (P^*_{700} in figure 11.32) donate electrons to a short ETC that ultimately transfers electrons to $NADP^+$, yielding NADPH. Thus when both photosystems work together, both ATP and NADPH are produced. Because the electrons flow in a noncyclic pattern, this is called noncyclic electron flow, and the ATP produced is said to be made by **noncyclic photophosphorylation.**

At this point the cell has a problem. When P700 was oxidized, its electrons were replaced by electrons arising from the oxidation of P680. But if P680 is to continue to absorb light, its electrons must be replaced. Notice in figure 11.32 that the standard reduction potential of P680 is more positive than that of the $\frac{1}{2}O_2/H_2O$ conjugate redox pair. This means that H_2O can be used to donate electrons to P680, resulting in the release of oxygen. This is accomplished by a metal-containing enzyme called the oxygen-evolving complex (OEC) or the water-splitting complex, which is closely associated with photosystem II. As photosystem II absorbs light, Mn atoms in the OEC become increasingly oxidized, enabling them to extract electrons from H_2O and pass them to a protein in the photosystem, which then passes them to P680. In the process, protons are moved across the membrane in which the photosynthetic machinery is located. Thus the evolution of oxygen from water also contributes to the PMF.

Photosystem I Working Alone Yields Only ATP

Now let us return our attention to photosystem I functioning alone. Notice in figure 11.32 that when electrons are donated from P^*_{700} to the photosystem I–associated ETC, they reach a branch point at ferredoxin (Fd). When photosystem I functions alone, the electrons follow the dashed line to an electron carrier in the photosystem II–associated ETC (cyt b_6). From there, the electrons eventually return to P700. In this case the electrons have flowed in a cyclic manner. This still generates a PMF, which can be used to synthesize ATP, and this is referred to as **cyclic photophosphorylation.** Also note that because the electrons were diverted at Fd, no electrons were available to reduce $NADP^+$ to NADPH. Thus when photosystem I functions alone, only ATP is made available to the cell.

You might be asking why an organism needs to use both photosystems at certain times, and only photosystem I at others. The answer is related to the requirements of the dark reactions. The dark reactions of oxygenic phototrophs use three ATP and two NADPH to reduce one CO_2 to carbohydrate (CH_2O).

$$CO_2 + 3ATP + 2NADPH + 2H^+ + H_2O \rightarrow CH_2O + 3ADP + 3P_i + 2NADP^+$$

Oxygenic phototrophs use noncyclic and cyclic electron flow to supply these needs. The noncyclic system generates one NADPH and one ATP per pair of electrons; therefore four electrons passing through the system produce two NADPH and two ATP. A total of 8 quanta of light energy (4 quanta for each photosystem) is needed to propel the four electrons from water to $NADP^+$. Because the ratio of ATP to NADPH required for CO_2 fixation is 3:2, at least one more ATP must be supplied. Cyclic photophosphorylation probably operates independently to generate the extra ATP. This requires absorption of another 2 to 4 quanta. It follows that around 10 to 12 quanta of light energy are needed to reduce and incorporate one molecule of CO_2 during photosynthesis. ▶| *CO_2 fixation: reduction and assimilation of CO_2 carbon (section 12.3)*

It is worth reemphasizing that although light is the source of energy for chlorophyll-based phototrophy, the process used to make ATP is virtually the same as seen for chemotrophs: Oxidation-reduction reactions occurring in ETCs generate a PMF used by ATP synthase to make ATP. Furthermore, just as is true of mitochondrial electron transport, photosynthetic electron transport takes place within a membrane. Chloroplast granal membranes contain both photosystems and their antennas. **Figure 11.33** shows a thylakoid membrane carrying out noncyclic photophosphorylation. Protons move to the thylakoid interior during photosynthetic electron transport and return to the stroma when ATP is formed. It is thought that stromal lamellae possess only photosystem I and are involved in cyclic photophosphorylation alone. In cyanobacteria, photosynthetic light reactions are located in thylakoid membranes within the cell. *Photosynthetic Electron Transport and ATP Synthesis*

Light Reactions in Anoxygenic Photosynthesis

Certain bacteria carry out a second type of photosynthesis called **anoxygenic photosynthesis.** Although the term *photosynthesis*

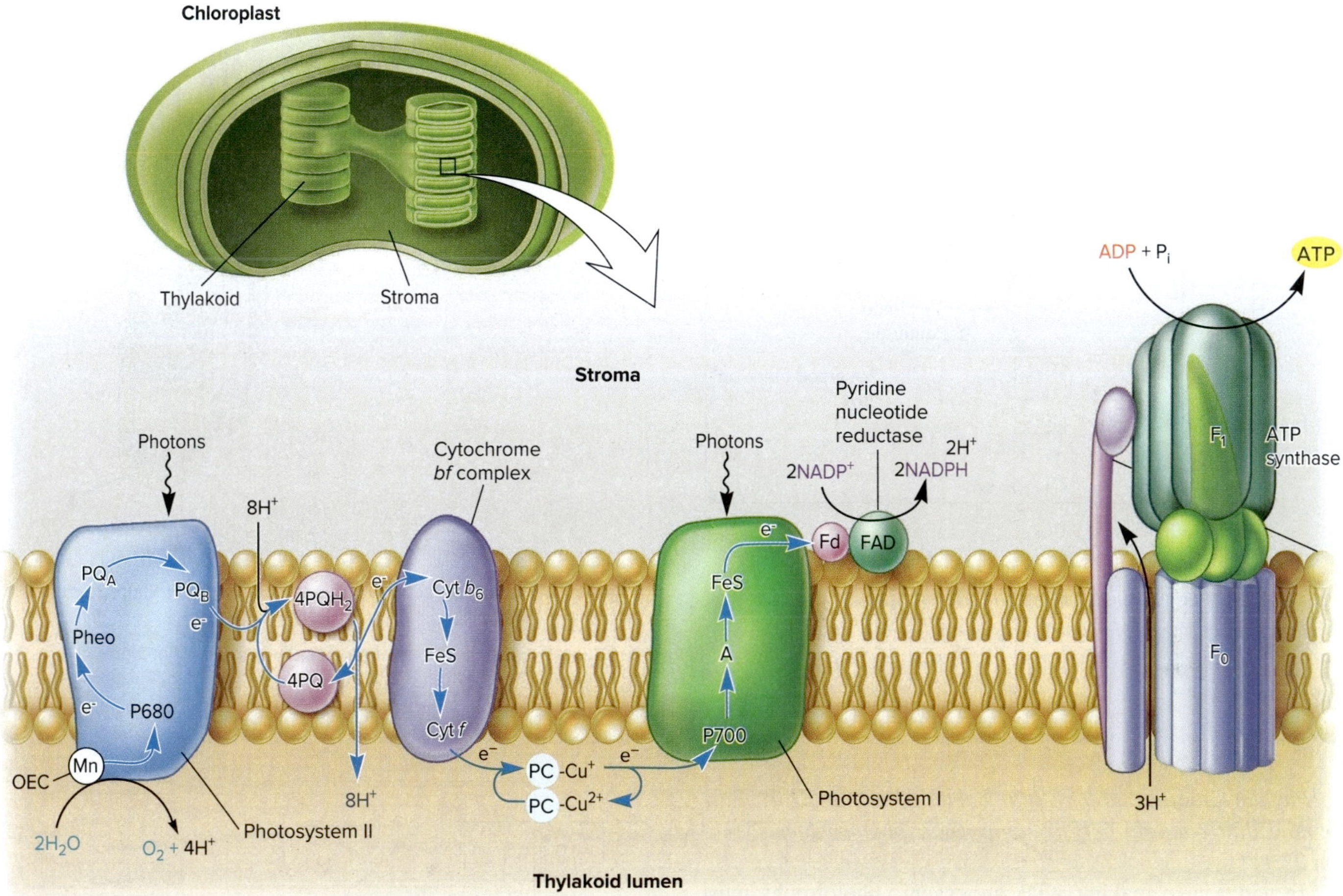

Figure 11.33 The Light Reactions of Photosynthesis. An illustration of the chloroplast thylakoid membrane showing photosynthetic ETC function and noncyclic photophosphorylation. The chain is composed of three complexes: photosystem (PS) I, the cytochrome *bf* complex, and PS II. Two diffusible electron carriers connect the three complexes. Plastoquinone (PQ) connects PS II with the cytochrome *bf* complex, and plastocyanin (PC) connects the cytochrome *bf* complex with PS I. The light-driven electron flow pumps protons across the thylakoid membrane and generates a PMF, which can then be used to make ATP. Water is the source of electrons and the oxygen-evolving complex (OEC) produces oxygen. A in PS I represents both A_0 and A_1, as shown in figure 11.32.

is used, many of these bacteria can function as photoheterotrophs and don't always use the energy gained from light to fix CO_2. *Anoxygenic phototrophy* is a better term; it derives its name from the fact that molecules other than water are used as an electron source and therefore O_2 is not produced. The process also differs in terms of (1) the pigments used, (2) the participation of just one photosystem, and (3) the mechanisms used to generate reducing power. Members of five bacterial phyla carry out anoxygenic phototrophy: *Proteobacteria* (purple sulfur and purple nonsulfur bacteria), *Chlorobi* (green sulfur bacteria), *Chloroflexi* (green nonsulfur bacteria), *Firmicutes* (heliobacteria), and *Acidobacteria*. The biology and ecology of these organisms are described in more detail in chapters 21, 22, and 23.

Anoxygenic phototrophs have light-absorbing pigments called **bacteriochlorophylls** (figure 11.30). Although bacteriochlorophylls are structurally very similar to the chlorophylls used in oxygenic photosynthesis, they are very sensitive to oxygen so are found only in bacteria that perform anoxygenic phototrophy. In some bacteria, bacteriochlorophylls are located in membranous vesicles called chlorosomes (*see figure 21.7*). The absorption maxima of bacteriochlorophylls are at longer wavelengths than those of chlorophylls, reaching into the infrared wavelengths. This better adapts anoxygenic phototrophic bacteria to their ecological niches. ▶| *Photosynthetic bacteria are diverse (section 21.4)*

The key difference between anoxygenic and oxygenic phototrophs is that anoxygenic phototrophs have only a single photosystem. Because they are restricted to cyclic photophosphorylation, they do not produce O_2 from H_2O. Indeed, almost all anoxygenic phototrophs are strict anaerobes. The schemes for the phototrophic ETCs of two purple nonsulfur bacteria are given in **figures 11.34** and **11.35**. When the reaction-center bacteriochlorophyll P870 is excited, it donates an electron to an ETC that transfers the electron back to P870 while generating sufficient PMF to drive ATP synthesis by ATP synthase.

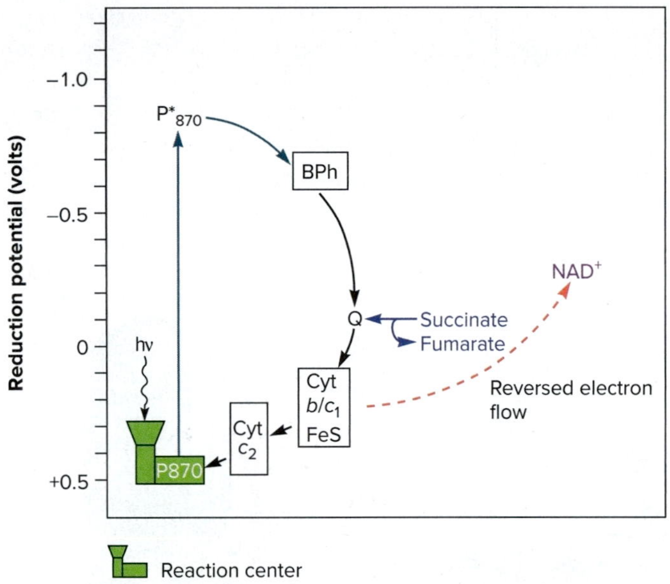

Figure 11.34 Purple Nonsulfur Bacterial Phototrophy. The ETC of the purple nonsulfur bacterium *Rhodobacter sphaeroides.* Ubiquinone (Q) is very similar to coenzyme Q. The electron source succinate is in blue. BPh, Bacteriopheophytin.

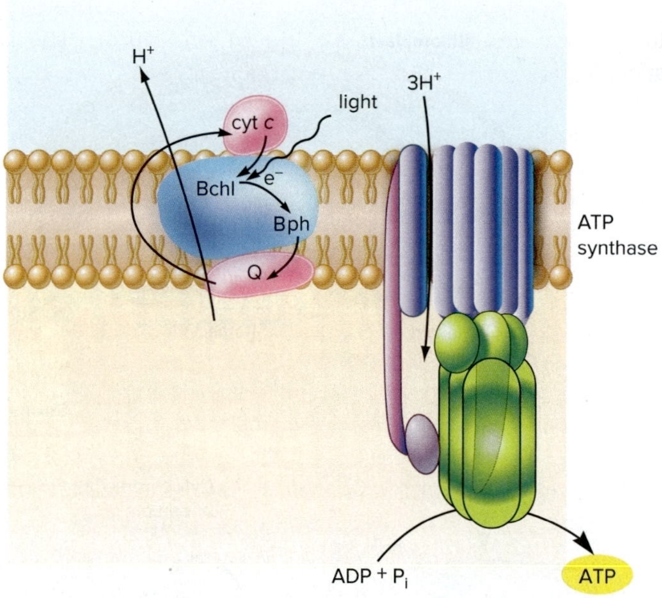

Figure 11.35 Cyclic Electron Flow During Anoxygenic Phototrophy. The photosystem of *Blastochloris viridis* is illustrated. Bchl, bacteriochlorophyll; Bph, bacteriophaeophytin; cyt c, cytochrome c; Q, quinones.

▶| *Phylum* Chlorobi: *green sulfur bacteria (section 21.4); Purple nonsulfur bacteria perform anoxygenic photosynthesis (section 22.1)*

In addition to making ATP, anoxygenic phototrophs must also generate reducing power (NAD[P]H or reduced ferredoxin) for CO_2 fixation (photoautotrophs) and other biosynthetic processes (photoautotrophs and photoheterotrophs). They are able to generate reducing power in at least three ways, depending on the bacterium. (1) Some have hydrogenases that are used to reduce $NAD(P)^+$ to NAD(P)H directly by donating electrons from hydrogen gas. This is possible because hydrogen gas has a more negative reduction potential than NAD^+ (figure 11.10). (2) Others, such as the photosynthetic purple bacteria, use reverse electron flow to generate NAD(P)H (figure 11.34). In this mechanism, electrons are drawn off the photosynthetic ETC and "pushed" to $NAD(P)^+$ using PMF. This process is similar to that seen for chemolithotrophs having inorganic energy sources with more positive reduction potentials than that of $NAD(P)^+$/NAD(P)H. (3) Phototrophic green bacteria and heliobacteria also draw off electrons from their ETCs. However, because the reduction potential of the carrier where this occurs is more negative than NAD^+, the electrons flow spontaneously to these electron acceptors. Thus these bacteria exhibit a simple form of noncyclic photosynthetic electron flow (**figure 11.36**). The diversion of electrons from the photosynthetic ETC means that continued generation of PMF will cease unless the electrons are replaced. Electrons from environmental electron donors such as hydrogen sulfide, elemental sulfur, and organic compounds replace the electrons removed from the ETC in this way. *Cyclic and Noncyclic Photophosphorylation*

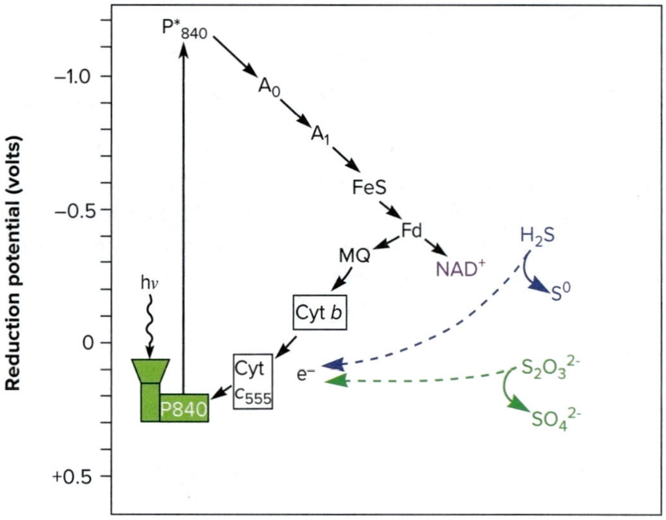

Figure 11.36 Green Sulfur Bacterial Light Reactions. Light energy is used to make ATP by cyclic photophosphorylation and to move electrons from thiosulfate ($S_2O_3^{2-}$) and H_2S (green and blue) to NAD^+. The ETC consists of A_0 (bacteriochlorophyll 663); A_1, a quinone-like molecule; FeS (iron sulfur molecules within the reaction center); Fd (ferredoxin); MQ (menaquinone); and two different cytochromes (Cyt).

Rhodopsin-Based Phototrophy

So far, we have discussed chlorophyll-based types of phototrophy; that is, the use of chlorophyll or bacteriochlorophyll to absorb light and initiate the conversion of light energy to chemical energy. Within the last decade, it has become apparent that many bacteria and archaea are capable of chlorophyll-independent phototrophy. These microbes rely on a form of microbial rhodopsin, a molecule similar to that found in the eyes of many multicellular organisms. The first microbial rhodopsin was discovered in certain archaea. However, it was first described so long ago that it was called bacteriorhodopsin; we now know that it is more correctly called **archaerhodopsin.** Since that time, a variety of rhodopsins have since been found in members of *Proteobacteria* (and thus called proteorhodospin), *Flavobacteria,* extremely halophilic bacteria (xanthorhodopsin), and some eukaryotes. The archaerhodopsin of the halophilic archaeon *Halobacterium salinarum* remains the best studied and is the focus of our discussion here. ▶| *Haloarchaea (section 20.3)*

H. salinarum normally depends on aerobic respiration for the release of energy from its organic energy source. It cannot grow anaerobically by anaerobic respiration or fermentation. However, under conditions of low oxygen and high light intensity, it synthesizes archaerhodopsin, a deep-purple pigment. Archaerhodopsin's color comes from retinal, a type of carotenoid. Retinal is covalently attached to a protein that is embedded in the plasma membrane such that retinal is in the center of the membrane.

Archaerhodopsin functions as a light-driven proton pump. When retinal absorbs light, a proton is released and the archaerhodopsin undergoes a sequence of conformation changes that transports the proton into the periplasmic space (*see figure 20.17*). This light-driven proton pumping generates a pH gradient that can be used to power the synthesis of ATP by chemiosmosis. This phototrophic capacity is particularly useful to *H. salinarum* because oxygen is not very soluble in concentrated salt solutions and may decrease to an extremely low level in its habitat. When the surroundings become temporarily anoxic, the archaeon uses light energy to synthesize sufficient ATP to survive until oxygen levels rise again.

The discovery that microbial rhodopsins are common in marine ecosystems blurs the line between metabolic strategies. To date, all rhodopsin-containing microbes are chemoorganoheterotrophs, using organic carbon as their source of carbon, electrons, and energy. Note that rhodopsin-based proton pumping generates a PMF without having to oxidize NADH via an ETC. What does this mean for these microorganisms? Many of these microbes live in nutrient-depleted environments where organic carbon is limited. With each organic molecule the organism takes up, it must "decide" to either oxidize it to make NADH and ATP or reduce it at the expense of NADPH and ATP to make more macromolecules for growth. The use of rhodopsin-based proton pumping enables the synthesis of ATP without having to catabolize as much precious organic carbon. That is, by performing rhodopsin-mediated phototrophy, these chemoorganoheterotrophs can shunt more organic carbon to growth rather than oxidizing it for energy. ▶| Microorganisms in the open ocean are adapted to nutrient limitation (section 30.2)

Comprehension Check

1. Define the following terms: *light reactions, dark reactions, chlorophyll, carotenoid, phycobiliprotein, antenna,* and *photosystems I* and *II*.
2. What happens to a reaction center chlorophyll pair, such as P700, when it absorbs light?
3. What is the function of accessory pigments?
4. What is photophosphorylation? What is the difference between cyclic and noncyclic photophosphorylation?
5. Compare and contrast anoxygenic phototrophy and oxygenic photosynthesis. How do these two types of phototrophy differ from rhodopsin-based phototrophy?
6. Suppose you isolated a bacterial strain that carried out oxygenic photosynthesis. What photosystems would it possess, and what group of bacteria would it most likely belong to?

Key Concepts

11.1 Metabolic Diversity and Nutritional Types

- All organisms require a source of carbon, energy, and electrons. Biologists have devised terms to describe how these needs are fulfilled (**table 11.1**).
- Heterotrophs use reduced organic molecules as their source of carbon for biosynthetic purposes. Autotrophs use CO_2 as their primary or sole carbon source.
- Phototrophs use light energy, and chemotrophs obtain energy from the oxidation of chemical compounds.
- Electrons are extracted from reduced inorganic substances by lithotrophs and from reduced organic compounds by organotrophs.
- Most organisms can be placed into one of five nutritional types: photolithoautotroph, photoorganoheterotroph, chemolithoautotroph, chemolithoheterotroph, and chemo-organoheterotroph (**table 11.2**).

11.2 There Are Three Chemoorganotrophic Fueling Processes

- Chemoorganotrophic microorganisms can use three kinds of electron acceptors during energy metabolism (**figure 11.2**). Electrons from the oxidized nutrient can be accepted by an endogenous electron acceptor (fermentation), by oxygen (aerobic respiration), or by another external electron acceptor (anaerobic respiration).

- Chemoorganotrophs use a variety of molecules as their energy, electron, and carbon source. These molecules are degraded by pathways that funnel into pathways used for glucose catabolism (**figure 11.3**).
- Some of the pathways used by chemoorganotrophs are amphibolic, having both catabolic and anabolic functions (**figure 11.4**).

11.3 Aerobic Respiration Can Be Divided into Three Steps

- Aerobic respiration of glucose begins with glycolytic pathways, which catabolize glucose to pyruvate. Pyruvate is fed into the TCA cycle for completion of catabolism.
- The glycolytic pathways and the TCA cycle generate numerous NADH and $FADH_2$ molecules. These are oxidized by an electron transport chain, using oxygen as the terminal electron acceptor. Electron flow generates a proton motive force (PMF), which is used to synthesize ATP by oxidative phosphorylation.

11.4 Glucose to Pyruvate: The First Step

- Glycolysis, used in its broadest sense, refers to all pathways used to break down glucose to pyruvate.
- The Embden-Meyerhof pathway has a net production of two NADH and two ATP, the latter being produced by substrate-level phosphorylation. It also produces several precursor metabolites (**figure 11.5**).
- In the Entner-Doudoroff pathway, glucose is catabolized to pyruvate and glyceraldehyde 3-phosphate (**figure 11.6**). The latter product can be oxidized by enzymes of the Embden-Meyerhof pathway to provide ATP, NADH, and another molecule of pyruvate.
- In the pentose phosphate pathway, glucose 6-phosphate is oxidized twice and converted to pentoses and other sugars. It is a source of NADPH, ATP, and several precursor metabolites (**figure 11.7**).

11.5 Pyruvate to Carbon Dioxide (Step 2) Is Accomplished by the Tricarboxylic Acid Cycle

- Pyruvate from the glycolytic pathways is fed into the tricarboxylic acid cycle by a reaction catalyzed by the enzyme pyruvate dehydrogenase. It converts pyruvate to acetyl-CoA. In the process, one of pyruvate's carbons is released in the form of carbon dioxide and one NADH is produced.
- The tricarboxylic acid cycle oxidizes acetyl-CoA to CO_2 and forms one GTP, three NADH, and one $FADH_2$ per acetyl-CoA (**figure 11.8**). It also generates several precursor metabolites.

11.6 Electron Transport and Oxidative Phosphorylation (Step 3) Generate the Most ATP

- The NADH and $FADH_2$ produced from the oxidation of reduced organic molecules can be oxidized by an electron transport chain (ETC). Electrons flow from carriers with more negative reduction potentials to those with more positive potentials (**figures 11.9–11.11**). As electrons flow through an ETC, protons are transported across the membrane, generating proton motive force (PMF). PMF is used for ATP synthesis by oxidative phosphorylation.
- Bacterial and archaeal ETCs are often different from eukaryotic chains with respect to such aspects as carriers and branching (**figures 11.12** and **11.13**).
- ATP synthase catalyzes the synthesis of ATP (**figure 11.16**). In eukaryotes, it is located on the inner surface of the inner mitochondrial membrane. Bacterial and archaeal ATP synthases are on the inner surface of the plasma membrane. ATP synthase uses the PMF for ATP synthesis.
- In eukaryotes, the P/O ratio for NADH is about 2.5 and that for $FADH_2$ is around 1.5; P/O ratios are usually much lower in bacterial and archaeal chains. Aerobic respiration in eukaryotes can theoretically yield a maximum of 32 ATP (**figure 11.17**).

11.7 Anaerobic Respiration Uses the Same Three Steps as Aerobic Respiration

- Anaerobic respiration uses an exogenous molecule other than O_2 as the terminal electron acceptor for electron transport. The most common acceptors are nitrate, sulfate, and CO_2 (**figure 11.18** and **table 11.3**).
- For some microorganisms, the same pathways used to aerobically respire an organic energy source are also used for anaerobic respiration. However, less energy is provided by anaerobic respiration than aerobic respiration because the alternate electron acceptors have reduction potentials that are less positive than the reduction potential of oxygen.

11.8 Fermentation Does Not Involve an Electron Transport Chain

- During fermentation, an endogenous electron acceptor is used to reoxidize any NADH generated by the catabolism of glucose to pyruvate (**figure 11.19**). Flow of electrons from the electron donor to the electron acceptor does not involve an ETC, and in most organisms, ATP is synthesized only by substrate-level phosphorylation.
- There are many different fermentation pathways. These are of practical importance in clinical and industrial settings (**figure 11.20**).

11.9 Catabolism of Organic Molecules Other Than Glucose

- Microorganisms catabolize many extracellular carbohydrates. Monosaccharides are taken in and phosphorylated; disaccharides may be cleaved to monosaccharides by either hydrolysis or phosphorolysis. External polysaccharides are degraded by hydrolysis and the products are actively transported. Intracellular glycogen and starch are converted to glucose 1-phosphate by phosphorolysis (**figure 11.21**).

- Triglycerides are hydrolyzed to glycerol and fatty acids by enzymes called lipases. Fatty acids are usually oxidized to acetyl-CoA in the β-oxidation pathway (**figure 11.23**).
- Proteins are hydrolyzed to amino acids that are then deaminated (**figure 11.24**). The carbon skeletons produced by deamination can be fermented or fed into the TCA cycle (**figure 11.3**).

11.10 Chemolithotrophy: "Eating Rocks"

- Chemolithotrophs synthesize ATP by oxidizing inorganic compounds—usually hydrogen, reduced nitrogen and sulfur compounds, or ferrous iron—with an ETC. O_2 is the usual electron acceptor (**figure 11.25** and **table 11.5**). The PMF produced is used by ATP synthase to make ATP.
- Many of the energy sources used by chemolithotrophs have a more positive standard reduction potential than the NAD^+/NADH conjugate redox pair. These chemolithotrophs must expend energy (PMF) to drive reverse electron flow and produce the NADH they need for CO_2 fixation and other processes (**figure 11.27**).

11.11 Flavin-Based Electron Bifurcation

- Some anaerobic bacteria and archaea conserve energy by using a cytoplasmic flavoenzyme that transfers electrons such that an endergonic redox reaction is possible because it is coupled to an exergonic one (**figure 11.28**).

11.12 Phototrophy

- In oxygenic photosynthesis, eukaryotes and cyanobacteria trap light energy with chlorophyll and accessory pigments, and move electrons through photosystems I and II to make ATP and NADPH (the light reactions). The ATP and NADPH are used in the dark reactions to fix CO_2.
- Cyclic photophosphorylation involves the activity of photosystem I and generates ATP only. In noncyclic photophosphorylation, photosystems I and II operate together to move electrons from water to $NADP^+$, producing ATP, NADPH, and O_2 (**figure 11.32**). In both cases, electron flow generates PMF, which is used by ATP synthase to make ATP.
- Anoxygenic phototrophs differ from oxygenic phototrophs in possessing bacteriochlorophyll and having only one photosystem (**figures 11.34–11.36**). Cyclic electron flow generates a PMF, which is used by ATP synthase to make ATP (i.e., cyclic photophosphorylation). They are anoxygenic because they use molecules other than water as an electron donor for electron flow and the production of reducing power.
- Some bacteria and archaea use a type of phototrophy that involves a rhodopsin, a proton-pumping pigment. This type of phototrophy generates PMF but does not involve an ETC. It supplements the energy needs of heterotrophs.

Active Learning

1. How would you isolate a thermophilic chemolithotroph that uses sulfur compounds as a source of energy and electrons? What changes in the incubation system would be needed to isolate bacteria using sulfur compounds in anaerobic respiration? How would you tell which process is taking place through an analysis of the sulfur molecules present in the spent medium?
2. Certain chemicals block ATP synthesis by allowing protons and other ions to "leak across membranes," disrupting the charge and proton gradients established by electron flow through an ETC. Does this observation support the chemiosmosis hypothesis? Explain your reasoning.
3. Two flasks of *E. coli* are grown in batch culture in the same medium (2% glucose and amino acids; no nitrate) and at the same temperature (37°C). Culture #1 is well aerated. Culture #2 is anoxic. After 16 hours the following observations are made:
   - Culture #1 has a high cell density; the cells appear to be in stationary phase, and the glucose level in the medium is reduced to 1.2%.
   - Culture #2 has a low cell density; the cells appear to be in logarithmic phase, although their doubling time is prolonged (over 1 hour). The glucose level is reduced to 0.2%.

   This phenomenon is known as the Pasteur effect. Like Pasteur did, figure out why culture #2 has so little glucose remaining relative to culture #1, even though culture #2 displayed slower growth and has less biomass. (Hint: It may be helpful to think about gas mileage obtained with a sports car versus an old clunker.)
4. Review the description of the Berkeley Pit Lake in this chapter's opening case study. In addition to heavy metals, the water has a high concentration of sulfate. Predict the nutritional types that might be isolated from the lake. What might they use as energy and electron sources? What might any chemotrophs in the lake use as terminal electron acceptors for their ETCs?
5. The South Pacific Gyre is a large oceanic region between South America and Australia. The sediments in this deep-water region are limited in the amount of

organic carbon that is easily assimilated but appear to be well oxygenated. In an effort to understand how microbes survive this harsh environment, which may be similar to about one-third of the ocean floor, scientists sequenced the genomes of the sediment microbial community. They found genes encoding enzymes involved in (1) iron oxidation, (2) manganese oxidation, and (3) carbon fixation linked to nitrification. What metabolic strategies are these three types of microbes using? Is this expected given the environmental conditions in this ecosystem?

Read the original paper: Tully, B. J., and Heidelberg, J. F. 2016. Potential mechanisms for microbial energy acquisition in oxic deep-sea sediments. *Appl Environ Microbiol.* 82:4232–4243. doi:10.1128/AEM.01023-16.

12

Anabolism: The Use of Energy in Biosynthesis

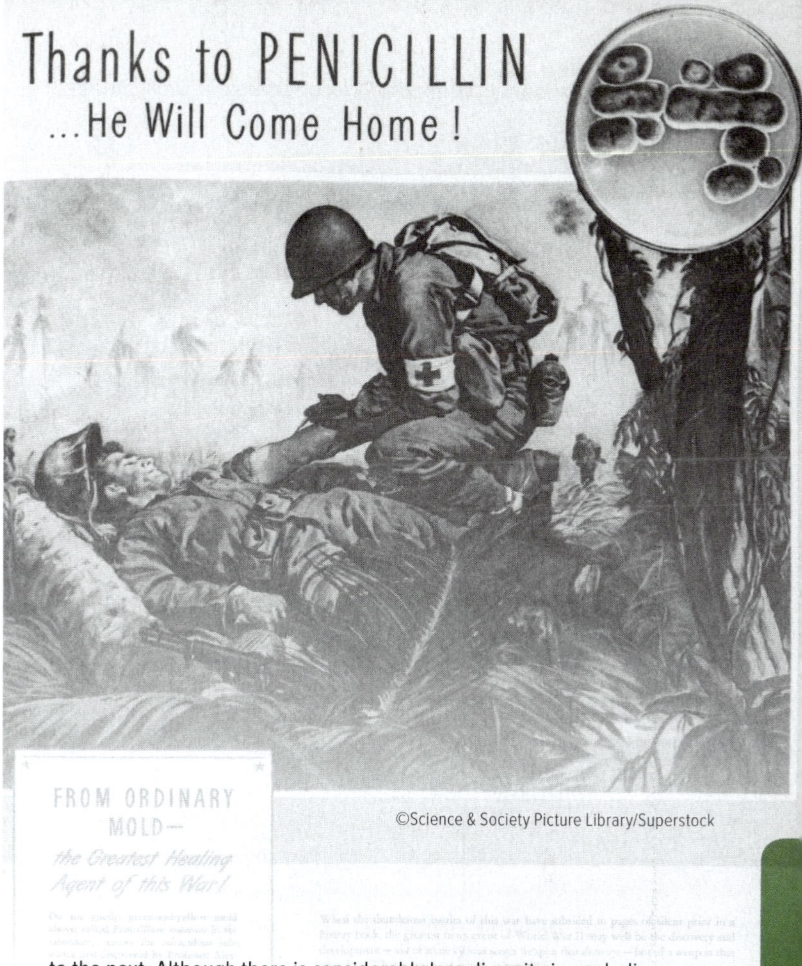

©Science & Society Picture Library/Superstock

Building Penicillin

The story of Alexander Fleming's rediscovery of penicillin in 1929 is well known. However, Fleming had little to do with developing penicillin as a treatment for infectious disease because after some initial experiments, he abandoned the study of its activity. Fortunately, Ernst Chain developed an interest in the chemical and eventually purified it. In 1940, as many nations battled in World War II, Howard Florey and his assistants began clinical trials with penicillin. By that time, Selman Waksman had discovered several other antibacterial agents and had coined the term *antibiotic* to mean an antibacterial agent produced by a microbe. The penicillin trials were successful, and the need for antibiotics during World War II to treat wound infections was the impetus for finding and developing more antibiotics for clinical use. A few years later a ready supply was available. ▶| *Antimicrobial chemotherapy evolved from antisepsis efforts (section 9.1)*

Antibiotics are interesting molecules for several reasons. One is that they are made at all. Why would a microbe expend materials and energy to make an antibiotic? What selective advantage do they provide? How do they contribute to the ecology of the microbe? Do they serve a purpose other than killing other microbes (*see chapter 23 opening story*)? These questions are still not fully answered. Another reason these molecules are interesting is that they are structurally complex. Despite this, they are built using the same precursor metabolites and many of the same biochemical processes used in the anabolic pathways that sustain life.

During anabolism, an organism begins with simple inorganic molecules and a carbon source, and constructs ever more complex molecules until new organelles and cells arise (**figure 12.1**). Thus anabolism is the creation of order. Because a cell is highly ordered and immensely complex, much energy is required for biosynthesis. This is readily apparent from estimates of the biosynthetic capacity of rapidly growing *Escherichia coli* cells (**table 12.1**). Furthermore, even nongrowing cells need energy for the biosynthetic processes they carry out. This is because nongrowing cells continuously degrade and resynthesize cellular molecules during a process known as turnover. Thus cells are never the same from one instant to the next. Although there is considerably less diversity in anabolic processes as compared to catabolic processes, anabolism is amazing in its own right. Using just 12 precursor metabolites, a cell is able to manufacture the myriad of molecules from which it is constructed. In this chapter, we discuss the synthesis of some of the most important types of cell constituents.

Readiness Check:

Based on what you have learned previously, you should be able to:

- ✓ Distinguish metabolism, catabolism, and anabolism
- ✓ List the products of the fueling reactions (section 11.1)
- ✓ Describe how carbon, hydrogen, oxygen, nitrogen, phosphorus, and sulfur are used by cells
- ✓ Differentiate between heterotrophy and autotrophy (section 11.1)
- ✓ Describe the components of a biochemical pathway and how they are organized (section 10.5)
- ✓ Discuss the function of ATP and how that function is related to its phosphate transfer potential (section 10.2)
- ✓ State the function of enzymes and ribozymes (section 10.6)

12.1 Principles Governing Biosynthesis

After reading this section, you should be able to:

a. Describe in general terms the steps organisms use to convert a carbon source and inorganic molecules to cells
b. Discuss the principles that govern biosynthesis

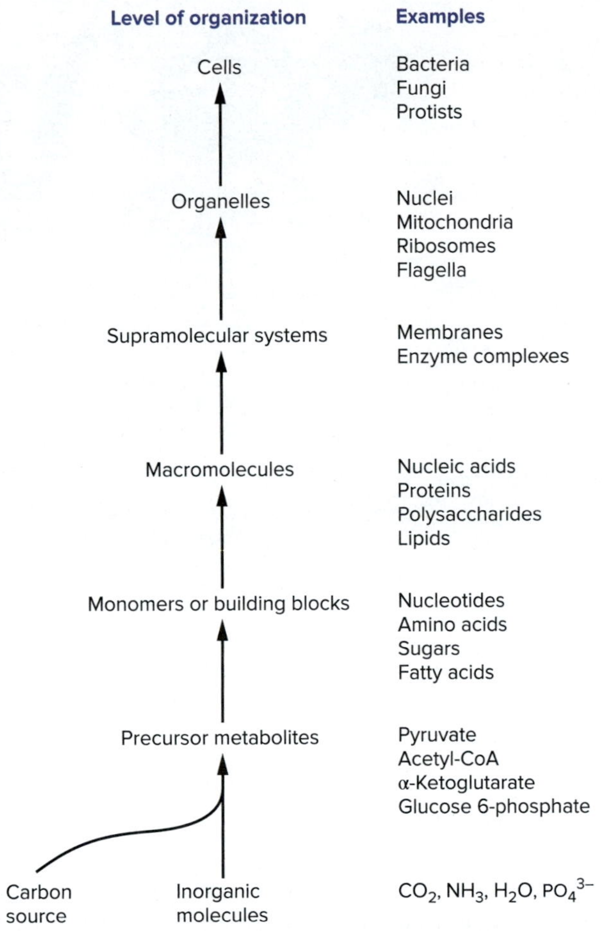

Figure 12.1 The Construction of Cells. The biosynthesis of cells and their constituents is organized in levels of ever greater complexity.

The problem faced by all cells is how to make the molecules they need as efficiently as possible. They have solved this problem by carrying out biosynthesis using a few basic principles, described here.

1. **Large molecules are made from small molecules.** The construction of **macromolecules** (complex molecules) from a few simple structural units (monomers) linked together by a single type of covalent bond saves much genetic storage capacity, biosynthetic raw material, and energy, making their synthesis highly efficient.
2. **Many enzymes do double duty. However, some enzymes function in one direction only.** As discussed in chapter 11, many enzymes are used for both catabolic and anabolic processes in pathways that are said to be amphibolic (*see figure 11.4*), saving additional materials and energy. Although many steps of amphibolic pathways are catalyzed by enzymes that act reversibly, some are not. These steps require the use of separate enzymes: one to catalyze the catabolic reaction, the other to catalyze the anabolic reaction. The use of two enzymes allows independent regulation of catabolism and anabolism. Thus catabolic and anabolic pathways are never identical, even though they share many enzymes. Although both types of pathways can be regulated by their end products as well as by the concentrations of ATP, ADP, AMP, and NAD^+/NADH, end-product regulation generally assumes more importance in anabolic pathways.
3. **Anabolism consumes energy.** Many reactions in anabolic pathways are endergonic and will not proceed in the direction of biosynthesis. Cells solve this problem by connecting endergonic biosynthetic reactions to the hydrolysis of ATP and other nucleoside triphosphates. When these two processes are coupled, the free energy made available during nucleoside triphosphate breakdown drives the biosynthetic reaction to completion. ◀ *ATP: the major energy currency of cells (section 10.2)*
4. **Catabolism and anabolism can be physically separated.** Catabolic and anabolic pathways can be localized into distinct cellular compartments—a process called compartmentation. In eukaryotes, this

Table 12.1 Biosynthesis of Macromolecules in *E. coli*

Cell Constituent	Number of Molecules per Cell	Rate of Synthesis (sec^{-1})
DNA	2	750 nucleotides
RNA	2×10^5	20–90 nucleotides
Polysaccharides	1×10^6	32.5 molecules
Lipids	2×10^7	12,500 molecules
Proteins	3×10^6	8 amino acids

is easily done because eukaryotic cells have membrane-bound organelles that carry out specific functions. Compartmentation also occurs in bacterial and archaeal cells. For instance, carboxysomes separate CO_2 fixation from other processes in bacteria. Compartmentation makes it easier for catabolic and anabolic pathways to operate simultaneously yet independently.

5. **Catabolism and anabolism often use different cofactors.** Usually catabolic oxidations produce NADH to transport electrons. In contrast, when an electron donor is needed during biosynthesis, NADPH often serves as the donor.
6. **After macromolecules have been constructed from simpler precursors, they are assembled into larger, more complex structures such as supramolecular systems and organelles by a process called self-assembly** (figure 12.1). This is possible because many macromolecules contain the necessary information to build multimolecular complexes without enzymatic help.

Comprehension Check

1. Explain why most enzymes that operate in amphibolic pathways function in both the anabolic and catabolic direction, but a few function only in either anabolism or catabolism.
2. How might the ratio of NADH to NADPH indicate the balance between anabolism and catabolism?
3. Provide an example of compartmentation in a eukaryotic cell and a bacterial or archaeal cell.
4. What is self-assembly? What biological entity also exhibits self-assembly?

12.2 Precursor Metabolites: Starting Molecules for Biosynthesis

After reading this section, you should be able to:

a. List the central metabolic pathways, noting which precursor metabolites are generated by each pathway
b. Draw a simple diagram that lists all the precursor metabolites and illustrates how they are used in biosynthesis

Generation of **precursor metabolites** is critical to anabolism because they give rise to all other molecules made by the cell. Precursor metabolites are carbon skeletons (i.e., carbon chains) used as the starting substrates for the synthesis of monomers and other building blocks needed to make macromolecules (*see figure 11.1*). During biosynthesis, functional moieties such as amino and sulfhydryl groups are added to the carbon chains. The precursor metabolites and how they are used in biosynthesis are shown in **figures 12.2** and **12.3**. Several things should be noted in figure 12.2. First, all the precursor metabolites are intermediates of the glycolytic pathways (Embden-Meyerhof, Entner-Doudoroff, and the pentose phosphate pathways) and the tricarboxylic acid (TCA) cycle. Therefore these pathways play a central role in metabolism and are often referred to as the **central metabolic pathways.** Note, too, that most of the precursor metabolites are used for synthesis of amino acids and nucleotides. ◀ *Glucose to pyruvate: the first step (section 11.4); Pyruvate to carbon dioxide (step 2) is accomplished by the tricarboxylic acid cycle (section 11.5)* ▶ *Common metabolic pathways (appendix II)*

If an organism is a chemoorganotroph using glucose as its energy, electron, and carbon source (either aerobically or anaerobically), it generates precursor metabolites as it generates ATP and reducing power. But what if the chemoorganotroph is using an amino acid as its sole source of carbon, electrons, and energy? And what about autotrophs? How do they generate precursor metabolites from CO_2, their carbon source? Heterotrophs growing on something other than glucose convert that carbon source into one or more intermediates of the central metabolic pathways. From there, they can generate the remaining precursor metabolites. Autotrophs must first convert CO_2 into reduced organic carbon from which they can generate the precursor metabolites. Many of the reactions that autotrophs use to synthesize the precursor metabolites are reactions of the central metabolic pathways, operating in either the catabolic direction or the anabolic direction. Thus the central metabolic pathways are important to the anabolism of both heterotrophs and autotrophs.

We begin our discussion of anabolism by first considering CO_2 fixation by autotrophs. Once CO_2 is converted to organic carbon, the synthesis of other precursor metabolites, amino acids, nucleotides, and additional building blocks is very similar in both autotrophs and heterotrophs. Recall that the precursor metabolites provide the carbon skeletons for the synthesis of other important organic molecules. In the process of transforming a precursor metabolite into an amino acid or a nucleotide, the carbon skeleton is modified in a number of ways, including the addition of nitrogen, phosphorus, and sulfur. Thus as we discuss the synthesis of monomers from precursor metabolites, we also address the assimilation of nitrogen, sulfur, and phosphorus. *A Biochemical Pathway*

12.3 CO_2 Fixation: Reduction and Assimilation of CO_2 Carbon

After reading this section, you should be able to:

a. List the pathways used by microbes to fix CO_2, noting the types of organisms (eukaryotes, bacteria, archaea) that use each pathway
b. Describe in general terms the three phases of the Calvin-Benson cycle
c. Identify the steps in the Calvin-Benson cycle that consume ATP and NADPH

Autotrophs use CO_2 as their sole or principal carbon source, and the reduction and incorporation of CO_2 requires much energy. Photoautotrophs obtain energy by trapping light during the light

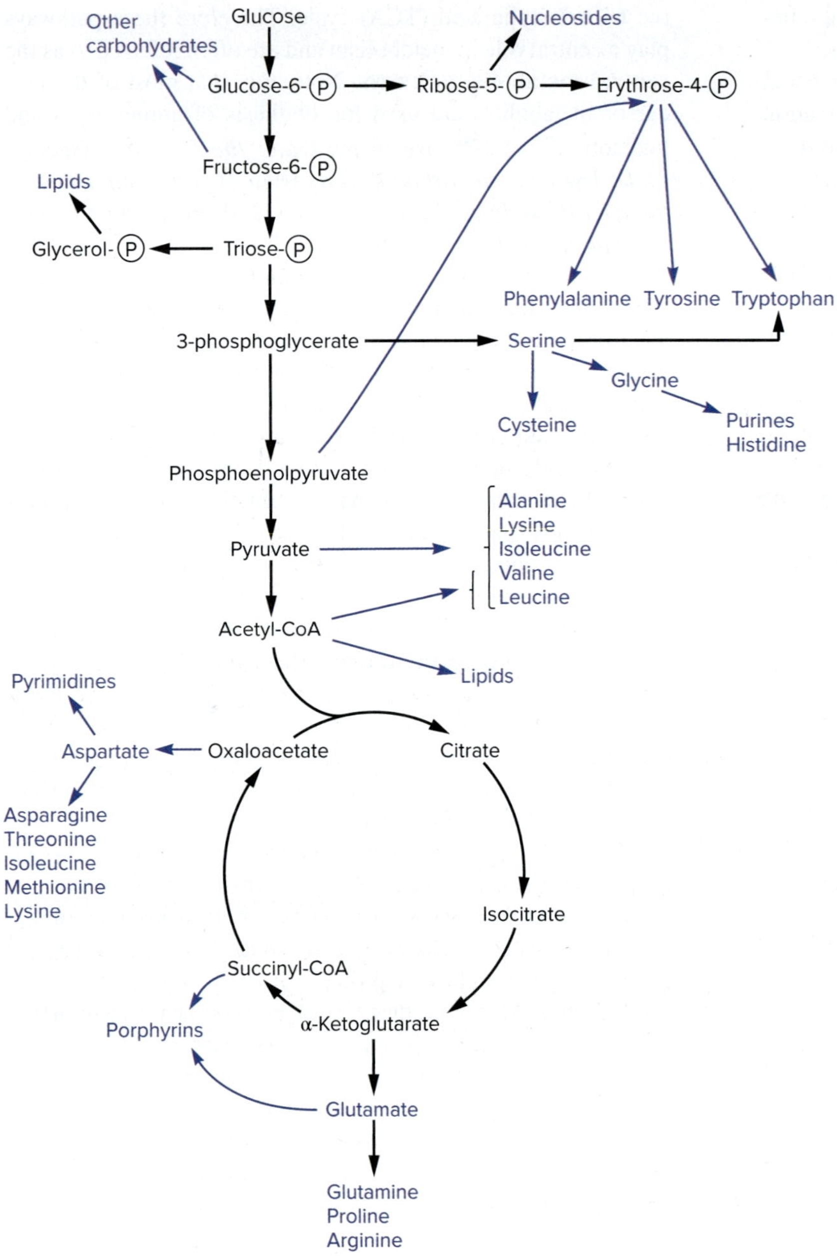

Figure 12.2 The Organization of Anabolism. Biosynthetic products (in blue) are derived from precursor metabolites. Porphyrin synthesis in nonplant eukaryotes and alphaproteobacteria begins with succinyl-CoA. For simplicity, the role of glycine in the succinyl-CoA pathway is not illustrated. In plants and all other bacteria, porphyrin synthesis begins with glutamate (section 12.5).

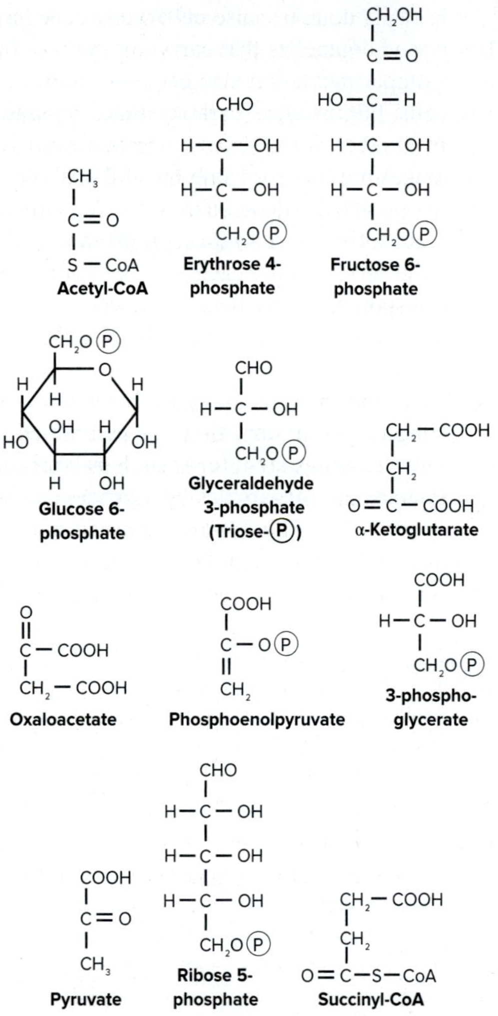

Figure 12.3 The 12 Precursor Metabolites. In chemoorganoheterotrophs, these are produced by the central metabolic pathways. Autotrophs use CO_2-fixation pathways and other pathways (e.g., gluconeogenesis) to make precursor metabolites.

reactions of photosynthesis, and chemolithoautotrophs derive energy from the oxidation of inorganic electron donors. Autotrophic CO_2 fixation is crucial to life on Earth because it provides the organic matter on which heterotrophs depend. ◀ *Chemolithotrophy: eating rocks (section 11.10); Phototrophy (section 11.12)* ▶ *Carbon cycle (section 28.1)*

Six different CO_2-fixation pathways have been identified in microorganisms. Eukaryotic autotrophs and most aerobic bacterial autotrophs use the **Calvin-Benson cycle,** also called the Calvin cycle. Other pathways are used by some obligatory anaerobic and microaerophilic bacteria and archaea. We consider the Calvin-Benson cycle first and then briefly introduce the other CO_2-fixation pathways.

Calvin-Benson Cycle

The Calvin-Benson cycle is also called the reductive pentose phosphate cycle because it is essentially the reverse of the pentose phosphate pathway. Thus many of the reactions are similar, in particular the sugar transformations. The reactions of the Calvin-Benson cycle occur in the chloroplast stroma of eukaryotic autotrophs. In cyanobacteria, some nitrifying bacteria, and thiobacilli (sulfur-oxidizing chemolithotrophs), the cycle is associated with inclusions called **carboxysomes.** These polyhedral structures contain the enzyme critical to the Calvin-Benson cycle and are the site of CO_2 fixation. ◀ *Pentose phosphate pathway: a major producer of reducing power for anabolic reactions (section 11.4); Inclusions (section 3.6)*

The Calvin-Benson cycle is divided into three phases: carboxylation phase, reduction phase, and regeneration phase (**figure 12.4** and *appendix II*). During the carboxylation phase, the enzyme **ribulose 1,5-bisphosphate carboxylase/oxygenase** (RuBisCO) catalyzes the addition of CO_2 to the 5-carbon molecule ribulose 1,5-bisphosphate (RuBP), forming a 6-carbon intermediate that rapidly and spontaneously splits into two molecules of 3-phosphoglycerate (PGA). PGA is an intermediate of the Embden-Meyerhof pathway (EMP), and in the reduction phase of the Calvin-Benson cycle, PGA is reduced to glyceraldehyde 3-phosphate by two reactions that reverse two EMP reactions. The EMP reactions differ from the Calvin-Benson cycle reactions in that the Calvin cycle enzyme glyceraldehyde 3-phosphate dehydrogenase uses NADPH rather than NADH (compare figures 12.4 and 11.5). Finally, in the regeneration phase, RuBP is reformed, so that the cycle can repeat. In addition, this phase produces carbohydrates such as fructose 6-phosphate and glucose 6-phosphate, which are precursor metabolites (figures 12.2 and 12.3). This portion of the cycle is similar to the pentose phosphate pathway and involves transketolase and transaldolase reactions. *The Calvin Cycle*

To synthesize fructose 6-phosphate or glucose 6-phosphate from CO_2, the cycle must operate six times to yield the desired hexose and reform the six RuBP molecules.

$$6\text{RuBP} + 6CO_2 \rightarrow 12\text{PGA} \rightarrow 6\text{RuBP} + \text{fructose-6-p}$$

The incorporation of one CO_2 into organic material requires three ATP and two NADPH. The formation of glucose from CO_2 may be summarized by the following equation.

$$6CO_2 + 18\text{ATP} + 12\text{NADPH} + 12H^+ + 12H_2O \rightarrow$$
$$\text{glucose} + 18\text{ADP} + 18P_i + 12\text{NADP}^+$$

Other CO_2-Fixation Pathways

The reactions of the Calvin-Benson cycle were determined in the 1940s and 1950s. At the time, it was thought that this pathway was used by all autotrophic organisms. This was shown not to be the case in the mid-1960s, when a second CO_2-fixation pathway was discovered. Over the years, and at an increasingly rapid rate, four more pathways were discovered. Many microbiologists think that there are still others. Why so many pathways have evolved is still an unanswered question. However, evidence suggests that certain pathways correlate with certain lifestyles. In other words, the ecological niche occupied by a microbe has shaped the evolution of its CO_2-fixation pathway.

The first alternative CO_2-fixation pathway discovered was the **reductive TCA cycle** (**figure 12.5**), which is used by some autotrophs in the bacterial phyla *Aquificae, Proteobacteria, Nitrospirae,* and *Chlorobi*. The reductive TCA cycle is so named because it runs in the reverse direction of the normal, oxidative TCA cycle (compare figures 12.5 and 11.8). ▶| Aquificae *and* Thermotogae *are ancient bacterial lineages (section 21.1); Phylum* Chlorobi *(section 21.4); Class* Epsilonproteobacteria *ranges from pathogens to deep-sea bacteria (section 22.5)*

The remaining CO_2-fixation pathways are used by members of the bacterial phylum *Chloroflexi* (3-hydroxypropionate bi-cycle; *see figure 21.8*): The reductive acetyl-CoA pathway (*see figure 20.2*) is used by some methanogens; the 3-hydroxypropionate/4-hydroxybutyrate pathway (*see figure 20.3*) is used by members of the order *Sulfolobales;* and the dicarboxylate/4-hydroxybutyrate cycle (*see figure 20.3*) was discovered in members of orders *Thermoproteales* and *Desulfurococcales.* A seventh pathway has been proposed that links CO_2 fixation to dissimilatory phosphite oxidation. This proposal is supported by genomic evidence, but has not yet been established biochemically.

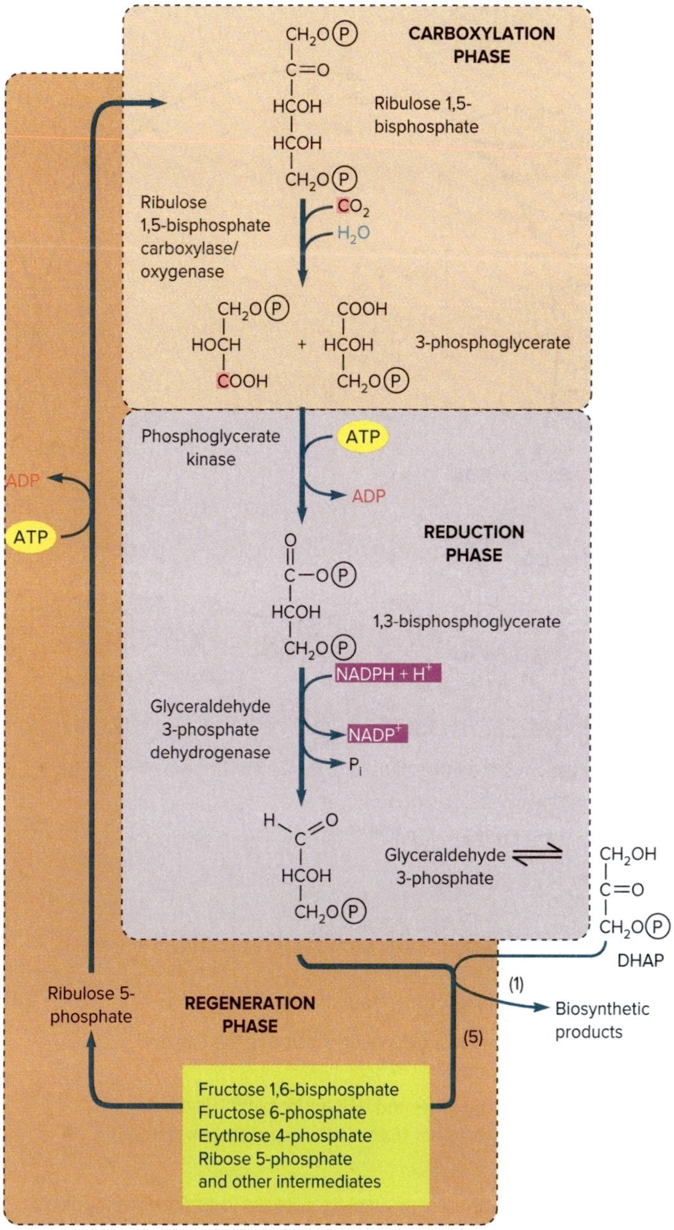

Figure 12.4 The Calvin-Benson Cycle. This overview of the cycle shows only the carboxylation and reduction phases in detail. Three ribulose 1,5-bisphosphates are carboxylated (the incorporated carbon atom is shaded in pink) to give six 3-phosphoglycerates in the carboxylation phase. These are converted to six glyceraldehyde 3-phosphates, which can be converted to dihydroxyacetone phosphate (DHAP). Five of the six trioses (glyceraldehyde phosphate and dihydroxyacetone phosphate) are used to reform three ribulose 1,5-bisphosphates in the regeneration phase. The remaining triose is used in biosynthesis. The numbers in parentheses at the lower right indicate this carbon flow.

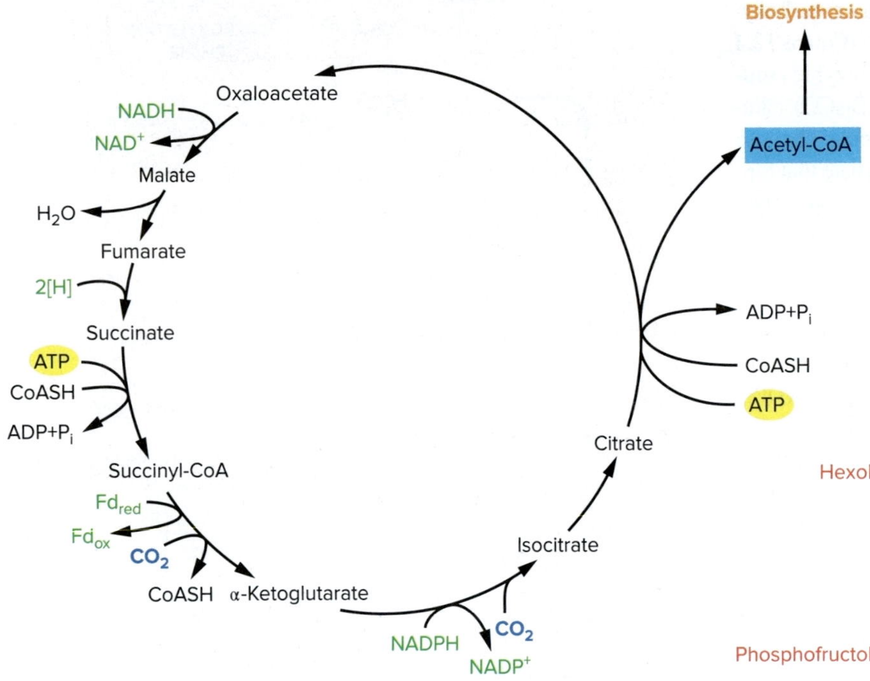

Figure 12.5 The Reductive TCA Cycle. This cycle is used by green sulfur bacteria and some chemolithotrophic bacteria to fix CO_2. The cycle runs in the opposite direction as the TCA cycle. ATP and reducing power (NADH, NADPH, and [H]) drive the reversal. The product of the pathway is acetyl-CoA.

MICRO INQUIRY *How might acetyl-CoA be used by the cell?*

Comprehension Check

1. Briefly describe the three phases of the Calvin-Benson cycle. What other pathways are used to fix CO_2?
2. Which two enzymes are specific to the Calvin cycle?

12.4 Synthesis of Carbohydrates

After reading this section, you should be able to:

a. Compare and contrast gluconeogenesis and the Embden-Meyerhof pathway
b. Describe the role of ATP and UTP in the synthesis of monosaccharides (other than glucose) and polysaccharides
c. Outline the major steps in peptidoglycan synthesis
d. Evaluate the effectiveness of targeting antibiotics to peptidoglycan synthesis

Many microbes need to synthesize glucose for anabolic purposes. The synthesis of glucose from noncarbohydrate precursors is called **gluconeogenesis.** The gluconeogenic pathway shares six enzymes with the Embden-Meyerhof pathway. However, the two pathways are not identical (**figure 12.6**). Four reactions are catalyzed by enzymes that are specific for gluconeogenesis. Two of these enzymes are involved in the conversion of pyruvate to phosphoenolpyruvate. The third enzyme catalyzes the formation of fructose 6-phosphate from fructose 1,6-bisphosphate, and the fourth removes the phosphate from glucose 6-phosphate to generate glucose.

Synthesis of Monosaccharides and Polysaccharides Often Involves Nucleoside Diphosphate Carriers

As can be seen in figure 12.6, gluconeogenesis synthesizes fructose 6-phosphate, glucose 6-phosphate, and glucose. Once these sugars have been formed, other common sugars can be manufactured. For

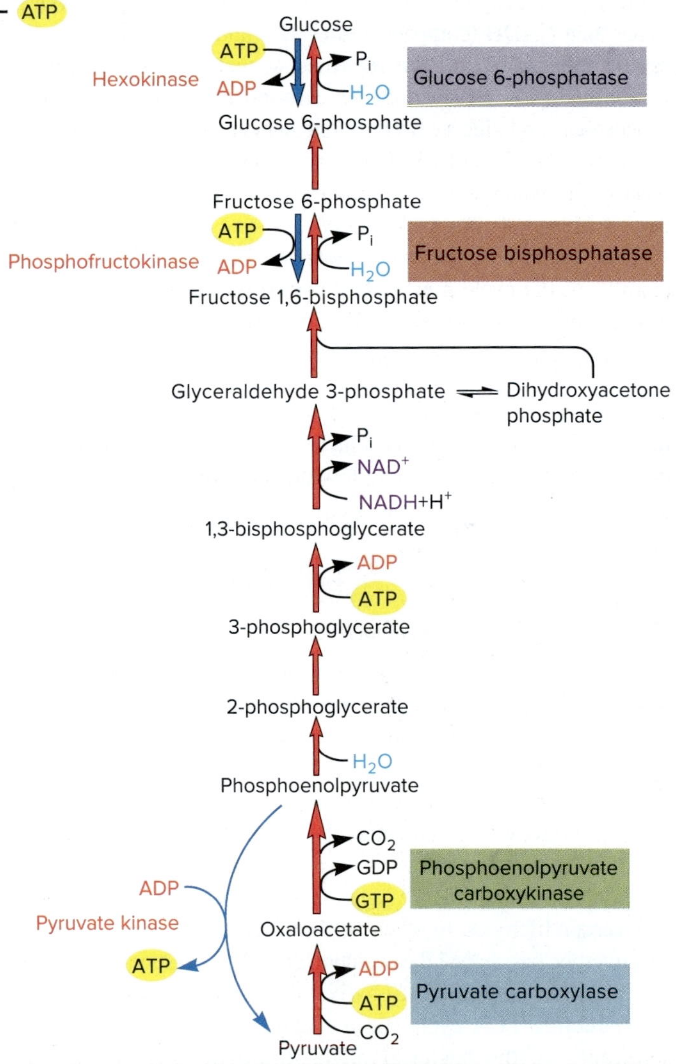

Figure 12.6 Gluconeogenesis Pathway Used by Many Microorganisms. The names of the four enzymes catalyzing irreversible reactions that are different from those found in the Embden-Meyerhof pathway (EMP) are in colored boxes. EMP steps are shown with blue arrows for comparison.

MICRO INQUIRY *Why is it important that the enzymes dedicated to gluconeogenesis (rather than those used in both catabolic and anabolic reactions) are found in both the 3-carbon and 6-carbon stages?*

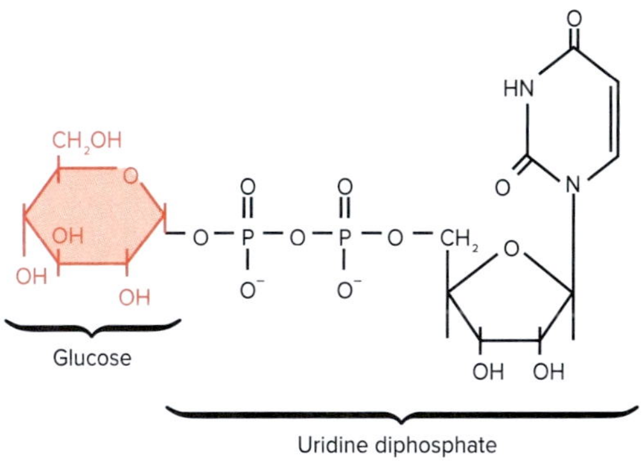

Figure 12.7 **Uridine Diphosphate Glucose (UDPG).**

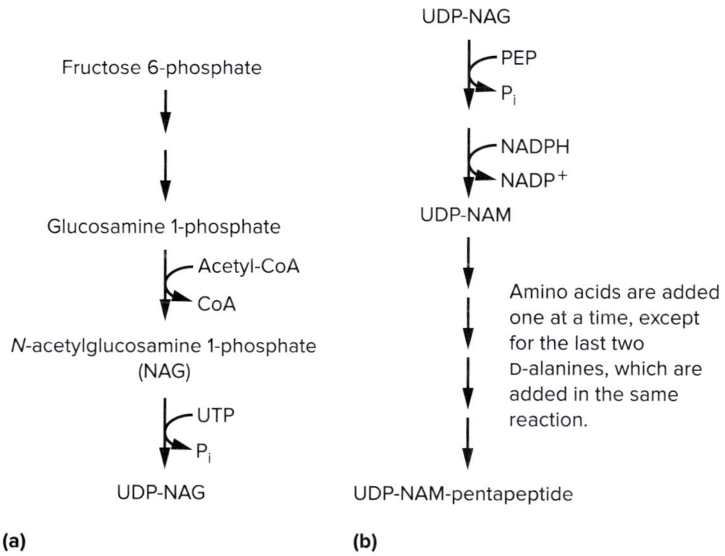

Figure 12.8 **Synthesis of UDP-NAG and UDP-NAM-Pentapeptide.** These are the initial steps in peptidoglycan synthesis. NAG, *N*-acetylglucosamine; NAM, *N*-acetylmuramic acid; PEP, phosphoenolpyruvate.

example, mannose comes directly from fructose 6-phosphate by a simple rearrangement of a hydroxyl group (*see figure AI.9*). Several sugars are synthesized while attached to a **nucleoside diphosphate.** One of the most important nucleoside diphosphate sugars is uridine diphosphate glucose (UDPG), which is formed when glucose reacts with uridine triphosphate (**figure 12.7**). Other important uridine diphosphate sugars are UDP-galactose and UDP-glucuronic acid. Uridine diphosphate (UDP) carries sugars around the cell for participation in enzyme reactions much like ADP bears phosphate in the form of ATP.

Nucleoside diphosphate sugars also play a central role in the synthesis of polysaccharides such as starch and glycogen, both of which are long chains of glucose. Again, biosynthesis is not simply a direct reversal of catabolism. For instance, during glycogen and starch synthesis in bacteria and protists, adenosine diphosphate glucose (ADP-glucose) is formed from glucose 1-phosphate and ATP. It then donates glucose to the end of growing glycogen and starch chains.

$$\text{ATP} + \text{glucose 1-phosphate} \rightarrow \text{ADP-glucose} + \text{PP}_i$$

$$(\text{Glucose})_n + \text{ADP-glucose} \rightarrow (\text{glucose})_{n+1} + \text{ADP}$$

Synthesis of Peptidoglycan Occurs in the Cytoplasm, at the Plasma Membrane, and in the Periplasmic Space

Nucleoside diphosphate sugars also participate in the synthesis of peptidoglycan. Recall that peptidoglycan is a large, complex molecule consisting of long polysaccharide chains made of alternating *N*-acetylmuramic acid (NAM) and *N*-acetylglucosamine (NAG) residues. The NAM groups bear stem peptides consisting of five amino acids (a pentapeptide). During peptidoglycan synthesis, adjacent polysaccharide chains are cross-linked by bonds formed between the stem peptides. In Gram-negative bacteria and many Gram-positive bacteria, the pentapeptide stems are directly linked. In some Gram-positive bacteria, the stems are connected by an interbridge consisting of one or more amino acids (*see figures 3.19 and 3.21*). ◀ *There are two main types of bacterial cell walls (section 3.4)*

Not surprisingly, such an intricate structure requires an equally intricate biosynthetic process, especially because some reactions occur in the cytoplasm, others at the plasma membrane, and others in the periplasmic space. Peptidoglycan synthesis involves two carriers. The first, uridine diphosphate, functions in the cytoplasmic reactions. The second carrier, bactoprenol phosphate, is needed for reactions that occur first on the cytoplasmic and then the periplasmic side of the plasma membrane.

In the first step of peptidoglycan synthesis, UDP derivatives of NAM and NAG are formed (**figure 12.8**). Amino acids are then added sequentially to UDP-NAM to form the pentapeptide stem. NAM-pentapeptide is then transferred to bactoprenol phosphate (also called undecaprenyl phosphate), which is located at the cytoplasmic side of the plasma membrane (**figure 12.9**). The resulting intermediate is called Lipid I. Bactoprenol is linked to NAM by two phosphates (pyrophosphate; **figure 12.10**). Next UDP transfers NAG to Lipid I, generating Lipid II. This creates the peptidoglycan repeat unit. If the peptidoglycan unit requires an interbridge, it is thought to be added while the repeat unit is at the cytoplasmic face of the plasma membrane. Once Lipid II is formed, the repeat unit is flipped across the plasma membrane so that it is now on the periplasmic side of the plasma membrane, while still attached to bactoprenol. This is catalyzed by the enzyme MurJ, often referred to as a "flippase." Enzymes called glycotransferases then attach the repeat unit to a growing nascent peptidoglycan strand, which is still attached to bactoprenol. Glycotransferases are one type of the penicillin-binding proteins (PBPs) that function during peptidoglycan synthesis. The growing chain is removed from bactoprenol during the final step in peptidoglycan synthesis: **transpeptidation** (**figure 12.11**), which creates the cross-links between the peptidoglycan chains. Enzymes called transpeptidases catalyze this reaction. They too are PBPs. During transpeptidation the terminal D-alanine is removed as the cross-link is formed. *Peptidoglycan Biosynthesis*

The result of the process we have just described is a peptidoglycan sacculus that is thicker by one layer. However, the peptidoglycan of any bacterium usually has a typical thickness that does not increase or decrease dramatically. Furthermore, for rod-shaped cells to elongate, and for any bacterial cell to divide, the structure of the sacculus must be sculpted and modified to allow for maintenance of

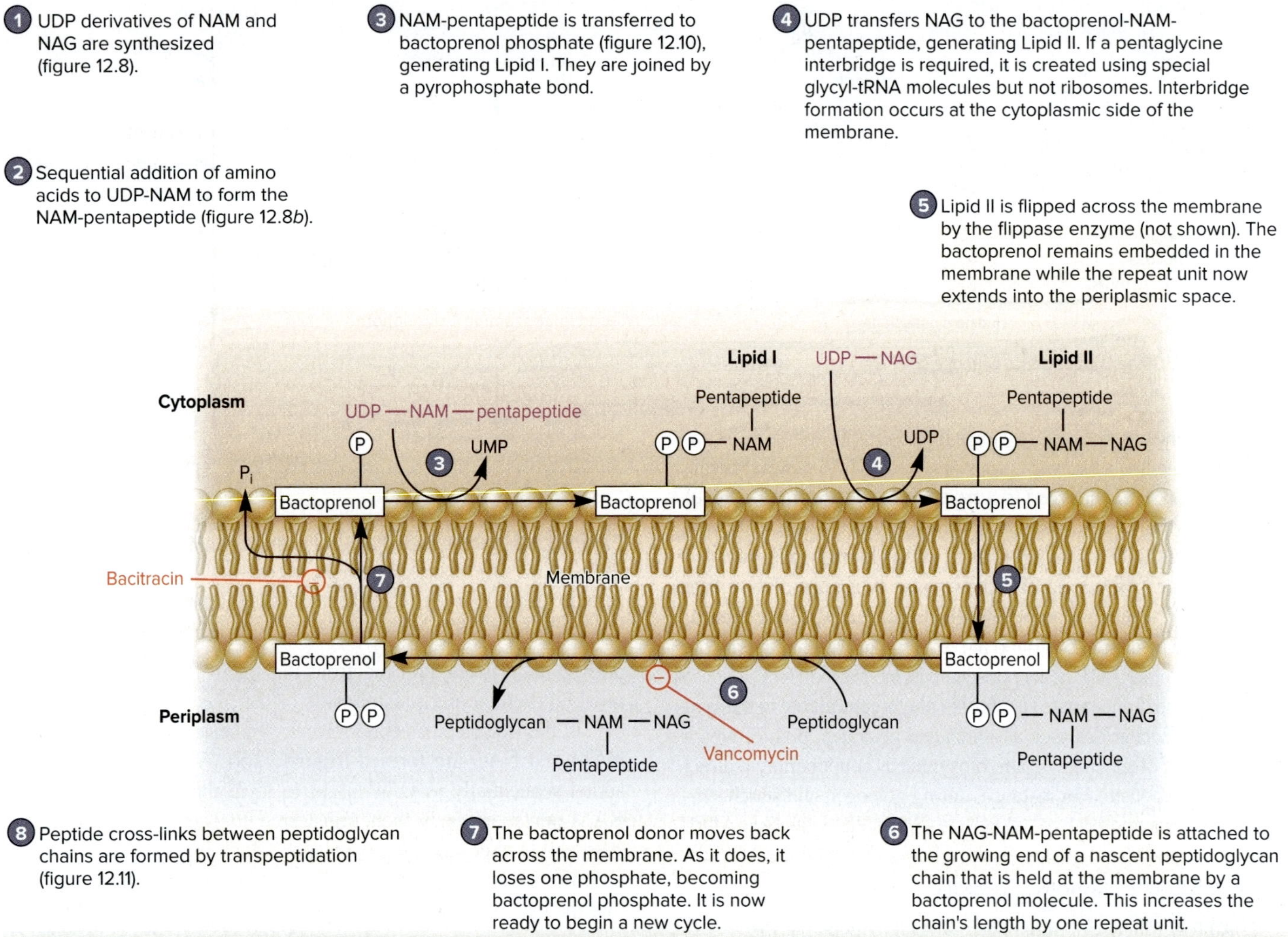

Figure 12.9 Peptidoglycan Synthesis. NAM is *N*-acetylmuramic acid and NAG is *N*-acetylglucosamine. Inhibition by bacitracin and vancomycin is shown.

MICRO INQUIRY *What is the difference between Lipid I and Lipid II?*

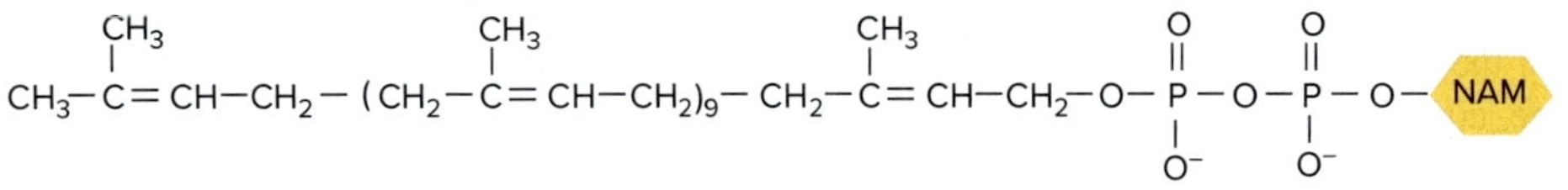

Figure 12.10 Bactoprenol-NAM. Bactoprenol is a 55-carbon, isoprene-derived alcohol (section 12.7) that is connected to *N*-acetylmuramic acid (NAM) by pyrophosphate.

the proper thickness, elongation, and separation during cytokinesis. This is accomplished by hydrolytic enzymes often called **autolysins.** These enzymes are also PBPs and they function in a precise and well-regulated way while maintaining wall shape and integrity in the presence of high osmotic pressure (*see figure 7.8*). Some autolysins attack the polysaccharide chains, while others hydrolyze the peptide cross-links. Autolysin activity is tightly controlled because unregulated activity would lead to weakening of the peptidoglycan sacculus. How autolysin activity is controlled is still poorly understood. ◀ *Cellular growth and determination of cell shape (section 7.2)*

Peptidoglycan synthesis is a particularly effective target for antimicrobial agents because of its importance to bacterial cell wall structure and function and because it is absent in animal cells. Inhibition of any stage of synthesis weakens the cell wall and can lead to lysis. For example, penicillin inhibits the transpeptidation reaction (figure 12.11), and bacitracin blocks the dephosphorylation of bactoprenol pyrophosphate (figure 12.9). ◀ *Inhibitors of cell wall synthesis (section 9.4)*

Comprehension Check

1. What is gluconeogenesis? Why is it important?
2. Describe the formation of mannose, galactose, starch, and glycogen. What are nucleoside diphosphate sugars? How do microorganisms use them?
3. Suppose a microorganism is growing on a medium that contains amino acids but no sugars. In general terms, how would it synthesize the pentose and hexose sugars it needs?
4. Diagram the steps involved in the synthesis of peptidoglycan and show where they occur in the cell. What are the roles of bactoprenol and UDP?
5. What would happen to a cell if it did not produce any autolysins? What if it produced too many in an unregulated fashion?

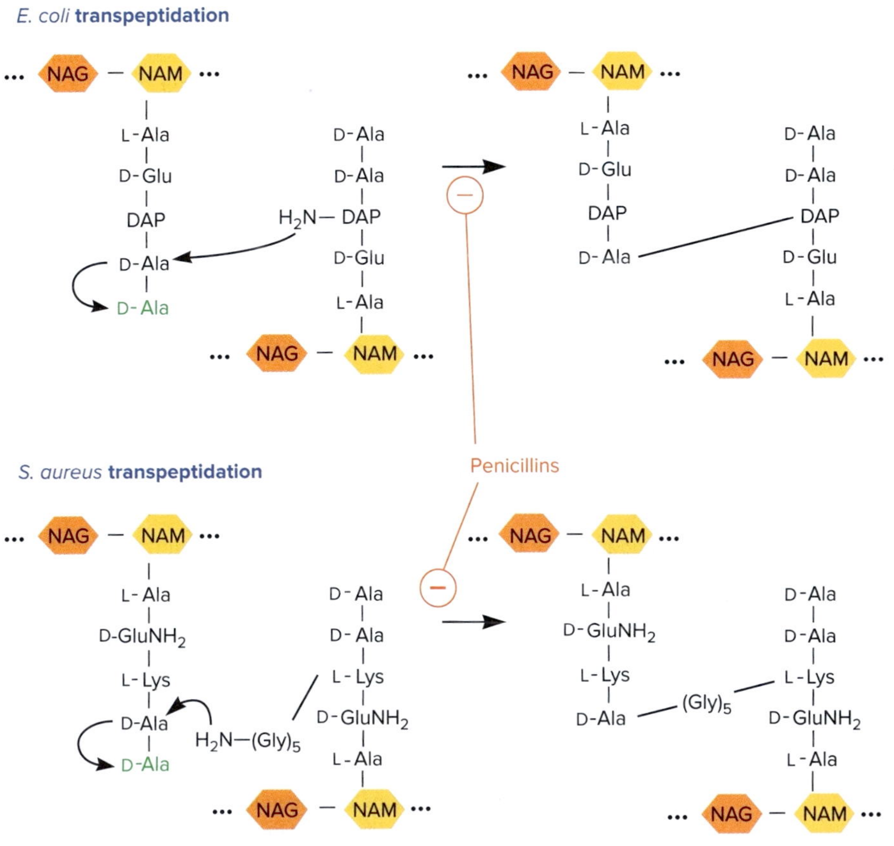

Figure 12.11 Transpeptidation. The transpeptidation reactions in the formation of the peptidoglycans of *E. coli* and *Staphylococcus aureus*. Note that the terminal D-alanine is removed during transpeptidation. The disruption of transpeptidation by penicillins is indicated. Other β-lactam antibiotics (e.g., cephalosporins) also inhibit this reaction. DAP, diaminopimelic acid.

12.5 Synthesis of Amino Acids Consumes Many Precursor Metabolites

After reading this section, you should be able to:

a. Discuss the three mechanisms microorganisms use to assimilate inorganic nitrogen and the role of transaminases in each
b. Describe the two methods microbes use to assimilate sulfur
c. Differentiate assimilatory nitrate reduction from dissimilatory nitrate reduction and assimilatory sulfate reduction from dissimilatory sulfate reduction
d. Evaluate the efficiency of using branched pathways for synthesizing amino acids
e. List the major anaplerotic reactions and explain their importance
f. Describe the two pathways for porphyrin synthesis and identify the types of organisms that use each

Many of the precursor metabolites serve as starting substrates for the synthesis of amino acids (figure 12.2). In the amino acid biosynthetic pathways, the carbon skeleton is remodeled and an amino group, and sometimes sulfur, is added. In this section, we first examine the mechanisms by which nitrogen and sulfur are assimilated and incorporated into amino acids. This is followed by a brief consideration of the organization of amino acid biosynthetic pathways.

Inorganic Nitrogen Assimilation

Nitrogen is a major component not only of proteins but also of nucleic acids, coenzymes, and many other cell constituents. Thus the cell's ability to assimilate inorganic nitrogen is critical. Although nitrogen gas is abundant in the atmosphere, only a few bacteria and archaea can reduce the gas and use it as a nitrogen source. Most incorporate either ammonia or nitrate. We examine ammonia and nitrate assimilation first and then discuss nitrogen assimilation in microbes that fix N_2.

Ammonia Incorporation Occurs in Two Ways

Ammonia can be incorporated into organic material relatively easily and directly because it is more reduced than other forms of inorganic nitrogen. Ammonia is initially incorporated into carbon skeletons by one of two mechanisms: reductive amination or the glutamine synthetase–glutamate synthase system. Once incorporated, the nitrogen can be transferred to other carbon skeletons.

The **reductive amination pathway** forms glutamate from α-ketoglutarate. It is catalyzed in many bacteria and fungi by glutamate dehydrogenase when the ammonia concentration is high (**figure 12.12**).

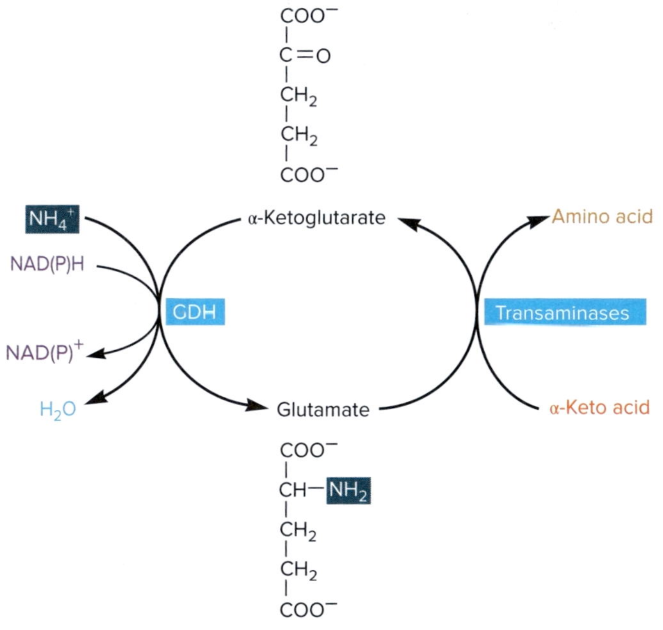

Figure 12.12 Ammonia Assimilation by Reductive Amination and Transaminases. Either NADPH- or NADH-dependent glutamate dehydrogenases may be involved. This route is most active at high ammonia concentrations. GDH, glutamate dehydrogenase.

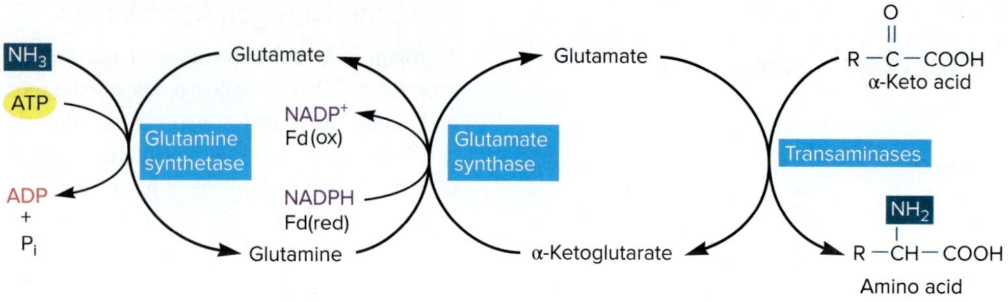

Figure 12.13 Ammonia Incorporation Using Glutamine Synthetase, Glutamate Synthase, and Transaminases. This route is effective at low ammonia concentrations. Fd, ferredoxin in either its reduced (Fd_{red}) state or oxidized (Fd_{ox}) form.

Once glutamate has been synthesized, the newly formed α-amino group can be transferred by enzymes called **transaminases** to other carbon skeletons. Microorganisms have a number of transaminases, each of which catalyzes the formation of several amino acids.

Many bacteria use the **glutamine synthetase–glutamate synthase (GS-GOGAT) system** to assimilate ammonia when its levels are low (**figure 12.13**). Incorporation of ammonia begins with synthesis of glutamine from glutamate in a reaction catalyzed by glutamine synthetase (**figure 12.14**). In this reaction, the carboxyl group found in the side chain of glutamate reacts with ammonia, and the incorporated nitrogen exists as an amide in glutamine. Glutamate synthase then transfers the amide nitrogen to α-ketoglutarate to generate a new glutamate molecule (figure 12.14). Because glutamate acts as an amino donor in transaminase reactions, ammonia may be used to synthesize all common amino acids when suitable transaminases are present.

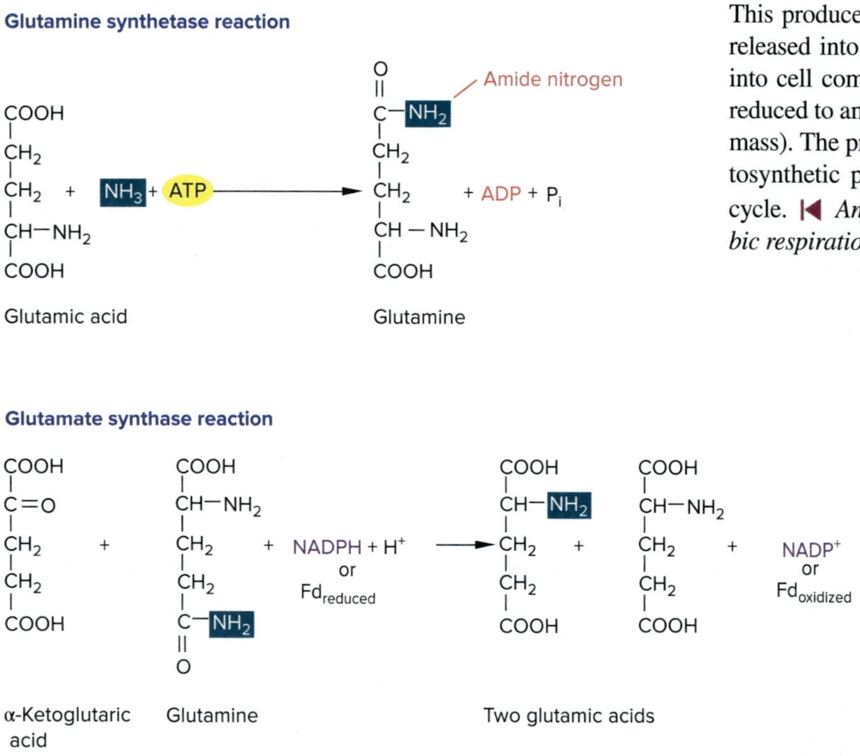

Figure 12.14 Glutamine Synthetase and Glutamate Synthase Reactions Involved in Ammonia Assimilation. Some glutamate synthases use NADPH as an electron source; others use reduced ferredoxin (Fd).

MICRO INQUIRY *What purpose is served for the cell when these reactions are used to produce two glutamic acids rather than a single glutamine?*

Assimilatory Nitrate Reduction: NO_3 to NH_3

The nitrogen in nitrate (NO_3^-) is much more oxidized than that in ammonia (NH_3). Therefore nitrate must first be reduced to ammonia before the nitrogen can be converted to an organic form. This reduction of nitrate is called **assimilatory nitrate reduction.** It is important to clearly distinguish assimilatory nitrate reduction from the type of nitrate reduction discussed in chapter 11—dissimilatory nitrate reduction. In dissimilatory nitrate reduction, nitrate serves as the terminal electron acceptor for anaerobic respiration. This produces reduced forms of nitrogen (e.g., N_2, N_2O) that are released into the environment and therefore are not incorporated into cell components. In assimilatory nitrate reduction, nitrate is reduced to ammonia, which is incorporated into cell material (biomass). The process is widespread among bacteria, fungi, and photosynthetic protists, and it is an important step in the nitrogen cycle. ◀ *Anaerobic respiration uses the same three steps as aerobic respiration (section 11.7)* ▶ *Nitrogen cycle (section 28.1)*

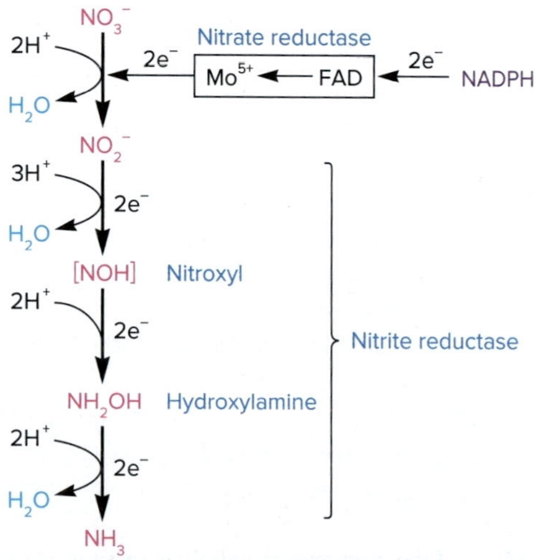

Figure 12.15 Assimilatory Nitrate Reduction. This sequence is thought to operate in bacteria that reduce and assimilate nitrate nitrogen. Here we show the electron donor as NADPH. However, some bacteria use other electron donors such as ferredoxin or NADH.

Assimilatory nitrate reduction takes place in the cytoplasm in bacteria. The first step in nitrate assimilation is its reduction to nitrite by **nitrate reductase,** an enzyme that contains both FAD and molybdenum (**figure 12.15**). Nitrite is next reduced to ammonia through a series of two electron additions catalyzed by nitrite reductase. The ammonia is then incorporated into amino acids by reductive amination or the GS-GOGAT system.

Nitrogen Fixation: N_2 to NH_3

The reduction of atmospheric gaseous nitrogen to ammonia is called **nitrogen fixation.** Only a few bacteria and archaea carry out nitrogen fixation: (1) free-living chemotrophic bacteria (e.g., *Azotobacter, Klebsiella,* and *Clostridium* spp.) and archaea (e.g., methanogens), (2) bacteria living in symbiotic associations with plants such as legumes (e.g., *Rhizobium* spp., *see figure 31.9*), and (3) cyanobacteria (e.g., *Nostoc, Anabaena,* and *Trichodesmium* spp.). These microbes play a critical role in the nitrogen cycle. The biological aspects of nitrogen fixation are discussed in chapters 28 and 31. The biochemistry of nitrogen fixation is the focus of this section.

During nitrogen fixation, nitrogen is reduced by sequential two-electron additions as illustrated in **figure 12.16**. The reduction of molecular nitrogen to ammonia is quite exergonic, but the reaction has a high activation energy because molecular nitrogen is an unreactive gas with a triple bond connecting the two nitrogen atoms. Therefore nitrogen reduction requires a large ATP expenditure—at least 8 electrons and 16 ATP molecules (4 ATP per pair of electrons); symbiotic nitrogen-fixing bacteria can consume almost 20% of the ATP supplied by their host plants. Note in figure 12.16 that two of the electrons and four of the ATP molecules are expended to reduce 2 protons to hydrogen gas at the start of the process. This is often thought of as a wasteful expenditure of electrons and energy. However, some bacteria have evolved mechanisms for capturing the hydrogen gas and using it as a source of electrons for nitrogen reduction.

The reduction of nitrogen to ammonia is catalyzed by the enzyme **nitrogenase** (figure 12.16). The best-studied nitrogenase is a complex enzyme consisting of two major protein components, a MoFe protein joined with a Fe protein. The MoFe protein has an organic cofactor that contains molybdenum, iron, and sulfur atoms, whereas the Fe protein associates with iron atoms. Fe protein is reduced by ferredoxin and then binds ATP. ATP binding changes the conformation of the Fe protein and lowers its reduction potential (−0.29 to −0.40 V), enabling it to reduce the MoFe protein. ATP is hydrolyzed when this electron transfer occurs. Finally, reduced MoFe protein donates electrons to protons

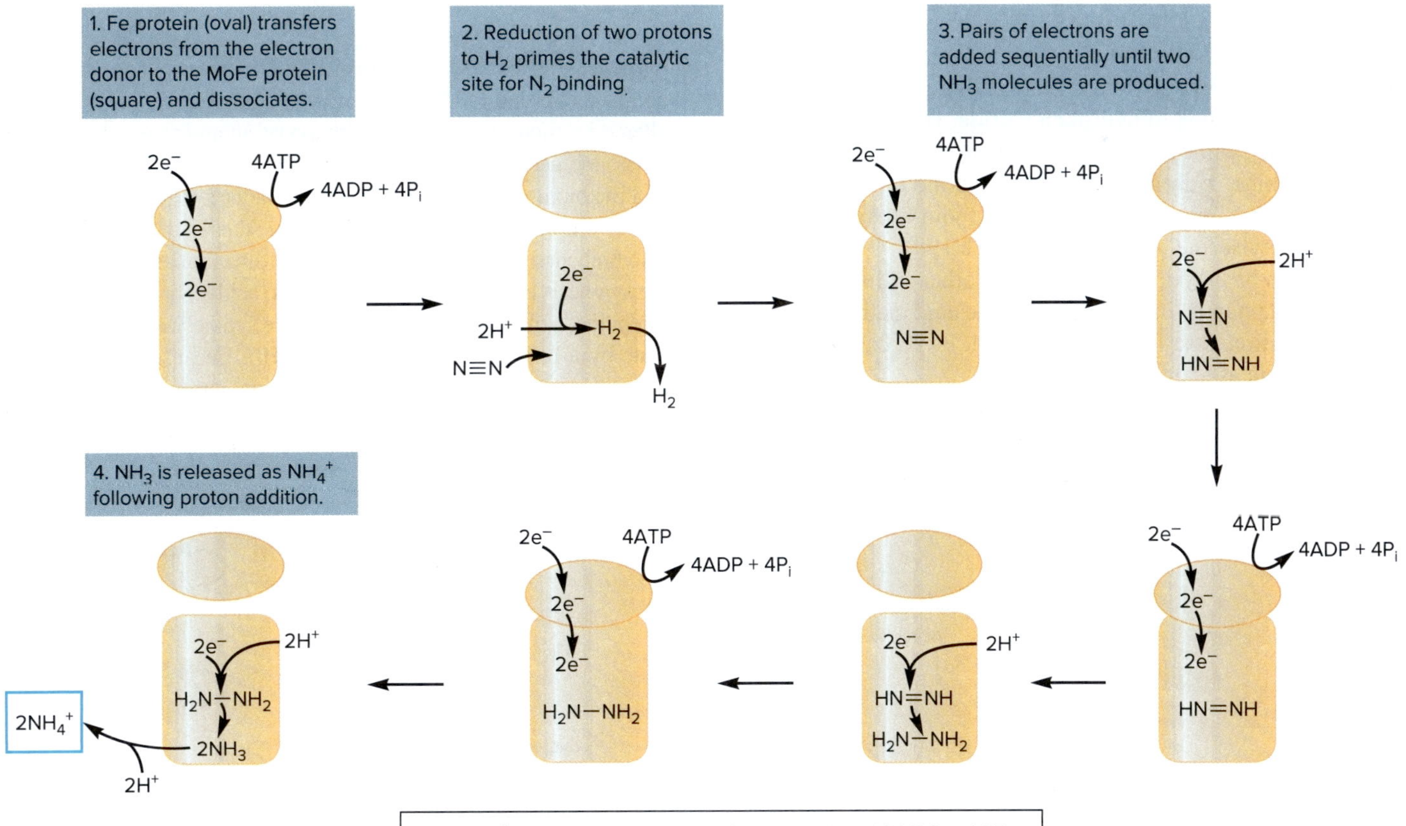

$$N_2 + 8H^+ + 8e^- + 16ATP \rightarrow 2NH_3 + H_2 + 16ADP + 16P_i$$

Figure 12.16 Nitrogen Reduction by Nitrogenase. Ferredoxin is often the electron donor for nitrogen fixation.

(the first two reactions of the process) or sequentially to atomic nitrogen (the series of reactions outlined by number 3 in figure 12.16). Nitrogenase is sensitive to O_2 and must be protected from O_2 inactivation. Microbes use a variety of strategies to protect nitrogenase, as we discuss more fully in chapters 22 and 31.

Sulfur Assimilation: SO_4 to S-Bearing Molecules

Sulfur is needed for the synthesis of the amino acids cysteine and methionine. It is also needed for synthesis of several coenzymes (e.g., coenzyme A and biotin). Sulfur is obtained from two sources: (1) the amino acids cysteine and methionine, and (2) sulfate. Many microorganisms use cysteine and methionine obtained from either external sources or intracellular amino acid reserves. Because the sulfur is already in a reduced organic form, it is readily incorporated into other organic molecules. In contrast, the sulfur atom in sulfate is more oxidized than it is in cysteine and methionine; thus sulfate must be reduced before it can be assimilated. This process is known as **assimilatory sulfate reduction** to distinguish it from the dissimilatory sulfate reduction that takes place when sulfate acts as an electron acceptor during anaerobic respiration. ◀ *Anaerobic respiration uses the same three steps as aerobic respiration (section 11.7)* ▶ *Sulfur cycle (section 28.1)*

Assimilatory sulfate reduction involves sulfate activation through the formation of phosphoadenosine 5′-phosphosulfate (PAPS), another modified nucleotide that functions as a carrier (**figure 12.17**). Sulfate activation prepares the sulfate for reduction. As shown in **figure 12.18**, sulfate is first reduced to sulfite (SO_3^{2-}), then to hydrogen sulfide. Hydrogen sulfide is then used to synthesize cysteine in two ways (**figure 12.19**). Fungi combine hydrogen sulfide with serine to form cysteine (figure 12.19*a*), whereas many bacteria join hydrogen sulfide with O-acetylserine instead (figure 12.19*b*). Once formed, cysteine can be used in the synthesis of other sulfur-containing organic compounds, including the amino acid methionine. Sulfur assimilation is so integral to the cell, PAPS can be used by some bacteria as a signal for growth.

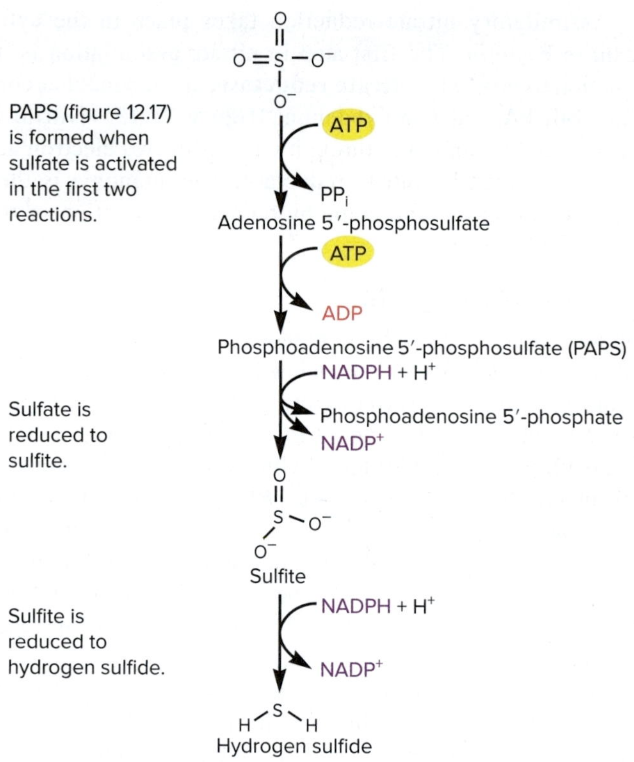

Figure 12.18 The Sulfate Reduction Pathway.

Amino Acid Biosynthetic Pathways

Some amino acids are made directly by transamination of a precursor metabolite. For example, alanine and aspartate are made from pyruvate and oxaloacetate, respectively, using glutamate as the amino group donor. However, most precursor metabolites must be altered by more than just the addition of an amino group. In many cases, the carbon skeleton must be reconfigured, and for cysteine and methionine, the sulfur must also be added. These pathways are more complex, and often involve many steps. These pathways are also branched; that is, a single precursor metabolite can be used for the synthesis of a family of related amino acids. For example, lysine, threonine, isoleucine, and methionine are synthesized from oxaloacetate by a branching route (**figure 12.20**). Because of the need to conserve nitrogen, carbon, and energy, amino acid synthetic

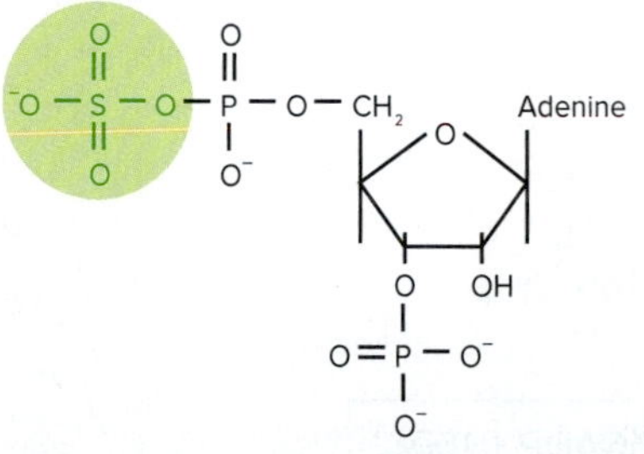

Figure 12.17 Phosphoadenosine 5′-phosphosulfate (PAPS). The sulfate group is shaded green.

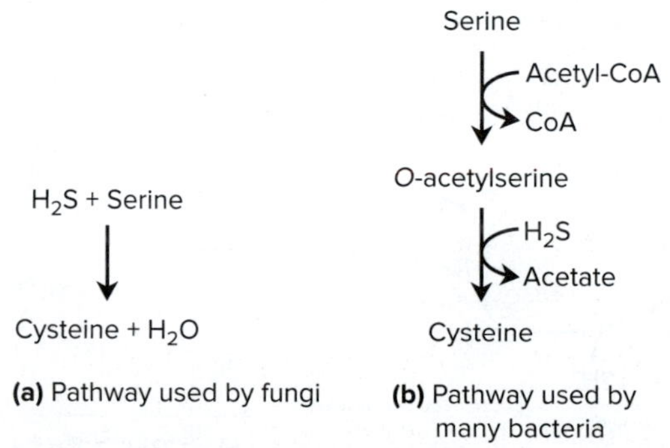

Figure 12.19 Incorporation of Sulfur from H_2S. H_2S is the product of sulfate reduction, as shown in figure 12.18.

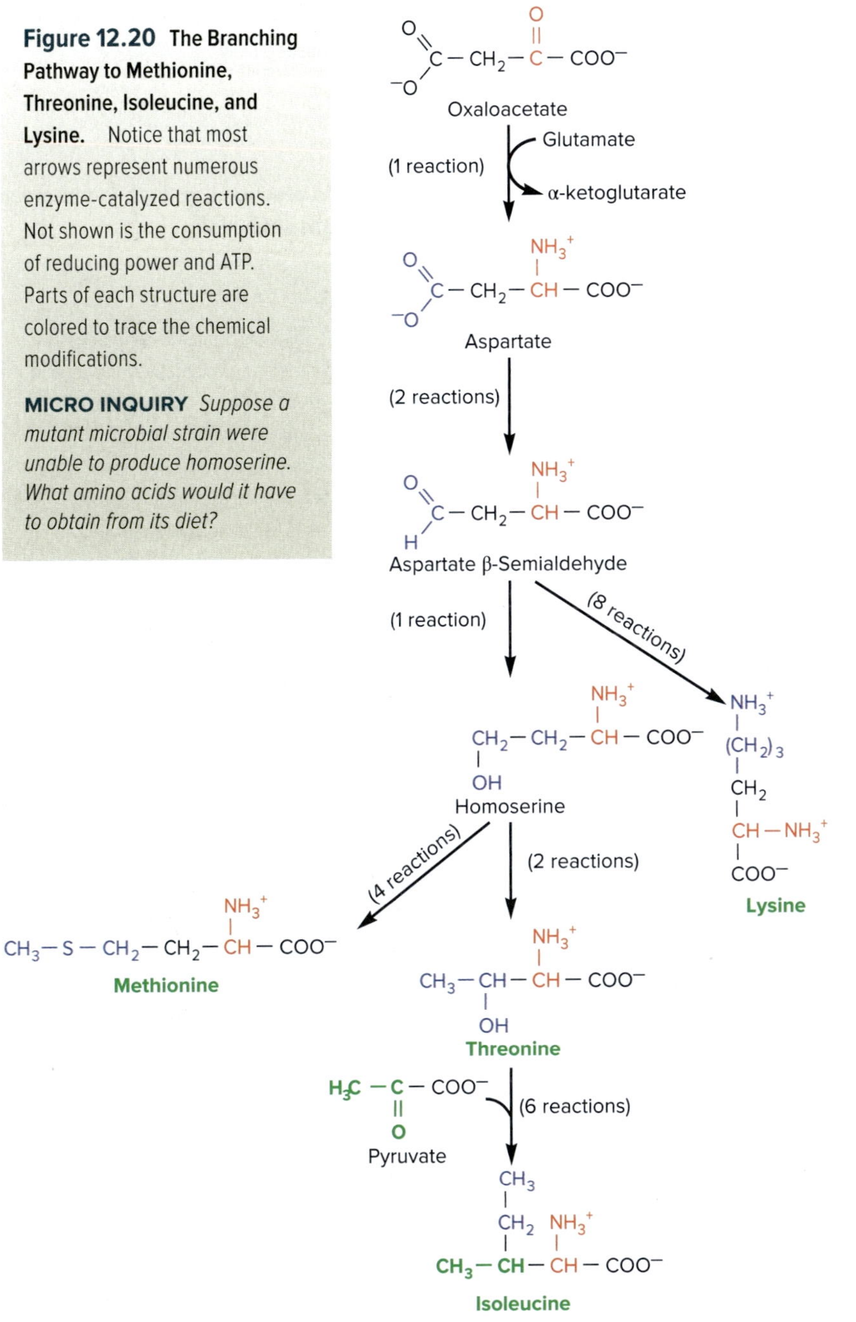

Figure 12.20 The Branching Pathway to Methionine, Threonine, Isoleucine, and Lysine. Notice that most arrows represent numerous enzyme-catalyzed reactions. Not shown is the consumption of reducing power and ATP. Parts of each structure are colored to trace the chemical modifications.

MICRO INQUIRY *Suppose a mutant microbial strain were unable to produce homoserine. What amino acids would it have to obtain from its diet?*

pathways are usually tightly regulated by feedback mechanisms. ◀ *Posttranslational regulation of enzyme activity (section 10.7)*

Anaplerotic Reactions Replace the Precursor Metabolites Used for Amino Acid Biosynthesis

When an organism actively synthesizes amino acids, heavy demand for precursor metabolites is placed on the central metabolic pathways, especially the TCA cycle. Therefore it is critical that TCA cycle intermediates be readily available. This is especially true for organisms carrying out fermentation, where the TCA cycle does not function to catabolize glucose. To ensure an adequate supply of TCA cycle–generated precursor metabolites, microorganisms use **anaplerotic reactions** (Greek *anaplerotic,* filling up) to replenish these intermediates.

Most microorganisms replace TCA cycle intermediates by generating oxaloacetate; this is done in two ways. One adds CO_2 as a carboxyl group to pyruvate (catalyzed by pyruvate carboxylase), and the other adds CO_2 to phosphoenolpyruvate (catalyzed by phosphoenolpyruvate carboxylase) (**figure 12.21**). The pyruvate carboxylase reaction is observed in yeasts and some bacteria. The phosphoenolpyruvate carboxylase reaction is used by other microorganisms (e.g., *E. coli* and *Salmonella* spp.).

Other anaplerotic reactions are part of the **glyoxylate cycle,** which functions in some bacteria, fungi, and protists (**figure 12.22**). This cycle is made possible by two unique enzymes, isocitrate lyase and malate synthase. The glyoxylate cycle is actually a modified TCA cycle. The two decarboxylations of the TCA cycle are bypassed, making possible the conversion of acetyl-CoA to oxaloacetate without the loss of the acetyl-CoA carbons as CO_2. ◀ *Pyruvate to carbon dioxide (step two) is accomplished by the tricarboxylic acid cycle (section 11.5)*

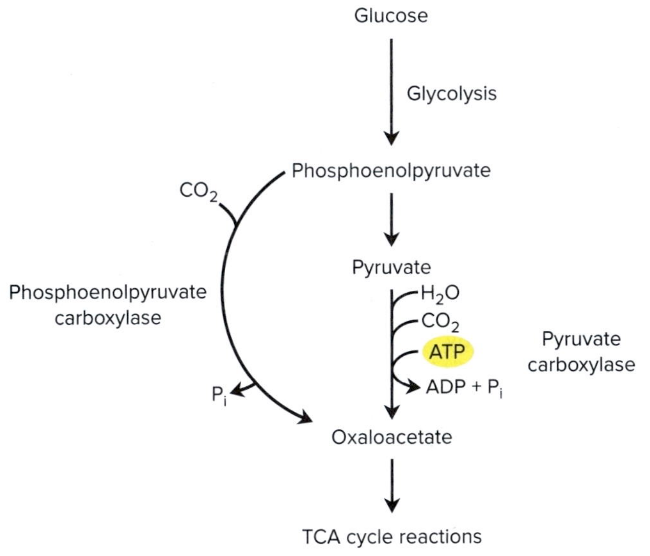

Figure 12.21 Anaplerotic Reactions That Replace Oxaloacetate, a TCA Cycle Intermediate. Pyruvate and phosphoenolpyruvate are formed by the Embden-Meyerhof glycolytic pathway.

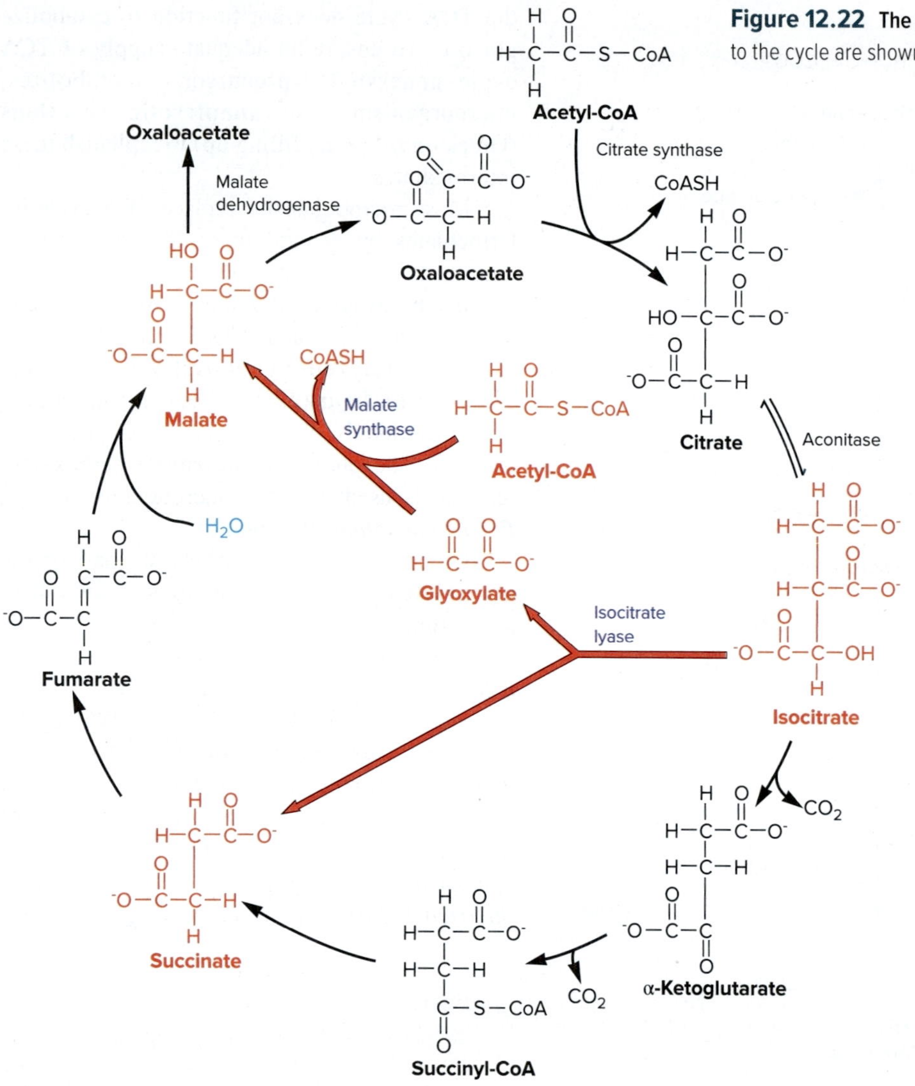

Figure 12.22 The Glyoxylate Cycle. The reactions and enzymes unique to the cycle are shown in red and blue, respectively.

The remaining reactions are highly conserved in all organisms studied thus far.

Comprehension Check

1. Describe the roles of glutamate dehydrogenase, glutamine synthetase, glutamate synthase, and transaminases in ammonia assimilation.
2. How does assimilatory nitrate reduction differ from dissimilatory nitrate reduction? What is the fate of nitrate following assimilatory nitrate reduction versus its fate following denitrification?
3. Briefly describe the structure and mechanism of action of nitrogenase. Why is nitrogen fixation so energetically expensive?
4. How do organisms assimilate sulfur? How does assimilatory sulfate reduction differ from dissimilatory sulfate reduction?
5. Why are branched pathways efficient mechanisms for synthesizing amino acids?
6. Define an anaplerotic reaction. Give three examples of anaplerotic reactions.
7. Describe the glyoxylate cycle. How is it similar to the TCA cycle? How does it differ?
8. List two porphyrin molecules. What is their common feature? Which amino acids and precursor metabolites are important in their synthesis?

Synthesis of Porphyrins

As we demonstrate in scction 12.6 and in figure 12.2, in addition to their role in protein synthesis, amino acids are required for the synthesis of other molecules. Here we consider their role in the synthesis of porphyrins (*see figures 10.10 and 11.30*). Porphyrins are found in a number of critically important molecules, including hemoglobin, cytochromes, phycobilins, chlorophylls, and vitamin B_{12}. There are two pathways leading to these important products: the succinyl-CoA/glycine pathway, used by all nonphotosynthetic eukaryotes and α-proteobacteria; and the glutamate pathway, used by all other bacteria, photosynthetic eukaryotes, and archaea (**figure 12.23**). Both pathways converge at the intermediate 5-aminolevulinic acid.

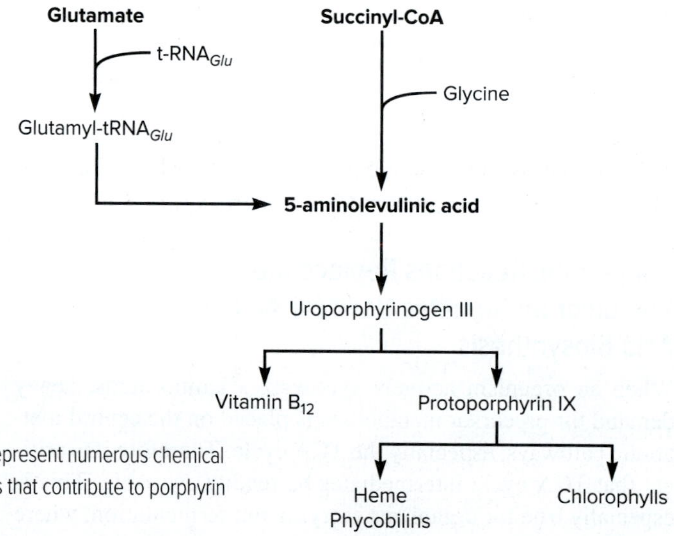

Figure 12.23 A Simplified Illustration of Porphyrin Biosynthesis. Most arrows represent numerous chemical reactions. Not shown are the many electron donors, ATP molecules, and other molecules that contribute to porphyrin synthesis.

12.6 Synthesis of Purines, Pyrimidines, and Nucleotides

After reading this section, you should be able to:

a. Identify which nitrogenous bases are purines and which are pyrimidines
b. Draw a simple diagram that illustrates the chemical moieties found in nucleosides and nucleotides
c. Discuss in general terms how phosphorus is assimilated
d. Compare and contrast purine and pyrimidine biosynthesis
e. Discuss the methods used to convert ribonucleotides to deoxyribonucleotides

Purine and pyrimidine biosynthesis is critical for all cells because these molecules are used to synthesize ATP, several cofactors, ribonucleic acid (RNA), deoxyribonucleic acid (DNA), and other important cell components. Nearly all microorganisms can synthesize their own purines and pyrimidines as these are crucial to cell function. ▶| *DNA replication in bacteria (section 13.3); Transcription in bacteria (section 13.5)*

Purines and **pyrimidines** are cyclic nitrogenous bases with several double bonds. **Adenine** and **guanine** are purines, consisting of two joined rings, whereas pyrimidines (**uracil, cytosine,** and **thymine**) have only one ring. A purine or pyrimidine base joined with a pentose sugar, either ribose or deoxyribose, is a **nucleoside.** A **nucleotide** is a nucleoside with one or more phosphate groups attached to the sugar.

Amino acids participate in the synthesis of nitrogenous bases and nucleotides in a number of ways, including providing the nitrogen that is part of all purines and pyrimidines. The phosphorus present in nucleotides is provided by other mechanisms. We begin this section by examining phosphorus assimilation. We then examine the pathways for synthesis of nitrogenous bases and nucleotides.

Phosphorus Assimilation

In addition to nucleic acids, phosphorus is found in proteins (i.e., phosphorylated proteins), phospholipids, and coenzymes such as $NADP^+$. The most common phosphorus sources are inorganic phosphate and organic molecules containing a phosphoryl group. Inorganic phosphate is incorporated through the formation of ATP in one of three ways: (1) photophosphorylation, (2) oxidative phosphorylation, and (3) substrate-level phosphorylation. |◀ *Glucose to pyruvate: the first step (section 11.4); Electron transport and oxidative phosphorylation (step 3) generate the most ATP (section 11.6); Phototrophy (section 11.12)*

Microorganisms may obtain organic phosphates from their surroundings in dissolved or particulate form. **Phosphatases** hydrolyze organic phosphate-containing molecules to release inorganic phosphate. Gram-negative bacteria have phosphatases in the periplasmic space, which allows phosphate to be taken up immediately after release. On the other hand, protists can directly use organic phosphates after ingestion or hydrolyze them in lysosomes and then incorporate the phosphate.

Purine Biosynthesis

The biosynthetic pathway for purines is a complex, 11-step sequence (*see appendix II*) in which seven different molecules contribute parts to the final purine skeleton (**figure 12.24**). The pathway begins with ribose 5-phosphate, and the purine skeleton is constructed on this sugar. Therefore the first purine product of the pathway is the nucleotide inosinic acid, not a free purine base. The cofactor folic acid is very important in purine biosynthesis because folic acid derivatives contribute two carbons to the purine skeleton.

Once inosinic acid has been formed, relatively short pathways synthesize adenosine monophosphate and guanosine monophosphate (**figure 12.25**), and produce nucleoside diphosphates and triphosphates by phosphate transfers from ATP. At this point, the cell has synthesized ribonucleotides. However, it needs deoxynucleotides for DNA synthesis. In deoxyribonucleotides there is an H atom rather than a hydroxyl group on carbon two of the sugar. Thus to convert a ribonucleotide into a deoxyribonucleotide, that carbon must be reduced. Deoxyribonucleotides arise from the reduction of nucleoside diphosphates or nucleoside triphosphates by two different routes. Some microorganisms reduce the ribose in nucleoside triphosphates with a system requiring vitamin B_{12} as a cofactor. Others reduce the ribose in nucleoside diphosphates. Both systems employ a small, sulfur-containing protein called thioredoxin as their reducing agent.

Pyrimidine Biosynthesis

Pyrimidine biosynthesis is equally as complex as purine synthesis. However, the two processes differ in an important way: Purines are synthesized by adding chemical moieties to ribose 5-phosphate and never exist as free bases; construction of pyrimidine rings is completed before ribose is attached.

Pyrimidine biosynthesis begins with aspartic acid and carbamoyl phosphate, a high-energy molecule synthesized from bicarbonate and ammonia provided by the amino acid glutamine (**figure 12.26**). Aspartate carbamoyltransferase catalyzes the joining (condensation) of these two substrates to form carbamoylaspartate, which is then converted to the initial pyrimidine product, orotic acid. The high-energy molecule 5-phosphoribosyl

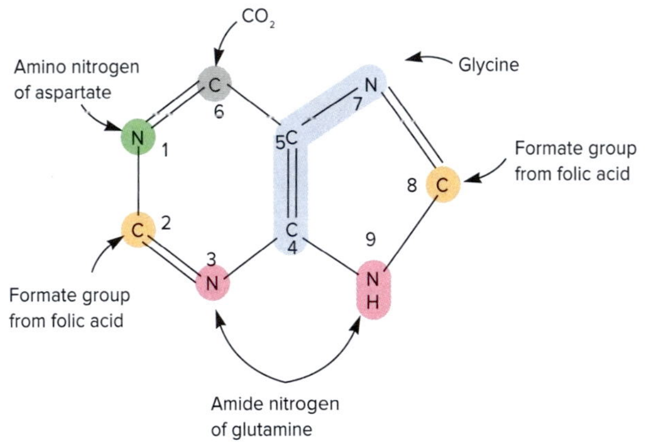

Figure 12.24 Sources of the Nitrogen and Carbon in Purines. The nitrogen numbered 9 is linked to the sugar ribose 5-phosphate, and the base is built while attached to the sugar.

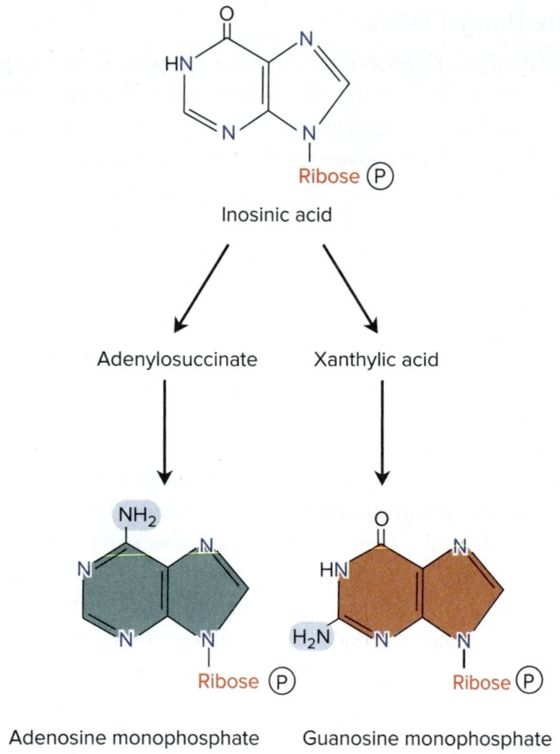

Figure 12.25 Synthesis of Adenosine Monophosphate and Guanosine Monophosphate. The blue highlighted groups differ from those in inosinic acid.

1-pyrophosphate (PRPP) provides the ribose 5-phosphate that is attached to orotic acid to form a nucleotide.

Orotidine monophosphate is sequentially converted to uridine monophosphate, uridine triphosphate, and cytidine triphosphate. These ribonucleotides are reduced to deoxy-forms in the same way that purine ribonucleotides are. That leaves the cell with only one deoxynucleotide yet to be made: deoxythymidine monophosphate. It is made when deoxyuridine monophosphate is methylated with a folic acid derivative (**figure 12.27**).

Comprehension Check

1. How is phosphorus assimilated? What roles do phosphatases play in phosphorus assimilation? Why can phosphate be directly incorporated into cell constituents, whereas nitrate, nitrogen gas, and sulfate cannot?
2. Explain the difference between a purine and a pyrimidine, and between a nucleoside and a nucleotide.
3. Outline the way in which purines and pyrimidines are synthesized. How is the deoxyribose component of deoxyribonucleotides made?

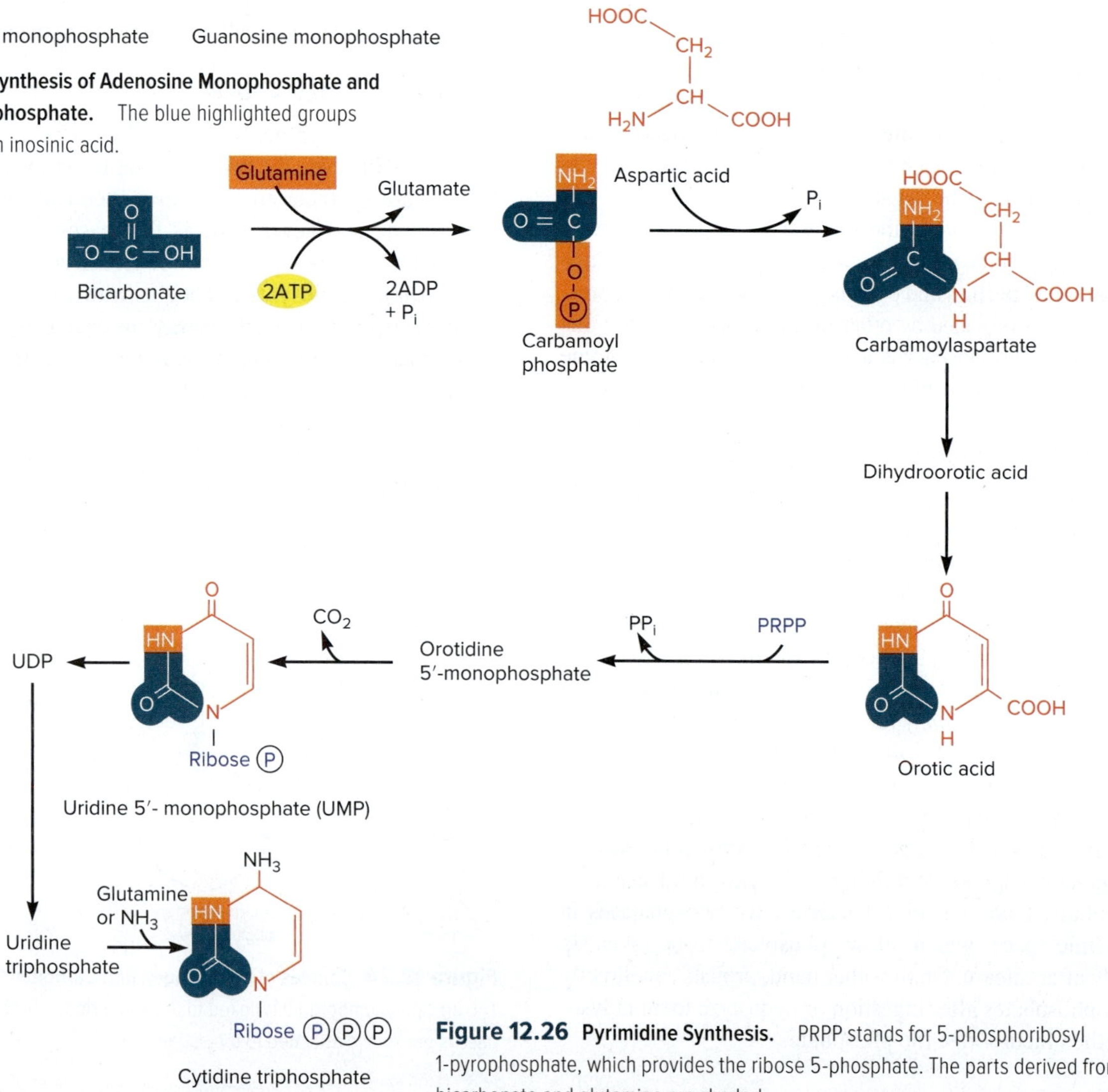

Figure 12.26 Pyrimidine Synthesis. PRPP stands for 5-phosphoribosyl 1-pyrophosphate, which provides the ribose 5-phosphate. The parts derived from bicarbonate and glutamine are shaded.

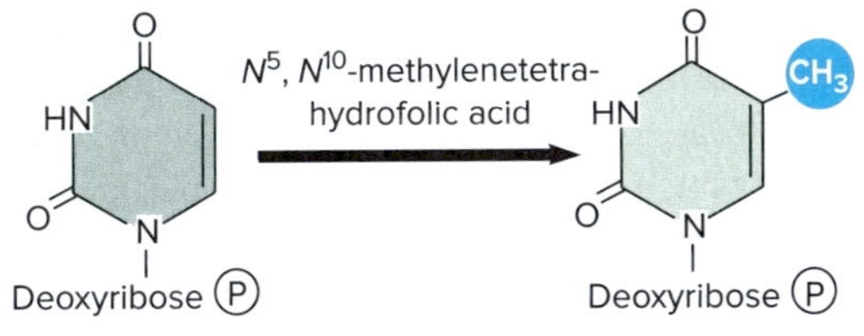

Figure 12.27 Deoxythymidine Monophosphate Synthesis. Deoxythymidine differs from deoxyuridine in the presence of the shaded methyl group.

acetyl-CoA molecules is further activated by the addition of carbon dioxide. This generates a 3-carbon molecule called malonyl-ACP. Of the three carbons, the two derived from acetyl-CoA are added to acetyl-ACP, while the remaining carbon (from CO_2) and the ACP carrier are released. In the next stage, the β-keto group arising from the initial condensation reaction is removed in a three-step process involving two reductions and a dehydration. The fatty acid is then ready for the addition of two more carbon atoms.

Unsaturated fatty acids are synthesized in two ways. Eukaryotes and many aerobic bacteria use a pathway that

12.7 Lipid Synthesis

After reading this section, you should be able to:

a. Describe the role of acetyl-CoA, acyl carrier protein (ACP), and fatty acid synthase in fatty acid synthesis
b. Distinguish saturated fatty acids from unsaturated fatty acids
c. Describe the roles of dihydroxyacetone phosphate and fatty acids in the synthesis of triacylglycerol
d. Describe the roles of diacylglycerol and CTP in the synthesis of phospholipids
e. Summarize lipopolysaccharide synthesis
f. Describe the Lpt pathway
g. Compare sterol synthesis to archaeal isoprenoid lipid synthesis

Lipids are absolutely required by cells, as they are the major components of cell membranes. Most bacterial and eukaryal lipids contain fatty acids or their derivatives. Archaeal lipids are built using an isoprene building block. In this section we consider the pathways leading to these molecules. We also consider the synthesis of lipopolysaccharides, major components of the outer membrane of typical Gram-negative bacteria.

Fatty Acids and Phospholipids

Fatty acids consist of a carboxyl group attached to a long chain of carbons (alkyl chains); the chains usually have an even number of carbons (the average length is 18 carbons). Some fatty acids have alkyl chains that are unsaturated; that is, they have one or more double bonds. Most microbial fatty acids are straight chained, but some are branched and some have one or more cyclopropane rings. Their synthesis is complex and is the target of some antimicrobial agents used to treat infectious disease caused by eukaryotic pathogens. ▶| *Lipids (appendix I)* |◀ *Antifungal drugs (section 9.6)*

Synthesis of saturated fatty acids is catalyzed by the **fatty acid synthase** complex. The process begins with two molecules of acetyl-CoA; both must be complexed with acyl carrier protein (ACP) in place of coenzyme A (CoA) (**figure 12.28**). One of the

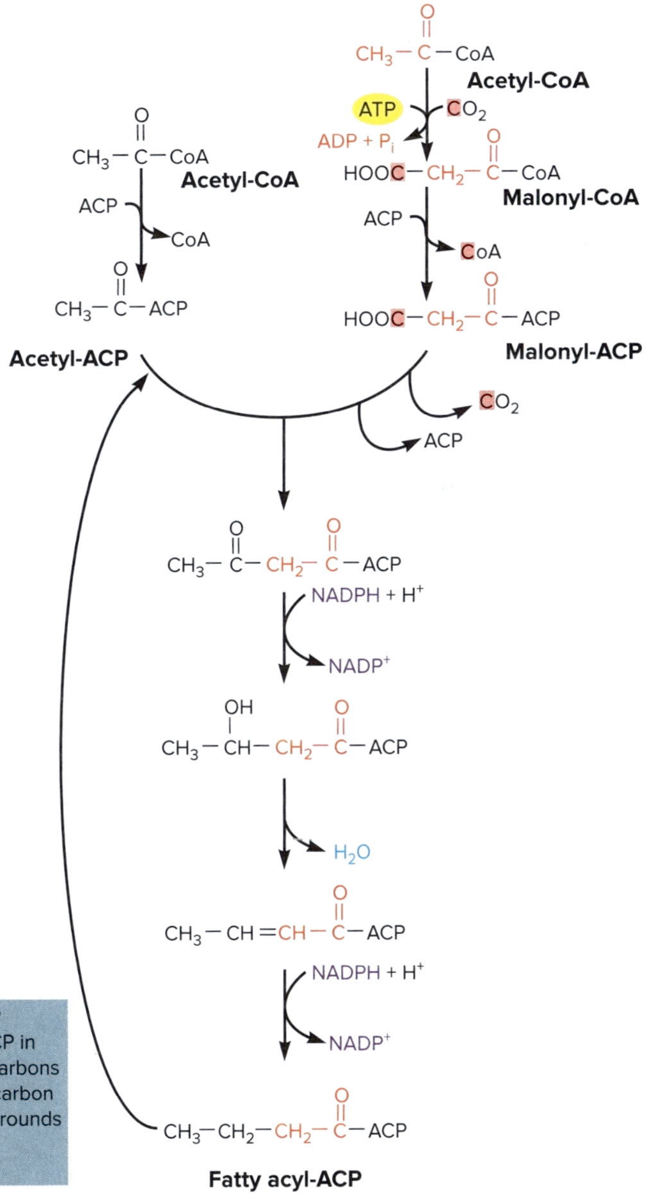

The 4-carbon fatty acyl-ACP takes the place of acetyl-ACP in the next round. Two more carbons are added, resulting in a 6-carbon fatty acyl-ACP. Subsequent rounds add 2-carbon units until the final length is reached.

Figure 12.28 Fatty Acid Synthesis. The cycle is repeated until the proper chain length has been reached. Carbon atoms derived from carbon dioxide are boxed in red; atoms derived from malonyl CoA are in red text. ACP, acyl carrier protein.

introduces double bonds into a saturated fatty acid using both NADPH and O_2 as shown in the following equation.

$$R{-}(CH_2)_9{-}\overset{\overset{\displaystyle O}{\|}}{C}{-}SCoA + NADPH + H^+ + O_2 \rightarrow$$

$$R{-}CH{=}CH(CH_2)_7{-}\overset{\overset{\displaystyle O}{\|}}{C}{-}SCoA + NADP^+ + 2H_2O$$

A double bond is formed between carbons nine and ten, and O_2 is reduced to water with electrons supplied by both the fatty acid and NADPH. Anaerobic bacteria and some aerobes create double bonds during fatty acid synthesis by dehydrating hydroxy fatty acids. Oxygen is not required for double-bond synthesis by this pathway.

Fatty acids are used to make other lipids such as triacylglycerol and phospholipids. Both of these molecules have glycerol linked to fatty acids by ester bonds and both are synthesized using the same branched pathway (**figure 12.29**). Glycerol arises from the reduction of dihydroxyacetone phosphate, an intermediate of the Embden-Meyerhoff pathway, to glycerol 3-phosphate. Glycerol 3-phosphate is then esterified with two fatty acids to give phosphatidic acid. Phosphatidic acid is the branchpoint of the pathway. The branch to triacylglycerol begins with the hydrolysis of phosphate from phosphatidic acid, giving a diacylglycerol; the third fatty acid is then attached to yield a triacylglycerol. The branch to a phospholipid begins when phosphatidic acid is attached to a cytidine diphosphate (CDP) carrier that plays a role similar to that of uridine and adenosine diphosphate carriers in carbohydrate biosynthesis. For example, bacteria synthesize phosphatidylethanolamine, a major cell membrane component, through the initial formation of CDP-diacylglycerol (figure 12.29). This CDP derivative then reacts with serine to form the phospholipid phosphatidylserine, and decarboxylation yields phosphatidylethanolamine. In this way, a complex membrane lipid is constructed from the products of glycolysis, fatty acid biosynthesis, and amino acid biosynthesis.

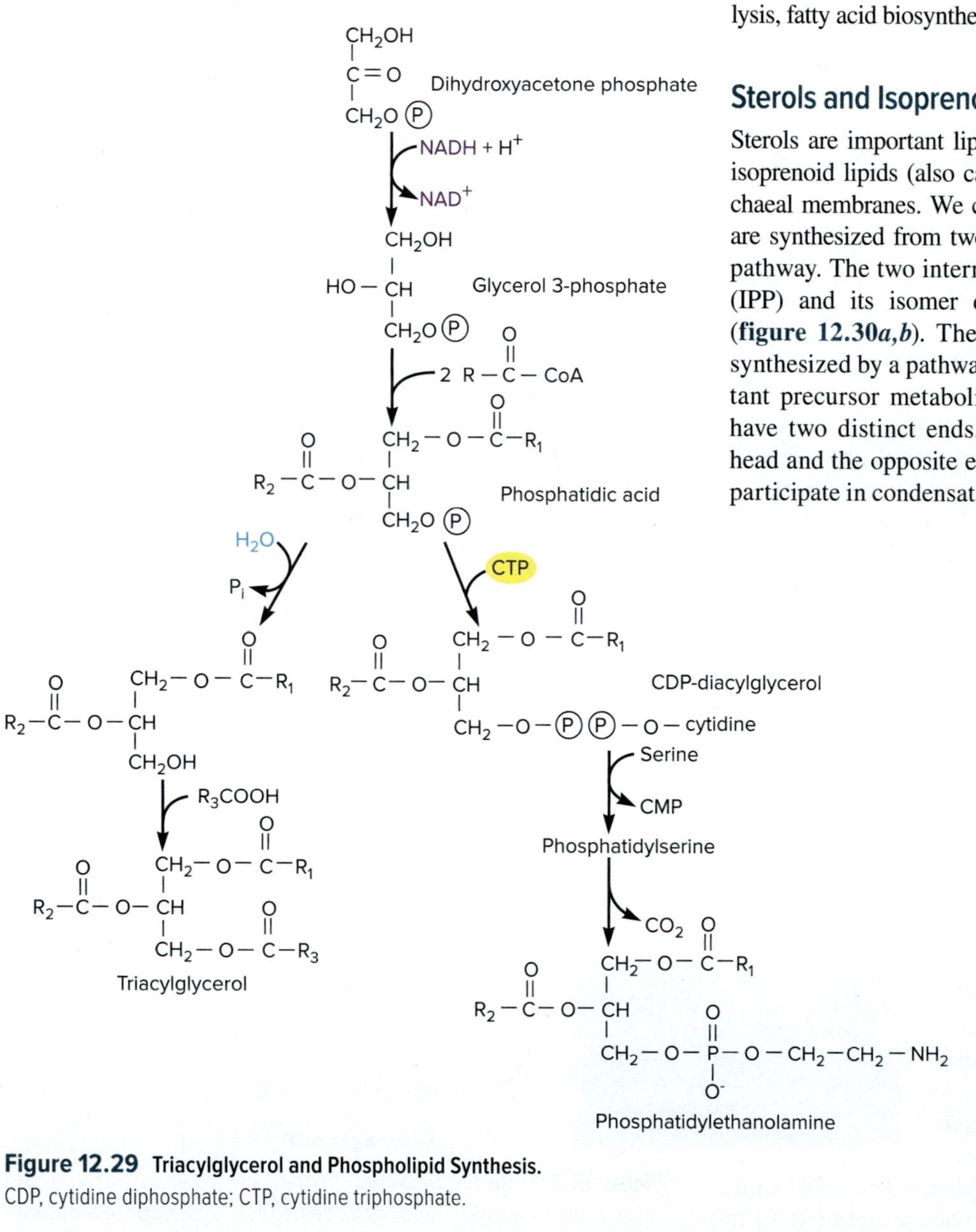

Figure 12.29 Triacylglycerol and Phospholipid Synthesis. CDP, cytidine diphosphate; CTP, cytidine triphosphate.

Sterols and Isoprenoid Lipids

Sterols are important lipids found in eukaryotic membranes, and isoprenoid lipids (also called isoprenes) are the type found in archaeal membranes. We consider them together because isoprenes are synthesized from two intermediates of the sterol biosynthetic pathway. The two intermediates are isopentenyl pyrophosphate (IPP) and its isomer dimethylallyl pyrophosphate (DMAPP) (**figure 12.30*a,b***). These two molecules are most commonly synthesized by a pathway that starts with acetyl-CoA, an important precursor metabolite. Notice that both IPP and DMAPP have two distinct ends. The pyrophosphate end is called the head and the opposite end is called the tail. The two molecules participate in condensation reactions that link molecules head to tail (figure 12.30*c*), head to head, or tail to tail, leading to the production of a variety of molecules, including intermediates that give rise to cholesterol and steroid hormones, carotenoids, and archaeal lipids. The pathways to the sterols found in eukaryotic cells are well understood and involve numerous reactions and enzymes. Although archaeal lipid synthesis pathways have not been fully established, there are themes in common with bacterial lipid synthesis pathways (**table 12.2**). As in the bacterial fatty acid synthesis pathway, carbon units are added repetitively until the final hydrophobic chain length is reached. The archaeal building blocks are IPP and DMAPP (figure 12.30). Two full-length fatty acids are bound to the glycerolphosphate backbone sequentially, in ether linkages to form

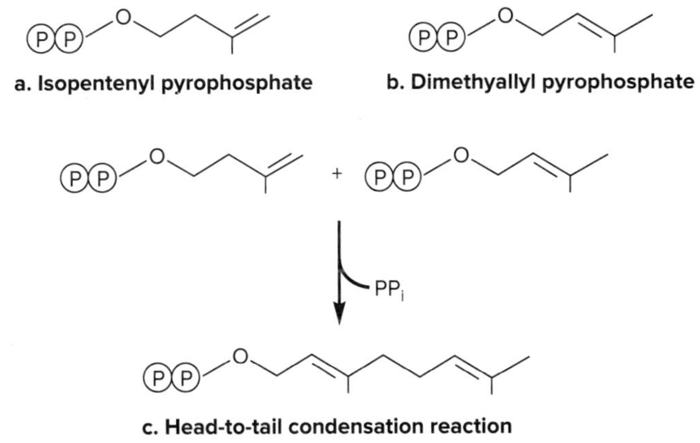

Figure 12.30 Isoprenoid Lipid Building Blocks. (a) Isopentenyl pyrophosphate and (b) dimethylallyl pyrophosphate are shown in (c) undergoing a head-to-tail condensation reaction.

archaeol, the compound analogous to phosphatidic acid. As a prelude to polar head group addition, both pathways activate the molecule by the addition of CDP. The polar head group addition reactions are catalyzed by enzymes that are highly conserved between bacteria and archaea. A final step in the archaeal pathway involves reduction of the unsaturated bonds in the isoprenoid chains. Still unclear are the pathways to synthesize archaeal tetraether lipids (*see figure 4.4*) and the mechanisms used by the cell to regulate lipid biosynthesis in response to changes in environmental conditions.

Lipopolysaccharides

Lipopolysaccharides are important components of the outer membrane of typical Gram-negative bacteria. Recall from chapter 3 that these molecules consist of a lipid component (lipid A), an oligosaccharide core, and the O polysaccharide (O antigen). We consider the synthesis of LPS here because its dual nature (lipid and carbohydrate) requires an understanding of the processes used to synthesize both types of molecules. Furthermore, many of the biosynthetic strategies we described earlier are apparent in the LPS biosynthetic pathway (**figure 12.31**). This is a complex pathway; most arrows in our rendition of the pathway represent multiple reactions, each catalyzed by a different enzyme. The pathway consists of two branches: one for the synthesis of lipid A attached to the core oligosaccharide (lipid A-core), and one for the synthesis of the O-antigen repeat unit and its polymerization into the final full-length O antigen.

The lipid A-core branch begins with UDP-*N*-acetylglucosamine (UDP-GlcNAc; abbreviated UDP-NAG in figure 12.8) and, using numerous reactions, generates the acyldisaccharide (i.e., a disaccharide linked to fatty acids) platform on which sugars that make up the core oligosaccharide are added. The fatty acids are delivered by acyl carrier protein (ACP), which was introduced in our discussion of fatty acid biosynthesis (figure 12.28). The process occurs mainly at the cytoplasmic face of the plasma membrane, where the fatty acids are inserted into the plasma membrane; the developing core oligosaccharide

Table 12.2 Comparison of Bacterial and Archaeal Lipid Synthesis

	Bacterial	Archaeal
Fatty Acid Synthesis		
Carrier for growing fatty acid chain	Malonyl-ACP	None
Number of carbon units added per cycle	2	5
Source of carbon units	Acetyl-CoA	Isopentenyl pyrophosphate
Hydrocarbon length	16–18	20–25
Phospholipid Synthesis		
Glycerophosphate backbone	Glycerol-3-phosphate	Glycerol-1-phosphate
Linkage between backbone and fatty acids	Ester	Ether
Compound with fatty acids and glycerophosphate	Phosphatidic acid	Archaeol
Activated carrier	CDP-diacylglycerol	CDP-archaeol
Common polar head groups	Serine, ethanolamine	Serine, ethanolamine

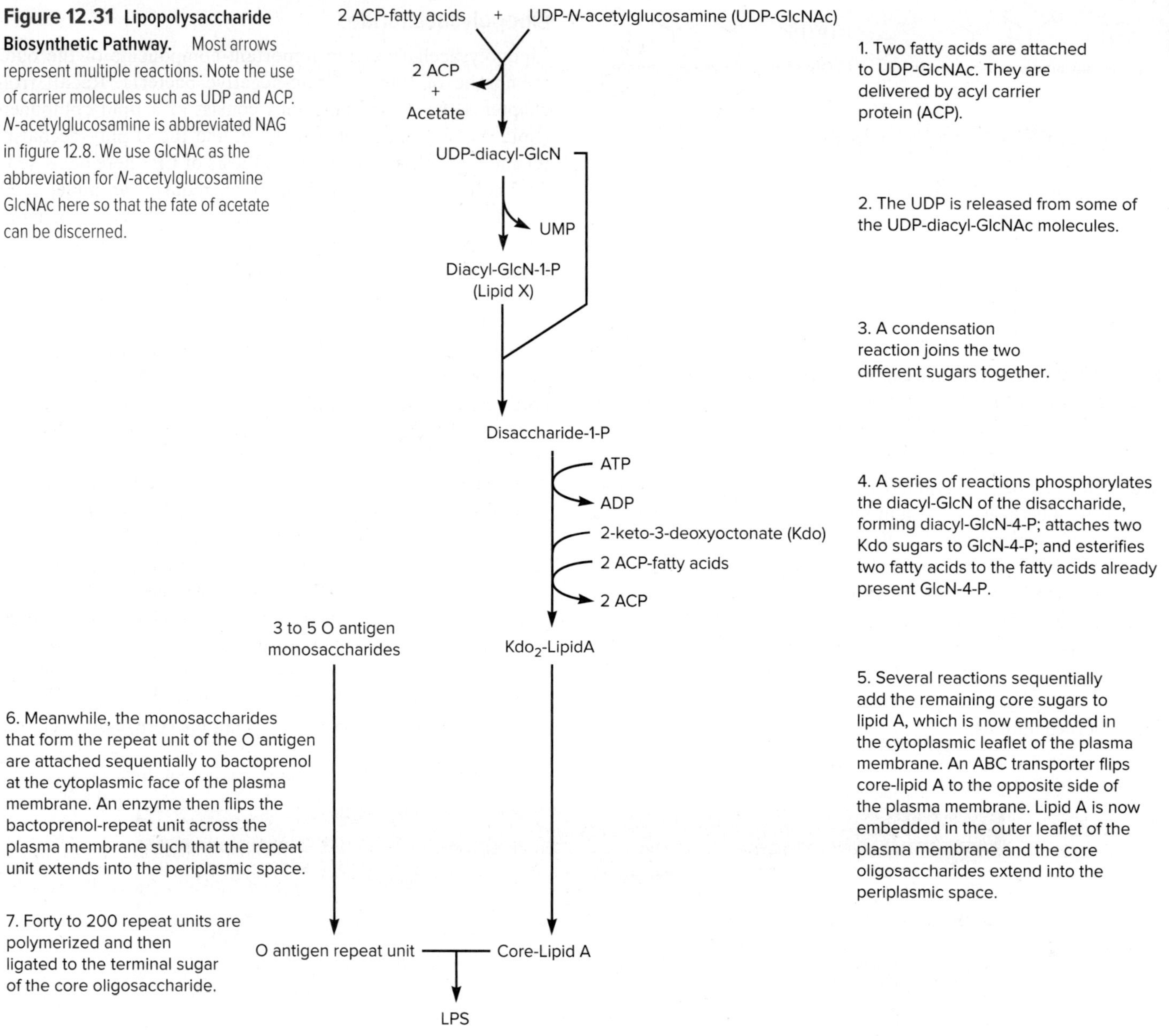

Figure 12.31 Lipopolysaccharide Biosynthetic Pathway. Most arrows represent multiple reactions. Note the use of carrier molecules such as UDP and ACP. *N*-acetylglucosamine is abbreviated NAG in figure 12.8. We use GlcNAc as the abbreviation for *N*-acetylglucosamine GlcNAc here so that the fate of acetate can be discerned.

extends into the cytoplasm. Once synthesis of the lipid A-core portion of LPS is finished, it is flipped to the periplasmic face of the plasma membrane, with the core oligosaccharides in the periplasmic space.

The O antigen branch of the pathway begins with the three to five sugars (each carried by a nucleoside diphosphate) that form the repeat unit of the O antigen. These are added sequentially to bactoprenol on the cytoplasmic face of the plasma membrane; the repeat-unit sugars are in the cytoplasm (recall the use of bactoprenol in peptidoglycan synthesis; figure 12.9). The bactoprenol-repeat unit is next flipped to the periplasmic side of the plasma membrane. The repeat units, which are now extending into the periplasmic space, are polymerized as a block, yielding the O antigen. Finally, the O antigen is ligated to the lipid A core, generating the final LPS molecule.

Figure 12.31 illustrates synthesis of LPS, but does not show how the mature molecule is inserted into the outer leaflet of the outer membrane. The problem faced by the cell is twofold. First, the carbohydrate portion of LPS is hydrophilic, making its transport across the outer membrane difficult. Second, the lipid A portion of LPS is hydrophobic, making its transport across the periplasmic space difficult. Current evidence supports a model in which seven proteins form a "walking path" along which the LPS moves (**figure 12.32**). The proteins are called Lpt proteins (for *l*ipo*p*olysaccharide *t*ransport) and the path they create is called the Lpt pathway. Lpt proteins embedded in the plasma membrane

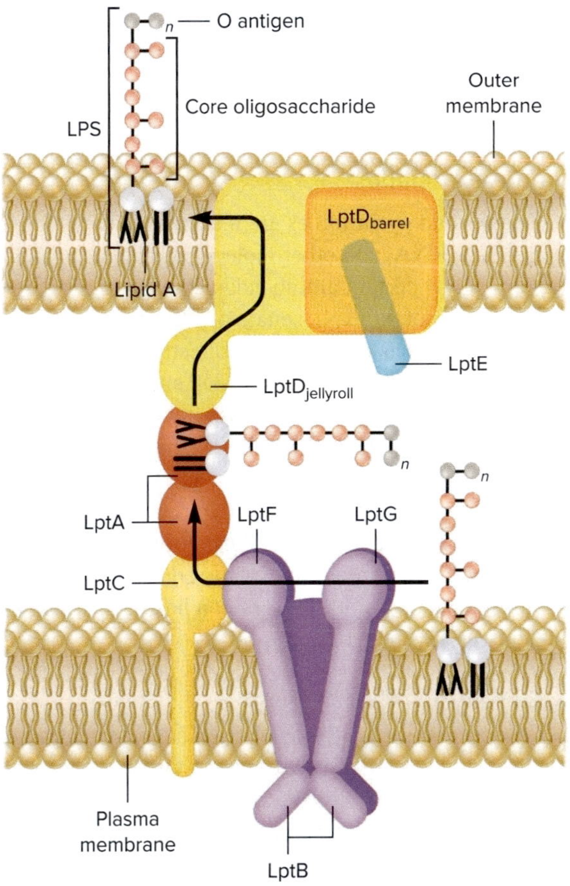

Figure 12.32 The Lpt Pathway Moves LPS from the Plasma Membrane to the Outer Membrane. The arrows track the path of LPS molecules through the proteins of the Lpt pathway.

transfer LPS from the plasma membrane to a protein that bridges the periplasm. The bridge transports lipid A across the aqueous periplasm, using a β-jelly roll structure. The β-jelly roll is a hydrophobic portion of a protein (often described as "greasy"), and readily interacts with lipid A. β-jelly rolls of LptC, LptA, and LptD form a continuous hydrophobic track to guide LPS. The core oligosaccharides and O antigen extend out from the bridge. This positions the O antigen near the next part of the pathway, a complex of two proteins. One of the proteins (LptD) forms a β-barrel. Recall from chapter 3 that β-barrels are common in transport proteins (e.g., porin proteins) that translocate molecules across membranes. The second protein, LptE, resides inside the LptD β-barrel. LptE directs LPS to enter the LptD lumen where it is reoriented and inserted into the outer leaflet of the outer membrane.

Comprehension Check

1. What is a fatty acid? Describe in general terms how fatty acid synthase manufactures a fatty acid.
2. How are unsaturated fatty acids made?
3. Briefly describe the pathways for triacylglycerol and phospholipid synthesis. Of what importance are phosphatidic acid and CDP-diacylglycerol?
4. What two building blocks are important for both sterol and isoprene biosynthesis? What pathway is most commonly used to provide them?
5. Activated carriers participate in carbohydrate, peptidoglycan, lipid, and LPS synthesis. Briefly describe these carriers and their roles. Are there any features common to all the carriers? Explain your answer.

Key Concepts

12.1 Principles Governing Biosynthesis

- Many important cell constituents are macromolecules, large polymers constructed of simple monomers.
- Although many catabolic and anabolic pathways share enzymes, making metabolism more efficient, some enzymes are separate and independently regulated.
- Macromolecular components often undergo self-assembly to form the final supramolecular system.

12.2 Precursor Metabolites: Starting Molecules for Biosynthesis

- Precursor metabolites are carbon skeletons used as the starting substrates for biosynthetic pathways. They are intermediates of glycolytic pathways and the TCA cycle (i.e., the central metabolic pathways) (**figures 12.2** and **12.3**).
- Most precursor metabolites are used for amino acid biosynthesis; others are used for synthesis of porphyrins, purines, pyrimidines, and lipids.

12.3 CO_2 Fixation: Reduction and Assimilation of CO_2 Carbon

- Six different CO_2-fixation pathways have been identified in autotrophic microorganisms: the Calvin-Benson cycle, the reductive TCA cycle, the acetyl-CoA pathway, the 3-hydroxypropionate bi-cycle, the 3-hydroxypropionate/4-hydroxybutyrate pathway, and the dicarboxylate/4-hydroxybutyrate cycle.
- The Calvin-Benson cycle is used by most aerobic autotrophs to fix CO_2. It can be divided into three phases: the carboxylation phase, the reduction phase, and the regeneration phase (**figure 12.4**). Three ATP and two NADPH are used during the incorporation of one CO_2.
- The reductive TCA cycle is used by a number of autotrophs belonging to the bacterial phyla *Aquificae, Chlorobi, Proteobacteria,* and *Nitrospirae* (**figure 12.5**).

12.4 Synthesis of Carbohydrates

- Gluconeogenesis is the synthesis of glucose and related sugars from nonglucose precursors.
- Glucose, fructose, and mannose are gluconeogenic intermediates or are made directly from them (**figure 12.6**); galactose is synthesized with nucleoside diphosphate derivatives (**figure 12.7**). Bacteria and protists synthesize glycogen and starch from adenosine diphosphate glucose.
- Peptidoglycan synthesis is a complex process involving both UDP derivatives and the lipid carrier bactoprenol, which helps transport NAG-NAM-pentapeptide units across the plasma membrane. Cross-links are formed by transpeptidation (**figures 12.8–12.11**).

12.5 Synthesis of Amino Acids Consumes Many Precursor Metabolites

- The addition of nitrogen to the carbon chain provided by a precursor metabolite is an important step in amino acid biosynthesis. Ammonia, nitrate, or N_2 can serve as the source of nitrogen.
- Ammonia can be directly assimilated by the activity of transaminases and either glutamate dehydrogenase or the glutamine synthetase–glutamate synthase system (**figures 12.12–12.14**).
- Nitrate is incorporated through assimilatory nitrate reduction catalyzed by the enzymes nitrate reductase and nitrite reductase (**figure 12.15**).
- Nitrogen fixation is catalyzed by nitrogenase. Atmospheric molecular nitrogen is reduced to ammonia, which is then incorporated into amino acids (**figure 12.16**).
- Microorganisms can use cysteine, methionine, and inorganic sulfate as sulfur sources. Sulfate must be reduced to sulfide before it is assimilated (**figures 12.18** and **12.19**).
- Some amino acids are made directly by the addition of an amino group to a precursor metabolite, but most amino acids are made by pathways that are more complex. Many amino acid biosynthetic pathways are branched. Thus a single precursor metabolite can give rise to several amino acids (**figure 12.20**).
- Anaplerotic reactions replace TCA cycle intermediates to keep the cycle in balance while it supplies precursor metabolites. The anaplerotic reactions include the glyoxylate cycle (**figures 12.21** and **12.22**).
- Amino acids are also used in the synthesis of porphyrins, molecules consisting of four pyrrole rings. Two pathways have been identified for porphyrin synthesis: glutamate pathway and succinyl-CoA pathway (**figure 12.23**).

12.6 Synthesis of Purines, Pyrimidines, and Nucleotides

- Purines and pyrimidines are nitrogenous bases found in DNA, RNA, and other molecules. The nitrogen is supplied by certain amino acids. Phosphorus is provided by either inorganic phosphate or organic phosphate.
- Phosphorus can be assimilated directly by phosphorylation reactions that form ATP from ADP and P_i. Organic phosphorus sources are the substrates of phosphatases that release phosphate from the organic molecule.
- The purine skeleton is synthesized beginning with ribose 5-phosphate and initially produces inosinic acid (**figures 12.24** and **12.25**). Pyrimidine biosynthesis starts with carbamoyl phosphate and aspartate, and ribose is added after the skeleton has been constructed (**figures 12.26** and **12.27**).

12.7 Lipid Synthesis

- Fatty acids are synthesized from acetyl-CoA, malonyl-CoA, and NADPH by fatty acid synthase. During synthesis, the intermediates are attached to the acyl carrier protein (**figure 12.28**). Double bonds can be added in two different ways.
- Triacylglycerols are made from fatty acids and glycerol phosphate. Phosphatidic acid is an important intermediate in this pathway (**figure 12.29**).
- Phospholipids can be synthesized from phosphatidic acid by forming CDP-diacylglycerol, then adding an amino acid.
- Sterols and isoprenes are synthesized from two building blocks called isopentenyl pyrophosphate and dimethylallyl pyrophosphate. Condensation reactions using these molecules in various ways build the ring structures of sterols and the glycerol diethers and diglycerol tetraethers found in archaeal membranes (**figure 12.30**).
- Lipopolysaccharides are synthesized by a complex branched pathway that uses one branch to make the lipid A core portion and the other to synthesize the O antigen (**figure 12.31**). These two are joined in the periplasmic space. LPS moves from the plasma membrane to the outer leaflet of the outer membrane via the Lpt pathway (**figure 12.32**).

Active Learning

1. What would happen if a microorganism that depended on the glyoxylate cycle no longer produced the enzyme isocitrate lyase? Could this defect be corrected by nutrients in the media? If so, which nutrients?
2. Intermediary carriers are in a limited supply: When they cannot be recycled because of a metabolic block, serious consequences ensue. Think of some examples of these consequences.
3. Bacteria in the genus *Rickettsia* are parasites that live in the cytoplasm of host cells. It is believed that *Rickettsia* ancestors invaded a eukaryotic cell and, over time, lost genes that encode enzymes in biosynthetic pathways. This forced *Rickettsia* to remain in the cytoplasm as a parasite, importing the precursors it requires to synthesize its macromolecules. Researchers have explored *Rickettsia* genomes to assess its metabolic capability. After examining figure 12.2, which biochemical intermediates would be most efficient for *Rickettsia* to import? Based on your understanding of the differences between bacterial and eukaryotic cells, which pathways would you expect to be fully retained in *Rickettsia*?

Read the original paper: Driscoll, T. P., et al. 2017. Wholly *Rickettsia*! Reconstructed metabolic profile of the quintessential bacterial parasite of eukaryotic cells. *mBio* 8(5):e00859-17.

13

Bacterial Genome Replication and Expression

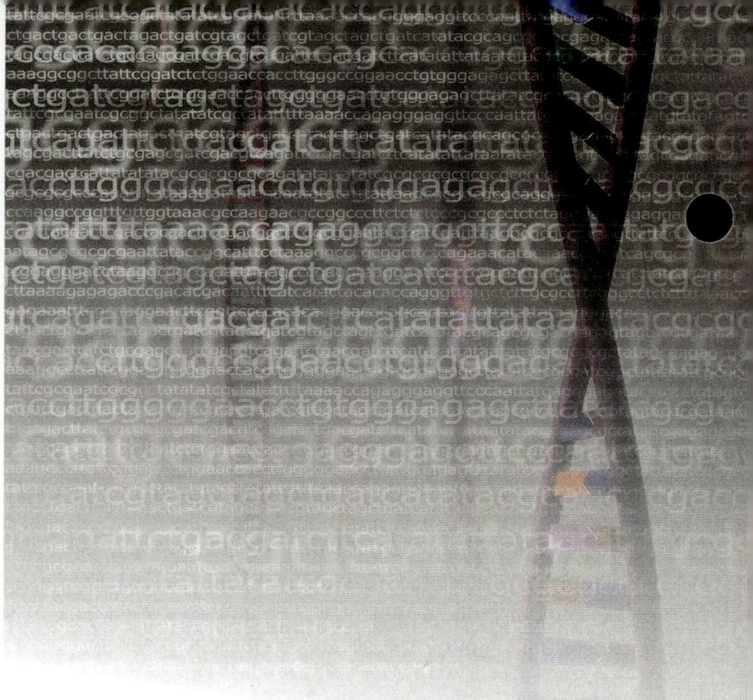

©Chad Baker/Getty Images

Making Code

Since ancient times, humans have sent important messages in code. Julius Caesar sent coded messages to his generals. Mary Queen of Scots used code to detail her plans to overthrow Queen Elizabeth the First. George Washington regularly communicated to his troops in code, as have commanders in all wars. Today work continues to create codes that are harder to break and more content rich.

Code writing is related to another field: storage of information, which often involves converting the information using a code—a process termed *encryption*. The history of information storage parallels the history of communicating in code. Initially messages were coded in language, song, and dance. Later, coded messages were hand-written on paper; manuscripts and books stored valuable data. As codes became more sophisticated and their content larger, the form of coded messages evolved to include radio communications, recordings, and binary digits in computers. Similar methods were developed to store data sets.

In the early years of this century, scientists began exploring a new method of encrypting and storing data called biostorage or bioencryption. The earliest approaches stored coded data in DNA inserted into a bacterial chromosome. Scientists turned to DNA because they estimated that the information in about 1,000 books having 500 pages of small print is the equivalent of the information stored in one strand of human DNA. Among the advantages of storing data in DNA is its chemical stability in cold, dry conditions over extended time periods, and its ability to be compacted into a small volume.

To accomplish data storage in DNA, binary code of zeroes and ones is converted to the four bases of the DNA alphabet, A, G, C, and T. Anything stored as digital files, like music, movies, or this textbook, can be converted to nucleic acid, and later retrieved and decoded. In a stunning demonstration of the principle, a team from Harvard encoded a short GIF of a horse and rider into short synthetic DNA pieces. These were introduced into a population of *Escherichia coli,* where they were stably maintained in the chromosome. Rapid, efficient DNA sequencing technology enabled data retrieval.

For millennia cellular organisms have used DNA to store genetic information, and they have devised methods for replicating, recoding, and decoding genetic information. In this chapter, we explore the polymerization reactions that generate DNA, RNA, and protein, reactions central to the inheritance of traits.

Throughout the twentieth century, microorganisms served as model experimental systems for genetic studies and helped to elucidate the nature of genetic information, gene structure, the genetic code, and mutation. Thus we begin by considering bacterial genetics in this chapter and the next. Because of the distinctive characteristics of eukaryotes and archaea, the replication and expression of their genomes are discussed in chapter 15.

Readiness Check:

Based on what you have learned previously, you should be able to:

- ✓ Define the terms *genome, chromosome, plasmid, haploid, diploid, genotype,* and *phenotype*
- ✓ Outline the flow of genetic information, identifying the processes used
- ✓ Draw a simple diagram that illustrates the chemical moieties found in nucleotides (section 12.6)
- ✓ State the base-pairing rules
- ✓ Describe the structure of bacterial genomes (section 3.6)

13.1 Experiments Using Bacteria and Viruses Demonstrated that DNA Is the Genetic Material

After reading this section, you should be able to:

a. Summarize Griffith's experiments on transformation
b. Relate how the contributions of Avery, MacLeod, McCarty, Hershey, and Chase confirmed that DNA stores genetic material

Although it is now hard to imagine, it was once thought that DNA was too simple a molecule to store genetic information. With only four different nucleotides, it seemed that a molecule of much greater complexity must house the genetic information of a cell. It was argued that proteins, being composed of 20 different amino acids, were the better candidate for this important cellular function.

The early work of Fred Griffith in 1928 on the transfer of virulence (i.e., the ability to cause disease) in the pathogen *Streptococcus pneumoniae*, commonly called pneumococcus, set the stage for research showing that DNA was indeed the genetic material. Griffith found that if he boiled virulent bacteria and injected them into mice, the mice were not affected and no pneumococci could be recovered from the animals (**figure 13.1**). When he injected a combination of killed virulent bacteria and a living nonvirulent strain, the mice died; moreover, he could recover living virulent bacteria from the dead mice. Griffith called this change of nonvirulent bacteria into virulent pathogens transformation.

Oswald Avery and his colleagues then set out to discover which constituent in the heat-killed virulent pneumococci was responsible for transformation. These investigators used enzymes to selectively destroy DNA, RNA, or protein in purified extracts of virulent pneumococci (S strain). They then exposed nonvirulent pneumococcal strains (R strains) to the treated extracts. Transformation of the nonvirulent bacteria was blocked only if DNA was destroyed, suggesting that DNA was carrying the

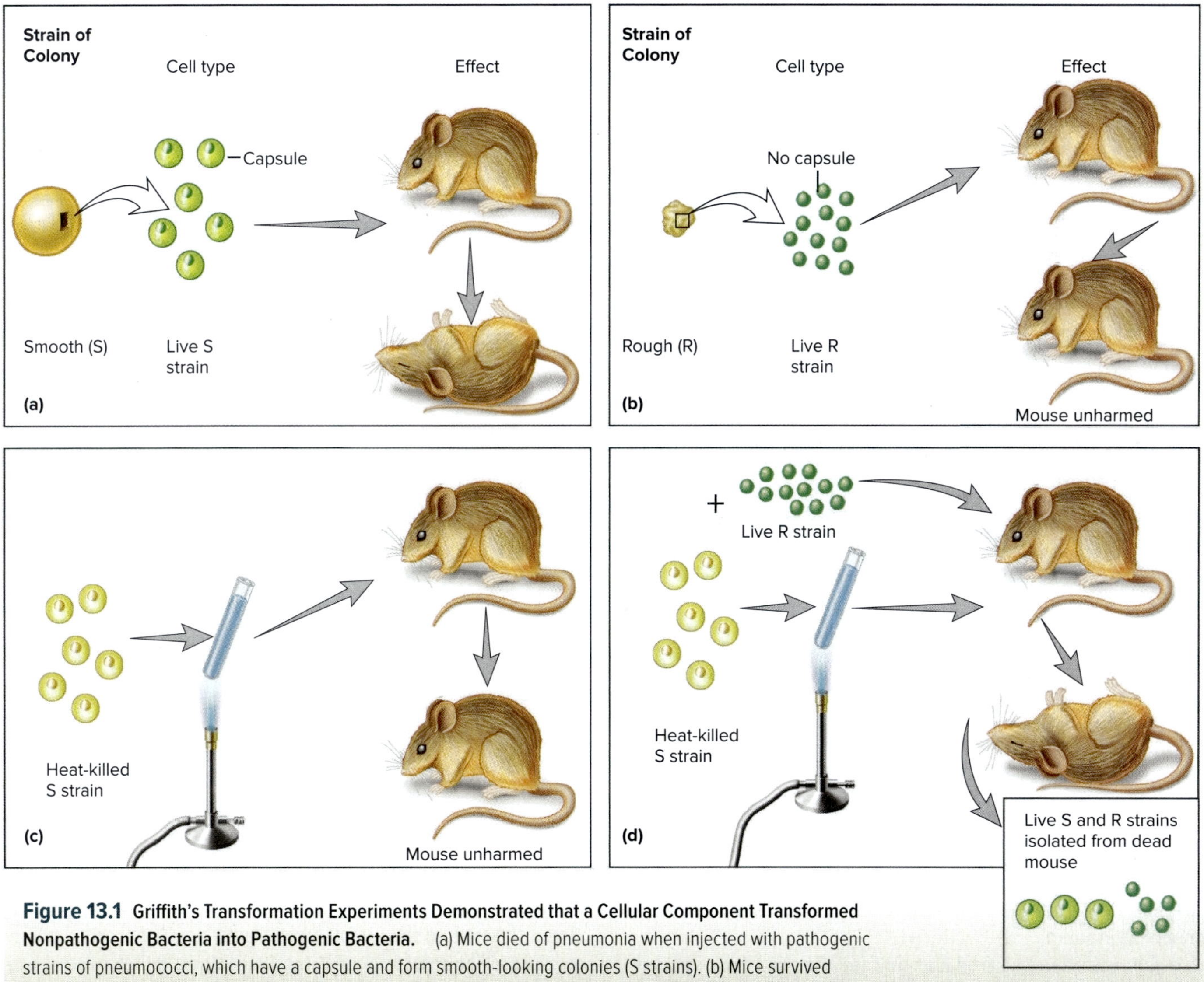

Figure 13.1 Griffith's Transformation Experiments Demonstrated that a Cellular Component Transformed Nonpathogenic Bacteria into Pathogenic Bacteria. (a) Mice died of pneumonia when injected with pathogenic strains of pneumococci, which have a capsule and form smooth-looking colonies (S strains). (b) Mice survived when injected with nonpathogenic strains of pneumococci, which lack a capsule and form rough colonies (R strains). (c) Injection with heat-killed S strains had no effect on the mice. (d) Injection with a live R strain and a heat-killed S strain caused pneumonia, and live S strain pneumococci could be isolated from the dead mice.

MICRO INQUIRY *Based on what we now know about proteins, why can we conclude from this experiment that genetic information was unlikely to be carried by proteins?*

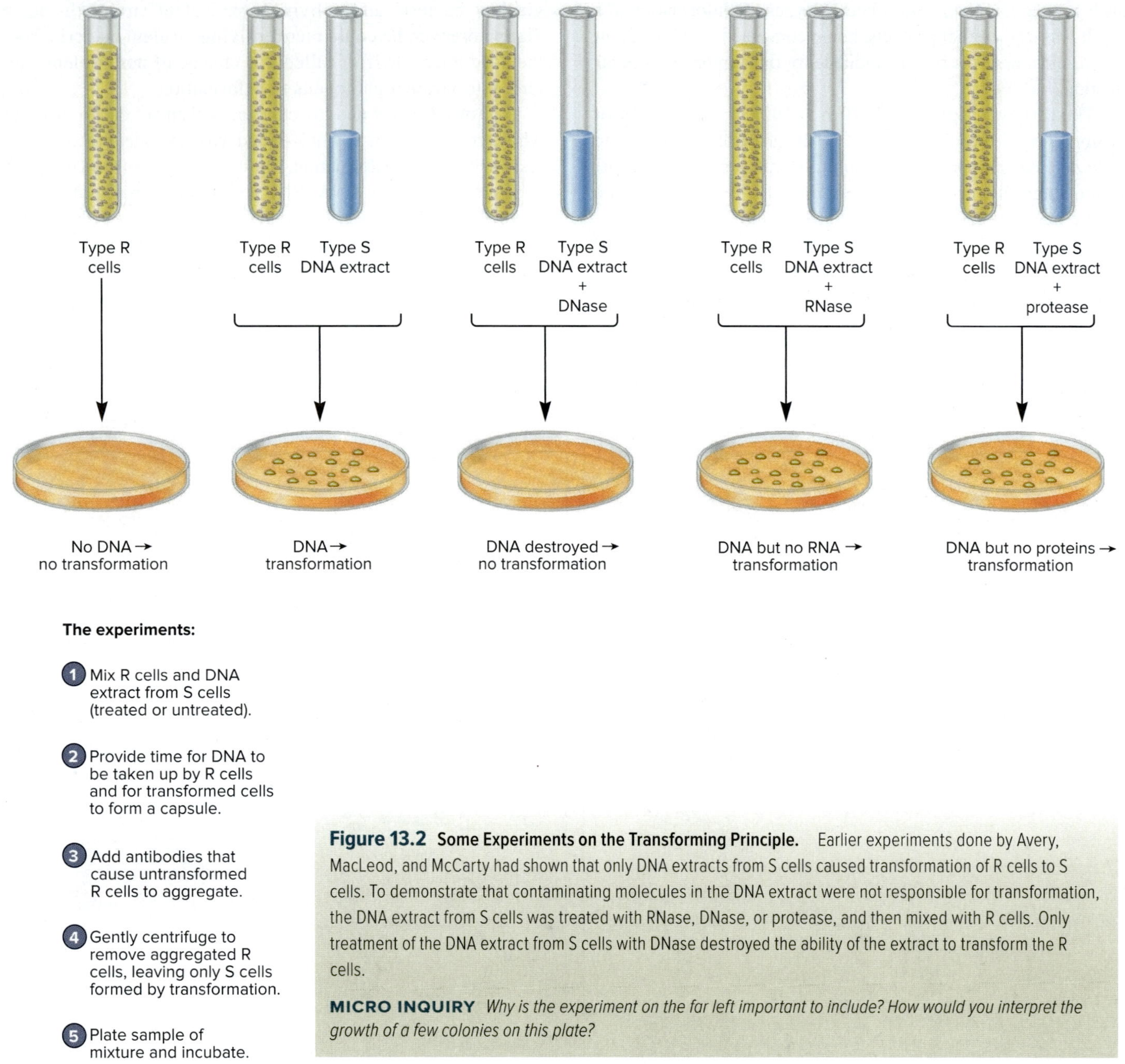

Figure 13.2 Some Experiments on the Transforming Principle. Earlier experiments done by Avery, MacLeod, and McCarty had shown that only DNA extracts from S cells caused transformation of R cells to S cells. To demonstrate that contaminating molecules in the DNA extract were not responsible for transformation, the DNA extract from S cells was treated with RNase, DNase, or protease, and then mixed with R cells. Only treatment of the DNA extract from S cells with DNase destroyed the ability of the extract to transform the R cells.

MICRO INQUIRY *Why is the experiment on the far left important to include? How would you interpret the growth of a few colonies on this plate?*

information required for transformation (**figure 13.2**). The publication of these studies by Avery, C. M. MacLeod, and M. J. McCarty in 1944 provided the first evidence that DNA carried genetic information.

Eight years later, Alfred Hershey and Martha Chase wanted to know if protein or DNA carried the genetic information of a bacterial virus called T2 bacteriophage. They performed experiments in which they made the virus's DNA radioactive with ^{32}P, or they labeled its protein coat with ^{35}S. They mixed radioactive virions (either ^{32}P-labeled or ^{35}S-labeled) with *Escherichia coli* and incubated the mixture for a few minutes. This allowed the virions to attach to *E. coli* and initiate the infection process (*see figure 6.16*). The culture was then centrifuged to separate the infected cells from any unadsorbed phage particles. The infected cells in the resulting pellet were resuspended and the suspension was agitated violently in a blender. The blender treatment sheared off bacteriophage particles adsorbed to *E. coli* cells (**figure 13.3**). Some of the suspension was used in a plaque assay (*see figure 6.20*) to determine if the blender treatment affected the ability of the phage to multiply

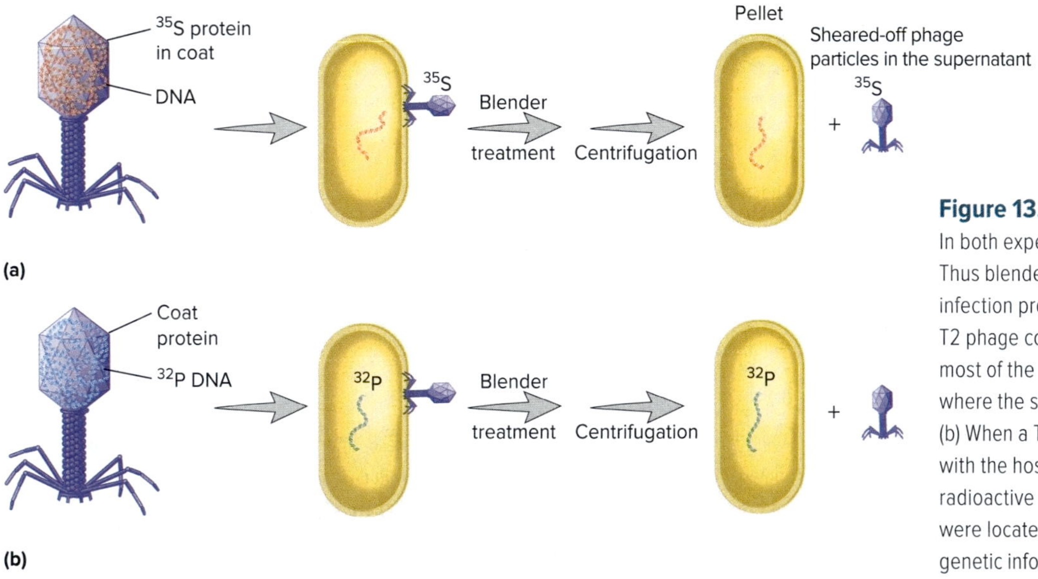

Figure 13.3 The Hershey-Chase Experiment. In both experiments, phage progeny were produced. Thus blender treatment did not interfere with the infection process. (a) When *E. coli* was infected with a T2 phage containing ^{35}S protein, after centrifugation most of the radioactivity remained in the supernatant where the sheared-off phage particles were located. (b) When a T2 phage containing ^{32}P DNA was mixed with the host bacterium, after centrifugation the radioactive DNA was in the pellet where the cells were located. Thus DNA was carrying the virus's genetic information.

within host cells. The remaining suspension was centrifuged to separate cells from the sheared-off phage particles. After centrifugation, radioactivity in the supernatant (where the phage particles remained) versus that in the bacterial cells in the pellet was determined.

The results of their experiments showed that blender treatment did not disrupt the infection process, because progeny phages were produced. They also demonstrated that when ^{35}S-labeled T2 was used in the experiment, the majority of the radioactive protein was in the supernatant; whereas when ^{32}P-labeled T2 was used, the radioactive DNA was in the bacterial cells that formed the pellet. Because DNA entered the cells and the protein did not, the phage DNA must have been carrying the genetic information needed to complete the infection process. Some luck was involved in their discovery, for the genetic material of many viruses is RNA and the researchers happened to select a DNA virus for their studies. Imagine the confusion if T2 had been an RNA virus! The controversy surrounding the nature of genetic information might have lasted considerably longer than it did.

Subsequent studies on the genetics of viruses and bacteria were largely responsible for the rapid development of molecular genetics. Furthermore, much of the DNA technology described in chapter 17 has arisen from studies of bacterial and viral genetics.

Comprehension Check

1. Briefly summarize the experiments of Griffith; Avery, MacLeod, and McCarty; and Hershey and Chase.
2. Explain how protein was ruled out as the molecule of genetic information storage in each of the experiments performed by these important microbiologists.

13.2 Nucleic Acid and Protein Structure

After reading this section, you should be able to:

a. Draw schematic representations of DNA, RNA, and amino acids that show their major features
b. Compare and contrast the structure of DNA and RNA
c. Identify the covalent bonds that are used to link nucleotides together to form a nucleic acid and amino acids together to form a polypeptide

DNA, RNA, and proteins are often called informational molecules. The information exists as the sequence of monomers from which they are built. Here we describe the monomers and how they are linked together to form these important macromolecules.

DNA Is a Polymer of Deoxyribonucleotides

Deoxyribonucleic acid (DNA) is a polymer of deoxyribonucleotides (**figure 13.4**) linked together by phosphodiester bonds (**figure 13.5*a***). It contains the bases adenine, guanine, cytosine, and thymine. DNA molecules are very large and are usually composed of two polynucleotide chains coiled together to form a double helix 2.0 nm in diameter (figure 13.5). The monomers of DNA are called deoxyribonucleotides because the sugar found in them is deoxyribose (figure 13.4*b*). The bond that links the monomers together to form the polymer is called a phosphodiester bond because it consists of a phosphate that forms ester linkages between the 3′-hydroxyl of one sugar and the 5′-hydroxyl of an adjacent sugar. Purine and pyrimidine bases are attached to the 1′-carbon of the deoxyribose sugars, and the bases extend toward the middle of the cylinder formed by the two chains. (The numbers designating the carbons in the sugars are given a prime notation to distinguish them from the numbers designating the carbons and

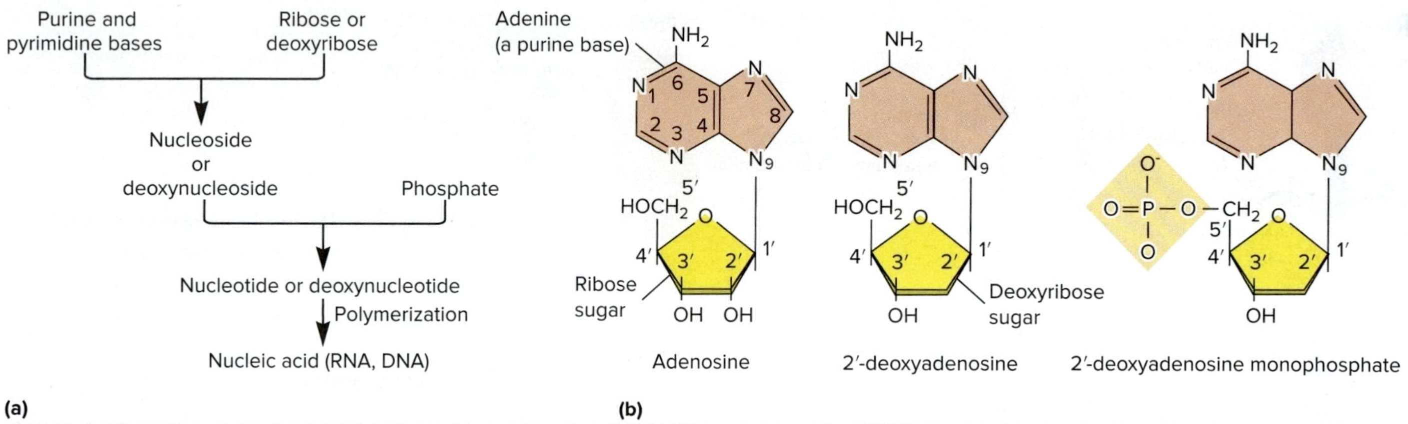

Figure 13.4 The Composition of Nucleic Acids. (a) A diagram showing the relationships of various nucleic acid components. Combination of a purine or pyrimidine base with ribose or deoxyribose gives a nucleoside (a ribonucleoside or deoxyribonucleoside). A nucleotide contains a nucleoside and one or more phosphates. Nucleic acid polynucleotide chains form when nucleotides are connected by phosphodiester bonds (figure 13.5*a*). (b) Examples of nucleosides—adenosine and 2′-deoxyadenosine—and the nucleotide 2′-deoxyadenosine monophosphate. The carbons of nucleoside and nucleotide sugars are indicated by numbers with primes to distinguish them from the carbons in the bases.

MICRO INQUIRY *To which carbon of ribose (deoxyribose) is each of the following bonded: adenine and the hydroxyls of adenosine and deoxyadenosine?*

nitrogens in the nitrogenous bases.) The bases from each strand interact with those of the other strand, forming base pairs. The base pairs are stacked on top of each other in the center, one base pair every 0.34 nm. The purine adenine (A) of one strand is always paired with the pyrimidine thymine (T) of the opposite strand by two hydrogen bonds. The purine guanine (G) pairs with cytosine (C) by three hydrogen bonds. This AT and GC base pairing means that the two strands in a DNA double helix are **complementary.** In other words, the bases in one strand match up with those of the other according to specific base-pairing rules. Because the sequences of bases in these strands encode genetic information, considerable effort has been devoted to determining the base sequences of DNA and RNA from many organisms, including hundreds of microbes. ▶| *Microbial genomics (chapter 18)*

The two polynucleotide strands of DNA fit together much like the pieces in a jigsaw puzzle. Inspection of figure 13.5*b,c* shows that the two strands are not positioned directly opposite one another. Therefore when the strands twist about one another, a wide major groove and narrower minor groove are formed by the backbone. There are 10.5 base pairs per turn of the helix, and each turn of the helix has a vertical length of 3.4 Å. The helix is right-handed; that is, the chains turn counterclockwise as they approach a viewer looking down the longitudinal axis. The two backbones are antiparallel, which means they run in opposite directions with respect to the orientation of their sugars. Thus the 5′ end of one strand is paired with the 3′ end of the complementary strand; or stated another way, one strand is oriented 5′ to 3′ and the other, 3′ to 5′ (figure 13.5*b*).

The structure of DNA just described is that of the B form, the most common form in cells. Two other forms of DNA have been identified. The A form primarily differs from the B form in that it has 11 base pairs per helical turn, rather than 10.5, and a vertical length of 2.6 Å, rather than 3.4. Thus it is wider than the B form. The A form DNA is found in bacterial endospores, and is partially responsible for protecting the DNA from UV damage. In addition, some virus genomes have A form DNA. The Z form of DNA is dramatically different, having a left-handed helical structure. The Z form has 12 base pairs per helical turn and a vertical rise of 3.7 Å. Thus it is more slender than the B form. Short sequences within eukaryotic chromosomes can be in the Z form. It is thought that they may relieve torsional stress placed on the DNA molecule when it is being actively transcribed. *DNA Structure*

Supercoiling is another property of DNA. DNA is helical; that is, it is a coil. Whenever the rotation of a coil is restrained in some way, it causes it to coil on itself. The coiling of a coil is supercoiling. Recall that most bacterial chromosomes are closed, circular double-stranded DNA molecules. In this state, the two strands are unable to rotate freely relative to each other, and the molecule is said to be constrained. The strain is relieved by supercoiling. There are two types of supercoiling: positive and negative. For DNA, these are defined by the change in number of base pairs per turn in the double helix. As just discussed, the B form of DNA has 10.5 base pairs per turn of the helix. Supercoiling that decreases the number of base pairs per turn is said to be negative supercoiling. Likewise, supercoiling that increases the number of base pairs per turn is called positive supercoiling.

Supercoiling is another level of DNA structure that is critical for its function. Although bacterial chromosomes are negatively supercoiled overall, smaller regions can vary depending on how the chromosome is used at a given time. Importantly for this chapter, supercoiling "loosens" up the DNA, making it easier to separate the two strands. Strand separation is an important early step in both DNA replication and transcription, as we discuss in sections 13.3 and 13.5, respectively.

RNA Is a Polymer of Ribonucleotides

Ribonucleic acid (**RNA**) is a polymer of ribonucleotides containing the sugar ribose and the bases adenine, guanine, cytosine, and uracil, instead of thymine (figure 13.4). The nucleotides are joined by a phosphodiester bond, just as they are in DNA. In cells, RNA molecules are single stranded. However, an RNA strand can coil back on itself to form secondary structures such as hairpins with complementary base pairing and helical organization. The formation of double-stranded regions in RNA is often critical to its function. ▶| *Attenuation and riboswitches can stop transcription prematurely (section 14.3); Translational riboswitches (section 14.4)*

Proteins Are Polymers of Amino Acids

Proteins are polymers of amino acids linked by peptide bonds; thus they are also called polypeptides. An amino acid is defined by the presence of a central carbon (the α carbon) to which are attached a carboxyl group, an amino group, and a side chain (**figure 13.6**). Twenty amino acids are normally used to form proteins. However, two unusual amino acids are found in some proteins (section 13.6). Amino acids differ in their side chains. Depending on the chemical structure of the side chain, the amino acid is described as nonpolar, polar, or charged. The peptide bonds linking the amino acids are formed by a reaction between the carboxyl group of one amino acid and the amino group of the next amino acid in the protein (**figure 13.7**). A polypeptide has polarity just as DNA and RNA

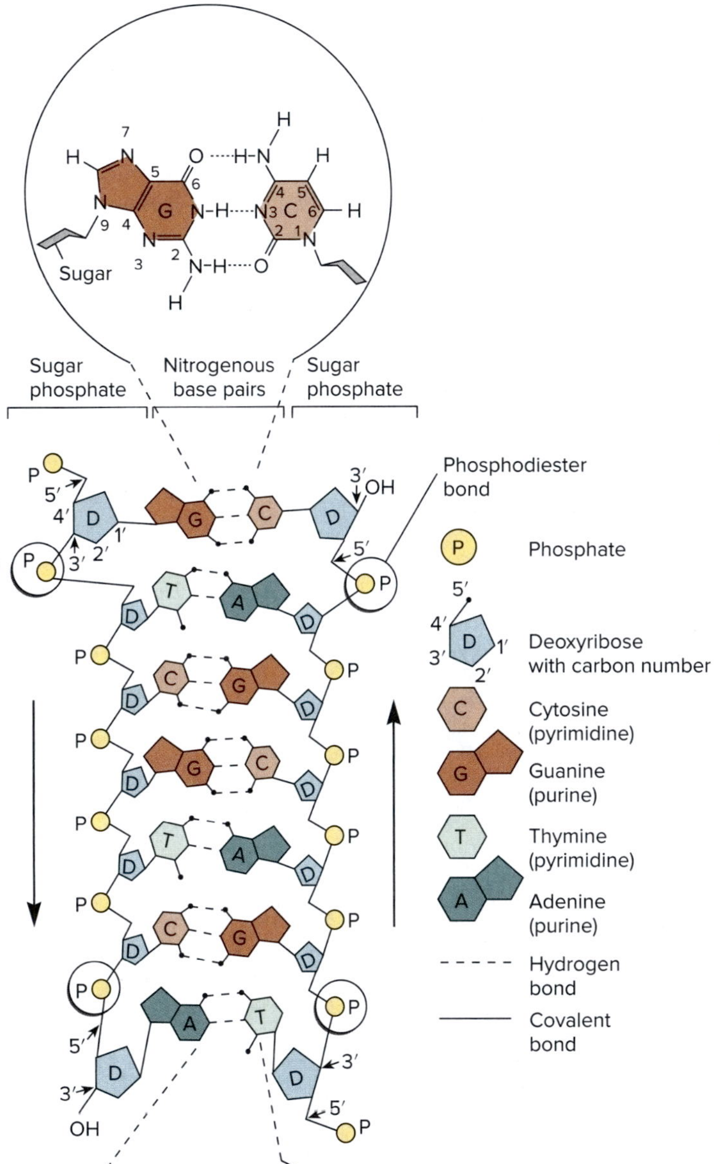

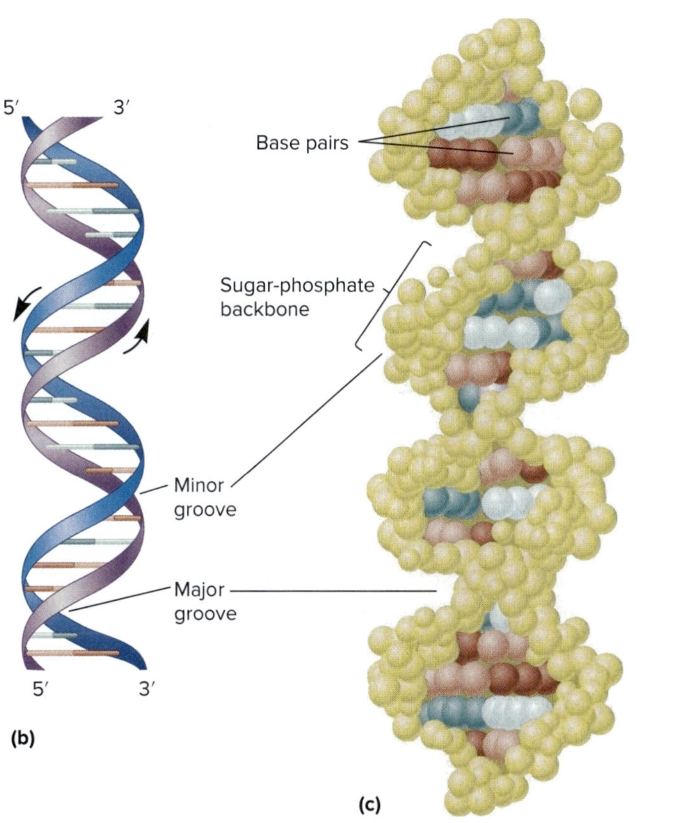

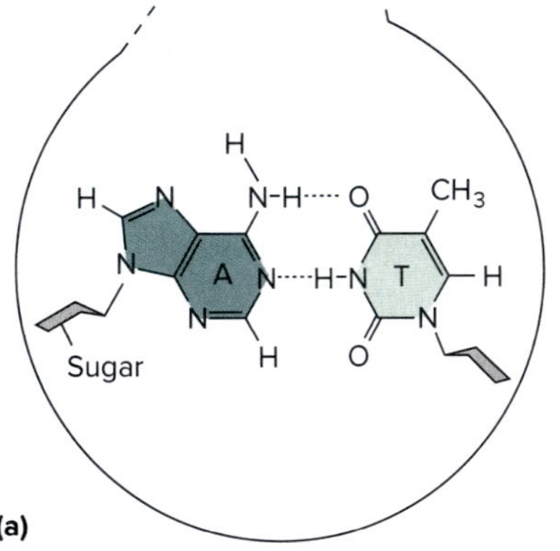

(a)

Figure 13.5 DNA Structure—B Form of DNA. DNA Is Usually a Double-Stranded Molecule. (a) A schematic, nonhelical model. In each strand, phosphates are esterified to the 3′-carbon of one deoxyribose sugar (blue) and the 5′-carbon of the adjacent sugar. The two strands are held together by hydrogen bonds (dashed lines). Because of the specific base pairing, the base sequence of one strand determines the sequence of the other. The two strands are antiparallel; that is, the backbones run in opposite directions, as indicated by the two arrows, which point in the 5′ to 3′ direction. (b) A simplified model that highlights the antiparallel arrangement and the major and minor grooves. (c) A space-filling model of the B form of DNA. Note that the sugar-phosphate backbone spirals around the outside of the helix and the base pairs are embedded inside.

MICRO INQUIRY *How many H bonds are there between adenine and thymine, and between guanine and cytosine?*

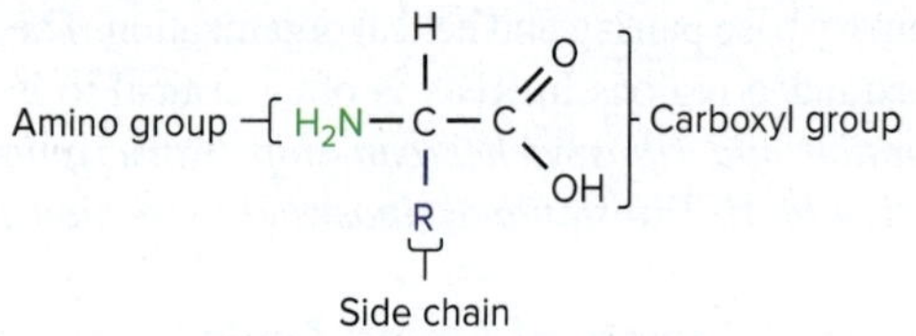

Figure 13.6 The General Structure of an Amino Acid. All amino acids have a central carbon (the α carbon) to which are attached a carboxyl group (COOH), an amino group (NH_2), and a side chain (R). Amino acids differ in terms of their side chains, which may be nonpolar, polar, negatively charged (acid), or positively charged (basic).

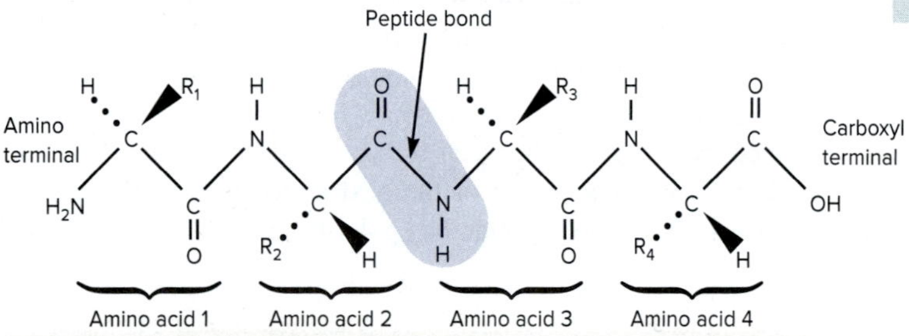

Figure 13.7 Peptide Bonds Link Amino Acids Together in Peptide Chains. A tetrapeptide chain is shown. One of the peptide bonds linking the four amino acids together is highlighted in blue. At one end of the peptide is an amino group (amino or N terminal); at the other end is a carboxyl group (carboxyl or C terminal).

MICRO INQUIRY *Identify the two other peptide bonds of the tetrapeptide chain.*

do. At one end of the chain is an amino group, and at the other end is a carboxyl group. Thus a polypeptide has an amino or N terminus and a carboxyl or C terminus.

Proteins do not typically exist as extended chains of amino acids. Rather, they fold to form three-dimensional structures, often more or less spherical in shape. The final shape is determined to a large extent by the sequence of amino acids in the polypeptide. This sequence is called the primary structure. Secondary and tertiary structures result from the folding of the chain. Finally, two or more polypeptide strands can interact to form the final, functional protein. This level of structure is called quaternary structure. These higher levels of structure are stabilized by intra- (and inter-) chain bonds. Protein structure is described in more detail in appendix I.

Comprehension Check

1. What are nucleic acids? How do DNA and RNA differ in structure?
2. What does it mean to say that the two strands of the DNA double helix are complementary and antiparallel? Examine figure 13.5*b* and explain the differences between the minor and major grooves.
3. Amino acids are described as nonpolar, polar, and charged, depending on the molecular makeup of their side chains. Which type of amino acid might be found in the transmembrane portion of a polypeptide located in a cell's plasma membrane? Explain your answer.

13.3 DNA Replication in Bacteria

After reading this section, you should be able to:

a. Describe a bacterial replicon
b. Summarize the events that occur during the three phases of DNA replication
c. Create a table or concept map that illustrates the function of the major proteins found in a bacterial replisome
d. List the enzymatic and structural elements needed by DNA polymerases for DNA synthesis
e. Outline the major events that occur at the replication fork

DNA replication is an extraordinarily important and complex process upon which all life depends. During DNA replication, the two strands of the double helix are separated; each then serves as a template for the synthesis of a complementary strand according to the base-pairing rules. Each of the two progeny DNA molecules consists of one new strand and one old strand, and DNA replication is said to be semiconservative (**figure 13.8**). DNA replication is also extremely accurate; *E. coli* makes errors with a frequency of only 10^{-9} or 10^{-10} per base pair replicated (or about one in a million [10^{-6}] per gene per generation). Despite its

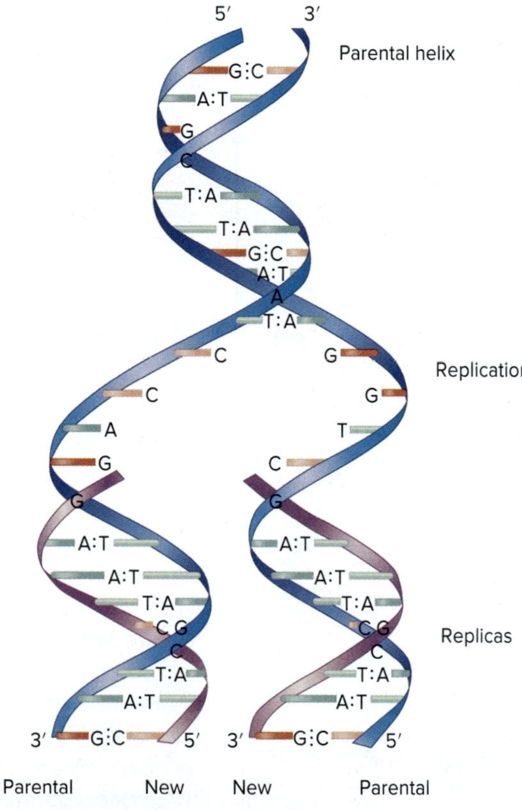

Figure 13.8 Semiconservative DNA Replication. The replication fork of DNA showing the synthesis of two progeny strands. Newly synthesized strands are purple. Each replica contains one new and one old strand.

complexity and accuracy, replication is very rapid. In bacteria, replication rates approach 750 to 1,000 base pairs per second. Most of our discussion in this section is based on studies of chromosomal DNA replication in *E. coli.*

Bacterial DNA Replication Initiates from a Single Origin of Replication

The replication of chromosomal DNA begins at a single point, the **origin of replication.** Synthesis of DNA occurs at the **replication fork,** the place at which the DNA helix is unwound and individual strands are replicated. Two replication forks move outward from the origin until they have copied the whole **replicon**—the portion of the genome that contains an origin that is replicated as a unit. When the replication forks move around the circular chromosomes observed in most bacteria, a structure shaped like the Greek letter theta (θ) is formed (**figure 13.9**). Because the bacterial chromosome is a single replicon, the forks meet on the other side and two separate chromosomes are released. *DNA Replication Fork; Bidirectional DNA Replication*

Replisomes Contain DNA Polymerases and Other Proteins Needed for DNA Replication

DNA replication is essential to organisms, and a great deal of effort has been devoted to understanding its mechanism. The replication of *E. coli* DNA requires at least 10 proteins that form a huge complex on the DNA called the **replisome** (**table 13.1**). As we describe in chapter 15, many of the archaeal and eukaryotic replication enzymes and proteins differ from those used by bacterial cells. In general, the archaeal replication machinery is more similar to the eukaryotic machinery than it is to the

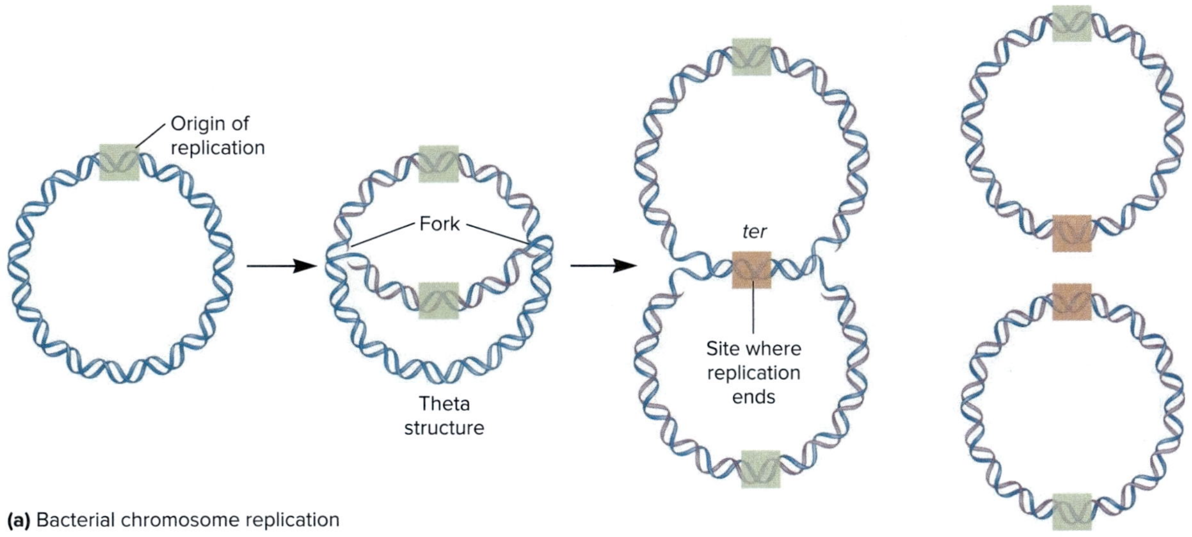

(a) Bacterial chromosome replication

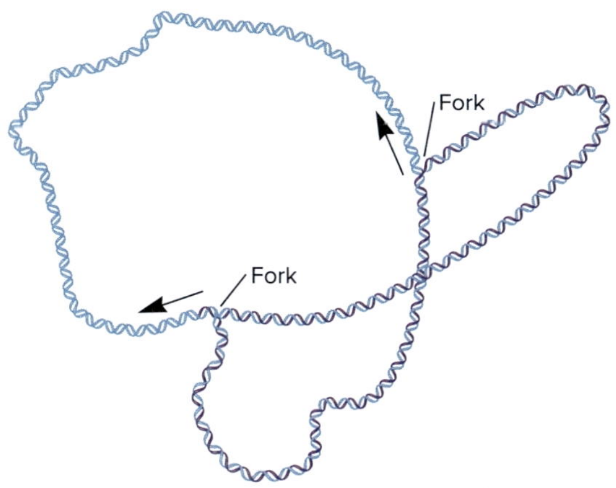

(b) Diagram of a theta structure during chromosome replication

Figure 13.9 Bidirectional Replication of the *E. coli* Chromosome. (a) Replication begins at one site on the chromosome, called the origin of replication. Two replication forks proceed in opposite directions from the origin until they meet at a site called the replication termination site (*ter*). (b) A schematic representation of the chromosome, illustrating the theta structure commonly observed as an intermediate. Parental DNA is blue; new DNA strands are purple; arrows represent direction of fork movement.

Table 13.1 Components of the *E. coli* Replication Machinery

Protein	Function
DnaA (initiator protein)	Initiation of replication; binds origin of replication (*oriC*)
Helicase (DnaB)	Breaks hydrogen bonds holding two strands of DNA together; promotes primase activity; involved in primosome assembly
Helicase loader (DnaC)	Together with DnaA, directs helicase to DNA template
DNA gyrase	Relieves supercoiling of DNA ahead of helicase; separates daughter molecules in final stages of replication
SSB proteins	Bind single-stranded DNA after strands are separated by helicases
Primase (DnaG)	Synthesis of RNA primer; component of primosome
DNA polymerase III holoenzyme	Catalyzes most of the DNA synthesis during replication; has 3′ → 5′ exonuclease (proofreading) activity
DNA polymerase I	Removes RNA primers and fills the resulting gaps in DNA
DNA ligase	Synthesizes phosphodiester bonds to join Okazaki fragments
Tus	Acts at the termination of replication
Topoisomerase IV	Separates daughter molecules in final stages of replication

bacterial system. Despite these differences, the overall process of DNA replication is similar in all organisms.

Within the replisome, enzymes called **DNA polymerases** catalyze DNA synthesis. All known DNA polymerases catalyze DNA synthesis in the 5′ to 3′ direction, and the nucleotide to be added is a deoxyribonucleoside triphosphate (dNTP; often referred to simply as deoxynucleoside triphosphates or deoxynucleotides). Deoxynucleotides are linked by phosphodiester bonds formed by a reaction between the hydroxyl group at the 3′ end of the growing DNA strand and the phosphate closest to the 5′ carbon (the α-phosphate) of the incoming deoxynucleotide (**figure 13.10**). The energy needed to form the phosphodiester bond is provided by release of the terminal two phosphates as pyrophosphate (PP_i) from the nucleotide that is added. The PP_i is subsequently hydrolyzed to two separate phosphates (P_i). Thus the deoxynucleoside triphosphates dATP, dTTP, dCTP, and dGTP serve as DNA polymerase substrates while deoxynucleoside monophosphates (dNMPs: dAMP, dTMP, dCMP, dGMP) are incorporated into the growing chain.

DNA polymerase needs three things to catalyze DNA synthesis. The first is a template, which is read in the 3′ to 5′ direction and is used to direct synthesis of a complementary DNA strand. The second is a primer (e.g., an RNA strand or a DNA strand) to provide a free 3′-hydroxyl group to which nucleotides can be added (figure 13.10). The third is a pool of the four dNTPs.

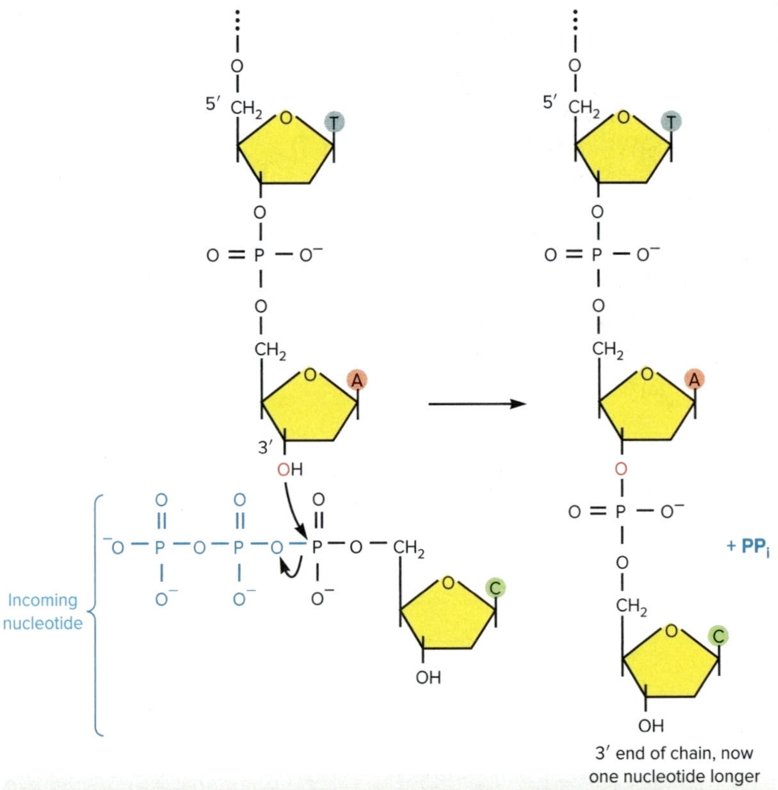

Figure 13.10 The DNA Polymerase Reaction. The hydroxyl of the 3′ terminal deoxyribose makes a nucleophilic attack on the α-phosphate of the nucleotide substrate (in this example, adenosine attacks cytidine triphosphate).

MICRO INQUIRY *What provides the energy to fuel this reaction?*

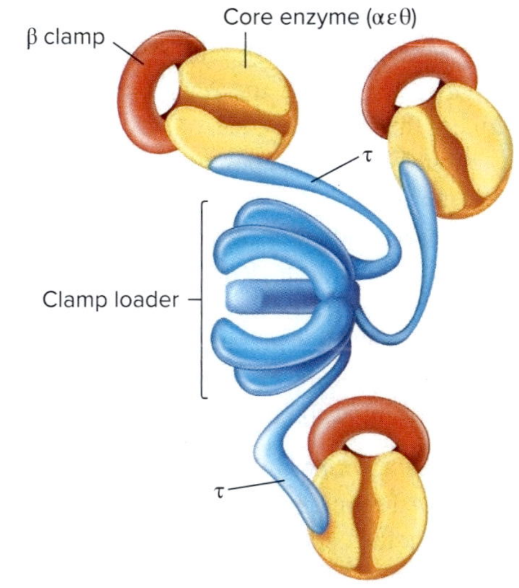

Figure 13.11 DNA Polymerase III Holoenzyme. The holoenzyme consists of three core enzymes (each composed of three different proteins; α, ε, θ, in gold) and several other proteins. Each core enzyme is associated with a β clamp, in red, which tethers a DNA template to the core enzyme. The clamp loader, in blue, is a complex of six proteins that load β clamps onto DNA.

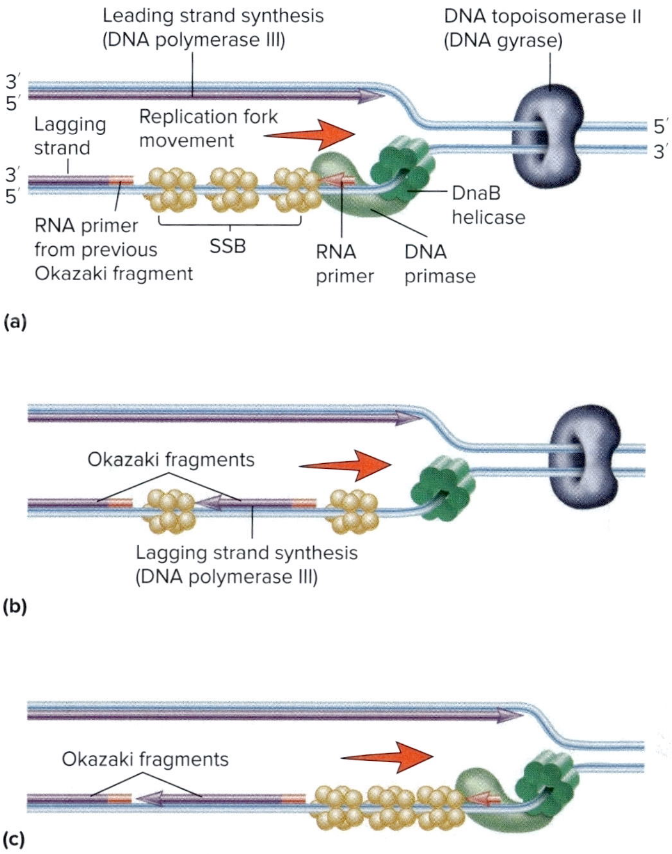

Figure 13.12 Other Replisome Proteins. A single replication fork showing the activity of replisome proteins other than DNA polymerase III holoenzyme (not shown) is illustrated. Helicase is responsible for separating the two strands of parental DNA. The strands are kept apart by single-stranded DNA binding proteins (SSB), which allows for synthesis of an RNA primer by primase. DNA gyrase eases the strain introduced into the DNA double helix by helicase activity. Both leading strand and lagging strand synthesis are illustrated. The lagging strand is synthesized in short fragments called Okazaki fragments. A new RNA primer is required for the synthesis of each Okazaki fragment.

MICRO INQUIRY *What is the difference between helicase and gyrase? Which is a topoisomerase?*

E. coli has five different DNA polymerases (DNA polymerase I–V). DNA polymerase III plays the major role in replication, although it is assisted by DNA polymerase I. DNA polymerase III holoenzyme is a multifunctional enzyme complex within the replisome (**figure 13.11**). Within the holoenzyme, there are three core DNA polymerases. Each core polymerase binds a strand of DNA and is responsible for catalyzing DNA synthesis and proofreading the product to ensure fidelity of replication. Associated with each core enzyme is a subunit called the β clamp that tethers it to the template DNA. At the center of the holoenzyme, and represented by an octopus-like structure in figure 13.11, is the clamp loader, a protein complex responsible for loading the β clamp onto DNA. Two of the core enzymes replicate one of the DNA strands, and the third replicates the other strand.

Other proteins found in the replisome include helicases, single-stranded DNA binding proteins, and topoisomerases (**figure 13.12**). **Helicases** are responsible for separating (unwinding) the DNA strands just ahead of the replication fork, using energy from ATP hydrolysis. **Single-stranded DNA binding proteins (SSBs)** keep the strands apart once they have been separated, and **topoisomerases** relieve the tension generated by the rapid unwinding of the double helix (the replication fork may rotate as rapidly as 75 to 100 revolutions per second). This is important because rapid unwinding leads to positive supercoiling in the helix ahead of the replication fork, which can impede replication if not removed. Topoisomerases like **DNA gyrase** change the structure of DNA by transiently breaking one or two strands without altering the nucleotide sequence of the DNA (e.g., a topoisomerase might tie or untie a knot in a DNA strand).

Once the template is prepared, the primer needed by DNA polymerase III is synthesized by an enzyme called **primase.** The primer is a short RNA strand, usually around 10 nucleotides long and complementary to the DNA (figure 13.12). RNA is used as the primer because unlike DNA polymerase, RNA polymerases (such as primase) can initiate RNA synthesis without an existing 3′-OH. Primase and several accessory proteins form a complex called the **primosome**. The primosome is another important component of the replisome.

As noted, DNA polymerase enzymes synthesize DNA in the 5′ to 3′ direction. Therefore one of the DNA polymerase core enzymes is able to move in the same direction as the replication fork and synthesize DNA continuously as the DNA unwinds ahead of the core enzyme. This strand is called the

leading strand (figure 13.12). The other strand, called the **lagging strand,** cannot be extended in the same direction as the movement of the replication fork because there is no free 3′-OH to which a nucleotide can be added. As a result, the lagging strand is synthesized discontinuously in the 5′ to 3′ direction (i.e., in the direction opposite of the movement of the replication fork) and produces a series of fragments called **Okazaki fragments,** after their discoverer, Reiji Okazaki (1930–1975). Discontinuous synthesis occurs as primase makes many RNA primers along the template strand. DNA polymerase III then extends these primers with DNA, and eventually the Okazaki fragments are joined to form a complete strand. Thus while the leading strand requires only one RNA primer to initiate synthesis, the lagging strand has many RNA primers that must eventually be removed. Okazaki fragments are about 1,000 to 3,000 nucleotides long in bacteria and approximately 100 nucleotides long in eukaryotic cells. *DNA Replication; Structural Basis of DNA Replication*

The Replication Fork Is Where the Complex Events of DNA Synthesis Occur

The details of DNA replication are outlined in **figure 13.13**. We present replication as a series of discrete steps, but in the cell these events occur quickly and simultaneously on both the leading and lagging strands. Although the DNA replication process has been described as though the replisome moves along the DNA, it is actually the opposite. Each of the two replication forks in the cell is stationary, while the parent DNA is threaded through during replication.

In *E. coli,* DNA replication is initiated at specific nucleotides called the *oriC* locus (for *ori*gin of *c*hromosomal replication). This site is AT rich. Recall that adenines pair with thymines using only two hydrogen bonds, so AT-rich segments of DNA become single stranded more readily than do GC-rich regions. This is important for initiation of replication.

1. The bacterial initiator protein DnaA is responsible for triggering DNA replication. DnaA proteins bind regions in *oriC* throughout the cell cycle, but to initiate replication, DnaA proteins must bind a few particular *oriC* sequences. The presence of DnaA at these sites recruits a helicase (usually DnaB helicase) to the origin.
2. The helicase unwinds the helix with the aid of topoisomerases such as DNA gyrase (figure 13.13, step 1). The strands are kept separate by SSBs.
3. Primase synthesizes RNA primers as needed (figure 13.13, step 1). DNA polymerase III holoenzyme catalyzes both

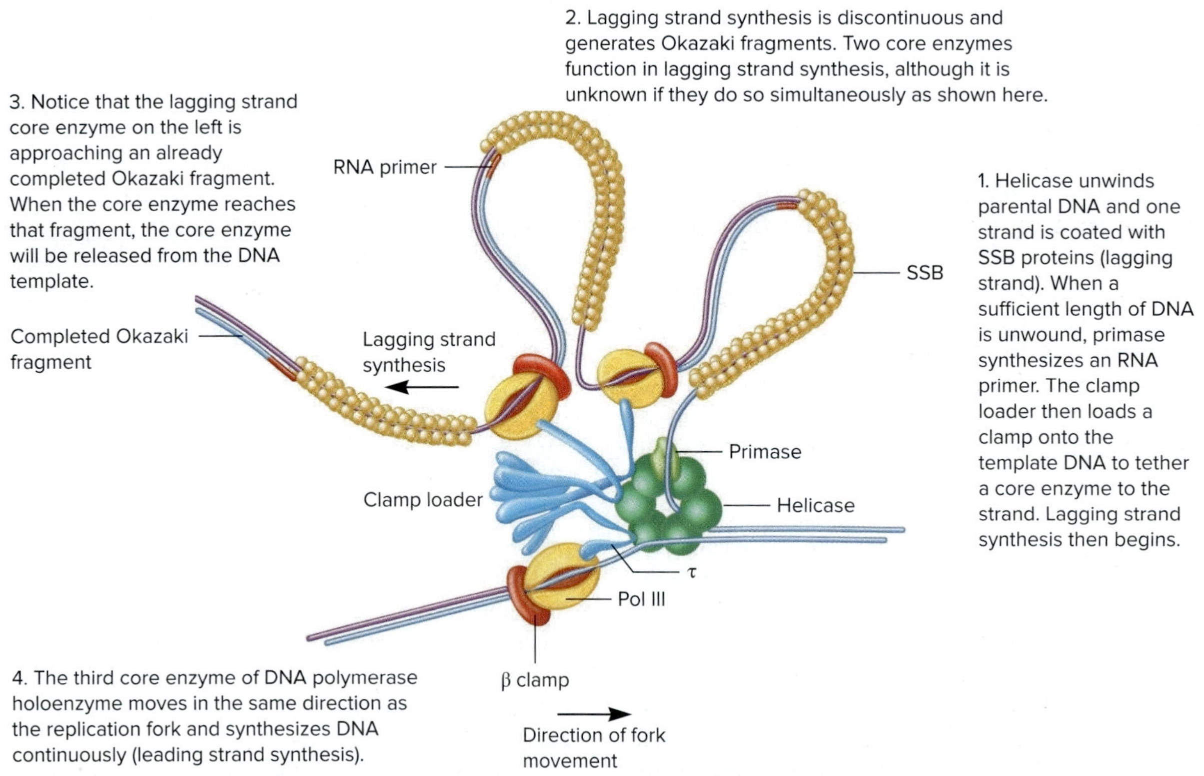

Figure 13.13 Events at the Replication Fork. Events are numbered and move in a counterclockwise direction. Components of the replisome are responsible for synthesis of both leading and lagging strands. The arrows show the direction of movement of lagging strand synthesis and the replication fork. Okazaki fragments are eventually joined together after removal of the RNA primers and synthesis of DNA to fill the gap, both catalyzed by DNA polymerase I (figure 13.14); DNA ligase seals the nick and joins the two fragments (figure 13.15).

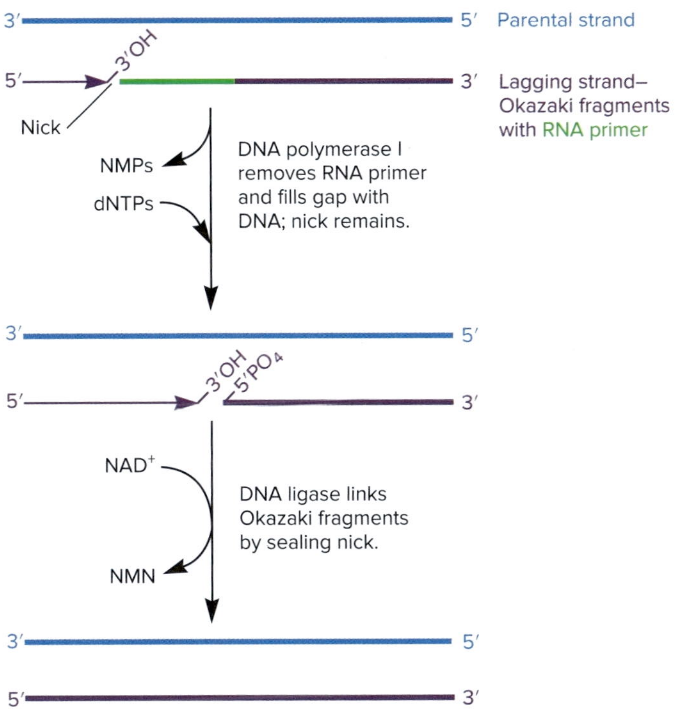

Figure 13.14 DNA Polymerase I Completes Lagging Strand Synthesis. dNTPs, deoxynucleoside triphosphates; NAD^+, nicotinamide adenine dinucleotide; NMN, nicotinamide mononucleotide; NMPs, nucleoside monophosphates.

leading strand and lagging strand synthesis from the RNA primers. Lagging strand synthesis is particularly amazing because of the "gymnastic" feats performed by the holoenzyme. It must discard old β clamps, load new β clamps, and tether the template to the core enzyme with each new round of Okazaki fragment synthesis. All of this occurs as DNA polymerase III is synthesizing DNA.

4. After most of the lagging strand has been synthesized by the formation of Okazaki fragments, DNA polymerase I removes the RNA primers using its unique ability to snip off nucleotides one at a time starting at the 5′ end while moving toward the 3′ end of the RNA primer. This ability is referred to as 5′ to 3′ exonuclease activity. DNA polymerase I begins its exonuclease activity at the free 5′ end of each RNA primer. With the removal of each ribonucleotide, the adjacent 3′-OH from the deoxynucleotide is used by DNA polymerase I to fill the gap between Okazaki fragments (**figure 13.14**).
5. Finally, the Okazaki fragments are joined by the enzyme **DNA ligase,** which forms a phosphodiester bond between the 3′-OH of the growing strand and the 5′-phosphate of an Okazaki fragment (**figure 13.15**).

Amazingly, DNA polymerase III, like all DNA polymerases (except DNA polymerases called reverse transcriptases), has an additional function that is critically important: **proofreading.** Proofreading is the removal of a mismatched base immediately after it has been added; its removal must occur before the next base is incorporated. The ε subunit of DNA polymerase III has 3′ to 5′ exonuclease activity. This activity enables the polymerase core enzyme to check each newly incorporated base to see that it forms stable hydrogen bonds. In this way, mismatched bases can be detected. If the wrong base has been mistakenly added, the exonuclease activity removes it, but only as long as it is still at the 3′ end of the growing strand (**figure 13.16**). Once removed, holoenzyme backs up and adds the proper nucleotide in its place. DNA proofreading is not 100% efficient, and as discussed in

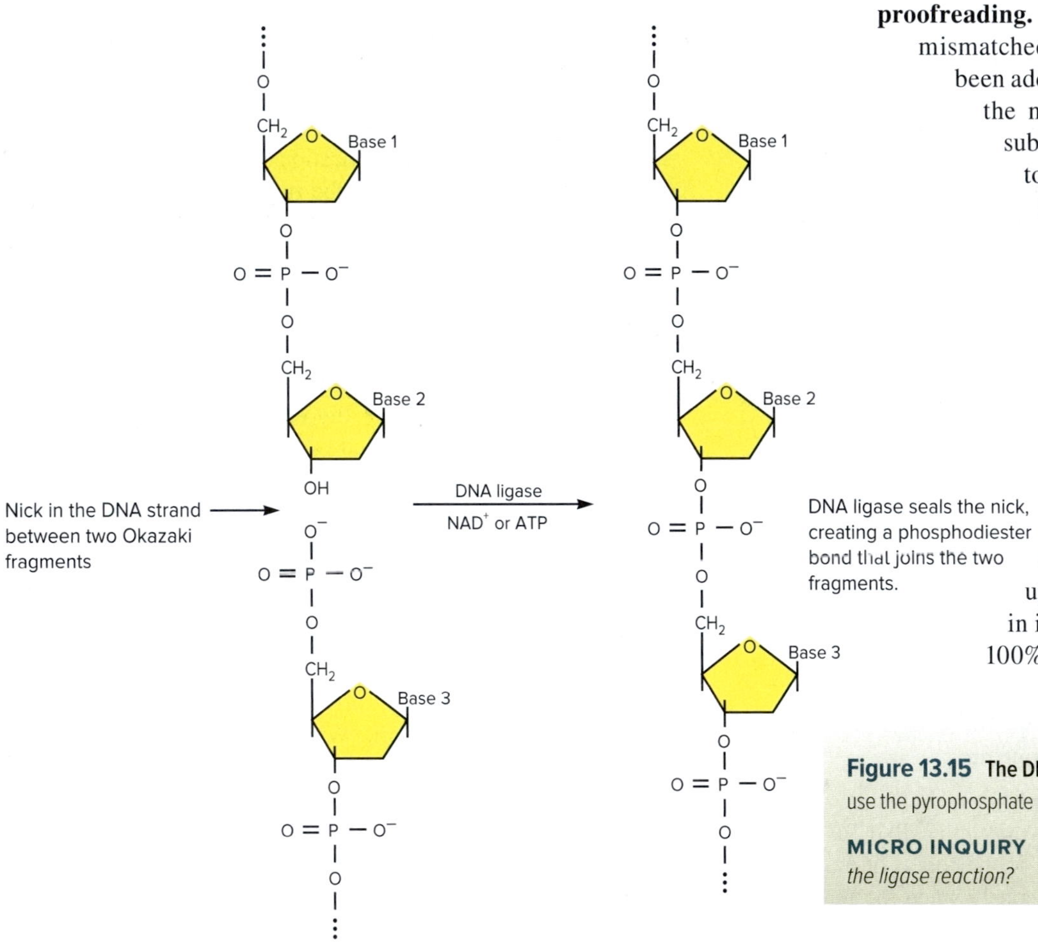

Figure 13.15 The DNA Ligase Reaction. Bacterial ligases use the pyrophosphate bond of NAD^+ or ATP as an energy source.

MICRO INQUIRY *Why can't DNA polymerase I perform the ligase reaction?*

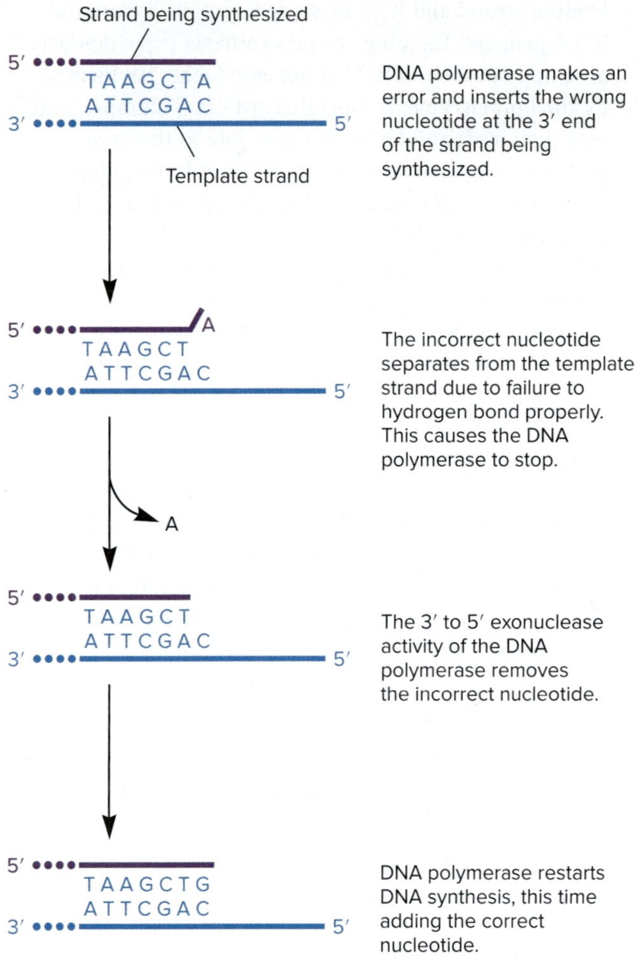

Figure 13.16 The Proofreading (Editing) Function of DNA Polymerase.

chapter 16, the mismatch repair system is the cell's second line of defense against the potential harm caused by the incorporation of the incorrect nucleotide. *Proofreading Function of DNA Polymerase*

As we have seen, DNA polymerase III is a remarkable multiprotein complex, with several enzymatic activities. In *E. coli,* the polymerase component is encoded by the *dnaE* gene. *Bacillus subtilis,* a Gram-positive bacterium that is another important experimental model, has a second polymerase gene called $dnaE_{Bs}$. Its protein product synthesizes the lagging strand. Thus while the overall mechanism of DNA replication is highly conserved, there can be variation in replisome components.

Termination of Replication Requires Steps to Separate Daughter Chromosomes

In *E. coli,* DNA replication stops when the replisome reaches a termination site (*ter*) on the DNA. A protein called Tus binds to the *ter* sites and halts progression of the forks. In many other bacteria, replication stops spontaneously when the forks meet. Regardless of how fork movement is stopped, there are two problems that must be solved by the replisome. The first is the formation of interlocked chromosomes called **catenanes** (**figure 13.17*a***). Catenanes are produced when topoisomerases break and rejoin DNA strands to ease supercoiling ahead of the replication fork. The two daughter DNA molecules are separated by the action of other topoisomerases that break both strands of one molecule, pass the other DNA molecule through the break, and then rejoin the strands. The other problem is a dimerized chromosome—two chromosomes joined together to form a single chromosome twice as long (figure 13.17*b*). Dimerized chromosomes result from DNA recombination that sometimes occurs between two daughter molecules during DNA replication. Recombinase enzymes (e.g., XerCD in *E. coli*) catalyze an intramolecular cross-over that separates the two chromosomes.

Linear Chromosome Ends Must Be Protected During Replication

All eukaryotic chromosomes and some bacterial chromosomes are linear. In both cases, this poses a problem during replication because all DNA polymerases synthesize DNA in the 5′ to 3′ direction from a primer that provides a free 3′-OH. When the RNA primer for the Okazaki fragment at the 5′ end of the daughter strand is removed, the daughter molecule is shorter than the parent molecule. Over numerous rounds of DNA replication and cell division, this leads to a progressively shortened chromosome. Ultimately the chromosome loses critical genetic information, which is lethal to the cell. This is called the "end replication problem," and a cell must solve it if it is to survive. Eukaryotic cells have solved the end replication problem with an enzyme called telomerase, as we describe in chapter 15. Bacteria have taken a different approach. ▶| *Telomeres and telomerases: protecting the ends of linear DNA molecules (section 15.2)*

Of those bacteria having linear chromosomes, the best understood mechanism for solving the end replication problem is that used by *Borrelia burgdorferi.* The approach *B. burgdorferi* cells use is to disguise the ends so well that they aren't really ends. How are the ends disguised? Consider a typical linear, double-stranded DNA molecule. At each terminus of the double helix is a strand having a 3′-OH and the complementary strand having a 5′-phosphate. Now examine **figure 13.18**, which shows that *B. burgdorferi* chromosomes do not have these free ends. Rather, a phosphodiester bond links the two complementary strands together. This is made possible by inverted repeats at each terminus and the formation of a hairpin. The origin of replication is located between the hairpins. Interestingly, only leading strand synthesis occurs at each replication fork. When replication is complete, a circular molecule has been formed that is twice the length of the parent chromosome. Thus it is a dimerized chromosome. An enzyme called telomere resolvase (ResT) cuts the two chromosomes apart as it forms hairpin telomeres for each daughter molecule.

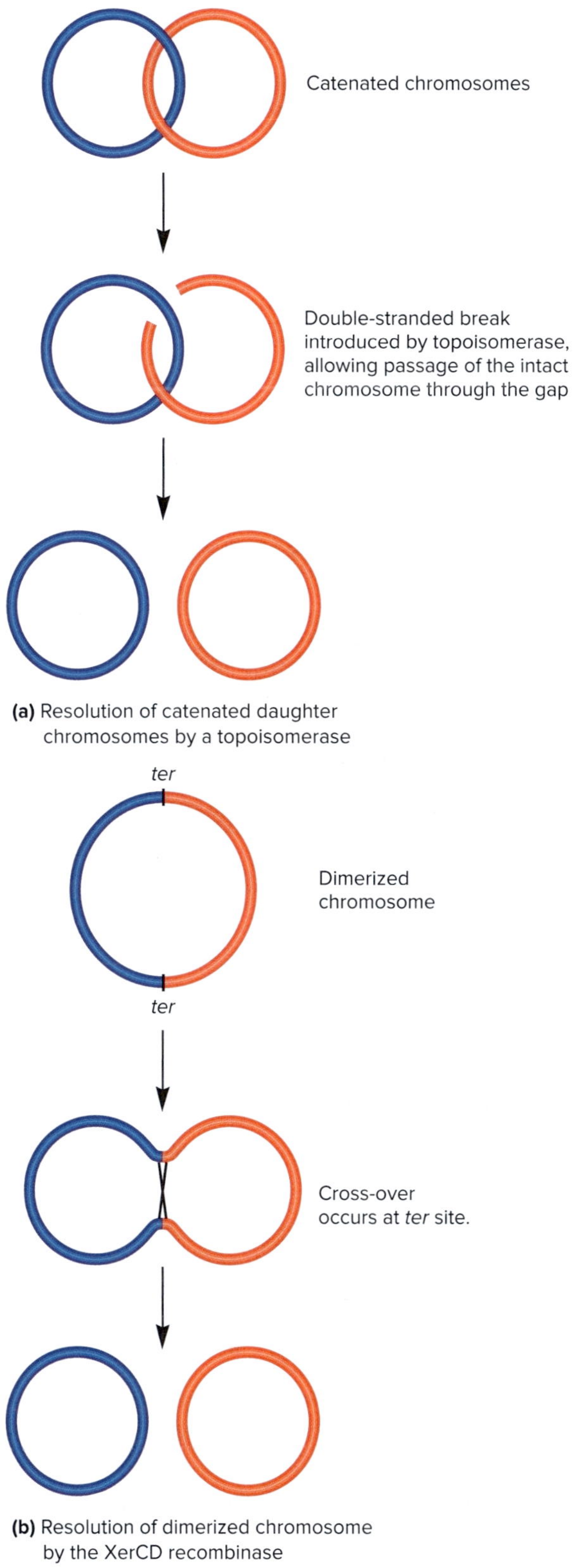

Figure 13.17 Resolving Catenated and Dimerized Chromosomes. (a) Catenated chromosomes arise from the activity of topoisomerases during chromosome replication. Accordingly, they are resolved by topoisomerases. (b) Dimerized chromosomes arise from recombination events that can occur during chromosome replication between the daughter chromosomes. They are resolved by recombinase enzymes such as XerCD of *E. coli.* These enzymes catalyze a cross-over that separates the two chromosomes.

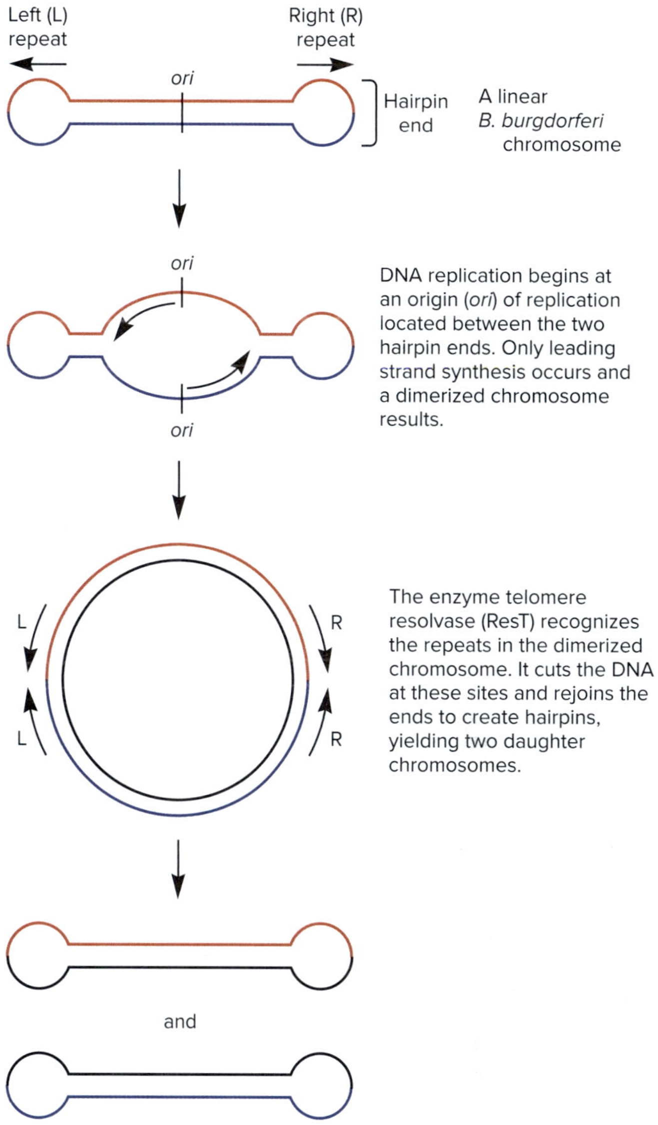

Figure 13.18 **Replication of a *Borrelia burgdorferi* Linear Chromosome.**

Comprehension Check

1. How many replicons do typical bacterial cells have (i.e., those having a single chromosome)? How many replication forks are used to replicate a circular chromosome? Is a primosome part of the replisome, or is the replisome part of the primosome? What is the function of each?
2. What is the difference between the replisome, DNA polymerase III holoenzyme, and core polymerase?
3. Describe the nature and functions of the following replication components and intermediates: DNA polymerases I and III, topoisomerase, DNA gyrase, helicase, single-stranded DNA binding proteins, Okazaki fragment, DNA ligase, leading strand, lagging strand, and primase.
4. Outline the steps involved in DNA synthesis at the replication fork. How do DNA polymerases correct their mistakes?
5. What is the end replication problem? How does *B. burgdorferi* solve it?

13.4 Bacterial Genes Consist of Coding Regions and Other Sequences Important for Gene Function

After reading this section, you should be able to:

a. Draw a typical bacterial protein-coding gene; label the important portions of the gene and the conventions for numbering base pairs in the gene
b. Draw typical tRNA- and rRNA-coding genes

DNA replication allows genetic information to be passed from one generation to the next. But how is the information used? To answer that question, we must first look at how genetic information is organized. The basic unit of genetic information is the gene. Genes have been regarded in several ways. At first, it was thought that a gene contained information for the synthesis of one enzyme—the one gene–one enzyme hypothesis. This was modified to the one gene–one polypeptide hypothesis because of the existence of enzymes and other proteins composed of two or more different polypeptide chains (subunits) coded for by separate genes. A segment of DNA that encodes a single polypeptide is sometimes termed a cistron. However, not all genes encode proteins; some code instead for ribosomal RNA (rRNA) and transfer RNA (tRNA), both of which function in protein synthesis (**figure 13.19**). In addition, it is now known that many eukaryotic genes encode more than one protein. Thus a **gene** might be defined as a polynucleotide sequence that codes for one or more functional products (i.e., a polypeptide, tRNA, or rRNA). In this section, we consider the structure of each of these three types of genes.

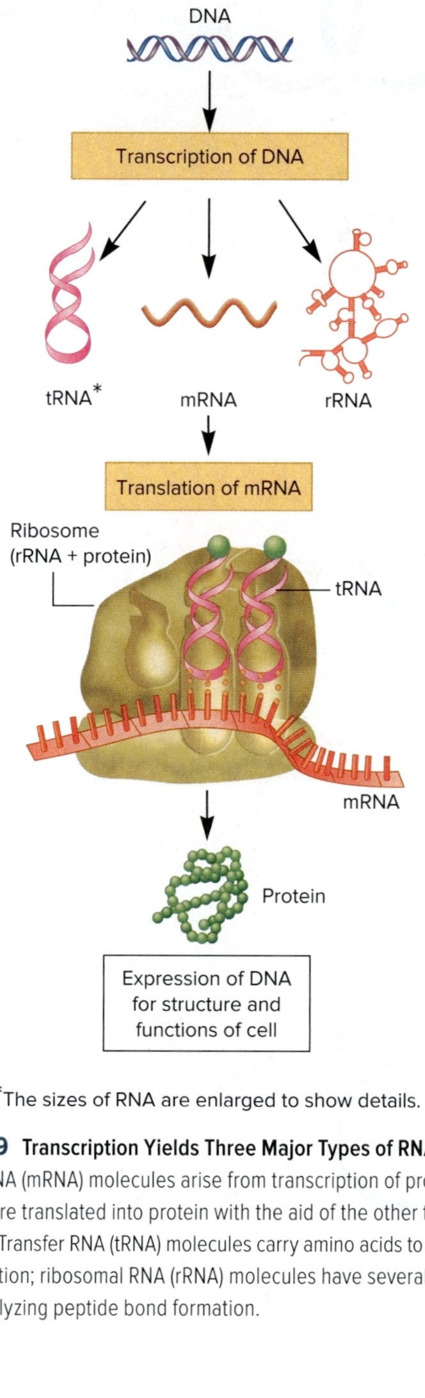

Figure 13.19 Transcription Yields Three Major Types of RNA Molecules. Messenger RNA (mRNA) molecules arise from transcription of protein-coding genes. They are translated into protein with the aid of the other two major types of RNA: Transfer RNA (tRNA) molecules carry amino acids to the ribosome during translation; ribosomal RNA (rRNA) molecules have several functions, including catalyzing peptide bond formation.

Protein-Coding Genes

Most of the genes found in bacterial genomes encode proteins. However, DNA does not serve directly as the template for protein synthesis. Rather, the genetic information in the gene is transcribed to give rise to a messenger RNA (mRNA), which is translated (section 13.7) into a protein (figure 13.19). For this to occur, protein-coding genes must contain signals that indicate where transcription should start and stop, and signals in the resulting mRNA that indicate where translation should start and stop. As we describe in more detail in section 13.5, during transcription only one strand of a gene directs mRNA synthesis. This strand is called the **template strand,** and the complementary DNA strand is known as the sense strand because it is the same nucleotide sequence as the mRNA, except in DNA bases (**figure 13.20**). Messenger RNA is synthesized from the 5′ to the 3′ end in a manner similar to DNA synthesis. Therefore the polarity of the DNA template strand is 3′ to 5′. In other words, the beginning of the gene is at the 3′ end of the template strand.

An important site called the **promoter** is located at the start of the gene. The promoter is the binding site for RNA polymerase, the enzyme that synthesizes RNA. The promoter is neither transcribed nor translated; it functions strictly to orient RNA polymerase so it is a specific distance from the first DNA nucleotide that will serve as a template for RNA synthesis. The promoter thus specifies which strand is to be transcribed and where transcription should begin. As we discuss in chapter 14, the sequences near the promoter often are very important in regulating when and at what rate a gene is transcribed. ▶| *Regulation of transcription initiation saves considerable energy and materials (section 14.2)*

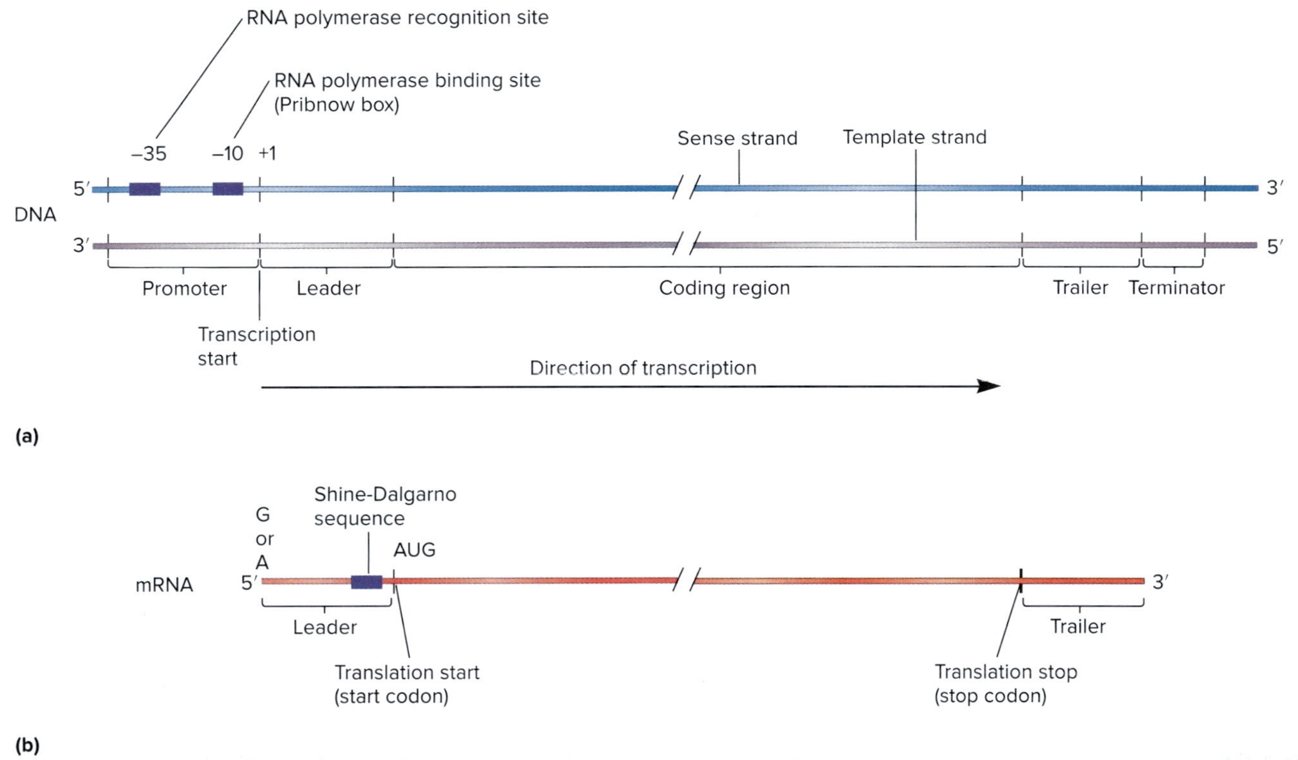

Figure 13.20 A Bacterial Structural Gene and Its mRNA Product. (a) The organization of a typical structural gene in a bacterial cell. Some genes lack leaders or trailers or both. Transcription begins at the +1 position in DNA and proceeds to the right, as shown. The numbering of nucleotides to the left of this spot is in a negative direction, while the numbering to the right is in a positive direction. For example, the nucleotide that is immediately to the left of the +1 nucleotide is numbered −1, and the nucleotide to the right of the +1 nucleotide is numbered +2. There is no zero nucleotide in this numbering system. In many bacterial promoters, sequence elements at the −35 and −10 regions play a key role in promoting transcription. During transcription, the template is read in the 3′ to 5′ direction. Regulatory sites are not shown but are usually upstream of the coding region and may overlap with the promoter. (b) Messenger RNA product of the gene shown in part a. The first nucleotide incorporated into mRNA is usually GMP or AMP. Translation of the mRNA begins with the AUG start codon.

MICRO INQUIRY *Why is the nontemplate strand called the "sense strand"?*

The transcription start site (labeled +1 in figure 13.20) represents the first nucleotide in the mRNA synthesized from the gene. However, the initially transcribed portion of the gene does not code for amino acids. Instead, it is a **leader** that is transcribed into mRNA but is not translated into amino acids. In bacteria, the leader includes a region called the **Shine-Dalgarno sequence,** which is important in the initiation of translation. The leader sometimes is also involved in regulation of transcription and translation. ▶| *Attenuation and riboswitches can stop transcription prematurely (section 14.3); Riboswitches and small RNAs can control translation (section 14.4)*

Immediately next to (and downstream of) the leader is the most important part of the gene, the **coding region** (figure 13.20). The coding region typically begins with the template DNA sequence 3′-TAC-5′. This is transcribed into the start codon, 5′-AUG-3′, which codes for the first amino acid of the polypeptide encoded by the gene. The remainder of the coding region is transcribed into a sequence of codons that specifies the sequence of amino acids for the rest of the protein. The coding region ends with a sequence that, when transcribed, is a stop codon. It signals the end of the protein and stops the ribosome during translation. The stop codon is immediately followed by the **trailer,** which is transcribed but not translated. The trailer contains sequences that prepare RNA polymerase for release from the template strand. Indeed, just beyond the trailer (and sometimes slightly overlapping it) is the **terminator.** The terminator is a sequence that signals RNA polymerase to stop transcription.

In bacteria, the coding region is usually continuous, unlike the coding regions of eukaryotic genes, which are often interrupted by noncoding sequences called **introns.** Those rare bacterial genes that do contain introns are transcribed into an intron-containing mRNA. The introns are eventually removed but by a mechanism different from that used to remove introns from eukaryotic mRNAs. ▶| *From pre-mRNA to mRNA: a unique eukaryotic process (section 15.3)*

tRNA and rRNA Genes

Actively growing cells need a ready supply of tRNA and rRNA molecules for protein synthesis. To ensure this, bacterial cells often have more than one gene for each of these molecules. Furthermore, it is important that the number of each tRNA or rRNA relative to other tRNAs or rRNAs be controlled. This is accomplished in part by having several tRNA or rRNA genes transcribed together from a single promoter.

In bacteria, genes for tRNA consist of a promoter, tRNA coding region, leader, and trailer. When more than one tRNA is transcribed from the promoter, the coding regions are separated by short spacer sequences (**figure 13.21*a***). Whether the gene encodes a single tRNA or multiple tRNAs, the initial transcript must be processed to remove the noncoding sequences (i.e., leader, trailer, and spacers, if present). This is called posttranscriptional modification, and it is accomplished by ribonucleases—enzymes (and in some cases ribozymes) that cut RNA. In addition, many bacterial tRNA genes contain introns that must be removed during tRNA maturation. |◀ *Ribozymes: catalytic RNA molecules (section 10.6)*

Bacterial cells usually contain more than one rRNA gene. Each gene has a promoter, trailer, and terminator (figure 13.21*b*). Thus, as seen for tRNA genes, the transcript from an rRNA gene is a single, large precursor molecule that is cut by ribonucleases to yield the final rRNA products. Interestingly, in many bacteria, the trailer regions and the spacers often contain tRNA genes. Thus the precursor rRNA encodes for both tRNA and rRNA.

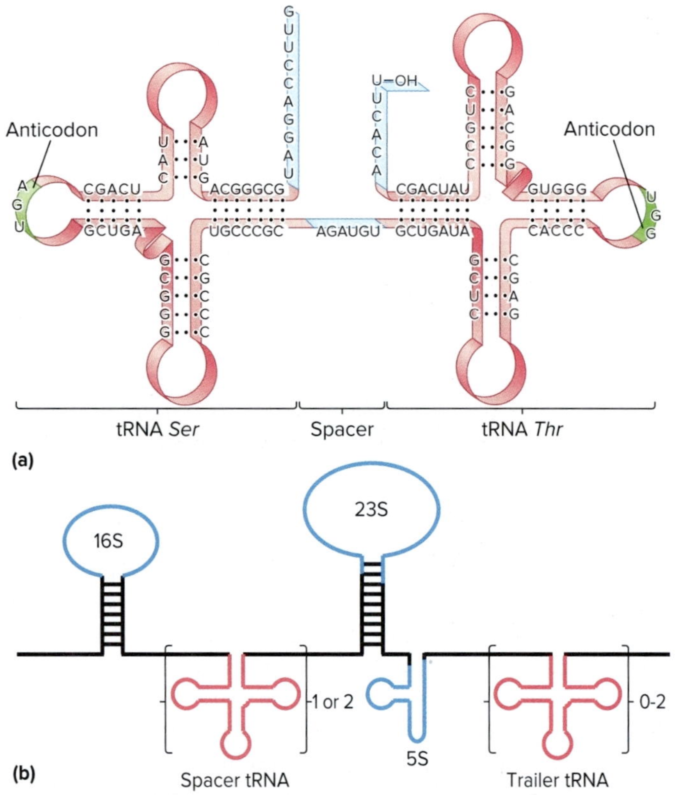

Figure 13.21 tRNA and rRNA Genes. (a) A tRNA precursor from *E. coli* that contains two tRNA molecules. The spacer and extra nucleotides at both ends are removed during processing. (b) The *E. coli* rRNA gene codes for a large transcription product that is cleaved into three rRNAs and one to three tRNAs. The 16S, 23S, and 5S rRNA segments are represented by blue lines, and tRNA sequences are placed in brackets. The seven copies of this gene vary in the number and kind of tRNA sequences.

Comprehension Check

1. The coding region of a gene is said to be "downstream" from the leader. Conversely, the leader is said to be "upstream" of the coding region. For each of the following portions of a gene, indicate whether it is downstream or upstream of the coding region: promoter, +1 nucleotide, trailer, terminator. Which portions of a gene are transcribed but not translated?
2. Which strand of a gene has sequences that correspond to the start (i.e., ATG rather than AUG) and stop codons in the mRNA product of the gene?
3. Briefly discuss the general organization of tRNA and rRNA genes. How does their expression differ from that of protein-coding genes with respect to posttranscriptional modification of the gene product?

13.5 Transcription in Bacteria

After reading this section, you should be able to:

a. Illustrate the organization of bacterial genes in a typical operon
b. Describe the structure of a typical bacterial RNA polymerase holoenzyme
c. Outline the events that occur during the three phases of transcription
d. Discuss the role of bacterial promoters and sigma factors in transcription initiation
e. Distinguish factor-independent termination of transcription from rho-dependent termination of transcription

Synthesis of RNA from a DNA template is called **transcription,** and the RNA product has a sequence complementary to the DNA template directing its synthesis. Although adenine directs the incorporation of thymine during DNA replication, it usually pairs with uracil during RNA synthesis. Transcription generates three major kinds of RNA. **Transfer RNA (tRNA)** carries amino acids during protein synthesis, and **ribosomal RNA (rRNA)** molecules are components of ribosomes. **Messenger RNA (mRNA)** bears the message for

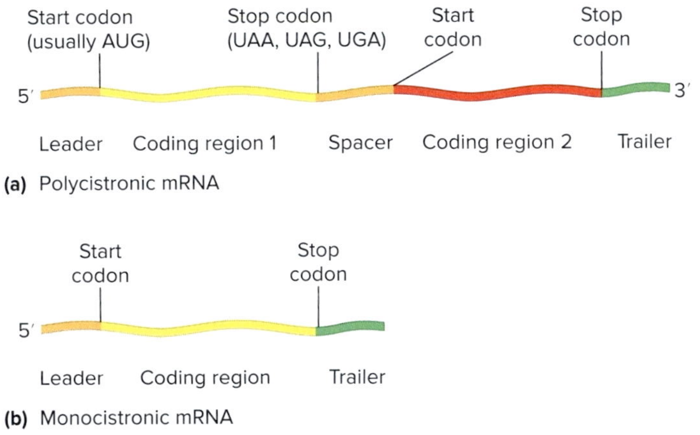

Figure 13.22 Polycistronic and Monocistronic mRNAs. Polycistronic mRNAs are commonly observed in bacteria and archaea. Eukaryotic genes usually give rise to monocistronic mRNAs.

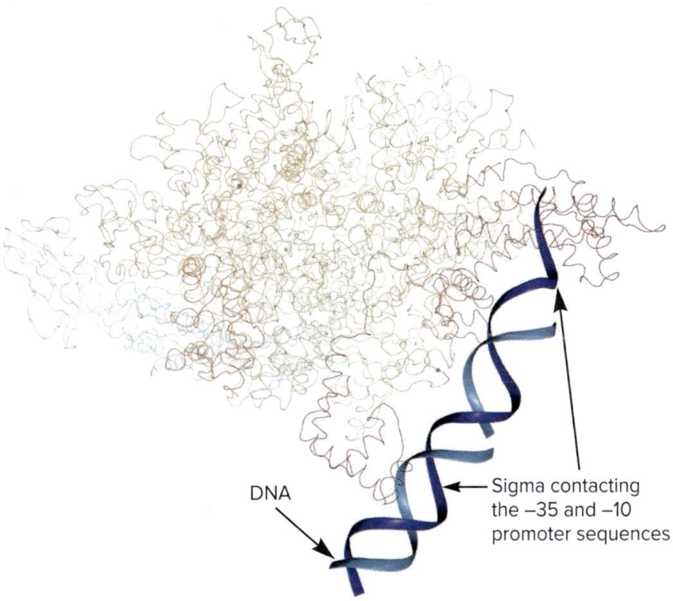

Figure 13.23 Structure of *Thermus aquaticus* RNA Polymerase on a Promoter. The promoter DNA is read by the σ subunit (red) at the −35 and −10 sequences. Other polypeptides in the enzyme are two copies of α (blue), β (olive), β′ (gold), and ω (orange).

protein synthesis. Bacterial genes encoding proteins involved in a related process (e.g., encoding enzymes for synthesis of an amino acid) are often located close to each other and are transcribed from a single promoter. Such a transcriptional unit is termed an **operon.** Transcription of an operon yields an mRNA consisting of a leader followed by one coding region, which is separated from the second coding region by a spacer, and so on, with the final sequence of nucleotides being the trailer. Such mRNAs are said to be **polycistronic mRNAs** (**figure 13.22*a***). Each coding region in the polycistronic mRNA is defined by a start and stop codon. Thus each coding region is translated separately to give rise to a separate polypeptide. Many archaeal mRNAs are also polycistronic. However, polycistronic mRNAs are rare in eukaryotes. Instead, their mRNAs are usually **monocistronic** (figure 13.22*b*), containing information of a single gene. ▶| *Transcription (section 15.3)*

Bacterial RNA Polymerases Consist of Five Different Proteins

RNA is synthesized by enzymes called **RNA polymerases.** In bacteria, a single RNA polymerase transcribes all genes. Most bacterial RNA polymerases contain five types of polypeptide chains: α, β, β′, ω, and σ (**figure 13.23**). The **RNA polymerase core enzyme** is composed of five polypeptides (two α subunits, β, β′, and ω) and catalyzes RNA synthesis. The **sigma factor** (σ) has no catalytic activity but instead functions as a transcription factor by helping the core enzyme recognize the promoter. When sigma is bound to the core enzyme, the six-subunit complex is termed **RNA polymerase holoenzyme.** Only holoenzyme can begin transcription, but the core enzyme completes RNA synthesis once it has been initiated.

Stages of Transcription: Initiation, Elongation, and Termination

Transcription involves three separate processes: initiation, elongation, and termination, which together are often referred to as the transcription cycle (**figure 13.24**). The transcription factor sigma is critical to the initiation process. As part of the RNA polymerase holoenzyme, it helps position the core enzyme at the promoter. Bacterial promoters have several characteristic features. Two of particular note are a sequence of six bases (often TTGACA) about 35 base pairs before (upstream of) the transcription starting point and a TATAAT sequence called the **Pribnow box,** usually about 10 base pairs upstream of the transcriptional start site (**figure 13.25**; also figure 13.20). These regions are called the −35 and −10 sites, respectively, because these are their approximate distances in nucleotides upstream of the first nucleotide to be transcribed (i.e., the +1 site). Sigma factor first recognizes the −35 sequence, allowing the holoenzyme to "settle down" on that region of the promoter (figure 13.23). Sigma and proteins in the core enzyme undergo conformational changes that cause the DNA strands in the AT-rich −10 region to separate, forming what will become the transcription bubble. Sigma then interacts with one of the strands and stabilizes the interaction of RNA polymerase with the unwound DNA. The resulting complex of RNA polymerase holoenzyme and unwound DNA

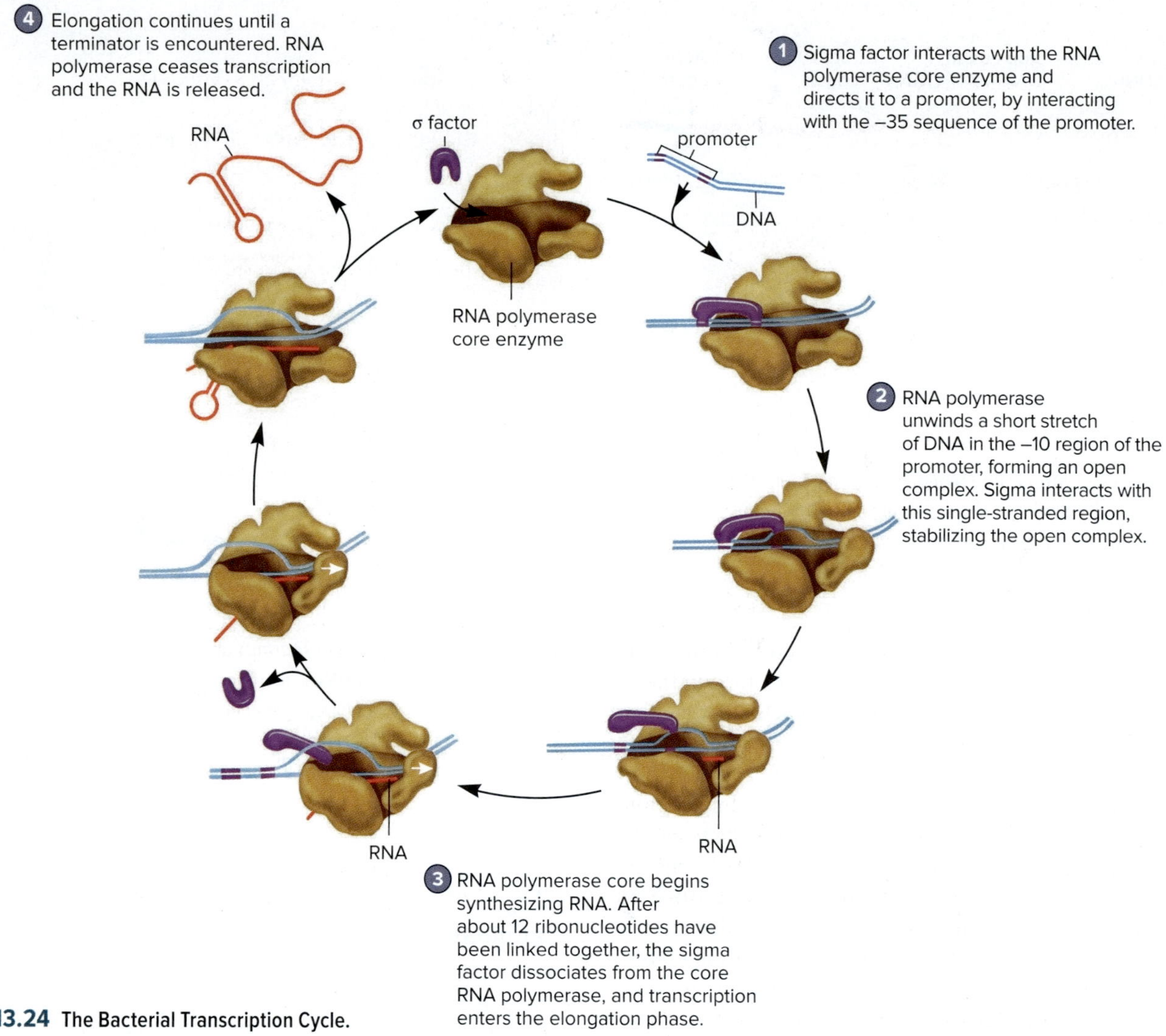

Figure 13.24 **The Bacterial Transcription Cycle.**

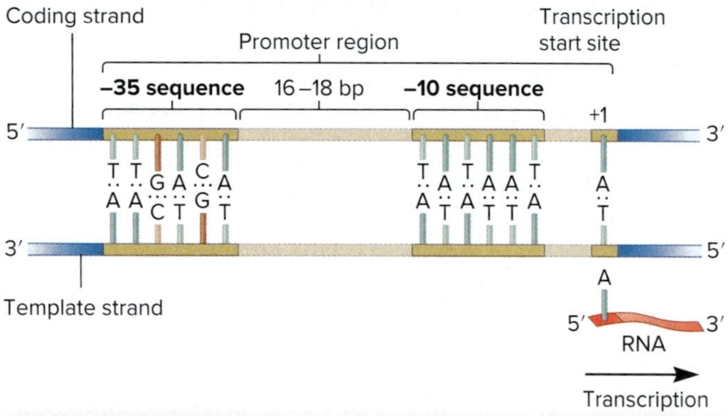

Figure 13.25 **A σ^{70} Promoter.** Many bacterial promoters are recognized by the sigma factor σ^{70}. Its promoters have a characteristic set of nucleotides centered at about –10 and –35. Shown are the consensus sequences for these two important promoter sites. Bacterial cells produce additional sigma factors that recognize different promoter sequences (table 13.2).

MICRO INQUIRY *Are the –35 and –10 regions considered "upstream" or "downstream" of the +1 nucleotide?*

is called the open complex (**figure 13.26**). *Stages of Transcription*

At this point in our discussion, it is worth noting that bacterial cells produce more than one type of sigma factor. Each sigma factor preferentially directs RNA polymerase to a distinct set of promoters. For instance, in *E. coli*, most genes have promoters recognized by a sigma factor called σ^{70}. This sigma factor recognizes promoters having the –10 and –35 sequences shown in figure 13.25 and **table 13.2**. These sequences are the **consensus sequences** for σ^{70}-recognized promoters. Promoters recognized by other sigma factors have different consensus sequences. The use of different sigma factors to initiate transcription is a common bacterial regulatory mechanism, as we describe in chapter 14. Our focus here is on transcription of genes recognized by σ^{70}.

Once the open complex is formed, transcription can begin. Within the open complex is a region of unwound DNA equivalent to about 16 to 20 base pairs. This is the "transcription bubble," and it moves with the RNA polymerase as it synthesizes mRNA from the template DNA strand during

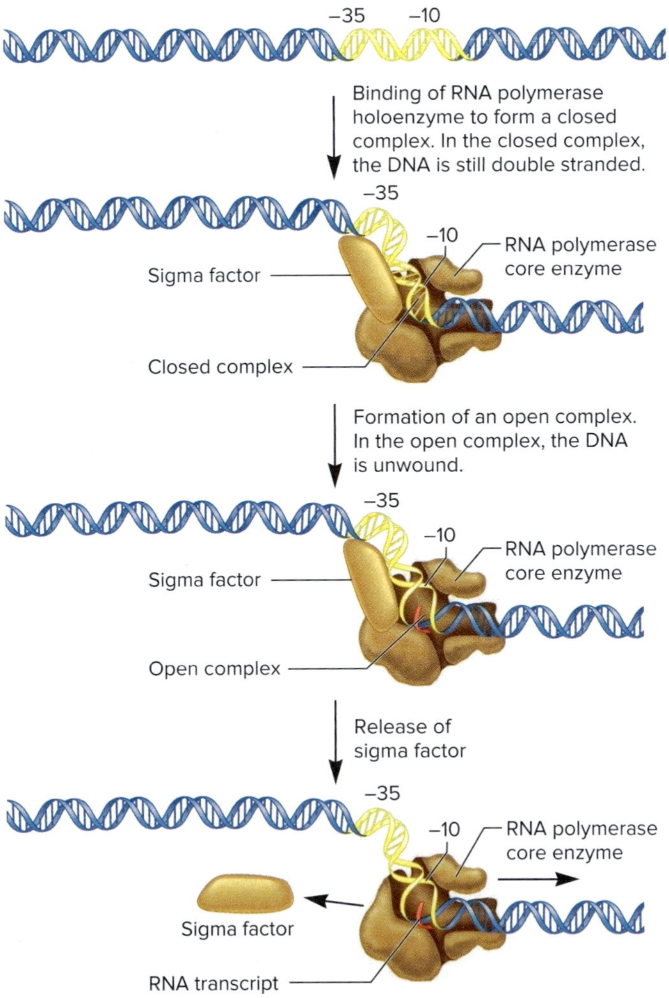

Figure 13.26 Initiation of Transcription in Bacteria. The sigma factor of the RNA polymerase holoenzyme is responsible for positioning the core enzyme properly at the promoter. Sigma factor recognizes two regions in the promoter, one centered at −35 and the other centered at −10. Once positioned properly, the DNA at the −10 region unwinds to form an open complex. The sigma factor dissociates from the core enzyme after transcription is initiated.

elongation (**figure 13.27**). Within the transcription bubble, a temporary RNA:DNA hybrid is formed. As RNA polymerase holoenzyme progresses along the DNA template, the sigma factor dissociates from the other subunits and can direct another RNA polymerase core enzyme to initiate transcription (figure 13.24).

The reaction catalyzed by RNA polymerase is quite similar to that catalyzed by DNA polymerase (figure 13.10). ATP, GTP, CTP, and UTP are used to produce RNA complementary to the DNA template, and pyrophosphate is produced as ribonucleoside monophosphates are incorporated into the growing RNA chain. Pyrophosphate is hydrolyzed to fuel the process. RNA synthesis also proceeds in a 5′ to 3′ direction with new ribonucleotides being added to the 3′ end of the growing chain, making the RNA complementary and antiparallel to the template DNA. As elongation of the mRNA continues, single-stranded mRNA is released, and the two strands of DNA behind the transcription bubble resume their double helical structure. As shown in figure 13.24, RNA polymerase is a remarkable enzyme capable of several activities, including unwinding the DNA, moving along the template, and synthesizing RNA. As is the case with DNA replication, DNA unwinding for transcription also introduces supercoiling. A transcribing RNA polymerase requires the activity of a topoisomerase to relieve the torsion introduced into the chromosome.

Termination of transcription occurs when the core RNA polymerase dissociates from the template DNA. This is brought about by the terminator. There are two kinds of terminators. The first type causes factor-independent termination (**figure 13.28**). This terminator consists of an inverted repeat followed by an A-rich nucleotide sequence. RNA polymerase transcribes the inverted repeat, but it pauses within the A-rich region. This allows the inverted repeat to fold back on itself, forming a hairpin-shaped stem-loop structure. The A-U base pairs holding the DNA and RNA together in the transcription bubble are too weak to hold the RNA:DNA duplex together and RNA polymerase falls off.

Table 13.2 *E. coli* **Sigma Factors and the Sequences They Recognize**

Sigma Factor	Consensus Promoter Sequences[1]		Genes Transcribed from Promoter
σ^{70}	TTGACA	TATAAT	Most genes
σ^{54}	YTGGCAC[2]	TTGC	Genes for nitrogen metabolism
σ^{38}	TTGACA	TCTATACTT	Genes for stationary phase and stress response
σ^{32}	TNTCNCCCTTGAA	CCCCATTTA	Genes for heat shock response
σ^{28}	TAAAG	GYCGATAR	Genes for motility
σ^{24}	GAACT	TCAAA	Genes for extreme heat shock response, membrane protein damage
σ^{19}	GAAAAT	TGTCCT	Genes for the transport of ferric citrate

1 With the exception of the σ^{54} promoters, all consensus sequences are located at −35 and −10, respectively. The σ^{54} consensus sequences are located at −24 and −12.
2 N indicates any nucleotide; Y indicates a pyrimidine, C or T; R indicates a purine, A or G.

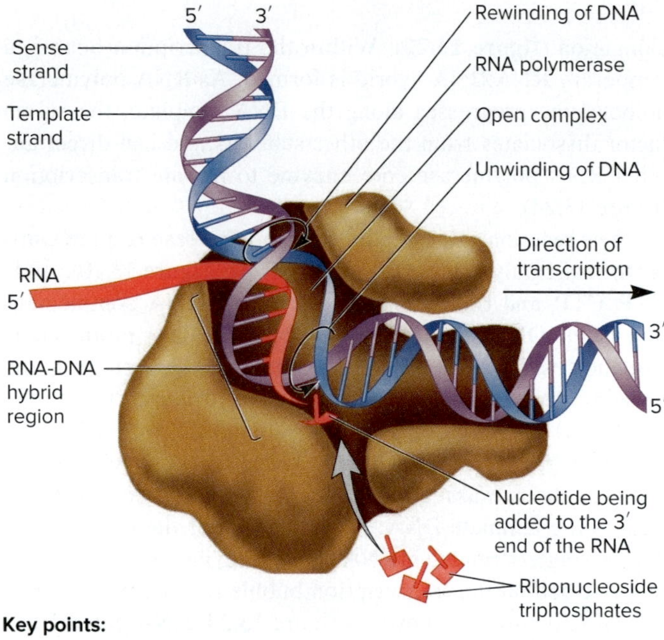

Key points:

- RNA polymerase slides along the DNA, creating an open complex as it moves.
- The template strand is used to make a complementary copy of RNA as an RNA-DNA hybrid.
- The RNA is synthesized in a 5′ to 3′ direction using ribonucleoside triphosphates as precursors. Pyrophosphate is released (not shown).
- The complementarity rule is the same as the AT/GC rule except that U is substituted for T in the RNA.

Figure 13.27 The Transcription Bubble.

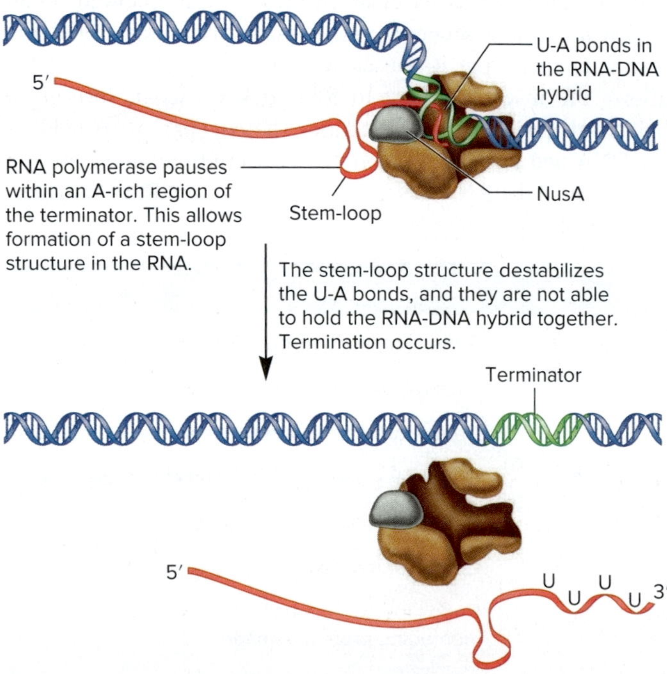

Figure 13.28 Factor-Independent Termination of Transcription. This type of terminator contains an inverted repeat and an A-rich sequence downstream from the repeat. A protein called NusA stimulates termination.

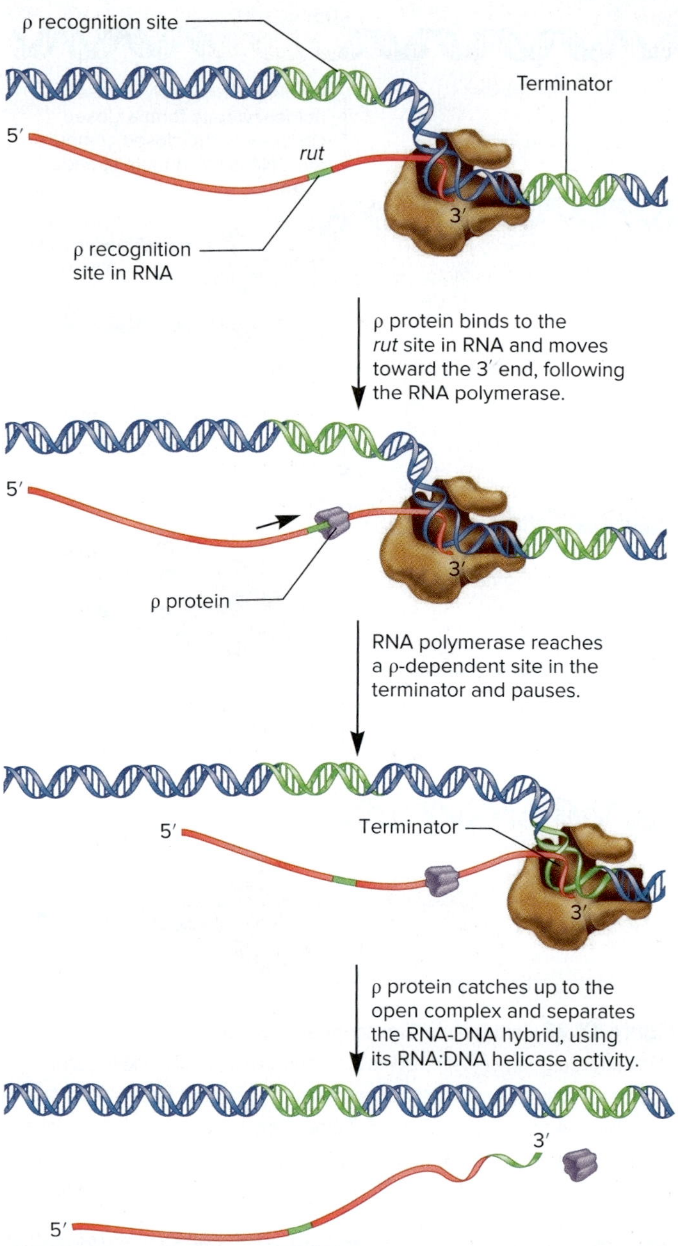

Figure 13.29 Rho-Factor (ρ)-Dependent Termination of Transcription. The *rut* site stands for *r*ho-*ut*ilization site.

The second kind of terminator is termed factor-dependent terminator because it requires the aid of a protein. The best-studied termination factor is rho factor (ρ). Rho factor can be involved in transcription termination of all types of genes, but its action is best studied for protein-coding genes. Current models propose that rho binds to mRNA at a site called *rut* for *r*ho-*ut*ilization site. For rho to bind, *rut* must be free of ribosomes, as shown in **figure 13.29**. Rho uses energy supplied by ATP hydrolysis to move along the mRNA, as it follows RNA polymerase. With a slower speed than RNA polymerase, rho can only catch up if the polymerase pauses at a rho-dependent pause site. If this occurs, rho causes RNA polymerase to dissociate from DNA. Rho has hybrid RNA:DNA helicase activity, and this activity causes unwinding of the mRNA-DNA complex and release of RNA polymerase from the template DNA strand.

Comprehension Check

1. Outline the transcription cycle observed in bacteria. In which steps do the following function: RNA polymerase core enzyme, sigma factor, and RNA polymerase holoenzyme?
2. What is a polycistronic mRNA? How are polycistronic mRNAs formed?
3. What is a consensus sequence? What is the importance of the existence of different sigma factors and consensus sequences in bacterial promoters?
4. Compare and contrast factor-dependent and factor-independent transcription termination.

13.6 The Genetic Code Consists of Three-Letter "Words"

After reading this section, you should be able to:

a. Explain the importance of the reading frame of a protein-coding gene
b. Describe the universal genetic code
c. List deviations from the universal genetic code that have been identified in some microorganisms
d. Explain how the wobble hypothesis enables organisms to encode fewer tRNA molecules

The final step in expression of protein-coding genes is translation. Protein synthesis is called translation because it is a decoding process. The information encoded in the language of nucleic acids must be rewritten in the language of proteins. During translation, the sequence of nucleotides is "read" in discrete sets of three nucleotides, each set being a **codon.** Each codon codes for a single amino acid. The sequence of codons is "read" in only one way—the **reading frame** (**figure 13.30**)—to give rise to the amino acid sequence of a polypeptide. Deciphering the genetic code was one of the great achievements of the twentieth century. Here we examine the nature of the genetic code. Translation is considered in section 13.7.

The genetic code, presented in RNA form, is summarized in **table 13.3**. Close inspection of the code reveals several features that are related not only to the way cells use DNA to store information but also to why it is valuable for storing data, as described in the chapter opening story. One feature is that the code words (codons) are three letters (bases) long; thus one small "word" conveys a significant amount of information. Each codon is recognized by an anticodon present on a tRNA molecule. Another feature is that the code has "punctuation." One codon, AUG, is almost always the first codon in the protein-coding portion of mRNA molecules. It is called the **start codon** because it serves as the start site for translation by coding for the initiator tRNA. Three other codons (UGA, UAG, and UAA) terminate translation and are called **stop** or **nonsense codons.** These codons do not encode an amino acid and therefore do not have a tRNA bearing their anticodon. Thus only 61 of the 64 codons in the code, the **sense codons,** direct amino acid incorporation into protein. Finally, the genetic code exhibits **code degeneracy** (also

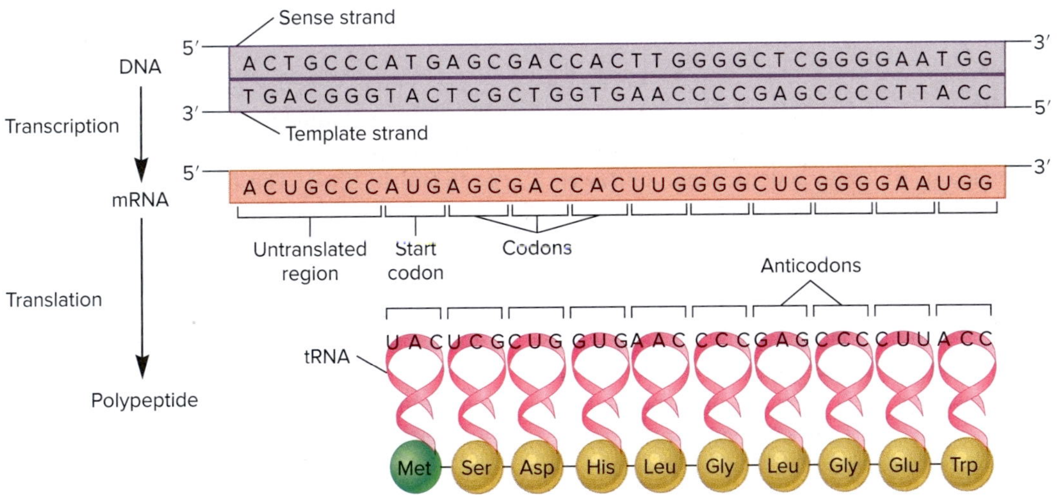

Figure 13.30 Reading Frame. During transcription, an mRNA complementary to the template strand of DNA is synthesized. The nucleotides in the mRNA are organized into groups of three, each group being a codon. The first codon translated into protein is the start codon. It establishes the reading frame and therefore the sequence of amino acids in the polypeptide chain that is made from the mRNA.

Table 13.3 The Genetic Code[1]

First Position (5′ End)[2]	SECOND POSITION: U		C		A		G		Third Position (3′ End)
U	UUU	Phe F	UCU	Ser S	UAU	Tyr Y	UGU	Cys C	U
	UUC		UCC		UAC		UGC		C
	UUA	Leu L	UCA		UAA	STOP	UGA	STOP	A
	UUG		UCG		UAG		UGG	Trp W	G
C	CUU	Leu L	CCU	Pro[3] P	CAU	His H	CGU	Arg R	U
	CUC		CCC		CAC		CGC		C
	CUA		CCA		CAA	Gln Q	CGA		A
	CUG		CCG		CAG		CGG		G
A	AUU	Ile I	ACU	Thr T	AAU	Asn N	AGU	Ser S	U
	AUC		ACC		AAC		AGC		C
	AUA		ACA		AAA	Lys K	AGA	Arg R	A
	AUG	Met M	ACG		AAG		AGG		G
G	GUU	Val V	GCU	Ala A	GAU	Asp D	GGU	Gly G	U
	GUC		GCC		GAC		GGC		C
	GUA		GCA		GAA	Glu E	GGA		A
	GUG		GCG		GAG		GGG		G

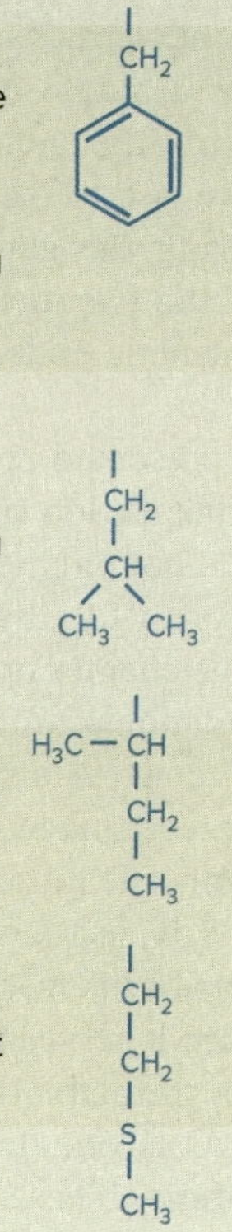
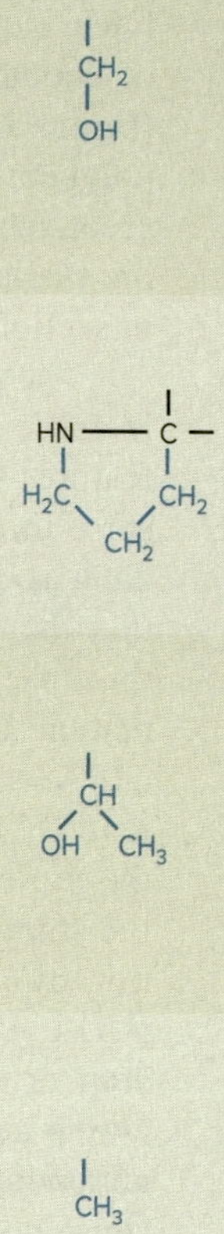
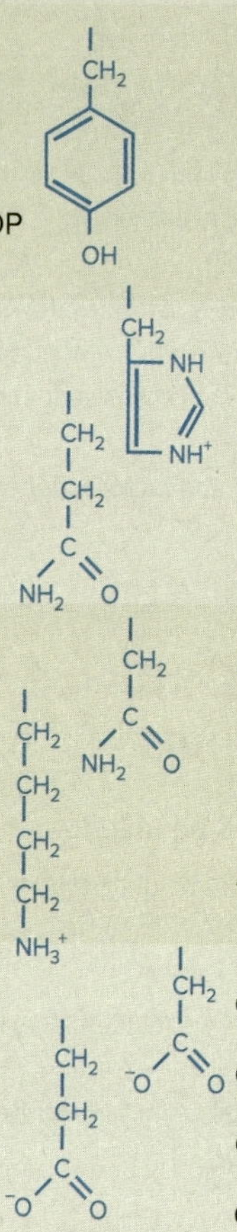
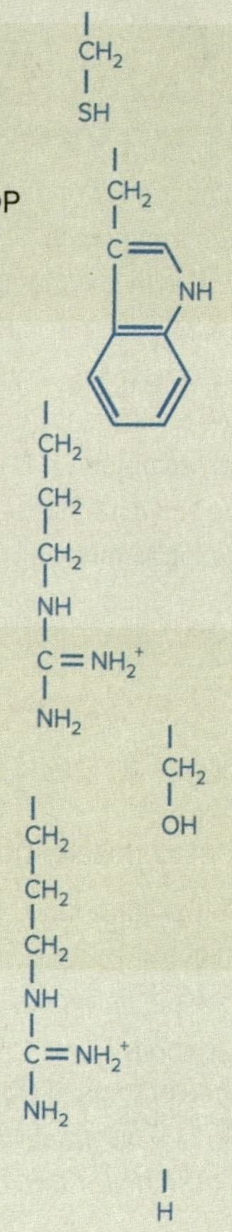

1 Next to each codon or set of codons is the R group characteristic of that amino acid drawn in blue.
2 The code is presented in RNA form. Codons run in the 5′ to 3′ direction. The 3- and 1-letter abbreviations for the amino acids designated by the codon are provided, as is the structure of the R chain of that amino acid.
3 Proline is an imino acid, rather than an amino acid.

called redundancy); that is, there are up to six different codons for a given amino acid.

Despite the existence of 61 sense codons, there are fewer than 61 different tRNAs. It follows that not all codons have a corresponding tRNA. Cells can successfully translate mRNA using fewer tRNAs because loose pairing between the 5′ base in the anticodon and the 3′ base of the codon is tolerated. Thus as long as the first and second bases in the codon correctly base pair with an anticodon, the tRNA bearing the correct amino acid will bind to the mRNA during translation. This is evident on inspection of the code. Note that the codons for a particular amino acid most often differ at the third position (table 13.3). This somewhat loose base pairing is known as **wobble,** and it relieves cells of the need to synthesize so many tRNAs (**figure 13.31**). Wobble also decreases the effects of some mutations. ▶| *Mutations: heritable changes in a genome (section 16.1)*

The description of the genetic code just provided is of the universal genetic code. However, there are exceptions to the code (**table 13.4**). The first exceptions discovered were stop codons that encoded one of the 20 amino acids. For instance, mycoplasma bacteria use the stop codon UGA to code for glutamine (Gln). More dramatic deviations from the code have also been discovered. Members of all three domains of life encode proteins containing the amino acid selenocysteine, the twenty-first amino acid (**figure 13.32*a***). Most selenocysteine-containing enzymes catalyze redox reactions. Pyrrolysine, the twenty-second amino acid, can be found in the proteins of several bacteria and methanogenic archaea (figure 13.32*b*). In the methanogens, pyrrolysine is found in methyltransferase enzymes, which function in methanogenesis. Selenocysteine is inserted at certain UGA codons, whereas pyrrolysine is inserted at UAG codons.

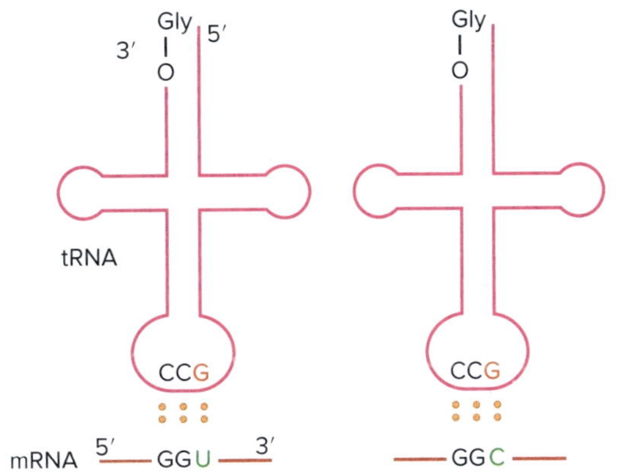

Figure 13.31 Wobble Decreases the Number of tRNAs Needed to Specify Glycine. Because of wobble, G in the 5′ position of the anticodon can pair with either C or U in the 3′ position of the codon. Thus two codons can be recognized by the same tRNA.

Comprehension Check

1. What is the difference between a codon and an anticodon?
2. List the "punctuation" functions of the genetic code and explain their importance.
3. What is meant by code degeneracy (redundancy)? How does wobble help alleviate the energy cost exacted on a cell due to code degeneracy?
4. Is the genetic code truly universal? Explain your answer.

Table 13.4 Some Exceptions to the Universal Genetic Code

Codon	Amino Acid Inserted	Where Observed
AGA and AGG	Stop	Mammalian mitochondria
AGA and AGG	Serine (Ser)	Invertebrate mitochondria
AUA	Methionine (Met)	Mammalian, invertebrate, and yeast mitochondria
CUA	Threonine (Thr)	Yeast mitochondria
CUG	Serine	Some fungi
UAA and UAG	Glutamine (Gln)	Some protists
UAG	Pyrrolysine	Some methanogens and bacteria
UGA	Selenocysteine	Members of all three domains
UGA	Tryptophan (Trp)	Mammalian, invertebrate, and yeast mitochondria
UGA	Glutamine	Mycoplasma bacteria

MICRO INQUIRY *For each of these codons, determine what it normally encodes. Choose one of the variants and suggest how it might have evolved.*

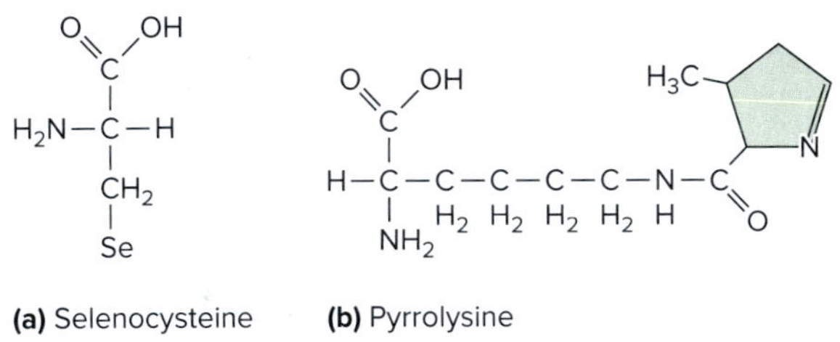

Figure 13.32 Selenocysteine and Pyrrolysine. These unusual amino acids are not coded for by DNA but are inserted in response to specific stop codons and stem-loop structures in mRNA.

13.7 Translation in Bacteria

After reading this section, you should be able to:

a. Relate the general structure of a tRNA molecule to its role in amino acid activation and translation
b. Summarize the formation of a translation initiation complex
c. Describe the structure of bacterial ribosomes
d. State the initiator tRNA used by bacteria
e. Outline the events that occur at the A, P, and E sites of the bacterial ribosome during the elongation phase of translation

Translation involves decoding mRNA and covalently linking amino acids together to form a polypeptide; this occurs at the ribosome. Translation begins when a ribosome binds mRNA and is positioned so that translation will yield the correct amino acid sequence in the polypeptide chain. Transfer RNA molecules carry amino acids to the ribosome so that they can be added to the polypeptide chain as the ribosome moves down the mRNA molecule. Just as DNA and RNA synthesis proceeds in one direction, so too does protein synthesis. Polypeptide synthesis begins with the amino acid at the end of the chain with a free amino group (the N-terminal) and moves in the C-terminal direction. Thus translation is said to occur in the amino terminus to carboxyl terminus direction. Protein synthesis is accurate and very rapid. In *E. coli,* synthesis occurs at a rate of at least 900 amino acids added per minute.

Cells that grow quickly must use each mRNA with great efficiency to synthesize proteins at a sufficiently rapid rate. To achieve rapid rates of protein synthesis, mRNAs often are simultaneously complexed with several ribosomes, each ribosome reading the mRNA message and synthesizing a polypeptide. At maximal rates of mRNA use, there may be a ribosome every 80 nucleotides along the mRNA or as many as 20 ribosomes simultaneously reading an mRNA that codes for a 50,000 dalton polypeptide. A complex of mRNA with several ribosomes is called a **polyribosome** or polysome (**figure 13.33**). Polysomes are present in all organisms. Bacteria can further increase the efficiency of gene expression by coupling transcription and translation (figure 13.33*b*). While RNA polymerase is synthesizing an mRNA, ribosomes can already be attached to the mRNA so that transcription and translation occur simultaneously. Recent work

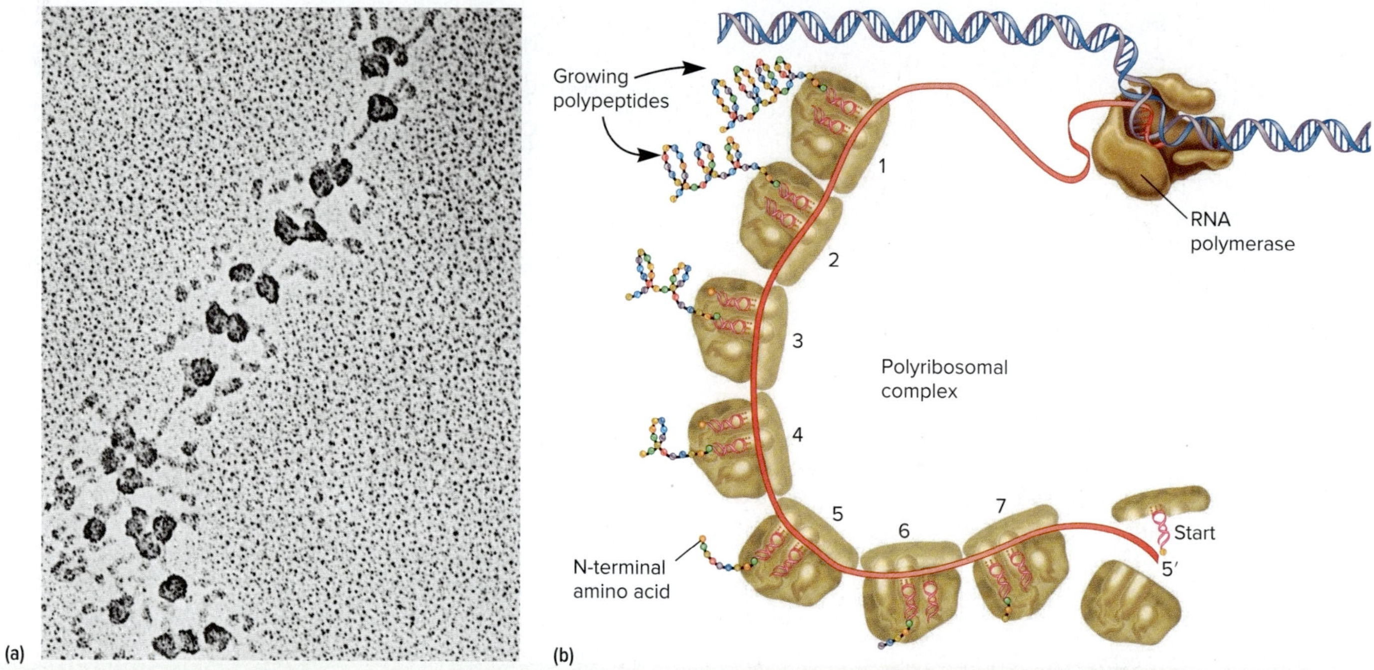

Figure 13.33 Coupled Transcription and Translation in Bacteria. (a) A transmission electron micrograph showing a polyribosome. (b) A schematic representation of coupled transcription and translation. As the DNA is transcribed, ribosomes bind the free 5′ end of the mRNA. Thus translation is started before transcription is completed. Note that there are multiple ribosomes bound to the mRNA, forming a polyribosome. The ribosomes are shown at different points in the translation process. Ribosomes 1, 2, 5, and 7 have completed the transpeptidation reaction, but translocation has not yet occurred. Ribosomes 3 and 4 have an A site containing an incoming aminoacyl-tRNA. Transpeptidation has not occurred. Ribosome 6 shows elongation upon completion of both transpeptidation and translocation. The tRNA bearing the growing polypeptide is in the P site and the empty tRNA is in the E site.
(a) ©Steven McKnight and Oscar L. Miller/University of Virginia

MICRO INQUIRY *Why is simultaneous transcription and translation impossible in eukaryotes?*

has demonstrated an alignment between the RNA exit tunnel of RNA polymerase and the small ribosomal subunit that directs the newly formed mRNA for efficient translation initiation. Coupled transcription and translation is possible in bacterial cells because a nuclear envelope does not separate the translation machinery from DNA, as it does in eukaryotes.

Amino Acid Activation: Attachment of an Amino Acid to a Transfer RNA

For translation to occur, a ready supply of tRNA molecules bearing the correct amino acid must be available. Thus a preparatory step for protein synthesis is **amino acid activation,** the process in which amino acids are attached to tRNA molecules. Before we discuss this process, we need to examine the structure of tRNA molecules.

Transfer RNA molecules are about 70 to 95 nucleotides long and possess several characteristic structural features. These features become apparent when the tRNA is folded so that base pairing within the tRNA strand is maximized. When represented two dimensionally, this base pairing causes the tRNA to assume a cloverleaf conformation (**figure 13.34*a***). However, the three-dimensional structure looks like the letter L (figure 13.34*b*). One important feature of tRNAs is the acceptor stem, which holds the activated amino acid. The 3′ end of all tRNAs has the same CCA sequence, and in all cases the amino acid is attached to the 3′-hydroxyl of the adenosine nucleotide. Another important feature of a tRNA is the **anticodon.** The anticodon is complementary to an mRNA codon and is located on the anticodon arm.

Enzymes called **aminoacyl-tRNA synthetases** catalyze amino acid activation (**figure 13.35**). As is true of DNA and RNA synthesis, the reaction is driven to completion when ATP is hydrolyzed to release pyrophosphate. The amino acid is attached to the tRNA by a high-energy bond. The storage of energy in this bond provides the fuel needed to generate the peptide bond when the amino acid is added to the growing peptide chain.

There are at least 20 aminoacyl-tRNA synthetases, each specific for a single amino acid and its tRNAs (cognate tRNAs). It is critical that each tRNA attach the corresponding amino acid because if an incorrect amino acid is attached to a tRNA, it will be incorporated into a polypeptide in place of the correct amino acid. The protein synthetic machinery recognizes only the anticodon of the

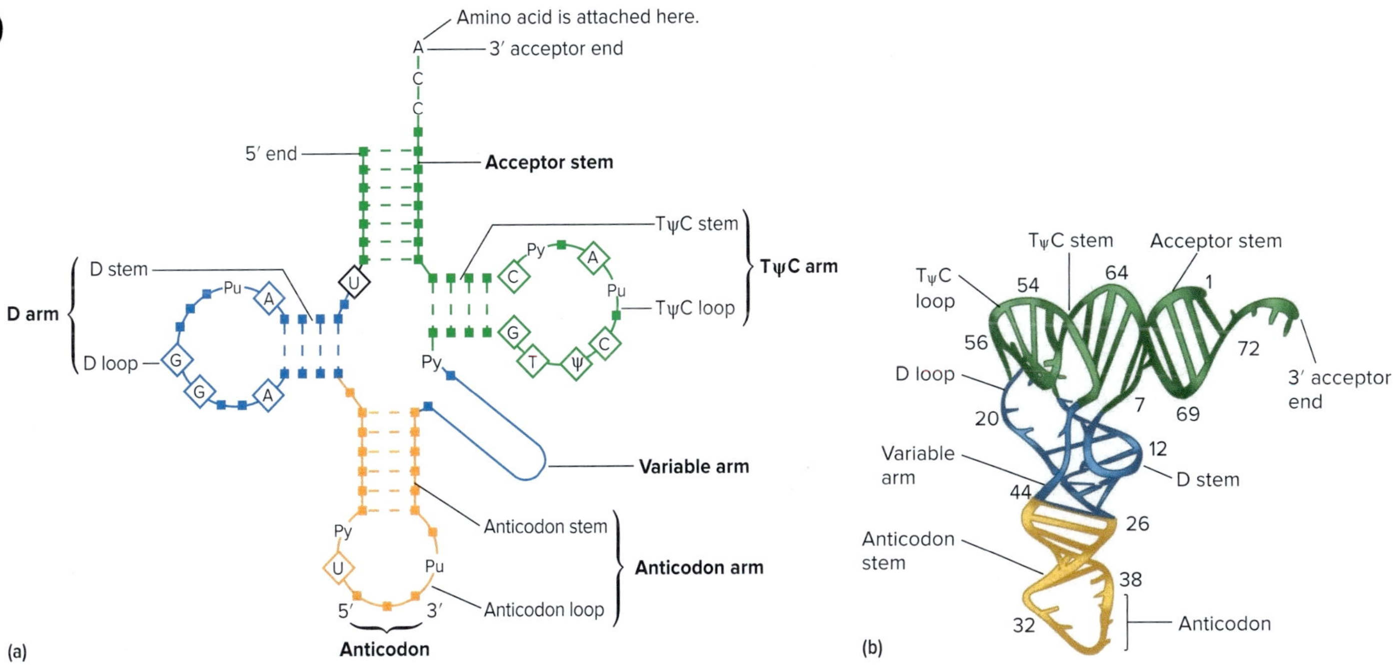

Figure 13.34 tRNA Structure. (a) The two-dimensional cloverleaf structure for tRNA. In addition to the anticodon arm, three other arms are readily observed: the D or DHU arm (blue), the T or TψC arm (green), and the variable arm (blue). The D and T arms are named because of the presence of unusual nucleotides. The variable arm is of different lengths depending on the tRNA; the other arms are fairly constant in size. Bases found in all tRNAs are in diamonds; purine and pyrimidine positions in all tRNAs are labeled Pu and Py, respectively. ψ represents pseudouridine, an isomer of uridine. (b) The three-dimensional structure of tRNA.

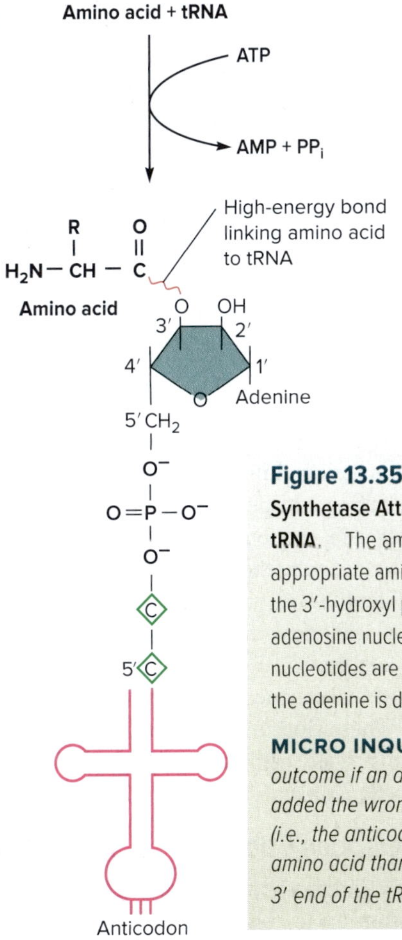

Figure 13.35 Aminoacyl-tRNA Synthetase Attaches an Amino Acid to a tRNA. The amino acid is attached by the appropriate aminoacyl-tRNA synthetase to the 3′-hydroxyl provided by the terminal adenosine nucleoside. The two cytosine nucleotides are indicated in diamonds and the adenine is drawn as atoms.

MICRO INQUIRY *What would be the outcome if an aminoacyl-tRNA synthetase added the wrong amino acid to a tRNA (i.e., the anticodon specified a different amino acid than that added to the 3′ end of the tRNA)?*

aminoacyl-tRNA and cannot tell whether the correct amino acid is attached. Some aminoacyl-tRNA synthetases proofread just like DNA polymerases do. If the wrong amino acid is attached to tRNA, the enzyme hydrolyzes the amino acid from the tRNA, rather than release the incorrect product. *Aminoacyl-tRNA Structure*

Ribosomes Have Three tRNA Binding Sites

Protein synthesis takes place on ribosomes that serve as workbenches, with mRNA acting as the blueprint. Recall that ribosomes are formed from two subunits, the large subunit and the small subunit, and each contains one or more rRNA molecules and numerous proteins. A bacterial ribosome and its components are shown in **figure 13.36**. The ribosome can be divided into two functional domains, the translational domain and the exit domain. Both subunits contribute to the formation of the translational domain, which interacts with tRNAs and is responsible for forming peptide bonds. The exit domain is located solely in the large subunit. Three sites are found within the translational domain for binding tRNAs: A, P, and E sites. The **A** (**aminoacyl** or **acceptor**) site receives tRNAs carrying an amino acid to be added to the protein being synthesized. The **P** (**peptidyl** or **donor**) site holds a tRNA attached to the growing polypeptide. The **E** (**exit**) site is the location from which empty tRNAs leave the ribosome. The growing peptide chain emerges from the large subunit at the exit domain.

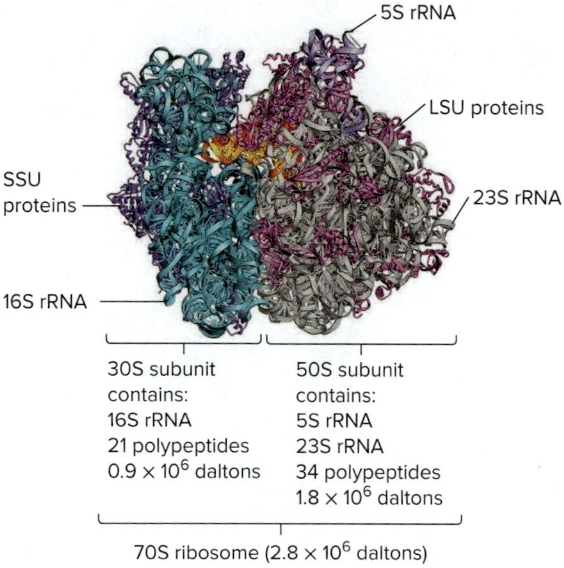

Figure 13.36 Bacterial Ribosome Structure. The *Thermus thermophilus* 70S ribosome viewed with the 30S subunit (SSU) on the left and the 50S subunit (LSU) on the right. The A site tRNA (gold) is visible in the interface cavity. 16S rRNA, cyan; 23S rRNA, gray; 5S rRNA, light blue; 30S proteins, dark blue; 50S proteins, magenta.

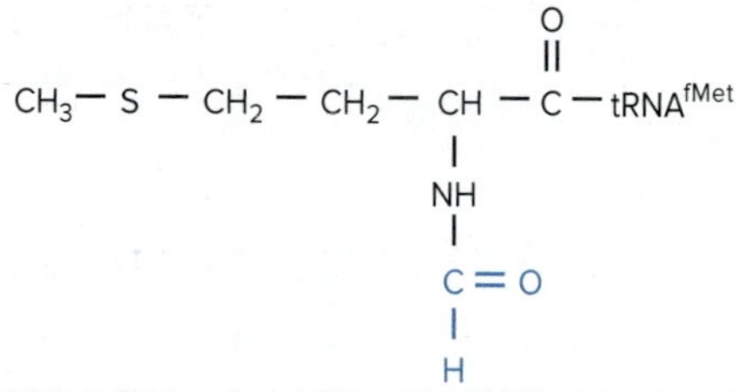

Figure 13.37 *N*-formylmethionyl-tRNA^{fMet} Is the Initiator tRNA Used by Bacteria. The formyl group is in color. Archaea and eukaryotes use methionyl-tRNA for initiation.

MICRO INQUIRY *Why would it be impossible for fMet-tRNA to initiate peptide bond formation with another amino acid? (Hint: Examine figure 13.40 closely.)*

Ribosomal RNA is thought to have three roles. (1) All three rRNA molecules contribute to ribosome structure. (2) The 16S rRNA of the 30S subunit is needed for initiation of protein synthesis because its 3′ end binds to a site on the leader of the mRNA called the Shine-Dalgarno sequence; thus the Shine-Dalgarno sequence is part of the **ribosome-binding site (RBS)**. This positions the mRNA on the ribosome. The 16S rRNA also binds a protein needed to initiate translation (initiation factor 3) and the 3′ CCA end of aminoacyl-tRNA. (3) The 23S rRNA is a ribozyme that catalyzes peptide bond formation.

Protein Synthesis Begins with Formation of the 70S Initiation Complex

Like transcription and DNA replication, protein synthesis is divided into three stages: initiation, elongation, and termination. Bacteria begin protein synthesis with a modified aminoacyl-tRNA, *N*-formylmethionyl-tRNA^{fMet} (fMet-tRNA), which is called the **initiator tRNA** and is coded for by the start codon AUG (**figure 13.37**). The amino acid of the initiator tRNA has a formyl group covalently bound to the amino group and can be used only for initiation because of the presence of the formyl group. When methionine is to be added to a growing polypeptide chain (i.e., at an AUG codon in the middle of the mRNA), a normal methionyl-tRNA^{Met} is employed. Although bacteria start protein synthesis with *N*-formylmethionine, the formyl group is not retained but is hydrolytically removed. In fact, one to three amino acids may be removed from the amino terminus of the polypeptide after synthesis.

Protein synthesis in bacteria begins with formation of the 30S initiation complex, consisting of the initiator tRNA, the mRNA to be translated, and the 30S ribosomal subunit; two **initiation factors** (IF-1 and IF-2) are involved (**figure 13.38**). Positioning of the initiator fMet-tRNA on the mRNA is crucial for proper translation of the mRNA. This is accomplished with the help of the 16S rRNA within the 30S subunit, which is complementary to and binds the Shine-Dalgarno sequence in the leader sequence of the mRNA. By aligning the Shine-Dalgarno sequence with the 16S rRNA, the start codon (AUG or sometimes GUG) specifically binds with the fMet-tRNA anticodon. This ensures that the start codon will be translated first.

Once the 30S initiation complex is formed, it binds the 50S ribosomal subunit, forming the 70S initiation complex. The fMet-tRNA is positioned at the peptidyl or P site. At this juncture, you may be wondering what kept the 30S and 50S subunits from binding each other earlier in the initiation stage. The answer is the third initiation factor (IF-3), as illustrated in figure 13.38. Also revealed in this figure is the energy cost of initiation. GTP, like ATP, is a high-energy molecule. Hydrolysis of GTP to GDP provides the energy needed to accomplish initiation.

The mechanism of translation initiation we have just described holds for mRNAs having a leader (i.e., a 5′ untranslated region) with a Shine-Dalgarno sequence. However, mRNAs lacking a Shine-Dalgarno sequence in the leader or lacking leaders altogether have been identified. Current evidence suggests that translation initiation of leaderless mRNAs is accomplished by 70S ribosomes. It appears that *E. coli* has a subset of ribosomes that lack the 16S rRNA sequence complementary to the Shine-Dalgarno sequence. These ribosomes may be more prevalent under stress conditions when nuclease activity results in many leaderless mRNAs. *Translation Initiation*

Elongation of the Polypeptide Chain

Every addition of an amino acid to a growing polypeptide chain is the result of an **elongation cycle** composed of three phases:

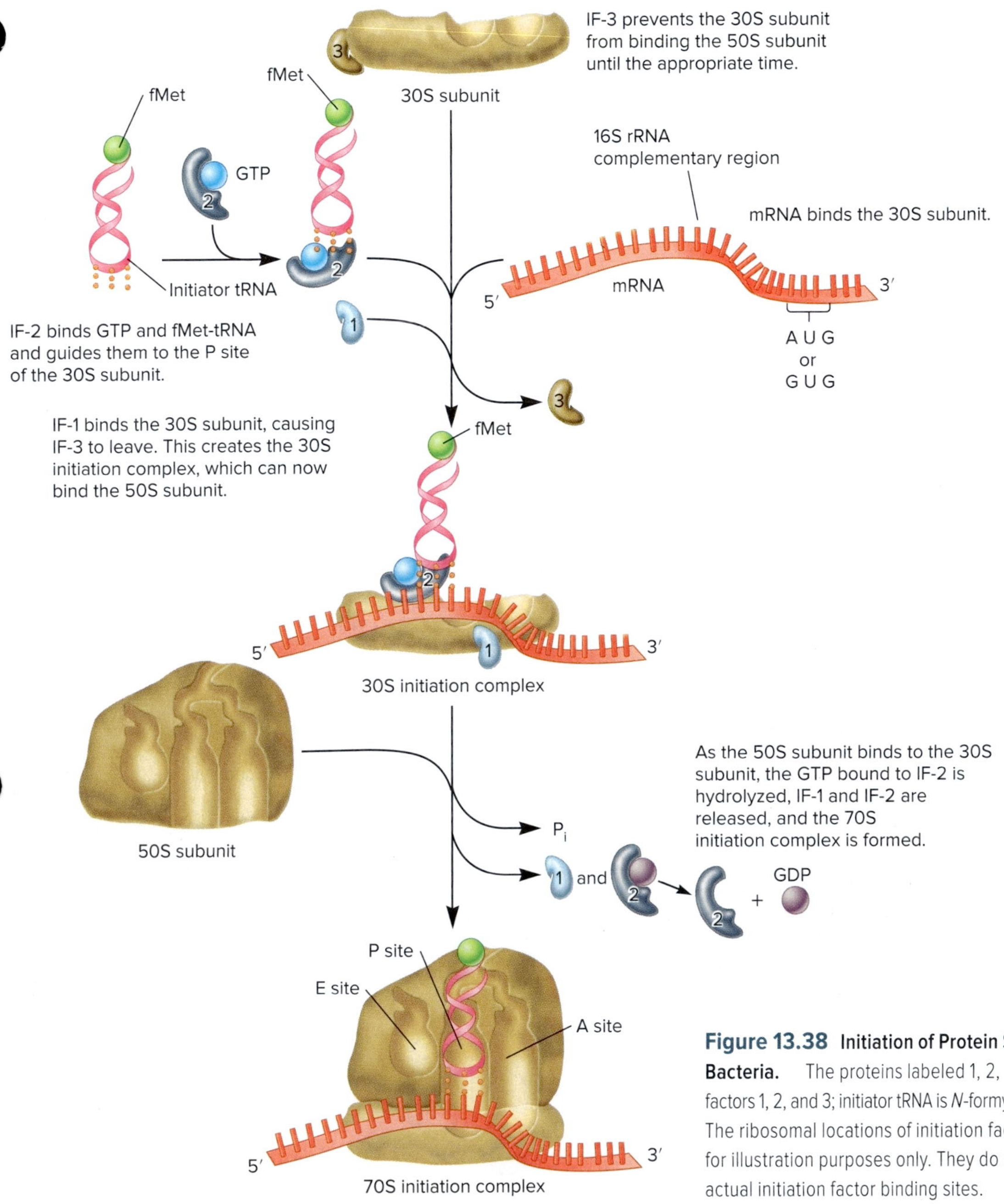

Figure 13.38 Initiation of Protein Synthesis in Bacteria. The proteins labeled 1, 2, and 3 are initiation factors 1, 2, and 3; initiator tRNA is *N*-formylmethionyl-tRNA^{fMet}. The ribosomal locations of initiation factors are depicted for illustration purposes only. They do not represent the actual initiation factor binding sites.

aminoacyl-tRNA binding, the transpeptidation reaction, and translocation. The process is aided by proteins called **elongation factors** (**EF**). In each turn of the cycle, an amino acid corresponding to the proper mRNA codon is added to the C-terminal end of the polypeptide chain as the ribosome moves down the mRNA in the 5′ to 3′ direction.

At the beginning of an elongation cycle, the P site is filled with either the initiator fMet-tRNA or a tRNA bearing a growing polypeptide chain (peptidyl-tRNA), and the A and E sites are empty (**figure 13.39**). Messenger RNA is bound to the ribosome in such a way that the proper codon interacts with the P site tRNA (e.g., an AUG codon for fMet-tRNA). The next codon is located within the A site and is ready to accept an aminoacyl-tRNA.

In the aminoacyl-tRNA binding phase, the first phase of the cycle, the aminoacyl-tRNA corresponding to the codon in the A site is inserted so its anticodon is aligned with the codon on the mRNA. In bacterial cells, this is aided by two elongation factors and requires the expenditure of one GTP (figure 13.39). Once the proper aminoacyl-tRNA is in the A

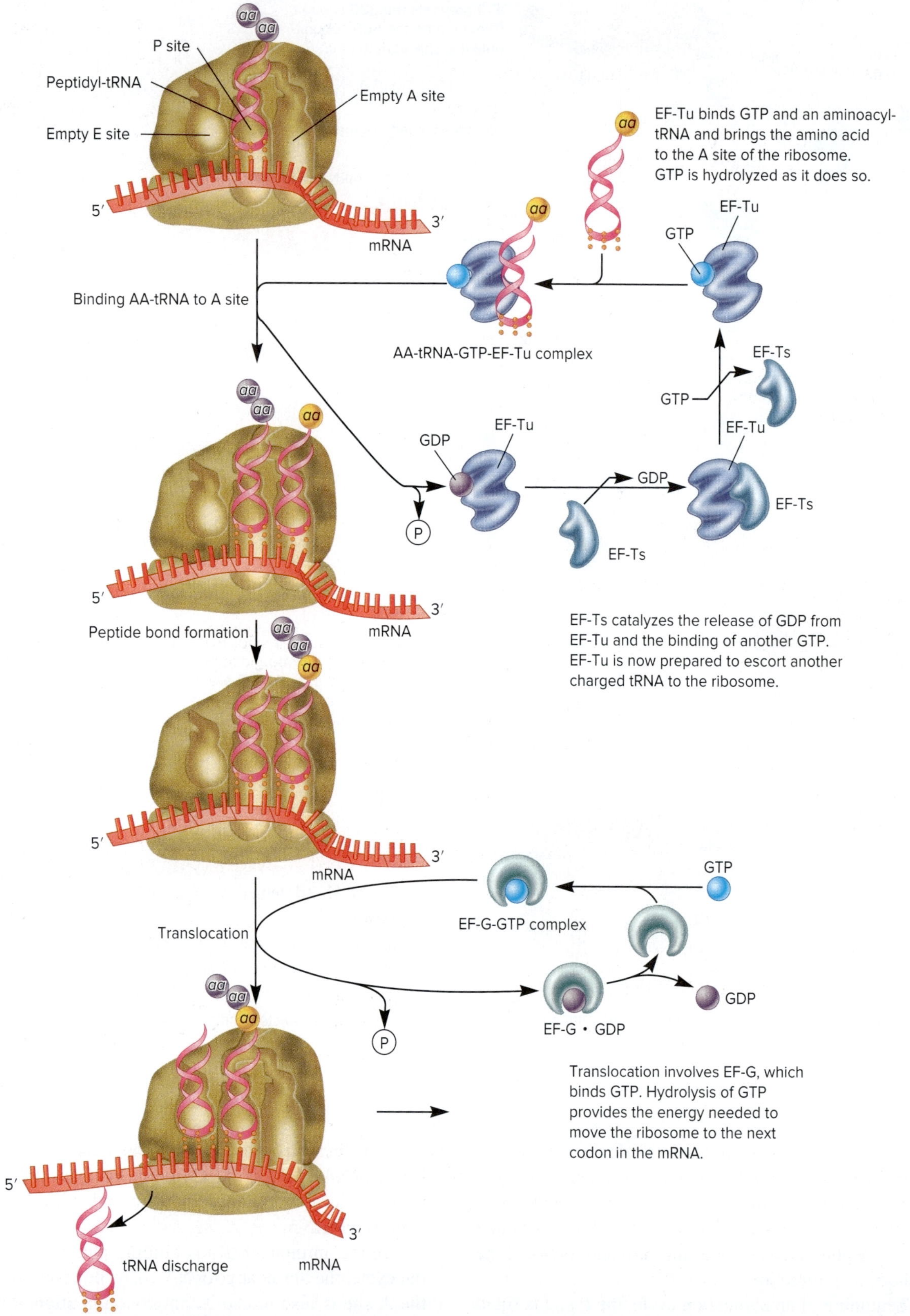

Figure 13.39 The Elongation Cycle of Protein Synthesis. The ribosome possesses three sites, a peptidyl or donor site (P site), an aminoacyl or acceptor site (A site), and an exit site (E site). EF-Tu, EF-Ts, and EF-G are elongation factors. The arrow to the right of the ribosome in the translocation step shows the direction of ribosome movement.

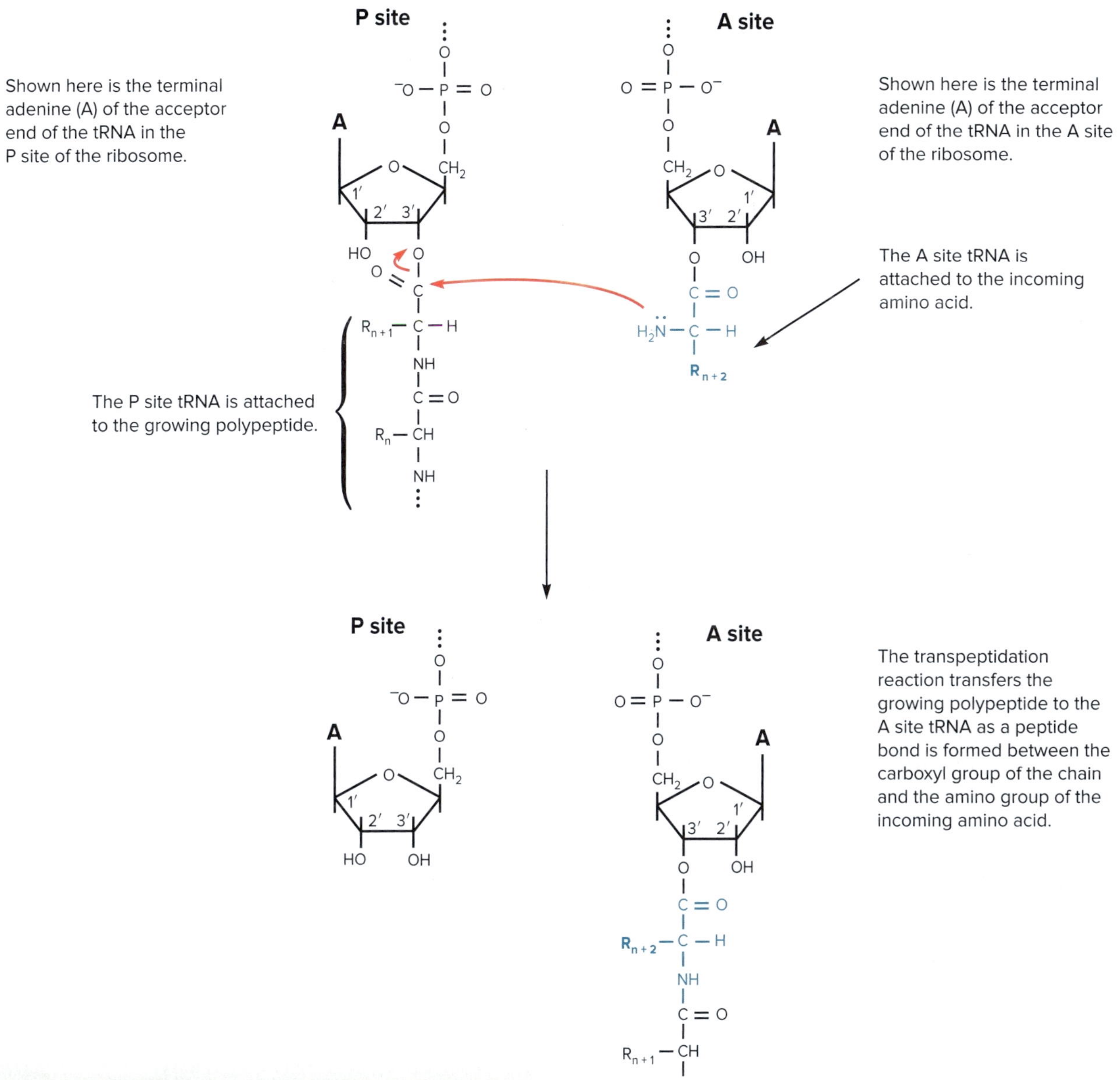

Figure 13.40 Transpeptidation. The peptidyl transferase reaction catalyzed by the 23S rRNA ribozyme. The peptide grows by one amino acid and is transferred to the A site.

MICRO INQUIRY *What provides the energy to fuel this reaction?*

site, the second phase of the elongation cycle, the transpeptidation reaction, occurs (figure 13.39 and **figure 13.40**). Transpeptidation is catalyzed by the **peptidyl transferase** activity of the 23S rRNA ribozyme, which is part of the 50S ribosomal subunit. In this reaction, the amino group of the A site amino acid reacts with the carboxyl group of the C-terminal amino acid on the P site tRNA (figure 13.40). This results in the transfer of the peptide chain from the tRNA in the P site to the tRNA in the A site, as a peptide bond is formed between the peptide chain and the incoming amino acid. No extra energy source is required for peptide bond formation because the bond linking an amino acid to tRNA is high in energy (figure 13.35).

The final phase in the elongation cycle is **translocation.** Three things happen simultaneously: (1) the peptidyl-tRNA moves from the A site to the P site; (2) the ribosome moves one codon along mRNA so that a new codon is positioned in the A site; and (3) the empty tRNA moves from the P site to the E site and subsequently leaves the ribosome. Translocation involves rotations of the 30S and 50S subunits relative to each other. In addition, the head portion of the 30S subunit swivels. These changes in ribosome structure move the tRNAs into their new locations; the codon-anticodon interactions between the tRNAs and the mRNA move the mRNA as the tRNAs move. One elongation factor participates and one GTP is hydrolyzed during this intricate process. *Translation Elongation*

Insertion of Selenocysteine and Pyrrolysine

Insertion of the unusual amino acids selenocysteine and pyrrolysine during translation occurs by two distinctive mechanisms. Selenocysteine is synthesized from serine after it has been attached to certain tRNAs. The enzyme catalyzing the conversion is selenocysteine synthase. Once formed, the amino acid is recognized by a specific elongation factor and is incorporated when a UGA stop codon is encountered in association with nucleotide sequences called cis-acting *se*lenocysteine *i*nsertion *s*equence elements (SECIS). In bacteria, SECIS are found immediately after the UGA stop codon.

Pyrrolysine insertion differs from that of selenocysteine in several ways. Pyrrolysine is synthesized from lysine before being attached to a tRNA. Organisms that use pyrrolysine make an unusual tRNA with a CUA anticodon; the pyrrolysine is attached by a specific aminoacyl-tRNA synthetase. Pyrrolysine is inserted at UAG stop codons located near a sequence element called *py*rro*l*ysine *i*nsertion *s*equence (PYLIS). Both SECIS and PYLIS form stem-loop structures that prevent cessation of translation.

Protein Synthesis Ends When the Ribosome Reaches a Stop Codon

Protein synthesis stops when the ribosome reaches one of three stop codons: UAA, UAG, and UGA (**figure 13.41**). The stop codon is found on the mRNA immediately before the trailer. Three release factors (RF-1, RF-2, and RF-3) aid the ribosome in recognizing these codons. Because there is no cognate tRNA for a stop codon, the ribosome halts. Peptidyl transferase hydrolyzes the bond linking the polypeptide to the tRNA in the P site, and the polypeptide and the empty tRNA are released. Ribosome recycling factor, a protein that is a structural mimic of tRNA, is responsible for disassembling the translational complex. The mRNA is released and the two ribosomal subunits separate. IF-3 binds the 30S subunit, which prepares it for the next round of protein synthesis. *Translation Termination*

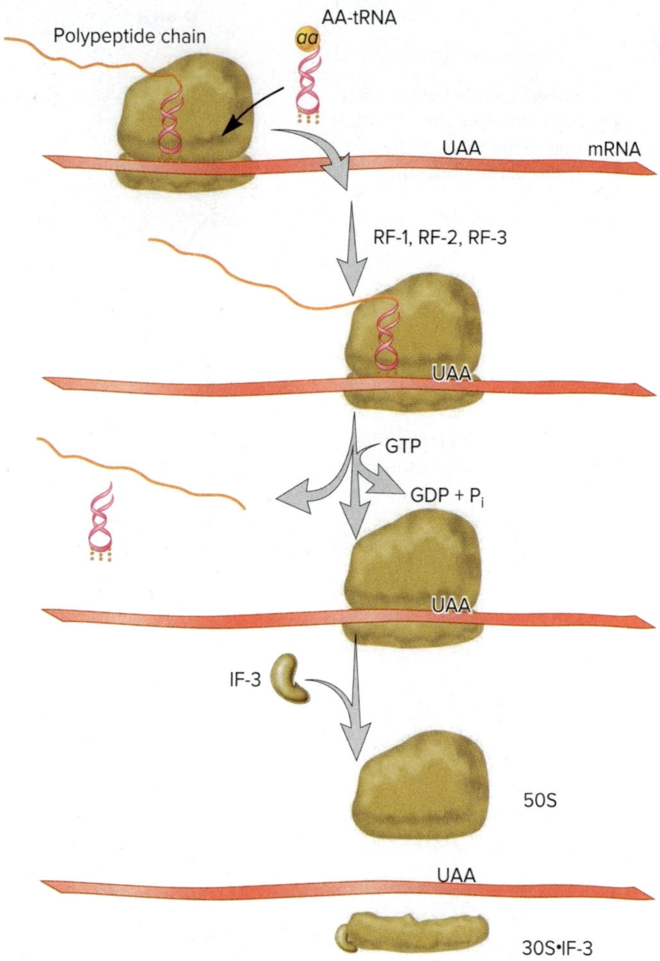

Figure 13.41 Termination of Protein Synthesis in Bacteria. Although three different nonsense codons can terminate chain elongation, UAA is most often used. Three release factors (RFs) assist the ribosome in recognizing nonsense codons and terminating translation and GTP is hydrolyzed. Binding of IF-3 to the 30S subunit prepares it for the next round of translation initiation. Transfer RNAs are in pink.

Ensuring Accuracy During Translation

Protein synthesis is a very expensive process. Two ATP high-energy bonds are required for amino acid activation, initiation consumes one GTP, two GTP molecules are used during each elongation cycle, and another GTP is hydrolyzed when protein synthesis terminates (figures 13.35, 13.38, 13.39, and 13.41). Presumably this large energy expenditure is required to ensure the fidelity of protein synthesis. Fidelity is assessed both before and after formation of the peptide bond. When an aminoacyl-tRNA enters the A site, correct pairing of the anticodon and codon causes conformational changes in components of the ribosome such that the aminoacyl-tRNA is "locked into place" in a manner that facilitates peptide bond formation. These conformational changes do not occur for an incorrect aminoacyl-tRNA and the tRNA is ejected. However, on rare occasions the incorrect aminoacyl-tRNA is selected and a peptide bond is formed between the growing polypeptide and the wrong amino acid. The presence of an incorrect amino acid is recognized by release factors. This leads to hydrolysis of the aberrant polypeptide from the tRNA, its release from the ribosome, and termination of translation.

Comprehension Check

1. In which direction are polypeptides synthesized? What is a polyribosome and why is it useful?
2. Briefly describe the structure of transfer RNA and relate this to its function. How are amino acids activated for protein synthesis, and why is the specificity of the aminoacyl-tRNA synthetase reaction so important?
3. What are the translational and exit domains of the ribosome? What roles do ribosomal RNAs have?
4. Describe the structure and function of the following: fMet-tRNA, start codon, initiation factors, elongation cycle, elongation factors, peptidyl and aminoacyl sites, transpeptidation reaction, peptidyl transferase, translocation, stop codon, and release factors.
5. How many ATP and GTP molecules would be hydrolyzed in the synthesis of a 125 amino acid protein? Explain why this is a good argument for careful regulation of gene expression (especially considering that most proteins are larger than 125 amino acids).

13.8 Protein Maturation and Secretion

After reading this section, you should be able to:

a. Describe the role of protein splicing in protein maturation
b. Discuss the role of molecular chaperones in protein folding, and list some important examples of chaperones
c. Distinguish protein translocation from protein secretion
d. List bacterial translocation and secretion systems, and indicate whether they function in Gram-positive, Gram-negative, or both types of bacteria
e. Describe the structure and function of outer membrane vesicles, and contrast them with secretion systems

For a polypeptide to assume its cellular function, more is required than simply having the correct sequence of amino acids linked together on the ribosome. Amino acids must be modified, and proteins must be properly folded and in some cases associated with other protein subunits to generate a functional enzyme (e.g., DNA and RNA polymerases are multimeric proteins). In addition, proteins must be delivered to the proper subcellular or extracellular site. Because these three events—maturation, folding, and secretion—are often linked in protein complexes, we discuss them together.

Modifications to the N-Terminus of a Polypeptide

The information for protein maturation is encoded in the protein's primary and secondary structures. Decoding that information occurs as the protein emerges from the ribosome. Because protein synthesis is initiated with N-formylmethionine, the N-terminus of a polypeptide matures with the removal of the N-formyl group. Some proteins require the additional removal of the methionine, while others remove several additional N-terminal amino acids as well. This is specified by the identity of the second amino acid in the polypeptide. The enzymes that catalyze these reactions associate with the ribosome near the polypeptide exit tunnel in the large subunit. These reactions occur **cotranslationally,** i.e., during protein synthesis.

Molecular Chaperones: Proteins That Help Proteins Fold

Polypeptides begin to assume their secondary structures, α-helices, β-strands, and coils, as they pass through the ribosome exit tunnel. Assembly of a functional tertiary structure requires the correct interaction of the secondary structural elements. The linear format of protein synthesis can sometimes result in improper folding or aggregation as polypeptides enter the crowded cytoplasm. Cells use proteins called **chaperones** to suppress incorrect folding and in some cases to correct any misfolding that may have occurred. Chaperones are so critical to protein structure that they are present in all three domains of life.

A chaperone called **trigger factor** (**TF**) associates with the ribosome and successfully folds many cytoplasmic proteins. TF binds to the growing polypeptide chain and is thought to mask hydrophobic regions so they don't interact with each other prematurely or with other proteins. Recall that hydrophobic regions are generally in the protein interior, so the presence of a hydrophobic surface region indicates that the polypeptide may need help to assume its functional shape. TF also acts as an isomerase on the peptide bond that precedes proline residues in the polypeptide. The proline cis/trans isomerization is a critical step in protein folding, and several different chaperones have this enzymatic activity.

About 30% of proteins need additional help folding, either cotranslationally or **posttranslationally.** For these proteins, other chaperones like DnaJ/DnaK/GrpE or GroES/GroEL complete the folding process. These chaperones are cytoplasmic and require ATP. Among its many roles, DnaJ/DnaK/GrpE assists proteins with iron-sulfur centers to insert and chemically coordinate the iron atom. GroES/GroEL forms a cage in which misfolded proteins are uncoiled and refolded, away from cytoplasmic crowding. Large proteins with complex topologies are those most likely to require GroES/GroEL for proper maturation.

Protein Translocation and Secretion in Bacteria

It has been estimated that more than one-third of the proteins synthesized leave the cytoplasm. They can have several destinations including the plasma membrane, the external milieu, or in the case of Gram negatives, the periplasm. Therefore it is not surprising that multiple systems for moving proteins have evolved. Some of these are found in all domains of life, while others are unique to bacterial cells, and still others are restricted to only Gram-negative or Gram-positive bacteria. When proteins are moved from the cytoplasm to or across the plasma membrane, the movement is called **translocation.** Protein **secretion** refers to the movement of proteins from the cytoplasm to the external environment. All protein translocation and secretion systems described here require the expenditure of energy in the form of ATP or the proton motive force. In addition, the protein complexes responsible for movement across the plasma membrane are localized to functional membrane microdomains. ◀ *Bacterial plasma membranes are dynamic (section 3.3); ATP (section 10.2); Electron transport and oxidative phosphorylation (step 3) generate the most ATP (section 11.6)*

The variations in cell envelope structure pose different challenges for protein secretion. In Gram-positive bacteria, the proteins must be translocated across the plasma membrane. Once across the plasma membrane, the protein either passes through the relatively porous peptidoglycan into the external environment or becomes attached to the peptidoglycan. Gram-negative bacteria also transport proteins across the plasma membrane, but then must move them across the outer membrane.

Translocation Systems Are Common to Both Gram-Positive and Gram-Negative Bacteria

Two translocation systems, the Sec system and the Tat system, are observed in both Gram-negative and Gram-positive bacteria. The **Sec system**, sometimes called the general secretion pathway, is highly conserved, having been identified in all three domains of life (**figure 13.42**). It translocates unfolded proteins across the plasma membrane or integrates them into the membrane itself.

Before proteins can be translocated, they must be sorted and targeted to their ultimate destination. The N-terminal polypeptide region, the first part to exit the ribosome, is termed the **signal peptide** or signal sequence. This region of the protein signals factors associated with the ribosome to direct it to one of several routes. **Signal recognition particle** (**SRP**) is a protein-RNA complex that surveys actively translating ribosomes at the same time as TF. Whereas TF folds and releases cytoplasmic proteins, SRP identifies membrane proteins with extremely hydrophobic signal peptides. Membrane proteins in particular have a transmembrane α-helix that SRP recognizes as it probes the ribosome exit tunnel. SRP escorts the entire ribosome to the membrane.

An alternative route to the membrane is via the SecA protein. SecA recognizes signal peptides that are less hydrophobic than those preferred by SRP. SecA transport to the membrane may occur cotranslationally or posttranslationally. One or more copies of SecA may bind the cargo protein, and SecA may be accompanied by other chaperones.

SecA does not function alone. Three proteins (SecY, SecE, and SecG) form a channel in the membrane. SecA threads the signal peptide and the preprotein passes through the channel (figure 13.42). SecA acts as a motor, using the energy from ATP hydrolysis, and two other proteins (SecDF) use the proton motive force to translocate the preprotein. The SecYEG channel is unique in that it opens in two ways, based on the delivery vehicle and the protein destination. SRP-ribosome complexes align the ribosome exit tunnel with the SecYEG channel, and SecYEG opens laterally for membrane proteins to be inserted. This process is similar to the cotranslational translocation that occurs in the eukaryotic Sec system that moves proteins across the membranes of the endoplasmic reticulum. ▶| *Protein localization and secretion in eukaryotes (section 15.4)*

The Tat system for protein secretion (**figure 13.43**) is also widespread in both Gram-positive and Gram-negative bacteria, although it translocates only a few dozen proteins. Tat is an acronym for *t*win-*a*rginine *t*ranslocase, referring to two consecutive arginines, a distinctive feature in the signal peptide of proteins translocated by this system. Tat-secreted proteins must completely fold in the cytoplasm; thus unlike the Sec system, the Tat system secretes only folded proteins. Some proteins have dedicated chaperones to insert cofactors and assist in folding. A protein docking complex recognizes the

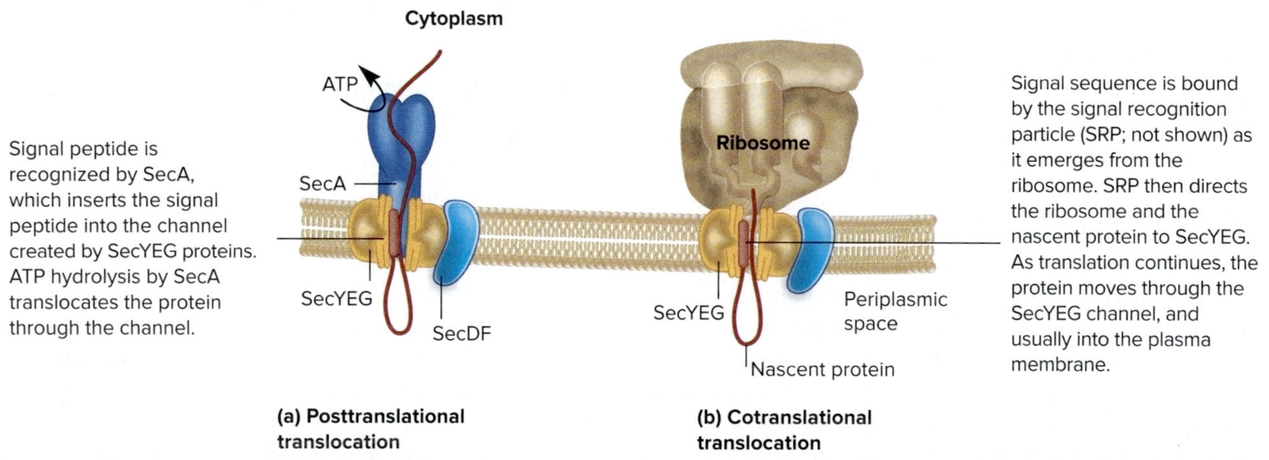

Figure 13.42 Posttranslational and Cotranslational Translocation by the Sec System Move Unfolded Proteins Across or Into the Plasma Membrane. (a) Posttranslational translocation uses chaperones to keep the protein unfolded until it is recognized by SecA. Here we show SecA existing as a dimer. There is also evidence that a monomer of SecA is sufficient for translocation. (b) In cotranslational translocation the protein has not had a chance to fold before it is bound by SRP and delivered, along with the ribosome, to SecYEG.

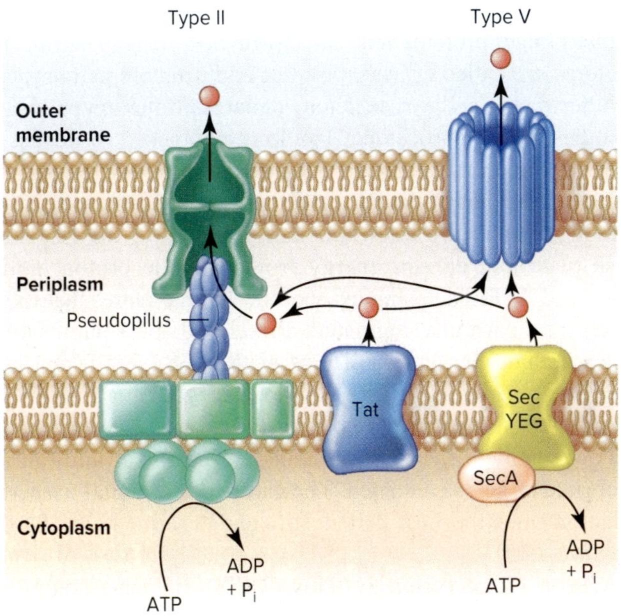

Figure 13.43 Two-Step Secretion Systems of Gram-Negative Bacteria. The Sec and Tat systems deliver proteins from the cytoplasm to the periplasmic space. Type II and Type V systems complete the secretion process. Each protein follows a specific path through one of the translocation systems and a secretion system.

signal peptide and escorts the protein to the pore complex in the membrane. Secretion by Tat relies on the proton motive force for energy.

Following polypeptide translocation, two additional maturation steps occur. Regardless of the secretion system used, the signal peptide must be removed from the preprotein. **Signal peptidase** is a membrane protein that recognizes the signal peptide and hydrolyzes the peptide bond just after it. Finally, for proteins that require disulfide bonds, the enzyme **protein disulfide isomerase** (**PDI**) oxidizes cysteine residues to form disulfides. Because the cytoplasm is a reducing environment, it does not support this reaction. In Gram-negative bacteria, it occurs in the periplasm. DsbA and DsbB act on proteins as they are released from SecYEG, and DsbA oxidizes cysteines in order from the N-terminus. Not all proteins fold in this pattern, and DsbC acts as both a chaperone and PDI to correct any misformed disulfides. Gram-positive bacteria, in general, do not rely on disulfide bond formation. However, some secreted proteins contain disulfides. Cysteine oxidation is catalyzed by oxidative enzymes, but the specific enzymes vary among Gram-positive organisms and their maturation pathways remain to be determined.

For Gram-positive bacteria, the Sec and Tat systems constitute the complete pathway for protein secretion. After the signal peptidase acts, the mature polypeptide is released to the cell's environment. A few proteins, however, are anchored to the cell surface by attachment to peptidoglycan. These proteins have a second signal peptide termed a **sorting signal** at the C-terminus that directs them to sortase enzymes embedded in the membrane. **Sortase** catalyzes a reaction between the sorting signal and Lipid II, the peptidoglycan precursor. These surface protein–Lipid II molecules are inserted into the growing cell envelope by penicillin-binding proteins. ◀ *Synthesis of peptidoglycan occurs in the cytoplasm, at the plasma membrane, and in the periplasmic space (section 12.4)*

Secretion in Gram-Negative Bacteria Requires Crossing Two Membranes

Proteins produced in Gram-negative bacteria destined for secretion or surface display must pass through the plasma membrane, the periplasmic space, and the outer membrane. Some proteins cross in a two-step process. The first step is translocation to the periplasm by the Sec or Tat system, as described, and the second step is secretion across the outer membrane. In contrast, one-step processes involve structures composed of multiple polypeptides that completely span the periplasm. Secretion systems are numbered (Type I secretion system, Type II secretion system, etc.) in order of discovery. Three of these secretion systems, Types II, V, and IX, comprise the second step in two-step processes; and the remainder, Types I, III, IV, VI, and VII, constitute one-step systems. Although Type VII secretion systems are found in Gram-positive bacteria, we discuss them here because several of these genera have an unusual cell wall structure that includes an outer membrane. Several secretion systems are best known for their roles in pathogenicity, where the proteins that they secrete, termed **effectors** contribute to the mechanisms of host injury.

Two-Step Secretion Systems

Type V secretion systems (**T5SS**) are mechanistically simple in that the protein to be exported passes from the periplasm through a barrel structure in the outer membrane (figure 13.43). The barrel may be either a part of the protein being translocated or a separate polypeptide that is coexpressed and cotranslocated to the periplasm. Chaperones accompany the barrel or barrel domain through the aqueous periplasm to the membrane for insertion and folding. T5SS secrete proteases involved in virulence and adhesins that mediate the attachment of bacterial cells to substrates.

Type II secretion systems (**T2SS**) are found only in *Proteobacteria*. A cell may have multiple T2SSs, as each system is dedicated to one or a few proteins. The T2SS is anchored in both membranes by complexes comprised of many proteins, and the two membrane complexes are connected with a pseudopilus (figure 13.43). Proteins for secretion enter from the periplasm and the pseudopilus acts like a piston to push the proteins through the outer membrane. Typical cargo for a T2SS includes degradative enzymes like proteases, lipases, and cellulases. Synthesis of the T2SS is often initiated by quorum sensing. ◀ *Cell-cell communication within microbial populations (section 7.6)* ▶ *Proteobacteria (chapter 22)*

Type IX secretion systems (**T9SS**) also have a narrow phylogenetic distribution. They are limited to the *Bacteroidetes*, where they are responsible for secreting the gliding motility proteins in *Flavobacterium* and the virulence proteases in the dental pathogen *Porphyromonas*. These proteins are translocated by the Sec system and fold in the periplasm. A C-terminal signal peptide tags the proteins for outer membrane localization or surface display. At the outer membrane, a sortase removes the signal peptide and attaches the protein to the lipopolysaccharide. ◀ *Twitching and gliding motility (section 3.8)*

One-Step Secretion Systems

Type I secretion systems (**T1SS**) are the simplest of the one-step secretion systems, and a bacterial cell may contain many of these systems for the export of different proteins (**figure 13.44*a***). Proteins exported by T1SS include S-layer subunits, degradative enzymes, and proteins important in pathogenesis. Each T1SS has three components: an ABC (*A*TP *b*inding *c*assette) transporter in the plasma membrane (*see figure 3.13*), a membrane fusion protein that bridges the periplasm, and a barrel structure that spans the outer membrane. The ABC component is specific for a single substrate, whereas the outer membrane barrel is common to multiple T1SS. The three components assemble only when the substrate binds to the ABC transporter.

Type III secretion systems (**T3SS**), also called **injectosomes**, are molecular syringes (*see figure 39.9*) used by pathogens to inject proteins into eukaryotic host cells. The injected proteins are called effector proteins and promote pathogenesis by altering the cell cytoskeleton, signal transduction pathways, or other cellular processes. T3SS are composed of dozens of proteins and extend through three membranes: the plasma and outer membranes of the bacterium and the host cell membrane. The T3SS structure is evolutionarily related to the flagellum base plate and derives energy from both the proton motive force and ATP. The complexity of the T3SS dictates that its assembly be carefully regulated. Proteins to be secreted by a T3SS include the distal portions of the T3SS itself, and later the effector proteins targeted to the eukaryotic host. Assembly is initiated in response to both bacterial and host factors, and once assembled, the T3SS remains poised to act until an appropriate target is contacted. The nature of the signal to secrete effectors into the host cell is still undetermined. ▶ *Plague (section 39.2)*

Type IV secretion systems (**T4SS**) are most commonly used for DNA transfer although some are dedicated to the secretion of DNA-protein complexes or proteins. They evolved from the conjugative machinery used to transfer DNA (*see figure 16.20*). Like T3SS, these structures transfer molecules directly to the cytoplasm of another cell. Assembly of the T4SS is believed to occur at the cell poles and includes integral membrane

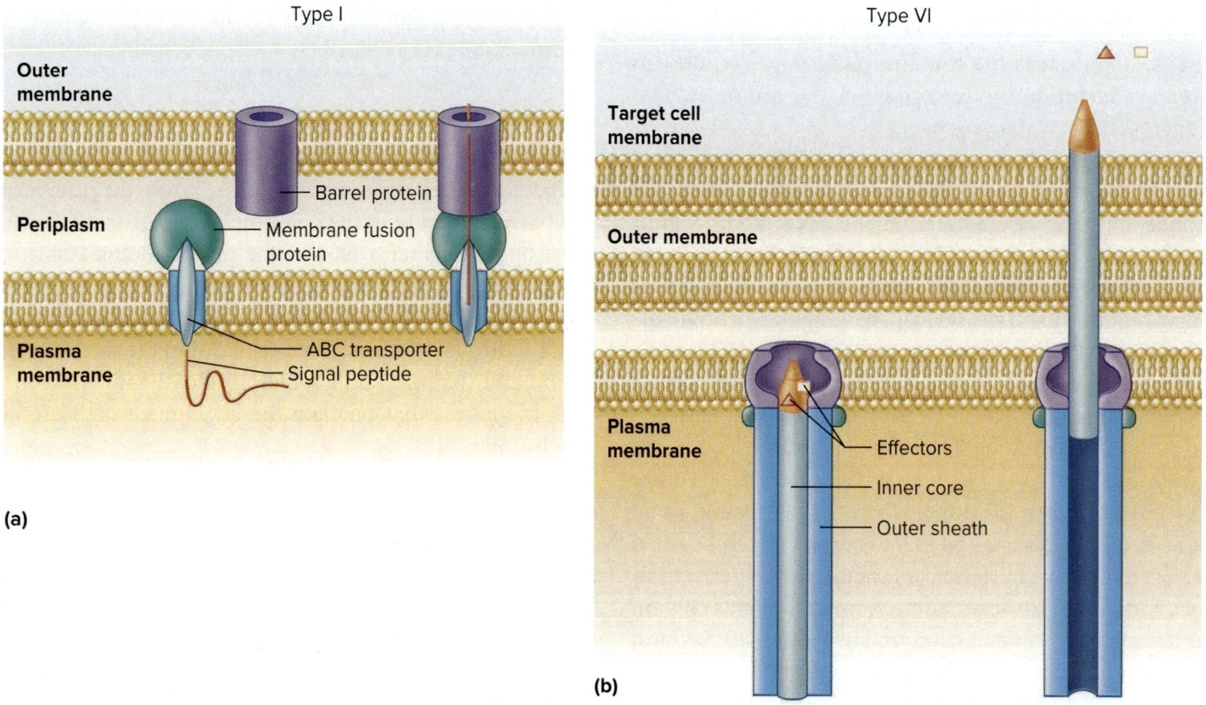

Figure 13.44 Selected One-Step Secretion Systems of Gram-Negative Bacteria. Each system is dedicated to secrete specific proteins. (a) The Type I secretion system assembles when the preprotein signal peptide interacts with the proteins in the plasma membrane. This complex recruits an outer membrane protein to provide a continuous channel for secretion. (b) The Type VI secretion system assembles in the cytoplasm and loads effector proteins to the tip of the needle. The system remains in the cytoplasm, ready to eject the needle into a target cell, upon a signal indicating contact.

proteins in both inner and outer membranes, ATPases in the cytoplasm to provide the energy for secretion, and a pilus. Pili are produced constitutively, but effector secretion requires a host cell for delivery. Understanding how pilus biogenesis and secretion of effector molecules are coupled is an active area of investigation.

Type VI secretion systems (**T6SS**) are contractile weapons used by bacteria to deliver toxins to neighboring cells, both bacterial and eukaryotic, and they have significant homology to the bacteriophage T4-cell penetration apparatus. T6SS are composed of a base plate in the plasma membrane and a membrane spanning complex in the periplasm. Two concentric cylindrical structures, an outer sheath and an inner tube, form in the cytoplasm and then dock onto the membrane complex (figure 13.44*b*). In response to a contact signal, the inner tube is ejected, puncturing the target cell and delivering multiple diverse effector proteins to the target cell's cytoplasm. Related bacteria have immunity proteins and are not affected by the secreted toxins. Thus a population of related cells can dominate a niche, as exemplified by *Vibrio cholerae*. Chitin in aquatic environments is a preferred nutrient, and *V. cholerae* competes with other bacteria and predatory amoeba by delivering toxins via the T6SS. Should humans ingest *V. cholerae* via contaminated water, the same T6SS is used to compete with the gut microbiome to establish the infection known as cholera. ▶| *Bacteriophage T4: a virulent bacteriophage (section 26.2); A functional core microbiome is required for human homeostasis (section 34.3); Cholera (section 39.4)*

Type VII secretion systems (**T7SS**) were initially described in *Mycobacterium* spp. Although these organisms are considered Gram positive, they have a unique cell envelope comprised of a plasma membrane and an outer mycomembrane that features mycolic acids (*see figure 23.7*). Protein secretion across both membranes necessitates a secretion system, as found in Gram-negative bacteria. *M. tuberculosis* uses a T7SS to deliver important virulence factors to host immune cells. Other Gram-positive organisms have related secretion systems. In *Staphylococcus aureus*, the T7SS is required for virulence and the establishment of persistent infections; and in other organisms T7SS have roles in DNA transfer, cell envelope integrity, sporulation, metal uptake, and competition for nutrients. ▶| Mycobacterium tuberculosis *(section 39.1)*

Membrane Vesicles Are an Alternative Secretion Mechanism

A final secretion system has no protein components but is rather composed of nanoparticles derived from cell membranes. Production of extracellular membrane vesicles is well documented in all cellular life forms, but few details are currently known about how these particles form in bacteria. Cell envelope structural differences dictate vesicle production. Outer membrane vesicles (OMVs) bud from the outer membrane of Gram-negative bacteria. OMVs are 20 to 200 nm in diameter and have a lipid bilayer composed of outer membrane with lipopolysaccharide (LPS) and integral outer membrane proteins. The vesicle lumen contains molecules typically found in the periplasm, specifically peptidoglycan, translocated proteins, and nucleic acids. OMV formation is believed to be essential because mutants unable to produce OMVs have not been uncovered. The formation rate and composition of OMVs can vary with environmental conditions, suggesting that the process may be regulated by the cell. Recent studies have demonstrated a difference in the phospholipid content in the outer leaflet of the outer membrane. As this affects local membrane curvature, it could represent a mechanism for OMV genesis.

In Gram-positive bacteria, extracellular membrane vesicles (EVs) pinch off from the plasma membrane, and must cross the thick peptidoglycan cell wall to be released to the environment. In *Staphylococcus aureus*, cytoplasmic surfactant peptides cause local membrane disruption that leads to EV formation. The ability of EVs to cross the cell wall is affected by the extent of peptidoglycan crosslinking. Enzymes called autolysins, which normally function to remodel peptidoglycan during cell division, facilitate transit of EVs to the cell exterior. Other Gram-positive microbes use bacteriophage-derived autolysins for this purpose.

Unlike proteins secreted by secretion systems, materials secreted via OMVs and EVs are more diverse and are more protected from the environment by the lipid bilayer. As a result, the vesicle contents can persist longer in the extracellular milieu than individual macromolecules. These vesicles deliver their contents to either bacterial or eukaryotic cells. OMVs fuse with the outer membrane of Gram-negative cells and release the contents into the periplasm. Vesicle contents are released in the cytoplasm of eukaryotic cells after either membrane fusion or endocytosis.

Molecules transferred via extracellular membrane vesicles are signals that once received may trigger cooperation or conflict. Among bacteria, documented functions of vesicle-delivered molecules include horizontal gene transfer, reinforcement of biofilm extracellular matrix, trafficking quorum sensing molecules, acting as decoy for bacteriophage, and as general modulators of intercellular interactions. For example, a diverse community of microbes may secrete multiple degradative enzymes to digest complex nutrients that any individual is incapable of catabolizing on its own. In addition, OMVs interact with host cells during pathogenesis. Their LPS surface represents a microbe-associated molecular pattern that will be recognized by the innate immune system and elicit a response. ▶| *Recognition of foreignness is key for survival (section 32.6)*

Comprehension Check

1. What are molecular chaperones? Describe their functions.
2. What is a signal peptide? Suggest why a protein's signal peptide is not removed until after the protein is translocated across the plasma membrane.
3. Give the major characteristics and functions of the protein secretion systems described in this section.
4. How does cell envelope structure determine the mechanism for extracellular membrane vesicle formation?

Key Concepts

13.1 Experiments Using Bacteria and Viruses Demonstrated that DNA Is the Genetic Material

- DNA is composed of only four different building blocks (nucleotides). Therefore it was originally thought to be too simple to function as an organism's genetic material.
- The knowledge that DNA is the genetic material for cells came from studies on transformation by Griffith and Avery and from experiments on T2 phage by Hershey and Chase (**figures 13.1–13.3**).

13.2 Nucleic Acid and Protein Structure

- DNA and RNA are polymers of deoxyribonucleotides and ribonucleotides, respectively. DNA differs in composition from RNA in having deoxyribose and thymine, rather than ribose and uracil.
- DNA is double stranded, with complementary AT and GC base pairing between the strands. The strands run antiparallel and are twisted into a right-handed double helix (**figure 13.5**).
- RNA is normally single stranded, although it can coil upon itself and base pair to form hairpin structures.
- Proteins are polymers of amino acids (**figure 13.6**) linked by peptide bonds (**figure 13.7**).

13.3 DNA Replication in Bacteria

- Circular bacterial DNAs have a single origin of replication. They are copied by two replication forks to form a theta-shaped (θ) figure (**figure 13.9**).
- The replisome is a complex of proteins responsible for DNA replication (**table 13.1** and **figures 13.11** and **13.12**).
- DNA polymerase enzymes catalyze the synthesis of DNA in the 5′ to 3′ direction while reading the DNA template in the 3′ to 5′ direction. The double helix is unwound by helicases with the aid of topoisomerases such as DNA gyrase. DNA binding proteins keep the strands separate (**figures 13.10, 13.12,** and **13.13**).
- DNA polymerase III holoenzyme, the replicative DNA polymerase in bacteria, synthesizes a complementary DNA copy beginning with a short RNA primer made by the enzyme primase. The leading strand is replicated continuously, whereas DNA synthesis on the lagging strand is discontinuous and forms Okazaki fragments (**figures 13.13** and **13.14**). DNA polymerase I excises the RNA primers and fills in the resulting gap. DNA ligase then joins the fragments together (**figures 13.14** and **13.15**).
- The linear chromosomes of *Borrelia burgdorferi* have hairpin ends and an origin of replication in the middle of the chromosome. DNA replication proceeds bidirectionally from the origin, but only leading strand synthesis occurs at each fork. The resulting molecule is a dimer that is then cut into the single linear molecules as hairpin ends are formed (**figure 13.18**).

13.4 Bacterial Genes Consist of Coding Regions and Other Sequences Important for Gene Function

- A gene is a nucleic acid sequence that codes for a polypeptide, tRNA, or rRNA.
- The template strand of DNA carries genetic information and directs the synthesis of an RNA transcript.
- A gene also contains a promoter, a coding region, and a terminator; it may have a leader and a trailer (**figure 13.20**).
- The genes for tRNA and rRNA often code for a precursor that is subsequently processed to yield several products (**figure 13.21**).

13.5 Transcription in Bacteria

- Genes of related function are often organized into a transcriptional unit called an operon. A single promoter directs transcription of the operon, yielding a polycistronic mRNA. Each coding region in the polycistronic mRNA has its own start and stop codons and gives rise to a single polypeptide (**figure 13.22**).
- RNA synthesis is catalyzed by enzymes called RNA polymerases. Bacterial RNA polymerases are composed of several protein subunits (**figure 13.23**).
- The transcription cycle (**figure 13.24**) includes initiation, elongation, and termination.
- A protein called sigma factor is a transcription factor that directs RNA polymerase to a promoter during transcription initiation. Sigma recognizes certain sequences in the promoter (**figure 13.25**). Once bound, RNA polymerase unwinds a short stretch of DNA (**figure 13.26**). It then begins moving down the template strand (**figure 13.27**), synthesizing RNA in a 5′ to 3′ direction (elongation). Termination of transcription is brought about by two mechanisms: factor-independent and factor-dependent termination (**figures 13.28** and **13.29**).

13.6 The Genetic Code Consists of Three-Letter "Words"

- Genetic information is carried in the form of 64 nucleotide triplets called codons (**table 13.3**); 61 sense codons direct amino acid incorporation, and three stop codons terminate translation. The code is degenerate (redundant); that is, there is more than one codon for most amino acids.
- Some proteins contain two rare amino acids: selenocysteine and pyrrolysine (**figure 13.32**).

13.7 Translation in Bacteria

- During translation, ribosomes attach to mRNA and synthesize a polypeptide beginning at the N-terminal end. A polysome or polyribosome is a complex of mRNA with several ribosomes (**figure 13.33**).
- Amino acids are activated for protein synthesis by attachment to the 3′ end of transfer RNAs. Activation requires ATP, and the reaction is catalyzed by aminoacyl-tRNA synthetases (**figure 13.35**).

- Ribosomes are large, complex organelles composed of rRNAs and many polypeptides (**figure 13.36**).
- Protein synthesis begins with the formation of the 30S initiation complex made up of the mRNA, initiator fMet-tRNA, two initiation factors, and the 30S subunit. The initiator tRNA is bound to the start codon on the mRNA. The 50S subunit then binds to form the 70S initiation complex (**figures 13.37** and **13.38**).
- In the elongation cycle, the proper aminoacyl-tRNA binds to the A site (**figure 13.39**). Then the transpeptidation reaction is catalyzed by peptidyl transferase (**figure 13.40**). Finally, during translocation, the peptidyl-tRNA moves to the P site and the ribosome moves down the mRNA by one codon. The empty tRNA leaves the ribosome by way of the exit site.
- Protein synthesis stops when a stop codon is reached. Bacteria require three release factors for codon recognition and ribosome dissociation from the mRNA (**figure 13.41**).

13.8 Protein Maturation and Secretion

- Chaperone proteins help ensure that proteins fold properly or are delivered to their destination site.
- Many proteins must be transported across the plasma membrane of cells. The most commonly used mechanisms are the Sec system, which is found in all organisms, and the Tat system (**figure 13.42**).
- Protein secretion in Gram-negative bacteria can be a two-step process using the Sec or Tat system and a secretion system located in the outer membrane. Alternately, secretion can occur in a single step when the cargo is passed directly from the cytoplasm through a single complex to the exterior of the cell (**figures 13.43** and **13.44**). In all cases, energy for secretion derives from the cytoplasm, either in the form of ATP or the proton motive force.
- Extracellular membrane vesicles are small nanoparticles that bud off the bacterial plasma membrane. A diverse set of molecules may be carried in these vesicles, and they can fuse with either bacterial cells or eukaryotic cells.

Active Learning

1. *Streptomyces coelicolor* has a linear chromosome. Interestingly, there are no genes that encode essential proteins near the ends of the chromosome in this bacterium. Why do you think this is the case?
2. You have isolated several *E. coli* mutants:

   Mutant #1 has a point mutation (a single base-pair change) in the −10 region of the promoter of a gene encoding an enzyme needed for synthesis of the amino acid serine.

   Mutant #2 has a mutation in the −35 region in the promoter of the same gene.

   Mutant #3 is a double mutant with mutations in both the −10 and −35 region of the promoter of the same gene.

   Only Mutant #3 is completely unable to make serine. Why do you think this is so?
3. *Desulfovibrio vulgaris* Hildenborough is a sulfate-reducing bacterium whose biofilms inflict corrosion on metals in anaerobic industrial settings. Its biofilm matrix is predominantly protein. Two strains believed to be identical were found to differ dramatically in their ability to form a biofilm. Comparison of the two genomes via whole genome sequencing identified a single nucleotide change as a good candidate to explain the observations. The affected gene encodes an ABC transporter, and the gene is adjacent to genes encoding a membrane fusion protein and an outer membrane pore protein. Suggest an explanation to relate this protein complex to the ability to form a biofilm in *D. vulgaris.*

   Read the original paper: DeLeon, K. B., et al. 2017. Unintended laboratory drive evolution reveals genetic requirements for biofilm formation by *Desulfovibrio vulgaris* Hildenborough. *mBio* 8(5):e01696-1717.
4. In 2017, researchers demonstrated that they could expand the genetic code by creating two novel nucleotides that would base-pair only with one another. This unnatural base pair was inserted in place of a serine codon in the gene that encodes green fluorescent protein. Recall from table 13.3 that serine is specified by six different codons, so repurposing one of these would not significantly affect the cell. What other modifications to the cell's gene expression systems would be required to accommodate this change?

   Read the original paper: Zhang, Y., et al. 2017. A semi-synthetic organism that stores and retrieves increased genetic information. *Nature* 551:644–647.

14

Regulation of Bacterial Cellular Processes

©Tetra Images/Getty Images

Light Show

If you've ever watched a light show, you know how magnificent and complex a series of electrical switches can be. Technicians use a series of regulators to control each light, making it brighten, dim, or pulse rhythmically. Gene expression, too, can be controlled and coordinated by manipulating promoters—the switches—on individual genes. In fact, this can even result in a bacterial light show.

Scientists have created sophisticated sensors by pairing promoters responsive to a specific stimulus with the gene that encodes green fluorescent protein (GFP). GFP is perfectly named: When made by the cell and excited by ultraviolet light, it fluoresces green. Fusing the gene that encodes GFP to a promoter effectively creates a switch to turn the green light on and off. The timing and intensity of the green light emitted are a measure of how the promoter is controlled.

In a remarkable demonstration of promoter specificity, scientists built a novel promoter-gene combination. They fused a promoter responsive to the explosive TNT with the gene for GFP and introduced it to *Escherichia coli*. This *E. coli* biosensor glows green after exposure to TNT. In the field, it is sensitive enough to detect the tiny amount of volatilized TNT that escapes from unexploded land mines buried in sand. This novel application of stimulating a promoter to turn on a light could save life and limb for people living near zones of armed conflict.

Despite the simplicity of this TNT-GFP biosensor, promoters are rarely as simple as an on/off switch. More commonly, cellular components interact with promoters and act as rheostats, or dimmers, to increase or decrease the level of transcription. Gene expression can also be controlled by adjusting the activity of the message, or its translation. In this chapter, we explore mechanisms of gene regulation and how they interact in cellular processes.

Readiness Check:

Based on what you have learned previously, you should be able to:

✓ Describe the mechanisms often used to regulate enzyme activity (section 10.7)

✓ Discuss genome structure and organization of genetic information in bacterial cells (sections 13.3 and 13.4)

✓ Summarize the events that occur during the three phases of transcription and translation (sections 13.5 and 13.7)

✓ Describe the phosphoenolpyruvate: sugar phosphotransferase system (section 3.3)

14.1 Bacteria Use Many Regulatory Options

After reading this section, you should be able to:

a. List when, during the flow of genetic information, bacterial cells can regulate gene expression
b. Speculate why microbial geneticists historically focused almost exclusively on the regulation of transcription initiation

Figure 14.1 summarizes the mechanisms that can be used to regulate the flow of genetic information. Regulation that occurs by controlling the processes of transcription and translation is often called regulation of gene expression. When gene expression is regulated by controlling mRNA synthesis, it is said to be "at the level of transcription" or at the "transcriptional level." Likewise, when gene expression is governed by controlling protein synthesis, it is said to be "at the level of translation" or at the "translational level." Translation yields a protein that may or may not be functional. The activity level of proteins may be altered by posttranslational modification.

Although the overall processes of transcription and translation are similar for all organisms, gene expression is somewhat different for members in each domain. Our focus in this chapter is on well-understood bacterial regulatory processes at the levels of transcription, translation, and posttranslation.

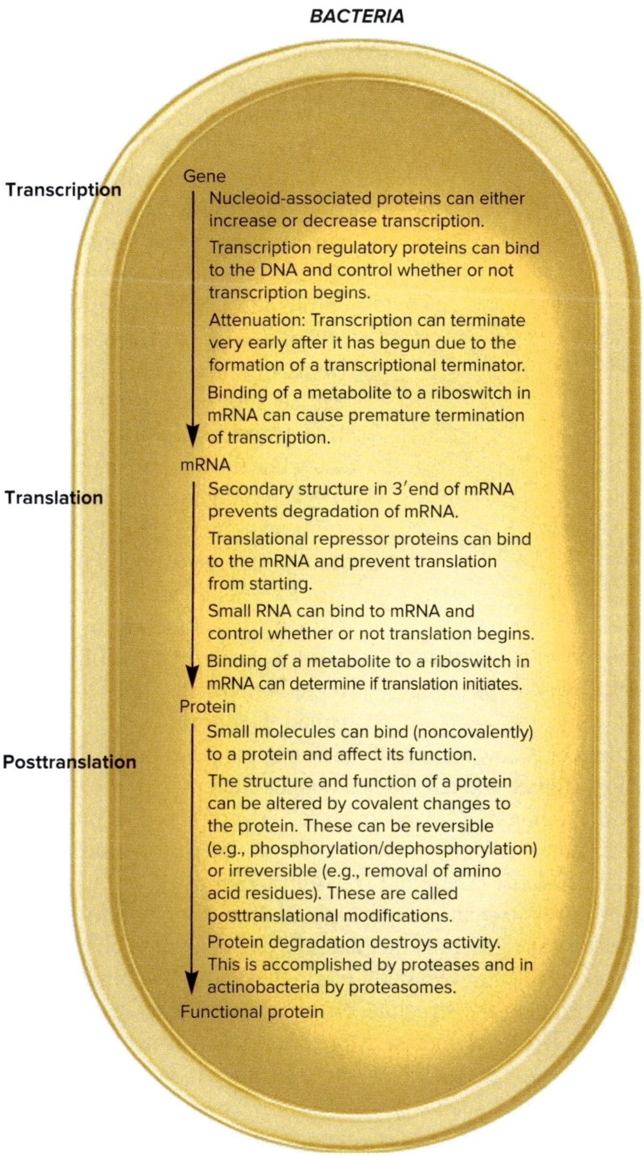

Figure 14.1 Some Common Regulatory Mechanisms in Bacteria.

Regulation at the level of transcription has been the focus of microbial geneticists for many years and is well understood. Transcription can be regulated by governing its initiation or its elongation, and these topics are covered in sections 14.2 and 14.3, respectively. We begin our discussion by introducing two phenomena: induction of enzyme synthesis and repression of enzyme synthesis. Induction and repression provided the first models for gene regulation. These models involve the action of regulatory proteins, and the notion that gene expression is regulated solely by proteins persisted for many years. Eventually it was demonstrated that RNA molecules also have regulatory functions.

14.2 Regulation of Transcription Initiation Saves Considerable Energy and Materials

After reading this section, you should be able to:

a. Compare and contrast housekeeping, constitutive, inducible, and repressible genes
b. Describe the helix-turn-helix motif in DNA-binding proteins
c. Summarize how negative transcriptional control and positive transcriptional control can be used to regulate both inducible and repressible genes
d. Outline the regulatory "decisions" made by cells

Induction and repression are historically important, as they were the first regulatory processes for gene expression to be understood in any detail. In this section, we first describe these phenomena and then examine the underlying regulatory mechanisms.

Induction and Repression of Enzyme Synthesis

Just as an automobile or a refrigerator does not work forever and eventually must be replaced, so too enzymes remain functional only for a certain time. The "old" enzymes are eventually degraded so that amino acids can be recycled and used to synthesize new proteins. However, this valuable process creates a problem for a cell because many enzymes catalyze reactions that are needed almost all the time (e.g., enzymes of the central metabolic pathways). Their functions are often referred to as "housekeeping functions," and the genes that encode them are often called **housekeeping genes.** To maintain the proper amount of housekeeping enzymes, many housekeeping genes are expressed continuously by the cell. Genes expressed continuously are said to be **constitutive genes.** Other enzymes are only needed at certain times and in certain environments. To conserve energy and cellular materials such as amino acids and nucleotides, genes encoding these enzymes are expressed only when needed; therefore their expression is regulated. The β-galactosidase gene is an example of a regulated gene.

The enzyme β-galactosidase catalyzes hydrolysis of the disaccharide sugar lactose to glucose and galactose (**figure 14.2**). When *E. coli* grows with lactose as its only carbon source, each cell contains about 3,000 β-galactosidase molecules, but it has less than three molecules in the absence of lactose. β-galactosidase is an inducible enzyme; that is, its level rises in the presence of a small effector molecule called an **inducer** (in this case, the lactose derivative allolactose). Likewise, the genes that encode inducible enzymes such as β-galactosidase are referred to as **inducible genes.**

β-galactosidase is an enzyme that functions in a catabolic pathway, and many catabolic enzymes are inducible enzymes. On the other hand, the genes for enzymes involved in

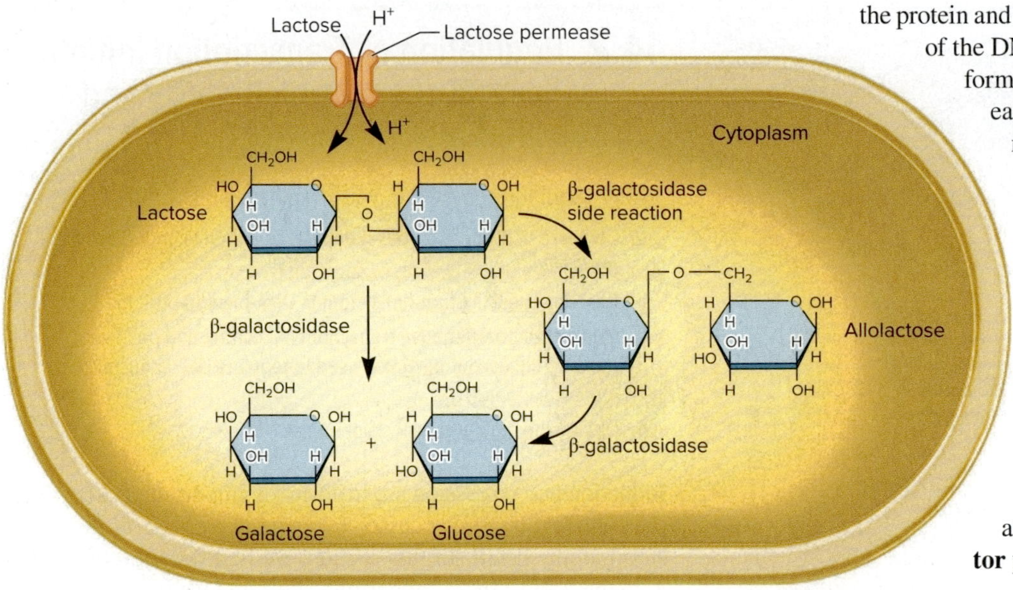

Figure 14.2 The Reactions of β-Galactosidase. The main reaction catalyzed by β-galactosidase is the hydrolysis of lactose, a disaccharide, into the monosaccharides galactose and glucose. The enzyme also catalyzes a minor reaction that converts lactose to allolactose. Allolactose acts as the inducer of β-galactosidase synthesis.

biosynthetic pathways are **repressible genes,** and their products are repressible enzymes. For instance, an amino acid present in the surroundings may inhibit the formation of enzymes responsible for its biosynthesis. This makes sense because the microorganism does not need the biosynthetic enzymes for a particular substance if it is already available. Generally, repressible enzymes are necessary for synthesis and are present unless the end product of their pathway is available. Inducible enzymes, in contrast, are required only when their substrate is available; they are missing in the absence of the inducer.

Regulatory Proteins Often Control Transcription Initiation

Proteins are often used to regulate transcription initiation, and therefore induction and repression. Many of these **transcriptional regulatory proteins** are DNA-binding proteins that form dimers and attach to short, inverted sequences of bases in the DNA called palindromes. Only a small portion of the proteins actually interacts with the palindromes, and these are referred to as DNA-binding domains. The other portion of the protein interacts with an effector molecule, like an inducer. This is an example of allosteric control. Binding of an effector to one domain changes the conformation of the DNA-binding domain, thus affecting its ability to interact with DNA. ◀ *Allosteric regulation (section 10.7)*

A common structure observed in the DNA-binding domain of transcriptional regulatory proteins is called a helix-turn-helix. Helix-turn-helix regions are generally about 20 amino acids in length and are folded into two α-helices (*see figure A1.14*) separated by a β-turn. One of the helices protrudes from the surface of the protein and is often positioned in the major groove of the DNA. Thus when the regulatory protein forms a dimer, two α-helices, one from each protein subunit, interact with the major groove of DNA.

Regulatory proteins can exert either negative or positive control. **Negative transcriptional control** occurs when the binding of the protein to DNA inhibits initiation of transcription. Regulatory proteins that act in this fashion are called **repressor proteins. Positive transcriptional control** occurs when the binding of the protein to DNA promotes transcription initiation. These proteins are called **activator proteins.** *Regulatory Proteins*

Repressor and activator proteins usually function by binding DNA at specific sites. In bacteria, repressor proteins attach to a region called the **operator,** which usually overlaps or is downstream of the promoter (i.e., closer to the coding region) (**figure 14.3*a,b***). When bound, the repressor protein either blocks binding of RNA polymerase to the promoter or prevents its movement. Activator proteins attach to **activator-binding sites** (figure 14.3*c,d*). These are often upstream of the promoter (i.e., farther away from the coding region). Binding of an activator to its regulatory site generally facilitates RNA polymerase binding. This is accomplished in two main ways. In some cases, binding of the activator alters the structure of the promoter, making it a better target for RNA polymerase. More often, the activator protein interacts with RNA polymerase's α or σ subunits to facilitate binding.

Figure 14.3 shows the four basic ways in which the interactions of an effector and a regulatory protein can affect transcription. (1) For negatively controlled inducible genes (e.g., those encoding enzymes needed for catabolism of a sugar), the repressor protein is active and prevents transcription when the substrate of the pathway is not available (figure 14.3*a*). It is inactivated by the inducer (e.g., the substrate of the pathway). (2) For negatively controlled repressible genes (e.g., those encoding enzymes needed for the synthesis of an amino acid), an inactive repressor protein, called the **aporepressor,** is initially synthesized. It is activated by the **corepressor** (figure 14.3*b*). For repressible enzymes that function in a biosynthetic pathway, the corepressor is often the product of the pathway (e.g., an amino acid). (3) The activator of a positively regulated inducible gene is activated by the inducer (figure 14.3*c*), whereas (4) the activator protein of a positively regulated repressible gene is inactivated by an inhibitor (figure 14.3*d*).

Before we continue our discussion, we must consider two general aspects of regulation. The first is that gene expression is rarely an all-or-nothing phenomenon; it is a continuum. Inhibition of transcription usually does not mean that genes are "turned off"

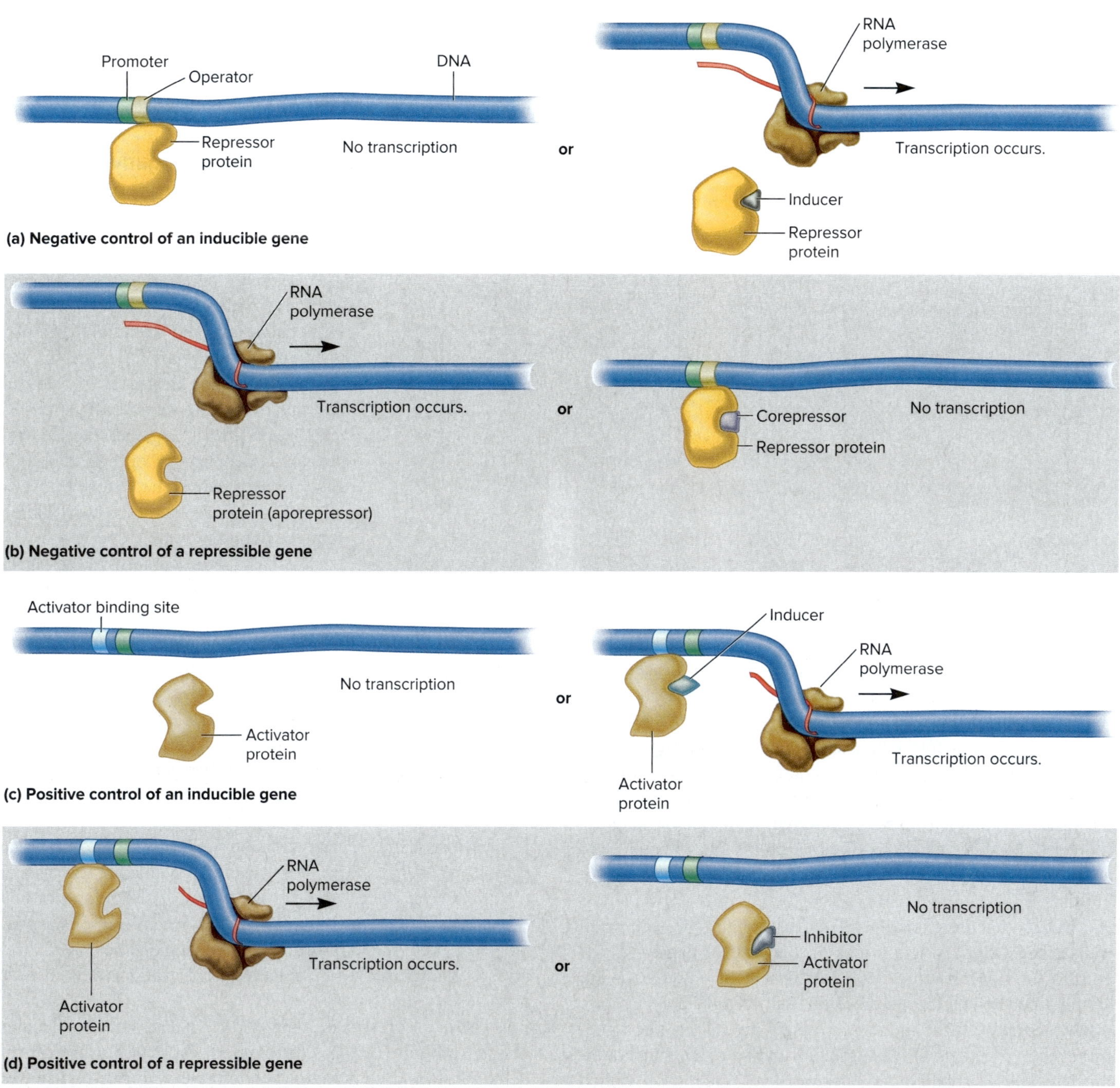

Figure 14.3 Action of Bacterial Regulatory Proteins. (a) In the absence of an inducer, the repressor protein blocks transcription. The presence of an inducer prevents the repressor from binding DNA and transcription occurs. (b) In the absence of a corepressor, the repressor is unable to bind DNA and transcription occurs. When the corepressor binds the repressor, the repressor attaches to DNA and blocks transcription. (c) The activator is only able to bind DNA and activate transcription when it interacts with the inducer. (d) The activator binds DNA and promotes transcription unless the inhibitor is present. When the inhibitor is present, the activator undergoes a conformational change that prevents it from binding DNA; this inhibits transcription.

MICRO INQUIRY *In what way is an inducer molecule that binds a repressor similar to an inhibitor molecule that binds an activator?*

(though this terminology is frequently used). Rather it means the level of mRNA synthesis is decreased significantly and in most cases is occurring at very low levels. In other words, many promoters of regulated genes and operons are considered "leaky" in that there is always some low, basal level of transcription. The second aspect of regulation is the "decision-making" process used by microbial cells. Although cells do not have thought processes, it is convenient to think of regulation in this way. Consider

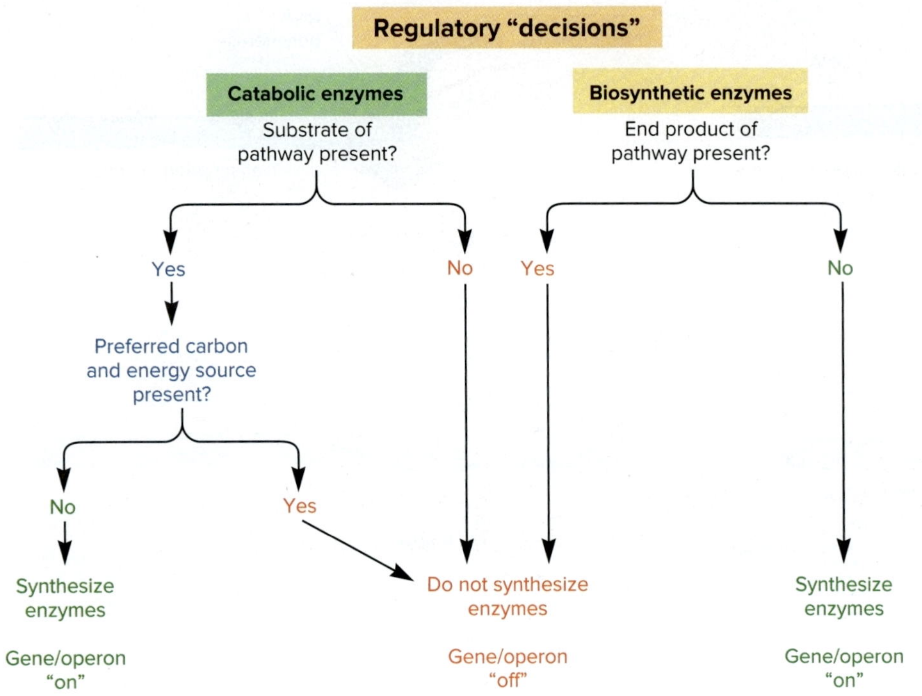

Figure 14.4 Examples of Regulatory "Decisions" Made by Cells.

the regulatory "decisions" made by an *E. coli* cell. It need only synthesize the enzymes of a specific catabolic pathway if the substrate of the pathway is present in the environment and a preferred carbon and energy source (e.g., glucose) is not (**figure 14.4**; *also see figure 22.4*). Preferred carbon and energy sources usually are more easily catabolized or yield more energy. Thus it is to the cell's advantage to use the preferred source before another source. Conversely, synthesis of enzymes for biosynthetic pathways is inhibited when the end product of the pathway is present.

Recall that functionally related bacterial genes are often transcribed from a single promoter. **Structural genes**—genes coding for nonregulatory polypeptides (e.g., enzymes)—are simply lined up together on the DNA, and a single, polycistronic mRNA carries all the messages (*see figure 13.22*). The structural genes, together with the promoter and operator or activator-binding sites, is called an **operon.** The three well-studied operons discussed next demonstrate different ways that regulatory proteins can be used to control gene expression at the level of transcription initiation.

Lactose Operon: Negative Transcriptional Control of Inducible Genes

In the late 1930s Jacques Monod (1910–1976) began examining bacterial growth and its regulation. He chose *E. coli* as a model bacterium and eventually focused his attention on genes involved in the growth of *E. coli* on lactose. He was joined in his studies about 15 years later by François Jacob (1920–2013). Together they developed the concept of an operon, and as a result of their studies, the lactose (*lac*) operon of *E. coli* is probably the best-studied negative control system.

The *lac* operon contains three structural genes controlled by the *lac* repressor (LacI). LacI is encoded by *lacI*, which is located upstream of the *lac* operon (**figure 14.5**). One gene in the operon codes for β-galactosidase; a second gene directs the synthesis of β-galactoside permease, the protein responsible for lactose uptake. The third gene codes for the enzyme β-galactoside transacetylase, whose function still is uncertain. The presence of the first two genes in the same operon ensures that the rates of lactose uptake and breakdown will vary together.

Lactose is one of many organic molecules *E. coli* uses as a carbon and energy source. It is wasteful to synthesize enzymes of the *lac* operon when lactose is not available. Therefore the cell expresses this operon at high levels only when lactose is available and a preferred carbon and energy source is not; the *lac* repressor is responsible for inhibiting transcription when there is no lactose.

The *lac* repressor is composed of four identical molecules of LacI protein (i.e., it is a tetramer), each with a helix-turn-helix DNA-binding domain. The tetramer is formed when two

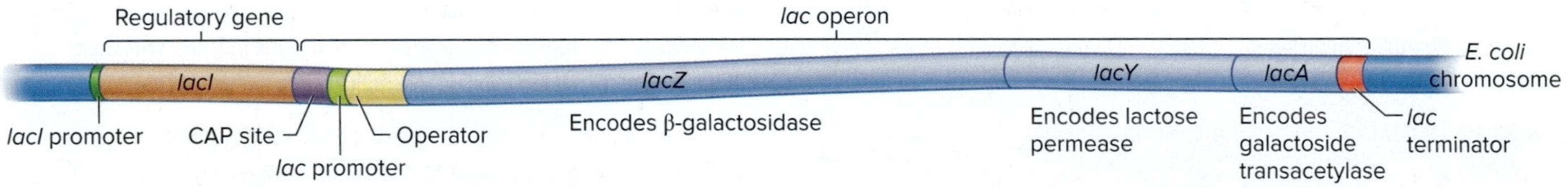

Figure 14.5 The *lac* Operon. The *lac* operon consists of three structural genes—*lacZ, lacY,* and *lacA*—that are transcribed as a single unit from the *lac* promoter. The operon is regulated both negatively and positively. Negative control is through the *lac* repressor, the product of the *lacI* gene. The operator is the site of *lac* repressor binding. Positive control results from the action of CAP. CAP binds the CAP site located just upstream from the *lac* promoter. CAP is partly responsible for a phenomenon called catabolite repression (section 14.5). For simplicity, the operator is represented as a single region. In reality, the *lac* operator consists of three distinct sites, as shown in figure 14.6.

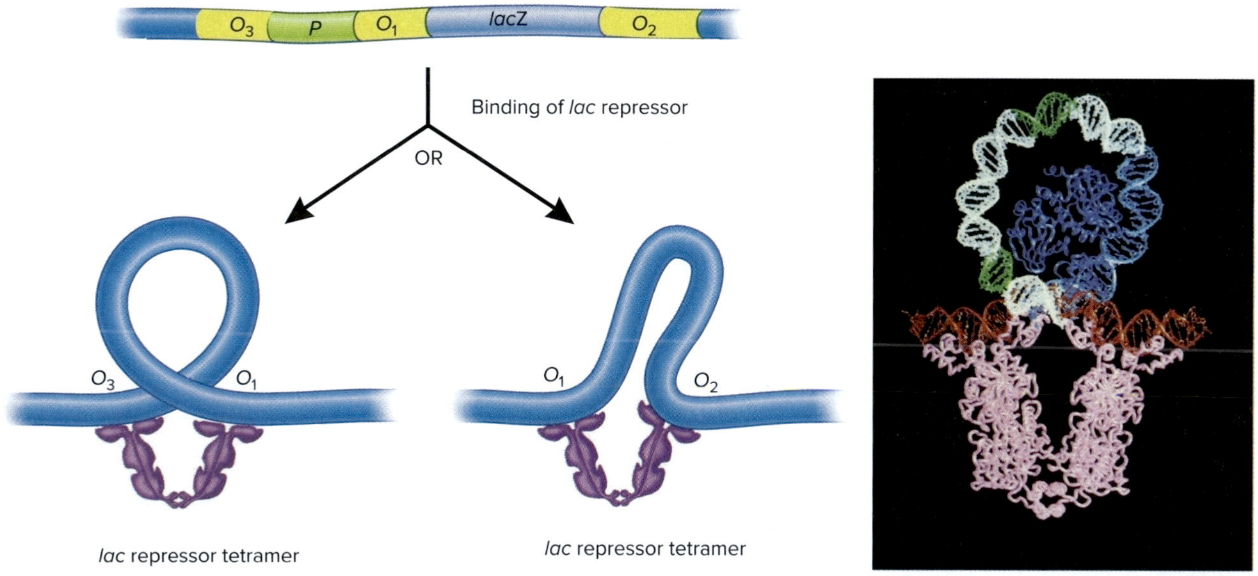

(a) DNA loops generated by binding of the *lac* repressor

(b) *lac* repressor bound to O_1 and O_3 (red)

Figure 14.6 The *lac* Operator Sites. The *lac* operon has three operator sites: O_1, O_2, and O_3 (a). As shown in (a) and (b), the *lac* repressor (violet) binds O_1 and one of the other operator sites, forming a DNA loop. The DNA loop contains the −35 and −10 binding sites (green) recognized by RNA polymerase holoenzyme. Thus these sites are inaccessible and transcription is blocked. The DNA loop also contains the CAP binding site, and CAP (blue) is shown bound to the DNA (b). When the *lac* repressor is bound to the operator, CAP is unable to activate transcription.

dimers interact. When lactose catabolism is not required, each dimer of the tetramer recognizes and tightly binds one of three different *lac* operator sites: O_1, O_2, and O_3 (**figure 14.6*a***). O_1 is the main operator site and must be bound by the repressor if transcription is to be blocked. When one dimer is at O_1 and another is at one of the two other operator sites, the dimers bring the two operator sites close together, with a loop of DNA forming between them. Attachment of the *lac* repressor is a two-step process. First, it binds nonspecifically to DNA. Then it rapidly slides along the DNA until it reaches an operator site. Two α-helices of the repressor fit into the major groove of operator-site DNA (figure 14.6*b*).

How does the repressor inhibit transcription? The *lac* promoter is located near the *lac* operator sites. When there is no lactose, the repressor binds O_1 and one of the other operator sites, bending the DNA in the promoter region. This prevents initiation of transcription either because RNA polymerase cannot access the promoter or because it is blocked from moving into the coding region (**figure 14.7*a***). When lactose is available, it is brought into the cell by lactose permease. Once inside the cell, β-galactosidase converts lactose to allolactose, the inducer of the operon (figure 14.2). This occurs because there is always a low level of permease and β-galactosidase synthesis. Allolactose noncovalently interacts with the *lac* repressor protein and causes it to change to an inactive shape that is unable to bind any operator sites. The inactivated repressor leaves the DNA and transcription of the operon occurs (figure 14.7*b*).

Close examination of figures 14.5 and 14.6 clearly shows that regulation of the *lac* operon is actually more complicated. That is because the *lac* operon is regulated by a second regulatory protein called *c*atabolite *a*ctivator *p*rotein (CAP). CAP functions in a global regulatory network that allows *E. coli* to use glucose preferentially over all other carbon and energy sources by a mechanism called catabolite repression. The use of two different regulatory proteins to control the expression of an operon illustrates another important point about regulatory processes—there are often layers of regulation. In the case of the lactose operon, the lactose repressor regulates gene expression in response to the presence or absence of lactose. CAP regulates the operon in response to the presence or absence of glucose. As described in section 14.5, the use of two regulatory proteins generates a continuum of expression levels. The highest levels of transcription occur when lactose is available and glucose is not; the lowest levels occur when lactose is not available and glucose is. For almost all of the examples described in this chapter, regulation occurs by more than one mechanism. lac *Operon Induction*

Tryptophan Operon: Negative Transcriptional Control of Repressible Genes

The tryptophan (*trp*) operon of *E. coli* consists of five structural genes that encode enzymes needed for synthesis of the amino acid tryptophan (**figure 14.8**). It is regulated by the *trp* repressor, which is encoded by the *trpR* gene. Because the enzymes

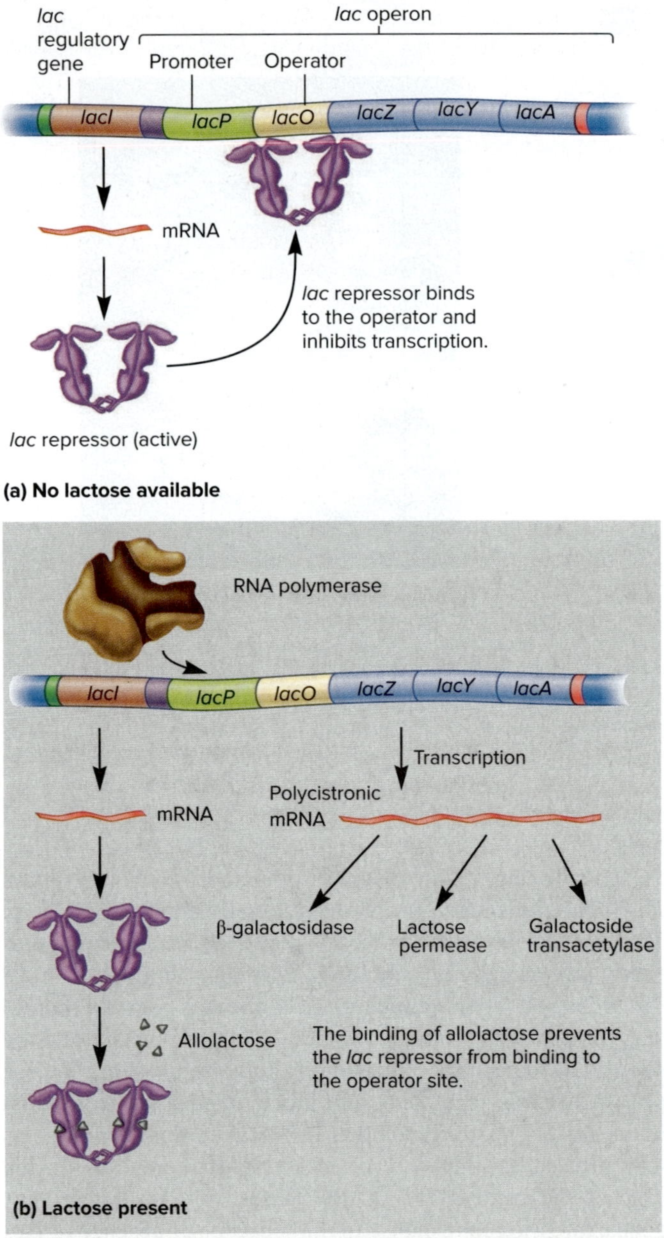

Figure 14.7 **Regulation of the *lac* Operon by the *lac* Repressor.** (a) The *lac* repressor is active and binds the operator when allolactose is not present. Repressor binding to the operator inhibits transcription. (b) When lactose is available, some of it is converted to allolactose by β-galactosidase. When sufficient amounts of allolactose are present, it binds and inactivates the *lac* repressor. The repressor leaves the operator and RNA polymerase is free to initiate transcription.

MICRO INQUIRY *Is allolactose a corepressor or inducer molecule? Explain your answer.*

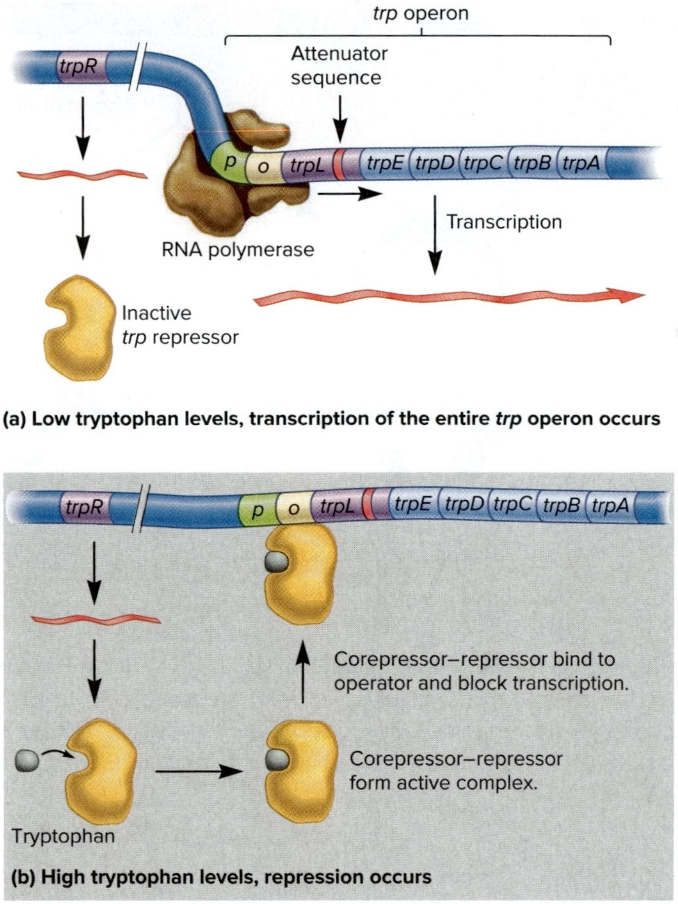

Figure 14.8 **Regulation of the *trp* Operon by Tryptophan and the *trp* Repressor.** When first synthesized, the *trp* repressor is unable to bind the operator. It is activated by the binding of tryptophan, the corepressor. (a) When tryptophan levels are low, the repressor is inactive and transcription occurs. Enzymes encoded by the operon catalyze the reactions for tryptophan biosynthesis. (b) When tryptophan levels are sufficiently high, it binds the repressor. The repressor-corepressor complex binds the operator, and transcription of the operon is inhibited.

encoded by the *trp* operon function in a biosynthetic pathway, it is wasteful to make the enzymes needed for tryptophan synthesis when tryptophan is readily available. Therefore the operon functions only when tryptophan is not present and must be made de novo from precursor molecules (figure 14.4). To accomplish this regulatory goal, the *trp* repressor protein is synthesized in an inactive form that cannot bind the *trp* operator as long as tryptophan levels are low (figure 14.8*a*). When tryptophan levels increase, tryptophan acts as a corepressor, binding the repressor thereby altering its DNA binding domain. The repressor-corepressor complex then attaches to the operator, blocking transcription initiation (figure 14.8*b*).

Like the *lac* operon, the *trp* operon is subject to another layer of regulation. In addition to being controlled at the level of transcription initiation by the *trp* repressor, expression of the *trp* operon is also controlled at the level of transcription elongation by a process called attenuation. This mode of regulation is discussed in section 14.3. *Tryptophan Repressor*

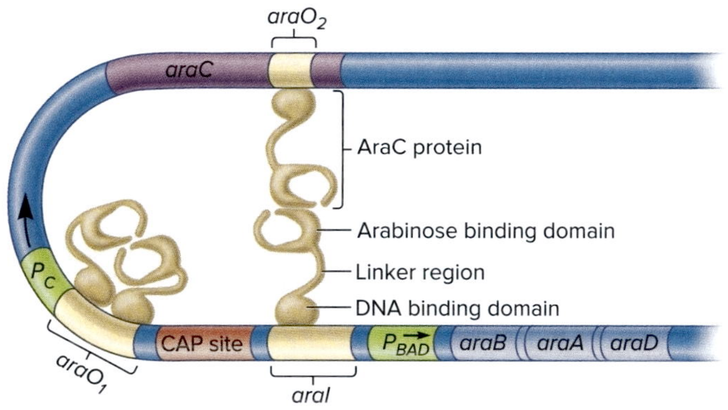

(a) Operon inhibited in the absence of arabinose

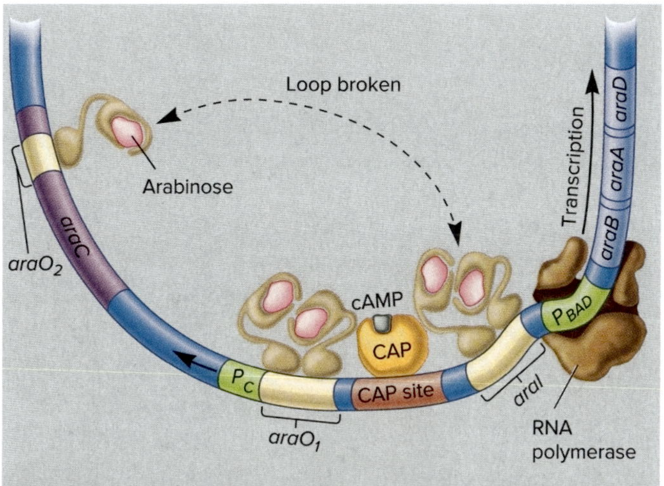

(b) Operon activated in the presence of arabinose

Figure 14.9 Regulation of the *ara* Operon by the AraC Protein. AraC can act both as a repressor and as an activator, depending on the presence or absence of arabinose. (a) When arabinose is not available, the protein acts as a repressor. Two AraC proteins are involved. One binds *araI* and the other binds *araO*$_2$. The two proteins interact to bend the DNA between the two sites, making it inaccessible to RNA polymerase. (b) When arabinose is present, it binds AraC, disrupting the interaction between the two AraC proteins. Subsequently, two AraC proteins, each bound to arabinose, form a dimer, which binds to *araI*. The AraC dimer functions as an activator.

Arabinose Operon: Transcriptional Control by a Protein That Acts Both Positively and Negatively

Regulation of the *E. coli* arabinose (*ara*) operon illustrates how the same protein can function either positively or negatively, depending on the environmental conditions. The *ara* operon encodes enzymes needed for catabolism of arabinose to xylulose 5-phosphate, an intermediate of the pentose phosphate pathway. The *ara* operon is regulated by AraC, which can interact with three different regulatory sequences: *araO*$_2$, *araO*$_1$, and *araI* (**figure 14.9**). When arabinose is not present, one molecule of AraC binds *araI* and another binds *araO*$_2$. The two AraC proteins interact, causing the DNA to bend. This prevents RNA polymerase from binding to the *ara* promoter, thereby blocking transcription. In these conditions, AraC acts as a repressor (figure 14.9*a*). However, when arabinose is present, it binds AraC and prevents AraC molecules from interacting. This breaks the DNA loop. Furthermore, binding of two AraC-arabinose complexes to *araI* promotes transcription. Thus when arabinose is available, AraC acts as an activator (figure 14.9*b*). The *ara* operon, like the *lac* operon, is also subject to catabolite repression (section 14.5). ◀ *Pentose phosphate pathway (section 11.4)*

Comprehension Check

1. Many genes and operons are regulated at the level of transcription initiation. Why is it advantageous to regulate gene expression at the transcriptional level, rather than during or after translation?
2. What is the difference between induction and repression? How do bacteria use them to respond to changing nutrient supplies?
3. Using figure 14.4 as a guide, trace the "decision-making" pathway of an *E. coli* cell growing in a medium containing arabinose, no glucose, and lacking tryptophan.
4. Suppose synthesis of an amino acid was accomplished by four enzymes encoded by four genes that formed an operon. Draw a model of the operon showing how it could be controlled by negative regulation of transcription initiation.

14.3 Attenuation and Riboswitches Stop Transcription Prematurely

After reading this section, you should be able to:

a. Describe in general terms how the *trp* operon of *E. coli* is regulated by attenuation
b. Compare and contrast regulation of transcription elongation by attenuation and riboswitches
c. Explain why the coupling of transcription and translation in bacterial cells is important to the regulatory mechanisms they use

The regulatory circuits described in the previous section control transcription by proteins binding to the promoter region of the DNA. RNA, the product of transcription, also plays a regulatory role. RNA adopts secondary and tertiary structures that allow it to interface with proteins and other molecules. In the regulatory mechanisms presented here, RNA folding patterns affect transcription, specifically at the level of transcription termination. In this type of regulation, transcription is initiated but prematurely stopped depending on the environmental conditions and the needs of the organism. Attenuation was the first example of this kind of regulation. It was discovered by Charles Yanofsky in his 1970s studies of the *trp* operon. More recently riboswitches have been described. These regulatory sequences in the mRNA leader both sense and respond to environmental conditions by either terminating transcription or blocking translation.

Attenuation: Ribosome Behavior Influences Transcription

As noted in section 14.2, the tryptophan (*trp*) operon of *E. coli* is under the control of a repressor protein, and excess tryptophan inhibits transcription of operon genes by acting as a corepressor of the repressor protein. Although the operon is regulated mainly by repression, the continuation of transcription also is controlled. That is, there are two decision points involved in transcriptional control: the initiation of transcription and the continuation of transcription into the operon's structural genes. This additional level of control serves to adjust transcription in a more subtle fashion, such that the two systems of control can decrease transcription more than either one alone. When the repressor protein is not active, RNA polymerase begins to synthesize the leader. However, it often does not progress to the first structural gene in the operon. Instead, transcription is terminated within the leader; this is called **attenuation.** ◀ *Protein-coding genes (section 13.4)*

The ability to attenuate transcription is based on two things: the nature of the leader and the fact that transcription and translation are coupled in bacteria (*see figure 13.33*). The *trp* operon leader contains a sequence of nucleotides called *trpL*. Interestingly, a small portion of *trpL* is transcribed and translated, giving rise to a small peptide called the leader peptide. The leader peptide has never been isolated, presumably because it is rapidly degraded. In addition to encoding the leader peptide, *trpL* contains **attenuator** sequences (**figure 14.10**). When transcribed, these sequences form stem-loop secondary structures in the newly formed mRNA. We define these sequences numerically (regions 1, 2, 3, and 4). When regions 1 and 2 pair (1:2; figure 14.10*a*), they form a structure called the pause loop, which slows RNA polymerase. The pause loop forms just prior to the formation of the terminator loop, which is made when regions 3 and 4 base-pair (3:4; figure 14.10*a*). A poly-U sequence follows the 3:4 terminator loop, just as it does in factor-independent transcriptional termination (*see figure 13.28*). However, in this case, the terminator is in the leader, rather than at the end of the operon. Alternatively, another stem-loop structure can be formed in the mRNA by pairing regions 2 and 3 (2:3, figure 14.10*b*). The formation of this antiterminator loop prevents generation of both the 1:2 pause and 3:4 terminator loops.

How do these various loops control transcription termination? Three scenarios illustrate the process. In the first, translation is not coupled to transcription because no ribosome is associated with the mRNA. In this scenario, the pause and terminator loops form, stopping transcription before RNA polymerase reaches the *trpE* gene (figure 14.10*a*).

In the next two scenarios, translation and transcription are coupled; that is, a ribosome initiates synthesis of the leader peptide. The distance between RNA polymerase and the nearest

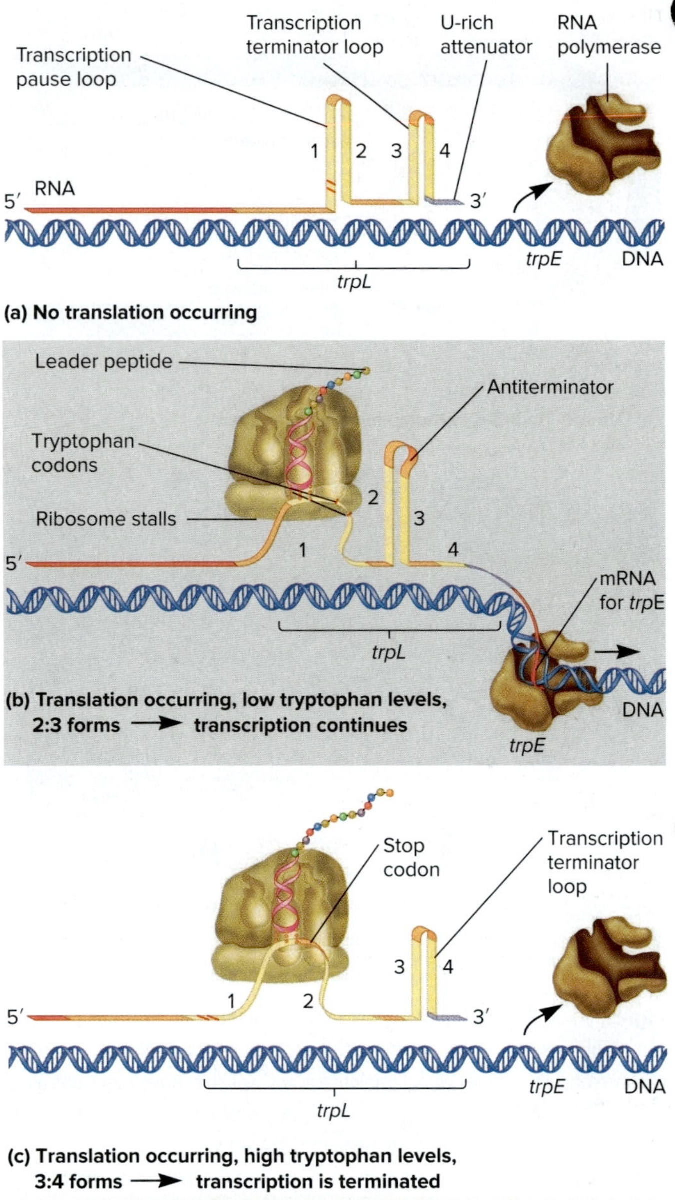

Figure 14.10 Attenuation of the *trp* Operon. (a) When protein synthesis slows, transcription and translation are not coupled. Under these conditions, the most stable form of the mRNA occurs when region 1 hydrogen bonds to region 2 (transcription pause loop) and region 3 hydrogen bonds to region 4 (transcription terminator loop). The formation of the transcription terminator causes transcription to stop within *trpL*. (b) When protein synthesis occurs, transcription and translation are coupled. When tryptophan levels are low, ribosome pausing blocks region 1 so that region 2 can hydrogen bond only with region 3. When the 2:3 loop forms, the 3:4 terminator loop cannot form and transcription proceeds. (c) When tryptophan levels are high, the ribosome progresses to the stop codon, blocking region 2. The 3:4 loop forms and transcription terminates.

MICRO INQUIRY *How does this attenuation respond to rates of protein synthesis in addition to the intracellular level of tryptophan?*

ribosome determines which stem-loop structures are formed. As a ribosome translates the leader mRNA, it follows the RNA polymerase. Among the first several nucleotides of region 1 are two tryptophan codons; this is unusual because normally there is only one tryptophan residue per 100 amino acids in *E. coli* proteins. If tryptophan levels are low, there will not be enough charged $tRNA^{trp}$ to fill the A site of the ribosome when the ribosome encounters the two tryptophan codons, and it will stall (figure 14.10*b*). Meanwhile RNA polymerase continues to synthesize mRNA, moving away from the stalled ribosome. The presence of the ribosome on region 1 prevents region 1 from base-pairing with region 2. As RNA polymerase continues, region 3 is transcribed, and the 2:3 antiterminator loop forms. This prevents the formation of the 3:4 terminator loop. Without the terminator loop, RNA polymerase continues into the *trp* structural genes. If, on the other hand, there is plenty of tryptophan in the cell, there will be an abundance of charged $tRNA^{trp}$, and the ribosome will translate the two tryptophan codons in the leader peptide without hesitation. Thus the ribosome remains close to the RNA polymerase. As RNA polymerase and the ribosome continue through *trpL*, regions 1 and 2 are transcribed and readily form a pause loop. Then regions 3 and 4 are transcribed, the terminator loop forms, and RNA polymerase is ejected from the DNA template. Finally, a UGA stop codon between regions 1 and 2 (i.e., the end of the leader peptide-coding region) causes termination of translation (figure 14.10*c*).

◀ *The genetic code consists of three-letter "words" (section 13.6)*

How is attenuation useful to *E. coli*? If the bacterium is deficient in an amino acid other than tryptophan, protein synthesis will slow and tryptophanyl-$tRNA^{Trp}$ (trp-tRNA) will accumulate. Transcription of the *trp* operon will be inhibited by attenuation. When the bacterium begins to synthesize protein rapidly, tryptophan may be scarce and the concentration of trp-tRNA may be low. This stimulates transcription, resulting in larger quantities of the tryptophan biosynthetic enzymes. Acting together, repression and attenuation can coordinate the rate of synthesis of amino acid biosynthetic enzymes with the availability of amino acid end products and with the overall rate of protein synthesis. When tryptophan is present at high concentrations, any RNA polymerases not blocked by the repressor protein probably will not get past the attenuator sequence. Repression decreases transcription about 70-fold and attenuation slows it another 8- to 10-fold; when both mechanisms operate together, transcription can be slowed about 600-fold.

Attenuation is important in regulating other operons that encode amino acid biosynthetic enzymes. Their leader peptide sequences resemble the tryptophan system in organization. For example, the leader peptide of the histidine operon has seven histidines in a row and is followed by an attenuator that is a terminator sequence.

Riboswitches: Effector–mRNA Interaction Regulates Transcription

Riboswitches (also called sensory RNAs) are regions in the 5′ untranslated leader of an mRNA to which a small effector molecule (i.e., not a protein) can bind. Binding of the effector determines whether or not its target mRNA continues to be synthesized. Other riboswitches function at the level of translation and are considered in section 14.4. For most transcriptional riboswitches, if the leader of an mRNA is folded one way, transcription continues; if folded differently, transcription terminates. The leader is called a riboswitch because it is analogous to a light switch turning lights on or off, depending on whether the switch is in the on position (one folding pattern of the mRNA) or the off position (another folding pattern of the mRNA). What makes riboswitches unique is that the mRNA folding is determined by effector binding directly to the mRNA—a capability previously thought confined to proteins.

The riboswitch that regulates the riboflavin (*rib*) biosynthetic operon of *Bacillus subtilis* serves as an example (**figure 14.11**). The production of riboflavin biosynthetic enzymes is regulated by the abundance of flavin mononucleotide (FMN), which is derived from riboflavin. Upon transcription of the *rib* operon, sequences in the leader mRNA fold into a structure called the *rfn* box. When FMN levels are low, it does not bind the *rfn* box and the leader folds so that transcription can proceed. Conversely, when FMN levels are high, it binds such

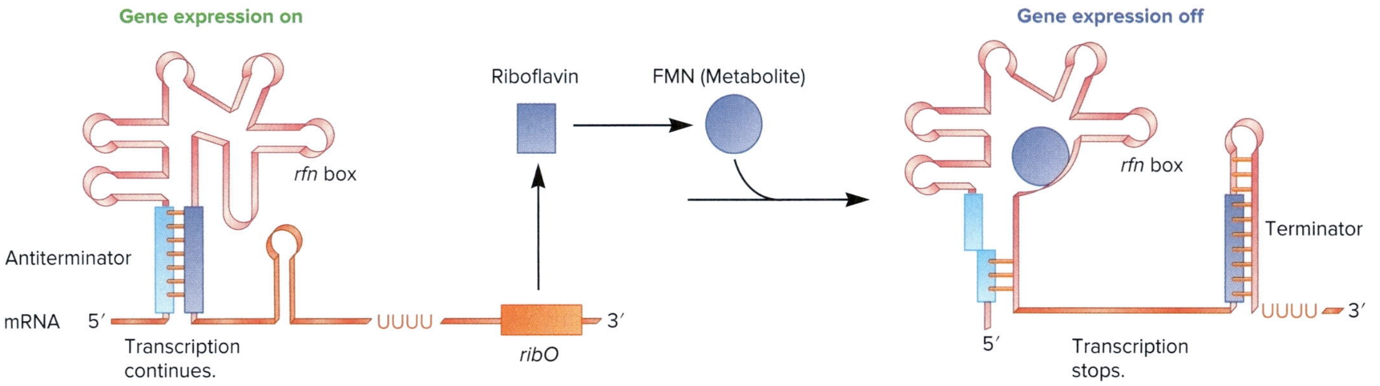

Figure 14.11 Riboswitch Control of the Riboflavin (*rib*) Operon of *Bacillus subtilis*. The *rib* operon encodes enzymes that catalyze riboflavin synthesis, a component of flavin mononucleotide (FMN). FMN binding to the *rfn* (*r*ibo*f*lavi*n*) box in the *rib* mRNA leader causes a transcription terminator to form.

Table 14.1 Regulation of Gene Expression by Riboswitches

System	Target genes encode:	Effector and Regulatory Response
T box	Amino acid biosynthetic enzymes	Uncharged tRNA; anticodon base-pairs to 5′ end of mRNA, preventing formation of transcriptional terminator.
Vitamin B_{12} element	Cobalamine biosynthetic enzymes	Adenosylcobalamine (AdoCbl) binds to *btuB* mRNA and blocks translation.
Cyclic-di-GMP	Self-splicing ribozyme	Binding of c-di-GMP alters splicing site.
Metal-sensing riboswitches (e.g., M box)	Metal transporter proteins	Intracellular levels of metal (e.g., Mg^{2+}) cause premature transcriptional termination or trigger mRNA degradation.
glmS riboswitch	GlmS enzyme	*glmS* mRNA is a self-cleaving ribozyme that encodes an enzyme in the biosynthetic pathway for glucosamine-6-phosphate (GlcN6P). GlcN6P binding to the riboswitch promotes self-cleavage of *glmS* mRNA to a smaller RNA that is degraded by RNase.

that a terminator loop forms, halting transcription. Transcription is stopped with the aid of rho protein, the RNA helicase that functions in factor-mediated transcription termination.

Examination of **table 14.1** reveals that riboswitches are often coordinated with other types of regulatory mechanisms discussed in sections 14.4 and 14.5. For instance, some respond to second messengers such as cyclic dimeric GMP (c-di-GMP), while others function in conjunction with small RNAs. This is another example of how bacteria use multiple regulatory mechanisms to control their activities.

14.4 Riboswitches and Small RNAs Control Translation

After reading this section, you should be able to:

a. Distinguish translational riboswitches from transcriptional riboswitches
b. Discuss regulation of translation by small RNAs

It appears that in general, the riboswitches found in Gram-positive bacteria function by transcriptional termination, whereas the riboswitches discovered in Gram-negative bacteria regulate the translation of mRNA. Translation is usually regulated by blocking its initiation. In addition, some small RNA molecules can control translation initiation of target mRNAs. Both riboswitch and small RNA regulatory mechanisms are described now.

Translational Riboswitches

Similar to the riboswitches described in section 14.3, riboswitches that function at the translational level contain effector-binding elements at the 5′ end of the mRNA. Binding of the effector molecule affects the folding pattern of the mRNA leader, which often results in occlusion of the Shine-Dalgarno sequence and other elements of the ribosome-binding site. This inhibits ribosome binding and initiation of translation (**figure 14.12**). An example of this type of regulation is observed for the thiamine biosynthetic operons of numerous bacteria. The mRNA leaders of thiamine operons contain a structure called the THI-element that binds thiamine pyrophosphate. High levels of thiamine pyrophosphate cause an association with the THI-element and result in a conformational change in the leader that sequesters the Shine-Dalgarno sequence and blocks translation initiation.

Small RNA Molecules

Many RNA molecules have been discovered that do not function as mRNAs, tRNAs, or rRNAs. Microbiologists often refer to them as **small RNAs (sRNAs)** or noncoding RNAs (ncRNAs); sRNAs range in size from around 25 to 500 nucleotides. Although some sRNAs have been implicated in the regulation of DNA replication and transcription, many function at the level of translation. Those that regulate translation do so by targeting mRNA molecules. These translation-controlling sRNAs are considered in this section.

Two main types of mRNA-targeting sRNAs can be described based on where they are encoded on the genome: cis-encoded and trans-encoded. Cis-encoded sRNAs are synthesized from the nontemplate strand (i.e., the sense strand) of the gene that gives rise to their mRNA target. Thus they are complementary to the target mRNA and are called **antisense RNAs.** Trans-encoded sRNAs are synthesized from genes located at sites distinct from the genes encoding their target mRNAs. Despite this, they have some regions of complementarity to their targets. Both work similarly, by binding to mRNA. As you might imagine, binding of an sRNA to an mRNA target can interfere with either ribosome binding or the ability of the mRNA to move through the ribosome. Thus

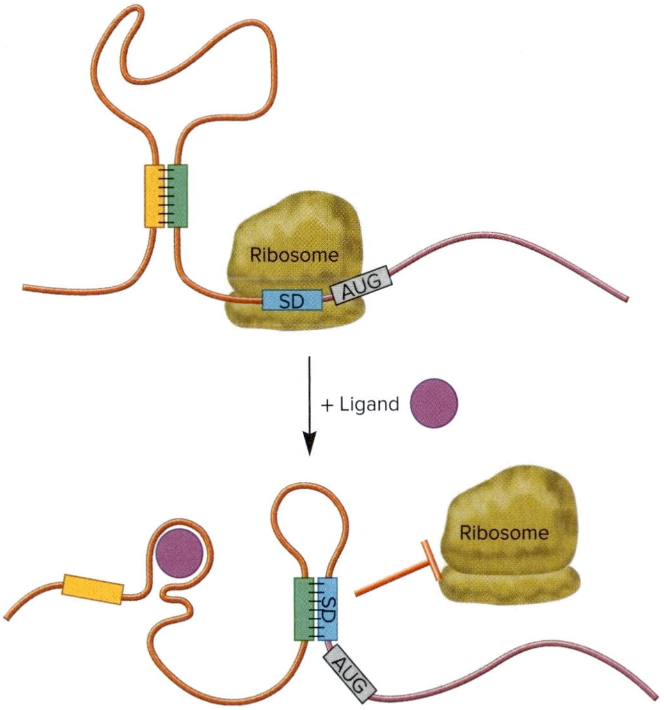

Figure 14.12 Inhibition of Translation by a Riboswitch. In the absence of a relevant metabolite, an effector binding site is formed in the leader of the mRNA (red) when complementary sequences (orange box and green box) hydrogen bond. This folding pattern exposes important sequences in the ribosome-binding site (e.g., the Shine-Dalgarno sequence; blue box) and translation occurs. When the appropriate effector molecule is present, it binds the leader, creating a structure containing the ribosome-binding site. Thus the ribosome-binding site becomes inaccessible and translation is blocked.

it seems logical that many sRNAs would inhibit translation. However, some sRNAs actually promote translation upon binding to the mRNA. Whether inhibiting or activating, most trans-encoded sRNAs work with the help of an RNA chaperone. Hfq, the RNA chaperone protein used by *E. coli*, binds and stabilizes many sRNAs. A short region of 6 to 10 nucleotides of sRNA protrudes from the Hfq-sRNA surface, primed to interact with complementary sequences that are the regulatory targets.

The regulation of the OmpF porin protein provides an example of translation control by a trans-encoded sRNA. Recall that the outer membrane of Gram-negative bacteria contains channels made of porin proteins (*see figure 3.26*). The two most important *E. coli* porins are OmpF and OmpC (Omp for *o*uter *m*embrane *p*rotein). OmpC pores are slightly smaller and are made when the bacterium grows at high osmotic pressures. It is the dominant porin when *E. coli* is in the intestinal tract. The larger OmpF pores are favored when *E. coli* grows in a dilute environment; OmpF allows solutes (e.g., nutrients) to diffuse into the cell more readily. The cell must maintain a constant level of porin protein in the membrane, with the relative levels of the two porins corresponding to the osmolarity of the medium. Expression of the *ompF* gene is regulated by MicF RNA, the product of the *micF* gene (*mic* for *m*RNA-*i*nterfering *c*omplementary RNA). The MicF RNA is complementary to *ompF* mRNA at the translation initiation site (**figure 14.13**). When MicF RNA base-pairs with *ompF* mRNA, translation is repressed. MicF RNA is produced under conditions such as high osmotic pressure or the presence of some toxic material, both of which favor *ompC* expression. Production of MicF RNA helps ensure that OmpF protein is not produced at high levels at the same time as OmpC protein. In addition to being regulated by MicF RNA, *ompF* (and *ompC*) is regulated by a two-component signal transduction system, as we describe in section 14.5.

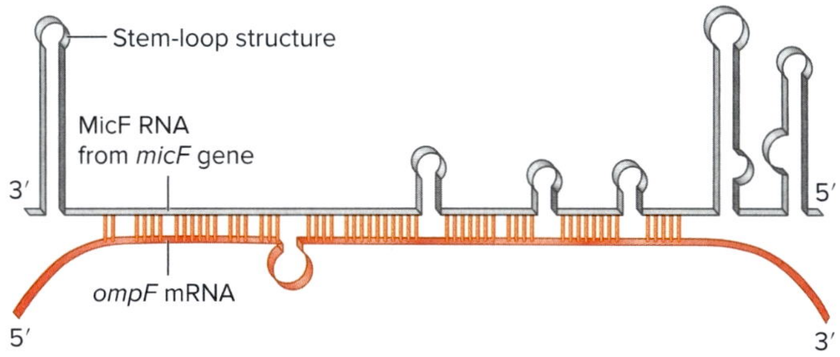

Figure 14.13 Inhibition of Translation by a Trans-Encoded sRNA. The *ompF* mRNA encodes the porin OmpF. Translation of this mRNA is regulated by the trans-encoded sRNA MicF, product of the *micF* gene. MicF is partially complementary to *ompF* mRNA and, when bound to it, prevents translation.

MICRO INQUIRY *How does inhibition of translation by sRNA differ from translational inhibition by riboswitches?*

Comprehension Check

1. What are the functions of *trpL* and the ribosome in attenuation?
2. How would attenuation in the *E. coli trp* operon change when a culture is shifted from a rich medium to one that is deficient in tryptophan?
3. Compare and contrast transcriptional and translational riboswitches.
4. How are translational riboswitches similar to attenuation as described for the *trp* operon? How do they differ? How do translational riboswitches differ from sRNAs?
5. Suggest a mechanism by which *micF* transcription might be regulated.

14.5 Mechanisms Used for Global Regulation

After reading this section, you should be able to:

a. Compare and contrast two-component signal transduction systems with phosphorelay signal transduction systems
b. Explain how alternate sigma factors can be used to initiate transcription of distinct sets of operons
c. List the nucleotides that are used by bacteria as second messengers

d. Summarize how CAP, cAMP, and the *lac* repressor work together to cause diauxic growth as well as other outcomes related to regulation of the lactose operon
e. Hypothesize how other operons encoding catabolic enzymes might be regulated by catabolite repression
f. Describe the stringent response
g. Describe the enzymes that synthesize and hydrolyze c-di-GMP, and summarize how its cellular concentration regulates biofilm formation

The types of regulation discussed thus far are responses to a single stimulus. A change in carbon source or nutrient concentration affects a few genes or operons. Microbes also need to respond to significant environmental changes, for example, when a pathogen encounters a susceptible host. In these situations, the microbe's response requires large-scale changes that coordinate sensory input with gene expression and cellular physiology. These rely on **global regulatory systems:** regulatory systems that affect many genes, operons, and pathways simultaneously. Perhaps the simplest approach to accomplish this is to use a single **global regulatory protein** to alter the expression of multiple operons (or genes). The collection of genes and operons controlled by the global regulator is called a **regulon.** A good example of global regulatory protein is CAP, which enables bacteria to use glucose preferentially regardless of the presence of other potential carbon sources. CAP's regulon includes the *lac* and *ara* operons.

Several themes recur in global regulation. The microbial sensory process involves converting an environment signal to a conformational change in the molecule that senses the signal. The sensor may be RNA or protein. The conformational change is then transmitted to effector molecules. Often the sensor is a large protein with multiple domains, so that the conformational change in one domain causes a change in a different domain. AraC is a good example of how detecting arabinose forces a change in DNA binding specificity. In other cases, a change in the sensory domain triggers enzymatic activity in a different domain. Such allosteric regulation plays an important role in global regulatory systems. In this section, we consider some ways bacteria control complex processes using global regulation.

Two-Component Signal Transduction and Phosphorelays

Responses to environmental stimuli are often controlled by regulatory proteins that make up **two-component signal transduction systems.** These systems link events occurring outside the cell to gene expression inside the cell.

Two-component signal transduction systems are named after the two proteins that govern the regulatory pathway. The first is an enzyme called a **sensor kinase** that spans the plasma membrane so that part of it is exposed to the extracellular environment (periplasm in Gram-negative bacteria), while another part is exposed to the cytoplasm. In this way, the sensor kinase can detect specific changes in the environment and communicate information to the cell's interior. It does so by first phosphorylating itself and then transferring the phosphate to the second component, the **response regulator.** Upon receiving the signal from its partner sensor kinase, the response regulator undergoes a conformational change to allow it to interact with other molecules. Some form homodimers that bind to DNA and either activate or repress transcription initiation. Other response regulators interact with enzymes or other proteins in the global network.

The EnvZ/OmpR system is an excellent example of a two-component signal transduction system that functions at the level of transcription; it is used to control porin production in *E. coli.* As we introduce in section 14.4, *E. coli* uses sRNA to control OmpF production. Thus two regulatory mechanisms are used to control the response to osmolarity. In fact, the two mechanisms are intertwined, with the two-component system also controlling the synthesis of the MicF sRNA. Two-component systems and sRNAs often operate together to form complex regulatory networks. In the EnvZ/OmpR system, EnvZ (Env for cell *env*elope) is the sensor kinase. EnvZ is a homodimer (i.e., two identical polypeptides interact to form the functional protein) that is an integral membrane protein anchored to the membrane by two membrane-spanning domains. EnvZ is looped through the membrane such that a central domain protrudes into the periplasm, while the amino and carboxyl termini are exposed to the cytoplasm (**figure 14.14**).

The response regulator OmpR is a cytoplasmic protein that controls transcription of the *ompF* and *ompC* structural genes. The N-terminal end of OmpR is called the receiver domain because it has an aspartic acid residue that accepts the signal (a phosphoryl group) from the sensor kinase EnvZ. Upon receipt of the signal, the C-terminal end of OmpR regulates transcription by binding DNA. When EnvZ senses high osmolarity in the periplasm, it phosphorylates itself (autophosphorylation) on a specific histidine residue. This phosphoryl group is quickly transferred to an aspartic acid of OmpR. Once OmpR is phosphorylated, it regulates transcription so that *ompF* is repressed and *ompC* is activated.

The effectiveness of two-component systems is illustrated by their abundance. Most bacterial cells use these systems to respond to an array of environmental stresses. For example, members of the soil bacterial genus *Streptomyces* have at least 80 such systems.

Two-component signal transduction systems involve a simple relay where the sensor kinase transfers a phosphoryl group directly to the response regulator. However, there are instances when more proteins participate in the transfer of phosphoryl groups. These more complex pathways are called **phosphorelay systems.** Phosphorelays are important control measures observed in quorum-sensing by *Vibrio harveyi* and endospore formation by *B. subtilis*, both described in detail later in this section.

Sigma Factors

Another way many genes can respond in coordinated fashion is through the use of **alternate sigma factors,** which immediately changes expression of many genes as they direct RNA polymerase to specific subsets of bacterial genes. This is possible because RNA polymerase core enzyme needs the

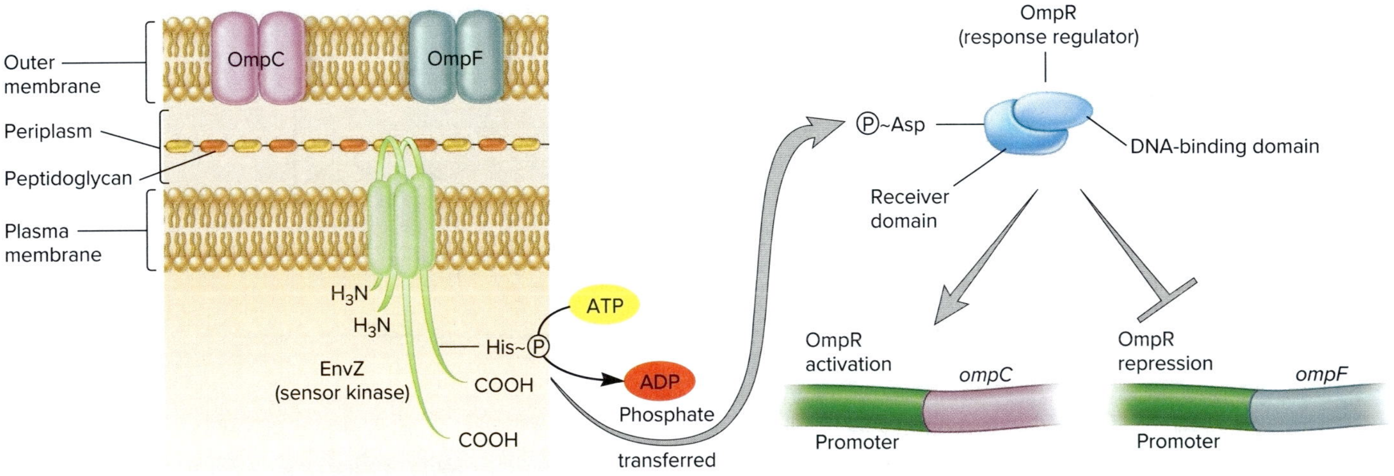

Figure 14.14 Two-Component Signal Transduction and the Regulation of Porins. The sensor kinase EnvZ loops through the cytoplasmic membrane so that both its C- and N-termini are in the cytoplasm. When EnvZ senses osmolarity change in the periplasm, it autophosphorylates a histidine residue. EnvZ then passes the phosphoryl group to the response regulator OmpR, which accepts it on an aspartic acid residue. This activates OmpR so that it binds DNA to repress *ompF* and induce *ompC* expression.

MICRO INQUIRY *Relative to each promoter, where would you predict phosphorylated OmpR would bind the* ompC *and* ompF *genes?*

assistance of a sigma factor to bind a promoter and initiate transcription. Each sigma factor recognizes promoters that differ in sequence (*see table 13.2*). ◀ *Transcription in bacteria (section 13.5)*

E. coli synthesizes multiple sigma factors (**table 14.2**). Under favorable growth conditions, a sigma factor called σ^{70} directs RNA polymerase activity. (The superscript number indicates the size of the sigma factor; 70 stands for 70,000 daltons.) When flagella and chemotactic proteins are needed, *E. coli* produces σ^{28}. σ^{28} then binds its consensus sequences in promoters of genes whose products are needed for flagella biosynthesis and chemotaxis. If the temperature rises too high, σ^{32} is produced and stimulates the formation of about 17 heat-shock proteins that protect the cell from thermal destruction.

Sigma factors known as *e*xtra*c*ytoplasmic *f*unction (ECF) sigmas deserve mention because they function in signal transduction. The best-studied system uses a plasma membrane-bound protein called an anti-sigma factor that has a periplasmic sensory domain. In the absence of the stimulus, the anti-σ binds the ECF σ, sequestering it so that it cannot bind its cognate promoters. Upon detecting its specific signal, the anti-σ is destroyed by proteolysis, and the ECF σ is released and activates transcription of its target genes. Other ECF sigmas are phosphorylated by plasma membrane-bound kinases (serine/threonine kinases) that detect the signal and phosphorylate and activate the ECF σ, which then initiates transcription.

Table 14.2 *E. coli* Sigma Factors

Sigma Factor	Genes Transcribed
σ^{70}	Genes needed during exponential growth
σ^{54}	Genes involved in nitrogen metabolism
σ^{38}	Genes needed during stationary phase and the general stress response
σ^{32}	Genes needed to protect against heat shock and other stresses
σ^{28}	Genes involved in flagellum assembly
σ^{24}	Genes needed for the proper folding of membrane proteins
σ^{19}	Genes that respond to iron starvation

Second Messengers

Many bacteria use **second messengers** in their global control systems. Second messengers are small molecules produced within a cell in response to a signal (i.e., the first messenger) that is outside the cell. The enzymes that synthesize the second messenger respond to environmental signals, and the second messenger relays that information to a larger set of effectors. Three bacterial second messengers were discovered many years ago: cyclic adenosine monophosphate (cAMP), guanosine tetraphosphate (ppGpp), and guanidine pentaphosphate (pppGpp) (**figure 14.15**). Newer additions to the bacterial second messenger family include **cyclic dimeric GMP (c-di-GMP)**, c-di-AMP, and cGMP. Each second messenger responds to a particular type of environmental signal, and redirects cellular physiology and gene expression to respond. We describe three second messenger regulons in detail.

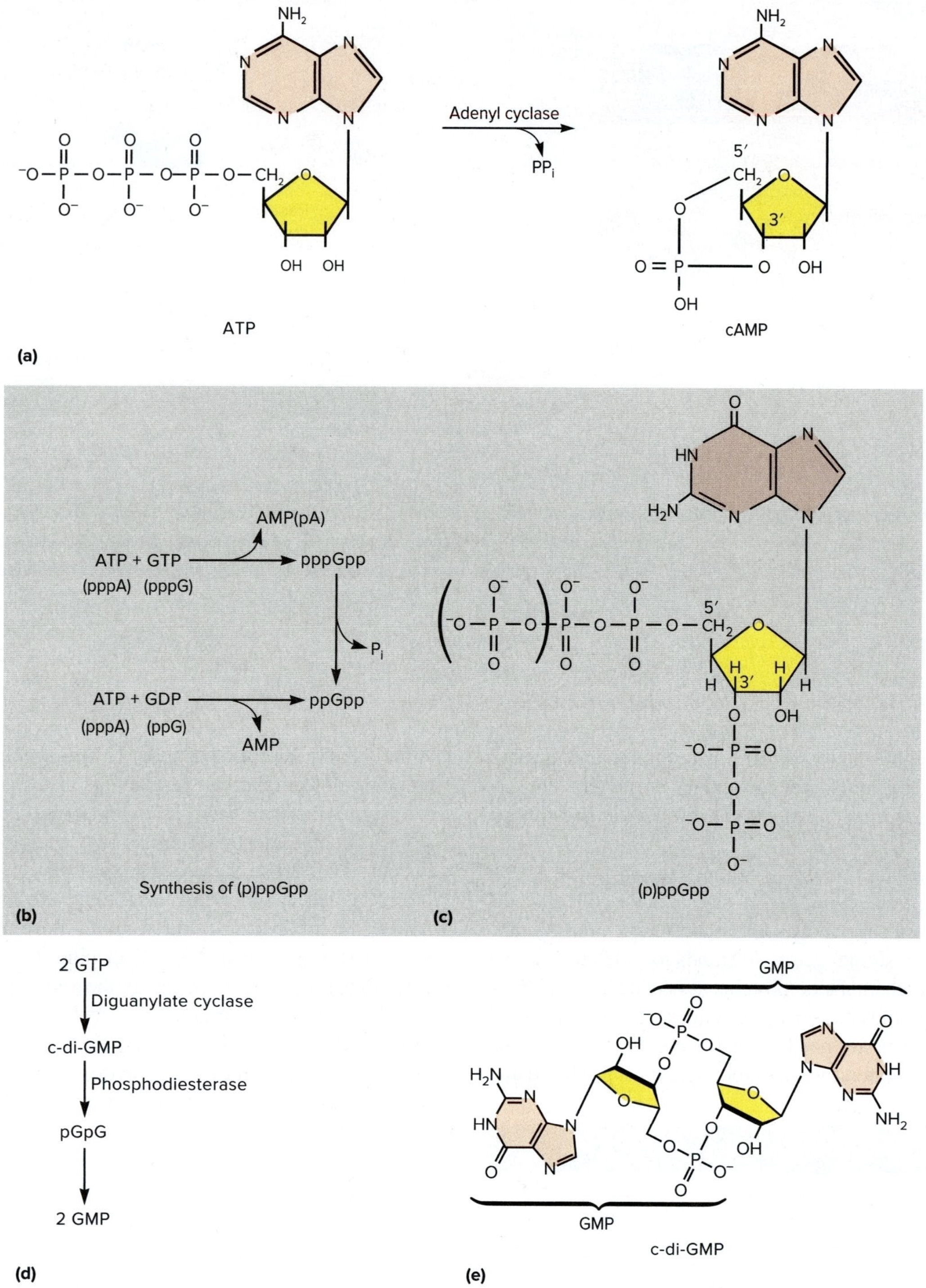

Figure 14.15 Second Messengers. (a) Cyclic adenosine monophosphate (cAMP) forms from ATP. The phosphate extends between the 3′ and 5′ hydroxyls of ribose. (b) Pathways to make guanosine tetraphosphate (ppGpp) and guanosine pentaphosphate (pppGpp). (c) The structure of (p)ppGpp. (d) Two enzymes control levels of cyclic dimeric GMP (c-di-GMP). Diguanylate cyclases form c-di-GMP and phosphodiesterases degrade c-di-GMP. (e) Structure of c-di-GMP, two GMP molecules linked by their phosphates.

Catabolite Repression in *E. coli*

If *E. coli* grows in a medium containing both glucose and lactose, it uses glucose preferentially until the sugar is exhausted. Then after a short lag, growth resumes at a slower rate with lactose as the carbon source (**figure 14.16**). This allows glucose, the more easily catabolized and more energy-yielding molecule, to be used first, followed by other energy sources that may be more difficult to degrade or yield less energy. This biphasic

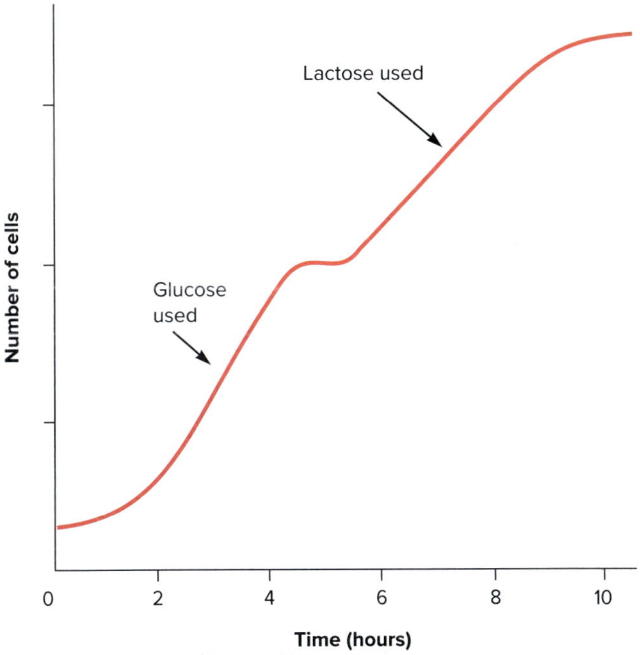

Figure 14.16 Diauxic Growth. The diauxic growth curve of *E. coli* grown with a mixture of glucose and lactose. Glucose is used first, then lactose. A short lag in growth occurs when glucose is exhausted, while the bacteria synthesize the enzymes needed for lactose use.

MICRO INQUIRY *For what other compounds would you also see this kind of diauxic growth?*

growth pattern is called **diauxic growth** and it is brought about by a global regulatory process called **catabolite repression.** In *E. coli,* two regulatory networks bring about catabolite repression. One network directly inhibits the transporters that take up the less preferred sugars. The other network dramatically decreases expression of the genes encoding the enzymes that catabolize those sugars. As might be expected, the two networks are linked, as we describe in a moment. We begin by looking at the latter network.

The enzymes for glucose catabolism are constitutive. However, operons that encode enzymes required for the catabolism of carbon sources that must first be modified before entering glycolysis, such as the *lac, ara, mal* (maltose), and *gal* (galactose) operons, are called catabolite operons, and their expression is coordinately repressed when glucose is plentiful. This is accomplished by **catabolite *a*ctivator *p*rotein (CAP)**, which is also called *c*yclic AMP *r*eceptor *p*rotein (CRP). CAP exists in two states. When the second messenger **3′, 5′-cyclic adenosine monophosphate (cAMP;** figure 14.15*a*) is bound, CAP binds DNA through a helix-turn-helix motif (**figure 14.17**). In the absence of cAMP, CAP cannot bind DNA. The enzyme adenyl cyclase synthesizes cAMP from ATP. Adenyl cyclase catalyzes cAMP synthesis only when little or no glucose is available. Thus the level of cAMP varies inversely with that of glucose. When glucose is unavailable and the catabolism of another sugar might be needed, the amount of cAMP in the cell increases, allowing cAMP to bind to and activate CAP.

All catabolite operons have promoters possessing a CAP recognition site, and cAMP-CAP must bind to this site before RNA polymerase can begin transcription efficiently. Upon binding, CAP interacts with the carboxyl terminal domain of the α subunits of RNA polymerase and stabilizes the interaction of RNA polymerase with the promoter; this stimulates transcription (figure 14.17). Thus all catabolite operons are

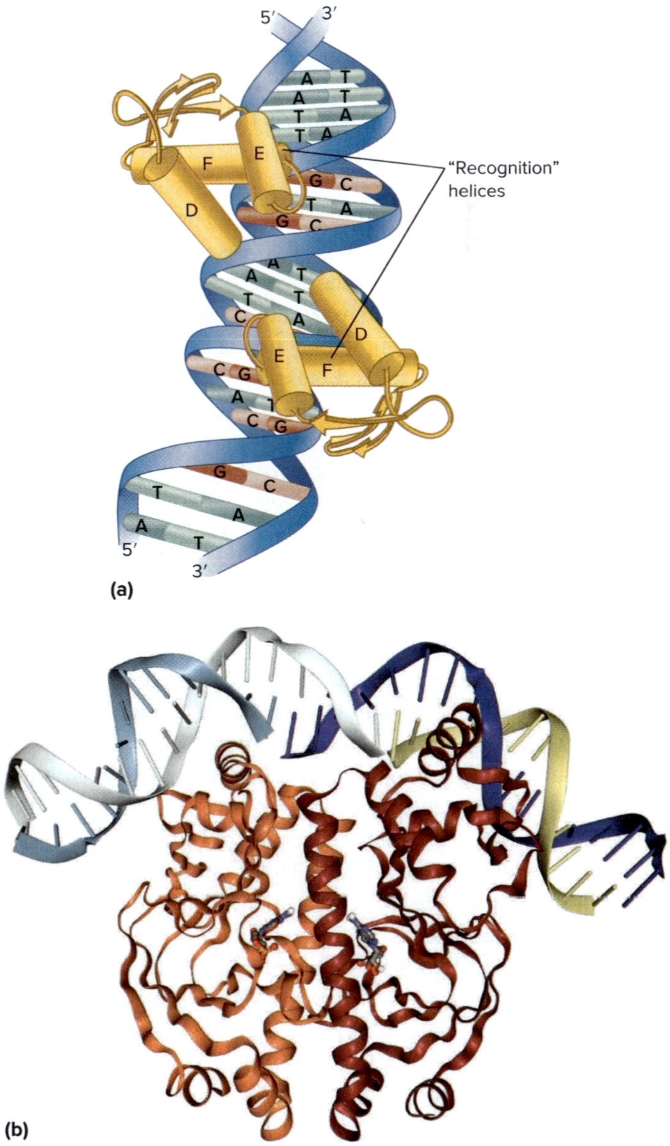

Figure 14.17 CAP Structure and DNA Binding. (a) The CAP dimer binding to DNA at the *lac* operon promoter. The recognition helices, labeled F, fit into two adjacent major grooves on the double helix. (b) A model of the *E. coli* cAMP-CAP-DNA complex.

controlled by two regulatory proteins: the regulatory protein specific to each operon (e.g., *lac* repressor and AraC protein) and CAP. In the case of the *lac* operon, if glucose is absent and lactose is present, the inducer allolactose will bind to and inactivate the *lac* repressor protein, CAP will be in the active form (with cAMP attached), and transcription will proceed (**figure 14.18*a***). However, if glucose and lactose are both in short supply, even though CAP binds to its site on the DNA, transcription will be inhibited by the presence of the repressor protein, which remains bound to the operator in the absence of inducer (figure 14.18*c*). Dual control ensures that the *lac* operon is expressed only when lactose catabolic genes are needed. *Combination of Switches: The* lac *Operon*

We have seen how CAP controls catabolite operons; now let's turn our attention to the regulation of cAMP and how glucose levels are sensed by the cell. Both depend on the state of a protein that functions in the phosphoenolpyruvate:sugar phosphotransferase system (PTS) (**figure 14.19**). Recall from chapter 3 that in the PTS, a phosphoryl group is transferred by a phosphorelay from phosphoenolpyruvate (PEP) to glucose, which then enters the cell as glucose 6-phosphate (*see figure 3.14*). The phosphorylation state of the PTS proteins is an indicator of glucose abundance. When glucose levels are low, enzyme IIB cannot accept the phosphate from enzyme IIA and transfer it to glucose. Instead, enzyme IIA phosphorylates adenyl cyclase, thus activating the enzyme and stimulating cAMP production. The cAMP second messenger, then, relays the information about decreasing glucose concentration to molecules that can respond by exploiting any alternative carbon sources that may be available. *Group translocation (section 3.3)*

Stringent Response in *E. coli*

Many bacteria regulate their growth rate with the second messengers guanosine tetraphosphate (ppGpp) and guanosine pentaphosphate (pppGpp) (figure 14.15*c*). These related molecules respond to nutritional stress and are therefore referred to as **alarmones.** Because they act similarly in the cell, they are designated together as (p)ppGpp. Under good growth conditions, the concentration of (p)ppGpp is low. Nutrient limitation or the presence of an antibiotic increases (p)ppGpp synthesis. This **stringent response** is a global mechanism to translate cellular stress into slowing the growth rate. This requires coordination of all major cellular processes: transcription, translation, DNA replication, and metabolism.

One of the best-studied stringent responses is the *E. coli* response to amino acid starvation. When *E. coli* grows in nutrient-rich conditions, RNA polymerase actively transcribes tRNA and rRNA genes, providing a ready supply of these molecules for protein synthesis. When cells are starved for amino acids, it's wasteful to keep synthesizing large quantities of tRNA and rRNA. Instead, the cell decreases synthesis and increases transcription of the appropriate amino acid biosynthetic genes.

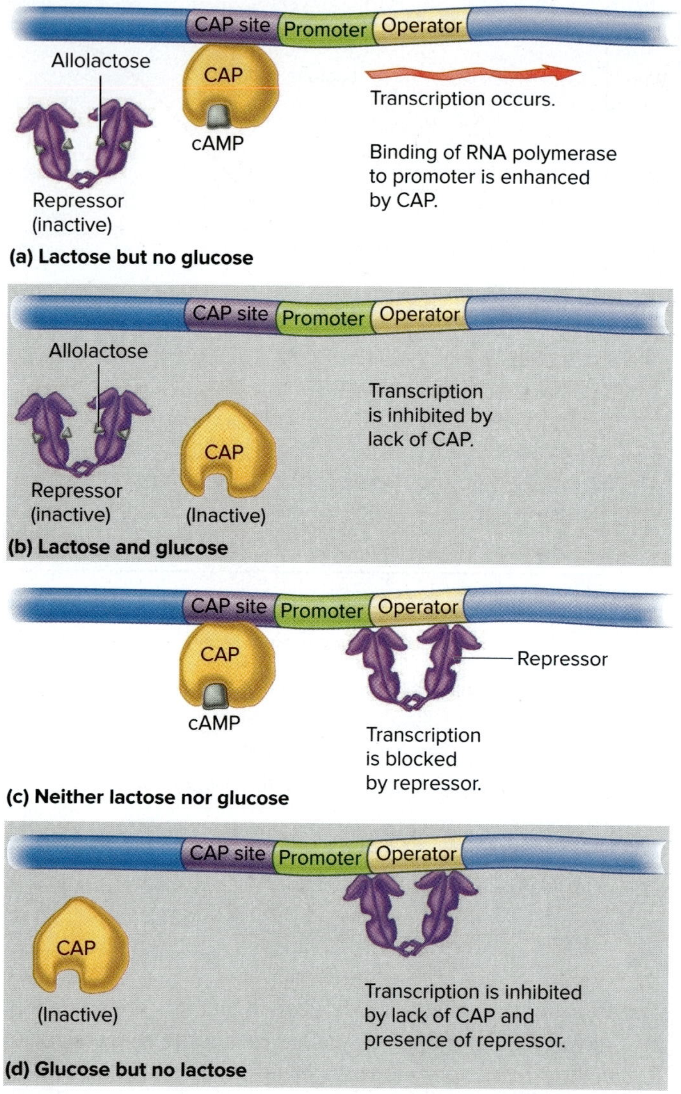

Figure 14.18 Regulation of the *lac* Operon by the *lac* Repressor and CAP. A continuum of *lac* mRNA synthesis is brought about by the action of CAP and the *lac* repressor. (a) When lactose is available and glucose is not, the repressor is inactivated and cAMP levels increase. Cyclic AMP binds CAP, and cAMP-CAP then binds the CAP binding site near the *lac* promoter and facilitates binding of RNA polymerase. Transcription occurs at maximal levels. (b) When both lactose and glucose are available, both CAP and the *lac* repressor are inactive. Transcription levels are low. (c) When neither glucose nor lactose is available, both CAP and the *lac* repressor are bound to their regulatory sites. CAP binding enhances RNA polymerase binding to the promoter. However, the repressor blocks transcription. Transcription levels are low. (d) When glucose is available and lactose is not, CAP is inactive and the *lac* repressor is active. Thus RNA polymerase binds inefficiently, and those polymerase molecules that do bind are blocked by the repressor. This condition results in the lowest levels of transcription observed for the *lac* operon.

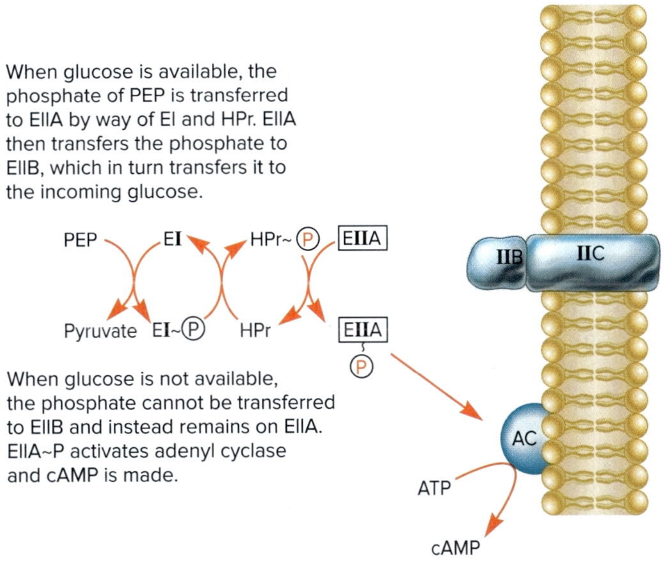

Figure 14.19 Activation of Adenyl Cyclase by the Phosphoenolpyruvate: Sugar Phosphotransferase System (PTS). PEP is phosphoenolpyruvate. EI and EII are enzymes I and II of the PTS, respectively. EII has three subunits: A, B, and C. HPr, heat-stable protein.

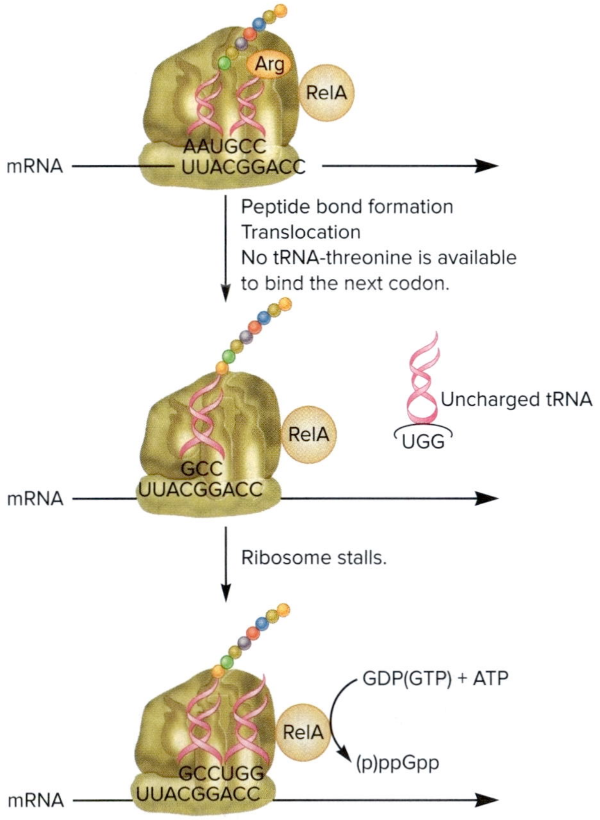

Figure 14.20 Activation of RelA in Response to Amino Acid Starvation.

When bacteria are starved of amino acids, tRNA molecules cannot attach their cognate amino acids. Thus during translation, there are no charged tRNAs to enter the ribosomal A site, and the ribosome stalls (**figure 14.20**). However, an uncharged tRNA eventually enters the A site. This triggers an enzyme called RelA, which is associated with the ribosome, to catalyze (p)ppGpp synthesis (figure 14.15*b*). The effects of (p)ppGpp are wide ranging and universally inhibitory. Because (p)ppGpp is structurally related to GTP, it is a competitive inhibitor of many translation factors (*see figures 13.38 and 13.39*). It also destabilizes RNA polymerase promoter open complexes (*see figure 13.26*) thereby blocking transcription, it inhibits the replication protein primase (*see figure 13.12*), and it inhibits several metabolic enzymes. Eventually (p)ppGpp is hydrolyzed, and the cell resets its level of metabolic activity to one better suited to the nutrients available. ◀ *DNA replication in bacteria (section 13.3); Transcription in bacteria (section 13.5); Translation in bacteria (section 13.7)*

Although the stringent response functions in a wide variety of bacteria, the precise triggers differ, reflecting the challenges unique to each microbe's lifestyle. Enzymes called small alarmone synthetases and small alarmone hydrolases set the cellular concentration of (p)ppGpp and are themselves regulated by global mechanisms like sigma factors. Some cells have these enzymes as domains within a larger protein to directly adjust (p)ppGpp levels upon activation from a sensory domain within the same protein. Recent studies have demonstrated high levels of (p)ppGpp in persister cells, suggesting a possible mechanism for their formation, as an extreme response to slowing the growth rate. ◀ *Microbial growth in natural environments (section 7.6)*

Signaling by c-di-GMP

The second messenger c-di-GMP is synthesized by enzymes called diguanylate cyclases (DGCs). Once formed, c-di-GMP functions until it is destroyed by enzymes called phophodiesterases (PDEs; figure 14.15*d*). Each type of enzyme includes a characteristic sequence of amino acids: GGDEF for DGCs, and EAL (or HD-GYP) for PDEs (*see table 13.3*). C-di-GMP was discovered in the 1980s but drew relatively little interest until the annotation of many bacterial genomes led to the discovery that an unexpectedly large number of proteins contained the amino acid sequences GGDEF and EAL. When the expression of these proteins was altered by genetic manipulation, a host of important and complex behaviors were affected. These included biofilm formation, motility, cell cycle events, and production of virulence factors. Interestingly, if a GGDEF-containing protein was overexpressed and affected a particular behavior, then overexpression of an EAL-containing protein had the opposite effect on the same behavior.

Since then, scientists have begun to unravel the mechanism of action of c-di-GMP. Both DGCs and PDEs have at least one sensory domain that detects environmental signals such as oxygen levels, decreasing nutrients, and light. When a DGC detects its signal, it catalyzes formation of c-di-GMP. When PDE receives its signal, it degrades c-di-GMP. When c-di-GMP is available, it interacts with an effector. Most effectors thus far identified are proteins, but as we note in section 14.3, some are riboswitches. Upon interacting with c-di-GMP, the effector is able to influence the activity of its target. Some effectors are transcriptional regulatory proteins and thus control transcription. Others act post-translationally, directly changing the activity of enzymes. Different effectors in the same cell have different affinities for c-di-GMP, allowing for responses at varying intracellular c-di-GMP concentrations.

The example of the transition from planktonic (motile) to biofilm (sessile) growth in the Gram-negative pathogen *Pseudomonas aeruginosa* demonstrates multiple roles of c-di-GMP signaling. Increasing cellular concentration of c-di-GMP exerts two effects that promote biofilm formation. First, c-di-GMP activates a transcription factor, resulting in the expression of biofilm-related genes, including the *cdrBA* operon. This operon encodes a type V secretion system (CdrB) and its cargo, CdrA (**figure 14.21*a***). CdrA is an adhesin displayed on the outer membrane that anchors the cell to the substrate in the developing biofilm. The CdrA C-terminus has two cysteine residues that are oxidized to a disulfide bond upon reaching the periplasm. The C-terminus remains in the periplasm in a shape that plugs the CdrB secretion system lumen and holds the adhesin to the cell. The second effect of c-di-GMP is on the multidomain protein LapD. LapD spans the plasma membrane with domains in both the cytoplasm and the periplasm. When c-di-GMP binds the LapD cytoplasmic domain, the conformational change propagates to the periplasmic domain, where LapD now binds and restrains a protease called LapG. Under these conditions, cells actively form a biofilm. As c-di-GMP levels fall, cells exit the biofilm by a reversal of the LapD-LapG interaction (figure 14.21*b*). As PDE activity eliminates c-di-GMP, the LapD cytoplasmic domain changes configuration, thereby releasing LapG to the periplasm. LapG cleaves CdrA near the C-terminus, releasing the cell from the adhesin, thereby allowing it to break free of the biofilm matrix. ◀ *Two-step secretion systems (section 13.8)*

Comprehension Check

1. What are global regulatory systems and why are they necessary? What are ECF sigma factors? How do they differ from other alternate sigma factors?
2. What is diauxic growth? Explain how catabolite repression causes diauxic growth.
3. Describe the events that occur with *E. coli* in each of the following growth conditions: in a medium containing glucose but not lactose; in a medium containing both sugars; in a medium containing lactose but no glucose; and in a medium containing neither sugar.
4. *E. coli* has two phosphate uptake systems (one with a high affinity for phosphate, the other with a low affinity). Describe how a two-component regulatory system might be used by *E. coli* to regulate phosphate transport.
5. Why is the stringent response typically active during nitrogen starvation but not in response to carbon sources?

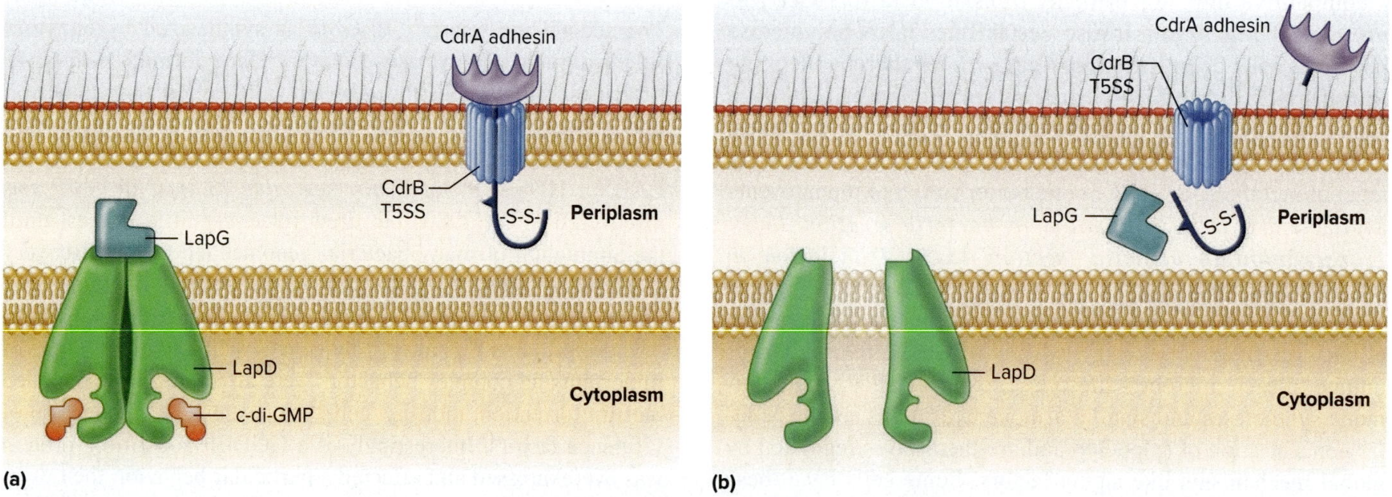

Figure 14.21 Sessile to Motile Transition in a *Pseudomonas* biofilm. (a) Surface adhesins attach cells to a substrate and are anchored to the cell through a type V secretion system. C-di-GMP binding to LapD promotes formation of a structure to sequester the LapG protease. (b) When c-di-GMP levels fall, LapG is released from LapD and cleaves the adhesin's periplasmic domain. The cell thus breaks free of the substrate and swims away from the biofilm.

14.6 Bacteria Combine Several Regulatory Mechanisms to Control Complex Cellular Processes

After reading this section, you should be able to:

a. Outline how the proteins of the *E. coli* chemotaxis system control flagellar rotation in the presence of a chemoattractant
b. Predict the events that would occur if *E. coli* cells were in a gradient of a chemorepellent
c. Differentiate quorum sensing observed in *Vibrio fischeri* from that in *Vibrio harveyi*
d. Outline the regulatory mechanisms used by *Bacillus subtilis* to control endospore formation
e. Compare the activities of restriction-modification and CRISPR-Cas in cellular responses to bacteriophage infection

In this section, we present four examples of complex processes that represent a significant change in bacterial lifestyle. These changes are energy-intensive, and the cell balances input from multiple sensors before committing its resources. These examples illustrate how global regulatory systems can be used to regulate a wide variety of behaviors, and how individual systems coordinate responses at the molecular level.

Chemotaxis in *E. coli*

Recall from chapter 3 that bacteria are able to sense chemicals in their environment and move either toward an attractant or away from a repellent. For simplicity, we only concern ourselves with movement toward an attractant. The best-studied chemotactic system is that of *E. coli,* which, like many other peritrichously flagellated bacteria, exhibits two movement modalities: a smooth swimming motion called a run, which is interrupted by the second type of movement called a tumble. A run occurs when the flagellum rotates in a counterclockwise direction (CCW), and a tumble occurs when the flagellum rotates clockwise (CW) (*see figure 3.44*). The cell alternates between these two types of movements, with the tumble establishing the direction of movement in the run that follows.

When *E. coli* is in an environment that is homogeneous—that is, the concentration of all chemicals in the environment is the same throughout its habitat—the cell moves randomly, with no apparent direction or purpose; this is called a random walk. However, if a chemical gradient exists in its environment, the frequency of tumbles decreases as long as the cell is moving toward the attractant. In other words, the length of time spent moving toward the attractant is increased, and eventually the cell gets closer to the attractant. The process is not perfect, and the cell must continually readjust its direction through a trial-and-error process mediated by tumbling. When one examines the path taken by the cell, it is similar to a random walk but is biased toward the attractant.

For many decades, scientists have been dissecting this complex behavior to understand how *E. coli* senses the presence of an attractant, how it switches from a run to a tumble and back again, and how it "knows" it is heading in the correct direction. These studies reveal that the chemotactic response of *E. coli* involves a two-component signal transduction system and covalent modification of chemotaxis receptors. As we noted previously, most two-component systems are used to regulate transcription initiation. However, some regulate protein activity. This is the case for chemotaxis. |◀ *Flagellar movement (section 3.8)*

For chemotaxis to occur, *E. coli* must determine if an attractant is present and then modulate the activity of the two-component system that dictates the rotational direction of the flagellum (i.e., either run or tumble). *E. coli* senses chemicals in its environment when they bind to chemoreceptors (**figure 14.22**). Numerous chemoreceptors have been identified. Here we focus on one class of receptors called methyl-accepting chemotaxis proteins (MCPs). The signal transduction system that controls the direction of flagellar rotation consists of the sensor kinase CheA and the response regulator CheY. When activated, CheA phosphorylates itself using ATP (figure 14.22*a*). This phosphoryl group is then quickly transferred to CheY. Phosphorylated CheY diffuses through the cytoplasm to the flagellar motor. Upon interacting with the motor, the direction of rotation is switched from CCW to CW, and a tumble ensues. When CheA is inactive, the flagellum rotates in its default mode (CCW), and the cell moves forward in a smooth run.

As implied by the preceding discussion, the state of the MCPs must be communicated to the CheA/CheY system. How is this accomplished? The MCPs are embedded in the plasma membrane with different parts exposed on each side of the membrane (figure 14.22). The periplasmic side of each MCP has a binding site for one or more attractant molecules. The cytoplasmic side of an MCP interacts with two proteins, CheW and CheA. The CheW protein helps attach the CheA protein to an MCP. Together with CheW and CheA, the MCP receptors form large receptor clusters at one or both poles of the cell (figure 14.22*b*). It is thought that smaller aggregations of the MCPs, CheA, and CheW function as signaling teams and are the building blocks of the receptor clusters (**figure 14.23**).

The MCPs in each signaling team appear to work cooperatively to modulate CheA activity. When any one of the MCPs in the signaling team is bound to an attractant, CheA autophosphorylation is inhibited, the flagellum continues rotating CCW, and the cell continues in its run. Because of this cooperation, the cell can respond to very low concentrations of attractant. Furthermore, it can integrate signals from all receptors in the team (figure 14.23). On the other hand, if attractant levels decrease, so that the level of attractant bound to the MCPs in a signaling team decreases, CheA is stimulated to autophosphorylate, the phosphate is then transferred to CheY, and the cell begins to tumble. However, tumbling does not continue indefinitely. About 10 seconds after the switch to CW rotation occurs, the phosphoryl group

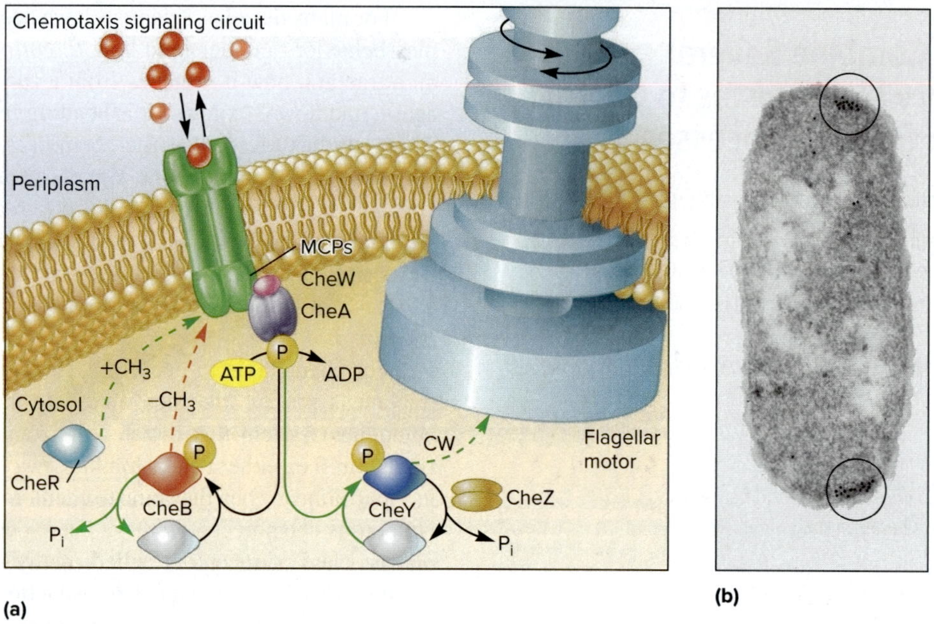

Figure 14.22 Proteins and Signaling Pathways of the Chemotaxis Response in *E. coli*. (a) The methyl-accepting chemotaxis proteins (MCPs) associate with CheA and CheW. CheA is a sensor kinase that, when activated, phosphorylates two different response regulators: CheB, a methylesterase, or CheY. Phosphorylated CheY interacts with the flagellar motor, causing rotation of the flagellum to switch from counterclockwise (CCW) to clockwise (CW). This results in a switch from a run (CCW rotation) to a tumble (CW rotation). The pathway that increases the probability of CCW is shown in red. CCW rotation is the default rotation. It is periodically interrupted by CW rotation, which causes tumbling. The pathways that lead to CW rotation are shown in green. Molecules in gray are unphosphorylated and inactive. (b) MCPs, CheW, and CheA complexes form large clusters of receptors at either end of the cell, as shown in this electron micrograph of *E. coli*. Gold-tagged antibodies were used to label the receptor clusters, which appear as black dots (encircled).

is removed from CheY by the CheZ protein, and CCW rotation is resumed.

But how does *E. coli* measure the concentration of attractant in its environment, and how does it know when it is moving toward the attractant? *E. coli* measures the concentration of an attractant every few seconds and determines if the concentration is increasing or decreasing over time. To compare concentrations of the attractant over time, *E. coli* must have a mechanism for "remembering" the previous concentration. *E. coli* accomplishes this by using covalent modification, specifically methylation of the MCPs. The CheR and CheB proteins compare the overall methylation level of the MCPs (on the cytoplasmic side) with the overall amount of attractant bound (on the periplasmic face). The cytoplasmic portion of each MCP has four to six sites that can be methylated. As long as the concentration of the attractant keeps increasing, the number of MCPs bound to attractant remains high, and the MCP methylation level remains high (i.e., most or all of the sites are methylated). However, if the attractant concentration decreases, the level of methylation will exceed the level of attractant bound. This disparity in methylation level and MCP-bound attractant stimulates CheA to autophosphorylate. As a result, the signal for CW flagellar rotation is initiated, and the cell tumbles in an attempt to reorient itself in the gradient so that it is moving up the gradient (toward the attractant), rather than down the gradient (away from the attractant). At the same time, some of the methyl groups are removed from the MCPs by the methylesterase CheB, establishing a methylation level that is commensurate with the number of MCPs bound to the attractant. A few seconds later, the number of MCPs bound to attractant will be compared to this new methylation level. Based on the correspondence of the two, the cell will determine if it is again moving up the gradient. If it is, tumbling will be suppressed and the run will continue.

Quorum Sensing by *Vibrio* spp.

Cell-to-cell communication among bacterial cells often occurs by the exchange of chemicals often termed signals or signaling molecules. The exchange of signaling molecules is essential in the coordination of gene expression in microbial populations. Cell communication plays an essential role in the regulation of genes whose products are needed for the establishment of virulence, symbiosis, biofilm production, and morphological differentiation in a wide range of bacteria. Quorum-sensing mechanisms vary among microbes. Here we focus on two of the best-studied systems.

The marine bioluminescent bacterium *Vibrio fischeri* uses quorum sensing to regulate light production within the light organ of its squid host; cells produce light only at a high

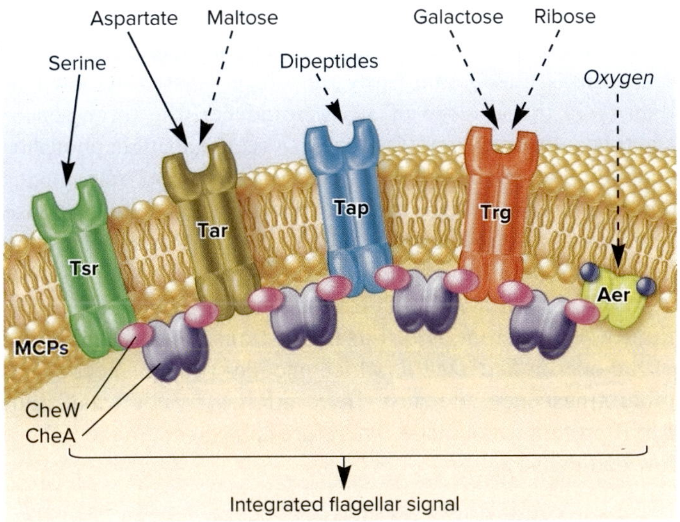

Figure 14.23 ***E. coli* Methyl-Accepting Chemotaxis Proteins.** The attractants sensed by each methyl-accepting chemotaxis protein (MCP) are shown. Some are sensed directly, when the attractant binds the MCP (solid lines). Others are sensed indirectly (dashed lines). Maltose, dipeptides, galactose, and ribose are detected by their interaction with periplasmic binding proteins. Oxygen is detected indirectly by the Aer chemoreceptor, which lacks a periplasmic sensing domain. Instead, the cytoplasmic domain has a binding site for FAD. FAD is an electron carrier found in many electron transport systems. The redox state of the MCP-bound FAD molecule is used to monitor the functioning of the electron transport system. This in turn mediates a tactic response to oxygen.

MICRO INQUIRY *Why doesn't the Aer receptor need a periplasmic domain?*

density. Quorum sensing is also used to control genes whose products are needed for maintenance of the symbiotic relationship between *V. fisheri* and its host. As a result, the squid/*V. fisheri* symbiosis has become an important model for understanding animal-bacterial associations. Our focus is on the regulation of a single operon, that involved with bioluminescence. However, it should be kept in mind that **quorum sensing** regulates multiple genes and operons. ◀ *Cell-cell communication within microbial populations (section 7.6)*

Quorum sensing in *V. fischeri* and many other Gram-negative bacteria uses an ***N*-acylhomoserine lactone** (**AHL**) signal (**figure 14.24**). Synthesis of this small molecule is catalyzed by an enzyme called AHL synthase, the product of the *luxI* gene. The *luxI* gene is subject to positive autoregulation; that is, transcription of *luxI* increases as AHL accumulates in the cell. This is accomplished through the transcriptional activator LuxR, which is active only when AHL binds to it. Thus a simple feedback loop is created. Without AHL-activated LuxR, the *luxI* gene is transcribed only at basal levels. When *V. fisheri* cell density within the squid light organ is low, the small amounts of AHL produced by the bacterial cells freely diffuse out of each cell and accumulate in the environment. As cell density increases, the concentration of AHL outside each cell eventually exceeds that inside the cell, and the concentration gradient is reversed. As the intracellular AHL concentration increases, it binds and activates LuxR. LuxR then increases transcription of *luxI* and the genes whose products are needed for bioluminescence (*luxCDABEG*). Quorum sensing is often called **autoinduction,** and the AHL signal is termed the **autoinducer** (**AI**) to reflect the autoregulatory nature of this system. *Quorum Sensing*

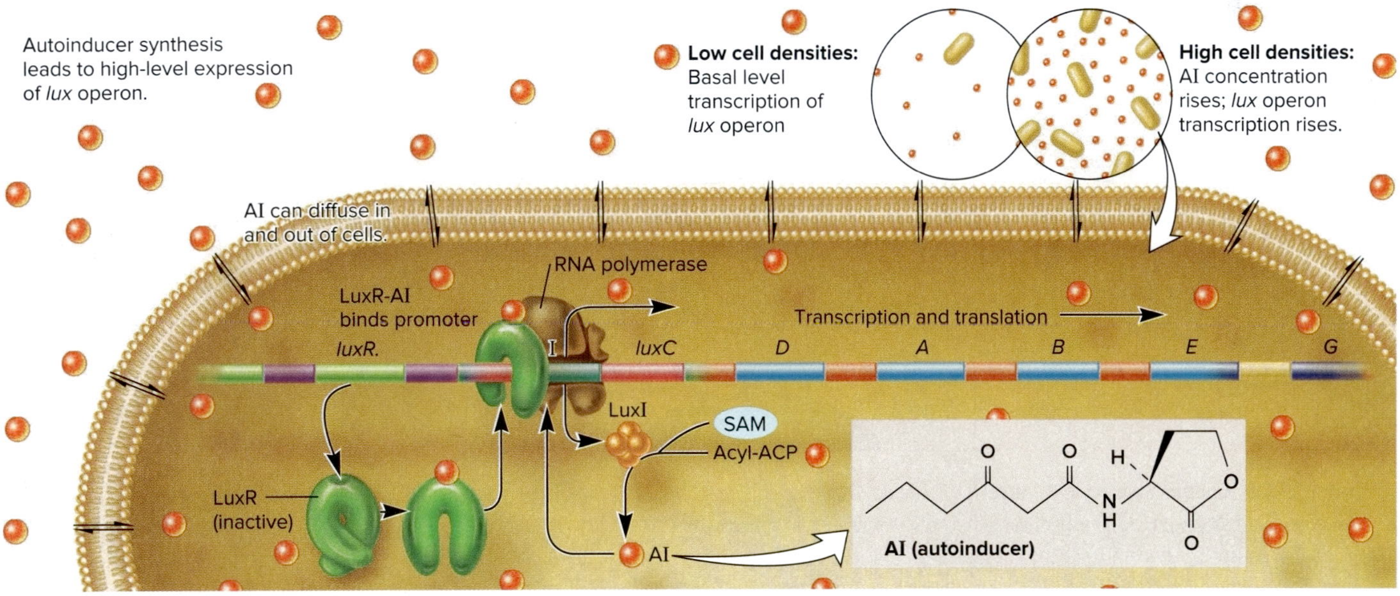

Figure 14.24 **Quorum Sensing in *V. fischeri*.** The AHL signaling molecule (AI) diffuses out of the cell; when cell density is high, AHL diffuses back into the cell, where it binds to and activates the transcriptional regulator LuxR. Active LuxR then stimulates transcription of the gene coding for AHL synthase (*luxI*), as well as the genes encoding proteins needed for light production.

Another kind of quorum sensing depends on an elaborate phosphorelay signal transduction system. It is found in both Gram-negative and Gram-positive bacteria, including *Staphylococcus aureus, Ralstonia solanacearum, Salmonella enterica, Vibrio cholerae,* and *E. coli*. It has been best studied in the bioluminescent bacterium *Vibrio harveyi*.

Unlike *V. fischeri, V. harveyi* responds to three autoinducer molecules: HAI-1, AI-2, and CAI-1. HAI-1 (harveyi autoinducer-1) is a homoserine lactone, and its synthesis depends on the *luxM* gene. AI-2 (autoinducer-2) is furanosylborate, a small molecule that contains a boron atom—quite an unusual component in an organic molecule (*see figure 7.24*). Its synthesis relies on the product of the *luxS* gene. CAI-1 (cholerae autoinducer-1) is the product of the enzyme CqsA. As shown in **figure 14.25**, HAI-1, AI-2, and CAI-1 are secreted by cells, which then use separate proteins called LuxN, LuxPQ, and CqsS, respectively, to detect their presence. LuxN, LuxQ, and CqsS are sensor kinases. At low cell density in the absence of any autoinducer, the three sensor kinases autophosphorylate and converge on a single phosphotransferase protein called LuxU. LuxU accepts phosphates from each sensor kinase and then phosphorylates the response regulator LuxO. Phosphorylated LuxO in turn activates the transcription of genes encoding several small RNAs that destabilize *luxR* mRNA. This is accomplished with the aid of the RNA chaperone Hfq. LuxR is a transcriptional activator of the operon *luxCDABE,* which encodes proteins needed for bioluminescence. Because *luxR* mRNA is not translated, LuxR protein is not made. Therefore cells do not make light at low cell density.

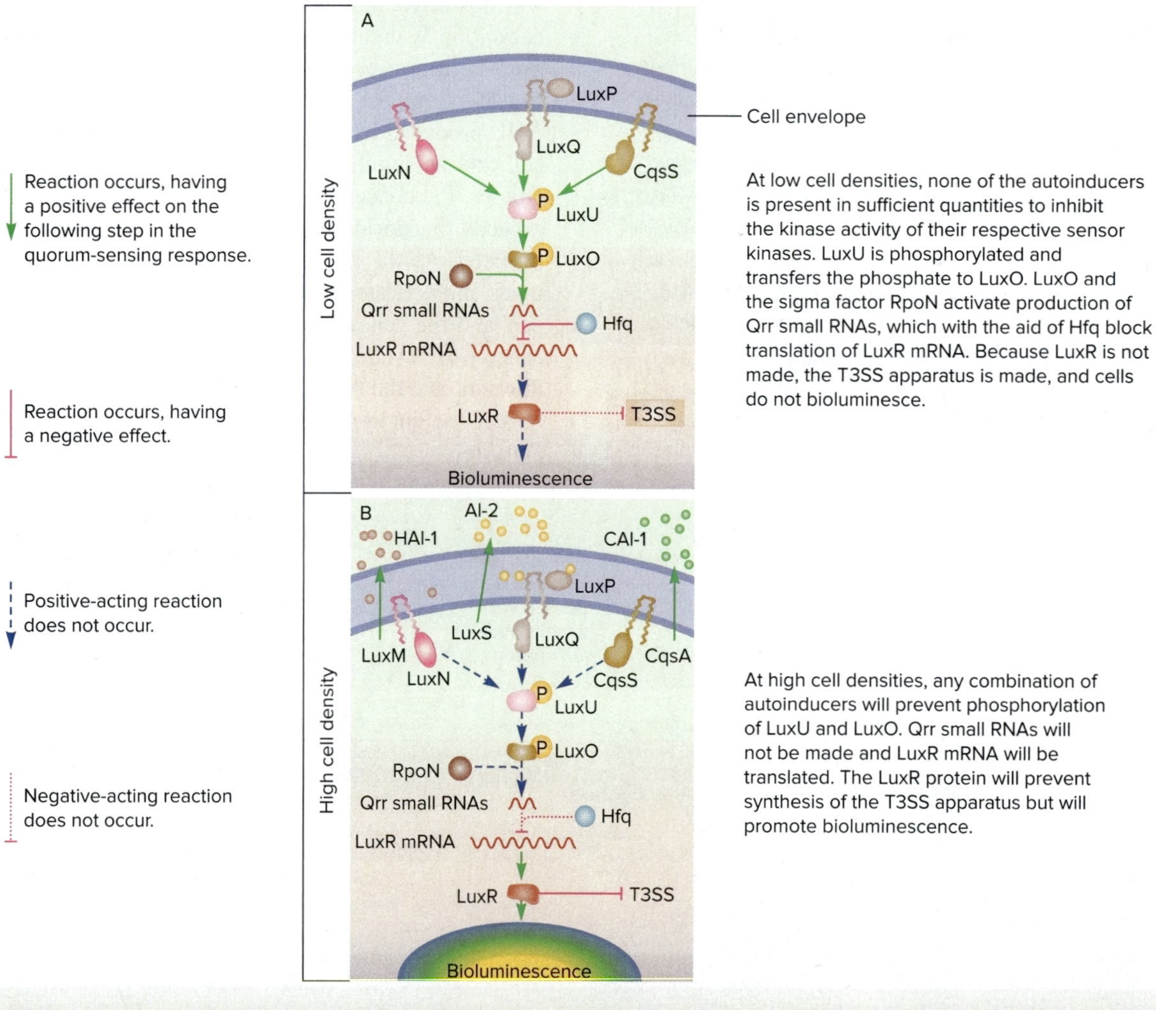

Figure 14.25 Quorum Sensing in *V. harveyi*. Three autoinducing signals—HAI-1, AI-2, and CAI-1—are detected. LuxN, LuxA, and CqsA are sensor kinases that detect HAI-1, AI-2, and CAI-1, respectively. Their kinase activity is inhibited by the presence of their respective autoinducer. All three signals converge at LuxU, which when phosphorylated transfers the phosphate to LuxO. LuxO working with the sigma factor RpoN activates transcription of the *qrr* (quorum regulatory RNA) genes. These gene products are small RNAs that block translation of the LuxR mRNA. This is accomplished in conjunction with the Hfq RNA chaperone. LuxR represses transcription of genes encoding components of a type III secretion system (T3SS) and activates transcription of genes required for bioluminescence.

MICRO INQUIRY *Why does* V. harveyi *make three separate signaling molecules?*

An interesting thing happens as the density of any one of these autoinducers increases: LuxN binds HAI-1, LuxPQ binds AI-2, and CqsS binds CAI-1. When this happens, the proteins switch from functioning as kinases to phosphatases, proteins that dephosphorylate, rather than phosphorylate, their substrates. The flow of phosphates is now reversed; LuxO is inactivated by dephosphorylation, the small RNAs are not made, and *luxR* mRNA is translated. LuxR now activates transcription of *luxCDABE* and light is produced. Careful inspection of figure 14.25 reveals that another set of genes is controlled by the quorum-sensing system of *V. harveyi*. In this microbe, genes for a type III protein secretion system (T3SS) are controlled in the opposite manner as those for bioluminescence.

At this point, we should pause and ask an important question: Why does *V. harveyi* need three different autoinducers? It appears that these molecules allow the bacterium to carry out three different kinds of conversations. HAI-1 is specific to *V. harveyi* and one other closely related species. Thus it is thought to allow *V. harveyi* to communicate with members of its own species. In essence, this autoinducer conveys the message, "There are many *V. harveyi* nearby." AI-2-like molecules are made by many Gram-negative and Gram-positive bacteria. Thus its message is thought to be, "There are many bacteria nearby." Finally, CAI-1 is produced by other members of the genus *Vibrio,* including *V. cholerae,* for which it is named. Its message is thought to be, "There are many *Vibrio* species nearby." As might be expected, each autoinducer has a different signal strength. HAI-1 is the strongest and CAI-1 is the weakest. The presence of all three autoinducers maximizes expression of the bioluminescence genes and fully represses expression of the type III secretion system genes.

Sporulation in *B. subtilis*

As we discuss in chapter 3, endospore formation is a complex process that involves asymmetric cell division to yield a large mother cell and a smaller forespore, engulfment of the forespore by the mother cell, and construction of additional layers of spore coverings (**figure 14.26*a***; *also see figure 3.47*). Sporulation takes approximately 8 hours. Numerous bacterial species form endospores, including *B. subtilis*, which has served as the model organism for understanding sporulation.

In *B. subtilis,* sporulation is controlled by phosphorelay, posttranslational modification of proteins, numerous transcription initiation regulatory proteins, and alternate sigma factors. The latter are particularly important. When growing vegetatively, *B. subtilis* RNA polymerase uses sigma factors σ^A and σ^H to recognize genes for normal growth. However, when cells sense a starvation signal, a cascade of events is initiated that results in the production of other sigma factors that are differentially expressed in the developing endospore and mother cell.

Initiation of sporulation is controlled by the protein Spo0A (Spo is short for sporulation, the number 0 represents the stage of sporulation at which the protein functions), a response regulator protein that is part of a phosphorelay system (figure 14.26*b*). Sensor kinases associated with this system detect environmental stimuli that trigger sporulation. One of the most important sensor kinases is KinA, which senses nutrient depletion and

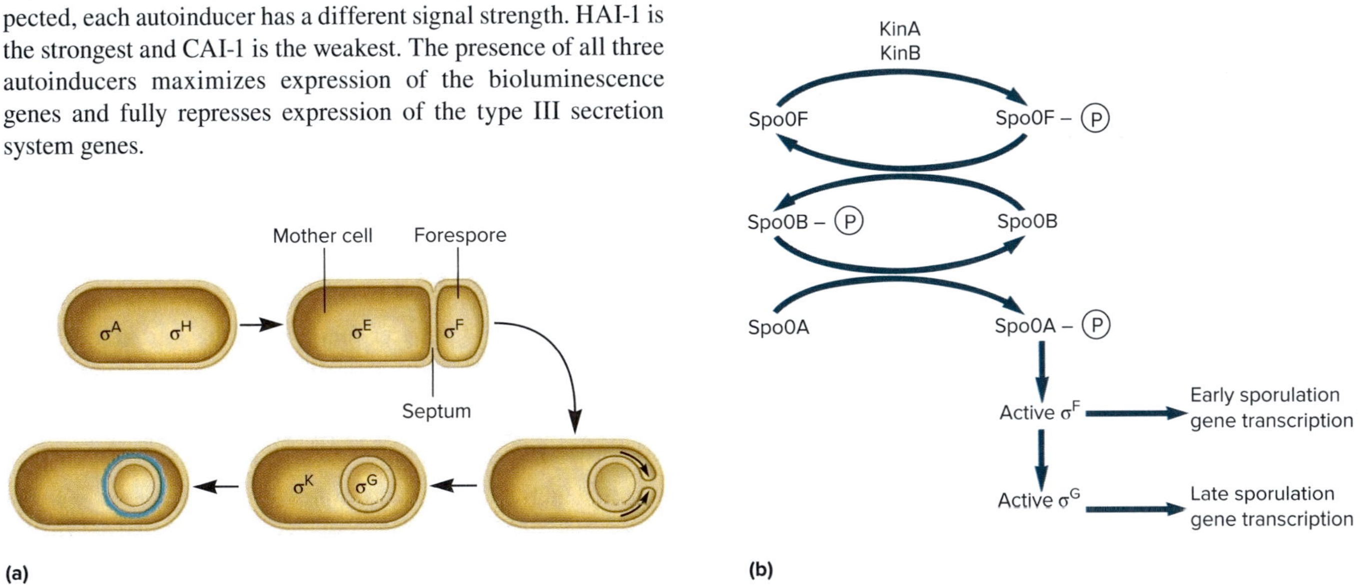

Figure 14.26 Regulation of Sporulation in *B. subtilis*. (a) The initiation of sporulation is governed in part by the activities of two spatially separated sigma factors. σ^F is located in the forespore, while σ^E is confined to the mother cell. These sigma factors direct the initiation of transcription of genes whose products are needed for early events in sporulation. Later, σ^G and σ^K are localized to the developing endospore and mother cell, respectively. They control the expression of genes whose products are involved in later steps of sporulation. (b) The activation of σ^F is accomplished through a phosphorelay system that is triggered by the activation of the sensor kinase protein KinA. When KinA senses starvation, it autophosphorylates a specific histidine residue. The phosphoryl group is then passed in relay fashion from Spo0F to Spo0B and finally to Spo0A.

MICRO INQUIRY *Compartmentation—the spatial localization of proteins as a means of regulation—is usually considered in the context of eukaryotic cells. How is the process of sporulation an example of compartmentation?*

autophosphorylates a specific histidine residue. The phosphoryl group is then transferred to an aspartic acid residue on Spo0F, the second member of the phosphorelay. Spo0F donates the phosphoryl group to a histidine on Spo0B. Spo0B in turn relays the phosphoryl group to Spo0A. Phosphorylated Spo0A positively controls genes needed for sporulation and negatively controls genes that are not needed. In response to Spo0A, the expression of over 500 genes is altered. This has earned Spo0A the name "master regulator." Among the genes whose expression is stimulated by Spo0A is *sigF,* the gene encoding σ^F, and *spoIIGB,* the gene encoding an inactive form of σ^E (pro-σ^E).

When sporulation starts, the chromosome has replicated, with one copy remaining in the mother cell and another to be partitioned in the forespore. Shortly after the formation of the spore septum, σ^F is found in the forespore, and pro-σ^E is localized to the mother cell. Pro-σ^E is cleaved by a protease to form active σ^E. The two sigma factors, σ^F and σ^E, bind to the promoters of genes needed in the forespore and mother cell, respectively. There they direct the expression of genes encoding products needed for the early steps of endospore formation. These genes are primarily responsible for the engulfment process. Another gene regulated by σ^F is one that encodes σ^G, which will replace σ^F in the developing endospore. Likewise, σ^E directs the transcription of a mother-cell–specific sigma factor, σ^K. σ^K ensures that genes encoding late-stage sporulation products are transcribed. These include genes for synthesis of the cortex and coat layers of the endospore. Overall, temporal regulation is achieved because σ^F and σ^E direct transcription of genes that are needed early in the sporulation process, whereas σ^G and σ^K are needed for the transcription of genes whose products function later. In addition, spatial control of gene expression is accomplished because σ^F and σ^G are located in the forespore and developing endospore, whereas σ^E and σ^K are found only in the mother cell.

Responses to Viral Infection

Our discussion of viruses and their life cycles in chapter 6 approaches the subject from the virus's perspective. Far from being passive victims, microbial cells have several systems to actively combat viral infection. The first system discovered was **restriction-modification** in the 1950s by Salvador Luria. He termed the system restriction because certain strains of *E. coli* restricted the growth of bacteriophage. We now know that microbes produce methylase enzymes that modify the nitrogenous bases in short stretches of DNA throughout the chromosome. Bacteria and archaea also produce enzymes called **restriction endonucleases** (**REs**) that recognize the same DNA sequence as the methylase and hydrolyze the phosphodiester bonds only if the DNA has not been methylated. This enables the microbe to protect its own DNA by adding methyl groups. When unmethylated viral DNA enters the cytoplasm, it is recognized by the RE and destroyed. However, when large populations are studied, as in Luria's experiments, a few bacteriophage have their DNA modified by the methylase, escape restriction, and become highly infective. REs are found in most bacteria and archaea, and have been adapted for use in genetic engineering. ▶| *Microbial DNA technologies (chapter 17)*

In an analogy with the immune system of higher organisms, bacterial responses to infection can be viewed as molecular mechanisms to distinguish self from nonself, and to destroy invaders. Restriction-modification is an innate system, encoded on the host chromosome and expressed constitutively at a low level. In addition, many microbes have another defense mechanism: the CRISPR/Cas system, best described as adaptive immunity. The similarity is based on the observation that the CRISPR/Cas system uses parts of the invading virus's genome to "remember" an attack and prepare for a possible future attack, just as the adaptive immune system uses parts of an invading microbe to remember and prepare for the next invasion. ▶| *Adaptive immunity (chapter 33)*

CRISPR/Cas was discovered when genome analyses identified sets of repeated nucleotide sequences separated by short spacers. These sets of nucleotides form a CRISPR array, CRISPR being short for *c*lustered *r*egularly *i*nterspersed *s*hort *p*alindromic *r*epeats. The sequence of base pairs in the repeats is identical (or nearly so) within each CRISPR array. The spacers, however, differ considerably in sequence (**figure 14.27**). Surprisingly, when the spacer sequences were used to probe nucleic acid databases, it was found that the spacers exhibit significant similarity to a variety of bacterial and archaeal virus genomes. CRISPR arrays were later shown to be associated with a set of genes encoding proteins called Cas proteins (short for *C*RISPR-*as*sociated *s*equences). Thus the designation: CRISPR/Cas system. CRISPR/Cas systems have been identified in about 50% of bacterial genomes and 90% of archaeal genomes.

The function of CRISPR/Cas systems is divided into three stages: adaptation, expression, and interference. The adaptation stage occurs upon infection by a virus. If the microbial cell survives (e.g., because the virus can't complete its life cycle, or because its DNA has been cut by REs), it adds pieces of the viral genome to its CRISPR array. The addition is made at the end of the array closest to the *cas* genes, and each new addition consists of a new repeat and a spacer consisting of DNA from the infecting virus. In any given bacterial strain, a CRISPR sequence inserted during the oldest infection is located farthest from the *cas* genes and the most recent is closest to the *cas* genes.

CRISPR response to infection involves the second and third stages of the CRISPR system, expression and interference. In the expression stage, the CRISPR region is transcribed to yield a large RNA containing all repeats and spacers. Cas proteins associate with this full-length RNA and process it into mature CRISPR RNAs (crRNAs). Each of these consists of one repeat and one spacer (figure 14.27). During viral infection, the Cas-crRNAs associate with either viral DNA or viral mRNA, leading to destruction of the molecule (interference). This prevents virus multiplication, and infection is thwarted. The interference stage bears significant similarity to silencing RNAs (siRNAs), as we describe in chapter 15.

CRISPR expression assimilates input from several global regulatory systems. The CRISPR array is believed to be

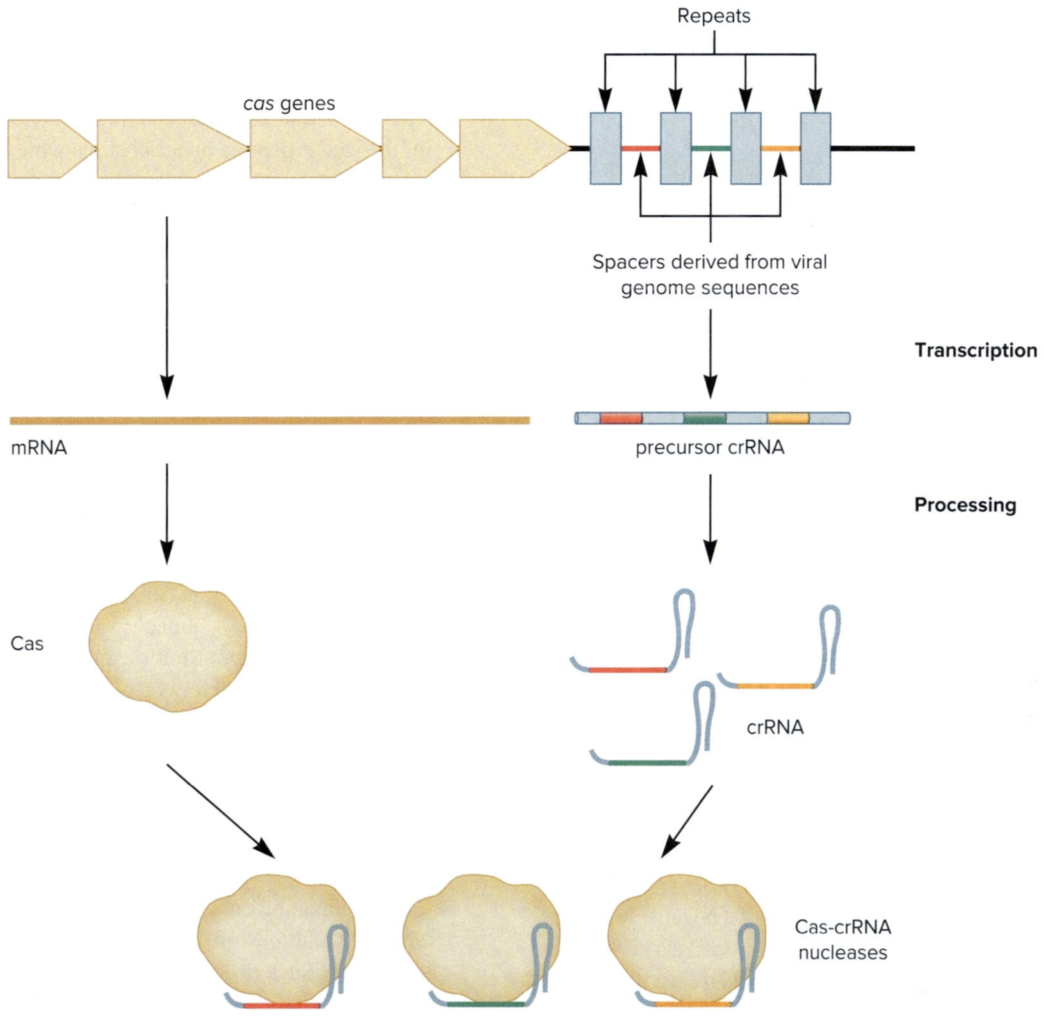

Figure 14.27 CRISPR/Cas System. Bacterial and archaeal genomes contain a series of repeats and spacer sequences in a CRISPR array that is usually adjacent to *cas* genes. Expression of these loci produces Cas proteins or a precursor CRISPR RNA (crRNA). The precursor crRNA is processed to a set of mature crRNAs that each associates with Cas to create a site-specific endonuclease.

constitutively transcribed at a low level, always primed to respond. Quorum sensing signals stimulate CRISPR/Cas expression, as high cell density renders a population of cells susceptible to infection. Although the detailed molecular mechanisms have not been completely established, one or more two-component systems that sense cell envelope stress can activate CRISPR transcription in response to viral contact. In addition, infection often disrupts host cell metabolism, and the CRISPR array responds to changes in cAMP-CAP levels.

Our focus has been on the use of CRISPR/Cas systems by bacteria to defend against viral attacks. However, there is considerably more to these systems. It is now clear that bacteria use CRISPR/Cas for the more general process of protecting their genomes from foreign nucleic acids that might enter via conjugation or natural transformation. Finally, like restriction endonucleases, CRISPR/Cas complexes have been developed into useful tools for genome engineering, as described in chapter 17. ▶| *Bacterial conjugation requires cell-cell contact (section 16.6); Bacterial transformation is the update of free DNA from the environment (section 16.7)*

Comprehension Check

1. Describe the MCP-CheW-CheA receptor complex. What two proteins are phosphorylated by CheA? What is the role of each?
2. How does an MCP regulate the rate of CheA autophosphorylation? How does this mediate chemotaxis?
3. What would be the phenotype of a *V. fischeri* mutant that could not regulate *luxI*, so that it was constantly producing autoinducer at high levels?
4. Why might bacteria use quorum sensing to regulate genes needed for virulence? How might this reason be related to the rationale behind using quorum sensing to establish a symbiotic relationship?
5. Briefly describe how a phosphorelay system and sigma factors are used to control sporulation in *B. subtilis*. Give one example of posttranslational modification as a means to regulate this process.
6. Briefly describe the three stages of CRISPR/Cas response to viral infection. Why is quorum sensing involved?

Key Concepts

14.1 Bacteria Use Many Regulatory Options

- Regulation of gene expression can be controlled at many levels, including transcription initiation, transcription elongation, translation, and posttranslation (**figure 14.1**).

14.2 Regulation of Transcription Initiation Saves Considerable Energy and Materials

- Induction and repression of enzyme levels are two important regulatory phenomena. They usually occur because of the action of regulatory proteins.
- Transcriptional regulatory proteins are DNA-binding proteins. When bound to DNA, they can either inhibit transcription (negative control) or promote transcription (positive control). Their activity is modulated by small effector molecules called inducers, corepressors, and inhibitors (**figure 14.3**).
- Repressors are responsible for negative control. They block transcription by binding an operator and interfering with the binding of RNA polymerase to its promoter or by blocking the movement of RNA polymerase after it binds DNA.
- Activator proteins are responsible for positive control. They bind DNA sequences called activator-binding sites and, in doing so, promote binding of RNA polymerase to its promoter.
- The *lac* operon of *E. coli* is an example of a negatively controlled inducible operon. When there is no lactose in the surroundings, the *lac* repressor is active and transcription is blocked. When lactose is available, it is converted to allolactose by the enzyme β-galactosidase. Allolactose acts as the inducer of the *lac* operon by binding the repressor and inactivating it. The inactive repressor cannot bind the operator and transcription occurs (**figure 14.7**).
- The *trp* operon of *E. coli* is an example of a negatively controlled repressible operon. When tryptophan is not available, the *trp* repressor is inactive and transcription occurs. When tryptophan levels are high, tryptophan acts as a corepressor and binds the *trp* repressor, activating it. The *trp* repressor binds the operator and blocks transcription (**figure 14.8**).
- The *ara* operon of *E. coli* is an example of an inducible operon that is regulated by the dual-function regulatory protein AraC. AraC functions as a repressor when arabinose is not available. It functions as an activator when arabinose, the inducer, is available (**figure 14.9**).

14.3 Attenuation and Riboswitches Stop Transcription Prematurely

- In the tryptophan operon, a region called *trpL* lies between the operator and the first structural gene (**figure 14.10**). It codes for synthesis of a leader peptide and contains an attenuator, a factor-independent termination site. Synthesis of the leader peptide by a ribosome while RNA polymerase is transcribing *trpL* regulates transcription. The tryptophan operon is expressed only when insufficient tryptophan is available. This mechanism of transcription control is called attenuation.
- The 5′ ends (leaders) of some mRNA molecules can bind metabolites and other effectors. Binding of the effector to the mRNA causes a change in the leader structure, which can terminate transcription. This regulatory mechanism is called a riboswitch (**figure 14.11**).

14.4 Riboswitches and Small RNAs Control Translation

- Some riboswitches regulate gene expression at the level of translation. For these riboswitches, the binding of an effector to specific sequences in the leader region of the mRNA alters leader structure and affects ribosome binding either negatively or positively (**figure 14.12**).
- Translation can also be controlled by small RNAs. These small RNA molecules are noncoding. They base-pair to the mRNA and usually inhibit translation (**figure 14.13**).

14.5 Mechanisms Used for Global Regulation

- Some cellular processes are so complex they require the action of multiple operons that must be simultaneously controlled. Such operon networks are global regulatory systems.
- Global regulatory systems often involve many layers of regulation. Mechanisms such as global regulatory proteins, two-component signal transduction systems, phosphorelay systems, alternate sigma factors, and second messengers are often used.
- Expression of the porin genes *ompF* and *ompC* is controlled by a two-component signal transduction system. These systems have a sensor kinase that detects an environmental change. The sensor kinase transduces the environmental signal to the response-regulator protein by transferring a phosphoryl group to it. The response regulator then activates genes needed to adapt to the new environmental conditions and inhibits expression of those genes that are not needed (**figure 14.14**).
- Core RNA polymerase can associate with different sigma subunits to control transcription of classes of genes. Each set of genes controlled by a sigma factor has a promoter sequence specifically recognized by that sigma.
- Diauxic growth is observed when *E. coli* is cultured in the presence of glucose and another sugar such as lactose (**figure 14.16**). This growth pattern is the result of catabolite repression, where glucose is used preferentially over other sugars. Operons that are part of the catabolite

repression system are regulated by the activator protein CAP. CAP activity is modulated by the second messenger cAMP, which is produced only when glucose is not available (**figure 14.19**). Thus when there is no glucose, CAP is active and promotes transcription of operons needed for the catabolism of other sugars (**figure 14.18**).

- The stringent response occurs when bacteria regulate their growth rate. It is mediated by the second messenger guanosine tetraphosphate (ppGpp) and guanosine pentaphosphate (pppGpp). The stringent response results in a decrease in transcription, translation, replication, and metabolism (**figure 14.20**). (p)ppGpp also serves as a general "alarm" for other stress conditions.
- Cyclic dimeric GMP (c-di-GMP) levels in bacterial cells depend on the activity of enzymes called diguanylate cyclases (DGCs), which synthesize c-di-GMP, and phosphodiesterases (PDEs), which degrade c-di-GMP. The DGCs and PDEs have sensory domains that detect environmental signals. Their response to the signal determines the level of c-di-GMP. C-di-GMP interacts with effector molecules that affect a target, bringing about the appropriate response to the environmental signal.

14.6 Bacteria Combine Several Regulatory Mechanisms to Control Complex Cellular Processes

- Chemotaxis in *E. coli* is regulated by covalent modification of chemoreceptor proteins and a two-component signal transduction system that controls the direction of flagellar rotation (**figure 14.22**). Methylation is used to measure the amount of a chemoattractant encountered over time. In this way, *E. coli* can determine if it is moving toward or away from an attractant.
- Quorum sensing is a type of cell-to-cell communication mediated by small signaling molecules such as *N*-acylhomoserine lactone (AHL). Quorum sensing couples cell density to regulation of transcription. Well-studied quorum-sensing systems include the regulation of bioluminescence in *Vibrio* spp. (**figures 14.24** and **14.25**). *V. harveyi* quorum sensing is an example of a phosphorelay signal transduction pathway.
- Endospore formation in *B. subtilis* is regulated by a phosphorelay system that is important in initiation of sporulation and the use of alternate sigma factors (**figure 14.26**).
- Bacteria respond to viral infection by deploying enzymes to degrade viral nucleic acids. Restriction endonucleases recognize unmethylated viral DNA and degrade it. CRISPR/Cas systems maintain a molecular database of viruses in a CRISPR array (**figure 14.27**). Following induction by viral contact, expression of the array results in Cas nucleases targeted to viral nucleic acid.

Active Learning

1. Attenuation affects anabolic pathways, whereas repression affects either anabolic or catabolic pathways. Provide an explanation for this.
2. Describe the phenotype of the following *E. coli* mutants when grown in two different media: glucose only and lactose only. Explain the reasoning behind your answer.
   a. A strain with a mutation in the gene encoding the *lac* repressor; the mutant repressor cannot bind allolactose.
   b. A strain with a mutation in the gene encoding CAP; the mutant form of CAP binds but cannot release cAMP.
   c. A strain in which the Shine-Dalgarno sequence has been deleted from the gene encoding adenyl cyclase.
3. What would be the phenotype of an *E. coli* strain in which the tandem tryptophan codons in the leader region were mutated so that they coded for serine instead?
4. What would be the phenotype of a *B. subtilis* strain whose gene for σ^G has been deleted? Consider the ability of the mutant to survive in nutrient-rich versus nutrient-depleted conditions.
5. Propose a mechanism by which a cell might sense and respond to levels of Na^+ in its environment.
6. Pathogens regulate their virulence factors (proteins that enable them to cause disease) so that they are only expressed when the pathogen is in a susceptible host. This regulation often responds to temperature, for example, when the pathogen is in the human body at 37°C. Several "RNA thermometers" have been described, in which the 5′ untranslated region (5′ UTR) of a gene encoding a virulence factor can detect temperature and affect gene regulation. Suggest a mechanism by which the 5′ UTR might function as a thermosensor. (HINT: Remember that the 5′ UTR is transcribed into mRNA but is not translated.)
7. Some genes whose products confer antibiotic resistance are regulated so that gene expression only occurs in the presence of that antibiotic. Consider an antibiotic that binds to the ribosome to slow translation. Diagram a regulatory circuit that would function through a leader peptide.

8. The sessile-to-motile transition for biofilm bacteria also occurs in Gram-positive organisms. Propose a mechanism to regulate the attachment of an adhesin to a microbe with a Gram-positive cell wall.
9. *Streptomyces coelicolor* encodes dozens of two-component regulatory systems (TCS), identified by genome inspection and comparison to known sensor kinases (SK) and response regulators (RR). Bioinformatic methods, however, cannot identify the activating signal for many of the TCS. A team of scientists created many mutant strains, each lacking a single TCS, and screened them for altered phenotypes. One mutant strain was dramatically different in its ability to produce antibiotics, and this phenotype varied with the concentration of glutamate. Coincidentally, the operon adjacent to the one mutated in this strain encodes a glutamate uptake system (*gluABCD*). This combination of phenotype and genomic location strongly hints at a glutamate response. To identify an interaction between the RR and the *gluABCD* operon, purified RR was mixed with DNA fragments derived from *gluABCD*. Where on the *gluABCD* operon do you think the RR binds? What modification to the RR was found to increase its binding affinity for *gluABCD*? The same experiment was performed with DNA fragments from operons encoding antibiotic biosynthesis enzymes, but no DNA binding was observed. What can be concluded about the regulation of antibiotic biosynthesis in *S. coelicolor*?

Read the original paper: Li, L., et al. 2017. A novel two-component system, GluR-GluK, involved in glutamate sensing and uptake in *Streptomyces coelicolor. J Bacteriol.* 199(8):e00097-117.

15

Eukaryotic and Archaeal Genome Replication and Expression

©AbbieImages/Getty Images

Plastics: Brought to You by Microbes

The next time you discard a plastic water bottle, consider this: That plastic bottle will take thousands of years to be converted back into the smaller chemical molecules from which it was made, assuming it ever degrades. Also consider that each year 140 million tons of equally recalcitrant plastics are purchased worldwide (and usually discarded). And what is often forgotten is that to make all that plastic, 150 million tons of fossil fuels must be processed each year, contributing additional waste, diversion of the fuels from refineries and power plants, and exacerbating geopolitical problems centered on oil and other fossil fuels. Could microbes solve this problem? Microbiologists are trying to answer that question.

Why could microbiologists suggest a solution to the plastics problem? Plastics are organic polymers, such as polyurethane, polystyrene, polyethylene, and polyester. Many bacteria and archaea store carbon as intracellular inclusions of poly-β-hydroxybutyrate (PHB) and other polyhydroxyalkanoates (PHAs). PHAs are polyesters; thus bacteria and archaea have evolved to synthesize a type of plastic. And even more important, because PHA-producing bacteria and archaea have a means to use PHA when carbon is in short supply, they produce the enzymes needed for PHA degradation.

For many years, microbiologists have worked to optimize production of PHAs by bacteria such as *Ralstonia eutropha*, and they have succeeded—up to 80% of cellular dry weight are PHAs in some strains. However, the cost of feeding these chemoorganoheterotrophic bacteria so that they produce PHAs makes the plastics they produce very expensive. This has limited the commercial success of bioplastics.

"If expense arises from feeding a chemoorganotroph," scientists thought, "why not move the bacterial genes encoding PHB-synthesizing enzymes into an organism that feeds itself?" The logical autotrophs were plants such as thale cress (*Arabidopsis thaliana*) and tobacco (*Nicotiana tabacum*). Though successful, scientists were then faced with another dilemma: the use of land to "grow" plastics rather than to grow food.

Once again, microbes may provide a solution. Microbiologists have introduced the genes for PHB synthesis into a microalga, the diatom *Phaeodactylum tricornutum*. The advantages of an alga-based system are that these photoautotrophs need only sunlight and CO_2, and they can be grown in areas (e.g., a desert) not used for food production. This greatly decreases cost and avoids the ethical problems associated with plant-based systems.

Although scientists have not yet maximized diatom PHB production, it is a "proof of concept" experiment. It shows that an alga-based system for plastic production may be successful. The next step is to maximize PHB production in a cost-effective manner. This will require that microbiologists take into account the differences between gene expression in bacteria and eukaryotes. They will need to modify the PHB-synthesis genes such that they are transcribed at a high rate in the diatom, producing mRNA molecules recognized and translated by eukaryotic ribosomes. Fortunately, microbiologists have a wealth of information provided by geneticists and molecular biologists studying plant systems and eukaryotic microbes such as the yeast *Saccharomyces cerevisiae* to guide them. What these geneticists and molecular biologists have learned is the focus of this chapter.

Readiness Check:

Based on what you have learned previously, you should be able to:

- ✓ Discuss bacterial genome structure and the organization of genetic information (sections 13.3 and 13.4)
- ✓ Describe the events that occur during the three phases of DNA replication and how the components of the bacterial replisome participate (section 13.3)
- ✓ Describe the structure of a typical bacterial RNA polymerase holoenzyme and how it functions during the three phases of transcription (section 13.5)

✓ List the major features of the universal genetic code and comment on any deviations from it (section 13.6)
✓ Describe the events that occur during the three phases of translation and the structure of bacterial ribosomes, and identify the initiator tRNA used by bacteria (section 13.7)
✓ Summarize the posttranslational events that result in a properly folded protein in its appropriate location within or outside a bacterial cell (section 13.8)
✓ Trace the movement of materials via the endocytic and secretory pathways of eukaryotic cells (section 5.4)
✓ Describe the structure of mitochondria and chloroplasts (section 5.6)
✓ Outline the steps of the eukaryotic cell cycle

15.1 Why Consider Eukaryotic and Archaeal Genetics Together?

If it looks like a duck, walks like a duck, and quacks like a duck, it must be a duck. Unfortunately, the logic of this saying fails when applied to archaea. For many years, microbiologists thought they looked like bacteria, did the things bacteria do, and therefore were just another group of bacteria. And why not? Archaea look like bacteria, and archaeal energy-conserving and biosynthetic processes and enzymes are similar to those of bacteria. However, archaea are not bacteria. Following Carl Woese's discovery that bacterial and archaeal rRNA molecules differ, other components of their genome replication and expression systems were also found to be distinct. These studies led to the realization that the archaeal machinery for genetic information processing is more similar to that of eukaryotes than of bacteria.

In this chapter, our goal is to accomplish three things: (1) to illustrate the similarities of archaea and eukaryotes with respect to how they replicate and express the information in their genomes, (2) to point out those aspects of the archaeal machinery that are distinct, and (3) to draw comparisons among the members of the three domains of life. To that end, we organize this chapter with sections that follow the steps of genetic information flow in cells. We begin each section with a brief summary of the most salient information about each process as it occurs in bacteria. We then describe the process in eukaryotes and finally consider archaea. As you will see, archaea are an interesting mix of bacterial and eukaryotic characteristics. It is this mix that has prompted further studies to elucidate the evolutionary history of microorganisms.

15.2 DNA Replication: Similar Overall, but with Different Replisome Proteins

After reading this section, you should be able to:

a. Create a table or concept map that illustrates the differences in eukaryotic, archaeal, and bacterial genome structure and organization
b. Distinguish the replicons of eukaryotes and archaeal cells from bacterial replicons
c. Compare and contrast the replisomes of eukaryotes, archaeal cells, and bacterial cells
d. Construct a model to demonstrate the role of telomerase in the replication of eukaryotic chromosomes

DNA replication is similar in all cellular organisms. It is accomplished by a huge complex of many different proteins called the **replisome.** Recall from chapter 13 that the DNA of most bacteria is circular and replication begins at a single origin of replication (*see figure 13.9*). The bacterial replicative DNA polymerase (DNA polymerase III; *see figure 13.11*) is recruited to the origin only after the initiator protein DnaA begins assembly of the bacterial replisome, which is composed of at least 30 proteins. The replication forks move bidirectionally from the origin until the entire circular chromosome is replicated. We now compare this to what occurs in eukaryotes.

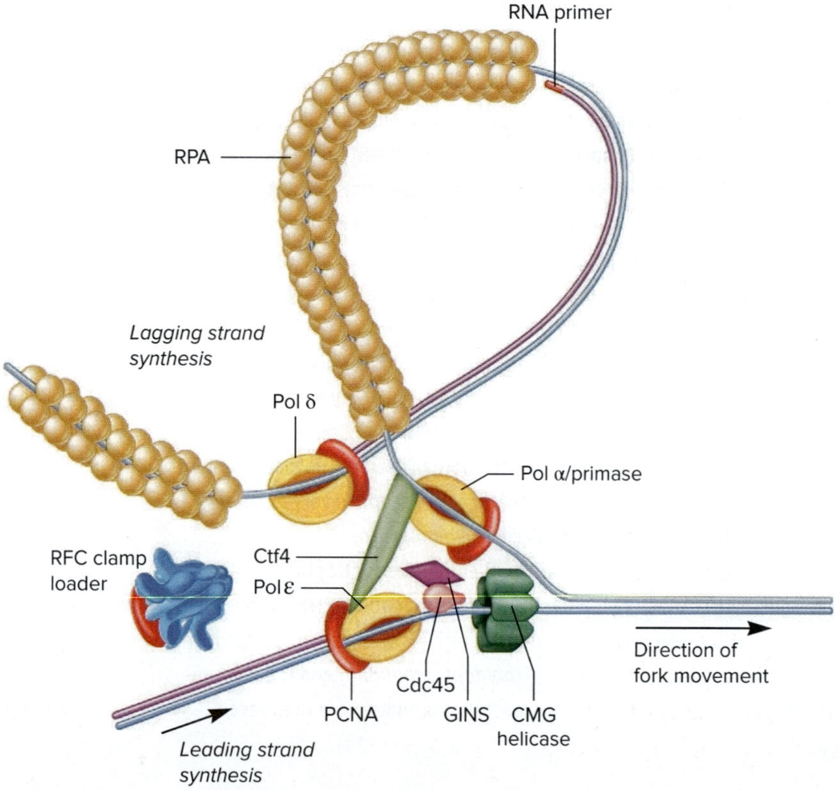

Figure 15.1 A Typical Eukaryotic Replication Fork. The replisome proteins and their functions are defined in table 15.1. Compare this figure to figure 13.13. Parental DNA strands are blue; newly synthesized DNA strands are purple.

Table 15.1 Major Replisome Proteins in Bacteria, Eukaryotes, and Archaea

Function	Bacteria	Eukaryotes	Archaea
Initiator; binds origin[1]	DnaA	ORC	Orc1/Cdc6
Helicase loader	DnaC	Cdc6/Cdt1	Orc1/Cdc6
Helicase	DnaB	CMG complex of Cdc45, *MCM*, GINS[2]	Complex of RecJ/Gan, *MCM*, GINS[2]
Single-strand DNA binding	SSB proteins	Replication protein A (RPA)	SSB or RPA, depending on the species
Primer synthesis	DnaG	Pol α-primase	PriS/PriL[3]
Replicative DNA polymerase	DNA polymerase III (C-family polymerase)	DNA polymerase (Pol) δ and DNA polymerase ε (B-family polymerases)	B-family polymerase (both crenarchaeotes and euryarchaeotes) and D-family polymerase (euryarchaeotes)
Clamp loader	τ complex	Replication factor C (RFC)-L and RFC-S (1-4)[4]	RFC-L and RFC-S[4]
Clamp	β clamp	Proliferating-cell nuclear antigen (PCNA)	PCNA
Primer removal	Ribonuclease (RNase) H[5]; DNA polymerase I	RNaseH[5]; Dna2; Flap endonuclease (FEN)-1	RNaseH[5]; FEN-1
Ligase	NAD^+-dependent DNA ligase; ATP-dependent DNA ligase[6]	ATP-dependent DNA ligase[6]	ATP-dependent DNA ligase; NAD^+-dependent DNA ligase[6]

1 Although we have included the initiators in this table, they are involved only in the formation of the replisome and do not remain in the replisome as it advances from the origin of replication.
2 In eukaryotes, MCM is heteromeric, composed of six different Mcm proteins (Mcm2-7). GINS is short for the first letters of the Japanese numbers representing the proteins composing GINS. In most archaea, MCM is homohexameric. The archaeal MCM is associated with GINS homologues. RecJ/Gan may be the archaeal version of Cdc45.
3 PriS and PriL are homologues of the primase subunits of Pol α-primase.
4 The eukaryotic clamp loader is a pentameric structure consisting of one large subunit (RFC-L) and four small subunits encoded by four different genes. The archaeal clamp loader is also pentameric consisting of RFC-L and four small subunits encoded by one or two genes, depending on the archaeon.
5 The bacterial RNase H is distinct, whereas the eukaryotic and archaeal enzymes are homologues.
6 All NAD^+-dependent DNA ligases belong to the same family of ligases and are evolutionarily related. Likewise ATP-dependent DNA ligases are related. NAD^+-dependent DNA ligases are primarily responsible for ligating Okazaki fragments in bacteria, whereas ATP-dependent DNA ligases play this role in eukaryotes and archaea.

DNA Replication in Eukaryotes

The distinctive features of eukaryotic DNA replication arise from differences in replication machinery and genome structure. Most eukaryotes use a similar set of proteins to replicate their DNA (**table 15.1** and **figure 15.1**). However, eukaryotic replisome proteins are given different names in part to reflect that they are generally unrelated to bacterial replisome proteins. The major features of eukaryotic genome structure relevant to DNA replication include having multiple chromosomes, each of which is usually much larger than a typical bacterial chromosome. Furthermore, the DNA in eukaryotic chromosomes is linear, which means that a mechanism for replicating chromosome ends is needed. Finally, eukaryotic DNA is wound around histones in nucleosomes. Nucleosomes must be deconstructed ahead of the replication fork (i.e., histones removed) and then formed on the parental and newly synthesized strands behind the replication fork.

DNA Replication Initiates from Multiple Origins

Replication of the large, eukaryotic chromosomes is initiated at multiple origins of replication (**figure 15.2**). This allows the

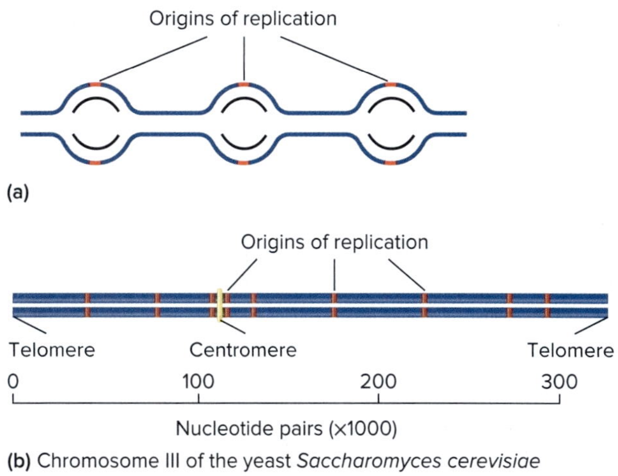

Figure 15.2 Eukaryotic Chromosomes Are Replicated from Multiple Origins of Replication. (a) Drawing of a eukaryotic chromosome being replicated. (b) Schematic of chromosome III from the yeast *Saccharomyces cerevisiae*. It has 180 genes and is replicated from nine origins of replication.

chromosome to be replicated much faster than it could be if there were only one origin per chromosome. Two replication forks move outward from each origin until they encounter replication forks that formed at adjacent origins. Thus eukaryotic chromosomes consist of multiple **replicons,** rather than the single replicon (i.e., the entire chromosome) observed in bacteria.

In eukaryotes, origins of replication are "marked" by a complex of proteins called the **origin recognition complex** (**ORC**), which remains bound to origins throughout much of the cell cycle (table 15.1). All ORCs consist of six different ORC proteins (Orc1 to Orc6), but the identity of these proteins varies among eukaryotic species. The ORC serves as a platform on which other proteins assemble in a cell cycle–dependent fashion. The first protein to associate with ORC is the helicase loader, either Cdc6 (*c*ell-*d*ivision-*c*ycle protein 6) or Cdt1 (*C*dc10-*d*ependent-*t*ranscript 1 protein), depending on the organism. This occurs in the late M/early G1 phases of the cell cycle. Together, ORC and the helicase loader recruit the helicase enzyme, a set of proteins called the MCM (*m*ini*c*hromosome *m*aintenance) complex, to the origin, thereby forming the pre-replicative complex (pre-RC). Pre-RCs are activated as the cell cycle transitions from the G1 phase to the S phase by phosphorylation of the helicase loader, which is then replaced by numerous other proteins, forming the complete replisome (table 15.1). Within the replisome, MCM joins with two coactivators to form the CMG helicase complex and unwinds DNA to initiate replication.

Table 15.2 DNA Polymerase Families

Family	Example(s)	Functions/Comments
A	Bacterial DNA polymerase I	Replacement of RNA primers present in Okazaki fragments
B	Archaeal DNA polymerase B (Pol B) Eukaryotic DNA polymerases α, δ, and ε	Replicative DNA polymerases in eukaryotes and some archaea
C	Bacterial DNA polymerase III	Replicative DNA polymerase
D	Archaeal DNA polymerase D (Pol D)	Replicative DNA polymerase in some archaea; unique to archaea
X	Eukaryotic DNA polymerase β	DNA repair
Y	Bacterial DNA polymerase IV (DinB)	DNA repair
Reverse transcriptase (RT)	Retroviral reverse transcriptases Telomerase RT	RNA-dependent DNA polymerase

Eukaryotic Primases and DNA Polymerases

DNA polymerases from many organisms have been analyzed and sorted into seven families based on their amino acid sequence and structure. As shown in **table 15.2**, eukaryotic DNA polymerases are found in different families than are the bacterial enzymes. In eukaryotes, three DNA polymerases function in DNA replication: DNA polymerases α-primase, ε, and δ. Primer synthesis is accomplished with the enzyme DNA polymerase α-primase (often called simply Pol α-primase). Pol α-primase is so named because it has two activities: DNA synthesis (Pol α) and RNA synthesis (primase). This interesting enzyme consists of four subunits: two for DNA synthesis (Pol 1 and Pol 2) and two for RNA synthesis (Pri1 and Pri2). The Pri1 subunit contains the catalytic site. The primer is made in two steps: The primase component of the enzyme makes a short RNA strand (~10 nucleotides), which is transferred to the active site of Pol α. Pol α adds an additional 20 or so deoxyribonucleotides to the RNA strand. Thus in eukaryotes, the single-stranded primer for DNA replication is an RNA-DNA hybrid molecule. Once the primer is formed, the other two DNA polymerases take over. DNA polymerase ε (Pol ε) catalyzes leading-strand synthesis, whereas DNA polymerase δ (Pol δ) carries out lagging-strand synthesis. Thus unlike most bacteria where DNA polymerase III synthesizes both leading and lagging strands, two distinct DNA polymerases carry out these functions in eukaryotes.

Telomeres and Telomerases: Protecting the Ends of Linear DNA Molecules

The fact that the DNA in eukaryotic chromosomes is linear poses several problems. Without protection, the ends are susceptible to degradation by enzymes called DNases. They are also able to fuse with the ends of other DNA molecules, thus generating aberrant chromosomes. Finally, linear chromosomes present a problem during replication because of DNA polymerase's need for a primer that provides a free 3′-OH. When the primer for the Okazaki fragment at the 5′ end of the daughter strand is removed, the daughter molecule is shorter than the parent molecule. Over numerous rounds of DNA replication and cell division, this leads to a progressively shortened chromosome. Ultimately the chromosome loses critical genetic information. This is called the "end replication problem," and a cell must solve it if it is to survive.

Eukaryotic cells have solved the end replication problem by forming complex structures called **telomeres** at the ends of their chromosomes and by using an enzyme called **telomerase**

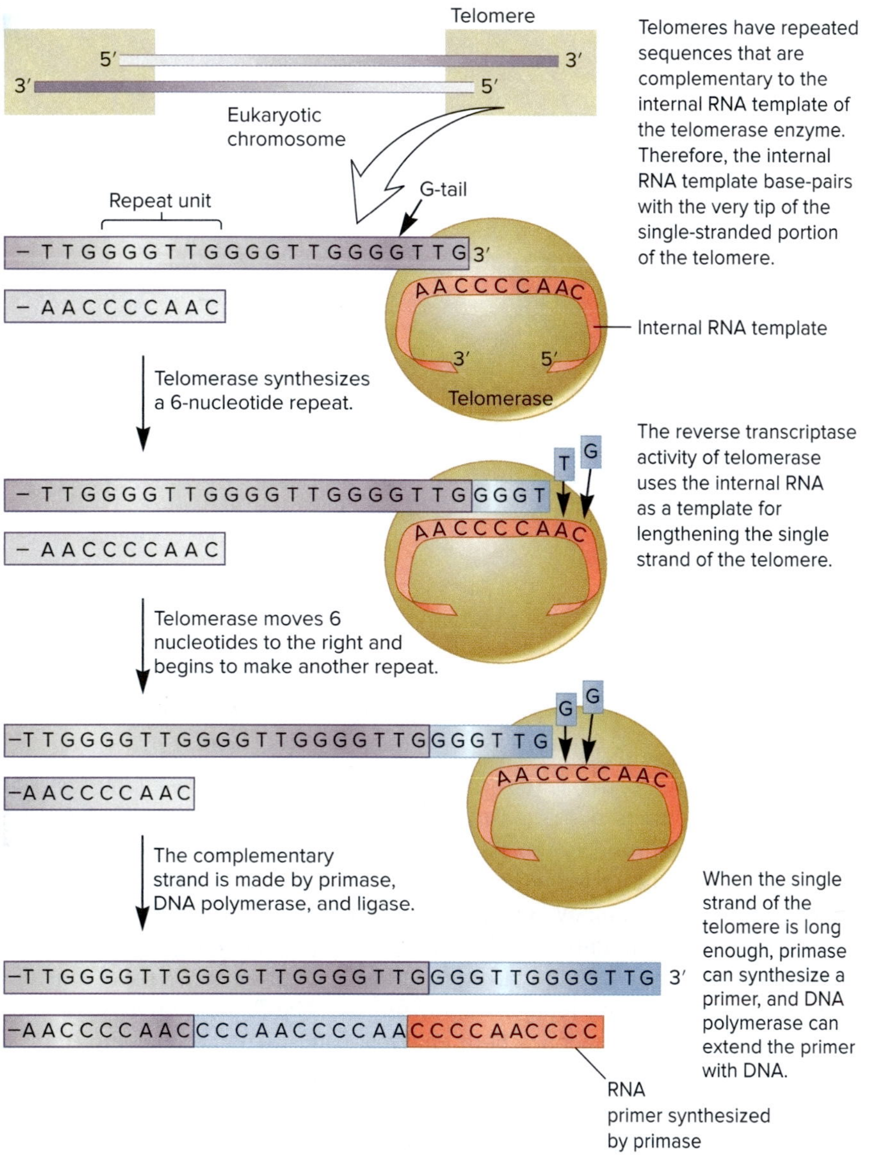

Figure 15.3 Telomerase Replicates the Telomeric DNA of Eukaryotic Chromosomes. Telomerase contains an RNA molecule that can base-pair with a small portion of the 3′ overhang (the G-tail). The RNA serves as a template for DNA synthesis catalyzed by the reverse transcriptase activity of the enzyme. The 3′-OH of the telomere DNA serves as the primer. The process is repeated many times until the 3′ overhang is long enough to serve as the template for synthesis of the complementary telomere DNA strand by the replisome.

(**figure 15.3**). Telomeres are protein-DNA complexes that protect the DNA within them from degradation and end fusion. The protein component of a telomere varies from species to species, as does the length of DNA covered. Telomeric DNA contains many copies of a particular sequence of nucleotides, placed one after the other (tandem repeats). Importantly, the DNA is single-stranded at the very end and is called the G-tail because it is rich in guanosine bases; telomerase uses the G-tail to maintain chromosome length.

Telomerase has two important components: an internal RNA template and an enzyme called telomerase reverse transcriptase (RT). The internal RNA is complementary to a portion of the G-tail and base-pairs with it (figure 15.3). The internal RNA provides the template for DNA synthesis, which is catalyzed by telomerase RT (i.e., the 3′-OH of the G-tail serves as the primer for DNA synthesis). After being lengthened sufficiently, there is room for synthesis of an RNA primer, and the single strand of telomere DNA can serve as the template for synthesis of the complementary strand. Thus the length of the chromosome is maintained.

Telomerase RT deserves additional comment. As described in the preceding paragraph, telomerase RT synthesizes DNA using an RNA template. Enzymes with this capability are termed RNA-dependent DNA polymerases. These enzymes are not unique to telomerases; certain viruses use RNA-dependent DNA polymerases to complete their life cycles (e.g., human immunodeficiency viruses and hepatitis B viruses). These interesting viruses are described more in chapters 26 and 38.

Archaea Use Homologues of Eukaryotic Replisome Proteins

The mixture of bacterial and eukaryotic features is obvious when archaeal genomes are examined. Archaeal chromosomes are circular and similar in size to those of bacteria. However, some archaea have histones associated with their chromosomes just like eukaryotes.

DNA replication in archaea also illustrates their "hybrid" nature. Because few archaea are amenable to laboratory culture, experimental studies on DNA replication are restricted to a small number of organisms, but some generalities can be made. The circular chromosomes of archaea are replicated by replisomal proteins that are similar to those of eukaryotes (**figure 15.4**; table 15.1). Some archaeal chromosomes have a single origin of replication, like bacterial chromosomes; however, others have two, three, or four origins. Multiple origins are believed to have resulted from genetic elements (plasmids, viruses) that have integrated into the chromosome. It is possible to delete one or more origins, and surprisingly, all replication origins could be deleted from *Haloferax volcanii* and *Thermococcus*

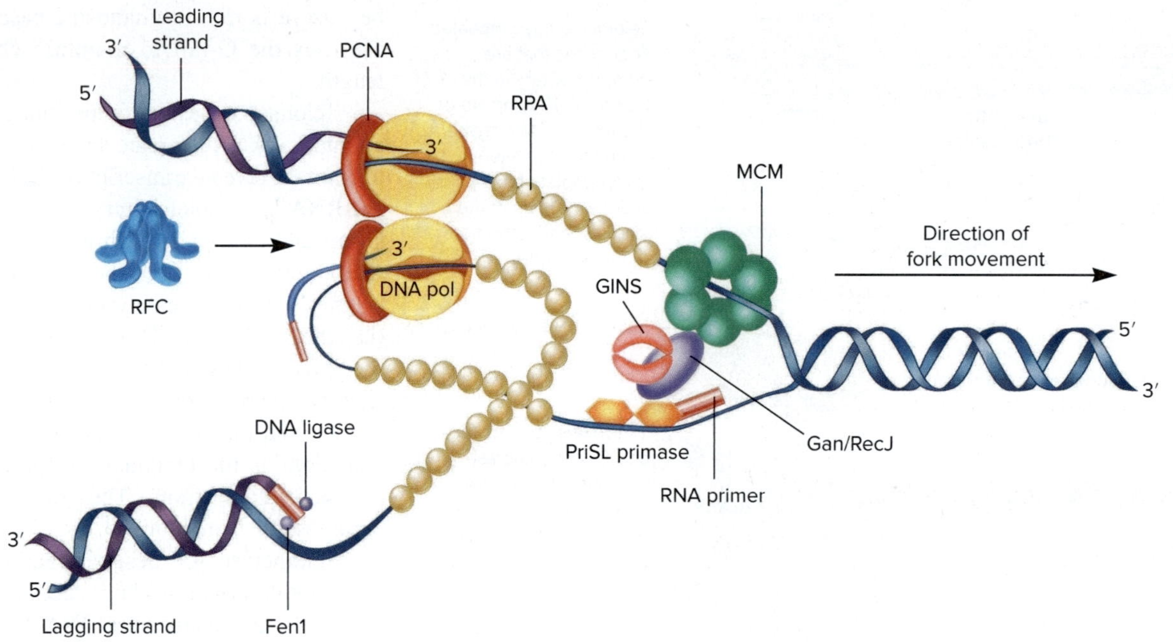

Figure 15.4 An Archaeal Replication Fork. The replisome proteins and their functions are defined in table 15.1. Compare this figure to figures 13.13 and 15.1. Blue, parental DNA strands; purple, newly synthesized DNA strands.

kodakarensis with no effect. These cells used a type of homologous recombination to initiate DNA replication, *r*ecombination-*d*riven DNA *r*eplication (RDR), a phenomenon also observed in mitochondria and chloroplasts.

Many archaeal cells are polyploid, and it is currently unknown if DNA replication and cell division are coupled in these cells. Polyploid cells may not have a defined cell cycle and chromosomes may not be evenly distributed to daughter cells upon cell division. Organisms shown to be capable of RDR are all polyploid, as RDR requires at least two genome copies. Interestingly, among polyploid genomes, many essential genes are located near replication origins. It is possible that this gene order ensures a greater number of copies per cell for these essential genes, as they are among the first replicated. This strategy may also be protective in environments where the cells are prone to DNA damage.

Analysis of archaeal primases and replicative DNA polymerases has led to some intriguing discoveries. Nearly all archaeal primases are heterodimeric; that is, they are composed of two different protein subunits, PriS and PriL, just like their eukaryotic counterparts. PriS is the catalytic subunit and PriL is thought to function as a regulatory subunit. Archaeal primases vary dramatically in terms of their activity. Some make RNA-DNA hybrid primers as in eukaryotes. Others make RNA primers. Furthermore, some archaeal primases have functions other than priming DNA synthesis, including filling gaps as part of DNA repair processes. Other archaea have monomeric primases consisting of both PriS and PriL sequences fused into a single polypeptide.

Archaeal DNA polymerases fall into two families. All archaea studied to date have DNA polymerases in the same family as the replicative eukaryotic DNA polymerases (B family polymerases; table 15.2) and are called Pol B. However, many archaea have a unique DNA polymerase called Pol D. Members of the phylum *Crenarchaeota* lack Pol D, however for other archaea, it has been proposed that Pol D is the primary replicative polymerase. The *T. kodakarensis* chromosome encodes both enzymes, and the genes encoding Pol B could be deleted with little effect on cell growth. Pol D, however, was essential for cell viability. How Pol D functions in organisms like *T. kodakarensis* that use RDR remains to be determined.

Comprehension Check

1. What is the advantage of having multiple origins of replication for a eukaryote?
2. Refer to figure 15.1 and draw the last replication fork present near the end of a chromosome. Use your drawing to illustrate the end replication problem.
3. Review the steps in mitosis and meiosis. Why would having unprotected ends of linear DNA disrupt these two important cellular processes?
4. Outline the process used by telomerase to maintain chromosome length.
5. The discovery of telomerase activity in an archaeon would be a surprise. Why?

15.3 Transcription

After reading this section, you should be able to:

a. Compare and contrast gene structure of eukaryotic, archaeal, and bacterial cells
b. Create a table to distinguish eukaryotic, archaeal, and bacterial promoters, RNA polymerases, and transcription factors
c. Discuss the modifications made to eukaryotic mRNA molecules during and after synthesis

Transcription is essentially the same in all cellular organisms: The template strand directs RNA synthesis, and RNA is made from the 5′ to the 3′ end by enzymes called RNA polymerases. Although this unifies the process of transcription across all domains of life, several key aspects differ. These arise in part from differences in cell and chromosomal structure, the nature of RNA polymerase, and the organization of protein-coding genes. Bacteria use a single RNA polymerase that consists of the catalytic core responsible for synthesizing mRNA and the intrinsic transcription factor subunit, sigma (σ; *see figure 13.23*). Bacterial genes with related function are often organized into operons, and are thereby transcribed from the same promoter. This results in polycistronic mRNA (*see figure 13.22*) from which individual proteins are translated. Some regions of a bacterial gene (e.g., leader and trailer) are transcribed but not translated, and for almost all protein-coding genes, the coding information within a gene is continuous. Now let's compare this to transcription in eukaryotes.

Transcription in Eukaryotes

In eukaryotes, transcription occurs in the nucleus, and the RNA products must move to the cytoplasm, where they function in translation. As in DNA replication, chromatin structure must be altered for RNA polymerases to be recruited to promoters and ahead of moving RNA polymerases as transcription proceeds. This is accomplished by chromatin-remodeling and chromatin-modifying enzymes described in section 15.5. Gene structure and organization in eukaryotes also is strikingly different from that seen in bacteria. For the vast majority of eukaryotes, each protein-coding gene has its own promoter, and the transcript synthesized from the gene is **monocistronic.** Furthermore, many eukaryotic protein-coding genes are composed of regions called exons and introns. An **exon** contains sequences that code for part of the polypeptide, whereas an **intron** is a stretch of noncoding sequences. Thus the coding portion of a eukaryotic gene is not continuous. Both exons and introns are transcribed, yielding an exon- and intron-containing transcript called a **primary** or **pre-mRNA.** The introns are spliced out of the primary transcript before translation occurs.

Eukaryotic RNA Polymerases Include Homologues of the Bacterial RNA Polymerase Core Subunits

Eukaryotes have three major RNA polymerases. RNA polymerase I catalyzes rRNA synthesis, RNA polymerase II is responsible for mRNA synthesis, and RNA polymerase III synthesizes all the tRNAs and one rRNA (5S rRNA). All five of the bacterial RNA polymerase core enzyme subunits are conserved in eukaryotic RNA polymerases (**figure 15.5**). However, eukaryotic RNA polymerases have several additional subunits.

Transcription Initiation in Eukaryotes

Most research on transcription in eukaryotes has been on the transcription of protein-coding genes, and that is our focus here. In eukaryotes, RNA polymerase II (often referred to as RNAPII) is responsible for transcribing protein-coding genes. This complex enzyme is at least 500,000 daltons in size, with about 10 subunits.

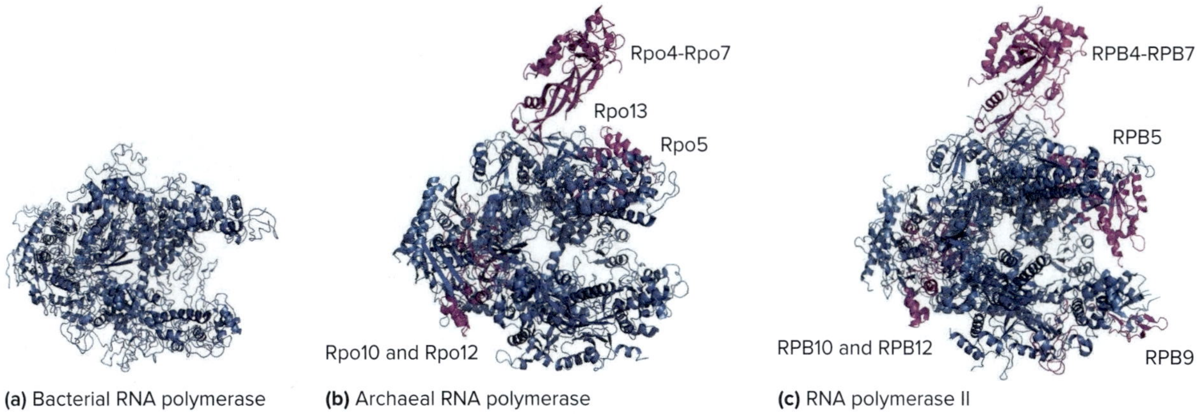

Figure 15.5 Bacterial, Archaeal, and Eukaryotic RNA Polymerases. Ribbon diagrams of the RNA polymerases of (a) the bacterium *Thermus aquaticus*, (b) the archaeon *Sulfolobus shibatae*, and (c) RNA polymerase II of eukaryotes. The subunits found in all three RNA polymerases are shown in blue. Those that are unique to archaea and RNA polymerase II are shown in magenta.
©Finn Werner

Table 15.3 Eukaryotic and Archaeal Basal Transcription Factor Homologues

Eukaryotic Transcription Factor	Archaeal Transcription Factor
TATA-binding protein (TBP)[1]	TBP
TFIIB[2]	TFB[2]
TFIIE	TFE
TFIIF	
TFIIB	
TFIIH	

1 Eukaryotic TBP is often part of a complex of proteins called TFIID.
2 TFII indicates that the protein is a *t*ranscription *f*actor for RNA polymerase *II*. TF indicates that the protein is a *t*ranscription *f*actor for archaeal RNA polymerase.

Transcription initiation requires that RNAPII be recruited to a promoter. This is accomplished by transcription factors called basal or general transcription factors, some of which bind the core promoter and then interact with other transcription factors to attract RNAPII (**table 15.3**). RNAPII-recognized promoters usually have two regions, the core promoter and a regulatory region. The core promoter is the minimal region needed for transcription to occur. One of the first features discovered in eukaryotic RNAPII core promoters was the conserved nucleotide sequence TATA; thus the region is named the TATA box (**figure 15.6**). Although it was initially thought that all RNAPII core promoters contain a TATA box, it is now known that many do not. TATA-less promoters are generally associated with housekeeping genes, that is, genes expressed under most conditions. Those that do contain a TATA box tend to be part of regulated genes.

Eukaryotic transcription factors are referred to as extrinsic factors because they are not part of the RNA polymerase. The TATA-binding protein (TBP) is one of the most important for transcription initiation. It is often part of a complex of proteins called TFIID (short for *t*ranscription *f*actor *D* for RNA polymerase *II*). Despite its name, TBP functions in transcription initiation for both TATA box-containing promoters and those that lack the TATA box.

While TFIID occupies the region around the core promoter, other large multisubunit protein complexes are required as co-activators. SAGA is recruited to upstream activating sequences, where it assembles with other transcription factors. Mediator stabilizes the entire pre-initiation complex (PIC) of RNAPII and transcription factors. These large assemblies coordinate input from transcription factors and chromatin structure to initiate transcription. Once the PIC (**figure 15.7**) is formed, TFIIE positions TFIIH to denature six base pairs of the DNA, forming an open complex (*see figure 13.26*). RNAPII can then begin synthesizing mRNA. However, transcription cannot continue to the elongation phase until the contacts between TFIIB and RNAPII are broken. This is accomplished by TFIIH, which phosphorylates the C-terminal domain of one of the RNAPII subunits. Once this occurs, RNAPII is released from the PIC and elongation proceeds. The transcription factors and co-activators remain on the promoter to recruit additional RNAPIIs.

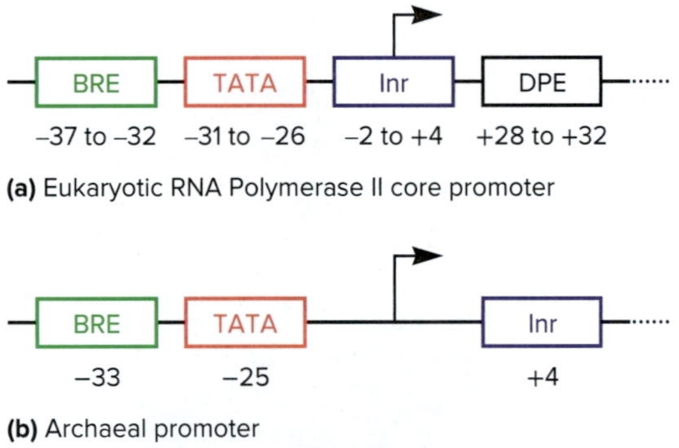

Figure 15.6 Eukaryotic and Archaeal Promoters. (a) Eukaryotic RNA polymerase II core promoters contain one or more of the elements shown. However, it is very rare for all of the elements to be present. BRE is the TFII*B* *r*ecognition *e*lement. Inr is the *in*itiato*r* element. DPE is the *d*ownstream *p*romoter *e*lement. The approximate positions of each element are indicated, relative to the transcriptional start site (indicated by the arrow). (b) An archaeal promoter. Promoter element designations are the same as those in eukaryotic promoters although the exact nucleotide sequences in the promoter elements differ.

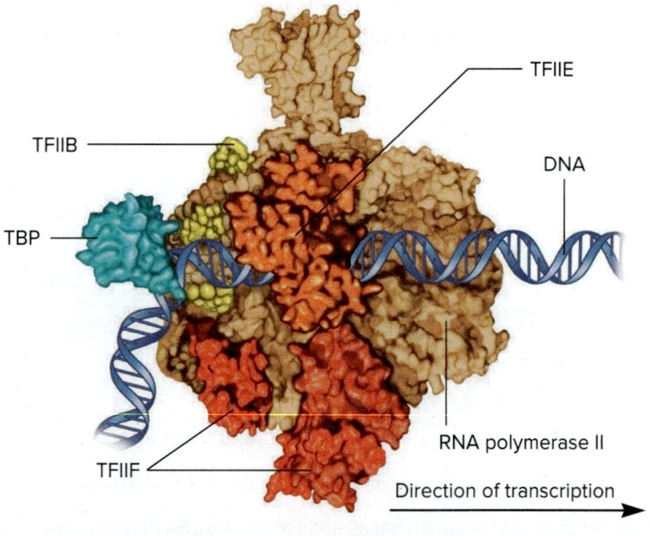

Figure 15.7 Initiation of Transcription in Eukaryotes. Schematic showing the location of some of the transcription factors in the pre-initiation complex (PIC). TFIIH and TFIIA are not visible in this view. Note how closely the transcription factors are associated with RNA polymerase II. The arrow shows the direction of transcription, once it initiates.

From Pre-mRNA to mRNA: A Unique Eukaryotic Process

Unlike nearly all bacterial mRNAs, the initial transcript synthesized from a eukaryotic protein-coding gene must be modified before giving rise to a translatable mRNA. Modification occurs in the nucleus and begins during the elongation phase of transcription; it is completed after transcription ends. Modification starts with the addition of an unusual nucleotide to the 5′ end of the transcript when the transcript is only about 25 ribonucleotides long (**figure 15.8**). The **5′ cap,** as it is called, is 7-methylguanosine. After the 5′ cap is added and a sufficiently long transcript is generated, removal of introns begins. After synthesis is completed, any remaining introns are removed, and the pre-mRNA is modified further by the addition of a **3′ poly-A tail** (figure 15.8). Both the 5′ cap and poly-A tail help protect the mRNA from enzymatic attack. Furthermore, they enable the cell to recognize intact mRNA ready to be transported from the nucleus to the cytoplasm. The 5′ cap has the additional important function of serving as a recognition signal for ribosome binding so that the mRNA can be translated.

Introns are removed from the pre-mRNA by a large complex of proteins and RNA molecules unique to eukaryotic cells. This large complex, called a **spliceosome,** is composed of protein splicing factors and several small nuclear ribonucleoproteins (snRNPs, pronounced "snurps"). The snRNPs consist of small nuclear RNA (snRNA) molecules (about 60 to 300 nucleotides long) associated with polypeptides. Some snRNPs recognize and bind exon-intron junctions (**figure 15.9**). The circular introns that result from splicing (figure 15.9, step 4) appear to persist in cells and there is evidence that they may be translated, although their function is unknown.

Sometimes a pre-mRNA is spliced so that different patterns of exons remain or the junction of two exons varies. This **alternative splicing** allows a single gene to code for more than one protein. The splice pattern determines which protein is synthesized. Splice patterns can be cell-type specific or determined by the needs of the cell. The importance of alternative splicing in multicellular eukaryotes was clearly demonstrated when it was discovered that the human genome has only about 20,000 genes, rather than the anticipated 100,000. Alternative splicing is one mechanism by which human cells produce a vast array of proteins using fewer genes. This mechanism for expanding the coding capacity of a genome is not available to bacteria and archaea because introns in their protein-coding genes are very rare. *Processing of Genetic Information: Prokaryotes Versus Eukaryotes*

Transcription in Archaea

An article describing archaeal transcription might have the headline: "Eukaryotic-like RNA polymerase functions in bacterial-like environment." This is because archaeal transcription occurs in the cytoplasm, as it does for bacteria. Many archaeal mRNAs are polycistronic like bacterial genes. Introns are rare in protein-coding archaeal genes (as they are in bacteria), and the few present are removed by a process different from that used by eukaryotes (again as they are in bacteria). Finally, archaeal cells use a single RNA polymerase to catalyze transcription.

The archaeal RNA polymerase is large and most similar to RNA polymerase II of eukaryotes (figure 15.5). Likewise, transcription initiation in archaea is very similar to that seen in eukaryotes, as evidenced by the similarity in promoters (figure 15.6) and the use of similar basal transcription factors (table 15.3). The archaeal process can be thought of as a simplified version of that in eukaryotes. As shown in **figure 15.10**, two archaeal transcription factors are important for recruiting the archaeal RNA polymerase to a promoter: TATA-binding protein (TBP) and transcription factor B (TFB). A third transcription factor TFE then associates with RNA polymerase and promotes DNA unwinding to form the open complex (*see figure 13.26*). RNA polymerase then moves down the gene, synthesizing

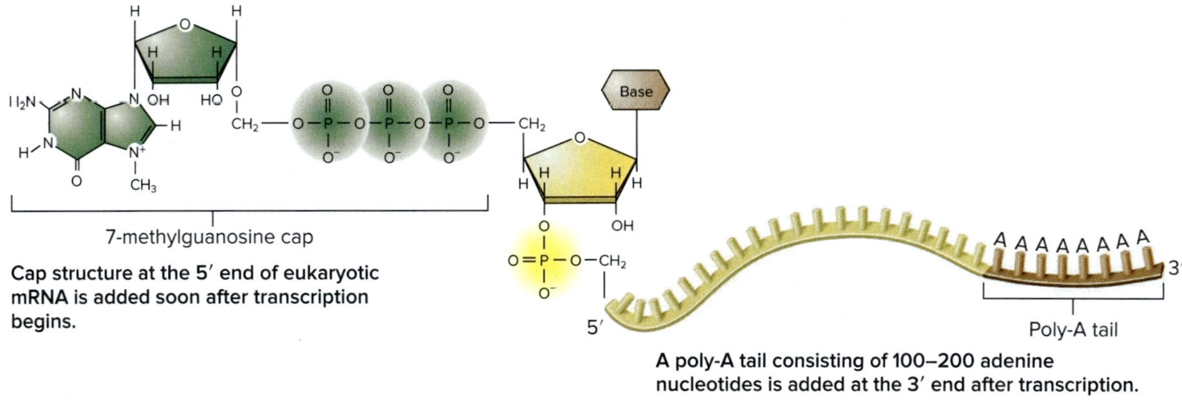

Figure 15.8 Modifications Made to the Ends of Eukaryotic mRNA.

MICRO INQUIRY *What functions are served by the 5′ cap and the 3′ poly-A tail?*

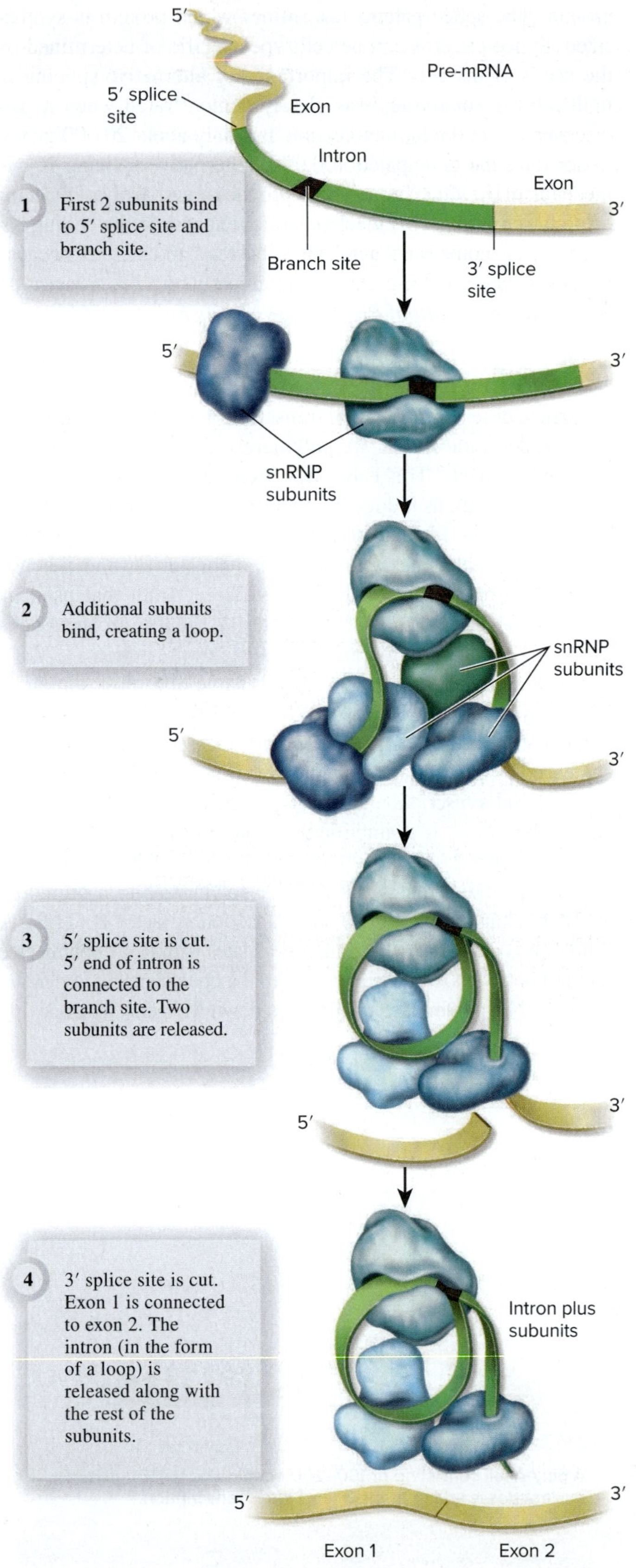

Figure 15.9 **Removal of an Intron by a Spliceosome.**

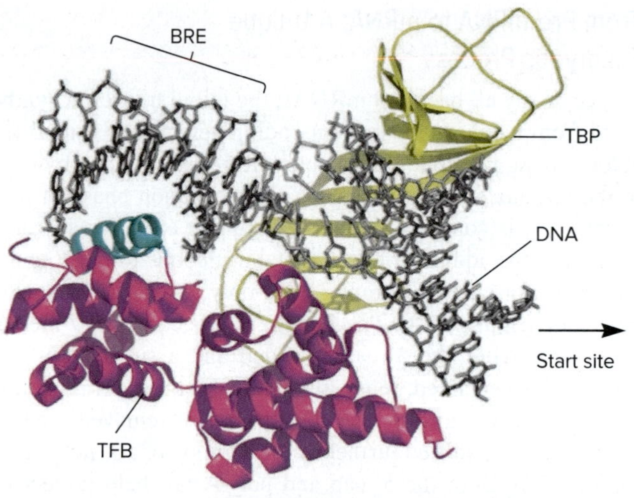

Figure 15.10 **Binding of TBP and TFB to an Archaeal Promoter.** The crystal structure of the ternary complex between TBP, the carboxyl terminus of TFB, and the region of an archaeal promoter containing the TATA box and BRE. The turquoise portion of TFB is an α-helix that recognizes BRE. TBP is TATA-box binding protein; TFB is transcription factor B; BRE is B recognition element.

RNA. TFB remains bound to the promoter, ready to initiate another round of transcription. TFE accompanies RNA polymerase until transcription terminates.

Comprehension Check

1. What elements in archaeal promoters are also observed in eukaryotic RNA polymerase II (RNAPII) core promoters? What additional elements are observed in RNAPII core promoters? How do bacterial promoters differ from those seen in eukaryotic and archaeal cells?
2. Compare the role of bacterial sigma factors in transcription initiation to that of eukaryotic and archaeal transcription factors. Why are bacterial sigma factors referred to as intrinsic transcription factors, whereas eukaryotic and archaeal transcription factors are extrinsic?
3. Outline the events that generate a mature, translatable mRNA in eukaryotes. Where does this processing occur?

15.4 Translation and Protein Maturation and Localization

After reading this section, you should be able to:

a. Differentiate the structure of eukaryotic, archaeal, and bacterial ribosomes
b. Identify the initiator tRNA used by eukaryotic and archaeal cells
c. Outline the steps in translation initiation observed in eukaryotes and compare them to those observed in bacteria and archaea
d. Discuss the role of molecular chaperones in eukaryotic and archaeal cells
e. Create a concept map or table that distinguishes the translocation and secretion systems observed in the three domains of life

For organisms to survive, they must be able to synthesize the proteins they need (translation), fold them into a functional conformation, and localize the proteins to the correct site within (or outside) the cell. In bacteria, translation initiation requires three protein initiation factors (IFs), the initiator tRNA (fMet-$tRNA_i^{Met}$), and the two ribosomal subunits (50S and 30S subunits) (*see figure 13.38*). The 16S rRNA and the Shine-Dalgarno sequence interact to position the ribosome properly at the start codon. Elongation is assisted by three elongation factors (EFs; *see figure 13.39*). Translation is tightly coupled with transcription, and polysomes are often observed (*see figure 13.33*). Translation termination involves three release factors (RFs). **Chaperones** help the protein fold, and some chaperones deliver proteins to protein translocation or secretion systems.

In this section, we focus on translation, and protein folding and localization as they occur in eukaryotes and archaea. As with DNA replication and transcription, the overall processes are similar in all organisms. Therefore we focus our attention primarily on those aspects of the eukaryotic and archaeal processes that differ from those observed in bacteria.

Translation in Eukaryotes Initiates in a Unique Manner but Proceeds As in Bacteria

There are many obvious differences between bacterial and eukaryotic translation. One is that eukaryotic ribosomes are larger than those in bacteria and require more initiation factors to be positioned properly on the mRNA. The 5′ cap and the 3′ poly-A tail of eukaryotic mRNAs are integral to translation initiation. The process begins with mRNA activation, when several eukaryotic initiation factors (eIFs) bind to the 5′ end of the mRNA and poly-A binding proteins (PABPs) attach to the 3′ end. The eIFs and PABPs interact, causing the mRNA to fold back on itself with the eIFs and PABPs forming a bridge between the two ends (**figure 15.11**). Meanwhile, the initiator tRNA (Met-$tRNA_i^{Met}$) is loaded onto the 40S ribosomal subunit (the small subunit, SSU) to form a structure termed the 43S complex. Formation of the 43S complex is facilitated by several additional eIFs. The 43S complex then binds the activated mRNA, generating the 48S complex.

As illustrated in figure 15.11, at this point in translation initiation, the SSU is not positioned at the start codon of the mRNA as it is in bacteria. Rather, it is at the 5′ end of the mRNA. The mRNA is then scanned for the start codon. Once the start codon is encountered, eIFs are released and the 60S ribosomal subunit (large subunit) joins with the 48S complex, forming the complete 80S ribosome. Translation can now commence.

Our description of translation initiation applies to mRNAs having a leader (i.e., a 5′ untranslated region) but some mRNAs lack a leader, notably all mRNAs produced by the protist *Giardia intestinalis*. At least four different mechanisms for translation initiation are available to leaderless mRNAs, to ensure that translation occurs in a wide range of cellular conditions.

Translation elongation and termination proceed in a fashion similar to that seen in bacteria. Three elongation factors function in both bacteria and eukaryotes, and the bacterial and eukaryotic factors are very similar structurally and functionally. Polysomes form during translation of eukaryotic mRNAs. However, because of the circularization of the mRNA during translation initiation, the polysomes have a characteristic structure (**figure 15.12**). Another difference is seen in the termination step. In eukaryotes, a single release factor (eRF) functions as opposed to three RFs in bacteria.

Protein Folding in Eukaryotes Is Similar to That Seen in Bacteria

Eukaryotes, like bacteria, use proteins called chaperones to help many proteins fold properly and restore proper folding if the protein is denatured by environmental conditions. Chaperones were first identified by their role in protecting the proteins of heat-stressed cells and are often called heat-shock proteins (HSPs). There are numerous HSPs, and these are often sorted into types of HSPs based on their size. For instance, the bacterial chaperones GroEL and GroES have a molecular weight of about 60,000 and are included with other HSPs of similar size. All are referred to as HSP60s. Representatives of all types of chaperones are found in eukaryotes.

Just as with bacteria, eukaryotes use several pathways to fold proteins properly (**figure 15.13**). Some chaperones perform their tasks cotranslationally, while others function posttranslationally. About 70% of proteins are successfully folded co-translationally by *r*ibosome-*a*ssociated *c*omplex (RAC) and *n*ascent-chain-*as*sociated *c*omplex (NAC), chaperones that interact with proteins in the ribosome exit tunnel. Proteins needing additional help to fold are bound by an HSP70 chaperone (similar to the bacterial DnaK protein) and an HSP40 cochaperone (similar to the bacterial DnaJ protein). HSP70 is a central folding hub that directs unfolded substrates to a protein called prefoldin, or to other classes of chaperones that may be specific for that substrate. Other proteins are folded by pathways that use HSP60 chaperones similar to the GroEL system of bacteria. In eukaryotes this is called TRiC/CCT [*t*ailless complex polypeptide-1 (TCP-1) *ri*ng *c*omplex/*c*haperonin-*c*ontaining *T*CP-1].

Protein Localization and Secretion in Eukaryotes

Proteins produced by eukaryotic cells have several possible destinations because these cells are highly compartmentalized, having membrane-bound organelles. Although mitochondria and chloroplasts have genomes that encode some of their proteins, most are encoded in the nucleus and translated in the cytosol. Although some proteins synthesized in the cytosol stay there, many are moved to compartments or are secreted. As described in chapter 5, proteins that are secreted or that move between the endoplasmic reticulum, Golgi apparatus, endosomes, and other membranous components of the endocytic and secretory pathways are delivered to these organelles or the plasma

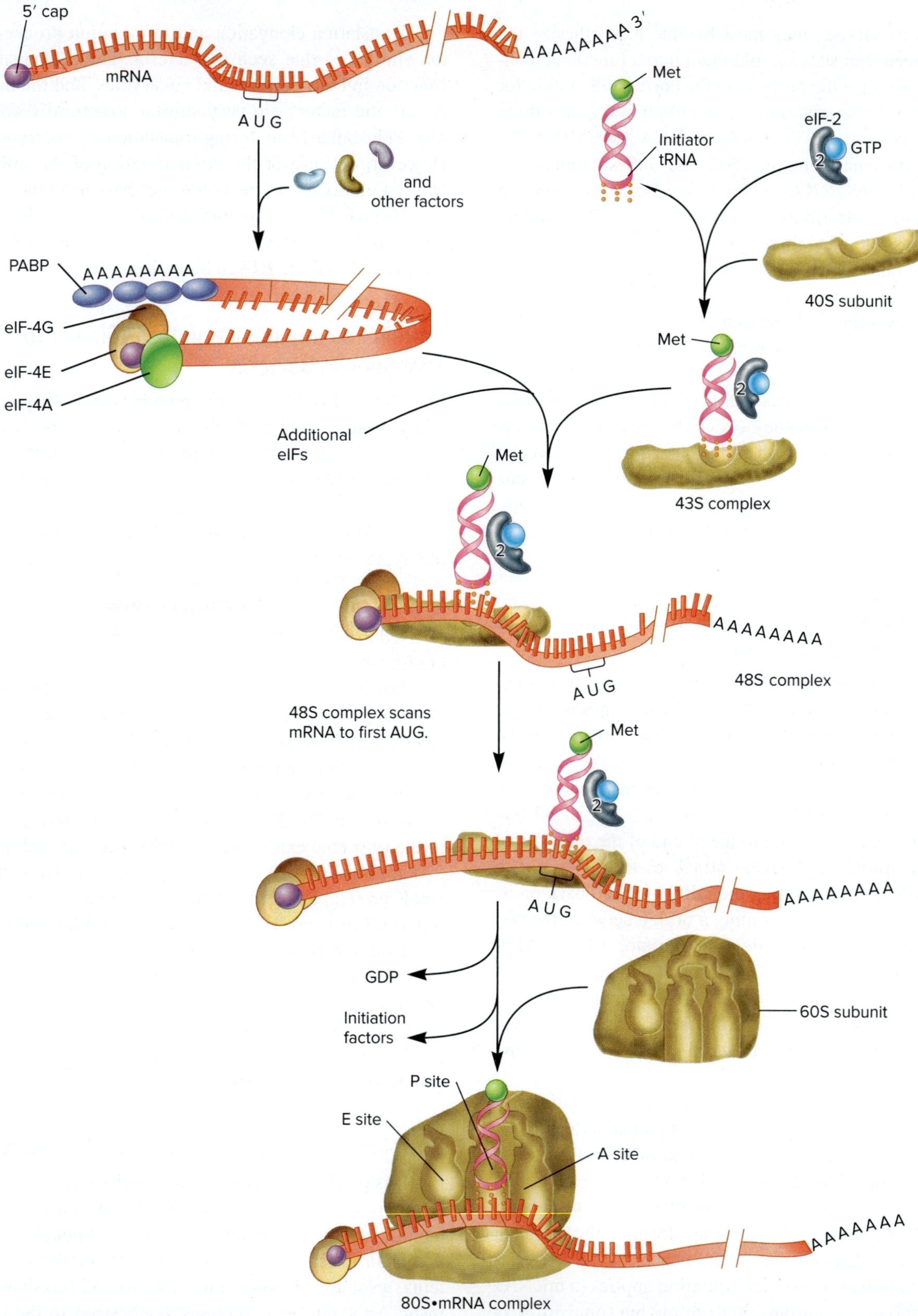

Figure 15.11 Translation Initiation in Eukaryotes. To simplify the drawing, many additional eukaryotic initiation factors (eIFs) are not shown. The upper left portion of the figure shows mRNA activation; the upper right shows the formation of the 43S complex. The 43S complex binds to the 5′ end of the activated mRNA to make the 48S complex, and then moves along the mRNA to find the first AUG. This positions the initiator tRNA carrying methionine in what will become the peptidyl (P) site of the ribosome after the addition of the 60S subunit. The acceptor (A) site is vacant and ready for the arrival of the tRNA carrying the second amino acid encoded by the mRNA. Met is methionine. PABP is poly-A binding protein. E site is exit site.

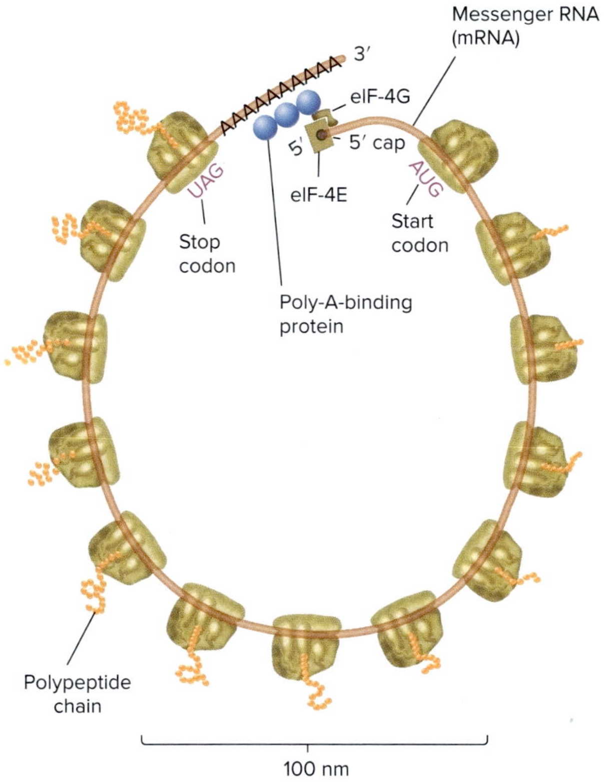

Figure 15.12 Eukaryotic Polysome. Schematic representation of a polysome. The distinctive circular form results from the interaction of the two ends of the mRNA brought about by the poly-A-binding proteins, eIF-4E, eIF-4G, and other initiation factors.

membrane by small vesicles that pinch off from the membrane of one organelle (e.g., the ER) and then fuse with a target membrane. This mechanism for localizing and secreting proteins is often called **vesicular transport,** and it does not involve the translocation of proteins across membranes. However, for proteins to move from the cytosol to the ER, mitochondria, and chloroplasts, the proteins must be translocated across membranes. This transmembrane transport involves systems similar to certain translocation and protein secretion systems used by bacteria, as well as other proteins found in bacterial membranes. ⏮ *Several cytoplasmic membranous organelles function in the secretory and endocytic pathways (section 5.4)*

The best-studied protein translocation system in eukaryotes is the **Sec system** (**figure 15.14**). Central to the functioning of this pathway in eukaryotes is the Sec61αβγ translocon. Several proteins that make up the Sec system in eukaryotes are homologous to Sec proteins in bacteria (**table 15.4**). In eukaryotes, the Sec system functions to move unfolded proteins from the cytosol into the membrane or lumen of the ER. It also functions in chloroplasts to move proteins from the stroma into the thylakoid membrane or the thylakoid lumen. Like the Sec system of bacteria, cotranslational translocation involves a signal recognition particle (table 15.4), which helps deliver the translating ribosome and nascent polypeptide to the Sec system. Likewise, posttranslationally translocated proteins are targeted to the Sec translocon by a **signal peptide** at their amino-terminus (*see figure 13.42*). In addition to the Sec system, the twin-arginine translocation (Tat) system is thought to function in the chloroplasts and mitochondria

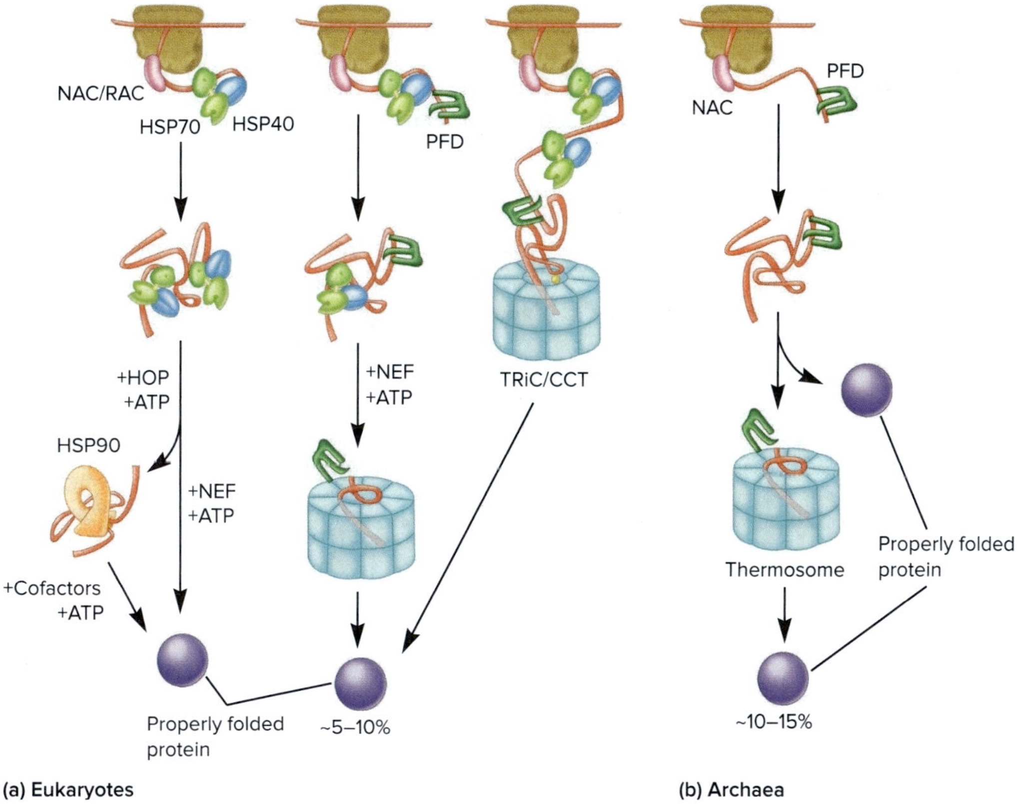

Figure 15.13 Protein Folding in Eukaryotes and Archaea. (a) The eukaryotic pathways all employ NAC (nascent-chain-associated complex), RAC (ribosome-associated complex), the cochaperone HSP40 and an HSP70 chaperone. Some proteins are properly folded by these chaperones alone. Other proteins require additional chaperones such as HSP90, prefoldin (PFD), and TRiC/CCT. The latter is an HSP60 chaperone as is GroEL of bacteria. HOP is HSP90-organizing protein, NEF is nucleotide exchange factor. (b) Most archaeal proteins are folded by NAC and PFD, as many archaea lack HSP70 chaperones. Other proteins require additional folding steps conferred by the thermosome, the archaeal homologue of TRiC/CCT. Although not shown, all HSP60 chaperones hydrolyze ATP.

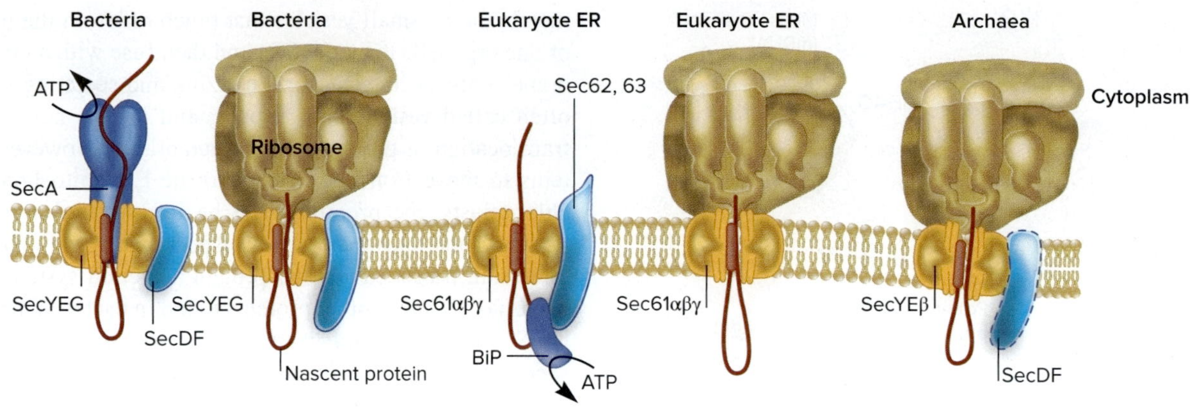

Figure 15.14 Comparison of the Sec Systems Found in Bacteria, Eukaryotes, and Archaea. In bacteria and eukaryotes, translocation can occur posttranslationally or cotranslationally. In archaea, cotranslational translocation is relatively well studied but posttranslational translocation is not. SecDF proteins are found only in some archaea.

Table 15.4 Comparison of Secretion Systems

Bacteria	Eukaryotes	Archaea
Sec Protein Homologues		
SecY	Sec61α	SecY[1]
SecE	Sec61γ	SecE
SecG	-	-
-	Sec61β[2]	Sec61β
Signal Recognition Particle Homologues		
4.5S RNA	7S RNA	7S RNA
Ffh (SRP54)	SRP54	SRP54
	SRP19	SRP19
	SRP72, SRP68, SRP14, SRP9	

1 Although given the same name as the bacterial protein, this archaeal protein's structure is more like the eukaryotic homologue.
2 All Sec systems have three proteins that form the pore through which proteins move. However, the Sec61β protein differs from SecG both in structure and in function.

of some eukaryotes. ◀ *Protein translocation and secretion in bacteria (section 13.8)*

Translation, Protein Maturation, and Protein Localization in Archaea

Most studies of archaeal translation have focused on translation initiation. These studies have yielded some surprising results. First, transcription and translation are tightly coupled, as in bacteria. Thus translation of mRNAs can begin before synthesis of the mRNA is finished. Some archaeal mRNAs have a Shine-Dalgarno sequence in the leader. Translation initiation of these mRNAs proceeds by a mechanism similar to that in bacteria. However, the initiator tRNA is Met-$tRNA_i^{Met}$, as in eukaryotes. Furthermore, more initiation factors (called archaeal IFs or aIFs) are involved than in bacteria. Some are similar to eIFs, some to bacterial IFs, and other aIFs are unique. Many archaeal mRNAs are either leaderless or lack a Shine-Dalgarno sequence. Leaderless mRNAs are translated using an intact 70S ribosome. For polycistronic mRNAs, each coding-region yields a separate protein, as occurs in bacteria.

Translation elongation and termination have been less well studied than translation initiation. However, several archaeal elongation factors have been characterized. They are similar to those in both bacteria and eukaryotes. Thus elongation factors, as opposed to initiation factors, are highly conserved in all cellular organisms. Finally, archaea use a single release factor during translation termination, as do eukaryotes.

Most archaeal proteins are folded using two chaperones called NAC and PFD, homologues of like-named chaperones in eukaryotes (figure 15.13*b*). Some proteins are folded with the additional help of an HSP60 chaperone called the thermosome, a homologue of TRiC/CCT. As you might imagine, archaeal chaperones have been of particular interest to microbiologists because many archaea are thermophilic. Like other HSP60s, the thermosome hydrolyzes ATP as it helps proteins fold. One well-studied thermosome derives from *Pyrodictium occultum,* an archaeon that grows at temperatures as high as 110°C. The *P. occultum* thermosome hydrolyzes ATP most rapidly at 100°C and makes up almost three-quarters of the cell's soluble protein when *P. occultum* grows at 108°C.

Archaeal protein translocation systems are less well studied than those of bacteria and eukaryotes. Thus far, two protein translocation systems have been identified: the Sec and the Tat systems (*see figure 13.43*). The archaeal Sec system is best studied and presents a clear example of the hybrid nature of archaea (figure 15.14 and table 15.4). In archaea, the Sec translocon is called the SecYEβ translocon because it consists of SecY and Sec E, which have homologues in both bacteria and eukaryotes but are most similar to those of eukaryotes (table 15.4), and Sec61β, which is found in eukaryotic cells. Co-translational translocation in archaea requires a signal recognition particle, and table 15.4 compares the signal recognition particle components from the three domains of life. *Protein translocation and secretion in bacteria (section 13.8)*

Comprehension Check

1. Is it surprising that polysomes are observed in all cellular organisms? Explain your answer.
2. How does placement of a ribosome at the start codon differ in bacteria and eukaryotes? What role do the 5′ cap and 3′ poly-A tail play in this process in eukaryotes?
3. List the aspects of translation in archaea that are similar to translation in bacteria. Create a similar list that compares archaeal translation to that seen in eukaryotes. What generalizations can be made from these comparisons?

15.5 Regulation of Cellular Processes

After reading this section, you should be able to:

a. List the steps at which eukaryotic and archaeal cells can regulate cellular processes
b. Explain why the separation in time and space of transcription and translation in eukaryotic cells is important to the possible regulatory mechanisms used
c. Write a paragraph comparing the mechanisms by which eukaryotic and bacterial regulatory proteins control transcription
d. Discuss the importance of small RNA molecules in regulating gene expression in eukaryotes

As is the case in bacteria, the regulation of cellular processes in eukaryotes and archaea can occur at transcriptional, translational, and posttranslational levels (**figure 15.15**; and *see figure 14.1*). Much of the research on regulation in eukaryotes and archaea has focused on transcription initiation. That is our focus here. *Control of Gene Expression in Eukaryotes*

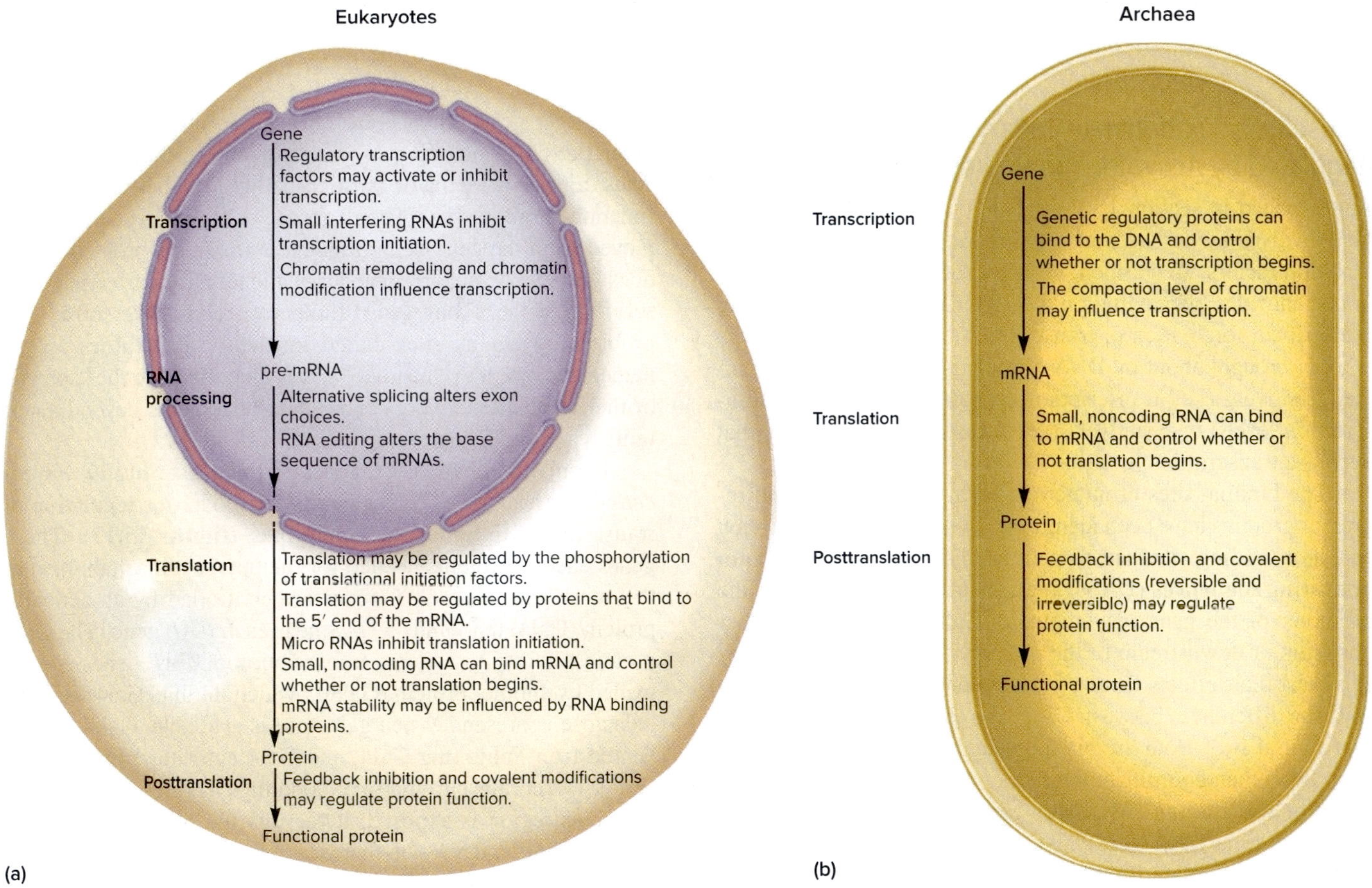

Figure 15.15 Some Common Regulatory Mechanisms in Eukaryotes and Archaea.

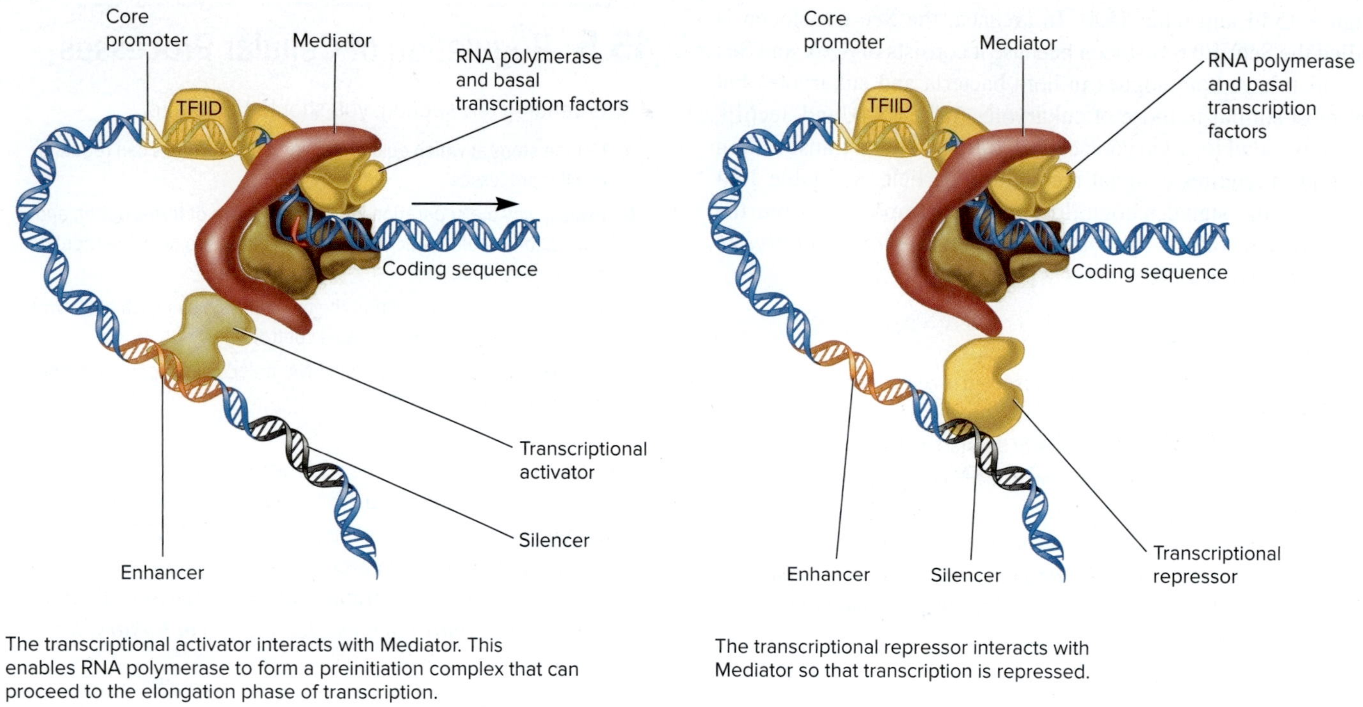

Figure 15.16 Enhancers and Silencers Exert Their Effects from a Distance. Enhancers and silencers are regulatory sites on DNA bound by activator proteins that increase the rate of transcription, or repressor proteins that inhibit transcription, respectively. Both sites can be quite far from the genes they regulate. When the regulatory protein binds its recognition site, it subsequently interacts with Mediator or a basal transcription factor (e.g., TFIID), causing a DNA loop to form. When Mediator functions in repression, additional subunits called the CDK8 (cyclin-*d*ependent *k*inase 8) module interfere with the assembly of the preinitiation complex.

Transcription Initiation in Eukaryotes Is Often Regulated by Activator and Repressor Proteins

As described in section 15.3, transcription initiation in eukaryotes involves numerous transcription factors called basal transcription factors. In addition, co-activators (e.g., Mediator) play important roles, depending on the organism and the gene being transcribed. Regulation of transcription initiation in eukaryotes is often brought about by DNA-binding proteins that can act either positively (i.e., transcription activators that increase transcription levels) or negatively (i.e., transcription repressors that decrease transcription levels).

The binding sites of numerous eukaryotic activator and repressor proteins have been identified. Two different transcription activator binding sites are sometimes distinguished: **upstream activating sequences** (**UASs**) and enhancers. UASs are near the promoter of the gene they regulate. **Enhancers** can be either upstream or downstream of the promoter, and importantly, they can exert their effects from a great distance (**figure 15.16**). Despite the differences between UASs and enhancers, they function in a similar way. Both are binding sites for transcription activators which then coordinate with other proteins to promote transcription, including Mediator and TFIID. Transcription repressors bind DNA sites called **silencers.** Silencers, like enhancers, can regulate transcription even when located far from their target gene. Eukaryotic repressor proteins interact with other proteins, including Mediator, to bring about repression. Thus, although Mediator is often called a co-activator, it also functions in repression. The existence of regulatory sites far removed from the gene being regulated presents a problem: How does the cell prevent an enhancer or silencer from affecting transcription of the wrong gene? Eukaryotic cells have solved this problem by having sites called insulators. **Insulators** are so named because they "insulate" genes by preventing the "spread" of the effects of activators and repressor proteins beyond a certain area of the chromosome.

Regulation of the *GAL* genes in the yeast *Saccharomyces cerevisiae* has served as an important model for regulation of transcription initiation in eukaryotes (**figure 15.17**). These genes encode proteins needed for the uptake and catabolism of the sugar galactose. *GAL* genes are controlled by an activator protein (Gal4) that binds a UAS near each *GAL* gene. The Gal4 protein's ability to activate transcription of *GAL* genes is controlled by another protein (Gal80), which inhibits Gal4 when no galactose is present. When galactose is available, Gal80 is prevented from inhibiting Gal4, and Gal4 activates transcription. Gal4 first recruits chromatin remodeling enzymes, and once the promoter region is exposed, it recruits the basal transcription factors. Finally, RNA polymerase II binds, setting the stage for transcription to start. The regulation of yeast *GAL* genes has

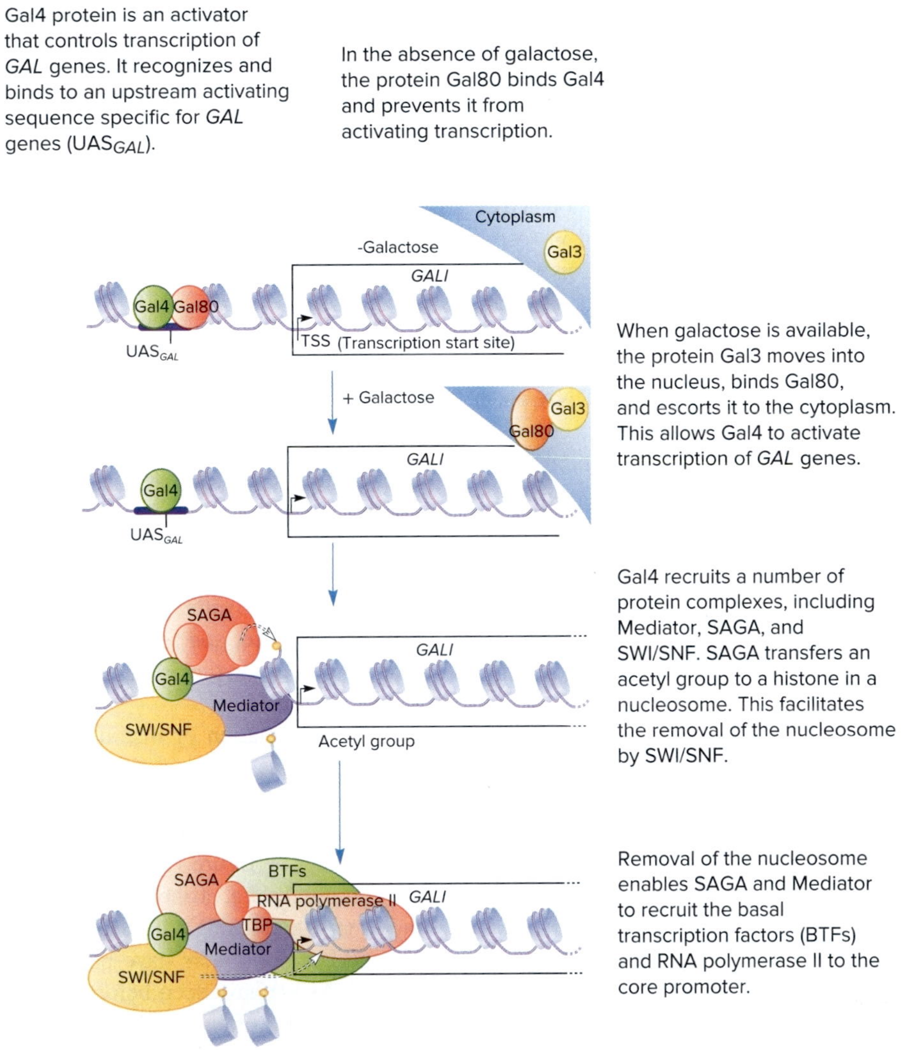

Figure 15.17 **Regulation of *GAL* genes in *Saccharomyces cerevisiae*.** TBP is TATA-binding protein, one of the basal transcription factors (BTFs) recruited to the promoter when galactose is available.

been an important model because it exemplifies many of the features of eukaryotic regulation. These include chromatin remodeling, chromatin modification, formation of activation complexes, and the indirect influence of activators and repressors on genes.

Chromatin remodeling and chromatin modification deserve additional comment. Both are important in regulation of gene expression because they facilitate changes in chromatin structure such that promoters can become either more or less accessible to the transcription machinery. Both chromatin remodeling and chromatin modification are catalyzed by enzymes that alter nucleosomes (figure 15.17). When **chromatin remodeling** occurs, the histones present in a nucleosome are changed, thus changing the histone-DNA interactions. Recall from chapter 5 that the nucleosome core is composed of DNA wound around two copies each of the major histone proteins H2A, H2B, H3, and H4. In addition to the major histone proteins, eukaryotic cells also produce small quantities of variant forms of histones H2A, H2B, and H3. These variant forms can replace a major histone, thereby altering chromatin structure. **Chromatin modification** refers to the covalent modification of the major histone proteins present in a nucleosome—often by addition or removal of acetyl, methyl, or phosphoryl groups. Variant histones and covalent modification of histones are used by cells as signals. For instance, acetylation of histones promotes transcription; methylation of histones often inhibits transcription.

Eukaryotes Use Several Other Regulatory Mechanisms

In addition to controlling transcription initiation with regulatory proteins, eukaryotes use a variety of other methods to regulate gene expression (figure 15.15*a*). Two-component signal transduction pathways have been identified in a number of eukaryotic microbes. The small nucleotide cAMP serves as a second messenger in eukaryotes, and c-di-GMP signaling functions in fruiting body formation by *Dictyostelium discoideum* and in the innate immune system. Eukaryotes produce cis-encoded small RNAs (sRNAs), often called antisense RNAs, that function at the level of translation. Riboswitches have been identified in some eukaryotes. However, they work by controlling whether or not RNA splicing occurs, rather than affecting transcription elongation or translation. Eukaryotes also produce RNA molecules that are important components of the spliceosome, where they contribute to the selection of splice sites used during mRNA processing (alternative splicing). In doing so, different proteins can be made at certain times in the life cycle of the organism by combining different exons. ◀ *Riboswitches: effector-mRNA interactions regulate transcription (section 14.3); Riboswitches and small RNAs control translation (section 14.4); Mechanisms used for global regulation (section 14.5); Supergroup* Amoebozoa *includes protists with pseudopodia (section 24.3)*

One of the most interesting regulatory mechanisms observed in eukaryotes is RNA interference (RNAi). RNAi is brought about by two types of very small RNAs: micro RNAs and small interfering RNAs. **Micro RNAs (miRNAs)** are formed when larger RNA molecules with double-stranded regions are cut by an RNase (usually an RNase called Dicer), yielding a tiny double-stranded miRNA. Once formed, the miRNAs interact with a protein called argonaute (AGO) to form a complex called the *R*NA-*i*nduced *s*ilencing *c*omplex (RISC). RISC surveys mRNAs present in the cell using one strand of the miRNA as the guide RNA. When RISC encounters an mRNA with sequences complementary to the guide RNA, the guide strand base-pairs with a target mRNA. This leads to inhibition of gene expression by preventing translation initiation, inhibiting translation elongation, and degrading mRNA. RNAi caused by miRNAs plays important roles in developmental processes in plants, animals, fungi, and some protists.

Small interfering RNAs (siRNAs) are formed in a similar manner as miRNAs. The difference lies in their source. Many naturally occurring siRNAs are formed from double-stranded RNAs (dsRNAs) that arise in cells during a viral infection. Once detected by the cell, an RNase (e.g., Dicer) cuts the dsRNA, forming the mature siRNA. This interacts with AGO to form a RISC, with one strand of the siRNA serving as the guide RNA as seen for miRNAs. siRNAs also differ from miRNAs in that they can function at the level of transcription. They do so by interacting with several proteins, including AGO, to form the *R*NA-*i*nduced *t*ranscriptional *s*ilencing (RITS) complex. The siRNA directs the RITS complex to a target on the chromosome by a mechanism that is not yet understood. Once bound to the chromosome, the RITS complex initiates chromatin remodeling that results in a highly compacted form of chromatin called heterochromatin. Heterochromatin is so compacted that promoters in it cannot be recognized by the basal transcription factors and RNA polymerase. Therefore transcription is prevented.

Regulation of Gene Expression in Archaea

The regulation of gene expression in archaea is of interest because archaeal transcription and translation machinery is most similar to that of eukaryotes, yet it functions in anucleate cells with circular chromosomes. Is archaeal regulation of gene expression more like bacterial regulation or more like eukaryotic regulation? Unfortunately, because of the challenges of culturing archaea in the laboratory, relatively little is known. For instance, some archaea use sRNAs and riboswitches to regulate gene expression, but little is known about how they function.

Some archaea use transcriptional regulatory proteins that are bacteria-like in their action. For example, repressor proteins have been identified that attach to operators in a fashion similar to bacterial repressors (*see figure 14.3*). However, other repressor proteins function in a unique manner: Instead of binding an operator, they associate with the TATA box in the promoter, thereby preventing the TATA-binding protein (TBP) from attaching. A few archaeal activator proteins have been studied. They bind near the promoter and recruit transcription factors (e.g., TATA-binding protein) to the promoter, thus facilitating RNA polymerase binding.

The association of histones with some archaeal chromosomes begs the question of whether chromosome architecture imposes obstacles for transcription. In eukaryotes, histone octamers wrap 146 base pairs (bp) around themselves, like thread around a spool (figure 15.17). Archaeal histone-DNA complexes are composed of a variable number of histone dimers, often homodimers, and a single dimer organizes about 30 bp. Archaeal histone homopolymers have been demonstrated to wrap up to 180 bp. As we have seen before, multiple protein complexes in eukaryotes are often simplified in archaea, where a single homologous protein can generate a comparable structure and perform the same function. Archaeal histones do not display posttranslational modifications (i.e., the acetylation and methylation used to activate or repress genes). However, archaeal promoter regions have fewer histones than other regions, and transcription elongation slows upon encountering histones on the DNA template. Clearly the ability of histones to organize a bacterial-sized chromosome located in the cytoplasm has important evolutionary implications. This modest platform for wrapping DNA has been expanded and embellished in eukaryotes.

Comprehension Check

1. List two similarities and two differences between bacterial and eukaryotic regulatory proteins.
2. What regulatory sequences in bacterial genomes are analogous to the enhancers and silencers observed in eukaryotic genomes?
3. How are cis-encoded RNAs, micro RNAs, and small interfering RNAs alike? How do they differ? What other non-coding RNAs can affect protein synthesis in eukaryotes?

Key Concepts

15.1 Why Consider Eukaryotic and Archaeal Genetics Together?

- Despite having simple cell architecture, archaea exhibit unique characteristics that distinguish them from bacteria.
- Examination of the machinery involved in processing and expressing archaeal genomes has revealed that it is more like that of eukaryotes than that of bacteria.

15.2 DNA Replication: Similar Overall, but with Different Replisome Proteins

- Eukaryotic DNA has many replicons and replication origins (**figure 15.2**). DNA replication is initiated when certain proteins are recruited to origins of replication by a set of proteins called the origin recognition complex (ORC; **table 15.1**).
- In eukaryotes, three different DNA polymerases are part of the replisome (**figure 15.1**). DNA polymerase α-primase synthesizes an RNA-DNA hybrid molecule that serves as the primer for DNA synthesis. DNA polymerase ε carries out leading strand synthesis, and DNA polymerase δ is responsible for lagging strand synthesis.
- Telomerase is responsible for replicating the ends of linear eukaryotic chromosomes (**figure 15.3**).
- All known archaeal cells have circular DNA. Their chromosomes are thought to be replicated in a manner similar to circular bacterial chromosomes.
- Some archaeal species have more than one origin of replication, while others have been shown to function without an origin. These organisms use a type of recombination to copy their chromosome.
- Archaeal replisomes of all species examined thus far contain a primase and a DNA polymerase (Pol B) similar to those in eukaryotic replisomes (**figure 15.4**). Some archaea have a second DNA polymerase (Pol D) that is unique to archaea (**table 15.2**).

15.3 Transcription

- Most eukaryotic protein-coding genes are not organized into operons. Therefore the mRNAs that arise from them are monocistronic.
- Three RNA polymerases transcribe genes in eukaryotes. RNA polymerase II transcribes protein-coding genes. RNA polymerase I transcribes rRNA genes, and RNA polymerase III transcribes the 5S rRNA and all tRNA genes. All three eukaryotic RNA polymerases have more subunits than bacterial RNA polymerases (**figure 15.5**).
- The RNA polymerase II core promoters of eukaryotes are distinctive, as is the mechanism of transcription initiation (**figure 15.6**). Several basal transcription factors are involved in recruiting RNA polymerase II to a promoter (**figure 15.7**).
- RNA polymerase II synthesizes pre-mRNA, which is modified by addition of a 3′ poly-A sequence and a 5′ cap (**figure 15.8**) and by removal of introns by spliceosomes (**figure 15.9**).
- Many archaeal genes are organized into operons (though perhaps not as many as in bacteria). The single archaeal RNA polymerase is similar to eukaryotic RNA polymerase II. The archaeal RNA polymerase requires two transcription factors to be positioned properly at a promoter (**figure 15.10**). Archaeal promoters have elements similar to those in the RNA polymerase II core promoters of eukaryotes, including a TATA box (**figures 15.6** and **15.10**).

15.4 Translation and Protein Maturation and Localization

- Translation initiation in eukaryotes requires more initiation factors than it does in bacteria (**figure 15.11**). Initiation involves an interaction of the 5′ and 3′ ends of the mRNA, followed by binding of the 43S complex (SSU of the ribosome and initiator tRNA) to the 5′ end of the message. The 48S complex then scans the mRNA until it encounters a start codon. Translation elongation is similar in eukaryotes and bacteria. Translation termination requires a single release factor, rather than the three used in bacteria.
- Protein folding in eukaryotes is aided by chaperone proteins (**figure 15.13**) and can occur cotranslationally or posttranslationally. Homologues of all types of chaperones are found in eukaryotes.
- In eukaryotes, proteins are translocated into the endoplasmic reticulum (ER) membrane by a Sec system (**figure 15.14**). The proteins are then trafficked between the ER, plasma membrane, Golgi, and other components of the endocytic and secretory pathways by vesicular transport. The Sec system is also used to move some proteins from the chloroplast stroma to the thylakoid membrane or the thylakoid lumen. The Tat system functions in some eukaryotes to translocate proteins into mitochondria and chloroplasts.
- In archaea, transcription and translation are coupled, as in bacteria. Some archaeal mRNAs have a Shine-Dalgarno sequence that is thought to function in positioning the ribosome as it does in bacteria. However, other archaea lack a Shine-Dalgarno sequence or are leaderless. It is thought that intact 70S ribosomes initiate translation of these mRNAs. Other aspects of archaeal translation are similar to those seen in eukaryotes, including the protein factors and initiator tRNA used.
- Archaea use chaperones to help fold proteins. They resemble those of eukaryotes.
- Archaea use the Sec and Tat pathways to translocate proteins into or across the plasma membrane.

The archaeal Sec system proteins are similar to the eukaryotic Sec system proteins (**figure 15.14**).

15.5 Regulation of Cellular Processes

- Eukaryotes have regulatory sites for the binding of transcription regulatory proteins. Activators bind either enhancers or upstream activating sites, and repressors bind sites called silencers (**figure 15.16**). The regulatory proteins exert their actions indirectly by affecting the activity of transcription factors, chromatin-modifying enzymes, and chromatin-remodeling enzymes (**figure 15.17**).
- Micro RNAs are used by eukaryotic cells to regulate translation. Micro RNAs are part of a complex that targets a particular mRNA and inhibits its translation.
- Small interfering RNAs are also used to control translation, in a manner similar to that used by micro RNAs. However, some small interfering RNAs regulate transcription. This is accomplished by causing the formation of heterochromatin, a highly compacted form of chromatin, in the region targeted by the small interfering RNA.
- The mechanism of action of archaeal proteins that regulate transcription initiation is best understood for repressor proteins. Some repressor proteins bind an operator sequence and are thought to work in a manner similar to that seen in bacteria. Other repressor proteins inhibit transcription by localizing to the TATA box, thereby preventing the archaeal transcription factors from attaching to the DNA and initiating transcription.

Active Learning

1. Budding yeasts such as *S. cerevisiae* exhibit telomerase activity throughout their life cycles, whereas most human somatic cells do not exhibit telomerase activity. Suggest why this is so.
2. All of the subunits in bacterial RNA polymerases have homologues in both archaeal and eukaryotic RNA polymerases. What does this suggest about the evolution of these enzymes?
3. Would you expect that one day microbiologists might discover small interfering RNAs in archaea? If so, how might they function? Explain your answer.
4. In the chapter opening story, it was stated that the next step in using algae to produce plastics is to maximize expression of the bacterial polyhydroxyalkanoate-synthesizing gene in algal cells. Suggest one genetic change that might be made to increase its expression.
5. Riboflavin biosynthesis in bacteria and archaea begins with the same precursor, but the enzymatic reactions to synthesize riboflavin occur in a different sequence and the enzymes that catalyze the reactions are not related. As described in chapter 14, regulation of the *Bacillus rib* operon occurs through a riboswitch. Researchers studying archaeal regulation found that, in contrast, riboflavin biosynthesis is controlled by a single protein with two functions, binding to DNA and catalyzing one step in the pathway. Propose a mechanism for the regulation of the archaeal *rib* genes, keeping in mind that archaeal regulators are most similar to bacterial regulators, but they operate on a eukaryotic-like gene expression platform.

   Read the original paper: Rodionova, I. A., et al. 2017. A novel bifunctional transcriptional regulator of riboflavin metabolism in Archaea. *Nucleic Acids Res.* 45:3785.

16 Mechanisms of Genetic Variation

Source: Photo by Jeff Venuga/USDA Natural Resources Conservation Service

Manure Happens

Consider this: A milk-producing dairy cow generates on average 150 pounds (68 kilograms) of manure each day and a beef cow will have produced about 9,800 pounds (4,445 kilograms) of manure during the finishing process that prepares it for slaughter. Multiply this by the number of dairy cows and beef cattle around the world and that's a lot of manure!

This much manure creates many problems: odor, release of volatile compounds into the atmosphere, water pollution, and the possible dissemination of pathogens. The latter is of considerable concern because it is coupled with another problem: the potential spread of antibiotic-resistant bacteria, pathogens and nonpathogens alike. Antibiotic resistance (AR) has been increasing since the beginning of antibiotic use. Currently, more than 30,000 people in Europe and the United States die each year from infections caused by microbes resistant to antimicrobial agents.

You may be wondering why there might be antibiotic-resistant bacteria in manure. The answer is simple. Livestock are exposed to antibiotics either through therapeutic use or at lower amounts (subtherapeutic levels) in their feed. These antibiotics are a selective pressure that acts on the resident microbiota, leading to the evolution of antibiotic-resistant bacteria. These bacteria pass their AR genes on to their progeny. But more worrisome is their ability to share AR genes with other, unrelated bacteria in their habitat by horizontal gene transfer (HGT). Bacteria use HGT to create genetic variability, which in turn helps at least some survive when encountering a new environment (e.g., exposure to an antibiotic).

Therapeutic use of antibiotics to cure infection is understandable, but what is less obvious is why livestock are given subtherapeutic levels of antibiotics in their feed. The major advantage is that feed is used more efficiently and in turn, more meat or milk is produced from each animal. This translates to greater profits for the rancher and lower costs for the consumer.

Do the benefits of subtherapeutic antibiotic use outweigh the risk of spreading AR genes from manure to other organisms? Studies using molecular approaches such as those described in chapter 17 have shown the presence of AR genes in manure. Many of the genes are only distantly related to clinically relevant AR genes, and some define an entirely new family of AR genes. This is either good news or bad news. It could be that most AR genes present in manure have not spread to humans and caused disease. However, it could also mean that manure is a reservoir of novel AR genes waiting to make their way into clinically important bacteria. Further complicating the issue is a recent report that manure from cows never exposed to antibiotics increases the number of soil bacteria with AR genes encoding β-lactamases (penicillin-destroying enzymes). Why this occurs is not clear. But what is clear is that bacteria are adept at altering their genomes, thereby generating more genetic diversity in a population.

The increase of antibiotic resistance in bacteria is only one example of microbial evolution. All microbes use a variety of mechanisms to help ensure their survival and to evolve into new species. HGT among and between bacteria and archaea plays a particularly important role. As more and more genomes are sequenced and annotated, it is clear that HGT has occurred many times. Furthermore, genes have also been transferred from bacteria and archaea to eukaryotes and from eukaryotes to bacteria and archaea.

How the genomes of microbes are altered to increase genetic diversity is the topic of this chapter. Our main focus is on how this is accomplished in the absence of sexual reproduction. Some of the mechanisms we discuss can create changes in the genome that are actually detrimental. Thus we also consider processes that help balance the stability of the genome with the capacity to introduce genetic variation.

Readiness Check:

Based on what you have learned previously, you should be able to:

- ✓ Draw a simple diagram that illustrates the three chemical moieties that define a nucleotide (section 13.2)
- ✓ State the base-pairing rules
- ✓ Explain the importance of the reading frame of a protein-coding gene (section 13.6)
- ✓ Describe the major types of plasmids (section 3.6)
- ✓ Create a concept map or table that distinguishes the secretion systems observed in Gram-negative bacteria (section 13.8)
- ✓ Describe the life cycle and structures of a virus (chapter 6)

16.1 Mutations: Heritable Changes in a Genome

After reading this section, you should be able to:

a. Distinguish spontaneous from induced mutations, and list the most common ways each arises
b. Construct a table, concept map, or picture to summarize how base analogues, DNA-modifying agents, and intercalating agents cause mutations
c. Discuss the possible effects of mutations

Perhaps the most obvious way genetic diversity can be created in the absence of sexual reproduction is by **mutations** (Latin *mutare,* to change), which are heritable changes in DNA sequence. Several types of mutations exist. Some arise from the alteration of single pairs of nucleotides and from the addition or deletion of one nucleotide pair in the coding regions of a gene. Such small changes in DNA are called **point mutations** because they affect only one base pair in a given location. Larger mutations are less common. These include large insertions, deletions, inversions, duplications, and translocations of nucleotide sequences.

Mutations occur in one of two ways. (1) **Spontaneous mutations** arise occasionally in all cells in the absence of any added agent. (2) **Induced mutations** are the result of exposure to a **mutagen,** which can be either a physical or a chemical agent. Mutations are characterized according to either the kind of genotypic change that has occurred or their phenotypic consequences. In this section, the molecular basis of mutations and mutagenesis is first considered; then the phenotypic effects of mutations are discussed.

Spontaneous Mutations

Spontaneous mutations result from errors in DNA replication, spontaneously occurring lesions in DNA, or the action of mobile genetic elements such as transposons (section 16.5). A few of the more prevalent mechanisms are described here.

Replication errors can occur when the nitrogenous base of a nucleotide shifts to a different form (isomer) called a tautomeric form. Nitrogenous bases typically exist in the keto form (*see figure 13.5*) but are in equilibrium with the imino and enol forms (**figure 16.1*a***). The shift from one form to another changes the hydrogen-bonding characteristics of the bases, allowing purine for purine (e.g., an A for a G) or pyrimidine for pyrimidine (e.g., a T for a C) substitutions that can eventually lead to a stable alteration of the nucleotide sequence (figure 16.1*b*). Such substitutions are known as **transition mutations** and are relatively common. On the other hand, **transversion mutations**—mutations where a purine is substituted for a pyrimidine (e.g., an A for a T) or a pyrimidine for a purine—are rarer due to the steric problems of pairing purines with purines and pyrimidines with pyrimidines.

Replication errors can also result in the insertion and deletion of nucleotides. These mutations generally occur where there is a short stretch of repeated nucleotides. In such a location, the pairing of template and new strand can be displaced,

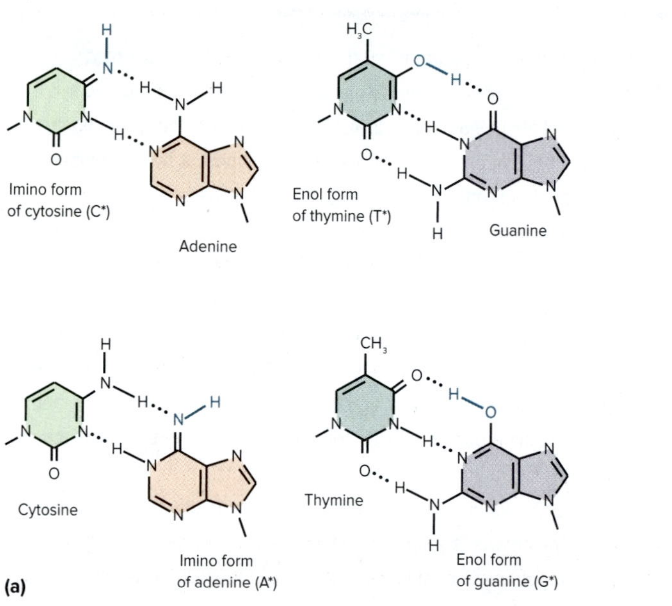

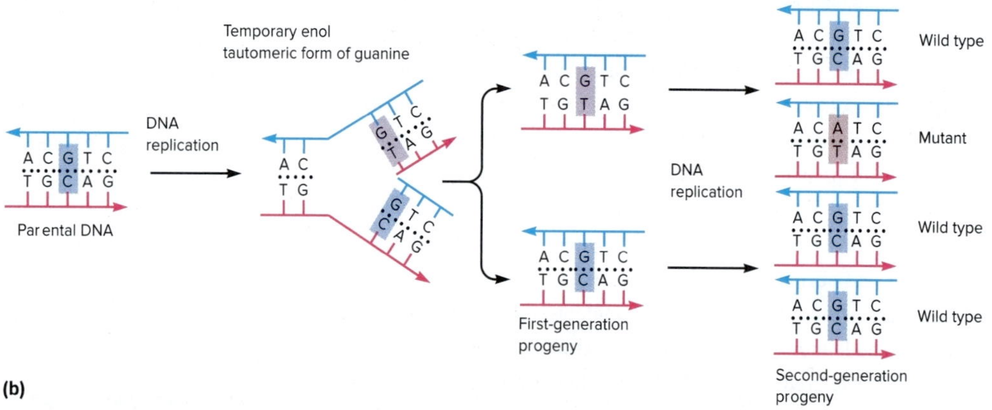

Figure 16.1 Tautomerization and Transition Mutations. Errors in replication due to base tautomerization. (a) Normally AT and GC pairs are formed when keto groups participate in hydrogen bonds. In contrast, enol tautomers produce AC and GT base pairs. The alteration in the base is shown in blue. (b) Mutation as a consequence of tautomerization during DNA replication. The temporary enolization of guanine leads to the formation of a GT base pair in one of the first-generation progeny molecules. Thus, the pyrimidine cytosine is replaced by the pyrimidine thymine. If this abnormal pairing is not detected by repair mechanisms, when that progeny DNA molecule is replicated, one of the second-generation progeny molecules will contain an AT base pair and a GC–to–AT transition mutation will have occurred. Note that the process requires two replication cycles. Wild type is the form of the gene before mutation occurred.

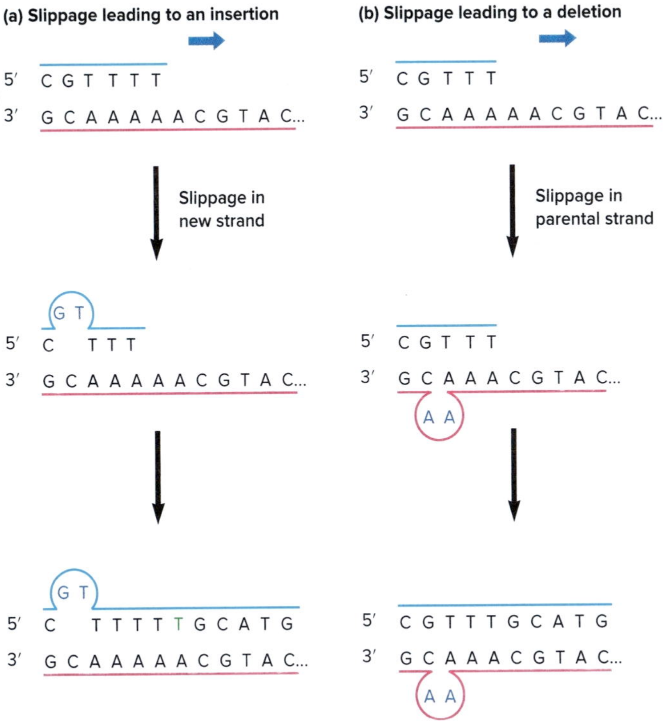

Figure 16.2 Insertions and Deletions. A mechanism for the generation of insertions and deletions during replication. The direction of replication is indicated by the blue arrow. In each case, there is strand slippage resulting in the formation of a small loop that is stabilized by hydrogen bonding in the repetitive sequence, the AT stretch in this example. (a) If the new strand slips, an addition of one T results. (b) Slippage of the parental strand yields a deletion (in this case, a loss of two Ts).

leading to insertions or deletions of bases in the new strand (**figure 16.2**).

Spontaneous mutations can originate from lesions in DNA as well as from replication errors. For example, it is possible for purine nucleotides (A and G) to be depurinated; that is, to lose their base. This results in the formation of an apurinic site, which does not base pair normally and may cause a mutation after the next round of replication (**figure 16.3**). Likewise, pyrimidine bases (T and C) can be lost, forming an apyrimidinic site. Other lesions are caused by reactive forms of oxygen such as oxygen free radicals and peroxides produced during aerobic metabolism. For example, guanine can be converted to 8-oxo-7,8-dihydrodeoxyguanine, which often pairs with adenine, rather than cytosine, during replication.

Induced Mutations Are Caused by Mutagens

Any agent that damages DNA, alters its chemistry, or in some way interferes with its functioning will probably induce mutations. Mutagens are classified according to their mode of action. Three common types of chemical mutagens are base analogues, DNA-modifying agents, and intercalating agents. Some physical agents (e.g., radiation) are mutagens that damage DNA.

Base analogues are structurally similar to normal nitrogenous bases and can be incorporated into the growing polynucleotide chain during replication (**table 16.1**). Once in place, these compounds typically exhibit base-pairing properties different from the bases they replace and can eventually cause a stable mutation. A widely used base analogue is 5-bromouracil, an analogue of thymine. It undergoes a tautomeric shift from the normal keto form to an enol much more frequently than a normal base. The enol tautomer hydrogen bonds like cytosine, pairing with guanine rather than adenine. The mechanism of action of other base analogues is similar to that of 5-bromouracil.

There are many **DNA-modifying agents**—mutagens that change a base's structure and therefore alter its base-pairing specificity. Some of these mutagens preferentially react with certain bases

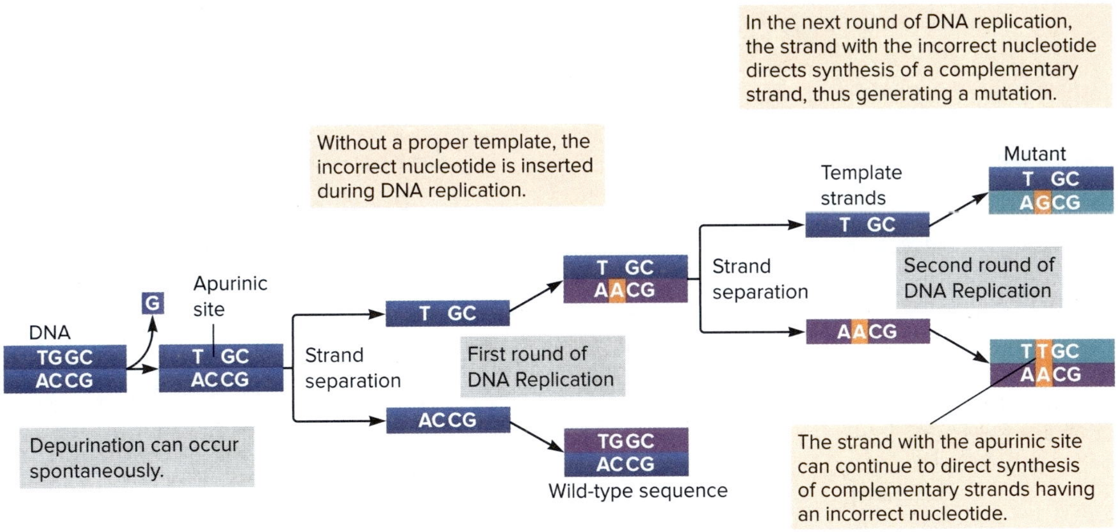

Figure 16.3 Mutation Due to Formation of an Apurinic Site. Apurinic sites arise spontaneously (depurination). They do not provide a template for DNA replication, and as a result, an incorrect nucleotide can be inserted. A similar process introduces mutations at sites where apyrimidinic sites arise.

Table 16.1 Examples of Mutagens

Mutagen	Effect(s) on DNA Structure
Chemical	
5-Bromouracil	Base analogue
2-Aminopurine	Base analogue
Ethyl methanesulfonate	Alkylating agent that adds an ethyl group to guanine
Hydroxylamine	Hydroxylates cytosine
Nitrogen mustard	Alkylating agent that adds a methyl group to guanine
Nitrous oxide	Deaminates bases
Proflavin	Intercalating agent
Acridine orange	Intercalating agent
Physical	
UV light	Promotes pyrimidine dimer formation
X rays	Cause base deletions, single-strand nicks, cross-linking, and chromosomal breaks

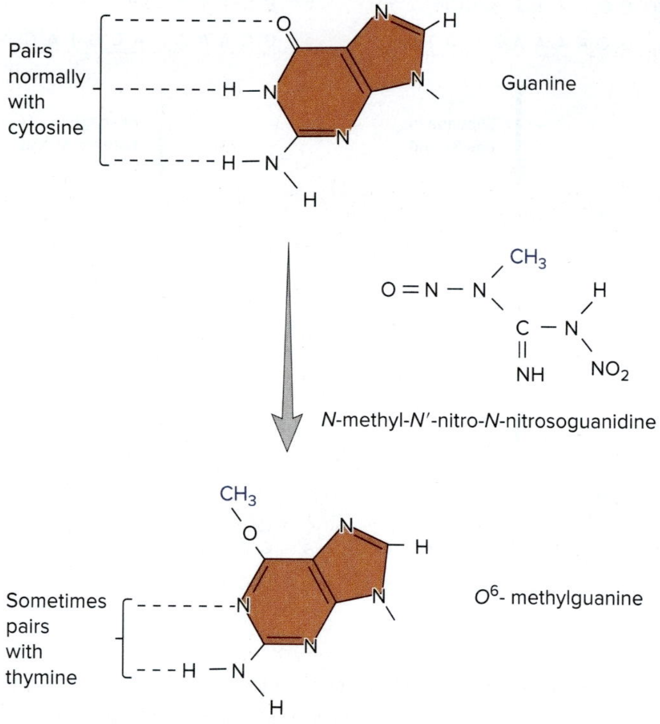

Figure 16.4 Methyl-Nitrosoguanidine Mutagenesis. Methyl-nitrosoguanidine methylates guanine.

and produce a particular kind of DNA damage. For example, methyl-nitrosoguanidine adds methyl groups to guanine, causing it to mispair with thymine (**figure 16.4**). A subsequent round of replication can then result in a GC-AT transition.

Intercalating agents distort DNA to induce single nucleotide pair insertions and deletions. These mutagens are planar and insert themselves (intercalate) between the stacked bases of the helix. This results in distortion of the helical structure, rendering the DNA inactive as a template for either replication or transcription.

Many mutagens, and indeed many carcinogens, damage bases so severely that hydrogen bonding between base pairs is impaired or prevented and the damaged DNA can no longer act as a template for replication. For instance, ultraviolet (UV) radiation often generates thymine dimers between adjacent thymines (**figure 16.5**).

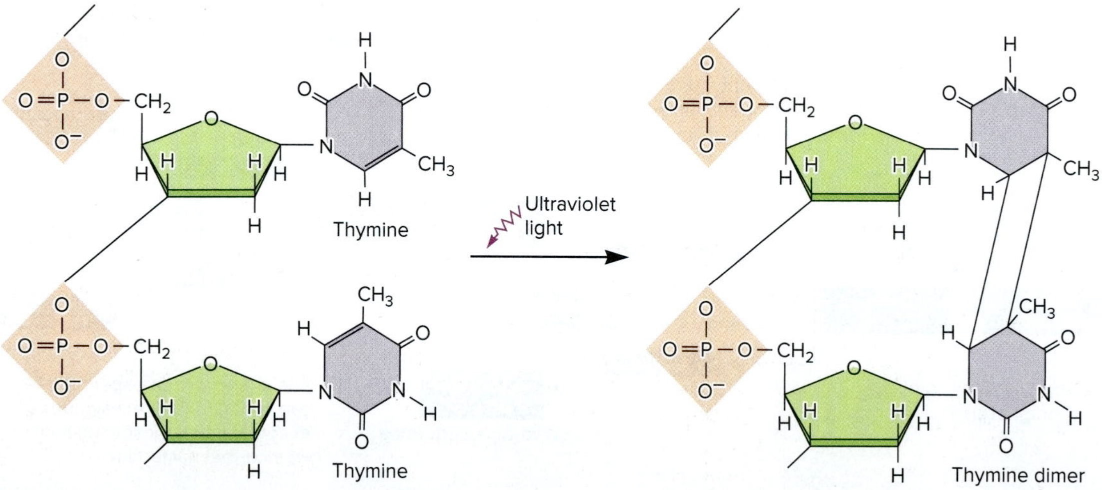

Figure 16.5 Thymine Dimers in a DNA Strand Are Formed by Ultraviolet Radiation. Thymine dimers are one type of cyclobutane dimer (i.e., a dimer formed between adjacent pyrimidines) generated by UV radiation.

Effects of Mutations

The effects of mutations can be described at the protein level and in terms of observed phenotypes. In all cases, the impact is readily noticed only if it produces a change in phenotype. In general, the more prevalent form of a gene and its associated phenotype is called the **wild type.** A change from wild type to a mutant form is a **forward mutation** (table 16.2). It is possible to restore a wild-type phenotype with a second mutation. If the

Table 16.2 Types of Point Mutations

Type of Mutation	Change in DNA	Example
Forward Mutations		
None	None	5′-A-T-G-A-C-C-T-C-C-C-C-G-A-A-A-G-G-G-3′ Met - Thr - Ser - Pro - Lys - Gly
Silent	Base substitution	5′-A-T-G-A-C-A-T-C-C-C-C-G-A-A-A-G-G-G-3′ Met - Thr - Ser - Pro - Lys - Gly
Missense	Base substitution	5′-A-T-G-A-C-C-T-G-C-C-C-G-A-A-A-G-G-G-3′ Met - Thr - Cys - Pro - Lys - Gly
Nonsense	Base substitution	5′-A-T-G-A-C-C-T-C-C-C-C-G-T-A-A-G-G-G-3′ Met - Thr - Ser - Pro - STOP!
Frameshift	Insertion/deletion	5′-A-T-G-A-C-C-T-C-C-G-C-C-G-A-A-A-G-G-G-3′ Met - Thr - Ser - Ala - Glu - Arg
Reverse Mutations		
	Base substitution	5′-A-T-G-A-C-C-T-C-C —forward→ A-T-G-C-C-C-T-C-C —reverse→ A-T-G-A-C-C-T-C-C Met - Thr - Ser → Met - Pro - Ser → Met - Thr - Ser
	Base substitution	5′-A-T-G-A-C-C-T-C-C —forward→ A-T-G-A-C-C-T-G-C —reverse→ A-T-G-A-C-C-A-G-C Met - Thr - Ser → Met - Thr - Cys → Met - Thr - Ser
	Base substitution	5′-A-T-G-A-C-C-T-C-C —forward→ A-T-G-C-C-C-T-C-C —reverse→ A-T-G-C-T-C-T-C-C Met - Thr - Ser → Met - Pro - Ser → Met - Leu - Ser (polar amino acid) → (nonpolar amino acid) → (polar amino acid) pseudo-wild type
Suppressor Mutations		
Frameshift of opposite sign (intragenic suppressor)	Insertion/deletion	5′-A-T-G-A-C-C-T-C-C-C-C-G-A-A-A-G-G-G-3′ Met - Thr - Ser - Pro - Lys - Gly ↓ Forward mutation (insertion) 5′-A-T-G-A-C-C-T-C-C-G-C-C-G-A-A-A-G-G-G-3′ Met - Thr - Ser - Ala - Glu - Arg ↓ Suppressor mutation (deletion) 5′-A-T-G-A-C-C-C-C-G-C-C-G-A-A-A-G-G-G-3′ Met - Thr - Pro - Pro - Lys - Gly
Extragenic suppressor Nonsense suppressor		Gene (e.g., for tyrosine tRNA) undergoes a mutation in its anticodon region that enables it to recognize and align with a nonsense mutation (e.g., UAG). Thus an amino acid (tyrosine) is inserted at the mutant stop codon and translation continues.
Physiological suppressor		A defect in one chemical pathway is circumvented by another mutation; for example, one that opens up another chemical pathway to the same product or one that permits more efficient uptake of a compound produced in small quantities because of the original mutation.

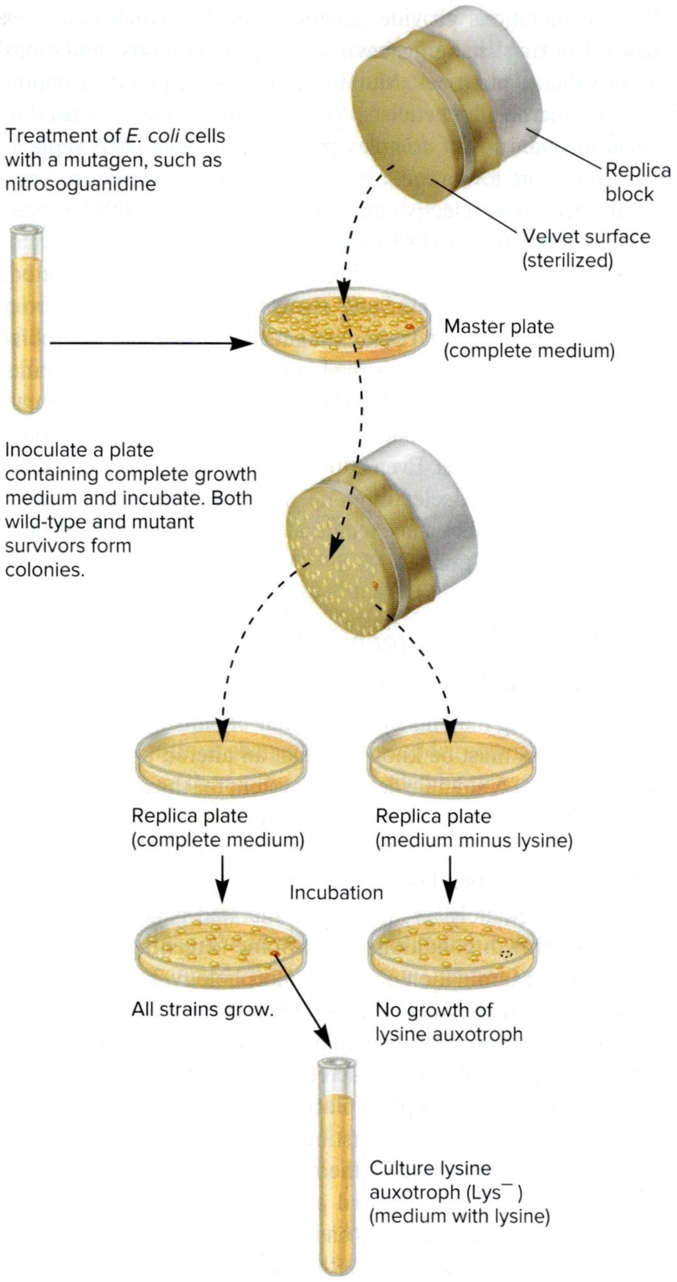

Figure 16.6 Replica Plating. The use of replica plating to isolate a lysine auxotroph. After growth of a mutagenized culture on a complete medium, a piece of sterile velvet is pressed on the plate surface to pick up bacteria from each colony. Then the velvet is pressed to the surface of other plates, and organisms are transferred to the same position as on the master plate. After the location of Lys^- colonies growing on the replica with complete medium is determined, the auxotrophs can be isolated and cultured.

MICRO INQUIRY *How would you screen for a tryptophan auxotroph? How would you select for a mutant that is resistant to the antibiotic ampicillin but sensitive to tetracycline (assume the parental stain is resistant to both antibiotics)?*

Mutant Selection: Using Conditions to Inhibit Growth of Wild-Type Organisms

Mutant selection techniques use incubation conditions under which the mutant grows because of properties conferred by the mutation, whereas the wild type does not. Selection methods often involve reversion or suppressor mutations that restore the wild-type phenotype. For example, if the intent is to isolate a prototroph from a lysine auxotroph (Lys^-), a large population of lysine auxotrophs is plated on minimal medium lacking lysine, incubated, and examined for colony formation. Only cells that have mutated to restore the ability to manufacture lysine (i.e., prototrophy) will grow on minimal medium. Several million cells can be plated on a single Petri dish, but only the rare revertant cells will grow. Thus many cells can be tested for mutations by scanning a few Petri dishes for growth.

Other selection methods identify mutants resistant to a particular environmental stress such as virus attack, antibiotic treatment, or specific temperatures. Therefore it is possible to grow the microbe in the presence of the stress and look for surviving organisms. Consider the example of an antibiotic-sensitive wild-type bacterium. When it is cultured in medium lacking the antibiotic and then plated on selective medium containing the antibiotic, any colonies that form are resistant to the antibiotic and carry a mutated gene that confers antibiotic resistance.

Mutant screening and selection methods are used for purposes other than understanding more about the nature of genes or the biochemistry of a particular microorganism. One very important role of mutant selection and screening techniques is in the study of carcinogens. One of the first and perhaps best known of the carcinogen testing systems is the Ames Test, developed by Bruce Ames. The Ames test is based on the observation that many carcinogens also are mutagens. The test determines if a substance increases the rate of mutation; that is, if it is a mutagen. If the substance is a mutagen, then it is also likely to be carcinogenic if an animal is exposed to it at sufficient levels. Note that the test does not directly test for carcinogenicity. This is because it uses a bacterium as the test organism. Carcinogenicity can only be directly demonstrated with animals, and testing is extremely expensive and takes much longer to complete than does the Ames test. Thus the Ames test serves as an inexpensive procedure to identify chemicals that may be carcinogenic and thus deserve further testing.

The Ames test is called a mutational reversion assay because it looks for histidine prototrophs that arise from histidine auxotrophs. The assay employs several "tester" strains of *Salmonella enterica* serovar Typhimurium. Each tester strain has a different mutation in the histidine biosynthesis operon. They also have mutations that make their cell walls more permeable to test substances and impair their ability to repair DNA.

In the Ames test, tester strains of *S.* Typhimurium are plated with the substance being tested and the number of visible colonies that form is determined (**figure 16.7**). To ensure that DNA

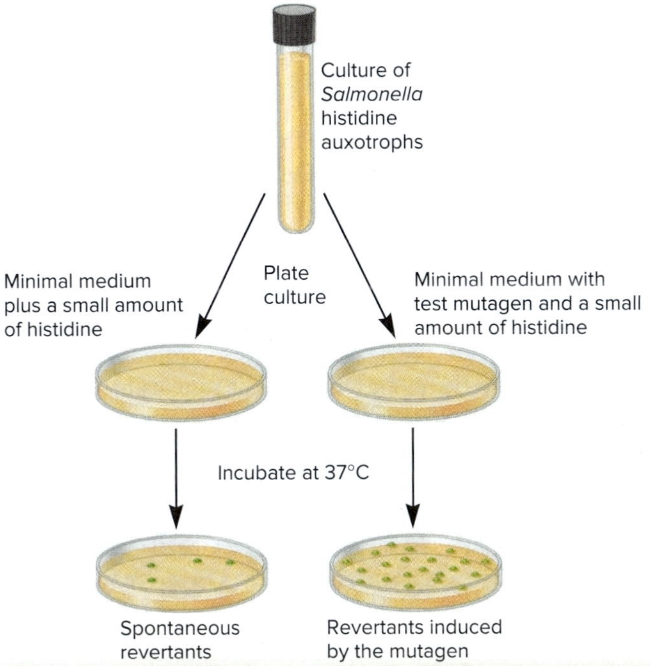

Figure 16.7 The Ames Test for Mutagenicity.

MICRO INQUIRY *Why is a small amount of histidine added to the test medium?*

replication can take place and allow for the development of a mutation in the presence of the potential mutagen, the bacteria and test substance are mixed in dilute molten top agar to which a trace of histidine has been added. This molten mix is then poured on top of minimal agar plates and incubated for 2 to 3 days at 37°C. All of the histidine auxotrophs grow for the first few hours in the presence of the test compound until the histidine is depleted. This initial growth does not produce a visible colony. Once the histidine supply is exhausted, only revertants that have mutated to regain the ability to synthesize histidine continue to grow and produce visible colonies. These colonies need only be counted and compared to controls to estimate the relative mutagenicity of the compound: The more colonies, the greater the mutagenicity.

Comprehension Check

1. Describe how replica plating is used to detect and isolate auxotrophic mutants.
2. Why are mutant selection techniques generally preferable to screening methods?
3. Briefly discuss how reversion mutations and resistance to an environmental factor (e.g., presence of an antibiotic) can be employed in mutant selection.
4. Describe how you would isolate a mutant that required histidine for growth and was resistant to penicillin. The wild type is a prototroph.
5. What is the Ames test and how is it carried out? What assumption concerning mutagenicity and carcinogenicity is it based upon?

16.3 DNA Repair Maintains Genome Stability

After reading this section, you should be able to:

a. Compare and contrast proofreading, excision repair, direct repair, mismatch repair, and recombinational repair
b. Propose a scenario that would elicit the SOS response and describe the response to those conditions

If there is a microbial equivalent of an extreme sport, *Deinococcus radiodurans*'s ability to repair its genome after it has been blasted apart by a high dose of radiation might be a contender. Surprisingly, this ability is primarily related to the resistance of *D. radiodurans* proteins and the structure of its genome, rather than to its DNA repair mechanisms. Following damage, its radiation-resistant proteins begin repairing the genome quickly. Repair is aided by the genome consisting of two chromosomes, each having numerous areas of homology. *D. radiodurans* is polyploid, that is, there are several copies of the chromosome. This allows the DNA fragments to anneal, allowing the shattered genome to be pieced together. Other than this, *D. radiodurans* uses pretty much the same DNA repair mechanisms as other organisms. Obviously mutations can have disastrous effects. Therefore it is imperative that a microorganism be able to repair changes. There are numerous repair mechanisms and these are well conserved across all domains of life. Repair in *E. coli* is best understood and is our focus in this section. ▶| Deinococcus-Thermus *includes radiation-resistant bacteria (section 21.2)*

Proofreading: The First Line of Defense

As we discuss in chapter 13, replicative DNA polymerases sometimes insert the incorrect nucleotide during DNA replication. However, these DNA polymerases have the ability to evaluate the hydrogen bonds formed between the newly added nucleotide and the template nucleotide, and correct any errors immediately; that is, before the next nucleotide is added. This ability is called **proofreading.** When a DNA polymerase detects that a mistake has been made, it backs up and removes the incorrect nucleotide with its 3′ to 5′ exonuclease activity (*see figure 13.16*). It then restarts DNA replication, this time inserting the correct nucleotide. Proofreading is efficient, but does not correct every error. Furthermore, it does not correct induced mutations. *E. coli* thus uses other repair mechanisms to help ensure the stability of its genome. |◀ *DNA replication in bacteria (section 13.3)*

Mismatch Repair

When proofreading by replicative DNA polymerases fails, mismatched bases are usually detected and repaired by the **mismatch repair** system (**figure 16.8**). In *E. coli* the enzyme MutS scans newly replicated DNA for mismatched pairs. Another enzyme, MutH, removes a stretch of newly synthesized DNA around the

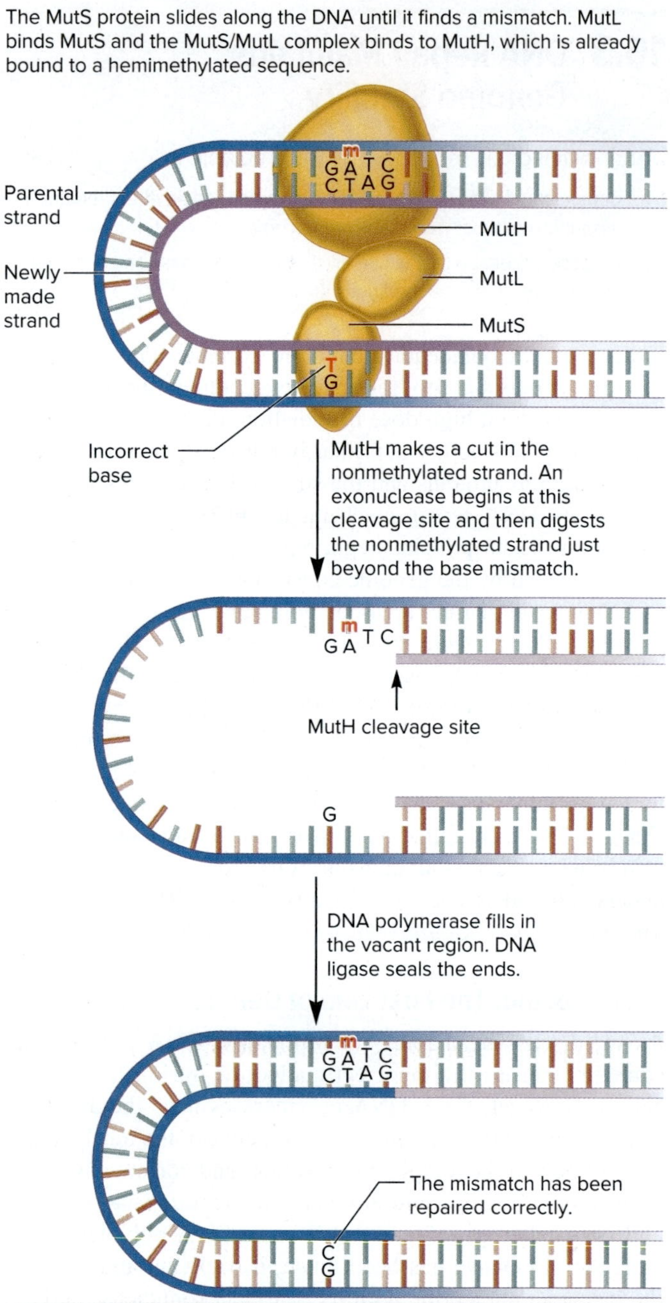

Figure 16.8 Methyl-Directed Mismatch Repair in *E. coli*. The role of MutH is to identify the methylated strand of DNA, which is the nonmutated parental strand. The methylated adenine is designated by an m.

MICRO INQUIRY *How is mismatch repair similar to DNA polymerase proofreading? How is it different?*

mismatch. A DNA polymerase then replaces the excised nucleotides, and the resulting nick is sealed by DNA ligase.

Successful mismatch repair depends on the ability of mismatch repair enzymes to distinguish between parental and newly synthesized DNA strands. This distinction is possible because newly synthesized DNA strands temporarily lack methyl groups on their bases, whereas the parental DNA is methylated. Thus for a short time after the replication fork has passed, the DNA is hemimethylated, that is, methylated only on one strand. The repair system cuts out the mismatch from the unmethylated strand. Eventually adenine bases in the GATC sequences of the newly synthesized strand are methylated by enzyme DNA adenine methyltransferase (Dam).

Excision Repair

Excision repair corrects damage that causes distortions in the DNA double helix. Two types of excision repair systems have been described: nucleotide excision repair and base excision repair. Both use the same approach to repair; they remove the damaged portion of a DNA strand and use the intact complementary strand as the template for synthesis of new DNA. The two repair systems are distinguished by the enzymes used to correct the damage.

In **nucleotide excision repair,** an *E. coli* enzyme called UvrABC endonuclease removes damaged nucleotides and a few nucleotides on either side of the lesion. The resulting single-stranded gap is filled by DNA polymerase I, and DNA ligase joins the fragments (**figure 16.9**). This system can remove thymine dimers (figure 16.5) and repair almost any other injury that produces a detectable distortion in DNA.

Base excision repair employs enzymes called DNA glycosylases. These enzymes remove damaged or unnatural bases yielding apurinic or apyrimidinic (AP) sites. Enzymes called AP endonucleases recognize the damaged DNA and nick the backbone at the AP site (**figure 16.10**). DNA polymerase I removes the damaged region, using its 5′ to 3′ exonuclease activity. It then fills in the gap, and DNA ligase joins the DNA fragments.

Direct Repair

Thymine dimers and alkylated bases (e.g., those with methyl or ethyl groups attached) often are corrected by **direct repair.** Thymine dimers (figure 16.5) are split apart by a process called **photoreactivation,** which requires visible light and is catalyzed by the enzyme photolyase. Methyl and some other alkyl groups that have been added to guanine can be removed with the help of an enzyme known as alkyltransferase or methylguanine methyltransferase. Thus damage to guanine from mutagens such as methyl-nitrosoguanidine (figure 16.4) can be repaired directly. *Direct Repair*

Recombinational Repair

Recombinational repair corrects damaged DNA in which both bases of a pair are missing or damaged, or where there is a gap opposite a lesion. In this type of repair, the protein **RecA** cuts a piece of template DNA from a sister molecule and puts it into the gap or uses it to replace a damaged strand (**figure 16.11**). Another copy of the damaged segment often is available because either it has recently been replicated or the cell contains more than one copy of its chromosome. Once the template is in place, the remaining damage can be corrected by another repair system.

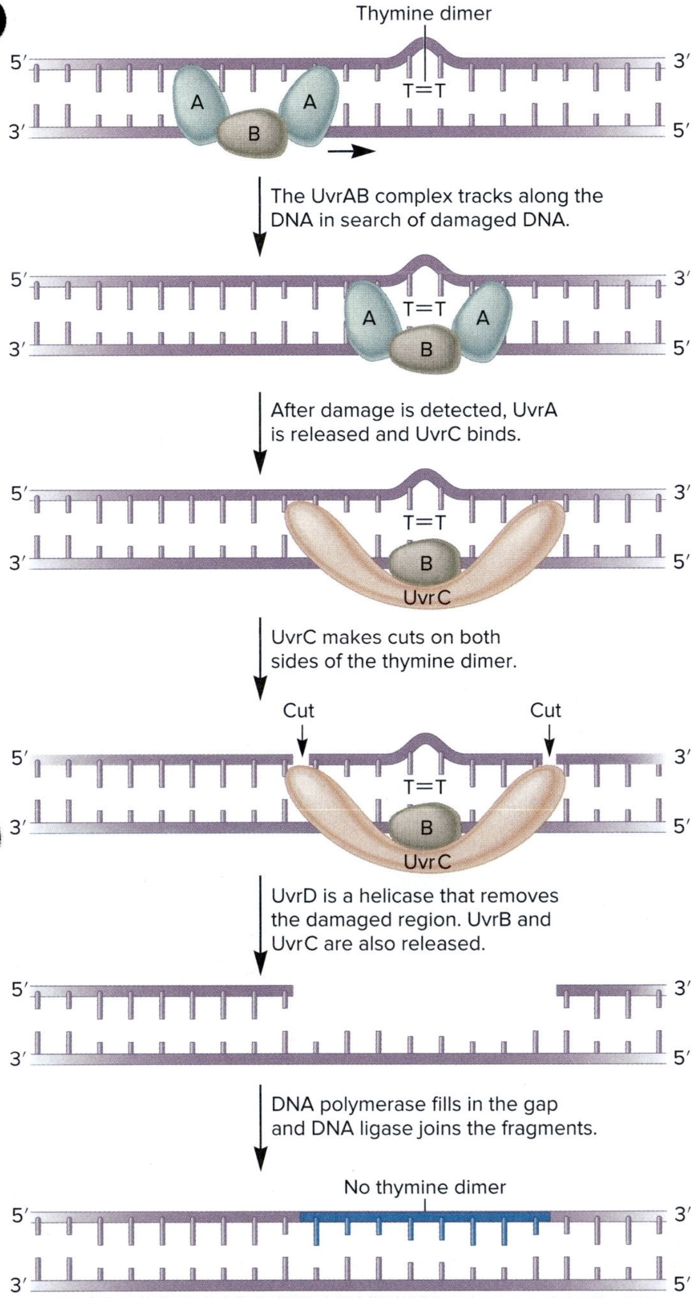

Figure 16.9 **Nutcleotide Excision Repair in *E. coli*.**

MICRO INQUIRY *How is damaged DNA recognized by the UvrAB complex?*

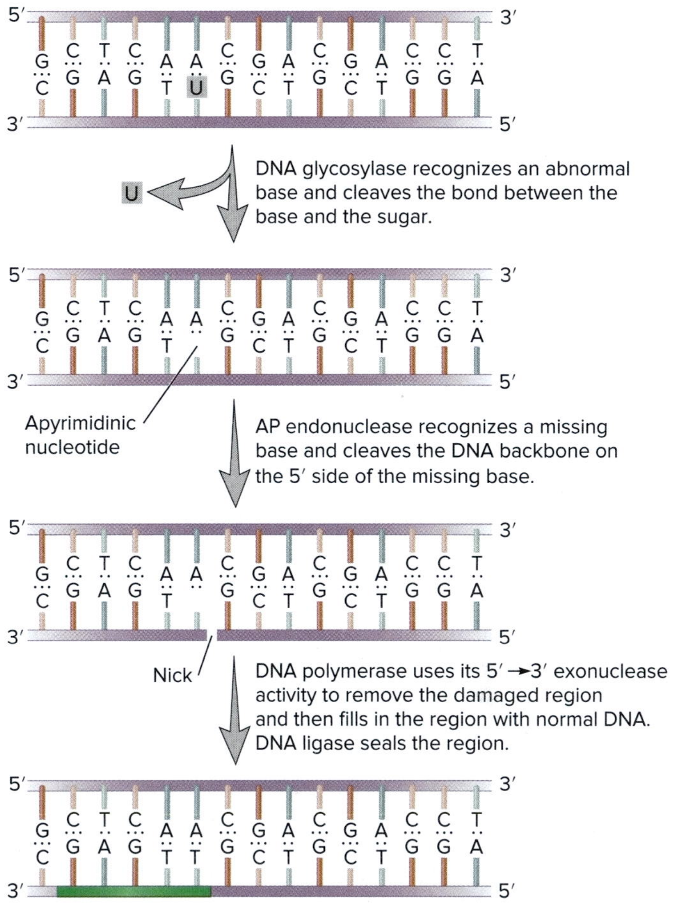

Figure 16.10 **Base Excision Repair.**

SOS Response

Despite having multiple repair systems, sometimes the damage to an organism's DNA is so great that the normal repair mechanisms just described cannot repair all the damage. As a result, DNA synthesis stops completely. In such situations, a global control network called the **SOS response** is activated. In this response, over 50 genes are activated when a transcriptional repressor protein called LexA is destroyed.

The SOS response, like recombinational repair, depends on the activity of RecA. RecA binds to single- or double-stranded DNA breaks and gaps generated by cessation of DNA synthesis. RecA binding initiates recombinational repair. Simultaneously RecA takes on coprotease function, causing LexA to destroy itself (autoproteolysis). The first genes transcribed in the SOS response encode the Uvr proteins needed for nucleotide excision repair (figure 16.9). Then expression of genes whose products are involved in recombinational repair is increased. To give the cell time to repair its DNA, the protein SfiA is produced; SfiA blocks cell division. Finally, if the DNA has not been fully repaired after about 40 minutes, a process called **translesion DNA synthesis** is triggered. In this process, DNA polymerase IV (also known as DinB; Din is short for *d*amage *in*ducible) and DNA polymerase V (also called UmuCD; Umu is short for *U*V *mu*tagenesis) synthesize DNA across gaps and other lesions (e.g., thymine dimers) that stopped DNA polymerase III. However, because an intact template does not exist, these DNA polymerases often insert incorrect bases. Furthermore, they lack proofreading activity. Therefore even though DNA synthesis continues, it is highly error prone and results in the generation of numerous mutations.

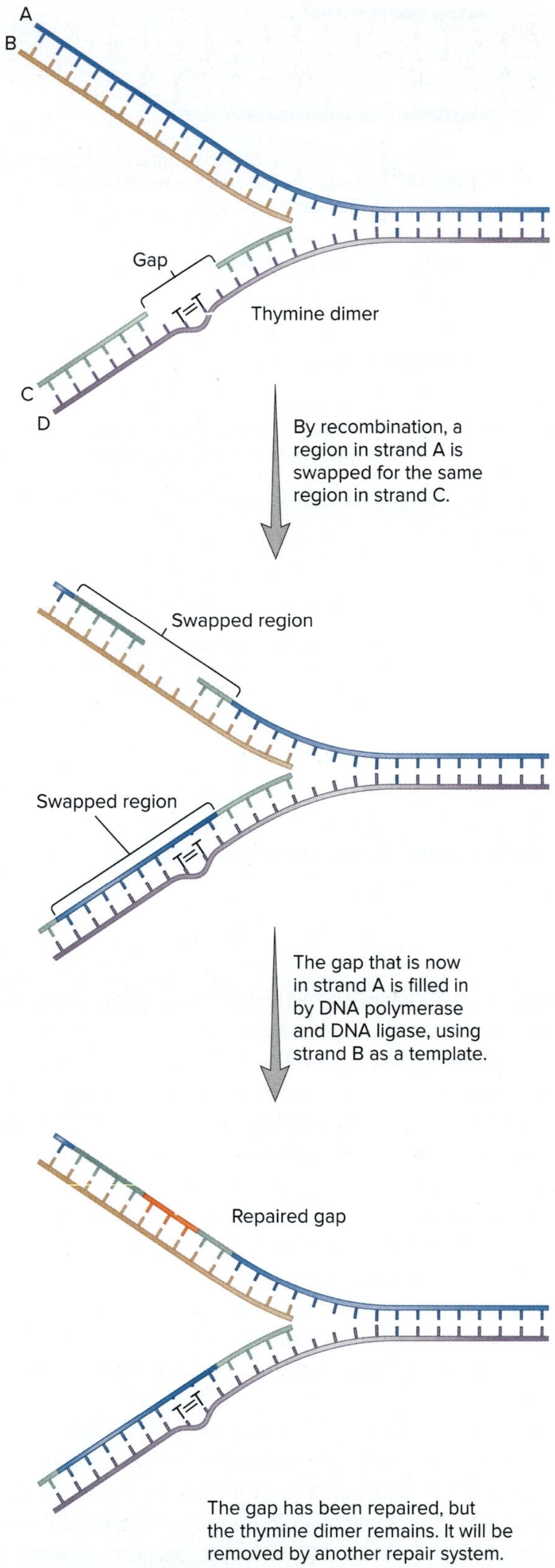

Figure 16.11 Recombinational Repair.

The SOS response is so named because it is made in a life-or-death situation. The response increases the likelihood that some cells will survive by allowing DNA synthesis to continue. For the cell, the risk of dying because of failure to replicate DNA is greater than the risk posed by the mutations generated by this error-prone process.

Comprehension Check

1. Compare and contrast the two types of excision repair.
2. What role does DNA methylation play in mismatch repair?
3. When *E. coli* cells are growing rapidly, they may contain up to four copies of their chromosomes. Why is this important if *E. coli* needs to carry out recombinational repair?
4. Explain how the following DNA alterations and replication errors would be corrected (there may be more than one way): base addition errors by DNA polymerase III during replication, thymine dimers, AP sites, methylated guanines, and gaps produced during replication.

16.4 Microbes Use Mechanisms Other than Mutation to Create Genetic Variability

After reading this section, you should be able to:

a. Distinguish vertical gene transfer from horizontal gene transfer
b. Summarize the four possible outcomes of horizontal gene transfer
c. Compare and contrast homologous recombination and site-specific recombination

In section 16.1, we discuss the consequences of mutations in terms of their effect on a protein and the phenotype of the organisms bearing the mutation. However, the effect also depends on the environment in which the organism lives. In other words, those mutations that don't immediately kill the organism are subject to selective pressure. Selective pressure determines if a mutation will persist in a population to become an alternate form of the gene. An alternate form of a gene is termed an **allele.** The existence of alleles for each gene in a genome means each organism in a population can have a distinctive (and probably a unique) set of alleles making up its genome, that is, its genotype. Each genotype in a population can be selected for or selected against. Organisms with genotypes, and therefore phenotypes, best suited to the environment survive and are most likely to pass on their genes. Shifts in environmental pressures lead to changes in the population and ultimately result in the evolution of new species. The mechanisms by which new combinations of genes are generated are the topic of this section. All involve **recombination,** the process in which one or more nucleic acid molecules are rearranged or combined to produce a new genotype. Geneticists refer to organisms produced following a recombination event as recombinant organisms or simply **recombinants.**

Sexual Reproduction and Genetic Variability

The transfer of genes from parents to progeny is called **vertical gene transfer.** This type of gene transfer is observed in all organisms. In eukaryotes capable of sexual reproduction, vertical gene transfer is accompanied by genetic recombination, which occurs in two ways. The first is from crossing-over between sister chromosomes during meiosis. This creates new mixtures of alleles on homologous chromosomes. The second type of recombination results from fusion of gametes. Since each gamete contains alleles from the parent, when the gametes fuse, the parental alleles are combined in the zygote. As a result of these two types of recombination events, progeny organisms are not identical to their parents or to each other.

Horizontal Gene Transfer: Creating Variability the Asexual Way

Bacteria and archaea do not reproduce sexually. This suggests that genetic variation in populations of these microbes should be relatively limited, only occurring with the advent of a new mutation and its passage to the next generation by vertical gene transfer. However, this is not the case. Bacteria and archaea have multiple mechanisms for creating recombinants collectively referred to as **horizontal (lateral) gene transfer (HGT)**. HGT is distinctive from vertical gene transfer because genes from one independent, mature organism are transferred to another mature organism, often creating a stable recombinant having characteristics of both the donor and the recipient. *Horizontal Gene Transfer*

The importance of HGT cannot be overstated. HGT has been important in the evolution of many species, and it continues to be commonplace in many environments. Indeed, HGT is the main mechanism by which bacteria and archaea evolve. Many examples are known of DNA transfer between distantly related genera and across domains of life, particularly among microbes sharing a habitat (e.g., the human gut or an Antarctic lake). HGT helps bacteria and archaea survive environmental stresses. For instance, some archaea respond to UV radiation by increasing their ability to carry out HGT so that they can obtain and use DNA from other cells to repair any UV-damaged DNA in their genomes. Another important example is the transfer of groups of genes encoding virulence factors from pathogens to other bacteria. These gene clusters are called pathogenicity islands, and their discovery has provided valuable insights into the evolution of pathogenic bacteria. Finally, in most cases, the acquisition of new traits by HGT enables microbes to rapidly expand their ecological niche. This is clearly seen when HGT results in the spread of antibiotic-resistance genes among pathogenic bacteria. ▶| *Evolutionary processes and the concept of a microbial species inspire debate (section 19.5); Pathogenicity islands encode virulence factors (section 35.4)*

During HGT, a piece of donor DNA, sometimes called the exogenote, enters a recipient cell. The transfer can occur in multiple ways. If the donor and recipient cells are closely related, the DNA may be transferred directly between the two cells in temporary physical contact (conjugation). Alternately, a bacteriophage may act as an intermediary, and transport DNA within the viral capsid (transduction). In both cases, cell surface structures dictate the range of organisms that may interact by these mechanisms. Two other ways that HGT occurs are through transfer of naked DNA (transformation) or transfer of DNA within a membrane vesicle. In these processes, the DNA is taken up directly from the environment, and the donor and recipient may not be closely related. All of these processes have been described in bacteria and archaea. In addition, membrane vesicle formation occurs in cells from all domains, and although the phenomenon of HGT by membrane vesicles is well documented, the mechanistic details remain obscure. Thus we focus on the well-established HGT processes of conjugation, transformation, and transduction. |◀ *Membrane vesicles are an alternative secretion mechanism (section 13.8)*

If the donor DNA contains genes already present in the recipient, the recipient will become temporarily diploid for those genes. This partially diploid cell is sometimes called a merozygote (Greek *meros*, part). The donor DNA has four possible fates in the recipient (**figure 16.12**). First, when the

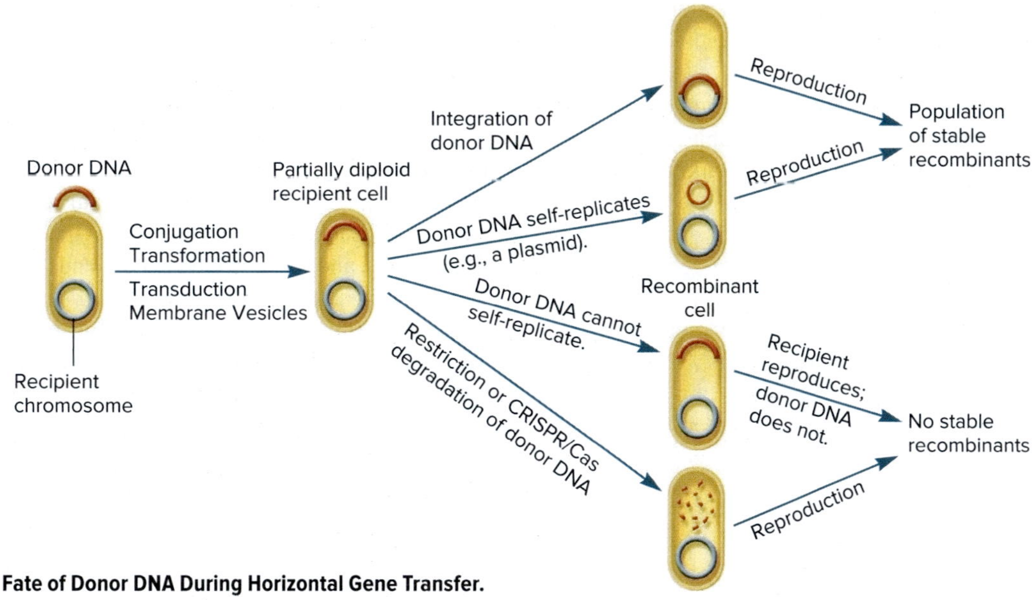

Figure 16.12 The Fate of Donor DNA During Horizontal Gene Transfer.

donor DNA has a sequence homologous to one in the recipient's chromosome (sometimes called an endogenote), integration may occur. That is, the donor's DNA may pair with the recipient's DNA and recombine. The recombinant then reproduces, yielding a population of stable genetic variants. Second, if the donor DNA is able to replicate (e.g., it is a plasmid), it may persist separate from the recipient's chromosome. When the recipient reproduces, the donor DNA replicates and a population of stable recombinants is formed. Third, the donor DNA remains in the cytoplasm but is unable to replicate. When the recipient divides, the donor DNA is eventually lost from the population. Finally, host restriction or CRISPR/Cas degradation of donor DNA may occur, thereby preventing the formation of a recombinant cell. ◀ *Membrane vesicles are an alternative secretion mechanism (section 13.8); Responses to viral infection (section 14.6)*

Molecular Recombination: Joining DNA Molecules Together

Crossing-over during meiosis and integration of donor DNA into the recipient's chromosome during HGT occur by similar mechanisms. **Homologous recombination** is the most common mechanism, and many of the enzymes used are similar for all organisms. Homologous recombination occurs wherever there are long regions of the same or similar nucleotide sequence in two DNA molecules (e.g., sister chromosomes in diploid eukaryotes or similar sequences on both a chromosome and plasmid). Homologous recombination results from DNA strand breakage and reunion leading to crossing-over. It is carried out by enzymes, many of which are also important for DNA repair (e.g., RecA; section 16.3).

The other major type of recombination is **site-specific recombination.** Site-specific recombination differs from homologous recombination in three significant ways. First, it does not require long regions of sequence homology. Second, recombination occurs at specific target sites in DNA molecules, and third, it is catalyzed by enzymes collectively called **recombinases.** Site-specific recombination is used by some plasmids and viral genomes to integrate into host chromosomes. It is also the type of recombination used by transposable elements to move from one place to another in DNA molecules, as we discuss in section 16.5.

Comprehension Check

1. An antibiotic-resistance gene located on a bacterium's chromosome is transferred from the bacterium to its progeny. What type of gene transfer is this?
2. What four fates can DNA have after entering a bacterium?
3. How does homologous recombination differ from site-specific recombination?

16.5 Transposable Elements Move Genes Within and Between DNA Molecules

After reading this section, you should be able to:

a. Differentiate insertion sequences from transposons
b. Distinguish simple transposition from replicative transposition
c. Defend this statement: "Transposable elements are important factors in the evolution of bacteria and archaea"

As more and more genomes are sequenced and annotated, it is apparent that all genomes are rife with genetic elements that are able to move into and out of the genomes where they reside. These elements are often referred to as "jumping genes," mobile genetic elements, and transposable elements. **Transposition** refers to the movement of a mobile genetic element.

Mobile genetic elements were first discovered in the 1940s by Barbara McClintock during her studies on maize genetics (a discovery for which she was awarded the Nobel Prize in 1983). As additional types of mobile genetic elements were discovered, nomenclature was developed to distinguish them. As is often the case when a scientific field expands and its language evolves, the nomenclature became unwieldy and confusing. In 2008 a group of scientists studying bacterial and archaeal mobile genetic elements proposed a new definition for **transposable elements:** "specific DNA segments that can repeatedly insert into one or more sites or into one or more genomes." This new definition encompasses elements that differ in structure, mechanisms of integration and excision, target sites, and ability to be transferred from one cell to another by HGT. *Transposition: Shifting Segments of the Genome*

The enzymes that function in transposition are collectively termed recombinases. However, the recombinase used by a specific mobile genetic element may be called an integrase, resolvase, or transposase. There is tremendous interest in these enzymes because their mechanisms of action are similar to those used to rearrange the gene segments that encode important immune system proteins such as antibodies and T-cell receptors. ▶ *Adaptive immunity (chapter 33)*

The simplest mobile genetic elements in bacteria are **insertion sequences,** or IS elements for short (**figure 16.13*a***). An IS element is a short sequence of DNA (around 750 to 1,600 base pairs [bp] in length). It contains only the gene for the enzyme transposase, and it is bounded at both ends by inverted repeats—identical or very similar sequences of nucleotides in reversed orientation. Inverted repeats are usually about 15 to 25 base pairs long and vary among IS elements so that each type of IS has a specific nucleotide sequence in its inverted repeats. **Transposase** recognizes the ends of the IS and catalyzes transposition. Each IS element is named by giving it the prefix IS followed by a number. IS elements have been observed in a variety of bacteria and some archaea.

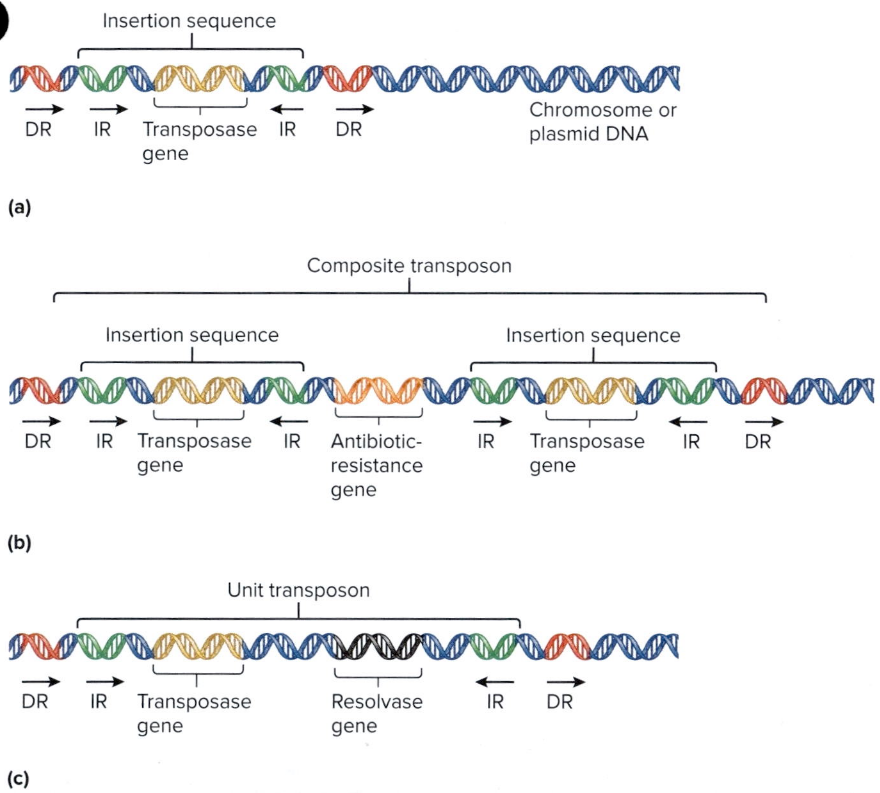

Figure 16.13 Transposable Elements. All transposable elements contain common features. These include inverted repeats (IRs) at the ends of the element and a recombinase (e.g., transposase) gene. (a) Insertion sequences (IS) consist only of IRs on either side of the transposase gene. (b) Composite transposons and (c) unit transposons contain additional genes (e.g., antibiotic-resistance genes) in addition to the recombinases that enable them to transpose. In composite transposons, the additional genes are flanked by insertion sequences, which supply the transposase. Unit transposons are not associated with insertion sequences. DRs, direct repeats in host DNA, flank a transposable element.

MICRO INQUIRY *What features are common to all types of transposable elements?*

Transposons are more complex in structure than IS elements. Some transposons (composite transposons) consist of a central region containing genes unrelated to transposition (e.g., antibiotic-resistance genes) flanked on both sides by IS elements that are identical or very similar in sequence (figure 16.13*b*). The flanking IS elements encode the transposase used by the transposon to move. Other transposons (unit transposons) lack IS elements and encode their own transposition enzymes (figure 16.13*c*). Most transposon names begin with the prefix Tn.

Two major transposition methods have been identified: simple transposition and replicative transposition (**figure 16.14**). **Simple transposition** is also called **cut-and-paste transposition.** In this method, transposase catalyzes excision of the transposable element, followed by cleavage of a new target site and ligation of the element into this site (**figure 16.15**). Target sites are specific sequences about five to nine base pairs long. When a mobile genetic element inserts at a target site, the target sequence is duplicated so that short, direct-sequence repeats flank the element's terminal inverted repeats. In **replicative transposition,** the original transposon remains at the parental site on the chromosome and a copy is inserted at the target DNA site (figure 16.14*b*). *Simple Transposition*

Transposable elements are of interest for many reasons. Their movement into a chromosome can alter gene function either by causing mutations in the genes into which they move or by providing regulatory sites that change the expression patterns of the adjacent genes. They also contribute to the evolution of an organism's chromosome, plasmids, and other mobile genetic elements. Of particular note is the role of mobile genetic elements in the spread of antibiotic resistance, as we discuss in section 16.9.

Comprehension Check

1. How does a transposon differ from an insertion sequence?
2. What is simple (cut-and-paste) transposition? What is replicative transposition? How do the two mechanisms of transposition differ? What happens to the target site during transposition?
3. What effect would you expect the existence of transposable elements to have on the rate of microbial evolution? Give your reasoning.

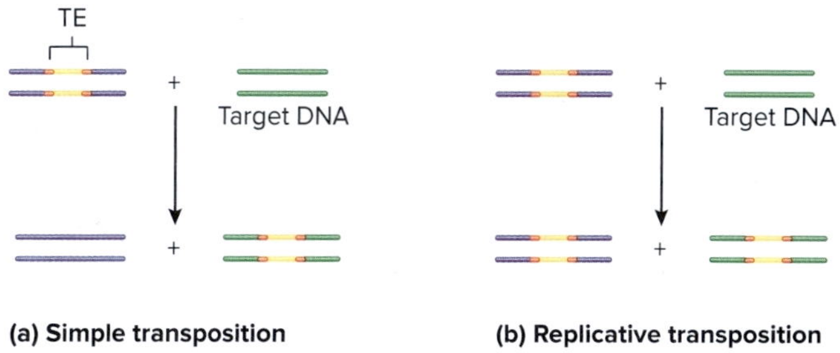

Figure 16.14 Comparison of Simple and Replicative Transposition. TE, transposable element.

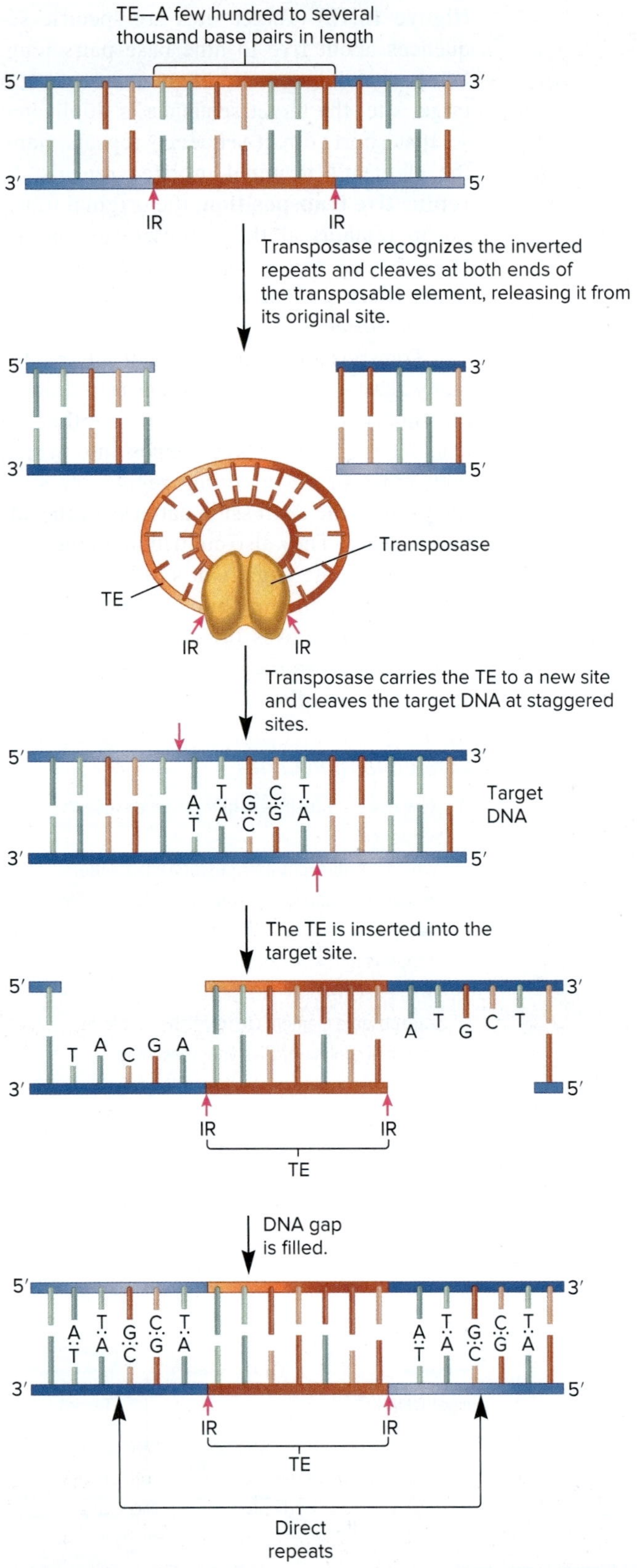

Figure 16.15 Simple Transposition. IR, inverted repeat; TE, transposable element.

16.6 Bacterial Conjugation Requires Cell-Cell Contact

After reading this section, you should be able to:

a. Identify the type of plasmids that are important creators of genetic variation
b. Describe the features of the F factor that allow it to (1) transfer itself to a new host cell and (2) integrate into a host cell's chromosome
c. Outline the events that occur when an F^+ cell encounters an F^- cell
d. Distinguish F^+, Hfr, and F′ cells from each other
e. Explain how Hfr cells arise
f. Outline the events that occur when an Hfr cell encounters an F^- cell

Conjugation, the transfer of DNA by direct cell-to-cell contact, depends on the presence of a **conjugative plasmid.** Recall from chapter 3 that **plasmids** are small, double-stranded DNA molecules that can exist independently of host chromosomes. They have their own replication origins, replicate autonomously, and are stably inherited. Some plasmids are **episomes,** plasmids that can exist either with or without being integrated into host chromosomes.

Perhaps the best-studied conjugative plasmid is the **F factor.** It plays a major role in conjugation in *E. coli,* and it was the first conjugative plasmid to be described (**figure 16.16**). The F factor is about 100,000 bases long and bears genes responsible for cell attachment and plasmid transfer between specific *E. coli* cells. Most of the information required for plasmid transfer is located in the *tra* operon, which contains at least 28 genes. Many of these direct the formation of sex pili that attach the F^+ cell (the donor cell containing an F plasmid) to an F^- cell (**figure 16.17**). Other gene products aid DNA transfer. In addition, the F factor has several IS elements that assist plasmid integration into the host cell's chromosome. Thus the F factor is an episome that can exist outside the bacterial chromosome or be integrated into it (**figure 16.18**).

The initial evidence for bacterial conjugation came from an elegant experiment performed in 1946 by Joshua Lederberg and Edward Tatum. They mixed two auxotrophic strains, incubated the culture for several hours in nutrient medium, and then plated it on minimal medium. To reduce the chance that their results were due to a reversion or suppressor mutation, they used double and triple auxotrophs on the assumption that two or three simultaneous reversion or suppressor mutations would be extremely rare. When recombinant prototrophic colonies appeared on the minimal medium after incubation, they concluded that the two auxotrophs were able to associate and undergo recombination.

Lederberg and Tatum did not directly prove that physical contact of the cells was necessary for gene transfer. This evidence was provided several years later by Bernard Davis, who constructed a U-tube consisting of two pieces of curved glass tubing fused at the base to form a U shape with a glass filter between the halves. The filter allowed passage of media but not bacteria. The U-tube was filled with a growth medium and each side inoculated with a different auxotrophic strain of *E. coli.* During incubation, the medium was pumped back and forth through the filter to ensure exchange.

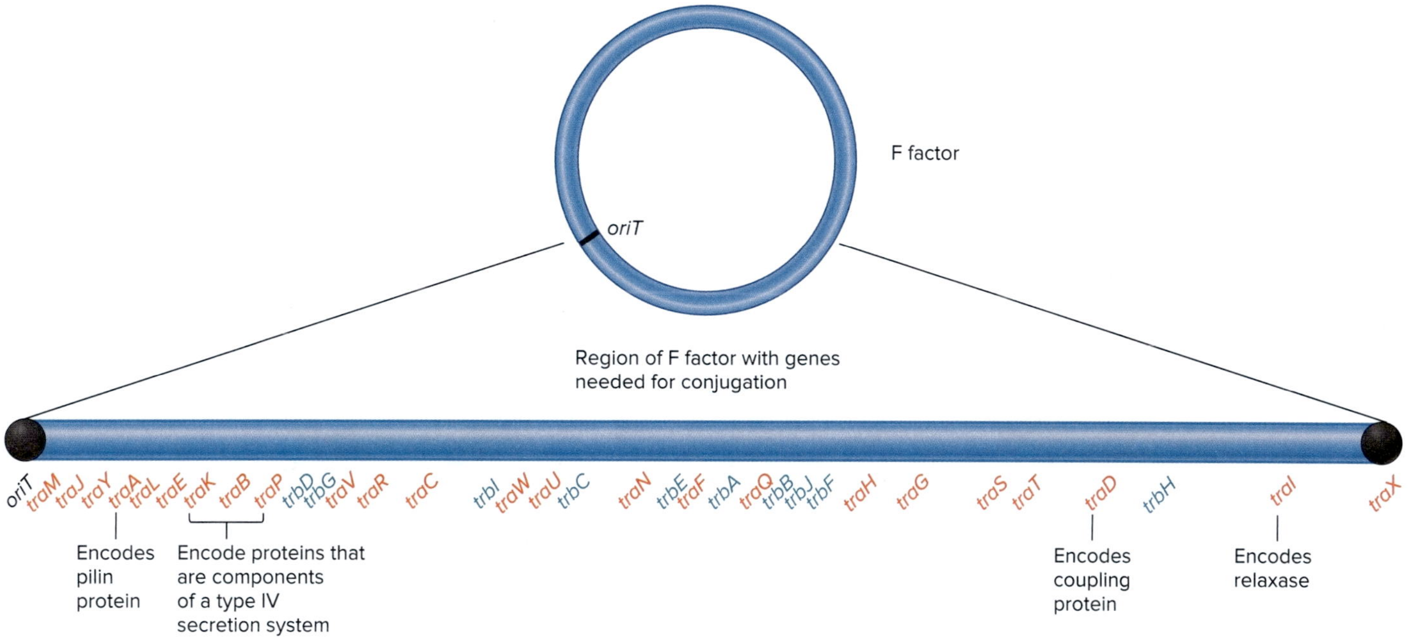

Figure 16.16 The F plasmid. Transfer (*tra*) genes are shown in red, and some of their functions are indicated. The plasmid also contains three insertion sequences and a transposon. The site for initiation of rolling-circle replication and gene transfer during conjugation is *oriT*.

When the bacteria were later plated on minimal medium, Davis discovered that if the two auxotrophic strains were separated from each other by the filter, gene transfer did not take place. Therefore direct contact was required for the recombination that Lederberg and Tatum had observed.

$F^+ \times F^-$ Mating: Plasmid Only Transfer

In 1952 William Hayes demonstrated that the gene transfer observed by Lederberg and Tatum was unidirectional. That is, there were definite donor (F^+, or fertile) and recipient (F^-, or nonfertile) strains, and gene transfer was nonreciprocal. He also found that in $F^+ \times F^-$ mating, the progeny were only rarely changed with regard to auxotrophy (i.e., chromosomal genes usually were not transferred). However, F^- strains frequently became F^+.

These results are now understood and readily explained in the following way. An F^+ strain contains an extrachromosomal F factor

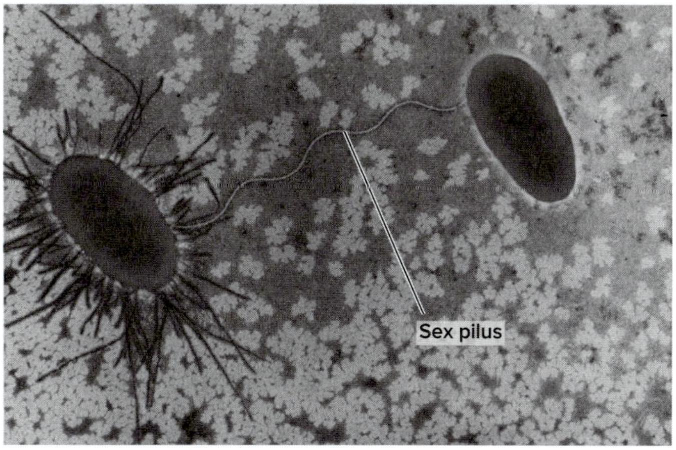

Figure 16.17 Bacterial Conjugation. An electron micrograph of two *E. coli* cells in an early stage of conjugation. The F^+ cell to the left is covered with fimbriae, and a sex pilus connects the two cells.

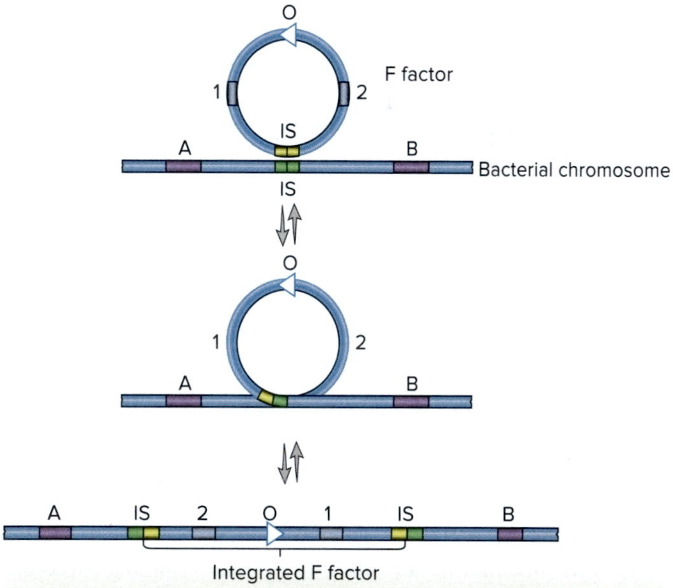

Figure 16.18 F Plasmid Integration. The reversible integration of an F plasmid or factor into a host bacterial chromosome. The process begins with association between plasmid and bacterial insertion sequences. Note that the ISs (insertion sequences) are used for recombination, not transposition. The O arrowhead (white) indicates the site at which oriented transfer of chromosome to the recipient cell begins. A, B, 1, and 2 represent genetic markers.

MICRO INQUIRY *What does the term* episome *mean?*

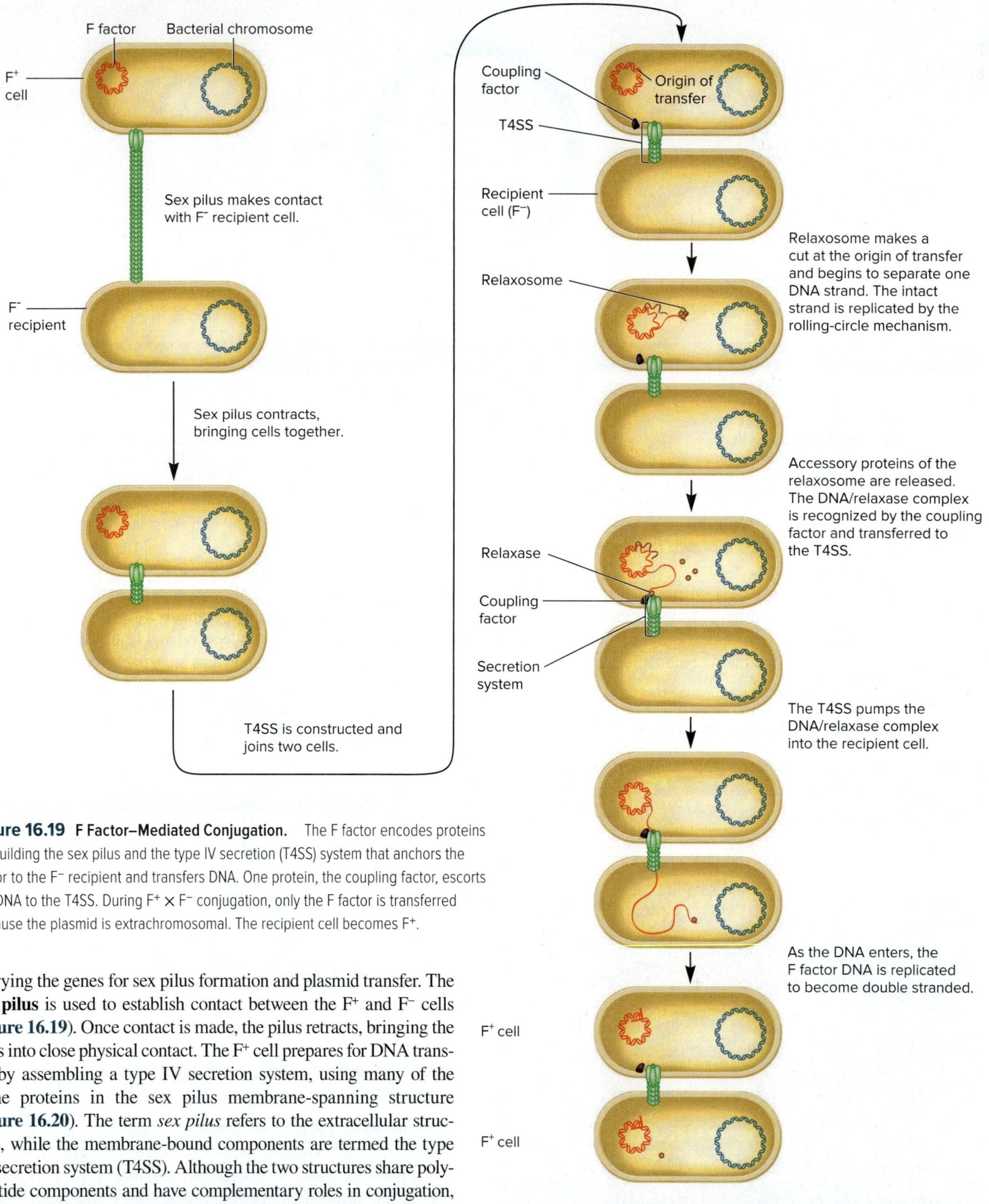

Figure 16.19 F Factor–Mediated Conjugation. The F factor encodes proteins for building the sex pilus and the type IV secretion (T4SS) system that anchors the donor to the F− recipient and transfers DNA. One protein, the coupling factor, escorts the DNA to the T4SS. During F+ × F− conjugation, only the F factor is transferred because the plasmid is extrachromosomal. The recipient cell becomes F+.

carrying the genes for sex pilus formation and plasmid transfer. The **sex pilus** is used to establish contact between the F+ and F− cells (**figure 16.19**). Once contact is made, the pilus retracts, bringing the cells into close physical contact. The F+ cell prepares for DNA transfer by assembling a type IV secretion system, using many of the same proteins in the sex pilus membrane-spanning structure (**figure 16.20**). The term *sex pilus* refers to the extracellular structure, while the membrane-bound components are termed the type IV secretion system (T4SS). Although the two structures share polypeptide components and have complementary roles in conjugation, pilus formation and DNA transfer are independent processes. *Conjugation: Transfer of the F Plasmid*

As the F plasmid is being transferred, it is also copied using a process called **rolling-circle replication.** One strand of the circular DNA is nicked, and the free 3′-hydroxyl end is extended by replication enzymes (**figure 16.21**). The 3′ end is lengthened while the growing point rolls around the circular template and the 5′ end of the strand is displaced to form an ever-lengthening tail, much like the peel of an apple is displaced by a knife as it is

Figure 16.20 The Conjugation Machinery Encoded by F Factor Is a Type IV Secretion System. The F factor–encoded type IV secretion system (T4SS) is composed of numerous Tra proteins, including TraA, which forms the sex pilus, and TraD, the coupling factor. The core complex is used for both pilus biogenesis and DNA transfer. LPS, lipopolysaccharide; OM, outer membrane; P, periplasm; PG, peptidoglycan; PM, plasma membrane.

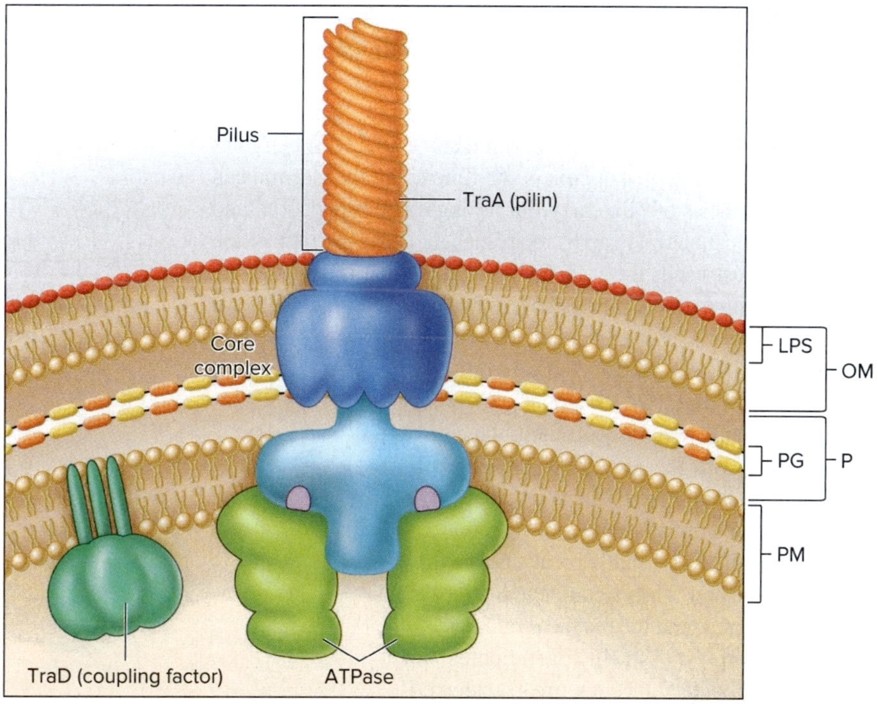

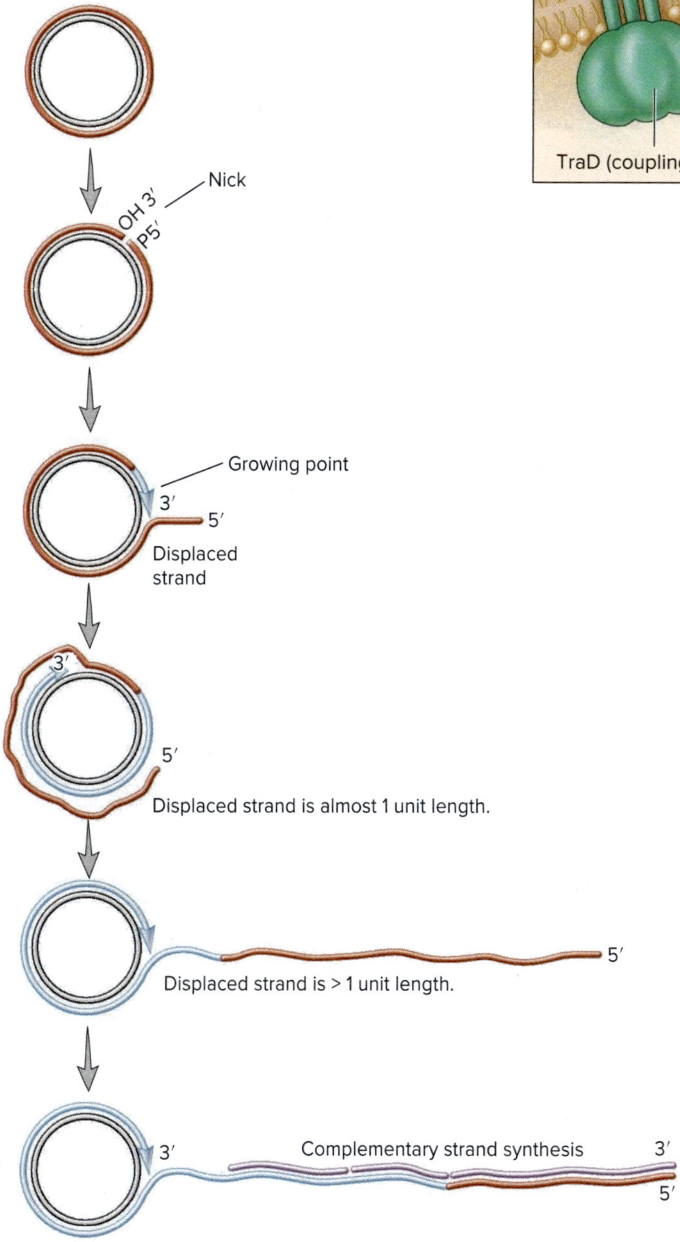

pared. We are concerned here with rolling-circle replication of a plasmid. However, rolling-circle replication is also observed during the replication of some viral genomes (e.g., phage lambda). *Rolling-Circle Replication*

During conjugation, rolling-circle replication is initiated by the relaxosome, a complex of proteins encoded by the F factor (figure 16.16). The relaxosome nicks one strand of the F factor at a site called *oriT* (for origin of transfer). The major component of the relaxosome is a protein called TraI. It has relaxase activity and remains attached to the 5′ end of the nicked strand. As the F factor is replicated, TraI guides the displaced strand through the T4SS to the recipient cell. It is believed that both ATP hydrolysis and the proton motive force provide the energy for DNA translocation. During plasmid transfer, the entering strand is copied to produce double-stranded DNA. When this is complete, the F^- recipient cell becomes F^+.

Hfr Conjugation Transfers Chromosomal DNA

By definition, an F^+ cell has the F factor separate from the chromosome, so in an $F^+ \times F^-$ mating, chromosomal DNA is not transferred. However, within this population, a few cells have the F plasmid integrated (i.e., recombined) into their chromosomes. This explains why not long after the discovery of $F^+ \times F^-$ mating, a second type of F factor–mediated conjugation was discovered. In this type of conjugation, the donor transfers chromosomal genes with great efficiency but does not change the recipient bacteria

Figure 16.21 Rolling-Circle Replication. A single-stranded tail, often composed of more than one genome copy, is generated and can be converted to the double-stranded form by synthesis of a complementary strand. The "free end" of the rolling-circle strand is bound to the relaxosome. OH 3′ is the 3′-hydroxyl and P 5′ is the 5′-phosphate created when the DNA strand is nicked.

into F+ cells. Because of the *h*igh *f*requency of *r*ecombinants produced by this mating, it is referred to as **Hfr conjugation** and the donor is called an **Hfr strain.**

Hfr strains contain the F factor integrated into their chromosome, rather than free in the cytoplasm (**figure 16.22**). When integrated, the F plasmid's *tra* operon is still functional; the plasmid can direct the synthesis of pili, carry out rolling-circle replication, and transfer genetic material to an F− recipient cell. However, rather than transferring just itself, the F factor also directs the transfer of the host chromosome. DNA transfer begins when the integrated F factor is nicked at *oriT*. As it is replicated, the F factor begins to move into the recipient (**figure 16.23**). Initially only part of the F factor is transferred, followed by the donor's chromosome. If the cells remain connected, the entire chromosome with the rest of the integrated F factor will be transferred; this takes about 100 minutes to accomplish. However, the connection between the cells usually breaks before this process is finished. Thus a complete F factor is rarely transferred, and the recipient remains F−.

When an Hfr strain participates in conjugation, bacterial genes are transferred to the recipient in either a clockwise or a counterclockwise direction around a circular chromosome, depending on the orientation of the integrated F factor. After the replicated donor chromosome enters the recipient cell, it may be degraded or incorporated into the F− genome by recombination. *Transfer of Chromosomal DNA*

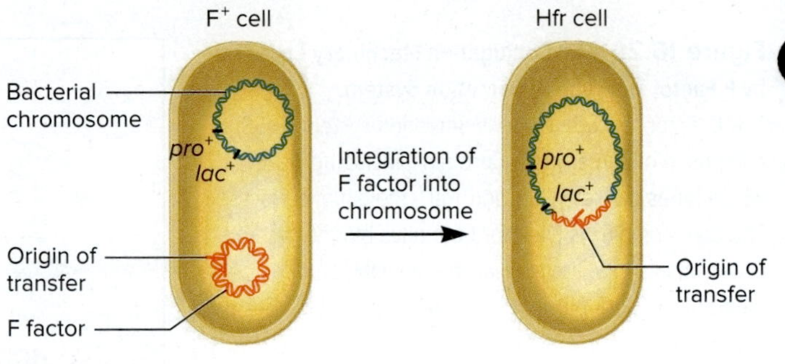

Figure 16.22 Insertion of an F Factor into the Donor Cell's Chromosome Creates an Hfr Cell.

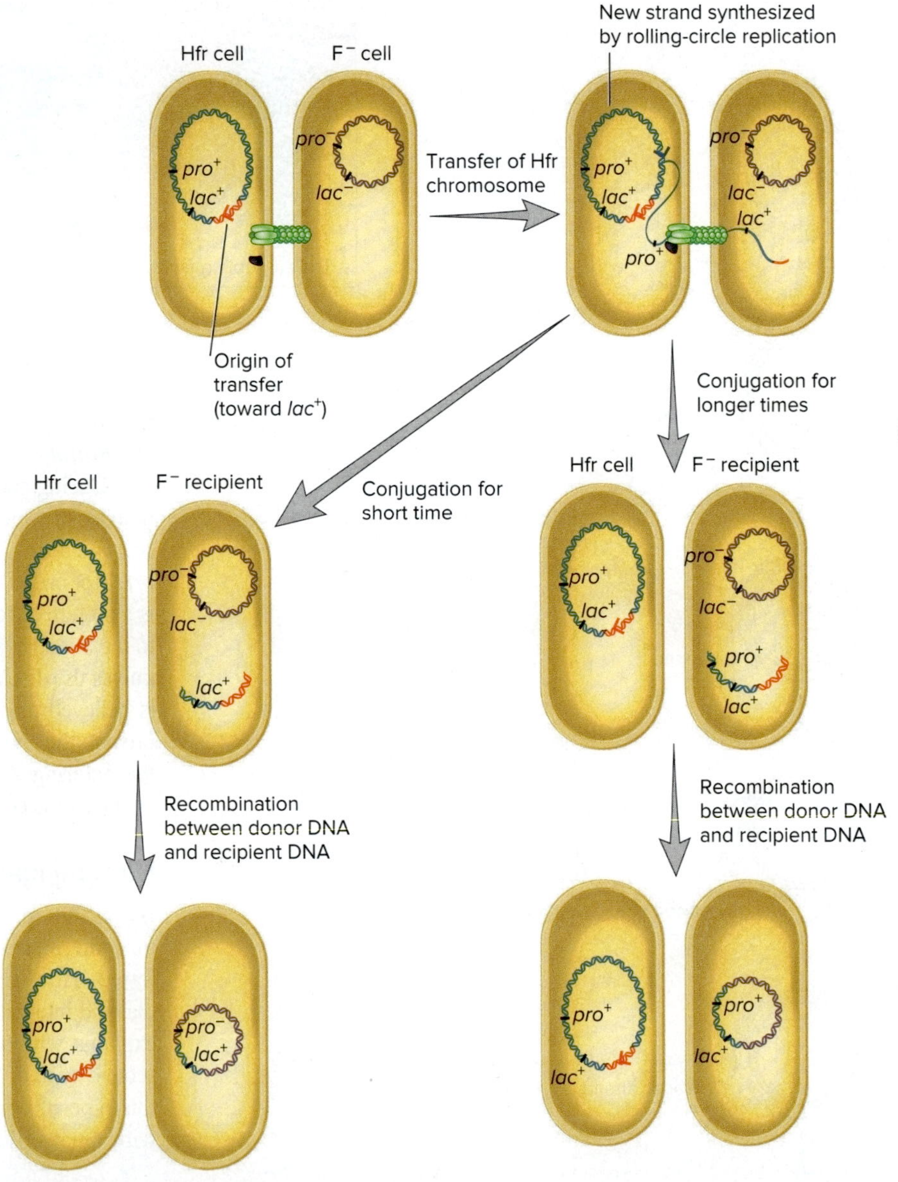

Hfr x F− conjugation

Figure 16.23 Hfr × F− Conjugation. Shown are the two cells after initial contact and elaboration of the T4SS. As illustrated, during Hfr × F− conjugation, some plasmid genes and some chromosomal genes are transferred to the recipient. Note that only a portion of the F factor moves into the recipient. Because the entire plasmid is not transferred, the recipient remains F−. In addition, the incoming DNA must recombine with the recipient's chromosome if it is to be stably maintained.

F′ Conjugation

Because the F factor is an episome, it can leave the bacterial chromosome and resume status as an autonomous plasmid. Sometimes during excision an error occurs and a portion of the chromosome is excised, becoming part of the F plasmid. Because this erroneously excised plasmid is larger and genotypically distinct from the original F factor, it is called an **F′ plasmid** (**figure 16.24*a***). A cell containing an F′ plasmid retains all of its genes, although some of them are on the plasmid. It mates only with an F− recipient, and F′ × F− conjugation is similar to an F+ × F− mating. Once again, the plasmid is transferred as it is copied by rolling-circle replication. Bacterial genes on the chromosome are not transferred (figure 16.24*b*), but bacterial genes on the F′ plasmid are transferred. These genes need not be incorporated into the recipient chromosome to be expressed. The recipient becomes F′ and is partially diploid because the same bacterial genes present on the F′ plasmid are also found on the recipient's

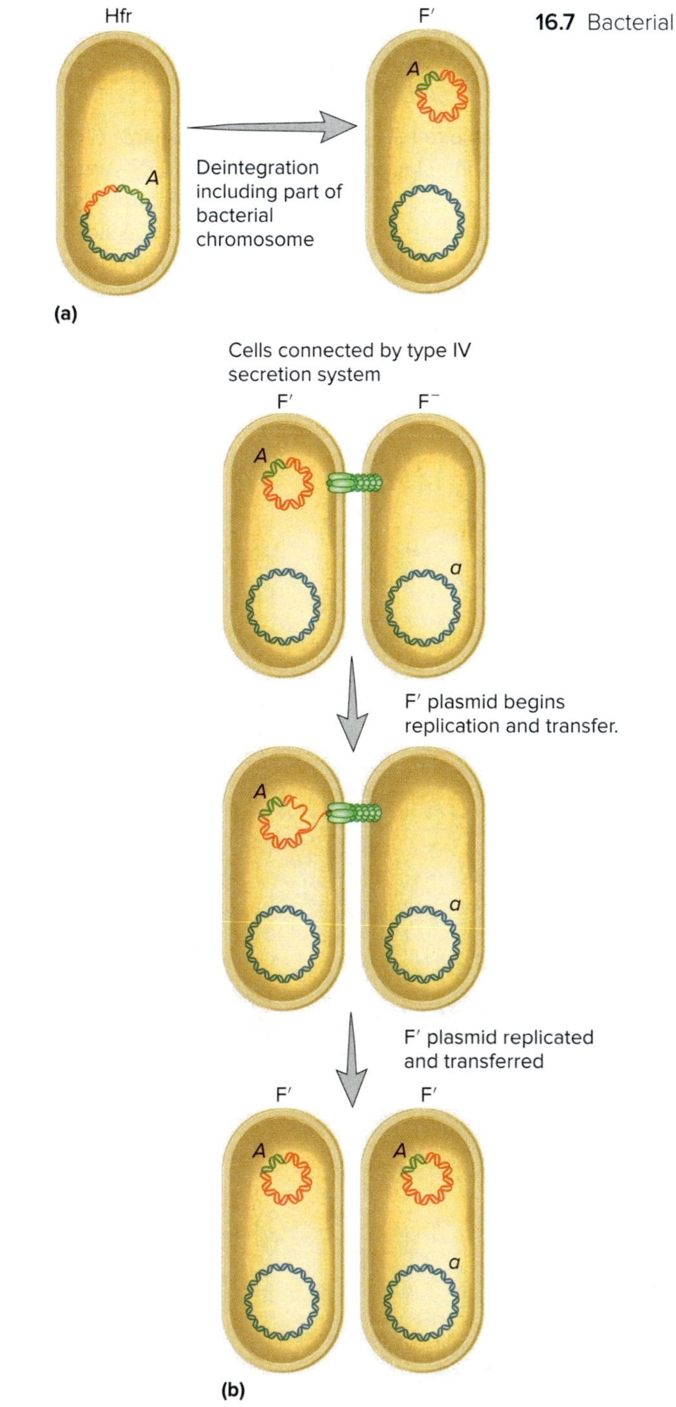

Figure 16.24 F′ Conjugation. (a) Due to an error in excision, the *A* gene of an Hfr cell is incorporated into the F factor. (b) During conjugation, the *A* gene is transferred to a recipient, which becomes diploid for that gene (i.e., Aa).

chromosome. In this way, specific bacterial genes may spread rapidly throughout a bacterial population.

Bacterial Conjugation in Gram-Positive Bacteria

Although most research on plasmids and conjugation has been done using *E. coli* and other Gram-negative bacteria, conjugative plasmids are present in numerous Gram-positive bacteria. With one exception (*Streptomyces* spp.), Gram-positive conjugative systems share similarities with those of Gram-negative bacteria. For instance, they have homologues of TraI (relaxase) and TraD (coupling factor), as well as other T4SS proteins. Unlike Gram-negative bacteria, which establish contact by way of the sex pilus, cell-to-cell contact in these Gram-positive bacteria is established by surface substances that enable cells to directly adhere to one another. One of the best-studied Gram-positive systems is that of *Enterococcus faecalis*. The recipient cells of this bacterium release short peptides that activate transfer genes in donor cells containing the proper plasmid. The conjugation systems of *Streptomyces* spp. differ from those of other Gram-positive bacteria in that T4SS, TraI homologues, and a specific nick site on the plasmid have not been identified.

Comprehension Check

1. What is bacterial conjugation and how was it discovered?
2. For F^+, Hfr, and F^- strains of *E. coli,* indicate which acts as a donor during conjugation, which acts as a recipient, and which transfers chromosomal DNA.
3. Describe how $F^+ \times F^-$ and Hfr conjugation processes proceed, and distinguish between the two in terms of mechanism and the final results.
4. Compare and contrast $F^+ \times F^-$ and $F' \times F^-$ conjugation.

16.7 Bacterial Transformation Is the Uptake of Free DNA from the Environment

After reading this section, you should be able to:

a. Describe the factors that contribute to a bacterium being naturally transformation competent
b. Predict the outcomes of transformation using a DNA fragment versus using a plasmid
c. Design an experiment to transform bacteria that are not naturally competent with a plasmid that carries genes encoding ampicillin resistance and the protein that generates green fluorescence

Another HGT mechanism is transformation, discovered by Fred Griffith in 1928. **Transformation** is the uptake of circular or linear DNA from the environment outside the cell and maintenance of the DNA in the recipient cell in a heritable form. Natural transformation is distinct from artificial transformation, a laboratory technique that induces cells to take up DNA. Natural transformation has been observed in some archaea and in members of several bacterial phyla. It occurs in soil and aquatic ecosystems, in vivo during infection, and in biofilm and other microbial communities.

Natural transformation is best studied in *Streptococcus pneumoniae, Bacillus subtilis,* and *Neisseria* spp. It occurs when bacteria lyse and release their DNA into the surrounding environment. These fragments may be relatively large and contain several genes. If a fragment contacts a **competent cell**—a cell that is able to take up DNA and be transformed—the DNA is bound to the cell and imported (**figure 16.25*a***). The transformation frequency of very competent cells is around 10^{-3} for most genera when an excess of DNA is used. That is, about one cell in every thousand will take up and integrate the gene.

Competency is a complex phenomenon that is induced in about 10 to 15% of a population, during a certain stage of growth. For instance *S. pneumoniae* becomes competent during early exponential phase, whereas *B. subtilis* becomes competent in stationary phase. Competence genes require activation through alternative sigma factors or transcription activator proteins. A few species are constitutively competent (e.g., *Neisseria* spp. and *Thermus* spp.).

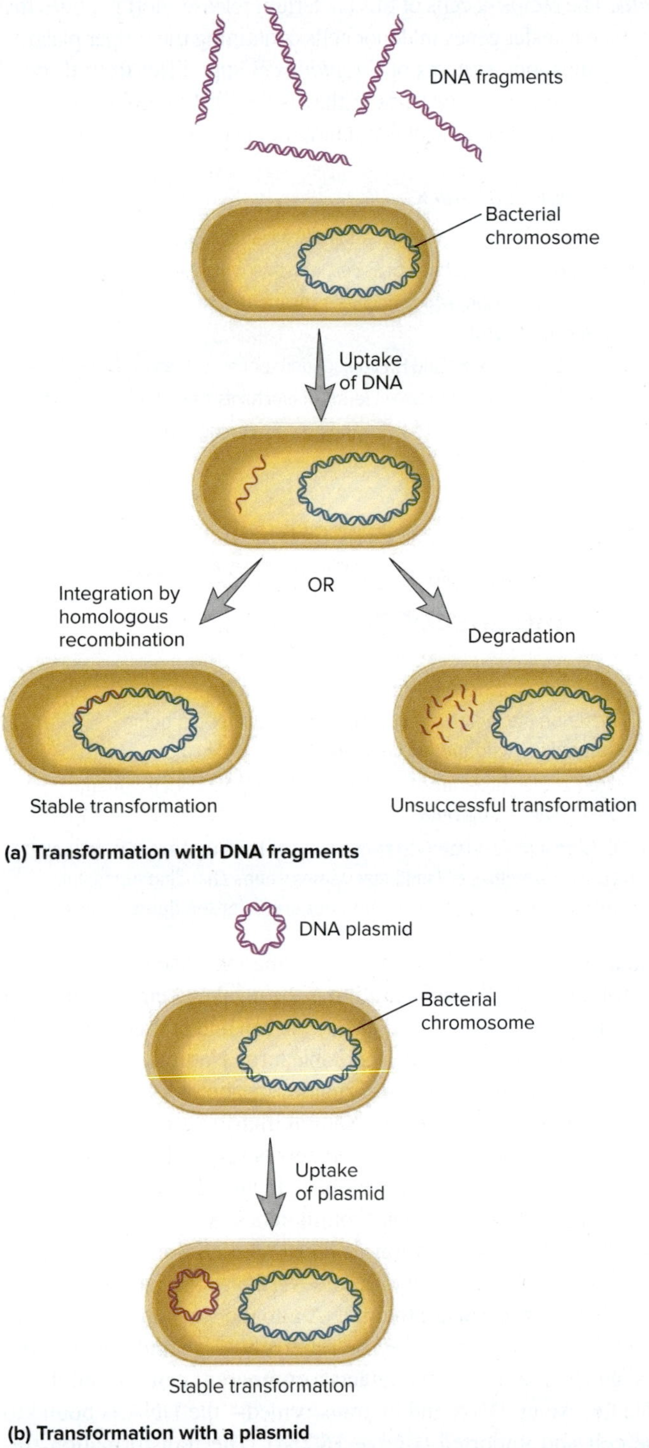

Figure 16.25 Bacterial Transformation. Transformation with (a) DNA fragments and (b) plasmids. Transformation with a plasmid often is induced artificially in the laboratory. The transforming DNA is in purple, and integration is at a homologous region of the genome.

Most bacteria are able to take up DNA from any source (i.e., other bacteria, archaea, and eukaryotes). Some take up DNA only from closely related species. For instance, *Haemophilus influenzae* only takes up DNA containing a specific 9-base-pair sequence that is repeated more than 1,400 times in the *H. influenzae* genome. ◀ *Mechanisms used for global regulation (section 14.5)*

Natural transformation is a multistep process (**figure 16.26**). Double-stranded (ds) DNA first binds a transformation pilus,

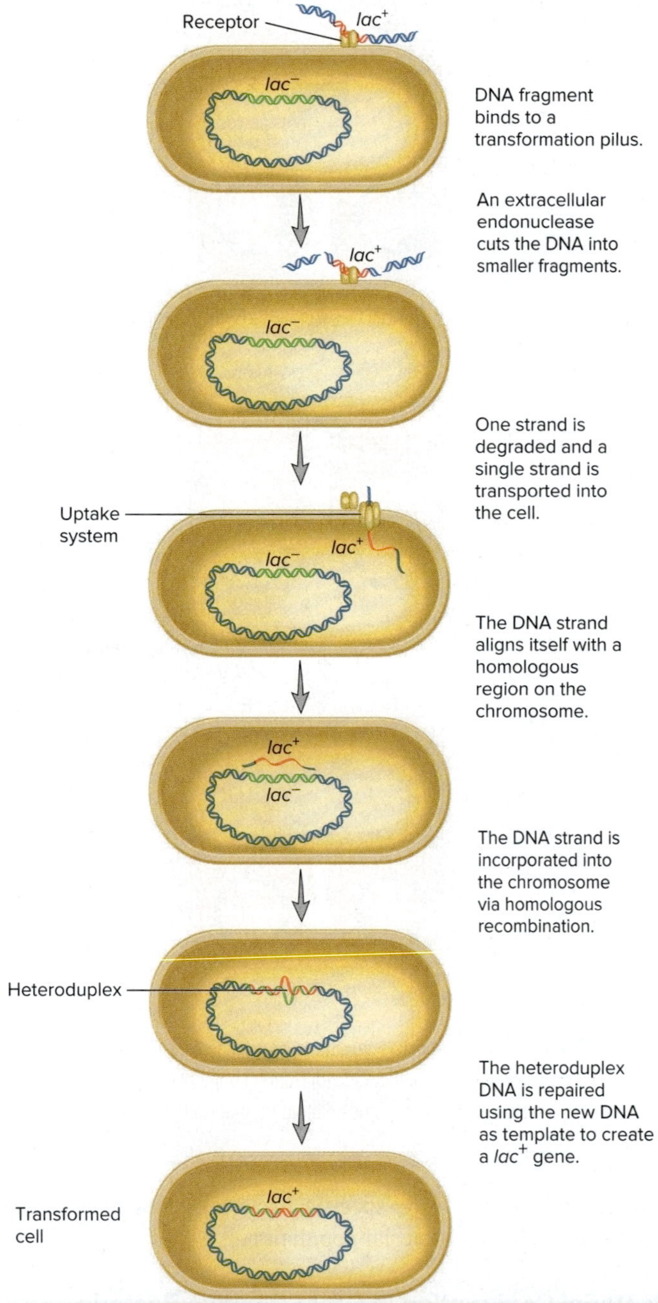

Figure 16.26 Bacterial Transformation in *S. pneumoniae*.

MICRO INQUIRY *According to this model, what would happen if DNA that lacked homology to the* S. pneumoniae *chromosome were taken into the cell?*

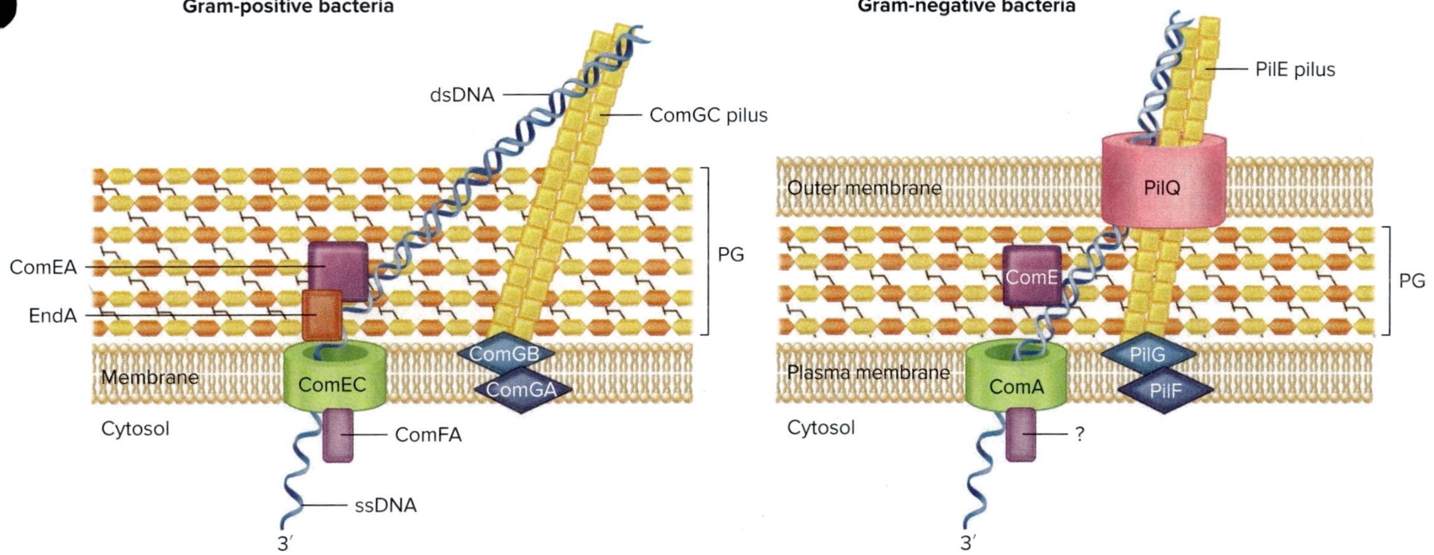

Figure 16.27 DNA Uptake Systems of Gram-Positive and Gram-Negative Bacteria. Competence (Com) proteins are related to type II secretion systems. Pilin (Pil) proteins are related to type IV pili. ComFA, ATPase; EndA, endonuclease; PG, peptidoglycan.

which directs the DNA to a membrane complex that includes a DNA receptor, an endonuclease, and a transmembrane pore. The nuclease degrades one strand and the remaining intact strand is translocated to the cytoplasm using energy from ATP hydrolysis. The incoming single-stranded (ss) fragment interacts with proteins including RecA that integrate the DNA into the recipient's chromosome by homologous recombination.

Despite the difference in cell envelope architecture, Gram-positive and Gram-negative bacteria use similar proteins as the main components of their DNA uptake machinery. These proteins are related to those used in type II secretion systems and type IV pili *(see figure 13.43)*. (Note that type IV pili are unrelated to type IV secretion systems.) The major differences arise from the need for Gram-negative bacteria to move dsDNA across the outer membrane prior to its translocation as ssDNA across the plasma membrane (**figure 16.27**). Despite knowing the basic machinery, scientists still have not determined the precise mechanisms by which DNA enters the cell or how each component interacts with the DNA. *Bacterial Transformation*

Microbial geneticists exploit transformation to move DNA (usually recombinant DNA) into cells. Because many species, including *E. coli,* are not naturally transformation competent, these bacteria must be made artificially competent by certain treatments. Two common techniques are electrical shock and exposure to calcium chloride. Both approaches render the cell membrane temporarily more permeable to DNA, and both are used with *E. coli*. To increase the transformation frequency with *E. coli,* strains unable to produce one or more nucleases are used. These strains are especially important when transforming the cells with linear DNA, which is vulnerable to attack by nucleases. It is easier to transform bacteria with plasmid DNA because plasmids replicate within the host and are not as easily degraded (figure 16.25*b*). ▶| *Introducing recombinant DNA into host cells (section 17.1)*

Comprehension Check

1. What is the relationship between transformation and competence?
2. Describe how transformation occurs in *S. pneumoniae*. How does the process differ in *H. influenzae* and *B. subtilis*?
3. Discuss two ways in which artificial transformation can be used to place functional genes within bacterial cells.

16.8 Transduction Is Virus-Mediated DNA Transfer

After reading this section, you should be able to:

a. Differentiate generalized transduction from specialized transduction
b. Correlate a phage's life cycle to its capacity to mediate generalized or specialized transduction
c. Draw a figure, create a concept map, or construct a table that distinguishes conjugation, transformation, and transduction

Transduction is mediated by viruses. It is a frequent mode of HGT in nature. Indeed evidence suggests that the number of genes moved by marine viruses from one host cell to another is huge (perhaps 10^{24} per year). Furthermore, viruses in marine environments and hot springs move genes between organisms in all three domains of life. ▶| *Marine viruses: mortality at sea (section 30.2)*

Recall from chapter 6 that virus particles are structurally simple, often composed of just a nucleic acid genome protected by a protein coat called the capsid. Viruses are unable to multiply autonomously. Instead, they infect and take control of a host cell, forcing it to make many copies of the virus. Viruses that infect bacteria are called bacteriophages, or phages for short. **Virulent bacteriophages** multiply in their bacterial host immediately after entry. After the progeny phage particles reach a certain number, they cause the host to lyse, so they can be released and infect new host cells (*see figure 6.16*). Thus this process is called the **lytic cycle. Temperate bacteriophages,** on the other hand, do not immediately kill their host. Instead, the phage establishes a relationship with their host called **lysogeny,** and bacteria that have been lysogenized are called **lysogens.** Many temperate phages establish lysogeny by inserting their genomes into the bacterial chromosome. The inserted viral genome is called a **prophage.** The host bacterium is unharmed by this, and the phage genome is replicated along with the host cell's genome. Temperate phages may remain inactive in their hosts for many generations. However, they can be induced to switch to a lytic cycle under certain conditions, including UV irradiation. When this occurs, the prophage is excised from the bacterial genome and the lytic cycle proceeds. ◀ *There are several types of viral infections (section 6.4)*

Transduction is the transfer of bacterial or archaeal genes by virus particles. It is important to understand that host genes are packaged in the virus particle because of errors made during the virus's life cycle. The virion containing these genes then transfers them to a new cell. Two kinds of transduction have been described: generalized and specialized.

Generalized Transduction Transfers Any Donor Gene

Generalized transduction most often occurs during the lytic cycle of virulent phages but sometimes happens during the lytic cycle of temperate phages. Any part of the bacterial genome can be transferred after being partially degraded as the virus takes control of its host (**figure 16.28**). During the assembly stage, viral genomes are packaged by the "headful mechanism"; that is, only genomes of a certain size (i.e., number of nucleotides in length) are packaged. During generalized transduction, a fragment of the host genome that happens to be about the same size as the phage genome is mistakenly packaged. Such a phage is called a generalized transducing particle, because once it is released, it may encounter a susceptible host cell and eject the bacterial DNA it carries into that cell. However, because it lacks viral genes this does not initiate a lytic cycle. As in transformation, once the DNA fragment has been released into the recipient cell, it must be incorporated into the recipient cell's chromosome to preserve the transferred genes. The DNA remains double stranded during transfer, and both strands are integrated into the recipient's chromosome. About 70 to 90% of the transferred DNA is not integrated but often is able to remain intact temporarily and be expressed. Abortive transductants are bacteria that contain this nonintegrated, transduced DNA and are partial diploids.

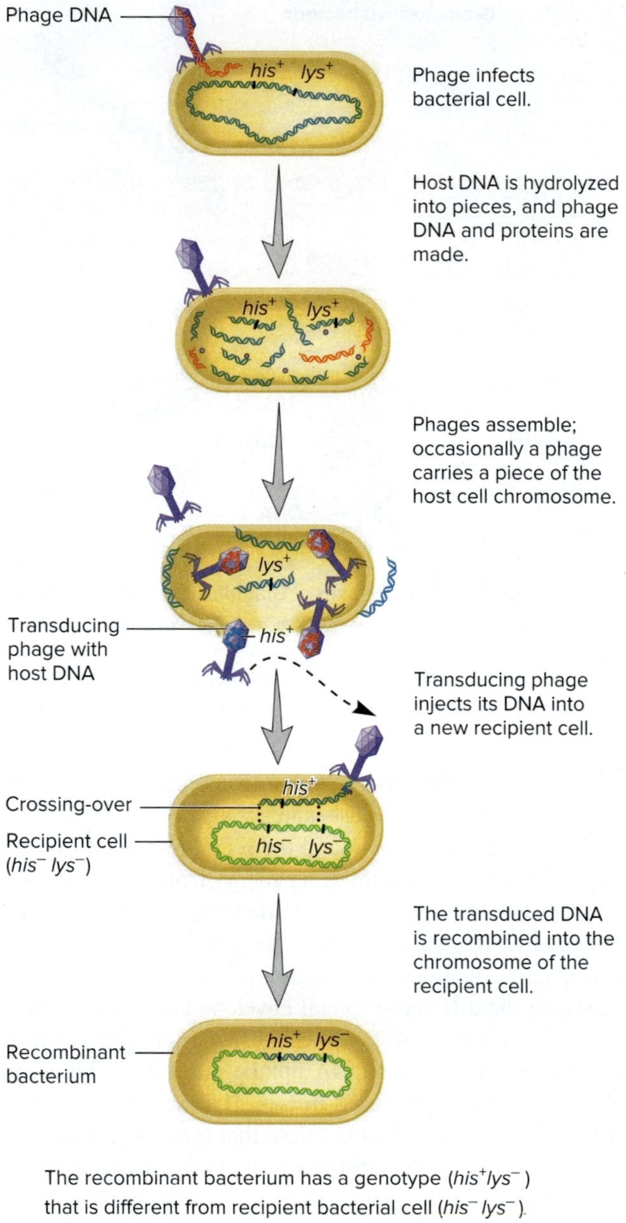

Figure 16.28 Generalized Transduction in Bacteria.

Specialized Transduction Transfers Specific Donor Genes

In **specialized transduction,** only specific portions of the bacterial genome are carried by transducing particles. Specialized transduction is made possible by an error in the lysogenic life cycle of temperate phages that insert their genomes into a specific site in the host chromosome. When a prophage is induced to leave the host chromosome, excision is sometimes carried out improperly. The resulting phage genome contains portions of the bacterial chromosome (about 5 to 10% of the bacterial DNA) next to the integration site, much like the situation with F′ plasmids (**figure 16.29**). However, the transducing particle is defective because it lacks some viral genes and cannot reproduce without assistance. In spite of this, it will inject the remaining viral genome and any bacterial genes it carries into another bacterium. The bacterial genes may become stably incorporated under the proper circumstances.

The best-studied example of specialized transduction is carried out by the *E. coli* phage lambda. The lambda genome inserts into the host chromosome at specific locations known as attachment or *att* sites (**figure 16.30**). The phage *att* sites and bacterial *att* sites are similar and can complex with each other. The *att* site for lambda is between the *gal* and *bio* genes on the *E. coli* chromosome; consequently when lambda excises incorrectly to generate a specialized transducing particle, these bacterial genes are most often present. The product of cell lysis (lysate) resulting from the induction of a population of lysogenized *E. coli* cells contains normal phage and a few defective transducing particles. These particles are called lambda *dgal* if they carry the galactose utilization genes or lambda *dbio* if they carry the *bio* from the other side of the *att* site (figure 16.30). ▶| *Bacteriophage lambda: a temperate bacteriophage (section 26.2)* *Specialized Transduction*

Comprehension Check

1. Describe generalized transduction and how it occurs. What is an abortive transductant?
2. What is specialized transduction and how does it come about?
3. How might one tell whether horizontal gene transfer was mediated by generalized or specialized transduction?
4. Why doesn't a cell lyse after successful transduction with a temperate phage?
5. Describe how conjugation, transformation, and transduction are similar. How are they different?

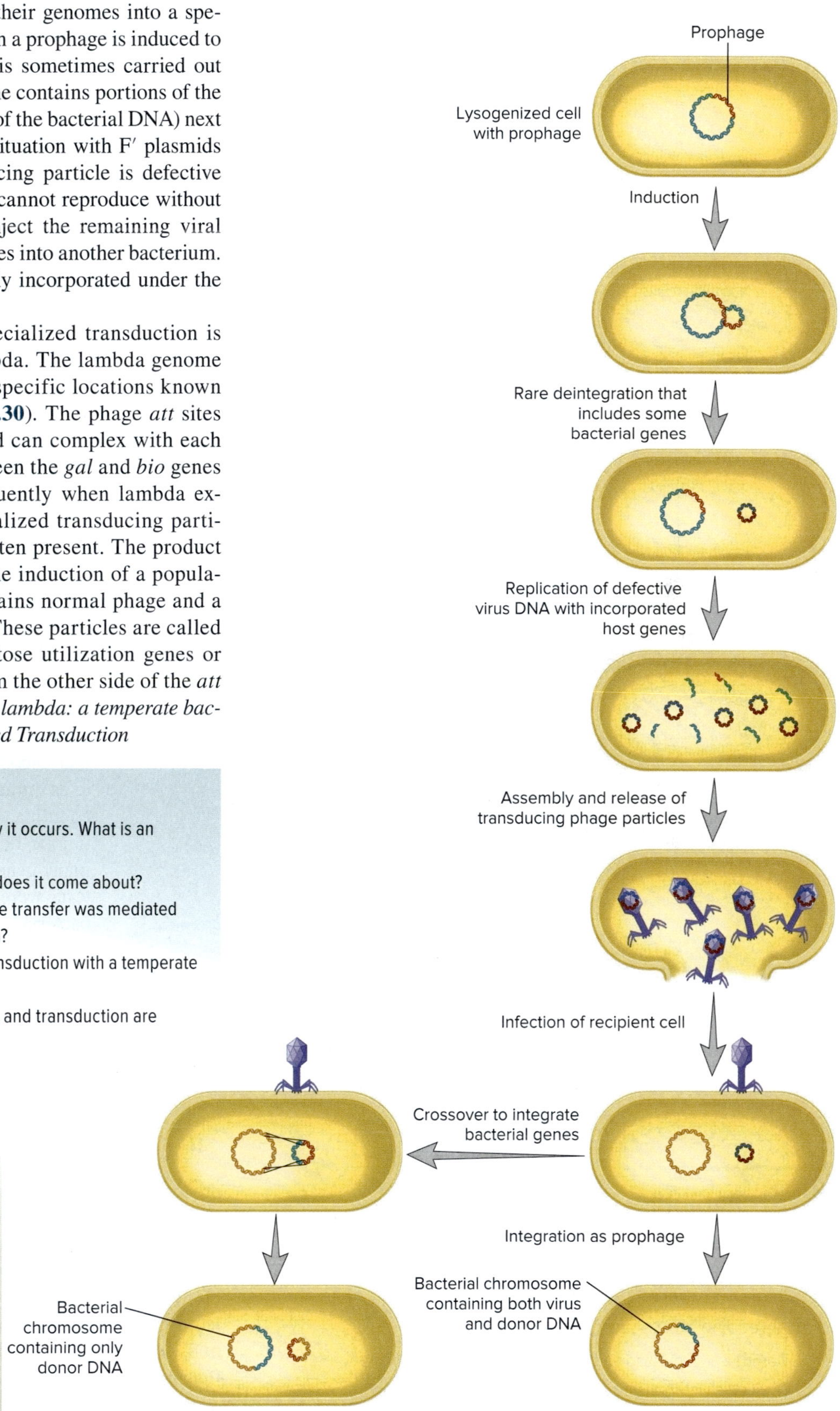

Figure 16.29 Specialized Transduction by a Temperate Bacteriophage. Recombination can produce two types of transductants.

MICRO INQUIRY *Compare the number of transducing particles that arise during generalized (figure 16.29) and specialized transduction. Why is there such a big difference?*

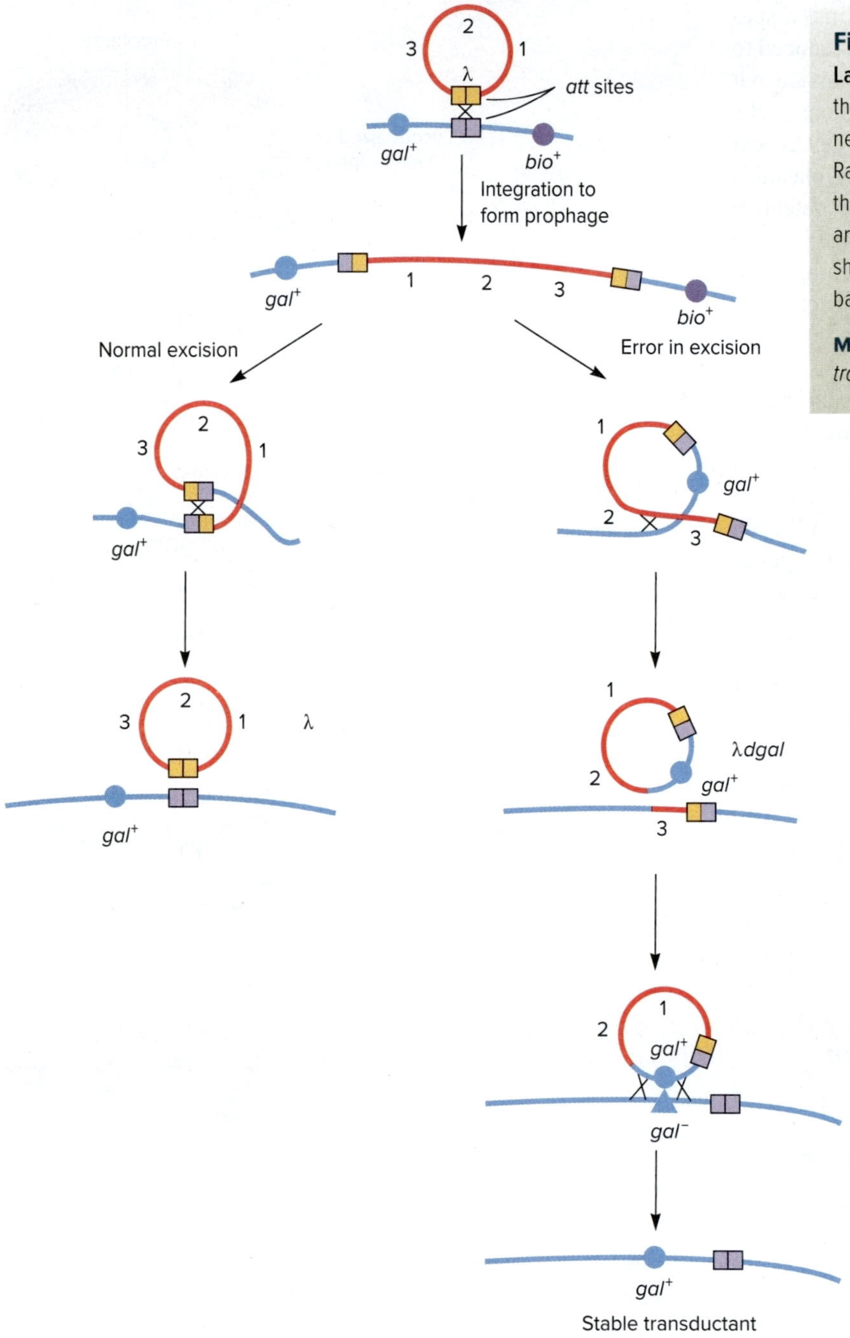

Figure 16.30 The Mechanism of Transduction for Phage Lambda and *E. coli*. Integrated lambda phage lies between the *gal* and *bio* genes. When it excises normally (top left), the new phage is complete and contains no bacterial genes. Rarely, excision occurs asymmetrically (top right), and either the *gal* or *bio* genes are picked up and some phage genes are lost (only aberrant excision involving the *gal* genes is shown). The result is a defective lambda phage that carries bacterial genes and can transfer them to a new recipient.

MICRO INQUIRY *Why can't the* gal *and* bio *genes be transduced by the same transducing particle?*

16.9 Evolution in Action: The Development of Antibiotic Resistance in Bacteria

After reading this section, you should be able to:

a. Describe an R plasmid and its associated genetic elements
b. Distinguish integrative conjugative elements, transposons, and conjugative plasmids
c. Describe how genetic elements mobilize portions of chromosomes

Within 3 years after the widespread use of penicillin began in the 1940s, penicillin-resistant bacteria were found in clinical specimens. To understand the origin of drug resistance, it is important to recall that in nature, antibiotic-producing microbes must also protect themselves from the antibiotics they secrete. In other words, microbes capable of synthesizing antibiotics must also have mechanisms to resist their effects—otherwise they would "commit suicide" by producing the antibiotic. In antibiotic-producing microorganisms, the genes that encode resistance proteins are often referred to as immunity genes. Immunity genes are usually coordinately regulated with genes that code for antibiotic biosynthetic enzymes. It is believed that many genes encoding antibiotic resistance in bacteria were "captured" from antibiotic-producing bacteria and moved by HGT to nonproducers, giving rise to a large pool of resistance-encoding genes outside the producing microorganisms. It is therefore not surprising that in nonproducing bacteria, genes for drug resistance may be present on bacterial chromosomes, plasmids, transposons, and other mobile genetic elements. Because they are often found on mobile genetic elements, they can easily exchange between bacteria. In addition to acquiring resistance genes by HGT, some nonproducers can become resistant due to spontaneous chromosomal mutations. Usually such mutations result in a change in the drug target; therefore the antibiotic cannot bind and inhibit growth. ◀ *There are several mechanisms of drug resistance (section 9.8)*

Frequently a bacterial pathogen is drug resistant because it has a plasmid bearing one or more resistance genes; such plasmids are called **R plasmids** (resistance plasmids; **figure 16.31**). Plasmid-borne resistance genes often code for enzymes that destroy or modify drugs. Plasmid-associated genes have been implicated in resistance to aminoglycosides, chloramphenicol, penicillins, cephalosporins, erythromycin, tetracyclines,

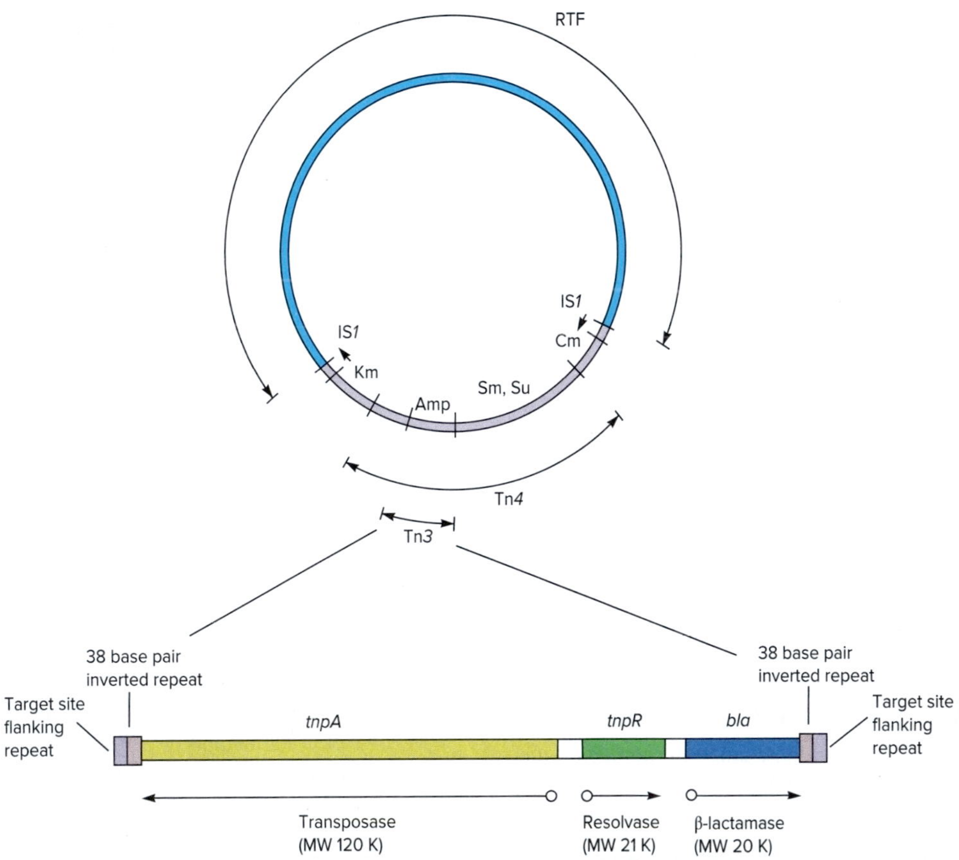

Figure 16.31 An R Plasmid. Plasmid R1 is an R plasmid that contains the replicative transposon Tn*3*. Tn*3* contains the gene for β-lactamase (*bla*), an enzyme that confers resistance to ampicillin (Amp). The arrows below the Tn*3* genes indicate the direction of transcription. Note that Tn*3* is inserted into another transposable element, Tn*4*. Tn*4* carries genes that provide resistance to streptomycin (Sm) and sulfonamide (Su). The R1 plasmid also carries resistance genes for kanamycin (Km) and chloramphenicol (Cm). The RTF region of R1 codes for proteins needed for plasmid replication and transfer.

MICRO INQUIRY *As a replicative transposon, what would happen if Tn3 "hopped" from this R1 plasmid into a different plasmid?*

sulfonamides, and others *(see table 9.1)*. Once a bacterial cell possesses an R plasmid, the plasmid (or its genes) may be transferred to other cells quite rapidly through HGT. Because a single plasmid may carry genes for resistance to several drugs, a pathogen population can become resistant to several antibiotics simultaneously, even though a patient may be treated with only one drug.

Antibiotic-resistance genes can be located on genetic elements other than plasmids. Many transposons contain genes for antibiotic resistance and can move rapidly between plasmids and through a bacterial population. As seen in figure 16.31, some resistance genes reside on a transposon. Tn*5*, a transposon active in many Gram-negative bacteria, is of particular note because it is a replicative transposon. Thus it not only moves but also leaves a copy of itself in its original location.

Antibiotic-resistance genes also reside in and move with genetic elements termed integrative conjugative elements (ICEs) and mobilizable genomic islands (MGIs). These elements share components and mechanisms with transposons and conjugative plasmids. ICEs, also referred to as conjugative transposons, carry genes for a type IV secretion system, a conjugation system, and a recombinase for chromosome integration and excision. These genes confer the ability to transfer the element to other organisms through conjugation, as described in section 16.6. A well-studied example of an ICE is Tn*916* from *Enterococcus faecalis*. Although Tn*916* cannot replicate autonomously, it can transfer itself from *E. faecalis* to a variety of recipients and integrate into their chromosomes.

MGIs are related to ICEs but they mobilize significantly larger DNA segments because of their recombination mechanism. They include an *att* site for recombination with a compatible site in the bacterial chromosome (*attC*). *attC* sites are often in tRNA genes. Recall that a genome must encode a large number of tRNAs, and that they are often redundant because of the genetic code degeneracy. Integration at tRNA genes may be less mutagenic than random integration. Imprecise integration and excision by MGIs results in the horizontal transfer of large chromosomal regions termed islands. One type, pathogenicity islands, are so named because the acquired genes enable the microbe to interact with its environment in a new way, conferring disease. Mutations in ICEs or MGIs that result in the loss of transfer functions can cement an island and its contents in place. ▶| *Pathogenicity islands encode virulence factors (section 35.4)*

Our focus here has been on the origin and transmission of resistance genes from antibiotic producers to nonproducers. What is important to recognize is that expression of these genes by nonproducers can be costly, expending considerable energy and metabolites. Resistance genes can be lost if the bacteria are not consistently exposed to the antibiotic, and in some cases, to biocides and heavy metals. Thus exposure to the antibiotic is a selective force that helps maintain the resistance gene in a population of bacteria. Of concern is the relatively low concentrations of antibiotics and heavy metals needed to maintain resistance genes. In most cases, sublethal levels of 1% of the minimum inhibitory concentration (MIC) of a single antibiotic is sufficient to maintain the resistant phenotype. Unfortunately, the levels of antibiotic and heavy metal contamination in the environment far exceed this amount. As we discuss in chapter 9, measures must be enforced to decrease the use (especially misuse) of antibiotics to help alleviate the antibiotic-resistance problem, including a multinational effort to reduce antibiotic use in agriculture.

Key Concepts

16.1 Mutations: Heritable Changes in a Genome

- A mutation is a stable, heritable change in the nucleotide sequence of the genetic material.
- Spontaneous mutations can arise from replication errors (transition, transversion, and insertion and deletion of nucleotides), and from DNA lesions (apurinic sites, apyrimidinic sites, oxidation of DNA) (**figures 16.1–16.3**).
- Induced mutations are caused by mutagens. Mutations may result from the incorporation of base analogues, specific mispairing due to alterations of a base caused by DNA-modifying agents, the presence of intercalating agents, and severe damage to the DNA caused by exposure to radiation (**table 16.1; figures 16.4** and **16.5**).
- Mutations are usually recognized when they cause a change from the more prevalent wild-type phenotype. A mutant phenotype can be restored to wild type by reversion mutations or suppressor mutations (**table 16.2**).
- There are four important types of point mutations: silent mutations, missense mutations, nonsense mutations, and frameshift mutations (**table 16.2**).
- Mutations can affect phenotype in numerous ways. Some mutations affect colony or cell morphology, viability (i.e., lethal mutations), biochemical pathways, and susceptibility to environmental stresses (e.g., resistance mutations). Some mutations are conditional, being expressed under some conditions but not others (e.g., high temperature but not low temperatures).

16.2 Detection and Isolation of Mutants

- A sensitive and specific screening method is needed for detecting and isolating mutants. An example is replica plating for the detection of auxotrophs (**figure 16.6**).
- One of the most effective techniques for isolating mutants is to select for mutants by adjusting environmental conditions so that the mutant will grow while the wild type does not.
- Because many carcinogens are also mutagenic, the Ames test can be used to measure mutagenicity and is therefore an indirect indication of carcinogenicity (**figure 16.7**).

16.3 DNA Repair Maintains Genome Stability

- Proofreading during DNA replication helps correct errors in replication. However, cells have multiple mechanisms for correcting mispaired and damaged DNA when proofreading is not effective.
- Mismatch repair replaces incorrect base pairs (**figure 16.8**).
- Excision repair systems remove damaged portions from a single strand of DNA (e.g., thymine dimers) and use the other strand as a template for filling in the gap (**figures 16.9** and **16.10**).
- Direct repair systems correct damaged DNA without removing damaged regions. For instance, during photoreactivation, thymine dimers are repaired by splitting the two thymines apart. This is catalyzed in the presence of light by the enzyme photolyase.
- Recombinational repair removes damaged DNA by recombination of the damaged DNA with a normal DNA strand elsewhere in the cell (**figure 16.11**).
- When DNA damage is severe, DNA replication is halted, triggering the SOS response. During the SOS response, genes of DNA repair systems are transcribed at a higher rate. In addition, special DNA polymerases are produced, which replicate damaged DNA. However, they do so without an intact template and therefore create mutations.

16.4 Microbes Use Mechanisms Other than Mutation to Create Genetic Variability

- In sexual organisms, the mixing of parental DNA molecules in the progeny, as well as crossing-over that occurs during gamete formation, generates genetic diversity.
- Horizontal gene transfer (HGT) is an important mechanism for creating genetic diversity, especially in bacteria and archaea. It is a one-way process in which DNA is transferred from a donor to a recipient. In many transfers, the donor DNA must be integrated into the recipient's chromosome to be stably maintained (**figure 16.12**).
- The formation of recombinant microbes by HGT is accomplished by homologous recombination, in which DNA molecules pair due to large areas of sequence similarity. Some viral genomes and all known transposable elements integrate into other DNA molecules by site-specific recombination. This type of recombination does not require extensive sequence homology.

16.5 Transposable Elements Move Genes Within and Between DNA Molecules

- Mobile genetic elements (transposable elements) are DNA segments that move about the genome in a process known as transposition.
- There are numerous types of transposable elements, including insertion sequences and several types of transposons (**figure 16.13**).
- Simple (cut-and-paste) transposition and replicative transposition are two distinct mechanisms of transposition (**figures 16.14** and **16.15**).

16.6 Bacterial Conjugation Requires Cell-Cell Contact

- Conjugation is the transfer of genes that depends upon cell-to-cell contact and is mediated by a plasmid. The F factor is one type of conjugative plasmid (**figure 16.16**). F factor conjugation in *E. coli* is accomplished by a sex pilus and a type IV secretion system (**figure 16.20**).
- In $F^+ \times F^-$ mating, the F factor remains independent of the chromosome and a copy is transferred to the F^- recipient; donor chromosomal genes are not usually transferred (**figure 16.19**).
- Hfr strains transfer bacterial genes to recipients because the F factor is integrated into the host chromosome (**figure 16.22**). A complete copy of the F factor is transferred very rarely (**figure 16.23**).
- When the F factor leaves an Hfr chromosome, it occasionally picks up some bacterial genes to become an F′ plasmid, which readily transfers these genes to other bacteria (**figure 16.24**).

16.7 Bacterial Transformation Is the Uptake of Free DNA from the Environment

- Transformation is the uptake of naked DNA by a competent cell and its incorporation into the genome (**figures 16.25** and **16.26**).
- Only certain bacterial species are naturally competent for transformation. Naturally competent bacteria have uptake machinery composed of proteins similar to those of type II secretion systems and type IV pili (**figure 16.27**). Other species can be made artificially competent.

16.8 Transduction Is Virus-Mediated DNA Transfer

- Bacterial viruses (bacteriophages) can multiply and destroy the host cell (lytic cycle) or become a latent prophage that remains within the host (lysogenic cycle).
- Transduction is the transfer of genes by viruses.
- In generalized transduction, any host DNA fragment can be packaged in a virus capsid and transferred to a recipient (**figure 16.28**).
- Certain temperate phages carry out specialized transduction by incorporating bacterial genes during prophage induction and then transferring those genes to another bacterium (**figures 16.29** and **16.30**).

16.9 Evolution in Action: The Development of Antibiotic Resistance in Bacteria

- Genes conferring resistance can be found on chromosomes, R plasmids (**figure 16.31**), integrative and conjugative elements (ICEs), and mobile genetic islands (MGIs). Thus resistance genes are readily moved from one cell to another by horizontal gene transfer.

Active Learning

1. Mutations are often considered harmful. Give an example of a mutation that would be beneficial to a microorganism. What gene would bear the mutation? How would the mutation alter the gene's role in the cell, and what conditions would select for this mutant allele?
2. Mistakes made during transcription affect the cell but are not considered mutations. Why not?
3. Suppose that transduction took place when a U-tube experiment was conducted. How would you confirm that a virus was passed through the filter and transduced the recipient?
4. Suppose that you carried out a U-tube experiment with two auxotrophs and discovered that recombination was not blocked by the filter but was stopped by treatment with deoxyribonuclease. What gene transfer process is

responsible? Why would it be best to use double or triple auxotrophs in this experiment?

5. What would be the evolutionary advantage of having a period of natural "competence" in a bacterial life cycle? What would be possible disadvantages?
6. Hydroxylamine is a mutagen that attaches a hydroxyl group to cytosine causing it to base-pair like thymine. Sketch a diagram like figure 16.1*b* to illustrate how point mutations occur following hydroxylamine treatment.
7. Pseudogenes are found in most genomes. These are genes with a mutation that renders them completely nonfunctional. Bacteria that are obligate intracellular symbionts contain many pseudogenes, often up to one-third of the genome. It is believed that some genes become dispensable in this niche, especially for functions that can be obtained from the host cell, like synthesis of biochemical precursors. Genomes of intracellular symbionts usually contain pseudogenes of DNA repair enzymes. Propose a mechanism to explain the evolutionary process of becoming an obligate intracellular symbiont.
8. Certain transposons have characteristic end sequences, as well as encoding a set of genes required for transposition. A search of a whole genome sequence will often turn up recognizable transposable elements. CRISPR/Cas systems are also recognizable by the CRISPR repeat array and adjacent Cas operon. (For a review of CRISPR/Cas systems, see section 14.6.) A survey of bacterial chromosomes revealed that many organisms have a DNA region that is a variant CRISPR/Cas system called Cas7f and a Tn7-like transposon. Cas7f contains a short CRISPR array and a subset of Cas genes. It lacks the gene encoding the interference nuclease, meaning that it directs an effector complex to a site on the chromosome specified by the array, but cannot cut the DNA at that site. Cas7f is located within a Tn7-like transposon. The transposon is also a variant, in that it has its characteristic repeats at each end of the element, and the genes encoding transposase, but it lacks the gene encoding the protein that directs transposition to plasmids and bacteriophages. Researchers realized that they were observing a DNA sequence recognition module with no enzymatic activity (Cas7f) and a transposon that had lost its ability to identify a target DNA sequence. They proposed that this new genetic element represents an RNA-guided transposon. Diagram how transposition might occur for this variant transposon.

   Read the original paper: Peters, J. E., et al. 2017. Recruitment of CRISPR-Cas systems by Tn7-like transposons. *Proc. Natl. Acad. Sci. USA* 114:e7358-e7366.

17

Microbial DNA Technologies

©David Planchet

Spinning Stronger Silk

Spider silk is an astounding fiber, lightweight and elastic, yet strong enough to suspend a dangling spider or capture prey for its next meal. Materials scientists have long sought to replicate spider silk in the laboratory to exploit its properties. Among the potential applications are lightweight bulletproof vests and medical devices like artificial ligaments, bandages, and sutures.

Spider silk fibers are composed of many copies of a single protein, spidroin. In the silk gland, spidroin is in a highly concentrated solution, rather than a fiber. As the protein solution moves through the spider's silk glands and spinning ducts, it is acidified and dehydrated. Individual protein molecules change conformation, aggregate, and elongate to become the silk fiber. Industrial production of fibers requires large quantities of spidroin, more than is feasible to extract from even an army of spiders.

By inserting the gene that encodes spidroin into bacteria, yeast, or cultured animal cells, large amounts of recombinant spidroin protein are produced. Even goats have been engineered to make spidroin in their milk! But the challenge in making spider silk is to spin the spidroin solution to a fiber.

It turns out that spidroin proteins from different types of spiders aren't exactly the same. They vary somewhat in their amino acid sequence, and as a consequence, their properties vary. Although several types can be prepared as recombinant proteins, all have some regions that are insoluble or less responsive to the pH changes that occur during fiber spinning. The result is a silk that is less elastic or weaker than the original spider-made version. So scientists designed the ideal spidroin consisting of the N-terminal and middle portions of the protein from *Euprosthenops australis,* the nursery web spider, and the C-terminal portion from *Araneus ventricosus,* an orb-weaving spider.

But how is a new hybrid, or chimeric, gene made? In this case, having the spidroin genes from both spider species already cloned made the task simpler. The polymerase chain reaction (PCR) makes many copies of a gene, or in this case, part of a gene. These parts can be assembled to build the complete gene, including a promoter "on" switch and an intact open reading frame that encodes the protein exactly as it has been designed. After introducing this hybrid gene to *E. coli*, flipping the "on" switch, and purifying the chimeric spidroin protein, 1 kilometer of spider silk fiber can be spun from the spidroin purified from just 1 liter of recombinant *E. coli* culture.

In this chapter, we introduce some of the DNA technologies that have made this type of project possible. From cloning a gene in *E. coli* (**figure 17.1**), to assembling a custom gene sequence, to genome editing, the field of biotechnology has been revolutionized by the ease and simplicity of in vitro DNA technologies.

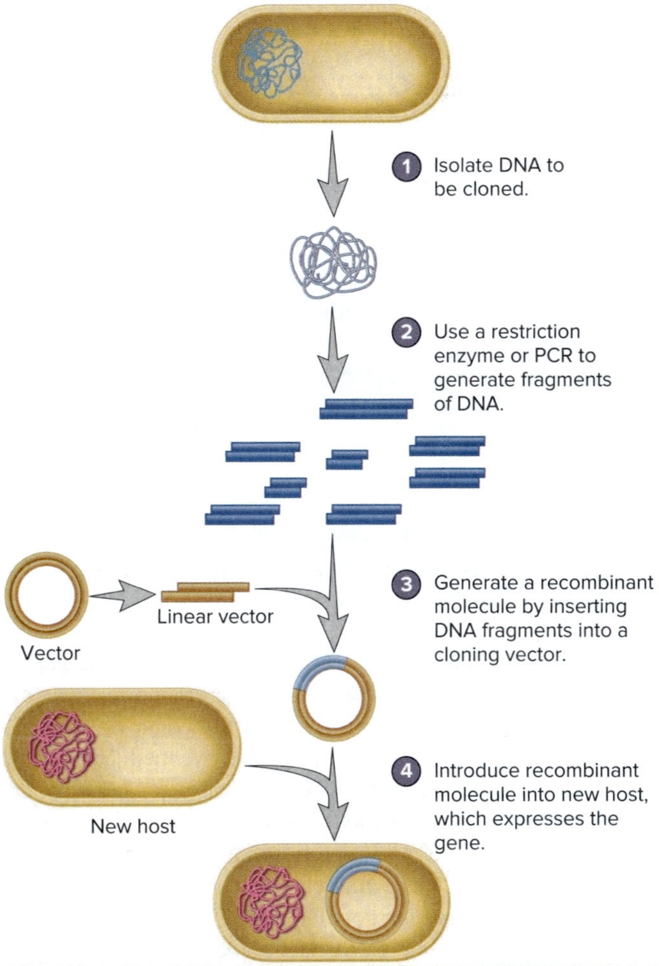

Figure 17.1 Steps in Cloning a Gene. Each step shown in this overview is discussed in more detail in this chapter.

MICRO INQUIRY *Which of the DNA molecules shown are recombinant?*

Readiness Check:

Based on what you have learned previously, you should be able to:

- ✓ Identify the key structural elements of DNA, RNA, and protein (section 13.2)
- ✓ Diagram the flow of cellular information from gene to protein as well as protein localization (e.g., subcellular localization or export) (sections 13.5, 13.7, and 13.8)
- ✓ Explain the general structural differences in a eukaryotic gene and a bacterial gene, including regulatory, coding, and noncoding sequences (sections 13.3, 13.5, and 15.3)
- ✓ Describe the capacity of proteins to recognize and bind to DNA (sections 13.2 and 14.2)
- ✓ List three differences between a plasmid and a chromosome (sections 3.6 and 16.6)
- ✓ Describe DNA repair and homologous recombination reactions (sections 16.3 and 16.4)

17.1 Key Discoveries Led to the Development of DNA Cloning Technology

After reading this section, you should be able to:

a. Explain how restriction enzymes recognize and digest DNA to create either blunt or sticky ends
b. Diagram the reaction catalyzed by reverse transcriptase and describe its application to biotechnology
c. List the three features of a cloning vector and why these elements are needed
d. Differentiate between plasmids, phage-based cloning vectors, cosmids, and artificial chromosomes in terms of structure and application
e. Explain how a piece of foreign DNA is recombined in vitro into a cloning vector
f. Identify host cells commonly used for transformation
g. Compare two common techniques by which recombinant DNA constructed in vitro is introduced into host cells

Genetic engineering refers to the modification of the genetic code of a living organism. For centuries, crop improvement and breeding animals for desirable traits constituted the only form of genetic engineering. But in the mid to late twentieth century, the identification of DNA as the chemical basis for heredity and advances in laboratory methods for manipulating DNA opened new avenues for scientists. Properties like rapid growth and easy laboratory handling made microbes like *E. coli* and the yeast *Saccharomyces cerevisiae* popular genetic models.

The application of DNA technologies to the study of microbes takes two forms. Microbes can be altered to allow scientists to ask direct questions about how they function in the environment and how the basic processes of life work. Additionally, microbes can be treated as factories for the production of biological molecules like insulin (a protein) and small molecules, such as vitamins, as discussed in chapter 42. The underlying principle behind these techniques is that the genetic code is universal. Whether it is from *E. coli,* a worm, or a tomato, DNA has the same structure and function. Genes can be moved from one organism to another and still encode the same protein. Similarly, enzymes that act on DNA can perform their catalytic role on DNA from any source.

The earliest methods for DNA manipulation used enzymes and bacterial cells to modify and amplify DNA in a process known as DNA cloning. DNA molecules with segments that originate from different organisms are termed recombinant. More recent methods bypass the need for cells and use purified enzymes to create recombinant DNA molecules in vitro. As researchers envision more complex engineered cells, the components of the cloner's toolbox (e.g., regulatory sequences, structural genes, and enzymes) are being standardized for widespread use.

Restriction Enzymes

One of the first breakthroughs leading to DNA cloning was the discovery by Werner Arber and Hamilton Smith in the late 1960s of bacterial enzymes that cut double-stranded DNA. These enzymes, known as **restriction enzymes** or restriction endonucleases, recognize and cleave specific sequences (**figure 17.2**). Each restriction enzyme has a four to eight base pair sequence as its substrate; and hundreds of different restriction enzymes have been purified and are commercially available. These enzymes can be used to prepare DNA fragments containing specific genes or portions of genes. For example, the restriction enzyme *Eco*RI, isolated by Herbert Boyer in 1969 from *Escherichia coli,* cleaves DNA between G and A in the base sequence 5′-GAATTC-3′ (**figure 17.3**). Because DNA is

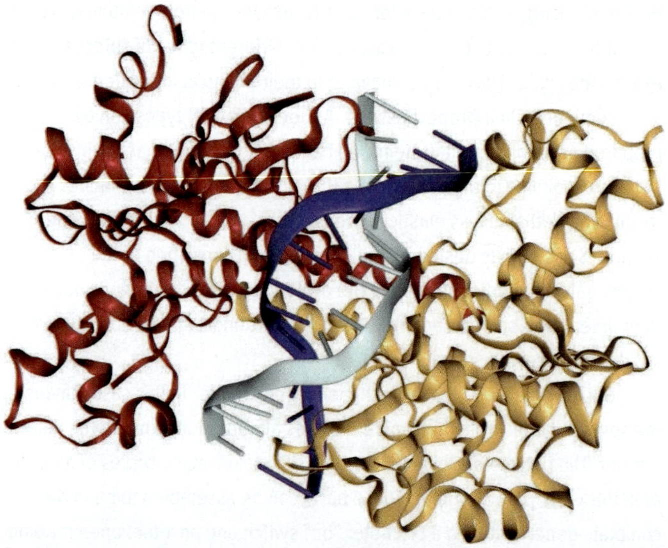

Figure 17.2 Restriction Endonuclease Bound to DNA. In this structure of *Hinc*II bound to DNA, the two strands of DNA are colored light and dark blue. The enzyme's two identical subunits are colored red and yellow. The symmetry of the palindromic DNA substrate is mirrored in the symmetry of the enzyme.

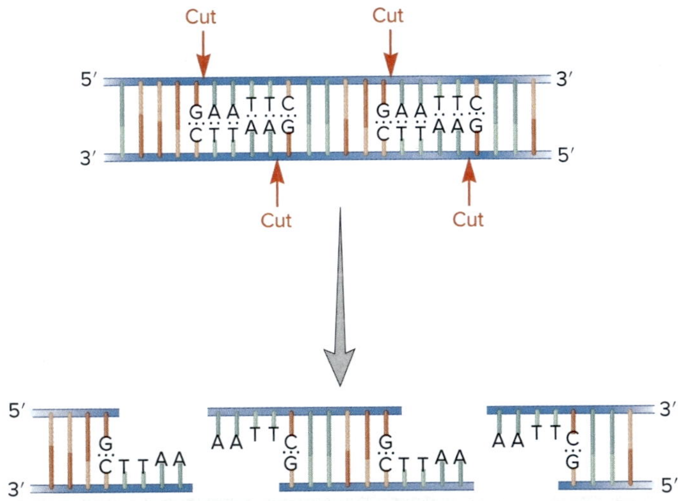

Figure 17.3 Restriction Endonuclease Action. The cleavage catalyzed by the restriction endonuclease *Eco*RI. The enzyme makes staggered cuts on the two DNA strands to form sticky ends.

MICRO INQUIRY *Examine the uncut piece of DNA shown in the upper half of this figure. Where would an exonuclease cleave?*

antiparallel, this sequence is also present on the complementary strand of DNA. When *Eco*RI hydrolyzes the phosphodiester bonds between the G and A residues, unpaired 5′-AATT-3′ remains at the end of each strand; such overhangs are termed **sticky ends.** In contrast, cleavage by restriction enzymes such as *Alu*I and *Hin*cII leaves blunt ends. Because the reaction occurs in the common structural component of DNA (the phosphate-sugar backbone), the information in DNA (the sequence of nitrogenous bases) is preserved. Analysis of restriction enzyme digests is typically done by agarose gel electrophoresis, as discussed in **Techniques & Applications 17.1**. A few restriction enzymes and the DNA sequence that they recognize are listed in **table 17.1**. Note that enzymes are named after the bacterium from which they originate. *Restriction Endonucleases*

Genetic Cloning and cDNA Synthesis

An important advance in cloning DNA came in 1972, when David Jackson, Robert Symons, and Paul Berg reported that they had successfully generated recombinant DNA molecules. They allowed the sticky ends of fragments to anneal—that is, to base pair with one another (**figure 17.4**), and then covalently joined the fragments with the enzyme DNA ligase (*see figure 13.15*). Within a year, cloning vectors into which foreign DNA fragments could be inserted had been developed. Such plasmids replicate within a microbial host, thereby maintaining the cloned fragment of DNA. *Early Genetic Engineering Experiment*

Once it was demonstrated that foreign genes could be inserted into cloning vectors, biologists sought to clone specific genes from various organisms. However, it was evident that cloning eukaryotic DNA into bacterial hosts would be problematic. This is because eukaryotic pre-mRNA must

Table 17.1 Some Restriction Endonucleases and Their Recognition Sequences

Enzyme	Microbial Source	Recognition Sequence[1]	Ends Produced
*Alu*I	*Arthrobacter luteus*	5′AG↓CT 3′ 3′ TC↑GA 5′	5′ AG CT 3′ 3′ TC GA 5′
*Eco*RI	*Escherichia coli*	5′ G↓AATTC 3′ 3′ CTTAA↑G 5′	5′ G AATTC 3′ 3′ CTTAA G 5′
*Hin*cII	*Haemophilus influenzae* c	5′ GTT↓AAC 3′ 3′ CAA↑TTG 5′	5′ GTT AAC 3′ 3′ CAA TTG 5′
*Not*I	*Nocardia otitidis-caviarum*	5′ GC↓GGCCGC 3′ 3′ CGCCGG↑CG 5′	5′ GC GGCCGC 3′ 3′ CGCCGG CG 5′
*Pst*I	*Providencia stuartii*	5′ CTGCA↓G 3′ 3′ G↑ACGTC 5′	5′ CTGCA G 3′ 3′ G ACGTC 5′

1 The arrows indicate the sites of cleavage on each strand.

MICRO INQUIRY: *Which enzymes yield blunt ends? Which yield sticky ends with a 5′ overhang? Which generate a 3′ overhang?*

TECHNIQUES & APPLICATIONS

17.1 Gel Electrophoresis

Agarose and polyacrylamide gel electrophoresis is routinely used to separate DNA fragments by size. When DNA molecules are placed at the negative end of an electrical field, they migrate toward the pole with a positive charge. Each fragment's migration rate is determined by its molecular weight so that the smaller a fragment is, the faster it moves through the gel. Migration rate is also a function of gel density. In practice, this means that high concentrations of gel material (agarose or acrylamide) provide better resolution of small fragments and vice versa. Agarose provides good resolution of DNA molecules larger than 500 base pairs (bp), while polyacrylamide can resolve fragments that differ by 1 base pair. Plasmid DNA that has not been digested with restriction enzymes is usually supercoiled and does not migrate to a position corresponding to its molecular weight. Comparison to molecular weight standards is therefore only valid for linear DNA.

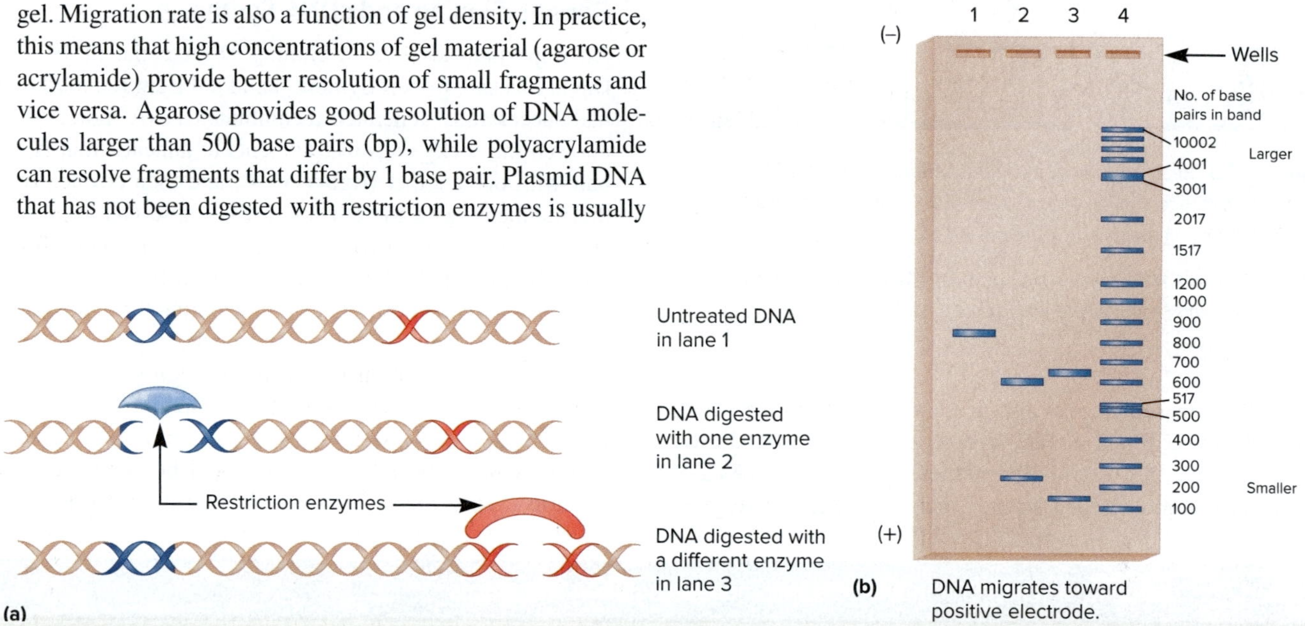

Gel Electrophoresis of DNA. (a) A DNA fragment with recognition sites for two restriction endonucleases. Each enzyme recognizes a specific sequence and cuts the DNA at that position, to create two fragments that can be separated and sized on an agarose gel. (b) DNA is loaded into wells on one end of an agarose gel. When an electrical current is passed through the gel, the negatively charged DNA migrates toward the positive pole. Larger fragments, measured in numbers of base pairs, migrate more slowly than the smaller (shorter) fragments. The same DNA fragment is used in lanes 1, 2, and 3. It is intact in lane 1, and cut with two different enzymes in lanes 2 and 3. The stained gel reveals the separation pattern of the DNA fragments. The size of a given DNA band can be determined by comparing it to a known set of molecular weight markers (lane 4) called a ladder.

MICRO INQUIRY *Approximate the sizes of the DNA fragments in lanes 1, 2, and 3 using the size standards in lane 4. Convince yourself that when the fragment in lane 1 is digested with restriction enzymes, no DNA is lost. The sizes of the fragments in each lane should add up to the size of the uncut fragment.*

be processed (e.g., introns spliced out), and bacteria lack the molecular machinery to perform this task. In 1970 Howard Temin and David Baltimore independently discovered the enzyme that solved this dilemma. They isolated the enzyme **reverse transcriptase** (**RT**) from retroviruses. These viruses have an RNA genome that is copied into DNA prior to replication. The mechanism by which reverse transcriptase accomplishes this is outlined in **figure 17.5**. Processed mRNA can be extracted from a eukaryotic cell and used as a template for **complementary DNA** (**cDNA**) synthesis in vitro. The resulting cDNA can then be cloned, without the need for RNA processing. ▶| *Retroviruses (section 26.7)* cDNA

Cloning Vectors

Cloning depends on the propagation of many copies of the nucleotide sequence of choice. To accomplish this, **cloning vectors**, DNA elements that replicate and can carry additional genetic cargo, had to be developed. The major types of vectors are plasmids, bacteriophages and other viruses, cosmids, and

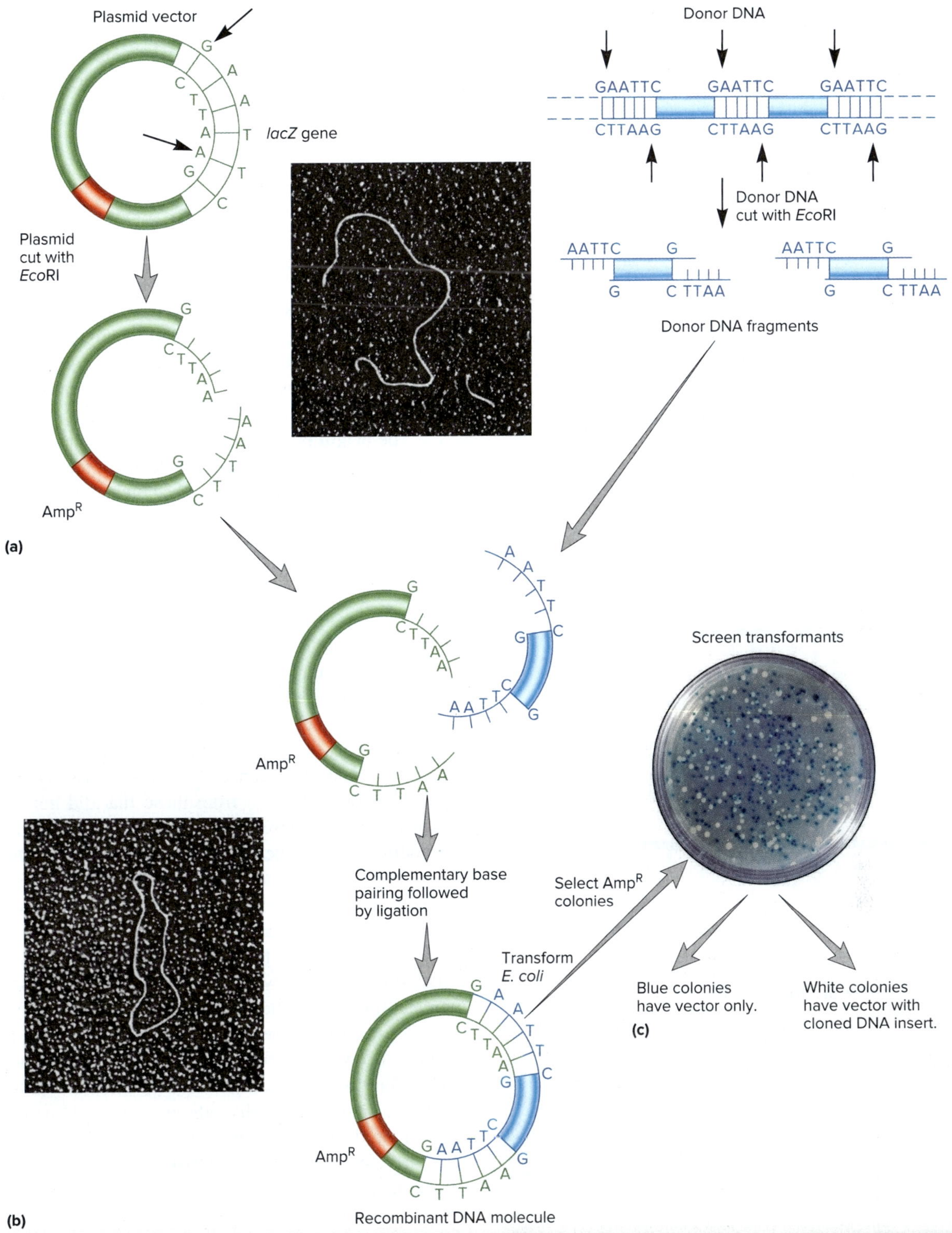

Figure 17.4 DNA Cloning. The construction of a recombinant plasmid. Treatment of both the cloning vector and the donor DNA with the same restriction endonuclease results in complementary sticky ends on both molecules. The scale of the sticky ends of the fragments and plasmid has been enlarged to illustrate base pairing. (a) The electron micrograph shows a plasmid that has been linearized by a restriction enzyme, and a donor DNA fragment. (b) The micrograph shows an intact plasmid. (c) After transformation, *E. coli* cells are plated on medium containing ampicillin and X-Gal. Ampicillin ensures that only ampicillin-resistant transformants grow; X-Gal enables the visualization of colonies that were transformed with recombinant vector (vector + insert, white colonies).

MICRO INQUIRY *What would you conclude if you obtained only blue colonies after cloning into a vector that enables blue/white screening of transformants? Why might such a result occur?*

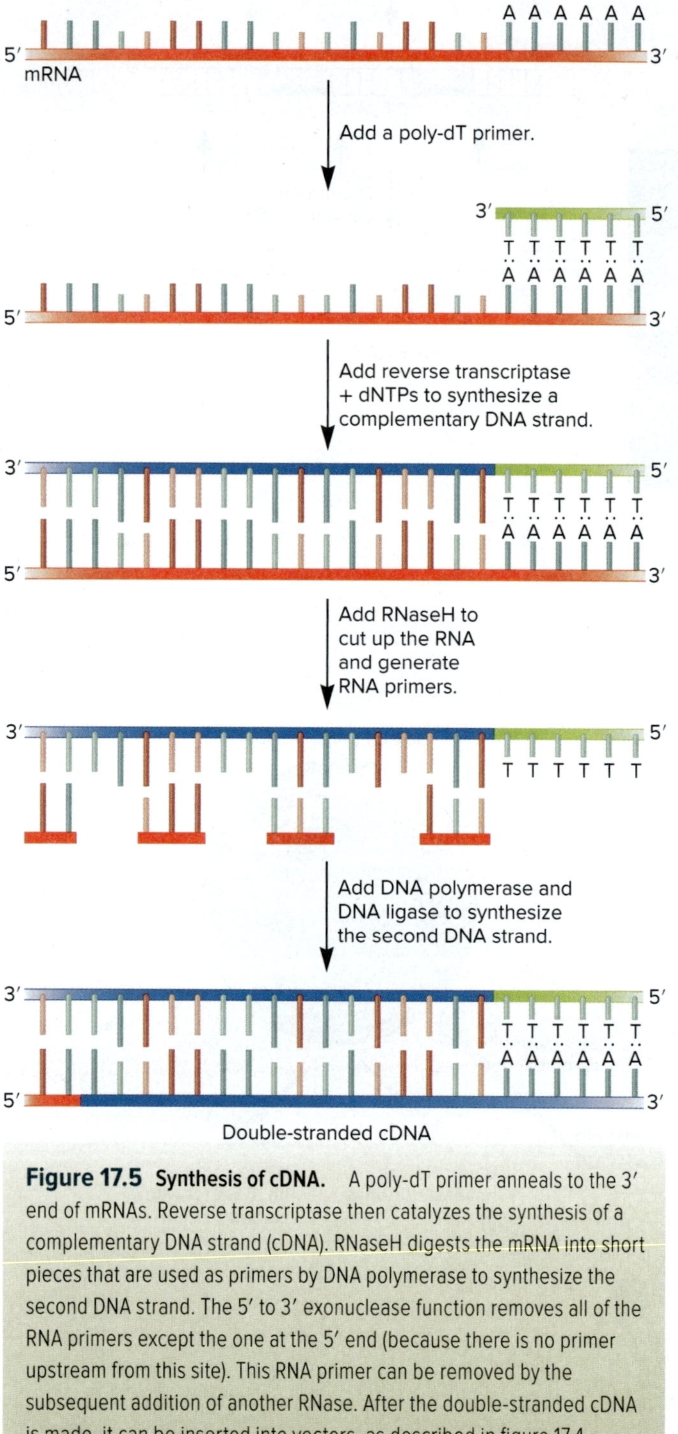

Figure 17.5 Synthesis of cDNA. A poly-dT primer anneals to the 3′ end of mRNAs. Reverse transcriptase then catalyzes the synthesis of a complementary DNA strand (cDNA). RNaseH digests the mRNA into short pieces that are used as primers by DNA polymerase to synthesize the second DNA strand. The 5′ to 3′ exonuclease function removes all of the RNA primers except the one at the 5′ end (because there is no primer upstream from this site). This RNA primer can be removed by the subsequent addition of another RNase. After the double-stranded cDNA is made, it can be inserted into vectors, as described in figure 17.4.

MICRO INQUIRY *Why must introns be removed from eukaryotic DNA before it can be expressed in a bacterium?*

artificial chromosomes. Each type has its own advantages and applications, so the selection of the proper cloning vector is critical to the success of any cloning procedure. Most cloning vectors share three important features: an origin of replication; a region of DNA that bears unique restriction sites, called a multicloning site; and a selectable marker. These elements are described next in the context of plasmids, the most frequently used cloning vectors. *Construction of a Plasmid Vector*

Plasmids

Plasmids make excellent cloning vectors because they replicate autonomously (i.e., independently of the chromosome) and are easy to purify. They can be taken up by microbes by conjugation, transformation, or electroporation. Many different plasmids are used in biotechnology, all derived from naturally occurring plasmids that have been genetically engineered (**figure 17.6**). *Plasmids (section 3.6); Bacterial conjugation requires cell-cell contact (section 16.6); Bacterial transformation is the uptake of free DNA from the environment (section 16.7)*

The **origin of replication** (*ori*) allows the plasmid to replicate in the microbial host independently of the chromosome, and it dictates the **copy number,** or the number of plasmid molecules present in a cell. High copy number facilitates plasmid purification; an *E. coli* high copy number plasmid like pUC19 is present at hundreds of copies per cell. Some plasmids have two origins of replication, each recognized by different host organisms. These plasmids are called **shuttle vectors** because they can be transferred, or "shuttle," from one host to another. YEp24 is a shuttle vector that can replicate in yeast (*Saccharomyces cerevisiae*) and in *E. coli* because it has the 2μ circle yeast replication element and the *E. coli* origin of replication (figure 17.6).

Following the uptake of vector by host cells, it is necessary to discriminate between cells that successfully imported the plasmid (transformants) from those that did not (nontransformants). This is achieved by the presence of a gene on the plasmid encoding a protein needed for the cell to survive under certain conditions. Such a gene is called a **selectable marker.** In the case of pUC19, the selectable marker, Amp^R, encodes the ampicillin-resistance enzyme. Because *E. coli* is normally susceptible to ampicillin, only those cells that have taken up plasmid (i.e., transformants) will grow when plated on agar containing ampicillin. The shuttle vector YEp24 bears both the Amp^R gene to be used when selecting *E. coli* transformants and URA3, which encodes a protein essential for uracil biosynthesis in yeast. Therefore this plasmid must be used in strains of *S. cerevisiae* that are uracil auxotrophs, otherwise it would be impossible to select, or identify, transformants.

To clone a fragment of DNA into a plasmid, both the fragment and plasmid are cut with the same restriction enzyme so that compatible sticky ends are generated. It is essential that these restriction enzymes cut at only one place in the plasmid. Cleavage at a unique restriction site generates a linear plasmid. Alternatively, two different, unique sites may be cleaved and the DNA sequence between the two sites replaced with cloned DNA. Cloning vectors have been designed with many restriction sites clustered in a single region called the **multicloning site (MCS)** (figure 17.6). When the plasmid and the DNA to be cloned are cut with one of these enzymes, the compatible sticky ends hydrogen bond. Incubation in the presence of DNA ligase

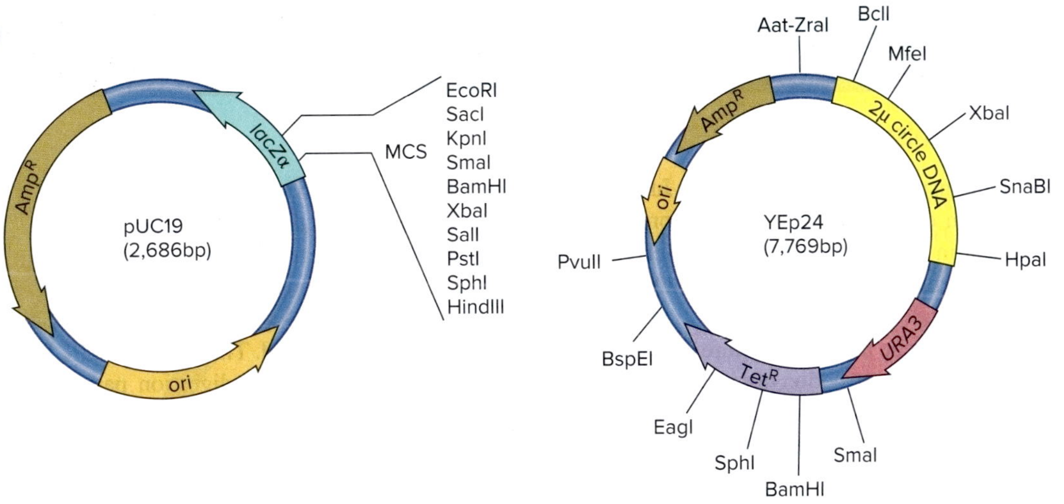

Figure 17.6 The Cloning Vectors pUC19 and YEp24. Restriction sites that are present only once in each vector are shown. pUC19 replicates only in *E. coli*, while YEp24 replicates in both *E. coli* and *S. cerevisiae*. MCS is multiple cloning site.

MICRO INQUIRY *Which plasmid is a shuttle vector? Why?*

results in the synthesis of phosphodiester bonds between the cloned DNA fragment and the vector (figure 17.4).

Unfortunately, the ligation of foreign DNA into a vector is never 100% efficient. When the ligation mixture is introduced into host cells, cells that carry a plasmid without the insert DNA must be distinguished from those that carry a plasmid into which DNA was successfully cloned. To accomplish this, the MCS in plasmids such as pUC19 is engineered within the 5′ end of the *lacZ* gene. *lacZ* encodes β-galactosidase (β-gal), an enzyme that cleaves the disaccharide lactose into galactose and glucose (*see figure 14.2*). An alternative substrate for the enzyme is X-Gal (5-bromo-4-chloro-3-indolyl-β-D-galactopyranoside), a chemical that is colorless until hydrolyzed by β-gal, when it turns blue. When DNA has been inserted into the MCS, the *lacZ* gene is interrupted and a functional β-gal is not produced. Plating the cells on a selective-differential growth medium permits a rapid visual determination of whether the cloning procedure has been successful. Ampicillin in the medium requires that the cells carry the plasmid, and X-Gal in the medium turns colonies blue if they can synthesize β-gal. Colonies carrying plasmids with an insert in the MCS, then, will be white, as the insert prevents β-gal synthesis (figure 17.4*c*). There are a number of other clever ways in which cells with vector can be distinguished from those with vector plus insert; the detection of blue versus white colonies, though, is a common approach.

Phage and Viral Vectors

Phage vectors are phage genomes that have been genetically modified to include useful restriction enzyme recognition sites for the insertion of foreign DNA. Once DNA has been inserted, the recombinant phage genome is packaged into viral capsids and used to infect host cells. The resulting phage lysate consists of thousands of phage particles that carry cloned DNA as well as the genes needed for host lysis. Common vectors are derived from the *E. coli* bacteriophage lambda (λ), which has a double-stranded DNA genome. ▶| *Bacteriophage lambda: a temperate bacteriophage (section 26.2)*

Genetic engineering of animal cells is accomplished by the use of viral vectors. These recombinant viruses are shuttle vectors that replicate in *E. coli* as bacterial plasmids, but they also direct the assembly of mature infective virus particles to shuttle the DNA to an animal cell. Commonly used viruses for mammalian cells are retroviruses, lentiviruses, and adenoviruses.

Cosmids

DNA inserts larger than about 15,000 bp to 25,000 bp cannot be stably maintained in plasmids and phage vectors, respectively. Instead, **cosmids** can be used to clone such large fragments of DNA. These engineered vectors have a *cos* site from phage and a selectable marker, origin of replication, and MCS from plasmids (thus the term "cos-mid"). These hybrid vectors replicate as plasmids within the host cell, but the presence of the *cos* sites means that the vector can be packaged into phage capsids and transferred to new host cells by transduction. |◀ *Transduction is virus-mediated DNA transfer (section 16.8)*

Artificial Chromosomes

Artificial chromosomes are used when particularly large fragments of DNA must be cloned, as when constructing a genomic library (section 17.3). In fact, the genomic library containing the human genome was constructed in **bacterial artificial chromosomes (BACs)**, which then enabled nucleotide sequencing. BACs were also used in the construction of a synthetic microbial genome (**Techniques & Applications 17.2**). Like natural chromosomes, artificial chromosomes replicate only once per cell cycle. **Yeast artificial chromosomes (YACs)** were developed first and consist of a yeast telomere (TEL) at each end, a centromere sequence (CEN), a yeast origin of replication (ARS, *a*utonomously *r*eplicating *s*equence), a selectable marker such as URA3, and an MCS to facilitate the insertion of foreign DNA (**figure 17.7*a***). YACs are used when extraordinarily large DNA pieces (up to 1,000 kb) are to be cloned. BACs were developed, primers are single-stranded DNA molecules, generally between 15 and 30 nucleotides long, that are made with a DNA synthesizer. Because primers are based on flanking nucleotide sequences, only DNA with known (or nearly known) sequence can be amplified by PCR. The reaction mixture, then, includes the template DNA (often copies of an entire genome), two primers, a thermostable DNA polymerase, and dNTPs.

PCR is a series of DNA replication reactions, called cycles. Each cycle precisely executes three steps in a machine called a thermocycler. In the first step, the double-stranded template DNA is denatured to single-stranded DNA by raising the temperature to about 95°C. Next the temperature is lowered to about 50°C so that the primers can hydrogen bond (anneal) to their complementary region on the template strands. Primers are included in the reaction at a huge molar excess because they must compete with the other template strand for binding. In the final step, the temperature is raised, usually to 68° to 72°C, so that DNA polymerase can extend the primers and synthesize copies of the target DNA sequence using dNTPs. Only sturdy DNA polymerases that can tolerate the repeated temperature fluctuations can be used in PCR. The most commonly used thermostable enzyme is ***Taq* polymerase** from the thermophilic bacterium *Thermus aquaticus*.

At the end of one PCR cycle, the targeted sequences on both strands have been copied. When the three-step cycle is repeated (figure 17.8), the two strands from the first cycle are copied to produce four DNA molecules. These are amplified in the third

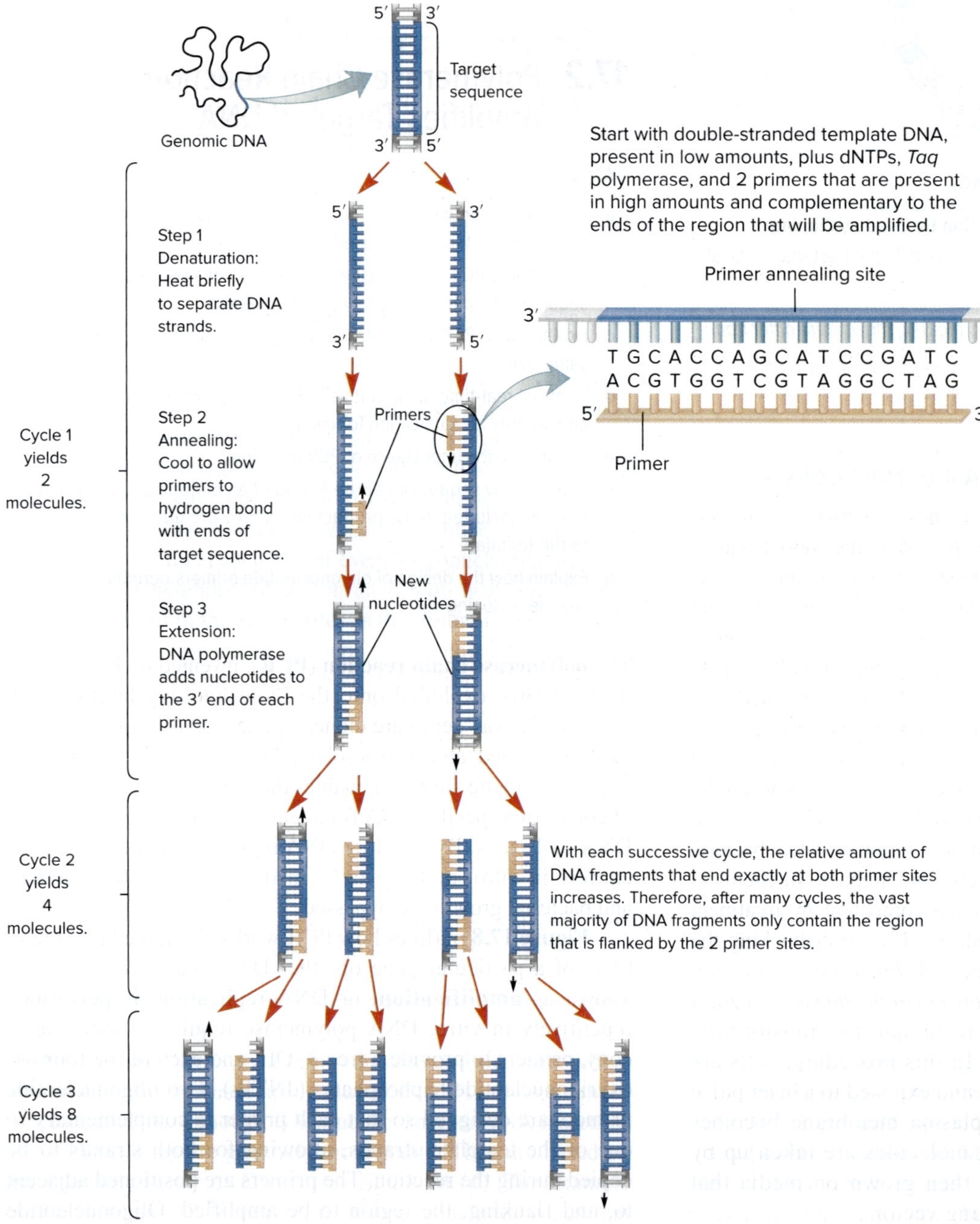

Figure 17.8 The Polymerase Chain Reaction (PCR) Technique. During each cycle, oligonucleotides that are complementary to the ends of the targeted DNA sequence bind to the DNA and act as primers for the synthesis of this DNA region. The primers are usually 15 to 30 nucleotides in length. The region between the two primers is typically hundreds of nucleotides in length, not just several nucleotides as shown here. The net result of PCR is synthesis of many copies of DNA in the region flanked by the two primers.

MICRO INQUIRY *Why, after three cycles, are the majority of amplified DNA molecules (i.e., PCR products) the size defined by the distance between the primers?*

cycle to yield eight double-stranded products. Thus, each cycle increases the number of target DNA molecules exponentially. Depending on the initial concentration of the template DNA and other parameters such as the G + C content of the DNA to be amplified, it is theoretically possible to produce about 1 million copies of targeted DNA sequence after 20 cycles and over 1 billion after 30 cycles. Pieces ranging in size from less than 100 base pairs to several thousand base pairs in length can be amplified, and the initial concentration of target DNA can be as low as 10^{-20} to 10^{-15} M. ◀ *DNA replication in bacteria (section 13.3)* *Polymerase Chain Reaction*

PCR can be performed in either of two ways. If large quantities of a specific piece of DNA are needed, the reaction products are collected and purified at the end of a designated number of cycles. This is sometimes called **end-point PCR,** and the final number of DNA fragments amplified is not quantitative, meaning that the amount of final product does not reflect the amount of template DNA present at the start of the reaction. End-point PCRs are useful when looking for the presence or absence of a particular DNA, for example, in detecting a pathogen in a clinical laboratory. In contrast, **real-time PCR** is quantitative; in fact, it is referred to as qPCR. That is, the amount of DNA or RNA template (which is converted to DNA with reverse transcriptase prior to starting PCR) present in a given sample can be determined. This is accomplished by adding a fluorescently labeled probe to the reaction mixture and measuring its signal during the initial cycles. This is when the rate of DNA amplification is exponential. However, as the PCR cycles continue, substrates are consumed and polymerase efficiency declines. So although the amount of product increases, its rate of synthesis is no longer exponential (this is why end-point collection of PCR products is not quantitative). Specially designed thermocyclers record the amount of PCR product generated as it occurs, thus the term real-time PCR. Gene expression studies often rely on real-time PCR because mRNA transcripts can be copied by reverse transcriptase to cDNA, which is then quantified. Therefore the procedure monitors transcription of the gene targeted by the primers.

PCR is an essential tool in many areas of molecular biology, medicine, and biotechnology. Because the primers used in PCR target specific DNA, PCR can isolate particular fragments of DNA (e.g., genes) from materials that contain many different genomes, such as soil, water, and blood. This explains why PCR has become an essential part of certain diagnostic tests, including those for *Chlamydia*, hepatitis, human papillomavirus infections, and other infectious agents and diseases. The tests are rapid, sensitive, and specific. PCR is also employed in forensic science, where it is used in criminal cases as part of DNA fingerprinting technology. ▶ *Metagenomics provides access to uncultured microbes (section 18.3); Molecular characteristics (section 19.3)*

Seamless Cloning

One limitation to conventional cloning is that it relies on the presence of naturally occurring restriction sites in DNA. A modification to PCR allows for the addition of custom restriction sites onto the ends of DNA to be cloned, thereby simplifying DNA assembly tasks. Remember that in PCR, the purpose of the oligonucleotide primers is to base-pair with the template and provide a free 3′ OH group for the addition of deoxyribonucleotides as the new DNA strand is synthesized. It is imperative that the primer hybridizes to the template precisely for 15 to 20 bases at the 3′ end. However, the 5′ end of the oligonucleotide primer is not required to bind to the template for replication to proceed (**figure 17.9*a***). Scientists have exploited this feature of DNA replication to add DNA sequences to the 5′ end of the primer to include, for example, a restriction endonuclease recognition site. Initially the primer will bind and copy the template as illustrated in figure 17.8. In subsequent cycles, when the initial copies are used as templates, those additions on the 5′ end will be copied. Because of the exponential nature of PCR, at the end point of the reaction, all of the product will include the restriction enzyme recognition sequences at the 5′ end of each strand. The result is a DNA fragment copied exactly from the template, with the addition of convenient restriction sites at each end.

Taking this idea one step further, Daniel Gibson developed a method in 2009 to eliminate the use of restriction endonucleases entirely. His method, termed Gibson assembly, uses the 5′ ends of oligonucleotide primers to specify the desired junction between the molecules. In a series of reactions diagrammed in figure 17.9*b*, a vector and insert fragment can be joined efficiently. The insert DNA molecule is prepared by PCR performed with primers that will add DNA that matches the vector. The addition of T5 exonuclease to both the linear vector and the insert cleaves the end of each molecule, leaving extended 3′ sticky ends. Just as in cloning with restriction enzymes, these single-stranded regions can hybridize. Two additional enzymes are added to the reaction, a DNA polymerase and a DNA ligase, to reform a circular molecule. Because the resulting product does not retain any scars or seams from restriction sites in the final product, this approach is sometimes called seamless cloning.

With the advent of seamless cloning techniques, scientists are free to envision a DNA molecule and create it in vitro. The procedure diagrammed in figure 17.9*b* can be expanded to assemble multiple fragments by amplifying each fragment with primers that specify the junction between it and its intended neighbor. An entire operon can be synthesized from promoter and regulatory sequences and genes. Indeed, entire genomes have been synthesized by Gibson assembly, as illustrated in Techniques & Applications 17.2.

Comprehension Check

1. Briefly describe the polymerase chain reaction. What is the difference between a step and a cycle?
2. Why is PCR used to detect infectious agents that are often hard to diagnose?
3. How would you use PCR to measure the concentration of a specific gene's mRNA?
4. Why is it possible to visualize a PCR product on an agarose gel even if the template genome is present at such a low concentration that it cannot be seen?
5. Why is it technically challenging to assemble a large operon using restriction endonucleases?

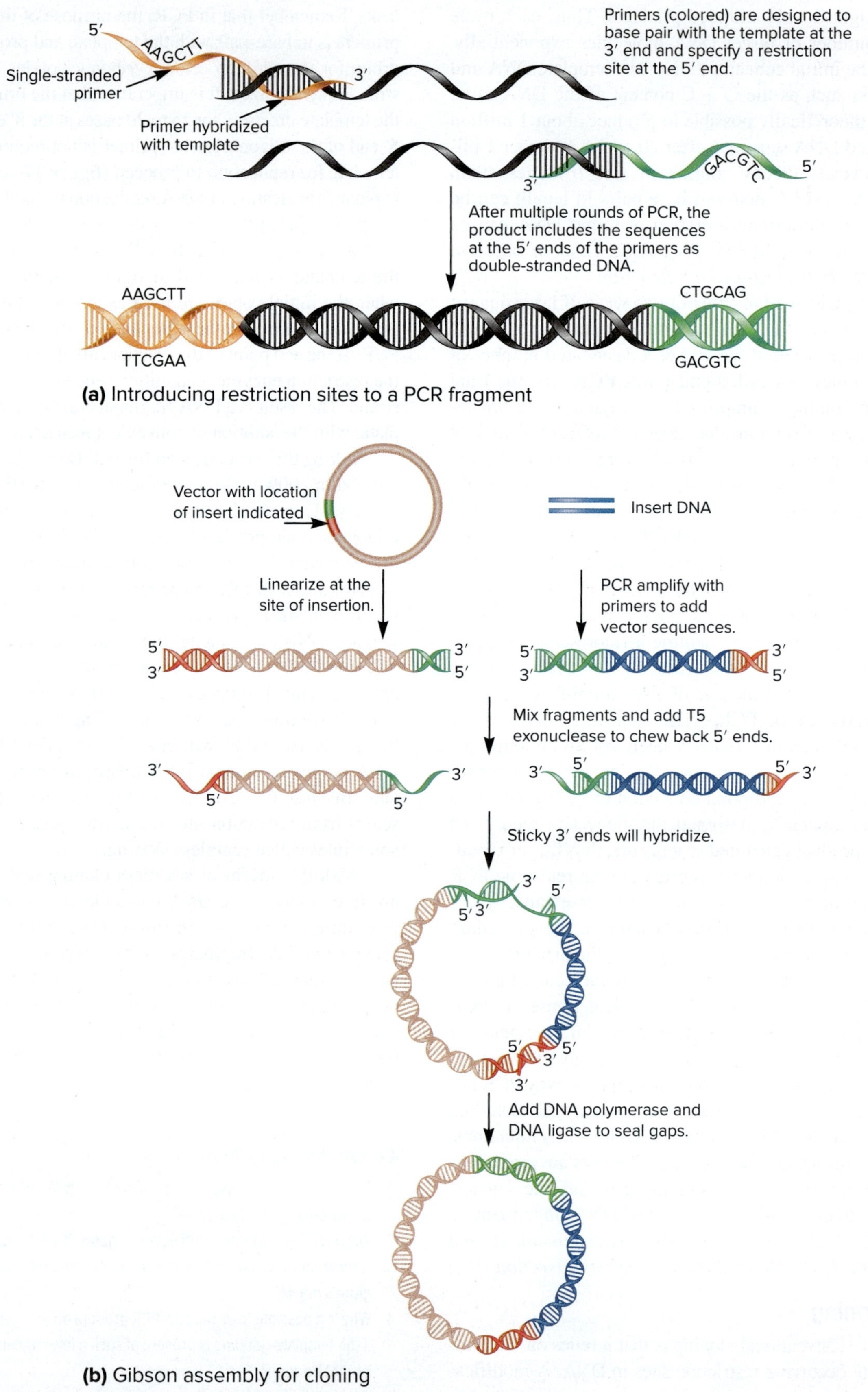

Figure 17.9 Polymerase Chain Reaction Using Primers with a 5′ Extension.

17.3 Genomic Libraries: Cloning Genomes in Pieces

After reading this section, you should be able to:

a. Explain why genomic libraries are useful
b. Outline the construction of a genomic library and how a clone of the gene of interest might be selected

When cloning, the DNA to be inserted into a vector can be obtained in several ways. It can be synthesized by PCR using the microbe's chromosome as template, as discussed in section 17.2, or it can be amplified from DNA extracted from a natural environment, like soil or water. PCR requires knowledge of the nucleotide sequence of the DNA to be cloned. However, what if researchers want to clone a gene based on the function of its product but have no idea what its DNA sequence might be? A genomic library must then be constructed and screened.

The goal of **genomic library** construction is to have all the DNA derived from a single organism or isolated from a single environment inserted into a set of cloning vectors. When cloning DNA directly from the environment in this way, a **metagenomic library** is produced. Genomic and metagenomic library construction starts by fragmenting the source DNA to a size compatible with the particular vector being used. **Figure 17.10** illustrates the process with recombinant plasmids, and comparable strategies may be used for bacterial artificial chromosomes and viral vectors. Each fragment is cloned into a separate vector so that each vector now carries a separate piece of chromosome. When a collection of these recombinant vectors is introduced (e.g., by transformation) into a population of host cells, each transformant carries a vector that has a different piece of foreign chromosome. Ideally the entire organism's genome or all the DNA from the sample environment is represented in the collection of recombinant vectors. In this way, isolated groups of genes can be analyzed and isolated (figure 17.10). ▶| *Metagenomics provides access to uncultured microbes (section 18.3)*

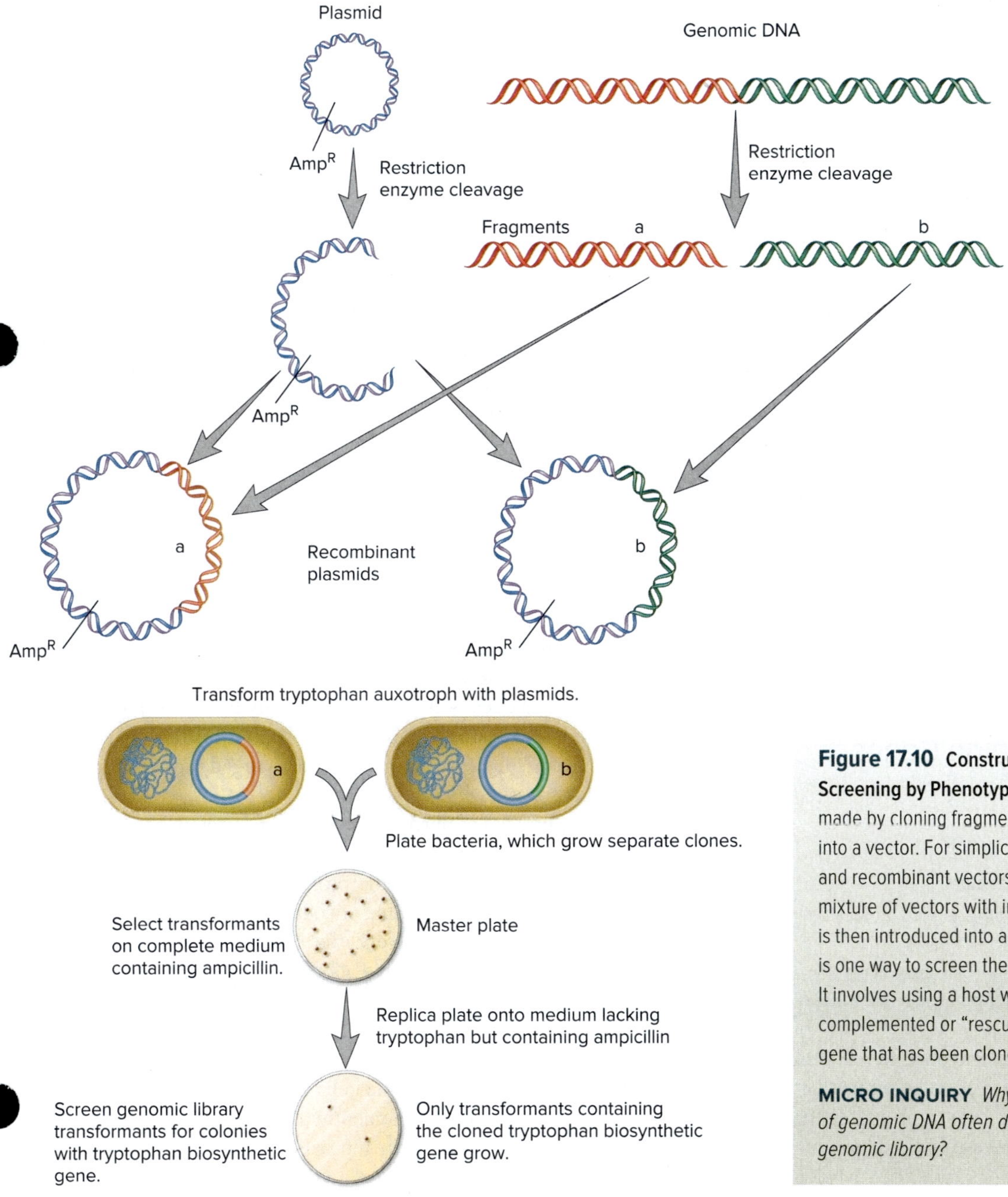

Figure 17.10 Construction of a Genomic Library and Screening by Phenotypic Rescue. A genomic library is made by cloning fragments of an organism's entire genome into a vector. For simplicity, only two genomic fragments and recombinant vectors are shown. In reality, a large mixture of vectors with inserts is generated. This mixture is then introduced into a suitable host. Phenotypic rescue is one way to screen the colonies for the gene of interest. It involves using a host with a genetic defect that can be complemented or "rescued" by the expression of a specific gene that has been cloned.

MICRO INQUIRY *Why are long fragments (e.g., 20,000 bp) of genomic DNA often desired when constructing a genomic library?*

Genomic libraries are often constructed to find a gene or genetic element (e.g., small noncoding RNA) with a specific function. In this case, it is necessary to know something about the function of the target gene or genetic element. If the genomic library has been introduced into a microbe that expresses the foreign gene, it may be possible to assay each clone for a specific protein or phenotype. For example, to find genes that encode enzymes needed for the biosynthesis of the amino acid tryptophan from a newly isolated bacterium, the library could be expressed in an *E. coli* tryptophan auxotroph (figure 17.10). Recall that the growth of tryptophan auxotrophs requires the addition of this amino acid to a defined medium. Following introduction of the genomic library into host cells, those that now grow without tryptophan would be good candidates for the genomic library fragment that possesses the tryptophan biosynthetic genes. Success with this approach depends on the assumption that the activity of the cloned gene product is similar in both organisms. If this is not the case, the host must be the same species from which the library was prepared. In this example, a bacterial mutant lacking the trait in question (e.g., a tryptophan auxotroph) is used as the genomic library host. The **genetic complementation** of a deficiency in the host cell is sometimes called **phenotypic rescue** because the mutant phenotype is (at least partially) restored, or "rescued." ◀ *Detection and isolation of mutants (section 16.2)*

If a genomic library is prepared from a eukaryote, a cDNA library is usually constructed. This removes introns and only the protein-coding regions of the genome are cloned. cDNA is prepared (figure 17.5) and cloned into a suitable vector. After the library is introduced into the host microbe, host cells carrying the gene of interest are typically identified by phenotypic rescue or PCR amplification of the gene of interest after the vector has been isolated from candidate transformants. In some cases these approaches are not feasible. Then the researcher must develop a novel approach that suits the particular set of circumstances to screen the genomic library.

17.4 Expressing Foreign Genes in Host Cells

After reading this section, you should be able:

a. Explain the utility of expression vectors
b. Outline the procedure by which a His-tagged protein is generated in vivo and purified in vitro
c. Summarize the role of GFP in protein analysis, and differentiate between a transcriptional and a translational GFP fusion

When a gene from one organism is cloned into another, it is said to be a **heterologous gene.** However, to be transcribed, a recombinant, heterologous gene must have a promoter that is recognized by the host RNA polymerase. Translation of its mRNA depends on proper ribosome binding. For instance, if the host is a bacterium and the gene has been cloned from a eukaryote, a bacterial promoter and ribosome binding site must be provided and introns removed.

The problems of expressing recombinant genes in host cells are largely overcome with the help of specially constructed cloning vectors called **expression vectors.** These vectors enable the high-level transcription of a gene because it is cloned adjacent to a particularly "strong" promoter. Often expression vectors contain regulatory regions of the *lac* operon (or another inducible promoter) so that expression of the cloned genes can be controlled by the addition of inducer molecules. This is important because sometimes high-level expression of a heterologous protein can be toxic to the host cell. The presence of an inducible promoter allows the biotechnologist to grow cells to a certain density before inducing the expression of the cloned gene. ◀ *Regulation of transcription initiation saves considerable energy and materials (section 14.2)*

Purification and Study of Recombinant Proteins

It is often necessary to isolate the protein product of the cloned gene so that its structure and activity can be studied. In addition, it is sometimes desirable to determine the subcellular localization of a protein. Finally, as discussed in chapter 42, heterologous genes are often expressed for the production of commercially prepared recombinant proteins. Here we discuss some clever ways in which proteins can be purified and visualized in living cells.

Protein Purification

A common method by which proteins are purified is called **polyhistidine tagging,** or simply **His-tagging.** His-tagging involves adding a series of histidine amino acid residues to either the N- or C-terminus of a protein. Most often six residues are added; this is called a 6xHis-tag. Histidine is used because it has a high affinity for metal ions, so a protein bearing multiple consecutive histidine residues will bind preferentially to a solid-phase material, called a resin, which has exposed nickel or cobalt atoms. This permits the separation of the His-tagged protein from other cellular constituents.

The process of purifying a His-tagged protein begins by adding the histidine residues to the protein. This is most commonly accomplished by cloning the protein into an expression vector that already has the histidine-encoding sequence present (**figure 17.11**, step 1). Alternatively, the histidine-encoding sequence can be amplified by PCR and added to the gene of interest. Once the protein-coding nucleotide sequence is fused to the histidine codons, the vector is introduced into the microbe of choice, usually *E. coli* (step 2). As *E. coli* grows, the His-tagged protein is produced (step 3). When the cells reach a sufficiently high density, they are lysed by treatment with detergent or enzymes, and all the soluble proteins in the cell—including the His-tagged

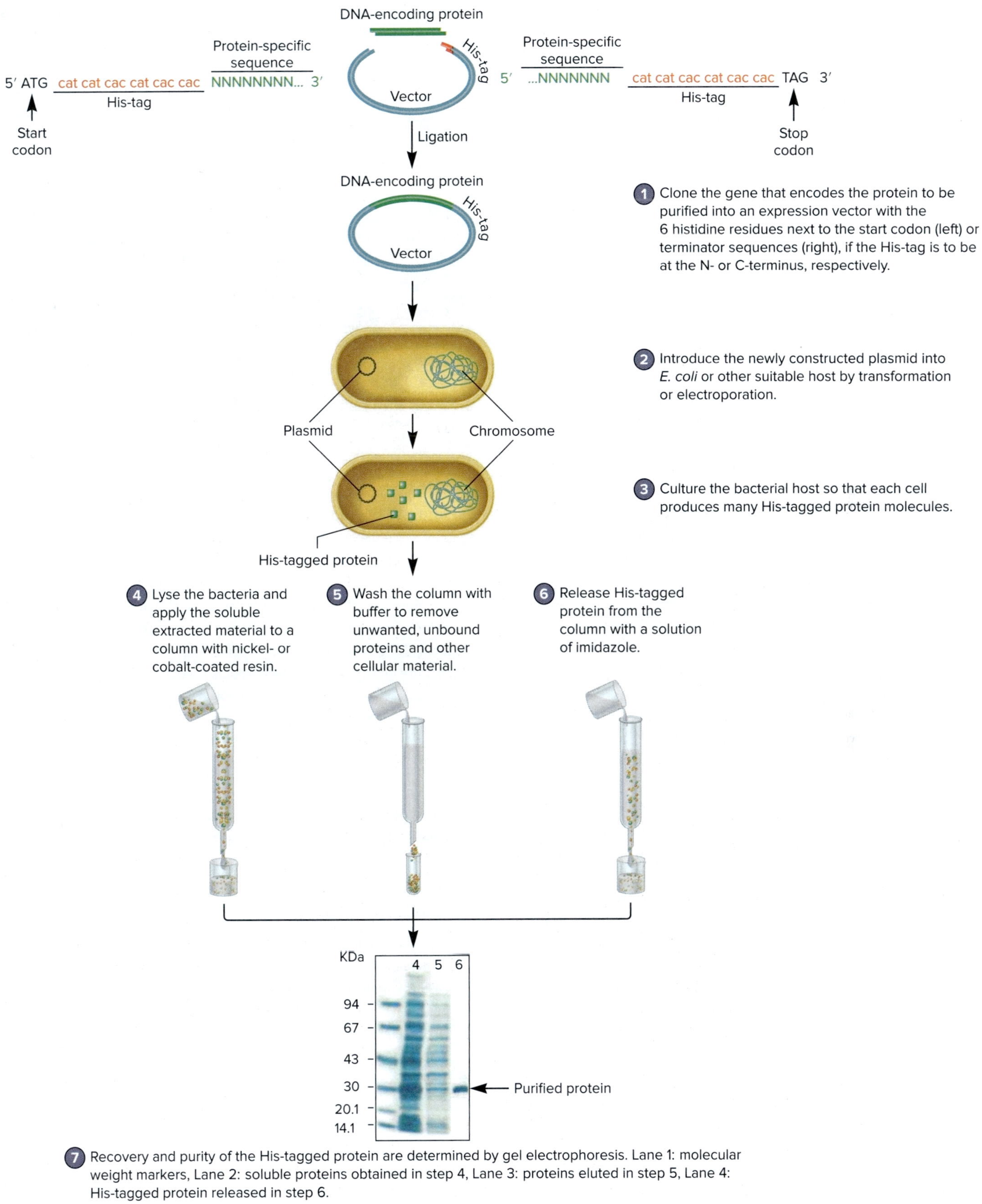

Figure 17.11 **Construction and Purification of a Polyhistidine-Tagged Protein.**

protein—are collected. These proteins are mixed with the metal resin that is bound to a column (step 4) so that only the His-tagged protein sticks to the resin. The remaining, unwanted material is washed from the resin (step 5). The His-tagged protein is then removed by passing the aromatic compound imidazole through the column (step 6). Depending on the purification system used, the His-tag can be cleaved from the protein while it is attached to the resin, or it can be removed following release of the protein. Often the protein is functional with the His-tag, so it may not be removed. His-tagging is used if the protein is soluble and is found in the cytoplasm. While His-tagging is certainly the most popular approach to purifying recombinant proteins, other approaches exist, especially for purifying membrane-associated proteins.

Fluorescence Labeling

What if the goal of cloning a gene is to study the regulation or function of the protein product in vivo? It is possible to visualize the activity of a specific promoter (i.e., when the promoter is on or off) as well as observe the localization of a protein by fluorescence microscopy. The development of fluorescent labeling of living cells began when it was discovered that the jellyfish *Aequorea victoria* produces a protein called **green fluorescent protein (GFP)**. GFP is encoded by a single gene that, when translated, undergoes self-catalyzed modification to generate a strong, green fluorescence. This means that it is easily cloned and expressed in the organism of choice. An entire palette of fluorescent labels is now available. These include mutant versions of the GFP protein that glow throughout the blue-green-yellow spectrum.

There are two ways in which the amino acid sequence that encodes fluorescent peptides can be attached to a gene. A transcriptional fusion is used to study gene regulation by replacing the entire protein coding sequence of the gene of interest with the fluorescence-encoding gene (**figure 17.12*a***). This places the expression of the fluorescence gene under the control of the promoter of the replaced gene. Thus, when the promoter is turned "on," it directs the transcription of the fluorescence gene and the cells glow. If the transcriptional fusion has been made in a multicellular organism, then not only is the timing of promoter activity seen but the specific cell type in which the gene is normally expressed is also determined. On the other hand, a translational fusion is used to determine where a particular protein is localized in the cell. Translational fusions add the GFP-coding sequence to the end of the structural gene of interest. This creates a chimeric protein—a protein that consists of two parts: the protein being studied and GFP. Now whenever that protein is made, and wherever it goes in the cell, the protein will fluoresce. Of course, care must be taken to ensure that the protein fusion still functions like the original protein. One way to do this is to test for phenotypic rescue of a mutant lacking the structural gene of interest. Once it has been determined that the chimeric protein is fully functional, the cellular "address" of the protein can be determined (figure 17.12*b*).

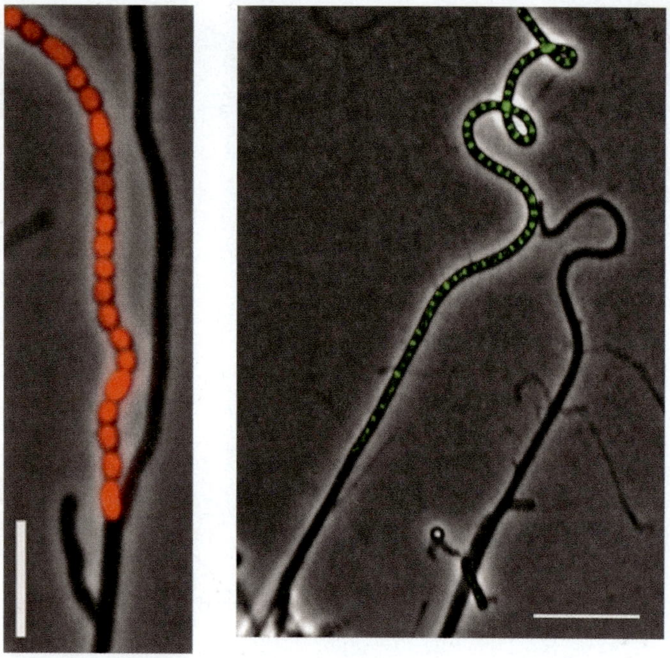

(a) (b)

Figure 17.12 Fluorescent Protein Labeling. (a) A transcriptional fusion places the gene encoding a fluorescent peptide under the control of a promoter, replacing the coding sequence of the gene. In this case, the mCherry fluorescence gene is controlled by the promoter that normally governs transcription of a gene in the actinomycete *Streptomyces coelicolor* that is expressed only when aerial hyphae form cross walls (septa) that give rise to a chain of spores. Thus red fluorescence is seen uniformly throughout the chain of spores but not in hyphae without septae. (b) A translational fusion generates a chimeric protein consisting of the normal protein attached to the GFP label. Here the FtsZ gene has been fused to GFP. This shows that FtsZ localizes at the septa. Notice that the filament to the right is not forming septae, so FtsZ is not present.
(a) ©Mark Buttner (b) ©Klas Flärdh

MICRO INQUIRY *What special considerations are necessary if one is constructing a translational fusion of a protein known to be exported?*

Comprehension Check

1. Describe how a genomic library might be screened for a gene that confers the production of an extracellular enzyme that degrades casein, a protein found in milk. (Hint: Agar with skim milk loses its opacity when casein is degraded.)
2. List several reasons why a cloned gene might not be expressed in a heterologous host cell.
3. Think of a situation in which you might want to keep the His-tag on a recombinant protein and a situation in which it would be important to remove it.
4. You are studying chemotaxis proteins in a newly described bacterium. You have cloned a gene that encodes the CheA protein. You need to be sure that this protein localizes to the inside face of the plasma membrane (*see figure 14.22*). Will you use a transcriptional or translational fusion? Explain your choice.

17.5 Cas9 Nuclease Is a Precise Tool for Genome Editing

After reading this section, you should be able to:

a. Distinguish the DNA recognition features of restriction endonucleases and Cas9 nuclease
b. Explain how Cas9 nuclease can be directed to cut at a unique site in a genome
c. Diagram how a new gene may be inserted into a chromosome by homologous recombination

Cas9 genome editing has rapidly become one of the most widely used tools for altering genomes in vivo. Cas9 genome editing is often referred to as CRISPR or CRISPR-Cas9, referencing the bacterial genome element from which it was developed, but in fact, only the Cas9 component is used in editing. Cas9 nucleases are encoded in the genomes of most bacteria and archaea, where they are usually adjacent to a CRISPR locus, *c*lustered *r*egularly *i*nterspaced *s*hort *p*alindromic *r*epeats. As the mechanistic details of Cas9 function were discovered, two research groups, one led by Jennifer Doudna and Emmanuelle Charpentier and the other by Feng Zhang, sought to adapt Cas9 for genome editing. In this process, genomic DNA can be directly modified and the procedures are general enough to be used for any cell into which DNA can be introduced and expressed. ◀ *Responses to viral infection (section 14.6)*

Like restriction enzymes, Cas9 is an endonuclease that cuts both strands of a target DNA. However, there is an important difference between the two types of nucleases, in terms of how they recognize their target sequence in double-stranded DNA. Restriction enzymes recognize four to eight base pairs through contacts between the DNA molecule and amino acid side chains in the enzyme (figure 17.2). Cas9, however, is a ribonucleoprotein consisting of a polypeptide and a **guide RNA** (**gRNA**). Recognition of target DNA for cleavage occurs by hybridization of about 20 bases between the gRNA and its complementary DNA sequence in the genome (**figure 17.13**).

In microbes, the CRISPR locus is the source of the gRNA (*see figure 14.27*), and the Cas9 nuclease protects the cell from viral attack. Sequences in the CRISPR locus derive primarily from mobile genetic elements (bacteriophage and plasmids), so the Cas9 nuclease in a microbial cell specifically targets invading DNA for destruction. The extreme specificity conferred by the gRNA is the key to genome editing because each 20-base target sequence is almost certainly unique, even in a large eukaryotic genome. In contrast, a restriction enzyme that recognizes a few nucleotides will cut the genome, on average, every few thousand bases.

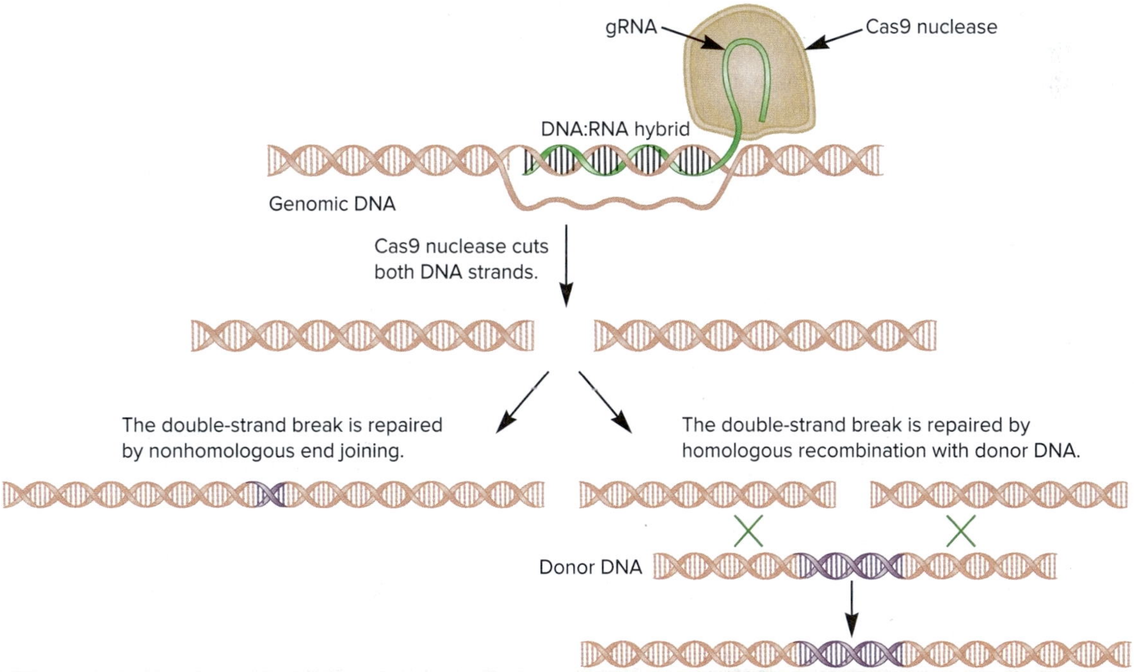

Figure 17.13 Genome Editing with Cas9 Nuclease.

MICRO INQUIRY *How could you assemble the donor DNA molecule for homologous recombination?*

Cas9 enzymes can be engineered to carry gRNAs with specified nucleotide sequences, thereby programming the recognition sequence for the nuclease. The gRNA directs Cas9 to hybridize with a single, desired site in a genome, making it the most precise mechanism available for targeting and cutting DNA. In eukaryotes, all of which lack a CRISPR/Cas system, the editing process begins by introducing the two components of the mature Cas9 endonuclease, the apoenzyme and the gRNA, to the host cells. These molecules may be added directly, or they may be added as cloned DNA regulated by an inducible promoter. In the latter case, upon induction, the Cas9-gRNA complex assembles and performs its DNA cutting function.

Figure 17.13 illustrates how Cas9 recognizes and hydrolyzes a specific DNA sequence. A portion of the gRNA protrudes from the enzyme, available for hybridization. Upon locating its complement, the gRNA induces a conformational shift in the nuclease (protein) portion of Cas9, which then hydrolyzes phosphodiester bonds in both DNA strands. In the simplest case, a point mutation is created as the cell scrambles to repair the damage. Some microbes and all eukaryotes have a **nonhomologous end joining** (**NHEJ**) system that can rejoin the two chromosome pieces. But because it's not a perfect repair, the fused DNA ends often have a deletion or insertion of a few base pairs. The consequence is usually a frameshift mutation in the gene that will result in an inactive protein. A limitation to this method is that the outcome differs in each cell.

A precise Cas9 genome edit changes the nucleotides to a predetermined sequence (e.g., a mutated gene could be replaced with the wild-type allele). In this case, a donor sequence must be constructed in vitro using seamless cloning techniques. The donor DNA includes chromosomal regions flanking the gRNA binding site to provide regions of homology for recombination with the chromosome. The donor sequence is introduced to the cell at the same time as the Cas9/gRNA molecules. **Homologous recombination** between the broken chromosome and the donor sequence will repair the double-stranded break and at the same time incorporate the DNA carried on the donor molecule into the genome. Selection and screening procedures to identify mutants are not required with Cas9 gene editing, unlike traditional cloning. Cells into which the Cas9-gRNA complex has been introduced can survive and reproduce only if the original target sequence is modified. Unrepaired double-stranded chromosome breaks are lethal. ◀ *Molecular recombination: joining DNA molecules together (section 16.4)*

Although the molecular components of Cas9 editing originally derived from the bacterium *Streptococcus pyogenes,* this technique has been used most frequently in eukaryotic systems. There are several reasons for this. First, many bacteria (~45%) and most archaea (~85%) have homologues of the Cas9 editing system, complicating its use in these organisms. Second, other tools for genome editing already exist for many microbes (*see figure 42.5*). Third, most bacteria lack a NHEJ repair pathway for repairing the double-stranded break generated by Cas9 nuclease and cannot survive this treatment, but homologous repair can be used efficiently. Despite these challenges, Cas9 editing has been used successfully in some microorganisms that are difficult to manipulate genetically.

Researchers have developed other applications for Cas9 nuclease that take advantage of its ability to bind chromosomal DNA in a single genomic location. An altered version of Cas9, termed dead Cas9 (dCas9), retains the DNA localization function conferred by the gRNA, but no longer cuts DNA. Instead, a dCas9-gRNA molecule bound to DNA can serve as a platform for other enzymes. In the simplest example, dCas9-gRNA bound to a promoter acts as a repressor to inhibit expression of target genes. Conversely, to activate transcription, dCas9-gRNA has been fused to the ω subunit of RNA polymerase so it can bind to the promoter of a target gene and recruit RNA polymerase. Genome editing has been demonstrated using dCas9-gRNA fused to additional DNA modifying enzymes. Because the dCas9 variant is used, there is no nuclease activity to break the phosphate-sugar backbone, but the other enzymes directly modify the base, for example, from C to T. The outcome is a precise change to the DNA sequence. Experiments like these have demonstrated the versatility of the dCas9-gRNA module as a homing device to direct a range of effector molecules to a specific chromosomal location. ◀ *Transcription in bacteria (section 13.5); Regulatory proteins often control transcription initiation (section 14.2)*

Cas9 gene editing is a relatively new technique that holds much promise, and scientists are actively exploring modifications and improvements. Projects as varied as removing allergens from peanuts to curing genetic diseases in humans have been proposed as examples of what might be achievable. Importantly, Cas9 technology opens the prospect of engineering the human genome in germ line cells (e.g., egg cells or sperm cells), thereby inviting discussions on many fronts to address ethical concerns.

Comprehension Check

1. Commercially grown mushrooms are sensitive to handling during harvest and may rapidly discolor, a reaction catalyzed by a single enzyme. Describe a strategy to extend the shelf life of mushrooms using Cas9 genome engineering.
2. A dinucleotide (e.g., GA or CC) appears in DNA with an average frequency of 1 in 16 (4^2) bases, and a trinucleotide appears with an average frequency of 1 in 64 (4^3) bases. Calculate how frequently you would expect a 20 bp sequence to occur. Compare this number to the size of the human genome (3×10^9 bp).

Key Concepts

17.1 Key Discoveries Led to the Development of DNA Cloning Technology

- Genetic engineering became possible after the discovery of restriction enzymes and reverse transcriptase, and the development of essential methods in nucleic acid chemistry.
- Restriction enzymes are important because they cut DNA at specific sequences, thereby releasing fragments of DNA that can be cloned or otherwise manipulated (**figure 17.3** and **table 17.1**).
- There are four types of cloning vectors: plasmids, viruses, cosmids, and artificial chromosomes. Cloning vectors generally have at least three components: an origin of replication, a selectable marker, and a multicloning site (**figures 17.6** and **17.7**).
- A common approach to cloning is to digest both vector and DNA to be inserted with the same restriction enzyme or enzymes so that compatible sticky ends are generated. The vector and DNA to be cloned are then incubated in the presence of DNA ligase, which catalyzes the formation of phosphodiester bonds once the DNA fragment inserts into the vector.
- Once the recombinant plasmid has been introduced into host cells, cells carrying vector must be selected. This is often accomplished by allowing the growth of only antibiotic-resistant cells because the vector bears an antibiotic-resistance gene. Cells that took up vector with inserted DNA must then be distinguished from those that contain only vector. Often a blue-versus-white colony phenotype is used; this is based on the presence or absence, respectively, of a functional *lacZ* gene (**figure 17.4**).
- The bacterium *E. coli* and the yeast *S. cerevisiae* are the most common host species.
- DNA can be introduced into microbes by transformation or electroporation.

17.2 Polymerase Chain Reaction Amplifies Targeted DNA

- The polymerase chain reaction (PCR) allows small amounts of specific DNA sequences to be amplified, or increased in concentration thousands of times (**figure 17.8**).
- PCR consists of multiple cycles of three steps each: DNA denaturation, primer annealing, and DNA synthesis.
- PCR has numerous applications. It often is used to obtain genes for cloning and in diagnostic and forensic science.
- Including additional sequences at the 5′ end of an oligonucleotide primer in PCR results in the presence of any desired sequence, for example a restriction site, in the final product (**figure 17.9**).
- Gibson assembly is a series of reactions that permits seamless cloning, that is, overlap of specific fragments by enzymatic reactions that do not include restriction enzymes.

17.3 Genomic Libraries: Cloning Genomes in Pieces

- It is sometimes necessary to find a gene without the knowledge of the gene's DNA sequence. A genomic library is constructed by cleaving an organism's genome into many fragments, each of which is cloned into a vector to make a unique recombinant plasmid.
- Genomic libraries are often screened for the gene of interest by phenotypic rescue (genetic complementation) (**figure 17.10**).

17.4 Expressing Foreign Genes in Host Cells

- An expression vector has the necessary features to express in high levels any recombinant gene it carries.
- If a eukaryotic gene is to be expressed in a bacterium, cDNA is used because it lacks introns; a bacterial ribosome binding site must also be added to the 5′ end of the gene.
- Purification of recombinant proteins is often accomplished by fusing the coding sequence of a protein to six consecutive histidine residue codons found on some expression vectors. When introduced and expressed in bacteria, the His-tagged protein can be selectively purified (**figure 17.11**).
- Fluorescent proteins can be used to study the regulation of gene expression (transcriptional fusions) and protein localization (translational fusions) (**figure 17.12**).

17.5 Cas9 Nuclease Is a Precise Tool for Genome Editing

- The Cas9-gRNA nuclease is a precise DNA cutting tool. Because the localization component of the enzyme is an RNA molecule, it can be modified to direct the enzyme to any specified location in a genome.
- Repair of the double-stranded DNA break catalyzed by Cas9-gRNA allows for the nearby DNA to be altered by substitution or deletion when the desired change is introduced to the cell as a donor sequence.
- A variant of Cas9 that lacks nuclease activity but retains the guide RNA (dCas9-gRNA) is a convenient means to direct a range of other enzymes to a specific chromosomal target.

Active Learning

1. You are performing PCR to amplify a gene encoding a tRNA from a bacterium that has recently been grown in pure culture. You are expecting a product of 95 base pairs (bp). However, following PCR, you observe three bands on an agarose gel and only one is 95 bp. List at least two possible explanations (excluding experimental error).
2. You have cloned a structural gene required for riboflavin synthesis in *E. coli.* You find that an *E. coli* riboflavin auxotroph carrying the cloned gene on a vector makes less riboflavin than the wild-type strain. Why might this be the case?
3. Suppose you transformed *E. coli* with a plasmid vector carrying a gene that encodes human interferon, but none of the transformants produce interferon. Give as many plausible reasons as possible for this result.
4. You are interested in the activity and regulation of a protease made by the Gram-positive bacterium *Geobacillus stearothermophilus.* What would be the purpose of constructing each of the following: a His-tagged protease, a transcriptional GFP fusion to the protease gene promoter, and a translational GFP fusion to the protease gene coding sequence?
5. The genome sequence of a recently isolated bacterium includes a gene that is predicted to encode a restriction endonuclease. List the steps you could take to express the gene in *E. coli* and purify the enzyme with a His-tag. How could you determine if the His-tag affects the activity of the enzyme? You'd also like to perform some experiments that involve introducing DNA to this new organism, and although it is naturally transformable, the restriction enzyme appears to be very active in degrading foreign DNA. Describe a strategy to develop a strain of this isolate that has lost the restriction enzyme activity.
6. Many *Streptomyces* spp. have genomic regions (gene clusters) that are predicted to encode a biosynthetic pathway. As members of this genus often produce antibiotics, there is much interest in exploring these cryptic genes. They are often not expressed under laboratory conditions. Cloning and expressing the genes in another organism have not been generally successful. Propose a CRISPR-based strategy to activate such a gene cluster directly in its native organism.
7. *Mesoplasma florum* is a wall-less bacterium related to *Mycoplasma* spp. There is considerable interest in developing a minimal genome host for applications like those described in Techniques & Applications 17.2. *M. florum* is attractive because it is not pathogenic to humans, it has a rapid growth rate, and has a very small genome: only 793 kb. However, genetic engineering tools do not exist for this organism, and no plasmids have ever been isolated. This prompted a group of scientists from the Université de Sherbrooke to build an *M. florum-E. coli* shuttle vector. What genetic elements are required for such a construct? Without a known plasmid, where could they get an origin of replication for *M. florum*? How do you think they could determine the copy number of the shuttle vector in *E. coli* and in *M. florum*?

   Read the original paper: Matteau, D., et al. 2017. Development of plasmids for *Mesoplasma florum*. *Appl. Environ. Microbiol.* 83:e02274-16.

18

Microbial Genomics

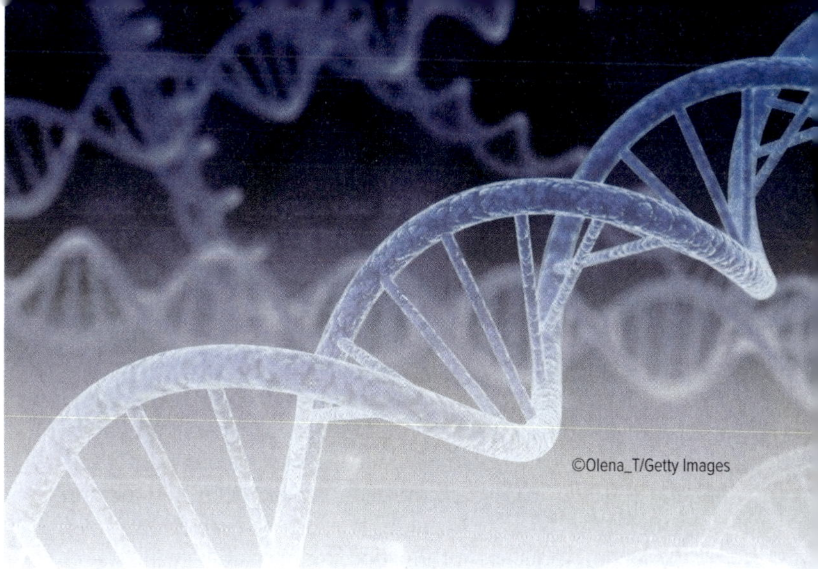

©Olena_T/Getty Images

"Synthetic Life": Oxymoron or the Future?

"To live, to err, to fall, to triumph, to recreate life out of life!" When James Joyce wrote these words in 1916 in *A Portrait of the Artist as a Young Man,* he never imagined that they would be encrypted in the genetic code of a synthetic genome in a novel bacterium. But 94 years later, Daniel Gibson and his co-workers at the J. Craig Venter Institute (JCVI) did just that. Building on 15 years of investigation, the JCVI researchers used a computer to design a genome and a DNA synthesizer to make short stretches of DNA. These were stitched together to create a 1.08 million base pair genome based on the chromosome of *Mycoplasma mycoides* (*see Techniques & Applications 17.2*). The genome was transplanted into *M. capricolum,* which after a few rounds of replication consisted entirely of molecules whose synthesis was directed by a chromosome that started as computer code and four bottles of deoxyribonucleotides. The new microbe is known as *M. mycoides* JCVI-Syn 1.0, or just Syn 1.0.

Why construct this microbe? Craig Venter views it as a $30 million proof of concept leading to the construction of genomes specifically designed for the development of vaccines, pharmaceuticals, clean water and food products, and biofuels. To reach this goal, Venter and his team took the 901 genes in the Syn 1.0 genome and, through trial and error, whittled the genome down to just 473 genes, thereby creating the smallest self-replicating organism, called Syn 3.0. The idea is that this leaves room to add DNA sequence needed to engineer microbes capable of special tasks.

But is synthetic life ethical? Depends on whom you ask. Some environmental groups say synthetic genome research should stop until specific regulations are in place. Most scientists and bioethicists take a more measured view, envisioning two ways such a microbe might be released: bioterror and bioerror. Protection against bioterror relies on a host of security measures, many of which are already in place. Bioerror brings us back to James Joyce. The Syn 1.0 genome includes a "watermark"—sequences unique to this microbe so that if it were to escape the lab, its identity could be determined. They developed a cipher to convert the genetic code so that when decoded, the watermark spells out the Joyce quote, as well as the words of the late physicist Richard Feynman, "What I cannot create, I do not understand."

To comprehend how we have arrived in a world where a computer is the starting point for a new organism, we need to review the individual technologies required. This involves the integration of a number of genetic and biotechnology principles; be sure you succeed in the readiness check, because the world of genomics is a vast and amazing place well worth understanding.

Readiness Check:

Based on what you have learned previously, you should be able to:

- ✓ Identify the key structural elements of DNA, RNA, and protein (section 13.2)
- ✓ Discuss what is meant by the term *genome*
- ✓ Explain the process and utility of PCR (section 17.2)
- ✓ Diagram the fundamental principles of gel electrophoresis (Techniques & Applications 17.1)
- ✓ Summarize the purpose, construction, and screening of a genomic library (section 17.3)

18.1 DNA Sequencing Methods

After reading this section, you should be able to:

a. Explain how DNA is sequenced by the Sanger chain termination method
b. Contrast and compare the advantages and disadvantages of the Sanger method with next-generation sequencing

The genomic era is a 21st-century phenomenon. However, one of the most crucial steps leading to the genomic revolution was figuring out how to sequence DNA. This was accomplished in 1977 by Alan Maxam and Walter Gilbert (collaborating on one technique), and Frederick Sanger. Sanger's method became the most commonly used and is now discussed.

Sanger DNA Sequencing

The Sanger method involves the synthesis of a new strand of DNA using the DNA to be sequenced as a template. The reaction begins when single strands of template DNA are mixed with an oligonucleotide primer (DNA about 12 to 20 nucleotides complementary to the region to be sequenced; it primes the initiation of new strand synthesis), DNA polymerase, the four deoxynucleoside triphosphates (dNTPs), and dideoxynucleoside triphosphates (ddNTPs). ddNTPs differ from dNTPs in that the 3′ carbon lacks a hydroxyl group (**figure 18.1**). In such a reaction mixture, DNA synthesis will continue until a ddNTP, rather than a dNTP, is added to the growing chain. Without a 3′-OH group to attack the 5′-PO_4 of the next dNTP to be incorporated, synthesis stops (*see figure 13.10*). Indeed, Sanger's technique is frequently referred to as a **chain-termination DNA sequencing** method.

To obtain sequence information, four separate synthesis reactions must be prepared, one for each ddNTP (**figure 18.2**). The ddNTP is mixed with all four normal dNTPs, and as DNA synthesis proceeds, only sometimes will the ddNTP be incorporated into the growing DNA strand, rather than its dNTP cousin. This

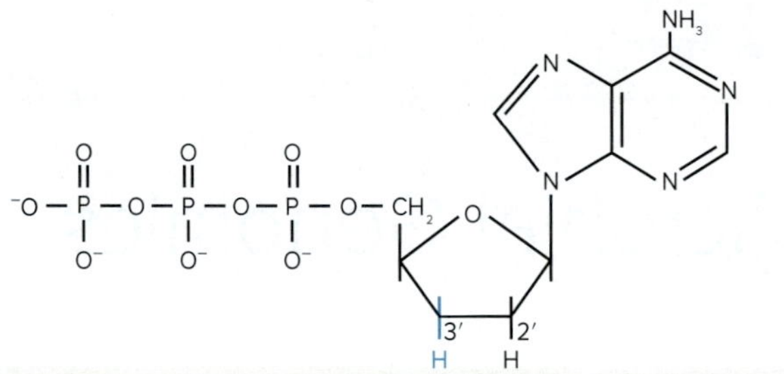

Figure 18.1 Dideoxyadenosine Triphosphate (ddATP). Note the lack of a hydroxyl group on the 3′ carbon, which prevents further chain elongation by DNA polymerase.

MICRO INQUIRY *What is the function of the 3′-OH during DNA synthesis?*

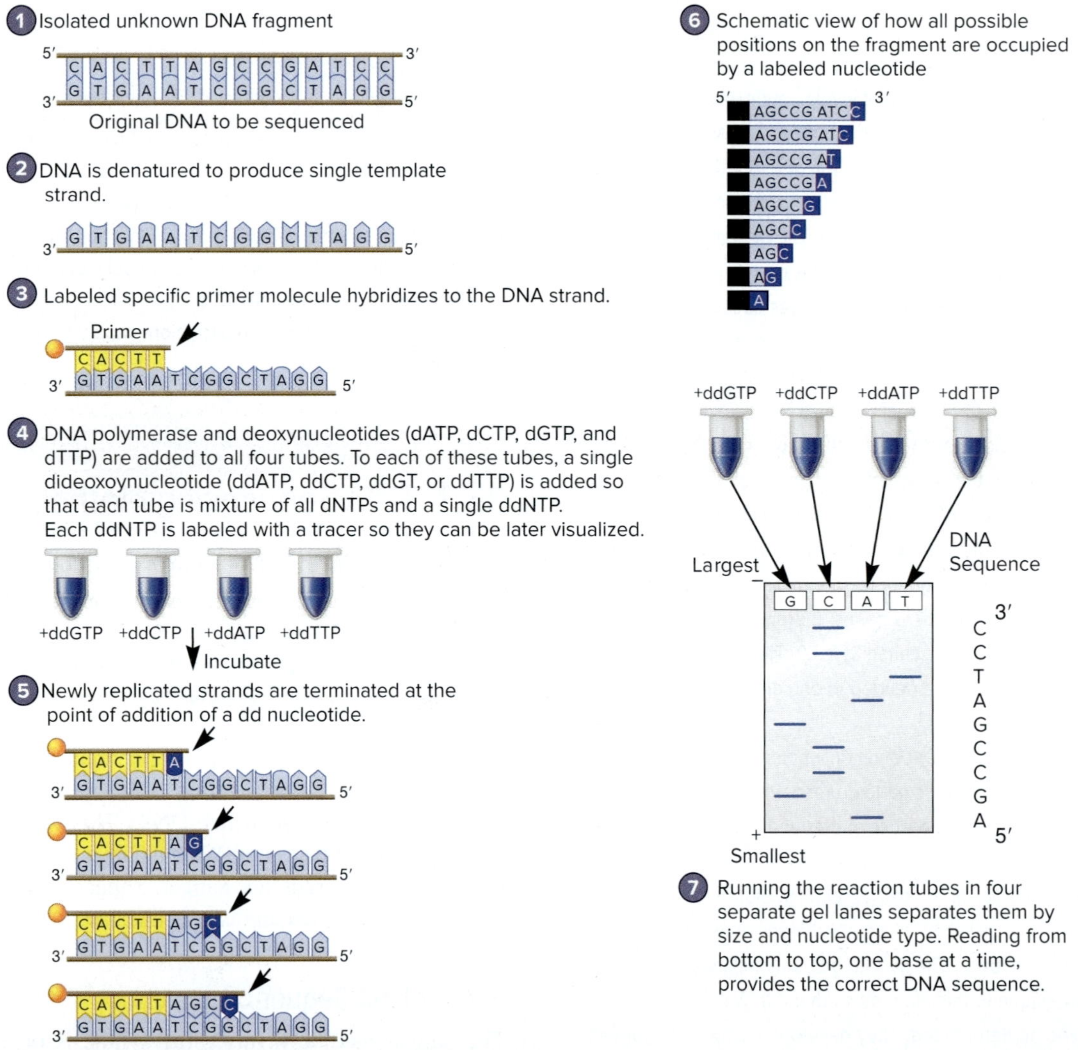

Figure 18.2 The Sanger Method of DNA Sequencing. Steps 1–6 are used for both manual and automated sequencing. Step (7) shows preparation of a gel for manual sequencing in which radiolabeled ddNTPs are used. Although manual sequencing is rarely used, it is instructive in understanding how DNA sequence is determined.

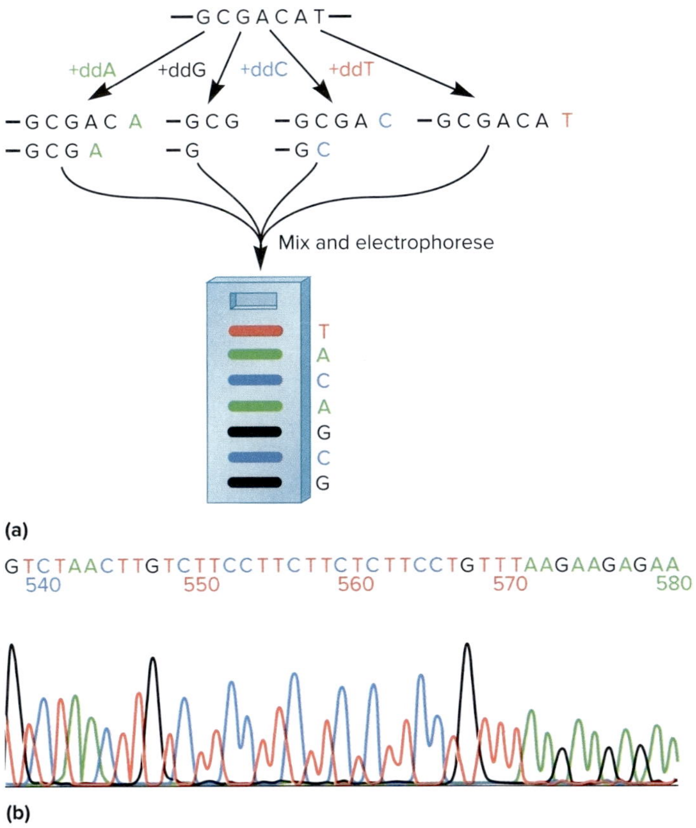

Figure 18.3 Automated Sanger DNA Sequencing. (a) Part of an automated DNA sequencing run. Here the ddNTPs are labeled with fluorescent dyes. (b) Data generated during an automated DNA sequencing run. Bases 538 to 580 are shown.

results in a collection of DNA fragments of varying lengths, each ending in the same ddNTP. For example, a reaction prepared with ddATP + dATP, dTTP, dGTP, and dCTP produces fragments ending with an A, those with ddTTP produce fragments with T termini, and so forth. After DNA synthesis is completed, the DNA is made single stranded, usually by heating. DNA is usually prepared for automated sequencing (**figure 18.3**). Here each ddNTP is labeled with a different colored fluorescent dye. The resulting fragments are then separated by electrophoresis. Recall that each fragment's migration rate is inversely proportional to the log of its molecular weight. Simply put, the smaller a fragment is, the faster it moves through the gel. Because synthesis always adds a nucleotide to the 3′-OH of the growing strand, the ddNTP at the end of the shortest fragment is assigned to the 5′ end of the DNA sequence, while the largest fragment is the 3′ end. In this way, the DNA sequence can be read directly from the gel from the smallest to the largest fragment. ◀ *Gel electrophoresis (Techniques & Applications 17.1)*

Sanger Sequencing

Automated sequencing uses a laser beam to detect DNA fragments as they exit the bottom of the gel from smallest to largest. This is possible because each ddNTP is labeled with a different fluorescent color. This is recorded on a graph called a chromatogram in which the amplitude of each spike represents the fluorescent intensity of each particular fragment (figure 18.3*b*). The corresponding DNA sequence is listed above the chromatogram. Typically, an automated Sanger chain termination sequencing system can accurately read 500 to 800 bp in a single electrophoresis run.

Next-Generation DNA Sequencing

Sanger's chain termination method was used to complete the first human genome sequence in 2001. This cost about $300 million and took about a decade to finish. Although this was an amazing feat at the time, it illustrates two limitations of Sanger sequencing: It is expensive and time consuming. Because scientists want to sequence genomes faster and more cheaply, innovative, newer DNA sequencing techniques have been invented.

This desire stimulated the development of **next-generation sequencing** (NGS) technologies. The setup is quite different from the Sanger approach; rather than long genomic strands in solution, NGS uses short, sheared pieces of DNA templates with oligonucleotides called adapters attached to each end (**figure 18.4**, step 1). The adapter at one end of each DNA fragment attaches it to a solid substrate. The adapter at the other end also attaches the DNA fragment to the substrate, but by annealing it to a primer used to initiate the polymerase chain reaction (PCR; figure 18.4, steps 2 and 3). PCR results in the production of many copies of the same fragment, which are sequenced simultaneously. Because thousands of identical DNA fragments are sequenced at the same time, these methods are sometimes called **massively parallel sequencing techniques.** In addition to making genomic sequencing faster and cheaper, as we discuss later in this chapter, NGS avoids the need to insert (i.e., clone) individual DNA fragments into vectors. This is important because it is almost impossible to clone every DNA fragment into any given genomic library. ◀ *Polymerase chain reaction amplifies targeted DNA (section 17.2); Genomic libraries: cloning genomes in pieces (section 17.3)*

Although several NGS techniques have been marketed, currently **reversible chain termination sequencing** is most frequently used. Here, the DNA is sequenced as each nucleotide is incorporated. This is called **sequencing by synthesis.** This takes place in a flow cell, which is a glass slide into which grooves have been cut. Flow cells allow reagents to be added, flushed, and new reagents added without the loss of the attached DNA. PCR amplification of DNA templates within the flow cell is called "bridge amplification" and it creates clusters of double-stranded fragments of DNA scattered over the surface of the slide (figure 18.4, step 4). As we will see, it is critical that fragments with identical nucleotide sequences are clustered together in the flow cell. In the next step, the double-stranded fragments are denatured, and the flow cell is flushed. This leaves bundles of single-stranded linear fragments ready for sequencing (figure 18.4, steps 5 and 6).

Similar to Sanger sequencing, sequencing by synthesis uses a modified fluorescent nucleotide that, when introduced into the growing strand, stops the reaction because the 3′-OH is blocked (**figure 18.5**). However, rather than using a dideoxynucleotide as in Sanger sequencing (figure 18.1), here a small chemical group is added that can be enzymatically removed. Another big difference is that the modified nucleotide does not fluoresce until it is incorporated into the growing DNA strand. Because sequencing proceeds in a flow cell, reagents needed for DNA synthesis can be introduced to catalyze the incorporation of a modified dNTP, which is then detected by the onset of fluorescence. These synthesis reagents are then flushed from the cell, and the enzyme cocktail that removes the fluorescent tag and blocking group is infused. The cycle can then repeat (figure 18.4, steps 7 through 9).

In summary, once the templates have been amplified in clusters, the incorporation of each nucleotide determined by laser optics involves the following events: (1) The DNA polymerase,

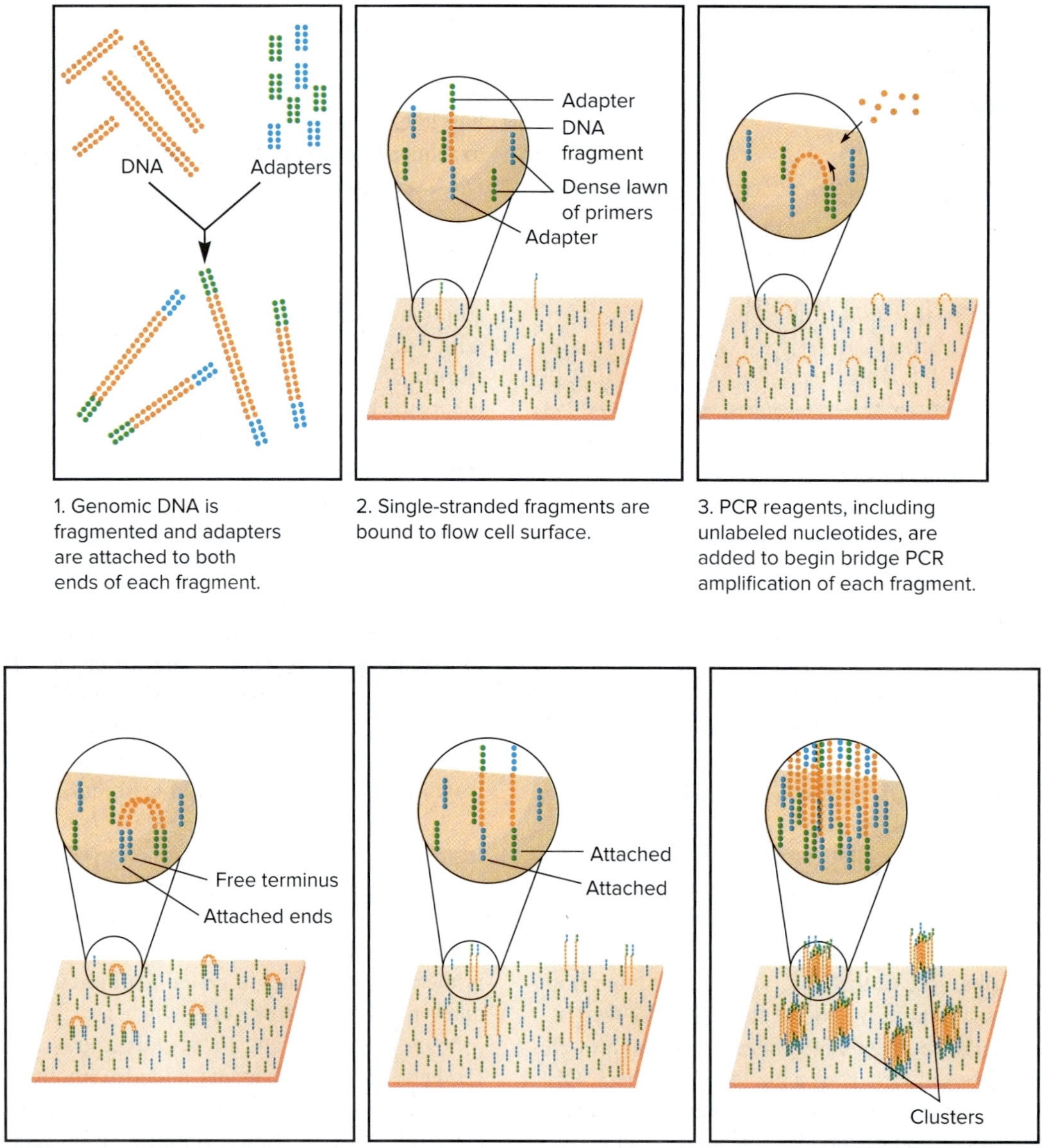

1. Genomic DNA is fragmented and adapters are attached to both ends of each fragment.

2. Single-stranded fragments are bound to flow cell surface.

3. PCR reagents, including unlabeled nucleotides, are added to begin bridge PCR amplification of each fragment.

4. Bridge PCR amplification generates double-stranded fragments bound to the flow cell surface at both ends.

5. Fragments are denatured to become single stranded.

6. Each fragment serves as PCR template to generate millions of clusters of identical fragments.

Figure 18.4 **Reversible Chain Termination Sequencing.**

MICRO INQUIRY *Why is it important that identical fragments of DNA to be sequenced are clustered together?*

(Continued)

Figure 18.4 *(Continued)*

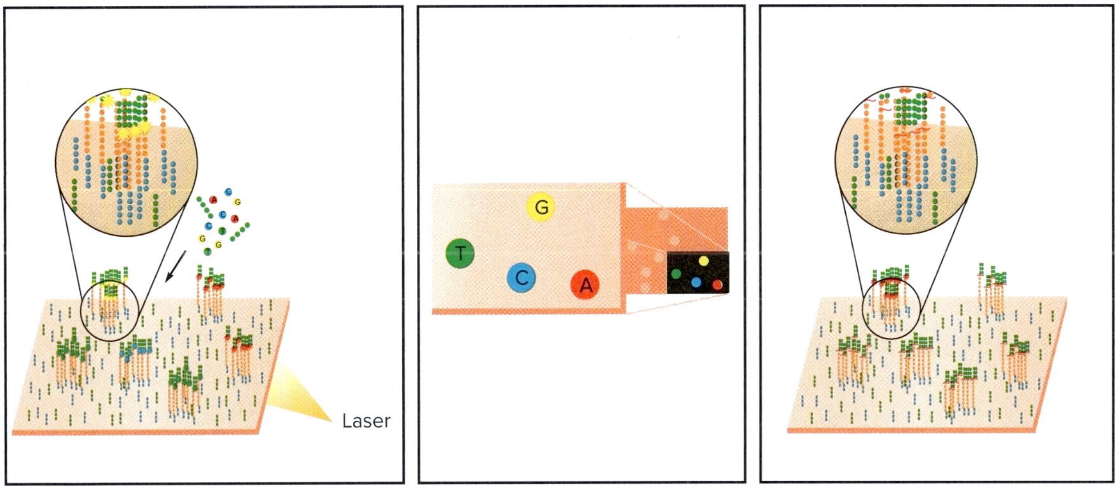

7. The first sequencing cycle begins when reagents, including primers and labeled nucleotides with reversible terminators, are added.

8. Laser excitation is followed by image capture of fluorescent signal from each cluster of fragments. This identifies the first base, which is recorded.

9. Enzymatic cleavage of the reversible terminator enables the second sequencing cycle to begin.

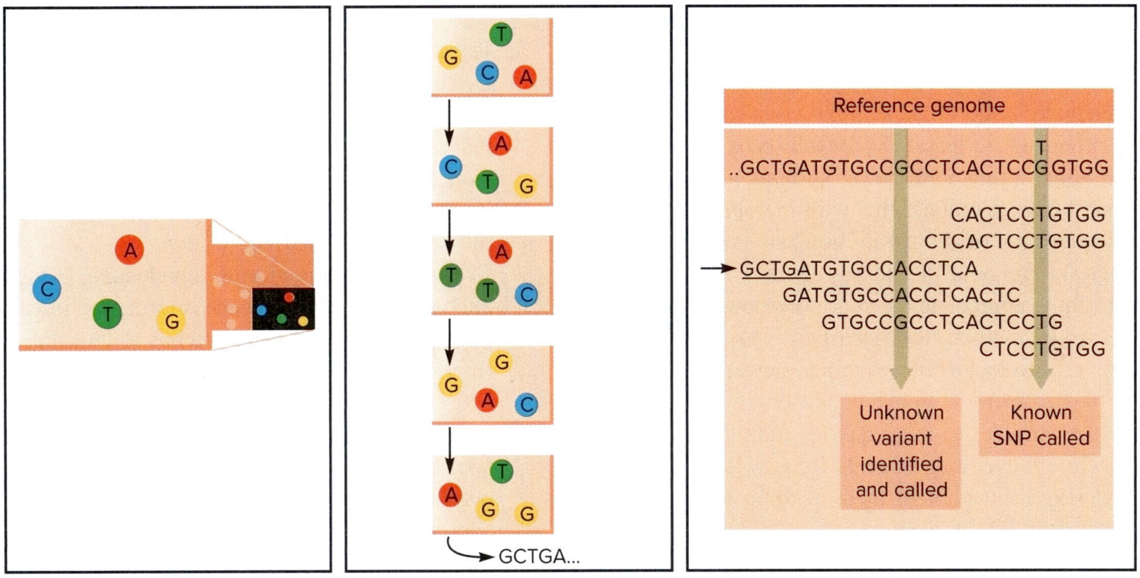

10. Laser excitation generates the signal that identifies the second base in each cluster of fragments.

11. Cleavage of terminator and sequencing cycles are repeated to determine base sequence of each cluster of fragments.

12. Sequences are aligned and compared with reference data base. SNP, single nucleotide polymorphism.

all four modified dNTPs, and other reagents needed for DNA synthesis, using the tethered fragments as template, are added to the flow cell. (2) The incorporation of a modified nucleotide as determined by the template strand occurs, causing its fluorescent tag to emit light while at the same time stalling synthesis. This pause allows the identity of the incorporated base to be determined by the color of the fluorescent tag. (3) Reagents needed for synthesis are then flushed from the flow cell, and (4) cleavage enzyme reagents are introduced. (5) The fluorescent label and the blocking molecule are removed, exposing the 3′-OH. Finally, (6) the enzyme cocktail is flushed, and synthesis reagents are added once again and the cycle repeats. Because the light emitted by any single nucleotide is too dim to accurately record, each cluster of identical fragments must grow synchronously so that a large enough signal is generated. The incorporation of the same nucleotide at the same time into each identical fragment in a cluster generates a signal with sufficient amplitude for detection (figure 18.4, steps 8 and 10). This explains why the length of each

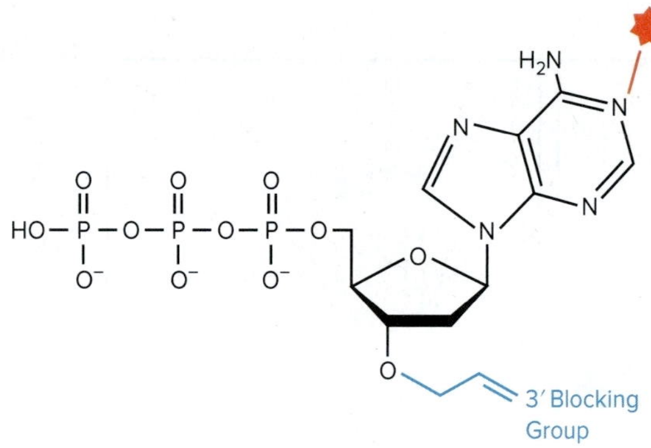

Figure 18.5 Modified Base Used in Reversible Chain Termination Sequencing. This example shows modified dATP, with a 3′-OH blocking group and a fluorescent tag attached to the adenine. The base only fluoresces when the PP_i is removed during incorporation into the growing strand. The 3′ blocking group is then enzymatically removed exposing the 3′-OH needed for continued DNA synthesis.

read (the number of nucleotides determined in a single run) is much shorter than it is for Sanger sequencing: After about 150 to 300 bases have been incorporated, the synchronicity of nucleotide incorporation deteriorates. That is, the synthesis of complementary strands in a cluster of identical templates becomes "out of sync." Without a strong synchronous signal, the data become ambiguous. Nonetheless, short read length is compensated by the volume of data produced: At least 1.5 GB (billion bases) are read per flow cell in about a day, with newer systems reading up to 120 GB in about 2 days. Compare this with the output of automated Sanger sequencing, of about 1 million bases per day. With the average bacterial genome measuring about 4 million base pairs, one can begin to appreciate the enormous power of NGS.

In addition to the method described here, various new approaches are being developed. For example, newer sequencing platforms seek to streamline sample preparation, improve signal detection, optimize surfaces to more rapidly attain high reagent concentrations while preventing reagents from sticking, and attain longer read lengths. Although there is room for improvement, next-generation DNA sequencing technologies have already revolutionized genomics by drastically reducing time and cost, and increasing accuracy.

18.2 Genome Sequencing

After reading this section, you should be able to:

a. List the steps used in whole-genome shotgun cloning
b. Compare and contrast genome assembly using Sanger and next-generation sequencing (NGS) approaches
c. Describe the multiple strand displacement method and how this technique is used

In 1995 J. Craig Venter, Hamilton Smith, and their collaborators were the first to sequence a bacterial genome. Prior to that, only the small genomes of viruses had been sequenced. To reduce bacterial genomes to manageable sizes, they developed **whole-genome shotgun sequencing** and the computer software needed to assemble sequence data into a complete genome. They used their new method to sequence the genomes of the bacteria *Haemophilus influenzae* and *Mycoplasma genitalium*. With this accomplishment, Venter and Smith ushered in the genomic era. Within 20 years the number of complete genomes published grew from two to thousands of sequenced genomes spanning all three domains of life.

Whole-Genome Shotgun Sequencing

Although revolutionary when it was introduced, sample preparation is now considered one of the biggest disadvantages of whole-genome shotgun sequencing. This is because it requires cloning all the DNA to be sequenced as genomic fragments inserted into cloning vectors (i.e., the construction of a genomic library; **figure 18.6**). The entire sequencing process can be broken into four stages starting with library construction, which is discussed in section 17.3. Here we briefly describe the subsequent stages: random sequencing, fragment alignment and gap closure, and editing.

1. **Random sequencing.** The vectors carrying the cloned DNA are purified and thousands of DNA fragments are sequenced using Sanger sequencing, employing primers that recognize the plasmid DNA sequences adjacent to the cloned, chromosomal insert. Usually all stretches of the genome are sequenced between 8 and 10 times to increase the accuracy of the final results.
2. **Fragment alignment and gap closure.** Using computer analysis, the DNA sequence information of each fragment is assembled into longer stretches of sequence. Two fragments are joined together to form a larger stretch of DNA if the sequences at their ends overlap and match. This comparison process results in a set of larger, contiguous nucleotide sequences called **contigs.** Sometimes an overlapping sequence is missing, generating gaps between contigs. There are several strategies to obtain the missing sequences. Ultimately, however, the contigs are aligned in the proper order to form the complete genome sequence. The term *scaffold* is used to describe sequence data with gaps that persist between contigs.
3. **Editing.** The sequence is then carefully proofread to resolve any ambiguities or frameshifts in the sequence. Proofreading is accomplished by ensuring that all reads of the same sequence are identical and the sequences of the two DNA strands are complementary.

Next-Generation Genomic Sequencing

Sanger sequencing ushered in the genomic era. However, the advent of NGS techniques has made genomic sequencing, particularly of microorganisms, much more practical in terms of

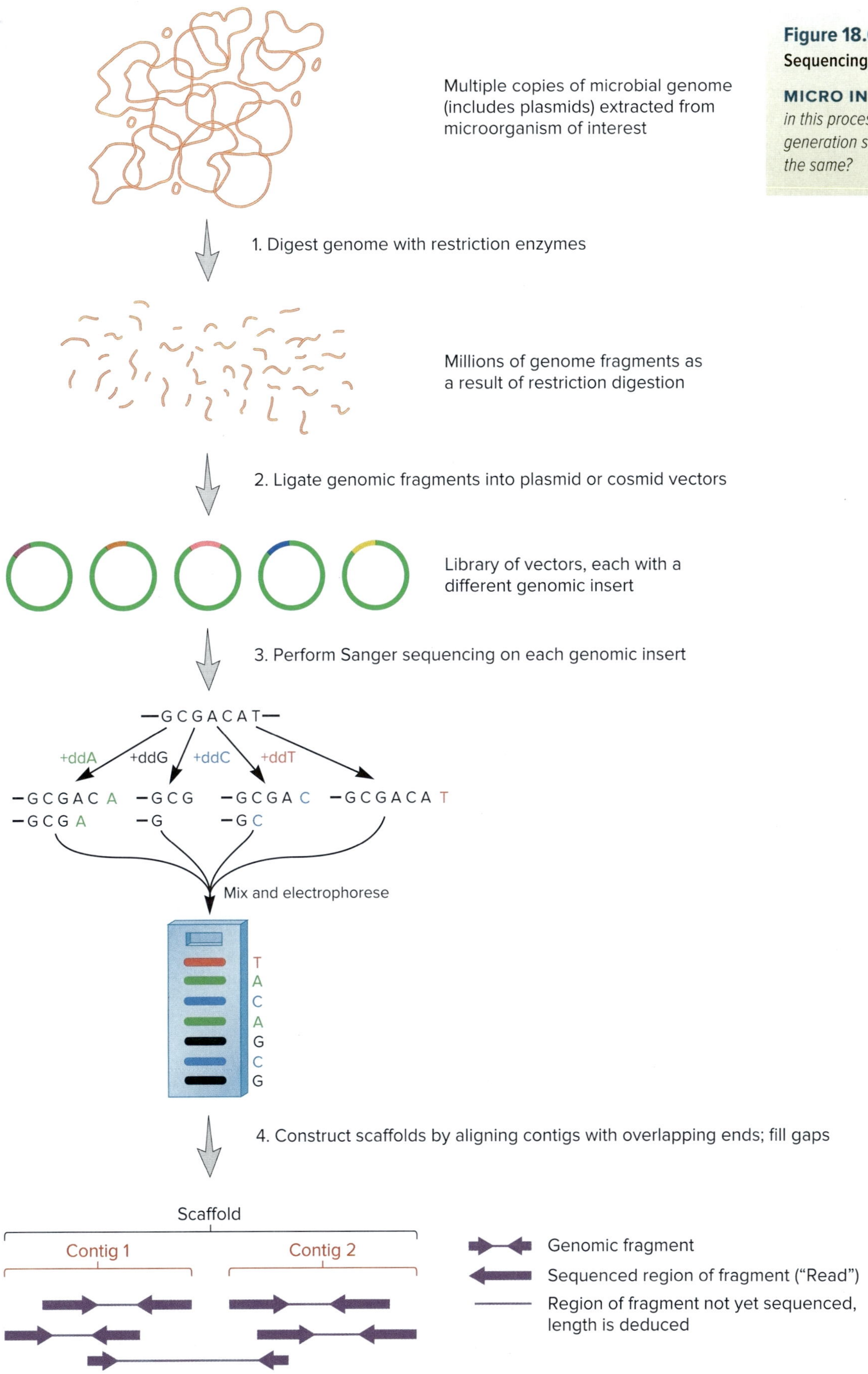

Figure 18.6 Whole-Genome Shotgun Sequencing.

MICRO INQUIRY *Which step (or steps) in this process is (are) not used in next-generation sequencing? Which are the same?*

time, money, and improved outcome. Like Sanger sequencing, preparing genomic DNA for sequencing by synthesis involves shearing the DNA into smaller pieces. The important difference is that, rather than inserting genomic fragments into cloning vectors as is the case with Sanger sequencing, adapters are added to the genomic DNA fragments, which are then attached to a solid substrate (figure 18.4). This is vastly more efficient; typically close to 100% of the genomic fragments with adapters bind to the solid substrate (e.g., flow cell surface), whereas cloning more than 80% of the genomic fragments into a genomic library requires a lot of skill and luck.

The advent of NGS has had a significant impact on the quality of the final genomic sequence, as assessed by two specific factors: depth of coverage and breadth of coverage. **Coverage** refers to the average number of times each nucleotide is sequenced in a genome (or other sequencing project). This redundancy of coverage is also referred to as the depth of sequencing, so "depth" and "coverage" can be used interchangeably. Depending on the technique used, a single nucleotide might be read (sequenced) 18 times or 80 times. The latter is described as deep sequencing; each nucleotide has been sequenced a very high average number of times. Breadth of coverage refers to how much of the entire genome was sequenced. A coverage of 100% means the genome sequence is complete, without gaps between contigs. Ideally one wants deep sequencing with 100% breadth of coverage. Because NGS does not involve the construction of a genomic library, it almost always yields higher breadth of coverage. The depth of coverage is also quite different. When Sanger genomic sequencing is used, any given region of the genome is typically sequenced no more than ten times. By contrast, NGS results in the same genomic fragment being read 30 to 100 times, greatly increasing accuracy. This is important because the DNA polymerase used in sequencing reactions lacks proofreading capability, so it cannot correct mismatched bases. Deep NGS overcomes the limitations of the DNA polymerase reaction by re-reading the same stretch of genome many times. A base at a given position that is not in agreement with the majority of the reads is recognized as a mistake or a rare variant worthy of further study.

One feature that is not terribly different between the two types of approaches is the assembly of a complete genome. Recall that NGS generates much shorter reads (about 150 bases rather than 500 to 800 bases for Sanger); nonetheless, the reads are aligned by overlapping ends (figure 18.6). A great deal of computing power is required to align and assemble the thousands of short, overlapping sequences.

Single-Cell Genomic Sequencing

It is now possible to amplify the few femtograms (10^{-15} gram) of DNA present in a single microbial cell to the several micrograms (10^{-6} gram) needed for sequencing. This is an important breakthrough because the vast majority of microbes cannot be grown axenically (i.e., in pure culture). The capacity to sequence the DNA from a single cell extracted from its natural environment is prompting new research strategies in microbial genetics, ecology, and infectious disease.

The process of **single-cell genomic sequencing** requires DNA amplification, but rather than PCR, a method called **multiple displacement amplification** (**MDA**) is used (**figure 18.7**). Unlike PCR DNA amplification, MDA occurs at a single temperature and uses the DNA polymerase from bacteriophage phi29 to synthesize new strands of DNA from the genome template. This polymerase is used because it does not readily dissociate from ("fall off") the template strand; the importance of this feature will be made clear shortly. In addition, phi29 polymerase rarely incorporates the wrong base; that is to say, it has higher fidelity than most thermostable DNA polymerases used in PCR. Another difference between PCR and MDA is the use of a collection of primers with random sequences, six bases in length (hexamers), to

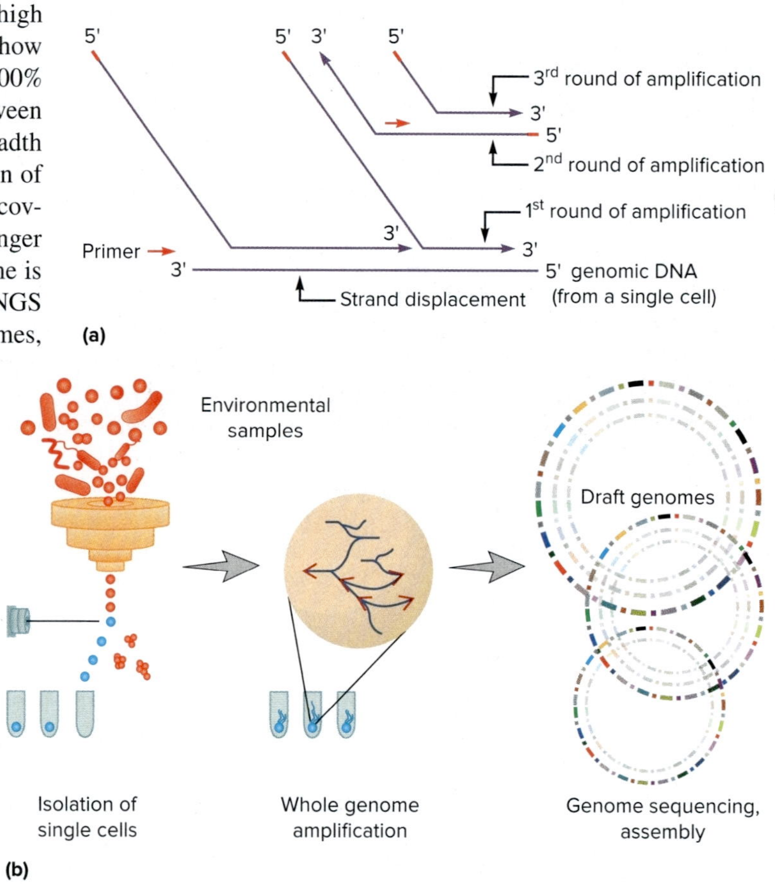

Figure 18.7 Single-Cell Genomic Sequencing. (a) Many copies of DNA extracted from a cell are generated by multiple displacement amplification. Random hexamer primers (red) bind to complementary template sequences, and DNA polymerase from bacteriophage phi29 is used to catalyze synthesis in the 5′ to 3′ direction (purple arrowheads). When the end of a newly synthesized strand meets double-stranded DNA, one strand is displaced by the growing DNA. (b) Single-cell genomic sequencing can be used to discover uncultivated bacteria and archaea from natural samples.

initiate synthesis. The primers hydrogen bond to complementary sequences scattered throughout the genome. As DNA synthesis proceeds from each primer, the growing 3′ end of one newly made strand will eventually bump into and then displace the 5′ end of another newly growing strand. Recall that the phi29 polymerase does not easily dissociate once bound to the DNA, so both new strands will continue to grow. In this way, many new strands are rapidly synthesized. The new strands have an average length of about 12,000 bases (12 kb) but can be as long as 100 kb. This makes them suitable for DNA sequencing. NGS is then performed to avoid the requirement of genomic library construction and to ensure higher breadth and depth of coverage.

Microbiologists have used single-cell genomics to sequence the genomes of hundreds of uncultivated microbes representing multiple undefined bacterial and archaeal taxa sampled from many different environments (figure 18.7*b*). These microbes constitute what the researchers call "microbial dark matter," because there was no prior knowledge of their existence. Although the average breadth of coverage is typically only about 40%, reflecting a major limitation of single-cell genomics, a variety of discoveries have been made, including at least 20,000 new hypothetical protein families, evidence for new superphyla, and the surprising presence of genes encoding sigma factors in archaea.

Comprehension Check

1. Why is the Sanger technique of DNA sequencing also called the chain-termination method?
2. Explain the difference between a dideoxynucleotide used in Sanger sequencing and the modified bases used in reversible chain termination sequencing.
3. Why does reversible chain termination sequencing yield short reads?
4. How would one recognize a gap in the genome sequence following nucleotide sequencing?
5. Suggest a medical and an ecological application of single-cell genomic sequencing.

18.3 Metagenomics Provides Access to Uncultured Microbes

After reading this section, you should be able to:

a. Differentiate between the construction and screening of a genomic library and a metagenomic library
b. List two applications of metagenomics in any field of microbiology

DNA sequencing is a powerful tool, but its impact would be diminished if it were applied only to the small minority of microbes currently held in cultures. Fortunately this is not the case. **Metagenomics**—the study of microbial genomes based on DNA extracted directly from the environment—has emerged as a key technology across a number of biological disciplines, including ecology, environmental microbiology, infectious disease, and immunology. Metagenomics samples the entire pool of nucleic acids found in any given ecosystem (e.g., soil, water, feces) and it is most frequently used to determine the members of the microbial community living there. Before the development of NGS, DNA extracted from the environment was typically used as a template to amplify small subunit rRNA genes or other target genes using PCR. This approach continues to be used, but many microbiologists now use shotgun metagenomics instead. The term *shotgun* is used because it involves sequencing all the DNA extracted from an environment, rather than using PCR primers to target a specific gene. In addition to taxonomic information, shotgun metagenomics also catalogs most of the genes present in the sample, providing important clues about microbial activity (**figure 18.8**). ▶| *Microbial taxonomy and phylogeny are largely based on molecular characterization (section 19.3)*

To illustrate the impact of coupling direct DNA extraction with NGS, consider a microbial census based only on the microorganisms that can be cultured from an environmental sample (e.g., soil). Based on the average recovery of microbes in culture, this approach will miss about 98% of the microbial species present. Such an approach would be like taking a census of a big city by randomly sampling a few individuals. Next best would be to base a microbial census on DNA extracted from soil and cloned into bacterial vectors that are then sequenced by Sanger chain termination, as this would certainly improve representation, but by an unknown amount. Keeping with our city census analogy, this would be like taking a census by sending out a questionnaire to each household, hoping everyone would respond, but understanding that only an unknown fraction of residents will. Metagenomic analysis using next-generation sequencing, on the other hand, is the equivalent of sending a team of census takers into the community to ensure that each person is counted: The team probably won't reach every single resident, but depth and breadth of coverage will be very high. Similarly, without the need to construct a genomic library and by sequencing deeply, metagenomic analysis coupled with NGS can detect organisms present in low numbers.

Once nucleotide sequences are obtained from DNA extracted directly from the environment, partial or full genomes can be detected in two ways. First, they can be assembled as previously discussed, by aligning overlapping sequences at the ends of reads (figure 18.6). However, this approach can be difficut with shotgun metagenomics, because a collection of reads 100 to 300 bp in length is generated; thus the number of gapped sequences tends to be large. Instead, one can attempt to align these short reads to previously sequenced genomes in an effort to identify similar sequences from known microorganisms. This too has the major drawback that many taxa obtained by metagenomics are uncultured and are therefore unrepresented in existing genomic databases; they too are microbial dark matter. This problem prompted the establishment of the Genomic Encyclopedia of Bacteria and Archaea Project. The goal of this program is to improve reference databases by sequencing the genomes of a wide diversity of

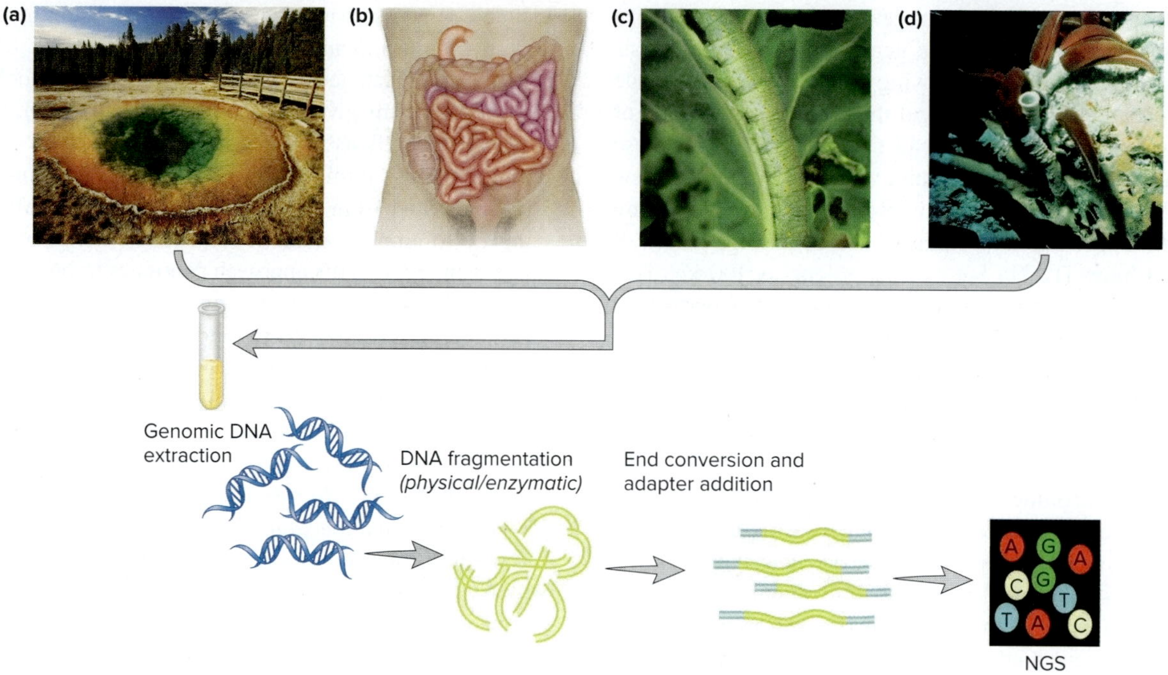

Figure 18.8 Metagenomic Analysis. DNA has been extracted directly from environments such as (a) bacterial mats at Yellowstone National Park, (b) the human colon, (c) cabbage white butterfly larvae, and (d) tube worms from hydrothermal vents. The DNA can be used to construct a metagenomic library and NGS is used to obtain nucleotide sequence information.

(a) ©Yi Xiang Yeng/iStock/Getty Images; (b) ©McGraw-Hill Education; (c) ©Nigel Cattlin/Science Source; (d) Source: OAR/National Undersea Research Program (NURP)/College of William & Mary/NOAA

cultured microorganisms. Metagenomic sampling of natural environments has so far led to the discovery of hundreds of new bacterial and archaeal phyla.

Comprehension Check

1. NGS techniques are considered well suited for metagenomics because a genomic library does not have to be constructed. Apart from convenience, explain why this is important for metagenomic analysis.
2. Examine figure 18.8. How might metagenomics be used to isolate genes encoding a potentially new peptide antibiotic?

18.4 Bioinformatics: What Does the Sequence Mean?

After reading this section, you should be able to:

a. Explain how a potential protein-coding gene is recognized within a genome sequence
b. Compare the meaning of the terms *orthologue* and *paralogue*
c. Differentiate between a conserved hypothetical protein and a putative protein of unknown function
d. Describe the construction of a physical genome map

As you might imagine, sequencing entire genomes generates a huge volume of information. The field of **bioinformatics** combines biology, mathematics, computer science, and statistics to convert raw nucleotide data into the location and potential function of genes or presumed genes on sequenced genomes using a complex process called **genome annotation.** Once genes have been identified, bioinformaticists can perform computer or **in silico analysis** to further examine the genome.

Obviously, obtaining nucleotide sequences without understanding the location and function of individual genes would be a pointless exercise. The goal of genome annotation is to identify every potential (putative) protein-coding gene as well as each rRNA- and tRNA-coding gene. A protein-coding gene is usually recognized as an **open reading frame** (**ORF**); to find all ORFs, both strands of DNA must be analyzed in all three reading frames (**figure 18.9**). A bacterial or archaeal ORF is generally defined as a sequence of at least 100 codons (300 base pairs) that is not interrupted by a stop codon and has terminator sequences at the 3′ end. The 5′ end of the gene should also bear a ribosome-binding site. Only if these elements are present is an ORF considered a putative protein-coding gene. This process is performed by gene prediction programs designed to find genes that encode proteins or functional RNA products. Ideally, computer-identified genes are then manually inspected by bioinformaticists to verify the computer-generated gene assignments. This process is called genome curation.

ORFs that appear to encode proteins are called ***coding sequences*** (**CDS**). Bioinformaticists have developed algorithms to compare the sequence of predicted CDS with those in large databases containing nucleotide and amino acid sequences of known proteins. The base-by-base comparison of two or more

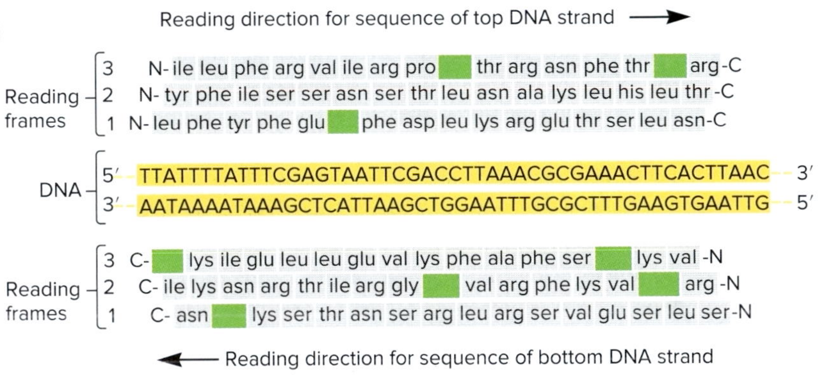

Figure 18.9 Finding Potential Protein-Coding Genes. Annotation of genomic sequence requires that both strands of DNA be translated from the 5′ to 3′ direction in each of three possible reading frames. Stop codons are shown in green.

gene sequences is called **alignment.** Alignments can also be performed by comparing amino acid sequences between two proteins. Scientists most often use **BLAST (*b*asic *l*ocal *a*lignment *s*earch *t*ool)** programs to perform this task. These programs compare the nucleotide (or amino acid) sequence of interest, called the query sequence, to all other sequences entered in the database. The results ("hits") are ranked in order of decreasing similarity. An E-value is assigned to each alignment; this value measures the possibility of obtaining the alignment by chance; thus highly homologous sequences have very low E-values.

Considerable information can often be inferred from translated amino acid sequences of potential genes. Often a short pattern of amino acids, called a motif or domain, will represent a functional unit within a protein, such as the active site of an enzyme. For instance, **figure 18.10** shows the C-terminal domain of the cell division protein MinD from a number of microbes. Because these amino acids are found in such a wide range of organisms, they are considered phylogenetically well conserved. In this case, the conserved region is predicted to form a coil needed for proper localization of the protein to the membrane. Finding this high level of conservation (similarity) allows the genome curator to confidently assign a function to the domain. ◀ *Cytokinesis (section 7.2)*

Genes from different organisms with such similar ORFs are called **orthologues.** Sometimes there appear to be duplicated genes on the same genome. This is discovered when two or more genes have very similar nucleotide sequences. Such genes are called **paralogues.**

As the number of sequenced genomes has expanded, so has the need to carefully define how new genes and proteins are named. The use of a structured vocabulary is called ontology, and a standard **gene ontology (GO)** has been adopted as the means by which proteins, or motifs within proteins, are commonly named. This is based on the similarities of amino acid sequences among orthologous proteins. A GO term not only reflects protein function but also defines the cellular process in which the protein participates (e.g., motility) and the cellular location of the protein (e.g., flagellum).

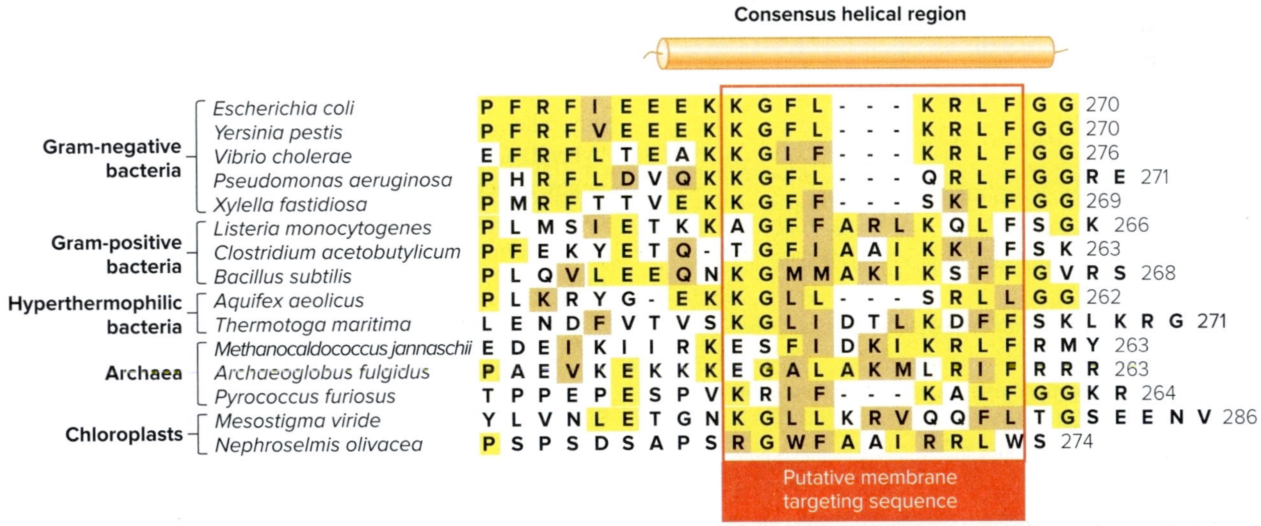

Figure 18.10 Analysis of Conserved Regions of Phylogenetically Well-Conserved Proteins. C-terminal amino acid residues of MinD from 15 organisms and chloroplasts are aligned to show strong similarities. Amino acid residues identical to *E. coli* are boxed in yellow, and conservative substitutions (e.g., one hydrophobic residue for another) are boxed in orange. Dashes indicate the absence of amino acids in those positions; such gaps may be included to maintain an alignment. The number of the last residue shown relative to the entire amino acid sequence is shown at the extreme right of each line.

MICRO INQUIRY *Which amino acids are most highly conserved?*

Proteins that do not align with known amino acid sequences fall into two classes: (1) **Conserved hypothetical proteins** are encoded by genes that have matches in the database but no function has yet been assigned to any of the sequences. (2) **Proteins of unknown function** are the products of genes unique to that organism. On the one hand, as more genomes are published, the likelihood of finding a match in another organism is increased. But on the other hand, as more metagenomic data become available, so too does the pool of putative proteins with unknown functions (i.e., conserved hypothetical proteins).

Once all the genes have been annotated, a **physical map** representing the entire genome may be drawn. A physical map is typically drawn as concentric circles depicting a bacterial or archaeal chromosome with each gene described by functional class (e.g., energy metabolism), which is color coded. Deviation from the mean % G + C is often indicated an the inner circle. Such deviations are common when DNA has been acquired by horizontal gene transfer. If the microbe has multiple chromosomes or plasmids, a genome map will show each chromosome and plasmid. Recall that the term *genome* includes all of the cell's DNA.

Comprehension Check

1. What is the goal of genome annotation? Why does it require knowledge of mathematics, statistics, biology, and computer science?
2. What three elements must be present in a sequence of DNA for it to be considered a potential open reading frame?
3. Why is conservation of nucleotide or amino acid sequence important in annotating a putative gene?

18.5 Functional Genomics Links Genes to Phenotype

After reading this section, you should be able to:

a. Explain how genome annotation can be used to graphically represent the metabolism, transport, motility, and other key features of a microbe
b. Contrast and compare microarray analysis with RNA-Seq in the study of transcriptomes
c. Explain how 2-D gel electrophoresis is able to separate proteins of identical molecular weight
d. Summarize the importance of mass spectrometry in analyzing protein structure
e. Explain why DNA-protein interactions are of interest and how they can be experimentally identified

Functional genomics seeks to place genomic information in a biological context. For instance, the careful annotation of a microbe's genome can be used to piece together metabolic pathways, transport systems, and potential regulatory and signal transduction mechanisms. A common outcome of a genome project is to infer the functional metabolic pathways, transport mechanisms, and other physiological features of the microbe (**figure 18.11**). One of the first microbes for which such analysis was performed was the causative agent of syphilis, *Treponema pallidum*. Because *T. pallidum* cannot be grown in pure culture, genomics is key to learning about its metabolism and the ways it avoids host defenses. The sequencing and annotation of the *T. pallidum* genome revealed that *T. pallidum* lacks genes encoding key enzymes in the TCA cycle and oxidative phosphorylation. *T. pallidum* also lacks many biosynthetic pathways and must rely on molecules supplied by its host. This inference is supported by the observation that about 5% of its genes code for transport proteins. Genomic analysis is a common approach to understanding microbial physiology. It is particularly rewarding when, like *T. pallidum*, the microbe cannot be cultured. ▶| *Phylum* Spirochaetes: *bacteria with a corkscrew morphology (section 21.6)*

Transcriptome Analysis

Once the identity and function of the genes that comprise a genome have been analyzed, the key question remains, "Which genes are expressed at any given time?" The entire collection of RNAs produced at any one time by an organism is its **transcriptome,** which is studied in the field of transcriptomics. Prior to the genomic era, researchers could identify only a limited number of genes whose expression was altered under specific circumstances. There are two technologies that allow scientists to look at the expression level of a vast collection of genes: DNA microarrays and RNA-Seq. The first to be developed, **DNA microarrays,** consists of solid supports, usually of glass or silicon, upon which DNA is attached in an organized grid fashion. Each spot of DNA, called a **probe,** represents a single gene. The probe may be a PCR product generated from genes of interest or complementary DNA (cDNA). The location and identity of each probe on the grid are carefully recorded.

The analysis of gene expression using microarray technology, like many other molecular genetic techniques, is based on hybridization between the single-stranded probe DNA (i.e., the genes attached to the microarray) and the target nucleotides from the microbe of interest. Rather that using mRNA, which is easily degraded, cDNA is made from all the mRNA molecules extracted from the microbe at a particular time of interest (for example before and after infecting a host organism). The target cDNAs are labeled with fluorescent dyes and incubated with the microarray under conditions that ensure proper binding of target cDNA to its complementary probe (gene). Unbound target is washed off and the microarray is then scanned with laser optics. Fluorescence at each spot or probe indicates that cDNA hybridized to its corresponding gene. Recall that each cDNA is derived from an mRNA molecule, so the color and intensity of each probe indicates the relative level of gene expression. However, microarray technology is difficult to standardize and large changes in gene expression cannot be detected. Furthermore, one can only detect changes in transcription level of those genes represented as probes. For these and other reasons, RNA-Seq is now the transcriptomic method of choice. *DNA Probe*

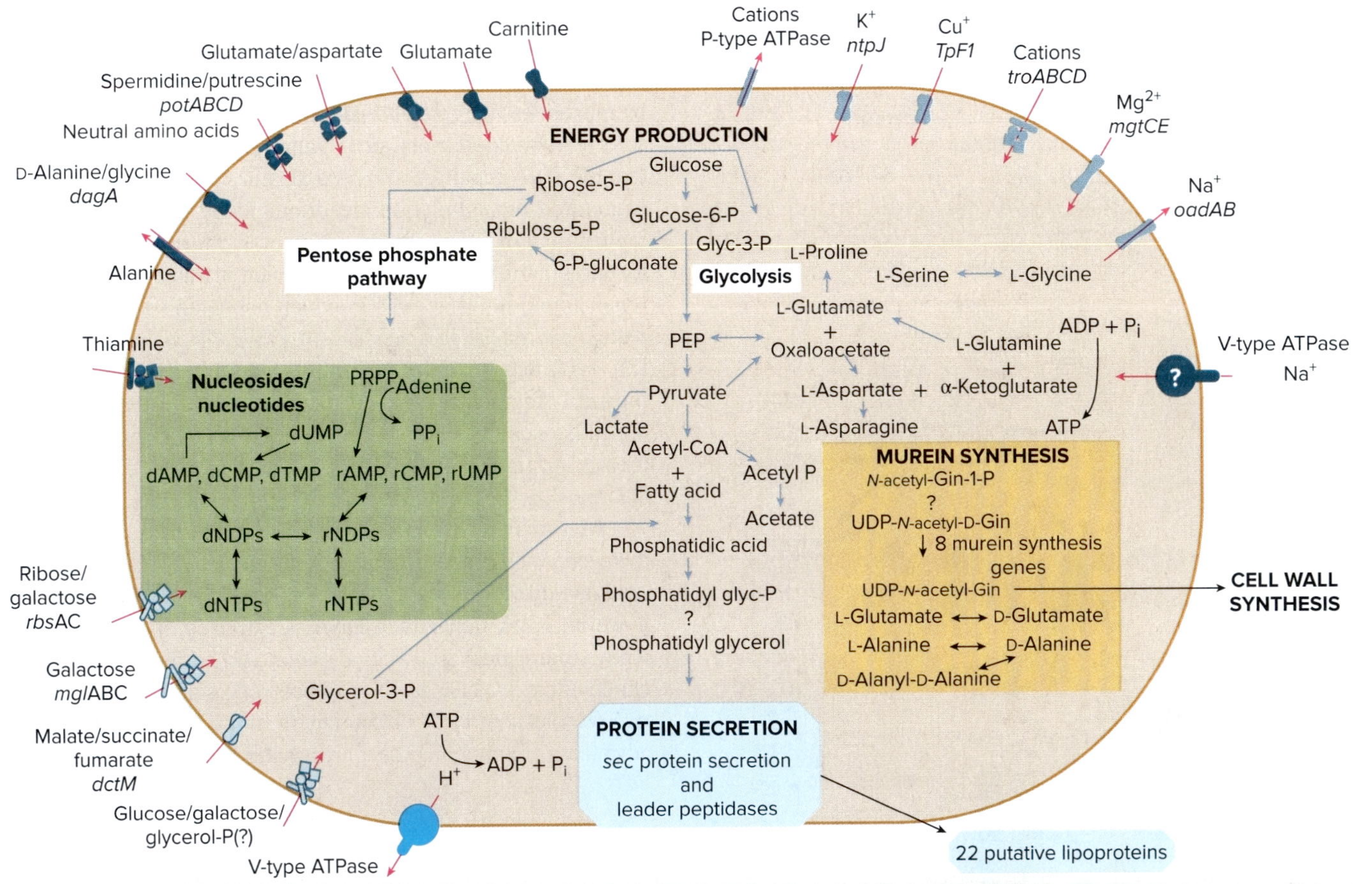

Figure 18.11 Metabolic Pathways and Transport Systems of *Treponema pallidum*. This depicts *T. pallidum* metabolism as deduced from genome annotation. Note the limited biosynthetic capabilities and extensive array of transporters. Although glycolysis is present, the TCA cycle and respiratory electron transport are lacking. Question marks indicate where uncertainties exist or expected activities have not been found.

MICRO INQUIRY *Based on this genomic reconstruction, can you determine if* T. pallidum *has a respiratory or fermentative metabolism?*

The advent of NGS techniques has made it possible to directly sequence total cellular mRNA, a process known as **RNA-Seq.** RNA-Seq has replaced microarray technology for many microbial transcriptomic applications. Quantification of mRNA by RNA-Seq is accomplished by measuring the number of reads (sequenced products) matching each gene.

The process of RNA-Seq begins with extracting all the RNA from the microbe of interest and converting all the cellular mRNA into cDNA by incubating the mRNA in the presence of the enzyme reverse transcriptase and dNTPs (**figure 18.12**; *also see figure 17.5*). The cDNA is then prepared for NGS by adding adapter sequences to each cDNA fragment. Each cDNA, representing an individual mRNA, is then sequenced using NGS. The resulting nucleotide sequences can be analyzed in two ways. Typically, the microorganism being studied has a sequenced genome, so the cDNA sequence is identified by alignment with the known genome sequence, called a reference genome. If no reference genome is available, the nucleotide sequences are converted to amino acid sequences and compared with databases that house millions of protein sequences.

RNA-Seq generates a wealth of information. Because the number of reads is not limited, there is no upper limit to the level of gene induction that can be detected. For instance, under specific growth conditions, a 9,000-fold increase in *Saccharomyces cerevisiae* (baker's yeast), gene expression was estimated for several genes when 16 million mapped reads (that is, sequences that could be aligned to the reference genome) were analyzed. Also, because RNA-Seq yields nucleotide sequence, information is revealed about the transcripts themselves. For instance, variations in transcribed sequences can be detected, as can the nucleotides where transcription starts and stops.

Transcriptomic experiments yield a vast amount of information that must be organized in some meaningful fashion. One common approach is to use **hierarchical cluster analysis,** which groups genes according to function or in terms of level of expression: Up-regulated genes (red spots) are grouped separately from

Figure 18.12 RNA-Seq.

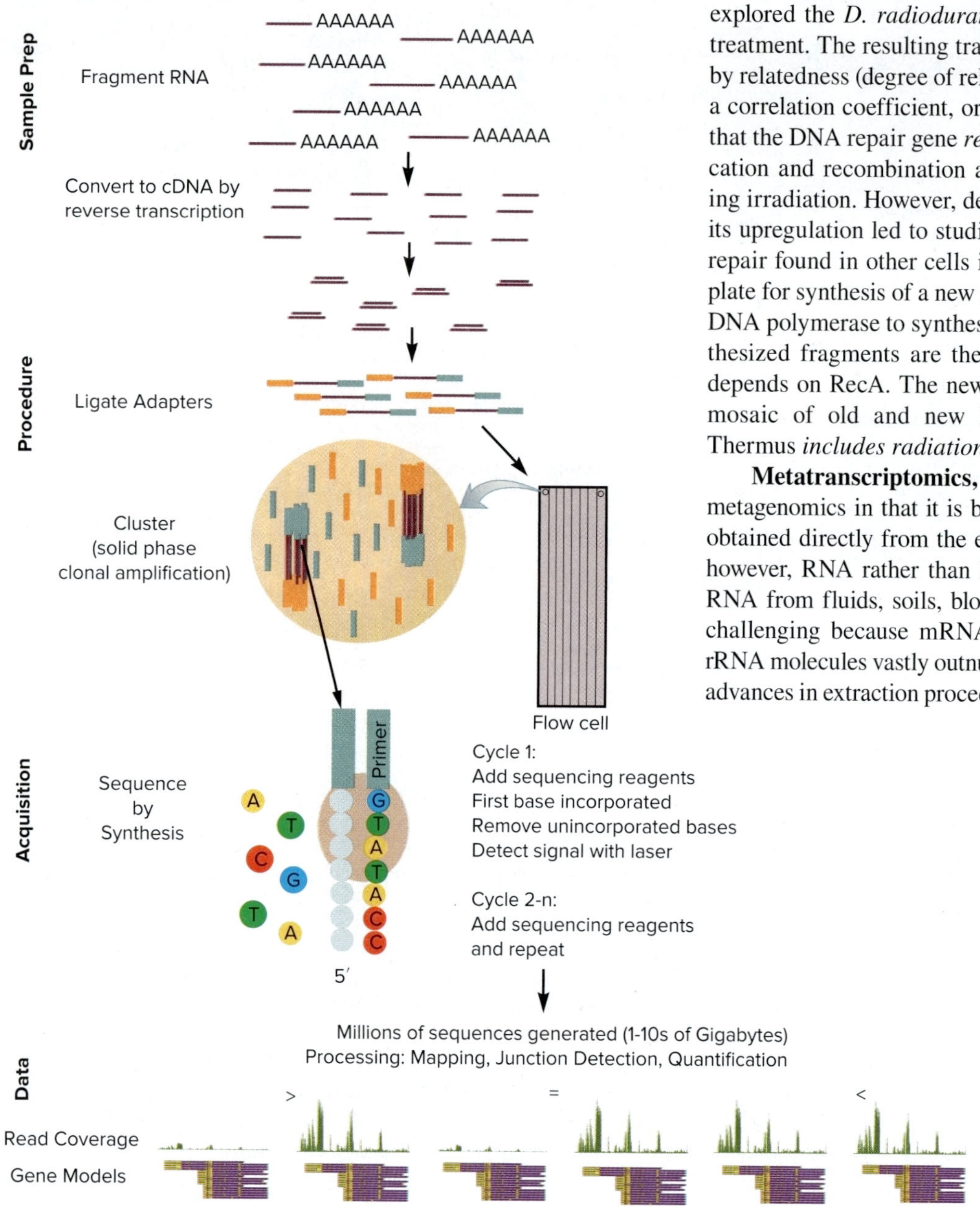

down-regulated genes (green spots); genes whose expression remains unaltered are shown in black. The hierarchical cluster analysis shown in **figure 18.13** was generated when *Deinococcus radiodurans* was exposed to γ-radiation. This bacterium has the unusual ability to withstand very high levels of ionizing radiation, which causes double-stranded breaks in DNA—the most lethal form of DNA damage. Remarkably, *D. radiodurans* reassembles its genome after it has been fragmented into thousands of pieces. It was thought that sequencing this bacterium's genome would reveal that *D. radiodurans* has many genes involved in DNA repair. But, surprisingly, *D. radiodurans* was discovered to have fewer DNA repair genes than *E. coli*.

To understand these unexpected findings, microbiologists explored the *D. radiodurans* transcriptome following radiation treatment. The resulting transcripts were identified and grouped by relatedness (degree of relatedness is statistically quantified by a correlation coefficient, or "r value"). Such analysis confirmed that the DNA repair gene *recA* and genes involved in DNA replication and recombination are dramatically upregulated following irradiation. However, detection of another gene identified by its upregulation led to studies in which it was found that unlike repair found in other cells in which one strand serves as a template for synthesis of a new strand, *Deinococcus* uses an unusual DNA polymerase to synthesize two new strands. The newly synthesized fragments are then pieced together in a process that depends on RecA. The newly repaired genome thus becomes a mosaic of old and new DNA fragments. ▶| Deinococcus-Thermus *includes radiation-resistant bacteria (section 21.2)*

Metatranscriptomics, as its name suggests, is similar to metagenomics in that it is based on the analysis of nucleic acids obtained directly from the environment. In metatranscriptomics, however, RNA rather than DNA is extracted. The extraction of RNA from fluids, soils, blood, stool, and so forth is technically challenging because mRNA is not very stable and tRNA and rRNA molecules vastly outnumber the more labile mRNA. Recent advances in extraction procedures enable the isolation of sufficient quantities of mRNA from diverse microorganisms to assess gene expression at the community level. Once obtained, the environmental mRNA is processed like that described for RNA-Seq from a single, cultured microorganism. However, because metatranscriptomics describes the transcriptome of an entire ecosystem, it will yield transcripts that map to multiple reference genomes as well as those that fail to align with any reference genomes. These novel transcripts represent newly discovered genes.

Proteomics Explores Total Cellular Protein

Genome function can be studied at the level of translation (i.e., proteins) as well as transcription (i.e., mRNA). The entire collection of proteins produced by an organism is called its **proteome.** Thus **proteomics** is the study of the proteome or the collection of proteins an organism produces at a given point in time. Proteomics provides information about genome function that mRNA studies cannot because a direct correlation between mRNA and the pool of cellular proteins does not always exist. Proteomics strives to determine the function of different cellular proteins, how they interact with one another and other molecules within the cell, and the ways in which they are regulated.

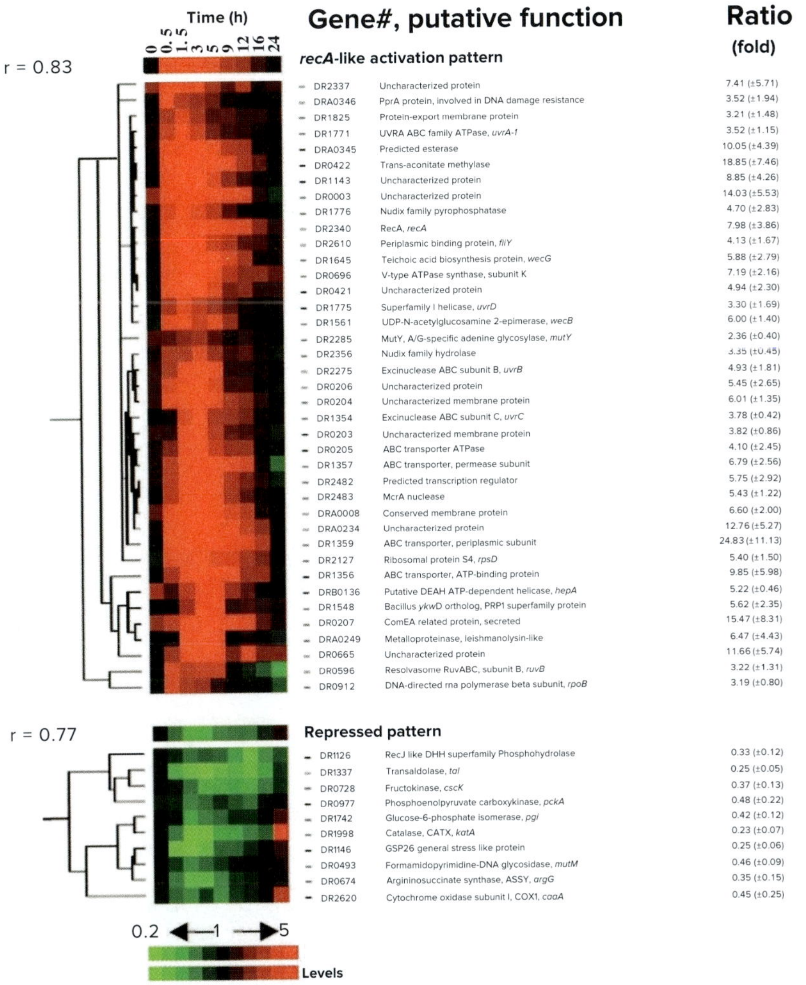

Figure 18.13 Hierarchical Cluster Analysis of Gene Expression of *D. radiodurans* Following Exposure to γ-Radiation. Each row of colored strips represents a single gene, and the color indicates the level of expression over nine time intervals. The far-left column is the control and thus is black (at control levels of expression). The level of induction or repression relative to the control value is indicated as the ratio (fold). Each group of genes has been scored for relatedness, and a "tree" has been generated on the far left of the clusters, with the indicated correlation coefficient (r value). Many genes encoding DNA repair, synthesis, and recombination proteins are induced upon radiation. These are grouped separately from repressed genes involved in other aspects of metabolism.

Liu Zhou, et al. "Transcritomedyamics," PNAS, April 2003,Vol. 100: 4191–4196. Copyright (2003) National Academy of Sciences, U.S.A.

MICRO INQUIRY *Based on your interpretation of the data, identify two genes that are strongly induced and two genes that are strongly repressed.*

Although new techniques in proteomics are continuously being developed, we focus briefly only on the most common approaches, including **two-dimensional gel electrophoresis.** In this procedure, a mixture of proteins is separated using two different electrophoretic procedures. This permits the visualization of thousands of proteins that would not otherwise be separated based on molecular weight, because many proteins have roughly the same mass. As shown in **figure 18.14*a***, the first gel involves **isoelectric focusing,** in which proteins move through a pH gradient (e.g., pH 3 to 10). In isoelectric focusing, a protein mixture is applied to an acrylamide gel in a tube with a pH gradient and electrophoresed. Each protein moves along the pH gradient until the protein's net charge is zero and the protein stops moving. The pH at this point is equal to the protein's **isoelectric point.** Thus the first gel separates proteins based on the content of ionizable amino acids. The second gel is SDS *p*oly*a*crylamide *g*el *e*lectrophoresis (SDS-PAGE). SDS (sodium dodecyl sulfate) is an anionic detergent that denatures proteins and coats them so that now all the proteins have a negative charge. After the isoelectric gel has been completed, the tube gel is soaked in SDS buffer and placed at the edge of an SDS-PAGE gel. A voltage is then applied. Because all the proteins are negatively charged thanks to SDS treatment, they migrate from the negative pole of the gel to the positive. In this way, polypeptides already separated along the horizontal length of the tube gel by charge are now separated by their molecular weight in the vertical direction; that is, the smallest polypeptide will travel fastest and farthest. Two-dimensional gel electrophoresis can resolve thousands of proteins; each protein is visualized as a spot of varying intensity, depending on its cellular abundance (figure 18.14*b*). Radiolabeled proteins are typically used, enabling greater sensitivity.

Two-dimensional gel electrophoresis is even more powerful when coupled with **mass spectrometry** (**MS**). To perform MS, a protein spot is cut from the gel and the extracted protein is cleaved into fragments by treatment with proteolytic enzymes. Then the fragments are analyzed by a mass spectrometer and the mass of the fragments is plotted. This mass fingerprint can be used to estimate the probable amino acid composition of each fragment and tentatively identify the protein. Sometimes proteins or collections of fragments are run through two mass spectrometers in sequence, a process known as **tandem MS** (**figure 18.15**). The first spectrometer separates proteins and fragments, which are further

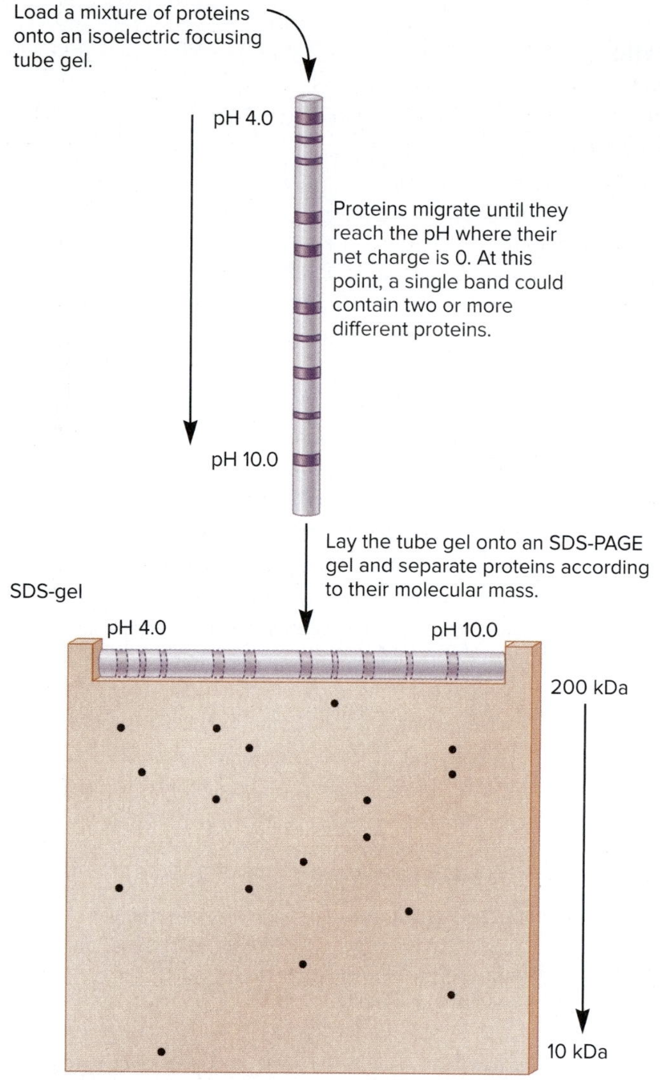

(a) The technique of two-dimensional gel electrophoresis

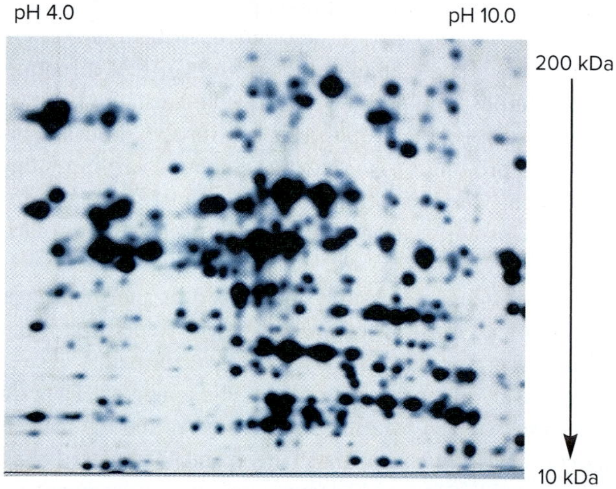

(b) An autoradiograph of a two-dimensional gel. Each protein is a discrete spot.

Figure 18.14 Two-Dimensional Gel Electrophoresis.

(b) ©Tyne/Simon Fraser/Science Source

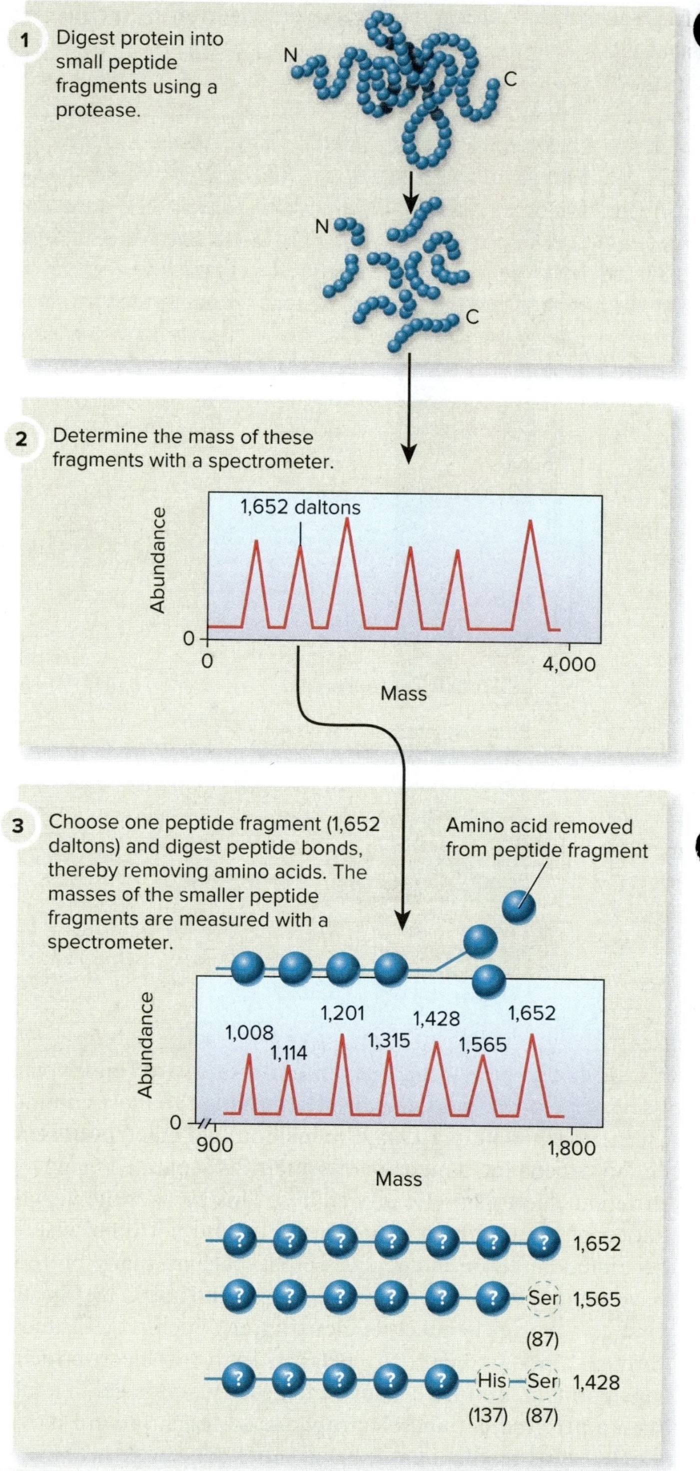

Figure 18.15 The Use of Tandem Mass Spectrometry to Determine the Amino Acid Sequence of a Peptide.

MICRO INQUIRY *How would determining the amino acid sequence of a 30 amino acid residue peptide differ from determining the amino acid sequence of a 200,000 dalton protein?*

fragmented. The second spectrometer then determines the amino acid sequence (based on molecular weight) of each smaller fragment. The sequence of a whole protein often can be determined by analysis of fragment sequence data. Alternatively, if the genome of the organism has been sequenced, only a partial amino acid sequence is needed. The partial amino acid sequence is aligned to the predicted translated sequences of all the annotated protein-coding genes on the organism's genome. In this way, both the protein and the gene that encodes it can be identified.

In **structural proteomics** the focus is on determining the three-dimensional structures of many proteins and using these to predict the structures of other proteins. The assumption is that proteins fold into a limited number of shapes and can be grouped into families of similar structures. When a number of protein structures are determined for a given family, the patterns of protein structure organization or protein-folding rules will be known. Computational biologists use this information to predict the most likely shape of a newly discovered protein, a process known as **protein modeling.**

Not surprisingly, a variety of other "-omics" are studied. For instance, **lipidomics** investigates a cell's lipid profile at a particular time, and **glycomics** is the systematic study of the cellular pool of carbohydrates. The goal of **metabolomics** is to identify all the small-molecule metabolites present in the cell at a given point in time. Because many metabolites are used in multiple pathways, the overview presented by a complete metabolome enables a researcher to discern which pathways are functional under a given set of conditions. Thus metabolomics provides a high-resolution means to evaluate physiological status. These "-omics" rely heavily on the use of chromatography and as such are beyond the scope of this chapter. Nonetheless, the development of these fields illustrates the dynamic and ever-changing nature of microbiology and the need to take a holistic view of cells and their communities if one seeks to understand the structure, function, and behavior of microbes.

Probing DNA-Protein Interactions

As described in chapters 13 and 15, many proteins directly interact with DNA; for example, regulatory proteins and proteins involved in replication and transcription function by binding to DNA. Until recently, the most common way to study protein-DNA interactions was through a procedure known as electrophoretic mobility shift assays (EMSA). This technique is also called a gel shift assay because when a DNA-binding protein is added to a mixture of purified target DNA, the mobility of the DNA is slowed in an agarose gel. When compared to the same DNA without protein, the DNA-protein complex is shifted to a higher molecular weight. The magnitude of the shift is usually related to the protein:DNA ratio such that when sufficient protein is added, the largest shift is seen. To perform EMSA, the target protein must be purified and the DNA to which it binds must be amplified. ◀ *Gel electrophoresis (Techniques & Applications 17.1)*

What if an investigator suspects a DNA-binding protein binds to more than one target DNA? He or she would turn to **chromatin immunoprecipitation** (**ChIP**) coupled with next-generation sequencing. In contrast to EMSA, which is entirely in vitro and uses one protein and one amplified fragment of DNA, ChIP surveys the living cell. First, cells are treated with a cross-linking agent such as formaldehyde, which "fixes" any DNA-binding proteins to their genomic targets (figure 18.16). The cells are then broken open and the DNA (with bound proteins) is cut into small pieces. Antibodies, which are molecules that bind to specified proteins, are used to tag the protein of interest bound to the DNA, and the antibody-protein-DNA complexes are precipitated. Next, the

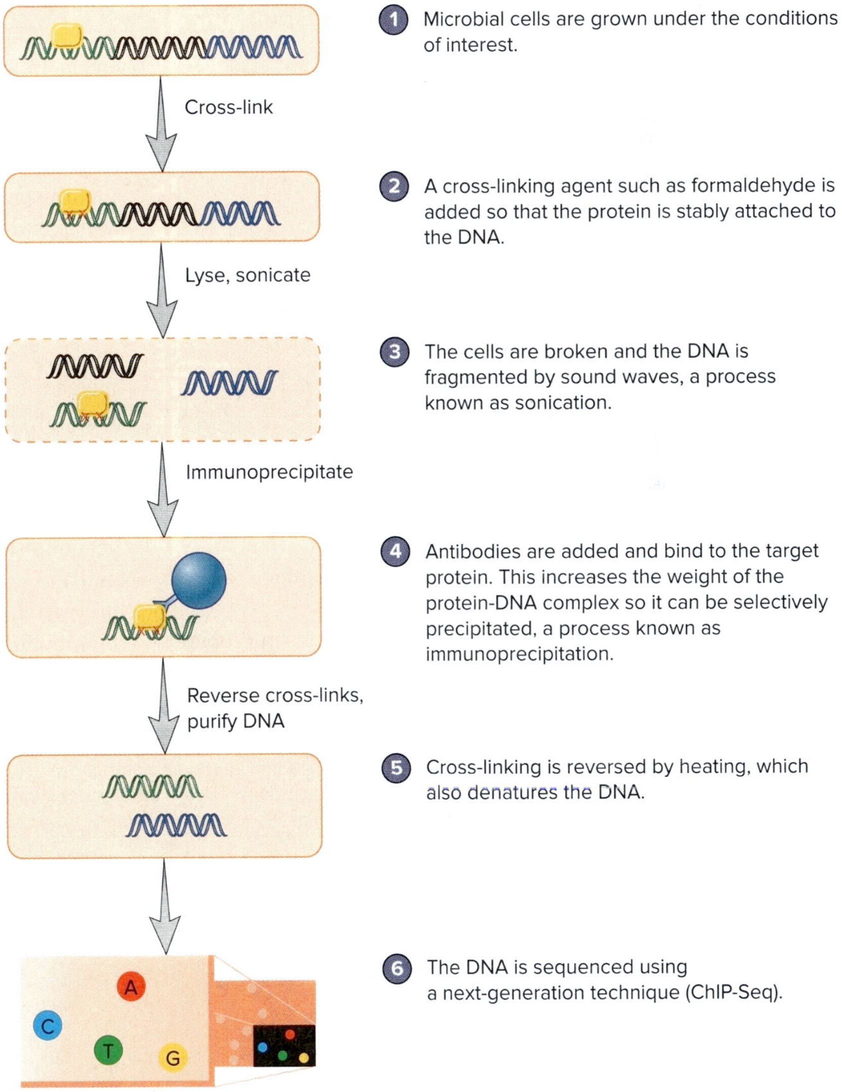

Figure 18.16 ChIP-Seq Analysis. This technology can be used to identify the DNA to which a protein binds.

proteins are removed and the DNA to which the proteins were bound can be identified by shearing the DNA so that a next-generation sequencing technique can be applied. This is called ChIP-Seq. ChIP-Seq enables the identification of DNA-protein interactions under different conditions. The same microbial strain can be treated with cross-linking agents when grown in different situations or at different phases of growth, so the DNA-binding profile of a protein can be followed. ▶| *Antibodies bind specific 3-D antigens (section 33.7)*

Comprehension Check

1. What is the role of NGS in RNA-Seq?
2. How might the following scientists use transcriptomics in their research?
   (a) A microbial ecologist who is interested in how the soil microbe *Rhodopseudomonas palustris* degrades the toxic compound 3-chlorobenzene
   (b) A medical microbiologist who wants to learn how the pathogen *Salmonella* survives within a host cell
3. Compare and contrast metagenomics with metatranscriptomics.
4. Why does two-dimensional gel electrophoresis allow the visualization of many more cellular proteins than seen in single-dimension electrophoresis (i.e., SDS-PAGE)?
5. What is the difference between mass spectrometry and tandem mass spectrometry?
6. Describe a ChIP-Seq experiment that will enable the identification of promoters to which the stationary phase sigma factor RpoS binds in *E. coli.* How could you also determine when RpoS binds to DNA?

18.6 Systems Biology: Making and Testing Complex Predictions

After reading this section, you should be able to:

a. Explain the role of genomics in shaping systems biology
b. Compare and contrast systems biology and synthetic biology

Imagine the following experiment: You have created a microbial strain in which you have deleted, or "knocked out," a single gene; it doesn't really matter what the gene encodes. You want to compare the transcriptome and proteome of the knockout mutant with its parent strain, which is identical to it in every way except the presence of this particular gene. To compare the transcriptomes of wild-type and mutant strains, you perform RNA-Seq, and to compare their proteomes, you run two-dimensional gels followed by mass spectroscopy. You find that the loss of just one gene has huge ramifications with regard to gene expression and protein synthesis. Not only is the expected pathway altered, but the expression of over a hundred other genes and proteins has also been affected. A single mutation has altered the function of the system, which in this case is a microbial cell.

This scenario has been repeated many times for numerous microorganisms. Life scientists now realize that most of the genes and proteins that have been widely investigated have a vastly broader role in cellular physiology than previously thought. Gone are the days of reductionist biology, when a single metabolic or regulatory pathway could be considered in isolation. **Systems biology** seeks to integrate the "parts list" of cells—mRNA, proteins, small molecules—with the molecular interactions that become pathways for catabolism, anabolism, regulation, behavior, responses to environmental signals, etc. To accomplish this, systems biology is interdisciplinary, requiring the expertise of physiologists, biochemists, bioinformaticists, mathematicians, geneticists, ecologists, and others.

One area where systems biology has been applied to microbiology is the study of metabolism (**figure 18.17**). Here genome sequencing and annotation lead to one set of predictions that can be tested by transcriptomics, proteomics, and metabolomics. The interactions between the individual products—RNA, protein, metabolites—can then be linked in a predictive network and used to construct testable hypotheses. Like most scientific endeavors, systems biology is an iterative process involving multiple rounds of data collection, analysis, and hypothesis building. What is unique, however, is the volume of data, the depth of sampling, and the use of sophisticated predictive models to guide the development of new hypotheses.

As systems biology matured, scientists realized that understanding cellular regulatory systems could lead to a new level of metabolic engineering such that a microbe could be constructed with novel genetic networks. The introduction of artificial regulatory and metabolic pathways into a microorganism so that it produces specific products is called **synthetic biology.** Examples of synthetic biology include biofuel production in *E. coli,* accomplished by rerouting amino acid biosynthetic pathways, and the commercial production of the important antimalaria drug artemisinin from *Saccharomyces cerevisiae.* ▶| *Synthetic biology (section 42.4)*

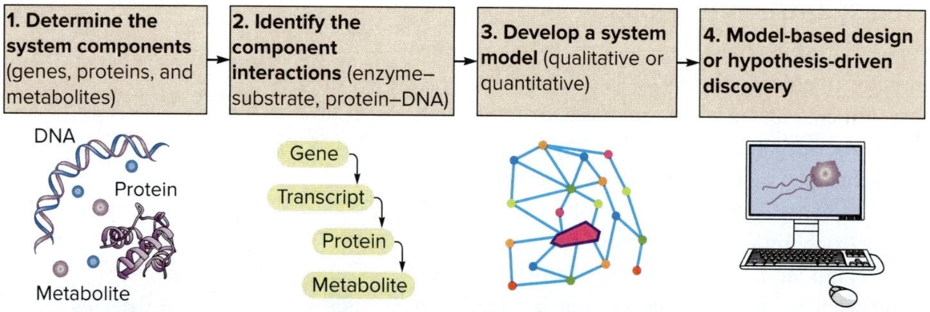

Figure 18.17 Systems Biology. Microbial metabolism is an active area of systems biology research. It can involve (1) identifying the system components, (2) recognizing their interactions, and (3) developing a model that predicts how they interact, which then (4) informs the synthesis of hypotheses that can be tested either by further modeling or by experimentation.

18.7 Comparative Genomics

After reading this section, you should be able to:

a. Predict the relative genome size of an organism given its domain and its morphologic and metabolic complexity

b. Describe the genomic differences that distinguish intracellular parasites from free-living microbes

A natural consequence of the tremendous number of sequenced genomes is the ability to make meaningful genome comparisons. **Comparative genomics** focuses on similar nucleotide and amino acid sequences among organisms to infer gene function and evolutionary relationships. Perhaps the most fundamental comparison is the relative sizes of microbial genomes. Several generalizations can be deduced from such analysis. Inspection of **figure 18.18** reveals that the smallest genomes published to date, across all three domains, belong to parasitic microbes. These organisms depend on metabolites from their hosts, with whom they have co-evolved. Parasitic microbes have therefore lost the genes encoding enzymes needed for the assembly of these molecules. For example, one of the smallest genomes belongs to *Candidatus* Carsonella *ruddii*, which lives within the sap-sucking insect *Pachypsylla venusta*. This bacterium, which cannot be grown without its host, has a genome of only about 160,000 (160 Kb) base pairs and 182 genes. This is roughly equivalent in size to a large virus (e.g., cytomegalovirus with about 160 proteins encoded on 230 Kb genome).

Generalizations about genome size can also be made across the three domains of life. Bacterial and archaeal genomes tend to be smaller than eukaryotic genomes, in part because they have more genes per Kb; that is, they have a high gene density. This is because bacteria and archaea lack introns and much of the other noncoding sequences found in eukaryotes. Furthermore,

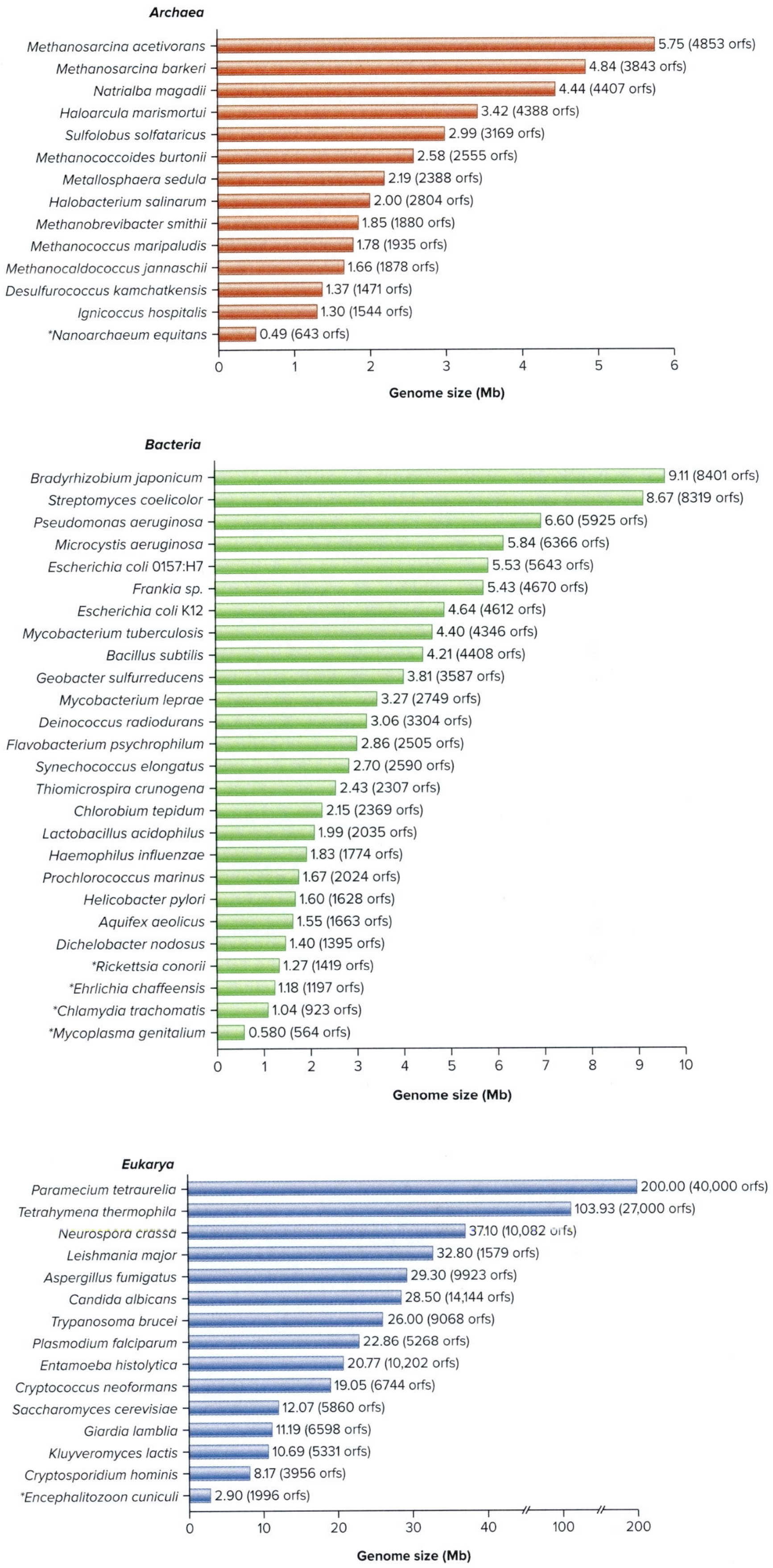

Figure 18.18 Microbial Genome Sizes. The distribution of genome sizes in millions of base pairs (Mb) among some archaea, bacteria, and eukaryotic microbes. Asterisk denotes that the microbe is parasitic.

MICRO INQUIRY *Two* E. coli *strains are shown in this figure. One is a pathogen and one is not. Which do you think is pathogenic and why?*

archaea tend to have smaller genomes than many bacteria (figure 18.18). The largest known archaeal genome, that of the methanogen *Methanosarcina acetivorans,* falls in the middle of the genome size range for bacteria. The genomes of free-living bacteria range from small (about a million base pairs) to over 9 million base pairs (Mb); genome size generally reflects metabolic and morphological complexity. Although the majority of eukaryotic microbe genomes are bigger than bacterial and archaeal genomes, figure 18.18 demonstrates that eukaryotic microbes have the widest range of genome sizes. The genome of the intracellular pathogenic microsporidian fungus *Encephalitozoon cuniculi* is smaller than many bacterial and archaeal genomes, yet the ciliated protist *Paramecium tetraurelia* has a tremendously large genome, with about twice as many genes as humans.

Comparing genome sequences between genera, species, and strains has revealed a high level of **horizontal gene transfer** (**HGT**). As detailed in chapter 16, HGT is the movement of genetic material between organisms that need not be of similar evolutionary lineages. Surprisingly, many genes involved in a myriad of cellular functions reside on mobile genetic elements, including transposons, plasmids, and phages. Genome analysis has revealed that HGT is frequently mediated by viruses and that lysogeny may be the rule, rather than the exception. In fact, some bacteria and archaea carry multiple proviruses. Some temperate phages carry lysogenic conversion genes, so called because they change the phenotype of their hosts. For instance, the virulence genes for some important pathogens, including the bacteria that cause diphtheria and cholera, are encoded by phages. *Lytic and lysogenic infections are common for bacterial and archaeal cells (section 6.4)*

Mobile genetic elements permanently integrated into a microbial genome are called **genomic islands.** If these new genes encode proteins that confer virulence, they are known as **pathogenicity islands.** The term *island* reflects the fact that these imported DNA sequences are different than the rest of the organism's genome. For example, genomic and pathogenicity islands often have a G + C content (measured as percentage) unlike the remainder of the genome. Genomic islands also sometimes retain inverted repeat sequences or other elements that mark their transposon or viral origin. It has become clear that HGT is a major evolutionary force in short-term microbial evolution and long-term speciation. *Microbes use mechanisms other than mutation to create genetic variability (section 16.4)* *Evolutionary processes and the concept of a microbial species inspire debate (section 19.5)*

It follows that evolutionary relationships between microbes can be explored by comparative genomics. One approach is to examine the organization of orthologous genes (i.e., homologous genes found in different organisms) in the genomes of different microbes. This is called **synteny,** and just as syntax refers to the order of words in a sentence, synteny describes the order of genes on a given genome and compares this order between the genomes of two or more organisms. Closely related microbes share a more recent common evolutionary history so generally have a higher degree of synteny than more distantly related organisms. It may be easiest to understand synteny by first considering the order of genes in similar operons found in two different microbes. For instance, one would not be surprised to learn that the order of tryptophan biosynthetic genes is the same (conserved) in the *E. coli* and *Salmonella* spp. genomes. Analysis of synteny takes this comparison to the level of whole genomes. As shown in **figure 18.19**, when the order of genes in the genome of the plant symbiont *Sinorhizobium meliloti* is compared with that of the plant pathogen *Agrobacterium tumefaciens,* it is clear that the gene order is very similar. This is in contrast to *E. coli* and *S. meliloti.* Although these two bacteria belong to the same phylum, *Proteobacteria,* they are not sufficiently related to show synteny. Proteobacteria *(chapter 22)*

Comparative genomics can also provide great insight into genes and gene products that are associated with virulence. Such insights are particularly needed in understanding *Mycobacterium tuberculosis,* the causative agent of tuberculosis (TB).

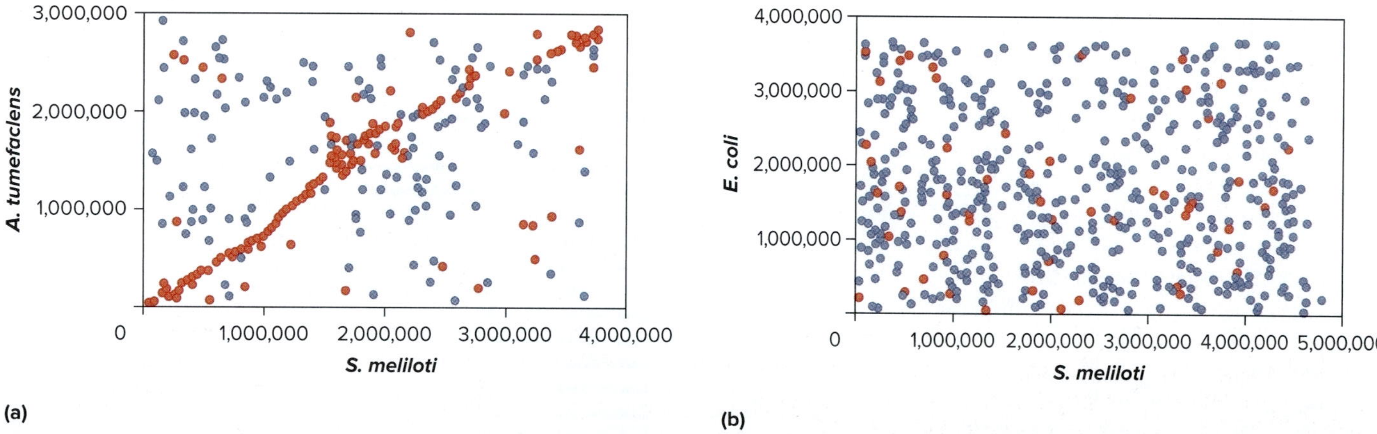

Figure 18.19 Synteny. Each axis on the graphs maps the genes on each organism's genome. For instance, each number on the X axis in both (a) and (b) indicates the position of genes (blue dots) on the 3.7 Mb *S. meliloti* genome. (a) When the genomes of *S. meliloti* and *A. tumefaciens* are compared, it is evident that the distribution of orthologous genes on the genome is quite similar, as shown by the red line indicating similar placement on the genome (red dots represent *A. tumefaciens* genes). (b) *E. coli* and *S. meliloti* are not closely related; thus their genomes show no synteny (red dots represent *E. coli* genes).

After establishing residence in immune cells in the lung, *M. tuberculosis* often remains in a dormant state until the host's immune system is compromised. It can then cause active disease, killing about 2 million people annually. The *M. tuberculosis* genome has been compared to the genomes of two relatives—*M. leprae,* which causes leprosy, and *M. bovis*, the causative agent of TB in a wide range of animals, including cows and humans. The genomes of *M. bovis* and *M. tuberculosis* are most similar—about 99.5% identical at the sequence level. However, the *M. bovis* genome is missing 11 separate regions, making its genome slightly smaller (4.3 Mb versus 4.4. Mb). The sequence dissimilarities involve the inactivation of some genes, leading to major differences in the way the two bacteria respond to environmental conditions. This may account for the host range differences between these two closely related pathogens. ▶| *Order* Corynebacteriales *includes important human pathogens (section 23.1); Mycobacterium* spp. *infections (section 39.1)*

The divergence between *M. tuberculosis* and *M. leprae* is even more striking. The *M. leprae* genome is a third smaller than that of *M. tuberculosis.* About half the genome is devoid of functional genes. Instead, there are over 1,000 degraded, nonfunctional genes called **pseudogenes** (**figure 18.20**). In total, *M. leprae* seems to have lost more than 1,000 genes during its career as an intracellular parasite. One hope from genomic analysis is that critical surface proteins can be discovered and used to develop a sensitive test for early detection of leprosy. This would allow immediate treatment of the disease before nerve damage occurs. ▶| *Leprosy (section 39.3)*

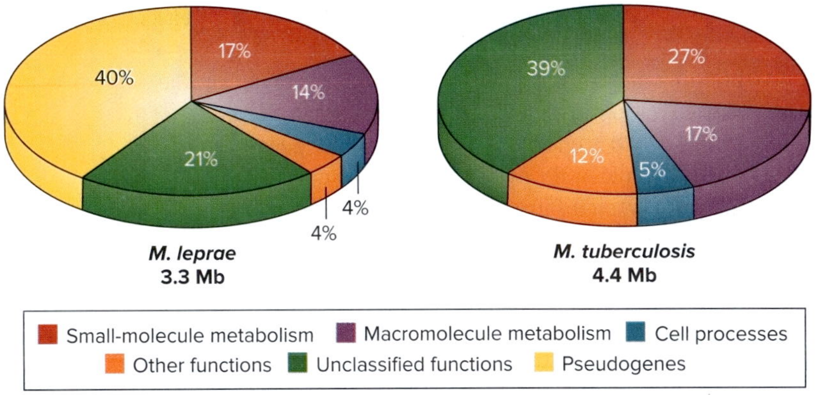

Figure 18.20 Comparative Genomic Analysis Between Two *Mycobacterium* Species. The genome of *M. leprae* is significantly smaller than that of *M. tuberculosis*. Over one-third of its genes are nonfunctional pseudogenes that will presumably be lost as genome reduction continues.

Comprehension Check

1. Cite an infectious disease for which you think a systems biology approach would be appropriate. List three specific questions you would like to explore regarding this host-pathogen interaction.
2. For what types of microorganisms is extensive gene loss common? What is the most likely explanation for this phenomenon?
3. How would you infer that a toxin gene was recently acquired by HGT in a food pathogen such as *E. coli* O157:H7?

Key Concepts

18.1 DNA Sequencing Methods

- Genomics is the study of the molecular organization of genomes, their information content, and the gene products they encode.
- DNA fragments have traditionally been sequenced using dideoxynucleotides and the Sanger chain termination technique (**figures 18.2** and **18.3**).
- Next-generation sequencing (NGS) technologies are faster and cheaper than the Sanger method and are based on sequencing by synthesis, employing reversible chain termination (**figures 18.4** and **18.5**).

18.2 Genome Sequencing

- When using the Sanger sequencing technique, the whole-genome shotgun technique is used. Four stages are involved: library construction, sequencing of randomly produced fragments, fragment alignment and gap closure, and editing the final sequence (**figure 18.6**).
- Genome sequencing using NGS techniques has accelerated the volume and pace of microbial genome sequencing.
- It is now possible to use the multiple strand displacement technique to copy the genome of a single microbial cell so there is sufficient DNA for sequencing. In this way, the genomes of uncultivated microbes can be studied (**figure 18.7**).

18.3 Metagenomics Provides Access to Uncultured Microbes

- Metagenomics enables the study of the biodiversity and metabolic potential of microbial communities without culturing individual microbes (**figure 18.8**).
- Metagenomics involves sequencing DNA extracted directly from the environment. NGS approaches are used to avoid genomic library construction.
- Metagenomics has been applied to many natural microbial habitats, including the human body.

18.4 Bioinformatics: What Does the Sequence Mean?

- Analysis of vast amounts of genome data requires sophisticated analytical procedures; this is the basis of the discipline of bioinformatics.

- An open reading frame is a putative protein-coding gene; the identity of the gene and its product are inferred by aligning the translated amino acid sequence with similar sequences in the database (**figures 18.9** and **18.10**).
- Bioinformatics enables the comparison of genes within genomes to identify paralogues and the comparison of genes between different organisms to identify orthologues.
- The physical map of a microbe's genome is often displayed so the location of genes with related functions are easily identified.

18.5 Functional Genomics Links Genes to Phenotype

- The presence of genes whose products are known to function in specific metabolic processes such as catabolic pathways and transport enables the genomic reconstruction of the organism—a type of map that illustrates these processes within the cell (**figure 18.11**).
- Transcriptomics is the study of all the RNA an organism makes at a given time. It is best studied with RNA-Seq, which assesses gene expression as a measure of individual gene transcripts (mRNA). Gene expression can be determined for mutant versus wild-type strains or for a given organism under specific environmental conditions (**figure 18.12**).
- The entire collection of proteins that an organism can produce is its proteome, and its study is called proteomics.
- The proteome is often analyzed by two-dimensional gel electrophoresis, in which the total cellular protein pool can be visualized. In many cases, the amino acid sequence of individual proteins is determined by mass spectrometry; if this is coupled to genomics, both a protein of interest and the gene that encodes it can be identified (**figures 18.14** and **18.15**).
- Structural proteomics seeks to model the three-dimensional structure of proteins based on computer analysis of amino acid sequence data.
- Protein-DNA interactions can be analyzed by chromatin immunoprecipitation, which enables the identification of both the protein and its target DNA (**figure 18.16**).

18.6 Systems Biology: Making and Testing Complex Predictions

- Systems biology combines experimentally derived data sets with predictive modeling to study an organismal process (**figure 18.17**).
- Systems biology is the opposite of reductionist biology because it integrates molecular networks with the goal of understanding the function of the cell as a whole.

18.7 Comparative Genomics

- Comparing genomes reveals information about genome structure and evolution, including the importance of horizontal gene transfer.
- Comparative genomics is an important tool in discerning how microbes have adapted to particular ecological niches and in developing new therapeutic agents.

Active Learning

1. You are annotating the genomic sequence of a new soil bacterium. You believe you have found a gene that encodes nitrate reductase. How did you arrive at this conclusion, and how will you strengthen this annotation using only in silico analysis?
2. You have discovered a new DNA-binding protein. You show by EMSA that it binds near the promoter of an operon that encodes enzymes required for carotenoid biosynthesis. You wonder if any other genes are controlled by this new protein. How will you find other regions of DNA to which the protein might bind? How will you determine if the new protein increases or decreases transcription of the genes it regulates?
3. You are developing a new vaccine for a pathogen. You want your vaccine to recognize specific cell-surface proteins. Explain how you will use genome analysis to identify potential protein targets. What functional genomics approach will you use to determine which of these proteins is produced when the pathogen is in its host?
4. A variety of potentially useful natural molecules can be extracted from the marine sponge *Theonella swinhoei.* However, these products are actually produced by a bacterial symbiont that has not yet been grown in the lab. Because the bacterium has not yet been isolated, it has candidate species status: *Candidatus* Entotheonella gemina. When a European team of scientists performed genomic analysis, they discovered that the bacterium is a chemoorganotrophic facultative anaerobe that requires specific vitamins and amino acids for growth. Proteomic analysis revealed the amino acid sequence of some peptide products that may have industrial or medical importance. Refer to sections 18.4 and 18.5 and describe the techniques these scientists most likely used to perform these experiments. Explain how they were able to conclude the specific metabolic details listed and how these might be used to help isolate and grow the microorganism in culture.

Read the original paper: Lackner, G., et al. 2017. Insights into the lifestyle of uncultured bacterial natural product factories associated with marine sponges. *Proc. Nat. Acad. Sci. USA.* 114:e347-e356. doi:10.1073/pnas.1616234114.

19

Microbial Taxonomy and the Evolution of Diversity

Source: NASA, ESA, R. O'Connell (University of Virginia), and the Hubble Heritage Team

Scientists Query: "Is the Microbial Universe Expanding?"

Should the audacious question "Does life exist that cannot be classified as archaeal, bacterial, or eukaryotic?" be taken seriously? After all, the elegant work of Carl Woese, who first described archaea, has survived the test of time to emerge stronger and more widely accepted. Yet this is the question that some microbiologists and evolutionary biologists ask when they consider "microbial dark matter." Just as astronomical dark matter represents the majority of mass in the observable universe, most of the biomass in microbial ecosystems (including the human body) is also difficult to define.

One way to view all microbes is to sort them into one of three categories: explored, unexplored, and undiscovered. Explored microbes are those we can culture. Here we see an overrepresentation of pathogens and microbes with medical, industrial, or food applications. The unexplored are those we know exist based on gene sequences obtained by metagenomic studies, but have not yet been grown in the lab. These microbes contribute to nutrient cycling and host-microbe physiology in ways that are defined in aggregate but not at the level of a single microbial taxon (e.g., genus or species). The undiscovered, however, are most intriguing. These are the microbes yet to be found—the real microbial dark matter. Do they exist? If so, how much bigger is the microbial universe than previously thought?

The only way to answer these questions is to explore. As discussed in chapter 18, metagenomic and single-cell genome sequencing harness the power of next-generation sequencing to probe genes and genomes that belong to uncultured microorganisms. The results have been startling: Hundreds of new archaeal and bacterial phyla have recently been proposed. Even more exciting is the fact that some newly discovered gene sequences don't match any previously identified archaeal, bacterial, or eukaryotic nucleotide sequences. For example, very large and unusual viruses have been found to possess genes whose evolutionary history places them somewhere between archaeal and eukaryotic domains. If cellular counterparts to these viruses exist, would they represent a new domain?

As we discuss in this chapter, phyla (s., phylum) are higher-level taxa to which a group of related organisms belong. The recent, sudden increase in the number of phyla has raised concern among taxonomists. While the criteria used to define a particular bacterial or archaeal species have long been a subject of discussion, many taxonomists feel the explosion in metagenomic and single-cell genomic data warrants a discussion regarding what truly constitutes higher-level taxa, including the recently minted term "superphylum," as well as the traditionally held phylum, class, order, family, and perhaps even domain. It seems the expansion of the microbial universe is a messy business that may require not only new nomenclature, but a new way of thinking about life itself.

In this chapter, we explore microbial taxonomy and evolution—how and why microorganisms are classified using an evolutionary framework. This was not always the case, and the move to an evolution-based system resulted in the reclassification of some genera. Indeed, the fields of microbial taxonomy and evolution continue to present challenges and controversy. But ultimately, microbial taxonomy and phylogeny open the door to a degree of biological diversity unmatched by multicellular organisms.

Readiness Check:

Based on what you have learned previously, you should be able to:

- ✓ Describe the structure of DNA (section 13.2) and summarize methods used in DNA sequencing and metagenomics (sections 18.1 to 18.3)
- ✓ Outline in general terms how small subunit rRNA has been used to define the three domains of life (section 1.2)
- ✓ Summarize the mechanisms of horizontal gene transfer (chapter 16)
- ✓ Describe the rationale for the RNA world hypothesis of the origin of life (section 1.2)
- ✓ Contrast and compare the structural differences between bacterial, archaeal, and eukaryotic microbes (chapters 3–5)
- ✓ Explain the principles of next-generation sequencing and metagenomics (sections 18.1–18.3)

19.1 Microbial Taxonomy Is Based on the Comparison of Multiple Traits

After reading this section, you should be able to:

a. Explain the utility of taxonomy and systematics
b. Draw a concept map illustrating the differences between phenetic and genotypic classification

Microbiologists are faced with the daunting task of understanding the diversity of life forms that cannot be seen with the naked eye but can live anywhere on Earth. Obviously a reliable classification system is paramount. The science of classifying living things is called **taxonomy** (Greek *taxis,* arrangement or order; *nomos,* law, or *nemein,* to distribute or govern). In a broader sense, taxonomy consists of three separate but interrelated parts: classification, nomenclature, and identification. A taxonomic scheme is used to arrange organisms into groups called taxa (s., **taxon**) based on mutual similarity. **Nomenclature** is the branch of taxonomy concerned with the assignment of names to taxonomic groups in agreement with published rules. Identification is the practical side of taxonomy—the process of determining if a particular isolate belongs to a recognized taxon and, if so, which one. The term **systematics** is often used for taxonomy, although it sometimes infers a more general scientific study of organisms with the ultimate objective of arranging them in an orderly manner. Thus systematics encompasses morphology, ecology, epidemiology, biochemistry, genetics, molecular biology, and physiology.

One of the oldest classification systems, called **natural classification,** arranges organisms into groups whose members share many characteristics. The Swedish botanist Carl von Linné, or Carolus Linnaeus as he often is called, developed the first natural classification in the eighteenth century. It was based largely on anatomical characteristics and was a great improvement over previously employed systems because natural classification provided information about many biological properties. For example, classification of humans as mammals denotes that they have hair, self-regulating body temperature, and milk-producing mammary glands in the female.

When natural classification is applied to higher organisms, evolutionary relationships become apparent simply because the morphology of a given structure (e.g., wings) in a variety of organisms (ducks, songbirds, hawks) suggests how that structure might have been modified to adapt to specific environments or behaviors. However, the traditional taxonomic assignment of microbes was not rooted in evolutionary relatedness. For instance, bacterial pathogens and microbes of industrial importance were historically given names that described the diseases they cause or the processes they perform (e.g., *Vibrio cholerae, Clostridium tetani,* and *Lactococcus lactis*). Although these labels are of practical use, they do little to guide the taxonomist concerned with the vast majority of microbes that are neither pathogenic nor of industrial consequence. Our present understanding of the evolutionary relationships among microbes now serves as the theoretical underpinning for taxonomic classification.

In practice, determining the genus and species of a newly isolated microbe is based on **polyphasic taxonomy.** As the term "polyphasic" suggests, this encompasses many aspects that describe the microorganism. These include phenotypic and genotypic features. To understand how all of these data are incorporated into a coherent profile of taxonomic criteria, we must first consider the individual components.

Phenetic Classification

For a very long time, microbial taxonomists had to rely exclusively on a **phenetic system,** which classifies organisms according to their phenotypic similarity **(Table 19.1)**. This system succeeded in bringing order to biological diversity and clarified the function of morphological structures. For example, because motility and flagella are always associated in particular microorganisms, it is reasonable to suppose that flagella are involved in at least some types of motility. Although phenetic studies can reveal possible evolutionary relationships, this is not always the case. For example, not all flagellated bacteria belong to the same phylum. This is why the best phenetic classification is one constructed by comparing as many attributes as possible.

Genotypic Classification

As the name suggests, **genotypic classification** seeks to compare the genetic similarity between organisms. Individual genes or whole genomes can be compared. Since the 1970s it has been widely accepted that bacteria and archaea whose genomes are at least 70% homologous belong to the same species. However, there is now a consensus building to replace this metric with a genomics-based assay that measures the average nucleotide identity between organisms. The means by which microbes are genotypically classified is discussed further in section 19.3.

Table 19.1 Components of Polyphasic Taxonomy

Classification System	Basis of Classification	Traits	Microbial Example
Phenetic	Phenotype	Morphology Physiology	*Clostridium* spp. (anaerobic, endospore-forming, Gram-positive rod) compared to *Bacillus* spp. (aerobic, endospore-forming, Gram-positive rod)
Genotypic	Genes and genomes	Average nucleotide identity rRNA genes	*E. coli* strain K12 compared to *E. coli* O157:H7 *E. coli compared* to *Salmonella enterica*

Phylogenetic Classification

©Pixtal/age fotostock

With the publication in 1859 of Charles Darwin's *On the Origin of Species,* biologists began developing **phylogenetic** or **phyletic classification systems** that sought to compare organisms on the basis of evolutionary relationships. The term **phylogeny** (Greek *phylon,* tribe or race; *genesis,* generation or origin) refers to the evolutionary development of a species. Scientists realized that when they observed differences and similarities between organisms as a result of evolutionary processes, they also gained insight into the history of life on Earth. However, for much of the twentieth century, microbiologists could not effectively employ phylogenetic classification, primarily because of the lack of a good fossil record. When Carl Woese and George Fox proposed using small subunit (SSU) rRNA nucleotide sequences to assess evolutionary relationships among microorganisms, the door opened to the resolution of long-standing inquiries regarding the origin and evolution of the majority of life forms on Earth—microbes. Whole genome sequencing is the most recent analytical tool to elucidate phylogenetic relationships among microbes.

Comprehension Check

1. What is a natural classification? What microbial features might have been considered when devising a natural classification scheme?
2. What is polyphasic taxonomy, and what types of data does it consider? Should each type of data be of equal weight? Why or why not?
3. Consider the finding that bacteria capable of anoxygenic photosynthesis belong to several different phyla. How do you think these bacteria might have been originally classified, and what types of data do you think were key in making the most recent taxonomic assignments?

19.2 Taxonomic Ranks Provide an Organizational Framework

After reading this section, you should be able to:

a. Outline the general scheme of taxonomic hierarchy
b. Explain how the binomial system of Linnaeus is used in microbial taxonomy

The definition of a bacterial or archaeal species is widely debated, as discussed in section 19.5. Nonetheless, for practical reasons it is essential that the established rules of taxonomy are followed. Microbes are placed in taxonomic levels arranged in a nonoverlapping hierarchy so that each level includes not only the traits that define the rank above it but also a new set of more restrictive traits (**figure 19.1**). Thus within each domain—*Bacteria, Archaea,* or *Eukarya*—each organism is assigned (in descending order) to a phylum, class, order, family, genus, and species epithet or name. Some microbes are also given a subspecies designation. Microbial groups at each level have a specific suffix that indicates rank or level.

The application of Linnaeus's classification system to bacteria began in Pasteur and Koch's time, about 1872. However, within 20 years microbiologists became dissatisfied, considered it haphazard, and called for more uniform criteria in organizing phyla, classes, orders, families, genera, and species. With the ongoing explosion in metagenomic analysis, some would argue that history is repeating itself. Within the last few decades, thousands of 16S rRNA genes and protein-coding genes have been sequenced that do not belong to any previously defined taxa. This data explosion has led to the recent development of the taxonomic classification **superphylum,** below domain and above phylum (e.g., in figure 19.1, superphylum would be placed between *Bacteria* and *Proteobacteria*). Ideally a superphylum includes organisms of several phyla that share a number of distinctive characteristics, such as unusual morphological or metabolic features. However, some feel that the term is being loosely applied based on insufficient data—for instance, to SSU rRNA sequences alone.

At the other end of the classification scheme is the most fundamental definition of a bacterial or archaeal **species.** A species is a collection of strains that share many stable properties and differ significantly from other groups of strains. A **strain** consists of the descendants of a single, pure microbial culture. Strains within a species may be described in a number of different ways. **Biovars** are variant strains characterized by biochemical or physiological differences, **morphovars** differ morphologically, and **serovars** have distinctive antigenic (immunologically reactive) properties. Because changes in species assignment are not uncommon, one strain is designated as the **type strain** for each species. The type strain is the holder of the species name. This ensures permanence of names when nomenclature revisions occur because the type species must remain within the original species. It is usually one of the first strains studied and often is more fully characterized than others; however, it does not have to be the most representative member. Only those strains very similar to the type strain or type species are included in a species. Each species is assigned to a **genus,** the next rank in the taxonomic hierarchy. A genus is a well-defined group of one or more species that is clearly separate from other genera. In practice,

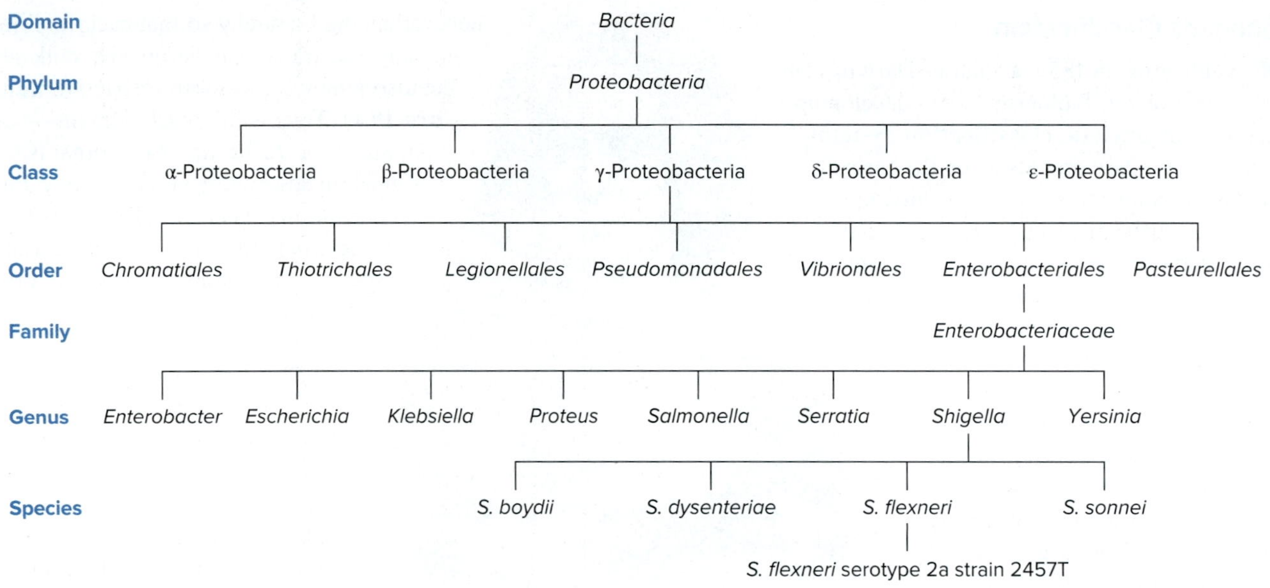

Figure 19.1 Hierarchical Arrangement in Taxonomy. In this example, members of the genus *Shigella* are placed within higher taxonomic ranks. Not all classification possibilities are given for each rank to simplify the diagram. Note that -*ales* denotes order and -*ceae* indicates family.

considerable subjectivity occurs in assigning species to a genus, and taxonomists may disagree about the composition of genera.

Microbiologists name microorganisms using the **binomial system** of Linnaeus. The Latinized, italicized name consists of two parts. The first part is the generic name (i.e., the genus), and the second is the species name (e.g., *Yersinia pestis,* the causative agent of plague). The species name is stable; the oldest epithet for a particular organism takes precedence and must be used. In contrast, a generic name can change if the organism is assigned to another genus. For example, some members of the genus *Streptococcus* were placed into two new genera, *Enterococcus* and *Lactococcus,* based on rRNA analysis and other characteristics. Thus *Streptococcus faecalis* is now *Enterococcus faecalis.* To be recognized as a new species, genomic, metabolic, morphological, reproductive, and ecological data must be accepted and published in the *International Journal of Systematic and Evolutionary Microbiology;* until that time, the new species name will appear in quotation marks. Microbes that have not been grown in pure culture but for which there is sufficient genetic characterization may be given a provisional genus and species name preceded by the term *Candidatus,* meaning candidate. For instance, *Candidatus* Mycoplasma girerdii is believed to be associated with preterm birth. It was first detected by metagenomic sequencing in the microbial community in a preterm infant's mouth. Although complete genomes are known for several related strains, none have been successfully cultured. Note that its provisional genus and species names are not italicized. *Bergey's Manual of Systematics of Bacteria and Archaea* contains only recognized bacterial and archaeal species as discussed in section 19.6.

Comprehension Check

1. What is the difference between a microbial species and a strain?
2. Why is it important to have a type strain for each species?
3. The genus *Salmonella* was once thought to contain five species. Most scientists now consider only two species valid: *S. bongori* and *S. enterica*. The latter contains six subspecies, and *S. enterica* subspecies *enterica* is further subdivided into eight serovars. Three of these, Enteritidis, Typhi, and Typhimurium, were once considered *Salmonella* species. (a) Inspect figure 19.1 and construct a similar lineage for *Salmonella*. (b) What criteria do you think were used to make these taxonomic changes?

19.3 Microbial Taxonomy and Phylogeny Are Largely Based on Molecular Characterization

After reading this section, you should be able to:

a. Construct a concept map describing the approaches commonly used to determine taxonomic classification
b. Assess the impact molecular methods have had on the field of microbial taxonomy and phylogeny
c. Compare and contrast nucleotide sequencing and nonsequencing-based molecular approaches used in microbial taxonomy and phylogeny
d. Select an appropriate technique to identify a microbial genus, species, and strain
e. Predict the basic biological as well as public health implications of microbial taxonomic identification

Many different approaches are used to classify and identify microorganisms that have been isolated and grown in pure culture. For clarity, we divide these approaches into two groups: classical and molecular. The most durable identifications are those that are based on a combination of approaches. Methods often employed in routine laboratory identification of pathogenic bacteria are covered in chapter 37, Clinical Microbiology and Immunology.

Classical Characteristics

Classical approaches to taxonomy make use of morphological, physiological, biochemical, and ecological characteristics. These characteristics have been employed in microbial taxonomy for many years and form the basis for phenetic (phenotypic) classification. When used in combination, they are quite useful in routine identification of well-characterized microbes.

Morphological Characteristics

Morphological features are important in microbial taxonomy for many reasons (**table 19.2**). Morphology is easy to study and analyze, particularly in eukaryotic microorganisms and more complex bacteria and archaea. In addition, morphological comparisons are valuable because structural features depend on the expression of many genes and are usually genetically stable. Thus morphological similarity often is a good indication of phylogenetic relatedness.

Physiological and Metabolic Characteristics

Physiological and metabolic characteristics are useful because they are directly related to the nature and activity of microbial proteins. For instance, the detection of specific end products of fermentation in a newly discovered microorganism reveals the presence of specific catabolic enzymes and the genes that encode them. Therefore analysis of characteristics, such as energy metabolism and nutrient transport, provides an indirect comparison of microbial genomes. **Table 19.3** lists some of the most important of these properties.

Biochemical Characteristics

Among the more useful biochemical characteristics used in microbial taxonomy are bacterial fatty acids, which can be analyzed using a technique called *f*atty *a*cid *m*ethyl *e*ster (FAME) analysis. A fatty acid profile reveals differences in chain length, degree of saturation, branched chains, and hydroxyl groups. Microbes of the same species will have identical fatty acid profiles, provided they are grown under the same conditions; this limits FAME analysis to only those microbes that can be grown in pure culture. Finally, because the identification of a species is done by comparing the results of the unknown microbe in question with the FAME profile of other, known microbes, identification is only possible if the species in question has been previously analyzed. Nonetheless, FAME analysis is particularly important in public health, food, and water microbiology. In these applications, microbiologists seek to identify specific pathogens. ▶| *Microbiology of food (chapter 41); Purification and sanitary analysis ensures safe drinking water (section 43.1)*

Advances in mass spectrometry (MS) have resulted in the fast and accurate identification of bacteria based on the presence of specific, highly abundant proteins. The specific type of MS used is called matrix-assisted laser desorption/ionization-time of flight (MALDI-ToF). MALDI-ToF enables the analysis of complex biomolecules that could not previously be studied by MS. The material to be analyzed is dried on a sample holder (called a target) and

Table 19.2 Some Morphological Features Used in Classification and Identification

Feature	Microbial Groups
Cell shape	All major groups[1]
Cell size	All major groups
Colony morphology	All major groups
Staining behavior	Bacteria, some fungi
Cilia and flagella	All major groups
Mechanism of motility	Gliding bacteria, spirochetes, protists
Spore morphology and location	Bacteria, protists, fungi
Cellular inclusions	All major groups

1 Used in classifying and identifying at least some bacteria, archaea, fungi, and protists.

Table 19.3 Some Physiological and Metabolic Characteristics Used in Classification and Identification

Carbon, nitrogen, and energy sources
Cell envelope constituents
Fermentation products
Motility
Oxygen requirements and tolerance
pH and temperature optimum and growth range
Photosynthetic pigments
Salt requirements and tolerance
Secondary metabolites formed
Storage inclusions

then mixed with a molecular film called a matrix. When a UV laser beam strikes the sample target, the matrix helps stimulate release of the sample from the surface; this is matrix-assisted desorption. Once the matrix-sample is released (desorbed), the matrix transfers protons to the sample, which then becomes ionized; this is matrix-assisted ionization. Once ionized, biomolecules are "flown" across a space within the instrument; the time taken for a molecule to fly from one side to the other is used to determine the mass of each molecule; this is time of flight.

In its simplest form, MALDI-ToF identification of bacteria involves the transfer of whole cells from a single colony grown under specific conditions to a sample target. Once dried, a matrix is deposited and the bacterial samples are analyzed. MALDI-ToF yields the masses of many highly abundant bacterial proteins. Like FAME analysis, each experimentally derived protein profile must be matched to a protein profile from a known bacterium. MALDI-ToF is becoming increasingly important in medical microbiology laboratories where the same strains of organisms are regularly encountered. Like FAME, microbes must be grown under very specific conditions and MALDI-ToF cannot be used to identify newly discovered microbes. ◀ *Proteomics explores total cellular proteins (section 18.5)*

Ecological Characteristics

The ability of a microorganism to colonize a specific environment is of taxonomic value. Some microbes may be very similar in many other respects but inhabit different ecological niches, suggesting they may not be as closely related as first suspected. Some examples of taxonomically important ecological properties are life cycle patterns; the nature of symbiotic relationships; the ability to cause disease in a particular host; and habitat preferences such as requirements for temperature, pH, oxygen, and osmotic concentration. Many growth requirements are considered physiological characteristics as well (table 19.3). ◀ *Environmental factors affect microbial growth (section 7.5)* ▶ *Many types of microbial interactions exist (section 27.1)*

Molecular Characteristics

It is hard to overestimate how the study of DNA, RNA, and proteins has advanced our understanding of microbial evolution and taxonomy. Evolutionary biologists studying plants and animals draw from a rich fossil record to assemble a history of morphological changes; in these cases, molecular approaches supplement such data. In contrast, microorganisms have almost no fossil record, so molecular analysis is the only feasible means of collecting a large and accurate data set that explores microbial evolution. When scientists are careful to make only valid comparisons, phylogenetic inferences based on molecular approaches provide the most robust analysis of microbial evolution.

Microbial genomes can be directly compared and taxonomic similarity can be estimated in many ways. DNA sequence can be determined for one or a few genes, or for an entire genome, depending on the degree of identification required. The recent rapid drop in cost and time to generate a genome sequence has made this approach viable for many taxonomic applications. Molecular techniques developed in the late twentieth century are now in transition to techniques based on whole-genome sequencing (WGS). In many instances, the original nomenclature remains, although the technique is performed in silico rather than in vitro. As with phenetic approaches, comparison to standards and type strains forms the basis for identification and phylogenetic placement. Many classical properties used in strain identification are now inferred from genome sequences.

Signature Sequences

The rRNAs from small ribosomal subunits (16S from bacterial and archaeal cells and 18S from eukaryotes) have become the molecules of choice for inferring microbial phylogenies and making taxonomic assignments at the genus level. The **small subunit rRNAs (SSU rRNAs)** are widely applicable in studies of microbial evolution, relatedness, and genus identification for several important reasons (**figure 19.2**). First, they play the same role in all microorganisms. In addition, because ribosomes are absolutely necessary for survival and ribosomes cannot function without SSU rRNAs, the genes encoding these rRNAs cannot tolerate large mutations. Thus these genes change very slowly with time. Finally, rRNA genes are rarely subject to horizontal gene transfer, an important factor in comparing sequences for phylogenetic purposes. The utility of SSU rRNAs is extended by the presence of certain sequences within SSU rRNA genes that are variable among organisms as well as other regions that are quite similar. The variable regions enable comparison between closely related microbes, whereas the stable sequences allow the comparison of distantly related microorganisms.

Comparative analysis of SSU rRNA sequences from thousands of organisms has demonstrated the presence of **oligonucleotide signature sequences.** These are short, conserved nucleotide sequences that are specific for phylogenetically defined groups of organisms. Thus the signature sequences found in bacterial rRNAs are rarely or never found in archaeal rRNAs and vice versa. Likewise, the 18S rRNA of eukaryotes bears signature sequences that are specific to the domain *Eukarya*. In addition, signature sequences for a lower taxon, for example, the genus *Pseudomonas*, may be found in the variable regions of a higher taxon like *Bacteria* (figure 19.2*b*).

The ability to amplify regions of rRNA genes (rDNA) by the polymerase chain reaction (PCR) and sequence the DNA using next-generation techniques has greatly increased the efficiency by which SSU rRNA sequences can be obtained. PCR can amplify rDNA from the genomes of different organisms because conserved nucleotide sequences flank the nucleotides that can be used to reveal the microbe's identity (figure 19.2*b*). In practice, this means that PCR primers can be generated to amplify rDNA from both cultured and uncultured microbes. Any uncultivated microorganism that is identified solely on its nucleic acid sequence (or other observable, quantifiable phenotype) is termed a **phylotype**.

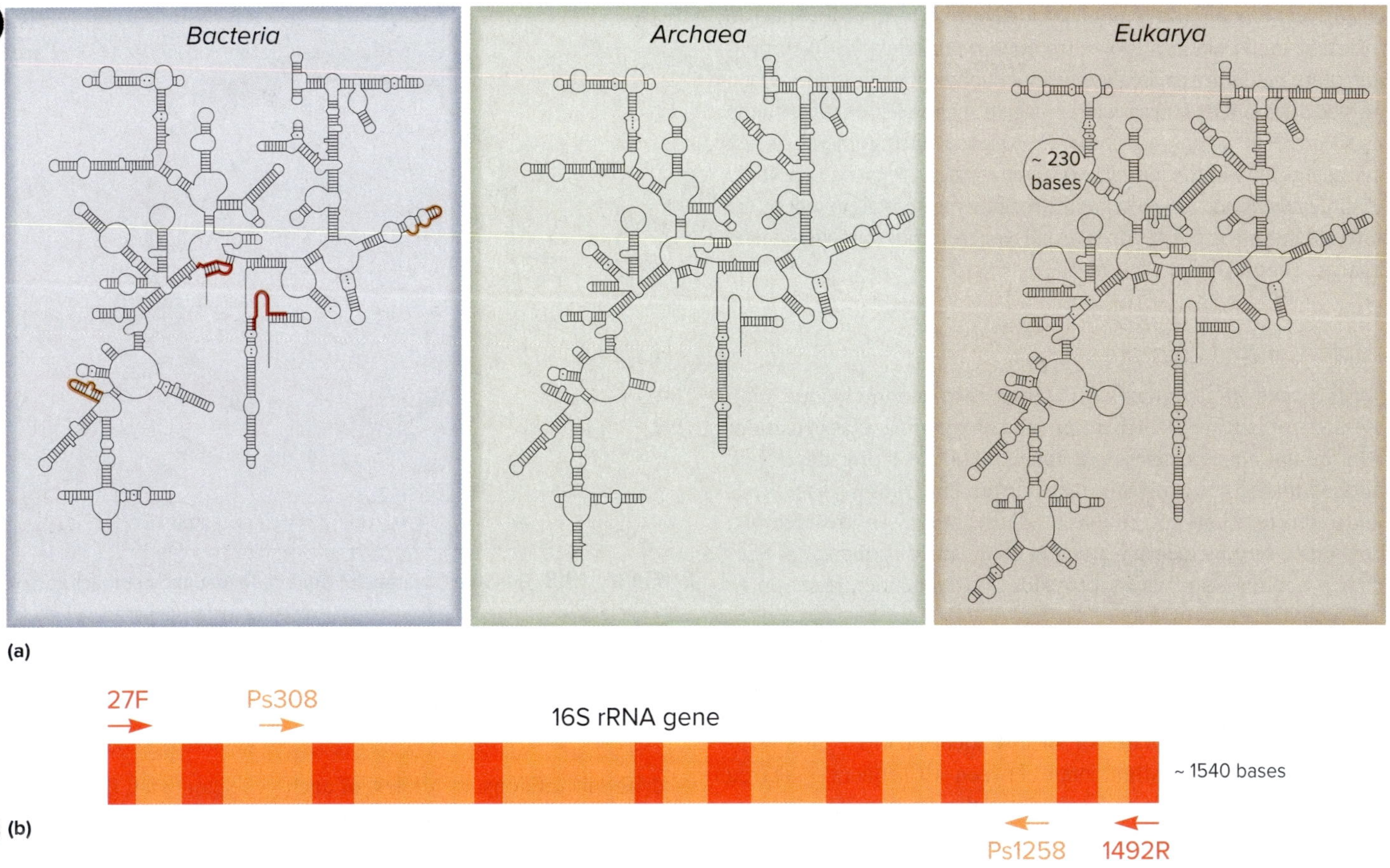

	Amino acid number		203																			221
Lyme disease *Borrelia*	***B. burgdorferi***		G	G	G	I	K	G	L	K	G		V	F	L	L	T	L	L	S	G
	B. afzelii		G	G	G	I	K	G	L	K	G		V	F	L	L	T	L	L	S	G
	B. garinii		G	G	G	I	K	G	L	K	G		V	F	**V**	L	T	L	L	S	G
Relapsing fever *Borrelia*	***B. hermsii***		G	G	G	I	K	**A**	L	K	G	**G**	**M**	F	**I**	L	T	L	L	S	G
	B. recurrentis		G	G	G	I	K	**S**	L	K	G	**G**	**M**	F	**I**	L	T	L	L	S	G
	B. parkeri		G	G	G	I	K	G	L	K	G	**G**	**M**	F	**I**	L	T	L	L	S	G
	B. miyamotoi		G	G	G	I	K	**A**	**L**	**Q**	**G**	**G**	**I**	**I**	**I**	L	T	L	L	S	G

(c)

Figure 19.2 Signature rRNA and Amino Acid Sequences Help to Identify Microbes. (a) Representative examples of rRNA secondary structures from the three domains: *Bacteria (Escherichia coli), Archaea (Methanococcus vannielii),* and *Eukarya (Saccharomyces cerevisiae).* (b) Diagram of a bacterial 16S rRNA gene. Conserved regions (red) are present in all bacteria, and bacterial-specific PCR primers 27F and 1492R bind where indicated. These sites are highlighted in red in (a). Variable regions (gold) contain genus-specific sequence, and PCR primers Ps308 and Ps1258 indicate the positions of signature sequences specific to *Pseudomonas*. These sites are highlighted in gold in (a). (c) Alignment of amino acids 203-221 in L-lactate permease from *Borrelia species*. Shaded boxes are amino acids identical to the *B. burgdorferi* sequence. An amino acid indel (shaded blue) distinguishes species that cause Lyme disease from those that cause the condition known as relapsing fever.

The use of SSU rRNA sequences as a taxonomic marker has been validated as a molecular measure of relatedness. Correlating SSU rRNA sequence data with traditional measures of assigning species, organisms of the same species show at least 98.65% sequence identity in their SSU rDNA sequences. This means that when comparing the SSU rRNA sequences of two organisms, differences at more than 20 positions of a ~1,540 base rRNA indicate that they probably derive from different species. ◀ *Polymerase chain reaction amplifies targeted DNA (section 17.2)*

Signature sequences are present in genes other than those encoding rRNA. Many genes have nucleotide insertions or deletions of specific lengths and sequences at fixed positions. A particular nucleotide sequence that is inserted or deleted may be found exclusively among all members of a phylum. These taxon specific insertions and deletions are called conserved **indels** (for *in*sertion/*del*etion). Indels are particularly useful in phylogenetic studies when they are flanked by conserved regions. In such cases, changes in the signature sequence cannot be due to sequence misalignments. When an indel occurs in the coding region of a gene, it must be a multiple of three nucleotides to preserve the reading frame. Alignment of the inferred amino acid sequences reveals indels useful for phylogenetic analysis (figure 19.2*c*).

Whole-Genome Comparison

As the field of taxonomy shifts to whole-genome sequencing (WGS) for strain identification and taxonomic classification, new quantitative measures of relatedness are being developed and evaluated. As with any metric, standardization of methods and interpretations are required. These processes are ongoing. An initial step to identify a newly determined sequence is SSU rRNA comparison, which provides confident identification to the genus level.

Identification to the species level requires detailed gene-by-gene comparison of the new isolate and type strains of its closest relatives. **Average nucleotide identity (ANI)** is now widely considered to be the standard for species identification. This technique uses pairwise alignments between all sequences shared between two genomes and calculates the fraction of identical nucleotides. ANI values for two genomes of the same species should be at least 95 to 96%.

ANI is poised to replace a biochemical technique called **DNA-DNA hybridization (DDH)**. DDH is performed by mixing genomic DNA from two strains. The mixture is heated until denaturation occurs, and then slowly cooled to allow renaturation. Noncomplementary regions remain unpaired, and the degree of renaturation is calculated. This is a relatively complex assay and results depend on quality of the extracted DNA and other factors. By contrast, bioinformatics software quickly calculates digital DDH values using WGS data, although this technique is less commonly used than ANI.

Another biochemical technique gone digital is determination of **G + C content.** This metric is a simple percentage of the bases in DNA that are G + C, and it is readily determined computationally from WGS data. Organisms range from around 30% to 80% G + C. Despite the wide range of variation, the G + C content of strains within a species is constant and varies little within a genus.

Subspecies and Strain Identification

For many applications, identification to levels below species is required. Typically genes that evolve more quickly than those that encode rRNA must be analyzed. The technique called **multilocus sequence analysis (MLSA)** compares the sequences of several conserved housekeeping genes (**figure 19.3**). At least 5 genes are examined to avoid misleading results that can arise through horizontal gene transfer. Because many

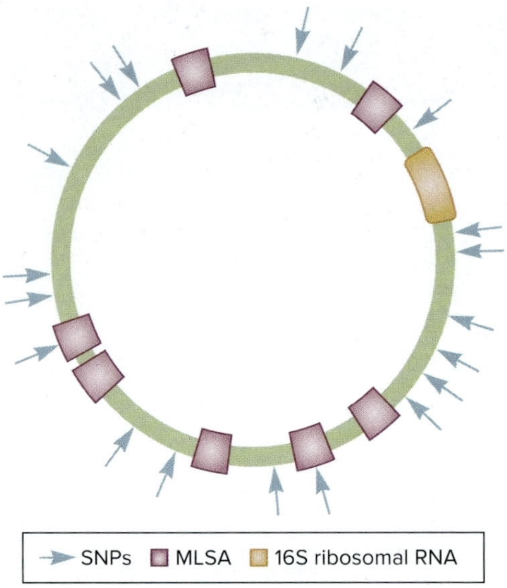

Figure 19.3 Genome Coverage of Genetic Taxonomic Approaches. A bacterial or archaeal genome contains at least one locus encoding 16S rRNA. MLSA samples multiple genes throughout the genome. SNP analysis compares more, but shorter, nucleotide sequences that span the entire genome.

different versions, or alleles, of each gene can exist, the finding that two microbial isolates share the same alleles for multiple genes is strong evidence that the two strains are closely related, perhaps even the same strain. MLSA is now often performed from *w*hole *g*enome sequences, where it is termed wgMLSA. The availability of wgMLSA data enables extended gene-by-gene comparison.

To survey a large fraction of the genome, **single nucleotide polymorphisms (SNPs**, pronounced "snips") are identified in specific genes, intergenic regions, or other noncoding regions (figure 19.3). Originally developed to analyze human DNA, SNP analysis targets specific regions because they are normally conserved, so single base pair differences reveal evolutionary change. SNP analysis shares features with analysis of **restriction fragment length polymorphism (RFLP)**. This technique identifies differences in restriction endonuclease digestion patterns, which reflect individual base pair changes. When this analysis is applied to the gene encoding the SSU rRNA, it is termed **ribotyping.** With the advent of WGS, these techniques are increasingly performed in silico.

Comprehension Check

1. What are the advantages of using each major group of characteristics (morphological, physiological and metabolic, biochemical, ecological, and molecular) in classification and identification? How is each group related to the nature and expression of the genome? Give examples of each type of characteristic.
2. Why is rRNA suitable for determining relatedness?
3. List the steps you would take to identify an organism just isolated in pure culture. What criteria would tell you if it is a new species?

19.4 Phylogenetic Trees Illustrate Evolutionary Relationships

After reading this section, you should be able to:

a. Paraphrase the rationale underpinning the construction of phylogenetic trees
b. Compare and contrast rooted and unrooted trees
c. Outline the general considerations used in building a phylogenetic tree
d. Characterize the challenges horizontal gene transfer introduces in the study of microbial evolution

Microbial taxa within *Bacteria* and *Archaea* form discrete, genealogically clustered groups that can be illustrated in phylogenetic trees. **Phylogenetic trees** show inferred evolutionary relationships in the form of multiple branching lineages connected by nodes (**figure 19.4**). Organisms whose nucleotide sequences have been analyzed are identified at the tip of each branch. Each node (branchpoint) represents a divergence event, and the length of the branches correlates with the number of molecular changes that have taken place between the two nodes.

Often sequences are obtained from well-classified microbes that have been grown in pure culture; however, this is not always the case. SSU rRNA sequences have become particularly important in both identifying microbes and constructing phylogenetic trees to describe their evolutionary relationships. An inclusive term for the organism at each branch tip, regardless of whether or not it has been cultured, is the **operational taxonomic unit** (**OTU**). The term *clade* refers to a group of organisms with a common ancestor. We now briefly describe how phylogenetic trees can be built with the intention of increasing an understanding of what they represent.

There are five steps in building a phylogenetic tree. First, the nucleotide or amino acid sequences must be aligned. Alignments are performed with software, although manual inspection of the alignment is also important (figure 19.4*a*). In addition to SSU rRNA gene sequences, protein-coding genes may be analyzed, in which case the alignment of amino acids is preferred. This is because the genetic code is degenerate, so even if a nucleotide sequence is not conserved, the amino acid sequence may be. For instance, if a protein-coding gene were mutated so that a codon changed from AGA to AGG, arginine would still be added to the growing polypeptide during translation. Next the alignment must be examined for a phylogenetic signal; this will determine if it is appropriate to continue with tree building (figure 19.4*b*). There are two extremes in this regard: At one end of the spectrum, the sequences align perfectly. The other extreme is 25% identity, or what would be expected if two random DNA sequences were aligned. (When there are four possible nucleotides, each position has a 1 in 4 chance of matching.) Phylogenetic analysis can only be performed on those sequences that fall in the middle, with a mixture of random and matched positions. The third step is the hardest: choosing which tree-building method to use. We briefly review some of the more popular methods. The last two steps involve the application of the selected method, performed by a computer, followed by manual examination of the resulting tree to confirm that it is logical. For example, a computer-generated tree

Nucleotide position	1	2	3	4	5	6	7	8	9	10	11	12	13	14	15
Microbe 1	A	C	U	G	A	C	U	C	A	U	A	G	A	U	C
Microbe 2	A	G	U	G	A	G	U	C	A	G	A	C	A	U	C
Microbe 3	U	C	U	G	G	G	U	C	A	G	A	C	A	U	C
Microbe 4	U	G	U	G	G	U	C	C	A	U	A	C	A	U	C

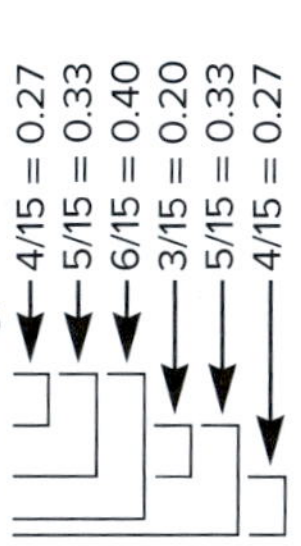

(a) Sequence alignment and analysis

Microbe	1	2	3	4
1	1.0	0.27	0.33	0.40
2		1.0	0.20	0.33
3			1.0	0.27
4				1.0

(b) Calculated evolutionary distance

Microbe	1	2	3	4
1	1.0	0.33	0.40	0.62
2		1.0	0.23	0.40
3			1.0	0.35
4				1.0

(c) Corrected evolutionary distance

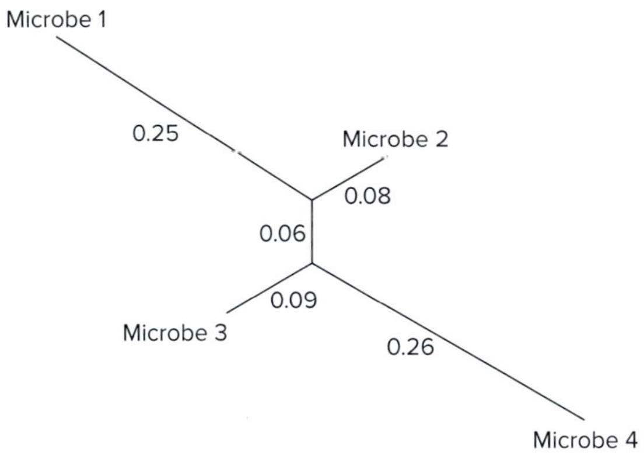

(d) Phylogenetic tree

Figure 19.4 Constructing a Phylogenetic Tree Using a Distance Method. (a) Nucleotide sequences are aligned, pairwise comparison is made, and the number of nonidentical nucleotides is scored. For example, when sequences from microbes 1 and 2 are compared, there are 4 mismatches out of 15 total nucleotides, yielding a calculated evolutionary distance (E_D) of 0.27. (b) The calculated E_D values are corrected to account for back mutation to the original genotype or other forward mutations that could have occurred at the same site before generating the observed genotype. (c) A tree-building method is then selected (in this case, a distance method is used), and computer analysis of the values generates a phylogenetic tree as shown in (d). E_D values are indicated for each branch. Numbers are not exact due to rounding for a short sequence.

that places a mammal and an archaeon on the same branch would certainly require correction.

Approaches to building a phylogenetic tree can be divided into two broad categories: a distance-based (phenetic) approach and a character-based (cladistic) approach. Distance-based approaches are the most intuitive. Here the differences between the aligned sequences are counted for each pair and summarized in a single statistic, which is roughly the percent difference between the two sequences (figure 19.4*b*,*c*). A tree is generated by serially linking pairs that are ever more distantly related (i.e., start with those with the least number of sequence differences and move to those with the most). This is called cluster analysis and should be carefully applied as it has the unattractive capability of generating trees even in the absence of evolutionary relationships. Neighbor joining is another distance-based method that uses a different matrix that attempts to avoid this problem by modifying the distance between each pair of nodes based on the average divergence from all other nodes.

Character-based methods for phylogenetic tree building are more complicated but generate more robust trees. These methods start with assumptions about the pathway of evolution, infer the ancestor at each node, and choose the best tree according to a specific model of evolutionary change. These methods include maximum parsimony, which assumes that the fewest number of changes occurred between ancestor and extant (living) organisms. Another approach is called maximum likelihood. This requires a large data set because for each possible tree that can be built, its probability (i.e., the likelihood) based on certain evolutionary and molecular information is determined so that the tree with the greatest probability based on these criteria is selected. All tree building methods have their advantages and disadvantages, so it is advisable to use several methods to analyze the same data set. Similar trees generated by different approaches is the desired outcome.

Importantly, a tree may be unrooted or rooted. An unrooted tree (**figure 19.5*a***) represents phylogenetic relationships but does not indicate which organisms are more primitive relative to the others. Figure 19.5*a* shows that A is more closely related to C than it is to either B or D, but it does not indicate which of the four species might be the oldest. In contrast, the rooted tree (figure 19.5*b*) includes a node that serves as the common ancestor and shows the development of the four species from this root. It is much more difficult to develop a rooted tree. For example, there are 15 possible rooted trees that connect four species but only three possible unrooted trees.

An unrooted tree can be rooted by adding data from an outgroup—a species known to be very distantly related to all the species in the tree (figure 19.5*c*). The root is determined by the point of the tree where the outgroup joins. This provides a point of reference to identify the oldest node on the tree, which is the node closest to the outgroup. So, for example, in figure 19.5*c*, organism Z is the outgroup and the oldest node on the tree is marked with an arrow.

Once a tree is constructed, it is important to get a sense of whether the placement of its branches and nodes is legitimate.

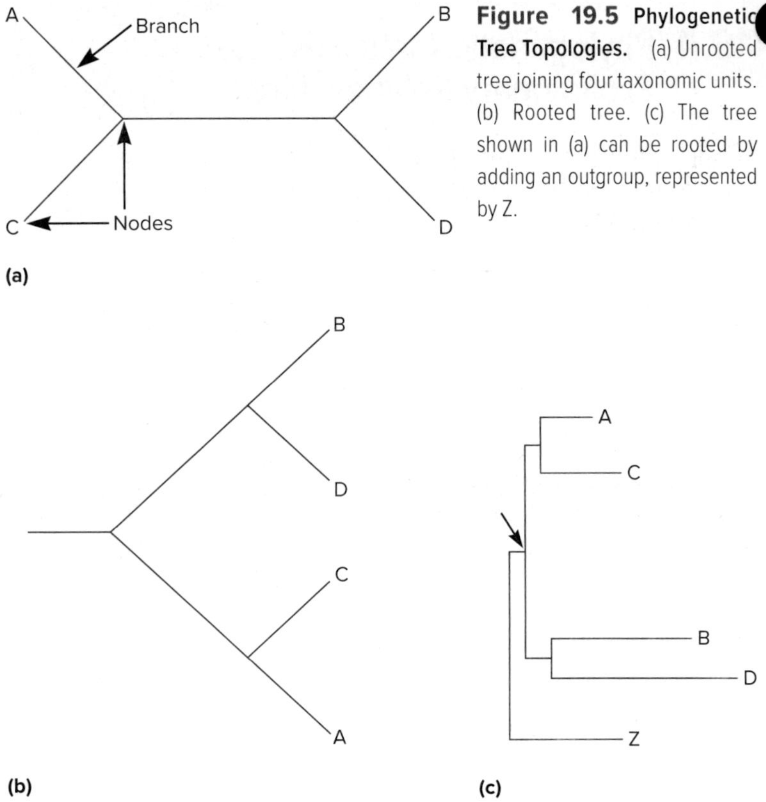

Figure 19.5 Phylogenetic Tree Topologies. (a) Unrooted tree joining four taxonomic units. (b) Rooted tree. (c) The tree shown in (a) can be rooted by adding an outgroup, represented by Z.

There are a variety of methods to assess the "strength" of a tree, but the most common is bootstrapping. Bootstrapping involves phylogenetic analysis of a randomly selected subset of the data presented on the tree. A bootstrap value is the percent of analyses in which that particular branch was found. Typically bootstrap values of 70% or greater are thought to support a tree. Another approach, called Bayesian inference, may also be used. Rather than looking at a single tree, Bayesian inference analyzes multiple potential trees and calculates the probability that each branch would appear based on this comparison. Although these values are also reported as percentages, they are not directly comparable to bootstrap values. Only values greater than 95% are acceptable when Bayesian inference is used.

An important feature in phylogenetic trees is the scale. Just as a scale bar on a road map indicates distance, the scale bar on a phylogenetic tree illustrates the evolutionary distance. This is usually measured in number of mutations per 100 nucleotides or amino acid substitutions per 100 amino acid residues. This may be expressed as a number without units, for example, 0.02 (2 per 100). To continue the map analogy, just as a map does not reveal how long it takes to get from one point to another (due to traffic and weather), the branches on a phylogenetic tree do not indicate the length of time it took for an ancestral microbe to give rise to an extant form. As discussed in section 19.5, the theory of punctuated equilibria is one important reason evolutionary distance, as measured by the similarity of genes or proteins in living organisms, provides little or no information regarding how long ago evolutionary divergence occurred.

One of the biggest challenges in constructing a satisfactory tree is widespread, frequent horizontal gene transfer (HGT). Although microbiologists are careful to exclude from their analysis genes and

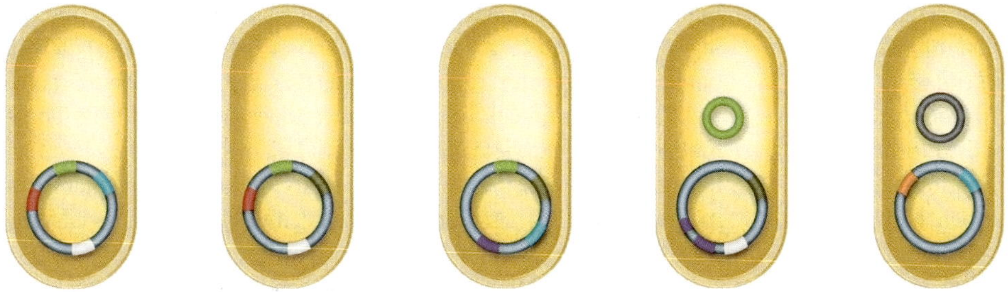

Figure 19.6. **Core and Pan-genomes.** The core and pan-genomes in cells comprising strains within a species. The core genome is common to all members (blue), while the pan-genome includes all sequences present in any member of the set.

proteins known to have been subject to HGT, the influence of HGT on phylogeny and evolution cannot be ignored. Indeed, there has been frequent gene transfer among all three domains. Clearly, the pattern of microbial evolution is not truly linear and treelike.

Assessing the impact of HGT on microbial evolution has been guided by genome sequence analysis. Genome comparisons of strains within a species and among species within a phylum have revealed that microbial genomes consist of an older core genome and a more recently acquired pan-genome (**figure 19.6**). The **core genome** is the set of genes found in all members of a species (or other monophyletic group). Thus it is thought to represent the minimal number of genes needed for the group of microbes to survive. In general, these genes encode informational proteins involved in DNA replication, transcription, and translation (e.g., rRNA genes). These genes are thought to have been present in the group's common ancestor. By contrast, the **pan-genome** consists of every gene in all strains of a species (or other taxonomic unit), so it includes the core genome plus every additional gene found in at least one strain. Genes outside the core genome are more recently acquired genes that enable microbial colonization of new niches. In general, genes unique to the pan-genome are considered to have been acquired by HGT. A comparison of the core genome size with the actual genome size of a particular strain thus indicates the evolution of new traits. For instance, current values of the core genome and pan-genome of *Bacillus anthracis* differ by only roughly 200 genes (about 3,600 versus 3,800, respectively), reflecting the limited genetic diversity within this species. By contrast, the *E. coli* core genome consists of about 2,800 genes, whereas some estimate that the pan-genome consists of roughly 37,000 genes. The broad genetic diversity among *E. coli* strains is illustrated by comparing the genomes of the nonpathogenic *E. coli* strain K12 and the pathogenic strain O157:H7. Their last common ancestor is estimated to have lived about 4.5 million years ago. During this time *E. coli* has mutated and exchanged genes, allowing the development of many strains. These have radiated to numerous habitats to which strains continue to adapt. Obviously, for any given species the estimated size of the core and pan-genome depends on the number of strains with sequenced genomes. Indeed, as more strains of each species are sequenced, the core genome tends to get smaller, as strains are discovered that lack genes once thought to be common to all genomes. It follows that as core genomes shrink, pan-genomes expand. ◀ *Mechanisms of genetic variation (chapter 16)*

Comprehension Check

1. Could a phylotype be considered an OTU? What about a species?
2. List the differences between distance-based and character-based methods for constructing a phylogenetic tree. Which type is maximum parsimony? Explain your answer.
3. What is the difference between a rooted and unrooted tree? Which provides more information?
4. You are building a tree based on 16S rRNA sequence alignments of a group of spirochetes. Suggest a possible outgroup so that you can build a rooted tree. Refer to chapter 21 for more information about these bacteria.
5. Is HGT involved in movement of genes in the core or pan-genome? Explain.

19.5 Evolutionary Processes and the Concept of a Microbial Species Inspire Debate

After reading this section, you should be able to:

a. Diagram the endosymbiotic theory of the origin of mitochondria and chloroplasts
b. Compare and contrast the two theories that address the origin of the nucleus
c. Explain why the concept of a microbial species is difficult to define
d. List the "gold standard" taxonomic methods currently applied to species designation
e. Explain the importance of adaptive mutations in giving rise to new ecotypes

It is our goal here to describe current models that seek to explain the evolution of new microbial species. Before we begin this discussion, however, it is helpful to consider the origin of all microbes—bacterial, archaeal, and eukaryotic. We then review the controversy that surrounds the word "species" as applied to bacteria and archaea. Only then will it be possible to understand and appreciate the evolutionary mechanisms that drive the development of new species of microorganisms.

Evolution of the Three Domains of Life

As we present in chapter 1, many scientists think that the first self-replicating entity was RNA. This is because RNA has the capacity to reproduce as well as to catalyze chemical reactions. It is thought that when RNA became enclosed in lipid spheres, the first primitive cell-like forms were generated (*see figures 1.5 and 1.6*). Considerable evidence indicates that by at least 3.5 billion years ago, such proto-cells (Greek *protos,* first) had evolved to form the ancestors of our extant microbes. Moreover, by 2.5 billion years ago, bacteria and archaea not only abounded, but each had evolved distinct taxonomic lineages. For instance, the Gram-positive bacterial phylum *Firmicutes* and Gram-negative phyla *Proteobacteria* and *Cyanobacteria* had developed. Indeed, the ancestors of modern cyanobacteria performed the oxygenic photosynthesis responsible for converting our anoxic planet to an oxygenated one.

A reexamination of the tree of life based on SSU rRNA (*see figure 1.2*) shows that the root of the tree is on the earliest region of the bacterial branch. As we discuss in chapter 1, the root is considered the *l*ast *u*niversal *c*ommon *a*ncestor, or LUCA. The placement of LUCA indicates that although bacteria and archaea share similar cellular construction, they are not phylogenetically linked. Because LUCA maps to the bacterial branch of the tree, it is thought that *Archaea* and *Eukarya* evolved independently of *Bacteria.* Recall that this was first suggested by Carl Woese and George Fox in the 1970s. Since that time, the distinction they inferred between bacterial and archaeal lineages has been confirmed by biochemical differences including membrane lipids, cell wall structure, and enzymes involved in gene transcription (**table 19.4**).

Although *Archaea* and *Eukarya* share a recent common ancestor, eukaryotes possess both archaeal and bacterial traits. There are several hypotheses that account for genes of both archaeal and bacterial ancestry on the eukaryotic nuclear genome. One hypothesis asserts that the first eukaryotic cell arose upon the fusion of an archaeon and a bacterium that lived in close association. Over time, archaeal genes involved in metabolism were lost while bacterial genes involved in information processing were also degraded. Contrary to this "single-step" hypothesis, others suggest a more multistep scenario involving a series of endosymbioses. Here an archaeal cell is thought to have engulfed a bacterium that donated genes that would eventually become the nuclear genome of an ancestral eukaryote. Most recently, evidence supporting the hypothesis that the first eukaryotic cell arose from within the archaeal lineage was reported. In 2017, archaeal DNA sequences recovered from marine sediments worldwide were found to encode a number of eukaryotic-like proteins, such as membrane proteins involved in phagocytosis and cytoskeletal components. These sequences were proposed to constitute a new archaeal superphylum, Asgard, that is more closely related to eukaryotes than to any other archaeal or bacterial lineage. The Asgard microbes have not been cultivated in the laboratory and are known only through DNA sequence fragments.

Unlike the uncertainty surrounding the origin of the first eukaryotic proto-cell, there is general agreement that mitochondria and chloroplasts arose by the incorporation of endosymbiotic bacteria. Like bacteria, most mitochondria and chloroplasts have a single, circular chromosome and undergo binary fission. In addition, mitochondria and chloroplasts have 70S, not 80S, ribosomes. These observations led to the development of the **endosymbiotic hypothesis** by Lynn Margulis (**figure 19.7**). This posits the following series of events. An ancestral eukaryotic cell lost its rigid cell wall but had evolved actin (or its precursor) that enabled amoeboid motility. This nucleated cell was thus able to develop endocytosis. These mobile proto-eukaryotes became predators of other cells, including bacteria. Predation imposed selection for cellular enlargement and increased motility. Engulfment without digestion of bacterial prey evolved because a smaller bacterial cell provided energy for the larger host cell, while the host protected and supplied nutrients to the bacterial cell. Because endocytosis and oxidative phosphorylation are not compatible on the same membrane, the endosymbiont was retained as a separate subcellular entity. The energy supplied by the endosymbiont conferred a growth advantage to the proto-eukaryote, enabling its dominance over and eventually eliminating other cells that lacked both cell walls and endosymbionts.

As the endosymbiont became more dependent on its host for nutrients and protection, there was little selective pressure for the retention of genes involved in these processes. Conversely, there was strong selective pressure to retain the genes involved in energy conservation. Thus genes whose products were redundant to the host were eventually lost. In fact, such genome reduction is the rule, rather than the exception, among obligate intracellular microbes (*see figure 18.18*).

Ultimately, the endosymbiont evolved into an energy-providing organelle like mitochondria, or other mitochondria-like organelles (e.g., hydrogenosomes and mitosomes). Hydrogenosomes are found in some protists where, like mitochondria, they take up pyruvate that results from glycolysis within the host cytoplasm. Unlike mitochondria, however, pyruvate in the hydrogenosome is reduced to acetate, H_2, and CO_2 with ATP generated (*see figure 5.15*). The similarity between certain key genes (and thus their protein products) supports the notion that hydrogenosomes and mitochondria evolved from a single common ancestor, most likely a proteobacterium. Mitosomes are found in some protists. Of these three organelles, mitochondria appear to be most highly derived (i.e., continued to evolve), since these organelles are the site of oxidative phosphorylation. ◀ *Mitochondria, related organelles, and chloroplasts are involved in energy conservation (section 5.6)* ▶ *The proteobacterial origin of mitochondria (section 22.1);* Microsporidia *are intracellular parasites (section 25.7)*

Chloroplasts arose when these new aerobic eukaryotes engulfed a cyanobacterium—probably an ancestor of *Prochlorococcus.* Again, this led to the development of a mutualistic relationship that evolved into our extant green plants and algae—organisms that possess both mitochondria and chloroplasts. Such endosymbioses exist today in certain protists that retain living cyanobacteria or the functional chloroplasts of their algal prey. Here it is thought a eukaryotic cell with mitochondrion-like organelles engulfed a photosynthetic cell that possessed both mitochondria and chloroplasts. Recent genomic data support this hypothesis, that is to say that plants arose from an ancestral protist that engulfed a cyanobacterium. The situation is more complicated for red algae, which are derived from a protist that engulfed an ancient photosynthetic eukaryote.

Table 19.4 Comparison of *Bacteria, Archaea,* and *Eukarya*

Property	*Bacteria*	*Archaea*	*Eukarya*
Membrane-Enclosed Nucleus with Nucleolus	Absent	Absent	Present
Complex Internal Membranous Organelles	Absent	Absent	Present
Cell Wall	Almost all have peptidoglycan containing muramic acid	Variety of types, no muramic acid	No muramic acid
Membrane Lipid	Ester-linked, straight-chained fatty acids	Ether-linked, branched isoprene-derived chains	Ester-linked, straight-chained fatty acids
Gas Vesicles	Present	Present	Absent
Transfer RNA			
Thymine	Present in most	No thymine in T or TΨC arm	Present
Amino acid carried by initiator tRNA	*N*-formylmethionine	Methionine	Methionine
Polycistronic mRNA	Present	Present	Present in some protists
mRNA Introns	Rare	Rare	Present
mRNA Splicing, Capping, and Poly-A Tailing	Absent	Absent	Present
Ribosomes			
Size	70S	70S	80S
Elongation factor 2 reaction with diphtheria toxin	Does not react	Reacts	Reacts
Sensitivity to chloramphenicol and kanamycin	Sensitive	Insensitive	Insensitive
Sensitivity to anisomycin	Insensitive	Sensitive	Sensitive
DNA-Dependent RNA Polymerase			
Number of enzymes	One	One	Three
Structure	Simple subunit pattern (6 subunits)	Complex subunit pattern similar to eukaryotic enzymes (8–12 subunits)	Complex subunit pattern (12–14 subunits)
Rifampicin sensitivity	Sensitive	Insensitive	Insensitive
RNA Polymerase II Type Promoters	Absent	Present	Present
Metabolism			
Similar ATP synthase	No	Yes	Yes
Methanogenesis	Absent	Present	Absent
Nitrogen fixation	Present	Present	Absent
Chlorophyll-based photosynthesis	Present	Absent	Present[1]
Chemolithotrophy	Present	Present	Absent

1 Present in chloroplasts (of bacterial origin).

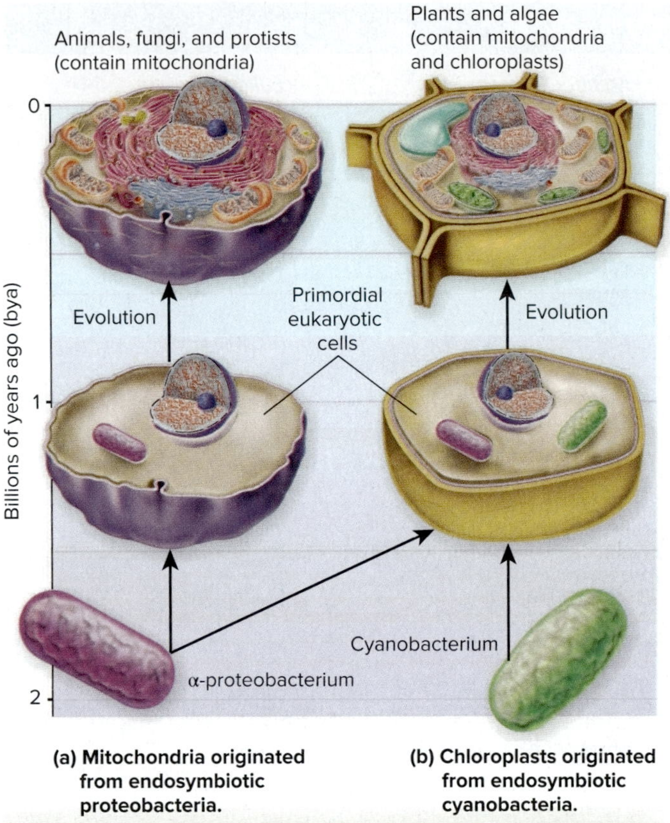

Figure 19.7 The Endosymbiotic Theory. (a) According to this hypothesis, mitochondria derived from a proteobacterium. (b) A similar phenomenon occurred for chloroplasts, which derived from cyanobacteria.

MICRO INQUIRY *On what evidence is this hypothesis based?*

▶| *Phylum* Cyanobacteria: *oxygenic photosynthetic bacteria (section 21.4); Protists (chapter 24)*

What Is a Microbial Species?

The term "species concept" describes a theoretical framework used to understand how and why certain organisms can be sorted into discrete taxonomic groups. We discussed "species definition" in sections 19.2 and 19.3, when we reviewed the criteria used to identify a microbial genus, species, or strain. From this, we can see that species definition is the application of the species concept.

Both species concept and definition have changed over time and continue to be difficult for microbiologists to agree upon. Because bacteria and archaea lack sexual reproduction, extensive morphological features, and a fossil record, microbiologists are at a distinct disadvantage when defining species, as compared to biologists studying other forms of life.

Historically, the application of different criteria in making species assignments has led to taxonomic confusion. In some cases, a single microbial species is so metabolically and genetically diverse that it seems probable that the group represents multiple species. On the other hand, some species are very narrowly defined, such that two species differ very little. For instance, *Bacillus anthracis* strains are so similar to *B. cereus,* many believe that all *B. anthracis* strains are really members of the *B. cereus* species. It is argued that only because *B. anthracis* causes anthrax does it have its own species designation.

In an effort to clarify and standardize microbial taxonomy, the ICSP recommends four criteria to meet a "gold standard" for species assignment: The microbe must be phenotypically similar to others in the group, whole genome similarity as determined by DNA-DNA hybridization must be at least 70%, the melting temperature of the DNA (a reflection of the G + C content) within 5°C, and less than 3% divergence in rRNA gene sequence. As noted previously, these techniques may soon be replaced with a genomic metric, such as average nucleotide identity. Even with these proposed updates, some remain uncomfortable with these criteria. They point out that two microorganisms with, for instance, only 75% similarity in DNA and 98% rRNA gene sequence identity can be considered the same species, but if these criteria were applied to eukaryotes, all primates (monkey, apes, you) would be lumped together as a single species! Indeed, it remains unresolved whether the species concept can be applied to microbes.

Microbial Evolutionary Processes

While the debate regarding the operational definition of a microbial species continues, microbiologists agree that the microbial species concept is grounded in natural selection and evolution. As the most ancient life forms on Earth, bacteria and archaea have evolved and adapted to virtually every habitat. While their diverse metabolic strategies and ability to tolerate extreme conditions explain *why* microbes display such enormous diversity, natural selection explains *how* this diversity came to be.

Recall that genetic diversity in members of *Archaea* and *Bacteria* must occur asexually. Thus heritable genetic changes in these organisms are introduced principally by two mechanisms: mutation and HGT, both of which are subject to natural selection. Generally speaking, it is thought that mutation drives initial speciation events, and HGT permits more rapid radiation thereafter. In other words, a new species must arise from its single, ancestral species. This makes sense if one assumes that the ancestral population of microbes was genetically homogeneous. Genetic variation can only arise within a population possessing identical (or nearly identical) genomes by mutation, gene loss and gain, and intragenomic recombination (**figure 19.8*a***). By definition, there is not enough genetic diversity to drive speciation among such a population by HGT.

Anagenesis, also known as **genetic drift,** refers to small, random genetic changes that occur over generations. It might

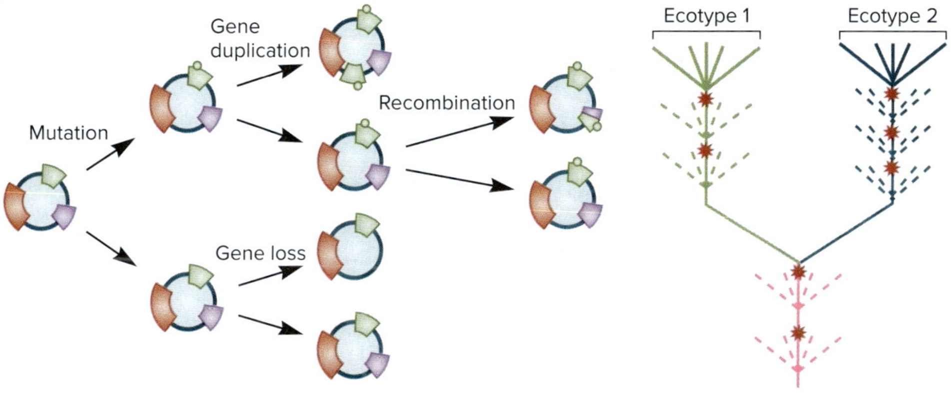

(a) Mechanisms of genetic variation within a homogeneous population **(b)** Stable ecotype model

Figure 19.8 Evolution of Microbial Diversity. (a) Mechanisms by which a single, genetically homogeneous population of microbes can develop genetic variation include mutation, gene loss and duplication, and recombination. (b) Genetic changes within a population can lead to the development of new ecotypes. Each ecotype is subject to periodic selection events (indicated by stars) that enable cells with adaptive mutations to outcompete other lineages, which are eventually driven to extinction (dotted lines). The solid lines represent successful populations or lineages; those at the top are extant.

MICRO INQUIRY *Construct a scenario in which each of the following factors leads to the establishment of two ecotypes from a single common ancestor, as shown in (b): the availability of carbon and nitrogen sources; terminal electron acceptor; and mean local temperature.*

seem that very small genetic differences within a microbial population would be of little evolutionary significance. However, model studies designed to assess competition between microbial populations has led to some surprising observations. When selection is applied, very small genetic differences can result in one population overtaking another. How does this happen when individuals within a population have similar mutation rates and most of these mutations are neutral and have no phenotypic effect? Only those rare mutations that confer a growth advantage, called **adaptive mutations,** are retained and passed from one generation to the next, in which case we say the mutation is fixed. The descendants of that individual continue to evolve through mutation and other intraspecific mechanisms.

Adaptive mutation is key to the ecotype model of microbial evolution. An **ecotype** is a population of microbes that is genetically very similar but ecologically distinct. Ecotypes arise when members of a microbial population living in a specific ecosystem undergo a genetic event (or series of events) that enables them to outcompete the remainder of the population. According to the ecotype model, the acquisition of adaptive mutations ultimately drives the remaining members of the population into extinction and reduces the amount of genetic diversity within the surviving population (figure 19.8*b*). The fossil record shows that evolution does not always occur at a constant rate but is periodically interrupted by rapid bursts of speciation driven by abrupt environmental changes. Niles Eldredge and Steven Jay Gould coined the term **punctuated equilibria** to describe this phenomenon. Certainly, the 3.5-billion-year history of microbial life on Earth affords the accumulation of many, many mutations; that in turn has resulted in vast speciation. ◀ *Mechanisms of genetic variation (chapter 16); Comparative genomics (section 18.7)*

Unlike variation introduced in the ecotype model, HGT-driven genetic variation requires genetically diverse groups of microbes. This is because HGT does not rely on replication but rather on the exchange of genetic material between microbes. The rate of HGT is extremely variable. Some microbes have very reduced genomes with no evidence of HGT. These microbes are generally highly adapted to a specific, stable ecological niche. The most extreme examples are obligate intracellular symbionts that can only grow within their host cells, where no other microbes exist with which to exchange genes. By contrast, some microbes appear to have high rates of HGT with more than half of their genome acquired from other organisms, as is the case for the members of the ancient phylum *Thermotogae*. In such microbes, genes acquired by HGT frequently expand metabolic capabilities, thereby enabling rapid adaptation to new environmental challenges. ▶ Aquaficae *and* Thermotogae *are ancient bacterial lineages (section 21.1)*

Finally, we pose the question: How many bacterial and archaeal species are there? There are two major obstacles to formulating an answer. First, most microbial species resist growth in the laboratory, so they can only be detected by metagenomic or other culture-independent approaches. Second, as we have seen, microbiologists cannot agree on a biological species concept. So we must resort to the operational definition of a species in terms of nucleic acid homologies, as well as similarities in physiology, morphology, and ecology. Based on these criteria, estimates range from 100,000 to 1,000,000 species in nature, with about 10^{30} individual cells. These estimates reveal that there are probably about a billion more microbes on Earth than stars in the universe.

Comprehension Check

1. Define ecotype. Do you think it is necessary to obtain microbes in pure culture before assigning different ecotypes? Explain.
2. What is the difference between the core genome and pan-genome? What might you infer if you compare two genera, one

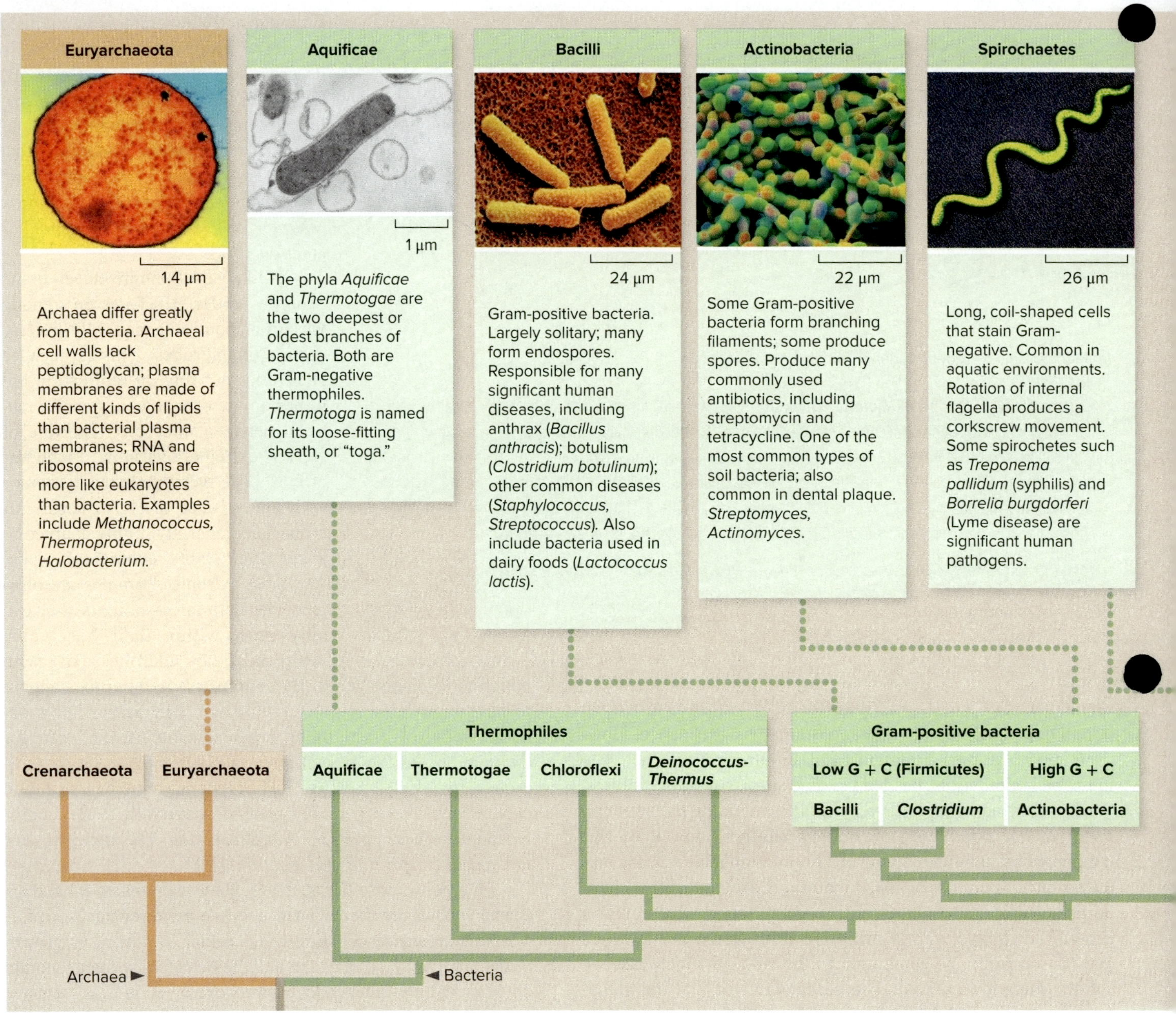

Figure 19.9 Some Major Clades of Bacteria and Archaea. This classification scheme is based on *Bergey's Manual of Systematics of Archaea and Bacteria.*
(a) ©SPL/Science Source; (b) ©Karl O. Stetter; (c) ©Andre Syred/SPL/Science Source; (d) ©Microfield Scientific Ltd/Getty Images; (e) ©Alfred Paseika/SPL/Science Source; (f) ©McGraw-Hill Education/Don Rubbelke photographer; (g) ©Dr. Kari Lounatmaa/Science Source; (h) Source: CDC/Janice Haney Carr; (i) ©Derek Lovley/Science Source

in which the size of the core genome and pan-genome are very similar, and one in which the core genome is much smaller than the pan-genome?

3. Of the following genes, which do you think are part of the pan-genome and which are part of the core genome: the genes for lactose catabolism in *E. coli*; the genes for heat-stable DNA polymerase in *Thermus aquaticus*; the genes for proteorhodopsin in marine bacteria; the genes for toxin production in *Vibrio cholerae*?
4. Would a protein encoded on the core genome or one encoded only on the pan-genome be best to use in constructing a phylogenetic tree? Explain your answer.

19.6 *Bergey's Manual of Systematics of Archaea and Bacteria*

After reading this section, you should be able to:

a. Employ *Bergey's Manual* to investigate the defining taxonomic elements used for a bacterium or archaeon that is unfamiliar to you

In 1923 David Bergey, professor of bacteriology at the University of Pennsylvania, and four colleagues published *Bergey's Manual of Determinative Bacteriology,* a classification of

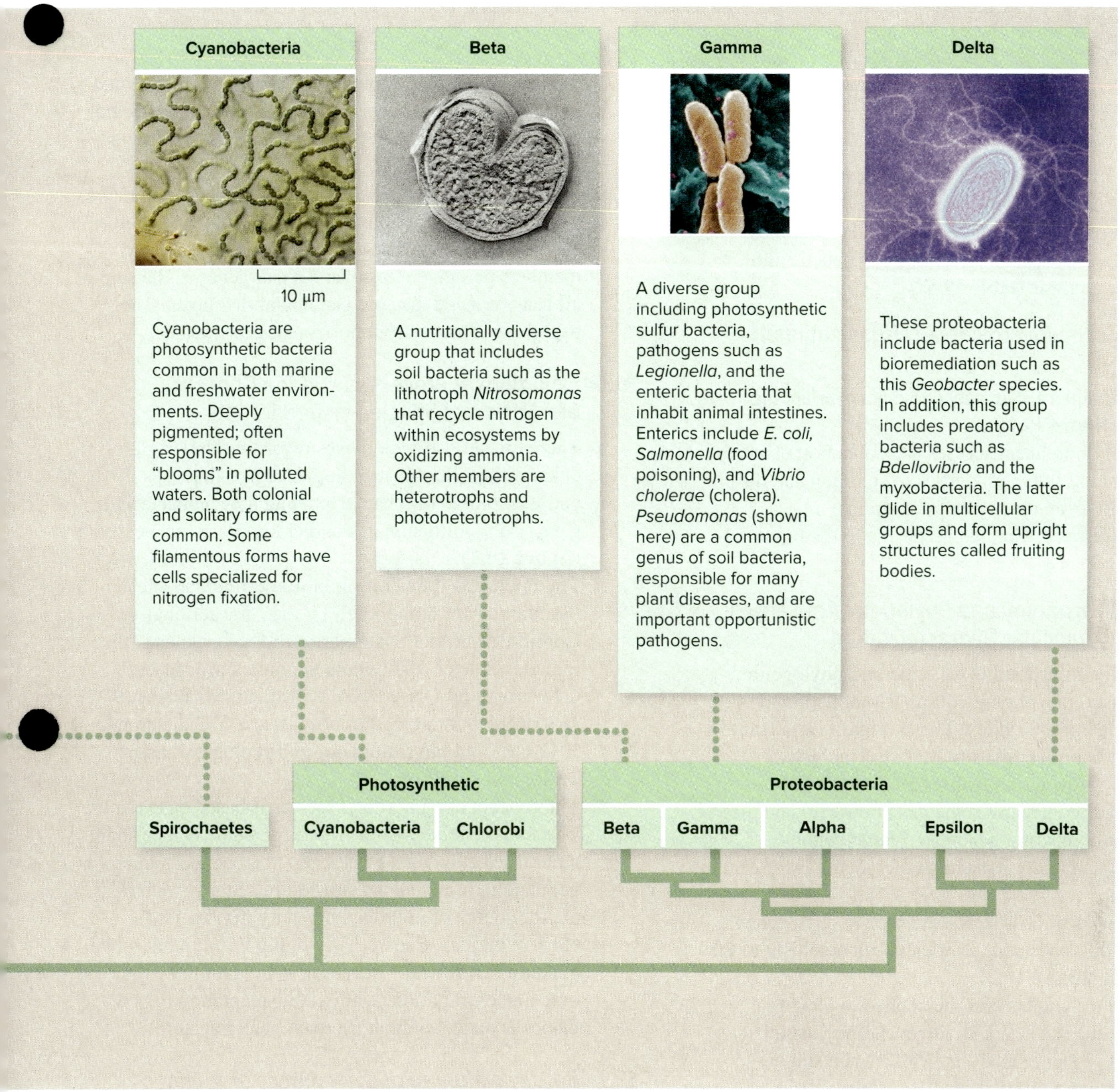

bacteria that could be used for the identification of many bacterial species. Nine editions of this manual were published in the twentieth century. Despite its age, this text continues to serve as a relatively brief reference guide in the identification of bacteria based on physiological and morphological traits. A related publication, *Bergey's Manual of Systematic Bacteriology* included the archaea for the first time in 1984. The final edition of this reference comprised five volumes, and included morphology, physiology, growth conditions, ecology, and other information. Classification was organized based on phylogeny, and clinically important bacteria were not discussed separately, but integrated into the phylogenetic scheme. Both references are referred to simply as *Bergey's Manual* and in the twenty-first century, the current manifestation is an online reference entitled *Bergey's Manual of Systematics of Archaea and Bacteria*. This electronic format allows for the resource to be frequently updated as new taxa are described and validated. **Figure 19.9** illustrates most of the groups covered in *Bergey's Manual* and in chapters 20–23.

Comprehension Check

1. *Bergey's Manual* is no longer based on phenetic classification. Why is this the case?
2. Describe two different situations in which it would be essential to identify the genus and species of a bacterium or archaeon.

Key Concepts

19.1 Microbial Taxonomy Is Based on the Comparison of Multiple Traits

- Taxonomy, the science of biological classification, is composed of three parts: classification, nomenclature, and identification.
- A polyphasic approach is used to classify microbes. This incorporates information gleaned from genetic and phenotypic analysis (**table 19.1**).

19.2 Taxonomic Ranks Provide an Organizational Framework

- Taxonomic ranks are arranged in a nonoverlapping hierarchy (**figure 19.1**).
- A bacterial or archaeal species is a collection of strains that have many stable properties in common and differ significantly from other groups of strains.
- Microorganisms are named according to the binomial system.

19.3 Microbial Taxonomy and Phylogeny Are Largely Based on Molecular Characterization

- Historically, microbial taxonomic and phylogenic analysis used morphological, physiological, and ecological characteristics. These remain important in building a complete picture that also includes molecular information (**tables 19.2 and 19.3**).
- Nucleic acid sequencing is the most powerful and direct method for comparing genomes. Whole-genome sequencing is rapidly replacing other methods for genome comparison.
- The sequences of SSU rRNA are used in phylogenetic studies of microbes and can identify an organism to the genus level (**figure 19.2**).
- Additional techniques must be applied to identify a microbe at the species or strain level. They include multilocus sequence analysis (MLSA) and single nucleotide polymorphism analysis (**figure 19.3**).
- Signature sequences and indels are conserved in various taxa and provide important information on phylogenetics.

19.4 Phylogenetic Trees Illustrate Evolutionary Relationships

- Phylogenetic relationships often are shown as branched diagrams called phylogenetic trees. Trees are based on pairwise comparison of amino acid or nucleotide sequences, followed by computer analysis (**figure 19.4**).
- Trees may be either rooted or unrooted and are created in several different ways. Unrooted trees can be rooted by including an outgroup when the tree is constructed (**figure 19.5**).
- Microbes have a long history of horizontal gene transfer, which confuses taxonomic analysis. Complete genome analysis has revealed a set of core genes found in all members of a given taxon, and a pan-genome, the sum of all genes outside the core genome of that taxon. The pan-genome arises from horizontal gene transfer.

19.5 Evolutionary Processes and the Concept of a Microbial Species Inspire Debate

- There are several hypotheses regarding the origin of eukaryotic cells. Most biologists agree that endosymbioses of a bacterium and a cyanobacterium gave rise to mitochondria and chloroplasts, respectively (**figure 19.7**).
- The operational definition of a microbial species is based on criteria approved by the International Committee on the Systematics of Prokaryotes. This includes at least 70% whole genome similarity as determined by DNA-DNA hybridization, at least 97% 16S rRNA homology, no more than a 5°C difference in % G + C, and physiological, morphological, and ecological similarity.
- The concept of a microbial species is based on evolution. The ecotype model describes the outcome of a periodic natural selection on a genetically homogeneous microbial population. Individuals that acquire adaptive mutations are the source of microbial diversity (**figure 19.8**).
- Horizontal gene transfer is also important in microbial evolution. However, speciation is thought to be the outcome of mutation, while rapid adaptation to new niches is mediated by horizontal gene transfer.

19.6 *Bergey's Manual of Systematics of Archaea and Bacteria*

- *Bergey's Manual of Systematics of Archaea and Bacteria* is based on the accepted system of prokaryotic taxonomy.
- Comparisons of nucleic acid sequences, particularly 16S rRNA, are the foundation of this classification.

Active Learning

1. Consider the fact that the use of 16S rRNA sequencing as a taxonomic and phylogenetic tool has resulted in tripling the number of bacterial phyla. Why has the advent of this genetic technique expanded the currently accepted number of microbial phyla?
2. You have recently established a pure culture of a new archaeon from soil. Describe the approaches you would use to identify your new microbe to the species level.
3. Discuss the problems in developing an accurate phylogenetic tree. Do you think it is possible to create a completely accurate universal phylogenetic tree? Explain your answer.
4. Why is the current classification system for *Bacteria* and *Archaea* likely to change considerably? How would one select the best features to use in the identification of unknown microbes and determination of relatedness?
5. How would you interpret a whole-genome sequence that is overall 45% G + C but shows a 20,000 bp region that is 55% G + C?
6. Horses were introduced to Iceland over 1,000 years ago and have remained an isolated population, not exposed to contagious diseases. A recent respiratory epidemic caused symptoms in almost all horses on the island. Viral infection could not be demonstrated, but *Streptococcus equi* subsp. *zooepidemicus* was recovered from nasal swabs of all infected animals. Although this bacterium was believed to be a commensal, whole-genome sequencing was performed on over 300 isolates. The population comprised four clades defined by multilocus sequencing analysis. The data table lists the distribution of these clades.

Clade	Number of horses infected	Number of locations	Number of SNPs in clade
1	45	21	25
2	14	10	153
3	14	8	151
4	8	1	44

Working on the hypothesis that a pathogenic strain of *S. zooepidemicus* was recently introduced to Iceland, which of the four clades is most likely to represent that strain? Why? Researchers compared the CRISPR spacers in different isolates to examine their relatedness. Why are CRISPR spacers a good measure of strain relatedness? (Review section 14.6 on CRISPR.)

Read the original paper: Björnsdóttir, S., et al. 2017. Genomic dissection of an Icelandic epidemic of respiratory disease in horses and associated zoonotic cases. *mBio.* 8:e00826-17.

7. *Achromatium oxaliferum* is an aquatic sulfur-oxidizing microbe with extremely large polyploid cells. Microscopy following DNA staining reveals that DNA is distributed around the cell, but not uniformly. This organism is not cultured in the lab, rather it is isolated directly from the environment; and because of its cell size, it can be manipulated to isolate single cells. Genomic sequencing of single cells revealed multiple SSU rRNA genes per cell with 93 to 95% similarity, less similarity than should exist within a single genome. Fluorescence in situ hybridization with probes specific for one SSU rRNA demonstrated that a single cell contains a mixture of SSU rDNA sequences. Comparable results were found for other genes, and the authors conclude that *Achromatium* doesn't have multiple copies of a single genome, but instead it appears to have a diverse population of genomes within a single cell. How would you define a species in this genus? How would you identify the core genome? Insertion sequences and transposases are abundant in this organism. How might these affect the observed genetic diversity in a cell?

Read the original paper: Ionescu, D., et al. 2017. Community-like genome in single cells of the sulfur bacterium *Achromatium oxaliferum*. *Nature Communications* 8:455.

20

Archaea

©Mike Groll/AP Images

Methanogenic Archaea Fuel Domestic Energy Debate

Long before gas prices surged and politicians began to argue about energy security, archaea were busy generating methane, otherwise known as natural gas. Although methane deposits are billions of years old, only recently have geologists recognized that at least 2,300 trillion cubic feet (Tcf) of natural gas could be recovered within the United States. This upward revision is largely due to the inclusion of methane trapped in underground shale. This domestic source of gas, much of it under Appalachian and Rocky Mountain states, provides over 65% of the total U.S. gas supply. This should be good news as the United States consumes about 27 Tcf annually. But there is a hitch. The methane has to be extracted from the earth.

Before we tackle the extraction problem, let's discuss how all that natural gas got there in the first place. There are two sources of methane: abiotic and biotic. Abiotic methane results principally from thermal decomposition of sedimentary rocks and is almost negligible compared to that produced by microbes. All biologically produced methane is generated by methanogenic archaea from either $CO_2 + H_2$ or acetate (section 20.3). Evidently, shale beneath the United States has been home to a vast and productive archaeal community for eons.

Although the abundance of natural gas in shale has been known for a long time, only recently has methane recovery become financially worthwhile. The price of extracting the gas trapped within the soft, porous shale far exceeded any later profits. But the recent development of horizontal drilling and hydraulic fracturing ("fracking") technologies changed the economic equation. Now a typical well can be drilled vertically about 3,000 meters and then extended just as far horizontally. Fracking involves high-pressure pumping of up to 9 million liters of water, sand, and chemicals into the well. This cracks the shale, while a "propping agent" keeps the pores open so the gas can be recovered.

However, some citizens in communities where fracking takes place are concerned about groundwater contamination, fracking-induced earthquakes, the escape of contaminated drilling fluids at the well surface, and methane leaking from the wells, thereby increasing the carbon footprint of natural gas usage. Federal regulations mandate that recovered drilling fluids be treated before they are discharged back into the environment. In some cases, drillers have used municipal wastewater treatment facilities for this, creating a tremendous burden on local infrastructure. Escaping methane is considered problematic because methane is a harmful greenhouse gas (*see section 28.2*). If the United States is to continue to use this tremendous natural resource, these concerns must be addressed to the satisfaction of producers and consumers.

In this chapter, we review archaea, which were once considered almost exclusively extremophiles—organisms that thrive in extreme conditions, where they do unusual things, such as make methane. However, we know now that they are physiologically and ecologically diverse, found in both mundane and challenging environments. Indeed, the third domain of life deserves our respect, especially if they are to provide us with energy to fuel our power grid.

Readiness Check:

Based on what you have learned previously, you should be able to:

- ✓ Draw the basic structure of the phylogenetic tree of life based on SSU rRNA sequences and explain the placement of *Archaea* (section 1.2)
- ✓ Describe an archaeal cell wall and membrane, including component parts (section 4.2)
- ✓ List environmental factors that influence microbial growth, including oxygen (section 7.5)
- ✓ Describe the difference between heterotrophy and autotrophy (section 11.1)
- ✓ Define chemolithotrophy and identify an example of an electron donor and an acceptor (section 11.10)
- ✓ Review archaeal DNA replication, gene transcription, and protein synthesis and secretion (chapter 15)
- ✓ Describe the use of metagenomics in the discovery of new microbial taxa (section 18.3)

20.1 Overview of *Archaea*

After reading this section, you should be able to:

a. List some common habitats in which archaea reside
b. Describe the challenges in archaeal taxonomy
c. Compare at least three key metabolic pathways that are central to archaeal physiology with those used by bacteria

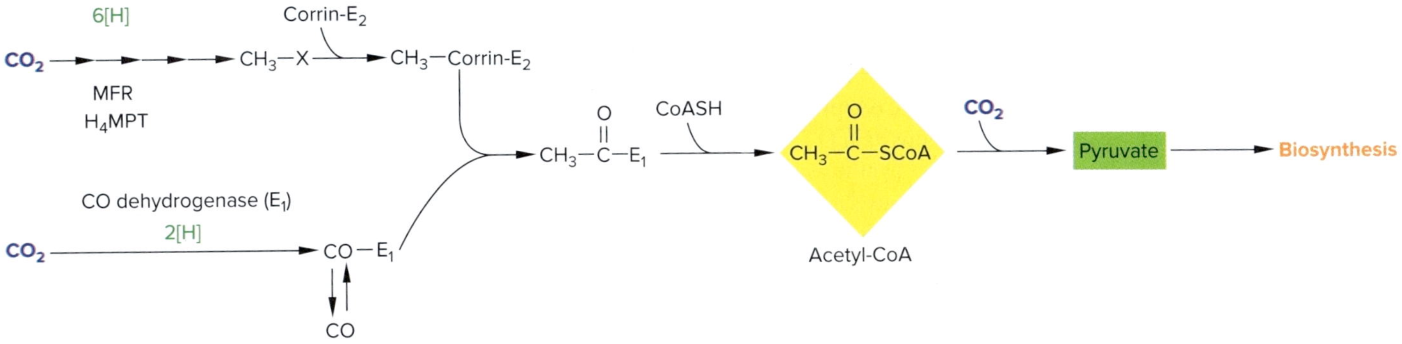

Figure 20.2 The Reductive Acetyl-CoA Pathway in *Archaea*. Two molecules of CO_2 are reduced, each by a different mechanism, and combined to form an acetyl group and then acetyl-CoA.

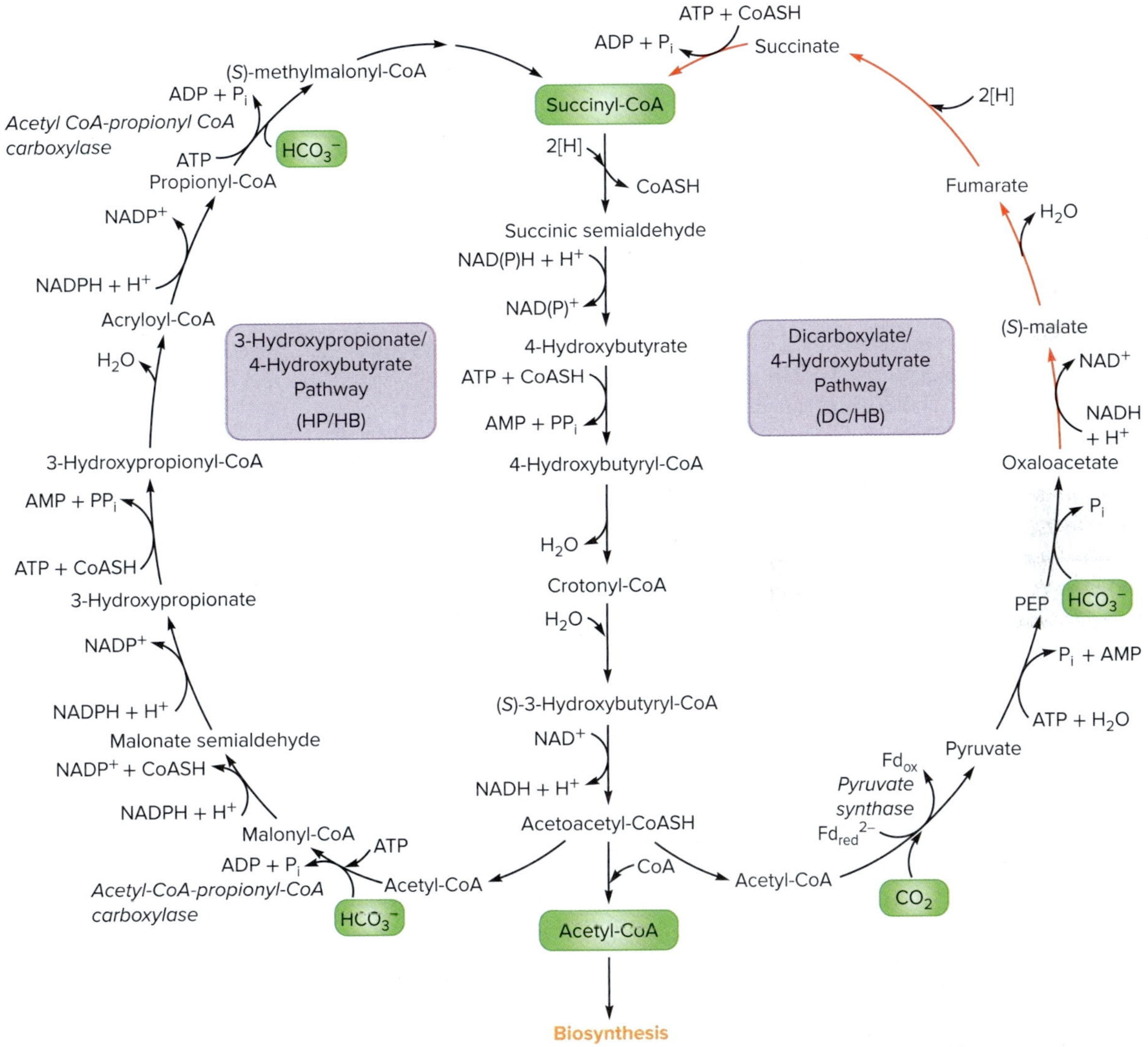

Figure 20.3 The 3-Hydroxypropionate/4-Hydroxybutyrate (HP/HB) and Dicarboxylate/4-Hydroxybutyrate (DC/HB) Cycles. Aerobic crenarchaea and some thaumarchaea use the HP/HB pathway (left) to convert two bicarbonate ions into acetyl-CoA. Some crenarchaea fix one molecule of CO_2 and one HCO_3^- using the anaerobic DC/HB pathway (right). The reactions shown with red arrows are also used in the reductive tricarboxylic acid cycle.

MICRO INQUIRY *Given that the DC/HB pathway uses far less ATP per pyruvate synthesized than the HP/HB pathway, why do some archaea use the HP/HB pathway instead?*

has other advantages. It has a low demand for metal cofactors, and unlike either the acetyl-CoA pathway or the DC/HB cycle, it can operate under aerobic conditions (table 20.1). The key carboxylase enzyme, a biotin-dependent acetyl-CoA/propionyl-CoA carboxylase, accepts two bicarbonate ions (HCO_3^-) in a series of steps that yields succinyl-CoA and net synthesis of one acetyl-CoA. Interestingly, bacteria and eukaryotes use acetyl-CoA carboxylase as the first step in fatty acid biosynthesis, but because archaea do not make fatty acids, this enzyme can be dedicated to carbon assimilation.

The last autotrophic pathway known to occur in archaea is the anaerobic DC/HB pathway (figure 20.3). It is found in the anaerobic and microaerophilic members of two crenarchaeote families, *Thermoproteales* and *Desulfurococcales*. The DC/HB pathway starts when an acetyl-CoA and a CO_2 molecule are condensed to form pyruvate, which is converted to phosphoenolpyruvate (PEP). PEP carboxylase adds HCO_3^- to PEP to form oxaloacetate (OAA). Through a series of reductive tricarboxylic acid cycle steps (i.e., backward TCA cycle steps), OAA is converted to succinyl-CoA, which is transformed through 4-hydroxybutyrate into two molecules of acetyl-CoA. As in the HP/HB pathway, one acetyl-CoA is then used for biosynthesis, and the other is used as a CO_2 acceptor in the next round of the cycle. The DC/HB cycle uses only five ATP per pyruvate made, but some of its enzymes are oxygen sensitive. Thus while the conversion of succinyl-CoA to acetyl-CoA is the same in both the HP/HB and DC/HB pathways, aerobes must perform the more expensive HP/HB pathway.

Inspection of figures 20.2 and 20.3 reveals that autotrophic pathways used by archaea result in the production of the 2-carbon compound (C2) acetyl-CoA that must be converted to the central metabolite pyruvate (C3) for anabolic purposes. The most common means by which this is accomplished is through the glyoxylate cycle (**figure 20.4**). Here the enzyme isocitrate lyase splits isocitrate (C6) into glyoxylate (C2) and succinate (C4); glyoxylate is condensed with acetyl-CoA (C2) to form malate (C4). While autotrophic archaea and haloarchaea such as *Haloferax volcanii* use the glyoxylate cycle, other haloarchaea use the methylaspartate pathway. This pathway also yields malate and succinate but requires many more steps. In the methylaspartate pathway (figure 20.4), α-ketoglutarate (C5) is converted to the amino acid glutamate, which is rearranged to form the nonstandard amino acid methylaspartate. The ammonia is next removed, and several steps are required before the cleavage of a C5 intermediate into propionyl-CoA (C3) and glyoxylate (C2). Just as in the glyoxylate pathway, acetyl-CoA condenses with glyoxylate to form malate, which is used for biosynthesis. Unlike the glyoxylate pathway, another CO_2 (as HCO_3^-) is accepted by propionyl-CoA to form succinate, which is also converted to malate. All this begs the question, why use a more complicated (and therefore costly) mechanism? The answer may be linked to the fact that so far, the methylaspartate pathway is found only among haloarchaea. These microbes are under constant osmotic stress in the high-salt environments they inhabit. Glutamate is used as an osmoprotectant, so it is hypothesized that haloarchaea may use the accumulation of intracellular glutamate as a signal to activate the

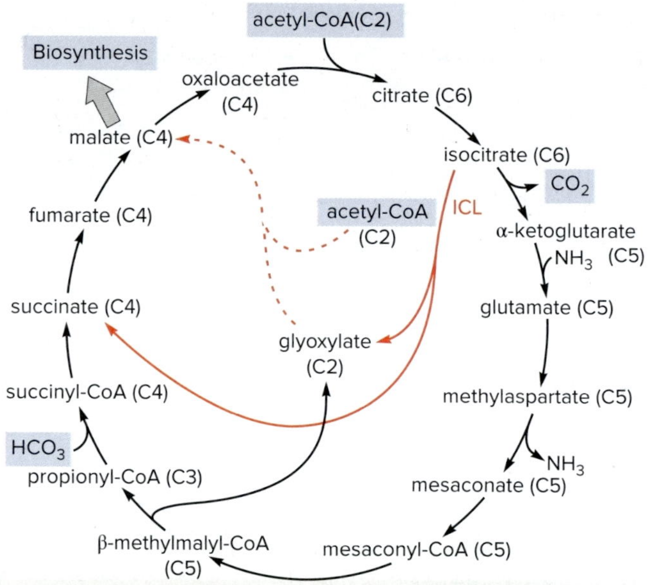

Figure 20.4 Glyoxylate and Methylaspartate Pathways Are Used for Acetyl-CoA Assimilation. Autotrophic archaea and some haloarchaea use the glyoxylate pathway (red and dashed arrows) to incorporate acetyl-CoA. Other haloarchaea lack the key enzyme isocitrate lyase (ICL) and assimilate acetyl-CoA via the longer methylaspartate pathway instead (black and dashed arrows).

MICRO INQUIRY *Why might the production of glutamate as an intermediate be beneficial for some haloarchaea?*

methylaspartate pathway when growing on acetate or compounds that produce acetate as an intermediate. ◀ *Solutes affect osmosis and water activity (section 7.5)*

Evidence suggests that archaeal pathways for amino acid, purine, and pyrimidine biosynthesis are similar to those in other organisms. Some methanogens can fix atmospheric dinitrogen, and this pathway also seems to be well conserved among bacteria and archaea. ◀ *Anabolism: the use of energy in biosynthesis (chapter 12)*

Chemoorganotrophic Pathways

Among chemoorganotrophic archaea, at least three novel carbohydrate catabolic pathways have been identified: a modified Embden-Meyerhof (EM) pathway and two variations of the Entner-Doudoroff pathway. The modified EM pathway used by hyperthermophilic euryarchaeotes involves several novel enzymes, including an ADP-dependent glucokinase and phosphofructokinase (**figure 20.5**). These kinases each result in the production of AMP; in fact, unlike the classical EM pathway, there is no net ATP produced. Another unique step in this modified EM pathway is the conversion of glyceraldehyde 3-phosphate directly to 3-phosphoglycerate. Some archaea use ferredoxin as the electron acceptor, while others use NAD^+.

The two Entner-Douderoff pathways include one called the semi-phosphorylative pathway and another dubbed the

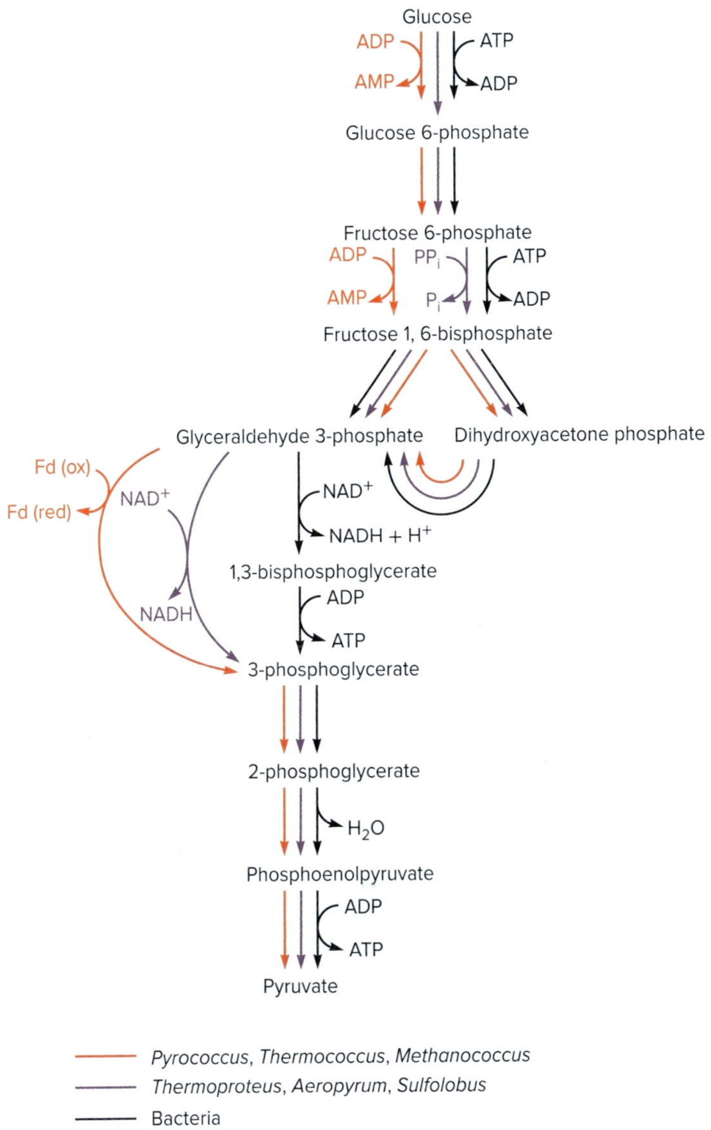

Figure 20.5 Modified Embden-Meyerhof Pathway. The pathway used by some archaea compared to that of bacteria. The only difference between the pathway used by *Thermoproteus* spp. and the other archaea shown is the use of NAD$^+$ as the electron acceptor rather than ferredoxin when glyceraldehyde 3-phosphate is oxidized to 3-phosphoglycerate. There is no net ATP produced in the archaeal pathway.

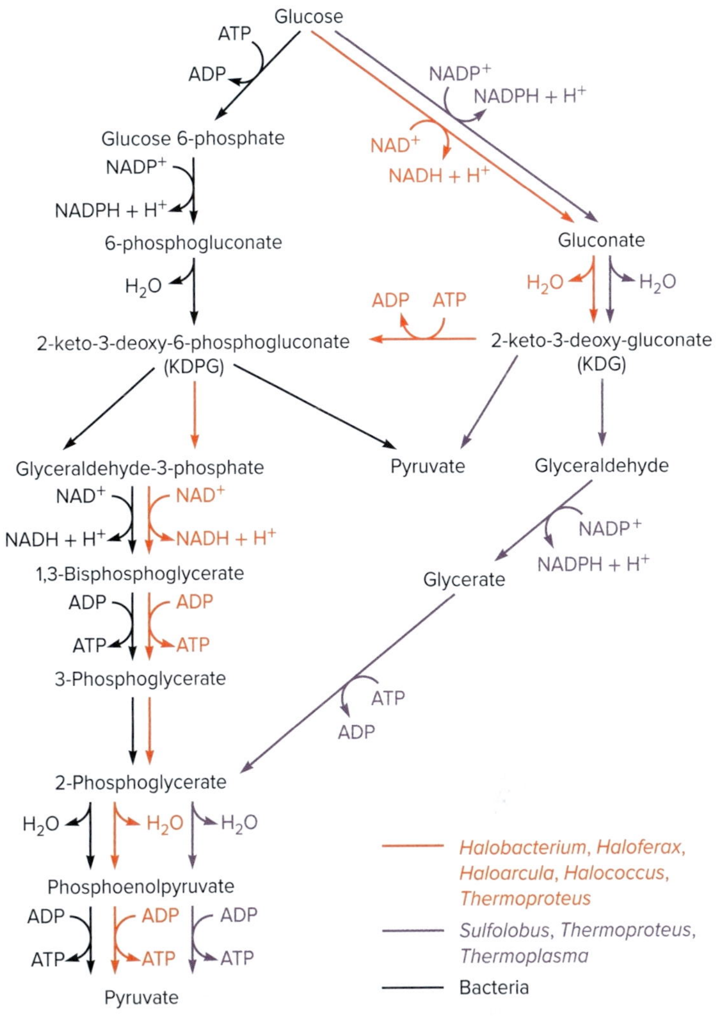

Figure 20.6 Two Alternate Versions of the Entner-Doudoroff Pathway. The semi-phosphorylative pathway (red) used by halophilic archaea involves the ATP-dependent phosphorylation of KDG to KDPG; the remainder of the pathway is like that of bacteria and an ATP is gained. The nonphosphorylative pathway (purple) used by *Sulfolobus* spp. and others joins the bacterial ED pathway when glycerate is converted to 2-phosphoglycerate, so no net ATP is produced.

nonphosphorylative pathway. Extreme halophiles and *Thermoproteus* spp. catabolize glucose using the semi-phosphorylative Entner-Doudoroff (ED) pathway. As shown in **figure 20.6**, glucose is oxidized to gluconate and then dehydrated to 2-keto-3-deoxygluconate (KDG). KDG is phosphorylated at the expense of ATP to KDPG, and the remainder of the ED pathway is like that of bacteria. This results in a net yield of one ATP. The nonphosphorylative ED pathway is found in *Sulfolobus, Thermoplasma,* and *Thermoproteus* spp. (Note that thermoprotei can utilize all three archaeal-specific pathways.) This pathway begins like that of the halophiles, but KDG is cleaved by KDG aldolase to pyruvate and glyceraldehyde. Glyceraldehyde is oxidized to glycerate, which is then phosphorylated to 2-phosphoglycerate. Pyruvate is generated as in the classical and halophilic ED pathway. Because these microbes bypass the energy-conserving step in which 1,3-bisphosphoglycerate is converted to 3-phosphoglycerate, there is no net yield of ATP. ◀ *Glucose to pyruvate: the first step (section 11.4)*

All archaea that have been studied can oxidize pyruvate to acetyl-CoA. However, they lack the pyruvate dehydrogenase complex present in eukaryotes and respiratory bacteria, and use the enzyme pyruvate ferridoxin oxidoreductase instead. Halophiles and the extreme thermophiles *Thermoplasma* spp. appear to have a functional tricarboxylic acid cycle. Methanogens do not catabolize glucose to any significant extent, so it is not surprising that they lack a complete tricarboxylic acid cycle. However, some store glycogen, which is catabolized via a modified EM pathway (figure 20.5). ◀ *Pyruvate to carbon dioxide (step 2) is accomplished by the tricarboxylic acid cycle (section 11.5)*

Comprehension Check

1. How are euryarchaeotes, crenarchaeotes, and thaumarchaeotes distinguished?
2. Why do methanogens use the reductive acetyl-CoA pathway for carbon fixation?
3. Why do haloarchaea use the methylaspartate rather than the glyoxylate cycle for the incorporation of acetate?
4. Compare the ATP and NADH yield of each of the archaeal glycolytic pathways with that used by bacteria.

20.2 *Proteoarchaeota*, or TACK Superphylum

After reading this section, you should be able to:

a. List the major physiological types in the TACK superphylum
b. Discuss hyperthermophilic and thermoacidophilic growth
c. Discuss the relationship of Asgard archaea to eukaryotes
d. Describe the importance of archaeal-specific lipids in the discovery of thaumarchaeotes
e. Explain, in basic terms, the metabolism of thaumarchaeotes
f. Predict where you might find members of this phylum
g. Discuss how metabolism of uncultured organisms can be inferred from partial genome sequence data

The *Proteoarchaeota* or TACK superphylum unites the well-studied *Crenarchaeota* and *Thaumarchaeota* with *Aigarchaeota* and *Korarchaeota*, two proposed phyla that lack cultivated members (figure 20.1). We now discuss representatives to sample the diversity in this superphylum.

Phylum *Crenarchaeota*

All crenarchaeotes are thermophilic and many are sulfur dependent. The sulfur may be used either as an electron acceptor in anaerobic respiration or as an electron donor by lithotrophs. They are often found in geothermally heated water or soils that contain elemental sulfur, which imparts a milky white appearance to water. These environments are sometimes called solfataras and are found all over the world (**figure 20.7**). Many hyperthermophiles have also been isolated from the waters surrounding submarine volcanoes (*see figure 27.4*). These archaea are classified as **hyperthermophiles** if their optimum growth temperature exceeds 85°C. The most extreme example is *Pyrolobus fumarii,* which was isolated from an active hydrothermal vent in the northeast Pacific Ocean. It grows best at about 105°C, but can survive autoclaving at 121°C for 1 hour. It is strictly anaerobic, using Fe(III) as a terminal electron acceptor and H_2 or formate as electron donors and energy sources.

Hydrothermal vents are also home to the parasitic archaeon, *Nanoarchaeum equitans*, the only cultured member of the DPANN superphylum. *N. equitans* relies on its crenarchaeal host, *Ignicoccus hospitalis,* for replication (**figure 20.8**). *I. hospitalis* is an anaerobic, chemolithoautotrophic hyperthermophile that uses H_2 as an electron donor and elemental sulfur as its electron acceptor. It uses the DC/HB pathway for carbon fixation (figure 20.3). Its cell envelope is unique as it lacks an S layer and possesses a double membrane (*see figure 4.6g*). The plasma membrane is composed of diethers and tetraethers that frequently give rise to small, bleblike vesicles. The outer membrane surrounds a wide periplasm-like compartment. These lipids, which should not be confused with fatty acids, are called caldarchaeol lipids. They are extremely thermostable tetraethers found in many hyperthermophilic archaea. They consist of two C40 isoprenoid units linked to glycerol at both ends (i.e., it has four ether bonds), thus generating a monolayer rather than the more familiar lipid bilayer seen in bacteria and eukaryotes (*see figures 4.4* and *4.5*). The outer membrane of *I. hospitalis* houses H_2:sulfur oxidoreductase and ATP synthase complexes, and is thereby energized. Its parasitic partner *N. equitans* is also quite interesting. It cannot be grown without its host and forms tiny cells of only 350 to 500 nm that attach to the

Figure 20.7 A Solfatara on Mt. Krafla, Iceland, Supports the Growth of Thermophilic Archaea.

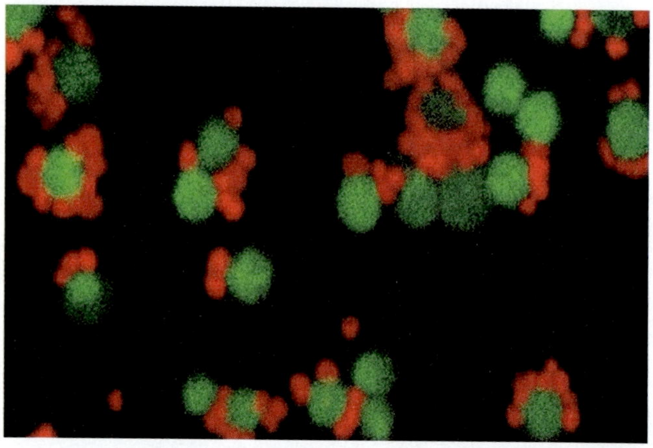

Figure 20.8 Parasitic Cells of *Nanoarchaeum equitans* Attached to the Surface of Their Host, *Ignicoccus hospitalis.* The crenarchaeote *I. hospitalis* has an outer membrane that may be important for *N. equitans* attachment. In this confocal laser scanning micrograph, *I. hospitalis* is stained green and *N. equitans* is red.

I. hospitalis outer membrane. Very little is known about the physiology of *N. equitans,* although genome analysis reveals that it lacks the capacity to synthesize most lipids, amino acids, nucleotides, and cofactors. At 490,000 bp, it seems *N. equitans* has taken genome reduction to the extreme. ◀ *Comparative genomics (section 18.7)*

Two of the best-studied members of *Crenarchaeota* belong to the genera *Sulfolobus* and *Thermoproteus*. Members of *Sulfolobus* are aerobic, and spherical, with a temperature optimum around 80°C and a pH optimum of only 2 to 3. For this reason, they are **thermoacidophiles.** Their cell walls contain carbohydrate and archaeal-specific lipoproteins. They grow chemoorganotrophically under oxic conditions, but they can also grow chemolithoautotrophically, using H_2, H_2S, and FeS_2 as electron donors and with oxygen as the terminal electron acceptor, although the use of ferric iron has been reported. When growing chemolithoautotrophically, they use the HP/HB cycle, and when growing chemoorganotrophically, they employ the nonphosphorylative version of the Entner-Doudoroff pathway (figure 20.6) and a complete TCA cycle. Unlike most organisms, *S. solfataricus* rarely uses NAD^+ as an electron acceptor; instead, it uses $NADP^+$ and ferredoxin-dependent oxidoreductases. In fact, archaea in general seem to lack NADH dehydrogenase, whereas it is integral to the function of bacterial and mitochondrial electron transport chains. The genomes of *Sulfolobus* spp. show a high level of plasticity; indeed, the *S. solfataricus* genome has 200 integrated insertion sequences, indicating a history of frequent horizontal gene transfer. Although *S. solfataricus* grows at pH 2 to 4, it maintains a cytoplasmic pH of about 6.5, thereby generating a large pH gradient across the plasma membrane. This energy is conserved in the formation of ATP by membrane-bound ATP synthases and at least 15 secondary transport systems that couple the transport of organic solutes (e.g., sugars, amino acids) with the movement of protons (**figure 20.9**).

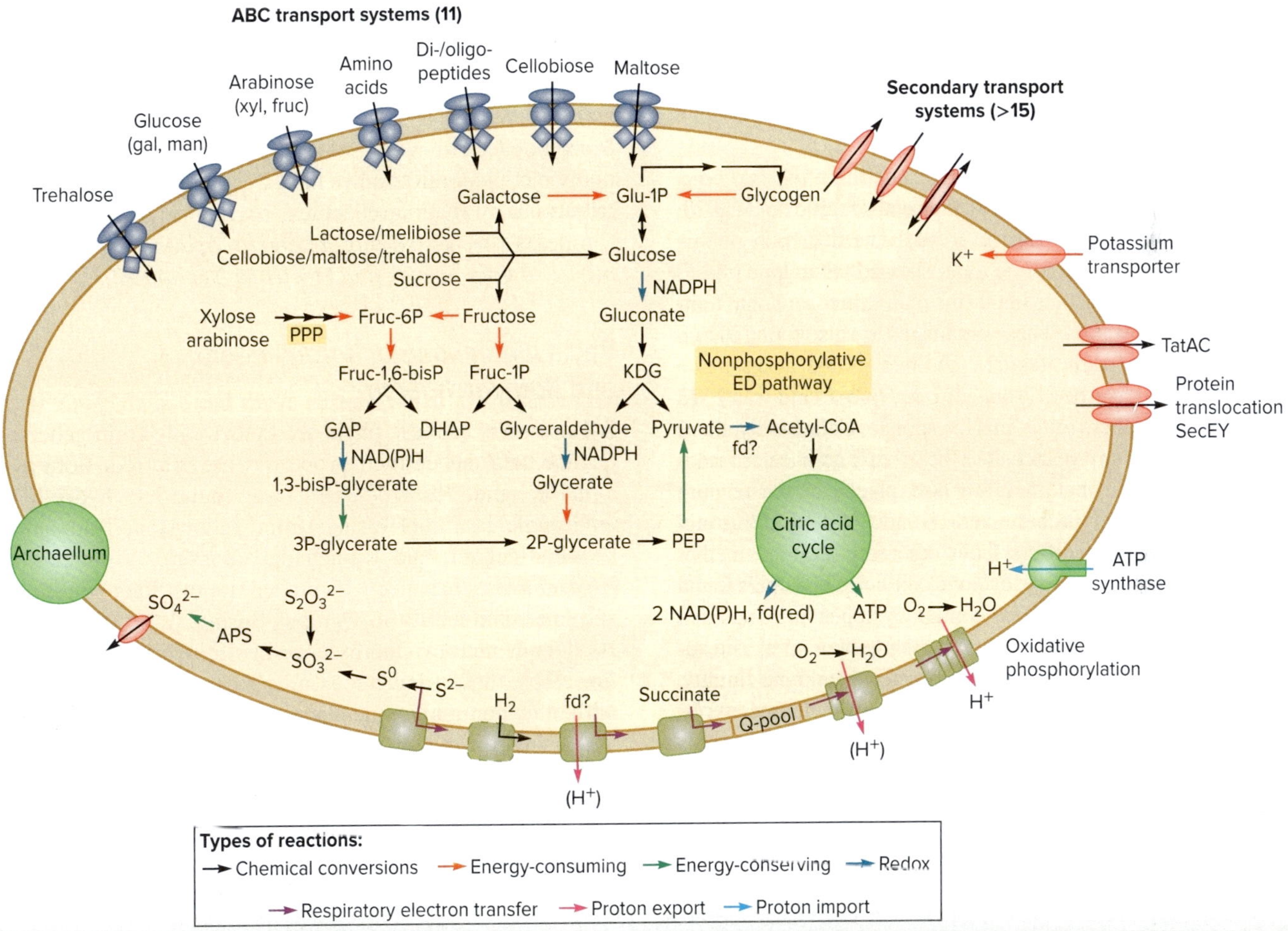

Figure 20.9 Genomic Reconstruction of *Sulfolobus solfataricus.* Each type of reaction is denoted by a different color arrow, as shown in the key. Components of the aerobic respiratory network include the reduction of a quinone (Q) pool, a putative ferredoxin dehydrogenase (fd?), and succinate dehydrogenase. Alternative electron donors include hydrogen and sulfide, which reduce the quinone pool through hydrogenase and sulfide reductase, respectively. Elemental sulfur (S^0) and thiosulfate ($S_2O_3^{2-}$) are converted to sulfate. The ABC transport systems import xylose (xyl), fructose (fruc), glucose (glu), galactose (gal), and mannose (man). Archaella and a gene encoding a methyl-accepting chemotaxis protein are present.

MICRO INQUIRY *Identify the potential electron donors to the electron transport chain. What is the source of energy for the potassium transporter shown on the right-hand side of the cell? Hint: Examine the arrow color code box.*

The ABC transporters also used for nutrient uptake have high substrate affinity, an important adaptation in low nutrient environments. ◀ *Electron transport and oxidative phosphorylation (step 3) generate the most ATP (section 11.6)*

Thermoproteus spp. are long, thin rods that can be bent or branched (*see figure 4.1*), with cell walls composed of glycoprotein. Thermoprotei are anaerobic and grow at temperatures from 75° to 100°C. Some are acidophiles, with optimum pH values between 3 and 4, while others are neutrophiles. They are found in hot springs and other hot aquatic habitats rich in sulfur. *Thermoproteus* spp. can grow organotrophically and oxidize glucose, amino acids, alcohols, and organic acids with elemental sulfur as the electron acceptor during anaerobic respiration. They both use alternate Entner-Doudoroff pathways (figure 20.6) to catabolize sugars. They also grow chemolithotrophically, oxidizing H_2 with S^0 as the electron acceptor. CO_2 is incorporated through the DC/HB cycle (figure 20.3).

Phylum *Thaumarchaeota*: Mesophilic Ammonia Oxidizers

The first evidence for what would become the phylum *Thaumarchaeota* was the discovery of an archaeal-specific lipid called, at that time, crenarchaeol (**figure 20.10**). The detection of this lipid, coupled with metagenomic data, showed what appeared to be mesophilic crenarchaeotes in marine plankton from polar, temperate, and tropical waters. These organisms are also found in rice paddies, soils, and freshwater lake sediments, and at least two symbiotic species have been isolated, one from a cold-water sea cucumber and the other from a marine sponge. Collectively, these microbes were for many years called the group I archaea or mesophilic *Crenarchaeota*, but they are now placed in the phylum *Thaumarchaeota,* and crenarchaeol is now called **thaumarchaeol.** Thaumarchaeol differs from other diglycerol tetraether lipids in the presence of a single cyclohexane ring. The lipids found in hyperthermophilic crenarchaea contain cyclopentane rings that confer additional membrane stability at high temperature. The addition of a cyclohexane ring promotes greater membrane fluidity, as required by thaumarchaeotes growing at lower temperatures. ◀ *Metagenomics provides access to uncultured microbes (section 18.3)*

The purification and growth of the marine sponge symbiont *Cenarchaeum symbiosum* and the free-living thaumarchaeote *Nitrosopumilus maritimus* confirmed that archaea are capable of nitrification. Indeed, ammonia oxidation is the metabolic feature that defines *Thaumarchaeota*. Genomic analysis of thaumarchaeotes reveals that they fix carbon via the HP/HB cycle. However, some thaumarchaeotes use organic carbon rather than CO_2 as their carbon source. These archaea are considered mixotrophic rather than heterotropic because they capture energy through the oxidation of ammonia to nitrite using oxygen as the terminal electron acceptor while using organic carbon for anabolic purposes. The first step in ammonia oxidation is its conversion to hydroxylamine (NH_2OH), a reaction catalyzed by ammonia monooxygenase (AMO; *see figure 22.13*). Bacterial AMO is composed of three subunits, AmoA, AmoB, and AmoC. The archaeal versions of the genes encoding these proteins are phylogenetically distinct from bacterial *amo* genes and a fourth subunit, AmoX, is found exclusively in archaeal AMO. This variation in *amo* genes has made it possible to assess the presence and, through reverse transcription PCR studies, the activity of archaeal nitrification in soils and waters. While it had long been thought that α- and β-proteobacteria were solely responsible for nitrification, many metagenomic studies have documented significant contributions by thaumarchaeotes, particularly in marine environments. ▶ *Nitrifying bacteria oxidize ammonium or nitrite to gain energy and electrons (section 22.1)*

Phyla *Aigarchaeota, Korarchaeota,* and *Bathyarchaeota*

All members of these phyla are known only from genome sequence data, hence their properties are inferred. Both *Aigarchaeota* and *Korarchaeota* are found in hydrothermal environments, in hot springs from Yellowstone National Park (YNP), Iceland, and Kamchatka (Russia). A population of *Aigarchaeota* has been identified in the long-studied pink streamer community in Octopus Spring, YNP (**figure 20.11**). As documented by fluorescence in situ hybridization (*see figure 29.6*), these filamentous microbes are in intimate contact with other community members. Their predicted anabolic pathways suggest a reliance on the import of biochemical precursors.

Bathyarchaeota, formerly known as the Miscellaneous *Crenarchaeota* Group, are found in sediments and soils in both marine and terrestrial environments. A metagenomic survey of an Australian coal bed methane well resulted in two genome sequence assemblies that include homologues of genes for methanogenesis and the reductive acetyl-CoA pathway. The amount of methane produced in this environment far exceeds the capacity of the small euryarchaeal methanogen population found at the site, and it appears that *Bathyarchaeota* share the ability to perform methanogenesis.

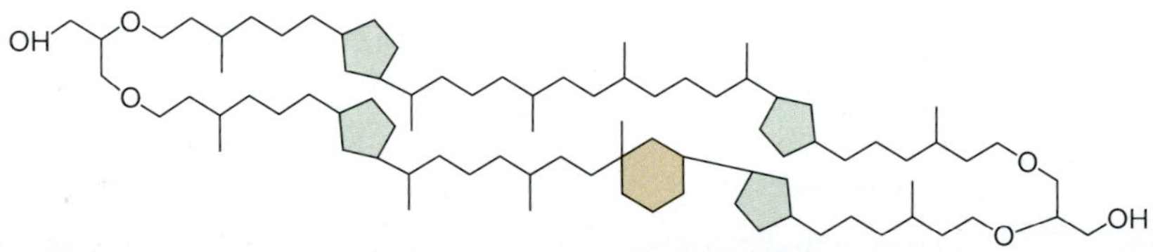

Figure 20.10 Thaumarchaeol. The presence of a cyclohexane ring (orange) is the molecular signature of thaumarchaeol.

Figure 20.11 Octopus Spring, Yellowstone National Park.

Asgard Archaea Are Close Relatives of Eukaryotes

The Asgard archaea comprise a proposed superphylum that encompasses four proposed phyla (figure 20.1), all named for characters who inhabit the kingdom of Asgard, one of the nine worlds described in Norse mythology. All Asgard archaea were discovered through metagenomics in surveys of aquatic sediments from geographically and chemically diverse sites worldwide. There are no representatives in culture.

Asgard genomes encode a significant number of **eukaryotic signature proteins,** that is, proteins characteristic of eukaryotes. These include cytoskeletal components (actin and tubulin), informational processing proteins (DNA polymerase and ribosome components), and components of intracellular trafficking and secretion pathways. The presence of these intracellular trafficking proteins in particular is puzzling, as archaeal cells are typically much smaller than eukaryotic cells and lack endoplasmic reticula and Golgi apparati where these proteins are known to function.

The existence of the Asgard archaea expands our understanding of microbial diversity, and allows an exploration of microbial evolutionary histories. We include the Asgard archaea in this section describing the TACK superphylum, but it is important to note that this placement is not universally accepted. The choice of conserved proteins used in phylogenetic tree construction can significantly influence the outcome, and it is likely that additional discoveries will spur modifications to currently proposed trees. ◀ *Phylogenetic trees illustrate evolutionary relationships (section 19.4)*

Comprehension Check

1. What are thermoacidophiles and where do they grow? In what ways do they use sulfur in their metabolism?
2. List three aspects of the *Ignicoccus-Nanoarchaeum* symbiosis that make it unusual.
3. Compare the carbon and sulfur metabolism of *Sulfolobus* and *Thermoproteus* spp.
4. Discuss the role of external pH on the magnitude of the proton motive force generated by *Sulfolobus* spp.
5. You have discovered a new microorganism that oxidizes ammonia. How will you determine if this microbe is a thaumarchaeote or a bacterium?

20.3 Phylum *Euryarchaeota:* Methanogens, Haloarchaea, and Others

After reading this section, you should be able to:

a. Outline the process of methanogenesis and discuss its importance in the flow of carbon through the biosphere as well as in the production of energy for human needs
b. Discuss the physiology and ecology of anaerobic methane oxidation
c. Explain the strategies halophiles have evolved to cope with osmotic stress and why these strategies are needed
d. Outline rhodopsin-based phototrophy as used by halophiles
e. Describe the habitats in which methanogens and halophiles reside
f. List one unique feature of *Thermoplasma, Pyrococcus,* and *Archaeoglobus* spp.

Euryarchaeota is a diverse phylum with many genera (figure 20.1). We discuss five major groups within the euryarchaeotes.

Methanogens and Methanotrophs

Late in the eighteenth century, Italian physicist Alessandro Volta discovered that he could ignite gas released from anoxic marshes. In this way, Volta discovered the biological production of methane and demonstrated the industrial importance of natural gas. The production of methane—**methanogenesis**—is the last step in the anaerobic degradation of organic compounds. Because the ΔG of methanogenesis compares unfavorably to other forms of respiration, methanogenesis occurs only when O_2 and most other electron acceptors are unavailable, making it a strictly anaerobic process (*see figure 28.1*). Two groups of archaea appear to be capable of methanogenesis. Within the *Euryarchaeota,* cultured methanogens generate methane from H_2 and CO_2 or from short-chain organic compounds, such as formate, acetate, and methanol. These compounds are usually the fermentation products of other microbes that live in the same community. When using H_2 and CO_2, their growth is autotrophic. These archaea use the reductive acetyl-CoA pathway to incorporate CO_2 (figure 20.2). A second group, the *Bathyarchaeota* (TACK superphylum), encode all the methanogenic enzymes and cofactors. The presence of methanogenesis in these two divergent lineages demonstrates the antiquity of this metabolism. Our discussion focuses on the well-established pathways in the euryarchaeal methanogens.

Methanogens are the largest group of cultured archaea. There are eight orders that differ greatly in overall shape, 16S rRNA sequence, cell wall chemistry and structure, membrane lipids, and other features. For example, methanogens construct five different types of cell walls. Several genera have walls with

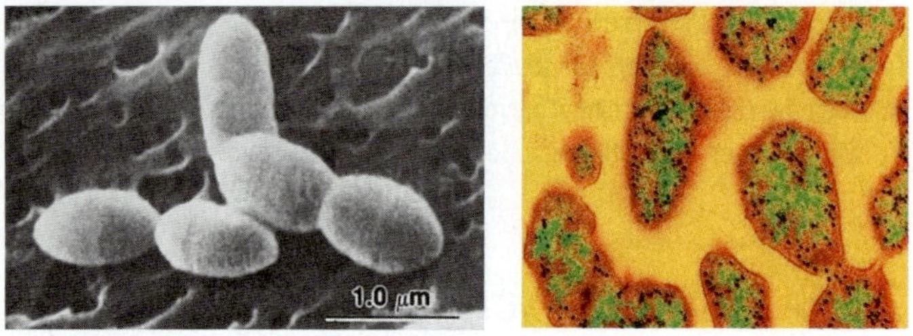

(a) (b)

Figure 20.12 Methanogens. (a) *Methanobrevibacter smithii;* SEM. (b) *Methanococcoides burtonii;* TEM.

(a) From J.T. Staley, M.P. Bryant, N. Pfenning and J.G. Holt (Eds), *Bergey's Manual of Systematic Bacteriology*, Vol. 3. © 1989 Williams and Wilkins Co., Baltimore; (b) ©Dr. M. Rohde,GBF/Science Source

pseudomurein; other walls contain either proteins or heteropolysaccharides (*see figure 4.6*). The morphology of two representative methanogens is shown in **figure 20.12**, and selected properties of genera within four orders are presented in **table 20.2**.

Methanogenesis

The production of methane (CH_4), methanogenesis, requires that the carbon atom be converted from its most oxidized state in CO_2 (+4) to its most reduced state (−4). Recall that carbon fixation in autotrophy takes a carbon atom from +4 to 0, a process that requires the hydrolysis of ATP. By contrast, methanogenesis is an energy-generating mechanism that requires several unique cofactors. Some are carriers for the carbon atom, whereas others are electron donors. As pictured in **figure 20.13**, these include *m*ethano*fur*an (MFR), tetrahydro*m*ethano*p*terin (H_4MPT), coenzyme F_{420}, *co*enzyme *M* (CoM; 2-mercaptoethanesulfonic acid), *co*enzyme *B* (CoB; 7-mercaptoheptanoylthreonine phosphate), and coenzyme F_{430}. These are used in the following series of reactions (**figure 20.14**): (1) Gaseous CO_2 is bound to an enzyme complex that first reduces it to formate at one active site. An internal tunnel in the enzyme complex directs formate to a second active site where it is further reduced by binding to MFR. This is an endergonic reaction, coupled to ferredoxin oxidation. The carbon atom is now activated and part of a formyl group (—HC═O). (2) The formyl group is transferred to H_4MPT and dehydrated. (3) Using F_{420} as the electron donor, the dehydrated formyl group (═HC—) is reduced to the methyl level (—CH_3) in a two-step process. (4) The methyl group is transferred to CoM in a membrane-bound enzyme that couples the reaction to Na^+ translocation across the membrane. This results in a sodium motive force to drive an ATPase. (5) The methyl group is now ready for final reduction to methane (CH_4) by coenzyme B (CoB). This is catalyzed by the F_{430}-containing *m*ethyl-*c*oenzyme M *r*eductase (MCR). H_2 is the electron donor and protons are released. In addition to methane, a second product is a disulfide of the two cofactors, CoM-S-S-CoB. The final step in the cycle (6) regenerates the two coenyzmes by reducing this disulfide in an exergonic reaction. The enzyme that catalyzes this reaction, heterodisulfide reductase, uses a flavin-based electron bifurcation mechanism to oxidize CoM-S-S-CoB and reduce the ferredoxin required for reaction (1). ◀ *Flavin-based electron bifurcation (section 11.11)*

It can be useful to compare methanogens that possess cytochromes with those that lack them. Only the *Methanosarcinales* have cytochromes and methanophenazine, a menaquinone-like, redox active molecule. These are the only methanogens with the capacity to grow on acetate, methanol, and methylamines, with a few species of *Methanosarcina* also growing on H_2 and CO_2. These archaea tend to have relatively high growth yields of around 7 grams of dried cellular

Table 20.2 Selected Characteristics of Representative Methanogens

Order and Genera	Morphology	Wall Composition	Motility	Methanogenic Substrates Used	Capable of N_2 Fixation
Order *Methanobacteriales*					
Methanothermobacter	Long rods or filaments	Pseudomurein	−	$H_2 + CO_2$, formate	Yes
Methanothermus	Straight to slightly curved rods	Pseudomurein with an S-layer	+	$H_2 + CO_2$	No
Order *Methanococcales*					
Methanococcus	Irregular cocci	Protein	+	$H_2 + CO_2$, formate	Yes
Order *Methanomicrobiales*					
Methanomicrobium	Short curved rods	Protein	+	$H_2 + CO_2$, formate	No
Methanogenium	Irregular cocci	Protein or glycoprotein	−	$H_2 + CO_2$, formate	No
Methanospirillum	Curved rods or spirilla	Protein	+	$H_2 + CO_2$, formate	Yes
Order *Methanosarcinales*					
Methanosarcina	Irregular cocci, packets	Protein sometimes with polysaccharide	−	$H_2 + CO_2$, methanol, methylamines, acetate	Yes

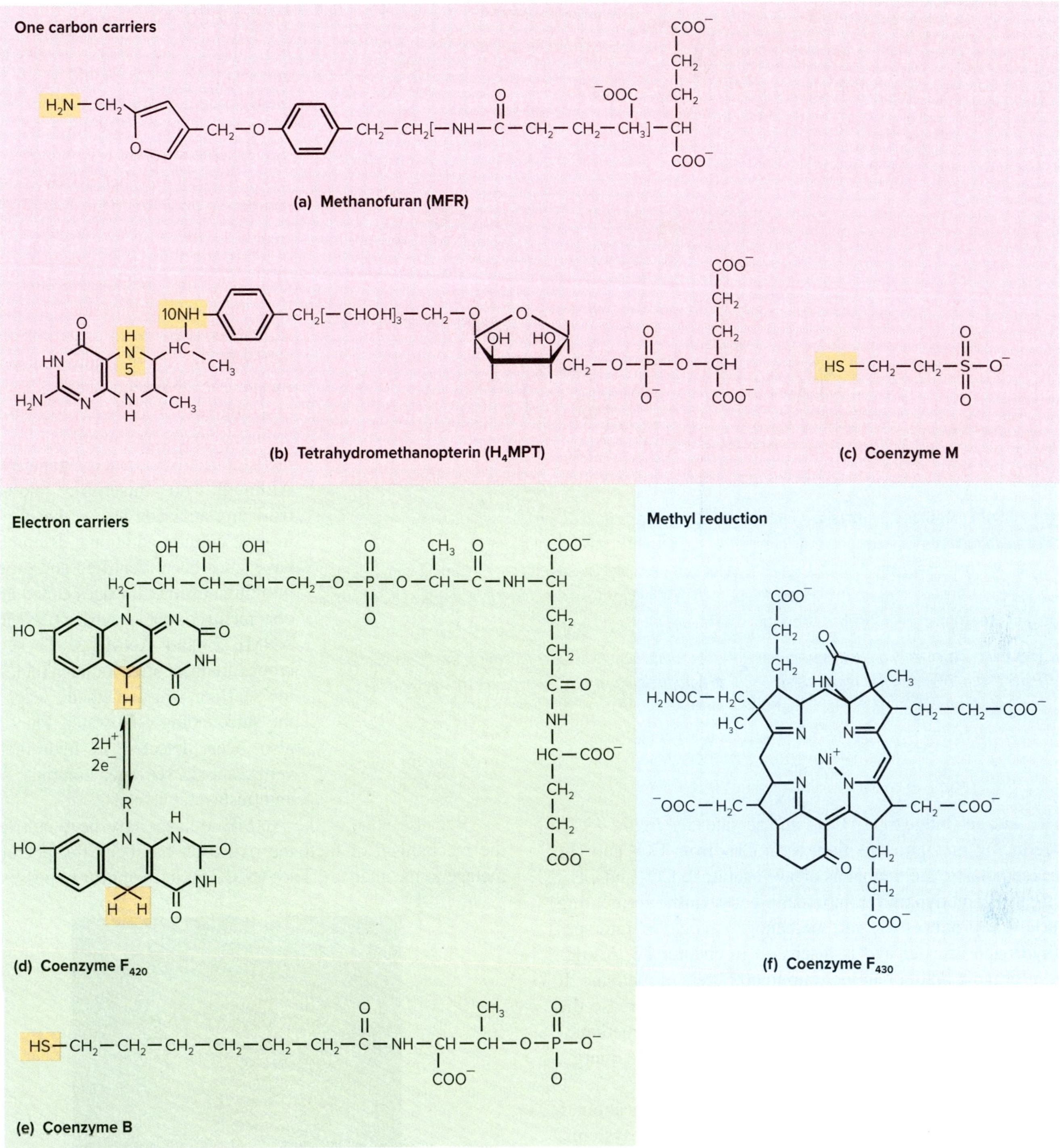

Figure 20.13 Methanogen Coenzymes. (a) Coenzymes MFR, (b) H_4MPT, and (c) coenzyme M carry 1-carbon units during methanogenesis. The portions of the coenzymes that carry the 1-carbon units are highlighted. H_4MPT carries carbon units on nitrogens 5 and 10, like the more common enzyme tetrahydrofolate. (d) Coenzyme F_{420} and (e) coenzyme B participate in redox reactions. The part of the molecule that is reversibly oxidized and reduced is highlighted. (f) Coenzyme F_{430} participates in reactions catalyzed by the methyl-CoM reductase.

material per mole of CH_4 produced. By contrast, most others convert H_2 and CO_2 to methane; a few can also use formate as a methanogenic substrate (table 20.2). These autotrophic methanogens have growth yields of between 1.5 and 3 grams per mole CH_4. Given their better growth yields, it is not surprising that members of *Methanosarcinales* account for about two-thirds of methanogenesis.

Methanogenic archaea are potentially of great practical importance, as discussed in the chapter opening story. Anaerobic digesters use fermentative bacteria to degrade particulate wastes

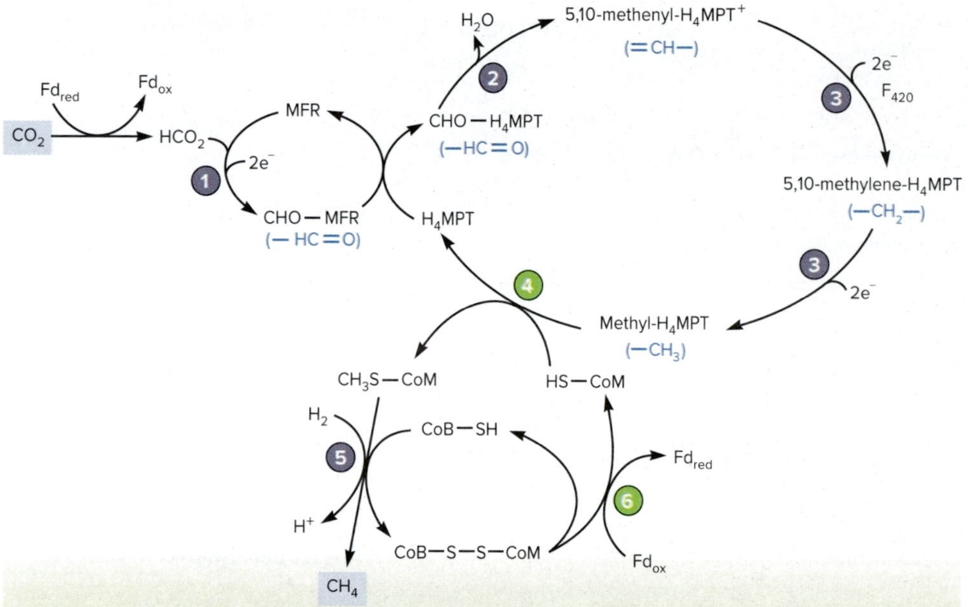

Figure 20.14 Methane Synthesis. Pathway for CH_4 synthesis from CO_2 in *M. thermautotrophicus.* Cofactor abbreviations: methanopterin (MPT), methanofuran (MFR), coenzyme M (CoM), coenzyme B (CoB). The nature of the carbon-containing intermediates leading from CO_2 to CH_4 is indicated in parentheses. The circled numbers correspond to steps described in text. Blue circles are endergonic reactions; green circles are exergonic reactions.

MICRO INQUIRY *What are the mechanisms by which methanogens are thought to couple CO_2 reduction to ATP generation? Why do you think this has been so hard to demonstrate?*

such as sewage sludge to H_2, CO_2, and acetate (*see figure 43.7*). CO_2-reducing methanogens then form CH_4 from CO_2 and H_2, while acetoclastic methanogens cleave acetate to CO_2 and CH_4. The tight thermodynamic relationship between the fermenting bacteria that produced H_2 and methanogens is called interspecies hydrogen transfer and is discussed in chapter 27. A kilogram of organic matter can yield up to 600 liters of methane. It is quite likely that future innovation will greatly increase the efficiency of industrial methane production and make methanogenesis from organic waste an important source of energy.
▶| *Biofuel production is a dynamic field (section 42.2)*

Methanogenic archaea are estimated to produce about 1 billion tons of methane annually. In freshwater ecosystems, rates of methane production can be so great that bubbles of methane sometimes rise to the surface of lakes and ponds. Rumen methanogens are so active that a cow can belch 200 to 400 liters of methane a day. Methanogenesis is also an environmental problem. Methane absorbs infrared radiation and is a more potent greenhouse gas than CO_2. Atmospheric methane concentrations have been rising over the last 200 years. Methane production significantly promotes global warming; this subject is discussed in chapter 28. ▶| *The rumen ecosystem (section 27.2)*

Methanotrophy

No doubt the levels of archaeal-generated methane would render this planet uninhabitable were it not for the oxidation of methane by microbes capable of methanotrophy. For many decades, it was thought that this process—the capacity to oxidize methane and use it as a carbon source—was restricted only to obligate aerobic proteobacteria. However, it is now recognized that about 90% of methane produced in marine sediments is oxidized anaerobically by archaea. The use of fluorescent probes for specific DNA sequences has revealed assemblages of archaea and sulfate-reducing bacteria inhabiting anoxic, methane-rich sediments (**figure 20.15**). Although no anaerobic methane-oxidizing archaeon has yet been grown in pure culture, additional assemblages have since been identified and grouped into three distinct lineages called anaerobic methanotrophic archaea (ANME)-1, ANME-2, and ANME-3. These microbes are most often found in regions of the seafloor where methane seeps into the surrounding cold water. They have also been detected in hydrothermal vents, anoxic seawater columns, soils, and freshwater habitats.

Because none of the ANME archaea have been cultivated, the mechanism of methane oxidation has been deciphered by metagenomic analysis. The ANME archaea appear to possess all

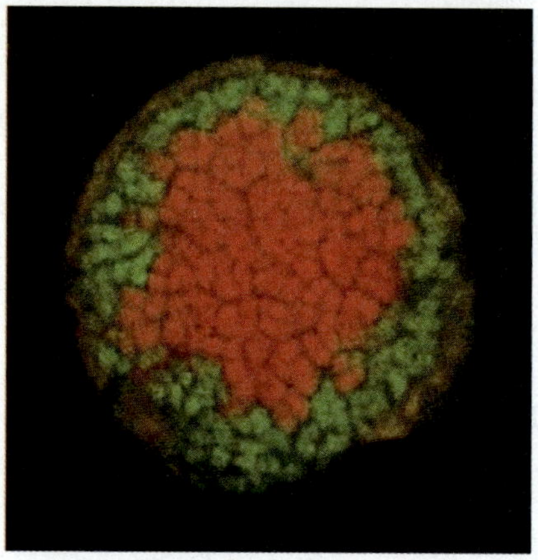

Figure 20.15 Methanotrophic Archaea Grow in Association with Sulfate-Reducing Bacteria. A cluster of methanotrophic archaea, stained red by an archaeal fluorescent 16S rRNA probe, surrounded by a layer of sulfate-reducing bacteria labeled by a green bacterial 16S rRNA fluorescent probe.

the genes needed for methanogenesis, implying that anaerobic methane oxidation is accomplished by running methanogenesis backward. The demonstration that purified enzyme methyl-coenzyme M reductase can catalyze the energetically unfavorable conversion of methane to methyl-coenzyme M supports the notion that methane is the electron donor. In the ANME consortia with sulfate-reducing bacteria, sulfate is the terminal electron acceptor:

$$CH_4 + SO_4^{2-} \rightarrow HCO_3^- + HS^- + H_2O \ (\Delta G'^\circ = -17 \text{ kJ/mol})$$

Electron transfer within these cell clusters appears to be mediated through nanowires, cell-to-cell connections that have been visualized by electron microscopy (*see figure 22.25b*). The archaea provide the bacteria with much of their organic carbon. In addition, ANME archaea also fix N_2 gas and provide the bacteria with organic nitrogen. We now know that sulfate is not the only electron acceptor used in the anaerobic oxidation of methane. One group of ANME, ANME-2d couples nitrate reduction by anammox bacteria to methane oxidation. Other ANME-2d members use Fe(III) and Mn(IV) minerals as electron acceptors, representing new aspects of biogeochemical cycling.

Haloarchaea

The **extreme halophiles** include some bacteria, eukaryotic microorganisms, and **haloarchaea,** which belong to the order *Halobacteriales.* This euryarchaeal group currently has 42 genera in one family, *Halobacteriaceae*. These microbes received this unfortunate name before *Archaea* and *Bacteria* were recognized as different taxa. Most are aerobic chemoorganotrophs with respiratory metabolism. Extreme halophiles demonstrate a wide variety of nutritional capabilities. When grown in defined media, these archaea use carbohydrates or simple compounds such as glycerol and organic or amino acids as their carbon source. Halophiles are found in a variety of cell shapes, including cubes and pyramids in addition to rods and cocci (**figure 20.16*a***). Some are motile by archaella.

The most obvious distinguishing trait of this family is its absolute dependence on a high concentration of NaCl, requiring at least 1.5 M NaCl (about 8%, wt/vol). Growth optima are usually 3 to 4 M NaCl (17 to 23%), but some will grow at salt concentrations approaching saturation (about 36%). The cell walls of most halophiles are so dependent on the presence of NaCl that they disintegrate when the NaCl concentration drops below 1.5 M. Thus they only grow in high-salinity habitats such as marine salterns and salt lakes such as the Dead Sea between Israel and Jordan, and the Great Salt Lake in Utah. Haloarchaea often have red-to-yellow pigmentation from carotenoids that are used as protection against the ultraviolet radiation in sunlight. They can reach such high population levels that salt lakes, called salterns, turn red (figure 20.16*b*).

Halophiles use two strategies to cope with osmotic stress. The first approach is to increase cytoplasmic osmolarity by accumulating small organic molecules called **compatible solutes.** These include glycine, betaine, polyols, ectoine, and amino acids. When this strategy is used, salt adaptation of cytoplasmic proteins is not necessary. Compatible solutes are found in extremely halophilic methanogenic archaea, as well as halophilic bacteria. By contrast, members of the orders *Halobacteriales* and *Haloanaerobiales* employ a "salt-in" approach. These archaea use Na^+/H^+ antiporters and K^+ symporters to concentrate KCl and NaCl to levels equivalent to the external environment. Thus proteins need to be protected from salt denaturation and dehydration. Proteins from these microbes have evolved to possess a limited number of hydrophobic amino acids and a greater number of acidic residues. The acidic amino acids tend to be located on the surface of the folded protein, where they attract cations, which form a hydrated shell around the protein, thereby maintaining its solubility.

(a) (b)

Figure 20.16 Halophilic Archaea. (a) A sample taken from a saltern in Australia viewed by fluorescence microscopy. Note the range of cell shapes includes cocci, rods, and cubes. (b) A solar evaporation pond is extremely high in salt and mineral content. The archaea that dominate this hot, saline habitat produce brilliant red pigments.

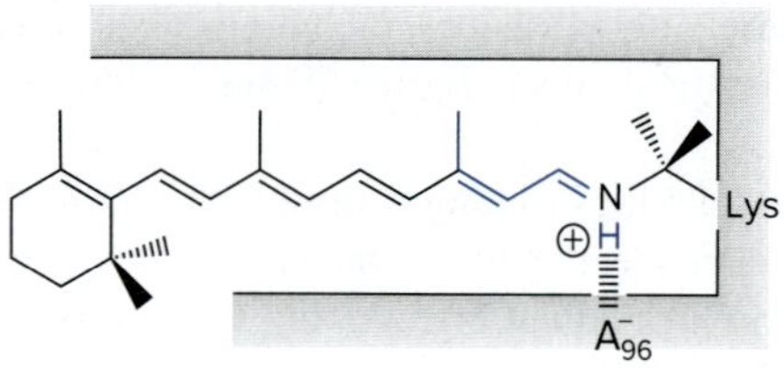

Figure 20.17 **The Chromophore of Archaerhodopsin Is a Retinal Schiff Base.**

Probably the best-studied member of the family is *Halobacterium salinarum*. This archaeon is unusual because it produces a protein called **archaerhodopsin** (originally termed bacteriorhodopsin) that can harvest light energy without the presence of chlorophyll. Structurally similar to the rhodopsin found in the mammalian eye, archaerhodopsin functions as a light-driven proton pump. Like all members of the rhodopsin family, archaerhodopsin has two distinct features: (1) a chromophore derivative of retinal (an aldehyde of vitamin A), which is covalently attached to the protein by a Schiff base with the amino group of lysine (**figure 20.17**); and (2) seven membrane-spanning domains connected by loops on either side of the membrane with the retinal resting within the membrane. Archaerhodopsin molecules form aggregates in a modified region of the plasma membrane called the **purple membrane.** As shown in **figure 20.18**, this light-driven proton pump generates a pH gradient without an electron transport chain. Nonetheless, this gradient can be used to power the synthesis of ATP by a chemiosmotic mechanism. ◀ *Rhodopsin-based phototrophy (section 11.12)*

H. salinarum has three additional rhodopsins, each with a different function. Halorhodopsin uses light energy to transport chloride ions into the cell and maintain a 4 to 5 M intracellular KCl concentration. The two additional rhodopsins are called sensory rhodopsin I (SRI) and SRII. **Sensory rhodopsins** act as photoreceptors; in this case, one for green-light and one for blue. They control archaellum activity to position the organism optimally in the water column. *Halobacterium* cells move to a location of high light intensity that optimizes archaerhodopsin proton pumping but limits the lethal effects of ultraviolet light.

Genomic analysis of *Halobacterium* NRC-1 illustrates many of these physiologic strategies (figure 20.18). The use of both archaerhodopsin-mediated proton motive force and oxidative phosphorylation to drive ATP synthesis is shown, as is the import of cations to increase cytoplasmic osmolarity. The signals from sensory rhodopsins converge on the archaellum, along with input from 17 methyl-accepting chemotactic proteins. *Halobacterium* NRC-1 uses a variety of ATP transporters for the uptake of cationic amino acids, peptides, and sugars. Carbohydrates are oxidized via the semi-phosphorylative variant of the Entner-Doudoroff pathway (figure 20.6) and the TCA cycle. Interestingly, genes for exporting the toxic heavy metals arsenite and cadmium are present, as well as genes homologous to nonspecific multi-drug-resistance exporters.

Thermoplasms

Archaea in the class *Thermoplasmata* are thermoacidophiles that lack cell walls and are therefore pleomorphic. At present, three genera—*Thermoplasma, Picrophilus,* and *Ferroplasma*—are known. They are sufficiently different from one another to be placed in separate families: *Thermoplasmataceae, Picrophilaceae,* and *Ferroplasmataceae.*

Members of the genus *Thermoplasma* grow in refuse piles of coal mines. These piles contain large amounts of pyrite (FeS), which is oxidized to sulfuric acid by chemolithotrophic bacteria. As a result, the piles become very hot and acidic. This is an ideal habitat for these archaea because they grow best at 55° to 59°C and pH 1 to 2. Although they lack cell walls, their plasma membranes are strengthened by large quantities of caldarchaeol, lipid-containing polysaccharides, and glycoproteins.

Picrophilus spp. are even more unusual than *Thermoplasma.* They have an S-layer outside the plasma membrane. The cells grow as irregularly shaped cocci, around 1 to 1.5 µm in diameter. They are aerobic and grow between 47° and 65°C with an optimum of 60°C. They have remarkable pH requirements, growing only below pH 3.5 with a growth optimum at pH 0.7. Growth even occurs at about pH 0. Both *Picrophilus* and *Thermoplasma* use secondary transporters for importing nutrients. This strategy contributes to maintaining the dramatic pH difference across the cell membrane.

Extremely Thermophilic S^0-Reducers

This physiological group contains the class *Thermococci,* with one order, *Thermococcales.* These archaea are strictly anaerobic chemoorganoheterotrophs that use a variety of carbon substrates (e.g., peptides and carbohydrates) and reduce sulfur to sulfide. They are motile by archaella and have optimum growth temperatures around 88° to 100°C. The order contains one family and three genera, *Thermococcus, Paleococcus,* and *Pyrococcus.*

Despite being thermophilic anaerobes, thermococci are among the easiest archaea to grow in the laboratory, as they tolerate brief room temperature handling. *Thermococcus kodakarensis* has become a model organism due to its ease of genetic transformation and the presence of plasmids to create shuttle vectors. The development of genetic techniques for an archaeal hyperthermophile has opened new avenues to explore the biology of these organisms. ◀ *Plasmids (section 17.1)*

Sulfate-Reducing *Euryarchaeota*

Although many euryarchaea reduce sulfur during anaerobic respiration, only a few can use sulfate as a terminal electron acceptor. The sulfate reducers are found in the class *Archaeoglobi* and the order *Archaeoglobales.* This order has only one family and three genera. *Archaeoglobus* spp. are irregular coccoid cells with cell walls consisting of glycoprotein subunits. They can extract electrons from a variety of electron donors (e.g., H_2, lactate, glucose) and reduce sulfate, sulfite, or thiosulfate to sulfide. Elemental sulfur is not used as an acceptor. They are thermophilic (optimum growth is about 83°C) and have been isolated from marine

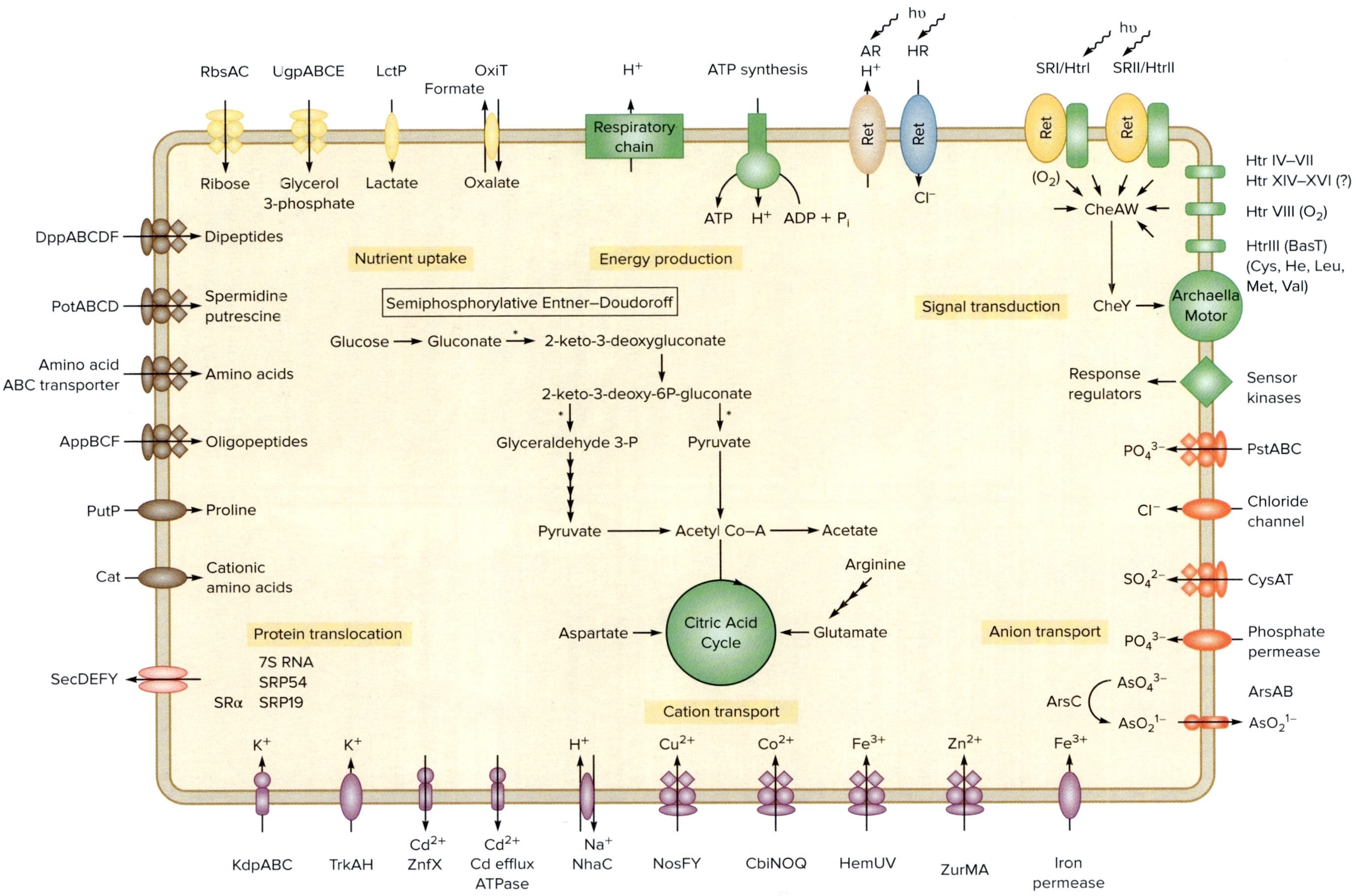

Figure 20.18 **Genomic Reconstruction of *Halobacterium* NRC-1.** ATP synthesis occurs through respiratory-dependent chemiosmosis and the ATP synthase, light-driven proton pumping by archaerhodopsin (AR, pink oval), and chloride transport by halorhodopsin (HR, blue oval). Nutrient uptake systems are shown in brown and yellow; names of transporters correspond with the gene(s) encoding them; the substrates transported are shown in the inside of the transporter. Cation transporters are shown in purple and anion transporters in red. Sensory rhodopsins (orange) and 17 light, oxygen, and amino acid receptors (Htr proteins) regulate chemotaxis and are thus shown linked to the archaella motor through the chemotaxis protein CheY.

hydrothermal vents. Some species are autotrophic and, like methanogens, use the reductive acetyl-CoA pathway to assimilate CO_2. *Archaeoglobus* spp. appear to have evolved from methanogens. Although members of this genus retain several methanogenic cofactors, they have lost the critical enzyme methyl reductase (figure 20.14, step 5), and do not produce methane.

Comprehension Check

1. Why are the growth yields of most methanogens much lower than those of microbes that grow aerobically?
2. Why do you think methanogenesis requires so many cofactors? (Hint: What would happen to CO_2 and CH_4 if they were not bound to larger molecules?)
3. What is the ecological and practical importance of methanogens?
4. Where are extreme halophiles found, and what is unusual about their cell walls and growth requirements?
5. What is the difference between sensory rhodopsin and archaerhodopsin? Find examples of each in figure 20.18.
6. How is *Thermoplasma* able to live in acidic, very hot coal refuse piles when it lacks a cell wall?
7. In what way are members of the genus *Archaeoglobus* similar to methanogens, and how do they differ from other extreme thermophiles?

Key Concepts

20.1 Overview of *Archaea*

- *Archaea* contains a highly diverse group of microbes with respect to morphology, reproduction, physiology, and ecology. Although best known for their growth in anoxic, hypersaline, and high-temperature habitats, they also inhabit marine Arctic, temperate, and tropical waters as well as the human body.
- *Archaea* comprises three recognized phyla, *Crenarchaeota*, *Euryarchaeota*, and *Thaumarchaeota*. At least a dozen other phyla are proposed based on culture-independent surveys (**figure 20.1**). In most cases, none of the organisms identified by metagenomics have been grown in culture.
- Autotrophic archaea generally use one of three pathways for CO_2 assimilation: the reductive acetyl-CoA pathway, the 3-hydroxypropionate/4-hydroxybutyrate (HP/HB) cycle, or the dicarboxylate/4-hydroxybutyrate (DC/HB) cycle. Acetate is incorporated either by the glyoxylate or methylaspartate cycle. Most other anabolic pathways appear to be similar to those found in bacteria (**table 20.1; figures 20.2** to **20.4**).
- Although much of archaeal catabolism appears similar to that of bacteria, archaea differ in glucose catabolism, using modified versions of the Embden-Meyerhof and Entner-Doudoroff pathways (**figures 20.5** and **20.6**).

20.2 *Proteoarchaeota,* or TACK Superphylum

- The extremely thermophilic S^0-metabolizers in the phylum *Crenarchaeota* depend on sulfur for growth and are frequently acidophiles. The sulfur may be used as an electron acceptor in anaerobic respiration or as an electron donor by chemolithotrophs. Many are strict anaerobes and grow in geothermally heated soil and water that are rich in sulfur.
- The discovery of an archaeal-specific lipid, now called thaumarchaeol, led to the discovery of globally distributed marine, mesophilic archaea.
- These mesophiles are placed in the phylum *Thaumarchaeota* and are united by their capacity to oxidize ammonia. They are found in many marine and terrestrial habitats, where they can be important contributors to nitrification.
- Asgard archaea encode multiple eukaryotic signature proteins, and are believed to resemble the earliest eukaryotes to evolve.

20.3 Phylum *Euryarchaeota:* Methanogens, Haloarchaea, and Others

- The phylum *Euryarchaeota* contains five major groups: methanogens, haloarchaea, thermoplasms, extremely thermophilic S^0-reducers, and sulfate-reducing archaea.
- Methanogenic archaea are strict anaerobes that obtain energy through the synthesis of methane. They have several unusual cofactors that are involved in methanogenesis (**figures 20.13** and **20.14**).
- Anaerobic methane-oxidizing archaea belong to *Methanomicrobiales* and *Methanosarcinales,* and commonly use sulfate or nitrate as electron acceptors in symbiotic relationships with other microbes. These methanotrophs oxidize methane by reversing the methanogenic pathway.
- Extreme halophilic archaea are aerobic chemoorganotrophs that require at least 1.5 M NaCl for growth. They are found in habitats such as salterns and salt lakes.
- *Halobacterium salinarum* carries out phototrophy without chlorophyll or bacteriochlorophyll by using archaerhodopsin, which employs retinal to pump protons across the plasma membrane (**figures 20.17** and **20.18**).
- The thermophilic *Thermoplasma* spp. grow in hot, acidic coal refuse piles and survive despite the lack of a cell wall. Another thermoplasm, *Picrophilus* spp., grows at pH 0.
- The class *Thermococci* contains extremely thermophilic organisms that can reduce sulfur to sulfide.
- The extreme thermophiles in the genus *Archaeoglobus* differ from other archaea in using a variety of electron donors to reduce sulfate. They also contain methanogenic cofactors F_{420} and methanopterin.

Active Learning

1. Some believe that archaea should not be separate from bacteria because both groups are prokaryotic. Do you agree or disagree? Explain your reasoning.
2. Explain why the fixation of CO_2 by *Thermoproteus* spp. using the DC/HB cycle is not photosynthesis.
3. When the temperature increases, some archaea change their shapes from elongated rods into spheres. Suggest one reason for this change.
4. Why are ether linkages more stable in membranes than ester lipids? How would the presence of tetraether linkages stabilize a thermophile's membrane?
5. Suppose you wished to isolate an archaeon from a hot spring in Yellowstone National Park. How would you go about it?
6. Do you think there will be a discovery of an archaeal pathogen of humans? Explain your reasoning.
7. The discovery that methanogenesis occurs in oxic environments has been termed the methane paradox, as several enzymes in the pathway are inactivated by oxygen. Scientists measured methane production in several wetland sites near Lake Erie, and established that methanogenesis occurs in both oxic and anoxic mud. One measure of methanogenesis is the quantity of *mcrA* transcripts, encoding a subunit of methyl coenzyme M reductase. Why is expression of this gene a good measure of methane production? Researchers found that methane production correlated with acetate concentration, but not formate, methanol, or methylamine concentrations. What taxonomic order is most likely prevalent in this ecosystem? Which carbon fixation pathway would you expect to be in use?

   Read the original paper: Angle, J. C., et al. 2017. Methanogenesis in oxygenated soils is a substantial fraction of wetland methane emissions. *Nature Comm.* 8:1567.

21

Nonproteobacterial Gram-Negative Bacteria

©Gary Alvis/Getty Images

Fuel From Food Waste

After finishing a meal, many people collect their food scraps and feed them to microbes. Home composters promote the degradation of organic waste, and the resulting material is recycled in gardens. But what if instead of tilling the compost into the soil, it could be fed to the electric grid? Microbes like *Thermotoga* perform dark fermentation, an anaerobic process that converts complex carbohydrates like those found in food waste to acetate, CO_2, and most importantly, hydrogen gas (H_2). Rather than a composter outside the kitchen, you might someday have a bioreactor that captures H_2 and converts it to electricity to power your kitchen appliances.

Thermotoga has distinguished itself by producing so much hydrogen that it almost reaches the theoretical maximum yield. It does so by secreting abundant hydrolytic enzymes like cellulases, xylanases, and invertases. These enzymes act on plant biomass to release smaller sugar units. The NADH and $FADH_2$ produced by these oxidative reactions are used in an electron transport chain that uses protons as the terminal electron acceptor, thereby producing an abundance of H_2.

Hydrogen is an attractive alternative fuel, as its combustion generates water instead of greenhouse gases. It has a high energy density, meaning that a given weight of H_2 has more energy than a comparable weight of fuels like methane, ethanol, or gasoline. Scientists in Tunisia studied the growth of *Thermotoga maritima* using an unusual medium—the fruit and vegetable waste from the city of Tunis and seawater from a nearby bay. The food waste provided the carbon and energy sources with additional micronutrients provided in the seawater. Only nitrogen and sulfur sources had to be added for growth, and the culture produced abundant H_2. An important advantage to using *Thermotoga* for dark fermentation is that at its growth temperature of 80°C, few contaminants survive.

In this chapter, we discuss these microbes as well as 12 other phyla of Gram-negative bacteria. This includes a variety of morphologically and physiologically amazing microbes. From deinococci, which can withstand levels of ionizing radiation that would kill a human, to the spirochetes that cause syphilis, this chapter truly illustrates the depth of microbial diversity.

Readiness Check:

Based on what you have learned previously, you should be able to:

- ✓ Draw a Gram-negative bacterial cell wall and list its constituent parts (section 3.4)
- ✓ Differentiate between swimming and gliding motility (section 3.8)
- ✓ List environmental factors that influence microbial growth (section 7.5)
- ✓ Compare and contrast the growth and physiology of obligate aerobes, facultative anaerobes, and strict anaerobes (section 7.5)
- ✓ Describe the difference between heterotrophy and autotrophy (section 11.1)
- ✓ Define chemolithotrophy and identify an example of an electron donor and an acceptor (section 11.10)
- ✓ Compare and contrast oxygenic and anoxygenic phototrophy (section 11.12)

21.1 *Aquificae* and *Thermotogae* Are Ancient Bacterial Lineages

After reading this section, you should be able to:

a. Compare and contrast the physiological and structural differences between members of the phyla *Aquificae* and *Thermotogae*

Thermophilic microbes are found in both the bacterial and archaeal domains, but to date, most hyperthermophilic microbes (those with optimum growth temperatures above 80°C) belong to *Archaea*. The phyla *Aquificae* and *Thermotogae* are two examples of bacterial hyperthermophiles.

The phylum *Aquificae* is thought to represent the deepest, oldest branch of *Bacteria*. Two of the best-studied genera are *Aquifex* and *Hydrogenobacter*. *Aquifex* spp. are chemolithoautotrophs that capture energy by oxidizing hydrogen, thiosulfate, and sulfur with oxygen as the terminal electron acceptor. The reductive citric acid cycle is used to fix CO_2 (*see figure 12.5*). *Aquifex pyrophilus* is a microaerophilic rod that has been shown to grow with a temperature optimum of 85°C and a maximum of 95°C. Because these genera contain thermophilic chemolithoautotrophic species, it has been suggested that the original bacterial ancestor was probably thermophilic and chemolithoautotrophic. ◀ *Chemolithotrophy: "Eating Rocks" (section 11.10)*

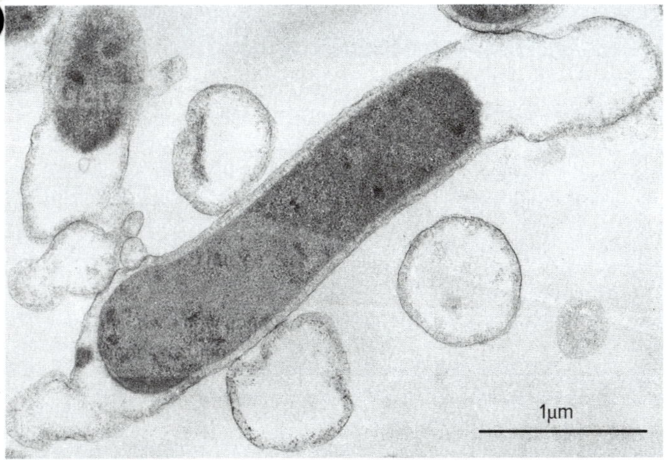

Figure 21.1 ***Thermotoga maritima.*** Note the loose sheath extending from each end of the cell.
©Karl O. Stetter

The second deepest branch is the phylum *Thermotogae.* Members of the genus *Thermotoga* (Greek *therme,* heat; Latin *toga,* outer garment) are thermophiles with a growth optimum of 80°C and a maximum of 90°C. They are rods with an outer sheathlike envelope (like a toga) that can extend or balloon out from the ends of the cell enclosing a large periplasmic space (**figure 21.1**). They grow in active geothermal areas in marine hydrothermal systems and terrestrial solfataric springs. In contrast to *Aquifex, Thermotoga* species are chemoorganotrophs with a functional glycolytic pathway that can grow anaerobically using protons, elemental sulfur, or ferric iron as terminal electron acceptors. Like other members of the phylum *Thermotogae,* the genomes of *Thermotoga* spp. have been subject to horizontal gene transfer (HGT). In this phylum, estimates of genes of archaeal origin range from 25% to 50%. *Horizontal gene transfer: creating genetic variation the asexual way (section 16.4)*

21.2 *Deinococcus-Thermus* Includes Radiation-Resistant Bacteria

After reading this section, you should be able to:

a. Describe habitats in which deinococci can be isolated
b. Discuss the unique capacity of deinococci to tolerate desiccation and high doses of radiation

The phylum *Deinococcus-Thermus* contains the orders *Deinococcales* (Greek *deinos,* unusual) and *Thermales.* Most are mesophilic, but *Thermus aquaticus* was one of the first thermophilic organisms discovered. It grows optimally at 70°C, and its thermostable DNA polymerase, *Taq,* is commonly used in polymerase chain reactions. Deinococci are spherical or rod-shaped and nonmotile, and they are often seen in pairs or tetrads (**figure 21.2*a***). They are aerobic heterotrophs producing acid from only a few sugars. Although they stain Gram positive, their cell wall includes an outer membrane (figure 21.2*b*). Almost all strains are extraordinarily resistant to oxidative stress, desiccation, and radiation; they can survive as much as 3 to 5 million rad of radiation. In contrast, exposure to 100 rad can be lethal to humans.

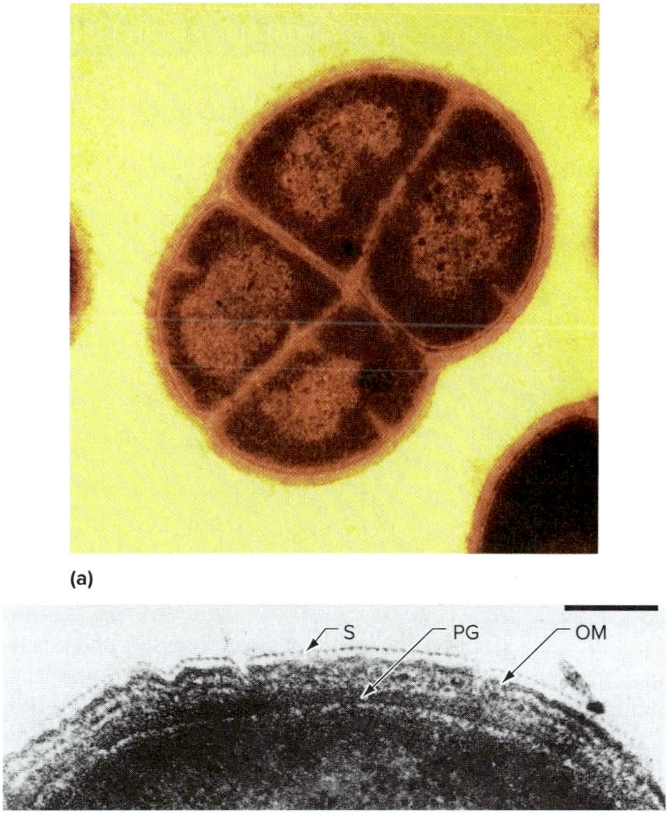

Figure 21.2 **Deinococci.** (a) A *Deinococcus radiodurans* tetrad, or cluster of four cells (average cell diameter 2.5 μm). (b) The cell wall of *D. radiodurans* with an S-layer (S), a peptidoglycan layer (PG), and an outer membrane (OM). Scale bar = 100 nm.
(a) ©Michael J. Daly/Science Source; (b) ©Dr. Robert G.E. Murray/University of Western Ontario

Deinococci are ubiquitous but not abundant; they can be isolated from ground meat, feces, air, freshwater, and other sources. Their resistance to radiation results from their ability to repair a severely damaged genome, which consists of a circular chromosome, a megaplasmid, and a small plasmid. High levels of radiation generate radical oxygen species that break the genome into hundreds of fragments. Amazingly, within 12 to 24 hours, it is pieced back together, ensuring viability. *Deinococcus* cells are polyploid, and it is unlikely that the genome breaks at the same place in each copy. Therefore each cell has the capacity to reassemble the genome from overlapping fragments. This process is termed extended synthesis-dependent strand annealing. Smaller pieces grow larger, and large fragments undergo RecA-mediated homologous recombination to reassemble. The genes responsible for this

Table 21.1 Properties of Some Members of the Class *Mollicutes*

Genus	No. of Recognized Species	Genome Size (Mb)	Sterol Requirement	Habitat	Other Distinctive Features
Acholeplasma	18	1.50–1.65	No	Vertebrates, some plants and insects	Optimum growth 30–37°C
Anaeroplasma	4	1.50–1.60	Yes	Bovine or ovine rumen	Strictly anaerobic
Mesoplasma	11	0.82–0.93	No	Insects, plants	Optimum growth 30°C; sustained growth in serum-free medium only with 0.04% detergent (Tween 80)
Mycoplasma	116	0.60–1.35	Yes	Humans, animals	Optimum growth usually 37°C
Spiroplasma	37	0.78–2.22	Yes	Insects, plants	Helical filaments; optimum growth at 30–37°C
Ureaplasma	7	0.75–1.20	Yes	Humans, animals	Urea hydrolysis

radiation and desiccation response are repressed under normal growth conditions by a transcriptional regulator that binds to their promoters. Radiation triggers the activity of a protease that destroys this repressor by an unknown mechanism, thereby inducing gene expression. ◂ *Transcriptome analysis (section 18.5)*

21.3 Class *Mollicutes*, Phylum *Tenericutes:* Bacteria That Lack Cell Walls

After reading this section, you should be able to:

a. Relate the structure of the mollicute cell envelope to its morphology
b. Outline how arginine can be catabolized to yield ATP by substrate-level phosphorylation
c. Draw the flux of protons that enables *Ureaplasma urealyticum* to generate a proton motive force despite the absence of an electron transport chain
d. Illustrate how gliding motility is accomplished by mycoplasmas
e. List several habitats in which mollicutes reside and relate this to their capacity to cause disease

Until recently the class *Mollicutes* was placed in the phylum *Firmicutes.* Although it is generally agreed that members of this class are related to certain clostridia (low G + C Gram-positive bacteria), *Mollicutes* are currently placed in the phylum *Tenericutes* (Latin *tener,* tender; *cutis,* skin). *Tenericutes* includes all bacteria that lack walls and do not synthesize peptidoglycan precursors. Members of *Mollicutes* are commonly called **mycoplasmas;** these bacteria are characterized by the absence of cell walls, small genomes, and simplified metabolic pathways. Their small genomes appear to be the result of genome reduction such that they now lack a variety of metabolic capabilities, including the ability to synthesize peptidoglycan precursors.

Some of the major characteristics of the best-studied mycoplasmas are summarized in **table 21.1**. Because they are bound only by a plasma membrane, they are pleomorphic, varying in shape from spherical or pear-shaped microbes from 0.3 to 0.8 μm in diameter, to slender branched or helical filaments (**figure 21.3*a***). Some mycoplasmas (e.g., *M. genitalium*) have a specialized terminal structure that projects from the cell and gives them a flask or pear shape. This structure aids in attachment to eukaryotic cells. They are among the smallest bacteria capable of self-reproduction. Most species are facultative anaerobes, but a few are obligate

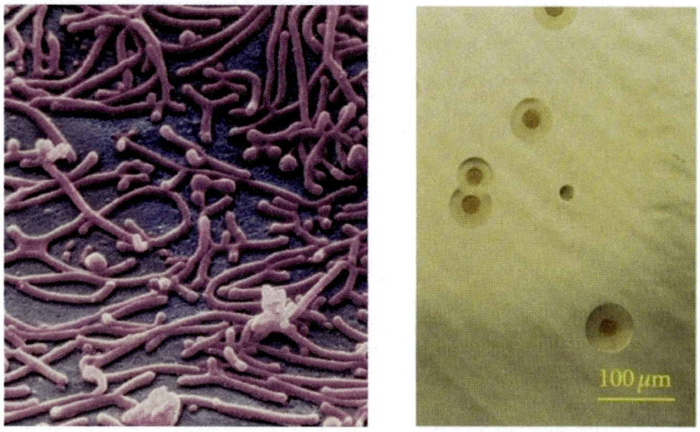

(a) (b)

Figure 21.3 Mycoplasma Cells and Colonies. (a) A scanning electron micrograph of *Mycoplasma pneumoniae* shows its pleomorphic nature. (b) Note the fried-egg appearance of mycoplasma colonies.

(a) ©Don W. Fawcett/Science Source; (b) From "*Isolation and Identification of Acholeplasma sp. from the Mud Crab, Scylla serrata*" by Chen JG, Lou D, Yang JF - (2011)

anaerobes. When grown on agar, most form colonies with a fried-egg appearance because they grow into the agar surface at the center while spreading outward on the surface at the colony edges (figure 21.3*b*).

Mollicute genomes are among the smallest in bacteria (table 21.1). The genomes of the human pathogens *Mycoplasma genitalium, M. pneumoniae,* and *Ureaplasma urealyticum* have fewer than 1,000 genes, suggesting a minimal genome size for free-living existence. Their small genome reflects their inability to synthesize a number of macromolecules. For instance, because *M. genitalium* relies so heavily on its host's biosynthetic capacity, it has lost genes that encode enzymes needed to synthesize amino acids, purines, pyrimidines, and fatty acids. In addition, most species require sterols for growth, which they obtain from the host as cholesterol. Sterols are an essential component of the mycoplasma plasma membrane, where they facilitate osmotic stability in the absence of a cell wall. Some mycoplasmas produce ATP by the Embden-Meyerhof pathway and lactic acid fermentation. The pentose phosphate pathway seems to be functional in at least some mycoplasmas; none appears to have the complete tricarboxylic acid cycle. Arginine can be catabolized to generate ATP in the following series of reactions:

$$\text{Arginine} + H_2O \rightarrow \text{citrulline} + NH_3$$
$$\text{Citrulline} + P_i \rightarrow \text{ornithine} + \text{carbamyl-P}$$
$$\text{Carbamyl-P} + \text{ADP} \rightarrow \text{ATP} + CO_2 + NH_3$$

The first reaction is catalyzed by the enzyme arginine deaminase, which is found in several firmicutes, including *Bacillus* and *Clostridium* species. ▶| *Gram-positive bacteria (chapter 23)*

The small mycoplasma genome has made it the model organism for development of a "synthetic" bacterium. As discussed in chapter 17, this has resulted in a bacterium with a genome that has been chemically synthesized. This work provides proof of principle that cells can be manufactured to encode only those products desired by the biotechnologist. |◀ *Techniques & Applications 17.2*

Ureaplasma urealyticum, a causative agent of urinary tract infections, has an interesting way of conserving energy: It hydrolyzes urea (found in urine) to generate an electrochemical gradient via the accumulation of ammonia/ammonium (NH_3/NH_4^+). This gradient is responsible for the chemiosmotic potential that drives ATP synthesis (**figure 21.4**).

Some mycoplasmas are capable of gliding motility that propels them across host cell surfaces. *Mycoplasma mobile* moves at a steady pace of 2 to 5 μm per second. This is mediated by cell surface adhesins that surround the "neck" of the cell (**figure 21.5*a***). In *M. mobile,* these proteins attach to a cytoskeleton-like structure and function like microscopic legs. Gliding is powered by ATP hydrolysis. By contrast, most other motile mycoplasmas move more like inchworms, at about 1 μm per second. Motility in *M. pneumoniae* accounts for its ability to move from bronchial cilia to the host cell surface. At the leading edge of a mycoplasma cell is an attachment organelle whose adhesins bind and release cell surface oligosaccharides. Alternating compression and extension of the organelle results in directional movement (figure 21.5*b*). |◀ *Twitching and gliding motility (section 3.8)*

Mycoplasmas are remarkably widespread and can be isolated from animals, plants, and soil. Although their complex growth requirements can make their growth in pure (axenic) cultures difficult, at least 10% of the mammalian cell cultures in use are probably contaminated with mycoplasmas. This seriously interferes with tissue culture experiments and has resulted in the development of rapid test kits for mycoplasma detection in cell culture.

In animals, mycoplasmas colonize mucous membranes and joints, and often are associated with diseases of the respiratory and urogenital tracts. Spiroplasmas have been isolated from insects, ticks, and a variety of agriculturally important plants. Arthropods often act as vectors and carry the spiroplasmas between plants. In humans, *U. urealyticum* and *M. hominis* are sexually transmitted bacteria that cause inflammation of the reproductive organs of men and women. *Ureaplasma urealyticum* is also associated with premature delivery of newborns, neonatal meningitis, and pneumonia. *M. pneumoniae* causes primary atypical pneumonia ("walking pneumonia") in humans. ▶| *Mycoplasmal pneumonia (section 39.1)*

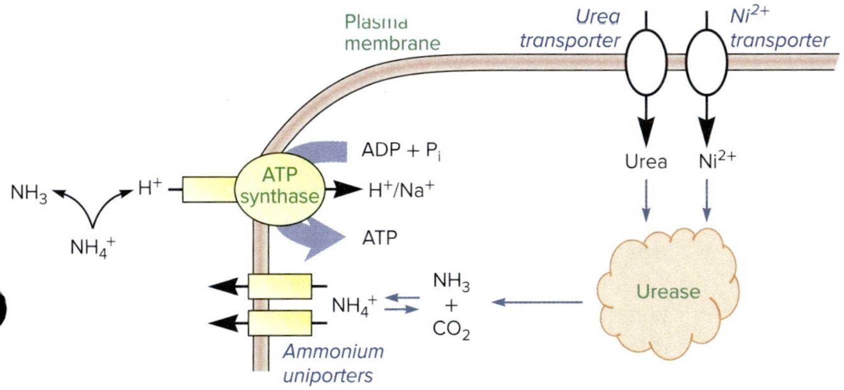

Figure 21.4 Energy Conservation in *Ureaplasma urealyticum.* To generate an electrochemical gradient through the accumulation of NH_3/NH_4^+, *U. urealyticum* imports urea through specific transporters. Once inside the cell, urease (a nickel-binding enzyme) catalyzes the hydrolysis of urea to NH_3 and CO_2. Ammonia accepts a proton to become NH_4^+, which is exported through NH_4^+ uniporters. Outside the cell, NH_4^+ is converted to NH_3 and H^+, providing protons to drive ATP synthesis via the membrane-bound ATP synthase.

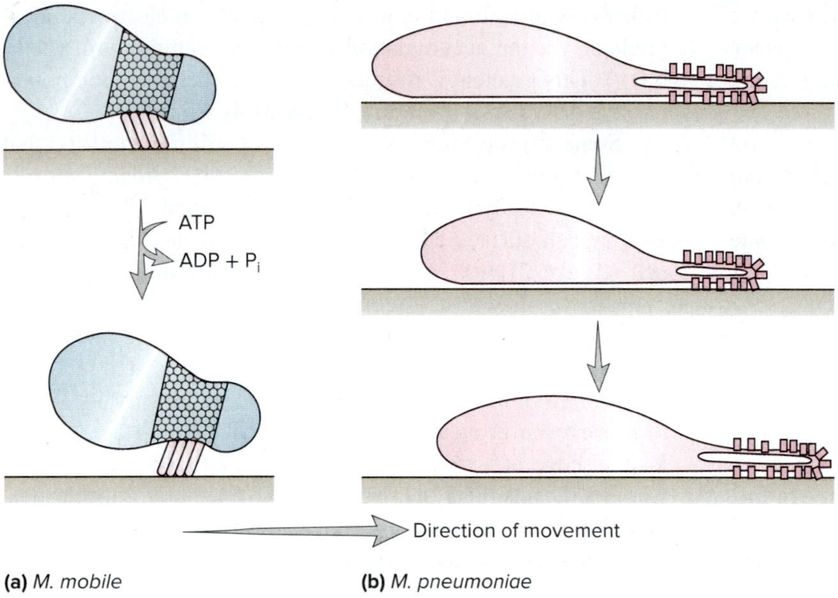

Figure 21.5 **Two Forms of Gliding Motility in Mycoplasmas.** (a) *M. mobile* uses cytoskeletal "leg" proteins to move. (b) *M. pneumoniae* has an attachment organelle that causes the cell to expand and contract, so that it moves at an uneven, inchwormlike pace.

Comprehension Check

1. In what ways are members of *Aquifex* and *Thermotoga* similar? How do they differ?
2. How do deinococci survive high doses of radiation?
3. Design an experiment to explore the role of sterols in mycoplasma structure and/or physiology.
4. Explain the relationship between genome size and growth requirements of mycoplasmas. Why is genome reduction common in intracellular microbes?

21.4 Photosynthetic Bacteria Are Diverse

After reading this section, you should be able to:

a. Assess the importance of photosynthetic pigments in the distribution of photosynthetic bacteria in nature
b. Draw a generic chlorosome and identify the function of its structural elements
c. Predict the habitat closest to your home where you might find members of the phyla *Chlorobi, Chloroflexi,* and *Cyanobacteria*
d. Draw a generic cyanobacterial cell and label its intracellular structures
e. List three types of specialized cells made by cyanobacteria and describe the function of each
f. Compare and contrast prochlorophytes with other cyanobacteria
g. Draw a concept map describing the structure and physiology of the six types of photosynthetic bacteria listed in table 21.2

Historically, the term *photosynthetic* has been applied to both photoheterotrophic and photoautotrophic microbes. We continue to use this term in this way, but care should be taken not to assume that the capture of light energy (phototrophy) is always linked with CO_2 fixation (autotrophy). As we shall see, unlike plants, photosynthetic microbes often have more than one metabolic strategy and may uncouple phototrophy from autotrophy and grow photoheterotrophically.

Bacteria capable of chlorophyll-based photosynthesis have been identified in seven phyla, with the most recent discovered in 2014 when a chlorophyll-containing photoheterotrophic, Gram-negative bacterium belonging to the phylum *Gemmatimonadetes* was isolated from a freshwater lake in the Gobi Desert. The phylum *Chloroflexi* contains green nonsulfur bacteria, and the phylum *Chlorobi* includes green sulfur bacteria. Cyanobacteria are placed in their own phylum, *Cyanobacteria.* Purple bacteria are divided between three groups. Purple sulfur bacteria are placed in γ-proteobacteria (families *Chromatiaceae* and *Ectothiorhodospiraceae*). Purple nonsulfur bacteria are distributed between α-proteobacteria and β-proteobacteria. Aerobic *An*oxygenic *P*hototrophs (AAnPs) are photoheterotrophs divided among α-, β-, and γ-proteobacteria. Finally, Gram-positive heliobacteria in the phylum *Firmicutes* are also photosynthetic. There appears to have been considerable horizontal transfer of photosynthetic genes among the phyla, with at least 50 genes related to photosynthesis common to all. In this chapter, we describe cyanobacteria and green bacteria; purple bacteria are discussed in chapter 22, while heliobacteria are discussed in chapter 23. Properties of the major groups are summarized in **table 21.2**.

Cyanobacteria differ fundamentally from the other photosynthetic bacteria because they perform **oxygenic photosynthesis:** They have photosystems I and II, use water as an electron donor, and generate oxygen during photosynthesis. In contrast, purple bacteria, green bacteria, and AAnPs have only one photosystem and use **anoxygenic photosynthesis.** Because they are unable to use water as an electron source, they employ other reduced molecules as electron donors. AAnPs and purple nonsulfur bacteria are photoorganoheterotrophs, so they use electrons obtained during the oxidation of organic substrates but make ATP from light energy. The purple and green sulfur bacteria use inorganic electron donors such as hydrogen sulfide, sulfur, and hydrogen, as well as organic matter, as an electron source for the reduction of $NAD(P)^+$ to NAD(P)H. Those that use sulfide as an electron source often form sulfur granules. Purple sulfur bacteria accumulate granules within their cells, whereas green sulfur bacteria deposit the sulfur granules outside their cells. ◀ *Phototrophy (section 11.12)*

Photosynthetic microbes are major contributors of fixed carbon, termed primary productivity, in a variety of habitats. The differences in photosynthetic pigments and oxygen requirements among photosynthetic bacteria have important ecological

Table 21.2 Characteristics of the Major Groups of Gram-Negative Photosynthetic Bacteria

Characteristic	Green Sulfur	Green Nonsulfur	Purple Sulfur	Purple Nonsulfur	Aerobic Anoxygenic Phototrophs	Cyanobacteria
Major photosynthetic pigments	Bacteriochlorophylls *a* plus *c*, *d*, or *e*	Bacteriochlorophylls *a* and *c*	Bacteriochlorophyll *a* or *b*	Bacteriochlorophyll *a* or *b*	Bacteriochlorophyll *a*	Chlorophyll *a* plus phycobiliproteins. *Prochlorococcus* has divinyl derivatives of chlorophyll *a* and *b*
Morphology of photosynthetic membranes	Photosynthetic system partly in chlorosomes	Chlorosomes present when grown anaerobically	Photosynthetic system contained in spherical or lamellar membrane complexes that are continuous with the plasma membrane	Photosynthetic system contained in spherical or lamellar membrane complexes that are continuous with the plasma membrane	Few, if any, intracytoplasmic membranes	Thylakoid membranes lined with phycobilisomes
Photosynthetic electron donors	H_2, H_2S, S^0	Photoheterotrophic donors—a variety of sugars, amino acids, and organic acids; photoautotrophic donors—H_2S, H_2	H_2, H_2S, S^0	Usually organic molecules: sometimes reduced sulfur compounds or H_2	A variety of sugars, amino acids, and organic acids	H_2O
Sulfur deposition	Outside of the cell	N/A[1]	Inside the cell[2]	Outside of the cell in a few cases	N/A	N/A
Nature of phototrophy	Anoxygenic	Anoxygenic	Anoxygenic	Anoxygenic	Anoxygenic	Oxygenic (some are also facultatively anoxygenic)
General metabolic type	Obligate anaerobic photolithoautotrophs	Usually photoheterotrophic; sometimes photoautotrophic or chemoheterotrophic (when aerobic and in the dark)	Obligate anaerobic photolithoautotrophs	Usually anaerobic photoorgano-heterotrophs; some facultative photolithoautotrophs; some are chemo-organoheterotrophs in the dark	Aerobic photoorgano-heterotrophs	Aerobic photo-lithoautotrophs
Motility	Many nonmotile; some glide; some have gas vesicles	Gliding	Motile with polar or peritrichous flagella	Motile with polar flagella or nonmotile; some have gas vesicles	Some are motile with a single or a few polar or subpolar flagella	Nonmotile, swimming motility without flagella; some glide; some have gas vesicles
Taxonomic distribution	*Chlorobi*	*Chloroflexi*	γ-proteobacteria	α-proteobacteria, β-proteobacteria (*Rhodocyclus*)	α-proteobacteria, β-proteobacteria, γ-proteobacteria	*Cyanobacteria*

1 N/A: Not applicable.
2 With the exception of *Ectothiorhodospira*.

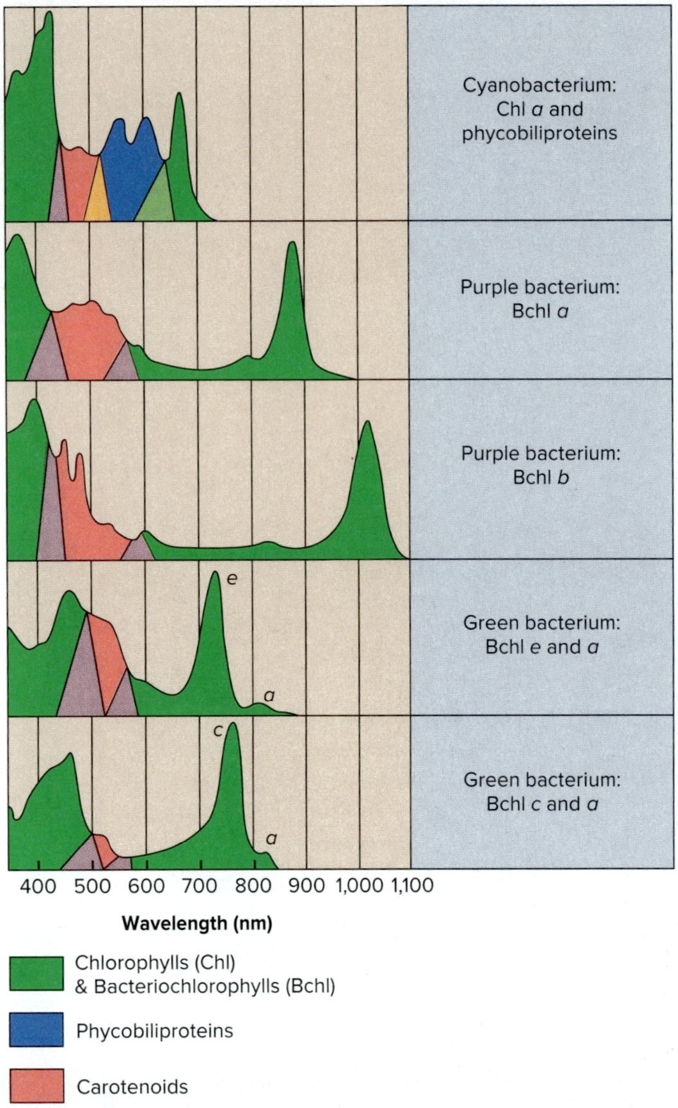

Figure 21.6 Photosynthetic Pigments. Absorption spectra of five photosynthetic bacteria showing the differences in absorption maxima and the contributions of various accessory pigments.

consequences. As shown in **figure 21.6**, chlorophylls (Chl), bacteriochlorophylls (Bchl), and their associated accessory pigments have narrow absorption spectra. Microbes therefore use accessory pigments to broaden their energy input. These pigments collect light of different wavelengths and transfer it to Chl or Bchl, where it is converted to chemical energy. Oxygenic cyanobacteria and diatoms dominate the aerated upper layers of freshwater and marine microbial communities, where they absorb large amounts of red and blue light. Anoxygenic purple and green photosynthetic bacteria inhabit the deeper anoxic zones rich in hydrogen sulfide and other reduced compounds that can be used as electron donors. Their Bchl and accessory pigments enable their use of light in the far-red spectrum (i.e., >750 nm) not used by other photosynthetic organisms (compare the Bchl *a* and Bchl *b* with Chl *a* in figure 21.6). Bchl absorption has additional peaks between 350 and 550 nm, allowing growth at greater depths because shorter-wavelength light can penetrate water farther.

Phylum *Chlorobi*: Green Sulfur Bacteria

Members of the phylum *Chlorobi* are commonly called **green sulfur bacteria.** This is a small group of obligately anaerobic photolithoautotrophs that use hydrogen sulfide, elemental sulfur, and hydrogen as electron sources. When sulfide is oxidized, elemental sulfur is deposited outside the cell. Photosynthetic pigments are located in ellipsoidal vesicles called **chlorosomes** that are attached to the plasma membrane by a proteinaceous baseplate that contains Bchl *a* molecules (**figure 21.7**). The chlorosome membrane is an unusual lipid monolayer, rather than bilayer. Within chlorosomes, bacteriochlorophyll

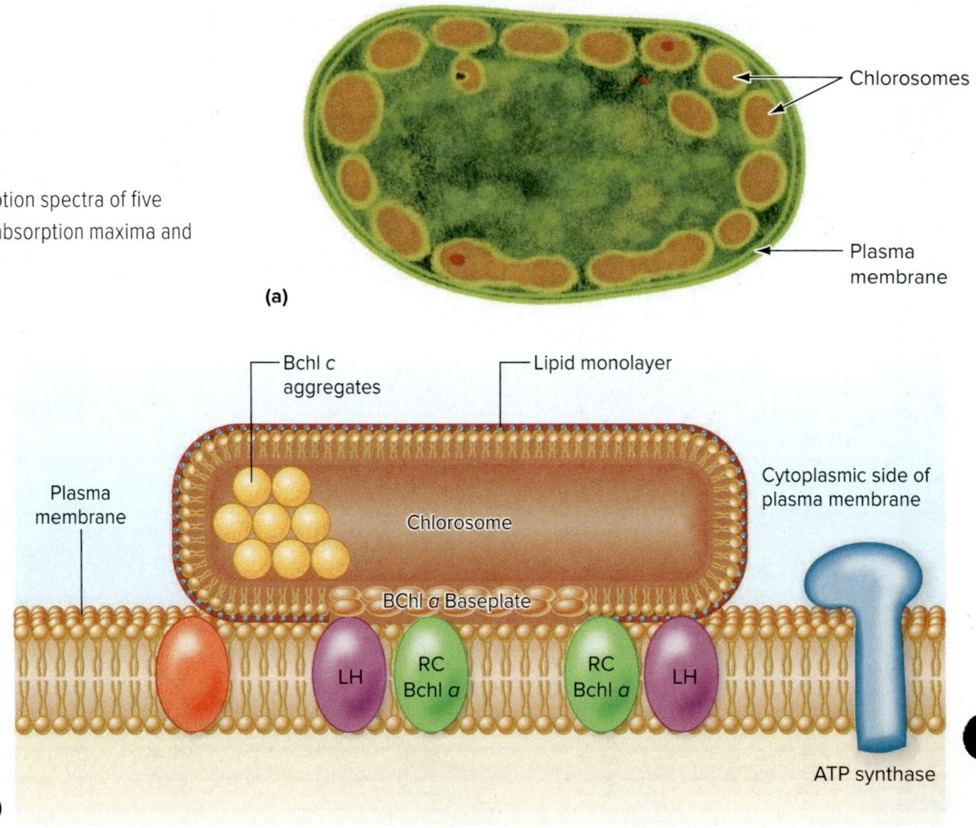

Figure 21.7 Green Sulfur Bacteria. (a) Light-harvesting chlorosomes surround the inside perimeter of the cytoplasmic membrane. Antenna bacteriochlorophylls (Bchl *c, d,* or *e*) transfer light energy to reaction center Bchl *a* in the plasma membrane. (b) A chlorosome is attached to the inner surface of the plasma membrane by a baseplate composed of protein, a lipid monolayer, and Bchl *a* molecules. The reaction center (RC) is located in the plasma membrane and contains only Bchl *a*. The light-harvesting complex (LH) and ATP synthase are also located in the plasma membrane.

molecules (Bchl *c, d,* or *e*) are grouped into rodlike structures held together by carotenoids and lipids. The light harvested by these pigments is transferred by baseplate Bchl *a* to reaction centers in the plasma membrane, which also contain Bchl *a*. The ATP produced by photophosphorylation is used to fuel CO_2 fixation by the reductive citric acid cycle (*see figure 12.5*); the more familiar Calvin-Benson cycle is used only by aerobes.

These bacteria flourish in anoxic, sulfide-rich zones of lakes. Although they lack flagella and most are nonmotile, members of at least one genus glides. Some species have gas vesicles to adjust their depth for optimal light and hydrogen sulfide (*see figure 3.33c*). Those without gas vesicles are found in sulfide-rich muds at the bottom of lakes and ponds.

Green sulfur bacteria are diverse morphologically. They may be rods, cocci, or vibrios; some grow singly, and others form chains and clusters. They are either grass-green or chocolate-brown in color. Representative genera are *Chlorobium, Prosthecochloris,* and *Pelodictyon.*

Phylum *Chloroflexi:* Green Nonsulfur Bacteria

The phylum *Chloroflexi* is a deep and ancient branch of the bacterial tree. It has two orders: One consists of thermophilic, facultatively photosynthetic bacteria (*Chlorflexales*), and the other consists of nonphotosynthetic thermophiles and mesophiles (*Herpetosiphonales*). Members of the genus *Chloroflexus* make up the majority of photosynthetic **green nonsulfur bacteria.** However, the term *green nonsulfur* is a misnomer because not all members of this group are green, and some can use sulfide. *Chloroflexus* spp. are filamentous, gliding bacteria often isolated from neutral to alkaline hot springs, where they grow in the form of orange-reddish mats, usually in association with cyanobacteria. They possess small chlorosomes with accessory Bchl *c*. The light-harvesting complexes contain Bchl *a* and are in the plasma membrane. *Chloroflexus* spp. are metabolically diverse. Some are facultative anaerobes, others are anaerobic, and while some are chemoorganotrophic, most are photoheterotrophic. They can carry out anoxygenic photosynthesis using sulfide or hydrogen as a source of electrons. Heterotrophic growth is supported by pyruvate, acetate, glycerol, and glucose oxidation. When growing autotrophically, they use an unusual pathway, the 3-hydroxypropionate bi-cycle to fix CO_2 (**figure 21.8**). An interesting aspect of this pathway is that the left half produces glyoxylate, which is then used in the right half to make pyruvate.

Members of the nonphotosynthetic, gliding, rod shaped, or filamentous genus *Herpetosiphon* are aerobic chemoorganotrophs with respiratory metabolism. They can be isolated from freshwater and soil habitats.

Comprehension Check

1. Which of the phototrophs discussed perform anoxygenic photosynthesis? Which are also capable of chemoorganoheterotrophy?
2. What is the function of chlorosomes? How does their unique structure contribute to this function?
3. Compare the capacity of green nonsulfur (*Chloroflexi*) and green sulfur bacteria (*Chlorobi*) to position themselves in light of optimum wavelength.

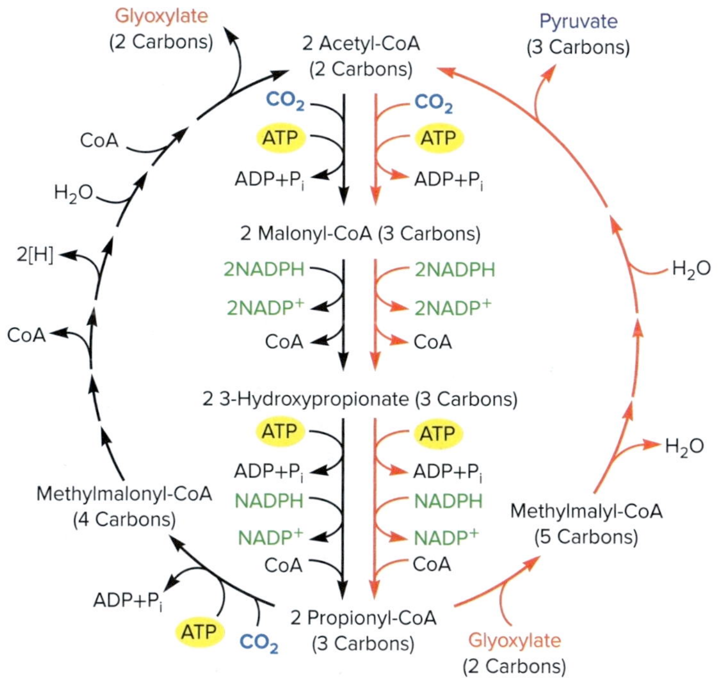

Figure 21.8 **The 3-Hydroxypropionate Bi-Cycle.** This pathway functions when photosynthetic members of the phylum *Chloroflexi* grow autotrophically; CO_2 exists as HCO_3^- when it is added to intermediates of the pathway.

Phylum *Cyanobacteria*: Oxygenic Photosynthetic Bacteria

Cyanobacteria are the largest and most diverse group of photosynthetic bacteria. Originally studied by botanists, they were classified as *Chlorophyceae,* erroneously known as blue-green algae. *Bergey's Manual* divides cyanobacteria into five subsections (**table 21.3**). At present all taxonomic schemes must be considered tentative, and while many genera have been assigned, species names have not yet been designated in most cases.

Recall that endosymbiotic cyanobacteria are thought to have evolved into chloroplasts. It follows that the cyanobacterial photosynthetic apparatus closely resembles that of eukaryotes; cyanobacteria have chlorophyll *a* and photosystems I and II, and thereby perform oxygenic photosynthesis. Light-harvesting and electron transport chain components are located in thylakoid membranes to which particles called **phycobilisomes** are attached (**figure 21.9**). Phycobilisomes are protein complexes made up of **phycobiliprotein** pigments, particularly **phycocyanin** and **phycoerythrin,** held together by linker proteins (figure 21.9*b*). Each phycobilisome consists of a core phycobiliprotein pigment, allophycocyanin, surrounded by other pigments arranged in stacklike structures. Each pigment absorbs light of different wavelengths, and this arrangement enables the efficient transfer of energy to photosystem II. Assembly of these complexes in the thylakoid membrane is aided by association with flotillins in functional membrane microdomains.

◀ *Bacterial plasma membranes are dynamic (section 3.3); Evolution of the three domains of life (section 19.5)*

Table 21.3 Characteristics of the Cyanobacterial Subsections

Subsection	General Shape	Reproduction and Growth	Heterocysts	Other Properties	Representative Genera
I	Unicellular rods or cocci; nonfilamentous aggregates	Binary fission, budding	–	Nonmotile or swim without flagella	*Chroococcus* *Prochlorococcus* *Synechococcus*
II	Unicellular rods or cocci; may be held together in aggregates	Multiple fission to form baeocytes	–	Some baeocytes are motile.	*Pleurocapsa* *Dermocarpella* *Chroococcidiopsis*
III	Filamentous, unbranched trichome with only vegetative cells	Binary fission in a single plane, fragmentation	–	Gliding motility	*Oscillatoria* *Prochlorothrix* *Spirulina*
IV	Filamentous, unbranched trichome may contain specialized cells	Binary fission in a single plane, fragmentation to form hormogonia	+	Often motile, may produce akinetes; symbiotic with fungi to form lichens	*Anabaena* *Cylindrospermum* *Nostoc*
V	Filamentous trichomes either with branches or composed of more than one row of cells	Binary fission in more than one plane, hormogonia formed	+	May produce akinetes; greatest morphological complexity and differentiation in cyanobacteria	*Fischerella* *Stigonema* *Geitleria*

Carbon dioxide is assimilated through the Calvin-Benson cycle (*see figure 12.4*). The enzymes needed for this process are found in microcompartments called **carboxysomes.** During daylight hours, when cyanobacteria fix carbon for anabolic purposes, they also synthesize **glycogen** as a reserve carbohydrate to supply electrons needed for respiration at night. Some cyanobacteria also store extra nitrogen as polymers of arginine and aspartic acid called **cyanophycin** granules. Phosphate is stored in polyphosphate granules. Because cyanobacteria lack the enzyme α-ketoglutarate dehydrogenase (α-KDH), it was long thought that they lack a fully functional TCA cycle. However, two enzymes that bypass the need for α-KDH were identified in a unicellular cyanobacterium belonging to the genus *Synechococcus*. In most cyanobacteria, the first enzyme converts α-ketoglutarate to succinic semi-aldehyde, which is then oxidized to succinate by the second enzyme; the intermediate succinyl-CoA is not produced (*see figure 11.8*). ◀ *Microcompartments (section 3.6)*

Although most cyanobacteria are obligate photolithoautotrophs, a few can grow in the dark as chemoheterotrophs by

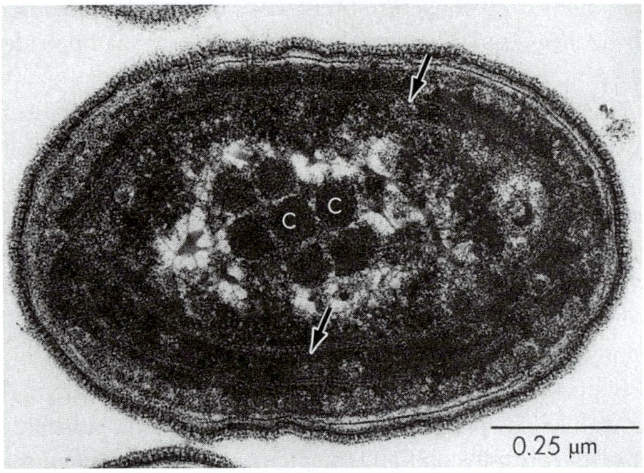

(a)

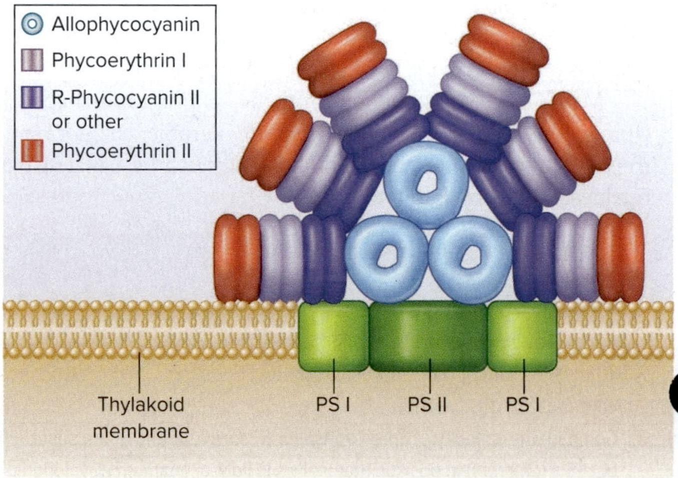

(b)

Figure 21.9 Cyanobacterial Thylakoids and Phycobilisomes. (a) Marine *Synechococcus* with thylakoids (arrows). Carboxysomes are labeled c. (b) Phycobilisomes, located on the thylakoid membrane, have an allophycocyanin core that transfers energy from the radially arranged pigment stacks to photosystem II. Phycobilisomes are found in all cyanobacteria except prochlorophytes.

Figure 21.10 Cyanobacteria Are Morphologically Diverse. (a) *Oscillatoria* trichomes seen with Nomarski interference-contrast optics (×250). (b) *Chroococcus turgidus* produces a gelatinous sheath that can surround one, two, or four cells. (c) *Nostoc* sp. produces heterocysts (arrows) in the absence of combined nitrogen. (d) The cyanobacterium *Anabaena spiroides* with heterocysts. (×1,000).

oxidizing glucose and a few other sugars. Under anoxic conditions, *Oscillatoria limnetica* oxidizes hydrogen sulfide instead of water and carries out anoxygenic photosynthesis much like green photosynthetic bacteria. *Oscillatoria* is one of several cyanobacterial genera whose members produce geosmins, volatile organic compounds that have an earthy odor. When populations of these cyanobacteria become dominant in lakes and rivers, this can be a nuisance, even changing the taste of drinking water. ▶| *Microorganisms in marine and freshwater ecosystems (chapter 30)*

Cyanobacteria vary greatly in shape and appearance. They range in diameter from about 1 to 10 μm and may be unicellular, exist as colonies of many shapes, or form filaments called trichomes (**figure 21.10*a***). A **trichome** is a row of bacterial cells that are in close contact with one another over a large area. Although many cyanobacteria appear blue-green because of phycocyanin, isolates from the open ocean are red or brown because their phycobilisomes contain the pigment phycoerythrin (figure 21.9*b*). Because each pigment evolved to absorb light of different wavelengths, cyanobacteria modulate the relative amounts of these pigments in response to available light in a process known as **chromatic adaptation.** When orange light is sensed, the production of phycocyanin (absorption peak at 618 nm) is stimulated, whereas blue and blue-green light promote the production of phycoerythrin (absorption peaks at 490, 546, and 576 nm). Chromatic adaptation means that cyanobacteria must be able to sense light quality, and they use a variety of strategies to do so, including sensory rhodopsins, plantlike proteins called phytochromes, and membrane-associated receptors that sense ultraviolet and blue light. Many cyanobacterial species use gas vacuoles to position themselves in optimum illumination in the water column—a form of **phototaxis.** Gliding motility is used by many filamentous cyanobacteria, including those found in dense microbial assemblages in sediments called microbial mats (*see figure 29.9*). Although cyanobacteria lack flagella, about one-third of the marine *Synechococcus* strains swim at rates up to 25 μm per second. Swimming motility is not used for phototaxis; instead, it appears to be used for chemotaxis toward simple nitrogenous compounds such as urea. The mechanisms for both gliding and swimming motility in cyanobacteria remain unresolved. |◀ *Chemotaxis (section 3.8)*

Cyanobacteria show great diversity with respect to reproduction and employ a variety of mechanisms: binary fission, budding, fragmentation, and multiple fission. In the last process, a cell enlarges and then divides several times to produce many smaller progeny called **baeocytes** (*see figure 7.2*), which are released upon rupture of the parental cell. Some baeocytes disperse through gliding motility. Fragmentation of filamentous cyanobacteria can generate small, motile filaments called **hormogonia.** Some filamentous species develop **akinetes,** specialized, dormant, thick-walled resting cells resistant to desiccation (**figure 21.11*a***). These later germinate to form new filaments.

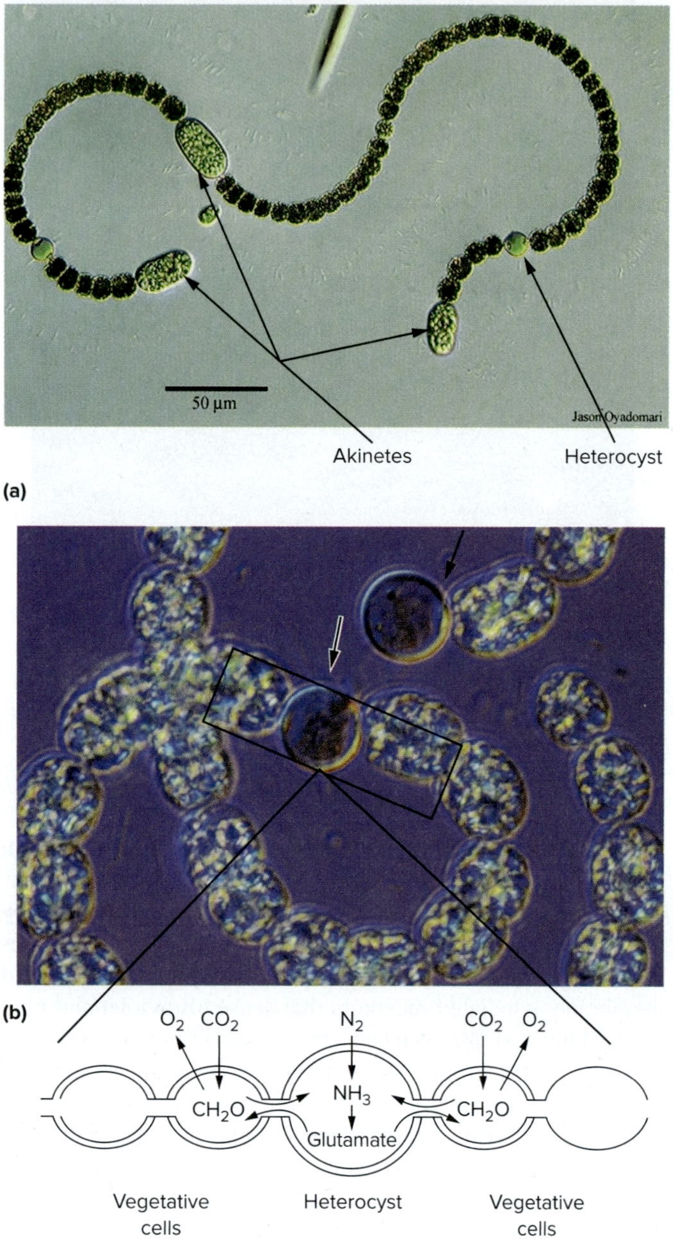

Figure 21.11 Cyanobacteria with Akinetes and Heterocysts. (a) The filamentous cyanobacterium *Anabaena* with akinetes and heterocysts. (b) *Anabaena* heterocysts (arrows) and vegetative cells. (c) Heterocysts fix dinitrogen to form glutamate, which is exchanged with adjacent vegetative cells for carbohydrate. Heterocysts use only photosystem I, so photosynthesis is anoxygenic in these differentiated cells.

MICRO INQUIRY *Why do you think NH_3 is converted to the amino acid glutamate in heterocysts before its transfer to neighboring vegetative cells?*

Many filamentous cyanobacteria fix atmospheric nitrogen by means of special cells called **heterocysts** (figure 21.11). Around 5 to 10% of the cells develop into heterocysts when these cyanobacteria are deprived of both nitrate and ammonia, their preferred nitrogen sources. Within these specialized cells, photosynthetic membranes are reorganized and the proteins that make up photosystem II and phycobilisomes are degraded. Photosystem I remains functional to produce ATP, but no oxygen is generated. Lack of O_2 production is critical because the N_2-fixing enzyme nitrogenase is extremely oxygen sensitive. Heterocysts develop thick cell walls, which slow or prevent O_2 diffusion into the cell, and any O_2 present is consumed during respiration. Heterocyst structure and physiology ensure that these specialized cells remain anaerobic; the heterocyst is dedicated to nitrogen fixation, not reproduction. Nutrients are obtained from adjacent vegetative cells, while the heterocysts contribute fixed nitrogen in the form of amino acids. Nitrogen fixation also is carried out by some cyanobacteria that lack heterocysts. Some fix nitrogen under dark, anoxic conditions in microbial mats. Planktonic forms such as those belonging to the genus *Trichodesmium* fix nitrogen and contribute significantly to the marine nitrogen budget *(see figure 30.8).* ◀ *Nitrogen fixation: N_2 to NH_3 (section 12.5)* ▶ *Nitrogen cycle (section 28.1)*

An important group of cyanobacteria is collectively referred to as **prochlorophytes** (genera *Prochloron, Prochlorococcus,* and *Prochlorothrix*). These microbes differ from other cyanobacteria by the presence of chlorophyll *b* as well as *a,* and the lack of phycobilins. This pigment arrangement imparts a grass-green color. As the only bacteria to possess chlorophyll *b,* the ancestors of unicellular prochlorophytes are considered the best candidates for the endosymbionts that gave rise to chloroplasts. ◀ *Evolution of the three domains of life (section 19.5)*

The three recognized prochlorophyte genera are quite different from one another. Members of the genus *Prochloron* were first discovered as symbionts growing either on the surface or within the cloacal cavity of marine colonial ascidian invertebrates (commonly called "sea squirts"). These bacteria are single-celled, spherical, and 8 to 30 µm in diameter. By contrast, those in the genus *Prochlorothrix* are free living, with cylindrical cells that form filaments growing in freshwater.

Prochlorococcus spp. are tiny coccobacilli less than 1 µm in diameter that flourish in marine plankton. They differ from other prochlorophytes in having divinyl chlorophyll *a* and *b,* and α-carotene, instead of chlorophyll *a* and β-carotene. Other cyanobacteria use monovinyl chlorophyll *a*; the difference between mono- and divinyl chlorophyll is the presence of a double bond between carbons attached to pyrrole ring II (*see figure 11.30*). *Prochlorococcus marinus* is globally distributed in the ocean between the latitudes 40°N and 40°S and to 200 meters in depth, making it the most abundant photosynthetic organism on Earth.

Figure 21.12 **Bloom of Cyanobacteria and Algae in a Eutrophic Pond.**

Prochlorococci have a profound effect on the global carbon cycle. Together with members of the genus *Synechococcus, Prochlorococcus* accounts for at least one-third of global CO_2 fixation. These microbes have coevolved to take maximum advantage of incident light. *Synechococcus* spp. occupy the upper 25 meters of the open ocean, where white and blue-green light can be harvested. In the waters beneath *Synechococcus,* two ecotypes of *Prochlorococcus* species exist: The high-light ecotype grows between about 25 to 100 meters, while the low-light ecotype lives from 80 to 200 meters, where the waters are penetrated by blue-violet light. This prevents the two genera of cyanobacteria from competing for the essential resource of light. ◀ *Microbial evolutionary processes (section 19.5)* ▶ *Microorganisms in the open ocean are adapted to nutrient limitation (section 30.2)*

In addition to those in coastal and open ocean waters, many other cyanobacteria inhabit a wide range of other habitats. A variety of cyanobacteria live in symbiotic association with protists and fungi. Thermophilic species may grow at temperatures up to 75°C in neutral to alkaline hot springs. Because these photoautotrophs are so hardy, they are primary colonizers of soils and surfaces devoid of plant growth. Some unicellular forms even grow in the fissures of desert rocks. In nutrient-rich warm ponds and lakes, filamentous cyanobacteria such as *Anacystis* and *Anabaena* spp. can reproduce rapidly to form blooms (**figure 21.12**). The release of large amounts of organic matter upon the death of the bloom microorganisms stimulates the growth of chemoheterotrophic bacteria. These microbes subsequently deplete available oxygen. This kills fish and other organisms. Some species can produce toxins that can kill animals that drink the water. Other cyanobacteria (e.g., *Oscillatoria* spp.) are so pollution resistant and characteristic of freshwater with high organic matter content that they are used as water pollution indicators. ▶ *Microorganisms in coastal ecosystems are adapted to a changeable environment (section 30.2)*

Comprehension Check

1. What are the major characteristics that distinguish cyanobacteria from other photosynthetic organisms?
2. What is a trichome and how does it differ from a simple chain of cells?
3. Briefly discuss the ways in which cyanobacteria reproduce.
4. Describe how a vegetative cell, a heterocyst, and an akinete are different. How are heterocysts modified to carry out nitrogen fixation?
5. Compare prochlorophytes with other cyanobacteria. Why do they occupy stratified layers in the ocean?
6. Describe some important positive and negative impacts cyanobacteria have on humans and the environment.

21.5 Superphylum *Planctomycetes-Verrucomicrobia-Chlamydiae:* Atypical Cell Division

After reading this section, you should be able to:

a. List unusual features that unite some members of *Planctomycetes, Verrucomicrobia,* and *Chlamydiae*
b. Identify the electron donor and acceptor used in the anammox reaction
c. Explain why the anammox reaction is important to the global flux of nitrogen
d. Describe habitats where methane oxidation might occur
e. Explain the term *obligate intracellular parasite*
f. Diagram the chlamydial life cycle
g. Deduce why chlamydia can survive despite significant metabolic limitations

This superphylum includes seven phyla: *Planctomycetes, Verrucomicrobia, Chlamydiae,* and four others. The grouping is supported by multiple phylogenetic markers and the observation that some genera within each of these phyla synthesize sterols and lack the cell division protein FtsZ. Because these features are not uniformly present, it is believed that the last common ancestor of these microbes had all these characteristics, which were subsequently lost by some. For many years the presence of peptidoglycan was debated, because although the genes encoding peptidoglycan synthesis enzymes are found in the genomes, peptidoglycan could not be detected. A new approach to labeling D-amino acids demonstrated peptidoglycan in *Chlamydia trachomatis.* In most cases, it is present in small quantities, and not necessarily at all stages of the life cycle. Cell division in these bacteria occurs by budding and binary fission processes that use other cytoskeletal proteins. In other bacteria, FtsZ functions both in constriction of the membrane during cell division and in deposition of peptidoglycan during growth, so it is not surprising that the absence of FtsZ in the PVC superphylum correlates with anomalies in both processes. ◀ *Cytokinesis (section 7.2)*

Phylum *Planctomycetes:* Anammox Bacteria

Members of the genera *Brocadia, Kuenenia, Scalindua,* and *Anammoxoglobus* possess a compartment called the **anammoxosome** that occupies much of the cell volume. This is the site of *an*aerobic *amm*onia *ox*idation, called the **anammox reaction,** or simply anammox. This unique form of chemolithoautotrophy employs ammonium ion (NH_4^+) as the electron donor and nitrite (NO_2^-) as the terminal electron acceptor—it is reduced to nitrogen gas (N_2) (**figure 21.13*a***). Hydrazine (N_2H_4) is an important intermediate in anammox. It is normally a cellular toxin, but in this case its very negative reduction potential enables it to donate electrons to ferredoxin. The four high-energy electrons carried by reduced ferredoxin can then be used to produce a proton motive force (PMF) that drives ATP synthesis by an ATP synthase embedded in the anammoxosome membrane. Reduced ferredoxin is also used in CO_2 fixation.

One interesting feature of the physiology of anammox bacteria is that they use nitrite not only as the electron acceptor for ammonium oxidation and generation of a PMF but also as an electron donor for CO_2 fixation, which is accomplished using the reductive acetyl-CoA pathway (*see figure 20.2*). Therefore nitrite acts in both anabolic and catabolic processes, as shown in the following reactions:

An anabolic reaction:

$$2NO_2^- + CO_2 + H_2O \rightarrow CH_2O + 2NO_3^-$$

And a catabolic reaction:

$$NO_2^- + NH_4^+ \rightarrow N_2 + 2H_2O$$

The anammoxosome is bounded by a single bilayer membrane that contains ladderane lipids (figure 21.13*b*). These lipids are unique among planctomycetes and they are thought to form an impermeable barrier, preventing the escape of toxic anammox intermediates into the cytoplasm. It is important to note that anammox is estimated to contribute as much as 70% to the cycling of nitrogen in the world's oceans. ▶| *Nitrogen cycle (section 28.1)*

Members of the genus *Planctomyces* attach to surfaces through a stalk and holdfast. Most of these bacteria have life cycles in which sessile cells bud to produce motile swarmer cells. The swarmer cells are flagellated and swim for a while before attaching to a substrate and beginning budding reproduction.

Early work with *Planctomycetes* like *Gemmata obscuriglobus* revealed the presence of intracellular membranes, including one to surround the chromosome. The extensive internal membranes plus the presence of the anammoxosome led scientists to propose that these cells may be a precursor to a eukaryotic cell plan. However, more recent studies dispute this idea. They show that the internal membranes are simple extensions of the plasma membrane, and that these cells have a somewhat modified Gram-negative cell structure.

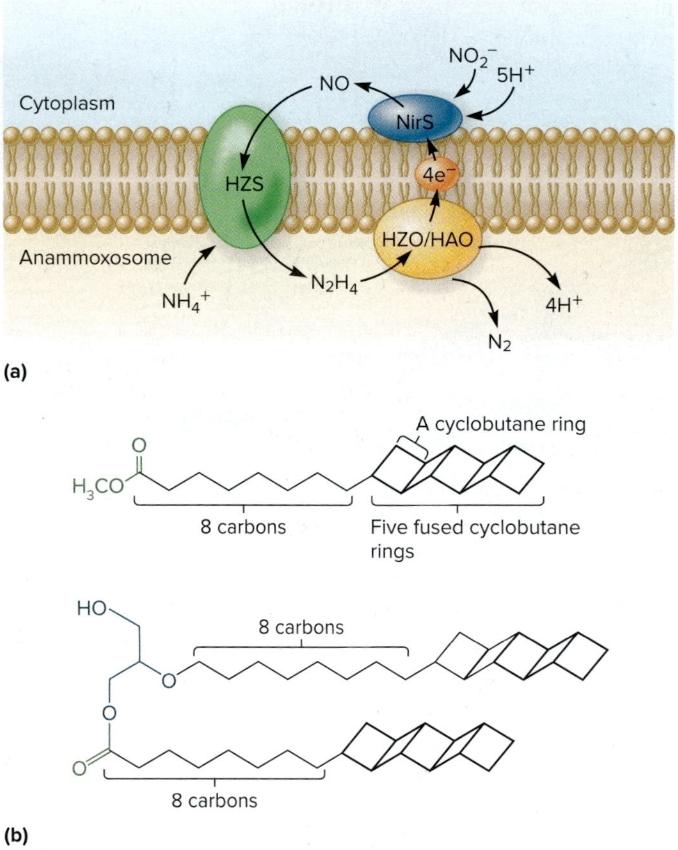

Figure 21.13 **Features of Anammox Bacteria.** (a) In the anammox reaction, ammonium (NH_4^+) combines with nitric oxide (NO) to form hydrazine (N_2H_4) in a reaction catalyzed by hydrazine synthase (HZS). Oxidation of hydrazine by a hydrazine/hydroxylamine oxidoreductase (HZO/HAO) results in the formation of one N_2 and four protons. The electrons are passed to NO_2^-, which is reduced by nitrite:nitric oxide reductase (NirS) to form another molecule of NO and the cycle continues. (b) Ladderane lipids, found in anammoxosomes, are unique to planctomycetes.

MICRO INQUIRY *How does a planctomycete use the anammox reaction to generate a proton motive force?*

Phylum *Verrucomicrobia* Includes Human Symbionts and Methylotrophs

Verrucomicrobia (Latin *verruco,* wart) are so named because of the wartlike structures that protrude from the cell surface. They comprise a significant fraction of soil microbes, and are found in a variety of aquatic and terrestrial environments. In particular, acidic geothermal environments appear to be a preferred niche for verrucomicrobial methane oxidizers. Methane oxidation is significant because methane is an important greenhouse gas. For many years, it was thought that only genera of aerobic α- and β-proteobacteria could oxidize methane. We now know that certain archaea anaerobically oxidize methane, usually when growing in association with sulfate-reducing bacteria. In all cases, methane oxidation ceases in acid environments below pH 4.2. However, it had been noted that methane is produced abiotically in certain geochemically active areas. These sites are also rich in H_2S, which microbes convert to sulfuric acid (H_2SO_4), so the pH can be as low as 1.0. Researchers hypothesized that despite the low pH, aerobic methane oxidation might occur in such environments. They used PCR to directly amplify from soil a methane monooxygenase gene conserved in proteobacterial methanotrophs. Enrichment cultures were then used to isolate *Acidimethylosilex fumarolicum.* This bacterium grows between pH 0.8 and 5.8 with a temperature optimum of 55°C. Since the discovery of *A. fumarolicum* in Italy, other *Verrucomicrobia* isolates have been found in similar sites across the globe. These bacteria demonstrate that methanotrophy is more widespread and that this phylum is more diverse than previously thought. Indeed, the discovery of additional novel *Verrucomicrobia* strains is anticipated as environmental clone libraries from a variety of ecosystems contain SSU rRNA of microbes that may be assigned to this phylum. |◀ *Methanotrophy (section 20.4)* ▶| *Order* Methylococcales *consists of C1-metabolizing bacteria (section 22.3)*

Another member of the phylum *Verrucomicrobia* has received a great deal of attention: *Akkermansia mucinophila.* This bacterium makes up 3 to 5% of the typical human gut microbial community and has been shown to diminish obesity-related systemic inflammation. As its name suggests, *A. mucinophila* consumes mucus and was initially grown in pure culture by providing gastric mucin as the sole carbon source. It is an obligate anaerobe with a fermentative metabolism, nonmotile and coccobacillus in shape. ▶| *A functional core microbiome is required for human homeostasis (section 34.3)*

Phylum *Chlamydiae:* Obligate Intracellular Parasites

Chlamydiae are Gram-negative obligate intracellular parasites. This means they must grow and reproduce within host cells. Although their ability to cause disease is widely recognized, many species grow within protists and animal cells without adverse effects. These hosts represent a natural reservoir for chlamydiae. Within the single class *Chlamydiae,* there are four families, of which *Chlamydiales* is most well known. This family was founded with the single genus, *Chlamydia.* In 1999 several species were split from this genus and placed in a new genus, *Chlamydophila.* Although controversial, this was based on molecular data. Since that time the complete genome sequences of many species in both genera have been published, fueling continued debate and recommendations to reunite the genera. Both genera include human and animal pathogens, including *Chlamydophila pneumoniae, Chlamydophila psittaci,* and *Chlamydia trachomatis.* In the human host, *C. trachomatis* is responsible for the most common sexually transmitted infection, nongonococcal urethritis. This microbe is also the most common cause of preventable blindness in the world. These diseases are discussed in chapter 39. |◀ *Microbial taxonomy and phylogeny are largely based on molecular characterization (section 19.3)*

Members of the genera *Chlamydia* and *Chlamydophila* are nonmotile, coccoid bacteria, ranging in size from 0.2 to 1.5 µm. Chlamydiae are extremely limited metabolically, relying on their host cells for key metabolites, including ATP. Although they have an ATP synthase, it is run in reverse such that the hydrolysis of ATP generates a proton motive force used to drive the uptake of nutrients from the host. Like other intracellular parasites, their genomes are small at 1.0 to 1.3 Mb. Comparative genomics reveals that some chlamydial genes were acquired by horizontal gene transfer (HGT) from free-living bacteria to a common chlamydial ancestor, and HGT accounts for the transfer of genes from chlamydiae to their protist hosts in the genera *Trypanosoma* and *Leishmania.* ▶| Euglenozoa *(section 24.2)*

Chlamydial reproduction is unique. It begins with the attachment of an **elementary body (EB)** to the host cell surface (**figure 21.14**). EBs are 0.2 to 0.6 µm in diameter, contain electron-dense nuclear material and a rigid cell wall, and are infectious. They achieve osmotic stability in the extracellular environment by cross-linking their outer membrane proteins, and possibly periplasmic proteins, with disulfide bonds. EBs stimulate host cell actin polymerization to enter the cell by endocytosis, and they reside in an inclusion. Bacterial proteins induce transport of the inclusion within the cytoplasm and hijack the host Golgi complex. Within the inclusion, the EB differentiates into a **reticulate body (RB)**. RBs have less dense nuclear material and more ribosomes than EBs; their walls are also more flexible. RBs divide by binary fission, eventually filling the inclusion with more than 1,000 chamydiae. RBs initiate cell division before doubling in size, resulting in a progressive decrease in cell size until the dimension of the EB is reached. Additional developmental processes result in mature, infectious EB formation. The host cell lyses and releases EBs 48 to 72 hours after infection.

Chlamydial metabolism is very different from that of other bacteria. It had been thought that chlamydiae cannot catabolize carbohydrates or synthesize ATP. *Chlamydophila psittaci,* one of the best-studied species, lacks both flavoprotein and cytochrome electron transport chain carriers but has a membrane translocase that acquires host ATP in exchange for ADP. The protist parasite *Chlamydia* UWE25 cannot synthesize NAD^+ but synthesizes a transport protein for the import of host NAD^+. Thus some chlamydiae are energy parasites, completely dependent on their hosts for ATP. However, in addition to two genes for ATP/ADP translocases, the *C. trachomatis* genome also has genes for substrate-level phosphorylation, electron transport, and oxidative phosphorylation. When supplied with precursors from the host, RBs can synthesize DNA, RNA, glycogen, lipids, and proteins. They also can synthesize at least some amino acids and coenzymes. The EBs have minimal metabolic activity and cannot take in ATP or synthesize proteins. They evolved exclusively for transmission.

Comprehension Check

1. Describe planctomycetes and their distinctive metabolic properties.
2. Explain how some planctomycetes produce N_2.
3. What might have been the culture conditions used to isolate *A. fumarolicum?*
4. Briefly describe the steps in the chlamydial life cycle. Why do you think chlamydiae differentiate into specialized cell types for infection and reproduction?
5. How does chlamydial metabolism differ from that of other bacteria?

21.6 Phylum *Spirochaetes:* Bacteria with a Corkscrew Morphology

After reading this section, you should be able to:

a. Draw a spirochete cell as it would be observed in a scanning electron micrograph and in cross section by transmission electron microscopy
b. Describe how the unusual flagellar arrangement is well suited for motility in common spirochete habitats

The phylum *Spirochaetes* (Greek *spira,* a coil; *chaete,* hair) contains Gram-negative, chemoorganotrophic bacteria distinguished by their cell shape and mechanism of motility. Carbohydrates, amino acids, long-chain fatty acids, and long-chain fatty alcohols may serve as carbon and energy sources. **Table 21.4** summarizes some of the more distinctive properties of selected genera.

(a)

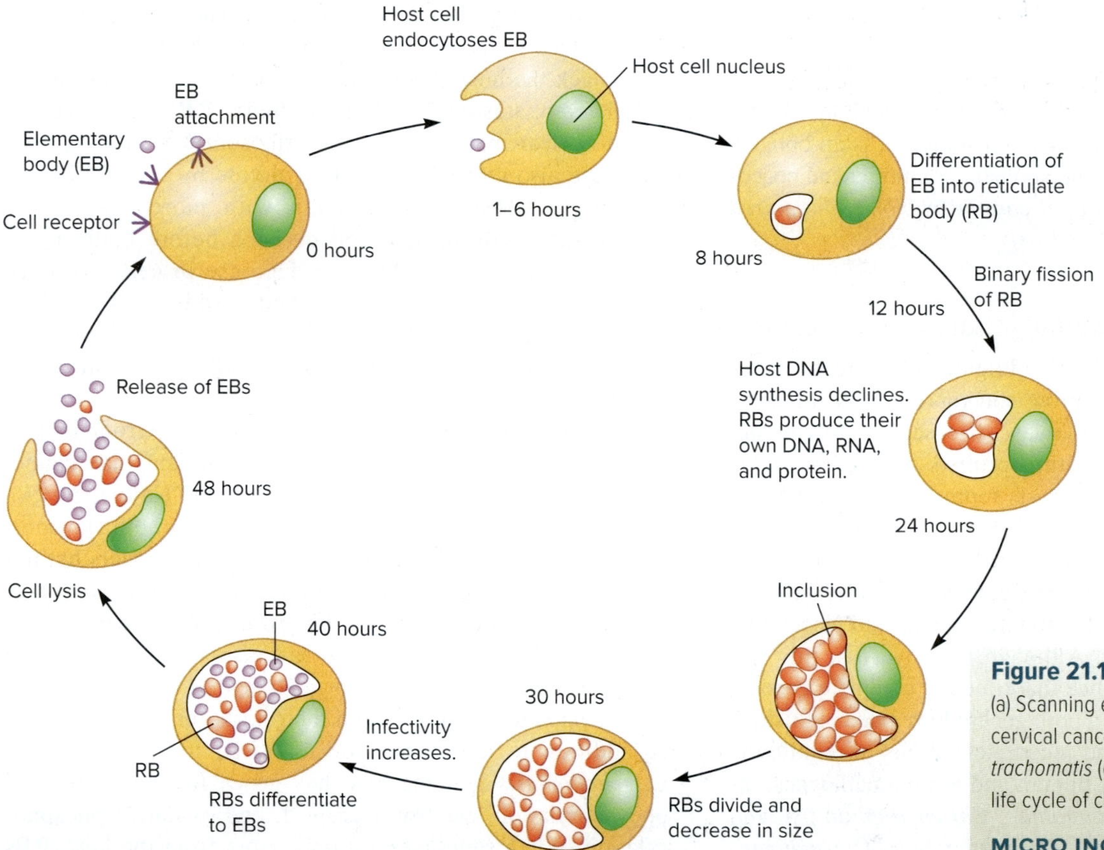

(b)

Figure 21.14 The Chlamydial Life Cycle. (a) Scanning electron micrograph of a human cervical cancer cell infected by *Chlamydia trachomatis* (green spheres). (b) Generalized life cycle of chlamydiae.

MICRO INQUIRY *Why do you think RBs have been so difficult to study?*

Table 21.4 Characteristics of Some Spirochete Genera

Genus	Dimensions (μm) and Flagella	Oxygen Relationship	Carbon + Energy Source	Habitats
Spirochaeta	0.2–0.75 × 5–250; usually 2 ppl[1] flagella	Facultatively anaerobic or anaerobic	Carbohydrates	Aquatic and free-living
Treponema	0.1–0.7 × 1–20; 2–16 ppl flagella	Anaerobic or microaerophilic	Carbohydrates or amino acids	Mouth, intestinal tract, and genital areas of animals; some are pathogenic (syphilis, yaws)
Borrelia	0.2–0.5 × 3–30; 15–20 ppl flagella	Microaerophilic	Carbohydrates	Arthropods; pathogens (relapsing fever, Lyme disease)
Leptospira	0.1 × 6–12; 2 ppl flagella	Aerobic or microaerophilic	Fatty acids and alcohols	Free-living or pathogens of mammals, usually located in the kidney (leptospirosis)
Leptonema	0.1–0.2 × 13–21; 2 ppl flagella	Aerobic	Fatty acids and alcohols	Free-living in aquatic and soil environments
Brachyspira	0.2 × 1.7–6.0; 8 ppl flagella	Anaerobic and aerotolerant	Carbohydrates	Mammalian intestinal tract; pathogens (swine dysentery)

[1]ppl, periplasmic.

Spirochetes are morphologically unique. They are slender, long bacteria with a flexible, helical shape (**figure 21.15**). The central protoplasmic cylinder, which contains cytoplasm and the nucleoid, is bounded by a plasma membrane and a Gram-negative cell wall. Two to more than a hundred **periplasmic flagella** extend from both ends of the cylinder and often overlap one another in the center third of the cell (*see figure 3.43*). The whole complex of periplasmic flagella lies inside a flexible outer membrane sheath. This outer sheath is made of lipid, protein, and carbohydrate, and varies in structure between different genera. Its precise function is unknown, but the sheath is essential because spirochetes die if it is damaged or removed. The outer membrane sheath of *Treponema pallidum,* the causative agent of syphilis, has few proteins exposed on its surface. This helps the spirochete avoid attack by host antibodies. *T. pallidum* is also discussed in chapter 18 (*see figure 18.11*). ▶| *Syphilis (section 39.3)*

Spirochete motility occurs by the rotation of the periplasmic flagella, like the external flagella of other bacteria. This results in the rotation of the outer membrane sheath in the opposite direction relative to the protoplasmic cylinder, and the cell moves in a corkscrewlike movement (**figure 21.16**). Flagellar rotation also enables flexing and crawling on solid substrates.

Spirochetes are exceptionally diverse ecologically and grow in habitats ranging from mud to the human mouth. Members of the genus *Spirochaeta* are free-living and often grow in anoxic and sulfide-rich freshwater and marine environments. Some species of the genus *Leptospira* grow in oxic water and moist soil. Some spirochetes form symbiotic associations with other organisms and are found in a variety of locations: the hindguts of termites and wood-eating roaches, the digestive tracts of mammals, and the oral cavities of animals (*Treponema*

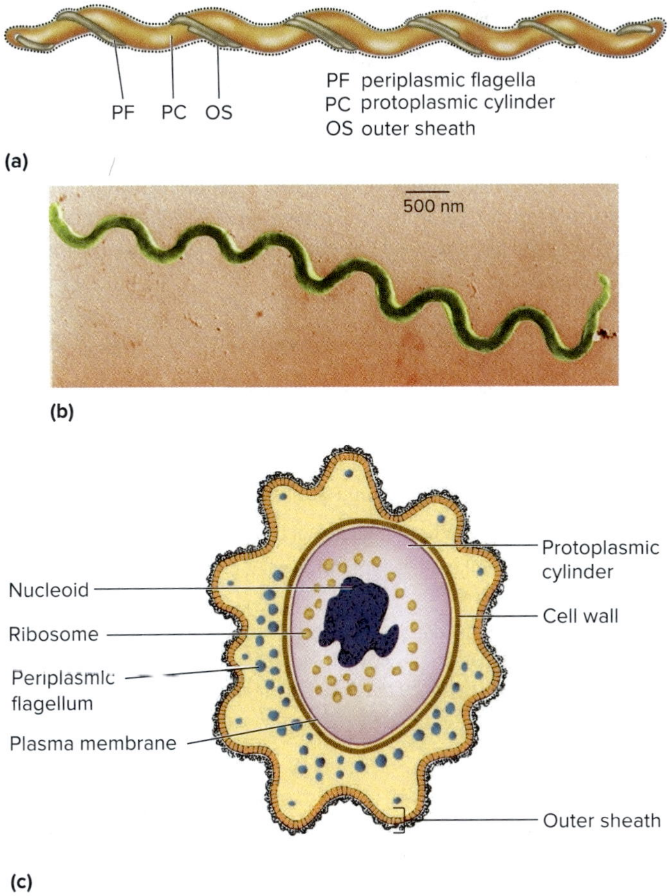

Figure 21.15 Spirochete Morphology. (a) A surface view of spirochete structure as interpreted from electron micrographs. (b) A longitudinal view of *Treponema pallidum*. (c) A cross section of a typical spirochete showing morphological details.

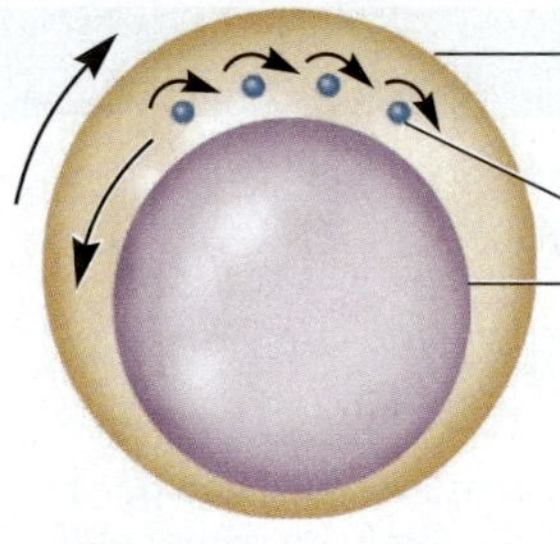

Figure 21.16 Spirochete Motility. A hypothetical mechanism for spirochete motility.

denticola, T. oralis). Spirochetes from termite hindguts and freshwater sediments can fix nitrogen. Spirochetes coat the surfaces of many protists that live in the guts of termites and wood-eating roaches. For example, the flagellate *Mixotricha paradoxa,* a termite symbiont, is covered with slender spirochetes (0.15 by 10 μm in length) that are firmly attached and help move the protist. ▶| *Microorganism-insect mutualisms (section 27.2)*

Borrelia burgdorferi is one of several *Borrelia* species that cause Lyme disease. It is an unusual bacterium in several respects. Its genome consists of a circular chromosome and 17 linear and circular plasmids. Metabolically, it is also unusual. For instance, *N*-acetyl glucosamine (NAG), a component of peptidoglycan, is required for growth. *B. burgdorferi* may also use it as an energy source. The microbe uses the Embden-Meyerhof pathway to oxidize glucose monomers to pyruvate. Consistent with the microaerophilic nature of the bacterium, *B. burgdorferi* lacks a TCA cycle and oxidative phosphorylation.

21.7 Phylum *Bacteroidetes* Includes Important Gut Microbiota

After reading this section, you should be able to:

a. Explain how humans host members of this phylum
b. Describe unusual carbon substrates used by some bacteroidetes
c. Explain the proposed mechanism for *Flavobacterium johnsoniae* gliding motility, and relate gliding motility to the microbe's capacity to degrade complex organic substrates

The phylum *Bacteroidetes* is diverse. All known members are chemoheterotrophs, and many contribute to the digestion of cellulose, agar (a component of seaweed), chitin (found in arthropod shells and fungal cell walls), proteins, and other macromolecules. They are common in terrestrial and marine environments, as well as sewage treatment plants where they contribute to the treatment process. ▶| *Wastewater treatment processes (section 43.2)*

The genus *Bacteroides* contains anaerobic, Gram-negative, nonspore-forming rods; motility is rare. They are important constituents of the human intestinal tract and the rumen of ruminants. About 30% of the bacteria cultured from human feces are members of the genus *Bacteroides,* and these organisms provide extra nutrition by degrading cellulose, pectins, and other complex carbohydrates. Although they often benefit their hosts, they are also involved in human disease. For example, *B. fragilis* is a particularly common anaerobic pathogen found in abdominal, pelvic, pulmonary, and blood infections. ▶| *The rumen ecosystem (section 27.2); The microbe-human ecosystem (chapter 34)*

Bacteroides chromosomes exhibit extreme genetic flexibility due to the presence of dozens of invertible DNA elements. These short (~250 base pair) sequences often contain promoters, and have inverted repeats at each end. The repeats are the substrate for recombination enzymes that flip the DNA by 180 degrees. Invertible promoters control the expression of adjacent genes, often cell surface proteins and capsular polysaccharides. The inversion event changes which adjacent gene is activated by the promoter, left or right. Multiple operons controlled by invertible promoters result in hundreds of possible cell surface combinations. It follows that *Bacteroides* can easily remodel its surface for rapid adaptation to new environmental niches, or to escape immune recognition.

Phenomena specific to individual genera within *Bacteroidetes* have recently been united by the discovery of a type IX secretion system (T9SS). The oral human pathogen *Porphyromonas gingivalis,* as its name suggests, is associated with gum disease. Strains use a T9SS to secrete a protease, gingipain, which accounts for much of its virulence. *Cytophaga* and *Sporocytophaga* secrete degradative enzymes that act on cellulose. Interestingly, microbes in these genera are responsible for damage to valuable artwork painted on canvas. *Flavobacterium* has a unique motility mechanism that enables gliding via long filamentous surface proteins. All of these phenomena rely on the secretion and anchoring of proteins to the outer membrane. |◀ *Two-step secretion systems (section 13.8)*

The **gliding motility** so characteristic of *Cytophaga, Sporocytophaga,* and *Flavobacterium* can be rapid; they typically travel 2 μm per second, but some are considerably faster. Low nutrient levels usually stimulate gliding. Unlike flagellar-mediated swimming motility, the mechanisms that enable gliding motility are not well understood. *Flavobacterium johnsoniae* depends on motility proteins called glide (Gld) proteins. These are thought to assemble into motors that cross the plasma membrane and extend into the periplasm (**figure 21.17**). Like the motor proteins in flagellar basal bodies, the Gld proteins probably harvest the energy of the proton motive force. Gld proteins propel large, filamentous surface proteins called SprB, resulting in cell movement along helical tracks. |◀ *Twitching and gliding motility (section 3.8)*

Gliding motility gives a bacterium many advantages. Many aerobic chemoheterotrophic gliding bacteria actively digest insoluble macromolecular substrates such as cellulose and chitin, and gliding motility is ideal for searching these out. Because

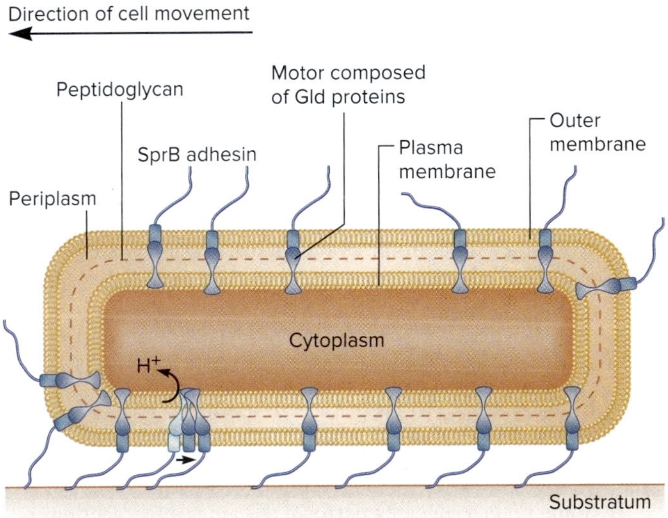

Figure 21.17 Proposed Mechanism for *Flavobacterium johnsoniae* Gliding Motility. Gld proteins within the cell envelope are thought to form motors that propel adhesins such as SprB along the cell surface.

MICRO INQUIRY *How would you test whether or not Gld proteins are necessary for gliding motility? Explain the experiment you would perform and what your controls and prediction would be.*

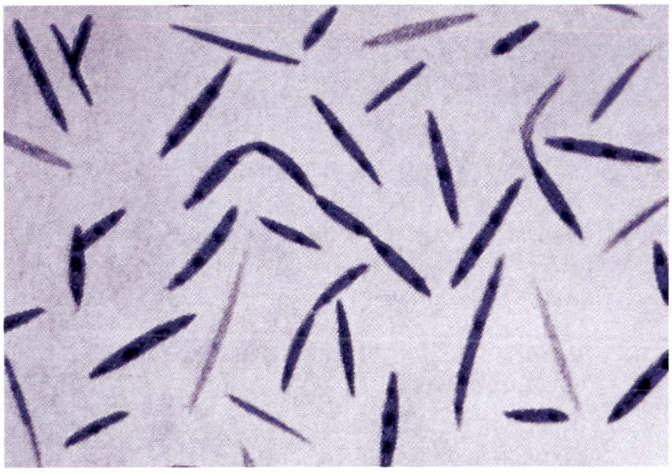

Figure 21.18 *Fusobacterium* Species Are Spindle-Shaped.
Source: CDC

the digestive enzymes are outer membrane-associated, the bacteria must be in contact with insoluble nutrient sources; gliding motility makes this possible. Gliding movement is well adapted to drier habitats and to movement within solid masses such as soil, sediments, and rotting wood that are permeated by small channels. Finally, gliding bacteria, like flagellated bacteria, can position themselves at optimal levels of oxygen, hydrogen sulfide, light intensity, temperature, and other factors that influence growth and survival.

Genomic analyses have led to proposals to unite *Bacteroidetes* with two other phyla, *Chlorobi* and *Fibrobacteres,* into what is termed the FCB superphylum. Supporting evidence includes the presence of conserved signature indels and genes encoding the unique T9SS. ◀ *Signature sequences (section 19.3)*

21.8 Phylum *Fusobacteria:* Commensal Anaerobes

After reading this section, you should be able to:

a. Describe amino acid metabolism in *Fusobacteria*
b. Describe habitats in the human body where *Fusobacteria* are found

Bacteria in the phylum *Fusobacteria* (Latin *fuso,* spindle) exhibit a characteristic spindle, or football, shape (**figure 21.18**). They are obligate or facultative anaerobes that inhabit both anoxic sediments and oral and intestinal habitats of mammals and insects. Amino acids and proteins are the primary energy source. The *Fusobacterium nucleatum* genome encodes biosynthetic pathways for only three amino acids. Amino acid fermentation results in end products that include butyric acid, which confers a rancid odor to cultures, and hydrogen sulfide, which promotes pathogenesis through damage to epithelial tissue.

As commensals, *Fusobacteria* are widespread in anaerobic niches in the human body, and can cause opportunistic infections. Oral biofilms feature *Fusobacteria* as bridge bacteria—that is, they physically interact with other species to support the overall biofilm architecture. Surveys of the human microbiome associated with specific tumors have revealed an overabundance of *F. nucleatum,* specificially with oral carcinomas and colorectal carcinomas. In vitro studies have shown that the presence of *Fusobacterium* in co-culture with cancer cells is associated with enhanced cancer cell proliferation, and in animal cancer models, *Fusobacterium* promotes a more rapid disease progression. It is not clear if *F. nucleatum* is the cause of these phenomena or if the tumor microenvironment is more permissive for its growth. Recent studies have established that this microbe is also involved in resistance to certain chemotherapeutic drugs. The presence of *F. nucleatum* activates cellular signaling pathways that prevent apoptosis, thereby allowing the survival of cancerous cells following chemotherapy. ▶ *Cancer (section 34.4)*

Comprehension Check

1. Draw and label a diagram of spirochete morphology, locating these structures: protoplasmic cylinder, periplasmic flagella, and outer membrane sheath.
2. List the major physiological properties of the genus *Bacteroides.*
3. List three advantages of gliding motility.
4. Why are cytophagas ecologically important?
5. What feature of *Fusobacteria* growth requirements is responsible for its ability to colonize tumors?

Key Concepts

21.1 *Aquificae* and *Thermotogae* Are Ancient Bacterial Lineages

- *Aquifex* and *Thermotoga* contain hyperthermophilic, Gram-negative rods that are the two deepest or oldest phylogenetic branches of *Bacteria*.
- *Aquifex* spp. are chemolithoautotrophs. *Thermotoga* spp. are chemoorganotrophs.

21.2 *Deinococcus-Thermus* Includes Radiation-Resistant Bacteria

- Members of the order *Deinococcales* are aerobic cocci and rods that have an unusual cell wall.
- *Deinococcus radiodurans* is distinctive because of its tremendous resistance to desiccation and radiation.

21.3 Class *Mollicutes*, Phylum *Tenericutes:* Bacteria That Lack Cell Walls

- Mycoplasmas stain Gram negative because they lack cell walls and cannot synthesize peptidoglycan, but many require sterols for growth.
- Many mycoplasmas are plant and animal pathogens. They are some of the smallest bacteria capable of self-reproduction.

21.4 Photosynthetic Bacteria Are Diverse

- Cyanobacteria carry out oxygenic photosynthesis; purple and green bacteria use anoxygenic photosynthesis.
- The four most important groups of purple and green photosynthetic bacteria are the purple sulfur bacteria, the purple nonsulfur bacteria, both of which are proteobacteria, green sulfur bacteria (phylum *Chlorobi*), and green nonsulfur bacteria in the phylum *Chloroflexi* (**table 21.2**).
- The bacteriochlorophyll pigments of purple and green bacteria enable them to live in deeper, anoxic zones of aquatic habitats (**figure 21.6**).
- Green sulfur bacteria are obligately anaerobic photolithoautotrophs that use hydrogen sulfide, elemental sulfur, and hydrogen as electron sources.
- Green nonsulfur bacteria include members of the genus *Chloroflexus*, gliding thermophilic filaments metabolically similar to purple nonsulfur bacteria. Both *Chlorobi* and *Chloroflexus* spp. have chlorosomes, which function to harvest light and transfer it to the reaction center in the plasma membrane (**figure 21.7**).
- Cyanobacteria carry out oxygenic photosynthesis by means of a photosynthetic apparatus similar to that of the eukaryotes. Phycobilisomes contain the light-harvesting pigments phycocyanin and phycoerythrin (**figure 21.9**).
- Cyanobacteria reproduce by binary fission, budding, multiple fission, and fragmentation by filaments to form hormogonia. Some produce dormant akinetes.
- Many nitrogen-fixing cyanobacteria form heterocysts, specialized cells in which nitrogen fixation occurs (**figure 21.11**).
- Members of the *Prochlorococcus* taxon are the most abundant photosynthetic organism on Earth. Together with other unicellular cyanobacteria in the genus *Synchcoccus,* they account for about one-third of global carbon fixation.

21.5 Superphylum *Planctomycetes-Verrucomicrobia-Chlamydiae:* Atypical Cell Division

- Members of the PVC superphylum undergo cell division by budding and other atypical processes.
- Some planctomycetes perform the anammox reaction (**figure 21.13**).
- *Verrucomicrobia* include methylotrophs and some human symbionts.
- Chlamydiae are nonmotile, coccoid, Gram-negative bacteria that lack peptidoglycan and must reproduce within cytoplasmic vacuoles of host cells. Their life cycle involves elementary bodies (EBs) and reticulate bodies (RBs) (**figure 21.14**).

21.6 Phylum *Spirochaetes:* Bacteria with a Corkscrew Morphology

- Spirochetes are slender, long, helical, Gram-negative bacteria.
- Spirochetes are motile by means of periplasmic flagella underlying an outer membrane sheath or outer membrane (**figures 21.15** and **21.16**).

21.7 Phylum *Bacteroidetes* Includes Important Gut Microbiota

- Members of the phylum *Bacteroidetes* are obligately anaerobic, chemoorganotrophic, nonsporing, motile, or nonmotile rods of various shapes.
- Some of these bacteria are important rumen and intestinal symbionts; others can cause disease. Many are important in degrading cellulose, chitin, and other macromolecules in terrestrial and aquatic environments.
- Gliding motility is present in a diverse range of bacteria; in members of *Bacteroidetes,* gliding is mediated by a unique group of transmembranous and surface proteins (**figure 21.17**).

21.8 Phylum *Fusobacteria:* Commensal Anaerobes

- Spindle-shaped *Fusobacteria* are anaerobes found in multiple environments including the human body. They are components of oral biofilms and have been associated with several types of carcinoma. It is not clear if they are causative or if they merely proliferate well in a tumor environment.

Active Learning

1. The cyanobacterium *Anabaena* grows well in liquid medium that contains nitrate as the sole nitrogen source. Suppose you transfer some of these filaments to the same medium except it lacks nitrate. Describe the morphological and physiological changes you would observe.
2. Compare the structural and functional differences between chlorosomes and thylakoid membranes. Do you think this is an example of convergent evolution—two structures evolving separately to fulfill the same function, or divergent evolution—two structures that arose from a single ancestor but adapted to meet the particular needs of the organism that bears them? What other data would you need (apart from structure and function) to answer this question?
3. Many types of movement are employed by bacteria discussed in this chapter. Review them and propose mechanisms by which energy (ATP or proton gradients) might drive the locomotion.
4. When *Chlamydia* was first identified in 1907, it was believed to be a virus. What aspects of its life cycle might lead to such a conclusion?
5. In an investigation into the evolutionary origin of the 3-hydroxypropionate (3HP) bi-cycle, scientists examined genome sequence data for the presence of genes encoding 3HP bi-cycle enzymes in a broad selection of *Chloroflexi*. Although all *Chloroflexi* have some of the enzymes, only the *Chloroflexales* have all of the enzymes, and thus could perform the complete set of reactions. All of the enzymes in the 3HP bi-cycle participate in other pathways of central carbon metabolism, only some of which are present in *Chloroflexi*. Propose a mechanism by which this pathway may have originated and evolved within this small group of organisms.

   Read the original paper: Shih, P., et al. 2017. Evolution of the 3-hydroxypropionate bicycle. *Proc. Natl. Acad. Sci. USA*. 114:10749-54.

22

Proteobacteria

©Ingram Publishing

Bison and Brucellosis Spark Controversy

Winter in Yellowstone National Park is said to be a "magical" time for visitors. But for the roughly 4,900 Yellowstone bison, winter brings the challenge of finding enough to eat. And as these iconic animals forage, some find themselves overstepping the bounds of the park. That's when things get complicated.

Brucellosis is a disease of ruminant animals caused by the bacterium *Brucella abortus.* As its name implies, it causes miscarriages in pregnant animals. Because Yellowstone bison may be asymptomatic carriers of *B. abortus,* ranchers fear they will infect cattle when they wander out of the park. Unfortunately, the first sign of infection is when an animal aborts a highly contagious stillborn calf.

Federal, state, and tribal agencies sought to address fears of disease transmission with a bison management plan aimed at maintaining a sustainable wild population without negative impact on the cattle industry. The basics include hazing and transport, screening animals for *B. abortus* exposure, and slaughter for those testing positive. Hazing—a process that can include anything from noisemakers to helicopters to rubber bullets—is used to discourage bison from leaving the park and to capture those that have. Bison found outside the park are transported to designated capture areas where they are tested for brucellosis. Those with *B. abortus* antibodies (seropositive) are sent to slaughter. Seronegative calves and yearlings are vaccinated and sent to a separate holding area along with seronegative adults. These animals are tagged and released back into the park in the spring.

There is only one problem with this management plan: No one likes it. Many ranchers fear that wandering bison will decimate their cattle. They recall a time when brucellosis cost the cattle industry $400 million annually. Others rebut that there has never been a documented case of brucellosis transmission from Yellowstone bison to cattle. They remind the public of the vast herds of bison that were so grossly overhunted that by 1902 only 50 remained in Yellowstone. Meanwhile, each winter hundreds of out-of-bounds bison are transported, tested, and either vaccinated and released or culled. At the same time, the public lands have been opened for regulated hunting of bison. It seems that the only thing the pro- and anti-bison factions can agree upon is that the current management plan has neither reduced the prevalence of brucellosis in the bison herd nor generated a sustainable population.

So what will be the fate of Yellowstone bison? The delicate balancing act between ranching, tourism, and environmental conservation will continue to drive any modifications to the bison management plan. It is in many ways odd to think that the future of these 1-ton animals rests upon a microbe that measures only 2 μm.

B. abortus is a member of the vast phylum ***Proteobacteria.*** This diverse group of microbes includes many pathogens as well as bacteria that play key roles in nutrient cycling. When reading the chapter, challenge yourself to find attributes of environmental microbes that are also found in proteobacterial pathogens.

Readiness Check:

Based on what you have learned previously, you should be able to:

- ✓ Draw a Gram-negative cell wall and list its constituent parts (section 3.4)
- ✓ Differentiate between swimming and gliding motility (section 3.8)
- ✓ List environmental factors that influence microbial growth with a particular emphasis on oxygen (section 7.5)
- ✓ Describe the difference between heterotrophy and autotrophy (section 11.1)
- ✓ Compare anaerobic and aerobic respiration and fermentation (sections 11.7 and 11.8)
- ✓ Draw an electron transport chain indicating the relative standard reduction potentials of the electron donor, acceptor, and electron transporters (sections 10.4 and 11.6)
- ✓ Define chemolithotrophy and identify an example electron donor and acceptor (section 11.10)
- ✓ Compare and contrast oxygenic and anoxygenic phototrophy (section 11.12)
- ✓ Summarize the function of the Calvin-Benson cycle and the enzyme nitrogenase (sections 12.3 and 12.5)

22.1 Class *Alphaproteobacteria* Includes Many Oligotrophs

After reading this section, you should be able to:

a. Describe the oligotrophic environments inhabited by some α-proteobacteria
b. List two unusual metabolic features of rickettsias and relate these to their intracellular lifestyle
c. Outline the metabolic flexibility of some purple photosynthetic bacteria
d. Draw the photosynthetic apparatus found in purple photosynthetic bacteria and label its constituent parts
e. Diagram the life cycles of *Hyphomicrobium* and *Caulobacter* spp.
f. Compare and contrast the relationship between *Rhizobium* and *Agrobacterium* spp. and plants
g. Identify the electron donors and the electron acceptor used in nitrification
h. List the physiological and taxonomic groups of microbes capable of nitrification

A wide variety of bacteria belong to the class *Alphaproteobacteria*, and **figure 22.1** illustrates a proposed phylogeny of the class. Several of these placements are controversial, but there is general agreement that the branch labeled "Protomitochondrion" belongs in the *Proteobacteria*. This branch represents the extinct lineage of bacteria that, through endosymbiosis, gave rise to eukaryotic mitochondria and related organelles. It has historically been positioned within the *Alphaproteobacteria*, although recent analyses suggest that it may be closer to the root of all *Proteobacteria*. The *Rickettsiales* have evolved to live within plant and animal cells as either mutualists, whereby both organisms benefit, or as pathogens. The remainder of the class includes free-living bacteria, many of which are oligotrophs, capable of growing at low nutrient levels. Some α-proteobacteria have unusual metabolic modes such as methylotrophy—the ability to grow using one-carbon compounds as their sole carbon source (*Methylobacterium*), chemolithotrophy (*Nitrobacter*), and the ability to fix nitrogen (*Rhizobium*). **Table 22.1** summarizes the general characteristics of many of the bacteria discussed in the following sections. ◀ *Evolution of the three domains of life (section 19.5)*

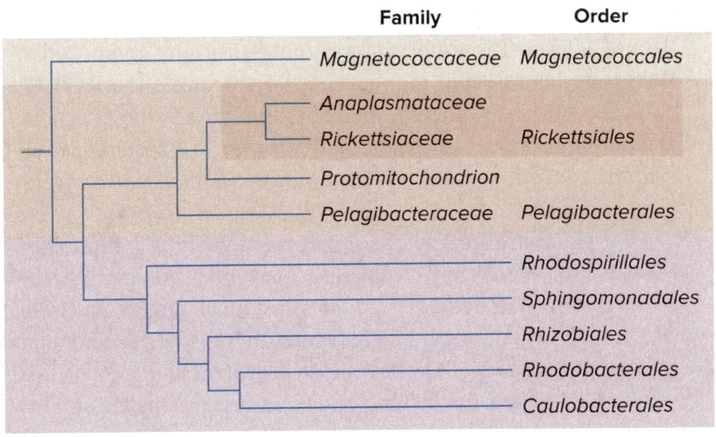

Figure 22.1 Proposed Phylogeny of *Alphaproteobacteria*.
Protomitochondrion is an extinct organism that is hypothesized to have given rise to the mitochondrial family of eukaryotic organelles.

Rickettsias Are Obligate Intracellular Bacteria

The genus *Rickettsia* contains rod-shaped, coccoid, or pleomorphic cells with typical Gram-negative walls. Although their size varies, they tend to be very small. All species are parasitic or mutualistic, living in association with eukaryotes from protozoa to vertebrates. The parasitic forms grow in erythrocytes, macrophages, and vascular endothelial cells. Others live in blood-sucking arthropods such as fleas, ticks, mites, or lice, which serve as vectors or primary hosts (**figure 22.2**).

Rickettsias are believed to have descended from a free-living, aerobic bacterium that became an intracellular parasite of an ancestral eukaryotic cell lacking organelles. In this way, the bacterium was able to assimilate many host substrates, making genes for the biosynthesis of these substrates unnecessary. Thus, as these genes mutated, there was no selection to prevent their loss and conversion to pseudogenes (genes that have lost the capacity to encode products due to mutation of critical elements, e.g., promoter, placement of an early stop codon). To decrease cellular energy burden, many pseudogenes were eventually lost. This gene reduction gave rise to a small genome that prevented the bacterium from living outside the host cell (*see figure 18.18*).

Rickettsias are therefore very different from most other bacteria in physiology and metabolism. They have lost the genes encoding glycolytic pathway enzymes, so they cannot use glucose as an energy source. Instead, they oxidize glutamate and tricarboxylic acid cycle intermediates such as succinate. The rickettsial plasma membrane has carrier-mediated transport systems for the assimilation of host cell nutrients and coenzymes. Its membrane also has a carrier that exchanges ADP for external ATP. An exception is *R. prowazekii*, which possesses a complete TCA cycle and a membrane-bound ATP synthase.

This genus includes important human pathogens including *R. prowazekii* and *R. typhi*, the causative agents of louse-borne and murine typhus, respectively, as well as *R. rickettsii*, which causes Rocky Mountain spotted fever. These microbes enter host cells by inducing phagocytosis but immediately escape the phagocytic vacuole to reproduce by binary fission in the cytoplasm. Eventually the host cell bursts, releasing new organisms. ▶ *Phagocytosis destroys invaders (section 32.6); Rocky mountain spotted fever (section 39.2)*

Purple Nonsulfur Bacteria Perform Anoxygenic Photosynthesis

All **purple nonsulfur bacteria** are α-proteobacteria, with the exception of *Rhodocyclus* (β-proteobacteria). These bacteria use anoxygenic photosynthesis and possess bacteriochlorophylls *a* or *b* (*see table 21.2*). Their photosynthetic apparatus consists of

Table 22.1 Characteristics of Selected α-Proteobacteria

Genus	Dimensions (μm) and Morphology	Genome Size (Mb)	Oxygen Requirement	Other Distinctive Characteristics
Agrobacterium	0.6–1.0 × 1.5–3.0; motile with peritrichous flagella, nonsporing rods	3.0–3.6	Aerobic; microaerophilic	Chemoorganotrophs that can invade plants and cause tumors
Caulobacter	0.4–0.6 × 1–2; comma-shaped with polar flagellum and prostheca with holdfast	4.0	Aerobic	Chemoorganotrophic and oligotrophic; asymmetric cell division
Hyphomicrobium	0.3–1.2 × 1–3; rod-shaped or oval with polar prosthecae	3.6–4.8	Aerobic	Reproduce by budding; methylotrophic
Magnetospirillum	0.2–0.7 × 3–4; spiral cells with a single polar flagellum; single chain of cytoplasmic magnetite crystals	4.8–5.4	Microaerophilic	Chemoorganotrophic; magnetotactic
Nitrobacter	0.5–0.9 × 1.0–2.0; rod- or pear-shaped, sometimes motile by flagella	3.4–4.4	Aerobic; microaerophilic	Chemolithotrophs, oxidize nitrite to nitrate
Pelagibacter	0.2 × 0.5; rod-shaped	1.3–1.5	Aerobic	Chemoorganotrophic; proteorhodopsin present
Rhizobium	0.5–1.0 × 1.2–3.0; motile rods with flagella	6.5–7.7	Aerobic	Invade leguminous plants to produce nitrogen-fixing root nodules
Rhodospirillum	0.8–1.0 × 7–10; spiral cells with polar flagella	4.4	Anaerobic, microaerophilic, aerobic	Anoxygenic photoheterotrophs under anoxic conditions
Rickettsia	0.3–0.5 × 0.8–2.0; short nonmotile rods	1.1–1.3	Aerobic	Obligate intracellular parasites

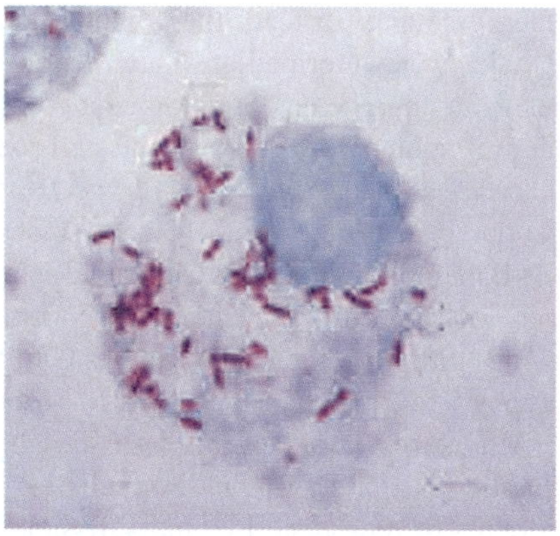

Figure 22.2 ***Rickettsia rickettsii.*** A tick hemolymph cell filled with *Rickettsia rickettsii*, the causative agent of Rocky Mountain spotted fever.
Source: CDC

invaginated plasma membrane, which makes many intracellular folds (**figure 22.3*a,b***). These *i*ntra*c*ytoplasmic *m*embranes (ICMs) increase the membrane surface area, thereby providing more space for *p*hoto*s*ynthetic *u*nits (PSUs). Each PSU includes a *r*eaction *c*enter (RC), consisting of a protein-bacteriochlorophyll (Bchl) complex that absorbs light in the infrared range. The RC-Bchl is responsible for donating electrons to the electron transport chain that results in a proton motive force sufficient to fuel ATP synthesis. The energy required for charge separation is supplied by the photons directly absorbed by the RC-Bchl or transferred from a light-harvesting complex. All purple bacteria (nonsulfur and sulfur) have a primary light-harvesting complex called LH-I closely associated with the RC. Most also have additional light-harvesting complexes, called LH-II (figure 22.3*c*). Light-harvesting complexes are composed of Bchl *a* and carotenoid pigments. LH-IIs encircle the RC/LH-I complex within the membrane. This maximizes the efficiency of energy transfer. ◀ *Light reactions in anoxygenic photosynthesis (section 11.12); Photosynthetic bacteria are diverse (section 21.4)*

Purple nonsulfur bacteria vary considerably in morphology. Most are motile by polar flagella. They may be spirals (*Rhodospirillum*), rods (*Rhodopseudomonas*), half circles or circles (*Rhodocyclus*), or they may even form prosthecae and buds (*Rhodomicrobium*). They are most prevalent in the mud and water of lakes and ponds with abundant organic matter and low sulfide levels.

Purple nonsulfur bacteria show exceptional metabolic flexibility. Normally they grow anaerobically as photoorganoheterotrophs, trapping light energy and employing organic molecules

(a) Intracytoplasmic membranes **(b)** *R. rubrum* **(c)** Integral light-harvesting complex

Figure 22.3 The Photosynthetic Apparatus of Purple Bacteria. (a) The arrangement of intracytoplasmic membranes (ICMs) varies, but in all cases, they are associated with the plasma membrane. (a_1) ICMs that resemble vesicles are found in the nonsulfur α-proteobacteria *Rhodospirillum rubrum* and *Rhodobacter capsulatus,* and the sulfur γ-proteobacteria *Chromatium vinosum* and *Thiocapsa roseopersicina.* (a_2) Tubelike ICMs are found in *T. pfennigii.* (a_3) ICMs arranged in small stacks are seen in *Rhodospirillum molischianum* and the γ-proteobacterium *Ectothiorhodospira mobilis.* (a_4) *Rhodopseudomonas palustris* has larger stacked ICMs. (b) A transmission electron micrograph of *Rhodospirillum* sp. grown anaerobically in the light shows vesicle-like ICMs. (c) Molecular model of light-harvesting complex-II (LH-II) from *Rhodopseudomonas acidophila.* LH-II contains bacteriochlorophyll *a* and carotenoids that transfer electrons to the centrally located reaction center and closely associated light-harvesting complex I; not shown.

as both electron and carbon sources (table 22.1). Although they are called nonsulfur bacteria, some species can oxidize very low, nontoxic levels of sulfide to sulfate, but they do not oxidize elemental sulfur. Some can also use hydrogen as an electron source. In the absence of light, most purple nonsulfur bacteria grow aerobically as chemoorganoheterotrophs, but some species carry out fermentation anaerobically. Oxygen inhibits bacteriochlorophyll and carotenoid synthesis so that cultures growing aerobically in the dark are colorless.

Rhodospirillum rubrum exemplifies the metabolic flexibility of this group. In a strategy analogous to *E. coli* catabolite repression, *R. rubrum* must assess its oxygen and light conditions and the available carbon sources, and adjust gene expression to optimize its metabolism for its environment. The "decision-making" process this microbe uses is displayed in **figure 22.4**. In oxic conditions, the photosynthetic apparatus is inhibited, and it grows chemoorganotrophically. However, when oxygen is not present, it can use one of three metabolic options, depending on the presence of light and organic carbon. These include fermentation, photoheterotrophy, or photoautotrophy. When growing phototrophically, *R. rubrum* makes photopigments that give it a rich red color. *R. rubrum* has sparked the interest of industrial microbiologists because it produces novel biodegradable plastics when grown on β-hydroxycarboxylic and *n*-alkanoic acids.

▶| *Biotechnology and industrial microbiology (chapter 42)*

In addition to the anaerobic anoxygenic photosynthesis displayed by the purple nonsulfur bacteria, some α-proteobacteria are capable of **aerobic anoxygenic photosynthesis (AAnP)**. These aerobic bacteria are photoheterotrophs that catabolize carbon and other organic materials using O_2-dependent respiration. They supplement this chemoorganotrophic source of energy using Bchl *a*–dependent phototrophy. This allows them to shunt more organic carbon to anabolic pathways, thereby enhancing their growth. They are found in marine and freshwater environments, and are often present in microbial mat communities. Important α-proteobacterial genera include *Erythromonas, Roseococcus, Porphyrobacter,* and *Roseobacter,* although metagenomic analysis reveals that a wider diversity of bacteria are capable of AAnP (*see table 21.2*). Their contribution to the global carbon cycle is an area of active investigation.

Rhodospirillum and *Azospirillum* (both in the family *Rhodospirillaceae*) are among several bacterial genera with species capable of forming **cysts** in response to nutrient limitation (**figure 22.5**). In addition to nutritional cues, cyst formation is

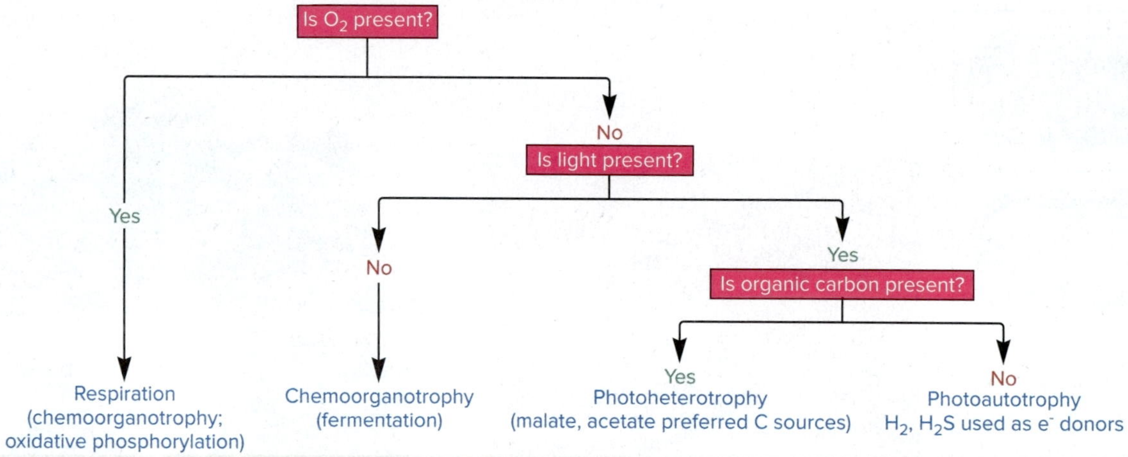

Figure 22.4 The Metabolic Flexibility of *Rhodospirillum rubrum*. This microbe can switch from one type of metabolism to another, depending on the availability of light and oxygen.

MICRO INQUIRY *What would* Rhodospirillum rubrum *use for carbon and energy sources if grown (a) anaerobically in the dark in the presence of malate, (b) anaerobically in the light without organic carbon, and (c) aerobically in the light in the presence of malate?*

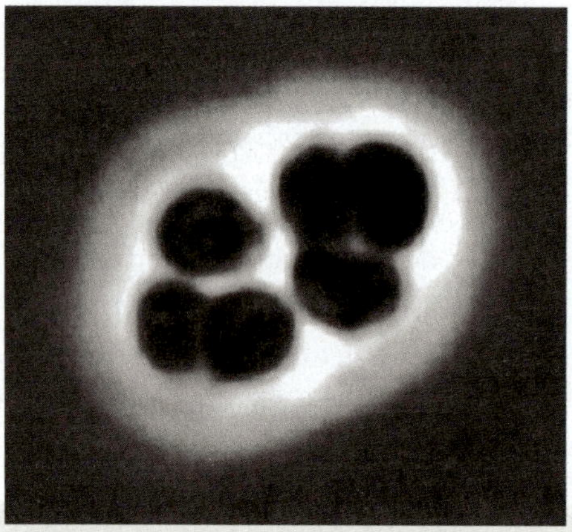

Figure 22.5 Cyst Formation in *Rhodospirillum centenum*. Cysts are most commonly found in clusters of four surrounded by a thick outer coat consisting of lipopolysaccharides and lipoproteins. Cysts are about 1 μm in diameter, nonmotile, and more spherical than growing cells.
Courtesy of James Berleman and Carl Bauer

MICRO INQUIRY *How are bacterial cysts different from the endospores of certain Gram-positive bacteria?*

regulated by the secretion and reuptake of cyclic GMP, a molecule better known for its role as a second messenger in eukaryotes. These resting cells differ from the well-characterized endospores made by Gram-positive bacteria, such as members of the genera *Bacillus* and *Clostridium*. Although cysts are also resistant to desiccation, they are less tolerant of other environmental stresses such as heat and ultraviolet light. They have a thick outer coat and store an abundance of poly-β–hydroxybutyrate. Cyst-forming bacteria are not limited to α-proteobacteria; for instance, members of *Azotobacter,* which are γ-proteobacteria, also form cysts. ◀ *Inclusions (section 3.6); Bacterial endospores are a survival strategy (section 3.9)*

The Proteobacterial Origin of Mitochondria

It has long been held that mitochondria originated from within the class α-proteobacteria. However recent evidence based on large metagenomic data sets suggest that mitochondria may have evolved from a group of bacteria related to, but more ancient than, the α-proteobacteria. These organisms have not yet been isolated in culture. Regardless, mitochondrial origin is like the chicken-or-egg dilemma: Did the protomitochondrion first become an intracellular entity (like rickettsiae) or did it first develop extensive intracytoplasmic membranes (like purple bacteria)?

Evidence in favor of the intracellular-first proposal includes phylogenetic trees based on 16S rRNA sequences (figure 22.1), and the diversity of rickettsiae. Many rickettsiae are symbionts of protozoa, associations that may closely resemble ancestral engulfment. Alternately, comparison of figures 22.3 and 5.14 illustrates the structural similarity between mitochondrial cristae and the intracytoplasmic membranes (ICMs) of purple nonsulfur bacteria. In both cases, these extended plasma membranes provide additional surface area for the membrane-embedded bioenergetic machinery. Beyond the visual similarities, the protein complex responsible for the growth and curvature of the ICMs appears to be conserved. The *mi*tochondrial contact site and *c*ristae *o*rganizing *s*ystem (MICOS) proteins are present in all eukaryotes. In addition, the gene encoding the MICOS core subunit is widespread among α-proteobacteria, but not other bacteria. Assuming that the proteins play the same role in these organisms, this strongly supports the proposal that ICMs are the functional counterparts to mitochondrial cristae. Interestingly, many other α-proteobacteria synthesize ICMs. As described in the following pages, nitrifying bacteria, methylotrophs, and magnetobacteria all contain extensive ICMs as architectural supports for their energy conservation systems.

Caulobacteraceae and *Hyphomicrobiaceae* Bacteria Reproduce in Unusual Ways

These two groups of α-proteobacteria have interesting life cycles that feature a prostheca and reproduction by budding. A **prostheca** (pl., prosthecae), also called a **stalk,** is an extension of the cell, including the plasma membrane and cell wall, that is narrower than the mature cell. **Budding** is a form of cellular reproduction that is distinctly different from binary fission. The bud first appears as a small protrusion at a single point on the mother cell. This new cell pinches free of the mother cell and is noticeably smaller than its parent. Most or all of the bud's cell envelope is newly synthesized. ◀ *Bacterial cell cycles can be divided into three phases (section 7.2)*

The families *Caulobacteraceae* and *Hyphomicrobiaceae* contain two of the best-studied prosthecate genera: *Hyphomicrobium* and *Caulobacter.* The genus *Hyphomicrobium* contains chemoorganotrophic, aerobic, budding bacteria that frequently attach to solid objects in freshwater, marine, and terrestrial environments. At the beginning of the reproductive cycle, the mature cell produces a prostheca (also called a hypha) 0.2 to 0.3 μm in diameter that grows to several μm in length (**figure 22.6**). The chromosome replicates, and a copy moves into the hypha while a bud forms at its end. As the bud matures, it produces one to three flagella, and a septum divides the bud from the hypha. The bud is finally released as an oval- to pear-shaped swarmer cell, which swims off, settles down, and begins budding. A mother cell may bud several times at the tip of its hypha.

Hyphomicrobium spp. also have distinctive physiology and nutrition. Sugars and most amino acids do not support robust growth; instead, these bacteria grow on ethanol and acetate, as well as 1-carbon (C1) compounds such as methanol, formate, and formaldehyde. This makes them facultative **methylotrophs** because they gain energy as well as carbon for biosynthesis from reduced C1 compounds, as well as other, larger organic substrates. The enzymes required for this metabolism are located in an extensive intracellular membrane system. Methylotrophs are so efficient at acquiring C1 molecules they can grow in medium without an added carbon source. Presumably the medium absorbs sufficient atmospheric carbon compounds. *Hyphomicrobium* spp. may comprise up to 25% of the total bacterial population in oligotrophic (nutrient-poor) freshwater habitats.

Bacteria in the genus *Caulobacter* form vibroid-shaped cells (*see figures 2.28 and 7.9*), which alternate between polarly flagellated rods and cells that possess a prostheca and a **holdfast,** by which they attach to solid substrata (**figure 22.7**). Incredibly, the material secreted at the end of the *Caulobacter crescentus* holdfast is the strongest biological adhesion molecule known—bacterial superglue. These bacteria often adhere to other microorganisms and may absorb nutrients released by those microbes. The prostheca differs from that of *Hyphomicrobium* spp. in that it lacks cytoplasmic components and is composed almost totally of the plasma membrane and cell wall. It grows longer in nutrient-poor media and can reach more than 10 times the length of the cell body. The prostheca may improve the efficiency of nutrient uptake from dilute habitats by increasing surface area.

The life cycle of *Caulobacter crescentus* is a model for the investigation of the bacterial cell cycle (**figure 22.8**). When ready to reproduce, the cell elongates and a single polar flagellum forms at the end opposite the prostheca. The cell then undergoes asymmetric transverse binary fission to produce a flagellated swarmer cell that swims away. The swarmer, which cannot reproduce, comes to rest, ejects its flagellum, and forms a new prostheca on the formerly flagellated end. The new stalked cell starts the cycle anew. This process takes about 2 hours to complete.

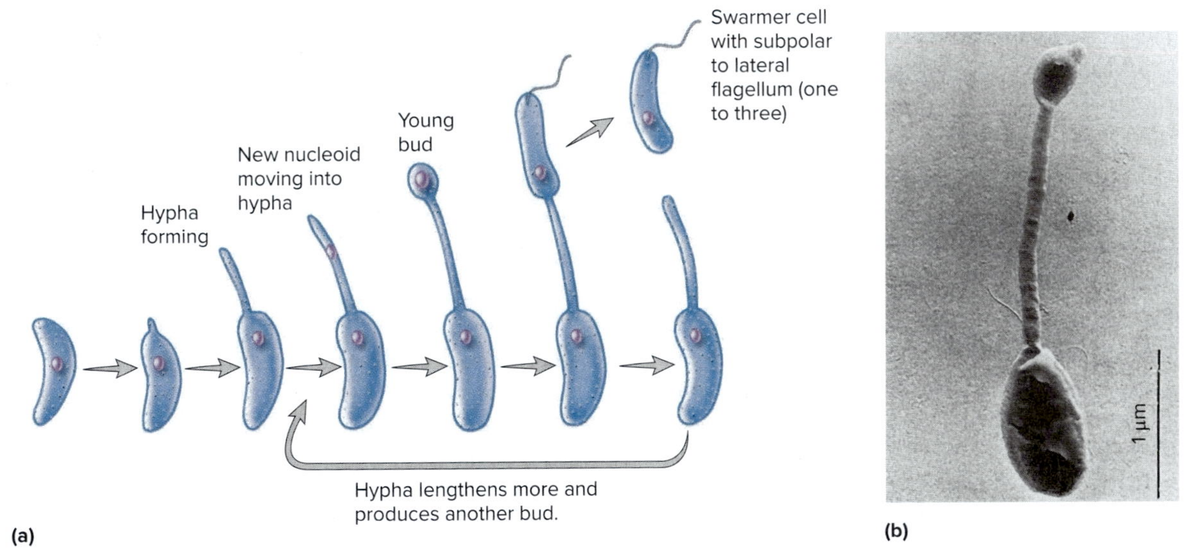

Figure 22.6 Budding in *Hyphomicrobium*. (a) The life cycle of *Hyphomicrobium*. (b) *Hyphomicrobium facilis* with hypha and young bud.
(b) From J.T. Staley, M.P. Bryant, N. Pfenning and J.G. Holt (Eds), *Bergey's Manual of Systematic Bacteriology*, Vol. 3. © 1989 Williams and Wilkins Co., Baltimore

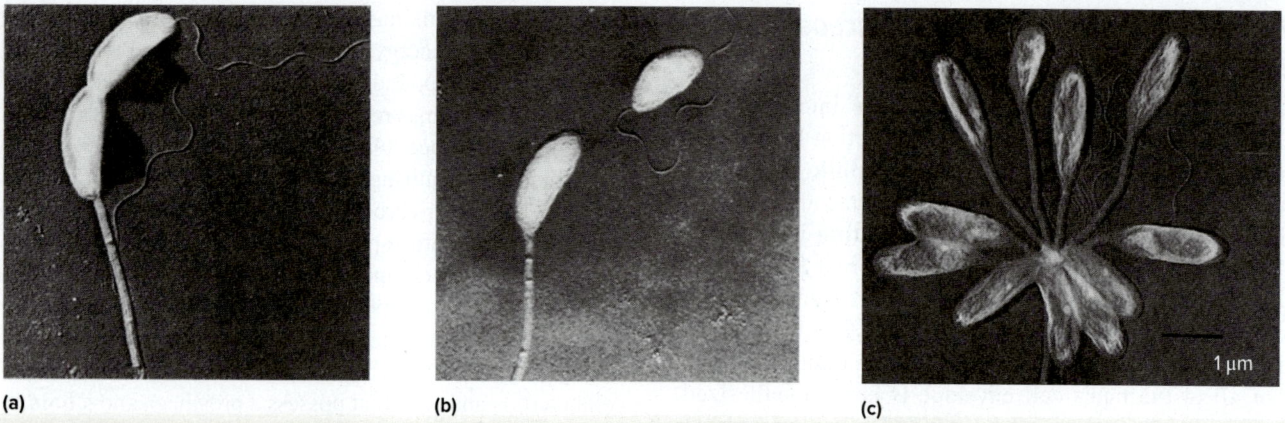

Figure 22.7 ***Caulobacter*** **spp. Morphology and Reproduction.** (a) A cell dividing to produce a swarmer (×6,030). Note prostheca and flagellum. (b) A stalked cell and a flagellated swarmer cell (×6,030). (c) A rosette of cells as seen in the electron microscope.
(a, b) ©Jeanne S. Poindexter/Barnard College; (c) From J.T. Staley, M.P. Bryant, N. Pfenning and J.G. Holt (Eds), *Bergey's Manual of Systematic Bacteriology*, Vol. 3. ©1989 Williams and Wilkins Co., Baltimore

MICRO INQUIRY *If* Caulobacter *bacteria were grown in a phosphate-limited broth, what would happen to the length of the stalks?*

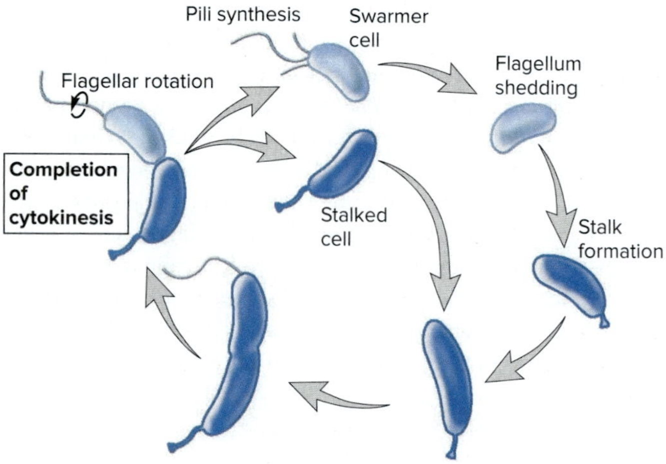

Figure 22.8 ***Caulobacter*** **Life Cycle.** Stalked cells attached to a substrate undergo asymmetric binary fission, producing a stalked and a flagellated cell, called a swarmer cell. The swarmer cell swims freely and synthesizes pili until it settles, ejects its flagella, and forms a stalk. Only stalked cells can divide.

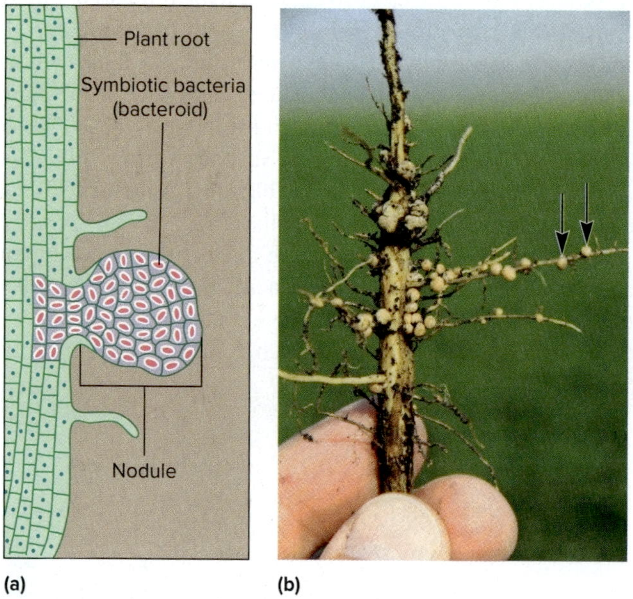

Figure 22.9 ***Rhizobium.*** (a) Diagram of differentiated bacteria, called bacteroids, in the nodule of a legume root. (b) Nodules on a legume root.
(b) ©DS70/Getty Images

Order *Rhizobiales* Includes Important N_2-Fixing Bacteria

The order *Rhizobiales* contains the family *Rhizobiaceae,* which includes aerobic bacteria in the genera *Rhizobium* and *Agrobacterium. Rhizobium* spp. are motile rods that become pleomorphic (having many shapes) under adverse conditions. They live symbiotically within root nodules of legumes as specialized nitrogen-fixing cells called bacteroids. Bacteroid physiology is optimized for nitrogen fixation within the nodule (**figure 22.9**; *see also figure 31.9*). The presence of rhizobium nodules has led to the success of leguminous plants likes soybeans. Indeed, *Leguminosae* is the most successful plant family on Earth, with over 18,000 species. Within the nodules, the microbes fix atmospheric nitrogen to ammonia, making it available to the plant host. The process by which bacteria perform this fascinating and important symbiosis is discussed in chapter 31. ◀ *Nitrogen fixation: N_2 to NH_3 (section 12.5)* ▶ *Rhizobia are symbionts of leguminous plants (section 31.3)*

The genus *Agrobacterium* differs from *Rhizobium* in that it contains plant pathogens. Although genetic analysis supports assigning many species of *Agrobacterium* to the genus *Rhizobium,* like many others, we continue to refer to the genus as *Agrobacterium*

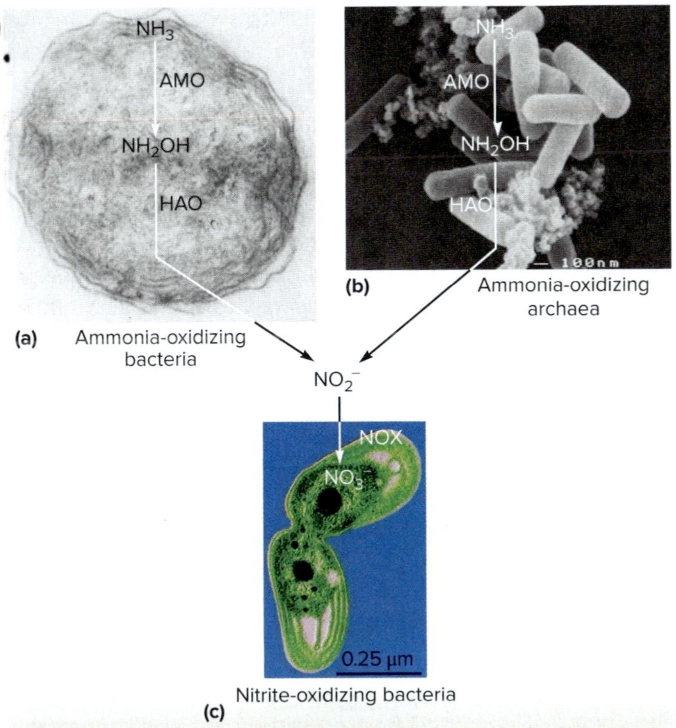

Figure 22.13 Nitrification. Ammonia-oxidizing bacteria (e.g., *Nitrosomonas europaea*) or archaea (*Nitrosopumilus maritimus*) use the enzymes ammonia monooxygenase (AMO) and hydroxylamine oxidoreductase (HAO) and other enzymes to convert ammonia to nitrite. Nitrite-oxidizing bacteria such as *Nitrobacter* sp. use nitrite oxidoreductase (NOX) to catalyze the oxidation of nitrite to nitrate.

(a) Courtesy of Luis Sayavedra-Soto and the Electron Microscopy Facility , Oregon State University; (b) ©Dr. Martin Konneke; (c) ©ISM/Phototake

MICRO INQUIRY *Why do nitrifying microbes possess intracellular membranes?*

class, *Nitrococcus* (*Ectothiorhodospiraceae* family). These microbes differ considerably in morphology and may be rod-shaped, ellipsoidal, spherical, spirillar, or lobate. They may possess either polar or peritrichous flagella. Often they have extensive intracytoplasmic membrane complexes that increase the surface area available for electron transport (*see figure 3.31*). Chemolithotrophy captures little free energy, so reduced inorganic compounds such as ammonia or nitrite must be oxidized at a great rate to conserve enough energy to support growth. This is analogous to an old car that gets bad gas mileage; it must burn more gas (electrons) to cover the same distance (do work) than an efficient, newer car.

Nitrifying microorganisms make important contributions to the nitrogen cycle. Nitrification occurs rapidly in oxic soil treated with fertilizers containing ammonium salts. By converting ammonium to nitrite and then to nitrate, nitrification increases the availability of nitrogen to plants because nitrate is readily used by plants. The discovery of ammonia-oxidizing archaea has triggered a reassessment of the relative contribution of bacteria to global nitrification. ▶| *Nitrogen cycle (section 28.1)*

Magnetospirillum Displays Cytoplasmic Magnetosomes

Magnetotactic bacteria are found in several phylogenetic groups, but the best characterized genera are *Magnetococcus* and *Magnetospirillum*. They occupy freshwater or marine sediments, and are highly motile, using their magnetosomes to orient in the planet's magnetic field. Studies of *Magnetospirillum* (**figure 22.14**) have identified enzymes and structural proteins involved in magnetosome formation. The magnetosome intracytoplasmic membrane forms from plasma membrane invagination but displays unique *ma*gnetosome *m*embrane (Mam) proteins. Iron is transported from the cytoplasm in an ATP-dependent process, and its nucleation to the growing crystal is driven by other Mam proteins. Each magnetosome contains a single magnetite crystal, and its size and structure are characteristic of a species. *Magnetospirillum* cells contain up to 60 magnetosomes, aligned across the center of the cell by their interaction with the cytoskeletal protein MamK. Genes encoding the magnetosome structure and assembly proteins are clustered on a genomic magnetosome island (GMI) of about 100 kilobases. Interestingly, the GMI can be transferred to the related α-proteobacterium *Rhodospirillum rubrum*, resulting in a "magnetized" microbe that forms ordered magnetosome chains in its cells.

SAR11 Clade, the Most Abundant Microorganisms on Earth

Candidatus Pelagibacter ubique was the first cultured representative of the world's most abundant group of bacteria, called the SAR11 clade. "SAR" references the *Sar*gasso Sea, where these marine bacteria were first discovered. All members of the SAR11 clade are chemoorganoheterotrophs, and

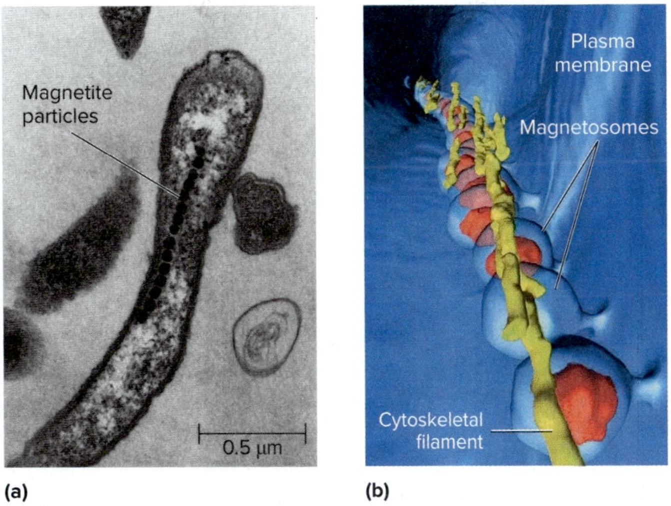

Figure 22.14 Magnetosomes. (a) Transmission electron micrograph of the magnetotactic bacterium *Magnetospirillum magnetotacticum*. (b) Three-dimensional reconstruction of the magnetosomes of *M. magneticum* (electron cryotomograph).

(a) ©Dennis Kunkel Microscopy, Inc./Phototake; (b) ©Grant J. Jensen information about their physiology comes from combined analysis of cultures, the complete genome of *Ca.* P. ubique, and metagenomic sequences of environmental bacteria. These microbes appear to reverse the Embden-Meyerhof pathway for gluconeogenesis and use a modified version of the Enter-Douderoff pathway for glucose oxidation (**figure 22.15**). In its oligotrophic marine environment, *Ca.* P. ubique is a superior competitor for low molecular weight dissolved organic matter, specifically compounds derived from eukaryotic algae and marine cyanobacteria.

The genome of *Ca.* P. ubique is 1.31 Mb, which is quite small for an independently growing microbe. As this bacterium evolved, it became optimized for a life with scant nutrients. The small size of its genome reflects the almost complete absence of intergenic nucleotides. There are no pseudogenes, phage genes, or duplicated genes. It encodes a relatively large number of high-affinity transporters, as it is energetically cheaper to transport macromolecules than to make them de novo. Even its small cell size maximizes the surface-to-volume ratio (*see figure 3.5*). The discovery of genes for proteorhodopsin perplexed microbiologists, as initial experiments failed to show that light enhanced rates of growth, as predicted if this light-driven proton pump contributed to the production of ATP. However, when such experiments were later performed on cells starved for carbon, there was an observable growth advantage for cells cultivated in the light. ◀ *Rhodopsin-based phototrophy (section 11.12)* ▶ *Microorganisms in the open ocean are adapted to nutrient limitation (section 30.2)*

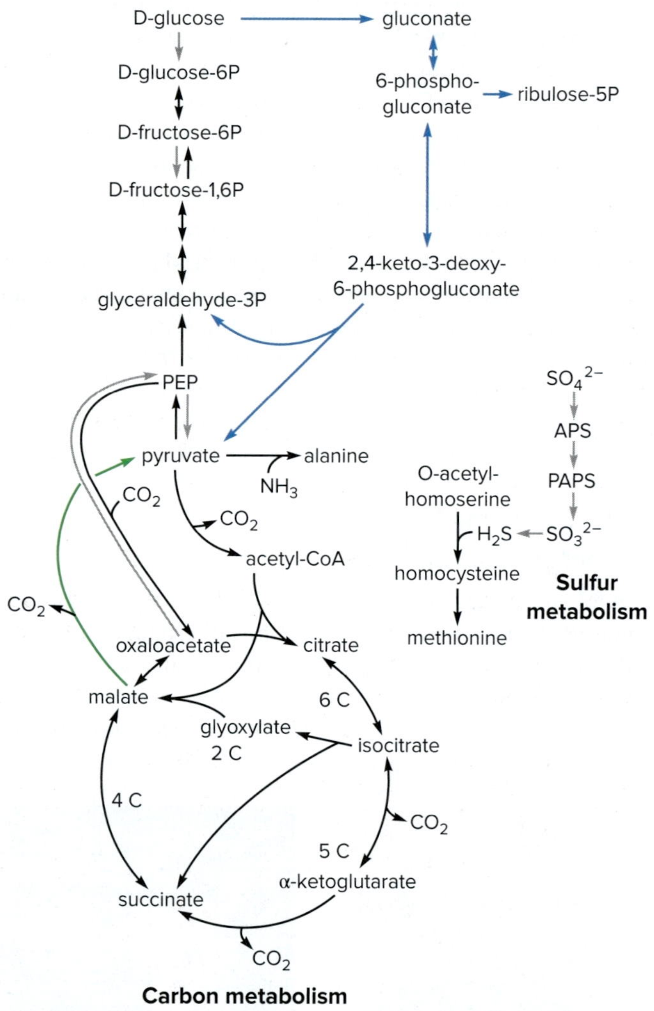

Figure 22.15 **Central Metabolism in *Ca.* P. ubique as Determined by Genome Analysis.** The upper half of the figure shows the Embden-Meyerhof pathway for glucose oxidation as used by many bacteria (gray arrows); *Ca.* P. ubique runs this pathway in reverse for gluconeogenesis (black arrows); glucose is oxidized by the modified Entner-Douderoff pathway shown with blue arrows. The bottom half of the figure shows the glyoxylate cycle; unlike other microbes that convert oxaloacetate to PEP (gray arrow), *Ca.* P. ubique starts gluconeogenesis by converting malate to pyruvate (green arrow). *Ca.* P. ubique lacks genes for assimilatory sulfate reduction (gray arrows), and requires reduced organic sulfur compounds such as methionine for growth.

MICRO INQUIRY *Why can oxaloacetate and glucose, but not acetate, replace the need for pyruvate for growth?*

Comprehension Check

1. What feature unites α-proteobacteria?
2. Why do rickettsias have very small genomes?
3. Why are purple nonsulfur bacteria called "nonsulfur" bacteria?
4. Compare and contrast the life cycles of members of the genera *Hyphomicrobium* and *Caulobacter.*
5. How do *Agrobacterium* and *Rhizobium* spp. differ in lifestyle?
6. What is a methylotroph? How do methane-oxidizing bacteria use methane as both an energy source and a carbon source? How do these microbes avoid poisoning themselves with the formaldehyde they produce as an intermediate in C1 metabolism?
7. Contrast and compare chemolithoautotrophic growth of a nitrifying bacterium with that of *Magnetococcus marinum*.
8. How is SAR11 adapted for life with limited nutrients?

22.2 Class *Betaproteobacteria* Includes Chemoheterotrophs and Chemolithotrophs

After reading this section, you should be able to:

a. List human diseases caused by *Neisseria* and *Bordetella* spp.
b. Discuss the function of the sheath in the physiology of *Leptothrix* spp.
c. Summarize the reactions used by *Nitrosomonas europaea* to oxidize ammonium to nitrite
d. Differentiate between nitrification and denitrification, and identify the electron donors and acceptors for each
e. Explain how sulfur-oxidizing bacteria obtain ATP from reduced sulfur species via substrate-level phosphorylation

Table 22.3 Characteristics of Selected β-Proteobacteria

Genus	Dimensions (μm) and Morphology	Oxygen Requirement	Other Distinctive Characteristics
Bordetella	0.2–0.5 × 0.5–2.0; nonmotile coccobacillus	Aerobic	Require organic sulfur and nitrogen; respiratory pathogen of vertebrates
Burkholderia	0.5–1.0 × 1.5–4; straight rods with single flagellum or a tuft at the pole	Aerobic, some capable of anaerobic respiration with NO_3^-	Poly-β-hydroxybutyrate as reserve; can be pathogenic to humans and animals
Leptothrix	0.6–1.5 × 2.5–15; straight rods in chains with sheath, free cells flagellated	Aerobic	Sheaths encrusted with iron and manganese oxides
Neisseria	0.6–1.9; cocci in pairs with flattened adjacent sides	Aerobic	Inhabitants of mammalian mucous membranes
Nitrosomonas	Size varies with strain; spherical to ellipsoidal cells with intracytoplasmic membranes	Aerobic	Chemolithotrophs that oxidize ammonia to nitrite
Sphaerotilus	1.2–2.5 × 2–10; single chains of cells with sheaths, may have holdfasts	Aerobic	Sheaths not encrusted with iron and manganese oxides
Thiobacillus	0.3–0.5 × 0.9–4; rods, often with polar flagella	Aerobic	All chemolithotrophic; oxidize reduced sulfur compounds to sulfate; some also capable of chemoorganotrophy

β-proteobacteria are similar to α-proteobacteria metabolically but tend to use substances released from organic decomposition in anoxic habitats. Some of these bacteria use hydrogen, ammonia, volatile fatty acids, and similar substances. As with α-proteobacteria, there is considerable metabolic diversity; β-proteobacteria may be chemoorganotrophs, photolithotrophs, and chemolithotrophs.

The class *Betaproteobacteria* includes 12 families. **Table 22.3** summarizes the general characteristics of many of the bacteria discussed in this section. Because sequence data suggest that the separation between *Betaproteobacteria* and *Gammaproteobacteria* is less distinct than once thought, *Betaproteobacteria* may one day be merged with *Gammaproteobacteria*.

Order *Neisseriales* Includes Important Pathogens

Neisseria is the best-studied genus in the order *Neisseriales*. Members of this genus are nonmotile, aerobic, Gram-negative cocci that most often occur in pairs with adjacent sides flattened. They may have capsules and pili. These chemoorganotrophs inhabit the mucous membranes of mammals, and some are human pathogens. *Neisseria gonorrhoeae* is the causative agent of gonorrhea; *Neisseria meningitidis* is responsible for some cases of bacterial meningitis. ▶| *Meningitis (section 39.1); Gonorrhea (section 39.3)*

Order *Burkholderiales* Includes Recently Evolved Pathogens

The order *Burkholderiales* has three well-known families. *Burkholderia* is placed in the family *Burkholderiaceae*. One of the most important species is *B. cepacia*, which can degrade over 100 different organic molecules and is very active in recycling organic materials in nature. Originally described as the plant pathogen that causes onion rot, it has emerged as a major nosocomial (hospital-acquired) pathogen. It is a particular problem for cystic fibrosis patients, where it forms biofilms in the lungs. Two other species, *B. mallei* and *B. pseudomallei*, are pathogens that could be misused as bioterrorism agents. Vaccines to both are under development. ▶| *Bioterrorism readiness is an integral component of public health microbiology (section 36.7)*

Surprisingly, two genera within the *Burkholderiaceae* family are capable of forming nitrogen-fixing symbioses with legumes much like the rhizobia (α-proteobacteria). Genome analysis of nitrogen-fixing *Burkholderia* and *Cupriavidus* (formerly *Ralstonia*) isolates reveals the presence of nodulation (*nod*) genes that are very similar to those of rhizobia. This suggests a common origin. It is thought these β-proteobacteria gained the capacity to form symbiotic, nitrogen-fixing nodules with legumes through horizontal gene transfer. |◀ *Horizontal gene transfer: creating genetic variation the asexual way (section 16.4)*

Some members of the order *Burkholderiales* have a **sheath**—a hollow, tubelike structure surrounding a chain of cells. Sheaths often are close fitting, but they are never in intimate contact with the cells they enclose. Some contain ferric or manganic oxides. They have at least two functions. Sheaths help bacteria attach to solid surfaces and acquire nutrients from slowly running water as it flows past, even if it is nutrient-poor. Sheaths also protect against predators such as protozoa.

Two well-studied sheathed genera are in the family *Comamonadaceae*. *Sphaerotilus* bacteria form long, sheathed chains of rods attached to submerged plants, rocks, and other solid objects. These chains release swarmer cells with a bundle of flagella into the water, where they attach to an object by means of a holdfast and form a new chain. They grow best in slowly running, polluted freshwater. In fact, they thrive in activated sewage sludge and sometimes form tangled masses of filaments that interfere with the proper settling of sludge. *Leptothrix* spp. characteristically deposit large amounts of iron and manganese oxides in the sheath. *Leptothrix* sheaths bind many chemicals in aqueous solutions, and are in development as a novel type of adsorbent and filter.

The family *Alcaligenaceae* contains the genus *Bordetella,* composed of Gram-negative, aerobic coccobacilli. They are chemoorganotrophs with respiratory metabolism and require organic sulfur and nitrogen (amino acids) for growth. As mammalian parasites, they multiply in respiratory epithelial cells. *B. bronchiseptica* is the causative agent of the canine condition kennel cough, and *B. pertussis* is a nonmotile, encapsulated species that causes pertussis (whooping cough); *B. parapertussis* is a closely related species that causes a milder form of the disease. *Pertussis (section 39.1)*

Order *Nitrosomonadales* Consists of Ammonia-Oxidizing Bacteria

A number of chemolithotrophs are found in the order *Nitrosomonadales* including the genera of nitrifying bacteria *Nitrosomonas* and *Nitrosospira* (figure 22.13). Both of these bacteria oxidize ammonia to nitrite; a metabolic and transport reconstruction based on the genome of *Nitrosomonas europaea* is shown in **figure 22.16**. This microbe uses the enzymes ammonia

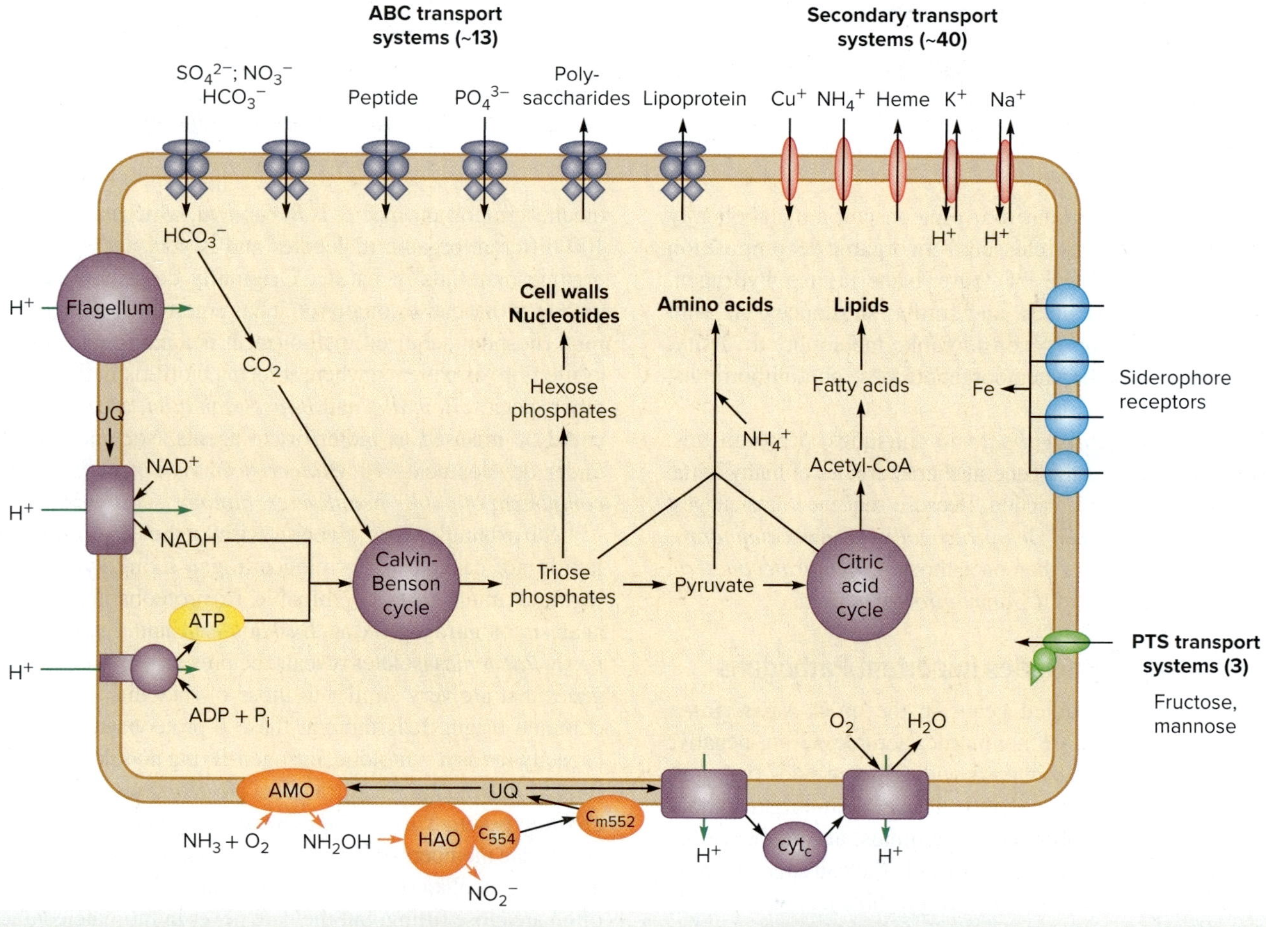

Figure 22.16 Genomic Reconstruction of *Nitrosomonas europaea*. This chemolithoautotroph uses the Calvin-Benson cycle to fix CO_2. Ammonia oxidation is catalyzed by a membrane-bound ammonia monooxygenase (AMO) and hydroxylamine oxidoreductase (HAO). Electrons from HAO flow through two cytochromes (c_{554} and c_{m552}) into an electron transport chain at the level of ubiquinone-cytochrome c oxidoreductase. O_2 is the terminal electron acceptor. Electrons from ammonia oxidation must also flow in reverse for NAD^+ and $NADP^+$ reduction. These enzymes, the flagellar motor, the ATP synthase, and secondary transport systems are fueled by the proton motive force established by ammonia oxidation.

MICRO INQUIRY *Suggest why* N. europaea *has PTS transport systems for the sugars fructose and mannose if it is autotrophic.*

monooxygenase (AMO) and hydroxylamine oxidoreductase (HAO) to oxidize ammonia to nitrite in the reactions:

$$NH_3 + O_2 + 2H^+ + 2e^- \rightarrow NH_2OH + H_2O \rightarrow$$
$$NO_2 + 5H^+ + 4e^-$$

Recent reports have demonstrated that nitric oxide (NO) is the product of HAO, and a third enzyme must be present to convert NO to NO_2. This enzyme and the precise chemical reaction catalyzed remain to be discovered. Of the four electrons released, two must return to AMO to continue ammonia oxidation, while the other two can be used either as reductants for biosynthesis or be donated to the electron transport chain to reduce the terminal electron acceptor, oxygen. Figure 22.16 also illustrates the autotrophic metabolism of this microbe, which uses the Calvin-Benson cycle to fix CO_2.

Order *Hydrogenophilales* Includes a Model Colorless Sulfur Bacterium

This small order contains the genus *Thiobacillus,* which are the best-studied **colorless sulfur bacteria,** so named because they use sulfur-containing compounds as energy sources but are not photosynthetic and therefore lack photosynthetic pigments. Many are unicellular, rod-shaped or spiral, sulfur-oxidizing bacteria that can be either nonmotile or flagellated (**table 22.4**). *Bergey's Manual* divides the colorless sulfur bacteria between two classes; for example, the genera *Thiobacillus* and *Macromonas* are β-proteobacteria, whereas the genera *Thiomicrospira, Thiobacterium, Thiospira, Thiothrix, Beggiatoa,* and others are γ-proteobacteria. Only some of these bacteria have been grown in pure culture.

Thiobacillus spp. metabolism has been intensely studied. The genus includes obligate aerobes and facultative denitrifiers. They use a variety of reduced, inorganic sulfur compounds as energy and electron sources and oxygen as electron acceptor. ATP is generated in two ways, both of which require that the cell first oxidize sulfide or thiosulfate to sulfite (SO_3^{2-}; **figure 22.17*a***). When sulfite donates its electrons to an electron transport chain, ATP is made by oxidative phosphorylation and sulfate is the end product. Alternatively, sulfite can be bound to AMP to form the molecule adenosine 5′-phosphosulfate (APS, figure 22.17*b*). The sulfate in APS is next enzymatically replaced with PP_i, yielding sulfate and ATP through substrate-level phosphorylation. Although all *Thiobacillus* spp. are autotrophic, fixing CO_2 via the Calvin-Benson cycle (*see figure 12.4*), *T. novellus* and a few other species can also grow heterotrophically. ◀ *Chemolithotrophy (section 11.10)*

The metabolic flexibility of *Thiobacillus denitrificans* has made it the focus of much investigation. Remarkably it can use either oxygen or nitrate as its terminal electron acceptor when growing chemolithotrophically, using hydrogen sulfide or thiosulfate to donate electrons to an electron transport chain. When using nitrate as a terminal electron acceptor, *T. denitrificans* can sequentially reduce nitrate to nitrogen gas—a process called **denitrification.** It can also use ferrous iron (Fe^{2+}) as an electron donor; this form of chemolithotrophy has been well documented in another species, *T. ferrooxidans*.

Sulfur-oxidizing bacteria have a wide distribution and great practical importance. *Thiobacillus* spp. are ubiquitous in soil, freshwater, and marine habitats. Many demonstrate great acid tolerance (e.g., *T. thiooxidans* grows at pH 0.5 and cannot grow above pH 6), and these bacteria prosper in habitats they have acidified by sulfuric acid production. Drainage from mining operations is arguably the most important high-sulfide, low-pH environment in the United States. *T. ferrooxidans* and *Leptospirillum ferrooxidans* control the levels of sulfide (as sulfuric acid and iron disulfide) in these waters. Acid mine drainage is inhospitable to life other than acidophilic bacteria and archaea and can pollute

Table 22.4 Colorless Sulfur-Oxidizing Genera

Genus	Cell Shape	Motility; Location of Flagella	Sulfur Deposit[1]	Nutritional Type
Thiobacillus	Rods	+; polar	Extracellular	Obligate or facultative chemolithotrophs
Thiomicrospira	Spirals, comma, or rod-shaped	– or +; polar	Extracellular	Obligate chemolithotrophs
Thiobacterium	Rods embedded in gelatinous masses	–	Intracellular[2]	Probably chemoorganoheterotrophs
Thiospira	Spiral rods, usually with pointed ends	+; polar (single or in tufts)	Intracellular	Unknown
Macromonas	Rods, cylindrical or bean-shaped	+; polar tuft	Intracellular[2]	Probably chemoorganoheterotrophs

1 When hydrogen sulfide is oxidized to elemental sulfur.
2 May use sulfur oxidation to detoxify H_2O_2.

MICROBIAL DIVERSITY & ECOLOGY

22.1 Acid Mine Drainage

Each year millions of tons of sulfuric acid flow to the Ohio River from the coal mines of the Appalachian Mountains. This sulfuric acid is of microbial origin and leaches enough metals from the mines to make the river reddish and acidic. The primary culprit is *Acidithiobacillus ferrooxidans,* a chemolithotrophic bacterium that derives its energy from oxidizing ferrous ion to ferric ion and sulfide ion to sulfate ion. The combination of these two energy sources is important because of the solubility properties of iron. Ferrous ion is somewhat soluble and can be formed at pH values of 3.0 or less in moderately reducing environments. However, when the pH is greater than 4.0 to 5.0, ferrous ion is spontaneously oxidized to ferric ion by O_2 in the water and precipitates as a hydroxide. If the pH drops below 2.0 to 3.0 because of sulfuric acid production by spontaneous oxidation of sulfur or sulfur oxidation by thiobacilli and other bacteria, the ferrous ion remains reduced, soluble, and available as an energy source. Remarkably, *A. ferrooxidans* grows well at such acidic pHs and actively oxidizes ferrous ion to an insoluble ferric precipitate. The water is rendered toxic for most aquatic life and unfit for human consumption.

The ecological consequences of this metabolic lifestyle arise from the common presence of pyrite (FeS_2) in coal mines. The bacteria oxidize both elemental components of pyrite for their growth and in the process form sulfuric acid, which leaches the remaining minerals.

Autoxidation or bacterial action

$$2FeS_2 + 7O_2 + 2H_2O \rightarrow 2Fe^{2+} + 4SO_4^{2-} + 4H^+$$

A. ferrooxidans

$$2Fe^{2+} + 1/2O_2 + 2H^+ \rightarrow 2Fe^{3+} + H_2O$$

Pyrite oxidation is further accelerated because the ferric ion generated by bacterial activity readily oxidizes more pyrite to sulfuric acid and ferrous ion. In turn, the ferrous ion supports further bacterial growth. It is difficult to prevent *A. ferrooxidans* growth as it requires only pyrite and common inorganic salts. Because *A. ferrooxidans* gets its O_2 and CO_2 from the air, the only feasible method of preventing its damaging growth is to seal the mines to render the habitat anoxic.

nearby streams and corrode natural and engineered structures (**Microbial Diversity & Ecology 22.1**). Conversely, sulfur-oxidizing bacteria can be beneficial. Other, neutrophilic species can increase soil fertility by oxidizing elemental sulfur to sulfate. Thiobacilli are used in processing low-grade metal ores because of their ability to leach metals from ore. There is much interest in using *T. denitrificans* to decontaminate water with high levels of nitrate. ▶| *Sulfur cycle (section 28.1)*

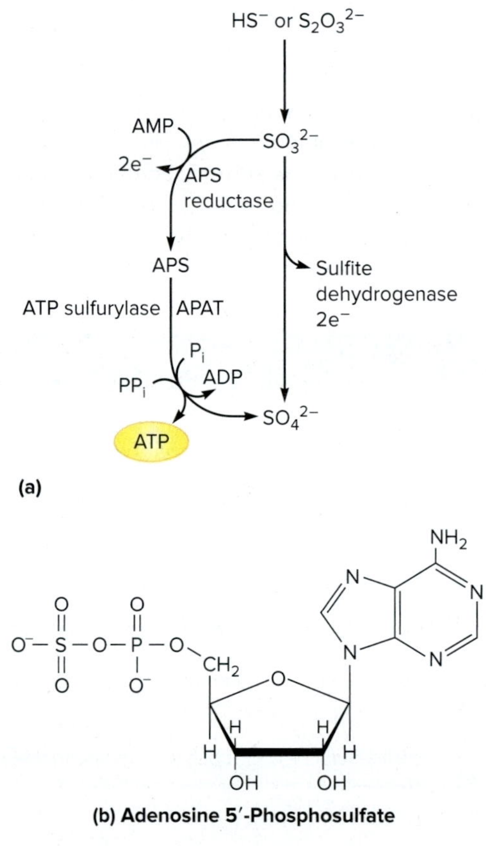

Figure 22.17 Sulfur Oxidation in *Thiobacillus ferrooxidans.* (a) Reduced sulfur compounds such as sulfide (HS⁻) or thiosulfate ($S_2O_3^{2-}$) are oxidized to sulfite (SO_3^{2-}). Electrons from sulfite can be donated to an electron transport chain via sulfite dehydrogenase and used to generate a proton motive force. Alternatively, the enzyme APS reductase catalyzes the oxidation of sulfite to sulfate, which is used to convert adenosine monophosphate (AMP) to adenosine 5′-phosphosulfate (APS). ATP sulfurylase and APS:phosphate adenylyltransferase (APAT) then catalyze the exchange of the sulfate on APS with PP_i to generate ATP via substrate-level phosphorylation. (b) The structure of APS.

MICRO INQUIRY *Identify the reactions that contribute to substrate-level phosphorylation and those that might contribute to ATP synthesis by a membrane-bound ATP synthase.*

Comprehension Check

1. Members of which β-proteobacterial genera make important contributions to the environment? Which are human or animal pathogens?
2. What is the end product of ammonium oxidation?
3. Examine figure 22.16. How does *N. europaea* generate a proton motive force (PMF), and what processes does this PMF fuel?
4. How do colorless sulfur bacteria obtain energy by oxidizing sulfur compounds?
5. What role does iron play in the redox metabolism of *Thiobacillus ferrooxidans*?

22.3 Class *Gammaproteobacteria* Is the Largest Bacterial Class

After reading this section, you should be able to:

a. Relate the electron donors used by a variety of γ-proteobacteria to the habitats in which they thrive
b. Describe the multicellular arrangements and unusual cellular morphologies displayed by some γ-proteobacteria
c. Outline the oxidation of methane by methanotrophs
d. Compare the intracellular life cycle of *Legionella* spp. to that of *Coxiella* spp.
e. Describe practical aspects of pseudomonads
f. List at least five electron acceptors used by *Shewanella* cells and explain the term *dissimilatory metal reduction*
g. Describe three strategies used by dissimilatory metal-reducing bacteria that enable their use of insoluble compounds as electron acceptors
h. Compare and contrast the physiology and importance of members of *Vibrionales* with those in *Enterobacteriales* and *Pasteurellales*
i. Summarize mixed acid fermentation
j. Describe the use of metabolic assays to distinguish between enteric bacteria

As the largest class of *Proteobacteria,* **γ-proteobacteria** demonstrate an extraordinary variety of physiological types. The diversity of these bacteria is evident from the fact that *Bergey's Manual* divides γ-proteobacteria into 14 orders and 27 families. **Table 22.5** outlines the general characteristics of some of the bacteria discussed in this section.

Purple Sulfur Bacteria Perform Anoxygenic Photosynthesis

Purple sulfur bacteria are strict anaerobes and usually photolithoautotrophic. *Bergey's Manual* places **purple sulfur bacteria** in the order *Chromatiales,* which has two families of interest: *Chromatiaceae*

Table 22.5 Characteristics of Selected γ-Proteobacteria

Genus	Dimensions (μm) and Morphology	Oxygen Requirement	Other Distinctive Characteristics
Beggiatoa	1–200 × 2–10; colorless cells form filaments, either single or in bundles	Aerobic or microaerophilic	Gliding motility; can form sulfur inclusions with hydrogen sulfide present
Chromatium	1–6 × 1.5–16; rod-shaped or ovoid, straight or slightly curved, polar flagella	Anaerobic	Anoxygenic photolithoautotrophs that can use sulfide; sulfur stored within the cell
Ectothiorhodospira	0.7–1.5 in diameter; vibrioid- or rod-shaped, polar flagella	Anaerobic, some aerobic or microaerophilic	Internal lamellar stacks of membranes; deposit sulfur granules outside cells
Escherichia	1.1–1.5 × 2–6; straight rods, peritrichous flagella or nonmotile	Facultatively anaerobic	Mixed acid fermenters; formic acid converted to H_2 and CO_2, lactose fermented, citrate not used
Haemophilus	<1.0 in width, variable lengths; coccobacilli or rods, nonmotile	Aerobic or facultatively anaerobic	Fermentative; require growth factors present in blood; parasites on mucous membranes
Leucothrix	Long filaments of short cylindrical cells, usually holdfast is present	Aerobic	Dispersal by gonidia, filaments don't glide; rosettes formed; chemoorganotrophic
Methylococcus	0.8–1.5 × 1.0–1.5; cocci with capsules, nonmotile	Aerobic	Can form cysts; use methane, methanol, and formaldehyde as sole carbon and energy sources; fix nitrogen nonsymbiotically
Photobacterium	0.8–1.3 × 1.8–2.4; straight, plump rods with polar flagella	Facultatively anaerobic	Several species can emit blue-green light; Na^+ needed for growth
Pseudomonas	0.5–1.0 × 1.5–5.0; straight or slightly curved rods, polar flagella	Aerobic or facultatively anaerobic	Respiratory metabolism with oxygen or nitrate as acceptor; degrade a broad range of organic compounds
Vibrio	0.5–0.8 × 1.4–2.6; straight or curved rods with sheathed polar flagella	Facultatively anaerobic	Fermentative or respiratory metabolism; sodium ions stimulate or are needed for growth; oxidase positive

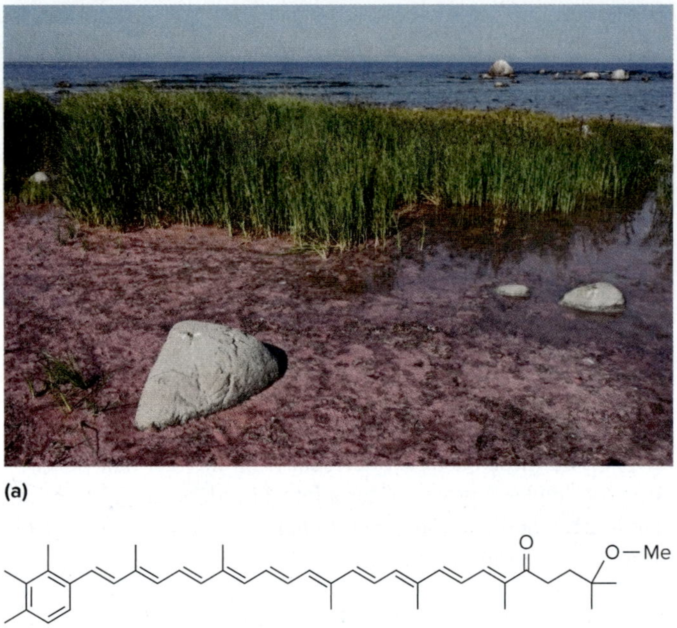

Figure 22.18 **Purple Photosynthetic Sulfur Bacteria.** (a) Purple photosynthetic sulfur bacteria growing in a salt marsh. (b) The chemical structure of okenone.
(a) ©Johnny Madsen/Alamy Stock Photo

Figure 22.19 ***Beggiatoa alba.*** A light micrograph showing part of a colony (×400). Note the dark sulfur granules within many of the filaments.
©Roger Burks/University of California at Riverside and Mark Schneegurt/Wichita State University, and Cyanosite (www-cyanosite.bio.purdue.edu)

and *Ectothiorhodospiraceae. Ectothiorhodospira* is composed of red, spiral-shaped, polarly flagellated cells that deposit sulfur globules externally. Internal photosynthetic membranes are organized as lamellar stacks (figure 22.3).

The majority of purple sulfur bacteria are in the family *Chromatiaceae*. Recall that during anoxygenic photosynthesis, electrons are required for the reduction of $NADP^+$. These microbes use hydrogen sulfide as an electron donor and deposit the resulting oxidized sulfur globules internally (*see figure 3.33*); often they further oxidize the sulfur to sulfate. Hydrogen also may serve as an electron donor. *Thiospirillum, Thiocapsa,* and *Chromatium* spp. are typical purple sulfur bacterial genera found in anoxic, sulfide-rich zones of lakes, marshes, and lagoons where large blooms can occur under certain conditions (**figure 22.18*a***). Okenone, a carotenoid pigment produced only in *Chromatiaceae,* is a useful geochemical biomarker (figure 22.18*b*). Its occurrence in sedimentary rock formations provides geochemists with an indicator of the environmental conditions at a specific historic time interval, thus aiding in climate reconstruction. ◀ *Light reactions in anoxygenic photosynthesis (section 11.12)*

Order *Thiotrichales* Includes Large Filamentous Bacteria

The largest family in the order *Thiotrichales* is *Thiotrichaceae.* This family has several genera whose members oxidize sulfur compounds. Morphologically both rods and filamentous forms are present, but the most striking members are filamentous.

Two genera in this family are *Beggiatoa* and *Leucothrix,* which contain well-studied gliding species. *Beggiatoa* spp. are microaerophilic and grow in sulfide-rich habitats such as sulfur springs, freshwater with decaying plant material, rice paddies, salt marshes, and marine sediments. Their filaments contain short, disklike cells (**figure 22.19**). These bacteria oxidize hydrogen sulfide to form large sulfur grains located in pockets formed by invaginations of the plasma membrane. *Beggiatoa* spp. also oxidize the sulfur to sulfate when donating electrons to the electron transport chain for energy production. Many strains grow heterotrophically with acetate as a carbon source, and some incorporate CO_2 autotrophically. Carbon fixation typically occurs using the Calvin-Benson cycle, but there is metagenomic evidence that some *Beggiatoa* from marine sediments perform the reductive TCA cycle as well (*see figure 12.5*). Freshwater species are much smaller than marine.

Members of three genera of *Thiotrichaceae—Beggiatoa, Thioploca,* and *Thiomargarita*—are notable because they form large mats on the seafloor near hydrothermal vents and adjacent to nutrient-rich coasts where methane seeps out of Earth's crust. These filamentous microbes are among the largest known bacteria. For example, *Thiomargarita* spp. can be over 100 µm in diameter (about twice as thick as a fingernail) and hundreds of centimeters long (**figure 22.20**). All three groups grow in bundles, making them even more conspicuous. These bacteria appear hollow because they form large intracellular vacuoles where high concentrations of nitrate (up to 500 mM) are stored. Sulfur inclusions are found in a thin layer of cytoplasm around the vacuoles.

Leucothrix spp. are aerobic chemoorganotrophs that form filaments or trichomes up to 400 µm long. Most are marine, and attach to solid substrates by a holdfast, a microscopic rootlike structure. These filamentous bacteria have a complex life cycle in which they are dispersed by the formation of small spherical cells called gonidia (**figure 22.21**). *Thiothrix* is a related genus whose members form sheathed filaments and release gonidia from the open end of the sheath. In contrast to *Leucothrix,*

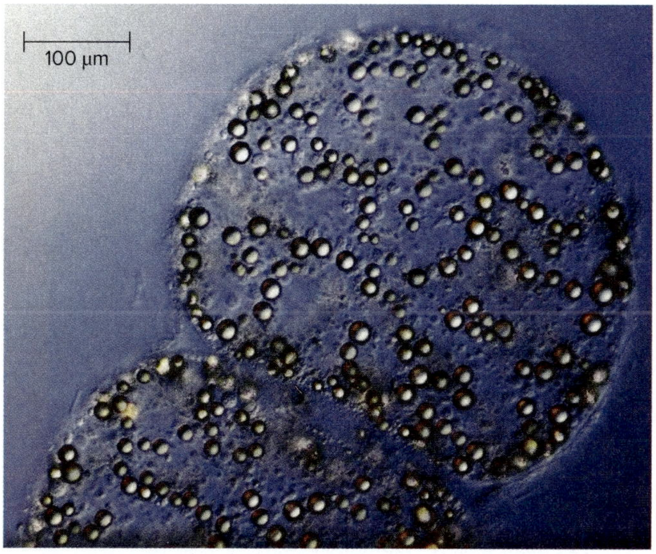

Figure 22.20 ***Thiomargarita namibiensis,* the World's Largest Known Bacterium.** This bacterium, usually 100–300 μm in diameter, occasionally reaches a size of 750 μm (larger than a period on this page). *T. namibiensis* uses sulfide from bottom sediments as an energy source, and nitrate from overlying waters as an electron acceptor. Vacuoles that store high concentrations of nitrate appear beadlike in this image.

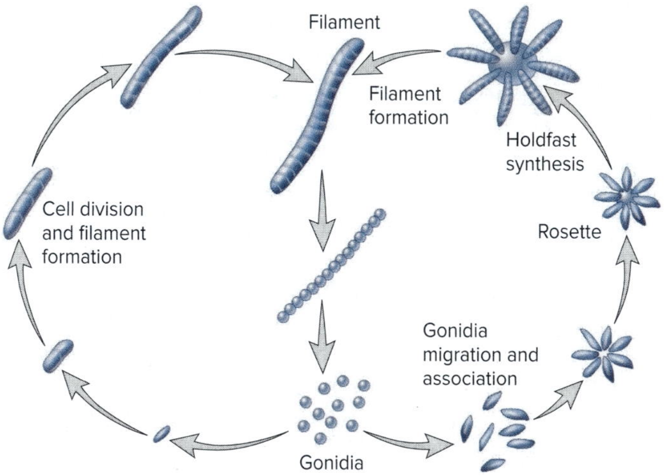

Figure 22.21 Life Cycle and Morphology of *Leucothrix mucor.*

Thiothrix includes species that are facultatively autotrophic, chemoorganotrophic or mixotrophic, oxidizing hydrogen sulfide and depositing sulfur granules internally. *Thiothrix* spp. grow in sulfide-rich flowing water and activated sludge sewage systems.

The genus *Thiomicrospira,* family *Piscirickettsiaceae,* has an interesting history. Originally it included all marine, spiral-shaped, sulfur-oxidizing bacteria. Later 16S rRNA analysis showed that this group was not monophyletic, and members were split between γ- and ε-proteobacterial classes.

Figure 22.22 shows a cell model based on the annotated genome of *Thiomicrospira crunogena* XCL-2. This deep-sea microbe was isolated from a hydrothermal vent (*see figure 27.4*). This environment is characterized by the mixture of hot, anoxic, highly reduced vent fluid that rapidly mixes with cold, oxic bottom water. Of importance to this autotroph is the difference in dissolved CO_2 concentrations that it encounters: Vent fluid is rich in CO_2, whereas bottom water is not. This has led to the evolution of a high affinity transporter for HCO_3^-. Also of note is the multienzyme complex capable of oxidizing a number of reduced sulfur compounds to sulfate. This "Sox" complex (for *s*ulfur *ox*idation) is conserved in a number of sulfur-dependent chemolithoautotrophs such as the β-proteobacteria *Thiobacillus* spp. Although *T. crunogena* can use a variety of reduced sulfur compounds as electron donors, it can use only oxygen as an electron acceptor. Interestingly, genome sequencing reveals the presence of what appears to be a complete prophage within the chromosome.

Order *Methylococcales* Consists of C1-Metabolizing Bacteria

The family *Methylococcaceae* in the order *Methylococcales* includes the genera *Methylomonas, Methylobacter,* and *Methylococcus,* all of which are obligate methylotrophs that oxidize methane and methanol as their only carbon and energy source. Like α-proteobacterial methanotrophs, these microbes also form extensive intracellular membranes; unlike α-proteobacterial methanotrophs, they form cysts instead of exospores (figure 22.5). They also differ in the use of the ribulose monophosphate pathway for the incorporation of C1 compounds (**figure 22.23**). ◀ *Methanogens and methanotrophs (section 20.3)*

Order *Legionellales* Includes Intracellular Pathogens

Two families make up the order *Legionellales.* The first is *Legionellaceae,* with its single genus, *Legionella.* The second family is *Coxiellaceae,* with two genera, *Coxiella* and *Rickettsiella* (not to be confused with the α-proteobacterial genus *Rickettsia*). These microbes are all intracellular pathogens that display a dimorphic (i.e., two cell types) lifestyle reminiscent of the chlamydial life cycle. ◀ *Phylum* Chlamydiae*: obligate intracellular parasites (section 21.5)*

Although there are over 58 species of *Legionella, L. pneumophila* has been most intensely studied because it causes a pneumonia in human hosts called Legionnaire's disease. In nature, *L. pneumophila* is an intracellular parasite of protozoa that live in moist soil and aquatic environments, including cooling towers, air conditioning units, and hot tubs. This explains why outbreaks of Legionnaire's disease occur in hotels, hospitals, and cruise ships—all of which have centralized cooling systems. *L. pneumophila* is transmitted by inhalation of aerosolized water droplets that enter the lungs where they infect macrophages. It is believed that the similarities between *Legionella*'s protozoan hosts and macrophages have made *L. pneumophila* an "accidental" pathogen of humans. Within a host cell (either protozoan or human), the bacteria are contained within a vacuole. The bacteria use a Type IV secretion system to transfer over 300 types of effector proteins to the host

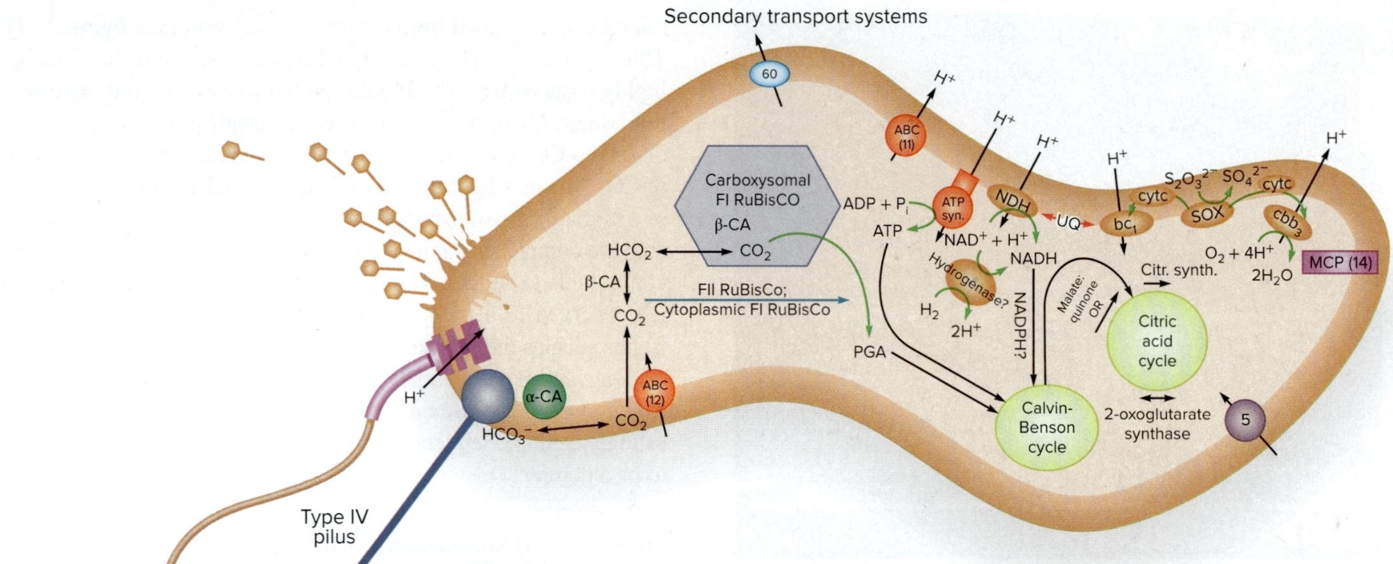

Figure 22.22 Model *Thiomicrospira crunogena* XCL-2 Cell. This chemolithoautotroph has high affinity carbonic anhydrases (α-CA and β-CA), enabling the accumulation of CO_2 to 100 times higher than extracellular concentrations. Three ribulose 1,5-bisphosphate carboxylase-oxygenase (RuBisCO) genes are present, and it is hypothesized that one RuBisCO protein is localized in carboxysomes. Energy and reducing power for carbon fixation is captured through Sox-dependent oxidation of reduced sulfur compounds. The electron transport chain includes cytochromes c (cytc), bc$_1$, and cbb$_3$ with O_2 as the terminal electron acceptor. A prophage is present within the genome; a suite of genes encoding lytic and lysogeny proteins suggests that the prophage can enter the lytic cycle, as indicated by the release of phage on the left side of the image. Transporters include over 60 secondary transporters (light blue) for metal, inorganic, and organic compounds, as well as a dozen ATP-binding cassette (ABC, red) uptake proteins and 11 ABC exporters (red). Other membrane proteins include 14 methyl-accepting proteins (MCP, purple) that help mediate chemotaxis and 5 unclassified transporters (lavender).

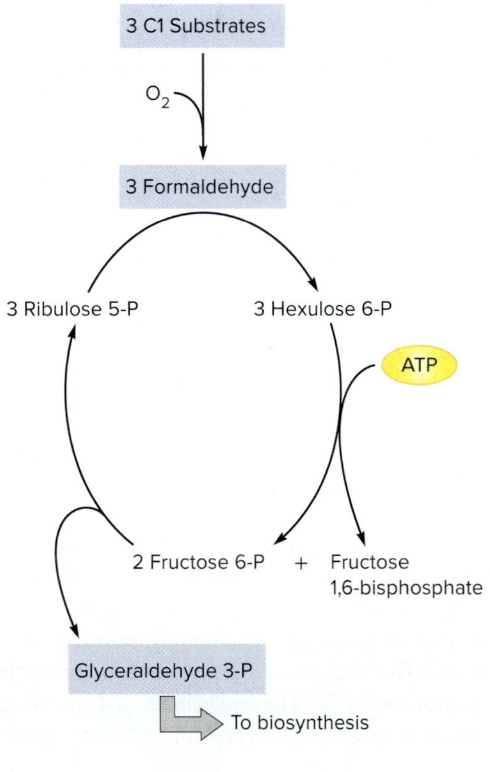

cytoplasm, thereby converting the vacuole to a *Legionella*-containing *v*acuole (LCV). Among the functions of these effectors is proteolysis to provide the amino acids required as the carbon and energy source for bacterial growth within the LCV. As amino acid nutrients become limited, *Legionella* initiates a stringent response by synthesizing (p)ppGpp. Accumulation of this secondary messenger results in the synthesis of the stress sigma factor RpoS, which in turn activates hundred of genes. The result is a metabolic shift to use other carbon sources and differentiation of *Legionella* cells to a cystlike transmissible cell called the *m*ature *i*ntracellular *f*orm (MIF). MIFs are nonreplicative; they exist only to infect the next host. They are heat tolerant and resistant to a variety of antibiotics. In aquatic environments, MIFs persist both in biofilms and as viable-but-not-culturable cells. ◀ *Stringent response in* E. coli *(section 14.5)* ▶ *Monocytes, macrophages, and dendritic cells (section 32.4); Legionnaire's disease (section 39.1)*

Figure 22.23 The Ribulose Monophosphate Pathway Is Used by γ-Proteobacterial Methylotrophs to Assimilate C1 Compounds. The net yield of this pathway is a glyceraldehyde 3-P molecule that can be used in central metabolism.

MICRO INQUIRY *How does the ribulose monophosphate pathway compare to the serine pathway (figure 22.12) in terms of energy and reducing power used? What might be the ecological consequence of this difference?*

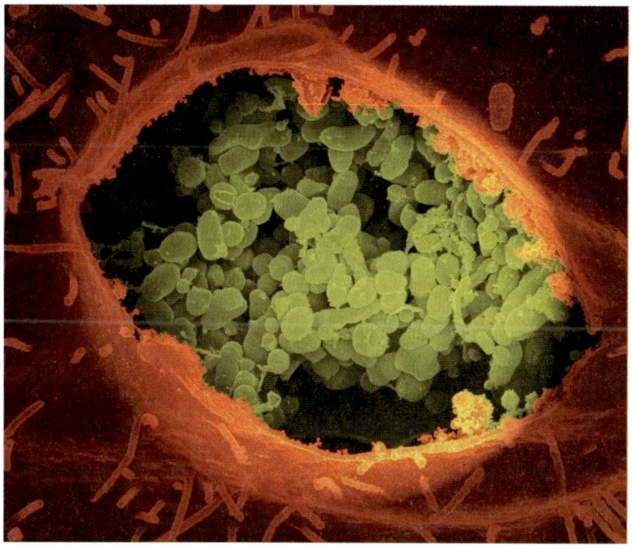

Figure 22.24 ***Coxiella burnetii*** **Within a Host Cell Vacuole.**
Source: National Institute of Allergy and Infectious Diseases (NIAID)/NIH/USHHS

Coxiella spp. also display two cell types within the host, although they have a much broader range of hosts, including insects, fish, birds, rodents, sheep, goats, and humans. As with *L. pneumophila,* humans become infected by inhaling contaminated aerosols, although in the case of the pathogen *C. burnetii,* the resulting Q fever is a flulike illness. The developmental life cycle of *C. burnetii* is similar to that of *L. pneumophila.* A small, infectious form called the *s*mall *c*ell *v*ariant (SCV, similar to MIFs and chlamydial elementary bodies) enters host cells by phagocytosis. Once the phagosome has fused with a lysosome, the low pH triggers SCVs to become metabolically active. The resulting vacuole is called the parasitophorous (parasite-eating) vacuole (**figure 22.24**). While they are in this vacuole, *C. burnetii* SCVs differentiate into the replicative form known as the *l*arge *c*ell *v*ariant (LCV). Both LCVs and SCVs undergo binary fission and infect host cells, so one might wonder why the microbe evolved the capacity to form SCVs. It is thought that SCVs, like MIFs, are responsible for long-term survival outside the host. It has been suggested that the compact and protected genome of SCVs makes these forms better suited for the initial respiratory burst and release of degradative enzymes when phagosomes and lysosomes initially fuse. The alternation of two cell types makes pathogens in the *Legionellales* distinct from the chlamydiae and rickettsiae, which cannot survive outside a host cell. ◀ *Phylum* Chlamydiae: *obligate intracellular parasites (section 21.5)* ▶ *Phagocytosis destroys invaders (section 32.6); Q fever (section 39.5)*

The Genus *Pseudomonas* Includes Many Species That Impact Human Activities

Pseudomonas is the most important genus in the order *Pseudomonadales,* family *Pseudomonaceae.* These bacteria are straight or slightly curved rods, and are motile by one or several polar flagella. They are chemoorganotrophs that usually carry out aerobic respiration, although some use nitrate as the terminal electron acceptor in anaerobic respiration. Most hexoses are degraded by the Entner-Doudoroff pathway; all pseudomonads have a functional tricarboxylic acid cycle. ◀ *Glucose to pyruvate (section 11.4); Pyruvate to carbon dioxide (step 2) is accomplished by the tricarboxylic acid cycle (section 11.5)*

The genus *Pseudomonas* is an exceptionally heterogeneous taxon, currently with about 150 species. As taxonomy and systematics have advanced, the composition of this genus has changed dramatically, yet it remains the bacterial genus with the largest number of species. Classification by 16S rRNA is insufficient to discriminate species, and multilocus sequence analysis using multiple housekeeping genes is the current standard. *Pseudomonas* is further categorized into 19 subgroups intermediate between genus and species. Three important subgroups are exemplified by *P. aeruginosa,* an opportunistic pathogen associated with cystic fibrosis; *P. putida,* a ubiquitous soil resident with broad degradative capacity; and *P. fluorescens,* a fluorescent pigment producer commonly associated with food spoilage. Bear in mind, however, that the large number of species reflects the ease of isolating and culturing *Pseudomonas* moreso than its dominance in the environment. ◀ *Subspecies and strain identification (section 19.3)*

A distinctive attribute of *Pseudomonas* species is their capacity to degrade an exceptionally wide variety of organic molecules. Thus they are important in the microbial breakdown of organic materials to inorganic substances (mineralization) in nature and in sewage treatment. Likewise, secreted proteases and lipases make *Pseudomonas* spp. common food spoilage organisms.

Several species (e.g., *P. aeruginosa*) are important experimental subjects. Many advances in microbial physiology and biochemistry have come from their study. For example, the study of *P. aeruginosa* has significantly advanced our understanding of how bacteria form biofilms and the role of extracellular signaling in bacterial communities and pathogenesis. The genome of *P. aeruginosa* has an unusually large number of genes for catabolism, nutrient transport, the efflux of organic molecules, and metabolic regulation, reflecting its ability to grow in many environments and resist antibiotics. ◀ *Biofilms are common in nature (section 7.6); Cell-cell communication within microbial populations (section 7.6)*

Order *Alteromonadales* Includes Anaerobes That Use a Range of Electron Acceptors

Many members of order *Alteromonadales,* including members of the genera *Marinomonas, Pseudoalteromonas,* and *Alteromonas,* are found in marine environments. *Shewanella* spp. have also been isolated from seawater, as well as lake sediments and salted foods. Bacteria in the genus *Shewanella* have generated much interest and are the focus of our discussion. This genus includes 66 facultatively anaerobic species that form straight or curved rods. It was first noted that *S. oneidensis* could use over 10 different electron acceptors and reduce thiosulfate and elemental sulfur to sulfide, a property usually seen only in strict anaerobes. Later it was discovered that all known *Shewanella* spp. show similar metabolic flexibility and can use a variety of metal electron acceptors, including uranium, chromium, iodate, technetium, neptunium, plutonium, selnite, tellurite, and vanadate, as well as nitroaromatic compounds. This wide range of metal electron acceptors makes *Shewanella* spp. excellent candidates as agents for bioremediation of environments contaminated with radionuclides. In addition, these bacteria can use a number of carbon substrates, and some can use H_2 as an energy and electron source. ▶| *Biodegradation and bioremediation (section 43.4)*

The use of metals as terminal electron acceptors is considered dissimilatory metal reduction because the metals are not incorporated into biomass. Dissimilatory metal reduction presents a problem for the microbe because many metals, such as iron and manganese, are insoluble in their oxidized forms, which is the form needed by the microbe for use as an electron acceptor. Dissimilatory metal-reducing bacteria have evolved several strategies that enable the use of solid metals such as Fe(III) or Mn(IV) as electron acceptors. The most direct approach is to localize cytochromes in the outer membrane, rather than the plasma membrane (**figure 22.25*a***). This enables the direct transfer of electrons from cytochromes to the extracellular metal. This approach is used by *Shewanella,* as well as by a number of Gram-negative dissimilatory metal-reducing bacteria such as the δ-proteobacterium *Geobacter sulfurreducens*. Another strategy used by members of the genera *Shewanella* and *Pseudomonas* is to transfer electrons to the metal via external, intermediary compounds such as humic acids found in the environment or their own secreted metabolites. These compounds are collectively known as **electron shuttles** because they pass (shuttle) electrons from the last point on the electron transport chain (ETC) through the growth substrate to the mineral surface. Species of *Shewanella, Geobacter,* and some cyanobacteria have evolved a third strategy: the production of electrically conductive **nanowires.** These appendages measure about 100 nm in diameter and tens of μm in length. In *Shewanella* spp., nanowires are chains of outer membrane vesicles containing cytochromes MtrC and OmcA. They facilitate electron transfer from the terminal point in the ETC to a metal surface (figure 22.25*b,c*). Nanowires enable the transport of electrons to solid-phase acceptors that are physically distant from the microbe. Interestingly, when near a redox active metal, *Shewanella* cells display electrokinesis—behavior that is analogous to chemotaxis. |◀ *Chemotaxis (section 3.8); Membrane vesicles are an alternative secretion mechanism (section 13.8)* ▶| *Microbial fuel cells: batteries powered by microbes (section 43.3)*

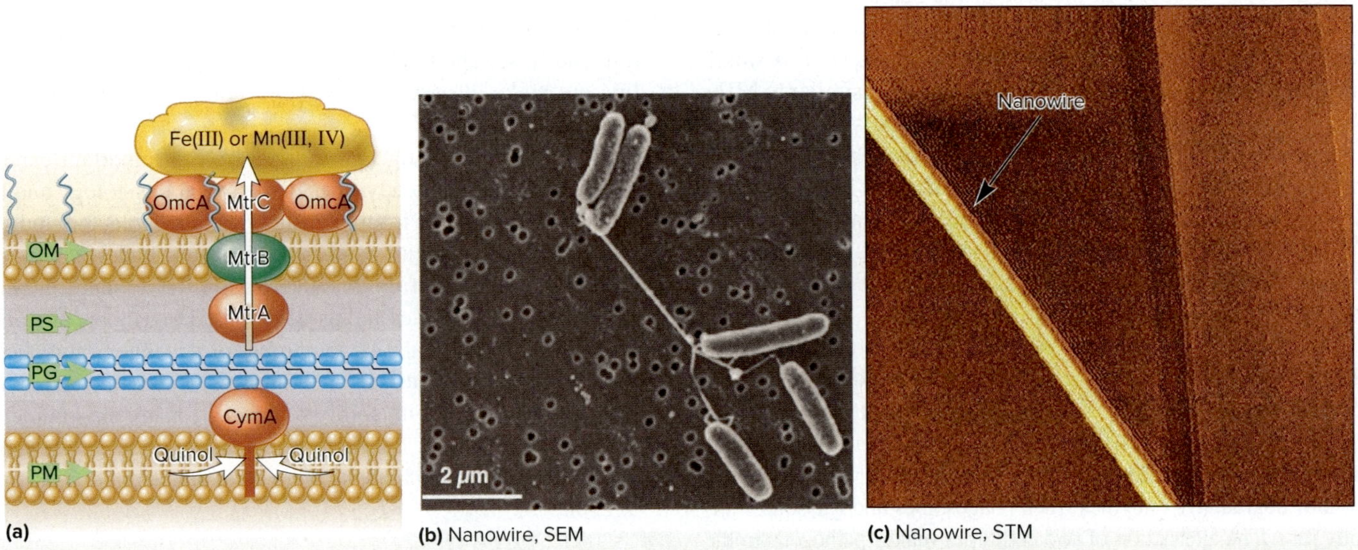

Figure 22.25 External Terminal Electron Acceptors. (a) *Shewanella oneidensis* MR-1 transfers electrons from the plasma membrane, through the outer membrane, to an insoluble metal. CymA is a quinol dehydrogenase that oxidizes quinol and passes the electrons to the c-type cytochrome MtrA. MtrB is not a cytochrome but helps move the electrons through the outer membrane to the surface-exposed c-type cytochromes, MtrC and OmcA. These redox proteins donate electrons to the oxidized, insoluble metals iron and manganese, which serve as terminal electron acceptors. (b) When electron acceptors are in short supply, *S. oneidensis* MR-1 makes nanowires along which electrons can travel to a terminal electron acceptor. This scanning electron micrograph shows the production of nanowires that occurs when cells are oxygen limited. (c) A scanning tunneling micrograph of a *S. oneidensis* MR-1 nanowire with a diameter of 100 nm. Note the ridges and troughs that run along the axis of the nanowire. OM, outer membrane; PG, peptidoglycan; PM, plasma membrane; PS, periplasmic space.

MICRO INQUIRY *Could a Gram-positive microbe use an insoluble metal as a terminal electron acceptor in the same way* S. oneidensis *does?*

Order *Vibrionales* Includes Aquatic Bioluminescent Bacteria and Pathogens

Three closely related orders of *Gammaproteobacteria* contain a number of important bacterial genera. Each order has only one family of facultatively anaerobic Gram-negative rods. **Table 22.6** summarizes the distinguishing properties of the families *Enterobacteriaceae, Vibrionaceae,* and *Pasteurellaceae.*

The order *Vibrionales* contains the family *Vibrionaceae.* We highlight two genera, *Vibrio* and *Photobacterium.*

Several vibrios are important pathogens. *Vibrio cholerae* causes cholera, and *V. parahaemolyticus* can cause gastroenteritis in humans following consumption of contaminated seafood. *V. anguillarum* and others are responsible for fish diseases, which can be especially problematic in fish farms. In recent years, the marine habitats of *Vibrio* spp. have changed. These bacteria prefer warm water, and climate change has allowed them to expand their range into ecosystems beyond the tropics. ▶| *Cholera (section 39.4)*

Some members of *Vibrionaceae* are unusual in being bioluminescent—that is, they can emit light of a blue-green color. The enzyme luciferase catalyzes the reaction and uses reduced flavin mononucleotide, molecular oxygen, and a long-chain aldehyde as substrates.

$$FMNH_2 + O_2 + RCHO \xrightarrow{\text{luciferase}} FMN + H_2O + RCOOH + \text{light}$$

Evidence suggests that an enzyme-bound, excited flavin intermediate is the direct source of luminescence. Because the electrons used in light generation are diverted from the electron transport chain and ATP synthesis, the bacteria expend considerable energy on luminescence. It follows that luminescence is regulated and can be turned off or on under the proper conditions.

There is much speculation about the role of bacterial luminescence and its value to bacteria, particularly because it is such an energetically expensive process. Luminescent bacteria do not emit light when they grow as free-living organisms in seawater; the light emitted by a single cell would not be visible. When free-living luminescent bacteria infect the light organ of young fish (**figure 22.26**) or squid (*see figure 7.25*), they reproduce and reach densities where light emission is visible. Bacterial luminescence was the first example of density-dependent gene regulation by the quorum-sensing molecule autoinducer. It remains unclear what advantage bacterial luminescence confers on some animal hosts. However, the mechanism by which autoinducer regulates light production in these marine bacteria is an important model for understanding quorum sensing in many Gram-negative bacteria, including a number of pathogens. |◀ *Cell-cell communication within microbial populations (section 7.6); Quorum sensing by* Vibrio *spp. (section 14.6)*

Order *Enterobacteriales* Includes Pathogens and Mutualists

The order *Enterobacteriales* has only one family, *Enterobacteriaceae,* which contains 42 genera. This family is the largest listed in table 22.6. It contains peritrichously flagellated or nonmotile, facultatively anaerobic, straight rods with simple nutritional requirements.

The metabolic properties (particularly during anaerobic growth) of members of *Enterobacteriaceae* are very useful in characterizing its constituent genera. These bacteria, often called **enterobacteria** or **enteric bacteria** (Greek *enterikos,* pertaining to the intestine), all degrade sugars by means of the Embden-Meyerhof pathway. Under microaerobic or anoxic conditions, the enzyme **pyruvate formate-lyase** (**PFL**) catalyzes the cleavage of

Table 22.6 Characteristics of Families of Facultatively Anaerobic Gram-Negative Rods

Characteristics	*Enterobacteriaceae*	*Vibrionaceae*	*Pasteurellaceae*
Cell dimensions	0.3–1.0 × 1.0–6.0 μm	0.3–1.3 × 1.0–3.5 μm	0.2–0.4 × 0.4–2.0 μm
Morphology and motility	Straight rods; peritrichous flagella or nonmotile	Straight or curved rods; polar flagella; lateral flagella may be produced on solid media	Coccoid to rod-shaped cells, sometimes pleomorphic; nonmotile
Physiology	Oxidase negative	Oxidase positive; all can use D-glucose as sole or principal carbon source	Oxidase positive; heme and/or NAD^+ often required for growth; organic nitrogen source required
Symbiotic relationships	Some are parasitic on mammals and birds; some species are plant pathogens	Most are not pathogens; several inhabit light organs of marine organisms	Parasites of mammals and birds
Representative genera	*Escherichia, Shigella, Salmonella, Citrobacter, Klebsiella, Enterobacter, Erwinia, Serratia, Proteus, Yersinia*	*Vibrio, Photobacterium*	*Pasteurella, Haemophilus*

(a)

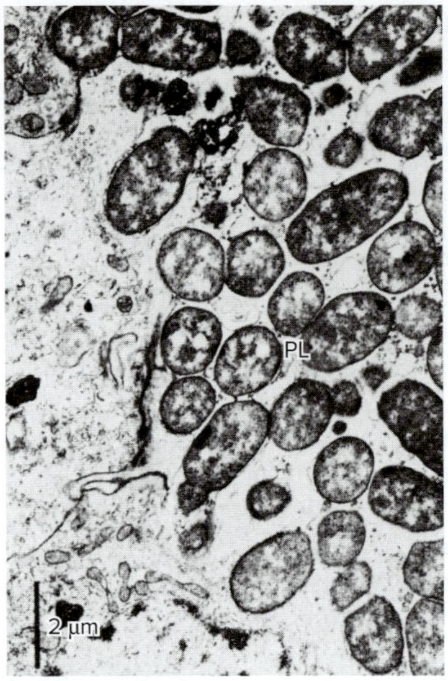

(b)

Figure 22.26 Bioluminescence. (a) A photograph of the Atlantic flashlight fish *Kryptophanaron alfredi.* The light area under the eye is the fish's luminous organ, which can be covered by a lid of tissue. (b) Ultrathin section of the luminous organ of a fish, *Equulities novaehollandiae,* with the bioluminescent bacterium *Photobacterium leiognathi,* PL.

(a) ©Frederick R. McConnaughey/Science Source; (b) ©James G. Morin/Cornell University

pyruvate to formate and acetyl-CoA. Those enteric bacteria that produce large amounts of gas during sugar fermentation, such as *Escherichia* spp., also have the enzyme **formate dehydrogenase (FDH)**, which splits formic acid to H_2 and CO_2 (**figure 22.27**). This enzyme complex is crucial to survival during fermentation, when one-third of the carbon derived from glucose is converted to formate. If this intermediate were allowed to accumulate within the cell, the internal pH would become far too acidic to support viability. In fact, *E. coli* has three different FDH enzymes. FDH-H (H for hydrogen production) is synthesized only during fermentative conditions. FDH-N is produced under anoxic conditions when nitrate serves as the terminal electron acceptor. FDH-O is produced under microaerobic conditions as well as when nitrate is the terminal electron acceptor.

The family can be divided into two groups based on their fermentation products. The majority (e.g., the genera *Escherichia, Proteus, Salmonella,* and *Shigella*) carry out mixed acid fermentation and produce mainly lactate, acetate, succinate, ethanol, formate, H_2, and CO_2. In contrast, *Enterobacter, Serratia, Erwinia,* and *Klebsiella* spp. are butanediol fermenters, producing butanediol, ethanol, and CO_2. ◀ *Fermentation does not involve an electron transport chain (section 11.8)*

Because enteric bacteria are so similar in morphology, biochemical tests are often used to identify them after a preliminary examination of their morphology, motility, and growth responses. Some more commonly used tests are those for the type of fermentation, lactose and citrate utilization,

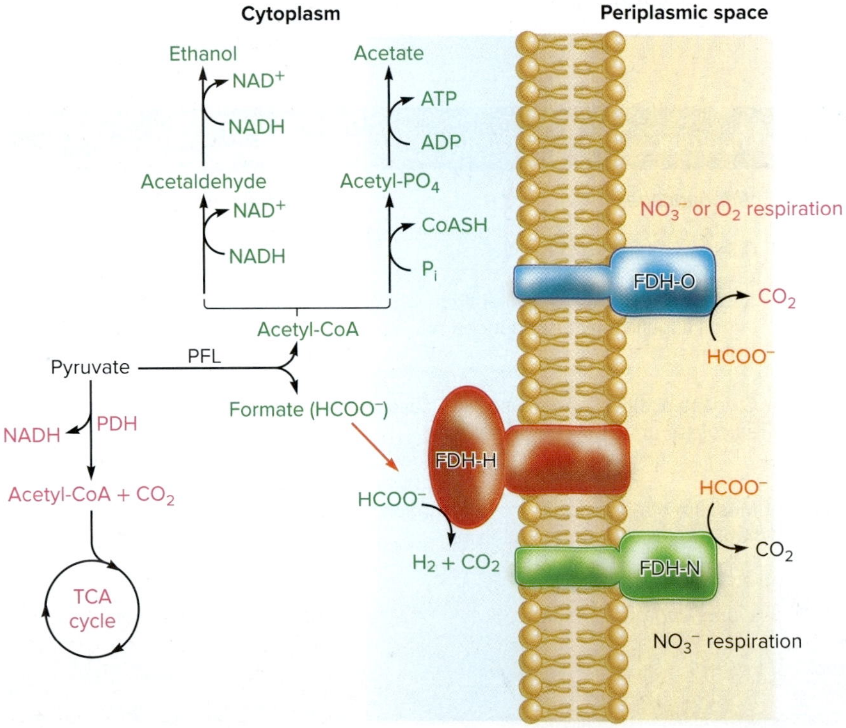

Figure 22.27 Metabolism of Pyruvate and Formate in *E. coli.* During aerobic growth, pyruvate is converted to CO_2 and acetyl-CoA by the pyruvate dehydrogenase complex (PDH) so that acetyl-CoA can enter the TCA cycle. Under low oxygen conditions, some pyruvate is also cleaved to formate and acetyl-CoA by pyruvate formate-lyase (PFL). Formate must then be converted into CO_2 and H_2 by one of three formate dehydrogenase (FDH) enzymes. FDH-O is active during aerobic respiration when oxygen levels are low. If respiration continues under anoxic conditions with nitrate as the terminal electron acceptor, both FDH-N and FDH-O split formate. These enzymes are located on the periplasmic face of the plasma membrane. During fermentative conditions, FDH-H is made, resulting in mixed acid fermentation: Formate is cleaved to yield CO_2 and H_2, whereas acetyl-CoA is converted to ethanol and acetate. FDH-H is a peripheral membrane protein, with its active site exposed to the cytoplasm. Aerobic metabolism is shown in pink, fermentative metabolism in green.

MICRO INQUIRY *Why do you think FDH-O and FDH-N are on the periplasmic face of the plasma membrane while FDH-H is on the cytoplasmic face?*

indole production from tryptophan, urea hydrolysis, and hydrogen sulfide production. For example, lactose fermentation occurs in some genera (e.g., *Escherichia, Klebsiella,* and *Enterobacter*) but not in others (e.g., *Shigella, Salmonella,* and *Proteus*). **Table 22.7** summarizes a few of the biochemical properties useful in distinguishing between genera of enteric bacteria. The mixed acid fermenters are located on the left in this table and the butanediol fermenters on the right. ▶| *Identification of microorganisms from specimens (section 37.2)*

Members of *Enterobacteriaceae* are so widespread and important that they are probably seen more often in laboratories than any other bacteria. *E. coli* is undoubtedly the best-studied bacterium and the experimental organism of choice for many microbiologists. It is an inhabitant of the colon of humans and other warm-blooded animals, and it is useful in detecting fecal contamination in water. Some strains cause gastroenteritis or urinary tract infections. Several other genera contain important human pathogens responsible for a variety of diseases: *Salmonella,* typhoid fever and gastroenteritis; *Shigella,* bacillary dysentery; *Klebsiella,* pneumonia; and *Yersinia,* plague. Members of the genus *Erwinia* are major pathogens of crop plants and cause blights, wilts, and several other plant diseases.

Order *Pasteurellales* Includes Important Pathogens

Bacteria in the family *Pasteurellaceae* (order *Pasteurellales*) differ from those in *Vibrionales* and *Enterobacteriales* in several ways (table 22.6). Most notably, they are smaller, nonmotile, have complex nutritional requirements, and are parasitic in vertebrates. The family contains seven genera: *Pasteurella, Haemophilus, Actinobacillus, Lonepinella, Mannheimia, Phocoenobacter,* and *Gallibacterium.*

Members of this family are best known for the diseases they cause in humans and animals. *Pasteurella multocida* is responsible for fowl cholera, which kills many chickens, turkeys, ducks, and geese each year. *P. haemolytica* is at least partly responsible for pneumonia in cattle, sheep, and goats (e.g., "shipping fever" in cattle). *H. influenzae* serotype b is a major human pathogen that causes a variety of diseases, including meningitis, sinusitis, pneumonia, and bronchitis. Fortunately, a sharp reduction in the incidence of *H. influenzae* serotype b infections began in the late 1980s due to administration of the *H. influenzae* type b ("Hib") vaccine. However, this bacterium still causes at least 3 million cases of serious disease and several hundreds of thousands of deaths each year globally. ▶| *Meningitis (section 39.1)*

Comprehension Check

1. What are the major characteristics of purple sulfur bacteria? Contrast the features that define the families *Chromatiaceae* and *Ectothiorhodospiraceae.*
2. Describe the metabolism of *Beggiatoa, Leucothrix,* and *Thiothrix* spp.
3. List three attributes that account for the importance of pseudomonads.
4. Why is metal reduction by *Shewanella* spp. considered dissimilatory? Describe two morphological adaptations that enable oxidation of insoluble metals outside the cell.
5. List the major distinguishing traits of bacteria in the families *Vibrionaceae, Enterobacteriaceae,* and *Pasteurellaceae.*
6. What two cofactors are required for bioluminescence?
7. Into what two groups can enteric bacteria be placed based on their fermentation patterns?

22.4 Class *Deltaproteobacteria* Includes Chemoheterotrophic Anaerobes and Predators

After reading this section, you should be able to:

a. Characterize the common physiological features of sulfate-reducing bacteria (SRB)
b. Identify common carbon substrates used by SRB
c. Draw an electron transport chain capable of dissimilatory sulfate reduction
d. Relate the physiology of SRB to their common environmental niches
e. Compare and contrast the physiology and ecology of *Shewanella* and *Geobacter* spp.
f. Draw the life cycle of a bdellovibrio and explain why it is considered a bacterial predator
g. Relate the life cycle of myxobacteria to their habitat and means of acquiring nutrients
h. Describe the two types of gliding motility displayed by myxobacteria, and sketch their respective mechanistic models

Although **δ-proteobacteria** are not a large assemblage of genera, they show considerable morphological and physiological diversity. These bacteria can be divided into two general groups, all of them chemoorganoheterotrophs. Some are predators, such as bdellovibrios and myxobacteria. Others are anaerobes that use sulfate and sulfur as terminal electron acceptors while oxidizing organic nutrients. The class has eight orders and 19 families. **Table 22.8** summarizes the defining properties of some representative genera.

Orders *Desulfovibrionales, Desulfobacterales,* and *Desulfuromonales*: Anaerobic Sulfate/Sulfur Reducers

Desulfovibrionales, Desulfobacterales, and *Desulfuromonales* are a diverse group of **sulfate-** or **sulfur-reducing bacteria (SRB)** that are united by their ability to reduce elemental sulfur or sulfate and other oxidized sulfur compounds to hydrogen sulfide during anaerobic respiration. Recall that anaerobic respiration is coupled with the oxidation of reduced organic compounds and in general the SRB can be grouped according to carbon metabolism. Many SRB oxidize lactate, formate, butyrate, propionate, pyruvate, and aromatic compounds to

Table 22.7 Some Characteristics of Selected Genera in *Enterobacteriaceae*

Characteristics	*Escherichia*	*Shigella*	*Salmonella*	*Citrobacter*	*Proteus*
Methyl red	+	+	+	+	+
Voges-Proskauer	–	–	–	–	d
Indole production	(+)	d	–	d	d
Citrate use	–	–	(+)	+	d
H_2S production	–	–	(+)	d	(+)
Urease	–	–	–	(+)	+
β-galactosidase	(+)	d	d	+	–
Gas from glucose	+	–	(+)	+	+
Acid from lactose	+	–	(–)	d	–
Motility	d	–	(+)	+	+
Genome size (Mb)	4.6–5.5	4.6	4.5–4.9	4.9–5.3	4.1
Other characteristics	1.1–1.5 × 2.0–6.0 μm; peritrichous flagella when motile	1–3 X 0.7–1.0 μm; no gas from sugars	0.7–1.5 × 2–5 μm; peritrichous flagella	1.0 × 2.0–6.0 μm; peritrichous flagella	0.4–0.8 × 1.0–3.0 μm; peritrichous flagella

1 (+) Usually present.
2 (–) Usually absent.
3 d, Strains or species vary in possession of characteristic.

Table 22.8 Characteristics of Selected δ- and ε-Proteobacteria

Class Genus	Dimensions (μm), Morphology, and Motility	Oxygen Requirement	Other Distinctive Characteristics
δ-Proteobacteria			
Bdellovibrio	0.2–0.4 × 0.5–1.4; comma-shaped rods with a sheathed polar flagellum; also capable of gliding motility	Aerobic	Prey on other Gram-negative bacteria and grow in the periplasm; alternate between predatory and intracellular reproductive phases
Desulfovibrio	0.5–1.5 × 2.5–10; curved or sometimes straight rods, motile by polar flagella	Anaerobic	Oxidize organic compounds to acetate and reduce sulfate or sulfur to H_2S
Desulfuromonas	0.4–0.9 × 1.0–4.0; straight or slightly curved or ovoid rods, lateral or subpolar flagella	Anaerobic	Reduce sulfur to H_2S, oxidize acetate to CO_2; form pink or peach-colored colonies
Myxococcus	0.8 × 3–6; fusiform, gliding motility	Aerobic	Form fruiting bodies with myxospores not enclosed in a sporangium
Stigmatella	0.7–0.8 × 4–8; fusiform, gliding motility	Aerobic	Stalked fruiting bodies with sporangioles containing myxospores (0.9–1.2 × 2–4 μm)
ε-Proteobacteria			
Campylobacter	0.2–0.8 × 0.5–5; spirally curved cells with a single polar flagellum at one or both ends	Microaerophilic	Carbohydrates not fermented or respired; oxidase positive and urease negative; found in intestinal tract, reproductive organs, and oral cavity of animals
Helicobacter	0.2–1.2 × 1.5–10; helical, curved, or fusiform; multiple, sheathed flagella	Microaerophilic	Catalase and oxidase positive; urea rapidly hydrolyzed; found in the gastric mucosa of humans and other animals

Yersinia	*Klebsiella*	*Enterobacter*	*Erwinia*	*Serratia*
+	(+)[1]	(–)[2]	+	d[3]
– (37°C)	(+)	+	(+)	+
d	d	–	(–)	(–)
(–)	(+)	+	(+)	+
–	–	–	(+)	–
d	(+)	(–)	–	–
+	(+)	+	+	+
(–)	(+)	(+)	(–)	d
(–)	(+)	(+)	d	d
– (37°C)	–	+	+	+
4.6	5.3–5.5	4.9–5.3	5.1	5.1
0.5–0.8 × 1.0–3.0 µm; peritrichous flagella when motile	0.3–1.0 × 0.6–6.0 µm; capsulated	0.6–1.0 × 1.2–3.0 µm; peritrichous flagella	0.5–1.0 × 1.0–3.0 µm; peritrichous flagella; plant pathogens and saprophytes	0.5–0.8 × 0.9–2.0 µm; peritrichous flagella; colonies often pigmented

acetate, which is not further metabolized. Other SRBs oxidize acetate completely to CO_2. ◀ *Anaerobic respiration uses the same three steps as aerobic respiration (section 11.7)*

The use of sulfate as a terminal electron acceptor is a multistep process. Sulfate cannot be directly reduced. Because sulfate is a stable molecule, it must first be activated by reacting with ATP to form adenosine 5′-phosphosulfate, or APS (**figure 22.28**; figure 22.17*b*). The fate of APS during anaerobic respiration differs from that seen in sulfide-oxidizing chemolithoautotrophs (figure 22.17*a*). During sulfate reduction, APS accepts two electrons, which results in the production of sulfite (SO_3^{2-}) and adenosine monophosphate (AMP). Six additional electrons are needed to reduce sulfite completely to hydrogen sulfide. These are captured by a hydrogenase enzyme located on the periplasmic face of

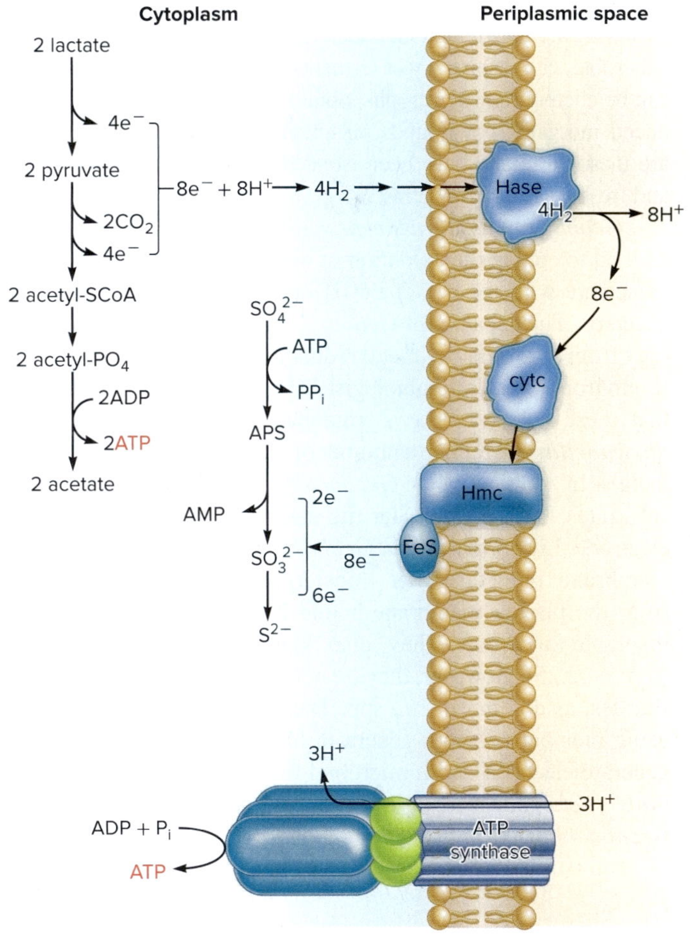

Figure 22.28 Dissimilatory Sulfate Reduction by *Desulfovibrio*. These microbes make ATP by substrate-level phosphorylation during acetate fermentation and by ATP synthase. Synthesis of ATP by ATP synthase is driven by a proton motive force generated by a periplasmic hydrogenase, which oxidizes H_2 to protons and electrons. The electrons are passed back into the cell, where they are used to reduce sulfate completely to sulfide. As detailed in the text, the oxidation of sulfate to sulfide requires eight electrons and the activation of sulfate by its covalent attachment to AMP to form APS.

the plasma membrane. The hydrogenase donates the electrons from molecular hydrogen to cytochrome c_3, which in turn transfers them to the integral membrane protein Hmc. Finally, an iron-sulfur protein delivers the electrons to sulfite reductase, and hydrogen sulfide is produced. Note that hydrogenase releases protons in the periplasm; the resulting proton motive force is sufficient to drive the synthesis of ATP.

SRB are very important in sulfur cycling within ecosystems. Because significant amounts of sulfate are present in almost all aquatic and terrestrial habitats, SRB are widespread and active in locations made anoxic by microbial digestion of organic materials. The SRB thrive in habitats such as muds and sediments of polluted lakes and streams, sewage lagoons and digesters, and waterlogged soils. *Desulfuromonas* spp. are most prevalent in anoxic marine and estuarine sediments. They also can be isolated from methane digesters and anoxic hydrogen sulfide–rich muds of freshwater habitats. Often sulfate and sulfur reduction are apparent from the smell of hydrogen sulfide (a rotten-egg odor) and the blackening of water and sediment by iron sulfide. Hydrogen sulfide production in waterlogged soils can kill animals, plants, and microorganisms. SRB negatively impact industry because of their primary role in the anaerobic corrosion of iron in pipelines, heating systems, and other structures. ▶| *Sulfur cycle (section 28.1)*

The order *Desulfuromonales* encompasses two families, *Desulfuromonaceae* and *Geobacteraceae.* They are strictly anaerobic with respiratory or fermentative metabolism. They can be chemolithoheterotrophs, obtaining electrons from reduced inorganic compounds, or chemoorganotrophs. They are mesophilic and have been isolated from anoxic marine and freshwater environments.

Geobacter metallireducens was the first microbe described to couple the oxidation of organic substrates such as acetate with the use of Fe(III) as a terminal electron acceptor. The capacity of *Geobacter* spp. to conserve energy from dissimilatory metal reduction is of great interest to environmental microbiologists. They have discovered that these bacteria, like γ-proteobacteria in the genus *Shewanella,* can reduce a number of toxic and radioactive metals. In addition, when *Geobacter* spp. oxidize organic substrates, they can transfer the electrons directly to an electrode. Like *Shewanella, Geobacter* spp. have an outer membrane reductase that transfers electrons received from the plasma membrane-bound ETC to an external, insoluble acceptor. They also synthesize nanowires (figure 22.25). However, they do not use external electron shuttles, as do *Shewanella* spp. There is great interest in using members of both genera in bioremediation and to generate electricity with microbial fuel cells. ▶| *Microbial fuel: batteries powered by microbes cells (section 43.3)*

Order *Bdellovibrionales* Includes Predators That Invade Other Gram-Negative Bacteria

The order *Bdellovibrionales* has only the family *Bdellovibrionaceae* and four genera. The genus *Bdellovibrio* (Greek *bdella,* leech) contains aerobic, Gram-negative, curved rods with polar flagella. The flagellum is unusually thick due to the presence of a sheath that is continuous with the cell wall. Bdellovibrios have a distinctive lifestyle: They prey on other Gram-negative bacteria and alternate between a nongrowing predatory phase and an intracellular reproductive phase.

The life cycle of a bdellovibrio is complex, although it requires only 1 to 3 hours for completion (**figure 22.29**). The free bacterium swims rapidly (about 100 cell lengths per second) until it collides violently with its prey. It attaches to the prey bacterium's surface, begins to rotate as fast as 100 revolutions per second, and bores a hole through the host cell wall in a process involving type IV pili and the release of several hydrolytic

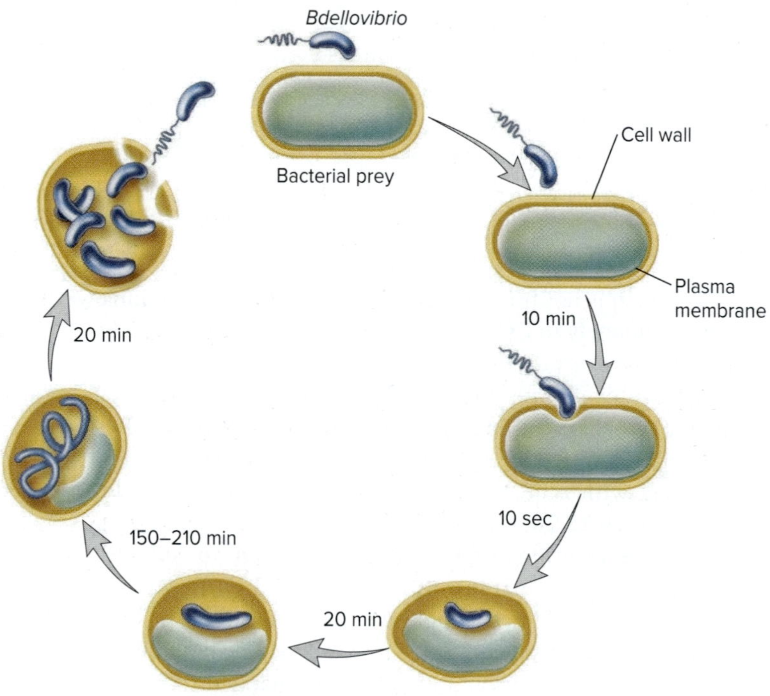

(a)

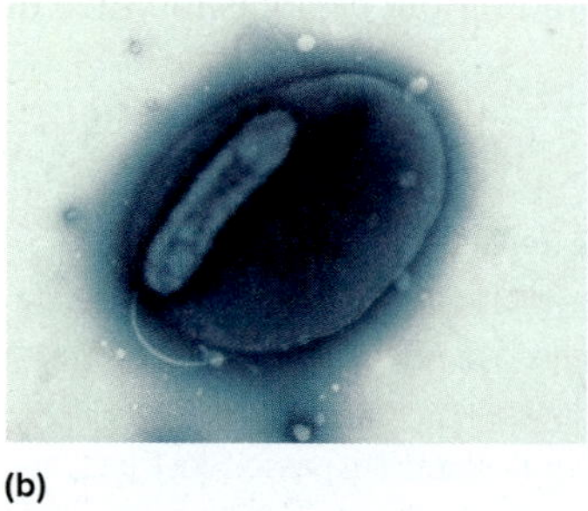

(b)

Figure 22.29 The Life Cycle of *Bdellovibrio*. (a) A general diagram showing the complete life cycle. (b) *Bdellovibrio bacteriovorus* encapsulated in the periplasm of *E. coli.*

(b) ©Katy Evans, Chi Aizawa and Liz Sockett

enzymes. Its flagellum is lost during penetration of the cell. These bacteria can also scout for bacterial prey by gliding slowly over solid substrates. ◀ *Bacterial pili and fimbriae (section 3.7)*

After entry, the bdellovibrio takes control of the host cell and grows in the periplasm while the host cell loses its shape and becomes spherical. The predator inhibits host DNA, RNA, and protein synthesis, and disrupts the host's plasma membrane so that cytoplasmic constituents leak out of the cell. The growing bdellovibrio uses host amino acids as its carbon, nitrogen, and energy source. It employs host fatty acids and nucleotides directly in biosynthesis, thus saving carbon and energy. The bdellovibrio grows into a long filament that septates into many smaller, flagellated progeny, which escape upon host cell lysis. Such multiple fission is rare in bacteria and archaea. *Bdellovibrio*

Order *Myxococcales*: Bacteria with Morphological Complexity and Multicellularity

Myxobacteria are Gram-negative, aerobic soil bacteria characterized by gliding motility and a complex life cycle that includes production of multicellular structures called fruiting bodies and the formation of spores called myxospores. Myxobacterial cells are rods, which may be either slender with tapered ends (fusiform) or stout with rounded, blunt ends. The order *Myxococcales* is divided into six families based on the shape of vegetative cells, myxospores, and sporangia.

Most myxobacteria are predators of other microbes but, unlike bdellovibrios, do not depend on predation for reproduction. Myxobacteria secrete lytic enzymes and antibiotics to kill their prey. The digestion products, primarily small peptides, are absorbed. Most myxobacteria use amino acids as their major source of carbon, nitrogen, and energy. All are aerobic chemoorganotrophs with respiratory metabolism.

The myxobacterial life cycle is quite distinctive and in many ways resembles that of cellular slime molds (**figure 22.30**) discussed in chapter 24. In the presence of a food supply, myxobacteria glide along a solid surface, feeding and leaving slime trails. During this stage, the cells often form a swarm and move in a coordinated fashion. Some species congregate to produce a sheet of cells that moves rhythmically to generate waves or ripples. When their nutrient supply is exhausted, the myxobacteria aggregate and differentiate into a fruiting body, a structure that is largely made up of individual cells that have differentiated into exospores called myxospores. The life cycle of the species *Myxococcus xanthus* serves as a model system for bacterial differentiation and behavior. Development in this microbe is induced by nutrient limitation and requires at least five different extracellular signaling molecules that allow the cells to communicate with one another.

Fruiting bodies range in height from 50 to 500 μm and often are colored red, yellow, or brown by carotenoid pigments. Each species forms a characteristic fruiting body. They vary in complexity from simple globular objects made of about 100,000 cells (*Myxococcus*) to the elaborate, branching, treelike structures formed by members of the genera *Stigmatella* and *Chondromyces*. In some species, cells develop into dormant myxospores enclosed in walled structures called sporangioles or sporangia. As might be expected for such complex structures, a variety of new proteins are synthesized during fruiting body formation, and *Myxococcus xanthus* has one of the largest bacterial genomes sequenced to date.

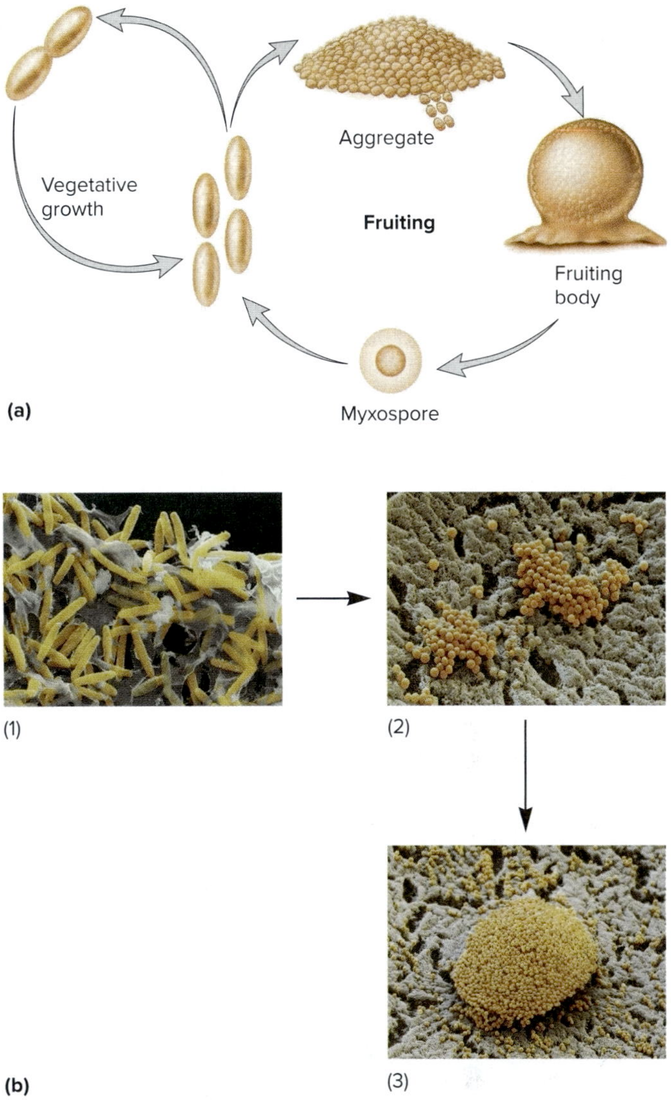

Figure 22.30 **Life Cycle of *Myxococcus xanthus*.** (a) When nutrients are plentiful, *M. xanthus* grows vegetatively. However, when nutrients are depleted, the exchange of extracellular signaling molecules triggers cells to aggregate and form fruiting bodies. Most of the cells within a fruiting body become resting myxospores that will not germinate until nutrients are available. (b) Scanning electron micrographs taken during (1) vegetative growth, (2) aggregation, and (3) fruiting body formation.

Myxospores are dormant and desiccation-resistant; they may survive up to 10 years under adverse conditions (compare this with thousands of years for Gram-positive endospores). The use of fruiting bodies provides further protection for myxospores and assists in their dispersal. Because myxospores are kept together within a fruiting body, a colony of myxobacteria automatically develops when myxospores are released and germinate. This communal organization may be advantageous because myxobacteria obtain nutrients by secreting hydrolytic enzymes and absorbing soluble digestive products. A mass of myxobacteria can produce enzyme concentrations sufficient to digest their prey more easily than an individual cell.

Fruiting body development requires gliding motility. Gliding motility in *M. xanthus* is much slower (2–4 mm/min) than in flavobacteria (*see figure 21.17*). Two types of motility have been characterized in *M. xanthus*. Social (S) motility is a kind of twitching that is governed by the production of retractable type IV pili from the front end of the cell (**figure 22.31*a***). When the pili retract, the cell creeps forward. This type of motility was originally called social motility because it is only observed when cells are in a group. It is now known that cell-to-cell contact is required for S motility because cells share outer membrane lipoproteins involved in pili secretion. The second type of motility is called adventurous (A) motility because it is exhibited by single cells on the periphery of the group, perhaps scouting for prey.

Adventurous motility requires a multiprotein motility machine that also assembles at the forward cell pole, coordinated by a protein called MglA (figure 22.31*b*). The molecular motor is made up of Agl proteins, which have several functions. AglQRS proteins in the plasma membrane couple the translocation of protons with gliding motility, much like Mot proteins in the flagellar basal body in swimming bacteria (*see figure 3.41*). AglZ coordinates the interaction of MglA with the cytoskeletal protein MreB and a complex of Glt (*gl*iding *t*ransducer) proteins spanning the cell wall. Critically, Glt proteins generate thrust when the entire multiprotein complex comes into contact with the substrate at specific areas in the cell envelope called focal adhesions. Each motility machine is anchored to the substrate by an unknown mechanism, and disassembles when it reaches the trailing pole of the cell.

It is intriguing that a single microbe has evolved two motility systems, and while S motility appears to be similar to twitching motility in *Pseudomonas aeruginosa,* neither system seems to be similar to the mechanism by which *Bacteroidetes,* cyanobacteria, or *Mycoplasma* spp. glide. This suggests that gliding motility has evolved independently multiple times. ◀ *Bacterial cytoskeleton (section 3.6); Twitching and gliding motility (section 3.8); Class* Mollicutes, *Phylum* Tenericutes *(section 21.3); Phylum* Bacteroidetes *includes important gut microbiota (section 21.7)*

Myxobacteria are found in soils worldwide. They are most commonly isolated from neutral soils or decaying plant material such as leaves and tree bark, and from animal dung. Although they grow in habitats as diverse as tropical rain forests and the Arctic tundra, they are most abundant in warm areas.

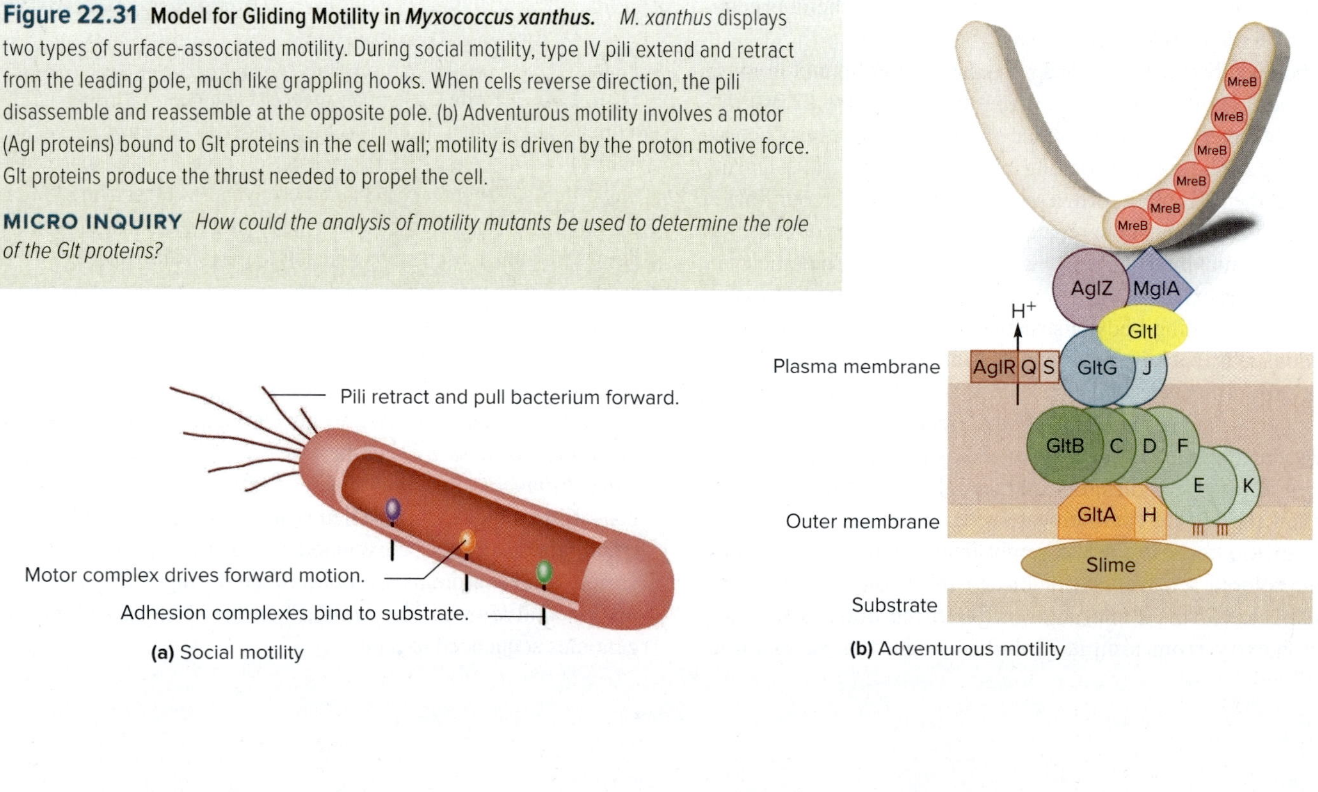

Figure 22.31 Model for Gliding Motility in *Myxococcus xanthus*. *M. xanthus* displays two types of surface-associated motility. During social motility, type IV pili extend and retract from the leading pole, much like grappling hooks. When cells reverse direction, the pili disassemble and reassemble at the opposite pole. (b) Adventurous motility involves a motor (Agl proteins) bound to Glt proteins in the cell wall; motility is driven by the proton motive force. Glt proteins produce the thrust needed to propel the cell.

MICRO INQUIRY *How could the analysis of motility mutants be used to determine the role of the Glt proteins?*

Comprehension Check

1. What general features are common to most δ-proteobacteria?
2. What is the role of fermentation in dissimilatory sulfate reduction? How do SRB make ATP?
3. What mechanisms do *Geobacter* spp. use to transfer electrons to extracellular insoluble metals? How does this compare with *Shewanella* spp?
4. Characterize *Bdellovibrio* spp. and outline the life cycle in detail. Why can they prey only on Gram-negative bacteria?
5. Compare the growth of a myxobacterium in nutrient-rich and nutrient-poor conditions. Why are extracellular signaling molecules required for fruiting body formation?

22.5 Class *Epsilonproteobacteria* Ranges from Pathogens to Deep-Sea Bacteria

After reading this section, you should be able to:

a. List two common human illnesses caused by ε-proteobacteria
b. Describe the unusual metabolism of *Helicobacter* spp. and explain how *H. pylori* can tolerate the acidic conditions within the stomach
c. Describe some of the unusual habitats that ε-proteobacteria inhabit

ε-proteobacteria are slender, Gram-negative rods, which can be straight, curved, or helical. Although *Bergey's Manual* places ε-proteobacteria as a class within the phylum *Proteobacteria,* their phylogenetic position as the most deeply branching clade has led to a proposed new phylum for these organisms. A taxonomic revision based on recent phylogenetic analysis that includes sequences from newly cultured and uncultured organisms may be on the horizon. Epsilonproteobacteria are physiologically and ecologically diverse, inhabiting extreme environments such as the acidic gastric mucosa, hydrothermal vents, and sulfidic caves. Two genera, *Campylobacter* and *Helicobacter,* contain members that are microaerophilic, motile, helical or vibrioid, Gram-negative rods. Table 22.8 summarizes some of the characteristics of these two genera.

The genus *Campylobacter* contains both nonpathogens and species pathogenic for humans and other animals. *C. foetus* causes reproductive disease and abortions in cattle and sheep. Other campylobacters are associated with conditions in humans ranging from septicemia (pathogens or their toxins in the blood) to enteritis (inflammation of the intestinal tract). Infection with some species frequently precedes the development of Guillain-Barré syndrome (GBS), the most common form of flaccid paralysis. It is thought that these bacteria trigger GBS by a phenomenon known as molecular mimicry. In this case, the structure of *Campylobacter* spp. lipopolysaccharide resembles components found on nerve cells. This causes the immune system to mistakenly attack the host's peripheral nervous system even after infection has cleared.

C. jejuni is found in the intestinal tract of animals. It causes an estimated 2.4 million human cases of *Campylobacter* gastroenteritis—inflammation of the stomach and intestine with subsequent diarrhea—in the United States each year. *C. jejuni* has a preferred niche in the lower intestinal tract, and virulence gene expression is induced by short chain fatty acids derived from the host gut microbiota. Studies with chickens, turkeys, and cattle have shown that as much as 50 to 100% of a flock or herd of these birds or other animals excrete *C. jejuni.* Campylobacteria are transmitted to humans by contaminated food and water, contact with infected animals, or anal-oral sexual activity.

The other ε-proteobacterial genus that contains pathogens is *Helicobacter.* There are at least 23 species of *Helicobacter,* all isolated from the stomachs and upper intestines of humans, dogs, cats, and other mammals. In developing countries, 70 to 90% of the population is infected; the rate in developed countries ranges from 25 to 40%. The major human pathogen is *Helicobacter pylori,* which causes gastritis, peptic ulcer disease, and gastric cancer. *H. pylori* is an obligate microaerophile. The only carbohydrate the bacterium can use is glucose, which it metabolizes by both respiratory and fermentative pathways. It uses the pentose phosphate pathway, which provides NADPH for biosynthesis, and the Entner-Doudoroff pathway to produce pyruvate. The microbe does not have a complete TCA cycle and appears to run the cycle "forward" (i.e., oxidatively) from oxaloacetate to α-ketoglutarate and "backward" (i.e., in the reductive direction) from oxaloacetate to succinate. Fumarate may act as the terminal electron acceptor during anaerobic respiration. It can also carry out aerobic respiration. Pyruvate is fermented to lactate, acetate, and formate during fermentation. As a resident of the gastric mucosa, it would seem that *H. pylori* is exposed to very acidic conditions, yet in the laboratory, it cannot be cultured below pH 4.5. The presence of urease enzymes in the cytoplasm and the cell surface helps explain this paradox. *H. pylori* burrows within the gastric mucosa, which is not as acidic as the lumen. There it depends on urease activity to convert urea to CO_2 and NH_3. The ammonia drives the local pH up while the microbe uses chemotaxis to maintain its position within the mucous membranes. Indeed, motility is required for colonization of the host. ▶| *Peptic ulcer disease and gastritis (section 39.3)*

ε-proteobacteria are also found in a variety of unusual ecosystems. For instance, filamentous microbial mats in anoxic, sulfide-rich cave springs are dominated by ε-proteobacteria (**figure 22.32**). Many have been cultured from marine hydrothermal vents; terrestrial forms have been isolated from groundwater, oil-field brines, limestone caves, and sulfidic springs. These include thermophilic, chemolithoautotrophic, chemoorganotrophic, and chemolithoheterotrophic species.

Chemolithoheterotrophic organisms use organic carbon for biosynthesis but inorganic compounds, such as sulfide and molecular hydrogen, as electron donors for energy conservation. The autotrophic species use the reductive TCA cycle for CO_2 fixation (*see figure 12.5*). Many ε-proteobacteria that

Figure 22.32 ε-Proteobacteria Dominate Filamentous Microbial Mats in a Wyoming Sulfidic Cave Spring.

inhabit hydrothermal vents are capable of nitrate respiration, that is, anaerobic respiration using nitrate as the terminal electron acceptor. This discovery suggests that microbial metabolism is more diverse in these interesting ecosystems usually thought of as sites of sulfur cycling. ▶| *Sulfide-based mutualisms (section 27.2)*

Comprehension Check

1. What property unifies ε-proteobacteria?
2. How does *H. pylori* survive the acidic conditions of the stomach?
3. Compare the use of sulfur and nitrate in the metabolism of ε-proteobacteria.

Key Concepts

22.1 Class *Alphaproteobacteria* Includes Many Oligotrophs

- Rickettsias are obligate intracellular parasites. They have numerous transport proteins in their plasma membranes and make extensive use of host cell nutrients, coenzymes, and ATP.
- The purple nonsulfur bacteria can grow anaerobically as photoorganoheterotrophs and aerobically as chemoorganoheterotrophs. When growing photosynthetically, purple bacteria localize the light-harvesting proteins and reaction centers on intracytoplasmic membranes (**figure 22.3**).
- The symbiont that ultimately evolved to form mitochondria was a proteobacterium. A conserved membrane organizing system points to a mechanism by which intracytoplasmic membranes developed into cristae as the site of energy conservation enzymes.
- Many proteobacteria have prosthecae or stalks or reproduce by budding. Two examples of budding or appendaged bacteria are *Hyphomicrobium* spp. (budding bacteria that produce swarmer cells) and *Caulobacter* spp. (bacteria with prosthecae and holdfasts) (**figures 22.6–22.8**).
- *Rhizobium* spp. carry out nitrogen fixation, whereas agrobacteria cause the development of plant tumors. Both are in the family *Rhizobiaceae* (**figures 22.9** and **22.10**).
- Proteobacteria that use 1-carbon compounds for carbon and energy are called methylotrophs; some of these microbes can use methane and are called methanotrophs. In bacteria, this is an aerobic process that produces the intermediate formaldehyde, which is either oxidized for energy or assimilated for anabolic purposes (**figures 22.11, 22.12,** and **22.23**).
- Nitrifying bacteria are chemolithotrophic aerobes that oxidize either ammonia to nitrite or nitrite to nitrate and are responsible for nitrification (**figure 22.13** and **table 22.2**).
- *Magnetospirillum* spp. assemble magnetosomes from iron on invaginated plasma membrane surfaces (**figure 22.14**).
- Members of the globally distributed marine SAR11 clade are the most abundant microorganisms on Earth. They are uniquely adapted to life in oligotrophic waters.

22.2 Class *Betaproteobacteria* Includes Chemoheterotrophs and Chemolithotrophs

- The genus *Neisseria* contains nonmotile, aerobic, Gram-negative cocci that usually occur in pairs. They colonize mucous membranes and cause several human diseases, including meningitis and gonorrhea.
- *Sphaerotilus, Leptothrix,* and several other genera contain species with sheaths, hollow tubelike structures that surround chains of cells without being in intimate contact with the cells.
- The order *Nitrosomonadales* includes two genera of nitrifying bacteria, including the ammonia-oxidizing microbe *Nitrosomonas europaea* (**table 22.2; figure 22.16**).
- Colorless sulfur bacteria such as *Thiobacillus* spp. oxidize elemental sulfur, hydrogen sulfide, and thiosulfate to sulfate while generating energy chemolithotrophically (**figure 22.17**).

22.3 Class *Gammaproteobacteria* Is the Largest Bacterial Class

- γ-proteobacteria are the largest subgroup of proteobacteria and have great variety in physiological types (**table 22.5**).
- Purple sulfur bacteria are anaerobes and usually photolithoautotrophs. They oxidize hydrogen sulfide to sulfur and deposit the granules internally (**figure 22.18**).
- Bacteria such as *Beggiatoa* and *Leucothrix* spp. grow in long filaments or trichomes. They have gliding motility. *Beggiatoa* spp. are primarily chemolithotrophs and *Leucothrix* spp. are chemoorganotrophs.
- Marine *Beggiatoa, Thioploca,* and *Thiomargarita* spp. are among the largest microbes. Members of the genus *Thiomargarita* form large intracellular vacuoles in which nitrate is stored (**figure 22.20**).
- The marine bacterium *Thiomicrospira crunogena* is a chemolithoautotroph that uses reduced sulfur compounds as a source of energy and electrons and O_2 as the terminal electron acceptor (**figure 22.22**).
- Members of the genus *Pseudomonas* are straight or slightly curved, Gram-negative, aerobic rods that are motile by one or several polar flagella and do not have prosthecae or sheaths.
- Pseudomonads participate in natural mineralization processes, are major experimental subjects, cause many diseases, and often spoil refrigerated food.
- The genus *Shewanella* is notable for species with metabolic flexibility and the ability to use extracellular, insoluble metals as terminal electron acceptors. They are among several microbes that have evolved strategies to accomplish this, including the use of electron shuttles and formation of nanowires (**figure 22.25**).
- The most important facultatively anaerobic, Gram-negative rods are found in three families: *Vibrionaceae, Enterobacteriaceae,* and *Pasteurellaceae* (**table 22.6**).
- *Enterobacteriaceae,* often called enterobacteria or enteric bacteria, are peritrichously flagellated or nonmotile, facultatively anaerobic, straight rods with simple nutritional requirements (**table 22.7**).
- Enteric bacteria are usually identified by a variety of physiological tests. *E. coli* performs a mixed acid fermentation, which depends on the enzymes pyruvate formate-lyase and formate dehydrogenase (**figure 22.27**).

22.4 Class *Deltaproteobacteria* Includes Chemoheterotrophic Anaerobes and Predators

- δ-proteobacteria are aerobic chemoorganotrophic bacteria that can use elemental sulfur and oxidized sulfur compounds as electron acceptors in anaerobic respiration. Other δ-proteobacteria are predatory aerobes (**table 22.8**).
- Sulfate-reducing bacteria are very important in sulfur cycling in ecosystems. *Desulfovibrio* spp. conserve energy by fermentation and a proton motive force (**figure 22.28**).
- *Bdellovibrio* spp. are aerobic curved rods with a sheathed polar flagellum that consume other Gram-negative bacteria and reproduce within the periplasmic space of their prey (**figure 22.29**).
- *Myxobacteria* spp. are Gram-negative, aerobic soil bacteria with gliding motility and a complex life cycle that leads to the production of dormant myxospores held within fruiting bodies (**figures 22.30** and **22.31**).

22.5 Class *Epsilonproteobacteria* Ranges from Pathogens to Deep-Sea Bacteria

- ε-proteobacteria are the smallest of the proteobacterial classes and include two important genera with pathogenic species: *Campylobacter* and *Helicobacter.* These bacteria are microaerophilic, motile, helical or vibroid, Gram-negative rods (**table 22.8**).
- Other ε-proteobacteria are chemolithoautotrophs from deep-sea hydrothermal vent ecosystems.

Active Learning

1. The advantages of metabolic flexibility in *Rhodospirillum rubrum* must offset the energetic burden of carrying all the structural and regulatory genes involved in these processes. What makes this trade-off possible?
2. Compare nitrification with denitrification: Which is found in heterotrophs and which is found in autotrophs? Are they both dissimilatory processes?
3. Why might the ability to form dormant cysts be of great advantage to an agrobacterium but not as much to a rhizobium?
4. Examine figures 22.17 and 22.28. Note that in both cases APS is formed, but in the case of *T. ferrooxidans*, it is involved in sulfide oxidation, whereas for *Desulfovibrio*, it is used during sulfate reduction. What other similarities and differences are present? Why must APS be used in both cases?
5. How might electron shuttles facilitate growth of bacteria in a biofilm that has become anoxic in the interior?
6. Intercellular communication has been studied in detail in *Myxococcus xanthus*. Review the life cycle of this developmentally complex microbe and indicate at what points during this process intercellular signaling molecules might be exchanged and the message they would convey.

7. *Bdellovibrio* preys on Gram-negative bacteria, including the well-studied *E. coli*. Since the morphology and biosynthesis of *E. coli* surface structures (lipopolysaccharide, capsule, flagella, pili) are well known, propose experiments to dissect the process of *Bdellovibrio* recognition and infection of *E. coli*. How would you determine which surface structures are recognized by *Bdellovibrio*?
8. Cold seeps are found near the ocean floor in regions where the Earth's tectonic plates meet. Chemicals seeping out of the Earth's crust form the basis of ecosystems that support a diverse community of microbes. Methane (natural gas) is common, and methanotrophs grow abundantly here. Methane concentrations can vary significantly over time in this environment. Methanotrophs must have mechanisms to adapt to starvation conditions when methane is limited, and mechanisms to recover when methane levels rise.

   A study of *Methyloprofundus sedimenti* showed that this microbe exhibited structural changes in its membranes and storage granules upon methane starvation. Reactions to capture methane occur in intracellular membranes. How does methane cross membranes?

   Storage granules that appear to contain glycogen were lost from the cells during starvation. In contrast, granules that resemble polyphosphate granules were not affected. Why would cellular stores of these two molecules respond differently to methane starvation?

   Read the original paper: Tavormina, P. L., et al. 2017. Starvation and recovery in the deep-sea methanotroph *Methyloprofundus sedimenti*. *Mol. Microbiol.* 103:242.
9. Nitrification requires the sequential action of ammonia-oxidizing bacteria (AOB) and nitrite oxidizing bacteria (NOB). Scientists exploring the coordination between these organisms looked for evidence of quorum sensing (QS) genes and signals. In a genome sequence, which genes are most likely to indicate the presence of a QS system? (You may wish to review figure 14.24.)

   Organisms that carried QS-related genes were screened for the production of acyl-homoserine lactones (AHLs). Culture supernatants of AOB or NOB were chemically extracted and concentrated, and applied to a strain of *Agrobacterium* that acted as a detector by changing color in the presence of AHLs. Only some organisms whose genomes contained QS-related genes responded positively in this assay. Why might this detection scheme fail to detect AHLs from some organisms?

   Read the original paper: Mellbye, B. L., et al. 2017. Acyl-homoserine lactone production in nitrifying bacteria of the genera *Nitrosospira, Nitrobacter,* and *Nitrospira* identified via a survey of putative quorum-sensing genes. *Appl. Env. Microbiol.* 83:e01540-17.

23

Gram-Positive Bacteria

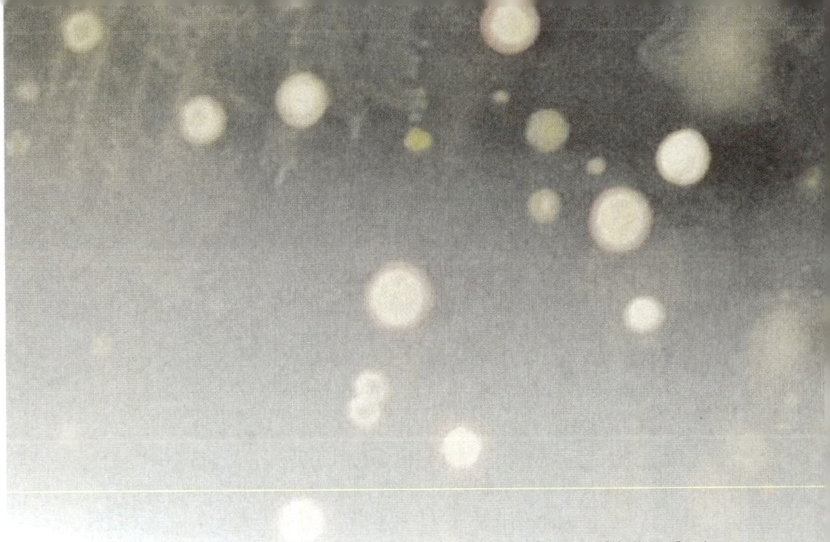

Courtesy of Kathleen Sandman

Antibiotic Production: Is It Actually Bacterial Chitchat?

Among bacteria, it is hard to find a genus whose constituents have given more to humankind than *Streptomyces.* When growing in their natural soil habitat, streptomycetes are key players in a microbial community that eats decaying plant and animal material. And like diners in a crowded restaurant, these microbes are "talking" up a storm, using a vocabulary of small molecules.

In fact, some microbiologists hypothesize that antibiotics are the language of inter- and intraspecies communication. If this is the case, then streptomycetes are very talkative indeed. About two-thirds of all antibiotics prescribed are made by these bacteria, including tetracycline and neomycin. Noting that streptomycetes produce and secrete such a wide range of natural products, pharmaceutical companies have screened countless isolates, discovering that they also make anticancer compounds (e.g., doxorubicin), antifungals (e.g., nystatin), and ivermectin, used to treat parasitic infections in humans and to prevent heartworm in dogs. All told, *Streptomyces* species produce about 10,000 bioactive products, but models suggest there may be a million yet undiscovered!

How can it be that after 75 years of research, so few streptomycete compounds are known? And how do we know this? One stark piece of evidence was revealed when the *S. coelicolor* genome was sequenced. This species was known to make four antibiotics, yet the genes for the production of 20 additional compounds were found. These genes are tightly regulated; to find and test new antibiotics, their regulatory "off" switch must be unlocked. This is where the ability of streptomycetes to communicate gets interesting. About 75 years ago, streptomycin, the first drug to treat tuberculosis, was purified from *S. griseus*. Ten years later, researchers discovered that streptomycin production depended on the release of tiny amounts of a small molecule called A-factor, a gamma-butyrolactone. If *S. griseus* cells fail to exchange A-factor, no streptomycin is produced. Much later it was discovered that a different group of chemicals, 2-alkyl-4-hydroxymethylfuran-3-carboxylic acids (or AHFCAs), controls the release of other antibiotics. And studies with natural isolates show that when cultured alone, many *Streptomyces* spp. lack discernible antibiotic activity. But when grown in the company of other streptomycetes, and allowed to exchange secreted molecules, new antibiotics are released.

What does all this bacterial chitchat mean? To the microbes it probably provides a way to ensure that enough of their kind are present. After all, the amount of antibiotic secreted by an individual isn't a very effective defense. To the world of drug discovery, it means that screening for new natural products is more complicated than simply growing a *Streptomyces* isolate in pure culture. New and creative methods of culturing and screening must be developed.

In this chapter, we survey the Gram-positive bacteria. Phylogenetic relationships among Gram-positive bacteria show that they can be divided into a low G + C group and a high G + C group based on the G + C content of their genomic DNA. By definition, high G + C Gram-positive bacteria have a DNA base composition above approximately 50 mol% G + C; these bacteria belong to the phylum *Actinobacteria*. Based on 16S rRNA sequence data, there are six classes in the phylum. Many actinobacteria, also called actinomycetes, are aerobic bacteria that are distinctive because most have filamentous cells called hyphae that differentiate to produce asexual spores. A summary of the actinobacteria described in this chapter is presented in **table 23.1**. The low G + C Gram-positive bacteria are placed in the phylum *Firmicutes* and divided into three classes: *Bacilli, Clostridia,* and *Negativicutes.* These microbes include endospore formers, bacteria notable as pathogens, and bacteria important for food production.

Readiness Check:

Based on what you have learned previously, you should be able to:

- ✓ Draw a Gram-positive cell wall and list its constituent parts (section 3.4)
- ✓ Differentiate between swimming and gliding motility (section 3.8)
- ✓ Outline the process of endospore formation and germination (section 3.9)
- ✓ Distinguish heterotrophy from autotrophy (section 11.1)
- ✓ List environmental factors that influence microbial growth (section 7.5)
- ✓ Compare and contrast the growth and physiology of obligate aerobes, facultative anaerobes, microaerophiles, and strict anaerobes (section 7.5)
- ✓ Explain how biofilms form and why they are important (section 7.6)
- ✓ Compare and contrast anaerobic and aerobic respiration, oxygenic and anoxygenic phototrophy, and substrate-level phosphorylation and oxidative phosphorylation (sections 11.4, 11.5, and 11.12)
- ✓ Draw an electron transport chain indicating the relative standard reduction potentials of the electron donor, acceptor, and electron transporters (sections 10.4 and 11.6)
- ✓ Summarize the generation of ATP and the fate of pyruvate in fermentation (section 11.8)

Table 23.1 Characteristics of Actinobacteria

Genus	Dimensions (μm), Morphology and Motility	Genome Size Range (Mb)	Oxygen Relationship	Other Distinctive Characteristics
Actinoplanes	Nonfragmenting, branching mycelium with little aerial growth; sporangia formed; motile spores with polar flagella	7.6–9.5	Aerobic	Hyphae often in palisade arrangement; highly colored; type II cell walls; found in soil and decaying plant material
Arthrobacter	0.8–1.2 × 1.0–8.0; young cells are irregular rods, older cells are small cocci; usually nonmotile	3.9–5.2	Aerobic	Have rod-coccus growth cycle; respiratory metabolism; catalase positive
Bifidobacterium	0.5–1.3 × 1.5–8; rods of varied shape, usually curved; nonmotile	1.9–2.8	Anaerobic	Cells can be clubbed or branched, pairs often in V arrangement; ferment carbohydrates to acetate and lactate but no CO_2; catalase negative
Corynebacterium	0.3–0.8 × 1.5–8.0; straight or slightly curved rods with tapered or clubbed ends; nonmotile	2.3–3.4	Aerobic or facultative	Cells often arranged in a V formation or in palisades of parallel cells; catalase positive and fermentative; polyphosphate granules
Frankia	0.5–2.0 in diameter; vegetative hyphae with limited-to-extensive branching and no aerial mycelium; multilocular sporangia formed	5.3–9.8	Aerobic to microaerophilic	Sporangiospores nonmotile; usually fix nitrogen; type III cell walls; most strains are symbiotic with angiosperm plants and induce root nodules
Micrococcus	0.5–2.0 in diameter; cocci in pairs, tetrads, or irregular clusters; usually nonmotile	2.3–2.6	Aerobic	Colonies usually yellow or red; catalase positive with respiratory metabolism; primarily on mammalian skin and in soil
Mycobacterium	0.2–0.6 × 1.0–10; straight or slightly curved rods, sometimes branched; acid-fast; nonmotile and nonsporing	3.3–7.0	Aerobic	Catalase positive; can form filaments that are readily fragmented; walls have high lipid content; in soil and water; some parasitic
Nocardia	0.5–1.2 in diameter; vegetative hyphae that fragment into rod-shaped and coccoid forms; nonmotile	6.2–9.4	Aerobic	Aerial hyphae formed; catalase positive; type IV cell walls; widely distributed in soil
Propionibacterium	0.2–1.5 × 1–5; pleomorphic nonmotile rods, may be forked or branched; nonsporing; nonmotile	2.5–2.8	Anaerobic to microaerophilic	Fermentation produces propionate and acetate, and often gas; catalase positive
Streptomyces	0.5–2.0 in diameter; vegetative mycelium extensively branched; aerial mycelium forms chains of three to many spores	6.5–11.9	Aerobic	Form discrete lichenoid or leathery colonies that often are pigmented; respiratory metabolism; use many organic compounds as nutrients; soil organisms

23.1 Class *Actinobacteria*

After reading this section, you should be able to:

a. Draw the life cycle of at least two different actinobacteria
b. Distinguish an exospore from an endospore
c. Discuss the taxonomic importance of cell wall structure
d. List common environments in which these microbes can be found
e. List one veterinary and two human diseases caused by actinobacteria
f. Explain snapping division and how this results in the palisade arrangement of cells
g. Relate the structure of mycolic acids to the properties of mycobacterial cell walls
h. Relate the unusual carbon and energy substrates used by bacteria in the family *Nocardiaceae* to their importance in bioremediation
i. Describe the unusual morphology of actinoplanete spores

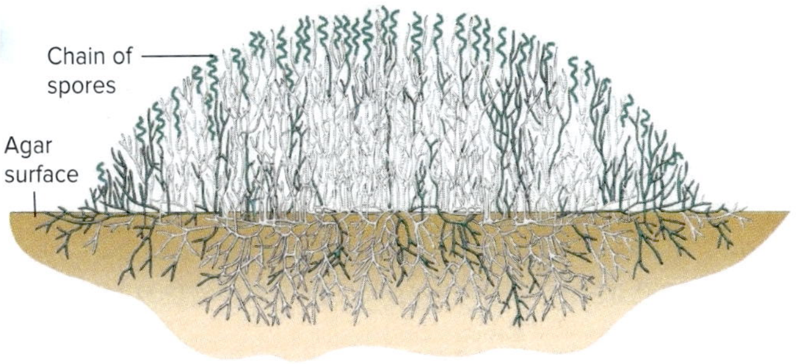

Figure 23.1 Diagram of an Actinobacterial Colony. The colony consists of actively growing aerial hyphae that cannibalize the substrate hyphae to obtain nutrients. Live hyphae are blue; dead are white.

j. Name two reasons propionibacteria are important to humans
k. Explain how streptomycete morphology helps this group thrive in soils
l. Discuss the origin of the antibiotic streptomycin and list five other medicinal agents made by streptomycetes
m. Describe the morphology of *Frankia* spp., relate this to their colonization of plants, and explain the advantage these bacterial symbionts confer to plants such as alder trees
n. Describe the morphology of members of *Bifidobacteriaceae*

Actinobacteria are fascinating microorganisms, of which table 23.1 lists characteristics of several genera. Although all are chemoorganoheterotrophs, several have been discovered to use atmospheric H_2 as an electron donor to maintain electron flow into the electron transport chain when growing in nutrient-depleted aerobic conditions. As described in the chapter opening story, they also produce metabolites used as antibiotics, anticancer drugs, antihelminthics, and drugs that suppress the immune system in patients who have received organ transplants. This practical aspect of actinobacteria is linked very closely to their mode of growth, which in many cases involves a complex life cycle. Many actinobacteria make filaments, called **hyphae,** instead of more familiar rods or cocci. Septae (crosswalls) usually divide the hyphae into long cells (20 μm and longer), each containing several nucleoids. When growing on a solid substratum such as soil or agar, filamentous actinobacteria develop a branching network of hyphae (**figure 23.1**). Hyphae grow on and into the substratum to form a dense mat called a **substrate mycelium.** In nature, the substratum is soil or other sediments. In the laboratory, substrate hyphae grow into agar, where the tangle of substrate hyphae imparts a leathery texture to the colonies of many actinomycetes. In many actinobacteria, substrate hyphae go on to differentiate into upwardly growing hyphae to form an **aerial mycelium** that gives the colony a fuzzy appearance. It is at this time that medically useful compounds are formed. Because the physiology of the actinobacteria has switched from actively growing vegetative cells into this special, secondary cell type, these compounds are often called **secondary metabolites.**

Over time, aerial hyphae septate to form chains of spores (*see figure 7.2*). When the septum completes constriction, individual exospores are released into the environment. These spores are considered **exospores** because they do not develop within a mother cell as do endospores. If the spores are located in a sporangium (a sheathlike structure), they may be called **sporangiospores.** The spores can vary greatly in shape (**figure 23.2**). Like spore formation in other bacteria, actinobacterial sporulation is in response to nutrient deprivation and chemical signaling between microbes. In general, actinobacterial spores are not particularly heat resistant but withstand desiccation well, so they are well suited to the sediment environment. Most actinobacteria are nonmotile, and spores dispersed by wind or animals may find a new habitat to provide needed nutrients. Interestingly, among the few motile actinobacteria, movement is confined to flagellated spores. ◀ *Bacterial endospores are a survival strategy (section 3.9)*

Because the sequences of the actinobacterial 16S rRNAs do not distinguish genera, other molecules are used for taxonomic

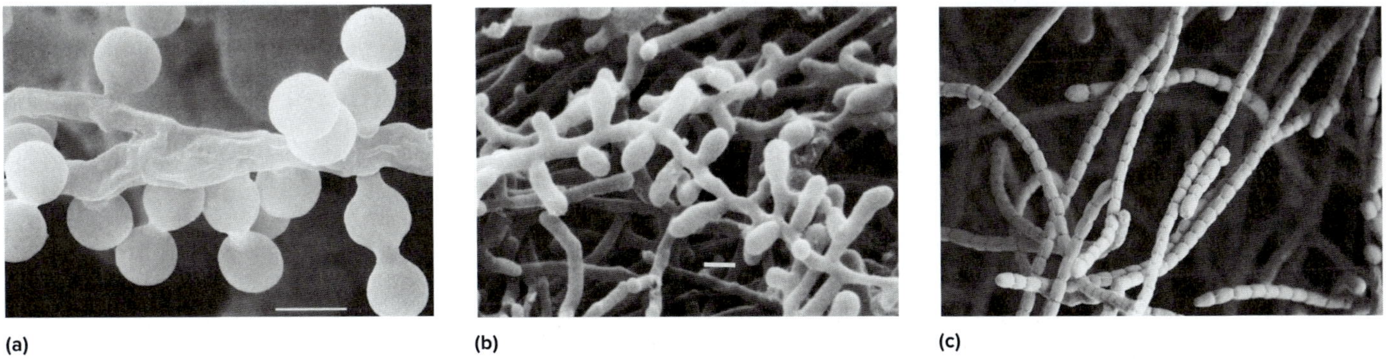

Figure 23.2 Examples of Actinomycete Spores as Seen in the Scanning Electron Microscope. (a) Spores of *Pseudonocardia yunnanensis* (bar = 1 μm). (b) Single spores of *Saccharomonospora viridis* are closely spaced along aerial hyphae. This organism causes farmer's lung. (c) Spore chain of *Streptomyces venezuelae*, which produces the antibiotic chloramphenicol.

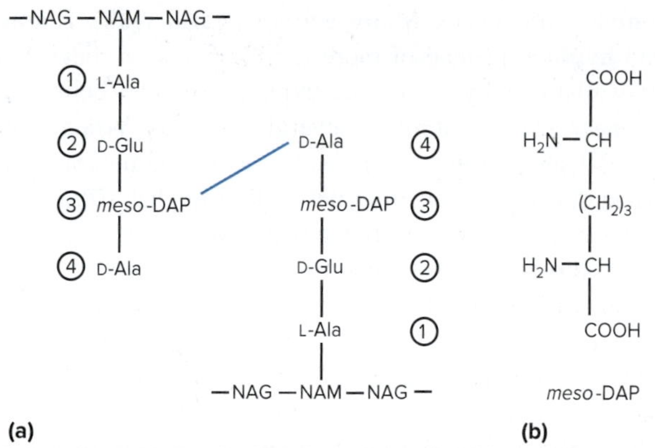

Figure 23.3 Actinobacterial Peptidoglycan Structure. (a) Peptidoglycan with *meso*-diaminopimelic acid (DAP) in position 3 and a direct cross-link between positions 3 and 4 of the peptide subunits is shown. NAM, N-acetylmuramic acid; NAG, N-acetylglucosamine. (b) One stereoisomer of DAP.

MICRO INQUIRY *Refer to table 23.2 and determine how the cell walls of* Streptomyces *and* Actinoplanes *spp. differ from that shown here.*

purposes. Actinobacterial cell wall composition varies greatly among different groups and remains of taxonomic importance. Eight major cell wall types can be distinguished based on the structure of the peptidoglycan and cell wall sugar content (**figure 23.3** and **table 23.2**; *also see figure 3.17*). Peptidoglycan containing unusual sugars can be used to categorize actinomycetes. In addition, the identity of the amino acids in the interpeptide bridge is a helpful discriminator. Besides the amino acids used in protein synthesis, other amino acids may be present. These include ornithine, diaminobutyric acid, and diaminopimelic acid (DAP) which occurs in several chiral forms. Some other taxonomically valuable properties include the color of the mycelium and sporangia, the surface features and arrangement of spores, the phospholipid composition of cell membranes, the type of isoprenoid quinones, cellular fatty acid patterns, and spore heat resistance. ◀ *Peptidoglycan structure (section 3.4)*

Order *Actinomycetales*

The order *Actinomycetales* includes the genera *Actinomyces, Actinobaculum, Arcanobacterium,* and *Mobiluncus.* Most are irregularly shaped, nonsporing, Gram-positive rods with aerobic, facultative, or microaerophilic metabolism. The rods may be straight or slightly curved, and usually have swellings, club shapes, or other deviations from normal rod-shape morphology. The *Actinomycetales* make a significant contribution to degrading complex polysaccharides in soil. They synthesize and secrete enzymes to hydrolyze molecules like chitin, pectin, and cellulose.

Members of the genus *Actinomyces* are either straight or slightly curved rods, or slender filaments with true branching (**figure 23.4**). The rods and filaments may have swollen or clubbed ends. They are either facultative or strict anaerobes that require CO_2 for optimal growth. *Actinomyces* species are normal inhabitants of mucosal surfaces of humans and other warm-blooded animals; the oral cavity is their preferred habitat. *A. bovis* causes lumpy jaw in cattle. *A. israelii* is most commonly responsible for actinomycoses, ocular infections, and lumpy jaw in humans.

Table 23.2 Actinobacterial Cell Wall Types

Cell Wall Type	Amino Acids in Interpeptide Bridge	Characteristic Sugars	Representative Genera
I	L, L-DAP[1], glycine	none	*Streptomyces, Nocardioides*
II	*meso*-DAP, glycine	arabinose, xylose	*Actinomyces, Actinoplanes*
III	*meso*-DAP	madurose[3]	*Actinomadura, Frankia*
IV	*meso*-DAP	arabinose, galactose	*Micropolyspora, Nocardia*
V	lysine, ornithine	none	*Actinomyces*
VI	aspartic acid (v)[2]	galactose	*Actinomyces, Microbacterium*
VII	diaminobutyric acid, glycine, lysine (v)	none	*Agromyces, Clavibacter*
VIII	ornithine	none	*Aureobacterium*

1 DAP, diaminopimelic acid.
2 v, presence is variable.
3 Madurose is 3-*O*-methyl-D-galactose.

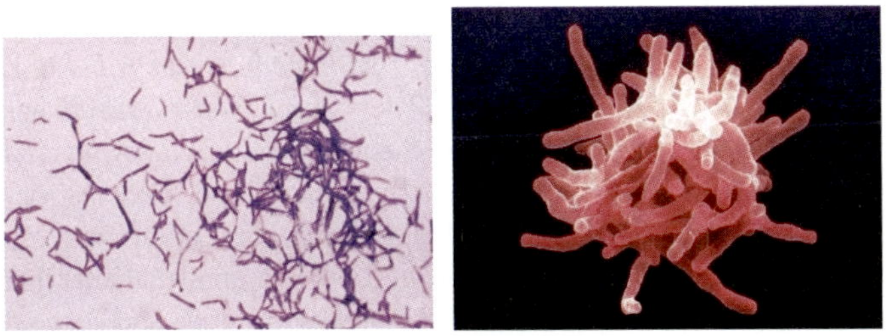

(a) (b)

Figure 23.4 Representatives of the Genus *Actinomyces*. (a) *Actinomyces* sp.; Gram stain shows branching forms (×400). (b) *Actinomyces* sp.: color-enhanced scanning electron micrograph (×2,500).
(a) ©Biophoto Associates/Science Source; (b) ©David M. Phillips/Science Source

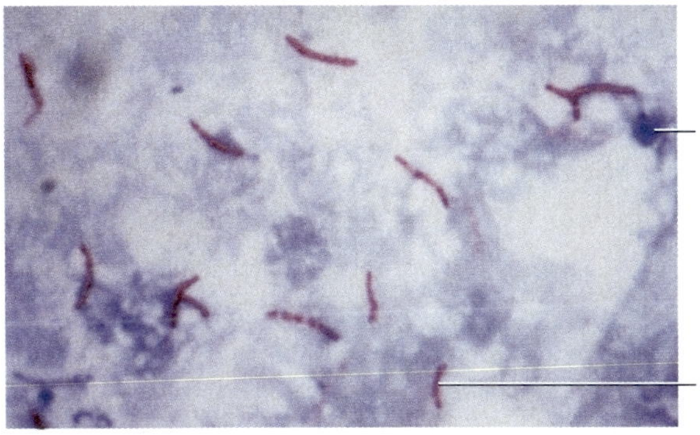

Figure 23.6 Mycobacteria. *Mycobacterium tuberculosis*, acid-fast stain.
Source: CDC/Dr. George P. Kubica

Order *Corynebacteriales* Includes Important Human Pathogens

The order *Corynebacteriales* includes three important genera: *Corynebacterium, Mycobacterium,* and *Nocardia*. Species in these genera produce long-chain fatty acids, called mycolic acids. These lipids are a virulence factor but are essential for viability only in mycobacteria.

The genus *Corynebacterium* includes aerobic and facultative, catalase-positive, straight to slightly curved rods. Club-shaped cells are common. Both corynebacteria and arthrobacters have a two-layered cell wall that leads to **snapping division** (**figure 23.5*a***). This occurs because during cell division, only the inner layer grows inward to generate a septum dividing new cells. As the septum thickens, tension rises on the outer wall layer, which still holds the two cells together. Eventually increasing tension ruptures the outer layer at its weakest point, and a snapping movement tears the outer layer apart around most of its circumference. The new cells now rest at an angle to each other and are held together by the remaining portion of the outer layer, which acts as a hinge. This results in angular arrangements of the cells, called a **palisade arrangement** in which rows of cells are lined up side by side (figure 23.5*b*). Although some species are harmless saprophytes, many corynebacteria are plant or animal pathogens. For example, *C. diphtheriae* is the causative agent of diphtheria in humans. ▶| *Diphtheria (section 39.1)*

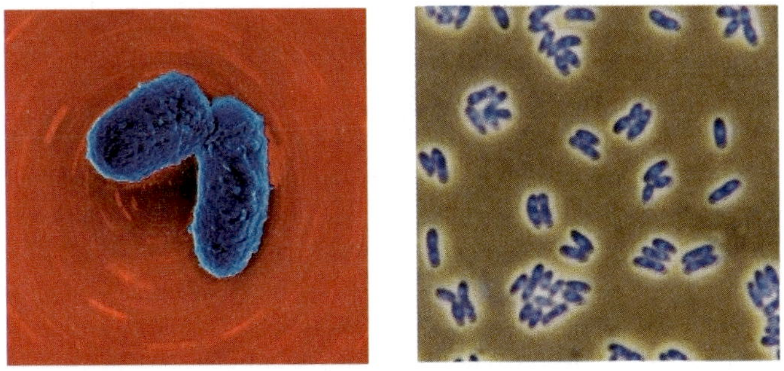

(a) (b)

Figure 23.5 *Corynebacterium diphtheriae* Undergoes Snapping Division, Giving Rise to Cells in Palisades. (a) Following snapping division, daughter cells remain attached and at an angle (×28,000). (b) This angular association of pairs of cells results in a palisade arrangement (×1,000).
(a) ©Dr. Gary Gaugler/Science Source; (b) ©M. Wachi/Society for Actinomycetes Japan

The family *Mycobacteriaceae* contains the genus *Mycobacterium,* which is composed of slightly curved or straight rods that sometimes branch or form filaments (**figure 23.6**). Mycobacterial filaments readily fragment into rods and coccoid bodies. They are aerobic and catalase positive. Mycobacteria grow very slowly and must be incubated for 2 to 40 days after inoculation on a complex medium to form a visible colony.

Although mycobacteria are considered Gram positive, their cell walls possess an outer membrane, and their peptidoglycan is held within the periplasm (**figure 23.7*a***). The outer membrane is constructed of tightly packed, covalently linked **mycolic acids** and noncovalently attached shorter lipids. Mycolic acids are complex fatty acids with an invariant C26 fatty acid that is attached to a longer, variable fatty acid (figure 23.7*c*). Mycolic acids in the outer membrane are covalently linked to the peptidoglycan by arabinogalactan, a polysaccharide made of arabinose and galactose subunits. The cell wall is extremely hydrophobic and impenetrable to most organic molecules, including antibiotics. In fact, mycobacteria have evolved porins that enable the entry of small molecules (*see figure 3.26*). In addition, the mycobacterial outer membrane possesses a type VII secretion system. The lipid content makes the cell wall **acid-fast** (carbol fuchsin dye cannot be removed from the cell by acid alcohol treatment). |◀ *Differential staining (section 2.3); Protein maturation and secretion (section 13.8)*

Although some mycobacteria are free-living saprophytes, they are best known as animal pathogens. *M. bovis* causes tuberculosis in cattle, other ruminants, and primates. Because this bacterium can produce tuberculosis (TB) in humans, dairy cattle are tested for the disease yearly; milk pasteurization kills the pathogen. Prior to widespread milk pasteurization, contaminated milk was a significant source of transmission. Today *M. tuberculosis* is

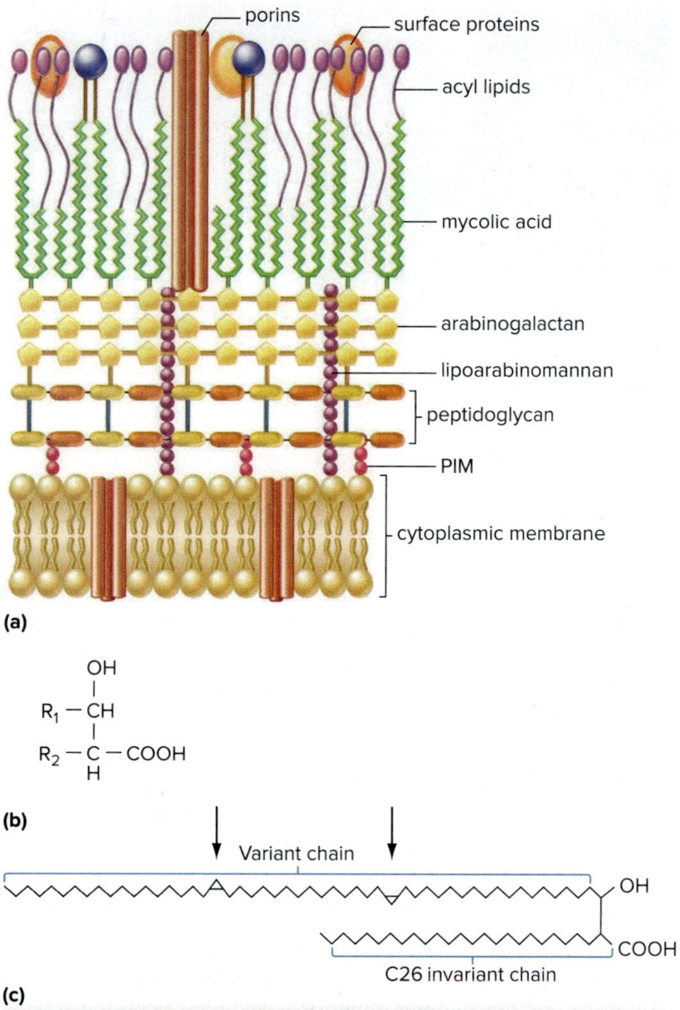

Figure 23.7 Mycobacterial Cell Wall. (a) The peptidoglycan is tethered to the plasma membrane by the glycolipids lipoarabinomannan and phosphatidylinositol mannosides (PIM). Arabinogalactan (D-arabinose and D-galactose), which is linked to mycolic acids, is also attached to the peptidoglycan, and is covered by a lipid bilayer. Like Gram-negative bacteria, mycobacteria have porins to regulate the passage of small molecules across the outer membrane. (b) The generic structure of mycolic acids, a family that includes over 500 different types. R1 is the variant chain; R2 is the C26 invariant chain. (c) An example mycolic acid with two cyclopropane rings (arrows).

MICRO INQUIRY *Why do* Mycobacterium *spp. have porins?*

the chief source of TB in humans. It is estimated that about one-third of the world's population is infected with *M. tuberculosis* (*see figure 39.5*). Leprosy is also a mycobacterial human disease; it is caused by *M. leprae*. Finally, *M. avium* and the related species *M. intracellulare* comprise *M. avium* complex, which has become a common mycobacterial disease in developed countries, particularly among individuals with compromised immunity. ◀ *Comparative genomics (section 18.7)* ▶ Mycobacterium *infections (section 39.1); Mycobacterial skin infections (section 39.3)*

Figure 23.8 ***Rhodococcus.***
Source: CDC/Dr. David Berd

The family *Nocardiaceae* has two notable genera, *Nocardia* and *Rhodococcus.* These bacteria develop a substrate mycelium with hyphae of 0.5 to 1.2 µm in diameter that readily breaks into rods and coccoid fragments (**figure 23.8**). Almost all are strict aerobes. These and related bacteria that resemble members of the genus *Nocardia* (named after Edmond Nocard [1850–1903], French bacteriologist and veterinary pathologist) are collectively called **nocardioforms.**

Nocardia spp. are distributed worldwide in soil and aquatic habitats. They are involved in the degradation of hydrocarbons and waxes, and contribute to the biodeterioration of rubber joints in water and sewage pipes. Although most are free-living saprophytes, some species, particularly *N. asteroides,* are opportunistic pathogens that cause nocardiosis in humans and other animals. People with impaired immune responses are most at risk. The lungs are typically infected; other organs and the central nervous system may be invaded as well.

Rhodococcus contains species widely distributed in soils and aquatic habitats. They are of considerable interest because they degrade an enormous variety of molecules such as petroleum hydrocarbons, detergents, benzene, polychlorinated biphenyls (PCBs), and various pesticides.

Order Micrococcales

Of the wide variety of genera within the order *Micrococcales*, two of the best known are *Micrococcus* and *Arthrobacter.* The genus *Micrococcus* contains aerobic, catalase-positive cocci that occur mainly in pairs, tetrads, or irregular clusters and are usually nonmotile (**figure 23.9**). Unlike many other actinobacteria, micrococci do not undergo morphological differentiation. Their colonies often are yellow, orange, or red. They are widespread in soil, water, and on mammalian skin.

The genus *Arthrobacter* contains aerobic, catalase-positive rods with respiratory metabolism. Their most distinctive feature is a rod-coccus growth cycle (**figure 23.10**). In exponential phase, these bacteria are irregular, branched rods that undergo snapping division. As they enter stationary phase, the cells change to a coccoid form. Upon transfer to fresh medium, the coccoid cells differentiate to form actively growing rods. Although arthrobacters often are isolated from fish, sewage, and plant surfaces, like many other actinobacteria, their most important habitat is soil, where they constitute a significant component of the culturable microbial community. They are well adapted to this niche because they are resistant to desiccation and nutrient deprivation, even though they do not form spores. Arthrobacters are unusually flexible nutritionally and can even degrade some herbicides and pesticides.

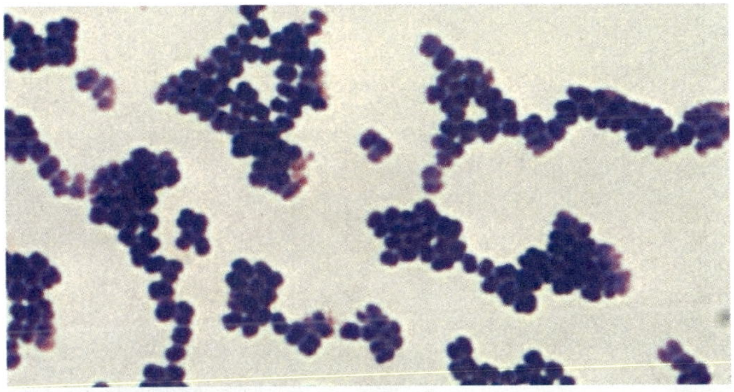

Figure 23.9 ***Micrococcus.***
Source: CDC/Dr. Richard Facklam

MICRO INQUIRY *What makes the morphology of* Micrococcus *distinct from most other actinobacteria discussed in this chapter?*

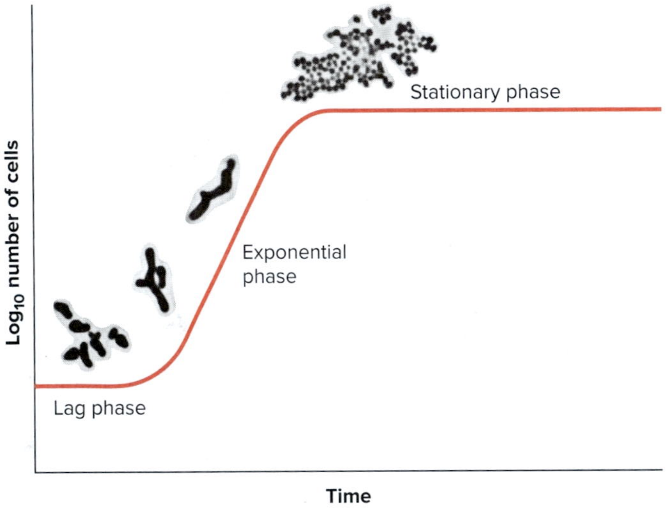

Figure 23.10 **The Rod-Coccus Growth Cycle of *Arthrobacter* spp.** When inoculated into fresh medium at 25°C, coccoid cells of about 1 μm in diameter (like those in stationary phase) swell and elongate to form branching rods of about 7–10 μm in length in exponential phase. As nutrients are depleted and cells enter stationary phase, the cells become spherical and form clusters.

Comprehension Check

1. Describe the function of the substrate mycelium, aerial mycelium, and exospore. Explain how these structures confer a survival advantage.
2. How can cell wall structure and sugar content be used to classify actinobacteria?
3. Why are actinobacteria of such practical interest?
4. List two genera that are commonly associated with humans.
5. Compare the morphology and cell wall structure of corynebacteria with that of mycobacteria.
6. What is snapping division? How does it lead to palisade formation?
7. What implications does the complex hydrophobic mycobacterial cell wall have in terms of treating diseases caused by these bacteria?
8. What is a nocardioform? List two reasons these microbes are relevant to humans.

Order *Micromonosporales* Includes Bacteria That Produce Morphologically Diverse Spores

The order *Micromonosporales* contains the genera *Micromonospora, Dactylosporangium, Pilimelia,* and *Actinoplanes.* These bacteria are often called actinoplanetes (Greek *actinos,* a ray or beam; *planes,* a wanderer). They have an extensive substrate mycelium but usually lack all but a rudimentary aerial mycelium. Often the hyphae are highly colored, and diffusible pigments may be produced. Spores are usually formed within a sporangium raised above the surface of the substratum at the end of a special hypha called a sporangiophore. The spores are quite unusual because in some species, they demonstrate flagellated motility. Flagella attach to a new substrate, where spores remain until triggered to germinate. The morphology and arrangement of spores are also variable. For instance, *Actinoplanes and Pilimelia* spp. have sporangia that may have only a few spores, while other species produce thousands of spores per sporangium (**figure 23.11**). The spores are arranged in coiled or parallel chains. *Dactylosporangium* spp. form club-shaped, finger-like, or pyriform sporangia with one to six spores (figure 23.11*c*). *Micromonospora* bacteria bear single spores, which often occur in branched clusters of sporophores.

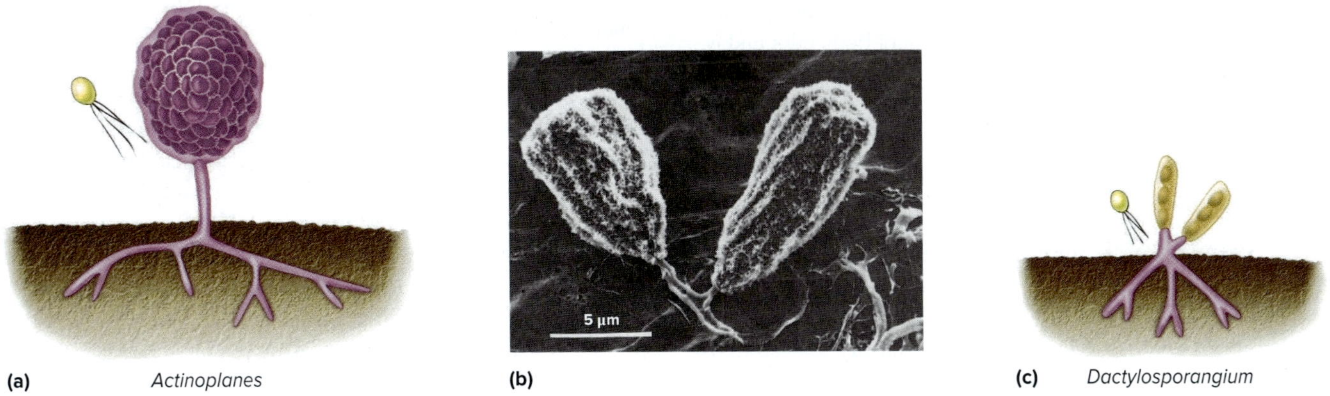

Figure 23.11 ***Micromonosporaceae.*** Actinoplanete morphology. (a) *Actinoplanes* structure. (b) A scanning electron micrograph of mature *Actinoplanes capillaceus* sporangia. (c) *Dactylosporangium* structure.
(b) ©S. Amano & S. Miyadoh/Society for Actinomycetes Japan

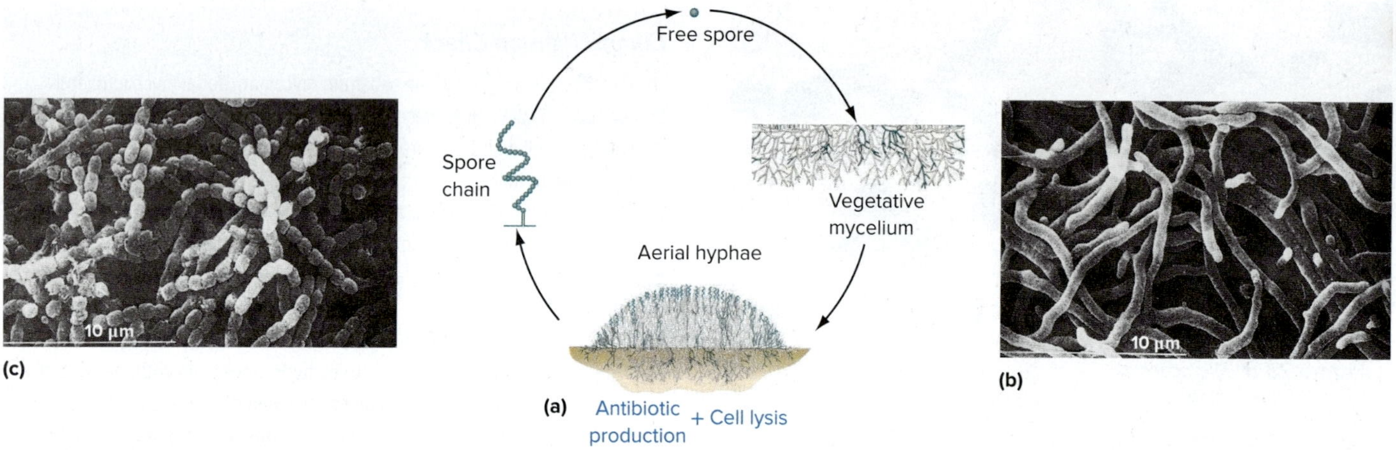

Figure 23.12 ***Streptomyces* Life Cycle.** (a) Streptomycetes form three types of cells: Exospores germinate to produce vegetative (also called substrate) hyphae, which then differentiate into aerial hyphae. These give rise to chains of exospores. Scanning electron micrographs are shown of (b) *S. coelicolor* vegetative hyphae and (c) chains of exospores.

Actinoplanetes grow in almost all soil habitats, ranging from forest litter to beach sand. Some flourish in freshwater, particularly in streams and rivers (probably because of abundant oxygen and plant debris), others are marine. *Actinoplanes* spp. are the source of the drug acarbose used to treat type 2 diabetes. The soil-dwelling species have an important role in the decomposition of plant and animal material. *Pilimelia* spp. grow in association with keratin. *Micromonospora* spp. actively degrade chitin and cellulose, and produce antibiotics such as gentamicin.

Order *Propionibacteriales*

The order *Propionibacteriales* contains the genus *Propionibacterium,* pleomorphic, nonmotile, nonsporing rods that are often club-shaped with one end tapered and the other end rounded. The most famous members of this genus are used to produce Swiss cheese, where its gaseous products of fermentation are responsible for the holes. Cells also may be coccoid or branched. They can be single, in short chains, or in clumps. They are facultatively anaerobic or aerotolerant; lactate and sugars are fermented to produce large quantities of propionic and acetic acids, and often carbon dioxide. In animals, these bacteria are found growing in the digestive tract and on the skin. *P. acnes* is involved with the development of body odor and is thought to cause acne vulgaris. It is also an emerging opportunistic pathogen associated with implant infections.

Order *Streptomycetales:* An Important Source of Antibiotics

The order *Streptomycetales* has only one family, *Streptomycetaceae,* and three genera, the most important of which is *Streptomyces.* This is a large genus; there are over 500 species. All are considered strict aerobes. Members of this family and similar bacteria are often called **streptomycetes** (Greek *streptos,* bent or twisted; *myces,* fungus).

Like Gram-negative myxobacteria, streptomycetes are a good example of true multicellularity among bacteria. This is illustrated by their life cycle (**figure 23.12**). Single, nonmotile exospores germinate to produce long substrate hyphae. These form a dense substrate mycelium. Upon receipt of cues that indicate nutrient limitation and intercellular signals, the substrate hyphae grow up into the air to form aerial hyphae. Aerial hyphae obtain nutrients by cannibalizing the filaments within the substrate mycelium. It is at this point in the life cycle that the stressed but still surviving substrate hyphae produce antibiotics and other secondary metabolites to compete with other soil microbes. As aerial hyphae continue to grow, they curl and septate in regular intervals. Ultimately the aerial hyphae become chains of spores that pinch off into single exospores to begin the cycle anew. Interestingly, the process of exospore formation is dependent on the cytoskeletal proteins FtsZ and MreB, but vegetative growth is not, because substrate hyphae grow by extending their tips rather than by binary fission. This results in long filaments with only occasional septae. Because multiple nucleoids are enclosed in the same cell, this is called syncytial growth. ◀ *Cytokinesis (section 7.2)* ▶ *Antibiotics (section 42.1)*

Streptomycetes are very important both ecologically and medically. The natural habitat of most streptomycetes is the soil, where they may constitute from 1 to 20% of the culturable population. In fact, the odor of moist earth is largely the result of streptomycete production of volatile substances such as geosmin. Streptomycetes play a major role in mineralization—the degradation of organic matter to inorganic elements. As substrate hyphae grow into the earth, they secrete a variety of exoenzymes enabling the degradation of resistant substances such as pectin, lignin, chitin, keratin, latex, agar, and aromatic compounds. ▶ *Diverse microorganisms inhabit soil (section 31.2)*

Selman Waksman's discovery that *S. griseus* (**figure 23.13*a***) produces streptomycin, the first drug to cure TB, set off a massive search resulting in the isolation of new *Streptomyces* species that produce other compounds of medicinal importance. Since that time, streptomycetes have been found to produce over 10,000 bioactive compounds; a few are listed in **table 23.3**.

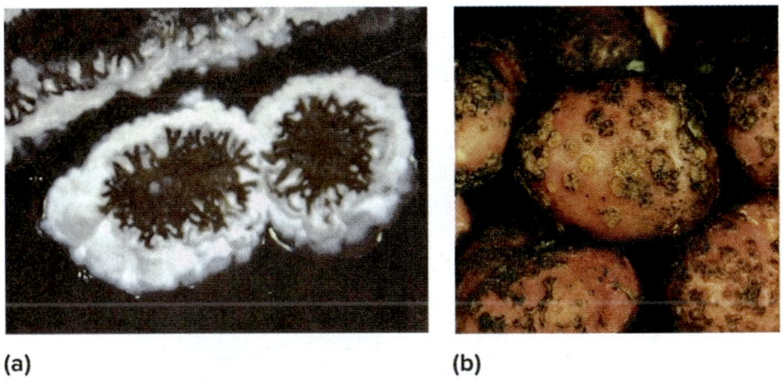

Figure 23.13 Streptomycetes of Practical Importance. (a) *Streptomyces griseus*. Colonies of the actinomycete that produces streptomycin. Each colony is about 0.5 cm in diameter. (b) *S. scabies* growing on a potato.
(a) ©Roger Greenwell; (b) ©Nigel Cattlin/Alamy Stock Photo

Table 23.3	**Examples of Natural Products Made by Streptomycetes**	
Microbe	**Natural Product**	**Application**
S. mediterranei	Rifamycin	Antibiotic: transcription inhibitor
S. rimosus	Tetracycline	Antibiotic: protein synthesis inhibitor
S. venezuelae	Chloramphenicol	Antibiotic: protein synthesis inhibitor
S. clavuligerus	Clavulanic acid	β-lactamase inhibitor
S. nodosus	Amphotericin B	Antifungal
S. noursei	Nystatin	Antifungal
S. peucetius	Daunorubicin, doxorubicin epirubicin	Anticancer
S. verticillus	Bleomycin	Anticancer

Hundreds of these natural products are now used in medicine and industry; about two-thirds of the antimicrobial agents used in human and veterinary medicine are derived from streptomycetes. Antibiotic-producing bacteria have "immunity" genes that encode proteins (or small noncoding RNA) that make them resistant to the antibiotics they synthesize. ◀ *Antimicrobial chemotherapy (chapter 9)*

The genome of *Streptomyces coelicolor,* which produces four antibiotics, is one of the largest bacterial genomes (8.67 Mbp). Its large number of genes (7,825) no doubt reflects the number of proteins required to undergo a complex life cycle. Many genes are devoted to regulation, with an astonishing 65 predicted RNA polymerase sigma transcription factors and about 80 two-component regulatory systems. The ability to exploit a variety of soil nutrients is also demonstrated by the presence of a large number of ABC transporters, the Sec and two Tat protein translocation systems, and over 50 secreted degradative enzymes. ◀ *Protein maturation and secretion (section 13.8); Regulation of transcription initiation saves considerable energy and materials (section 14.2)*

Although most streptomycetes are nonpathogenic saprophytes, a few are associated with plant and animal diseases. *Streptomyces scabies* causes scab disease in potatoes and beets (figure 23.13*b*). *S. somaliensis* and *S. sudanensis* are the only streptomycetes known to be pathogenic to humans. They are associated with actinomycetoma, a rare infection of subcutaneous tissues that causes swelling, abscesses, and even bone destruction if untreated. *S. albus* and other species have been isolated from patients with various ailments and may be pathogenic.

Order *Streptosporangiales*

The order *Streptosporangiales* includes maduromycetes, so named because they produce the sugar derivative **madurose** (3-*O*-methyl-D-galactose). Aerial hyphae bear pairs or short chains of spores, and the substrate hyphae are branched (**figure 23.14**). Members of some genera form sporangia; spores are not heat resistant. *Thermomonospora* spp. produce single spores on aerial hyphae or on both aerial and substrate hyphae. They have been isolated from moderately high-temperature habitats such as compost piles and hay and can grow at 40° to 48°C.

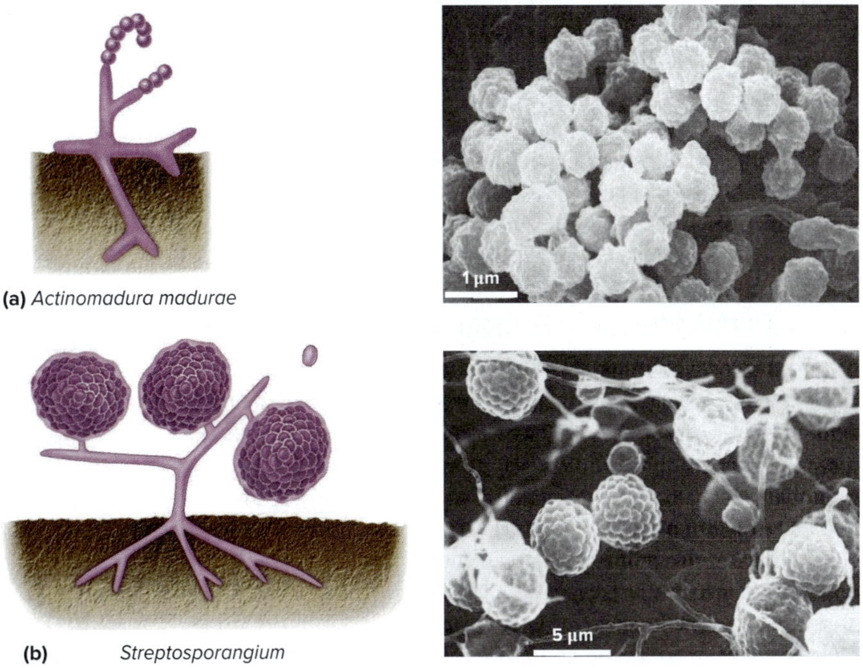

Figure 23.14 Maduromycetes. (a) *Actinomadura madurae* morphology; illustration and electron micrograph of a spore chain. (b) *Streptosporangium roseum* morphology; illustration and micrograph of sporangia and hyphae; SEM.
(a) ©R.H. Berg/Society for Actinomycetes Japan; (b) ©S. Amano, S. Miyadoh & T. Shomura/Society for Actinomycetes Japan

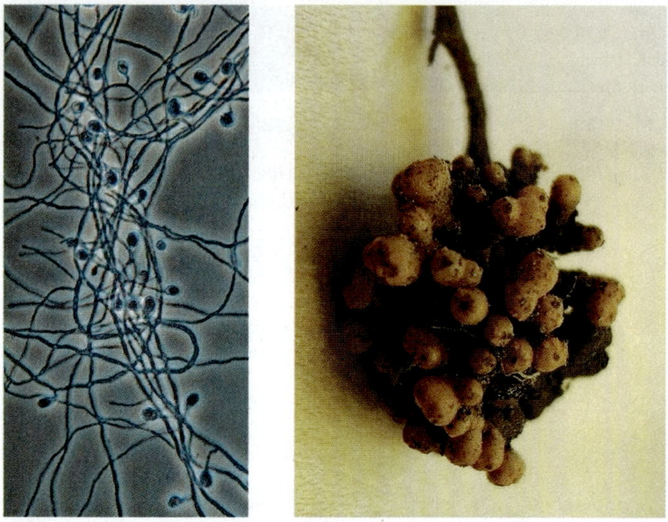

(a) (b)

Figure 23.15 ***Frankia.*** (a) A phase contrast micrograph showing hyphae and multilocular spores. (b) Nodules of the alder *Alnus rubra*.

Order *Frankiales:* Bacteria That Fix N_2 in Trees

The order *Frankiales* includes bacteria in the genus *Frankia*, which form multilocular sporangia, characterized by clusters of spores when hyphae divide both transversely and longitudinally. (Multilocular means having many cells or compartments.) *Frankia* spp. form nonmotile spores in a sporogenous body (**figure 23.15**). They grow in symbiotic association with the roots of at least eight families of higher nonleguminous plants (e.g., alder trees). They are microaerophilic and can fix atmospheric nitrogen.

Frankia spp. fix nitrogen so efficiently that a plant such as an alder can grow in the absence of combined nitrogen (e.g., NO_3^-) (figure 23.15*b*). Within the nodule, these bacteria form branching hyphae with globular vesicles at their ends, which may be the sites of nitrogen fixation. Their nitrogenase resembles that of rhizobia in that it is oxygen sensitive and requires molybdenum and cobalt. ▶| Frankia *form actinorhizae (section 31.3)*

Order *Bifidobacteriales:* Pathogens and Probiotics

The order *Bifidobacteriales* includes members of the genera *Falcivibrio* and *Gardnerella,* which are found in the human genital/urinary tract. *Gardnerella* spp. are thought to be a major cause of bacterial vaginosis. Bifidobacteria have gained recent attention as more is learned about the importance of human gut microflora. They are nonmotile, nonsporing, Gram-positive rods of varied shapes that are slightly curved and clubbed; often they are branched (**figure 23.16**). The rods can be single or in clusters and V-shaped pairs. They are anaerobic and actively ferment carbohydrates to produce acetic and lactic acids but not carbon dioxide. They are found in the mouth and intestinal tract of warm-blooded vertebrates, in sewage, and in insects. *B. bifidum* is a pioneer colonizer of the human intestinal tract, especially when babies are breast fed. *Bifidobacterium* spp. are sold as

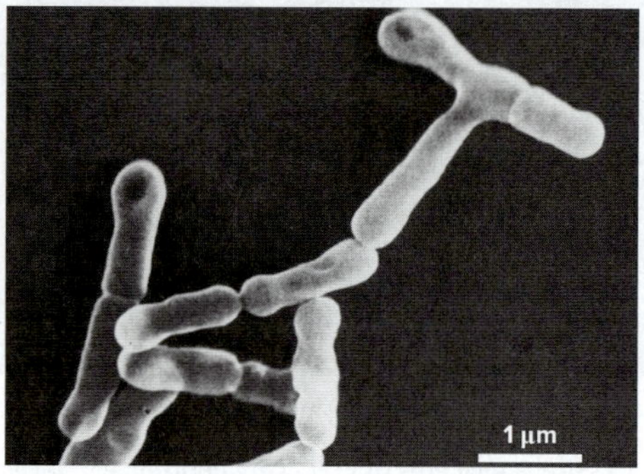

Figure 23.16 ***Bifidobacterium.*** *Bifidobacterium bifidum*; scanning electron photomicrograph.

probiotic agents (particularly in yogurt) because they are thought to impart a health benefit. ▶| *The microbiome develops from birth to adulthood (section 34.2); Microbiome manipulation can be therapeutic (section 34.5)*

Comprehension Check

1. List the distinguishing properties of actinoplanetes. What other actinomycetes form motile spores?
2. Where would you most likely find a member of the genus *Propionibacterium* in your home or dormitory?
3. List three types of natural products made by members of the genus *Streptomyces.*
4. Describe three ways in which *Streptomyces* spp. are of ecological importance.
5. Briefly describe the defining properties of genera in the order *Streptosporangiales.* What is madurose?
6. Why are *Frankia* spp. considered ecologically important?
7. Examine the ingredients list on a commercial yogurt cup; are bifidobacteria present?

23.2 Class *Bacilli:* Aerobic Endospore-Forming Bacteria

After reading this section, you should be able to:

a. List terminal electron acceptors and fermentation products produced by *Bacillus* spp. and staphylococci
b. Discuss reasons that make *Bacillus subtilis* an important model organism
c. List three reasons *Bacillus* spp. are of practical importance
d. Summarize the evolution and emergence of MRSA, and list at least two diseases caused by *Staphylococcus aureus*
e. Describe the structure and medical importance of *Listeria monocytogenes*
f. Identify the genera considered lactic acid bacteria, and discuss their importance in the food industry
g. Distinguish between enterococci and streptococci
h. Name bacterial genera with species capable of hemolysis, and differentiate between α-hemolysis and β-hemolysis

There are two classes of *Firmicutes* that form endospores: *Bacilli* and *Clostridia*. Recall that endospores are made within a mother cell upon nutrient depletion. Once released, these endospores are remarkably resistant to environmental challenges, sometimes surviving thousands of years. ◀ *Bacterial endospores are a survival strategy (section 3.9)*

The class *Bacilli* is very large but includes only two orders, *Bacillales* and *Lactobacillales*. These include genera representing cocci, endospore-forming rods and cocci, and nonsporing rods. We first describe the biology of some members of the order *Bacillales* and then important representatives of the order *Lactobacillales*. Characteristics of selected genera are summarized in **table 23.4**.

Order *Bacillales*

The genus *Bacillus,* family *Bacillaceae,* is the largest in the order *Bacillales*. The genus contains endospore-forming, chemoheterotrophic rods that are usually motile with peritrichous flagella (**figure 23.17**). Members of this genus are aerobic or facultative, and catalase positive. *Bacillus subtilis* is the most well-studied Gram-positive bacterium. It is a facultative anaerobe that can use nitrate as a terminal electron acceptor or perform mixed acid fermentation with lactate, acetate, and acetoin as major end products. It is nonpathogenic and a popular model organism for the study of gene regulation, cell division, quorum sensing, and cellular differentiation. Its 4.2 Mb genome was one of the first genomes to be

Table 23.4 Characteristics of Members of the Class *Bacilli*

Genus	Dimensions (μm), Morphology, and Motility	Genome Size (Mb)	Oxygen Relationship	Other Distinctive Characteristics
Bacillus	0.4–1.8 × 0.9–10; straight rods, peritrichous flagella, endospore-forming	4.2–5.5	Aerobic or facultative	Catalase positive; chemoorganotrophic
Enterococcus	0.6–2.0 × 0.6–2.5; spherical or ovoid cells in pairs or short chains, nonsporing, sometimes motile	2.6–3.2	Facultative	Ferment carbohydrates to lactate with no gas; complex nutritional requirements; catalase negative; gut microbiota of mammals and birds
Lactobacillus	0.5–1.2 × 1.0–10; usually long, regular rods, nonsporing, rarely motile	1.8–3.4	Facultative	Fermentative, at least half the end product is lactate; require rich, complex media; catalase and cytochrome negative
Lactococcus	0.5–1.2 × 0.5–1.5; spherical or ovoid cells in pairs or short chains, nonsporing, nonmotile	2.0–2.6	Facultative	Chemoorganotrophic with fermentative metabolism; lactate without gas produced; catalase negative; complex nutritional requirements; in dairy and plant products
Leuconostoc	0.5–0.7 × 0.7–1.2; cells spherical or ovoid, in pairs or chains; nonmotile and nonsporing	1.7–3.2	Facultative	Require fermentable carbohydrate and nutritionally rich medium for growth; fermentation produces lactate, ethanol, and gas; catalase and cytochrome negative
Listeria	0.4–0.5 × 1–2; rods occurring singly or in short chains; motile with peritrichous flagella only at <30°C	2.9–3.0	Aerobic or facultative	Widely distributed in nature; food-borne contaminant that can grow at refrigeration temperatures and in food preserved with salt and sodium nitrate
Staphylococcus	0.5–1.5; spherical cells occurring singly and in irregular clusters, nonmotile and nonsporing	2.5–2.9	Facultative	Chemoorganotrophic with both respiratory and fermentative metabolism; usually catalase positive; associated with skin and mucous membranes of vertebrates
Streptococcus	0.5–2.0; spherical or ovoid cells in pairs or chains, nonmotile and nonsporing	1.6–2.4	Facultative	Fermentative, producing mainly lactate and no gas; catalase negative; commonly attack red blood cells (α- or β-hemolysis); complex nutritional requirements; commensals or parasites on animals
Thermoactinomyces	0.4–1.0 in diameter; branched, septate mycelium resembles those of actinomycetes	2.6–3.4	Aerobic	Usually thermophilic; true endospores form singly on hyphae; numerous in decaying hay, vegetable matter, and compost

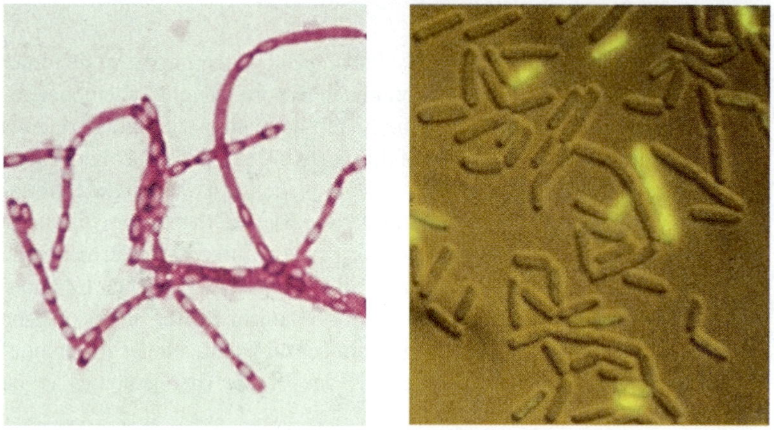

(a) *Bacillus anthracis* (b) *B. cereus*

Figure 23.17 *Bacillus.* (a) *B. anthracis* endospores are elliptical and central (×1,600). (b) *B. cereus* stained with SYTOX Green nucleic acid stain and viewed by epifluorescence and differential interference contrast microscopy. The cells that glow yellow are dead.
(a) ©Michael Abbey/SPL/Science Source; (b) ©Thermo Fisher Scientific

(a) Biofilm formation (b) Fruiting body formation

Figure 23.18 Biofilm Formation by *Bacillus subtilis.* (a) When grown for several days without agitation in nutrient-limiting medium, *B. subtilis* forms a hydrophobic biofilm at the aqueous-air interface. (b) *B. subtilis* forms multicellular structures called fruiting bodies, when grown on minimal-salts agar for several days. These cells are labeled so that spores turn blue, demonstrating that spores are abundant at the tips of these structures.
Branda et al., PNAS 25 Dept. 2001, vol. 98, fig 1A, p. 11623 Copyright (2001) National Academy of Sciences, U.S.A.

MICRO INQUIRY *What growth or survival advantage might this type of multicellular behavior confer to* B. subtilis?

completely sequenced and reveals a number of interesting elements. For instance, several families of genes have been expanded by gene duplication; the largest such family encodes ABC transporters—the most abundant class of protein in *B. subtilis.* There are 18 genes that encode sigma factors. Recall that the use of alternative sigma subunits of RNA polymerase is one way in which bacteria regulate gene expression. In this case, many of the sigma factors govern spore formation and other responses to stressful conditions. ◀ *Bacterial endospores are a survival strategy (section 3.9); Protein maturation and secretion (section 13.8); Sporulation in* Bacillus subtilis *(section 14.6)*

B. subtilis is a soil-dwelling microbe whose endospores are easily extracted from diverse soil types. Interestingly, it is often found in its vegetative state (i.e., not as an endospore) when growing in association with actively decaying material and in close association with plant roots. In addition to forming endospores, *B. subtilis* also develops biofilms comprised of vegetative cells and endospores that are spatially distinct (**figure 23.18**). ◀ *Biofilms are common in nature (section 7.6)*

Many species of *Bacillus* are of considerable importance. They produce the antibiotics bacitracin, gramicidin, and polymyxin. *B. cereus* (figure 23.17*b*) can cause food poisoning. Several species produce insecticides. For example, *B. thuringiensis* and *B. sphaericus* form a solid protein crystal, the **parasporal body,** during endospore formation (**figure 23.19**). As discussed in chapter 42, the *B. thuringiensis* parasporal body contains protein toxins that can kill over 100 species of moths when activated in the alkaline gut of caterpillars thereby destroying the epithelium. The *B. sphaericus* parasporal body contains proteins toxic for mosquito larvae and may be useful in controlling the mosquitoes that carry the malaria parasite *Plasmodium. B. anthracis* is the causative agent of the disease anthrax, which can kill both farm animals and humans. ▶ *Anthrax (section 39.5); Agricultural biotechnology relies on a plant pathogen (section 42.5)*

Members of the genus *Thermoactinomyces* were originally classified as actinomycetes because, like these microbes, their soil-associated substrate hyphae

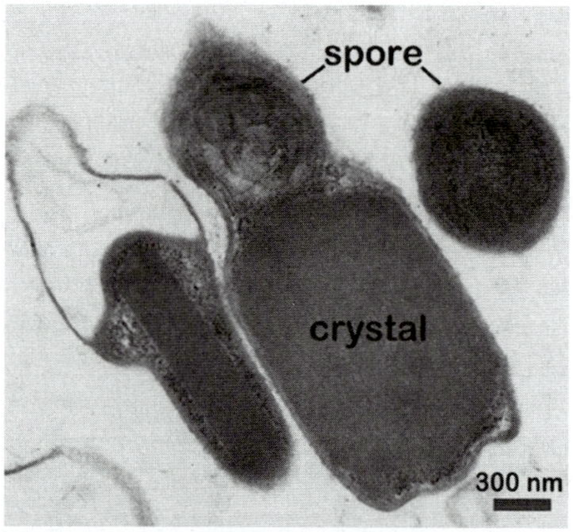

Figure 23.19 The Parasporal Body. An electron micrograph of a *B. thuringiensis* sporulating cell containing a parasporal crystal just beneath the endospore. A free spore is also shown.
From: "Protein crystal structure obtained at 2.9 Å resolution from injecting bacterial cells into an X-ray free-electron laser beam" Michael R. Sawaya, Duilio Cascioa, Mari Gingerya, Jose Rodrigueza, Lukasz Goldschmidt, Jacques-Philippe Colletier, Marc M. Messerschmidt, Sébastien Boutet, Jason E. Koglin, Garth J. Williams, Aaron S. Brewster, Karol Nassh, Johan Hattne, Sabine Bothh, R. Bruce Doak, Robert L. Shoeman, Daniel P. DePonte, Hyun-Woo Park, Brian A. Federici, Nicholas K. Sauter, Ilme Schlichting, and David S. Eisenberg, PNAS vol. 111 no. 35, fig 1, p. 12769–12774. Copyright (2014) National Academy of Sciences, U.S.A.

(filamentous cells) differentiate into upwardly growing aerial hyphae. However, phylogenetic analysis places them in the order *Bacillales,* family *Thermoactinomycetaceae.* Their G + C content (52–55 mol%) is considerably lower than that of actinobacteria. Also unlike actinobacteria, *Thermoactinomyces* species form true endospores within both the aerial and substrate hyphae. These bacteria are thermophilic and grow between 45° and 60°C. *Thermoactinomyces* spp. are commonly found in damp haystacks and compost piles. In fact, *T. vulgaris* from haystacks, grain storage silos, and compost piles is a causative agent of farmer's lung, an allergic disease of the respiratory system. Spores from *T. vulgaris* recovered from the mud of a Minnesota lake were found to be viable after about 7,500 years of dormancy.

One other genus within the order *Bacillales* bears mentioning: *Sporosarcina.* This genus is 1 of 10 genera in the family *Planococcaceae,* and it contains the only known endospore-forming bacterium that has a coccoid rather than rod shape. *S. ureae* is particularly interesting because it is tolerant of very alkaline conditions (up to pH 10), which occur when it degrades urea to CO_2 and NH_3. Interestingly, it is easily isolated from agricultural soils on which animals urinate.

Family *Staphylococcaceae*

The family *Staphylococcaceae* contains four genera, the most important of which is *Staphylococcus.* Members of this genus are facultatively anaerobic, nonmotile cocci that divide in more than one plane to form irregular clusters (**figure 23.20**). They are usually catalase positive and oxidase negative. They respire using oxygen as the terminal electron acceptor, although some can reduce nitrate to nitrite. They are also capable of fermentation and convert glucose principally to lactate.

Staphylococci are normally associated with the skin and mucous membranes of warm-blooded animals. They are responsible for many human diseases. *S. aureus* has an impressive collection of virulence factors to evade the immune system and subvert the host (figure 23.20*b*). These include Protein A, a surface-displayed protein that binds host antibodies by the conserved Fc region, rendering them ineffective. Several additional proteins aid in colonization and dissemination; clumping factor (also called coagulase) enables bacteria to attach to platelets and wounded tissue, thereby promoting initial colonization. By contrast, the enzyme staphylokinase dissolves fibrin clots that would otherwise keep the bacterium from invading beyond the site of initial colonization. The presence of coagulase is used as a marker for the identification of *S. aureus.* Growth and hemolysis patterns on blood agar are also useful in distinguishing these bacteria. The staphylococcal β-hemolysin lyses cells by forming solvent-filled channels in target plasma membranes. ▶| *Staphylococcal diseases (section 39.3); Staphylococcal food poisoning (section 39.4)*

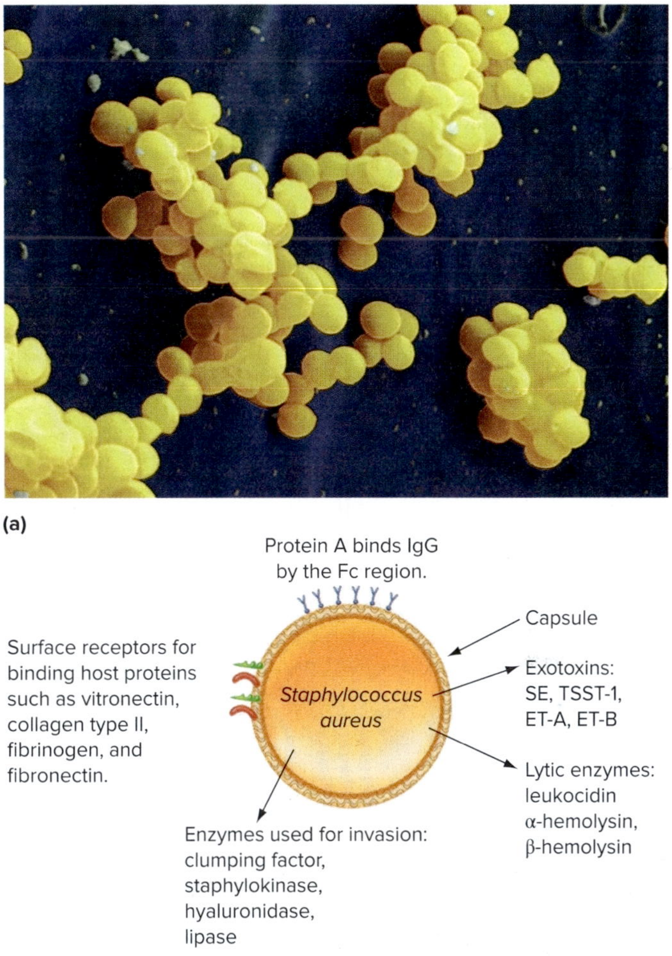

Figure 23.20 ***Staphylococcus.*** (a) *S. aureus* cocci arranged in grapelike clusters; color-enhanced scanning electron micrograph. Each cell is about 1 μm in diameter. (b) *S. aureus* produces many virulence factors that account for its success as a pathogen. Staphylokinase catalyzes the degradation of host fibrin, facilitating microbial invasion. Hyaluronidase degrades hyaluronic acid, which is found in connective and epithelial tissues.

Family *Listeriaceae*

The family *Listeriaceae* includes two genera: *Brochothrix* and *Listeria. Brochothrix* spp. are most commonly found in meat but are not pathogenic. *Listeria* is the medically important genus in this family. It contains short rods that are facultatively anaerobic, catalase positive, and motile by peritrichous flagella. In addition to aerobic respiration, these microbes ferment glucose mainly to lactate. *Listeria* spp. are widely distributed in nature, particularly in decaying matter. *L. monocytogenes* is a pathogen of humans and other animals, and causes listeriosis, an important food-borne infection. ▶| *Food-borne disease outbreaks (section 41.3)*

Order *Lactobacillales*

Streptococcus, Enterococcus, Lactococcus, Lactobacillus, and *Leuconostoc* are all genera in the order *Lactobacillales.* These bacteria do not form endospores and are usually nonmotile. They are strictly fermentative, depending on sugars for carbon and energy. They lack cytochromes and an electron transport chain so

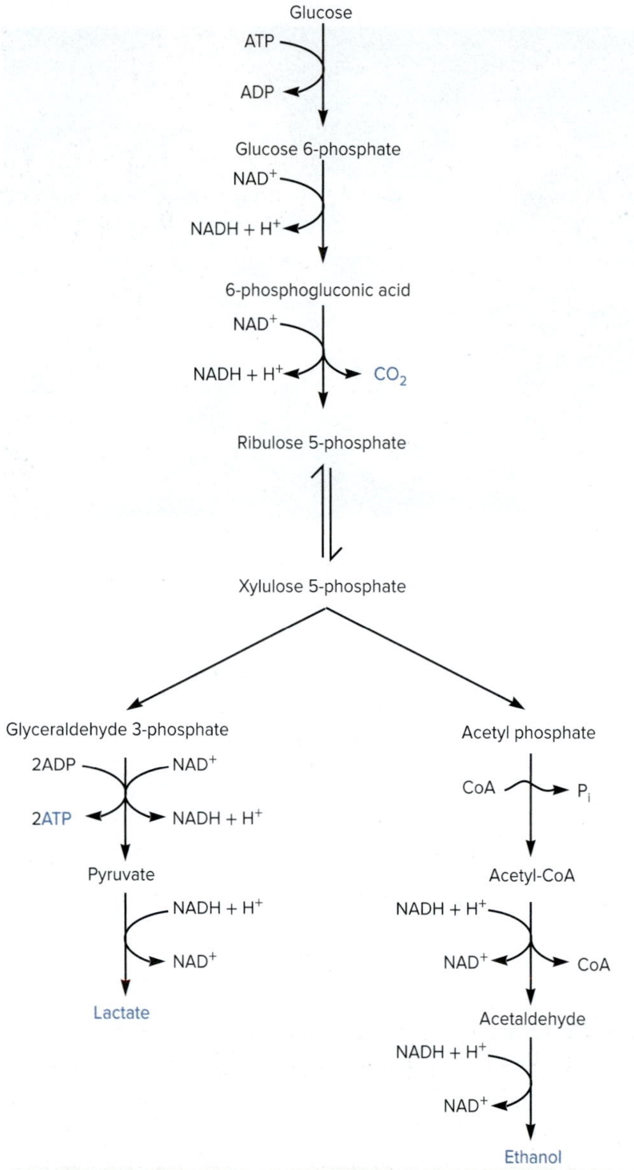

Figure 23.21 Heterolactic Fermentation and the Phosphoketolase Pathway. The phosphoketolase pathway converts glucose to lactate, ethanol, and CO_2.

MICRO INQUIRY *What is the net yield of ATP for each glucose that enters the phosphoketolase pathway?*

they obtain energy only by substrate-level phosphorylation. Those that produce lactic acid as their major or sole fermentation product are sometimes collectively called **lactic acid bacteria.** Nutritionally, they are fastidious, and many vitamins, amino acids, purines, and pyrimidines must be supplied because of their limited biosynthetic capabilities. Members of this order usually are categorized as facultative anaerobes but can be considered aerotolerant. ◀ *Oxygen concentration (section 7.5); Fermentation does not involve an electron transport chain (section 11.8)*

The largest genus in this order is *Lactobacillus,* with about 100 species. Lactobacilli include rods and some coccobacilli. All lack catalase and cytochromes, and produce lactic acid as their main or sole fermentation product (*see figure 11.20*). They carry out either **homolactic fermentation** using the Embden-Meyerhof pathway or **heterolactic fermentation** with the phosphoketolase pathway (**figure 23.21**). They grow optimally under slightly acidic conditions, at a pH between 4.5 and 6.4, and are found in dairy products, meat, water, sewage, beer, fruits, and many other materials. Lactobacilli also are part of the normal flora of the human body; in particular they are important normal vaginal flora.

Lactobacillus species are indispensable to the food and dairy industry. They are used in the production of fermented vegetable foods (e.g., sauerkraut, pickles), sourdough bread, Swiss and other hard cheeses, yogurt, and sausage. Yogurt is probably the most popular fermented milk product in the United States, and *Streptococcus thermophilus* and *Lactobacillus delbrueckii* subspecies *bulgaricus* are used in its production. At least one species, *L. acidophilus,* is sold commercially as a probiotic agent that may provide some health benefits for the consumer. On the other hand, some lactobacilli are problematic. They can be responsible for spoilage of beer, wine, milk, and meat. ▶ *Fermented milks (section 41.5)*

Leuconostoc, family *Leuconostocaceae,* contains facultative Gram-positive cocci, which may be elongated or elliptical and arranged in pairs or chains (**figure 23.22**). Leuconostocs carry out heterolactic fermentation by converting glucose to D-lactate and ethanol or acetic acid by means of the phosphoketolase pathway (figure 23.21). They can be isolated from plants, silage, and milk. They are used in wine production, in the fermentation of vegetables, and in the manufacture of buttermilk, butter, and cheese. *L. mesenteroides* synthesizes dextrans from sucrose and is important in industrial dextran production. *Leuconostoc* species are involved in food spoilage and tolerate high sugar concentrations so well that they grow in syrup and can be a major problem in sugar refineries.

Bacteria in the important families *Enterococcaceae* and *Streptococcaceae* occur in pairs or chains when grown in liquid media (**figure 23.23** and **table 23.5**). They are unable to respire and have a strictly fermentative metabolism. They ferment sugars to produce lactic acid but no gas; that is, they carry out homolactic fermentation. A few species are strict anaerobes rather than

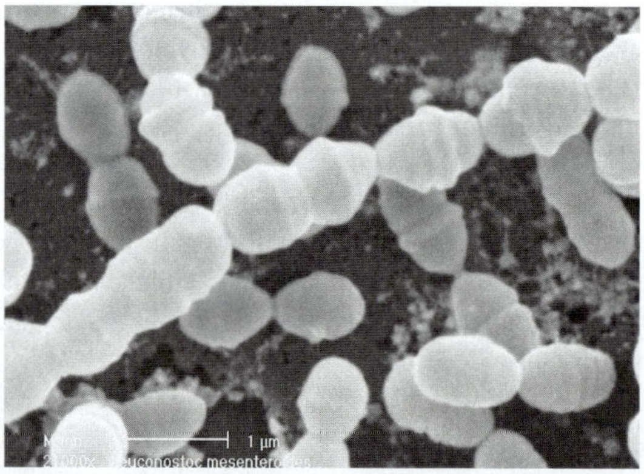

Figure 23.22 *Leuconostoc.* *Leuconostoc mesenteroides;* scanning electron micrograph.

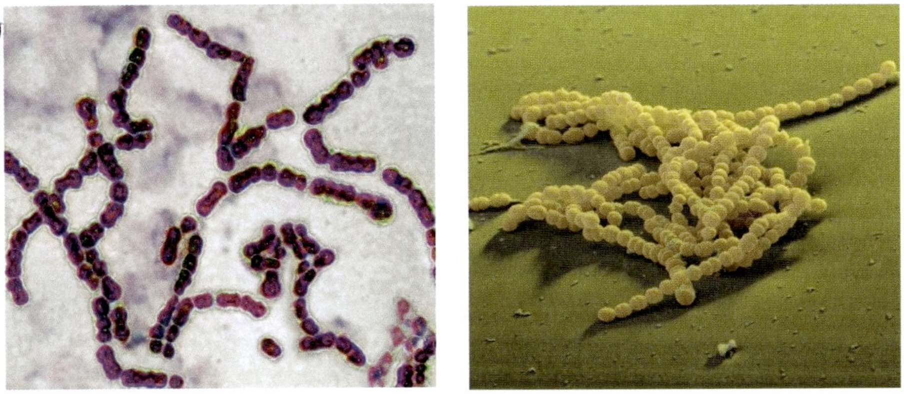

(a) *Streptococcus pyogenes* **(b)** *S. agalactiae*

Figure 23.23 ***Streptococcus.*** (a) *Streptococcus pyogenes*. (b) *Streptococcus agalactiae,* the cause of group B streptococcal infections. Note the long chains of cells; color-enhanced scanning electron micrograph.
(a) ©James Cavallini/Science Source; (b) ©Science Source

aerotolerant. Enterococci such as *E. faecalis* are normal residents of the intestinal tracts of humans and most other animals. *E. faecalis* is an opportunistic pathogen that can cause urinary tract infections and endocarditis. Enterococci are major agents in the horizontal transfer of antibiotic-resistance genes.

The family *Streptococcaceae* includes only two genera of note: *Lactococcus* and *Streptococcus*. *Lactococcus* spp. ferment sugars to lactic acid and can grow at 10°C but not at 45°C. *L. lactis* is widely used in the production of buttermilk and cheese because it can curdle milk and add flavor through the synthesis of diacetyl and other products. ▶| *Fermented milks (section 41.5)*

The genus *Streptococcus* is large and complex. All species in the genus are facultatively anaerobic and catalase negative. One of the most important characteristics to identify these cocci is the ability to lyse red blood cells when grown on blood agar, an agar medium containing 5% sheep or horse blood (**figure 23.24**). In **α-hemolysis,** a 1- to 3-mm greenish zone of incomplete hemolysis forms around the colony; **β-hemolysis** is characterized by a clear zone of complete lysis. Serological studies are also very important in identification because streptococci often have distinctive cell wall antigens. These bacteria produce surface carbohydrates that are important in species identification through the **Lancefield grouping system**, which divides the streptocoocci into 20 groups. For example, *S. pyogenes* is often referred to as GAS, for group A streptococcus (figure 23.23). Biochemical and physiological tests, and sensitivity to bacitracin, sulfa drugs, and optochin (a quinine analogue) also are used to identify particular species. ▶| *Clinical microbiology and immunology (chapter 37)*

The streptococci are divided into three groups based on pathogenicity: pyogenic streptococci, oral streptococci, and other streptococci. Pyogenic streptococci are pathogens associated with pus formation. Most species are β-hemolytic on blood agar and form chains of cells. The major human pathogen in this group is

Table 23.5 Classification of Streptococci, Entercocci, and Lactococci

Characteristics	*Streptococcus*	*Enterococcus*	*Lactococcus*
Predominant arrangement (most common first)	Chains, pairs	Pairs, chains	Pairs, short chains
Capsule/slime layer	+	–	–
Habitat	Mouth, respiratory tract	Gastrointestinal tract	Dairy products
Growth at 45°C	Variable	+	–
Growth at 10°C	Variable	Usually +	+
Growth at 6.5% NaCl broth	Variable	+	–
Growth at pH 9.6	Variable	+	–
Hemolysis	Usually β (pyogenic) or α (oral)	α, β, –	Usually –
Serological group (Lancefield)	Variable (A–O)	Usually D	Usually N
Representative species	Pyogenic streptococci: *S. agalactiae, S. pyogenes, S. equi, S. dysgalactiae;* Oral streptococci: *S. gordonii, S. salivarius, S. sanguis, S. oralis, S. pneumoniae, S. mitis, S. mutans*; Other streptococci: *S. bovis, S. thermophilus*	*E. faecalis, E. faecium, E. avium, E. durans, E. gallinarum*	*L lactis, L. raffinolactis, L. plantarum*

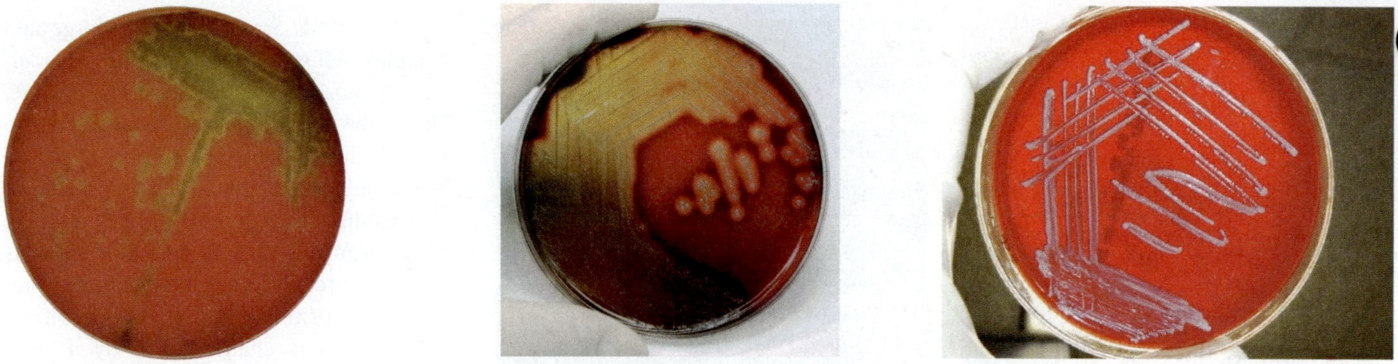

(a) Alpha-hemolysis (b) Beta-hemolysis (c) No hemolysis

Figure 23.24 Staphylococcal and Streptococcal Hemolytic Patterns. (a) *Staphylococcus aureus* on blood agar, illustrating α-hemolysis. (b) *Streptococcus pyogenes* on blood agar, illustrating β-hemolysis. (c) *Staphylococcus epidermidis* on blood agar with no hemolysis, also referred to as γ-hemolysis.
(a) ©Matt Mezydlo; (b) ©Linde Stewart/Getty Images; (c) ©ksass/iStock/Getty Images

MICRO INQUIRY *Why is there a green zone around α-hemolytic colonies?*

S. pyogenes, which causes streptococcal sore throat, acute glomerulonephritis, and rheumatic fever. The normal habitat of oral streptococci is the oral cavity and upper respiratory tract of humans and other animals. In other respects, oral streptococci are not necessarily similar. *S. mutans* is associated with the formation of dental caries. *S. pneumoniae* is α-hemolytic and grows as pairs of cocci. It is associated with lobar pneumonia and otitis media (inflammation of the middle ear) in young children. ▶| *Streptococcal diseases (sections 39.1 and 39.3)*

Comprehension Check

1. You are studying what you think is a new *Bacillus* isolate. Without performing molecular analyses, what morphological and physiological traits would you assay?
2. What makes the genus *Thermoactinomyces* unique? What disease does *T. vulgaris* cause?
3. List three environments where you would find *Staphylococcus* spp. On what basis are pathogenic and nonpathogenic species and strains distinguished?
4. List the major properties that define the genus *Lactobacillus*. In what ways are species in this genus important in the food and dairy industries?
5. How are various species of *Streptococcus* identified?
6. Of what practical importance are leuconostocs? What are lactic acid bacteria?
7. What is the difference between α-hemolysis and β-hemolysis?

23.3 Class *Clostridia:* Anaerobic Endospore-Forming Bacteria

After reading this section, you should be able to:

a. Identify several *Clostridia* spp. that form endospores
b. Outline the reaction by which some clostridia ferment amino acids, and relate this carbon substrate to their environmental distribution and pathogenic potential
c. Identify the terminal electron acceptors used by *Desulfotomaculum* spp.
d. Distinguish phototrophy in heliobacteria from that of other anoxygenic phototrophic bacteria

The class *Clostridia* contains a diverse group of bacteria. The characteristics of some of the more important genera are summarized in **table 23.6**. With over 150 species, *Clostridium* is the largest genus in the class. It includes obligately anaerobic, fermentative, Gram-positive bacteria that form endospores (**figure 23.25**). As we noted for the *Actinobacteria*, 16S rRNA sequences are insufficient to discriminate among strains of *Clostridia*. Additional conserved gene sequences reveal that there is sufficient diversity within the genus to split it into several additional genera. *C. difficile*, a human pathogen responsible for severe gastrointestinal disease, has been transferred to the new genus *Clostridioides* and retains its abbreviated name "C. diff."
We can expect major revisions within the class *Clostridia* in the near future.

Clostridium spp. have great practical impact. Because they are anaerobic and form heat-resistant endospores, they are responsible for many cases of food spoilage, even in canned foods. Clostridia often ferment amino acids via the **Stickland reaction** (**figure 23.26**). In this process, one amino acid is oxidized to produce pyruvate, which then is fermented to yield acetate and ATP. A different amino acid serves as an electron acceptor, so it is reduced, thereby reoxidizing NADH to NAD^+ and forming two additional acetate molecules. This reaction also generates ammonia, hydrogen sulfide, and fatty acids during the anaerobic decomposition of proteins. These products are responsible for unpleasant odors arising during putrefaction.

Although some clostridia are industrially valuable (e.g., *C. acetobutylicum* is used to manufacture butanol), the pathogenic species produce well-known toxins. For instance, *C. botulinum* produces the botulism toxin and *C. tetani* causes tetanus through the release of tetanus toxin. *C. perfringens* secretes a

Table 23.6 Characteristics of Selected Members of the Class *Clostridia*

Genus	Dimensions (μm), Morphology, and Motility	Genome Size (Mb)	Oxygen Relationship	Other Distinctive Characteristics
Clostridium	0.3–2.0 × 1.5–20; rod-shaped, often pleomorphic, nonmotile or peritrichous flagella	3.0–5.0	Anaerobic	Usually chemoorganotrophic, fermentative, and catalase negative; form oval or spherical endospores
Desulfotomaculum	0.3–2.5 × 2.5–15; straight or curved rods, peritrichous or polar flagella	2.7–4.9	Anaerobic	Reduce sulfate to H_2S, form subterminal to terminal endospores; stain Gram negative but have Gram-positive wall; catalase negative
Heliobacterium	0.8–1 × 2.5–9; rods that are frequently bent, gliding motility or peritrichous flagella	3.1	Anaerobic	Chemoheterotrophic or photoheterotrophic with bacteriochlorophyll *g;* stain Gram negative but have Gram-positive wall; some form endospores; fix nitrogen

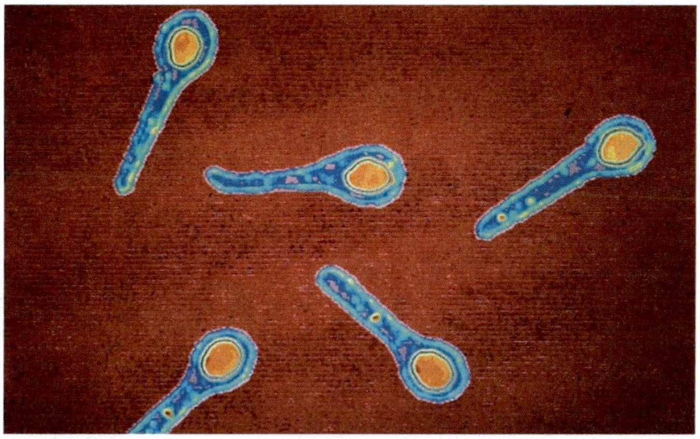

Figure 23.25 *Clostridium tetani.* This pathogen makes endospores that are round and terminal. When the mother cell surrounding each endospore lyses, the released spores remain viable for hundreds, if not thousands of years (×3,000).
©Alfred Pasieka/Getty Images

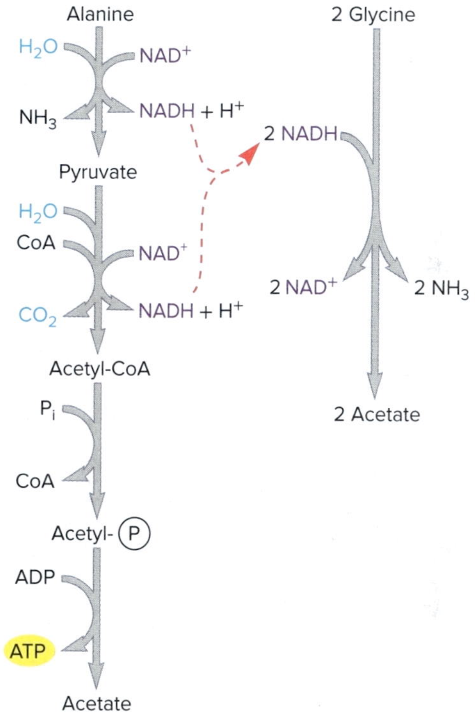

Figure 23.26 The Stickland Reaction. Alanine is oxidized to acetate, and glycine is used to reoxidize the NADH generated during alanine fermentation. The fermentation also produces some ATP.

variety of protein toxins and, depending on the strain, causes gas gangrene or food poisoning. Although it lacks a TCA cycle and makes ATP only by fermentation, it has an extraordinary doubling time of only 8 to 10 minutes when in a human host. This is due to the ready supply and assimilation of host macromolecules. ▶| *Botulism (section 39.4); Tetanus (section 39.3)*

C. tetani ferments a number of amino acids to lactate, butyrate, and H_2 (**figure 23.27**). It lacks the genes to catabolize sugars. By contrast, *C. acetobutylicum* and *C. perfringens* can catabolize both carbohydrates and amino acids. *C. tetani* uses an ATP-binding cassette transporter and a V-type ATPase to couple ATP hydrolysis to the flow of sodium ions out of the cell. V-type ATPases pump protons, or in this case sodium ions, out of the cell at the expense of ATP made by substrate-level phosphorylation during fermentation. This is not a futile exercise, because it creates a sodium motive force, which in turn drives amino acid and sugar uptake systems. Although *C. tetani* also exports protons via a pyrophosphatase, external protons seem to serve only as the counter ion in Na^+-H^+ antiport. Surprisingly, *C. tetani* possesses genes for a membrane-bound electron transport chain; microbes that make ATP exclusively by fermentation do not generally possess electron transport chains. In this case, it appears that this chain links Na^+-extrusion with oxidation of NADH generated during fermentation. |◀ *Primary and secondary active transport (section 3.3)*

Desulfotomaculum is a genus that contains endospore-forming species that reduce sulfate and sulfite to hydrogen

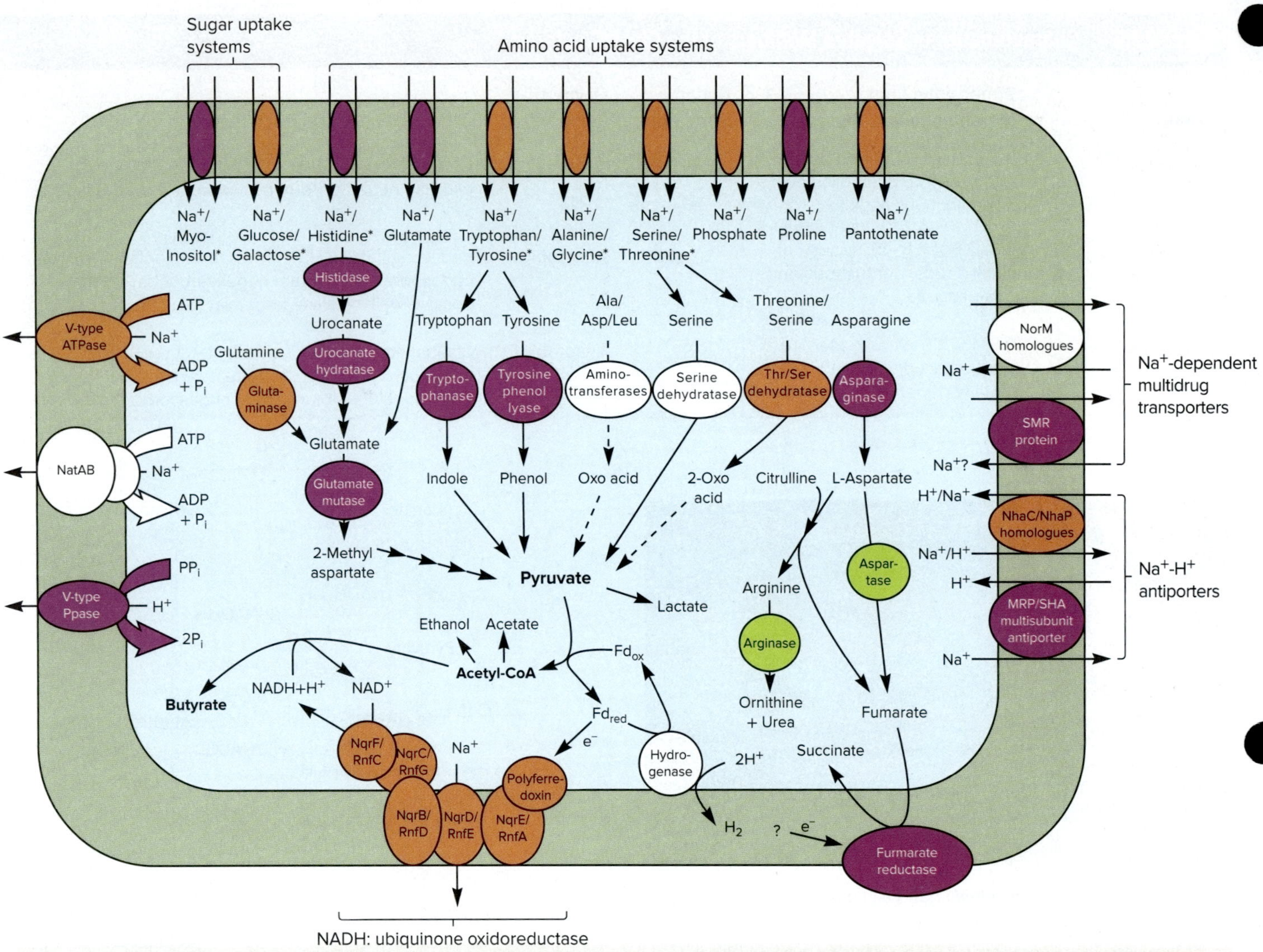

Figure 23.27 Amino Acid Fermentation and Na$^+$ Flux in *Clostridium tetani*. The fermentation of amino acids and the generation of a Na$^+$ motive force is central to survival of this pathogen. Note that amino acids are transported into the cell by Na$^+$-dependent uptake systems. Sugars are also transported in this manner but are used in anabolic reactions only. Several Na$^+$-dependent transporters with unknown substrates are present. A V-type ATPase and an ATP-binding cassette transporter (NatAB) pump Na$^+$ out of the cell at the expense of ATP hydrolysis. A proton translocating pyrophosphatase (V-type Ppase) is used to export H$^+$. Components of a membrane-bound electron transport chain are present, but it is thought that the NADH dehydrogenase participates in amino acid fermentation. The lack of quinones suggests that the ETC is not used to establish a proton motive force.

MICRO INQUIRY *The presence of an electron transport chain in a strictly fermentative microbe is unusual. Describe how this microbe uses its ETC and how this differs from a respiratory microbe.*

sulfide during anaerobic respiration (**figure 23.28**). Although they stain Gram negative, electron microscopy studies show that these bacteria have Gram-positive cell walls, and phylogenetic studies place the genus with the *Firmicutes.*

Heliobacteria further illustrate the diversity of this class. The genera *Heliobacterium* and *Heliophilum* contain unusual anaerobic, phototrophic species characterized by the presence of bacteriochlorophyll *g.* They have a photosystem I–type reaction center like green sulfur bacteria but have no intracytoplasmic photosynthetic membranes; pigments are contained in the plasma membrane (**figure 23.29**). They also differ from many other anoxygenic phototrophic bacteria because they cannot grow autotrophically. Pyruvate is used during photoheterotrophic growth in light; it is fermented in dark, anoxic conditions. Like clostridia, heliobacteria are capable of nitrogen fixation. Although they have a Gram-positive cell wall, they have a low peptidoglycan content, and they stain Gram negative. Some heliobacteria form endospores. ◀ *Light reactions in anoxygenic photosynthesis (section 11.12); Photosynthetic bacteria are diverse (section 21.4)*

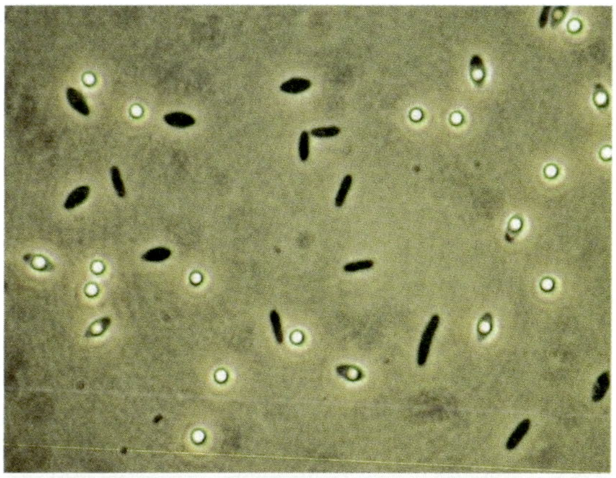

Figure 23.28 ***Desulfotomaculum.*** *Desulfotomaculum acetoxidans* with endospores (bright spheres); phase contrast.
©R. John Parkes/Cardiff University

MICRO INQUIRY *In addition to its Gram-staining characteristics, why do you think* Desulfotomaculum *spp. were originally thought to be Gram-negative microbes?*

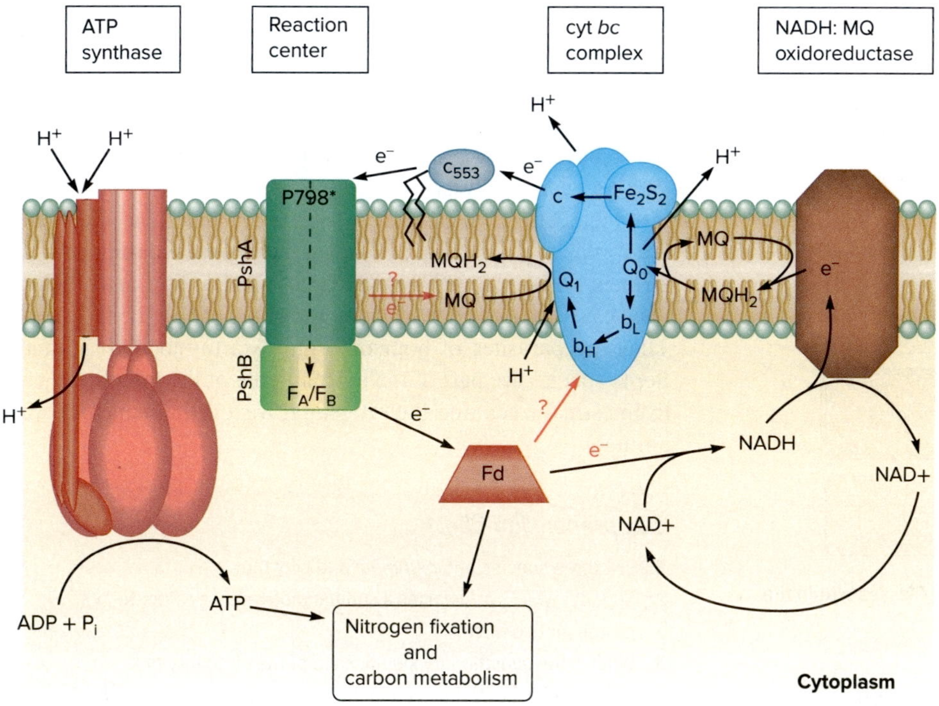

Figure 23.29 Electron Transfer in *Heliobacterium modesticaldum*. Genes encoding all subunits of NADH:quinone oxidoreductase are present on the *H. modesticaldum* genome. This suggests that electrons flow from NADH to the NADH:menaquinonone (MQ) oxidoreductase, to cytochrome (cyt) *bc* complex, which in turn reduces a 2Fe-2S subunit and cytochrome b_L. Electrons from cyt *bc* can be transferred via cyt c_{553} to the reaction center (RC) primary electron donor P798. Based on similarities between the heliobacterial RC and photosystem I in other anoxygenic photosynthetic bacteria, it is hypothesized that a proton motive force is generated by cyclic electron flow.

MICRO INQUIRY *What are the two sources of electrons for the* H. modesticaldum *electron transport chain?*

23.4 Class *Negativicutes:* Gram-Positive Bacteria with Outer Membranes

After reading this section, you should be able to:

a. Explain why these bacteria are considered Gram positive despite possessing outer membranes
b. Discuss why new taxonomic ranks must be developed
c. Identify the location where *Veillonella* sp. might be found on your body

In 2010 three bacterial isolates were cultured from human skin and soft-tissue wounds. What happened next illustrates the ever-changing world of microbial taxonomy and systematics. Further investigation of these bacteria revealed that although they had an inner and outer membrane, like Gram-negative bacteria, 16S rRNA sequences showed that they were closely related to microbes in the Gram-positive family *Veillonellaceae*. Because the three isolates were distinct from any other bacteria in this family, they were assigned to a new genus and species: *Negativicoccus succinicivorans*. The name reflects their unusual cell envelope and succinate fermentation.

The discovery of *N. succinicivorans* triggered a reexamination of the class *Clostridia*, and more broadly, the phylum *Firmicutes*. It was determined that bacteria that have peptidoglycan between the inner and outer membranes and therefore stain Gram negative, but are otherwise genetically *Firmicutes*, require their own class. Therefore the new class *Negativicutes* and order *Selenomonadales* were approved (**table 23.7**). Two existing families, *Acidaminococcaceae* and *Veillonellaceae*, were then assigned to this new order. Such changes are always controversial, and this one is no exception.

Electron microscopy confirmed that bacteria in the *Negativicutes* and *Halanaerobiales* exhibit two distinct membranes, that is, they are diderm. Bacteria in other classes in the phylum *Firmicutes* have the typical Gram-positive cell envelope. Phylogenetic analyses using 16S rRNA sequence and conserved protein sequences indicate that these two classes do not group together (**figure 23.30**). Although there are only a limited number of genome sequences from these microbes, genes encoding the components of the cell envelope are present. These include enzymes involved in

Table 23.7 Characteristics of Selected Members of the Class *Negativicutes*

Genus	Dimension (µm), Morphology, Motility	Genome Size (Mb)	Oxygen Relationship	Other Distinctive Characteristics
Veillonella	0.3–0.5; cocci in pairs, short chains, and masses; nonmotile	2.1–2.8	Anaerobic	Pyruvate and lactate fermented but not carbohydrates; parasitic in mouths, intestines, and respiratory tracts of animals
Megasphaera	1.7–2.6; cocci; nonmotile	1.8–3.2	Anaerobic	Ferment a variety of carbohydrates, including lactate; isolated from mouths, intestines of animals, and spoiled beer
Negativicoccus	0.4; cocci; nonmotile	1.45	Anaerobic and microaerophilic	Isolated from human skin and soft-tissue wounds; ferments succinate to acetate and propionate

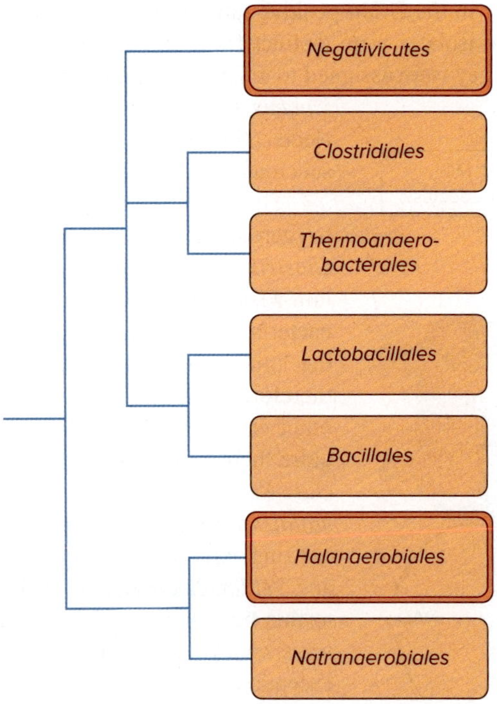

Figure 23.30 Presence of an Outer Membrane in Classes Within the Phylum ***Firmicutes*.**

lipopolysaccharide synthesis and outer membrane proteins. Detailed comparisons of these genes establish that they are related, and it has been proposed that the ancestral *Firmicutes* were diderm, and possessed an outer membrane (OM) that was lost several times during the evolution of these microbes to the present-day *Firmicutes*. ◀ *Gram-positive and Gram-negative or monoderms and diderms? (Microbial Diversity and Ecology 3.1)*

How might loss of the OM be tolerated in bacteria? A mutant microbe that synthesized and deposited excess peptidoglycan on its surface, thereby disrupting the OM, might survive and replicate. Note that the endospore-forming members of the genera *Bacilli* and *Clostridia* do not have an OM. The intricate membrane folding processes during endospore engulfment (*see figure 3.51*) may be facilitated with only a single membrane. The evolutionary success of endospore formation may have driven the stabilization of microbes lacking the OM. ◀ *Sporulation: making endospores (section 3.9)*

Members of the genus *Veillonella* are anaerobic, chemoheterotrophic cocci. All have complex nutritional requirements and ferment carbohydrates, organic acids, and amino acids to produce gas (CO_2 and often H_2) plus a mixture of volatile fatty acids. They are parasites of homeothermic (warm-blooded) animals. Some species are part of the normal biota of the mouth, the gastrointestinal tract, and the urogenital tract of humans and other animals.

Comprehension Check

1. List three species of *Clostridium* and why they are notable.
2. Suggest why *C. tetani* uses a sodium motive force rather than a proton motive force.
3. What is the evolutionary significance of the discovery of a phototrophic bacterium within the low G + C Gram-positive bacteria? Do you think the phototrophic apparatus in *Heliobacterium* spp. evolved independently of those found in Gram-negative anoxygenic phototrophs? Explain.
4. What kind of genetic evidence, in addition to 16S rRNA sequences, might have demonstrated that *Negativicutes* should be considered Gram positive?

Key Concepts

23.1 Class *Actinobacteria*

- Members of this class that form branching, usually nonfragmenting hyphae and asexual spores are collectively called actinobacteria or actinomycetes (**figure 23.1**).
- The asexual spores borne on aerial hyphae are called exospores because they do not develop within a mother cell. Some produce sporangiospores.
- Actinobacteria have several distinctively different types of cell walls and often also vary in terms of the sugars present in cell extracts. Properties such as color and morphology are also taxonomically useful (**tables 23.1** and **23.2; figure 23.3**).
- Members of the genus *Actinomyces* are irregularly shaped, nonsporing rods that can cause disease in cattle and humans.
- Corynebacteria form irregular and club-shaped cells that are often arranged in palisades. *C. diphtheriae* is the cause of diphtheria (**figure 23.5**).
- Mycobacteria form either rods or filaments that readily fragment. Their cell walls have a high lipid content and mycolic acids, which makes them acid-fast (**figures 23.6** and **23.7**).
- Nocardioform actinobacteria have hyphae that readily fragment into rods and coccoid elements (**figure 23.8**).
- Members of the genus *Micrococcus* are unusual members of this class because they form simple coccoid cells. Some species are normal inhabitants on human skin.
- Arthrobacters have an unusual rod-coccus growth cycle and, like corynebacteria, carry out snapping division (**figure 23.10**).
- *Micromonospora* and *Actinoplanes* spp. have an extensive substrate mycelium and form aerial sporangia and are present in soil, freshwater, and the ocean (**figure 23.11**).
- Members of the genus *Propionibacterium* are common skin and intestinal inhabitants, and are important in cheese manufacture and the development of acne vulgaris.
- The order *Streptomycetales* includes the genus *Streptomyces*. Members of this genus produce substrate hyphae that differentiate into aerial hyphae-bearing chains of 3 to 50 or more exospores (**figure 23.12**).
- Streptomycetes are important in the degradation of resistant organic material in the soil and produce many useful antibiotics. A few cause diseases in plants and animals. However, they are most notable for the variety of antibiotics and other medically and industrially important compounds they produce (**table 23.3**).
- The genera *Frankia* and *Geodermatophilus* are placed in the order *Frankiales*. *Frankia* spp. grow in symbiotic association with nonleguminous plants and fix nitrogen (**figure 23.15**).
- The genus *Bifidobacterium* is placed in the order *Bifidobacteriales*. Bifidobacteria are irregular, anaerobic rods and one of the first colonizers of the intestinal tract in nursing babies.
- Members of the genus *Bifidobacterium* are often sold for their probiotic activity.

23.2 Class *Bacilli*: Aerobic Endospore-Forming Bacteria

- The class *Bacilli* is divided into two orders: *Bacillales* and *Lactobacillales*.
- The genus *Bacillus* contains aerobic and facultative, catalase-positive, endospore-forming, chemoorganotrophic rods that are usually motile with peritrichous flagella. *Bacillus subtilis* is a model organism that forms endospores and multicellular structures such as biofilms and fruiting bodies (**figures 23.17** and **23.18**).
- *Thermoactinomyces* spp. are thermophilic and form a mycelium and true endospores. *T. vulgaris* causes allergic reactions that lead to farmer's lung.
- Members of the genus *Staphylococcus* are facultatively anaerobic, nonmotile, Gram-positive cocci that form irregular clusters. They carry out aerobic respiration; some also ferment and use nitrate as a terminal electron acceptor. *S. aureus* is a pathogen that possesses an array of virulence factors (**figure 23.20**).
- Several important genera such as *Lactobacillus* and *Listeria* contain regular, nonsporing rods. *Lactobacillus* spp. carry out lactic acid fermentation and are extensively used in the food and dairy industries. *Listeria monocytogenes* is an important agent in food-borne disease. *Leuconostoc* spp. carry out heterolactic fermentation using the phosphoketolase pathway (**figure 23.21**); they are involved in the production of fermented vegetable and dairy products.
- The genera *Streptococcus, Enterococcus,* and *Lactococcus* contain cocci arranged in pairs and chains that are usually facultative anaerobes that carry out fermentation (**table 23.5**). Some important species are the pyogenic coccus *S. pyogenes, S. pneumoniae,* the enterococcus *E. faecalis,* and the lactococcus *L. lactis.*

23.3 Class *Clostridia*: Anaerobic Endospore-Forming Bacteria

- Members of the genus *Clostridium* are anaerobic, Gram-positive rods that form endospores and do not carry out dissimilatory sulfate reduction. They are responsible for food spoilage, botulism, tetanus, and gas gangrene. They have a fermentative metabolism (**figures 23.25–23.27**).

- *Desulfotomaculum* contains anaerobic, endospore-forming species that reduce sulfate to sulfide during anaerobic respiration.
- Heliobacteria are anaerobic, phototrophic bacteria with bacteriochlorophyll *g*. Some form endospores (**figure 23.29**).

23.4 Class *Negativicutes*: Gram-Positive Bacteria with Outer Membranes

- The discovery of several new Gram-negative staining bacteria that are phylogenetically *Firmicutes* led to the development of the class *Negativicutes*. It has been proposed that the ancestral *Firmicutes* had an outer membrane and that it may have been lost to facilitate endospore formation.

Active Learning

1. Even though actinobacteria are high G + C organisms, there are regions of the genome that are AT-rich. Suggest a few such regions and explain why they must be more AT-rich.
2. Given that mycolic acids are essential for mycobacterial viability, how would you test their importance in mycobacterial infection?
3. *Streptomyces coelicolor* is studied as a model system for cellular differentiation. Some of the genes involved in sporulation contain an AT-rich leucine codon not used in vegetative genes. Suggest how *Streptomyces* might use this rare codon to regulate sporulation.
4. Suppose that you discovered a nodulated plant that could fix atmospheric nitrogen. How might you show that a bacterial symbiont was involved and that a *Frankia* sp. rather than a *Rhizobium* sp. was responsible?
5. Some species of *Clostridium* demonstrate much faster doubling times when in the host than when grown in vitro. Design a demonstration that would explore this phenomenon.
6. What physiological properties might account for the ease with which anaerobic clostridia can be isolated from soil and other generally aerobic niches?
7. *S. aureus* strains resistant to methicillin (MRSA) are also resistant to many other antibiotics and are more virulent than non-MRSA strains. Refer to chapters 16 and 18, and develop a testable hypothesis that predicts how these many traits came to be in a single group of strains.
8. Actinomycete genomes often contain biosynthetic gene clusters (BGCs) that encode enzymes in a pathway to assemble secondary metabolites like antibiotics. The challenge is to determine the conditions under which the genes are expressed and the identity and activity of the compound synthesized. Scientists exploring the phenomenon of bacterial cross-talk inducing gene expression have begun to coculture organisms to discover new metabolites. Coculture of different microbes with a *Micromonospora* strain carrying multiple cryptic BGCs revealed metabolite production by the presence of red pigmentation not observed in monoculture. The compound, keyicin, was most active against *Rhodococcus*. A two-flask system was devised in which the two pure cultures were separated by a 0.2-μm filter that did not permit passage of either microbe. The presence of *Rhodococcus* elicited keyicin production without direct contact, and it exhibited susceptibility to keyicin. Draw growth curves for each strain growing in this system to illustrate keyicin production by *Micromonospora* and its effects on *Rhodococcus*. To confirm the correlation of a specific gene cluster with keyicin, under what conditions would you expect the BGC to be expressed?

   Read the original paper: Adnani, N., et al. 2017. Coculture of marine invertebrate-associated bacteria and interdisciplinary technologies enable biosynthesis and discovery of a new antibiotic, keyicin. *ACS Chem. Biol.* 12:3093.

24 Protists

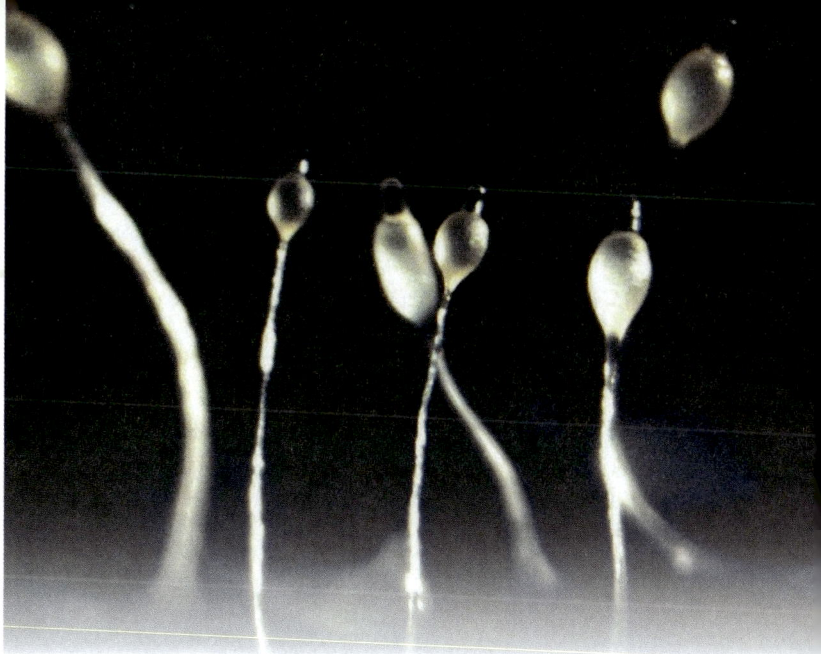

©Dr. Owen Gilbert

Sustainable Farming Practiced by Amoebae

The advent of farming changed the course of human evolution; without the need to hunt and gather, our ancestors could form larger, more cohesive social groups. However, we are not the first farmers. For decades, scientists have known of beetles, termites, and ants that farm fungi (*see figure 27.8*). Now an even more primitive kind of farming has been discovered. It is performed by an amoeba called *Dictyostelium discoideum* (Dicty for short).

Dicty lives in soil where it grows as solitary, single-celled microbes preying on bacteria. If Dicty runs short of bacterial food, cells join forces and form a multicellular slug that migrates to a new, food-rich area (section 24.3). Once there, the slug forms a fruiting body consisting of two parts: a spherical sorus that houses cells that have become spores and a supporting stalk, also made of Dicty cells. While it was well established that fruiting body formation is a means of dispersal to a site with a better food supply, about a third of Dicty fruiting bodies bring their food with them.

When grown in the presence of bacterial food, the farming Dicty stop eating before those that do not farm. Dicty farmers then pack the bacterial leftovers in each sorus. The sorus of nonfarming Dicty contains only amoebae. When slugs arrive at a spot deemed suitable for germination, the farmers have a ready food supply. So regardless of where they move, they are sure to have their "favorite" species for consumption.

But there are a couple of drawbacks with this system. First, if farmer and nonfarmer slugs are placed in a head-to-head migration race, the nonfarmers go farther (but not necessarily faster). This means the nonfarmer is more likely to escape the poor conditions that triggered slug formation in the first place. Furthermore, if both types of Dicty settle in regions of equal food supply, the farmers have no advantage; farmers and nonfarmers grow equally well. Nonetheless, bacterial husbandry can be a lifesaver, so why is it found in only a third of all isolates? The answer may rest with the fact that Dicty eats a varied diet, consuming a wide range of soil bacteria. This presumably has prevented the kind of co-evolution between prey and predator seen in other systems ranging from viruses and their cellular hosts to superpredators and their quarry on the African plains.

So far, our discussion of microbial diversity has focused on *Bacteria* and *Archaea*. We now survey the vast diversity of the eukaryotic microbial world, starting with protists and moving on to fungi in chapter 25. The general structure of eukaryotic cells is reviewed in chapter 5; here we focus on adaptations that make these microorganisms successful competitors.

Eukaryotic microorganisms suffer from a confused taxonomic history. The kingdom Protista, as presented in many introductory biology texts, is an artificial grouping of over 64,000 different single-celled life forms that lack common evolutionary heritage; that is, they are not monophyletic (*see figure 1.2*). In this text, we refer to these organisms simply as **protists.** Protists are unified only by what they lack: Absent is the level of tissue organization found in the more evolved fungi, plants, and animals. The term **protozoa** (s., protozoan; Greek *protos,* first; *zoon,* animal) refers to chemoorganotrophic protists. The term **algae** can be used to describe photosynthetic protists that, like plant cells, possess a cell wall. The study of photosynthetic protists (algae) is often referred to as **phycology** and is the realm of both botanists and protistologists. The study of all protists, regardless of their metabolic type, is called **protistology,** although the term **protozoology** is more commonly used.

Readiness Check:

Based on what you have learned previously, you should be able to:

- ✓ Draw a eukaryotic cell envelope and identify the structure and function of its component layers (section 5.2)
- ✓ List the three filaments that make up the cytoskeleton of eukaryotic cells and describe how they are used (section 5.3)
- ✓ Identify the structure and function of eukaryotic cellular organelles (sections 5.4–5.6)
- ✓ Describe the endocytic pathway in general terms (section 5.4)
- ✓ Explain the difference between a mitochondrion and a hydrogenosome (section 5.6)
- ✓ Describe the structure and function of eukaryotic flagella and cilia (section 5.7)

24.1 Protist Diversity Reflects Broad Phylogeny

After reading this section, you should be able to:

a. Describe three nutritional strategies used by protists
b. Draw a typical protist cell membrane and underlying features, labeling the plasmalemma, ecto- and endoplasm, and pellicle
c. Explain how contractile vacuoles are used to maintain osmotic stability
d. Trace the passage of a food particle as it is consumed by a protozoan
e. Compare binary fission as it occurs in a protist with that in a bacterium or archaeon
f. Outline the process of sexual reproduction and how it can vary in protists

Most eukaryotes are microbes. It is therefore not surprising that the vast diversity of protists is a function of their capacity to thrive in a wide variety of habitats. Their one common requirement is moisture because all are susceptible to desiccation. Most protists are free living and inhabit freshwater or marine environments. Many terrestrial chemoorganotrophic forms can be found in decaying organic matter and soil. Whether terrestrial or aquatic, protists play an important role in nutrient cycling. ▶| *Biogeochemical cycling sustains life on Earth (section 28.1)*

Some protozoa, or chemoorganoheterotrophic protists, are **saprophytes,** securing nutrients from dead organic material by releasing degradative enzymes into the environment. They then absorb the soluble products—a process sometimes called **osmotrophy.** Other protozoa employ **holozoic nutrition,** in which solid nutrients are acquired by phagocytosis. Photoautotrophic protists are strict aerobes and, like cyanobacteria, use photosystems I and II to perform oxygenic photosynthesis. It is difficult to classify the nutritional strategies of some protists because they simultaneously use reduced organic molecules and CO_2 as carbon sources. This strategy is called **mixotrophy.**

Protist Morphology

Despite their diversity, protists share many common features. In many respects, their morphology and physiology are the same as the cells of multicellular plants and animals. However, because many protists are unicellular, all of life's various functions must be performed within a single cell. Those that are multicellular lack highly differentiated tissues. Therefore the structural complexity observed in protists arises at the level of specialized organelles.

The protist cell membrane is called the **plasmalemma** and is identical to that of multicellular organisms (**figure 24.1**). In some protists, the cytoplasm immediately under the plasmalemma is divided into an outer gelatinous region called the **ectoplasm** and an inner fluid region, the **endoplasm.** The ectoplasm imparts rigidity to the cell body. Many protists also have a supportive layer or cuticle called the **pellicle.** The pellicle consists of the plasmalemma and a relatively rigid layer just beneath it. The pellicle may be simple in structure. For example, *Euglena* spp. are protists with

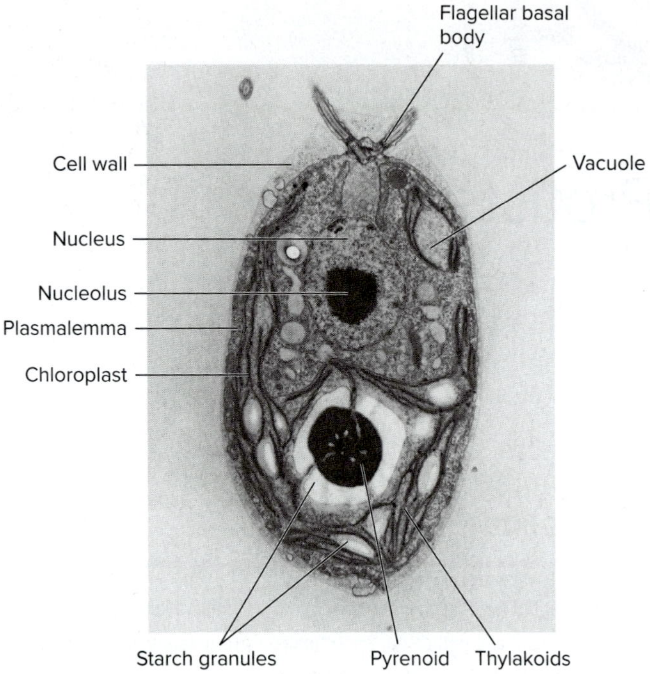

Figure 24.1 Protist Structure. A transmission electron micrograph of the photosynthetic protist *Chlamydomonas reinhardtii*, which measures about 10 μm in diameter.

a series of overlapping strips with a ridge at the edge of each strip fitting into a groove on the adjacent one, much like how the "tongue-and-groove" boards of a hardwood floor fit together. In contrast, the pellicles of ciliate protists are exceptionally complex with two membranes and a variety of associated structures. Although pellicles are not as strong and rigid as cell walls, for those that possess them, pellicles serve a protective function and impart the characteristic shape associated with that particular species.

One or more vacuoles are usually present in the cytoplasm of protozoa. These are differentiated into contractile, secretory, and food or phagocytic vacuoles. **Contractile vacuoles** function as osmoregulatory organelles in protists that live in hypotonic environments, such as freshwater lakes. Contractile vacuoles maintain osmotic balance by continuous water expulsion. Secretory vacuoles, or vesicles, are discussed in section 5.4. **Phagocytic vacuoles** are conspicuous in protists that ingest food by phagocytosis (holozoic protists). Phagocytic vacuoles are the sites of food digestion. In some organisms, they may occur anywhere on the cell surface, while others have a specialized structure for phagocytosis called the **cytostome** (cell mouth). Once the cytostome has engulfed a food particle, digestion commences when the phagocytic vacuole becomes acidic. As digestion proceeds, the vacuole membrane forms small blebs that pinch off and carry nutrients throughout the cytoplasm. The undigested contents of the original phagocytic vacuole are expelled from the cell either at a random site on the cell membrane or at a designated position called the **cytoproct** (Greek *proktos,* anus).

The diversity of energy-conserving organelles observed in protists reflects their evolutionary history and lifestyle. Most aerobic chemoorganotrophic protists have mitochondria, while photosynthetic forms have chloroplasts. A dense proteinaceous area, the **pyrenoid,** which is associated with the synthesis and storage of starch, may be present in chloroplasts. The majority of anaerobic chemoorganotrophic protists lack mitochondria; some of these organisms have hydrogenosomes. ◀ *Mitochondria, related organelles, and chloroplasts are involved in energy conservation (section 5.6)*

Many protists feature cilia or flagella at some point in their life cycle. Their formation is associated with a basal bodylike organelle called the kinetosome. In addition to aiding in motility, these organelles may be used to generate water currents for feeding and respiration.

Encystment and Excystment

Actively growing and replicating protists are called **trophozoites,** which roughly means "growing little animals." Many are also capable of forming metabolically dormant cysts, in a process called **encystment.** The cyst is morphologically simpler than the trophozoite and has a cell wall even if the protist lacks one when growing vegetatively. Cyst formation is particularly common among aquatic protists and parasitic forms. Cysts serve three major functions: (1) They protect against adverse changes in the environment, such as nutrient deficiency, desiccation, adverse pH, and low levels of O_2; (2) they are sites for nuclear reorganization and cell division (reproductive cysts); and (3) in parasitic species they are the infectious stage between hosts. Protists escape from cysts by a process called **excystment** (**figure 24.2**). Although the exact stimulus for excystment is unknown for most protists, it is generally triggered by a return to favorable environmental conditions. For example, cysts of parasitic species excyst after ingestion by the host.

Protist Reproductive Cells and Structures

Most protists have both asexual and sexual reproductive phases in their life cycles. The most common method of asexual reproduction is **binary fission.** During this process, the nucleus first undergoes mitosis and then the cytoplasm divides by cytokinesis to form two identical individuals (**figure 24.3**). Multiple fission is also common, as is budding. Some filamentous, photosynthetic protists undergo fragmentation so that each piece of the broken filament contains a nucleus and grows independently.

Sexual reproduction involves the formation of gametes. Protist cells that produce gametes are termed **gamonts.** The fusion of haploid gametes is called **syngamy.** Among protists, syngamy can involve the fusion of two morphologically similar gametes (isogamy) or two morphologically different types (anisogamy). Meiosis may occur before the formation and union of gametes, as in most animals, or just after fertilization, as is the case with lower plants. Furthermore, the exchange of nuclear material may occur in the familiar fashion—between two different individuals (**conjugation**)—or by the development of a genetically distinct nucleus within a single individual (autogamy).

With this level of reproductive complexity, perhaps it is not surprising that the nuclei among protists show considerable diversity. Most commonly, a vesicular nucleus is present. This is 1 to 10 μm in diameter, spherical, and has a distinct nucleolus and uncondensed chromosomes (figure 24.1). The uncondensed chromosomes give these nuclei a lacy or vesicular appearance when viewed microscopically. Ovular nuclei are up to 10 times this size and possess many peripheral nucleoli. Still others have chromosomal nuclei, in which the chromosomes remain condensed throughout the cell cycle. Finally, many ciliated forms have two types of nuclei: a large macronucleus with distinct nucleoli and condensed chromatin, and a smaller, diploid micronucleus with dispersed chromatin but

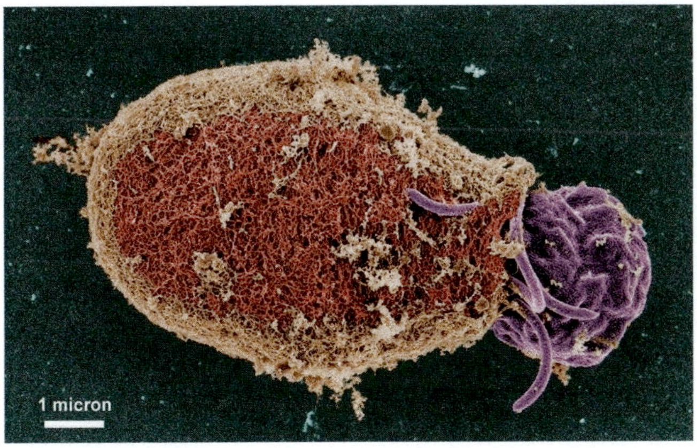

Figure 24.2 Excystment. This colorized scanning electron micrograph shows the protist *Giardia* sp. undergoing excystment. The flagellated vegetative cell, called a trophozoite, is emerging from the right of the cyst.
Source: CDC/Dr. Stan Erlandsen

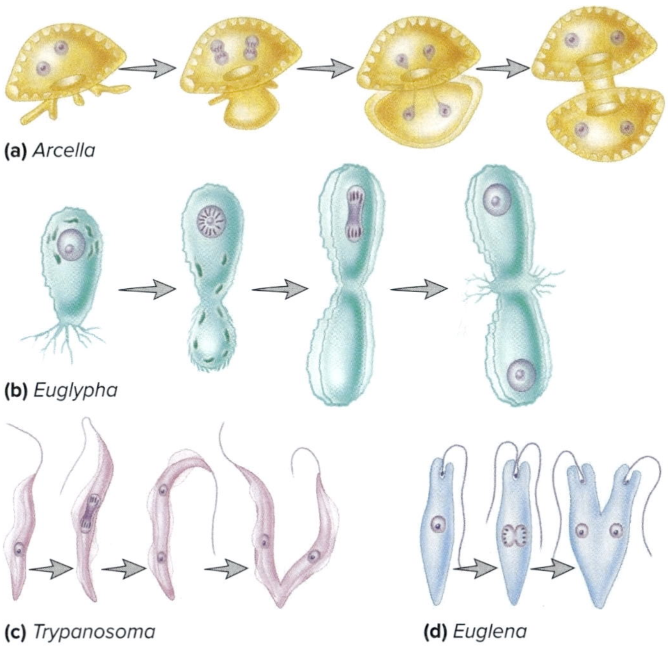

Figure 24.3 Binary Fission in Protists. (a) The two nuclei of a testate (shelled) amoeba in the genus *Arcella* divide as some of its cytoplasm is extruded and a new test for the daughter cell is secreted. (b) Members of another testate amoeba genus, *Euglypha,* begin to secrete new platelets before cytoplasm begins to move out of the aperture. The nucleus divides while the platelets are used to construct the test for the daughter cell. For many protists, all organelles must be replicated before the cell divides. This is also the case with species of (c) *Trypanosoma* and (d) *Euglena.*

lacking nucleoli. Genes in the macronucleus encode proteins involved in housekeeping functions and binary fission, whereas genes on the micronucleus function only in sexual reproduction.

Protist Taxonomy Is Controversial

Ever since Antony van Leeuwenhoek described the first protozoan "animalcule" in 1674, the taxonomic classification of protists has remained in flux. During the twentieth century, classification schemes were based on morphology and protists were typically classified into four major groups based on their means of locomotion: flagellates (*Mastigophora*), ciliates (*Infusoria* or *Ciliophora*), amoebae (*Sarcodina*), and stationary forms (*Sporozoa*). Although these terms may still be encountered, they are without evolutionary context. While it is now agreed that the old classification system is best abandoned, little agreement exists on what should take its place. Here we use a modified version of the higher-level classification system for the eukaryotes based on morphological, biochemical, and phylogenetic analyses proposed by the International Society of Protistologists in 2005. This scheme does not use formal hierarchical rank designations such as class and order, reflecting the fact that protist taxonomy remains an area of active research.

Comprehension Check

1. Describe the pellicle. What is its function?
2. Trace the path of a food item from cytostome to cytoproct.
3. What functions do cysts serve for a typical protist? What causes excystment to occur?
4. What is a gamont? What is the difference between anisogamy, isogamy, and syngamy?
5. How does conjugation differ from autogamy?
6. Describe vesicular, ovular, and chromosomal nuclei. Which is most like the nuclei found in higher eukaryotes?
7. Why has taxonomic classification of protists been problematic?

24.2 Supergroup *Excavata:* Primitive Eukaryotes

After reading this section, you should be able to:

a. Differentiate a mitosome from a mitochondrion
b. Identify which members of *Excavata* have hydrogenosomes
c. Identify a location in your community where you might find protists in the taxa *Fornicata, Parabasalia,* and *Euglena*
d. List one human and one veterinary disease caused by species of *Fornicata* and *Parabasalia*
e. Draw a typical euglenoid cell and explain the function of the stigma and pyrenoid
f. Discuss the etiology of Chagas' disease and African sleeping sickness

Excavata includes some of the most primitive eukaryotes. Most possess a cytostome characterized by a suspension-feeding groove with a posteriorly directed flagellum used to generate a feeding current. This enables the capture of small particles. Those that lack this morphological feature are presumed to have had it at one time during their evolution; that is, it is thought to have been secondarily lost.

Fornicata

Ever curious, van Leeuwenhoek described *Giardia lamblia* (**figure 24.4*a***) from his own diarrheic feces. Over 300 years later, this species, now known as *G. intestinalis,* continues to be a public health concern. Today this microaerophilic protist most often infects campers and other individuals who

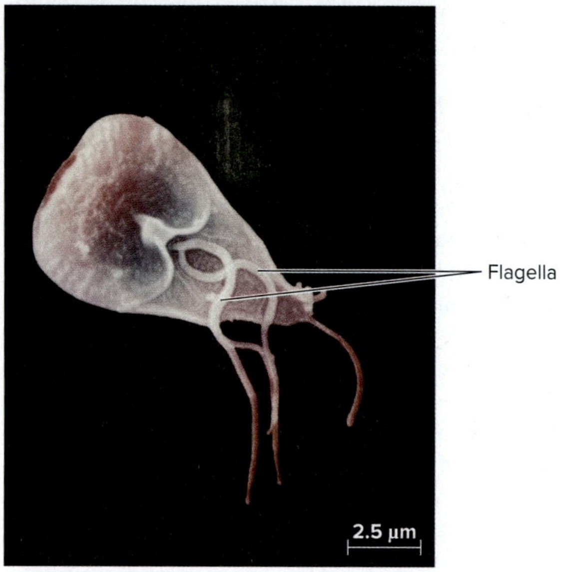

(a) *Giardia intestinalis*

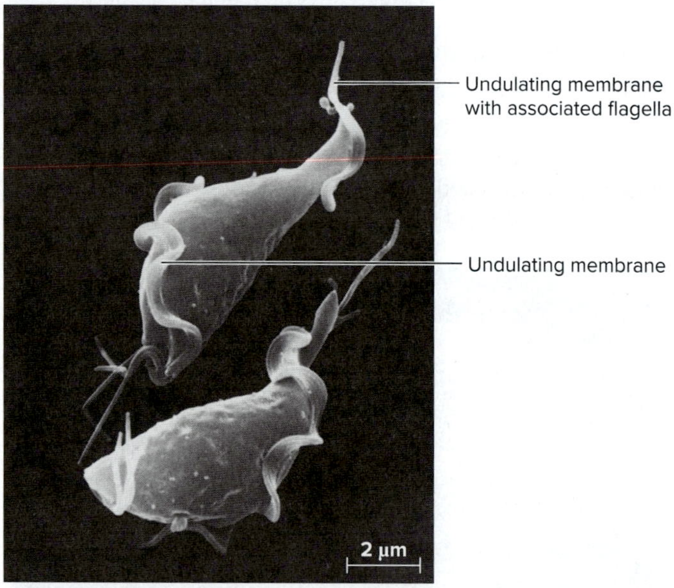

(b) *Trichomonas vaginalis*

Figure 24.4 Parasitic Members of the Supergroup *Excavata.* (a) *Giardia intestinalis.* (b) *Trichomonas vaginalis.* These specialized parasitic flagellates absorb nutrients from living hosts.
(a) ©Science Source; (b) ©David M. Phillips/The Population Council/Science Source

MICRO INQUIRY *How are these two protists morphologically similar but physiologically different?*

unwittingly consume contaminated water. It is the most common cause of epidemic waterborne diarrheal disease in the United States (about 30,000 cases yearly). Like all members of *Fornicata, G. intestinalis* bears flagella and lacks mitochondria. However, it possesses mitochondria-like double-bounded membrane organelles called **mitosomes.** Although these primitive organelles share ancestry with mitochondria and hydrogenosomes, they are not involved in ATP synthesis. Instead the mitosomes of *G. intestinalis* are involved in producing FeS proteins and energy is produced by substrate level phosphorylation in the cytoplasm. ▶| *The aerobic mitochondrion belongs to the family of mitochondrial-related organelles (section 5.6)*

Cyst formation is key to *G. intestinalis* survival. Ingestion of only about 10 cysts is needed for infection. Within the small intestine, bile and slight alkalinity trigger excystment (figure 24.2). Trophozoites then attach to gut cells called enterocytes. When these enterocytes die and are sloughed from the gut, the microbe must "decide" whether to attach to another enterocyte or encyst and exit to find a new host. ▶| *Giardiasis (section 40.5)*

Unlike *Giardia* spp. most members of *Fornicata* are harmless symbionts. The few free-living forms are most often found in waters that are heavily polluted with organic nutrients (a condition known as eutrophication). Only asexual reproduction by binary fission has been observed. Other pathogenic species include *Hexamida salmonis,* a fish parasite found in hatcheries and fish farms, and *H. meleagridis,* a turkey pathogen that is responsible for the annual loss of millions of dollars in poultry revenue.

Parabasalia

Members of *Parabasalia* are flagellated; most are endosymbionts of animals. They use phagocytosis to engulf food items, although they lack a distinct cytostome. Here we consider two subgroups: *Trichonymphida* and *Trichomonadida*. Trichonymphids are obligate mutualists in the digestive tracts of wood-eating insects such as termites and wood roaches. They secrete the enzyme cellulase needed for the digestion of wood particles, which they entrap with pseudopodia. One species, *Trichonympha campanula,* can account for up to one-third of the biomass of an individual termite (*see figure 27.2*). Although asexual reproduction is the norm, a hormone called ecdysone produced by the host when molting triggers sexual reproduction. ▶| *Microorganism-insect mutualisms (section 27.2)*

Trichomonadida spp., or simply trichomonads, do not require oxygen and possess hydrogenosomes rather than mitochondria (*see figure 5.15*). They undergo asexual reproduction only. They are symbionts of the digestive, reproductive, and respiratory tracts of many vertebrates. *Tritrichomonas foetus* is a cattle parasite and an important cause of spontaneous abortion in these animals. Four species infect humans: *Dientamoeba fragilis, Pentatrichomonas hominis, Trichomonas tenax,* and *T. vaginalis* (figure 24.4*b*). *D. fragilis* has recently been recognized as a cause of diarrhea, while *T. vaginalis* has long been known to be pathogenic. It is found in the genitourinary tract of both men and women, and causes the sexually transmitted infection trichomoniasis. Because the protozoan does not live outside the host, it does not have a cyst stage in its life cycle. ▶| *Trichomoniasis (section 40.4)*

Euglenozoa

Members of *Euglenozoa* are commonly found in freshwater, although a few species are marine. About one-third of euglenids are photoautotrophic; the remaining are free-living chemoorganotrophs. Most chemoorganotrophic forms are saprophytic, although a few parasitic species have been described. The representative genus *Euglena* contains photoautotrophic euglenids. A typical *Euglena* spp. cell (**figure 24.5**) is elongated and bounded by a plasmalemma. The pellicle consists of proteinaceous strips and microtubules; it is elastic enough to enable turning and flexing of the cell, yet rigid enough to prevent excessive alterations in shape. *Euglena* spp. contain chlorophylls *a* and *b* together with carotenoids. The large nucleus contains a prominent nucleolus. Euglenids store carbon as paramylon (a polysaccharide composed of β(1→3) linked glucose molecules), which is unique to euglenids. A red eyespot called a **stigma** helps the organism orient to light and is located near an anterior reservoir. A contractile vacuole near the reservoir continuously collects water from the cell and empties it into the reservoir, thus regulating the osmotic pressure within the organism. Two flagella arise from the base of the reservoir, although only one may emerge from the canal and actively beat to move the cell. Reproduction in euglenids is by longitudinal mitotic cell division.

Trypanosomes are important members of *Euglenozoa*. These microbes exist only as parasites of plants and animals, and have global significance. For example, members of the genus *Leishmania,* which cause a group of conditions collectively termed leishmaniasis, are typanosomes that affect some 12 million people. Chagas' disease is caused by *Trypanosoma cruzi,* which is transmitted by "kissing bugs" (*Triatominae*), so called because they bite the face of sleeping victims (*see figures 40.10*

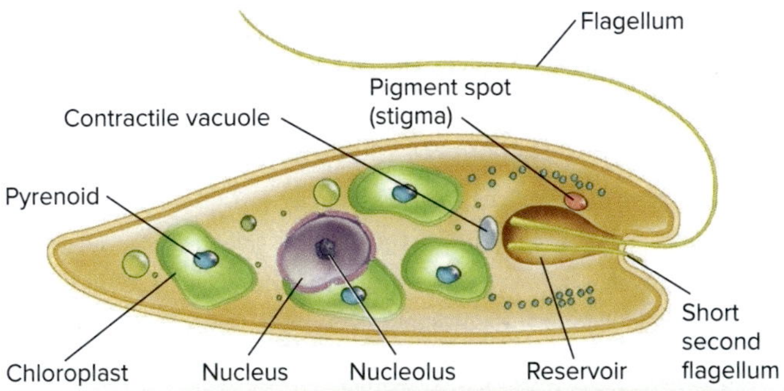

Figure 24.5 ***Euglena:* Principle Structures Found in This Euglenoid.** Notice that a short second flagellum does not emerge from the anterior invagination. In some euglenids, both flagella are emergent.

MICRO INQUIRY *What are the functions of the stigma and the reservoir?*

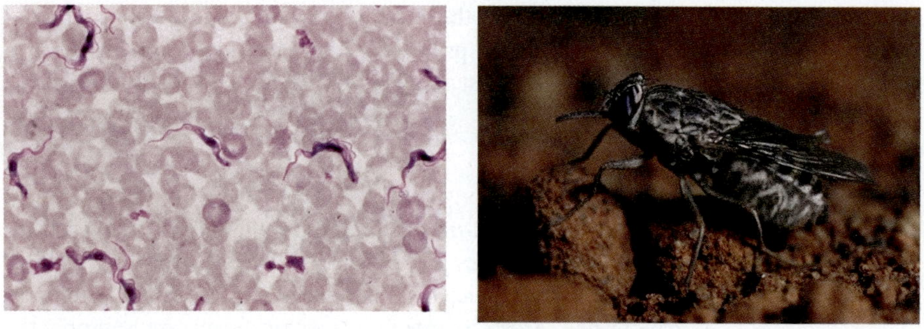

(a) (b)

Figure 24.6 The Euglenozoan *Trypanosoma* and Its Insect Host. (a) *Trypanosoma* among red blood cells. Note the dark-staining nuclei, anterior flagella, and undulating, changeable shape. (b) The tsetse fly is an important vector of the *Trypanosoma* species that causes African sleeping sickness.
(a) Source: CDC/Dr. Mae Melvin; (b) ©Oxford Scientific/Getty Images

to 40.12). ▶| *Leishmaniasis (section 40.3); Trypanosomiasis (section 40.3)*

The trypanosomes *T. brucei* subspecies *rhodesiense* and subspecies *gambiense* cause African sleeping sickness (**figure 24.6**). Ingestion of these parasites by the blood-sucking tsetse fly triggers a complex cycle of development and reproduction, first in the fly's gut and then in its salivary glands (*see figure 40.11*). From there the parasite is easily transferred to a vertebrate host, where it often causes a fatal infection. It is estimated that about 50,000 people die annually of sleeping sickness. The presence of this dangerous parasite prevents the use of about 11 million square kilometers of African grazing land.

Comprehension Check

1. List at least three features that distinguish *Parabasalia* from *Euglenozoa*.
2. What is the function of the plasmalemma and pellicle in euglenid cells? How are these structures similar to the bacterial cell wall?
3. Members of which *Euglenozoa* genera cause disease?

24.3 Supergroup *Amoebozoa* Includes Protists with Pseudopodia

After reading this section, you should be able to:

a. Draw three different types of pseudopodia
b. Differentiate between a naked and a testate amoeba
c. Compare and contrast the structure, motility, and life cycles of acellular and cellular slime molds
d. Explain the importance of molecular signaling in the morphological development of *Dictyostelium discoideum*
e. Identify a nearby location where you might find members of the genera *Tubulinea, Entamoebida,* and *Eumycetozoa*

Because the amoeboid form arose independently numerous times from various flagellated ancestors, some amoebae are placed in the supergroup *Amoebozoa,* while others are in *Rhizaria*. One of the morphological hallmarks of amoeboid motility is the use of **pseudopodia** (meaning "false feet") for both locomotion and feeding (**figure 24.7**). Pseudopodia can be rounded (**lobopodia**), long and narrow (**filopodia**), or form a netlike mesh (**reticulopodia**). Amoebae that lack a cell wall and are surrounded only by a plasma membrane are called **naked amoebae.** In contrast, the plasma membrane of a **testate amoeba** is covered by material that is either made by the protist itself or collected by the organism from the environment. Binary fission is the usual means of asexual division, although some amoebae form cysts that undergo multiple fission.

Tubulinea

Members of *Tubulinea* inhabit almost any environment where they will remain moist; this includes glacial meltwater, marine environments, tide pools, lakes, and streams. Free-living forms are known to dwell in ventilation ducts and cooling towers, where they feed on microbial biofilms. Others are endosymbionts, commensals, or parasites of invertebrates, fishes, and

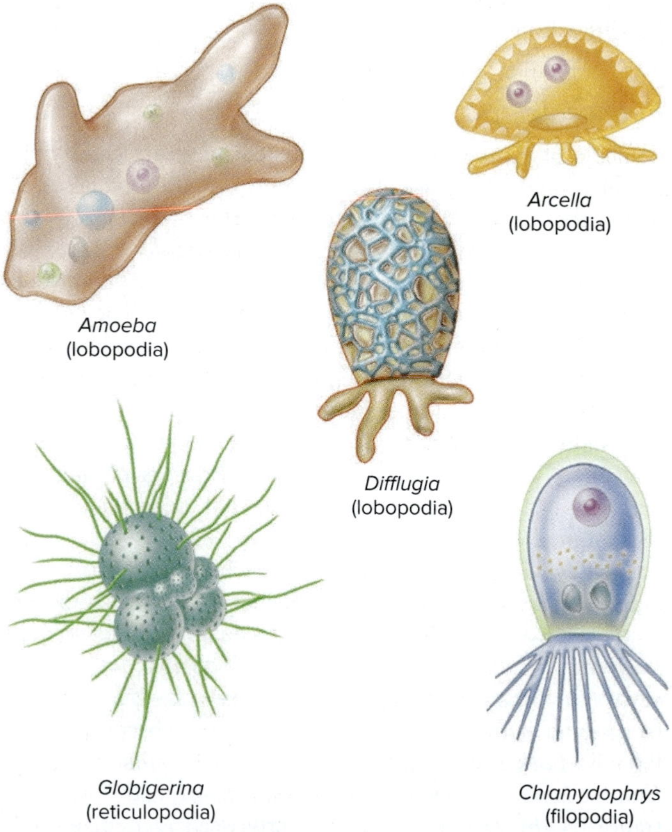

Figure 24.7 Pseudopodia.

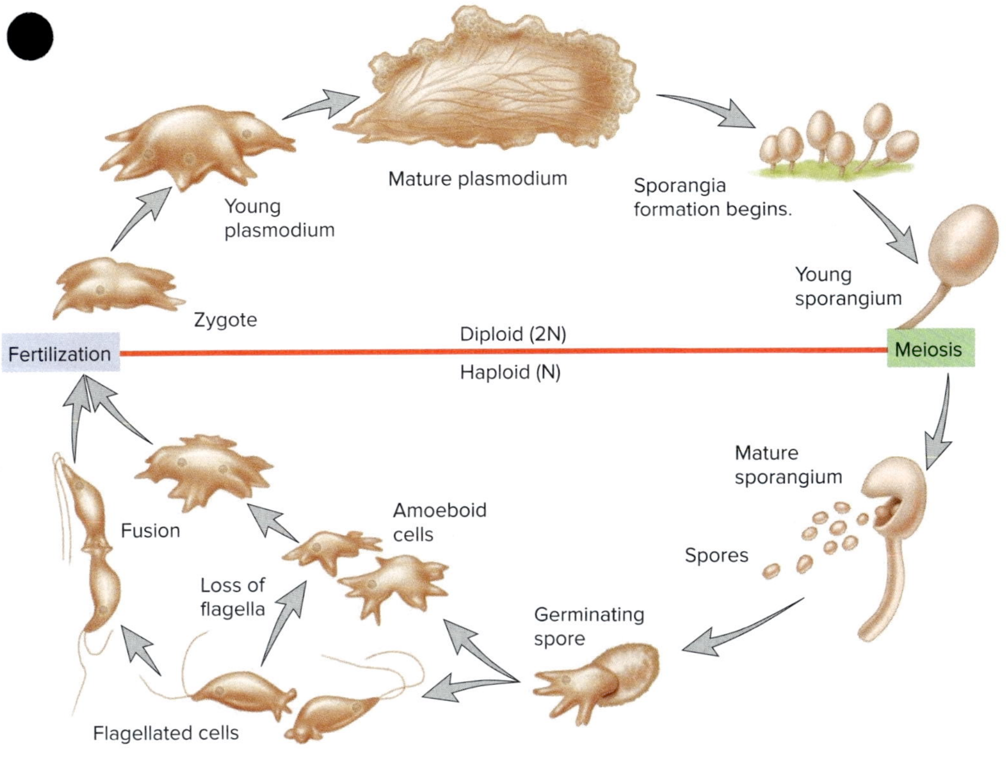

(a)

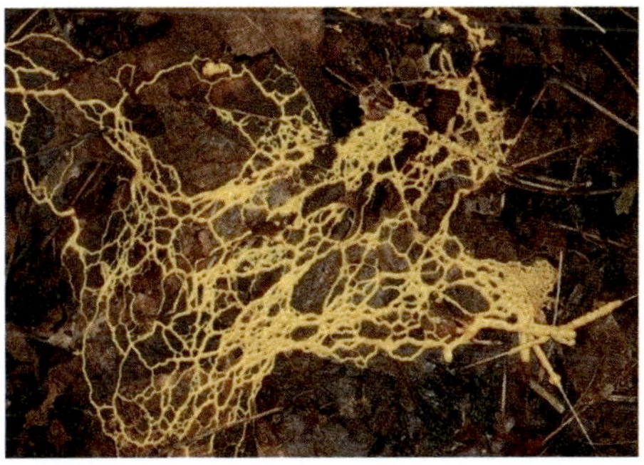
(b) *Physarum* sp.

(c) *Physarum polycephalum*

(d) *Stemonitis*

Figure 24.8 Acellular Slime Molds. (a) The life cycle of a plasmodial slime mold includes sexual reproduction; when conditions are favorable for growth, the adult diploid forms sporangia. Following meiosis, the haploid spores germinate, releasing haploid amoeboid or flagellated cells that fuse. (b) Plasmodium of the slime mold *Physarum* sp. Sporangia of (c) *Physarum polycephalum,* and (d) *Stemonitis* sp.

MICRO INQUIRY *How does a plasmodium differ from a single cell?*

mammals. Some harbor intracellular symbionts, including algae, bacteria, and viruses, but the nature of these relationships is not well understood. *Amoeba proteus,* a favorite among introductory biology laboratory instructors, is included in this group.

Entamoebida

Like *Excavata* protists, entamoebid protists lack mitochondria but contain mitosomes. However, the identification of about 20 putative mitochondrial proteins in the free-living amoeba *Mastigamoeba balamuthi* suggests that these mitosomes are more advanced than those found in *Giardia* spp. *M. balamuthi* is a relative of *Entamoeba histolytica,* the cause of amoebic dysentery—the third-leading cause of parasitic death worldwide. In fact, all *Entamoeba* spp. are either parasites like *E. histolytica* or commensals—that is, living in, but causing no harm to, their hosts. ▶| *Amebiasis (section 40.5)*

Eumycetozoa

First described in the 1880s, members of *Eumycetozoa* or "slime molds" have been classified as plants, animals, and fungi. As we examine their morphology and behavior, the source of this confusion should become apparent. Recent analysis of certain proteins (e.g., elongation factor EF-1, α-tubulin, and actin) as well as physiological, behavioral, biochemical, and developmental data point to a monophyletic group. *Eumycetozoa* includes *Myxogastria* and *Dictyostelia*.

The life cycle of **acellular slime mold,** exemplified by those in the genus *Myxogastria*, includes a distinctive stage when the organisms exist as streaming masses of colorful protoplasm. The protoplasm creeps along in amoeboid fashion over moist, rotting logs, leaves, and other organic matter, which it degrades (**figure 24.8**). Acellular slime molds derive their name from the stage in their life cycle when a large, multinucleate mass called a plasmodium is formed; there can be as many as 10,000 synchronously dividing nuclei within a single plasmodium (figure 24.8*b*). When starved or dried, the plasmodium develops ornate fruiting bodies. As these mature, they form stalks with cellulose walls that are resistant to environmental stressors (figure 24.8*c,d*). When conditions improve, spores germinate and release haploid amoeboflagellates. These fuse, and as the resulting zygotes feed by endocytosis, nuclear divisions give rise to the multinucleate plasmodium.

Cellular slime molds (*Dictyostelia*) use endocytosis to feed on bacteria and yeasts. The species *Dictyostelium discoideum* is an attractive model organism. During its life cycle individual cells cooperate to form different morphologies. Multicellularity is demonstrated in the pseudoplasmodium, which consists of thousands of individual vegetative cells moving together as a mass (**figure 24.9**, step 2). The pseudoplasmodium forms when starved cells release the signaling molecules, cyclic AMP (cAMP) and a specific glycoprotein. Other *D. discoideum* cells sense these compounds and respond by forming an aggregate around the signal-producing cells. In this way, large, motile, multicellular slugs develop and serve as precursors to fruiting body formation (figure 24.9, step 3). Fruiting body morphogenesis commences when the slug stops and cells pile on top of each other. Cells at the bottom of this vertically oriented structure form a stalk by secreting cellulose, while cells at the tip differentiate into spores (figure 24.9, steps 4 and 5). Germinated spores become vegetative amoebae to start this asexual cycle anew (figure 24.9, step 6).

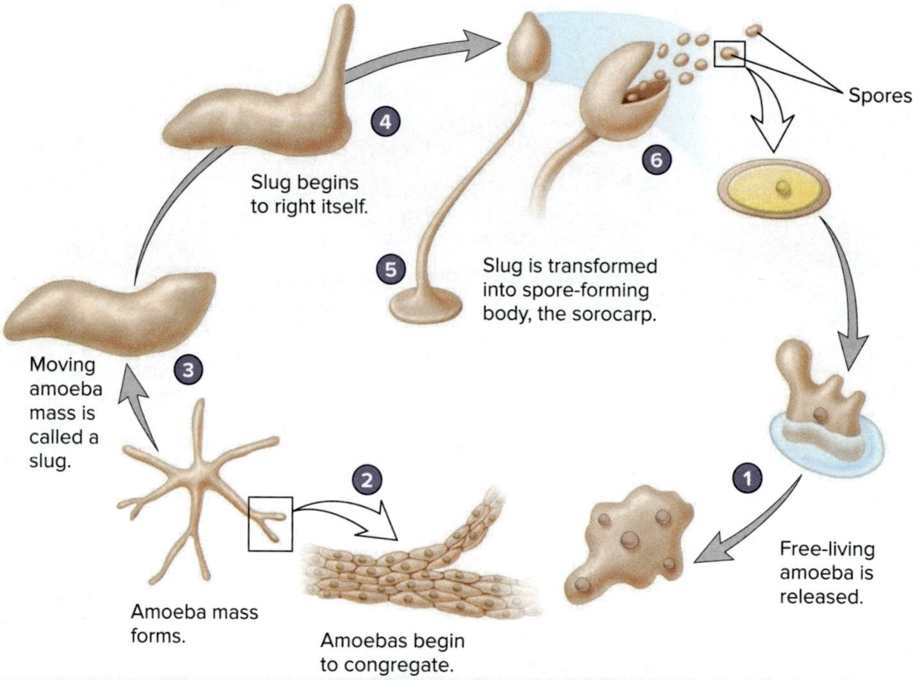

Figure 24.9 Life Cycle of *Dictyostelium discoideum*, a Cellular Slime Mold.

MICRO INQUIRY *How would a mutant strain unable to secrete cAMP behave if mixed with a wild-type strain? What if it were mixed with another cAMP mutant?*

Dictyostelium spp. display complex behaviors. In addition to farming, described in the chapter-opening story, they also differentiate to resemble primitive immune cells. During slug formation, some cells become "sentinel cells" and vanquish harmful bacteria. Sentinel cells accomplish this by producing proteins similar to those involved in immune responses in higher organisms. They roam within the slug patrolling for pathogenic bacteria such as *Legionella pneumophila,* which infect *Dictyostelium* spp. Immunologists are not too surprised, noting that all multicellular organisms need protection against bacterial pathogens.

Sexual reproduction in *D. discoideum* involves the formation of special spores call macrocysts. These arise by a form of conjugation that has some unusual features. First, a group of amoebae become enclosed within a wall of cellulose. This brings cells into close contact so that conjugation can occur between members of different mating types. There are three mating types determined by the sequence of a single gene, as well as those capable of self-fertilization. Following conjugation, a large amoeba forms and cannibalizes the remaining amoebae. The now-giant amoeba matures into a macrocyst. Macrocysts can remain dormant within their cellulose walls for extended periods. Vegetative growth resumes after the diploid nucleus undergoes meiosis to generate haploid amoebae.

Comprehension Check

1. Describe filopodia, lobopodia, and reticulopodia form and function.
2. Why have slime molds been so hard to classify?
3. Describe the life cycle of *Dictyostelium discoideum*. Predict the behavior of a mutant that fails to secrete cAMP.

24.4 Supergroup SAR: Protists of Global Importance

After reading this section, you should be able to:

a. Describe the morphology of the filopodia and axopodia found in members of *Rhizaria*
b. Explain the morphology, feeding mechanisms, and reproduction of members of *Radiolaria* and *Foraminifera*
c. Describe the morphology, motility, and feeding behaviors of dinoflagellates
d. Explain the importance of dinoflagellates to coral biology and the generation of "red tides"
e. Diagram how a ciliate captures and consumes a food particle

f. Compare and contrast the morphology and function of the micro- and macronucleus and trace their fate during ciliate conjugation
g. Identify the unique structures found in an apicomplexan cell and explain their function
h. Outline the apicomplexan life cycle
i. List at least two human diseases caused by apicomplexans
j. Explain the importance of heterokont flagella in the taxonomy of the *Stramenopila*
k. Review the structure, morphology, reproduction, and ecological importance of diatoms
l. Deduce why the peronosporomycetes and *Labyrinthulids* were erroneously considered fungi

The supergroup SAR, which stands for stramenopiles, alveolates, and *Rhizaria,* has only recently been recognized. Phylogenomic evidence strongly supports the view that these organisms form a monophyletic clade. As such, it elevates stramenopiles and alveolates to the level of supergroup, and eliminates *Chromalveolata,* to which they were formerly assigned. The SAR supergroup encompasses some of the world's most important protists, including those that fix as much carbon as the rain forests (diatoms) and those that infect over 200 million people each year worldwide (*Plasmodium* spp., which cause malaria).

Rhizaria

Members of *Rhizaria* include those in *Radiolaria* and *Foraminifera.* These organisms are amoeboid in morphology and can be distinguished by their filopodia, which can be simple, branched, or connected. Filopodia supported by microtubules are known as **axopodia** (s., axopodium). Axopodia protrude from a central region of the cell called the axoplast and are primarily used in feeding (**figure 24.10*a***).

Radiolaria

Radiolaria are notable for their beautifully intricate, microscopic skeletons. Most have an internal skeleton made of siliceous material; however, members of the subgroup *Acantharia* have endoskeletons consisting of strontium sulfate. A few genera contain species with an exoskeleton of siliceous spines or scales, while some lack a skeleton completely. Skeletal morphology is highly variable and often includes radiating spines that help the organisms float, as does the storage of oils and other low-density fluids (figure 24.10*b*).

Radiolaria feed by endocytosis, using mucus-coated filopodia to entrap prey including bacteria, other protists, and even small invertebrates. Large prey items are partially digested extracellularly by the secretion of degradative enzymes before becoming encased in a food vacuole. Many surface-dwelling radiolarians have algal symbionts that enhance their net carbon assimilation. Acantharians reproduce only sexually by consecutive mitotic and meiotic divisions that release hundreds of biciliated cells. Asexual reproduction (binary or multiple fission, or budding) is most common in other radiolaria, but sexual reproduction can be triggered by a heavy feeding or nutrient limitation. In this case, two haploid nuclei fuse to form a diploid zygote encased in a cyst from which it is released when survival conditions improve.

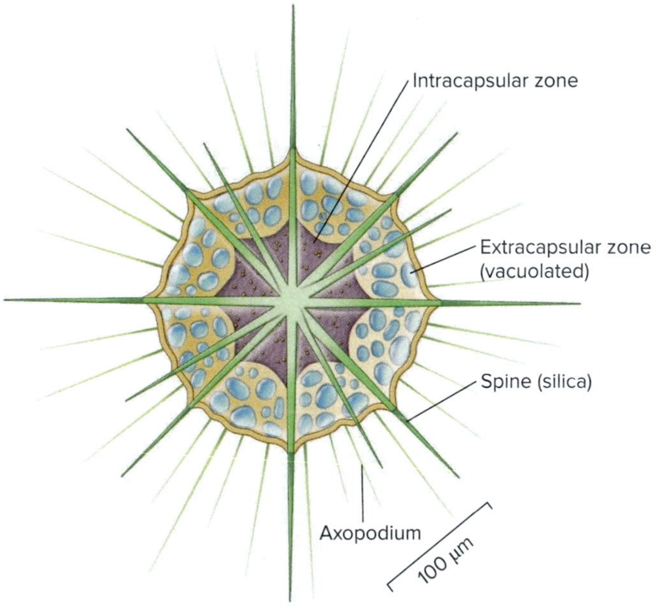

(a)

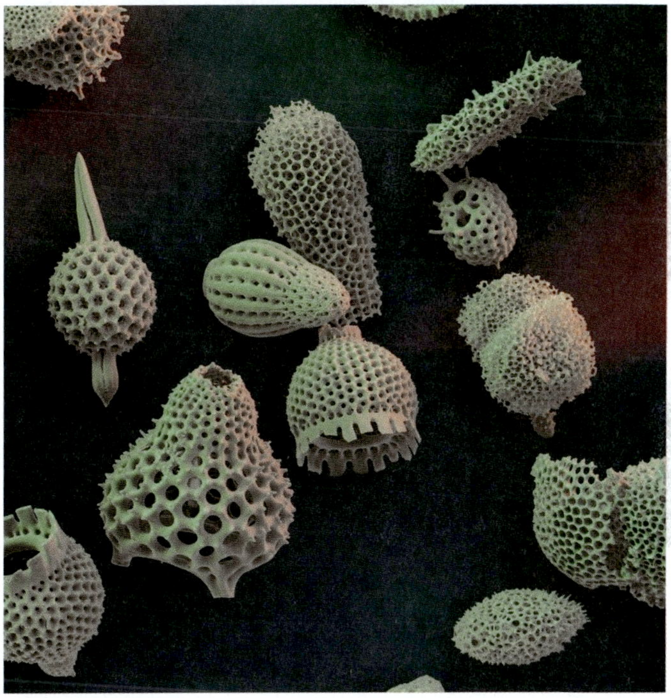

(b) Radiolarian shells

Figure 24.10 Radiolaria. (a) The radiolarian *Acanthometra elasticum* demonstrates the internal skeleton. (b) Radiolarian shells made of silica.
(b) ©Eye of Science/Science Source

Foraminifera

Members of *Foraminifera* (also called forams) range in size from roughly 20 µm to several centimeters. Their filopodia are arranged in a branching network called reticulopodia (Latin

Figure 24.11 A Foraminiferan. The reticulopodia project through pores in the calcareous test, or shell, of this protist.
©David Caron/Science Source

Figure 24.12 White Cliffs of Dover. The limestone that forms these cliffs is composed almost entirely of fossil shells of protists, including foraminifera.
©Elena Castaldi/123RF

reticulum, little net; reticulopodia thus means "little net feet"). These bear vesicles at their tips that secrete a sticky substance used to trap prey. Many species harbor endosymbiotic algae that migrate out of reticulopodia (without being eaten) to expose themselves to more sunlight. These algae contribute significantly to foram nutrition. Forams have characteristic tests (plates) arranged in multiple chambers that are sequentially added as the protist grows (**figure 24.11**).

Some forams are free-floating planktonic forms but most are benthic, living within the sediments. A few benthic forams inhabiting anoxic sediments perform denitrification, reducing nitrate completely to dinitrogen gas. Foraminifera were the first example of a denitrifying eukaryote; it is now known that some diatoms and fungi are also capable of this process. ▶| *Nitrogen cycle (section 28.1)*

Foraminiferan life cycles can be complex. While some smaller species reproduce only asexually by budding or multiple fission, larger forms frequently alternate between sexual and asexual phases. During the sexual phase, flagellated gametes pair, fuse, and generate asexual individuals (agamonts). Meiotic division of agamonts gives rise to haploid gamonts. There are several mechanisms by which gamonts become diploid again. For instance, a variety of forams release flagellated gametes that become fertilized in the open water. In others, two or more gamonts attach to one another, enabling gametes to fuse within the chambers of the paired tests. When the shells separate, newly formed agamonts are released.

Forams are found in marine and estuarine habitats. Some are planktonic, but because most are benthic, their tests accumulate on the seafloor, where they create a fossil record dating back to the Early Cambrian (543 million years ago). This is helpful in oil exploration. Their remains, or ooze, can be up to hundreds of meters deep in some tropical regions. Foram tests make up most modern-day chalk, limestone, and marble, and formed the familiar white cliffs of Dover in England (**figure 24.12**). The stones used to build the great pyramids of Egypt are also made of forams.

Alveolata

Alveolata is a large group that includes three imporant taxa: *Dinoflagellata* (dinoflagellates), *Ciliophora* (ciliates), and *Apicomplexa* (apicomplexans). We begin our discussion with **dinoflagellates**—a large group most commonly found in marine plankton, where some species are responsible for the phosphorescence sometimes seen in seawater. Their nutrition is complex; photoautotrophy, heterotrophy, and mixotrophy are all observed. Most of the heterotrophic forms are saprophytic (using osmotrophy to assimilate nutrients), but some also use endocytosis. All dinoflagellates bear two distinctively placed flagella: One is wrapped around a transverse groove (the girdle), and the other is draped in a longitudinal groove (the sulcus; **figure 24.13**). The orientation and beating patterns of these flagella cause the cell to spin as it is propelled forward; the name dinoflagellate is derived from the Greek *dinein,* to whirl. Many dinoflagellates are covered with cellulose plates secreted within alveolar sacs that lie just under the plasma membrane. These forms are said to be thecate or armored; those with empty alveoli are called athecate or naked. Like some other protists, dinoflagellates have specialized compressed proteins called **trichocysts.** When attacked, trichocysts shoot out from the cell as a means of defense.

Most dinoflagellates are free living, although some form important associations with other organisms. Endosymbiotic dinoflagellates that live as undifferentiated cells occasionally send out motile cells called **zooxanthellae.** The most well-known

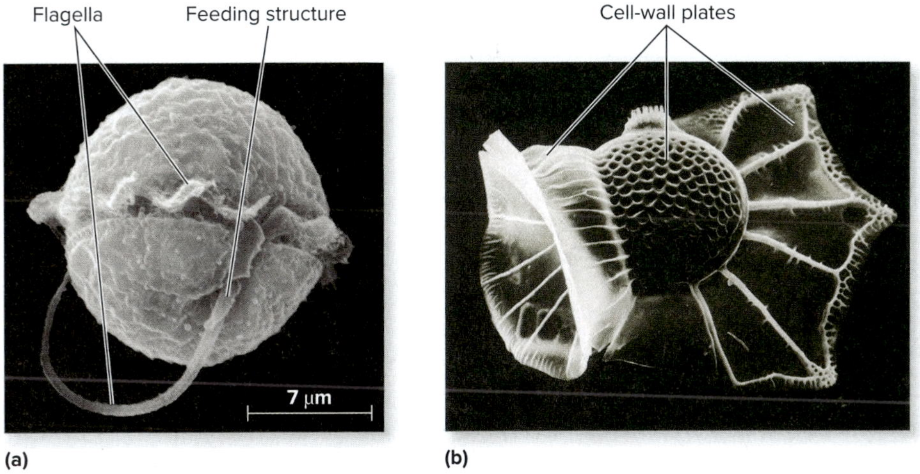

Figure 24.13 Dinoflagellates. (a) The surface of *Peridiniopsis berolinensis* is smooth because the alveoli appear empty. Two types of flagella can be seen in this freshwater dinoflagellate. One flagellum coils around the girdle; as it moves, this flagellum causes the cell to spin. The straight flagellum that extends from the cell acts as a rudder to determine the direction of movement. (b) The alveoli of a marine dinoflagellate in the genus *Ornithocercus* contains cellulose cell-wall plates. The cell wall extends to form sail-like structures.
(a) ©Lee W. Wilcox, Ph.D.; (b) ©Claude Taylor, III

MICRO INQUIRY *What do you think is the function of the "sails" in* Ornithocercus *sp.?*

zooxanthellae belong to the genus *Symbiodinium*. These are photosynthetic endosymbionts of reef-building coral (*see figure 27.3*). They provide fixed carbon to the coral animal and help maintain the internal chemical environment needed for the coral to secrete its calcium carbonate exoskeleton. Dinoflagellates are also responsible for toxic "red tides" that harm other organisms, including humans. Common red tide dinoflagellates include members of the genera *Alexandrium* and *Lingulodinium* (formerly called *Gonyaulax; see figure 30.3*). ▶| *Zooxanthellae (section 27.2); Microorganisms in coastal ecosystems are adapted to a changeable environment (section 30.2)*

Ciliates (*Ciliophora*) include about 12,000 known species. All are chemoorganotrophic and range from about 10 μm to 4.5 mm long. They inhabit both benthic and planktonic communities in marine and freshwater systems, as well as moist soils. As their name implies, they employ many cilia for locomotion and feeding. The cilia are generally arranged either in longitudinal rows (**figure 24.14**) or in spirals around the body of the

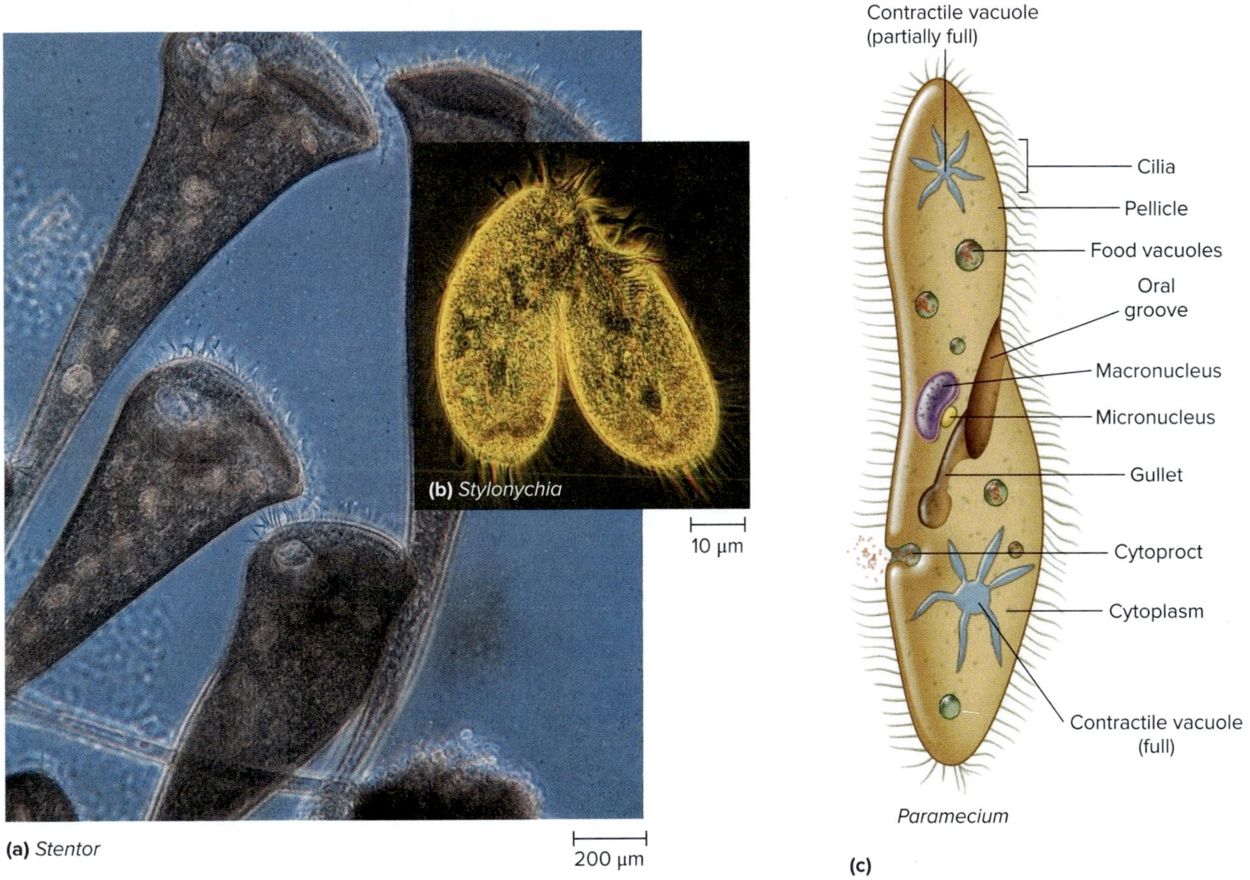

Figure 24.14 *Ciliophora*. (a) *Stentor*, a large, vase-shaped, freshwater protozoan. (b) Two *Stylonychia* cells conjugating. (c) Structure of *Paramecium* spp.
(a) ©Eric Grave/Science Source; (b) ©Eric Grave/Science Source

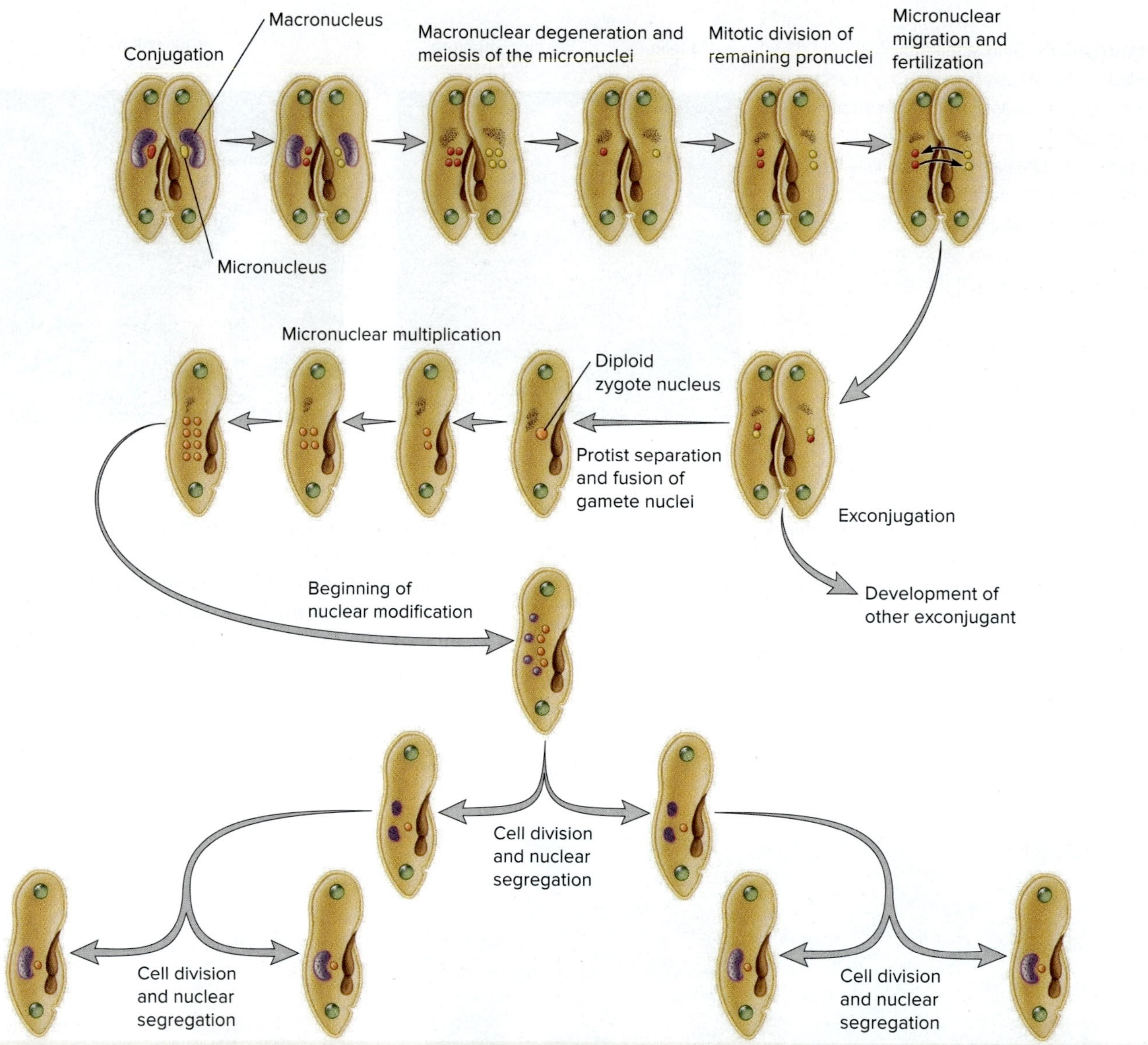

Figure 24.15 Conjugation in *Paramecium caudatum*. After the conjugants separate, only one of the exconjugants is shown; however, a total of eight new protists result from each conjugation.

MICRO INQUIRY *What are the functions of the micronucleus and macronucleus?*

organism. They beat with an oblique stroke, causing the protist to revolve as it swims. Ciliary beating is so precisely coordinated that ciliates can go both forward and backward. There is great variation in shape, and most ciliates do not look like the slipper-shaped members of the genus *Paramecium.* Some species, including *Vorticella* spp., attach to substrates by a long stalk. *Stentor* spp. attach to substrates and stretch out in a trumpet shape to feed. A few species have tentacles for the capture of prey. Some can discharge toxic, threadlike darts called toxicysts to capture prey. A striking feature of ciliates is their ability to quickly entrap many particles by the action of the cilia. Food first enters the cytostome and passes into phagocytic vacuoles that fuse with lysosomes after detachment from the cytostome and the vacuole's contents are digested. After the digested material has been absorbed into the cytoplasm, the vacuole fuses with the cytoproct and waste material is expelled.

Most ciliates have two types of nuclei: a large **macronucleus** and a smaller **micronucleus.** The micronucleus is diploid and contains the normal somatic chromosomes. It divides by mitosis and transmits genetic information through meiosis and sexual reproduction. Macronuclei gene expression is needed to maintain routine cellular functions and control normal cell metabolism. Macronuclei are derived from micronuclei by a complex series of steps. Within the macronucleus are many chromatin bodies, each containing many copies of only one or two genes. Macronuclei are thus polyploid and divide by elongating and then constricting.

The fate of the macro- and micronuclei can be followed during conjugation of *Paramecium caudatum* (**figure 24.15**). In this

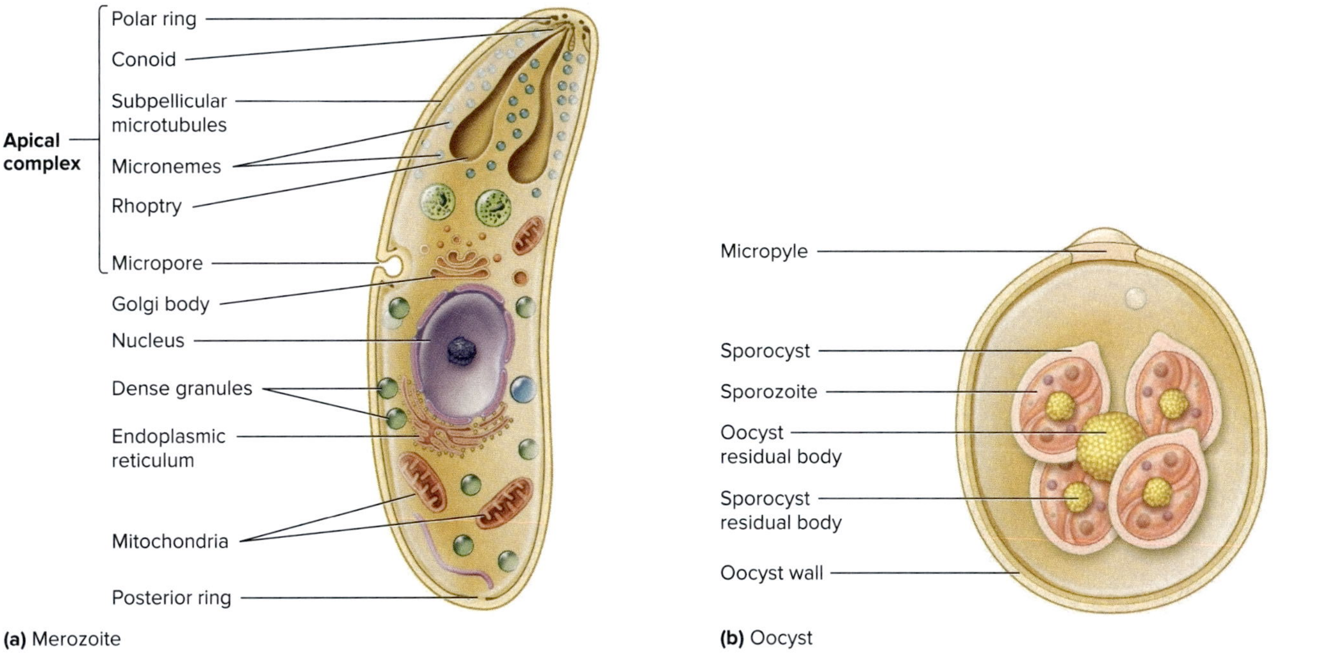

Figure 24.16 The Apicomplexan Cell. (a) The vegetative cell, or merozoite, illustrating the apical complex, which consists of the polar ring, conoid, rhoptries, subpellicular microtubules, and micropore. (b) The infective oocyst of *Eimeria* sp. The oocyst is the resistant stage and has undergone multiple fission after zygote formation (sporogony).

process, there is an exchange of gametes between paired cells of complementary mating types (conjugants). At the beginning of conjugation, two ciliates fuse their pellicles at the site of contact. The macronucleus is no longer needed, so it is degraded in each cell. The individual micronuclei undergo meiosis to form four haploid pronuclei, three of which disintegrate. The remaining pronucleus divides again mitotically to form two gametic nuclei—a stationary one and a migratory one. The migratory nuclei pass into the respective conjugants. Then the ciliates separate, the gametic nuclei fuse, and the resulting diploid zygote nucleus undergoes three rounds of mitosis. The eight resulting nuclei have different fates: One nucleus is retained as a micronucleus; three others are destroyed; and the four remaining nuclei develop into macronuclei. Each separated conjugant now undergoes cell division. Eventually progeny with one macronucleus and one micronucleus are formed.

Although most ciliates are free living, symbiotic forms also exist. Some live as harmless commensals; for example, *Entodinium* spp. are found in the rumen of cattle, and *Nyctotherus* protists occur in the colon of frogs. Other ciliates are strict parasites; for example, *Balantidium coli* lives in the intestines of mammals, including humans, where it can cause dysentery. *Ichthyophthirius* spp. live in freshwater, where they can attack many species of fish, producing a disease known as "ick."

All **apicomplexans** are parasites of animals and are distinguished by a unique arrangement of fibrils, microtubules, vacuoles, and other organelles, collectively called the **apical complex,** which is located at one end of the cell (**figure 24.16*a***). There are three secretory organelles—the rhoptries, micronemes, and dense granules—that release enzymes and calcium needed to penetrate host cells. In addition, apicomplexans (with the exception of *Cryptosporidium* spp.) contain plastids, called **apicoplasts.** Although these arose by endosymbiosis of an ancient cyanobacterium, unlike chloroplasts, apicoplasts are not the site of photosynthesis. In the malaria-causing pathogen *Plasmodium* spp., apicoplasts are known to be the site of acetyl-CoA and isoprenoid biosynthesis. In all apicomplexans studied to date, the apicoplast is essential for viability. This makes them attractive targets for the development of drugs to inhibit the growth of pathogenic apicomplexans. ◀ *Sterols and isoprenoid lipids (section 12.7)*

Apicomplexans have complex life cycles in which certain stages sometimes occur in one host and other stages occur in a different host. The life cycle has both asexual (clonal) and sexual phases, and is characterized by an alternation of haploid and diploid generations. The clonal and sexual stages are haploid, except for the zygote. The motile, infective stage is called the **sporozoite.** When this haploid form infects a host, it differentiates into a gamont; male and female gamonts pair and undergo multiple fission, which produces many haploid cells called merozoites. Merozoites may continue to infect new host cells or pair and fuse to form zygotes. Each zygote secretes a protective covering and is then considered a spore. Within the spore, the nucleus undergoes meiosis (restoring the haploid condition) followed by mitosis to generate eight sporozoites ready to infect a new host (figure 24.16*b*). Motility (flagellated or amoeboid) is confined to the gametes and zygotes of a few species.

Among the most important human pathogens are the five species in the apicomplexan genus *Plasmodium,* which cause malaria: *P. falciparum, P. malariae, P. vivax, P. ovale,* and *P. knowlesi.* Other *Plasmodium* spp. cause malaria in animals such as monkeys, birds, bats, lizards, and antelopes. The life cycle of the human pathogen *P. vivax* is shown in figure 40.6; it infects both liver and red blood cells. Prevention measures have resulted in about a 50% decline in malaria since 2000; however, the disease still causes about 665,000 deaths per year. *P. falciparum* is the most

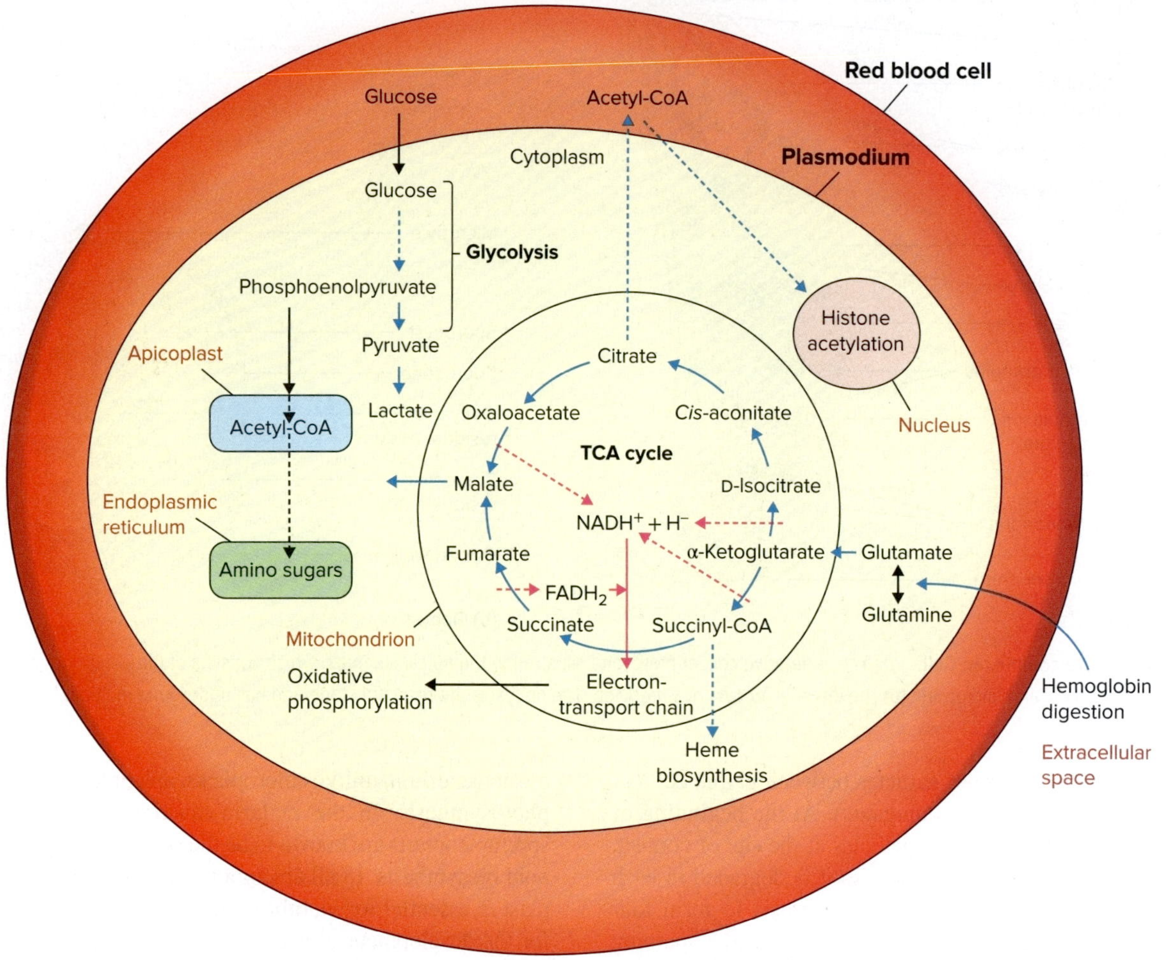

Figure 24.17 ***Plasmodium falciparum*** **Drives the TCA Cycle in Two Directions.** Glutamate is deaminated to yield α-ketoglutarate from which reactions proceed in both the oxidative and reductive direction. Pyruvate is fermented in the parasite's cytosol to lactate rather than oxidized to acetyl-CoA. Phosphoenolpyruvate is converted to acetyl-CoA in the apicoplast and used to synthesize amino sugars. The host may convert citrate produced by the protist to acetyl-CoA, which then re-enters the microbe, where it is thought to be used to acetylate histones. The uncertainty of these steps is denoted by a dashed line.

deadly of the five species, accounting for at least one-third of malaria deaths. The energy metabolism of this microbe is unusual. Despite possessing genes for a mitochondrial ATP synthase, it is fermentative. Like all apicomplexans, *P. falciparum* cannot convert pyruvate to acetyl-CoA in the mitochondrion. Instead, phosphoenolpyruvate generated by the Embden-Meyerhof pathway in the cytosol is transported to the apicoplast where it is converted to acetyl-CoA. Without acetyl-CoA in the mitochondrion, the TCA cycle begins with the deamination of glutamate to α-ketoglutarate. Reactions then proceed in both the reductive and oxidative directions. Reductive TCA steps result in citrate and malate, which are exported from the mitochondrion, while in the oxidative direction, succinyl-CoA, a key precursor for heme biosynthesis, is generated (**figure 24.17**). ▶| *Malaria (section 40.3)*

A number of other apicomplexans are important infectious agents. Several species of *Eimera* are the causative agents of cecal coccidiosis in chickens, a condition that costs hundreds of millions of dollars in lost animals each year in the United States. *Theileria parva* and *T. annulata* are tick-borne parasites that cause a fatal disease in cattle called East Coast fever. This kills over a million animals annually in sub-Saharan Africa. Toxoplasmosis, caused by members of the genus *Toxoplasma,* is transmitted either by consumption of undercooked meat or by fecal contamination from a cat's litter box. Cryptosporidia are responsible for cryptosporidiosis, an infection that begins in the intestines but can disseminate to other parts of the body. Cryptosporidiosis and members of another apicomplexan genus, *Cyclospora,* most commonly infect immunocompromised individuals. ▶| *Toxoplasmosis (section 40.5); Cryptosporidiosis (section 40.5); Cyclosporiasis (section 40.5)*

Stramenopila

The large and diverse *Stramenopila* group (informally known as stramenopiles) includes photosynthetic protists such as diatoms, brown and golden algae (*Chrysophyceae*), as well as chemoorganotrophic (saprophytic) genera such as öomycetes (*Peronosporomycetes*), labyrinthulids (slime nets), and members of *Hyphochytriales.* Stramenopiles also include brown seaweeds and kelp that form large, rigid structures and macroscopic forms that were once considered fungi and plants. One unifying feature

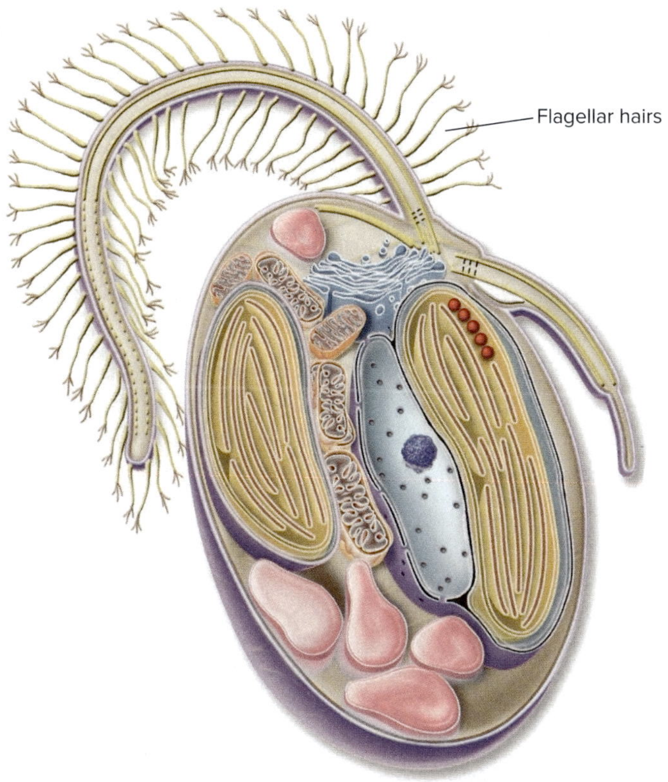

Figure 24.18 A Stramenopile Cell. Note the two flagella, one with "straw hairlike" projections.

of these very diverse protists is the presence of tubular christae in their mitochondria. Another is the possession of **heterokont flagella** at some point in the life cycle. This is characterized by two flagella—one extending anteriorly and the other posteriorly. These flagella bear small hairs with a unique, three-part morphology; stramenopila means "straw hair" (**figure 24.18**).

Diatoms (*Bacillariophyta*) possess chlorophylls *a* and c_1/c_2, and the carotenoid fucoxanthin. When fucoxanthin is the dominant pigment, cells have a golden-brown color. Their major carbohydrate reserve is chrysolaminarin, a polysaccharide storage product composed principally of $\beta(1\rightarrow3)$ and $\beta(1\rightarrow6)$–linked glucose residues in a ratio of 11:1. Diatoms have a distinctive, two-piece cell wall of silica called a **frustule** composed of crystallized silica $[Si(OH)_4]$ with very fine markings. Diatom frustules are composed of two halves or thecae that overlap like those of a Petri dish (**figure 24.19**). The larger half is the epitheca, and the smaller half is the hypotheca. Frustule morphology is very useful in diatom identification, and diatom frustules have a number of practical applications. The fine detail and precise morphology of these frustules have made them attractive for nanotechnology. ▶| *Some microbes are products (section 42.6)*

Although the majority of diatoms are strictly photoautotrophic, some are facultative chemoorganotrophs, absorbing carbon-containing molecules through the holes in their walls. The vegetative cells of diatoms are diploid and can be unicellular, colonial, or filamentous. They lack flagella and have a single, large nucleus and smaller plastids. Reproduction consists of the organism dividing asexually, with each half then constructing a new hypotheca (**figure 24.20*a***). Because the epitheca and hypotheca are of different sizes, each time the parental hypotheca is used as an epithecal template to construct a new hypotheca, the diatom gets smaller. This cannot go on forever without the diatom disappearing altogether, so when a cell has diminished to about 30% of its original size, sexual reproduction is usually triggered. The diploid vegetative cells undergo meiosis to form gametes,

(a) *Cyclotella meneghiniana* **(b)**

Figure 24.19 Diatoms. (a) The siliceous epitheca and hypotheca of the diatom *Cyclotella meneghiniana* fit together like halves of a Petri dish. (b) A variety of diatoms show the intricate structure of the silica cell wall.

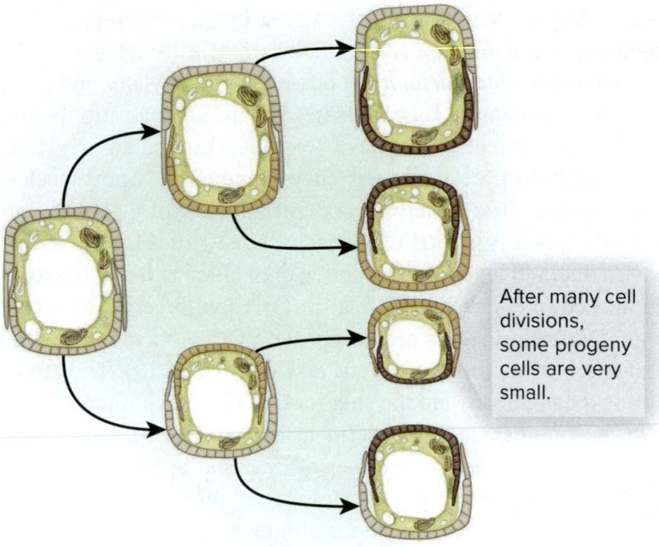

(a) Asexual reproduction in diatoms

Figure 24.20 Gametic Life Cycle as Illustrated by Diatoms. (a) Diatom asexual reproduction involves repeated mitotic divisions. Because a new lower cell-wall piece is always synthesized, asexual reproduction eventually causes the mean cell size to decline in a diatom population. (b) Small cell size triggers sexual reproduction, which regenerates maximal cell size.

MICRO INQUIRY *How does the trigger for sexual reproduction among diatoms compare with more common stimuli that induce sexual reproduction? (Hint: Compare to radiolaria.)*

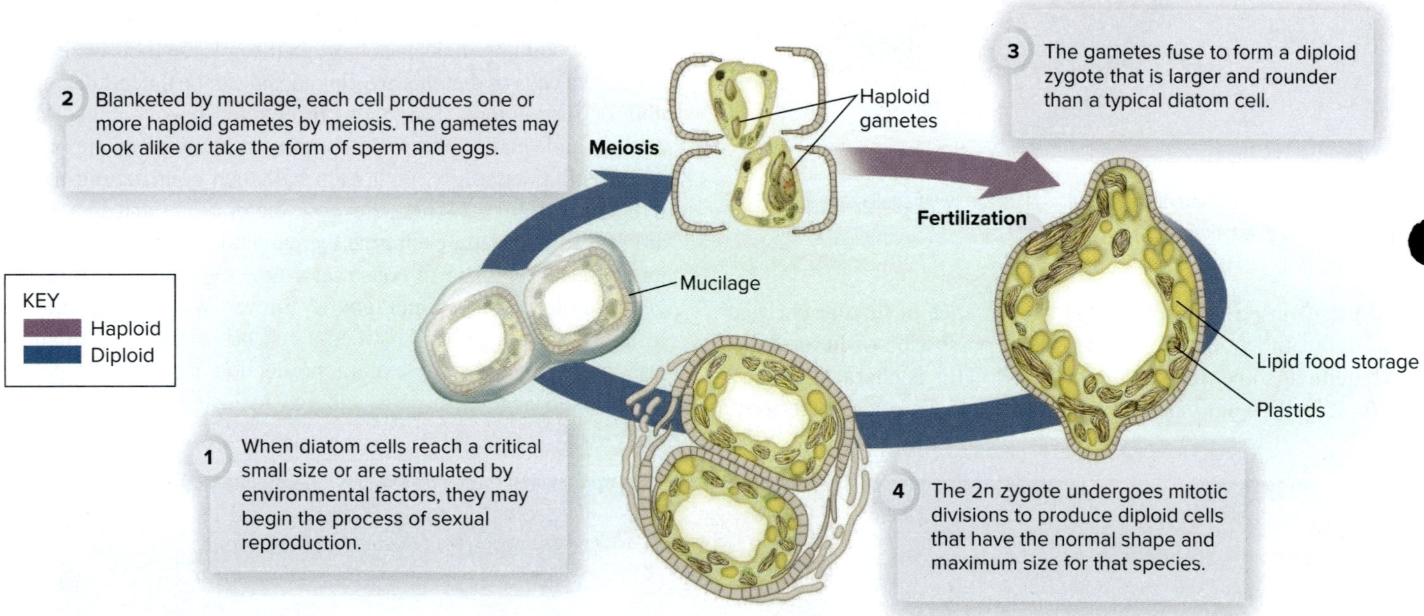

(b) Sexual reproduction in diatoms

which then fuse to produce a zygote. The zygote develops into an auxospore, which increases in size again and forms a new wall. The mature auxospore eventually divides mitotically to produce vegetative cells with frustules of the original size (figure 24.20*b*).

Diatoms are found in freshwater lakes, ponds, streams, and throughout the world's oceans. Marine planktonic diatoms produce about 50% of the organic carbon in the ocean; they are therefore critical in global carbon cycling. This is why some call them "the rain forests of the oceans." ▶| *Carbon cycle (section 28.1); Microorganisms in the open ocean are adapted to nutrient limitation (section 30.2)*

A group of protists once considered fungi and traditionally called **öomycetes,** meaning "egg fungi," are now called **peronosporomycetes.** They differ from fungi in a number of ways, including cell walls made of cellulose and the fact that they are diploid throughout their life cycle, except for the short period of time when they are haploid gametes. When undergoing sexual reproduction, they form a relatively large egg cell (öogonium) that is fertilized by either a sperm cell or a smaller gametic cell (called an antheridium) to produce a zygote. When the zygote germinates, the asexual zoospores display heterokont flagellation.

Some peronosporomycetes help degrade plant and animal detritus, while others are pathogenic. For instance, species of *Saprolegnia* and *Achlya* are saprophytes that form cottony masses on dead algae and animals, mainly in freshwater environments. Certainly the most famous peronosporomycete is *Phytophthora infestans,* which attacked the European potato crop in the mid-1840s, spawning the Irish famine. The original classification of *P. infestans* as a fungus was misleading, and for decades farmers attempted to control its growth with fungicides, to which it is (of course) resistant. This protist continues to take its toll; potato blight costs some $6.7 billion annually worldwide. *P. sojae* is another agriculturally important species; it infects soybeans and costs this industry millions of dollars per year. Finally, *P. ramorum* is a plant pathogen that causes a disease called sudden oak death. It kills oak trees and a variety of woody shrubs.

Labyrinthulids also have a complex taxonomic history: Like peronosporomycetes, they were formerly considered fungi. However, molecular phylogenetic evidence combined with the observation that they form heterokont flagellated zoospores places them in *Stramenopila*. The nonflagellated stage of the life cycle features spindle-shaped cells that form complex colonies that glide rapidly along an ectoplasmic net made by the organism. This net is actually an external network of calcium-dependent contractile fibers made up of actinlike proteins that facilitate the movement of cells. Their feeding mechanism is like that of fungi: osmotrophy aided by the production of extracellular degradative enzymes. In marine habitats, members of the genus *Labyrinthula* grow on plants and algae, and are thought to play a role in the wasting disease of eelgrass, an important intertidal plant.

Haptophyta

The phylogenetics of *Haptophyta* remains unsettled; it may be a sister group to the supergroup SAR. It includes the *Coccolithales,* photosynthetic protists that bear ornate calcite scales called coccoliths (**figure 24.21**). Together with forams, **coccolithophores** precipitate calcium carbonate ($CaCO_3$) in the ocean, thereby removing soluble CO_2 from the ocean and influencing Earth's carbon budget. Cells are usually biflagellated and possess a unique organelle called a haptonema, which is somewhat similar to a flagellum but differs in microtubule arrangement. One species, *Emiliania huxleyi,* has been studied extensively. Like all coccolithophores, it is planktonic and can cause massive blooms in the open ocean (*see figure 30.10*). These high concentrations of *E. huxleyi* can significantly alter nutrient flux as well as alter the weather by emitting sulfur (as dimethylsulfide or DMS) to the atmosphere. The release of DMS causes clouds to grow larger. Other coccolithophores are known to cause toxic blooms. ▶| *Water is the largest microbial habitat (section 30.1)*

Picozoa

The phylogenetic origin of another marine lineage, *Picozoa,* is very mysterious. Evidence for *Picozoa* (*pico* Greek, tiny; *zoa,* animal) emerged in the early 2000s when 18S rRNA metagenomic data revealed the presence of eukaryotic organisms small enough to fit through filters of 2 to 3 μm. Originally thought to be photosynthetic, they were given the tentative name "phycobiliphytes." Further analysis showed that these protists are not photosynthetic. They are broadly distributed in high numbers throughout the world's oceans.

In 2013 the first cultured picozoan was reported. These protists, named *Picomonas judraskeda,* were isolated from marine coastal waters and consist of two roughly spherical halves divided by a deep cleft. Total cell size ranges from 2 × 2.5 μm to 2.5 × 3.8 μm. They have two flagella and demonstrate an unusual kind of motility that has been described as cycles of "jump-drag-skedaddle." They consume colloidal organic materials by endocytosis, and it is thought that their unusual motility assists in feeding.

Figure 24.21 The Haptophyte *Emiliania huxleyi*. The ornamental scales are made of calcite.

Comprehension Check

1. What adaptations do planktonic radiolaria have to help them float?
2. Compare how radiolaria use axopodia with the way forams use reticulopodia.
3. What are alveoli? Describe the difference between a naked and an armored dinoflagellate.
4. What is the morphology of a typical ciliate? Why do you think ciliates are the fastest-moving protists?
5. Describe conjugation as it occurs in ciliates. What is the fate of the micronucleus and the macronucleus during this process?
6. What function does the apical complex serve in apicomplexans?
7. Explain the unique structural features of diatoms. How does their morphology play a role in the alternation between asexual and sexual reproduction?
8. Why do you think öomycetes and labyrinthulids were formerly considered fungi?
9. What is the ecological importance of coccolithophores?

24.5 Supergroup *Archaeplastida* Includes "Green Algae"

After reading this section, you should be able to:

a. List at least three different morphologies found among the *Chloroplastida*
b. Describe the structure and life cycle of the model organism, *Chlamydomonas*

Archaeplastida includes all organisms with a photosynthetic plastid that arose through an ancient endosymbiosis with a cyanobacterium. It thus includes all higher plants as well as many protist

Figure 24.22 ***Chlorophyta*** **(Green Algae); Light Micrographs.** (a) *Chlorella,* a unicellular nonmotile chlorophyte (×160). (b) *Volvox,* which demonstrates colonial growth. (c) *Spirogyra;* each filament contains ribbonlike, spiral chloroplasts. (d) *Acetabularia,* the mermaid's wine goblet.

species. ◀ *Mitochondria, mitochondria-like organelles, and chloroplasts evolved from endosymbionts (section 1.2)*

Chloroplastida

Members of *Chloroplastida* are often referred to as green algae (Greek *chloros,* green). These phototrophs grow just about anywhere: in freshwater and saltwater, in soil, and on and within other organisms. They have chlorophylls *a* and *b* along with specific carotenoids, and they store carbohydrates such as starch. Many have cell walls made of cellulose. They exhibit a wide diversity of body forms, ranging from unicellular to colonial, filamentous, membranous or sheetlike, and tubular types (**figure 24.22**). Some species have a holdfast structure that anchors them to a substratum. Both asexual and sexual reproduction are observed.

Chlamydomonas spp. are members of the subgroup *Chlorophyta* (**figure 24.23** and 24.1). Individuals have two flagella of equal length at the anterior end by which they move rapidly in water.

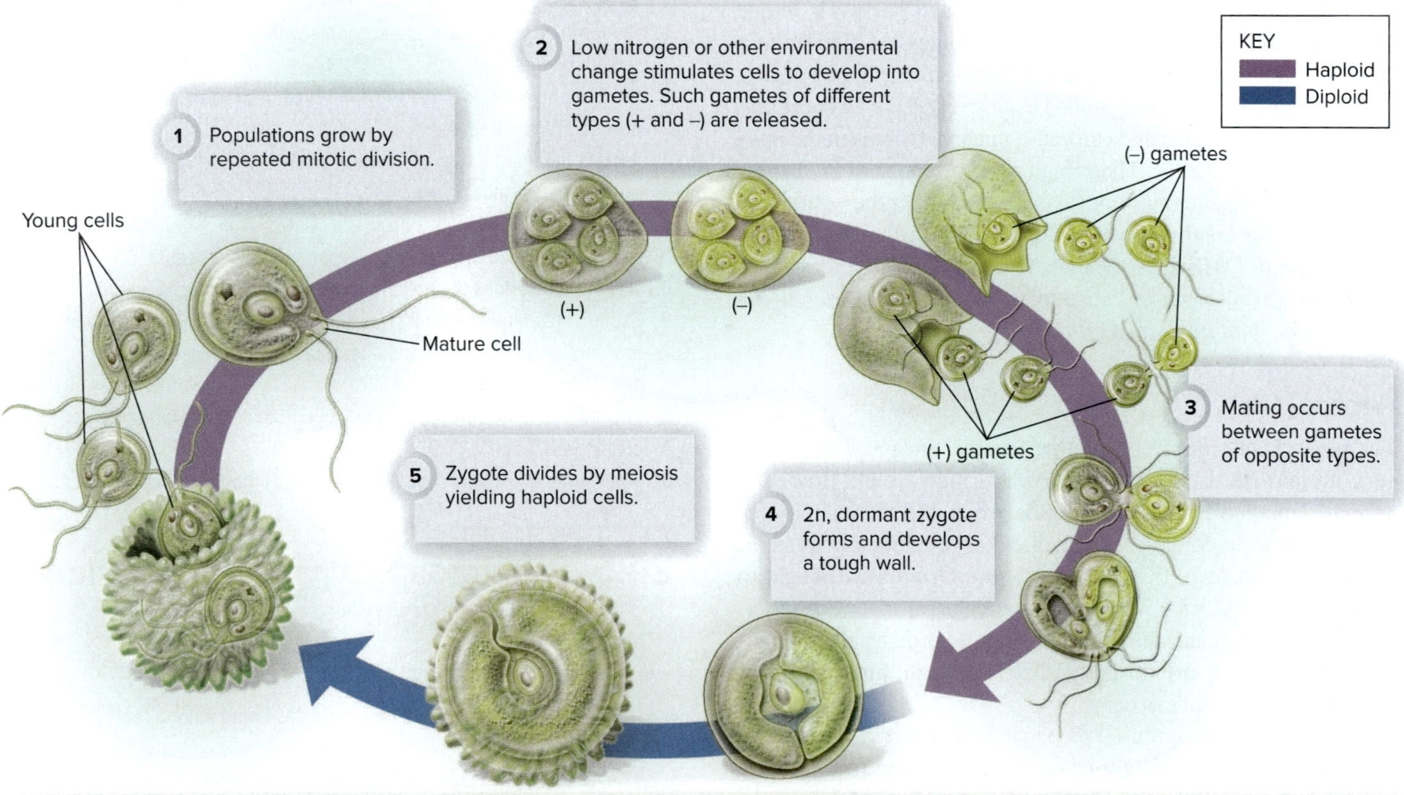

Figure 24.23 ***Chlamydomonas reinhardtii:*** **The Structure and Life Cycle of This Motile Green Alga.** During asexual reproduction, all structures are haploid; during reproduction, only the zygote is diploid.

MICRO INQUIRY *Why do you think environmental stress stimulates gamete development?*

Each cell has a single haploid nucleus, a large chloroplast, a conspicuous pyrenoid, and a stigma that aids the cell in phototactic responses. Two small contractile vacuoles at the base of the flagella function as osmoregulatory organelles. *Chlamydomonas* spp. reproduce asexually by producing zoospores through mitotic division. Sexual reproduction occurs when some cells act as gametes and fuse to form a four-flagellated, diploid zygote that ultimately loses the flagella and enters a resting phase. Meiosis occurs at the end of this resting phase and produces four haploid cells that give rise to adults.

The chlorophyte *Prototheca moriformis* causes the disease protothecosis in animals and, rarely, in humans. *P. moriformis* cells are fairly common in the soil, and it is from this site that most infections occur. Severe systemic infections, such as massive invasion of the bloodstream, have been reported in animals.

Comprehension Check

1. Compare the morphology of members of the genus *Chlamydomonas* with those belonging to *Euglena* (figures 24.5 and 24.23).
2. Like many protists, sexual reproduction in *C. reinhardtii* involves individuals from different mating types. From an evolutionary point of view, why is this the case?

Key Concepts

24.1 Protist Diversity Reflects Broad Phylogeny

- Protists are widespread in the environment, found wherever water, suitable nutrients, and an appropriate temperature occur.
- Protists are important components of many terrestrial, aquatic, and marine ecosystems, where they contribute to nutrient cycling. Many are parasitic in humans and animals, and some have become useful in the study of molecular biology.
- The protistan cell membrane is called the plasmalemma, and the cytoplasm can be divided into the ectoplasm and the endoplasm. A pellicle lies beneath the plasma membrane of some protists. The cytoplasm contains contractile, secretory, and food (phagocytic) vacuoles.
- Energy metabolism occurs within mitochondria, hydrogenosomes, or chloroplasts.
- Some protists can enter a resting stage (encystment) called a cyst. Cysts protect the organism against adverse environments, function as a site for nuclear reorganization, and serve as a means of transmission in parasitic species (**figure 24.2**).
- Most protists reproduce asexually by binary or multiple fission or budding (**figure 24.3**).
- Protists also use a variety of sexual reproductive strategies, including syngamy and autogamy.
- Protist phylogeny is the subject of active research and debate. The classification scheme presented here is based on that of the International Society of Protistologists.

24.2 Supergroup *Excavata:* Primitive Eukaryotes

- The supergroup *Excavata* includes *Fornicata, Parabasalia,* and *Euglenozoa*. Most have a cytoproct and use a flagellum for suspension feeding.
- The human pathogen *Giardia intestinalis* is a member of *Fornicata*. It is considered one of the most primitive eukaryotes (**figure 24.4**).
- Most parabasalids are flagellated endosymbionts of animals. They include the obligate mutualists of wood-eating insects such as *Trichonympha* spp. and the human pathogen *Trichomonas vaginalis.*
- Many members of *Euglenozoa* are photoautotrophic (**figure 24.5**). The remainder are chemoorganotrophs; most of these are saprophytic. Important human pathogens in this group cause leishmaniasis, Chagas' disease, and African sleeping sickness.

24.3 Supergroup *Amoebozoa* Includes Protists with Pseudopodia

- Amoeboid forms use pseudopodia, which can be lobopodia, filopodia, or reticulopodia (**figure 24.7**). Amoebae that bear external plates are called testate; those without plates are called naked or atestate amoebae.
- The *Amoebozoa* subgroup *Eumycetozoa* includes acellular and cellular slime molds. Acellular slime molds form a large, multinucleate mass of protoplasm (**figure 24.8**).
- Cellular slime molds produce a pseudoplasmodium, consisting of many individual cells moving together as a mass. *Dictyostelium discoideum* is a cellular slime mold that is used as a model organism in the study of chemotaxis, cellular development, and cellular behavior (**figure 24.9**).

24.4 Supergroup SAR: Protists of Global Importance

- *Rhizaria* encompasses amoeboid forms, including members of *Radiolaria,* which have a siliceous internal skeleton and filopodia that are used to entrap prey (**figure 24.10**).
- Members of *Foraminifera* bear netlike reticulopodia and tests. Most foraminifera are benthic, and their tests accumulate on the ocean floor, where they are biomarkers used in oil exploration (**figure 24.11**).

- *Alveolata* consists of dinoflagellates, ciliates, and apicomplexans.
- Dinoflagellates are a large group of nutritionally complex protists. Most are marine and planktonic. They are known for their phosphorescence and for causing toxic blooms (**figure 24.13**).
- Ciliates are chemoorganotrophic protists that use cilia for locomotion and feeding. In addition to asexual reproduction, conjugation is used in sexual reproduction (**figures 24.14** and **24.15**).
- Apicomplexans are parasitic with complex life cycles. The motile, infective stage is called the sporozoite. This group includes a number of pathogens, including five species of *Plasmodium,* which cause malaria in humans (**figure 24.16**).
- Stramenopiles are extremely diverse and include diatoms, golden and brown algae, peronosporomycetes, and labyrinthulids. Diatoms are found in freshwater and saltwater, and are important components of marine plankton (**figures 24.18–24.20**).
- Haptophytes include coccolithophores, planktonic photosynthetic protists that contribute to the global carbon budget by precipitating calcium carbonate for their ornate scales (**figure 24.21**).
- *Picozoa* include very small, heterotrophic, marine protists with a global distribution.

24.5 Supergroup *Archaeplastida* Includes "Green Algae"

- *Archaeplastida* includes *Chlorophyta,* also known as green algae. All are photosynthetic with chlorophylls *a* and *b* along with specific carotenoids. They exhibit a wide range of morphologies (**figure 24.22**).
- *Chlamydomonas* spp. are model protists, which undergo sexual and asexual reproduction (**figure 24.23**).

Active Learning

1. Protist encystment is usually triggered by changes in the environment. Compare and contrast this with endospore formation in Gram-positive bacteria.
2. Vaccine development for diseases caused by protists (e.g., malaria, Chagas' disease) has been much less successful than for bacterial or viral diseases. Discuss one biological reason and one geopolitical reason for this fact.
3. Most apicomplexans bear the unpigmented chloroplast remnant called the apicoplast. Only the recently described apicomplexan *Chromera velia* has been shown to use the apicoplastid as the site of photosynthesis. What genetic and physiological features do you think were characterized to show that this organism is truly photosynthetic? How would you determine if this protist is photoautotrophic, photoheterotrophic, or mixotrophic?
4. Benthic foraminifera inhabit marine sediments. It was discovered that the benthic foram *Globobulimina pseudospinescens* stores intracellular nitrate, which it reduces to nitrogen gas. How would you go about designing a comparative analysis of eukaryotic and bacterial denitrification? What physical factors and physiological and enzymatic features would you characterize?
5. Protist taxonomy was once based entirely on morphology and ecology. This has been replaced by molecular characterization. In an effort to unite classical and molecular taxonomic approaches, an international group of scientists has started an open-access platform called *UniEuk* to build a common community of expert-driven taxonomy. Select a protist discussed in this chapter and list the kinds of information that would be helpful in exploring its taxonomy. You can then compare your proposed inputs with those described on the *UniEuk* website.

   Read the original paper: Berney, C., et al. 2017. *UniEuk*: time to speak a common language in protistology! *J. Euk. Microbiol.* 64:407-11. doi:10.1111/jeu.12414.

25

Fungi

©Y H Lim/Alamy Stock Photo

The Complex Story of Caterpillar Fungus

Q*uestion:* What do caterpillars, athletic performance, the economy of the Tibetan autonomous region (TAR), and global climate change have in common? *Answer:* The fungus *Ophiocordyceps sinensis,* more commonly known as cordyceps or *yartsa gunbu,* which means "summer grass winter worm." This fungus is one of many entomopathogenic fungi, which means it kills insects. *O. sinensis* is found in the TAR where it parasitizes caterpillars of more than 30 types of ghost moths (*Thitarodes* spp.). Each spring, when the fungal spores germinate, the resulting hyphae infect these soil-dwelling caterpillars. The following spring, fruiting bodies called stoma have formed within the heads of the caterpillars, which poke through the soil surface. In May and June, local people comb the hillsides at altitudes between 3,000 and 5,000 meters looking for fungal stomas poking through the earth, like mushroom stems. Once found, the mummified caterpillar with its attached fungus, as shown, is harvested, dried, and sold for its purported medical qualities.

What are these presumed medical benefits? The first written accounts of caterpillar fungus date from the fifteenth century, when it was credited as an aphrodisiac with the ability to improve a myriad of medical conditions from asthma to cancer. For centuries, cordyceps remained a local medicinal product. However, in the early twenty-first century, caterpillar fungus was credited with the outstanding performance of Chinese track athletes. This triggered a run on the cordyceps market, and between 2000 and 2011, the value of caterpillar fungus increased about 1,000%.

The tremendous popularity of cordyceps had a profound effect on the TAR economy as yak herders and subsistence farmers turned to caterpillar fungus as a means of income supplementation. By about 2005, the caterpillar fungus harvest was so profitable it accounted for about 8.5% of the gross national product in the TAR. So-called "spore money" provided economic opportunities that had been out of reach for centuries. However, as the number of people harvesting cordyceps grew, the yield per person declined. However, reasons for the decline in harvest yields extend beyond supply and demand. First, there was a lack of understanding of the fungal life cycle. The most valuable form of caterpillar fungus occurs in early spring, when the size of the fungal stoma is about equal to the caterpillar, as shown. The fungus at this point is all hyphae without spores; not until the end of the 6-week harvesting season does the fungus mature and form spores. At this point, the size of the caterpillar has greatly diminished, so the product fetches the lowest price. But because late-season caterpillar fungus is often harvested anyway, spores are removed from the environment, eliminating the fungal spore dissemination that ensures sustainable harvests. Another reason harvest yields are declining may be global climate change. For reasons not fully understood, the rate at which temperatures rise is accelerated at high altitude. As recorded temperatures in the TAR plateaus have climbed, there has been a decline in precipitation. Fungal growth is sensitive to temperature and humidity, so these two factors also appear to be limiting harvest size. Nonetheless in 2016, 500 grams of cordyceps had a street value of about US $40,000. Hopefully, the implementation of sustainable harvesting techniques will continue to allow economically marginalized people the opportunity to profit from *yartsa gunbu.*

In this chapter, we introduce ***Fungi,*** sometimes referred to as true fungi or *Eumycota* (Greek *eu*, true; *mykes*, fungus). The term *fungus* (pl., fungi; Latin *fungus*, mushroom) describes eukaryotic organisms that are spore-bearing, have absorptive nutrition, lack chlorophyll, and reproduce both sexually and asexually. Scientists who study fungi are **mycologists,** and the scientific discipline devoted to fungi is called **mycology.** The study of fungal toxins is called **mycotoxicology,** and the diseases caused by fungi in animals are known as **mycoses** (s., mycosis).

Fungi is an enormous group of organisms; about 100,000 fungal species have been described, although some estimate the true number of species to be in the millions. It is thus not surprising that the taxonomy of these organisms has been revised numerous times and remains uncertain. The application of molecular techniques combined with morphological and ecological considerations has led to a better understanding of the phylogenetic relationships between various fungal groups. In deciphering fungal taxonomy, it is helpful to know that a term describing a phylum ends with *-mycota,* a subphylum with *-mycotina*, a class with *-mycetes,* and an order with *-ales*. For example, neither the traditionally regarded phylum *Zygomycota* nor *Chytridiomycota* is monophyletic and no longer an accepted phylum. As shown in **figure 25.1**, their members have been distributed among other taxa. Most recently, metagenomic analysis from a wide range of soils revealed a large clade of environmental RNA sequences. This has led to the discovery of a new phylum called *Cryptomycota*, meaning "hidden fungi" because they may constitute almost half the fungal biomass in some locations. Here we present the six major fungal groups that have representatives grown in culture: Chytridiomycetes, Zygomycetes, *Glomeromycota, Ascomycota, Basidiomycota,* and *Microsporidia,* shown in **figure 25.2** and summarized in **table 25.1**.

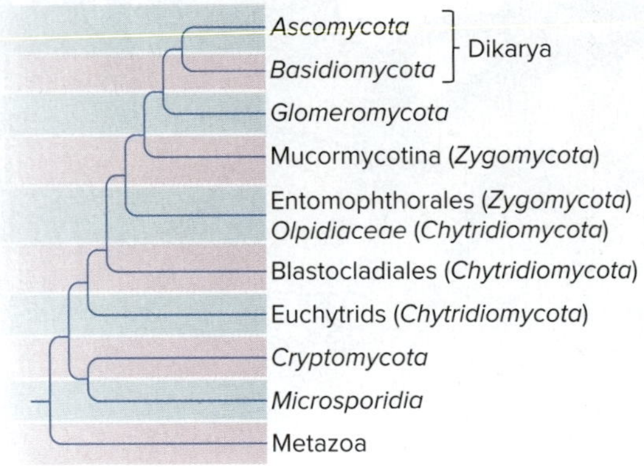

Figure 25.1 The Major Branches of *Fungi*. Phylogenetic analysis of fungal groups. The reclassification of members of the formerly accepted phyla *Zygomycota* and *Chytridiomycota* are indicated parenthetically.

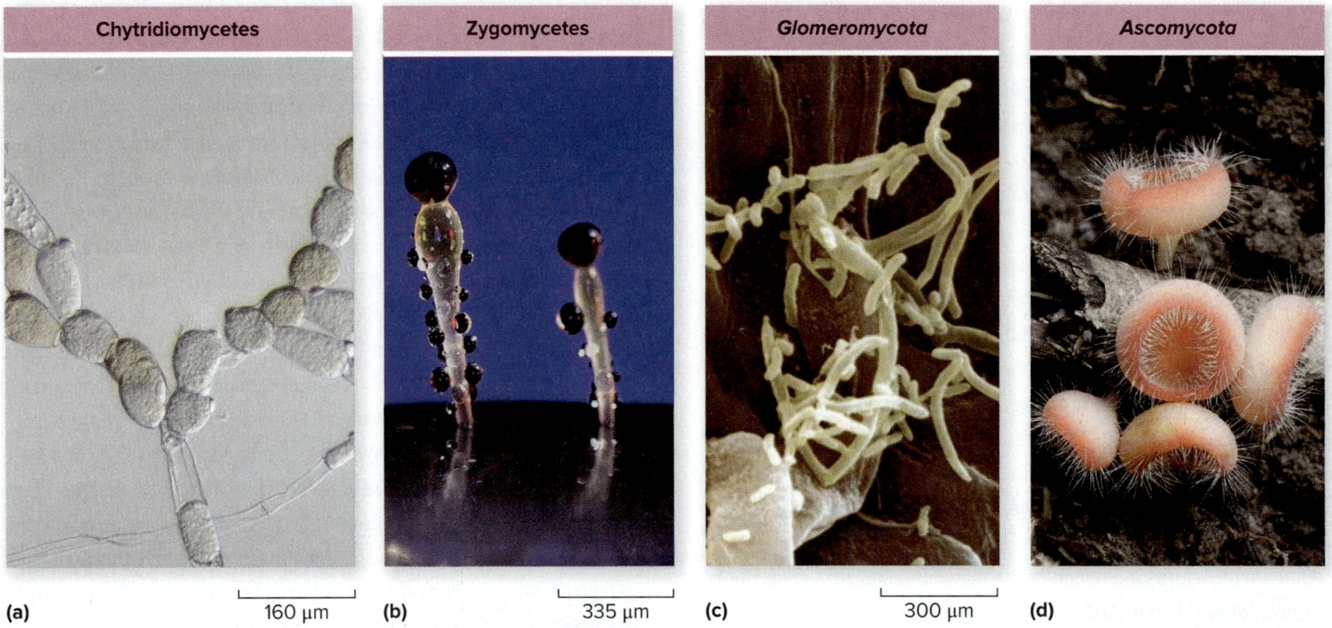

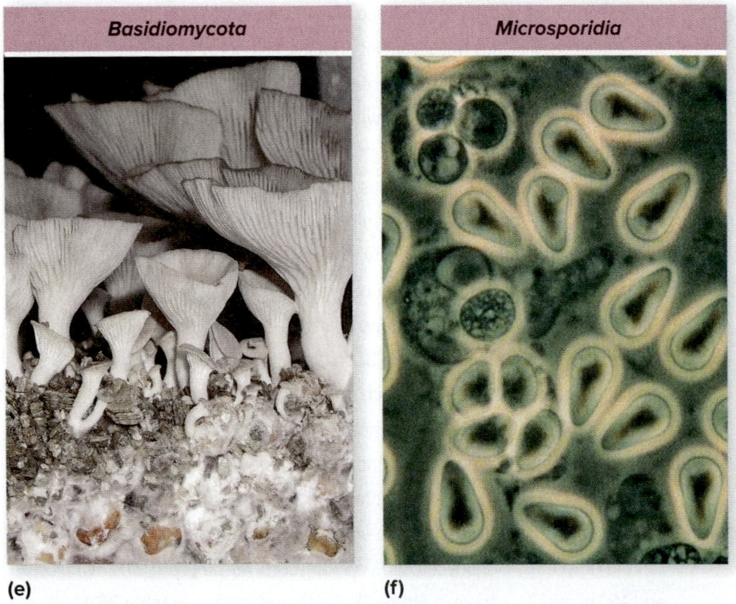

Figure 25.2 Representatives of the Fungal Groups. (a) The free-living chytrid *Allomyces macrogynus,* which lives in mud. (b) *Pilobolus* sp., a zygomycete, grows on animal dung and on culture medium. Stalks about 1 mm long contain dark, spore-bearing sacs. (c) Mycorrhizae are glomeromycetes associated with roots. (d) The cup fungus *Cookeina tricholoma* is an ascomycete from the rain forest of Costa Rica. In cup fungi, the spore-producing structures line the cup; in basidiomycetes that form mushrooms such as *Amanita* sp. they line the gills beneath the cap of the mushroom. (e) *Pleurotus ostreatus,* the oyster mushroom, is a basidiomycete. All visible structures of fungi arise from an extensive network of hyphae that penetrate and are interwoven with the substrate as they grow. (f) By contrast, the unusual fungi that belong to *Microsporidia* form spores that germinate in host cells. Shown here is the mosquito pathogen *Edhazardia aedis.*

(a, b) ©Carolina Biological Supply Company/Phototake; (c) ©Eye of Science/Science Source; (d) ©Sinclair Stammers/Science Source; (e) ©Robert W. Seagull; (f) Source: Dr. James Becnel, USDA ARS Gainesville, FL and Society of Invertebrate Pathology

Table 25.1 Abbreviated Classification of *Fungi*

Subclass	Characteristics	Examples
Chytridiomycetes	Flagellated cells in at least one stage of life cycle; may have one or more flagella. Cell walls with chitin and β-1,3 / 1,6-glucan; glycogen is used as a storage carbohydrate. Sexual reproduction often results in a zygote that becomes a resting spore or sporangium; saprophytic or parasitic. Chytrid subdivisions include *Blastocladiales, Monoblepharidales, Neocallimastigaceae, Spizellomycetales,* and *Chytridiales.*	*Allomyces* *Blastocladiella* *Coelomomyces* *Physoderma* *Synchytrium*
Zygomycetes	Thalli usually filamentous and nonseptate; sexual reproduction gives rise to thick-walled zygospores that are often ornamented. Includes seven subdivisions: *Basidiobolus, Dimargaritales, Endogonales, Entomophthorales, Harpellales, Kickxellales, Mucorales,* and *Zoopagales.* Human pathogens found in *Mucorales* and *Entomophthorales.*	*Mucor* *Phycomyces* *Rhizopus* *Thamnidium*
Ascomycota	Sexual reproduction involves meiosis of a diploid nucleus in an ascus, giving rise to haploid ascospores; most also undergo asexual reproduction with the formation of conidiospores with specialized aerial hyphae called conidiophores. Many produce asci within complex fruiting bodies called ascocarps. Includes saprophytic, parasitic forms; many form mutualisms with phototrophic microbes to form lichens. Four monophyletic subdivisions: *Saccharomycetes, Pezizomycotina, Taphrinomycotina,* and *Neolecta.* Includes recently discovered soil fungi *Archaeorhizomycetes.*	*Aspergillus* *Candida* *Neurospora* *Penicillium* *Pneumocystis* *Saccharomyces*
Basidiomycota	Includes many common mushrooms and shelf fungi. Sexual reproduction involves formation of a basidium (small, club-shaped structure that typically forms spores at the ends of tiny projections) within which haploid basidiospores are formed. Usually four spores per basidium but can range from one to eight. Sexual reproduction involves fusion with opposite mating type resulting in a dikaryotic mycelium with parental nuclei paired but not initially fused. No subdivisions recognized. Also includes plant pathogens (rusts and smuts) belonging to *Urediniomycota* and *Ustilaginomycota.*	*Agaricus* *Boletus* *Dacrymyces* *Lycoperdon* *Polyporus* *Uromyces* *Ustilago*
Glomeromycota	Filamentous, most are endomycorrhizal arbuscular; form asexual spores outside of host plant; lack centrioles, conidia, and aerial spores. No subdivisions recognized.	*Acaulospora* *Entrophospora* *Glomus*
Microsporidia	Obligate intracellular parasites usually of animals. Lack mitochondria, peroxisomes, kinetosomes, cilia, and centrioles; spores have an inner chitin wall and outer wall of protein; produce a tube for host penetration. Subdivisions currently uncertain.	*Amblyospora* *Encephalitozoon* *Enterocytozoon*
Cryptomycota	3–5 μm roughly spherical cells, flagellated, osmotrophic. Cell wall does not possess chitin. Broadly distributed in soil, freshwater, and aquatic sediments.	*Rozella*

Source: Adl, S. M., et al. "The new higher level classification of eukaryotes with emphasis on the taxonomy of protists." *Journal of Eukaryotic Microbiology*, vol. 52, 2005, 399–451.

25.1 Fungal Biology Reflects Vast Diversity

Readiness Check:

Based on what you have learned previously, you should be able to:

- ✓ Draw a eukaryotic cell envelope and identify the structure and function of its component layers (section 5.2)
- ✓ List the three filaments that make up the cytoskeleton of eukaryotic cells and describe how they are used (section 5.3)
- ✓ Identify the structure and function of eukaryotic cellular organelles (sections 5.4–5.6)
- ✓ Describe the endocytic pathway in general terms (section 5.4)
- ✓ Describe the structure and function of eukaryotic flagella and cilia (section 5.7)

After reading this section, you should be able to:

a. Discuss the importance of fungi in the environment as well as their practical importance
b. Differentiate between a yeast and a mold, and between hypha and mycelium
c. Compare and contrast different spore morphologies
d. Explain how mating types govern sexual reproduction
e. Infer what is meant by a "dikaryotic fungus"
f. Assess the role of fungal decomposition on global carbon flux

Fungal Distribution and Importance

Unlike protists, fungi are primarily terrestrial organisms. They have a global distribution from polar to tropical regions. Fungi are **saprophytes,** securing nutrients from dead organic material by releasing degradative enzymes into the environment. This enables their absorption of the soluble products—a process sometimes called **osmotrophy** (Greek *osmos,* pushing; perhaps a misnomer, as nutrients are transported rather than pushed into the cell). Fungi are important decomposers. They degrade complex organic materials in the environment to simple organic compounds and inorganic molecules. In this way, carbon, nitrogen, phosphorus, and other critical constituents of dead organisms are released and made available for living organisms. Many fungi are pathogenic, with over 5,000 species known to attack economically valuable crops and many other plants. About 20 new human fungal pathogens are documented each year, although most human mycoses are caused by a limited number of species. Conversely, fungi also form beneficial relationships with other organisms. For example, the vast majority of vascular plant roots form important associations with fungi called mycorrhizae. ▶| *Mycorrhizae facilitate the growth of most plants (section 31.3)*

Fungi, especially yeasts (single-celled fungi), are essential to many industrial processes involving fermentation. Examples include the making of bread, wine, beer, cheeses, and soy sauce. They are also important in the commercial production of many organic acids (citric, gallic) and certain drugs (ergometrine, cortisone), and in the manufacture of many antibiotics (penicillin, griseofulvin) and the immunosuppressive drug cyclosporine. In addition, fungi are important research tools in the study of fundamental biological processes. Cytologists, geneticists, biochemists, biophysicists, and microbiologists regularly use fungi in their research. ▶| *Microbiology of fermented foods: beer, cheese, and much more (section 41.5); Microbes are the source of many products of industrial importance (section 42.1)*

Fungal Structure

The body or vegetative structure of a fungus is called a **thallus** (pl., thalli). It varies in complexity and size. Single-cell microscopic fungi are referred to as yeasts, while multicellular masses are called molds. Fungi also include macroscopic puffballs and mushrooms. Like most bacteria, fungi possess cell walls; however, fungal cell walls are made of glucans, mannans, glycoproteins, trehalose, and chitin. Chitin is a strong but flexible nitrogen-containing polysaccharide consisting of *N*-acetylglucosamine residues.

A **yeast** is a unicellular fungus with a single nucleus that reproduces either asexually by budding and transverse division or sexually through spore formation. Generally yeast cells are larger than bacteria and are commonly spherical to egg-shaped. They lack flagella and cilia but have most other eukaryotic organelles (*see figures 5.2 and 5.3a*). Unlike molds, yeasts grow as smooth colonies on agar plates much like bacteria.

The thallus of a **mold** consists of long, branched, threadlike filaments called **hyphae** (s., hypha; Greek *hyphe,* web) that form a tangled mass called a **mycelium** (pl., mycelia). In some fungi, protoplasm streams lengthwise through the hyphae, uninterrupted by cross walls. These hyphae are called **coenocytic** or **aseptate hyphae** (**figure 25.3*a***). The hyphae of other fungi (figure 25.3*b*) have cross walls called **septa** (s., septum) with either a single pore (figure 25.3*c*) or multiple pores (figure 25.3*d*) that enable cytoplasmic streaming. These hyphae are termed **septate hyphae.** When grown in the laboratory, molds form fuzzy colonies on agar plates.

Hyphae are composed of an outer cell wall and an inner lumen, which contains the cytosol and organelles. A plasma membrane surrounds the cytoplasm and lies next to the cell wall. The filamentous nature of hyphae results in a large surface area relative to the volume of cytoplasm. This makes nutrient absorption by osmotrophy possible.

Fungal Reproduction

Reproduction in fungi can be either asexual or sexual. Asexual reproduction is accomplished in several ways: (1) A parent cell undergoes mitosis and divides into two daughter cells by a central constriction and formation of a new cell wall (**figure 25.4*a***), or (2) mitosis in vegetative cells may be concurrent with budding to produce a smaller daughter cell. This is very common in yeasts. The formation of asexual spores often accompanies asexual reproduction and is usually used as a means of dispersal. There are many types of asexual spores, each with its own name. **Arthroconidia** (**arthrospores**) are formed when hyphae fragment (figure 25.4*b*). **Sporangiospores** develop within a sac called a **sporangium** (pl., sporangia) at a hyphal tip (figure 25.4*c*).

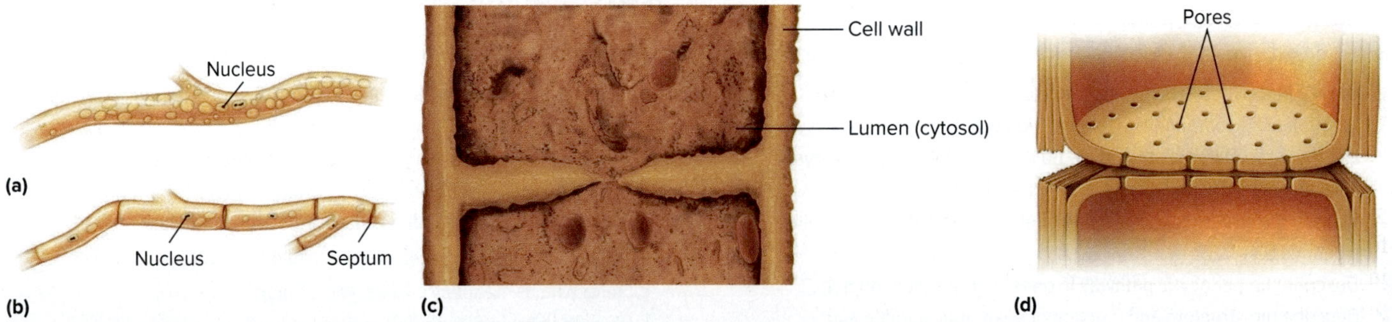

Figure 25.3 Hyphae. Drawings of (a) coenocytic hyphae (aseptate) and (b) hyphae divided into cells by septa. (c) A section of a representative fungal filament demonstrates that in some species, cellular compartments communicate by a single pore. (d) Drawing of a multiperforate septal structure.

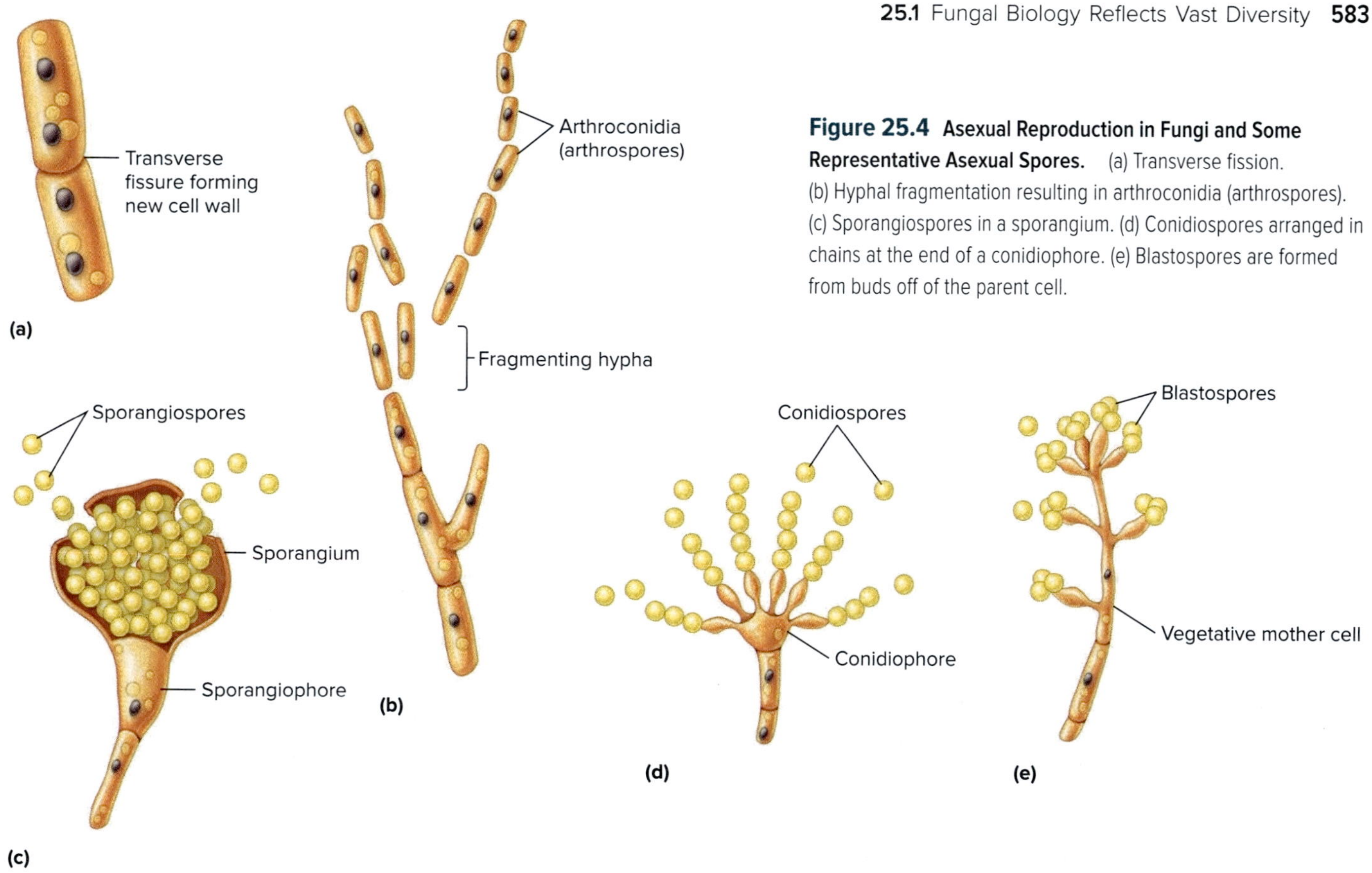

Figure 25.4 Asexual Reproduction in Fungi and Some Representative Asexual Spores. (a) Transverse fission. (b) Hyphal fragmentation resulting in arthroconidia (arthrospores). (c) Sporangiospores in a sporangium. (d) Conidiospores arranged in chains at the end of a conidiophore. (e) Blastospores are formed from buds off of the parent cell.

Conidiospores are spores that are not enclosed in a sac but produced at the tips or sides of the hypha (figure 25.4*d*). **Blastospores** are produced from a vegetative mother cell budding (figure 25.4*e*).

Sexual reproduction in fungi involves the fusion of compatible nuclei. Homothallic fungal species are self-fertilizing and produce sexually compatible gametes on the same mycelium. Heterothallic species require outcrossing between different but sexually compatible mycelia called mating types. Depending on the species, sexual fusion may occur between haploid gametes, gamete-producing bodies called **gametangia,** or hyphae. Sometimes both the cytoplasm and haploid nuclei fuse immediately to produce the diploid zygote, as seen in higher eukaryotes. More often, cytoplasmic fusion occurs before nuclear fusion. This produces a **dikaryotic stage** in which a single cell contains two separate haploid nuclei (N + N), one from each parent (**figure 25.5**). After a period of dikaryotic existence, the two nuclei fuse and undergo meiosis to yield haploid spores. This is seen in both ascomycetes and basidiomycetes, so these are sometimes referred to as dikaryotic fungi.

Fungal spores are important for several reasons. They enable fungi to survive environmental stresses such as desiccation, nutrient limitation, and extreme temperatures, although they are not as stress resistant as bacterial endospores. They aid in fungal dissemination, which helps explain their wide distribution. Because spores are often small and light, they can remain suspended in air for long periods and are often spread by adhering to the bodies of insects and other animals. The bright colors and fluffy textures of many molds often are due to their aerial hyphae and spores. Finally, the size, shape, color, and number of spores are useful in the identification of fungal species.

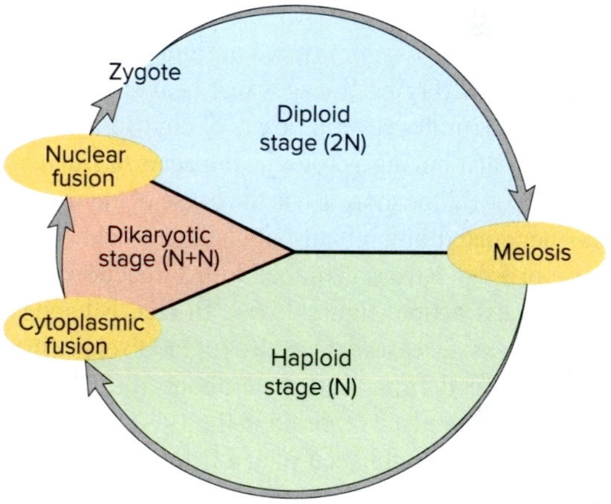

Figure 25.5 Reproduction in Fungi. A drawing of the generalized life cycle for fungi showing the alternation of haploid and diploid stages. Some fungal species do not pass through the dikaryotic stage indicated in this drawing. The asexual (haploid) stage is used to produce spores that aid in the dissemination of the species. The sexual (diploid) stage involves the formation of spores that survive adverse environmental conditions (e.g., cold, dryness, heat).

Comprehension Check

1. How can fungi be defined? What is the difference between a yeast and a mold?
2. What is the distribution of these microbes?
3. What organelles would you expect to find in the cytoplasm of a typical fungus?
4. Compare and contrast the following types of asexual fungal spores: sporangiospore, arthrospore, conidiospore, and blastospore.
5. What does the term *dikaryotic* mean? How does this differ from sexual reproduction in other organisms?

25.2 Chytridiomycetes Produce Motile Spores

After reading this section, you should be able to:

a. Explain why chytrids are unique among fungi
b. Describe general chytrid morphology
c. Discuss the role of chytridriomycosis in the loss of amphibian diversity worldwide

The polyphyletic **chytridiomycetes,** commonly called **chytrids,** includes free-living and parasitic genera. Free-living chytrids are saprophytic, living on plant or animal matter in freshwater, mud, or soil. Parasitic forms infect aquatic plants and animals, including insects (**figure 25.6*a***). A few are found in the anoxic rumen of herbivores. Chytrids display a variety of life cycles involving both asexual and sexual reproduction. Members of this group are microscopic in size and may consist of a single cell, a small multinucleate mass, or a true mycelium.

Chytrids are unique among fungi in the production of a spore with a single, posterior, whiplash flagellum. It has been hypothesized that all fungi were once flagellated. The loss of flagella as more advanced groups evolved may have coincided with the development of other means of spore distribution, such as aerial dispersal from mycelial mats. Chytrids are not considered monophyletic (note three clades of *Chytridiomycota* in figure 25.1). Instead, they derive from early diverging lineages that bear spores. The largest clade, *Euchytrids,* unites several orders of chytrids.

A chytrid that is having a profound impact is *Batrachochytrium dendrobatidis,* or Bd for short. Bd is the cause of chytridiomycosis, which has decimated amphibian populations worldwide. At least 200 species of frogs have succumbed to Bd infections; some have been driven to extinction (figure 25.6*b*). In fact, Bd appears to be causing the greatest disease-induced loss of biodiversity in recorded history. Bd emerged from East Asia about 100 years ago, and deadly infections were first reported in the 1990s. It is unclear why chytridiomycosis has established such a firm grip on some species of frogs but not others. Those that are unaffected often serve as carriers. Bd affects the skin of infected frogs, which becomes overly thick (hyperkeratinized). Because gas exchange occurs through amphibian skin, infected frogs are unable to regulate the concentration of blood electrolytes and die in cardiac arrest. As if this situation were not dire enough, the closely related species *Batrachochytrium salamandrivorans* (Bsal) was introduced to Europe from Asia in 2010 via the pet trade. Whereas salamanders and newts in Asia are immune to Bsal, those in Europe are exquisitely sensitive and large die-offs of salamander populations have been reported. Bsal is particularly hearty in the environment because it produces an encysted spore that is covered by a thick, water-resistant cell wall.

25.3 Zygomycetes: Fungi with Coenocytic Hyphae

After reading this section, you should be able to:

a. Recognize zygomycetes in nature
b. Outline the life cycle of a *Rhizopus* species
c. Discuss the benefits and harm attributed to zygomycetes

Like chytrids, **zygomycetes** are also polyphyletic. Most live on decaying plant and animal matter in the soil; a few are parasites of plants, insects, and other animals, including humans. The hyphae of zygomycetes are coenocytic (aseptate), with many haploid nuclei. Asexual spores develop in sporangia at the tips of aerial hyphae and are usually wind dispersed. Sexual reproduction produces tough, thick-walled zygotes called **zygospores** that can remain dormant when the environment is too harsh for growth of the fungus.

The mold *Rhizopus stolonifer* is a common member of this division. This fungus grows on the surface of moist, carbohydrate-rich foods, such as breads, fruits, and vegetables. Hyphae called rhizoids extend into the substrate and absorb nutrients. Other hyphae (stolons) become erect, then arch back into the substratum, forming new rhizoids. Still others remain erect and produce at their tips asexual sporangia filled with hundreds of black spores, giving the mold its characteristic color. Each spore, when liberated, can germinate to start a new mycelium.

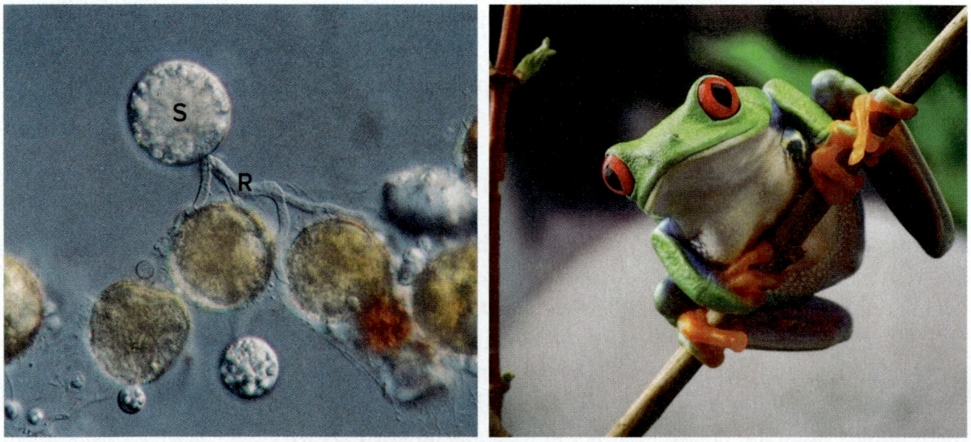

(a) (b)

Figure 25.6 Chytrids and a Chytridiomycosis Susceptible Host. (a) A chytrid with a single round sporangium (S) that is about 50 μm in diameter. In this image, the fungus is growing between brown pollen grains, which it will eventually penetrate with its branched rhizoids (R). (b) Chytridiomycosis caused by *Batrachochytrium dendrobatidis* (Bd) is killing amphibians. Shown here is the red-eyed tree frog (*Agalychnis callidryas*), whose population has succumbed to Bd infection.
(a) ©Martha Powell; (b) ©Irina Kozorog/Shutterstock

MICRO INQUIRY *What is the function of chytrid rhizoids? How might rhizoids contribute to the ability of Bd to cause infection?*

Rhizopus spp. usually reproduce asexually, but if food becomes scarce or environmental conditions unfavorable, sexual reproduction occurs (**figure 25.7**). Sexual reproduction requires compatible strains of opposite mating types (called + and −). When + and − hyphae are close, each produces a different hormone, called a pheromone, that causes their hyphae to form projections called progametangia; these mature into gametangia. First the gametangia fuse, a process called plasmogamy. Then, the nuclei of the two gametes fuse (karyogamy), forming a zygote. The zygote develops a thick, rough, black coat and becomes a dormant zygospore. Meiosis often occurs at the time of germination; the zygospore then splits open and produces a hypha that bears an asexual sporangium filled with haploid spores that begin the cycle again.

One important zygomycete is *Rhizopus chinensis*, the cause of the rice disease known as seedling blight. If one considers that rice feeds more people on Earth than any other crop, the implications of this disease are obvious. It was thought that the fungus secreted a toxin that kills rice seedlings, so scientists set about isolating the toxin and the genes that produce it. Much to everyone's surprise, the toxin is produced by an α-proteobacterium, *Burkholderia* sp., growing within the fungus.

Zygomycetes also contribute to human welfare. For example, *Rhizopus oligosporus* is used in Indonesia to produce tempeh from boiled soybeans. Other zygomycetes (*Mucor* spp.) are used with soybeans in Asia to make a curd called sufu. Others are employed in the commercial preparation of some anesthetics, birth control agents, industrial alcohols, meat tenderizers, and the yellow coloring used in margarine and butter substitutes. ▶| *Microbiology of fermented foods: beer, cheese, and much more (section 41.5)*

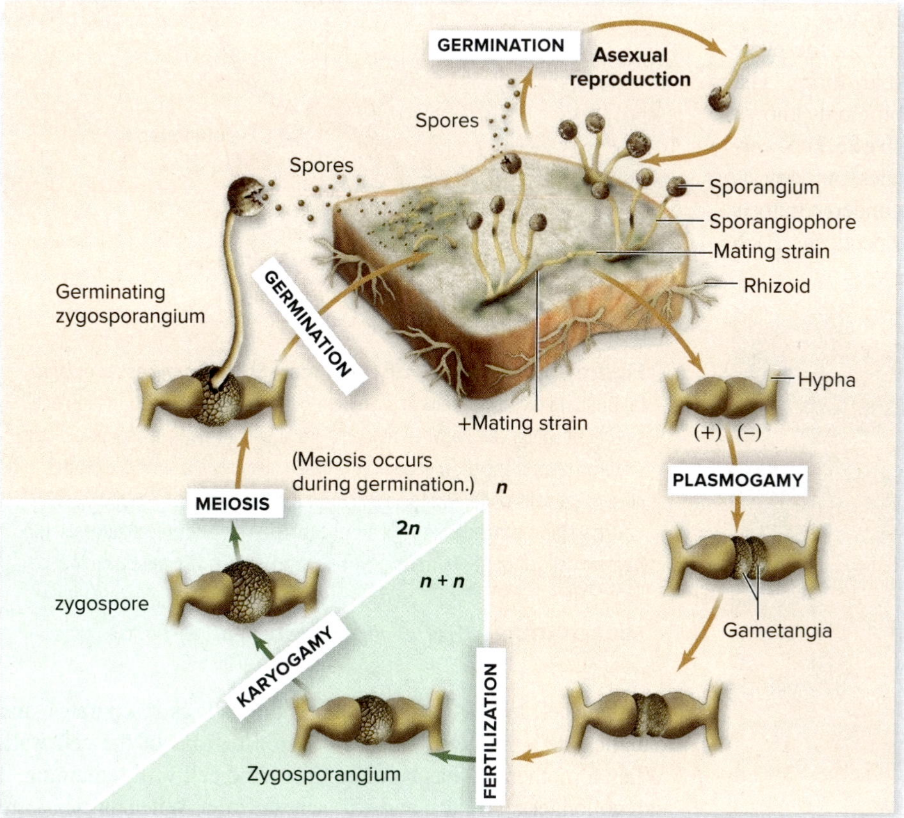

Figure 25.7 ***Rhizopus stolonifer*, a Zygomycete That Grows on Simple Sugars.** This fungus is often found on moist bread or fruit.

25.4 *Glomeromycota* Are Mycorrhizal Symbionts

After reading this section, you should be able to:

a. Describe the functional importance of glomeromycetes
b. Discuss the contributions of each partner in a mycorrhizal fungus-plant symbiosis

Glomeromycetes are of critical ecological importance because they are symbionts of vascular plants. All glomeromycetes are **mycorrhizal fungi** that grow in association with the roots of their plant hosts, which include herbaceous plants and tropical trees. This is a mutualistic relationship because both the host plant and the fungus benefit: The fungus delivers soil nutrients to the plant, and the plant in turn provides carbohydrate to the fungus. Among the six types of mycorrhizal fungi discussed in chapter 31, only the arbuscular mycorrhizal hyphae are glomeromycetes; the other groups are basidiomycetes or ascomycetes. Arbuscular mycorrhizae are important because of the number and diversity of plants they colonize. ▶| *Mycorrhizae facilitate the growth of most plants (section 31.3)*

Glomeromycetes have aseptate hyphae and produce large, multinucleate spores. Only asexual reproduction has been documented but there is evidence that sexual reproduction may occur as well. Spores germinate when in contact with the roots of a suitable host plant. Glomeromycetes have evolved specialized flat hyphae called **appressoria** (s., appressorium) to penetrate their host plants.

Comprehension Check

1. What morphological feature makes the chytrids unique among fungi?
2. What is the difference between hyphae called rhizoids and those called stolons?
3. How do different mating types of *Rhizopus stolonifer* recognize each other?
4. Given that mycorrhizal fungi obtain carbohydrates from their plant partners, what specific nutrients are the fungi likely to provide to plants?
5. Which fungal class would you most likely find in each of the following environments: 2-week-old cake, amphibian skin, tree roots?

25.5 *Ascomycota* Includes Yeasts and Molds

After reading this section, you should be able to:

a. Describe ecological, economic, and medical importance of ascomycetes
b. Diagram the *Saccharomyces cerevisiae* life cycle
c. Outline the mechanisms used by typical filamentous ascomycetes to reproduce, survive unfavorable conditions, and disperse
d. Explain the function of sclerotia
e. Identify the origin of St. Anthony's fire, LSD, and "sick building syndrome"

Members of *Ascomycota,* or **ascomycetes,** commonly known as **sac fungi,** are named for their characteristic reproductive structure, the saclike **ascus** (pl., asci; Greek *askos,* sac), which holds **ascospores.** Ascomycetes are ecologically important in freshwater, marine, and terrestrial habitats because they degrade many chemically stable organic compounds, including lignin, cellulose, and collagen. Many species are quite familiar and economically important (**figure 25.8**). For example, most of the red, brown, and blue-green molds that cause food spoilage are ascomycetes. The powdery mildews that attack plant leaves and the fungi that cause chestnut blight and Dutch elm disease are ascomycetes. Most fungi that infect humans are ascomycetes. On the other hand, many yeasts as well as edible morels and truffles are also ascomycetes. The pink bread mold *Neurospora crassa* is an important research tool in genetics and biochemistry.

Some ascomycetes are yeasts, while others have a life cycle that alternates between yeast and filamentous (mold) forms. The life cycle of the yeast *Saccharomyces cerevisiae,* commonly known as brewer's or baker's yeast, is well understood (**figure 25.9**). *S. cerevisiae* alternates between haploid and diploid states. As long as nutrients remain plentiful, haploid and diploid cells undergo mitosis to produce haploid and diploid daughter cells, respectively. Each daughter cell leaves a scar on the mother cell as it separates, and daughter cells bud only from unscarred regions of the cell wall. When a mother cell has no more unscarred cell wall remaining, it can no longer reproduce and will senesce (die). When nutrients are limited, diploid *S. cerevisiae* cells undergo meiosis to produce four haploid cells that remain bound within a common cell wall, the ascus. Upon the addition of nutrients, two haploid cells of opposite mating types (a and α) come into contact and fuse to create a

(a) *Morchella esculenta* **(b)** *Tuber brumale*

Figure 25.8 ***Ascomycota.*** (a) The common morel, *Morchella esculenta,* is one of the choicest edible fungi. It fruits in the spring. (b) The black truffle, *Tuber brumale,* is highly prized for its flavor.
(a) ©Juniors Bildarchiv GmbH/Alamy Stock Photo; (b) ©Kelly Cline/E+/Getty Images

(a) *Saccharomyces cerevisiae:* budding division

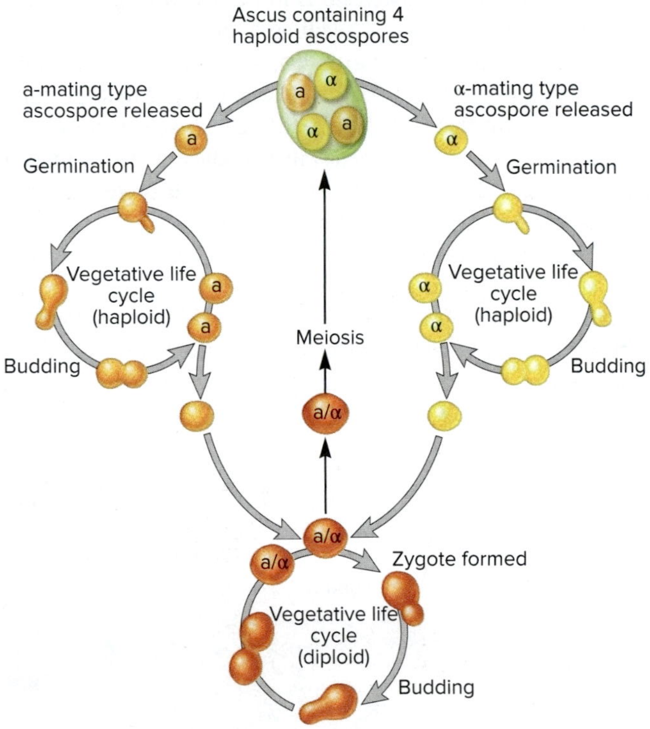

(b) *S. cerevisiae* life cycle

Figure 25.9 The Life Cycle of the Yeast *Saccharomyces cerevisiae.* (a) Budding division results in asymmetric septation and the formation of a smaller daughter cell. (b) When nutrients are abundant, haploid and diploid cells undergo mitosis and grow vegetatively. When nutrients are limited, diploid *S. cerevisiae* cells undergo meiosis to produce four haploid cells that remain bound within a common cell wall, the ascus. Upon the addition of nutrients, two haploid cells of opposite mating types (a and α) fuse to create a diploid cell.
(a) ©SCIMAT/SPL/Science Source

MICRO INQUIRY *What determines when a yeast cell can no longer bud?*

Figure 25.10 Asexual Reproduction in Filamentous Ascomycetes. Characteristic conidiospores of *Aspergillus* spp. as viewed with the scanning electron microscope.

diploid. Typically only cells of opposite mating types can fuse; this process is tightly regulated by the action of pheromones.

S. cerevisiae is a valuable model organism. Research on this organism has revealed the importance of many cellular processes. For instance, it is a favorite model system for studying cell cycling and the events during mitosis. This research is critical not only for our understanding of normal cell division, but also for exploring the loss of cell cycle control that occurs in cancerous cells.

Unlike zygomycetes, filamentous ascomycetes form hyphae with septae (crosswalls). Asexual reproduction is common and is associated with the production of asexual spores called **conidia** (**figure 25.10**). Sexual reproduction involves two strains of opposite mating types. Hyphae from one mating type form a specialized structure called an ascogonium, and hyphae of the other mating type form an antheridium (**figure 25.11**). The nuclei from the antheridium migrate to the ascogonium during the exchange of cytoplasm (plasmogamy). The nuclei do not yet fuse; rather nuclei from each mating type pair and replicate in unison during the ensuing dikaryotic phase. The ascogonium is said to be fertilized and the hyphae are called **ascogenous hyphae.** Ascogenous hyphae, supported by sterile hyphae, next develop into a cuplike structure

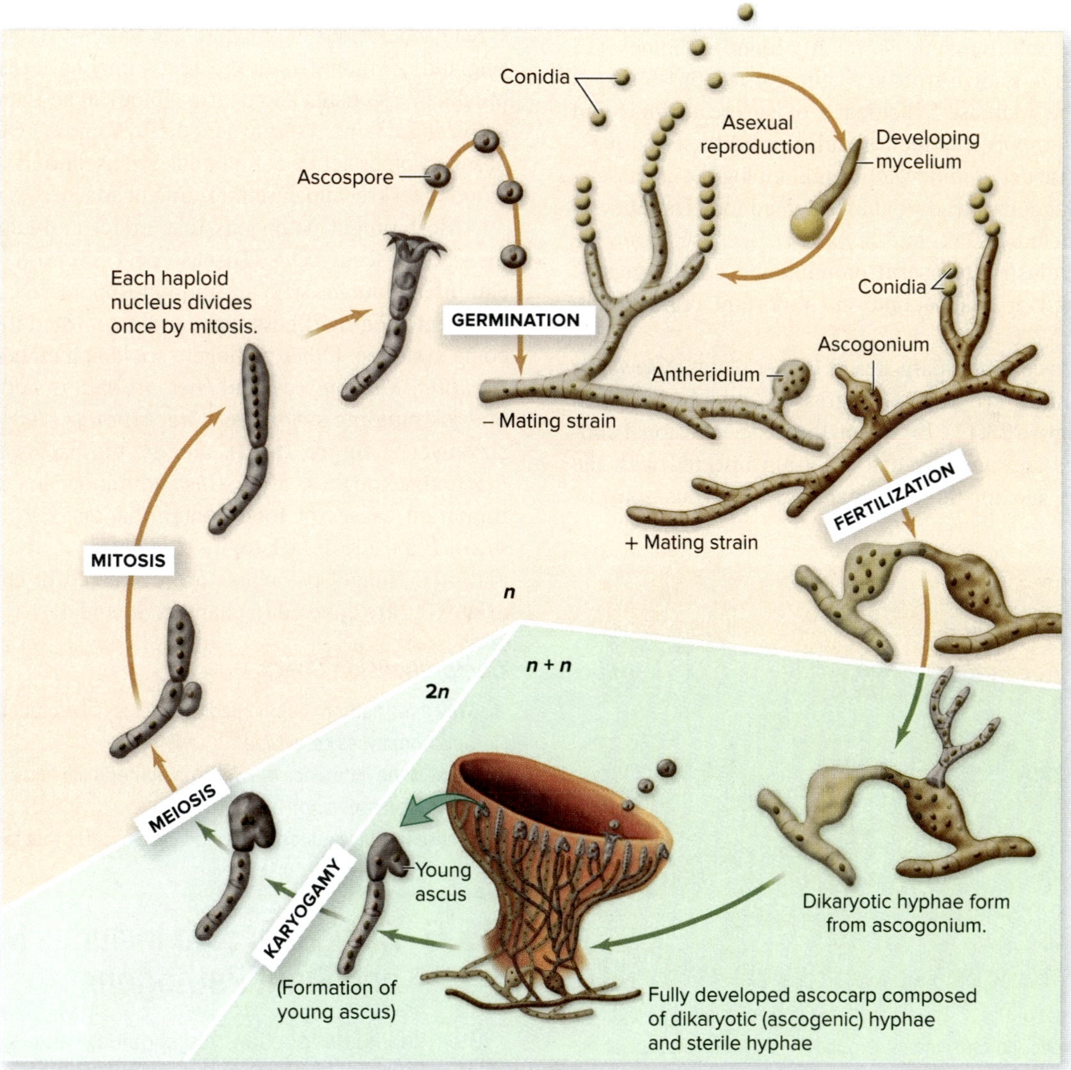

Figure 25.11 The Typical Life Cycle of a Filamentous Ascomycete. Sexual reproduction involves the formation of asci and ascospores. The events pictured on the left side of the figure (karyogamy, meiosis, mitosis, and release of ascospores) occur within the ascocarp.

MICRO INQUIRY *In what ways is the life cycle of filamentous ascomycetes similar to that of the zygomycete shown in figure 25.7? How do the two life cycles differ?*

called an **ascocarp.** This is where ascogenous hyphae mature, and nuclear fusion (karyogamy) takes place. The diploid zygote nucleus then undergoes meiosis, and the resulting four haploid nuclei divide mitotically again to produce a row of eight nuclei in each developing ascus. These nuclei are walled off from one another and each will develop into an **ascospore.** Thousands of asci may be packed together in an ascocarp. When ascospores mature, they often are released from the asci with great force. If the mature ascocarp is jarred, it may appear to belch puffs of "smoke" consisting of thousands of ascospores. Upon reaching a suitable environment, the ascospores germinate and start the cycle anew.

Although conidia are the major form of dissemination, structures called **sclerotia** are compact masses of dormant hyphae that permit survival in stressful conditions. Sclerotia confer a competitive advantage to the fungi that produce them. For instance, some species of *Aspergillus* that infect plants form sclerotia to remain viable in the soil during cool months. They then take advantage of nutrient resources when the temperature rises.

Several *Aspergillus* species are noteworthy. *A. fumigatus* is ubiquitous in the environment, commonly found in homes and the workplace. It is known to trigger allergic responses and is implicated in the increased incidence of severe asthma and sinusitis. *A. nidulans* is a model organism used to study questions of eukaryotic cell and developmental biology. *A. oryzae* is used in the production of traditional fermented foods and beverages in Japan, including saki and soy sauce. Because *A. oryzae* secretes many industrially useful proteins and can be genetically manipulated, it has become an important organism in biotechnology.

Many ascomycetes are parasites of higher plants. *Claviceps purpurea* parasitizes rye and other grasses, causing the plant disease **ergot** (**figure 25.12**). Ergotism, the toxic, delusional condition in humans and animals that eat grain infected with the fungus, is often accompanied by gangrene, nervous spasms, abortion, and convulsions. During the Middle Ages, ergotism, then known as St. Anthony's fire, killed thousands of people. For example, over 40,000 deaths from ergot poisoning were recorded in France in the year 943. It has been suggested that the widespread accusations of witchcraft in Salem Village and other New England communities in the 1690s may have resulted from outbreaks of ergotism. The pharmacological activities are due to an active ingredient, lysergic acid diethylamide (LSD). In controlled dosages, other active compounds are used to induce labor, lower blood pressure, and ease migraine headaches.

Figure 25.12 **Ergot of Rye.** The ascomycete *Claviceps purpurea* infects rye and other grasses, producing hard masses of hyphae known as ergots in place of some of the grains (fruits). Ergots produce alkaloids that cause psychotic delusions in humans and animals when products made with infected grain are consumed. Ergots have been used to treat migraine headaches and hasten childbirth.

Figure 25.13 **Fungal "Sick Building Syndrome" Can Be Caused by an Ascomycete.** Scanning electron micrograph of dense mycelia and conidiophores characteristic of *Stachybotrys chartarum*.

Most fungal pathogens that infect animals, including humans, are ascomycetes (**Disease 25.1**). Several *Aspergillus* spp. can infect humans who are immunosuppressed. Aspergillosis is a life-threatening illness that can be confined to the lung or become systemic. Other pathogens include members of the genera *Candida, Blastomyces,* and *Histoplasma*. In addition, one cause of "sick building syndrome," *Stachybotrys chartarum* is also an ascomycete (**figure 25.13**). Finally, the *Aspergillus* and *Fusarium* toxins known as aflatoxins and fumonisins, respectively, are important causes of food contamination (*see figures 41.5 and 41.6*). Exposure to aflatoxins can result in liver cancer. These and other fungal pathogens are discussed in chapter 40, while aflatoxins are covered in chapters 40 and 41.

Comprehension Check

1. Why does nutrient deprivation trigger sexual reproduction in *Saccharomyces cerevisiae*?
2. What is the difference between an antheridium and an ascogonium? What is an ascocarp?
3. Where would you most likely find ascomycetes that form sclerotia?

25.6 *Basidiomycota* Includes Mushrooms and Plant Pathogens

After reading this section, you should be able to:

a. Describe the life cycle of a typical basidiomycete
b. Identify the fungi that cause plant rusts
c. Describe a plant disease caused by a basidiomycete

DISEASE

25.1 White-Nose Syndrome Is Decimating North American Bat Populations

Bats evoke all kinds of images. Some people immediately think of vampire bats and are repulsed. Others think of large fruit bats often called flying foxes. If you have spent a summer evening outdoors in North America, mosquitoes and the small bats that eat them may come to mind. A new scene can now be added to these: bats with white fungal hyphae growing around their muzzles (**box figure**). This is the hallmark of white-nose syndrome (WNS), and if its rate of infection continues unchecked, it is projected to eliminate the most common bat species in eastern North America (*Myotis lucifugus*) by 2026.

WNS was first spotted in 2006 among bats hibernating in a cave near Albany, New York. Scientists quickly became alarmed for two reasons. First, it spreads rapidly—it's known to occur in at least 11 bat species and is now found in 31 states in the United States, including Texas and Washington, and five Canadian provinces. Second, it is deadly. The population of bats declines from 30 to 99% in any given infected hibernacula (the place where bats hibernate, which unfortunately rhymes with Dracula).

***Pseudogymnoascus destructans* causes WNS.** A little brown bat (*Myotis lucifugus*) with the white fungal hyphae (arrow) for which WNS is named. Lower figure shows damage to wing as a result of fungal infection.
(Top) Source: U.S. Fish and Wildlife Service/Marvin Moriatry; (Bottom) Source: USGS/Carol Uphoff Meteyer

WNS is caused by the ascomycete *Pseudogymnoascus destructans*. It colonizes a bat's wings, muzzle, and ears where it erodes the epidermis before invading the underlying skin and connective tissue. Despite the name WNS, the primary site of infection is the wing. Wings provide a large surface area for colonization, and once infected, the thin layer of skin is easily damaged, leading to adverse physiological changes during hibernation. These in turn result in premature awakening, loss of essential fat reserves, and strange behavior.

Where did this pathogen come from and why does it infect bats? The best hypothesis regarding its origin is that humans inadvertently brought it from Europe, where it causes mild infection in at least one hibernating bat species. This makes *P. destructans* an apparent case of pathogen pollution—the human introduction of invasive pathogens of wildlife and domestic animal populations that threaten biodiversity and ecosystem function.

The capacity of *P. destructans* to sweep through bat populations results from a "perfect storm" of host- and pathogen-associated factors. *P. destructans* is psychrophilic, with a growth optimum around 12°C; it does not grow above 20°C. All infected bat species hibernate in cold and humid environments such as caves and mines. Because their metabolic rate is drastically reduced during hibernation, their body temperature reaches 2 to 7°C. Thus WNS is only seen in hibernating bats or those that have just emerged from hibernation. When metabolically active, the bat's body temperature is too warm to support pathogen growth.

While it is too late to save the millions of bats that have already succumbed to WNS, microbiologists, conservationists, and government agencies are trying to limit the continued decline in bat populations. Caves have been closed to human traffic, and protocols for decontamination after visiting hibernacula have been developed to limit the spread from cave to cave. Although we cannot cure sick bats, it is our responsibility to stop the continued spread of this pathogen.

Fungi in *Basidiomycota* are commonly known as **basidiomycetes,** or **club fungi.** Examples include jelly fungi, rusts, shelf fungi, stinkhorns, puffballs, toadstools, mushrooms, and bird's nest fungi. Most are saprophytes that decompose plant debris, especially cellulose and lignin. For example, the common fungus *Polyporus squamosus* forms large, shelflike structures that project from the lower portion of dead trees, which they help decompose. Many mushrooms are used as food, and the cultivation of the mushroom *Agaricus campestris* is a multimillion-dollar business. Of course, not all mushrooms are edible. Many mushrooms produce alkaloids that act as either poisons or hallucinogens. One such example is the "death angel" mushroom, *Amanita phalloides* (**figure 25.14**). Two toxins isolated from this species are phalloidin and α-amanitin. Phalloidin primarily attacks liver cells, where it binds to and ruptures plasma membranes. Alpha-amanitin attacks the cells lining the stomach and small intestine, causing severe gastrointestinal symptoms associated with mushroom poisoning. *A. muscaria* is both poisonous and hallucinogenic.

Basidiomycetes are named for their characteristic structure or cell, the **basidium,** which is involved in sexual reproduction

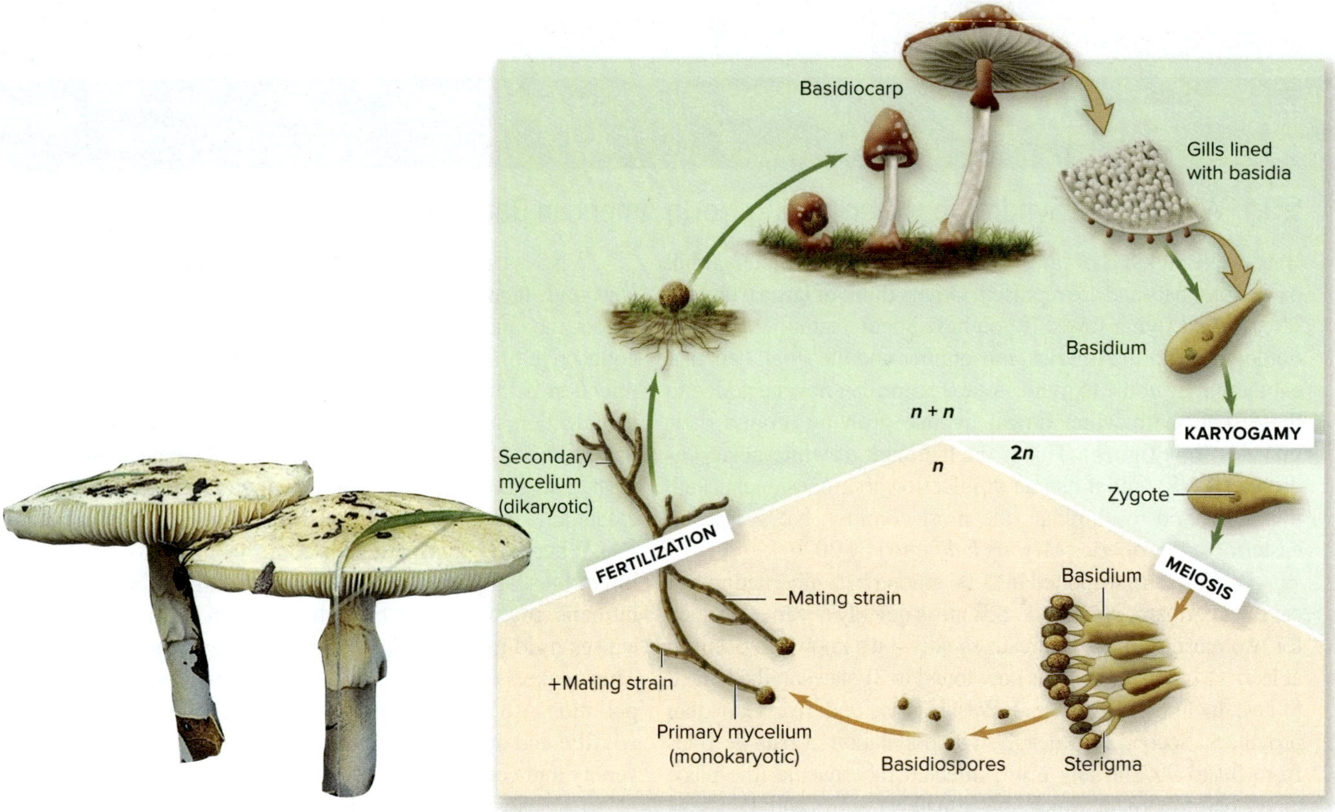

Figure 25.14 ***Basidiomycota.*** (a) The death cap mushroom *Amanita phalloides* is usually fatal when eaten. (b) The life cycle of a typical soil basidiomycete starts with a basidiospore germinating to produce a monokaryotic mycelium (a single nucleus in each septate cell). The mycelium grows and spreads throughout the soil. When this primary mycelium meets another monokaryotic mycelium of a different mating type, the two fuse to initiate a dikaryotic secondary mycelium. Hyphae in the secondary mycelium are divided by septa into cells, each of which contains two nuclei, one of each mating type. This dikaryotic mycelium is eventually stimulated to produce basidiocarps. A solid mass of hyphae forms a button that pushes through the soil, elongates, and develops a cap. The cap contains many platelike gills, each of which is coated with basidia. The two nuclei in the tip of each basidium fuse to form a diploid zygote nucleus, which immediately undergoes meiosis to form four haploid nuclei. These nuclei push their way into the developing basidiospores, which are then released at maturity.

(figure 25.14*b*). A basidium (Greek *basidion,* small base) is produced at the tip of hyphae and normally is club-shaped. Two or more **basidiospores** are produced by the basidium, and basidia may be held within fruiting bodies called **basidiocarps.** Basidiocarps can reach 60 cm in diameter and have many pores, each lined with basidia that produce basidiospores. Thus a single fruiting body can produce millions of spores. Like ascomycete hyphae, the hyphae of basidiomycetes are septate.

Basidiomycetes known as urediniomycetes and ustilaginomycetes include important plant pathogens causing "rusts" and "smuts," respectively, that destroy millions of dollars worth of crops annually. These fungi do not form large basidiocarps. Instead, small basidia arise from hyphae at the surface of the host plant. The hyphae grow extracellularly in plant tissue. In addition, some urediniomycetes are human pathogens.

The ustilaginomycete *Ustilago maydis* is a common corn pathogen that has become a model organism for plant smuts (**figure 25.15**). It is dimorphic; plant-associated fungi grow in the mycelial form but are yeastlike in the external environment. Yeastlike saprophytic *U. maydis* cells can be easily grown in the laboratory. In nature, the yeast (haploid sporidia) must mate to produce infectious, filamentous dikaryons that depend on the host plant for continued development (**figure 25.16**). Once a plant is infected, *U. maydis* forms appressoria within the host. This triggers the plant to form tumors in which the fungus proliferates and eventually produces diploid spores called

Figure 25.15 ***Ustilago maydis.*** This pathogen causes tumor formation in corn. Note the enlarged tumors in place of normal corn kernels.

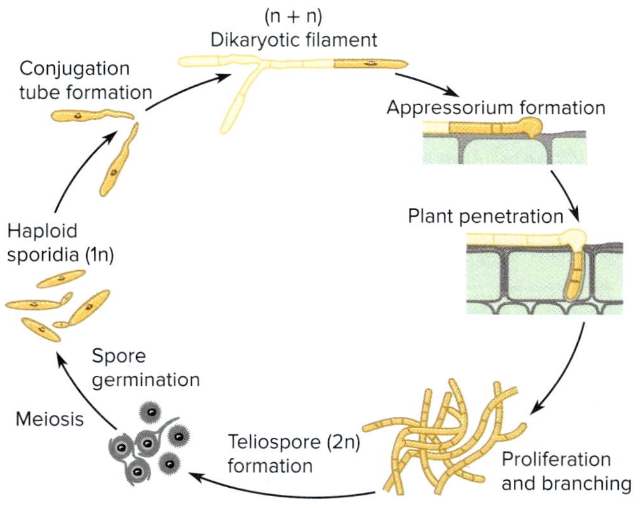

Figure 25.16 **Life Cycle of the Plant Pathogen *Ustilago maydis*.** Haploid sporidia mate to form infectious dikaryons, which depend on the host plant for continued development. The formation of fungal appressoria within the host induces tumor formation.

MICRO INQUIRY *What adaptations do you think* U. maydis *has evolved that has led to its pathogenicity? Contrast this with mycorrhizae.*

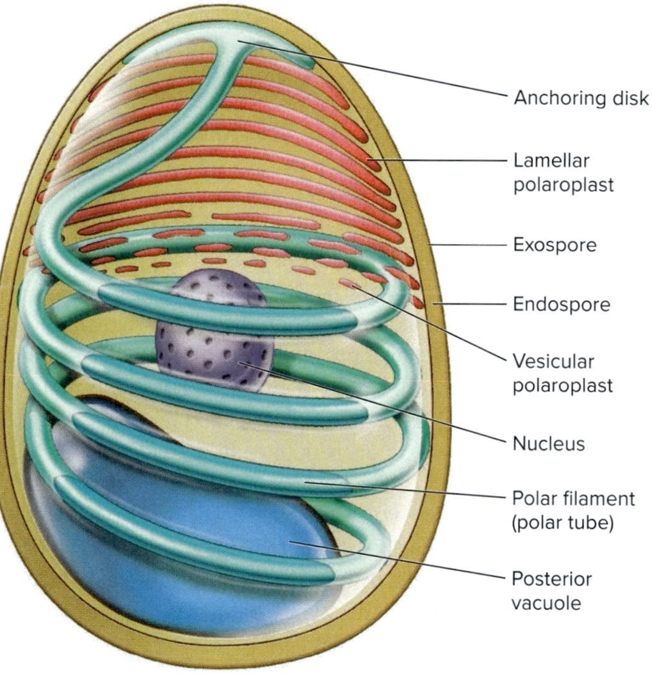

Figure 25.17 **The Unique Structure of a Microsporidian Spore.** Upon germination in a host cell, the polar tube is ejected with enough force to pierce the host cell membrane, allowing the fungus to gain entry.

teliospores. Upon germination, cells undergo meiosis and haploid sporidia are released. These sporidia can spread from plant to plant, where they start the growth and infection cycle again.

25.7 *Microsporidia* Are Intracellular Parasites

After reading this section, you should be able to:

a. Explain why microsporidia are no longer considered protists
b. Describe the microsporidian life cycle and relate its unique morphology to host invasion
c. List at least two human diseases caused by microsporidia

Of all the fungi, members of ***Microsporidia*** have the most confused taxonomic history. These obligately endoparasitic microbes are sometimes still considered protists. However, analysis of ribosomal RNA and specific proteins such as α- and β-tubulin shows that they are fungi. However, unlike other fungi, they lack mitochondria, peroxisomes, and centrioles but possess mitosomes, like some protists. Their genomes are highly reduced in size and complexity. But like most fungi, microsporidian cell walls contain chitin and trehalose.

Microsporidial morphology is also unique among eukaryotes. It is distinct in the possession of an organelle called the polar tube that is essential for host invasion. Small spores of 1 to 40 μm are viable outside the host. Depending on the species, spores may be spherical, rodlike, or egg- or crescent-shaped. Spore germination occurs only in the host, where it expels the tightly packed polar tube or filament (**figure 25.17**). The polar tube is ejected with enough force to pierce the host cell membrane, permitting parasite entry. Once inside the host cell, the microsporidian undergoes a developmental cycle that differs among the various microsporidian species. However, in all cases, more spores are produced and eventually take over the host cell.

Microsporidia are important obligate intracellular parasites that infect insects, fish, and humans. They are particularly problematic for immunocompromised individuals. Human pathogens include *Enterocytozoon bieneusi,* which causes diarrhea and pneumonia (depending on whether it was acquired through ingestion or inhalation, respectively), and *Encephalitozoon cuniculi,* which causes encephalitis and nephritis (kidney disease). Aquatic birds are an important source of pathogenic microsporidia in nature; indeed, a single flock of waterfowl can deposit over 10^8 microsporidian spores in surface waters after one visit. ▶| *Microsporidiosis (section 40.6)*

Comprehension Check

1. What is the difference between a basidiocarp and a basidium?
2. How do cryptococci differ from most basidiomycetes?
3. How does *Ustilago maydis* trigger tumor formation? How is tumor formation advantageous to the fungi?
4. Why do you think members of *Microsporidia* are still sometimes considered protists? What structure is essential for the germination of a microsporidian spore? Describe its function.

Key Concepts

25.1 Fungal Biology Reflects Vast Diversity

- Fungi are important decomposers that break down organic matter; live as parasites on animals, humans, and plants; play a role in many industrial processes; and are used as research tools in the study of fundamental biological processes. Fungi secrete enzymes and absorb the digested food.
- A fungus is a eukaryotic, spore-bearing organism that has absorptive nutrition and lacks chlorophyll; reproduces asexually, sexually, or by both methods; and normally has a cell wall containing glucans and chitin.
- Hyphae are long, branched, threadlike filaments that form a tangled mass called a mycelium. Hyphae may be either septate or coenocytic (nonseptate). The mycelium can produce reproductive structures (**figure 25.3**).
- Asexual reproduction in the fungi often leads to the production of specific types of spores that are easily dispersed (**figure 25.4**).
- Sexual reproduction is initiated in fungi by the fusion of hyphae (or cells) of different mating strains. In some fungi, the nuclei in the fused hyphae immediately combine to form a zygote. In dikaryotes, the two genetically distinct nuclei remain separate, forming pairs that divide synchronously. Eventually some nuclei fuse (**figure 25.5**).

25.2 Chytridiomycetes Produce Motile Spores

- Fungi belonging to *Chytridiomycota* produce motile spores. Most are saprophytic; chytridiomycosis is causing massive, global amphibian die-offs (**figure 25.6**).
- Chytrids are microscopic and display both sexual and asexual reproduction.

25.3 Zygomycetes: Fungi with Coenocytic Hyphae

- Most zygomycetes are saprophytic. One example is the common bread mold, *Rhizopus stolonifer.* Sexual reproduction occurs through conjugation between strains of opposite mating type (**figure 25.7**).
- Zygomycetes are used in food and pharmaceutical manufacturing.

25.4 *Glomeromycota* Are Mycorrhizal Symbionts

- *Glomeromycota* includes arbuscular mycorrhizal fungi that grow in association with plant roots. They serve to increase plant nutrient uptake.
- Arbuscular mycorrhizae grow between the plant cell wall and plasma membrane.

25.5 *Ascomycota* Includes Yeasts and Molds

- Members of *Ascomycota* are known as sac fungi because they form a sac-shaped reproductive structure called an ascus. Ascomycetes can have either a yeast morphology (**figure 25.9**) or a mold morphology (**figure 25.11**).
- Important ascomycetes include the yeast *Saccharomyces cerevisiae, Aspergillus* spp., *Claviceps purpurea* (the cause of ergotism), and the indoor mold *Stachybotrys chartarum.*

25.6 *Basidiomycota* Includes Mushrooms and Plant Pathogens

- Members of *Basidiomycota* are called club fungi. They are named after their basidium that produces basidiospores (**figure 25.14**).
- Genera from urediniomycetes and ustilaginomycetes include important plant pathogens called smuts and rusts (**figure 25.15**).

25.7 *Microsporidia* Are Intracellular Parasites

- Members of *Microsporidia* have a unique morphology and are obligate intracellular pathogens.
- They include opportunistic human pathogens; some infect other vertebrates and insects (**figure 25.17**).

Active Learning

1. Some fungi can be viewed as coenocytic organisms that exhibit little differentiation. When differentiation does occur, such as in the formation of reproductive structures, it is preceded by septum formation. Why does this occur?
2. Both bacteria and fungi are major environmental decomposers. Obviously competition exists in any given environment, but fungi usually have an advantage. Discuss characteristics specific to fungi that provide this advantage.
3. *Neurospora crassa* and several other filamentous fungi demonstrate a circadian rhythm such that conidiation occurs in roughly 21.5-hour cycles, depending on temperature and the light/dark regime. This can be observed in nature as bands of conidia alternating with bands of undifferentiated fungi. Suggest why these fungi evolved a circadian rhythm. If you were to study this phenomenon, what questions would you want to explore?

4. Analysis of sclerotia and conidia formation in the filamentous ascomycete *Aspergillus flavus* suggests that formation of these structures is cell density dependent. Specifically, high cell density cultures yield conidia, whereas low cell density triggers sclerotia formation. Review the structure and function of conidia and sclerotia, and formulate a hypothesis to explain why their development would be regulated in a density-dependent manner.
5. *Ustilago maydis* causes corn smut disease (figure 25.15). Although it is known to produce a variety of molecules that suppress host plant defenses and modulate plant metabolism, not much is known about these effectors. Two European investigators explored one such protein, called ApB73. They found that ApB73 is needed for host tumor (gall) formation but the protein remains stuck to the fungal cell wall during infection. How do you think it was determined that ApB73 was required for gall formation? Given that it does not enter host plants but remains at the fungus-host interface, discuss possible roles ApB73 might have in mediating infection.

   Read the original paper: Stirnberg, A., and Djamei, A. 2016. Characterization of ApB73, a virulence factor important for colonization of *Zea mays* by the smut *Ustilago maydis*. *Mol. Plant Pathol.* 17:1467–79. doi:10.1111/mpp.12442.

26 Viruses

Courtesy of Kathleen Sandman

Disrupting the Viral Life Cycle

Cats and humans have cohabited for thousands of years, and as every cat owner knows, they live in intimate contact. From the cat's perspective, all food and household goods, especially beds, are communal. As with all intimate relationships, partners also share microbiomes. While we know for sure that you can catch a cold or the flu from another person in the household, we also know that it is uncommon to get a virus from a pet in the household. Why? As we discuss in chapter 6, for a virus to complete its life cycle it must bind to a specific receptor on the surface of a cell, enter, and subvert cell mechanisms to ensure its replication and release.

Feline leukemia virus (FeLV) causes an immune deficiency in cats (FeLeuk; pronounced "fee-luke") with symptoms comparable to AIDS in humans. Both diseases are caused by retroviruses, and FeLV infects several cell types in cats. Transmission between cats occurs by contact with an infected animal's secretions (saliva, urine, milk), and is correlated with the frequency of interaction with other animals. As cats groom with a coating of saliva, simply touching a cat exposes humans to the virus. How does the FeLV life cycle work at a molecular level, and can we learn from that how humans avoid infection, despite repeated exposure over millennia?

The first step in infection is critical—virus attaching to a host cell receptor. In animals, each tissue has a unique and characteristic set of surface structures. A virus typically infects only a single tissue type, because that's where its compatible receptor is located. This tissue tropism explains why we avoid touching our face during flu season—virus on our hands won't infect, but transfer to eyes or nose introduces the influenza virus to our susceptible respiratory epithelium.

FeLV gains access to cat cells through a membrane protein that normally transports phosphate. This common and widespread protein is highly conserved between cats and humans, and indeed, phosphate transporters in human plasma membranes are the point of entry for FeLV. Most human cells are susceptible to FeLV infection. Importantly, though, the virus and host cell must have an intimate and prolonged molecular conversation for the virus to succeed in infecting the host. Any breakdown in communication can stall viral transport to the nucleus, viral gene expression, virion assembly, or virion release. Several of these later steps in the viral life cycle are where FeLV is halted in human cells.

In this chapter, we examine the varied strategies used by viruses. In many ways, this discussion is a continuation of topics introduced in chapters 14 and 15, where we consider the regulation of cellular processes, and is continued in chapter 38 where we discuss diseases caused by human viruses.

Readiness Check:

Based on what you have learned previously, you should be able to:

- ✓ Describe common virion structures and morphology (section 6.2)
- ✓ Summarize the five steps common to the life cycle of most viruses (section 6.3)
- ✓ Describe the events that occur during the three phases of DNA replication, transcription, and translation (sections 13.3, 13.5, 13.7, 15.2, 15.3, and 15.4)
- ✓ Draw typical bacterial, archaeal, and eukaryotic protein-coding genes (sections 13.4, 13.5, and 15.3)

26.1 Virus Phylogeny Is Difficult to Establish

After reading this section, you should be able to:

a. Distinguish between the Baltimore system of grouping viruses and the official taxonomy of viruses proposed by the International Committee on Taxonomy of Viruses
b. Determine if a virus has a positive- or negative-strand genome
c. Differentiate the Baltimore groups of viruses

In 1971 the International Committee on Taxonomy of Viruses (ICTV) developed a uniform classification system for viruses. Since then the number of viruses and taxa has continued to expand. In its 2018 report, the ICTV describes over 2,000 virus species assigned to 9 orders, 131 families, 46 subfamilies, and 803 genera. The committee considers many viral characteristics but places greatest weight on the following properties to define families: nucleic acid type, presence or absence of an envelope, symmetry of the capsid, and dimensions of the virion and capsid. Virus order names end in *-virales;* virus family names in *-viridae;* subfamily names in *-virinae;* and genus names in *-virus*. An example of this nomenclature scheme is shown in **figure 26.1**. ◀ *Virion structure is defined by capsid symmetry and presence or absence of an envelope (section 6.2)*

As we discuss in chapter 19, taxonomy classifies organisms based on their evolutionary history. To piece together the phylogenetic relationships of viruses, virologists often use two major approaches: comparisons of genome sequences and comparisons of protein folds observed in their major capsid proteins. Both approaches are challenging, in part because horizontal gene transfer among unrelated viruses and between viruses and their host cells is clearly evident. Despite this, evidence suggests that retroviruses (section 26.7) and reverse transcribing DNA viruses (section 26.8) share a common evolutionary history. All of these viruses use an enzyme called reverse transcriptase in their life cycles. Evidence also suggests that double-stranded (ds) DNA viruses can be divided into two lineages based on folding of their capsid proteins. These viruses include nucleocytoplasmic large DNA (NCLD) viruses (section 26.2), which have garnered considerable interest in part because they seem to blur the distinctions between cells and viruses.

Although ICTV reports are the official authority on viral taxonomy, many virologists find it useful to group viruses using a scheme devised by Nobel laureate David Baltimore. The Baltimore system complements the ICTV system but focuses on the viral genome and the process used to synthesize viral mRNA. Recall from chapter 6 that all four nucleic acid types can be found in viruses: dsDNA, single-stranded (ss) DNA, dsRNA, and ssRNA. The characterization of the genome of a ssRNA virus is further differentiated by the sense of the ssRNA—that is, whether the genome is identical to or complementary to the mRNA produced by the virus. ssRNA viruses with an RNA genome that is identical in base sequence to that of the mRNA it produces are said to have **plus-strand** or **positive-strand RNA** genomes (**figure 26.2**). Other ssRNA viruses have genomes that are complementary to the mRNA they produce. These viruses are said to have **minus-strand** or **negative-strand RNA.** Baltimore's system organizes viruses into seven groups (**table 26.1**). This system helps virologists (and microbiology students) simplify the vast array of viral life cycles into a relatively small number of basic types; thus we use it here.

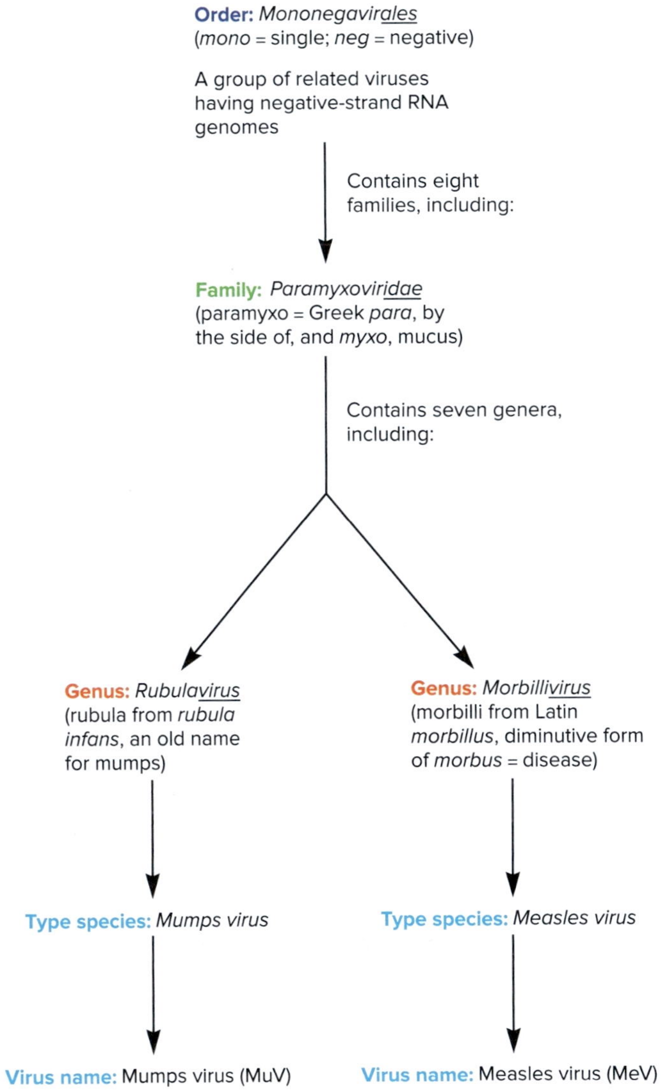

Figure 26.1 Virus Taxonomy. The taxa names are derived from various aspects of the biology and history of the members of the taxa, including features of their structure, diseases they cause, and locations where they were first identified or recognized. The suffixes used for each type of taxon (e.g., order, family, genus) are underlined.

Viral mRNA	5′...GAC UCG AGC...3′
Plus-strand RNA	5′...GAC UCG AGC...3′
Negative-strand RNA	3′...CUG AGC UCG...5′
Plus-strand DNA	5′...GAC TCG AGC...3′
Negative-strand DNA	3′...CTG AGC TCG...5′

Figure 26.2 Plus-Strand and Negative-Strand Viral Genomes. The genomes and replication intermediates of viral genomes can be either plus strand or negative strand. This designation is relative to the sequence of nucleotides in the mRNA of the virus. Plus-strand genomes have the same sequence as the mRNA, either using DNA nucleotides if a DNA genome or RNA nucleotides if an RNA genome. Negative-strand genomes are complementary to the viral mRNA.

Table 26.1 The Baltimore System	
Group	**Description**
Double-stranded (ds) DNA viruses	Genome replication: dsDNA → dsDNA mRNA synthesis: dsDNA → mRNA
Single-stranded (ss) DNA viruses	Genome replication: ssDNA → dsDNA → ssDNA mRNA synthesis: ssDNA →dsDNA →mRNA
Double-stranded RNA viruses	Genome replication: dsRNA → ssRNA → dsRNA mRNA synthesis: dsRNA → mRNA
Plus-strand RNA (+RNA) viruses	Genome replication: +RNA → −RNA → +RNA mRNA synthesis: +RNA = mRNA → −RNA → mRNA
Negative-strand RNA (−RNA) viruses	Genome replication: −RNA → +RNA → −RNA mRNA synthesis: −RNA → mRNA
Retroviruses	Genome replication: ssRNA → dsDNA → ssRNA mRNA synthesis: ssRNA → dsDNA → mRNA
Reverse transcribing DNA viruses	Genome replication: dsDNA → ssRNA → dsDNA mRNA synthesis: dsDNA → mRNA

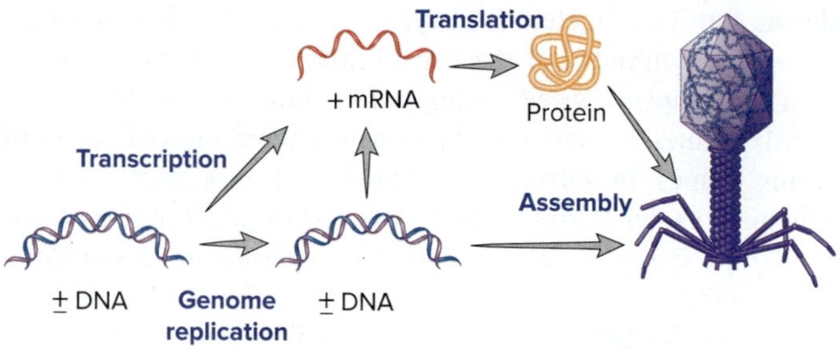

Figure 26.3 Multiplication Strategy of Double-Stranded DNA Viruses. Because the genome of dsDNA viruses is similar to that of the host, genome replication and mRNA synthesis closely resemble that of the host cell and can involve host polymerases, viral polymerases, or both. The DNA genome serves as the template for DNA replication and mRNA synthesis. Translation of the mRNA by the host cell's translation machinery yields viral proteins, which are assembled with the viral DNA to make mature virions. These are eventually released from the host.

Comprehension Check

1. List some characteristics used in classifying viruses. Which seem to be the most important?
2. Consider these terms: *Equine torovirus, Nidovirales,* Equine torovirus (EToV), *Coronaviridae*, and *Torovirus*. Construct a figure similar to figure 26.1 using these terms.

26.2 Double-Stranded DNA Viruses Infect All Cell Types

After reading this section, you should be able to:

a. Describe in general terms the strategy used by double-stranded (ds) DNA viruses to synthesize their nucleic acids and proteins
b. Describe in general terms how bacteriophage lambda regulates the switch between lytic and lysogenic cycles
c. Choose one specific dsDNA virus and outline the major events in its life cycle, noting, when possible, the specific mechanisms used to accomplish each step

Perhaps the largest group of known viruses is the double-stranded (ds) DNA viruses; most bacteriophages and archaeal viruses have dsDNA genomes, as do several insect viruses and a number of important vertebrate viruses, including herpesviruses and poxviruses. The pattern of multiplication for dsDNA viruses is shown in **figure 26.3**. The synthesis of DNA and RNA is similar to what occurs in cellular organisms; therefore many dsDNA viruses can rely entirely on their host's DNA and RNA polymerases.

Bacteriophage T4: A Virulent Bacteriophage

The life cycle of T4 bacteriophage (family *Myoviridae,* species *Enterobacteria phage T4*) serves as our example of a virulent (lytic) dsDNA phage. Cell lysis is the outcome of an infection with a lytic bacteriophage. As with most viruses (except plant viruses), the first step of viral infection is attachment (adsorption) to the host cell surface. T4 attachment begins when a long tail fiber contacts either the lipopolysaccharide or certain proteins in the outer membrane of its *Escherichia coli* host (**figure 26.4*a***). As more long tail fibers make contact, the baseplate settles down on the surface (figure 26.4*b*). Both baseplate and sheath change shape, and the tail sheath shortens (figure 26.4*c,d*). As the sheath becomes shorter and wider, the central tube located within the sheath is pushed through the bacterial cell wall. Finally, the linear DNA is extruded from the head into the host cell (figure 26.4*e,f*).

Within 2 minutes after entry of T4 DNA into an *E. coli* cell, the *E. coli* RNA polymerase starts synthesizing T4 mRNA (**figure 26.5**). This mRNA is called early mRNA because it is made before viral DNA is made. One of the earliest T4 genes to be expressed encodes a protein that binds to the host enzyme RNaseE and directs it to degrade host mRNA. The consequence of host mRNA degradation is that ribonucleotides and ribosomes are now free for transcription and translation of T4 genes. Within 5 minutes, viral DNA synthesis commences, catalyzed by a virus-encoded **DNA-dependent DNA polymerase.** DNA replication is initiated from several origins of replication and proceeds bidirectionally from each. Viral DNA replication is followed by the synthesis of late mRNAs, which are important in later stages of the infection.

As with most viruses, expression of T4 genes is carefully timed, and the transcripts and protein products of the genes are named to reflect the time of appearance. Thus depending on the virus, terms such as immediate-early, middle, and late may be used. T4 controls the expression of its genes by regulating the

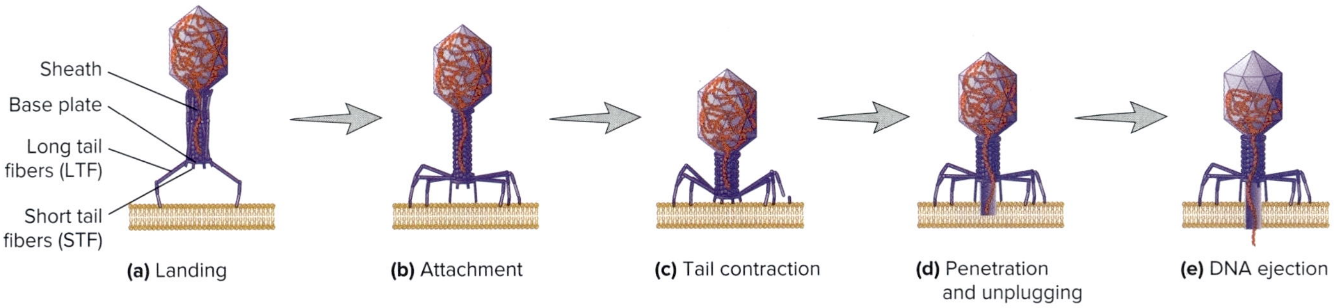

Attachment begins when one or two LTFs contact the cell surface reversibly. This allows the virion to "walk" to locate its preferred surface site. As more LTFs contact the surface, the STFs extend downward. Interaction between STFs and their receptors is irreversible and the virion is now committed to attachment. The baseplate settles onto the cell surface.

After the baseplate is firmly seated, it expands to create a central hole and the tail sheath contracts and shortens. These movements push the central tube through the outer membrane.

Baseplate protein gp5 is released into the periplasm, where its lysozyme activity is needed for further movement through the cell wall. Another protein, gp29, plays two roles. It connects the baseplate to the sheath, and when these structures change shape upon irreversible binding, gp29 travels to the plasma membrane to form a transmembrane channel through which DNA is ejected.

Cell wall

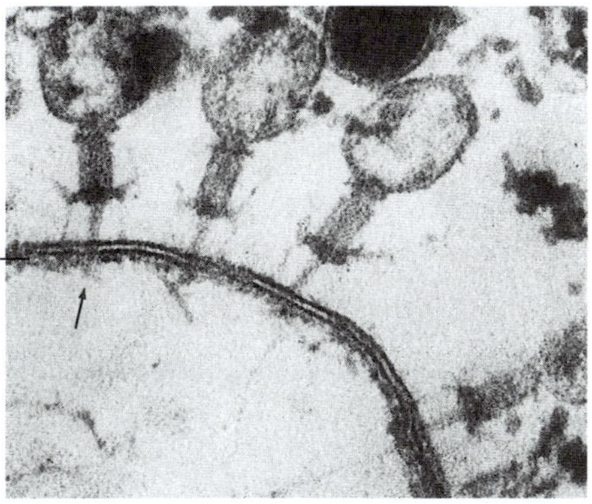

(f)

Figure 26.4 **T4 Phage Adsorption and DNA Entry.**
(a–e) Adsorption and viral DNA entry into the *E. coli* host are mediated by the phage's tail fibers and base plate. (f) An electron micrograph of an *E. coli* cell being infected by T-even phages. These phages have released their dsDNA into the cell and have empty capsids.
(f) ©Lee D. Simon/Science Source

MICRO INQUIRY *What proteins found in the T4 baseplate facilitate penetration through the cell wall?*

activity of *E. coli* RNA polymerase. Initially T4 genes are transcribed by the host RNA polymerase and the housekeeping sigma factor σ^{70} (*see table 14.2*). After a short interval, a viral enzyme catalyzes addition of the chemical group ADP-ribose to one of the α-subunits of RNA polymerase (*see figure 13.23*). This modification helps inhibit transcription of host genes and promotes viral gene expression. Later the second α-subunit receives an ADP-ribosyl group. This halts transcription of some early T4 genes but not before the product of one early gene (*motA*) stimulates transcription of somewhat later genes. One of these later genes encodes the sigma factor gp55. This viral sigma factor directs the host cell's RNA polymerase core enzyme to viral late promoters to transcribe the late genes, around 10 to 12 minutes after infection. ◀ *Bacteria combine several regulatory mechanisms to control complex cellular processes (section 14.6)*

The tight regulation of T4 gene expression is aided by the organization of the T4 genome. Genes whose products have related functions—such as the genes for phage head or tail fiber construction—are clustered together. Early and late genes are clustered separately on the genome; they are even transcribed in different directions. A considerable portion of the T4 genome encodes products needed for replication, including the protein subunits of its replisome and enzymes needed to prepare for DNA synthesis. Some of these enzymes synthesize hydroxymethylcytosine (HMC), a modified nucleotide that replaces cytosine in T4 DNA (**figure 26.6**). Once HMC is synthesized, replication ensues by a mechanism similar to that seen in bacteria. After T4 DNA synthesis, it is modified by the addition of glucose (a process called glycosylation) to HMC residues. Glycosylated HMC residues protect T4 DNA from attack by *E. coli* endonucleases called **restriction enzymes,** which would otherwise destroy the viral DNA. This bacterial defense mechanism is called **restriction.** Other phages also chemically modify their DNA to protect against host restriction. ◀ *Responses to viral infection (section 14.6); Restriction enzymes (section 17.1)*

The linear dsDNA genome of T4 is generated in an interesting manner involving the formation of long DNA molecules called **concatemers,** which are composed of several genome units linked together in the same orientation (**figure 26.7**). How does this occur? As we discuss in chapter 15, the ends of linear DNA molecules cannot be replicated without mechanisms such as the enzyme telomerase. T4 and its *E. coli* host do not have telomerase enzyme activity. Therefore each progeny viral DNA molecule has single-stranded 3′ ends. These ends participate in homologous

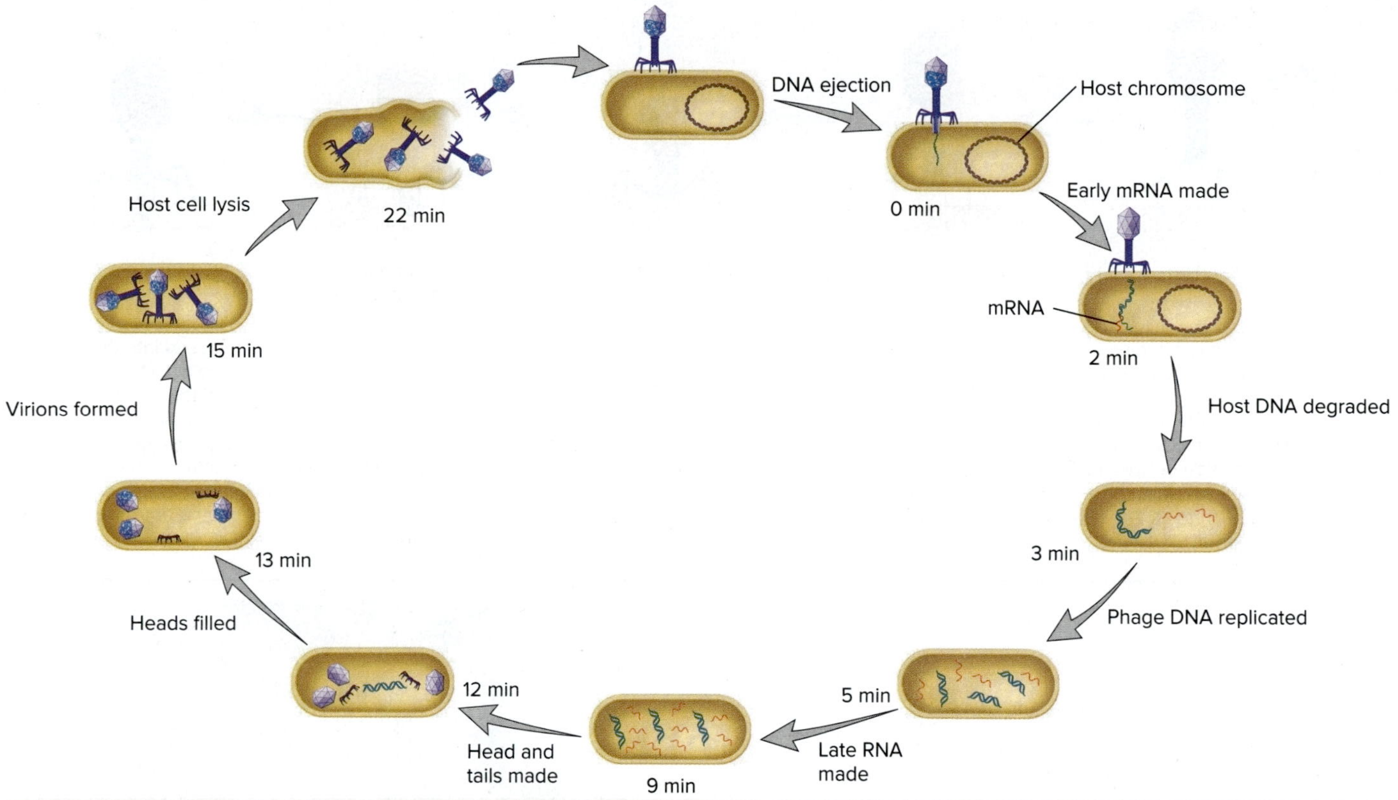

Figure 26.5 The Life Cycle of Bacteriophage T4. A diagram depicting the life cycle with the minutes after DNA ejection given for each stage.

MICRO INQUIRY *Why do you think T4 evolved to initiate DNA replication from multiple origins, rather than from a single origin of replication?*

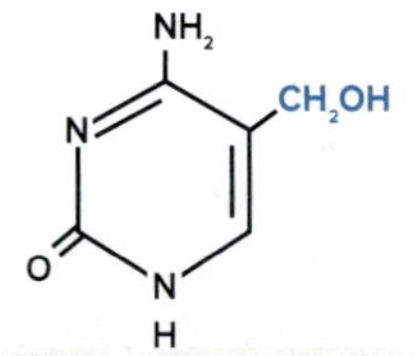

Figure 26.6 5-Hydroxymethylcytosine (HMC). In T4 DNA, HMC often has glucose attached to its hydroxyl.

MICRO INQUIRY *What function does HMC glycosylation serve?*

recombination with double-stranded regions of other progeny DNA molecules, generating concatemers. During assembly, concatemers are cleaved such that the genome packaged in the capsid is slightly longer than the T4 gene set. Thus each progeny virus has a genome unit that begins with a different gene and ends with the same set of genes. If each viral genome was circularized, the sequence of genes in each virion would be the same. Therefore the T4 genome is said to be terminally redundant and circularly permuted, and the genetic map of T4 is drawn as a circular molecule.

The formation of new T4 phage particles is an exceptionally complex self-assembly process that involves viral proteins and some host cell factors (*see figure 6.13*). Late mRNA transcription begins about 9 minutes after entry of T4 DNA into *E. coli* (figure 26.5). Late mRNA products include phage structural proteins, proteins that help with phage assembly but do not become part of the virion, and proteins involved in cell lysis and phage release. These proteins are used in four fairly independent subassembly lines that ultimately converge to generate a mature T4 virion.

A critical step in T4 virion construction is filling the head portion of the virion with the T4 genome. This is no simple matter—the dsDNA genome is somewhat rigid and has many negatively charged moieties. Therefore the dsDNA must be crammed into the capsid. This is accomplished by a complex of proteins sometimes called the "packasome." The T4 packasome has more power than an automobile engine.

The packasome includes a protein called terminase, which has two functions: to cut the concatemers formed during T4 genome replication and to coordinate the insertion of DNA into the T4 head. Terminase threads the end of the T4 genome through a portal, where it enters the phage head using energy supplied by ATP hydrolysis. The packasome has been proposed to move the DNA by causing a transition from B-form DNA to A-form DNA, and then reversing back to B-form DNA. This compresses the helix like a spring. When the phage head is filled with a DNA molecule roughly 3% longer than the length of one set of T4 genes, terminase then makes a second cut, and the packaging process for that head is complete. Terminase then leaves the

Figure 26.7 The Terminally Redundant, Circularly Permuted Genome of T4. The formation of concatemers during replication of the T4 genome is an important step in phage multiplication. During assembly of virions, the phage head is filled with DNA cleaved from the concatemer. Because slightly more than one set of T4 genes is packaged in each head, each virion contains a different DNA fragment (note that the ends of the fragments are different). However, if each genome were circularized, the sequence of genes would be the same.

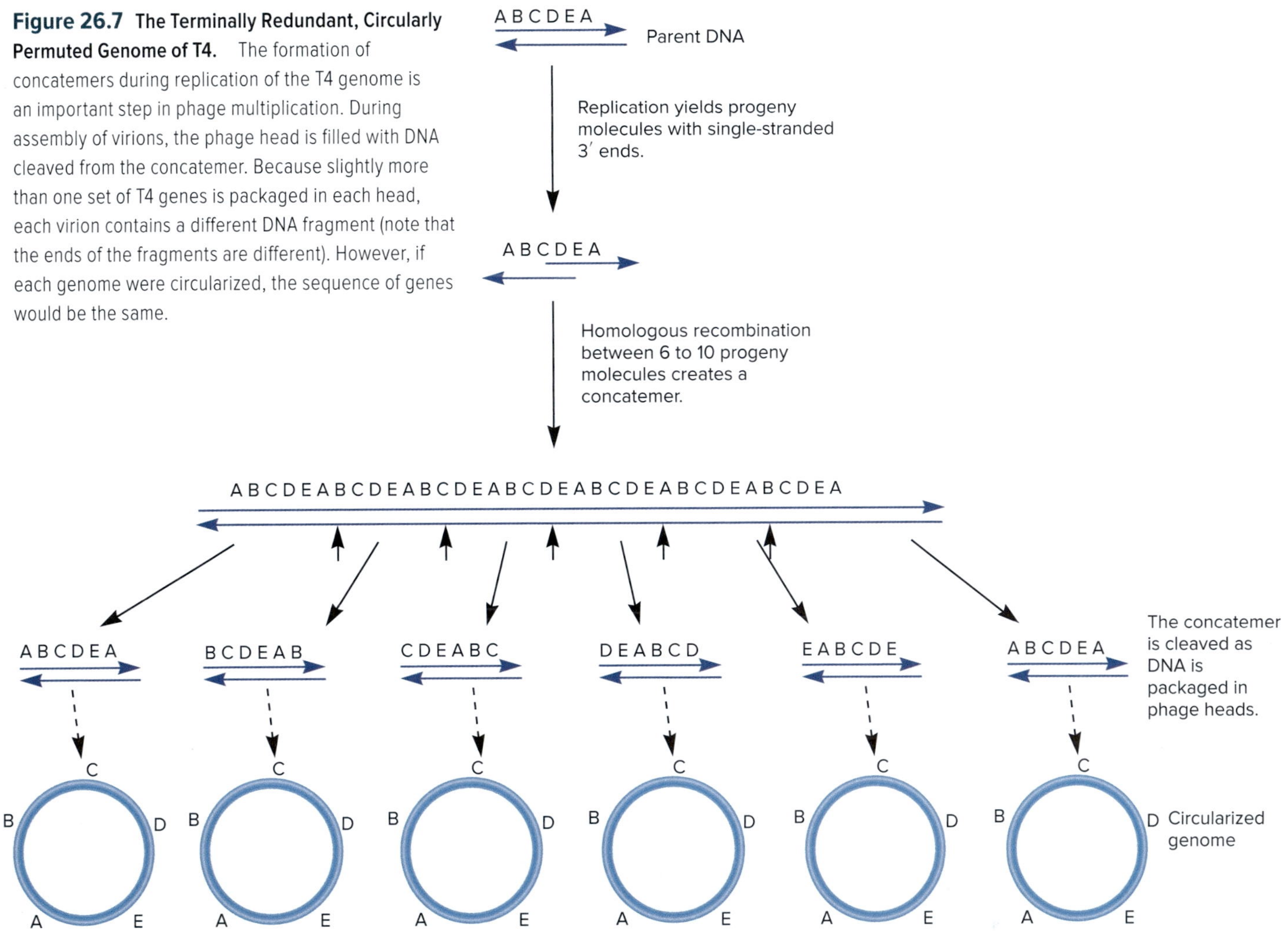

head, and several other viral proteins bind at the portal through which the DNA entered. This seals the head and prepares it for addition of the tail and tail fibers.

Finally, virions are released so that they can infect new cells and begin the cycle anew. T4 lyses *E. coli* when about 150 virus particles have accumulated in the host cell. T4 encodes two proteins to accomplish this. The first, called holin, creates holes in the *E. coli* plasma membrane. The second, an endolysin called T4 lysozyme (distinct from the gp5 lysozyme used for bacteriophage entry), degrades peptidoglycan in the host's cell wall. Thus the activity of holin enables T4 lysozyme to move from the cytoplasm to the peptidoglycan so that both the plasma membrane and the cell wall are destroyed.

Bacteriophage Lambda: A Temperate Bacteriophage

Unlike bacteriophage T4, temperate bacteriophages, such as phage lambda (family *Siphoviridae,* species *Enterobacteria phage lambda*), can enter either the lytic or lysogenic cycle upon infecting a host cell. If a temperate virus enters the lysogenic cycle, its dsDNA genome often is integrated into the host's chromosome, where it resides as a prophage until conditions for induction occur (*see figure 6.16*). Upon induction, the viral genome is excised from the host genome and the lytic cycle is initiated. ◀ *Lytic and lysogenic infections are common for bacterial and archaeal cells (section 6.4)*

How does a temperate phage "decide" which cycle to follow? The process by which phages make this decision is best illustrated by bacteriophage lambda (λ), which infects *E. coli.* As shown in **figure 26.8**, λ has an icosahedral head 55 nm in diameter and a noncontractile tail with a thin tail fiber at its end. Its DNA genome is a linear molecule with cohesive ends—single-stranded stretches, 12 nucleotides long, that are complementary to each other and can base-pair.

Like most bacteriophages, λ attaches to its host and then releases its genome into the cytoplasm, leaving the capsid outside. Once inside the cell, the linear genome is circularized when the two cohesive ends base-pair with each other; the breaks in the strands are sealed by the host cell's DNA ligase (**figure 26.9**). The λ genome has been carefully mapped, and over 70 genes have been identified (**figure 26.10**). Most genes are clustered according to their function, with separate groups involved in

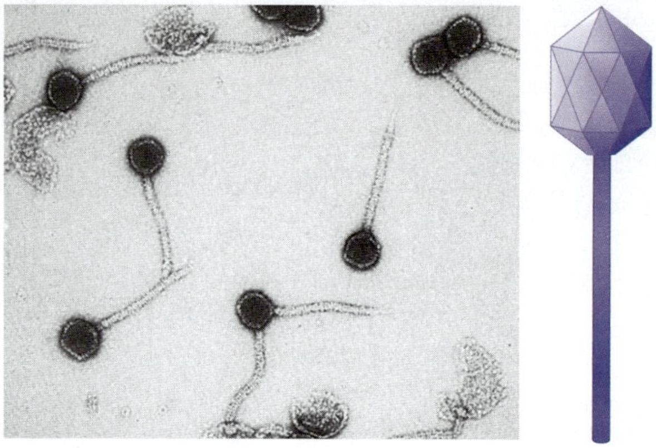

Figure 26.8 Bacteriophage Lambda (λ). The icosahedral head illustrated in the drawing is not obvious in the photomicrograph at this resolution.

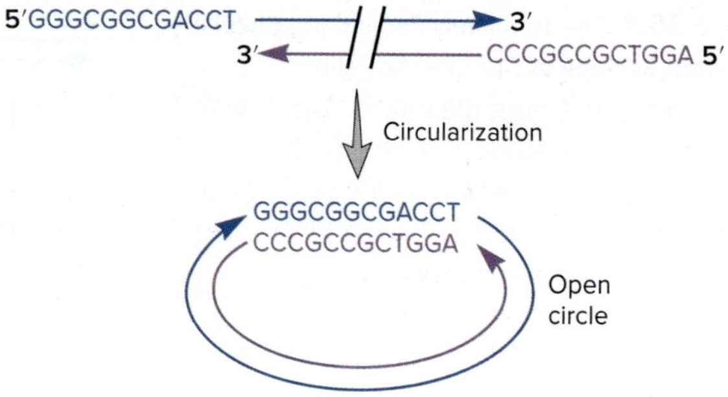

Figure 26.9 Lambda Phage DNA. A diagram of λ phage DNA showing its 12-base, single-stranded cohesive ends.

head synthesis, tail synthesis, lysogeny, DNA replication, and cell lysis. This organization is important because once the genome is circularized, a cascade of regulatory events occurs that determine if the phage pursues a lytic cycle or establishes lysogeny. Regulation of appropriate genes is facilitated by clustering and coordinated transcription from the same promoters.

The cascade of events leading to either lysogeny or the lytic cycle serves as a model for complex regulatory processes. It involves the action of several regulatory proteins that function as transcriptional repressors or activators or both, proteins that regulate transcription termination, and antisense RNA molecules (**figure 26.11**). The protein cII is an activator protein that plays a pivotal role in determining if λ will establish lysogeny or follow a lytic pathway. If the cII protein reaches high enough levels early in the infection, lysogeny will occur; if it does not reach this critical level, the lytic cycle will occur.

Transcription of the λ genome is catalyzed by the host cell's **DNA-dependent RNA polymerase** (an RNA polymerase that uses DNA as the template for RNA synthesis), and the cII protein is synthesized relatively early in the infection. If the cII protein levels are high enough, it will increase transcription of the *int* gene, which encodes the enzyme integrase. **Integrase** catalyzes integration of the λ genome into the host cell's chromosome, thus establishing lysogeny. The cII protein also increases transcription of the *cI* gene. This gene encodes a regulatory protein often called the λ repressor because it represses the transcription of all bacteriophage genes (except its own). This repression maintains the lysogenic state. As just noted, the cI

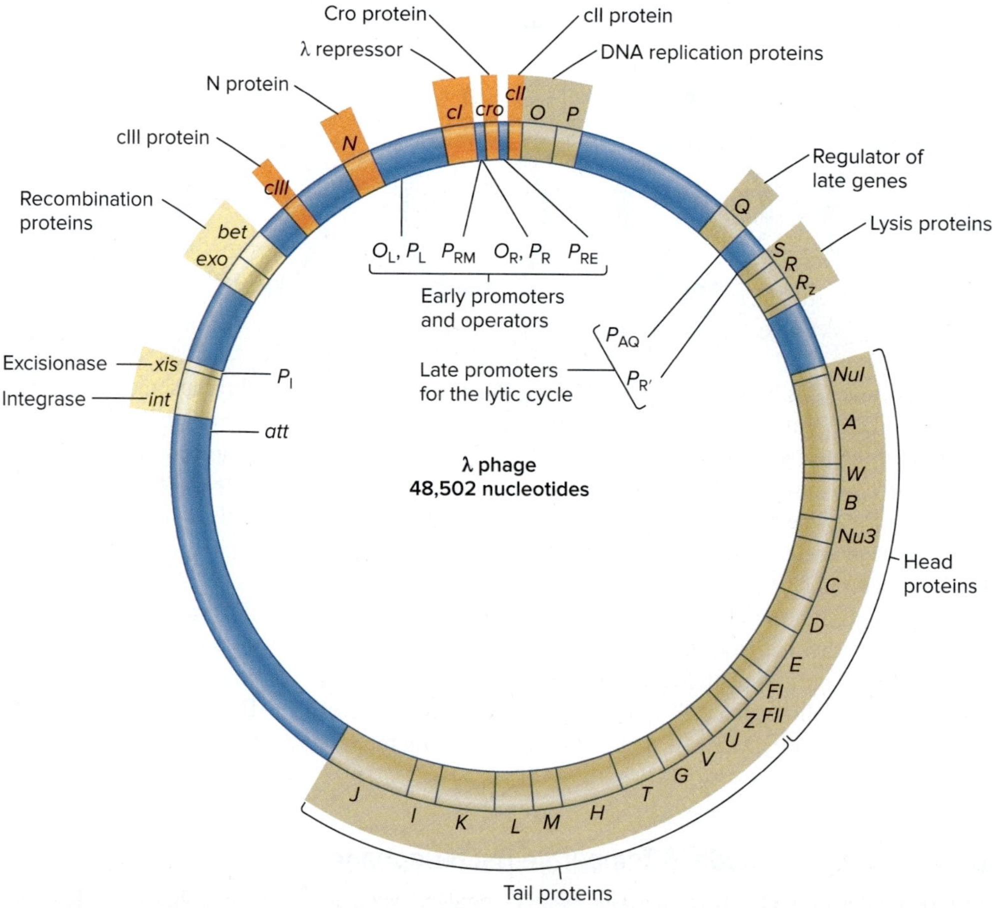

Figure 26.10 The Genome of Phage λ. The genes are color-coded according to function. Genes shaded in orange encode regulatory proteins that determine if the lytic or lysogenic cycle will be followed. Genes involved in establishing lysogeny are shaded in yellow. Those required for the lytic cycle are shaded in tan. Important promoters and operators are noted inside the circle.

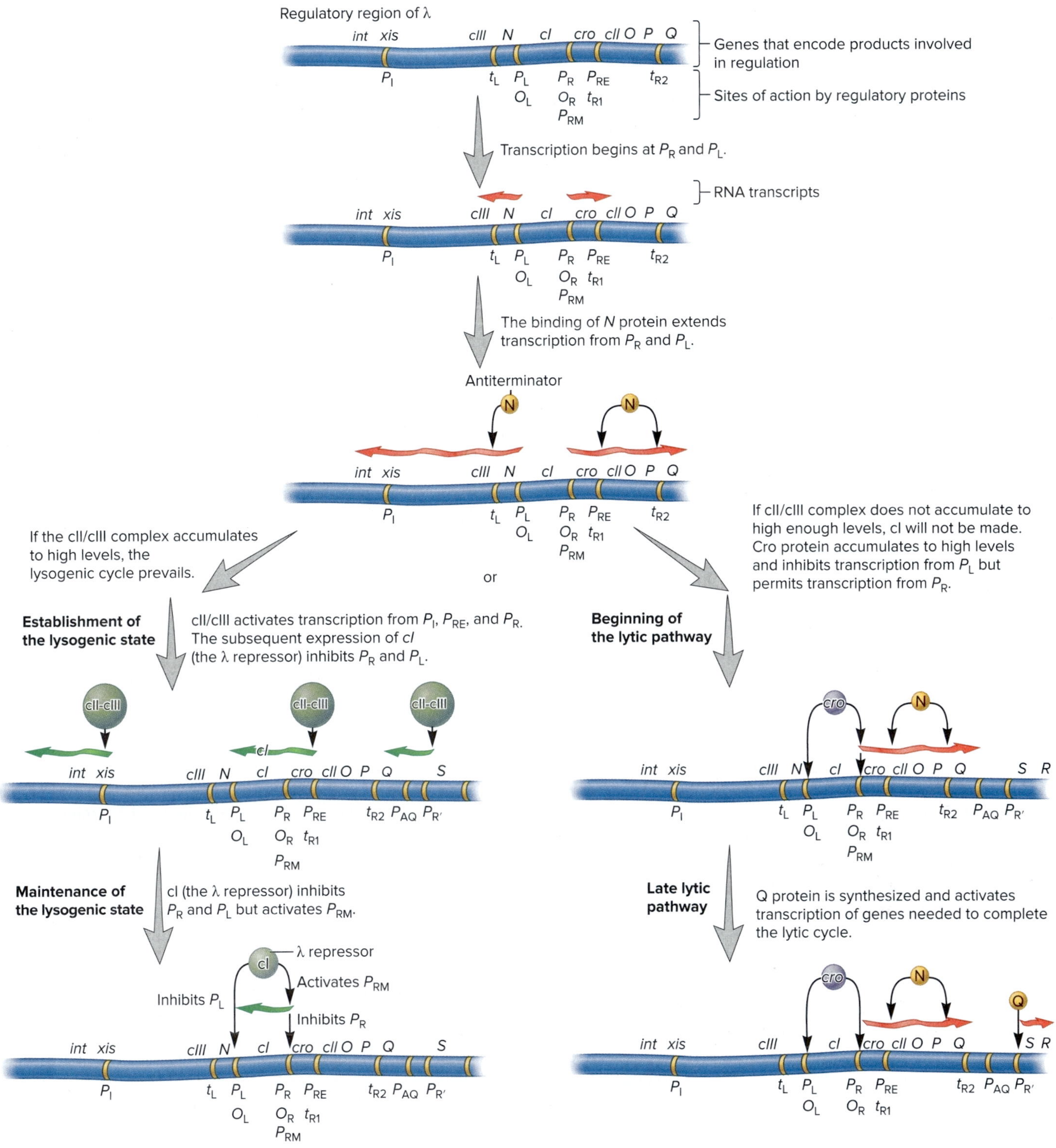

Figure 26.11 The Decision-Making Process for Establishing Lysogeny or the Lytic Pathway. The initial transcripts are synthesized by the host RNA polymerase. These encode the N protein and the Cro protein. The N protein is an antiterminator that allows transcription to proceed past the terminator sequences t_L, t_{R1}, and t_{R2}. This allows transcription of other regulatory genes, as well as the *xis* and *int* genes. The latter genes encode the enzymes excisionase and integrase, respectively. The left side of the figure illustrates what occurs if lysogeny is established. The right side of the figure shows the lytic pathway.

protein (λ repressor) allows transcription of its own gene. This is because cI functions as an activator protein when it binds to the P_{RM} promoter from which the *cI* gene can be transcribed (**table 26.2**).

Integration of the λ genome into the host chromosome takes place at a site in the host chromosome called the attachment site (*att*). A corresponding site is present on the phage genome, and the λ integrase catalyzes site-specific recombination between the two sites. The bacterial site is located between the galactose (*gal*) and biotin (*bio*) operons, and as a result of integration, the circular λ genome becomes a linear stretch of DNA located between these two host operons (*see figure 16.30*). The prophage can remain

Table 26.2 Functions of Lambda Promoters and Operators

Promoter or Operator	Name Derivation	Function
P_L	**P**romoter **L**eftward	Promoter for transcription of *N, cIII, xis,* and *int* genes; important in establishing lysogeny
O_L	**O**perator **L**eftward	Binding site for lambda repressor and Cro protein; binding by lambda repressor maintains lysogenic state; binding by Cro protein prevents establishment and maintenance of lysogeny
P_R	**P**romoter **R**ightward	Promoter for transcription of *cro, cII, O, P,* and *Q* genes; Cro, O, P, and Q proteins are needed for lytic cycle; cII protein helps establish lysogeny
O_R	**O**perator **R**ightward	Binding site for lambda repressor and Cro protein; binding by lambda repressor maintains lysogenic state; binding by Cro allows transcription to occur
P_{RE}	**P**romoter for Lambda **R**epressor **E**stablishment	Promoter for *cI* gene (lambda repressor gene); recognized by cII protein, a transcriptional activator; important in establishing lysogeny
P_I	**P**romoter for **I**ntegrase Gene	Transcription from P_I generates mRNA for integrase protein but not excisionase; recognized by the transcriptional activator cII; important for establishing lysogeny
P_{AQ}	**P**romoter for **A**nti-**Q** mRNA	Transcription from P_{AQ} generates an antisense RNA that binds Q mRNA, preventing its translation; recognized by the transcriptional activator cII; important for establishing lysogeny
P_{RM}	**P**romoter for **R**epressor **M**aintenance	Promoter for transcription of lambda repressor gene (*cI*); activated by lambda repressor; important in maintaining lysogeny
$P_{R'}$	**P**romoter **R**ightward′	Promoter for transcription of viral structural genes; activated by Q protein; important in lytic cycle

integrated indefinitely, being replicated as the bacterial genome is replicated. ◀ *Molecular recombination: joining DNA molecules together (section 16.4)*

Because the cII protein is made early in the infection, it might seem that it would accumulate quickly and ensure that lysogeny always occurs. However, cII is degraded by a host protease unless it is protected by a viral protein called cIII. The cIII protein is synthesized at the same time as cII, and as long as its levels remain high enough, cII will be protected. However, if cII is not protected from proteolysis, the level of a protein called Cro (product of the *cro* gene) will increase. Cro protein is both a repressor protein and an activator protein. It inhibits transcription of the *cIII* and *cI* genes, further decreasing the amount of cII and λ repressor. However, it increases its own synthesis as well as the synthesis of another regulatory protein called Q. When Q protein accumulates at a high enough level, it activates transcription of genes required for the lytic cycle. When that occurs, the infection process is committed to the lytic cycle. Ultimately the host is lysed and new virions are released.

We have now considered the regulatory processes that dictate whether lysogeny is established or the lytic cycle is pursued. However, how does induction reverse lysogeny? Induction usually occurs in response to environmental factors such as ultraviolet light or chemical mutagens that damage DNA. This damage alters the activity of the host cell's **RecA** protein. As we describe in chapter 16, RecA plays important roles in recombination and DNA repair. When activated by DNA damage, RecA interacts with λ repressor, causing the repressor to cleave itself. As more and more λ repressor proteins destroy themselves, transcription of the *cI* gene is decreased, further lowering the amount of λ repressor in the cell. Eventually the level becomes so low that transcription of the *xis, int,* and *cro* genes begins. The *xis* gene encodes the protein excisionase. Excisionase binds integrase, causing it to reverse the integration process, and the prophage is freed from the host chromosome. As λ repressor levels continue to decline, the Cro protein levels increase. Eventually synthesis of λ repressor is completely blocked, protein Q levels become high, and the lytic cycle proceeds to completion.

Our attention has been on λ phage, but there are many other temperate phages. Most, like λ, exist as integrated prophages in the lysogen. However, not all temperate phages integrate into the host chromosome at specific sites. Bacteriophage Mu integrates randomly into the genome. It then expresses a repressor protein that inhibits lytic growth. Furthermore, integration is not an absolute requirement for lysogeny. The *E. coli* phage P1 is similar to λ in that it circularizes after infection and begins to manufacture repressor. However, it remains as an independent circular DNA molecule in the lysogen and is replicated at the same time as the host chromosome. When *E. coli* divides, P1 DNA is apportioned between the daughter cells so that all lysogens contain one or two copies of the phage genome.

The λ mechanism for the lysis-lysogeny decision is based on transcriptional regulatory circuits involving the cII protein. A different mechanism used by the *B. subtilis* phage ϕ3T is based on quorum sensing. Upon infection, a small phage protein is synthesized, proteolyzed, and exported as a six amino acid peptide. Other nearby cells import the peptide, where it influences the lytic-lysogeny decision. Low levels of the peptide, indicating few nearby infections, are likely to result in the lytic cycle. By contrast, high levels of the peptide promote lysogeny, indicating that there are few host cells available for further infection.

Comprehension Check

1. Explain why the T4 genome is circularly permuted.
2. Explain in molecular terms how a bacterial cell is made lysogenic by a temperate phage such as λ.
3. How is a prophage induced to become active again?
4. Describe the roles of cII, cIII, λ repressor (cI), Cro, Q, RecA, integrase, and excisionase in lysogeny and induction.
5. How do the temperate phages Mu and P1 differ from λ?

Herpesviruses

We now turn our attention to some important dsDNA viruses of eukaryotes, beginning with herpesviruses. As you will see, the life cycle of these viruses shares many features with that of bacteriophages such as T4. However, important distinctions in the life cycles exist because of the nature of their respective host cells.

Herpesviruses are members of the order *Herpesvirales*. This order includes herpes simplex virus type 1 (HSV-1; also called human herpesvirus 1) and herpes simplex virus type 2 (HSV-2; also called human herpesvirus 2), which cause cold sores and genital herpes, respectively (*see figures 38.14 and 38.17*); varicella-zoster virus (also called human herpesvirus 3), which causes chickenpox and shingles (*see figures 38.1 and 38.2*); Epstein-Barr virus (also called human herpesvirus 4), which causes infectious mononucleosis and has been implicated in some human cancers; cytomegaloviruses (species *Human herpesvirus 5*); HHV 6 and HHV 7, which infect almost the entire human population as children (species *Human herpesvirus 6*, species *Human herpesvirus 7*); and HHV 8 (species *Human herpesvirus 8*), the cause of Kaposi's sarcoma in AIDS patients. ▶| *Chickenpox (varicella) and shingles (herpes zoster) (section 38.1); Cold sores (section 38.3); Genital herpes (section 38.3); Cytomegalovirus inclusion disease (section 38.3); Mononucleosis (infectious) (section 38.3); Acquired immune deficiency syndrome (AIDS) (section 38.3)*

Herpesvirus virions are 150 to 200 nm in diameter, somewhat pleomorphic, and enveloped (**figure 26.12**). The envelope contains distinct viral projections (called spikes) that are regularly dispersed over the surface. The envelope surrounds a layer of proteins called the tegument (Latin *tegumentum,* to cover), which in turn surrounds the nucleocapsid. Herpesvirus genomes are linear, about 125 to 295 kilobase pairs (kb) long, and encode 70 to over 200 proteins. When herpesviruses target cells of vertebrate hosts, some bind to epithelial cells, others to neurons. Host cell selection is mediated by the binding of envelope spikes to specific host cell surface proteins termed receptors.

Herpesviruses cause both productive infections and latent infections. In a productive infection, the virus multiplies explosively; between 50,000 and 200,000 new virions are produced from each infected cell. As the virus multiplies, the host cell's metabolism is inhibited and the host's DNA is degraded. As a result, the cell dies. The first exposure to a herpesvirus usually causes this type of infection. Some of the cells infected in the initial infection develop a latent infection. During the latent infection, virions cannot be detected. However, the virus can be reactivated in the host cells, leading to a productive infection. The viral genome remains in the host cell after reactivation; thus once infected, the host can experience repeated productive infections.

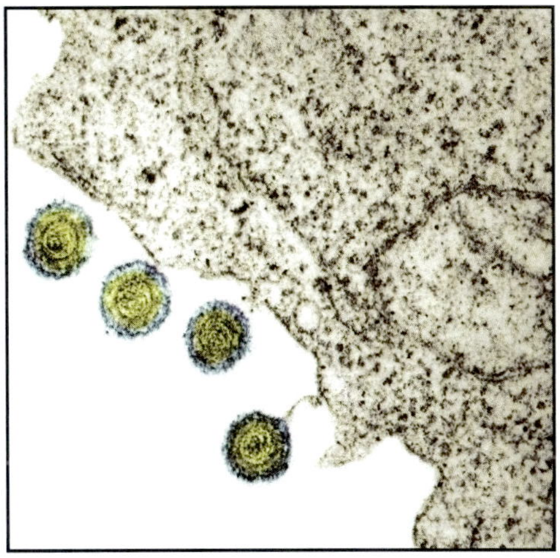

Figure 26.12 HHV-6. Human herpesvirus 6 at the surface of a T lymphocyte. This TEM has been colored to highlight the capsid (yellow) and glycoprotein spikes (blue). Virions measure 150–200 nm in diameter.

A productive infection caused by HSV-1 is shown in **figure 26.13**. It begins with receptor-mediated attachment followed by fusion of the viral envelope with the host cell membrane. The initial association is between proteoglycans of the epithelial cell surface and viral glycoproteins. This is followed by a specific interaction with one of several cellular receptors. The nucleocapsid and some associated tegument proteins are released into the cytoplasm and transported by the host cell's microtubule/dyenin system to the nucleus. The linear dsDNA and some of the tegument proteins then enter the nucleus by way of a nuclear pore complex. One of these proteins is a ribonuclease that degrades host mRNA thereby providing free nucleotides for viral gene expression. Immediately upon entry of the viral DNA into the host nucleus, the DNA circularizes and is transcribed by host DNA-dependent RNA polymerase to form mRNAs, which are translated to yield several immediate early and early proteins. These are mostly regulatory proteins and enzymes required for replication of viral DNA (figure 26.13, steps 1 and 2). Replication of the genome with a virus-specific DNA-dependent DNA polymerase begins in the cell nucleus within 4 hours after infection (figure 26.13, step 3). Viral structural proteins are the products of the late genes. These proteins enter the nucleus, where they are assembled into nucleocapsids. Capsid formation and genome packaging are similar to that seen for bacteriophage T4.

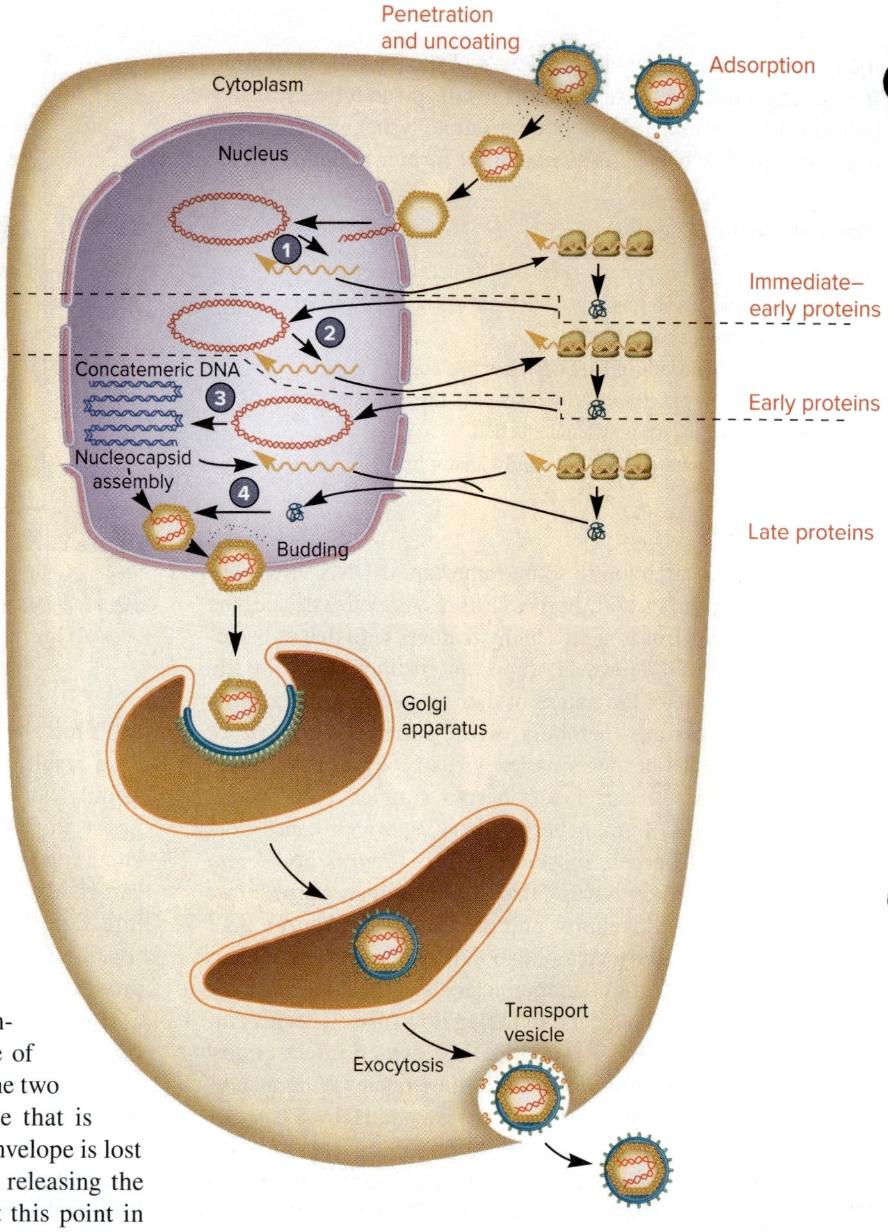

1. Circularization of genome and transcription of immediate-early genes
2. Immediate-early proteins (products of immediate-early genes) stimulate transcription of early genes.
3. Early proteins (products of early genes) function in DNA replication, yielding concatemeric DNA. Late genes are transcribed.
4. Late proteins (products of late genes) participate in virion assembly.

Figure 26.13 **Life Cycle for Herpes Simplex Virus Type 1.**

MICRO INQUIRY *How is the envelope of this virus formed? How does this differ from most enveloped viruses?*

The acquisition of the tegument and envelope, and exit from the host cell require several steps to complete. Once the nucleocapsid is assembled, it makes contact with the inner membrane of the nucleus. It then buds into the space between the two nuclear membranes. This generates an envelope that is called the primary viral envelope. The primary envelope is lost when it fuses with the outer nuclear membrane, releasing the herpesvirus nucleocapsid into the cytoplasm. At this point in the life cycle, some tegument proteins associate with the nucleocapsid. Additional tegument proteins are added when the developing virion is enveloped by membranes of either the Golgi apparatus or endosomes. This yields a mature enveloped virion that is transported to the plasma membrane in a membrane vesicle. The membrane vesicle fuses with the plasma membrane and releases the mature virion from the host cell. Thus unlike many other enveloped viruses, the source of the HSV-1 envelope is an organelle membrane, rather than the plasma membrane.

Just as the life cycle of λ has been of interest as a model of regulatory processes, so too has the establishment of latency by herpesviruses. Both viral and host proteins play a role in establishing a latent HSV-1 infection. In epithelial cells, where productive infections occur, an HSV-1 protein called VP16 and a host protein called host cell factor 1 (HCF-1) are both required for full expression of the virus's immediate early genes. HCF-1 is a nuclear protein that establishes the chromatin structure in the HSV genome for gene expression. In infected neurons, however, where the virus is latent, HCF-1 is cytoplasmic. In addition, small noncoding RNAs produced only in the neuron act to repress the viral lytic cycle. The inhibition of early gene expression helps establish latency. Interestingly, during the reactivation of the virus in neurons, one of the earliest events is the movement of HCF-1 to the nucleus and the production of VP16.

Nucleocytoplasmic Large DNA Viruses (Megaviruses)

Nucleocytoplasmic large DNA (NCLD) viruses are viruses of eukaryotic cells that are thought to have arisen from a common ancestor; a recently discovered virus order, *Megavirales,* has been proposed to encompass them. NCLD viruses, often referred to as megaviruses, are large enough to be visible by light microscopy *(see figure 6.1).* They have similar life cycles in which most, if not all, of the events occur in the cytoplasm of their hosts. Most have icosahedral capsids that enclose a lipid membrane and all have a large dsDNA genome of 500 to over 1,000 kilobases. Thus both the virion itself and the genome often rival the size of small bacteria and their genomes. In fact, the original NCLD virus discovered, Mimivirus, had been known to scientists for years as a bacterium called the Bradford coccus. Electron microscopy revealed not cocci, but icosahedral capsids. Megaviruses infect a variety of eukaryotic hosts, including amoebas, algae, insects, and humans. This extraordinarily broad host range is a consequence of the viral uptake mechanism by the host—phagocytosis. There is no viral envelope–cell surface receptor pairing to confer specificity.

The large genomes of megaviruses enable them to encode many proteins. Thus they are self-sufficient and rely on their hosts for much less than do other viruses. They generally encode all the proteins needed for DNA replication, enzymes involved in recombination, RNA polymerases and associated transcription factors, and chaperone proteins. In addition, they encode some components of the translational machinery (tRNAs and aminoacyl-tRNA synthetases). There are only three features that distinguish NCLD viruses and cells: the presence of ribosomes, a mechanism to generate ATP, and their method of reproduction.

Comprehension Check

1. Why do cold sores recur throughout the lifetime of an HSV-1–infected individual?
2. In what part of the host cell does a herpesvirus genome replicate? Where does the viral genome reside during a latent infection?
3. Many small DNA viruses rely on host enzymes for replication and transcription. Why are megaviruses able to use their own DNA-dependent RNA polymerases and DNA polymerases?

26.3 Single-Stranded DNA Viruses Use a Double-Stranded Intermediate in Their Life Cycles

After reading this section, you should be able to:

a. Describe in general terms the strategy used by single-stranded (ss) DNA viruses to synthesize their nucleic acids and proteins
b. Choose one specific bacterial and eukaryal ssDNA virus and illustrate the major events in their life cycles, noting, when possible, the specific mechanisms used to accomplish each step

Most DNA viruses are double-stranded, but several important viruses with single-stranded (ss) DNA genomes have been described. The life cycles of ssDNA viruses are similar to those of dsDNA viruses with one major exception. An additional step must occur in the synthesis stage because the ssDNA genome needs to be converted to a dsDNA molecule. A few ssDNA viruses are discussed next.

Bacteriophages ϕX174 and Pf1

The life cycle of ϕX174 (family *Microviridae,* species *Enterobacteria phage phiX174*) begins with attachment to the cell wall of its *E. coli* host. The circular ssDNA genome enters the cell with the help of a pilot protein called protein H. Protein H forms a tubelike structure through which the DNA passes; the protein capsid remains outside the cell. The ϕX174 ssDNA genome has the same base sequence as viral mRNA and is therefore **plus-strand DNA.** For either transcription or genome replication to occur, the phage DNA must be converted to a double-stranded **replicative form (RF)** (**figure 26.14**). This is catalyzed by the host's DNA polymerase. The RF directs the synthesis of more RF copies and plus-strand DNA, both by rolling-circle replication (*see figure 16.21*). Assembly of ϕX174 virions begins with formation of a procapsid. This involves scaffolding proteins that interact with capsid proteins, causing them to undergo conformational changes and form a procapsid with the proper morphology. Once the procapsid is generated, the ssDNA genome is inserted, creating the mature virion. After assembly of virions, the host is lysed by a viral enzyme (protein E) that blocks peptidoglycan synthesis. Protein E inhibits the activity of the bacterial protein MraY, which catalyzes the transfer of peptidoglycan precursors to lipid carriers (*see figure 12.9*). Blocking cell wall synthesis weakens the host cell wall, causing the cell to lyse and release the progeny virions.

We include *Pseudomonas* phage Pf1 (family *Inoviridae*) in our discussion because of one striking feature: its release from its host by a secretory process, without lysis. Assembly of the mature phage occurs in a type II secretion system (T2SS) composed of phage components (*see figure 13.43*). The circular ssDNA molecule is compressed and flattened so that the two sugar phosphate backbones contact one another and the bases protrude outward to contact the assembling capsid. Protomers embedded in the plasma membrane are added as the DNA passes through the T2SS. Complete coverage of the DNA with protomers results in bacteriophage release from the cell. Although the host cell is not lysed, it grows and divides at a reduced rate. ⏮ *Two-step secretion systems (section 13.8)*

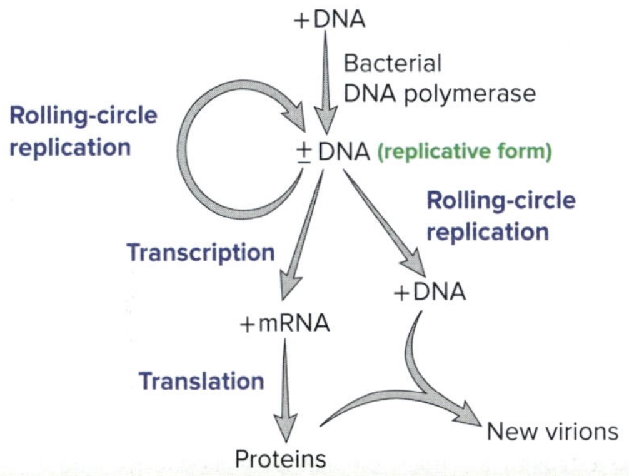

Figure 26.14 **The Multiplication Strategy of ϕX174, a Plus-Strand DNA Phage.**

MICRO INQUIRY *Why is the ϕX174 genome considered plus stranded?*

Parvoviruses

Parvoviruses (family *Parvoviridae*) infect numerous animal hosts, including crustaceans, dogs, cats, mice, and humans. Since its discovery in 1974, human parvovirus B19 (genus *Erythroparvovirus*) has emerged as a significant human pathogen, responsible for the childhood skin rash called fifth disease. Parvovirus virions are uniform, icosahedral, nonenveloped particles approximately 26 nm in diameter. Their genomes are composed of one ssDNA molecule of about 5,600 bases. Most of the genomes are **negative-strand DNA** molecules. That is, their sequence of nucleotides is complementary to that of the viral mRNA (figure 26.2). Parvoviruses are among the simplest of the DNA viruses. The genome is so small that it directs the synthesis of only three proteins and some smaller polypeptides. None has enzymatic activity. ▶| *Human parvovirus B19 infection (section 38.3)*

A productive parvovirus infection requires attachment to a particular cell type. Parvovirus B19 infects several human tissues but only red blood cell progenitors in the bone marrow support its replication. Virions enter host cells by receptor-mediated endocytosis. Each virion escapes the endosome and is transported to the nucleus by the host cell's microtubules, directed by a nuclear localization signal on its capsid. The nucleocapsid enters the nucleus, followed by release of viral DNA from the capsid.

Because the parvovirus genome does not code for any enzymes, the virus must use host cell enzymes for all biosynthetic processes. Viral infection causes arrest of the cell cycle in S phase, when the host cell can provide all of the components needed for viral replication. Because the genome is negative-strand DNA, it serves as the template for mRNA synthesis. Some of the RNA products encode polypeptides required for the interesting way the virus's genome is replicated. The ends of the parvovirus genome are palindromic sequences that can fold back on themselves. Formation of a complex composed of a viral protein and a hairpin at the 3′ end of the genome provides the primer needed for replication (**figure 26.15**). This is recognized as a substrate by the host DNA polymerase, and DNA replication ensues by a process that is somewhat similar to rolling-circle replication. The parvovirus version of this replication method is often called rolling-hairpin replication because DNA polymerase seems to shuttle back and forth as it synthesizes genomes. The process involves a dsDNA intermediate, much like ϕX174.

Comprehension Check

1. Why is it necessary for some ssDNA viruses to manufacture a replicative form?
2. From the point of view of the virus, compare the advantages and disadvantages of host cell lysis, as seen in ϕX174, with the continuous release of phage without lysis, as with Pf1 phage and other filamentous phages.
3. How do parvoviruses "trick" the host DNA polymerase into replicating their genomes? Why must this occur in the nucleus only during the S phase of the host's cell cycle?

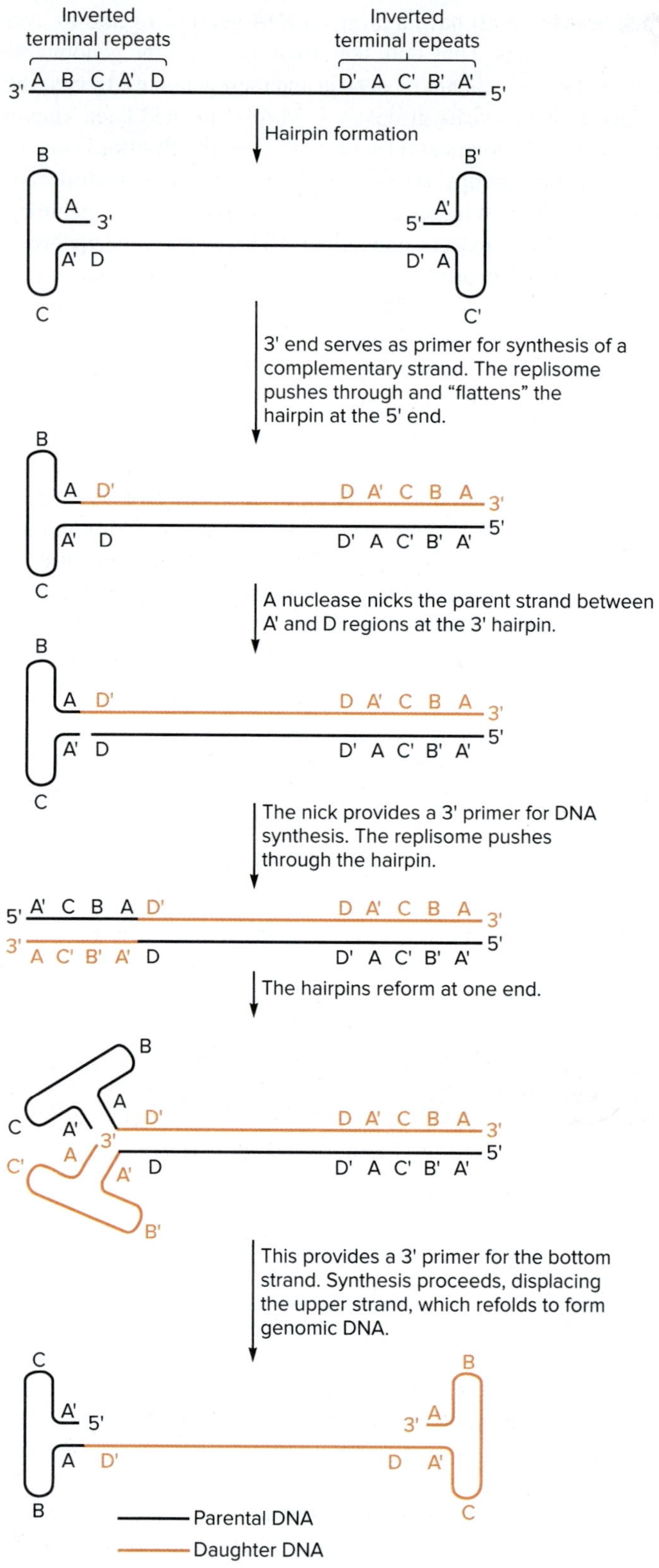

Figure 26.15 The Rolling-Hairpin Replication Used by Parvoviruses to Replicate Their DNA.

26.4 Double-Stranded RNA Viruses: RNA-Dependent RNA Polymerase Replicates the Genome and Synthesizes mRNA

After reading this section, you should be able to:

a. Distinguish RNA-dependent RNA polymerases from DNA-dependent RNA polymerases
b. Describe what an RNA-dependent RNA polymerase is doing when functioning as a replicase and when functioning as a transcriptase
c. Describe in general terms the strategy used by dsRNA viruses to synthesize their nucleic acids and proteins
d. Describe the major events in the life cycles of ϕ6 and rotaviruses, noting, when possible, the specific mechanisms used to accomplish each step

The Baltimore system divides viruses with RNA genomes into four groups (table 26.1). However, they all share the same dilemma: Their host cells have dsDNA genomes and thus lack a polymerase that can make RNA from an RNA template. Therefore RNA viruses classified in the Baltimore System as double-stranded, plus-strand, and negative-strand RNA viruses must produce an enzyme called **RNA-dependent RNA polymerase** (RdRp; an RNA polymerase that uses an RNA molecule as the template for RNA synthesis). When the RdRp is used to replicate the viral RNA genome, it is often referred to as a **replicase.** When it is used to synthesize mRNA, the RdRp is often said to have **transcriptase** activity. In most cases, the same enzyme carries out both functions. Viruses in the fourth group, retroviruses, use a very different approach: They use the enzyme reverse transcriptase to convert their ssRNA genomes into dsDNA, which is then recognized by the host cell's DNA-dependent RNA polymerase.

Among viruses with RNA genomes, those with dsRNA are uncommon. These viruses share a common multiplication strategy (**figure 26.16**). Here we discuss two representative dsRNA viruses: a bacteriophage and a vertebrate virus. We then turn our attention to other RNA viruses with other types of RNA genomes.

Bacteriophage ϕ6

Several dsRNA phages have been discovered and assigned to the family *Cystoviridae*. The best studied, *Pseudomonas* phage phi6 (ϕ6), is unusual for several reasons. One is that it is an enveloped phage. The envelope encloses the nucleocapsid, which consists of an outer protein shell, composed of a viral protein called P8, and an inner core sometimes called the polymerase complex. It is so named because it consists of the virus's RNA-dependent RNA polymerase (RdRp) and three other proteins, and encloses the genome. The ϕ6 genome illustrates one of the methods used by RNA viruses to ensure synthesis of all viral proteins: it is a **segmented genome,** composed of more than one distinct RNA molecule. For most viruses, each segment encodes a single protein. However, ϕ6 deviates from this because each of its three dsRNA molecules generates a polycistronic mRNA and several proteins. Many other viruses also generate polycistronic mRNAs, much like the polycistronic mRNAs observed in bacteria and archaea (*see figure 13.22*).

The life cycle of the virus also has unusual features. Like some other phages, ϕ6 attaches to a pilus via an envelope spike. Retraction of the pilus brings the phage into contact with the outer membrane of its Gram-negative host. The viral envelope then fuses with the cell's outer membrane, a process mediated by another envelope spike. Fusion of the two membranes delivers the nucleocapsid to the periplasmic space. A protein associated with the nucleocapsid digests the peptidoglycan, allowing the nucleocapsid to cross this layer of the cell wall. Finally, the intact nucleocapsid enters the host cell by a process that resembles endocytosis. An interaction between the shell protein P8 and plasma membrane phospholipids mediates this mechanism of entry.

Once inside the host, the viral RNA polymerase acts as a transcriptase, catalyzing synthesis of viral mRNA from each dsRNA segment while they are still inside the capsid. The mRNAs leave the capsid and some are translated to early proteins. These assemble to form new, empty polymerase complexes (i.e., lacking RNA). Meanwhile, the replicase synthesizes plus-strand RNA from each segment. The plus-strand RNAs become enclosed within the newly formed polymerase complexes, where they serve as templates for the synthesis of the complementary negative strand, regenerating the dsRNA genome. Proteins made later in the infection cycle are found in the outer shell. These associate with the polymerase complex, forming the nucleocapsid. Once the nucleocapsid is complete, a nonstructural viral protein called P12 surrounds the nucleocapsid with a plasma membrane–derived envelope while the nucleocapsid is within the host cytoplasm. Finally, additional viral proteins are added to the envelope and the host cell is lysed, releasing mature virions.

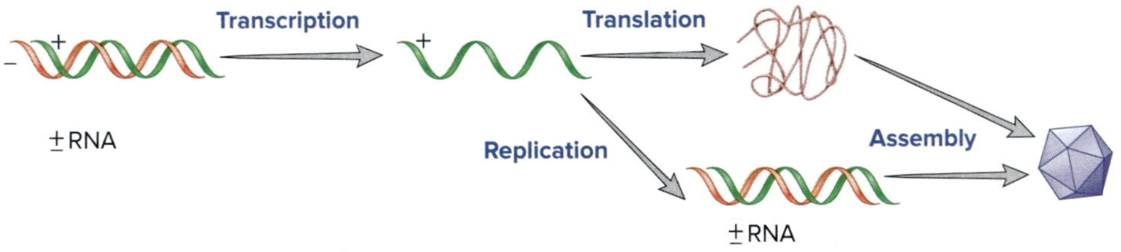

Figure 26.16 Multiplication Strategy of Double-Stranded RNA Viruses. Although we have designated plus and minus RNA strands, both strands may serve as templates for synthesis of mRNA by transcriptase, depending on the virus. For those viruses, each strand is said to be ambisense. The strand of the viral genome we have designated as the negative strand serves as the template for synthesis of positive strands by replicase. The positive strands are used by replicase to synthesize negative strands, thus replicating the dsRNA genome.

Rotaviruses

Human rotaviruses (family *Reoviridae*) are responsible for the deaths of over 200,000 children worldwide each year. They cause severe diarrhea, which results in dehydration and death if appropriate therapy is not provided promptly. Because of their impact on humans, rotaviruses have been studied intensely to better understand their life cycles and pathogenesis. ▶| *Gastroenteritis (viral) (section 38.4)*

Viewed by electron microscopy, rotavirus virions have a characteristic wheel-like appearance (Latin *rota,* wheel). Virions are nonenveloped and are composed of 11 segments of dsRNA surrounded by three concentric layers of proteins (*see figure 38.22*). The RNA segments code for six structural and six nonstructural proteins (NSPs).

When a rotavirus virion enters a host cell, it loses the outermost protein layer and is then referred to as a double-layered particle (DLP; **figure 26.17**). The genome is transcribed by the viral transcriptase while still inside the DLP. The mRNA passes through channels in the DLP and is released into the cytosol of the host cell. There the mRNAs are translated by the host cell's ribosomes. The newly formed proteins cluster together, forming an inclusion called a viroplasm. It is within the viroplasm that new DLPs are formed. Initially the DLPs contain plus-strand RNA, but this is soon used as a template for synthesis of the negative strand. Replicase activity of the RdRp requires that the enzyme be in complex with the core shell protein. The mechanism to coordinate replication of all 11 genome segments is not known. Thus the dsRNA molecules are synthesized within the developing DLP.

DLPs containing dsRNA are enveloped by membranes of the endoplasmic reticulum. While in the endoplasmic reticulum, the outermost layer is added to the DLP, converting it to the mature triple-layered virion. The mature virion is released by an unknown mechanism.

Comprehension Check

1. The rotavirus genome encodes 12 proteins. Suggest one or more strategies it might use to ensure that all are synthesized.
2. Describe the life cycle of ϕ6 phage. What makes this phage unusual when compared with other bacteriophages?
3. How does the rotavirus structure give rise to its name?
4. In what ways are the life cycles of ϕ6 and rotaviruses similar? How do they differ?

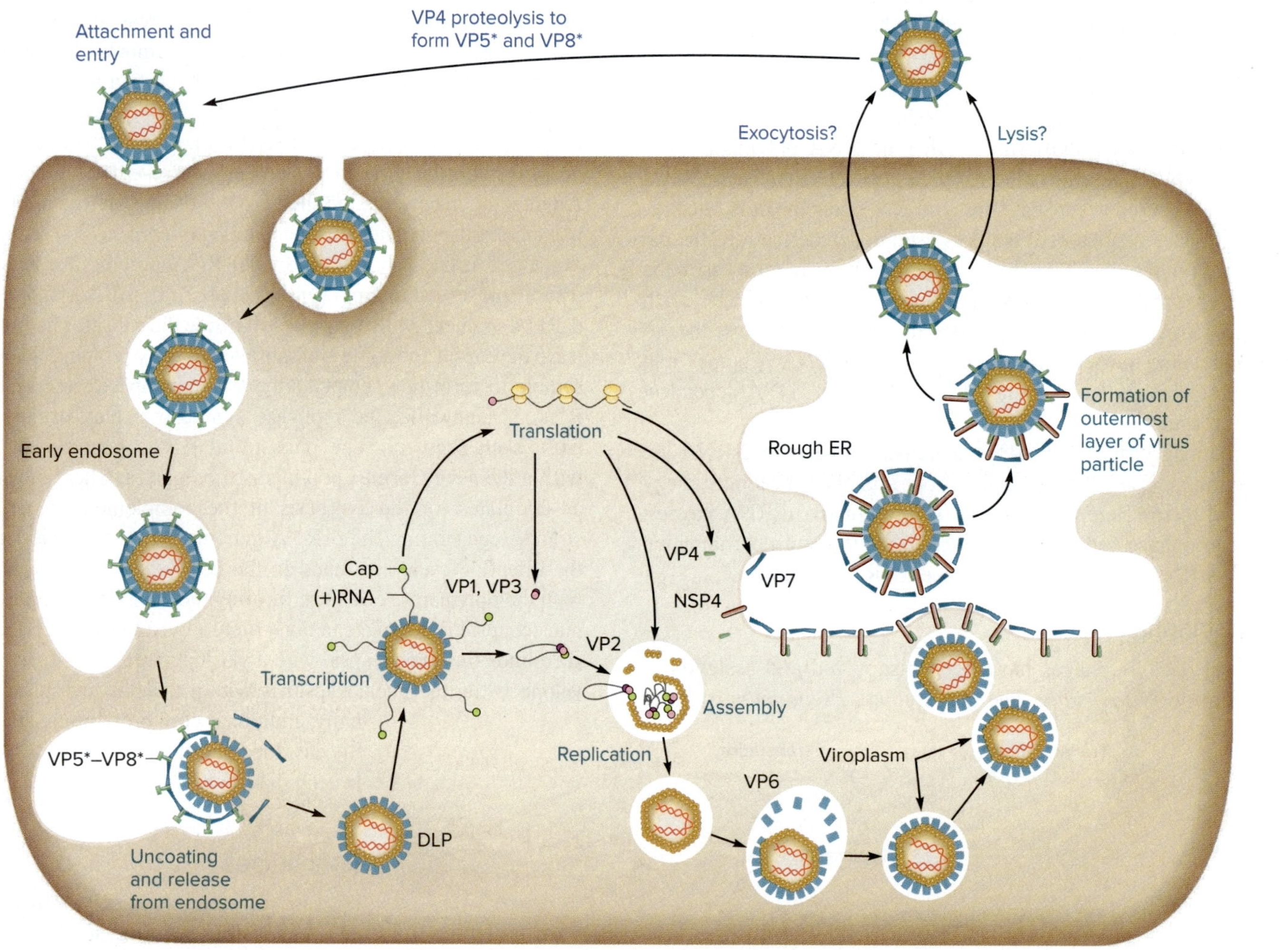

Figure 26.17 Life Cycle of a Rotavirus. Proteases in the human intestinal tract act on the rotavirus outer shell protein (VP4) to generate the infective form of the virus. VP5* and VP8* are fragments of VP4 that bind the cell surface receptors. Upon viral uncoating, the RdRp (VP1) transcribes the genome and VP3 caps the mRNA. Genome replication and virion assembly are coupled in the viroplasm. NSP4 is a viroporin protein that creates a membrane channel through which ions leak into the cytoplasm from the endoplasmic reticulum (ER). This leakage signals the outer shell proteins VP7 and VP4 to coat the virion as it enters the ER. The method of release of progeny virions has not been definitively determined.

26.5 Plus-Strand RNA Viruses: Genomes That Can Be Translated upon Entry

After reading this section, you should be able to:

a. Describe in general terms the strategy used by plus-strand RNA viruses to synthesize their nucleic acids and proteins
b. Outline the major events in the life cycles of poliovirus and tobacco mosaic virus, noting, when possible, the specific mechanisms used to accomplish each step

Plus-strand RNA viruses have genomes that can act as mRNA and be translated upon entry into the host cell. One of the first products is the RNA-dependent RNA polymerase, which catalyzes synthesis of negative-strand RNAs; these are then used to make more plus-strand RNAs (**figure 26.18**). In some cases, this occurs by way of a double-stranded replicative form (RF), as seen with ssDNA viruses (section 26.3). In plant and animal viruses, viral genome replication occurs in a structure formed within the cytoplasm called a replication complex (*see figure 6.12*). These are compartments formed in response to factors produced by the virus, and they are derived from the membranes of cell organelles (e.g., ER membrane); the source of the membrane depends on the virus. Assembly of progeny virions sometimes also occurs within the replication complex. There are a number of medically important positive-strand animal viruses (e.g., poliovirus, Zika virus, eastern equine encephalitis virus [**figure 26.19**], and hepatitis A virus), and most plant viruses have plus-stranded RNA genomes.

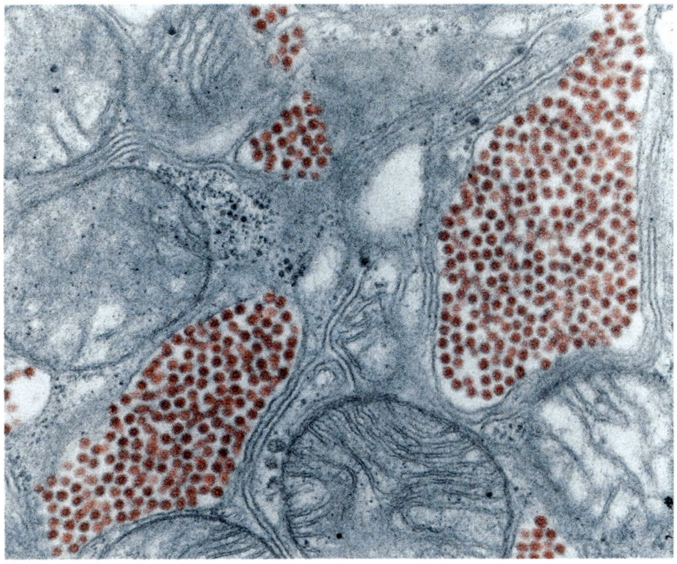

Figure 26.19 Eastern Equine Encephalitis Virus in the Cell Cytoplasm. This TEM has been colored to highlight the virions (red). Each virion measures 65-70 nm in width.

Source: CDC/Fred Murphy and Sylvia Whitfield

Poliovirus

Human *Enterovirus C* (commonly called poliovirus), the causative agent of poliomyelitis, has caused disease in humans for centuries (*see Historical Highlights 38.1*). It primarily targets children, with paralysis being the tragic result in some cases. The life cycle of poliovirus illustrates another strategy used by RNA viruses to ensure synthesis of needed proteins: synthesis of a polyprotein. ▶| *Poliomyelitis (section 38.4)*

The biology of polioviruses (family *Picornaviridae*) has been intensely studied for decades but many questions linger. This nonenveloped virus makes its way into a human host by ingestion. It attaches to a cell surface molecule called CD155, found on certain white blood cells and neurons of the central nervous system. The nucleocapsid enters the host cell, and the plus-strand RNA genome is released into the cytosol while the virion is at the cell periphery and held within an endocytic vesicle. The genome acts as mRNA and is translated by host cell ribosomes. This process has garnered considerable attention because the virus's RNA does not have the 5′ cap found on eukaryotic mRNAs, which is important for ribosome binding. Poliovirus "tricks" its host into translating its capless RNA using a 5′ region on the RNA called the internal ribosome entry site (IRES). In this region the ssRNA folds back on itself and forms extensive secondary structures (regions of dsRNA and numerous ssRNA loops), which are important for recognition of the RNA by ribosomes (**figure 26.20**).

Translation of the poliovirus genome yields a single polyprotein (**figure 26.21**). A **polyprotein** is a large protein that is cleaved into smaller proteins by proteases. The poliovirus polyprotein has protease activity and cleaves itself into three smaller proteins. Additional cuts in these proteins eventually yield all the structural proteins needed for capsid formation, as well as the virus's RdRp. Three viral proteins, 2B, 2C, and 3A, are membrane proteins that redirect cell membranes to form a new organelle, termed the replication organelle (RO). The RO both protects the viral RNAs from degradation and segregates viral and host proteins for replication. Within the RO, the RdRp generates negative-strand RNA molecules that serve as templates for synthesis of plus-strand RNAs. Some of these are translated, but eventually most will be used as new genomes and incorporated into capsids. Mature particles may be released from the cell

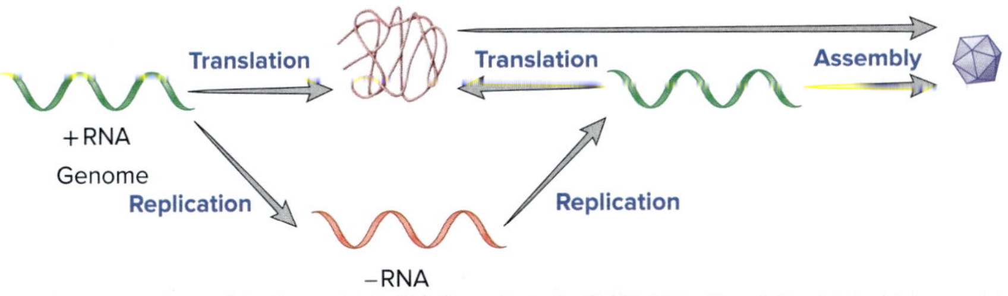

Figure 26.18 Multiplication Strategy of Plus-Strand RNA Viruses. The plus-strand RNA genome can serve directly as mRNA. One of the first viral proteins synthesized is replicase, which replicates the genome, sometimes via a double-stranded replicative form. Transcriptase is responsible for synthesizing additional mRNA molecules.

MICRO INQUIRY *Where in the host does the plus-strand RNA genome replicate?*

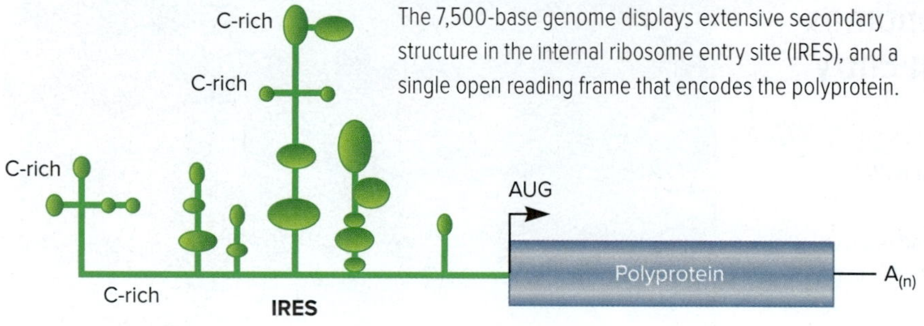

Figure 26.20 The Poliovirus Genome. The 7,500-base genome displays extensive secondary structure in the internal ribosome entry site (IRES), and a single open reading frame that encodes the polyprotein.

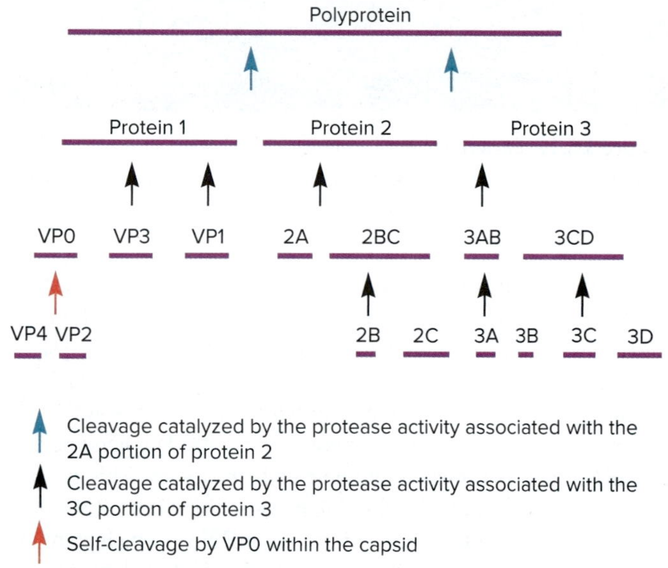

Figure 26.21 Cleavage of the Poliovirus Polyprotein. Proteins VP1-4 are capsid proteins. Protein 3D is the RNA-dependent RNA polymerase.

by lysis, although it has been suggested that virions may be released from the RO at the cell surface in vesicles. ◀ *Synthesis stage (section 6.3)*

Tobacco Mosaic Virus

Most plant viruses are RNA viruses, and of these, plus-strand RNA viruses are most common. Tobacco mosaic virus (TMV; family *Virgaviridae*) is the best-studied plus-strand RNA plant virus. TMV virions are filamentous with coat proteins arranged in a helical pattern. The TMV life cycle illustrates two new strategies to produce its needed proteins: subgenomic RNAs and readthrough. **Subgenomic RNAs** are mRNAs that are shorter than the RNA genome. They can be synthesized in several ways, depending on the virus. Usually, synthesis of subgenomic RNAs is dictated by internal transcription start sites, internal transcription termination sites, or both. **Readthrough** occurs during translation. It occurs when the ribosome ignores a stop signal and continues translation.

Plant viruses, including TMV, usually enter the host through an abrasion or wound on the plant; biting (e.g., sap sucking) insects are often involved in transmission of the virus. Following entry into its host, the TMV RNA genome is translated. Two proteins are produced, one about 125 kilodaltons (kD) and one about 185 kD. The larger protein is produced by readthrough of the stop codon at the end of the coding region for the 125-kD protein. The 185-kD protein has RdRp activity and functions as both a transcriptase and a replicase. TMV uses membranes of the endoplasmic reticulum (ER) to form a replication complex. There, the RdRp synthesizes negative-strand RNA using the plus-strand genome as the template. It is not clear whether a double-stranded RF is created in vivo, but RFs have been observed in vitro. The negative-strand RNA serves as a template for synthesis of new genomes. It also serves as the template for synthesis of subgenomic mRNAs that are translated into a variety of proteins needed by the virus.

After the coat protein and RNA genome of TMV have been synthesized, they spontaneously assemble into complete TMV virions in a highly organized process (**figure 26.22**). The protomers come together to form disks composed of two layers of protomers arranged in a helical spiral. Association of coat protein with TMV RNA begins at a specific assembly initiation site close to the 3′ end of the genome. The helical capsid grows by the addition of protomers, probably as disks, to the end of the rod. As the rod lengthens, the RNA passes through a channel in its center and forms a loop at the growing end. In this way, the RNA easily fits as a spiral into the interior of the helical capsid.

Multiplication of plant viruses within their host depends on the virus's ability to spread throughout the plant. Viruses can move to new sites in the plant through the plant vasculature; usually they travel in the phloem. They can also spread locally in nonvascular tissue. Because plant cells have tough cell walls, plant viruses move from cell to cell through plant cell plasmodesmata. These are slender bridges of cytoplasmic material, including extensions of the ER. Plasmodesmata extend through holes in cell walls to join adjacent plant cells. Viral "movement proteins" are required for transfer through the plasmodesmata. TMV movement proteins are closely associated with ER membranes and so are near the TMV replication complex where viral genomes are being synthesized. They bind the viral genomic RNA (vRNA) and interact with components of the cytoskeleton. Therefore they facilitate the movement of vRNA from the replication complex to the plasmodesmata. TMV movement proteins also cause the plasmodesmata to increase in diameter, allowing the vRNA to pass through to the adjacent cell.

Several cytological changes can take place in TMV-infected cells. These include microscopically visible intracellular inclusions similar to the replication complexes observed in animal cells infected with plus-strand viruses. Hexagonal crystals of almost pure TMV virions sometimes develop in TMV-infected cells. In addition, host cell chloroplasts become abnormal and often degenerate, while new chloroplast synthesis is inhibited.

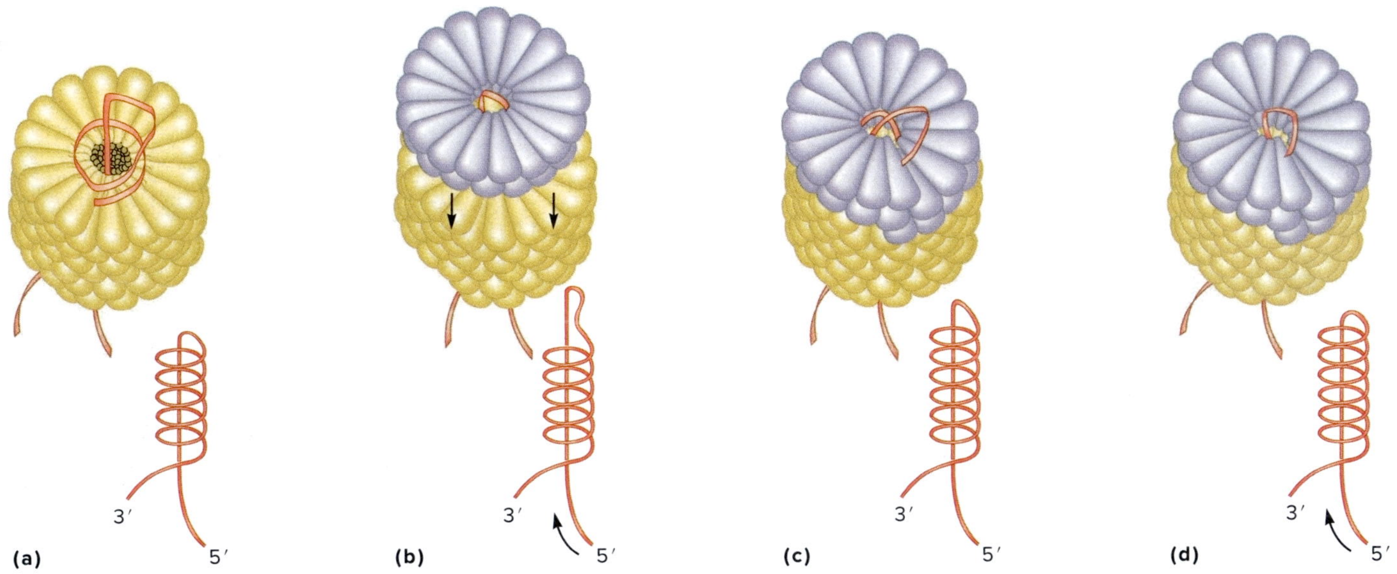

Figure 26.22 TMV Assembly. The elongation phase of tobacco mosaic virus nucleocapsid construction. The lengthening of the helical capsid through the addition of a protein disk to its end is shown in a sequence of four illustrations; line drawings depicting RNA behavior are included. The RNA genome inserts itself through the hole of an approaching disk and then binds to the groove in the disk as it locks into place at the end of the cylinder.

Comprehension Check

1. How do some plus-strand viruses use polyproteins to complete their life cycles? Describe two other strategies used by plus-strand viruses to generate all their proteins.
2. What is an IRES? Why is it needed?
3. How does TMV spread from one host cell to another?

26.6 Minus-Strand RNA Viruses: RNA-Dependent RNA Polymerase Is Part of the Virion

After reading this section, you should be able to:

a. Describe in general terms the strategy used by minus-strand RNA viruses to synthesize their nucleic acids and proteins
b. Explain how having a segmented genome impacts synthesis of viral mRNA and proteins and the generation of new strains of a virus
c. Create a flow chart that summarizes the life cycle of influenza virus, noting the specific mechanisms it uses to accomplish each step of its life cycle

Negative-strand viruses are found in many families, including *Rhabdoviridae* (e.g., rabies virus), *Filoviridae* (e.g., Ebola virus), *Paramyxoviridae* (e.g., measles and mumps virus), and *Orthomyxoviridae* (e.g., influenza viruses). Most negative-strand RNA viruses are enveloped viruses that infect plants and animals. They vary in morphology from spherical, to filamentous, rod-shaped, bullet-shaped, and pleomorphic. Members of eight families have nonsegmented linear genomes and are grouped into the order *Mononegavirales*. The remaining negative-strand RNA viruses have segmented genomes that range from two to eight segments, each encoding usually one protein.

The genomes of negative-strand RNA viruses cannot function as mRNA. Therefore these viruses must bring at least one RNA-dependent RNA polymerase (RdRp) into the host cell during entry. Initially the viral genome serves as the template for mRNA synthesis (**figure 26.23**). Later the virus switches from mRNA synthesis to genome replication, as the RdRp synthesizes a distinct plus-strand RNA for replication. During this phase of the life cycle, the plus-strand RNA molecules synthesized from the minus-strand genome serve as templates for the manufacture of new negative-strand RNA genomes.

Our focus in this section is on influenza viruses. Influenza virus virions are composed of an envelope enclosing seven or eight nucleocapsids, depending on the type of influenza virus. Each nucleocapsid consists of a single, negative-strand RNA that exists as a double helical hairpin (**figure 26.24*a***). Formation and maintenance of the double helix do not involve base pairing.

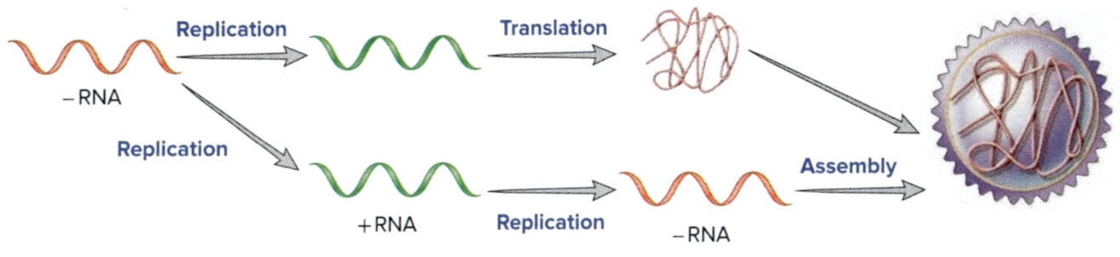

Figure 26.23 Multiplication Strategy of Negative-Strand RNA Viruses. An RNA-dependent RNA polymerase enters the host cell at the same time the negative-strand RNA genome enters. The genome serves as a template for synthesis of mRNA. Later in the infection, the negative-strand genome is used for plus-strand synthesis. These plus strands then act as templates for replication of the negative-strand genomes.

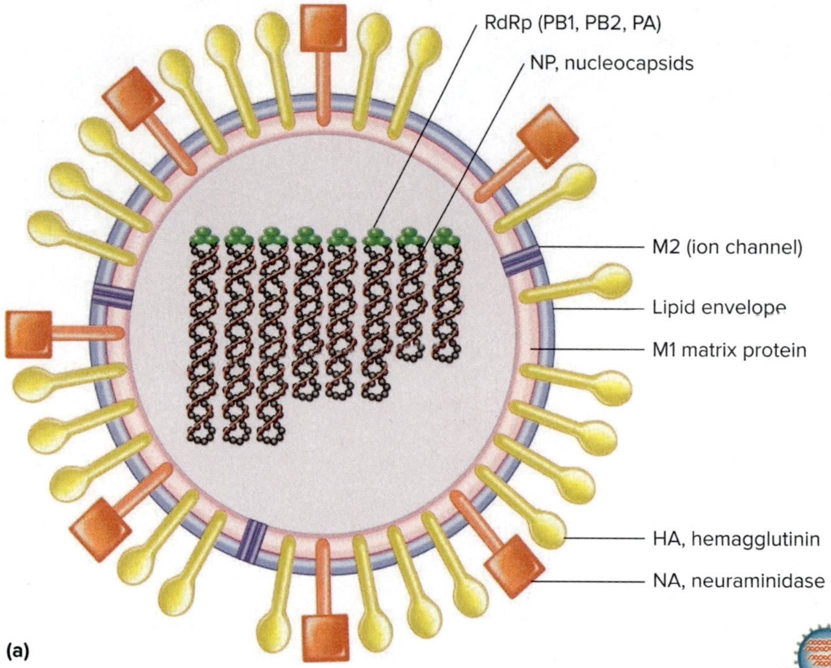

(a)

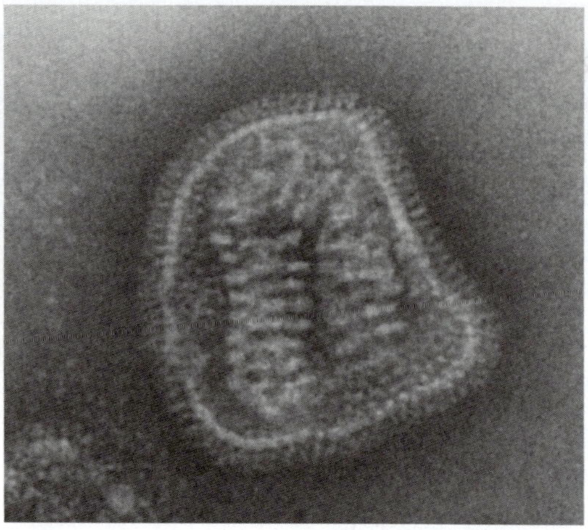

(b)

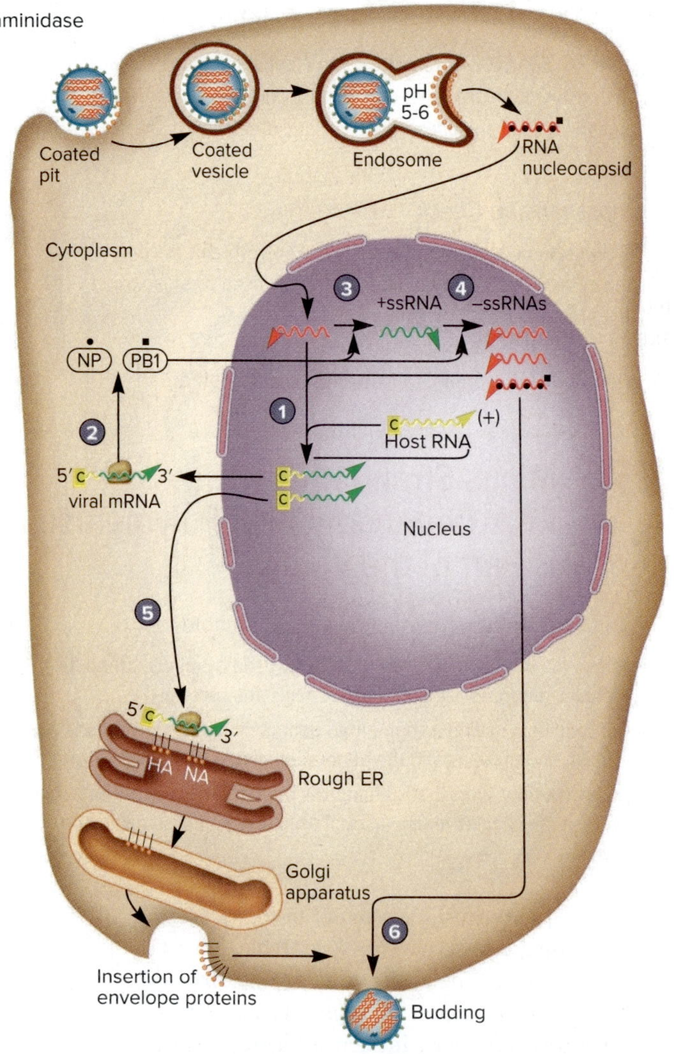

1 The PB2 protein cleaves the cap and about 10 nucleotides from the 5′ end of host mRNA (cap snatching). The fragment is used to prime viral mRNA synthesis by the transcriptase activity of the PB1 protein.

2 Viral mRNA is translated. Early products include more NP and PB1 proteins.

3 PB1 synthesizes +ssRNA from genomic –ssRNA molecules.

4 PB1 synthesizes new copies of the genome using +ssRNA made in step 3 as templates. Some of these new genome segments serve as templates for the synthesis of more viral mRNA. Later in the infection, they will become progeny genomes.

5 Viral mRNA molecules transcribed from other genome segments encode structural proteins such as hemagglutinin (HA) and neuraminidase (NA). These messages are translated by ER-associated ribosomes and delivered to the cell membrane.

6 Viral genome segments are packaged as progeny virions bud from the host cell.

(c)

Figure 26.24 An Influenza Virus and Its Life Cycle. (a) Schematic representation of an influenza A virus. (b) Negative-stained transmission electron micrograph (TEM) of an influenza virus particle, which measures 80–120 nm in width. (c) This simplified version of the influenza life cycle omits several steps and represents the viral RNA molecules as simple lines. Negative-strand RNA is red and positive-strand RNA is green Host-derived RNA is yellow. Steps 1 through 6 illustrate the steps of the virus's life cycle following endocytosis.

(b) Source: CDC/Fred Murphy, Cynthia Goldsmith, Dr. Erskine L. Palmer and Dr. M.L. Martin

Rather, this is accomplished by numerous copies of the nucleocapsid protein (NP), which coat the RNA. The three polypeptides of the influenza RdRp (PA, PB1, PB2) are attached to the open end of the hairpin.

An influenza virion enters a host cell by receptor-mediated endocytosis. This encloses the virus in an endosome (figure 26.24*c*). When the endosomal pH decreases, a conformational change in the virion facilitates contact with the endosome membrane, allowing fusion of the endosome membrane and viral envelope to occur, releasing the nucleocapsids into the cytosol. ▶| *Influenza (flu; section 38.1)*

To execute its life cycle, influenza components shuttle between the nucleus and cytoplasm using nuclear localization signals on proteins and nuclear export signals on RNA. Uncoated nucleocapsids are transported to the nucleus for mRNA synthesis, mRNAs move to the cytoplasm for translation, and the newly synthesized viral RdRps return to the nucleus for replication and capsid assembly. When RNA segments and associated RdRp enter the host cell nucleus, genome segments serve as templates for mRNA synthesis (figure 26.24*c*, step 1). Copies of the negative-strand genome destined for translation are distinguished from copies destined for replication by the presence of a 5′ cap structure. Both termini of each genome segment form a panhandle structure recognized by RdRp as the substrate for transcriptase activity. Once bound to the panhandle, the RdRp acquires a 5′ cap by a process termed cap snatching. Host mRNAs with a 5′ cap are cut by the RdRp about 8 to 14 bases after the cap, and that short primer is used for generating viral mRNA. Later the virus switches from mRNA synthesis to genome replication. During this phase of the life cycle, the plus-strand RNA molecules synthesized from the minus-strand genome segments serve as templates for new negative-strand RNA genomes (figure 26.24*c*, step 4). Virions exit the host cell by budding and thus acquire their envelope (figure 26.24*c*, step 6; *see also figure 6.15*).

Comprehension Check

1. How does the use of a segmented genome by influenza viruses differ from that used by bacteriophage ϕ6?
2. Trace the multiplication of an influenza virus starting with host cell attachment and ending with the exit of virions.

26.7 Retroviruses: Plus-Strand Viruses That Use Reverse Transcriptase in Their Life Cycles

After reading this section, you should be able to:

a. Describe in general terms the strategy used by retroviruses to synthesize their nucleic acids and proteins
b. Differentiate a segmented genome from the genome of a retrovirus
c. Describe the life cycle of HIV, noting the specific mechanisms it uses to accomplish each step
d. Defend the statement "all three enzymatic activities of reverse transcriptase are required for dsDNA to be synthesized from RNA"
e. Distinguish DNA-dependent DNA polymerases, DNA-dependent RNA polymerases, RNA-dependent RNA polymerases, and RNA-dependent DNA polymerases in terms of their templates, products synthesized, and proofreading activity, as discussed in this and previous sections

Retroviruses have positive-strand RNA genomes. However, their genomes do not function as mRNA (**figure 26.25**). Instead, retroviruses first convert their ssRNA genomes into dsDNA using a multifunctional enzyme called **reverse transcriptase.** The dsDNA is then integrated into the host's DNA, where it can serve as a template for mRNA synthesis and synthesis of the plus-strand RNA genome. The host cell's DNA-dependent RNA polymerase catalyzes both of these processes.

Numerous retroviruses have been identified and studied. However, human immunodeficiency virus (HIV), the cause of AIDS (acquired immune deficiency syndrome), is of particular interest. Because of its global importance, we focus exclusively on HIV in this section. ▶| *Acquired immune deficiency syndrome (AIDS) (section 38.3)*

HIV is a member of the genus *Lentivirus* within the family *Retroviridae*. In the United States, AIDS is caused primarily by HIV-1, while in many developing nations HIV-2 is prevalent. HIV is an enveloped virus. The envelope surrounds an outer shell, which encloses a cone-shaped core (**figure 26.26**). The core contains two copies of the 10,000-base HIV RNA genome and several enzymes, including reverse transcriptase (RT) and integrase. Thus far about 10 virus-specific proteins and dozens of host proteins have been discovered in the HIV virion.

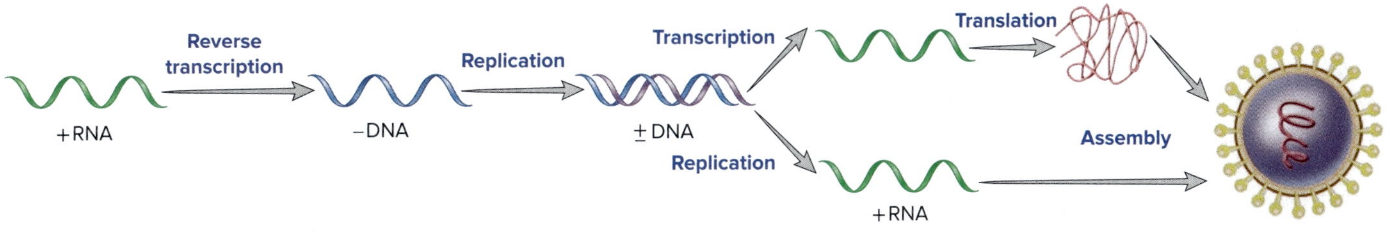

Figure 26.25 Multiplication Strategy of Retroviruses. Retroviruses have a plus-strand RNA genome that is first converted into dsDNA by the enzyme reverse transcriptase. The viral dsDNA integrates into the host chromosome, where it serves as the template for synthesis of viral mRNA and viral genomes. Both are synthesized using the host cell's DNA-dependent RNA polymerase.

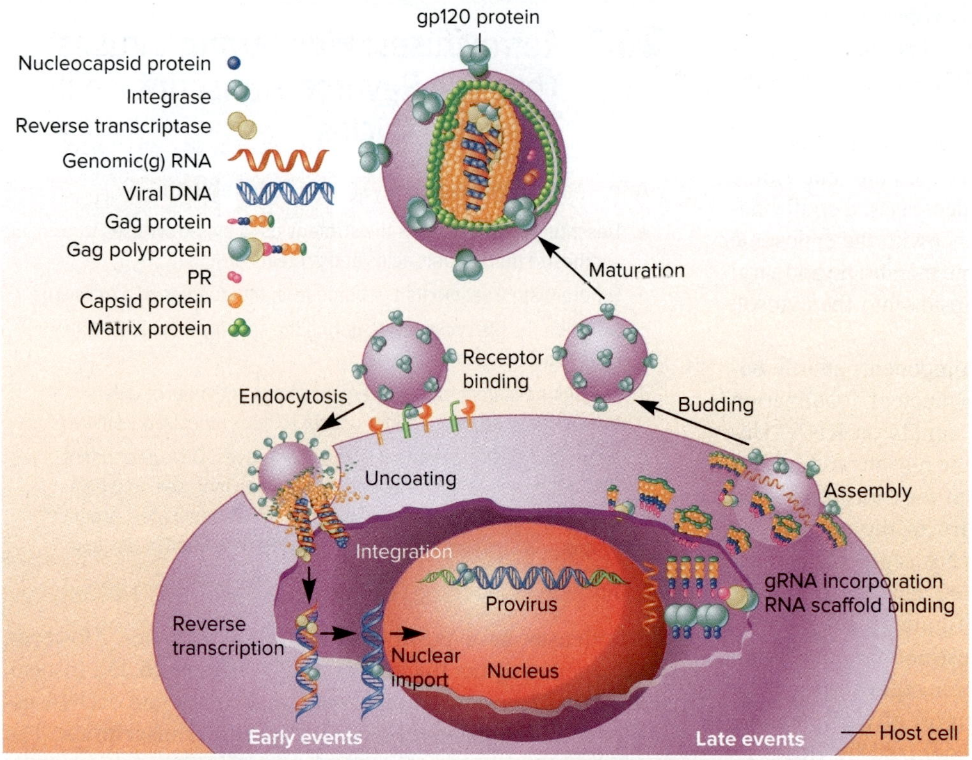

Figure 26.26 The HIV Life Cycle.

MICRO INQUIRY *What three enzymatic activities does reverse transcriptase display? When during the HIV life cycle is each activity used?*

After entering the body, the gp120 viral envelope protein binds to host cells that have a surface glycoprotein called CD4. The virus requires a coreceptor in addition to CD4, and this varies depending on the host cell infected (*see figure 38.11*). The process of viral entry has been actively debated. Both viral envelope fusion with the plasma membrane and endocytosis have been demonstrated. Some scientists suggest that virions may enter by either method. HIV can infect several different cell types, and subtle differences may affect the entry process. Inside the infected cell, the core protein dissociates from the RNA, and the RNA is copied into a single strand of DNA by the RT. The RNA is next degraded by RT, and the DNA strand is copied to form a double-stranded DNA version of the original RNA genome (figure 26.25).

Reverse transcription is a critical step in the life cycle of HIV, and RT is a remarkable enzyme with multiple activities. It is an **RNA-dependent DNA polymerase,** a DNA-dependent DNA polymerase, and a ribonuclease. Despite its versatility, RT lacks an important function observed in other DNA polymerases: proofreading. Thus it makes errors as it synthesizes DNA. For the initial steps in DNA synthesis, a host tRNA molecule is used as a primer (**figure 26.27**). In the process, a newly synthesized, small negative-strand DNA molecule is transferred from one end of the RNA template to the other to prime the synthesis of the rest of the minus-strand DNA. Later the full-length minus-strand DNA circularizes to allow completion of the plus-strand DNA.

Once dsDNA is formed, a complex of dsDNA, integrase, and other factors (including some host molecules) moves into the nucleus, where the proviral DNA is integrated into the cell's DNA by integrase. Once integrated, the provirus can force the cell to synthesize viral mRNA (figure 26.26). Several different mRNAs are transcribed using the host cell's DNA-dependent RNA polymerase. Some are full-length mRNAs and others are subgenomic mRNAs. The subgenomic mRNAs are formed by alternative splicing that brings different coding regions together, eliminating any intervening coding regions (rather than introns, as typically occurs during alternative splicing). Translation of two of the mRNAs yields polyproteins that are cleaved to give rise to numerous proteins. One of the polyproteins is synthesized by ribosomal frameshifting followed by read-through. **Ribosomal frameshifting** is often used when viral genes overlap but are in different reading frames. It occurs at specific sites at which the ribosome pauses and then either continues in the same reading frame or shifts to the alternative reading frame. In the case of HIV, the ribosome reaches a stop codon, pauses, and either stops to yield one protein or shifts reading frame and continues translating the mRNA to yield the larger polyprotein. Recall that full-length unspliced mRNAs are the same as genomic RNA: Both are copied from the same strand. It is not known how or if the two molecules are distinguished, or if there is a competition for this RNA between ribosomes and the genome packaging complex. Some of the earliest proteins synthesized are used to regulate cellular activities so that HIV genes are preferentially expressed. Proteins needed to form HIV virions are made later.

Eventually viral proteins and the complete HIV RNA genome are assembled into new nucleocapsids that bud from the infected host cell (figure 26.26). The viral particles released from the host cell are not infective, and require a maturation process. The viral Gag protein, a structural component of the virion, is cleaved by proteases. The result is a change in virion shape from a sphere to a cone, and RNA genome repackaging. Antiviral therapy with protease inhibitors targets maturation.

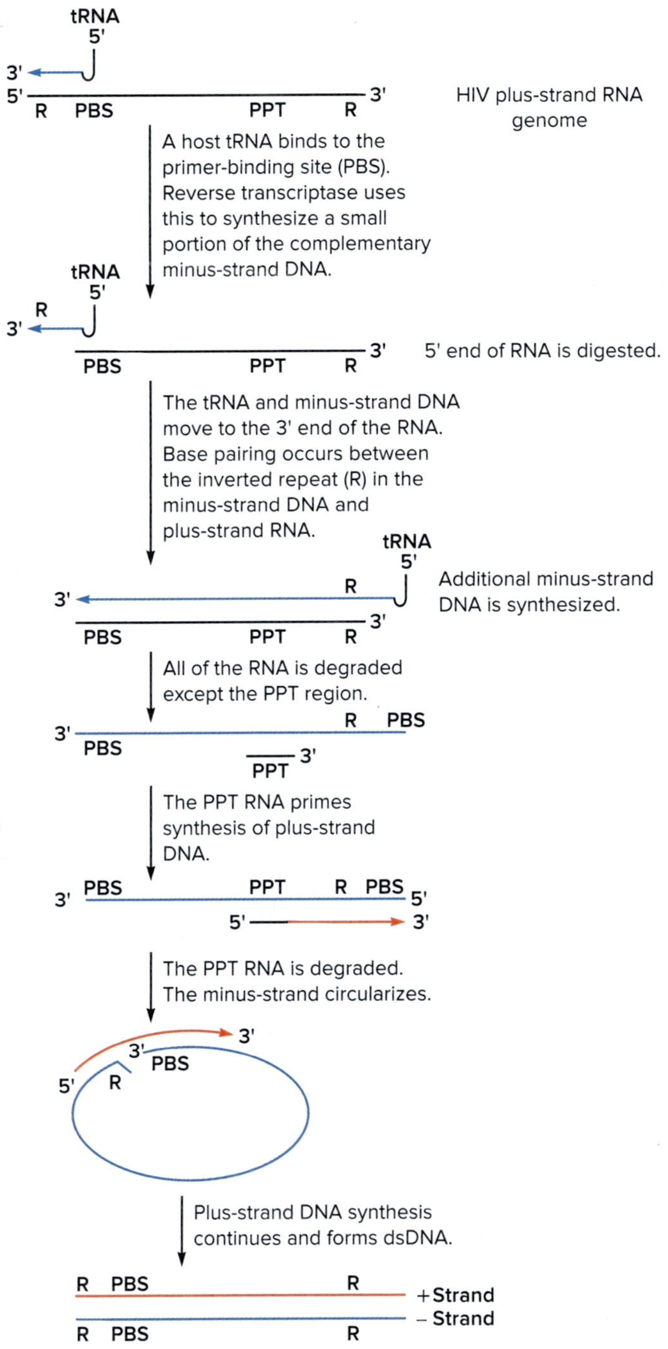

Figure 26.27 Reverse Transcription.

Comprehension Check

1. What is the function of each of the following HIV proteins: Gag, gp120, reverse transcriptase, and integrase?
2. CD4 is found on several different immune system cells. Consult chapter 33. Why is the presence of this molecule on the surface of many immune system cells important to the development of AIDS?
3. What role does alternative splicing play in the life cycle of HIV? For what purpose does the virus use ribosomal frameshifting and readthrough?

26.8 Reverse Transcribing DNA Viruses

After reading this section, you should be able to:

a. Describe in general terms the strategy used by reverse transcribing DNA viruses to synthesize their nucleic acids and proteins
b. Compare the role of reverse transcriptase in the life cycle of a retrovirus to that in the life cycle of a hepadnavirus

Reverse transcribing DNA viruses use reverse transcriptase (RT) to replicate their genomes. Members of four virus families fall into this category: *Caulimoviridae* (e.g., cauliflower mosaic virus and other plant viruses), *Metaviridae* (fungal, plant, and animal viruses), *Pseudoviridae* (fungal, plant, protist, and animal viruses), and *Hepadnaviridae* (e.g., hepatitis B virus and related animal viruses). Hepatitis B virus (HBV) is one of the best-studied reverse transcribing DNA viruses and is our focus here.

HBV virions are 42-nm spherical particles that contain the viral genome. The HBV genome is a circular dsDNA molecule that consists of one complete but nicked strand and a complementary strand that has a large gap; that is, it is incomplete (**figure 26.28**). The genome is 3.2 kb in length and consists of four partially overlapping, open reading frames that encode viral proteins. Production of new virions takes place predominantly in liver cells (hepatocytes).

After infecting a host cell, the viral-gapped DNA is released into the nucleus. There host repair enzymes fill the gap and seal the nick, yielding a covalently closed, circular DNA (cccDNA). Transcription of viral genes occurs in the nucleus using host RNA polymerase and yields several mRNAs, including a large 3.4-kb RNA known as the pregenome (a plus-strand RNA). The RNAs move to the cytoplasm, and the mRNAs are translated to produce

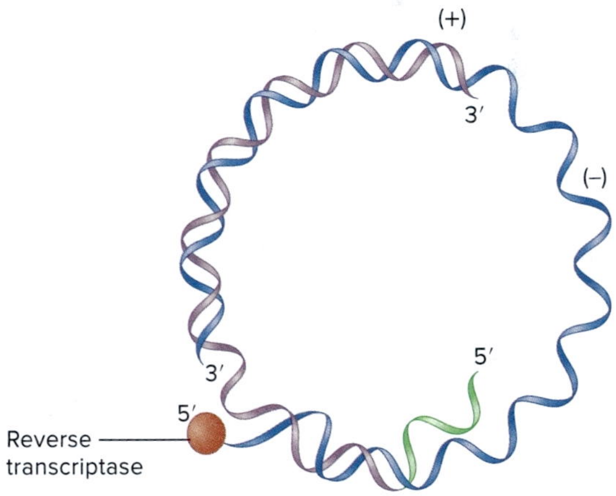

Figure 26.28 The Gapped Genome of Hepadnaviruses. The genomes of hepadnaviruses are unusual in several respects. The negative strand of the dsDNA molecule is complete but nicked (shown in blue). The enzyme reverse transcriptase is attached to its 5′ end. The positive strand is gapped; that is, it is incomplete (shown in purple). A short stretch of RNA is attached to its 5′ end (shown in green). Upon entry into the host nucleus, the nick in the negative strand is sealed and the gap in the positive strand is filled, yielding a covalently closed, circular dsDNA molecule.

viral proteins, including core protein and RT. RT then associates with the plus-strand RNA pregenome and core protein to form an immature core particle. RT subsequently reverse transcribes the RNA using a protein primer to form a minus-strand DNA from the pregenome RNA. After almost all the pregenome RNA has been degraded by the RNaseH activity of RT, the remaining RNA fragment serves as a primer for synthesis of the gapped dsDNA genome, using the minus-strand DNA as template. Finally, the nucleocapsid is completed and progeny virions are released. An important challenge to treating hepadnavirus infection is the persistence of cccDNA in the nucleus. Treatment with drugs that inhibit RT results in elimination of virions. However, when drug treatment is halted, viral mRNA can again be synthesized from cccDNA, resuming the viral life cycle.

Comprehension Check

1. Describe the HBV genome. How is it converted to covalently closed, circular DNA in the host?
2. Trace the HBV multiplication cycle, paying particular attention to localization within the host cell during the biosynthetic stage.

Key Concepts

26.1 Virus Phylogeny Is Difficult to Establish

- Currently viruses are classified with a taxonomic system placing primary emphasis on the type and strandedness of viral nucleic acids, and on the presence or absence of an envelope (**figure 26.1**).
- Increasingly, comparisons of viral genome sequences and the folding patterns of viral capsid proteins are being used to establish phylogenetic relationships among viruses.
- The Baltimore system is used by many virologists to organize viruses based on their genome type and the mechanisms used to synthesize mRNA and replicate their genomes (**table 26.1**).

26.2 Double-Stranded DNA Viruses Infect All Cell Types

- T4 is a virulent bacteriophage that causes lytic infections of *E. coli.* After attachment to a specific receptor site on the bacterial surface, T4 releases its dsDNA into the cell (**figures 26.4** and **26.5**). T4 DNA contains hydroxymethylcytosine (HMC) in place of cytosine, and glucose is often added to the HMC to protect the phage DNA from attack by host restriction enzymes (**figure 26.6**). T4 DNA replication produces concatemers, long strands of several genome copies linked together (**figure 26.7**).
- Lambda (λ) phage is a temperate bacteriophage. It can establish lysogeny, rather than pursuing a lytic infection. During lysogeny, the viral DNA, called a prophage, is replicated as the cell's genome is replicated. Lysogeny is reversible, and the prophage can be induced to become active again and lyse its host. This highly regulated process is an important model system for regulatory processes. The protein cII plays a central role in regulating the choice between lysogeny and a lytic cycle. If cII protein levels are high enough, lysogeny is established. If not, the lytic cycle is initiated (**figures 26.10** and **26.11**).
- Herpesviruses are a large group of dsDNA viruses. They cause acute infections such as cold sores, genital herpes, chickenpox, and mononucleosis. This is followed by a lifelong latent infection in which the virus genome resides within host cells. The virus can be reactivated at later times to cause a productive infection.
- Nucleocytoplasmic large DNA (NCLD) viruses, also called megaviruses, are members of several virus families. They are thought to share an evolutionary history and are notable for their large genome and virion size. Furthermore, their genomes contain genes for translation-related functions that are not usually observed in viral genomes.

26.3 Single-Stranded DNA Viruses Use a Double-Stranded Intermediate in Their Life Cycles

- ϕX174 is an example of a ssDNA bacteriophage. Its replication involves the formation of a dsDNA replicative form (RF) (**figure 26.14**).
- *Pseudomonas* Pf1 phage is filamentous phage that upon infection is continuously released by the host without causing lysis.
- Parvoviruses cause a spectrum of diseases in a wide variety of animals. The parvovirus genome is replicated by host DNA polymerase in the host cell's nucleus using a process that is similar to rolling-circle replication (**figure 26.15**).

26.4 Double-Stranded RNA Viruses: RNA-Dependent RNA Polymerase Replicates the Genome and Synthesizes mRNA

- Double-stranded RNA viruses use a viral enzyme called RNA-dependent RNA polymerase to synthesize mRNA (transcriptase activity) and replicate their genomes (replicase activity) (**figure 26.16**).
- Bacteriophage ϕ6 is an unusual phage in that it is enveloped and enters the host bacterium through a process that resembles endocytosis. It has a segmented genome, in

which each segment encodes a polycistronic mRNA that yields several proteins.

- Rotaviruses have segmented genomes, in which each segment codes for one or two proteins. They multiply in an inclusion called the viroplasm (**figure 26.17**).

26.5 Plus-Strand RNA Viruses: Genomes That Can Be Translated upon Entry

- The genomes of plus-strand RNA viruses serve directly as mRNA molecules. Among the first viral proteins synthesized is an RNA-dependent RNA polymerase that replicates the plus-strand RNA genome, sometimes by forming a dsRNA replicative intermediate. The negative RNA strands produced by the RNA-dependent RNA polymerase are used to make more genomes and mRNA (**figure 26.18**).
- Poliovirus genomic RNA is translated into a single polyprotein that is cleaved to form all the proteins needed by the virus during its life cycle (**figures 26.20** and **26.21**).
- TMV is like most known plant viruses in that it has a plus-strand RNA genome. Translation of TMV genomic RNA yields two proteins, the larger of which is produced by a readthrough mechanism. The RNA-dependent RNA polymerase encoded by the genome synthesizes minus-strand RNAs that serve as templates for synthesis of subgenomic mRNAs that are translated to yield other proteins needed by the virus. The minus-strand RNAs also serve as templates for synthesis of new plus-strand RNA genomes. The TMV nucleocapsid forms spontaneously when disks of coat protein protomers complex with the RNA (**figure 26.22**).

26.6 Minus-Strand RNA Viruses: RNA-Dependent RNA Polymerase Is Part of the Virion

- For minus-strand RNA viruses to synthesize mRNA and replicate their genomes, their virions must carry an RNA-dependent RNA polymerase, which enters the host cells as the genome does. The polymerase first synthesizes mRNA from the negative-strand genome. Later it is used to replicate the genome by way of a plus-strand intermediate (**figure 26.23**).
- Influenza viruses have segmented genomes; most segments encode a single protein. The viruses exit by budding (**figure 26.24**).

26.7 Retroviruses: Plus-Strand Viruses That Use Reverse Transcriptase in Their Life Cycles

- Retroviruses replicate their genomes and synthesize mRNA via a dsDNA intermediate. The dsDNA is formed by a multifunctional enzyme called reverse transcriptase (**figure 26.25**).
- HIV is an enveloped retrovirus with a cone-shaped core that contains two copies of its genome and several enzymes, including reverse transcriptase. Upon infection, its ssRNA genome is converted to dsDNA, which is then integrated into the host genome (**figure 26.26**). HIV uses alternative splicing to generate numerous different mRNA molecules. One encodes two polyproteins. One of the polyproteins arises by a combined ribosomal frameshifting, readthrough mechanism.

26.8 Reverse Transcribing DNA Viruses

- There are four groups of DNA viruses that use reverse transcriptase in their life cycles. One group, hepadnaviruses, includes hepatitis B virus, which has a dsDNA genome that consist of one complete but nicked strand and an incomplete (i.e., gapped) complementary strand (**figure 26.28**).
- Upon infection with hepatitis B virus, the host cell repairs the gap and seals the nick to generate a covalently closed, circular viral genome. This serves as the template for the synthesis of pregenome RNA, which is the template for reverse transcription. Reverse transcription produces the dsDNA, gapped genome.

Active Learning

1. No temperate RNA phages have yet been discovered. How might this be explained?
2. The choice between lysogeny and lysis is influenced by many factors. How would external conditions such as starvation or crowding be "sensed" and communicated to the transcriptional machinery to influence this choice?
3. The most straightforward explanation as to why the endolysin (i.e., the T4 lysozyme) of T4 is expressed so late in infection is that its promoter is recognized by the gp55 alternative sigma factor. Propose a different explanation.
4. You are studying RNA viruses and have discovered a new one that grows well in a culture of eukaryotic cells. You know that the virus is a single-stranded RNA virus, but you don't know if it is plus or minus stranded. Your lab-mate says, "Well, just treat your cell culture with cyclohexamide and see if the virus replicates its genome." You know that cyclohexamide inhibits protein elongation by binding to eukaryotic ribosomes. What is the basis of your lab-mate's suggestion?
5. Although they do not contain ribosomes, and therefore genes to encode small subunit rRNA, nucleocytoplasmic large DNA (NCLD) viruses do contain conserved genes for functions related to translation. When phylogenetic trees are created using these conserved genes, the NCLD viruses group as a fourth domain, beside Bacteria,

Archaea, and Eukaryotes. Defend or refute the creation of a fourth domain to classify these viruses.

6. Coronaviruses (CoV) that cause Middle East respiratory syndrome (MERS) are plus-strand RNA viruses. Upon infecting cells, they remodel intracellular membranes by membrane pairing, curvature, and fission to form double-membrane vesicles (DMVs). DMV formation is induced by viral proteins, and DMVs develop into replication organelles (ROs), a process observed by microscopy. The MERS-CoV genome encodes a single polyprotein that is proteolyzed to generate multiple functional polypeptides. Three peptides (nsp3, nsp4, and nsp6), predicted to have transmembrane domains, were identified as candidates for inducing DMV formation. Researchers inserted plasmids with various gene combinations into host cells and looked for DMV development. These cellular constructs and the results are summarized in the accompanying table. What sequence of events induces DMV formation?

   Read the original paper: Oudshoorn, D., et al. 2017. Expression and cleavage of Middle East Respiratory Syndrome Coronavirus polyprotein induce the formation of double-membrane vesicles that mimic those associated with coronaviral RNA replication. *mBio.* 8:e01658-17.

Components of the Polyprotein Introduced to Host Cells	Polypeptides Present	Observations of DMV Formation
nsp3	nsp3	no DMVs
nsp4	nsp4	no DMVs
nsp3 + nsp4 mixed	nsp3 + nsp4	DMVs present
nsp3-nsp4 polyprotein	nsp3 + nsp4	DMVs present
nsp3-nsp6 polyprotein	nsp3 + nsp456 (incomplete proteolysis)	aberrant membranes
nsp3-nsp5-GFP-nsp6 polyprotein	nsp3 + nsp4 + nsp5 + nsp6 + GFP	DMVs present
nsp3-nsp4 polyprotein with altered peptide cleavage site	nsp3-nsp4	paired membranes but no curvature
nsp3-nsp4 polyprotein with inactive protease domain	nsp3-nsp4	paired membranes but no curvature

27

Microbial Interactions

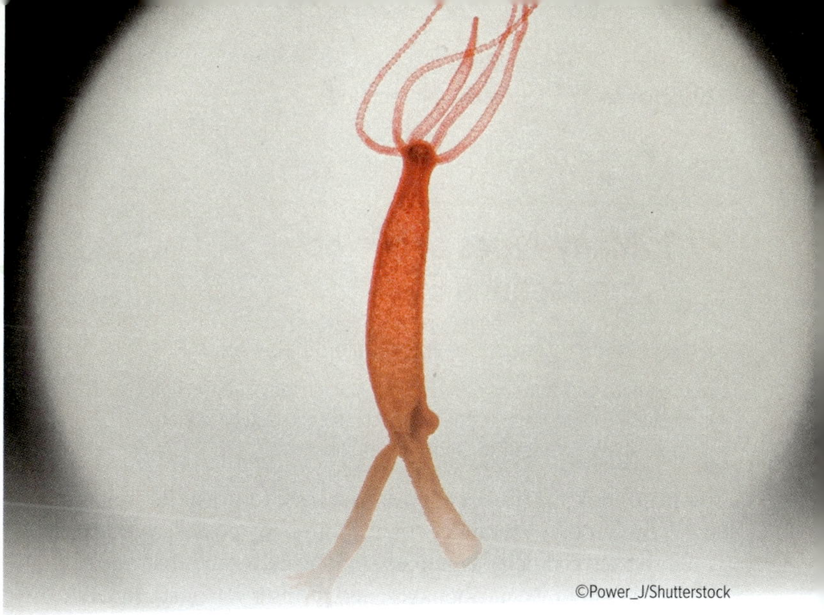

©Power_J/Shutterstock

Microbes in Community

It has been suggested that the only pure cultures of microorganisms are found in Petri dishes. Microbes elsewhere are in constant contact with organisms in other species, other genera, and other domains. Animals of all types have an external epithelial tissue, including that in their gut, genitourinary, and respiratory tracts, colonized by microbes specifically adapted to these habitats. These interactions are established and maintained as an ecosystem, and an animal and its microbiota have been termed metaorganism, holobiont, and even superorganism.

Hydra, as shown, is one of the simplest animals, with only a few differentiated tissues, but like other animals, its associated microbiota changes as the animal grows and matures. This makes it a good model system to study microbiome development and host-microbe interactions. *Hydra* embryogenesis proceeds through specific developmental stages. It produces distinct antimicrobial peptides (AMPs) at each stage, and these regulate the composition of its microbiota until a mature body plan and a stable microbiome are established in the adult. Gram-negative β-proteobacteria in the genus *Curvibacter* predominate in the adult *Hydra,* replacing the Gram-positive bacteria associated with the embryo. In a remarkable demonstration of molecular multitasking, the *Hydra* neuropeptide NDA-1 both promotes development of its nervous system and acts as an AMP that governs spatial structuring of the microbial community. *Curvibacter* is absent from the animal's glycocalyx (the dense carbohydrate-rich external surface layer) in the regions where NDA-1 is present, and in vitro is highly susceptible to the antibiotic activity of the purified NDA-1 protein.

Like a variety of other microbe-host interactions, many microbes within *Hydra* communicate with each other and with the host by secreting N-acyl homoserine lactone (AHL) signaling molecules. Interestingly, *Hydra* produces an oxidoreductase that specifically alters AHL structure. Therefore, in addition to controlling its microbiome composition, *Hydra* also controls microbial gene activity and the phenotype of certain associated microbiota. For example, *Curvibacter* gene expression varies dramatically in the presence of unmodified AHL and *Hydra*-modified AHL. While quorum sensing via AHLs promotes colonization of the *Hydra,* alternate gene expression induced by modified AHLs may do the opposite.

These examples of chemical communication illustrate just two aspects of the *Hydra-Curvibacter* interaction. The complexity of microbiome development rivals the complexity of embryogenesis, and it can be argued, is every bit as crucial to a successful outcome.

Microbes rarely live alone. They have evolved in specific ways with other organisms, including other microbes, plants, and animals. In this chapter we define types of microbial interactions and present illustrative examples.

Readiness Check:

Based on what you have learned previously, you should be able to:

- ✓ List environmental factors that influence microbial growth (section 7.5)
- ✓ Compare and contrast the growth and physiology of obligate aerobes, facultative anaerobes, and strict anaerobes (section 7.5)
- ✓ Explain the general morphology of protists and the structure and function of hydrogenosomes (sections 5.6 and 24.2)
- ✓ Summarize the mechanisms of horizontal gene transfer (section 16.4)
- ✓ Diagram, in general terms, the standard reduction potentials of electron donors as compared to electron acceptors and relate this to standard free energy change (section 10.3)
- ✓ Describe the difference between heterotrophy and autotrophy (section 11.1)
- ✓ Compare anaerobic and aerobic respiration (section 11.2)
- ✓ Define chemolithotrophy and identify an example electron donor and acceptor (section 11.10)
- ✓ Compare and contrast oxygenic and anoxygenic phototrophy (section 11.12)
- ✓ Explain carbon and nitrogen fixation (sections 12.3 and 12.5)
- ✓ Identify substrates used in methanogenesis (section 20.3)
- ✓ Relate genome reduction to obligate intracellular lifestyles (section 18.7)
- ✓ Describe the principles and application of metagenomics (sections 18.3 and 29.3)

27.1 Many Types of Microbial Interactions Exist

After reading this section, you should be able to:

a. Compare and contrast different types of microbial interactions, including mutualism, cooperation, commensalism, predation, parasitism, amensalism, and competition

You need look no further than in a mirror to grasp the importance of microbial-host interactions. Your body possesses at least as many bacterial cells as it does somatic cells (and that doesn't include the fungi, bacteriophages, and protists you host). Indeed, the vast majority of microbes live in complex communities that include organisms from other domains of life. The stable association of one organism with another is called **symbiosis,** regardless of whether the association is beneficial for both partners, as in mutualism, cooperation, and commensalism, or has a negative impact, described as predation, parasitism, amensalism, and competition (**figure 27.1**). These terms will be defined and exemplified in the following sections.

For many years bacterial symbioses were viewed as the relationship between one bacterial species and one host species. However, the wide application of metagenomics, single-cell sequencing, next-generation sequencing, and new approaches to culturing microbes have revealed that microbial symbioses can range from a single species of microorganism and its host to hosts with thousands of different kinds of microbes. In fact, there have been so many discoveries of new, and never imagined, symbiotic relationships that the vocabulary of microbial interactions is changing. For example, while the smaller organism in a symbiotic relationship has always been called the **symbiont** and the larger organism the host, these organisms are now sometimes regarded as partners. What was once considered a microbe-host relationship is now sometimes called a metaorganism, holobiont, or superorganism to reflect the critical nature of the interaction.

All symbionts must undergo the process of colonization, reproduction, and persistence within the host, as well as transmission to a new host. Because microbial symbioses are the result of thousands, if not millions, of years of coevolution, these processes have been optimized. However, their specifics differ and make each symbiotic relationship seem like a clever feat of natural history. For example, host specificity can range from very narrow to very broad. Symbionts may be transmitted horizontally from host to host or vertically during reproduction, or each host may have to acquire the symbiont from the environment. Symbioses can range from those that are very stable (e.g., aphids and their bacterial mutualists, as discussed next), to interactions that are temporary, depending on the life cycle of the host or other variables. Symbionts can live within the host as an **endosymbiont,** or on the surface of the host as an **ectosymbiont.** Finally, symbionts alter not only the health of their hosts, but their behavior, evolution, and reproductive success as well (**Microbial Diversity & Ecology 27.1**). ◀ *Metagenomics provides access to uncultured microbes (section 18.3)* ▶ *Microbial biology relies on cultures (section 29.1)*

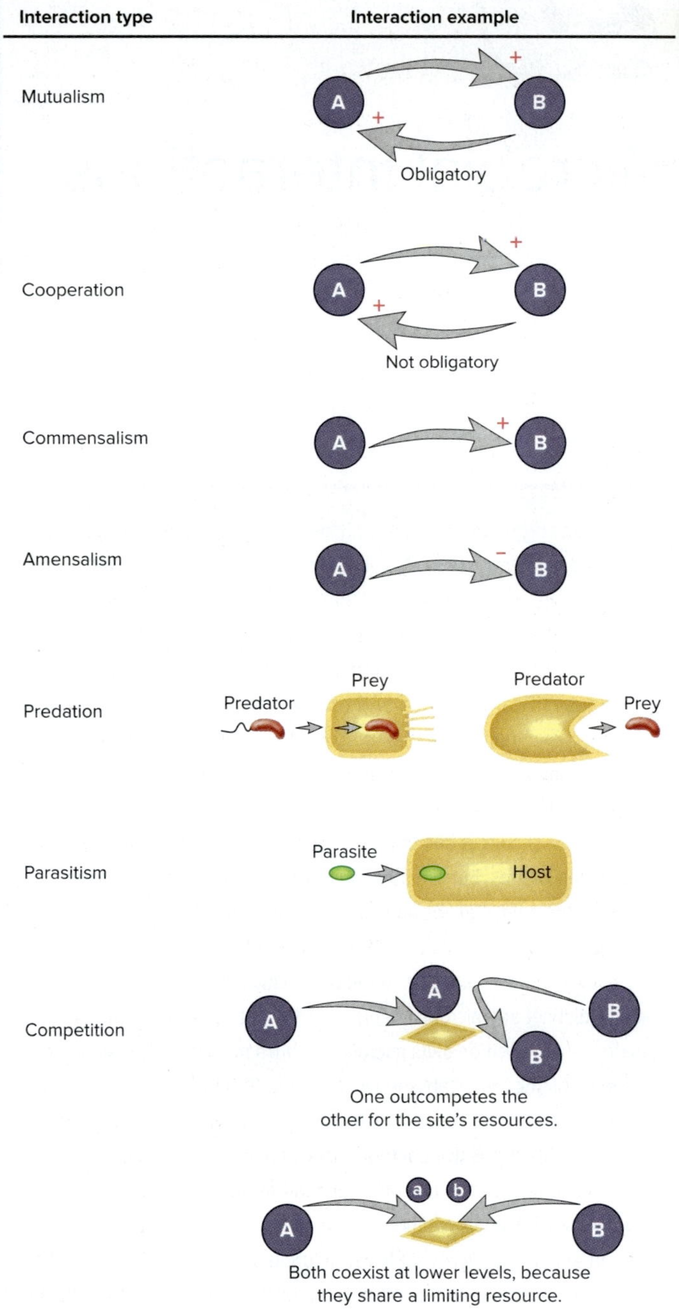

Figure 27.1 Microbial Interactions. Basic characteristics of symbiotic interactions that can occur between different organisms.

27.2 Mutualism and Cooperation Are Two-Way Interactions

After reading this section, you should be able to:

a. Distinguish mutualism and cooperation
b. Discuss the metabolic relationships between symbiotic partners
c. Relate coral bleaching to a symbiotic relationship
d. Describe the relationship between microbes and complex organisms, as illustrated in the tube worm and the mammalian rumen

MICROBIAL DIVERSITY & ECOLOGY

27.1 *Wolbachia pipientis:* The World's Most Infectious Microbe?

Most people have never heard of the bacterium *Wolbachia pipientis,* but this rickettsia infects more organisms than any other microbe. It is known to infect a broad range of crustaceans, spiders, mosquitoes, millipedes, and nematodes, and may infect more than 2 million insect species worldwide. To what does *W. pipientis* owe its extraordinary success? Quite simply, this endosymbiont is a master at manipulating its hosts' reproductive biology.

W. pipientis inhabits the cytoplasm of its host's cells and is transferred from one generation to the next through the eggs of infected females. To survive, *Wolbachia* must ensure the fertilization and viability of infected eggs while decreasing the likelihood that uninfected eggs survive. The mechanism by which this is accomplished depends on the host. In wasps and mosquitoes, *W. pipientis* causes cytoplasmic incompatibility, which means that embryonic development will be abnormal if only the male is infected. For instance, when infected sperm of the wasp *Nasonia vitripennis* fertilizes uninfected eggs, chromosomes from the *W. pipientis*–laden sperm prematurely try to align with the egg's chromosomes. These eggs then divide as if never fertilized. However, chromosomes behave normally when an infected female mates with an uninfected male. This yields a normal sex distribution, and all progeny are infected with the rickettsia.

In other infected insects, *W. pipientis* may simply kill all the male offspring and induce parthenogenesis in infected females; that is, the females simply clone themselves. This limits genetic diversity but allows 100% transmission of rickettsias to the next generation. In still other hosts, the microbe modifies male hormones so that the males become feminized and produce eggs.

Another effect of *Wolbachia* infection is interference with viral replication. Viruses normally spread by insects may not be transmissible if the insect is infected with *Wolbachia*. This has led to the notion that this infection could be used as a means of biological control. *Aedes* mosquitoes transmit numerous viruses, like Zika, dengue, chikungunya, and West Nile. In humans, these viral infections have no cure and no treatment beyond supportive care, so insect control is the best prevention.

However, the mosquito vectors of these viruses, which belong the genus *Aedes* (**box figure**), are not natural hosts to *Wolbachia*. Nonetheless, infection can be established by transinfection, the process of transferring the microbes from another insect species. Many *Wolbachia* isolates have been screened for those that have the most severe effects on *Aedes,* and several stable insect strains have been established with *Wolbachia* from *Drosophila*. Cytoplasmic interference is extensive in these strains of *Wolbachia*. To establish a stable population of *Aedes* outside the laboratory, it is important that the infection not impose a dramatic burden on the insect, as it must compete with and integrate into wild populations.

In a promising development, *Wolbachia*-infected *Aedes* mosquitoes are effective at blocking the transmission of both Zika and dengue viruses. Experiments in Australia have confirmed the ability of these insects to persist and spread in local mosquito populations, validating this approach to insect vector control.

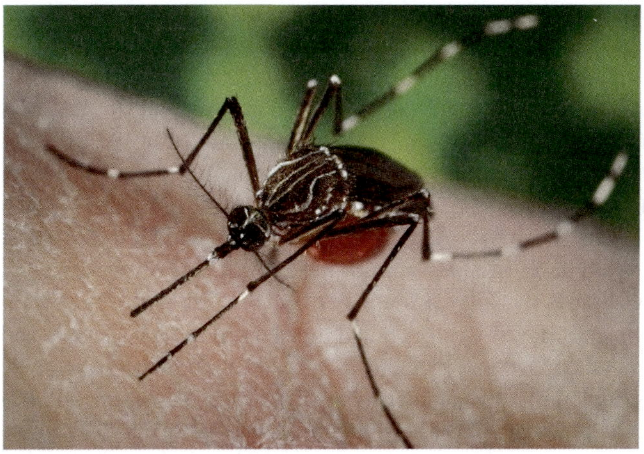

Female *Aedes aegypti* mosquito. Infection with *Wolbachia* may render these insects incapable of spreading arboviruses to humans.
Source: CDC/James Gathany

For most microbial ecologists, the nonobligatory aspect between host and symbiont differentiates **cooperation** from **mutualism** (Latin *mutuus,* borrowed or reciprocal) (figure 27.1). Unfortunately, it is often difficult to distinguish obligatory from nonobligatory because that which is obligatory in one habitat may not be in another (e.g., the laboratory). Nonetheless, the most useful distinction between cooperation and mutualism is the observation that cooperating organisms can grow independently, although they may not function as well.

Mutualism defines the relationship in which some reciprocal benefit accrues to both partners. This is an obligatory relationship in which the **mutualist** and the host are dependent on each other. In many cases, the individual organisms will not survive when separated. Several examples of mutualism are presented next.

Microorganism-Insect Mutualisms

Mutualistic associations are common between microbes and insects partly because insects often consume plant sap or animal fluids lacking essential vitamins and amino acids. These are provided by bacterial symbionts in exchange for a secure habitat and ample nutrients. Aphids and their γ-proteobacterium *Buchnera aphidicola* partner are a model system for the study of mutualistic symbioses. A mature aphid contains millions of these bacteria within specialized cells called **bacteriocytes.** Strains of *B. aphidicola* provide their hosts with 10 essential amino acids absent in sap. If the insect is treated with antibiotics, it dies. Likewise, *B. aphidicola* is an obligate mutualistic symbiont. The inability of either partner to grow without the other indicates that the two organisms underwent **coevolution;** that is, they have evolved together. In fact, the pea aphid *Acyrthosiphon pisum* and its *B. aphidicola* symbionts share the pathway for the biosynthesis of several amino acids such that certain steps occur only in one organism.

It is estimated that the *B. aphidicola*–aphid endosymbiosis was established about 150 million years ago. The genomes of different endosymbiotic *B. aphidicola* strains reveal that about 75% of the ancestral *B. aphidicola* genome has been eliminated. It follows that these genomes are very small: in most cases, only about 0.65 Mb each. Over 90% of the genes are common among these bacterial strains, and the genomes show no evidence of gene duplication, translocation, inversion, or horizontal transfer. This implies that although the initial acquisition of the endosymbiont by ancestral aphids enabled their use of an otherwise deficient food source (sap), the bacteria have not continued to expand the ecological niche of their insect host through the acquisition of new traits that might be advantageous elsewhere. ◀ *Comparative genomics (section 18.7)*

Termites are another good model system for the study of endosymbiosis. Some termites harbor protist, bacterial, and archaeal endosymbionts and eat only wood. The main structural polysaccharides of wood are cellulose (an unbranched glucose polymer) and hemicellulose (a smaller, branched glucose polymer), which combine with lignin *(see figure 31.2)* to form lignocellulose. A diet of wood poses two problems for the termite: how to degrade polysaccharides that may possess as many as 15,000 glucose monomers and where to get organic nitrogen needed for nucleotide and protein synthesis. Although these termites produce a cellulolytic enzyme, only their mutualistic protists can complete lignocellulose degradation. Nitrogen-fixing bacteria that live in the termite gut solve the problem of obtaining organic nitrogen. ◀ *Nitrogen fixation (section 12.5)*

Most termite-associated protists are members of the *Excavata* supergroup and as such are quite ancient, possessing hydrogenosomes rather than mitochondria. These protists ferment cellulose to acetate, CO_2, and H_2. Acetate is the termite's preferred carbon source, and termites rely on bacterial symbionts to convert the CO_2 and H_2 released by the protists to acetate by the reductive acetyl-CoA pathway (*see figure 20.2*). In addition, termites harbor methanogenic archaea that use these substrates to form CH_4, which is excreted. ◀ *Mitochondria related organelles and chloroplasts are involved in energy conservation (section 5.6)*; *Supergroup* Excavata *(section 25.2)*

Some termite protists also harbor endosymbionts (i.e., endosymbionts of endosymbionts). For instance, the protist *Trichonympha* sp. relies on a bacterial endosymbiont in the genus *Elusimicrobium* to convert glutamine to other amino acids and nitrogenous compounds (**figure 27.2**). In exchange, the protist supplies *Elusimicrobium* sp. with glucose 6-phosphate, which can directly enter glycolysis. In addition, motility is often conferred by spirochetes that cover the protist surface. Motility is essential to prevent protist expulsion by the termite gut and to acquire food.

Zooxanthellae

Many marine invertebrates (sponges, jellyfish, sea anemones, corals, ciliated protists) harbor endosymbiotic dinoflagellates called zooxanthellae within their tissues. Because the degree of host dependency on the dinoflagellate is variable, only one well-known example is presented here. ◀ *Alveolata (section 24.4)*

The hermatypic (reef-building) corals satisfy most of their energy requirements using their zooxanthellae, which include members of the dinoflagellate genus *Symbiodinium* (**figure 27.3*a***). These protists line the coral gastrodermal tissue at densities between 5×10^5 and 5×10^6 cells per square centimeter of coral animal. *Symbiodinium* cells give up to 95% of the carbon they fix in exchange for nitrogenous compounds, phosphates, CO_2, and protection from ultraviolet light from their coral hosts. In addition to dinoflagellates, corals host distinct bacterial communities in three habitats (figure 27.3*b*). The surface mucus layer supports the growth of nitrogen-fixing and chitin-degrading bacteria, while the animal tissue within the gastrodermal cavity features another population of bacteria (although little is known about these microbes). The coral skeleton hosts yet another group of bacteria. These microbial communities are responsible for extremely efficient nutrient cycling and tight coupling of trophic levels, which accounts for the stunning success of reef-building corals in developing vibrant ecosystems.

During the past several decades, the number of coral bleaching events has increased dramatically. **Coral bleaching** is defined as a loss of either the photosynthetic pigments from the zooxanthellae or the complete expulsion of the dinoflagellates from the coral (figure 27.3*c*). Damage to photosystem II of the zooxanthellae generates reactive oxygen species (ROS); these ROS appear to directly damage many corals (recall that photosystem II uses water as the electron source, resulting in the evolution of oxygen). Coral bleaching appears to be caused by a variety of additional stressors, but one important factor is temperature. Temperature increases as small as 2°C above the average summer maxima can trigger coral bleaching. Increased temperature has also been correlated with coral infectious diseases. Some disease-causing microbes are thought to reside in the coral mucus layer, where their numbers are normally kept in check. But during and following bleaching, the coral microbial community is in disequilibrium, and stressed corals become more susceptible to pathogenic attack. Conditions such as white plague and yellow blotch disease have been noted to follow bleaching events in the Caribbean, reducing corals by roughly a third. Finally, ocean acidification caused by increased CO_2 dissolution in

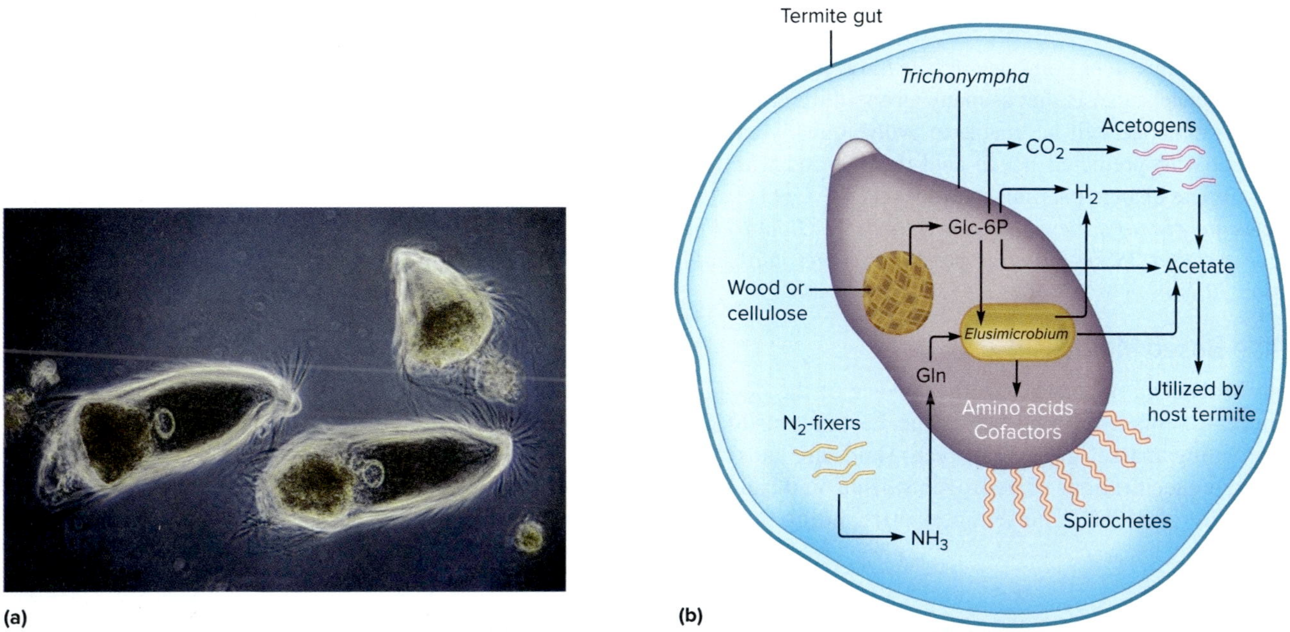

Figure 27.2 Mutualism. (a) Light micrograph of *Trichonympha* sp., a multiflagellated protozoan from the termite's gut (×135). The ability of *Trichonympha* to break down cellulose allows termites to use wood as a food source. (b) Interaction between the termite endosymbiotic protist *Trichonympha* and its bacterial endosymbiont *Elusimicrobium* sp. Within the termite gut, bacteria convert N_2 to ammonia, which is taken up and converted to glutamine (Gln) by *Trichonympha* sp. The protist relies on *Elusimicrobium* sp. to convert Gln to other amino acids, which are released with other cofactors that the protist cannot synthesize. Both the protist and *Elusimicrobium* sp. ferment glucose to acetate, H_2, and CO_2. These products are exported by the protist and made available to the termite, which also relies on free-swimming gut bacteria to boost acetate levels through the reductive acetyl-CoA pathway (*see figure 20.2*).

MICRO INQUIRY *What methanogenic substrates are available in the termite gut?*

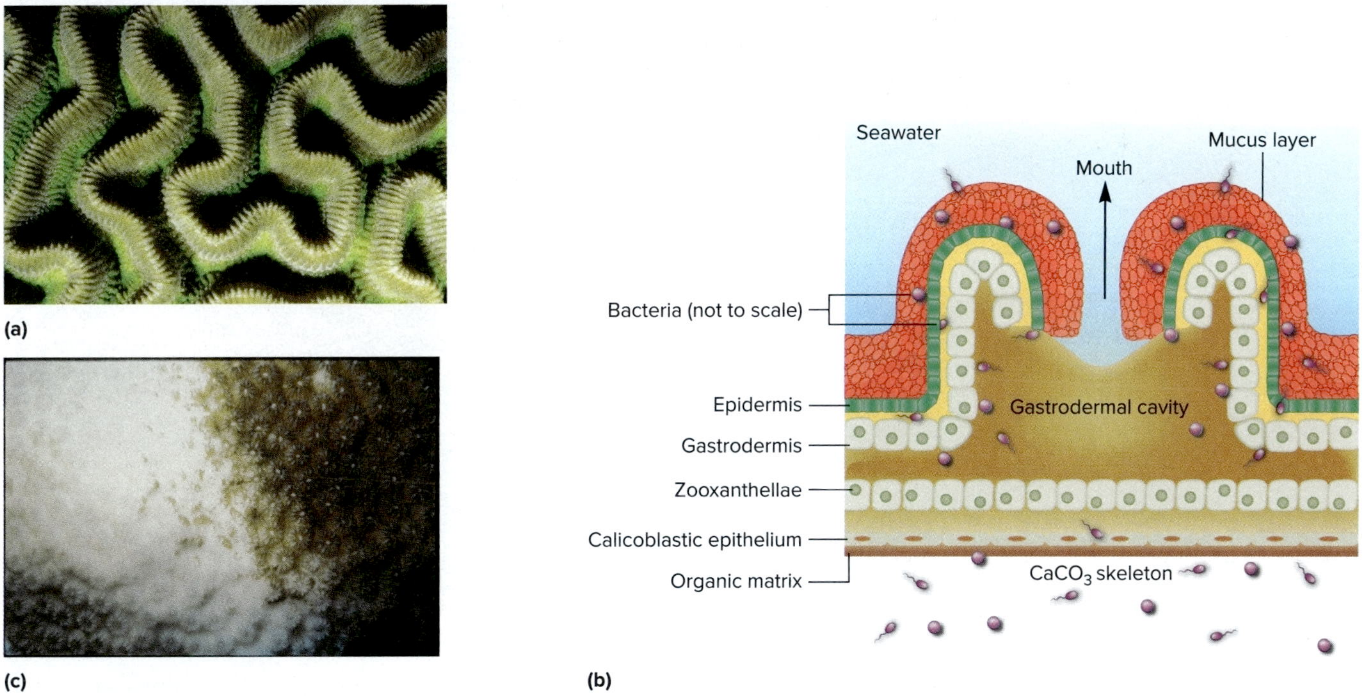

Figure 27.3 Zooxanthellae. (a) The green color of this rose coral (*Manicina* sp.) is due to the abundant zooxanthellae within its tissues. (b) Corals provide three distinct habitats for bacterial growth: the mucus layer at the surface, the gastrodermal cavity, and the skeleton. Calicoblastic epithelial cells precipitate calcium carbonate ($CaCO_3$) from the water to build the exoskeleton. Zooxanthellae inhabit gastrodermal cells and are shown in green. (c) Coral bleaching in star coral in the Caribbean. Golden-brown color indicates healthy coral tissue; white (bleached) area has lost zooxanthellae.

seawater also threatens coral health (*see figures 30.1 and 30.2*). Sadly, evidence suggests that as many as one-third of corals and their zooxanthellae will be unable to evolve quickly enough to keep pace with predicted ocean acidification and warming if global climate change continues unchecked. ◀ *Light reactions in oxygenic photosynthesis (section 11.12)* ▶ *Global climate change (section 28.2); Water is the largest microbial habitat (section 30.1)*

Sulfide-Based Mutualism

Several thousand meters below the surface of the ocean, hydrothermal vents are found where the Earth's crustal plates are spreading apart (**figure 27.4**). Vent fluids are anoxic, contain high concentrations of hydrogen sulfide, and can reach a temperature of 350°C. The seawater surrounding these vents has sulfide concentrations around 250 μM and temperatures 10° to 20°C above the ambient seawater temperature of about 2°C. Although the high pressure at such depths prevents the water from boiling, one might think these conditions are inhospitable to life. But nothing could be further from the truth.

Giant (>1 m in length), red, gutless tube worms (*Riftia* spp., **figure 27.5*a***) growing near hydrothermal vents exemplify a remarkably successful mutualism in which bacterial endosymbionts are maintained within specialized cells of the tube worm host (figure 27.5*b,c,d*). *Riftia* worms live at the interface between hot, anoxic, sulfide-containing fluids of the vents and cold, oxygenated seawater. The worm lacks a mouth and gut and relies on endosymbiotic bacteria to provide organic carbon. The bacteria are chemolithotrophic, using sulfide as an electron donor and

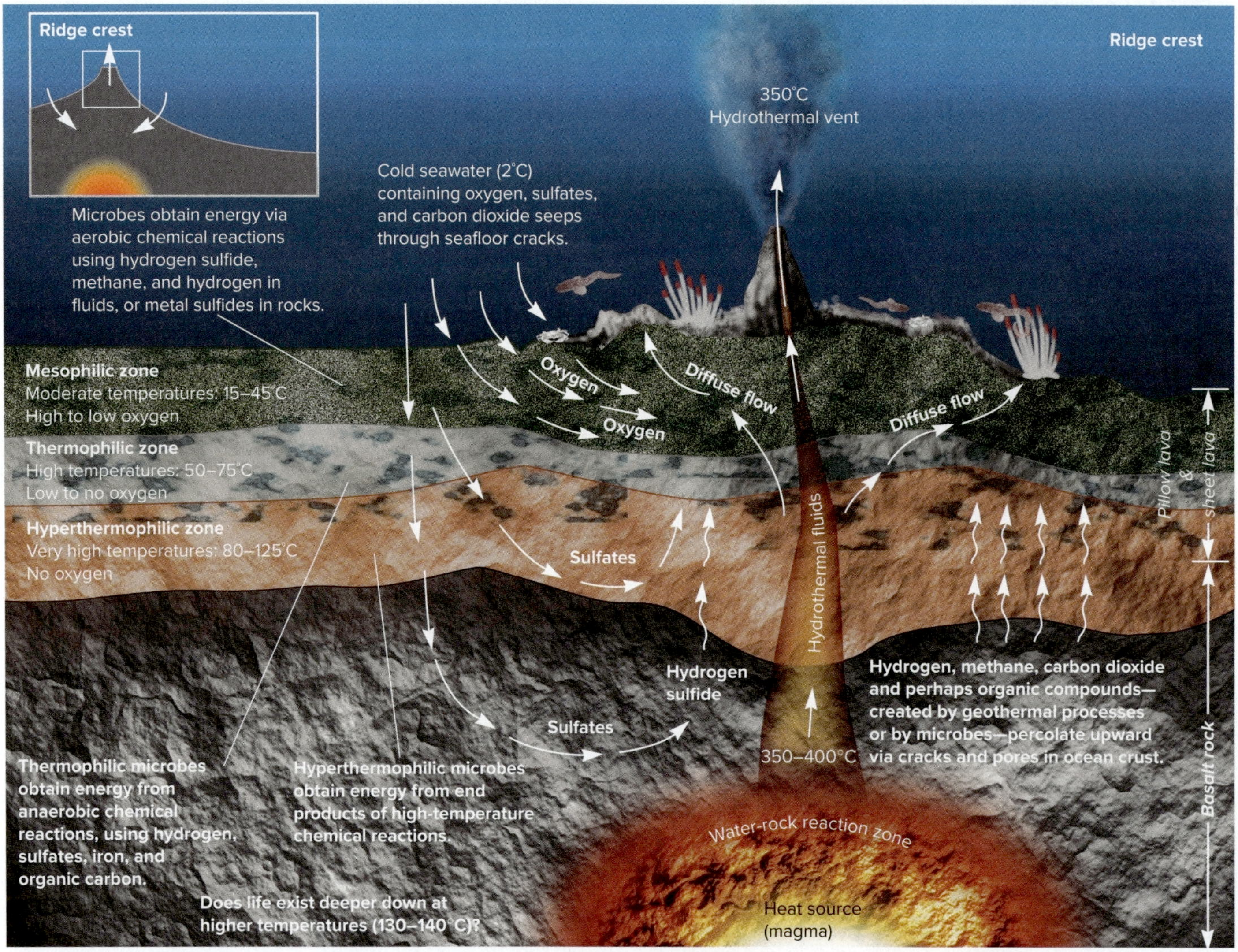

Figure 27.4 Hydrothermal Vents and Related Geological Activity. The chemical reactions between seawater and rocks that occur over a range of temperatures on the seafloor supply the carbon and energy that support a diverse collection of microbial communities in specific niches within the vent system.

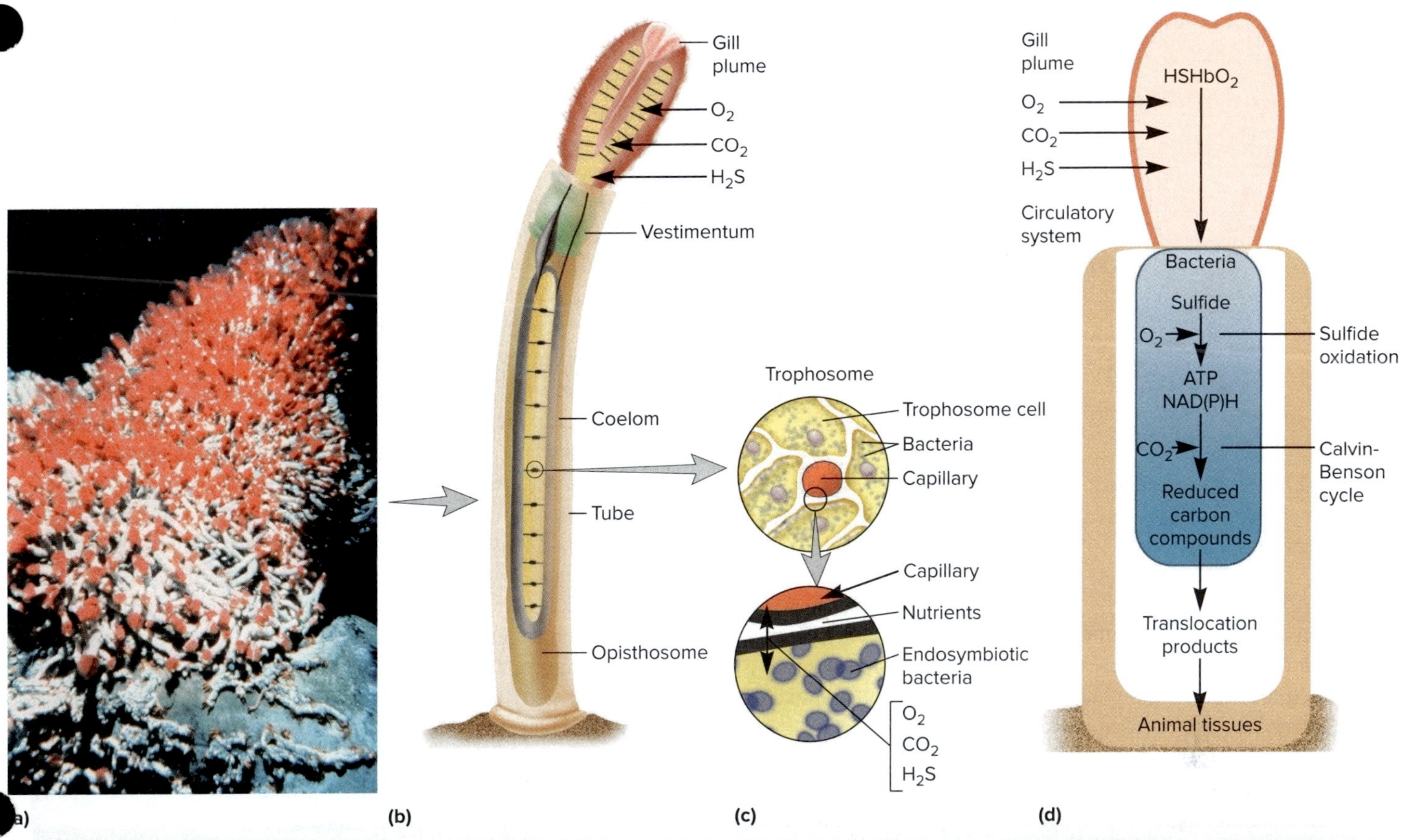

Figure 27.5 The Tube Worm–Bacterial Relationship. (a) A community of tube worms (*Riftia pachyptila*) at the Galápagos Rift hydrothermal vent site (depth 2,550 m). Each worm is more than a meter in length and has a 20-cm gill plume. (b, c) Schematic illustration of the anatomical and physiological organization of the adult tube worm. The animal's heart and brain is housed within the vestimentum. At its anterior end is a respiratory gill plume. Inside the trunk of the worm is a trophosome consisting primarily of endosymbiotic bacteria, associated cells, and blood vessels. At the posterior end of the animal is the opisthosome, which anchors the worm in its tube. (d) Oxygen, carbon dioxide, and hydrogen sulfide are absorbed through the gill plume and transported to the blood cells of the trophosome. Hydrogen sulfide is bound to the worm's hemoglobin ($HSHbO_2$) and carried to the endosymbiont bacteria. The bacteria oxidize the hydrogen sulfide and use some of the released energy to fix CO_2 using the Calvin-Benson cycle. Some of the reduced carbon compounds synthesized by the endosymbiont are translocated to the animal's tissues.
(a) Source: OAR/National Undersea Research Program (NURP)/College of William & Mary/NOAA

MICRO INQUIRY *What is the nutritional type of the endosymbiotic bacteria?*

oxygen as a terminal electron acceptor. In fact, *Riftia* worms have unique hemoglobin that provides both reduced sulfur and oxygen to the endosymbionts. Hydrogen sulfide (H_2S) and O_2 are removed from the seawater by the worm's hemoglobin and delivered to an organ called the trophosome. The trophosome is packed with bacteriocytes, specialized cells that contain *Candidatus* Endoriftia persephone. These chemolithotrophic endosymbionts reach densities of up to 10^{11} cells per gram of worm tissue. The endosymbionts use the Calvin-Benson cycle to fix CO_2 (*see figure 12.4*), which they reduce with electrons provided by H_2S. The CO_2 is carried to the endosymbionts in three ways: (1) freely in the bloodstream, (2) bound to hemoglobin, and (3) as organic acids such as malate and succinate. When these acids are decarboxylated, they release CO_2. This CO_2 is then fixed using the same pathway used by plants and cyanobacteria, but it occurs in the deepest, darkest reaches of the ocean. This mutualism enables *Riftia* worms to grow to an astounding size in densely packed communities.

The Rumen Ecosystem

Ruminants are the most successful and diverse group of mammals on Earth today. Examples include cattle, deer, elk, bison, water buffalo, camels, sheep, goats, giraffes, and caribou. These animals spend vast amounts of time chewing their cud—a small ball of partially digested grasses that the animal has consumed but not yet completely digested. It is thought that ruminants evolved an "eat now, digest later" strategy because their grazing can often be interrupted by predator attacks.

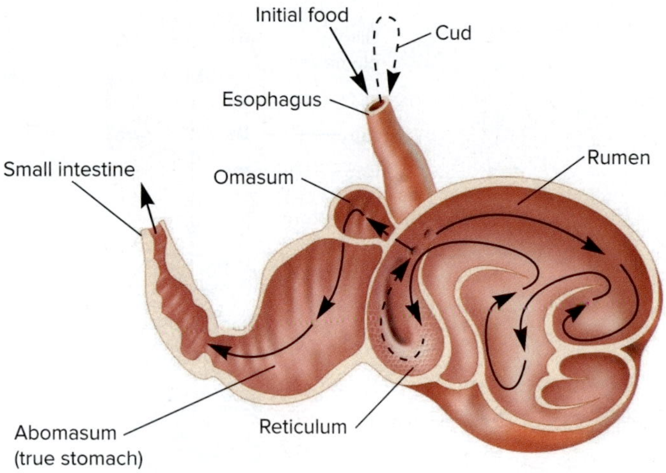

Figure 27.6 Ruminant Stomach. The stomach compartments of a cow. The microorganisms are active mainly in the rumen. Arrows indicate direction of food movement.

These herbivorous animals have stomachs that are divided into four chambers (**figure 27.6**). The upper part of the stomach is expanded to form a large pouch called the rumen and a smaller, honeycomb-like region, the reticulum. The lower portion is divided into an antechamber, the omasum, followed by the "true" stomach, the abomasum. The rumen is a highly muscular, anaerobic fermentation chamber where grasses eaten by the animal are digested by a diverse microbial community that includes bacteria, archaea, fungi, and protists. This microbial community is large—about 10^{12} organisms per milliliter of digestive fluid. When the animal eats plant material, it is mixed with saliva and swallowed without chewing to enter the rumen. Here, microbial action and further mixing coats the grass with microbes, reducing it to a pulpy, partially digested mass. At this point, the mass moves into the reticulum, where it is regurgitated as cud, chewed, and reswallowed by the animal. During this process, the grass becomes progressively more liquefied and flows out of the rumen into the omasum and then the abomasum. Here the nutrient-enriched grass meets the animal's digestive enzymes, and soluble organic and fatty acids are absorbed into the animal's bloodstream.

The microbial community in the rumen is extremely dynamic and, as a whole, forms a mutualistic relationship with the animal host. The rumen is slightly warmer than the rest of the animal, and with a redox potential of about −30 mV, all resident microorganisms must carry out anaerobic metabolism. Very specific interactions occur between microbes. One group of bacteria secretes cellulases that cleave the $\beta(1\rightarrow4)$ linkages between successive D-glucose molecules that form plant cellulose. The D-glucose is then fermented to organic acids such as acetate, butyrate, and propionate. These organic acids, as well as fatty acids, are the true carbon and energy sources for the animal. ◀ *Fermentation does not involve an electron transport chain (section 11.8)*

In some ruminants, organic matter processing stops with the production of short-chain fatty acids. In others, such as cattle, CO_2, H_2, and to a lesser extent acetate are used by methanogenic archaea to generate methane (CH_4). Methanogens provide two services: They synthesize many of the vitamins needed by their animal host, and they efficiently remove H_2. This is important because the microbial oxidation of a wide range of organic compounds is thermodynamically feasible within rumen and many other anoxic environments only if H_2 is kept at very low levels. The oxidation of these organic substrates, which results in the generation of H_2 quickly consumed by methanogens, is termed **syntrophy** (Greek *syn*, together; *trophe*, nourishment). Syntrophy is an association in which the growth of one organism either depends on, or is improved by, growth factors, nutrients, or substrates provided by another organism growing nearby.

In addition to the rumen ecosystem, syntrophic relationships occur in anoxic methanogenic ecosystems such as sludge digesters, anoxic freshwater sediments, and flooded soils. Obligate proton-reducing acetogens live in these environments. These microbes oxidize ethanol, butyrate, propionate, and other fermentation end products to H_2, CO_2, and acetate (hence the name "acetogen"). Although the enzymology is not well understood, it is clear that these reactions are thermodynamically unfavorable under standard conditions; that is, they have a positive ΔG. However, because methanogens consume the H_2 as it is generated, the reaction becomes thermodynamically possible. This type of syntrophy is called **interspecies hydrogen transfer.** Thus when fermentative bacteria produce low molecular weight fatty acids (e.g., propionate), anaerobic bacteria such as δ-proteobacteria in the genus *Syntrophobacter* produce H_2 as follows:

$$\text{Propionate}^- + 3H_2O \rightarrow \text{acetate}^- + HCO_3^- + H^+ + 3H_2 \quad (\Delta G^\circ = +76.1\ \text{kJ/mol})$$

Syntrophobacter spp. use protons ($2H^+ + 2e^- \rightarrow H_2$) as terminal electron acceptors. The products H_2 and CO_2 (HCO_3^-) are used by methanogenic archaea such as those in the genus *Methanospirillum*:

$$4H_2 + CO_2 \rightarrow CH_4 + 2H_2O \ (\Delta G^\circ = -25.6\ \text{kJ/mol})$$

By synthesizing methane, *Methanospirillum* spp. consume H_2, keeping it at a low concentration in the immediate environment of both microbes. As a consequence, the reaction becomes exergonic, with a ΔG between −20 and −25 kJ/mol (compare to +76.1 kJ/mol under standard H_2 concentrations). Continuous removal of H_2 promotes further fatty acid fermentation and H_2 production. Because increased H_2 production and consumption stimulate the growth of *Syntrophobacter* and *Methanospirillum* spp., all participants in the relationship benefit. The syntrophic nature of this relationship is further illustrated by the observation that the propionate-oxidizing bacterium *Pelotomaculum thermopropionicum* tethers itself to the methanogen *Methanothermobacter thermautotrophicus* with its flagellum. It is believed that *P. thermopropionicum* does not use the flagellum for motility, but rather to synchronize this syntrophic interaction. Remarkably, one flagellar protein, FliD, serves as an interspecies signal to increase the transcription of methanogenesis-related genes in *M. thermautotrophicus*. ◀ *Methanogens and methanotrophs (section 20.3)*

Cooperation: Nonobligatory Positive Interactions Between Host and Microbe

So far our examples have featured symbionts that promote the growth of hosts in exchange for a safe, nutrient-rich home. The example of the Gram-negative bacterium *Xenorhabdus nematophila* and its nematode (worm) host *Steinernema carpocapsae* is remarkable because the bacterium contributes directly to the reproductive success of its host. Juvenile nematodes harboring *X. nematophila* within their guts live in the soil. To mature, the juvenile must find an insect to infect and consume; this is when things get really interesting (**figure 27.7**). As the hungry juvenile consumes insect blood (haemolymph), *X. nematophila* is excreted in its feces. *X. nematophila* is now free-living, and as they replicate, they use a type III secretion system to secrete enzymes to kill the insect. Once the insect is dead, the bacteria switch to the production of different compounds that protect the insect from degradation by other bacteria and from attack by ants, thereby protecting the home of their host nematode. Amazingly, as the process continues, *X. nematophila* produces yet a different set of molecular signals that trigger *S. carpocapsae* development to adulthood. A single insect cadaver may host many adult nematodes that mate, yielding a cadaver that ultimately teems with *S. carpocapsae* eggs. As these eggs mature to the juvenile stage with *X. nematophila* symbionts, they leave to find new insect homes and start the cycle anew. Because the partners can be grown separately in the laboratory, it has been possible to dissect these events at the structural, molecular, and biochemical level, making this an important model system.

Comprehension Check

1. How could one test to see if an insect-microbe relationship is mutualistic?
2. What is the role of the *Riftia* tube worm's hemoglobin in the success of the tube worm–endosymbiont mutualistic relationship?
3. How is the *Riftia* tube worm endosymbiont similar to cyanobacteria? How is it different?
4. What is meant by "tight trophic coupling" between the coral animal and its microbial symbionts?
5. Why is it important that the rumen is a reducing environment?
6. Describe the mutualism between methanogenic archaea and their ruminant hosts and between methanogens and bacteria such as *Syntrophobacter* spp.
7. How does cooperation differ from mutualism? What might be some of the evolutionary implications of both types of symbioses?
8. Why is the *X. nematophila–S. carpocapsae* symbiosis considered cooperation? What other types of microbial interactions are in play during the nematode life cycle?

27.3 Commensalism and Amensalism Are One-Way Interactions

After reading this section, you should be able to:

a. Distinguish commensalism and amensalism
b. Provide specific examples of each of these types of interaction

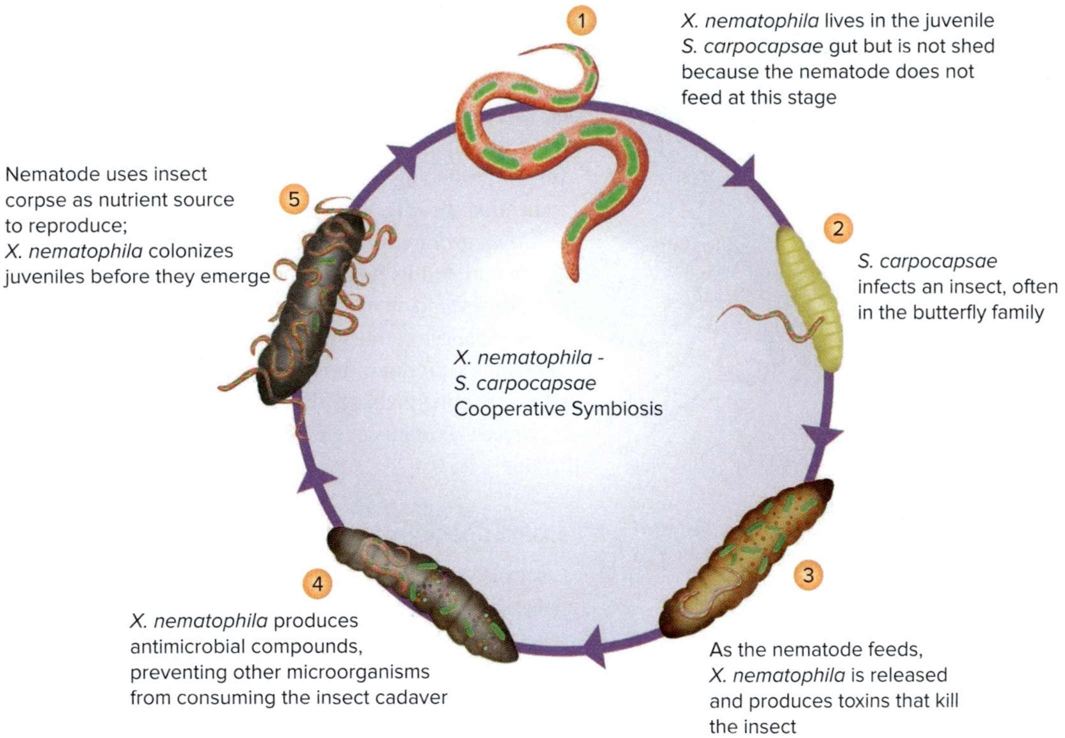

Figure 27.7 The *Xenorhabdus nematophila - Steinernema carpocapsae* Is an Intriguing Model Bacterial-Nematode System.

Unidirectional symbiotic associations benefit only one partner. These may have either little effect on the other (commensalism) or a negative effect on the other (amensalism).

Commensalism (Latin *com,* together; *mensa,* table) is a relationship in which one symbiont, the **commensal,** benefits, while the host is neither harmed nor helped, as shown in figure 27.1. The spatial proximity of the two partners permits the commensal to feed on substances captured or ingested by the host, and the commensal often obtains shelter by living either on or in the host. The commensal is not directly dependent on the host's metabolism, so when it is separated from the host experimentally, it can survive without the addition of factors of host origin.

Commensalistic associations also occur when one microbial group modifies the environment to make it better suited for another organism. For instance, the synthesis of acidic waste products during fermentation stimulates the proliferation of more acid-tolerant microorganisms, which may be only a minor part of the microbial community at neutral pH. A good example of this is the succession of microorganisms during milk spoilage (*see figure 41.3*). Biofilm formation provides another example. The colonization of a newly exposed surface by one type of microorganism (an initial colonizer) makes it possible for other microorganisms to attach to the microbially modified surface. This is a good example of the unidirectional interaction that defines commensalism. ◀ *Biofilms are common in nature (section 7.6)*

The microorganisms that live in and on humans, known as our microbiome, have traditionally been called commensals. Although this term is still applied broadly to these microbes, it is important to keep in mind that the more we learn about the vast microbial ecology of our bodies, the more we come to understand that we depend on our microbes for a myriad of important functions that we cannot achieve without them. These include the synthesis of vitamins, maturation of our immune systems, maintaining a healthy body weight, and prevention of and recovery from disease. The term commensals does not apply to our microbiome, and it is being replaced by the more appropriate term, mutualists. ◀ *The microbe-human ecosystem (chapter 34)*

Amensalism: Microbial Products Harm Other Microbes

Amensalism describes the adverse effect that one organism has on another organism (figure 27.1). This is a unidirectional process based on the release of a specific compound by one organism that has a negative effect on another. A classic example of amensalism is the production of antibiotics that can inhibit or kill other, susceptible microorganisms.

Amensalisms can be quite complex. Attine ants (ants belonging to a New World tribe) are perhaps the most fascinating example. These ants cultivate for their own nourishment a garden of fungi belonging to the genus *Leucocoprinus* (**figure 27.8**). Their ability to accomplish this advanced task depends upon actinobacteria in the genus *Pseudonocardia.* These bacteria produce inhibitory compounds that prevent the growth of *Escovopsis* spp., a predatory fungus that would otherwise destroy the "crop" *Leucocoprinus* fungi.

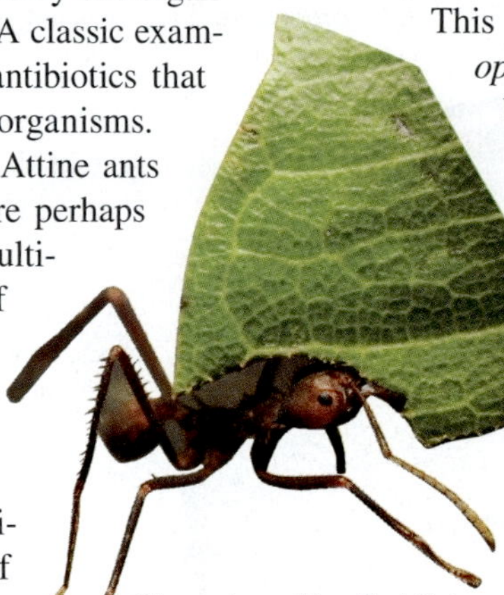

©Amazon-Images/Alamy Stock Photo

These unique interactions appear to have evolved 50 to 65 million years ago in South America. Millions of years of coevolution have resulted in particular groups of ants that cultivate specific strains of *Leucocoprinus* fungi, which are then subject to different groups of *Escovopsis* parasites. To promote the amensalistic interaction between *Pseudonocardia* and *Escovopsis,* the ants have developed intricate crypts within their exoskeletons for the growth of the antifungal-producing *Pseudonocardia.* These crypts have been modified through the ants' evolutionary

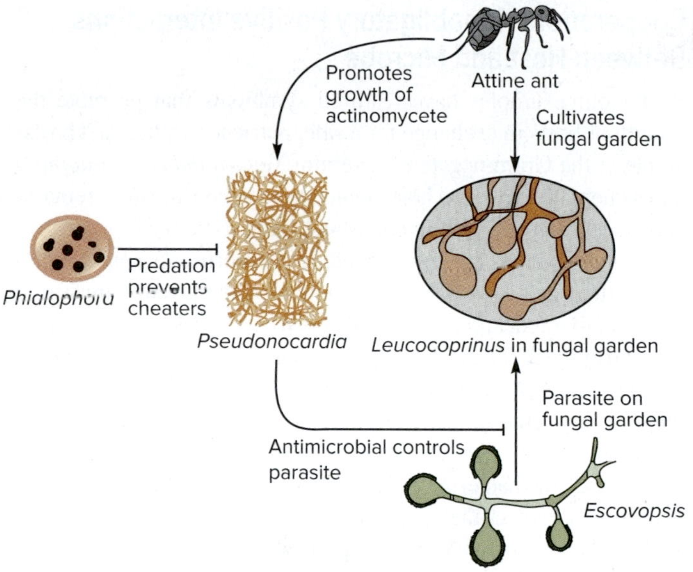

Figure 27.8 Amensalism: An Adverse Microbe-Microbe Interaction. A diagram describing the use of antimicrobial-producing bacteria by ants to control fungal parasites in their fungal garden.

history. The most primitive "paleoattine" ants carry the bacteria on their forelegs, "lower" and "higher" attines have evolved special plates on their ventral surfaces, while the entire surface of the most recent attines, leaf-cutter ants of the genus *Acromyrmex*, are covered with the bacterium. Related ants that do not cultivate fungal gardens (e.g., *Atta* spp.) do not host *Pseudonocardia* cells.

Surprisingly, this four-membered symbiosis is even more complicated, as a fifth member of the symbiosis is found in some cases. This member is a black ascomycete yeast of the genus *Phialophora.* It grows in association with the *Pseudonocardia* symbiont, on which it also preys. But why would attine ants evolve the capacity to tolerate a fungus that eats their antifungal-producing bacteria? The answer has to do with cheating. Briefly, it is energetically expensive for *Pseudonocardia* cells to produce an antimicrobial compound, but its production ensures a home on the ant. However, some members of the *Pseudonocardia* population could get a "free ride" by letting their neighbors produce the antifungal agent. It is thought that only antifungal-producing bacteria can keep the *Phialophora* fungus in check; this prevents cheaters from thriving.

Comprehension Check

1. How does commensalism differ from cooperation?
2. Why is nitrification a good example of a commensalistic process? (Hint: See figure 22.13.)
3. Describe the amensalism observed in attine ant communities. What other types of interactions are present in these communities?
4. The production of antimicrobial agents and their effects on target species have been called a "microbial arms race." Explain what you think this phrase means and its implications for the attine ant system.

27.4 Antagonistic Interactions Characterize Predation, Parasitism, and Competition

After reading this section, you should be able to:

a. Distinguish predation from parasitism
b. Provide examples of each of these forms of symbiosis

Antagonistic interactions include killing (predation), exploitation (parasitism), and competition. Knowing when the dominant microbe benefits is a helpful distinction. Those that obtain biochemical precursors and energy after the victim is killed are predators; parasites benefit while the victim is alive. Competition is best described as an uneasy truce between the microbes involved. We present examples of each.

As is the case with larger organisms, predation among microbes involves a predator species that attacks and kills its prey. Over the last several decades, microbiologists have discovered a number of fascinating microbes that survive by their ability to prey upon others.

Microbial predators have been classified according to their predation strategy. The first two classes describe individual cells; the third class acts as a population. **Epibiotic predators** attach to their prey organism's surface. *Vampirococcus* cells (**figure 27.9*a***) attach to the outer membrane and secrete degradative enzymes that result in lysis and the release of the prey's cytoplasmic contents. **Endobiotic predators** invade the victim's cytoplasm or periplasm where they consume the contents to obtain the energy and precursors needed for cell division. *Daptobacter* (figure 27.9*b*) and *Bdellovibrio (see figure 22.29)* use this predation strategy. ◀ *Order* Bdellovibrionales *includes predators that invade other Gram-negative bacteria (section 22.4)*

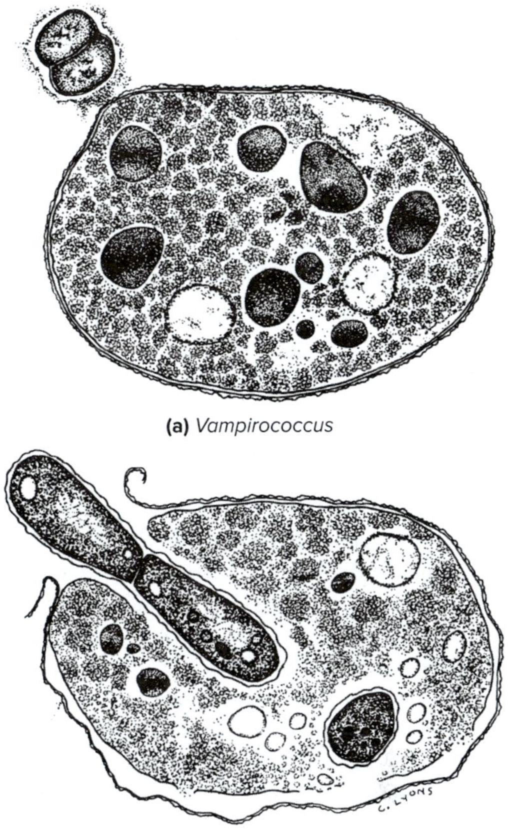

(a) *Vampirococcus*

(b) *Daptobacter*

Figure 27.9 Examples of Predatory Bacteria Found in Nature. (a) *Vampirococcus* has an epibiotic mode of attacking a prey bacterium; (b) *Daptobacter* enters the cytoplasm as it attacks a susceptible bacterium.

In contrast to the predatory bacteria shown in figure 27.9, some bacteria are facultative predators. Sometimes they consume organic matter released from dead organisms, whereas at other times, they actively prey on other microbes. Two examples are members of the γ-proteobacterial genus *Lysobacter* and the δ-proteobacterial genus *Myxococcus.* Both display predation whereby populations of cells use gliding motility to creep toward and cover their prey while releasing an arsenal of degradative enzymes. *M. xanthus* fruiting body formation can be triggered when its prey has been exhausted (**figure 27.10**).

Importantly, predation has many beneficial effects, especially when one considers predators and prey at the population and community levels. Ingestion

(a) (b) (c)

Figure 27.10 Predation by *Myxococcus xanthus.* (a) In this demonstration, a drop of *M. xanthus* cells is placed to the left of a colony of prey bacteria (*Escherichia coli*). (b) The *E. coli* cells lyse as *M. xanthus* glides over and consumes them; thus the colony becomes clear. (c) As *M. xanthus* cells swarm over the prey bacteria they move in waves. They will go on to form fruiting bodies (*see figure 22.30*).
Courtesy of James Berleman and Carl Bauer

MICRO INQUIRY *How does myxobacterial predation differ from that of* Vampirococcus?

and assimilation of a prey bacterium can lead to increased rates of nutrient cycling, critical for the functioning of the microbial loop (*see figure 30.5*). Although here we focus on bacterial predators, the impact of viral predation on communities, ecosystems, and even the global carbon budget is vast. As such, it is an active area of research. ◀ *Marine viruses: mortality at sea (section 30.2)*

Parasitism: Microbes Harm the Host

Parasitism is one of the most complex microbial interactions; **parasite** and host must coexist at least temporarily, so the microbe has enough time to reproduce and colonize a new host. Coexistence may involve nutrient acquisition, physical maintenance in or on the host, or both. But what happens if the host-parasite equilibrium is upset? If the balance favors the host (perhaps by a strong immune defense or antimicrobial therapy), the parasite loses its habitat and may be unable to survive. On the other hand, if the equilibrium is shifted to favor the parasite, the host becomes ill and, depending on the specific host-parasite relationship, may die.

On the other hand, a controlled parasite-host relationship can be maintained for long periods of time and the parasitic nature of the relationship may not be obvious. For example, lichens (**figure 27.11**) are the association between specific ascomycetes (fungi) and either green algae or cyanobacteria. The fungal partner is termed the **mycobiont** and the algal or cyanobacterial partner, the **phycobiont.** The fungus obtains nutrients from its partner by projections of fungal hyphae called haustoria, which penetrate the phycobiont cell wall. It also uses O_2 produced by the phycobiont for respiration. In turn, the fungus protects the phycobiont from high light intensities, provides water and minerals, and creates a firm substratum within which the phycobiont can grow protected from environmental stress. This is an ancient association, having developed prior to the evolution of vascular plants. Based on the above description, one could easily conclude that the relationship is mutualistic. However, some cyanobacteria and algae phycobionts grow more quickly when cultured without their mycobiont. These are considered to be in a relationship in which the mycobiont is the parasite. This is in contrast to cooperation, in which the symbiont often grows more slowly without its host. ◀ Ascomycota *includes yeasts and molds (section 26.5)*; *Phylum* Cyanobacteria: *oxygenic photosynthetic bacteria (section 21.4)*

(a) (b)

Figure 27.11 Lichens. (a) The flattened leaf-shaped genus *Umbilicaria* is a common foliose lichen. (b) A thin slice of *Umbilicaria* viewed with a compound microscope shows that the green algae occur in a thin upper layer and fungal hyphae make up the rest of the lichen.
©Lee W. Wilcox, Ph.D.

An important aspect of many symbiotic relationships, including parasitism, is that over time, the symbiont will discard excess, unused genomic information, a process called **genomic reduction** (*see figure 18.18*). This occurs when the symbiont has become dependent on the host for specific functions, such as synthesis of key metabolites. It is observed in the aphid endosymbiont *Buchnera aphidicola*; it has also occurred in the human pathogens *Mycobacterium leprae, Mycoplasma genitalia,* and *Encephalitozoon cuniculi,* a microsporidium. These microorganisms now can only survive inside their host cells. ◀ *Comparative genomics (section 18.7); Order* Corynebacteriales *includes important human pathogens (section 23.1);* Microsporidia *are intracellular parasites (section 25.7)*

Competition: Microbes Vying for Common Goods

Competition arises when different organisms within a population or community try to acquire the same resource, whether this is a physical location or a particular limiting nutrient (figure 27.1). If one of the two competing organisms can dominate the environment, whether by occupying the physical habitat or by consuming a limiting nutrient, it will overtake the other. This phenomenon was studied by E. F. Gause, who in 1934 described it as the **competitive exclusion principle.** In chemostats, competition for a limiting nutrient may occur among microorganisms with transport systems of differing affinity. This can lead to the exclusion of the slower-growing population under a particular set of conditions. If the dilution rate in the chemostat is changed, the previously slower-growing population may become dominant. Often two microbial populations that appear to be similar nevertheless coexist. In this case, they share the limiting resource (space, a nutrient) and coexist while surviving at lower population levels. Competitive exclusion in part accounts for the phenomenon of colonization resistance—the capacity of gut microflora to prevent pathogen growth. ◀ *Chemostats (section 7.9)* ▶ *A functional core microbiome is required for human homeostasis (section 34.3)*

Comprehension Check

1. Define predation and parasitism. How are these similar and different?
2. What is a lichen? Why is this considered an example of a parasite-host relationship? Why might you also consider it a mutualistic interaction?
3. What is the competitive exclusion principle? List another example where this principle is demonstrated in the natural world.

Key Concepts

27.1 Many Types of Microbial Interactions Exist

- Symbiotic interactions include mutualism (mutually beneficial and obligatory), cooperation (mutually beneficial, not obligatory), and commensalism (product of one organism can be used beneficially by another organism). Predation involves one organism (the predator) ingesting or killing a larger or smaller prey, parasitism (a longer-term internal maintenance of another organism or acellular infectious agent), and amensalism (a microbial product inhibiting another organism) (**figure 27.1**).

27.2 Mutualism and Cooperation Are Two-Way Interactions

- Mutual advantage is central to many organism-organism interactions. These interactions can be based on material transfers related to energetics or the creation of physical environmental changes that offer protection. With several important mutualistic interactions, chemolithotrophic microorganisms play a critical role in making organic matter available for use by an associated organism (e.g., endosymbionts in *Riftia* tube worms) (**figure 27.5**).
- The rumen is an excellent example of a mutualistic interaction between an animal and a complex microbial community. In this microbial community, complex plant materials are broken down to simple organic compounds that can be absorbed by the ruminant, as well as forming waste gases such as methane that are released to the environment.

27.3 Commensalism and Amensalism Are One-Way Interactions

- Commensalism or syntrophism simply means growth together. It does not require physical contact and involves a mutually positive transfer of materials, such as interspecies hydrogen transfer.
- Amensalism describes the adverse effects of one organism on another. Amensalism includes antibiotic production; and the system of attine ants using antibiotic-producing *Pseudonocardia* to control fungus in their gardens (**figure 27.8**) is a fascinating example.

27.4 Antagonistic Interactions Characterize Predation, Parasitism, and Competition

- Predation and parasitism are closely related. Predation has many beneficial effects on populations of predators and prey. These include recycling immobilized minerals in organic matter for reuse by chemotrophic and photosynthetic primary producers, and protection of prey from heat and damaging chemicals (**figures 27.9–27.11**).

Active Learning

1. Describe an experimental approach to determine if a plant-associated microbe is a commensal or a mutualist.
2. Some patients who take antibiotics for acne develop yeast infections of the mouth or genitourinary tract. Explain.
3. *Variovorax paradoxus* is a Gram-negative soil organism that has adapted to use N-acyl homoserine lactones as a carbon source. In so doing, *V. paradoxus* interferes with quorum sensing in other microbes in its habitat. Classify this interspecies interaction, referring to figure 27.1.
4. Scientists interested in the predatory relationship between a microalga (*Chlorella vulgaris*) and a Gram-negative bacterium (*Vampirovibrio chlorellavorus*) noted publications detailing extensive morphological studies of this symbiotic interaction published in the 1970s. However, the bacterial strain was not able to be revived from culture collections. Undeterred, they extracted DNA from the archival material, sequenced it, and assembled almost the entire genome of the bacterium in silico. Comparing the predicted proteins from the genome sequence with the decades-old studies, the scientists inferred a likely predatory sequence. For each step of predation in a freshwater or marine environment, what types of proteins would be required?
   - Detecting prey and moving toward it
   - Attachment and transferring enzymes to the prey
   - Consuming the prey
   - Binary fission within the prey

   Read the original paper: Soo, R. M., et al. 2015. Back from the dead; the curious tale of the predatory cyanobacterium *Vampirovibrio chlorellavorus*. *PeerJ*. 3:e968.

28

Biogeochemical Cycling and Global Climate Change

©VOISHMEL/AFP/Getty Images

Global Climate Change; Global Infectious Disease Change?

When most people think of the bubonic plague, they think of medieval Europe. Malaria brings to mind sub-Saharan Africa. And if they have even heard of dengue (aka breakbone) fever, they think of a jungle somewhere in the tropics. However, these are just three of the diseases that may threaten residents living in temperate climates thanks to global climate change, according to scientists and the World Health Organization.

The links between climate and infectious disease are complex, so the exact shifts in disease patterns are difficult to predict. But consider the following facts. Insects are the vectors for many of the world's worst scourges: plague, malaria, Lyme disease, yellow and dengue fevers (just to name a few). Warmer temperatures not only expand the geographical range of many insects, they also lengthen their breeding season, boost their reproductive rates, and drive them to feed (i.e., bite) more frequently. In addition to diseases carried by insects, about 600 of the known 14,000 infectious agents, such as the viruses causing Ebola and avian flu, are shared between humans and animals. Historically, changes in the environment have been associated with the emergence of new diseases or the reemergence of old diseases.

With these facts in mind, the effects of global climate change on infectious disease can be broadly divided into three categories: altered disease patterns that are already being witnessed, those that can be predicted with confidence, and those that will surprise us. Let's start with a disease whose range has already increased—plague. This bacterial disease is endemic in human and rodent populations in parts of New Mexico, where its prevalence is correlated with the timing and amount of rainfall. Plague has recently been diagnosed farther north than ever before, including Wyoming and Idaho. Similar temperature-dependent shifts in the number of rodents carrying plague bacteria have been documented in Central Asia.

The patterns of insect-borne diseases are also likely to change. Mosquitoes have brought malaria to Texas, Michigan, New Jersey, and New York. Dengue fever–bearing mosquitoes have spread as far north as Chicago and the Netherlands. Lyme disease, a tick-borne disease, has spread into Canada. And as drenching rain and flooding becomes more common, the rates of cholera and other diarrheal diseases are expected to rise.

This leaves us with diseases that could surprise us. These could be completely new diseases or diseases that were thought to be well controlled. But whether the infectious diseases are new or old, their prevention and treatment will require the expertise of a new generation of epidemiologists, health care professionals, and (of course) microbiologists.

It is obvious why microbiologists are involved in the study of the link between climate and infectious disease. What may be less obvious is the role of microbes in climate change. However, as this chapter illustrates, the cycling of all elements (including those involved in climate change) is mediated by microorganisms. The sum of the microbial, physical, and chemical processes that drive the flow of elements between sediments, waters, and the atmosphere is known as **biogeochemical cycling.** We explore biogeochemical cycling by focusing on the flux of carbon, nitrogen, sulfur, phosphorus, iron, and manganese. However, many other elements, such as arsenic, vanadium, uranium, and chromium, also support microbial growth and thus are cycled through the biosphere. We then turn our attention to global climate change, which is driven by changes in the carbon and nitrogen cycles. Collectively, this is the realm of **environmental microbiology,** which is concerned with microbial processes that occur in ecosystems, rather than the individual microorganisms dwelling there. However, it is critical to understand that bacteria, archaea, fungi, and protists are the agents of change.

Readiness Check:

Based on what you learned previously, you should be able to:

- ✓ Explain how nutrient availability and environmental factors influence microbial growth (sections 3.3, 7.5, and 7.6)
- ✓ Compare and contrast the growth and physiology of obligate aerobes, facultative anaerobes, and strict anaerobes (section 7.5)
- ✓ Explain oxidation-reduction (redox) reactions (section 10.3)
- ✓ Diagram, in general terms, the standard reduction potentials of electron donors as compared to electron acceptors and relate this to standard free-energy change (sections 10.3 and 10.4)
- ✓ Describe the difference between heterotrophy and autotrophy (section 11.1)
- ✓ Compare anaerobic respiration, aerobic respiration, and fermentation (sections 11.3, 11.7, and 11.8)
- ✓ Define chemolithotrophy and identify an example electron donor and acceptor (section 11.10)
- ✓ Compare and contrast oxygenic and anoxygenic phototrophy (section 11.12)

28.1 Biogeochemical Cycling Sustains Life on Earth

After reading this section, you should be able to:

a. Explain the term *redox potential* and how it influences the flux of elements found in specific habitats
b. Sketch generalized carbon, nitrogen, phosphorus, sulfur, iron, and manganese cycles
c. Identify, by metabolic type, the microorganisms responsible for carbon and nitrogen fixation, and the mineralization of organic carbon and nitrogen under oxic and anoxic conditions
d. List environments in which methanogenesis and methane oxidation occur
e. Trace the fate of an N atom during dissimilatory nitrate reduction, nitrification, denitrification, and the anammox reaction, and indicate whether each must occur in an oxic or anoxic environment
f. Contrast and compare dissimilatory nitrate reduction with dissimilatory sulfate reduction, and assimilatory nitrate reduction with assimilatory sulfate reduction
g. Identify, by metabolic type, microorganisms responsible for dissimilatory nitrate, sulfate, and iron reduction; nitrification; and sulfur and iron oxidation
h. Contrast and compare the iron and manganese cycles
i. Provide at least two examples of how the flux of one element influences the cycling of another
j. Summarize the importance of biogeochemical cycling in maintaining life on Earth

The long success and vast diversity of microbes reflect the variety of habitats found on Earth. In addition to differences in temperature, pressure, salinity, pH, and nutrient availability, variation in redox potential is also extremely important in determining microbial activity. **Redox potential,** or more formally, oxidation-reduction potential, is a measure of the tendency of molecules in a system to accept or donate electrons. In chapter 10, we discuss redox potential as it relates to specific conjugate redox pairs *(see table 10.2)*. Here our focus is on redox potential in the context of microbial environments, such as soil, sludge, or growth media. The redox potential of these environments can be quantified in volts or millivolts, but to do so, each environment must be compared to a standard. Because no "standard environment" exists, a hydrogen electrode set to zero volts is used. Thus the redox potential of an environment is determined by the electrical potential difference between the environment and a standard hydrogen electrode. The value is denoted as E_h. Environments composed principally of compounds with high redox potentials (more positive) are more likely to accept electrons (i.e., become reduced) when a new compound is added. The newly added compound that donates electrons then becomes oxidized (**table 28.1**).

Why do microbiologists care about redox potentials? The redox state of the environment plays a critical role in determining the types of microbes present because it determines (1) which oxidized compounds are available as terminal electron acceptors for anaerobic respiration and (2) what reduced molecules are present for use as electron donors in chemolithotrophy. In general, electron acceptors used in anaerobic respiration follow a continuum defined by the free energy yield of each reaction. The terminal electron acceptors and their vertical zonation are shown in **figure 28.1**; however, microbial metabolic versatility is far more amazing than illustrated here. For example, the recent discovery of cable bacteria—multicellular sulfide oxidizing deltaproteobacteria in the *Desulfobulbaceae*

Table 28.1 Elements Important in Biogeochemical Cycling and Their Redox States

		MAJOR FORMS AND VALENCES				
Cycle	**Significant Gaseous Component Present?**	**Reduced Form**	**Intermediate Oxidation State Forms**			**Oxidized Form**
C	Yes	Methane: CH_4 (−4)	Carbon monoxide: CO (+2)			CO_2 (+4)
N	Yes	Ammonium: NH_4^+; organic N (−3)	Nitrogen gas: N_2 (0)	Nitrous oxide: N_2O (+1)	Nitrite: NO_2^- (+3)	Nitrate: NO_3^- (+5)
S	Yes	Hydrogen sulfide: H_2S; SH groups in organic matter (−2)	Elemental sulfur: S^0 (0)	Thiosulfate: $S_2O_3^{2-}$ (+2)	Sulfite: SO_3^{2-} (+4)	Sulfate: SO_4^{2-} (+6)
Fe	No	Ferrous iron: Fe^{2+} (+2)				Ferric Iron: Fe^{3+} (+3)
P	No	Phosphonates (+3)	Phosphite: PO^{3-} (+3)			Polyphosphate, phosphate: PO_4^- (+5)
Mn	No	Manganese ion: Mn^{2+} (+2)	Manganese oxide: MnO (+2)			Manganese dioxide: MnO_2 (+4)

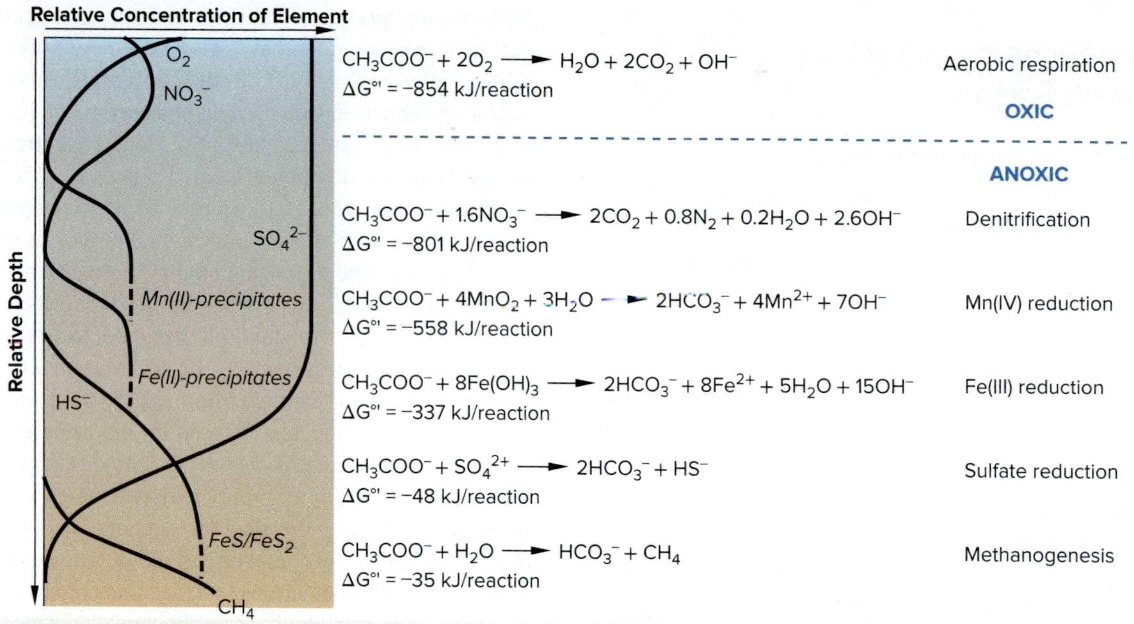

Figure 28.1 The Redox Potential, pH, and Oxygen Concentration Influence the Availability of Terminal Electron Acceptors. As oxygen concentration decreases with depth, other terminal electron acceptors are used in anaerobic respiration. The distribution of microbes in sediments or soils is dictated by the free energy yield for each respiratory process. The relative concentration of elements with depth is shown for marine sediments; freshwater sediments and soil have less sulfate thus methanogenic archaea may be more abundant.

MICRO INQUIRY *Why do sulfate reduction and methanogenesis rarely occur at the same depth or in the same microhabitat?*

family—demonstrates nature's ingenuity. Inspection of figure 28.1 reveals that sulfide and oxygen are spatially separated in the sediment column. Cable bacteria use gliding motility to position themselves vertically in marine sediments. This enables cells at the anoxic end of the filament to use sulfide as an electron donor. The electrons are passed from cell to cell until they reach those in oxic sediments, where O_2 is used as the terminal electron acceptor. ◀ *Redox reactions: reactions of central importance in metabolism (section 10.3)*

Carbon Cycle

Carbon is the element that defines life, and as such it is everywhere. The carbon cycle is often regarded in terms of sources, sinks, and reservoirs of CO_2. Sources, such as heterotrophic microbes, release carbon as CO_2, whereas sinks include plants, phytoplankton, and other autotrophic microbes that take up CO_2 from the atmosphere. Reservoirs store carbon for geological periods of time. The exchange of elements between sources and sinks is referred to as flux. Carbon is present in reduced forms, such as methane (CH_4) and other, more complex organic matter, and in the oxidized, inorganic forms, carbon monoxide (CO) and carbon dioxide (CO_2). Although carbon is continuously transformed from one form to another, for the sake of clarity, we shall say that the cycle "begins" with carbon fixation—the conversion of CO_2 into organic matter (**figure 28.2**). Plants are often regarded as the principal CO_2-fixing organisms, but at least half the carbon on Earth is fixed by microbes, particularly by marine cyanobacteria in the genera *Prochlorococcus* and *Synechococcus,* and diatoms. Importantly, microbes also fix carbon in anoxic environments using anoxygenic photosynthesis as well as by chemolithoautotrophy in the absence of light. In fact, recent evidence suggests that bacterial and archaeal chemolithoautotrophy in

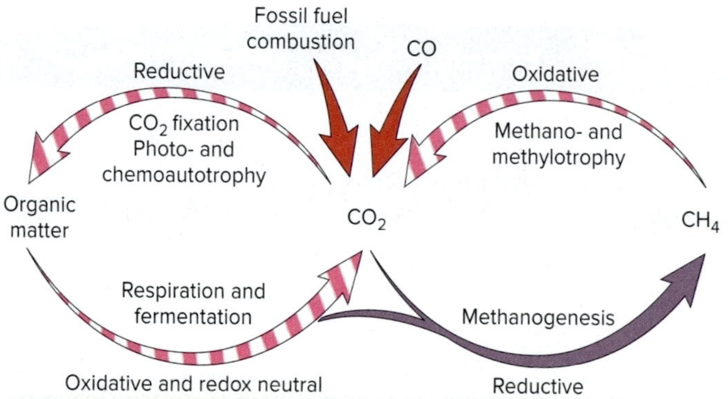

Figure 28.2 The Carbon Cycle. Carbon fixation occurs through the activities of photoautotrophic and chemoautotrophic microorganisms. Methane is produced from inorganic substrates ($CO_2 + H_2$) or from organic matter. Carbon monoxide (CO), produced by sources such as automobiles and industry, is returned to the carbon cycle as CO_2 by CO-oxidizing bacteria. Aerobic processes are indicated by red arrows, those that occur under both oxic and anoxic conditions are noted with striped arrows, and a purple arrow leads to methanogenesis, an anaerobic process.

MICRO INQUIRY *What microbes are capable of methanogenesis and methane oxidation?*

deep, dark subsurface sediments may constitute a significant fraction of global carbon fixation.

Alternatively, inorganic carbon (CO_2) can be reduced anaerobically to methane (CH_4). Recall that only archaea form methane from either $H_2 + CO_2$ or H_2 + acetate. As discussed in chapters 20 and 22, methane produced in sediments is then oxidized either aerobically by proteobacteria or anaerobically by archaea or denitrifying bacteria (**figure 28.3**). Large amounts of methane are also generated in the guts of ruminant animals. Globally, sediments found in rice paddies, coal mines, sewage treatment plants, landfills, marshes and mangrove swamps, and archaea found in the guts of ruminant animals and even termites are important sources of methane. ◀ *Methanogens and methanotrophs (section 20.3); Class* Alphaproteobacteria *includes many oligotrophs (section 22.1); Class* Gammaproteobacteria *is the largest bacterial class (section 22.3)*

All fixed carbon enters a common pool of organic matter that can then be oxidized back to CO_2 through aerobic or anaerobic respiration and fermentation (i.e., heterotrophy). In the carbon cycle depicted in figure 28.2, no distinction is made between the different types of organic matter formed and degraded. This is a marked oversimplification because organic matter varies widely in terms of elemental composition, structure of basic repeating units, and linkages between repeating units. Its bioavailability and degradation are also influenced by other factors, including (1) oxidation-reduction potential; (2) availability of competing nutrients; (3) abiotic conditions such as pH, temperature, O_2, and osmotic conditions; and (4) the microbial community present.

Many of the complex organic substrates used by microorganisms are summarized in **table 28.2**. Lignin, an important structural component in mature plant materials, is notoriously stable. Lignin is actually a family of complex polymers linked by carbon-carbon and carbon-oxygen (ether) bonds (*see figure 31.2*). Filamentous microorganisms—fungi and the streptomycetes—secrete hydrolytic enzymes that degrade lignin by oxidative depolymerization, a process that requires oxygen. A few microbes, such as the purple bacterium *Rhodopseudomonas palustris,* can degrade lignin anaerobically but very slowly. Diminished biodegradability under anoxic conditions results in accumulation of lignified materials, including the formation of peat bogs.

Two terms are helpful in considering the fate and availability of organic materials in nature: mineralization and immobilization. **Mineralization** describes the decomposition of organic matter to simpler, inorganic compounds (e.g., CO_2, NH_3, CH_4, H_2). These compounds may or may not be recycled within the same

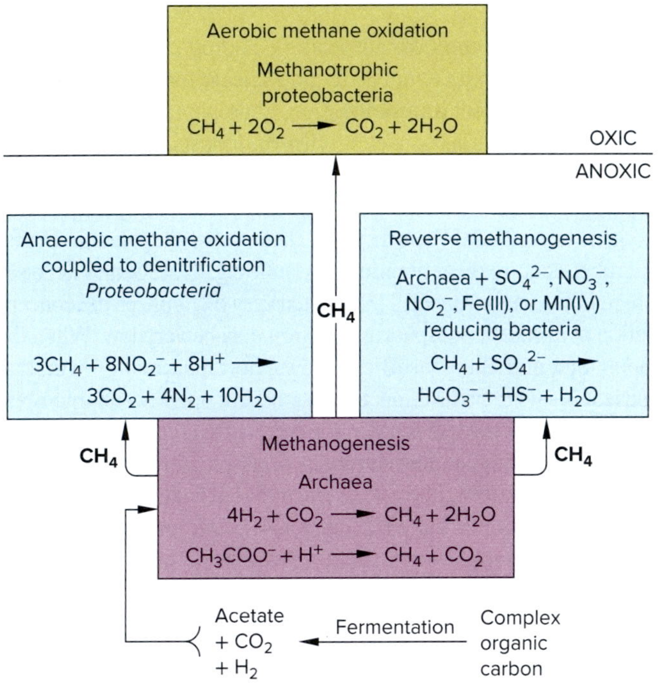

Figure 28.3 Methanogenesis and Methanotrophy. As methane is produced by archaea in anoxic sediments, it diffuses upward where it is oxidized by one of three general types of methanotrophs: proteobacteria that couple methane oxidation with denitrification (left blue box); anaerobic archaea that reverse the methanogenic pathway in association with sulfate (shown), nitrate, or Fe(III) reduction (right blue box); and aerobic proteobacteria.

Table 28.2 Complex Organic Substrate Characteristics That Influence Decomposition and Degradability

			ELEMENTS PRESENT IN LARGE QUANTITY					DEGRADATION	
Substrate	Basic Subunit	Linkages (If Critical)	C	H	O	N	P	With O_2	Without O_2
Starch	Glucose	α(1→4) α(1→6)	+	+	+	–	–	+	+
Cellulose	Glucose	β(1→4)	+	+	+	–	–	+	+
Hemicellulose	C6 and C5 monosaccharides	β(1→4), β(1→3), β(1→6)	+	+	+	–	–	+	+
Lignin	Phenylpropene	C—C, C—O bonds	+	+	+	–	–	+	+/–
Chitin	*N*-acetylglucosamine	β(1→4)	+	+	+	+	–	+	+
Hydrocarbon	Aliphatic, cyclic, aromatic		+	+	–	–	–	+	+/–

environment. By contrast, nutrients (including carbon) converted into biomass become temporarily unavailable for nutrient cycling; this is called nutrient **immobilization.** Saprophytes, which degrade dead organic material, as well as viruses, protists, and other predators are important in recycling immobilized organic compounds.

Many complex substrates contain only carbon, hydrogen, and oxygen (table 28.2). Growth using these substrates demands that microbes acquire the remaining nutrients (e.g., N, P, S, and Fe) elsewhere in the environment. This is often very difficult, as the concentration of nitrogen, phosphorus, and iron may be very low. When the supply of a nutrient is insufficient to support maximal growth, that nutrient is said to be limiting. For instance, in open-ocean microbial communities, growth of many microbes is often nitrogen limited. In other words, if higher concentrations of usable nitrogen (e.g., NO_3^-, NH_4^+) were available, the rate of microbial growth would increase. This phenomenon is called Leibig's law of the minimum, named after the nineteenth-century chemist Justus van Leibig.

Comprehension Check

1. In an anoxic environment containing equal concentrations of nitrate and sulfate, which terminal electron acceptor will support the growth of most anaerobic microbes? Explain your answer.
2. What is the microbial contribution to global carbon fixation?
3. Where will lignin degrade faster: in a shallow riverbed or in a rice paddy? Explain.
4. What kinds of microbes would be able to recycle the carbon atoms in CH_4?
5. Compare and contrast the processes that lead to mineralization and immobilization of carbon.

Nitrogen Cycle

Nitrogen exists in redox states from −3 to +5, and this is reflected in the transformations that drive nitrogen cycling (table 28.1). As shown in **figure 28.4**, depending on its level of oxidation, nitrogen species can serve as electron acceptors in

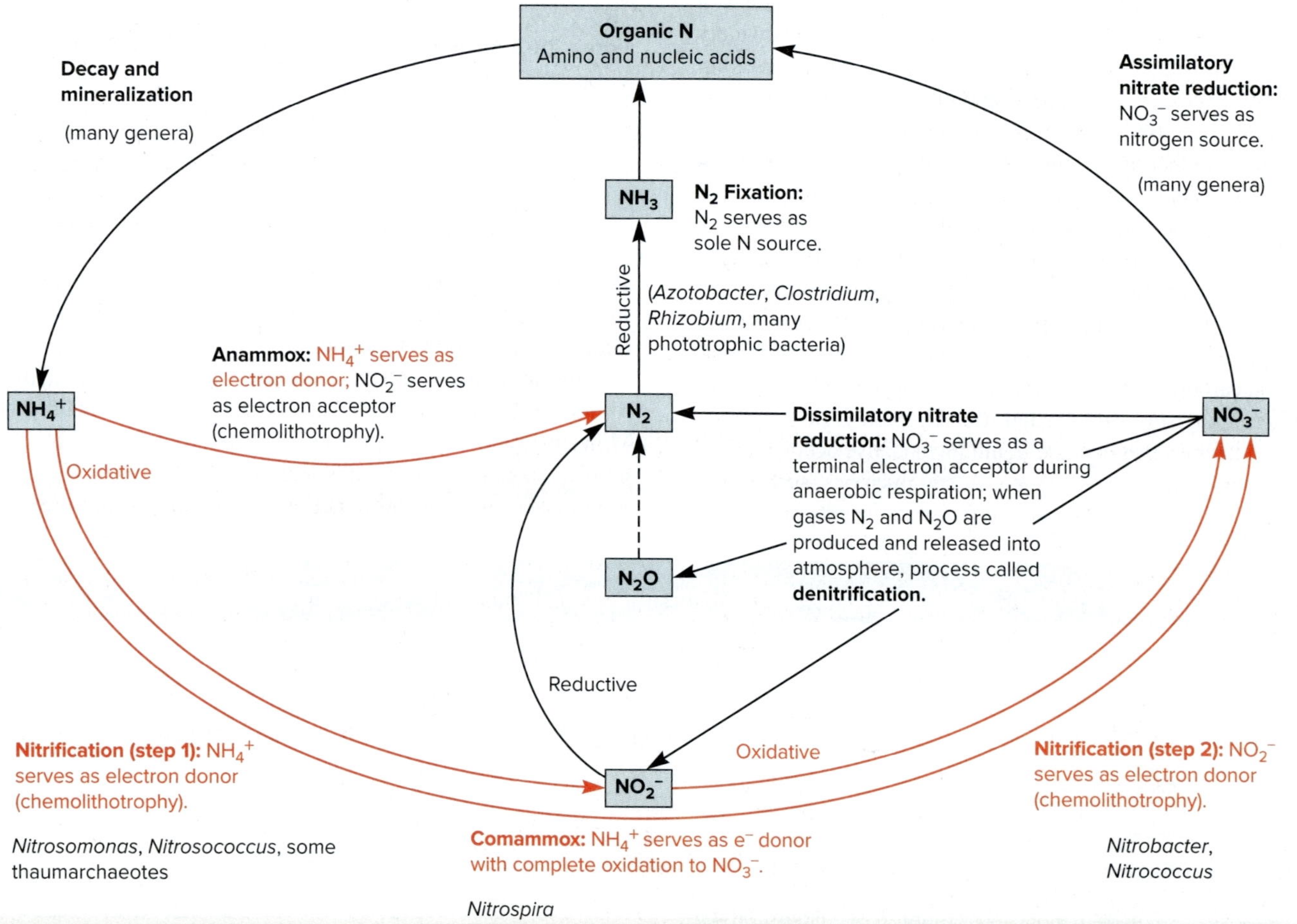

Figure 28.4 A Simplified Nitrogen Cycle. Depending on their oxidative state, some nitrogen species can function as either electron donors or electron acceptors and therefore play very different roles in the environment. NH_4^+ is fully reduced; therefore it can serve only as an electron donor. The fully oxidized species NO_3^- can function as a terminal electron acceptor, or plants and microbes can reduce it and incorporate it into biomass during assimilatory reduction. Oxidative reactions are labeled in red, reductive reactions in black.

MICRO INQUIRY *What are the products of denitrification, nitrification, dissimilatory nitrate reduction, and assimilatory nitrate reduction?*

anaerobic respiration (e.g., NO_3^- and NO_2^-) or as electron donors (e.g., NO_2^- and NH_4^+) in chemolithotrophy. This can lead to some confusion, so keep in mind the oxidation state of the nitrogen species: Is it fully oxidized (i.e., NO_3^-)? If so, it can only accept electrons. Conversely, if it is fully reduced (NH_4^+), its role is limited to electron donor. Nitrite (NO_2^-) is neither fully oxidized nor reduced, so it can function as either electron donor or acceptor, depending on the organism.

We begin our discussion of this cycle with **nitrogen fixation**—the reduction of the inorganic gaseous molecule N_2 to its organic form (e.g., amino acids, purines, and pyrimidines). Nitrogen fixation is performed only by some bacteria and archaea. Although the nitrogenase enzyme is sensitive to oxygen, nitrogen fixation can be carried out under both oxic and anoxic conditions. Microbes such as *Azotobacter* spp. and cyanobacteria in the genus *Trichodesmium* fix nitrogen aerobically, while free-living anaerobes such as *Clostridium* spp. also fix nitrogen. Perhaps the best-studied nitrogen-fixing microbes are the bacterial symbionts of leguminous plants, including rhizobia, their α-proteobacterial relatives, and a few β-proteobacteria. Other bacterial symbionts fix nitrogen as well. For instance, actinomycete species in the genus *Frankia* fix nitrogen while colonizing many types of woody shrubs, and heterocyst-forming cyanobacteria *Anabaena* spp. fix nitrogen when in association with the water fern *Azolla* spp. ◀ *Nitrogen fixation (section 12.5); Phylum* Cyanobacteria: *oxygenic photosynthetic bacteria (section 21.4); Order* Frankiales: *bacteria that fix N_2 in trees (section 23.1)* ▶ *Nitrogen-fixing bacteria are vital to agriculture (section 31.3)*

The product of N_2 fixation is ammonia (NH_3), which is immediately incorporated into organic matter as an amine. These amine N atoms are introduced into proteins, nucleic acids, and other biomolecules. When these organic molecules are degraded (dissimilated) and mineralized, nitrogen is released as ammonium (NH_4^+). Many microbial genera are capable of dissimilation of organic nitrogen substrates, but complete mineralization requires an assemblage of microbes.

Because ammonium is fully reduced, it can only donate electrons, and this is what happens during **nitrification** (figure 28.4). In the first step, ammonium is oxidized to nitrite (NO_2^-), and in the second step, NO_2^- is oxidized to nitrate (NO_3^-). For many years, it was thought that nitrification had to be performed by two different groups of bacteria. For example, some archaea and bacteria in the genera *Nitrosomonas* and *Nitrosococcus* play important roles in ammonia oxidation, while *Nitrobacter* spp. and related bacteria carry out nitrite oxidation. Both steps of nitrification are usually aerobic, with O_2 as the terminal electron acceptor. There are now two known exceptions. First is the β-proteobacterium *Nitrosomonas eutropha,* which oxidizes ammonium anaerobically to nitrite and nitric oxide (NO) using nitrogen dioxide (NO_2) as an acceptor in a denitrification-related reaction (*see figure 22.13*). Most recently, several *Nitrospira* species have been shown to perform both nitrification steps. This ability has been dubbed **comammox,** for *com*plete *amm*onia *ox*idation (figure 28.4). ◀ *Nitrifying bacteria oxidize ammonium or nitrite to gain energy and electrons (section 22.1)*

The production of nitrate is important because it can be reduced and incorporated into microbial and plant cell biomass in a process known as **assimilatory nitrate reduction.** Alternatively, some microorganisms use nitrate as a terminal electron acceptor during anaerobic respiration. When respired, the nitrogen is not incorporated into cellular material, so this is called **dissimilatory nitrate reduction.** A variety of microbes, including *Geobacter metallireducens* and *Desulfovibrio* spp., are capable of dissimilatory nitrate reduction. When nitrate is fully reduced to dinitrogen gas (N_2), nitrogen is removed from the ecosystem and returned to the atmosphere through a series of reactions collectively known as **denitrification.** This dissimilatory process is performed by a variety of heterotrophic bacteria, such as *Pseudomonas denitrificans.* The major products of dissimilatory nitrate reduction include nitrogen gas (N_2) and nitrous oxide (N_2O), although nitrite (NO_2^-) also can accumulate *(see figure 11.18)*. N_2O is an important greenhouse gas, as discussed in section 28.2. ◀ *Anaerobic respiration uses the same three steps as aerobic respiration (section 11.7); Inorganic nitrogen assimilation (section 12.5)*

The **anammox reaction** (*an*oxic *amm*onium *ox*idation) is an anaerobic reaction performed by chemolithotrophs in the phylum *Planctomycetes.* Here ammonium ion (NH_4^+) serves as the electron donor and nitrite (NO_2^-) as the terminal electron acceptor; it is reduced to nitrogen gas (N_2) (*see figure 21.13*). In effect, the anammox reaction is a shortcut to N_2, proceeding directly from ammonium and nitrite, without having to cycle first through nitrate (figure 28.4). The discovery that planctomycete bacteria oxidize considerable amounts of NH_4^+ to N_2, thereby removing ammonia from the environment, has sparked keen interest in the wastewater management field, where high levels of ammonia are undesirable. ◀ *Phylum* Planctomycetes: *anammox bacteria (section 21.5)*

Comprehension Check

1. Organic nitrogen introduced into an ecosystem directly by nitrogen fixation is sometimes called "new" nitrogen. Explain why.
2. Describe the two-step process that makes up nitrification. Could microbes performing nitrification and anammox live in the same environment? Why or why not?
3. What is the difference between assimilatory nitrate reduction and denitrification? Which reaction is performed by many microbes, and which is a more specialized metabolic capability?
4. Refer to section 21.5 and figure 21.13. What nitrogenous intermediates are used in the anammox reaction? Why don't they accumulate in the environment?

Phosphorus Cycle

Biogeochemical cycling of phosphorus is important for a number of reasons. All living cells require phosphorus for nucleic acids, and some lipids and polysaccharides. For many years, it was believed that all phosphorus is originally derived solely from the weathering of phosphate-containing rocks and assimilated as

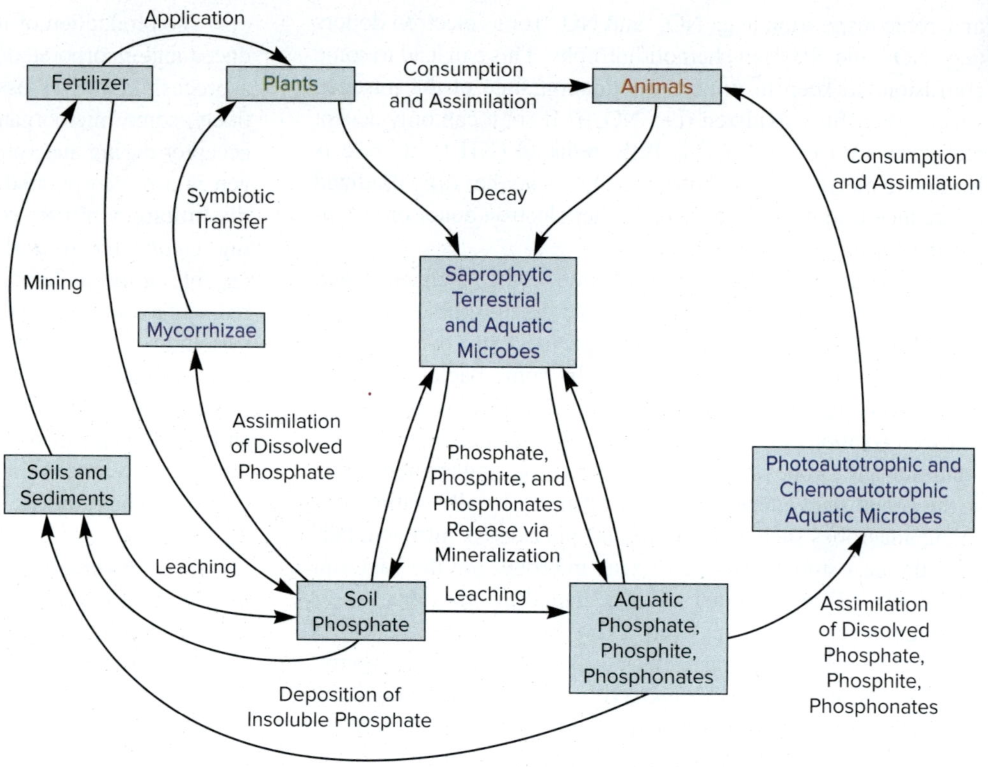

Figure 28.5 A Simplified Phosphorus Cycle. Phosphorus enters soil and water through the degradation of plants and animals, weathering of rocks, and fertilizer application. Phosphate in soil and phosphate, phosphite, and phosphonates in waters are consumed by terrestrial and aquatic microbes and passed on to larger organisms. However, much of the soil phosphorus can leach great distances or complex with cations to form relatively insoluble compounds. Terrestrial chemolithotrophic microbes also take up soil phosphate but are not shown.

phosphate (**figure 28.5**). However, recent evidence suggests that phosphonates, which bear a C—P bond and are thus organic, are significant sources of phosphorus for marine microorganisms.

In soil, phosphorus exists in both inorganic and organic forms. Organic phosphorus includes not only that found in biomass but also that in materials such as humus and other organic compounds. The phosphorus in such organic materials is readily recycled by microbial activity. Inorganic phosphorus is negatively charged, so it complexes with positively charged elements in the environment, such as iron, aluminum, and calcium. These compounds are relatively insoluble, and their dissolution is pH dependent such that phosphate is most available to plants and microbes between pH 6 and 7. Microbes transform simple orthophosphate (PO_4^{3-}) to more complex forms. These include the polyphosphates present as inclusions as well as more familiar macromolecules such as nucleotides and phospholipids. ◀ *Inclusions (section 3.6); Phosphorus assimilation (section 12.6)*

Sulfur Cycle

Microorganisms contribute greatly to the sulfur cycle; a simplified version is shown in **figure 28.6**. The sulfur cycle is similar to the nitrogen cycle in that, depending on the oxidation state of the sulfur species, it can serve as an electron acceptor, an electron donor, or both. Sulfate, the fully oxidized species, is reduced by plants and microbes for use in amino acid and protein biosynthesis; this is described as **assimilatory sulfate reduction.** By contrast, when sulfate diffuses into anoxic habitats, it provides an opportunity for microbial **dissimilatory sulfate reduction.** Here sulfate serves as a terminal electron acceptor during anaerobic respiration by a variety of microbes, including δ-proteobacteria such as *Desulfovibrio* and *Desulfonema* spp. and archaea belonging to the genus *Archaeoglobus*. This results in sulfide accumulation. Sulfide, which is fully reduced, can then serve as an electron source for anoxygenic photosynthetic microorganisms and chemolithoautotrophs, including members of the phylum *Chlorobi* and the genus *Thiobacillus,* respectively. These microbes convert sulfide to elemental sulfur and sulfate.

Other microorganisms have been found to carry out dissimilatory elemental sulfur (S^0) reduction. These include members of the genus *Desulfuromonas,* thermophilic archaea, and cyanobacteria in hypersaline sediments. Sulfite (SO_3^{2-}) is another critical intermediate that can be reduced to sulfide by a wide variety of microorganisms, including members of the genera *Alteromonas, Clostridium, Desulfovibrio,* and *Desulfotomaculum.*

Dimethylsulfoniopropionate (DMSP) is an important organic sulfur compound. It is produced by marine phytoplankton as a compatible solute. When these cells die and release DMSP, bacterioplankton (floating bacteria) use it as a sulfur and carbon source. In the process, it is metabolized to dimethylsulfide (DMS) and released into the atmosphere. There DMS is rapidly converted into a variety of sulfur compounds that serve as nuclei for water droplet formation, contributing to the formation of clouds. Because clouds help keep the Earth's surface cool, it is hypothesized that increased DMS production could help mitigate the effects of global climate change.

Iron Cycle

The iron cycle features the interchange of ferrous iron (Fe^{2+}) to ferric iron (Fe^{3+}) (**figure 28.7**). In environments that are fully aerated with a neutral pH, iron is present primarily as insoluble minerals of either oxidation state. The solubility of both reduced, ferrous iron (Fe^{2+}) and oxidized, ferric iron (Fe^{3+}) increases with

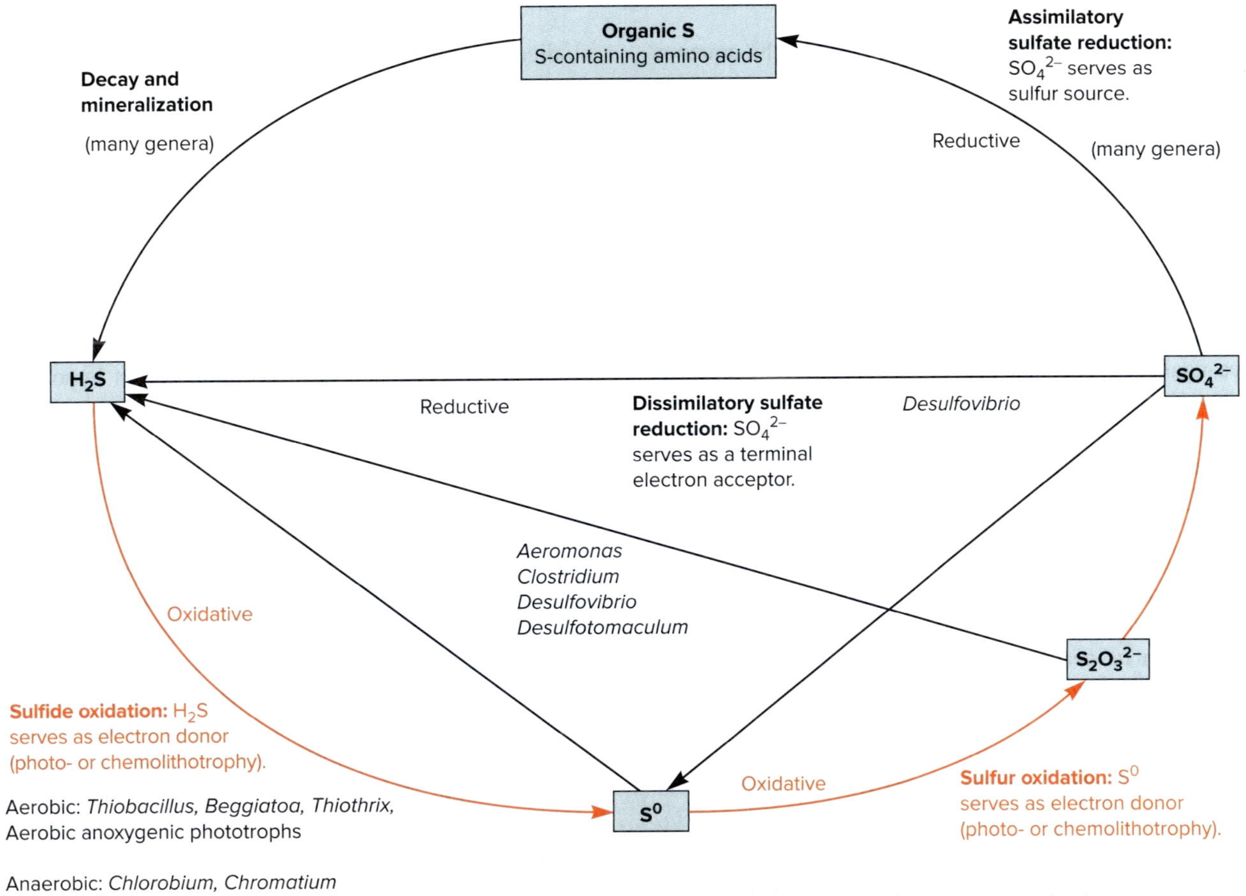

Figure 28.6 A Simplified Sulfur Cycle. Sulfate is fully oxidized thus it serves as the terminal electron acceptor in anaerobic respiration, termed dissimilatory sulfate reduction. Alternatively, plants and microbes can donate the S atom to organic compounds during assimilatory sulfate reduction. H_2S is fully reduced, so it can serve as an electron donor during chemo- and photolithotrophy. Because elemental sulfur and thiosulfate are neither fully oxidized nor reduced, they can serve as either electron donors or acceptors.

MICRO INQUIRY *Which sulfur species is gaseous?*

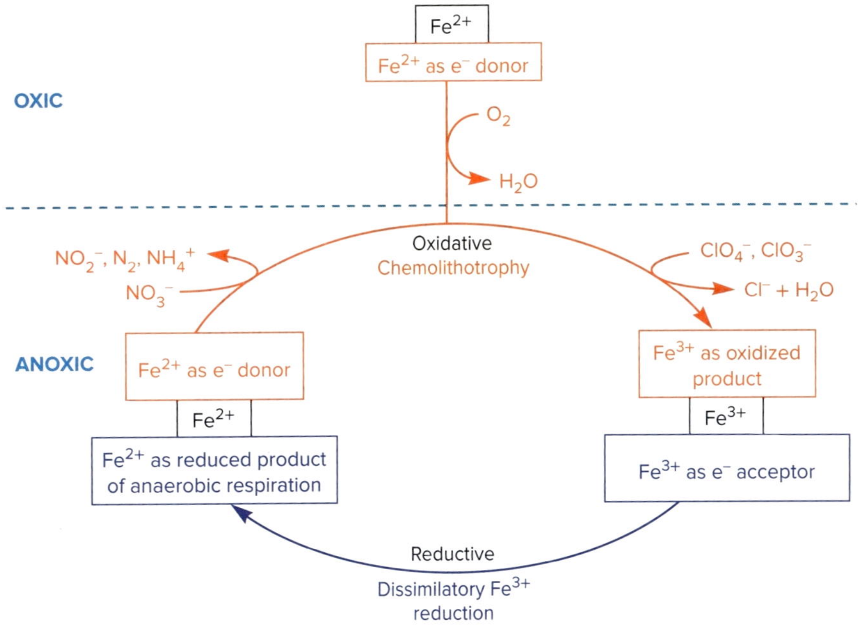

Figure 28.7 The Iron Cycle. Dissimilatory reduction of iron takes place when Fe^{3+} is used as a terminal electron acceptor during anaerobic respiration. This occurs only in anoxic environments. By contrast, oxidation of Fe^{2+} can occur under both oxic and anoxic conditions. Fe^{2+} can be used as an electron donor with O_2 as electron acceptor, although some lithotrophs use NO_3^- or even the environmental contaminants chlorate (ClO_3^-) and perchlorate (ClO_4^-) as electron acceptors.

MICRO INQUIRY *Why is Fe^{2+} used as an electron donor chiefly in acidic environments?*

compounds such as ammonium or ferrous iron as a source of electrons, they must oxidize large amounts of these compounds to generate enough ATP and reducing equivalents for carbon fixation. This reflects the relatively small amount of energy released during the oxidation of most inorganic chemicals. Thus chemolithotrophs not only link the carbon cycle to either the nitrogen, sulfur, or iron (or other metal) cycles, they also significantly impact the global flux of these elements.

In the case of nitrogen fixation, bacteria and archaea convert N_2 to ammonia, which is then incorporated into organic nitrogen molecules such as amino and nucleic acids. In many environments, the amount of carbon that is taken up by plants and microbes (either as CO_2 or as dissolved organic carbon) is dictated by how much nitrogen is available to support growth. Thus by introducing "new" nitrogen into the system, nitrogen-fixing microbes can also increase the rate of carbon flux.

Comprehension Check

1. Compare the products of dissimilatory sulfate reduction to those of dissimilatory nitrate reduction.
2. How might iron have been important in the evolution of early life?
3. What are some important microbial genera that contribute to manganese cycling?
4. How can microbial activity render some metals more or less toxic to warm-blooded animals?
5. Suggest how a chemolithoautotroph might link the carbon and sulfur cycles.

28.2 Global Climate Change: Biogeochemical Cycling Out of Balance

After reading this section, you should be able to:

a. Differentiate between the terms "global warming" and "global climate change" and identify greenhouse gases
b. Explain the link between greenhouse gas accumulation and microbial cycling of carbon and nitrogen
c. Discuss the origin of the greenhouse gases CO_2, methane, and nitrogen oxide species

Microbial activity is critical in maintaining the dynamic equilibrium that defines our biosphere. The capacity of microbes to thrive in seemingly every niche on Earth is the result of 3.5 billion years of microbial evolution. Importantly, changes in the physical and chemical environment to which microbes have adapted generally occurred over geological time scales. However, since the beginning of the twentieth century, the rate at which CO_2 and other so-called greenhouse gases have entered the atmosphere has been much faster than at any other time in the known history of life on Earth. Atmospheric gases such as CO_2, CH_4, and nitrogen oxides are called **greenhouse gases** because they trap heat reflected from Earth's surface in the atmosphere, rather than allowing it to radiate into space. The rate at which these gases are entering the atmosphere exceeds the rate by which the natural carbon and nitrogen cycles can recycle them. It is the ever-increasing atmospheric concentration of these gases that has resulted in what is often called global warming. However, the term *global climate change* more accurately reflects the current changes in patterns of wind, precipitation, and ocean and atmospheric temperatures.

The most abundant greenhouse gas is CO_2. Since the onset of the industrial era, about 150 years ago, global CO_2 levels have risen from 278 ppm (parts per million) to a current level over 400 ppm (**figure 28.11**). Most of this can be attributed to the burning of fossil fuel, which releases about 10 billion tons of CO_2

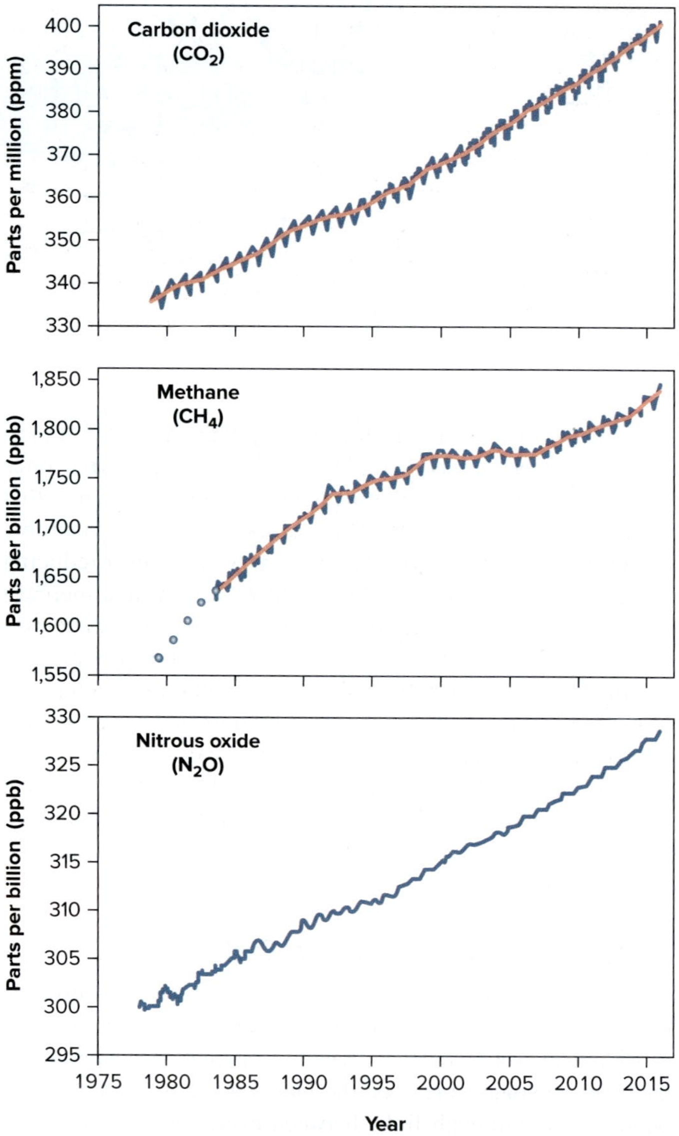

Figure 28.11 Global Atmospheric Concentrations of Greenhouse Gases.

Source: Butler, James, & Montzka, Stephen. "The NOAA Annual Greenhouse Gas Index (AGGI)." 2017.

each year. Changes in land use management, principally through deforestation, account for another 1.2 billion tons of CO_2 annually. To understand why these activities result in the release of CO_2, consider that fossil fuel (e.g., oil) is formed from decaying organic matter. This organic matter arose through millions of years of CO_2 fixation, which removed CO_2 from the atmosphere. It has been locked deep within the Earth for almost as long. When fossil fuels are burned, the energy of combustion oxidizes this material back to CO_2 in a tiny fraction of the time it took to accumulate. The equilibrium of carbon flux is lost because atmospheric inputs (CO_2 release) greatly exceed the rate of removal (CO_2 fixation).

The situation is different with deforestation. Forests are generally regarded as CO_2 sinks because the abundance of photoautotrophic plants pulls a great deal of CO_2 out of the atmosphere, converting it to biomass. Indeed, during the 1990s terrestrial ecosystems sequestered about 3 gigatons of carbon per year (a gigaton is 1 billion metric tons). Thus when forests are cut down for firewood or agriculture, the Earth loses a carbon sink. While some speculate that other, remaining plants will assimilate the extra CO_2 and simply grow faster, changes in the rate of soil microbial respiration complicate matters. Recall that as respiration rates increase, so does the quantity of CO_2, CH_4, and N_2O released. In fact, warming permafrost in subarctic regions has dramatically increased soil microbial respiration, and experiments in mineral soils show that CO_2 release increases with warming temperatures.

Methane is a greenhouse gas of increasing concern because it has about 30 times the global warming potential of CO_2. This means that a single molecule of CH_4 released into the atmosphere has the same thermal retention capacity as 30 CO_2 molecules. Based on analyses of gas bubbles in glacier ice cores, the levels of methane in the atmosphere remained essentially constant until about 150 years ago. Since then, methane levels have increased 2.5 times to the present level of about 1,890 parts per billion (ppb) (figure 28.11). Considering these trends, there is worldwide interest in understanding the factors that control methane production and oxidation by microorganisms.

The nitrogen oxide greenhouse gases NO and N_2O, collectively referred to as "Nox," have about 280 times the global warming potential of CO_2. To understand the origin of rising Nox, we need to examine food production. The "green revolution" of the mid-twentieth century introduced human-made fertilizer as a key means of ensuring that agricultural output could keep pace with population growth. However, manufacturing fertilizer is an energy-intensive process that uses hydrogen gas to reduce N_2 to NH_4^+ at high temperature and pressure. This process was first described by Fritz Haber and Karl Bosch in 1913, but it was not until the middle of the twentieth century that the Haber-Bosch reaction was used at an industrial level. Current fertilizer use is about 200 million metric tonnes per year, or about 500 times greater now than it was in 1940. In contrast to the positive outcome of increased agricultural output, the addition of so much ammonium to the soil has drastically altered the balance of the nitrogen cycle.

The problem is that not all the NH_4^+ is taken up by plants. The extra $NH4^+$ generally has one of two fates: runoff or nitrification followed by denitrification (**figure 28.12**). Ammonium runoff leaches into lakes and streams, frequently causing **eutrophication**—an increase in nutrient levels that stimulates the growth of a limited number of organisms, thereby disturbing the ecology of these aquatic ecosystems. By contrast, microbial nitrification results in the oxidation of ammonium to more nitrate than can be immobilized by plants and microbes, because organisms need a specific ratio of C:N:P. Thus the extra nitrate is reduced to N_2 and reactive greenhouse Nox gases through the process of

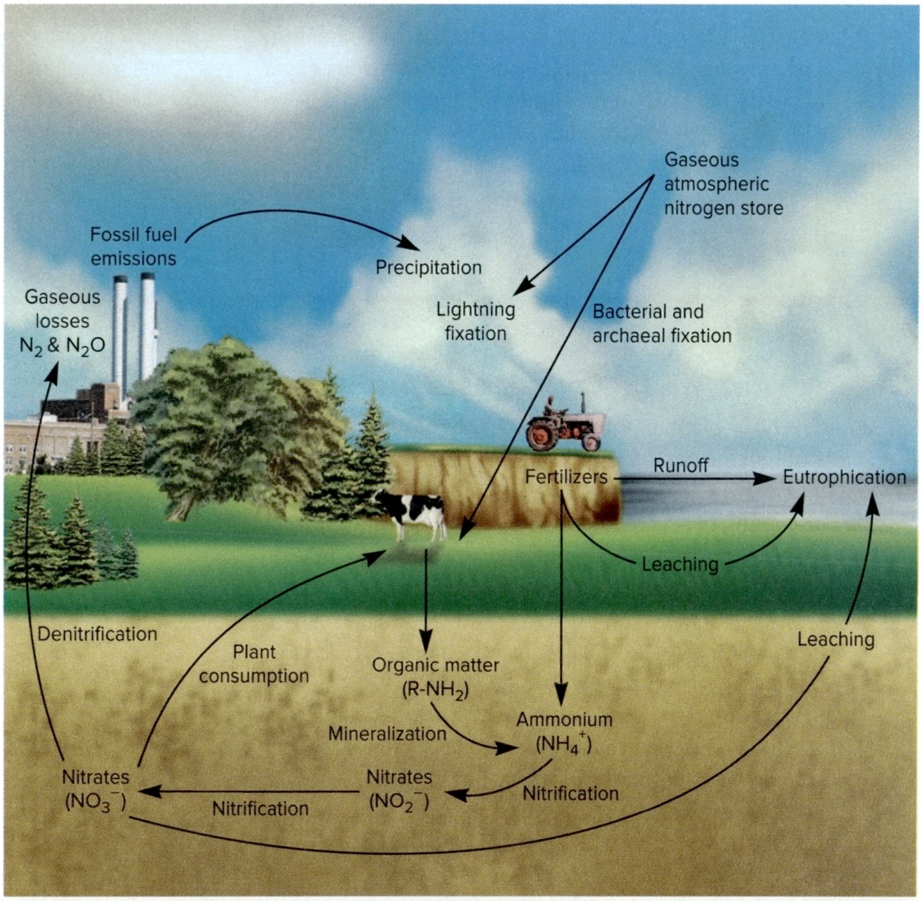

Figure 28.12 Natural and Human-Made Influences on the Nitrogen Cycle.

MICRO INQUIRY *What organisms benefit from nitrification?*

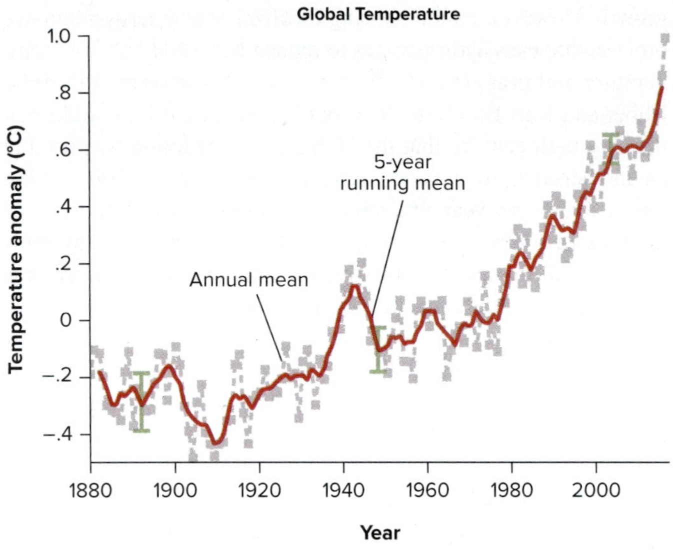

Figure 28.13 **Global Annual-Mean Surface Air Temperature Change.**

Source: NASA, "GISS Surface Temperature Analysis."

denitrification. Cycles of nitrification/denitrification fueled by NH_4^+ fertilizer are responsible for the highest N_2O levels in 650,000 years.

What are the consequences of disrupting the carbon and nitrogen cycles? Global climate change is the most obvious example. It is important to keep in mind that weather is not the same as climate. While North America has suffered some of the hottest summers on record in the past decade, a single day or week in July that is particularly hot is not, by itself, evidence of global climate change. Global climate change is measured over decades and includes many parameters such as surface temperature on land and sea, and in the atmosphere and troposphere; rates of precipitation; and frequency of extreme weather. Based on these analyses, the average global temperature has increased 0.8°C since 1880, with two-thirds of the increase occurring since 1975 (**figure 28.13**). Depending on the rate of continued increase in greenhouse gases, the average global surface temperature is predicted to rise as much as 6.4°C by 2100.

An important question is how will microbes respond to a changing world. For the vast majority of Earth's history, microorganisms have been the drivers of balanced elemental cycling, so changes in microbial activities will have a major impact on the rate and magnitude of greenhouse gas accumulation and global climate change. As introduced in chapters 30 and 31, the role microbes play in balancing carbon and nitrogen fluxes has opened new avenues of research and discovery in microbial ecology.

Comprehension Check

1. List three greenhouse gases. Discuss their origins.
2. Discuss the possible role of forests in controlling atmospheric concentrations of CO_2.
3. How do changes in the nitrogen cycle caused by fertilization influence the carbon cycle?
4. Given that each microbial group has an optimum temperature range for growth, how might you predict changes to a soil microbial community living in your geographic area?

Key Concepts

28.1 Biogeochemical Cycling Sustains Life on Earth

- The redox potential of the environment has a major impact on the type of metabolic processes that can occur (**figure 28.1**).
- The cycling between inorganic (CO_2) and organic carbon occurs aerobically and anaerobically by all types of microbes (**figure 28.2**). Only methanogenic archaea can produce methane, an anaerobic process. Methane can be oxidized aerobically or anaerobically (**figure 28.3**).
- Assimilatory processes involve incorporation of nutrients into the organism's biomass during metabolism; dissimilatory processes involve the release of nutrients to the environment after metabolism. Dissimilatory reduction involves the use of the oxidized compound (e.g., NO_3^- or SO_4^{2-}) as the terminal electron acceptor during anaerobic respiration (**figures 28.4** and **28.6**).
- Chemolithotrophs use reduced elements as electron donors and generate the oxidized form, which can then be used for assimilatory or dissimilatory processes (**figures 28.4** and **28.6**).
- Phosphorus is chiefly cycled in its oxidized state, phosphate. In marine systems, phosphite and phosphonates are also cycled (**figure 28.5**).
- Iron and manganese cycle between their oxidized and reduced forms without the accumulation of intermediates (**figures 28.7** and **28.8**).
- Mercury cycling is complex but when methylated can have significant effects on the health of higher organisms (**figure 28.9**).
- Although it is convenient to consider a single element cycling, in nature there are extensive and complex interactions between nutrient cycles (**figure 28.10**).

28.2 Global Climate Change: Biogeochemical Cycling Out of Balance

- Microbes have evolved slowly over time; this has given rise to the biogeochemical cycles that sustain life on Earth. However, since the beginning of the twentieth century, the rate at which CO_2, CH_4, and nitrogen oxides have been released into the atmosphere has outpaced the rate at which they can be recycled. Thus these greenhouse gases are accumulating in Earth's atmosphere (**figures 28.11** and **28.12**).

- The increase in greenhouse gases is correlated with an increase in the global annual-mean surface air temperature (**figure 28.13**).

Active Learning

1. Examine figure 28.12. Farmers in Argentina use a crop/cattle rotation system in which cattle graze on pastures for about 5 years. Cattle are then moved to a different area and crops are planted where they had grazed. Crops are grown for 3 years, and cattle are rotated back and the field reverts to a pasture. This dramatically reduces the amount of nitrogen fertilizer that must be added when crops are grown. Discuss why this practice minimizes fossil fuel emissions.
2. A bacterium isolated from sewage sludge was recently found to be capable of anoxygenic photosynthesis using nitrite as the electron donor, converting it to nitrate. Compare this form of nitrification to that which is well characterized. By removing nitrite from the environment, what other processes within the nitrogen cycle might this bacterium be influencing?
3. It has been shown that the addition of phosphorus to lake ecosystems promotes excess nitrogen removal. This has real-world implications, as many communities have implemented management plans to reduce phosphorus influx to natural waters. Explain how an increase in one element, phosphorus, could lead to the removal of another, nitrogen. Think about the carbon cycle. How could an increase in another element in a given ecosystem alter the loss of carbon as CO_2?
4. Microbial methane oxidation is an important process because it drastically limits the release of this dangerous greenhouse gas. A group of European microbiologists used enrichment cultures of archaea from Dutch canal sediments to obtain two isolates. One oxidized CH_4 to CO_2 using NO_3^- as an electron acceptor to generate N_2. The other isolate reduced Fe(III) to Fe(II) during methane oxidation. Refer to figures 28.2, 28.3, 28.4, and 28.7 to review the parts of the carbon, nitrogen, and iron cycles involved. If you mixed the two isolates with equal amounts of nitrate and Fe(III), which methane-oxidizing archaeon do you think would grow fastest? Why?

Read the original paper: Etwing, K. F. 2016. Archaea catalyze iron-dependent anaerobic oxidation of methane. *Proc. Nat. Acad. Sci. USA*. 113:12792–6. doi:10.1073/pnas.1609534113.

29

Methods in Microbial Ecology

Source: NOAA/OER

Scientists Search for Intraterrestrial Life—and Find It

In 1955 the "father of marine microbiology," Claude ZoBell, concluded that there was very little, if any, microbial life beneath the seabed. He based this assessment on the absence of bacterial growth in cultures from marine sediments deeper than about 10 meters. About a decade later, this notion was reinforced when the deep-sea research submarine *Alvin* broke free of its tethers and sank. Although the scientists in the sub were rescued, a sandwich was not. Ten months later, *Alvin* was recovered and the sandwich was still intact. "Aha!" said scientists, "the ocean depths are too extreme to support microbial life!" Not surprisingly, this approach to studying deep-sea microbial ecology has not stood the test of time.

The modern age of subsurface microbiology was launched in the mid-1980s, when the U.S. Department of Energy drilled wells 200 m deep and a variety of microorganisms from subsurface aquifers were cultured. This was followed by a 1994 report of the growth of microorganisms from sediments collected 500 m below the seabed. In both cases, novel culture conditions were employed that more closely resembled the pressure, temperature, and redox potential of subsurface environments. Nonetheless, it was hard to convince the scientific community that these microbes weren't contaminants.

In the twenty-first century, advances in the tools available to microbial ecologists have propelled the "deep hot biosphere" to one of the most dynamic fields in microbiology and ecology. Technological advances including the advent of culture-independent methods have been crucial in probing what may be half of all microbial life on Earth. These techniques are designed to ask not just "Is there life?" but "Who is there?" "How fast are they growing?" and "What is their food?" Obtaining microorganisms in pure culture remains the most powerful tool available for understanding the taxonomy, genetics, and physiology of a single microbial strain. But just as it is impossible to learn everything about animals by studying only those in zoos, microbiologists cannot fully understand microbes without exploring them in nature. In the past, investigations were often limited by technology; now this rarely seems to be the case.

Recent estimates suggest that most microbial communities have between 10^{10} and 10^{17} individuals representing at least 10^7 different taxa. How can such huge groups of organisms exist and, moreover, survive together in a productive fashion? To answer this question, one must know which microbes are present and how they interact with each other and the environment. In other words, one must study **microbial ecology**—an interdependent series of inquiries that involves traditional, laboratory-based analyses, metagenomics, and in situ (Latin, in place) biogeochemical assessments (**figure 29.1**). The application of these techniques has brought an explosion of recent advances generating new appreciation of the vast diversity of microbes and their role in nutrient cycling and the health of their symbiotic partners.

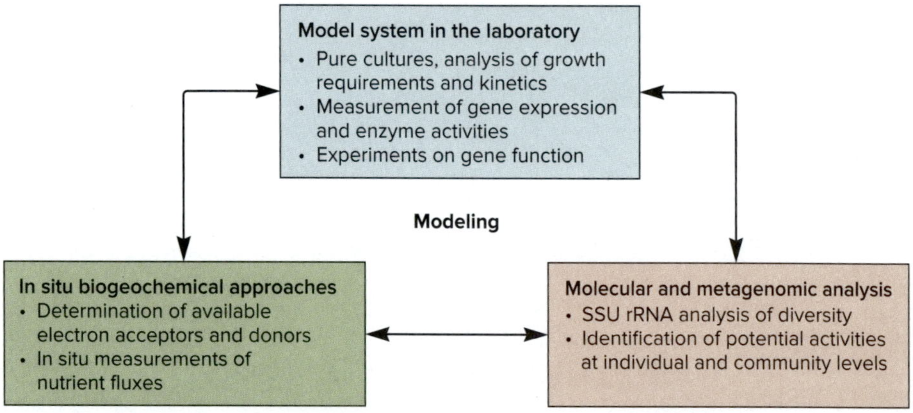

Figure 29.1 Microbial Ecology Is an Interdisciplinary Science. It relies on the interplay between a variety of approaches, which can be broadly classified as shown here.

Readiness Check:

Based on what you have learned previously, you should be able to:

- ✓ Differentiate between defined and complex culture media (section 7.7)
- ✓ Describe conventional techniques used to isolate single colonies and establish a pure culture (section 7.7)
- ✓ List environmental factors that influence microbial growth (section 7.5)
- ✓ Explain why and how microbes frequently grow in biofilms (section 7.6)
- ✓ Evaluate the importance of intercellular molecular signaling to microbial growth in nature (section 7.6)
- ✓ Explain why PCR results in the amplification of a specific DNA sequence despite many competing sequences (section 17.2)
- ✓ Summarize the principle and application of gel electrophoresis (section 17.1)
- ✓ Review the theory and practice of metagenomics and functional genomics (sections 18.3 and 18.5)
- ✓ Compare and contrast different molecular approaches to microbial taxonomy and phylogeny (section 19.3)
- ✓ Explain why SSU rRNA is used as a taxonomic marker (section 19.3)
- ✓ Explain how the oxidation-reduction (redox) state of an environment impacts the types of microbial metabolic processes that might be found there (section 28.1)

29.1 Microbial Biology Relies on Cultures

After reading this section, you should be able to:

a. Explain why, to the best of our knowledge, most microorganisms resist culturing in the laboratory
b. Describe the use of enrichment cultures
c. State when and how to apply the most probable number (MPN) method
d. Describe at least two innovative approaches used to coax growth in the laboratory of previously uncultured microbes
e. Explain the fundamentals of flow cytometry

It is now well understood that less than 5% of microbes have been grown under laboratory conditions. There are many reasons a particular microbe may resist culturing; a few are listed in **table 29.1**. The discrepancy between the number of microbial cells observed by microscopic examination and the number of colonies cultivated from the same natural sample has been called the **great plate count anomaly.** This has led to the description of potentially viable microbes as being "nonculturable" (*see figure 7.34*). A microbe is deemed **viable but nonculturable** (**VBNC**) if it fails to be cultivated but is determined to be viable by one of several methods. A common technique uses dyes that discriminate between live and dead cells that can be visualized microscopically (**figure 29.2**). A molecular approach, called viability polymerase chain reaction (vPCR) also uses a dye to differentiate between live and dead cells. In this case, the dye enters dead cells whose membranes have become damaged. Upon photoactivation, the dye crosslinks the DNA, thereby preventing subsequent PCR amplification of target genes (e.g., SSU rRNA). In this way, the DNA of only viable cells is present in the pool of PCR templates and is therefore detected as amplified DNA product.

Although the "gold standard" is an **axenic culture** (i.e., a pure culture) of the microbe of interest, one typically has to purify the microbe from a natural microbial assemblage. In such cases, it is often best to increase the number of microbes of interest relative to others in the same sample. **Enrichment culture** techniques are based on creating a microenvironment in the laboratory to promote the growth of a microorganism formerly restricted to a small ecological niche. At the same time, growth of other microbial types is inhibited. This approach often plays a central role in finding new microbes. As described in chapter 7, the success of an enrichment culture depends on a thorough understanding of the specific niche the microbe of interest inhabits and the physiological features that set that microbe apart from others. ◀ *Enrichment cultures (section 7.7)*

Table 29.1 Reasons Microbes May Be "Unculturable"

Potential Problem	Example Method Designed to Overcome
Microbe is slow growing.	Incubation times of weeks to months
Microbe is present in very low abundance.	Extinction cultures, using many replicates
Different microbes in same habitat are physiologically very similar. OR Inhibition by other microbes in a mixed culture	Remove other microbes from the sample by physical methods such as filtration or density-gradient centrifugation, or use extinction cultures.
Fastidious growth requirement	Assess growth requirements of similar microbes, if known; use annotation of metagenomic sequences to infer nutritional capabilities and requirements; grow in diffusion chambers that allow influx of small molecules from natural environmental samples without microbial contamination.
Cross-feeding or communication signals from other microbes are needed.	Co-cultivate strains; use diffusion chambers; grow in conditioned spent medium of "helper" microbe.
Triggers for growth or exit from a dormant state are not present.	Add known growth triggers such as N-acetylmuramic acid.

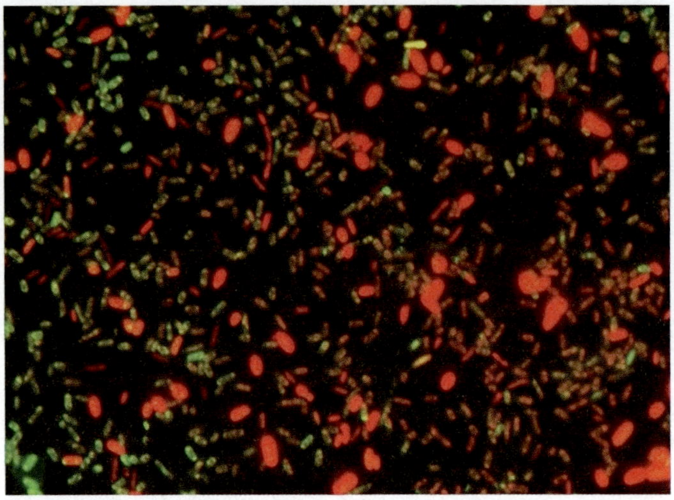

Figure 29.2 Assessment of Microbial Viability by Use of Direct Staining. By using differential staining methods, it is possible to estimate the portion of cells in a given population that are viable. In the LIVE/DEAD *Bac*Light Bacterial Viability procedure, two stains are used: a membrane-permeable green fluorescent nucleic acid stain and propidium iodide (red), which penetrates only cells with damaged membranes. In this *Escherichia coli* culture, living cells stain green, while dead and dying cells stain red.

A traditional approach to quantifying the presence of a particular microbial genus or species is the **most probable number (MPN)** technique. To perform MPN, the microbe must be culturable. This technique is still used to quantify specific types of microbes in natural samples, enrichment cultures, or, as is often the case, in food or water samples. The MPN technique involves establishing serial, 10-fold dilutions of the sample. Triplicate tubes of each dilution are usually established, although sometimes five or six replicates of each dilution are used for more accurate results. The theoretical basis of the MPN approach is that even if only one cell is inoculated into a tube, growth will occur. As shown in **figure 29.3**, after incubation at the appropriate temperature and duration, the tubes are examined for growth. Ideally, beyond a certain dilution, tubes fail to show growth because no cells were introduced into the medium. This is seen in the 10^{-5} and 10^{-6} triplicates in figure 29.3. In this case, the first dilution without growth would be marked as the last dilution in a set of three that includes the two dilutions that precede it. The number of tubes in the first set (10^{-3}, in this case) and the middle set is recorded. In our example, we see that this gives us a pattern of three tubes with growth in the 10^{-3} dilution tubes, one tube with growth in 10^{-4}, and zero tubes in 10^{-5}. A table based on statistical and theoretical considerations is consulted (figure 29.3*b*), and we see that the pattern 3-1-0 gives a

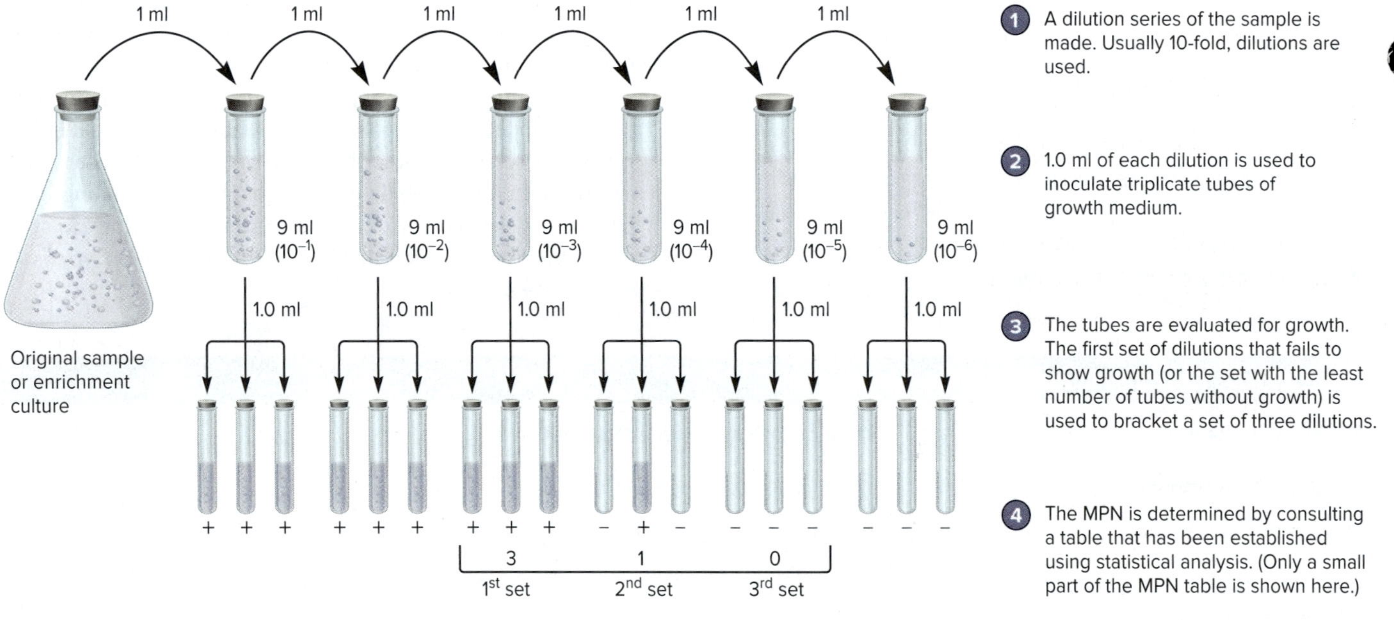

(a) Procedure for MPN analysis

Number of positive tubes in:			MPN
1st set	2nd set	3rd set	
3	0	3	0.95
3	1	0	0.43
3	1	1	0.75

MPN equals 4.3×10^3 microbes per 1.0 ml of sample

(b) MPN table

Figure 29.3 Most Probable Number Technique.

MPN of 0.43. What does this mean? The most probable outcome is that an average of 0.43 organisms were inoculated into each of the tubes of the middle set (in this case, 10^{-4}). From this, we conclude that the most-probable number of organisms per 1 milliliter of the original, undiluted sample would be 0.43×10^4 or 4.3×10^3.

Inspection of the MPN table in figure 29.3 reveals that the results do not always show the expected lack of growth only in the lowest dilution tubes. This can happen if the microbes were not randomly distributed within the sample or if they grew in clumps or chains. When this occurs, less than ideal results are obtained, and it may be best to use more replicates or a plating technique so that individual colonies can be counted.

Clearly, the MPN technique is only valid if the microbe of interest can be grown in the laboratory. Although most microorganisms found in nature have not yet been grown in the laboratory, it would be premature to claim that microbes currently unculturable will never be cultivated. Traditional growth media are now giving way to novel media and incubation conditions. One approach is to use unusual electron donors and acceptors. For instance, a novel isolate from Mono Lake (California) sediments was cultivated only when anoxic conditions with arsenite as the electron donor and nitrate as the electron acceptor were provided. It has also become clear that growing "unculturable" microbes requires patience. Microbiologists are most familiar with doubling times in the order of hours or, at the most, a few days, when in fact, this is not always the case. For example, newly described anaerobic soil isolates were discovered only when cultures on solid media were incubated over a 3-month period. Furthermore, microbiologists have been trained to equate turbidity in liquid culture with growth. This is not always true, as some bacteria stop dividing (i.e., reach stationary phase) at cell densities too low to appear turbid. Lastly, in recognition that growth of one microbe might depend on the product(s) of another, techniques have been developed that allow bacteria to exchange secreted metabolites and signaling molecules through filters or gels.

A variety of new culture techniques have been applied that address specific needs of microorganisms growing in diverse environments. Efficiency can be greatly improved by using high-throughput methods. One popular approach is the **extinction culture technique,** so named because natural samples are first microscopically examined and then diluted to a density of 1 to 10 cells, that is, diluted to extinction, much like in the MPN assay. Multiple 1-milliliter cultures are established in 48-well microtiter dishes. After incubating for the desired time under appropriate conditions, the samples are stained and examined microscopically to screen for growth. In many cases, cultures that do not appear turbid will nonetheless show signs of viable microbial growth and can be transferred to fresh medium. It is important that the initial culture has fewer than 10 cells; additional transfers and dilutions can ensure that a culture arising from a single cell (i.e., clonal growth) eventually is obtained.

Recent progress in cataloging microbes that make up the human microbiome has impelled the development of the high-throughput cultivation technique called culturomics. **Culturomics** involves at least three steps: First, samples of a mixed assemblage of microbes are used as inoculum for hundreds of different culture media incubated under a variety of physical conditions (e.g., temperature, pH, and oxygen concentration). Microorganisms from colonies that grow are then identified by mass spectrometry or small subunit (SSU; 16S for bacteria and archaea, 18S for eukaryotic microorganisms) rRNA gene sequencing. From these initial cultures, larger cultures can be started—that is, cultures are scaled up and in some cases organisms are grown in pure culture. Even if pure culture is not possible, these microbes can be the source for genome sequencing by single-cell genomics. ◀ *Single-cell genomic sequencing (section 18.2); Proteomics explores total cellular proteins (section 18.5)*

Two different techniques are commonly used to isolate single cells from a mixed population: flow cytometry and optical tweezers. In both cases, the isolated cells can be used to start cultures. In **flow cytometry,** cells are tagged with a fluorescent dye and injected into a flowing stream of fluid. As shown in **figure 29.4**, as the diameter of the stream narrows, one cell at a time is forced through a thin tube. Each cell is detected by a laser beam. The laser excites the fluorescent label and light emitted by each cell is collected by a detector that converts the signal made by the cell into a voltage pulse proportional to the size of the cell. Some of the cell's emitted light scatters to the side; this is a function of morphological complexity of the cell. This is also detected and quantified. Thus based on cell size and morphology, flow cytometry can detect

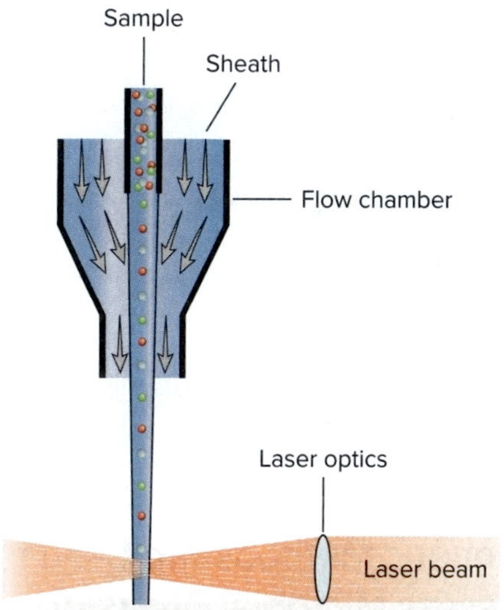

Figure 29.4 Flow Cytometry. Fluorescently labeled cell samples are injected into a fluid stream that narrows using a technique called hydrodynamic focusing. This forces only one cell at a time to pass a laser beam and scatter light in all directions. Forward-scattering light is a function of cell size, while the nature of laterally scattered light is determined by morphological complexity. This information is detected, measured, and converted to an electrical signal that reports the number and types of cells present.

MICRO INQUIRY *What happens to the flow stream in fluorescence activated cell sorting (FACS)?*

different populations of cells. This enables an additional function, **fluorescence activated cell sorting (FACS)**. Here, as the stream leaves the detector, vibration causes the fluid to break into single droplets with one cell per drop. Cells become electrostatically charged as they pass the detector. The charge is used to guide the cell into a collection vessel. Thousands of cells per second pass through the laser beam, making "flow" an extremely efficient means to count and sort cells.

A less efficient but more visually rewarding approach of isolating single cells is accomplished by **optical tweezers.** This technique uses a highly focused laser beam to manipulate microbial cells. As the laser beam is focused, it passes through a very strong electric field gradient. Cells can be polarized in an electric field (i.e., they are dielectric), so they are attracted to the center of the light beam within the gradient. Because light is a collection of particles (photons) that interact with matter in its path, the traveling laser beam impinges on cells as they travel through the electric current, moving the cells toward the narrowest point of the focused beam. In this way, an optical trap is made that moves microbial cells (and other small, dielectric particles, including DNA and proteins) in a directed fashion. When coupled with micromanipulation, a desired cell or cellular organelle can be drawn up into a micropipette. As with flow cytometry, once the microbe is isolated, it can be used as inoculum for the development of an axenic culture. Alternatively, PCR amplification of DNA from the individual cell can provide sequence data for phylogenetic analysis or single cell genomic sequencing.

Comprehension Check

1. When might a microbe be considered VBNC?
2. How would you determine the ratio of dead to living microbial cells in a soil sample?
3. Describe the rationale upon which the extinction culture technique is based.
4. Would you use fluorescence activated cell sorting or optical tweezers to obtain cells for analysis from (a) lake water, (b) a biofilm, (c) soil? Explain your choices.

29.2 Genetic Methods Are Used to Assess Microbial Diversity

After reading this section, you should be able to:

a. Explain why microbial communities are often investigated in situ
b. Describe why, when, and how FISH and CARD-FISH are used
c. Summarize how PCR is used to take a microbial census
d. Explain why and how DGGE is used
e. Describe the use of phylochips to assess microbial communities
f. Predict which techniques would be appropriate for assessing the microbial community in a gram of forest soil and in a milliliter of seawater

Microbial ecologists study natural microbial communities in soils or waters, or in association with other organisms, including humans. In all cases, scientists seek to assess microbial diversity. Diversity can be a measure of the number of genera, species, or ecotypes in an ecosystem. Alternatively, diversity can reflect the various types of metabolic strategies used in a single habitat: What carbon sources and electron donors and acceptors are used by the different microbes? A microbial **population** is a group of microorganisms within a single species or other taxon in an ecosystem. When a group of microbes is defined by its physiological activity, it is called a **guild,** in the same way that a guild of people often represents their occupation, for example, a carpenter's guild. A microbial example would be a guild of nitrifying microbes in a soil sample.

Staining Techniques

The most direct way to assess microbial diversity is to observe microbes in nature. This can be carried out in situ using immersed microscope slides or electron microscope grids placed in a location of interest, which are then recovered later for observation. Samples taken from the environment often are examined in the laboratory using classical cellular stains or fluorescent dyes that stain specific molecules. Individual cells can then be counted and their concentration calculated. The fluorescent stains DAPI (4′,6-*dia*mido-2-*p*henyl*i*ndole) and acridine orange are commonly used to visualize microbes in environmental, food, and clinical samples (**figure 29.5**). These dyes specifically label

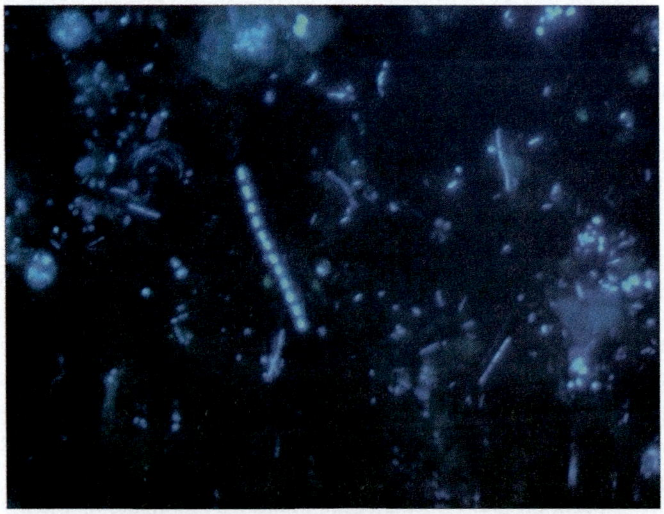

Figure 29.5 DAPI Stains Nucleic Acids. A microbial assemblage from the North Sea off the Belgian coast as visualized following DAPI staining. ©David C. Gillan/Mons University, Belgium

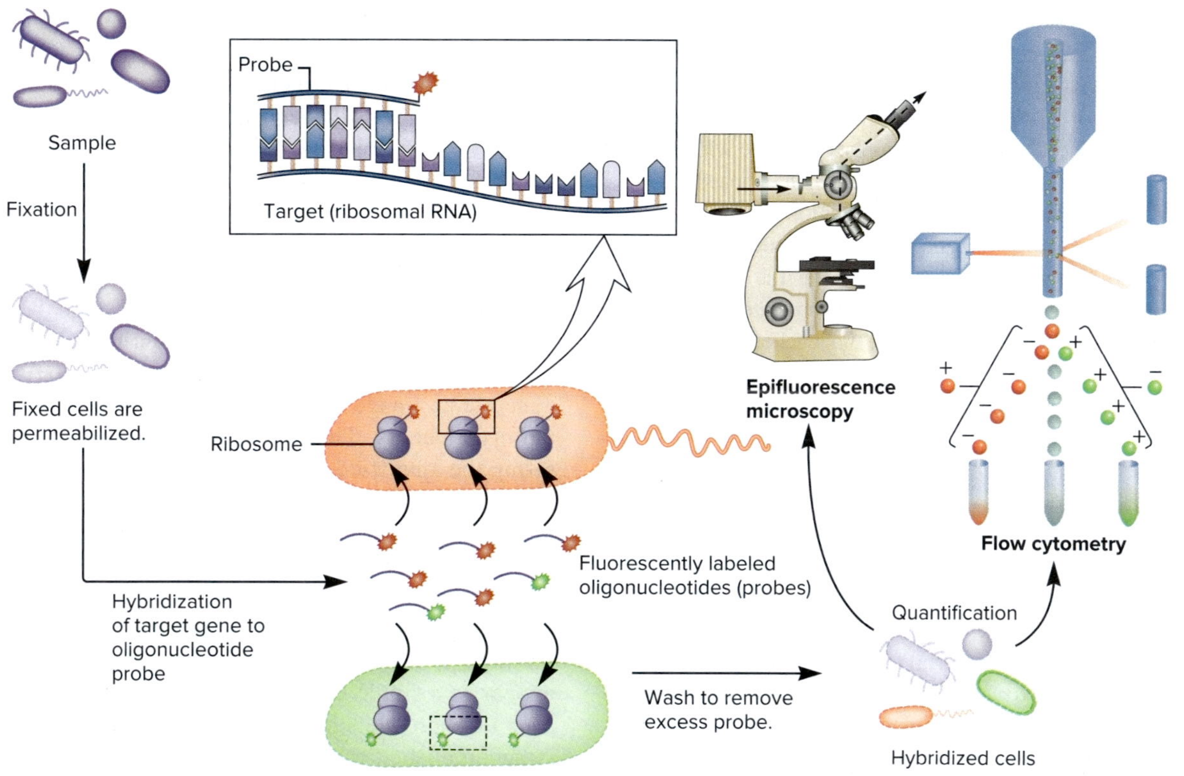

Figure 29.6 Fluorescence in Situ Hybridization (FISH).

nucleic acids (i.e., both DNA and RNA), and little sample preparation is required prior to their application. The nucleic acid dye SYBR Green is also used, in part because the signal is so bright virus particles can be seen.

Nucleic acid staining allows one to count the total number of microbes, but microbiologists often want to enumerate a specific genus or type of microbe. In these cases, a technique called ***f*luorescent *in s*itu *h*ybridization (FISH)** can be used. Here fluorescently labeled, single-stranded oligonucleotides (small lengths of DNA) that are known to be specific to the microbe of interest are used to label natural samples (**figure 29.6**). These oligonucleotides are called **probes.** Sequences within the gene encoding SSU rRNA are commonly used probes (*see figure 19.2*). To perform FISH, microbial cell membranes are made permeable so the probe can be internalized. Once inside the cell, it hybridizes to complementary nucleotide sequences. Probe that does not bind is washed off, and probe fluorescence reveals the organisms of interest. Probe fluorescence is visualized by **epifluorescence microscopy** (**figure 29.7**). Alternatively, individual cells can be separated and counted (but not visualized) by flow cytometry (figure 29.4). ◀ *Fluorescence microscopes use emitted light to create images (section 2.2)*

Depending on the probe used, FISH can identify a particular species, strain, or ecotype. This has made it a popular tool in clinical diagnostics and food microbiology, as well as microbial ecology. For example, FISH was used to discover the association between the archaea *Nanoarchaeum equitans* and *Ignicoccus hospitalis* (*see figure 20.8*).

Sometimes when FISH is used in natural samples, the fluorescent signal is not bright enough to be detected microscopically.

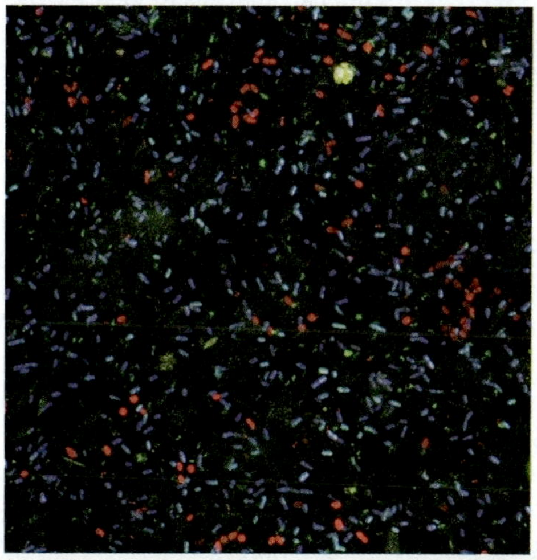

Figure 29.7 The Use of FISH to Study Microbial Populations. In this study, 16S rRNA probes tagged with different color fluorescent labels were used to study the microbial diversity within an earthworm gut. ©Seana Davidson

MICRO INQUIRY *Under what circumstances would one use epifluorescence microscopy, rather than flow cytometry, and vice versa?*

This is because microbes living in these environments often grow quite slowly. Thus each cell has relatively few ribosomes and the amount of probe that binds is below detectable limits. A clever modification of the FISH technique amplifies the signal produced by each cell. The trick is to label the fluorescent probe with an enzyme that makes lots of fluorescent product when exposed to substrate. This is called *ca*talyzed *r*eported *d*eposition-FISH, or simply **CARD-FISH.** The enzyme attached to the oligonucleotide probe is usually horseradish peroxidase (HRP). After the oligonucleotides are hybridized to the sample, the HRP substrate tyramide is added. Tyramide is oxidized by HRP and the level of fluorescence is amplified. In this way, a single oligonucleotide probe bound to just one molecule of rRNA can generate an "enzyme boosted" fluorescent signal.

Molecular Techniques

Because so few microbes can be grown easily in culture, the importance of molecular techniques to identify and quantify microorganisms is paramount. Typically, DNA from samples of soil, water, or other natural material (e.g., blood in a clinical setting) is extracted. The extracted DNA can then be used as template for amplification of specific genes by PCR or as template for shotgun metagenomics. Metagenomic analysis using high-throughput, next-generation sequencing is an important tool in studying microbial diversity. Computer software has been developed that analyzes shotgun metagenomic reads and quantifies the output such that the relative abundance of different microbial taxa can usually be assessed. ◀ *Next-generation DNA sequencing (section 18.1); Metagenomics provides access to uncultured microbes (section 18.3)*

When targeting specific genes, microbial ecologists regularly use SSU rRNA sequence analysis to identify the microbes that populate a community, although the **internal transcribed spacer region** (**ITS**) between the 16S and 23S rRNA genes may also be used *(see figure 13.21)*. Sometimes other genes besides SSU rRNA analysis are used. When protein-coding genes are used to define phylotypes, the genes must meet three important criteria: They need to be found in all known microbes, they need to have only one copy in the genome, and they need to rarely show evidence of horizontal gene transfer. Regardless of the target gene(s), PCR can be used to amplify DNA directly from the environment. Quantitative PCR can be used to assess the relative abundance of taxa. ◀ *Polymerase chain reaction amplifies targeted DNA (section 17.2)*

Despite the power of PCR, its use under these circumstances can be problematic because the rRNA genes of some microbes are more readily amplified than others. There are a variety of reasons for this, including the G + C content of a microbe's genome, the amount of nucleic acid degradation that might have occurred, and the relative ease with which some microbes can be lysed and thus their nucleic acids extracted (e.g., spores vs. vegetative cells). The observation that certain nucleotide templates are more readily amplified than others is termed **PCR bias.**

Once SSU rRNA nucleotide sequences are obtained, they can be compared with sequences isolated from other microbes that have been deposited in databases such as the Ribosomal Database Project. In this way, microbial ecologists can get a reasonable idea of the identity of the microbes that occupy a specific niche. Because whole organisms are not isolated and studied, it is said that specific **phylotypes** have been identified.

It is sometimes desirable to avoid nucleic acid sequencing and obtain a DNA fingerprint instead. DNA fingerprints generate an image much like a barcode that represents community diversity. This is most commonly achieved by separating the amplified rRNA gene fragments on an agarose gel (*see Techniques & Applications 17.1*). One microbial community can then be compared with another (differing in either time or space). Often a single pair of primers is used to amplify the same region of a SSU rRNA gene from a mixture of genomes. Despite different nucleotide sequences, the PCR products will have very similar molecular weights and thus appear as a single band on an agarose gel. To be of any use, such PCR products must be separated from one another. This is accomplished through **denaturing gradient gel electrophoresis** (**DGGE**).

DGGE relies on the fact that although the PCR fragments are about the same size, they differ in nucleotide sequence. Recall that the rate at which double-stranded DNA becomes single stranded varies with the G + C content; that is, it varies with DNA sequence. DGGE takes advantage of the observation that DNA of different nucleotide sequences will denature at varying rates and uses a gradient of chemicals (usually urea and formamide) to denature the DNA. In this technique, DNA fragments generated by PCR are loaded in a single well in a gradient gel. As electrophoresis proceeds, fragments will migrate until they become denatured. What appeared to be a single DNA fragment (band) on a nongradient gel will resolve into separate fragments (multiple bands) by DGGE. Because DGGE can be limited by poor gel-to-gel reproducibility, its popularity has declined in recent years. ◀ *DNA sequencing methods (section 18.1)*

Specialized microarrays called **phylochips** (**figure 29.8**) can also be used to assess microbial diversity in natural samples without nucleotide sequencing. Microarrays are glass chips upon which single-stranded DNA probes are attached in a grid-like pattern. Each DNA probe is a single gene (or part of a gene), and its identity and location on the slide are recorded. Single-stranded nucleotides from an environmental sample are incubated with the phylochip so that complementary probe and sample sequences hybridize. Hybridization to each probe (gene) is measured by laser beams. Analysis of DNA hybridization thus reveals the presence or absence of the specific genes in the environment.

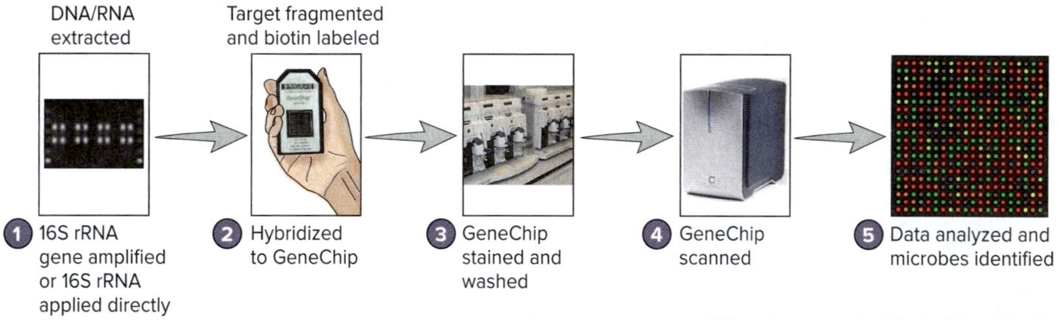

Figure 29.8 Phylochip Analysis. This form of microarray analysis enables the rapid evaluation of microbial diversity. (1) 16S rRNA genes or 16S rRNA molecules are obtained from the natural sample and labeled. (2) The 16S nucleic acids are hybridized to probes on a microarray called a phylochip. The probes consist of short regions of 16S rRNA genes obtained from preselected microbes. (3) The chip is washed to remove labeled 16S rRNA nucleic acids that could not hybridize because their complement is not found among the probes. (4) Laser beams scan the phylochip and (5) generate a series of spots of various colors and intensities that indicate the presence, absence, and, to a lesser extent, relative abundance of each microbe represented with an rRNA probe. These data are computer analyzed.

Comprehension Check

1. What macromolecules does DAPI stain? What are its advantages and disadvantages?
2. What is FISH? Describe an example in which the application of CARD-FISH might yield better results than traditional FISH.
3. Compare and contrast analysis of DNA extracted from a natural sample by shotgun metagenomics and targeted PCR amplification of specific genes. What kinds of questions can be asked with each method?
4. How might phylochip analysis be useful in food safety investigations?

29.3 Assessment of Microbial Community Activity Relies on Biochemistry and Genetics

After reading this section, you should be able to:

a. Explain how microbial ecologists measure community activity
b. Describe why microelectrode measurements are important tools in the study of microbial community activity
c. Compare the application of traditional stable isotope analysis with stable isotope probing
d. Assess the advantages and disadvantages of measuring in situ mRNA abundance
e. List the types of data that can be generated by MAR-FISH and compare this with functional gene arrays
f. Differentiate metagenomics from metaproteomics

Many microbial ecologists want to understand how microbes drive and respond to the biogeochemical cycles essential for life on Earth. Thus they seek to identify and measure the flux of nutrients, the interactions between microbes that make these fluxes possible, and the influences of abiotic factors. Here we divide methods into biogeochemical and molecular techniques, but it should become clear that the results of one type of approach are often used to formulate hypotheses that will be tested by another (figure 29.1). ◀ *Biogeochemical cycling sustains life on Earth (section 28.1)*

Biogeochemical Approaches

Microbial growth rates in complex systems can be measured directly. Changes in microbial numbers are followed over time, and the frequency of dividing cells is also used to estimate production. Alternatively, the incorporation of radiolabeled components such as thymidine (usually labeled with ^3H) into microbial biomass provides information about growth rates and microbial turnover. This approach is most easily applied to aquatic microorganisms; it is more difficult to directly observe terrestrial microbial communities.

Aquatic microorganisms can be recovered directly by filtration; their volume, dry weight, or chemical content can then be measured. It is sometimes not possible to directly measure cell density. In that case, carbon, nitrogen, phosphorus, or an organic constituent of the cells (such as protein) can be determined. This will give a single-value estimate of the microbial community. Such chemical measurements are typically expressed as microbial biomass.

To determine community activity, it is necessary to have a good understanding of the microenvironment inhabited by the microbial community in question. In many ecosystems, these microhabitats are studied by **microelectrodes**—electrodes capable of measuring pH, oxygen tension, H_2, H_2S, or nitrogen species. The tips of these electrodes range from 2 to 5 μm wide and 100 to 200 μm long, enabling measurements to be taken as the tip penetrates the sample in increments of a millimeter or less. Because the concentration and distribution of electron donors (e.g., sulfide, ammonium) and acceptors (e.g., oxygen, nitrate, sulfate) are measured, microbial activities can be inferred.

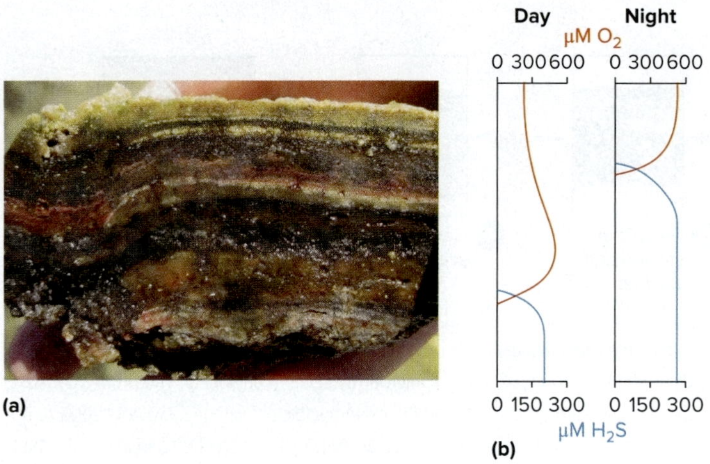

Figure 29.9 Microelectrode Analysis of Microbial Mats. (a) Microorganisms, through their metabolic activities, can create environmental gradients resulting in layered ecosystems. A vertical section of a hot spring (55°C) microbial mat, showing the various layers of microorganisms. (b) The measurement of oxygen and H_2S with microelectrodes demonstrates the diurnal shift in gas production: During the day, oxygen produced by cyanobacteria in the upper layers inhibits anaerobic respiration, so sulfate reduction is confined to the lower (dark) strata. At night, when oxygenic photosynthesis is diminished, atmospheric oxygen has a limited range of diffusion into the mat, and H_2S diffuses upward, where it can be used as an electron donor by purple photosynthetic bacteria.

(a) ©Reut S. Abramovich/Bar-Ilan University, Israel

Microelectrodes have been widely used in the study of **microbial mats** (figure 29.9). These highly striated communities of photoautotrophs and chemotrophs develop in saline lakes, marine intertidal zones, and hot springs.

In determining community function, it is often instructive to explore the cycling of a particular elemental species. In aquatic and marine systems, the amount of carbon fixed by microbes is important. This is most commonly accomplished by measuring how much radioactive bicarbonate ($H^{14}CO_3$) is taken up by microbes. As we discuss in chapter 30, when bicarbonate dissolves in water, it dissociates to CO_2. Thus when added to a sample of microbes, its incorporation into particulate carbon (i.e., biomass) represents the gross primary production; that is, the total amount of carbon fixation without accounting for CO_2 lost through respiration. Specialized CO_2 traps can be added to the incubation vessel to measure its loss by respiration. When CO_2 respired is subtracted from the amount of CO_2 fixed into biomass, net primary production has been assessed.

In addition to radioactive isotope incorporation, **stable isotope analysis** may be used. Isotopes are those forms of elements that differ in atomic weight because they bear different numbers of neutrons. Isotopes can be stable or unstable; only radioactive isotopes are unstable and decay. As shown in table 29.2, stable isotopes are rare when compared to their lighter counterparts. For example, both ^{14}N and ^{15}N occur in nature, but ^{14}N is found in much greater abundance. Organisms discriminate between stable isotopes, and for any given element, the lighter (in this case, ^{14}N) is preferentially incorporated into biomass, a phenomenon known as **isotope fractionation.** This discrimination can be used to follow the fate of a specific element by labeling the compound of interest with the heavier isotope. For instance, to explore the fate of NO_3^- in soil, samples of the soil would be incubated in the presence of $^{15}NO_3^-$. One would then measure the amount of ^{15}N incorporated into biomass, released as $^{15}N_2$, and reduced to $^{15}NH_4^+$. However, because the absolute value of ^{15}N is so small, it cannot be directly measured. Instead, the ratio of ^{15}N to ^{14}N in the sample is compared to the standard ratio found in inorganic material. An organism that assimilates $^{15}NO_3^-$ would contain more ^{15}N when compared to the standard. This difference, called delta (δ), is calculated as follows:

$$[(R_{sample} - R_{standard})/(R_{standard})] \times 1{,}000 = \delta_{sample\text{-}standard}$$

where R_{sample} is the ratio of heavy to light isotope in the sample and $R_{standard}$ is the ratio of heavy to light isotope in the standard. For instance, if our hypothetical soil community is found to have a $^{15}N/^{14}N$ ratio greater than the standard by 3 parts per thousand (ppt), this value is reported as $\delta\ ^{15}N = +3\ \delta$ ‰ (the symbol ‰ represents ppt). The true value of stable isotope fractionation rests in the fact that different microbes discriminate between heavy and light isotopes differently. For instance, nitrate production can be discerned from N_2 formation because the isotopic enrichment factor for denitrification is less than that for nitrification (about −26 ‰ vs. −34 ‰, respectively).

Another technique, called **stable isotope probing,** can be used to examine nutrient cycling in microbial communities as well as identify the microbes taking up the element of interest. As an example, this technique was employed to explore

Table 29.2 Natural Abundances of Some Biologically Relevant Isotopes

Hydrogen	Carbon	Nitrogen	Oxygen	Sulfur
^{1}H 99.984%	^{12}C 98.89%	^{14}N 99.64%	^{16}O 99.763%	^{32}S 95.02%
^{2}D 0.0156% (deuterium)	^{13}C 1.11%	^{15}N 0.36%	^{17}O 0.0375%	^{33}S 0.75%
			^{18}O 0.1995%	^{34}S 4.21%
				^{36}S 0.02%

methanogenesis in a rice paddy soil ecosystem. Researchers built **microcosms**—small incubation chambers with conditions that mimic the natural setting. In this case, they simulated natural rice paddies and introduced $^{13}CO_2$. Gaseous $^{13}CH_4$ was collected and RNA was isolated from the soil. The ^{12}C-containing RNA was separated from the ^{13}C-RNA on the basis of differing densities. The ^{13}C-containing rRNA, which could only be synthesized by those bacteria that assimilated the $^{13}CO_2$, was then used to identify these archaea; recall that all methanogenic microbes are archaea. This study demonstrates the value and importance of culture-independent approaches that combine biogeochemistry and molecular biology to understand microbial community ecology and physiology. ◀ *Methanogens and methanotrophs (section 20.3)*

Molecular Approaches

As introduced in section 18.3, the use of metagenomics to investigate natural populations has revolutionized the depth to which community dynamics can be assessed. Metagenomics applied to diverse natural habitats can have several outcomes, including the identification of new genes and gene products from uncultured microbes, the assembly of whole or partial genomes, and comparisons of community gene content from microbial assemblages of different origin. Researchers are finding large numbers of new genes from metagenomic analysis, and it is agreed that the discovery of new genes is far from complete. In fact, the number of metagenomic sequences obtained from uncultivated microbes in mixed assemblages outnumbers those obtained from microbes grown in pure culture. Here we discuss a few metagenomic approaches most commonly used by microbial ecologists when assessing community dynamics.

Recall that shotgun metagenomics involves the nucleotide sequencing of community DNA fragments. So far we have focused on using these nucleotide sequences to identify community members, but metagenomic sequence can also be used to infer community function. Metagenomes from a variety of biomes (e.g., mines, hypersaline ponds, estuaries) reflect the biogeochemical conditions of a habitat. For instance, community metagenomes from a well-oxygenated coral reef are rich in genes encoding respiration proteins, while the metagenomes from the colons of terrestrial animals have a high number of genes coding for enzymes involved in fermentation.

mRNA present in the environment can also be monitored, a technique called **metatranscriptomics.** Here mRNA extracted from the natural environment is reverse transcribed to cDNA (*see figure 17.5*) and directly analyzed by next-generation sequencing. The presence of mRNA demonstrates that a gene is actively transcribed and the gene product is probably active. Thus one can assess community activity at the time the sample was collected. However, rather than surveying the entire genetic landscape, a researcher may be interested in a particular metabolic function. In this case, a specific gene or mRNA encoding enzymes of interest would be targeted by specific primers. Example genes are listed in **table 29.3**. Such studies have shown that both dissimilatory nitrate reduction and the anammox reaction account for loss of nitrogen in specific habitats. ◀ *Transcriptome analysis (section 18.5)*

Table 29.3 Example Genes Used to Assess Community Function and Metabolic Diversity

Process	Gene	Gene Product
Ammonium oxidation	*amoA, amoC, amoB*	Ammonia monooxygenase
Anammox	*hzf*	Hydrazine hydrolase
Dissimilatory nitrate reduction	*narG* *nosZ* *nir*	Nitrate reductase Nitrous oxide reductase Nitrite reductase
Nitrogen fixation	*nifH, nifD, nifK*	Different subunits of nitrogenase
Methanogensis	*mcrA*	Methyl coenzyme M reductase
Methane oxidation	*mmoA*	Methane monooxygenase
Sulfate reduction	*apsA*	Adenosine phosphosulfate reductase

Unfortunately, most mRNAs are unstable, making recovery technically challenging. Instead of measuring mRNA, a **metaproteomics** approach can be used to identify each protein present at the time of sampling. There are two general approaches by which all the proteins in a community can be sampled and identified. The more labor-intensive technique uses 2D-polyacrylamide gel electrophoresis (2D-PAGE) so that each protein recovered from the environment is visualized (*see figure 18.14*). Individual proteins can be eluted from the gel and analyzed by mass spectroscopy (MS). Total peptide mass and amino acid sequence information is then used to search databases to identify the protein. This approach is best if the relative concentration of individual proteins is considered important. If a census of proteins is the goal, a more automated, high-throughput approach is taken. Here community proteins are extracted and digested with a protease to generate protein fragments called peptides. This mixture of peptides is separated by a technique known as 2D nano-liquid chromatography (nano-LC). Just as 2D gels separate proteins first by charge and then by molecular weight, 2D nano-LC separates peptides by two different properties: Peptides are first separated by charge and then by hydrophobicity. The peptide masses are determined by advanced mass spectrometry, and finally, tandem MS is used to obtain amino acid sequence information (*see figure 18.15*). In both cases, the protein sequences obtained are compared to metagenomic data obtained from the same environment. Metaproteomics is limited by the complexity of the environment being assessed. Complex environments

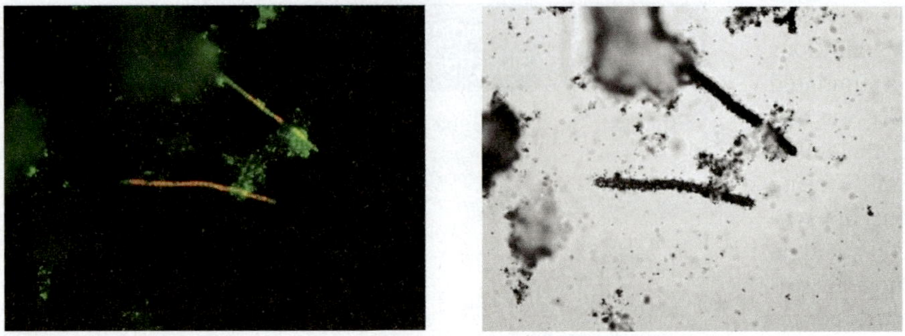

(a) (b)

Figure 29.10 MAR-FISH Technique Combines Microautoradiography with FISH in a Single Sample. Cells of interest are triple-labeled for identification and measurement of metabolic activity even if present in a complex microbial community. (a) The filamentous microbe of interest identified using a FISH probe that generates an orange signal under fluorescence microscopy. The green fluorescence is the result of a dye (YO-PRO-1) that binds to double-stranded nucleic acid such as DNA. (b) MAR resulted in black staining of the filamentous bacteria when viewed under bright-field microscopy, indicating that the microbes took up (assimilated) ^{3}H-labeled amino acids.

such as a pristine soil may have as many as a billion or more different proteins per gram. Even the most advanced and automated MS approaches are not yet able to resolve such a large and diverse group of proteins. |◀ *Proteomics explores total cellular proteins (section 18.5)*

Both metatranscriptomics and metaproteomics seek to assess activities that are taking place in the environment. Another method that also measures active processes uses FISH to identify specific mRNAs (figure 29.7). This approach is called **in situ reverse transcriptase (ISRT)-FISH.** Recall that in situ means "in place," so this technique allows the microbial ecologist to examine specimens and determine if a gene of interest is being expressed, as well as visualize the microbes expressing it. Let's say that one is interested in recently identified archaea capable of nitrification. These microbes express the archaeal *amoA* (ammonia monooxygenase) gene to oxidize ammonium to nitrite (table 29.3). One could design an *amoA* DNA probe—a short piece of DNA that is complementary to the *amoA* mRNA. Once the DNA probe has hybridized to the *amoA* mRNA in the natural sample, reverse transcriptase is used to generate a cDNA, which is then amplified by PCR. One now has an abundance of nucleotide fragments that can be fluorescently tagged in FISH. In our example of the archaeal nitrifier, we might want to determine where these archaea are metabolically active in the water column of an oceanic sampling station. Water samples of varying depths would be collected and filtered onboard a research vessel. The filters would most likely be fixed with preservative for later analysis (it is difficult to perform microscopy on a moving boat). The data collected by ISRT-FISH would indicate where archaeal nitrification was most active. This could then be correlated with physical parameters such as light penetration, oxygen concentration, and levels of nitrate, nitrite, ammonia, and dissolved organic nitrogen. In this way, the microbial ecologist can begin to build a hypothesis about the importance and distribution of these interesting microbes.

But what if one wanted to identify the microbe responsible for a specific metabolic activity? Recall that ISRT-FISH enables the visualization of the microbes expressing a specific gene but does not identify the microbe. To simultaneously assess metabolic activity and phylogenetic identity, a technique that combines an older method, **microautoradiography (MAR)**, with FISH can be employed. This method is thus called MAR-FISH. To understand MAR-FISH, it is necessary to first explain MAR. In MAR, a radioactive substrate is mixed with the sample and incubated under specific conditions, usually closely resembling those in the field. At the completion of the incubation period, the sample is washed to remove any radioactive substrate not taken up by the cells. A sample attached to a microscope slide is treated with a photographic emulsion. The slide is carefully wrapped and stored in complete darkness. As the incorporated radioactive material decays, the silver grains in the photographic emulsion are exposed. When developed, this leaves a series of black dots around the cells that assimilated the radioactive substrate (**figure 29.10**). When MAR is combined with FISH, one can analyze the phylogenetic identity and the specific substrate uptake patterns of microbes of interest at the level of a single cell in complex microbial communities.

Comprehension Check

1. How do radioactive and stable isotopes differ? Design an experiment to determine whether or not denitrification is taking place in a soil ecosystem.
2. How might microelectrodes be used to gather data in preparation for a stable isotope study?
3. Develop a question that could be answered by metatranscriptome analysis.
4. What is metaproteomics? How is it similar to metagenomics? How does it differ?
5. What is the difference between MAR and MAR-FISH? Suggest a microbial ecology question that could be answered by MAR and one that would require MAR-FISH.

Key Concepts

29.1 Microbial Biology Relies on Cultures

- Less than 5% of microbial taxa have been cultured. The discrepancy between the number of microbial taxa observed and that grown is termed the great plate count anomaly and reflects the large proportion of microbes that are viable but nonculturable.
- Although most microbes cannot be grown in pure culture, novel approaches such as extinction culture and culturomics have resulted in the isolation of previously uncultured microorganisms.
- Flow cytometry is a fast and efficient method of counting and sorting cells; optical tweezers can be used to isolate single cells. Both methods can be used to obtain cells for culture.

29.2 Genetic Methods Are Used to Assess Microbial Diversity

- A variety of staining techniques are used to observe microbes in natural environments. Fluorescent in situ hybridization (FISH) labels specific microbes that possess a specific nucleotide sequence, usually a region of the SSU rRNA gene (**figures 29.5–29.7**).
- Shotgun metagenomics provides the most complete information about the microbial population because all genes present in DNA extracted from natural samples are sequenced and analyzed.
- Amplification of SSU rRNA genes from DNA extracted from natural samples is a common approach for assessing microbial diversity.
- Microarrays that have SSU rRNAs as probes, called phylochips, can quickly evaluate population and community diversity (**figure 29.8**).

29.3 Assessment of Microbial Community Activity Relies on Biochemistry and Genetics

- Microelectrodes can be used to determine physical and biological parameters such as pH, O_2, and H_2S concentrations in microniches (**figure 29.9**).
- Stable isotope analysis is based on the fact that organisms discriminate between heavy and light naturally occurring isotopes. It is often used to monitor nutrient flux through ecosystems.
- Metatranscriptomics assesses community activity by measuring microbial gene expression.
- Metaproteomics can be used to identify the proteins that are expressed by a microbial community. Because the number of proteins is so large, traditional 2D gel analysis is usually not practical and high-throughput chromatographic techniques are used instead.

Active Learning

1. How might you attempt to grow in the laboratory a chemolithoautotroph that uses ferrous iron as an electron donor and oxygen as an electron acceptor?
2. Both acetate and CO_2 can be used by methanogenic archaea to generate methane (CH_4). How would you determine the source of carbon for methanogenesis in a waterlogged peat?
3. There are two types of microbial mats in Shark Bay, Australia: smooth and pustular (bumpy). Noting that these mats seem to have a higher number of archaea than other mats, a team of U.S. and Australian scientists sought to identify and enumerate archaeal genera, determine their metabolic strategies (e.g., methanogenesis), and determine their spatial distribution relative to the physical features of the mats, such as oxygen penetration. Discuss the techniques that might have been used to answer these questions.

Read the original paper: Wong, H. L., et al. 2017. Dynamics of archaea at fine spatial scales in Shark Bay mat microbiomes. *Sci. Rep.* 7:46160. doi:10.1038/srep46160.

30

Microorganisms in Marine and Freshwater Ecosystems

©Explorer/Science Source

Ocean Death Coming Soon to a Coast Near You

If you live near a coast, you're probably not far from an ocean dead zone. Ocean dead zones are regions in which all plant and animal life dwelling below the surface has suffocated. These regions have been increasing in size and frequency since the 1960s, and could occupy as much as 20% of the world's oceans by 2100. Dead zones result when phytoplankton respond to increases in nutrients and temperature by growing (and growing and growing) and then sinking. Initially, phytoplankton on the ocean floor means more food for heterotrophic organisms. But as the number of oxygen-consuming animals and microbes increases, available oxygen dwindles. Invertebrates and fish die. Plants cannot grow because there is not enough light. The only organisms left are anaerobic chemolithoautotrophs—microbes that make a living fixing CO_2 and extracting energy from reduced inorganic compounds.

Of the more than 500 ocean dead zones globally covering over 245,000 square kilometers, the Baltic Sea is the largest, while the second largest dead zone is the mouth of the Mississippi River. This mighty river dumps fertilizer collected from watersheds ranging from Montana to Louisiana. The addition of literally tons of extra nitrogen and phosphorus to the Gulf of Mexico has resulted in a dead zone that is roughly the size of Connecticut and Rhode Island combined. The presence of such a large dead zone has huge and obvious implications for fisheries, sea birds, tourism, and the overall health of the Gulf.

Are dead zones reversible? In some cases, yes. But like most problems, prevention is the best approach. Prudent use of fertilizer and limiting surface runoff are good places to start. Another area of concern is the deposition of animal waste into local waterways. Finally, proper sewage treatment (*section 43.2*) and the restoration of natural wetlands to filter and consume much of the nitrogen and phosphorus before it flows to the coast are projects worth pursuing.

In this chapter, we turn our attention to the microbial ecology of marine and freshwater environments. **Oceanography** is the study of marine systems and the biological, physical, geological, and chemical factors that impact biogeochemical cycling, water circulation, and climate. Microbial oceanography seeks to understand marine microorganisms at the level of the individual taxon and at the same time elucidate how they influence biogeochemical cycling at a global level. **Limnology** is the investigation of aquatic systems within continental boundaries, including glaciers, groundwater, rivers, streams, and wetlands. All terrestrial life depends on the sustained provision of clean freshwater from these ecosystems. This chapter provides a brief introduction to these dynamic and interdisciplinary fields.

Readiness Check:

Based on what you have learned previously, you should be able to:

- ✓ Explain the global impact that microbes have on carbon and nitrogen cycling (section 28.1)
- ✓ Diagram the viral life cycle (section 6.3)
- ✓ List environmental factors that influence microbial growth (section 7.5)
- ✓ Compare and contrast the growth and physiology of obligate aerobes, facultative anaerobes, and strict anaerobes (section 7.5)
- ✓ Explain why and how microbes frequently grow in biofilms (section 7.6)
- ✓ Explain how the oxidation-reduction (redox) state of an environment impacts the types of microbial metabolic processes that might be found there (section 28.1)
- ✓ Identify, by nutritional type, microorganisms responsible for carbon fixation, nitrogen fixation, nitrification, dissimilatory nitrate reduction, denitrification, nitrification, the anammox reaction, and sulfur reduction (sections 20.2, 21.4, 21.5, 22.1, 22.3, and 22.4)
- ✓ Review innovative methods for growing microbes isolated from nature in the laboratory (section 29.1)

30.1 Water Is the Largest Microbial Habitat

After reading this section, you should be able to:

a. Explain the carbonate equilibrium system
b. Discuss the cause and consequences of ocean acidification

Microbes drive marine and freshwater ecosystems. They are the most diverse and abundant organisms in these habitats. The importance of microbial activities in oceans, rivers, and lakes has been brought into sharp focus by global climate change. To understand the capacity of microbes to influence the worldwide flux of nutrients and greenhouse gases, we must first consider marine and freshwater habitats.

The nature of water as a microbial habitat depends on a number of physical factors such as temperature, pH, and light penetration. One of the most important factors is dissolved oxygen. The flux rate of oxygen through water is about 10,000 times less than its rate through air. However, in some aquatic habitats, the limits to oxygen diffusion can be offset by the increased solubility of oxygen at colder temperatures and increasing atmospheric pressures. For instance, in the very deep ocean, the dissolved oxygen concentration actually increases with depth, even though the air-water interface can be literally miles away. On the other hand, tropical lakes and summertime temperate lakes may become oxygen limited only meters below the surface.

As seen in the development of dead zones (chapter opening story), light is also critical for the health of marine and freshwater ecosystems. Like all life on Earth, all organisms in these environments depend on **primary producers**—autotrophic organisms—to fix CO_2, thereby providing organic carbon. In streams, lakes, and coastal marine systems, macroscopic algae and plants are the chief primary producers. Organic carbon also enters these systems in terrestrial runoff. The situation is very different in the open ocean, where all organic carbon is the product of microbial autotrophy. In fact, about half of all **primary production** (carbon fixation) on Earth is the result of this microbial carbon fixation.

Solar radiation warms the water, and this can lead to thermal stratification. Warm water is less dense than cool water, so as the sun heats the surface in tropical and temperate waters, a **thermocline** develops. A thermocline can be thought of as a mass of warmer water floating on top of cooler water. These two water masses remain separate until there is either a substantial mixing event, such as a severe storm, or in temperate climates, the onset of autumn. As the weather cools, the upper layer of warm water becomes cooled and the two water masses mix. This is often associated with a pulse of nutrients from the lower, darker waters to the surface. This pulse of nutrients can trigger a sudden and rapid increase in the population of certain microbes, and a bloom may develop. This is considered more fully in the discussion of microorganisms in coastal marine systems and lakes.

Because CO_2 is critical in many chemical and biological processes, it is important to understand its fate in freshwater and marine ecosystems. The pH of unbuffered distilled water is determined by dissolved CO_2 in equilibrium with the air and is typically 5.0 to 5.5. The pH of freshwater systems such as lakes and streams, which are usually only weakly buffered, is therefore controlled by terrestrial input (e.g., minerals that may be either acidic or alkaline) and the rate at which CO_2 is removed by photoautotrophic organisms such as diatoms; CO_2 fixation can increase the pH of the water.

In contrast, seawater is strongly buffered by the balance of CO_2, bicarbonate (HCO_3^-), and carbonate (CO_3^{2-}). Atmospheric CO_2 enters the oceans and either is converted to organic carbon by photosynthesis or reacts with seawater to form carbonic acid (H_2CO_3), which quickly dissociates to form bicarbonate and carbonate (**figure 30.1**):

$$CO_2 + H_2O \rightleftharpoons H_2CO_3 \rightleftharpoons H^+ + HCO_3^- \rightleftharpoons 2H^+ + CO_3^{2-}$$

The oceans are effectively buffered between pH 7.6 and 8.2 by this **carbonate equilibrium system.** Much like the buffer one might use in a chemistry experiment, the pH of seawater is determined by the relative concentrations of bicarbonate and carbonate.

The reactions of the carbonate equilibrium system have taken on new importance as increases in atmospheric CO_2 result in higher rates of CO_2 dissolution in seawater. In fact, the ocean is already absorbing about 25% of the anthropogenically produced CO_2. This has slowed the rate of global climate change, but has also led to **ocean acidification.** The average pH of the ocean has dropped by 0.1 unit (to 8.1) since the pre-industrial age. It is expected to decline another 0.35 to 0.50 by 2100, unless effective means of limiting CO_2 emissions are implemented. This is significant because as CO_2 increases, the carbonate equilibrium shifts away from $CaCO_3$, potentially limiting its availability for the formation of shells

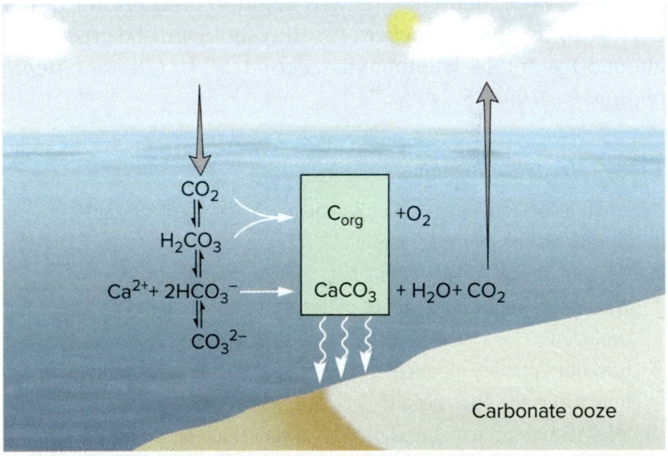

Figure 30.1 The Carbonate Equilibrium System. Atmospheric CO_2 enters seawater and is converted to organic carbon (C_{org}) or to carbonic acid (H_2CO_3) that spontaneously dissociates into the weak acids bicarbonate (HCO_3^-) and carbonate (CO_3^{2-}). Calcium carbonate ($CaCO_3$), a solid, precipitates to the seafloor, where it helps form a carbonate ooze. This system keeps seawater buffered at about pH 8.0.

MICRO INQUIRY *How can the production of $CaCO_3$ serve to reduce atmospheric CO_2 concentrations?*

Figure 30.2 **Coccolithophores Respond to Seawater Acidification.** *Emiliania huxleyi,* an abundant species in the open ocean.

and skeletons of marine organisms. Because protists known as coccolithophores produce about a third of the marine $CaCO_3$, they have been the focus of investigations seeking to predict the effects of ocean acidification. While responses are species specific, it is clear that these seemingly slight shifts in pH have an impact on their ability to form their shells, or coccoliths (**figure 30.2**). Ocean acidification is also having a negative impact on corals. Although it is difficult to predict, some models estimate that when atmospheric CO_2 reaches about 550 ppm, many corals will begin to erode faster rather than grow. If left unchecked, this threshold will be crossed in about 50 years. ◀ Haptophyta *(section 24.4); Global climate change (section 28.2)*

Comprehension Check

1. What factors influence oxygen solubility? How is this important in considering aquatic environments?
2. Describe the buffering system that regulates the pH of seawater. What might be the implications of this stable buffering on microbial evolution?
3. How does primary production in lakes and coastal ecosystems differ from that in the open ocean?
4. What features of a thermocline make it similar to hot and cool masses of air?

30.2 Microorganisms in Marine Ecosystems

After reading this section, you should be able to:

a. Explain the ecological and economic importance of estuaries and harmful algal blooms
b. Diagram carbon and sulfur flux in a Winogradsky column
c. Assess the importance of marine microbial primary productivity in the global carbon budget
d. Illustrate the microbial loop and evaluate its importance in the open ocean
e. Evaluate the importance of members of the SAR11 clade in marine ecosystems
f. Describe at least three mechanisms microbes have evolved to adapt to the low nutrient conditions found in the open ocean
g. Explain the role of viruses in maintaining and influencing marine microbial populations, as well as their contribution to nutrient cycling
h. Explain the distribution of microbial types that inhabit marine subsurface sediment

As terrestrial organisms, we must remind ourselves that about 96% of Earth's water is in marine environments. From a microbiological perspective, surface ocean waters have been most intensely studied; only recently have scientists begun to probe the benthos—deep-sea sediments and the subsurface. We begin our discussion of marine ecosystems with estuaries, then consider the microbial communities that inhabit the open ocean and, finally, the dark, cold, high-pressure benthos.

Microorganisms in Coastal Ecosystems Are Adapted to a Changeable Environment

An estuary is a semi-enclosed coastal region where a river meets the sea. Estuaries are defined by tidal mixing between freshwater and saltwater. Most estuaries undergo large-scale tidal flushing; this forces organisms to adapt to changing salt concentrations on a daily basis. Microbes that live under such conditions combat the resulting osmotic stress by adjusting their intracellular osmolarity to limit the difference with that of the surrounding water. Most protists and fungi produce osmotically active carbohydrates for this purpose, whereas bacteria and archaea regulate internal concentrations of potassium or compounds such as proline and betaine. Salt is another stressor, and most microbes that inhabit estuaries are halotolerant, which is distinct from halophilic. **Halotolerant** microbes can withstand significant changes in salinity; halophilic microorganisms have an absolute requirement for high salt concentrations. ◀ *Solutes affect osmosis and water activity (section 7.5)*

Estuaries are unique in many respects. Their calm, nutrient-rich waters serve as nurseries for juvenile forms of many commercially important fish and invertebrates. However, despite their importance to the commercial fishing industry, estuaries are among the most polluted marine environments. Estuaries and the rivers that feed them are the receptacles of toxins discharged during industrial processes as well as agricultural and street runoff. These pollutants include organic materials, particularly fertilizer, that can be used as nutrients.

If too much organic matter is consistently added, the system is said to be **eutrophic,** and the rate of respiration greatly exceeds that of photosynthesis. Thus the water may become anoxic (chapter opening story). By contrast, if nutrients are introduced in a pulselike fashion, a single photoautotrophic species, either algal or cyanobacterial, grows at the expense of all other organisms in the community. This phenomenon, called a **bloom,** often results from the introduction of nutrients combined with changes in light, temperature, and mixing of water masses. If the microbes produce a toxic product or are in themselves poisonous to other organisms such as shellfish or fish, the term **harmful algal bloom (HAB)** is used. Some HABs are called **red tides** because the microbial density is so great that the water becomes red or pink, the color of the algae (**figure 30.3**).

The number of HABs has dramatically increased in the last decade or so. It is predicted that global climate change may trigger more HABs, in part because warmer water forms a more stable thermocline that lasts longer into the autumn. HABs are sometimes responsible for killing large numbers of fish or marine mammals. For instance, off the coast of California, periodic HABs are responsible for sudden, large-scale deaths of sea lions. These blooms are caused by diatoms in the genus *Pseudonitzschia.* Anchovies consume the diatoms, and a potent neurotoxin, domoic acid, accumulates in the fish. The mammals are poisoned after they ingest large quantities of the fish. ◀ Stramenopila *(section 25.4)*

In addition to diatoms, HABs are often caused by dinoflagellates. Some bloom-causing dinoflagellates produce potent neurotoxins. Dinoflagellates of the genus *Alexandrium* produce a toxin responsible for paralytic shellfish poisoning (PSP), which affects humans as well as other animals in coastal, temperate North America. The toxin produced by the dinoflagellate *Karenia brevis* killed a number of endangered manatees and bottlenose dolphins in 2002 in a bloom in Florida. ◀ Alveolata *(section 25.4)*

Salt marshes generally differ from estuaries because their freshwater input is from multiple sources, rather than a single river. The microbial communities within salt marsh sediments are very dynamic. These ecosystems can be modeled in **Winogradsky columns** (**figure 30.4**), named after the pioneering microbial ecologist Sergei Winogradsky (1856–1953).

A Winogradsky column is easily constructed using a glass cylinder into which sediment is placed and then overlaid with seawater. When the top of the column is sealed, much of the cylinder becomes anoxic. The addition of shredded newspaper introduces a source of organic carbon that is fermented by members of the genus *Clostridium.* With these fermentation products available as electron donors, *Desulfovibrio* spp. and other microbes use sulfate as a terminal electron acceptor to produce hydrogen sulfide (H_2S). The H_2S diffuses upward, creating a stable H_2S gradient where photolithoautotrophs of the genera *Chlorobium* and *Chromatium* develop as visible olive green and purple zones, respectively.

Figure 30.3 Harmful Algal Blooms Are Also Called Red Tides. Blooms often color the water with the pigment of the microbe responsible. This HAB occurred in Puget Sound, Washington, USA.

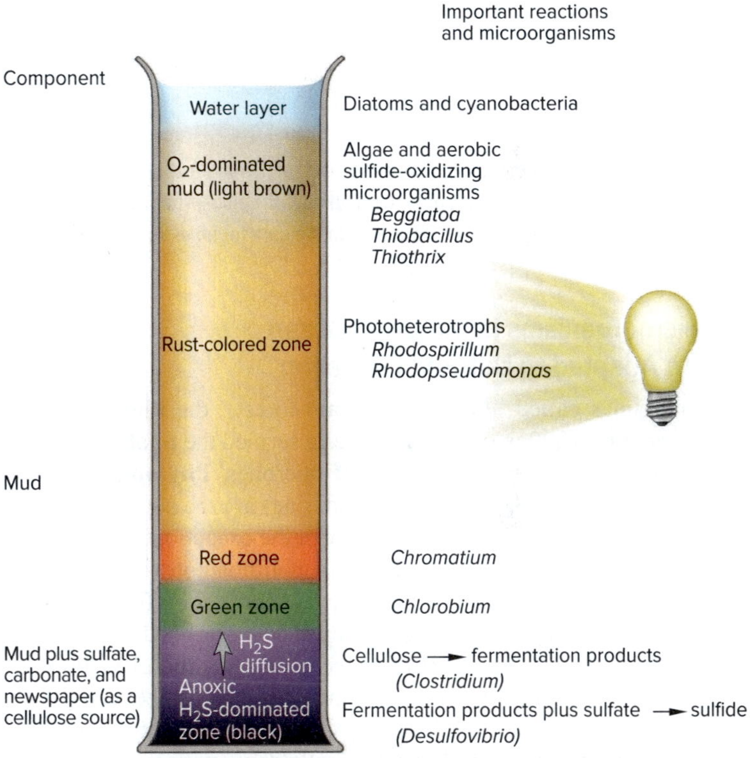

Figure 30.4 The Winogradsky Column. Microorganisms and nutrients interact over a vertical gradient in these microcosms. Fermentation products and sulfide migrate up from the reduced lower zone, and oxygen penetrates from the surface. This creates conditions similar to those in a lake or salt marsh with nutrient-rich sediments. Light simulates the penetration of sunlight into the anoxic lower region, which stimulates photosynthetic microorganism growth.

MICRO INQUIRY *How is sulfur cycled between the anoxygenic photosynthetic microbes belonging to the genera* Chlorobium *and* Chromatium, *and anaerobic heterotrophic* Desulfovibrio *spp.?*

These green and purple sulfur bacteria use H_2S as an electron source and CO_2, from sodium carbonate, as a carbon source. Above this region, purple nonsulfur bacteria of the genera *Rhodospirillum* and *Rhodopseudomonas* grow. These photoorganotrophs use organic matter as an electron donor under anoxic conditions and function in a zone where the sulfide level is lower. Both O_2 and H_2S may be present higher in the column, promoting the growth of other microorganisms. These include the chemolithotrophs in the genera *Beggiatoa* and *Thiothrix,* which use reduced sulfur compounds as electron donors and O_2 as an acceptor. In the upper, aerated portion of the column, diatoms and cyanobacteria may be visible. ◀ *Anaerobic respiration uses the same three steps as aerobic respiration (section 11.7); Phototrophy (section 11.12); Photosynthetic bacteria are diverse (section 21.4); Purple nonsulfur bacteria perform anoxygenic photosynthesis (section 22.1); Class* Gammaproteobacteria *is the largest bacterial class (section 22.3); Class* Clostridia *(section 23.2)*

Comprehension Check

1. Name two groups of protists known to cause HABs in marine ecosystems. What hazards do HABs pose for other marine organisms and humans?
2. Describe the ecosystem that develops within a Winogradsky column. Which of these organisms are chemolithoautotrophic? What would happen if the column were incubated in the dark for an extended period of time?

Microorganisms in the Open Ocean Are Adapted to Nutrient Limitation

Sometimes called "the invisible rain forest," the illuminated upper 200 to 300 m of the open ocean, termed the **photic zone,** is home to a diverse collection of microbes. **Phytoplankton** (Greek *phyto,* plant; *planktos,* wandering) are the source of all organic matter in the open ocean, with the exception of the hydrothermal vent and subsurface chemolithoautotrophic communities (*see figure 27.4*). Roughly half the carbon fixation in the open ocean is performed by cyanobacteria in the genera *Prochlorococcus* and *Synechococcus.* These microbes help constitute the **picoplankton** (planktonic microbes between 0.2 and 2.0 µm in size), which can represent 20 to 80% of the total phytoplankton biomass. Larger unicellular cyanobacteria (*Crocosphera watsonii*) and eukaryotic autotrophs, especially diatoms, contribute the other 50% of fixed carbon. ◀ *Phylum* Cyanobacteria: *oxygenic photosynthetic bacteria (section 21.4);* Stramenopila *(section 24.4); Sulfide-based mutualisms (section 27.2)*

In contrast to coastal regions, nutrient levels are extremely low in the open ocean, which is unaffected by rivers, streams, and terrestrial runoff. The key to survival in the photic zone is the tight recycling of nutrients, rather than allowing them to sink to the seafloor; this is the essence of the **microbial loop** (**figure 30.5**). Within the microbial loop, heterotrophic microbes consume **dissolved organic matter (DOM)** released as exudates by other microbes as well as larger plants and animals. The DOM released by photoautotrophic microbes is sometimes called **photosynthate.** Marine viruses, present at concentrations up to 10^8 per milliliter, are also a source of DOM as the lysis of their host cells contributes significantly to the return of nutrients into the microbial loop. This generates a common pool of DOM that serves as a nutrient source for all osmotrophic organisms. Heterotrophic protists, including flagellates and ciliates, consume smaller microbes, which may include other protists (figure 30.5*b,c*). Microbial cells, insoluble detritus, and other solid organic material are called **particulate organic matter (POM)**. Because protists are consumed by zooplankton, both DOM and POM are recycled at a number of trophic levels. The net effect of the microbial loop is the reuse of essential nutrients such as organic carbon, nitrate, and phosphate within the photic zone.

The fate of all the carbon fixed by marine phytoplankton has practical importance. Global climate change has focused intense scrutiny on determining how much CO_2 phytoplankton can remove from the atmosphere and sequester in the benthos. In other words, if the microbial loop could be overloaded, organic carbon that started out as atmospheric CO_2 might be pumped to the seafloor. This is sometimes called the "biological carbon pump," and it describes the export of carbon to the seafloor.

To understand how the microbial loop prevents most carbon from escaping to the seafloor, we need to consider the fate of organic matter that leaves the photic zone and falls through the depths as **marine snow** (**figure 30.6*a***). This material gets its name from its appearance, sometimes seen in video images of mid- and deep-ocean water, where drifting flocculent particles resemble snow. Marine snow consists primarily of fecal pellets, diatom frustules, and other materials not easily degraded. During its fall to the bottom, marine snow is colonized by microbes so that nutrient recycling and mineralization continue. Bacterial population densities associated with marine snow can reach 10^8 to 10^9 per milliliter, which is orders of magnitudes higher than the surrounding water. These microbes secrete a variety of exoenzymes to break down organic polymers so the resulting monomers can be assimilated. As the marine snow sinks, it leaves a plume of DOM, nitrogen, phosphorus, and iron. These nutrients attract chemotactic free-swimming microbes, which help to deplete the nutrient pool even further (figure 30.6*b*). By the time a piece of marine snow reaches the seafloor, less than 1% of photosynthetically derived organic carbon remains unaltered. Once there, only a tiny fraction will be buried in the sediments; most will return to the surface in upwelling regions (**figure 30.7**).

An urgent question is whether or not the oceans can increase CO_2 drawdown (remove and bury) as atmospheric

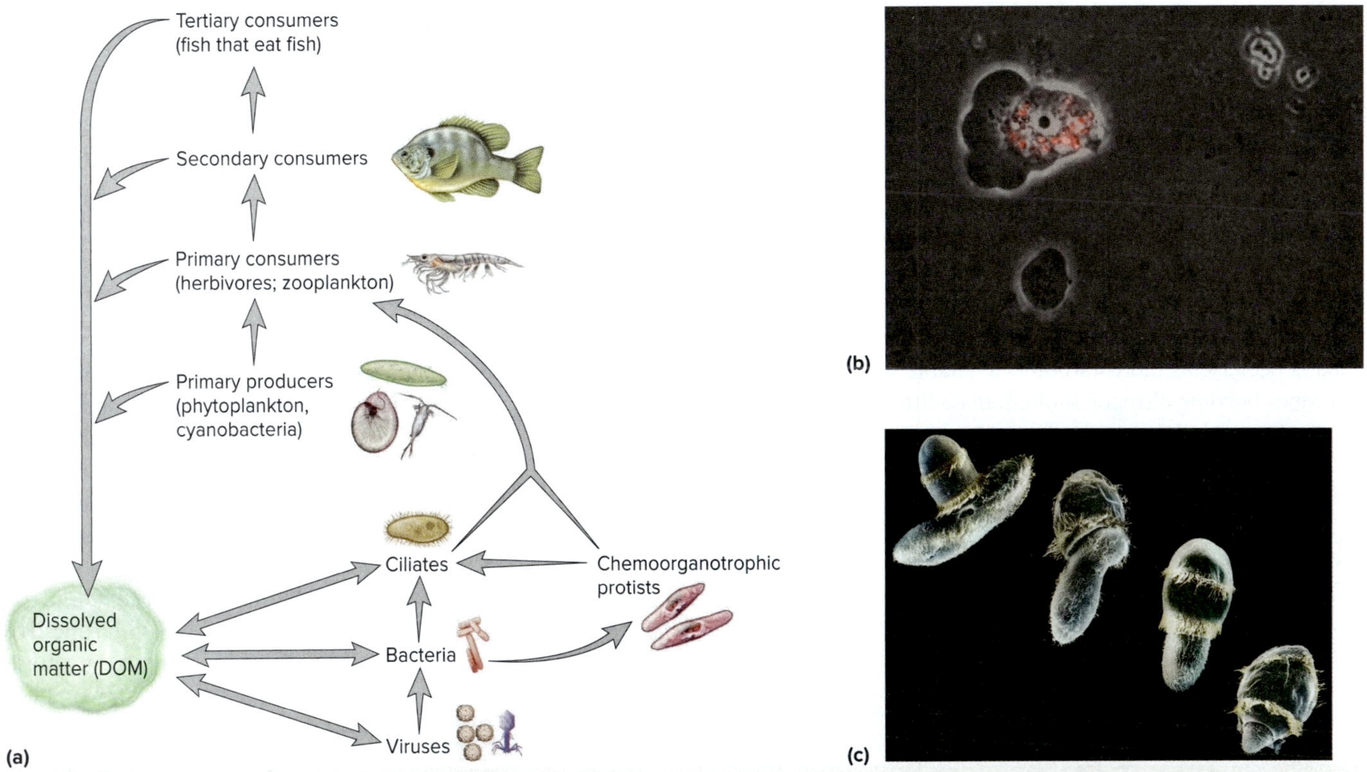

Figure 30.5 The Microbial Loop. (a) Microorganisms play vital roles in ecosystems as primary producers, decomposers, and primary consumers. All organisms contribute to a common pool of dissolved organic matter (DOM) consumed by microbes. Viruses contribute DOM by lysing their hosts, and bacterial and archaeal cells are consumed by protists, which also consume other protists. These microbes are then consumed by herbivores that often select food items by size, thereby ingesting both heterotrophic and autotrophic microbes. Thus nutrient cycling is a complex system driven in large part by microbes. (b) Protists consume bacteria; in this case, an amoeba is consuming the cyanobacterium *Synechococcus* sp., which fluoresces red. (c) Protists consume protists; in this time-lapse image, the ciliate *Didinium* sp. (rounded organism with two rows of cilia) is preying upon another ciliate, *Paramecium* sp.
(b) ©Jackie Parry; (c) ©Greg Antipa/Biophoto Associates/Science Source

MICRO INQUIRY *How do heterotrophic microbes contribute to the pool of DOM? How does this compare with the manner by which microbial autotrophs contribute to the DOM pool?*

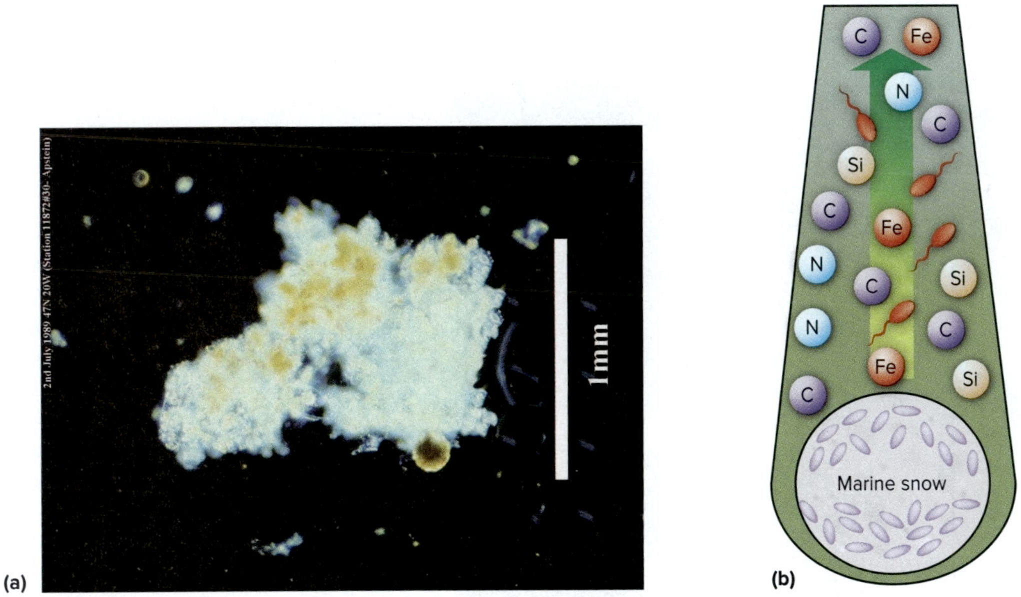

Figure 30.6 Marine Snow. The export of organic matter out of the photic zone to deeper water occurs through the sinking of marine snow. (a) Marine snow includes fecal pellets, planktonic tests (white objects), transparent snail-like pteropod shells, radiolarians, and diatoms. (b) The hydrolytic activity of the exoenzymes secreted by the microbes that colonize marine snow generates a plume of nutrients during sinking. This in turn attracts more microbes.
(a) Courtesy of Prof. Richard Lampitt/National Oceanography Centre

concentration continues to climb. Presently it is estimated that about 3 billion tons of carbon per year are buried (sequestered) in deep-ocean sediments. Can increases in atmospheric CO_2 be offset by increasing the rate and volume of carbon sequestration? Could this in turn slow global warming?

Although no one is sure of the answers to these questions, by piecing together empiric and modeling data, an understanding of how phytoplankton respond to climate change is beginning to emerge. For example, as atmospheric CO_2 increases, there appears to be selection for microorganisms that can fix N_2. This is because as more carbon is available for photosynthesis, microbes become nitrogen limited. In addition to increased CO_2 input, our oceans are also responding to warming temperatures (*see figure 28.13*). Long-term growth experiments in temperate waters measuring the growth of the cyanobacteria in the genus *Synechococcus* suggest the annual spring bloom will occur 4 to 5 days earlier with each 1°C increase in surface temperature. However, it is unclear if this response will be shared by other primary producers or if this will significantly increase CO_2 drawdown.

As suggested by the selection for nitrogen-fixing bacteria (diazotrophs) as CO_2 levels climb, microbial growth in the open ocean is often limited by nitrogen. Because there is no terrestrial runoff, the only source of "new" organic nitrogen at sea is biological nitrogen fixation. There appear to be at least four significant sources of N_2 fixation—all of which are cyanobacterial. For many years, it was thought that the filamentous cyanobacterium *Richelia intracellularis* was the only microbe that fixed significant amounts of N_2. This microbe lives in association with certain diatoms. However, filamentous cyanobacteria of the genus *Trichodesmium* are now known to be much more important in terms of the amount of N_2 fixed (**figure 30.8**). This microbe drifts on the surface in rafts that cover many square meters in tropical and subtropical oceans. It protects its oxygen-sensitive nitrogenase enzyme by partitioning it to regions of trichomes (filaments) called diazocytes. Because the region of a filament that functions as a diazocyte is not separated from the rest of the trichome by a septum, these structures differ from heterocysts seen on other

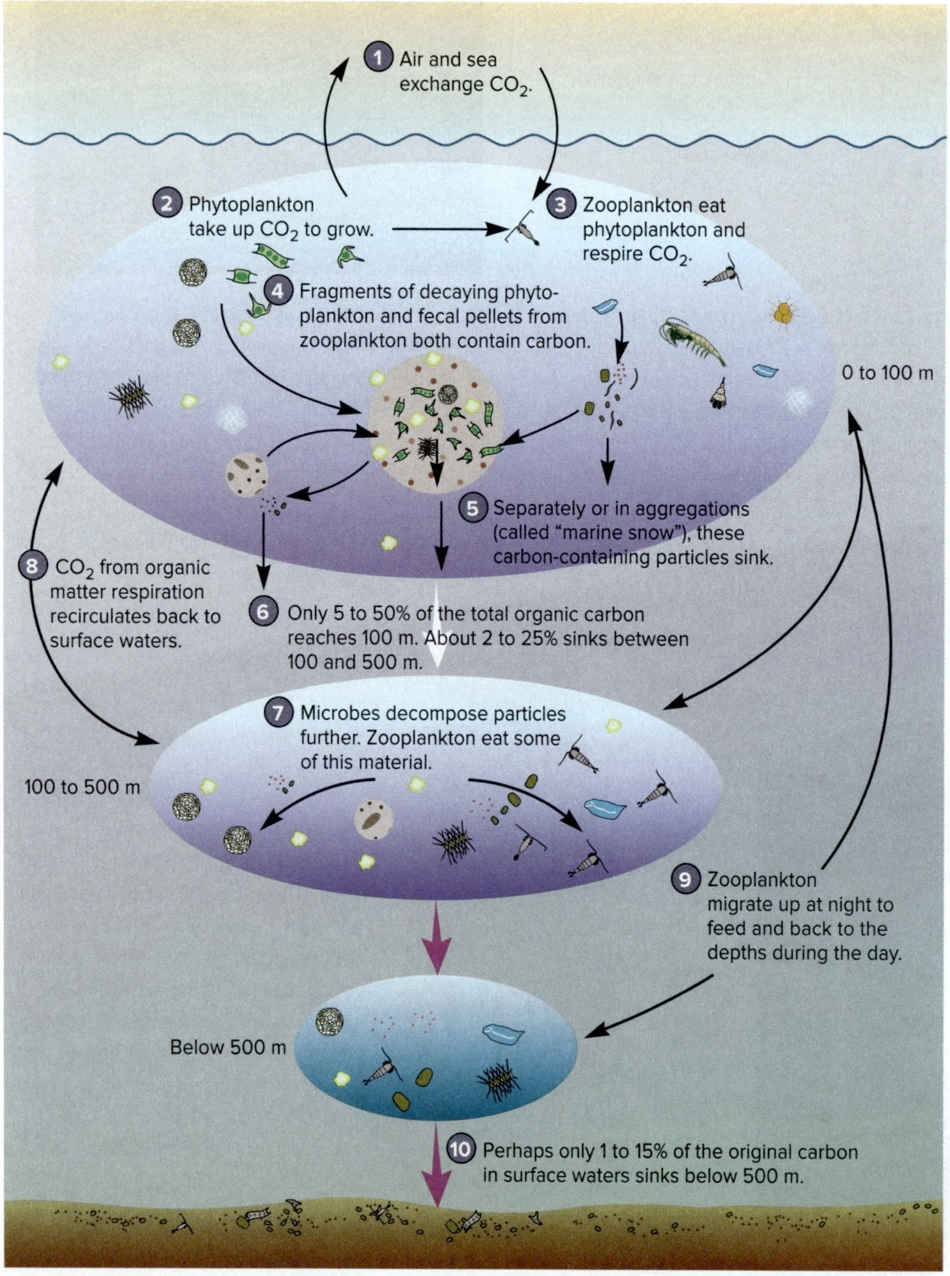

Figure 30.7 Fate of Carbon from the Photic Zone to the Deep Benthos. The vast majority of the carbon fixed by microbial autotrophs at the surface of the open ocean remains in the photic zone. Although a small fraction reaches the seafloor, much of it returns to the surface in regions where deep and surface waters mix (upwelling regions; not shown). The flux of CO_2 in and out of the world's oceans is in equilibrium, but this equilibrium may be upset by the increase in atmospheric CO_2, the hallmark of global climate change.

Figure 30.8 **The Filamentous Cyanobacterium *Trichodesmium* sp. Fixes N_2 in the Open Ocean.**
©Image Quest 3-D/NHPA/Photoshot/Newscom

filamentous cyanobacteria (*see figure 21.11*). Another source of "new" nitrogen is the N_2-fixing, unicellular cyanobacterium *Crocosphera watsonii*. It measures 3 to 8 µm in diameter, making it much bigger than the other unicellular cyanobacteria in the genera *Synechococcus* or *Prochlorococcus*. *C. watsonii* is found in tropical surface waters.

The last source of "new" nitrogen is an unusual cyanobacterium first reported in 2008. This cyanobacterium fixes N_2 but lacks photosystem II and is photoheterotrophic. Organic carbon is fermented and because it uses only photosystem I no O_2 is produced. Thus, it has no trouble fixing N_2 during the day. This microbe has not yet been cultured and is known only as UCYN-A. When observed in natural samples, it is a little less than 1 µm in diameter. It is both free-living and a symbiont of haptophyte algae. It inhabits temperate latitudes and is found in coastal waters as well as the open ocean. ◀ *Nitrogen fixation (section 12.5); Phylum* Cyanobacteria: *oxygenic photosynthetic bacteria (section 21.4);* Haptophyta *(section 24.4)*

In addition to nitrogen fixation, other components of the nitrogen cycle occur in the open ocean. Much of this cycling occurs below the photic zone, where oxygen concentrations reach a minimum before increasing with depth. Here two types of microbial-mediated reactions result in the loss of nitrogen as N_2 gas. The first is denitrification—anaerobic respiration using nitrate (NO_3^-) as a terminal electron acceptor and reducing it to N_2. The second is the anammox reaction—the anaerobic oxidation of ammonium (NH_4^+) to N_2 using nitrite (NO_2^-) as the terminal electron acceptor (*see figure 21.13*). Anammox is performed by members of the phylum *Planctomycetes*. Nitrogen that escapes denitrification or anammox in the oxygen minimum zone and sinks further is mostly NH_3/NH_4^+. However, most of the nitrogen that accumulates at depth is nitrate (NO_3^-). This is because the sinking NH_3/NH_4^+ is oxidized by microbes performing nitrification—the oxidation of NH_3 first to NO_2^- and then to NO_3^-. Most marine nitrification involves archaea in the phylum *Thaumarchaeota*. These mesophilic archaea oxidize NH_3 to NO_2^- which is then oxidized to NO_3^- by bacteria. The urgency to understand oceanic nitrogen cycling reflects concern about release of the greenhouse gas N_2O, principally by ammonia oxidizing archaea. ◀ *Phylum* Thaumarchaeota: *mesophilic ammonia oxidizers (section 20.2); Phylum* Planctomycetes: *anammox bacteria (section 21.5); Nitrogen cycle (section 28.1)*

Given the enormous volume of the ocean, it is probably not surprising that the most abundant clade (monophyletic organisms) on Earth are marine α-proteobacteria. Members of the clade called **SAR11** (named after the *Sar*gasso Sea, where they were first discovered) have been detected from almost all open ocean samples taken worldwide. In addition, they have been found at depths of 3,000 m, in coastal waters, and in some freshwater lakes. Using fluorescence in situ hybridization (FISH; *see figure 29.6*), SAR11 bacteria have been found to constitute 25 to 50% of the microbial cells in surface waters of both coastal and open-ocean samples. Indeed, SAR11 bacteria are estimated to constitute about 25% of all microbial life on the planet. ◀ *SAR11 clade, the most abundant microorganisms on Earth (section 22.1)*

The physiology of SAR11 illustrates a recurring metabolic theme that has emerged in the study of marine microorganisms: Many have evolved strategies to supplement their energy reserves. This metabolic diversity blurs the trophic divisions among marine microbes and further contributes to the microbial loop (figure 30.5). For example, genes for rhodopsin, first identified in halophilic archaea, are found in α-proteobacteria (i.e., SAR11), as well as β-proteobacteria, bacteroidetes, and in at least one marine dinoflagellate. Rhodopsin found in proteobacteria is termed proteorhodopsin (PR). None of these microorganisms is autotrophic; it appears that they use the light-driven rhodopsin pump to supplement ATP pools in nutrient-depleted waters. Given that 50% of heterotrophic marine bacteria and archaea in the photic zone have the genes needed for rhodopsin biosynthesis, we can infer that this form of phototrophy is important to survival. This notion is supported by the fact that PR-containing bacteria at the surface have a form of rhodopsin that absorbs more abundant green light (525 nm), while those living in deeper waters often bear PR tuned to blue light (490 nm), which penetrates deeper. ◀ *Rhodopsin-based phototrophy (section 11.12); Haloarchaea (section 20.3)*

Another group of marine microbes that supplement their energy reserves are aerobic anoxygenic phototrophs (AAnPs). Bacteria capable of **aerobic anoxygenic phototrophy** are known to possess bacteriochlorophyll *a* (Bchl *a*); this photosynthetic pigment is widely distributed in surface ocean waters. A combination of cultivation and metagenomic techniques demonstrates that some α-proteobacteria (the *Roseobacter* clade), as well as β- and γ-proteobacteria, are capable of this form of phototrophy (*see table 21.2*). Like rhodopsin-containing microbes, AAnPs are photoheterotrophs. That is, they use light to drive the synthesis of ATP (photophosphorylation) but meet their carbon needs by the assimilation of organic carbon. By meeting some of their ATP demands phototrophically, photoheterotrophs divert more organic carbon to biosynthesis and less to catabolic processes. However, open ocean bacteria that possess Bchl *a* are much less abundant than those that produce proteorhodopsin. This may reflect the fact that genes for proteorhodopsin have been widely subject to horizontal gene transfer, as only six genes are required for a functional light-driven proton pump. This is in contrast to the roughly 50 genes needed to assemble the anoxygenic light-harvesting complex, reaction center, Bchl *a*, and carotenoids found in AAnPs.

The α-proteobacterium *Silicobacter pomeroyi* appears to be capable of a third interesting metabolic strategy to increase its ATP yield: **lithoheterotrophy** (or chemolithoheterotrophy). Like photoheterotrophy, in which light is used for energy and electrons while organic carbon is consumed, lithoheterotrophy uses inorganic chemicals as a source of energy and electrons but requires organic carbon for anabolic purposes. *S. pomeroyi* is one of a number of marine carbon monoxide (CO) oxidizing heterotrophs that have been cultured. These bacteria produce CO dehydrogenase, thereby oxidizing CO to CO_2, but unlike other CO oxidizing bacteria, they do not have the enzymes necessary for the conversion of CO_2 into biomass. Instead, they use CO as a supplemental electron donor for their electron transport chain. Thus they need not oxidize as much organic carbon to generate NADH for this purpose. Instead, more organic carbon can be used for biosynthesis. This hypothesis is supported by thermodynamic considerations: The oxidation of one molecule of CO should release enough energy to drive the production of one ATP. Measurements of CO lithoheterotrophs show that they make up about 7% of the bacterioplankton in the open ocean and coastal waters.

It is not difficult to hypothesize that these alternatives to "traditional" organotrophy would be beneficial to individual cells in a nutrient-stressed environment. What is more problematic is estimating how much these processes contribute to the carbon budget of marine habitats. Theoretically, the advantage these processes confer must more than offset the energetic cost of carrying the genes required for their expression. But it remains to be seen if these processes actually make these microbes better competitors. Likewise, the impact these metabolic strategies have on carbon flux at the community, ecosystem, and global levels awaits more investigation. It is questions like these that make microbial oceanography such a dynamic and exciting field.

Marine Viruses: Mortality at Sea

As mentioned in our discussion of the microbial loop (figure 30.5), viruses are important members of marine and freshwater microbial communities. In fact, **virioplankton** are the most numerous members of marine ecosystems. However, quantifying viruses is tricky: The traditional method of plaque formation requires knowledge of both the virus and its host, and the ability to grow the host in the laboratory. Because so few microbes have been cultured, this prevents the measurement of virus diversity by demonstrating actual virus infection. Instead, virus particles may be visualized directly. There are at least two approaches for this. Electron microscopy is the most rigorous, but fluorescence microscopy is a common and convenient method used to quantify viruses. When seawater is filtered, viruses can be collected on filters with a pore size of 0.02 μm. Viral nucleic acids are stained with a fluorescent dye, such as YO-PRO or SYBR Green for fluorescence microscopy. Because microscopy reveals the presence of virions outside the host cell, it does not prove that they can infect a host cell. Therefore, viruses enumerated in this way are called ***virus*like particles** (**VLPs**). Using this approach, the average VLP density in seawater is between 10^6 and 10^7 per milliliter (although in some cases, it may be closer to 10^8 per milliliter); their numbers decline to roughly 10^6 below about 250 m. Marine viruses are so abundant that VLPs are now recognized as the most abundant microbes on Earth (**figure 30.9**).

Viral diversity is vast, including single- and double-stranded RNA and DNA viruses that infect archaea, bacteria, and protists. Metagenomic analysis of viral genomes has revealed that as many as 90% of the genomes have no recognizable match on any database. Thus not only is marine viral genetic diversity immense, it is largely unexplored. ◀ *Viruses (chapter 26)*

The abundance of marine viruses indicates that they are major agents of mortality in the sea. Indeed, viruses are thought

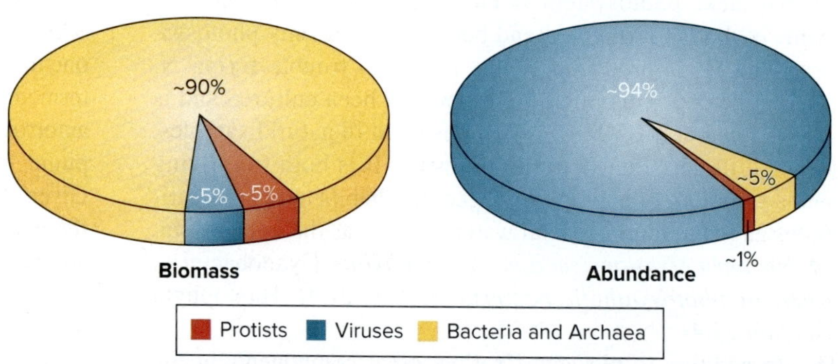

Figure 30.9 Relative Biomass and Abundance of Marine Microbes. (a) The biomass of bacteria and archaea far exceeds that of both protists and viruses. (b) When individuals are counted, however, the vast majority of marine microbes are viruses.

Figure 30.10 Viral Lysis and the Termination of a Massive Plankton Bloom. This satellite image shows the demise of a 500-km-long bloom (light blue region) of the coccolithophore *Emiliania huxleyi* in the Barents Sea. While the exhaustion of available nutrients also contributes to the death of such an enormous bloom, viruses specific to this protist are principal agents in the bloom's demise.
Source: NASA image courtesy Jeff Schmaltz, MODIS Rapid Response Team at NASA GSFC

to kill on average about 20% of the marine microbial biomass every day. However, measurement of virus-induced microbial mortality shows that it is highly variable. Virus abundance frequently corresponds to the microbial host that is most active in the community. One model predicts that as one microbial species (or strain) becomes numerically dominant, it soon will be targeted by lytic viruses and thus its population will decline. This permits another microbial species (or strain) to flourish, which then becomes subject to intense viral lysis, and so on. This "kill the winner" model has been challenged by the "piggyback the winner" model based on observations of higher than expected rates of lysogeny in marine bacterial populations. In this model, virus infection ends in insertion of the viral genome in the host chromosome (lysogeny) rather than lysis. ◀ *Bacteriophage lambda: a temperate bacteriophage (section 26.2)*

Perhaps the most compelling example of marine virus infection occurs during blooms of the coccolithophore *Emiliania huxleyi*. Such blooms are so intense they can be imaged from space, yet their collapse is thought to be (at least in part) the result of viral lysis (**figure 30.10**). In this case, the "Cheshire Cat" strategy serves to illustrate the complex relationship between virus and host. The *E. huxlei* life cycle involves both diploid and haploid forms and only diploid cells are susceptible to virus infection. During a bloom, a third population of *E. huxlei* cells emerges that are morphologically similar to haploid cells (and virus resistant), but remain diploid. More broadly speaking, however, by lysing their hosts, viruses accelerate the rate at which large populations of microbes are converted to POM and DOM. This "short-circuits" the microbial loop and contributes significantly to nutrient cycling (**figure 30.11**).

Microbial Communities in Benthic Marine Sediments Are Enormous and Mysterious

Sixty-seven percent of the Earth's surface lies under the sea, which means that of all the world's microbial ecosystems, we know the least about the largest. However, the combination of deep-ocean drilling projects and the exploration of geologically active sites, such as submarine volcanoes and hydrothermal vents, has revealed that the study of ocean sediments, or benthos, is both rewarding and surprising. Marine sediments range from the very shallow to the deepest trenches, from dimly illuminated to completely dark, and from the newest sediment on Earth to material that is millions of years old. The temperature and age of marine sediments depend on their proximity to geologically active areas. Hydrothermal vent communities with large and diverse invertebrates, some of which depend on endosymbiotic chemolithotrophic bacteria, have been

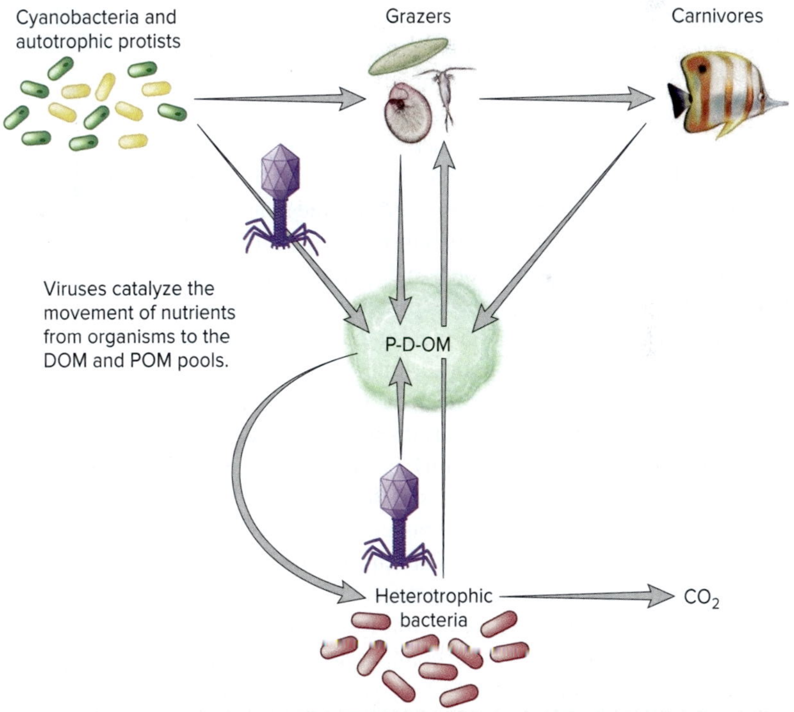

Figure 30.11 The Role of Viruses in the Microbial Loop. Viral lysis of autotrophic and heterotrophic microbes accelerates the rate at which these microbes are converted to particulate and dissolved organic matter (P-D-OM). This is thought to increase net community respiration and decrease the efficiency of nutrient transfer to higher trophic levels.

MICRO INQUIRY *When might viruses choose lysogeny rather than lysis and how would this affect carbon flow (also see figure 30.5)?*

intensely investigated since their exciting discovery in the late 1970s. These microbes are discussed in chapter 27. Because the vast majority of Earth's crust lies at great depth far from geothermally active regions, most benthic marine microbes live under high pressure, without light, and at temperatures between 1° and 4°C.

Deep-ocean sediments were once thought to be devoid of all life and thus not worth the considerable effort it takes to study them. But sediment samples from water depths up to 8,200 m (at its deepest, the ocean is about 11,000 m) reveal the presence of a vast "piezosphere" (Greek *piezein,* to squeeze or press). The piezosphere describes the biome at depths with pressures exceeding 100 atm (pressure increases about 1 atm per 10 m depth). Microbes that grow at such great pressures are termed **piezophilic** and their abundance is surprisingly high, typically exceeding that of surface waters. Reasons for this difference are complicated but include energy availability and limited predation.

In addition to being piezophilic, most deep-sea microbes are psychrophilic. The majority recovered so far are Gram-negative, facultative anaerobic bacteria. Common genera include γ-proteobacteria in the genera *Shewanella, Photobacterium, Colwellia, Moritella,* and *Psychromonas.* Interestingly, it appears that the subterranean depth limit of microbial growth is not governed by pressure; rather, it is determined by temperature. It seems unlikely that the maximum temperature at which life can be sustained has been identified. ◀ *Environmental factors affect microbial growth (section 7.5)*

It is now accepted that the biosphere below the seafloor is roughly equal to the microbial biomass in all the world's oceans. As might be predicted, most of these microbes are not amenable to laboratory culture. In fact, these microbes "breathe" about a million times slower than surface microorganisms, making their metabolism so slow that doubling times are measured in hundreds of years *(see chapter 7 opening story)*. Subsurface microbes truly stretch the limits of what we have come to define as life.

Global surveys show that the deep subsurface biosphere within relatively nutrient-rich continental margin sediments differs from that in the organic-poor distant abyss. Seawater is rich in sulfate, which is used as an electron acceptor in anaerobic respiration. In the continental margin, sulfate in sediments is depleted within 100 m depth, at which point methanogenesis becomes dominant. However, in the abyss, sulfate penetrates much more deeply, and distinct zonation between sulfate-reducing bacteria and methanogenic archaea is absent. Furthermore, the sediment surface and subsurface appear to be teeming with viruses; it is estimated that each year viral lysis in the benthos releases up to 630 million tons of carbon that had been sequestered by marine snow and other falling particulates (figure 30.6). These exciting findings suggest that a significant fraction of Earth's biomass is amazingly active yet largely uncharacterized.

Exploration of deep-sea subsurface microbiology has turned our understanding of bacterial energetics literally upside down. As discussed in chapters 11 and 28, it is generally understood that anaerobic respiration occurs such that there is preferential use of available terminal electron acceptors (*see figure 28.1*). Acceptors that yield the most energy (more negative Δ*G*) from the oxidation of NADH or reduced inorganic compounds (e.g., H_2, H_2S) will be used before those that produce a less negative Δ*G*. Thus following oxygen depletion, nitrate will be reduced, then manganese, iron, sulfate, and, finally, carbon dioxide. This is demonstrated in sediment cores collected in sediments 420 m deep off the coast of Peru (**figure 30.12**). However, the profile of

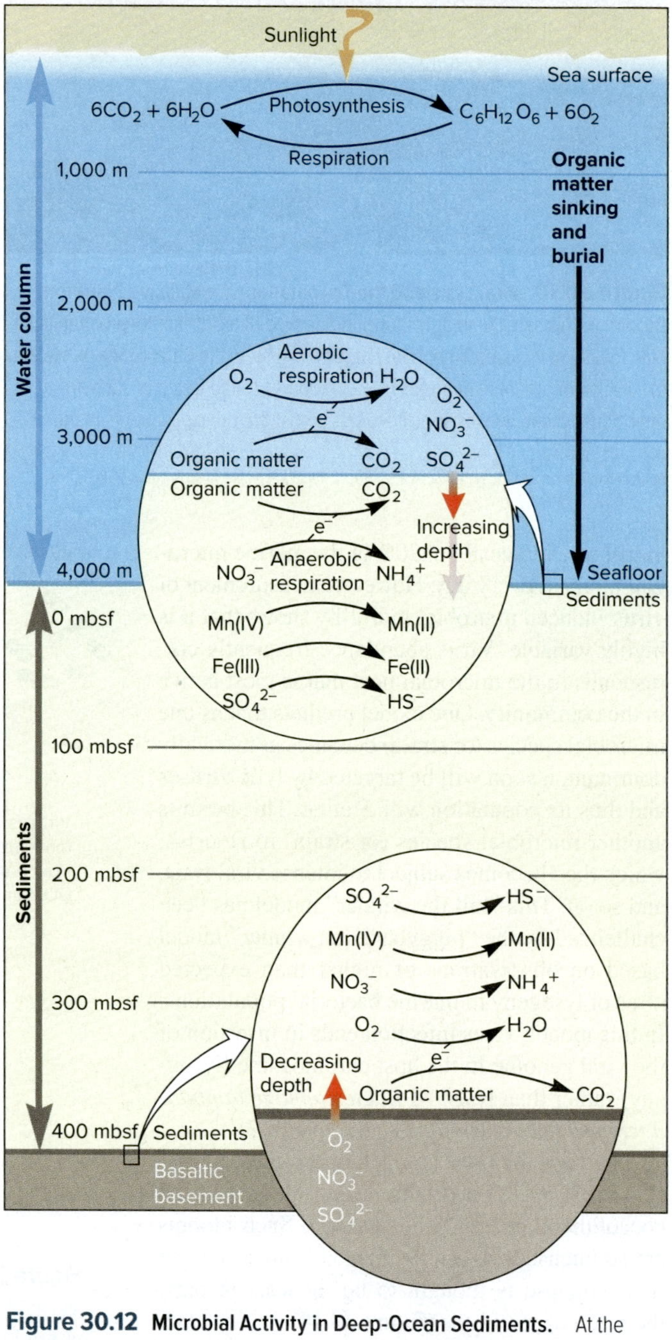

Figure 30.12 Microbial Activity in Deep-Ocean Sediments. At the surface of the seafloor, reduction of oxidized substrates that serve as electron acceptors in anaerobic respiration follows a predictable stratification based on thermodynamic considerations. This sequence reverses in very deep subsurface sediments, suggesting a source of electron acceptors from deep within Earth's crust. mbsf, meters below seafloor

these signature chemical compounds is reversed at depths greater than about 400 m. This suggests the presence of unknown sources of these electron acceptors at subsurface depths of more than 420 m. In addition, contrary to our long-held notion of thermodynamic limits, methane formation (methanogenesis) and iron and manganese reduction co-occur in sediments with high sulfate concentrations. Although the identity of the microbes that make up these communities awaits further study, it is clear that with densities of 10^8 cells per gram of sediment at the seafloor surface and 10^4 cells per gram in deep subsurface sediments, these communities are important. ◀ *Redox reactions (section 10.3); Anaerobic respiration uses the same three steps as aerobic respiration (section 11.7); Chemolithotrophy (section 11.10)*

Another outcome of deep-ocean sediment exploration has been the discovery of microbial communities on continental margins fueled by the release of hydrocarbons at depths between 200 and 3,500 m. Depending on the surface topology and the rate at which hydrocarbons are emitted, they are called pockmarks, gas chimneys, mud volcanoes, brine ponds, and oil and asphalt seeps. A variety of microbes use hydrocarbons as a sole carbon source, oxidizing them with oxygen or sulfate as the electron acceptor. Some regions are the sites of microbial mats, which can sometimes cover several hundred meters. These mats are dominated by the giant, vacuolated sulfur-oxidizing bacteria in the genera *Beggiatoa* and *Thiomargarita* (*see figures 22.19 and 22.20*). ◀ *Class* Gammaproteobacteria *is the largest bacterial class (section 22.3)*

Perhaps the most industrially important of these hydrocarbon-fueled communities are those found at **methane hydrates.** Methane hydrates are latticelike ice structures in which methane is trapped. The methane is produced by both geothermal activity and methanogenic archaea. The lack of associated methane-oxidizing microbes has enabled pools of natural gas to accumulate in latticelike cages of crystalline water 500 m or more below the sediment surface in many regions of the world's oceans. This discovery is significant because there may be up to 10^{13} metric tons of methane hydrate worldwide. Thus 80,000 times the world's current known natural gas reserve awaits the development of the technology to access it. ◀ *Phylum* Euryarchaeota: *methanogens, haloarchaea, and others (section 20.3)*

Comprehension Check

1. What is marine snow? Why is it important in CO_2 drawdown?
2. Explain how new organic nitrogen is introduced into the open ocean.
3. Explain the means by which proteorhodopsin contributes to the success of bacteria in the SAR11 clade.
4. Compare and contrast the use of lithoheterotrophy and photoheterotrophy by bacteria in oligotrophic waters.
5. How might community dynamics change in the absence of viral lysis?
6. Explain what is meant by "upside-down microbial energetics," as described for some deep-subsurface ocean sediments.
7. What are methane hydrates? Why are they important?

30.3 Microorganisms in Freshwater Ecosystems

After reading this section, you should be able to:

a. Appreciate the scarcity of freshwater on Earth
b. Contrast and compare the microbial communities found in lotic systems with those in lentic systems
c. Describe the differences between the microbial communities found in a deep, stratified lake with those of a shallow lake that lacks stratification
d. Describe how seasonal changes influence lake microbial communities and how these changes can lead to blooms

While the vast majority of water on Earth is in marine environments, freshwater is crucial to our terrestrial existence. As shown in **figure 30.13**, of the 3.9% of Earth's water that is not saline, only a tiny fraction represents the familiar streams, rivers, and wetlands that we associate with the term *freshwater*. We begin our discussion with the largest fraction of freshwater—that which is frozen at the Earth's polar regions. We then move to streams, rivers, and lakes.

Glaciers and Permanently Frozen Lakes Support Active Microbial Communities

Much of the ice on Earth has remained frozen for millions of years. Indeed, it is important to note that a majority of the Earth's surface never exceeds 5°C. This includes polar regions, the deep ocean, and high-altitude terrestrial locations. Surprisingly, microbes within glaciers are not dormant. Rather, evidence that active microbial communities exist in these environments has been growing over the last few decades. In fact, this is an interesting time in glacial microbiology; determining the diversity in these systems and assessing the role of these microorganisms in biogeochemical cycling as glaciers melt can involve novel and creative techniques. In addition, since the discovery of ice on Mars and on Jupiter's moon Europa, astrobiologists have become very interested in ice-dwelling psychrophilic microbes. ◀ *Temperature (section 7.5).*

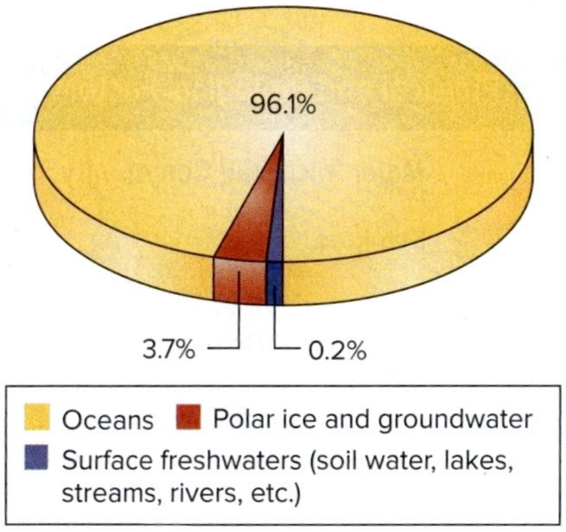

Figure 30.13 Relative Distribution of Freshwater on Earth.

The biology of lakes associated with glaciers has gained importance as it has been recognized that these ecosystems function as sentinels of climate change. Glacial lakes are constrained by low temperature and reduced solar radiation, and in the Antarctic, lakes are separated from macrofauna as well. Here phytoplankton, bacteria, and protists form simple food webs almost completely devoid of multicellular organisms (**figure 30.14**). Microbial growth rates are slow, so it not surprising that primary productivity within these lakes must be supplemented by the deposition of organic carbon from the surrounding environment. Indeed, phototrophic flagellated and ciliated protists are also capable of consuming bacteria and dissolved organic carbon. These mixotrophs are not the only consumers of bacteria; most of the viruses are bacteriophages. Viral lysis of bacteria in Antarctic lakes is almost 10 times that found in similar temperate and tropical lakes.

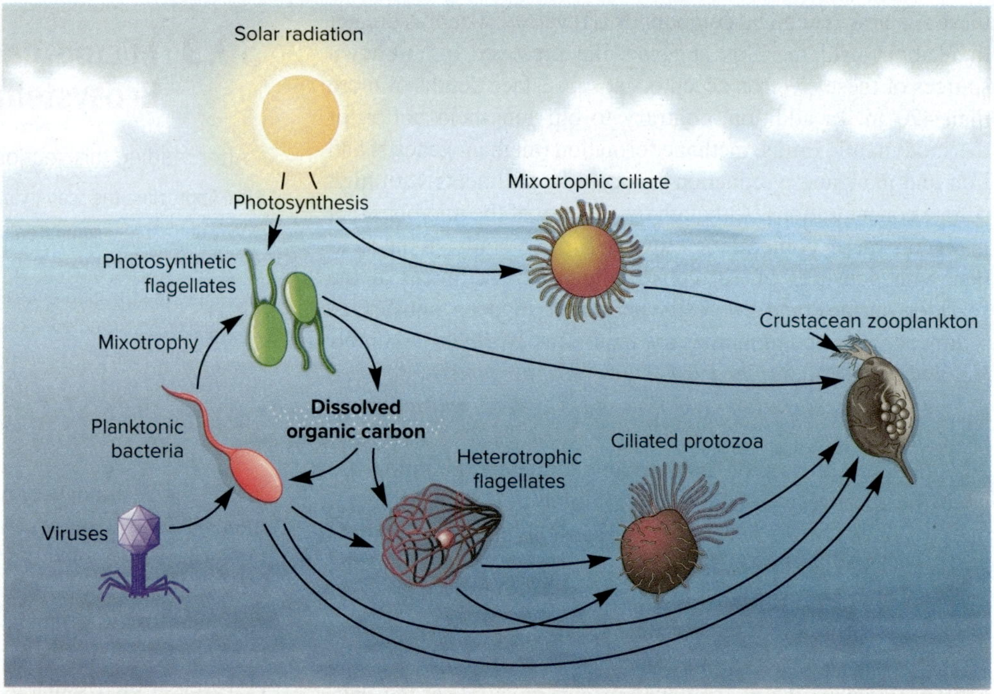

Figure 30.14 Nutrient Cycling in Antarctic Lakes Is Performed Entirely by Microbes. Antarctic glacial lakes lack multicellular organisms, so all nutrient cycling is between various types of microbes, including bacteriophages. The surrounding environment provides some organic carbon.

MICRO INQUIRY *List three major differences between this microbial loop and that found in the open ocean (figure 30.5).*

Microbial Communities in Streams and Rivers Differ

Streams and rivers are much more dynamic than glacial lakes. Some of the physical and biological factors that help regulate the microbial communities of these freshwater systems are listed in **table 30.1**. Lakes, rivers, and streams can be broadly classified as either **lotic** (Latin *lotus,* washing) systems with free-running waters (streams, most rivers) or **lentic** (Latin *lentus,* calm or slow) systems with free-standing waters (e.g., ponds; lakes; marshes; slow-moving rivers). In large, lentic rivers and lakes, phytoplankton thrive and are responsible for much of the fixed carbon. The biologically available carbon produced within the system is called **autochthonous.** By contrast, the constant flow of water in streams and all but the largest rivers prevents the development of significant planktonic communities. In addition, phytoplankton growth can be further limited by the amount of available light due to overhanging foliage, turbidity, and rapid mixing of the water. In most streams and rivers, the source of nutrients therefore comes from the surrounding land. Such nutrients are called **allochthonous.** Thus lentic systems often have a net autotrophic metabolism, whereas lotic systems are generally heterotrophic.

Table 30.1 Physical and Biological Factors Influencing Microbial Communities in Freshwater Habitats

Habitat	Major Microbial Community	Major Physical Characteristic	Major Biological Constraints
Lake	Plankton	Stratification, wind-generated turbulence	Nutrient competition, grazing, parasitism
Floodplain	Plankton, biofilms attached to plants	Periodic desiccation	Competition between algae and macrophytes
Rivers and streams	Benthic microbes, biofilms	Flow-generated turbulence	Colonization competition, biofilm grazing
Estuary—mudflats	Biofilms	Desiccation, high light, salt exposure	Competition and grazing at mud surface
Estuary—outflow	Plankton	Mixing with saltwater, turbidity	Grazing in water column

The allochthonous carbon that enters rivers and streams includes both dissolved and particulate organic carbon (DOC and POC, respectively). The amount of carbon entering freshwater systems depends on surrounding vegetation, proximity to human activity, and local geology. In addition, hydrological factors such as how long water stays in the soil before entering the river or stream also influence the nature of the organic carbon. Thus DOC ranges from material that can be readily assimilated to refractory carbon that only a few microbial groups can slowly degrade. Much of the particulate organic carbon includes large items such as leaf litter, which provide substrates for fungi and bacteria to colonize. Initially, the activity of these microbes releases a pulse of DOC; this is followed by the slow leaching of nutrients that may not be as easy to assimilate.

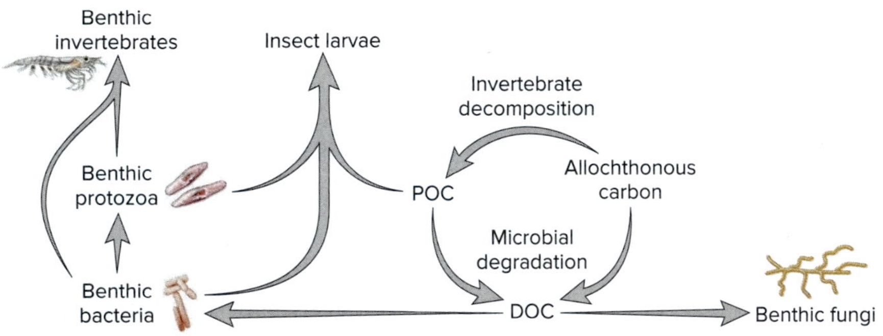

Figure 30.15 A Benthic Microbial Loop in Lotic Systems. In contrast to the microbial loop in planktonic systems (figure 30.5), the benthos of a lotic system, such as a fast-moving river or stream, features small eukaryotes such as invertebrates and insect larvae. These organisms play an important part in nutrient cycling. At the same time, macrofauna such as fish and other top carnivores may be absent or play relatively minor roles.

Most rivers and streams have healthy benthic communities of microbes that form biofilms consumed by invertebrates and other grazing organisms. In shallow rivers and streams, biofilms are often dominated by diatoms and other photosynthetic protists that contribute to the total primary production of the system. Biofilms that consist principally of heterotrophic microbes are important sources of degradation and mineralization. The role of benthic microbes in rivers and streams can be generally characterized by the microbial loop shown in **figure 30.15**.

As in any microbial community, when river and stream chemoorganotrophic microorganisms metabolize available organic material, they recycle nutrients within the ecosystem. Autotrophic microorganisms require minerals released from organic matter. This leads to the production of O_2 during daylight hours, when the rates of photosynthesis and respiration are in dynamic equilibrium. When the amount of organic matter added to streams and rivers does not upset this balance, so that heterotrophy and autotrophy are roughly equal, productive streams and rivers are maintained. However, as seen with ocean dead zones, when the supply of organic matter supports a vast heterotrophic population whose respiration rate outpaces oxygenic photosynthesis, the water may become anoxic. This can occur in streams and rivers adjacent to urban and agricultural areas. The release of inadequately treated municipal wastes and other materials from a specific location along a river or stream represents point source pollution. Such additions of organic matter can produce distinct and predictable changes in the microbial community and oxygen concentration, creating an oxygen sag curve in which oxygen is depleted just downstream of the pollution source (**figure 30.16**). Runoff from agriculturally active fields and feedlots is an example of nonpoint source pollution. This can also cause disequilibrium in the microbial community, leading to algal or cyanobacterial blooms.

Microorganisms in Lakes Can Cause Seasonal Blooms

Lakes differ from rivers and streams because they are lentic systems, thus dominated by planktonic microbes and invertebrates.

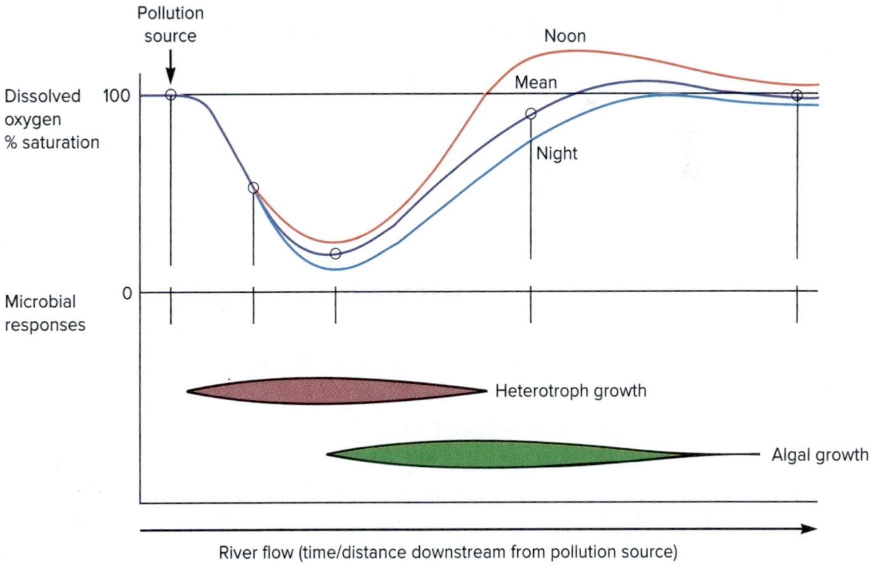

Figure 30.16 The Dissolved Oxygen Sag Curve. Microorganisms can create gradients over distance and time when nutrients are added to rivers. When organic wastes are added at a single site to a clean river system, the rate of O_2 consumption directly downstream increases resulting in a dissolved O_2 sag curve. Further downstream, the phototrophic community will again become dominant, resulting in diurnal changes in river oxygen levels.

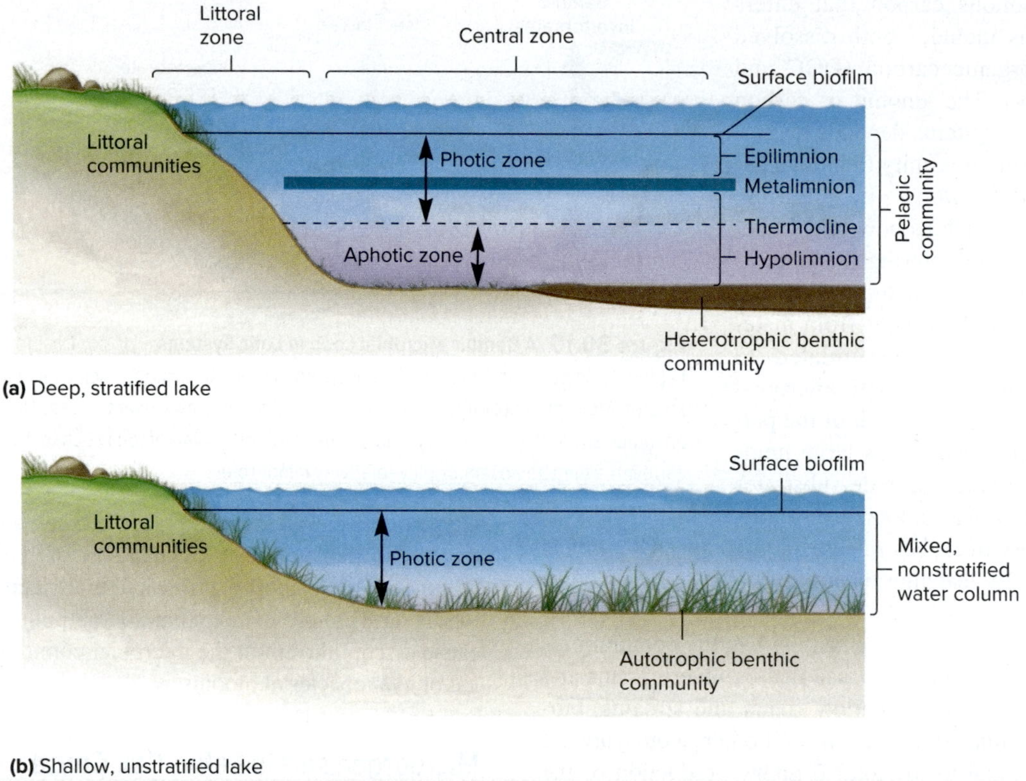

Figure 30.17 The Depth of a Lake Is a Key Factor in Shaping the Microbial Community. (a) Deep lakes have autotroph-dominated littoral and epilimnion microbial communities. Waters below the thermocline and photic zone form the hypolimnion, which can become oxygen depleted. Here a heterotrophic benthic community thrives. (b) Shallow lakes have little mixing, and the photic zone reaches the lake bottom. Thus autotrophs dominate the water column and benthic regions throughout the lake.

MICRO INQUIRY *How does the contribution of benthic autotrophs differ in deep lakes compared to that in shallow lakes? How does this affect the relative contribution of phytoplankton in these two aquatic systems?*

Because much of the water in a lake comes from the surrounding land, the geology and use of the land that feeds the lake largely determine the chemistry of the inflow water. For instance, mountain lakes are fed water from relatively infertile land, resulting in oligotrophic conditions within the lake. This explains why mountain lakes tend to be so clear: Microbial growth is limited to those species that tolerate low nutrient levels. On the other hand, low-lying lakes generally catch water from fertile, cultivated soils and so are nutrient rich. This supports a high level of planktonic growth, and the lakes may appear turbid. The picoplankton often include freshwater species of *Synechococcus,* whereas filamentous cyanobacteria such as those in the genera *Anabaena* and *Microcystis* constitute the macroplankton (microbes larger than 200 μm).

Deep lakes can be divided into two main regions: the **littoral zone,** or shoreline, and the central, **pelagic zone** (these terms also apply to marine systems). As shown in **figure 30.17*a*,** the littoral region is also defined by the depth of the photic zone. The two zones have distinct microbial communities. In the littoral zone, allochthonous DOC introduces organic substrates and nutrients such as phosphate and nitrate. Because the water is shallow, light penetrates throughout the water column, enabling the development of larger plants. Biofilms frequently form on these plants and in the sediments.

The microbial community within the pelagic zone is partly determined by the depth of the lake. Light penetrates the entire water column in shallow lakes (less than about 10 m), so benthic autotrophs, including diatoms and cyanobacteria, may be the principle primary producers (figure 30.17*b*). In deeper lakes, the central, pelagic region can become thermally stratified. This is the case in temperate regions (**figure 30.18**). In the summer, the upper layer, called the **epilimnion,** becomes warmed and oxygen is released into the atmosphere. The thermocline spans a region called the **metalimnion;** this region acts as a barrier to mixing of upper and lower water masses. The lower, colder, dark region, known as the **hypolimnion,** can become anoxic because oxygenic photosynthesis is confined to

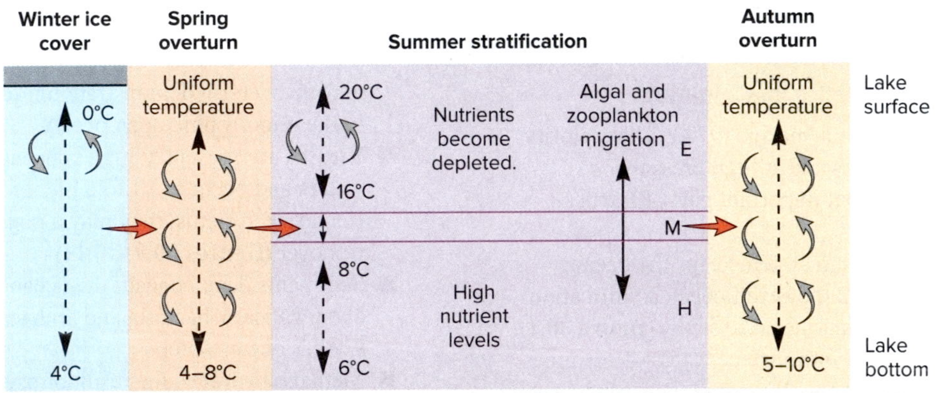

Figure 30.18 **Seasonal Changes in a Temperate Lake Are Associated with Blooms.** In the spring, the water begins to warm, and lack of thermal stratification ensures ample nutrients for microbes that can grow in the relatively warmer surface waters. During the summer as the surface waters warm, thermal stratification separates the cool, deep, nutrient-rich waters from the warm, illuminated waters. Because microbial growth is favored in the warmer surface water, this water becomes depleted of nutrients. During the autumn, waters cool and storms mix the layers, bringing the bottom nutrients to the surface. This pulse of nutrients can support a bloom of microbes or algae. Water turbulence is indicated by arrows; E, epilimnion; H, hypolimnion; M, metalimnion.

MICRO INQUIRY *Why does water turbulence play only a minor role in mixing the water column in the summer?*

the epilimnion. Because these two water masses do not mix (or mix very little), the upper layer supports a rich diversity of primary producers and consumers, which can deplete nutrients. Just the opposite is true for the hypolimnion, where dark, anoxic, cold conditions limit growth. Nutrients accumulate and the microbial community primarily consists of heterotrophs. When autumn weather cools the surface and storms physically mix the two layers, a short-lived bloom can occur as nutrients from the bottom of the lake are mixed into the photic zone. Much longer blooms may occur in the spring, when thermal stratification occurs but nutrients in the epilimnion have not yet been depleted by microbial growth.

Tropical lakes lack significant temperature changes during the year but have seasonal differences in rainfall. Although many aspects of tropical lakes are similar to those of a summer temperate lake, there are several key differences. The warmer, year-round temperatures support higher productivity, and plankton blooms may occur at any time during the year. Cyanobacteria capable of nitrogen fixation may dominate, reflecting the low N:P ratios frequently found in tropical lakes. Nutrient cycling in temperate lakes is largely driven by physical mixing, whereas biological processes are more important in tropical lakes. In this regard, the microbial loop shown in figure 30.5 can be applied to the epilimnion of larger tropical lakes that receive few allochthonous nutrients.

Lakes vary in nutrient status. Some are oligotrophic and others are eutrophic. Nutrient-poor lakes remain oxic throughout the year, and seasonal temperature shifts usually do not result in distinct oxygen stratification. In such oligotrophic lakes, cyanobacteria capable of nitrogen fixation may bloom. Members of several genera, notably *Anabaena, Nostoc,* and *Cylindrospermum,* can fix nitrogen under oxic conditions. In contrast, eutrophic lakes usually have bottom sediments rich in organic matter. If both nitrogen and phosphorus are present, cyanobacteria compete with algae. Cyanobacteria function more efficiently in alkaline waters (pH 8.5 to 9.5) and higher temperatures (30° to 35°C). By using CO_2 at rapid rates, cyanobacteria also increase the pH, making the environment less suitable for protists. Cyanobacteria have additional competitive advantages. Many produce siderophores to bind iron, making this important trace nutrient less available for protists. Some cyanobacteria also resist predation because they produce toxins. In addition, some synthesize odor-producing compounds that affect the quality of drinking water. These problems become magnified when cyanobacteria or protists (algae) produce massive blooms in strongly eutrophied lakes. ◀ *Phylum* Cyanobacteria: *oxygenic photosynthetic bacteria (section 21.4)*

Comprehension Check

1. Why is mixotrophy suited for survival in Antarctic glacial lakes?
2. What is an oxygen sag curve? What changes in a river cause this effect?
3. What are point and nonpoint source pollution? Can you think of examples in your community?
4. Compare and contrast the types of microbes found in the epilimnion and hypolimnion of a deep lake.
5. What are some important effects of eutrophication on lakes?
6. Why do cyanobacteria often dominate waters that have been polluted by the addition of phosphorus?

Key Concepts

30.1 Water Is the Largest Microbial Habitat

- Most water on the Earth is marine (97%). The majority of this is cold (2° to 5°C) and at high pressure. Freshwater is a minor but important part of Earth's biosphere.
- The carbonate equilibrium system keeps the oceans buffered at pH 7.6 to 8.2; however, ocean acidification threatens the health of marine ecosystems (**figure 30.1**).
- The penetration of light into surface waters determines the depth of the photic zone. Warming the surface waters can lead to the development of a thermocline.

30.2 Microorganisms in Marine Ecosystems

- Tidal mixing in estuaries causes daily swings in osmolarity. Thus microbes within estuaries have evolved mechanisms to cope with rapid changes in salinity.
- Coastal regions such as estuaries and salt marshes can be the sites of harmful algal blooms (**figure 30.3**).
- Autotrophic microbes in the photic zone within the open ocean account for about one-half of all the carbon fixation on Earth.
- The microbial loop describes the transfer of nutrients between trophic levels while taking into account the multiple contributions of microbes to recycling nutrients. Nutrients are recycled so efficiently, very little organic carbon is buried in the seabed (**figures 30.5–30.7**).
- The carbon and nitrogen budgets of the open ocean are intensely studied because of their implications for the future of climate change.
- Bacteria in the α-proteobacterial clade SAR11 are the most abundant organisms on Earth and like many marine bacteria demonstrate unique adaptations to life in the oligotrophic open ocean.
- Viruses are present at high concentrations in many waters and occur at 10-fold higher levels than bacteria. In marine systems, they play a major role in nutrient turnover (**figures 30.9–30.11**).
- Sediments deep beneath the ocean's surface are home to about as many bacteria and archaea as found in the world's ocean waters.
- Methane hydrates, the result of psychrophilic archaeal methanogenesis under extreme atmospheric pressure, may contain more natural gas than is currently found in known reserves.

30.3 Microorganisms in Freshwater Ecosystems

- Glaciers and permanently frozen lakes are sites of active microbial communities. Antarctic lakes lack multicellular organisms and have truncated food webs (**figure 30.14**).
- Nutrient sources for streams and rivers may be autochthonous or allochthonous. Often allochthonous inputs include urban, industrial, and agricultural runoff.
- The depth of a lake determines the microbial communities it will support. Deep lakes have a planktonic, autotrophic community in the epilimnion, and shallow lakes support large autotrophic communities (**figure 30.17**).
- Seasonal mixing of temperate lakes can introduce conditions that support microbial blooms in the spring and fall (**figure 30.18**).

Active Learning

1. The unicellular cyanobacterium *Prochlorococcus* sp. is the most abundant photosynthetic microbe in tropical and subtropical oceans. At least two ecotypes exist: One is adapted to high light and the other to lower light intensities. How does the presence of these two ecotypes contribute to their physiological success and their numerical success? How would you determine the amount of fixed carbon each ecotype contributes to open-ocean ecosystems?
2. How might it be possible to cleanse a eutrophic lake? Consider chemical, biological, and physical approaches as you formulate your plan.
3. It is well known that bacterivory (the consumption of bacteria) supports the growth of marine protists that lack plastids. However, measurements of bacterivory among plastid-containing planktonic algae show that small (less than 5 μm) algae carry out between about 40 and 95% of the bacterivory in the photic zone of the North Atlantic. Discuss how this level of mixotrophy impacts the concept of the microbial loop (figure 30.5).
4. A large international group of scientists used a variety of sampling and genomic techniques in a research program designed to better understand how microbial community dynamics in the photic zone impact the export of carbon to the seafloor (i.e., the biological pump). Discuss three specific questions you would ask if you were on the research team. Refer to chapters 18 and 29 to suggest methods you might use to answer these questions.

Read the original paper: Guidi, L., et al. 2016. Plankton networks driving carbon export in the oligotrophic ocean. *Nature.* 532:465–70. doi:10.1038/nature16942.

31

Microorganisms in Terrestrial Ecosystems

Source: Photo by Yue Jin/USDA-ARS

Bread for a Hungry World

Aristotle wrote about them. The Romans had a ceremony to prevent them. Over 2 million Indians starved in the 1940s because of them. They are responsible for the loss of more than a million metric tons of wheat in the United States every year. Norman Borlaug won a Nobel Peace Prize fighting them. What are they? Fungal plant pathogens: the cause of more than 70% of all major crop diseases.

The worst offenders among fungal plant pathogens are the "rusts," which are basidiomycetes (*see section 25.6*). In North America, probably the most important rust is *Puccinia graminis* subsp. *tritici*, the causative agent of wheat stem rust. Other important rusts include barley stem rust, soybean rust, cedar-apple rust, and poplar leaf rust. The introduction of fungi from Europe triggered chestnut blight (*Cryphonectria parasitica*), Dutch elm disease (*Ophiostoma ulmi*), and white pine blister rust (*Cronartium ribicola*). With over 6,000 different rust species known to cause plant diseases, it would seem the only approach to saving millions of dollars of valuable agricultural products would be to spray heavily with a fungicide and hope for the best.

Norman Borlaug knew that expensive fungicides were not an option for inhabitants of developing countries. Shortly after earning his Ph.D. in plant pathology at the University of Minnesota, he moved to Mexico to breed new, disease-resistant wheat strains. He later turned his attention to the Indian subcontinent, where he helped develop hardy strains of rice. In 1970 the Nobel committee noted, "More than any other single person of this age, he has helped provide bread for a hungry world." It is estimated that each day at least half the global population consumes grain that arose through Borlaug's plant breeding efforts. Borlaug was an active scientist until his death at age 91 in 2009.

Most microbes that associate with plants, whether that association is good, bad, or neutral, live in soil. Soil is a hugely underappreciated resource. It is a vast reservoir of nutrients, wastes, energy, and even pharmaceuticals. Although much is known about the geology and chemistry of soil, our understanding of the microbial communities that drive the development and continued health of soils is far from complete. However, the introduction of new, innovative culture and genomic techniques has made the study of terrestrial microbiology an exciting and dynamic field.

Readiness Check:

Based on what you learned previously, you should be able to:

- ✓ List environmental factors that influence microbial growth (section 7.5)
- ✓ Compare and contrast the growth and physiology of obligate aerobes, facultative anaerobes, and strict anaerobes (section 7.5)
- ✓ Evaluate the importance of intercellular molecular signaling to microbial growth in nature (section 7.6)
- ✓ Recall the basic enzymology of nitrogenase (section 12.5)
- ✓ Explain why SSU rRNA is used as a taxonomic marker (section 19.3)
- ✓ Identify microorganisms responsible for nitrogen fixation (sections 22.1 and 23.1)
- ✓ Explain the global impact that microbes have on carbon and nitrogen cycling (section 28.1)

31.1 Soils Are an Important Microbial Habitat

After reading this section, you should be able to:

a. Differentiate between a mineral and an organic soil, and describe the role of microbes in generating soil organic matter
b. Explain why lignin degradation is confined to only a few types of microbes
c. Predict how a specific carbon to nitrogen ratio will influence the microbial community within a given soil

The level of microbial diversity in soil exceeds that of any other habitat on Earth, with up to 50,000 species at densities of 10^7 to 10^{10} cells per gram of soil. The soil environment supports active microbial growth that contributes to global elemental cycling and makes nutrients available to plants. Simply put, without soil microbes, there would be no plants, which would eliminate herbivores, and so on up the food chain.

Soil microbial communities are supported by complex physical and chemical parameters, which provide a vast array of microhabitats. These include soil particles and the pore space between them, which is critical for the movement of water and

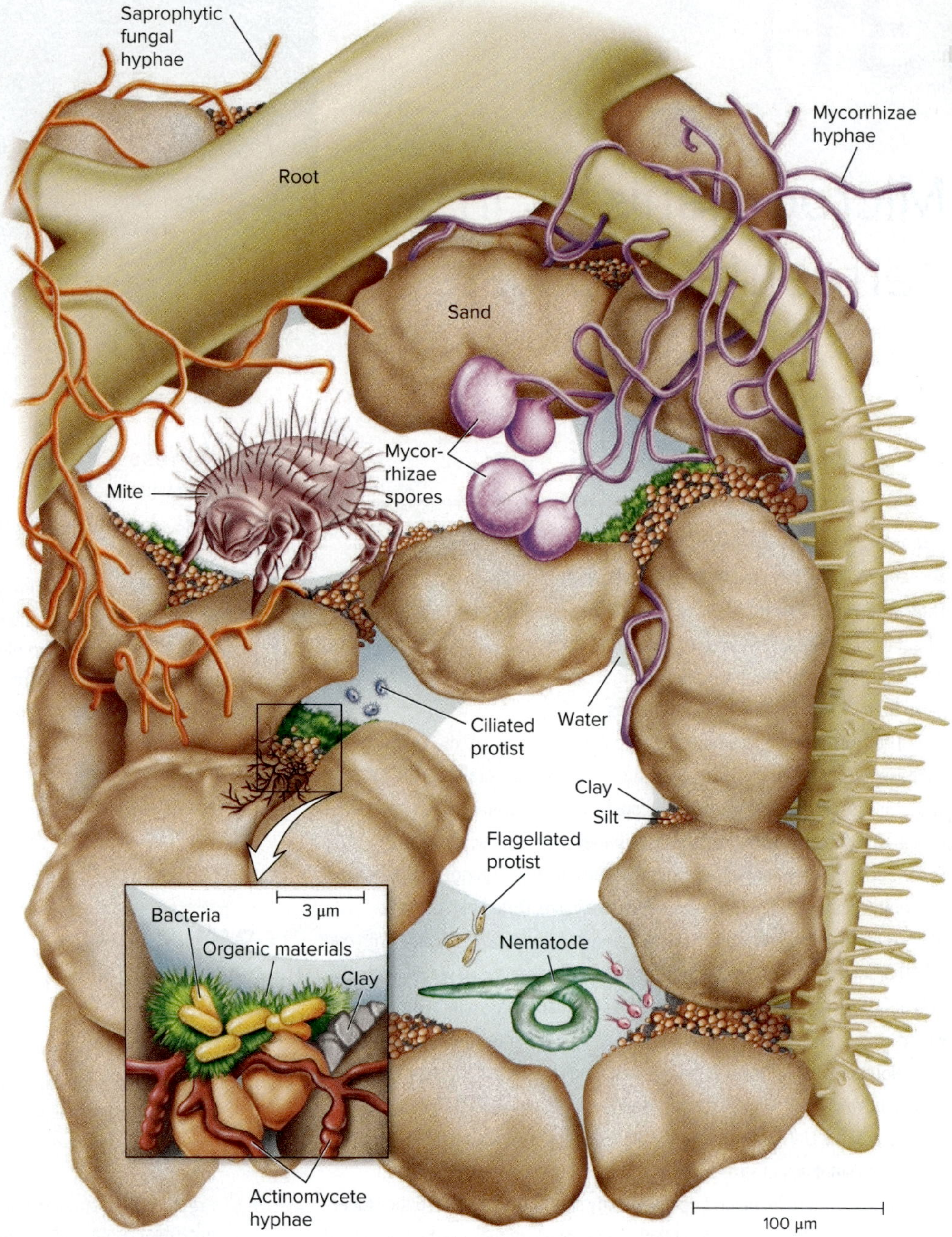

Figure 31.1 The Soil Habitat. A typical soil habitat contains a mixture of clay, silt, and sand along with soil organic matter. Roots, animals (e.g., nematodes and mites), and chemoorganotrophic bacteria consume oxygen, which is rapidly replaced by diffusion within the soil pores where the microbes live. Two types of fungi are present: mycorrhizal fungi, which derive their organic carbon from their symbiotic partners, plant roots; and saprophytic fungi, which contribute to the degradation of organic material. In addition to heterotrophs, chemolithotrophic microorganisms are important in elemental cycling within the soil environment.

MICRO INQUIRY *How are filamentous microbes especially well suited to the soil ecosystem?*

gases (**figure 31.1**). Total pore space, and thus gas diffusion, is determined by soil texture. For instance, sandy soils have larger pore spaces than do clay soils, so sandy soils tend to drain more quickly. Pores are also critical because they provide the optimum environment for microbial growth. Here microbes are within thin water films on particle surfaces where oxygen is present at high levels and can be easily replenished by diffusion.

Soil environments can change rapidly. For instance, rainfall or irrigation may quickly change a well-aerated soil to one with isolated pockets of water, generating miniaquatic habitats. If oxygen consumption exceeds that of oxygen diffusion, waterlogged soils can become anoxic. Shifts in water content and gas fluxes also affect the concentrations of CO_2 and other gases. These changes are accentuated in the smaller pores, where many bacteria are found. The roots of plants growing in aerated soils also consume oxygen and release CO_2, influencing the concentrations of these gases in the root environment.

Soils can be divided into two general categories: A **mineral soil** contains less than 20% organic carbon, whereas an **organic soil** possesses at least this amount. By this definition, the vast majority of Earth's soils are mineral. The importance of organic matter within soils cannot be overstated. **Soil organic matter** (**SOM**) helps retain nutrients, maintain soil structure, and hold water for plant use. SOM is subject to gains and losses, depending on changes in environmental conditions and agricultural practices. Plowing and other disturbances expose SOM to more oxygen, leading to extensive microbiological degradation of organic matter. Irrigation causes periodic wetting and drying, which can also lead to increased degradation of SOM, especially at higher temperatures.

Microbial degradation of plant material results in the release of CO_2 and the incorporation of plant carbon into microbial biomass. However, a small fraction of the decomposed plant material remains in the soil as SOM. This material can be divided into humic and nonhumic fractions. Nonhumic SOM has few, if any, biochemical degradation products. It can represent up to about 20% of the soil organic matter. Humic SOM, or humus, results when plant material has been incompletely degraded by microbes. Although it has no defined chemical composition, humus can be described as a complex blend of phenolic compounds, polysaccharides, and proteins.

Microbial degradation of plant material and the development of SOM can be thought of as a three-step process. First, easily degraded compounds such as soluble carbohydrates and proteins are broken down by a variety of heterotrophic microorganisms. About half the carbon is respired as CO_2, and the remainder is rapidly incorporated into new biomass. During the second stage, complex carbohydrates, such as cellulose, are degraded. Cellulose, an important plant cell wall structural polysaccharide, represents a vast store of carbon and it is the most abundant organic compound on Earth. Brown rot fungi and filamentous bacteria in the genera *Coniophora* and *Streptomyces,* respectively, and nonfilamentous bacteria such as *Cytophaga* and *Bacillus* spp. produce extracellular cellulase enzymes that break down cellulose into two and three glucose units called cellobiose and cellotriose, respectively. These smaller compounds are readily degraded and assimilated as glucose monomers.

©Pixtal/age fotostock

Lignin, an important structural component of woody plants, is the soil component most resistant to degradation. While its exact structure differs among plant species, the common building block is the phenylpropene unit. This consists of a hydroxylated benzene ring and a three-carbon linear side chain (**figure 31.2**). A single lignin molecule can consist of up to 600 cross-linked phenylpropene units. It is therefore not surprising that degradation of lignin is much slower than that of cellulose. Basidiomycete fungi commonly termed white rot fungi (e.g., *Phanerochaete* and *Phlebia* spp.) and actinobacteria (e.g., *Streptomyces* spp.) secrete extracellular phenoloxidase enzymes that degrade lignin aerobically. Lignin can be degraded anaerobically, but this process is very slow, so lignin accumulates in wet, poorly oxygenated soils, such as peat bogs. Regardless of the presence or absence of oxygen, no more than 10% of lignin is converted to biomass.

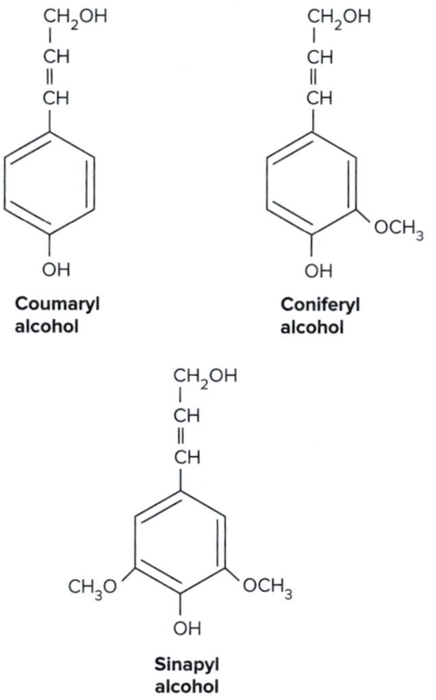

Figure 31.2 Phenylpropene Units. Hundreds of these molecules are polymerized to form lignin, an irregular branched polymer.

Nitrogen is another important element in soil ecosystems. Nitrogen in soil is often considered in relation to the soil carbon content as the organic **carbon to nitrogen ratio (C/N ratio)**. As a general rule, decomposition is thought to be maximal at a C/N ratio of 30 (i.e., 30 times more carbon than nitrogen). A ratio of less than 30 can result in the loss of soluble nitrogen from the system. Many agricultural soils have ratios above 30 and are therefore nitrogen limited. This is why each year tons of nitrogen fertilizer are added to agricultural soils. Most fertilizers include ammonium ion because it will bind to the negatively charged clays in a soil until used as a nutrient by plants. However, when ammonium is in excess, nitrifying microorganisms in a soil can oxidize the ammonium ion to nitrite and nitrate, and these anions can be leached from the plant environment and pollute surface waters and groundwater. ◀ *Nitrifying bacteria oxidize ammonium or nitrite to gain energy and electrons (section 22.1); Nitrogen cycle (section 28.1); Global climate change (section 28.2)*

Phosphorus is also commonly in short supply in soils, so it is also added to fertilizers. The binding of anionic phosphate molecules to soils depends on several factors, including soil pH. When the soil has bound all the phosphorus it can, the excess, together with phosphorus released by soil erosion, moves to lakes, streams, and estuaries, where it can stimulate the growth of microorganisms, particularly nitrogen-fixing cyanobacteria, resulting in eutrophication. ◀ *Microorganisms in lakes can cause seasonal blooms (section 30.3)*

Comprehension Check

1. List three reasons why soil organic matter (SOM) is important.
2. In what order will these compounds be degraded in soils: peptides, lignin, starch, cellulose, fructose?
3. What microbes commonly degrade cellulose and lignin? Why is their filamentous morphology advantageous?
4. Which of the following soils would a farmer be most likely to fertilize: a soil with a C:N ratio of 23, 28, or 37?
5. Why is most nitrogen fertilizer added as ammonium ion?

31.2 Diverse Microorganisms Inhabit Soil

After reading this section, you should be able to:

a. Estimate the relative microbial populations in several soil types
b. List the microbial genera easily cultured from soil
c. Describe the different ecological niches occupied by microbes in soil and how microbes have evolved to take advantage of the niches

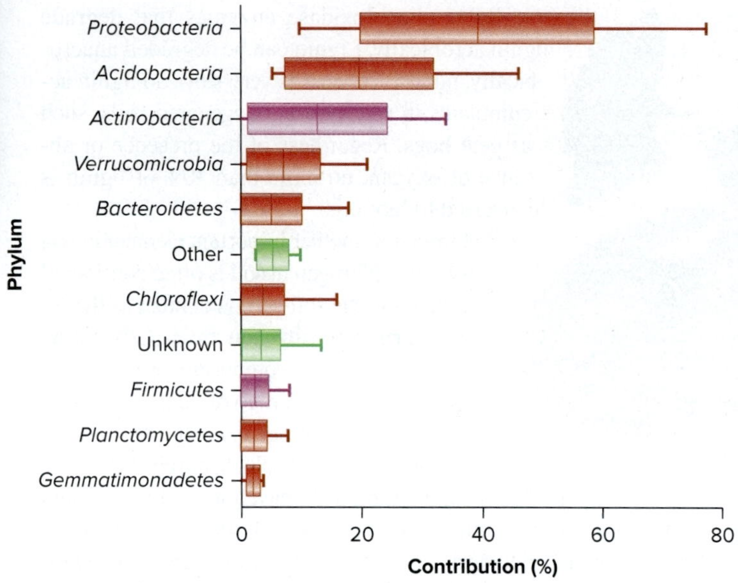

Figure 31.3 Bacterial Phyla Identified from 16S rRNA and 16S rRNA Genes Extracted from Soil. Data from 21 soil metagenomic libraries were analyzed to assess soil bacterial diversity. The horizontal line in the middle of each block represents the mean. The block represents one standard deviation around the mean. The vertical lines extending above and below indicate the maximum and minimum values for each phylum. Gram-negative phyla are shown in red, Gram-positive in violet, unknown and other phyla found in very small numbers in green.

Table 31.1 Easily Cultured Gram-Positive Soil Bacteria

Phylum	Representative Genera	Comments and Characteristics
Actinobacteria	*Arthrobacter*	Rod-coccus cycle
	Cellulomonas	Important in degradation of cellulose
	Corynebacterium	Club-shaped cells
	Mycobacterium	Acid-fast
	Nocardia	Rudimentary branching
	Streptomyces	Aerobic filamentous bacteria
Firmicutes	*Thermoactinomyces*	Higher temperature growth
	Bacillus	Aerobic, endospore-forming

An average gram of forest bulk soil—soil not close to plant roots—contains about 4×10^7 bacteria and archaea, while a gram of grassland or cultivated soil may possess 2×10^9 cells. It is difficult to determine how many different taxa are represented, but estimates based on DNA reassociation range from 2,000 to 18,000 distinct genomes per gram of soil. In addition, metagenomic surveys from different soils show some major trends. Although most rRNA gene sequences can be assigned to well-studied bacterial phyla, many (about 40%) cannot be placed in any known genera. Overall, bacteria belonging to at least 35 different phyla have been identified, with over 90% of soil bacterial 16S rRNA assigned to the nine phyla displayed in **figure 31.3**. Interestingly, this includes *Acidobacteria* and *Verrucomicrobia,* phyla about which little is known. Other important phyla not shown in figure 31.3 include *Chlamydiae, Chlorobi, Cyanobacteria, Deinococcus-Thermus,* and *Nitrospirae*. Many *Actinobacteria* are easily cultured and therefore well-characterized members of the soil community (**table 31.1**). These actinobacteria play a major role in the degradation of hydrocarbons, older plant materials, and soil humus. In addition, some members of these groups actively degrade pesticides. Filamentous bacteria, primarily of the genus *Streptomyces,* produce an odor-causing compound called geosmin, which gives soils their characteristic earthy odor. ◀ *Class* Actinobacteria *(section 23.1)*

Of course, bacteria are not the only microbes present in soils. Among archaea, ammonia-oxidizing members of the phylum *Thaumarchaeota* often prevail, and may constitute up to 10% of the microorganisms identified by SSU rRNA sequencing. The number of archaea appears to be related to the C/N ratio, with larger archaeal populations occurring in soils with higher concentrations of nitrogen. The relative contribution of bacterial and archaeal nitrification in soils is an active area of research *(see figure 22.13).* ◀ *Nitrogen cycle (section 28.1)*

Fungi are also very important members of the soil microbial community. Fungi that grow in bulk soil are saprophytic, helping to degrade complex plant molecules. Fungi can extend up to several hundred meters of hyphae per gram of soil. Although we tend to think of soil fungi as the mushrooms sprouting on our lawns, the vast majority of fungal biomass is below ground. There filaments bridge open areas between soil particles. These fungi form darkly pigmented resting structures called sclerotia and hyphal cords. Hyphal cords consist of parallel fungal filaments that move nutrients and water over great distances, including across air spaces, a unique part of their functional strategy. They often look similar to plant roots so are also called **rhizomorphs** (Greek *rhiza,* root; *morphe,* form).

Fungal biomass can be enormous. In fact the largest organism on Earth is found mostly underground in the mountains of Oregon. There a clone of the fungus *Armillaria ostoyae,* whose rhizomorphs infect the roots of many conifers, has been determined to cover over 4 square miles. Estimated to be over 2,000 years old, it is one of the oldest living organisms as well.

The flux of nutrients through soil in nearly all terrestrial ecosystems depends on plants for primary production. Soil microbes rapidly recycle organic material derived from plants and animals, including the many nematodes and insects, which function to reduce the size of solid organic matter. In turn, microbes themselves are preyed upon by viruses, other bacteria (e.g., species within the genera *Lysobacter* and *Myxococcus*), and soil protists, whose numbers can reach 100,000 per gram of soil. This makes microbial organic matter available to other trophic levels.

Comprehension Check

1. What are the differences in preferred soil microhabitats between bacteria and filamentous fungi?
2. What types of archaea have been detected in soils?
3. Outline the contribution of plants, bacteria, fungi, protists, and viruses to the flux of carbon in soil.

31.3 Microbe-Plant Interactions Can Be Positive, Negative, or Neutral

After reading this section, you should be able to:

a. Describe mechanisms by which plants and microbes communicate
b. Differentiate between the phyllosphere, rhizosphere, and rhizoplane environments and the microbes that dwell there
c. Explain the importance of mycorrhizae to the growth of vascular plants
d. Contrast and compare the ectomycorrhizal life cycle and that of arbuscular fungi
e. Diagram the steps that result in the formation of rhizobium nodules on leguminous plants
f. Compare the process of stem nodulation with root nodulation
g. Discuss other microbes, in addition to rhizobia, that fix nitrogen
h. Explain how *Agrobacterium tumefaciens* causes tumors in plants
i. List at least five plant pathogens other than *Agrobacterium tumefaciens* and the diseases they cause
j. Identify a plant disease that is likely to be found near your home or school

The evolution of aquatic plants into terrestrial organisms roughly 470 million years ago was made possible by cooperation with soil microbes, and many of these microbe-plant interactions remain. Some are commensalistic—no harm is done to the plant, but the microbe gains some advantage. Many others are beneficial to both the microorganism and the plant (e.g., they are mutualistic). Finally, other microbes are plant pathogens that parasitize their plant hosts. In all cases, the microbe and the plant have established the capacity to communicate. The microbe detects and responds to plant-produced chemical signaling molecules. This generally triggers the release of microbial compounds recognized by the plant, thereby beginning a two-way "conversation" that employs a molecular lexicon. Once a microbe-plant relationship is initiated, microbes and plants continue to monitor the physiology of their partner and adjust their own activities accordingly. The nature of the signaling molecules and the mechanisms by which both plants and microbes respond have stimulated exciting multidisciplinary research in soil microbiology, ecology, molecular biology, genetics, and biochemistry. ◀ *Cell–cell communication within microbial populations (section 7.6); Many types of microbial interactions exist (section 27.1)*

Microbes that interact with plants can be broadly divided into two classes: microbes that live on the surface of plants, called **epiphytes,** and those that colonize internal plant tissues, called **endophytes.** We can further classify microbes into those that live in the aboveground, or aerial, surfaces of plants separately and those that inhabit below-ground plant tissues. We begin our discussion of microbe-plant interactions by introducing the microbial communities associated with aerial regions of plants. We then turn our attention to two important microbe-root symbioses—mycorrhizal fungi and nitrogen-fixing rhizobia. Finally, we consider several important microbial plant pathogens.

Phyllosphere Microorganisms Are Stress Resistant

The environment of the aerial portion of a plant (e.g., stems and leaves), called the **phyllosphere,** undergoes frequent and rapid changes in humidity, ultraviolet (UV) light exposure, and temperature. This causes fluctuations in the leaching of organic material (primarily simple sugars) that supports a diverse assortment of microbes, including bacteria, filamentous fungi, yeasts, and photosynthetic and heterotrophic protists. Bacterial densities are estimated to be 10^6 to 10^7 cells per cm^2 of plant tissue. Culturable bacteria include the γ-proteobacteria *Pseudomonas syringae, Erwinia* spp., and *Pantoea* spp. In addition, methylotrophic α-proteobacteria such as *Methylobacterium* spp. are supported by plant release of methanol. Members of another abundant bacterial genus, *Sphingomonas,* produce pigments that function like sunscreen so they can survive the UV radiation striking plant surfaces. Microbial diversity in the phyllosphere is positively correlated with the productivity of the entire ecosystem. In addition, the phyllosphere microbial community contributes to global carbon and nitrogen cycling, the removal of airborne pollutants, and the decomposition of leaf litter.

The Rhizoplane and Rhizosphere Support Complex Microbial Communities

Plant roots receive between 30 and 60% of the net photosynthesized carbon. Of this, an estimated 40 to 90% enters soils as a wide variety of materials, including alcohols, ethylene, sugars, amino and organic acids, vitamins, nucleotides, polysaccharides, and enzymes. The plant root surface, termed the **rhizoplane** (Greek *rhiza,* root; Latin *planum,* flat surface), provides an exceptional environment for microorganisms, as these gaseous, soluble, and particulate materials move from plant to soil.

The **rhizosphere** is the region surrounding a plant root rich in plant exudates more easily degraded than lignin and cellulose. This in turn supports bacterial and fungal populations that can be as much as 100 times greater than surrounding bulk soil. Despite the seeming abundance of nutrients, microbes in rhizosphere communities must compete with one another, and they do so by the secretion of antimicrobial agents, lytic enzymes, and even hydrogen cyanide. Other microbes counteract these inhibitors by efflux pumps, detoxification enzymes, and secretion of molecules that alter competitor gene expression. Importantly, rhizosphere and rhizoplane microorganisms provide nutrients for other organisms, including plants and other microbes, thereby playing critical roles in organic matter synthesis and degradation.

Microbes in the rhizoplane and rhizosphere enhance plant growth by producing a variety of compounds. These include molecules that regulate plant growth called phytohormones (e.g., auxin and gibberellin) and compounds that stimulate host plant production of phytohormones. The production of volatile organic compounds including acetoin and phenazines—a diverse class of heterocyclic compounds—stimulates plant responses to stress. In addition, some bacteria secrete compounds that suppress the growth of plant fungal pathogens. The use of these compounds in agriculture is an active area of research.

Associative nitrogen fixation—N_2 fixation by epiphytic bacteria—is a critical process performed in the rhizoplane as well as in the rhizosphere. It is carried out by members of a number of genera, including *Azotobacter, Azospirillum,* and *Acetobacter.* These bacteria enhance the growth of agriculturally important crops such as wheat, potato, and barley. In addition to fixing N_2, many of these bacteria produce growth-promoting hormones that increase root hair development, thereby enhancing plant nutrient uptake.

Comprehension Check

1. Define rhizosphere, rhizoplane, and associative nitrogen fixation.
2. What unique stresses does a microorganism on a leaf contend with that a microbe in the soil does not?
3. List two ways in which compounds produced by rhizosphere and rhizoplane microbes promote plant growth.
4. What important genera are involved in associative nitrogen fixation?

Mycorrhizae Facilitate the Growth of Most Plants

Mycorrhizae (Greek, *myco,* fungus; *rhizo,* root) are mutualistic relationships that develop between about 80% of all land plants and a limited number of filamentous fungal species. Mycorrhizae are so abundant that up to 100 m of mycorrhizal filaments can be found in a gram of soil. Unlike most fungi, **mycorrhizal fungi** are not saprophytic; that is, they do not obtain organic carbon from the degradation of organic material. Instead, they use photosynthetically derived carbohydrate from their host. In return, they provide a number of services for their plant hosts. In moist environments, they increase the availability of nutrients, especially phosphorus. In arid environments, mycorrhizae aid in water uptake, increasing transpiration rates in comparison with nonmycorrhizal plants. ▶| *Mutualism and cooperation are two-way interactions (section 27.2)*

Mycorrhizae can be broadly classified as **endomycorrhizae**—those with fungi that enter root cells, or as **ectomycorrhizae**—those that remain extracellular, forming a sheath of interconnecting filaments (hyphae) around roots. Although six types of mycorrhizae are detailed in **table 31.2** and **figure 31.4**, we focus primarily on the two most important types: ectomycorrhizae and arbuscular mycorrhizae.

Table 31.2 Mycorrhizal Associations

Mycorrhizal Classification	Fungi Involved	Plants Colonized	Fungal Structural Features	Fungal Function
Ectomycorrhizae	Basidiomycetes, including those with large fruiting bodies (e.g., toadstools); some ascomycetes	~90% of trees and woody plants in temperate regions; fungal-plant colonization is often species specific	Hartig net, mantle or sheath; rhizomorphs; root hair development is usually limited	Nutrient (N and P) uptake and transfer
Arbuscular	Glomeromycetes, in particular six genera of glomeromycetes	Wild and crop plants, tropical trees; fungal-plant colonization is not highly specific	Arbuscules: hyphae-filled invaginations of cortical root cell	Nutrient (N and P) uptake and transfer; facilitate soil aggregation; promote seed production; reduce pest and nematode infection; increase drought and disease resistance
Ericaceous	Ascomycetes Basidiomycetes	Low evergreen shrubs, heathers	Some intracellular, some extracellular	Mineralization of organic matter
Orchidaceous	Basidiomycetes	Orchids	Hyphal coils called pelotons within host tissue	Some orchids are non-photosynthetic and others produce chlorophyll when mature; these organisms are almost completely dependent on mycorrhizae for organic carbon and nutrients.
Ectendomycorrhizae	Ascomycetes	Conifers	Hartig net with some intracellular hyphae	Nutrient uptake and mineralization of organic matter
Monotropoid mycorrhizae	Ascomycetes Basidiomycetes	Flowering plants that lack chlorophyll (*Monotropaceae;* e.g., Indian pipe)	Hartig net one cell deep in the root cortex	Nutrient uptake and transfer

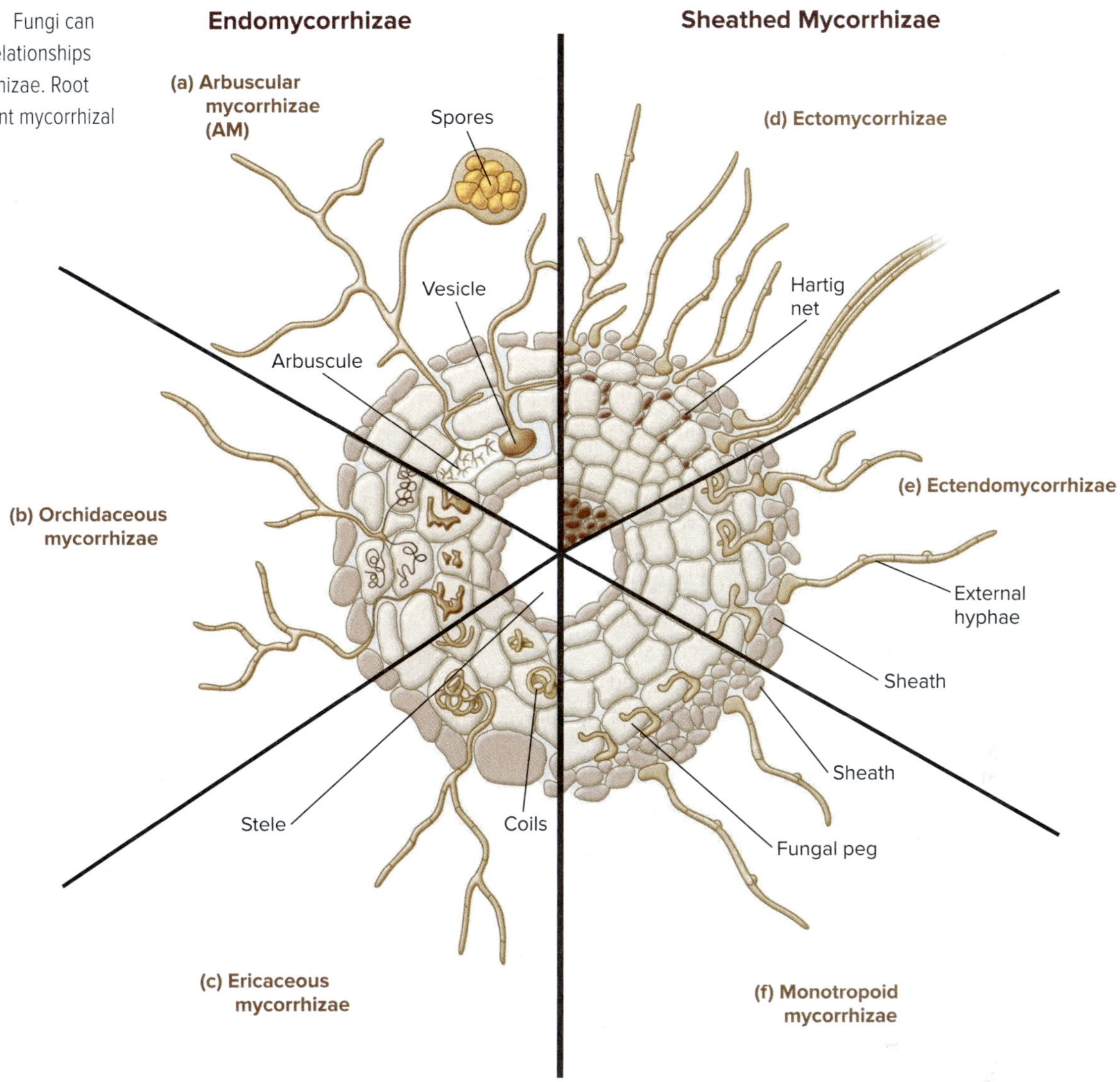

Figure 31.4 Mycorrhizae. Fungi can establish mutually beneficial relationships with plant roots, called mycorrhizae. Root cross sections illustrate different mycorrhizal relationships.

Ectomycorrhizae (ECM) are formed by both ascomycete and basidiomycete fungi. ECM colonize almost all trees in cooler climates. Their importance arises from their ability to transfer essential nutrients, especially phosphorus and nitrogen, to the root. The development of ECM starts with the growth of a fungal mycelium around the root. As the mycelium thickens, it forms a sheath or mantle so that the entire root may be covered by the fungal mycelium (**figure 31.5**). Most ECM produce signaling molecules that limit the growth of root hairs; thus ECM-colonized roots often appear blunt and covered in fungi. From the root surface, the fungi extend hyphae into the soil; these filaments may aggregate to form rhizomorphs visible to the naked eye. Hyphae within rhizomorphs are enclosed in a sheath and those on the inner side of the sheath penetrate between (but not within) root cells, forming a characteristic meshwork of hyphae called the **Hartig net** (figure 31.4*d*). Soil nutrients taken up by rhizomorphs first pass through the hyphal sheath and then into the Hartig net filaments, which form numerous contacts with root cells. This results in efficient transfer

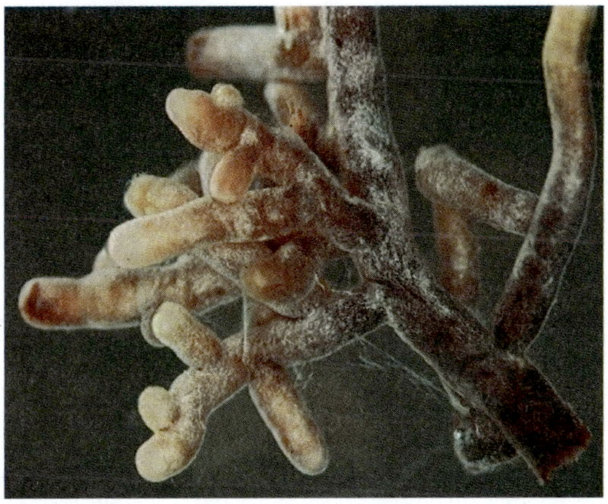

Figure 31.5 Ectomycorrhizae as Found on Roots of a Pine Tree. Typical irregular branching of the white, smooth mycorrhizae is evident.

of soil nutrients to the plant and carbohydrates to the fungus. This relationship has evolved to the point that some plants synthesize sugars solely for the use of their fungal partners, such as mannitol and trehalose.

Arbuscular mycorrhizae (AM) are the most common type of mycorrhizae and make up about half the fungal biomass in most soils. They are found in association with many tropical plants and, importantly, with most crop plants. AM fungi belong to the taxon *Glomeromycota* and can be cultured only with roots of their plant hosts. These microbes enter root cells between the plant cell wall and invaginations in the plasma membrane (figure 31.4*a*). So, although AM are endomycorrhizae, they do not breach the root cell membrane. Instead, treelike hyphal networks called **arbuscules** develop within folds of the plasma membrane. Individual arbuscules are transient, lasting at most 2 weeks. Spores released by nearby root-associated AM germinate and colonize roots. This requires a number of specific steps, as shown in **figure 31.6**. AM can be vigorous colonizers: A 5-cm segment of root can support the growth of as many as eight species, and hyphae from a single germinated spore can simultaneously colonize multiple roots of unrelated plant species, forming a complex underground highway for nutrient transport.

◀ Glomeromycota *are mycorrhizal symbionts (section 25.4)*

Like the α-proteobacteria belonging to the rhizobium family (discussed next), AM establish symbioses with legumes (among other plants). The secretion of so-called "**Nod factors**" by rhizobia has been well established, and analogous "**Myc factors**" have been purified from AM fungi (**figure 31.7**). Both Nod and Myc factors are lipochitooligosaccharides that prompt legume roots to accept symbionts. Myc factors have a simpler structure than Nod factors. Interestingly, the rhizobium-legume symbiosis is believed to have arisen only about 60 million years ago (Ma), while AM were present with the development of land plants 407 Ma. This suggests that this

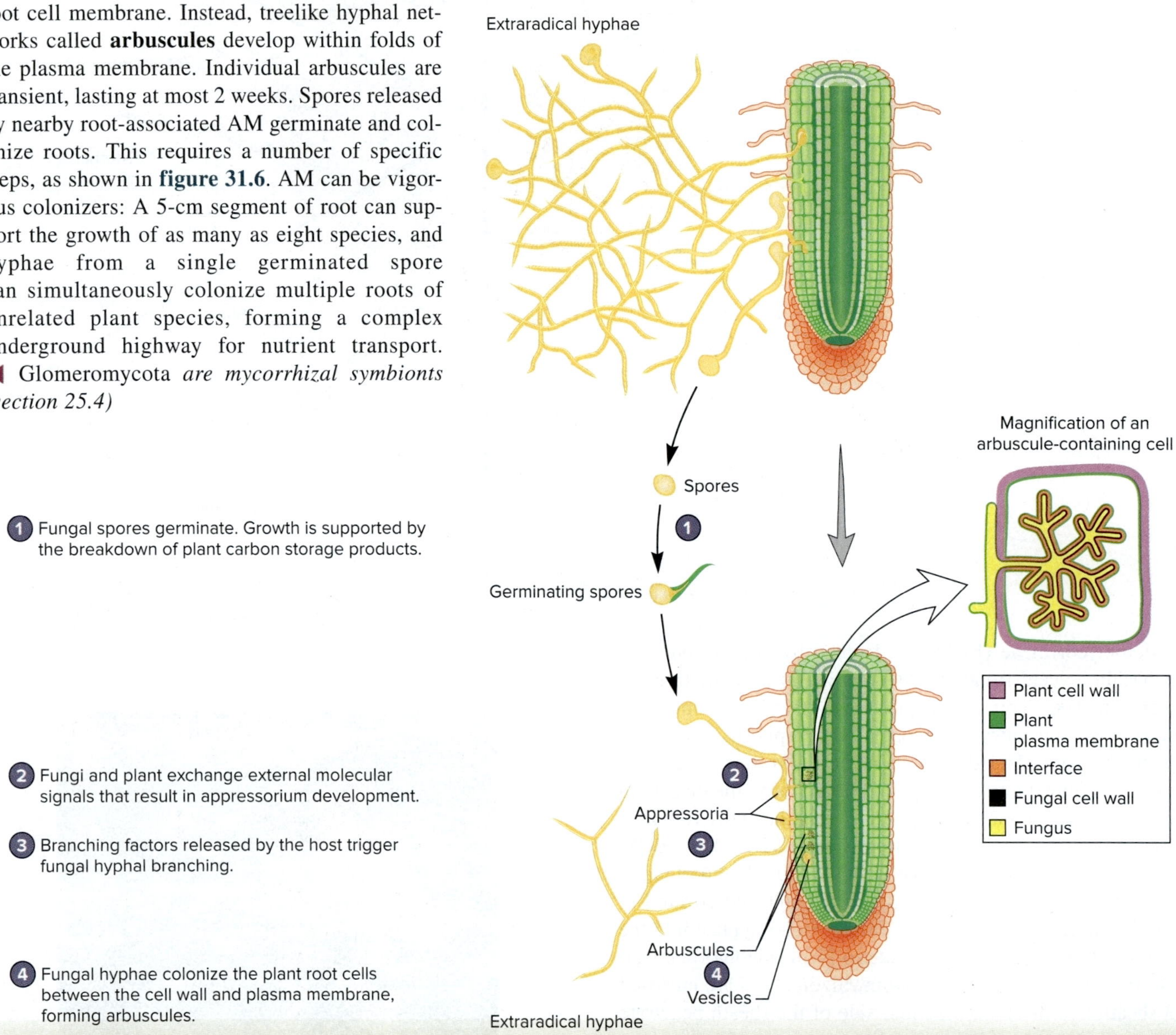

Figure 31.6 The Process of Root Colonization by Arbuscular Mycorrhizal Fungi.

MICRO INQUIRY *When does the fungus grow saprotrophically? How does the proportion of the organism growing saprotrophically change during host colonization?*

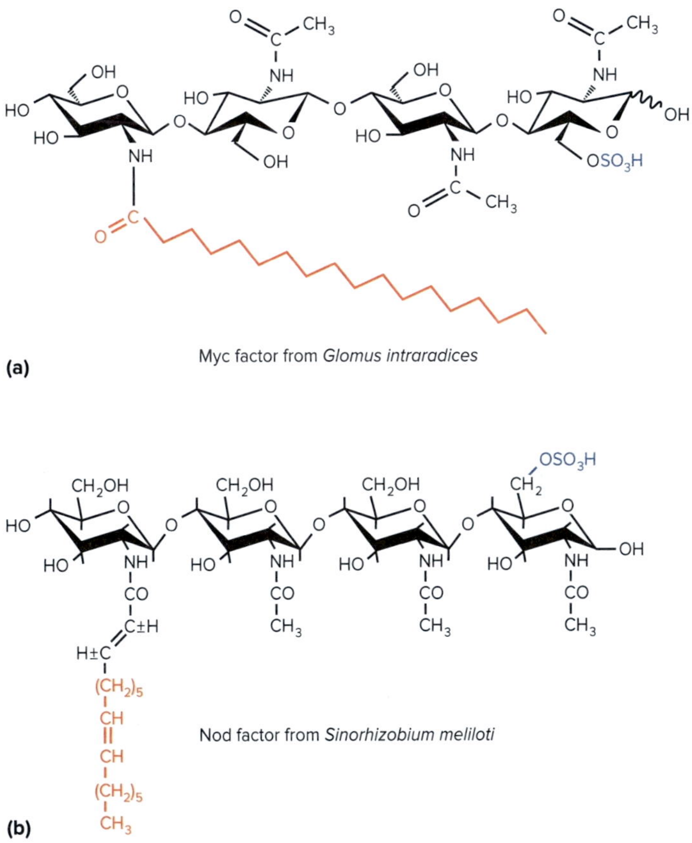

Figure 31.7 Myc and Nod Factors. Both arbuscular mycorrhizae and rhizobia form symbioses with legumes initiated when the microbes secrete a lipochitooligosaccharide. (a) AM produce Myc factors. (b) Bacteria secrete Nod factors. The precise structure of individual Myc and Nod factors depends on the microbial species, but all consist of four to five units of β–1,4 linked *N*-acetyl-*D*-glucosamine bearing a species-specific acyl chain (red) at the nonreducing terminal residue. All Nod factors have a sulfate attached to the reducing end (blue), while not all Myc factors are sulfonated.

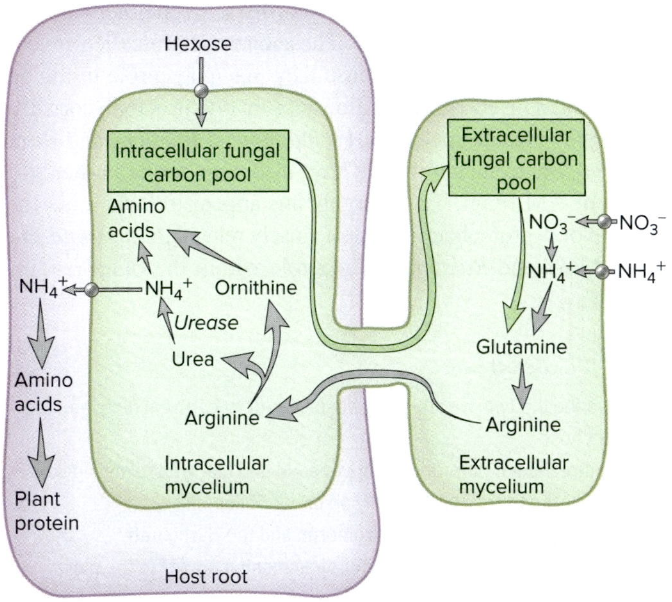

Figure 31.8 Nitrogen and Carbon Exchange Between Arbuscular Mycorrhizal Fungi and Host Plant. Nitrate and ammonium are taken up by fungal hyphae outside the host plant cell (extracellular mycelium) and converted to arginine. This amino acid is transferred to the mycelium within the host plant cell and broken down so that only the ammonium enters the plant.

signaling mechanism was adopted by rhizobia, which then further modified it.

AM are believed to provide a number of services to their plant hosts, including protection from disease, drought, nematodes, and other pests. Their capacity to transfer phosphorus to roots has been well documented. AM also provide nitrogen for their plant partners. Interestingly, AM assimilate NO_3^- and NH_4^+ through the glutamine synthetase–glutamate synthase (GOGAT) pathway (*see figure 12.13*), and therefore incorporate nitrogen into glutamine, which they then convert to arginine. But rather than transferring arginine to host cells, intracellular fungal hyphae degrade the amino acids, transferring only the NH_4^+ (**figure 31.8**). Thus while the fungus provides its host with much needed ammonium, it retains the carbon skeleton required for nitrogen uptake. In contrast, less is known about how mycorrhizal fungi take up carbon from their hosts. Interestingly, plants can detect, discriminate, and reward the best fungal partners with more carbohydrates and in at least one case, fatty acids as well. In return, the fungi transfer more nutrients to the plants that "feed" them best. This tight plant-microbe coupling probably accounts for the evolutionary stability of this partnership. ◀ *Inorganic nitrogen assimilation (section 12.5)*

In addition to these two most abundant mycorrhizal types, the orchidaceous mycorrhizae are of particular interest (figure 31.4*b*). Orchids are unusual plants because many never produce chlorophyll, while others only do so after they have matured past the seedling stage. Therefore all orchids have an absolute dependence on their endomycorrhizal partners for at least part of their lives. Indeed, orchid seeds will not germinate unless first colonized by a basidiomycete orchid mycorrhiza. Because orchids produce very little or no photosynthetically derived organic carbon, orchid mycorrhizal fungi are saprophytic. They must degrade soil organic matter to obtain carbon, which the orchids then also consume. In this case, the orchid functions as a parasite.

The nature of some mycorrhizae-host relationships is currently being reevaluated. It is now known that in addition to orchids, some green plants such as serrated wintergreen (*Orthilia secunda*) also parasitize their mycorrhizal partners. Likewise, the nuisance plant spotted knapweed (*Centaurea maculosa*) may obtain up to 15% of its organic carbon from mycorrhizal fungi. Finally, mycorrhizae may limit plant species diversity in some ecosystems, as the growth of certain seedlings is promoted by mycorrhizae that will only associate with a particular plant species.

Just as bacteria associate with plant roots, they also interact with mycorrhizal fungi. As the external hyphal network radiates out into the soil, a mycorrhizosphere is formed due to the flow of carbon from the plant into the mycorrhizal hyphal network and then into the surrounding soil. These bacteria are called mycorrhization helper bacteria because they may play a role in the development of mycorrhizal relationships involving ectomycorrhizal fungi; they include species of *Pseudomonas, Bacillus*, and *Burkholderia*. Bacterial symbionts have also been found in the cytoplasm of AM fungi. These organisms appear to form a distinct lineage of β-proteobacteria, most closely related to *Burkholderia, Pandoraea*, and *Ralstonia*. Their role within the fungi remains unclear.

Comprehension Check

1. Describe the two-way relationship between mycorrhizal fungi and the plant host.
2. List three major differences between AM and EM. What plants in your neighborhood do you think rely on either AM or EM?
3. What is the function of the rhizomorph and the Hartig net?
4. Describe the uptake and transfer of ammonium by AM to the plant host. Why is only the ammonium transferred?
5. Propose two potential functions for mycorrhization helper bacteria.

Nitrogen-Fixing Bacteria Are Vital to Agriculture

The conversion of gaseous nitrogen (N_2) to ammonia (NH_3) often occurs as part of a symbiotic relationship between bacteria and plants. These symbioses produce more than 100 million metric tons of fixed nitrogen annually and are a vital part of the global nitrogen cycle. Indeed, without microbial nitrogen fixation life on Earth could not have evolved. Currently, symbiotic nitrogen fixation accounts for more than half of the nitrogen used in agriculture (the remainder is applied as fertilizer). The provision of fixed nitrogen enables the growth of host plants in soils that would otherwise be nitrogen limiting. At the same time, it reduces loss of nitrogen by denitrification and leaching. ◀ *Nitrogen cycle (section 28.1)*

Rhizobia Are Symbionts of Leguminous Plants

Several α-proteobacterial genera contain species able to form nitrogen-fixing nodules with legumes. These include *Allorhizobium, Azorhizobium, Bradyrhizobium, Mesorhizobium, Sinorhizobium*, and *Rhizobium*. Collectively these bacteria are often called **rhizobia.** The phylogenetic diversity of rhizobia has been extended by the discovery that β-proteobacteria in the genera *Burkholderia* and *Cupriavidus* (formerly *Ralstonia*) also form nitrogen-fixing nodules on legumes. There are also reports of γ-proteobacteria nodulating wild legumes. Here we discuss the general process of nodulation, including some molecular details that have been revealed largely through studies of bacteria in the genus *Rhizobium*. ◀ *Class* Alphaproteobacteria *includes many oligotrophs (section 22.1)*

Rhizobia live freely in soil. In nitrogen-sufficient soils, plants secrete ethylene-mediated compounds that inhibit nodulation. But even in nitrogen-starved soils, when rhizobia approach the plant root, they are assumed to be invaders. The plant responds with an oxidative burst, producing a mixture of compounds that can contain superoxide radicals, hydrogen peroxide, and N_2O. This oxidative burst is critical in determining the fate of the infection process. Only rhizobia with sufficient antioxidant defenses can proceed to the next step in the infection process.

Infection starts with colonization, which is initiated by the exchange of signaling molecules between the plant and rhizobia in the rhizosphere. Plant roots release flavonoid inducer molecules (2-phenyl-1,4-benzopyrone derivatives) that stimulate rhizobial colonization of the root surfaces (**figure 31.9*a***). Flavonoids bind the bacterial protein NodD, which functions as a transcriptional regulator. NodD activates transcription of *nod* genes encoding the biosynthetic enzymes needed for the production of Nod factor signaling compounds (figure 31.9*b*). Conversely flavonoids from nonhost plants inhibit the production of Nod factors. Upon receipt of the Nod factor signal, gene expression in the outer (epidermal) cells of the roots is altered, resulting in oscillations in the levels of intracellular calcium. This calcium spiking triggers root hairs to curl, thereby entrapping bacteria (figure 31.9*c,d*). Nod factors trigger root cortex cells to initiate cell division, and these cells eventually form the nodule primordium that accepts the invading rhizobium.

In addition to Nod factors, the bacteria now produce an exopolysaccharide that induces changes in the plant cell wall, and the plant plasma membrane invaginates. This leads to the development of a bacteria-filled, tubelike structure called the **infection thread** (figure 31.9*e,f*). Plant cell division drives the growth of the infection thread. When bacteria penetrate the base of a root hair cell, they stimulate extension of the infection thread through additional plant cell walls and membranes. This is dependent on the plant hormone cytokinin in concert with Nod factor stimulation of plant cell mitosis. Finally, the infection thread filled with bacteria reaches the inner plant cortex, where each bacterial cell is endocytosed by a plant cell in a discrete, unwalled membrane compartment that arises from the infection thread. This unit, consisting of an individual bacterium and the surrounding endocytic membrane, is called the **symbiosome** (figure 31.9*g*). It is here that each bacterial cell differentiates into the nitrogen-fixing form called a **bacteroid.** Depending on the plant species, bacteroids may be terminally differentiated: They can neither divide nor revert back to the nondifferentiated state. Terminal differentiation is controlled by the production of nodule-specific cysteine-rich (NCR) peptide signals by the plant host; interestingly, these signals are similar in structure to antimicrobial peptides produced by humans called defensins.

The low levels of oxygen within the symbiosome triggers a regulatory cascade that controls nitrogen fixation and the switch to microaerobic respiration needed to provide ATP and reducing equivalents to the nitrogenase enzyme. These regulatory proteins include an oxygen-sensing, two-component signal transduction system (FixL and FixJ) and an alternative sigma factor, σ^N. Overall, most metabolic processes within bacteroids diminish with the exception of nitrogen fixation and respiration, which are upregulated. The assembly of many symbiosomes gives rise to the **root nodule** (figure 31.9*h,i*). *Nitrogen fixation: N_2 to NH_3 (section 12.5); Bacteria combine several regulatory processes to control complex cellular processes (section 14.6)*

The reduction of atmospheric N_2 to ammonia by the differentiated bacteroids depends on the production of a protein called **leghemoglobin.** Recall that the nitrogenase enzyme is inactivated by oxygen. To protect the nitrogenase, leghemoglobin binds oxygen and maintains microaerobic conditions within the mature nodule. Leghemoglobin is similar in structure to the hemoglobin found in animals; however, it has a higher affinity for oxygen. Interestingly, the protein

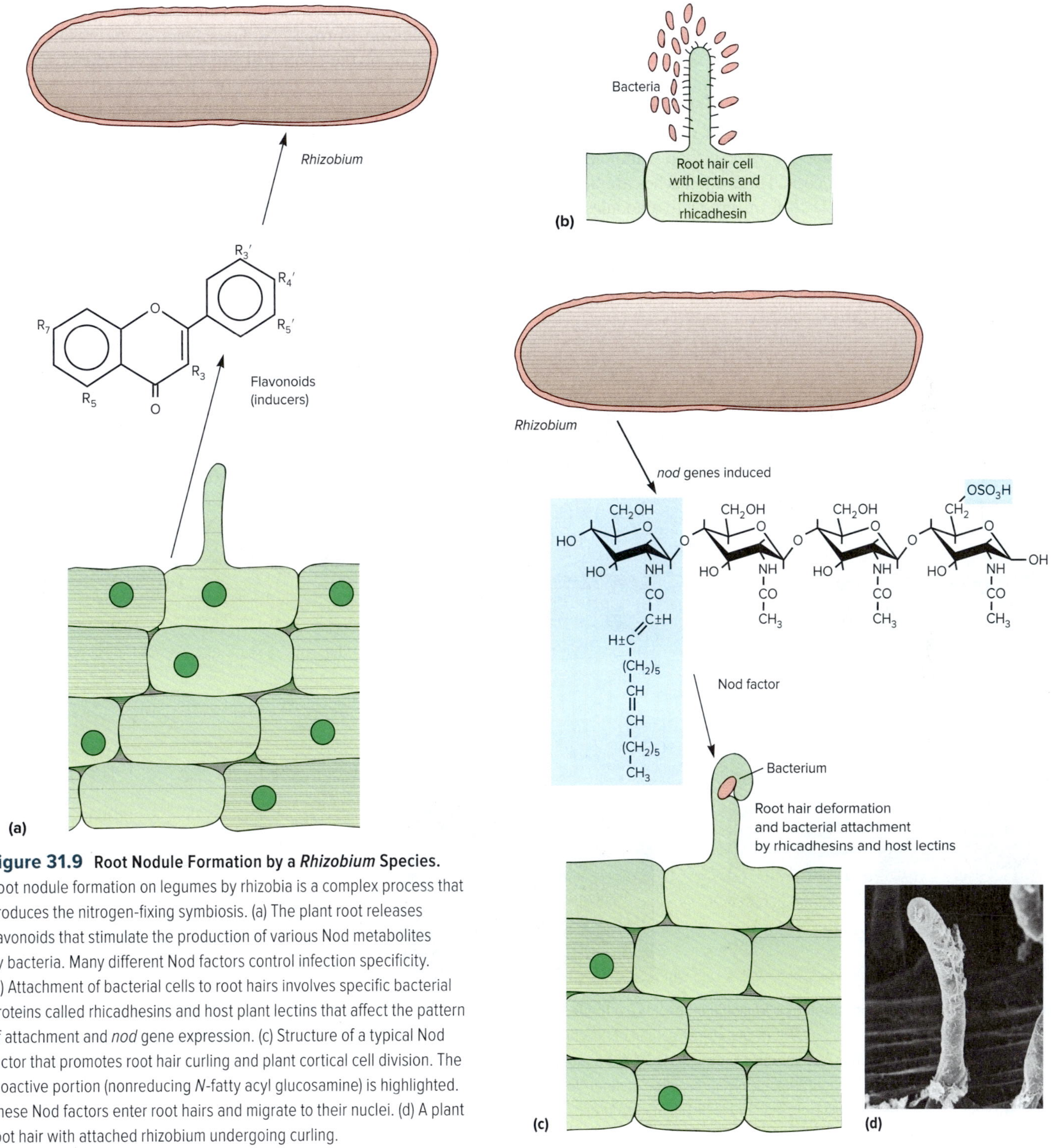

Figure 31.9 Root Nodule Formation by a *Rhizobium* Species. Root nodule formation on legumes by rhizobia is a complex process that produces the nitrogen-fixing symbiosis. (a) The plant root releases flavonoids that stimulate the production of various Nod metabolites by bacteria. Many different Nod factors control infection specificity. (b) Attachment of bacterial cells to root hairs involves specific bacterial proteins called rhicadhesins and host plant lectins that affect the pattern of attachment and *nod* gene expression. (c) Structure of a typical Nod factor that promotes root hair curling and plant cortical cell division. The bioactive portion (nonreducing *N*-fatty acyl glucosamine) is highlighted. These Nod factors enter root hairs and migrate to their nuclei. (d) A plant root hair with attached rhizobium undergoing curling.

(d) ©Dr. Jeremy Burgess/SPL/Science Source

moiety is encoded by plant genes, whereas the heme group is a bacterial product.

The transfer of ammonia from the bacteroid to the host plant involves a complex cycling of amino acids. The plant provides certain amino acids to the bacteroids so that they do not need to assimilate ammonia. In return, the bacteroids shuttle amino acids (which bear the newly fixed nitrogen) back to the plant. This creates an interdependent relationship, providing selective pressure for the evolution of mutualism. The bacteria also receive carbon and energy in the form of dicarboxylic acids from their host legume.

The molecular mechanisms by which both the legume host and the rhizobial symbionts establish productive nitrogen-fixing bacteroids within nodules is an intense area of research. A major goal of biotechnology is to enable nitrogen fixation within cereal plants such as rice, wheat, and corn. Three approaches are being pursued: (1) modifying rhizobia to form productive nodules on these nonleguminous plants; (2) creating strains of non-nodulating nitrogen-fixing bacteria that will colonize cereal plants; and (3) introducing nitrogen-fixation genes into plants that are not rhizobium hosts. Although progress has been made using each approach, nitrogen fixation on these important crops has yet to be realized, and the use of human-made fertilizer, with all its environmental problems, continues. ◀ *Global climate change (section 28.2)*

Stem-Nodulating Rhizobia

Some leguminous plants support rhizobia that nodulate stems, often in addition to roots. Nodules form at the base of roots branching out of the stem just above the soil surface. These plants are generally found in waterlogged soils and riverbanks.

Stem nodulation has not been as well studied as root nodulation, but it is clear that the two processes differ. Stem nodulation starts with the entry of bacteria through ruptures in stem epidermal cells that occur during emergence of the root tip. Unlike root nodulation, stem nodulation does not appear to involve the

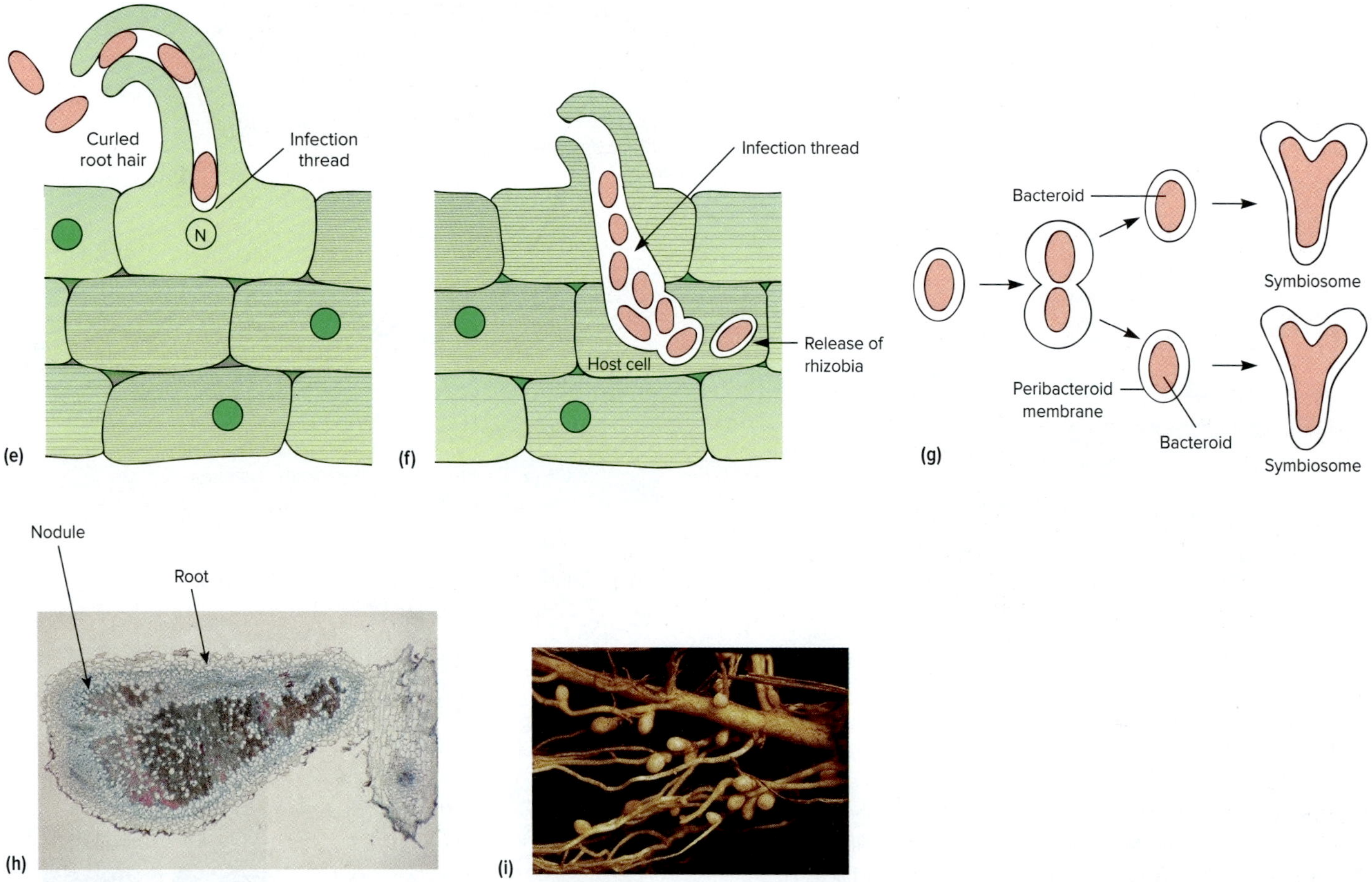

Figure 31.9 Root Nodule Formation by a *Rhizobium* Species (*continued*). (e) Initiation of bacterial penetration into the root hair cell and infection thread growth coordinated by the plant nucleus "N." (f) Cell-to-cell spread of rhizobia through transcellular infection threads followed by release of rhizobia and infection of host cells. (g) Formation of bacteroids surrounded by plant-derived peribacteroid membranes and differentiation of bacteroids into nitrogen-fixing symbiosomes. The bacteria change morphologically and enlarge around 7 to 10 times in volume. The symbiosome contains the nitrogen-fixing bacteroid, a peribacteroid space, and the peribacteroid membrane. (h) Light micrograph of two nodules that develop by cell division (×5). This section is oriented to show the nodules in longitudinal axis and the root in cross section. (i) Nodules on bur clover (*Medicago* sp.).

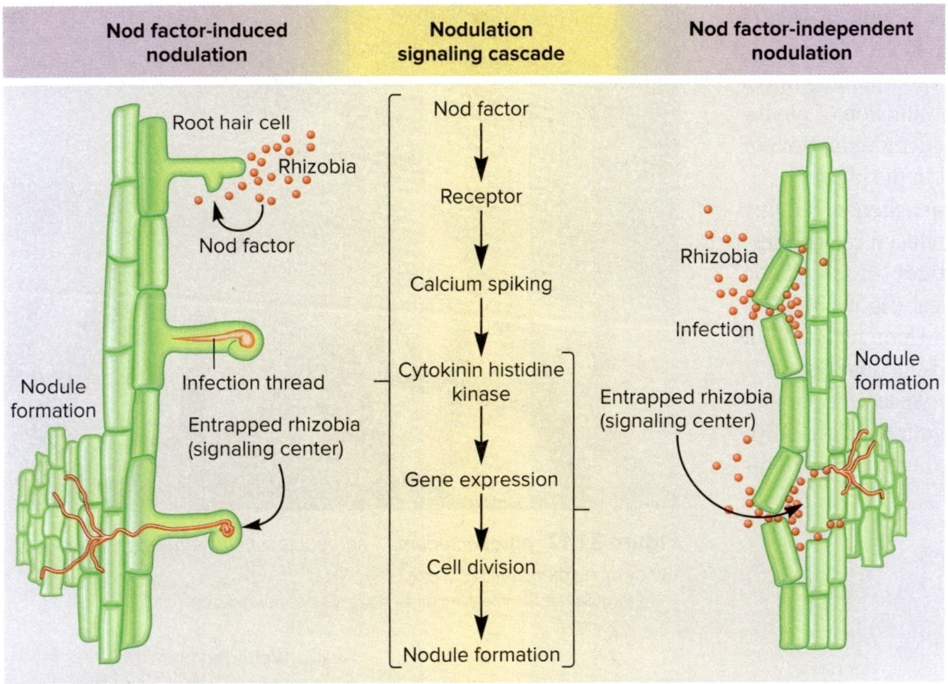

Figure 31.10 Root and Stem Nodulation. Root nodulation (left) involves the production of Nod factors by rhizobium and the signaling molecule cytokinin by the host plant. By contrast, Nod factor and plant cytokinin release is not necessary for stem nodulation by at least two rhizobia (right). It is not yet known if this is generally the case.

production of Nod factors or plant calcium spikes (**figure 31.10**). Nonetheless, as with root nodulation, bacteria within stem nodules are protected from the external environment and can obtain plant-derived energy substrates.

Frankia form Actinorhizae

Another example of symbiotic nitrogen fixation occurs between actinobacteria of the genus *Frankia* and eight nonleguminous host plant families. These bacterial associations with plant roots are called **actinorhizae** or actinorhizal relationships (**figure 31.11**). *Frankia* spp. are important particularly in trees and shrubs. The nodules of some plants (*Alnus, Ceanothus* spp.) can be as large as baseballs. The nodules of *Casuarina* spp. (Australian pine) approach soccer ball size.

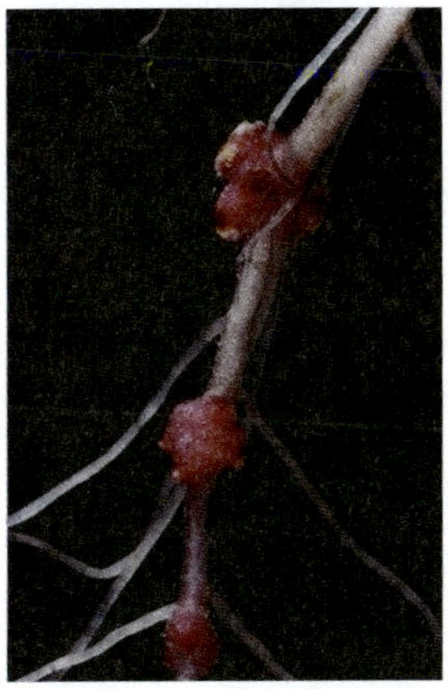

Figure 31.11 Actinorhizae. *Frankia*-induced actinorhizal nodules.

MICRO INQUIRY *How do you think* Frankia *spp. protect nitrogenase from oxygen damage?*

Depending on the host plant, the infection process proceeds either by root hair infection or by direct penetration of the matrix between root cells. When infecting through root hairs, the bacteria induce plant cell division, leading to the development of a prenodule. *Frankia* cells then enter the cortex of the root and form a nodule that resembles a lateral root. When infection is independent of root hair invasion, the bacteria are encapsulated in a shell made of pectin that then extends through the plant intercellular spaces. Although *Frankia* nodules vary in morphology, they all bear the general appearance of lateral root clusters that grow by repeated branching of the tips. Some nodules are found just below the surface of the soil, whereas others—particularly those forming on plants that grow in dry, rocky soil—develop very deep below the surface. ◀ *Order* Frankiales: *bacteria that fix N_2 in trees (section 23.1)*

Comprehension Check

1. List several bacteria that are considered rhizobia.
2. What is the function of the infection thread?
3. What does the term *terminally differentiated* mean?
4. How does nitrogen transfer between a rhizobium and its host differ from that between AM and its host?
5. What is unusual about leghemoglobin production and what is its function?
6. What are the two general mechanisms by which *Frankia* spp. nodulate their plant hosts? Which types of plants are infected?

Agrobacterium Is a Model Plant Pathogen

Clearly some microbe-plant interactions are beneficial for both partners. However, others involve microbial pathogens that harm or even kill their host. *Agrobacterium tumefaciens* is an α-proteobacterium that has been studied intensely for several decades. Initially research focused on how this microbe causes crown gall disease, which results in the formation of tumorlike growths in a wide variety of plants (**figure 31.12**). A better understanding of this mechanism led to *A. tumefaciens* becoming one of biotechnology's most important tools, because the molecular genetics by which this pathogen infects its host are the basis for plant genetic engineering. ▶ *Agricultural biotechnology relies on a plant pathogen (section 42.5)*

The genes for plant infection and virulence are encoded on an *A. tumefaciens* plasmid called the **Ti** (**tumor-inducing**) **plasmid.** These genes include 21 *vir* genes (*vir* stands for virulence), found in six separate operons. Two of these genes, *virD1* and *virD2,* encode proteins that excise a specific region of the Ti plasmid, called **T DNA.** After excision, the T DNA fragment is integrated into the host plant's genome. Once incorporated into a plant cell's genome, T DNA directs the overproduction of phytohormones that cause unregulated growth and reproduction of plant cells, thereby generating a tumor or gall in the plant.

The *vir* genes are not expressed when *A. tumefaciens* is living saprophytically in the soil. Instead, they are induced by the presence of plant phenolics and monosaccharides in an acidic (pH 5.2–5.7) and cool (below 30°C) environment (**figure 31.13*a***). The microbe infects its host through a wound. Upon reception of the plant signal, a two-component signal transduction system is activated: VirA is a sensor kinase that, in the presence of a phenolic signal, phosphorylates the response regulator VirG. Activated VirG then induces transcription of the other *vir* genes. This

Figure 31.12 ***Agrobacterium.*** *Agrobacterium tumefaciens*-caused tumor on a *Kalanchoe* sp. plant.

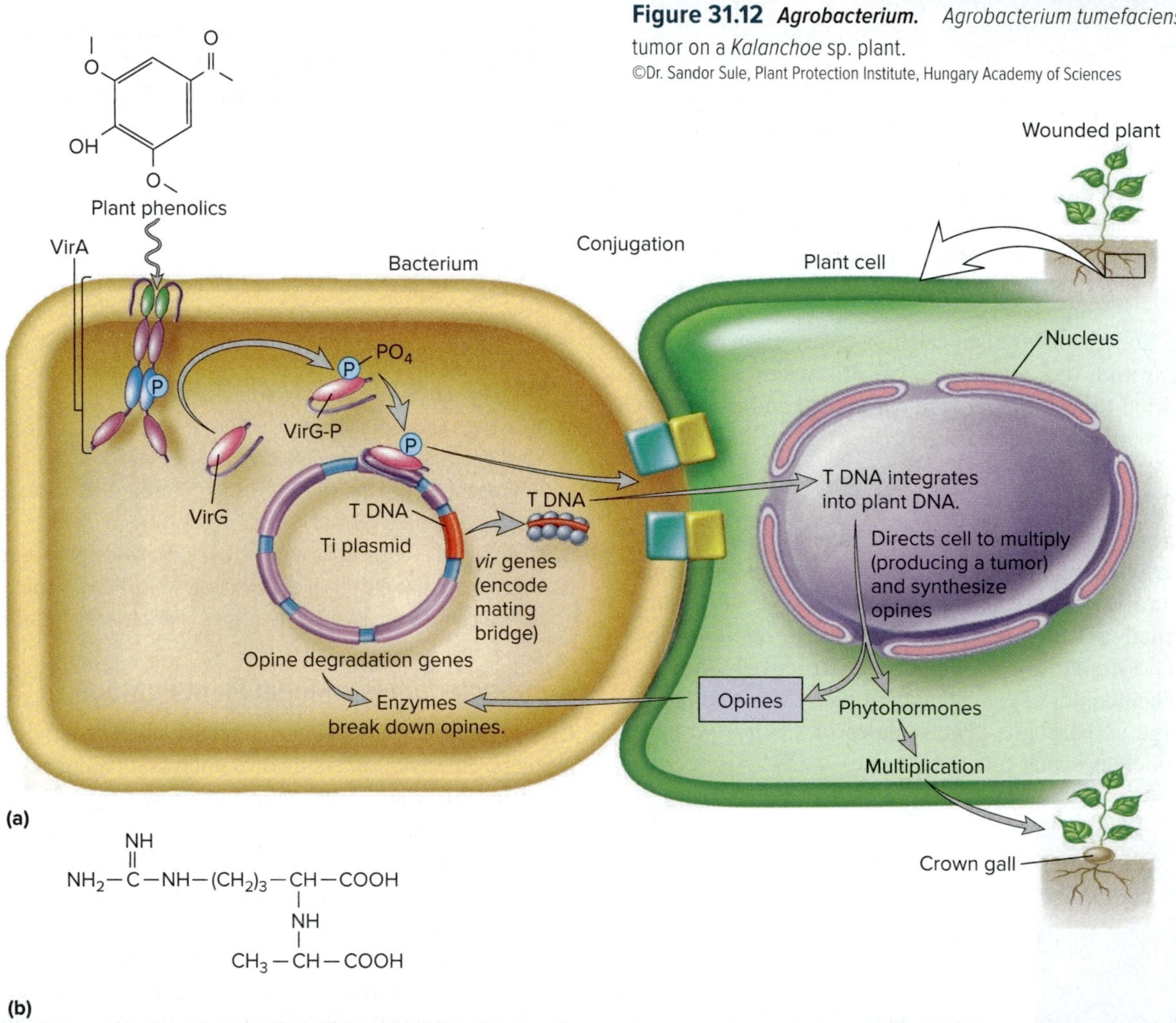

Figure 31.13 Functions of Genes Carried on the *Agrobacterium tumefaciens* Ti Plasmid. (a) Genes carried on the Ti plasmid of *A. tumefaciens* control tumor formation by a two-component regulatory system that stimulates formation of the mating bridge and excision of the T DNA. The T DNA is moved into the plant cell, where it integrates into the plant DNA. T DNA induces the production of plant hormones that cause the plant cells to divide, producing a tumor. (b) Tumor cells produce opines that can serve as a carbon source for the infecting *A. tumefaciens*. Ultimately a crown gall is formed on the stem of the wounded plant above the soil surface.

MICRO INQUIRY *How does the production of opines by the plant accelerate and amplify the* A. tumefaciens *infection process?*

enables the bacterial cell to become adequately positioned relative to the plant cell, at which point the *virB* operon expresses the apparatus that will transfer the T DNA. This transfer is similar to bacterial conjugation and involves a type IV secretion system. After VirD1 and VirD2 excise the T DNA from the Ti plasmid, the T DNA, with the VirD2 protein attached to the 5′ end, is delivered to the plant cell cytoplasm. The protein VirE2 is also transferred, and together with VirD2, the T DNA is shepherded to the plant cell nucleus, where it is integrated into the host's genome.

Once it is part of the host cell genome, the T DNA has two specific functions. First, it directs the host cell to overproduce phytohormones that cause tumor formation. Second, it stimulates the plant to produce special amino acid and sugar derivatives called opines (figure 31.13*b*). Opines are not metabolized by the plant, but *A. tumefaciens* is attracted to opines; chemotaxis of bacteria from the surrounding soil population will further advance the infection because the bacterium can use opines as sources of carbon, energy, nitrogen, and, in some cases, phosphorus. ◀ *Bacteria combine several regulatory processes to control complex cellular processes (section 14.6); Bacterial conjugation requires cell-cell contact (section 16.6)*

Microbial Plant Pathogens Are Diverse

In terms of both diversity and economic costs, fungi are the most important plant pathogens, causing about 70% of plant disease (**table 31.3**; also see chapter-opening story). Fungal plant pathogens can be classified by their relationship with their

Table 31.3 Major Plant Diseases Caused by Bacteria (B) and Fungi (F)

Symptoms	Examples	Pathogen
Spots and blights	Wildfire (tobacco)	*Pseudomonas syringae* pv.[1] *tabaci* (B)
	Blight (rice)	*Xanthomonas campestris* pv. *oryzae* (B)
	Citrus blast	*P. syringae* pv. *syringae* (B)
	Leaf spots and blights (many plants)	*Alternaria alternata* (F)
	Head blight (wheat)	*Fusarium graminearum* (F)
	Leaf spot (alfalfa)	*Pseudopeziza trifolii* (F)
	Leaf spot (cereals, grasses)	*Mycosphaerella graminicola* (F)
	Brown spot (corn)	*Physoderma maydis* (F)
Vascular wilts	Wilt (tomato)	*Clavibacter michiganensis* pv. *michiganensis* (B)
	Stewart's wilt (corn)	*Erwinia stewartii* (B)
	Fire blight (apples)	*E. amylovora* (B)
	Verticullium wilt (potato)	*Verticullium dahliae* (F)
Soft rots	Black rot (crucifers)	*X. campestris* pv. *campestris* (B)
	Soft rots (numerous)	*E. carotovora* pv. *carotovora* (B)
	Black leg (potato)	*E. carotovora* pv. *atroseptica* (B)
Smuts	Loose smuts (oats, barley, wheat)	*Ustilago avenae, U. nuda, U. tritici* (F)
	Covered kernel smut (sorghum)	*Sproisorium sorghi* (F)
Cankers	Canker (stone fruit)	*P. syringae* pv. *syringae* (B)
	Hyoxylon canker (oak trees)	*Hyoxylon atropunctatum* (F)
Rusts	Asian soybean rust	*Phakopsora pachyrrhizi* (F)
	Leaf rust (cereals)	*Puccinia recondita* (F)
	White pine blister rust	*Cronartium ribicola* (F)

1 pv., pathovar, a variety of microorganisms with phytopathogenic properties.

Source: From Lengler, J. W., et al. 1999. Biology of the prokaryotes. *Malden, MA: Blackwell Science, table 34.4.*

plant hosts: They consume either living or dead plant material. **Biotrophic fungi** assimilate living plant material but do not kill their hosts. To do so, they must overcome the plant immune hormone salicylic acid. Parasitic biotrophic fungi include powdery mildew and rust fungi; these microorganisms have evolved a special feeding apparatus called the haustorium to take up nutrients. Haustoria develop by the expansion of fungal hyphae that penetrate plant host cells. By contrast, necrotrophic fungi kill their host plant cells by releasing toxins or enzymes prior to consuming plant nutrients. These pathogens have evolved to evade the plant immunity hormones jasmonic acid and ethylene. **Necrotrophic fungi** infect (and kill) a broader range of hosts. *Sclerotinia sclerotiorum* (white mold fungus) and *Botrytis cinerea* (grey mold fungus) are two notorious necrotrophic fungi that each parasitize over 200 different plant species and are responsible for the loss of millions of pounds of produce both pre- and post-harvest annually.

Many bacteria are also plant pathogens (table 31.3). Like *A. tumefaciens,* some plant pathogenic bacteria infect through a wound in the plant. Others, like the soft-rot bacteria *Erwinia chrysanthemi* and *E. carotovora,* digest plant tissue by secreting enzymes that degrade pectin, cellulose, and proteins. Another group of important bacterial plant pathogens includes the wall-less bacterial phytoplasms that infect vegetable and fruit crops such as sweet potatoes, corn, and citrus.

As discussed in chapters 24 and 26, protists, viruses, and viroids can also be devastating plant pathogens. Examples include the öomycete *Phytophthora infestans,* which was responsible for the Irish potato famine (and continues to extract an economic burden), and tobacco mosaic virus (TMV), the first virus to be characterized. A virus of particular interest in terms of plant-pathogen interactions is a hypovirus (family *Hypoviridae*) that infects the fungus *Cryphonectria parasitica,* the cause of chestnut blight. Based on studies carried out in Europe, workers in the United States noted that if they infected the fungus with this hypovirus, the rate and occurrence of blight were decreased. Infection of trees with the less lethal virus strains is the basis of biological control efforts for chestnut blight. The goal is to vaccinate trees against the indigenous lethal strains of *C. parasitica*.

Comprehension Check

1. What is the difference between the Ti plasmid and T DNA?
2. What functions do the members of the two-component system play in infection of a plant by *A. tumefaciens*? What are the roles of phenolics and opines in this infection process?
3. What are the two general ways by which plant pathogens gain entry into the host plant tissue? What are the advantages and disadvantages of each mechanism?
4. How are plant pathologists attempting to control chestnut blight?

31.4 The Subsurface Biosphere Is Vast

After reading this section, you should be able to:

a. Relate the metabolic processes that occur in the subsurface biosphere to the free energy released during the use of each successive compound as a terminal electron acceptor
b. Describe the distribution of subsurface methane production
c. Explain at least one new discovery that has come from exploring the "deep hot biosphere"

For many years, it was thought that life could exist only in the thin veneer of Earth's surface and that any microbes recovered from sediments hundreds of meters deep were contaminants obtained during sampling. This view was drastically altered in the 1980s when the U.S. Department of Energy started looking for novel ways to store toxic waste. The agency began funding studies that applied modern technologies to sample the deep subsurface biosphere. Subsequent reports of microbes at great depth gained credibility, and international teams of geologists and microbiologists have since recovered microbes from thousands of meters below Earth's surface. The application of culture-independent techniques to quantify the density, diversity, and activity of microbes has revealed that subsurface microbes constitute about one-third of Earth's living biomass. This realization has made deep subsurface microbiology an exciting and active field.

Microbial processes take place in different subsurface regions, including (1) the shallow subsurface where water flowing from the surface moves below the plant root zone; (2) regions where organic matter, originally from the Earth's surface, has been transformed by chemical and biological processes to yield coal (from land plants), kerogens (from marine and freshwater microorganisms), and oil and gas; (3) zones where methane is being synthesized as a result of microbial activity; and (4) the energy-starved deep biosphere.

In the shallow subsurface, surface waters often move through aquifers—porous geological structures below the plant root zone. In a pristine system with an oxic surface zone, the electron acceptors used in catabolism are distributed from the most oxidized and energetically favorable (oxygen) near the surface to the least favorable (in which CO_2 is used in methanogenesis) in lower zones (*see figures 28.1 and 30.12*).

In subsurface regions where organic matter from the Earth's surface has been buried and processed by thermal and biological processes, kerogens and coals break down to yield gas and oil. These mobile products move upward into more porous geological structures where microorganisms are active. Chemical signature molecules from plant and microbial biomass are present in these petroleum hydrocarbons. Microorganisms also appear to be growing in intermediate-depth, oil-bearing structures. Active bacterial and archaeal assemblages are present in high-temperature (60° to 90°C) oil reservoirs, including members of such genera as *Thermotoga, Thermoanaerobacter,* and *Thermococcus.* Archaeal species are dominated by methanogens.

The "deep hot biosphere" is a term suggested by Thomas Gold to describe the largest and most mysterious biosphere on Earth: Bacteria and archaea living kilometers beneath the Earth's crust. The lack of apparent substrates to support growth and the calculated generation times in centuries to even 1,000 years seem alien to our understanding of microbial life. Anaerobic chemolithoautotrophy, independent of photosynthetically derived compounds, fuels these microbes; abiotically produced H_2 is believed to the most important energy and electron source. Genomic studies suggest the presence of N_2-fixation and denitrification form a subterranean nitrogen cycle. A notable example is found in 2.8-km-deep, high-pressure water within a crustal fracture in a South African gold mine. Surprisingly, at least 99% of the microbes belong to a single firmicute species, as determined by 16S rRNA analysis. DNA collected from this unusual habitat was assembled into a complete genome to reveal a novel bacterium provisionally called *Candidatus* Desulforudis audaxviator. It appears to be a motile, spore-forming, sulfate-reducing chemoautotroph that is moderately thermophilic. This microbe most likely infiltrated the fracture zone 3 to 25 million years ago, and its growth has apparently been sustained by hydrogen as electron donor and sulfate as terminal electron acceptor.

The discovery of deep subsurface microbes has a variety of implications. For example, the presence of primary production in the absence of sunlight suggests that autotrophy might also be possible in the subsurface sediments of Mars. Exploring this possibility, along with investigating the level of metabolic activity among Earth's deep subsurface microbes, remains the next challenge. ◀ *Microbial communities in benthic marine sediments are enormous and mysterious (section 30.2)*

Comprehension Check

1. Compare and contrast the metabolism (specifically, electron donors and acceptors) of microbes in the shallow subsurface with that of microbes in the very deep subsurface.
2. What happens in terms of microbiological processes when organic matter leaches from the surface into the subsurface?
3. What microbial genera have been observed in oil field materials?

Key Concepts

31.1 Soils Are an Important Microbial Habitat

- Soil is a complex mixture of clay, sand, silt, organic matter, and water. This provides many microenvironments that can support a diverse microbial community (**figure 31.1**).
- Soil organic matter (SOM) helps retain nutrients and water, and maintain soil structure. It can be divided into humic and nonhumic material.

31.2 Diverse Microorganisms Inhabit Soil

- Soils support the highest levels of microbial diversity measured.
- Archaea are not as abundant as bacteria in soils; ammonia-oxidizing archaea contribute to nitrification in soils.
- Saprophytic and mycorrhizal fungi constitute about half the soil microbial biomass in most soils.

31.3 Microbe-Plant Interactions Can Be Positive, Negative, or Neutral

- The phyllosphere includes all parts of a plant that are aboveground. These structures offer a unique but changeable environment for diverse microbial growth.
- The rhizosphere is the region around plant roots into which plant exudates are released. The rhizoplane is the plant root surface. A variety of microbes growing in these regions promote plant growth.
- Mycorrhizal relationships (plant-fungal associations) are varied and complex. Six basic types are observed, including endomycorrhizal and sheathed/ectomycorrhizal types. The hyphal network of the mycobiont can lead to the formation of a mycorrhizosphere (**figures 31.4–31.8; table 31.2**).
- The rhizobium-legume symbiosis is one of the best-studied examples of plant-microorganism interactions. Rhizobia form nodules on the roots of host plants in which specialized bacterial cells (bacteroids) fix nitrogen (**figure 31.9**).
- The interaction between rhizobia and host plants is mediated by complex chemicals that serve as communication signals (**figures 31.7–31.10**).
- *Actinobacteria* in the genus *Frankia* form nitrogen-fixing symbioses with some trees and shrubs.
- *Agrobacterium tumefaciens* establishes a complex communication system with plant hosts into which they transfer a fragment of DNA. Genes on this DNA encode proteins that promote bacterial growth and result in the formation of plant tumors or galls (**figures 31.12** and **31.13**).

31.4 The Subsurface Biosphere Is Vast

- The subsurface includes at least four zones: the shallow subsurface; the zone where gas, oil, and coal have accumulated; intermediate depths where oil and methane are found; and the very deep, hot subsurface.
- Microbial activity in the deep subsurface suggests extraordinarily slow growth fueled by substrates of geological, rather than photosynthetic, origin. These compounds support anaerobic chemolithoautotrophy.

Active Learning

1. Considerable research is currently devoted to developing a better understanding of the "rhizobiome," defined as all microbes within the rhizosphere, particularly of commercial crops. The goal is to increase the use of biocontrol microbes that will reduce the need for pesticides and fertilizers, limit plant pathogens, and provide larger yields. Select a crop and suggest microbes that might be candidates to be used in this way. Be sure to explain what each microbe would contribute to the microbe-plant relationship.
2. The Nod factor signals produced by the root nodulating bacterium *Sinorhizobium meliloti* are essential for biofilm formation as well as N_2 fixation. In what ways are nodulation and biofilm formation similar, and why do you think this microbe evolved to use the same signal for the development of both?
3. The ectomycorrhizal basidiomycete *Laccaria bicolor* produces a large number of small, secreted proteins (SSPs), some of which are only expressed in hyphae that colonize the host root. Discuss the potential role these proteins might have in the life cycle of *L. bicolor* as it colonizes host plants and moves from one host to another.
4. Although the role of viruses and bacteriophages in marine and freshwater systems has been studied for a number of years, little is known about viruses in soils. Stable isotope probing (*see section 29.3*) shows that carbon in rice plants is incorporated into the capsid of a soil bacteriophage. What does this imply about the trophic transfer of carbon in this soil? Formulate two questions that might increase our understanding of viruses in soil ecosystems.
5. Many have predicted that rising atmospheric CO_2 levels will drive increased carbon fixation by plants. However, others argue that this CO_2 fertilization effect will be limited by nitrogen availability. An international team of scientists hypothesized that mycorrhizae symbionts would provide the necessary nitrogen to their host plants. They found that the biomass of plants with ectomycorrhizal symbionts increased with rising CO_2 levels, but those with arbuscular mycorrhizae did not. What can we infer about the relationship between EM and AM and their plant hosts? Consider the types of plants colonized by these mycorrhizae and develop a hypothesis regarding plant growth rate as CO_2 levels continue to rise.

Read the original paper: Terrer, C., et al. 2016. Mycorrhizal association as a primary control of the CO_2 fertilization effect. *Science.* 353:72–4. doi:10.1126/science.aaf4610.

32

Innate Host Resistance

©Lightspring/Shutterstock

Supersize Me!

About 60% of the U.S. population is overweight; of these about half are considered obese. Although the causes of the obesity epidemic are complex, obesity rates correlate to diets high in saturated fats, as found in many fast-food products. The fats in foods are not the only problem, though. Lack of exercise and increasing portion size have added dramatically to the obesity issue. In many ways, the typical American diet is supersized. Surprisingly, obesity and the innate arm of the immune system are linked.

Innate resistance is an elegant system of sensory and response molecules that detect invasion by microorganisms. Capture of microbial pieces and parts by cells of the innate immune system triggers an inflammatory process featuring the release of chemical messengers (cytokines) that recruit an army of proteins and cells that help defend the host. An equally elegant system has recently gained attention whereby innate immune cells detect and respond to host cell stress by recognizing damage-associated molecular patterns (DAMPs) released from stressed or dying host cells. Recognition of DAMPs likewise stimulates cytokine expression to facilitate cellular repair. Both systems activate the same inflammatory process but usually with very different outcomes. But what does this have to do with obesity?

Originally thought to just be an energy reserve, visceral (belly) fat is a specialized tissue composed of lipid-laden adipocytes (fat cells), preadipocytes, phagocytes called macrophages, and other immune cells held in a web of extracellular matrix proteins. Normal adipocytes make pro- and anti-inflammatory cytokines, striking a delicate balance between opposing inflammatory functions to maintain homeostasis needed for tissue repair and regeneration functions.

We now know there is a strong correlation between diet, gut microbes, visceral adipose (obese conditions), and increased serum levels of pro-inflammatory cytokines. This is, in part, because saturated fatty acids are recognized as DAMPs by visceral adipocytes and their associated macrophages. One of the pro-inflammatory cytokines, interleukin (IL)-18, and a macrophage marker of inflammation, C-reactive protein (CRP), are especially elevated in the obese. These and other data show that the healthy balance of pro- and anti-inflammatory cytokines seen in normal adipose tissue is skewed toward inflammation when individuals become obese. This promotes chronic and systemic, low-grade inflammation with severe consequences for the cardiovascular system. Importantly, return to a leaner mass reverses this effect. ▶| *Many diseases are linked to dysbiosis (section 34.4)*

The integrity of any host organism depends on its ability to recognize foreign substances and microorganisms and provide an appropriate response to prevent harm. In this chapter, we explore the mechanisms by which mammals (principally humans) express and use innate resistance factors to defend themselves against invasion by foreign substances.

Readiness Check:

Based on what you have learned previously, you should be able to:

- ✓ Explain basic eukaryotic cell biology (sections 5.1–5.7)
- ✓ Diagram the flow of information in eukaryotes from DNA to protein and regulation of this flow (sections 15.2–15.5)

32.1 Immunity Arises from Innate Resistance and Adaptive Defenses

After reading this section, you should be able to:

a. Define the term *antigen*
b. In general terms, differentiate the role of the innate and adaptive immune system

To establish an infection, an invading microorganism must first overcome many surface barriers, such as skin and mucous membranes, that have either direct antimicrobial activity or inhibit attachment of the microorganism to the host. Because none of these barriers is an ideal environment for the vast majority of microorganisms, most **pathogens,** or disease-causing microorganisms, must breach these surfaces to reach underlying tissues and cause disease. However, any microorganism that penetrates these barriers encounters two levels of host defenses: other innate resistance mechanisms and an adaptive immune response.

Animals (including humans) are continuously exposed to microorganisms that can cause disease. Fortunately animals have evolved an **immune system** composed of widely distributed proteins, cells, tissues, and organs that recognizes foreign substances, including microorganisms. Together they act to neutralize or destroy substances recognized as foreign, or non-self.

Immunity

The term **immunity** (Latin *immunis,* free of burden) refers to the general ability of a host to resist infection or disease. **Immunology** is the study of immune responses and how they protect the host. It includes the distinction between "self" and "non-self" and all the biological, chemical, physiological, metabolic, and physical aspects of the immune response.

There are two different yet complementary components of the mammalian immune response. **Innate immunity** is the first line of defense against any foreign material, including microorganisms, encountered by the host. It includes general mechanisms such as skin, mucus, and constitutively produced antimicrobial chemicals such as lysozyme. Innate resistance mechanisms are always prepared to defend against foreign invaders to the same maximal extent each time a foreign invader is encountered. In contrast, the **adaptive immune response** must be activated ("turned on") by cells and chemicals of innate immunity. Unlike innate immunity, each adaptive immune response is tailored to a particular foreign agent (e.g., virus, bacterium, parasite, fungus, or toxin). Adaptive immune responses coordinate and amplify a variety of cellular and chemical responses that evolved to meet the challenge of each specific invader. The effectiveness of adaptive immune responses increases if the host encounters the specific foreign agent again; that is to say, adaptive immunity has a "memory." We learn about adaptive immunity in chapter 33.

Any substance recognized by the immune system—either the innate or adaptive arm—is called an **antigen.** An absolutely critical feature of the immune system is that it can differentiate between self-antigen and non-self (foreign) antigen. As discussed later in this chapter and in the next, discriminating self from non-self enables the immune system to rid the host of foreign invaders (bacteria, viruses, other infectious agents), as well as eliminate cancer cells. However, when this discrimination goes awry, it is the basis for allergic responses and autoimmune diseases such as rheumatoid arthritis and multiple sclerosis.

The distinction between innate and adaptive immunity is somewhat artificial as there are many bridges that interconnect the innate and adaptive immune responses. Although this may present challenges for students of immunology, the complexity of the mammalian immune system and its capacity to prevent infection from the billions of microorganisms encountered every day is truly remarkable. We begin our discussion with an introduction to the physical and chemical barriers that microbes face when encountering a host. We then present the various cells, tissues, and organs of innate defenses, discussing the mechanism by which they detect and respond to foreign invaders. In this discussion, reference is made to processes and chemicals that are described in detail either later in this chapter or in chapter 33 (adaptive immunity). To help you navigate this information, cross-references to the detailed discussion are provided.

Comprehension Check

1. List two key differences between innate and adaptive immunity.
2. What is an antigen?

32.2 Innate Resistance Starts with Barriers

After reading this section, you should be able to:

a. Identify the barriers that help prevent microbial invasion of the host
b. Explain how physical and mechanical barriers function to prevent microbial invasion of the host
c. Relate host anatomy and secretions to the success of innate resistance strategies

With few exceptions, a microbial pathogen attempting to invade a human host immediately confronts a vast array of innate defense mechanisms (**figure 32.1**). All vertebrate hosts have specific physical and mechanical barriers. These barriers, along

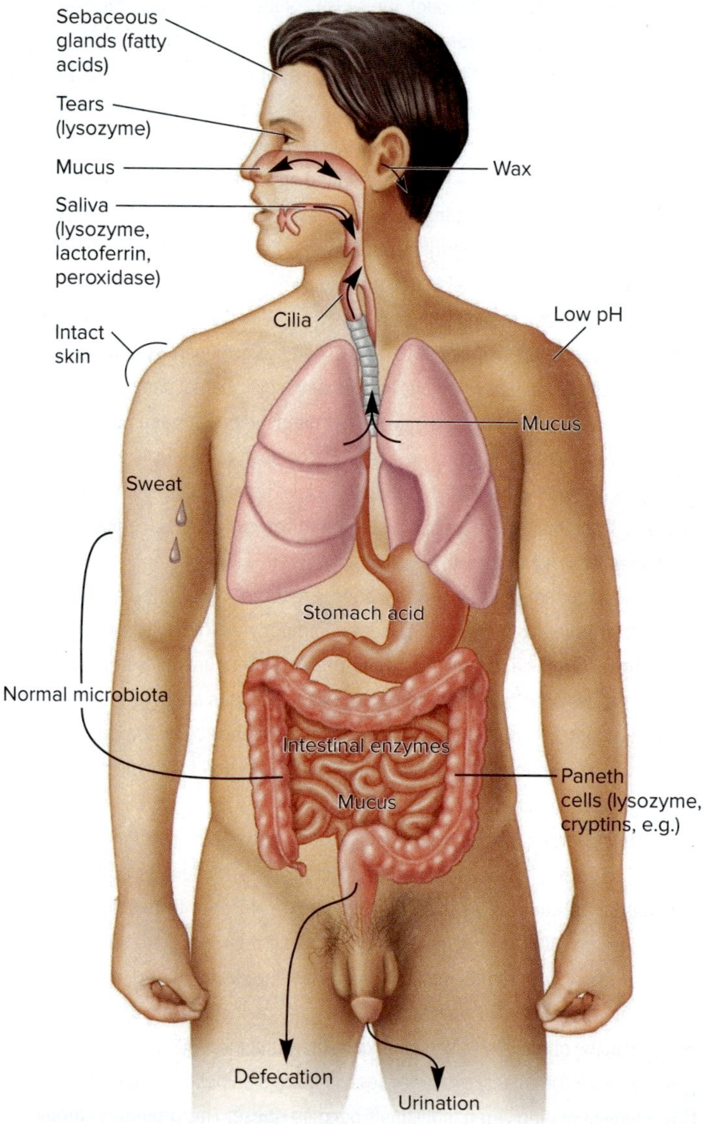

Figure 32.1 Host Defenses. Some nonspecific host defense mechanisms that help prevent entry of microorganisms into the host's tissues.

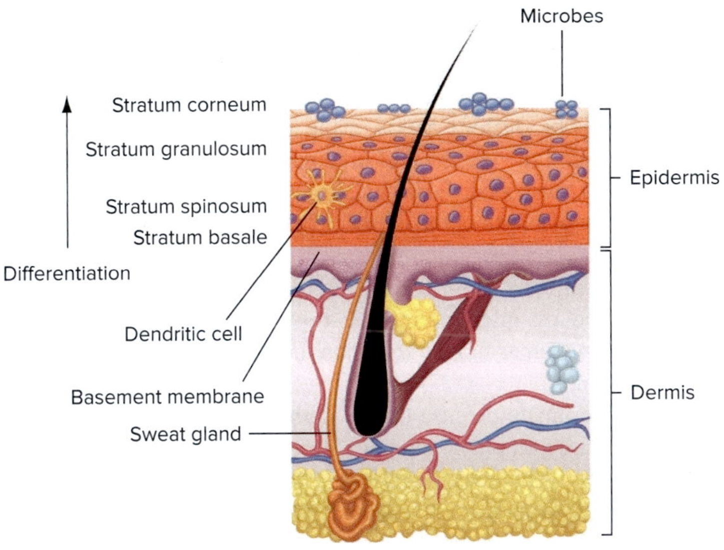

Figure 32.2 Human skin. Skin is composed of stratified cell layers, epidermis and dermis. This schematic identifies the interlocking tiers that create the antimicrobial barrier of skin.

with the host's flushing mechanisms (coughing, urinating, etc.), are the first line of defense against chemical, microbiological, and physical assaults. Protection of body surfaces and the mucous membrane interfaces between the host and its environment is essential in preventing microbial access to the host.

Skin

Skin consists of stratified, cornified (converted into keratin) epithelial cells (**figure 32.2**). Intact skin contributes greatly to innate host resistance because it is a very effective mechanical barrier, and its associated microbiota defend against invaders. Its outer layer consists of thick, closely packed cells called **keratinocytes.** These produce keratins, insoluble proteins that are the main components of hair, nails, and the outer skin cells. These outer skin cells shed continuously. Select areas of the skin are bathed in sebum, a lubricating oil produced by the sebaceous glands. The skin surface is topographically, ecologically, and physiologically diverse; this yields numerous microenvironments, each with distinct moisture, pH, temperature, and sebum content. These niches are colonized by resident microbial populations that evolved to thrive in the specific skin environment and help prevent the colonization of pathogens (*see figure 34.3*).

Mucous Membranes

The mucous membranes line internal structures connected to the environment, such as the gastrointestinal, genitourinary, and respiratory tracts. Also called mucosa, mucous membranes form a protective covering that resists penetration and traps microbial invaders. These membranes consist of a layer of epithelial cells and several types of specialized cells. For example, cells called goblet cells produce mucus (a glycoprotein-rich watery fluid), which bathes the membrane surface. Other specialized cells called Paneth cells secrete antimicrobial peptides and the antibacterial enzyme lysozyme. These compounds make mucosal secretions like tears and prostatic fluid toxic to many microorganisms. In addition to epithelial, goblet, and Paneth cells, molecules and cells of both innate and adaptive immunity are found in three types of specialized immune structures: Peyer's patches, isolated lymphoid follicles, and diffuse lymphoid tissue (section 32.5).

Respiratory System

The average person inhales at least eight microorganisms a minute, or 10,000 each day. Once inhaled, a microorganism must first survive the defenses of the mucous membranes that line the upper and lower respiratory tracts. Because the airflow in these tracts is very turbulent, microorganisms are deposited on the moist, sticky mucosal surfaces. Microbes larger than 10 μm usually are trapped by hairs and cilia lining the nasal cavity. The cilia in the nasal cavity beat toward the pharynx, moving mucus with trapped microorganisms toward the mouth to be expelled (**figure 32.3**). Humidification of the air within the nasal cavity causes many microorganisms to swell, and this aids phagocytosis. Microbes smaller than 10 μm (i.e., most bacterial cells) pass through the nasal cavity and are trapped by the **mucociliary blanket** that coats the mucosal surfaces of lower portions of the respiratory system. The trapped microbes are transported by ciliary action—the mucociliary escalator—that moves them away from the lungs toward the mouth. Coughing and sneezing reflexes also clear the respiratory system of microorganisms. Salivation washes microorganisms from the mouth and nasopharyngeal areas into the stomach. Microorganisms that succeed in reaching the alveoli of the lungs encounter a population of specialized phagocytic cells called **alveolar macrophages.** These cells ingest and kill most inhaled microorganisms by phagocytosis (sections 32.4 and 32.6).

Gastrointestinal Tract

The mucous membranes of the mouth and esophagus present a formidable barrier to penetration. However, with every snack, meal, and drink, we ingest thousands of microorganisms that reach the stomach. There, many are killed by acidic gastric secretions (pH 2 to 3), containing a mixture of hydrochloric acid, proteolytic enzymes, and mucus. However, some microorganisms and their products (e.g., protozoan cysts, *Helicobacter pylori, Clostridium* spp., and staphylococcal toxins) can survive the stomach acidity. Furthermore, organisms embedded in food particles may be protected and reach the small intestine. There, these microorganisms may be damaged by various pancreatic enzymes, bile, enzymes in intestinal secretions, and cells and molecules within the gut-associated lymphoid tissue (GALT) (section 32.5). **Peristalsis** (Greek *peri,* around; *stalsis,* contraction) and the normal shedding of columnar epithelial cells also act in concert to purge intestinal microorganisms. As discussed in chapter 34, the microbiota that inhabit the GI tract are essential for human health and homeostasis. These microbes are required for maturation of the immune system and proper immune function. In addition,

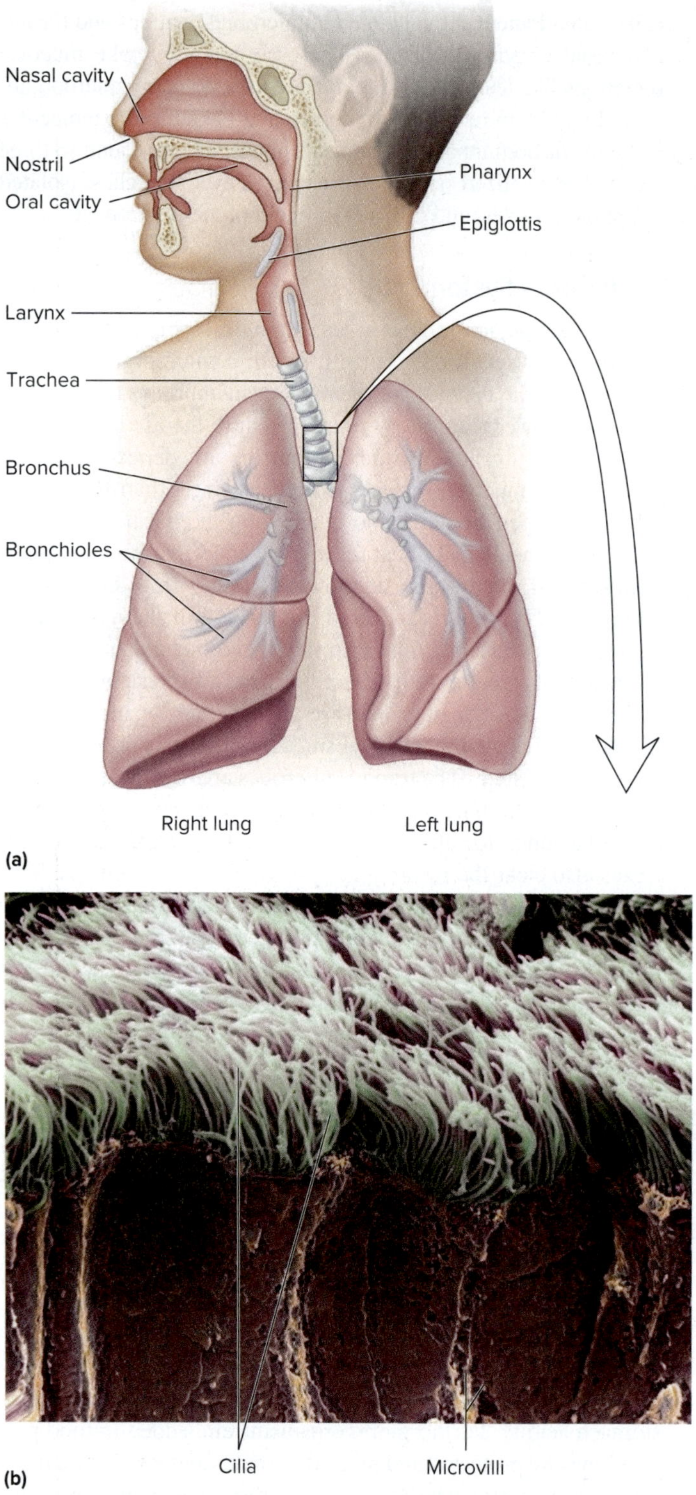

Figure 32.3 Respiratory Tract. (a) A schematic of the respiratory tract showing its components. (b) The respiratory tract is lined with a mucous membrane of ciliated epithelial cells. SEM of cillia; these specialized structures sweep particles toward the mouth to be expectorated or swallowed (X5,000).
(b) ©SPL/Science Source

gut microbiota help prevent the growth of potential pathogens. ◀ *A functional core microbiome is required for human homeostasis (section 34.3)*

Genitourinary Tract

The normal sterility of the kidney, ureters, and urinary bladder is maintained by a complex set of factors. In addition to removing microbes by flushing action, urine kills some bacteria due to its low pH and the presence of urea and other metabolic end products (e.g., uric acid, fatty acids, mucin, enzymes). In males, the anatomical length of the urethra (20 cm) provides a distance barrier that excludes microorganisms from the urinary bladder. Conversely, the short urethra (5 cm) in females is more readily traversed by microorganisms; this explains why women are about four times more likely than men to develop a urinary tract infection.

The innate defense of the vagina consists primarily of the normal microbiota combined with the protective measures provided by the mucous membrane. The vaginal microbiota of nonpregnant women is dominated by *Lactobacillus* spp. These lactic-acid-forming bacteria generate an acidic environment that is inhospitable to pathogenic organisms. Cervical mucus also has antibacterial activity from defensins, lysozyme, lactoferrin, and natural antimicrobial peptides, as discussed next.

Comprehension Check

1. Why is the skin such a good first line of defense against pathogenic microorganisms?
2. Describe one specific antimicrobial defense mechanism that operates within the respiratory, gastrointestinal, and genitourinary systems of mammals. Explain how it protects against pathogenic microorganisms.

32.3 Innate Resistance Relies on Chemical Mediators

After reading this section, you should be able to:

a. Discuss host molecules that have antimicrobial actions
b. Describe in general terms the activation of the host complement system and its three outcomes
c. List three categories of cytokines and discuss their major functions
d. Correlate host protection from microbial invasion with specific mediators

Mammalian hosts have a chemical arsenal with which to combat the continuous onslaught of microorganisms. Importantly, the normal microbiota also assist in host resistance by contributing their own chemicals to fend off competitors, as discussed in chapter 34. How chemical defenses produced by the host protect against microbial invasion is the next topic.

Lysozyme (muramidase) is an enzyme that lyses bacteria by hydrolyzing the bond connecting *N*-acetylmuramic acid and *N*-acetylglucosamine of the bacterial cell wall peptidoglycan—especially in Gram-positive bacteria (**figure 32.4**). Tears and other mucous secretions also contain significant amounts of the iron-binding protein lactoferrin. **Lactoferrin,** found in breast milk and mucous secretions, is similar to **transferrin,** which is found in blood. Both proteins sequester iron, reducing its

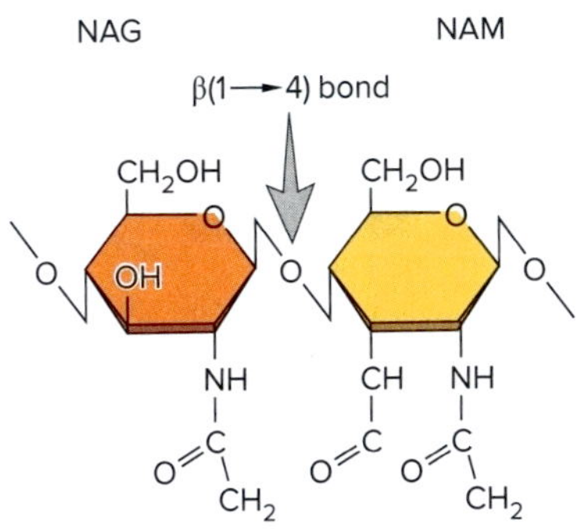

Figure 32.4 **Action of Lysozyme on Peptidoglycan.** This enzyme degrades the cell wall peptidoglycan by hydrolyzing the β(1→4) bonds that connect alternating *N*-acetylglucosamine (NAG) with *N*-acetylmuramic acid (NAM) residues.

availability to invading microbial pathogens thereby limiting their ability to multiply. Mucous membranes also produce lactoperoxidase, an enzyme that catalyzes the production of superoxide radicals, a reactive oxygen species toxic to many microorganisms. ◀ *Peptidoglycan structure (section 3.4); Oxygen concentration (section 7.5)*

Antimicrobial Peptides

Peptides that exhibit broad-spectrum antimicrobial activity are generally considered to be the most ancient primary defense mechanism of animals. Many are positively charged, and therefore electrostatically attracted to microbial cell surfaces. Antimicrobial peptides are amphipathic; that is, they have hydrophobic and hydrophilic components so they are soluble in both aqueous and lipid-rich environments. Antimicrobial peptides kill target cells through diverse mechanisms after interacting with their membranes. Two major types of antimicrobial peptides are cationic peptides and bacteriocins.

Cationic Peptides

Humans produce three generic classes of cationic peptides whose biological activity is related to their ability to damage bacterial plasma membranes. Most form pores, thereby altering membrane permeability, often leading to lysis.

The first group of cationic peptides includes linear, α-helical peptides. Important examples are the **cathelicidins.** There are about 30 members of this family found in many vertebrates, including humans. They are secreted as large proteins that become activated by proteolytic cleavage. Cathelicidins have broad antimicrobial activity against bacteria, enveloped viruses, and fungi. In humans, cathelicidin peptides are produced by a variety of cells including neutrophils, respiratory and urogenital epithelial cells, and alveolar macrophages.

A second group, called **defensins,** are similarly produced from precursor proteins that mature through protease activity. However, unlike cathelicidins, defensins are found in invertebrates as well as vertebrates. Humans have two types of defensins—α and β. α defensins tend to be slightly smaller than β defensins and are found in the granules of neutrophils (phagocytic cells), intestinal Paneth cells, and intestinal and respiratory epithelial cells, where they assist in bacterial degradation. β defensins are typically found in epithelial cells and are known to stimulate release of immune system chemical mediators by immune cells called mast cells (section 32.4). Like cathelicidins, they are active against many bacteria, fungi, and enveloped viruses.

A third group of cationic peptides are the **histatins.** Histatins are larger peptides found in saliva of humans and other higher primates, where they have antifungal activity. Unlike cathelicidins and defensins that disrupt microbial cell membranes, histatins compromise mitochondrial function and generate oxidative and osmotic stress resulting in fungal death.

Bacteriocins

In addition to host cells releasing antimicrobial peptides, some members of the bacterial microbiome release toxic proteins called bacteriocins that kill closely related bacteria. Bacteriocins are peptides that may give their producers, which are immune to their bacteriocins, an adaptive advantage against other bacteria. However, they sometimes increase bacterial virulence by damaging host immune cells. An example of a bacteriocin-producing bacterium is *Escherichia coli.* It synthesizes bacteriocins called **colicins,** encoded on several different Col plasmids (*see table 3.3*). Some colicins bind to specific receptors on the cell envelope of sensitive target bacteria and cause cell lysis; others attack specific intracellular sites such as ribosomes or disrupt energy production. Other examples include the lantibiotics produced by genera such as *Streptococcus, Bacillus, Lactococcus,* and *Staphylococcus.* ▶ *A functional core microbiome is required for human homeostasis (section 34.3)*

Complement Is a Cascading Enzyme System

The **complement system** is an ancient form of innate immunity found in primitive invertebrates as well as mammals. It is so named because when it was discovered, this collection of over 30 heat-labile serum proteins was found to "complement" the antimicrobial effects of other components of the immune system. The complement system can be initiated by three different types of signals, but all lead to three common final outcomes including (1) stimulating an inflammatory response by promoting the recruitment of white blood cells (leukocytes), (2) penetrating microbial cell membranes, thereby causing cell lysis, and (3) defending against microbial invasion by promoting phagocytosis through a process called opsonization. **Opsonization** (Greek *opson,* to prepare victims for) is the term used when microorganisms or other non-self antigens are coated with certain complement proteins, antibodies, or both. In this context, these proteins are called

Phagocytic cell–microbe interaction	Degree of binding	Opsonin
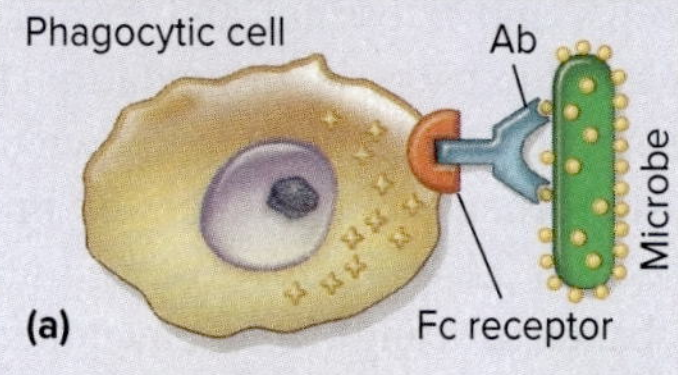	+	Antibody
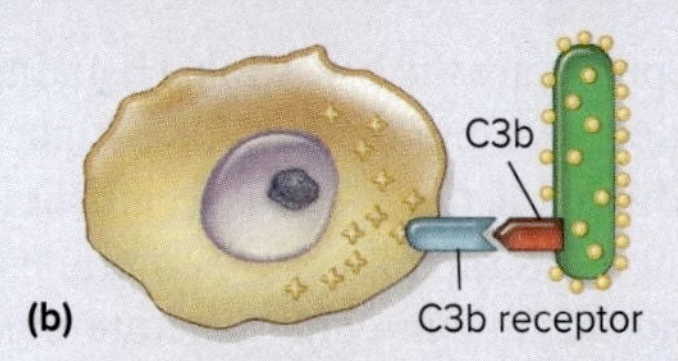	+ +	Complement C3b
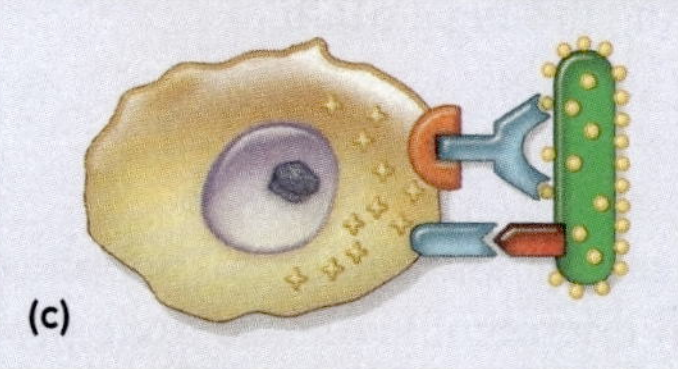	+ + + +	Antibody and complement C3b

Figure 32.5 Opsonization. (a) The ability of a phagocyte to bind to a microorganism is enhanced if the microorganism elicits the formation of antibodies (Ab) that act as a bridge to attach the microorganism to the Fc antibody receptor on the phagocytic cell. (b) If the activated complement (C3b) is bound to the microbe, the degree of binding is further increased by the C3b receptor. (c) If both antibody and C3b opsonize, binding is maximal.

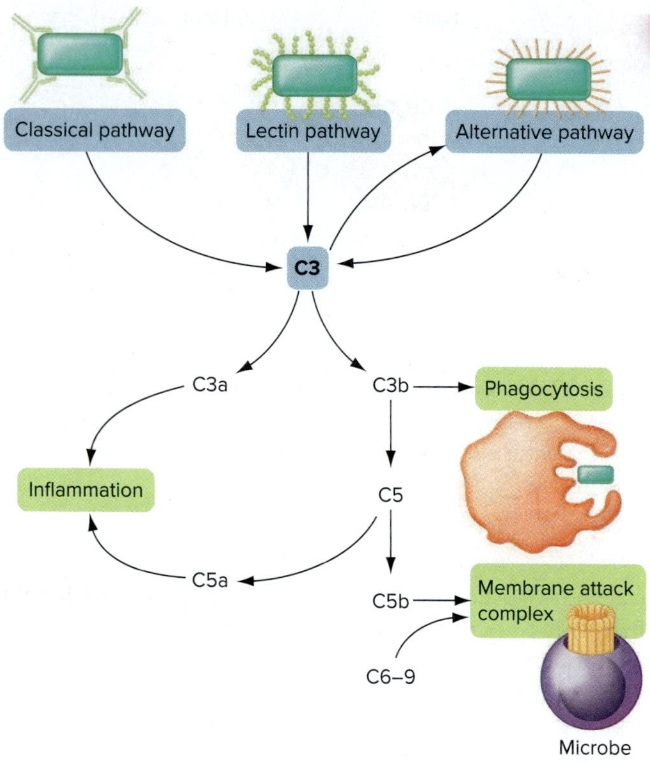

Figure 32.6 Complement Activation and Action. Complement activation involves a series of enzymatic reactions that lead to the formation of a C3 convertase, which cleaves complement component C3 into C3a and C3b. C3a is a peptide mediator of local inflammation. C3b is an opsonin that binds covalently to bacterial cells, thereby enhancing phagocytosis. C5a and C5b are generated by the cleavage of C5 by a C5 convertase. C5a functions together with C3a to mediate inflammation. C5b promotes assembly of a membrane attack complex from the terminal C6–9 complement components.

opsonins and the microorganism is said to be **opsonized.** The presence of opsonins on the surface of antigens acts like a beacon or target that promotes recognition by phagocytic cells. This is because phagocytic cells have surface receptors that specifically recognize opsonins. Importantly, pathogen clearance by phagocytes increases dramatically when they are coated with opsonins (**figure 32.5**). Complement proteins are the first opsonins to be made, so they play a vital role in the phagocytic removal of microorganisms.

Complement proteins are produced in an inactive form; complement protein activity is regulated by a cascade of proteolytic cleavage. Initial activation of key complement proteins results in their proteolytic activity, which is used to convert other complement proteins to the active state. Because activation pathways are complex, we provide a simplified overview. The three pathways of complement activation include the alternative, lectin, and classical pathways (**figure 32.6**). VIDEO ICON: COMPLEMENT

The **alternative complement pathway** (figure 32.6) is initiated in response to bacterial molecules with repetitive structures, such as the lipopolysaccharide (LPS) from Gram-negative bacteria. These molecules activate complement (C) protein 3 (C3) to cleave itself into two fragments, C3a and C3b (**table 32.1**). C3b is normally labile, but it becomes stable when it binds to microbial cell surfaces. Membrane-bound C3b acts as an opsonin, enhancing phagocytosis (figure 32.5). To bring about the other two outcomes (inflammation and microbial cell lysis), C3b also interacts with another complement protein called factor B, which once bound to C3b is cleaved by the complement protease factor D. The smaller cleavage product is released but factor Bb remains bound to C3b, thereby forming a complex called C3bBb. The C3bBb functions as a protease called the C3 convertase of the alternative complement pathway. This C3 convertase amplifies the pathway by cleaving more C3 to C3a and C3b, thereby increasing the amount of C3b coating microbial structures.

Not only does C3bBb serve to amplify C3b, but it also proteolytically activates another complement protein, C5. Because C3bBb cleaves both C3 and C5, it is both a C3 and C5 convertase. As a C5 convertase, it leads to the production of the cleavage products C5a and C5b. C5b rapidly complexes with two additional complement proteins, C6 and C7, forming a C5b67 complex that is stabilized by binding to microbial

Table 32.1 Key Complement Proteins and Their Functions

	Protein	Function
COMPLEMENT ACTIVATION (initiation through three unique pathways)		
Alternative Pathway	C3	C3 is activated by repeating patterns within microbial structures and then is spontaneously cleaved into C3a and C3b; C3b binds to nearby membrane.
	Factor D	Factor D is activated by C3b to become an active enzyme, which activates Factor B.
	Factor B	Factor Bb adsorbs to C3b, forming C3 convertase (protease) C3bBb.
	C3bBb	C3bBb (the active C3 convertase) cleaves additional C3 into its a and b fragments; C3b binds to the nearby membrane.
	Properidin	Properidin stabilizes the C3 convertase.
Lectin Pathway	Mannose-binding protein (MBP)	MBP binds to mannose on microorganisms, then recruits and binds plasma esterase to become mannose-associated serine protease (MASP).
	MASP-1, 2, 3	MASP proteins cleave C3 into C3a and C3b; C3b binds to nearby microbial membranes.
Classical Pathway	Antibody (Ab) formed by adaptive immune system upon exposure to antigen (Ag)	Ab binds to Ag on microorganism, causing three-dimensional shift in Ab structure, which reveals cryptic amino acids near carboxy-terminus (Fc region); the newly revealed amino acids attract plasma C1 protein.
	C1 (trimer of polypeptides q, r, and s)	C1q binds to Fc region of Ab-Ag complex; C1r,s binds plasma calcium.
	Ag-Ab-C1q,r,s complex	Ag-Ab-C1 complex is an activated enzyme (called the C2/C4 esterase) that cleaves plasma C2 and C4 into their a and b fragments, respectively.
	C2a	C2a binds the Ag-Ab-C1 complex.
	C4b	C4b binds the Ag-Ab-C1-C2a complex, forming a C3 convertase (protease).
	Ag-Ab-C1-C2a-C4b	Ag-Ab-C1-C2a-C4b (the active C3 convertase) cleaves plasma C3 into its a and b fragments; C3b binds to the nearby membrane.

membranes. C8 and C9 then bind, forming the **membrane attack complex** (**MAC**), which creates a pore in the plasma membrane or outer membrane of the target cell (**figure 32.7**). MAC is very effective against Gram-negative bacteria, less so against Gram-positive bacteria and fungi, whose membranes are protected by thick cell walls. Importantly, host cells have several mechanisms to protect their membranes from complement attack. Chief among these is the protein factor H and *d*ecay *a*ccelerating *f*actor (DAF) on the surface of host cells. Factor H inactivates Bb while DAF binds to factors C4b and C3b and displaces C2a and Bb. As we will learn, this disarms the lectin and classical, as well as the alternative pathways, respectively.

The production of both C3a and C5a leads to important pro-inflammatory effects. For example, binding of C3a and C5a to receptors on specific host cells induces the release of

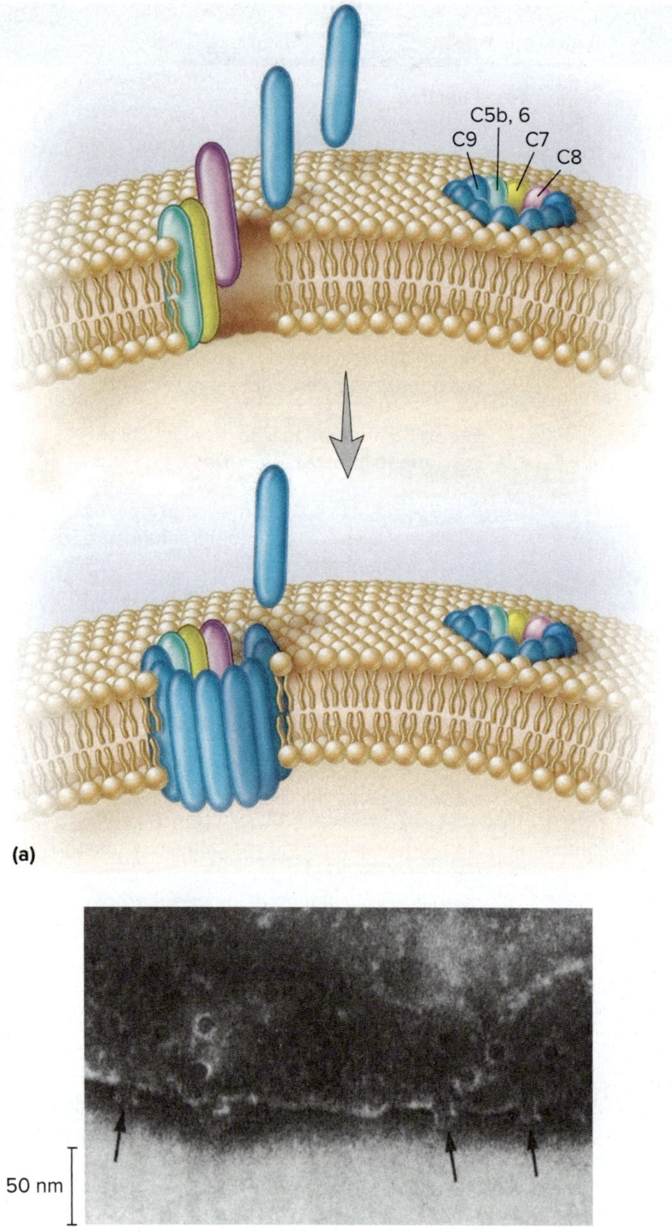

Figure 32.7 The Membrane Attack Complex. The membrane attack complex (MAC) is a barrel-shaped structure that forms a transmembrane pore in the target cell's membrane. (a) This representation shows the subunit architecture of the MAC. The transmembrane channel is formed by a C5b678 complex and 10 to 16 polymerized molecules of C9. (b) MAC pores appear as craters or donuts by electron microscopy.
(b) ©Sucharit Bhakdi

MICRO INQUIRY *What strategies could a Gram-negative bacterium use to protect itself against lysis by complement?*

other biological mediators. These amplify the inflammatory signals by dilating vessels, stimulating nerves, and increasing permeability of local blood vessels so that phagocytic cells, particularly neutrophils, can be recruited to the site of infection. Neutrophils are voracious phagocytes critical to the removal of extracellular pathogens. The inflammatory signals released as a result of C3a and C5a activation also stimulate other innate immune cells (dendritic cells and macrophages), which in turn help activate an adaptive immune response (section 32.4).

The **lectin complement pathway** also requires the activation of a C3 convertase. However, in this case, a soluble host lectin—a protein that binds to specific carbohydrates—initiates the proteolytic cascade. A lectin of major importance is **mannose-binding protein (MBP)**. This protein binds mannose, a major sugar component of bacterial and some fungal cell walls, as well as some virus envelopes. When MBP binds to mannose on pathogens, it is able to form a complex with other blood proteins known as *m*annose-binding lectin-*a*ssociated *s*erine *p*rotease-1 (MASP-1), MASP-2, and MASP-3. The MASPs are then activated to serve as proteases. MASP proteases first cleave the complement protein C4 to generate C4a and C4b. C4b is much like C3b because it too binds to microbial surfaces. Once C4b is bound to the target microbe, another complement protein called C2 complexes with it. C2 is then cleaved by a MASP to release C2a and C2b. C2a then binds to C4b to form a C3 convertase. Like the C3 convertase of the alternative pathway, this C3 convertase now generates C3a and C3b. In addition to acting as an opsonin, some of the C3b joins forces with C4b and C2a to form a C5 convertase (C3b plus C4bC2a) so that MAC components can be assembled and C5a and C3a can help trigger inflammation. Thus although different C3 and C5 convertases are used, the three outcomes are achieved as seen in the alternative pathway (figure 32.6).

Activation of the **classical complement pathway** typically depends on the presence of antibody specific to the pathogen. Antibodies are glycoproteins produced by activated B cells as part of an adaptive response (*section 33.7*). Each antibody binds a specific antigen. When this happens in the context of complement activation, the antibody-antigen complex binds C1, a complement component composed of three proteins (q, r, s). This forms a trimolecular complex (C1qrs + antibody + antigen), which has protease activity. The activated C1s subcomponent cleaves C4 and C2 to generate the C3 convertase C4b2a as described for the lectin pathway. The major difference is that C4b2a is bound to the antigen-antibody-C1 complex rather than directly to a microbial cell or other antigenic particle. However, as we saw in the lectin pathway, C4b2a serves as a C3 convertase to cleave C3 into a pathogen-bound subcomponent C3b and a C3a soluble component. Similarly, the C5 convertase of the classical pathway is the same as that for the lectin pathway (C3b + C4bC2a). Thus the major differences between the three pathways are (1) the types of molecules that trigger initial activation, (2) the means by which the C3 and C5 convertases are activated, and (3) the identity of the C3 and C5 convertases, although these are the same for lectin and classical pathways.

It is important to note that the three pathways are not simultaneously activated. Microbes arriving at a local tissue site for the first time interact with components of the alternative pathway and then the lectin pathway, resulting in the generation of C3a and C5a to help trigger inflammation, opsonization of the microorganisms with C3b, and initiation of the lytic sequence by MAC. If the pathogens persist or if they invade the host a second time, antibody responses also will activate the classical complement pathway.

Cytokines Are Chemical Messages Between Cells

A functional, coordinated immune system requires a highly specialized communication system. The "language" used by the mammalian immune system is based on small, soluble proteins that when bound to their cellular receptors initiate a signal transduction pathway that regulates the transcription of genes and the activation of proteins needed for an appropriate immune response. Because each cytokine triggers a specific cellular response or set of responses, the resulting immune activity depends on which signals are received, how many receptors become occupied, and interference from either external or internal sources.

Cytokine (Greek *cyto*, cell; *kinesis*, movement) is a generic term for any soluble, low-molecular-weight protein or glycoprotein released by one cell population that acts as an intercellular (between cells) mediator or signaling molecule. Cytokines are produced by many different types of cells. Once released, cytokines impact a broad range of cells, resulting in a variety of cellular responses. Cytokine synthesis is tightly regulated and is induced by stimuli such as a viral, bacterial, or parasitic infection, malignancy, and inflammation. In addition, the release of certain cytokines by one kind of immune cells can induce the production of other cytokines by other cells.

Cytokines produce biological actions only when bound to specific receptors on the surface of target cells. Most cells have hundreds to a few thousand cytokine receptors, but cytokines are very potent molecules and a maximal cellular response results when only a small percentage of receptors are occupied by cytokine molecules. Although there are many ways to group cytokines, we can place them in three simplified categories: (1) cytokines that principally regulate innate immunity, (2) cytokines that principally regulate adaptive immunity, and (3) cytokines that stimulate hematopoiesis (**table 32.2**).

The naming of cytokines has been one of those unfortunate stories with the left hand not knowing what the right has done. Historically, each newly discovered protein would be named based on its function. As protein biochemistry became more exact, scientists grouped proteins based on structure. This resulted in a list of proteins with different names yet the same function and proteins with identical structures but different names. For simplicity, cytokine nomenclature can be divided into four functional groups: (1) Cytokines that stimulate cell migration (i.e., chemotaxis and chemokinesis) are called **chemokines,** (2) cytokines produced by one leukocyte acting on another leukocyte are called **interleukins,** (3) regulatory cytokines produced in response to infection are called **interferons,** and (4) cytokines that stimulate the growth and differentiation of immature leukocytes in the bone marrow are called **colony-stimulating factors** (CSFs). The last important class of cytokines produced by some immune cells is called **tumor necrosis factors** (**TNF**), based on early observations of their in vitro activity; their immune function is to stimulate an inflammatory response. Although these terms are a helpful shorthand method of recognizing function, the groups are not mutually exclusive. For instance interferons produced by one type of leukocyte will influence the behavior of others, that is, interferons also serve as interleukins.

Acute-Phase Proteins

In addition to chemical mediators from leukocytes, a number of biologically active molecules from other somatic cells can cause dramatic physiological changes in the host. An important example occurs when macrophages respond to sites of injury. They release pro-inflammatory cytokines that stimulate the liver to rapidly produce **acute-phase proteins.** These mediators assist in the prevention of blood loss and ready the host for microbial invasion. They include C-reactive protein (CRP), mannose-binding protein (MBP), and surfactant proteins A (SP-A) and D (SP-D), all of which bind bacterial surfaces and act as opsonins. In addition, CRP interacts with C1q to activate the classical complement pathway and MBP binds to bacteria and fungi to activate complement through the lectin pathway. SP-A, SP-D, and C1q are also known as collectins—proteins composed of a collagen-like motif connected by α-helices to globular binding sites. These proteins (along with others) police host tissues by binding to ("collecting") foreign materials, thereby assisting in the removal of microorganisms and their products (section 32.6).

Comprehension Check

1. How do lysozyme and transferrin limit bacterial growth?
2. Describe the production and function of cationic peptides and bacteriocins.
3. Compare and contrast the means by which the alternate, lectin, and classical pathways are activated and the activities of the resulting C3 and C5 convertases.
4. Describe the three outcomes of complement activation.
5. Based on the information in table 32.2, which cytokines trigger fever? Which have antiviral activity?
6. Why is it important that some cytokines have anti-inflammatory activities?
7. How do acute-phase reactants assist in pathogen removal?

Table 32.2 Examples of Cytokines Grouped by Function

	Cytokine[1]	Source	Role
Innate Resistance	IL-1	Macrophages, endothelial and epithelial cells	Upregulates inflammatory response, including fever; stimulates release of acute-phase proteins
	IL-6	Macrophages, T cells, endothelial cells	Upregulates acute-phase response, including fever; stimulates neutrophil maturation; induces differentiation of T cells into T helper cells (TH_{17})
	IL-8	Neutrophils, macrophages, dendritic cells, endothelial and epithelial cells	Activates neutrophils, induces chemotaxis, enhances inflammation
	IL-15	Macrophages	Upregulates NK cell production
	IL-23	Macrophages and dendritic cells	Upregulates inflammatory response by inducing T cells to release IL-17, which stimulates neutrophil activity
	TGF-β	Dendritic cells, T helper cells	Induces differentiation of T cells into T helper cells (TH_{17}, Treg); stimulates antibody production
	TNF-α	Macrophages, NK cells, and T cells	Upregulates inflammatory response, including fever; stimulates acute-phase protein synthesis; stimulates neutrophil activity
	Type I IFN (IFN α/β)	Virus-infected host cells; plasmacytoid dendritic cells, macrophages, and NK cells	Induces broad-spectrum antiviral and anti-tumor activity; enhances other cytolytic function; promotes inflammation
Adaptive Immunity	IL-2	T cells	Stimulates growth and differentiation of T cells and NK cells
	IL-4	T helper cells (TH_2), dendritic cells	Induces differentiation of T cells into T helper cells (TH_2 cells); stimulates antibody production
	IL-5	T-cell subset and mast cells	Stimulates growth of B cells; enhances antibody secretion; activates eosinophils
	IL-10	T regulatory cells, macrophages	Anti-inflammatory effects largely mediated by inhibiting release of pro-inflammatory cytokines
	IL-12	T helper cells (TH_1), dendritic cells, macrophages	Induces differentation of T cells into T helper cells (TH_1 cells); stimulates activity of NK cells
	IL-17	T helper cells (TH_{17})	Monocyte and neutrophil chemokine; induces pro-inflammatory cytokines IL-6, TNF-α, IL-1, chemokines, and prostaglandins from various cells
	Type II IFN (IFN γ)	T helper cells (TH_1), dendritic cells, NK cells	Induces differentiation of T cells into T helper cells; stimulates antibody production; upregulates macrophage and NK cell activity
Hematopoiesis	IL-3	Basophils and activated T cells	Stimulates pluripotent hematopoietic stem cells to become myeloid progenitor cells; stimulates myeloid cell proliferation
	IL-7	Bone marrow and thymic stromal cells, dendritic and epithelial cells, and hepatocytes	Stimulates pluripotent hematopoietic stem cells to become lymphoid progenitor cells; stimulates lymphoid cell proliferation
	CSF-1	Osteoblasts	Induces hematopoietic stem cells to proliferate and differentiate into monocytes/macrophages; promotes monocyte survival
	CSF-2	Macrophages, T cells, endothelial and mast cells, and fibroblasts	Induces hematopoietic stem cells to proliferate and differentiate into granulocytes and monocytes
	CSF-3	Numerous cells and tissues	Induces hematopoietic stem cells to proliferate and differentiate into neutrophils; stimulates neutrophil function and survival

1 Cytokine: CSF, colony-stimulating factor; IFN, interferon; IL, interleukin; TNF, tumor necrosis factor

32.4 Each Type of Innate Immune Cell Has a Specific Function

After reading this section, you should be able to:

a. Recognize the different types of leukocytes involved with innate resistance
b. Outline the leukocyte response to microbial invasion
c. Relate leukocyte distribution within the host to host resistance

Blood cell development occurs in the **bone marrow** of mammals during the process of **hematopoiesis.** All white blood cells, or **leukocytes** (Greek *leukos,* white; *kytos,* cell), originate from hematopoietic precursor cells in the fetal liver and after birth, in the bone marrow and thymus (**figure 32.8**). When stimulated to undergo further development, some leukocytes become residents within tissues, where they respond to local trauma. These cells sound the alarm that signals invasion by foreign organisms. Other leukocytes circulate in body fluids and are recruited to the sites of infection after the alarm has been raised.

The average adult has between 4,500 and 10,000 leukocytes per cubic millimeter of blood. This average value shifts substantially during an immune response. For example, in most bacterial and fungal infections the white blood cell count may increase as leukocytes migrate from the bone marrow to the blood, then to the site of invasion. This transient increase in circulating leukocytes (leukocytosis) is a helpful biomarker for clinicians in diagnosing an infectious process. We now introduce these cells and their specialized functions.

Mast Cells

Mast cells are bone marrow–derived cells that differentiate in connective tissue. Mast cells (figure 32.8) have an indented nucleus and a cytoplasm filled with specialized organelles called granules. Mast cells are not phagocytic, but when stimulated, they rapidly release the contents of their granules into the extracellular environment, a process called degranulation. Mast cell granules contain histamine, prostaglandins, serotonin, heparin, dopamine, platelet-activating factor, and leukotrienes. Because these compounds influence the tone and diameter of blood vessels, they are termed **vasoactive mediators.** Additionally, mast cells possess high-affinity receptors for the type of antibody associated with allergic responses (IgE). When IgE is bound in sufficient quantity to these receptors the release of preformed vasoactive mediators occurs. As discussed in chapter 33, this plays a major role in certain allergic responses such as eczema, hay fever, and asthma. ▶| *Antibodies bind specific 3-D antigens (section 33.7); Hypersensitivities (section 33.10)*

Granulocytes

Granulocytes also have irregularly shaped nuclei with two to five lobes (figure 32.8). Their cytoplasm has granules that contain reactive substances that kill microorganisms and enhance inflammation. Three types of granulocytes exist: basophils, eosinophils, and neutrophils.

Basophils (Greek *basis,* base; *philein,* to love) have irregularly shaped nuclei with two lobes (figure 32.8). The granules contain histamine and other substances similar to those in mast cells. Like mast cells, basophils are also important in the development of allergies and hypersensitivities, because they possess high-affinity receptors for the type of antibody associated with allergic responses and release preformed vasoactive mediators.

Eosinophils (Greek *eos,* dawn; *philein*) have a two-lobed nucleus connected by a slender thread of chromatin (figure 32.8). Eosinophil granules contain hydrolytic enzymes (e.g., nucleases and glucuronidases), in addition to peroxidases and a caustic protein called major basic protein. Eosinophils circulate in low numbers and migrate from the bloodstream into tissue spaces, especially mucous membranes, when recruited by soluble chemotactic mediators. They defend against protozoan and helminth (worm) parasites. These organisms are too large to be phagocytosed, so when eosinophils release hydrolytic enzymes, cationic peptides, and reactive oxygen species into extracellular fluid, the parasite's plasma membrane is damaged and the organism is (hopefully) killed. Eosinophils also play a role in allergic reactions, as they have granules that contain histaminase. Thus their numbers in the bloodstream often increase during parasite infections and allergic reactions. ▶| *Type I hypersensitivity (section 33.10)*

Neutrophils (Latin *neuter,* neither; *philein*) are also called ***polymorphonuclear* cells** (PMNs) because their nuclei have three to five lobes connected by slender threads of chromatin (figure 32.8). Neutrophils can be described as "phagocytic machines" that harbor inconspicuous organelles known as primary and secondary granules. Primary granules contain peroxidase, lysozyme, defensins, and various hydrolytic enzymes, whereas the smaller secondary granules have collagenase, lactoferrin, cathelicidin, and lysozyme. These enzymes and other molecules help digest foreign material after it is phagocytosed (section 32.6).

Mature neutrophils leave the bone marrow and circulate in blood so they can rapidly migrate to a site of tissue damage and infection, where they become the principal phagocytic responders. Neutrophils migrate to sites where complement C3a and C5a have been released, where they find opsonized particles, which are more readily phagocytosed. Neutrophils have a limited life span of about a week. Neutrophils and their antimicrobial compounds are described in more detail in the context of phagocytosis (section 32.6) and the inflammatory response (section 32.7).

Monocytes, Macrophages, and Dendritic Cells

Monocytes (Greek *monos,* single; *cyte,* cell) are leukocytes with an ovoid- or kidney-shaped nucleus and granules in the cytoplasm (figure 32.8). They are produced in the bone marrow

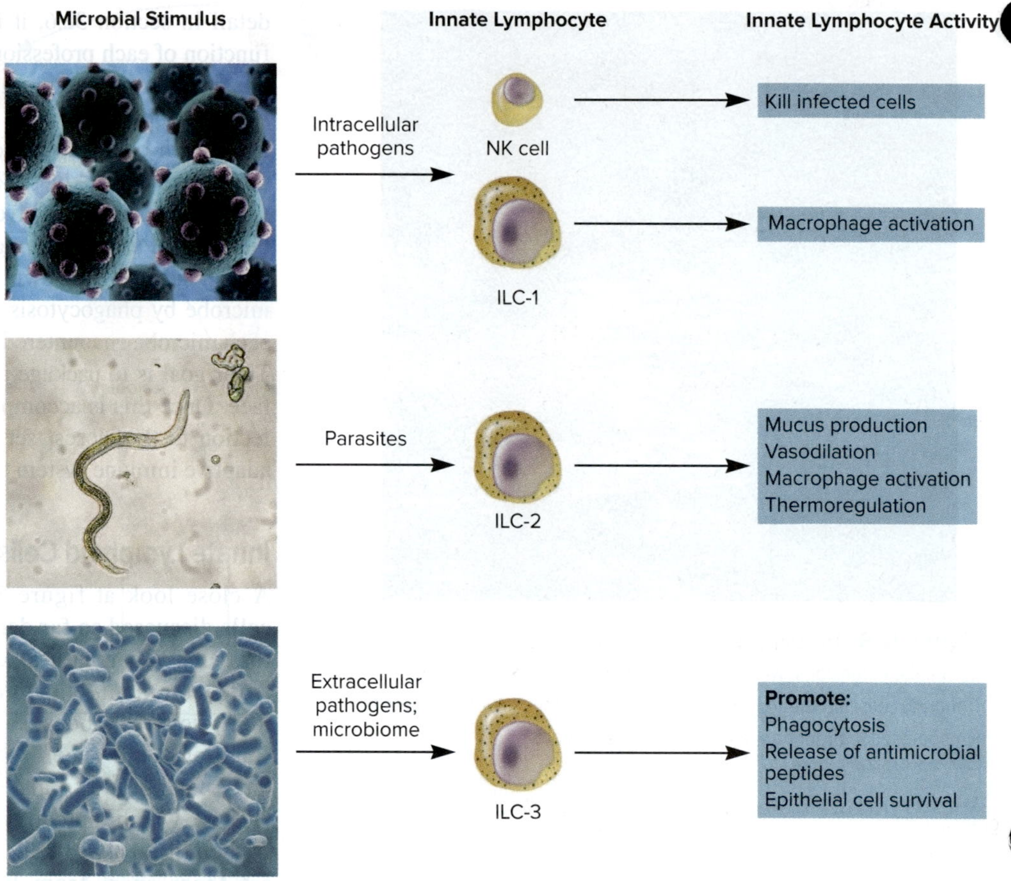

Figure 32.10 Innate Lymphocyte Stimuli and Responses.
(Left top) ©Nixxphotography/Getty Images; (Left middle) ©olgaru79/Shutterstock; (Left bottom) ©iStockphoto/Getty Images

of antimicrobial peptides, preventing an immune response to normal gut microbiota, and promoting phagocytosis and the survival of epithelial cells that line the mucosa. ▶| *A functional core microbiome is required for human homeostasis (section 34.3)*

Natural killer (NK) **cells** are sometimes also referred to as cytotoxic ILCs (figure 32.10). These large, granular cells detect and destroy stressed, malignant, or virally infected cells (**figure 32.11**). NK cells survey adjacent cells by contacting them via several surface receptors (figure 32.11*a*). These interactions provide both kill and don't kill signals to the NK cell. It is the balance of these signals that determines the fate of the target cell. If the target cell is found to be defective (e.g., malignant) or infected, the kill signals outweigh the survival signals and the NK cell becomes deadly.

Once activated to kill the target cell, the NK cell mobilizes its cytoskeletal proteins to closely adhere to the target cell and releases a cargo of deadly molecules (figure 32.11). These include a pore-forming protein called perforin and enzymes called granzymes. Together, these proteins trigger the target cell to commit suicide (apoposis).

In addition to responding to the balance of kill versus survival signals transmitted by the target cell, NK cells have another way to sense cells that need to be cleared. NK cells have receptors for antibodies, and therefore can attack cells that are opsonized by antibodies. Recall that self cells should not be antigenic, and should therefore not be coated with antibody. However, malignant cells may produce mutant proteins perceived to be non-self by the adaptive immune system. In addition, host cells infected with viruses may attach viral proteins to their surfaces. Both cases lead to antibody formation and attachment. The process by which NK cells recognize and kill such cells is called **antibody-dependent cell-mediated cytotoxicity** (ADCC) (**figure 32.12**).

Comprehension Check

1. Describe the structure and function of each of the following blood cells: basophil, eosinophil, neutrophil, monocyte, macrophage, and dendritic cell. Which cells are phagocytic?
2. How are macrophages and dendritic cells different? How are they similar?
3. How does a dendritic cell bridge the innate and adaptive arms of immunity?
4. How do innate lymphoid cells differ from other innate immune cells and other lymphocytes?
5. Compare and contrast the function of ILCs and LTi cells.
6. Discuss the role of NK cells in protecting the host. What are the two mechanisms NK cells use to detect altered cells?

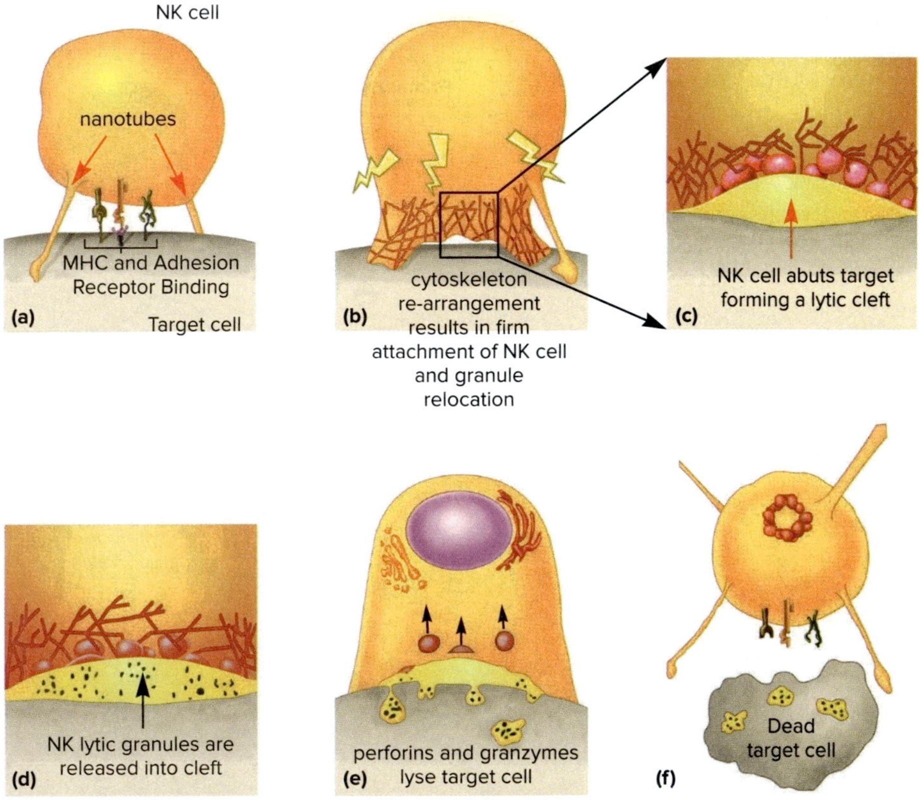

Figure 32.11 The System Used by Natural Killer (NK) Cells to Recognize and Destroy Abnormal Cells. (a) NK cell evaluates target cell using membrane nanotubes. (b) Microtubules organize lytic granules toward the target cell. (c) A lytic cleft forms and (d) granules exit the NK cell into the cleft. (e) Lytic perforins and granzymes lyse the target cell. (f) The NK cell releases the dead target cell and migrates toward another cell.

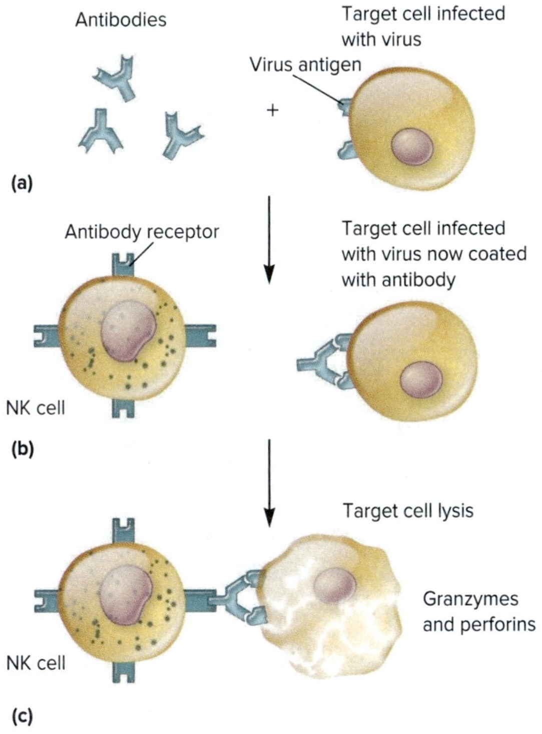

Figure 32.12 Antibody-Dependent Cell-Mediated Cytotoxicity. (a) In this mechanism, antibodies bind to a malignant cell or a host cell infected with a virus. (b) NK cells have specific antibody receptors on their surface. (c) The antibody bridges the infected cell with the NK cell so that the target is close enough for enzymatic attack.

32.5 Organs and Tissues of the Immune System Are Sites of Host Defense

After reading this section, you should be able to:

a. Differentiate between primary and secondary lymphoid tissues and organs in terms of structure and function
b. Explain the role of the spleen in host defense
c. Rationalize the cellular architecture of skin- and mucosal-associated lymphoid tissues

The organs and tissues of the immune system can be divided into primary or secondary lymphoid organs and tissues based on function (**figure 32.13**). The primary organs are where hematopoietic stem cells differentiate into the adaptive immune cells B and T lymphocytes. As discussed in chapter 33, B and T cells are considered naive until they encounter the antigen to which they will respond and remember. The thymus is the primary lymphoid organ for T cells, and the bone marrow is the primary lymphoid tissue for B cells. The secondary organs and tissues are areas where B and T lymphocytes encounter and bind antigen, which triggers their differentiation, activation, and proliferation into antigen-specific effector cells. The spleen is a secondary lymphoid organ, and lymph nodes and mucosal-associated tissues (GALT—gut-associated lymphoid tissue, and SALT—skin-associated lymphoid tissue) are secondary lymphoid tissues. These are now discussed in more detail.

Figure 32.13 Anatomy of the Lymphoid System. Lymph is distributed through a system of lymphatic vessels, passing through many lymph nodes. Lymph enters the lymph node through afferent lymphatic vessels, percolates through and around the follicles in the node, and leaves through the efferent lymphatic vessels. The lymphoid follicles are the sites of cellular interactions and extensive immunologic activity.

MICRO INQUIRY *In which of these organs and tissues are macrophages and dendritic cells found in abundance?*

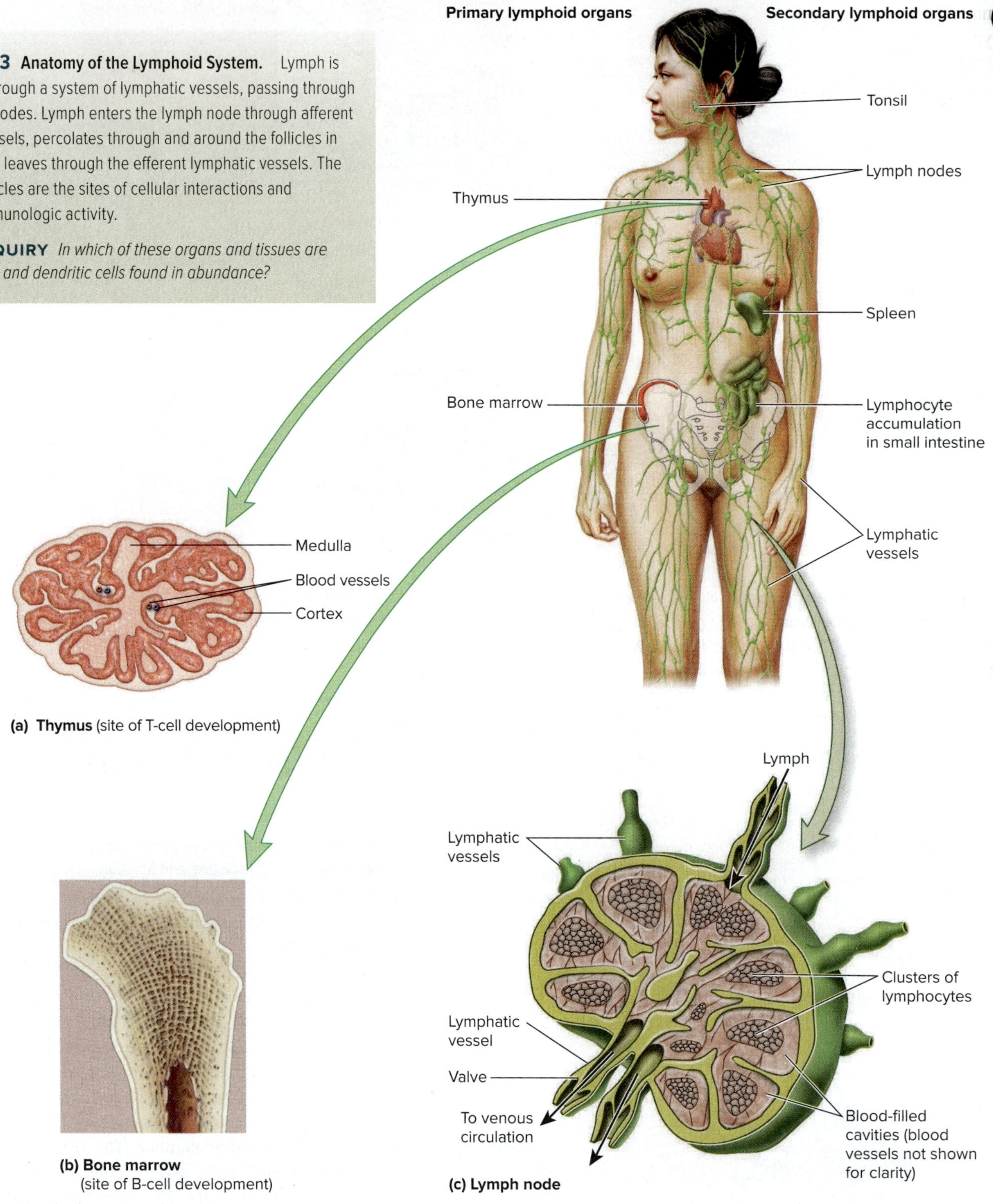

Primary Lymphoid Organs and Tissues

The **thymus** is a highly organized lymphoid organ located above the heart. Precursor cells from the bone marrow migrate into the outer layers of the thymus (cortex), where they proliferate. As they mature, about 98% are directed to undergo apoptosis because they cannot distinguish self from non-self antigen. This process is known as negative selection. The remaining 2% move deeper into the thymus (medulla), where they mature and subsequently enter the bloodstream and secondary lymphoid tissues where they await activation by innate immune cells. ▶| *T-cell development (section 33.5)*

In mammals, the bone marrow (figure 32.13*b*) is the site of B-cell maturation. Like T-cell maturation, a selection process within the bone marrow eliminates B cells that recognize self. Remaining B cells mature in the bone marrow and subsequently enter the bloodstream. Like T cells, they migrate to lymph nodes and the spleen, where they await introduction to the antigen to which they will make targeted antibody. ▶| *B cells make antibodies (section 33.6)*

Secondary Lymphoid Organs and Tissues

The **spleen** is the most highly organized secondary lymphoid organ. It is a large organ located in the abdominal cavity. It functions to filter blood and trap blood-borne particles to be assessed for foreignness by phagocytes and B cells (figure 32.13). Macrophages and dendritic cells are present in abundance, and serve to phagocytose blood-borne pathogens. Antigens captured by DCs are presented to T cells, activating a specific immune response.

Lymph nodes are placed in strategic locations throughout the body so that DCs that have captured antigen by phagocytosis can quickly migrate to them and present antigen to T cells that have not yet been activated (figure 32.13*c*). Lymph nodes also are rich in macrophages, which can remove antigens in the lymphatic fluid by phagocytosis. The outer region of lymph nodes (cortex) is rich in B cells that can bind antigen directly from the blood or lymph. The inner region of lymph nodes (medulla) is the T-cell zone. There T cells await interaction with DCs or B cells. ▶| *T cell activation (section 33.5)*

Lymphoid tissues are found throughout the body either as highly organized or as loosely associated cellular complexes. Highly organized lymphoid tissues feature DCs and macrophages surrounded by specific areas of B and T lymphocytes. Loosely associated lymphoid tissue lacks the cellular partitioning found in highly organized lymphoid tissue. The primary role of both types of lymphoid tissues is to sample the host's microbial environment in a region where leukocytes are localized in such a way that interaction between the innate and the adaptive arms of the immune response is optimized. We now discuss these tissues in more detail.

Despite the skin's defenses, at times pathogenic microorganisms gain access to the tissue under the skin surface. Here they encounter the highly organized lymphoid tissue called the ***s**kin-**a**ssociated **l**ymphoid **t**issue* **(SALT)** (**figure 32.14**). The major function of SALT is to confine microbial invaders to the area immediately underlying the epidermis thereby preventing access to the bloodstream. One type of SALT cell is the **Langerhans cell,** a dendritic cell that phagocytoses microorganisms that penetrate the skin. Once the Langerhans cell has internalized a foreign particle or microorganism, it migrates from the epidermis to nearby lymph nodes, where it presents antigen to activate lymphocytes, inducing a specific immune response to that antigen. This dendritic cell–lymphocyte interaction is an important bridge between innate resistance and adaptive immunity.

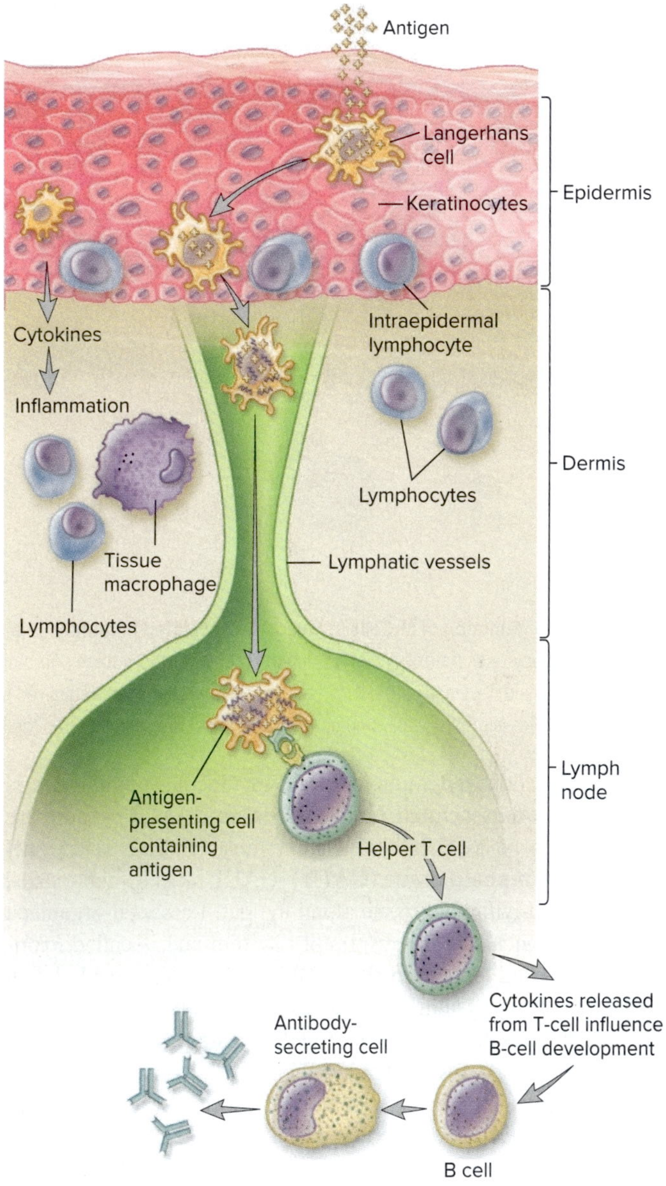

Figure 32.14 Skin-Associated Lymphoid Tissue (SALT). Keratinocytes make up 90% of the epidermis. They can secrete cytokines that cause an inflammatory response to invading pathogens. Langerhans cells are specialized DCs that internalize antigen and move to a lymph node, where they present antigen to helper T cells. Intraepidermal lymphocytes respond to specific microbial invaders that have previously been encountered by the host.

The epidermis also contains another type of SALT cell called the **intraepidermal lymphocyte** (figure 32.14), a specialized T cell having potent cytolytic and immunoregulatory responses to antigen. These cells are strategically located in the skin so that they can intercept antigens that breach the first line of defense. Most of these specialized SALT cells have limited receptor diversity and have likely evolved to recognize molecules on the surface of common skin microbes.

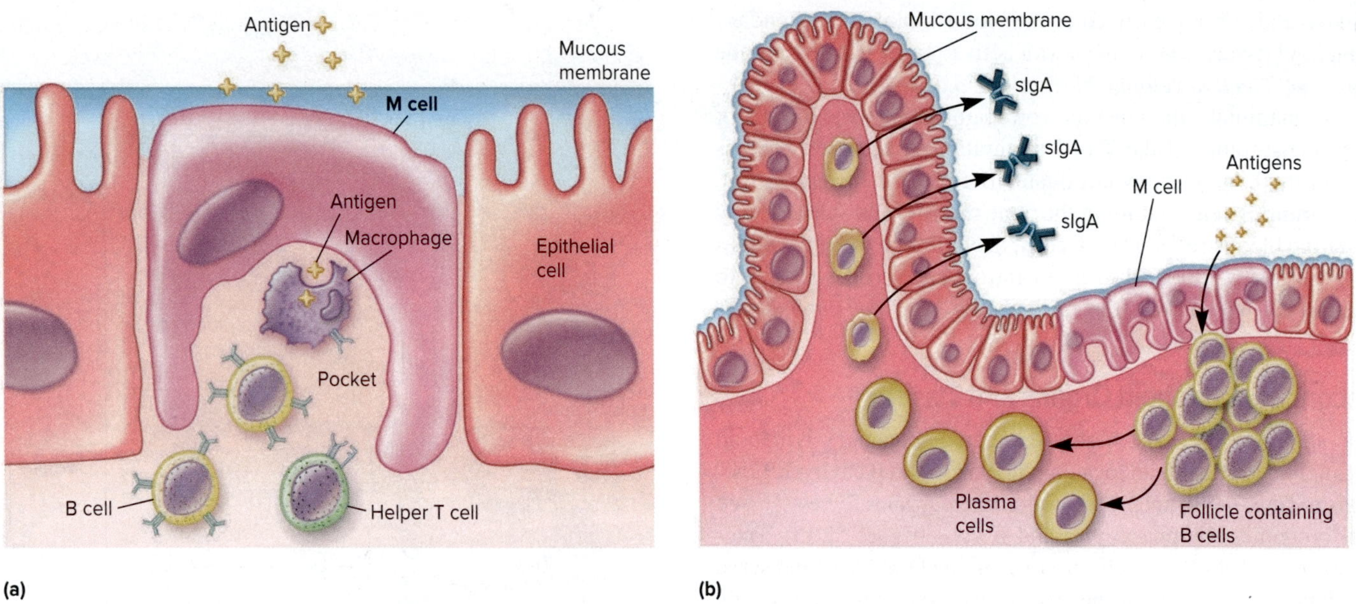

Figure 32.15 Function of M Cells in Mucosal-Associated Immunity. (a) Structure of an M cell located between two epithelial cells in a mucous membrane. The M cell endocytoses pathogens and releases them into the pocket containing helper T cells, B cells, and macrophages. It is within the pocket that pathogens often are destroyed. (b) Once in the M-cell pocket, antigen enters an organized lymphoid follicle containing B cells. The activated B cells mature into plasma cells, which produce antibodies (sIgA; *s*ecretory *i*mmuno*g*lobulin *A*) that are released into the lumen, where they react with the antigen that triggered their production.

The specialized lymphoid tissue in mucous membranes is called ***m**ucosal-**a**ssociated **l**ymphoid **t**issue* (**MALT**). There are several types of MALT. The system most studied is the ***g**ut-**a**ssociated **l**ymphoid **t**issue* (**GALT**). GALT includes the tonsils, adenoids, and lymphoid tissue along the gut. Less well-organized MALT also occurs in the respiratory system and is called ***b**ronchial-**a**ssociated **l**ymphoid **t**issue* (**BALT**); the diffuse MALT in the genitourinary system does not have a specific name.

All mucous membranes are lined by a single layer of epithelial cells that separate the environment (or lumen) from the underlying tissue. The lumen is subject to constant interaction with microbes, so it is imperative that the MALT defend any breach of the mucosal barrier. Within the GALT, **Peyer's patches** (in the small intestine) and **isolated lymphoid follicles** (ILFs; throughout the gut) have specialized cells called **M cells,** in between epithelial cells. M cells are flattened on the lumen side of the gut, and have a large pocket facing the interior (**figure 32.15*a***). When a foreign particle contacts an M cell, it is endocytosed and released into the pocket. There macrophages and DCs engulf the particle or pathogen and degrade it. In addition to macrophages, T and B cells are clustered in what are termed *follicles* within Peyer's patches and ILFs. Follicles are regions with a high density of B cells surrounded by a zone of T cells. As microbes are introduced into the M-cell pocket, they can be captured by B cells and stimulate the production of antibody that can be secreted into the gut lumen (secretory IgA; *section 33.7*). DCs activate T cells within the follicles or migrate to local lymph nodes (called mesenteric lymph nodes) to activate T cells there. Similar to SALT, GALT intra- and interepithelial lymphocytes are strategically distributed so that the detection of foreign particles is increased, should the intestinal membrane be breached. In addition to Peyer's patches and ILFs, diffuse lymphoid tissue is widely distributed in the large intestine, as well as the respiratory and genitourinary tracts, breast, eye, salivary glands, and skin. These areas typically have a high density of immune cells but do not have T and B cells in an organized follicle. ▶| *Antibodies bind specific 3-D antigens (section 33.7)*

As seen in our discussion thus far, a microbe attempting to invade a potential host encounters innate mechanisms including physical, chemical, and cellular barriers designed to kill the invader, digest the carcass into small antigens, and assist lymphocytes in developing long-term protection against the next invasion. We now examine the phagocytic processes in more detail and consider how the host integrates many of the innate resistance activities into the coordinated activities that define an inflammatory response.

Comprehension Check

1. Compare and contrast the function of primary and secondary lymphoid tissues and organs, citing examples of each.
2. Injury to the spleen can lead to its removal. What impact would this have on host defenses?
3. Describe SALT and MALT functions in innate resistance.

32.6 Phagocytosis Destroys Invaders

After reading this section, you should be able to:

a. Explain the methods by which pathogens are recognized by phagocytes
b. Compare and contrast the processes of autophagy and phagocytosis
c. Correlate the biochemical activities within the phagolysosome with pathogen destruction

In an average lifetime, humans and other vertebrates encounter thousands of microbial species, but only a few of these are pathogenic. Phagocytic cells (macrophages, dendritic cells, and neutrophils) are an important early defense against invading microorganisms. These cells recognize, ingest, and kill cellular microbes by the process of **phagocytosis** (Greek *phagein*, to eat; *cyte*, cell; *osis*, a process) and intracellular pathogens by the process of **autophagocytosis** (or autophagy). In addition to recognizing cells that have been opsonized, as discussed in section 32.3, phagocytes detect microorganisms on the basis of some of their unique molecules, as we now discuss.

Recognition of Foreignness Is Key for Survival

Professional phagocytes (neutrophils, macrophages, and DCs) and other leukocytes detect microbial macromolecules by the presence of repetitive structural patterns. These unique microbial signatures are called ***m*icrobe-*a*ssociated *m*olecular *p*atterns** (**MAMPs**), although they are referred to as pathogen-associated molecular patterns (PAMPs), too. We use the more inclusive term, MAMP, because the immune system responds to many microorganisms that are not pathogens. MAMPs are specific regions within common microbial macromolecules, such as lipopolysaccharide (LPS), peptidoglycan, fungal cell wall components, viral nucleic acids, and other microbial structures. MAMPs do not identify a specific microorganism; rather they inform the host that a microbe is present, sounding the alarm of potential infection.

Microbial invaders can be either extracellular, like most fungi and many bacteria, or intracellular, like viruses and some bacteria, and a few fungi. It therefore follows that cells of the innate immune system recognize extracellular infectious agents with receptors on their surface, while intracellular pathogens are detected by host receptors in the cytosol. All receptors that recognize MAMPs, regardless of cellular location, are called ***p*attern *r*ecognition *r*eceptors** (**PRRs**). We have already introduced soluble PRRs; they include the acute-phase proteins, mannose-binding protein (of the lectin complement pathway), and C-reactive protein (CRP). The membrane-bound PRRs are found on phagocytic cells, enabling them to respond by ingesting and degrading the source of the MAMP. As we now take a closer look at these receptors, keep in mind they are critical in distinguishing between self and non-self.

C-Type Lectin Receptors

C-type lectin receptors (CLRs) are a large group of calcium-dependent membrane-bound proteins having one or more lectin-binding domains specific for one type of carbohydrate (lectin) found on MAMPs. Ligands include mannose, fucose, or glucan carbohydrates found on some bacteria (*Mycobacterium tuberculosis* and *Helicobacter pylori)*, fungi (e.g., *Candida albicans* and *Aspergillus fumigatus)*, helminths (worms), and some viruses. CLRs also bind the *N*-acetylglucosamine components of peptidoglycan from a number of different bacteria, and the capsular polysaccharides of *Streptococcus pneumoniae* and *Klebsiella pneumoniae*. CLRs are concentrated on macrophages and dendritic cells. When a CLR on the surface of a phagocyte binds a MAMP, a signaling cascade is triggered that induces cytokine gene expression. This is mediated by the CLR cytoplasmic domain, which is called the immunoreceptor tyrosine-based activation motif (ITAM). Binding also results in receptor-MAMP internalization, its degradation, and subsequent antigen presentation to lymphocytes, as described in chapter 33.

Toll-Like Receptors

Toll-like receptors (**TLRs**) are an important class of transmembrane receptors expressed by a variety of cells that participate in innate immunity, especially macrophages and dendritic cells. They are found attached to the plasma membrane as well as on membranes of endosomes, lysosomes, endolysosomes, and endoplasmic reticulum, where they detect MAMPs that have gained access to the host cell cytosol. TLRs have an extracellular (or extra-organelle) domain that binds MAMPs, and a cytoplasmic Toll/interleukin-1 receptor (TIR) domain that activates downstream transcription signaling (**figure 32.16**). Binding of a MAMP to a TLR causes the TIR to recruit a specific adapter protein, either MyD88 or TRIF, which in turn activates a signal transduction cascade that ends in the activation of transcription factors required for expression of genes encoding cytokines. For example, activation through the MyD88-dependent adapter results in the activation of the transcription factor NF-κB, which activates transcription of pro-inflammatory cytokine genes. There are at least 10 distinct human TLRs that recognize distinct but sometimes overlapping MAMPS and subsequently initiate an immune response.

NOD-Like Receptors

Unlike the transmembranous TLRs, **nucleotide-binding and oligomerization domain (NOD)-like receptors** are located only in the host cell cytosol. There NOD-like receptors (NLRs) can sense intracellular MAMPs (e.g., viral RNA, intracellular bacterial toxins, and intracellular bacteria), along with host molecules called "damage-associated molecular patterns" (DAMPs), such as uric acid and heat-shock proteins. DAMPs are a diverse group of molecules that arise from cell injury and stress. Upon detection of one of these foreign or native signals, oligomerization of several NLRs occurs. Then cofactors and the enzyme caspase-1 are recruited to the oligomerized unit. This NLR-caspase complex is called the **inflammasome,** because once assembled, it triggers the release of pro-inflammatory cytokines. This is accomplished when caspase-1 cleaves and thereby activates the cytokines IL-1 and IL-18, which are also produced in response to signals transduced by other PRRs, such as TLRs. Thus NLRs (inside host cells) and TLRs (on the surface of host cell plasma and some intracellular membranes) work together to provide a strong inflammatory response.

RIG-I-Like Receptors

Retinoic-acid-inducible gene I (RIG-I)–like receptors (RLRs) are yet another family of cytoplasmic receptors. These receptors

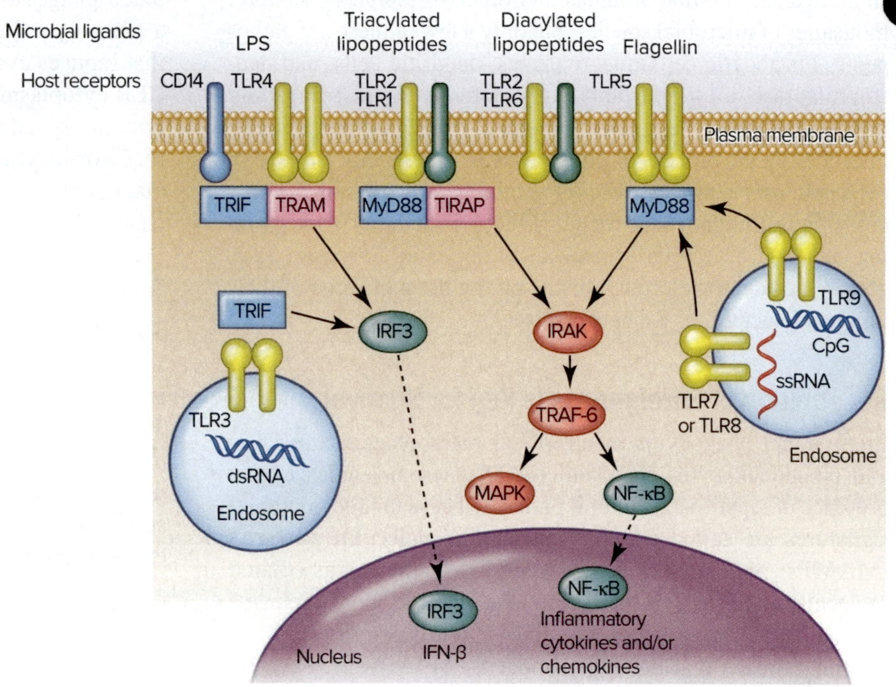

Figure 32.16 Recognition of Microbe-Associated Molecular Patterns (MAMPs) by Pattern Recognition Receptors (PRRs). MAMPs bind PRRs, such as toll-like receptors (TLRs) found on both the plasma membrane and endosomal membranes. PRR binding results in signaling that upregulates cytokine gene expression through common signal transduction pathways, like the transcription factor NF-κB. Interferon-β (IFN-β) is an antiviral cytokine. TRIF and TIRAP are proteins that, like MyD88, help mediate intracellular signaling. In addition to TLR4, CD14 also binds lipopolysaccharide (LPS). TLR10 is not shown; its MAMP ligand has not yet been identified.
Source: "Introduction to the Immune System." What-When-How.

evolved to sense viral RNA. Replication of viruses with RNA genomes within the cytosol leads to the production of dsRNA and triphosphate-capped ssRNA. RLRs recognize these forms of virus-specific RNA and stimulate an antiviral cytokine response. |◀ *Viruses and other acellular infectious agents (chapter 6)*

Intracellular Digestion

Two parallel processes, phagocytosis and autophagy, accomplish ingestion and digestion of foreign materials. During phagocytosis, extracellular particles are engulfed by pseudopodia (Latin *pseudo,* false; *podium,* foot) formed by the extension of the cell membrane, which surround the particle to form an internal phagosome. This is often opsonin and receptor-mediated (**figure 32.17, steps 1 and 2**). Autophagy begins when intracellular microorganisms, or their products, are captured by intracellular membranes through PRRs, or are coated with ubiquitin, the molecule used to tag proteins for recycling by proteasomes (*see figure 5.8*). In this case, rather than targeting proteins, ubiquitin "labels" microorganisms for capture by a phagophore (free-floating, open membrane in the cytosol). The phagophore then encircles the microorganism, creating the double-membrane autophagosome (**figure 32.18**). Thus, whereas the phagosome is derived by pinching off a bit of cell membrane, the autophagosome arises from an intracellular membrane. |◀ *There are several endocytic pathways (section 5.4)*

Once microorganisms have been enclosed in a phagosome or autophagosome, lysosomes fuse to the phagosome (figure 32.17, step 3) or autophagosome (figure 32.18) creating a phagolysosome or autolysosome, respectively. **Lysosomes** deliver a variety of hydrolases such as lysozyme, phospholipase A2, ribonuclease, deoxyribonuclease, and proteases. The activity of these degradative enzymes is enhanced by the acidic vacuolar pH. Collectively these enzymes participate in the destruction of the entrapped microorganisms (figure 32.17, step 4). In addition, toxic **reactive oxygen species (ROS)** such as the superoxide radical ($O_2^{\bar{\cdot}}$), hydrogen peroxide (H_2O_2), singlet oxygen ($^1O_2$), and hydroxyl radical (·OH) are produced. Neutrophils also contain the heme-protein enzyme, myeloperoxidase, which catalyzes the production of hypochlorous acid (bleach). The generation of ROS begins with the recruitment of an NADPH oxidase to the phagosome or autosome membrane. This enzyme uses molecular oxygen to oxidize NADPH to $NADP^+$ and generate superoxide radical (**table 32.3**). The consumption of O_2 and release of ROS is called the **respiratory** or **oxidative burst.** |◀ *Oxygen concentration (section 7.5)*

Macrophages, neutrophils, and mast cells have also been shown to form **reactive nitrogen species** (RNS). These molecules include nitric oxide (NO) and its oxidized forms, nitrite (NO_2^-) and nitrate (NO_3^-). RNS are very potent cytotoxic agents; nitric oxide is probably the most effective RNS. It can block cellular respiration by complexing with iron in electron transport proteins. Macrophages use RNS in the destruction of a variety of infectious agents as well as to kill tumor cells.

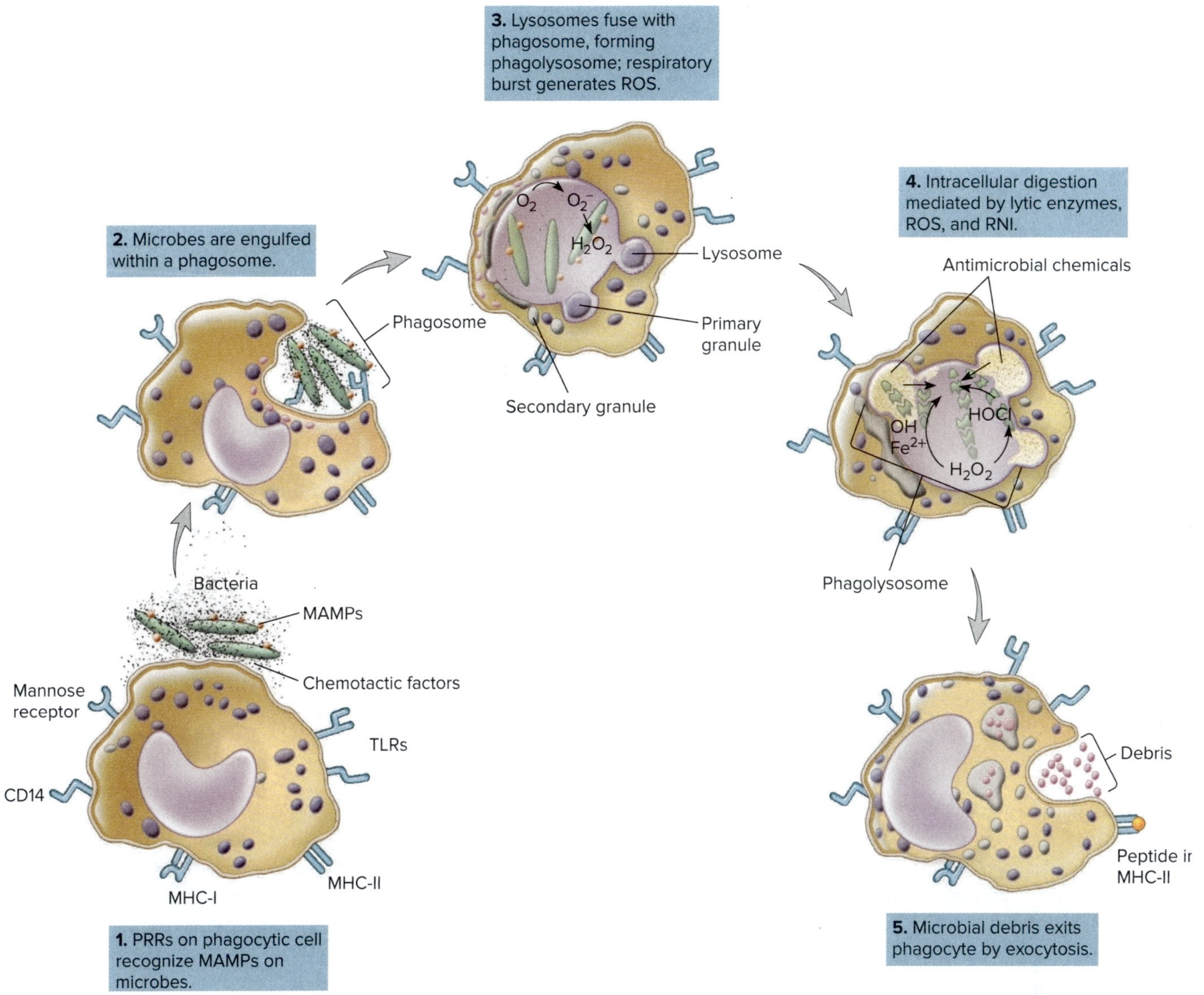

Figure 32.17 Phagocytosis. CD14 is a lipopolysaccharide receptor that associates with TLR-4; MAMPs: microbe-associated molecular patterns; MHC-I: class I major histocompatibility protein; MHC-II: class II major histocompatibility protein (MHC proteins are discussed in chapter 33); RNI: reactive nitrogen intermediates; ROS: radical oxygen species; TLRs: toll-like receptors

Neutrophil granules contain a variety of other microbicidal substances such as cationic peptides, the bactericidal permeability-increasing protein (BPI), and broad-spectrum antimicrobial peptides, including defensins (section 32.3). These substances are compartmentalized for extracellular secretion or delivery to phagocytic vacuoles. Susceptible microbial targets include a variety of Gram-positive and Gram-negative bacteria, yeasts and molds, and some viruses.

Exocytosis

Once the microbial invaders have been killed and digested into small antigenic fragments, the phagocyte may do one of two things. The cell may expel the microbial fragments by the process of **exocytosis** (figure 32.17, step 5). This is essentially a reverse of the phagocytic process whereby the phagolysosome unites with the cell membrane, resulting in the extracellular release of the microbial fragments. Neutrophils are dedicated to phagocytosis and exocytose their cargo until they become exhausted and die. By contrast, macrophages and dendritic cells become **antigen-presenting cells** (*see figure 33.4*). This is accomplished by passing microbial fragments from the phagolysosome to the endoplasmic reticulum. Here the peptide components of the microbial fragments are united with glycoproteins, called major histocompatibility (MHC) proteins *(see figure 33.3)*, destined for the cell membrane. The purpose is to place a

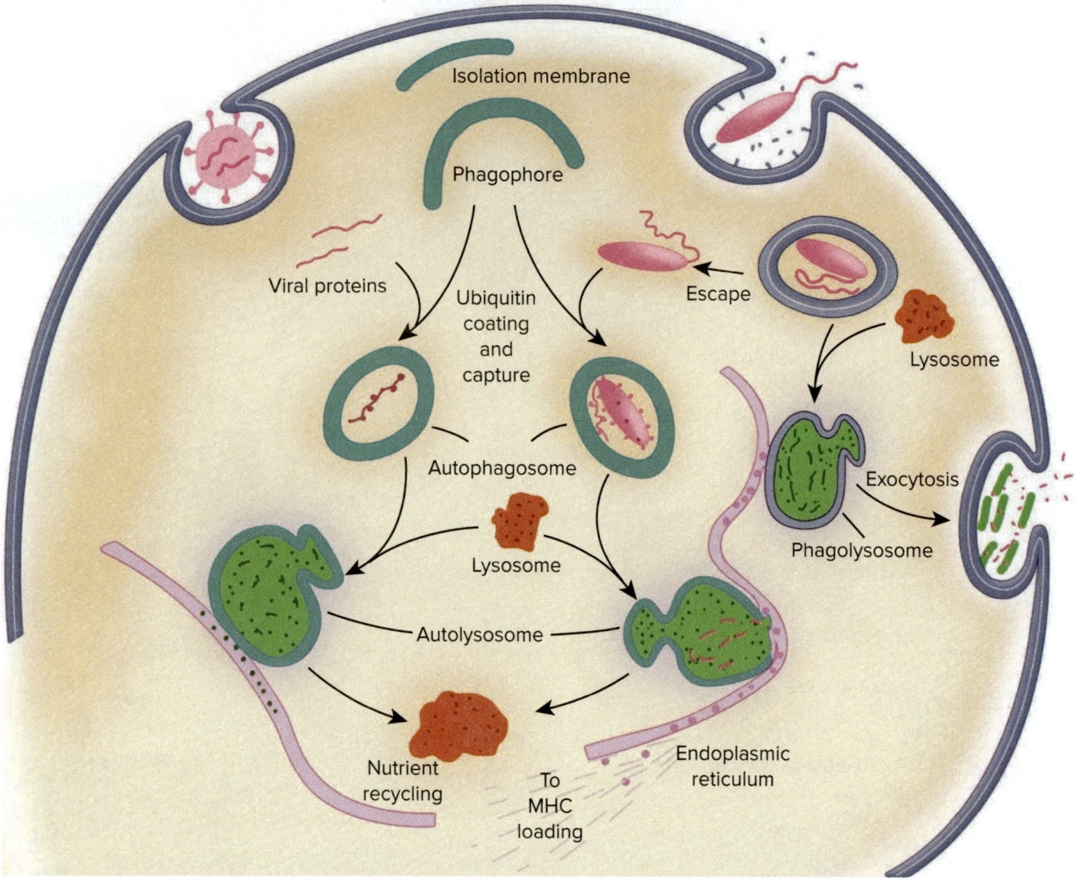

Figure 32.18 Autophagy Is Similar to Phagocytosis. Intracellular infectious agents within the cytosol are captured by free-floating membranes that form a phagophore, often through membrane-bound toll-like receptors. Cytosolic ubiquitin molecules coat the intracellular invaders and their macromolecules (such as viral proteins). The ubiquitinated particles are then surrounded by the phagophore to form an autophagosome. Autophagosome fuses to lysosomes forming an autolysosome and its contents degraded. Both phagolysosome and autolysosome can donate some of their debris to the endoplasmic reticulum, where antigenic peptides are further processed to bind MHC molecules destined for the phagocyte's cell membrane. The remaining debris is recycled or released to the environment.

Table 32.3 Formation of Reactive Oxygen Intermediates

Oxygen Intermediate	Reaction
Superoxide ($O_2^{\bar{\cdot}}$)	$NADPH + 2O_2 \xrightarrow{\text{NADPH oxidase}} 2O_2^{\bar{\cdot}} + H^+ + NADP^+$
Hydrogen peroxide (H_2O_2)	$2O_2^{\bar{\cdot}} + 2H^+ \xrightarrow{\text{Superoxide dismutase}} H_2O_2 + O_2$
Hypochlorous acid (HOCl)	$H_2O_2 + Cl^- \xrightarrow{\text{Myeloperoxidase}} HOCl + OH^+$
Singlet oxygen ($^1O_2$)	$ClO^- + H_2O_2 \xrightarrow{\text{Peroxidase}} {}^1O_2 + Cl^- + H_2O$
Hydroxyl radical ($\cdot OH^-$)	$O_2^{\bar{\cdot}} + H_2O_2 \xrightarrow{\text{Peroxidase}} 2{\cdot}OH^- + O_2$

microbe-specific peptide in the MHC so that when the MHC-peptide complex is attached to the DC or macrophage plasma membrane, the peptide is facing the exterior (figure 32.17, step 5). This starts the process of antigen presentation, which is critical because it enables DCs and macrophages to show or "present" microbial antigens to lymphocytes. This in turn triggers activation of the adaptive immune response. ▶| *Recognition of foreignness is critical for a strong defense (section 33.4); T-cell activation (section 33.5)*

Comprehension Check

1. How are intracellular and extracellular microbes recognized by leukocytes? Consider both the microbial and host components.
2. What is the purpose of the respiratory burst that occurs within phagocytic cells? Describe the nature and function of reactive oxygen and nitrogen species.
3. Of the three professional phagocytes, which are capable of antigen presentation and what is the purpose of this process?

32.7 Inflammation Unites All Components of Immunity

After reading this section, you should be able to:

a. Outline the sequence of innate host responses that result in inflammation
b. Distinguish acute and chronic inflammation in terms of the host responses involved in each
c. Construct a concept map relating host cells and processes that remove pathogens

Inflammation (Latin, *inflammatio,* to set on fire) is an important defense reaction to tissue injury, such as that caused by a pathogen or wound. Acute inflammation is the immediate response of the body to injury or cell death. The gross features were described over 2,000 years ago and are still known as the cardinal signs of inflammation: redness (*rubor*), warmth (*calor*), pain (*dolor*), swelling (*tumor*), and altered function (*functio laesa*). While often thought to be a negative event, wounds do not heal without inflammation. *Inflammatory Response*

The goal of an **acute inflammatory response** is to deliver immune cells to the site of injury or infection. It begins when injured tissue cells release chemical signals (chemokines) that activate the inner lining of nearby capillaries (**figure 32.19**). **Selectins,** a family of cell adhesion molecules, are displayed on activated endothelial cells that line blood vessels nearby. Selectins cause circulating neutrophils to slow down and roll along the endothelium, where they encounter pro-inflammatory signals (figure 32.19*b*). These signals activate selectin receptors called **integrins** on the surface of neutrophils. The integrins then attach tightly to the selectins, causing the neutrophils to stop rolling (margination). The neutrophils now undergo dramatic shape changes, squeeze through the endothelial wall (diapedesis) into the interstitial tissue fluid, migrate to the site of injury (extravasation), and attack the pathogen or other cause of the tissue damage. Neutrophils and other leukocytes

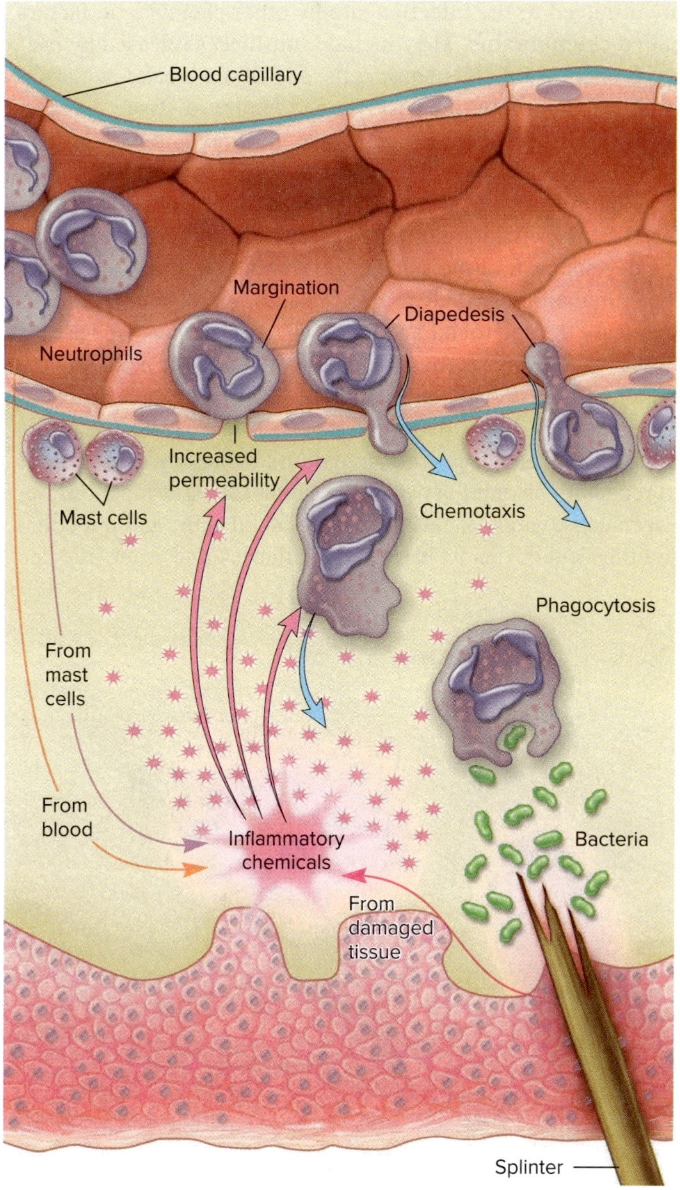

(a)

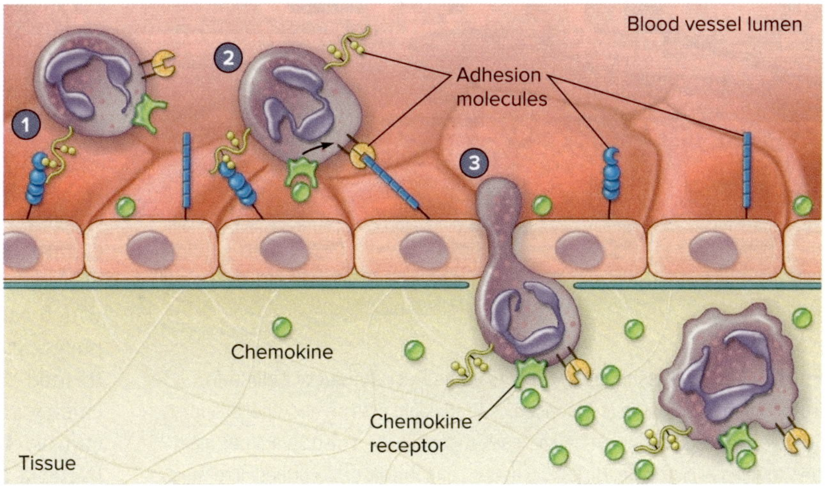

(b)

Figure 32.19 Physiological Events of the Acute Inflammatory Response. (a) At the site of injury (splinter), chemical messengers are released from the damaged tissue, mast cells, and the blood plasma. These inflammatory mediators stimulate neutrophil migration, diapedesis, chemotaxis, and phagocytosis. (b) Neutrophil integrins interact with endothelial selectins (1) to facilitate margination (2) and diapedesis (3).

MICRO INQUIRY *How does the movement of leukocytes in response to chemokines contribute to the cardinal signs of inflammation?*

are attracted to the infection site by other chemotactic factors called **chemotaxins.** They include substances released by bacteria, endothelial cells, mast cells, and tissue breakdown products. Depending on the severity and nature of tissue damage, other types of leukocytes (e.g., lymphocytes, monocytes, and DCs) may follow the neutrophils.

The release of inflammatory mediators from cells in injured tissue sets into motion a cascade of events that results in the development of the signs of inflammation. One response is the stimulation of resident macrophages and distant liver cells to release antimicrobial and acute-phase proteins, respectively. In the local response, these mediators increase the acidity in the surrounding extracellular fluid, which activates the extracellular enzyme **kallikrein** (**figure 32.20**). Cleavage of kallikrein releases the smaller peptide bradykinin. Bradykinin then binds to receptors on capillary walls, opening the tight junctions between cells, which enables fluid, red blood cells, and infection-fighting leukocytes to leave the capillaries and enter infected tissue. Simultaneously, bradykinin binds to mast cells in connective tissue. This activates mast cells by causing an influx of calcium ions, which leads to degranulation and release of preformed mediators such as histamine. Histamine in turn also makes intercellular junctions in the capillary wall wider so that more fluid, leukocytes, kallikrein, and bradykinin move out, causing swelling or edema. Bradykinin also stimulates the production of prostaglandins (PGE_2 and PGF_2) by endothelial cells to promote tissue swelling in the infected area. Prostaglandins also bind free nerve endings, making them fire and start a pain impulse. At the same time, liver cells release additional complement proteins, the iron-binding glycoprotein lactoferrin, and collectins.

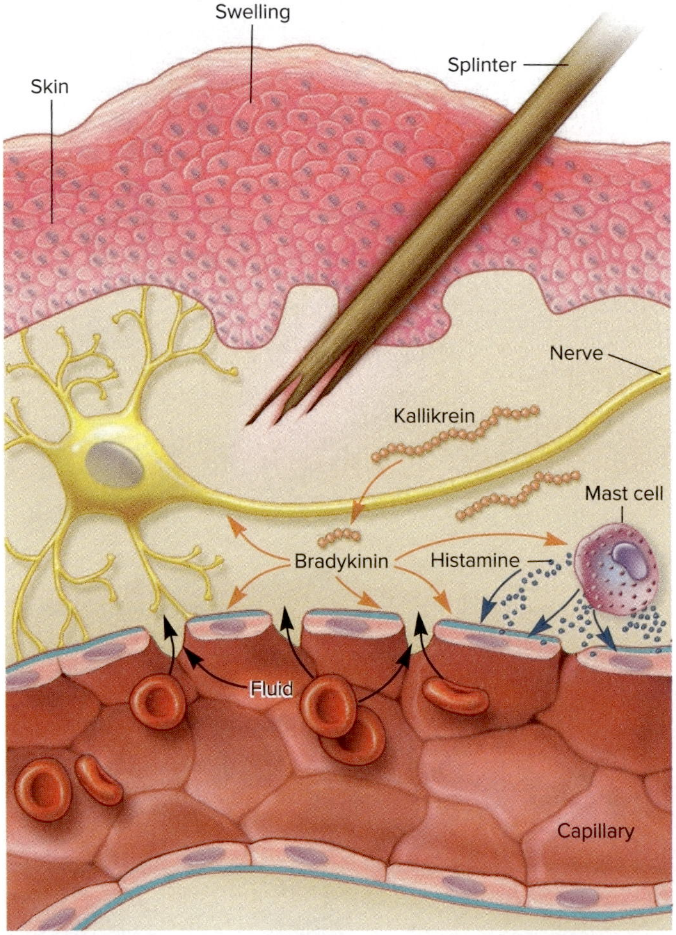

Figure 32.20 Tissue Injury Results in the Recruitment of Kallikrein, from Which Bradykinin Is Released. Bradykinin acts on endothelial and nerve cells, resulting in edema and pain, respectively. It also stimulates mast cells to release histamine. Histamine acts on endothelial cells, further increasing blood leakage into injured tissue sites.

As the inflammatory response continues, more phagocytes are recruited from the bloodstream to the inflamed area. To keep up with this demand, chemicals stimulate the bone marrow to release neutrophils and increase the rate of granulocyte production. Another important step in resolving a local infection is the formation of a fibrin clot to limit the spread of the infectious agents to the bloodstream or other parts of the body.

An acute inflammatory response resolves in days to a few weeks. Prolonged stimulation of the inflammatory pathways leads to progressive changes in cellular response and outcome, known as **chronic inflammation.** Whereas tissue healing and repair mark the end of acute inflammation, chronic inflammation is characterized by infiltration of lymphocytes and macrophages into the affected site and the formation of new connective tissue. This can cause permanent tissue damage. As cycles of cellular infiltration and resolution continue, degradative enzymes from macrophages destroy more underlying tissue than is replaced. Furthermore, if innate responses are unable to protect the host from ongoing infection, chronic tissue injury, or the persistence of poorly degradable materials (sutures, implants, etc.), the body attempts to wall off and isolate the site by forming a **granuloma** (Latin, *granulum,* a small particle; Greek, *oma,* to form). A granuloma is a well-organized mass of macrophages, eosinophils, multinucleated giant cells (two or more cells fused into one large cell), T lymphocytes, fibroblasts, and collagen (**figure 32.21**). Together, the cells and extracellular matrix proteins form a spherical mass that can bind calcium and form a recalcitrant nodule, making the granuloma harder to break down.

Importantly, chronic inflammation can occur as a distinct process without much acute inflammation, as evidenced by the chapter opening story about visceral fat causing chronic inflammation. The persistence of bacteria can also stimulate chronic inflammation. For example, mycobacteria, some of which cause tuberculosis and leprosy, have evolved to survive phagocytosis and reside within macrophages. Granulomas formed around mycobacteria-containing macrophages often calcify and appear on chest X rays, indicating potential tuberculosis. Importantly, the mycobacterial granuloma keeps the bacteria from spreading throughout the body. However, if immune function declines, granuloma integrity will fail and

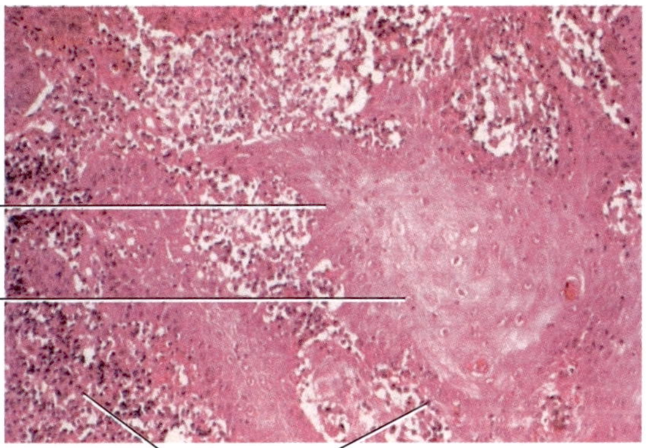

Figure 32.21 Photomicrograph of a Granuloma. A granuloma is a well-organized mass of cells and proteins resulting from a chronic inflammatory response. Note the epithelioid macrophages in the granuloma as contrasted to the large number of normal macrophages surrounding it.
Source: CDC/Dr. Cornelio Arevalo, Venezuela

bacteria will be released into the alveoli, causing active tuberculosis. ▶| *Order* Corynebacteriales *includes important human pathogens (section 23.1); Tuberculosis (section 39.1)*

Comprehension Check

1. What major events occur during an inflammatory reaction, and how do they contribute to pathogen destruction?
2. Compare and contrast acute and chronic inflammation.

Key Concepts

32.1 Immunity Arises from Innate Resistance and Adaptive Defenses

- There are two interdependent components of the immune response to invading microorganisms and foreign material: innate resistance and adaptive immune responses.
- Innate resistance mechanisms offer substantial host defense against any microorganism or foreign material. It has no immunological memory and innate responses are always ready for host defense.
- The adaptive immune response resists a particular foreign agent; moreover, adaptive immune responses improve on repeated exposure to the agent but must be activated by innate immune mechanisms.

32.2 Innate Resistance Starts with Barriers

- Many direct factors (age, nutrition) contribute in some degree to all host-microbe relationships. At times, they favor the establishment of the microorganism; at other times, they provide the host some measure of general defense (**figure 32.1**).
- Physical and mechanical barriers along with host secretions are the host's first line of defense against pathogens. Examples include the skin and mucous membranes and the epithelia of the respiratory, gastrointestinal, and genitourinary systems (**figures 32.2–32.4**).

32.3 Innate Resistance Relies on Chemical Mediators

- Mammalian hosts have specific chemical barriers that help combat the continuous onslaught of pathogens. Examples include cationic peptides, bacteriocins, cytokines, acute-phase proteins, and complement.
- The complement system is composed of a large number of serum proteins that play a major role in an animal's immune response. There are three pathways of complement activation: the alternative, lectin, and classical pathways (**figure 32.6** and **table 32.1**). All pathways lead to opsonization, enhanced inflammation, and development of the membrane attack complex (**figure 32.7**)
- Cytokines are required for regulation of innate and adaptive immune responses. Cytokines have a broad range of actions on eukaryotic cells (**table 32.2**).
- Acute-phase proteins are released from liver cells to bind microbial surfaces. They act as opsonins.

32.4 Each Type of Innate Immune Cell Has a Specific Function

- The cells responsible for both nonspecific and adaptive immunity are white blood cells, which are also called leukocytes (**figure 32.8**). Examples include monocytes and macrophages, dendritic cells, granulocytes, and mast cells.
- There are five kinds of innate lymphoid cells (ILCs): Lymphoid tissue inducing (LTi) cells stimulate the production of secondary lymphoid tissue, ILC-1s have specific functions in host defense in response to intracellular pathogens, ILC-2s respond to parasites, ILC-3s are stimulated by extracellular microbes including normal microbiota, and natural killer cells are ILCs that destroy cancer cells and cells infected with microorganisms (**figures 32.10–32.12**).

32.5 Organs and Tissues of the Immune System Are Sites of Host Defense

- The secondary lymphoid organs and tissues serve as areas where lymphocytes may encounter and bind antigens, then proliferate and differentiate into fully mature, antigen-specific effector cells. The spleen is a secondary lymphoid organ, and the lymph nodes and mucosal-associated tissues (GALT and SALT) are the secondary lymphoid tissue (**figures 32.13–32.15**).

32.6 Phagocytosis Destroys Invaders

- Phagocytes use pattern recognition receptors (PRRs) to detect microbial-associated molecular patterns (MAMPs) and damage-associated molecular patterns (DAMPs) on microorganisms and host cells. Membrane-bound and cytoplasmic PRRs transduce signals via the same molecular pathway to activate the inflammatory response. Toll-like receptors and NOD-like receptors are distinct classes of pathogen recognition receptors (**figure 32.16**).
- Phagocytosis involves the recognition, ingestion, and destruction of pathogens by lysosomal enzymes, reactive oxygen species, defensins, reactive nitrogen species, and metallic ions. A more conserved method of detection and digestion, called autophagy, uses intracellular membranes and lysosomes to kill and degrade intracellular pathogens (**figure 32.17** and **32.18**).

32.7 Inflammation Unites All Components of Immunity

- Inflammation is a nonspecific defense mechanism to tissue injury and infection.
- Inflammation can be either acute or chronic (**figures 32.19–32.21**).

Active Learning

1. Develop a list of four attributes a bacterium would need to evade or subvert innate immune responses. For each attribute, suggest a mechanism that might have evolved to accomplish this trait.
2. Age-related macular degeneration (AMD) is the principal cause of blindness in industrial nations. Vision is lost as a result of destruction of the macula (i.e., the central region of the retina). Around 2005, it was noted that mutation of the gene encoding complement factor H (*CFH*) led to a 7.4-fold increased risk of AMD development. Two hypotheses were suggested to explain how the *CFH* mutation might contribute to AMD. In the first, it is proposed that the combination of the mutant factor H and a triggering event, such as an infection, gives rise to AMD. The second predicts that chronic inflammation contributes to the development of AMD. Why would a defect in factor H lead to an aberrant response to infection? Why might chronic inflammation cause cellular damage? Assuming you had an excellent animal model for AMD, discuss how you could test one or both hypotheses.
3. Think about the last time you had a cold, flu, skin infection, or a stomach "bug." Which innate barrier was breached by the infectious agent? Refer to the cytokines listed in table 32.2; which cytokines were part of your immune response? Consider the effect the release of these cytokines may have had on how healthy (or unhealthy) you felt at the time.
4. TLR-10 is the only human toll-like receptor for which the ligand is unknown. Evidence now suggests that it may function very differently from other TLRs. TLR-10 receptor activity appears to suppress pro-inflammatory responses and monocyte differentiation into fully functional dendritic cells and macrophages. Discuss why the host cell may have evolved a strategy to downregulate innate immunity. Refer to figure 32.16 and suggest how TLR-10 receptor activity could diminish a pro-inflammatory response.

   Read the original paper: Hess, N. J., et al. 2016. TLR10 is a negative regulator of human monocyte innate immune responses. *J. Immunol.* 196(1):S126.3.

33 Adaptive Immunity

©AVAVA/iStockphoto/Getty Images

Killing Cancer, Immunologically

This year, about 90,000 people in the United States will be diagnosed with malignant melanoma. About 10,000 people will die of this cancer, which has been notoriously hard to treat. It is one of the most common cancers in people under 30 years of age. But these grim truths are beginning to be countered by reason for optimism. The field of immunology is changing the way we regard, and treat, melanoma and other cancers.

The idea that the immune system could mediate cancer cell growth is not new. For many years, it has been understood that chronic inflammation, as occurs with hepatitis B and C infections, drives the development of liver cancer. And for over 35 years, we have known that the immune system can be harnessed to treat superficial bladder cancer. Treatment of this cancer involves instilling the noninfectious vaccine strain of *Mycobacterium tuberculosis* into the patient's bladder to stimulate a robust local immune response that eliminates cancer cells.

In the last 15 years however, scientists and clinicians have come to understand much more about both cancer cells and other types of cells in the tumor microenvironment. Importantly, they have described how these cells interact with the immune system. This in turn has led to the development of two different approaches with real promise for how cancer is treated, and cured. The goal of both is to use the patient's adaptive immune response to specifically target, and kill, cancer cells.

The first approach is termed immunomodulation, and it changes how the immune system responds to cancerous cells. Malignant cells display unique proteins on their surfaces called neoantigens. Neoantigen simply means new antigen; thus it has not previously been detected by the patient's immune cells, and so triggers an immune response. If the response is vigorous enough, cancer cells bearing the neoantigen are killed. The goal of immunomodulatory cancer therapy is to ensure this kind of robust immune response occurs. The general approach is to design monoclonal antibodies (mAbs) that bind to immune cell receptors or other regulatory proteins to either prevent T-cell inhibition function or incite T-cell stimulation. In the treatment of melanoma, the combination of two mAbs that prevent T-cell inhibition has proven successful in prolonging the lives in many but not all patients, and some with late-stage disease have even been cured.

The second approach is called adoptive cellular therapy (ACT). Here a patient's T cells are engineered to express *c*himeric *a*ntibody *r*eceptors (CARs). As we discuss in this chapter, antibodies bind antigen with high specificity and one group of T cells, called cytotoxic T lymphocytes (CTLs), kills host cells expressing neoantigens. The idea behind CARs is to construct superkiller-CTLs. To do this, the domain of the T-cell receptor (TCR) that normally binds antigen is replaced with an antibody domain, while the part of the TCR that spans the membrane and instructs the T cell to kill remains intact. The antibody replacement domain has astonishingly high affintity and avidity for the cancer cell neoantigen, so when CARs bind neoantigens the CTLs truly become superkillers. This approach has been tremendously successful in treating B-cell acute lymphocytic lymphoma and leukemia. By targeting a single antigen, complete remission can be achieved in 70 to 95% of patients who otherwise would have only a 7% chance of living another 5 years. Now the challenge is to design modifications of these promising immunotherapies to treat other cancers.

In chapter 32 we learned that the innate resistance system responds to a foreign substance in the same manner and to the same magnitude each time it encounters an antigen. We also learned that if the innate immune response is not sufficient to contain a foreign invader, it activates an adaptive immune response. We now continue our discussion of the immune response by describing adaptive immunity.

Readiness Check:

Based on what you have learned previously, you should be able to:

- ✓ Explain basic eukaryotic cell biology (sections 5.1–5.7)
- ✓ Diagram the flow of information in eukaryotes from DNA to protein and regulation of this flow (sections 15.2–15.4)
- ✓ Discuss the importance of the host complement response (section 32.3)
- ✓ Describe the function of host leukocytes (section 32.4)
- ✓ Outline the events that occur during phagocytosis and inflammation (sections 32.6 and 32.7, respectively)

33.1 Adaptive Immunity Relies on Recognition and Memory

After reading this section, you should be able to:

a. Contrast host innate resistance with adaptive immunity
b. Define the two branches of adaptive immunity by their cell types and general function

Although the separation of innate and adaptive functions of immunity is artificial, it allows us to distinguish the preprogrammed, innate responses from the more diverse and long-lasting adaptive responses. In reality the two sides of this coin we call immunity represent a single, fully integrated host defense system.

The adaptive immune system of animals has three major functions: (1) recognize anything foreign to the body ("non-self"); (2) defend the host from this foreign material; and (3) remember the foreign invader. Like innate immunity, adaptive immunity must differentiate between self and non-self, but in contrast to innate immunity, recognition of non-self antigens by adaptive cells is highly specific. After recognizing non-self, the adaptive system responds by activating and amplifying **T and B lymphocytes.**

Most B and some types of T lymphocytes that respond to an antigen are induced to elicit an **effector response.** A successful effector response either eliminates the foreign material or renders it harmless to the host. As a consequence of initial activation, some B and T lymphocytes enter a state of limited activity and are referred to as **memory cells** because they will "remember" the foreign material that initially activated them. If the same substance is encountered at a later time, B and T memory cells are already programmed to mount a more intense and rapid response to eliminate the non-self material to once again protect the host.

Thus four characteristics distinguish adaptive immunity:

1. **Discrimination between self and non-self.** While both innate and adaptive immune cells must discriminate between self and non-self, the adaptive immune system responds much more selectively to non-self.
2. **Specificity.** Recognition of non-self antigen is directed against one particular pathogen or foreign substance (among trillions); the immunity to one pathogen or substance usually does not confer immunity to others.
3. **Diversity.** To mount these very specific responses, the system is able to generate an enormous diversity of cellular receptors and antibodies that recognize trillions of unique foreign substances.
4. **Memory.** When reexposed to the same pathogen or substance, the host reacts so quickly there is usually no noticeable illness. By contrast, the reaction time for innate defenses is just as long for later exposures to a given antigen as it was for the initial one.

Two branches or arms of adaptive immunity are recognized: humoral immunity and cellular (cell-mediated) immunity. B cells constitute **humoral (antibody-mediated) immunity.** The older term *humoral* reflects the fact that antibodies were discovered in the blood plasma (serum), so the term refers to the fluids or "humors" of the body. Circulating antibodies bind to microorganisms, toxins, and extracellular viruses, neutralizing them or "tagging" them for destruction by phagocytes and other mechanisms described in section 33.8. T cells constitute **cellular (cell-mediated) immunity.** Broadly speaking, there are two types of T cells. T lymphocytes that differentiate into cytotoxic T cells (CTLs) directly attack target cells infected with intracellular pathogens such as viruses as well as cancer cells. The other type of T cells, T helper cells, are so named because they can direct CTLs to target cell lysis, interact wih B cells to stimulate antibody production, as well as up- or downregulate innate defenses, such as phagocytosis and inflammation. Because the activity of the acquired immune response is so potent, it is imperative that T and B cells consistently discriminate between self and non-self with great accuracy.

33.2 Antigens Elicit Immunity

After reading this section, you should be able to:

a. Predict the types of molecules that can serve as antigens
b. Describe the relationship between haptens and antigens

While cells of the innate arm of immunity are limited to detecting molecular patterns as non-self, the adaptive arm of immunity uses a more sophisticated mechanism to recognize foreignness. The term **antigen** refers to any self or non-self substance that elicits an immune response and is therefore also termed immunogen (*immu*nity *gen*erating). Antigens include microbe-associated molecular patterns (MAMPS, described in chapter 32) as well as other molecules, including proteins, nucleoproteins, polysaccharides, and some glycolipids (**figure 33.1**). The ability of a molecule to function as an antigen depends on its size, structural complexity, chemical nature, and degree of foreignness to the host. Most antigens are large, complex molecules with a molecular mass generally greater than 10,000 daltons (Da).

Each antigen can have several **antigenic determinant sites,** or **epitopes** (figure 33.1). Epitopes are regions or sites of the antigen that bind to a specific antibody. Chemically, epitopes include sugars, organic acids and bases, amino acid side chains, hydrocarbons, and aromatic groups. The number of epitopes on the surface of an antigen is its **valence.** The valence determines the number of antibody molecules that can combine with the antigen at one time. If one determinant site is present, the antigen is monovalent. Most antigens, however, have more than one copy of the same epitope and so are multivalent. Multivalent antigens generally elicit a stronger immune response than monovalent antigens. Thus structurally repetitive molecules are multivalent and said to be strongly immunogenic. **Antibody affinity** relates to the strength with which an antibody binds to its antigen at a given antigen-binding site. Affinity tends to increase during the course of an

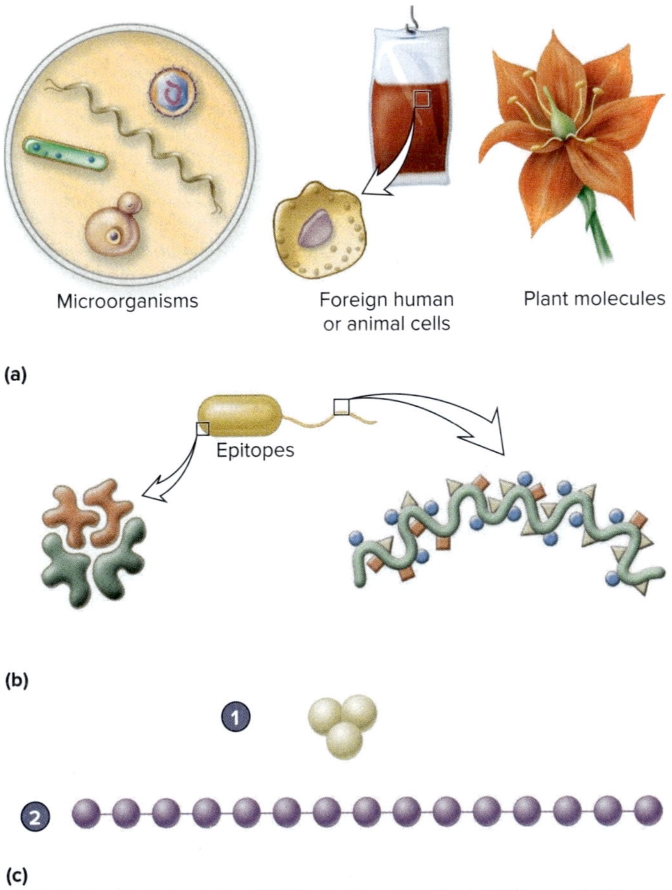

Figure 33.1 Antigen Characteristics Are Numerous and Diverse. (a) Whole cells, viruses, and some plant molecules are immunogenic. (b) The specific regions of an antigen recognized by antibodies are called epitopes. (c) Small molecules that are poorly immunogenic include (1) small molecules (haptens) not attached to a carrier molecule. (2) By contrast, large repetitive molecules such as flagellin and peptidoglycan tend to be very immunogenic.

MICRO INQUIRY *What does the term* valence *mean and how does an antigen's valence influence an immune response?*

immune response (section 33.7). The **avidity** of an antibody relates to its overall ability to bind antigen at all antigen-binding sites.

Haptens

As shown in figure 33.1*c,* some small antigens may not trigger an immune response—that is, they are not immunogenic. However, such molecules may become immunogenic if they become attached to a larger host protein, called a carrier protein. Small molecules that become immunogenic when combined with a carrier protein are called **haptens** (Latin *haptein,* to grasp). One example of a hapten is penicillin. By itself, penicillin is a small molecule that is not antigenic. However, when combined with certain serum proteins in sensitive individuals, it becomes immunogenic, activates lymphocytes, and can initiate a severe allergic immune reaction. As in the case of penicillin, the hapten is often not a protein and lacks a repetitive structure.

Comprehension Check

1. What structural features make an antigen immunogenic?
2. What is an antigenic determinant site or epitope?
3. What are the distinguishing features of a hapten? How does it become immunogenic?

33.3 Adaptive Immunity Can Be Earned or Borrowed

After reading this section, you should be able to:

a. Compare and contrast methods by which immunity occurs by natural and artificial means
b. Differentiate between the active and passive forms of natural and artificial immunity

So far, we have learned that adaptive immunity is very specific, but must be activated before enacting its defense. Humans evolved the natural ability to mount an adaptive immune response, but we now also use technology to artificially stimulate such a response to protect the host. Here we discuss these two types of adaptive immunity: natural and artificial, and the means by which they can be either passively or actively developed (**figure 33.2**).

Naturally Acquired Immunity

Naturally acquired active immunity occurs when an individual's immune system contacts an antigen such as a pathogen that causes an infection. The adaptive immune system responds by producing antibodies and activated T lymphocytes that neutralize or destroy the pathogen. The immunity produced can be either lifelong, as with measles or chickenpox, or last for only a few years, as with influenza. **Naturally acquired passive immunity** involves the transfer of antibodies from mother to child. For example, some of a pregnant woman's antibodies pass across the placenta to her fetus. If the woman is immune to diseases such as polio or diphtheria, this placental transfer also gives her fetus and newborn temporary immunity to these diseases. Certain other antibodies can pass from a mother to her offspring when the child is breastfed. These maternal antibodies are essential for providing immunity to the newborn for the first few weeks or months of life, as the child's own immune system matures. Naturally acquired passive immunity generally lasts only a short time (about 6 months).

Artificially Acquired Immunity

Artificially acquired active immunity results upon vaccination; that is, the host is intentionally exposed to a foreign material to stimulate the formation of antibodies and activated lymphocytes. A **vaccine** consists of antigens that can be prepared in several different ways including killed microorganisms; living, weakened (attenuated) microorganisms; genetically engineered organisms or their products; or inactivated bacterial toxins (toxoids). Vaccines and immunizations are discussed in detail in chapter 36.

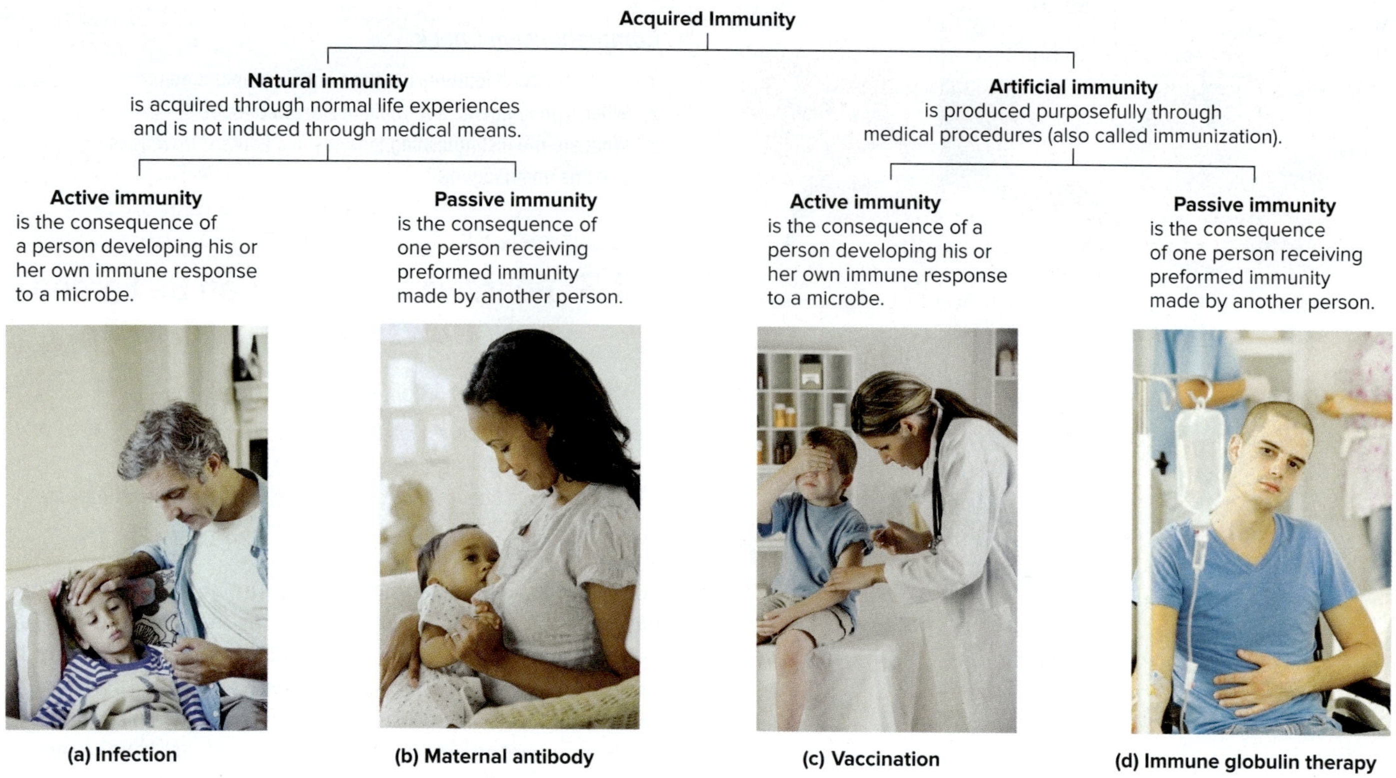

Figure 33.2 Immunity Can Be Acquired by Various Means.
(a) ©Paul Bradbury/Getty Images; (b) ©JGI/Blend Images LLC; (c) ©Phil Nealey/Getty Images; (d) ©Miodrag Gajic/Getty Images

Artificially acquired passive immunity results when antibodies or lymphocytes produced by one host are introduced into another. Although this type of immunity is immediate, it is short-lived, lasting only a few weeks to a few months. An example of artificially acquired passive immunity would be botulinum antitoxin produced in a horse and given to a human suffering from botulism food poisoning.

Comprehension Check

1. What distinguishes natural from artificial immunity?
2. What are ways that active immunity is different from passive immunity?
3. Of the four types of acquired immunity, which do you think your immune system has undergone?

33.4 Recognition of Foreignness Is Critical for a Strong Defense

After reading this section, you should be able to:

a. Diagram the host cell receptors that distinguish self from non-self
b. Compare the processes by which MHC class I and class II receptors recognize foreignness
c. Identify cells that function as antigen-presenting cells (APCs)

Distinguishing between self and non-self is essential in maintaining host integrity. This distinction must be highly specific and selective not only so invading pathogens are eliminated but also to prevent destruction of host tissue. An important extension of the capacity to distinguish self from non-self is seen in the modern use of tissue transplantation, where organs, tissues, and cells from unrelated persons are matched to the self-recognition markers on the recipient's cells. While entire textbooks are devoted to this subject, we discuss those aspects of foreignness recognition that assist us in understanding why and how lymphocytes respond.

It is not enough for the immune system to recognize foreign antigens as non-self; there must also be a mechanism that allows the immune system to recognize host cells as self. To accomplish this, each host cell expresses surface proteins that mark it as a resident of that host. This self signal must be broadcast to all cells of the immune system so that a selective defense is raised only against non-self antigens. This system has evolved in mammals and is called the major histocompatibility complex.

The Major Histocompatibility Complex

The ***m**ajor **h**istocompatibility **c**omplex* (**MHC**) is a collection of genes encoding proteins that enable the host to distinguish between self and non-self. The term *histocompatibility* is derived from the Greek word for tissue (*histo*) and the ability to get along (*compatibility*). Human MHC is located on chromosome 6 and is called the ***h**uman **l**eukocyte **a**ntigen* (**HLA**) complex. HLA proteins can be divided into three classes: Class I are found on all types of nucleated body cells; class II appear only on cells that can process antigens and present them to T lymphocytes. As we will learn, macrophages, dendritic cells (DCs), and B cells have the ability to present antigen in this way. Finally, class III molecules include various secreted proteins that have immune functions. Class III molecules are not required for the discrimination between self and non-self and will not be discussed further.

Class I MHC molecules include HLA types A, B, and C, and identify all nucleated cells of the body as "self." They consist of a complex of two protein chains, a larger alpha chain (45,000 Da), and a smaller chain called β_2-microglobulin (12,000 Da). As shown in **figure 33.3*a***, only the alpha chain spans the plasma membrane. Both chains interact to form a pocket that projects from the cell surface. This pocket can bind either self antigen, thereby marking the cell as a host cell, or it can bind a peptide extracted from an intracellular pathogen, signaling the immune system that it is an infected host cell.

HLA proteins differ among individuals; the closer two people are related, the more similar are their HLA molecules. This is because many forms of each HLA gene exist (i.e., HLA genes are polymorphic). Throughout evolution, multiple alleles of each gene have arisen by gene mutation, recombination, and other mechanisms. Variation among individuals is further amplified because HLA genes are codominant; thus each person expresses alleles from both mom and dad at each A, B, and C locus. Thus a person produces six different class I MHC proteins. Because MHC class I proteins are found on all nucleated cells (i.e., only red blood cells lack them), they stimulate an immune response when cells from one host are introduced into another host with different class I molecules. This is the basis for tissue typing when a patient is being prepared for an organ or bone marrow transplant.

Class II MHC molecules are produced only by certain immune cells, such as DCs, activated macrophages, mature B cells, and some innate lymphocyte cells (ILCs). As we will learn, class II molecules on DCs and B cells are key to initiating an adaptive immune response. Like class I MHC proteins, class II

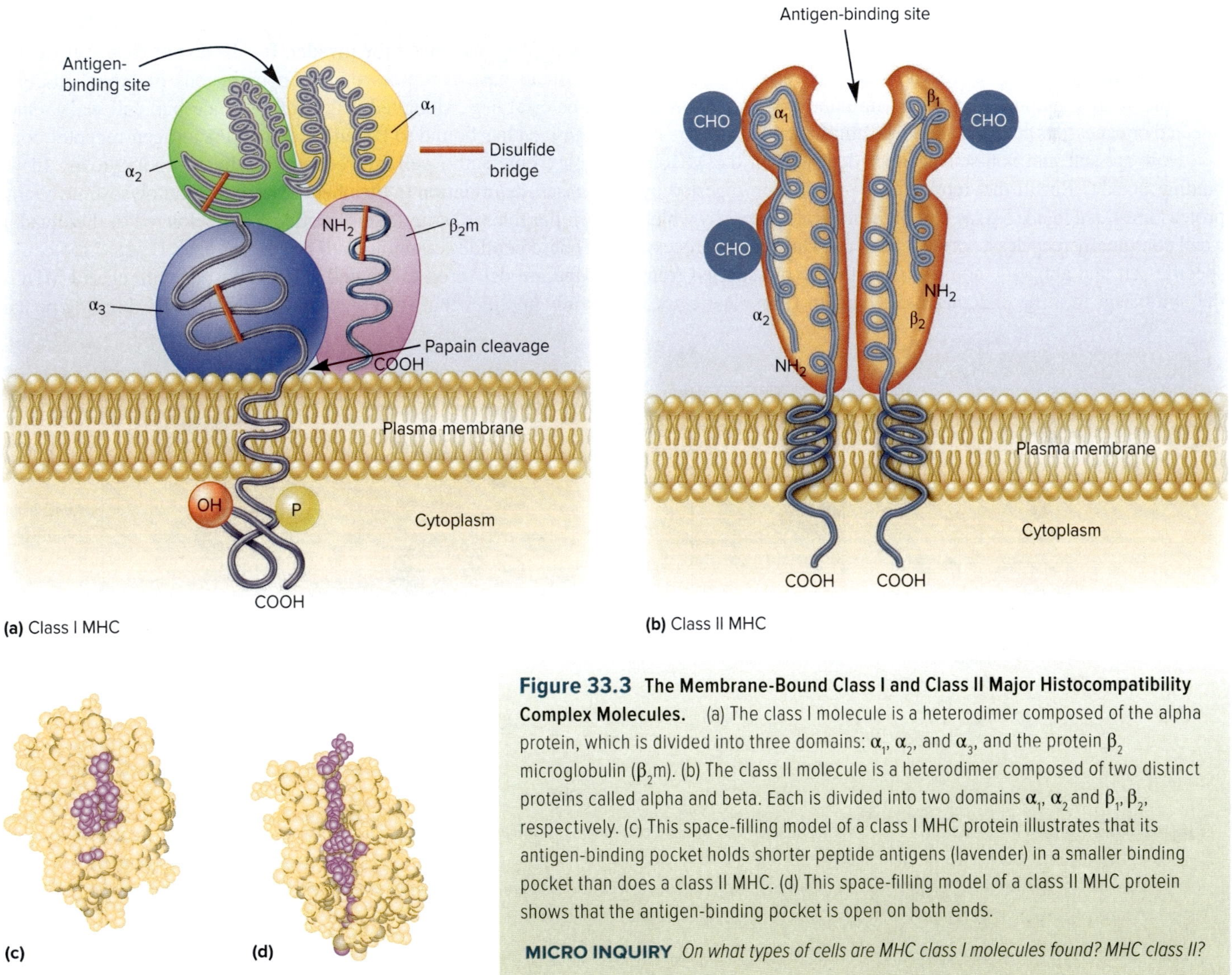

Figure 33.3 The Membrane-Bound Class I and Class II Major Histocompatibility Complex Molecules. (a) The class I molecule is a heterodimer composed of the alpha protein, which is divided into three domains: α_1, α_2, and α_3, and the protein β_2 microglobulin (β_2m). (b) The class II molecule is a heterodimer composed of two distinct proteins called alpha and beta. Each is divided into two domains α_1, α_2 and β_1, β_2, respectively. (c) This space-filling model of a class I MHC protein illustrates that its antigen-binding pocket holds shorter peptide antigens (lavender) in a smaller binding pocket than does a class II MHC. (d) This space-filling model of a class II MHC protein shows that the antigen-binding pocket is open on both ends.

MICRO INQUIRY *On what types of cells are MHC class I molecules found? MHC class II?*

MHC molecules are transmembrane proteins consisting of two distinct chains (figure 33.3*b*). However, MHC class II molecules fold to form a three-dimensional antigen-binding pocket into which only a non-self peptide fragment can be inserted for presentation to other cells of the immune system. MHC class II molecules also differ from MHC class I because they have a much deeper groove into which a longer peptide derived from a foreign substance can bind (figure 33.3*c,d*). The important differences between the two classes are therefore that MHC I molecules have smaller binding pockets capable of tethering self and non-self peptides, while MHC II molecules bind only larger, non-self peptides. As discussed in section 33.5, non-self peptide antigen fragments in the MHC groove must be present to activate T cells, which in turn activate other immune cells.

The mechanism by which peptides become bound to MHC molecules prior to the translocation of the MHCs to the host cell surface is known as **antigen processing** (**figure 33.4**). Class I molecules bind to peptides that originate in the cytoplasm. In healthy cells, self peptide obtained from the cytosol is bound as a way to communicate to the immune system, "I am a healthy self cell—please leave me alone!" However, if a host cell is infected with an intracellular pathogen, such as a virus, or the cell has become cancerous, a foreign peptide is placed in the binding groove of the MHC I molecule. This communicates to the immune system, "I am an infected or cancerous cell, so I should be eliminated!"

How are self and non-self peptides deposited in the MHC I binding pocket? Recall that intracellular proteins are digested by proteasomes (*see figure 5.8*) as part of the natural process by which a cell continually recycles its protein contents. During this process, specific self and antigenic non-self peptides are transported from the cytoplasm into the endoplasmic reticulum (ER). The class I MHC alpha chain and the β_2-microglobulin associate within the lumen of the ER. Only peptides that fit perfectly within the binding cleft of one of the six different MHC I molecules are attached. This process, known as **endogenous antigen processing,** enables the host cell to present peptide antigen to a subset of T cells called $CD8^+$, or cytotoxic T lymphocytes (CTLs). $CD8^+$ T cells bear a receptor specific for class I MHC molecules presenting antigen. If the host cell is presenting self antigen in its MHC I molecules, the T cells "ignore" it. However, if the MHC class I molecule is presenting a non-self antigen (e.g., from an intracellular pathogen or a new peptide found only in cancerous cells), these T cells bind and kill the host cell (section 33.5).

Class II MHC molecules differ from class I MHC molecules because they bind fragments that initially come from antigens outside the cell, thus they undergo **exogenous antigen processing.** Only foreign particles (e.g., bacteria, viruses, toxins) that have been taken up by endo- or phagocytosis can be bound to MHC class II molecules. Immune cells that place non-self antigen on MHC class II molecules are called **antigen-presenting cells (APCs)**, and these include macrophages, DCs, and B cells. Exogenous antigen processing by macrophages and DCs starts with phagocytosis of the invader. B cells do not phagocytose, but instead have receptors that bind pathogens or other antigenic material (e.g., sloughed fragments of microbial cell wall). Once antigen has bound to B-cell receptors, the antigen-receptor complex enters the cell by receptor-mediated endocytosis. In all cases, degradation in the phagolysosome or autolysosome results in the release of antigenic peptides that belonged to the invader. These peptides combine with preformed class II MHC molecules and are delivered to the cell surface. Just as with class I MHCs, only peptides that fit snugly in the class II MHC binding pocket

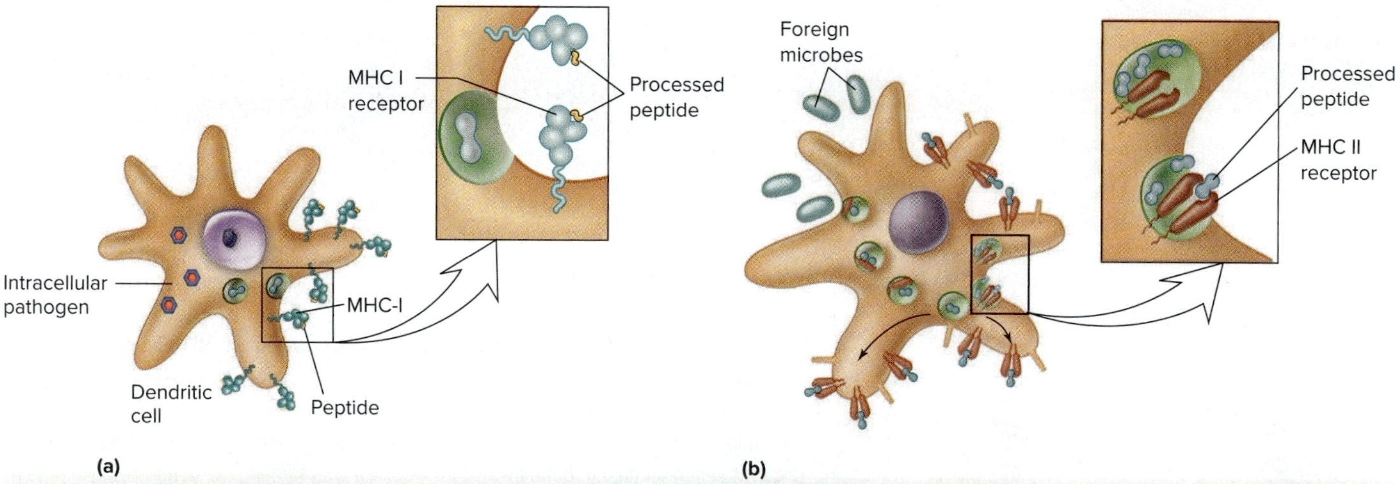

Figure 33.4 Antigen Presentation. (a) Antigens arising from within a cell (including self antigen and those from intracellular pathogens) are degraded by proteasomes and inserted into the antigen-binding pocket of the MHC class I proteins for presentation on the cell surface to a variety of immune cells. (b) Antigens arising from outside a phagocytic cell are captured by phagocytosis and degraded by the lysosomal process and inserted into the antigen-binding pocket of MHC class II proteins for presentation on the cell surface to T lymphocytes.

MICRO INQUIRY *What cells recognize and bind to MHC class I molecules that carry antigens? What about MHC class II molecules with antigen?*

are bound. This peptide can now be recognized by T cells called $CD4^+$ T-helper cells. DCs are particularly adept at presenting foreign peptides to T cells and stimulating them to become activated T cells. Unlike $CD8^+$ T cells, $CD4^+$ T cells do not directly kill target cells. Instead, they secrete cytokines to orchestrate an integrated innate and adaptive immune response. ◀ *Several cytoplasmic membranous organelles function in the secretory and endocytic pathways (section 5.4); Cytokines are chemical messages between cells (section 32.3); Phagocytosis destroys invaders (section 32.6)*

Comprehension Check

1. Construct a list of the differences between MHC class I and class II cells.
2. Why is it important that MHC class I molecules bind and present self antigen?
3. Compare and contrast the mechanism by which MHC class I and class II molecules bind antigen.

33.5 T Cells Are Critical for Immune Function

After reading this section, you should be able to:

a. Discuss the difference between T-cell development and T-cell activation
b. Contrast the biological functions of T-cell subsets
c. Describe T-cell receptor structure and function
d. Diagram antigen presentation within MHC receptors interacting with the appropriate T-cell type
e. Compare and contrast the activities of T_H1, T_H2, T_H17, Treg cells, and CTLs

Adaptive immunity involves the coordinated activity of both B and T cells. However, while B cells have a single function, to produce antibody, there are multiple subsets of T cells that work together to initiate, orchestrate, and carry out an adaptive immune response. We therefore begin our discussion by learning about T cells.

T-Cell Development

Common lymphoid progenitor (CLP) cells from the bone marrow migrate to the thymus, a primary lymphoid organ located above the heart (**figure 33.5**). The sole purpose of the thymus is to serve as a site for T-cell development. During a process known as **thymic selection,** two major events occur. One mechanism determines what kind of T cell any given immature cell will become. This is defined first by the structure of the **T-cell receptor** (**TCR**) and then by the structure of the T-cell coreceptor. The TCR consists of two interacting proteins. Most TCRs are made up of an α and a β protein, and are called **αβ T cells.** A small number of T cells bear TCRs made of γ and δ proteins. These are called **γδ T cells** and they leave the thymus shortly after development and migrate to the lymphoid tissue in the skin and mucous membranes. They do not become further selected on the basis of coreceptor structure. By contrast, the vast majority of cells are αβ T cells, which remain in the thymus where the second type of receptor specificity—that of coreceptor—is determined. T-cell coreceptors are members of the family of **cluster of differentiation (CD)** molecules. Cells of the immune system bear many different CDs on their cell surface; we are concerned with CD4 and CD8. Mature T lymphocytes will bear one but not both CDs. T cells selected to bear only the CD4 coreceptor interact only with MHC class II molecules on APCs (i.e., DCs, macrophages, and B cells). Other αβ T cells are selected to become CD8 cells; the CD8 coreceptor mediates interaction between the TCR and MHC class I molecules found on all host cells except red blood cells. Importantly, before leaving the thymus, all T cells are screened by a mechanism that forces any T cell that strongly recognizes self-antigen to undergo cell death (apoptosis). The entire process of T-cell development removes about 98% of the cells that enter the thymus as CLP cells. T cells that leave the thymus and migrate to lymph nodes and the spleen are called mature but naïve. As we will learn, this reflects the fact that they have fully developed (i.e., have matured) but cannot respond to antigen (i.e., are naïve) until instructed by antigen presented on a MHC molecule. By analogy, consider a 19-year-old person who has fully developed (matured) but still requires instruction (is naïve) before launching a career.

T-cell receptors (TCRs) play an essential role in both T-cell development and activation. TCRs evolved to detect antigen fragments presented by MHC molecules on the surface of other cells. Each TCR is composed of two parts: a heterodimeric polypeptide receptor, and six accessory polypeptides, collectively referred to as CD3 (**figure 33.6*a***). Each heterodimer forms a transmembrane receptor stabilized by disulfide bonds. The antigen recognition sites of TCRs form when the extracellular regions of the α and β or γ and δ chains interact to create a three-dimensional pocket. The amino acid sequence that contributes to each pocket is variable, resulting in the ability to bind only the specific antigen fragment that perfectly fits. In this way, TCR binding pockets are similar to MHC antigen-binding sites. However, there is one very important distinction between MHC molecules and TCRs. Recall that each person inherits and expresses six different alleles for each MHC class. By contrast, during T-cell maturation in the thymus, TCRs undergo a series of genetic events that make the binding pocket highly variable among T cells (section 33.5). Ultimately, this allows each T-cell receptor to recognize a different, specific antigenic peptide when it is bound to a MHC molecule. When a mature T cell is able to bind to a peptide in a MHC molecule, it is no longer naïve because it now "knows" the identity of the antigenic peptide that fits its particular binding cleft and its career path has been determined. This is the first step in T-cell activation, which is discussed next.

T-Cell Activation

Activation of a naïve T cell begins with antigen presentation in lymphoid tissue. In this process, the antigen held by the MHC of the

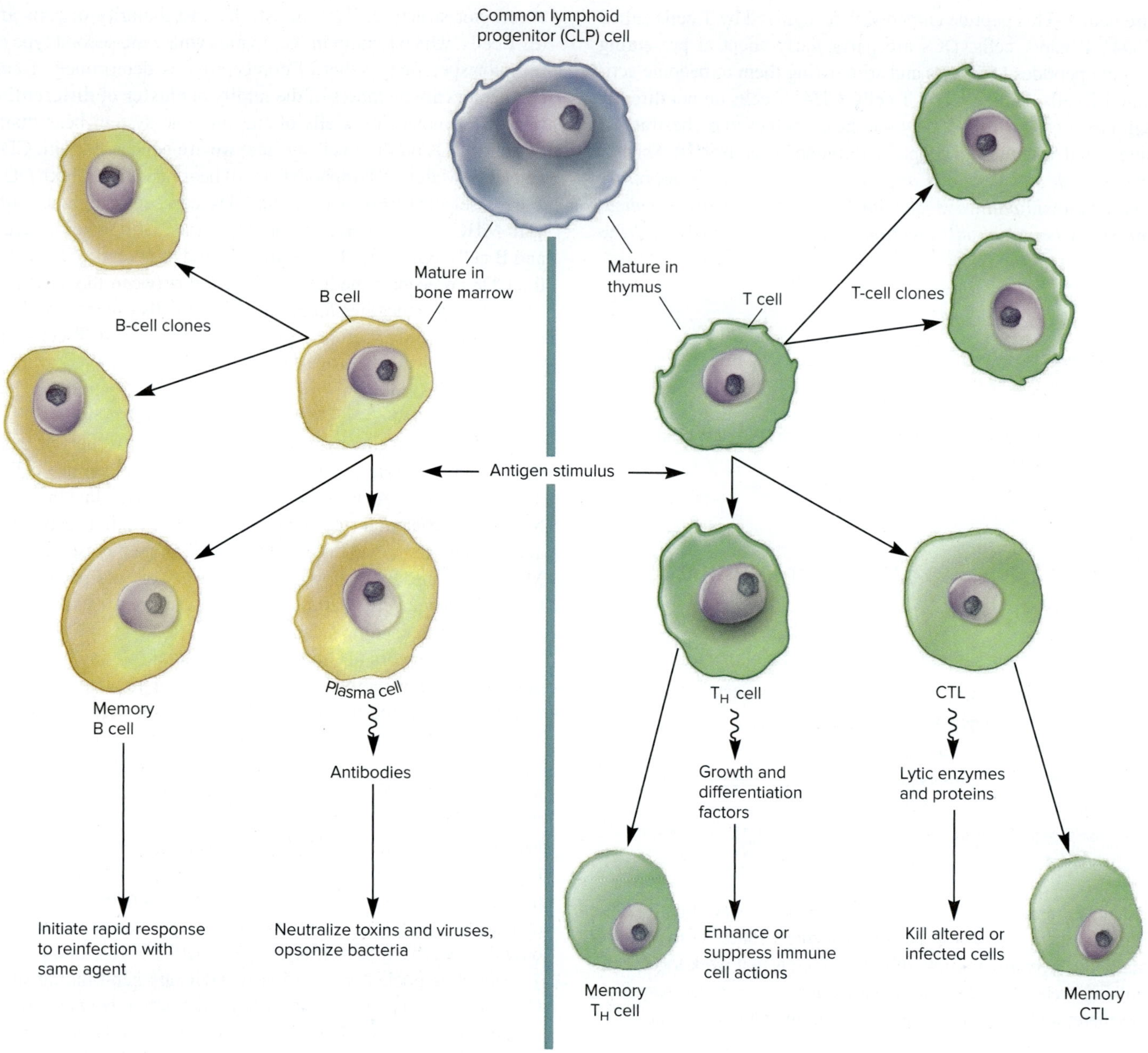

Figure 33.5 Development and Function of B and T Lymphocytes. Common lymphoid progenitor (CLP) cells in the bone marrow develop into B cells, which then migrate to secondary lymphoid tissue and the spleen where they can be activated to multiply (B-cell clones) and differentiate into antibody-producing plasma cells or long-lasting memory cells. CLP cells in the thymus differentiate into T cells that become two main types of cells: T-helper cells (T_H cells) and cytotoxic T lymphocytes (CTLs). Mature, naïve T cells leave the thymus and migrate to secondary lymphoid tissues where, upon antigenic stimulation, they replicate and become activated T_H cells and CTLs. T_H cells mediate the responses of other immune cells while CTLs kill host cells infected with intracellular pathogens and malignant host cells. Like B cells, T_H cells and CTLs can also differentiate into memory cells.

antigen-presenting cell (APC) is recognized by the TCR of the T cell because the peptide binds snugly to the TCR binding pocket. This TCR-peptide-MHC interaction forms a bridge between the APC and T cell called an **immune synapse** (figure 33.6*b*). Immune synapse formation causes a three-dimensional change in the TCR that is communicated to the T-cell nucleus so the appropriate genes can be expressed. However, as inspection of figure 33.6*b* shows, very little of the TCR extends into the cytoplasm. Therefore, to transmit the message that antigen has been detected, the cytoplasmic tails of the receptors associate with CD3 accessory polypeptides to transduce the intracellular signal. However, this is only the first step in T-cell activation. Full T-cell activation occurs only when two additional signals are communicated along with the antigen. Thus all naïve T cells require three signals to be activated.

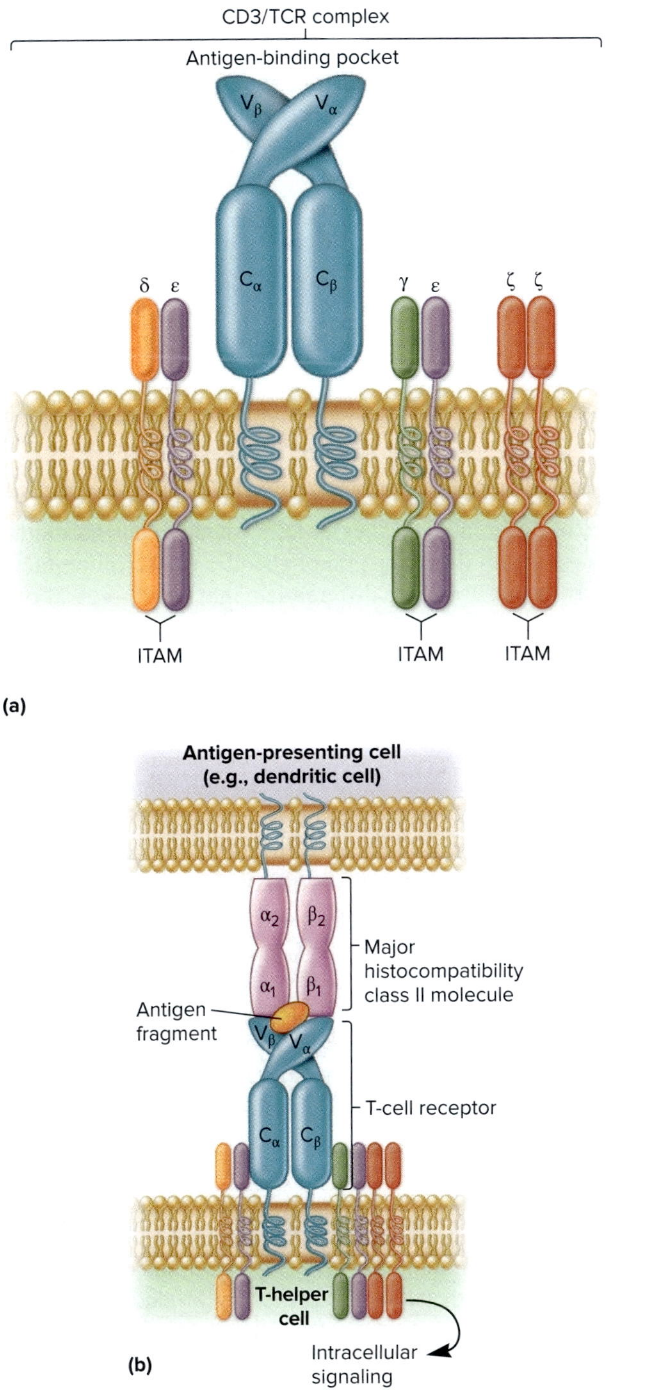

Figure 33.6 T-Cell Receptor and Immune Synapse Formation Activation. (a) The overall structure of the antigen receptor complex on a T-cell plasma membrane. Short, transmembrane CD3 accessory proteins have immunoreceptor tyrosine-based activation motifs (ITAMs), which upon immune synapse formation start the intracellular signaling that will result in T-cell activation. (b) An APC begins the activation process by displaying an antigenic peptide in the binding cleft of its MHC. A T cell is activated after the variable region of its receptor (designated V_α and V_β) also binds the antigen fragment in the MHC molecule on the APC surface. Shown here is a T-helper cell, as noted by its interaction with peptide bound to a class II MHC molecule. For simplicity, the CD4 coreceptor is not shown.

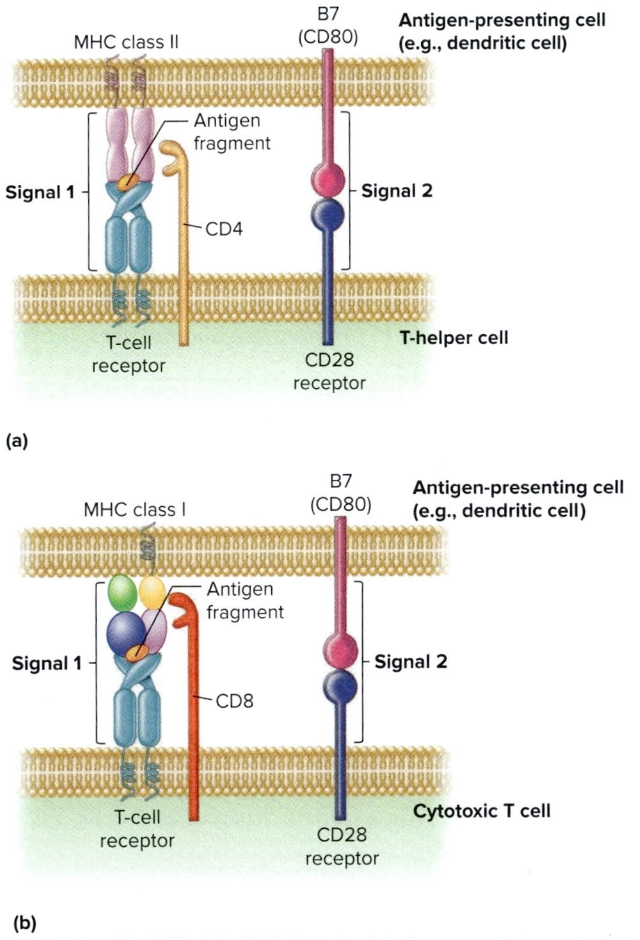

Figure 33.7 Three Signals Are Essential for T-Cell Activation. The first signal is the presentation of the antigen fragment by a dendritic cell. (a) The CD4 coreceptor on a T-helper cell mediates recognition of a MHC class II molecule on an APC. (b) The MHC class I molecule is recognized by the CD8 coreceptor on cells that will be activated to become CTLs. The second signal occurs when the dendritic cell presents the B7 (CD80) protein to the T helper or CTL with its CD28 protein receptor. The third signal is the release of cytokines by the APC and the T cells themselves, which drives the T cells to differentiate into memory and effector cells, which proliferate and become fully activated. Note that CD3 complexes are not shown for simplicity.

MICRO INQUIRY *How can T-helper cells facilitate $CD8^+$ cell activation?*

We have just described **signal 1,** which occurs when an antigen fragment, presented in an MHC molecule of an APC is recognized and bound by the appropriate T-cell receptor. In the case of T-helper (T_H) cells, CD4 coreceptors direct the interaction with MHC class II molecules (**figure 33.7*a***). For $CD8^+$ cells to be activated, CD8 coreceptors restrict interaction to MHC class I molecules on the APC. So while both CD8 and CD4 cells use the TCR to form an immune synapse, it is the CD8 or CD4 coreceptor that limits interaction with MHC class I or class II molecules, respectively.

In addition to signal 1, naïve CD4+ and CD8+ cells require co-stimulatory **signal 2** to become activated. Signal 2 occurs when the **B7 (CD80) protein** on the surface of an APC binds to the CD28 receptor on the T cell (figure 33.7*b*). Dendritic cells stimulate naïve T cells because they express high levels of B7 at all times. Other APCs do not produce high levels of B7 unless stimulated to do so. They therefore can upregulate previously activated cells (e.g., memory cells) but cannot activate naïve T cells. T cells that receive signal 1 without signal 2 become **anergic,** that is, they can recognize "their" antigen, but are unable to respond to it. ◀ *Monocytes, macrophages, and dendritic cells (section 32.4)*

Signal 3 is not as well defined as signals 1 and 2. Signal 3 involves the release of key cytokines by immune cells including the APC and the T cell making the immune synapse. These cytokines stimulate the target T cells to differentiate into effector or memory cells, which then proliferate, and become fully active cells. Thus while signals 1 and 2 involve receptors on an APC interacting with receptors on T cells, signal 3 involves soluble molecules binding to cognate receptors on the target T cell. All three signals trigger a change in T-cell gene expression enabling a specific, adaptive immune response.

Types of T Cells

Recall that T cells include a small number of γδ cells and a vast number of αβ T cells (**figure 33.8**). Whereas γδ T cells reside in the skin and mucous membranes, αβ T cells migrate to secondary lymphoid organs and tissues. However, αβ cells are not stationary. Instead, they periodically migrate to the bloodstream as they travel from lymph node to lymph node as well as to the spleen. Mature, naïve T cells in these tissues are awaiting activation by dendritic cell antigen presentation. Upon antigen stimulation, αβ T cells assume their active roles as effector cells and memory cells (figure 33.5). Effector T cells carry out specific functions to protect the host against foreign antigen. Effector and memory CD4+ T cells include T-helper and T regulatory cells, whereas activated CD8+ T cells function as CTLs and memory cells.

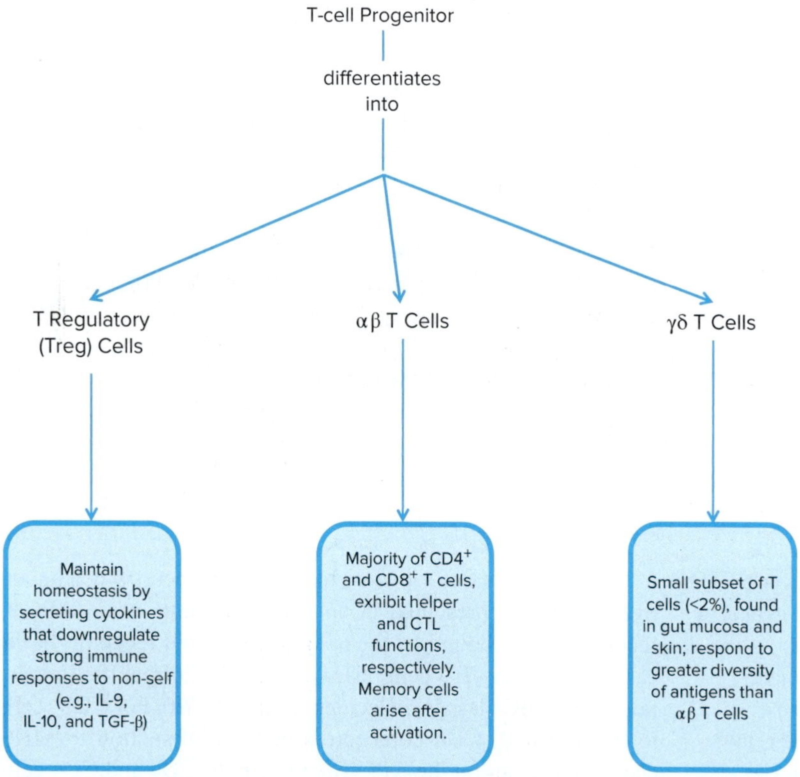

Figure 33.8 T-Cell Populations of Cell-Mediated Adaptive Immunity. T-cell subsets differentiate from their progenitor cells in the thymus. They leave the thymus upon maturation; however, they are still naïve because they have not been activated by their specific antigen.

T-Helper Cells

T-helper (T_H) cells are also known as **CD4+ T cells.** The release of different cytokines by DCs (signal 3) during immune synapse triggers the activation of specific transcription factors that drive naïve T cells to differentiate into one of several types of CD4+ T-helper cells. The most important types include T_H type 1 (T_H1) cells, T_H type 2 (T_H2) cells, T_H type 17 (T_H17) cells, and regulatory T cells (Treg) (**figure 33.9*a***). **T_H0** is often used to designate mature, naïve cells that have not yet been activated.

The release of interferon-γ (IFN-γ) and interleukin (IL)-12 by DCs induces the activity of the transcription factor T-bet in T_H0 cells. This results in their differentiation into **T_H1 cells.** T_H1 cells promote cytotoxic T lymphocyte activity, activate macrophages, and mediate inflammation by producing interleukins, interferon (IFN)-γ, and tumor necrosis factor-α (TNF-α) and TNF-β. Because DCs stimulate T_H1 cells to produce IFN-γ, the effects of this potent cytokine are amplified. These cytokines make T_H1 cells particularly important during infections by intracellular microbes such as viruses.

If during immune synapse, DCs release IL-4, the resulting induction of the T-cell transcription factor GATA-3 drives T_H0 cells to become **T_H2 cells.** T_H2 cells stimulate antibody responses and defend against helminth (worm) parasites by producing cytokines IL-4, IL-5, IL-6, and IL-13.

The secretion of two cytokines, transforming growth factor (TGF)-β and IL-6, results in the activation of the T-cell transcription factor RORγt. This triggers T_H0 differentiation into **T_H17 cells.** These cells respond to bacterial invasion by secreting IL-17 and defensins, recruiting neutrophils, and inducing a strong inflammatory response.

Regulatory T cells (Tregs) are derived from approximately 10% of CD4+ T cells and 2% of CD8+ T cells. When DCs release TGF-β during immune synapse, it induces the T-cell transcription factor FOXP3, which is responsible for the development of Tregs. Tregs are important because they dial back an

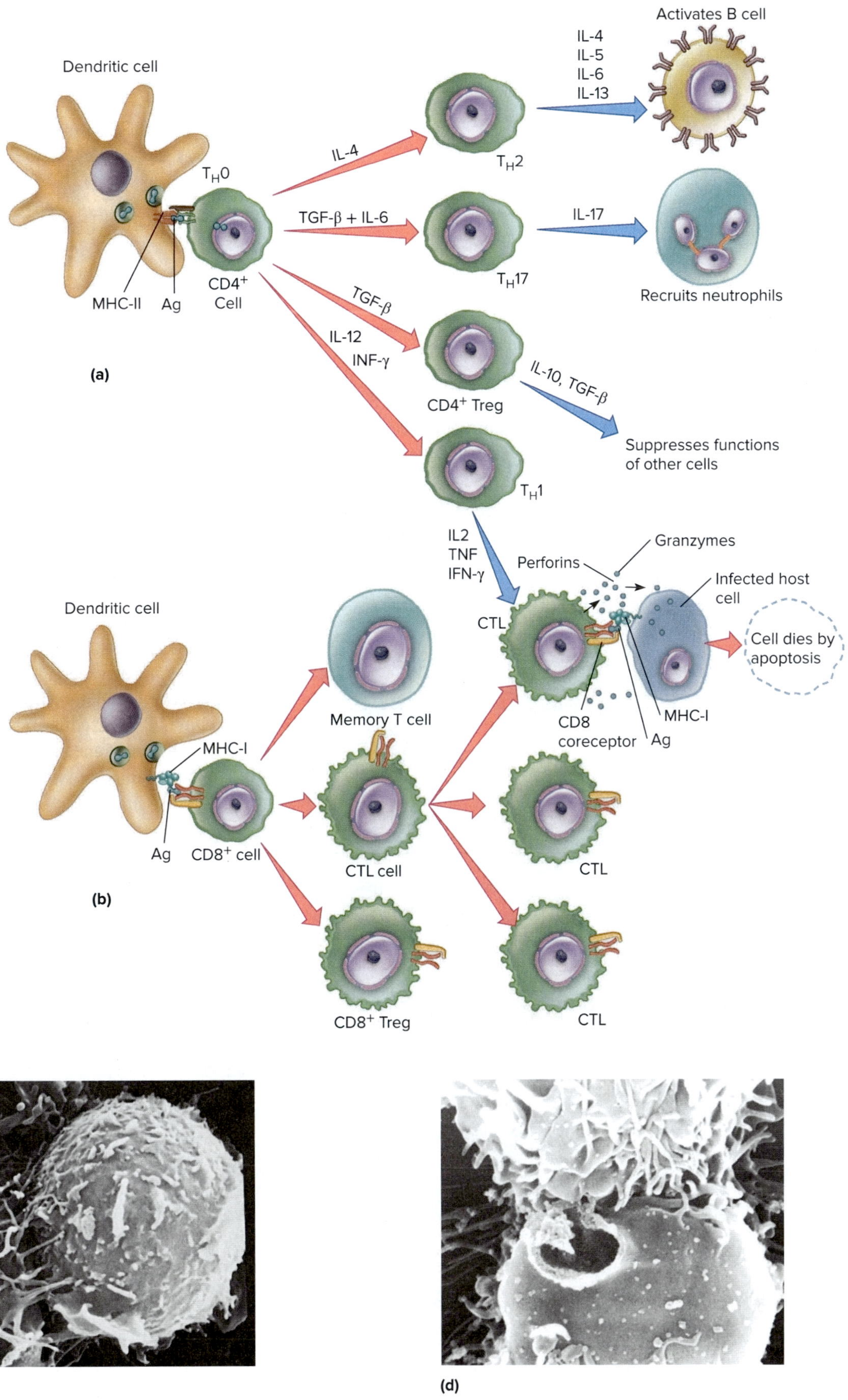

Figure 33.9 T-Cell Activation. (a) During activation, a CD4+ T_H0 cell may differentiate into several kinds of T cells, including T_H1, T_H2, T_H17, or Treg cells. Cytokines near red arrows are produced by DCs during immune synapse and trigger differentiation of T_H0 cells while cytokines near blue arrows are produced by activated T cells to communicate with other types of cells. Memory T cells for each T-cell class are also produced (not shown). (b) CD8+ T cells are activated by DCs presenting non-self antigen on MHC class I molecules. This drives CD8+ T cells to differentiate into activated CTLs. T_H1 cells can also stimulate CTLs by releasing specific cytokines. CTLs induce apoptosis in target cells. A small population of CD8+ cells differentiate into Tregs, but the role of these cells is not well understood. (c) A cytotoxic T cell (left) contacts a target cell (right) (×5,700). (d) The CTL secretes perforin that forms pores in the target cell's plasma membrane. These pores allow the contents of the target cell to leak out and granzymes to enter and induce apoptosis.

(c, d) ©Dr. Gilla Kaplan, PHRI, UMDNJ

immune response by producing the anti-inflammatory cytokines IL-10 and TGF-β. In addition to damping down acute inflammation and adaptive immune responses, Tregs are critical for maintaining mucosal immunity homeostasis. ◀ *Inflammation unites all components of immunity (section 32.7)* ▶ *A functional core microbiome is required for human homeostasis (section 34.3)*

Cytotoxic T Lymphocytes

Naïve **CD8⁺ T cells** are activated upon binding non-self antigens presented in MHC class I molecules of dendritic cells (figure 33.9*b*). Activation by its specific antigen causes a CD8⁺ T cell to mature into a **cytotoxic T lymphocyte (CTL)** that functions to destroy host cells that have been infected by intracellular pathogens, such as a virus, are presenting cancer neoantigens, or have altered MHC surface proteins. Any host cell that presents the same antigen in its class I MHC receptors as that presented by the DC that activated the CTL is thus targeted for destruction. Once activated, CTLs kill target cells in at least two ways: A cytolytic pathway, using **perforins,** granzymes, and granulysins, similar to NK cell–mediated lysis (*see figure 32.11*); and an apoptotic (programmed cell death) pathway triggered when the apoptotic protein Fas ligand (FasL) engages it receptor, Fas (also called CD95) expressed on the surface of target cells.

Comprehension Check

1. Describe the selection processes that result in the release of mature, naïve T cells from the thymus.
2. Describe the formation and structure of an immune synapse.
3. What are the three signals needed for T-cell activation? Describe the function of each.
4. Contrast and compare the activation of CD4⁺ and CD8⁺ T cells.
5. Outline the functions of a T-helper cell. How do T_H1, T_H2, T_H17, and Treg cells differ in function?
6. Briefly describe the two ways in which CTLs destroy target cells.

33.6 B Cells Make Antibodies

After reading this section, you should be able to:

a. Describe the B-cell receptor structure and function
b. Diagram the steps required for a B cell to become an APC
c. Compare T-cell-dependent and T-cell-independent B-cell activation in terms of mechanism and outcome

Stem cells in the bone marrow produce B-cell precursors (figure 33.5). Like thymic selection during T-cell maturation, a selection process within the bone marrow eliminates B cells that recognize self antigens. Upon leaving the bone marrow, B cells (like T cells) migrate between lymph nodes and spleen. Within these secondary lymphoid tissues and organs, B cells meet the specific antigen that activates them. Activation results in proliferation and differentiation into mature, antibody-secreting **plasma cells.** Not surprisingly, this activation process requires a B-cell receptor that recognizes and binds specific antigen. Interestingly, B-cell receptors are made of membrane-bound antibodies. Mature, naïve B cells attach a type of antibody (IgM or IgD) to their cell membrane so that the part of the antibody that binds to antigen is facing outward, away from the cell (**figure 33.10**).

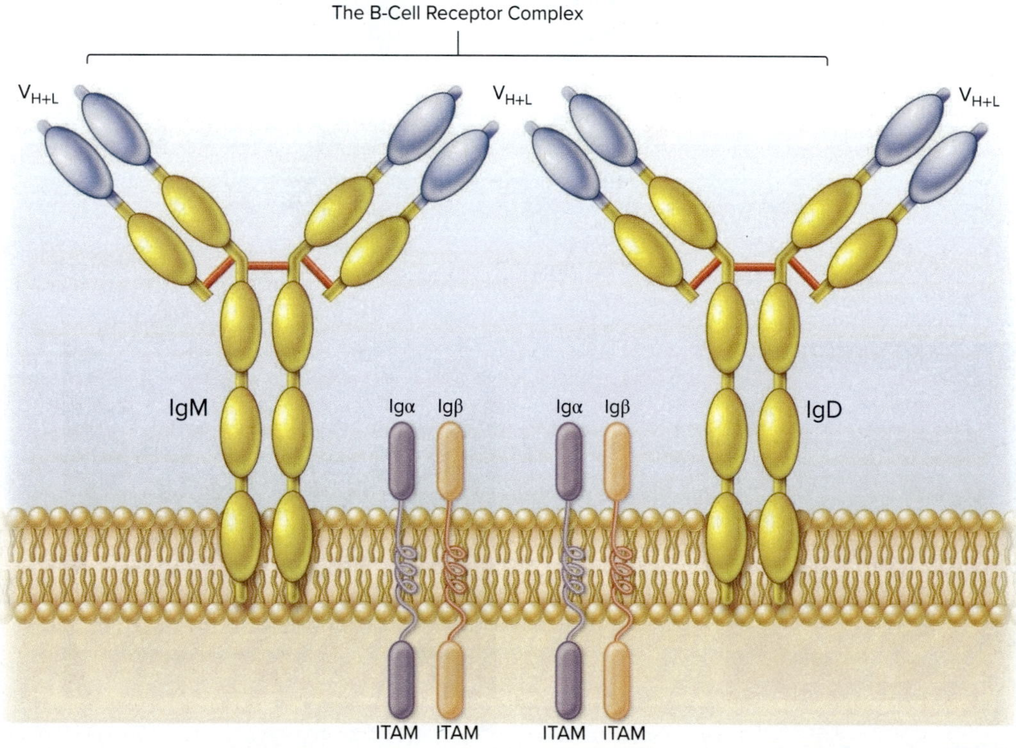

Figure 33.10 The Membrane-Bound B-Cell Receptor (BCR) Complex. The BCR is composed of monomeric IgM and IgD antibody, and the coreceptors Igα and Igβ. A B cell is activated after the variable region of one receptor (designated V_{H+L}) binds an antigen. The antigen is internalized, processed, and a peptide portion can be presented by the B cell's class II MHC molecule, to a T_H cell that has been activated by the same antibody after immune synapse formation with a DC (T-cell-dependent activation). Activation can also occur after the variable regions of multiple receptors are bridged by an antigen, causing the BCR to cross-link on the B-cell surface (T-cell-independent activation).

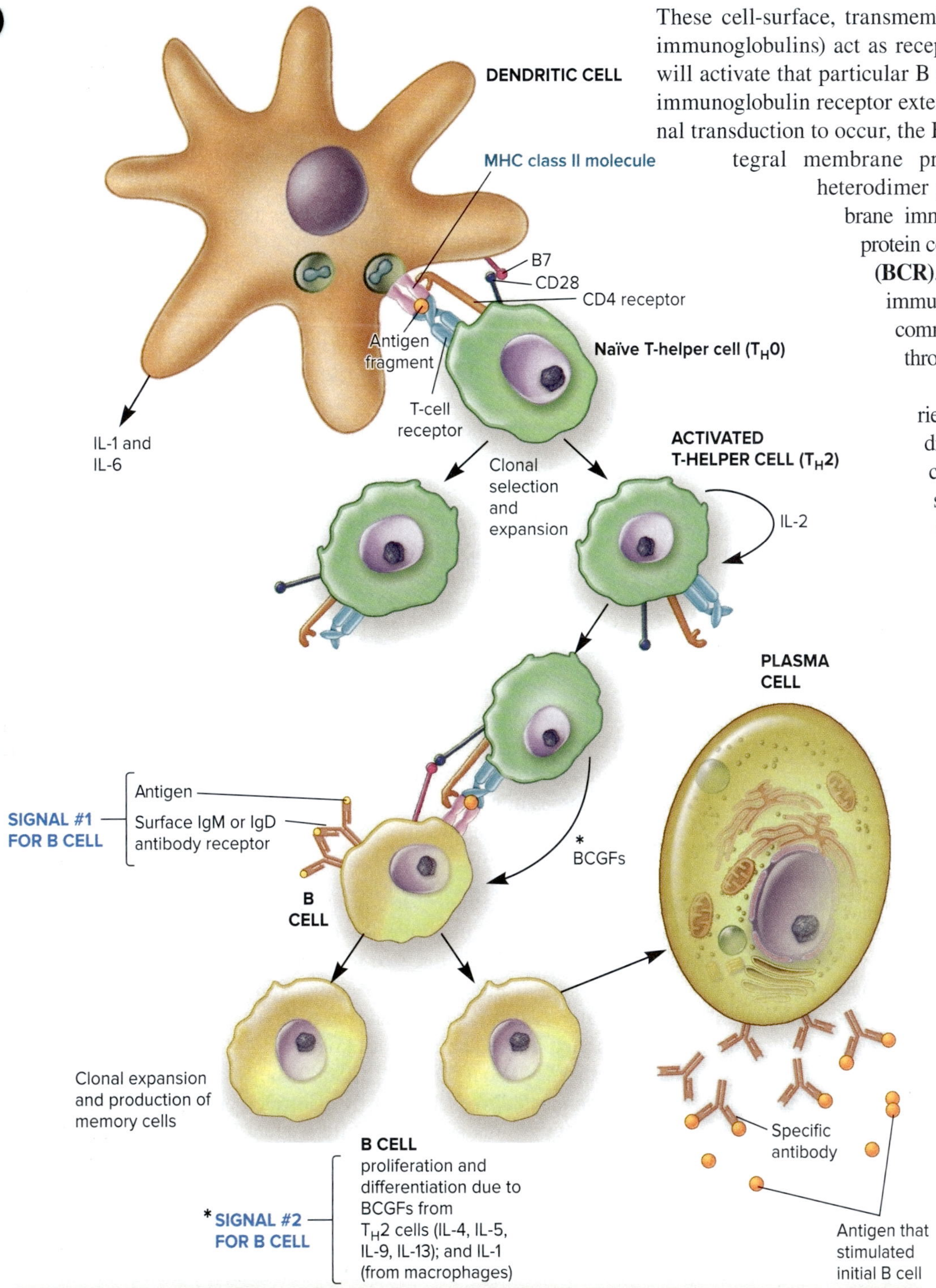

Figure 33.11 T-Dependent Antigen Activation of a B Cell. Schematic diagram of the events occurring in the interactions between macrophages, T-helper cells, and B cells. Many cytokines act as B-cell growth factors (BCGFs) to stimulate B-cell proliferation, while others stimulate B-cell differentiation into plasma cells.

MICRO INQUIRY *Which cells function as APCs in this process?*

These cell-surface, transmembrane antibodies (also known as immunoglobulins) act as receptors for the specific antigen that will activate that particular B cell. Like TCRs, very little of the immunoglobulin receptor extends into the cytoplasm, so for signal transduction to occur, the BCR must associate with other integral membrane proteins, in this case Ig-α/Ig-β heterodimer proteins. Together the transmembrane immunoglobulin and the heterodimer protein complex is called the **B-cell receptor (BCR)**. When an antigen is captured by the immunoglobulin receptor, the receptor communicates this capture to the nucleus through a signal transduction pathway.

While each individual human carries BCRs specific for as many as 10^{13} different antigens, each individual B cell possesses BCRs specific for a single, unique antigen and one particular epitope on that antigen, each human carries BCRs specific for about 10^{13} different antigens. Therefore a host produces at least 10^{13} different, mature, naïve B cells. This astounding level of diversity enables the host to respond to antigens for an entire lifetime.

So far we have introduced the B-cell receptor and the fact that activated B cells secrete antibody. The fate of antigen bound to the BCR is critical for the transition from naïve B cell to antibody-secreting plasma cell (**figure 33.11**). In most cases, B-cell activation involves interaction with T cells, and is thus called **T-cell dependent B-cell activation.** This process involves the simultaneous activation of both T and B cells with the same antigen, as is now described.

Upon binding antigen snugly in their BCR, B cells internalize the antigen-bound receptor by receptor-mediated endocytosis. Within the B cell, the antigen, which may be large and complex, is broken down and a peptide from the antigen is transferred to a MHC class II molecule. This peptide-MHC class II complex then moves to the B-cell membrane to display the antigen on the B-cell surface. Within the same lymph node and at the same time the B cell is undergoing antigen processing by receptor-mediated endocytosis, T_H0 cells are activated by forming an immune synapse with a DC presenting the same antigenic peptide as that presented by the B cell. Thus an identical peptide is presented in the MHC class II of the B cell and the DC. The immune synapse between the DC and the T cell results in T-cell activation and differentiation into a T_H1 or T_H2 cell. Because the B cell is now an APC, displaying the same antigen in its MHC class II molecule as the DC, the B cell presents the antigen to an activated T_H cell (figure 33.11). Upon formation of the immune synapse between the MHC class II on the B cell and TCR on the DC-activated T_H cell, a specific suite of cytokines from the T_H cell is released. These cytokines stimulate the expression of genes in the B cell that results in B-cell proliferation followed by differentiation into many antibody-secreting **plasma cells** or relatively fewer memory B cells. For example, if IFN-γ secreting T_H1 cells interact with B cells, the resulting plasma cells produce a type of antibody called IgG, while T_H2 cells secrete lots of IL-4 that stimulates B cells to differentiate into plasma cells that produce IgE. The release of TGF-β results in the development of IgA secreting plasma cells.

Antigens that trigger T-cell-dependent B-cell activation are called **T-dependent antigens.** This complex process of B-cell activation, with its two signals, antigen-BCR interaction (signal 1) and T-cell cytokines (signal 2), ensures that a B cell does not differentiate into a potent plasma cell that produces antibody directed against a self antigen. Indeed, one plasma cell can synthesize more than 10 million antibody molecules per hour!

As you might imagine, the cellular choreography needed for T-cell-dependent B-cell activation takes time—typically at least a week. Humans have therefore evolved a "shortcut" to antibody formation. However, unlike the vast diversity of molecules that can trigger T-cell-dependent B-cell activation, only a few specific antigens can trigger B cells to produce antibody without T-cell cooperation. These are called **T-independent antigens,** and their stimulation of B cells is known as **T-cell-independent B-cell activation.** Examples include bacterial lipopolysaccharides and certain polysaccharides. T-independent antigens tend to be polymeric, composed of repeating sugars or amino acids. The resulting antibody generally has a relatively low affinity for antigen.

The mechanism for activation by T-independent antigens relies on their polymeric structure. Such molecules present a wide array of identical epitopes to a B-cell receptor. The repeating epitopes cross-link membrane-bound BCRs such that cell activation occurs and B cells differentiate into antibody-secreting plasma cells. However, without T-cell help, the B cell cannot produce high avidity, high affinity antibodies, and no (or few) memory B cells are formed. Thus T-cell-independent B-cell activation is less effective than T-dependent B-cell activation. However, antibody is produced much more quickly and can be a good immunological stop-gap measure until T-cell-dependent B-cell activation is complete.

Comprehension Check

1. What is the structure of B-cell receptors? How are they involved in B-cell activation?
2. Why must T-cell activation by a DC and B-cell antigen presentation during T-cell-dependent B-cell activation occur in the same lymph node?
3. How does T-cell-independent antigen triggering of B cells differ from T-cell-dependent triggering?

33.7 Antibodies Bind Specific 3-D Antigens

After reading this section, you should be able to:

a. Compare and contrast the structure and function of the five classes of antibody
b. Explain how primary and secondary antibody responses differ
c. Discuss the genetic diversity of BCRs and soluble antibodies that results from recombination, alternative splicing, and somatic hypermutation

So far we have discussed how B cells are activated to become antibody-secreting plasma cells. We now turn our attention to antibody structure and how a seemingly infinite diversity of antibodies can be synthesized by a single individual. An **antibody** or **immunoglobulin (Ig)** is a glycoprotein. There are five classes of human antibodies; all share a basic structure that varies, to accomplish specific biological functions.

Antibody Structure

All antibody molecules have a basic structure composed of four polypeptide chains: two identical heavy and two identical light chains (**figure 33.12**). Each light-chain polypeptide consists of about 220 amino acids and has a mass of approximately 25,000 Da. Each heavy chain consists of about 440 amino acids and has a mass of about 50,000 to 70,000 Da. Heavy and light chains are connected to each other by disulfide bonds (figure 33.12*b*). The heavy chains are structurally distinct for each immunoglobulin class or subclass. Both light (L) and heavy (H) chains contain two different regions. The **constant (C) regions** (**C_L** and **C_H**) have amino acid sequences that do not vary significantly between antibodies of the same class. The **variable (V) regions** (V_L and V_H) have different amino acid sequences and these regions fold together to form the antigen-binding sites. Structural diversity in the V region accounts for the ability of antibodies to bind a seemingly infinite number of antigens.

The four chains are arranged in the form of a Y with a hinge region. This hinge allows the antibody molecule to be more flexible, adjusting to the different spatial arrangements of epitopes on antigens. The stalk of the Y is termed the **crystallizable fragment (Fc)** and is recognized by host immune cells that bear surface Fc receptors. The top of the Y consists of two **antigen-binding fragments (Fab)** that bind with compatible epitopes.

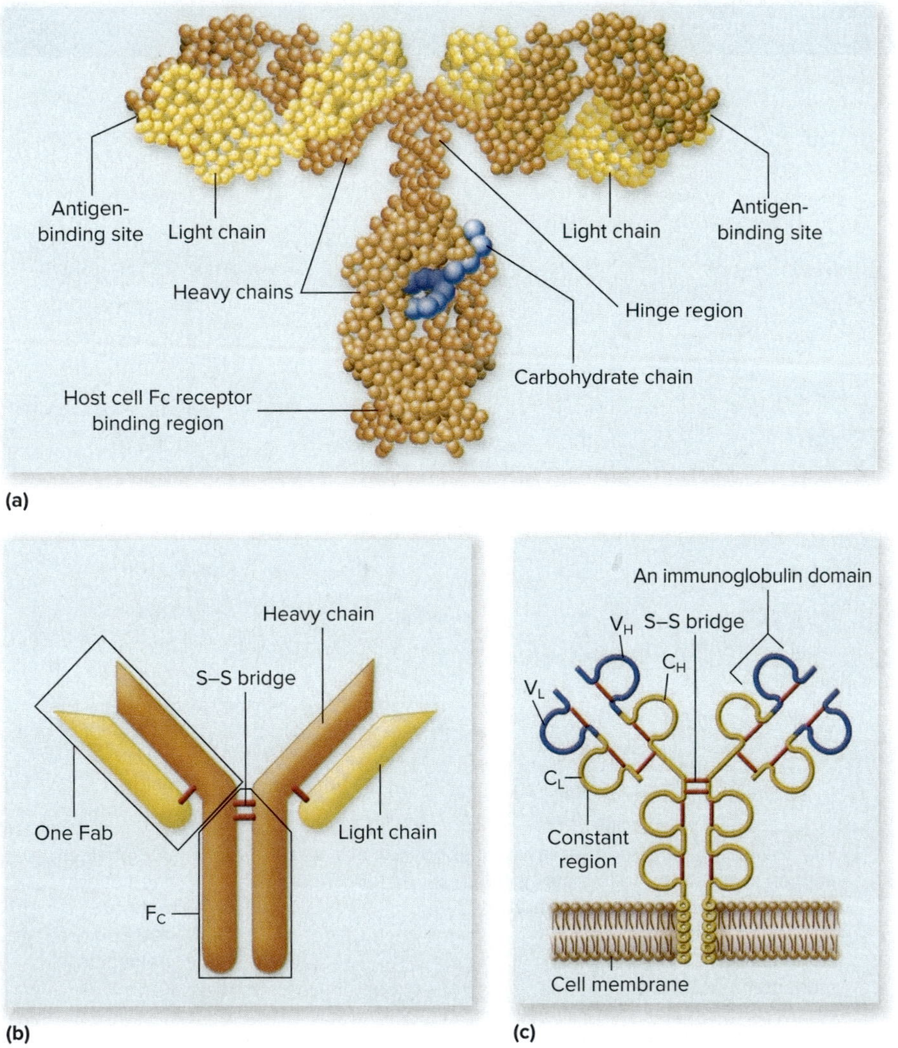

Figure 33.12 Immunoglobulin (Antibody) Structure. (a) A model of immunoglobulin structure showing the arrangement of the four polypeptide chains. (b) Heavy and light chains are connected to each other by disulfide bridges (bonds). Variable regions of heavy- and light-chains (V_H, V_L) form an antigen-binding fragment (Fab). The disulfide-linked heavy chains form the crystalizable fragment (Fc). (c) Both heavy and the light chains contain several homologous domains of about 100 to 110 amino acids. Within each domain, disulfide bonds form a loop of approximately 60 amino acids. Interchain disulfide bonds also link heavy and light chains together. All light chains contain a single variable domain (V_L) and a single constant domain (C_L). Heavy chains contain a variable domain (V_H) and either three or four constant domains (C_H1, C_H2, C_H3, and C_H4).

In humans, the constant regions of all light chains are identical. Each antibody molecule produced by a sole B cell will contain one of two types of light chains from two different genetic loci: kappa (κ) or lambda (λ). The light-chain variable (V) domain contains hypervariable regions, called complementarity-determining regions (CDRs), that differ in amino acid sequence more frequently than the rest of the variable domain. These regions are essential in conferring the ability to bind a specific antigen.

Like the light chain, the heavy chain also has a variable region (VH). The other domains of the heavy chains are termed constant (C) domains. The constant domains of the heavy chain form the constant (C_H) region. The amino acid sequence of this region determines the classes of heavy chains. In humans, there are five classes of heavy chains designated by lowercase Greek letters: gamma (γ), alpha (α), mu (μ), delta (δ), and epsilon (ε), and usually written as G, A, M, D, or E. The properties of these heavy chains determine, respectively, the five immunoglobulin (Ig) classes: IgG, IgA, IgM, IgD, and IgE (**table 33.1**). Each immunoglobulin class differs in its specific structure, half-life, distribution in the body, and interaction with other components of the host's defensive systems.

Antibody Function

An antibody binds to an antigen at the antigen-binding site within the Fab region of the antibody. Within this region, a pocket is formed by the folding of the V_H and V_L regions (figure 33.12). At this site, specific amino acids contact the antigen's epitope and form multiple noncovalent bonds between amino acids of the binding site and the epitope. Because binding is due to weak, noncovalent bonds such as hydrogen bonds and electrostatic attractions, the antigen's shape must exactly match that of the antigen-binding site. If the shapes of the epitope and binding site are not truly complementary, the antibody will not effectively bind the antigen.

Whereas the Fab region is concerned with binding antigen, the Fc region binds to receptors (called Fc receptors) found on various immune cells such as neutrophils, DCs, and macrophages. Recall that antibody is also important in opsonizing microbes and initiating the classical complement cascade. The Fab region of some antibodies binds to the first component of this complement pathway. The binding of an antibody to an antigen usually does not destroy the antigen. Rather, the antibody opsonizes, or marks, the non-self agent as a target for immunological attack. Other cells, such as neutrophils and NK cells, destroy antibody-coated cells through phagocytosis and antibody-dependent cell-mediated cytotoxicity, respectively (*see figures 32.12 and 32.17*). ◀ *Complement is a cascading enzyme system (section 32.3)*

Table 33.1 Physicochemical Properties of Human Immunoglobulin Classes

	IMMUNOGLOBULIN CLASSES				
Property	**IgG**[1]	**IgM**	**IgA**[2]	**IgD**	**IgE**
Heavy chain	γ_1	μ	α_1	δ	ε
Mean serum concentration (mg/ml)	9	1.5	3.0	0.03	0.00005
Percent of total serum antibody	80–85	5–10	5–15	<1	<1
Valency	2	5(10)	2(4)	2	2
Mass of entire molecule (kDa)[3]	146	970	160[3]	184	188
Placental transfer	+	–	–	–	–
Half-life in serum (days)[4]	23	5	6	3	2
Complement activation					
Classical pathway	++	+++	–	–	–
Alternative pathway	–	–	+	–	–
Induction of mast cell degranulation	–	–	–	–	+
% carbohydrate	3	7–10	7	12	11
Major characteristics	Most abundant Ig in body fluids; neutralizes toxins; opsonizes bacteria	First to appear after antigen stimulation; activates classical complement pathway; agglutinator; monomer functions as part of BCR complex	Secretory antibody; protects mucous membranes	Serves as part of BCR complex	Anaphylactic-mediating antibody; resistance to helminths

1 Properties of IgG subclass 1.
2 Properties of IgA subclass 1.
3 sIgA = Secretory IgA, 360 – 400 kDa.
4 Time required for half of the antibodies to disappear.

Immunoglobulin (Antibody) Classes

Immunoglobulin γ, or **IgG,** is the major immunoglobulin in human serum, where it accounts for 80% of the immunoglobulin pool (**figure 33.13**). IgG acts against bacteria and viruses by opsonizing them. It also neutralizes toxins and viruses (section 33.8). It is one of the two immunoglobulin classes that activate complement by the classical pathway (the other is IgM). IgG is the only immunoglobulin molecule able to cross the placenta and provide natural immunity in utero and to infants at birth.

There are four human IgG isotypes (IgG1, IgG2, IgG3, and IgG4); these vary chemically in their heavy-chain composition and the number and arrangement of interchain disulfide bonds. Differences in biological function have been noted in these isotypes. For example, IgG2 and IgG4 antibodies bind and neutralize toxins. IgG1 and IgG3 are important for opsonization.

Immunoglobulin μ, or **IgM,** accounts for about 5 to 10% of the serum immunoglobulin pool. In the serum, it is a polymer of five monomeric units (pentamer), each composed of two heavy chains and two light chains (**figure 33.14*a***). The monomers are arranged in a pinwheel array with the Fc ends in the center, held together by disulfide bonds and a protein called the J (joining) chain. Soluble, pentameric IgM is the immunoglobulin made upon initial antigen stimulation. It is the only antibody class produced following T-cell-independent B-cell activation. Pentameric IgM tends to remain in the bloodstream, where it agglutinates (or clumps) bacteria, activates complement by the classical pathway, and enhances the ingestion of pathogens by phagocytic cells. In contrast to pentameric IgM, individual IgM monomers are expressed on B-cell surfaces where they serve as the antibody component of the BCR.

Immunoglobulin α, or **IgA,** accounts for about 12% of serum immunoglobulin, but is actually the most abundant antibody.

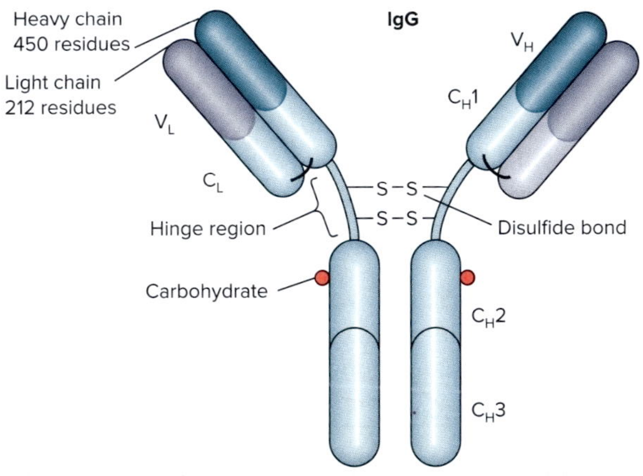

Figure 33.13 Immunoglobulin G. The basic structure of human IgG.

MICRO INQUIRY *What makes this class of immunoglobulin unique?*

This is because most IgA is found in mucous secretions. Whereas serum IgA is monomeric, when secreted into mucous membranes it forms a dimer held together by a J chain (figure 33.14*b*). Transport of IgA from mucosal-associated lymphoid tissue (MALT) to mucosal surfaces requires the attachment of a second protein, the secretory component (*see figure 32.15*). **Secretory IgA (sIgA)**, as the modified molecule is called, is the primary immunoglobulin of MALT. Secretory IgA is therefore found in any secretion associated with mucous membranes such as saliva, tears, and intestinal mucus, where it plays a major role in protecting surface tissues against microorganisms by the formation of an immune barrier. For example, when sIgA binds to viruses, bacteria, and protozoa, these microorganisms are blocked from attaching to mucosal surfaces and invading host tissues, a phenomenon known as immune exclusion. In addition, sIgA is found in breast milk, thereby providing a measure of mucosal immunity to breastfed infants. ◀ *Secondary lymphoid organs and tissues (section 32.5)*

Immunoglobulin δ, or **IgD,** has a monomeric structure (figure 33.14*c*). IgD antibodies are abundant in combination with IgM on the surface of B cells and thus are part of the B-cell receptor complex. Therefore their function is to help signal the B cell to start antibody production upon initial antigen binding.

Immunoglobulin ε, or **IgE** (figure 33.14*d*), makes up only a small percent of the total immunoglobulin pool. A number of different cell types, especially mast cells, eosinophils, and basophils, bind IgE molecules with their IgE-specific Fc receptors. When cell-bound IgE molecules become cross-linked by binding to the same antigen, these cells degranulate. Degranulation releases preformed mediators of inflammation (e.g., histamine). It also stimulates production of an excessive number of eosinophils in the blood (eosinophilia) and increases the rate of movement of the intestinal contents (gut hypermotility). This aids in the elimination of helminthic parasites, which is presumably why we evolved this antibody class. However the release of these mediators, particularly histamine, is also responsible for allergic reactions and anaphylaxis (section 33.10).

Antibody Kinetics

The rate of synthesis and secretion of antibody changes over time. The relatively slow antibody response when an antigen is encountered for the first time is called the primary antibody response.

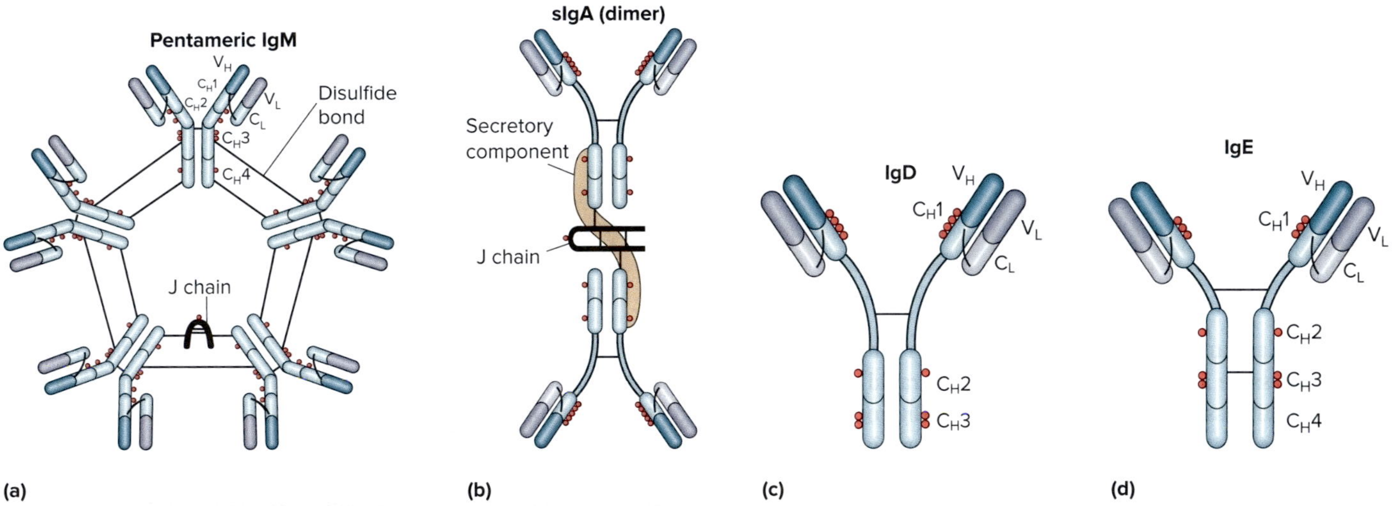

Figure 33.14 Immunoglobulins M, A, D, and E. (a) The pentameric structure of human IgM. The disulfide bonds linking peptide chains are shown in black. Note that 10 antigen-binding sites are present. (b) The dimeric structure of human secretory IgA. Notice the secretory component (tan) wound around the IgA dimer and attached to the constant domain of each IgA monomer. (c) The structure of human IgD showing the disulfide bonds that link protein chains (shown in black). (d) The structure of human IgE. Carbohydrate side chains are in red.

MICRO INQUIRY *Where is monomeric IgM commonly found? What about secretory IgA?*

Table 33.2	Theoretical Antibody Diversity Resulting from Combinatorial Joining of Germ-Line Genes[1]
λ light chains	V regions = 2 J regions = 3 Combinations = 2 × 3 = 6
κ light chains	V_κ regions = 250 – 350 J_κ regions = 4 Combinations = 250 × 4 = 1,000 = 350 × 4 = 1,400
Heavy chains	V_H = 250 – 1,000 D = 10–30 J_H = 4 Combinations = 250 × 10 × 4 = 10,000 = 1,000 × 30 × 4 = 120,000
Diversity of antibodies	κ-containing: 1,000 × 10,000 = 10^7 1,400 × 120,000 = 2×10^8 λ-containing: 6 × 10,000 = 6×10^4 6 × 120,000 = 7×10^5

1 Approximate values.

cutting and joining the heavy-chain counterparts of V and J DNA, as well as the D (diversity) sequences (**figure 33.17**).

During the process of joining either VJ or VDJ sequences in light- and heavy-chain rearrangement, respectively, there is an additional opportunity to introduce genetic diversity. Because combinatorial joining occurs between different nucleotides different codons can result in a process known as **splice site variability.** For example, one VJ splicing event can join the V sequence CCTCCC with the J sequence TGGTGG in two ways: CCTCCC + TGGTGG = CCGTGG, which codes for proline and tryptophan. Alternatively, the VJ splicing event can eliminate nucleotides, giving rise to the sequence CCTCGG, which codes for proline and arginine. Thus the same VJ joining could produce polypeptides differing in a single amino acid. Importantly, because combinatorial joining, tdt-dependent nucleotide addition, and splice site variability involve changing DNA sequences (as opposed to RNA splicing), each B cell produced has a new and different genotype and its replication will give rise to genetically distinct clones.

Initially, all heavy chains have the μ type of constant region (figure 33.17*b*). This explains why IgM is produced upon initial antigen stimulation (figure 33.15). Upon T-cell-dependent B-cell activation in lymphoid tissue, **antibody class switching** occurs when the VDJ region is joined with a new constant region that encodes a different class of antibody, usually IgG or IgA (figure 33.17*c*).

A unique editing process is used to switch from IgM to another antibody class. Activated B cells in lymphoid tissue synthesize the enzyme *a*ctivation-*i*nduced cytidine *d*eaminase (AID). AID replaces cytosine residues with uracil. This stimulates DNA repair enzymes to remove uracil residues from the DNA. The resulting abasic nucleotides are then excised, nicking one DNA strand. When nicking occurs adjacent to μ and γ switch regions, recombination occurs; the μ constant region is competely removed, and the γ constant region is joined to the VDJ region. Thus once class switching has occurred, the B cell will produce IgG and never be able to produce IgM again.

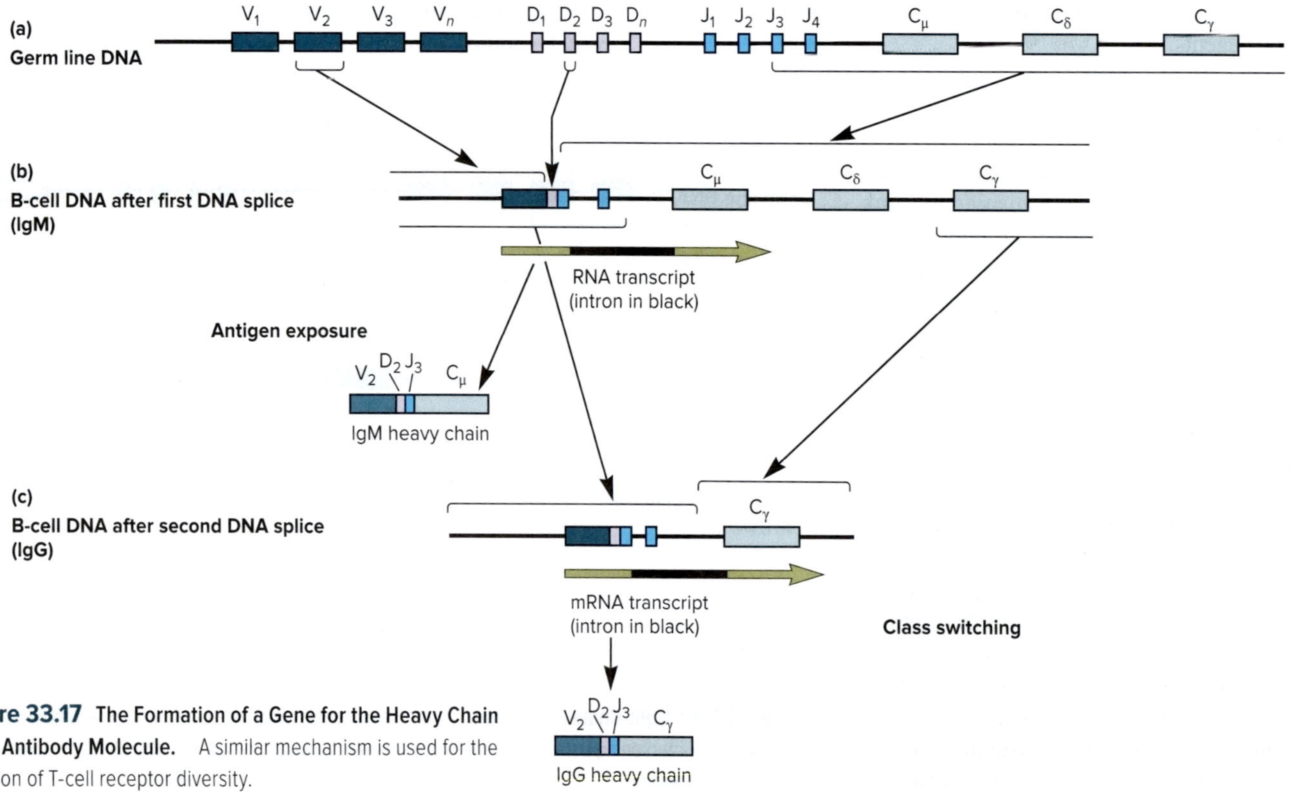

Figure 33.17 The Formation of a Gene for the Heavy Chain of an Antibody Molecule. A similar mechanism is used for the creation of T-cell receptor diversity.

In addition to class switching, T cells help trigger B cells to increase antibody diversity through **somatic hypermutation.** This mutational process occurs because the rearranged variable heavy and light genes possess "hotspots" that are more likely to be altered by point mutations than other areas of the variable regions. The resulting point mutations are recognized by DNA repair enzymes that attempt to correct the altered bases. AID, the same enzyme that mediates antibody class switching, is one of the enzymes involved. Because AID deaminates DNA at cytosine, base pair mismatches and transition-type mutations occur. Furthermore, repair of the base pair mismatch is error-prone, thereby resulting in even more sequence diversity. Accumulation of the altered bases in the variable regions of B-cell antibody coding genes following T-cell activation promotes significant antibody diversity. Because mutations are random, these variant antibodies must be tested within the lymphoid tissue for affinity and avidity. Only B cells producing "new and improved" antibody survive this selection process and are allowed to differentiate into mature, antibody-secreting plasma cells.

Clonal Selection

Each of the processes that lead to antibody diversity involves changes to a cell's DNA and all are therefore irreversible. Thus when each genetically distinct B cell is stimulated by a specific antigen to reproduce, it gives rise to a collection of clones; this process is known as clonal selection. T-cell receptor diversity is generated by similar VDJ recombination of the β chain and VJ recombination of the lighter, α chain. T cells are therefore genetically distinct and also undergo clonal selection when stimulated by antigen presentation. Clonal selection is important because it accounts for immunological specificity.

The clonal selection theory has four components or tenets. The first tenet is that there exists a pool of lymphocytes that can bind to a tremendous range of epitopes (**figure 33.18**). Because some of the B and T cells in the pool react with self-epitopes, the second tenet is that these self-reactive cells are eliminated at an early stage of development. This process, called negative selection, occurs in the bone marrow for B cells, while self-reacting T cells are eliminated in the thymus. The third tenet is that once a lymphocyte has been released from either the bone marrow or thymus and is exposed to its specific antigen, it proliferates to form a clonal population of identical cells derived from a single parent cell. Note that each genetically distinct B or T cell has been selected by its unique ability to interact with a specific antigen. The final tenet states that all clonal cells react with the same antigenic epitope that stimulated its formation. Clonal selection theory has also led to the development of monoclonal antibody (mAb) technologies that have been exploited to treat disease (**Techniques & Applications 33.1**).

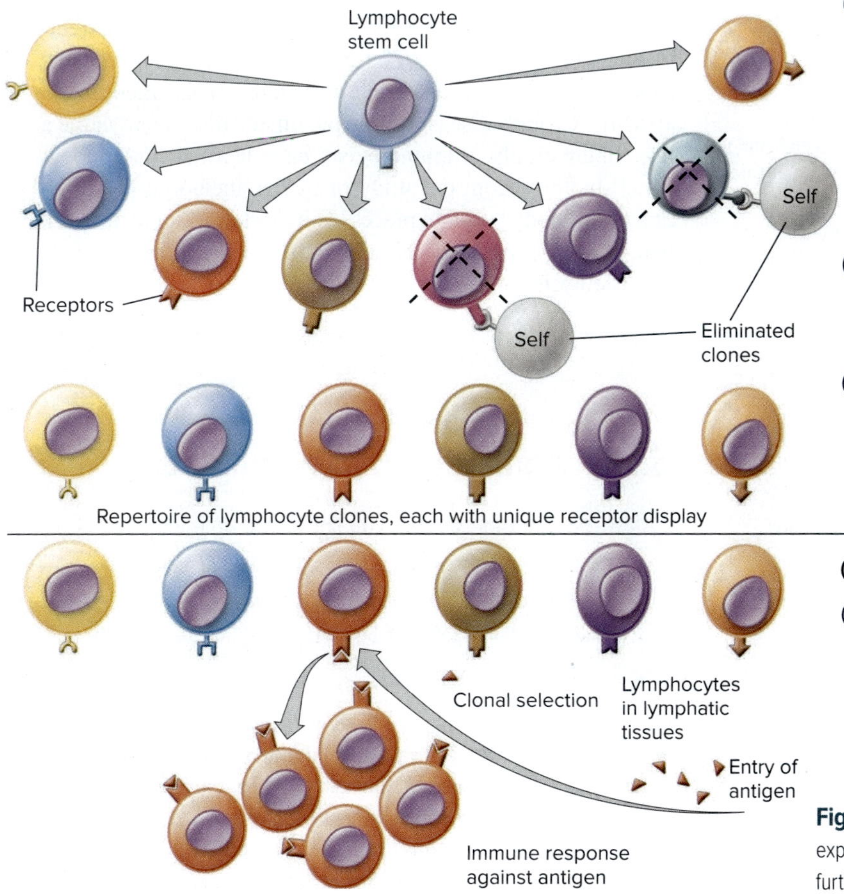

(a) Antigen-Independent Period

1. During development of early lymphocytes from stem cells, a given stem cell undergoes rapid cell division to form numerous progeny.

   During this period of cell differentiation, random rearrangements of the genes that code for the TCR or BCR occur. The result is a large array of genetically distinct cells, called clones; each clone bears a different receptor that can react with only a single antigen.

2. At the same time, any lymphocyte clones with receptors that recognize self molecules and could be harmful are eliminated from the pool of immature cells.

3. The specificity for a single antigen molecule is set for the life of a given clone. The end result is an enormous pool of mature but naïve lymphocytes ready to further differentiate under the influence of immune stimuli.

(b) Antigen-Dependent Period

4. Lymphocytes will then encounter antigens. These antigens will become the stimulus for the lymphocytes' final activation and immune function. A specific antigen selects only the lymphocyte clone or clones that carry matching TCR or BCR. This will trigger an immune response, which varies according to the type of lymphocyte involved.

Figure 33.18 Lymphocyte Clonal Expansion. (a) Cell populations expand and are restricted based on their MHC expression. (b) They are further expanded when activated by a specific antigen.

TECHNIQUES & APPLICATIONS

33.1 Monoclonal Antibody Therapy

The value of antibodies as tools for locating or identifying antigens is well established. For many years, antiserum extracted from human or animal blood was the main source of antibodies for tests and therapy, but most antiserum is problematic. It contains polyclonal antibodies, meaning it is a mixture of different antigen-specific antibodies, reflecting thousands of immune reactions experienced by the animal. An important advance in research and medicine has been the use of cell culture technology to produce a pure preparation of monoclonal antibodies (mAbs) that originate from a single clone and therefore have a single specificity for the same epitope. Monoclonal antibodies are used therapeutically when they have been modified to target a protein that contributes to disease (see chapter opener). Thus mAbs against specific antigens on cancer cells, infectious agents, and disease-associated proteins can facilitate their eradication without affecting other proteins, cells, or tissues.

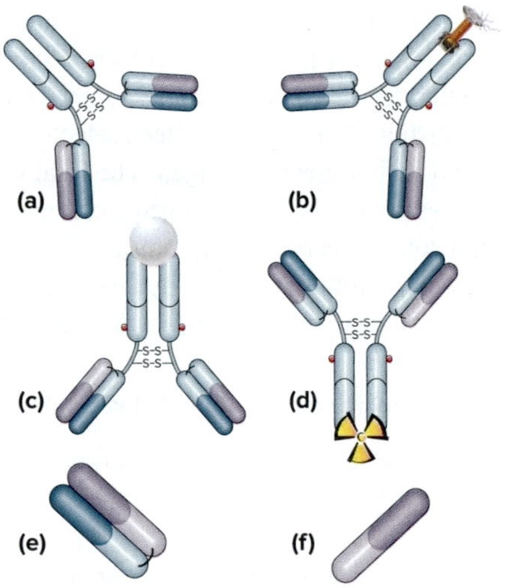

Monoclonal Antibody (mAb) Therapeutics. mAbs that retain their antigen binding specificity can be used (a) as the naked glycoprotein to stimulate antigen recognition and removal, or they can be altered to deliver (b) cytokines, (c) drugs, or (d) radionucleotides to kill their targets. Additionally, antibody pieces such as (e) the single domain and (f) the single-chain fragments can also be used.

The technology for producing monoclonal antibodies involves hybridizing human cancer cells and activated rabbit or mouse B cells in tissue culture systems. Mouse-human chimeric mAbs (mouse variable chains on human constant chains) and humanized (mouse hypervariable amino acids in human antibodies) are available. In addition, transgenic mice containing human antibody genes can be immunized with specific antigens to produce completely human B cells and mAbs. Additionally, human mAbs can be made in *Escherichia coli* from bacteriophage containing human B-cell DNA.

It is estimated that there are over 500 mAbs presently licensed or in development for the specific diagnosis or treatment of human disease. The mAbs target various types of cancer, autoimmune diseases, infection, sepsis, rheumatoid arthritis, Crohn's disease, and Alzheimer's disease, to name a few. Interestingly, immunologists have learned much about B-cell biology from the study of mAbs, including the function of antibody "orphan pieces" (single chain, single domain, Fab, $F(ab')_2$), previously thought to be nonfunctional products of proteolysis.

Comprehension Check

1. Describe the structure and function of the variable region, the complementarity-determining region, and the constant region of an antibody.
2. Explain the different functions of IgM when it is bound to B cells versus when it is soluble in serum.
3. Describe the major functions of each immunoglobulin class.
4. Which immunoglobulin can cross the placenta? Which is found in mucous membranes and breast milk?
5. Compare and contrast combinatorial joining of heavy- and light-chain antibody gene segments.
6. What events lead to B-cell diversity during development in the bone marrow? Compare these with the events that occur only after T-cell activation of B cells in lymphoid tissue.

33.8 Antibodies Doom Antigens

After reading this section, you should be able to:

a. Explain the consequences of antibody binding of antigen
b. Assess the effectiveness of antigen removal by antibody
c. Predict which antigens will be most susceptible to antibody action

As we have seen, the antigen-antibody interaction exhibits exquisite specificity. The interactions that occur in animals are absolutely essential in protecting the animal against the continuous onslaught of microorganisms and their products, as well as cancer cells. The mechanisms by which antibodies achieve this are now discussed.

Neutralization

Binding of antibody to biologically active materials (such as bacteria, toxins, or viruses) causes their inactivation or **neutralization** (**figure 33.19**). The capacity of bacteria to colonize mucosal surfaces of mammalian hosts depends in part on their ability to adhere to mucosal epithelial cells. Secretory IgA (sIgA) antibodies block certain bacterial adherence factors, preventing their attachment to host cells.

Antibody can also neutralize extracellular toxins made by pathogens. Such toxins must attach to target host cells before exerting their bioactivity. When antibodies bind to toxins, the toxin-antibody complex either is unable to attach to receptor sites on host target cells and is unable to enter the cell, or it is ingested by phagocytes. An antibody capable of neutralizing a toxin or antiserum containing neutralizing antibody against a toxin is called **antitoxin.** ▶| *Toxins are biological poisons (section 35.4)*

In a similar fashion, IgG, IgM, and IgA antibodies can bind to some viruses during their extracellular phase and block their capacity to bind to host cells. This antibody-mediated viral inactivation is called **viral neutralization.**

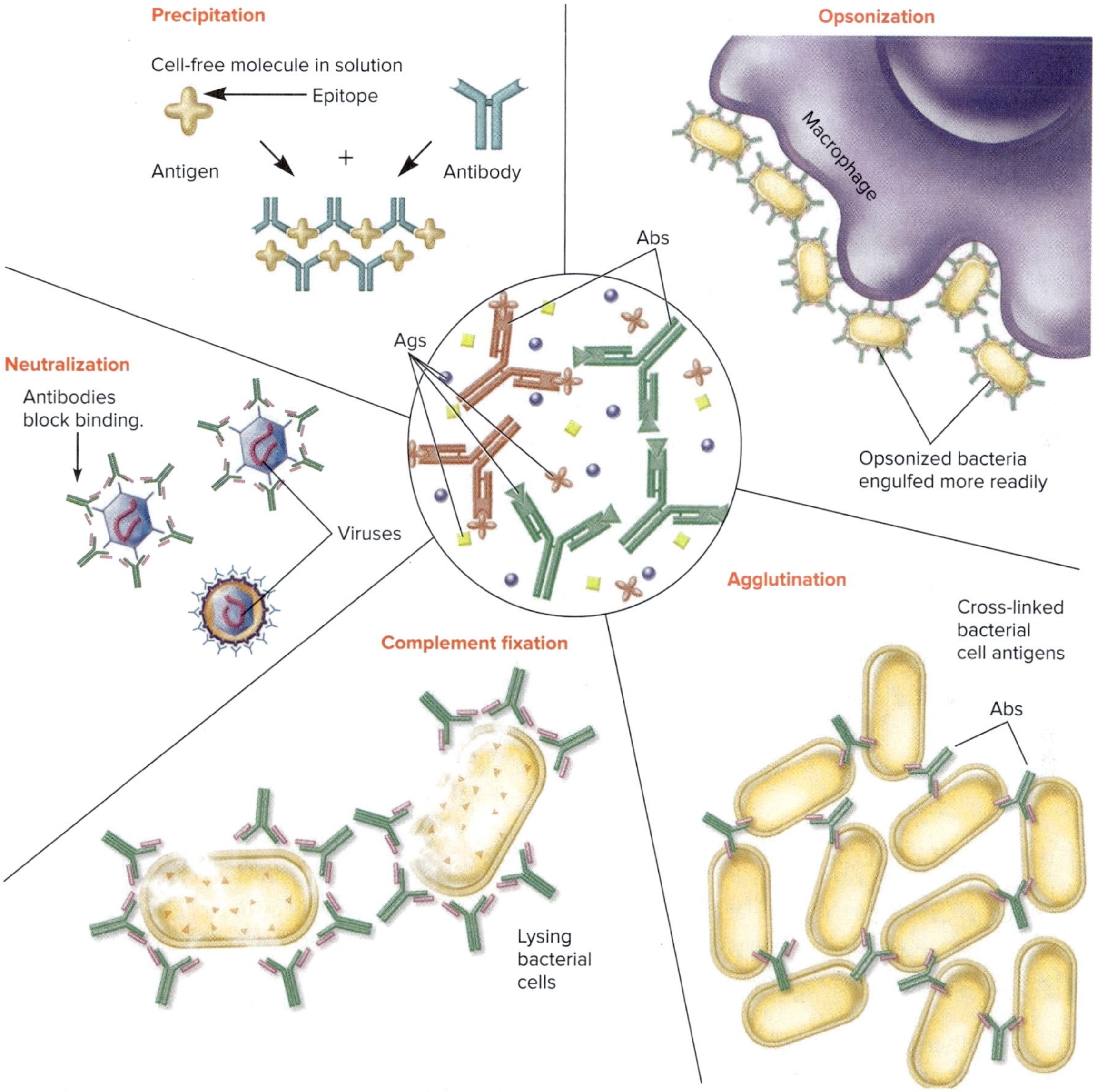

Figure 33.19 Consequences of Antigen-Antibody Binding. Immune complexes form when soluble antigens (Ag) bind soluble antibody (Ab), resulting in precipitation. Opsonization occurs when antibody binds to antigens on larger molecules or cells, enhancing their recognition by phagocytic cells. Agglutination results when antigens are cross-linked by antibody. The resulting complexes are insoluble. The classical complement cascade can be activated by immune complexes; this is called complement fixation. Neutralization results when antibody binds to toxins or viruses, preventing these antigens from binding to host cells.

MICRO INQUIRY *What is the difference between a precipitation and an agglutination reaction?*

Opsonization

The phagocytic process (*see figure 32.17*) can be greatly enhanced by opsonization. As noted in *section 32.3,* opsonization is the process by which complement and antibody attach to microorganisms or other foreign particles making them more easily recognized by phagocytic cells. Opsonizing antibodies include IgM and IgG, which trigger the classical complement cascade. These antibodies bind to Fc receptors on the surface of dendritic cells, macrophages, and neutrophils. This process dramatically increases the efficiency of antigen removal (*see figure 32.5*).

Immune Complex Formation

Because antibodies have at least two antigen-binding sites (the variable region at the end of each arm) and most antigens have at least two epitopes, cross-linking occurs when a single antibody binds two different antigens—one with each arm. When a large number of antigens are bound in this way by multiple antibodies, aggregates termed **immune complexes** are formed (figure 33.19). If the antigens are soluble molecules and the complex becomes large enough to settle out of solution, a **precipitation** (Latin *praecipitare,* to cast down) or **precipitin reaction** occurs. When the immune complex involves the cross-linking of cells or particles, an **agglutination reaction** occurs. These immune complexes are more rapidly phagocytosed than free antigens. However, immune complex formation can harm the host, as occurs in some types of kidney and vascular pathologies. Importantly, formation of immune complexes is the basis of some diagnostic tests. ▶| *Immune responses can be exploited to detect infections (section 37.3)*

Comprehension Check

1. What is the difference between neutralization and opsonization?
2. How does toxin neutralization occur? Viral neutralization?
3. Draw an immune complex (antigens are circles, antibodies are Y-shaped).

33.9 Immune Tolerance Is a Must

After reading this section, you should be able to:

a. Contrast the outcome of B-cell and T-cell exposure to self antigens
b. Compare and contrast central and peripheral tolerance

The process of T-cell and B-cell development can be described as a series of exquisite mechanisms designed to weed out self-reactive cells. Collectively, these mechanisms give rise to **acquired immune tolerance.** Acquired immune tolerance is more formally defined as the body's ability to produce T cells and antibodies against non-self antigens such as microbial antigens, while "tolerating" (not responding to) self antigens. Three general tolerance mechanisms protect the host from an immune response to self antigens: (1) negative selection, (2) the induction of anergy, and (3) inhibition of the immune response by the action of Treg cells.

As we have discussed, during T-cell and B-cell development, **negative selection** removes lymphocytes that recognize self antigens by apoptosis. Most negative selection of T cells occurs in the thymus and removal of self-reactive B cells takes place in the bone marrow. This is called **central tolerance** because it occurs in the primary sites of lymphocyte development (figure 33.18*a*).

Mechanisms occurring elsewhere in the body are collectively referred to as **peripheral tolerance.** These supplement central tolerance by limiting the activity of self-reactive lymphocytes released from the bone marrow or thymus (i.e., negative selection is not perfect). Whereas central tolerance removes self-reactive lymphocytes by apoptosis, peripheral tolerance involves incomplete activation signals given to T lymphocytes when they encounter self antigens in the periphery of the body. This mechanism leads to a state of unresponsiveness called **anergy.** Peripheral tolerance among self-reactive B cells is less robust and is governed by the T-cell population. This is sufficient to sustain tolerance because it denies the help essential for antibody production by self-reactive B cells that would produce high affinity and high avidity antibody as well as memory cells. By contrast, T-cell-independent B-cell activation results in the short-lived, low affinity and low avidity IgM.

Comprehension check

1. Describe the three ways acquired immune tolerance develops in the vertebrate host.
2. Why is anergy important?

33.10 The Immune System Can Malfunction

After reading this section, you should be able to:

a. Explain the differences between hypersensitivity, autoimmunity, tissue rejection, and immunodeficiency
b. Explain the immunologic basis for donated tissue rejection
c. Predict the impact immunosuppression or immunodeficiency has on the host

As in any body system, disorders also occur in the immune system. Immune disorders can be categorized as hypersensitivities, autoimmune diseases, transplantation (tissue) rejection, and immunodeficiencies. Each of these is now discussed.

Hypersensitivities

Hypersensitivity is an exaggerated adaptive immune response that occurs in an individual on a second or subsequent contact with an antigen. Hypersensitivity reactions can be immediate to delayed, but the main difference between them is the nature of the immune response to the antigen. Realizing this fact in 1963, Peter Gell (1914–2001) and Robert Coombs (1921–2006) developed a classification system for reactions responsible for hypersensitivities. Their system correlates clinical symptoms with the underlying immunologic events that occur during hypersensitivity reactions. The **Gell-Coombs classification** system divides hypersensitivities into four types: I, II, III, and IV.

Type I Hypersensitivity

An **allergy** (Greek *allos,* other; *ergon,* work) is one of the most common kinds of **type I hypersensitivity** reaction. Although IgE is important in the defense against helminths, it also mediates allergic responses. Allergic reactions occur when an individual who has produced IgE antibody in response to the initial exposure to an antigen (in this case called an **allergen**) subsequently encounters the same allergen. Upon initial exposure to a soluble allergen, B cells are stimulated to differentiate into plasma cells and produce IgE antibodies against the allergen with the help of T_H2 cells (**figure 33.20**). The initial exposure to an allergen is called sensitization, and there is a hereditary predisposition for an individual to be sensitized by a given allergen. Once synthesized, IgE binds irreversibly to the Fc receptors of mast cells, and to a lesser extent, basophils and eosinophils. When a subsequent exposure to the allergen occurs, the allergen attaches to the surface-bound IgE on the sensitized mast cells. The resulting IgE-mediated receptor cross-linking stimulates mast cell degranulation.

Degranulation releases physiological mediators such as histamine, leukotrienes, heparin, prostaglandins, PAF (*p*latelet-*a*ctivating *f*actor), ECF-A (*e*osinophil *c*hemotactic *f*actor of *a*naphylaxis), and proteolytic enzymes. These mediators trigger

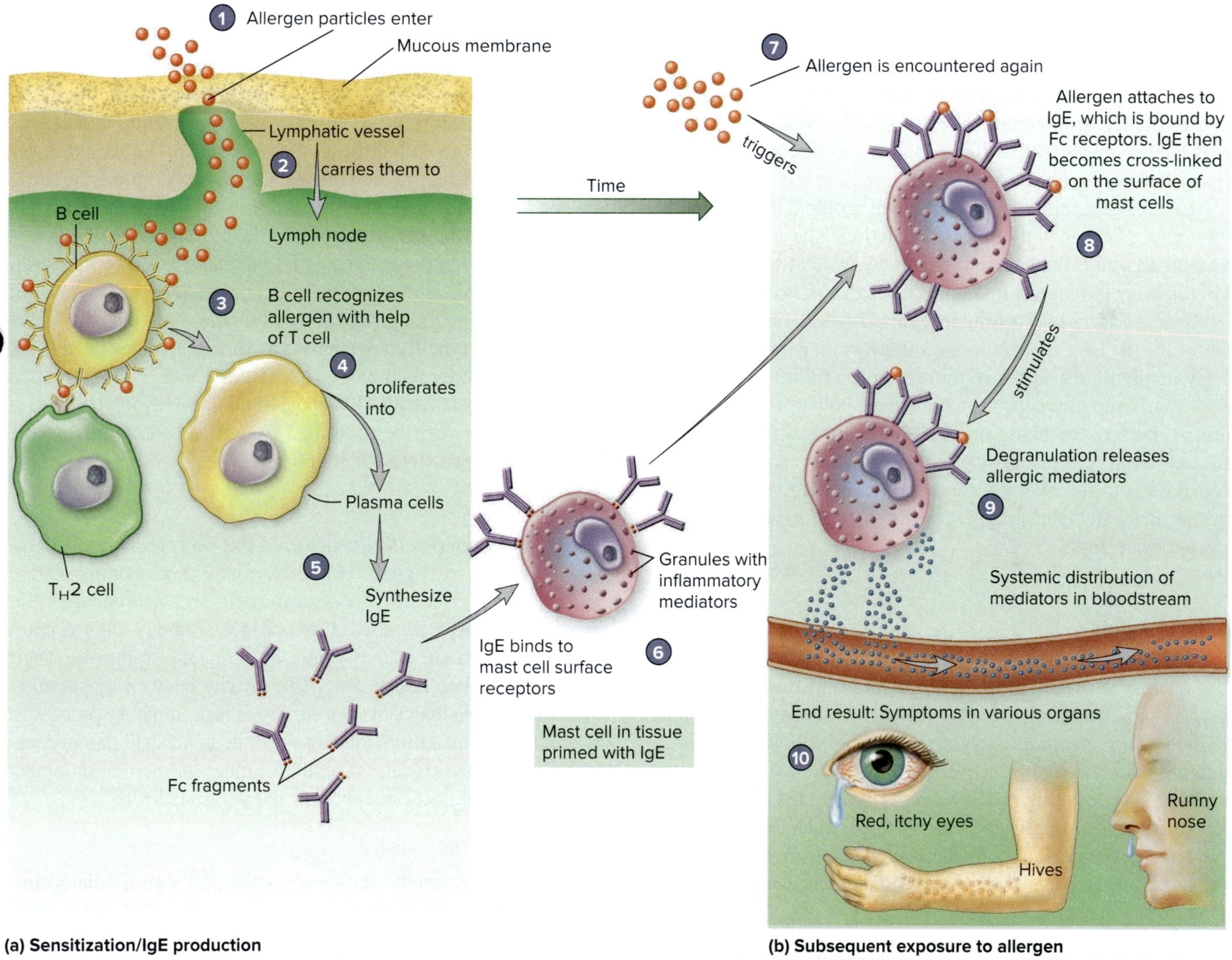

Figure 33.20 Type I Hypersensitivity (Allergic Response). (a) The initial contact (sensitization) of lymphocytes by small-protein allergens at mucous membranes results in T_H2 cell–assisted antibody class switching; plasma cells secrete IgE antibody. The IgE binds to its receptor on tissue mast cells (1–6). (b) Subsequent exposure to the same allergens results in their capture by the cell-bound IgE (7), triggering mast cell degranulation (8, 9). Characteristic signs and symptoms (hives, swelling, itching, etc.) of allergy ensue (10).

MICRO INQUIRY *What are some of the allergic mediators released during mast cell degranulation?*

smooth muscle contractions, vasodilation, increased vascular permeability, and mucus secretion (figure 33.20). The inclusive term for these responses is **anaphylaxis** (Greek *ana,* up, back again; *phylaxis,* protection). Anaphylaxis can be divided into systemic and localized reactions.

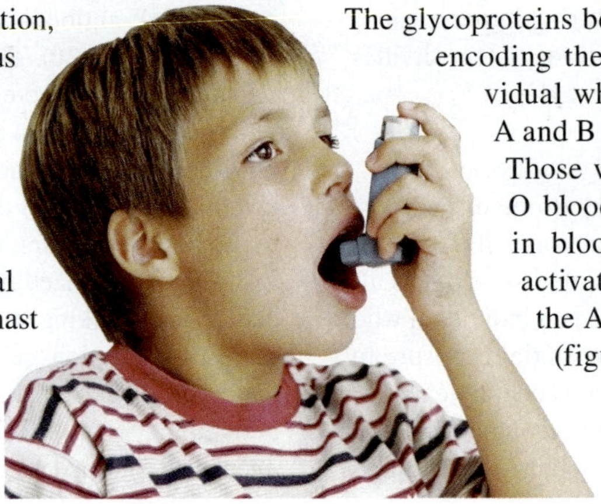
©Pixtal/age fotostock

Systemic anaphylaxis is a medical emergency caused by a sudden burst of mast cell mediators. It is characterized by respiratory distress caused by smooth muscle constriction in the bronchioles. The arterioles dilate, which greatly reduces arterial blood pressure and increases capillary permeability with rapid loss of fluid into the tissue spaces. Systemic anaphylaxis can be fatal within a few minutes due to asphyxiation, reduced blood flow, reduced blood pressure, and circulatory shock. Common examples of allergens that can produce systemic anaphylaxis include drugs (penicillin), peanuts, and insect venom from wasps, hornets, or bees.

Localized anaphylaxis is called an **atopic** (Greek, *a,* without; *topos,* place or "out of place") **reaction.** The symptoms that develop depend primarily on the route by which the allergen enters the body. Allergens that enter through the skin stimulate a localized reaction that results in a raised (wheal) and reddened (flare) lesion. Hay fever (allergic rhinitis) is a good example of atopy involving the upper respiratory tract. Initial exposure involves airborne allergens—such as plant pollen, fungal spores, animal dander, and house dust mites—that sensitize mast cells located within the mucous membranes of the respiratory tract. Reexposure to the allergen causes a localized anaphylactic response: itchy and tearing eyes, congested nasal passages, coughing, and sneezing.

Type II Hypersensitivity

Type II hypersensitivity is called a cytolytic or cytotoxic reaction because it results in the destruction of host cells, either by lysis or toxic mediators. In type II hypersensitivity, IgG or IgM antibodies are inappropriately directed against host cell-surface antigens. This stimulates activation of the classical complement pathway. The resulting opsonization of target host cells leads to activation of neutrophils and macrophages when their Fc receptors bind opsonizing antibody. Cell death occurs by phagocytosis and lysis, similar to that seen when complement exerts its effects on microbial membranes (*see figure 32.7*).

Mismatched blood transfusions are an example of a type II hypersensitivity. Blood transfusion was often fatal prior to the discovery of distinct blood types by Karl Landsteiner in 1904. Landsteiner's observation that sera from one person could agglutinate the blood cells of another person led to his identification of four distinct types of human blood. The red blood cell (RBC) types were later shown to result from cell-surface glycoproteins, now called the **ABO blood groups** (**figure 33.21*a***). The glycoproteins belong to group A or B. Because the genes encoding the glycoproteins are codominant, an individual who inherits both alleles will produce both A and B glycoproteins and have an AB blood type. Those who do not express either gene have type O blood. Type II hypersensitivity reaction seen in blood transfusion occurs as complement is activated by cross-linking antibodies to either the A or B glycoprotein on the suface of RBCs (figure 33.21*b*).

Blood typing can be accomplished by a slightly more sophisticated method of Landsteiner's process whereby blood from one host is mixed with antibodies specific for type A or type B blood. Agglutination of red cells by antibody (figure 33.21*c*) is used as a diagnostic tool to determine blood type.

After the ABO blood types were discovered, another RBC antigen was discovered as a result of experiments with rhesus monkeys. The so-called rhesus (Rh) factor (or D antigen) is determined by the expression of two alleles, one dominant (coding for the factor) and one recessive (not coding for the factor). Thus an individual who is homozygous or heterozygous for the dominant Rh allele has the antigen and is indicated as Rh^+. Expression of two recessive alleles results in the designation of Rh^- (no Rh factor). Incompatibility between Rh^- mothers and their Rh^+ fetus can result in a type II hypersensitivity in which maternal anti-Rh antibodies destroy fetal blood cells (**figure 33.22**). This is called **erythroblastosis fetalis.** This potentially fatal hemolytic disease of the newborn can be prevented by passively immunizing the mother with anti-Rh factor antibodies.

Type III Hypersensitivity

Type III hypersensitivity involves the overproduction of immune complexes (**figure 33.23**). Normally these complexes are phagocytosed effectively by dendritic cells and macrophages. But when an excessive number of IgG or IgM antibody-antigen complexes are formed, they may not be efficiently removed. Their accumulation can lead to a hypersensitivity reaction by stimulating complement that triggers a variety of inflammatory processes. The resulting inflammation can result in significant damage, especially of blood vessels (vasculitis), kidney glomerular basement membranes (glomerulonephritis), and joints (arthritis).

Type IV Hypersensitivity

Type IV hypersensitivity involves delayed, cell-mediated immune reactions. Therefore, unlike types I, II, and III hypersensitivities, type IV is mediated by T cells rather than by antibodies. Both T_H and CTL cells can elicit type IV hypersensitivity reactions, depending on the pathway in which the antigen is processed and presented. These events usually take a day or more to cause signs and symptoms, so type IV hypersensitivities are also called **delayed-type hypersensitivities** (**DTH**). Two important DTHs are tuberculin hypersensitivity and allergic contact dermatitis. Other examples of type IV

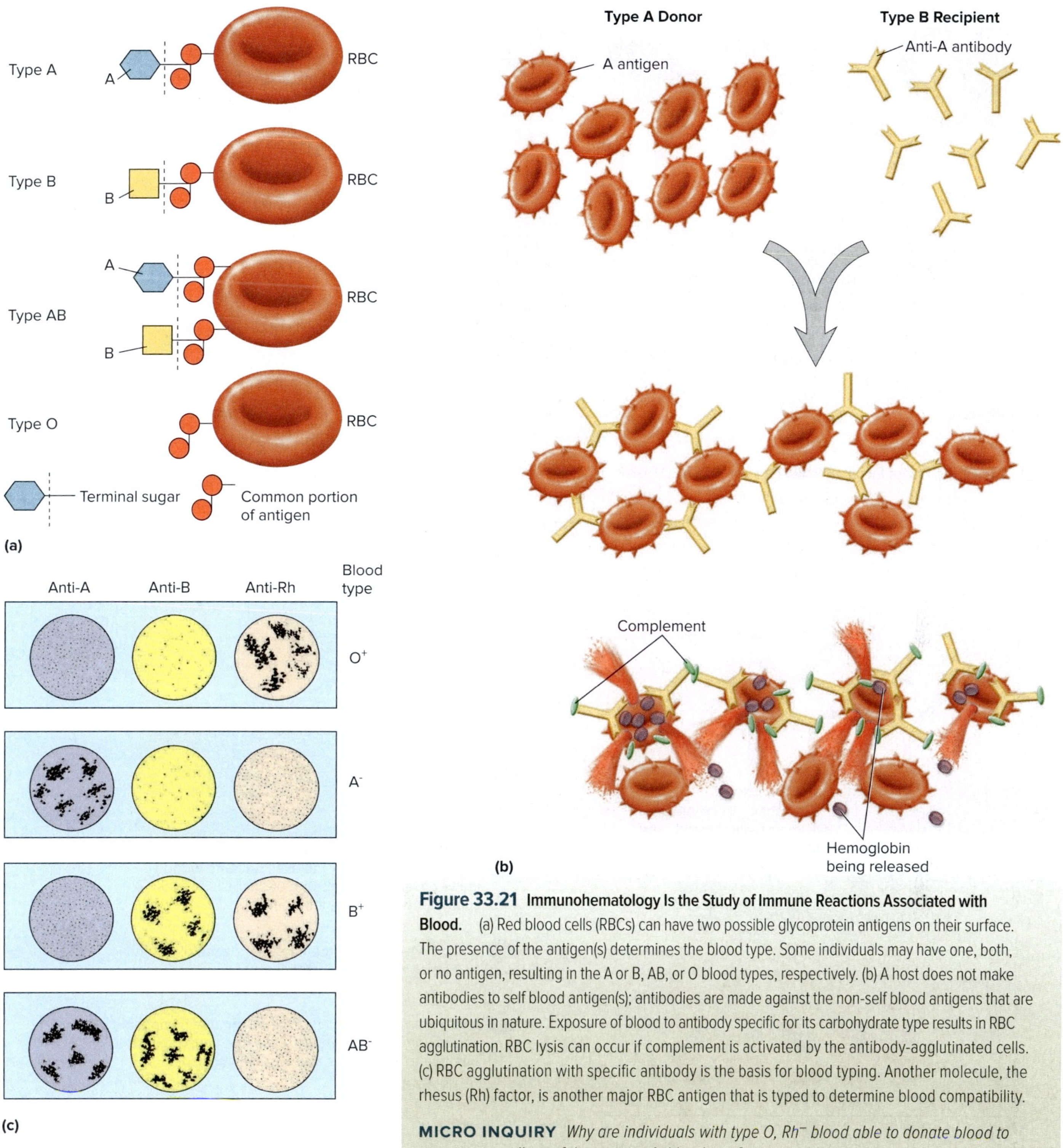

Figure 33.21 Immunohematology Is the Study of Immune Reactions Associated with Blood. (a) Red blood cells (RBCs) can have two possible glycoprotein antigens on their surface. The presence of the antigen(s) determines the blood type. Some individuals may have one, both, or no antigen, resulting in the A or B, AB, or O blood types, respectively. (b) A host does not make antibodies to self blood antigen(s); antibodies are made against the non-self blood antigens that are ubiquitous in nature. Exposure of blood to antibody specific for its carbohydrate type results in RBC agglutination. RBC lysis can occur if complement is activated by the antibody-agglutinated cells. (c) RBC agglutination with specific antibody is the basis for blood typing. Another molecule, the rhesus (Rh) factor, is another major RBC antigen that is typed to determine blood compatibility.

MICRO INQUIRY *Why are individuals with type O, Rh⁻ blood able to donate blood to anyone, regardless of the recipient's blood type?*

hypersensitivity reactions include cell and tissue destruction caused by leprosy, tuberculosis, leishmaniasis, candidiasis, and herpes simplex lesions. These infectious diseases are discussed in chapters 38–40.

Type IV hypersensitivity reactions occur when antigens, especially those binding to serum proteins or tissue cells, are processed and presented to T cells. If the antigen is phagocytosed, it will be presented to T_H cells by class II MHC molecules on the antigen presenting cell. This activates T_H1 cells, causing proliferation and pro-inflammatory cytokine secretion. If the antigen is lipid soluble, it can cross the cell membrane to be processed within the cytosol and be presented to CTL cells by

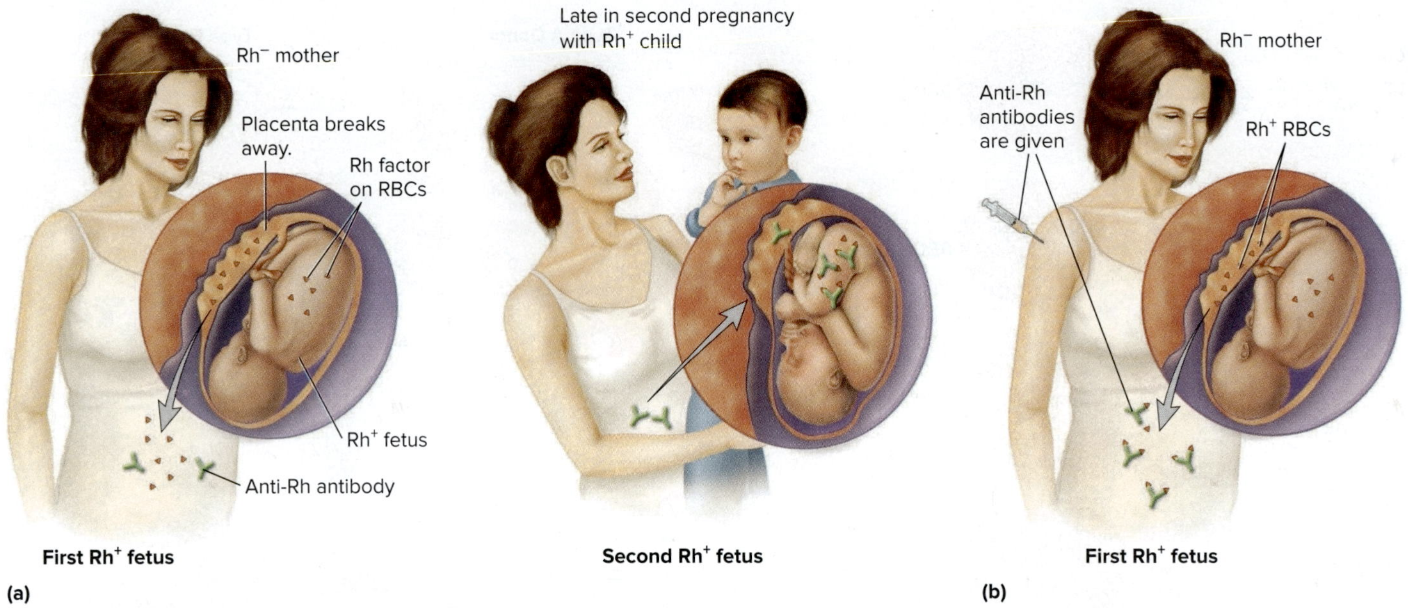

Figure 33.22 Rh Factor Incompatibility Can Result in RBC Lysis. (a) A naturally occurring blood cell incompatibility results when an Rh^+ fetus develops within an Rh^- mother. Initial sensitization of the maternal immune system occurs during birth when the placenta detaches and fetal blood passes through the placental barrier. In most cases, the fetus develops normally. However, a subsequent pregnancy with an Rh^+ fetus results in a severe, fetal hemolysis because the mother has made abundant anti-Rh antibody. (b) Anti-Rh antibody administered to Rh^- mothers during first and subsequent pregnancies bind, inactivate, and remove Rh factor that is transferred from the fetus, thereby eliminating the inciting antigen.

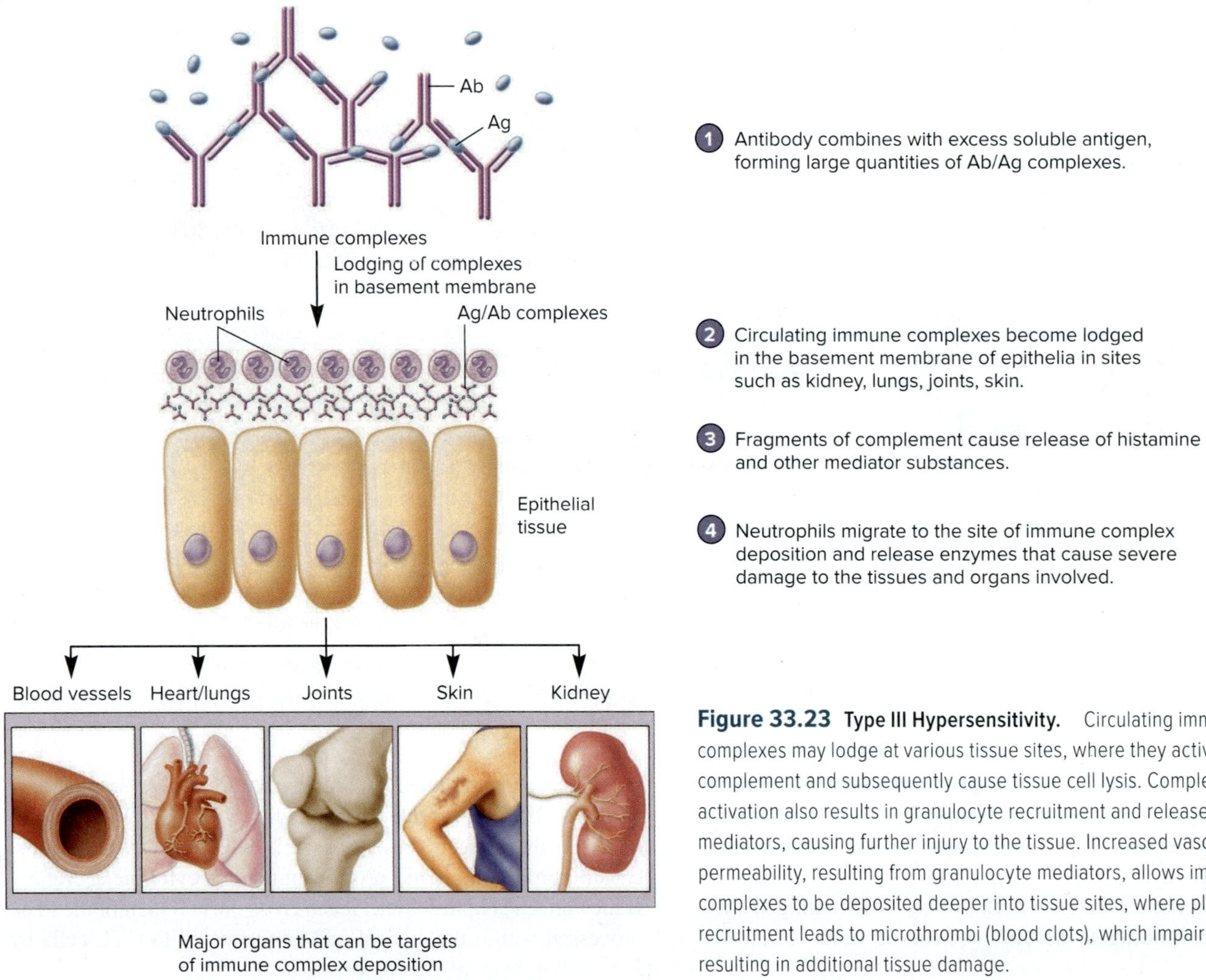

Figure 33.23 Type III Hypersensitivity. Circulating immune complexes may lodge at various tissue sites, where they activate complement and subsequently cause tissue cell lysis. Complement activation also results in granulocyte recruitment and release of their mediators, causing further injury to the tissue. Increased vascular permeability, resulting from granulocyte mediators, allows immune complexes to be deposited deeper into tissue sites, where platelet recruitment leads to microthrombi (blood clots), which impair blood flow, resulting in additional tissue damage.

class I MHC molecules. CTLs are thus stimulated to kill the cell presenting the antigen. T_H cell and CTL signals also stimulate the expression of adhesion molecules on local blood vessel endothelium to increase vascular permeability. This enables fluid and cells to enter the tissue space. The cytokines attract and recruit lymphocytes, macrophages, and basophils, exacerbating the inflammatory response. Extensive tissue damage may result.

In tuberculin hypersensitivity, partially purified proteins collectively called tuberculin are obtained from the bacterium that causes tuberculosis (TB). The proteins are injected into the skin of the forearm. If by 48 to 72 hours the area becomes reddened, firm, and hard (indurated), the test is considered positive. T_H1 cells previously activated by exposure to the tuberculosis bacterium migrate to the injection site and are responsible for the induration. If a person has never been exposed to the bacteria that cause TB, the injection site may initially become inflamed but does not enlarge and become indurated.

Allergic contact dermatitis is a type IV reaction caused by haptens that combine with carrier proteins in the skin to form the allergen that elicits the immune response (**figure 33.24**). Examples of these haptens include cosmetics, plant materials (catechol molecules from poison ivy and poison oak), topical chemotherapeutic agents, metals, and jewelry (especially jewelry containing nickel).

Comprehension Check

1. Discuss the mechanism of type I hypersensitivity reactions and how these can lead to systemic and localized anaphylaxis.
2. Why are type II hypersensitivity reactions called cytolytic or cytotoxic?
3. What characterizes a type III hypersensitivity reaction? Give an example.
4. How does the TB skin test work? Do you think the TB skin test would be accurate in someone who is immunosuppressed? Why or why not?

Autoimmunity and Autoimmune Diseases

The importance of the immune system's ability to distinguish between self and non-self antigens is dramatically illustrated by autoimmune diseases. In these conditions, the body loses tolerance and mounts an abnormal immune attack, either with antibodies or T cells, against its own self antigens.

It is important to distinguish between autoimmunity and autoimmune disease. Autoimmunity is benign, whereas autoimmune disease can be fatal. **Autoimmunity** is characterized by the presence of serum antibodies that react with self antigens. These antibodies are called autoantibodies. The formation of autoantibodies is a normal consequence of aging; is readily inducible by infectious agents or drugs; and is typically reversible (it disappears when the triggering antigen is removed). **Autoimmune disease** results from the activation of self-reactive T and B cells that, following stimulation by genetic or environmental triggers, cause chronic tissue damage. Examples of some autoimmune diseases are listed in **table 33.3**.

There is no single etiology for autoimmune disease. Rather, each disease has its own set of unique immune pathologies (table 33.3). Some autoimmune diseases are largely T-cell driven, others involve large quantities of autoantibodies, and some involve both mechanisms. In many cases, an infectious trigger, such as a viral or bacterial infection, may be involved. Evidence suggests that a change in the normal host-microbiota relationship (dysbiosis) results in immune dysfunction (*see figures 34.5 and 34.6*). An important component of autoimmune disease is the inability of regulatory T cells to limit immune activities.

Transplantation (Tissue) Rejection

Transplants between genetically different individuals within a species are termed **allografts** (Greek *allos,* other). Some

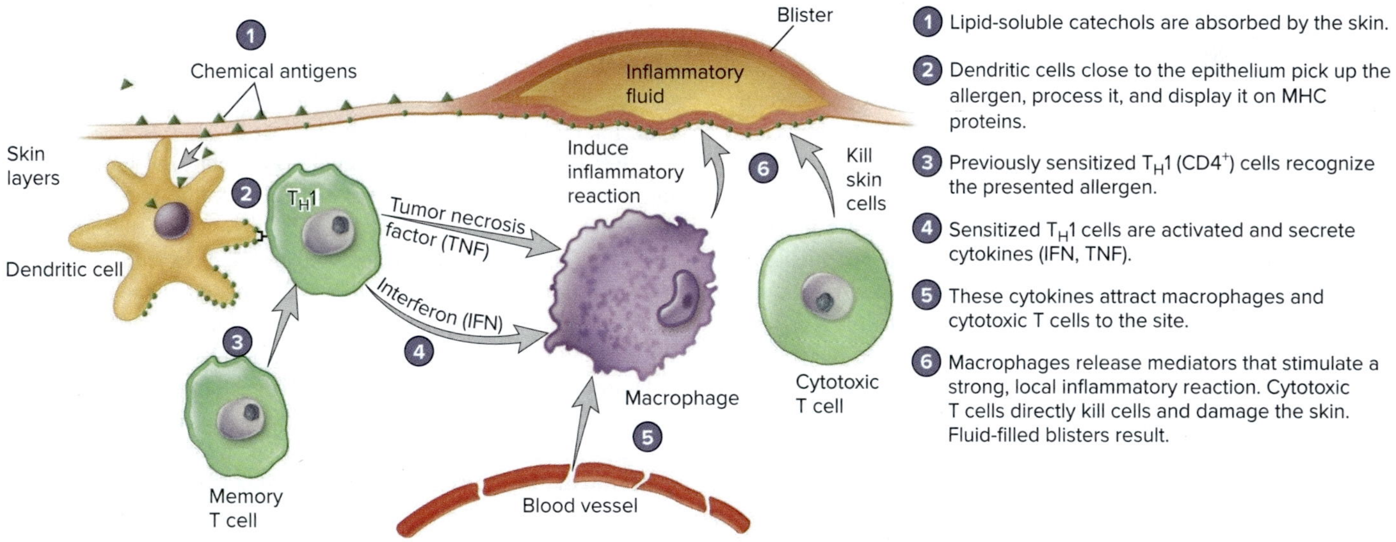

Figure 33.24 Contact Dermatitis. In contact dermatitis to poison ivy, a person initially becomes exposed to the allergen, predominantly 3-n-pentadecyl-catechol found in the resinous sap produced by the leaves, fruit, stems, and bark of the poison ivy plant. The catechol molecules, acting as haptens, combine with high-molecular-weight skin proteins. After 7 to 10 days, sensitized T cells are produced and give rise to memory T cells. Upon second contact, the catechols bind to the same skin proteins, and the memory T cells become activated in only 1 to 2 days, leading to an inflammatory reaction (contact dermatitis).

Table 33.3 Some Autoimmune Diseases in Humans

Disease	Autoantigen	Pathophysiology
Acute rheumatic fever	Streptococcal cell wall antigens mimic self antigens and induce antibodies that cross-react with antigens on heart muscle and other cells.	Type II hypersensitivity leads to myocarditis, heart valve scarring, and arthritis.
Autoimmune hemolytic anemia	Rh blood group antigen induces antibody to Rh antigen on red blood cells.	Type II hypersensitivity leads to anemia when red blood cells are destroyed by complement and phagocytosis.
Goodpasture's syndrome	Damage to kidney basement membrane exposes cryptic collagen protein, inducing anti-collagen antibody.	Type II hypersensitivity results in glomerulonephritis and pulmonary hemorrhage.
Graves' disease	Antibody to thyroid-stimulating hormone (TSH) receptor mimics TSH.	Overstimulation of TSH receptor leads to hyperthyroidism.
Multiple sclerosis	Antibody and activated T cells to several nervous system antigens	Types II and IV hypersensitivities alter nerve communications, leading to numbness, weakness, spasm, and loss of motor and cognitive function.
Myasthenia gravis	Antibody to acetycholine receptor in skeletal muscle	Antibody blockade of neurotransmitter receptors results in progressive muscular weakness.
Rheumatoid arthritis	IgG antibody (to synovial joint cartilage antigen) recognized as foreign	Type III hypersensitivity from immune complexes of antibodies to antibodies results in joint inflammation and destruction.
Systemic lupus erythematosus	Antibodies to various cellular (DNA, nucleoprotein, cardiolipin) and blood clotting components	Type III hypersensitivity results in immune complex-induced arthritis, glomerulonephritis, vasculitis, and rash.
Type I diabetes mellitus	Antibody and activated T cells to pancreatic beta cell antigens	Types II and IV hypersensitivities destroy beta cells, resulting in insulin deficiency.

transplanted tissues do not stimulate an immune response. For example, a transplanted cornea is rarely rejected because lymphocytes do not circulate into the anterior chamber of the eye. This site is considered an immunologically privileged site. Another example of a privileged tissue is the heart valve, which can be transplanted from a pig to a human without stimulating an immune response. Such a graft between different species is termed a **xenograft** (Greek *xenos,* strayed).

However, most transplants involve tissues that are not immunologically privileged. Unless all MHC proteins are identical between organ donor and recipient (as in identical twins), the recipient's cells will recognize the donor's tissues as foreign. This triggers the recipient's innate and adaptive immune responses, which may destroy the donor tissue. Such a response is called **host-versus-graft disease,** or organ rejection. Organ rejection can occur by two different mechanisms. First, foreign MHC molecules on transplanted tissue, or the "graft," are recognized as non-self, which results in activation of cytotoxic T cells (**figure 33.25**). Cytotoxic T cells then recognize the graft because it bears foreign class I MHC molecules. This response is much like the activation of CTLs by virally infected host cells, and results in destruction of the donated tissue. A second mechanism involves T-helper cells reacting to the graft and releasing cytokines. The cytokines stimulate macrophages to enter, accumulate within the graft, and destroy it. MHC molecules therefore play a dominant role in tissue rejection reactions because of their unique association with the recognition of self and non-self and their interaction with T cells.

Because class I MHC molecules are central to the development of host-versus-graft disease, it follows that the greater the antigenic difference between class I molecules of the recipient and donor tissues, the more rapid and severe the rejection reaction is likely to be. Class II MHC molecule mismatch can also result in rejection. Thus the goal in matching donor to recipient is to have as many MHC alleles in common as possible. Most recipients are not 100% matched to their donors, so immunosuppressing drugs are used to prevent host-mediated rejection of the graft.

Bone marrow (i.e., stem cell) transplant recipients may develop **graft-versus-host disease,** which is distinct from host-versus-graft seen in organ rejection. Graft-versus-host disease occurs when the transplanted bone marrow contains immune cells that recognize host MHC molecules as foreign and attack the host's cells. Because bone marrow transplant involves the introduction of new immune cells, it requires more rigorous matching between donor and

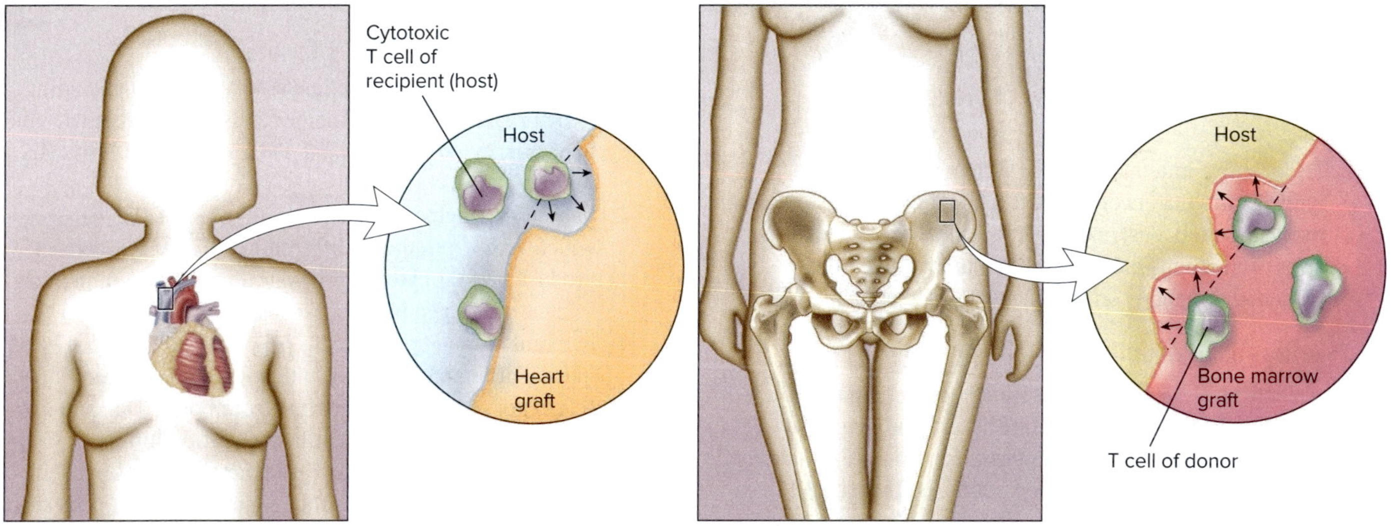

(a) Host-versus-graft disease **(b) Graft-versus-host disease**

Figure 33.25 Potential Transplantation Reactions. (a) Donated tissues that are not from an identical twin contain cellular MHC proteins that are recognized as foreign by the recipient host (host-versus-graft disease). The tissue is then attacked by host CTLs, resulting in its damage and rejection. (b) Donated stem cells may also contain immune cells that react against host antigens. Recognition of a foreign host by donor CTLs results in graft-versus-host disease.

recipient. Some level of mismatch is typical, however, and is controlled by immunosuppressive drugs.

Immunodeficiencies

Defects in one or more components of the immune system can result in its failing to recognize and respond properly to antigens. Such **immunodeficiencies** can make a person more prone to infection than people capable of a complete and active immune response. Despite the increase in knowledge of functional derangements and cellular abnormalities in many immunodeficiency disorders, the fundamental biological errors responsible for them remain largely unknown. To date, most genetic errors associated with these immunodeficiencies are located on the X chromosome and produce primary or congenital immunodeficiencies (**table 33.4**). Other immunodeficiencies can be acquired because of infections by immunosuppressive microorganisms, such as HIV.

Comprehension Check

1. What is an autoimmune disease and how might it develop?
2. What is an immunologically privileged site and how is it related to transplantation success?
3. Why are most patients given immunosuppressive drugs prior to receiving an organ transplant, as well as afterwards?
4. Describe an immunodeficiency. How might immunodeficiencies arise?

Table 33.4 Some Congenital Immune Deficiencies in Humans

Condition	Symptoms	Cause
Chronic granulomatous disease	Defective monocytes and neutrophils leading to recurrent bacterial and fungal infections	Failure to produce reactive oxygen intermediates due to defective NADPH oxidase
X-linked agammaglobulinemia	Plasma cell or B-cell deficiency and inability to produce adequate specific antibodies	Defective B-cell differentiation due to loss of tyrosine kinase
DiGeorge syndrome	T-cell deficiency and very poor cell-mediated immunity	Lack of thymus or a poorly developed thymus
Severe combined immunodeficiency disease (SCID)	Both antibody production and cell-mediated immunity impaired due to a great reduction of B- and T-cell levels	Various mechanisms (e.g., defective B- and T-cell maturation because of X-linked gene mutation; absence of RAG enzymes in lymphocytes)

Key Concepts

33.1 Adaptive Immunity Relies on Recognition and Memory

- The adaptive immune response system consists of lymphocytes that recognize and respond to specific foreign molecules (antigens) and remember them. Two branches or arms of immunity are recognized: humoral (antibody-mediated) immunity and cellular (cell-mediated) immunity.

33.2 Antigens Elicit Immunity

- An antigen is a substance that stimulates an immune response. Each antigen can have several antigenic determinant sites called epitopes that stimulate production of and combine with specific antibodies (**figure 33.1**).
- Haptens are small molecules that are not antigenic by themselves but can become antigenic if bound to a larger carrier molecule.

33.3 Adaptive Immunity Can Be Earned or Borrowed

- Immunity can be acquired by natural means—actively through infection or passively through receipt of preformed antibodies, as through breast milk (**figure 33.2**).
- Immunity can be acquired by artificial means—actively through immunization or passively through receipt of preformed antibodies, as with antisera.

33.4 Recognition of Foreignness Is Critical for a Strong Defense

- MHC molecules are cell-surface proteins coded by a group of genes termed the major histocompatibility complex. Class I MHC proteins are found on all nucleated cells of mammals. Class II MHC proteins are expressed on macrophages, dendritic cells, and B cells. The human MHC gene products are called the human leukocyte antigens (HLAs) (**figure 33.3**).
- Class I MHC proteins collect self and non-self peptides processed by the proteasome and present them to cytotoxic T cells, which respond only to non-self peptides (**figure 33.4**).
- Class II MHC proteins collect foreign peptides processed by the phagosome (macrophages and dendritic cells) or endosome (B cells) and present them to T-helper cells.

33.5 T Cells Are Critical for Immune Function

- T-cell development in the thymus involves selection for the CD4 or CD8 coreceptor and against T cells that recognize self antigens (**figure 33.5**).
- Antigen-presenting cells—macrophages, dendritic cells, and B cells—take in foreign antigens or pathogens, process them, and present antigenic peptides complexed with MHC class II molecules to T-helper cells ($CD4^+$ cells). Antigens presented in MHC class I are presented to $CD8^+$ T cells (**figure 33.6**).
- T cells control the development of other cells, including effector B and T cells. T-helper cells ($CD4^+$) regulate cell behavior; cytotoxic T cells ($CD8^+$) kill altered host cells directly.
- Subsets of T-helper cells include T_H1, T_H2, T_H17, and Treg, which arise from the undifferentiated T_H0. T_H1 cells produce various cytokines involved in cellular immunity. T_H2 cells also produce cytokines but are involved in humoral immunity. T_H17 cells primarily produce IL-17 and induce inflammation by recruiting neutrophils (**figure 33.9**).
- Regulatory T (Treg) cells arise from both $CD4^+$ and $CD8^+$ T cells and downregulate specific immune reactions, especially those rare reactions attacking self.
- Cytotoxic T lymphocytes recognize target host cells such as virus-infected cells that have foreign antigens within the class I MHC molecules on their surface. The CTLs then attack and destroy the target cells using the perforin pathway, the Fas-FasL apoptotic pathway, or both.

33.6 B Cells Make Antibodies

- B cells defend against antigens by differentiating into plasma cells that secrete antibodies into the blood and lymph, providing humoral or antibody-mediated immunity.
- B cells can be stimulated to divide and differentiate into antibody-secreting plasma cells when triggered by the appropriate signals.
- B cells have receptor immunoglobulins on their plasma membrane surface that are specific for three-dimensional epitopes on an antigen (**figure 33.10**).
- B cell differentiation is usually T-cell dependent, which yields plasma cells and memory cells. T-cell-independent B-cell activation results in the formation of only plasma cells that produce low affinity antibodies (**figure 33.11**).

33.7 Antibodies Bind Specific 3-D Antigens

- Antibodies (immunoglobulins) are a group of glycoproteins present in the blood, tissue fluids, and mucous membranes of vertebrates. All immunoglobulins have a basic structure composed of four polypeptide chains (two light and two heavy) connected to each other by disulfide bonds (**figure 33.12**). In humans, five immunoglobulin classes exist: IgG, IgA, IgM, IgD, and IgE (**figures 33.13** and **33.14** and **table 33.1**).
- The primary antibody response occurs following initial exposure to the antigen. This response has lag, log, plateau, and decline phases. Upon secondary antigen challenge, the B cells mount a heightened and accelerated response (**figure 33.15**).
- Antibody diversity results during B-cell development by the rearrangement and splicing of the individual gene segments on the antibody-coding chromosomes, the

generation of different codons during splicing, and the independent assortment of light- and heavy-chain genes. Antibody diversity further increases upon T-cell interaction, when activated B cells undergo somatic hypermutation (**figures 33.16** and **33.17**).

- Clonal selection results in large populations of genetically distinct B and T cells, each of which recognizes a unique antigen (**figure 33.18**).

33.8 Antibodies Doom Antigens

- The complement system can be activated by IgG or IgM, leading to cell lysis, phagocytosis, chemotaxis, or stimulation of the inflammatory response.
- Other defensive antigen-antibody interactions include toxin neutralization, viral neutralization, adherence inhibition, opsonization, and immune complex formation (**figure 33.19**).

33.9 Immune Tolerance Is a Must

- Central tolerance occurs during negative selection of lymphocytes that respond to self antigen. It occurs in the bone marrow for B cells and in the thymus for T cells.
- Peripheral tolerance occurs in the tissues and predominantly prevents T cells from responding to self.
- During peripheral tolerance, T cells may become anergic, which is a state of immune unresponsiveness.

33.10 The Immune System Can Malfunction

- When the immune response occurs in an exaggerated form and results in tissue damage to the individual, the term *hypersensitivity* is applied. There are four types of hypersensitivity reactions, designated as types I through IV (**figures 33.20–33.24**).
- Autoimmune diseases result when self-reactive T and B cells attack the body and cause tissue damage. A variety of factors can influence the development of autoimmune disease (**table 33.3**).
- The immune system can reject tissue transplants. Xenografts involve transplants of privileged tissue between different species, and allografts are transplants between genetically different individuals of the same species (**figure 33.25**).
- Immunodeficiency diseases are a diverse group of conditions in which an individual's susceptibility to various infections is increased; several severe diseases can arise because of one or more defects in the adaptive or innate immune responses (**table 33.4**).

Active Learning

1. A rare immunodeficiency disease called bare lymphocyte syndrome-2 is caused by a genetic mutation that prevents the production of MHC class II molecules. Predict the changes in development and function seen in T_H, CTLs, and B cells in a patient with this condition.
2. The diagnosis of mononucleosis, which is caused by the Epstein-Barr virus (EBV), is sometimes made by looking for anti-EBV antibodies, a process called EBV serology. You receive EBV serology reports for two patients: One has a high anti-EBV IgM titer, the other has a high anti-EBV IgG titer. Are you confident that both patients have mononucleosis? Explain your answer. (Hint: Refer to figure 33.15.)
3. Very young children have a large pool of mature, naïve T and B cells but very few memory cells; elderly adults have just the opposite: very few mature, naïve T and B cells but many memory cells. Discuss how this affects the ability of young children and the elderly to respond to infectious agents.
4. Discuss reasons that might explain why three signals are required for T-cell activation but only one signal is required for activation of an APC.
5. The differentiation of $CD8^+$ cells to CTLs requires activation by DCs presenting non-self antigenic peptides in their MHC class I molecules. It has long been known that for many antigens, this is sufficient to trigger $CD8^+$ activation and subsequent CTL development. However, for other antigens, immune synapse formation with a DC is not sufficient to drive differentiation of $CD8^+$ cells. In these cases, T_H1 help is also required. The nature of this T_H1 help has been debated. Immunologists explored this question in a model using herpes simplex virus and showed that rather than directly stimulating $CD8^+$ cells, T_H1 cells triggered DCs to increase the level of cytokines they released, which then resulted in $CD8^+$ activation and CTL development. Discuss how this experimental system might have been designed and the implications of this finding in terms of developing a topical agent to limit herpes outbreaks.

   Read the original paper: Greyer, M., et al. 2016. T cell help amplifies innate signals in $CD8^+$ DCs for optimal $CD8^+$ T cell priming. *Cell Reports.* 14:586–97. doi:10.1016/j.celrep.2015.12.058.

34

The Microbe-Human Ecosystem

©Radius Images/Alamy Stock Photo

Embrace Your Gut Flora

Samantha D. was desperate. Over the course of 8 months, she had been battling diarrhea so severe she had lost 60 pounds, was confined to a wheelchair, and was wearing adult diapers. For a period of 6 months, Vanessa R. ran high fevers and had diarrhea about once an hour on her worst days. Her better days were spent sleeping. Both these women suffered from *Clostridioides* (formerly *Clostridium*) *difficile* infection. Both were eventually cured by a treatment once considered crazy: fecal microbiome transplant (FMT).

First, the microbe: *C. difficile* (C. diff for short) is a Gram-positive, endospore-forming firmicute (*see sections 23.2 and 39.6*) that resides in low numbers in the human gut. It is normally outcompeted by nonpathogenic bacteria. Many C. diff infections arise in people undergoing antibiotic therapy. This leads to the inhibition of normal gut microbes, enabling the overgrowth of C. diff and its release of toxins triggering disabling diarrhea accompanied by sloughing of the gut epithelium. It is most commonly seen in hospitalized patients, but community-acquired C. diff infections are on the rise.

Next the cure: Antibiotic therapy frequently fails because C. diff has acquired resistance to multiple antimicrobials and antibiotic treatment continues to suppress the growth of gut microorganisms needed to outcompete this virulent pathogen. A growing number of physicians are now treating patients with recurrent C. diff infections with FMT. Feces collected from healthy donors without a recent history of antibiotic use is first tested to be sure it is free of pathogenic viruses, bacteria, and protozoa. The stool is then mixed with sterile saline and infused into the patient from one end (via a nasogastric tube) or the other (via a colonoscope). About 90% of patients report a complete cure within 48 hours to a week. FMT is so successful for recurrent C. diff infections that efforts are now under way to commercialize encapsulated fecal microbiota as a more acceptable and standardized treatment.

Apart from the gross factor, what does FMT teach us? For a very long time, gut microorganisms have been termed commensals—symbiotic organisms that neither injure nor benefit their host. Clearly, the severity of C. diff infections and its cure by FMT argue that this term should no longer be considered accurate for many of the microbes that live in and on us. In this chapter, we explore the relationship microbes have with their human hosts. As we will see, it is more complex than anyone previously imagined.

Readiness Check:

Based on what you have learned previously, you should be able to:

- ✓ Compare and contrast the growth and physiology of obligate aerobes, facultative anaerobes, and strict anaerobes (section 7.5)
- ✓ Relate the advances in metagenomic analysis with the development of next-generation sequencing (sections 18.1 to 18.3)
- ✓ Explain how small subunit ribosomal RNA is used to identify microbes (section 19.3)
- ✓ Differentiate between commensalism, mutualism, and parasitism (section 27.1)
- ✓ Identify and describe the function of professional phagocytes (section 32.4)
- ✓ Describe the structure and function of mucosal-associated lymphoid tissue (sections 32.5 and 32.6)
- ✓ Discuss the process of inflammation (section 32.7)
- ✓ Compare and contrast the major functional differences between T cells and B cells (sections 33.5, 33.6, and 33.7)

34.1 Humans Are Holobionts

After reading this section, you should be able to:

a. Define the word *microbiome*
b. Explain why humans are considered holobionts

In 2001, renowned microbiologist and Nobel laureate Joshua Lederberg coined the term **microbiome** to describe all the microbes "that literally share our body space." Very few scientists at that time would have predicted the profound impact the discoveries made over the next two decades would have on our understanding of human physiology, pathophysiology, and what it means to be human. We have learned that humans cannot live a normal life without their microbial partners—that we are holobionts: hosts and microbes living together, and importantly, evolving together. After all, the average adult carries about three times more microbial cells (10^{14}) than human somatic cells (average about 4×10^{13}). With a combined microbial contribution of over 8 million protein-encoding genes (as compared to our roughly 22,000 genes), we have come to realize that many of these gene products serve essential functions our genome cannot provide.

Microorganisms commonly associated with the human body are also known as **normal microbiota** (the term "normal flora" was once common). The acquisition of normal microbiota is a selective process. In humans, each microbial niche is related to a variety of factors such as body location, age, gender, diet, nutrition, and environment. For instance, many microorganisms revealed by 16S ribosomal RNA sequencing (e.g., members of *Clostridiales* and *Bacteroidales*) can survive only within the anoxic and nutritionally unique environments established in or on the human host, thus preventing their culture in the lab. Furthermore, some organisms long thought to be normal microbiota turn out to be transient associates or laboratory contaminants.

Without technological advances associated with next-generation nucleotide sequencing and metagenomics, we would still know very little of our microbiomes. Through the application of these techniques, the Human Microbiome Project has produced over 2,300 metagenomes derived from multiple body sites from hundreds of healthy volunteers. Similar human microbiome surveys have been performed worldwide. Together these data demonstrate that there is no single healthy microbiome. Rather, each person harbors a unique collection of microorganisms that assemble from a lifetime of interaction with their environment, diet, medications, and many other factors. We thus begin the chapter with the story of human colonization by their microbial partners. ◀ *Next-generation DNA sequencing (section 18.1); Next-generation genomic sequencing (section 18.2); Metagenomics provides access to uncultured microbes (section 18.3)*

Comprehension Check

1. Define the term *microbiome*.
2. How have advances in molecular biology techniques contributed to our definition of the human microbiome?

34.2 The Microbiome Develops from Birth to Adulthood

After reading this section, you should be able to:

a. Discuss the role of bifidobacteria in the infant microbiome
b. Predict the environmental host conditions that favor the association of acidophilic, anaerobic, lactic acid fermenting, and halophilic bacteria
c. Discuss the factors that help mediate microbial diversity in and on the human body

Development of a Stable Microbiome

The normal community of mutualistic and commensal microbiota is not static. It begins developing at birth and changes as we age. It is perhaps not surprising that during infancy our microbiota are most sensitive to the environment. This early time in our lives also appears to be a "window of opportunity" for the development of the kinds of microbes that will be most beneficial for the maturation of our immune systems. By the time we are 3 years old, we have adopted an adultlike community of microbes (**figure 34.1**).

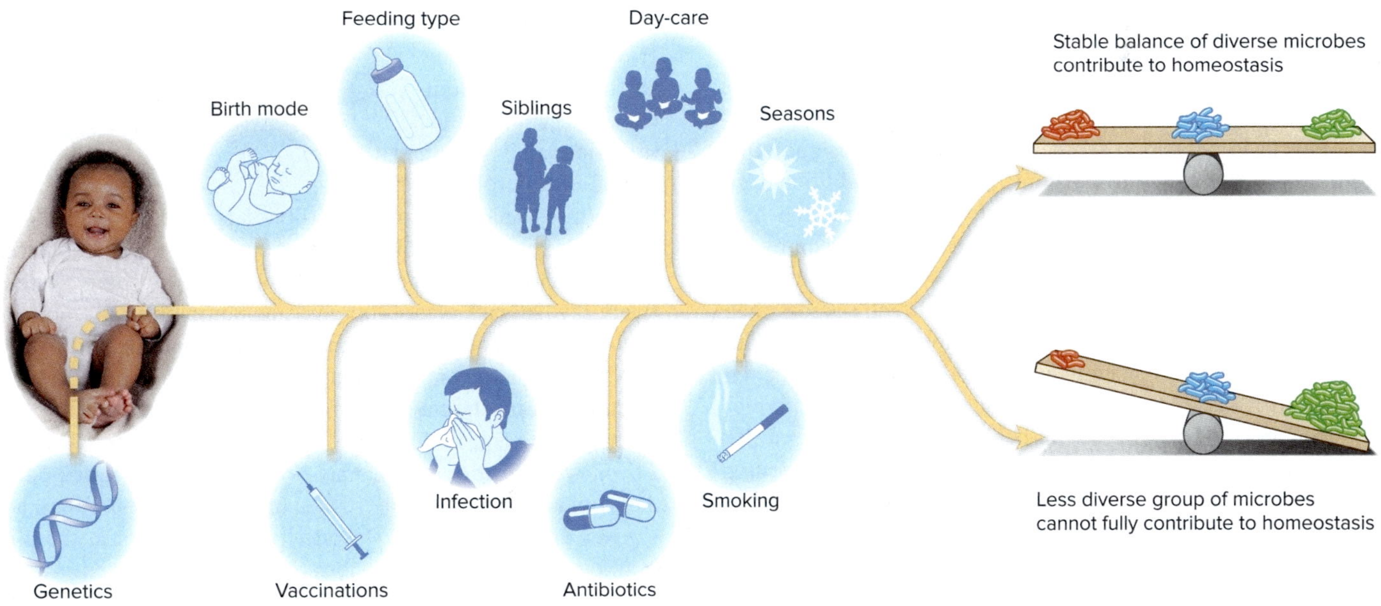

Figure 34.1 Many Factors Contribute to the Development of a Stable Microbiome. A diverse microbiome is linked to better health.
(photo) ©michel74100/Getty Images. (Text) Source: Man, Wing Ho, Piters,Wouter A.A. de Steenhuijsen and Bogaert, Debby. "The microbiota of the respiratory tract: gatekeeper to respiratory health," *Nature Reviews*, March 20, 2017.

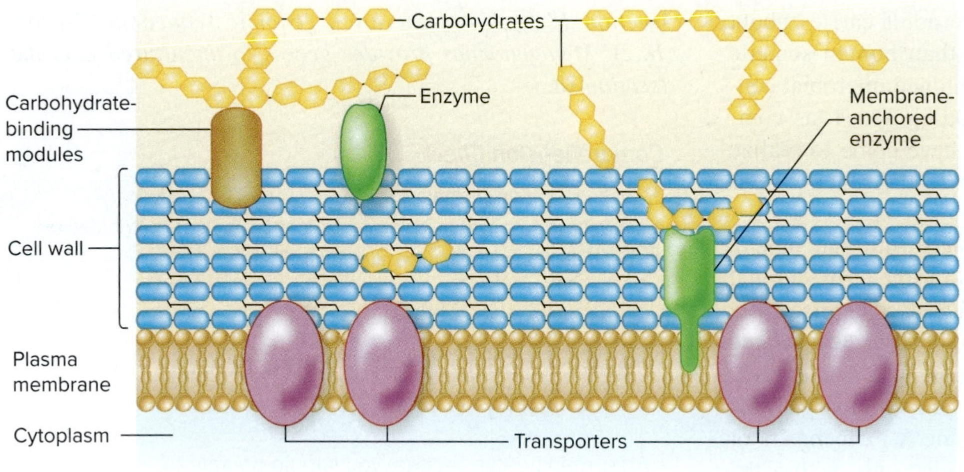

Figure 34.2 Sugar Acquisition Strategy of Bifidobacteria. Carbohydrates are bound to the bifidobacterial cell by use of an enzyme "docking station." Anchored to the cell surface, modular glycanases degrade complex carbohydrates to oligosaccharides, which are then transported across the plasma membrane. Bacteroidetes use a similar strategy to bind glycans to the outer membrane of their Gram-negative cell walls.

As we further explore, it is important to develop a microbiome rich in diversity.

Newborns are initially colonized by microorganisms from the immediate environment. For example, babies born vaginally acquire most of their microorganisms from their mother's birth canal, whereas babies born by cesarean delivery acquire microorganisms from the skin of their initial caretakers (nurses, doctors, midwives, parents, etc.). The importance of newborn colonization is now understood to be so critical that some health-care providers swab the mouths of newborns with maternal vaginal secretions after cesarian delivery.

The early colonization of the infant intestinal tract by *E. coli* and streptococci appears to establish a reducing environment for the growth of anaerobes such as bifidobacteria and bacteriodetes. In addition, human milk acts as a selective medium for nonpathogenic bacteria, as bottle-fed babies have a much smaller proportion of intestinal bifidobacteria. Bifidobacteria appear to be able to synthesize all amino acids and other required growth factors from simple nutrients. In addition, bifidobacteria and bacteroidetes have surface proteins that bind sugars, called glycans, present in the diet and produced by the host (**figure 34.2**). Bifidobacteria transport some of the polymeric sugars found in human breast milk directly across their plasma membranes—a remarkable feat for a bacterium. This helps explain its dominance in a breastfed infant's gut. The fermentation of these sugars produces acetate and lactate, which provide calories for the growing baby, and lowers the gut pH, helping to limit the growth of certain pathogens. An interesting discovery is that a robust immune response to vaccination is linked to the presence of bifidobacteria in an infant's GI tract. Switching to cow's milk or solid food (mostly polysaccharide) appears to result in the loss of bifidobacteria predominance; proteobacteria, firmicutes, and bacteroidetes—specifically enterobacteria, enterococci, lactobacilli, clostridia, and *Bacteroides* spp.—increase and outcompete bifidobacteria.

The initial microbiome changes with host age and gender, along with associated physiological and anatomical differences, until a stable resident (normal) microbiota becomes established (**table 34.1**). Each individual cultivates his or her own microbial population that is heavily influenced by genetics, developmental changes (e.g., puberty, menopause), diet, personal hygiene, anatomical site, and "life events" (antimicrobial use, travel, occupation, sexual partners, etc.).

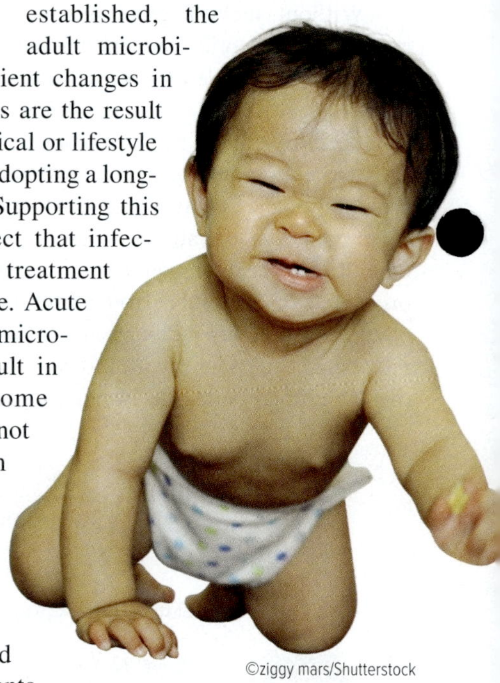

Interestingly, once established, the adult microbiota permits only transient changes in diversity. These changes are the result of substantial host physical or lifestyle changes (e.g., puberty, adopting a long-term vegetarian diet). Supporting this observation is the effect that infectious disease and its treatment have on the microbiome. Acute infection and brief antimicrobial chemotherapy result in short-term microbiome shifts but typically do not alter the long-term composition of the adult microbiota. However, long-term shifts in microbiota can occur with chronic infection and sustained use of antimicrobial agents.

While the adult human microbiota is relatively stable over time, it is highly variable from person to person and at different sites within the same person. In other words, there are relatively few species of microorganisms common to all humans; each human has a relatively unique microbiota. Based on 16S rRNA and whole genome metagenomic data, bacteria common to human skin, the intestinal tract, and the other mucosal surfaces include five major phyla: *Actinobacteria, Bacteroidetes, Firmicutes, Fusobacteria*, and *Proteobacteria* (**figure 34.3**). Additionally, a number of archaea (euryarchaeotes), fungi, and viruses are also present. Despite limited diversity at the phylum level, an enormous number of species are present. In fact the average adult gut is host to 500 to 1,000 different microbial species.

Table 34.1 Innate and Environmental Factors That Impact Microbial Diversity and Human Immunity

Relative Age	Innate Factors[1]	Environmental Factors	Microbial Diversity[2]
Birth	Genetics	Multiple caregivers, breast milk/formula, immunizations	Mother's vaginal and intestinal flora, transient species from caregivers
Infant	Developing immunity	Solid foods, crawling, mouth contact, first infections, first antibiotics	40–100 species, 7–8 phyla, transient fungi and virus
Child	Body growth	Changing diet, socialization/contact with peers, play, adventure	40–100 species, 7–8 phyla, rare fungi, rare virus
Teen	Puberty, sexual activity	Travel, sports, fitness	Species change, 7–8 phyla, increasing fungi, increasing virus
Adult	Weight gain, pregnancy, metabolic disease diagnosis, stress	Stable cohabitation, job changes, shift work, relocations, children, medications	Stable population ~500–1,000 species, 7–8 phyla, 1–10% fungi, 10–40% virus
Senior Citizen	Muscle atrophy, declining immunity, system failures	Day care, diet change, menopause, declining fitness, loss of mobility, incontinence, cardiopulmonary disorders, age-related infections, hospital, nursing care	Less stable population, 7–8 phyla, variable fungi, variable virus

1 Each factor included in subsequent age groups.
2 Type and relative frequency of taxa.
Sources: Spor, A., O. Koren and R. Ley, "Unravelling the effects of the environment and host genotype on the gut microbiome." *Nature Rev Microbiol* 9, 2014, 279–290; and Belkaid, Y., and J. A. Segre, "Dialog between skin microbiota and immunity." *Science* 346, 2011, 954–959.

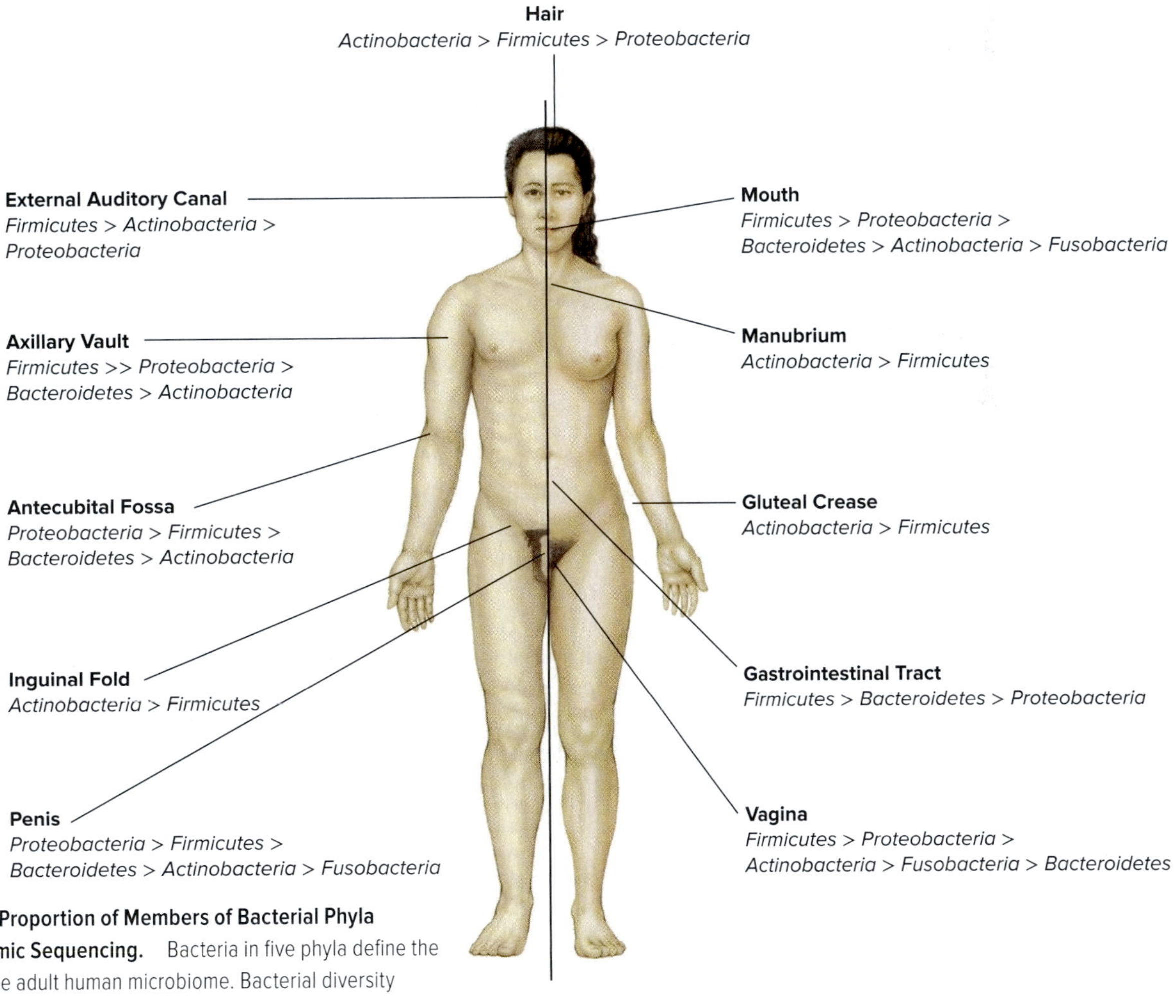

Figure 34.3 Relative Proportion of Members of Bacterial Phyla Identified by Metagenomic Sequencing. Bacteria in five phyla define the majority of microbes in the adult human microbiome. Bacterial diversity occurs at the species level.

The Microbiome Varies by Body Site

In a healthy human, regardless of age, internal tissues (e.g., brain, blood, cerebrospinal fluid, muscles) are normally free of microorganisms. Conversely, surface tissues (e.g., skin and mucous membranes) are constantly in contact with the environment and are colonized by various microbes. An overview of the culturable microbiota typically native to different regions of the body is presented in **table 34.2**. Because bacterial species make up most of the microbiome, they are emphasized over archaea, fungi, protists, viruses (including phages), and arthropods that are also part of the microbial assemblage.

Skin

The skin surface or epidermis has a slightly acidic pH, a high concentration of sodium chloride, a lack of moisture in many areas, and the oily lubricant sebum and antimicrobial peptides in other areas. It has been estimated that this surface area supports about 10^{10} bacteria dwelling in different microenvironments with distinct moisture, pH, sebum content, temperature, and topography (figure 34.3). Microorganisms that are temporarily present are transient and are typically unable to multiply on the skin.

The skin surface can be divided into three environmental niches: dry, moist, and sebaceous (containing sebum). In general, bacterial diversity is greatest at dry sites (e.g., forearm, buttocks, hands), harboring a mixture of Gram-positive and Gram-negative organisms from the phyla *Actinobacteria*, *Bacteroidetes*, *Firmicutes*, and *Proteobacteria*. Moist areas (e.g., umbilicus, underarm, inguinal and gluteal creases) exhibit less diversity, supporting mostly members of *Firmicutes* and *Actinobacteria* (e.g., *Staphylococcus* and *Corynebacterium* spp., respectively), in addition to low numbers of other bacteria. The skin surfaces with the lowest bacterial diversity are oily sebaceous sites (e.g., forehead, behind the ear, and the back), where *Cutibacterium* (formerly *Propionibacterium*) spp. and other *Actinobacteria* dominate (figure 34.3).

Most skin bacteria are found on superficial cells, colonizing dead cells, or closely associated with the oil and sweat glands. Secretions from these glands provide the water, amino acids, urea, electrolytes, and fatty acids that serve as nutrients primarily for resident bacteria such as *S. epidermidis*. The oil glands secrete complex lipids that may be partially degraded by enzymes from certain Gram-positive bacteria, in particular *Cutibacterium acnes*. These bacteria usually are harmless; however, they are associated with the skin condition acne vulgaris. It is not clear how *C. acnes* causes acne. However, it is known that these bacteria convert lipids secreted by oil glands to unsaturated fatty acids such as oleic acid that have strong antimicrobial activity against Gram-negative bacteria and some fungi. Some of these fatty acids are volatile and may have a strong odor. This is why many deodorants contain antibacterial substances that act selectively against Gram-positive bacteria to reduce the production of volatile unsaturated fatty acids and body odor.

Respiratory Tract

The respiratory tract is divided into the upper respiratory tract (URT)—the nostrils (also called nares), sinuses, pharynx, and oropharynx; and the lower respiratory tract (LRT)—the larynx below the vocal cords, trachea, bronchi, and lungs. While it has long been understood that the URT is colonized by a diverse group of microbes, we are only beginning to explore the LRT as a site that is not sterile, as previously thought.

The regions of the URT closest to the environment are the anterior nares, which like skin include host cells that produce sebum. Thus like skin, the anterior nares are colonized by the Gram-positive, lipophilic genera *Staphylococcus, Corynebacterium,* and *Cutibacterium*. Deeper into the nasal cavity we find additional microbial genera, including members of the Gram-positive genera *Streptococcus* and *Dolosigranulum* and Gram-negative *Moraxella* and *Haemophilus* spp. Finally, the oropharynx is most diverse, where in addition to bacteria belonging to genera found in the nares and nasopharynx, there are representatives of the genera *Neisseria, Rothia, Veillonella, Prevotella,* and *Leptotrichia*. The URT is also home to a robust diversity of viruses, some of which are pathogenic. In fact, many healthy people host pathogenic viruses without becoming ill. Perhaps not surprisingly, children are a richer source of viruses than adults.

The microbiome of the LRT has only recently been considered worthy of exploration. This is partly because there was strong consensus that lungs are sterile. Also, sampling the lungs without contamination with microbes from the URT is technically challenging and invasive, so not likely to be performed on healthy individuals. Nonetheless, a picture is emerging whereby the lungs function much like a revolving door: Microbes are introduced principally from the oropharynx but their stay is temporary as they are expelled and replaced by new transients.

Eye and External Ear

At birth and throughout human life, a small number of bacteria are found on the conjunctiva of the eye. The predominant bacterium is *S. epidermidis*, followed by *S. aureus, Haemophilus* spp., and *S. pneumoniae*. By contrast, the normal microbiota of the external ear resembles that of the skin, with coagulase-negative staphylococci and *Corynebacterium* spp. predominating.

Gastrointestinal Tract

The GI tract starts at the mouth and ends at the anus. Along the way, there are several distinct environments. Soon after an infant is born, the mouth is colonized by microorganisms from the surrounding environment. Most microorganisms that initially invade the oral cavity are aerobes and obligate anaerobes. When the first teeth erupt, anaerobes (*Porphyromonas, Prevotella,* and *Fusobacterium* spp.) become dominant due to the anoxic nature of the space between the teeth and gums. As teeth grow, *Streptococcus parasanguis* and *S. mutans* attach to enamel surfaces; *S. salivarius* attaches to the buccal (e.g., inside the cheeks) and gum epithelial surfaces and colonizes the saliva. These streptococci produce a glycocalyx and various other adherence factors that enable them

Table 34.2 Normal Human Microbiota Routinely Cultured from Various Body Sites

Body System	Body Site	Microorganisms	
Eye	Conjunctiva	Coagulase-negative staphylococci *Hemophilus* spp.	*Staphylococcus aureus* *Streptococcus* spp.
	Outer ear	Coagulase-negative staphylococci Diphtheroids	*Pseudomonas* spp. Enterobacteria (occasionally)
Skin	Nonmucous membrane surfaces	Coagulase-negative staphylococci Diphtheroids *Cutibacterium acnes* *Staphylococcus aureus* *Streptococcus* spp.	*Bacillus* spp. *Malassezia furfur* *Candida* spp. *Mycobacterium* spp. (occasionally)
Respiratory tract	Nose	Coagulase-negative staphylococci *Streptococcus* spp. (including *S. pneumoniae*) *Staphylococcus aureus*	*Neisseria* spp. *Haemophilus* spp.
Gastrointestinal tract	Mouth and oropharynx	*Streptococcus* spp. (including *S. pneumoniae*) Coagulase-negative staphylococci *Veillonella* spp. *Fusobacterium* spp. *Treponema* spp. *Porphyromonas* spp. *Prevotella* spp. *Neisseria* spp.	*Branhamella* spp. *Hemophilus* spp. Diphtheroids *Candida* spp. *Actinomyces* spp. *Eikenella corrodens* *Staphylococcus aureus*
	Stomach	*Streptococcus* spp. *Staphylococcus* spp.	*Lactobacillus* spp. *Peptostreptococcus* spp.
	Small intestine	*Lactobacillus* spp. *Bacteroides* spp. *Clostridium* spp.	*Mycobacterium* spp. *Enterococcus* spp. Enterobacteria
	Large intestine	*Bacteroides* spp. *Fusobacterium* spp. *Clostridium* spp. *Peptostreptococcus* spp. *Escherichia coli* *Klebsiella* spp. *Proteus* spp. *Lactobacillus* spp.	*Enterococcus* spp. *Streptococcus* spp. *Pseudomonas* spp. *Acinetobacter* spp. Coagulase-negative staphylococci *Staphylococcus aureus* *Mycobacterium* spp. *Actinomyces* spp.
Genitourinary tract	Distal urethra	Coagulase-negative staphylococci Diphtheroids *Streptococcus* spp. *Mycobacterium* spp.	*Bacteroides* spp. *Fusobacterium* spp. *Peptostreptococcus* spp.
	Vagina	*Lactobacillus* spp. *Peptostreptococcus* spp. Diphtheroids *Streptococcus* spp.	*Clostridium* spp. *Bacteroides* spp. *Gardnerella vaginalis* *Candida* spp.

to attach to oral surfaces. The presence of these bacteria contributes to the eventual formation of dental plaque, caries, gingivitis, and periodontal disease. ▶| *Dental disease (section 39.6)*

From the mouth, food and microbes travel through the esophagus to the stomach. The very acidic pH (2 to 3) of the stomach kills most microorganisms. As a result, the stomach usually contains less than 10 viable bacteria per milliliter of gastric fluid. These are mainly *Streptococcus, Staphylococcus, Lactobacillus, Peptostreptococcus* spp., and yeasts such as *Candida* spp. Microorganisms may survive if they pass rapidly through the stomach or if the organisms ingested with food are particularly resistant to gastric pH.

The small intestine is divided into three regions: duodenum, jejunum, and ileum. The duodenum (the first 25 cm of the small intestine) contains few microorganisms because of the combined influence of the stomach's acidic juices and the inhibitory action of bile and pancreatic secretions that enter here. Of the bacteria present, Gram-positive cocci and rods comprise most of the microbiota. *Enterococcus faecalis,* lactobacilli, diphtheroids, and the yeast *Candida albicans* are occasionally found in the jejunum. In the distal portion of the small intestine (ileum), the microbiota begins to take on the characteristics of the colon microbiota. It is within the ileum that the pH becomes more alkaline. For example, this is where anaerobic Gram-negative bacteria and members of the family *Enterobacteriaceae* become established.

The human large intestine is one of the most densely packed microbial ecosystems on Earth, with 10^{12} cells/gram of stool in the colon (large intestine). Over a thousand different species have been detected; most of them are not yet cultured. At the phylum level, *Firmicutes* and *Bacteroidetes* dominate healthy gut microbiome (figure 34.3). By contrast, at the genus level there is tremendous individual variability. This has led to the development of the notion of a core microbiome, that is, a group of taxa (usually at the genus level) consistent among people who share some common feature. A core microbiome predicts that rather than possessing a particular group of microbes, the microbiome collectively accomplishes specific metabolic and regulatory functions. For example, metagenomic sequencing studies reveal that people who follow the low-fiber, high-fat diet typical of most industrialized nations have a different core microbiome from that of individuals who consume a plant-based (high-fiber, low-fat) diet. In essence, these two groups of people are feeding their microbiomes differently. Interestingly, while the microbiome rapidly responds to large changes in diet, these population changes are highly variable and endure only as long as the new dietary habits are maintained.

Analyses of large metagenomic data sets have attempted to define the core gut microbiome, at least within Western populations who consume low-fiber, high-protein diets. Among people living in industrialized nations, genera that appear to be part of the core microbiome include (but certainly are not limited to) *Bacteroides, Faecalibacterium, Clostridia, Prevotella, Coprococcus,* and *Ruminococcus*. Phylogenetic analysis of gut bacteria that have only recently been cultured has led to the reassignment of some former members of the genus *Bacteroides* to the genera *Alistipes, Prevotella, Paraprevotella, Parabacteroides,* and *Odoribacter*. These investigations also identify human traits that vary with microbiome genera. Some of these traits we have already mentioned, such as age, gender, and antibiotic use. Other traits are somewhat more surprising, like red blood cell count, smoking history, and amount of sick time taken from work and school. Finally, some characteristics linked to gut microbiome population diversity may not be appropriate for discussion outside of a scientific or medical context, such as stool consistency and amount of time since last bowel movement.

Genitourinary Tract

The upper urinary tract (kidneys, ureters, and urinary bladder) is usually free of microorganisms. In both males and females, a few common bacteria (*S. epidermidis, E. faecalis,* and *Corynebacterium* spp.) can be cultured from the distal portion of the urethra. In males, metagenomic sequencing also suggests several anaerobic Gram-negative bacteria are present.

In contrast, the adult female genital tract, because of its large surface area and mucous secretions, has a complex microbiota that changes with a woman's menstrual cycle. The culturable microorganisms are acid-tolerant lactobacilli, including *Lactobacillus acidophilus*, also called Döderlein's bacillus. They ferment glycogen produced by the vaginal epithelium, forming lactic acid. As a result, the pH of the vagina and cervix is maintained between 4.4 and 4.6, inhibiting other microorganisms. In addition to *L. acidophilus,* metagenomic sequencing has identified a variety of anaerobic Gram-negative and Gram-positive bacteria. Some of these include those that normally inhabit moist skin areas (figure 34.3 and table 34.2).

Comprehension Check

1. Why are bifidobacteria important members of an infant's microbiome?
2. Discuss three factors that have affected the development of your microbiome.
3. What is the role of pH in determining the microbiota of the stomach, and what is the role of microbiota in determining the pH of the vagina?
4. What is a core microbiome? Why is it difficult for scientists to define?

34.3 A Functional Core Microbiome Is Required for Human Homeostasis

After reading this section, you should be able to:

a. Explain the contribution the gut microbiome makes to human metabolism
b. Analyze the importance of germ-free mice to our understanding of the human microbiome
c. Differentiate between direct and indirect mechanisms responsible for colonization resistance
d. List three specific reasons the immune system depends on the microbiome to maintain homeostasis
e. Compare and contrast three routes by which gut microbes communicate with the central nervous system (CNS)

The person-to-person variability in the species composition of our microbiomes makes it very hard to establish broad generalizations about which microbes contribute to good health. But one theme that repeatedly emerges is that a taxonomically diverse microbiome is essential for well-being. As discussed with regard to the GI tract, rather than trying to define a specific group of microbial genera that constitute the "healthy microbiome," a more realistic accounting is the concept of functional core microbiome. A **functional core microbiome** includes those microbes that provide the host with a suite of activities (to be described) required for health and homeostasis. Some of these activities, like the provision of vitamin K by *Escherichia coli*, have been known for many years. Other activities, like the emerging role of gut microbiota in human behavior, are surprising and just beginning to be understood. The notion of a functional core microbiome means that in addition to enumerating common genera, we need to turn our attention to characterizing microbial genes and metabolites. Here we discuss some of the functions of a healthy microbiome with the caveat that this is a rapidly advancing field, and we present just some of the highlights.

Host Metabolism

Body weight is most closely related not to the number of calories we eat, but the number of calories we absorb. Our cells are very good at taking up simple sugars and a few polysaccharides, but for the most part, we rely on our gut microbiota to convert our food into calories we can use. This was shown in the early 2000s in a series of experiments that used germ-free mice. **Germ-free (GF)** mice are born by cesarian delivery, reared in sterile enclosures, and fed sterile food and water. Although GF animals are nowhere near representative of normal life, comparing GF mice with those bearing a normal microbiome (in this context, known as conventional animals) allows scientists to test many of the complex relationships between microorganisms, hosts, and environmental factors in a controlled fashion.

These early microbiome studies showed that GF mice could eat more but gain less weight than conventional mice. For example, in one 8-week study, GF mice gained 50% less than conventional mice after both groups consumed a diet of 40% fat. While this may sound like an ad for a weight-loss plan ("Eat more! Weigh less!"), this study and others like it showed that gut microbiota convert much of the food eaten into degradation products that contribute to total caloric intake of the host. The causal nature of the microbiome population and weight gain has been demonstrated by fecal microbiome transplant (FMT) experiments (see chapter opener). In one study, FMT was performed using conventional, obese mice as donors and GF mice as recipients. Following FMT, the formerly GF recipient mice became obese as well, despite no change to their diet or exercise regime. When these types of experiments are repeated in a variety of different settings and contexts, they consistently demonstrate that obesity can be transferred by FMT. Additional experiments have shown that FMT from lean to obese mice can trigger weight loss in the recipients.

Table 34.3 Trends in Gut Microbiota Population Dynamics

Bacteria in Functional Core Microbiome	Microbiome Changes Associated with Obesity
Firmicutes	Increased *Firmicutes*
Bacteroidetes	Decreased *Bacteroidetes*
Verrucomicrobia	Decreased *Verrucomicrobia*
Actinobacteria	Increased *Actinobacteria*
Faecalibacterium prausnitzii	Decreased *F. prausnitzii*

It is frequently (but not always) observed that obese mice, like overweight and obese people, turn out to have higher relative concentrations of gut bacteria belonging to the phylum *Firmicutes* compared to bacteria belonging to the phylum *Bacteroidetes*, along with several other changes in gut microbiota population (**table 34.3**). Why is this important? Simply put, we cannot break down fiber very well but our bacteria can. Although the term *fiber* is not very specific, all fiber consists of a variety of linear and branch chained sugars. Because we can digest only a few complex polysaccharides, we feed most of them to our microbiome. Unlike our own cells, the typical gut microbiome produces thousands of enzymes that evolved to break down complex polysaccharides into their constituent monomers. Importantly many bacteria ferment these monomers into short-chain fatty acids (SCFAs). These SCFAs include butyrate, propionate, and acetate. Butyrate is the principle source of calories for intestinal epithelial cells (called colonocytes) that line the gut. In fact, butyrate increases colonocyte mitochondrial activity so that many carbon substrates are oxidized to CO_2—thus calories consumed are literally vaporized. Propionate travels to the liver where it inhibits cholesterol synthesis, and along with butyrate helps regulate weight by increasing the release of intestinal hormones that suppress hunger. On the other hand, acetate is readily absorbed by a variety of host cells and acts as a precursor for lipid synthesis in liver and fat cells. For this reason, acetate in this context is sometimes called "obesogenic." The shift in the population of gut microbiota observed in obese individuals results in more bacteria that produce acetate, which promotes weight gain and fewer bacteria that produce butyrate and propionate, which protect against obesity. The role in SCFAs in weight regulation and maintaining host homeostasis has refocused attention from the individual species that constitute the gut microbiome to the products these microorganisms secrete, called the **metabolome.** As we now discuss, SCFAs and the microbes that produce them are key players in immune modulation as well.

Immunity

The first clue that the microbiome was critical to host immunity was gleaned from an observation made over 50 years ago. At that time it was noted that following antibiotic treatment, people had a much higher risk of GI tract infection by members of the *Enterobacteriaceae,* such as *E. coli* or *Salmonella enterica*. It was rightly assumed that antibiotics disrupted the gut microbial community, and this phenomenon was explained by what was dubbed **colonization resistance.** Colonization resistance is based on the ecological principle of competitive exclusion. Fifty years ago, it was assumed that the commensal bacteria in the healthy colon outcompete pathogens for nutrients and space. Although this is true, like every other aspect of the microbiome, colonization resistance has turned out to be much more complex, involving not just direct microbiome-pathogen interactions, but indirect microbiome-host-pathogen interactions as well.

Among well characterized direct microbiome-pathogen interactions is the release of toxic peptides by one group of bacteria that specifically targets others, while the producing microbes remain unharmed (**figure 34.4**). These peptides include bacteriocins, microcins, and colicins. Bacteriocins are produced by Gram-positive bacteria and target closely related strains or species. These amphipathic peptides kill target bacteria by forming pores and depolarizing plasma membranes. Microcins are made by a wide range of *Enterobacteriaceae;* colicins are *E. coli* products. These two classes of peptides also employ a wider arsenal of toxic mechanisms including inhibition of cell wall synthesis, RNA polymerase, and tRNA synthetase, as well as possessing nuclease activity. Like bacteriocins, microcins and colicins target bacteria taxonomically related to the producing microbes. Some Gram-negative bacteria deliver bacteriocidal toxins directly through the cell envelope of competing bacteria through the needle of their type VI secretion system (T6SS). ◀ *Protein translocation and secretion in bacteria (section 13.8); Bacteriocins (section 32.3)*

Examples of indirect, microbiome-host-pathogen interaction include the ability of peptidoglycan and SCFAs released by microbiota to induce host cell production of a C-type lectin (figure 34.4, left side of figure). C-type lectins are a class of carbohydrate-binding proteins that require calcium (thus the "C" designation). In this case, the C-type lectin selectively kills Gram-positive microbes, like pathogenic strains of *C. difficile* and *Enterococcus faecalis*. Gut microbiota also chemically alter bile acids produced by the host. Bile acids when initially produced are called primary bile acids; after chemical modification (e.g., dehydroxylation, dehydrogenation) by gut microbiota, they are termed secondary bile acids. Secondary bile acids serve as signals to the host to limit inflammation and pathogen growth. Finally the fermentation product butyrate stimulates host colonocytes to consume oxygen through β-oxidation of fatty acids. This in turn limits oxygen diffusion into the lumen. By keeping the oxygen tension low, the growth of anaerobes is favored over other, less desirable facultative anaerobes such as *E. coli* and *S. enterica*. ◀ *Lipid catabolism (section 11.9)*

While the initial observations that led to an appreciation of colonization resistance were based on antibiotic therapy in people, sophisticated experiments using GF mice have been essential in advancing our understanding. Among the early results that signaled the importance of the microbiome were observations that the intestinal cellular architecture of GF mice differed from conventional mice. Specifically, the fingerlike projections (villi) that line the intestine were found to be longer, while the gulfs, or crypts, between villi were less developed in GF mice. To explore this change in morphology, scientists measured levels of antimicrobial peptides (AMPs) normally secreted by host Paneth cells located in the crypts. They discovered GF mice produced very low levels of AMPs, suggesting AMP production requires bacteria in the gut. In the early days of microbiome research, this result was notable because it suggested that the host had to detect, rather than ignore, gut microbes.

Scientists were further surprised when they examined tissue outside the intestine of GF mice, specifically lymphoid tissue in the spleen, thymus, and lymph nodes. These tissues were quite underdeveloped, indicating that microorganisms in the lumen of the gut were influencing immune cell and tissue

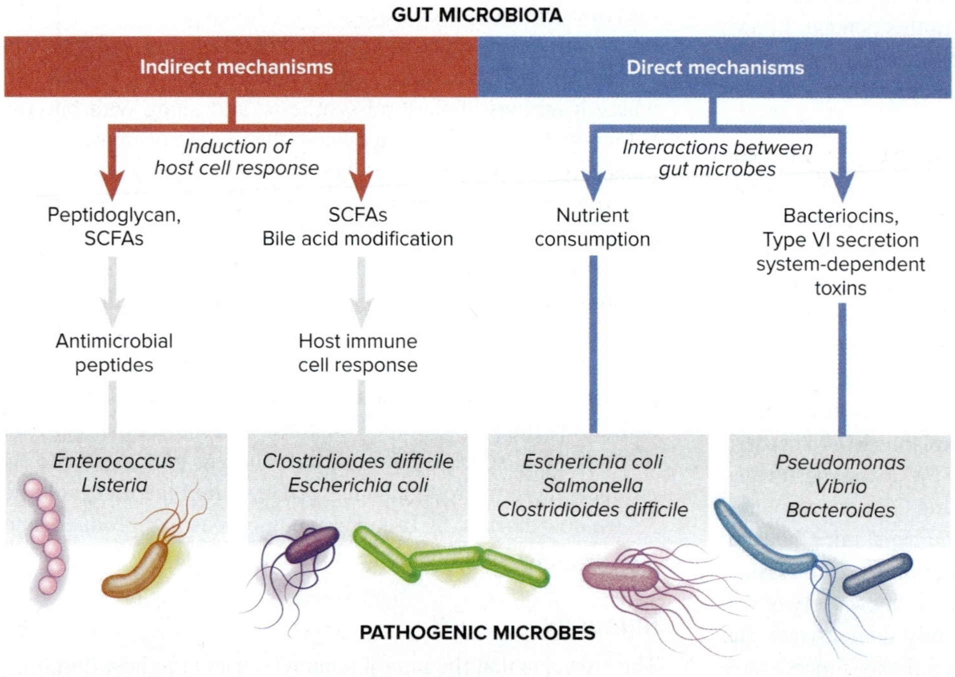

Figure 34.4 **Colonization Resistance Involves Direct and Indirect Mechanisms.**

Source: Pamer, Eric G. "Resurrecting the intestinal microbiota to combat antibiotic-resistant pathogens," *Science*, 352, April 29, 2016, 535–538.

development from afar. Additional studies showed that GF mice have limited development of white blood cells of the innate immune system (**figure 34.5**). These cells include neutrophils, monocytes, macrophages, and dendritic cells—all cells that mature in the bone marrow (*see figure 32.8*). This prompted investigators to look for soluble factors from the microbiome in the bloodstream that could drive immune cell maturation. In what has now become a recurrent theme, soluble SCFAs produced by gut microbiota as well as cytokines released by intestinal immune cells are needed for immune cell maturation throughout the body. Interestingly, cytokine release is triggered by stimulation of toll-like receptors (TLR) when gut immune cells detect microbial-associated molecular patterns (MAMPs) found on our normal microbiota (*see figure 32.16*). ◀ *Recognition of foreignness is key for survival (section 32.6)*

However, immune cell maturation is not the end of the story. Even after immune cells leave the bone marrow and enter the bloodstream of GF mice, they do not function normally. For example, neutrophils do not age in the same way as they do in conventional mice. Neutrophils are normally short-lived, with a lifespan of about 7 days. Their aging and removal ensure proper phagocytic function when they migrate to an infection site, so changes in aging and clearance of neutrophils in GF mice compromise this important innate immune response. Macrophages residing in tissues such as the CNS, skin, and gut also require a functional microbiome to maintain normal morphology and migration to sites of infection. Although we are still learning about the mechanisms responsible for these effects, it is clear that the microbiome has a profound effect on the development of the immune system. ◀ *Each type of innate immune cell has a specific function (section 32.4)*

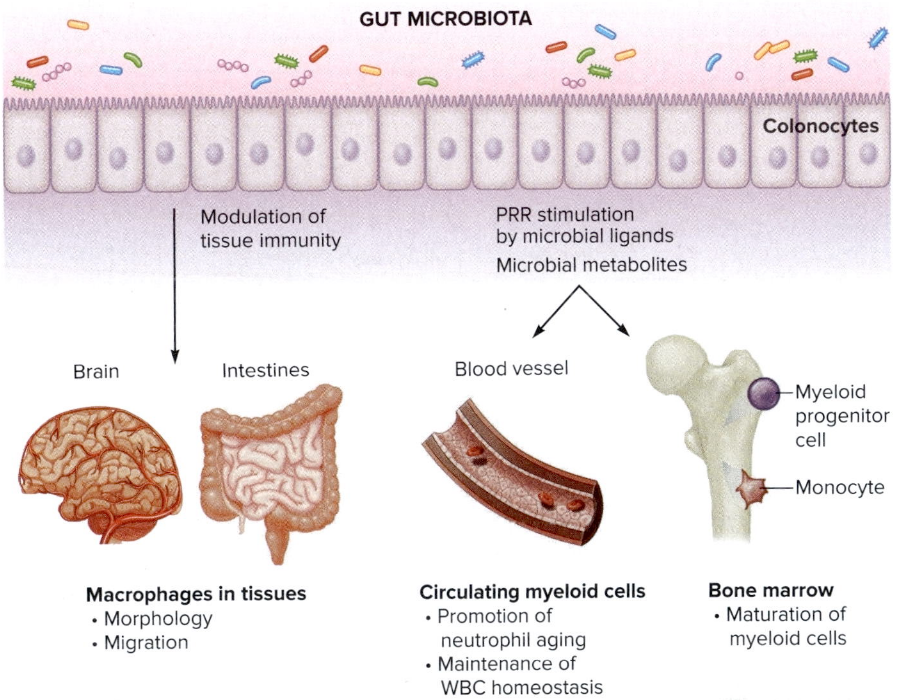

Figure 34.5 Microbiome Signals Influence Immune Cell Function at Sites Distant from the Gut. Studies in GF mice demonstrate that the maturation and function of innate immune cells depend on signals from gut microbiota. PRR, pattern recognition receptor.
(left) ©GraphicsRF/Shutterstock. Source: Thaiss, C. A., Zmora, N., Levy, M., and Elinav, E. "The microbiome and innate immunity." *Nature*, 535(7610), July 7, 2016, 65–74.

MICRO INQUIRY: *How might frequent antibiotic treatment in infants and toddlers affect their immune systems?*

These observations argue strongly against the long-held notion that the single layer of host colonocytes that lines the intestinal wall forms an impenetrable barrier that only pathogens can cross. This convenient belief helped explain the concept of **mucosal tolerance,** or how we could host so many microbes and foreign food antigens in our guts without responding with an overwhelming inflammatory response. However, the discovery of host cell pattern recognition receptors (PRRs), such as TLRs on colonocytes, combined with GF mice experiments and a plethora of metagenomic microbiome data led to a true paradigm shift: Colonocytes and the immune cells beneath and between them recognize and respond to microorganisms of the microbiome in addition to pathogens. While our understanding of mucosal tolerance is incomplete, it has been well documented that SCFAs, especially butyrate, inhibit the release of pro-inflammatory cytokines by colonocytes and macrophages, in part by inhibiting the host enzyme histone deacetylase (HDAC), thereby regulating host cell gene transcription.

Mucosal tolerance is now viewed as a balance between the activity of pro-inflammatory T_H17 cells and Treg cells, which produce the anti-inflammatory cytokine IL-10 (**figure 34.6*a***). But how is this balance determined and maintained? One of the early, exciting discoveries was that the culturable bacterium *Bacteroides fragilis* produces a molecule called polysaccharide A (PSA) that reverses inflammation by suppressing the release of the pro-inflammatory cytokine IL-17 by T_H17 cells and promoting the production of IL-10 by binding to a PRR (toll-like receptor 2) on Treg cells (figure 34.6*b*). Another early discovery was the capacity of SCFAs to increase both the number and activity of intestinal Treg cells. Remarkably, butyrate directly signals Treg cells by binding to their G-protein coupled receptors. In contrast to SCFAs, long-chain fatty acids appear to promote T_H1 and T_H17 differentiation and proliferation. Based on microbiome studies in model mouse systems and patients with inflammatory bowel disease, it would appear that microbiomes with a disproportionate number of bacteria belonging to the phylum *Proteobacteria* promote inflammation at least in part through the production of long-chain fatty acids. ◀ *Types of T cells (section 33.5)*

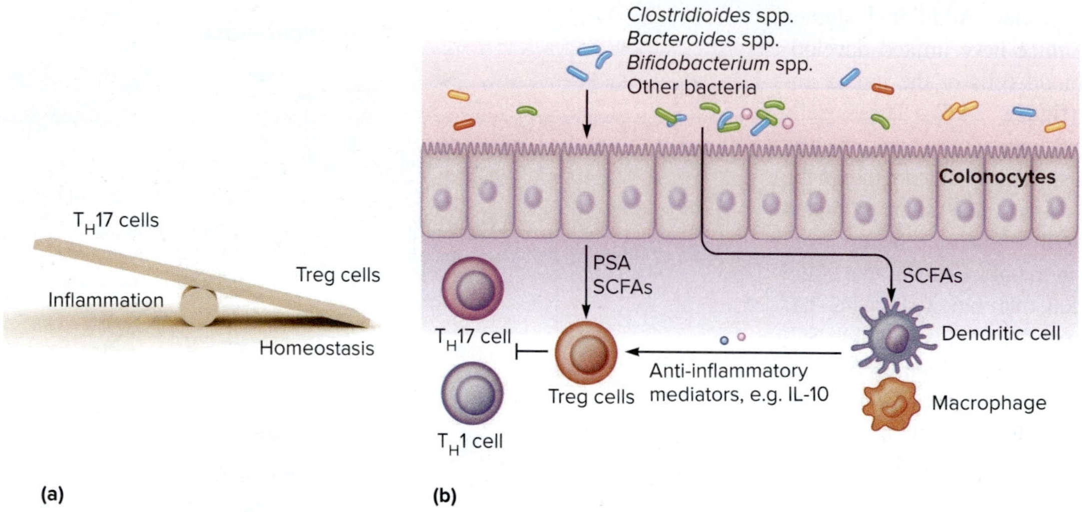

Figure 34.6 Gut Microbiota Metabolites Regulate Inflammation. (a) Inflammation reflects the balance of pro- and anti-inflammatory cytokines released by T_H17 cells and Treg cells, respectively. (b) Bacterial metabolites such as PSA from *B. fragilis* and SCFAs from clostridia and other fermentative organisms directly promote Treg activity. SCFAs also limit inflammation by interacting with macrophages and dendritic cells, which in turn stimulate Treg cells. PSA, polysaccharide A; SCFAs, short-chain fatty acids.

(a) ©ayagiz/iStockphoto/Getty Images. (b) Source: Thaiss, C. A., Zmora, N., Levy, M., and Elinav, E. "The microbiome and innate immunity." *Nature*, 535(7610), July 7, 2016, 65–74.

As discussed in chapter 32, the gut mucosa is lined with gut-associated lymphoid tissue (GALT)—regions rich in immune cells, including Peyer's patches and isolated lymphoid follicles (*see figure 32.15*). When stimulated by microbes or their products, these regions develop into lymphoid follicles, where cells of the innate and adaptive immune system are activated. Among these cells are B cells that when stimulated to differentiate into plasma cells produce antigen-specific secretory IgA (sIgA). Recall that sIgA molecules coat mucosal surfaces where they bind their microbial targets, preventing attachment and promoting clearance of foreign antigens. B cells must be activated to produce sIgA, a process that depends, in part, on innate lymphoid cells (ILC3s). Although ILCs develop normally in GF mice, they fail to function properly. Indeed, the production of sIgA depends on a functional microbiome (**figure 34.7**). Therefore, like certain host molecules and cells, the microbiome serves to link innate and adaptive immunity. ◀ *Innate lymphoid cells (section 32.4); Immunoglobulin (antibody) classes (section 33.7)*

Gut-Brain Axis

One of the most intriguing areas of microbiome investigation concerns how our gut microbiota may affect our central nervous system (CNS). A variety of studies have shown that specific behavioral traits (inquisitiveness, sociability, anxiety, depression) differ when comparing GF mice and conventional mice. Amazingly, when GF mice are fed bacteria, these traits are normalized. In addition, some behavioral traits are reproducibly transferred between mice via FMT. The influence of the microbiome is heritable as well, because the gut microbiome of pregnant mice influences the neurodevelopment of their offspring. Intriguingly, bacteria that stimulate maternal secretion of the pro-inflammatory cytokine IL-17 give rise to pups (baby mice) that display repetitive and antisocial behaviors.

Such results have prompted much research into the molecular relationships between microbiome and the CNS. Neurotransmitters are small molecules required for nerve function. Relative to conventional mice, GF mice have higher turnover rates (i.e., synthesis and degradation) of the neurotransmitters dopamine, norepinephrine, and serotonin in the brain. Serotonin produced in the CNS is involved in mood, appetite, sleep, learning, and memory. Because GF mice have lower blood levels of serotonin, it is a bit easier to draw a connection between gut microbiota and these levels of serotonin. This is because roughly 90% of the body's serotonin is produced in the GI tract, where it regulates intestinal movement or motility. Gut bacteria that produce the SCFA butyrate induce host intestinal cells to produce serotonin, and serotonin levels are restored in GF mice when butyrate-fermenting microbiota are administered. Going one step further, simply feeding butyrate to GF mice restores serotonin to normal levels.

However, these observations do not provide a complete explanation of how gut bacteria interact with the brain. In this regard, we can predict at least three ways the microbiome can influence the CNS (**figure 34.8**). First as we have seen, the microbiome has a profound impact on the immune system, which in turn modulates the CNS. One documented way in which this occurs is through microbial regulation of the balance between pro- and anti-inflammatory cytokines (figure 34.6). When pro-inflammatory cytokines reach the brain, they alter the function of neurons and CNS macrophages, called microglia. For example when lipopolysaccharide (LPS) crosses the gut, it triggers the production of pro-inflammatory cytokines, which in turn are responsible for the behaviors we associate with being sick: loss

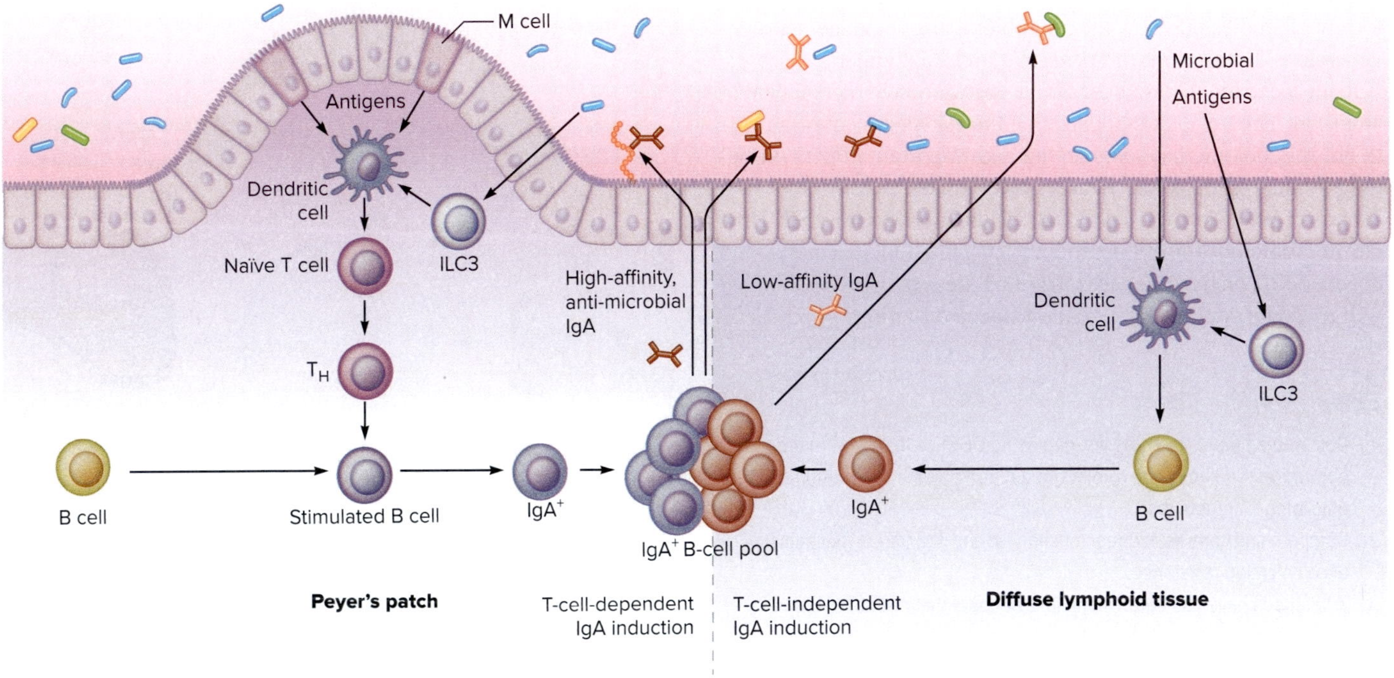

Figure 34.7 Production of sIgA Requires ILC3 and Microbial Stimulation. B-cell stimulation can be T-cell dependent (left, Peyer's patch) or T-cell independent (right, isolated lymphoid follicle). In both cases, innate lymphoid cells-3 (ILC3s) optimization of dendritic cell activity is needed. ILC3 function requires interaction with host microbes or their products.

Source: Honda, K., and Littman, D. R. "The microbiota in adaptive immune homeostasis and disease." *Nature*, 535(7610), July 7, 2016, 75–84.

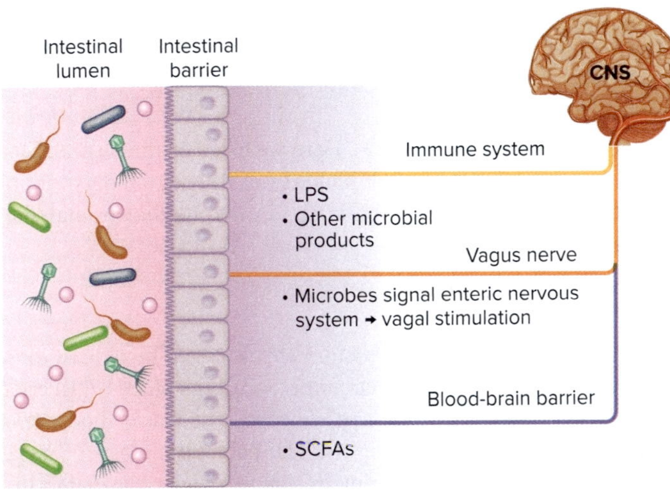

Figure 34.8 The Gut-Brain Axis. Gut microbes influence the central nervous system using at least three mechanisms. LPS, lipopolysaccharide; SCFAs, short-chain fatty acids.

(right top) ©GraphicsRF/Shutterstock. Source: Sampson, R. R., and Mazmanian, S. K. "Control of brain development, function, and behavior by the microbiome." *Cell Host Microbe* 17(5), May 13, 2015, 565–576.

of appetite, decreased motor activity, loss of sociability, and reduced cognition ("trouble thinking").

Another mechanism that helps mediate the gut-brain axis is a direct pathway from gut to brain. This is possible because the GI tract is lined with a network of nerves, called the enteric or intrinsic nervous system, that is connected to the CNS via the vagus nerve. Gut bacteria directly stimulate neurons within the enteric nervous system, which then communicate with the vagus nerve. The vagus nerve in turn transmits signals to the brain. The vagus nerve can differentiate between stimuli from normal microbiota and pathogens even in the absence of inflammation, although how this fascinating feat is possible remains under investigation.

Finally, as suggested by experiments in which butyrate and FMT had similar effects on the behavior of GF mice, soluble microbial products can impact the CNS. To understand this, we must first consider the CNS as a site that is set apart from the remainder of the body by the blood-brain barrier (BBB). The BBB is lined with cells that protect the CNS from normal blood components as well as toxins and infectious agents. This is largely due to impermeable connections, called tight junctions, between the cells that make up the BBB. Once again, a flurry of research was triggered by an observation made in GF mice: Their BBBs have the undesirable quality of being more permeable than conventional mice. And once again, it is SCFAs, particularly butyrate, that help maintain the BBB as an impenetrable barricade. Because SCFAs are most abundant when mice or humans eat a high-fiber diet, the broader implication is that one's diet can govern the entry of metabolites and other blood-borne compounds to the CNS.

The unexpected influence gut microbiota has on behavior, documented in animals and some human trials, has unleashed a torrent of studies into how the microbiome may mediate neurodegenerative diseases like Parkinson's and Alzheimer's diseases. Autism spectrum disorder (ASD) has also garnered a great deal of attention. This reflects evidence that the establishment of a functional microbiome has long-lasting and important effects on behavior and perhaps on neural development within the first 3 years of life, when most children are diagnosed with ASD. How our microbiomes impact neurodevelopment on the one hand, and degeneration on the other are intriguing areas of investigation that will no doubt continue to reveal unexpected findings.

Comprehension Check

1. Describe an experiment performed using GF mice that broadened our understanding of the role of the gut microbiome in maintaining physiologic homeostasis.
2. Compare and contrast a direct mechanism and an indirect mechanism of colonization resistance.
3. What foods have you eaten in the last 24 hours that promoted the microbial production of short-chain fatty acids and why is this desirable?
4. What is mucosal tolerance and how does the microbiome contribute to maintaining it?
5. Describe one way the microbiome impacts innate immunity and one way it impacts adaptive immunity.
6. Define the blood-brain barrier and how it may be involved in the gut-brain axis.

34.4 Many Diseases Have a Connection with Dysbiosis

After reading this section, you should be able to:

a. Describe metabolic syndrome and the metabolic endotoxemia hypothesis
b. Discuss the microbial link between a diet rich in red meat and atherosclerosis
c. Discuss in broad terms the relationship between the microbiome and cancer

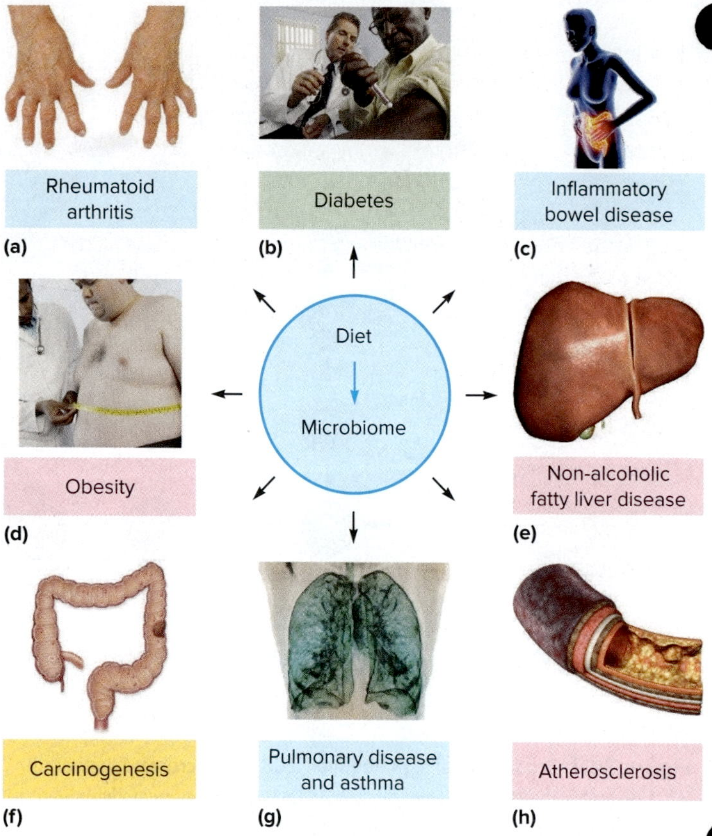

Figure 34.9 Loss of Microbiome Diversity Leads to Dysbiosis and a Variety of Diseases That Involve Inflammation.

(a) ©Suze777/Getty Images; (b) ©Adam Gault/Science Source; (c) ©Roger Harris/Science Source; (d) ©Adam Gault/Science Source; (e) ©MedicalRF.com; (f) ©McGraw-Hill Education; (g) ©Science Photo Library/Alamy Stock Photo; (h) ©MedicalRF.com. Source: Thaiss, C. A., Zmora, N., Levy, M., and Elinav, E. "The microbiome and innate immunity." *Nature*, 535(7610), July 7, 2016, 65–74.

A casual review of recent literature might lead one to think that all human disease is linked to the microbiome. A quick internet search reveals hundreds of reports describing the role of our microbial partners in the development of a variety of illnesses such as autoimmune disease (e.g., multiple sclerosis, lupus erythematosus, type 1 diabetes), asthma and allergies, inflammatory bowel disease, and neurologic conditions ranging from Parkinson's disease to autism spectrum disorder. Some of these conditions are inflammatory disorders and it has become clear within the last decade or so that conditions involving chronic inflammation are influenced and in some cases caused by an altered microbiome. The contribution of the microbiome to disease is called **dysbiosis** (**figure 34.9**). Here we focus on three conditions that together account for the majority of illness and death in industrialized nations: metabolic syndrome, cardiovascular disease, and cancer.

Metabolic Syndrome

Metabolic syndrome is a condition characterized by at least three of the following: large waist circumference, high blood triglyceride level, high blood pressure, and elevated low-density lipoprotein and fasting blood glucose levels. The pathophysiology of metabolic syndrome involves insulin resistance, which is the inability of the body's cells to take up insulin from the bloodstream as they should. Because insulin is needed to move glucose from the bloodstream into cells, insulin resistance leads to high blood sugar levels—the hallmark of diabetes mellitus. Insulin resistance precedes the development of type 2 diabetes and both conditions are associated with obesity (by contrast, type 1 diabetes is an autoimmune disease whereby host immunity attacks the pancreas, the site of insulin production). Metabolic syndrome also increases the risk of stroke and heart disease. Importantly the World Health Organization reports that global obesity has more than doubled since 1980, and metabolic syndrome is now a global epidemic.

Metabolic syndrome and obesity are associated with chronic, low-level inflammation, and this appears to be linked to the microbiome as explained by a phenomenon known as **metabolic endotoxemia** (figure 34.10). Metabolic endotoxemia is caused by a series of events that begins with a diet-induced change in the microbiome leading to limited microbial diversity and dysbiosis. Dysbiosis in turn results in an increase in the permeability of tight junctions between colonocytes, making the mucosal barrier "leaky." This enables LPS, which is sloughed from Gram-negative bacteria as they grow, to breach the mucosal barrier. In fact, blood levels of LPS in obese persons can be two to three times higher than that measured in lean individuals. LPS is a very potent immune modulating molecule that when detected by immune and other cells including adipocytes (fat cells), readily stimulates the release of pro-inflammatory cytokines. Over time, this is thought to result in the low-level, chronic inflammation that characterizes obesity and metabolic syndrome. In addition, LPS in the liver promotes insulin resistance, accumulation of triglycerides, and the development of fatty liver disease. LPS in the blood also induces the expression of pro-inflammatory and adhesion molecules by cells that line blood vessels, promoting the development of cardiovascular disease through atherosclerosis. Atherosclerosis is marked by the accumulation of fat, cholesterol, calcium, and other compounds collectively called plaques on the insides of blood vessels.

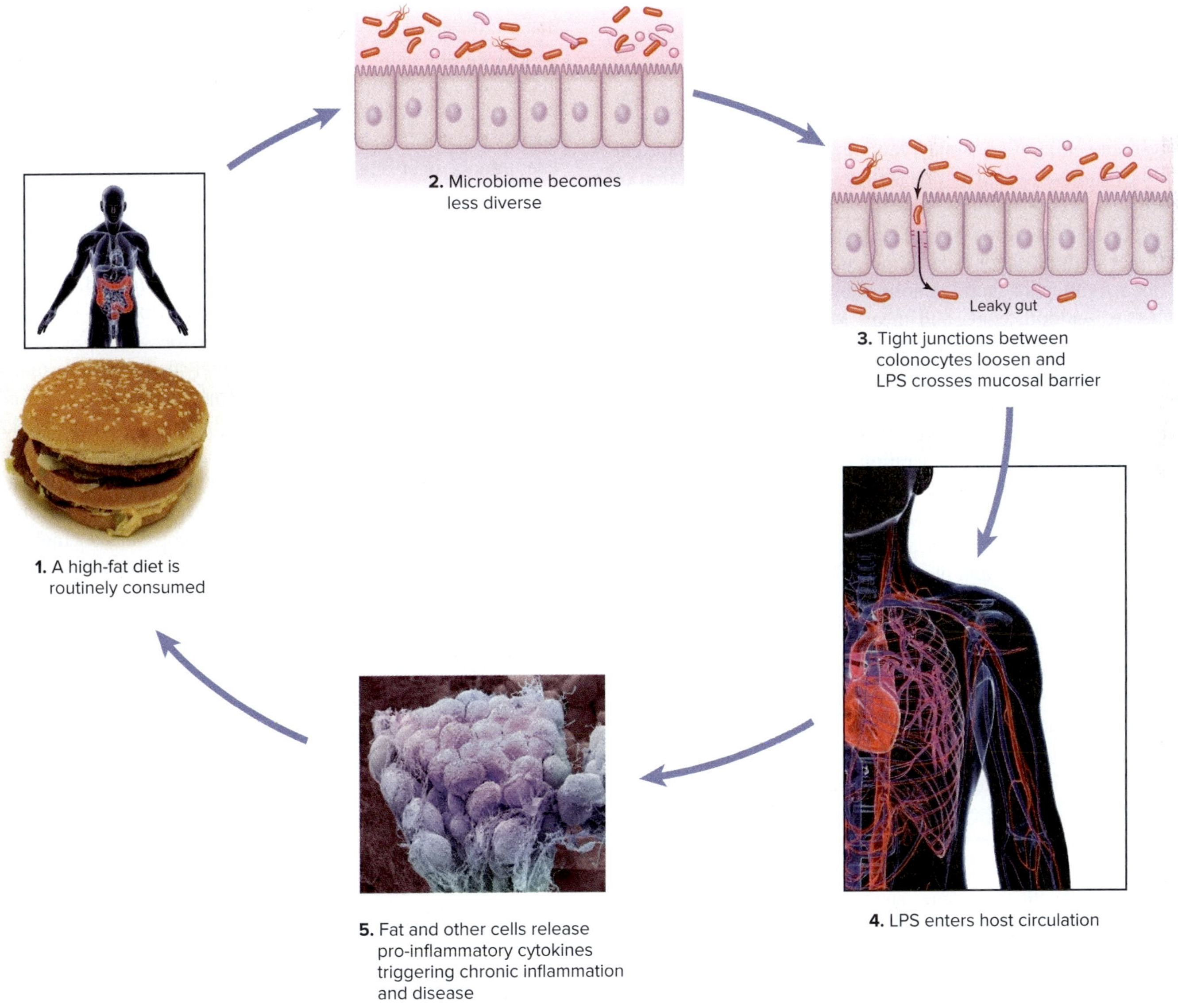

Figure 34.10 The Proposed Mechanism of Metabolic Endotoxemia. LPS, lipopolysaccharide.
(1 top) ©Eraxion/Getty Images; (1 bottom) ©McGraw-Hill Education/Mark Dierker, photographer; (4) ©Science Photo Library - SCIEPRO/Getty Images; (5) ©Science Photo Library/Alamy Stock Photo

While metabolic endotoxemia may explain the chronic inflammation associated with obesity and metabolic syndrome, it does not address the composition of the microbiome itself. Compelling evidence for the role of the microbiome in metabolic syndrome is obtained by the reproducible observation that insulin resistance can be transferred by FMT. Although this is not the whole story, the microbes that constitute dysbiosis result in the increased production of acetate over butyrate. Acetate diminishes the positive effects butyrate has on the immune system and high acetate levels promote the production of the appetite-stimulating hormone ghrelin. Because both animals and humans with insulin resistance also have elevated levels of branched-chain amino acids (i.e., leucine, isoleucine, and valine) and certain triglycerides, it is believed that a high-fat diet may select for a less diverse microbiome that produces these products as well as acetate.

Cardiovascular Disease

Although cardiovascular disease and metabolic syndrome are linked by the common finding of atherosclerosis, let's now consider the stereotypical diet consumed by a person with cardiovascular disease: lots of red meat and high-fat foods. From a microbiome point of view, such individuals are feeding their gut microbes little of the fiber that will result in the production of anti-inflammatory SCFAs. Instead, such a diet promotes the growth of a "meat-eating" microbial population. These microbes encounter high levels of L-carnitine, an amino acid enriched in red meat, and phosphatidylcholine, a phospholipid abundant in cheese, seafood, eggs, and meat. Microbial metabolism of L-carnitine and phosphatidylcholine results in the production of trimethylamine (TMA). Once TMA is absorbed into the bloodstream, it travels to the liver where it is enzymatically oxidized by liver cells to trimethylamine *N*-oxide (TMAO). TMAO has been associated with the acceleration of atherosclerosis in mice. The production of TMA and ultimately TMAO is a good example of how our diets influence our microbiome, as the microbiota of vegetarian animals produce very little TMA even when fed a diet rich in L-carnitine and phosphatidylcholine.

Cancer

The relationship between the microbiome and cancer is particularly complex. Although we commonly think of environmental factors and lifestyle habits like smoking or chewing tobacco as drivers of cancer, microorganisms are estimated to be involved in about 20% of all malignancies. There are several known mechanisms by which microorganisms can cause or contribute to cancer in humans.

We turn our attention first to specific microbes known to cause cancer. It has been known for decades that certain human viruses cause their host cells to become cancerous by two mechanisms operating simultaneously: altering the host cell cycle to favor proliferation and preventing host cells from repairing DNA damage, thereby increasing mutation rates. These two mechanisms also account for how bacterial products promote carcinogenesis. Bacteria can cause DNA damage either directly or indirectly. For example, *E. coli* in the phylogenetic group B2 harbor a genomic island that directs the synthesis of a DNA-damaging compound called colibactin. When taken up by intestinal cells, colibactin results in double-stranded DNA breaks. Colibactin-producing *E. coli* have been found in human colorectal cancers and stimulate intestinal tumor formation in mice. A more indirect approach to DNA damage is taken by enterotoxin-producing strains of *Bacteroides fragilis* that secrete a metalloprotease called Bft. Bft stimulates the production of high levels of radical oxygen species by host cells, which in turn are mutagenic. Its role in carcinogenesis has been confirmed by experiments in which purified Bft causes tumor formation in a mouse model of colorectal cancer. *Helicobacter pylori* provides a good example of the second mechanism of microbial-mediated carcinogenesis: bacterial dysregulation of host cell cycling. *H. pylori* is well known for its ability to cause stomach ulcers, but certain strains also produce a protein called CagA, which it injects into host cells. Inside host stomach cells, CagA leads to upregulation of genes that control cell proliferation, survival, migration, and blood vessel formation; these are activities required for tumor formation. In addition to causing cancer in a specific site (primary tumorigenesis), bacteria can be involved in metastasis of tumors to distant sites. For example, *Fusobacterium* and related microbiome species are frequently isolated from human colorectal tumors as well as metastatic liver tumors. Interestingly, in a mouse model of this type of cancer, the administration of antibiotics that target these anaerobic bacteria limits tumor growth. This highlights ongoing research that focuses on inhibiting the growth of cancer-causing microorganisms. ◀ *Viruses and cancer (section 6.4); Phylum* Bacteroidetes *includes important gut microbiota (section 21.7); Phylum* Fusobacteria: *commensal anaerobes (section 21.8)*

These examples of single microbial species known to cause cancer are well documented. By contrast, many cancers linked to microorganisms are driven by the inflammatory state associated with dysbiosis (figure 34.9). Low-level chronic inflammation provoked by obesity and metabolic endotoxemia is associated with an increased risk to a variety of cancers. Epidemiology studies reveal that some of the cancers for which obese people are at a higher risk include esophageal, liver, kidney, pancreatic, colorectal, endometrial, breast, ovarian, and thyroid cancers. In fact, about one in five of all cancer deaths in the United States is linked to obesity.

Interestingly, the microbiome even impacts the host response to chemotherapy. As discussed, a functional microbiome is needed for optimal immunity. Because response to both traditional chemotherapy and newer immunotherapies requires an active, functional immune system within the tumor environment, the microbiome can have a significant effect on treatment. In addition, there is evidence suggesting that consumption of *Bifidobacterium* spp. promotes immunity to tumors. This microbe is abundant in yogurt and other probiotic foods and products, as discussed next.

Comprehension Check

1. What is metabolic endotoxemia and how is it related to dysbiosis?
2. Compare the microbial products released after eating a fiber-rich salad to that after a hamburger and fries.
3. Describe the relationship between microbes, obesity, and cancer.

34.5 Microbiome Manipulation Can Be Therapeutic

After reading this section, you should be able to:

a. Describe how a product is classified as a probiotic
b. Predict uses of probiotics in the prevention of illness

Many years ago, a television commercial for yogurt featured a town full of centenarians (at least they were advertised as 100-year-old people) living in Eastern Europe who attributed their longevity to eating yogurt of a specific brand. It turns out that in 1908, microbiologist and Nobel laureate Élie Metchnikov also knew about these villagers and postulated that lactic acid–producing bacteria promoted their health. In the microbiome era, the probiotic industry has blossomed into a US$30 billion enterprise that continues to expand.

Despite this burgeoning market, many consumers are not really certain what the term "probiotic" means. In an effort to standardize the use of the term **probiotic,** the Food and Agricultural Organization of the United Nations–World Health Organization (FAO-WHO) has defined it as "live microorganisms, which, when administered in adequate amounts, confer a health benefit to the host." FAO-WHO guidelines provide specific requirements that should be met for foods and supplements to meet this standard (**figure 34.11**).

Further confusing the situation is the fact that the U.S. Food and Drug Administration (FDA) does not regulate probiotic foods and supplements, so the health benefits the manufacturers of such products claim have not been rigorously tested. By contrast, if viable microbes are to be used as a medicine, the FDA must approve their use. Examples include the use of *Mycobacterium bovis* BCG for superficial bladder cancer and live, attenuated vaccines (e.g., measles and rubella vaccines). These products have a long track record of safe, therapeutic success and are regulated as drugs. The next phase of probiotic development will likely include products consisting of microorganisms used to treat a specific condition rather than to convey unspecified health benefits. These newer probiotics will require approval by the FDA (or similar agencies in other countries) after extensive safety and efficacy clinical trials. For example, bacterial mixtures marketed to prevent or treat *C. difficile* infection will be classified as a drug; this contrasts to fecal microbiome transfer, which is currently considered a procedure and is (at the time of this writing) permitted only for treatment of recurrent infection (see chapter opening story).

Another confounding factor has been that until recently it has been difficult to evaluate whether probiotic bacteria actually colonize the gut or are retained long enough to have an effect. The discovery that many so-called probiotic microorganisms do not effectively colonize the intestines has led to the development of synbiotics. **Synbiotics** are foods or supplements that include both a prebiotic and a probiotic. The prebiotic is a compound(s) added to enhance the colonization and positive health benefits of probiotic microbes.

Although it is much harder to control the uniformity and effectiveness of a microbial product than a drug with a unique chemical structure, the recognition that the microbiome is, in essence, a newly discovered organ needed for human homeostasis is propelling the unstoppable development of probiotics for preventative medicine. It is likely that hospitalized patients will routinely be given an oral dose of bacterial consortia to protect against the development of infection with antibiotic-resistant pathogens. Similarly, outpatients requiring antibiotic therapy may also be prescribed a probiotic cocktail for microbiome reconstitution following treatment. Tourists may soon be packing probiotics to prevent travelers' diarrhea along with their sunscreen and flip-flops.

Until researchers, pharmaceutical companies, and regulatory agencies synchronize their approach to probiotics, it is instructive to consider the successful use of probiotics in animals. For example, *Lactobacillus acidophilus* is used in beef cattle feed. When these bacteria are sprayed on feed, cattle appear to have markedly lower (as much as 60%) carriage of the pathogenic *E. coli* strain O157:H7. This can make it easier to produce beef that meets current standards for microbiological quality at the time of slaughter. Probiotics are also used successfully with poultry. For instance, a probiotic *Bacillus subtilis* strain is fed to chickens to increase body weight. There is also a reduction in coliforms and *Campylobacter* spp. in the processed carcasses. This probiotic may decrease the need for antibiotics in poultry production and pathogen levels on farms. *Salmonella enterica* can be controlled by spraying a patented blend of 29 bacteria, isolated from the chicken cecum, on day-old chickens. As they

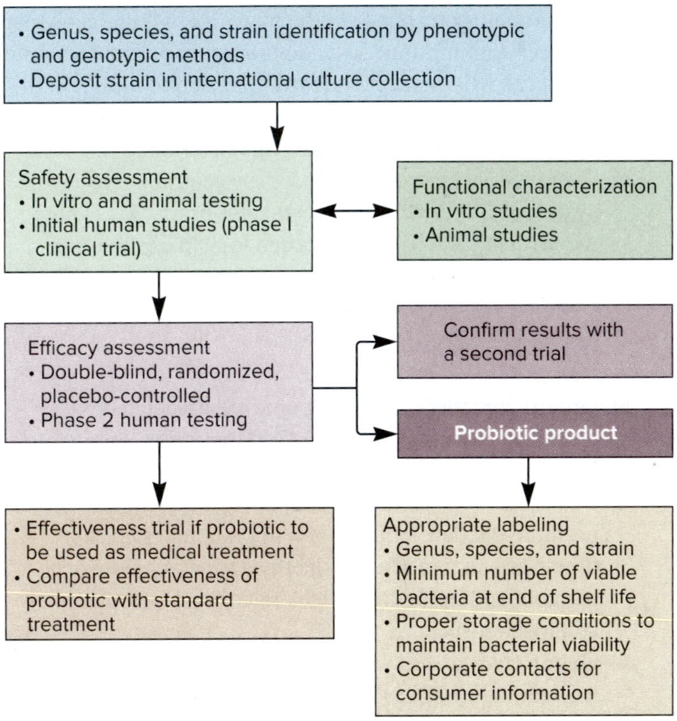

Figure 34.11 The Development of a Probiotic Product.

preen themselves, the chicks ingest the bacterial mixture, establishing a functional microbiome in the cecum that limits *Salmonella* colonization of the gut. In an age of increasing antimicrobial resistance, the use of probiotics may prove to be valuable for both humans and the animals they consume.

Comprehension Check

1. Describe the specific use of a currently used probiotic product and one under development.
2. How are probiotics used in agriculture?

Key Concepts

34.1 Humans Are Holobionts

- The average human adult carries about three times more microbial cells than human cells. Microorganisms that take up residence on a human host are called the normal microbiota.
- The Human Microbiome Project and similar projects worldwide are evaluating the role microbes, their genes, and their metabolites have in maintaining homeostasis in the host.
- Metagenomic analysis of the human microbiome has revealed an enormous diversity of microorganisms living on and in the human body. These microbes are essential in establishing and maintaining the health of the host.

34.2 The Microbiome Develops from Birth to Adulthood

- Various microbes have adapted to specific niches found on the human host. These niches are uniquely able to support microbial growth (**figure 34.3**).
- Microorganisms living on or in the skin can be characterized as either transients or residents. The normal microbiota have evolved the ability to tolerate vastly different conditions found on skin (e.g., oily, moist, and dry).
- The oral cavity provides a nutrient-rich habitat but subjects microbes to physical processes that can dislodge them. The normal microbiota of the oral cavity is composed of those microorganisms able to resist this mechanical removal.
- The stomach contains very few microorganisms due to its acidic pH.
- The distal portion of the small intestine and the entire large intestine have the largest microbial community in the body. Over 400 species have been cultured, the vast majority of them anaerobic; metagenomic studies suggest that approximately 12,000 different genera inhabit the large intestine (**table 34.2**).
- The upper urinary tract is usually free of microorganisms. The tip of the urethra (a mucous membrane) harbors numerous skin bacteria. In contrast, the adult female genital tract has a more complex microbiota that prevents other microorganisms from colonizing it.

34.3 A Functional Core Microbiome Is Required for Human Homeostasis

- A functional core microbiome provides the host with many enzymes capable of digesting complex polysaccharides that would otherwise be indigestible, as well as a variety of metabolites that help maintain homeostasis.
- Colonization resistance is the process by which the normal microbiota competitively exclude pathogens using both direct and indirect mechanisms (**figure 34.4**).
- Germ-free mice do not develop a mature, functional immune system, demonstrating the need for microbial colonization for optimal host immunity (**figure 34.5**).
- Fiber-rich foods promote the development of a microbiome that includes microorganisms that produce the short-chain fatty acids acetate, butyrate, and propionate. Butyrate and propionate have positive effects on innate and adaptive immune cells.
- Mucosal tolerance is marked by the absence of an inflammatory response to the many foreign antigens that interact with our mucous membranes, in particular in our GI tracts. Microbial products are needed to maintain the appropriate balance of pro-inflammatory T_H17 cells and anti-inflammatory Treg cells (**figure 35.6**).
- The gut microbiome influences the central nervous system in a system known as the gut-brain axis. Three mechanisms have been proposed to help explain this interesting phenomenon (**figure 34.8**).

34.4 Many Diseases Have a Connection with Dysbiosis

- Obesity and metabolic syndrome are associated with dysbiosis and involve chronic inflammation, thought to be caused by metabolic endotoxemia (**figure 34.9**).
- Foods rich in L-carnitine and phosphatidylcholine promote the development of a microbiome rich in organisms that produce trimethylamine, which when metabolized in the liver accelerates the development of atherosclerosis.
- One in five cancers is linked to obesity and the inflammatory state it causes through dysbiosis.

34.5 Microbiome Manipulation Can Be Therapeutic

- Foods containing microorganisms thought to confer a health benefit have been consumed for hundreds of years and are termed probiotics. The recent appreciation of human gut microbiome has helped engender more interest in probiotic foods, supplements, and medicines.
- Probiotic products are also used in agriculture, where it is hoped they will diminish the reliance on antibiotics.

Active Learning

1. A group of obese mice and a group of lean mice are fed an identical amount of the same chow each day. The mice are carefully monitored so there is no difference in the activity level in the two groups. Discuss why neither group of mice gains or loses weight.
2. There have been conflicting reports regarding the contribution of methanogenic archaea to dysbiosis. Some studies suggest the number of these microorganisms is elevated in obese as opposed to lean mice and humans while other research shows no such difference. Design an experiment that tests the hypothesis that methanogenic archaea are more prevalent in the gut microbiomes of obese individuals.
3. Butyrate is known to increase mitochondrial activity in colonocytes, priopionate is known to contribute to gluconeogenesis in the liver, while acetate contributes to lipid biosynthesis in the liver and other tissues. Refer to chapters 11 and 12 and trace the biochemical pathways that each of these fatty acids follows (these pathways are conserved across domains). Based on your review, why does acetate serve as a lipid substrate, thereby making it more "obesogenic" than either butyrate or propionate?
4. Yo-yo dieting is all too common; after losing weight, about 80% of people regain every kilogram within a year or less. Then comes another diet, and another round of weight loss and gain. A group of Israeli scientists explored this problem in mice and discovered that while all metabolic markers of obesity (e.g., blood triglyceride levels) were abolished by weight loss, the community of gut microbiota lagged far behind in returning to that of lean controls. This persistence of dysbiosis contributed to faster weight gain when mice resumed a high-fat diet. The investigators suggest this research should contribute to the development of "microbiome-targeted approaches" to help limit yo-yo dieting. Discuss the information you need to develop such a new approach to weight loss.

   Read the original paper: Thaiss, C.A., et al. 2016. Persistent microbiome alterations modulate the rate of post-dieting weight regain. *Nature*. 540:544–51. doi:10.1038/nature20796.

35

Infection and Pathogenicity

©Fuse/Getty Images

Dodging the Bullet

Just as we dispose of our garbage and recycling at home to prevent an unhealthy buildup of waste, eukaryotic cells also have a remarkable mechanism to clear out their debris and recycle key nutrients. The molecular study of this process, known as autophagy, earned Dr. Yoshinori Ohsumi the 2016 Nobel Prize in Medicine and Physiology. Autophagy, or "self-eating," involves the formation of a phagophore, a double-membraned structure that engulfs cytoplasmic material and delivers it to the lysosome for digestion. It is important to point out that this process differs from phagocytosis, whereby the phagosome is formed following endocytosis of an exogenous pathogen; here, the autophagosome originates *within* the cytoplasm. In addition to removing damaged or obsolete organelles, autophagy also recycles nonessential parts of the cell during periods of starvation, and removes cell components to make room for new ones. During infection, it is an important process that clears intracellular pathogens and toxins from the cytoplasm. Not surprisingly, microorganisms that have evolved to elude autophagy have enhanced pathogenicity.

Before Dr. Ohsumi's research, autophagy was recognized but not well understood. Once he elucidated the mechanism, doors were opened for others to study the process in greater depth. Microbiologists were interested in the role autophagy plays in pro-inflammatory signaling, in secretion of immune mediators, and in coordinating the adaptive immune system. They discovered that certain pathogens, including *Salmonella enterica* serovar Typhimurium and *Mycobacterium tuberculosis,* can evade or suppress autophagy by producing specific virulence factors.

The *S.* Typhimurium genome contains two gene clusters called "pathogenicity islands." Proteins encoded in the first island mediate invasion of the host cell, while genes in the second island code for proteins important in intracellular survival of the pathogen. Researchers in Germany showed that these virulence factors degrade the host proteins required for autophagy, allowing *S.* Typhimurium to obtain nutrients and replicate within the cell.

Following autophagy, *M. tuberculosis* can survive within macrophages by preventing fusion of the phagosome with the lysosome; moreover, the organism promotes the accumulation of lipid bodies in the host cell. These accumulations are typically degraded by autophagy during times of stress and starvation as a survival mechanism. By preventing host lipid metabolism, *M. tuberculosis* maintains these molecules as a nutrient source that enhances pathogen survival in the host.

Avoidance of autophagy is not limited to bacteria; researchers at the University of Texas Southwestern Medical Center demonstrated that the herpes simplex virus type 1 (HSV-1) encodes a neurovirulence protein that binds to a host-cell protein integral to autophagy, inhibiting its function. A mutant HSV-1 lacking the ability to bind to the protein was incapable of causing lethal encephalitis in mice, further solidifying the link between autophagy suppression and pathogenesis. In this chapter, we look at microbial virulence factors and their role in pathogenesis by mediating transmission, invasion, and avoidance of host defense mechanisms.

Readiness Check:

Based on what you have learned previously, you should be able to:

- ✓ Explain symbiosis and differentiate the true definition of parasitism from several common uses of the word (sections 27.1 and 27.4)
- ✓ Relate the principles of symbiosis to human-microbe interactions (chapter 34)
- ✓ Describe the structure and function of the main cellular components of bacteria, viruses, protists, and fungi as well as their replication processes and energy requirements (chapters 3, 5, 6, 7, 10, and 12)

35.1 The Process of Infection

After reading this section, you should be able to:

a. Differentiate between pathogenicity and virulence
b. Compare opportunistic, extracellular, and intracellular pathogens and discuss their requirements for survival
c. Relate the infectious disease process to time, identifying events associated with each stage of the process
d. Create a concept map to summarize the infectious process

When a microorganism is growing and multiplying on or within a larger organism, the latter, known as the **host,** is said to have an **infection.** The nature of an infection can vary widely with respect to severity, location, and type of organisms involved. An infection may or may not result in overt disease. Any organism that causes disease is known as a **pathogen** (Greek *patho,* disease; *gennan,* to produce), and its ability to cause disease is called **pathogenicity.** An organism's **virulence** is the *degree* of harm (pathogenicity) inflicted on its host. An **infectious disease** is any change from a state of health in which part or all of the host is incapable of carrying on its normal functions due to the presence of a pathogen or its products (e.g., toxins). Even resident microbiota, such as those associated with the gut or skin, can become **opportunistic pathogens** if they infect a host away from their typical niche, especially in a host with a weakened immune system (a compromised host). Host and tissue specificity offer microorganisms shelter and access to resources; however, relationships between organisms can be complex. Microorganisms actively seek host-specific sites that meet their unique growth and replication requirements.

To cause disease, microorganisms must not only contact the host but be able to survive in it. Bacteria, viruses, protists, and fungi need a suitable environment in which they can grow and reproduce. The range of factors required—temperature, pH, moisture, oxygen concentration, etc.—must be provided by the human host. These conditions can be met outside or inside host cells, depending on the pathogen. Infectious microorganisms are considered **extracellular pathogens** if they remain in tissues and fluids but never enter host cells during the course of disease. For instance, some bacteria and fungi can actively grow and multiply in the blood or tissue spaces. *Yersinia pestis,* which causes plague (also called Black Death), is a good example of an extremely virulent extracellular pathogen. Microbes that grow and multiply within host cells are called **intracellular pathogens;** they can be further subdivided into two groups. **Facultative intracellular pathogens** are those organisms that reside within the cells of the host or in the environment but can also be grown in pure culture without host-cell support. *Brucella abortus* is a bacterium that grows and replicates within macrophages, neutrophils, and trophoblast cells (cells that surround the developing embryo) and *Histoplasma capsulatum* is a fungus that grows within phagocytes, yet both can be grown in vitro. In contrast, **obligate intracellular pathogens** are incapable of growth and multiplication outside a host cell. By definition, all viruses are obligate intracellular pathogens in that they require a host cell for replication, often to the detriment of that cell. Some bacteria are also obligate intracellular bacterial pathogens, such as *Chlamydia* spp. and rickettsias (which cause diseases such as Rocky Mountain spotted fever and typhus). These microbes cannot be grown in the laboratory outside of their host cells. Malarial parasites are examples of protozoa that require host cells for growth and differentiation.

Clinically, the course of an infectious disease usually has a characteristic pattern and can be divided into several phases (**figure 35.1**). The **incubation period** is the time between pathogen entry and development of signs and symptoms. The pathogen is reproducing but has not reached a sufficient level to cause clinical manifestations. This period's length varies with pathogen. Next, the **prodromal stage** occurs with an onset of signs and symptoms that are not yet specific enough to make a clear

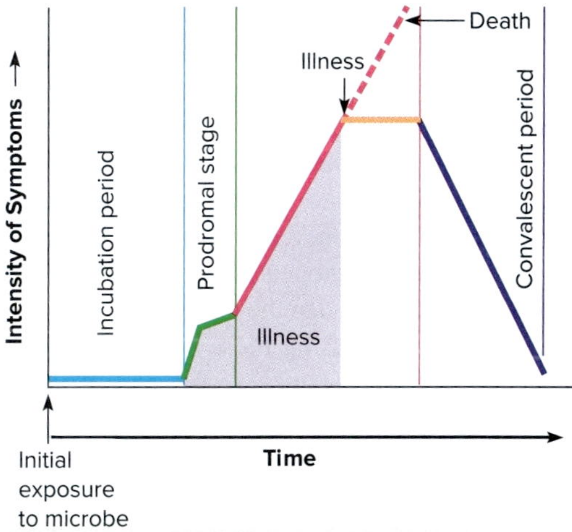

Figure 35.1 The Course of an Infectious Disease. Most infectious diseases occur in four stages. The duration of each stage is a characteristic feature of each disease. The shaded area represents when many diseases are communicable. Should the host be unable to overcome illness, death results.

MICRO INQUIRY *During which stages does the host exhibit signs and symptoms? How does this correlate to when the disease is communicable?*

diagnosis. However, the patient often is contagious. This is followed by the **illness period,** when the disease is most severe and displays characteristic signs and symptoms. The host immune response is typically triggered at this stage. Finally, during the period of decline, the signs and symptoms begin to disappear. The recovery stage often is referred to as **convalescence.**

Infectious diseases are typically not detected until distinct physical and/or physiological events alert the host of cellular battle. Infectious diseases often have characteristic signs and symptoms. **Signs** are objective changes in the body, such as a fever or rash, that can be directly observed. **Symptoms** are subjective changes, such as pain and loss of appetite, that are experienced by the patient. The term symptom is often used in a broader scope to include the clinical signs. A **disease syndrome** is a set of signs and symptoms that are characteristic of the disease.

The fundamental process of infection is essentially a competition for resources. The host is a source of protection, nutrients, and energy for the pathogen to use. Infectious agents must therefore develop mechanisms to access and exploit their hosts. Furthermore, to continue surviving, pathogens must also devise methods to move on to better environments once the immediate one declines in value, at which point dissemination to another location or a new host must occur.

For a pathogen to infect a host and subsequently cause disease, specific events must take place. First, the organism must be transmitted from the previous host or reservoir and enter the new host. Certain factors will affect the success of this process: the virulence of the organism, the number of invading organisms, and the presence of adhesion and invasion factors. Second, the organism must be able to outcompete the resident microbiota for resources and to survive the onslaught of the host defense mechanisms by producing a variety of molecules (e.g., antiphagocytic factors) as a means of counterattack. Third, disease ensues when the organism either produces molecules that directly damage host cells, stimulates the immune cells to destroy infected tissue, or alters the host-cell genome, which affects normal function. It is important to note that often the factors that help the pathogen avoid the host defenses are the same molecules that damage the host, hence there is a great deal of overlap when discussing these stages of the infection process.

Comprehension Check

1. Define infection, infectious disease, pathogenicity, virulence, and opportunistic pathogen.
2. Why are psychrophiles unable to infect humans?
3. What are some important characteristics of a pathogen?
4. What is an obligate intracellular pathogen? How might it survive?
5. Differentiate between the signs and symptoms of a disease.
6. Which phase of an infectious disease usually coincides with spread of the disease? Suppose an individual is contagious in the incubation period or in the early prodromal stage. How would that alter the ease with which the pathogen is spread to new hosts?

35.2 Transmission and Entry into the Host

After reading this section, you should be able to:

a. List and describe the means by which microorganisms access human cells and tissues to cause disease
b. Correlate initial microbial numbers and replication rates to infection and lethality
c. Compare the molecular mechanisms by which microorganisms adhere to and invade human cells and tissues

Before a microbe can cause disease in a human, the two organisms must come into contact. Pathogens are transmitted from a **source,** the location from which the pathogen is transferred to the host, either directly or indirectly through an intermediate agent. The source can be either animate (e.g., humans or animals) or inanimate (e.g., water or food). The period of infectivity is the time during which the source—such as a human or contaminated water—is disseminating the pathogen.

A **reservoir** is the natural environmental location in which the pathogen normally resides. It is also the site from which a source acquires the pathogen or where direct infection of the host can occur. Thus a reservoir sometimes functions as a source. Reservoirs also can be animate or inanimate. Examples include birds as the reservoir of West Nile virus, and contaminated drinking water as the reservoir of *Vibrio cholerae.* The increasing impingement of humans on the environment and increased exposure to antibiotics and mutagens have played significant roles in the substantive change in reservoirs over the last 100 years.

When infectious agents are transmitted from animals to humans, the diseases that ensue are termed **zoonoses** (Greek *zoon,* animal; *nosos,* disease); thus animals can serve as reservoirs, for example when a rabid dog transmits the rabies virus while biting a human. In addition, being bitten by arthropod **vectors** (organisms that spread disease from one host to another), such as mosquitoes, ticks, fleas, mites, or biting flies, can lead to infection (e.g., West Nile virus, malaria, Lyme disease, Rocky Mountain spotted fever, and plague).

Transmission and Virulence Are Interrelated

Transmission is a critical component of the infection process. It occurs either directly or indirectly by four main routes; airborne, contact, vehicle, or vector-borne. Pregnant women can also pass on a pathogen to their unborn child via **vertical transmission.** A pathogen's virulence may be strongly influenced by its mode of transmission and ability to live outside its host (**Historical Highlights 35.1**).

When a pathogen can survive for long periods outside its host, it can afford to leave the host and simply wait for a new one to come along. This seems to promote increased virulence. Host health is not critical, but extensive multiplication within the host will increase the efficiency of transmission. Good examples are tuberculosis, diphtheria, and measles caused by *Mycobacterium tuberculosis, Corynebacterium diphtheriae,* and measles viruses, respectively. These microbes survive for a relatively long time, from hours to days and weeks, outside human hosts. Many bacterial virulence factors that facilitate survival are encoded in unique sequences of DNA that are readily transferred between bacteria through conjugation, transduction, or transformation (horizontal gene transfer). ◀ *Horizontal gene transfer: creating genetic variation the asexual way (section 16.4); Microbial evolutionary processes (section 19.5)*

Exposure alone is not sufficient for infection to occur. Rather, the pathogen must also make contact with the appropriate host tissue. For instance, rhinoviruses are spread by airborne transmission from one host to another. Once the virus makes contact with the upper respiratory tract epithelium, the virus invades to cause disease only in the upper respiratory tract; rhinoviruses do not usually infect any other cell type in the host. This specificity is called a **tropism** (Greek *trope,* turning). Many pathogens exhibit cell, tissue, and organ specificities. A tropism by a specific microbe usually reflects the presence of specific cell surface receptors for that microbe on the host cell. ◀ *Attachment (Adsorption) (section 6.3)*

Airborne Transmission

Because air is not a suitable medium for the growth of pathogens, any pathogen that is airborne must have originated from a source such as humans, other animals, plants, soil, food, or water. In **airborne transmission,** the pathogen is suspended in the air in droplets, droplet nuclei, or dust.

Droplets form from saliva, mucus, and other body fluids. They can measure up to 2 mm in diameter and are produced when liquids are placed under force. Because of their size, droplets require close proximity for transfer between hosts (less than 1 meter)

HISTORICAL HIGHLIGHTS

35.1 The First Indications of Person-to-Person Spread of an Infectious Disease

In 1773 Charles White, an English surgeon and obstetrician, published his "Treatise on the Management of Pregnant and Lying-in Women." In it, he appealed for surgical cleanliness to combat childbed or puerperal fever, an acute febrile condition that can follow childbirth and is caused by streptococcal infection of the uterus or adjacent regions. In 1795 Alexander Gordon, a Scottish obstetrician, published his "Treatise on the Epidemic Puerperal Fever of Aberdeen," which demonstrated for the first time the contagiousness of the disease. In 1843 Oliver Wendell Holmes, a noted physician and anatomist in the United States, published a paper entitled "On the Contagiousness of Puerperal Fever" and also appealed for surgical cleanliness to combat this disease.

However, the first person to realize that a pathogen could be transmitted from one person to another was the Hungarian physician Ignaz Philipp Semmelweis. Between 1847 and 1849, Semmelweis observed that women who had their babies at the hospital with the help of medical students and physicians were four times as likely to contract puerperal fever as those who gave birth with the help of midwives. He concluded that the physicians and students were infecting women with material remaining on their hands after autopsies and other activities (amazingly, these individuals did not wash their hands after such procedures!). Semmelweis thus began washing his hands with a calcium chloride solution before examining patients or delivering babies. This simple procedure led to a dramatic decrease in the number of cases of puerperal fever and saved the lives of many women. As a result, Semmelweis is credited with being the pioneer of antisepsis in obstetrics. Unfortunately, in his own time, most of the medical establishment refused to acknowledge his contribution and adopt his procedures. After years of rejection, Semmelweis had a nervous breakdown in 1865. He died a short time later of a wound infection. It is very probable that it was a streptococcal infection, arising from the same pathogen he had struggled against his whole professional life.

and are therefore considered to be transmitted directly. **Droplet nuclei** are smaller particles, 1 to 5 μm in diameter, that result from the evaporation of the larger droplets. In contrast to droplets, the much smaller droplet nuclei can remain airborne for hours or days and travel long distances. **Dust** is an important factor in airborne transmission. When dust is resuspended by some disturbance, microorganisms, adhering to the particles, become airborne. If the dust carries pathogens that can survive for relatively long periods outside a host cell, this creates an epidemiological problem, particularly in clinical environments where hospital-acquired infections can arise. Because of the distance droplet nuclei and dust particles can travel (greater than 1 meter), these are considered transmission by indirect means. Pathogens that can be transmitted indirectly through aerosolized droplet nuclei or dust can also be transmitted directly via respiratory droplets. The reverse is not always true, for example the influenza virus can be transmitted either by direct (droplet) or indirect (aerosolized droplet nuclei) means; however, the Ebola virus can only be transmitted through droplets if the droplets directly contact the potential host's mucous membranes.

Contact Transmission

Contact transmission implies the coming together or touching of the source or reservoir of the pathogen and the host. Contact can be direct or indirect. Direct contact implies an actual physical interaction with the infectious source (**figure 35.2**). This route is frequently called person-to-person contact. Person-to-person transmission occurs primarily by touching, kissing, or sexual contact; by contact with oral secretions or body lesions (e.g., herpes and boils); by nursing mothers (e.g., HIV, staphylococcal infections); and through the placenta (e.g., syphilis; see Vertical Transmission) or during birth). Some pathogens also can be transmitted by direct contact with animals or animal products (e.g., members of the genera *Salmonella* and *Campylobacter*). Indirect contact transmission occurs when an inanimate object called a **fomite** transfers the infectious agent between hosts.

Vehicle Transmission

Inanimate materials that indirectly transmit pathogens are called **vehicles** (**figure 35.3**). In contact **vehicle transmission,** the vehicle, often a fomite, spreads the pathogen to multiple hosts. Examples of fomites include surgical instruments, drinking vessels, stethoscopes, bedding, eating utensils, and clothing. A single source containing pathogens (e.g., blood, drugs, intravenous fluids) can contaminate a common vehicle that causes multiple infections. Food, water, and biological materials (fluids and tissues) are important common vehicles for many human diseases; moreover they often support pathogen reproduction. Recall that aerosolized agents traveling greater than 1 meter to reach a new host are using the air as a vehicle.

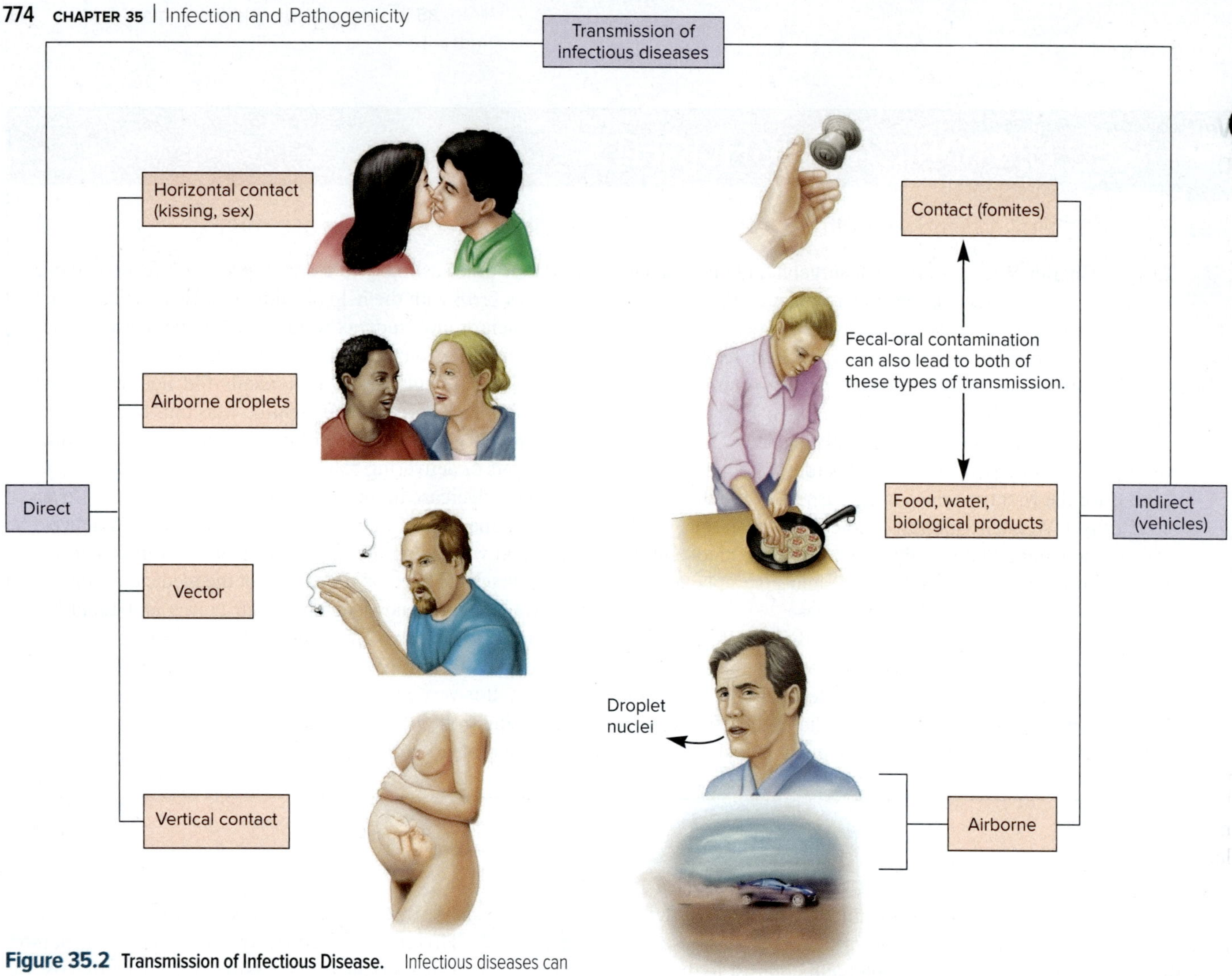

Figure 35.2 Transmission of Infectious Disease. Infectious diseases can be transmitted by various direct and indirect methods.

Figure 35.3 Common Vehicles of Microbial Contamination. Any inanimate object that transmits infectious agents to humans is referred to as a vehicle.

Vector-Borne Transmission

Direct living transmitters of a pathogen are called vectors. Most vectors are arthropods (e.g., insects, ticks, mites, fleas) or vertebrates (e.g., dogs, cats, skunks, bats). **Vector-borne transmission** often benefits the pathogen by permitting extensive reproduction and spread within the host. If pathogen levels are very high in the host, a vector such as a biting insect has a better chance of picking up the pathogen and transferring it to a new host. Indeed, pathogens transmitted by biting arthropods often are highly virulent, causing diseases such as malaria, typhus, and sleeping sickness. It is important that such pathogens do not harm their vectors and the vector generally remains healthy, at least long enough for pathogen transmission.

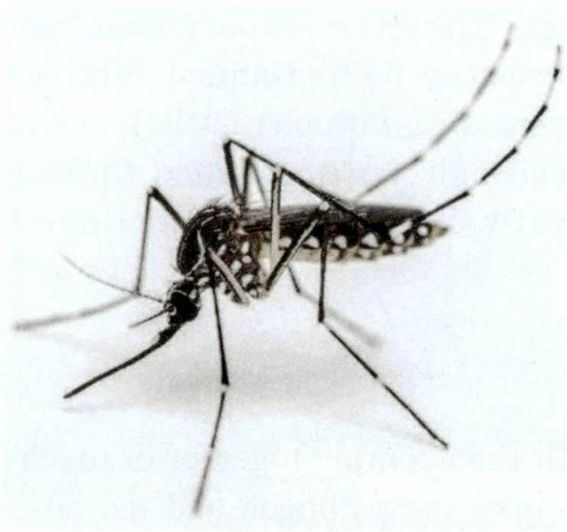

©Akil Rolle-Rowan/Shutterstock

Vertical Transmission

Direct transmission of pathogens is not only mediated by horizontal, person-to-person passage but by vertical passage as well. **Vertical transmission** of pathogens occurs when the unborn child acquires a pathogen from an infected mother. While vertical transmission does not occur as frequently as horizontal transmission, several microorganisms exploit this route, extending their host range. Babies born with an infectious disease are said to have a congenital (Latin, with birth) infection. Examples of congenital infections include gonorrhea, syphilis, herpes, German measles, and toxoplasmosis. These diseases are discussed in chapters 38 through 40.

The Number of Invading Organisms Differs Among Infections

The rate at which an infection proceeds (the timing of the incubation period and prodromal stage, for example) and its severity are functions of the initial inoculum of microorganisms and their virulence (in addition to the host's resistance). Some microorganisms are sufficiently adept at entering a host that very few are required to establish an infection. Conversely, other microorganisms are required in high numbers to establish an infection. The inocula can be measured experimentally to determine the **infectious dose 50 (ID_{50})**—that is, the number of microorganisms required to cause disease in 50% of the inoculated hosts (**figure 35.4**). One example of this is found in the oral transmission of *Salmonella* spp., which are shed in the feces of infected hosts and often contaminate hands. During food preparation, the bacteria are transferred to foods, where they can then be ingested. The ID_{50} for *Salmonella* spp. ingested in contaminated food is approximately 1×10^5 bacteria. Suffice it to say, the reason for thorough handwashing as a means of maintaining public health is to decrease the number of pathogens that can be transmitted. In general, the lower the infectious dose (i.e., the fewer organisms needed for infection), the higher the risk of infection.

Another measure of infectivity, and thus virulence, is the experimental determination of the number of microorganisms required to be lethal to a host species. This is known as the **lethal dose 50 (LD_{50})** and refers to the dose or number of pathogens that kills 50% of an experimental group of hosts within a specified period. Highly virulent pathogens kill at a lower dose when compared to less virulent organisms.

It is important to note that the ID_{50} and LD_{50} for some organisms can also be influenced by the susceptibility of the host. If the potential host has a robust immune system, the likelihood of developing a disease may be diminished. Factors such as nutrition, cleanliness, emotional health, and genetic predisposition play a role; thus, those with a weakened immune system due to medications, stress, or lack of nutrients, for example, may be more likely to contract an infectious disease.

Adhesion and Invasion Factors Initiate Disease

The need to adhere to a host cell, invade the underlying tissues, and colonize is a major obstacle that must be overcome by a potential pathogen. To accomplish these feats, the microbial cell wall is often covered with molecules, typically proteins (or glycoproteins), that are anchored to the peptidoglycan. These surface molecules serve a range of functions to facilitate access to the host. In addition to mediating adhesion and invasion, microbial surface proteins also allow the pathogen to evade the host immune responses and aid in the formation of biofilms.

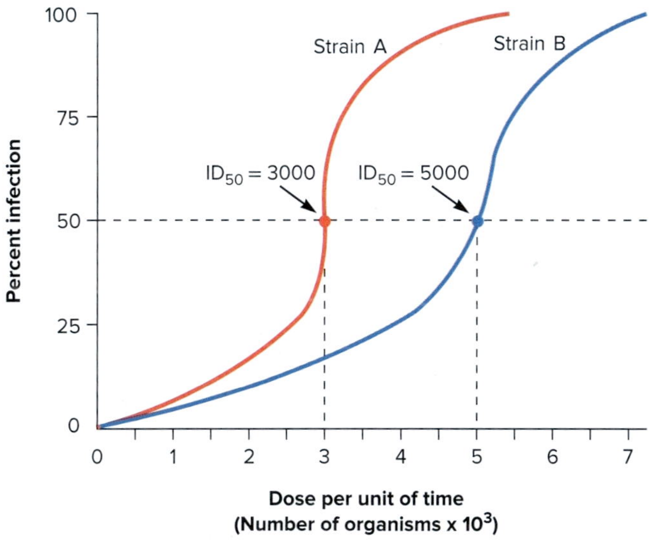

Figure 35.4 Determination of the ID_{50} of a Pathogenic Microorganism. Various doses of a specific pathogen are introduced into experimental host animals. Infections are recorded and a graph constructed. In this example, the graph represents the susceptibility of host animals to two different strains of a pathogen—strain A and strain B. For strain A, the ID_{50} is 3000, and for strain B, it is 5000. Hence strain A is more virulent than strain B.

Entry and Adhesion Establish Colonization

Once a pathogen has been transmitted from the reservoir or source to a susceptible host in sufficient numbers, access to the host is gained through the appropriate **portal of entry:** the skin, respiratory system, gastrointestinal system, urogenital system, or the conjunctiva of the eye. The first step in the infectious disease process following entry is the attachment of the microorganism to a target cell.

After gaining access to the host, the pathogen must be able to adhere to and colonize host cells or tissues. In this context, colonization means the establishment of a site of microbial replication on or within a host. It does not necessarily result in tissue invasion or damage. Adherence can be thought of as a two-step process. Step 1 is nonspecific and often reversible. It results from hydrophobic, electrostatic, and vibrational forces that encourage "docking." Step 2 is specific and permanent, as complementary molecules (cognate receptors) fulfill a lock-in-key style bonding. This latter process is an example of coevolution, permitting pathogen exploitation of the characteristics of host-cell surfaces. Thus pathogens adhere with a high degree of specificity to target tissues. Adherence structures, also known as adhesins, such as pili and fimbriae (**table 35.1**), membrane and capsular materials, and specialized adhesion

Table 35.1 Examples of Microbial Attachment Mechanisms

Microbe	Disease	Adhesion Mechanism	Host Receptor
Neisseria gonorrheae	Gonorrhea	Type IV fimbriae	Mannose on urethral epithelium
Escherichia coli	Diarrhea	Type I fimbriae	Sugar residue on intestinal epithelium
	Hemolytic uremic syndrome	P pili	Sugar residue on kidney cell
	Urinary tract infection	P pili	Sugar residue on urethral epithelium
Treponema pallidum	Syphilis	Outer membrane protein	Fibronectin on mucosal cell
Mycoplasma pneumoniae	Pneumonia	Membrane protein	Sialic acid on lung cell
Streptococcus pyogenes	Sore throat	Protein F	Fibronectin on upper respiratory tract cell
Streptococcus mutans	Dental caries	Sugar residue	Salivary glycoprotein on tooth
Influenza virus	Influenza	Hemagglutinin spike protein	Protein residue on upper respiratory tract cell
HIV-1	AIDS	gp120 protein	CD4 receptor on T cells
Polio virus	Poliomyelitis	Capsid protein VP1	CD 155 protein on intestinal and nerve cells

molecules on the surface of the invading microbe, bind to host sites (**figure 35.5**). These microbial products and structural components facilitate infection, and since they also contribute to the virulence of the organism, they are classified as virulence factors. Examples of pathogen adherence are also discussed in the context of specific diseases in chapters 38 through 40.

Invasion Disseminates Pathogens

Pathogens can be described in terms of their infectivity and invasiveness. **Infectivity** is the ability of the organism to establish a discrete focal point of infection. Attachment to specific host cell surface molecules permits microbe-mediated infection but may also induce cell-mediated microbe internalization. Viruses are adept at these methods of infectivity. **Invasiveness** is the ability of the organism to spread to adjacent or other tissues. For some pathogens, a localized infection is sufficient to cause disease. However, many pathogens invade other tissues.

Pathogens can either actively or passively penetrate the host's surfaces. These include mucous membranes in the respiratory, gastrointestinal, and genitourinary tracts, and other surfaces made up of epithelial cells, also referred to as epithelium. Active penetration may be accomplished through production of lytic substances that alter the host tissue by (1) attacking the extracellular matrix and basement membranes of body surfaces (integuments) and intestinal linings, (2) degrading carbohydrate-protein complexes between cells or on the cell surface (the glycocalyx), or (3) disrupting the host cell surface. Passive mechanisms of penetration are not related to the pathogen itself. Examples include (1) small breaks, lesions, or ulcers in a mucous membrane that permit initial entry; (2) wounds, abrasions, or burns on the skin's surface; (3) arthropod vectors that create small wounds while feeding; (4) separation of tight junctions between cells during inflammation, and (5) tissue damage caused by other organisms (e.g., a dog bite).

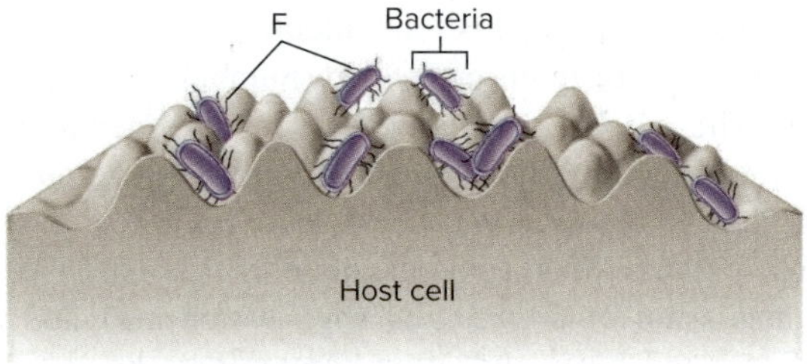

(a) Fimbriae (F)

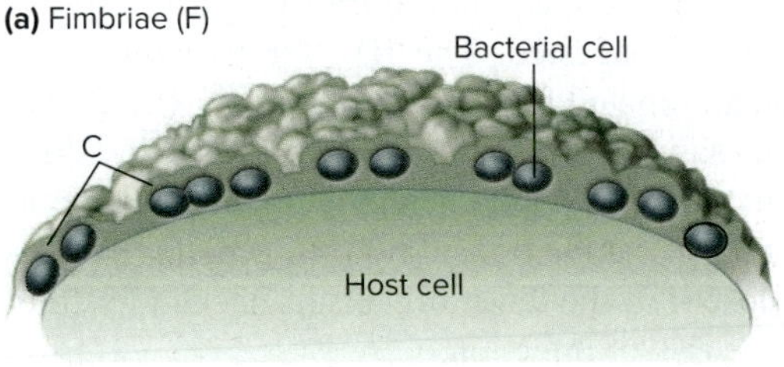

(b) Capsules (C)

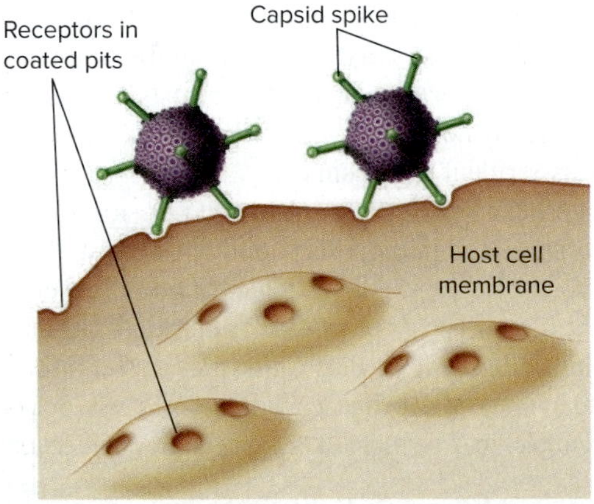

(c) Spikes

Figure 35.5 Microbial Adherence Mechanisms. (a) Bacterial fimbriae. (b) Bacterial capsules. (c) Viral protein spikes.

Once under the mucous membrane or other integument, a pathogen may penetrate to deeper tissues and continue disseminating throughout the body of the host. One way the pathogen accomplishes this is by producing specific structures or enzymes that promote spreading (**table 35.2**). These products are also considered virulence factors because they contribute to pathogen success within the host. Pathogens may also enter the small terminal lymphatic capillaries that surround epithelial cells. These capillaries merge into large lymphatic vessels that eventually drain into the circulatory system. Once the circulatory system is reached, the microbes have access to all organs and systems of the host. The presence of viable bacteria in the bloodstream is called **bacteremia.** The infectious disease process caused by bacterial or fungal toxins in the blood is termed **septicemia** (Greek *septikos,* produced by putrefaction; *haima,* blood).

Invasiveness varies greatly among pathogens. For example, *Clostridium tetani* (the cause of tetanus) is considered noninvasive because it does not spread from one tissue to another, but its toxin becomes blood-borne, thereby causing disease. *Bacillus anthracis* (the cause of anthrax) and *Yersinia pestis* (the cause of plague) also produce toxins; moreover the bacteria are also highly invasive. Members of the genus *Streptococcus* span the spectrum of virulence factors and invasiveness with some causing strep throat and others causing severe skin and soft-tissue infections ("flesh-eating" infections). Another method of invasion is seen with *Listeria monocytogenes, Shigella* spp., and *Rickettsia* spp. These bacteria use actin-based motility to move within mammalian host cells and spread between them. They activate the assembly of an actin tail using host cell actin and other cytoskeletal proteins (**figure 35.6**). The actin tails propel the bacteria through

Table 35.2 Bacterial Virulence Factors Involved in Bacterial Pathogen Invasion and Dissemination

Product	Organism Involved	Mechanism of Action
Coagulase	*Staphylococcus aureus*	Coagulates (clots) the fibrinogen in plasma. The clot protects the pathogen from phagocytosis and other host defenses.
Collagenase	*Clostridium* spp.	Breaks down collagen that forms the framework of connective tissues; allows the pathogen to spread
Deoxyribonuclease	Group A streptococci, staphylococci, *Clostridium perfringens*	Lowers viscosity of exudates, giving the pathogen more mobility
Elastase and alkaline protease	*Pseudomonas aeruginosa*	Cleaves laminin associated with basement membranes
Hemolysins	Staphylococci, streptococci, *Escherichia coli, Clostridium perfringens*	Lyse erythrocytes; make iron available for microbial growth
Hyaluronidase	Groups A, B, C, and G streptococci, staphylococci, clostridia	Hydrolyzes hyaluronic acid, a constituent of the extracellular matrix that cements cells together. Its hydrolysis renders the intercellular spaces amenable to passage by the pathogen
Hydrogen peroxide (H_2O_2) and ammonia (NH_3)	*Mycoplasma* spp., *Ureaplasma* spp.	These metabolic wastes are toxic and damage epithelia in respiratory and urogenital systems.
Immunoglobulin A protease	*Streptococcus pneumoniae*	Cleaves immunoglobulin A into Fab and Fc fragments
Lecithinase or phospholipase	*Clostridium* spp.	Destroys the lecithin (phosphatidylcholine) component of plasma membranes, allowing pathogen to spread
Leukocidins	Staphylococci, pneumococci, and other streptococci	Pore-forming exotoxins that kill leukocytes; cause degranulation of lysosomes within leukocytes, which decreases host resistance
Porins	*Salmonella enterica* serovar Typhimurium	Inhibit leukocyte phagocytosis by activating the adenylate cyclase system
Protein A Protein G	*Staphylococcus aureus* *Streptococcus pyogenes*	Located on cell wall. Immunoglobulin G (IgG) binds to either protein A or protein G by its Fc end, thereby preventing complement from interacting with bound IgG.
Pyrogenic exotoxin B (cysteine protease)	Group A streptococci (*Streptococcus pyogenes*)	Degrades proteins
Streptokinase (fibrinolysin, staphylokinase) Staphylokinase	Groups A, C, and G streptococci, staphylococci *S. aureus*	A protein that binds to plasminogen and activates the production of plasmin, thus digesting fibrin clots; this allows the pathogen to move from the clotted area

Figure 35.6 **Formation of Actin Tails by Intracellular Bacterial Pathogens.** Colorized scanning electron micrograph of *Listeria monocytogenes* in a host macrophage. The bacterium (pink) has polymerized host actin into a long tail (green) at only one pole that it uses for intracellular propulsion and to move from one host cell to another.

the cytoplasm of the infected cell to its surface, where they push out against the plasma membrane and form protrusions. The protrusions are engulfed by adjacent cells, and the bacteria enter the unsuspecting neighboring host cell, thereby spreading throughout the tissue. *Microsporidia* spp. use a novel polar tubule to bore into host cells (*see figure 25.17*), whereas many other fungi use hydrolytic enzymes to invade cells and tissues. These virulence factors are discussed in more detail in chapters 39 and 40.

Comprehension Check

1. Define source and reservoir. How are they related?
2. Describe the four main types of infectious disease transmission methods and give examples of each.
3. Define droplet nuclei, vehicle, fomite, and vector.
4. Why are pathogens that are transmitted person-to-person often less virulent than those transmitted by arthropod vectors?
5. How would you determine the ID_{50} of a suspected pathogen?
6. Describe several specific adhesins by which bacterial pathogens attach to host cells.

35.3 Surviving the Host Defenses

After reading this section, you should be able to:

a. Compare and contrast competition between microbial species with competition between microbial and human cells
b. Describe features that allow microorganisms to overcome host resistance and immunity
c. Discuss strategies microorganisms have evolved to exploit human cells and tissues as resources for their survival
d. Differentiate between the characteristics of microbes in a biofilm and their planktonic counterparts

The vast majority of microorganisms do not cause disease, in part because they are antagonized by the host's normal microbiota or eliminated by its immune system before infection begins. In these circumstances, competing for space and nutrients with the resident microbiota and innate defense mechanisms come into play upon initial host exposure, followed by specific immune responses that become paramount in preventing further proliferation of infectious agents. However, some microorganisms have evolved mechanisms that enable them to overcome the competition and elude the initial host responses, as well as those governed by the adaptive immune system.

Before bacteria can establish an infection, they must be able to overcome resident microbiota in direct competition for resources. Certain pathogens produce a Type VI secretion system (T6SS)—a molecular nanoweapon that delivers lethal effector molecules from its cytoplasm to a competing organism through direct contact. These secretion systems are especially effective in Gram-negative enteric pathogens, for whom competition for resources is especially fierce. Survival of the microorganism in the host also depends on its ability to "find shelter," thereby avoiding recognition by the defense cells. Some microorganisms survive and replicate inside host cells; in some cases, they reside in the very cells that are supposed to destroy them. Other organisms squeeze between host cells, make capsules to avoid phagocytosis, burrow under mucus, and secrete exopolysaccharides to form communal shelters within biofilms. Some microorganisms produce enzymes that inactivate innate resistance mechanisms. Still other organisms have evolved specialized protein secretion systems to selectively kill host cells. These physical and chemical characteristics reflect microbial virulence and facilitate microbial survival. ◀ *One-step secretion systems (section 13.8); A functional core microbiome is required for human homeostasis (section 34.3)*

Pathogens have evolved a variety of ways to *evade* the host's immune response. For example, cells infected with the hepatitis B virus produce large amounts of proteins not associated with the complete virus particle. These proteins serve as decoys, binding the available antimicrobial proteins so that the virus goes unnoticed. Some microbes change their surface proteins by mutation or recombination (e.g., the influenza virus) or downregulate the level of expression of cell surface proteins. Some bacteria (e.g., *Streptococcus pyogenes*) produce **capsules** that resemble host tissue components and thus remain covert. The capsules of some other bacteria (e.g., *Haemophilus influenzae*) prevent the deposition of host complement, thereby protecting the pathogen from host defenses. *Neisseria gonorrhoeae* can evade the specific immune response by two mechanisms: (1) It makes changes in the gene encoding the pili protein through a process called phase variation, resulting in altered pilin protein sequence and expression. This renders specific host responses useless against the new pili so that adherence to host tissue is not obstructed. (2) It produces proteases that degrade host proteins such as immunoglobulin A (IgA). Other bacteria produce special proteins (such as staphylococcal protein A, and the G and M proteins of *S. pyogenes*) that interfere with the host's ability to detect and remove them (*see figure 23.11*b). Human immunodeficiency virus (HIV), the measles virus, and cytomegalovirus are examples of pathogens that cause host cell fusion. This allows these viruses to spread from an infected cell to an uninfected cell with minimal exposure to host defenses. |◀ *Capsules and slime layers (section 3.5); Phylum* Bacteroidetes *includes important gut microbiota (section 21.8); Complement is a cascading enzyme system (section 32.3)*

Pathogens have evolved a variety of ways to *suppress* the host's immune response. Some viruses, most notably HIV, infect cells of the immune system and diminish their function while ensuring their own survival (*see chapter 38*). Some bacteria such as *Streptococcus pneumoniae, Neisseria meningitidis,* and *Haemophilus influenzae* produce a slippery mucoid capsule that prevents the host immune cells from effectively capturing the bacterium. Others such as *N. gonorrhoeae* have eliminated the O-antigen on their lipopolysaccharide (thus called lipooligosaccharide) thereby diminishing immune recognition and clearance. |◀ *Typical Gram-negative cell walls include additional layers besides peptidoglycan (section 3.4); Antimicrobial peptides and proteins (section 33.3)*

Biofilms Provide Microbial Protection

Many bacteria live within stable, sessile communities surrounded by a hydrated polymeric matrix of their own production. These **biofilm** bacteria are protected from nutrient deprivation, predators, environmental shifts, antimicrobial agents, and host immune cells. In contrast to free-swimming (planktonic) bacteria of the same species, biofilm bacteria are phenotypically different, expressing genes that favor survival when attached and in poor nutrient conditions. However, the role of the biofilm is much more complex, especially for biofilms that cause chronic infections in humans. Using high throughput sequencing, fluorescent in situ hybridization, and global gene analysis, studies have revealed that some pathogenic bacteria within biofilms exchange plasmids, nutrients, and quorum-sensing molecules, altering their behavior compared to planktonic forms. The differences are striking in that biofilm bacteria coordinate gene expression to upregulate mechanisms that make the biofilm community less sensitive to antibiotics and more resistant to host defense mechanisms. For example, studies on *Pseudomonas aeruginosa* biofilms reveal that shared communication signals activate global response regulators, which increase the activity of multidrug efflux pumps, increase starvation or stress responses, stimulate exopolysaccharide secretion, increase resistance to host cell attack, suppress host cell killing mechanisms, and change rapidly growing, antibiotic-sensitive cells into slower-growing, antibiotic-resistant persister cells. |◀ *Microbial growth in natural environments (section 7.6)*

Importantly, one remarkable activity of biofilms formed by pathogenic *P. aeruginosa* is the ability to suppress host immune responses (**figure 35.7**). This is evident in studies evaluating *P. aeruginosa* mixed with antibiotics, antibodies, and the phagocytic white blood cells (macrophages and neutrophils). Antibiotics, antibodies, and the phagocytes readily attack and kill planktonic *P. aeruginosa*. However, these agents are ineffective at killing *P. aeruginosa* in biofilms. *P. aeruginosa* in biofilms appears to prevent penetration of antibiotics and antibodies, and to cause phagocytes to downregulate killing mechanisms. In addition, efficacy of host

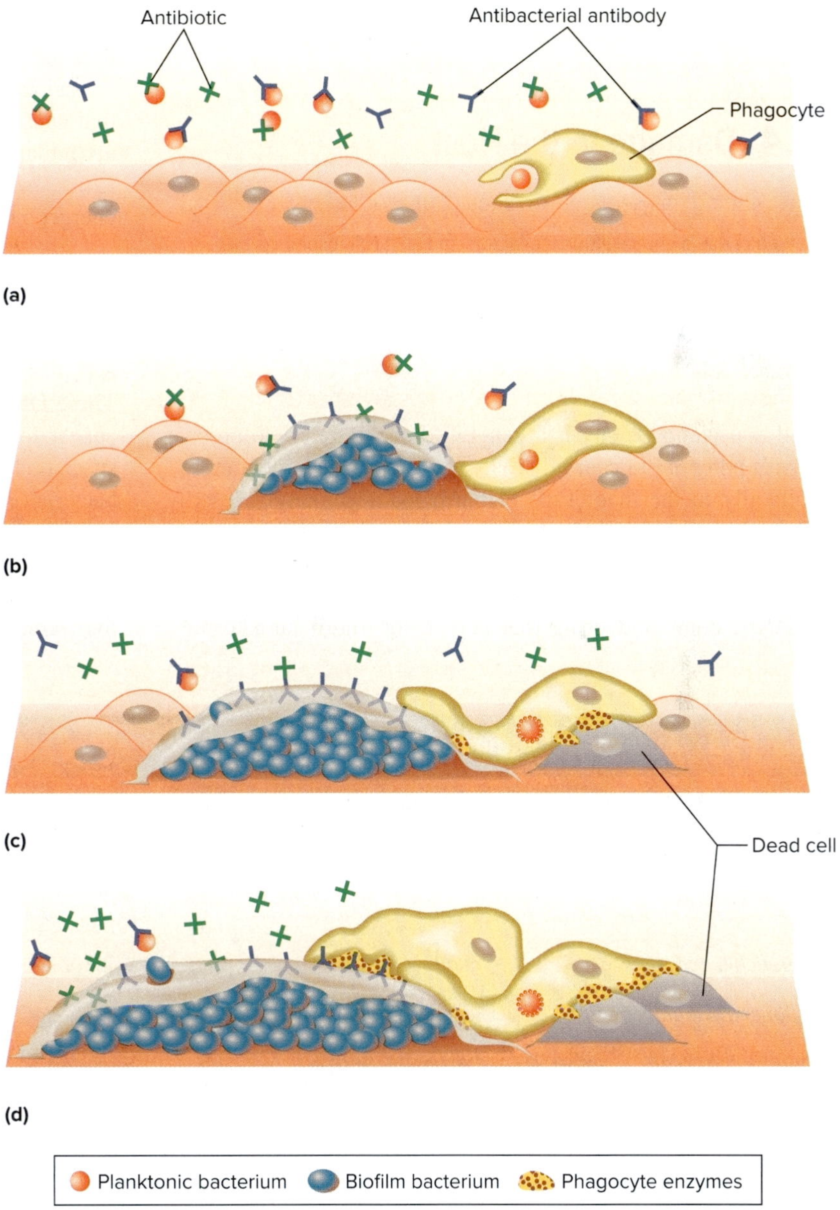

Figure 35.7 Frustrated Phagocytosis. (a) Planktonic cells are sensitive to antibiotics and are detected by host phagocytes and antibodies. (b) Planktonic cells settle to form a biofilm and become biofilm cells. (c) Biofilm cells resist antibiotics and antibody detection. Host phagocytes detect and attempt to destroy biofilm cells. (d) Unable to capture biofilm cells, phagocytes release antimicrobial products that kill host cells but not biofilm cells.

hydrolytic enzymes is decreased, while enzymes that create super- and nitric oxides are inhibited. Simply put, the *P. aeruginosa* biofilms "frustrate" phagocytes by reducing the magnitude of some antibacterial responses and preventing others altogether. ◀ *Phagocytosis destroys invaders (section 32.6)*

Comprehension Check

1. How does a capsule enhance pathogenicity?
2. What is a secretion system and how does it aid pathogen survival?
3. Give examples of organisms that can evade and suppress host immune responses.
4. What is a biofilm? How is the ability to form a biofilm a virulence factor?

35.4 Damage to the Host

After reading this section, you should be able to:

a. Predict the microbial virulence factors and host cell responses that result in host damage
b. Discuss the origin and nature of pathogenicity islands
c. Illustrate the mechanisms by which microbial toxins impact human cells

Recall that pathogenicity is a general term that refers to an organism's potential to cause disease, while virulence refers more specifically to the magnitude of harm (pathogenicity) caused. Some microorganisms are pathogenic due to their rapid replication and explosive departure from host cells. Others are pathogens because they integrate parts of their genome into the host cell DNA, commandeering that cell in the future or altering normal function. Many microorganisms, particularly bacteria, use toxins to poison host cells. In many cases of infection, tissue damage is caused indirectly when the pathogen stimulates an excessive or inappropriate immune response by the host. This is an extremely important point because it means that pathogenicity is not a trait inherent in microorganisms but rather a consequence of the interplay between microbe and host. Whatever the outcome, any structural or soluble product that increases pathogenicity is a virulence factor.

Virulence factors can be encoded in the microbial chromosome or as extrachromosomal elements. As mentioned earlier in the chapter, there is a pronounced overlap between (1) those factors that enable the microorganism to establish an infection (adherence and colonization factors), spread to other tissues (invasion factors), and avoid host defenses and (2) those factors that cause damage to the host. Synthesis of virulence factors is upregulated by specific host triggers, enabling the microorganism to better survive its residency in the host and avoiding the energetic cost of factor production when in the environment.

Pathogenicity Islands Encode Virulence Factors

Large segments (10 to 200 kilobases) of bacterial chromosomal and plasmid DNA have been found to encode virulence factors. These DNA segments are called **pathogenicity islands,** and they appear to have been inserted into the existing DNA during the speciation of pathogens from their nonpathogenic ancestors. Many bacteria (e.g., *Yersinia* spp., *P. aeruginosa, Shigella flexneri, Salmonella* spp., enteropathogenic *Escherichia coli*) carry at least one pathogenicity island. Pathogenicity islands generally increase bacterial virulence and are absent in nonpathogenic members of the same genus or species (**table 35.3**). For example, the

Table 35.3 Examples of Pathogenicity Islands and the Products They Encode

Organism	Pathogenicity Island[1]	Gene Product	Function
Helicobacter pylori	PAI-III	Type 1 pili Secreted protein	Attachment Hemolysin, cytotoxic necrosing factor, uropathogenic protein
Escherichia coli	cagPI	Type IV secretion proteins	Cytotoxin
Legionella pneumophila	icm/dot	Type IV secretion proteins	Intracellular survival
Rhodococcus equi	PAI-vap	Secreted proteins	Intracellular survival
Salmonella enterica	SPI-1, SPI-2	Type III secretion proteins	Cytotoxin
Shigella flexneri	SHI-1, SHI-2	Type III secretion proteins	Cytotoxin
Staphylococcus aureus	SaPI	Secreted proteins	Superantigens
Vibrio cholerae	VPI	Secreted protein	Toxin
Yersinia pestis	HPI-1	Siderophore synthesis	Iron uptake and storage

1 PI and PAI are both common abbreviations for pathogenicity island.

Staphylococcus aureus pathogenicity island encodes several superantigen genes, including the gene for the toxin that causes toxic shock syndrome. The movement of genes from one bacterial cell to another is discussed in chapter 16, while the detection of pathogenicity islands within a microbe's genome is explained in chapter 18. ◀ *Comparative genomics (section 18.7)*

Toxins Are Biological Poisons

A **toxin** (Latin *toxicum,* poison) is a substance that disrupts the normal metabolism of host cells with deleterious effects on the host. **Toxigenicity** is the pathogen's ability to produce toxins, and **intoxications** are diseases that result from a specific toxin produced by the pathogen. Intoxication diseases do not require the presence of the actively growing pathogen—just its toxin, as in the case of botulism. Bacteria produce two structurally different types of toxins—exotoxins (proteins) and endotoxin (lipopolysaccharide)—and fungi produce potent mycotoxins.

Exotoxins

Exotoxins are soluble, heat-labile proteins (inactivated at 60° to 80°C) usually released into host tissues as the bacterial pathogen metabolizes. Often exotoxins travel from the site of infection to other body tissues or target cells, where they exert their effects (**figure 35.8**). Exotoxins are often encoded by genes carried on plasmids or prophages within certain bacteria. They are associated with specific diseases and often are named for the disease they produce (e.g., the diphtheria toxin). Some are among the most lethal substances known—toxic in nanogram-per-kilogram of body weight concentrations (e.g., the botulinum toxin).

Exotoxins exert their biological activity by specific mechanisms and are grouped by either mechanism of action (e.g., a cytotoxin kills cells) or their protein structure. A common type is the **AB toxin,** which gets its name from the fact that it has two distinct toxin subunit types, an "A" (or active) component and a "B" (or binding) component. The B portion of the toxin binds to a host-cell receptor and triggers endocytosis of the toxin, which is eventually uncoupled. The A component has enzyme activity that causes toxicity (figure 35.8*a*). Thus the B component determines the cell type the toxin will affect, whereas the A component exerts the deleterious effect. AB toxins act on cells by different mechanisms. Many A subunits have ADP ribosylation activity, which catalyzes the transfer of adenosine diphosphate and ribose moieties of host NAD^+ to target host molecules *(see figure 10.7)*. Another type of exotoxin disrupts membranes. Examples of this type of exotoxin are the channel (pore)-forming toxins (figure 35.8*b*). They destabilize membrane integrity so that the host cell lyses. The general properties of several exotoxins are presented in **table 35.4**. As proteins, toxins are easily recognized by the host immune system, which produces antitoxin—antibodies that recognize

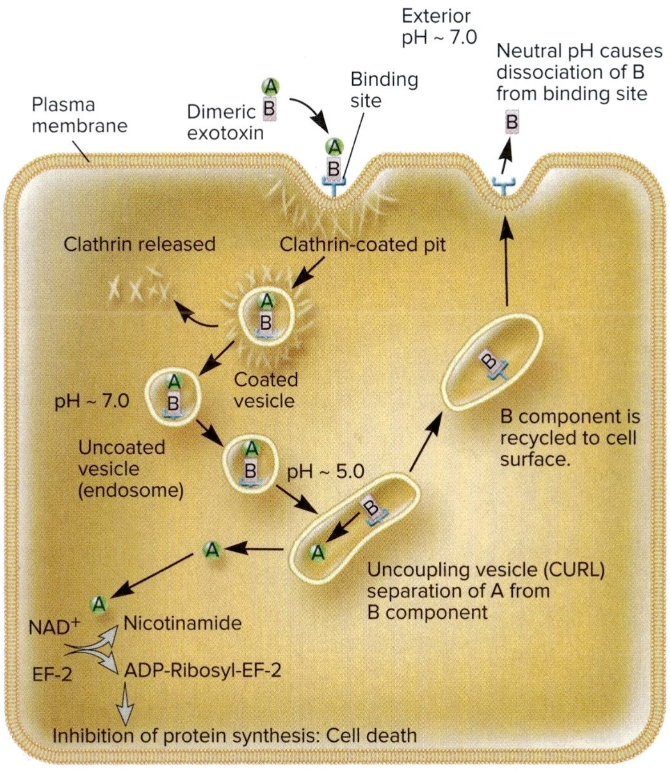

(a)

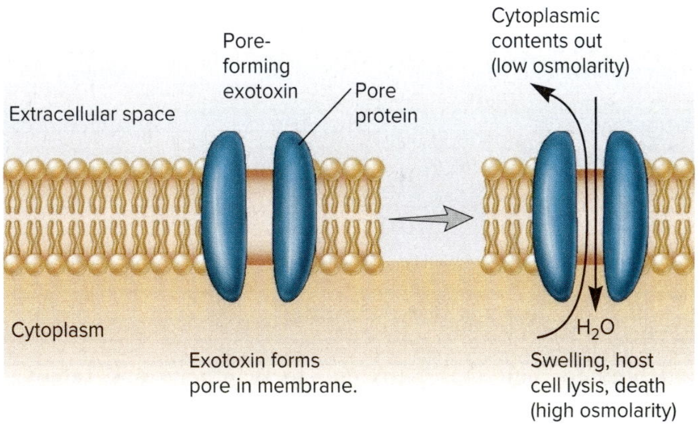

(b)

Figure 35.8 Two Examples of Exotoxin Mechanisms. (a) This AB toxin binds to the cell receptor within clathrin-coated pits by the B subunit. The intact toxin is endocytosed, and the pH change within the endosome causes the subunits to separate. An endosome in which this separation occurs is sometimes called a compartment of uncoupling of receptor and ligand (CURL). The B subunit is then recycled. The active toxin (A) subunit exerts its effect on its target, in this case the elongation factor-2 (EF-2) of protein synthesis, by enzymatically attaching an ADP-ribose to EF-2. The altered EF-2 is inhibited from assisting in protein synthesis, leading to cell death. (b) This channel-forming (pore-forming) toxin inserts itself into the host cell membrane, forming a channel (or pore). Multiple membrane pores result in an osmolarity shift, as water enters the cell and cytoplasmic contents move out. The resulting effect of this toxin is cell lysis.

Table 35.4 Exotoxins Produced by Human Bacterial Pathogens

Toxin	Organism	Gene Location	Toxin Type	Mechanism of Action
Edema factor (EF) Lethal factor (LF) Protective antigen (PA)	*Bacillus anthracis*	Plasmid	Tripartite AB	EF causes edema. LF is a cytotoxin. PA is a B component.
Pertussis toxin	*Bordetella pertussis*	Chromosome	AB	↓ATP, ↑cAMP alters cell function, leading to death.
Botulinum toxin	*Clostridium botulinum*	Prophage	AB	Blocks acetylcholine release, leading to flaccid paralysis
CPE enterotoxin	*Clostridium perfringens*	Chromosome	Cytotoxin	Hemolysis
Tetanospasmin	*Clostridium tetani*	Plasmid	AB	Blocks release of inhibitory neurotransmitter, leading to spastic paralysis
Diphtheria toxin	*Corynebacterium diphtheriae*	Phage	AB	Alters translation, leading to protein synthesis inhibition
Enterotoxin Shiga-like toxin	*Escherichia coli* *E. coli* O157:H7	Plasmid Phage gene integrated into chromosome	AB AB	↑cAMP, leading to water secretion from cell Inhibits protein synthesis leading to death
Cytolysin	*Salmonella* spp.	Chromosome	Cytotoxin	↑cAMP, leading to water secretion from cell
Shiga toxin	*Shigella dysenteriae*	Chromosome	AB	Inhibits protein synthesis, leading to death
Exfolative toxin Toxic shock syndrome toxin-1 Panton-Valentine leukocidin	*Staphylococcus aureus*	Chromosome Chromosome Phage	Protease Superantigen Cytotoxin	Skin peeling Cytokine-induced shock Lyses white blood cells
Streptolysin O Erythrogenic toxin	*Streptococcus pyogenes*	Chromosome Phage	Cytolysin Superantigen	Hemolysis Cytokine-induced shock
Cholera toxin	*Vibrio cholerae*	Phage	AB	↑cAMP, leading to water secretion from cell

and bind to their target toxins—to inactivate and remove them.

Finally, exotoxins called **superantigens** act by stimulating as many as 30% of host T cells to overexpress and release massive amounts of pro-inflammatory cytokines from host immune cells in the absence of a specific antigen (**figure 35.9**). The excessive concentration of cytokines causes multiple host organs to fail, giving the pathogen time to disseminate. By triggering this "cytokine storm," superantigens cause life-threatening disease; fever, fluid loss, and low blood pressure result in shock and death. Superantigens are so potent because they bind antigen-presenting cells nonspecifically to T cells, which then become activated. The host response is very similar to that described later in this section for endotoxin-mediated septic shock. Examples of each specific toxin type are discussed as part of the disease process in chapters 38 through 40. ◀ *Cytokines are chemical messages between cells (section 32.3)*

Endotoxin

The lipopolysaccharide (LPS) in the outer membrane of Gram-negative bacteria is toxic to humans. Thus LPS is called an **endotoxin** because it is bound to the bacterium and is released when the microorganism lyses, although some may also be released during cell division. The toxic component of LPS is the lipid portion, called lipid A (*see figure 3.25*). **Lipid A** is not a single macromolecular structure; rather, it is a complex array of lipid residues. Lipid A is heat stable and toxic in nanogram amounts but only weakly immunogenic. ◀ *Typical Gram-negative cell walls include additional layers besides peptidoglycan (section 3.4)*

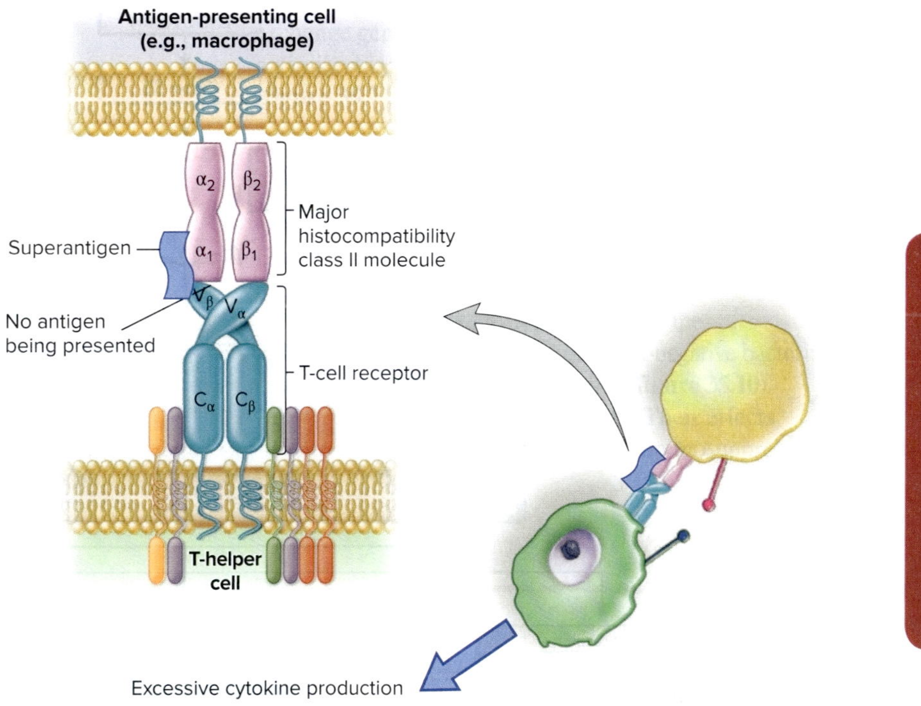

Figure 35.9 Superantigen. Nonspecific activation of T cells by a superantigen, in the absence of antigen, stimulates massive amounts of cytokines that overwhelm the host and result in shock and death.

Unlike the structural and functional diversity of exotoxins, the lipid A of various Gram-negative bacteria produces similar systemic effects regardless of the microbe from which it is derived. These include fever (i.e., endotoxin is pyrogenic), shock, damage to the lining of blood vessels, weakness, diarrhea, inflammation, intestinal hemorrhage, and fibrinolysis (enzymatic breakdown of fibrin, the major protein component of blood clots).

Lipid A is the primary initiator of **septic shock,** a culmination of systemic responses to the toxin. The biological effect of lipid A is an indirect one, mediated by host molecules and systems rather than by lipid A itself. Endotoxin stimulates the release of the endogenous mediators of sepsis (e.g., cytokines like TNF-a, IL-1, IL-6, and IL-18) from damaged endothelial cells, monocytes, macrophages, neutrophils, and plasma cell precursors. The ensuing "cytokine storm" causes rapid dilation of the blood vessels accompanied by a drastic drop in blood pressure that has profound physiological effects on the heart, vasculature, and other body organs. Endotoxin also activates a protein called the Hageman factor (blood clotting factor XII), which in turn results in unregulated blood clotting within capillaries (disseminated intravascular coagulation) and multiorgan failure (**figure 35.10**). Endotoxins also indirectly induce a fever in the host by causing macrophages to release endogenous pyrogens that reset the hypothalamic thermostat. One important endogenous pyrogen is the cytokine interleukin-1 (IL-1). Other cytokines released by macrophages, such as tumor necrosis factor, also produce fever. Other than antimicrobial agents to control infection and vasopressor medications to stabilize blood pressure, there is no specific drug therapy to stem the septic cascade. Thus the consequences of septic shock are either recovery or death. Death usually ensues if one or more organ systems fail completely. To prevent inadvertently causing septic shock, all drugs, especially those given intravenously or intramuscularly, must be free of endotoxin and are therefore routinely screened.

Mycotoxins

Mycotoxins are toxins produced by fungi. For example, *Aspergillus flavus* and *A. parasiticus* produce aflatoxins, and *Stachybotrys* spp. produce satratoxins. These fungi commonly contaminate food crops and water-damaged buildings, respectively. An estimated 4.5 billion people in developing countries may be exposed chronically to aflatoxins through their diet. Exposure to aflatoxins causes both chronic and acute liver disease and liver cancer. Aflatoxins are extremely carcinogenic, mutagenic, and immunosuppressive. Approximately 18 different types of aflatoxins exist; these are classified in two broad groups according to their chemical structure (*see figures 41.5 and 41.6*). *Stachybotrys trichothecene* mycotoxins are potent inhibitors of DNA, RNA, and protein synthesis. They induce inflammation, disrupt surfactant phospholipids in the lungs, and may lead to pathological changes in tissues.

The fungus *Claviceps purpurea* also produces toxic substances. The products are generically referred to as ergots, reflecting the name of the tuberlike structure of the fungi (*see figure 25.12*). Ergots from various *Claviceps* spp. produce alkaloids that have varying physiological effects on humans. One such alkaloid is lysergic acid, the psychotropic hallucinogen LSD.

Comprehension Check

1. What are virulence factors?
2. What are pathogenicity islands and why are they important?
3. Describe some general characteristics of exotoxins.
4. Discuss the role of lipid A in the development of septic shock.

36

Epidemiology and Public Health Microbiology

Practice What You Preach

The years 2011 and 2014 marked a practical opportunity for instructors of microbiology laboratories to practice what we preach. Using DNA fingerprinting techniques, investigators from the Centers for Disease Control and Prevention tracked two multistate outbreaks of *Salmonella* infection in a total of 150 people. The two groups were approximately 70% female, 30% male, with a median age of 21.5 years. The original illnesses were identified in employees of clinical microbiology laboratories and students in microbiology teaching laboratories who reported working with *Salmonella* bacteria. The outbreak strain associated with several of the ill people was indistinguishable from a commercially available strain of *Salmonella enterica* serovar Typhimurium found in several clinical and teaching laboratories associated with ill employees and students. Sadly, several children known to reside in households of individuals who work or study in a microbiology laboratory using this "lab strain" became seriously ill. Thus we teachers of microbiology were schooled in infectious disease tracking, case definitions, incidence reporting, laboratory testing, disease control, and disease prevention. Certainly, in addition to all the other lessons learned from this outbreak, proper hand washing upon leaving the laboratory should be a best practice that is never overlooked (ever).

In this chapter, we describe the practical goal of epidemiology: to establish effective disease recognition, control, prevention, and eradication measures for a given population. Because emerging and reemerging diseases and pathogens, healthcare-acquired (nosocomial) infections, and bioterrorism are major concerns for public health, these topics are also covered here.

Readiness Check:

Based on what you have learned previously, you should be able to:

- ✓ Identify specific chemotherapeutic treatments to control infection (chapter 9)
- ✓ Describe innate human host resistance mechanisms (chapter 32)
- ✓ Discuss the human adaptive immune response to pathogens (chapter 33)
- ✓ Explain pathogenicity and the infectious disease process (chapter 35)

36.1 Epidemiology Is an Evidence-Based Science

After reading this section, you should be able to:

a. Describe the agencies that are responsible for disease prevention and control
b. Define the basic vocabulary and processes used in the science of epidemiology

By definition, **epidemiology** (Greek *epi,* upon; *demos,* people or population; *logy,* study) is the science that evaluates the occurrence, determinants, distribution, and control of health and disease in a defined human population (**figure 36.1**). An individual who practices epidemiology is an **epidemiologist.** Epidemiologists are, in effect, disease detectives. Their major concerns are the discovery of the factors essential to disease occurrence and the development of methods for disease prevention. In the United States, the **Centers for Disease Control and Prevention (CDC)**, headquartered in Atlanta, Georgia, serves as the national agency for developing and carrying out disease prevention and control, environmental health, and health promotion and education activities (**Historical Highlights 36.1**). Its international counterpart is the **World Health Organization** (WHO), located in Geneva, Switzerland.

The science of epidemiology originated and evolved in response to great epidemic diseases such as cholera, typhoid fever, smallpox, influenza, and yellow fever (**Historical Highlights 36.2**). More recent epidemics of Ebola, HIV/AIDS, cryptosporidiosis, enteropathogenic *Escherichia coli*, severe acute respiratory syndrome (SARS), salmonellosis, and influenza have underscored the importance of epidemiology

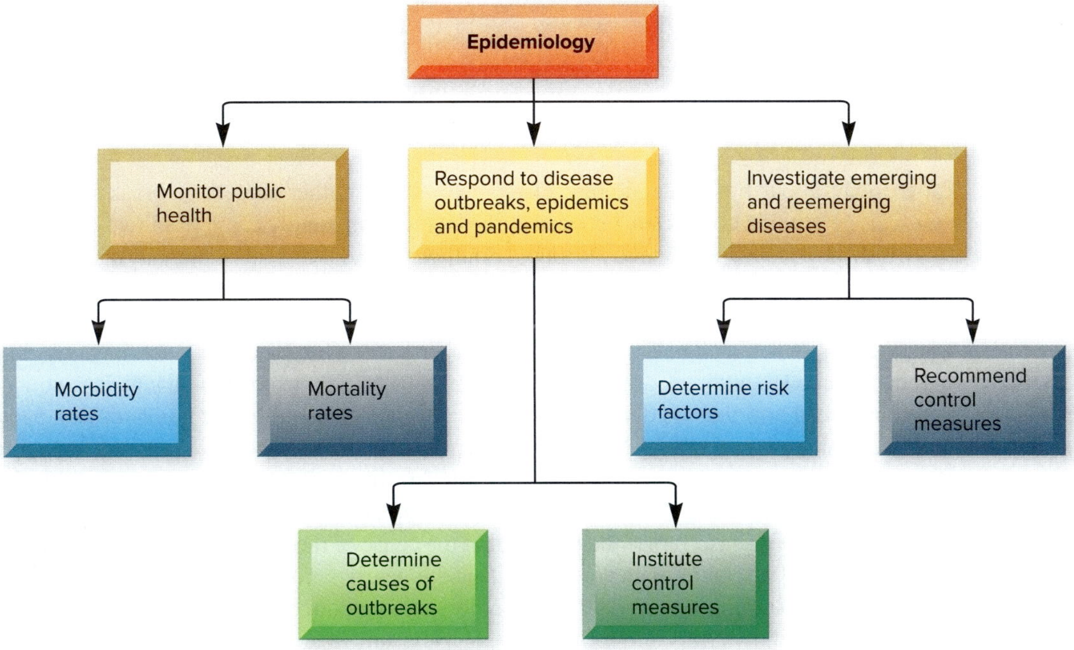

Figure 36.1 Epidemiology. Epidemiology is a multifaceted science that investigates diseases to discover their origin, evaluates diseases to assess their risk, and controls diseases to prevent future outbreaks.

HISTORICAL HIGHLIGHTS

36.1 The History of Public Health in the United States

The U.S. Public Health Service was born in 1798 out of the need to keep sailors from becoming sick while at sea. As defenders of the new republic, sailors, or seamen as they were called, could not find adequate healthcare at port cities of the day. Public hospitals were few in number and often did not have the resources to care for the large numbers of sick seamen that could arrive. As the seamen came from all over the country, their healthcare was determined to be a national responsibility. Thus the Marine Hospital Act of 1798 established a network of federal hospitals, called the Marine Hospital Service (MHS). The oversight responsibility was assigned to the Revenue Marine Division of the Treasury Department. The act permitted the taxation (20 cents per month) of the seamen, creating the first medical insurance program in the United States.

In 1878 the MHS was reorganized by congressional action from individually operated hospitals into a centrally controlled, national agency headquartered in Washington, D.C. The new federal agency became a separate bureau of the Treasury Department overseen by a central administrator, the "Supervising Surgeon," who was appointed by the Secretary of the Treasury. The administrative title of Supervising Surgeon was changed to Supervising Surgeon General in 1875. In 1902 the MHS became the Public Health and Marine Hospital Services, and the administrator's title was changed to Surgeon General. A national "hygienic" laboratory located on Staten Island, New York, was established in 1887. The Pure Food and Drugs Act was established in 1906. In 1912 the Public Health and Marine Hospital Services was renamed the Public Health Service and authorized to investigate communicable disease.

In 1930 the Hygienic Lab moved to Washington to become the National Institute of Health, but it was not until 1939 that all health, education, and welfare agencies and services created by Congress were combined, under the Federal Security Agency. The Public Health Service opened the Communicable Disease Center in Atlanta, Georgia, in 1946. In 1953 the Federal Security Agency became the Department of Health, Education, and Welfare, and in 1970 the Communicable Disease Center became the Center for Disease Control. Ten years later, the Department of Health, Education, and Welfare became the Department of Health and Human Services, and the Center for Disease Control was renamed the Centers for Disease Control. It was renamed again in 1992 to become the Centers for Disease Control and Prevention (although it is still abbreviated as the CDC).

HISTORICAL HIGHLIGHTS

36.2 John Snow, the First Epidemiologist

Much of what we know today about the epidemiology of cholera is based on the classic studies conducted between 1849 and 1854 by the British physician John Snow. During this period, a series of cholera outbreaks occurred in London, England, and Snow set out to find the source of the disease. Some years earlier when he was still a medical apprentice, Snow was sent to help during a cholera outbreak among coal miners. His observations convinced him that the disease was spread by unwashed hands and shared food, not by "bad" air or casual direct contact.

Thus when the outbreak of 1849 occurred, Snow continued to support the germ theory of the disease versus the popular "miasma" theory, suspecting that contaminated water was the source of the infection. He examined official death records and discovered that most of the victims in the Broad Street area lived close to the Broad Street water pump or were in the habit of drinking from it. He deduced that cholera was spread by drinking water from the Broad Street pump, which was contaminated with raw sewage containing the disease agent. When the pump was disabled by removing its handle, the number of cholera cases dropped dramatically.

In 1854 another cholera outbreak struck London. Part of the city's water supply came from two different suppliers: the Southwark and Vauxhall Company and the Lambeth Company. Snow interviewed cholera patients and found that most of them purchased their drinking water from the Southwark and Vauxhall Company that obtained its water from the Thames River below locations at which Londoners discharged their sewage. In contrast, the Lambeth Company took its water from the Thames before the river reached the city. The death rate from cholera was over eightfold lower in households supplied with Lambeth Company water. Snow deduced that the cause of the disease must be able to multiply in water. Thus he nearly recognized that cholera was caused by a microorganism, though Robert Koch did not discover the causative bacterium (*Vibrio cholerae*) until 1883.

To commemorate these achievements, the John Snow Pub now stands at the site of the old Broad Street pump. Those who complete the Epidemiologic Intelligence Program at the Centers for Disease Control and Prevention receive an emblem bearing a replica of a barrel of Watney's Ale, the brew dispensed at the John Snow Pub.

in preventing global catastrophes caused by infectious diseases. Today, epidemiology's scope encompasses all public health issues: infectious diseases, genetic abnormalities, metabolic dysfunction, malnutrition, neoplasms, psychiatric disorders, obesity, and aging. This chapter discusses only infectious disease epidemiology.

Epidemiology studies the factors determining and influencing the frequency and distribution of health-related events. When a disease occurs occasionally or at irregular intervals in a human population, it is a **sporadic** disease (e.g., bacterial meningitis). When it maintains a steady, low-level frequency at a moderately regular interval, it is an **endemic** (Greek *endemos,* dwelling in the same people) **disease** (e.g., the common cold). Hyperendemic diseases gradually increase in frequency beyond the endemic level but not to the epidemic level (e.g., the common cold during winter months). An **outbreak** is the sudden, unexpected occurrence of a disease, usually in a limited segment of a population (e.g., the 2014 measles outbreak that occurred at Disneyland California). An **epidemic** (Greek *epidemios,* upon the people), on the other hand, is an outbreak affecting many people at once (i.e., there is a sudden increase in the occurrence of a disease above the expected level). Influenza is an example of a disease that may occur suddenly and unexpectedly in a family and often achieves epidemic status in a community. The influenza virus resides in a **reservoir host** (usually birds or pigs) until it is transmitted to a human host. The first case in an epidemic is called the **index case.** A **pandemic** (Greek *pan,* all) is an increase in disease occurrence within a large population over at least two countries around the world. Usually pandemic diseases spread among continents. The global H1N1 influenza outbreaks of 1918 and 2009 are good examples. The **attack rate** is the proportional number of cases that develop in a population exposed to an infectious agent.

36.2 Epidemiology Is Rooted in Well-Tested Methods

After reading this section, you should be able to:

a. Evaluate the effectiveness of public health surveillance methods
b. Deduce the impact of public health methods on the quality of U.S. life over the last 100 years
c. Describe the use of geographical information systems to track diseases remotely
d. Use standardized surveillance data to measure infectious disease frequency
e. Calculate rates of incidence, prevalence, morbidity, and mortality

Public health is the science of protecting populations and improving the health of human communities through education, promotion of healthy lifestyles, and prevention of disease and injury.

Public health practitioners use a methodical approach to identify population health issues to prevent or correct negative outcomes. The methodical approach mirrors the scientific method in that it identifies a population health problem, determines the cause, proposes preventative or corrective action, implements that action, and assesses the outcomes of the action.

Public Health Surveillance

The identification of a population health issue is rooted in public health surveillance. Public health surveillance involves the proactive evaluation of genetic background, environmental conditions, human behaviors and lifestyle choices, emerging infectious agents, and microbial responses to chemotherapeutic agents to monitor the health of a population. In other words, public health practitioners look for cause-and-effect relationships to determine risk. The impact of public health surveillance is obvious when one considers the change in the leading causes of death in the United States from 1900 to 2016 (**table 36.1**). The public health landscape of 1900 was a valley of death due to infectious disease. Surveillance identified infectious disease problems and their associated risks. These were remedied by water treatment, strict sanitation guidelines, and later by the use of antimicrobial agents and vaccines. The result changed the death-by-infection landscape of the early twentieth century to one of very few deaths due to infectious disease by the 1960s. Ironically, the modernization that helped prevent the leading causes of mortality in the 1900s also led to a longer life expectancy and a more sedentary lifestyle in the twenty-first century. This changed the landscape of 2016 to one of a mountain of metabolic disease.

Public health surveillance has its roots in the fourteenth century as attempts were made to control the spread of bubonic plague. However, it was not until after the germ theory of disease was widely accepted (late 1800s) that a scientifically based monitoring process was used as a means to track communicable disease. This was not the confidential monitoring that we expect today. The first disease tracking practices were intrusive, as people were inspected for signs of clinical disease; those found to be infected were quarantined. The change from inspecting people to monitoring disease occurred around 1940. The emphasis on watching disease progress in a population and disseminating information in advance of the disease radically improved the impact on public health by changing the role of public health practitioners from disease informants to healthcare allies (**figure 36.2**). Importantly, disease data come from a number of sources, including mandated clinical reports, laboratory specimens, records and databases, sentinel surveillance, disease registries, and field surveys. These data are used most often for the generation of

Table 36.1 Ten Leading Causes of Death in the United States, 1900[1] and 2016[2]

Cause	Percentage	
	1900	2016*
Heart diseases	8.0	23.1
Cancer	3.7	21.8
Pneumonia and influenza	11.8	1.9
Tuberculosis	11.3	ND
Gastrointestinal diseases	8.3	ND
Cerebrovascular diseases	6.2	5.2
Senility (Alzheimer's disease and other dementias)	6.8	4.2
Pulmonary diseases	ND	5.6
Kidney diseases	4.7	1.8
Accidents	4.2	5.9
Diabetes mellitus	ND	2.9
Diphtheria	2.3	ND
Suicide	ND	1.6
All other causes	32.7	25.9
Total	100.0	100.0

1 *Source: National Office of Vital Statistics, 1954.*
2. *Source: National Center for Health Statistics*, National Vital Statistics System, Mortality Data Brief 293: Mortality in the United States, 2016: ND = Not determined.
*Latest available data

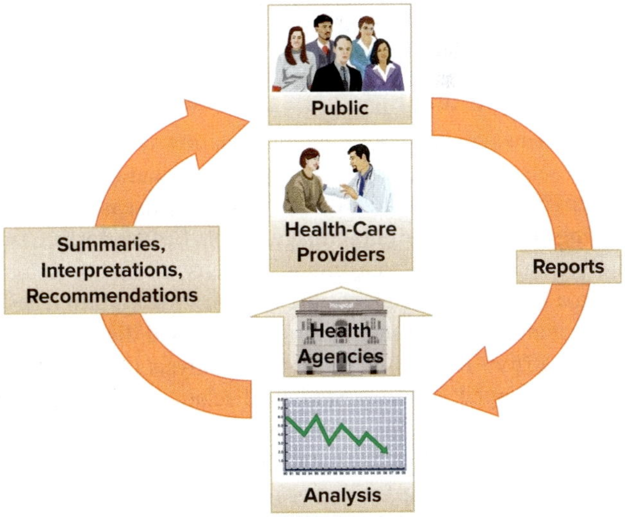

Figure 36.2 Information Loop of Public Health Surveillance. The Centers for Disease Control and Prevention use a number of different data sources to maintain surveillance of the infectious diseases that may erupt into localized epidemics or more widespread pandemics.

MICRO INQUIRY *How might biased data collection impact public health recommendations? Explain.*

reports informing public health officials of (1) population morbidity and mortality, (2) disease effect on school attendance, (3) impact of epidemics on employee absenteeism, and (4) animal and vector control policies.

In 1878 Congress authorized the U.S. Marine Hospital Service (the forerunner of the Public Health Service; Historical Highlights 36.1) to collect illness data associated with cholera, smallpox, plague, and yellow fever. These data were used to quarantine the sick and prevent the spread of these diseases into the United States. The following year, a congressional appropriation directed the collection and publication of these "notifiable diseases." Congress subsequently directed the Surgeon General to compile and publish the data. The CDC assumed this responsibility in 1961. As of 2017, there were 72 infectious diseases on the list of notifiable diseases; it is updated periodically to reflect emerging infectious diseases (e.g., Zika virus disease was added in 2016 and carbapenemase-producing carbapenem-resistant *Enterobacteriaceae* (CP-CRE) was new in 2018). Importantly, effective reporting of notifiable (and other) diseases is contingent on reliable surveillance data collection systems.

Effective use of public health surveillance data stems from high-integrity data collection (i.e., documented first-hand information of the data origin and the methods used to collect them). Surveillance of health issues is typically accomplished by two methods: population surveys and case reporting. Population surveys tend to be more of a sentinel surveillance method in that they look for trends and outliers within a given population (e.g., immunization records revealing infection control success). Case reports tend to be retrospective comparisons of specific conditions that assign people to risk groups (e.g., influenza test results to rule out SARS).

Remote Sensing and Geographic Information Systems: Charting Infectious Disease Data

Remote sensing and geographic information systems are map-based tools that can be used to study the distribution, dynamics, and environmental correlates of microbial diseases. **Remote sensing (RS)** gathers digital images of the Earth's surface from satellites and output from biological sensors, and transforms the data into maps. A **geographic information system (GIS)** is a data management system that organizes and displays digital map data from RS and facilitates the analysis of relationships between mapped features. Statistical relationships often exist between mapped features and diseases in natural host or human populations. Examples include the location of the habitats of the malaria parasite and mosquito vectors in Mexico and Asia, ticks that transmit Lyme disease in the United States, and African trypanosomes in both humans and livestock in Africa. RS and GIS are most useful if disease dynamics and distributions are clearly related to mapped environmental variables. For example, if a microbial disease is associated with certain vegetation types or physical characteristics (e.g., elevation, precipitation), RS and GIS can identify regions in which risk is relatively high.

Measuring Infectious Disease Frequency

To determine if an outbreak, epidemic, or pandemic is occurring, epidemiologists measure disease frequency at single time points and over time. They then use statistics to analyze the data and determine risk factors and other factors associated with disease.

To determine the frequency or rate of an event, an accurate count of the total population, the population exposed, and the number of affected people needs to be made. Disease surveillance practices enable the tracking of infections. Many infectious diseases (e.g., food- and waterborne diseases) must, by law, be reported within a specific time frame. This permits public health agencies to act swiftly to contain the outbreak and mobilize control measures. Timely notification of "reportable" diseases to public health agencies has stemmed numerous epidemics such as cholera and typhoid.

Measures of frequency usually are expressed as fractions. The numerator is the number of individuals experiencing the event—infection—and the denominator is the number of individuals in whom the event could have occurred, that is, the population at risk. The fraction is a proportion or ratio but is commonly called a rate because a time period is specified. (A rate also can be expressed as a percentage.) In population statistics, rates usually are stated per 1,000 individuals, although other powers of 10 may be used for particular diseases (e.g., per 100 for very common diseases and per 10,000 or 100,000 for uncommon diseases). Disease **incidence** is a measure of the number of diseased people during a defined time period, as compared to the total (healthy) population. The incidence of a disease reports both the rate of occurrence and the relative risk. While the incidence of a disease reflects a rate of infection over time, the **prevalence** refers to the total number of individuals infected in a population at a specific point in time, no matter when the disease began. The prevalence rate depends on both the incidence rate and the duration of the illness. Prevalence is calculated as follows:

$$\text{Prevalence} = \frac{\text{Total number of cases in population}}{\text{Total population}} \times 100$$

Since disease reflects a change in health status over time, a **morbidity rate** is used. The rate is commonly determined when the number of new cases of illness in the general population is known from clinical reports. It is calculated as follows:

$$\text{Morbidity rate} = \frac{\text{Number of new cases of a disease during a specified period}}{\text{Number of individuals in the population}}$$

For example, if in 1 month, 700 new cases of influenza per 100,000 individuals occur, the morbidity rate would be expressed as 700 per 100,000, or 0.7%.

The **mortality rate** is the relationship between the number of deaths from a given disease and the total number of cases of the disease. The mortality rate is a simple statement of the proportion of all deaths that are assigned to a single cause. It is calculated as follows:

$$\text{Mortality rate} = \frac{\text{Number of deaths due to a given disease}}{\text{Size of the total population with the same disease}}$$

For example, if 15,000 deaths occurred due to hepatitis C in a year and the total number of people infected was 30,000, the mortality rate would be 15,000 per 30,000, or 50%.

The determination of prevalence, morbidity, and mortality rates helps public health personnel direct healthcare efforts to control the spread of infectious diseases. A sudden increase in the morbidity rate of a particular disease may indicate a need for implementation of preventive measures designed to reduce the disease.

Comprehension Check

1. What is epidemiology?
2. Define and distinguish epidemic, endemic, and pandemic.
3. What types of surveillance data are most useful in determining infectious disease penetration into a population?
4. How might remote sensing be used to track Rocky Mountain spotted fever (*section 39.2*)?
5. Define morbidity rate, prevalence rate, and mortality rate.

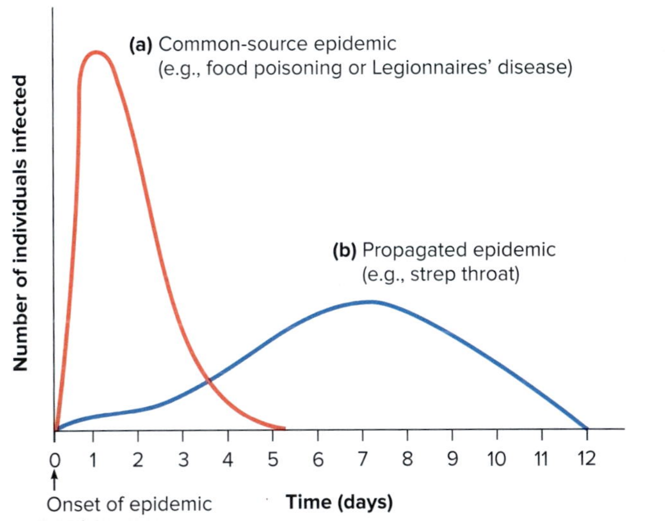

Figure 36.3 Epidemic Curves. (a) In a common-source epidemic, there is a rapid increase up to a peak in the number of individuals infected and then a rapid but more gradual decline. Cases usually are reported for a period that equals approximately one incubation period of the disease. (b) In a propagated epidemic, the curve has a gradual rise and then a gradual decline. Cases usually are reported over a time interval equivalent to several incubation periods of the disease.

MICRO INQUIRY *Is a flu epidemic an example of a common-source or a propagated epidemic? Explain your answer.*

36.3 Infectious Disease Is Revealed Through Patterns Within a Population

After reading this section, you should be able to:

a. Compare and contrast communicable and noncommunicable diseases
b. Interpret population infection data to define epidemic and pandemic events
c. Apply the concept of herd immunity to public health in your community
d. Predict potential infectious disease outbreaks from molecular data revealing antigenic drift and antigenic shift events

An **infectious disease** results from an infection by microbial agents such as viruses, bacteria, fungi, protozoa, and helminths (worms). A **communicable disease** is an infectious disease that can be transmitted from person to person. Not all infectious diseases are communicable; for example, rabies is an infectious disease acquired only through contact with a rabid animal. An epidemiologist studying an infectious disease is concerned with the causative agent, the source or reservoir of the disease agent, how it is transmitted, what host and environmental factors can aid development of the disease within a defined population, and how best to control or eliminate the disease. These factors describe the natural history or cycle of an infectious disease.

Two major types of epidemics are recognized: common source (noncommunicable) and propagated (communicable). A **common-source epidemic** is characterized as having reached a peak level within a short period of time (1 to 2 weeks), followed by a moderately rapid decline in the number of infected patients (**figure 36.3*a***). This type of epidemic usually results from a single, common contaminated source such as food (food poisoning) or water (cholera). By contrast, a **propagated epidemic** is characterized by a relatively slow and prolonged rise, and then a gradual decline in the number of individuals infected (figure 36.3*b*). This type of epidemic usually results from the introduction of a single infected individual into a susceptible population. The initial infection is then propagated to others in a gradual fashion until many individuals within the population are infected. An example is the increase in influenza cases that coincides with the return of college students after their winter break. Droplet and contact transmission of flu virus spreads rapidly among susceptible, and unvaccinated, students. Whereas one infected student is sufficient to initiate the epidemic, it is the person-to-person transmission of the flu virus that extends the epidemic. After an infectious disease has been recognized in a population, epidemiologists correlate the disease outbreak with a specific organism; its exact cause must be discovered (**Historical Highlights 36.3**). At this point, the clinical or diagnostic microbiology laboratory enters the investigation. Its purpose is to isolate and identify the organism responsible for the disease.

▶| *Clinical microbiology and immunology (chapter 37)*

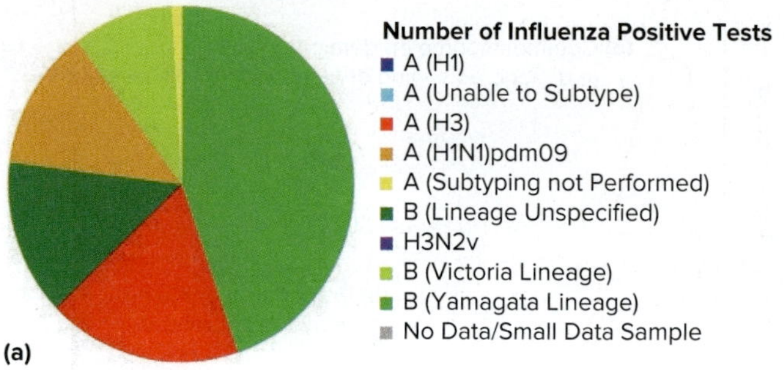

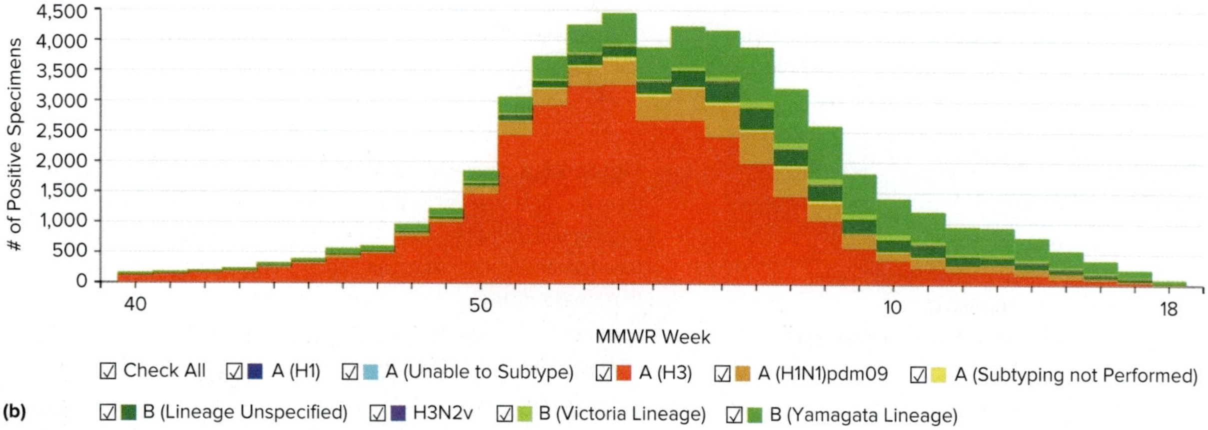

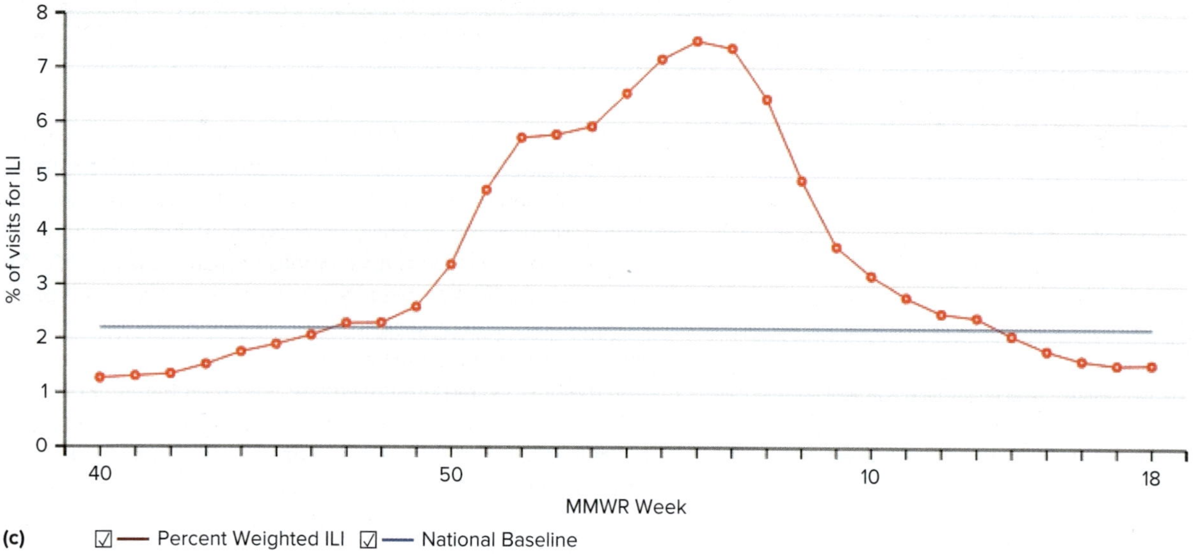

Figure 36.4 Graphic Representation of Epidemiological Data. The Centers for Disease Control and Prevention collect and evaluate various parameters related to human health and disease. These data reflect the cumulative number of confirmed human cases of influenza for the 2016–2017 flu season by (a) numbers of influenza subtypes, (b) strain number and percent per week, and (c) number of hospitalizations. ILI, influenza-like illness; MMWR, mortality and morbidity weekly report

As mentioned, epidemiologists recognize an infectious disease in a population by using various surveillance methods. Surveillance is a dynamic activity that includes gathering information on the development and occurrence of a disease, collating and analyzing the data, summarizing the findings, and using the information to select control methods (**figure 36.4**). Surveillance may not always require the direct examination of cases. However, to accurately interpret surveillance data and study the course of a disease in individuals, epidemiologists and other medical professionals must be aware of the pattern of infectious diseases that reflects how the disease spreads through a population or at least to one person (*see figure 35.3*). Epidemiologists study disease patterns so they can rapidly identify new outbreaks before substantial morbidity and mortality occur. ◀ *The process of infection (section 35.1)*

To understand how epidemics are propagated, consider **figure 36.5**. At time 0, all individuals in this population are susceptible to a hypothetical pathogen. The introduction of an infected individual initiates the epidemic outbreak (lower curve), which spreads to reach a peak by day 15. As individuals recover from the disease, they become immune and no longer

HISTORICAL HIGHLIGHTS

36.3 "Typhoid Mary"

In the early 1900s there were thousands of typhoid fever cases, and many people died of the disease. Most of these cases arose when people drank water contaminated with sewage or ate food handled by or prepared by individuals who were shedding the typhoid fever bacterium (*Salmonella enterica* serovar Typhi). The most famous carrier of the typhoid bacterium was Mary Mallon.

Between 1896 and 1906, Mallon worked as a cook in seven homes in New York City. Twenty-eight cases of typhoid fever occurred in these homes while she worked in them. As a result, the New York City Health Department had Mallon arrested and admitted to an isolation hospital on North Brother Island in New York's East River. Examination of her stools showed that she was shedding large numbers of typhoid bacteria, though she exhibited no external symptoms of the disease. An article published in 1908 in the *Journal of the American Medical Association* referred to Mallon as "Typhoid Mary," an epithet by which she is still known today. She was released when she pledged not to cook for others or serve food to them. Mallon changed her name and began to work as a cook again. For 5 years she managed to avoid capture while continuing to spread typhoid fever. Eventually the authorities tracked her down. She was held in custody for 23 years, until she died in 1938. As a lifetime carrier, Mary Mallon was positively linked with 10 outbreaks of typhoid fever, 53 cases, and 3 deaths.

transmit the pathogen (upper curve). The number of susceptible individuals therefore decreases. The decline in the number of susceptible individuals to the threshold density (the minimum number of individuals necessary to continue propagating the disease) coincides with the peak of the epidemic wave, and the incidence of new cases declines because the pathogen cannot propagate itself.

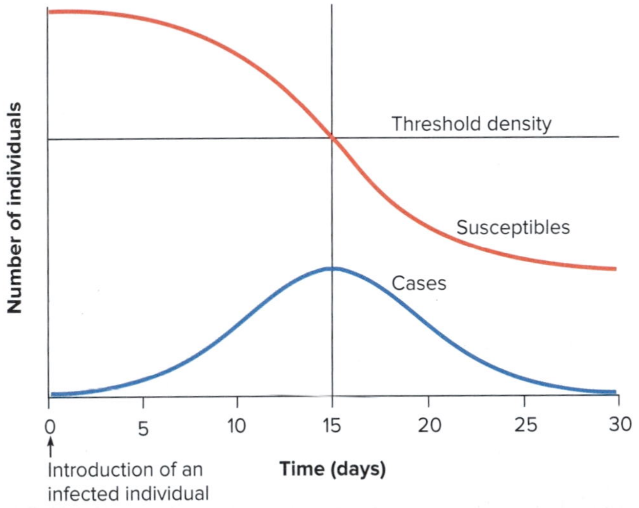

Figure 36.5 The Spread of an Imaginary Propagated Epidemic. The lower curve represents the number of cases, and the upper curve, the number of susceptible individuals. Notice the coincidence of the peak of the epidemic wave with the threshold density of susceptible people.

MICRO INQUIRY *Why does the number of susceptible individuals decline after the epidemic has reached its peak?*

Herd immunity is the resistance of a population to infection and pathogen spread because of the immunity of a large percentage of the population. The larger the proportion of those immune, the smaller the probability of effective contact between infective and susceptible individuals; that is, many contacts will be immune, and thus the population will exhibit a group resistance. A susceptible member of such an immune population enjoys an immunity that is not of his or her own making but instead arises because of membership in the group.

At times public health officials immunize large portions of a susceptible population in an attempt to maintain a high level of herd immunity. Any increase in the number of susceptible individuals may result in an endemic disease becoming epidemic. The proportion of immune to susceptible individuals must be constantly monitored because new susceptible individuals continually enter a population through migration and birth.

Pathogens can cause endemic diseases because infected humans continually transfer them to others (e.g., sexually transmitted diseases) or because the pathogens continually reenter the human population from animal reservoirs (e.g., West Nile virus, transmitted by mosquito vectors; birds serve as the reservoir). Other pathogens continue to evolve and may produce epidemics (e.g., HIV, influenza virus [A strain], and *Legionella* bacteria). As described in chapter 38, the influenza virus evolves in two distinct ways. **Antigenic drift** involves an accumulation of small genetic changes that, over time, result in a phenotype that is antigenically different from its predecessors. As a result, flu vaccines are continually reviewed and updated, and individuals should be reimmunized annually to maintain immunity to the constantly evolving viruses. A more abrupt genetic change that can occur in influenza A viruses is **antigenic shift** (*see figure 38.3*). Two or more different strains can exchange genetic fragments resulting in a major change in the hemagglutinin and/or neuraminidase composition of the viral genome. When a new subtype forms as a result of antigenic shift, often the human population has little or no immunity to the novel virus. ◀ *Minus strand RNA viruses: RNA-dependent RNA polymerase is part of the virion (section 26.6)* ▶ *Influenza (flu) (section 38.1)*

If the percentage of susceptible people is above the threshold density (figure 36.5), the level of protection provided by herd immunity will decrease and the morbidity rate will increase. For example, the morbidity rates of influenza among schoolchildren may reach epidemic levels if the number of susceptible individuals rises above 30% for the whole population.

One of the primary goals of public health is to ensure that the general public is sufficiently protected to withstand epidemics. This has resulted in the recommendation that at least 70% of the population be immunized against common infectious diseases, providing the herd immunity necessary for the protection of those who are not immunized. In other words, by keeping the population sufficiently immunized, the chain of infectious disease transmission is broken, limiting the reach of the disease. This is especially true for infectious diseases that quickly arise by antigenic shift, for example, the 2014–2015 influenza season's shift to an H3N2 viral strain.

Comprehension Check

1. How do epidemiologists recognize an infectious disease in a population?
2. Provide an example of a common-source and a propagated epidemic.
3. Explain herd immunity. How does this protect the community?
4. What is the significance of antigenic shift and drift in epidemiology?

36.4 Infectious Diseases and Pathogens Are Emerging and Reemerging

After reading this section, you should be able to:

a. Report on recently emerged and reemerging global infectious diseases
b. Organize geographic data to reveal changes in microbial drug resistance and vector spread of infection

Only a few decades ago, a grateful public trusted that science had triumphed over infectious diseases by building a fortress of health protection. Antibiotics, vaccines, and aggressive public health campaigns had yielded a string of victories over old enemies such as whooping cough, pneumonia, polio, and smallpox. In developed countries, people were lulled into believing that microbial threats were a thing of the past. Trends in the number of deaths caused by infectious diseases in the United States from 1900 through 2016 supported this conclusion (table 36.1). Sadly, despite this overall downward trend, the incidence of infectious disease resulting from emerging and reemerging microbial pathogens has claimed the lives of millions of people. The world has seen the global spread of HIV/AIDS, the resurgence of tuberculosis, the H1N1 influenza virus, and the appearance of new enemies such as Zika virus, hepatitis C and E viruses, pandemic emergence of SARS and MERS, Ebola virus, the Lyme disease spirochete, *Cryptosporidium* spp., and several strains of shiga toxin–producing *Escherichia coli*, not to mention the rise of extended-spectrum drug-resistant bacteria (**table 36.2**). In addition, during this same time period:

1. A "bird flu" virus that had never before attacked humans began to kill people in Southeast Asia. ▶ *Influenza (flu) (section 38.1)*
2. A new variant of a fatal prion disease of the brain, Creutzfeldt-Jakob disease, was identified in the United Kingdom, transmitted by beef from animals with bovine spongiform encephalopathy, nicknamed "mad cow disease." ▶ *Prion proteins transmit diseases (section 38.6)*
3. Methicillin-resistant species of *Staphylococcus* were seen for the first time. ◀ *Evolution in action development of antibiotic resistance in bacteria (section 16.9)*

Table 36.2 National Institute of Allergy and Infectious Diseases List of Emerging and Reemerging Infectious Diseases

Pathogens Newly Recognized in the Past Three Decades	Reemerging Pathogens
BACTERIUM	BACTERIUM
Bartonella henselae	*Clostridium (Clostridioides) difficile*
Ehrlichia spp.	*Streptococcus pyogenes*
Helicobacter pylori	*Staphylococcus aureus*
Borrelia burgdorferi	*Bordetella pertussis*
FUNGUS	
Encephalitozoon spp.	
Cryptococcus gattii strains	
PROTIST	
Acanthamoeba spp.	
Babesia spp.	
VIRUS	VIRUS
Australian bat lyssavirus	Ebola virus
Hendra virus	Measles virus
Hepatitis C virus	Mumps virus
Hepatitis E virus	Enterovirus A71
Human herpesvirus 8	Enterovirus D68
Human herpesvirus 6	Polio virus
Parvovirus B19	Dengue virus
MERS-CoV	Zika virus
Avian influenza A (H7N9)	
Bourbon virus	
Chikungunya virus	
Nipah virus	

4. Several major multistate food-borne outbreaks occurred in the United States and Germany, including those caused by *Salmonella* spp. in peanut butter and sprouts, *Cyclospora* protists on raspberries, viruses on strawberries, and various bacteria in produce, ground beef, cold cuts, and breakfast cereal. ▶| *Food-borne disease outbreaks (section 41.3)*
5. New strains of the tuberculosis bacterium resistant to many drugs and occur most often in people infected with HIV arose across the globe. ▶| *Mycobacterium tuberculosis (section 39.1)*
6. The resurgence of vaccine-preventable, childhood diseases, most notably whooping cough, measles, and mumps, erupted in several cities and towns across the United States and Europe.
7. In May 2015, the first confirmed Zika virus infection reached Brazil. Since then the first cases of locally transmitted Zika virus were confirmed in Florida in July 2016, followed by Texas in November that same year.

By the 1990s the idea that infectious diseases no longer posed a serious threat to human health was obsolete. In the twenty-first century, it is clear that humans will continually be faced with both new infectious diseases and the reemergence of older diseases once thought to be conquered. The CDC has defined these diseases as "new, reemerging, or drug-resistant infections whose incidence in humans has increased within the past three decades or whose incidence threatens to increase in the near future." Of note is the fact that while most emerging infectious diseases arise from the "Old World tropics," with today's capacity to travel anywhere in the world within 24 hours, infectious diseases can erupt just about anywhere. In addition, newer data suggest that emergence and reemergence of infectious disease stem from zoonotic sources. Areas that are dense with livestock are highly correlated with disease eruption that then disseminates globally. The increased importance of emerging and reemerging infectious diseases has stimulated the establishment of a field called **systematic epidemiology,** which focuses on the ecological and social factors that influence the development of these diseases. Coupled with systematic epidemiology and remote sensing techniques, the growing field of public health informatics integrates surveillance activities, defines data elements, and disseminates information via the internet so as to better track emerging and reemerging infectious diseases in real time.

Why are pathogens posing such a problem, despite dramatic advances in medical research, drug discovery, technology development, and sanitation? Many factors characteristic of the modern world undoubtedly favor the development and spread of these microorganisms and their diseases (**figure 36.6**).

As population density increases in cities, the dynamics of microbial exposure and evolution increase in humans. Urbanization often crowds humans and increases exposure to microorganisms. Crowding leads to unsanitary conditions and hinders the effective implementation of adequate medical care, enabling more widespread transmission and propagation of pathogens. In modern societies, crowded workplaces, communal-living settings, day-care centers, large hospitals, and public transportation all facilitate microbial transmission. In this millennium, the speed and volume of international travel are major factors contributing to the global emergence of infectious diseases. The spread of a new disease often used to be limited by the travel time needed to reach a new host population. If the travel time was sufficiently long, as when a ship crossed the ocean, the infected travelers would either recover or die before reaching a new population. Because travel by air has essentially eliminated time between exposure and disease outbreak, a traveler can spread virtually any infectious disease in a matter of hours. The H1N1 influenza pandemic of 2009 is one example of how one virus-infected individual (in Mexico) can lead to a global pandemic as asymptomatic carriers transport the virus to others during normal activities.

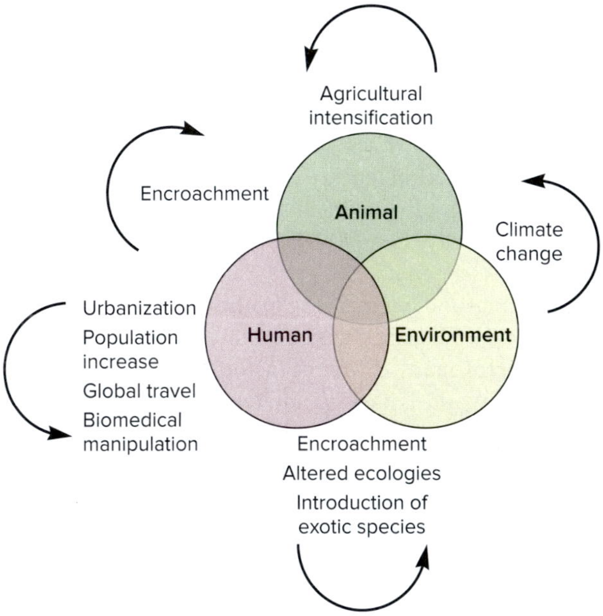

Figure 36.6 Disease Emergence and Reemergence Involve Multiple Factors and Mechanisms. The emergence and reemergence of disease-causing pathogens involve a dynamic, multifactorial interplay between humans, animals, and the environment.

Source: Daszak, P., Cunningham, A. A., and Hyatt, A. D. "Emerging infectious diseases of wildlife—threats to biodiversity and human health." *Science* 287(5452), January 2000, 443–449.

When changes in climate or ecology occur, it should not be surprising to find changes in both beneficial and detrimental microorganisms. Climate change also affects microorganism survival. Mass migrations of refugees, workers, and displaced persons have led to a steady growth of urban centers at the expense of rural areas. Furthermore, land development and the exploration and destruction of natural habitats have increased the likelihood of human exposure to new pathogens and may select for pathogens better able to adapt to new hosts and changing environments. The introduction of pathogens to a new environment or host can alter transmission and exposure patterns, leading to sudden proliferation of disease. For example, the spread of Ebola virus disease in 2013–2015 across West Africa clearly resulted when a suspected bat pathogen jumped into the encroaching human population. Migration of infected persons along established roadways facilitated viral transmission. Here, a virus with a week-long incubation period quickly exploited cultural practices to reemerge as an insidious and deadly foe. ◀ *Global climate change (section 28.2)*

Emerging and reemerging pathogens and their diseases are therefore the outcome of many different factors. Because the world is now so interconnected, we cannot isolate ourselves from other countries and continents. Changes in the disease status of one part of the world or the misuse of antibiotics in another part may well affect health around the planet. As Nobel laureate Joshua Lederberg so eloquently stated, "The microbe that felled one child in a distant continent yesterday can reach yours today and seed a global pandemic tomorrow."

Comprehension Check

1. What factors influence the definitions of emerging or reemerging infectious diseases?
2. What are some of the factors in your community that may lead to the emergence or reemergence of pathogens?
3. Describe how global climate change can lead to new human infectious diseases.

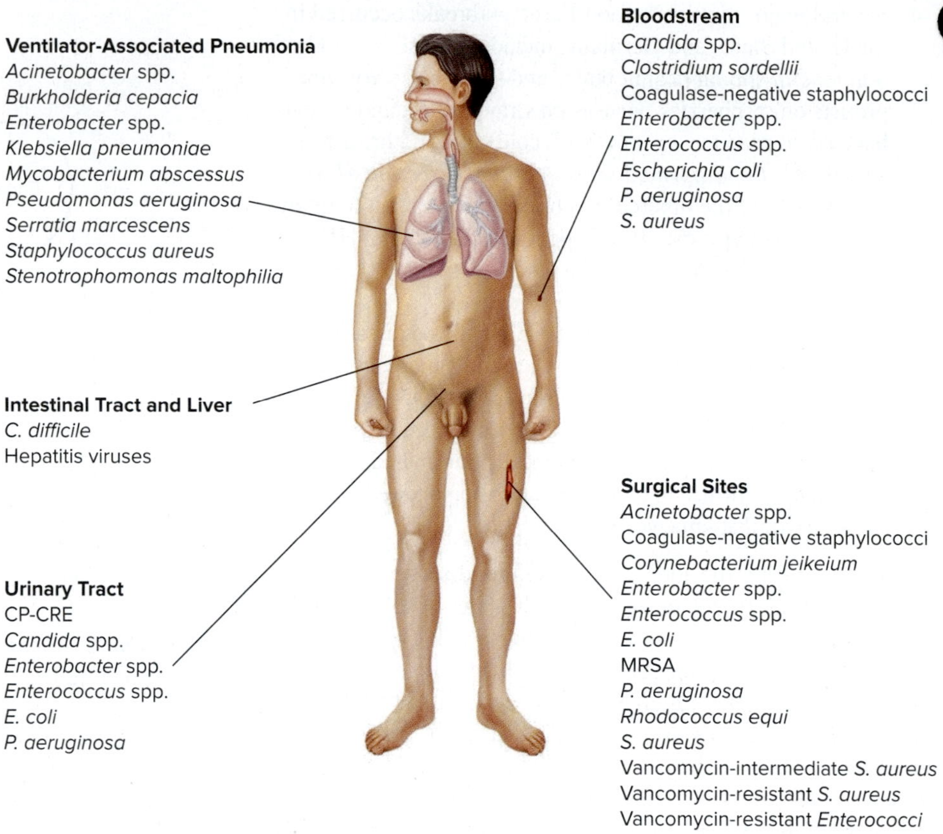

Figure 36.7 Healthcare-Associated Infections (HAI). The most common HAI are caused by bacteria, fungi, and viruses that are transferred to a patient from a healthcare worker, contaminated instruments, plants and flowers, fresh foods, and so on. CP-CRE: Carbapenemase-producing carbapenem-resistant *Enterobacteriaceae;* MRSA: methicillin-resistant *Staphylococcus aureus*

36.5 Healthcare Facilities Harbor Infectious Agents

After reading this section, you should be able to:

a. Report the major causes of healthcare-associated infection in the United States
b. Distinguish community-acquired pathogens from healthcare-associated pathogens, and explain why community-acquired pathogens are of concern to hospital personnel
c. Recommend measures for preventing and controlling healthcare-associated infections

Healthcare-associated, sometimes called **healthcare-acquired** or **nosocomial infections** (Greek *nosos*, disease; *komeion*, to take care of), result from infections acquired by patients while in a hospital or other clinical care facility. Besides harming patients, **healthcare-associated** infections (HAI) can affect nurses, physicians, aides, visitors, salespeople, delivery personnel, custodians, and anyone else who has contact with the hospital. Most HAI become clinically apparent while patients are still hospitalized; however, disease onset can occur after patients have been discharged. Infections that are incubating when patients are admitted to a hospital are not healthcare associated; they are community acquired. The CDC estimates that between 5% and 10% of hospital patients acquire some type of HAI. Thus healthcare-associated infections represent a significant proportion of infectious diseases acquired by humans, averaging approximately 1.7 million infections per year in the United States alone.

Healthcare-associated diseases are usually caused by bacteria, most of which are noninvasive and part of the normal human microbiota. **Figure 36.7** summarizes the most common types of healthcare-associated infections and pathogens. Interestingly, throughout most of the twentieth century, HAI were dominated by penicillin-sensitive staphylococci. During subsequent years, the emergence of methicillin-resistant *Staphylococcus aureus* (MRSA) increased dramatically. First reported in the late 1980s, vancomycin-resistant enterococci (VRE) are now seen in U.S. hospitals. Similar patterns are emerging for penicillin-resistant *Streptococcus pneumoniae.*

Today the most common HAI are (1) catheter-associated urinary tract infections, (2) surgical site infections, (3) central line

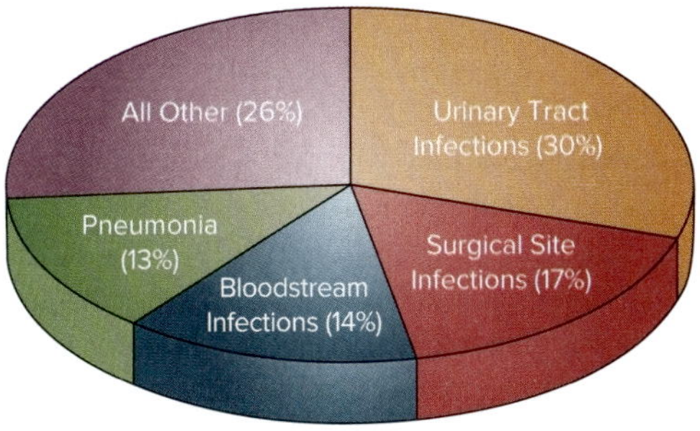

Figure 36.8 Leading Types of Healthcare-Associated Infections in the United States. These data reflect national statistics as of March 2015 (most recent data).

(catheters)–associated bloodstream infections, and (4) ventilator-associated pneumonias (**figure 36.8**). In addition to *C. difficile*, recently recognized healthcare-associated, Gram-positive species include *Clostridium sordellii, Corynebacterium jeikeium,* and *Rhodococcus equi*. The incidence of infections by the Gram-negative pathogens *Pseudomonas aeruginosa, Acinetobacter* spp., *Burkholderia cepacia,* and *Stenotrophomonas maltophilia* have also increased. Many Gram-negative bacilli are resistant to β-lactam antibiotics. These bacteria include *Klebsiella pneumoniae, E. coli,* other *Klebsiella* spp., *Proteus* spp., *Morganella* spp., *Citrobacter* spp., *Salmonella* spp., and *Serratia marcescens,* which are resistant to penicillins; many older cephalosporins; and even some newer cephalosporins such as cefotaxime, ceftriaxone, ceftazidime, and aztreonam. The β-lactamase-resistant members of *Enterobacteriaceae* are collectively referred to as extended spectrum β-lactamase (producing) *Enterobacteriaceae*. ◀ *Antimicrobial drug resistance is a public health threat (section 9.8)*

Many potential exogenous sources of infection exist in a hospital. Animate sources are the hospital staff, other patients, and visitors. Some examples of inanimate exogenous sources are food, plants and flowers, computer keyboards, intravenous and respiratory therapy equipment, and water systems (e.g., softeners, dialysis units, and hydrotherapy equipment).

In the United States, HAI prolong hospital stays by 4 to 14 days, result in an additional \$28 to \$33 billion per year to overall direct healthcare costs, and lead to approximately 99,000 deaths annually. The enormity of this problem has led hospitals to allocate substantial resources to developing methods and programs for the surveillance, prevention, and control of HAI.

Comprehension Check

1. Describe a healthcare-associated infection, and explain why such infections are significant.
2. What two general sources are responsible for healthcare-associated infections? Give some specific examples of each.
3. What can an infection control practitioner do to control healthcare-associated infections?

36.6 Coordinated Efforts Are Required to Prevent and Control Epidemics

After reading this section, you should be able to:

a. Discuss the three control measures employed by epidemiologists to break the chain of infection
b. Explain how the weakest link in the chain of infection can be exploited to control or prevent disease
c. Advise vaccine use to prevent infectious disease transmission in a population
d. Explain the role of disinfection, sanitation, and chemotherapy to control infectious disease transmission in a population

The control of an infectious disease relies heavily on a well-defined network of clinical microbiologists, nurses, physicians, epidemiologists, and infection control personnel who supply epidemiological information to a network of local, state, national, and international organizations. These individuals and organizations comprise the public health system. Each state has a public health laboratory responsible for overseeing infection surveillance and control. The communicable disease section of a state laboratory includes specialized laboratory services for the examination of specimens or cultures submitted by physicians, local health departments, hospital laboratories, sanitarians, epidemiologists, and others. These groups share their findings with other health agencies in the state, the CDC, and the WHO.

Control Starts with Breaking the Chain of Infection

Just as the development of an infectious disease is a complex process involving many factors, so too is the design of specific epidemiological control measures. Epidemiologists must consider available resources and time constraints, adverse effects of potential control measures, and human activities that might influence the spread of the infection. Often control activities reflect compromises among alternatives. To proceed intelligently, one must identify components of the infectious disease chain that are primarily responsible for a particular epidemic. Control measures should be directed toward that part of the chain that is most susceptible to control—the weakest link in the chain.

There are three types of control measures. The first type is directed toward reducing or eliminating the source or reservoir of infection: social distancing and isolation of carriers, destruction of an animal or arthropod reservoirs of infection, treatment of water and sewage to reduce contamination (*see chapter 43*), and therapy that reduces or eliminates infectivity of the individual.

The second type of control measure is designed to break the connection between the source of the infection and susceptible individuals. Examples include general sanitation measures: chlorination of water supplies, pasteurization of milk and other beverages, supervision and inspection of food and food handlers, and destruction of vectors (e.g., spraying to eliminate mosquitoes).

The third type of control measure reduces the number of susceptible individuals and raises the general level of herd immunity by immunization. Examples include passive immunization (*see figure 33.2*) to give temporary immunity following exposure to a pathogen or when a disease threatens to become an epidemic, active immunization to protect the host population, and prophylactic treatment to prevent infection (e.g., use of chloroquine when traveling to malaria-endemic countries). Antimicrobial chemotherapy is discussed in chapter 9. We discuss the use of vaccines and immunization in the next section.

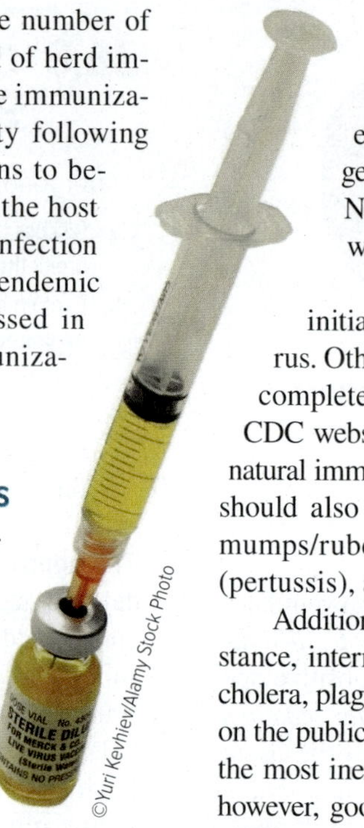

©Yuri Kevhiev/Alamy Stock Photo

Vaccines Immunize Susceptible Populations

A **vaccine** (Latin *vacca,* cow) is a preparation of one or more microbial antigens that induce protective immunity in the host. **Immunization** occurs when the host's immune system has successfully responded to the vaccine. The specific goal of vaccination is to induce antibodies and activated T cells to protect a host from future infection. Many epidemics have been stayed by mass prophylactic immunization (**table 36.3**). Vaccines have eradicated smallpox, pushed polio to the brink of extinction, and spared countless individuals from hepatitis A and B, influenza, measles, rotavirus disease, tetanus, typhus, and other dangerous diseases. **Vaccinomics,** the application of genomics and bioinformatics to vaccine development, is bringing a fresh approach to the Herculean problem of making vaccines against various microorganisms and helminths. ▶| *Vaccines (section 42.1)*

To promote a more efficient immune response, antigens in vaccines can be mixed with an **adjuvant** (Latin *adjuvans,* aiding), which enhances the rate and degree of immunization. Adjuvants can be any nontoxic material that prolongs antigen interaction with immune cells, assists in processing of antigens by antigen-presenting cells (APCs), or otherwise nonspecifically stimulates the immune response to the antigen. Several types of adjuvants can be used, including oil in water emulsions (Freund's incomplete adjuvant), aluminum hydroxide salts (alum), beeswax, and various combinations of bacteria (live or killed). In most cases, the adjuvant materials trap the antigen, thereby promoting a sustained release as APCs digest and degrade the preparation. In other cases, the adjuvant activates APCs so that antigen recognition, processing, and presentation are more efficient. |◀ *Recognition of foreigness is critical for a strong defense (section 33.4)*

The modern era of vaccines and immunization began in 1798 with Edward Jenner's use of cowpox as a vaccine against smallpox (**Historical Highlights 36.4**) and in 1881 with Louis Pasteur's rabies vaccine. Vaccines for other diseases did not emerge until later in the nineteenth century when, largely through a process of trial and error, methods for inactivating and attenuating microorganisms were improved and vaccines were produced. Vaccines were eventually developed against most of the epidemic diseases that plagued Western Europe and North America (e.g., diphtheria, measles, mumps, pertussis, German measles, and polio). Indeed, by the late twentieth century, it seemed that the combination of vaccines and antibiotics had solved the problem of microbial infections in developed nations. Such optimism was cut short by the emergence of new and previously unrecognized pathogens and antibiotic resistance among existing pathogens. Nevertheless, vaccination is still the most cost-effective weapon for preventing microbial disease (table 36.2).

Vaccination of most children should begin at birth, initiating protection against the blood-borne hepatitis B virus. Other vaccines should be given at about age 2 months; a complete schedule of childhood vaccines is available on the CDC website. Before that age, children are protected by passive natural immunity from maternal antibodies. Adults (over 50 years) should also be vaccinated/revaccinated for influenza, measles/mumps/rubella, streptococcal pneumonia, whooping cough (pertussis), and shingles, as determined by a physician.

Additional or booster vaccines are suggested for many. For instance, international travelers may need to be immunized against cholera, plague, polio, typhoid, typhus, and yellow fever, depending on the public health policies of the country visited. Immunization is the most inexpensive method to prevent some infectious diseases; however, good hand hygiene and food/water cautions are still advised as some countries lack public health basics such as clean water. Veterinarians, forest rangers, and others whose jobs involve contact with animals may be vaccinated against rabies, plague, and anthrax. Vaccine recommendations for U.S. citizens are updated annually; travelers, veterinarians, and healthcare providers should consult the CDC website, which lists vaccine and booster recommendations for all countries outside the United States, as well as recommendations for at-risk individuals. The role of immunization as a protective measure cannot be overstated: Immunizations save lives.

Whole-Cell Vaccines

Many of the current vaccines effective against human viral and bacterial diseases consist of whole microorganisms, termed **whole-cell vaccines.** These are either inactivated (killed) or attenuated (live but avirulent). The major characteristics of these vaccines are compared in **table 36.4**. **Inactivated vaccines** are effective, but they are less immunogenic, so they often require several boosters and normally do not adequately stimulate cell-mediated immunity or secretory IgA production. In contrast, **attenuated vaccines** usually are given in a single dose and stimulate both humoral and cell-mediated immunity. |◀ *Innate host resistance (chapter 32); Adaptive immunity (chapter 33)*

Even though whole-cell vaccines are considered the "gold standard" of vaccines, they are not without their drawbacks. In the 1990s the combined whole-cell pertussis vaccine (DTwP) was replaced with combined acellular pertussis (DTaP) due to the risk of serious adverse reactions from the whole-cell preparation. Attenuated vaccines can also cause full-blown illness in an individual whose immune system is compromised (e.g., AIDS patients, cancer patients undergoing chemotherapy, the elderly). These same individuals may also contract the disease from healthy people who have been vaccinated recently. Moreover,

Table 36.3 Examples of Vaccines to Prevent Viral and Bacterial Diseases in Humans

Disease	Vaccine	Booster*	Recommendation
Viral Diseases			
Chickenpox	Attenuated Oka strain	4–6 years	Children 12–18 months, older children who have not had chickenpox
Hepatitis A	Inactivated virus	6–18 months	Children 12 months
Hepatitis B	Viral antigen	1–4 months 6–18 months	High-risk medical personnel, children, birth to 18 months and 11–12 years of age
Human papillomavirus infections	Recombinant protein subunits	2–3 months 6 months	Children 11–12 years of age
Influenza A/B	Inactivated virus or live attenuated	Yearly	All persons
Measles, Mumps, Rubella	Attenuated viruses (combination MMR vaccine)	None	First dose 12–15 months, 2nd dose 4–6 years
Poliomyelitis	Attenuated (oral poliomyelitis vaccine, OPV) or inactivated virus (inactivated polio vaccine, IPV)	Adults as needed	First dose at 2 months, 2nd at 4 months, 3rd at 16–18 months, 4th at 4–6 years
Rabies	Inactivated virus	None	Individuals in contact with wildlife, animal control personnel, veterinarians
Respiratory disease	Live attenuated adenovirus	None	Military personnel
Shingles	Recombinant varicella surface glycoproteins	None	Individuals over 50 years
Smallpox	Live attenuated vaccinia virus	None	Laboratory, healthcare, and military personnel
Yellow fever	Attenuated virus	10 years	Military personnel and individuals traveling to endemic areas
Bacterial Diseases			
Anthrax	Extracellular components of unencapsulated *Bacillus anthracis*	None	Agricultural workers, veterinary, and military personnel
Cholera	Fraction of *Vibrio cholerae*	6 months	Individuals in endemic areas, travelers
Diphtheria, Pertussis, Tetanus	Diphtheria and tetanus toxoids, and acellular *Bordetella pertussis* vaccine (DTap) or tetanus toxoid, reduced diphtheria toxoid, and acellular pertussis vaccine (Tdap)	10 years	Children from 2–3 months old to 12 years, and adults; children 10–18 years, at least 5 years after DPT series, should receive Tdap
Haemophilus influenzae type b	Polysaccharide-protein conjugate (HbCV) or bacterial polysaccharide (HbPV)	None	First dose at 2 months, 2nd at 4 months, 3rd at 6 months, 4th at 12–15 months
Meningococcal infections	*Neisseria meningitidis* polysaccharides of serotypes A/C/Y/W-135	None	First dose at 11–12 years, 2nd dose at 16 years
Plague	Fraction of *Yersinia pestis*	Yearly	Individuals in contact with rodents in endemic areas
Pneumococcal pneumonia	Purified *S. pneumoniae* polysaccharide of 23 pneumococcal types or pneumococcal conjugate of 13 strains	None	Adults over 50 with chronic disease
Q fever	Killed *Coxiella burnetii*	None	Workers in slaughterhouses and meat-processing plants
Tuberculosis	Attenuated *Mycobacterium bovis* (BCG vaccine)	3–4 years	Individuals exposed to TB for prolonged periods of time; used in some countries, not licensed in the United States
Typhoid fever	*Salmonella enterica* Typhi Ty21a (live attenuated or polysaccharide)	None	Residents of and travelers to areas of endemic disease
Typhus	Killed *Rickettsia prowazekii*	Yearly	Scientists and medical personnel in areas where typhus is endemic

* Subsequent immunization dose(s) after the initial immunization.

HISTORICAL HIGHLIGHTS

36.4 The First Immunizations

Since the time of the ancient Greeks, it has been recognized that people who have recovered from plague, smallpox, yellow fever, and various other infectious diseases rarely contract the diseases again. The first scientific attempts at artificial immunizations were made in the late eighteenth century by Edward Jenner (1749–1823), who was a country doctor from Berkeley, Gloucestershire, England. Jenner investigated the basis for the widespread belief of the English peasants that anyone who had vaccinia (cowpox) never contracted smallpox. Smallpox was often fatal—10 to 40% of the victims died—and those who recovered had disfiguring pockmarks. Yet most English milkmaids, who were routinely infected with cowpox (a relatively mild infection), had clear skin because they rarely contracted smallpox.

In 1796, in an experiment that by today's standards was completely unethical, Jenner extracted the contents of a pustule from the arm of a cowpox-infected milkmaid, Sarah Nelmes, and injected it into the arm of 8-year-old James Phipps. As Jenner expected, immunization with the cowpox virus caused only mild symptoms in the boy. When he subsequently inoculated the boy with smallpox virus, the boy showed no symptoms of the disease. Jenner then inoculated large numbers of his patients with cowpox pus, as did other physicians in England and on the European continent (**box figure**). By 1800 the practice known as variolation had begun in America, and by 1805 Napoleon Bonaparte had ordered all French soldiers to be vaccinated.

Further work on immunization was carried out by Louis Pasteur (1822–1895). Pasteur discovered that if cultures of chicken cholera bacteria were allowed to age for 2 or 3 months, the bacteria produced only a mild attack of cholera when inoculated into chickens. Somehow the old cultures had become less pathogenic (attenuated). He then found that fresh cultures of these attenuated bacteria failed to produce cholera in chickens that had been previously inoculated with old, attenuated cultures. To honor Jenner's work with cowpox, Pasteur gave the name *vaccine* to any preparation of a weakened pathogen that was used (as was Jenner's "vaccine virus") to immunize against infectious disease.

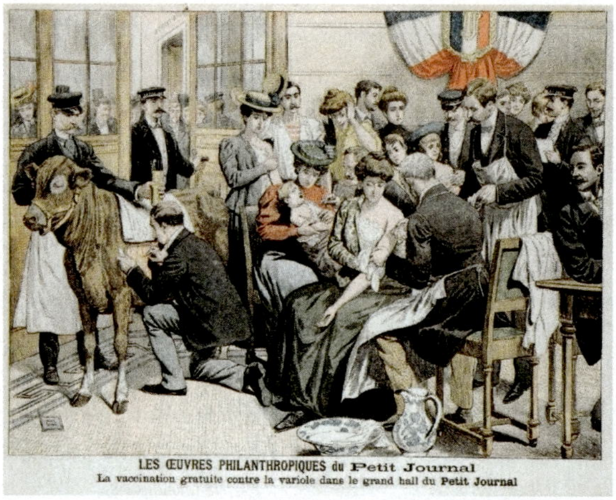

Nineteenth-Century Physicians Performing Vaccinations on Children.

Table 36.4 A Comparison of Vaccine Types[1]

Major Characteristic	Whole Cell Vaccines		Acellular or Subunit Vaccines	Recombinant or DNA Vaccines	
	Live, attenuated	Inactivated		Recombinant	DNA
Typical Immunogen	Virus	Bacteria	1–20 antigens from microorganism	Virus or bacterial vector	Antigen(s) encoded by DNA
T Response	Strong	Weak	Very strong	Very strong	Very strong
B-cell Response	Strong	Strong	Very strong	Very strong	Very strong
Stability	Decreased without refrigeration	Longer shelf life, freeze-dried	Longer shelf life, freeze-dried	Longer shelf life, freeze-dried	
Potential Problems	Reversion to pathogen possible, not given to immunocompromised	Weaker response, requires booster doses	Identification of antigens is costly and research intensive	Only effective against protein immunogens; DNA vaccines have efficacy issues	

[1]Source: US Department of Health & Human Services. "Vaccine Types."

attenuated viruses can at times mutate in ways that restore virulence. In developing countries, the polio vaccine is in the form of a live, attenuated virus. If the virus co-circulates with Coxsackie virus, the two viruses can recombine to form pathogenic vaccine-derived poliovirus. Several outbreaks of poliomyelitis have been attributed to this recombination. Children in the United States and other developed nations receive the killed polio vaccine, which is incapable of such genetic events.

Acellular or Subunit Vaccines

A few of the common risks associated with whole-cell vaccines can be avoided by using only specific, purified macromolecules derived from pathogenic microorganisms. There are three general forms of **subunit vaccines:** (1) capsular polysaccharides, (2) recombinant surface antigens, and (3) inactivated exotoxins called **toxoids.** The purified microbial subunits or their secreted products can be prepared as nontoxic antigens to be used in the formulation of vaccines (**table 36.5**).

Recombinant-Vector and DNA Vaccines

Genes isolated from a pathogen that encode major antigens can be inserted into nonvirulent viruses or bacteria. Such recombinant microorganisms serve as vectors, replicating within the host and expressing the gene product of the pathogen-encoded antigenic proteins. The antigens elicit humoral immunity (i.e., antibody production) when they are released by the vector, and they also elicit cellular immunity when they are broken down and properly displayed on the cell surface (just as occurs when host cells harbor an active pathogen). Several microorganisms, such as rotavirus and attenuated *Salmonella enterica,* have been used in the production of these **recombinant-vector vaccines.**

Table 36.5 Example Subunit Vaccines

Microorganism or Toxin	Vaccine Subunit
Capsular polysaccharide	
Haemophilus influenzae type b	Polysaccharide-protein conjugate (HbCV) or bacterial polysaccharide (HbPV)
Neisseria meningitidis	Polysaccharides of serotypes A/C/Y/W-135
Streptococcus pneumoniae	23 distinct capsular polysaccharides
Surface antigen	
Hepatitis B virus	Recombinant surface antigen (HbsAg)
Human papillomavirus	Recombinant protein subunits
Toxoids	
Corynebacterium diphtheriae toxin	Inactivated exotoxin
Clostridium tetani toxin	Inactivated exotoxin

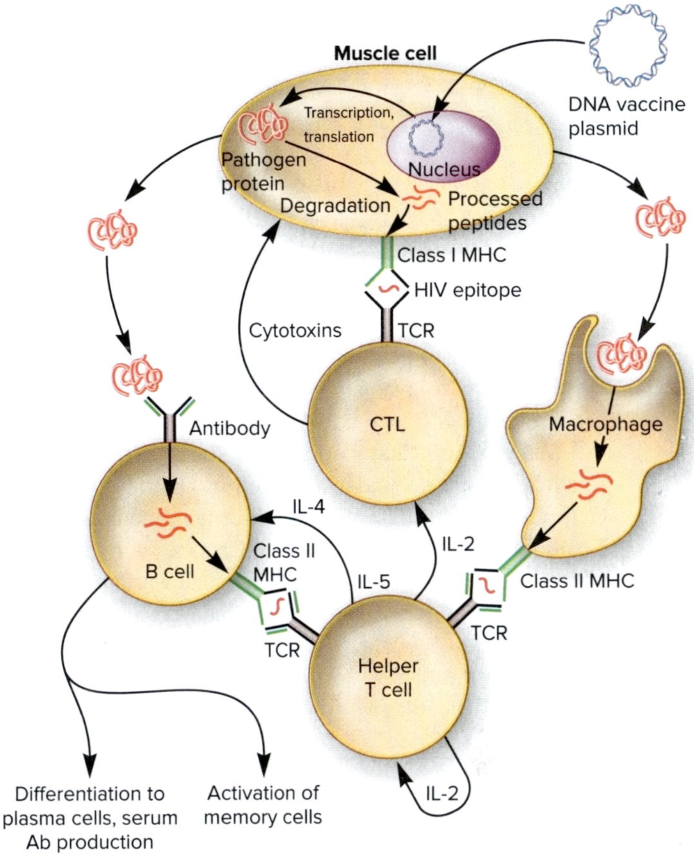

Figure 36.9 DNA Vaccine. DNA coding for a pathogen's proteins is inserted into host cells and expressed. The foreign proteins elicit both cell-mediated and antibody (Ab) responses.

On the other hand, **DNA vaccines** introduce fragments of pathogen DNA directly into the host cell. When injected into muscle cells, the DNA is taken into the nucleus and the pathogen's DNA is transiently expressed, generating foreign proteins to which the host's immune system responds (**figure 36.9**). DNA vaccines are very stable; refrigeration is often unnecessary. At present, human trials are under way with several different DNA vaccines against malaria, HIV, influenza, hepatitis B, rabies, HPV, and Zika. *Constructing Vaccines*

36.7 Bioterrorism Readiness Is an Integral Component of Public Health Microbiology

After reading this section, you should be able to:

a. Discuss the readiness efforts necessary to be prepared for a bioterrorism attack
b. Prioritize microorganisms based on public health threat
c. Construct tables of prevention and control information relative to an infectious agent
d. Identify appropriate authorities that respond to suspected acts of bioterrorism

HISTORICAL HIGHLIGHTS

36.5 1346—Early Biological Warfare Attack

The Black Death, which swept through Europe, Asia, and North Africa in the mid-fourteenth century, was probably the greatest public health disaster in recorded history. Europe alone lost an estimated quarter to a third of its population. This is not only of great historical interest but also relevant to current efforts to evaluate the threat of military or terrorist use of biological weapons. ▶I *Plague (section 39.2)*

Some believe that evidence for the origin of the Black Death in Europe is found in a memoir by Gabriele de' Mussi of Genoa. According to this fourteenth-century memoir, the Black Death reached Europe from the Crimea (a region of the Ukraine) in 1346 as a result of a biological warfare attack. The Mongol army hurled plague-infected cadavers into the besieged Crimean city of Caffa (now Feodosija, Ukraine), thereby transmitting the disease to the inhabitants; fleeing survivors then spread the plague from Caffa to the Mediterranean Basin. Such transmission was especially likely at Caffa, where cadavers would have been badly mangled by being hurled and the defenders probably suffered cut or abraded hands from coping with the bombardment. Because many cadavers were involved, the opportunity for disease transmission was greatly increased. Disposal of victims' bodies in a major disease outbreak is always a problem, and the Mongol army used their hurling machines as a solution to limited mortuary facilities. It is possible that thousands of cadavers were disposed of this way; de' Mussi's description of "mountains of dead" might have been quite literally true. Indeed, Caffa could be the site of the most spectacular incident of biological warfare ever, with the Black Death as its disastrous consequence. It is a powerful reminder of the horrific consequences that can result when disease is successfully used as a weapon.

Bioterrorism is defined as "the intentional or threatened use of viruses, bacteria, fungi, or toxins from living organisms to produce death or disease in humans, animals, and plants." The use of biological agents to effect personal or political outcomes is not new, and the modern use of biological agents is a reality (**Historical Highlights 36.5**). The most notable intentional uses of biological agents for criminal or terror intent are (1) the use of *Salmonella enterica* serovar Typhimurium in 10 restaurant salad bars (by the Rajneeshee religious cult in The Dalles, Oregon, 1984); (2) the intentional release of *Shigella dysenteriae* in a hospital laboratory break room (perpetrator[s] unknown; Texas, 1996); and (3) the use of *Bacillus anthracis* spores delivered through the U.S. postal system (perpetrator[s] unknown, although the FBI named Bruce Ivins, now deceased, as the likely suspect; five eastern U.S. states, 2001). The *S. enterica*–contaminated salads resulted in 751 documented cases and 45 hospitalizations due to salmonellosis. The *S. dysenteriae* release resulted in eight confirmed cases and four hospitalizations for shigellosis. The *B. anthracis* spores infected 22 people (11 cases of inhalation anthrax and 11 cases of cutaneous anthrax) and caused five deaths. The list of biological agents that could pose the greatest public health risk in the event of a bioterrorism attack is relatively short and includes viruses, bacteria, and toxins that, if acquired and properly disseminated, could become a difficult public health challenge in terms of limiting the numbers of casualties and controlling panic (**table 36.6**).

Among weapons of mass destruction, biological weapons can be as destructive as chemical weapons, including nerve gas. In certain circumstances, biological weapons can be as devastating as a nuclear explosion: A few kilograms of anthrax spores could kill as many people as a Hiroshima-size nuclear bomb. Biological agents are likely to be chosen as a means of localized attack (biocrime) or mass casualty (bioterrorism) for several reasons. They are mostly invisible, odorless, tasteless, and difficult to detect. Furthermore, perpetrators may escape undetected as it may take hours to days before signs and symptoms of their use become evident. Additionally, the general public is not likely to be protected immunologically against agents that might be used in bioterrorism. Ultimately the use of biological agents in terrorism results in fear, panic, and chaos.

In 1998 the U.S. government launched the first national effort to create a biological weapons defense. The initiatives included (1) the first-ever procurement of specialized vaccines and medicines for a national civilian protection stockpile; (2) invigoration of research and development in the science of biodefense; (3) investment of more time and money in genome sequencing, new vaccine research, and new therapeutic research; (4) development of improved detection and diagnostic systems; and (5) preparation of clinical microbiologists and clinical microbiology laboratories as members of first-responder teams in the event of a bioterror attack. In 2002 Congress enacted the Public Health Security and Bioterrorism Preparedness and Response Act, which identified "select" agents whose use is now tightly regulated. A final rule that governs the possession, use, and transport of select agents was issued in 2005. The select

Table 36.6 Some CDC-Defined Biological Select Agents and Toxins (BSATs)

Bacteria	Virus	Toxin
*Bacillus anthracis**	Variola major*	*Clostridium botulinum* toxin*
*Yersinia pestis**	Ebola*	*Clostridium perfringens* toxin
*Francisella tularensis**	Marburg*	Staphylococcal enterotoxin B
Brucella spp.	Lassa*	
Salmonella spp.	Machupo*	
Escherichia coli 0157:H7	Equine encephalitis	
Shigella spp.	Nipah	
Burkholderia mallei	Hantavirus	
Chlamydia psittaci		
Coxiella burnetii		
Rickettsia prowazekii		
Vibrio cholerae		

* Tier 1 agents. The documented risk of causing a high-consequence event is greater for Tier 1 BSATs than for the other BSATs.
Source: CDC, "Select Agents and Toxins List."

agent definition, however, was revised in 2012 to identify a "Tier 1" subset that (1) could produce a mass casualty event or economic devastation, (2) were highly communicable, (3) had a low infectious dose, and (4) might be readily weaponized (table 36.6).

In 2003 Congress established the Department of Homeland Security to coordinate the defense of the United States against terrorist attacks. As one of many duties, the secretary of Homeland Security is responsible for maintaining the National Incident Management System to monitor large-scale hazardous events. Bioterrorism and other public health incidents are managed within this system. The Department of Health and Human Services responds and deploys assets as needed within the areas of its statutory responsibility (e.g., the Public Health Service Act and the Federal Food, Drug, and Cosmetic Act), while apprising the secretary of Homeland Security of the nature of the response. The secretary of Health and Human Services directs the CDC to effect the necessary integration of public health activities.

The events of September and October 2001 in the United States changed the world. Global efforts to prevent terrorism, especially using biological agents, evolved from cautious planning to proactive preparedness. In the United States, the CDC partners with academic institutions across the country to educate, train, and drill public health employees, first responders, and numerous environmental and healthcare providers. Centers for Public Health Preparedness were established to bolster the overall response capability to bioterrorism. Another CDC-managed program, the Laboratory Response Network (LRN) serves to ensure an effective laboratory response to bioterrorism by helping to improve the nation's public health laboratory infrastructure through its partnership with the FBI and the Association of Public Health Laboratories (APHL). The LRN maintains an integrated network that links state and local public health, federal, military, and international laboratories so that a rapid and coordinated response to bioterrorism or other public health emergencies (including veterinary, agriculture, military, and water- and food-related) can occur.

In the absence of overt terrorist threats and without the ability to rapidly detect bioterrorism agents, it is likely that a bioterrorism act will be defined by a sudden spike in an unusual (nonendemic) disease reported to the public health system. Also, suddenly increased numbers of zoonoses, diseased animals, or vehicle-borne illnesses may indicate bioterrorism. Important guidelines and standardized protocols prepared for all sentinel laboratories to assist in the management of clinical specimens containing select agents have been established by the American Society for Microbiology in coordination with the CDC and the APHL. A summary of the rule-out tests for four Tier 1 bacterial agents is presented in **table 36.7**. The diseases and microbiology associated with specific select agents are discussed in chapters 38 to 40.

Comprehension Check

1. In what three general ways can epidemics be controlled? Give one or two specific examples of each type of control measure.
2. Name some of the microorganisms that can be used to commit biocrimes. From this list, which pose the greatest risk for causing large numbers of casualties?
3. Why are biological weapons more destructive than chemical weapons?
4. What is the Public Health Security and Bioterrorism Preparedness and Response Act designed to do?

Table 36.7 Criteria for Presumptive Identification of Four Bacterial Select Agents

Pathogen	Gram Morphology	Colonial Morphology	Biochemical Results
Bacillus anthracis	Gram-positive, endospore-forming rod	Gray, flat, "Medusa head" irregularity, nonhemolytic on sheep's blood agar	Catalase positive, oxidase positive, urea negative, Voges-Proskauer (VP) positive, phenylalanine (Phe) deaminase negative, NO_3^- to NO_2^- positive
Clostridium botulinum	Gram-positive, endospore-forming rod	Creamy, irregular, rough, broad, nonhemolytic	Catalase negative, urea negative, gelatinase positive, indole negative, VP negative, Phe deaminase negative, NO_3^- to NO_2^- negative
Francisella tularensis	Gram-negative rod (tiny)	Gray-white, shiny, convex, pinpoint after 72 hr, nonhemolytic	Catalase positive (weak), oxidase negative, β-lactamase positive, urea negative
Yersinia pestis	Gram-negative rod (bipolar staining)	Gray-white, "fried-egg" irregularity, nonhemolytic, grows faster and larger at 28°C	Catalase positive, oxidase negative, urea negative, indole negative, VP negative, Phe deaminase negative

Key Concepts

36.1 Epidemiology Is an Evidence-Based Science

- Epidemiology is the science that evaluates the determinants, occurrence, distribution, and control of health and disease in a defined population.
- Specific epidemiological terminology is used to communicate disease incidence in a given population. Frequently used terms include sporadic disease, endemic disease, hyperendemic disease, epidemic, index case, outbreak, and pandemic.

36.2 Epidemiology Is Rooted in Well-Tested Methods

- Public health surveillance is necessary for recognizing a specific infectious disease within a given population. This consists of gathering data on the occurrence of the disease, collating and analyzing the data, summarizing the findings, and applying the information to control measures.
- Population-based and case-based surveillance data are used to track infections within a population.
- Remote sensing and geographic information systems are used to digitally gather data from natural environments.
- Statistical analysis is an important tool used in the study of modern epidemiology.
- Epidemiological data can be obtained from such factors as morbidity, prevalence, and mortality rates.

36.3 Infectious Disease Is Revealed Through Patterns Within a Population

- A common-source epidemic is characterized by a sharp rise to a peak and then a rapid but not as pronounced decline in the number of individuals infected (**figure 36.3**). A propagated epidemic is characterized by a relatively slow and prolonged rise and then a gradual decline in the number of individuals infected.
- Herd immunity is the resistance of a population to infection and pathogen spread because of the immunity of a large percentage of the individuals within the population.

36.4 Infectious Diseases and Pathogens Are Emerging and Reemerging

- Humans will continually be faced with both new infectious diseases and the reemergence of older diseases once thought to be conquered (**table 36.2**).
- The CDC has defined these diseases as "new, reemerging, or drug-resistant infections whose incidence in humans has increased within the past two decades or whose incidence threatens to increase in the near future."
- Many factors characteristic of the modern world undoubtedly favor the development and spread of these microorganisms and their diseases (such as transportation, commerce, housing development).

36.5 Healthcare Facilities Harbor Infectious Agents

- Healthcare-associated (nosocomial) infections are infections acquired within a hospital or other clinical care facility and are produced by a pathogen acquired during a patient's stay. These infections come from either endogenous or exogenous sources (**figures 36.7** and **36.8**).
- Healthcare-associated diseases are usually caused by bacteria, most of which are noninvasive and part of the normal human microbiota.

36.6 Coordinated Efforts Are Required to Prevent and Control Epidemics

- The public health system consists of individuals and organizations that function in the control of infectious diseases and epidemics.
- Vaccination is one of the most cost-effective weapons for preventing microbial disease, and vaccines constitute one of the greatest achievements of modern medicine.
- Many of the current vaccines in use for humans (**table 36.3**) consist of whole organisms that are either inactivated (killed) or attenuated (live but avirulent).
- Some of the risks associated with whole-cell vaccines can be avoided by using only specific purified macromolecules derived from pathogenic microorganisms. Currently there are three general forms of subunit or acellular vaccines: capsular polysaccharides, recombinant surface antigens, and inactivated exotoxins (toxoids) (**table 36.5**).
- A number of microorganisms have been used for recombinant-vector vaccines. The attenuated microorganism serves as a vector, replicating within the host and expressing the gene product of the pathogen-encoded antigenic proteins. The proteins can elicit humoral immunity when the proteins escape from the cells and cellular immunity when they are broken down and properly displayed on the cell surface.
- DNA vaccines elicit protective immunity against a pathogen by activating both branches of the immune system: humoral and cellular (**figure 36.9**).
- Epidemiological control measures can be directed toward reducing or eliminating infection sources, breaking the connection between sources and susceptible individuals, or isolating susceptible individuals and raising the general level of herd immunity by immunization.

36.7 Bioterrorism Readiness Is an Integral Component of Public Health Microbiology

- Among weapons of mass destruction, biological weapons can be more destructive than chemical weapons. The list of biological agents that could pose the greatest public health risk in the event of a bioterrorism attack is short and includes viruses, bacteria, parasites, and toxins (**table 36.6**).
- Specific laboratory tests define the process for hospital and other sentinel labs to rule out a select agent.

Active Learning

1. Why is international cooperation a necessity in the field of epidemiology? What specific problems can you envision if there were no such cooperation?
2. What common sources of infectious disease are found in your community? How can the etiologic agents of such infectious diseases spread from their source or reservoir to members of your community?
3. How could you prove that an epidemic of a given infectious disease was occurring?
4. How can changes in herd immunity contribute to an outbreak of a disease on an island?
5. College dormitories are notorious for outbreaks of flu and other infectious diseases. These are particularly prevalent during final exam weeks. Using your knowledge of the immune response and epidemiology, suggest practices that could be adopted to minimize the risks at such a critical time.
6. Why does an inactivated vaccine induce only a humoral response, whereas an attenuated vaccine induces both humoral and cell-mediated responses?
7. Why is a DNA vaccine delivered intramuscularly and not by intravenous or oral routes?
8. In 2018 the CDC upgraded three carbapenemase-producing carbapenem-resistant *Enterobacteriaceae* (CP-CRE) to its list of national notifiable conditions. *Escherichia coli*, *Klebsiella* spp., and *Enterobacter* spp. were considered an emerging and epidemiologically important threat due to their resistance against multiple antibiotics and their role as healthcare-associated infectious agents. Carbapenemase is a formidable enzyme because it hydrolyzes not only the unique B-lactam moiety present in carbapenems, but also the structures of penicillins and cephalosporins (*see figure 9.20*). Organisms with carbapenem resistance were rare until the early 1990s. At that time organisms with chromosomal genes encoding the hydrolytic enzymes were discovered. The more recent discovery of plasmid-encoded carbapenemases marked a change in the dissemination of CREs.

   How do you think the rate of CRE dissemination would have proceeded if the genes encoding the carbapenemases remained localized to the chromosomes, rather than recombining with plasmids? What mechanisms guarantee the dissemination in both cases?

   How are the prevalence and incidence of CP-CRE calculated? Discuss steps healthcare institutions take to limit the spread of this dangerous bacterium.

   Read the original paper: Martirosov, D. M., and Lodise, T. P. 2016. Emerging trends in epidemiology and management of infections caused by carbapenem-resistant Enterobacteriaceae. *Diag. Microbiol. Infect. Dis.* 85:266.

37

Clinical Microbiology and Immunology

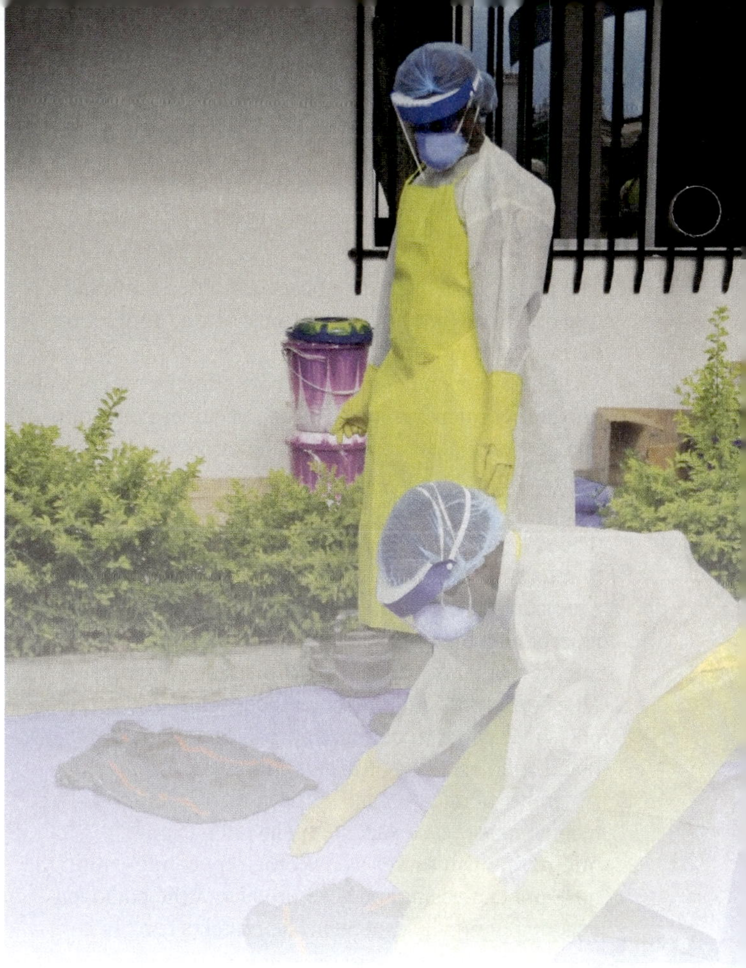

Source: CDC/Dr. Heidi Soeters

Ebola and Global Health Security

The 2014–2016 Ebola virus disease (Ebola) epidemic in West Africa was the longest and largest outbreak of its kind. Unlike previous Ebola epidemics, which were largely rural, this outbreak rapidly spread via close contacts in urban environments, leading to massive death tolls in Guinea, Liberia, and Sierra Leone. The global community was shocked as horrifying scenes of death and grief flooded the media. But as the epidemic began to wind down, it was clear that there was much to be learned from this tragic event. At the Centers for Disease Control and Prevention (CDC), Beth Bell, then-director of the National Center for Emerging and Zoonotic Infectious Diseases, pointed out, "This outbreak is a case study in why the Global Health Security Agenda is so important."

The Ebola epidemic put a spotlight on the need for rapid, reliable laboratory testing of samples from patients suspected of disease and those already infected. Without timely, accurate diagnostic testing, it was impossible to identify patients in the early stages of disease and those who could be safely discharged to the community. Moreover, persons who may have been exposed could not be traced. As you might imagine, rapid and reliable testing of a highly contagious, deadly virus in the developing world is no simple task. Indeed, very early in the Ebola epidemic, fragile local health agencies were quickly overwhelmed. Early response to the outbreak, when it might have been contained, was hobbled by a lack of diagnostic tests that could be rapidly deployed on a large scale, a shortage of skilled workers, mechanisms for sample collection and transport, limited understanding of biosafety, and an insufficient supply chain for personal protective equipment.

In response the CDC, the World Health Organization (WHO), and other governmental and nongovernmental organizations coordinated efforts with African governments to establish field laboratories and personnel training. Before the outbreak, few assays had been developed to detect and diagnose Ebola virus infection. Why this was the case is debatable, as many smaller Ebola outbreaks had occurred since it was first discovered in 1976. But during this outbreak, new assays were developed, which was a little like building a boat while already in the water. Nonetheless, the deployment of two types of assays—one that amplifies viral RNA and another that detects anti-Ebola IgM and IgG—were used to diagnose patients, determine when they were safe to return to their families, and trace contacts. Although initial tests were complex, limiting the turnaround time, by March 2015, a lateral flow antigen capture assay (the same technology used in home pregnancy tests) was approved, easing the burden on personnel and accelerating response time. In addition to the development of new tests, training protocols were codified ensuring that laboratory personnel could accurately perform these tests, and do so using appropriate biosafety precautions.

These tests enabled the WHO to declare the end of the Ebola transmission in Guinea in December 2015, in Liberia in January 2016, and in Sierra Leone in March 2016. However, the WHO and others continue to stress that sporadic transmission of Ebola continues. Therefore health agencies must remain on high alert, continue monitoring, and be ready to respond.

The Ebola epidemic points out how important both laboratory testing and laboratory safety are to effective health care. In this chapter we review some basic microbiological and immunological techniques that drive the diagnosis of infectious disease. But first, we discuss the common denominator in all diagnostic testing: biosafety.

Readiness Check:

Based on what you have learned previously, you should be able to:

- ✓ Describe the infectious disease process (chapter 35)
- ✓ Discuss the uses of tools and techniques used to culture and characterize microorganisms (chapters 2, 7, 17, 18, and 29)
- ✓ Outline important aspects of microbial structure, function, survival requirements, and replication strategies (chapters 3, 4, 5, 6, 7, 13, and 15)
- ✓ Describe the role of complement and antibodies in host defense (sections 32.3, 33.7, and 33.8)

37.1 The Clinical Microbiology Laboratory Detects Infectious Agents and Protects Its Workers

After reading this section, you should be able to:

a. Describe the various scientific disciplines involved in clinical microbiology
b. Describe the risk assessment process leading to a safe laboratory
c. Paraphrase the "standard microbiological practices" that guide lab safety
d. Differentiate between biosafety levels and provide examples of microorganisms assigned to each level

The clinical microbiology laboratory is the primary site for the diagnosis of human infectious disease. The clinical microbiology lab also supports public health infrastructure, acting as a sentinel in the rapid detection of naturally occurring outbreaks of infection and bioterrorism. The major goals of the **clinical microbiologist** are (1) rapid and accurate identification of pathogens from clinical specimens, and (2) antimicrobial susceptibility testing of these organisms (*see figures 9.2 to 9.4*). Clinical microbiology is interdisciplinary, and the clinical microbiologist must have working knowledge of microbial morphology, metabolism, growth, reproduction, biochemistry, and physiology. Additionally, the clinical microbiologist needs to understand the principles of microscopy, immunology, molecular biology, genomics, aseptic technique, sterilization, disinfection, and the dynamics of host-parasite relationships (**figure 37.1**).

Testing a clinical sample must follow a prescribed process to avoid contamination and waste. Thus, a workflow protocol is often established in the laboratory to rapidly move the specimen through testing procedures while maintaining its integrity. Throughout this process, the safety of clinical laboratory workers is paramount, as we now discuss.

The Need for Safety Guidance

Although appalling today, not so long ago mouth pipetting, smoking, and eating in the microbiology laboratory were not uncommon. Over the years, several key events have changed laboratory culture to promote safety. Pioneering biosafety techniques developed by Arnold G. Wedum, M.D., Ph.D., at the Army Biological Research Laboratory, Fort Detrick, Maryland, during the 1950s and 1960s set a standard for physical containment strategies and best practices to keep microbiologists from acquiring laboratory infections. The ground-breaking methods for manipulating DNA in the 1970s sparked an international discussion about the possible hazards of recombinant DNA technology and drove the requirement for the systematic evaluation of safe practices in the laboratory. Subsequent concern over standards needed to safely study cancer-causing viruses prompted the National Institutes of Health to establish criteria for physical containment and specific procedures to prevent human contamination. Together, these best methods and work practices were assembled into the *Biosafety in Microbiological and Biomedical Laboratories,* or BMBL. The current edition of the BMBL (2009) is prefaced with the caveat that "the intent was and is to establish a voluntary code of practice, one that all members of a laboratory community will together embrace to safeguard themselves and their colleagues, and to protect the public health and environment."

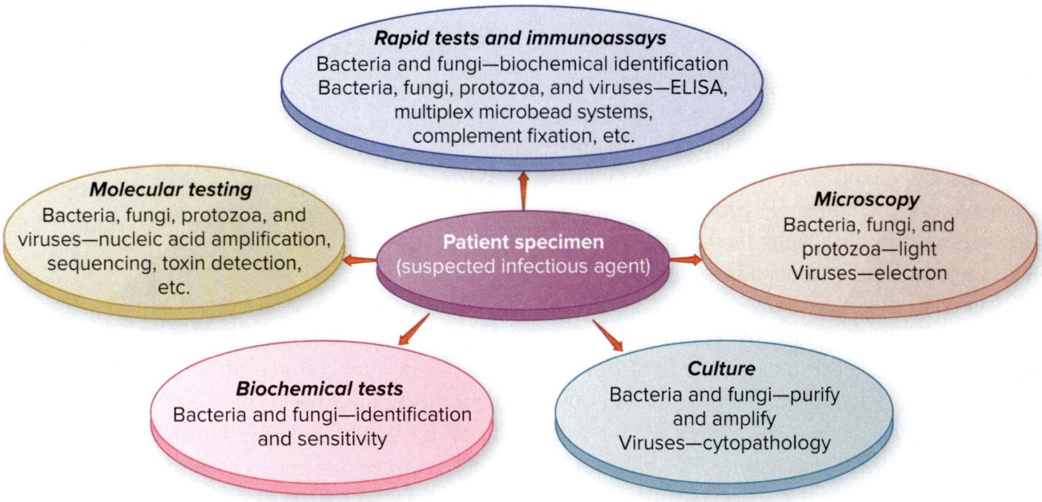

Figure 37.1 The Patient Specimen Is Evaluated by Various Techniques. The specimen source and patient history guide the decision for use of tests and techniques. Not all tests are performed on all specimens, but all laboratories need to be able to perform these tests or refer the specimen to other laboratories.

Table 37.1 Standard Microbiological Practices

1.	Persons must wash their hands after working with potentially hazardous materials and before leaving the laboratory.
2.	Eating, drinking, smoking, handling contact lenses, applying cosmetics, and storing food for human consumption must not be permitted in laboratory.
3.	Mouth pipetting is prohibited; mechanical pipetting devices must be used.
4.	Policies for the safe handling of sharps (such as needles, scalpels, pipettes, and broken glassware) must be developed and implemented.
5.	Procedures must be developed and implemented to minimize the creation of splashes and/or aerosols.
6.	Work surfaces must be decontaminated with an appropriate disinfectant after completion of work and after any spill or splash of potentially infectious material.
7.	All cultures, stocks, and other potentially infectious materials must be decontaminated using an effective method before disposal.
8.	A sign incorporating the universal biohazard symbol must be posted at the entrance to the laboratory when infectious agents are present. The sign may include the biosafety level, name of the agent(s) in use, and the name and phone number of the laboratory supervisor or other responsible personnel. Information should be posted according to institutional policy.
9.	Persons must report all injuries incurred in the laboratory to the laboratory supervisor as soon as possible.

Adapted from: Chosewood, L. C., and Wilson, D. E., 2009. *Biosafety in Microbiological and Biomedical Laboratories*, 5th ed. Washington, D.C.: U.S. Department of Health and Human Services, CDC, National Institutes of Health, 2009; and Emmert, E. "Biosafety guidelines for handling microorganisms in the teaching laboratory: Development and rationale." *J. Microbiol. & Biol. Educ.* 14(1), 2013. doi:10.1128/jmbe.v14i1.531.

To help prevent laboratory-acquired infections (LAIs), microbiologists today follow strict procedures, often called standard microbiological practices, codified in the BMBL (**table 37.1**). In addition, some events have led to legislation affecting how microbiologists practice laboratory safety and security. For example (1) the recognition that the human immunodeficiency virus (HIV) and other viruses are transmitted by blood led to the 1991 Bloodborne Pathogens Act, (2) the use of anthrax as a bioterrorism agent led to the Public Health Security and Bioterrorism Preparedness and Response Act of 2002, and (3) the response to emerging viruses with high infectivity or high lethality (such as avian influenza) resulted in the 2005 Pandemic and All-Hazards Preparedness Act.

Table 37.2 Microbiology Risk Group (RG) Characterization with Example Organisms

1.	Not known to consistently cause disease in healthy adults (e.g., *Lactobacillus casei, Vibrio fischeri*)
2.	Associated with human disease; potential hazard if percutaneous injury, ingestion, or mucous membrane exposure occurs (e.g., *Salmonella typhi, Escherichia coli* O157:H7, *Staphylococcus aureus*)
3.	Indigenous or exotic agents with potential for aerosol transmission; disease may have serious or lethal consequences (e.g., *Coxiella burnetii, Yersinia pestis,* herpesviruses)
4.	Dangerous/exotic agents that pose high risk of life-threatening disease, aerosol-transmitted lab infections; or related agents with unknown risk of transmission (e.g., smallpox virus, Ebola virus, hemorrhagic fever viruses)

Adapted from: Chosewood, L. C., and Wilson, D. E., 2009. *Biosafety in Microbiological and Biomedical Laboratories*, 5th ed. Washington, D.C.: U.S. Department of Health and Human Services, CDC, National Institutes of Health, 2009.

Practicing Biosafety

Safety planning starts with laboratory design (e.g., airflow, sink access, decontamination processes) and easy access to biosafety cabinets to minimize hazard (*see figure 8.6*). Following this initial survey, potential risks associated with manipulation of microorganisms are considered, including (1) procedure(s) used with each agent, (2) experience of those using the laboratory, and (3) infectious agent risk group. Each pathogen is assigned to one of four risk groups (RGs) reflecting increasing danger to humans (**table 37.2**). A pathogen's RG is determined by a number of criteria, including pathogenicity, virulence factors, mode of transmission, availability of protective vaccines, and effectiveness of chemotherapeutic agents.

The rest of the risk assessment determines the biosafety level (BSL). The BSL designation indicates that a particular risk is acknowledged and that a specific set of work practices, safety equipment, and facility engineering controls must be used in an assigned laboratory space to keep risk low. The four biosafety levels reflect an appropriate containment level used to prevent LAI and environmental contamination (**table 37.3**). Each subsequent level builds on the previous. The common component to all levels is the assumption that standard microbiology practices will be followed. These represent minimum guidelines for laboratory safety.

In addition to accepted laboratory behaviors, specific safety equipment is often recommended for each BSL. Personal

Table 37.3 Biosafety Recommendations for BSL 1-4 Microbiology Laboratories

Biosafety Level	Work Practices	Safety Equipment	Facilities
BSL-1	Standard microbiology practices	No barriers required. Personal protective equipment (PPE): laboratory coats and gloves; eye, face protection, as needed	Laboratory bench and sink required
BSL-2	BSL-1 practice plus: Limited lab access Biohazard warning signs "Sharps" precautions Biosafety manual defining any needed waste decontamination or medical surveillance policies	Biosafety cabinets (BSCs) or other physical containment devices used for all manipulations of agents that cause splashes or aerosols of infectious materials PPE: Laboratory coats, gloves, face and eye protection, as needed	BSL-1 facilities plus autoclave available
BSL-3	BSL-2 practice plus: Controlled lab access All waste decontaminated Lab clothing decontaminated before laundering Baseline serum testing	BSCs or other physical containment devices used for all open manipulations of agents PPE: Protective lab clothing, gloves, respiratory protection as needed	BSL-2 facilities plus self-closing double door, negative air pressure, room exhaust not recirculated, away from access corridors
BSL-4	BSL-3 practice plus: Different clothing to enter Shower to exit Decontaminate all material on exit	Class III (glove box) cabinet-based work or Class II (laminar flow) BSC with full-body, positive pressure suit	BSL-3 facilities plus separate building or zone, dedicated heat, ventilation, and air conditioning system, vacuum, and decontamination systems

Adapted from: Chosewood, L. C., and Wilson, D. E., 2009. *Biosafety in Microbiological and Biomedical Laboratories*, 5th ed. Washington, D.C.: U.S. Department of Health and Human Services, CDC, National Institutes of Health, 2009.

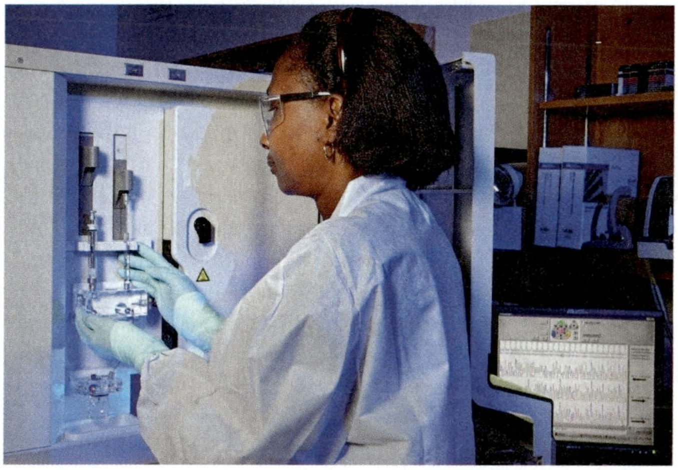

Figure 37.2 Personal Protective Equipment. This person is wearing a lab coat, safety glasses, and gloves to prevent contamination of the laboratory and herself from infectious agents.
Source: CDC/Hsi Liu, Ph.D., MBA; James Gathany

protective equipment (PPE) includes (1) protective coverings (coats, gowns, or uniforms), (2) protective eyewear, and (3) gloves. PPE prevents transmission of infectious agents by blocking their access to entry portals (**figure 37.2**). Additional PPE (e.g., face shield, respiratory protection) may be necessary as determined by the risk assessment.

37.2 Identification of Microorganisms from Specimens

After reading this section, you should be able to:

a. Construct a generalized work plan for the identification of a microorganism from a clinical specimen
b. Provide examples of appropriately collected specimens
c. Describe the use of microscopy to identify microorganisms in patient specimens
d. Explain common methods for culturing bacteria and fungi frequently isolated from patient specimens
e. Interpret microscopy, culture, biochemical, and molecular assays used to identify microorganisms from patient specimens

The clinical specimen is an unknown, and microbiology laboratory scientists are the detectives who identify the microorganism(s) it contains. In doing so, they determine the cause of a patient's infection. There are two types of clinical specimens: those from normally sterile body sites, such as cerebrospinal fluid or bone, and those from nonsterile sites, such as skin, mucous membranes, or feces. An important consideration for the clinical microbiologist is to remember that any microorganism found in a sterile-site specimen is a potential problem for the patient and thus should be rapidly identified. Assessment of nonsterile-site specimens is no less important but requires a working

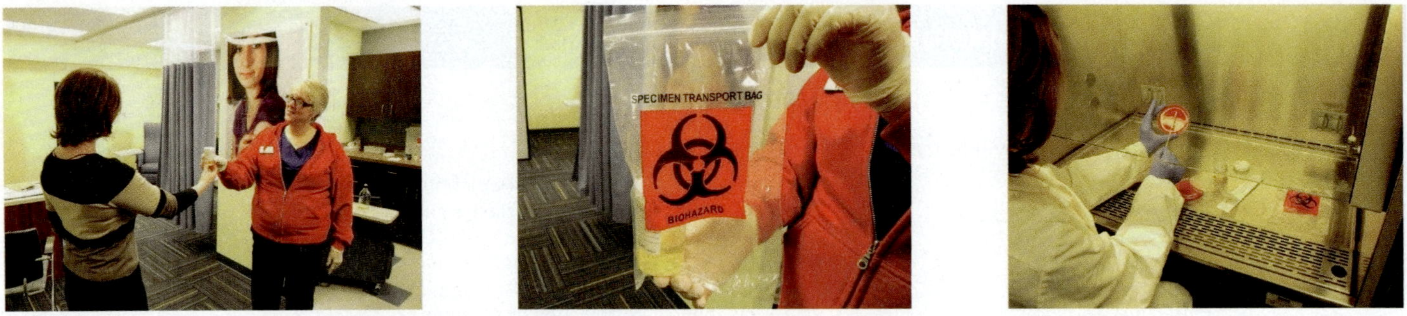

(a) (b) (c)

Figure 37.3 Collection of Patient Specimen for Microorganism Testing. (a) The identification of the microorganism begins with proper instruction for specimen collection. (b) The specimen is protected by transport media and transported in a biohazard containment bag to prevent contamination and spillage. (c) The specimen is processed in a biosafety cabinet. In this case, the specimen is cultured to identify the infectious agent and determine its sensitivity to antimicrobial drugs.

knowledge of the normal microbiota associated with that site. The challenge for the microbiologist is to identify the source of the infection amid the background "noise" of the normal microbiota.

The Clinical Specimen

A clinical specimen represents a portion or quantity of human material that is tested, examined, or studied to determine the presence or absence of particular microorganisms (**figure 37.3**). In general, the clinical specimen should:

1. Adequately represent the diseased area and may include additional sites (e.g., urine and blood specimens) to isolate and identify potential agents of the particular disease process.
2. Be of sufficient quality and quantity to allow a variety of diagnostic tests, should they be necessary.
3. Avoid preventable contamination from the normal microbiota and environment.
4. Be collected in appropriate containers, kept at an appropriate temperature, and forwarded promptly to the clinical laboratory.
5. Be obtained before antimicrobial agents have been administered to the patient, if possible.
6. Be accompanied by a putative diagnosis permitting laboratory personnel to contact the physician suggesting tests necessary to rule out other pathogens.

Choosing the appropriate test or procedure is determined by the specimen and the suspected pathogen(s) in the specimen based on the patient's signs and symptoms. Most laboratories have prescribed algorithms that dictate how each specimen is processed when it enters the laboratory. In some cases, the specimen is evaluated for microorganisms by direct techniques, including methods that first require growth of the microbe prior to testing. Indirect techniques, such as detection of antibody against a specific pathogen or amplification of pathogen genes, are used to identify a pathogen that is difficult to culture; or a molecular test can provide sensitive and specific results in a short period of time. We now highlight some of the microbiological methods commonly used.

Culture-Based Methods

For many bacteria and fungi, the process of obtaining pure cultures involves their growth on selective or differential media that also provide the first clues to the identity of the microbe. The choice of media used depends in part on the source of the specimen and knowledge of likely causes of the infection. Some examples of media used are described in **table 37.4** (*also see table 22.7*). In many cases, classical enrichment culture has been replaced by immunomagnetic bead (IMB) technology. IMBs are small particles coated with antibody targeting a specific microbe. When a clinical sample is mixed with IMBs specific for a suspected pathogen, antibody mediates binding of the pathogen to the beads. A magnet is then used to pull the beads and their attached microbes out of solution. Beads are then washed of debris and nonspecifically bound microbes. By using IMB-bound microbes to inoculate culture medium, a pure culture is more rapidly attained than with enrichment cultures. IMB technology can be used to separate microbes for other applications as well.

Once isolated in pure culture, growth characteristics of the bacterium may be examined by specific biochemical tests. Some of the most common biochemical tests used to identify bacterial isolates are listed in **table 37.5**. Classic dichotomous keys are coupled with the biochemical tests for the identification of bacteria from specimens. Generally, fewer than 20 tests are required to identify clinical bacterial isolates to the species level (**figure 37.4**). ◀ *Culture media (section 7.7)*

Miniaturization into kits and robotic automation of biochemical tests have made identification of pathogens in specimens much more efficient. Some laboratories use biochemical kits, such as the API 20E, for the identification of members of the

Table 37.4 Media Commonly Used for Isolation of Pure Bacterial Cultures from Clinical Specimens

Selective Media
Salmonella-Shigella agar (SS) is used to isolate *Salmonella* and *Shigella* species. Its bile salt mixture inhibits many groups of coliforms. Both *Salmonella* and *Shigella* spp. produce colorless colonies because they are unable to ferment lactose. Lactose-fermenting bacteria produce pink colonies.
Mannitol salt agar (MS) is used for the isolation of staphylococci. The selectivity is obtained by the high (7.5%) salt concentration that inhibits growth of many groups of bacteria. The mannitol helps differentiate pathogenic from nonpathogenic staphylococci, as the former ferment mannitol to form acid, whereas the latter do not. Thus this medium is also differential.
Bismuth sulfite agar (BS) is used for the isolation of *Salmonella enterica* serovar Typhi, especially from stool and food specimens. *S. enterica* serovar Typhi reduces the sulfite to sulfide, resulting in black colonies with a metallic sheen.
Hektoen enteric agar is used to increase the yield of *Salmonella* and *Shigella* spp. relative to other microbiota. The high bile salt concentration inhibits the growth of Gram-positive bacteria and retards the growth of many coliform strains.
Differential Media
Eosin methylene blue agar (EMB) differentiates between lactose fermenters and nonlactose fermenters. EMB contains lactose, salts, and two dyes—eosin and methylene blue. *E. coli,* which is a lactose fermenter, produces a dark colony or one that has a metallic sheen. *S. enterica* serovar Typhi, a nonlactose fermenter, appears colorless.
MacConkey agar is used for the selection and recovery of members of *Enterobacteriaceae* and related Gram-negative rods. The bile salts and crystal violet in this medium inhibit the growth of Gram-positive bacteria and some fastidious Gram-negative bacteria. Because lactose is the sole carbohydrate, lactose-fermenting bacteria produce colonies that are various shades of red, whereas nonlactose fermenters produce colorless colonies.
Blood agar: addition of citrated blood to tryptic soy agar makes possible variable hemolysis, which permits differentiation of some species of bacteria. Three hemolytic patterns can be observed on blood agar (*see figure 23.24*). 1. α-hemolysis—greenish to brownish halo around the colony (e.g., *Streptococcus gordonii* and *S. pneumoniae*) 2. β-hemolysis—complete lysis of blood cells resulting in clearing around the colony (e.g., *Staphylococcus aureus* and *Streptococcus pyogenes*) 3. Nonhemolytic—no change in medium (e.g., *Staphylococcus epidermidis* and *S. saprophyticus*)

family *Enterobacteriaceae* and other Gram-negative bacteria. The API 20E consists of a plastic strip with 20 microtubes containing biochemical substrates, each of which detects a different biochemical activity (**figure 37.5**). The biochemical substrates are inoculated with bacteria previously isolated in pure culture. After 5 to 12 hours of incubation, the 20 test results are converted to a seven- or nine-digit profile. This profile number is entered into the computerized *API Profile Index* to identify the bacterium.

In addition to traditional techniques and kits, larger laboratories also use semiautomated robotic systems to detect microorganisms. Some of these are culture-independent systems, reporting infectious agents by directly identifying microbial protein (e.g., toxins) or DNA within the patient specimen. Several of these technologies couple pure culture isolates with electrospray ionization-mass spectroscopy, peptide nucleic acid fluorescence *in situ* hybridization (PNA-FISH), or matrix-assisted laser desorption/ionization–time of flight (MALDI-TOF)–mass spectrometry to provide high-throughput, high-reliability identification capabilities. ◀ *Polymerase chain reaction amplifies targeted DNA (section 17.2); Genomic libraries: cloning genomes in pieces (section 17.3); Molecular characteristics (section 19.3); Molecular techniques (section 29.2)*

One example of an independent identification system is the automated blood culture system. It measures CO_2 concentration and gas pressure changes as glucose is consumed as a presumptive indicator of microorganism growth. Bacterial and fungal antimicrobial susceptibility is determined by evaluating organism growth when exposed to various antimicrobial agents and reported by similar means. The "gold standard" for antimicrobial susceptibility testing is still the agar dilution method. However, the most common practice is automated microbroth dilution antibiotic susceptibility tests, and the manual Etest and Kirby-Bauer tests (*see figures 9.2–9.4*). Together, the organism's identity and its sensitivity to a panel of antibiotics are determined, often within hours.

Cultures remain the standard for the recovery of fungi from patient specimens; however, the time needed to culture fungi varies anywhere from a few days to several weeks, depending on the organism. Cultures of fungi are evaluated for growth rate and appearance on at least one selective and one nonselective agar medium, with careful examination of colony morphology, color, and dimorphism. Typically, the isolation of fungi is accomplished by concurrent culture of the specimen on media that are respectively supplemented and unsupplemented with antibiotics and cycloheximide. Antibiotics inhibit bacteria that may be in the specimen, and

Table 37.5 Some Common Biochemical Tests Used in the Identification and Differentiation of Bacteria

Biochemical Test	Description	Laboratory Application
Carbohydrate fermentation	Acid and/or gas are produced during fermentative growth with sugars or sugar alcohols.	Fermentation of specific sugars used to differentiate enteric bacteria as well as other genera or species
Casein hydrolysis	Detects caseinase, an enzyme that hydrolyzes the milk protein casein. Bacteria that produce caseinase appear as colonies surrounded by a clear zone.	Used to cultivate and differentiate aerobic actinomycetes based on casein utilization. For example, *Streptomyces* uses casein and *Nocardia* does not.
Catalase	Detects the presence of catalase, which converts hydrogen peroxide to water and O_2	Used to differentiate *Streptococcus* (–) from *Staphylococcus* (+) and *Bacillus* (+) from *Clostridium* (–)
Citrate utilization	When citrate is used as the sole carbon source, this results in alkalinization of the medium.	Used in the identification of enteric bacteria. *Klebsiella* (+), *Enterobacter* (+), *Salmonella* (often +); *Escherichia* (–), *Edwardsiella* (–)
Coagulase	Detects the presence of coagulase. Coagulase causes plasma to clot.	This is an important test to differentiate *Staphylococcus aureus* (+) from *S. epidermidis* (–).
Decarboxylases (arginine, lysine, ornithine)	The decarboxylation of amino acids releases CO_2 and amine.	Used in the identification of enteric bacteria
Esculin hydrolysis	Tests for the cleavage of a glycoside	Used to differentiate *Staphylococcus aureus, Streptococcus mitis,* and others (–) from *S. bovis, S. mutans,* and enterococci (+)
β-galactosidase (ONPG) test	Demonstrates the presence of an enzyme that cleaves lactose to glucose and galactose	Used to differentiate enterics (*Citrobacter* +, *Salmonella* –) and to identify pseudomonads
Gelatin liquefaction	Detects whether or not a bacterium can produce proteases that hydrolyze gelatin and liquefy solid gelatin medium	Used in the identification of *Clostridium, Serratia, Pseudomonas,* and *Flavobacterium*
Hydrogen sulfide (H_2S)	Detects the formation of hydrogen sulfide from the amino acid cysteine due to cysteine desulfurase	Important in the identification of *Edwardsiella, Proteus,* and *Salmonella*
IMViC (indole; methyl red; Voges-Proskauer; citrate)	The indole test detects the production of indole from the amino acid tryptophan. Methyl red is a pH indicator to determine whether the bacterium carries out mixed acid fermentation. Voges-Proskauer (VP) detects the production of acetoin. The citrate test determines whether or not the bacterium can use sodium citrate as a sole source of carbon.	Used to separate *Escherichia* (MR+, VP–, indole+) from *Enterobacter* (MR–, VP+, indole–) and *Klebsiella pneumoniae* (MR–, VP+, indole–); also used to characterize members of the genus *Bacillus*
Lipid hydrolysis	Detects the presence of lipase, which breaks down lipids into simple fatty acids and glycerol	Used in the differentiation of clostridia
Nitrate reduction	Detects whether a bacterium can use nitrate as an electron acceptor	Used in the identification of enteric bacteria, which are usually +
Oxidase	Detects the presence of cytochrome *c* oxidase which is able to reduce O_2 and artificial electron acceptors	Important in distinguishing *Neisseria* and *Moraxella* spp. (+) from *Acinetobacter* (–) and enterics (all –) from pseudomonads (+)
Phenylalanine deaminase	Deamination of phenylalanine produces phenylpyruvic acid, which can be detected colorimetrically.	Used in the characterization of the genera *Proteus* and *Providencia*
Starch hydrolysis	Detects the presence of the enzyme amylase, which hydrolyzes starch	Used to identify typical starch hydrolyzers such as *Bacillus* spp.
SIM (sulfide, indole, motility) medium	Detects sulfides, indole (a metabolic product of tryptophan utilization), and motility	Used to differentiate enteric organisms colorimetrically
TSI (triple sugar iron) medium	Detects metabolism of glucose, lactose or sucrose, and sulfides from ammonium sulfate or sodium thiosulfate	Used to differentiate enteric organisms colorimetrically
Urease	Detects the enzyme that splits urea to NH_3 and CO_2	Used to distinguish *Proteus, Providencia rettgeri,* and *Klebsiella pneumoniae* (+) from *Salmonella, Shigella,* and *Escherichia* (–)

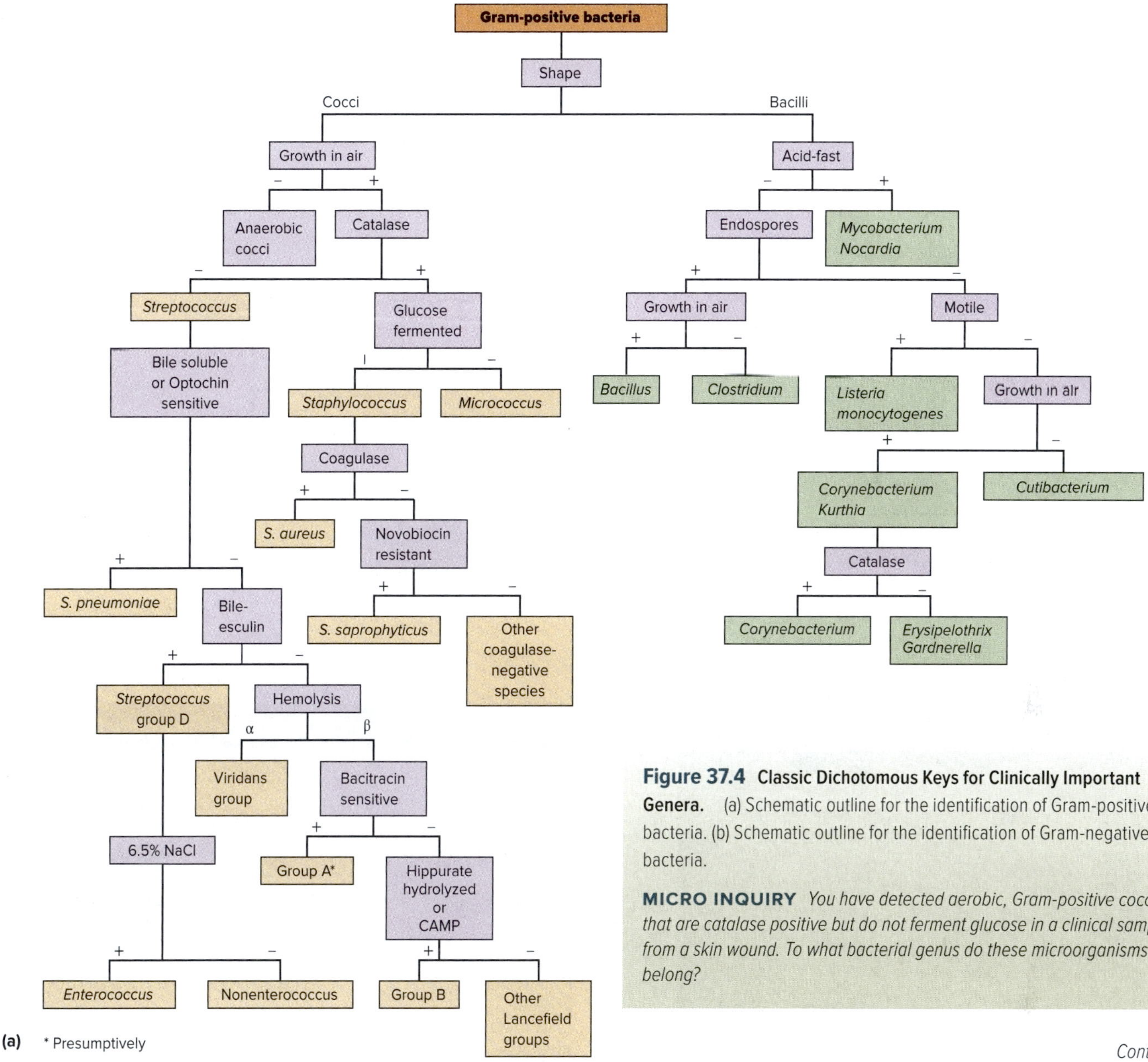

Figure 37.4 Classic Dichotomous Keys for Clinically Important Genera. (a) Schematic outline for the identification of Gram-positive bacteria. (b) Schematic outline for the identification of Gram-negative bacteria.

MICRO INQUIRY *You have detected aerobic, Gram-positive cocci that are catalase positive but do not ferment glucose in a clinical sample from a skin wound. To what bacterial genus do these microorganisms belong?*

Continued

cycloheximide inhibits saprophytic molds. However, a number of media formulations are routinely used to culture specific fungi (**table 37.6**). If pathogenic fungal infections are suspected, serological tests for antibodies to specific fungal pathogens may also be performed. Culture of viruses and protozoa from clinical specimens is not routine. Their detection and identification is by microscopy, molecular techniques, or both.

Microscopy

Sometimes microscopic visualization of microorganisms in a specimen (e.g., urine or sputum sample) or in culture can aid the identification process. Wet-mount, heat-fixed, or chemically fixed specimens are examined with bright-field or phase-contrast microscopy. Among the many stains used to examine specimens, the Gram stain is the most common (*see figure 2.19*). Although it does not identify bacteria, preliminary reports indicate the Gram-morphology, such as "Gram-positive cocci in pairs and chains." Such information is often helpful in ensuring appropriate antibiotic therapy. Note in figure 37.4 that the Gram reaction and shape of a bacterium are critical pieces of information used to decide which biochemical tests are appropriate to perform.

Definitive identification of most fungi is based on the morphology of hyphae or conidia (spores). Calcofluor white is a

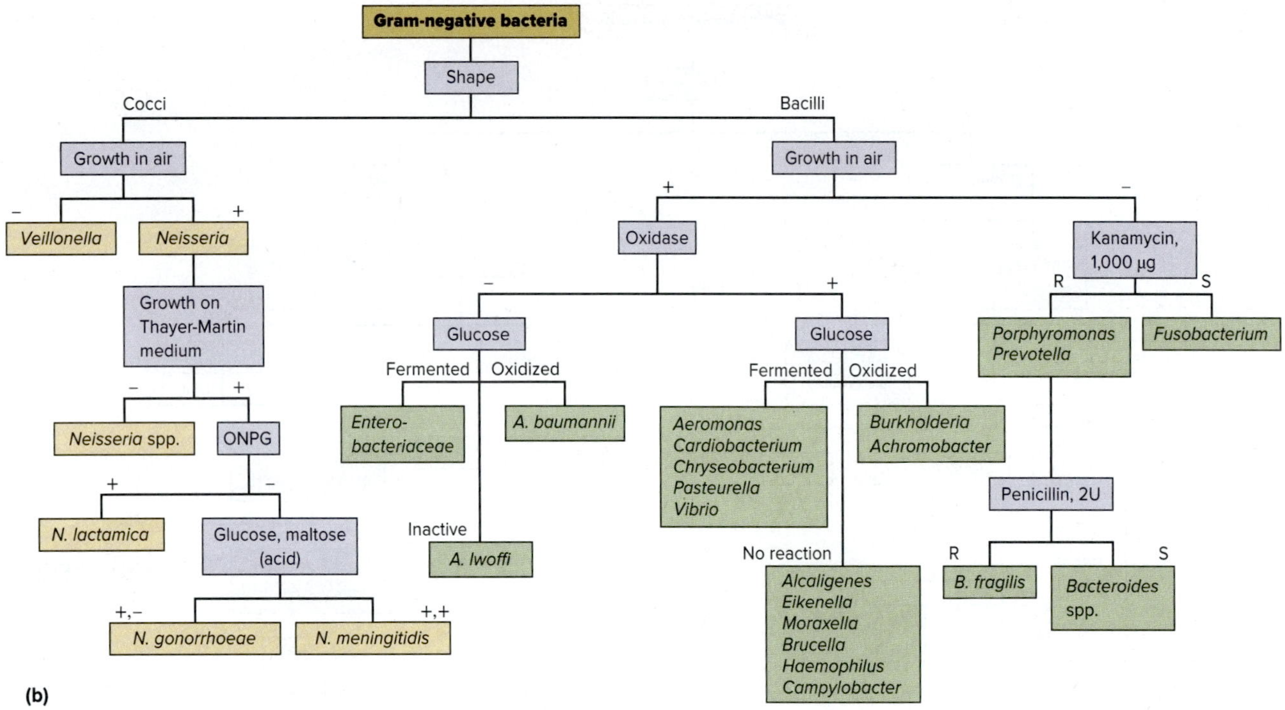

(b)

Figure 37.4 *Continued*

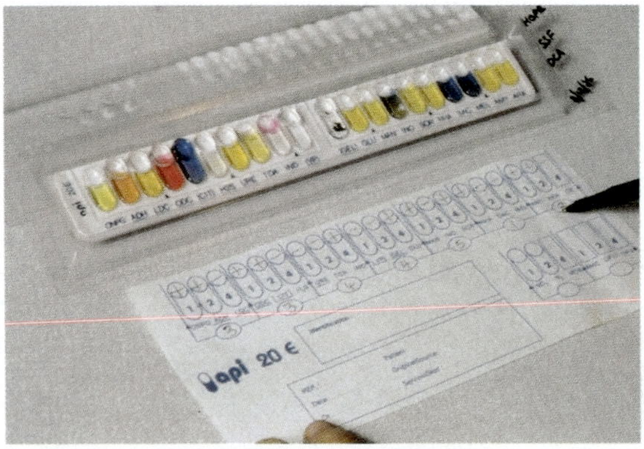

Figure 37.5 A Kit Approach to Bacterial Identification. The API 20E manual biochemical system for microbial identification.

Table 37.6 Examples of Media Used to Isolate Fungi

Culture Medium	Target Fungi
Antibiotic agar	Fungi in polymicrobial specimens
Caffeine agar	*Aspergillus, Rhizopus,* and others
Cornmeal agar	Most fungi, including pathogens
Malt agar	Ascomycetes
Malt extract agar	Basidiomycetes
Sabouraud dextrose agar (SAB)	Most fungi
SAB + chloramphenicol + cycloheximide	Dimorphic fungi and dermatophytes

fluorescent stain often used to detect fungi because it binds to the chitin in their cell walls. Protozoa are most often identified based on the morphology of their ova, trophozoites, and cysts. This is typically accomplished by direct microscopic evaluation of the clinical specimen. Helminths (worms) are often visualized in concentrated wet mounts of stool or urine specimens examined microscopically for the presence of eggs, larvae, or adult worms.

The fluorescence microscope is used to visualize microorganisms after staining with fluorochromes such as acridine orange, which stains nucleic acids. Viruses are too small to visualize with the light microscope, but some can be diagnosed using fluorescence microscopy as discussed next. ◀ *Antibodies bind specific 3-D molecules (section 33.7)*

In contrast to nonspecific stains, antibodies with an attached fluorescent marker can be used to identify microbes at the species level. There are two main kinds of fluorescent antibody techniques for microscopy: direct, which detects antigen, and indirect, which detects patient antibody to the pathogen. **Direct immunofluorescence** involves fixing the specimen containing the pathogen to be detected onto a slide (**figure 37.6*a***). Fluorochrome-labeled antibodies are then added to the slide and incubated. The slide is washed to remove any unbound antibody and examined with the

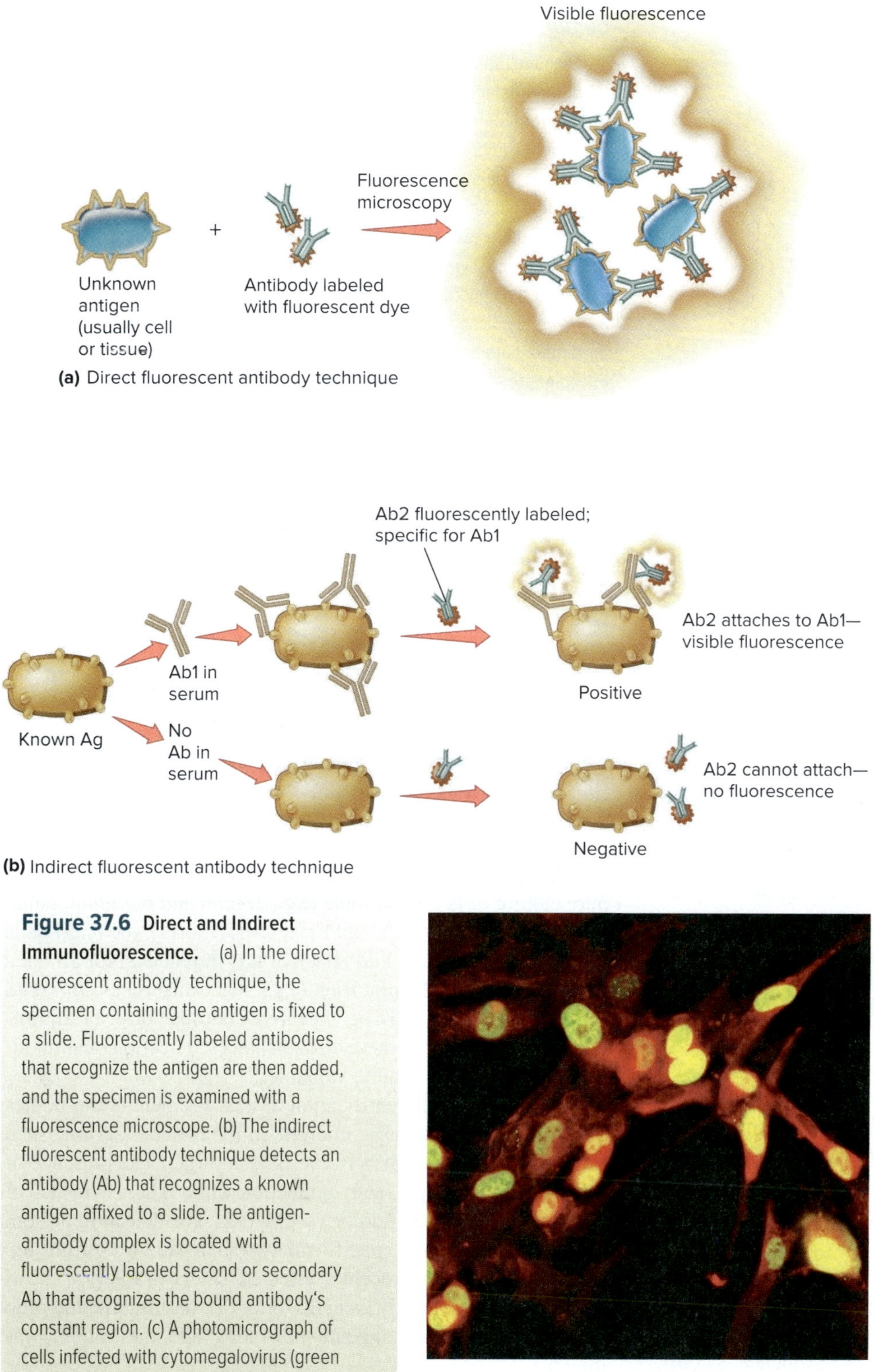

Figure 37.6 Direct and Indirect Immunofluorescence. (a) In the direct fluorescent antibody technique, the specimen containing the antigen is fixed to a slide. Fluorescently labeled antibodies that recognize the antigen are then added, and the specimen is examined with a fluorescence microscope. (b) The indirect fluorescent antibody technique detects an antibody (Ab) that recognizes a known antigen affixed to a slide. The antigen-antibody complex is located with a fluorescently labeled second or secondary Ab that recognizes the bound antibody's constant region. (c) A photomicrograph of cells infected with cytomegalovirus (green fluorescence) and adenovirus (yellow fluorescence) using an indirect immunofluorescent staining procedure.
(c) ©Millipore Corporation

MICRO INQUIRY *Which type of immunofluorescence would most likely be used to test a clinical sample for the presence of an intracellular pathogen such as hepatitis C virus?*

fluorescence microscope (*see figures 2.12 and 2.13*). If fluorescence is seen, the presence of the pathogen is confirmed. Example uses of direct immunofluorescence are the identification of the bacteria *Treponema pallidum* and *Rickettsia rickettsii,* the fungus *Pneumocystis jiroveci,* as well as many viruses including adenovirus, rabies virus, and herpes simplex viruses.

Indirect immunofluorescence (figure 37.6*b*) is used to detect the presence of antibodies in serum following an individual's exposure to microorganisms. In this technique, a known antigen (e.g., a virus) is fixed onto a slide. The patient serum suspected of containing (antiviral) antibodies is then added, and if the specific antibody is present, it reacts with antigen to form a complex. When a second, commercially prepared fluorescein-labeled antibody directed against the constant region of the antibody is added, it binds the fixed antibody. After incubation and washing, the slide is examined with fluorescence microscopy. The occurrence of fluorescence shows that antibody specific to the (viral) antigen is present in the serum as reported by the fluorescence of the secondary antibody that has attached to the serum antibody (figure 37.6*c*).

Molecular Methods

The identification of microorganisms through the analysis of specific proteins and nucleic acids is a rapid, accurate, and sensitive approach. Examples include the detection of bacterial exotoxin and viral capsid proteins; the detection of specific genes; and nucleic acid sequencing (*see figures 18.2–18.3*). Other molecular methods widely used are nucleic acid probes, real-time PCR amplification of DNA, and DNA fingerprinting. Details of techniques that use microbial DNA, RNA, or protein are presented in chapters 17–19.

Nucleic acid–based diagnostic methods for the detection and

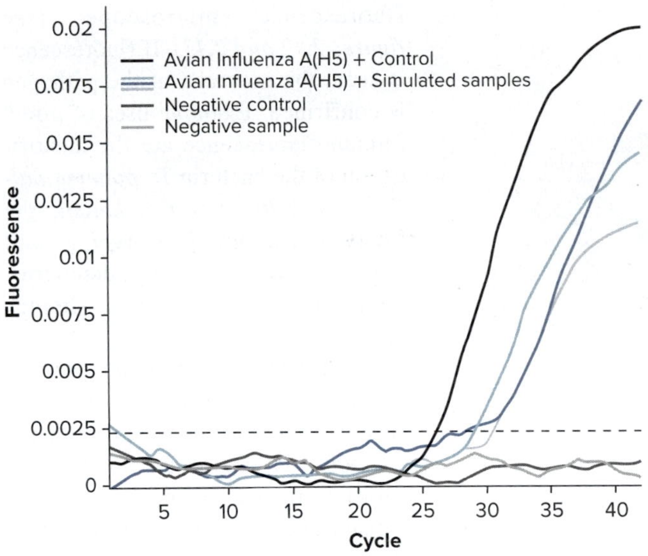

Figure 37.7 PCR Kit Results. Real-time detection of avian influenza is possible using a PCR kit. Note that sufficient DNA amplification occurs after 25 to 30 cycles.

identification of microorganisms have become routine in clinical microbiology laboratories. For example, most viruses are detected by PCR. If the virus has an RNA genome, it is reverse-transcribed to DNA prior to PCR amplification. In all cases, real-time PCR is used to provide rapid, accurate, and quantitative results (**figure 37.7**; also *see figure 17.8*). Often a single sample will be used in a multiplex PCR assay. Multiple primers targeting different viral genomes are used in this application of multiplex PCR enabling a single sample to be screened for all suspected viruses. For example, multiplex PCR for respiratory viruses assays a patient sample for all common respiratory viral pathogens (e.g., adenovirus, influenza, enteroviruses, parainfluenza virus, etc.). Another use of multiplex PCR is the analysis of blood cultures. Recall that blood is normally sterile and blood-borne infections can cause septic shock, so rapid and accurate identification of a blood-borne pathogen is imperative. A multiplex PCR approach enables the rapid identification of over 20 different pathogens and the genes for resistance to certain antibiotics. ◀ *Polymerase chain reaction amplifies targeted DNA (section 17.2)*

The gene most commonly used to identify bacteria in blood and other types of cultures is the 16S ribosomal RNA (rRNA) (*see figure 19.2*). Typically the rRNA gene or gene fragment is amplified by PCR. After nucleotide sequencing, the sequence is compared with those in international databases. This method of bacterial identification, called **ribotyping,** is based on the high level of 16S rRNA conservation among bacteria. As discussed in chapter 19, multiple housekeeping genes, instead of 16S rRNA genes, may be sequenced and analyzed in a technique called **multilocus sequence typing** (**MLST**). ◀ *Signature sequences (section 19.3)*

Comprehension Check

1. Examine table 37.4. What is the difference between selective and differential media? Why are some media both selective and differential?
2. Describe in general how biochemical tests are used in the API 20E system to identify bacteria.
3. What is the advantage of visualizing a bacterium using direct immunofluorescence as compared to Gram staining?
4. Why are most viruses identified by molecular methods, such as PCR, rather than microscopy methods?
5. Using the dichotomous key shown in figure 37.4, how would you differentiate between *Staphylococcus saprophyticus* and *S. aureus?*
6. Compare and contrast direct and indirect immunofluorescence. What is the difference in terms of what is detected in the patient's sample?

37.3 Immune Responses Can Be Exploited to Detect Infections

After reading this section, you should be able to:

a. Diagram the methods by which antibodies are used to identify microorganisms
b. Interpret immunologic data used to identify microorganisms from patient specimens
c. Design an antibody-based test for the identification of a specific microorganism from a patient specimen

It may not be possible to culture certain microorganisms from clinical specimens because the organism is difficult to grow in pure culture or is unculturable (e.g., *Treponema pallidum*; hepatitis A, B, and C viruses; and Epstein-Barr virus), the microbe is unsafe (e.g., rickettsias and HIV), or it is impractical for all but a few microbiology laboratories (e.g., strict anaerobes, *Borrelia* spp.). Cultures also may be "negative" because of prior antimicrobial therapy. Under these circumstances, detection of antibodies or antigens can be quite valuable.

Serology is the identification or measurement of antibodies directed against a specific microbe in the patient's serum. Such assays often distinguish between IgM and IgG. The presence of IgM indicates a recent, active infection while IgG is evidence of a past or chronic infection or vaccination. The detection of both IgM and IgG against a specific microbe in a patient's serum suggests the patient was recently infected and is in the process of class switching antibody (*see figure 33.15*). In this case, the ratio of IgM to IgG may indicate how long ago the patient was infected.

Dramatic advances in clinical immunology have given rise to a marked increase in the number, sensitivity, and specificity of serological tests. This reflects the availability of highly specific monoclonal antibodies with very high affinity and avidity, and the development of sensitive antibody-binding reporter systems (*see Techniques & Applications 33.1*). Although most of these tests are relatively simple to use, the validity of test results depends on proper test selection, use, and timing of specimen collection. For instance, individual immunologic responses to a microorganism can be quite variable, making the interpretation of immunologic tests potentially difficult. Unfortunately, lack of a measurable antibody titer does not always mean the absence of infection. Rather it may reflect an organism's lack of immunogenicity or insufficient time for an antibody response to develop following the onset of the infectious disease. Some patients are also immunosuppressed due to other disease processes or treatment procedures (e.g., cancer and AIDS patients) and therefore do not respond. In this section, some of the more common antibody-based techniques employed in the diagnosis of microbial and immunological diseases are discussed. |◀ *Antibody structure (section 33.7)*

Serotyping

Serotyping is used to differentiate strains (serovars or serotypes) of microorganisms that differ in the antigenic composition of a structure or product. This has traditionally been performed by testing the microorganism's response to different antigen-specific antibodies. For example, there are about 100 serotypes of *Salmonella enterica* that cause human disease. Serotype identification can help trace the source of food-borne disease outbreaks.

Agglutination

An **agglutination reaction** occurs when an immune complex is formed by cross-linking cells or particles with specific antibodies (*see figure 33.19*). Agglutination reactions usually form visible aggregates or clumps, called **agglutinates,** that can be seen with the unaided eye. Direct agglutination reactions are very useful in the diagnosis of certain diseases. For example, the Rapid Plasma Reagin test is used to screen for syphilis. Antigenic cardiolipin released from spirochete-damaged host cells stimulates the production of antibodies. These antibodies, collected from syphilis-suspect patients, will agglutinate syphilis bacteria in vitro, yielding a positive test. |◀ *Immune complex formation (section 33.8)*

Techniques have also been developed that employ microscopic synthetic latex spheres coated with antigens. These coated microspheres bind antibodies in a patient's serum specimen to identify viral disease rapidly when cultures are not feasible. Latex agglutination tests are also used to detect antibodies that develop during certain mycotic (fungal), helminthic (worm), and bacterial infections, as well as in drug testing. Microspheres can also be coated with monoclonal antibodies to capture antigens from patient specimens.

Hemagglutination assays use antibodies to cross-link red blood cells (RBCs) by attaching to surface antigens. Hemagglutination is routinely used in blood typing (*see figure 33.21*). When viruses mediate RBC clumping, it is termed **viral hemagglutination.** For example, if a person has a certain viral disease, such as measles, antibodies present in the serum react with the measles virus particles and neutralize them. Normally hemagglutination occurs when measles virus particles and red blood cells are mixed. However, if a person's serum is mixed first with virions, the antibodies bind all the virions. When this is followed by the addition of RBCs, hemagglutination is prevented. Therefore, in contrast to the formation of clumped particles in blood typing, here the lack of visible agglutination is a positive test result for the presence of virus-specific antibodies (**figure 37.8**).

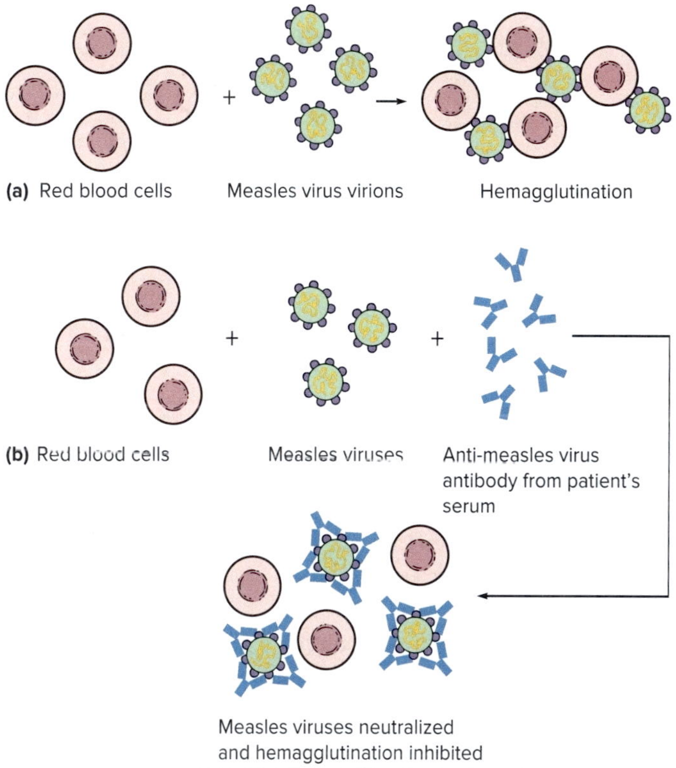

Figure 37.8 Viral Hemagglutination. (a) The virions of certain viruses bind to red blood cells, causing hemagglutination. (b) If patient serum containing specific antibodies to the virus is mixed with the red blood cells, the antibodies will neutralize the virus and inhibit hemagglutination (a positive test).

MICRO INQUIRY *Is the presence of cellular clumping a positive or negative result in this case?*

Agglutination tests are also used to measure antibody titer. In tube or well agglutination tests, a specific amount of antigen is added to a series of tubes or shallow wells in a microtiter plate (**figure 37.9**). Serial dilutions of serum (1/20, 1/40, 1/80, 1/160, etc.) containing the antibody are then added to each tube or well. The greatest dilution of serum showing an agglutination reaction is determined, and the reciprocal of this dilution is the serum antibody titer.

Complement Fixation

When complement binds to an antigen-antibody complex, it is removed from solution and becomes fixed, meaning it is attached to the antigen-antibody complex. Complement fixation tests are very sensitive and can be used to detect extremely

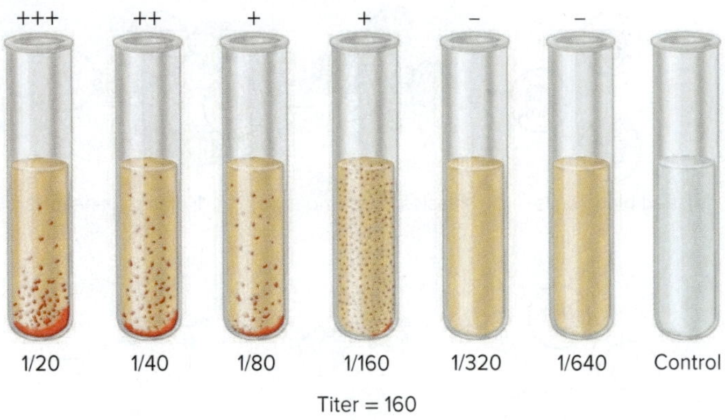

Figure 37.9 Tube Agglutination Test. Agglutination tests for determining antibody titers can be performed in tubes or in microtiter plates. The principle by which the test works is the same in both. The titer in this example is 160 because there is no agglutination in the next tube in the dilution series (1/320). The buff color in the dilution tubes indicates the presence of the patient's serum.

small amounts of an antibody directed against a suspect microorganism in an individual's serum or cerebral spinal fluid (CSF). A known antigen is mixed with test serum or CSF lacking complement (**figure 37.10*a***). After antibody-antigen complexes have had time to form, complement is added to the mixture (figure 37.10*b*). If antibody-antigen complexes are present, they fix complement, removing it from solution. Next, sensitized indicator cells, usually sheep RBCs previously coated with complement-fixing antibodies, are added to the mixture. If all the complement has already been consumed by the immune complexes, complement will be unable to lyse the indicator cells. On the other hand, if the serum or CSF does not have antigen-specific antibodies, complement remains in solution and lyses the indicator cells (figure 37.10*c*). Thus the absence of lysis shows that specific antibodies are present in the test serum or CSF and is thus considered a positive result. Complement fixation is used as a rapid, inexpensive screening method in the diagnosis of certain fungal, chlamydial, and protozoan diseases. *Complement Fixation Test*

Enzyme-Linked Immunosorbent Assay

The ***e**nzyme-**l**inked **i**mmuno**s**orbent **a**ssay* (**ELISA**) is one of the most widely used serological tests. The science behind the technique is similar to that described for fluorescence microscopy and involves the detection of proteins with antibodies that have been labeled with a molecule that can react with substrate to yield a color change (a chromogenic substrate) or light (a chemiluminescent substrate). There are two types of ELISAs: direct and indirect. Direct ELISAs detect protein antigens (e.g., virus capsid proteins) in patient samples; indirect ELISAs detect specific human antibodies in patient samples. Thus indirect ELISAs identify pathogens by the presence of antibodies directed against a specific pathogen, not by the presence of the pathogen itself.

The indirect immunosorbent assay starts with the addition of patient serum (with its soluble antibodies) to the wells of a microtiter plate. Because the wells of the microtiter plate have been previously coated with the antigen of interest, if the patient's serum contains antibody against this antigen, the antibody will bind, or be captured by, the antigen (**figure 37.11*a***). Excess serum, containing antibody to other antigens, is removed; and labeled antibody directed against the constant region of human

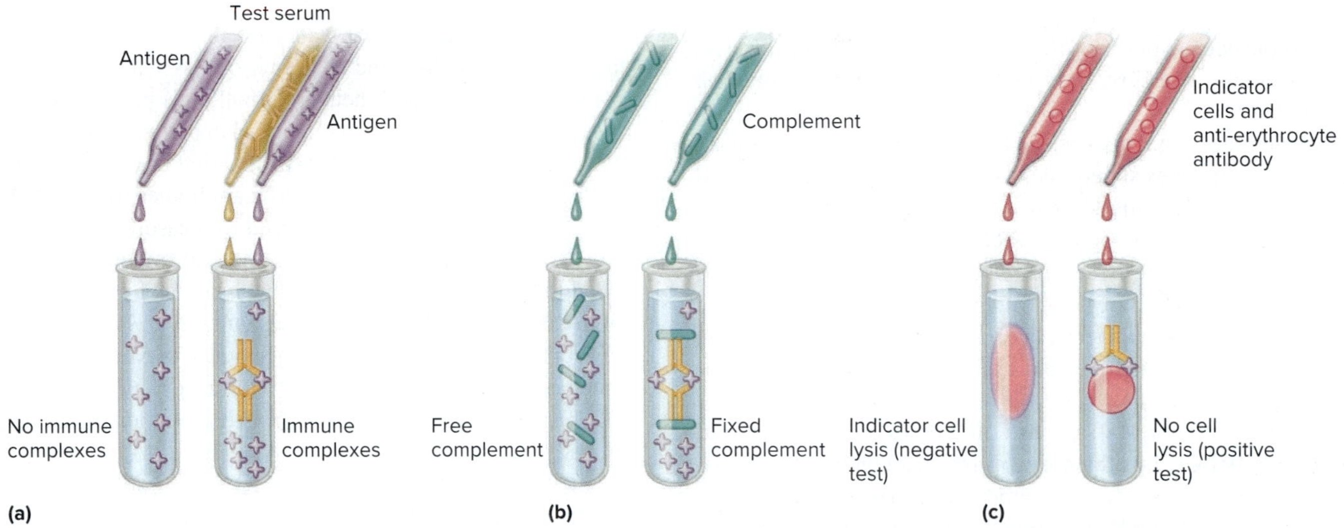

Figure 37.10 Complement Fixation. (a) Test serum is added to one test tube. A fixed amount of antigen is then added to both tubes. If antibody is present in the test serum, immune complexes form. (b) When complement is added, if complexes are present, they fix complement and consume it. (c) Indicator cells and a small amount of anti-erythrocyte antibody are added to the two tubes. If there is complement present, the indicator cells will lyse (a negative test); if the complement has been consumed by the immune complex, no lysis will occur (a positive test).

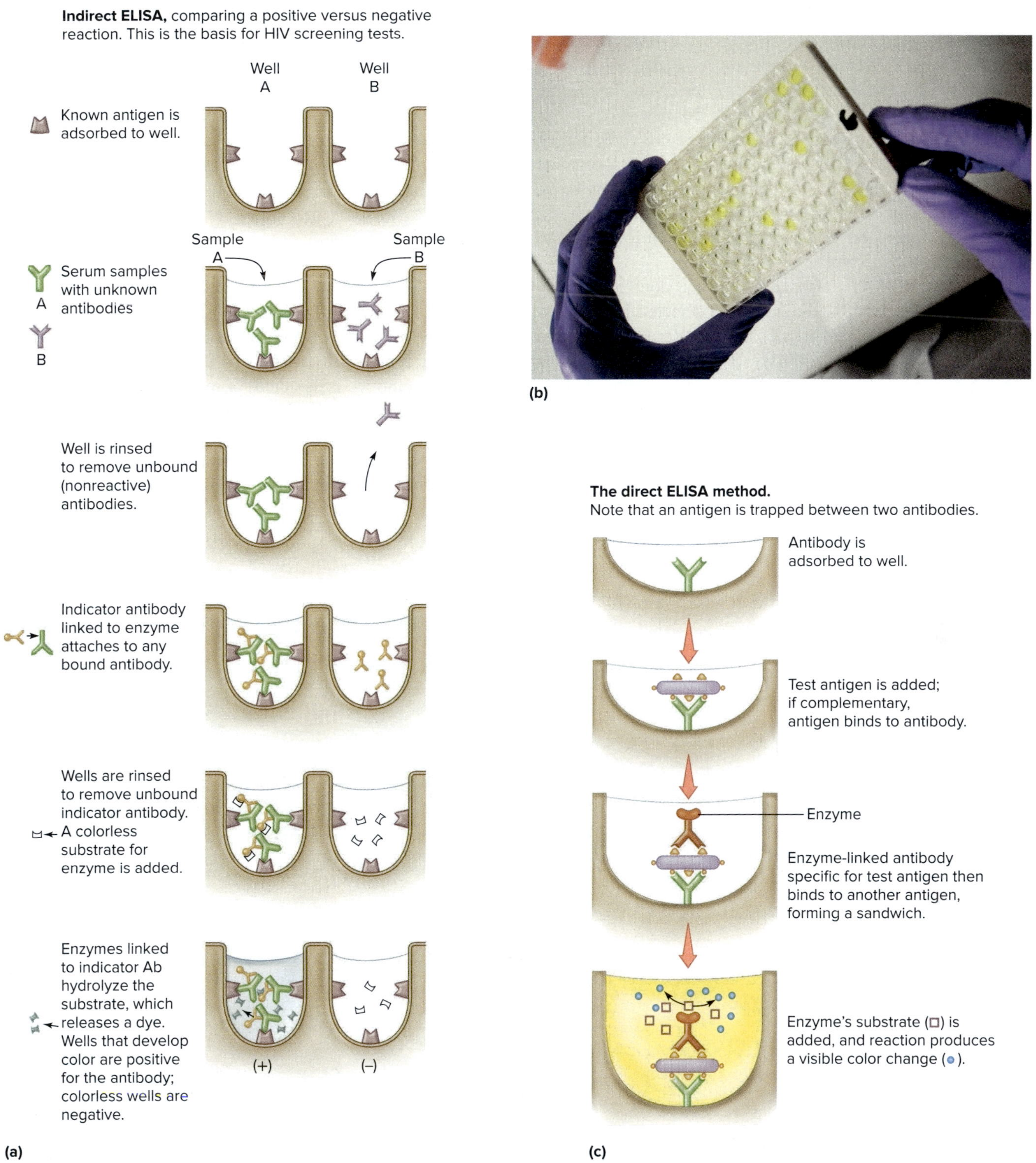

Figure 37.11 The ELISA Test. (a) The indirect immunosorbent assay for detecting antibodies. (b) ELISA test results. (c) The direct or sandwich method for the detection of antigens.

antibodies is added to each well. Only if patient antibody has been captured by the known antigen will the labeled, or secondary, antibody bind. Secondary antibody is needed to detect a positive result; it is formulated with an enzyme or small molecule (e.g., biotin) attached, or conjugated, to it. The presence of secondary antibody (and thus patient antibody) is detected by adding the substrate that reacts with the labeling molecule to generate either color or light (figure 37.11*b*). Because secondary antibody, typically made by injecting rabbits or mice with human antibody, can be formed against either human IgM or IgG, it is possible to assay for each antibody type.

The direct ELISA is also known as the double antibody sandwich assay because the antigen to be detected is first captured by antibody that coats the wells of the microtiter plate and then secondary (labeled) antibody. Thus patient antigen is wedged between these two types of antibodies (figure 37.11*c*). Microtiter plates with wells coated with commercially prepared monoclonal antibody (Mab) enable the high affinity capture of antigen from patient blood or other body fluid. After the removal of excess patient sample (which contains other antigens not bound by the Mab), antibody directed against the antigen of interest is added. As with indirect ELISAs, this antibody has an enzyme or small molecule conjugated to it. Thus addition of the appropriate substrate generates either a color change or the production of light. In both direct and indirect ELISAs, the presence and magnitude of signal are determined by placing the microtiter plate in a plate reader, an instrument designed to measure and record color or light in microtiter plate wells in units relative to standard reactions. *ELISA: Enzyme-Linked Immunosorbent Assay*

The principle of the ELISA methodology has been applied to a multitude of quick tests collectively called **lateral flow assays.** These are simple, low cost, rapid, and portable detection devices that you may have seen if you have ever been screened for strep throat. Lateral flow assays, like ELISAs, can be direct (antigen is detected) or indirect (antibody is detected). Patient sample is placed at one end of a filter paper, typically mounted in a plastic or cardboard case. The fluid sample is wicked or drawn up the filter paper (i.e., flowing laterally) where antibody-antigen interaction takes place. As outlined in

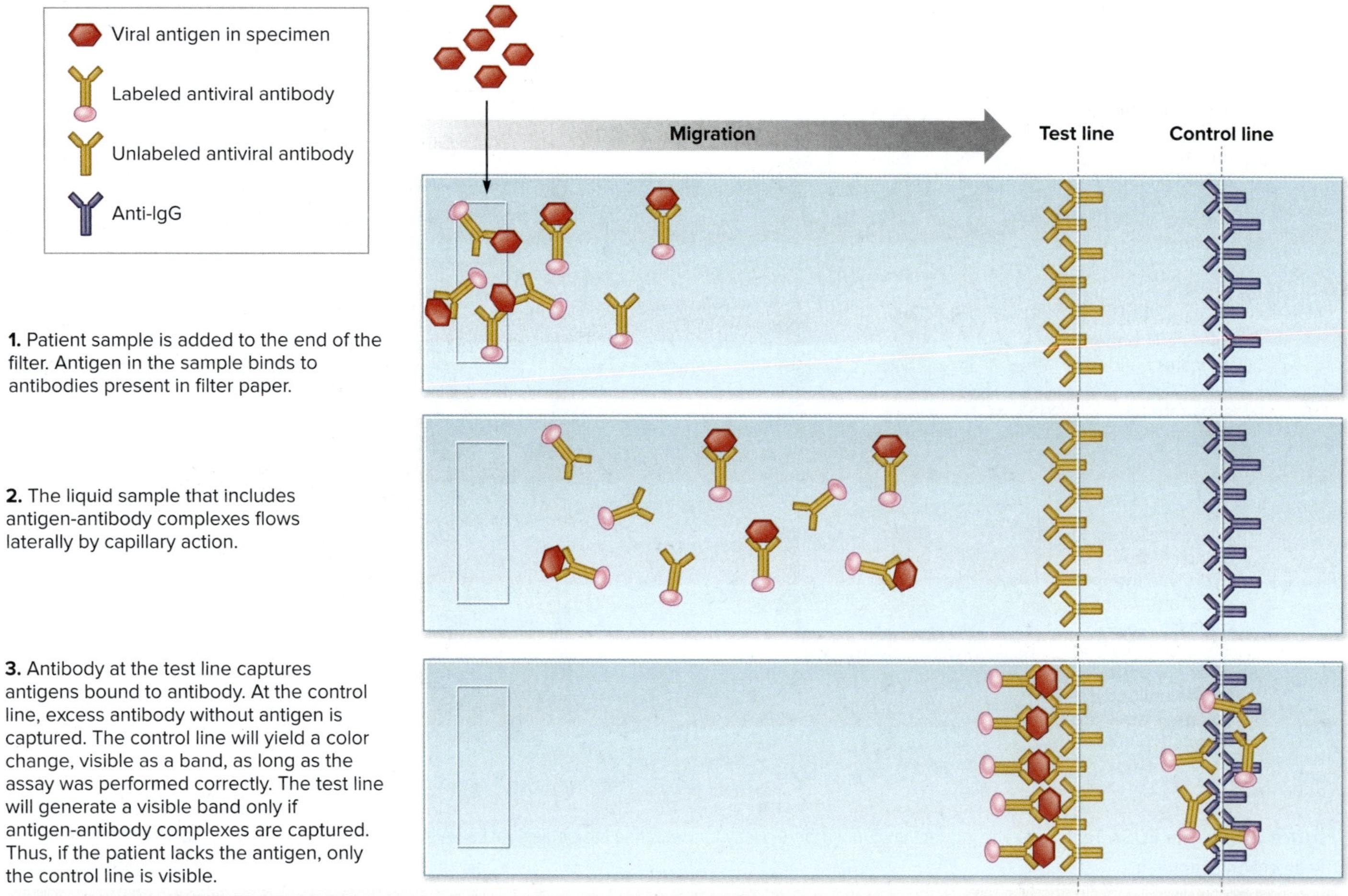

Figure 37.12 Lateral Flow Assay. In this case, patient antigen is detected. Other lateral flow assays detect patient antibody by providing antigen at the site where sample is added.

MICRO INQUIRY *Why must an excess of antibody be present on the filter?*

figure 37.12, the detection lines include chromogenic substrates. In addition to detecting *Streptococcus pyogenes* (strep throat), examples of other direct lateral flow assays include those for Rotavirus, *Giardia intestinalis,* and toxins produced by *Clostridioides* (formerly *Clostridium*) *difficile* in stool samples. Indirect lateral flow assays are used to detect antibodies to dengue and chikungunya viruses and *Anaplasma* bacteria in serum.

Immunoblotting (Western Blotting)

Another immunologic technique used in the microbiology laboratory is immunoblotting, also known as Western blotting. **Western blotting** involves polyacrylamide gel electrophoresis to separate the proteins in a patient specimen. The proteins are then transferred to sheets of filter paper (nitrocellulose or polyvinyl difluoride). This immobilizes the proteins on a sturdy substrate. Protein bands are then visualized by treating the filter papers with solutions of enzyme-tagged antibodies. This procedure was considered more specific than ELISA and lateral flow assays for many applications. However, the use of monoclonal antibodies in these other assays for antigen capture or antibody detection, along with other technological advances, has reduced the need for Western blotting, a more complex and time-consuming process. Western blotting is still used to confirm certain infections, such as Lyme disease.

Immunoprecipitation

The **immunoprecipitation** technique detects antibodies that when bound to antigens are called **precipitins.** The precipitin reaction occurs when antibodies and antigens are mixed in the proper proportions. The antibodies link the antigen to form a large antibody-antigen network or lattice that settles out of solution when it becomes sufficiently large (**figure 37.13*a***). Immunoprecipitation reactions occur at what is known as the zone of equivalence. This describes conditions within the assay tube or plate when there is an optimal ratio of antigen to antibody so that almost all antigen is bound by antibody and an insoluble lattice forms. If the precipitin reaction takes place in a test tube (figure 37.13*b*), a precipitation ring forms in the area in which the optimal ratio or equivalence zone develops. The principle of immunoprecipitation forms the basis of the immunodiffusion techniques discussed next. ◀ *Immune complex formation (section 33.8)*

Immunodiffusion

Immunodiffusion refers to a precipitation reaction that occurs between an antibody and antigen in an agar gel. Two techniques are routinely used: single radial immunodiffusion and double diffusion in agar.

The **single *r*adial *i*mmuno*d*iffusion (RID) assay** or Mancini technique quantitates antigens. Antibody is added to agar and the mixture is poured onto slides and allowed to set. Wells are cut in the agar and known amounts of standard antigen added. The unknown test antigen is added to a separate well. The slide is incubated for 24 hours or until equilibrium has been reached, during which time the antigen diffuses out of the wells to form insoluble complexes with antibodies. The size of the resulting precipitation ring surrounding various dilutions of antigen is proportional to the amount of antigen in the well (the wider the ring, the greater the antigen concentration). This is because antigen concentration drops as it diffuses farther out into the agar. The antigen forms a precipitin ring in the agar when its level has decreased sufficiently to reach equivalence and combine with the antibody to produce a large, insoluble network. This method can be used to quantitate serum immunoglobulins, complement proteins, and other substances.

The **double diffusion agar assay (Öuchterlony technique)** is based on the principle that diffusion of both antibody and antigen (hence double diffusion) through agar form stable and easily observable immune complexes. Test solutions of antigen and antibody are added to separate wells punched in agar. The solutions diffuse outward, and when antigen and the appropriate antibody meet, they combine and precipitate at the equivalence zone, producing an indicator line (or lines)

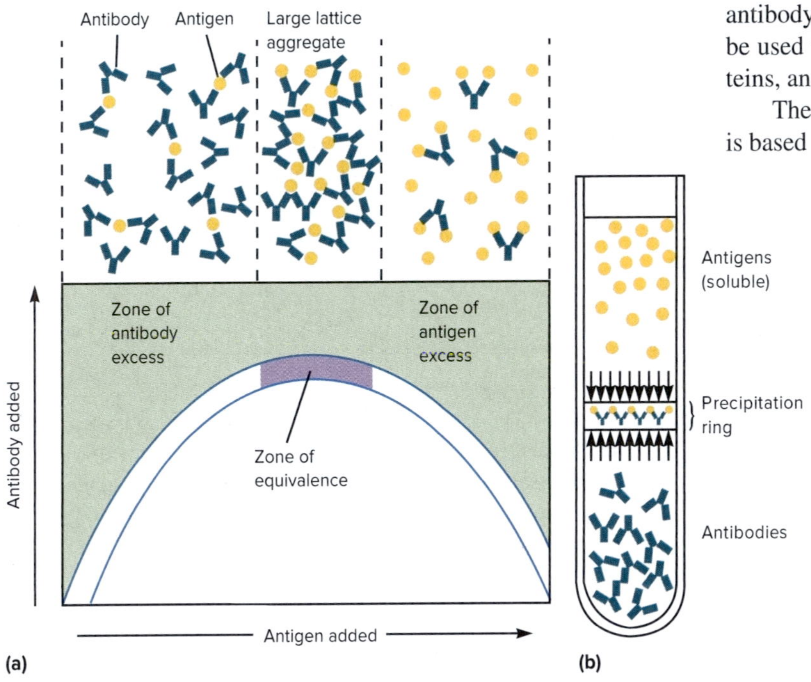

Figure 37.13 Immunoprecipitation. (a) Graph showing that a precipitation curve is based on the ratio of antigen to antibody. The zone of equivalence represents the optimal ratio for precipitation. (b) A precipitation ring test. Antibodies and antigens diffuse toward each other in a test tube. A precipitation ring is formed at the zone of equivalence.

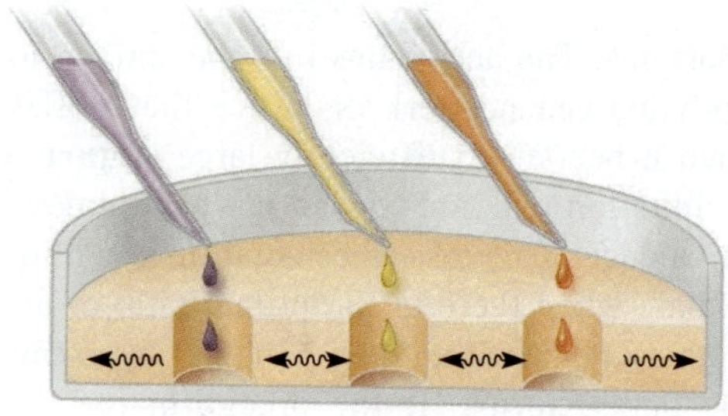

(a) In one method of setting up a double-diffusion test, wells are punctured in soft agar, and antibodies and antigens are added to separate wells in a pattern. As the contents of the wells diffuse toward each other, a number of reactions can result, depending on whether antibodies meet and precipitate antigens.

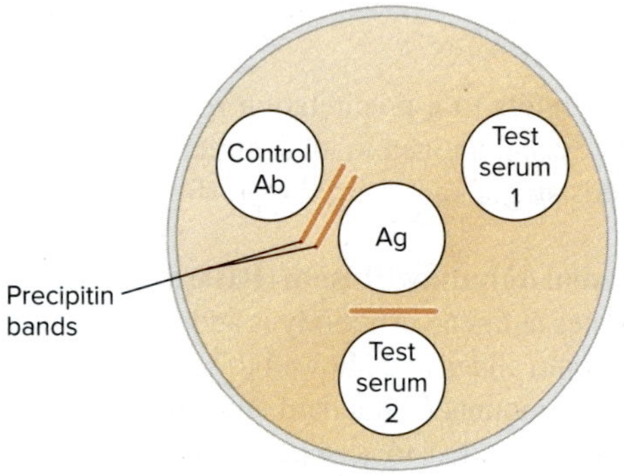

(b) Example of test pattern and results. Antigen (Ag) is placed in the center well and antibody (Ab) samples are placed in outer wells. The control contains known Abs to the test Ag. Note bands that form where Ab and Ag meet. The other wells (1, 2) contain unknown test sera. One is positive and the other is negative. Double bands indicate more than one antigen and antibody that can react.

Figure 37.14 Immunodiffusion. Double diffusion agar assay showing characteristics of identity (test serum 2) and a reaction of nonidentity (test serum 1).

(**figure 37.14**). The visible line of precipitation permits a comparison of antigens for identity (same antigenic determinants), partial identity, or nonidentity against a given selected antibody. Immunodiffusion assays are most commonly used to confirm certain fungal infections.

Flow Cytometry

Flow cytometry allows detection of specific lymphocyte subsets ($CD4^+$ T cells, for example, in HIV-infected blood) in an easy, reliable, fast way. Flow cytometry is described in chapter 29 (*see figure 29.4*). Here we emphasize that it has become a routine technique for counting $CD4^+$ and $CD8^+$ lymphocytes, as well as lymphocytes in varying stages of development as needed in the diagnosis of leukemia and lymphoma. This technique has also enabled the development of quantitative methods to assess antimicrobial susceptibility and drug cytotoxicity in a rapid, accurate, and highly reproducible way.

Radioimmunoassay

The ***radioimmunoassay*** (**RIA**) technique is an important tool in biomedical research and clinical practice (e.g., in cardiology, blood banking, diagnosis of allergies, and endocrinology). Indeed, Rosalyn Yalow (1921–2011) won the 1977 Nobel Prize in Physiology or Medicine for its development. RIA uses a purified antigen that is radioisotope-labeled. It competes for antibody with unlabeled standard antigen or test antigen in experimental samples. The radioactivity associated with the antibody is then detected by means of radioisotope analyzers and autoradiography (photographic emulsions that show areas of radioactivity). If there is a lot of antigen in an experimental sample, it will compete with the radioisotope-labeled antigen for antigen-binding sites on the antibody, and little radioactivity will be bound. A large amount of bound radioactivity indicates that little antigen is present in the experimental sample.

Comprehension Check

1. Why is serology an indirect means of detecting infectious disease?
2. What is the difference between viral and antibody-mediated hemagglutination?
3. You perform a complement fixation test and find that the test cells do not lyse. Is this a positive or negative result? Explain.
4. What does an indirect ELISA detect? How does this differ from a direct ELISA?
5. What is the zone of equivalence and why is it important in immunoprecipitation reactions?
6. Name two types of immunodiffusion tests and describe how they operate.

Key Concepts

37.1 The Clinical Microbiology Laboratory Detects Infectious Agents and Protects Its Workers

- Clinical microbiologists provide the rapid and accurate identification of disease-causing microorganisms from clinical specimens and the accurate antimicrobial susceptibility testing of isolated organisms.
- A variety of laboratory tests and techniques are used (**figure 37.1**). Many traditional bench tests have been miniaturized, automated, or replaced by faster and more sensitive tests.
- Over 50 years of pioneering work led to protocols and equipment to protect laboratory workers.

- To prevent laboratory-associated infections, scientists follow standard microbiology practices and wear personal protective equipment (**figure 37.2** and **tables 37.1–37.3**).

37.2 Identification of Microorganisms from Specimens

- Detection of disease-causing organisms begins with the collection of patient specimens. The correct selection of body fluid or tissue, and subsequent collection, transport, and preanalytical processing of the specimen are paramount for accurately identifying infecting agents.
- The clinical microbiology laboratory can provide preliminary or definitive identification of microorganisms from sterile site and nonsterile site specimens. A thorough knowledge of the normal human microbiota is necessary to identify pathogens from host organisms.
- The type of suspected microorganisms dictates which tests are performed. Bacteria and fungi are often grown in culture to purify and amplify the pathogen for identification. Viruses and protozoa are typically identified by immunologic or molecular tests.
- Identification of pathogens is accomplished by (1) direct recovery and (2) indirect or circumstantial evidence of a pathogen. Direct recovery of infectious agents is by culture and biochemical analysis, as well as molecular techniques. The culturing of many types of bacteria and fungi (filamentous and yeast forms) is routine; however, a few require special consideration.
- Identification of bacteria, fungi, and protozoa often can be made by concentrating and staining the organisms. Gram stain for bacteria and calcofluor white for fungi permit detection using light microscopy.
- The initial identity of a bacterial organism may be suggested by (1) the source of the culture specimen; (2) its microscopic appearance; (3) its pattern of growth on selective, differential, enrichment, or characteristic media; and (4) its hemolytic, metabolic, and fermentative properties. A simple dichotomous key is used to facilitate the identification (**figure 37.4**).
- Immunofluorescence is a process in which fluorochromes are excited with ultraviolet, violet, or blue light to make them fluoresce. These dyes can be coupled to an antibody. There are two main kinds of fluorescent antibody assays: direct and indirect (**figure 37.6**).
- Various molecular methods are mostly used to identify microorganisms directly from specimens. Complementary methods like real-time PCR, nucleic acid fingerprinting, PNA-FISH, and MALDI-TOF are used on culture isolates to speed identification.

37.3 Immune Responses Can Be Exploited to Detect Infections

- Serotyping refers to procedures used to differentiate strains (serovars or serotypes) of microorganisms that have differences in surface antigenic composition.
- In vitro agglutination reactions usually form aggregates or clumps (agglutinates) visible with the naked eye. Tests have been developed, such as the Rapid Plasma Reagin test for syphilis, latex microsphere agglutination reaction, hemagglutination, and viral hemagglutination, to detect antigen as well as to determine antibody titer (**figures 37.8** and **37.9**).
- The complement fixation test can be used to detect a specific antibody for a suspect microorganism in an individual's serum (**figure 37.10**).
- The enzyme-linked immunosorbent assay (ELISA) involves linking various enzymes to either antigens or antibodies. Two basic methods are involved: the double antibody sandwich method and the indirect immunosorbent assay (**figure 37.11**). The first method detects antigens and the latter, antibodies.
- Lateral flow assays are inexpensive easy tests that use antigen-antibody reactions to rapidly detect either antibody or antigen in patient samples (**figure 37.12**).
- Immunoblotting involves polyacrylamide gel electrophoresis of a protein specimen followed by transfer of the separated proteins to filter paper and identification of specific bands by labeled antibodies.
- Immunoprecipitation reactions occur only when there is an optimal ratio of antigen and antibody to produce a lattice at the zone of equivalence, which is evidenced by a visible precipitate (**figure 37.13**).
- Immunodiffusion refers to a precipitation reaction that occurs between antibody and antigen in an agar gel medium. Two techniques are routinely used: single radial diffusion and double diffusion in agar (**figure 37.14**).
- Flow cytometry enables detection of specific cell types, such as $CD4^+$ T cells in an HIV-infected patient, based on their cytometric parameters or by means of fluorochromes.

Active Learning

1. ELISA tests usually use a primary and secondary antibody. Why? What are the necessary controls one would need to perform to ensure that the antibody specificities are valid (i.e., no false-positive or false-negative reactions)?
2. *Legionella pneumophila* is a bacterium that is often found in water systems (e.g., shower heads, air-cooling towers). The dispersal of these bacteria can lead to the development of Legionnaires' disease, a particularly virulent pneumonia in the elderly and immunocompromised. Water systems are routinely monitored for *L. pneumophila* contamination using GVPC medium, which contains glycine and the antibiotics vancomycin, polymyxin B, and cyclohexamide. However, this method is known to underestimate the number of *L. pneumophila* cells in the environment.

   Suggest another method to monitor *L. pneumophila* in the environment? Discuss how would you collect your samples and compare your results to those obtained by plating on GVPC medium. Remember to discuss your positive and negative controls.
3. *Coccidioides immitis* is a common cause of community acquired fungal pneumonia in the southwestern United States. Complement fixation assays are used by some labs to detect infection, but a more rapid and direct assay is needed. A group of clinical microbiologists designed an experiment to determine the accuracy of IgG and IgM detection using a novel, semiautomated anti-*Coccidioides* antibody enzyme immunoassay. They wanted to determine the sensitivity and specificity of this immunoassay. What is meant by sensitivity and specificity? Discuss how you would design an experiment to answer these questions. Keep in mind your selection of positive and negative controls.

   Read the original paper: Malo, J., et al. 2017. Enhanced antibody detection and diagnosis of coccidioidomycosis with the MiraVista IgG and IgM detection enzyme immunoassay. *J. Clin. Microbiol.* 55:893–901. doi.org/10.1128/JCM.01880-16.

38

Human Diseases Caused by Viruses and Prions

Source: Photographs in the Carol M. Highsmith Archive, Library of Congress, Prints and Photographs Division

Remembering HIV/AIDS

If you are young, you do not remember the early days of human immunodeficiency virus (HIV) when large swaths of communities died. You do not remember how a young hemophiliac named Ryan White had to fight, and then move to a new community, to attend middle school. You do not remember highly visible public statements that HIV/AIDS was God's punishment. You do not remember high-level U.S. government officials who would not mention the term "HIV/AIDS" or the South African president who told his citizens that HIV did not cause AIDS and denied access to drugs that would prevent maternal-fetal transmission. You probably don't remember the quilt that toured the United States (shown here), each patch telling the story of a life cut short.

That's because you were probably born after 1996, when highly effective drug cocktails were introduced and the Joint United Nations Program on HIV/AIDS (UNAIDS) was formed. UNAIDS approached (and continues to approach) HIV education, screening, and treatment as a human right—a first for a disease. If one compares the early days of the HIV pandemic, which started in 1981 with the first reports of gay men suffering from an unusual fungal pneumonia caused by *Pneumocystis jiroveci* and a cancer called Kaposi's sarcoma, to the post-1996 era, it is obvious that a lot has changed. Public fear of HIV is (mostly) a thing of the past, and HIV can now be treated as a chronic disease.

HIV has also brought changes that are not so obvious. For example, HIV research led to much of our current understanding of the immune system, which in turn is yielding new and promising cancer treatments. HIV disrupted the pharmaceutical industry as developing nations began manufacturing their own lifesaving antiretroviral drugs that were still under patent protection and thus far too expensive to provide to their citizens. The global commitment to address HIV grew from $250 million in 1996 to over $10 billion by 2007; and in 2013, the UNAIDS reported a 30% decline in new HIV cases since its peak in 2005. In 2016 over 18 million people were treated, including almost 1 million children.

The HIV pandemic tells us a lot about viruses that seemingly emerge from nowhere. As new viral illnesses emerge, some will be caused by known viruses, as was the case with West Nile virus and Zika, but others will be entirely novel, like severe acute respiratory syndrome (SARS) and HIV. Most, like the SARS virus, will "burn out" through a combination of preventative measures and mutation. Others, like Zika virus, will cause devastating illness before they are brought under control. And others, like HIV, will cause pandemics and global crises. In all cases, the destruction viruses cause seems incongruent with their size and relative simplicity.

Chapters 6 and 26 review the general biology of viruses and introduce basic virology. In chapter 38, we continue this coverage by discussing some of the most important viruses that are human pathogens. We group viral diseases according to their mode of acquisition and transmission; viral diseases that occur in the United States are emphasized.

Readiness Check:

Based on what you have learned previously, you should be able to:

- ✓ Review basic virology (sections 6.1–6.6) and prion biochemistry (section 6.7)
- ✓ List the major features of each group of viruses in the Baltimore system of viral classification (chapter 26)
- ✓ Explain pathogenicity and the infection process (chapter 35)

38.1 Viruses Can Be Transmitted by Airborne Routes

After reading this section, you should be able to:

a. Discuss the viruses that cause common diseases spread by airborne transmission
b. Identify typical signs and symptoms of viral diseases spread by airborne transmission
c. Correlate airborne viral infection and disease severity with viral virulence factors

Because air does not support virus multiplication, any virus that is airborne originated from a living source. When humans are the source of the airborne virus, it usually is propelled from the respiratory tract by coughing, sneezing, or vocalizing.

Chickenpox (Varicella) and Shingles (Herpes Zoster)

Chickenpox (varicella) is a highly contagious disease primarily of children 2–7 years of age. Humans are the reservoir and source for this virus, which is acquired by droplet inhalation into the respiratory system (**figure 38.1*a***). The virus is highly infectious, with transmission rates in susceptible household contacts of 65 to 86%. In the prevaccine era, about 4 million cases of chickenpox occurred annually in the United States, resulting in approximately 11,000 hospitalizations and 100 to 150 deaths.

The causative agent is the enveloped, double-stranded (ds) DNA varicella-zoster virus (VZV; species *Human herpesvirus 3*), a member of the family *Herpesviridae.* The virus produces at least six glycoproteins that play a role in viral attachment to specific receptors on respiratory epithelial cells. Following an incubation period of 10 to 23 days, small vesicles erupt on the face or upper trunk, fill with pus, rupture, and become covered by scabs (figure 38.1*b*). Healing of the vesicles occurs in about 10 days. During this time, intense itching often occurs.

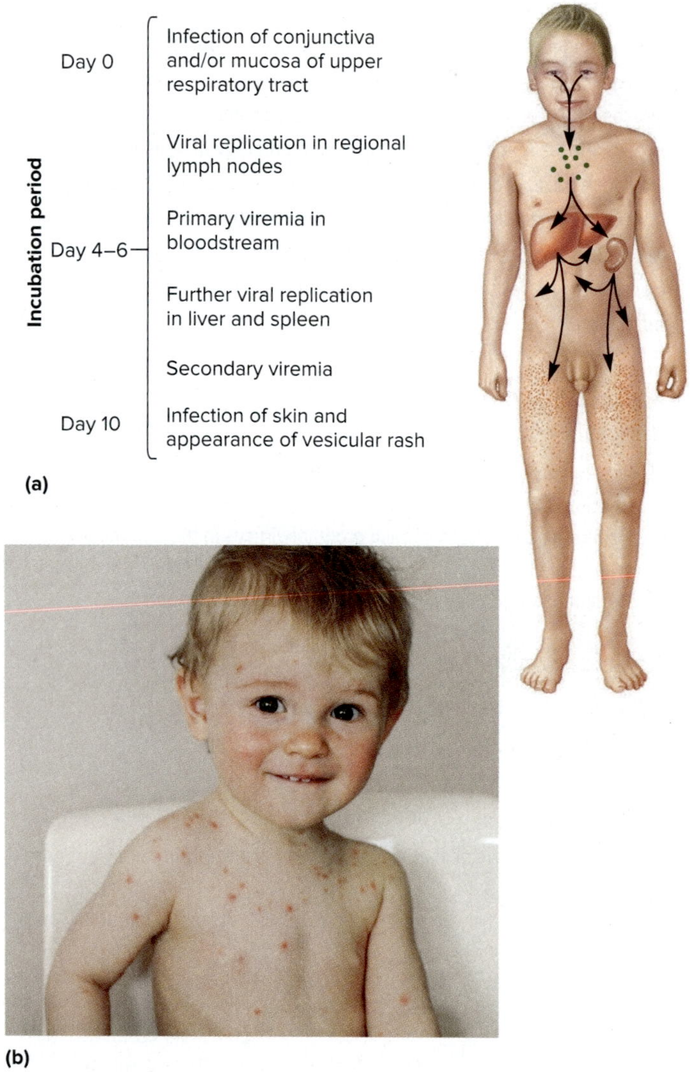

Figure 38.1 Chickenpox (Varicella). (a) Course of infection. Primary viremia occurs when virions first enter the bloodstream from the initial site of infection (in this case, lungs). Secondary virema refers to the spread of virions to other body sites. (b) Typical vesicular skin rash. This rash occurs all over the body but is heaviest on the trunk.

(b) ©Asisyaj/Shutterstock

Laboratory testing for VZV is not normally required, as the diagnosis of chickenpox is typically made by clinical assessment. Laboratory confirmation is recommended, though, to confirm the diagnosis of severe or unusual cases of chickenpox. The introduction of the attenuated varicella vaccine (*see table 36.3*) in the 1990s has drastically reduced the incidence of chickenpox. However, because fewer cases are seen, the likelihood of misdiagnosis is increased. Furthermore, in persons who have previously received varicella vaccination, chickenpox is still possible, as immunity wanes. In these individuals, mild or atypical chickenpox symptoms may occur, posing challenges for clinical diagnosis.

Individuals who recover from chickenpox are subsequently immune to this disease; however, they are not free of the virus, as viral DNA resides as a circular episome in a dormant (latent) state within the nuclei of cranial nerves and sensory neurons in the dorsal root ganglia. Although viral DNA is maintained in infected cells, virions cannot be detected (**figure 38.2*a***). When the infected person becomes immunocompromised by such factors as age, cancer, organ transplants, HIV, or psychological or physiological stress, the viruses may become activated and begin replication (figure 38.2*b*). They migrate down sensory nerves, replicate further, and produce painful vesicles and sensory nerve damage (figure 38.2*c*). This condition, known as **shingles** or **herpes zoster,** is characterized by vesicles in a stripe around the torso or one half of the face. This pattern reflects the infected nerve and is described as a dermatomal distribution. Shingles is most common in people over age 50. In some cases, pain persists even after the vesicles have healed, a disabling complication called **postherpetic neuralgia.**

The antiviral drug acyclovir or its derivative valacyclovir can be used to shorten the duration of a shingles outbreak. Antiviral medication is used to treat chickenpox only in severe cases or in immunocompromised patients. A vaccine for shingles is recommended for people 60 years and older. More than 1 million cases of herpes zoster occur annually in the United States with about 95 deaths.

Influenza (Flu)

Influenza (Italian *un influenza di freddo,* to be influenced by the cold), or the flu, was described by Hippocrates in 412 BCE. The first well-documented global epidemic of influenza-like disease occurred in 1580. Since then, 31 possible influenza pandemics have been documented, with four occurring in the twentieth century. The worst pandemic on record occurred in 1918 and killed at least 50 million people around the world. This disaster, traced to the Spanish influenza virus, was followed by pandemics of Asian flu (1957), Hong Kong flu (1968), Russian flu (1977), and swine flu (2009). These names reflect popular impressions of where or from what the episodes began, although all are now thought to have originated in China. ◀ *Minus-strand RNA viruses: RNA-dependent RNA polymerase is part of the virion (section 26.6)*

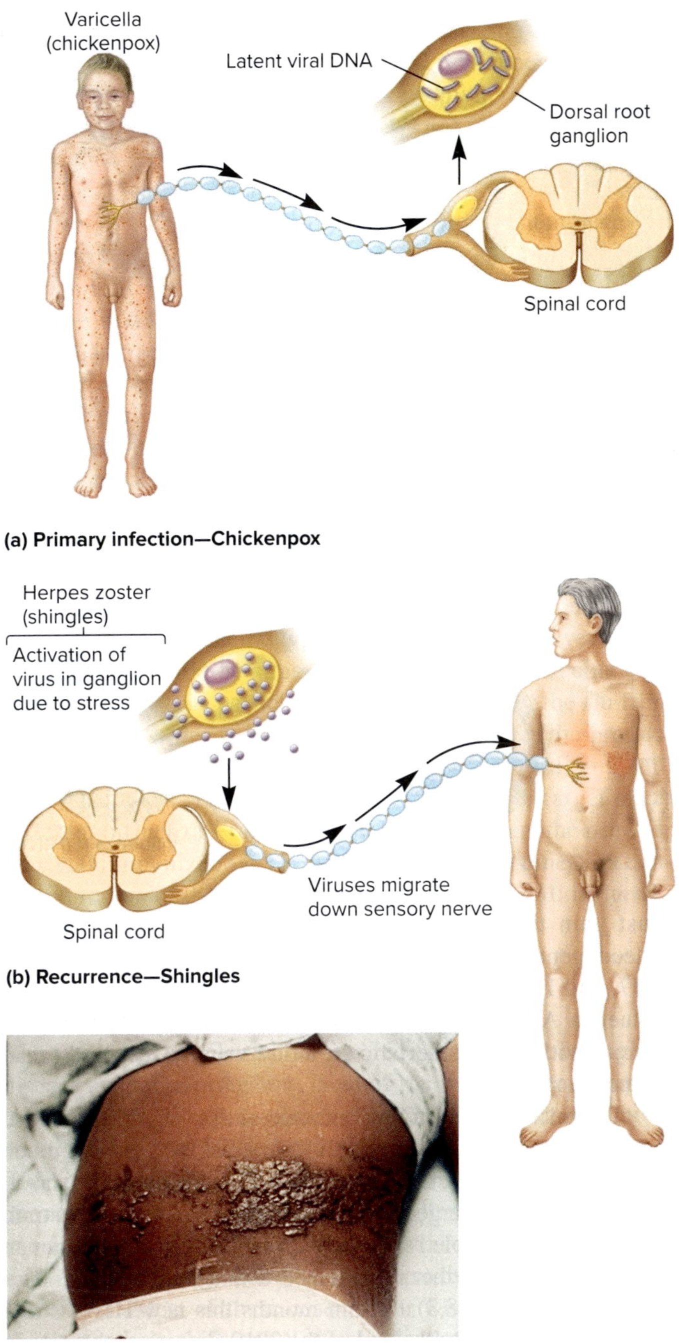

Figure 38.2 Pathogenesis of the Varicella-Zoster Virus. (a) After initial infection with varicella (chickenpox), the viruses migrate up sensory peripheral nerves to the dorsal root ganglia, producing a latent infection. (b) When a person becomes immunocompromised or is under psychological or physiological stress, the viruses may be activated. (c) They migrate down sensory nerve axons, replicate further, and produce painful vesicles. Because these vesicles usually appear around the trunk of the body, the name *zoster* (Greek for girdle) is used.

(c) Source: CDC

MICRO INQUIRY *Where does viral DNA reside during latent infection? Can virions be detected at this time?*

Influenza is a respiratory system disease caused by negative-strand RNA viruses that belong to the family *Orthomyxoviridae* (*see figure 26.24*). The three genera are *Influenzavirus A, Influenzavirus B,* and *Influenzavirus C*; only types A and B cause significant human disease. Influenza viruses are widely distributed in nature, infecting a variety of mammal and bird hosts. All influenza viruses are acquired by inhalation or ingestion of virus-contaminated respiratory secretions. During an incubation period of 1 to 2 days, viral particles adhere to host respiratory epithelium and initiate the replication cycle (*see figure 26.24c*). Two envelope spikes, hemagglutinin (HA) and neuraminidase (NA) play critical roles. HA is the adherence factor that binds host cells and thereby triggers receptor-mediated endocytosis. This is facilitated by NA, which is thought to hydrolyze the mucus that covers the epithelium. Once within the cytoplasm enclosed in an endosome, HA again functions, this time to bring about release of the virus's seven to eight nucleocapsids into the cytosol. To do this, HA undergoes a dramatic conformational change when the endosomal pH decreases. The hydrophobic ends of the hemagglutinin spring outward and extend toward the endosomal membrane. After they contact the membrane, fusion occurs and the RNA nucleocapsids are released into the cytoplasm. Each influenza virus nucleocapsid consists of a single negative-strand RNA molecule associated with nucleoprotein (NP) and three subunits of the RNA-dependent RNA polymerase (PA, PB1, and PB2; **figure 38.3**). NA also has an additional role in the replication cycle. It aids in the release of newly assembled influenza virus virions from infected cells by cleaving sialic acid residues between virus and host membranes.

Influenza viruses are classified into subtypes based on their HA and NA. There are 18 HA (H1–H18) and 11 NA (N1–N11) antigenic forms known; they recombine to produce the various HA/NA subtypes. Host specificity can be explained in part by the difference in receptor binding. The HA proteins of human influenza viruses preferentially attach to receptors on human tracheal epithelial cells. In contrast, the HA proteins of avian influenza viruses preferentially bind receptors found on intestinal epithelial cells of waterfowl, the main viral replication site of influenza viruses in birds.

The genomes of influenza viruses are linear RNA segments that change frequently due to point mutations and reassortment (figure 38.3). As RNA viruses, influenza viruses must use an RNA-dependent RNA polymerase (RdRp) to replicate their genomes. This gives rise to a high mutation rate because RdRps lack proofreading capabilities. These smaller genetic changes often result in subtle protein (antigen) changes, known as **antigenic drift.** Antigenic drift results from the accumulation of mutations in a single strain of flu virus and is occasionally responsible for a mismatch between one of the flu strains used in the seasonal vaccine and the circulating strain. This occurs about every 3 to 5 years, causing local increases in the number of flu infections. **Antigenic shift,** however, is a larger protein sequence change made possible by the linear, segmented nature of the RNA

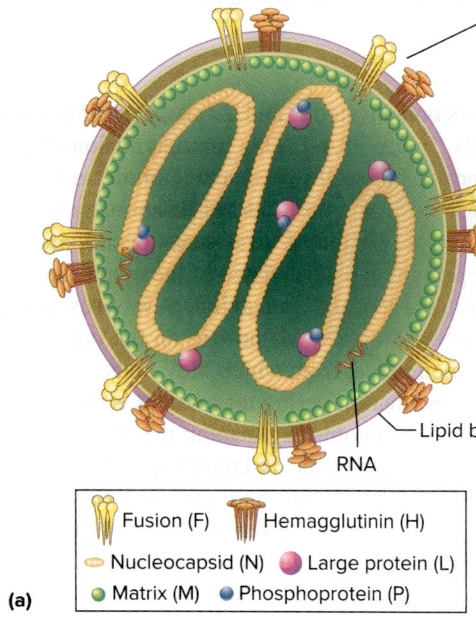

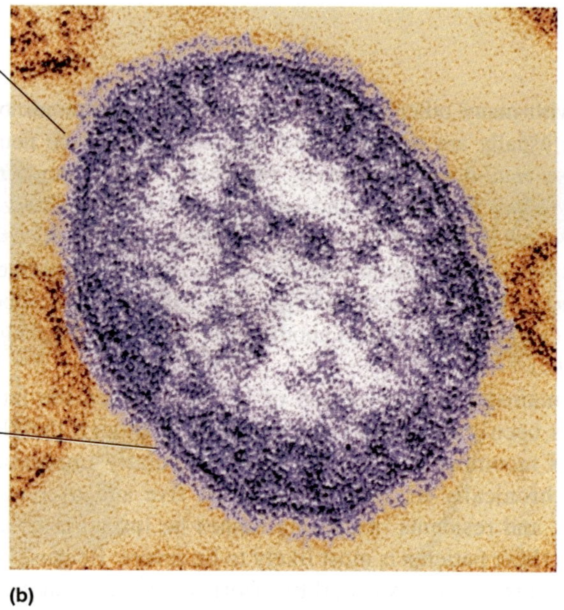

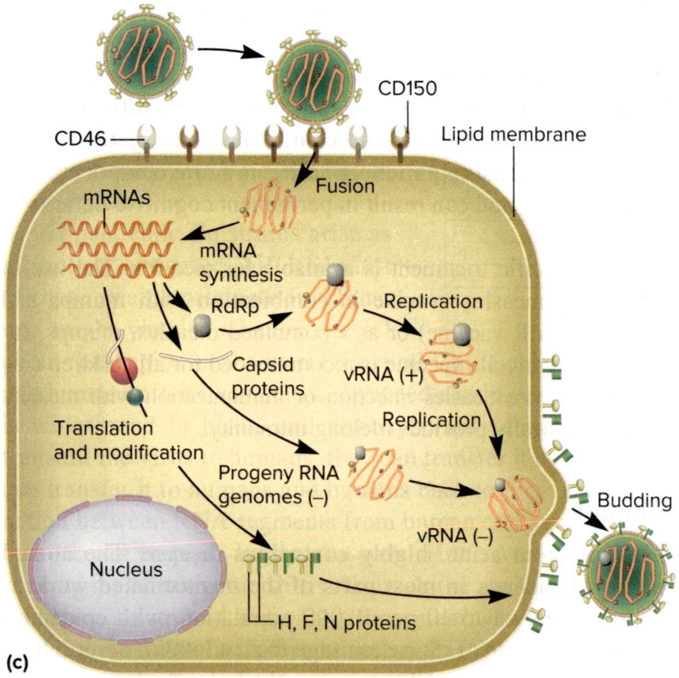

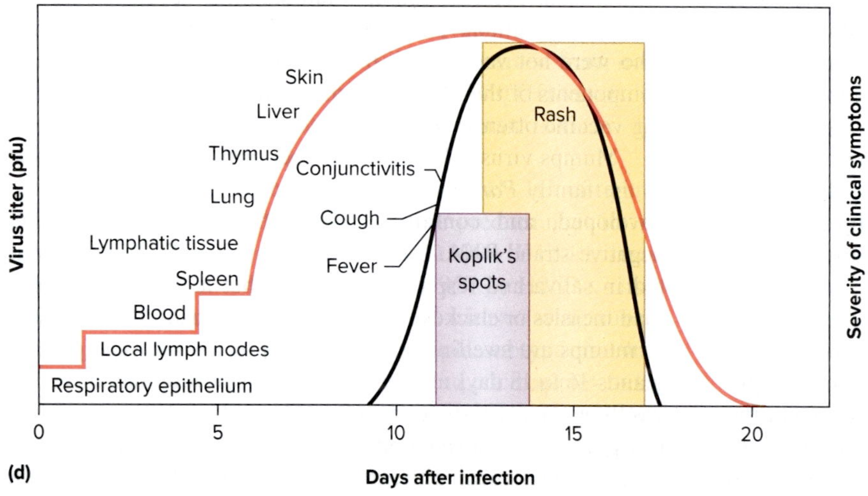

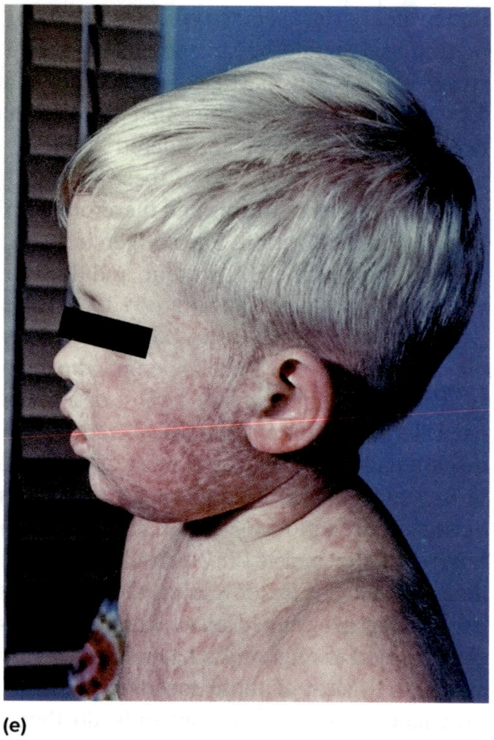

Figure 38.4 Measles (Rubeola). (a) Schematic of the spherical, RNA morbillivirus that causes measles. (b) Colorized transmission electron micrograph of a measles virus (MeV) virion (virion is 100 to 200 nm in diameter). (c) Life cycle of MeV. MeV virions exit the host by budding. (d) Severity of measles symptoms (black line) corresponds to the viral load (red line) in the tissue sites. (e) The rash of small, raised spots is typical of measles. The rash usually begins on the face and moves downward to the trunk.

(b) Source: CDC/Cynthia Goldsmith; (e) Source: CDC

epididymis and testes (orchitis) sometimes leading to sterility can occur, especially in postpubescent males. Therapy is limited to symptomatic and supportive measures.

Respiratory Syndromes and Viral Pneumonia

Acute viral infections of the respiratory system are among the most common causes of human disease. The infectious agents are called acute respiratory viruses, and they collectively produce a variety of clinical manifestations, including rhinitis (inflammation of the mucous membranes of the nose), tonsillitis, laryngitis, and bronchitis. Adenoviruses, coronaviruses, picornoviruses (rhinoviruses, coxsackieviruses A and B, and echovirus), influenza viruses, parainfluenza viruses, respiratory syncytial virus, and human metapneumovirus can all cause respiratory illness. Most of these viruses have many strains, so immunity is not complete, and reinfection is common. The best treatment is rest with symptom management.

Some viral infections of the respiratory system culminate in pneumonia. The clinical picture of viral pneumonia is nonspecific. Symptoms may be mild, or there may be severe illness and death. Importantly, secondary (subsequent) bacterial infections may also contribute to respiratory illness initiated by viruses. We highlight two coronaviruses and human respiratory syncytial virus as examples of these pathogens.

Coronaviruses are positive-strand RNA viruses named for the crownlike spikes on their surface. There are four main subtypes of coronaviruses: alpha, beta, gamma, and delta. Coronavirus virions are relatively large (120 to 150 nm), with a helical nucleocapsid surrounded by an envelope. Large spikes protrude from the envelope to aid in attachment and entry into host cells. The protruding envelope spikes extend from the oval-to-spherical virion to give the illusion of a halo, or corona, around the virus (**figure 38.5**). Most human coronaviruses are associated with common cold symptoms. However, two relatively new coronaviruses cause more severe illness: SARS-CoV and MERS-CoV.

Severe acute respiratory syndrome (SARS) was a highly contagious viral disease caused by the SARS betacoronavirus (SARS-CoV). SARS-CoV was first recognized in China in 2002, where it infected 8,098 people; it killed 774 worldwide in about a year. Since 2004 there have been no known cases of SARS-CoV. We report on it here as a great example of how a novel virus can emerge, wreak havoc on society, and burn itself out. The virus caused a febrile (38°C [100.4°F]) lower respiratory tract illness. SARS patients developed a dry cough after a few days, and most developed pneumonia. SARS was transmitted by close contact with respiratory secretions (droplet spread). No specific treatment is currently approved and no vaccine is licensed for preventative use. Molecular evidence suggests that amino acid substitution within the receptor-binding domain of the SARS membrane spike protein was likely the reason for reduced virulence of SARS-CoV infections late in 2004. This suggests that this mutation may be responsible for the apparent eradication of the SARS virus.

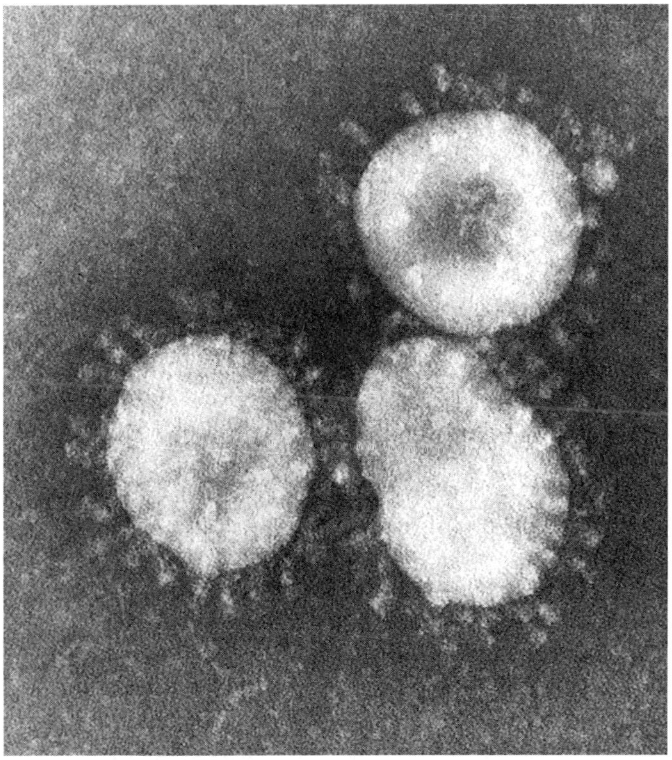

Figure 38.5 SARS Coronavirus (SARS-CoV). Coronaviruses are named for the crownlike (corona) appearance seen by electron microscopy.
Source: CDC/Dr. Fred Murphy

Middle East respiratory syndrome (MERS) is a relatively new viral respiratory illness of humans caused by the betacoronavirus MERS-CoV. It was first reported in Saudi Arabia in 2012 and has since spread to several other countries. Although it can be transmitted from person to person, MERS is also zoonotic, being transmitted to humans from dromedary camels. Both modes of transmission are responsible for illness in hundreds of people from several countries in and near the Arabian Peninsula. In 2015, an individual traveling from the Middle East sparked an outbreak in South Korea that eventually resulted in about 35 deaths. Most people infected with MERS-CoV develop severe acute respiratory illness, including fever, cough, and shortness of breath 4 to 5 days after exposure. Many of the infected also develop diarrhea. MERS has a 30 to 40% mortality rate. There is no currently approved antiviral treatment. Prevention is through diligent respiratory protection and thorough hand hygiene when exposure is possible.

Respiratory syncytial virus (RSV) often is described as the most dangerous cause of lower respiratory infections in infants (less than one year of age) and the elderly (age sixty-five and older). The virus is found worldwide and causes seasonal (November to March) outbreaks. In the United States, over 2.1 million children (under five years old) require outpatient visits, and between 100,000 and 126,000 infants are hospitalized each year for RSV. Additionally, there are 177,000 hospitalizations and 14,000 deaths among those sixty-five years or older.

RSV is a member of the negative-strand RNA virus family *Paramyxoviridae*. The virion is variable in shape and size (average diameter 120 to 300 nm). It is enveloped with two virus-specific glycoproteins as part of the structure. One of the two glycoproteins, G, is responsible for binding the virus to the host cell; the other, the fusion or F protein, permits fusion of the viral envelope with the host cell plasma membrane, leading to entry of the virus. The F protein also induces fusion of the plasma membranes of infected cells. Thus RSV gets its name from the resulting formation of a syncytium or multinucleated mass of fused cells. The syncytia are responsible for inflammation, alveolar thickening, and fluid accumulation in alveolar spaces. RSV is transmitted by direct contact with respiratory secretions of humans. Virions are unstable in the environment (surviving only a few hours on environmental surfaces) and are readily inactivated with soap and water, and disinfectants.

Clinical manifestations of RSV infection include acute onset of fever, cough, rhinitis, and nasal congestion. In infants and young children, this can progress to severe respiratory distress. Diagnosis is by either RSV rapid antigen test kit or RSV detection by reverse transcriptase PCR. Treatment is supportive and includes fluid replacement and fever reducers. Because premature infants are more likely to develop severe respiratory distress, the monoclonal antibody palivizumab that targets the F protein is given as a preventative measure to these infants.

Rubella (German Measles)

Rubella (Latin *rubellus*, little red) was first described in Germany in the 1800s giving rise to its moniker German measles. It is a contagious disease that occurs primarily in children five to nine years of age. It is caused by rubella virus, an enveloped, positive-strand RNA virus that is a member of the family *Togaviridae*. Rubella is spread in droplets shed from the respiratory secretions of infected individuals. Once the virus is inside the body, the incubation period ranges from 12 to 23 days. A rash of small red spots (**figure 38.6**), usually lasting no more than 3 days, and a low-grade fever are the normal symptoms. The rash appears as immunity develops and the virus disappears from the blood, suggesting that the rash is immunologically mediated and not caused by the virus infecting skin cells.

Although rubella is typically a mild disease, it can be disastrous if contracted by a woman in her first trimester of pregnancy. It can lead to fetal death, premature delivery, or a wide array of congenital defects called **congenital rubella syndrome.** These defects include damage to the heart, eyes, and ears, and brain damage that can lead to cognitive disabilities. Because there is no effective treatment for rubella, all children and women of childbearing age who have not been previously exposed to rubella should be vaccinated. The United States was declared rubella free in 2009. The live attenuated rubella vaccine (part of MMR or MMRV; *see table 36.3*) continues to be recommended.

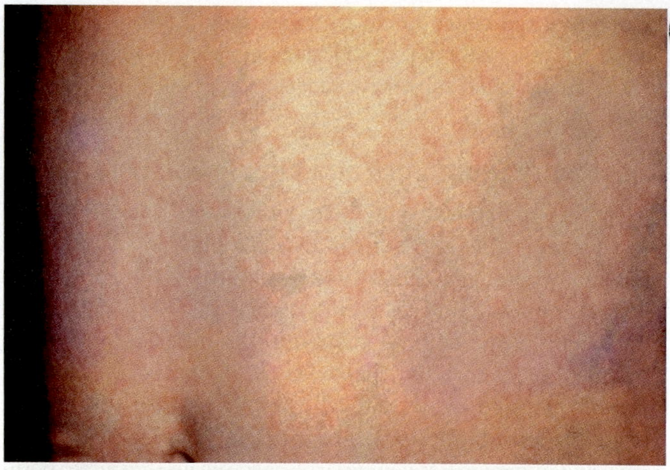

Figure 38.6 German Measles (Rubella). This disease is characterized by a rash of red spots. Spots are not raised above the surrounding skin as in measles (rubeola; see figure 38.4e).
Source: CDC

MICRO INQUIRY *Why is the characteristic rash not caused by viral infection of skin cells?*

Smallpox (Variola)

Smallpox (**variola**) is a highly contagious disease of antiquity. We highlight it because it is the only example of vaccine eradication of disease, and it is of concern should it be used in bioterrorism. Both eradication and bioterrorism are linked to the fact that smallpox is uniquely an illness of humans. It is caused by orthopoxviruses belonging to the species *Variola virus*. Variola viruses belong to the family *Poxviridae*, which also includes the species *Monkeypox virus*, *Molluscum contagiosum virus*, and *Vaccinia virus*. Vaccinia virus is the virus that serves as the antigen in the smallpox vaccine. Poxviruses are large, brick-shaped, and contain a dumbbell-shaped core (*see figure 6.6*). The genome consists of a single, linear molecule of double-stranded DNA that is replicated in the host cell's cytoplasm. The genome is composed of approximately 186 kilobase (kb) pairs with covalently closed ends (**figure 38.7*a***). Variola virus isolates have approximately 200 closely spaced and nonoverlapping genes. Interestingly, the size of the virion (300 by 250 to 200 nm) is slightly larger than that of some of the smallest bacteria (e.g., *Chlamydia* spp.).

Smallpox was once one of the most prevalent of all diseases. It was a universally dreaded scourge for more than 3 millennia, with case fatality rates of 20 to 50%; it disabled and disfigured those who survived. The disease begins with a prodrome of systemic aches. Characteristic symptoms of infection also include acute onset of fever, followed by a rash that features firm, deep-seated vesicles or pustules in the same stage of development (figure 38.7*b*). First subjected to some control in tenth-century India and China, it was gradually suppressed in the industrialized world after Edward Jenner's 1796 landmark discovery that infection with the harmless cowpox virus renders humans immune to the smallpox virus (*Historical Highlights 36.5*).

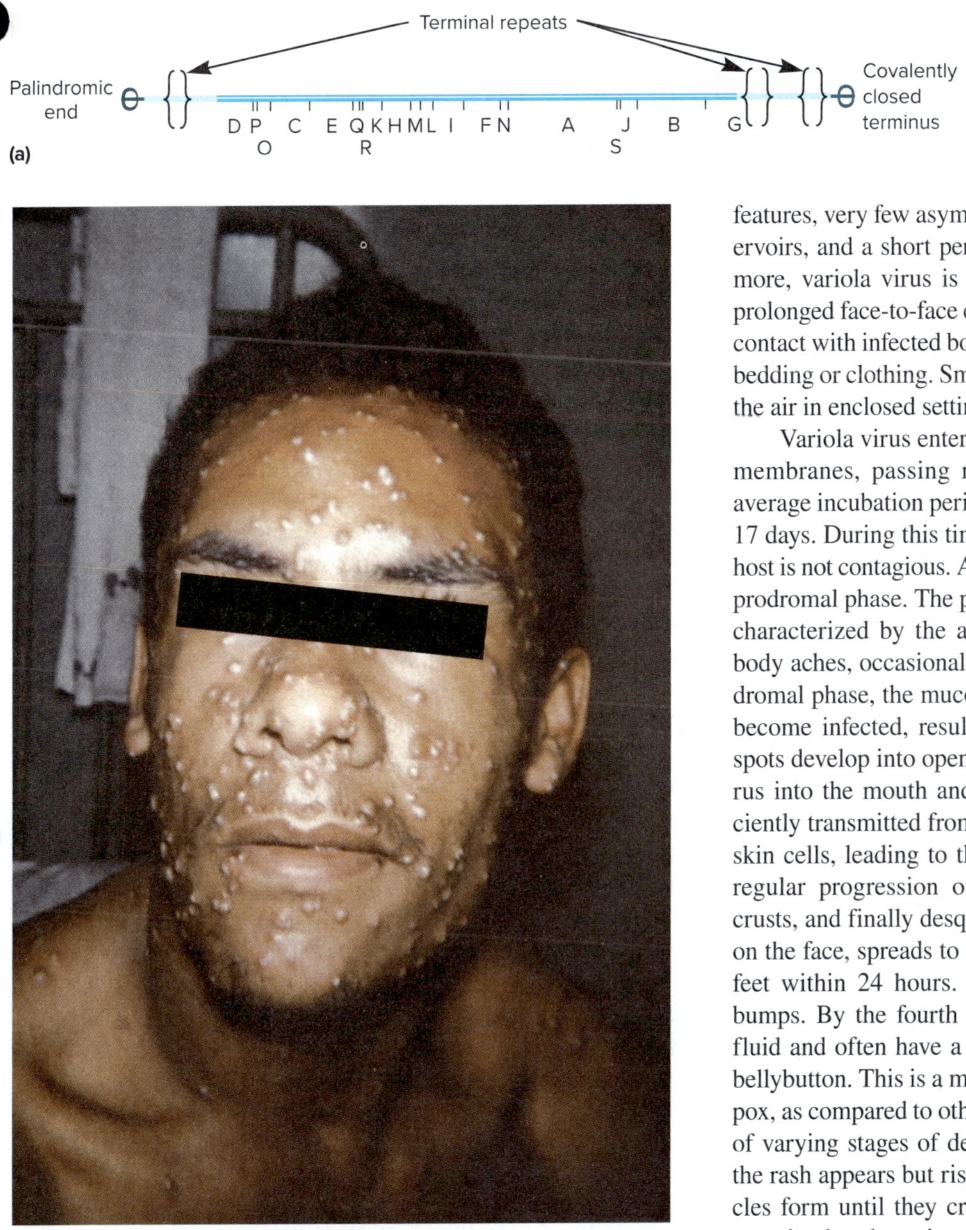

Figure 38.7 Variola Virus (VARV) Causes Smallpox.
(a) Genome architecture of VARV showing its linear, 186-kb pair, covalently closed DNA. (b) Smallpox rash demonstrating the virus-filled pustules. Pustules erupt at the same time all over the body.
(b) Source: CDC/Dr. Noble

MICRO INQUIRY *What specific clinical and biological features of smallpox made its eradication possible?*

Protection from smallpox was through vaccination. The smallpox vaccine contains live vaccinia virus. Through the concerted efforts by the WHO, smallpox was eradicated throughout the world—the greatest public health achievement ever. The last case from a natural infection occurred in Somalia in 1977. The infected man was part of the vaccination program, but he failed to make neutralizing antibodies when he was vaccinated (i.e., his vaccine did not "take"). Fortunately he survived the disease and died in 2013 of other causes. Eradication was possible because smallpox has obvious clinical features, very few asymptomatic carriers, only human hosts as reservoirs, and a short period of infectivity (3 to 4 weeks). Furthermore, variola virus is generally transmitted by direct and fairly prolonged face-to-face contact. It also can be spread through direct contact with infected bodily fluids or contaminated objects such as bedding or clothing. Smallpox has been reported to spread through the air in enclosed settings such as buildings, buses, and trains.

Variola virus enters the respiratory tract, seeding the mucous membranes, passing rapidly into regional lymph nodes. The average incubation period is 12 to 14 days but can range from 7 to 17 days. During this time, the virus multiplies in the host, but the host is not contagious. Another brief period of viremia precedes the prodromal phase. The prodromal phase lasts for 2 to 4 days and is characterized by the abrupt onset of malaise, severe head and body aches, occasional vomiting, and high fever. During the prodromal phase, the mucous membranes in the mouth and pharynx become infected, resulting in a rash of small, red spots. These spots develop into open sores that spread large amounts of the virus into the mouth and throat. This allows the virus to be efficiently transmitted from person to person. The virus then invades skin cells, leading to the development of lesions that undergo a regular progression of eruptions, papules, vesicles, pustules, crusts, and finally desquamation (figure 38.7*b*). The rash appears on the face, spreads to the arms and legs, and then the hands and feet within 24 hours. By the third day, the rash forms raised bumps. By the fourth day, the bumps fill with a thick, opaque fluid and often have a depression in the center that looks like a bellybutton. This is a major distinguishing characteristic of smallpox, as compared to other diseases that exhibit rashes with lesions of varying stages of development. The fever usually declines as the rash appears but rises and is sustained from the time the vesicles form until they crust over. Oropharyngeal and skin lesions contain abundant virus particles, particularly early in the disease process. Death from smallpox is due to toxemia associated with immune-mediated blood clots and elevated blood pressure.

Routine immunization for smallpox was discontinued in the United States in 1972. An accidental or deliberate release of the virus would be catastrophic in an unimmunized population and could cause a major pandemic. Thus if an outbreak occurred, prompt recognition and institution of control measures would be paramount as there is no FDA-approved treatment for smallpox.
◀ *Bioterrorism readiness is an integral component of public health microbiology (section 36.7)*

Comprehension Check

1. Compare and contrast chickenpox and shingles.
2. What is the difference between antigenic shift and antigenic drift?
3. What are some common symptoms of measles? When are individuals most contagious?

4. What is one side effect that mumps can cause in a young postpubescent male?
5. Describe some clinical manifestations caused by acute respiratory viruses.
6. When is rubella most dangerous and why?
7. Both VZV and smallpox are called "pox diseases." Compare the biology of these viruses and their disease manifestations and explain why you think this is the case.

38.2 Arthropods Can Transmit Viral Diseases

After reading this section, you should be able to:

a. Discuss the viruses that cause common arthropod-borne diseases
b. Identify typical signs and symptoms of two arthropod-borne viral diseases
c. Correlate arthropod-borne virus infection with geography and time of year

*Ar*thropod-*bo*rne viruses (arboviruses) are transmitted by bloodsucking arthropods (e.g., ticks, mosquitoes, certain types of flies) from one vertebrate host to another. They multiply in the tissues of their arthropod hosts without producing disease. Approximately 150 arboviruses cause illness in humans. Diseases produced by arboviruses can be divided into three clinical syndromes: (1) fevers of an undifferentiated type with or without a rash; (2) encephalitis (inflammation of the brain), often with a high fatality rate; and (3) hemorrhagic fevers, also frequently severe and fatal. No antiviral therapy is effective in treating arbovirus diseases, although supportive treatment is beneficial. Detection of these viruses is by nucleic acid amplification and serology.

Chikungunya

Chikungunya virus, a member of *Togaviridae*, is a positive-strand, enveloped RNA virus. Until recently, chikungunya virus was contained in Africa. However in 2007, it spread to Europe and in 2013 it was found on Caribbean Islands; by 2015, chikungunya virus disease cases had been reported in 32 U.S. states, Puerto Rico, and the U.S. Virgin Islands. The virus is transmitted to humans by *Aedes aegypti* and *Aedes albopictus* mosquitoes.

Acute chikungunya virus infection in humans often includes fever and joint pain. Other symptoms may include headache, muscle pain, joint swelling, or rash. These symptoms often resolve in 7 to 10 days. However, many patients report prolonged joint pain persisting for months to years. Rare complications include eye inflammation, myocarditis, hepatitis, nephritis, hemorrhage, encephalitis, and Guillain-Barré syndrome (GBS). GBS is thought to arise when the immune system mistakenly attacks peripheral nerves instead of the infecting pathogen (it is also seen after Zika infection and food poisoning with the bacterium *Campylobacter jejuni*).

There is no medicine to treat, or vaccine to prevent, chikungunya virus infection. Travelers to endemic areas should follow guidelines to prevent mosquito-borne infectious disease; they should use insect repellent, wear light-colored clothes with long sleeves and pants, and stay in places with mosquito netting, window and door screens, and fans.

Dengue and Dengue Hemorrhagic Fever

Dengue virus (DENV) is a member of the *Flaviviridae* family, and as such is an enveloped, positive-strand RNA virus. Its distribution is worldwide and dengue is endemic in over 100 countries where its mosquito vector, *Aedes aegypti,* is found. It can also be transmitted by *A. albopictus,* although not as efficiently. The WHO estimates there are about 400 million cases of dengue annually worldwide, of which roughly 3.2 million progress to dengue hemorrhagic fever (DHF). Each year, DHF takes the lives of approximately 9,000 people, mostly children.

There are four antigenically distinct serotypes: DENV-1 through DENV-4. All have a genome of about 110,000 bases encoding a single polypeptide that is cleaved into seven nonstructural proteins and three structural proteins—the capsid protein, the envelope glycoprotein (E), and the membrane protein. The E protein is responsible for attachment of the virion to host cells, which is followed by endocytosis and entry into the host. Following replication and packaging, the virus buds from the host cell's endoplasmic reticulum prior to exiting the cell.

The phenomenon of **antibody-dependent enhancement (ADE)** is responsible for development of DHF, which occurs only if a person has previously been infected by a DENV of a different serotype. This is because the E protein is antigenically distinct among each DENV serotype. For example, about 4 to 7 days after being bitten by a mosquito carrying DENV-1, the individual will develop fever, rash, joint and bone pain (earning it the moniker "breakbone fever"), muscle aches, and perhaps mild bleeding of the gums and easy bruising. This will last between 3 and 10 days. The patient recovers thanks to the formation of neutralizing antibodies that bind to DENV-1 E protein, thereby preventing infection of newly made virions into additional host cells. However, if this same individual is bitten anytime afterward with a mosquito carrying any of the other DENV serotypes (e.g., DENV-2), anti-DENV-1 antibodies will recognize the E protein on DENV-2 even though it is structurally distinct (**figure 38.8**). Unfortunately in this case rather than neutralizing the virus, the bound antibodies serve to opsonize (tag) the virus for phagocytosis by white blood cells. All would be well if these cells could destroy DENV, but that is not the case. Rather these immune cells, including dendritic cells, monocytes, and macrophages, serve as productive hosts for DENV. Because these cells are now so efficient at taking up DENV, very high levels of virus are made and DHF may ensue. DHF begins like dengue, but after the initial weeklong febrile illness, the patient experiences severe abdominal pain and persistent vomiting. The patient may have difficulty breathing as fluid leaks from capillaries into the lung (pleural) cavity. This crisis lasts 1 to 2 days and can lead to circulatory system failure, shock, and death unless the patient is given supportive care. In 2015, a dengue vaccine was introduced and is recommended for children older than 9 years of age in dengue endemic countries. ◀ *Monocytes, macrophages, and dendritic cells (section 32.4); Antibodies doom the antigen (section 33.8)*

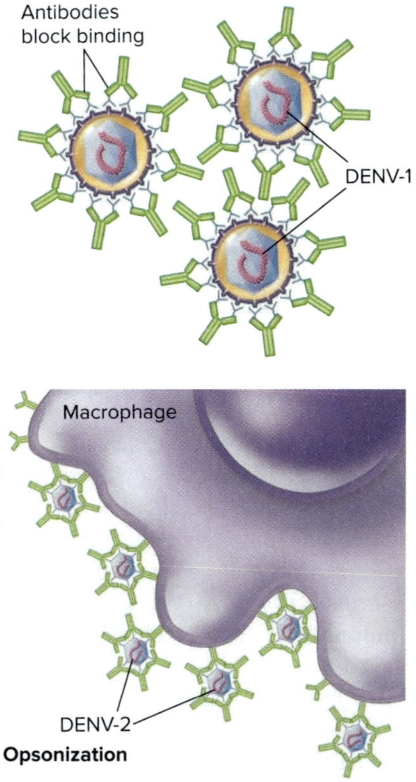

Figure 38.8 Antibody-Dependent Enhancement Leads to Dengue Hemorrhagic Fever.

Equine Encephalitis

Equine encephalitis is caused by viruses in the genus *Alphavirus,* family *Togaviridae*. They are positive-strand, enveloped RNA viruses. As the name implies, this is a disease of horses, where it can be fatal. However, in humans, the disease can present as a spectrum from fever and headache to meningitis and encephalitis. Human disease can progress to include seizures, paralysis, coma, and death. The virus is transmitted to humans by *Aedes* and *Culex* spp. mosquitoes. Various geographic descriptors are used to define the disease caused by genetically distinct strains of these arboviruses. These include eastern equine encephalitis (EEE), which occurs along the eastern Atlantic coast from Canada to South America; western equine encephalitis (WEE) is found from Canada to South America along the western coast; and Venezuelan equine encephalitis occurs in central and southern parts of the United States into South America. Reservoir hosts are important in the replication, maintenance, and dissemination of these arboviruses. Treatment consists of the supportive care of symptoms. Currently no human vaccine is available to prevent disease. A vaccine for horses can help reduce transmission of the virus.

West Nile Fever (Encephalitis)

West Nile fever (encephalitis) is caused by a positive-strand RNA flavivirus (West Nile virus) that once occurred primarily in the Middle East, Africa, and Southwest Asia. In 1999 the virus appeared unexpectedly in the United States, most likely via a mosquito (or mosquitoes) that hitched a ride on a jet from an endemic area that landed at New York's LaGuardia airport. That year it caused seven deaths among 62 confirmed human encephalitis cases and extensive mortality in a variety of domestic and exotic birds.

By 2015, 47 U.S. states and the District of Columbia reported West Nile virus (WNV) infections. In 2016, there were 2,038 confirmed cases and 94 deaths in the United States. WNV is transmitted to humans predominantly by *Culex* spp. mosquitoes that feed on infected birds (crows and sparrows). Humans and other mammals (e.g., horses) infected with WNV are considered "dead end hosts" because the level of virus in the bloodstream is not sufficient to transmit the virus to uninfected mosquitoes. Mosquitoes harbor the greatest concentration of virus in the late summer and early fall, and corresponding to this, there is a peak of disease in late August to early September. The risk of disease then decreases as the mosquitoes die when the weather becomes colder. Although many people are bitten by WNV-infected mosquitoes, most do not know they have been exposed because they remain asymptomatic or exhibit only mild, flulike symptoms.

There is no treatment other than hospitalization and intravenous fluids. There is no human vaccine to prevent WNV.

Zika

Zika virus (ZIKV) took the world by surprise in the spring of 2015 when people became ill with a virus about which little was known. In the roughly 75 years since its discovery, it had caused only mild febrile illness in fewer than 3,000 people in Asia and Africa. Like dengue, it is a member of the *Flaviviridae*, so it has an enveloped capsid and a positive-strand RNA genome. It is transmitted by many species of *Aedes* mosquitoes including *A. africanus, A. aegypti,* and *A. albopictus*. It can also be transmitted from person to person in body fluids, specifically breast milk and semen; men are advised to use condoms for at least 6 months after infection. There are two genetically distinct virus lineages. In both, the genome is about 10,800 bases and encodes a single polypeptide of about 3,400 amino acids that is subsequently cleaved into three structural and seven nonstructural proteins. Human neurons, skin cells, and immature dendritic cells are productive hosts in culture; more in vivo studies are needed to reveal how exactly the virus causes the devastating birth defects, including microcephaly, that have come to define ZIKV infection.

ZIKV has an incubation period of about 10 days and is symptomatic in only about 18% of people. When it does cause illness, it manifests as a mild fever, rash, joint pain, and conjunctivitis ("pink eye"), which resolves in 3 to 7 days. It has been associated with the development of GBS in a small minority of patients. Microcephaly seen in babies exposed to ZIKV in utero is characterized by diminished brain tissue with calcifications; the skull of some infants is also collapsed. These children also often have damage to the eyes, abnormally formed brain structures, and hydrocephalus ("water on the brain"). Although there is now consensus in the scientific and medical community that ZIKV causes these problems, there is no treatment, only prevention. Several vaccines are currently in clinical trials.

The rise and fall of the prevalence of Zika infection illustrates how quickly a virus can expand in an immunologically naïve population, that is, people who have never been exposed to the infectious agent. Zika infections peaked in the spring of 2016 with over 35,000 verified infections per week in South and Central America and the Caribbean. Since that time, the prevalence of infections has declined precipitously. Although no one knows for sure why this is the case, herd immunity certainly comes into play. Because roughly 80% of people who are exposed are unaware of infection, yet they nonetheless develop immunity, the transmission cycle from mosquito to human and back to mosquito becomes interrupted. Although epidemiologists use computer models to predict the rate and location of future infection, we will have to wait and see what actually happens.

◀ *Epidemiology is an evidence-based science (section 36.1)*

38.3 Direct Contact Diseases Can Be Caused by Viruses

After reading this section, you should be able to:

a. Discuss the viruses that cause common diseases spread by direct contact
b. Identify typical signs and symptoms of common viral diseases spread by direct contact
c. Correlate direct contact virus infection and disease severity with viral virulence factors

Transmission of disease from one person to another often requires close personal contact. For example, transfer of microorganisms between people occurs by direct contact with the infected person through touching, kissing, sexual contact, contact with body fluids and secretions, or contact with open wounds. We now consider several viral diseases that are transmitted by direct contact.

Acquired Immune Deficiency Syndrome (AIDS)

It is hard for those who were not adults in the 1980s and 1990s to imagine the devastating toll **AIDS (acquired immune deficiency syndrome)** took on infected men and women across the globe. Fortunately the 2015 report issued by the Joint United Nations Programme on HIV/AIDS and the WHO indicates that the number of new HIV infections has decreased 35% since 2000. At the end of 2014, 37 million people were infected and living with HIV/AIDS worldwide and about 41% of infected people were being treated with antiretroviral therapy. This represents a marked improvement, as only 23% were treated in 2010 (**figure 38.9**). First described in 1981, AIDS is the result of an infection by the **human immunodeficiency virus** (**HIV**), a positive-strand, enveloped RNA virus within the family *Retroviridae*. In the United States, AIDS is caused primarily by HIV-1; the HIV-2 strain predominates in Africa.

HIV is transmitted from one person to another when infected blood, semen, vaginal secretions, or breastmilk come in contact with an uninfected person's mucous membranes. In the developing world, AIDS affects as many women as men. In the United States, the groups most at risk for acquiring HIV are (in descending order) men who have unprotected sex with other men; intravenous drug users; heterosexuals who have

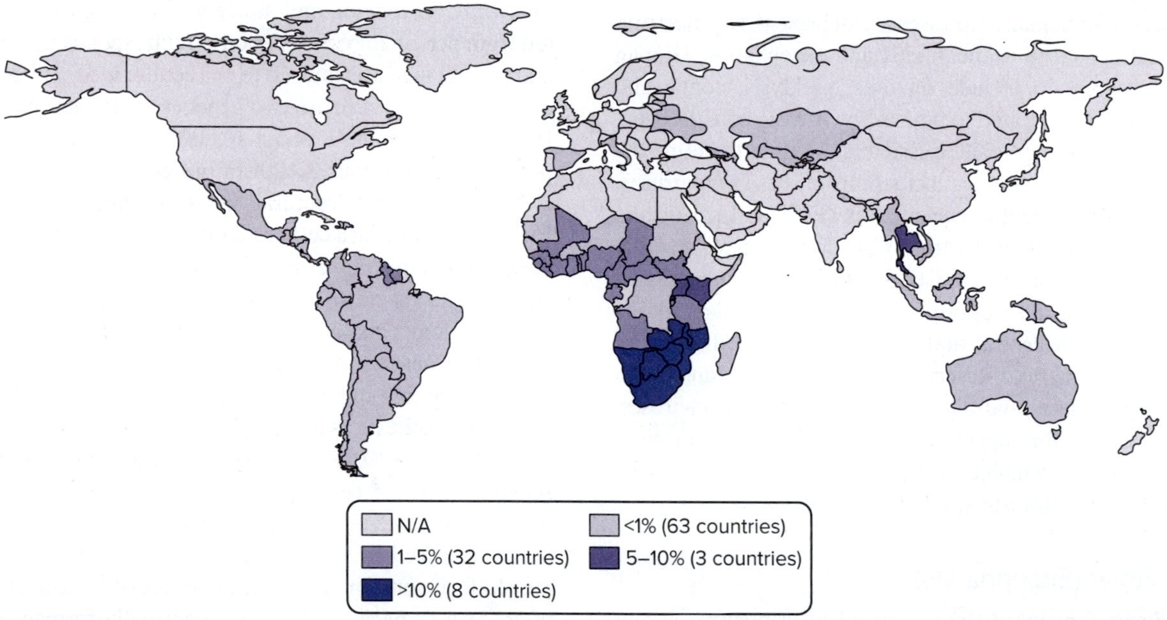

Figure 38.9 The Global HIV/AIDS View. This figure shows adult HIV prevalence from the 2015 UNAIDS report on the global AIDS epidemic.

Source: KFF, "The Global HIV/AIDS Epidemic," November 29, 2017.

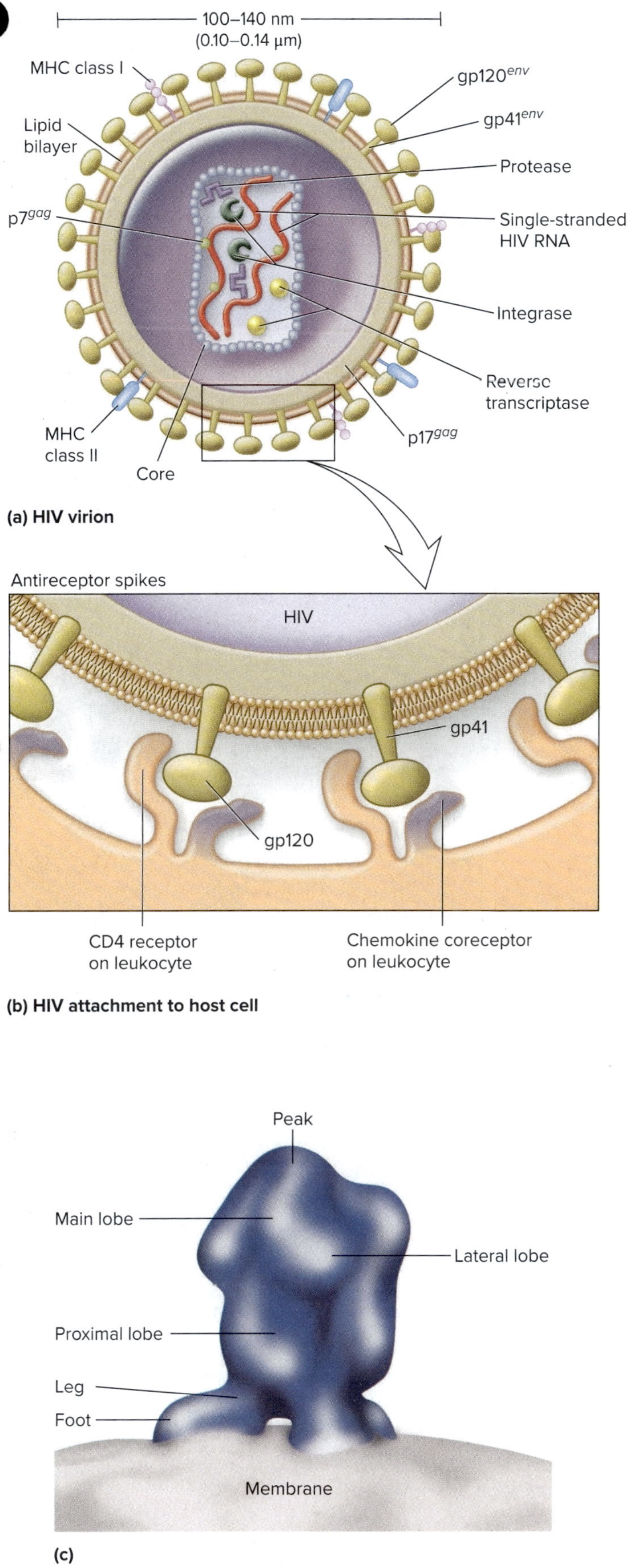

Figure 38.10 Schematic Diagram of the HIV Virion. (a) The HIV virion is enveloped and contains 72 external spikes. These spikes are formed by the two major viral envelope proteins, gp120 and gp41 (gp stands for glycoprotein—proteins linked to sugars—and the number refers to the mass of the protein, in kilodaltons). The HIV lipid bilayer is also studded with various host proteins, including class I and class II major histocompatibility complex molecules, acquired during virion budding. The cone-shaped core of HIV contains four nucleocapsid proteins (p24, p17, p9, p7), each of which is proteolytically cleaved from a 53 kDa Gag precursor by the HIV protease. The phosphorylated p24 polypeptide forms the chief component of the inner shell of the nucleocapsid, whereas the p17 protein is associated with the inner surface of the lipid bilayer and stabilizes the exterior and interior components of the virion. The p7 protein binds directly to genomic RNA and together with p9 forms the core. The retroviral core contains two copies of the single-stranded HIV genomic RNA to which preformed viral reverse transcriptase, integrase, and protease enzymes are attached. (b) The snug attachment of HIV glycoprotein molecules (gp41 and gp120) to their specific receptors on a human cell membrane. These receptors are CD4 and chemokine coreceptors CCR5 and CXCR4 (sometimes called fusin) that permit docking with the host cell and fusion with the cell membrane. (c) Surface-rendered model of the gp120 ENV protein spike of human and simian immunodeficiency viruses.

unprotected sex with infected partners; and children born of infected mothers. Transmission through semen increases HIV virulence as much as 100,000-fold relative to transmission by other routes. If untreated, HIV progresses to AIDS and is almost 100% fatal 9 to 11 years post infection. The use of combined antiviral medications has significantly reduced the morbidity and mortality of AIDS in developed nations. Modern medications can provide good quality of life, making HIV infection a chronic disease.

HIV-1 virions are approximately 110 nm in diameter, have a cylindrical core capsid, and their membrane envelopes are peppered with viral spike proteins (**figure 38.10*a***). The core contains two identical copies of its single-stranded RNA genome and several enzymes. Among the 10 virus-specific proteins, the gp120 envelope protein facilitates attachment to a susceptible host cell (e.g., T-helper cell). The gp120 proteins form trimeric spikes; that is, three lobes form the final three-dimensional spike structure, which sits on three membrane-bound feet (figure 38.10*c*).

To infect a host cell, the virus's gp120 envelope protein (figure 38.10*b*) binds the CD4 glycoprotein in the plasma membranes of $CD4^+$ T cells, macrophages, dendritic cells, and monocytes. Because dendritic cells are present throughout the body's mucosal surfaces and bear the CD4 protein, these are likely the first cells infected by HIV in sexual transmission. The virus requires a coreceptor in addition to CD4. Macrophage-tropic strains, which predominate early in the disease and infect both macrophages and T cells, require CCR5 (a chemokine coreceptor) as well as CD4. A second chemokine coreceptor, called CXCR-4, is used by T-cell–tropic strains active at later stages of infection. These strains induce the formation of syncytia (multinucleated masses of fused cells). Rare individuals with two defective copies of the CCR5 gene do not develop AIDS because the virus cannot infect their macrophages and T cells. People with one good copy of the CCR5 gene are infected but have an altered disease presentation than those with no mutation. ◀ *Retroviruses: plus-strand viruses that use reverse transcriptase in their life cycles (section 26.7); T cells are essential for immune function (section 33.5)*

HIV enters host cells by endocytosis as well as membrane fusion. When gaining entrance by endocytosis, the virus uncoats from within the endocytic compartment as the viral envelope fuses with the endosome membrane, releasing the virion contents into the cytosol (**figure 38.11*a***). When HIV uses membrane fusion to enter host cells, a conformational change in gp120 permits another protein, gp41, to extend, coil, and insert into the host cellular membrane. This process pulls the viral and cellular membranes together, allowing them to fuse. Inside the cell, the core protein remains associated with the RNA as it is copied into a single strand of DNA by the RNA-dependent DNA polymerase activity of the reverse transcriptase enzyme (*see figure 26.25*). The RNA template is next degraded by another reverse transcriptase component, ribonuclease H. The newly synthesized DNA strand now serves as template to form a double-stranded DNA copy of the original RNA genome also performed by reverse trancriptase. A complex of the double-stranded DNA (provirus) and integrase enzyme moves into the nucleus. Integrase catalyzes the integration of viral DNA into the cell's DNA; once integrated the viral genome is known as a provirus. The provirus can remain latent, giving no clinical sign of its presence.

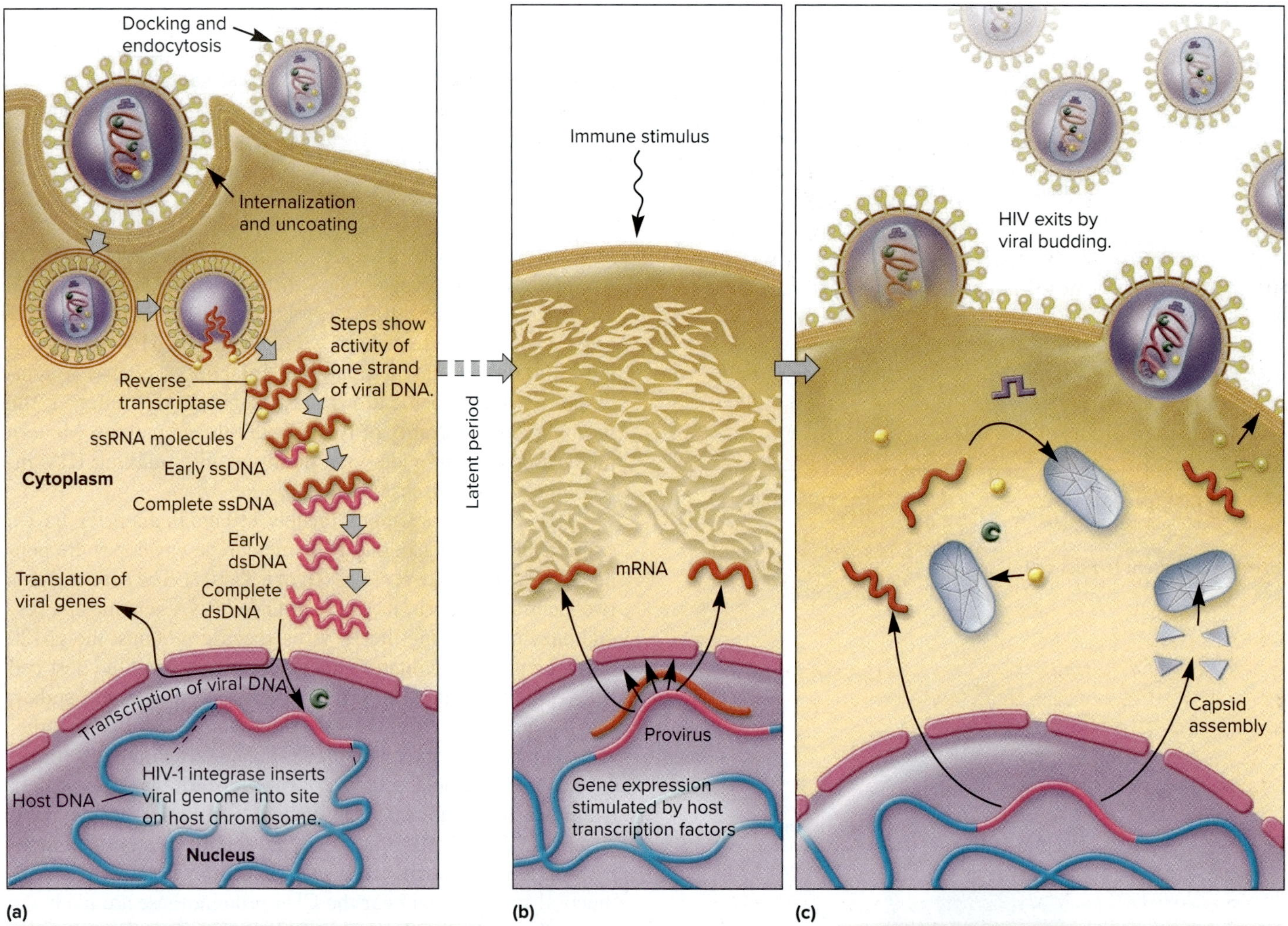

Figure 38.11 HIV Life Cycle. (a) The HIV virion binds to the $CD4^+$ host cell and can enter the host cell by endocytosis. The viral envelope fuses with the endocytic compartment releasing the virion contents into the cytosol. Reverse transcriptase catalyzes the synthesis of a single complementary strand of DNA from each of the twin RNA molecules. This DNA serves as a template for synthesis of double-stranded (ds) DNA. The dsDNA is inserted into the host chromosome as a provirus, establishing latency. (b) The provirus genes are transcribed. (c) Viral mRNA is translated into virion components (capsid, reverse transcriptase, spike proteins), and the virion is assembled. Mature virus particles bud from the host cell, taking host membrane as their envelope.

MICRO INQUIRY *What is the function of integrase and how is its activity related to latency?*

Alternatively the provirus can force the cell to synthesize viral mRNA (figure 38.11*b*). Some of the RNA is translated by host cell ribosomes to produce a long polypeptide that is cleaved into individual proteins by a viral protease. These proteins complete the assembly of new virions that bud from the infected host cell (figure 38.11*c*). Eventually the host cell dies.

Once a person becomes infected with HIV, the course of disease can vary greatly. The acute infection stage occurs 2 to 8 weeks after HIV infection. About 70% of individuals in this stage experience a brief illness referred to as acute retroviral syndrome, with symptoms that may include fever, malaise, headache, macular (small, red, spotty) rash, weight loss, and lymph node enlargement (lymphadenopathy). During this stage, the virus multiplies rapidly and disseminates to lymphoid tissues throughout the body, until an acquired immune response (antibodies and cytotoxic T cells) is generated to reduce virus multiplication (**figure 38.12**). During the acute infection stage, blood levels of HIV increase rapidly to 10^5 to 10^6 copies of viral RNA per milliliter of plasma. It is believed that the extent to which the immune response is able to control this initial burst of virus multiplication may determine the amount of time required for progression to the next clinical stage.

The asymptomatic stage of HIV infection may last from 6 months to 10 years (or longer in some individuals). During this stage, the levels of detectable HIV in the blood decrease but are still detectable because the virus continues to replicate, particularly in lymphoid tissues. Even before any decline in $CD4^+$ T cells (T-helper cells) can be detected, the virus may compromise certain immune functions (e.g., memory cell responses to common antigens such as tetanus toxoid or *Candida albicans*).

During the chronic symptomatic stage, which can last for months to years, virus multiplication continues and the number of $CD4^+$ T cells begins to significantly decrease. Because T-helper cells are critically important in the generation of acquired immunity, individuals at this stage develop a variety of symptoms including fever, weight loss, malaise, fatigue, anorexia, abdominal pain, diarrhea, headaches, and lymphadenopathy. Paradoxically, some patients develop increased serum antibody production during this stage, perhaps as a result of generalized immune dysfunction. These antibodies, however, do little to protect the host from infection. As $CD4^+$ T-cell numbers continue to decline, patients develop opportunistic infections (**Table 38.1**). Certain infections are correlated with specific levels of $CD4^+$ T cells (figure 38.12*b*). These include oral candidiasis (thrush), tuberculosis, and infections rarely seen in other patients, including *Mycobacterium avium* complex (MAC), cryptococcal pneumonia or meningitis, cytomegalovirus (CMV) infections, encephalitis caused by toxoplasmosis, and *Pneumocystis jiroveci* pneumonia (PCP).

In addition to its devastating effects on the immune system, HIV infection can also lead to disease of the central nervous system when virus-infected macrophages cross the blood-brain barrier. The classical symptoms of central nervous system disease include headaches, fevers, subtle cognitive changes, abnormal reflexes, and ataxia (irregularity of muscular action). Dementia and severe sensory and motor changes characterize more advanced stages of the disease. Another potential complication of HIV infection is cancer. Individuals infected with HIV-1 have an increased risk of three types of tumors: (1) Kaposi's sarcoma, caused by activation of latent human herpesvirus 8 (**figure 38.13**), (2) carcinomas of the mouth and rectum, and (3) B-cell lymphomas or lymphoproliferative disorders. It seems likely that the depression of the initial immune response enables other tumor-causing agents to initiate the cancers.

Laboratory diagnosis of HIV infection is most commonly accomplished through the detection of specific anti-HIV antibodies and antigens in the blood. Once a patient is confirmed to

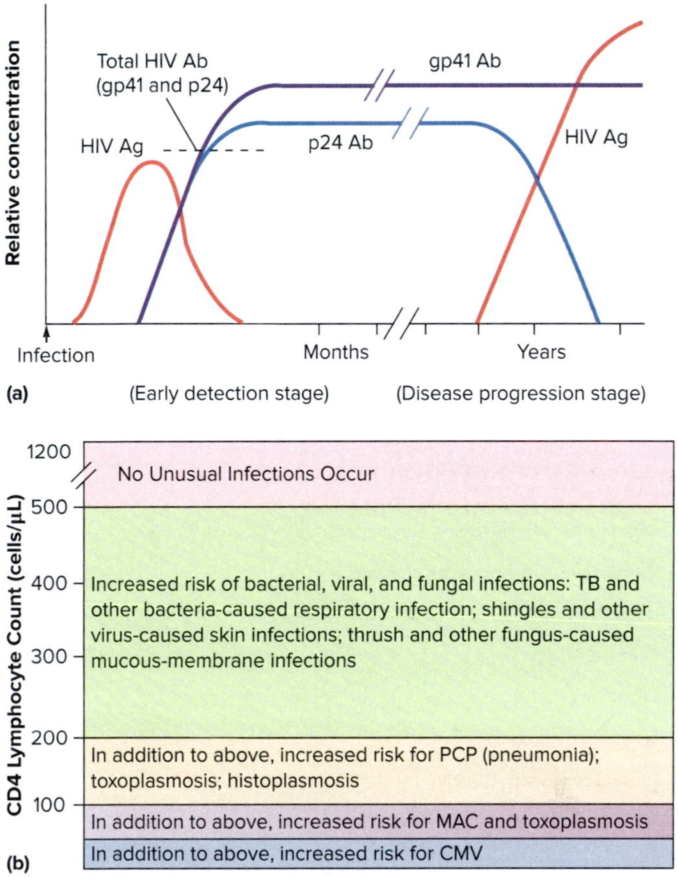

Figure 38.12 The Typical Immunological Pattern in an HIV-1 Infection. (a) HIV-1 antigen (HIV Ag) is detectable as early as 2 weeks after infection and typically occurs before IgG antibodies against gp41 and p24 develop. Seroconversion occurs when HIV-1 antibodies have risen to detectable levels. This usually takes place several weeks to months after infection. The period between HIV-1 infection and seroconversion often is associated with an acute flulike illness. HIV-1 antigen usually disappears following seroconversion but reappears in later stages of the disease. The reappearance of antigen usually indicates impending clinical deterioration. HIV-2 progresses more slowly and may not progress to AIDS. (b) As CD4 lymphocytes are destroyed by HIV, their declining numbers result in increased host susceptibility to specific infections.

Table 38.1 Disease Processes Associated with AIDS
Bacterial infection, multiple or recurrent
Candidiasis of bronchi, trachea, or lungs
Candidiasis, esophageal
Cervical cancer, invasive
Coccidioidomycosis, disseminated or extrapulmonary
Cryptococcosis, extrapulmonary
Cryptosporidiosis, chronic intestinal (>1 month's duration)
Cytomegalovirus disease (other than liver, spleen, or lymph nodes)
Cytomegalovirus retinitis (with loss of vision)
Encephalopathy, HIV-related
Herpes: chronic ulcer(s) (>1 month's duration); or bronchitis, pneumonitis, or esophagitis
Histoplasmosis, disseminated or extrapulmonary
Isosporiasis, chronic intestinal (>1 month's duration)
Kaposi's sarcoma
Lymphoma, Burkitt's
Lymphoma, immunoblastic
Lymphoma, primary, of brain
Mycobacterium avium complex or *M. kansasii*
Mycobacterium tuberculosis infection, any site
Mycobacterium infection, other species or unidentified species
Pneumocystis pneumonia
Pneumonia, recurrent
Progressive multifocal leukoencephalopathy
Salmonella septicemia, recurrent
Toxoplasmosis of brain
Wasting syndrome

be HIV positive, the viral load (number of viral copies per milliliter of blood) is determined by real-time polymerase chain reaction (RT-PCR). Viral load is particularly significant because the level of virions in the blood, as well as the concentration of $CD4^+$ cells, is predictive of the clinical course of the infection. These values can be used to determine the likelihood of specific opportunistic infections (figure 38.12*b*) as well as monitor the patient's response to therapy. ◀ *Polymerase chain reaction amplifies targeted DNA (section 17.2)*

There is no cure for HIV infection. Even when viral loads are undetectable, the virus cannot be eradicated from long-lived resting memory $CD4^+$ T cells. Thus the goal of treatment is to reduce viral load to undetectable levels, which eliminates disease burden on the patient and prevents transmission. Importantly, if a patient stops taking antiviral medications, the reservoir of quiescent $CD4^+$ T cells will become productive hosts and the viral load will rebound to high levels.

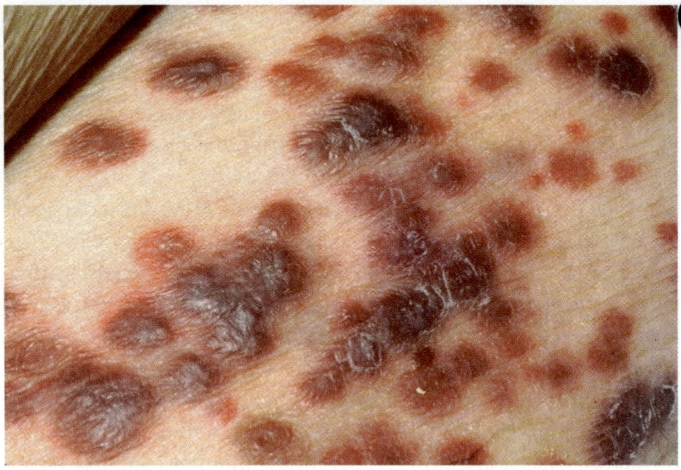

Figure 38.13 Kaposi's Sarcoma on the Arm of an AIDS Patient. The flat purple tumors can occur in almost any tissue.

The antiviral drugs currently approved for use in HIV disease are described in detail in chapter 9. The most successful treatment approach in controlling HIV is to use a combination of three antiviral agents simultaneously. This limits the development of drug resistant viral strains and reduces viral loads to undetectable levels.

The development of a vaccine for AIDS has been a long-sought goal. Among the many problems encountered in developing an HIV vaccine is the fact that the viral envelope proteins continually change their antigenic properties; HIV has a high mutation rate because reverse transcriptase lacks proofreading capabilities. One new experimental approach has been to synthesize hybrid antibodies to block the highly conserved gp120 protein from docking with host cells.

Prevention and control of HIV are achieved primarily through education. Understanding risk factors and practicing strategies to reduce risk are essential in the fight against HIV. Barrier protection from blood and body fluids greatly limits risk of HIV infection. HIV-positive mothers need not transmit the virus to their infant if treated with antiretroviral therapy prenatally and are then educated about breastmilk transmission. Individuals at high risk of acquiring HIV should be offered the opportunity to take preexposure prophylaxis (PrEP), a combination of two antiviral agents that dramatically reduces the probability of transmission. Education to prevent the sharing of intravenous needles and syringes is also very important. The public should also understand that blood and blood products are screened to prevent HIV transmission.

Cold Sores

A cold sore, also known as a fever blister, is a fluid-filled lesion typically found on the lips or inside the nostrils. Cold sores cause pain, itching, or a burning sensation because they reside on or near nerve endings. They are caused by infection with the double-stranded DNA virus species *Human herpesvirus 1,* HSV-1, which has an icosahedral capsid and is enveloped. The genome encodes approximately 100 proteins.

Clinical descriptions of herpes lesions date back to the time of Hippocrates (circa 400 BCE). Transmission is through direct contact of epithelial tissue surfaces with the virus. HSV-1 was initially thought to infect oral mucosal epithelium to cause cold sores or fever blisters, and herpes simplex virus type 2 (HSV-2; species *Human herpesvirus 2*) was thought to infect genital epithelium to cause genital herpes. However, both viruses can infect either tissue site. The herpesvirus genome must first enter an epithelial cell for the initiation of infection. The initial association is between glycoproteins of the epithelial cell surface and viral glycoproteins. This is followed by a specific interaction with one of several cellular receptors collectively termed HVEMs for *h*erpes*v*irus *e*ntry *m*ediators. The capsid, along with some associated proteins, then migrates along the cellular microtubule transport machinery to nuclear envelope pores. This "docking" is thought to result in viral DNA being injected through nuclear envelope pores while the capsid remains in the cytoplasm.

Active and latent phases occur within an infected host. After an incubation period of about 1 week, the active phase begins. During the active phase, the virus multiplies explosively; between 50,000 and 200,000 new virions are produced from each infected cell. During this replication cycle, herpesvirus inhibits its host cell's metabolism and degrades host DNA, inducing apoptosis. As a result, the cell dies, releasing viral progeny to infect other cells. Such an active infection may be symptom-free, or painful blisters may arise as a result of a local inflammatory response to cellular lysis. Blisters contain fluid and infectious virions. Importantly, apparently healthy people can transmit HSV to other hosts. Blisters involving the epidermis and surface mucous membranes of the lips, mouth, and gums are referred to as herpes labialis (**figure 38.14**). Primary and recurring HSV infections also may occur in the eyes, causing **herpetic keratitis** (inflammation of the cornea)—the most common cause of infectious blindness in developed nations.

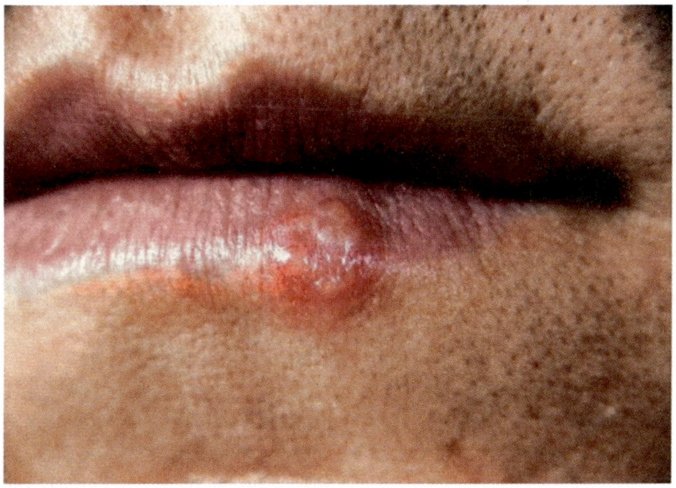

Figure 38.14 Cold Sores. Herpes simplex fever blisters on the lip, caused by herpes simplex type 1 virus.
Source: CDC/Dr. Herrmann

Blisters usually heal within 1 week, and then a very interesting virulence factor initiates the long-term survival of HSV-1 in infected neurons, inducing the latent phase of infection. As the active infection is curtailed by a vigorous cellular immune response, viruses released from infected oral epithelia travel to the trigeminal nerve ganglion (cluster of nerve cells in the face and jaw). There the virus will remain in a latent state for the lifetime of the infected host. However, stressful stimuli such as excessive sunlight, fever, trauma, chilling, emotional stress, and hormonal changes can reactivate the virus. Once reactivated, the virus travels from the nerve ganglion down a peripheral nerve to epithelial cells to produce another active phase. Interestingly, not all infected neurons die from a productive infection. Latently infected neurons and some neurons with productive HSV infection survive. This is possible because an HSV gene product is produced that protects infected neurons from host immune cells. The latency-associated transcript (LAT) is an 8.5-kb RNA spliced out of a transcript similar to that of an mRNA precursor, but no consistently expressed protein appears to be translated. Instead, a long-lived 2-kb microRNA (miR-LAT) remaining in the nucleus promotes degradation of two proteins that inhibit cell proliferation and initiate apoptosis. In other words, miR-LAT prevents the infected cell from self-destruction, ensures neuron survival, and provides a hiding place for the virus.

The drugs acyclovir, valacyclovir, and famciclovir are effective against cold sores. Idoxuridine and trifluridine are used to treat herpes infections of the eye. By adulthood, 70 to 90% of people in the United States have been infected and have anti-HSV-1 antibodies. Diagnosis of HSV-1 infection is typically by clinical evaluation of the lesions. However, in rare cases, HSV-1 can cause infection of the brain (encephalitis). When this is suspected, it is typically detected by nucleic acid amplification from cerebral spinal fluid.

Common Cold

The **common cold** is one of the most frequent infections experienced by humans. The incidence of infection is greater during cooler months, likely due to increased exposure to others while indoors and the effect of dry air on mucous membranes. About 50% of cases are caused by rhinoviruses (Greek *rhinos*, nose), which are nonenveloped, positive-strand RNA viruses in the family *Picornaviridae*. There are over 115 distinct serotypes, and each of these antigenic types has a varying capacity to infect the nasal mucosa and cause a cold. In addition, immunity to many of these viruses is not lifelong. Several other respiratory viruses are also associated with colds (e.g., coronaviruses and parainfluenza viruses). Thus colds are common because of the diversity of rhinoviruses, the involvement of other respiratory viruses, and the lack of a durable immunity. ◀ *Plus-strand RNA viruses: genomes that can be translated upon entry (section 26.5)*

Rhinoviruses provide an excellent example of the medical relevance of research on virion morphology (**figure 38.15**). The complete rhinovirus capsid structure has been elucidated with the use of X-ray diffraction techniques. The results help explain

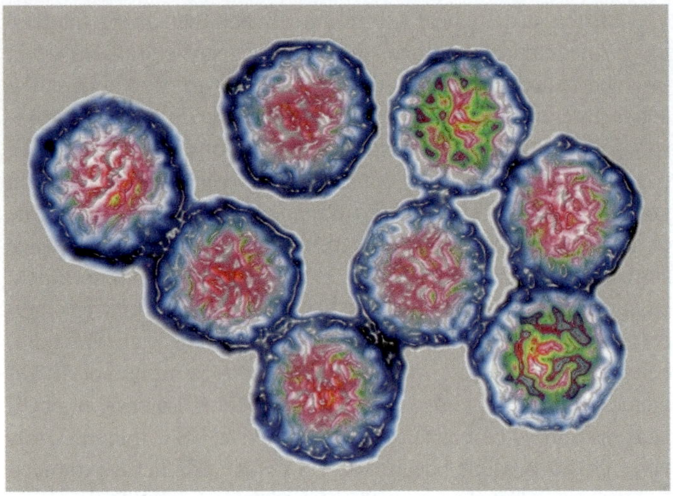

Figure 38.15 Rhinovirus Virion. One of over 100 types of virus that cause the common cold. Each virion is about 30 nm in diameter. Colorized transmission electron micrograph.

rhinovirus resistance to human immune defenses. The capsid protein that recognizes and binds to host cell surface molecules during infection is hidden from the immune system at the bottom of a relatively deep surface cleft (sometimes called a "canyon") making it a difficult target for antibodies.

Viral invasion of the upper respiratory tract occurs by binding to cellular adhesion molecules. The clinical manifestations include the familiar nasal stuffiness, sneezing, scratchy throat, and watery discharge from the nose. General malaise is commonly present. Fever is usually absent in uncomplicated colds, although a low-grade fever may occur in infants and children. The disease usually runs its course in about 1 week. Diagnosis of the common cold is made from observations of clinical symptoms. There are no routine procedures for direct examination of clinical specimens or for serological diagnosis. However, rhinovirus and other cold viruses are often detected when multiplex PCR is used to diagnose other suspected respiratory infections.

Sources of the cold viruses include the nasal secretions of infected individuals, airborne transmission over short distances by way of moisture droplets, and transmission on contaminated hands or fomites (contaminated inanimate objects). Epidemiological studies of rhinovirus colds have shown that the familiar explosive, noncontained sneeze may not be as effective in spreading virus as hand-to-hand contact between a rhinovirus "donor" and a susceptible "recipient." This emphasizes the need for good hand washing during cold season, which has two peaks, spring and early autumn. Infection is most common early in life and generally decreases with an increase in age. Treatment for the common cold includes rest, extra fluids, and the use of anti-inflammatory agents for alleviating local and systemic discomfort.

Cytomegalovirus Inclusion Disease

Cytomegalovirus inclusion disease is caused by cytomegalovirus (CMV). As a member of the family *Herpesviridae* (species *Human herpesvirus 5*), CMV is an enveloped, double-stranded DNA virus with an icosahedral capsid. Most people become infected with this virus at some time during their life; in the United States, about 80% of individuals older than 35 years have been exposed and carry a lifelong infection. Although most CMV infections are asymptomatic, certain groups are at risk for serious illness and long-term effects. For example, CMV remains the leading cause of congenital virus infection in the United States, a significant cause of transfusion-acquired infections, and a frequent contributor to morbidity and mortality among organ transplant recipients and immunocompromised individuals. Because the virus persists in the body, it is shed for several years in saliva, urine, semen, and cervical secretions.

CMV can infect any cell of the body, where it multiplies slowly and causes the host cell to swell in size; hence the prefix *cytomegalo-*, which means "enlarged cell." CMV is well known for its ability to interfere with many host immune functions, such as antigen presentation, cytokine production, and natural killer cell activity. Infected cells contain unique **intranuclear inclusion bodies** and cytoplasmic inclusions (**figure 38.16**). In fatal cases, cell damage is seen in the gastrointestinal tract, lungs, liver, spleen, and kidneys. In less-severe cases, cytomegalovirus inclusion disease symptoms resemble those of infectious mononucleosis.

Laboratory diagnosis is by the detection of anti-CMV IgM and IgG in patient serum or amplification of CMV nucleic acid by PCR. The virus may also be cultured.

The virus has a worldwide distribution, especially in developing countries, where infection is universal during childhood. The prevalence of this disease increases with poor hygiene or as socioeconomic status declines. The antivirals ganciclovir, valganciclovir, and foscarnet are used to treat high-risk patients. The nucleotide analogue cidofovir is used to treat CMV retinitis in HIV-positive patients. Transmission by blood transfusion or organ transplantation can be avoided by using blood or organs from seronegative donors. However, because the prevalence of CMV infection is so high, this is not always possible.

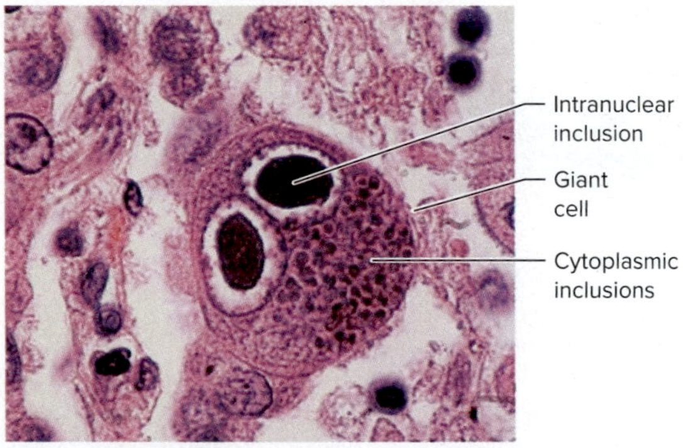

Figure 38.16 Cytomegalovirus Inclusion Disease. Light micrograph of a giant cell in a lung section infected with cytomegalovirus (×480). Cells containing intranuclear inclusion bodies have a typical "owl-eyed" appearance because of the clear halo that surrounds the inclusion. Sometimes inclusions also are visible in the cytoplasm.

Genital Herpes

Genital herpes is a lifelong infection caused by herpes simplex virus, predominantly type 2 (HSV-2) (**figure 38.17*a***). These viruses have a very short replication cycle. The core DNA is linear and double stranded. The HSV-2 envelope contains at least eight glycoproteins. HSV-2 is usually sexually transmitted, although it can be transmitted nonsexually as well. Infection begins when the virus is introduced into a break in the skin or mucous membranes. The virus infects the epithelial cells of the external genitalia, the urethra, and the cervix (figure 38.17*b,c*). Rectal and pharyngeal (oral) herpes are also transmitted by sexual contact.

The details of HSV-2 infection are very similar to those of HSV-1 discussed in the section on cold sores (HSV-1 infection). In the case of genital HSV-2, viruses retreat to nerve cells in the sacral plexus of the spinal cord, where they remain latent. As described for HSV-1, during the latent phase, the host cell does not die. Because viral genes are not expressed, the infected person is symptom-free. It should be noted that both primary infection and reactivation can occur without any symptoms, and apparently healthy people can transmit HSV-2 to their sexual partners or their newborns during vaginal delivery, the latter leading to **congenital (neonatal) herpes.** Congenital herpes is a life-threatening infection in newborns, affecting approximately 30 of every 100,000 babies per year in the United States. It can result in neurological involvement, as well as blindness. As a result, any pregnant woman who has active or new genital herpes should have a cesarian section instead of delivering vaginally. HSV-2 is also associated with a higher-than-normal rate of cervical cancer and miscarriages.

Diagnosis of HSV-2 infection is by screening of blood or serum for anti-HSV-2 antibodies, direct fluorescent antibody testing of tissue, and/or PCR. Although there is no cure for genital herpes, oral use of the antiviral drugs acyclovir, valacyclovir, or famciclovir has proven to be effective in ameliorating recurrent blister outbreaks. Topical acyclovir is more effective than famciclovir in reducing virus shedding.

In the United States, the incidence of genital herpes has increased during the past several decades. Currently one of six individuals aged fourteen to forty-nine is infected with HSV-2.

Human Herpesvirus 6 Infection

Human herpesvirus 6 (HHV-6) is the etiologic agent of **exanthem subitum** (Greek *exanthema*, rash), also called roseola infantum, which occurs in infants and toddlers. As a member of the family *Herpesviridae*, it is an enveloped virus with an icosahedral capsid and a core containing double-stranded DNA that establishes latency in host cells. Exanthem subitum is a short-lived disease characterized by a high fever of 3 to 4 days' duration, after which the temperature suddenly drops to normal and a macular rash appears on the trunk and then spreads to other areas of the body. HHV-6 infects over 95% of the U.S. infant population, and most children are seropositive for HHV-6 by three years of age. $CD4^+$ T cells are the main site of viral replication, whereas monocytes are in an infected, latent state. In adults, HHV-6 is commonly found in peripheral-blood mononuclear cells and saliva. Since the salivary glands are the major site of latent infection, transmission is probably by way of saliva.

Because HHV-6 produces latent infections, it occasionally reactivates in immunocompromised hosts, leading to pneumonitis. Diagnosis is by immunofluorescence, enzyme immunoassay, or PCR. There is no antiviral therapy or prevention.

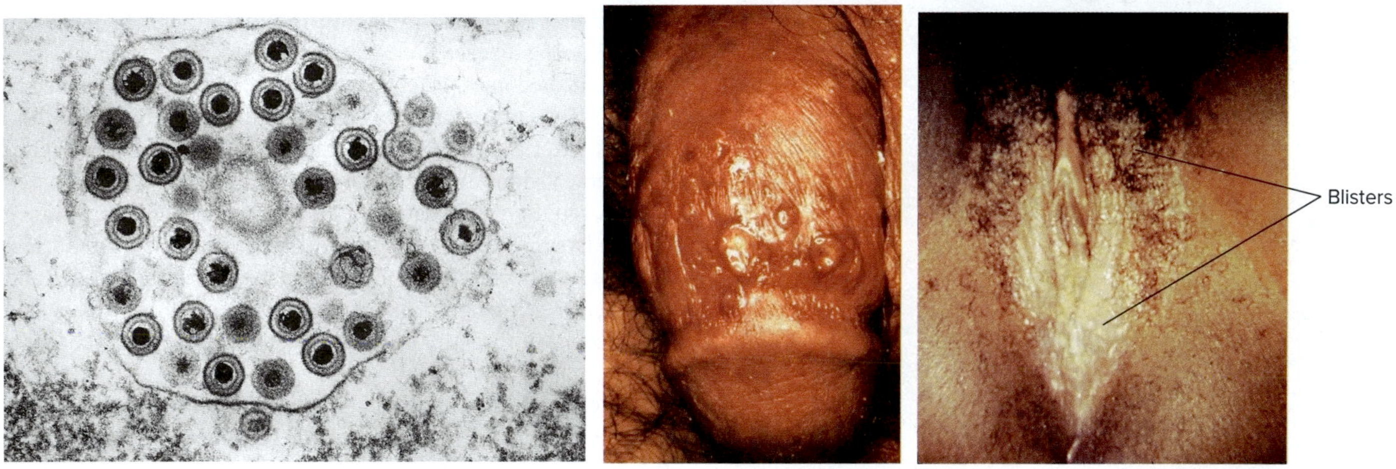

(a) (b) (c)

Figure 38.17 Genital Herpes. (a) Herpes simplex virus virions inside an infected cell. Each virion measures 150-200 nm in diameter. (b) Herpes vesicles on the penis. The vesicles contain fluid that is infectious. (c) Herpes blistering around the vaginal opening.

(a) Source: CDC/Dr. Fred Murphy & Sylvia Whitfield; (b) Source: CDC/Dr. Fiumara and Dr. Gavin Hart; (c) Source: CDC/Susan Lindsley

MICRO INQUIRY *Where does HSV-2 reside during latency?*

Human Parvovirus B19 Infection

Since its discovery in 1974, **human parvovirus B19** (family *Parvoviridae*, genus *Erythrovirus*) has emerged as a significant human pathogen. B19 virions are uniform, icosahedral, naked particles approximately 23 nm in diameter. Parvoviruses have a genome composed of one ssDNA molecule of about 5,000 bases. Parvoviruses are among the simplest of the DNA viruses. The genome is so small that it must resort to the use of overlapping genes to encode the very few proteins it encodes. The genome does not code for any enzymes, and the virus uses host cell enzymes for all biosynthetic processes. Thus viral DNA can only be replicated in the nucleus during the S phase of the cell cycle, when the host cell replicates its own DNA. Because the viral genome is single stranded and linear, host DNA polymerase must be tricked into copying it. By using a self-complementary sequence at the ends of the viral DNA, the parvovirus genome folds back on itself to form a primer for replication (*see figure 26.15*). This is recognized by the host DNA polymerase and DNA replication ensues. |◀ *Parvoviruses (section 26.3)*

A spectrum of disease is caused by parvovirus B19 infection, ranging from mild symptoms (fever, headache, chills, malaise) in normal persons and **erythema infectiosum** (**figure 38.18**) in children (also called **fifth disease**), to a joint disease syndrome in adults. Because the virus can infect bone marrow and cause a temporary cessation of red blood cell development, it can cause more serious complications, including aplastic crisis in persons with sickle cell disease or autoimmune hemolytic anemia, and pure red cell aplasia in immunocompromised individuals. The B19 parvovirus can also infect a fetus, resulting in anemia, fetal hydrops (the accumulation of fluid in tissues), and spontaneous abortion. This has prompted a grass-roots awareness of B19 complications among school teachers who may be pregnant. It is assumed that the mode of infection is by the respiratory route. The average incubation period is 4 to 14 days. Approximately 20% of infected individuals are asymptomatic, and a smaller percentage of infected individuals have symptoms for up to 3 weeks. Infection typically results in a lifelong immunity to B19.

B19 infection in otherwise healthy children is made clinically (based on signs and symptoms). In high-risk patients, it is detected by serology and/or PCR. These patients may be treated with commercial immunoglobulins containing anti-B19 antibodies. As with other diseases spread by contact with respiratory secretions, frequent hand washing is the best prevention of the disease.

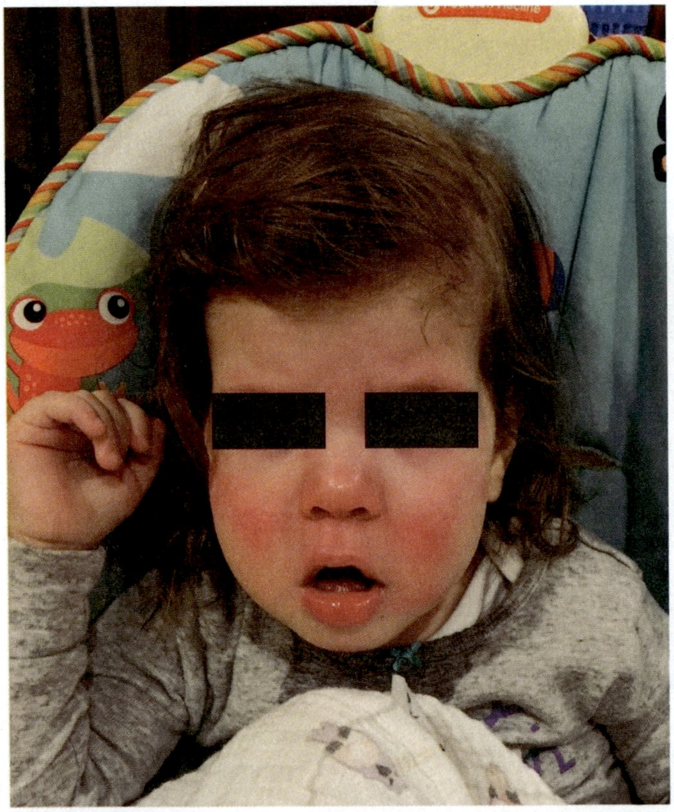

Figure 38.18 Erythema Infectiosum. The facial rash of fifth disease, sometimes called "slapped face syndrome."

Mononucleosis (Infectious)

Epstein-Barr virus (EBV; species *Human herpesvirus 4*) is a member of the family *Herpesviridae*. EBV exhibits the characteristic herpesvirus morphology with an enveloped, icosahedral capsid. Like all members of *Herpesviridae*, its dsDNA genome exists as a linear form in the mature virion and a circular episomal form in latently infected host cells. EBV is the etiologic agent of **infectious mononucleosis,** a disease whose symptoms closely resemble those of cytomegalovirus-induced mononucleosis. Because EBV occurs in oropharyngeal secretions, it can be spread by mouth-to-mouth contact (hence the term "kissing disease") or shared drinking bottles and glasses. A person becomes infected when the virus from someone else's saliva makes its way into epithelial cells lining the throat. After a brief bout of multiplication in the epithelial cells, new virions are shed and infect memory B cells. Infected B cells rapidly proliferate and take on an atypical appearance (called Downey cells) that is useful in diagnosis (**figure 38.19**). The disease is manifested by enlargement of lymph nodes and spleen, sore throat, headache, nausea, general weakness and tiredness, and a mild fever that usually peaks in the early evening.

Treatment of infectious mononucleosis is largely supportive and includes plenty of rest. Diagnosis is usually confirmed by demonstration of an increase in circulating mononuclear cells, along with a serological test for nonspecific (cross-reactive) antibodies, specific anti-viral antibodies, or identification of viral nucleic acid. Several rapid tests are available.

The peak incidence of mononucleosis occurs in people fifteen to twenty-five years of age. College students, particularly

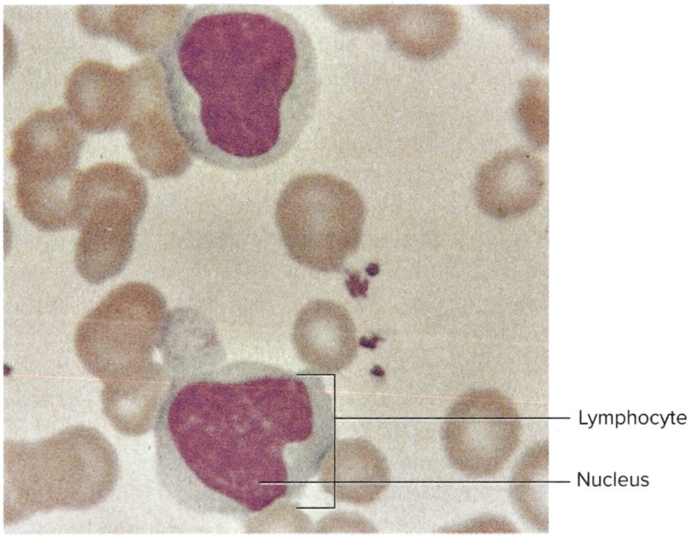

Figure 38.19 Evidence of Epstein-Barr Virus Infection in the Blood Smear of a Patient with Infectious Mononucleosis. Note the abnormally large lymphocytes (called Downey cells) containing indented nuclei with light discolorations.

those in upper socioeconomic classes, have a high incidence of the disease. Many individuals acquire immunity to the disease following early childhood infection, which is typically asymptomatic. EBV may well be the most common virus in humans, as it infects 80 to 90% of all adults worldwide. EBV infections are associated with the cancers Burkitt lymphoma in tropical Africa and nasopharyngeal carcinoma in Southeast Asia, East and North Africa, and in Inuit populations. Although these associations are well documented, the pathophysiological mechanism remains unclear.

Viral Hepatitides

Hepatitis (pl., hepatitides; Greek *hepaticus*, liver) is a general term for infection and inflammation of liver cells. Currently 11 viruses are recognized as causing hepatitis. Two are herpesviruses (CMV and EBV, which infect a variety of other cells) and nine are hepatotropic viruses that specifically target liver cells, or hepatocytes. EBV and CMV cause mononucleosis syndromes that can sometimes lead to mild, self-resolving forms of hepatitis with no permanent liver damage. Of the nine human hepatotropic viruses, only five are well characterized (**table 38.2**); hepatitis G and TTV (transfusion-transmitted virus) were only recently discovered. Hepatitis A and hepatitis E are transmitted by fecal-oral contamination and discussed in the section on food-borne and waterborne diseases (section 38.4). The other major types include hepatitis B, hepatitis C, and hepatitis D.

Hepatitis B is caused by **hepatitis B virus** (HBV), an enveloped virus with a partially double-stranded circular DNA genome of complex structure. HBV is classified as an *Orthohepadnavirus* within the family *Hepadnaviridae*. HBV is a reverse-transcribing DNA virus, so named because it uses a reverse transcriptase enzyme to replicate its genome. ◀ *Reverse transcribing DNA viruses (section 26.8)*

Serum from individuals infected with HBV contains three distinct antigenic particles, only one of which is the mature virion, sometimes also called a **Dane particle**. The two other antigenic particles, a smaller spherical and larger filamentous one, serve as decoys for host antibodies (**figure 38.20**). The viral genome is 3.2 kb in length, consisting of four partially overlapping, open-reading frames that encode viral proteins. Viral multiplication takes place predominantly in hepatocytes. The HBV envelope is coated with hepatitis B surface antigen (HBsAg). The inner nucleocapsid core antigen (HBcAg) encloses a single molecule of double-stranded HBV DNA and polymerase. HBsAg in body fluids is (1) an indicator of active hepatitis B infection, (2) used in the large-scale screening of

Table 38.2 Characteristics of Hepatitides Caused by Hepatotropic Viruses

Disease	Genome	Classification	Transmission	Characteristic	Prevention
Hepatitis A	RNA	*Picornaviridae, Hepatovirus*	Fecal-oral	Subclinical, acute infection	Killed HAV vaccine
Hepatitis B	DNA	*Hepadnaviridae, Orthohepadnavirus*	Blood, needles, body secretions, placenta, sexually	Subclinical, acute, or chronic infection; cirrhosis; primary hepatocarcinoma	Recombinant HBV vaccine
Hepatitis C	RNA	*Flaviviridae, Hepacivirus*	Blood, sexually	Subclinical, acute, or chronic infection; primary hepatocarcinoma	Routine screening of blood
Hepatitis D	RNA	Satellite	Blood, sexually	Coinfection with HBV required for satellite replication	HBV vaccine
Hepatitis E	RNA	*Hepeviridae, Hepevirus*	Fecal-oral	Subclinical, acute infection (but high mortality in pregnant women)	Improve sanitary conditions

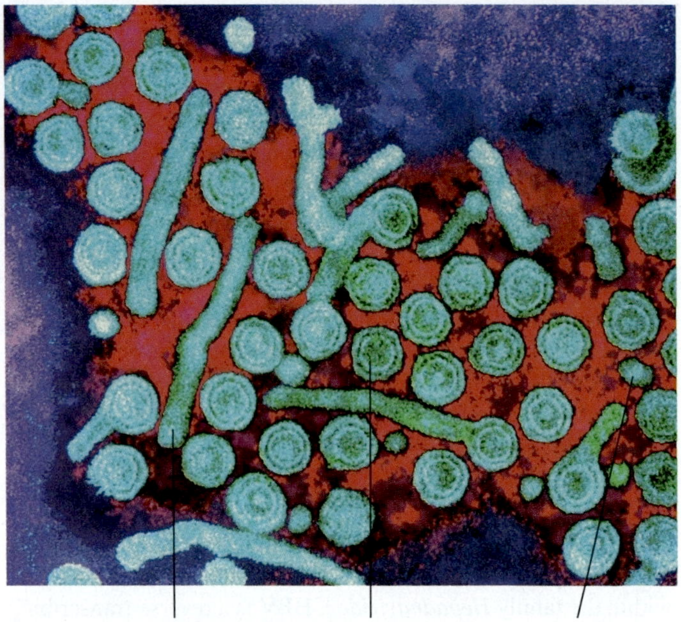

Figure 38.20 Hepatitis B Virus in Serum. Electron micrograph (×210,000) showing the three distinct types of hepatitis B antigenic particles. The spherical particles and filamentous forms are small spheres or long filaments without an internal structure, and only two of the three characteristic viral envelope proteins appear on their surface. Dane particles are the complete, infectious virion.

blood for HBV, and (3) the basis for the HBV vaccine, which was the first vaccine for human use developed by recombinant DNA technology. Diagnosis of HBV is made by serology, which can be complex. Most straightforward is the detection of HBsAg in unimmunized individuals; the presence of IgM or IgG anti-HBcAg antibody helps determine if the infection is acute or chronic. Viral load is determined using PCR to measure HBV nucleic acid.

Hepatitis B virus is normally transmitted through blood or other body fluids (saliva, sweat, semen, breast milk, urine, feces) and body-fluid–contaminated equipment (including shared intravenous needles). The virus can also pass through the placenta to the fetus of an infected mother. It is estimated that there are 257 million persons worldwide infected with HBV. The number of children becoming infected has dramatically declined thanks to vaccination campaigns. In the United States, there are currently about 850,000 confirmed chronically infected individuals. However, this prevalence most likely underestimates the true burden because many people do not get tested who are at high risk of contracting HBV. About 1,800 persons die yearly from HBV-related complications.

The clinical signs of hepatitis B vary widely. Most cases are asymptomatic. However, sometimes fever, loss of appetite, abdominal discomfort, nausea, fatigue, and other symptoms gradually appear following an incubation period of 1 to 3 months. The virus causes liver tissue degeneration and the release of liver-associated enzymes (transaminases) into the bloodstream. This is followed by jaundice, the accumulation of bilirubin (a breakdown product of hemoglobin) in the skin and other tissues with a resulting yellow appearance. Chronic hepatitis B infection also causes the development of hepatocellular carcinoma (liver cancer).

General measures for prevention include (1) vaccination; (2) preventing contact with HBV-infected blood and secretions, and minimizing accidental needlesticks; and (3) passive prophylaxis with intramuscular injection of hepatitis B immune globulin within 7 days of exposure. Recommended antiviral chemotherapy for HBV include entecavir (ETV) and tenofovir disoproxil fumarate (TDF).

Hepatitis C is caused by **hepatitis C virus** (HCV), which has a 50-nm enveloped virion that contains a single strand of linear RNA. HCV is a member of the family *Flaviviridae*. It is transmitted by contact with virus-contaminated blood, by the fecal-oral route, by in utero transmission from mother to fetus, sexually, or through organ transplantation. Screening for HCV is performed by detection of anti-HCV antibodies. Positive results are then confirmed by HCV nucleic acid amplification to determine viral load. Worldwide, hepatitis C has reached epidemic proportions, with about 68 million people infected. In the United States, HCV is the most common chronic bloodborne infection; approximately 9.2 million persons are chronically infected and 30,000 new cases occur annually. Like HBV, HCV infection can cause hepatocellular carcinoma. HCV is currently the leading reason for liver transplantation in the United States. There are six major genotypes with genotype 1 most common, followed by genotype 3. Treatment for genotype 1 disease is with the 2014 FDA-approved drug combination of sofosbuvir (a nucleotide analogue polymerase inhibitor) and ledipasvir (an HCV replication inhibitor). A 12-week treatment with this drug combination results in complete cure of the virus from the host. Other genotypes are now also usually treated for 12 weeks wtih sofosbuvir, coupled with velpatasvir, which has a similar mechanism to ledipasvir. Newer drugs and drug combinations designed to treat HCV infection caused by less prevalent genotypes are becoming available.

Hepatitis D virus (HDV) was discovered in 1977 and the disease hepatitis D was designated. It is now named **hepatitis delta virus.** There are eight known genotypes of HDV: Genotype 1 has a worldwide distribution; genotypes 2 and 4 exist in East Asia; genotype 3 is found in South America; and genotypes 5 through 8 are in Africa. All genotypes can cause acute as well as chronic liver disease. HDV is a satellite virus that is dependent on hepatitis B virus to provide the envelope protein (HBsAg) for its own replication. HDV's dependence on HBV means that HDV only replicates in liver cells coinfected with actively replicating HBV. HDV is spread only to persons who are already infected with HBV (superinfection) or to individuals who get HBV and the satellite at once (coinfection). The negative-strand RNA of HDV is smaller than the RNA of the smallest picornaviruses, and its circular conformation differs from the linear structure typical of animal negative-strand RNA viruses.

The primary laboratory tools for diagnosis of an HDV infection are serological tests for antidelta antibodies. Treatment of patients mirrors that of HBV-infected patients. Worldwide, there are approximately 15 million individuals infected with HDV. Prevention and control involve the widespread use of the hepatitis B vaccine. ◀ *Viroids and satellites: nucleic acid-based subviral agents (section 6.6)*

Warts

Warts, or verrucae (Latin *verruca*, wart), are horny projections on the skin caused by human papillomaviruses. Human papillomavirus (HPV) is the name given to a group of DNA viruses that includes more than 100 different strains, some of which are oncogenic (cancer-causing) (**figure 38.21*a***). They differ in the types of epithelium they infect; some infect cutaneous sites, whereas others infect mucous membranes. More than 30 of these viruses are sexually transmitted and cause genital warts—soft, pink cauliflower-like growths that occur on genitalia and around the rectum (figure 38.21*b*, *c*). Genital warts are the most common sexually transmitted infection in the United States today.

Papillomaviruses are placed in the family *Papillomaviridae*. These viruses have nonenveloped icosahedral capsids with a double-stranded, supercoiled, circular DNA genome. At least eight distinct genotypes produce benign epithelial tumors that vary in respect to their location, clinical appearance, and histopathologic features. Warts occur principally in children and young adults, and are limited to the skin and mucous membranes. The viruses are spread between people by direct contact; autoinoculation occurs through scratching. Four major kinds of warts are **plantar warts, verrucae vulgaris, flat** or **plane warts,** and **anogenital condylomata (venereal warts)**. Treatment includes physical destruction of the wart by electrosurgery, cryosurgery with liquid nitrogen or solid CO_2, laser fulguration (drying), direct application of the drug podophyllum to the wart, or injection of IFN-α.

Anogenital condylomata are sexually transmitted. Once HPV enters the body, the incubation period is 1 to 6 months. Infection with specific types of genital HPV plays a major role in the pathogenesis of epithelial cancers of the male and female genital tracts. HPV strains are designated by numbers and can be divided into "high-risk" (oncogenic) and "low-risk" (nononcogenic) strains. Four high-risk strains (6, 11, 16, and 18) are responsible for 70% of cervical cancers and 90% of genital warts. Specifically they have been found in

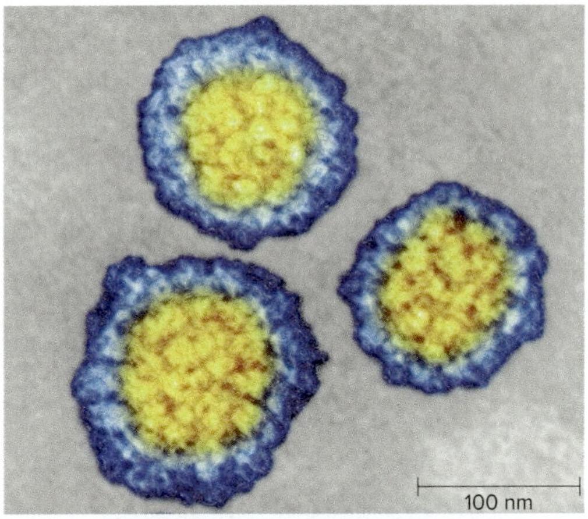

(a)

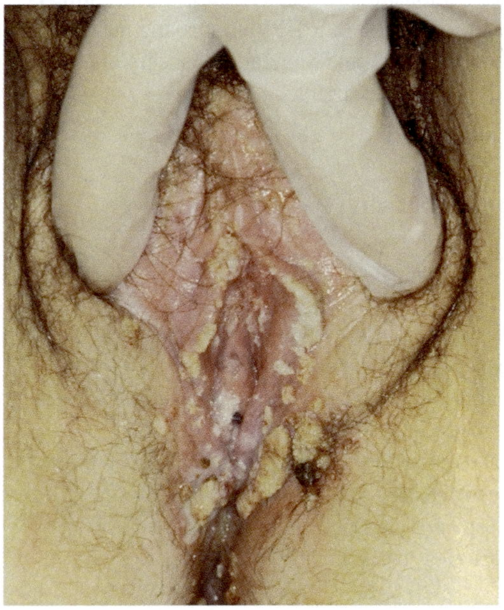

(b)

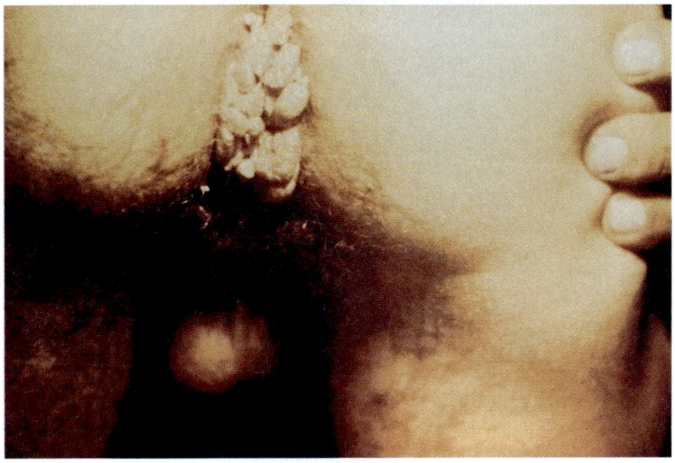

(c)

Figure 38.21 Human Papillomavirus (HPV). (a) Colorized transmission electron micrograph of HPV virions. Each virion is about 60 nm in diameter. (b) Genital warts of the vaginal labia. (c) Genital warts in the anal region.
(a) ©James Cavallini/Science Source; (b) Source: CDC/Joe Miller; (c) Source: CDC

association with invasive cancers of the cervix, vulva, penis, or anus (and other sites). Other high-risk strains have been found to be associated with cancers no more than 1% of the time. These other strains cause benign or low-grade cervical cell changes and genital warts but are rarely, if ever, associated with invasive cancer.

About 110 million Americans are currently infected with HPV and roughly 20 million become newly infected annually. Frequent sexual contact is the most consistent predictor of HPV infection; the number of sexual partners is proportionately linked to the risk of HPV infection. HPV is typically transmitted through direct contact, especially during vaginal or anal sex. Other types of genital contact in the absence of penetration (oral-genital, manual-genital, and genital-genital contact) can also lead to infection with HPV, but infection by these routes is less common. Most HPV infections are asymptomatic, transient, and clear on their own without treatment. However, in some individuals, HPV infections result in genital warts weeks or months after infection.

A quadrivalent vaccine against the four high-risk HPV strains (6, 11, 16, and 18) was licensed in the United States in 2006. The vaccine is made from noninfectious HPV-like particles and is recommended for prophylactic use in females eleven to twenty-six years of age and males eleven to twenty-one years of age. The vaccine has been found to be safe and to cause no serious side effects. Clinical trials have demonstrated 100% efficacy in preventing cervical precancers caused by the four HPV strains and nearly 100% efficacy in preventing vulvar and vaginal precancers and genital warts caused by the same four strains.

Comprehension Check

1. Describe HIV and how it cripples the immune system. What types of pathological changes can result?
2. Why do people periodically get cold sores? Describe the causative agent.
3. Describe two major ways in which herpes simplex virus type 2 is spread. Why do herpes infections become active periodically?
4. Describe the causative agent and some symptoms of mononucleosis and exanthem subitum.
5. What are the different causative viruses of hepatitis, and how do they differ from one another? How can one avoid hepatitis?
6. What viruses cause warts? Describe the formation of venereal warts and a method to prevent infection.

38.4 Food and Water Are Vehicles for Viral Diseases

After reading this section, you should be able to:

a. Describe the viruses that cause common food-borne and waterborne diseases
b. Identify typical signs and symptoms of common food-borne and waterborne viral diseases
c. Correlate food-borne and waterborne virus infection and disease severity with viral virulence factors

Food and water have been recognized as potential carriers (vehicles) of disease since the beginning of recorded history. Collectively more infectious diseases are transmitted by these two routes than any other. A few of the many human viral diseases that are food- and waterborne are now discussed. ▶| *Purification and sanitary analysis ensure safe drinking water (section 43.1)*

Gastroenteritis (Viral)

Acute viral gastroenteritis, inflammation of the stomach or intestines, is caused by six major categories of viruses: adenoviruses, astroviruses, bocaparvoviruses (bocaviruses and parvoviruses), caliciviruses (noroviruses and sapoviruses), parechoviruses, and rotaviruses (**figure 38.22**). The medical importance of these viruses is summarized in **table 38.3**. Acute viral gastroenteritis is a common illness, with a global incidence of 3 to 5 billion cases and 1.4 million deaths annually. Viral diarrhea is an especially common cause of mortality among children less than five years of age in developing countries, with over 800,000 deaths occurring annually. In the United States, acute diarrhea in children accounts for more than 3.5 million cases, resulting in more than 500,000 office visits, 55,000 hospitalizations, and 30 deaths per year. The viruses responsible for gastroenteritis are transmitted by the fecal-oral route.

Two groups of viruses cause the majority of viral gastroenteritis cases: noroviruses in children and adults, and rotaviruses in children. Noroviruses are estimated to cause about 20 million U.S. cases of acute gastroenteritis (at least 50% of all food-borne outbreaks of gastroenteritis). Furthermore, noroviruses are responsible for 1.9 million outpatient visits, 400,000 emergency department visits (primarily in young children), about 56,000 to 71,000 hospitalizations, and 570 to 800 deaths (mostly of young children and the elderly). At least another 20% of global cases and the majority of severe cases of diarrhea in children under age five years are due to rotaviruses, which are the leading cause of severe diarrhea in infants and children

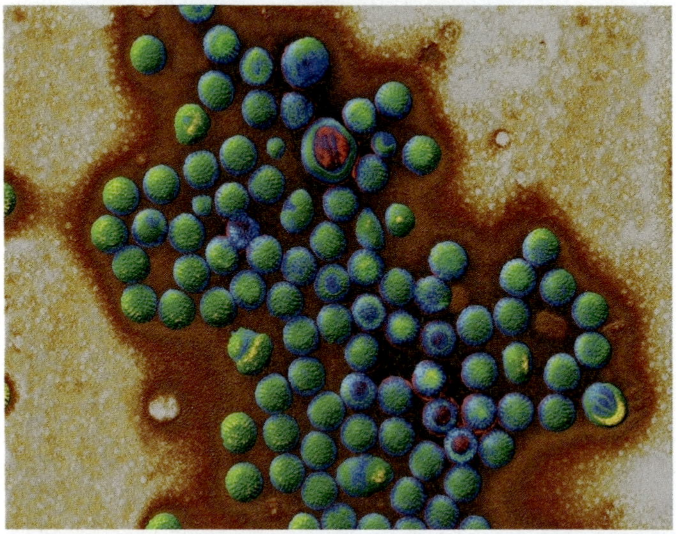

Figure 38.22 Rotavirus Virions. Colorized electron micrograph of rotavirus virions (family *Reoviridae*) in a human gastroenteritis stool filtrate (×90,000). Each virion is about 80 nm in diameter.

Table 38.3 Medically Important Gastroenteritis Viruses

Family	Virus	Characteristics
Adenoviridae	Adenoviruses (subgroup F, serotypes 40 & 41)	Primarily pediatric cases, outbreaks on military bases, more severe in the immunocompromised
Astroviridae	Human Astroviruses	Primarily pediatric cases and outbreaks in nursing homes, lasts 1–3 days
Caliciviridae	Noroviruses (genotypes I and II)	Pediatric epidemics, outbreaks common in older children and adults (families, communities, social events); often associated with shellfish, food handlers with poor hand hygiene, and cruise ships
	Sapoviruses (genogroups 1, 2, 4, & 5)	Primarily pediatric cases with mild diarrhea, rarely adult cases associated with shellfish and other foods
Parvoviridae	Human bocavirus 2	Primarily pediatric cases; mild to severe diarrhea, rarely with acute flaccid paralysis
Picornaviridae	Human parechoviruses	Primarily pediatric cases; mild diarrhea, rarely with myocarditis and encephalitis complications
Reoviridae	Rotaviruses (Groups A-C)	Primarily pediatric cases, large outbreaks in adults; dehydrating diarrhea (5–7 days) with cramps, fever, nausea, and vomiting in pediatric cases, 3–5 days diarrhea in adult cases

worldwide. Prior to the widespread use of vaccine to counter rotavirus, there were more than 400,000 doctor visits, more than 200,000 emergency room visits, 55,000 to 70,000 hospitalizations, and 20 to 60 deaths in the United States annually. Vaccination has reduced yearly rotavirus hospitalizations to fewer than 20,000, and deaths to single digits. Adenoviruses, sapoviruses, and astroviruses are the other major viral agents responsible for gastroenteritis. Genomic testing of patients with gastroenteritis has revealed two additional viral causes: human bocavirus 2 and parechoviruses (table 38.3).

The average incubation period for most of these viral diseases is 1 to 2 days. Viral gastroenteritis is seen most frequently in infants one to eleven months of age, where viruses attack the epithelial cells of the upper intestinal villi, causing malabsorption, impairment of sodium transport, and diarrhea. The clinical manifestations typically range from asymptomatic to a relatively mild diarrhea with headache and fever, to a severe, watery, nonbloody diarrhea with abdominal cramps. Fatal dehydration is most common in young children. Vomiting is often present, especially in children. Viral gastroenteritis can be self-limiting. Treatment is designed to provide relief through the use of oral fluid replacement with isotonic liquids, analgesics, and antiperistaltic agents. Symptoms usually last for 1 to 5 days, and recovery often results in protective immunity to subsequent infection.

Hepatitis A

Hepatitis A (infectious hepatitis) usually is transmitted by fecal contamination of food or drink, or shellfish that live in contaminated water. The disease is caused by hepatitis A virus (HAV) of the genus *Hepatovirus* in the family *Picornaviridae*. Recall that while all hepatitis viruses can cause liver disease, they are not taxonomically related (table 38.2). HAV is an icosahedral, linear, positive-strand RNA virus that lacks an envelope. Once in the digestive system, the virus multiplies within the intestinal epithelium. Usually only mild intestinal symptoms result. Occasionally viruses are found in the blood (viremia) and may spread to the liver. The virus reproduces in the liver, enters the bile, and is released into the small intestine. This explains why feces are so infectious. After about a 4-week incubation period, symptoms develop that include anorexia, general malaise, nausea, diarrhea, fever, chills, and jaundice. Most cases resolve in 4 to 6 weeks and yield a strong immunity, although some patients relapse, exhibiting symptoms for 6 months or more. The mortality rate is low (less than 1%), and about 40 to 80% of the U.S. population has antibodies, though few are aware of having had the disease. Control of infection is by simple hygienic measures, the sanitary disposal of excreta, and the HAV vaccine. The number of new cases has been dramatically reduced since the introduction of the hepatitis A vaccine in the 1990s. This vaccine is recommended for travelers (*see table 36.3*) going to regions with high rates of hepatitis A.

Hepatitis E

Hepatitis E caused by HEV genotypes 1 and 2 is implicated in many epidemics in developing countries, while genotypes 3 and 4 are found in Europe and, to a lesser extent, the United States. The single, positive-strand RNA viral genome (7,900 nucleotides) is linear. The virion is spherical, nonenveloped, and 32 to 34 nm in diameter (table 38.2).

Infection with HEV genotypes 1 and 2 usually is associated with feces-contaminated drinking water, whereas genotypes 3 and 4 are transmitted by contaminated food (particularly undercooked pork). HEV enters the blood from the gastrointestinal tract, replicates in the liver, is released from hepatocytes into the bile, and is subsequently excreted in the feces. Like hepatitis A, an HEV infection usually runs a benign course and is self-limiting. The incubation period varies from 15 to 60 days, with an average of 40 days. The disease is most often seen in patients who are fifteen to forty

HISTORICAL HIGHLIGHTS

38.1 A Brief History of Polio

Like many other infectious diseases, polio is probably of ancient origin. Various Egyptian hieroglyphics dated approximately 2000 BCE depict individuals with wasting, withered legs and arms (**box figure**). In 1840 the German orthopedist Jacob von Heine described the clinical features of poliomyelitis and identified the spinal cord as the problem area. Little further progress was made until 1890, when Oskar Medin, a Swedish pediatrician, portrayed the natural history of the disease as epidemic in form. He also recognized that a systemic phase, characterized by minor symptoms and fever, occurred early and was complicated by paralysis only occasionally. Major progress occurred in 1908, when Karl Landsteiner and William Popper successfully transmitted the disease to monkeys. In the 1930s much public interest in polio occurred because of the polio experienced by U.S. President Franklin D. Roosevelt. This led to the founding of the March of Dimes campaign in 1938; the sole purpose of the March of Dimes was to collect money for research on polio. In 1949 John Enders, Thomas Weller, and Frederick Robbins discovered that poliovirus could be propagated in vitro in cultures of human embryonic tissues of nonneural origin. This was the keystone that later led to the development of vaccines.

In 1952 David Bodian recognized that there were three distinct serotypes of poliovirus. Jonas Salk successfully immunized humans with formalin-inactivated poliovirus in 1952, and this vaccine (IPV) was licensed in 1955. The live attenuated poliovirus vaccine (oral polio vaccine, OPV) developed by Albert Sabin and others was employed in Europe in 1960 and was licensed for U.S. use in 1962. Both the Salk and Sabin vaccines led to a dramatic decline of paralytic poliomyelitis. Polio is likely to be the second disease eradicated through global vaccination efforts.

Ancient Egyptian with Polio. Note the withered leg.
©Bettmann/Getty Images

years of age. Children are typically asymptomatic or present mild signs and symptoms, including abdominal pain, anorexia, dark urine, fever, enlarged liver, jaundice, malaise, nausea, and vomiting. Case fatality rates are low (1 to 3%), except for pregnant women (15 to 25%), who risk death from fulminant hepatic failure. Diagnosis of HEV infection is by detection of anti-HEV antibodies or reverse transcriptase PCR. There are no specific measures for preventing HEV infections, other than those aimed at improving the level of health and sanitation in affected areas.

Poliomyelitis

Poliomyelitis (Greek *polios*, gray; *myelos*, marrow or spinal cord), **polio,** or **infantile paralysis** is caused by poliovirus (species *Human enterovirus C*) and was first described in England in 1789 as a "leg-wasting" disease of children, although a pictograph from ancient Egypt clearly depicts a man with what appears to be polio (**Historical Highlights 38.1**). The virus is an enterovirus—a transient inhabitant of the gastrointestinal tract—and a member of the family *Picornaviridae* and as such is a nonenveloped, positive-strand RNA virus. Three different poliovirus subtypes have been identified. Strain 2 was declared eradicated in 2015 and strain 3 has not been detected since 2012. Three vaccines are used: One tripartite vaccine (containing all three strains) and two monovalent vaccines (one with strain 1, the other contains strain 3). Poliovirus virions are very stable, especially at acidic pH, and can remain infectious for relatively long periods in food and water, its main routes of transmission. The incubation period ranges from 6 to 20 days.

Once ingested, the virus multiplies in the mucosa of the throat or small intestine. From these sites, the virus invades the tonsils and lymph nodes of the neck and terminal portion of the small intestine. Generally there are either no symptoms or a brief illness characterized by fever, headache, sore throat, vomiting, and loss of appetite. The virus sometimes enters the bloodstream, causing viremia. In more than 99% of cases, the viremia is transient and clinical disease does not result. However, in a minority of cases, the viremia persists and the virus enters the central nervous system and causes paralytic polio. The virus has a high affinity for anterior horn motor nerve cells of the spinal cord. Once inside these cells, it multiplies and destroys the cells, resulting in motor and muscle paralysis. Since the licensing of the tripartite Salk vaccine (1955) and the tripartite Sabin vaccine (1962), the incidence of polio has decreased markedly. No wild polio viruses exist in the United States. An ongoing global polio eradication effort has been very successful and only 22 cases were reported in 2017. Three polio-endemic countries remain: Afghanistan, Nigeria, and Pakistan. Substantial efforts are being made to eradicate polio and certify these remaining countries as polio-free. However, war and misinformation diminish vaccination efforts. Nonetheless, it is likely that polio, like smallpox, will be completely eradicated.

In addition to polio, there are over 100 other enteroviruses that all have the potential to invade the human central nervous system. But like poliovirus, this is very rare. In 2014 there was an outbreak of one such virus, enterovirus D68, that targeted young children with respiratory problems, such as asthma, causing life-threatening respiratory distress and in a few cases paralysis. Occasional, sporadic cases of enterovirus D68 continue to be diagnosed.

Comprehension Check

1. What virus groups are associated with acute viral gastroenteritis? How do they cause the disease's symptoms?
2. Describe some symptoms of hepatitis A.
3. Compare and contrast measures that prevent transmission of hepatitis A and E.
4. At what specific sites within the body can the poliomyelitis virus multiply? What is the usual outcome of an infection?

38.5 Zoonotic Diseases Arise from Human-Animal Interactions

After reading this section, you should be able to:

a. Discuss the viruses that cause common diseases spread by contact with animals
b. Identify typical signs and symptoms of zoonotic viral diseases
c. Correlate zoonotic virus infection and disease severity with viral attachment factors

The diseases discussed here are caused by viruses that are normally zoonotic (animal-borne). Members of the RNA virus families *Arenaviridae, Bunyaviridae, Flaviviridae, Filoviridae,* and *Picornaviridae* are notable examples of viruses found in animal reservoirs before transmission to and between humans. Although many different animals harbor viruses that also infect people, bats appear to be the most common zoonotic reservoir.

Ebola Virus and Marburg Diseases

Ebola virus disease (EVD) is caused by Ebola viruses, first recognized near the Ebola River in the Democratic Republic of the Congo in Africa. They are members of the genus *Ebolavirus* in the family *Filoviridae*, a group of filamentous, negative-strand RNA viruses (**figure 38.23**). Five Ebola species are known: *Sudan ebolavirus, Tai Forst ebolavirus* (formerly *Côte d'Ivoire ebolavirus), Bundibugyo ebolavirus, Zaire ebolavirus,* and *Reston ebolavirus*. Members of the first four species cause disease in humans. The fifth, *Reston ebolavirus*, was first discovered in 1989, and to date Reston viruses have caused disease only in nonhuman primates and pigs. Only Reston virus is spread by aerosol transmission. All five types of Ebola virus are transmitted by contact with body fluids. Convalescing patients continue to harbor virus in body fluids and in rare instances there has been sexual transmission through infected semen up to 17 months after recovery. Bats are thought to be the natural reservoir for Ebola viruses, although the exact species is not known.

Viral hemorrhagic fever (VHF) is the term used to describe the severe, multisystem syndrome seen in Ebola epidemics. Because the host vascular system is damaged, vascular leakage (hemorrhage) accompanied by the blood's diminished ability to clot (coagulopathy) occurs. The incubation period for EVD ranges from 2 to 21 days and is characterized by abrupt fever, headache, joint and muscle aches, sore throat, and weakness, followed by diarrhea, vomiting, and stomach pain. Signs of infection include fever, rash, red eyes, bleeding, and hiccups—symptoms alerting of internal hemorrhage. Minor hemorrhage is symptomatic in all ebolavirus infection. Severe hemorrhage is late stage and may be minimized if patients are resuscitated in time. There is no standard treatment for Ebola infection. Patients receive supportive therapy consisting of balancing patients' fluids and electrolytes, maintaining their oxygen status and blood pressure, and treating them for any complicating

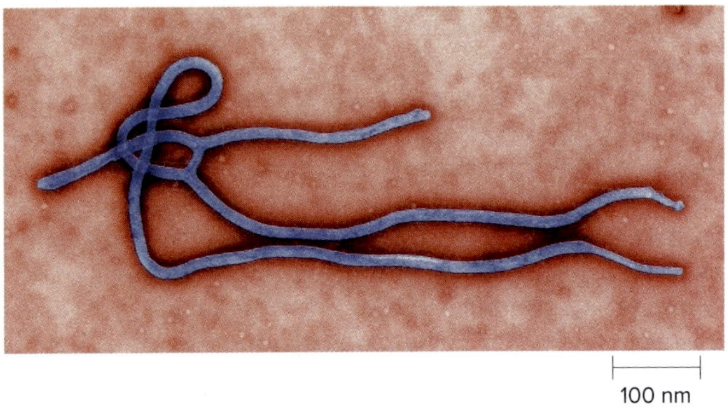

Figure 38.23 Ebola Virus Virions.
Source: CDC/Cynthia Goldsmith

infections. Convelscent sera, collected from Ebola survivors, and experimental antibody cocktails demonstrated promising success in the 2014 treatments of ebolavirus patients. In 2017, a vaccine shown to be highly efficacious in clinical trials was approved for use during an Ebola outbreak in the Democratic Republic of the Congo.

Marburg disease is caused by a genetically unique RNA virus also in the *Filoviridae* family. Marburg virus (MARV) similarly causes a hemorrhagic fever. It is a rare, severe type of hemorrhagic fever that affects both humans and nonhuman primates. MARV was first recognized in 1967, when outbreaks of hemorrhagic fever occurred simultaneously in laboratories in Marburg and Frankfurt, Germany, and in Belgrade, Yugoslavia (now Serbia). The first people infected had been exposed to African green monkeys or their tissues. In Marburg, the monkeys had been imported for research and to prepare polio vaccine. MARV is indigenous to Africa, and the reservoir host of MARV is the African fruit bat, *Rousettus aegyptiacus*. The average incubation period for Marburg hemorrhagic fever is 5 to 10 days. The disease symptoms are abrupt, marked by fever, chills, headache, and myalgia. A maculopapular rash (i.e., discolored with bumps), most prominent on the chest, back, and stomach, typically appears around the fifth day after onset. Nausea, vomiting, chest pain, sore throat, abdominal pain, and diarrhea may also occur in infected patients. Symptoms become increasingly severe and may include jaundice, delirium, liver failure, pancreatitis, severe weight loss, shock, and multiorgan dysfunction. A specific treatment for this disease is unknown and is limited to supportive hospital therapy.

Hantavirus Pulmonary Syndrome

Hantavirus pulmonary syndrome (HPS) is a disease caused by a negative-strand RNA virus of the family *Bunyaviridae*. HPS is typically transmitted to humans by inhalation of viral particles shed in urine, feces, or saliva of infected rodents. Because it does not cause disease in the rodent hosts, they are considered the reservoir. Although rare in the United States, HPS is potentially deadly. Rodent control in and around the home remains the primary strategy for preventing hantavirus infection. HPS in the United States is not transmitted from person to person, nor is it known to be transmitted by rodents purchased from pet stores.

Viral transfer may occur between rodents through biting, as field studies suggest that viral transmission in rodent populations occurs horizontally and more frequently between males. In the environment, virus particles can remain infective for 2 to 3 days. Hantaviruses have lipid envelopes susceptible to most disinfectants, so prevention involves cleaning areas that may be contaminated with rodent droppings. A specific treatment for HPS is unknown. Supportive therapy is used to treat symptoms, including balancing the patient's fluids and electrolytes, maintaining oxygen status and blood pressure, replacing lost blood and clotting factors, and treating for other complicating infections.

Rabies

Rabies (Latin *rabere*, rage or madness) is caused by a number of different strains of highly neurotropic viruses. Most belong to a single serotype in the genus *Lyssavirus* (Greek *lyssa*, rage or rabies), family *Rhabdoviridae*. The enveloped, bullet-shaped virion contains a negative-strand RNA genome (**figure 38.24**). Rabies has been the object of human fascination, torment, and fear since the disease was first recognized. Prior to Pasteur's development of a rabies vaccine, few words were more terrifying than the cry of "mad dog!" Improvements in prevention during the past 50 years have led to almost complete elimination of indigenously acquired rabies in the United States, where rabies is primarily a disease of feral animals and domestic pets that contact feral animals. Many wild mammals can become infected with rabies, but susceptibility varies according to species. Foxes, coyotes, and wolves are the most susceptible; intermediate are skunks, raccoons, insectivorous bats, and bobcats. Worldwide, almost all cases of human rabies are attributed to dog bites. In the more than 150 developing countries and territories, where canine rabies is still endemic, rabies accounts for about 59,000 deaths per year. In the United States, fewer than 10 cases are reported annually. Feral raccoons, skunks, foxes, and bats account for the majority of cases. Rabid animals may exhibit signs of agitation and aggression (known as furious rabies), although paralysis (dumb rabies) is the more common sign exhibited by rabid animals.

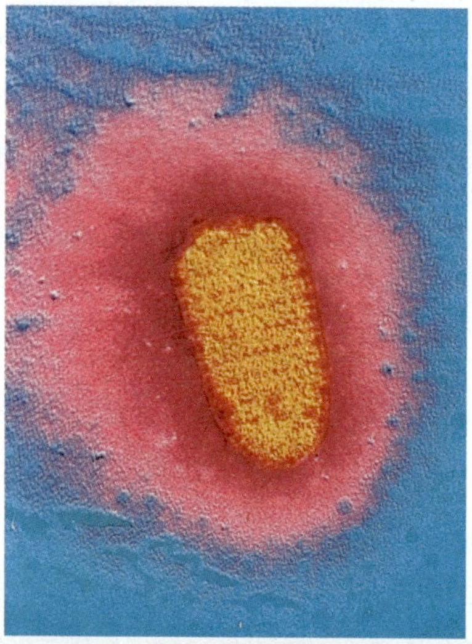

Figure 38.24 Rabies. Electron micrograph of a rabies virus virion (yellow) (×36,700). Note the bullet shape. The virus envelope contains spikelike glycoprotein projections that bind specifically to cellular receptors. The virus measures about 180 nm long and 75 nm wide.

MICRO INQUIRY *What type of host cell does the rabies virus initially infect? To where does the virus then travel?*

The virus multiplies in the salivary glands of an infected host. It travels to the brain, where the virus may change the animal's behavior. The virus also damages the pharyngeal nerve, making swallowing difficult. This causes the animal to "foam at the mouth" with saliva laden with virions. It is transmitted to humans or other animals by biting; by aerosols of the virus that can be spread in caves where bats dwell; or by contamination of scratches, abrasions, open wounds, and mucous membranes with saliva from an infected animal. After inoculation, the virion's envelope spike attaches to the plasma membrane of nearby skeletal muscle cells, which the virus enters. Multiplication of the virus then occurs. When the concentration of the virus in the muscle is sufficient, the virus enters the nervous system through unmyelinated sensory and motor terminals; the binding site is the nicotinic acetylcholine receptor.

The virus spreads by retrograde axonal flow at 8 to 20 mm per day until it reaches the spinal cord, when the first specific symptoms of the disease—pain or paresthesia at the wound site—may occur. A rapidly progressive encephalitis develops as the virus quickly disseminates through the central nervous system. The virus then spreads throughout the body along the peripheral nerves, including those in the salivary glands, where it is shed in the saliva.

Within brain neurons, the virus produces characteristic Negri bodies, masses of virions or unassembled viral subunits that are visible in the light microscope. In the past, diagnosis of rabies consisted solely of examining tissue for the presence of these bodies. Today diagnosis is based on *d*irect immuno*f*luorescent *a*ntibody (DFA), serology for antirabies antibody, and by PCR testing. ◀ *Identification of microorganisms from specimens (section 37.2)*

Symptoms of rabies in humans usually begin 2 to 16 weeks after exposure and include anxiety, irritability, depression, fatigue, loss of appetite, fever, and a sensitivity to light and sound. The disease quickly progresses to paralysis. In about 50% of all cases, intense and painful spasms of the throat and chest muscles occur when the victim swallows liquids. The mere sight, thought, or smell of water can set off spasms. Consequently, rabies has been called hydrophobia (fear of water). Once symptoms of rabies develop in a human, the disease has a 100% mortality rate. Death results from destruction of the regions of the brain that regulate breathing. Safe and effective vaccines (*h*uman *d*iploid-*c*ell rabies *v*accine [HDCV] or *r*abies *v*accine *a*dsorbed [RVA]) against rabies are available. Prevention and control involve preexposure vaccination of dogs and cats, and preexposure vaccination of humans at special risk, including veterinarians and persons spending a month or more in countries where rabies is common in dogs.

Individuals suspected of rabies exposure are given rabies immune globulin for passive immunity and rabies vaccine for active immunity to exploit the relatively long incubation period of the disease. Rabies immune globulin is typically administered directly into the wound after cleaning. The vaccine is administered as four doses given on days 3, 7, and 14 after the initial dose. This is usually recommended for anyone bitten by one of the common reservoir species (raccoons, skunks, foxes, and bats), unless it is proven that the animal was uninfected.

A global initiative to make rabies vaccines more available and less expensive is under way. The goal is to vaccinate at least 70% of dogs in developing nations by 2030 in an attempt to break the infection cycle. Because rabies has a variety of animal hosts, it cannot be eradicated.

Comprehension Check

1. What factors make the Ebola and Marburg hemorrhagic fever diseases so contagious and deadly?
2. What precautions can be taken to prevent hantavirus and rabies virus transmission to humans?
3. What is the rationale in treating suspected rabies with vaccination?

38.6 Prion Proteins Transmit Disease

After reading this section, you should be able to:

a. Describe diseases caused by prions
b. Differentiate prion disease by route of infection

Prion diseases, also called **transmissible spongiform encephalopathies (TSEs)**, are fatal neurodegenerative disorders that have attracted enormous attention not only for their unique biological features but also for their impact on public health. Prions consist of abnormally folded proteins (PrP^{Sc}) that induce normal forms of the protein (PrP^{C}) to fold abnormally (*see figure 6.24*). This group of diseases includes scrapie, kuru, Creutzfeldt-Jakob disease (CJD), variant Creutzfeldt-Jakob disease (vCJD), Gerstmann-Sträussler-Scheinker disease (GSD), and fatal familial insomnia (FFI). Scrapie occurs primarily in sheep. The first of these diseases to be studied in humans was kuru, discovered in the Fore tribe of New Guinea. Carlton Gadjusek and others showed that the disease was transmitted by ritual cannibalism (especially where brains and spinal cords were eaten). The primary symptom of the human disorders is dementia, usually accompanied by manifestations of motor dysfunction such as cerebral ataxia (inability to coordinate muscle activity) and myoclonus (shocklike contractions of muscle groups). FFI is also characterized by dysautonomia (abnormal functioning of the autonomic nervous system) and sleep disturbances. These symptoms appear insidiously in middle to late adult life and last from months (CJD, FFI, and kuru) to years (GSD) prior to death. ◀ *Prions are composed only of proteins (section 6.7)* *Prion Diseases*

Neuropathologically these disorders produce a characteristic spongiform degeneration of the brain, as well as deposition of amyloid plaques (**figure 38.25**). Prion diseases thus share important clinical, neuropathological, and cell biological features with another, more common cerebral amyloidosis, Alzheimer's disease. Prion diseases are organized into three subcategories: familial (inherited) types, sporadic types, and acquired types. Familial

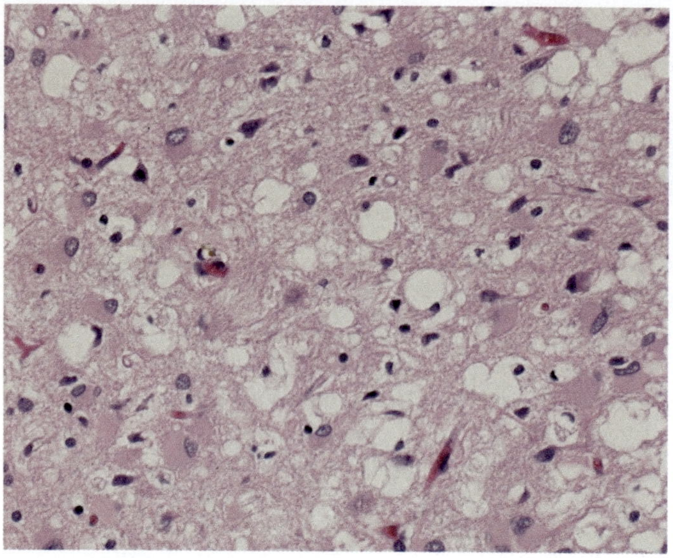

Figure 38.25 Variant Creutzfeldt-Jakob Disease (vCJD). vCJD is a prion-dependent disease that results in prominent spongiotic (holey) changes in the cortex, and loss of neurons. Note the spongiotic appearance of this brain tissue photomicrograph (magnified 100X and stained with hematoxylin and eosin).
Source: CDC/Teresa Hammett

CJD is the result of spontaneous mutation within the prion-encoding gene. Certain residues when mutated are known to give rise to CJD, but it is not clear why these amino acid substitutions confer inducible misfolding to the (PrPSc) conformation. Familial CJD is very rare, accounting for 5 to 15% of all CJD cases. It progresses rapidly and is always fatal, usually within 1 year of onset of illness.

Sporadic CJD (sCJD) is caused by a spontaneous mutation in the PrP^{C} protein coding gene. It has a global incidence of one case per million people per year and accounts for 75% of global CJD cases. There are no known gender, environmental, or lifestyle associations that would induce sCJD. This form of the disease is also rapidly progressive; symptoms move from depression and loss of coordination and walking, to loss of vision and speech, to incontinence and immobility, over an average of 4 to 5 months. Seventy percent of patients die within 6 months of symptom onset.

Variant CJD (vCJD) is acquired when a human consumes meat products from a BSE cow. Such transmission occurred during the 1980s and early 1990s in the United Kingdom, resulting in 175 cases of vCJD reported between 1996 and 2011. These data suggest a 5-year incubation period. Iatrogenic CJD is acquired from a physician or surgeon, a medical treatment, or diagnostic procedures. The most common source of iatrogenic CJD has been prion-contaminated human growth hormone, corneal grafts, and grafts of dura mater (tissue surrounding the brain). Donor screening and more thorough testing of grafts have decreased the frequency of prion transmission. Iatrogenic CJD represents less than 5% of global CJD cases (*see chapter 8 opening story*).

Comprehension Check

1. How are prions different from viruses? How are they similar?
2. In what way are spongiform encephalopathies commonly acquired?

Key Concepts

38.1 Viruses Can Be Transmitted by Airborne Routes

- Chickenpox and shingles are caused by the same herpesvirus, varicella zoster virus (VZV). Like all members of the *Herpesviridae,* VZV establishes a latent infection, as it can give rise to shingles many years later. Both chickenpox and shingles are vaccine preventable (**figures 38.1** and **38.2**).
- Two genera cause the flu, *Influenzavirus A* and *Influenzavirus B,* and are distinguished by their spike proteins, hemagglutinin (HA) and neuraminidase (NA). Antigenic drift is the steady accumulation of point mutations. Large changes in host specificity and virulence occur when two different strains infect the same host cell and swap RNA genome segments in a process called antigenic shift (**figure 38.3**).
- Measles virus is highly contagious; measles is a vaccine-preventable disease that kills over 130,000 children each year. It is characterized by a rash, fever, conjunctivitis, and congestion; it can also cause encephalitis that may result in permanent brain damage or death (**figure 38.4**).
- A number of viruses cause respiratory syndromes that range from the common cold (rhinovirus, adenovirus, coronavirus) to pneumonia and other respiratory syndromes. Respiratory syncytial virus (RSV) causes a respiratory syndrome in infants and toddlers that can be severe. There is currently no vaccine.
- Rubella is a vaccine-preventable disease that typically causes mild illness except when contracted during the first trimester of pregnancy, when it can cause congenital rubella syndrome (**figure 38.6**).
- Smallpox is caused by variola virus and was eradicated in 1977 (**figure 38.7**). Until that time smallpox extracted a large global burden with a 20 to 50% mortality rate; those who survived suffered lifelong disfigurement with scars from the pustular rash.

38.2 Arthropods Can Transmit Viral Diseases

- The arthropod-borne viral diseases are transmitted by arthropod vectors from human to human or animal to human. In many cases, the majority of infections are mild or go unnoticed. However, for each disease discussed, a small population of patients develops severe, sometimes deadly, disease.

- Chikungunya, dengue, and Zika are mosquito (*Aedes* spp.) borne diseases that cause rash, headache, and muscle and joint aches. An individual with a history of dengue who is infected with a different serotype may develop dengue hemorrhagic fever as a result of antibody-dependent enhancement of virus uptake (**figure 38.8**). Zika was almost unknown until 2015 when it invaded Brazil and other American countries. If contracted during pregnancy, it can cause devastating birth defects, most notably microcephaly.
- Since its introduction to New York in 1999, West Nile virus has spread across the continental United States. West Nile fever is mosquito (*Culex* spp.) borne; birds are its natural host. Humans are considered dead end hosts. There is no vaccine or antiviral treatment.

38.3 Direct Contact Diseases Can Be Caused by Viruses

- HIV is a sexually transmitted retrovirus. If infection is left untreated AIDS will develop after a variable period of latency. HIV targets immune cells by using the host's CD4 protein as the viral receptor. AIDS is characterized by the loss of immune function and the patient is overwhelmed with opportunistic infections. HIV cannot be cured and there is no vaccine. A combination of antiretroviral drugs has made living with HIV a chronic condition (**figures 38.9–38.13; table 38.1**).
- Both cold sores (**figure 38.14**) and genital herpes (**figure 38.17**) are caused by herpesviruses: Herpes simplex virus (HSV)-1 and HSV-2, respectively. They cause painful vesicular rashes when active and reside in neurons when latent. Physical and psychological stress can trigger latent viruses to become active.
- Viruses that cause hepatitis are not taxonomically related (**table 38.2**). Those transmitted by direct contact include hepatitis B virus (HBV, a DNA virus), hepatitis C virus (HCV, an RNA virus), and hepatitis D virus (HDV, a satellite virus dependent on HBV). Hepatitis caused by HBV (with or without HDV) or HCV can be acute or chronic and can lead to hepatocellular carcinoma. Hepatitis caused by HBV (and therefore HDV) is vaccine preventable; recently developed antiviral drugs can cure most hepatitis C infections.
- Cytomegalovirus (CMV), human herpesvirus 6 (HHV-6) infections, and human parvovirus B19 are typically infections of childhood, while Epstein-Barr virus (EBV) infects children asymptomatically. In young adults EBV causes infectious mononucleosis. CMV, HHV-6, and EBV are members of the *Herpesviridae,* and thus are enveloped DNA viruses that develop latency in all those who have been infected.

38.4 Food and Water Are Vehicles for Viral Diseases

- Viruses transmitted in food and water usually grow in the intestinal system and leave the body in feces (**table 38.3**).
- Viral gastroenteritis is the most common form of food-borne illness. Norovirus is most common in adults and children, rotavirus is common in young children globally; it is vaccine preventable (**figure 38.22**).
- Hepatitis acquired by fecal contamination of food or water is caused by hepatitis A, which is vaccine preventable, and hepatitis E.
- Poliovirus is an enterovirus and like other enteroviruses, usually causes mild gastrointestinal upset. In a small minority of cases, poliovirus and other enteroviruses infect the central nervous system. A global polio eradication program has made great strides in eliminating polio.

38.5 Zoonotic Diseases Arise from Human-Animal Interactions

- Diseases transmitted from animals are zoonotic.
- Examples of viral zoonoses include Ebola (**figure 38.23**) and Marburg diseases, hantavirus pulmonary syndrome, and rabies (**figure 38.24**).
- These viruses cause severe disease with high mortality rates. Rabies can be treated with a combination of active and passive immunization.

38.6 Prion Proteins Transmit Disease

- A prion disease is a pathological process caused by a transmissible protein (a prion) that remains clinically silent for a prolonged period, after which the clinical disease becomes apparent.
- Examples of prion diseases include Creutzfeldt-Jakob disease, variant Creutzfeldt-Jakob disease, kuru, Gerstmann-Sträussler-Scheinker disease, and fatal familial insomnia. These diseases are chronic infections of the central nervous system that result in progressive degenerative changes and eventual death.

Active Learning

1. You are a county health commissioner who has been informed of a new mosquito-borne disease. Mosquitoes in your county just tested positive for the virus. Most people are asymptomatic if infected but about 2% of individuals may develop encephalitis (brain infection). Discuss the efforts you will put into place to protect and educate the citizens of your county.
2. Flaviviruses like West Nile and dengue viruses are often referred to as arboviruses. Hepatitis C virus is also in the genus *Flavivirus* but is not considered an arbovirus. Why is this the case?
3. Infection with some viruses, like measles virus and hepatitis A virus, confers strong lifelong immunity. However, this is not the case with many other viruses such as dengue virus, rhinovirus, coronavirus, and influenza virus. Discuss the factors that contribute to this difference.
4. Discuss the biological, sociological, and political factors that made it possible to eradicate smallpox. Consider these factors in the context of polio eradication; how do they compare? Do you predict eradication efforts for any other viruses?
5. In 2005 a reconstructed influenza virus containing eight genes from the 1918 Spanish influenza virus was introduced into human lung epithelial cells and mice. This virus was able to rapidly reproduce and cause sudden illness and death in mice. In 2008 these genes were replaced on an individual basis with homologues from a contemporary H1N1 influenza virus and tested for virulence. Substitution of the 1918 gene with the H1N1 hemagglutinin (HA), neuraminidase (NA), or polymerase PB1 subunit severely diminished the virulence of the 1918 virus.

   Explain why each of these genes is so critical for infection and virulence. Do you predict an additive effect if more than one gene were replaced? Explain your answer.
6. Avian influenza viruses (AIVs) pose a public health threat if they gain the capacity to infect humans. The H5N6 AIV emerged in China in 2013 where it has caused several human deaths. To understand how this virus is evolving, a group of Chinese scientists collected specimens from poultry markets in central China. The complete genomes of 175 H5N6 AIVs were sequenced, revealing six different genotypes that arose through genetic reassortment. Refer to figure 38.3 and discuss how the scientists were able to arrive at this conclusion. Will all changes make the virus more likely to infect humans? What changes do you think would be most alarming?

   Read the original paper: Du, Y., et al. 2017. Molecular evolution and emergence of H5N6 avian influenza virus in central China. *J. Virol.* 91:e00143-17. doi.org/10.1128/JVI.00143-17.

39

Human Diseases Caused by Bacteria

©VCG Wilson/Corbis via Getty Images

The Plague Family Tree

Almost everyone has heard of the Black Death—the outbreak of plague in Europe from 1347 to 1351 when as much as 40% of the population died. Caused by the Gram-negative bacterium *Yersinia pestis*, it was called Black Death because victims bled under the skin and experienced necrosis (cell death) of the extremities, which turned skin purple-black.

For almost as long as the Black Death has been studied, it has been believed that the outbreak in 1347 and other waves of plague that followed over the next several hundred years were independent introductions of *Y. pestis* from Asia. But new evidence suggests that this may not be the case. Europe may have reciprocated and provided Asia with a more virulent *Y. pestis* strain.

How do you study a disease outbreak that happened over 500 years ago? It helps to have a team of archaeologists, who can find, document, and exhume graves, and a team of forensic anthropologists and bacteriologists, who can extract and sequence *Y. pestis* DNA from teeth and bones. Studies of plague victims in Spain, England, Germany, and Russia now give a more complete, albeit complex, story of *Y. pestis* evolution. It is broadly agreed that the story starts about 1320 with a plague outbreak in the Gobi Desert of Mongolia; from there it spread to China in the 1330s. It came to Europe in 1347 on a dozen trade ships that docked in Sicily, where accounts from the time describe dead sailors covered in black boils. From Sicily, the disease quickly migrated north, reaching Russia by 1351.

After the plague epidemic receded in the mid-1350s, it reemerged every few generations for the next three centuries, as depicted in the adjacent painting by Josse Lieferinxe dated 1497. If Europeans were the victims of independent waves of *Y. pestis* introduction from Asia, bacterial DNA from victims of each outbreak should demonstrate genetically different strains. However, *Y. pestis* DNA from victims spanning the fourteenth to seventeenth centuries is highly similar to earlier victims, suggesting it was the same strain that evolved right up until the last outbreak in France in 1722. Some scientists believe that the Black Death *Y. pestis* strain became more virulent and made its way back to China, where it caused outbreaks in the nineteenth and twentieth centuries.

Why does anyone even care about a disease that happened so long ago? Diseases emerge and move. We need to understand the past to respond in the future. In this chapter we learn about bacteria that cause human disease. Although it is good to remember that only a tiny fraction of bacteria are pathogens, it is wise to become acquainted with those that impact humans.

Readiness Check:

Based on what you have learned previously, you should be able to:

- ✓ Review basic bacterial cell biology (sections 3.2–3.9)
- ✓ Explain how key pathogens cause infection (chapter 35)
- ✓ Compare and contrast the general principles of innate and adaptive immunity (chapter 32; sections 33.1–33.8)
- ✓ Differentiate between different types of vaccines (section 36.6)
- ✓ Review methods by which pathogenic bacteria are identified (chapter 37)

39.1 Bacteria Can Be Transmitted by Airborne Routes

After reading this section, you should be able to:

a. Discuss bacteria that cause common airborne diseases
b. Identify typical signs and symptoms of airborne bacterial diseases
c. Correlate airborne bacterial infection and disease severity with bacterial virulence factors

Most airborne bacterial diseases infect the respiratory tract. Inhalation of infectious particles or self-inoculation can cause disease of the sinuses, throat, bronchus, or lungs. From the lungs, some bacteria can disseminate into the bloodstream, particularly in elderly and immunocompromised individuals. Some of the most common of these diseases are now discussed.

Chlamydial Pneumonia

Chlamydial pneumonia is caused by *Chlamydophila pneumoniae.* The Gram-negative chlamydiae are obligate intracellular parasites, therefore they must grow and reproduce within host cells. Although their ability to cause human disease is widely recognized, many species grow within protists and animal cells as their natural reservoir. Infection begins when bacteria in the elementary body (EB) morphology are inhaled (*see figure 21.14*). EBs evolved exclusively for transmission and infection. Host cells phagocytose the EBs, which remain in inclusion bodies, where they reorganize to form reticulate bodies (RBs). About 8 to 10 hours after infection, RBs begin binary fission, and replicate for 20 to 25 hours. They then differentiate into infectious EBs. The host cell lyses and releases EBs 48 to 72 hours after infection. ◀ *Phylum* Chlamydiae: *obligate intracellular parasites (section 21.5)*

Clinically, infections are generally mild; pharyngitis, bronchitis, and sinusitis commonly accompany some lower respiratory tract involvement. Symptoms include fever, cough that may be productive (respiratory secretions brought up by coughing), sore throat, hoarseness, and pain on swallowing. Infections with *C. pneumoniae* are common but sporadic. Approximately 5 to 10% of all cases of adult pneumonia and bronchitis are caused by *C. pneumoniae,* and about 50% of adults have antibodies against this pathogen, indicating prior exposure. *C. pneumoniae* appears to be primarily a human pathogen directly transmitted from human to human by droplet (respiratory) secretions. Diagnosis of chlamydial pneumonia is based on symptoms and serology to detect IgM antibody or nucleic acid amplification. Macrolides are the first line of treatment; tetracycline and fluoroquinolones are also effective.

In seroepidemiological studies, *C. pneumoniae* infections have been linked with coronary artery disease, as well as vascular disease at other sites. Although *C. pneumoniae* genes and antigens have been detected in artery plaque, the bacterium itself is inconsistently recovered in cultures of artery plaque. Thus a possible etiologic role of *C. pneumoniae* infection in coronary artery disease and systemic atherosclerosis remains unclear.

Diphtheria

Diphtheria (Greek *diphthera,* membrane; *-ia,* condition) is a vaccine-preventable disease caused by the Gram-positive bacterium *Corynebacterium diphtheriae* (*see figure 23.5*). *C. diphtheriae* is well adapted to airborne transmission by way of nasopharyngeal secretions because it is very resistant to drying. Interestingly, the *tox* gene that encodes diphtheria toxin is found on a prophage, so only *C. diphtheriae* lysogens that carry this prophage cause disease. This is an example of lysogenic conversion.

Diphtheria toxin is absorbed into the circulatory system and distributed throughout the body, where it may cause destruction of cardiac, kidney, and nervous tissues by inhibiting protein synthesis (**figure 39.1**). The toxin is an AB toxin encoded by a single gene that gives rise to a protein with two domains, A and B (*see figure 35.8*). The B domain binds to the heparin-binding epidermal growth factor receptor on the surface of various eukaryotic cells. Once bound, the toxin enters the cytoplasm by endocytosis. The B domain contains a transmembrane region that embeds itself in the endosome membrane. Acidification of the endosome breaks a disulfide bond holding the A and B domains together and releases the A domain into the cytoplasm. There it catalyzes the attachment of ADP-ribose (from NAD^+) to translation elongation factor-2 (EF-2). A single enzyme (i.e., catalytic domain) can exhaust the entire supply of cellular EF-2 within hours, resulting in protein synthesis inhibition and cell death. ◀ *Order* Corynebacteriales *includes important human pathogens (section 23.1); Exotoxins (section 35.4)*

Typical signs and symptoms of diphtheria include a thick nasal discharge containing both mucus and pus, pharyngitis, fever, cough, and paralysis. The cardinal sign of diphtheria is a thick, grayish pseudomembrane on the pharynx and respiratory mucosa consisting of dead host cells and *C. diphtheriae* bacteria. Most current diphtheria cases involve people over thirty years of age who have a weakened immunity and live in tropical areas. The vaccine consists of inactivated toxin, making it a toxoid vaccine. Between 1980 and 2016, fewer than 10 diphtheria cases were reported annually in the United States, and most occurred in nonimmunized individuals. ◀ *Vaccines immunize susceptible populations (section 36.6)*

Legionnaires' Disease

In 1976 the term **Legionnaires' disease,** or **legionellosis,** was coined to describe an outbreak of pneumonia that occurred at an American Legion convention in Philadelphia. The bacterium responsible for the outbreak was *Legionella pneumophila*, a nutritionally fastidious, aerobic, Gram-negative rod (**figure 39.2**). A variety of free-living amoebae and ciliated protozoa serve as the natural reservoir for *L. pneumophila.* In addition to living in freshwater ecosystems, these host protists inhabit cooling water found in air-conditioning systems and are an important part of the infection cycle. In fact, most outbreaks are traced to contaminated cooling water systems. ◀ *Order* Legionellales *includes intracellular pathogens (section 22.3)*

Legionella spp. multiply intracellularly within protozoa, just as they do within human macrophages. There is no human-to-human spread of legionellosis. Whereas most healthy people

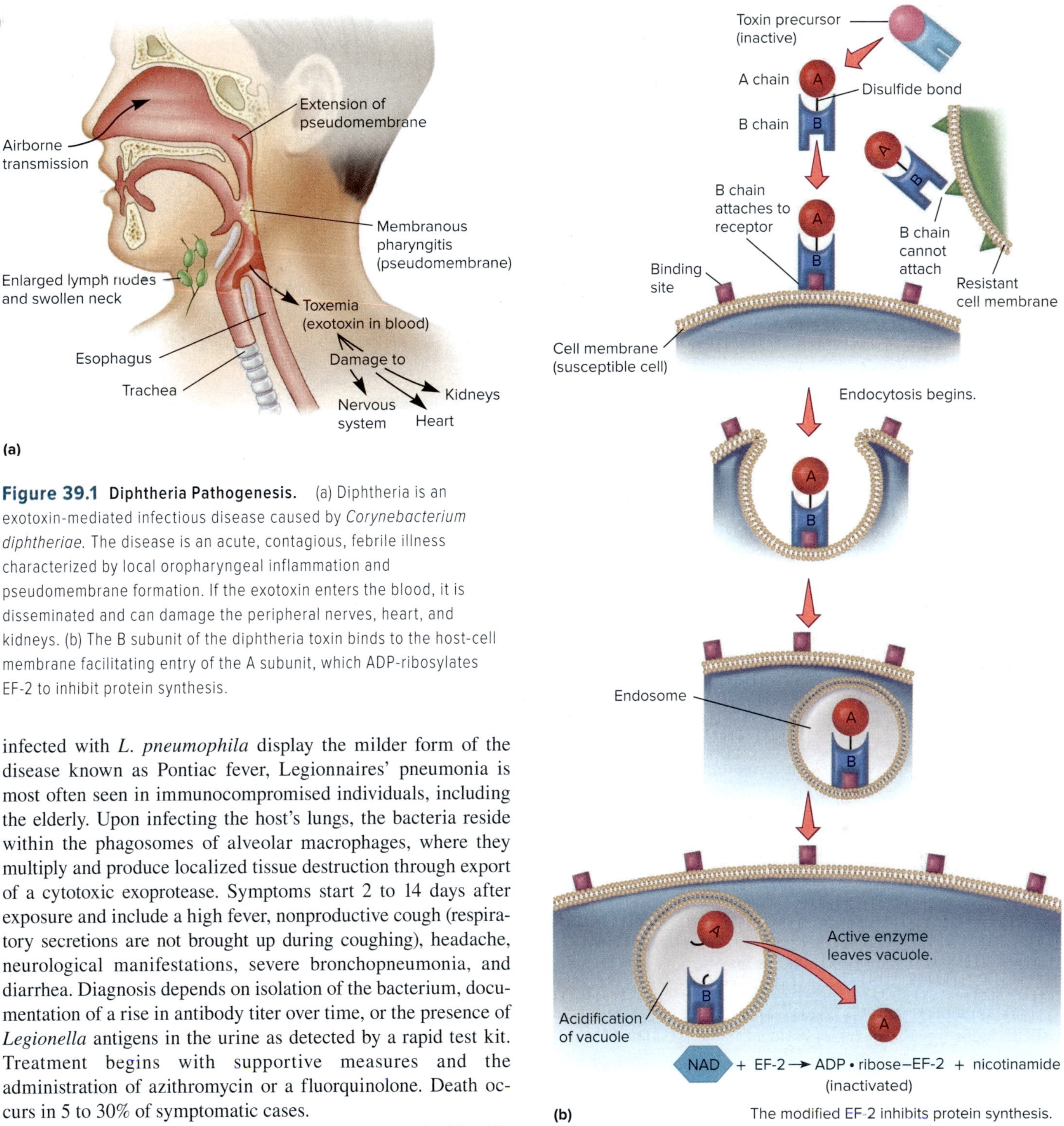

Figure 39.1 Diphtheria Pathogenesis. (a) Diphtheria is an exotoxin-mediated infectious disease caused by *Corynebacterium diphtheriae*. The disease is an acute, contagious, febrile illness characterized by local oropharyngeal inflammation and pseudomembrane formation. If the exotoxin enters the blood, it is disseminated and can damage the peripheral nerves, heart, and kidneys. (b) The B subunit of the diphtheria toxin binds to the host-cell membrane facilitating entry of the A subunit, which ADP-ribosylates EF-2 to inhibit protein synthesis.

infected with *L. pneumophila* display the milder form of the disease known as Pontiac fever, Legionnaires' pneumonia is most often seen in immunocompromised individuals, including the elderly. Upon infecting the host's lungs, the bacteria reside within the phagosomes of alveolar macrophages, where they multiply and produce localized tissue destruction through export of a cytotoxic exoprotease. Symptoms start 2 to 14 days after exposure and include a high fever, nonproductive cough (respiratory secretions are not brought up during coughing), headache, neurological manifestations, severe bronchopneumonia, and diarrhea. Diagnosis depends on isolation of the bacterium, documentation of a rise in antibody titer over time, or the presence of *Legionella* antigens in the urine as detected by a rapid test kit. Treatment begins with supportive measures and the administration of azithromycin or a fluorquinolone. Death occurs in 5 to 30% of symptomatic cases.

Prevention of Legionnaires' disease depends on the identification and elimination of the environmental source of *L. pneumophila*. Chlorination, ozonation, the heating of water, and the cleaning of water-containing devices can help control the multiplication and spread of *L. pneumophila*. A variety of regulatory control measures are now in place to reduce the incidence of Legionnaires' disease in hospitals, hotels, and so on. However, use of private hot tubs, fountains, and other water systems that aerosolize water can also be sources of bacteria. Since the initial outbreak of this disease in 1976, many outbreaks during summer months have been recognized in all parts of the United States. About 8,000 to 18,000 people are hospitalized each year with Legionnaires' disease, and another 18,000 or more mild or subclinical cases are thought to occur.

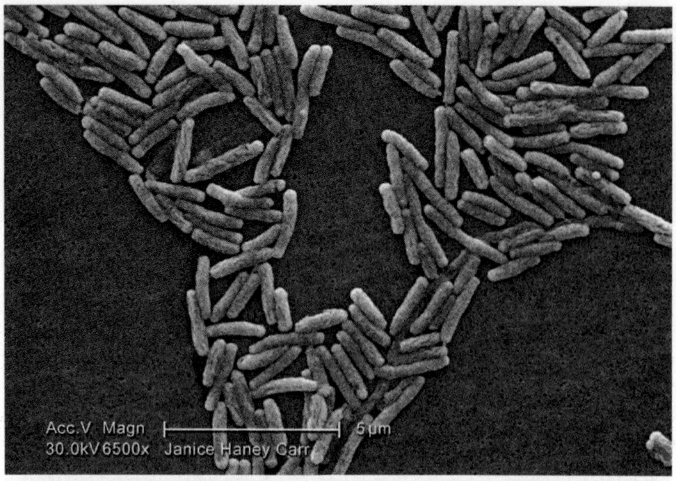

Figure 39.2 ***Legionella pneumophila.*** This peritrichously flagellated Gram-negative bacterium is the causative agent of Legionnaires' disease; SEM (×10,000).
Source: CDC/Margaret Williams, PhD; Claressa Lucas, PhD; Tatiana Travis, BS

MICRO INQUIRY *What organisms serve as the environmental reservoir for* L. pneumophila?

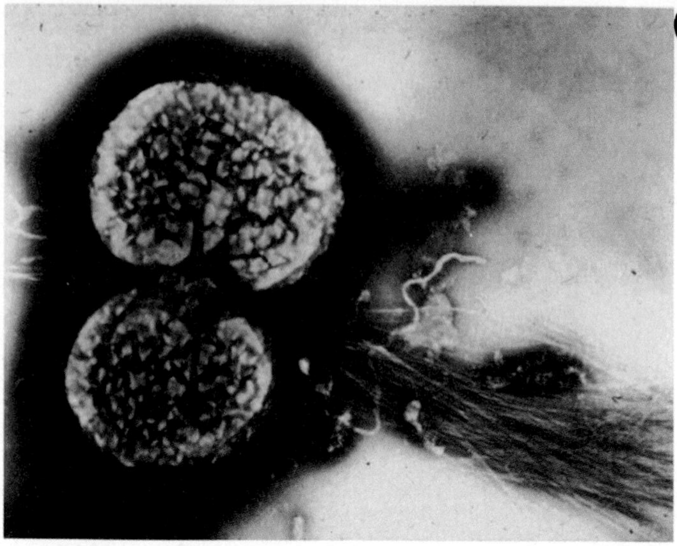

Figure 39.3 ***Neisseria meningitidis.*** Type IV pili are present on the cell surface as bundled filamentous appendages, facilitating attachment. Each cell is about 0.6 to 1.0 µm in diameter
©Tone Tonjum

Meningitis

Meningitis (Greek *meninx*, membrane; *-itis*, inflammation) is an inflammation of the brain or spinal cord meninges (membranes). In addition to many viruses, fungi, and protozoa, a variety of bacteria can cause meningitis. In adults and children over the age of about two years, the most common bacterial agents are *Streptococcus pneumoniae*, *Neisseria meningitidis*, and *Haemophilus influenzae* serotype b. Meningitis in infants is most frequently caused by group B streptococci (*S. agalactiae*), *Listeria monocytogenes*, and *E. coli*. Bacterial meningitis can also be the result of *Mycobacterium tuberculosis*, *Nocardia asteroides*, and *Staphylococcus aureus* infection.

Accurate identification of the causative agent is essential for proper treatment of meningitis. However, this can sometimes be difficult because meningitis symptoms may be present without a microbial agent in cerebral spinal fluid (CSF) as determined by Gram staining and culture. In such cases, the diagnosis often is called aseptic (meaning lack of an identifiable agent) meningitis. This can be caused by a virus or protozoan. Aseptic meningitis is typically more difficult to treat as therapy targeting a specific microorganism is not possible.

The immediate sources of the bacteria responsible for meningitis are respiratory secretions from carriers. *N. meningitidis*, often referred to as meningococcus, is a normal inhabitant of the human nasopharynx (5 to 10% of humans carry nonpathogenic serotypes). Most disease-causing *N. meningitidis* strains belong to serotypes A, B, C, Y, and W-135. In general, serotype A, and to a lesser extent W-135, strains are the cause of epidemic disease in Africa, whereas serotype B, C, Y, and W-135 strains are responsible for meningitis outbreaks in other countries. The disease process is initiated by pili-mediated colonization of the nasopharynx by pathogenic bacteria (**figure 39.3**). The bacteria cross the nasopharyngeal epithelium and the blood-brain barrier (typically through endocytosis) to enter the central nervous system. Often, *N. meningitidis* also invades the bloodstream (meningococcemia), where they proliferate.

Symptoms caused by *N. meningitidis* are variable, depending on whether meningococcemia is established. The usual symptoms of meningitis include one or more of the meningeal syndromes: vomiting, headache, lethargy, confusion, photophobia (sensitivity to light), and stiff neck. If the bacteria have spread to the bloodstream, a characteristic, rapidly progressing, purple rash develops along with septic shock. Once bacterial meningitis is suspected, ceftriaxone, a third-generation cephalosporin, is administered immediately. Penicillin may be used after it is established that the causative strain has not acquired penicillin resistance. Antibiotics (ciprofloxicin or rifampin) should also be administered prophylactically to patient contacts; if left untreated, meningococcal meningitis is fatal.

Control of *N. meningitidis* infection is with vaccination and antibiotics. Two vaccine types, meningococcal polysaccharide vaccine and meningococcal conjugate vaccine, are available. Both vaccine types are effective against serotypes A, C, Y, and W-135. Two vaccines that use recombinant proteins (antigen) are available for protection against serogroup B.

Another agent of meningitis is *H. influenzae* serotype B (Hib), a Gram-negative bacterium. Transmission is by inhalation of droplet nuclei shed by infectious individuals or carriers. Hib infects mucous membranes, resulting in sinusitis, pneumonia, bronchitis, and epiglottitis in young, unvaccinated children. In addition to gaining access to the CSF, it can disseminate to the bloodstream and cause bacteremia. *H. influenzae* disease has declined to fewer than one case per 100,000 children annually in the United States. Globally, it is estimated that *H. influenzae* serotype b causes at least 3 million cases of serious disease and almost 400,000 vaccine-preventable deaths each year.

Three to 6% of all *H. influenzae* infections are fatal. Furthermore, up to 20% of surviving patients have permanent hearing loss or other long-term sequelae. Therefore it is important that all children begin vaccination with the Hib conjugate vaccine at the age of two months.

Mycobacterium Infections

Mycobacteria are high G + C Gram-positive bacteria that normally inhabit soil, water, and house dust. They cause a number of human infections that are difficult to treat. Recall that mycobacteria have cell walls containing mycolic acids. These complex fatty acids have 60 to 90 carbon atoms with a shorter β-hydroxy chain and a longer α-alkyl side chain (*see figure 23.7*). The presence of mycolic acids and other lipids outside the peptidoglycan layer makes mycobacteria acid-fast (*see figure 2.19*). ◀ *Differential staining (section 2.3)*

M. avium Complex

Three mycobacteria are noteworthy pathogens in the United States: *Mycobacterium avium, Mycobacterium intracellulare,* and *M. chimaera*. They are so closely related that they are referred to as the *M. avium* complex (MAC). MAC has become the most common cause of mycobacterial infections in the United States. ◀ *Order* Corynebacteriales *includes important human pathogens (section 23.1)*

MAC is found worldwide and infects a variety of insects, birds, and animals. Both the respiratory and the gastrointestinal tracts have been proposed as entry portals for MAC; however, person-to-person transmission is not very efficient. MAC causes a pulmonary infection in humans similar to that caused by *M. tuberculosis*. Pulmonary MAC is more common in elderly persons with preexisting pulmonary disease and individuals with suppressed immune systems. The gastrointestinal tract is the most common site of colonization and dissemination in HIV-positive patients. It produces disabling symptoms, including fever, malaise, weight loss, night sweats, and diarrhea.

MAC can be isolated from sputum, blood, and aspirates of bone marrow. Cultures and acid-fast stains are of value in making a diagnosis. Like all mycobacterial infections, MAC is treated with a combination of drugs. Because no drug combination is currently approved by the U.S. Food and Drug Administration (FDA) for the therapy of MAC, each isolate needs to be tested for antibiotic sensitivity. Often a macrolide and an aminoglycoside along with ethambutol paired with rifamycin are given for an extended period of time.

Mycobacterium tuberculosis

More than a century ago, Robert Koch identified *Mycobacterium tuberculosis* as the causative agent of **tuberculosis (TB)**. We now know TB can also be caused by *M. bovis* (which can be transmitted by drinking unpasteurized milk) and *M. africanum*. Together these three species make up *M. tuberculosis* complex. TB is a global health problem of enormous dimension. It is estimated that one-third of the world's human population is infected, with at least 2 billion people with latent infections. There were 10.4 million people infected with TB in 2015, and 1.8 million deaths. In 2016, there were 9,287 new TB cases in the United States, where TB occurs most commonly among the homeless, elderly, malnourished, alcoholic males, minorities, immigrants, prison populations, and Native Americans. It appears that about one-fourth to one-third of active TB cases in the United States may be due to recent transmission. Thus the majority of active cases result from the reactivation of latent infections.

Infection results when the bacteria are inhaled. The infectious dose is quite low, around 10 bacterial cells. The incubation period is about 4 to 12 weeks, and the disease develops slowly. The symptoms of TB are fever, fatigue, night sweats, and weight loss. A cough, which is characteristic of pulmonary involvement, may result in expectoration of bloody sputum.

Inhaled *Mycobacterium tuberculosis* (Mtb) cells infect resident macrophages. Although innate immune defenses are raised, mycobacteria survive macrophage phagocytosis by blocking lysosome fusion with the phagosome. Infected macrophages fuse and a focus of infection begins. Upon activation of the adaptive arm of the immune system, T_H1 and cytotoxic T lymphocytes (CTLs) are recruited to the site of infection. These cells typically do little to clear the infection, but their additional cytokines help maintain a stalemate between pathogen and host in the form of **granulomas,** whereby infected macrophages are surrounded by other immune cells (**figure 39.4**). As granulomas persist, two things happen. First, infected macrophages in the center die, imparting a cheesy appearance and the granuloma is said to be caseating. Second, calcium is deposited in and around the granuloma. In new infections, these appear on X rays and are called **Ghon complexes.** As the granulomas enlarge, the degree of calcification also increases, and these regions are called **tubercles.** Tubercles are commonly seen in latent infections. Sometimes the tubercle lesions liquefy and form air-filled **tuberculous cavities.** From these cavities, the bacteria can be disseminated to new foci throughout the body (**figure 39.5**). This spreading, seen most commonly in children and immunocompromised people, is called **miliary tuberculosis** due to the development of many tubercles the size of millet seeds. Mycobacteria can disseminate anywhere in the host, including the brain, bones, and gut. Fortunately, most people with TB have the latent form of the disease, in which bacteria are confined to granulomas and are held in a relatively dormant state by the host's immune defenses. If the host immune system is stressed, **reactivation tuberculosis** may result. In this case, bacteria at the initial site of infection (i.e., the lungs) begin dividing, and the patient develops the signs or symptoms of TB.

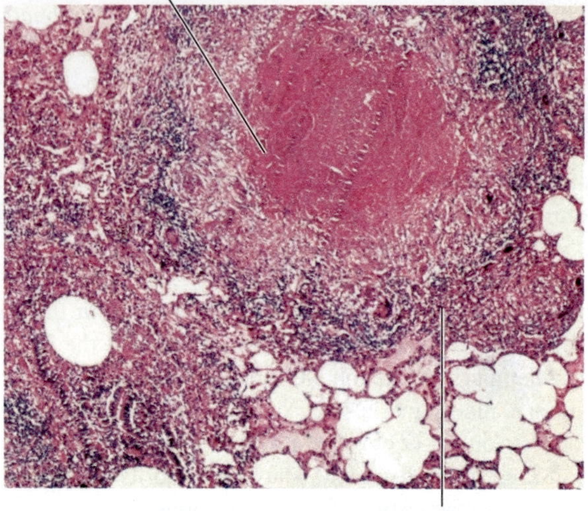

Figure 39.4 ***Mycobacterium tuberculosis* Resides in Granulomas.** In the lungs, tuberculosis is identified by the tubercle, a granuloma of white blood cells, bacteria, fibroblasts, and epithelioid cells. The center of the tubercle contains caseous (cheesy) pus and bacteria.
©Biophoto Associates/Science Source

Mtb does not produce typical virulence factors such as exotoxins, capsules, and fimbriae. The unique lipids and glycolipids in the cell envelope of Mtb are directly toxic to eukaryotic cells and create a hydrophobic barrier around the bacterium that facilitates impermeability and resistance to antimicrobial agents. It also protects against killing by acidic and alkaline compounds, osmotic lysis, and lysozyme. Furthermore, cell wall glycolipids associate with mannose, giving Mtb control over entry into macrophages by exploiting the macrophage mannose receptors. |◀ *Phagocytosis destroys invaders (section 32.6)*

TB must be treated with combination antimicrobial therapy. Four drugs are initially administered simultaneously: isoniazid (INH), rifampin (Rif), ethambutol, and pyrazinamide. After 6 to 8 weeks, only two drugs, INH and Rif, are administered for a total of 6 months as a way of decreasing the possibility that the bacterium develops drug resistance. However, ***multidrug-resistant strains of tuberculosis* (MDR-TB)** have developed and are spreading. A multidrug-resistant strain is an Mtb strain resistant to INH and rifampin, with or without resistance to other drugs. Inadequate therapy is the most common means by which resistant bacteria are acquired, and patients who have previously undergone therapy are presumed to harbor MDR-TB until proven otherwise. In addition to MDR-TB strains, other Mtb strains have emerged that are e*x*tensively *d*rug-*r*esistant (XDR); they are resistant to INH, rifampin, as well as second-line antibiotics. XDR-TB is a unique challenge to public health as first-line and second-line drug resistance severely limits treatment options. For this reason, MDR-TB and XDR-TB have a higher mortality rate than drug-sensitive TB.

Drug-resistant TB arises because Mtb cells within granulomas have spontaneous, predictable rates of chromosomal mutations that confer resistance to drugs. These mutations are unlinked; hence resistance to one drug is not associated with resistance to an unrelated drug. In the circumstances of monotherapy, erratic drug compliance, omission of one or more drugs, suboptimal dosage, poor drug absorption, or an insufficient number of active drugs in a regimen, a susceptible strain of *M. tuberculosis* may become resistant to multiple drugs within a matter of months. This is why many public health organizations worldwide practice *d*irectly *o*bserved *t*reatment *s*hort course (DOTS), in which each dose is taken in the presence of a health-care worker.

Persons infected with Mtb develop cell-mediated immunity because the bacteria stimulate a strong T_H1-cell response. These sensitized T cells are the basis for the tuberculin skin test. In this test, a *p*urified *p*rotein *d*erivative (PPD) of Mtb is injected intradermally. If the person has had TB or has been exposed to Mtb, sensitized T cells react with these proteins and a delayed hypersensitivity reaction occurs within 48 hours. This positive skin reaction appears as an induration (hardening) and reddening of the area around the injection site. A positive skin test cannot differentiate between active or latent tuberculosis, or prior immunization with the BCG vaccine (not used in the United States). Therefore, a positive PPD test needs to be confirmed with a chest X ray. Laboratory tests include culture, assays that measure cytokine release by a patient's immune cells in response to MTb antigens, microscopy of the acid-fast bacterium, culture, and commercially available DNA probes. |◀ *Type IV hypersensitivity (section 33.10)*

Mycoplasmal Pneumonia

Bacterial pneumonia is most frequently caused by *Streptococcus pneumoniae*, which presents with fairly consistent signs and symptoms (section 39.6). If the symptoms of pneumonia are different from what is usually observed, the disease is called **atypical pneumonia.** One cause of atypical pneumonia is *Mycoplasma pneumoniae*, a bacterium that lacks a cell wall and has worldwide distribution. Transmission involves close contact and airborne droplets. It is seen principally in older children and young adults. Because these otherwise healthy people are not incapacitated by the infection, it has acquired the nickname "walking pneumonia." |◀ *Class* Mollicutes, *phylum* Tenericutes: *bacteria that lack cell walls (section 21.3)*

M. pneumoniae lacks a cell wall so cells vary in shape and stain Gram negative. Furthermore, their lack of peptidoglycan makes them resistant to all β-lactam antibiotics (e.g., penicillin). Mycoplasmas typically colonize the upper respiratory tract and subsequently move to the lower respiratory tract, where they attach to respiratory mucosal cells. *M. pneumoniae* releases a toxin called community acquired respiratory distress syndrome toxin (CARDS). This toxin triggers the formation of vacuoles in host cells, as well as the release of high levels of cytokines that mediate allergic responses (IL-4, IL-13; *see table 32.2*). The activity of the CARDS toxin accounts for the accumulation of eosinophils and wheezing seen in walking pneumonia. Symptoms also

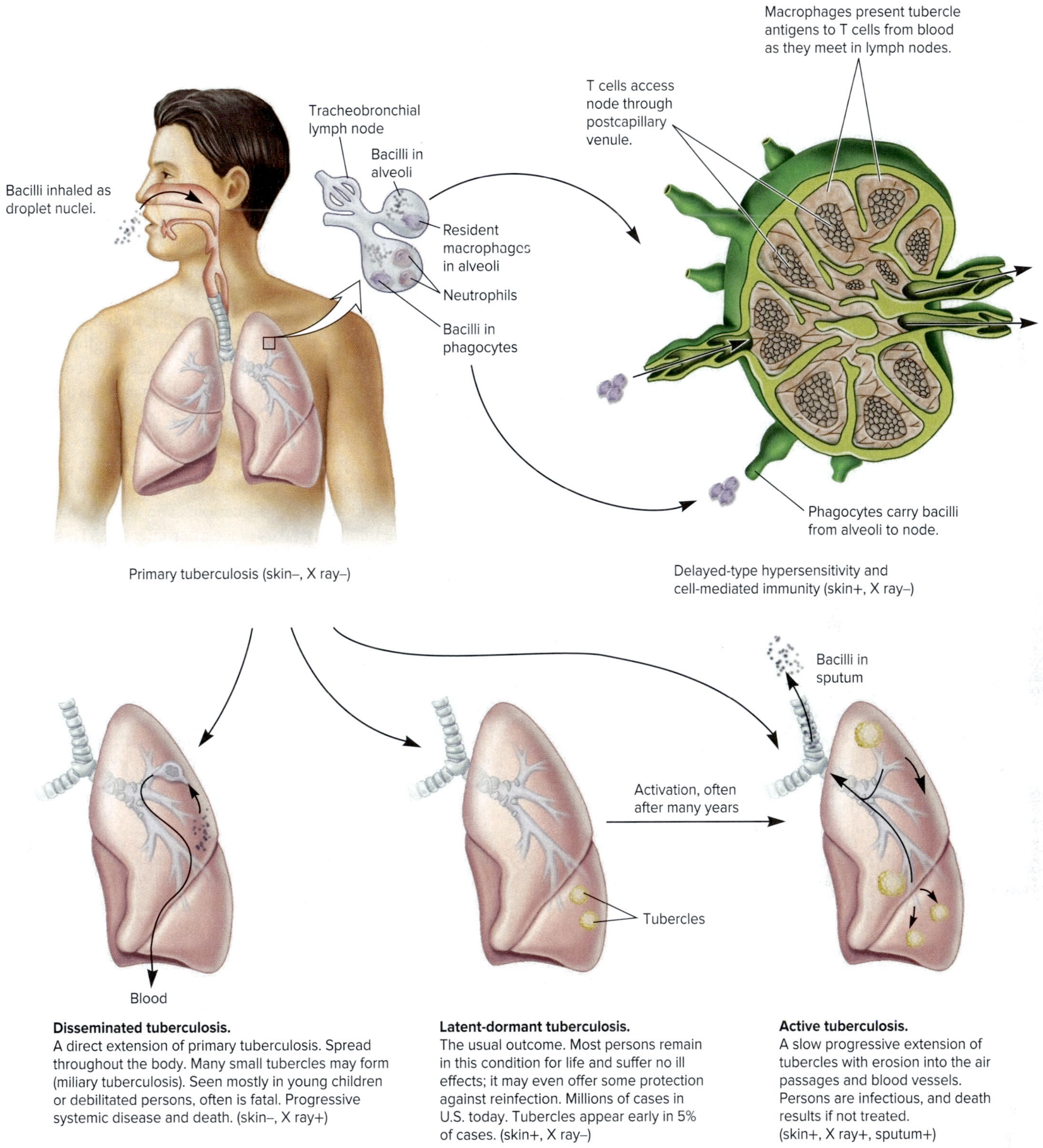

Figure 39.5 Tuberculosis.

include headache, weakness, a low-grade fever, and a predominant, characteristic cough. The disease and its symptoms usually persist for weeks. The mortality rate is less than 1%.

Several rapid tests that detect *M. pneumoniae* antibodies are available for diagnosis of mycoplasmal pneumonia; it can also be detected by nucleic acid amplification. When isolated from respiratory secretions, some mycoplasmas form distinct colonies with a "fried-egg" appearance on agar (*see figure 21.3*). During the acute stage of disease, diagnosis must be made by clinical observations. Tetracycline or macrolides are effective in treatment.

Pertussis

Pertussis (Latin *per*, intensive; *tussis*, cough), also called "whooping cough," is caused by the Gram-negative bacterium *Bordetella pertussis. B. parapertussis* is a closely related species that causes a milder form of the disease. Pertussis bacteria colonize the respiratory epithelium to produce a disease characterized by fever, malaise, uncontrollable cough, and cyanosis (bluish skin color resulting from inadequate tissue oxygenation). Whooping cough gets its name from the characteristically prolonged and paroxysmal coughing that ends in an inspiratory gasp, or whoop. Pertussis is a highly contagious, vaccine-preventable disease that primarily affects children. The World Health Organization (WHO) estimates there are approximately 16 million global cases of pertussis per year, with 195,000 deaths, mostly of infants less than six months old. The overall number of pertussis cases had been increasing in the United States since the 1980s, reaching a peak of 48,277 cases (and 20 deaths) in 2012. The number of cases has since declined to about 20,000 per year, with fewer than 10 deaths annually. The increase in pertussis in the United States is principally a result of the replacement of a whole-cell killed vaccine, which caused an unacceptable number of allergic responses, to an acellular subunit vaccine. Although the acellular vaccine has fewer short-term complications, immunity wanes by late adolescence. Because the tetanus, diphtheria, and pertussis vaccines are formulated in a single dose (DTaP or TDaP), adults receive booster shots when tetanus vaccines are renewed every 10 years. ◀ *Order* Burkholderiales *includes recently evolved pathogens (section 22.2); Vaccines immunize susceptible populations (section 36.6)*

Disease transmission occurs by inhalation of the bacterium in droplets released from an infectious person. The incubation period is 7 to 14 days. Once inside the upper respiratory tract, the bacteria colonize the cilia of the mammalian respiratory epithelium through fimbria-like structures called filamentous hemagglutinins that bind to complement receptors on phagocytes. Additionally, some of the components of the pertussis toxin assist in adherence to cilia by bridging bacterial and host cells.

B. pertussis produces several toxins, of which the pertussis toxin (PTx) is most important. PTx is an AB exotoxin secreted by a type IV secretion system (*see figure 35.8*). The B portion of the toxin is composed of five polypeptides that bind to specific carbohydrates on host cell surfaces and mediate transport of the A subunit from the plasma membrane to the cytoplasm. The A subunit is an ADP-ribosyl transferase, similar to the diphtheria toxin. However, PTx adds ADP-ribose moieties to a regulator of the host cell's G protein coupled receptor, Gi. This disables Gi, which normally functions to inhibit host adenyl cyclase. Without this molecular brake, adenyl cyclase is locked in the active state and ATP is continuously converted to cyclic AMP (cAMP). The resulting increase in intracellular levels of cAMP accelerates mucin secretion and alters water transport, affecting electrolyte balance. *B. pertussis* also produces several other toxins, including one with its own adenyl cyclase activity, which further increases intracellular cAMP. Other toxins include tracheal cytotoxin and dermonecrotic toxin, which destroy epithelial tissue. Working together, Ptx, adenyl cyclase toxin, and tracheal cytotoxin provoke secretory cells in the respiratory tract to produce nitric oxide, which kills nearby ciliated cells, inhibiting removal of bacteria and mucus. The secretion of thick mucus also impedes ciliary action. Breathing becomes so difficult, coughing is followed by the whooping sound as patients try to catch their breath.

Pertussis is divided into three stages: (1) The catarrhal (Latin *catarrhus,* flow down) stage, so named because the mucous membrane inflammation resembles the common cold. (2) The paroxysmal stage is characterized by prolonged coughing sieges. During this stage, the infected person tries to cough up mucous secretions by making 5 to 15 rapidly consecutive coughs followed by the characteristic whoop. Coughing can be so violent it is not uncommon for patients to fracture a rib. The catarrhal and paroxysmal stages last about 6 weeks. (3) The convalescent stage, when final recovery occurs, may take several months.

Laboratory diagnosis of pertussis is by culture of the bacterium and nucleic acid amplification. The development of a strong, lasting immunity takes place after infection. Treatment is with a macrolide. Treatment ameliorates clinical illness when begun during the catarrhal phase and may also reduce the severity of the disease when begun within 2 weeks of the onset of the paroxysmal cough. After this time, antibiotic therapy has only minimal impact.

Streptococcal Diseases (Group A)

Pathogenic streptococci, commonly called strep, are a heterogeneous group of Gram-positive bacteria. *Streptococcus pyogenes* are in group A. These β-hemolytic streptococci are of major importance. The different serotypes of **group A streptococci (GAS)** produce a number of virulence factors, including (1) extracellular enzymes that break down host molecules; (2) streptokinases, enzymes that activate a host blood factor that dissolves blood clots; (3) the cytolysins streptolysin O and streptolysin S, which kill host leukocytes; and (4) capsules and M protein, which limit phagocytosis (**figure 39.6**). M protein, an alpha helical coiled-coil protein anchored in the streptococcal cell membrane, facilitates attachment to host cells and prevents opsonization by complement. The amino acid sequence of M proteins varies among serotypes of *S. pyogenes*, and different M proteins mediate interaction with different host receptors. ◀ *Order* Lactobacillales *(section 23.2); Complement is a cascading enzyme system (section 32.3)*

In addition to the M protein, the surface of *S. pyogenes* is covered with a capsule made of hyaluronic acid, a molecule also found in humans. There is no vaccine for *S. pyogenes*. In general, the best antigen for a vaccine against streptococci is the capsule, as is the case with *S. pneumoniae*. However, because hyaluronic acid is also found in the host, it is not immunogenic. The best control measure is prevention of transmission. Individuals with a known infection should be isolated and treated.

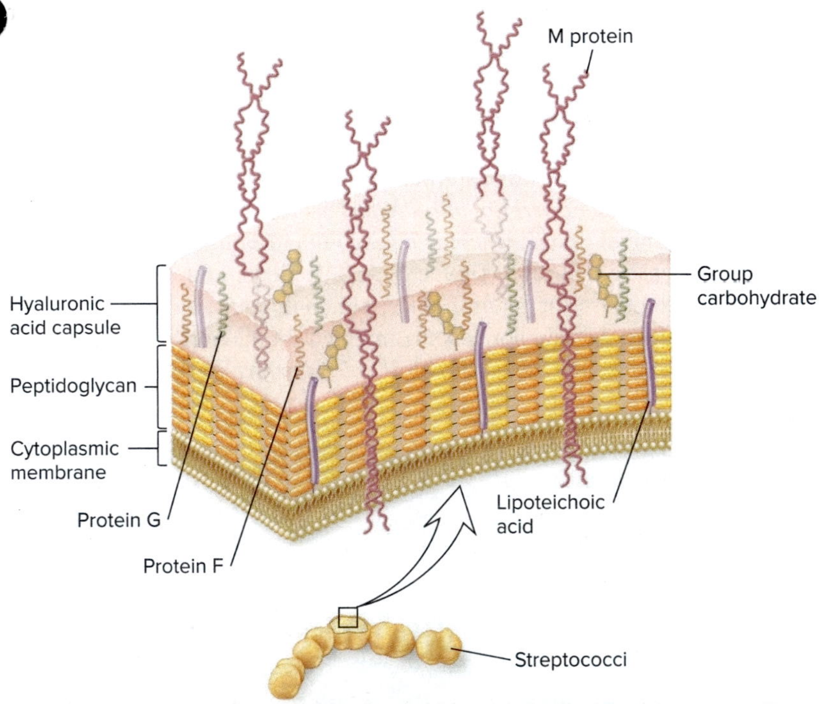

Figure 39.6 ***Streptococcus pyogenes*** **Cell Envelope.** The M protein is a major virulence factor for group A streptococci. It facilitates bacterial attachment to host cells and has antiphagocytic activity. Protein G prevents attack by antibodies because it binds to the Fc portion of antibody molecules, preventing the antigen-binding site from interacting with the bacterium. Protein F is an epithelial cell attachment factor. The Lancefield group is based on the antigenicity of the group carbohydrate.

MICRO INQUIRY *How is the M protein also thought to be involved in poststreptococcal diseases?*

Streptococcal Pharyngitis

Streptococcal pharyngitis is one of the most frequent bacterial infections of humans and is commonly called strep throat (**figure 39.7**). GAS are spread by droplets of saliva or nasal secretions. The incubation period in humans is 2 to 4 days.

The action of the strep bacteria in the throat (**pharyngitis**) or on the tonsils (**tonsillitis**) stimulates an inflammatory response and lysis of white and red blood cells. An inflammatory exudate consisting of cells and fluid is released from blood vessels and deposited in the surrounding tissue in about 50% of patients with strep pharyngitis. This is accompanied by a general feeling of discomfort or malaise, fever, and headache. Prominent physical manifestations include redness, edema, and lymph node enlargement in the throat. Signs and symptoms alone are not diagnostic because they are manifested by several other infections as well. Rapid test kits are commonly used to diagnose strep throat. In the absence of complications, the disease can be self-limiting and may disappear within 1 week. However, antibiotic treatment (penicillin, cephalosporins, or a macrolide antibiotic) can shorten the infection and clinical syndromes, and is especially important in children for the prevention of complications such as glomerulonephritis and rheumatic fever, as discussed next. Infections in older children and adults tend to be milder and less frequent due in part to the immunity developed against the many serotypes encountered in early childhood.

Poststreptococcal Diseases

The two most common poststreptococcal diseases caused by GAS are glomerulonephritis and rheumatic fever. They occur 1 to 4 weeks after an acute streptococcal infection. **Glomerulonephritis,** also called **Bright's disease,** is an inflammatory disease (type II hypersensitivity reaction) of the renal glomeruli—membranous structures within the kidney where blood is filtered. Damage results from the deposition of antigen-antibody complexes in the glomeruli. Although the causative antigen(s) remain in question, evidence suggests the production of antibody against *S*treptococcal *p*yrogenic *e*xotoxin *B* (SpeB) is involved. The complexes cause destruction of the glomerular membrane, allowing proteins and blood to leak into the urine. Clinically, the affected person exhibits edema, fever, hypertension, and hematuria (blood in the urine). Diagnosis is based on clinical history, blood and urine analyses, renal biopsy, and confirmatory evidence of prior streptococcal infection. Penicillin G or erythromycin can be given for any residual streptococci. However, there is no specific therapy once kidney damage has occurred. About 80 to 90% of all cases undergo slow, spontaneous healing of the damaged glomeruli, whereas others develop a chronic form of the disease. The latter may require a kidney transplant or lifelong renal dialysis. ⏴ *Type II hypersensitivity (section 33.10)*

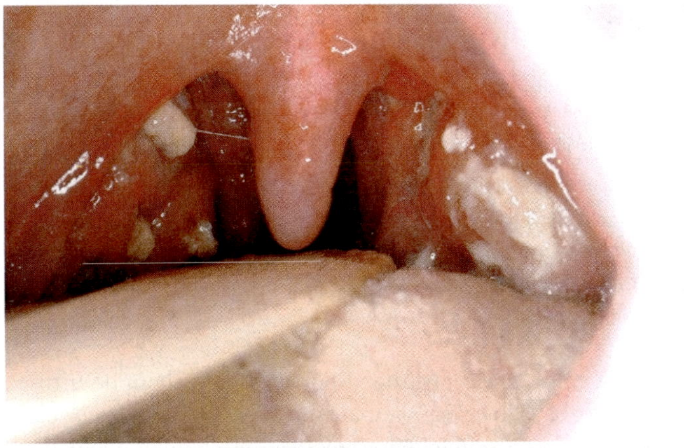

Figure 39.7 **Streptococcal Pharyngitis (Strep Throat).**
©Dr. Allan Harris/Phototake

MICRO INQUIRY *What causes the exudate seen in strep throat? Is it seen in all cases?*

Rheumatic fever (RF) is an autoimmune disease characterized by inflammatory lesions involving the heart valves, joints, subcutaneous tissues, and central nervous system. RF is associated with prior streptococcal pharyngitis with specific M protein strains. It is believed these M proteins trigger the production of autoantibodies that target cardiac proteins. The disease occurs most frequently among children six to fifteen years of age and manifests as a variety of signs and symptoms, making diagnosis difficult. In the United States, rheumatic fever has become very rare (less than 0.05% of streptococcal infections), but it occurs more frequently in tropical countries. Therapy is directed at decreasing inflammation and fever, and controlling cardiac failure. Salicylates and corticosteroids are the mainstays of treatment. Although rheumatic fever is rare, it is still the most common cause of permanent heart valve damage in children.

Comprehension Check

1. Why do you think chlamydiae differentiate into specialized cell types for infection and reproduction?
2. Compare and contrast the means by which people contract chlamydial pneumonia with Legionnaires' disease.
3. What causes the typical symptoms of diphtheria? How are individuals protected against this disease?
4. What are the three major causes of bacterial meningitis? Why is it important to determine which type a person has?
5. What immune cells would you find in a granuloma? Why are children and immunocompromised individuals more likely to develop miliary TB?
6. How do multidrug-resistant strains of tuberculosis develop?
7. Why is pneumonia caused by *M. pneumoniae* also called walking pneumonia?
8. What is the mechanism by which PTx kills host cells?
9. Describe the three stages of pertussis.
10. Which GAS disease is most prevalent? Describe the two major complications of a strep infection.

39.2 Arthropods Can Transmit Bacterial Diseases

After reading this section, you should be able to:

a. Discuss bacteria that cause common arthropod-borne diseases
b. Identify typical signs and symptoms of arthropod-borne bacterial diseases
c. Correlate arthropod-borne bacterial infection and disease severity with bacterial virulence factors

Bacteria cause fewer arthropod-borne diseases than do viruses and protozoa. However, they are of interest either historically (plague) or because they have emerged into human populations (ehrlichiosis, Lyme disease).

Lyme Disease

Lyme disease (Lyme borreliosis) was first documented among residents of Old Lyme, Connecticut. It has become the most common tick-borne zoonosis in the United States. Annually, an average of approximately 30,000 confirmed cases of Lyme disease are reported to the CDC, with 95% of these located in only 14 states. The disease is also present in Europe and forested areas of Asia.

At least three spirochete species are responsible for Lyme disease. *Borrelia burgdorferi* is the primary cause in the United States, whereas *B. garinii* and *B. afzelii* cause the disease in Europe (**figure 39.8*a***). Rodents and other wild animals are the natural hosts; although adult deer ticks feed on deer, deer do not become infected. In the northeastern United States, the reservoir of *B. burgdorferi* is the white-footed

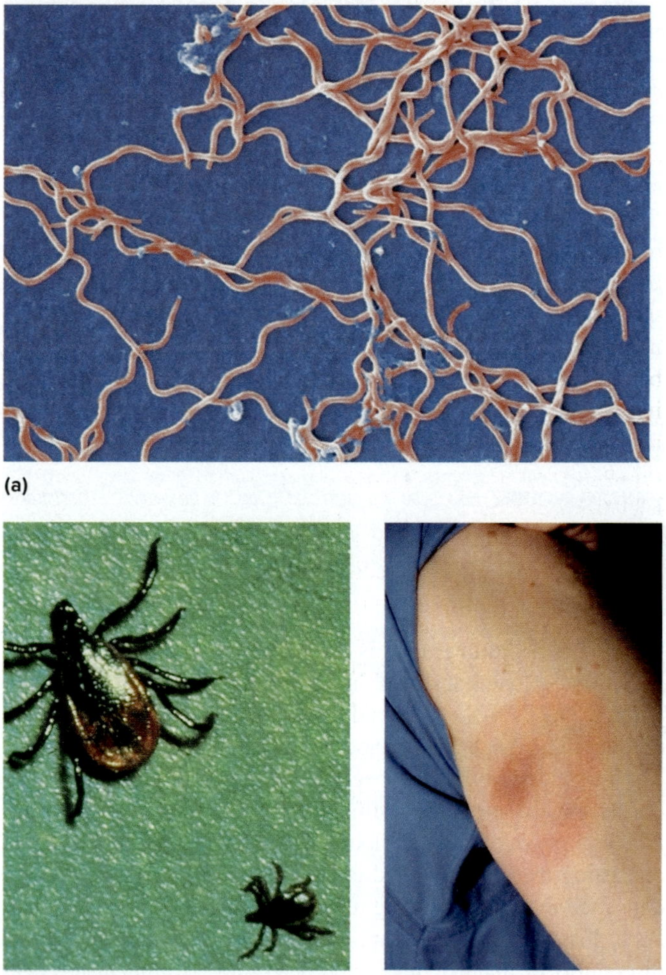

(a) (b) (c)

Figure 39.8 Lyme Disease. (a) The spirochete *Borrelia burgdorferi.* (b) An unengorged adult tick (*Ixodes scapularis*; bottom) is about the size of a pinhead, and an engorged adult (top) can reach the size of a jelly bean. (c) The typical rash (erythema migrans) showing concentric rings around the initial site of the tick bite.

(a)CDC/Janice Haney Carr; (b) Source: CDC/Michael L. Levin, Ph. D; (c) Source: CDC/James Gathany

MICRO INQUIRY *What are the three stages of Lyme disease? What other spirochete disease does this resemble?*

mouse. Spirochetes are transmitted to humans by the bite of infected black-legged (or deer) ticks (*Ixodes scapularis*; figure 39.8*b*). On the Pacific Coast, especially in California, the reservoir is a dusky-footed woodrat and the tick *I. pacificus.* ◀ *Phylum Spirochaetes: bacteria with a corkscrew morphology (section 21.6)*

Clinically, Lyme disease is a complex illness, which if untreated has three major stages. The initial, localized stage occurs 7 to 10 days after an infectious tick bite. The illness sometimes begins with erythema migrans, an expanding, ring-shaped skin lesion with a red outer border and partial central clearing (figure 39.8*c*). This is accompanied by flulike symptoms (malaise and fatigue, headache, fever, and chills). However, often the tick bite is unnoticed, or the skin lesion is missed due to skin coloration or its obscure location, such as on the scalp. Thus treatment may be delayed because the illness is assumed to be "just the flu."

The second, disseminated stage appears weeks or months after the initial infection. It consists of several symptoms such as neurological abnormalities, heart inflammation, and bouts of arthritis, usually in major joints such as elbows or knees. Current research indicates that Lyme arthritis might be an autoimmune response to major histocompatibility (MHC) molecules on cells in the synovium (joint) that are similar to bacterial antigens. Finally, like the progression of syphilis—another disease caused by a spirochete—the late stage may appear years later. Infected individuals may develop neuron demyelination with symptoms resembling Alzheimer's disease and multiple sclerosis. Behavioral changes can also occur. ◀ *Major histocompatibility complex (section 33.4)*

Laboratory diagnosis of Lyme disease is based on serological testing for IgM or IgG antibodies to the pathogen and Western blot for specific bacterial proteins or detection of *Borrelia* DNA in patient specimens (especially synovial fluid) after amplification by PCR. Treatment with doxycycline early in the illness results in prompt recovery and prevents arthritis and other complications.

Prevention and control of Lyme disease involve environmental modifications such as clearing tick habitat and the application of acaricidal compounds, which destroy mites and ticks. An individual's risk of acquiring Lyme disease may be greatly reduced by education and personal protection.

Plague

Plague (Latin *plaga*, pest) is caused by the Gram-negative bacterium *Yersinia pestis*. Between 1,000 and 2,000 cases of plague are reported to the WHO each year; U.S. cases average seven annually. Initially spread by contact with flea-infested animals such as rats, mice, and prairie dogs, or (more rarely) by direct contact with infected animals, *Y. pestis* can then spread among people by airborne transmission (**figure 39.9*a***). Once in the human body, the bacteria multiply in the blood and lymph. Symptoms include subcutaneous hemorrhages, fever, chills, headache, extreme exhaustion, and dramatically enlarged lymph nodes called **buboes** (hence the old name, **bubonic plague**) (figure 39.9*b*). In 50 to 70% of the untreated cases, death follows in 3 to 5 days from toxic conditions caused by the large number of bacteria in the blood.

Like many other Gram-negative pathogens, *Y. pestis* uses a type III secretion system to deliver effector proteins directly into targeted host cells. *Y. pestis* secretes plasmid-encoded yersinial outer membrane proteins (YOPS) into phagocytic cells to counteract natural defense mechanisms and help the bacteria multiply and disseminate in the host (**figure 39.10**). Additionally, when *Y. pestis* is phagocytosed by immune cells, it is able to survive and proliferate. ◀ *Protein maturation and secretion (section 13.8); Order* Enterobacteriales *includes pathogens and mutualists (section 22.3); Pathogenicity islands encode virulence factors (section 35.4)*

Pneumonic plague arises (1) from primary exposure to infectious respiratory droplets from a person or animal with respiratory plague or (2) secondary to blood-borne spread in a patient with bubonic or septicemic plague. The mortality rate for bubonic plague is about 50%. It is almost 100% for septicemic plague.

Because *Y. pestis* is listed as a potential bioterrorism agent, sentinel laboratory identification is restricted to initial staining and culture tests. Confirmation of the identity of *Y. pestis* by PCR, advanced serological studies, and other approved practices is restricted to national reference laboratories. Treatment is with streptomycin and recovery from the disease provides immunity. Prevention and control involve flea and rodent control, isolation of human patients, prophylaxis therapy of exposed persons, and vaccination of persons at high risk (e.g., park rangers, veterinarians).

Rocky Mountain Spotted Fever

Rocky Mountain spotted fever (RMSF) is caused by the rickettsia, *Rickettsia rickettsii*. Although this disease was originally detected in the Rocky Mountain area, most cases occur east of the Mississippi River. The disease is transmitted by two principal tick vectors: *Dermacentor andersoni*, the wood tick, which is active during the spring and early summer in the Rocky Mountain states; and *D. variabilis*, the dog tick, which has assumed greater importance and is almost exclusively confined to the eastern half of the United States. Unlike other rickettsias, *R. rickettsii* can pass from generation to generation of ticks through their eggs in a process known as transovarian passage. Therefore, no animals are needed as reservoirs for the continued propagation of this rickettsia.

The rickettsial bacteria are transmitted to humans from the saliva of an infected tick during feeding. The rickettsias induce phagocytosis by endothelial cells of small blood vessels, in which they multiply. This results in the characteristic rash said to be nonblanching because it does not fade when pressed upon. This reflects damage to blood vessels, with blood

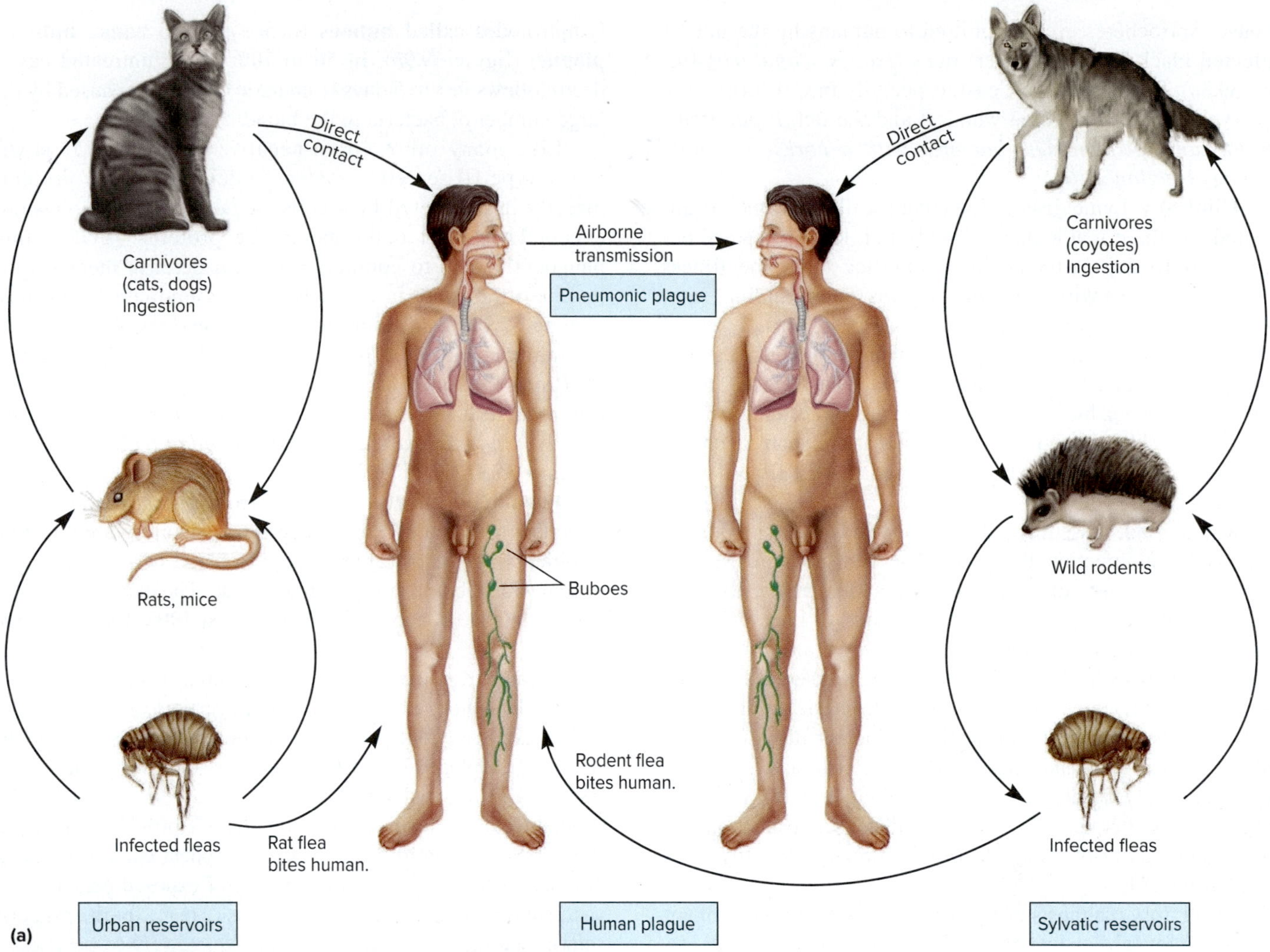

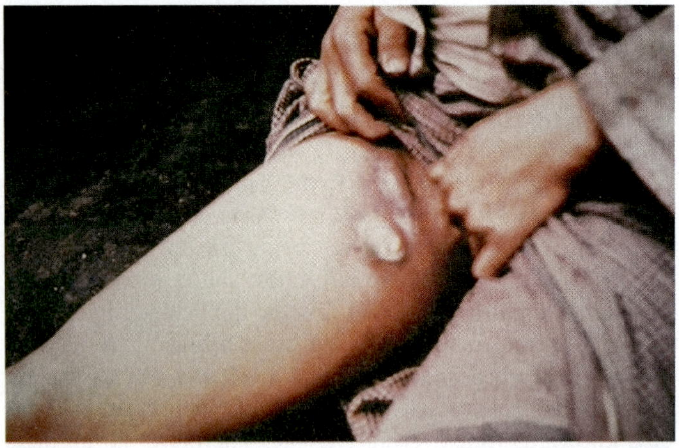
(b)

Figure 39.9 Plague. (a) Plague is spread to humans through (1) the urban cycle and rat fleas, (2) the sylvatic cycle, centering on wild rodents and their fleas, or (3) by airborne transmission from an infected person, leading to pneumonic plague. Dogs, cats, and coyotes can also acquire the bacterium by ingestion of infected animals. (b) Enlarged lymph nodes called buboes are characteristic of *Yersinia* infection.
Source: CDC

leaking into tissues. As rickettsia reproduce within the cytosol of endothelial cells, they polymerize host actin at one pole to make a tail that propels bacteria from one endothelial cell to the next (*see figure 35.6*). This limits rickettsial exposure to immune cells and host antibodies. ◀ *Rickettsias are obligate intracellular bacteria (section 22.1)*

RMSF is characterized by headache, high fever, chills, and a skin rash (**figure 39.11**) that initially appears on the hands and feet (including palms and soles, respectively), and then spreads to the trunk of the body. The symptoms reflect the accumulation of lysed host cells and toxic effects of rickettsial cell walls. If the disease is not treated, the rickettsias destroy blood vessels in the heart, lungs, or kidneys and can cause death about 8 days after infection. Severe pathological changes are avoided by antibiotic therapy (doxycycline), development of immune resistance, and supportive therapy. Diagnosis is made through observation of symptoms and signs such as the characteristic rash, by serological and nucleic acid amplification tests. The best means of prevention remains avoidance of tick-infested habitats and animals.

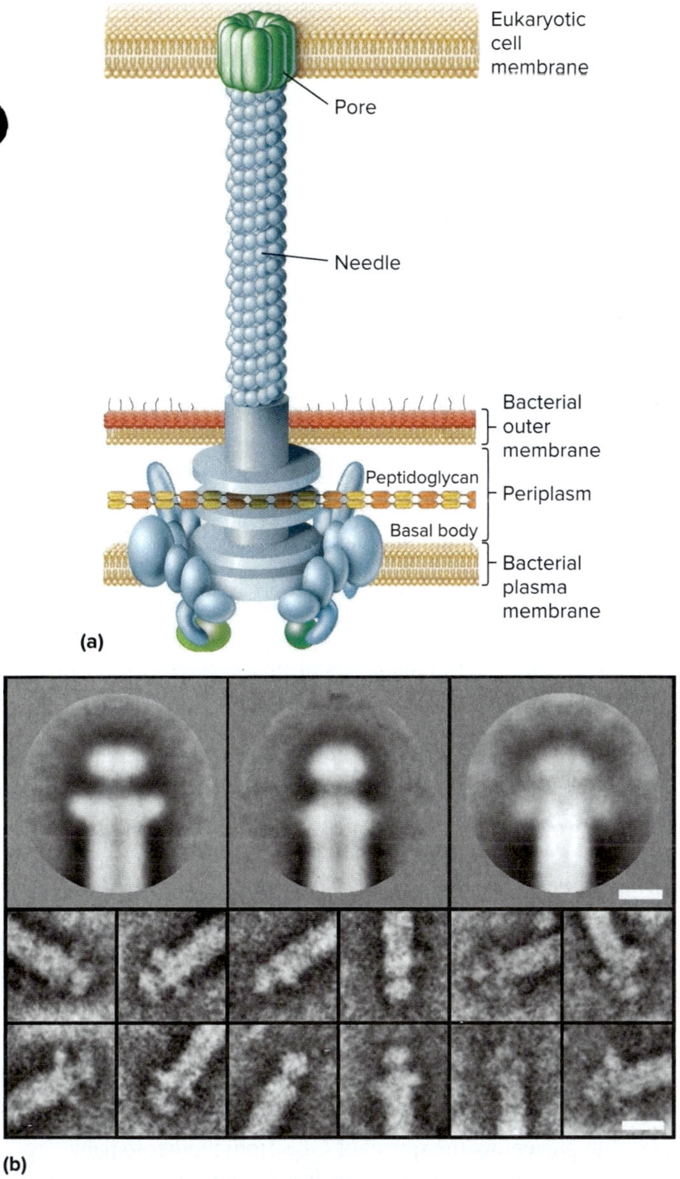

Figure 39.10 Type III Secretion System. (a) X-ray fiber diffraction studies resolve the injectisome as a helical structure. (b) Scanning tunneling electron microscopy reveals the injectisome tip, indicating how it may lock into the translocator pore on the target cell.

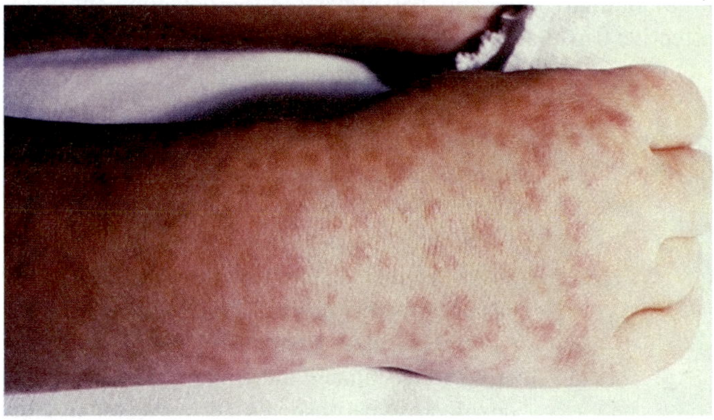

Figure 39.11 Rocky Mountain Spotted Fever. Typical nonblanching rash occurs on the extremities, including palms of the hands and soles of the feet.
Source: CDC

RMSF cases are reported year-round, with the greatest incidence between May and August. The CDC reports approximately 2,000 cases and several deaths each year in the United States.

Comprehension Check

1. What is the causative agent of Lyme disease and how is it transmitted to humans? Describe the three stages of Lyme disease.
2. Distinguish between bubonic and pneumonic plague. Why is pneumonic plague a more severe public health crisis?
3. Compare the rash seen in Lyme disease with that seen in Rocky Mountain spotted fever.
4. What type of host cell does *Rickettsia rickettsii* infect and how is this related to the signs, symptoms, and severity of Rocky Mountain spotted fever?

39.3 Direct Contact Diseases Can Be Caused by Bacteria

After reading this section, you should be able to:

a. Describe bacteria that cause common diseases spread by direct contact
b. Identify typical signs and symptoms of bacterial diseases spread by direct contact
c. Correlate direct contact bacterial infection and disease severity with bacterial virulence factors

Most of the direct contact bacterial diseases involve the skin, underlying tissues, or mucous membranes. Occasionally, bacteria penetrate beyond these regions to cause disseminated disease. We now discuss some of the better-known bacterial diseases transmitted by direct contact.

Gas Gangrene (Clostridial Myonecrosis)

Clostridium perfringens, C. novyi, and *C. septicum* are Gram-positive, endospore-forming rods termed the histotoxic (tissue damaging) clostridia. All can produce a necrotizing infection of skeletal muscle called **gas gangrene** (Greek *gangraina*, an eating sore) or **clostridial myonecrosis** (Greek *myo*, muscles; *necrosis*, death); however, *C. perfringens* is the most common cause. Analysis of the *C. perfringens* genome sequence reveals that the microbe possesses the genes for fermentation with gas production but lacks genes encoding enzymes for the TCA cycle or a respiratory chain. Nonetheless, *C. perfringens* has an extraordinary doubling time of only 8 to 10 minutes in the human host where all its nutritional needs are met. This makes prompt diagnosis and treatment of gangrene critical. ◀ *Class* Clostridia *(section 23.3)*

Histotoxic clostridia occur in the soil worldwide and also are part of the normal microbiota of the human large intestine. Contamination of injured tissue with spores from soil containing these clostridia is the usual means of transmission. Infections are commonly associated with wounds resulting from abortions, automobile accidents, frostbite, gunshot wounds, and other types of trauma. If

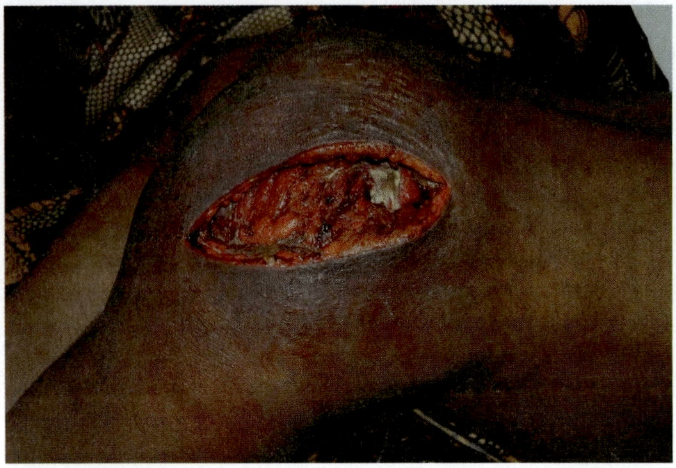

Figure 39.12 **Gas Gangrene (Clostridial Myonecrosis).** Necrosis of muscle and other tissues results from the numerous toxins produced by *Clostridium perfringens.*
©Scott Camazine/Phototake

spores germinate in anoxic tissue, the vegetative bacteria secrete several toxins, including α-toxin (a lecithinase), which breaks down muscle tissue by disrupting cell membranes thereby causing cell lysis. These bacteria also secrete collagenase, which as its name suggests, degrades host collagen to release amino acids, which it consumes. Additionally, *C. perfringens* secretes an enzyme that enables the use of sialic acid, found on the surface of many host cells, as a carbon and nitrogen source. These exoenzymes facilitate its rapid growth and spread in the host. Growth often results in accumulation of gas (mainly H_2 and CO_2 as a result of fermentation) and toxic breakdown products of skeletal muscle tissue (**figure 39.12**). ◀ *Exotoxins (section 35.4)*

Gas gangrene is a medical emergency! Clinical manifestations of gas gangrene include severe pain, edema, drainage, and muscle necrosis. Diagnosis is through culture of the causative species of clostridia and evaluation of patient signs and symptoms. Treatment requires extensive surgical removal of all dead tissue and antimicrobial therapy with a β-lactam and a macrolide. Prevention and control include debridement of contaminated traumatic wounds plus antimicrobial prophylaxis and prompt treatment of all wound infections. Interestingly, laboratory-raised, sterile maggots can also be used to remove dead and infected tissue. Amputation of limbs may be necessary to prevent further spread of the disease.

Group B Streptococcal Disease

Streptococcus agalactiae, or **group B streptococcus (GBS)**, is a Gram-positive bacterium that causes illness in newborn babies, pregnant women, the elderly, and immunocompromised adults. In these adults, GBS causes sepsis, pneumonia, and skin and bone infections. However, it is also prevalent in newborns, and is a frequent cause of newborn meningitis, bacteremia, and pneumonia. Nearly 75% of infected newborns present symptoms in the first week of life; this is called early-onset disease. GBS can also develop in infants who are one week to three months old (late-onset disease). Although early-onset disease is known to be transmitted maternally (mothers can be asymptomatic carriers), only about half of the infants with late-onset GBS disease are associated with a mother who is a GBS carrier. The source of infection for many newborns with late-onset GBS disease is unknown. Infant mortality due to GBS disease is about 5%. Babies who survive GBS disease, particularly those with meningitis, may have permanent disabilities, such as hearing or vision loss or developmental disabilities. The CDC reports approximately 20,000 total cases of invasive GBS annually in all age groups. The number of infants with GBS has dropped sharply since physicians began routine screening of all pregnant women in the mid-1990s.

GBS disease is diagnosed by culturing the Gram-positive, β-hemolytic streptococcus. An FDA-approved DNA test for GBS is also available if rapid diagnosis of disease is necessary or culture results are equivocal.

Many people are asymptomatic GBS carriers—they are colonized by GBS but do not become ill from it. Adults can carry GBS in the bowel, vagina, bladder, or throat. Twenty-five percent of pregnant women carry GBS in the rectum or vagina. Thus a fetus may be exposed to GBS before or during birth if the mother is a carrier. GBS carriage is temporary; that is, carriers do not harbor the bacteria for life. GBS carriage can be detected during pregnancy by the presence of the bacterium in culture specimens of either the vagina or the rectum. The CDC recommends that pregnant women have vaginal and rectal specimens cultured for GBS in late pregnancy (35 to 37 weeks' gestation). A positive culture means that the mother carries GBS—not that she or her baby will definitely become ill. Administration of antibiotics during labor is effective in preventing the spread of GBS from mother to baby.

GBS infections in both newborns and adults are usually treated with intravenous penicillin, unless resistance or hypersensitivity is indicated. A GBS vaccine based on genomic analysis has been under development. However, GBS vaccine development is problematic due to multiple serotypes and a capsule containing sialic acid, a carbohydrate also found on many human cells. This has prevented the development of an effective capsular vaccine. ▶ *Vaccines (section 42.1)*

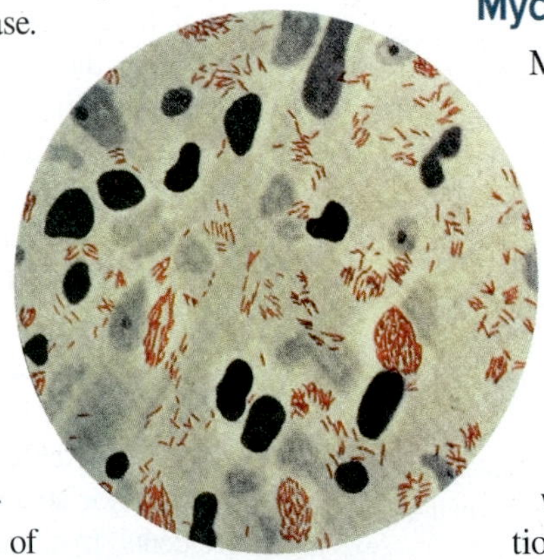

Source: CDC

Mycobacterial Skin Infections

Many pathogenic species of nontuberculous mycobacteria cause skin or soft-tissue infections. These diseases often are hard to diagnose as they may cause diverse and misleading signs and symptoms. Two skin-associated diseases caused by nontuberculous mycobacteria are presented here. Historically, people with leprosy were shunned rather than helped. We discuss leprosy because it is a complicated disease that still often stigmatizes its victims. In contrast, we also discuss an emerging mycobacterial infection associated with tattoo art.

Leprosy

Leprosy (Greek *lepros*, scaly, scabby, rough), a skin disease caused by *Mycobacterium leprae,* can be severely disfiguring. The only reservoirs of proven significance are humans and armadillos. The disease most often occurs in tropical countries, and the global leprosy burden is approximately 180,000 cases with less than 1,000 new cases reported each year. Transmission of leprosy is most likely to occur when individuals are exposed for prolonged periods to infected individuals who shed large numbers of *M. leprae*. Nasal secretions probably are the infectious material for family contacts. ◀ *Order* Corynebacteriales *includes important human pathogens (section 23.1)*

The incubation period for leprosy is about 3 to 5 years but may be even longer, and the disease progresses slowly. The bacterium invades peripheral nerve and skin cells, and reproduces as an obligately intracellular parasite. It is most frequently found in Schwann cells that surround peripheral nerve axons and in mononuclear phagocytes. The earliest symptom is usually a slightly pigmented skin eruption several centimeters in diameter. Approximately 75% of all individuals with this early, solitary lesion heal spontaneously because they mount an effective cell-mediated immune response. However, in some individuals, this immune response fails to clear the mycobacteria and one of two distinct forms of the disease occurs: tuberculoid or lepromatous leprosy.

Tuberculoid (neural) leprosy is a mild, nonprogressive form of leprosy associated with a delayed-type hypersensitivity reaction to antigens on the surface of *M. leprae*. It is characterized by damaged nerves and regions of the skin that have lost sensation and are surrounded by a border of nodules (**figure 39.13**). Afflicted individuals who do not develop hypersensitivity have a relentlessly progressive form of the disease, called **lepromatous (progressive) leprosy,** in which large numbers of *M. leprae* develop in skin cells, particularly in extremities where the temperature is slightly cooler. The bacteria kill skin tissue, leading to a progressive loss of facial features, fingers, toes, and other structures. Moreover, disfiguring nodules form all over the body. Nerves are also infected but usually are less damaged than in tuberculoid leprosy. ◀ *Type IV hypersensitivity (section 33.10)*

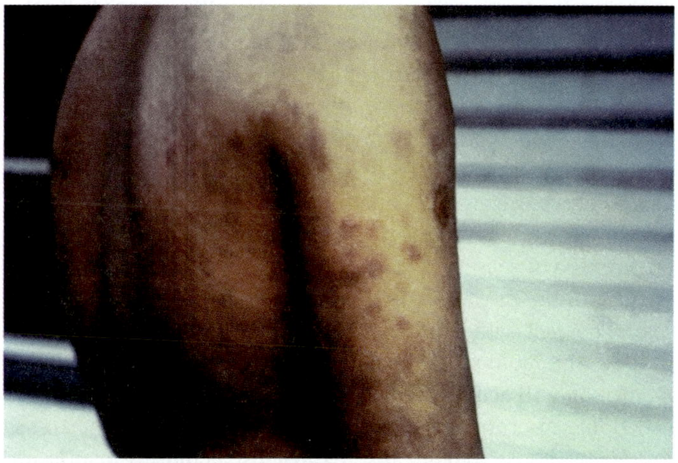

Figure 39.13 Leprosy. In tuberculoid leprosy, the skin manifests shallow, painless lesions.
Source: CDC/Arthur E. Kaye

MICRO INQUIRY *What are the chief differences between tuberculoid and lepromatous leprosy?*

Because *M. leprae* cannot be cultured, laboratory diagnosis is supported by the demonstration of the bacterium in biopsy specimens using direct fluorescent antibody staining. Serodiagnostic methods, such as the fluorescent leprosy antibody absorption test and ELISA, have been developed.

Treatment of tuberculous leprosy is accomplished with rifampicin combined with dapsone. Rifampicin with clofazimine is combined with dapsone to treat lepromatous leprosy (*see figure 9.12*). Identification and treatment of patients with leprosy are key to its control. Close contacts (e.g., family members) should be given chemoprophylactic drugs.

Tattoo-Associated Mycobacterial Infections

Sporadic cases and outbreaks of tattoo-associated mycobacterial infections remind us that emerging diseases do not have to be exotic. Rapidly growing mycobacteria (RGM) are a group of non-tuberculous mycobacteria commonly found in water, soil, and dust. RGM have been identifed as the cause of tattoo-associated skin infections. The index (first) case was a 49-year-old man who had over 100 small pimplelike pustules and reddened skin blotches within 2-week-old tattoos. Microscopic examination of biopsy specimens were negative for bacteria, including acid-fast bacilli. However, culture of the specimen revealed *M. chelonae*. Additional reported cases had a range of skin lesions, including pink, red, and purple papules (pimples); scaly papules; granulomatous papules; lichenoid papules; pustules; and plaques. The common link to all was the inking of tattoos by the same artist; all of the lesions were found to be concentrated in "gray wash" (areas of shading using gray tones) within the tattoos. Other tattoo-associated infections, including some caused by *M. haemophilum*, have also been reported. In most cases, it takes 6 to 9 months of multicourse antibiotic regimens to eradicate the infection.

Laboratory diagnosis for rapidly growing mycobacterial organisms is by culture and routine bacterial staining. Antibiotic sensitivity testing is also important. As with other mycobacterial infections, multiple antibiotics are used concurrently to prevent the development of drug resistance. Treatment of rapidly growing mycobacterial skin infections is successful often with macrolide antibiotics. Topical steroids may also be indicated to control the lesions.

Peptic Ulcer Disease and Gastritis

Peptic ulcer disease is caused by the Gram-negative, spiral bacterium *Helicobacter pylori*. Found in gastric biopsy specimens from patients with gastritis, *H. pylori* is responsible for most cases of chronic **gastritis** not associated with another known primary cause. Importantly, untreated *H. pylori* infection can lead to gastric cancer and it has been classified as a Class I carcinogen by the WHO. ◀ *Class* Epsilonproteobacteria *ranges from pathogens to deep-sea bacteria (section 22.5)*

H. pylori colonizes gastric mucus-secreting cells, beneath the gastric mucous layers using fimbriae as adhesins (**figure 39.14**). It binds to Lewis b antigens, which are on the surface of red blood cells. *H. pylori* also binds to sialic acid, a common sugar component of glycoproteins on the surface of gastric epithelial cells. *H. pylori* produces urease, which creates a localized alkaline environment when hydrolysis of urea produces ammonium. The increased pH may protect the bacterium from gastric acid as it grows under the layer of mucus in the stomach. The potential virulence factors responsible for epithelial cell damage and inflammation include proteases, phospholipases, cytokines, and cytotoxins.

Laboratory identification of *H. pylori* is by culture of gastric biopsy specimens, examination of stained biopsies for the presence of bacteria, detection of antigen in stool, and a test that detects urea in the patient's breath. The goal of *H. pylori* treatment is the complete elimination of the organism. Treatment is two-pronged: use of drugs to decrease stomach acid (proton pump inhibitors) and antibiotics (metronidazole and tetracycline, amoxicillin, or clarithromycin) to kill the bacteria.

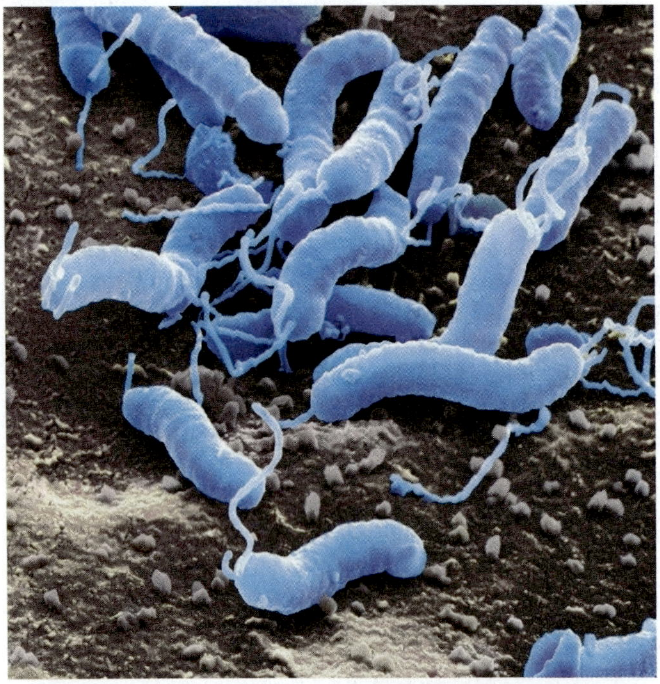

Figure 39.14 ***Helicobacter pylori*, the Causative Agent of Peptic Ulcer Disease.** Scanning electron micrograph of *H. pylori* adhering to gastric cells. Each cell is about 3 μm long.

MICRO INQUIRY *How does* H. pylori *increase the local pH in its gastric microenvironment?*

Approximately 50% of the world's population is estimated to be infected with *H. pylori*. It is most likely transmitted from person to person (usually acquired in childhood), although infection from a common exogenous source cannot be ruled out, and some hypothesize that it is also spread by food or water. Support for person-to-person transmission comes from evidence of clustering within families and from reports of higher than expected prevalences in residents of custodial institutions such as nursing homes.

Sexually Transmitted Infections

Sexually transmitted infections (STIs) represent a worldwide public health problem. There are over 30 different bacteria, viruses, and parasites transmitted between humans through sexual contact. The various viruses that cause STIs are presented in chapter 38, and the yeasts and protozoa in chapter 40. The most common bacterial conditions include chlamydial infections, gonorrhea, and syphilis. These and other sexually transmitted bacteria, and the diseases they cause, are presented in **table 39.1**.

The spread of most STIs is currently at an all-time high. The WHO estimates that 1 million new STI cases occur each day, with most infections occurring in fifteen- to thirty-year-old individuals. In the United States, the CDC reports that while substantial progress has been made in preventing, diagnosing, and treating certain STIs, over 20 million new infections occur each year, nearly half of them among people aged fifteen to twenty-four. STIs also exact a tremendous economic toll, with direct medical costs estimated at over $16 billion annually in the United States alone. Many cases go undiagnosed; others are not reported at all. This is particularly problematic as women with untreated chlamydia or gonorrhea infections risk infertility, and both men and women with papillomavirus infection have a greater risk of cancer.

STIs were formerly called venereal diseases (from Venus, the Roman goddess of love) and are also referred to as sexually transmitted diseases (STDs). Although they occur most frequently in the most sexually active age group, anyone who has sexual contact with an infected individual is at increased risk. In general, the more sexual partners a person has, the more likely the person will acquire an STI.

Some microorganisms that cause STIs can also be transmitted by nonsexual means. Examples include transmission by contaminated hypodermic needles and syringes shared among intravenous drug users and transmission from infected mothers to their infants. Preventive measures are based mainly on education, barrier protection, and treatment of infected individuals and their partners with chemotherapeutic agents.

Table 39.1 Summary of the Major Sexually Transmitted Infections (STIs) Caused by Bacteria

Microorganism	Disease	Comments	Treatment
Chlamydia trachomatis	Nongonococcal urethritis (NGU); cervicitis, pelvic inflammatory disease (PID), lymphogranuloma venereum	Serovars D-K cause most of the STIs in the U.S.; lymphogranuloma venereum rare in the U.S.	Doxycycline and azithromycin
Haemophilus ducreyi	Chancroid ("soft chancre")	Open sores on the genitals can lead to scarring without treatment; on the rise in the U.S.	Azithromycin and ceftriaxone
Helicobacter cinaedi, H. fennelliae	Diarrhea and rectal inflammation in men who have sex with men	Common in immunocompromised individuals	Metronidazole, macrolides
Mycoplasma genitalium	Implicated in some cases of NGU	Only recently described as an STI	Azithromycin
Mycoplasma hominis	Implicated in some cases of PID	Widespread, often asymptomatic but can cause PID in women	Azithromycin
Neisseria gonorrhoeae	Gonorrhea, PID	Usually symptomatic in men and asymptomatic in women; antibiotic-resistant strains are increasing in prevalence	Ceftriaxone with either azithromycin or doxycycline
Treponema pallidum subsp. *pallidum*	Syphilis, congenital syphilis	Manifests many clinical syndromes	Benzathine penicillin G or doxycycline if allergic
Ureaplasma urealyticum	Urethritis	Widespread, often asymptomatic but can cause PID in women and NGU in men; premature birth	Azithromycin or doxycycline

Chlamydial Diseases

Chlamydia is the most frequently reported sexually transmitted bacterial disease. Over 1.5 million cases of chlamydia were reported to the CDC in 2015. However, this number is thought to be grossly underestimated as most people with chlamydial infections have no symptoms and thus go undiagnosed. The most frequently isolated species is *C. trachomatis*. It is transmitted through anal, oral, and vaginal sex, and can be transmitted from mother to child during vaginal childbirth. Symptoms of chlamydial disease vary widely. Men may have no or few manifestations (including a urethral discharge, burning during urination, and itching); however, complications such as inflammation of reproductive structures can occur. Women may be asymptomatic or have a severe infection of the cervix and urethra. Symptoms include abnormal vaginal discharge and burning on urination. The infection may spread from the cervix to the fallopian tubes, leading to **pelvic inflammatory disease (PID)**, with symptoms of abdominal pain, back pain, fever, nausea, and unusual vaginal bleeding (**figure 39.15**). Furthermore, chlamydial infections of the throat can occur in men or women who have oral sex with infected partners. In pregnant females, chlamydial infections are especially serious because they are directly related to miscarriage, stillbirth, inclusion conjunctivitis, and infant pneumonia. Chlamydia is effectively treated with a macrolide or doxycycline; because it is frequently co-transmitted with *N. gonorrheae*, patients treated for gonorrhea are also treated for chlamydia.

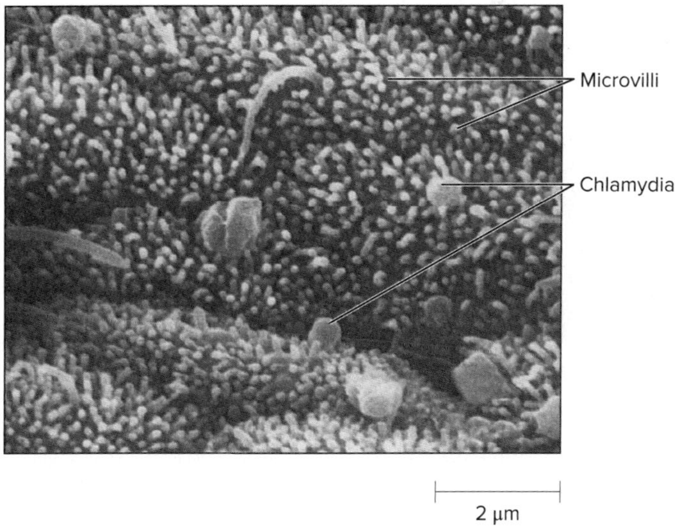

Figure 39.15 ***Chlamydia trachomatis*** **Attached to the Mucosa of a Fallopian Tube.**

Gonorrhea

Gonorrhea (Greek *gono*, seed; *rhein*, to flow) is an acute, infectious, sexually transmitted disease of the mucous membranes of the genitourinary tract, eye, rectum, and throat (table 39.1). It is caused by the Gram-negative, diplococcus *Neisseria gonorrhoeae*. These bacteria are also referred to as **gonococci** and have a worldwide distribution. It is estimated that there are over 400,000 cases of gonorrhea in the United States each year, but fewer than half are reported. ◀ *Order* Neisseriales *includes important pathogens (section 22.2)*

Once inside the body, gonococci use type IV pili and a surface molecule called protein II as adhesins to attach to the microvilli of mucosal cells. This attachment prevents the bacteria from being washed away by normal cervical and vaginal discharges or by the flow of urine. They are often endocytosed by mucosal cells and may even be transported through the cells to the intercellular spaces and subepithelial tissue, a process known as transcytosis. Neutrophils phagocytose the bacterium but gonococci survive because they can withstand the harsh phagosomal environment (**figure 39.16*a***). When gonococci are intracellular, the host's defenses have little effect on the bacteria.

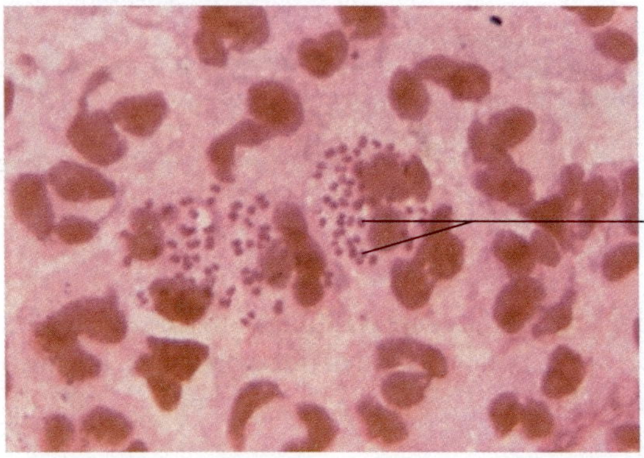

(a)

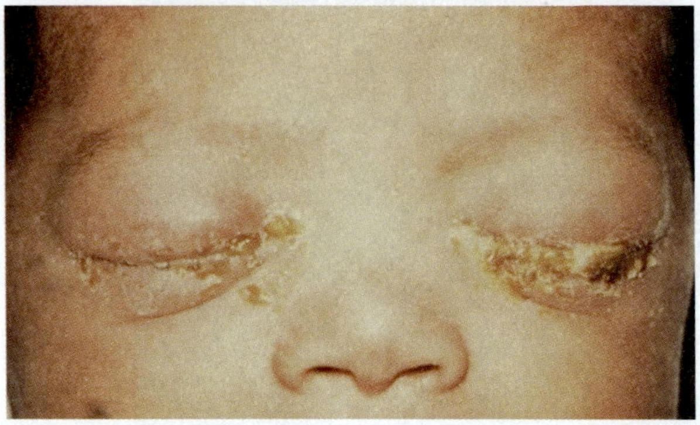

(b)

Figure 39.16 Gonorrhea. (a) Gram stain showing *Neisseria gonorrhoeae* (diplococci) inside neutrophils; light micrograph (×500). Although the presence of Gram-negative diplococci in exudates is a probable indication of gonorrhea, the bacterium should be isolated, identified, and tested for antibiotic sensitivity. (b) Gonococcal ophthalmia neonatorum in a one-week-old infant.
(a) Source: CDC/Bill Schwartz; (b) Source: CDC/J. Pledger

Extracellular gonococci evade detection by altering their adhesin proteins (a process known as antigenic variation), forcing immune cells to repeatedly re-identify and re-respond to the bacteria. Neutrophils are later replaced by fibrous tissue that may lead to scarring and stricture in the genitourinary tract. ◀ *Granulocytes (section 32.4); Phagocytosis destroys invaders (section 32.6); Surviving host defenses (section 35.3)*

In males, the incubation period is 2 to 8 days. Signs and symptoms include urethral discharge of yellow, creamy pus and frequent, painful urination accompanied by a burning sensation. In females, the cervix is the principal site infected and the disease is more insidious, with few women experiencing symptoms. In those who do, vaginal discharge occurs 1 to 3 weeks after infection. The gonococci also can infect the fallopian tubes and surrounding tissues, leading to PID. This occurs in 10 to 20% of infected women. Gonococcal PID is a major cause of sterility and ectopic pregnancies because scar formation in the fallopian tubes prevents passage of the egg to the uterus. Gonococci disseminate most often during menstruation, when there is an increased concentration of free iron available to the bacteria. In both genders, disseminated gonococcal infection with bacteremia may occur. This can lead to involvement of the joints (gonorrheal arthritis), heart (gonorrheal endocarditis), or pharynx (gonorrheal pharyngitis).

Gonorrheal eye infections can occur in newborns as they pass through an infected birth canal. The resulting disease is called **ophthalmia neonatorum,** or **conjunctivitis of the newborn** (figure 39.16*b*). This was once a leading cause of blindness in many parts of the world. To prevent this disease, as well as chlamydial conjunctivitis, erythromycin is placed in the eyes of all newborns. This treatment is required by law in many nations.

Laboratory diagnosis of gonorrhea relies on the successful growth of *N. gonorrhoeae* in culture to determine oxidase reaction, Gram-stain reaction, and colony and cell morphology. Because gonococci are very sensitive to adverse environmental conditions and survive poorly outside the body, specimens should be plated directly; when this is not possible, special transport media are necessary. *N. gonorrhoeae* can also be identified by nucleic acid amplificaton tests.

Gonorrhea is currently treated with a single injection of ceftriaxone. Alarmingly, strains resistant to all classes of antibiotics have been isolated, and some question whether (or when) gonorrhea will become untreatable. Because *N. gonorrheae* and *C. trachomatis* are often co-transmitted, current treatment for gonorrhea also includes a macrolide or doxycycline to treat chlamydia as well. Later stages of gonorrhea are more difficult to treat and may require larger drug doses over a longer period of time.

Syphilis

Syphilis (Greek *syn*, together; *philein*, to love) is an ulcerative disease (table 39.1) caused by the spirochete *Treponema pallidum* subsp. *pallidum* (*T. pallidum*). Infection occurs through intimate contact with the infected lesion of a sexual partner.

Syphilis has many nicknames including the "great imitator" as its many signs and symptoms mimic those of other diseases. The signs and symptoms of syphilis vary depending on the stage of infection (primary, secondary, latent, and tertiary). **Congenital syphilis** occurs when the disease is transmitted in utero from mother to child. ◀ *Phylum* Spirochaetes: *bacterial with a corkscrew morphology (section 21.6)*

As a spirochete, *T. pallidum* is actively motile, allowing it to easily enter the body through mucous membranes and minor breaks or abrasions of the skin. It replicates locally, inducing an ulceration, before it migrates to the regional lymph nodes, where it rapidly spreads throughout the body. Three recognizable stages of syphilis occur in untreated individuals. In the primary stage, after an incubation period of about 10 days to 3 weeks or more, the initial symptom is a small, painless, reddened ulcer, or **chancre** (French from Latin, *cancer*, creeping ulcer) with a hard ridge that appears at the infection site and contains spirochetes (**figure 39.17*a***). Contact with the chancre during sexual contact may result in disease transmission. In about one-third of the cases, the disease does not progress further and the chancre disappears. Serological tests are positive in about 80% of the infected individuals during this stage (**figure 39.18**).

Within 2 to 10 weeks after the primary lesion appears, the disease may enter the secondary stage, characterized by a highly variable skin rash (figure 39.17*b*). By this time, 100% of infected individuals are serologically positive. Other symptoms during this stage include the loss of hair in patches, malaise, and fever. Like chancres, rash lesions are infectious.

After several weeks, the disease becomes latent. During the latent period, the disease is rarely infectious, except for possible transmission from mother to fetus. After many years, tertiary stage disease develops in about 40% of untreated individuals. During this stage, masses of dead, fibrous tissue called **gummas** (figure 39.17*c*) form in the skin, bone, and nervous system as the result of hypersensitivity reactions. This stage also is characterized by a great reduction in the number of spirochetes in the body. Involvement of the central nervous system may result in tissue loss that can lead to cognitive deficits, blindness, a "shuffle" walk (tabes), or insanity. Many of these symptoms have been associated with such well-known people as Al Capone, Henry VIII, Adolf Hitler, and Friedrich Nietzsche. Sadly much of what

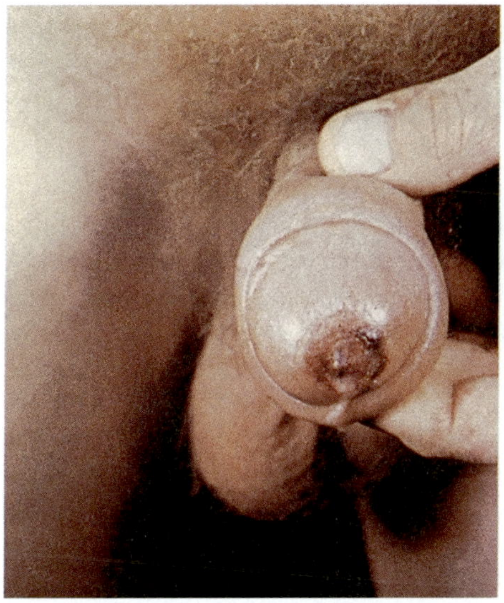

(a)

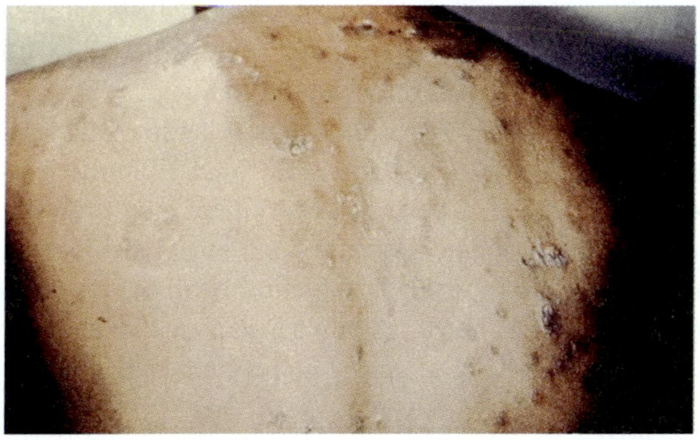

(b)

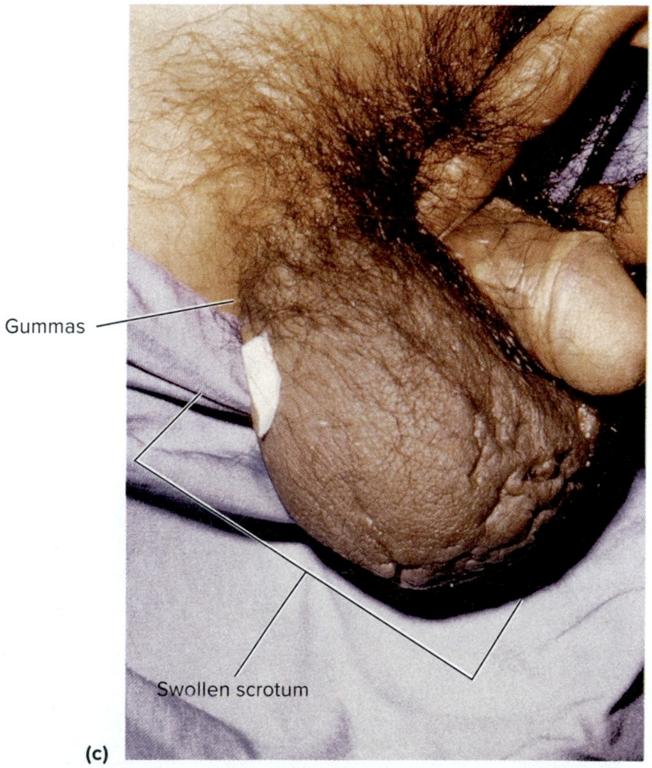

(c)

Figure 39.17 Syphilis. (a) Primary syphilitic chancre of the penis. (b) Whole body rash of secondary syphilis. (c) Late-stage tertiary syphilis marked by masses of swollen and dead, fiberlike tissue.
(a) Source: CDC; (b) Source: CDC; (c) Source: CDC/Donated by Brian Hill, New Zealand

MICRO INQUIRY *Which of these clinical manifestations are associated with syphilis transmission? Explain.*

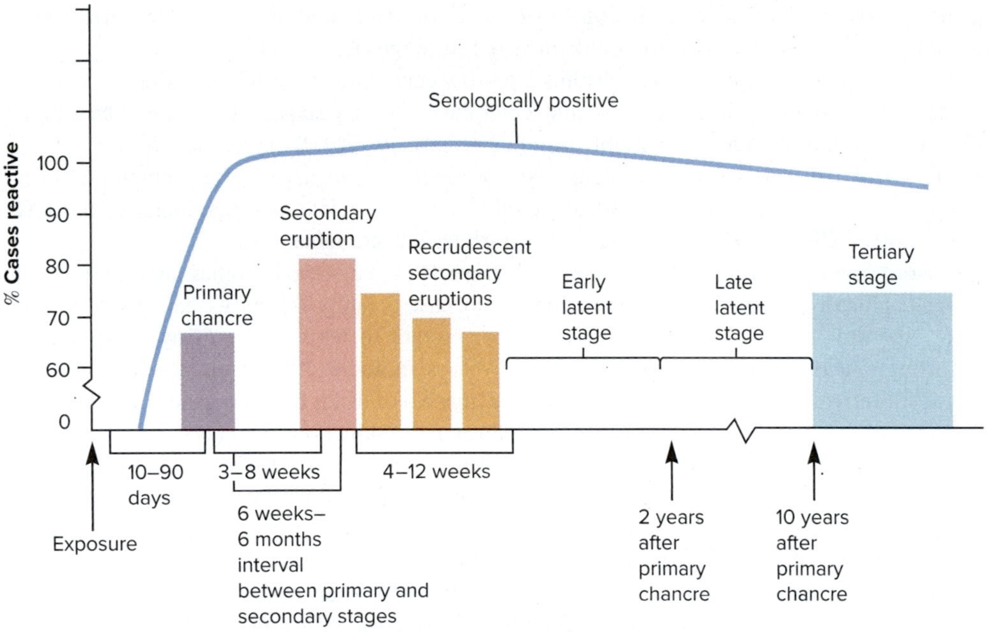

Figure 39.18 The Course of Untreated Syphilis.

we know about the progression and prognosis of syphilis is based on shameful studies performed in the United States during the twentieth century (**Disease 39.1**).

Diagnosis of syphilis is through clinical history, physical examination, and immunoassay. Two tests, the venereal disease research laboratory (VDRL) and rapid plasmin reagin (RPR), measure the production of antibody made in response to host cells damaged by *T. pallidum*. These are called nontreponemal tests because they do not detect spirochete-specific antibody. Diagnosis is made by pairing a nontreponemal assay (i.e., VDRL or RPR) with a treponemal specific test, such as the *T. pallidum* particle agglutination (TPPA) test, which

DISEASE

39.1 Syphilis and the Tuskegee Study

A research investigation named "Tuskegee Study of Untreated Syphilis in the Negro Male" would be unthinkable today. But it was the reality in 1932 Macon County, Alabama, when the federal Public Health Service began the study on 600 black men (399 with syphilis, 201 without the disease). The tale of this study and its participants is a stain on the history of U.S. public health, for which President Bill Clinton formally apologized in 1997.

Source: CDC

The "Tuskegee Study," as it was known, started with a racist objective—to develop syphilis treatment for black people. The enrollees were provided free medical checkups, meals, and burial assistance, but were not told the study had anything to do with syphilis. Rather, they were informed that they were being treated for "bad blood" (**box figure**). Except they weren't being treated. Even after penicillin was shown to be a highly effective cure for syphilis in 1947, treatment was withheld. The project was supposed to last 6 months; it went on for 40 years.

Finally in July 1972, a newspaper story broke the news that men were unknowingly enrolled in this highly unethical study and a government panel confirmed that study participants had been misled and appropriate medical treatment had been withheld. At this time, it was also revealed that the men were never given the opportunity to quit the study. Three months later, the panel shut down the study.

The following year, a class-action lawsuit was filed on behalf of the study participants. In 1974, a $10 million settlement was reached. The U.S. government also promised to provide lifetime medical benefits and burial services for all enrollees. In 1975 the wives and children of the men participants were added. The last study participant died in 2004 and there are currently 12 offspring receiving benefits.

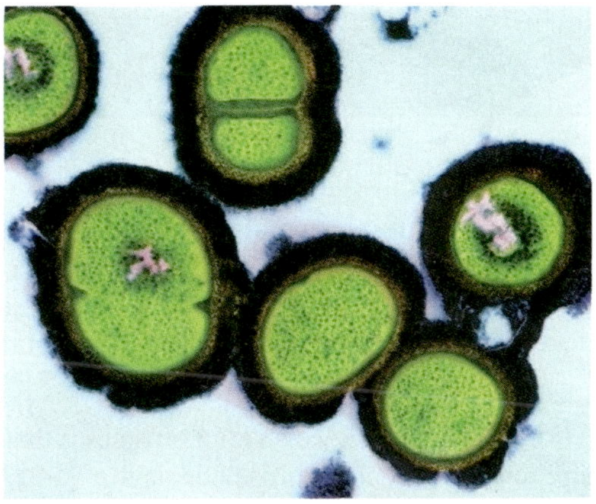

(a)

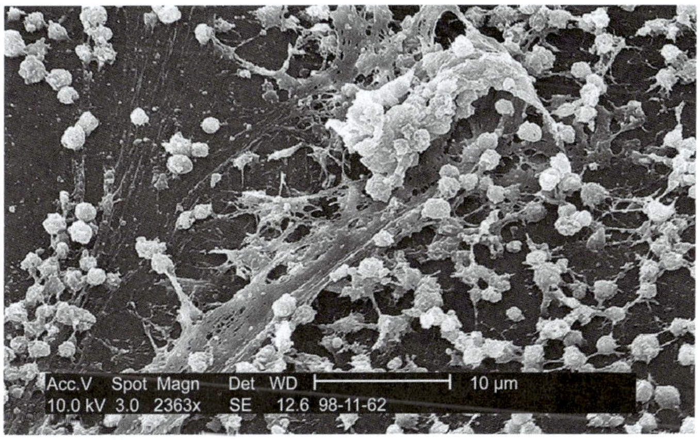

(b)

Figure 39.19 Slime and Biofilm. (a) Cells of *S. aureus* producing slime, seen as the black halo surrounding each cell (colorized transmission electron microscopy, ×10,000). (b) A biofilm on the inner surface of an intravenous catheter. Extracellular polymeric substances, mostly polysaccharides, surround and encase the staphylococci (scanning electron micrograph).
(a) ©Dr. Kari Lounatmaa/Science Source; (b) Source: CDC/Janice Carr

detects antibodies made against the pathogen. ◀ *Immune complex formation (section 33.8)*

Treatment in the early stages of the disease is easily accomplished with a single dose of long-acting benzathine penicillin G or a 2-week course of doxycycline, for penicillin-allergic individuals. Later stages of syphilis are more difficult to treat and require much larger doses over a longer period. Immunity to syphilis is not complete, and subsequent infections can occur.

Prevention and control of syphilis depend on (1) public education, (2) prompt and adequate treatment of all new cases, (3) follow-up on sources of infection and contacts so they can be treated, and (4) prophylaxis (barrier protection) to prevent exposure. At present, the incidence of syphilis, as well as other sexually transmitted infections, is rising in most parts of the world. An estimated 12 million new cases continue to occur each year. The latest available U.S. data report 76,015 total cases of syphilis in 2015, about an 18% increase over the previous year.

Staphylococcal Diseases

Staphylococci are among the most important bacteria that cause disease in humans. They are normal inhabitants of the upper respiratory tract, skin, intestine, and vagina. Staphylococci are members of a group of invasive Gram-positive bacteria known as the **pyogenic** (pus-producing) cocci. These bacteria cause various suppurative (pus-forming) diseases in humans (e.g., boils, carbuncles, abscesses, impetigo). ◀ *Family* Staphylococcaceae *(section 23.2)*

Staphylococci can be divided into pathogenic and relatively nonpathogenic strains based on synthesis of the enzyme coagulase. The coagulase-positive species *Staphylococcus aureus* is the most important human pathogen in this genus, while coagulase negative staphylococci (CoNS) are less virulent. However, they have increasingly been associated, as opportunistic pathogens, with serious nosocomial infections.

Staphylococci are further classified into slime producers and nonslime producers. The ability to produce slime has been proposed as a marker for pathogenic strains of both coagulase positive and negative staphylococci (**figure 39.19*a***). Slime is a viscous, extracellular glycoconjugate that allows these bacteria to adhere to smooth surfaces such as prosthetic medical devices and catheters. Scanning electron microscopy clearly demonstrates that biofilms (figure 39.19*b*) consisting of staphylococci encased in a slimy matrix are formed in association with biomaterial-related infections (**Disease 39.2**). Slime also inhibits neutrophil chemotaxis, phagocytosis, and antimicrobial agents *(see figure 35.7)*. ◀ *Biofilms are common in nature (section 7.6); Biofilms provide microbial protection (section 35.3)*

Staphylococci are common in nature. About one-third of people harbor *S. aureus* on their nasal mucosa as asymptomatic carriers. Staphylococci can be spread by the hands, expelled from the respiratory tract, or transported in or on animate and inanimate objects. *S. aureus* can produce disease in almost every organ and tissue of the body.

S. aureus produces disease through the ability to multiply and spread widely in tissues and produce many virulence factors (**table 39.2**). Some of these factors are exotoxins, and others are enzymes thought to be involved in invasiveness *(see figure 23.20b)*. Many toxin genes are carried on plasmids; in some cases, genes responsible for pathogenicity reside on both a plasmid and the bacterial chromosome. The pathogenic capacity of a particular *S. aureus* strain is due to the combined effect of extracellular factors and toxins. At one end of the disease spectrum is staphylococcal food poisoning, a brief, self-limiting condition caused solely by the ingestion of preformed enterotoxin (table 39.2). At the other end of the spectrum are staphylococcal bacteremia and disseminated abscesses in most organs of the body.

DISEASE

39.2 Biofilms

Biofilms consist of microorganisms embedded in an organic polymer matrix of microbial origin. They develop on virtually all surfaces immersed in natural aqueous environments, including both biological (aquatic plants and animals) and abiological (concrete, metal, plastics, stones). Biofilms form particularly rapidly in flowing aqueous systems where a regular nutrient supply is provided to the microorganisms. Extensive microbial growth, accompanied by excretion of copious amounts of extracellular organic polymers, leads to the formation of visible slimy layers (biofilms) on solid surfaces. ◀ *Biofilms are common in nature (section 7.6)*

Insertion of a prosthetic device into the human body often leads to the formation of biofilms on the surface of the device. The microorganisms primarily involved are *Staphylococcus aureus, S. epidermidis,* and other coagulase-negative staphylococci, as well as some Gram-negative bacteria. These normal skin inhabitants possess the ability to tenaciously adhere to the surfaces of inanimate prosthetic devices. Within the biofilms, they are protected from the body's normal defense mechanisms and also from antibiotics; thus the biofilm also provides a source of infection for other parts of the body as bacteria detach during biofilm sloughing.

Some examples of biofilms of medical importance include:

- Infections of artificial knee and hip prostheses that can require replacement of the prosthetic device
- Cystic fibrosis patients harboring great numbers of *Pseudomonas aeruginosa* that produce large amounts of alginate polymers
- Teeth, where biofilm forms plaque that leads to tooth decay
- Contact lenses, where bacteria may produce severe eye irritation, inflammation, and infection
- Air-conditioning and other water retention systems where potentially pathogenic bacteria, such as *Legionella* species, may be protected by biofilms from the effects of chlorination

The classic example of a staphylococcal lesion is the localized abscess of a domed pustule (**figure 39.20**). When *S. aureus* becomes established in a hair follicle, tissue necrosis results. Coagulase is produced and forms a fibrin wall around the lesion that limits bacterial spread. Within the center of the lesion, liquefaction of necrotic tissue occurs. The abscess may be either a furuncle (boil) or a carbuncle (cluster of subdermal boils). The central necrotic tissue drains, and healing eventually occurs. However, the bacteria may spread from any focus through lymph or blood to other parts of the body.

Infants sometimes develop **staphylococcal scalded skin syndrome (SSSS)** caused by strains of *S. aureus* that produce **exfoliative toxin (exfoliatin)**. This protein is usually plasmid encoded. Exfoliative toxin clips proteins that keep epidermal cells attached to the underlying skin layer, so the epidermis peels off to reveal inflamed dermis underneath—thus the name of the disease.

Toxic shock syndrome (TSS) is a staphylococcal disease with potentially serious consequences. This syndrome may occur in women who use superabsorbent tampons. These tampons can trigger a change in the normal vaginal flora, thereby enabling the growth of toxin-producing *S. aureus*. Toxic shock syndrome is characterized by low blood pressure, fever, diarrhea, an extensive skin rash, and shedding of the skin (**figure 39.21**). TSS is caused by the TSS toxin (TSST-1), a **superantigen** that triggers massive overproduction of cytokines by T cells. As discussed in chapter 35, superantigens activate 5 to 30% of the total T-cell population, whereas specific antigens activate only 0.01 to 0.1% of the T-cell population. The net effect is cytokine overproduction leading to circulatory collapse, shock, and multiorgan failure. Tumor necrosis factor α (TNF-α) and interleukins (IL) 1 and 6 are strongly associated with superantigen-induced shock. Mortality rates for TSS are 30 to 70%. Since the initial outbreak of TSS linked to superabsorbent tampons in the 1970s, the manufacture of these products has changed so they can no longer absorb as much fluid, thereby diminishing the risk of acquiring TSS in this way. ◀ *Toxins Are Biological Poisons (section 35.4)*

The definitive diagnosis of staphylococcal disease can be made only by isolation and identification of the staphylococcus involved. This requires culture, catalase, coagulase, and other biochemical tests. Commercial rapid test kits also are available. There is no specific prevention for staphylococcal disease. The mainstay of treatment is the administration of specific antibiotics: Antistaphylococcal β-lactams (e.g., nafcillin) and first- and second-generation cephalosporins are first-line treatments. Because of the prevalence of drug-resistant strains, all staphylococcal isolates should be tested for antimicrobial susceptibility. Cleanliness, good hygiene, and aseptic management of lesions are the best means of control.

Table 39.2 Some of the Enzymes and Toxins Produced by Staphylococci

Product	Physiological Action
β-lactamase	Breaks down β-lactam antibiotics
Catalase	Converts hydrogen peroxide into water and oxygen and reduces killing by phagocytosis
Coagulase	Reacts with prothrombin to form a complex that can cleave fibrinogen and cause the formation of a fibrin clot; fibrin may also be deposited on the surface of staphylococci, which may protect them from destruction by phagocytic cells; coagulase production is synonymous with invasive pathogenic potential
DNase	Destroys DNA
Enterotoxins	Are divided into heat-stable toxins of six known types (A, B, C1, C2, D, E); responsible for the gastrointestinal upset typical of food poisoning
Exfoliative toxins A and B (superantigens)	Cause loss of the surface layers of the skin in scalded-skin syndrome
Hemolysins	Alpha hemolysin destroys erythrocytes and causes skin destruction. Beta hemolysin destroys erythrocytes and sphingomyelin around nerves.
Hyaluronidase	Also known as spreading factor; breaks down hyaluronic acid located between cells, allowing for penetration and spread of bacteria
Lipases	Break down lipids
Panton-Valentine leukocidin	Inhibits phagocytosis by granulocytes and can destroy these cells by forming pores in their phagosomal membranes
Phenol soluble modulins	Contribute to biofilm formation and survival within phagosome
Protein A	Antiphagocytic by competing with neutrophils for the Fc portion of specific opsonins
Proteases	Break down proteins
Toxic shock syndrome toxin-1 (a superantigen)	Associated with the fever, shock, and multisystem involvement of toxic shock syndrome

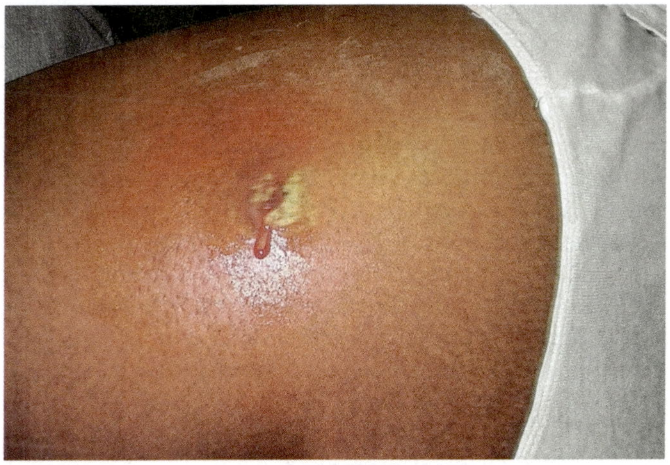

Figure 39.20 Staphylococcal Skin Infections. Draining abscess on the hip.
Source: CDC

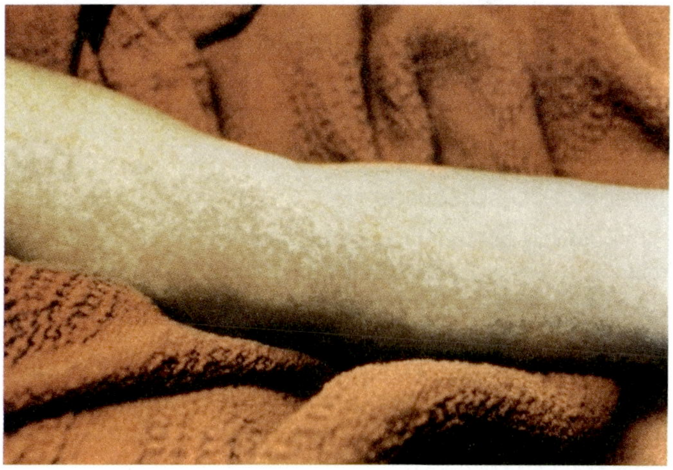

Figure 39.21 Effects of Toxic Shock Toxin. Toxic shock toxin produces a rash (shown here) that may resemble measles or a sunburn with peeling skin.
Source: CDC

Methicillin-Resistant *Staphylococcus aureus* (MRSA)

Isolates of *Staphylococcus aureus* resistant to all β-lactam antibiotics are known as ***methicillin-resistant Staphylococcus aureus*** **(MRSA)**. These include penicillins (e.g., penicillin, amoxicillin), "antistaphylococcal" penicillins (e.g., methicillin, oxacillin), cephalosporins (except the fifth-generation drug ceftaroline), and carbapenems. Antibiotic resistance is driven by the *mecA* (and in rare cases, the *mecC*) gene encoding an altered penicillin-binding protein called PBP2a. Otherwise healthy people can acquire MRSA infections, usually on or in the skin, such as abscesses, boils, and other pus-filled lesions. About 100,000 infections and 20,000 deaths due to MRSA occur annually in the United States. When the infection is acquired by people who have not recently been hospitalized or had a medical procedure (e.g., dialysis, surgery, catheters), the infection is referred to as a *c*ommunity-*a*ssociated (CA)-MRSA infection. However, MRSA is often a nosocomial infection; such infections are called hospital-acquired MRSA (HA-MRSA). Because CA-MRSA strains are now commonly reported inside health-care settings, the designation CA- and HA-MRSA is helpful only in distinguishing their general phenotypes. HA-MRSA strains are also resistant to several other classes of antibiotics. By contrast, CA-MRSA strains tend to remain sensitive to other classes of antibiotics but are often more virulent as they produce additional toxins, such as the Panton-Valentine leukocidin (table 39.2). ◀ *Antibacterial drugs (section 9.4)*

Similar to other infections caused by staphylococci, MRSA skin infections begin as a bump or reddened area that looks like a large pimple. But unlike carbuncles or boils caused by methicillin sensitive *S. aureus* (MSSA), MRSA infections result in the production of more pus (**figure 39.22**). MRSA also causes the same variety of diseases as MSSA.

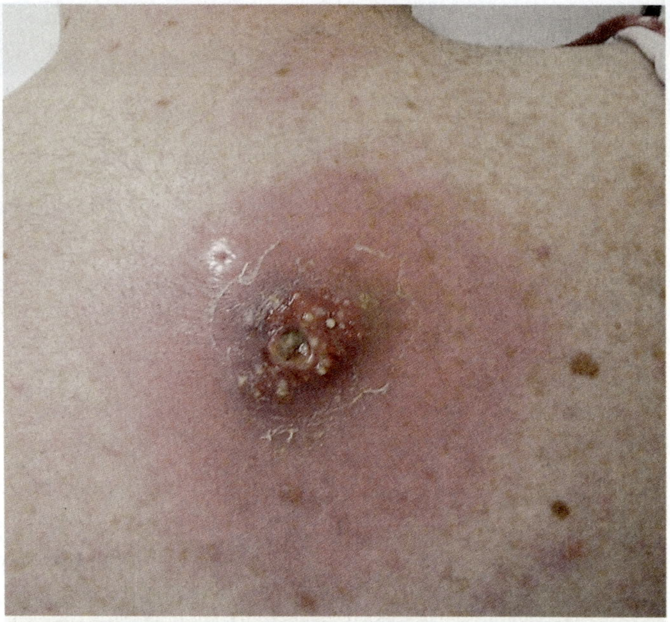

Figure 39.22 Methicillin-Resistant *Staphylococcus aureus* Wound.
Source: CDC/Gregory Moran, M.D.

MICRO INQUIRY *To what drugs, besides methicillin, are MRSA strains resistant?*

Treatment includes antibiotic therapy. Determining the antibiotic sensitivity of the cultured pathogen is a must. Doxycycline or trimethoprim plus sulfamethoxazole are common treatment options. Vancomycin is used for more severe MRSA infections; linezolid and daptomycin can also be used. Good aseptic technique is paramount in preventing the spread of MRSA.

Streptococcal Diseases

Skin and soft tissue infections can also be caused by group A streptococci (*S. pyogenes*). In addition, streptococcal disease can invasively infect the skin, spreading deeper into the tissue, often reaching the underlying muscle. This tissue is a very rich "growth medium" for streptococci, providing the nutrients for rapid replication. We discuss both the superficial and invasive streptococcal diseases next. Airborne transmission of *S. pyogenes* is discussed in section 39.1.

Cellulitis, Impetigo, and Erysipelas

Cellulitis is a diffuse, spreading infection of subcutaneous skin tissue, caused by either *S. pyogenes* or *S. aureus*. The resulting inflammation is characterized by a defined area of redness (erythema) and the accumulation of fluid (edema).

A frequently diagnosed skin infection caused by *S. pyogenes* and *S. aureus* is **impetigo.** Impetigo is a superficial infection, most commonly seen in children, usually located on the face, and characterized by crusty lesions and vesicles surrounded by a red border. It is contagious and can spread rapidly through a nursery, day care, or school. Treatment of impetigo is with topical mupirocin. In severe cases, systemic antibiotic therapy is also prescribed. Penicillin or a cephalosporin is the drug of choice; erythromycin is prescribed for individuals allergic to β-lactams.

Unlike cellulitis and impetigo, which can be caused by either *S. pyogenes* or *S. aureus*, **erysipelas** (Greek *erythros*, red; *pella*, skin) is almost always caused by *S. pyogenes*. It is an acute infection of the dermal layer of the skin. It occurs primarily in infants and people over thirty years of age; it sometimes follows a streptococcal sore throat. The skin often develops painful, reddish patches that enlarge and thicken with a sharply defined edge. Recovery usually takes a week or longer if no treatment is given. The drugs of choice for the treatment of erysipelas are penicillin or a cephalosporin. Erysipelas may recur periodically at the same body site for years.

Invasive Streptococcal Infections

The development of invasive disease initiated by *S. pyogenes* is linked to specific serotypes, as determined by M protein structure (figure 39.6), and predisposing host factors (surgical or

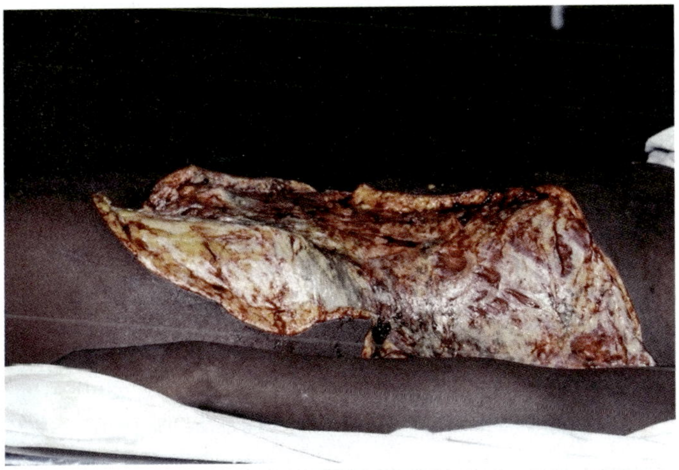

Figure 39.23 Necrotizing Fasciitis. Rapidly advancing streptococcal disease can lead to large necrotic sites that require surgical removal, called debridement, as shown here. Necrotizing fasciitis and myositis are sometimes called flesh-eating disease.

MICRO INQUIRY *What virulence factors enable the rapid invasion of certain strep A strains?*

nonsurgical wounds, diabetes, and other underlying medical problems). A life-threatening infection begins when GAS strains penetrate a mucous membrane or take up residence in a wound such as a bruise. This infection can quickly lead either to **necrotizing fasciitis** (Greek *nekrosis*, deadness; Latin *fascis*, band or bandage; *itis*, inflammation), which destroys the sheath covering skeletal muscles, or to **myositis** (Greek *myos*, muscle; *-itis*), the inflammation and destruction of skeletal muscle and fat tissue (**figure 39.23**).

It is estimated that approximately 9,000 to 11,500 cases of invasive GAS infections occur annually in the United States (1,000 to 1,800 deaths), and 6 to 7% of these are associated with necrotizing conditions. Rapid treatment is necessary to reduce the risk of death, and penicillin G remains the treatment of choice. In addition, surgical removal of dead and dying tissue is needed in more advanced cases.

One reason invasive GAS strains are so deadly is that about 85% of them carry the genes for the production of streptococcal pyrogenic exotoxins A and B (Spe exotoxins). Exotoxin B is a protease that hydrolyzes proteins found in the extracellular matrix of tissues, thereby facilitating bacterial invasion. Exotoxin A is a superantigen, so GAS skin and soft-tissue infections can lead to severe sepsis and septic shock, a condition called streptococcal toxic shock syndrome (STSS). STSS is very difficult to treat and has a mortality rate ranging from 20 to 40%. ◀ *Toxins are biological poisons (section 35.4)*

Tetanus

Tetanus (Greek *tetanos*, to stretch) is caused by *Clostridium tetani*, an anaerobic, Gram-positive, endospore-forming rod (*see figure 23.25*). The endospores of *C. tetani* are commonly found in hospital environments, in soil and dust, and in the feces of many farm animals and humans. Transmission to humans is associated with skin wounds as a break in the skin allows *C. tetani* endospores entry. If the oxygen tension is low enough, the endospores germinate and release the neurotoxin tetanospasmin. **Tetanospasmin** is an endopeptidase that selectively cleaves the synaptic vesicle membrane protein synaptobrevin. This prevents release of inhibitory neurotransmitters (γ-aminobutyric acid and glycine) at synapses within spinal cord motor nerves. The result is uncontrolled stimulation of skeletal muscles (spastic paralysis). A second toxin, **tetanolysin,** is a hemolysin that aids in tissue destruction. ◀ *Class Clostridia (section 23.3)*

Early in the course of the disease, tetanospasmin causes tension or cramping and twisting in skeletal muscles surrounding the wound and tightness of the jaw muscles. With more advanced disease, there is trismus ("lockjaw"), an inability to open the mouth because of the spasm of the masseter muscles. Facial muscles may go into spasms, producing the characteristic expression known as risus sardonicus. Spasms or contractions of the trunk and extremity muscles may be so severe that there is boardlike rigidity, painful convulsions, and backward bowing of the back (opisthotonus; **figure 39.24**). Death results from respiratory failure caused by spasms of the diaphragm and intercostal respiratory muscles.

Control measures for tetanus are not possible because the bacterium is widely disseminated in soil and its endospores remain viable for a remarkably long time. Therefore prevention is all important and depends on (1) active immunization with toxoid, which includes boosters every 10 years; (2) proper care of wounds contaminated with soil; (3) prophylactic use of antitoxin; and (4) administration of penicillin. In Western countries, tetanus is rare. However neonatal tetanus can occur through contamination of the umbilicus following non-sterile delivery. It is almost 100% fatal, causing about 61,000 infant deaths annually. Neonatal tetanus can be prevented by immunization of expectant mothers as maternally transmitted antibodies protect the newborn.

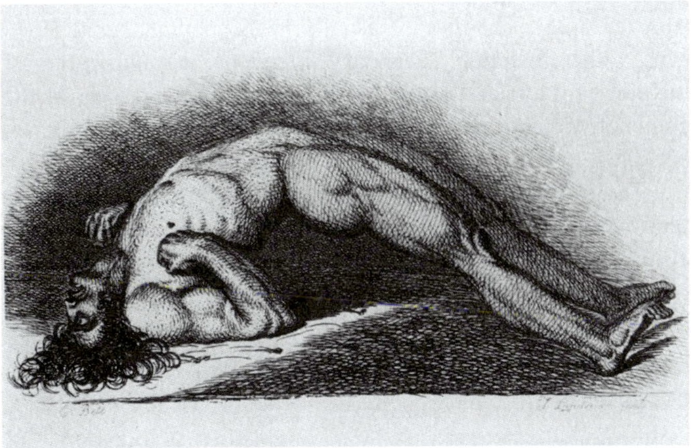

Figure 39.24 The Toll of Tetanus. Sir Charles Bell's portrait (c. 1821) of a soldier wounded in the Peninsular War in Spain shows opisthotonus resulting from tetanus.

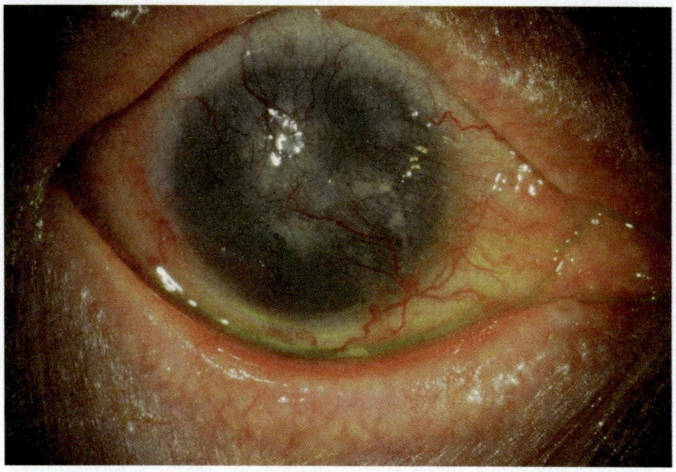

Figure 39.25 **Trachoma.** An active infection showing marked follicular hypertrophy of the eyelid. The inflammatory nodules cover the thickened conjunctiva of the eye.

Trachoma

Trachoma (Greek *trachoma*, roughness) is a highly contagious disease caused by *Chlamydia trachomatis* serotypes A–C. It is the greatest single cause of blindness throughout the world: Annually, 8 million people are infected worldwide, and about half develop irreversible blindness due to chronic or repeated infection. Active disease in adults over age twenty is three times more common in women than men because of mother-child contact. Although trachoma is uncommon in the United States, except among Native Americans in the Southwest, it is widespread in Asia, Africa, and South America. *C. trachomatis* can also cause "pink eye," a milder form of eye infection.

Trachoma is transmitted by contact with inanimate objects such as soap and towels, by hand-to-eye contact, or by flies. The disease begins abruptly with an inflamed conjunctiva. This leads to an inflammatory cell exudate and necrotic eyelash follicles (**figure 39.25**). The disease usually heals spontaneously. However, with reinfection, secondary infections, vascularization of the cornea, pannus (tissue flap) formation, and scarring of the conjunctiva can occur. If scar tissue accumulates over the cornea, blindness results.

Diagnosis of trachoma is by serological methods. The infection is treated with doxycycline, macrolides, or tetracycline. Prevention and control of trachoma depend on health education, personal hygiene, and sanitation including access to clean water.

Comprehension Check

1. How can humans acquire gas gangrene? Group B strep? Leprosy? Describe the major symptoms of each.
2. Why do you think the slow growth rate of *M. leprae* makes leprosy treatment more difficult?
3. Apart from gastric ulcers, what other conditions can *H. pylori* cause?
4. What is ophthalmia neonatorum and how is it transmitted?
5. Describe the lesions of syphilis. What do the symptoms indicate about the disease progression?
6. Infection with *N. gonorrhoeae* or *T. pallidum* does not confer lifelong protection from reinfection. Why is this the case?
7. Compare and contrast the diseases caused by the TSST and SSSS proteins of *S. aureus*.
8. What is the difference between *S. aureus* and MRSA?
9. What conditions are caused by superantigens? Why are these toxins so dangerous?
10. Explain why tetanus is potentially life-threatening. When was the last time you had a tetanus booster?
11. How is trachoma transmitted, and how does it cause blindness?

39.4 Food and Water Are Vehicles for Bacterial Diseases

After reading this section, you should be able to:

a. Describe bacteria that cause the common food-borne and waterborne diseases
b. Identify typical signs and symptoms of food-borne and waterborne bacterial diseases
c. Correlate food-borne and waterborne bacterial infection and disease severity with bacterial virulence factors

Many microorganisms that contaminate food and water can cause acute gastroenteritis—inflammation of the stomach and intestinal lining. When food is the source of the pathogen, the condition is often called **food poisoning.** Gastroenteritis can arise in two ways. Some bacteria cause a food infection. Alternatively, the pathogen may secrete an exotoxin that contaminates the food and is then ingested by the host. This is referred to as a **food intoxication** because the toxin is ingested but the presence of living microorganisms is not required. Because these toxins disrupt the functioning of the intestinal mucosa, they are called **enterotoxins.**

Worldwide, there are nearly 1.7 billion cases of childhood diarrheal disease annually. In fact, diarrheal diseases are the leading cause of childhood death, and in some parts of the world, they are responsible for more years of potential life lost than all other causes combined. For example, each year around 525,000 children (about 1,440 a day) die from diarrheal diseases. In the United States, the CDC estimates that there are about 76 million cases of food-borne illness, leading to 325,000 hospitalizations and 5,000 deaths each year.

This section describes several of the more common bacterial gastrointestinal infections, food intoxications, and waterborne diseases. **Table 39.3** summarizes many of the bacterial pathogens responsible for food poisoning, and **table 39.4** lists many important waterborne bacterial pathogens. The protozoa responsible for food-borne and waterborne diseases are covered in chapter 40. ▶| *Food-borne disease outbreaks (section 41.3)*

Botulism

Food-borne **botulism** (Latin *botulus*, sausage) is a food intoxication because disease results from ingesting preformed botulinum toxin, not the bacteria. A common source of disease is home-canned food contaminated with *Clostridium botulinum* endospores. The endospores germinate, and a toxin is produced during anaerobic vegetative growth. If the food is later eaten without adequate heating to denature all toxin molecules, the toxin causes disease.

Botulinum toxin is an AB neurotoxin that binds to the same target as tetanus toxin, the synaptobrevin of motor neurons, but with opposite effects (**figure 39.26**). It selectively cleaves the synaptic vesicle membrane protein synaptobrevin, thereby preventing release of the neurotransmitter acetylcholine. As a consequence, muscles do not contract in response to motor neuron activity, and flaccid paralysis results (**Techniques & Applications 39.3**). Symptoms of botulism occur within 12 to 72 hours of toxin ingestion and include blurred vision, difficulty swallowing and speaking, bilateral muscle weakness, nausea, and vomiting. Treatment relies on the adminstration of antitoxin and supportive care. If untreated, up to 5% of patients die of either respiratory or cardiac failure within a few days.

Infant botulism is the most common form of botulism in the United States with about 100 cases reported each year. Ingested endospores, which may be naturally present in honey or house dust, germinate in the infant's intestine. *C. botulinum* then multiplies and produces the toxin. The infant becomes constipated, listless, generally weak, and eats poorly. Death may result from respiratory failure. This is why it is recommended that infants not be fed raw honey.

Campylobacter jejuni Gastroenteritis

Campylobacter jejuni is the most commonly reported bacterial cause of food-borne illness in the United States. The CDC estimates there are between 2.1 and 2.4 million cases of ***Campylobacter* gastroenteritis** (**campylobacteriosis**) and subsequent diarrhea in the United States each year. Studies with chickens, turkeys, and cattle have shown that 50 to 100% of these animals excrete *C. jejuni*. It also can be isolated in high numbers from surface waters. It is transmitted to humans by contaminated food and water, contact with infected animals, and anal-oral sexual activity. ◀ *Class* Epsilonproteobacteria *ranges from pathogens to deep-sea bacteria (section 22.5)*

The incubation period for campylobacteriosis is 2 to 10 days. *C. jejuni* invades the epithelium of the small intestine, causing inflammation. It also secretes a cytolethal distending toxin that cleaves host DNA. Symptoms include diarrhea, high fever, severe inflammation, and ulceration of the intestine, which can cause bloody stools. The structural resemblance between *C. jejuni* surface polysaccharides and a sugar moiety found on nerve cell glycoproteins has been linked to the development of Guillain-Barré syndrome in rare cases following *C. jejuni* infection. This is a disorder in which the body's immune system attacks peripheral nerves, resulting in life-threatening paralysis.

Laboratory diagnosis is accomplished through stool cultures. The disease is usually self-limited, and treatment is supportive, including fluid and electrolyte replacement. Azithromycin should be reserved only for those with additional health problems. Recovery usually takes 5 to 8 days. Prevention and control involve good personal hygiene and food-handling precautions, pasteurization of milk, and thorough cooking of poultry.

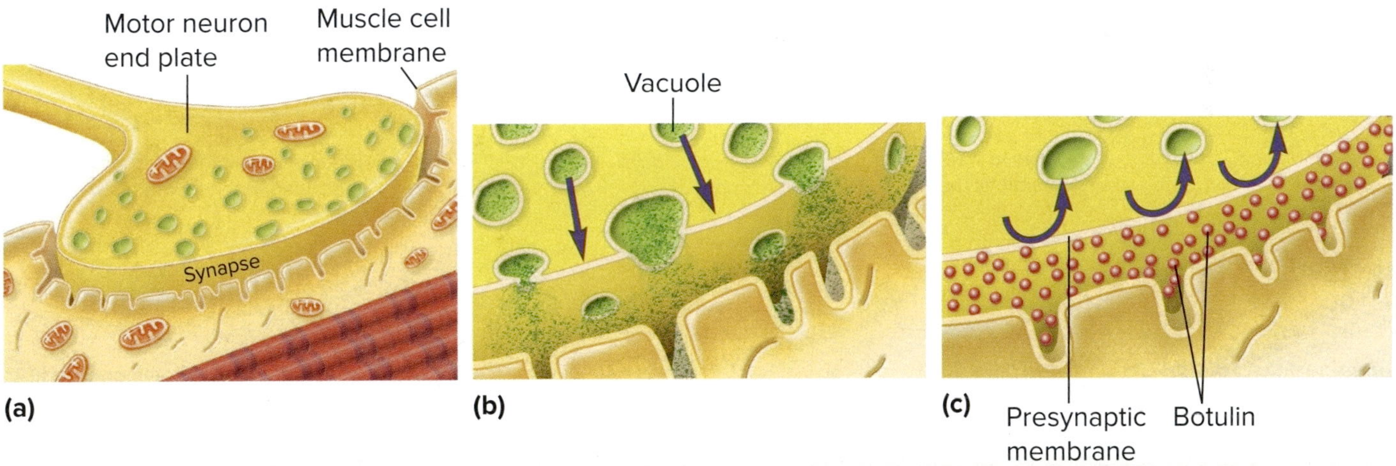

Figure 39.26 The Physiological Effects of *Botulinum* Toxin. (a) The relationship between the motor neuron and the muscle at the neuromuscular junction. (b) In the normal state, acetylcholine released at the synapse crosses to the muscle and creates an impulse that stimulates muscle contraction. (c) In botulism, the toxin enters the motor end plate and attaches to the presynaptic membrane, where it blocks release of the acetylcholine. This prevents impulse transmission and keeps the muscle from contracting.

MICRO INQUIRY *How do the neurological effects of botulinum toxin differ from those of tetanus toxin?*

Table 39.3 Bacteria That Cause Acute Diarrhea and Food Poisoning

Organism	Incubation Period (Hours)	Vomiting	Diarrhea	Fever	Common Source
Staphylococcus aureus	1–8 (rarely, up to 18)	+++	+	–	Meats, dairy and bakery products; produces enterotoxins
Bacillus cereus	2–16	+++	++	–	Reheated fried rice
Clostridium perfringens	8–16	±	+++	–	Rewarmed meat dishes
Clostridium botulinum	18–24	±	Rare	–	Anoxic foods (e.g., canned goods), raw honey
Escherichia coli (enterohemorrhagic)	3–5 days	±	++	±	Undercooked ground beef and unpasteurized fruit juices and cider
Escherichia coli (enterotoxigenic strain)	24–72	±	++	–	Raw fruits and vegetables, contaminated water
Vibrio parahaemolyticus	6–96	+	++	±	Seafood, particularly shellfish and crustaceans
Vibrio cholerae	24–72	+	+++	–	Fecally contaminated water; grows on exoskeleton of marine invertebrates (e.g., copepods)
Shigella spp.	24–72	±	++	+	Raw fruits and vegetables, contaminated water
Salmonella spp. (gastroenteritis)	8–48	±	++	+	Undercooked poultry, meats; egg, peanut butter
Salmonella enterica serovar Typhi (typhoid fever)	10–14 days	±	±	++	Fecally contaminated food; humans are the sole reservoir
Clostridioides (Clostridium) difficile	Days to weeks after antibiotic therapy	–	+++	+	Fecally contaminated food; spores are found in the envirnoment
Campylobacter jejuni	2–10 days	–	+++	++	Undercooked poultry, meats, seafood; eggs; unpasteurized milk
Yersinia enterocolitica	4–7 days	±	++	+	Food-borne (most commonly pork)

Source: Geo. F. Brooks, et al., *Medical Microbiology,* 21st ed. Norwalk, CT: Appleton & Lange, 1998.

Pathogenesis	Clinical Features and Treatment
Enterotoxins act on gut receptors that transmit impulses to medullary centers.	Abrupt onset (2–6 hours after consumption), intense vomiting for up to 24 hours, recovery in 24–48 hours. No treatment usually necessary except to restore fluids and electrolytes.
Enterotoxins formed in food or in gut from growth of *B. cereus*.	Incubation period of 8–16 hours, mainly vomiting or diarrhea.
Enterotoxins produced during sporulation in gut; causes hypersecretion.	Abrupt onset of profuse diarrhea; vomiting occasionally. Recovery usual without treatment in 1–4 days.
Toxin absorbed from gut and blocks acetylcholine release at neuromuscular junction.	Difficulty speaking or breathing. Treatment requires clearing the airway, ventilation, and polyvalent antitoxin. Mortality rate high.
Shiga toxins cause epithelial necrosis in colon; mild to severe complications.	Symptoms vary from mild to severe bloody diarrhea. The toxin can be absorbed, becoming systemic and producing hemolytic uremic syndrome, most frequently in children.
Heat-labile (LT) and heat-stable (ST) enterotoxins cause hypersecretion in small intestine.	Usually abrupt onset of diarrhea; vomiting rare. A serious infection in newborns. In adults, “travelers’ diarrhea” is usually self-limited in 1–3 days.
Toxin causes hypersecretion; vibrios invade epithelium; stools may be bloody.	Abrupt onset of diarrhea. Recovery is usually complete in 1–3 days. Food and stool cultures are positive.
Toxin causes hypersecretion in small intestine; infective dose $>10^5$ vibrios; cholera stool classically described as “rice-water” consistency.	Abrupt onset of liquid diarrhea in endemic area. Needs prompt replacement of fluids and electrolytes IV or orally. Tetracyclines shorten excretion of vibrios.
Organisms invade epithelial cells; blood, mucus, and neutrophils in stools; infective dose $<10^3$ organisms.	Abrupt onset of diarrhea, often with blood and pus in stools, cramps, and lethargy. Trimethoprim-sulfamethoxazole or ampicillin given in severe cases. Often self-limited.
Superficial infection of gut, little invasion; infective dose $>10^5$ organisms.	Gradual or abrupt onset of diarrhea and low-grade fever. Nausea, headache, and muscle aches common. Administer no antimicrobials unless systemic dissemination is suspected. Prolonged carriage is frequent.
Symptoms due to exotoxin; infective dose $\geq 10^7$ organisms.	Initially fever, headache, malaise, anorexia, and muscle pains. Fever may reach 40°C and lasts for 2 or more weeks. Diarrhea and abdominal pain may occur. Flat, red rash appears 2-4 weeks after illness begins. Antibiotic therapy indicated; may be fatal.
Toxins cause epithelial necrosis in colon	Especially after antibiotic therapy, abrupt bloody diarrhea and fever. Toxins in stool. Oral vancomycin useful in therapy. Recurrent cases treated with fecal microbiome transplant.
Invasion of mucous membrane; cytolethal distending toxin produced	Fever, diarrhea; PMNs and fresh blood in stool, especially in children. Usually self-limited. Special media needed for culture at 43°C. Erythromycin given in severe cases with invasion. Usual recovery in 5–8 days.
Gastroenteritis; occasional bacteremia; toxin produced occasionally.	Severe abdominal pain, diarrhea, fever; PMNs and blood in stool; polyarthritis, erythema nodosum, especially in children.

Table 39.4 Waterborne Bacterial Pathogens

Organism	Reservoir	Comments
Aeromonas hydrophila	Free-living	Sometimes associated with gastroenteritis, cellulitis, and other diseases
Campylobacter jejuni	Bird and animal reservoirs	Major cause of diarrhea; common in processed poultry; a microaerophile
Helicobacter pylori	Free-living	Can cause gastritis, peptic ulcers, gastric adenocarcinomas
Legionella pneumophila	Free-living and associated with protozoa	Found in cooling towers, evaporators, condensers, showers, and other water sources
Leptospira spp.	Infected animals	Hemorrhagic effects, jaundice
Mycobacterium spp.	Infected animals and free-living	Complex recovery procedure required
Pseudomonas aeruginosa	Free-living	Swimmer's ear and skin infections
Salmonella enteritidis	Animal intestinal tracts	Common water contaminant
Vibrio cholerae	Free-living	Found in many waters, including estuaries
Vibrio parahaemolyticus	Free-living in coastal waters	Causes diarrhea in shellfish consumers
Yersinia enterocolitica	Frequent in animals and in the environment	Waterborne gastroenteritis

Cholera

Cholera is an acute diarrheal disease caused by infection of the intestine with the Gram-negative, comma-shaped bacterium *Vibrio cholerae*. The disease is characterized by profuse watery diarrhea, vomiting, and leg cramps. These symptoms result from rapid loss of body fluids leading to dehydration and shock. Death can occur within hours if left untreated. Throughout recorded history, cholera (Greek *chole*, bile) has caused seven pandemics in various areas of the world, especially in Asia, the Middle East, and Africa. The disease has been rare in the United States since the 1800s. Unfortunately, 2010 will be remembered for the resurgence of cholera in Haiti. After Haiti had gone almost 100 years without cholera, well-intentioned relief workers responding to a devastating earthquake imported cholera bacteria to Haiti. Since that time, cholera has caused over 770,000 cases and at least 10,000 deaths. Cholera epidemics are a testament to how quickly bacteria can invade and infect.

Although there are many *V. cholerae* serogroups, only the toxigenic O1 and O139 (encapsulated) organisms have caused epidemics. *V. cholerae* O1 is divided into two biotypes, Classical and El Tor. Each biotype is divided into three serotypes, Inaba, Ogawa, and Hikojima, based on the structure of the O-antigen. Interestingly, the cholera toxin gene (*ctxAB*) is carried by the temperate CTX filamentous bacteriophage, which exists as a prophage within the genome of toxin-producing strains of *V. cholerae*. The receptor for the phage is the toxin coregulated pilus—the same structure used to colonize the host's gut. *V. cholerae* is an excellent example of how horizontal transfer of genes can confer pathogenicity.

Cholera is transmitted by ingesting food or water contaminated by fecal material from infected individuals. The incubation period is 3 to 72 hours and infection with serotypes O1 or O139 results in mild to severe diarrhea. Approximately 20% of those with severe diarrhea have watery diarrhea; 10 to 20% of those with watery diarrhea produce "rice-water-stool," so named because of the flecks of mucus floating in it. The bacteria adhere to the intestinal mucosa of the small intestine, where they are not invasive but secrete cholera toxin, an AB toxin (*see figure 35.8*). The A subunit enters the intestinal epithelial cells and adds an ADP-ribosyl group to the G protein coupled receptor, $G_{s\alpha}$, which becomes locked in the activated state. This results in constant stimulation of host adenyl cyclase and overproduction of cAMP. This leads to hypersecretion of water and chloride ions while inhibiting absorption of sodium ions. The infected person loses massive quantities of fluid and electrolytes, causing abdominal muscle cramps, vomiting, fever, and diarrhea. Diarrhea can be so profuse that a person can lose up to 20 liters of fluid per day (**figure 39.27**). Death results from dehydration and electrolyte imbalance.

Passage through the human host enhances infectivity, although the exact mechanism is unclear. Before *V. cholerae* exits the body, the intestinal environment stimulates the activity of certain bacterial genes. These genes, in turn, seem to prepare the bacteria for ever more effective colonization of their next victims, possibly fueling epidemics. *V. cholerae* can also be free-living in warm, alkaline, and saline environments.

Treatment is by oral rehydration therapy with NaCl plus glucose to stimulate water uptake by the intestine; doxycycline may

TECHNIQUES & APPLICATIONS

39.3 Clostridial Toxins as Therapeutic Agents: Benefits of Nature's Most Toxic Proteins

Some toxins are currently used for the treatment of human disease. Specifically, botulinum toxin, the most poisonous biological substance known, is used for the treatment of specific neuromuscular disorders characterized by involuntary muscle contractions. Since approval of type-A botulinum toxin (Botox) by the FDA in 1989 for three disorders (strabismus [crossing of the eyes], blepharospasm [spasmotic contractions of the eye muscles], and hemifacial spasm [contractions of one side of the face]), the number of neuromuscular problems treated has increased to include other tremors, migraine and tension headaches, and other maladies. In 2000 dermatologists and plastic surgeons began using Botox to temporarily eradicate wrinkles caused by repeated muscle contractions as we laugh, smile, or frown. The remarkable therapeutic utility of botulinum toxin lies in its ability to specifically and potently inhibit muscle activity for an extended duration. Overall, clostridia (currently one of the largest and most diverse groups of bacteria that includes about 130 species) produce more protein toxins than any other known bacterial genus and are a rich reservoir of toxins for research and medicinal uses. For example, research is under way to use clostridial toxins or toxin domains for drug delivery, prevention of food poisoning, and the treatment of cancer and other diseases. The remarkable success of botulinum toxin as a therapeutic agent has thus created a new field of investigation in microbiology.

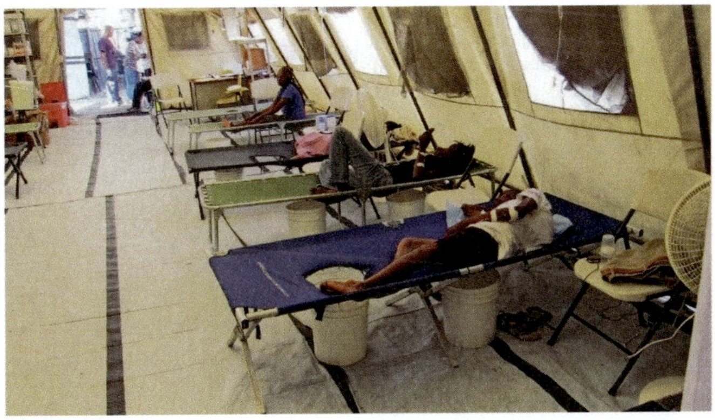

Figure 39.27 Cholera Cots. Cholera, caused by *V. cholerae*, causes such profuse diarrhea that special cots are used during outbreaks to keep suffering patients clean.
Source: CDC

also be given. The most reliable control methods are based on proper sanitation, especially of water supplies. The mortality rate without treatment is often over 50%, but with treatment and supportive care, it is less than 1%.

Escherichia coli Gastroenteritis

A major causative agent of food-borne diarrhea is *Escherichia coli*. The vast majority of *E. coli* strains are harmless and provide assistance in the metabolism of the food we eat. However, certain *E. coli* strains have become pathogenic by acquiring genes through horizontal gene transfer. Expression of virulence genes often alters the bacterial surface molecules (antigens). Thus, strains of *E. coli* are readily tracked by serotyping. Two of the most commonly used serotyping antigens are the H (flagellar) and the O (O-antigen portion of LPS) antigens.

Five categories (pathotypes) of diarrheagenic *E. coli* are recognized (**figure 39.28**): enterotoxigenic *E. coli* (ETEC), enteroinvasive *E. coli* (EIEC), enteropathogenic *E. coli* (EPEC), enteroaggregative *E. coli* (EAEC), and enterohemorrhagic *E. coli* (EHEC). All except EHEC are found in human feces; transmission is from poor hand hygiene or sanitation. Each pathotype uses a different mechanism to cause disease, as we now discuss. ◀ *Order* Enterobacteriales *includes pathogens and mutualists (section 22.3)*

Enterotoxic *E. coli* (ETEC) strains produce one or both of two distinct enterotoxins responsible for diarrhea. The toxins are distinguished by their heat stability: heat-stable enterotoxin (ST) and heat-labile enterotoxin (LT) (figure 39.28*a*). The genes for ST and LT production and for colonization factors are usually plasmid-borne and acquired by horizontal gene transfer. ST is a small protein that increases host cGMP concentrations. LT has a structure and activity very simlar to cholera toxin. Both toxins induce hypersecretion of electrolytes and water into the intestinal lumen, triggering the watery diarrhea. ETEC is the most common cause of travelers' diarrhea.

Enteroinvasive *E. coli* (EIEC) strains penetrate and multiply within the intestinal epithelial cells (figure 39.28*b*). EIEC and *Shigella* spp. are considered members of the same pathovar. Both *Shigella* spp. and EIEC strains produce an enterotoxin that stimulates diarrhea and dysentery symptoms as discussed in the section on shigellosis.

Enteropathogenic *E. coli* (EPEC) strains cause a specific kind of cell damage called effacing lesions by attaching to the brush border of intestinal epithelial cells using an adhesin known as intimin (figure 39.28*c*). Proteins encoded by the LEE (*l*ocus of *e*nterocyte *e*ffacement) region of a pathogenicity island cause effacing lesions or attaching-effacing (AE) lesions. This ultimately results in destruction of brush border microvilli (hemorrhagic colitis) adjacent to adhering bacteria. AE lesions result from the

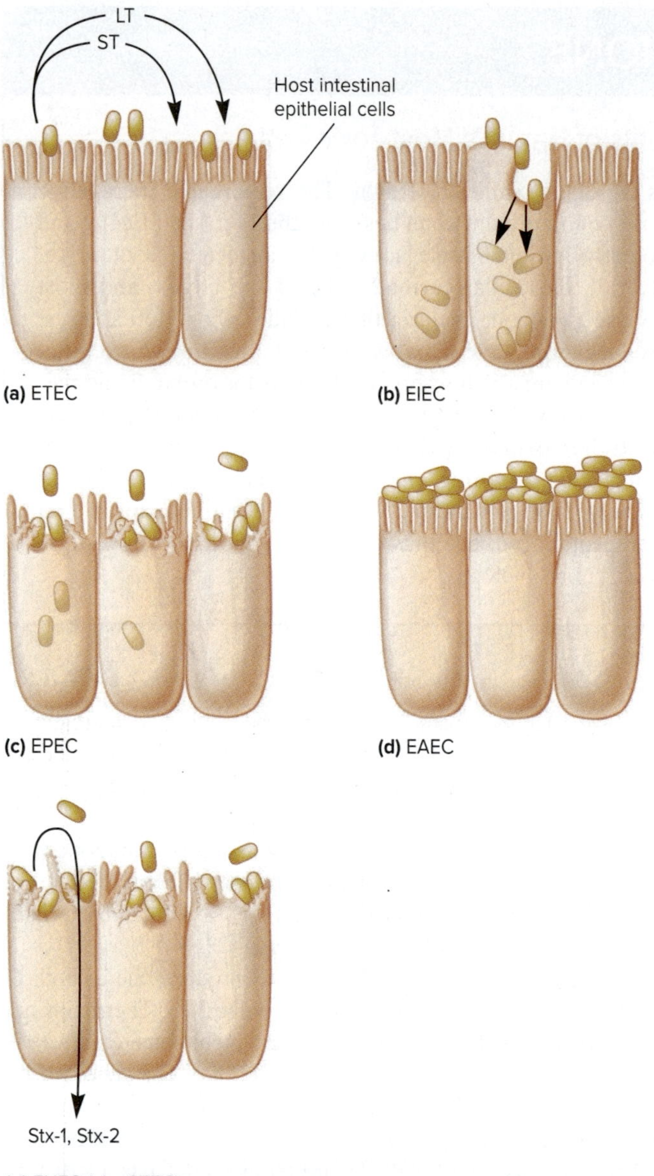

Figure 39.28 Five Strains of Diarrheagenic *E. coli.* Each strain of diarrhea-causing *E. coli* can be classified by the nature of its interaction with host intestinal epithelial cells.

MICRO INQUIRY *To which* E. coli *class does the strain O157:H7 belong?*

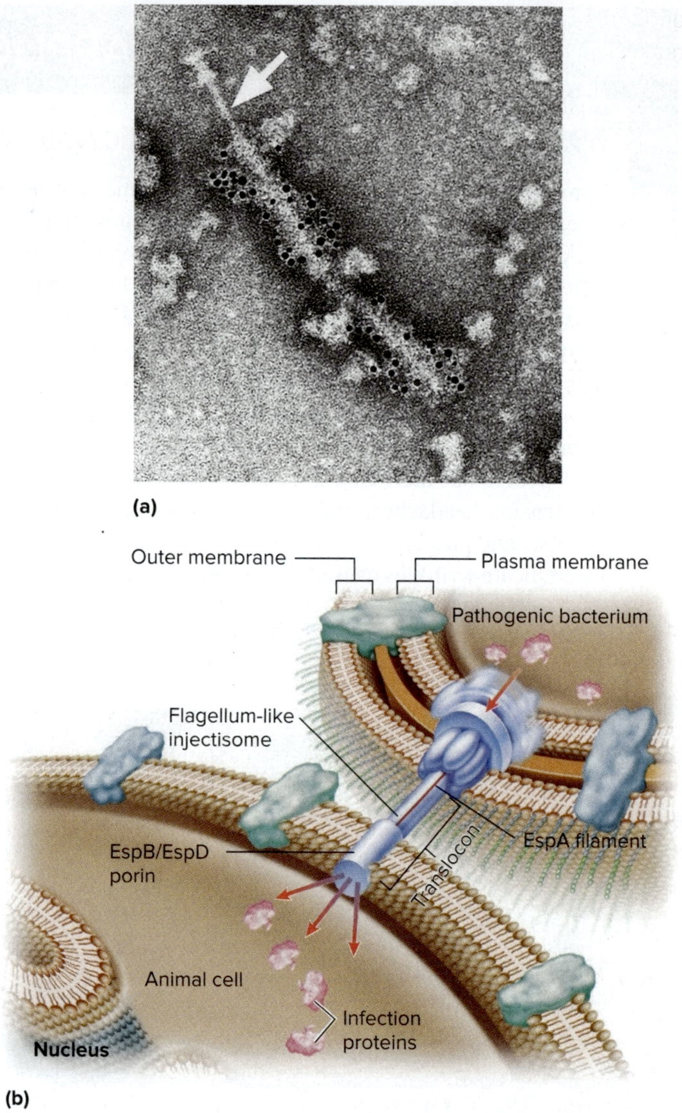

Figure 39.29 Type Three Secretion System (T3SS) of Enteropathogenic *E. coli* (EPEC). (a) Transmission electron micrographs of T3SS structures from EPEC cells. Sheathlike structures are labeled with 6-nm immunogold particles (black dots); the needle complexes, neck, and stem components (white arrows) remain unlabeled. (b) The T3SS proteins function like a syringe to inject virulence proteins into a host cell.

(a) From: Sekiya, K., et al., "Supermolecular structure of the enteropathogenic Escherichia coli type III secretion system and its direct interaction with the EspA-sheath-like structure," PNAS, Vol. 98, no. 20, September 25, 2001, p 11638–11643. ©2001 National Academy of Sciences, U.S.A.

delivery of seven different virulence proteins through a LEE-encoded type III secretion (T3SS) system, into host cell membranes (**figure 39.29**). These virulence proteins have a variety of activities that together limit host immune responses as well as trigger host cell destruction, leading to bloody diarrhea. EPEC is an important cause of potentially fatal diarrhea in infants and children living in countries without adequate sanitation infrastructure.

Enteroaggregative *E. coli* (EAEC) strains adhere to epithelial cells in localized regions, forming clumps of bacteria with a "stacked brick" appearance (figure 39.28*d*). EAEC can also produce several plasmid-encoded toxins that cause diarrhea, particularly in children in nations lacking good sanitation. It is the second most common cause of travelers' diarrhea.

Cattle serve as the reservoir for **enterohemorrhagic *E. coli* (EHEC)** or **Shiga toxin–producing *E. coli* (STEC)**, which is transferred to humans by water or food. There are 200 to 400 serotypes that carry the bacteriophage-encoded genetic determinants for Shiga toxin (also called Shiga-like toxin Stx-1 and/or Stx-2 proteins; figure 39.28*e*). Some EHEC strains also carry the LEE gene sequence to produce AE lesions, causing hemorrhagic colitis (HC) with severe abdominal pain and cramps followed by bloody diarrhea. Stx-1 and Stx-2 can cause the extraintestinal disease hemolytic uremic syndrome (HUS), a severe hemolytic anemia that leads to kidney failure. The A subunit of these toxins enters renal cells and depurinates 28S rRNA molecules, thereby disabling the ribosome, eliminating protein synthesis, and causing cell death.

A major form of STEC (EHEC) in North America is *E. coli* O157:H7 (STEC O157), which has caused many outbreaks of HC and HUS in the United States since it was first recognized in 1982. A variety of other STEC serotypes cause similar disease. Currently there are an estimated 265,000 STEC (36% STEC O157) cases in the United States each year, resulting in an average of 60 deaths. Despite the variety of STEC serotypes, most laboratories do not test for non-O157 strains, so the actual incidence of EHEC is underreported. Importantly, STECs are the only strains passed in nonhuman (animal) feces. Thus animal manure and farm contamination of foods are major sources of these bacteria. ▶I *Food-borne disease outbreaks (section 41.3)*

Diagnosis of diarrhea caused by *E. coli* is based on travel history and symptoms. Laboratory diagnosis is by culture of the specific type of *E. coli* from feces and immunoassays to detect Shiga toxins. Treatment for strains that do not produce Shiga toxin is with fluid and electrolytes plus doxycycline and trimethoprim-sulfamethoxazole. Antibiotics should not be given if the strain produces Shiga toxin as this can cause an increase in toxin release. Recovery can be without complications unless HUS develops. Prevention and control involve avoiding contaminated food and water.

Salmonellosis

Salmonellosis (*Salmonella* gastroenteritis) is caused by over 2,000 *Salmonella* serovars. Based on DNA homology studies, all known *Salmonella* serovars that cause human disease are classified as a single species, *Salmonella enterica*. The most frequently isolated serovars from humans are Typhimurium and Enteritidis. (Serovar names are not italicized and the first letter is capitalized.) An estimated 1.2 million cases of nontyphoidal salmonellosis occur annually in the United States; of these, approximately 19,000 result in hospitalization, and 380 result in death.

With the exception of *S. enterica* serovar Typhi (*S.* Typhi), for which humans are the sole reservoir, the initial source of the bacterium is the intestinal tracts of birds and other animals. Humans acquire the bacteria from contaminated water or foods such as beef products, poultry, eggs, or egg products. Once the bacteria are in the body, the incubation time is only about 12 to 72 hours. Disease results when the bacteria multiply and invade the intestinal mucosa. Abdominal pain, cramps, diarrhea, nausea, vomiting, and fever are the most prominent symptoms, which usually persist for 4 to 7 days but can last for several weeks. During the acute phase of the disease, as many as 1 billion *Salmonella* cells can be found per gram of feces. Most adult patients recover, but the loss of fluids can quickly dehydrate children and elderly people. Persons with severe diarrhea may therefore require rehydration therapy with intravenous fluids. Antibiotics are not usually given unless the infection spreads from the intestines.

Serovars including *S.* Agona, *S.* Heidelberg, *S.* Montevideo, *S.* Mbandaka, and *S.* Tennessee have been reported as contaminants of cereal products, chicken, tahini sesame paste, and peanut butter. The ongoing concern over food safety has prompted greater diligence for the detection and reporting of food contamination. Of note is the increasing concern regarding infection resulting from antibiotic-resistant *Salmonella* spp. *S.* Typhimurium is now resistant to at least five antimicrobial agents. Another serovar, *S.* Newport, is resistant to at least seven agents.

Typhoid (Greek *typhodes*, smoke) fever is caused by *S. enterica* serovar Typhi and is acquired by ingestion of food or water contaminated by feces of infected humans or person-to-person contact. Unlike nontyphoidal serovars of *S. enterica*, *S.* Typhi can result in multiorgan failure and death. *S.* Typhi is the only serovar known to produce a toxin. Specifically, it produces an unusual AB toxin with heterodimeric A (catalytic) subunits. One subunit possesses nuclease activity, so is considered a cytolethal distending toxin. Although the second A subunit is known to have ADP-ribosylating activity, its cellular target has not yet been discovered.

In earlier centuries, the disease occurred in great epidemics; it remains endemic in some regions that lack adequate sanitation. The incubation period for typhoid fever is about 5 to 21 days. Initially, fever is present with a paradoxically slow heart rate. This is followed by a characteristic rash that features small pink lesions called "rose spots." The bacteria colonize the small intestine, penetrate the epithelium, and spread to lymphoid tissue, blood, liver, and gallbladder. In some patients, infection is dire and can cause septic shock. In others, symptoms are less severe. Following acute illness, it typically takes about 3 months for most individuals to stop shedding bacteria in their feces. However, about 4% of individuals continue to shed the pathogen for extended periods but show no symptoms. In these carriers, the bacteria continue to grow in the gallbladder and reach the intestine through the bile duct. This was almost certainly the case for "Typhoid Mary" (*see Historical Highlights 36.4*). In the pre-antibiotic era, the mortality rate was at least 15%; it is now less than 1%. Treatment is with azithromycin, a fluoroquinolone, or a cephalosporin. Travelers to typhoid-endemic regions are advised to get the typhoid vaccine before leaving.

Shigellosis

Shigellosis, or bacillary dysentery, is a diarrheal illness resulting from an acute inflammatory reaction of the intestinal tract caused by the four species of the genus *Shigella*. Globally, *Shigella* spp. are estimated to cause up to 165 million cases of disease, with 600,000 deaths, per year. About 14,000 cases a year are reported in the United States. However, real numbers may be 20 times greater because mild cases of disease typically are not reported. The predominant disease-causing species in the United States is *S. sonnei*. *S. flexneri* predominates in many tropical countries. *S. boydii* and *S. dysenteriae* infections are uncommon.

The organism is transmitted by the fecal-oral route and is most prevalent among young children. The infectious dose is only around 10 to 100 bacteria in large part because the bacteria are very tolerant of stomach acidity. Infection is marked by fever, abdominal cramps, and a bloody, mucoidy diarrhea. In severe cases, the colon can become ulcerated.

Shigellae are intracellular parasites that penetrate the gut mucosal barrier. The bacteria are phagocytosed by macrophages within Peyer's patches. After being ingested by macrophages, the bacteria disrupt the phagosome membrane and are released into the cytosol, reproduce, and kill the macrophage. They then invade mucosal epithelial cells where they polymerize actin and use it to propel their intracellular motility, moving from one cell to the next. Both endotoxins and exotoxins (some secreted by a T3SS) participate in disease progression, but bacteria do not usually spread beyond the colonic epithelium. *S. dysenteriae* serotype 1 produces Shiga toxin (Stx). This is very similar to Shiga (or Shiga-like) toxins previously described EHEC. All these toxins can target the glomerular endothelium leading to hemolytic uremic syndrome (HUS) and kidney failure. ◀ *Phagocytosis destroys invaders (section 32.6); Exotoxins (section 35.4)*

The incubation period for shigellosis usually ranges from 1 to 7 days, and the organisms are shed over a period of 1 to 2 weeks. Unless the causative agent is *S. dysenteriae*, disease normally is self-limiting in adults and lasts an average of 4 to 7 days; in infants and young children, shigellosis may be fatal regardless of causative species. Fluid and electrolyte replacement are usually sufficient to treat disease; antibiotics may not be required in mild cases, but in severe cases azithromycin or a third-generation cephalosporin are indicated. Sometimes, particularly in infants and children, neurological complications occur. Prevention is a matter of good personal hygiene and the provision of a clean water supply.

Staphylococcal Food Poisoning

Staphylococcal food poisoning is the major type of food intoxication in the United States. It is caused by ingestion of improperly stored or cooked food (particularly foods such as ham, processed meats, chicken salad, pastries, ice cream, and creamy sauces) in which *S. aureus* has been allowed to reproduce and release enterotoxins. Because these enterotoxins are heat stable, even if food has been properly cooked and all bacteria killed, active toxin remains. Staphylococcal enterotoxins (SE) are primarily produced by *S. aureus,* although other staphylococci produce them. SE are structurally related, single-chain proteins but are antigenically distinct. Thirteen different SE have been identified; SE A, B, C1, C2, C3, D, and E are the most common. Typical symptoms include severe abdominal pain, cramps, diarrhea, explosive vomiting, and nausea. The onset of symptoms is rapid (usually 1 to 8 hours) and of short duration (usually less than 24 hours). The mortality rate of staphylococcal food poisoning is negligible among healthy individuals.

Treatment of staph food poisoning primarily involves maintaining hydration while letting the toxin flush from the body. The toxemia commonly resolves without medication. Prevention is through thorough hand washing during food preparation and the use of safe cooking and dining practices.

Comprehension Check

1. How can *Clostridium botulinum* cause disease without the presence of viable bacteria?
2. Compare the gastroenteritis caused by *Vibrio cholerae* with that of *Campylobacter jejuni*. How are they similar and different?
3. What are the mechanisms by which *E. coli* can cause disease?
4. Compare and contrast ETEC and EHEC in terms of toxin activity and severity of disease.
5. How do the mechanisms of pathogenicity differ between *Salmonella* Typhi and nontyphoidal *Salmonella* serovars?
6. Compare the intestinal diseases caused by *Salmonella* and *Shigella* spp. How are they the same? Different?
7. Why is staphylococcal food poisoning considered an intoxication? How is it treated?

39.5 Zoonotic Diseases Arise from Human-Animal Interactions

After reading this section, you should be able to:

a. Discuss the bacteria that cause common diseases spread by contact with animals
b. Identify typical signs and symptoms of zoonotic bacterial diseases
c. Correlate zoonotic bacterial infection and disease severity with bacterial virulence factors

Diseases transmitted from animals to humans are called zoonotic diseases, or zoonoses. As we now discuss, some pathogens that are normal microbiota or pathogens of animals can also cause disease in humans. Many are considered **select agents** indicating they represent a bioterrorism threat. ◀ *Bioterrorism readiness is an integral component of public health microbiology (section 36.7)*

Anthrax

Anthrax (Greek *anthrax*, coal) is a highly infectious animal disease caused by the Gram-positive, endospore-forming *Bacillus anthracis*. *B. anthracis* is found worldwide and can be transmitted to humans by direct contact with infected animals (cattle, goats, sheep) or their products, especially hides. *B. anthracis* endospores remain viable in soil and animal products for decades (*see figure 23.17a*). Although strains of *B. anthracis* have very little genetic variation, it is now possible to separate all known strains into five categories (providing some clues to their geographic sites of origin) based on the number of tandem repeats in various genes.

There are three forms of anthrax disease. When the bacterium enters through a cut or abrasion of the skin, **cutaneous anthrax** results. Inhaling endospores may result in **pulmonary anthrax,** also known as woolsorter's disease. If endospores reach the gastrointestinal tract, **gastrointestinal anthrax** may result. Discounting the 2001 bioterrorism events in the United States, more than 95% of anthrax is the cutaneous form. In pulmonary anthrax, the endospores (1 to 2 μm in diameter) are inhaled and lodge in the alveolar spaces, where they are engulfed by alveolar macrophages. This initially causes rapidly progressing flulike symptoms. Symptoms of gastrointestinal anthrax typically occur after the ingestion of undercooked meat containing endospores and include nausea, vomiting, fever, and abdominal pain. Cutaneous anthrax produces a lesion called an eschar, characterized by black, necrotic tissue with minimal inflammation of the surrounding tissue. *B. anthracis* bacteremia can develop from any form of anthrax, with pulmonary anthrax almost always resulting in bacteremia.

B. anthracis carries two plasmids encoding virulence factors. One plasmid has genes involved in the synthesis of its polyglutamyl capsule that inhibits phagocytosis, and the other bears genes for the synthesis of anthrax exotoxin. This exotoxin is composed of three proteins: protective antigen (PA), edema factor (EF), and lethal factor (LF). Macrophages bear two different receptors to which PA attaches. Attachment continues until seven PA-receptor complexes gather in a doughnut-shaped ring (**figure 39.30**). The ring acts like a syringe, boring through the plasma membrane of the macrophage. Toxin proteins EF and LF are then engulfed by the macrophage and shuttled to an endosome (*see figure 32.18*). Once there, PA molecules form a pore that pierces the endosomal membrane, and EF and LF enter the cytosol. Toxin activity results in fluid release, with the formation of edema. Additionally, LF prevents the transcription factor nuclear factor kappa B ($NF_{\kappa}B$) from regulating genes needed for an effective immune response. As thousands of macrophages die, they release their lysosomal contents, leading to fever, internal bleeding, septic shock, and rapid death. Without antibiotic treatment and supportive therapy, mortality rates approach 100% for inhalational and gastrointestinal anthrax, and between 20 and 50% for cutaneous anthrax.

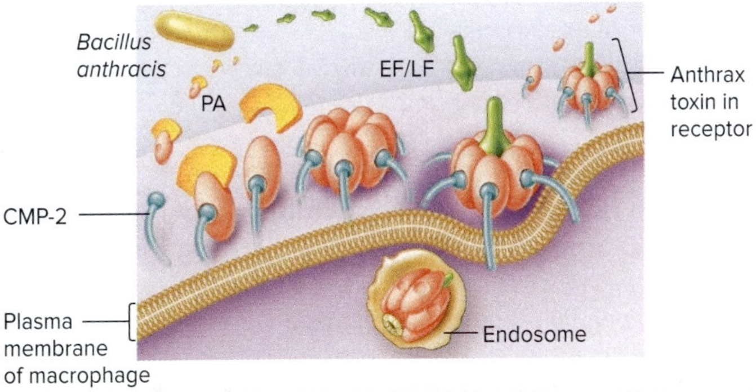

Figure 39.30 Anthrax. Protective antigen (PA) assembles in a barrel-shaped structure on target macrophages. Two proteins, edema factor (EF) and lethal factor (LF), are transported to an endosome. PA then delivers EF and LF from the endosome into the cytoplasm of the macrophage, where they exert their toxic effects.

MICRO INQUIRY *By what mechanisms do EF and LF kill macrophages?*

Between 2,000 and 20,000 cases of anthrax are estimated to occur worldwide annually. In the United States, the annual incidence was less than one case per year—a rate maintained for 20 years until 2001. The 2001 occurrence of 22 cases of anthrax spotlighted concern about anthrax as a weapon of bioterrorism.

Treatment of each form of anthrax is the same: Ciprofloxacin, penicillin, or doxycycline is successful only if begun before a critical concentration of toxin accumulates. Although antibiotics may kill the bacterium or suppress its growth, the exotoxin can still eventually kill the patient. Thus supportive care is essential in preventing death. Vaccination of animals, primarily cattle, is an important control measure. However, people with a high occupational risk, such as those who handle infected animals or their products, including hides and wool, should be immunized. Certain U.S. military personnel also receive the vaccine.

Brucellosis (Undulant Fever)

The genus *Brucella* contains important human and animal pathogens. **Brucellosis,** also called undulant fever, is caused by faintly staining Gram-negative coccobacilli of the species *B. abortus, B. melitensis, B. suis,* or *B. canis*. These bacteria are commonly transmitted through consumption of contaminated animal products or through abrasions of the skin while handling infected mammals (cattle, sheep, goats, pigs, and, rarely, dogs). The most common route is ingestion of contaminated or unpasteurized milk products. Direct person-to-person spread of brucellosis is extremely rare. However, infants may be infected through their mother's breast milk; sexual transmission of brucellosis has also been reported. A controversy in the northwestern United States over the transmission of brucellosis, endemic in wild bison and elk populations, to otherwise *Brucella*-free cattle has continued for many years. ◀ *Class* Alphaproteobacteria *includes many oligotrophs (section 22.1)*

Most human *Brucella* infections are caused by *B. melitensis*. Brucellosis is rare in the United States, with no more than 140 cases per year. In contrast to the United States, human brucellosis is the most common zoonosis worldwide. About 500,000 cases are reported annually, but it is suspected that the real incidence may be up to 25 times greater.

Brucellosis, in the acute (8 weeks from onset) form, presents as nonspecific, flulike symptoms, including fever, sweats, malaise, anorexia, headache, myalgia (muscle pain), and back pain. In the undulant (rising and falling) form, symptoms of brucellosis include cyclic fevers, arthritis, and testicular inflammation. Neurologic symptoms occur in up to 5% of the cases. In the chronic form (1 year from onset), brucellosis symptoms include chronic fatigue, depression, and arthritis. Mortality is low, less than 2%. However, it is known to trigger spontaneous abortion, premature delivery, and fetal death. Treatment is usually with doxycycline plus streptomycin or rifampin for 6 weeks to prevent recurring infection. The insidious disease, potential for food supply disruption, and significant morbidity associated with *Brucella* infection have prompted its restriction as a select agent; the public health impact, should it be used as a bioweapon, would be substantial.

Psittacosis (Ornithosis)

Psittacosis (ornithosis) is a worldwide infectious disease of birds that is transmissible to humans. Since 1996 fewer than 50 confirmed cases have been reported in the United States each year. Psittacosis was first described in association with parrots and parakeets, both of which are psittacine birds. The disease is now recognized in many other birds—among them, pigeons, chickens, ducks, and turkeys—and the general term ornithosis (Latin *ornis*, bird) is used.

Ornithosis is caused by *Chlamydophila psittaci*. Humans contract this disease either by handling infected birds or by inhaling dried bird excreta that contains viable *C. psittaci*. Incubation is typically 5 to 19 days. After entering the respiratory tract, the bacteria are transported to the cells of the liver and spleen. The bacteria multiply within these cells and then invade the lungs, where they cause inflammation, hemorrhage, and pneumonia. Endocarditis (inflammation of the heart muscle), hepatitis, and neurologic complications may occasionally occur. With doxycycline or tetracycline therapy, the mortality rate is about 2%.

Q Fever

The Q in **Q fever** stands for "query" because the cause of the fever was long unknown. The disease is now known to be caused by the γ-proteobacterium *Coxiella burnetii*. Although it is an intracellular pathogen, *C. burnetii* survives outside host cells by forming resistant, endospore-like bodies. This bacterium infects both wild animals and livestock. Cattle, sheep, and goats are the primary reservoirs. It does not need an arthropod vector for transmission, but in animals, many species of ticks can transmit *C. burnetii*. By contrast, human transmission is primarily by inhalation of dust contaminated with bacteria from dried animal feces or urine, or consumption of unpasteurized milk. Importantly, organisms are shed in high numbers within amniotic fluids and the placenta during birthing. The organisms are resistant to heat, drying, and many common disinfectants. These features enable the bacteria to survive for long periods in the environment. The disease can occur in epidemic form among slaughterhouse workers and sporadically among farmers and veterinarians. Each year, fewer than 30 cases of Q fever are reported in the United States. However, approximately 3% of U.S. adults and an additional 10 to 20% of high-risk populations (farmers, veterinarians, etc.) have antibodies suggesting prior exposure to *C. burnetii*. ◀ *Class* Gammaproteobacteria *is the largest bacterial class (section 22.3)*

In humans, Q fever is an acute illness characterized by the sudden onset of severe headache, malaise, confusion, sore throat, chills, sweats, nausea, chest pain, myalgia, and high fever. It is rarely fatal, but endocarditis occurs in about 10% of the cases. Five to 10 years may elapse between the initial infection and the appearance of endocarditis. During this interval, the bacteria reside in the liver and may cause hepatitis. The fatality rate is approximately 4%.

Diagnosis of Q fever is made by national reference laboratories because it is a select agent. PCR or culture of the bacterium and fluorescent antibody and agglutination tests are used. Treatment is with doxycycline and is most effective when initiated within the first 3 days of illness. Prevention and control involve public education, protective clothing, and vector control.

Tularemia

Another select agent is the Gram-negative bacterium *Francisella tularensis*, which causes the disease **tularemia.** It can be transmitted to humans by biting arthropods (ticks and deer flies), direct contact with infected tissue (rabbits), inhalation of aerosolized bacteria, or ingestion of contaminated food or water. However, tularemia is most often transmitted through contact with infected animals; it is called rabbit fever in the central United States because it can occur in hunters who skin their catch. After an incubation period of 2 to 10 days, it causes a disease that resembles plague but is not as severe. The patient develops a primary ulcerative lesion at the infection site, then lymph nodes enlarge, and a high fever develops. Without treatment, tularemia has a 30 to 60% mortality rate.

Diagnosis is made by national reference laboratories using PCR or culture of the bacterium and fluorescent antibody and agglutination tests. Treatment is with streptomycin, tetracycline, or aminoglycoside antibiotics for 10 to 21 days. An attenuated live vaccine is available from the U.S. Army for high-risk laboratory workers. Typically 150 to 200 cases of tularemia are reported annually in the United States.

Comprehension Check

1. How do humans acquire anthrax? Brucellosis?
2. Compare and contrast the symptoms of anthrax, brucellosis, and tularemia.
3. How is ornithosis transmitted?
4. Describe the disease of tularemia. Why is *F. tularensis* considered a potential agent of bioterrorism?

39.6 Opportunistic Diseases Can Be Caused by Bacteria

After reading this section, you should be able to:

a. Explain how disease can result from human microbiota that is normal to another body site
b. Cite specific environmental conditions that result from overgrowth of one bacterial species
c. Design a model of bacterial biofilm formation to explain its role in human disease

Opportunistic infections can be divided into two categories. The first involves microorganisms that are part of the normal microbiome, but cause disease if they overgrow other members of the microbial community or gain access to otherwise sterile sites in the host. The second are opportunistic microbes found in the environment. These cause disease only in immunologically compromised hosts who cannot mount the robust immune defenses that would otherwise eliminate these microbes. We discuss a few examples of opportunistic disease next.

Antibiotic-Associated Colitis

Antibiotic-associated colitis is a spectrum of disease caused by the overgrowth of *Clostridioides* (*Clostridium*) *difficile* that affects about 450,000 patients annually in the United States. *C. difficile* is sometimes part of the human intestinal microbiota but is normally kept at low numbers by other bacteria. The disease spectrum (*C. difficile* infection, or CDI) includes uncomplicated diarrhea, pseudomembranous colitis (a viscous collection of inflammatory cells, dead cells, necrotic tissue, and fibrin that obstructs the intestine), and toxic megacolon (inflammation that results in intestinal tissue necrosis), which is often lethal (**figure 39.31**). The most common symptom of CDI is watery diarrhea, consisting of three or more bowel movements per day for 2 or more days accompanied by fever, loss of appetite, nausea, and abdominal cramping. In severe cases, patients have as many as 20 diarrhea episodes per day, which leads to extreme weight loss, dehydration, and debilitation.

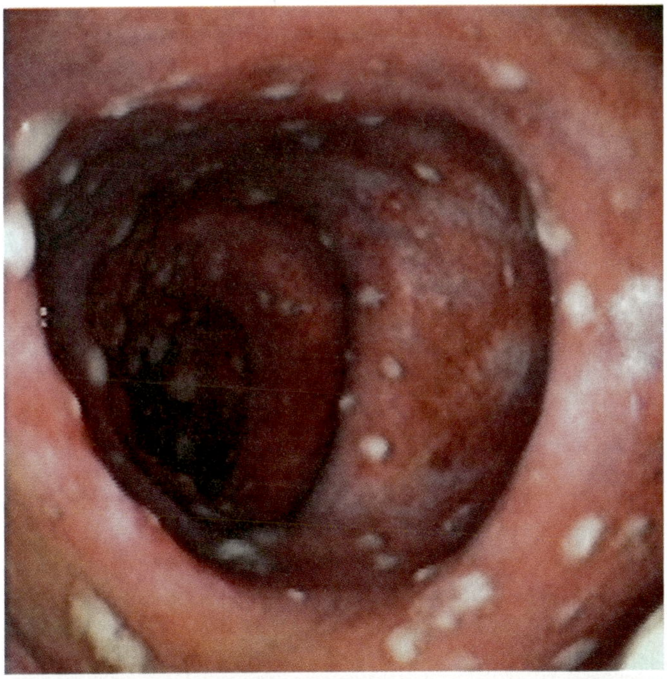

Figure 39.31 ***Clostridioides difficile*** **Infection.** Endoscopic view of *C. difficile*–associated colitis. The white patches represent inflamed tissue.

MICRO INQUIRY *What cells and tissues make up the pseudomembrane?*

C. difficile is an anaerobic, endospore-forming rod; endospores are often found in health-care facilities. Although therapy with any antibiotic can trigger CDI, the most common antibiotic associated with CDI is clindamycin, a protein synthesis inhibitor that targets anaerobes that normally compete with *C. difficile (see figure 34.4)*. When antibiotics inhibit the growth of normal microbiota, *C. difficile* bacteria multiply, produce toxins, and sporulate. All pathogenic strains of *C. difficile* produce two exotoxins: enterotoxin (TcdA) and a cytotoxin (TcdB), both of which are responsible for the inflammation and diarrhea. Some of the more aggressive strains also secrete binary toxin, which limits host innate immunity. Strains that do not produce TcdA and TcdB do not cause disease—that is, they are avirulent.

C. difficile is the most commonly recognized cause of diarrhea in hospitalized patients, as endospores from one patient can be transmitted to others. Antibiotic use is the main modifiable risk factor for CDI. In addition to suppressing other bacteria and thereby promoting *C. difficile* overgrowth, antibiotics increase the risk of selecting for antimicrobial-resistant strains. Ironically, treatment of CDI is by oral metronidazole or oral vancomycin. Fecal microbiome transplant (FMT), in which prescreened feces from a healthy donor are transplanted to the CDI patient's colon, can be used to treat recurrent cases. This reestablishes a healthy microbiome and has a cure rate of about 92%. Clinical trials involving treating patients with a limited number of bacterial species identified as critical to the success of FMT show promise. ◀ *The microbe-human ecosystem (chapter 34)*

Bacterial Vaginosis

Bacterial vaginosis is a polymicrobial infection, meaning it is an infection that involves multiple microbial species. It is most commonly associated with *Gardnerella vaginalis* (a Gram-negative to Gram-variable, pleomorphic, nonmotile rod), *Mobiluncus* spp. (Gram-negative, anaerobic rods), and various other anaerobic bacteria. These microorganisms inhabit the vagina and rectum of 20 to 40% of healthy women but their growth is inhibited by vaginal lactobacilli (**figure 39.32**). Bacterial vaginosis is typically a mild disease, although it is a risk factor for obstetric infections, adverse outcomes of pregnancy, pelvic inflammatory disease, and HIV infection.

Vaginosis is characterized by a copious, frothy, fishy-smelling discharge with varying degrees of pain or itching. Diagnosis is based on the microscopic observation of clue cells in the discharge. Clue cells are sloughed-off vaginal epithelial cells covered with bacteria, mostly *G. vaginalis* (figure 39.32). Treatment for bacterial vaginosis is with metronidazole, a drug that kills anaerobic bacteria and protozoa.

Dental Diseases

Only a few species of bacteria in the oral cavity can be considered true opportunistic dental pathogens, or odontopathogens. These few odontopathogens are responsible for the most common bacterial diseases in humans: tooth decay and periodontal disease.

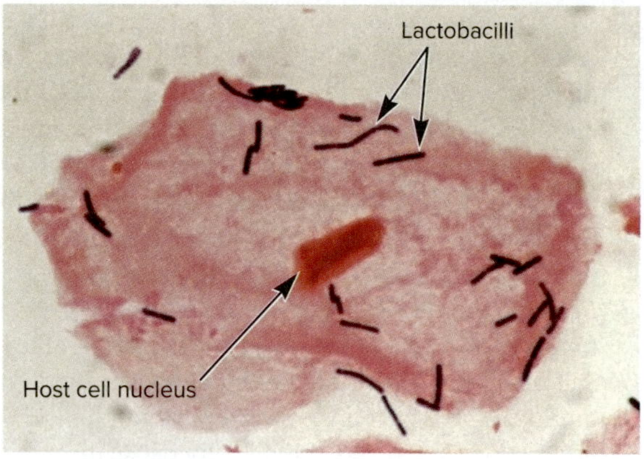

(a) Vaginal lactobacilli

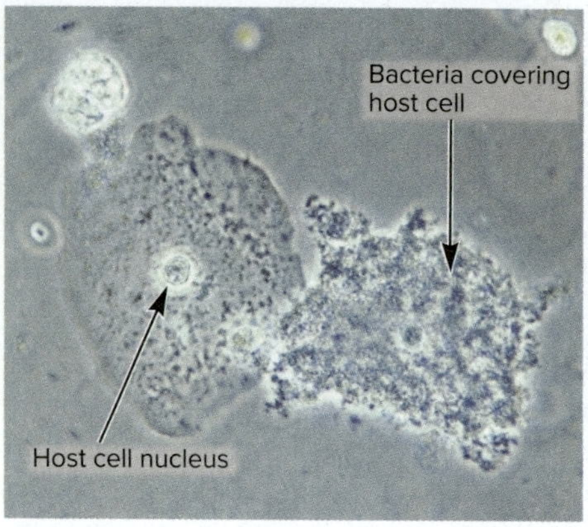

(b) Clue cell

Figure 39.32 Bacterial Vaginosis Is Diagnosed by the Presence of Clue Cells. (a) In the health host, Gram-positive lactobacilli keep the vaginal mucosa at an acidic pH, thereby preventing the growth of many other microorganisms. (Gram stain; lactobacilli are 2 to 4 μm in length.) (b) When vaginal lactobacilli growth is inhibited by antibiotic therapy, hormones, or other factors, Gram-negative anaerobes predominate. Host epithelial cells to which these bacteria attach are called clue cells. (Unstained, phase contrast light microscopy.)
(a) Source: CDC/Dr. Mike Miller; (b) Source: CDC/M. Rein

Dental Plaque

The human tooth has a natural defense mechanism against bacterial colonization that complements the protective role of saliva. The hard enamel surface selectively absorbs acidic glycoproteins (mucins) from saliva, forming a membranous layer called the acquired enamel pellicle. This pellicle, or organic covering, contains many sulfate (SO_4^{2-}) and carboxylate (—COO^-) groups that confer a net negative charge to the tooth surface. Because most bacteria also have a net negative charge, there is a natural repulsion between the tooth surface and bacteria in the oral cavity. Unfortunately, this natural defense mechanism breaks down when dental plaque forms.

Dental plaque is one of the densest collections of bacteria in the body—perhaps the source of the microorganisms seen under a microscope by Antony van Leeuwenhoek in the seventeenth century. Dental plaque formation begins with initial colonization of the pellicle by *Streptococcus gordonii*, *S. oralis,* and *S. mitis* (**figure 39.33**). These bacteria selectively adhere to the pellicle by specific ionic, hydrophobic, and lectinlike interactions. Once the tooth surface is colonized, subsequent attachment of other bacteria results from a variety of specific coaggregation reactions.

S. mutans and *S. sobrinus* become established on the tooth surface by attaching to these initial colonizers (figure 39.33). *S. mutans* and *S. sobrinus* produce extracellular enzymes that polymerize the glucose moiety of sucrose into a heterogeneous group of extracellular, water-soluble, and water-insoluble polysaccharides. These act like a cement that binds bacterial cells together, forming a plaque ecosystem that is also a biofilm. The fructose by-product can be used in fermentation. Once plaque becomes established, the surface of the tooth becomes anoxic. This leads to the growth of strictly anaerobic bacteria (*Bacteroides melaninogenicus, B. oralis,* and *Veillonella alcalescens*), especially between teeth and in the dental-gingival crevices (**figure 39.34**). In addition, bacteria produce lactic and possibly acetic and formic acids from sucrose and other sugars. Because plaque is not permeable to saliva, the acids are not diluted or neutralized, and they demineralize the enamel to produce a lesion on the tooth. It is this chemical lesion that initiates dental decay. ◂ *Biofilms are common in nature (section 7.6)*

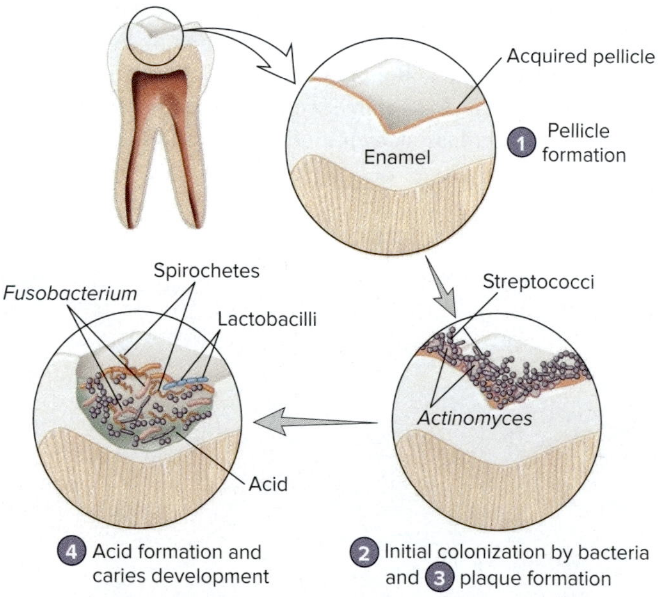

Figure 39.33 Stages in Plaque Development and Cariogenesis. A drawing of a microscopic view of pellicle and plaque formation, acidification, and destruction of tooth enamel.

Dental Decay (Caries)

As fermentation acids move below the enamel surface, they dissociate and react with the hydroxyapatite of the enamel to form

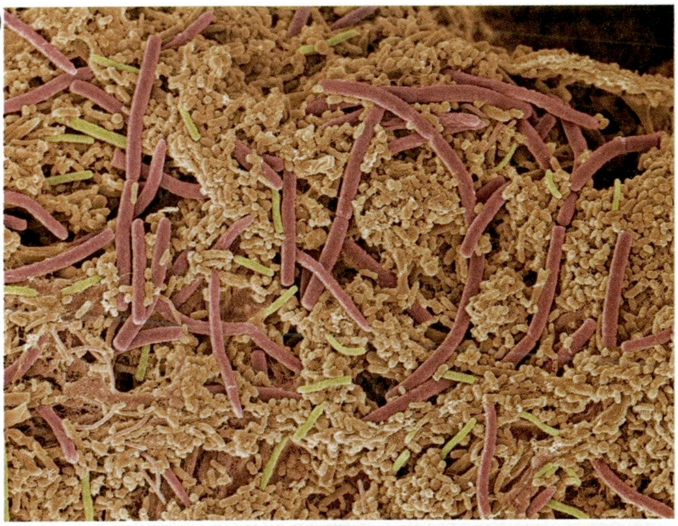

Figure 39.34 The Microscopic Appearance of Plaque. Colorized scanning electron micrograph of plaque shows a polymicrobial biofilm.

MICRO INQUIRY *How is tooth enamel demineralized by bacterial growth?*

soluble calcium and phosphate ions. As the ions diffuse outward, some reprecipitate as calcium phosphate in the tooth's surface layer to create an outer layer overlying a porous subsurface area. Between meals and snacks, the pH returns to neutrality and some calcium phosphate reenters the lesion and crystallizes. The result is a demineralization-remineralization cycle.

When an individual eats foods high in sucrose for prolonged periods, acid production overwhelms the repair process, and demineralization is greater than remineralization. This leads to dental decay or **caries**. Once the hard enamel has been breached, bacteria can invade the dentin and pulp of the tooth and cause its death.

Periodontal Disease

Periodontal disease refers to a diverse group of inflammatory diseases that affect the periodontium. It is the most common chronic infection in adults. The **periodontium** is the supporting structure of a tooth and includes the cementum, the periodontal membrane, the bones of the jaw, and the gingivae (gums). The gingiva is dense, fibrous tissue and its overlying mucous membrane that surrounds the necks of the teeth. The gingiva helps to hold the teeth in place.

Disease is initiated by the formation of plaque that forms at the dentogingival margin and extends down into the gingival tissue. A number of bacterial species contribute to tissue damage. The result is an initial inflammatory reaction known as **periodontitis,** which is caused by the host's immune response to both plaque bacteria and tissue destruction. This leads to swelling of the tissue and formation of periodontal pockets. Bacteria colonize these pockets and cause more inflammation, which leads to formation of a periodontal abscess, bone destruction (periodontosis), inflammation of the gingiva (gingivitis), and general tissue necrosis. If the condition is not treated, teeth may be lost.

The main strategies for preventing dental caries and periodontal disease include reduced consumption of sucrose; daily brushing, flossing, and rinsing with mouthwashes; and professional cleaning at least twice a year to remove plaque. The use of fluorides in toothpaste, drinking water, and mouthwashes or fluoride and sealants applied professionally to the teeth protects against lactic and acetic acids and reduces tooth decay.

Streptococcal Pneumonia

Streptococcal pneumonia is considered an opportunistic infection because it is typically contracted from the host's normal microbiota. It is caused by Gram-positive *Streptococcus pneumoniae*, normally found in the upper respiratory tract. Also called pneumococcal pneumonia, only those individuals with predisposing factors such as viral infections of the respiratory tract, physical injury to the tract, alcoholism, diabetes, or smoking are susceptible to disease. In addition to pneumonia, *S. pneumoniae* causes acute middle ear infection, bacteremia, and meningitis. About 60 to 80% of all respiratory diseases known as pneumonia are caused by *S. pneumoniae*. About 1 million cases of pneumococcal pneumonia occur in the United States annually, with a 5 to 7% mortality rate.

The primary virulence factor of *S. pneumoniae* is its capsule. Each of the over 90 serotypes of *S. pneumoniae* construct capsule from different polysaccharides. The capsule plays an important role in protecting the organism from ingestion and killing by phagocytes. Pathogenesis is due to the rapid multiplication of the bacteria in alveolar spaces. The bacteria also produce a variety of virulence factors including the toxin pneumolysin, which destroys host cells. As host cells are killed, alveoli fill with blood cells and fluid, and become inflamed. Blood in sputum imparts a rust color. The onset of clinical symptoms is usually abrupt, with chills; hard, labored breathing; and chest pain. Diagnosis is by chest X ray, Gram stain, culture, and tests for metabolic products. Treatment needs to be based on antibiotic susceptibility, because some strains are resistant to β-lactam and macrolide antibiotics.

Pneumococcal vaccines (PCV-13 and PPSV23) are available for people who are at risk for exposure (e.g., individuals with chronic diseases, college students, children, the elderly, and people in long-term care facilities). The pneumococcal conjugate vaccine (PCV) covers 13 pneumococcal serotypes, while the pneumococcal polysaccharide vaccine (PPSV23) covers another 23 serotypes of pneumococcal bacteria. These vaccines are effective because they generate antibodies to the strains that cause about 90% of pneumococcal disease. When these antibodies are deposited on the surface of the bacteria, they become opsonic and enhance phagocytosis (*see figure 32.5*).

Urinary Tract Infection and Pyelonephritis

Urinary tract infection (UTI), also called **cystitis**, is one of the most common infections, with an estimated global prevalence of about 150 million cases annually. In the United States, over

12 million people seek help for UTIs in doctor offices and emergency departments every year. The causative bacteria reside in the gut and when introduced into the urethra, ascend to reach the bladder. If left untreated, the bacteria may move up the ureters to the kidneys; kidney infections are called **pyelonephritis.** Women have a much shorter urethra, which accounts for the higher incidence of UTIs among women, but they are also not uncommon among infant boys and older men. A variety of bacteria cause UTIs including *Klebsiella pneumoniae, Staphylococcus saprophyticus, Enterococcus faecalis, Proteus mirabilis*, group B *Streptococcus*, and *Pseudomonas aeruginosa*, but roughly 90% of all UTIs are caused by uropathogenic *E. coli* (UPEC), which is our focus here.

UPEC is unique among *E. coli* strains in its ability to infect the bladder and avoid shedding during urination. UPEC localizes the protein FimH at the tip of type 1 pili, where it serves as an adhesin that attaches to receptors on bladder cells. Such binding triggers host-cell actin rearrangement so that UPEC cells are enveloped and internalized by these normally nonphagocytic cells. Inside bladder cells, UPEC replicates and establishes an intracellular biofilm community, called a pod. UPEC also produces the cytolytic toxin HylA, which contributes to host-cell invasion. As UPEC cells flux in and out of pods, neutrophils invade the bladder and cause acute inflammation, which accounts for the characteristic symptoms of frequency of urination, burning when urinating, abdominal cramping, and blood in the urine.

As long as the infection remains in the bladder, the patient will not develop a fever. However, if the bacteria ascend a ureter and reach a kidney (this is often not bilateral), pyelonephritis develops. This is a much more serious condition as it can cause permanent renal damage. UPEC regulates type 1 pili and flagella in a reciprocal fashion, so that when motility is needed to move up the ureter to the kidney, type 1 pili and FimH are not produced. Once UPEC has entered the kidney, it produces fimbriae to mediate attachment. In addition to fever and chills, pyelonephritis is characterized by pain over the affected kidney.

Uncomplicated UTIs are typically diagnosed clinically with the collection of urine. A simple dipstick test will reveal the presence of leukocyte esterase indicating neutrophil invasion of the bladder. If the causative agent is UPEC, the dipstick will also indicate the presence of nitrite in the urine. Nitrite is produced by UPEC cells that have switched to anaerobic respiration with the use of nitrate as the terminal electron acceptor. The causative agent of pyelonephritis must be determined by urine culture, as urine is normally sterile. The antibiotic sensitivity profile of the causative bacterium is essential. Uncomplicated UTIs are treated with antibiotics that reach a high level in the bladder including sulfamethoxazole/trimethoprim, nitrofurantoin, or fosfomycin. Individuals with pyelonephritis in the outpatient setting can be treated with a fluoroquinolone.

Comprehension Check

1. In both *C. difficile*–associated disease and bacterial vaginosis, the disease begins when the normal microbiota of the tissue site is altered. Why is this?
2. Name some common odontopathogens responsible for dental caries, dental plaque, and periodontal disease.
3. Discuss reasons cigarette smoking increases the risk of streptococcal pneumonia.
4. Discuss two adaptations used by UPEC to cause bladder and kidney infection.

Key Concepts

39.1 Bacteria Can Be Transmitted by Airborne Routes

- The vaccine-preventable diseases diphtheria and pertussis are toxin-mediated. Diphtheria toxin blocks host-cell protein synthesis by ADP-ribosylating elongation factor (EF)-2 (**figure 39.1**). *Bordetella pertussis* produces four exotoxins, including pertussis toxin, which stimulates host adenyl cyclase activity, and another toxin that has adenyl cyclase activity. Thus overproduction of cAMP is a key component of *B. pertussis* virulence.
- Atypical pneumonia can be caused by *Chlamydophila pneumoniae, Mycoplasma pneumoniae*, or *Legionella pneumophila*. Infections are considered atypical because they are not usually associated with a productive cough and other usual signs of pneumonia. *L. pneumophila* infection causes Legionnaires' disease, which is the most serious of these infections.
- About one-third of the global population has tuberculosis (TB) in its latent form. Caused by *Mycobacterium tuberculosis*, TB usually infects the lungs but can also infect any other body system. Strong cell-mediated immunity can keep TB in its latent form, which is why children, HIV-positive individuals, and others who are immunosuppressed often develop severe, life-threatening infections. Treatment is complex, involving multiple antibiotics administered for at least 6 months (**figure 39.4** and **39.5**).
- Vaccine-preventable meningitis is caused by *Neisseria meningitidis, Haemophilus influenzae* serotype b, and *Streptococcus pneumoniae*. However, meningitis in neonates is typically caused by group B streptococcus (*S. agalactiae*), *E. coli*, and *Listeria monocytogenes*. When *N. meningitidis* (meningococcus) causes bacteremia, it is termed meningococcemia. Aggressive

antibiotic therapy and supportive care are needed for this medical emergency.

- Streptococcal pharyngitis (strep throat) is caused by *S. pyogenes*, or group A streptococcus. Prompt antibiotic treatment prevents the development of post-streptococcal glomerulonephritis and rheumatic fever (**figure 39.7**).

39.2 Arthropods Can Transmit Bacterial Diseases

- Bacteria can be transmitted to humans by arthropod vectors, such as ticks, lice, and fleas.
- Lyme disease is a tick-borne disease caused by spirochetes in the genus *Borrelia*. In the northeast United States, the tick vector *Ixodes scapularis* transmits *B. burgdorferi*. Treatment with doxycycline prevents the progression of disease through three stages: flulike illness, arthritis and cardiac inflammation, and neurological deficits (**figure 39.8**).
- Plague, caused by *Yersinia pestis*, remains endemic in some regions. The flea vector feeds on rodents; humans are infected by flea bites or direct contact with infected animals. Humans can transmit plague by droplet transmission between individuals (pneumonic plague). Toxic proteins inserted into macrophages and other immune cells enable the survival and growth of *Y. pestis*. Lymph tissue forms bubos, while blood-borne *Y. pestis* causes mottling of skin and necrosis of extremities, thus the term Black Death (**figures 39.9** and **39.10**).
- The majority of Rocky Mountain spotted fever (RMSF) cases occur east of the Mississippi River. *Rickettsia rickettsii* is transmitted by the ticks *Dermacentor andersoni* and *D. variabilis*. *R. rickettsii* invades endothelial cells, causing vasculitis marked by a nonblanching rash. Without treatment (doxycycline), RMSF is usually fatal (**figure 39.11**).

39.3 Direct Contact Diseases Can Be Caused by Bacteria

- Direct contact of skin, mucous membranes, open wounds, or body cavities can lead to bacterial colonization and disease.
- *Staphylococcus aureus* and group A streptococcus (*S. pyogenes*) cause the superficial skin infections cellulitis and impetigo. *S. pyogenes* and *Clostridium perfringens* cause the life-threatening skin infections necrotizing fasciitis (**figure 39.22**) and gas gangrene (**figure 39.11**), respectively. These are medical emergencies that require surgical excision of infected tissue and aggressive antibiotic therapy.
- In infants, group B streptococcus (*S. agalactiae*) can cause meningitis, bacteremia, and pneumonia; in the elderly and immunocompromised adults, it can cause bacteremia, pneumonia, and skin and bone infections.
- *Mycobacterium leprae* causes leprosy, which in its tuberculoid form is mild and nonprogressive. Lepromatous leprosy results if cell-mediated immune responses are inadequate. It is marked by necrosis of skin and nerve damage (**figure 39.13**).
- *Helicobacter pylori* infection can result in peptic ulcer disease and gastric cancer. This bacterium survives stomach acidity by enclosing itself in the mucosa and producing ammonium via urease activity.
- *Chlamydia trachomatis* is sexually transmitted and can cause pelvic inflammatory disease (PID) in women and the eye infection trachoma in neonates. Trachoma that is not sexually transmitted is also seen in adults (**figure 39.25**).
- Gonorrhea, caused by *Neisseria gonorrhoeae*, is another cause of PID, partly because it is easier to diagnose in men than women, who typically have fewer symptoms. Like *C. trachomatis*, *N. gonorrhoeae* can cause eye infection (ophthalmia neonatorum) in newborns during birth (**figure 39.16**). Because of frequent coinfection, gonorrhea and chlamydia are treated simultaneously with ceftriaxone as well as either azithromycin or doxycycline.
- The spirochete *Treponema pallidum* causes syphilis, which progresses through three stages: painless chancre; skin rash, malaise, and fever; and neurosyphilis and the development of gummas (**figure 39.17**). Benzathine penicillin is an effective cure but like gonorrhea, infection with syphilis does not lead to lifelong immunity.
- Methicillin-resistant *Staphylococcus aureus* (MRSA) can be community or hospital acquired (CA and HA, respectively). MRSA can infect any organ system and is resistant to all β-lactam antibiotics. CA-MRSA is typically sensitive to a wider range of other antibiotics than HA-MRSA, but produces more toxins.
- Tetanus, caused by *Clostridium tetani*, is vaccine preventable. Tetanus toxin prevents the release of inhibitory neurotransmitters thereby causing spastic paralysis. Newborns delivered without sterile precautions are at risk for tetanus.

39.4 Food and Water Are Vehicles for Bacterial Diseases

- Food and water can serve as vehicles that transport bacteria to humans. Food-borne illnesses are classified as infections or intoxications, which are toxin mediated so presence of the bacterium is not needed (**tables 39.3** and **39.4**).
- *Clostridium botulinum* produces botulinum toxin, which blocks the release of the neurotransmitter acetylcholine, thereby causing flaccid paralysis (**figure 39.26**).
- *Campylobacter jejuni* is the most frequently reported cause of food-borne diarrhea in the United States. It produces a toxin with DNase activity, cytolethal distending toxin, but is not considered an intoxication. In rare cases, *C. jejuni* infection is associated with Guillain-Barré syndrome.
- Cholera is caused by the aquatic bacterium *Vibrio cholerae*. *V. cholerae* produces an AB toxin that ADP-ribosylates the stimulator of adenyl cyclase, which becomes locked in the active state. This leads to high levels of host cell cAMP. This results in the loss of ions

and water; patients can lose 20 liters of fluid in diarrhea daily. Rehydration with a salt solution prevents loss of life.

- There are five major strains of pathogenic, food-borne *E. coli*, each classified by the nature of its interaction with host intestinal epithelium (**figure 39.28**). Enterotoxigenic and enterohemorrhagic *E. coli* (ETEC and EHEC, respectively) produce toxins. EHEC also causes hemolytic uremic syndrome.
- Serovars of *Salmonella enterica* are classified as nontyphoidal or as causing typhoid fever. *S.* Typhi, the cause of typhoid fever, is the only serovar to produce an exotoxin. Non-typhoidal strains cause salmonellosis, a self-limiting gastroenteritis, while typhoid fever can be life threatening.
- Shigellosis is caused by one of four species of *Shigella*. *S. dysenteriae* produces Shiga toxin (Stx), which blocks protein synthesis by depurinating host 28S rRNA. Stx has been horizontally transferred to EHEC strains.
- Staphylococcal food poisoning, usually caused by *S. aureus*, is an intoxication with staphylococcal enterotoxins. Vomiting begins within hours of ingesting the tainted food and resolves within 24 hours.

39.5 Zoonotic Diseases Arise from Human-Animal Interactions

- Anthrax is a toxin-mediated disease that can manifest in one of three ways depending on mode of transmission: cutaneous, pulmonary, and gastrointestinal. The toxin consists of two effector proteins (lethal factor and edema factor) and protective antigen, which binds to host cells (**figure 39.30**).
- Human brucellosis is caused by several species of *Brucella*. Infection is most frequently transmitted by cattle and can be acute or insidious. Brucellosis is a nonspecific febrile illness that is particularly dangerous for pregnant women.
- Psittacosis is a rare pulmonary infection acquired from birds. The causative agent is the obligate intracellular pathogen *Chlamydophila psittaci*. Q fever is also rare. It is caused by *Coxiella burnetii*, also an intracellular pathogen transmitted by cows, sheep, and goats.
- Tularemia, caused by *Francisella tularensis*, can be transmitted by infected animals (e.g., rabbits), arthropods (ticks and deer flies), or direct inhalation of aerosolized bacteria. It causes an illness that is similar to plague, but less severe.

39.6 Opportunistic Diseases Can Be Caused by Bacteria

- Normal microbiota can cause disease especially when they invade normally sterile tissues. Opportunistic pathogens also include those that only infect individuals with compromised immune function.
- Opportunistic diseases that correspond to a loss of normal microbiota include antibiotic-associated colitis (*C. difficile;* **figure 39.31**) and bacterial vaginosis (*Gardnerella vaginalis*; **figure 39.32**).
- Dental plaque formation begins with the initial colonization of the acquired enamel pellicle by *Streptococcus gordonii*, *S. oralis*, and *S. mitis*. Other bacteria then become attached and form a plaque ecosystem (**figures 39.34** and **39.35**). The bacteria produce acids that cause a chemical lesion on the tooth and initiate dental decay or caries.
- Periodontal disease is a group of diverse clinical entities that affect the periodontium. Disease is initiated by the formation of subgingival plaque, which leads to tissue inflammation known as periodontitis and to periodontal pockets. Bacteria that colonize these pockets can cause an abscess, periodontosis, gingivitis, and general tissue necrosis.
- *Streptococcus pneumoniae*, a normal inhabitant of the nasopharynx, is a common cause of pneumonia particularly in the elderly. There are over 90 serotypes, each with a unique polysaccharide capsule. There are two vaccines protecting against over 30 serotypes that commonly cause disease.
- Urinary tract infections are one of the most common infections for which people seek treatment. The most common cause is uropathogenic *E. coli* (UPEC). Prompt antibiotic treatment prevents UPEC from ascending the ureters and causing kidney infection (pyelonephritis), a more severe condition.

Active Learning

1. Construct a list of pathogens discussed in this chapter that infect immune cells and compare the diseases they cause with several other diseases caused by pathogens that do not infect immune cells. Can you see any trends or make any inferences about these two types of pathogens and the diseases they cause?
2. Worldwide, pneumonia and diarrhea claim the lives of more children than any other reason. Review the UNICEF Pneumonia and Diarrhea Report (available online) and provide the three most important steps that can be taken to reverse these avoidable deaths.
3. The incidence of syphilis continues to climb worldwide, including in the United States. Review the three stages of syphilis and discuss why the disease may go unnoticed and untreated.

4. In 2011, Germany saw a large outbreak of EHEC that was eventually traced to contaminated bean sprouts. This outbreak was unusual in several ways. First, the rate of hemolytic uremic syndrome (HUS) was more than double that seen in previous EHEC outbreaks (20% rather than about 8%), and HUS was most commonly seen in adults rather than children. The causative strain, O104:H4, was found to be genetically complex, with elements seen in both EHEC and EAEC strains. Discuss features of both EHEC and EAEC that might have combined to make this strain so virulent.
5. When the Lyme disease bacterium *B. burgdorferi* is in the gut of its tick vector, its outer surface protein A (OspA) is expressed at high levels. Once it has been transmitted to a human, OspA production diminishes. However, to complete its infection cycle and return to the tick, *B. burgdorferi* cells must increase OspA synthesis. To accomplish this, *B. burgdorferi* binds the host neuroendocrine stress hormones epinephrine and norepinephrine, which results in upregulation of OspA.

   Discuss the role of coevolution in the development of *B. burgdorferi*'s ability to sense and respond to these host compounds. How would you show that among all the mediators of inflammation at the site of a tick bite, the spirochete responds specifically to epinephrine and norepinephrine? Your answer should include some carefully considered control experiments.
6. In investigating an outbreak of invasive group A streptococcus (iGAS) in Liverpool, UK, microbiologists first noted that the M protein found on these strains had not been seen previously in Liverpool. This was followed by whole-genome sequencing. When compared with iGAS strains with a different M protein, the outbreak strain had 19 unique genes, 5 of which were known to encode virulence factors. Given the types of diseases caused by iGAS, hypothesize what these virulence factors might be. What about the other 14 genes? How could gene products that are not bona fide virulence factors contribute to pathogenicity nonetheless?

   Read the original paper: Cornick, J. E., et al. 2017. Epidemiological and molecular characterization of an invasive group A streptococcus emm32.2 outbreak. *J. Clin. Microbiol.* 55:1837–46. doi:10.1128/JCM.00191-17.

40

Human Diseases Caused by Fungi and Protists

©RFStock/Alamy Stock Photo

Mushrooms of Death

Ask Kentuckian Mark Tatum who would win a battle of man versus mushroom and he'll likely bet on the fungus. In 2000 Tatum lost a portion of his face to a severe fungal infection called mucormycosis; such infections are often fatal. In fact, the capacity of certain fungi to degrade animal tissues has been leveraged by entrepreneur and artist Jae Rhim Lee to develop a "green" burial suit. Lee began this work about a decade ago, when she was a research fellow at the Massachusetts Institute of Technology. In what was known as the "Infinity Burial Project," Lee explored alternatives to the toxic formaldehyde-based burial process. Growing fungi in Petri dishes filled with human hair, skin, nails, and so forth, she selectively bred mushrooms that thrive on this macabre medium. Lee has since founded a company that sells the Infinite Burial Suit, which features a cocktail of mushroom spores that accelerate decomposition. Lee states that she is building an acceptance of natural processes, pointing out that cemetery soil already contains many fungal species and mushroom-based decomposition eliminates some of the more than 200 toxic chemicals that accumulate in the human body.

Ethicists will debate the religious and spiritual issues, while microbial ecologists consider the impact such a strain might have in the environment. For example, do such strains already exist in cemeteries? How will a "man (or woman)-eating mushroom" change the microbiota of soil? Will mourners be able to visit the cemeteries without fear of hitchhiking spores coming home to dinner? Regardless of the questions yet unanswered, one thing is certain: Fungi are masters at producing extracellular enzymes that digest complex foods into soluble nutrients. Just ask Mark Tatum.

In this chapter, we describe some of the fungi and protists that are pathogenic to humans and discuss the clinical manifestations, diagnosis, epidemiology, pathogenesis, and treatment of the diseases caused by them. The biology and diversity of these organisms are covered in chapters 5, 24, and 25. This chapter follows the format of the previous two chapters in categorizing their diseases by route of transmission (**figure 40.1**).

Readiness Check:

Based on what you have learned previously, you should be able to:

- ✓ Review basic eukaryote cell biology (chapter 5)
- ✓ Distinguish between yeast and mold (section 25.1)
- ✓ Identify the major factors supporting growth and reproduction of protists and fungi (sections 24.1 and 25.1)
- ✓ Explain pathogenicity and the infection process (chapter 35)
- ✓ Review methods by which pathogenic fungi and protozoa are identified (chapter 37)

40.1 Relatively Few Fungi and Protists Are Human Pathogens

After reading this section, you should be able to:

a. Describe the system used to categorize fungal diseases
b. Appreciate the burden of protozoan disease worldwide

While there are well over a million species of fungi on the Earth, only about 300 produce disease in humans. Medical mycology is the discipline that deals with fungi that cause human disease. These fungal diseases, known as **mycoses** (s., mycosis; Greek *mykes*, fungus), can be divided into five groups: superficial, cutaneous, subcutaneous, systemic, and opportunistic mycoses (**table 40.1**). Superficial, cutaneous, and subcutaneous mycoses are direct contact infections of the skin, hair, and nails. Systemic mycoses are fungal infections that have disseminated to visceral tissues. The fungi that cause systemic or deep mycoses are dimorphic; that is, they exhibit a yeast (single-cell) phase (Y) and a saprophytic mold or mycelial (filamentous) phase (M). The switch from growth as a mycelial form at external, cooler temperatures to growth as a yeast at human body temperatures is called the **Y-M shift** and contributes to virulence. Most systemic mycoses are acquired by inhalation of spores from soil. If a susceptible person inhales enough spores, infection begins as a lung infection that may reach the bloodstream and spread to other organs. Some fungi produce toxins

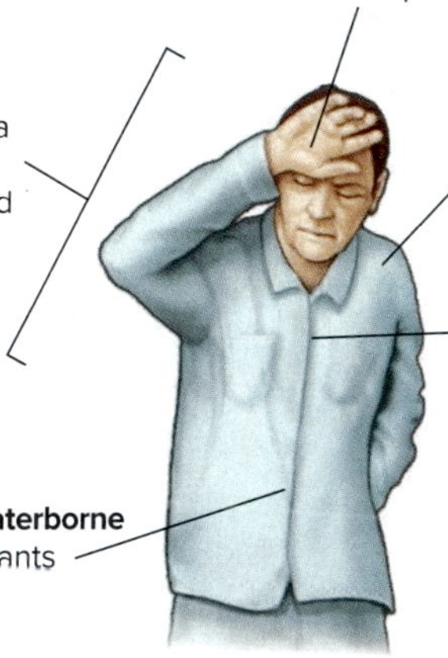

Figure 40.1 Common Routes of Fungal and Protozoan Disease Transmission. Examples of disease-causing fungi and protists are presented in this chapter according to route of transmission.

(mycotoxins) harmful to humans. We discuss them with other toxins in chapter 35 and in chapter 41 (*see figures 41.5 and 41.6*). *Fungi (chapter 25)*

Protozoa have become adapted to practically every type of habitat on Earth, including the human body. Pathogenic protozoa can be transmitted to humans by arthropod vectors, from contaminated food or water, or by direct contact. Although members of fewer than 20 genera of protists cause disease in humans (**table 40.2**), their impact is formidable. For example, in 2016, there were 216 million new cases of malaria and about 445,000 deaths, with children under the age of five bearing most of the burden. In the Americas, Chagas disease (caused by a trypanosome) affects 6 to 8 million people, with about 12,000 deaths annually. Outbreaks of the intestinal pathogens *Cryptosporidium* and *Cyclospora* caused by contaminated food and water supplies is a global problem (**table 40.3**). Elderly and immunosuppressed

Table 40.1 Examples of Some Medically Important Fungi

Disease Category	Pathogen	Location	Disease
Superficial mycoses	*Piedraia hortae*	Scalp	Black piedra
	Trichosporon beigelii	Beard, mustache	White piedra
	Malassezia furfur	Trunk, neck, face, arms	Tinea versicolor
Cutaneous mycoses	*Trichophyton mentagrophytes, T. verrucosum, T. rubrum*	Beard hair	Tinea barbae
	Trichophyton, Microsporum canis	Scalp hair	Tinea capitis
	Trichophyton rubrum	Smooth or bare parts of the skin	Tinea corporis
	Epidermophyton floccosum, T. mentagrophytes, T. rubrum	Groin, buttocks	Tinea cruris (jock itch)
	T. rubrum, T. mentagrophytes, E. floccosum	Feet	Tinea pedis (athlete's foot)
	T. rubrum, T. mentagrophytes, E. floccosum	Nails	Tinea unguium (onychomycosis)
Subcutaneous mycoses	*Phialophora verrucosa, Fonsecaea pedrosoi*	Legs, feet	Chromoblastomycosis
	Madurella mycetomatis	Feet, other areas of body	Maduromycosis
	Sporothrix schenckii	Puncture wounds	Sporotrichosis
Systemic mycoses	*Blastomyces dermatitidis*	Lungs, skin	Blastomycosis
	Coccidioides immitis	Lungs, other parts of body	Coccidioidomycosis
	Cryptococcus neoformans	Lungs, skin, bones, viscera, central nervous system	Cryptococcosis
	Histoplasma capsulatum	Within phagocytes	Histoplasmosis
Opportunistic mycoses	*Aspergillus fumigatus, A. flavus*	Respiratory system	Aspergillosis
	Candida albicans	Skin or mucous membranes	*Candidiasis*
	Pneumocystis jiroveci	Lungs, sometimes brain	*Pneumocystis* pneumonia
	Encephalitozoon, Pleistophora, Enterocytozoon, Microsporidium	Lungs, sometimes brain	Microsporidiosis

Table 40.2 Examples of Medically Important Protozoan Infections

Initial Infection Site	Pathogen	Disease
Blood	*Plasmodium* spp. *Trypanosoma brucei* *Trypanosoma cruzi*	Malaria African sleeping sickness American trypanosomiasis
Intestinal Tract	*Cryptosporidium* spp. *Cyclospora* spp. *Entamoeba histolytica* *Giardia intestinalis* *Toxoplasma gondii*	Cryptosporidiosis Cyclosporidiosis Amebiasis Giardiasis Toxoplasmosis
Lungs	*Balamuthia mandrillaris* *Pneumocystis jiroveci* *Trichomonas tenax*	*Balamuthia* amebic encephalitis *Pneumocystis* pneumonia Pulmonary trichomoniasis
Muscosal Membranes Ears, Eyes, Nose Genitals	*Acanthamoeba* spp. *Naegleria fowleri* *Trichomonas vaginalis*	Amebic meningoencephalitis Amebic meningoencephalitis Trichomoniasis
Skin	*Leishmania braziliensis* *Leishmania donovani* *Leishmania tropica*	Mucocutaneous leishmaniasis Visceral leishmaniasis Cutaneous leishmaniasis

Table 40.3 Water-Based Protozoan Reservoirs

Organism	Reservoir	Comments
Acanthamoeba spp.	Sewage sludge disposal areas	Can cause granulomatous amoebic encephalitis, keratitis, corneal ulcers
Cryptosporidium spp.	Many species of domestic and wild animals	Causes acute enterocolitis; important with immunologically compromised individuals; cysts resistant to chemical disinfection
Cyclospora cayetanensis	Waters—does not withstand drying; possibly other reservoirs	Causes long-lasting (43 days average) diarrheal illness; infection self-limiting in immunocompetent hosts; sensitive to prompt treatment with sulfonamide and trimethoprim
Giardia intestinalis	Livestock, dogs, cats, humans	Major cause of early spring diarrhea; important in cold mountain water
Naegleria fowleri	Warm ponds and lakes	Inhalation in nasal passages; central nervous system infection; causes primary amoebic meningoencephalitis

individuals are at increased risk for protozoan infections. ◀ *Protists (chapter 24)*

While seemingly different, fungi and protists share a number of phenotypic features that serve them in their ability to cause infection: microscopic size, eukaryotic physiology, cell walls or wall-like structures, and alternative life stages for survival outside of the host. We now discuss diseases caused by fungi and protists based on how they are acquired by the human host.

40.2 Fungi Can Be Transmitted by Airborne Routes

After reading this section, you should be able to:

a. Describe fungi that cause common diseases spread by airborne transmission
b. Identify typical signs and symptoms of common airborne fungal diseases

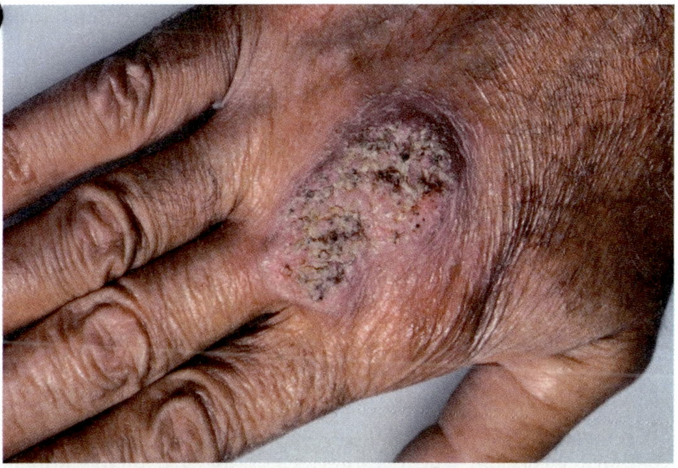

Figure 40.2 Systemic Mycosis: Blastomycosis. Blastomycosis of the hand caused by *Blastomyces dermatitidis*.

MICRO INQUIRY *What is the morphology of* B. dermatitidis *in the environment? In the human host?*

Eukaryotic pathogens transmitted to people through air are small enough to be carried by air currents and inhaled by unsuspecting hosts. Few, if any, human protozoan diseases are spread this way. We therefore present several examples of pathogenic fungi acquired through airborne transmission and the diseases they cause.

Blastomycosis

Blastomycosis is the systemic mycosis caused by *Blastomyces dermatitidis*, a dimorphic fungus that grows as a budding yeast in humans but as a mold on culture media and in the environment. It is found predominately in moist soil enriched with decomposing organic debris. *B. dermatitidis* is endemic in the Mississippi and Ohio river basins, and in some regions in Central and South America and Africa. The disease occurs in three clinical stages: pulmonary, disseminated, and cutaneous. The initial infection begins when blastospores are inhaled into the lungs. If it spreads systemically from the lungs, thereby becoming disseminated. It often spreads to the skin, where cutaneous ulcers and abscess formation occur (**figure 40.2**). *B. dermatitidis* can be isolated from sputum, wound exudates, and biopsy sections. Diagnosis requires the demonstration of thick-walled, yeastlike cells 8 to 15 μm in diameter. Fungal histology, serology, and antigen tests are also useful. Amphotericin B and itraconazole are the drugs of choice for treatment (*see figure 9.17*). Mortality is about 5%. There are no preventive or control measures. ◀ *Antifungal drugs (section 9.6)*

Coccidioidomycosis

Coccidioidomycosis is also known as valley fever, San Joaquin fever, or desert rheumatism because of the geographical distribution of the fungus. It is caused by *Coccidioides immitis,* which thrives in the semiarid, highly alkaline soils of southwestern United States and parts of Mexico and South America. In the United States about 30,000 people are infected annually, resulting in approximately 200 deaths. Endemic areas have been defined by massive skin testing with the antigen coccidioidin, where upward of 15% positive cases are reported; endemic areas include Arizona, California, Nevada, New Mexico, and Utah. In these areas coccidiodomycosis causes up to 30% of community-acquired pneumonia. In the soil and on culture media, this fungus grows as a mold that forms arthroconidia at the tips of hyphae (*see figure 25.4*). Because arthroconidia are so abundant, immunocompromised individuals can acquire the disease by inhalation as they simply move through these endemic areas. Wind turbulence and even construction of outdoor structures have been associated with infection. In humans, the fungus grows as thick-walled structures called spherules filled with spores about 2–5 μm in diameter (**figure 40.3**). Most cases of coccidioidomycosis are asymptomatic or indistinguishable from ordinary upper respiratory infections and resolve in a few weeks, with lasting immunity. A few infections result in progressive chronic pulmonary disease or disseminated infections of other tissues; the fungus can spread throughout the body, involving almost any organ or site.

Diagnosis is accomplished by identification of the large spherules (~60 μm in diameter) in pus, sputum, and aspirates. Culturing clinical samples in the presence of penicillin and streptomycin on the fungal selective medium Sabouraud agar also is diagnostic. Rapid confirmation methods include serology for IgM and IgG titers and skin testing. Fluconazole, itraconazole, ketoconazole, and amphotericin B are the drugs of choice for treatment. Prevention involves reducing exposure to dust (soil) in endemic areas. ◀ *Antifungal drugs (section 9.6).*

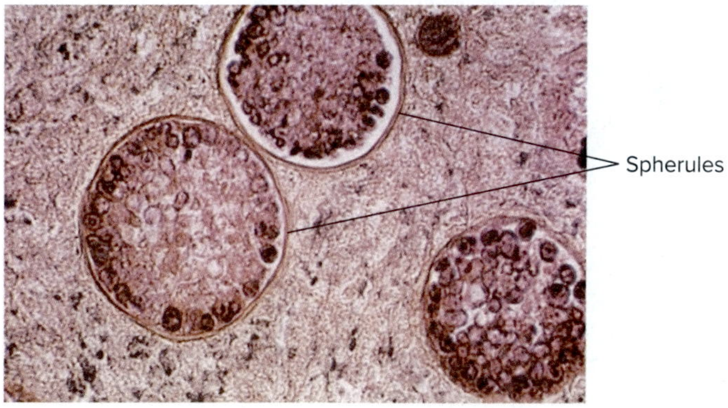

Figure 40.3 Systemic Mycosis: Coccidioidomycosis. *Coccidioides immitis* mature spherules filled with spores within a tissue section; light micrograph (×400).
Source: CDC/Lucille K. Georg

Histoplasmosis

Histoplasmosis is primarily a disease of the lungs caused by *Histoplasma capsulatum* var. *capsulatum*. Like *B. dermatitidis* and *C. immitis*, *H. capsulatum* can spread from the lungs to cause systemic disease. It differs in that it can grow intracellularly in host macrophages. Symptoms include flulike responses such as fever, cough, headaches, and muscle aches. However, infection is often asymptomatic. *H. capsulatum* is a small, budding yeast in humans and on culture media at 37°C. At 25°C it grows as a mold, with septate hyphae, microconidia, and macroconidia formed on conidiophores (**figure 40.4**; *also see figure 25.4*). Microconidia can become airborne when contaminated soil is disturbed. Microconidia are most prevalent where bird and bat droppings have accumulated. It is noteworthy that birds themselves are not infected; their droppings simply provide the nutrients for this fungus. Only bats and humans demonstrate the disease and harbor the fungus.

Infection ensues when a large dose of microconidia is inhaled. Histoplasmosis is found worldwide, with Africa, Australia, India, and Malaysia being endemic regions. Within the United States, histoplasmosis is endemic within the Mississippi, Kentucky, Tennessee, Ohio, and Rio Grande river basins. As many as 90% of people who reside in these areas have antibodies against the fungus. It has been estimated that in endemic areas of the United States, about 500,000 individuals are infected annually: 50,000 to 200,000 become ill; 3,000 require hospitalization; and about 50 die. The total number of infected individuals may be over 40 million in the United States alone. Histoplasmosis is a common disease among poultry farmers, spelunkers (people who explore caves), and bat guano miners (bat guano is used as fertilizer).

Because *H. capsulatum* grows within macrophages, the fungus can be disseminated to many organs and is often found in bone marrow. Lesions may appear in the lungs and show calcification; thus the disease may resemble tuberculosis. Laboratory diagnosis is accomplished by isolation of the fungus from tissue specimens, by calcofluor-white or other fungal stain, and serology. Most individuals with this disease exhibit a hypersensitive state that can be demonstrated by the histoplasmin skin test. Severe infections are treated with amphotericin B, whereas the less toxic drugs itraconazole and fluconazole are used to treat mild to moderate disease. Prevention and control involve wearing protective clothing and masks before entering or working in infested habitats.

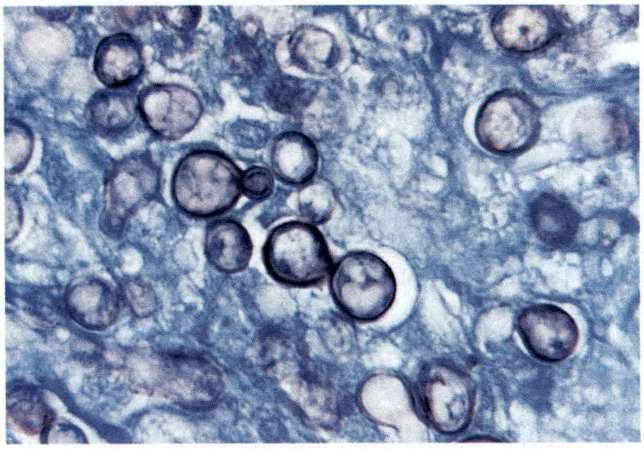
(a)

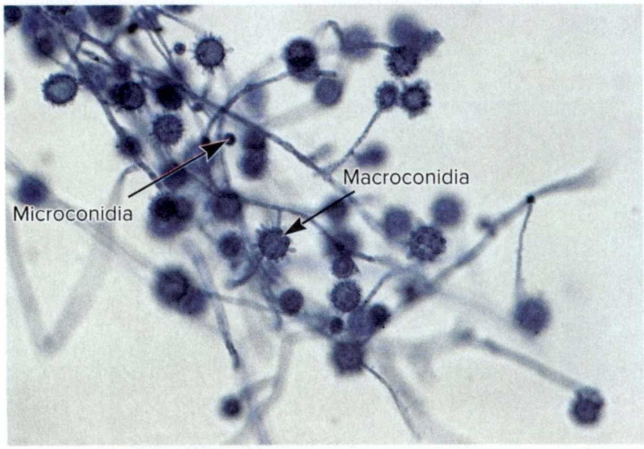

(b)

Figure 40.4 Morphology of *Histoplasma capsulatum* var. *capsulatum*. Light micrographs. (a) In the host, *H. capsulatum* grows as a yeast, with cells 4 to 6 μm in diameter. (b) Mycelia, microconidia (1–5 μm in diameter), and macroconidia (8–15 μm in diameter) as found in the soil.
Source: CDC/Dr. Libero Ajello

Comprehension Check

1. Apart from mode of transmission, how can diseases caused by fungi and protozoa be classified?
2. Compare and contrast the severity of disease caused by the three fungi discussed in this section.
3. Given your geographic location, are you at risk for any of these mycoses?

40.3 Arthropods Can Transmit Protozoal Disease

After reading this section, you should be able to:

a. Discuss the protozoa that cause common diseases spread by arthropod transmission
b. Identify typical signs and symptoms of arthropod-borne protozoal diseases
c. Correlate arthropod-borne protozoal disease severity with virulence factors used by the causative agents

Protozoa transmitted by insects interact with the human host in a manner that is different from arthropod-borne bacteria and viruses. The rapid replication of bacteria and viruses in the host

DISEASE

40.1 A Brief History of Malaria

No other single infectious disease has had more impact on humans than malaria. The first references to its periodic fever and chills can be found in early Chaldean (pre-Babylonian), Chinese, and Hindu writings. In the late fifth century BCE, Hippocrates described malaria symptoms. In the fourth century BCE, the Greeks noted an association between individuals exposed to swamp environments and the subsequent development of periodic fever and enlargement of the spleen. In the seventeenth century, the Italians named the disease *mal' aria* (bad air) because of its association with ill-smelling vapors from swamps near Rome. At about the same time, the bark of the quinaquina (cinchona) tree of South America was used to treat the intermittent fevers, although it was not until the mid-nineteenth century that quinine was identified as the active alkaloid. The major epidemiological breakthrough came in 1880, when French army surgeon Charles Louis Alphonse Laveran observed protozoan gametocytes in his patients' blood. Five years later, the Italian histologist Camillo Golgi observed the multiplication of the asexual blood forms. In the late 1890s, Patrick Manson postulated that malaria was transmitted by mosquitoes. Sir Ronald Ross, a British army surgeon in the Indian Medical Service, subsequently observed developing plasmodia in the intestine of mosquitoes, supporting Manson's theory. Using birds as experimental models, Ross definitively established the major features of the life cycle of plasmodia and received the Nobel Prize in 1902.

Human malaria is known to have contributed to the fall of the ancient Greek and Roman empires. Troops in both the U.S. Civil War and the Spanish-American War also were severely incapacitated by the disease. During World War II, malaria epidemics severely threatened both the Japanese and Allied forces in the Pacific. The same can be said for the military conflicts in Korea and Vietnam.

In the twentieth century, efforts were directed toward understanding the biochemistry and physiology of malaria, controlling the mosquito vector, and developing antimalarial drugs. In the 1960s it was demonstrated that resistance to *P. falciparum* infection among West Africans was associated with the presence of hemoglobin-S (Hb-S) in their erythrocytes. Hb-S differs from normal hemoglobin-A by a single amino acid, valine, in each half of the Hb molecule. Consequently these erythrocytes—responsible for sickle cell disease—have a low binding capacity for oxygen. Because the malarial parasite has a very active aerobic metabolism, it cannot grow and reproduce within these erythrocytes.

In 1955 the World Health Organization (WHO) began a worldwide malarial eradication program that collapsed by 1976. Among the major reasons for failure were the development of resistance to the insecticide DDT by mosquitoes and the development of resistance to chloroquine by strains of *Plasmodium*. In 2000, the United Nations Millennium Development Goals articulated ambitious objectives to reduce malaria that were echoed by the WHO, which targeted a 75% reduction in malaria incidence by 2015. Although this very aggressive goal was not realized, by 2015 the prevalence of malaria caused by the most virulent species, *Plasmodium falciparum,* was cut in half in Africa, averting an estimated 663 million cases. This was largely accomplished by the distribution of insecticide-treated bednets to protect children. Unfortunately during that same time, multidrug-resistant strains of *P. falciparum* emerged in Cambodia and its neighbors. While there is great hope that the coming years will continue to bring sharp declines in malaria in Africa, where the burden is highest, there is also fear that the drug-resistant strains in Southeast Asia will take hold in other regions. There is much work to be done to curb this ancient disease.

ensures further transmission. By contrast, protozoa grow more slowly and therefore have evolved an array of mechanisms to prevent clearance by the human host's immune system. Thus when considering arthropod-borne protozoan infections, it is important to keep in mind that the immune system may prevent clinical disease but not clear the protozoan from the host. This is seen in malaria; people who live in endemic areas may be referred to as "immune" when, in fact, most carry a low parasite load.

Malaria

The most important human pathogenic protozoa belong to the genus *Plasmodium* and cause **malaria** (**Disease 40.1**). A 2016 World Health Organization (WHO) report estimates 3.2 billion people (almost half of the world's population) across 95 countries are at risk for malaria (**figure 40.5**). Although mortality rates have fallen by more than 50% since 2000, 429,000 people died of malaria in 2015; 90% of those deaths were in Africa and most were children under the age of five. About 1,600 cases are reported each year in the United States, divided between returning U.S. travelers and non-U.S. citizens.

Human malaria is caused by five species of *Plasmodium.* In descending order of incidence, they include: *P.falciparum, P. vivax, P. malariae, P. ovale,* and *P. knowlesi.* The

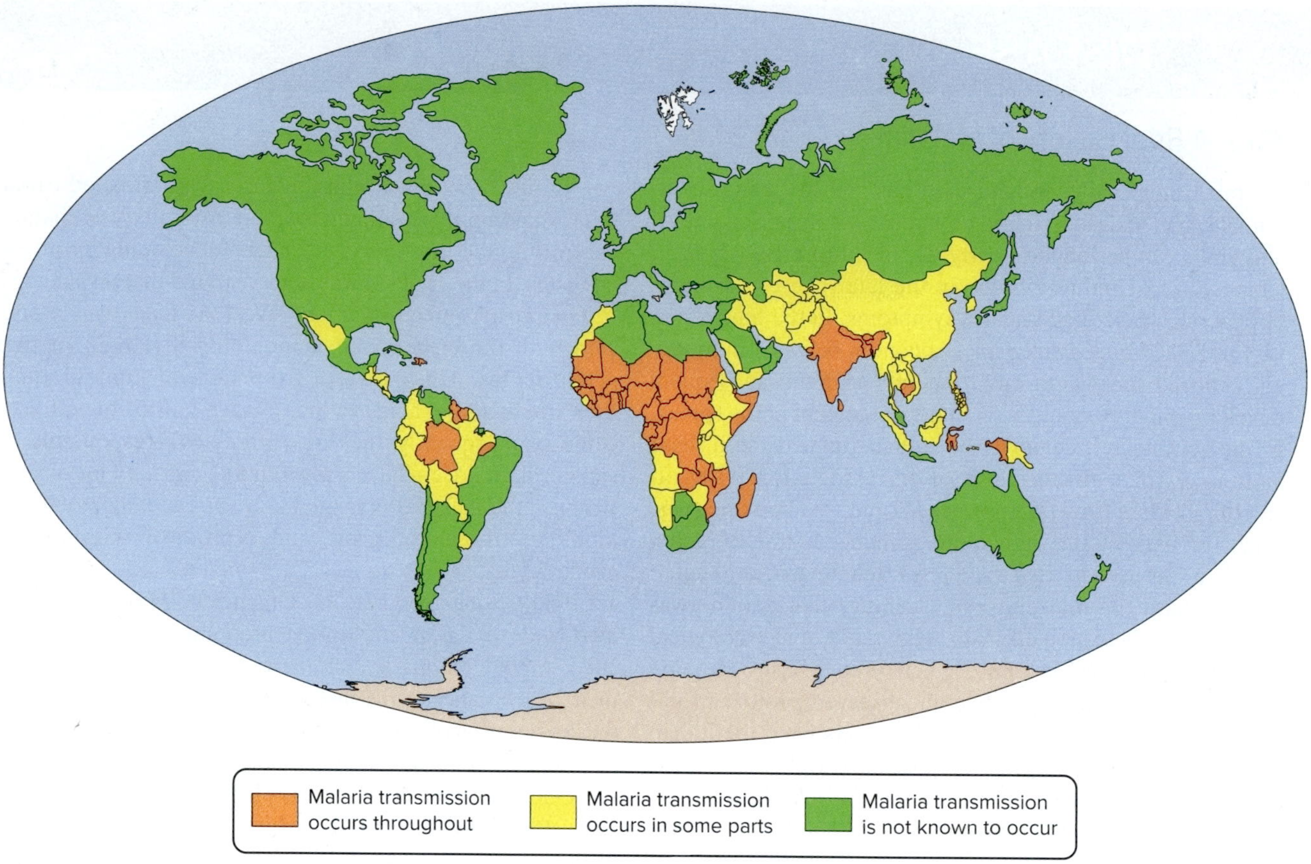

Figure 40.5 **Geographic Distribution of Malaria in 2017.**
Source: CDC, "Where Malaria, Occurs."

life cycle of plasmodial protists is shown in **figure 40.6**. The protozoan first enters the bloodstream through the bite of an infected female *Anopheles* mosquito. As she feeds, the mosquito injects a small amount of saliva containing an anticoagulant along with haploid sporozoites. The sporozoites in the bloodstream migrate to and enter hepatic cells of the liver. There they undergo multiple asexual fission (schizogony) and produce merozoites. After being released from hepatocytes, merozoites attach to and penetrate erythrocytes. The incubation period is as short as 7 days (frequently due to *P. falciparum*) and as long as 30 days (usually due to *P. malariae*). ◀ Alveolata *(section 24.4)*

Once inside the erythrocyte, the plasmodial cell begins to enlarge as a mononucleate cell termed a trophozoite. The trophozoite's nucleus divides asexually to produce a schizont that has 6 to 24 nuclei. The schizont divides and produces mononucleated merozoites that are released when infected erythrocytes lyse. These then infect other erythrocytes. This erythrocytic stage repeats approximately every 48 to 72 hours or longer, depending on the species of *Plasmodium*. Occasionally, merozoites differentiate into macrogametocytes and microgametocytes, which do not rupture the erythrocyte (**figure 40.7**). For sexual reproduction to occur, these must be ingested by a mosquito during a blood meal. In the mosquito's gut, the infected erythrocytes lyse and the gametes fuse to form a diploid zygote called the ookinete. The ookinete migrates to the mosquito's gut wall, penetrates, and forms an oocyst. In a process called sporogony, the oocyst undergoes meiosis and forms sporozoites, which migrate to the salivary glands of the mosquito. The cycle is now complete, and when the mosquito bites another human host, the cycle begins anew.

The pathological changes caused by malaria involve not only erythrocytes but also the spleen (where old red blood cells are normally removed from circulation) and other organs. Symptoms first develop with the release of merozoites and erythrocyte debris into the bloodstream, resulting in the malarial paroxysms—shaking chills, then burning fever followed by sweating caused partly by macrophage release of the cytokines TNF-α and interleukin-1. If *P. vivax* or *P. ovale* are the causative agents, there is a remission that lasts from a few weeks to several months while dormant protozoa called hypnozooites reside in the liver. This is followed by a relapse. Between paroxysms, the patient feels normal. Anemia results from the loss of erythrocytes, and the spleen and liver often become enlarged. Cerebral malaria occurs most frequently in children and nonimmune and immunocompromised individuals, including pregnant women. Its

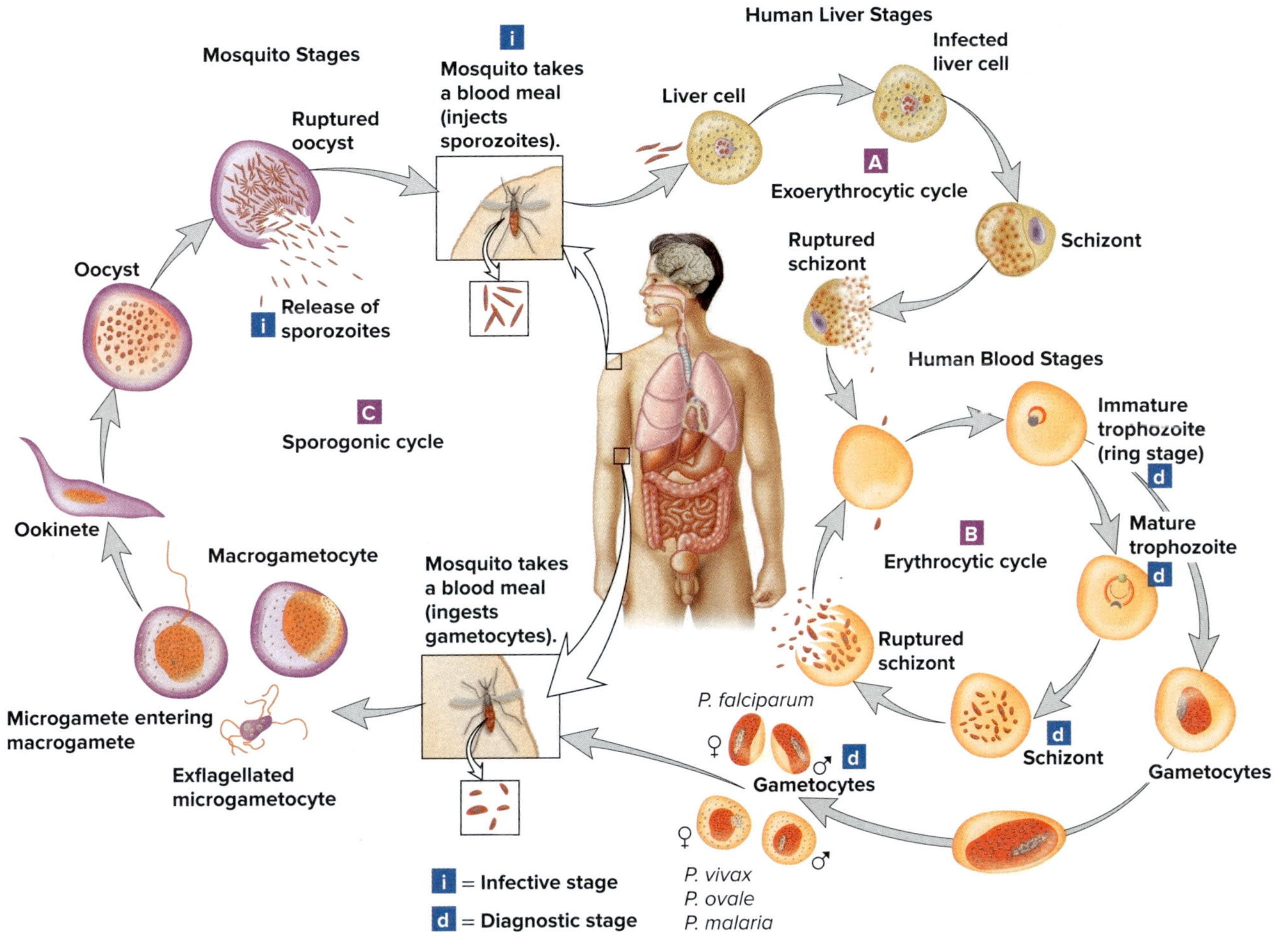

Figure 40.6 Malaria. Life cycle of *Plasmodium vivax.* Note the (A) exoerythrocytic cycle, (B) the erythrocytic cycle, and (C) the sporogonic cycle.

MICRO INQUIRY *How is a schizont formed? What is the cell type that is released from erythrocytes, and where does the protist go next?*

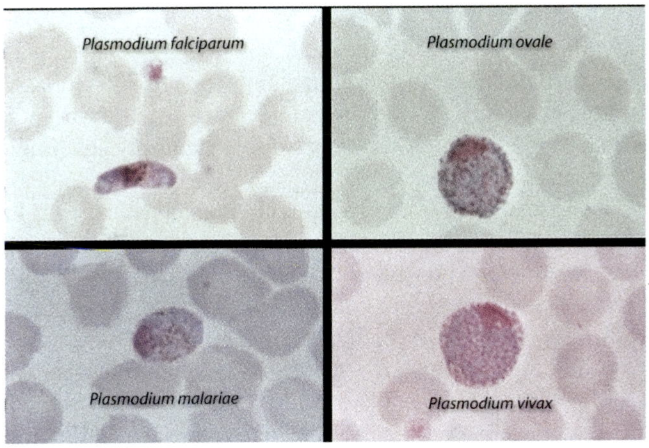

Figure 40.7 Macrogametocytes of *Plasmodium* spp. The morphology of gametocytes and trophozoites (not shown) in blood smears is used to diagnose malaria and determine the causative *Plasmodium* spp.; light micrograph (×1,200).
Source: CDC/Steven Glenn, Laboratory & Consultation Division

mortality rate is about 20%. ◀ *Cytokines are chemical messages between cells (section 32.3)*

Diagnosis of malaria is made by demonstrating the presence of parasites within Wright- or Giemsa-stained erythrocytes (figure 40.7). When blood smears are negative, serological testing can establish a diagnosis of malaria. Outside the United States, rapid diagnostic tests using species-specific antibodies are commonly used to diagnose malaria. PCR-based malarial identification and drug resistance testing may also be performed.

Specific recommendations for treatment of malaria depend on the geographic region and include administration of chloroquine, amodiaquine, or mefloquine. In the regions where *P. falciparum* predominates, artemisinin is used for treatment. In fact, artemisinin-based combination therapy (ACT)—combining artemisinin with another antimalarial drug—is recommended for the treatment of all *P. falciparum* malaria in endemic regions. These drugs suppress protozoan reproduction and are effective in eradicating erythrocytic asexual stages. Those infected with *P. vivax* or *P. ovale* must be treated with primaquine to eradicate hypnozooites. It is worth noting that individuals who are traveling to areas where malaria is endemic (figure 40.5) should receive chemoprophylactic treatment as recommended by the CDC.

Control of malaria is an urgent problem. The ideal approach would be to vaccinate all young children in endemic areas. However, to be effective, a vaccine must interfere with a critical malarial component or process (**figure 40.8**). In 2016, the WHO recommended the pilot use of the pre-erythrocytic, recombinant vaccine RTS,S in sub-Saharan Africa. This vaccine stimulates the host immune system to neutralize sporozoites as they transit from the skin to the liver. It reduces rates of malaria infection in children by 50%. This is part of a multipronged global effort to further the success in malaria control accomplished between 2000 and 2015 through the use of insecticide-treated bednets, artemisinin combination therapy, and indoor insecticide spraying. Research continues with the goal of developing a more effective vaccine.

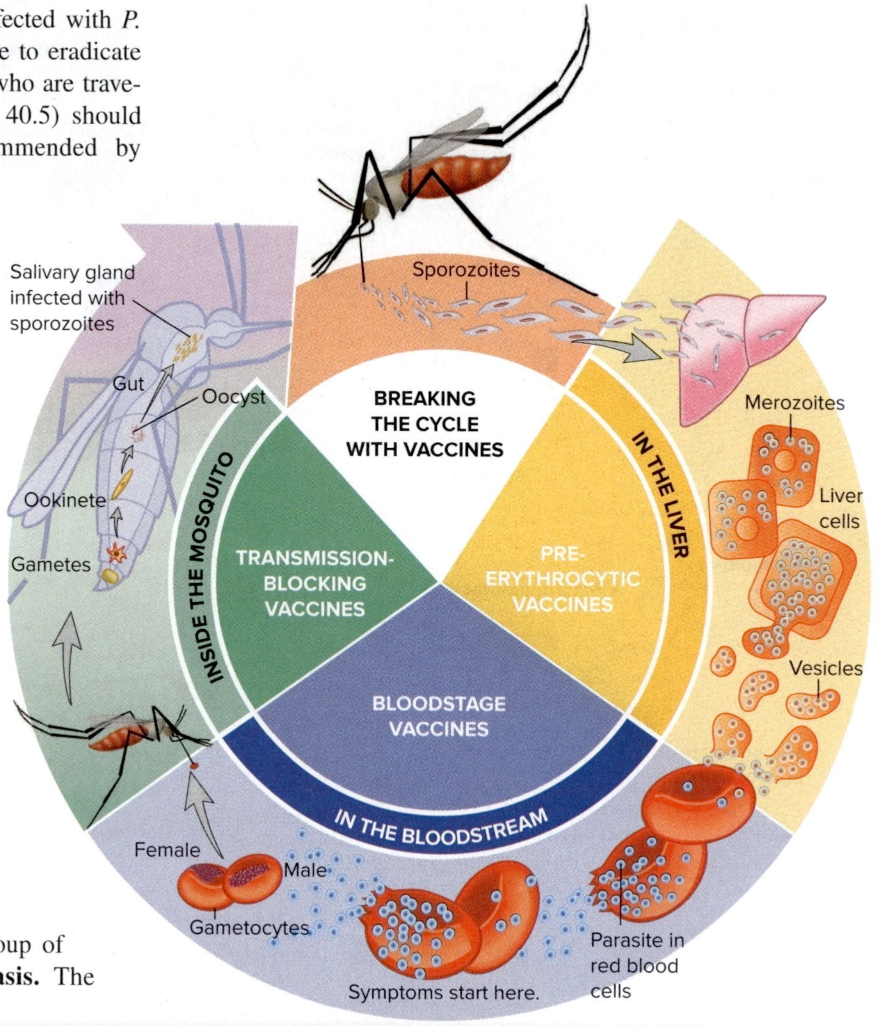

Figure 40.8 Malaria Vaccine Strategies.

MICRO INQUIRY *What step does the RTS,S vaccine target?*

Leishmaniasis

Leishmanias are trypanosomes that cause a group of human diseases collectively called **leishmaniasis.** The most common forms of leishmaniasis are cutaneous (skin associated ulcers), mucocutaneous (ulcers of the skin, mouth, nose, and throat), and visceral (affecting liver, spleen, and blood). Worldwide, leishmaniasis is a disease of poverty associated with malnutrition, displacement, poor housing, and discrimination. There are about 1 million new cases of the cutaneous disease and 300,000 cases of the visceral disease diagnosed around the globe each year. One billion people are at risk of infection across 98 countries in Asia, Africa, and Central and South America. About 30,000 people die from visceral leishmaniasis each year, and many more are permanently disfigured (**figure 40.9**).

The primary reservoirs of these parasites are canines and rodents. All species of *Leishmania* use female sand flies such as those of the genera *Lutzomyia* and *Phlebotomus* as intermediate hosts. Sand flies are about one-third the size of a mosquito, so they are hard to see and hear. The leishmanias are transmitted from animals to humans or between humans by sand flies. When an infected sand fly takes a human blood meal, it introduces flagellated promastigotes into the skin of the host (**figure 40.10**). The promastigotes are then engulfed by macrophages, in which they multiply by binary fission, and form small, nonmotile cells called amastigotes. These destroy the host cell and released amastigotes are engulfed by other macrophages in which they continue to multiply. |◀ Euglenozoa *(section 24.2)*

Over 20 species of *Leishmania* are human pathogens. Most cause cutaneous or mucocutaneous leishmaniasis. Diffuse cutaneous leishmaniasis begins with a relatively small, red papule at the site of each insect bite—the inoculation site. These papules are frequently found on the face and ears. They eventually develop into crustated ulcers (figure 40.9*b*). Healing occurs with scarring and a permanent immunity. *L. donovani* is the major cause of visceral leishmaniasis. It is endemic in large areas within northern China, eastern India, the Sudan, Mediterranean countries, and Latin America. During visceral leishmaniasis infected macrophages lodge in organs where the ongoing cycle of macrophage capture of promastigotes, amastigote development, macrophage death, and amastigote release damages host organs

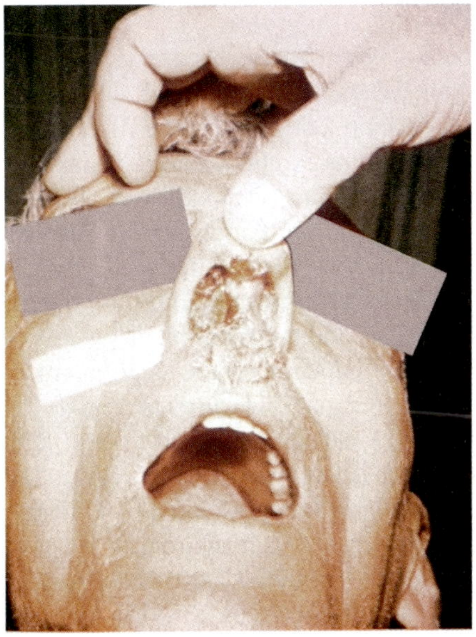

(a)

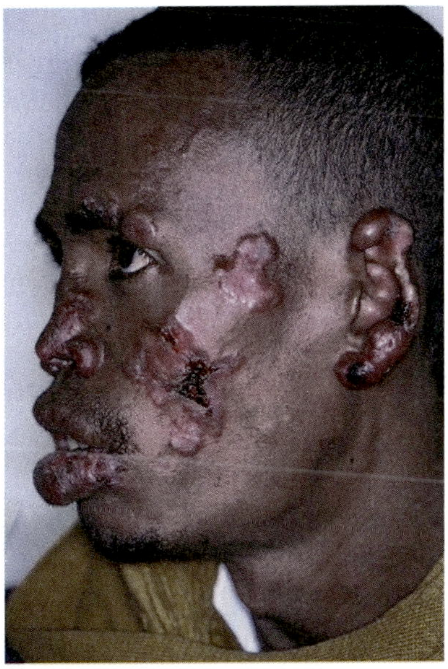

(b)

Figure 40.9 Leishmaniasis. (a) A person with mucocutaneous leishmaniasis, which has destroyed the nasal septum. (b) A person with diffuse cutaneous leishmaniasis.
(a) Source: CDC/Dr. Mae Melvin; (b) ©Andy Crump, TDR, World Health Organization/Science Source

MICRO INQUIRY *What is the primary reservoir of leishmanias? What is the intermediate host?*

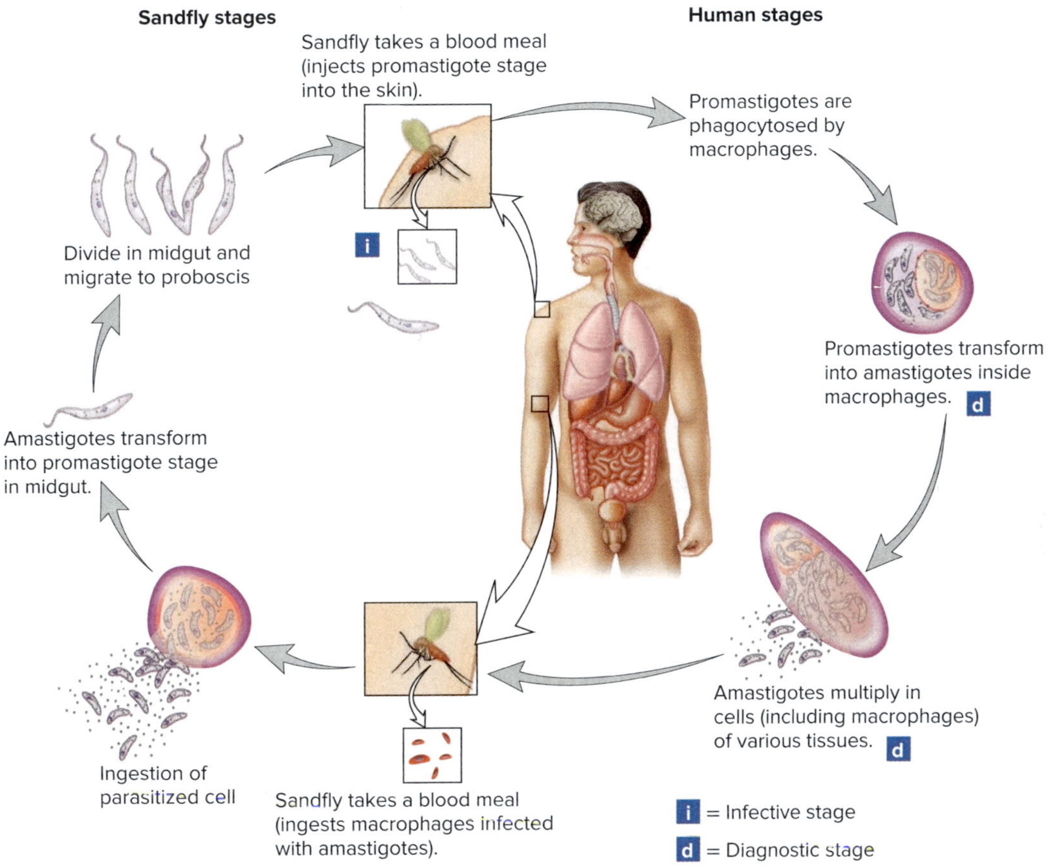

Figure 40.10 ***Leishmania*** **spp. Life Cycle.** Note the promastigote-to-amastigote cycle.

and tissues. Visceral leishmaniasis is called kala azar, meaning black fever, as infected individuals develop fever and in some cases hyperpigmented skin lesions.

Laboratory diagnosis of leishmaniasis is based on finding protozoa within infected macrophages in stained smears from lesions or infected organs. Serological tests are also available for diagnosis. Treatment includes paromomycin, liposomal amphotericin B, and the recently developed drug miltefosine. Vector and reservoir control and aggressive epidemiological surveillance are the best options for prevention and containment of this disease. Although there are no approved vaccines against leishmaniasis, several recombinant protein and DNA-based vaccines are in clinical trials.

Trypanosomiasis

Human African **Trypanosomiasis** (HAT), also known as African sleeping sickness, is caused by two subspecies of trypanosomes. Gambiense HAT is caused by *Trypanosoma brucei* subspecies *gambiense*, which is distributed over western and central countries and causes over 98% of reported cases. The remaining cases (rhodesiense HAT) are caused by the more aggressive *T. brucei* subspecies *rhodesiense*, found in the upland savannas of eastern and southern Africa. Reservoirs for both trypanosome subspecies are domestic cattle and wild animals, within which they cause severe malnutrition. Both species use tsetse flies (genus *Glossina*) as intermediate hosts.

Trypanosomiasis is found in 35 African countries, and in some, it is such a problem that millions of hectares are not fit for human habitation. Control efforts have reduced the incidence of HAT and 2009 was the first time in over 50 years that the number of cases fell below 10,000. In 2016, there were 2,184 reported cases across Africa.

The trypanosomes are transmitted to humans by the bite of the tsetse fly (**figure 40.11**; *also see figure 24.6*). The protists pass through the lymphatic system, enter the bloodstream, and replicate by binary fission as they pass to other body fluids. The tsetse fly bite is painful and can develop into a chancre. Patients may exhibit symptoms of fever, severe headaches, extreme fatigue, muscle and joint aches, irritability, and swollen lymph nodes. Some patients develop a skin rash. The disease gets its name from the fact that patients exhibit lethargy—characteristically lying prostrate, drooling from the mouth, and insensitive to pain; they also exhibit progressive confusion, slurred speech, seizures, and personality changes due to infection of the central nervous system. The protists cause interstitial inflammation and necrosis within lymph nodes and the small blood vessels of the brain and heart. Treatment depends on the stage of infection. Pentamidine isethionate and suramin are used to treat the hemolymphatic stages of HAT. Melarsoprol is used to treat late disease. If left untreated, HAT is fatal within 1 to 3 years.

T. cruzi causes **American trypanosomiasis** (**Chagas disease**), which occurs in the tropics and subtropics of the continental Americas (southern United States, Mexico, Central America, and northern South America). *T. cruzi* uses the triatomine (kissing) bug as a vector (**figure 40.12**). As the triatomine bug takes a blood meal, trypanosomes are discharged in the insect's feces, which contain an irritant that stimulates the victim to scratch. This enables trypanosomes to enter the bloodstream and invade the liver, spleen, lymph nodes, and central nervous system; disease occurs immediately after infection, lasting weeks or

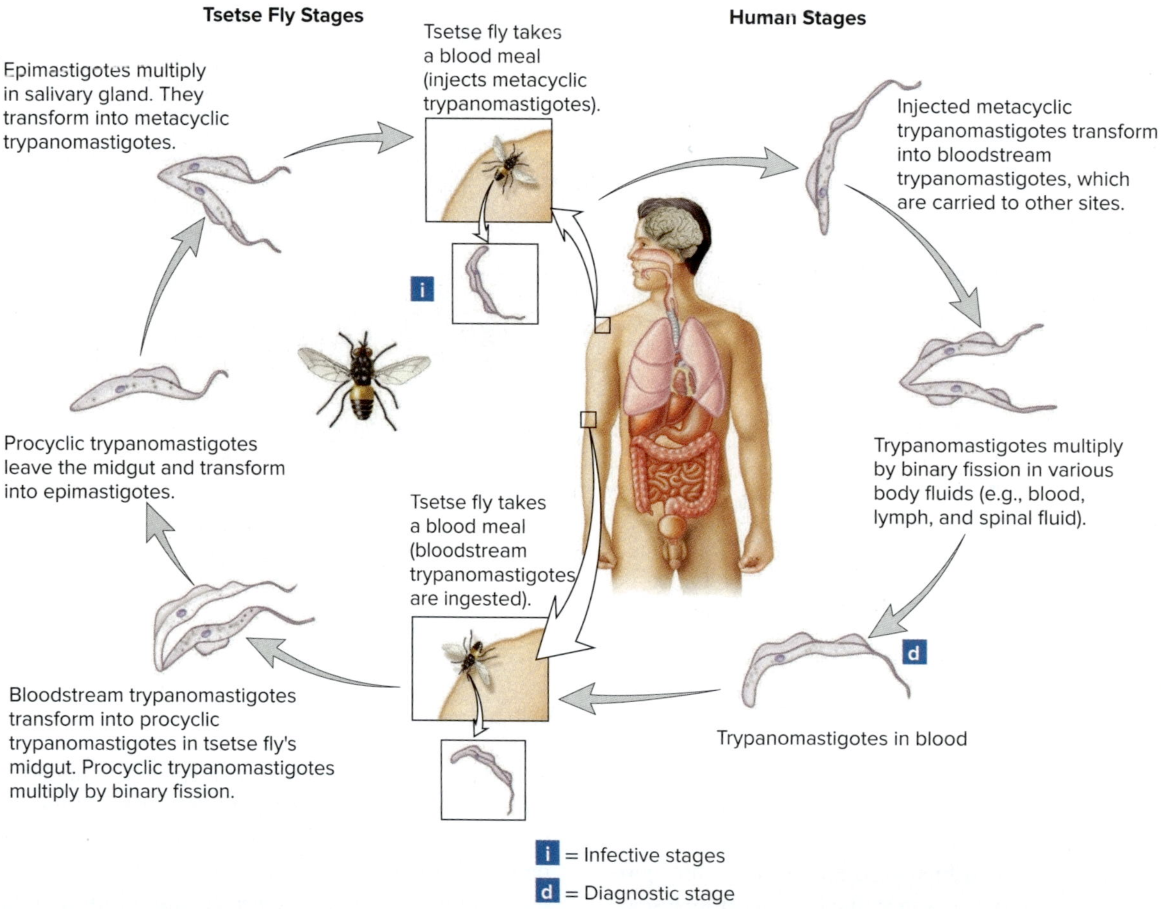

Figure 40.11 Life Cycle of *Trypanosoma brucei*.

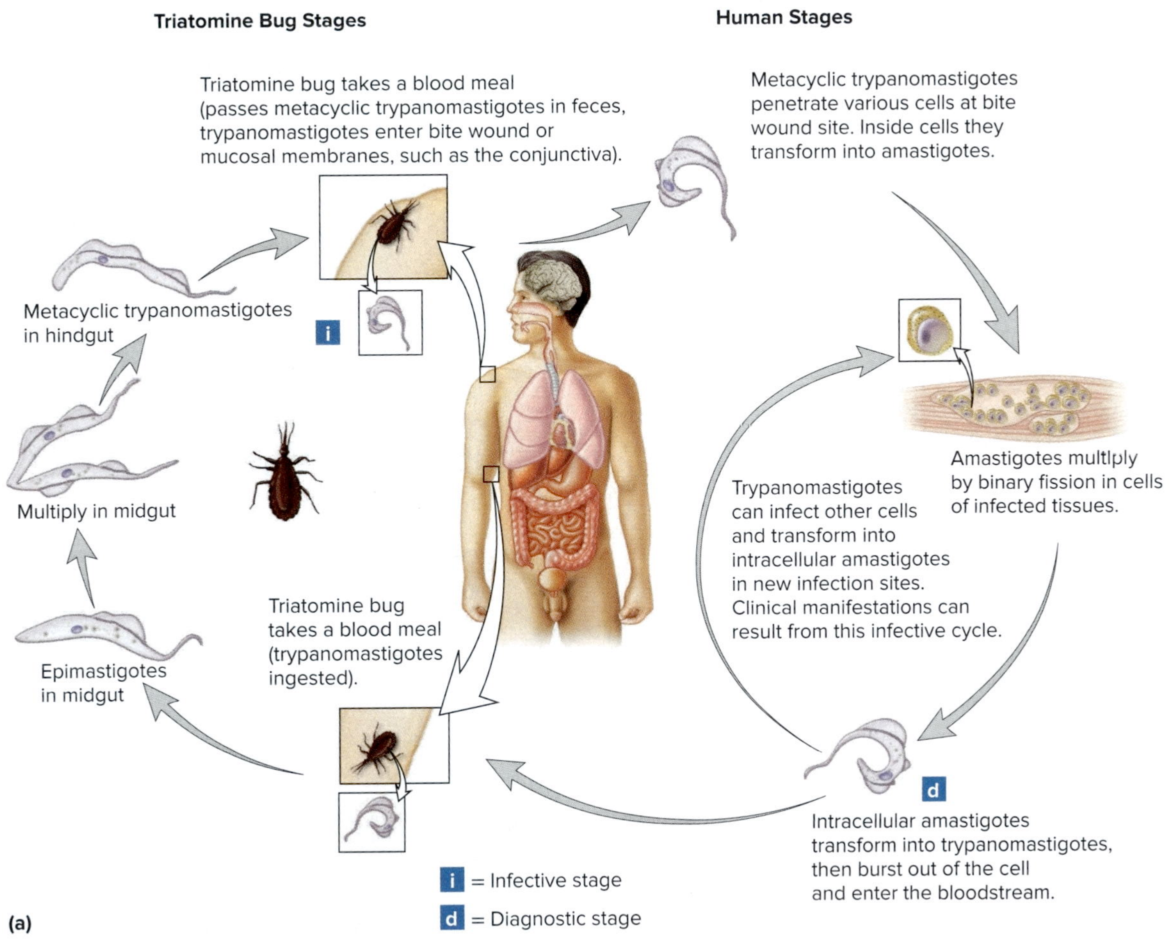

months (acute phase). Cell invasion stimulates the trypanosome's transformation into amastigotes, resulting in clinical manifestations of infection. The bloodstream trypanomastigotes do not replicate, however, until they enter a cell (becoming amastigotes) or are ingested by the arthropod vector (becoming epimastigotes). A chronic indeterminate (asymptomatic) phase of the disease follows the acute phase. During this phase, a high percentage of heart and gastrointestinal disease occurs due to parasitized cardiac and intestinal cells, respectively. An estimated 6 to 7 million people are infected, mostly in Latin America, with an annual incidence of about 28,000 new cases and 12,000 deaths each year.

Trypanosomiasis is diagnosed by finding motile protists in blood, spinal fluid, or skin biopsy and by serological testing. Acute Chagas is treated with benznidazole or nifurtimox; chronic Chagas is more difficult to treat as there is often significant cardiac and/or gastrointestinal pathology. Vaccine development for both HAT and Chagas has been stymied by the ability of these protozoa to change surface proteins **(antigenic variation)** to evade the immunologic response. ◀ Euglenozoa *(section 24.2)*

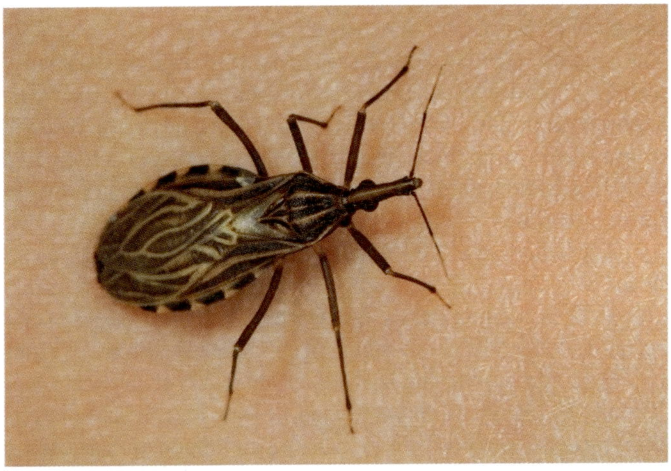

(b)

Figure 40.12 ***Trypanosoma cruzi*** **Infection.** (a) *T. cruzi* life cycle. (b) The triatomine bug.

Comprehension Check

1. Describe the *Plasmodium* spp. life cycle. What stages of the life cycle occur in humans? Where does sexual reproduction occur?
2. What are the three types of leishmaniasis? Compare the morphology of promastigotes and amastigotes. What is the location of each in the host?
3. African sleeping sickness and Chagas disease are both caused by trypanosomes. What similar measures can be taken to prevent both diseases? Why has vaccine development been difficult?

40.4 Direct Contact Diseases Can Be Caused by Fungi and Protists

After reading this section, you should be able to:

a. Discuss the microorganisms that cause common fungal and protozoal diseases spread by direct contact
b. Identify typical signs and symptoms of fungal and protozoal diseases spread by direct contact

Transmission of eukaryotic pathogens by direct contact results in a number of infectious diseases of the skin and reproductive organs. The direct contact diseases caused by fungi include cutaneous and subcutaneous mycoses. Diseases caused by protists can occur through hand-to-mouth/nose exchange, as well as their transfer through body fluids. We discuss fungal pathogens first and then a pathogenic protozoan.

Fungal Skin Infections

Fungi responsible for infections limited to the outer surface of hair and skin cause superficial mycoses. These fungi remain above the basement membrane of skin, consuming keratin as their primary carbon source. Infections of the hair shaft are collectively called **piedras** (Spanish for stone because their hyphae form hard nodules on the hair shaft). For example, black piedra caused by *Piedraia hortae* forms hard, black nodules on the hairs of the scalp. White piedra is caused by the yeast *Trichosporon beigelii* and forms light-colored nodules on the beard and mustache. Some superficial and cutaneous mycoses are called **tineas** and involve the outer layers of the skin, nails, and hair. The specific type of tinea is designated by a modifying term. Tinea versicolor is a common mycosis caused by the yeast *Malassezia furfur,* which forms brownish-red scales on the skin of the trunk, neck, face, and arms. These mycoses are treated with topical antifungal agents. Good personal hygiene prevents these infections.

Tinea barbae (Latin *barba*, the beard) is an infection of beard hair caused by *Trichophyton mentagrophytes* or *T. verrucosum*. It is predominantly a disease of men who live in rural areas and acquire the fungus from infected animals. Tinea capitis (Latin *capita*, the head) is an infection of the scalp hair (**figure 40.13*a***). It is characterized by hair loss, inflammation, and scaling. Tinea capitis is primarily a childhood disease caused by *Trichophyton* or *Microsporum* species. Person-to-person transmission of the fungus occurs frequently when poor hygiene and overcrowded conditions exist. The fungus also occurs in domestic animals, which can transmit it to humans. An ultraviolet [UV] light called a Wood's lamp can help with the diagnosis of tinea capitis because some species of fungi fluoresce when illuminated by UV radiation (figure 40.13*b*).

Members of three genera of fungi, or dermatophytes, are involved in cutaneous mycoses: *Epidermophyton, Microsporum,* and *Trichophyton*. Tinea corporis (Latin *corpus*, the body) is an infection that can occur on any part of the skin

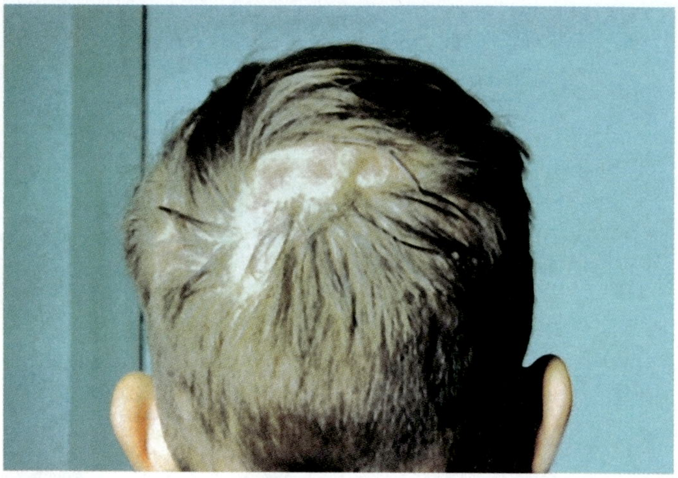

(a)

(b)

Figure 40.13 Superficial Mycosis: Tinea Capitis. (a) Fungal infection of the head caused by *Microsporum canis*. (b) Close-up using a Wood's lamp (a UV light).
(a) Source: CDC; (b) ©Medical-on-Line/Alamy Stock Photo

MICRO INQUIRY *What is the difference between a piedra and a tinea? Which is shown here?*

(**figure 40.14*a***). The disease is characterized by circular, red, well-demarcated, scaly, vesiculopustular lesions accompanied by itching. Tinea corporis, also called ringworm, is caused by the fungi *Trichophyton rubrum, T. mentagrophytes, T. concentricum,* and *Microsporum canis*. Fungi are transmitted by direct contact with infected animals and humans or by indirect contact through fomites (inanimate objects).

Tinea pedis (Latin *pes*, the foot), also known as athlete's foot, and tinea manuum (Latin *mannus*, the hand) are dermatophytic infections of the feet (figure 40.14*b*) and hands, respectively. Clinical symptoms vary from a fine scale to a pimplelike eruption. Itching is frequently present. Warmth, humidity, trauma, and occlusion increase susceptibility to infection. Most

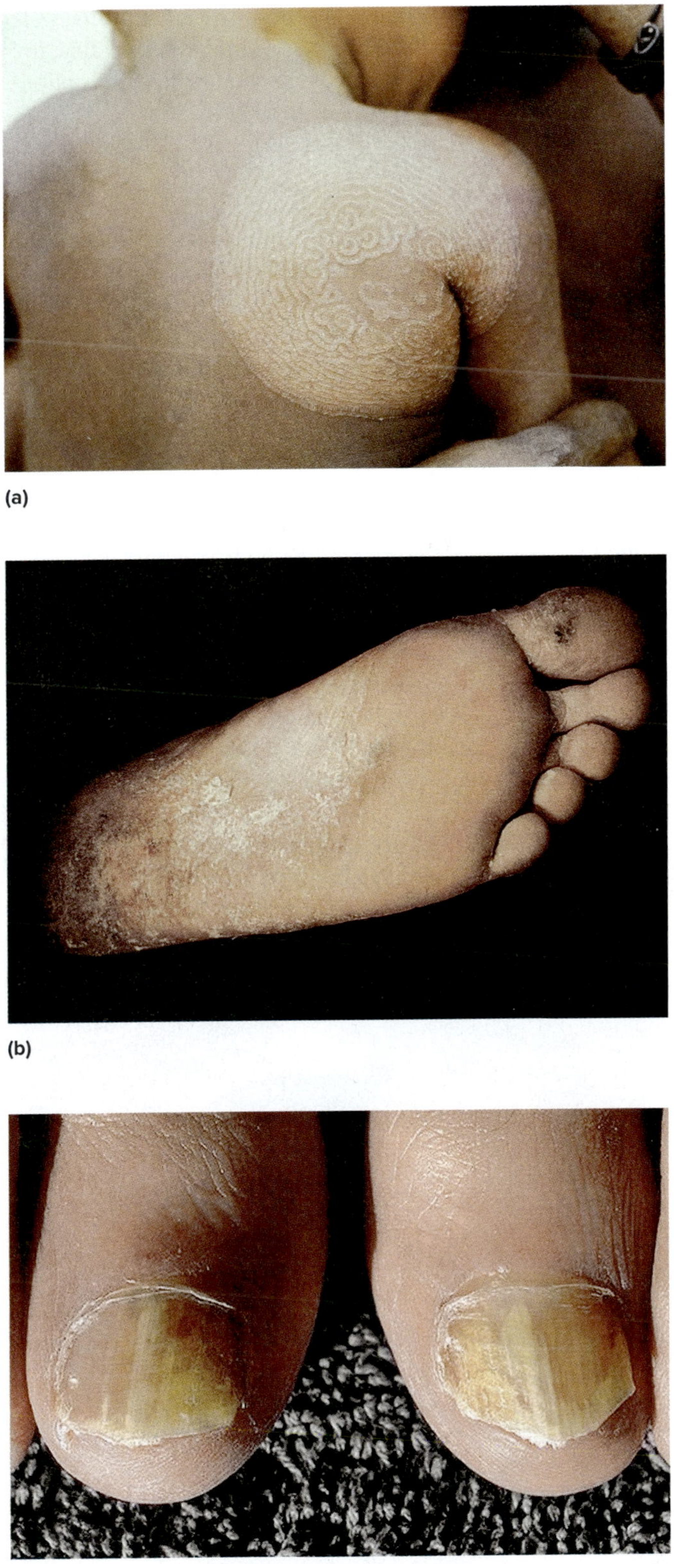

(a)

(b)

(c)

Figure 40.14 Cutaneous Mycosis. (a) Tinea corporis. Ringworm of the body (shoulder) caused by *Trichophyton concentricum*. (b) Tinea pedis caused by *Trichophyton rubrum*. (c) Tinea unguium of the nails caused by *Trichophyton rubrum*.

(a) Source: CDC/K. Mae Lennon, Tulane Medical School; Clement Benjamin; (b, c) Source: CDC

infections are caused by *T. rubrum, T. mentagrophytes*, or *E. floccosum*. Tinea pedis and tinea manuum occur throughout the world, are most commonly found in adults, and increase in frequency with age.

Tinea unguium (Latin *unguis*, nail) is a dermatophytic infection of the nail bed (figure 40.14*c*). In this condition, the nail becomes discolored and then thickens. The nail plate rises and separates from the nail bed. *Trichophyton rubrum* or *T. mentagrophytes* is the causative fungus.

Diagnosis is by microscopic examination of biopsied areas of the skin cleared with 10% potassium hydroxide and by culture on Sabouraud dextrose agar. Treatment is with topical ointments such as miconazole, tolnaftate, or clotrimazole for 2 to 4 weeks. Griseofulvin is used to treat nail infections. Dermatomycoses may be insidious; the desquamated (shed) skin cells can contain viable fungi for months or years. Thus any treatment plan should also include thorough cleaning of bedding, carpets, and any other substrates that could hold desquamated skin cells.

Dermatophytes that infect underneath the skin cause subcutaneous mycoses, and include sporotrichosis, chromoblastomycosis, and maduromycosis. These fungi are normally saprophytic inhabitants of soil and decaying vegetation. Because they are unable to penetrate the skin, they must be introduced into the subcutaneous tissue by a puncture wound. Most infections involve barefooted agricultural workers. Once in the subcutaneous tissue, the disease develops slowly—often over a period of years. During this time, the fungi produce a painless nodule that eventually ulcerates, and the organisms spread along lymphatic channels, producing more subcutaneous nodules. At times, the nodules drain to the skin surface. Diagnosis is accomplished by culture of the infected tissue.

Sporotrichosis is the subcutaneous mycosis caused by the dimorphic fungus *Sporothrix schenckii*. The disease occurs throughout the world and is the most common subcutaneous mycotic disease in the United States. The fungus can be found in the soil; on living plants, such as roses; or in plant debris, such as sphagnum moss, baled hay, and pine-bark mulch. Infection occurs by a puncture wound from a thorn or splinter contaminated with the fungus. The disease is an occupational hazard for florists, gardeners, and forestry workers. It is not spread from person to person. After an incubation period of 1 to 12 weeks, a small, red papule arises and begins to ulcerate. Although the infection usually remains localized, new lesions may appear along lymph channels (**figure 40.15**). In rare cases fungi spread throughout the body, producing extracutaneous sporotrichosis.

Sporotrichosis is typically treated by oral itraconazole for 3 to 6 months. Systemic or severe disease is treated with a lipid formulation of amphotericin B or ingestion of supersaturated potassium iodide until the lesions are healed, usually over several weeks.

Chromoblastomycosis, a disease in which dark brown nodules are formed, is caused by the black molds *Phialophora verrucosa* and *Fonsecaea pedrosoi*. These fungi exist worldwide,

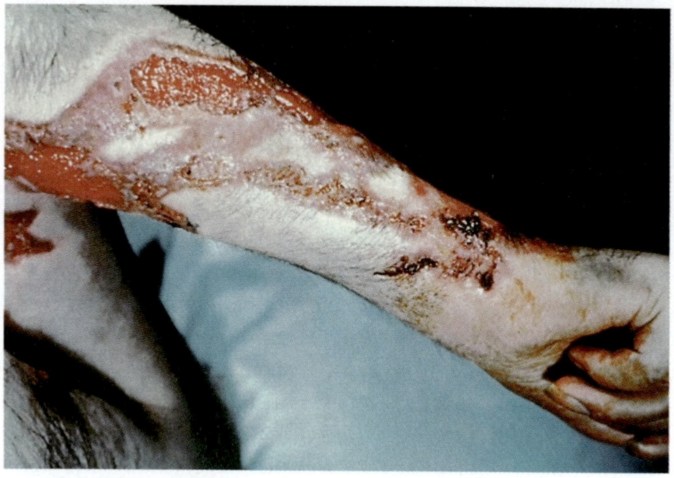

Figure 40.15 Subcutaneous Mycosis: Sporotrichosis. Sporotrichosis of the arm caused by *Sporothrix schenckii.*
Source: CDC/Dr. Lucille K. Georg

especially in tropical and subtropical regions. Most infections involve the legs and feet (**figure 40.16**). **Maduromycosis** is caused by *Madurella mycetomatis*, which is distributed worldwide and is especially prevalent in the tropics. Because the fungus destroys subcutaneous tissue and produces serious deformities, the resulting infection is often called a **eumycotic mycetoma,** or fungal tumor. One form of mycetoma, known as Madura foot, occurs through skin abrasions acquired while walking barefoot on contaminated soil. Griseofulvin, terbinafine, itraconazole, and fluconazole are common therapeutic options; however, these infections are very difficult to cure.

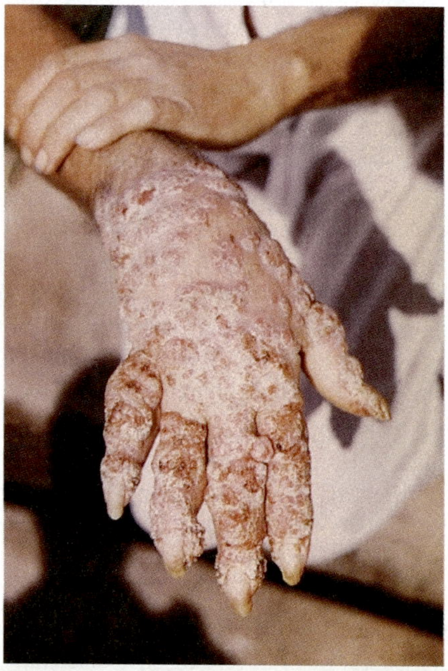

Figure 40.16 Subcutaneous Mycosis: Chromoblastomycosis. Chromoblastomycosis of the hand caused by *Phialophora verrucosa.*
Source: CDC/Dr. Lucille K. Georg

MICRO INQUIRY *How do fungi causing subcutaneous mycoses gain access to tissue below the skin surface?*

Trichomoniasis

Very few human pathogenic protists are transmitted through direct contact. Thus we only discuss one such disease, **trichomoniasis,** which is a sexually transmitted disease caused by the flagellated protozoan *Trichomonas vaginalis* (*see figure 24.4*). Humans are the only known host of *T. vaginalis.* Infection with this protist is often found with other sexually transmitted infections (STIs) and may increase the risk of acquiring other STIs. It is one of the most common STIs, with an estimated 2.3 million cases (women aged 14–49) annually in the United States and 143 million cases annually worldwide. In response to the protist, the body accumulates leukocytes at the site of infection. In women, this usually results in a profuse, purulent vaginal discharge that is yellowish to light cream in color and characterized by a disagreeable odor. The discharge is accompanied by itching. The cervical mucosa is covered with punctate hemorrhages, papules, and vesicles, referred to as "strawberry cervix" (**figure 40.17**). Males are generally asymptomatic because of the trichomonacidal action of prostatic secretions; however, a burning sensation may occur during urination. Diagnosis is made by microscopic identification of the protozoan. Treatment is by administration of metronidazole to all sexual contacts. ◀ Parabasalia *(section 24.2)*

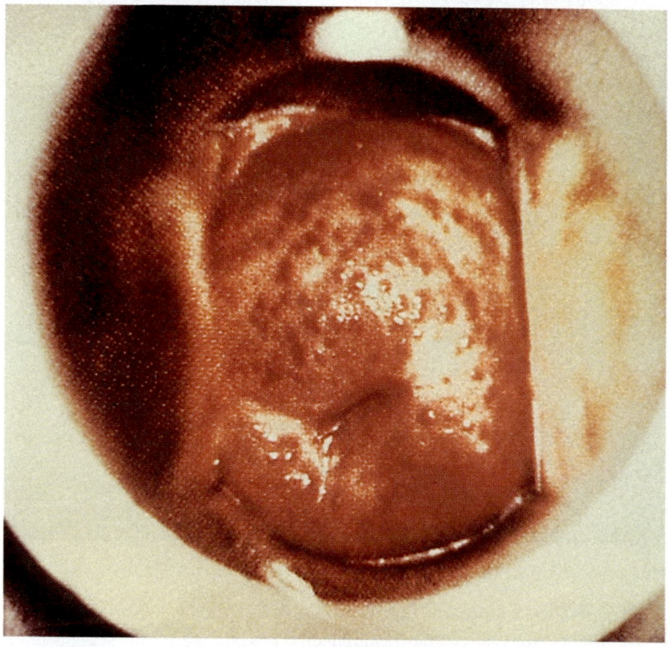

Figure 40.17 Trichomoniasis. *Trichomonas vaginalis* infection of the cervix causing hemorrhages, papules, and vesicles ("strawberry cervix").
Source: CDC

Comprehension Check

1. Describe two piedras that infect humans.
2. Briefly describe the major tineas that occur in humans.
3. Describe the three types of subcutaneous mycoses that affect humans.
4. Why do the signs and symptoms of trichomoniasis differ in men and women?

40.5 Food and Water Are Vehicles of Protozoal Diseases

After reading this section, you should be able to:

a. Discuss the microorganisms that cause common food-borne and waterborne fungal and protozoal diseases
b. Identify typical signs and symptoms of common food-borne and waterborne fungal and protozoal diseases
c. Explain in specific terms why access to clean water is considered a basic human right

An estimated 4% of deaths per year worldwide are caused by waterborne infectious diseases and many of them are caused by protozoa. Fungal and protozoal contamination of food, particularly as it is shipped great distances prior to consumption, is a problem that affects all nations. We now discuss a few of the more common protozoan diseases transmitted by ingesting tainted food or water.

Amebiasis

Amebiasis (amebic dysentery) is caused by the amoeba *Entamoeba histolytica,* which infects about 500 million people worldwide. While there is some evidence that *E. dispar* can also cause amebiasis, this has not been demonstrated unequivocally. Within the United States, about 3,000 to 5,000 cases are reported annually. However, worldwide as many as 100,000 die of amebiasis each year. ◀ Entamoebida *(section 24.3)*

Infection occurs by ingestion of mature cysts from fecally contaminated water, food, or hands, or from fecal exposure during sexual contact. After excystation in the lower region of the small intestine, the metacyst divides rapidly to produce eight small trophozoites (**figure 40.18**). These trophozoites move to the large intestine, where they can invade host tissue, live as commensals in the lumen of the intestine, or undergo encystation. In some hosts, the trophozoites remain in the intestinal lumen, resulting in an asymptomatic carrier state with cysts shed in feces.

If the trophozoites invade intestinal tissues, they multiply rapidly and spread laterally, while feeding on erythrocytes, bacteria, and yeasts. The invading trophozoites destroy the epithelial lining of the large intestine by producing a protease that plays a role in intestinal invasion by degrading the extracellular matrix and circumventing the host immune response by cleaving secretory immunoglobulin A (sIgA), IgG, and complement factors. Lesions (ulcers) are characterized by minute points of entry into the mucosa and extensive enlargement of the lesion after penetration into the submucosa. *E. histolytica* also may invade and produce lesions in other tissues, especially the liver, to cause hepatic amebiasis. All extraintestinal amebic lesions are secondary to those established in the large intestine. The symptoms of amebiasis are highly variable, ranging from an asymptomatic carriage to fulminating dysentery (exhaustive diarrhea accompanied by blood and mucus), appendicitis, and abscesses in the liver, lungs, or brain. ◀ *Antibodies bind specific 3-D molecules (section 33.7)*

Laboratory diagnosis of amebiasis is based on finding cysts in stools. Serological testing for *E. histolytica* is also available. Therapy for amebiasis depends on the location of the infection within the host and the host's condition. Asymptomatic carriers who are passing cysts are treated with paromomycin because they represent the most important reservoir of the protozoan. In symptomatic intestinal amebiasis, paromomycin and metronidazole are the drugs of choice. Prevention and control of amebiasis are achieved by practicing good personal hygiene and avoiding water or food that might be contaminated with human feces. Viable cysts in water can be destroyed by hyperchlorination or iodination. ◀ *Antiprotozoan drugs (section 9.7)*

Amebic Meningoencephalitis and Keratitis

Free-living amoebae of the genera *Naegleria* and *Acanthamoeba* directly invade the central nervous system. Therefore unlike other waterborne protozoal pathogens, they are not transmitted by ingestion. *Naegleria fowleri* causes primary amoebic meningoencephalitis (PAM), which is almost always fatal within 10 days. *N. fowleri* infection usually occurs when swimming in warm lakes. There have also been occasional deaths attributed to *Naegleria* infections acquired through the use of neti pots used to rinse the sinuses. In each case, tap water was reported to be contaminated with the protist. PAM results because *N. fowleri* is able to penetrate the cell layer between sinuses and the brain.

Several species of *Acanthamoeba* cause granulomatous amoebic encephalitis and infect the eye, causing a chronically progressive, ulcerative *Acanthamoeba* keratitis (inflammation of the cornea) that may result in blindness. Wearers of soft contact lenses may be predisposed to this infection and should take care to prevent contamination of lens-cleaning and soaking solutions. Diagnosis of these infections is by detection of amoebae in clinical specimens. Most freshwater amoebae are resistant to commonly used antimicrobial agents. These amoebae are reported in fewer than 100 human disease cases annually in the United States, although the incidence (especially of *Acanthamoeba* keratitis) is likely higher.

Cryptosporidiosis

A number of cryptosporidial species infect humans and other animals. *Cryptosporidium* (Greek *kruptos,* hidden) spp. are found in about 90% of sewage samples, in 75% of river waters, and in 28% of drinking waters. *C. parvum* is a common protist found in the intestine of many birds and mammals. When these animals

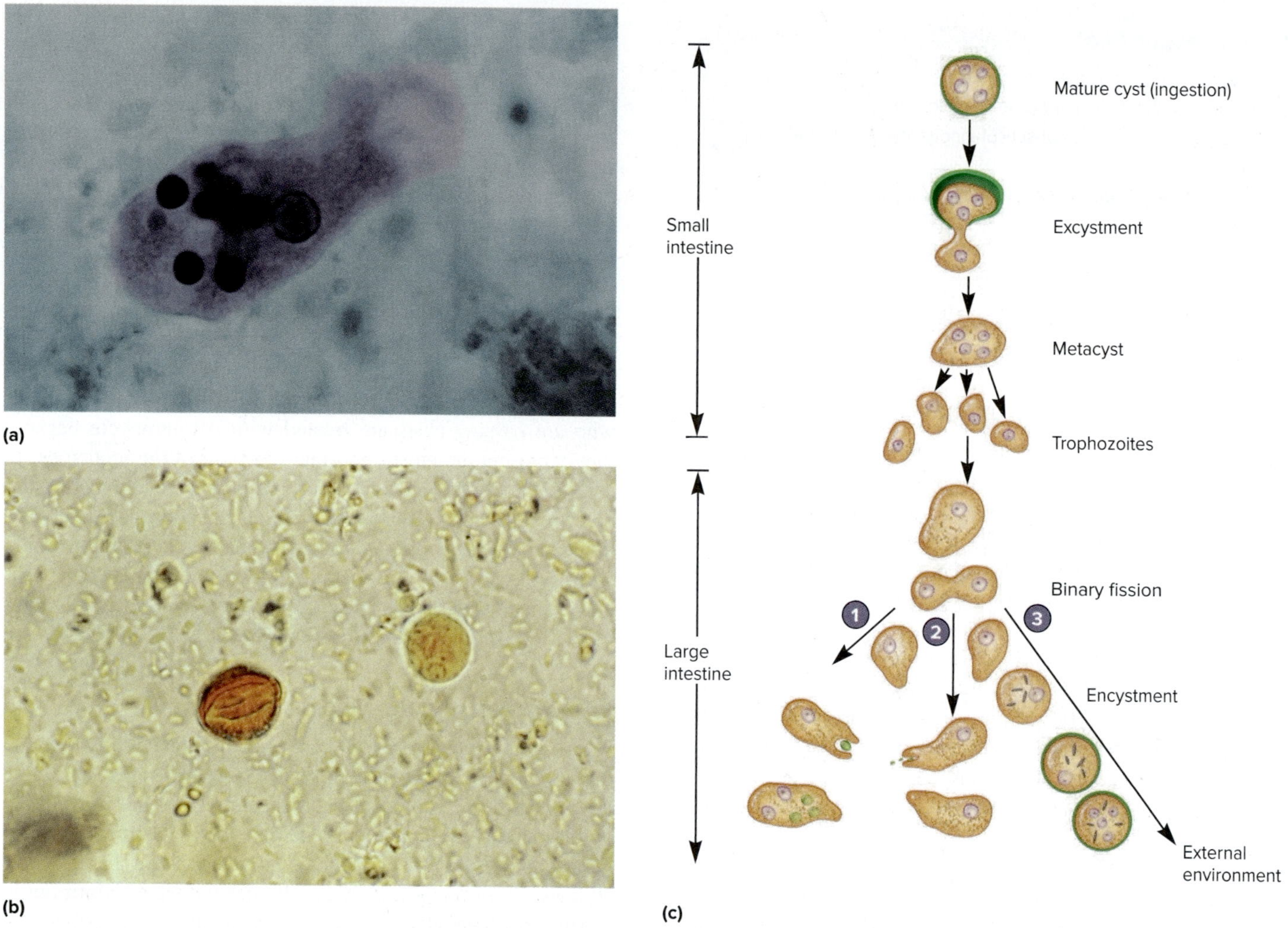

Figure 40.18 Amebiasis Caused by *Entamoeba histolytica.* Light micrographs of (a) a trophozoite (×1,000), which measure 15–20 µm, and (b) a cyst about 12 µm in diameter. (c) Life cycle. Infection occurs by ingestion of a mature cyst. Excystment occurs in the lower region of the small intestine, and the metacyst rapidly divides to give rise to eight small trophozoites (only four are shown). These enter the large intestine, undergo binary fission, and may (1) invade host tissues, (2) live in the lumen of the large intestine without invasion, or (3) undergo encystment and pass out of the host in feces.

(a) Source: CDC/Dr. Mae Melvin and Dr. Greene; (b) Source: CDC

defecate, oocysts are shed into the environment. If a human ingests food or water contaminated with oocysts, excystment occurs within the small intestine, and sporozoites enter epithelial cells and develop into merozoites. Some of the merozoites subsequently undergo sexual reproduction to produce zygotes, which differentiate into thick-walled oocysts. Oocysts released into the environment begin the life cycle again. ◀ Alveolata *(section 24.4)*

Major problems for public health arise from the fact that (1) oocysts are small, only 4 to 6 µm in diameter—too small to be easily removed by sand filters used to purify drinking water; (2) they have a low infectious dose, around 10 to 100 oocysts; and (3) *Cryptosporidium* spp. also are extremely resistant to disinfectants such as chlorine and oocysts may remain viable for 2 to 6 months in moist environments.

The incubation period for cryptosporidiosis ranges from 5 to 28 days. Diarrhea, which characteristically may be cholera-like (i.e., very watery, without blood), is the most common symptom. Other symptoms include abdominal pain, nausea, fever, and fatigue. The pathogen is routinely diagnosed by fecal concentration and acid-fast stained smears. Treatment of people with healthy immune systems is with nitazoxanide along with rehydration. Although the disease usually is self-limiting in healthy individuals, patients who are immunocompromised may develop prolonged, severe, and life-threatening diarrhea.

Cyclosporiasis

Cyclosporiasis is caused by the protist *Cyclospora cayetanensis*. The disease is most common in tropical and subtropical environments, although it has been reported in most countries. Each year, the CDC reports between 300 and 650 cases of cyclosporiasis in the United States. Many of these are the result of outbreaks linked to contaminated produce.

The disease presents with frequent, sometimes explosive, diarrhea. Often the patient exhibits cramps and bloating due to

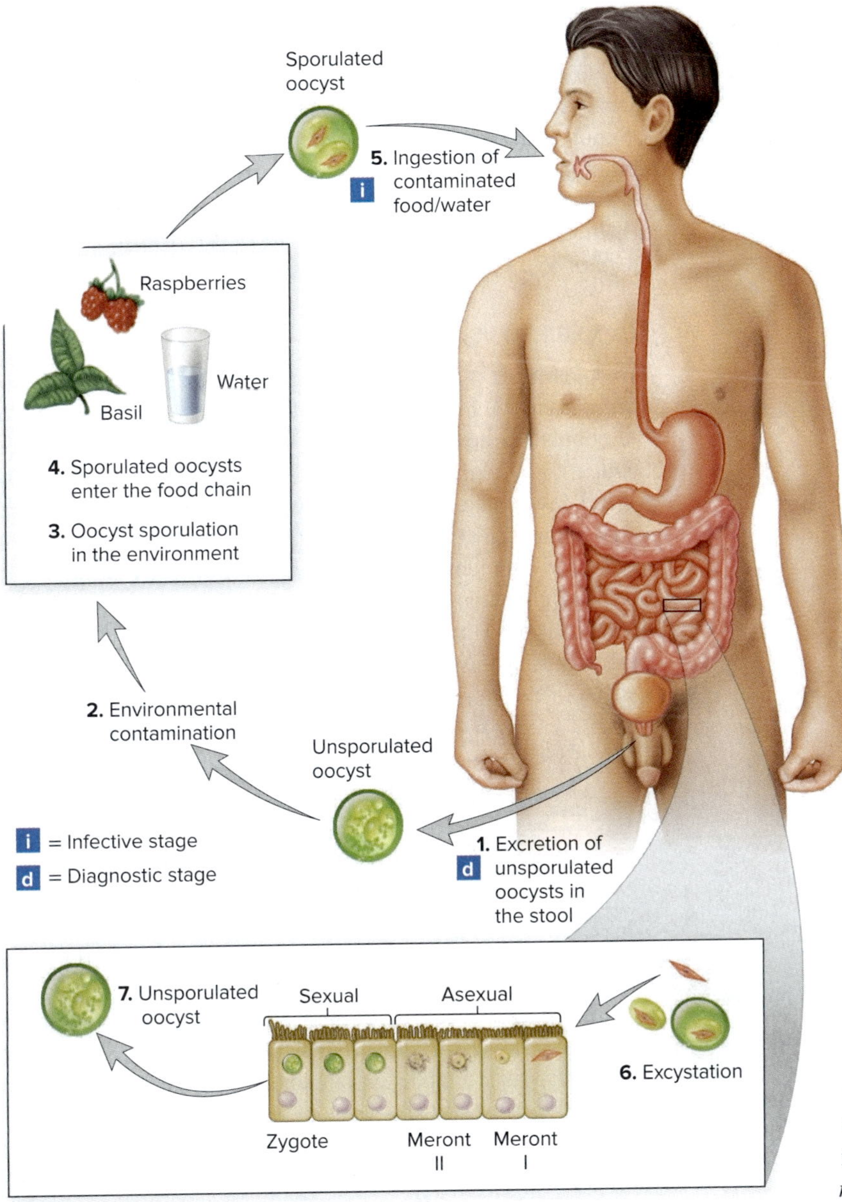

Figure 40.19 Life Cycle of *Cyclospora cayetanensis.* Unsporulated oocysts, shed in the feces of infected animals, sporulate and contaminate food and water. Once ingested, the sporulated oocysts excyst, penetrate host cells, and reproduce by binary fission (merogony). Sexual stages unite to form zygotes, resulting in new oocysts (unsporulated) that are shed in feces.

Source: CDC, "Biology."

substantial gas production, nausea, vomiting, fever, fatigue, and substantial weight loss. Patients may report symptoms for days to weeks with decreasing frequency and then relapses. The protozoa infect the small intestine with a mean incubation period of approximately 1 week. Cyclosporan oocysts in freshly passed feces are not infective (**figure 40.19**). Thus direct oral-fecal transmission does not occur. Instead, the oocysts must differentiate into sporozoites after days or weeks at temperatures between 22° and 32°C. Sporozoites often enter the food chain when oocyst-contaminated water is used to wash fruits and vegetables prior to their transport to market. Once ingested, the sporozoites are freed from the oocysts and invade intestinal epithelial cells, where they replicate asexually. Sexual development is completed when sporozoites mature into new oocysts and are released into the intestinal lumen to be shed with feces.

Laboratory diagnosis of cyclosporiasis is by the identification of oocysts in feces. Identification may require several specimens over several days. Treatment is with a combination of trimethoprim and sulfamethoxazole, and fluids to restore water lost through diarrhea. Prevention of cyclosporiasis is by avoidance of contaminated food and water.

Giardiasis

Giardia intestinalis is a flagellated protist that causes the very common intestinal disease **giardiasis,** which is worldwide in distribution. It has a very low infectious dose of about 10 cysts, and affects children more seriously than adults. In the United States, it is the most common cause of intestinal parasitic disease (between 15,000 and 20,000 cases yearly). *G. intestinalis* is endemic in child day-care centers in the United States. Transmission occurs most frequently by cyst-contaminated water supplies. Epidemic outbreaks have been recorded in wilderness areas, suggesting that humans may be infected from stream water with *G. intestinalis* harbored by livestock or household pets. Although this implies that infection also can be a zoonosis, it is estimated that about 7% of the U.S. population are asymptomatic carriers who shed cysts in their feces. As many as 200 million humans may be infected worldwide.

Following ingestion, *Giardia* cysts undergo excystment in the duodenum, forming trophozoites. The trophozoites inhabit the upper portions of the small intestine, where they attach to the intestinal mucosa by means of their sucking disks (**figure 40.20**; *also see figure 24.4*). The trophozoites feed on mucous secretions and reproduce to become such a large population that they interfere with nutrient absorption by the intestinal epithelium. Symptoms of giardiasis typically begin 1 week after infection and can last 6 to 8 weeks or longer if untreated. The disease can be acute or chronic. Acute giardiasis is characterized by severe diarrhea, epigastric pain, cramps, voluminous flatulence ("passing gas"), and anorexia. Chronic giardiasis causes intermittent diarrhea, with periodic appearance and remission of symptoms.

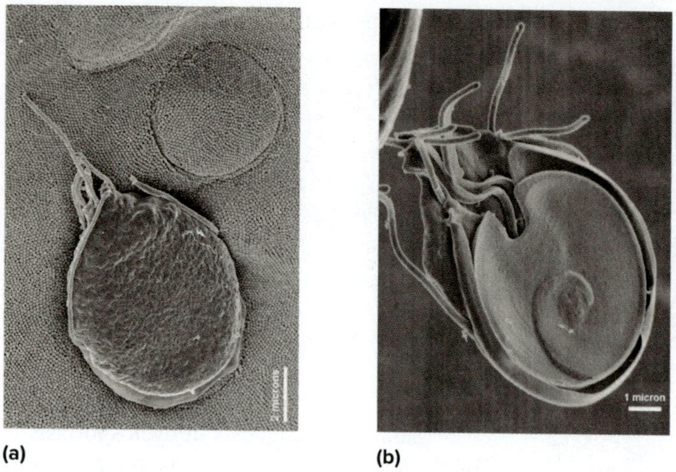

Figure 40.20 **Giardiasis.** (a) Scanning electron micrograph (SEM) depicting a *Giardia muris* protozoan adhering to the microvillus border of an intestinal epithelial cell. The circular lesion above the protozoan is an impression made by the ventral adhesive disk of another *G. muris* organism. (b) This SEM reveals the ultrastructural morphology of a *Giardia* protozoan's ventral adhesive disk. This disk acts like a suction cup, facilitating the organism's attachment to the intestinal surface.
Source: CDC/Dr. Stan Erlandsen

Laboratory diagnosis is based on the identification of trophozoites or cysts in stools. A commercial ELISA test is also available for the detection of *G. intestinalis* antigen in stool specimens. Tinidazole, nitazoxanide, and metronidazole are the drugs of choice. Prevention and control involve proper treatment of community water supplies, especially the use of slow sand filtration because cysts are highly resistant to chlorine treatment. ▶| *Purification and sanitary analysis ensure safe drinking water (section 43.1); Wastewater treatment (section 43.2)*

Toxoplasmosis

Toxoplasmosis is a worldwide disease caused by the protist *Toxoplasma gondii.* In some regions of the world, as much as 95% of the population has been infected, whereas in the United States, about 10% of people over the age of six have been infected. The natural reservoir of *T. gondii* is wild rodents, birds, and other small animals. Cats are the definitive host, meaning they are required for completion of the sexual cycle (**figure 40.21**). Cats (especially kittens) shed oocysts in their feces for several weeks after infection. Human infection can result through one of three modes of transmission: (1) zoonotic, through the accidental ingestion of oocytes while cleaning a litter box or handling a cat; (2) food-borne via the consumption of raw or undercooked infected meat; and (3) maternal-fetal (congenital) transmission (the mother may be unaware she is infected). Congenital transmission gained public notice when it was discovered that it can cause serious fetal defects or death. It follows that pregnant women are advised not to empty cat litter boxes.

In most cases, the host is unaware of *T. gondii* infection because a strong cell-mediated immune response is stimulated. Absent this immune response, acute toxoplasmosis occurs and may be accompanied by lymph node swelling (lymphadenopathy) with white blood cell enlargement. Pulmonary necrosis, myocarditis, and hepatitis caused by tissue necrosis may occur. Retinitis (inflammation of the retina of the eye) is associated with necrosis due to proliferation of the parasite within retinal cells. *T. gondii* in immunocompromised individuals can produce a unique encephalitis with necrotizing

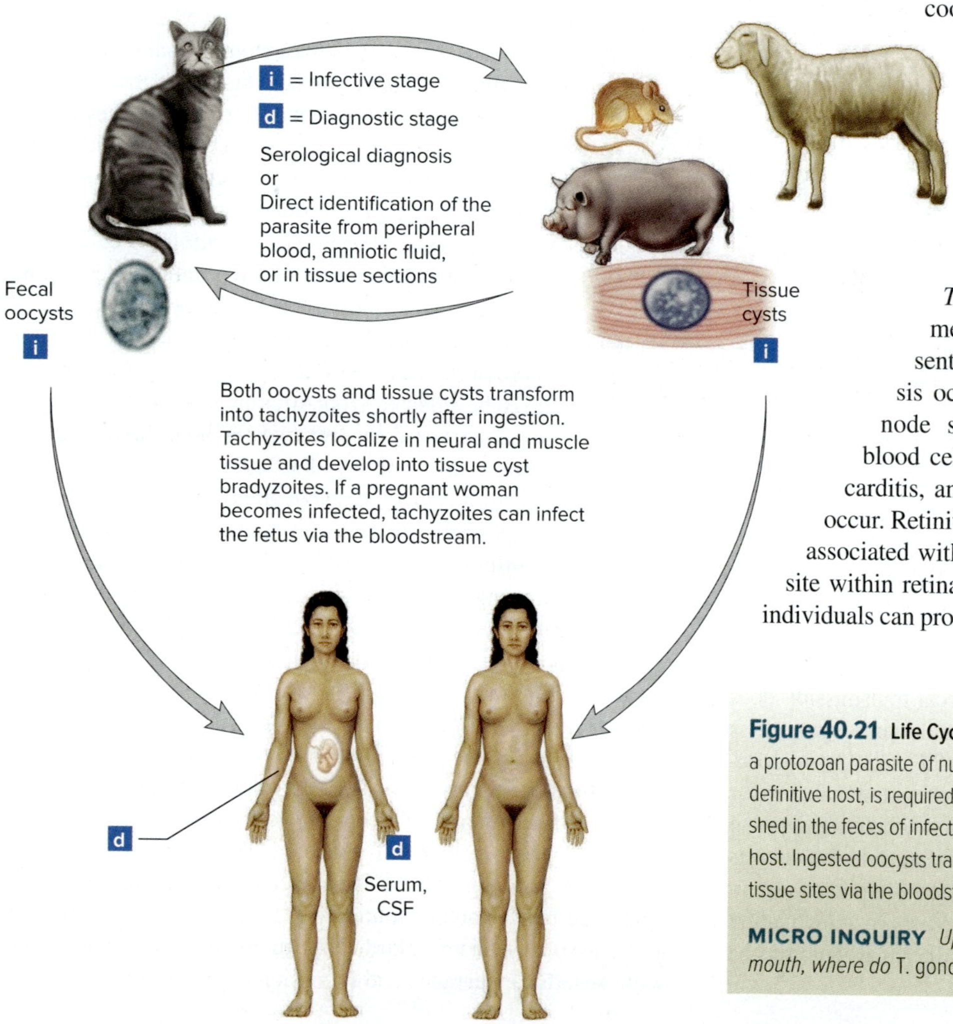

Figure 40.21 **Life Cycle of *Toxoplasma gondii.*** *Toxoplasma gondii* is a protozoan parasite of numerous mammals and birds. However, a cat, the definitive host, is required for completion of the sexual cycle. Oocysts are shed in the feces of infected animals, where they may be ingested by another host. Ingested oocysts transform into tachyzoites, which migrate to various tissue sites via the bloodstream.

MICRO INQUIRY *Upon entry into the human host by the nose or mouth, where do* T. gondii *colonize?*

lesions in the brain accompanied by inflammatory infiltrates. *T. gondii* causes about 400 congenital infections per year in the United States. Toxoplasmosis in newborns can result in a wide array of problems involving many different organ systems, but most commonly affects the central nervous system. ◀ Alveolata *(section 24.4)*

Laboratory diagnosis of toxoplasmosis is by serological tests. Epidemiologically, toxoplasmosis is ubiquitous in all higher animals. Treatment of toxoplasmosis is with a combination of pyrimethamine and sulfadiazine. Prevention and control require minimizing exposure by the following: avoiding eating raw or undercooked meat and eggs, washing hands after working in soil, cleaning cat litter boxes daily, keeping household cats indoors if possible, and feeding cats commercial food.

©Ingram Publishing

40.6 Opportunistic Diseases Can Be Caused by Fungi and Protists

After reading this section, you should be able to:

a. Predict the impact to the host when normal fungal microbiota are introduced to a new host environment
b. Distinguish growth and reproduction challenges experienced by normal fungal microbiota and ubiquitous environmental fungi when attempting to infect a healthy human
c. Cite specific host conditions that promote fungal infection

Certain fungi pose a specific risk for immunosuppressed individuals. The mycoses they cause are rarely seen in healthy people, but for those whose immune systems have become compromised, they can cause life-threatening infections. Immunosuppression has a myriad of causes, including: malnutrition, alcoholism, cancer, diabetes, another infectious disease (e.g., HIV/AIDS), trauma from surgery or injury, an altered microbiota from prolonged use of antibiotics, genetic deficiencies, cancer chemotherapy, and old age. Opportunistic mycoses may start as normal microbiota or ubiquitous environmental contaminants. Here we discuss a few common opportunistic fungal pathogens.

Aspergillosis

Of all the fungi that cause disease in human hosts, none is as widely distributed in nature as members of the genus *Aspergillus*. These filamentous fungi are omnipresent: They are found wherever organic debris occurs, especially in soil, decomposing plant matter, household dust, building materials, some foods, and water. In fact, it is almost impossible to avoid daily inhalation of *Aspergillus* spores (**figure 40.22**). However, infection results only when healthy immune functions deteriorate.

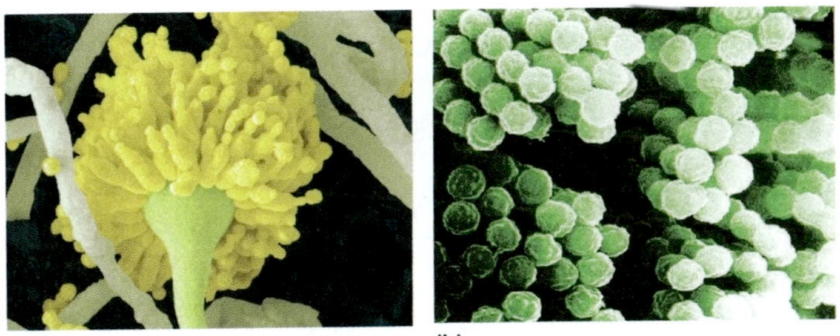

(a) (b)

Figure 40.22 ***Aspergillus* spp. Cause Opportunistic Mycoses.** (a) Conidiophores of *Aspergillus flavus*. (b) SEM of *Aspergillus* spores.
(a) ©Thierry Berrod, Mona Lisa Production/SPL/Science Source; (b) Source: CDC/Robert Simmons

The major portal of entry for *Aspergillus* spp. is the respiratory tract. Inhalation of conidiospores can lead to several types of aspergillosis. *Aspergillus fumigatus* is the species commonly involved in pulmonary **aspergillosis** marked by fever, chest pain, and cough. It also causes both superficial infection and invasive disease of immunosuppressed patients. Invasive disease typically results when pulmonary infection disseminates to the brain, kidney, liver, bone, or skin. Invasive aspergillosis has a mortality rate of 50 to 100%. ◀ Ascomycota *includes yeasts and molds (section 25.5)*

These fungi also cause allergic aspergillosis. Infected individuals may develop an immediate allergic response and suffer asthma attacks when exposed to fungal antigens. Another response is bronchopulmonary aspergillosis, in which patients with underlying pulmonary disease (tuberculosis, asthma, cystic fibrosis) suffer bronchitis resulting from both type I and type III hypersensitivities. Finally, the most common manifestation of pulmonary involvement is the occurrence of colonizing aspergillosis, in which fungal colonies form within the lungs and develop into "fungus balls" called aspergillomas. These consist of a tangled mass of hyphae growing in a circumscribed area. From the lung, the fungus may spread, producing disseminated aspergillosis in a variety of tissues and organs. In patients whose resistance is severely compromised, invasive aspergillosis may occur and fill the lung with fungal hyphae. ◀ *Hypersensitivities (section 33.10)*

Laboratory diagnosis of aspergillosis depends on identification either by direct examination of pathological specimens or by isolation and characterization of the fungus. An enzyme immunoassay that detects galactomannan (an antigen of *Aspergillus* spp.) can be used to screen suspected cases of aspergillosis. Treatment for invasive disease is with voriconazole, while allergic-type aspergillosis is treated with itraconazole and corticosteroids. However, successful therapy depends on treatment of the underlying disease so that host resistance increases. ◀ *Antifungal drugs (section 9.6)*

Candidiasis

Candidiasis is a mycosis caused by fungi in the genus *Candida*. *Candida* spp. disseminate within a human host depending on their need to invade through body fluids as yeast or tissues as

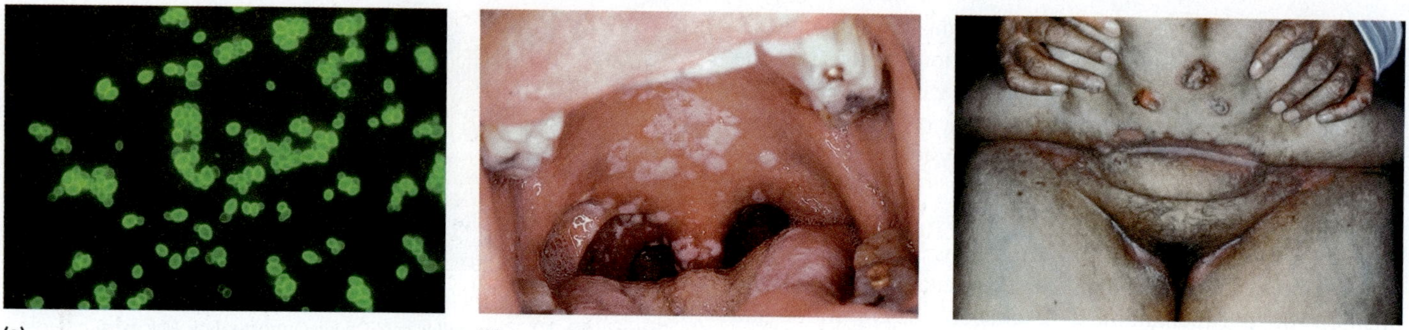

(a) (b) (c)

Figure 40.23 Opportunistic Mycoses Caused by *Candida albicans*. (a) Fluorescent antibody (FA) stain revealing the oval budding of the yeast form. (b) Thrush, or oral candidiasis, is characterized by the formation of white patches on the mucous membranes of the mouth and tongue. These patches form a pseudomembrane composed of spherical yeast cells, leukocytes, and cellular debris. (c) Intertriginous candidiasis involves those areas of the body, usually opposed skin surfaces, that cause irritation, inflammation, and areas of cutaneous ulceration.
(a) Source: CDC/Maxine Jalbert, Dr. Leo Kaufman; (b) Source: CDC/Sol Silverman, Jr., DDC; (c) Source: CDC/Dr. Hardin

pseudohyphae. As the term suggests, pseudohyphae are not true hyphae, rather they arise from incomplete budding. Cells elongate but remain attached after division. There are over 20 species of *Candida* that can cause infection in humans, most notably *C. albicans* (**figure 40.23*a***) and *C. glabrata*. In contrast to other opportunistic fungi, *Candida* spp. are members of the normal microbiota within the gastrointestinal tract, respiratory tract, vaginal area, and mouth. In healthy individuals, they do not produce disease because growth is suppressed by other microbiota and host resistance mechanisms. However, in the absence of competing microbiota, populations of *Candida* spp. grow. As they gain access to host epithelial cells, *C. albicans* secretes a peptide toxin called candidalysin that permeabilizes host-cell membranes. This provides additional nutrients, which further increases their numbers.

No other mycotic pathogens produce as diverse a spectrum of disease in humans as do *Candida* spp. **Oropharyngeal candidiasis,** or **thrush** (figure 40.23*b*), is a fairly common disease in newborns, persons with dentures, and people who use inhaled steroids. Thrush appears as many small, white flecks that cover the tongue and mouth. At birth, newborns do not have a normal microbiota in the oropharyngeal area. If the mother's vaginal area is heavily colonized with *Candida* spp., the upper respiratory tract of the newborn becomes colonized during passage through the birth canal. Once the newborn has developed his or her own normal oropharyngeal microbiota, thrush becomes uncommon. ▶| *Development of a stable microbiome (section 34.2)*

Intertriginous candidiasis involves those areas of the body, usually opposed skin surfaces, that are warm and moist: axillae, groin, skin folds (figure 40.23*c*). Napkin (diaper) candidiasis is typically found in infants whose diapers are not changed frequently and therefore are not kept dry. **Candidal vaginitis** can result as a complication of diabetes, antibiotic therapy, oral contraceptives, pregnancy, or any other factor that compromises the female host. Normally the omnipresent vaginal lactobacilli create a low pH, which limits fungal growth. However, if the numbers of lactobacilli are decreased by any of these factors, yeast may proliferate, causing a curdlike, yellow-white discharge from the vagina. *C. albicans* and *C. glabrata* can be transmitted to males during intercourse and lead to **balanitis.** Balanitis is an infection of the male glans penis and occurs primarily in uncircumcised males. In both sexes, candidiasis causes severe itching and burning. *Candida* spp. can also cause urinary tract infections and blood-borne infections.

Diagnosis of candidiasis is made by a combination of clinical observation and in more invasive cases, by direct microscopic observation. Mortality is almost 50% when the fungi invade the blood or disseminate to visceral organs.

There is no one satisfactory treatment for candidiasis. Cutaneous lesions can be treated with topical agents such as nystatin, miconazole, and trichomycin. Clotrimazole lozenges and nystatin swishing are used to treat oropharyngeal candidiasis. Similar drugs are used to treat genital/vulvovaginal and rectal candidiasis. Fluconazole can also be used for local infections while amphotericin B is used to treat systemic candidiasis.

Cryptococcosis

Cryptococcosis is a mycosis caused by *Cryptococcus gattii* and *C. neoformans*. These fungi grow as large, budding yeasts in the human host but as smaller yeast cells that undergo monokaryotic or dikaryotic reproduction in the environment, where they are saprophytes with a worldwide distribution. *C. gattii* is found in tropical and subtropical regions, as well as the U.S. Pacific Northwest and British Columbia. *C. neoformans* has a broader geographic distribution. Cryptococcosis manifests as pneumonia and/or meningitis with a mortality rate of about 33%. Typically, cryptococci cause disease in the immunocompromised and most often cause disease in people who are HIV positive, have cancer, or experienced a recent organ transplant.

C. neoformans is found in the droppings of pigeons and several other birds. It is not clear if these birds serve as reservoirs or if their excrement selectively enriches the growth of fungal spores in soil. In healthy people, fungi enter the body by the respiratory tract and are immediately cleared by innate immunity; or a minor, transitory pulmonary infection may occur. In contrast, if an

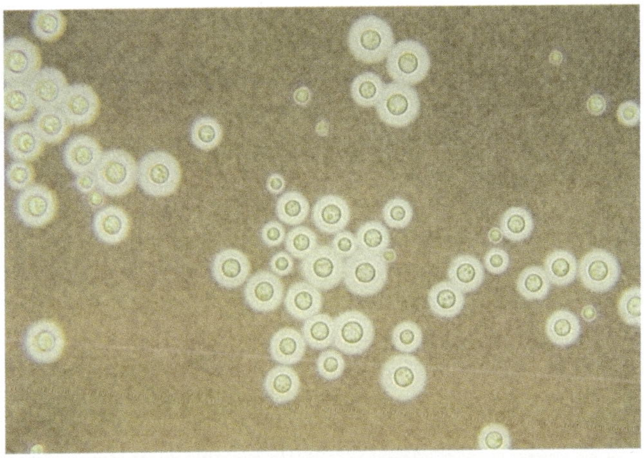

Figure 40.24 Systemic Mycosis: Cryptococcosis. India ink preparation showing *Cryptococcus neoformans.* Note the distinctly outlined capsules surrounding all cells and refractile inclusions in the cytoplasm; light micrograph; average cell size is 5–7 μm.
Source: CDC/Dr. Leanor Haley

MICRO INQUIRY *Once this microbe is acquired by inhalation, to what other body systems can it spread?*

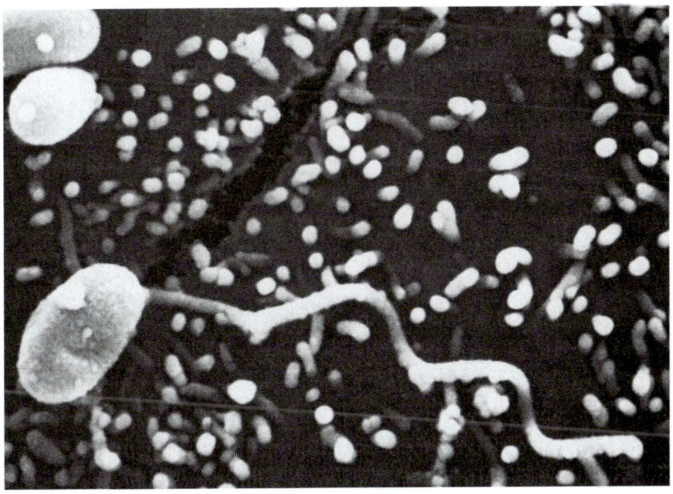

Figure 40.25 Microsporidia. Scanning electron micrograph of a microsporidian spore with an extruded polar tubule inserted into a eukaryotic cell. The spore injects the infective sporoplasms through its polar tubule.
Source: CDC-DPDx

immunosuppressed individual inhales the fungi, a pneumonia-like disease with coughing and fever occurs. Some pulmonary infections spread to the skin, bones, viscera, and central nervous system, causing headaches and mental status changes. Once the nervous system is involved, cryptococcal meningitis usually results. Diagnosis is accomplished by detection of the thick-walled, spherical yeast cells from blood, sputum, or cerebral spinal fluid using India ink, calcofluor-white, or fluorescent antibodies (**figure 40.24**). The fungi can be easily cultured on Sabouraud dextrose agar. Treatment for *C. neoformans* includes amphotericin B plus flucytosine. Immunosuppressed patients with a history of infection are treated prophylactically with fluconazole.

Microsporidiosis

More than 1,200 species of *Microsporidia* have been catalogued, in 143 genera; at least 14 species are known to be human pathogens causing **microsporidiosis.** Several domestic and feral animals appear to be reservoirs for species that infect humans. Increasingly recognized as opportunistic infectious agents, microsporidia infect a wide range of vertebrate and invertebrate hosts. One unifying characteristic of microsporidia is their production of a highly resistant spore, capable of surviving long periods of time in the environment. Spore morphology varies with species. Microsporidial spores recovered from human infections are oval to rodlike, measuring 1 to 4 μm. Microsporidia possess a unique organelle known as the polar tubule, which is coiled within the spore (**figure 40.25**; *also see figure 25.17*). ◀ Microsporidia *(section 25.7)*

Infection of a host cell results when a microsporidial spore extrudes its polar tubule, which bores through the target cell's plasma membrane. A sudden increase in spore calcium results in the injection of the sporoplasm (cytoplasm-like contents) through the polar tubule into the host cell. Inside the cell, the sporoplasm condenses and undergoes asexual multiplication. Multiplication of the sporoplasm is species-specific, occurring within the host cytoplasm or within a vacuole, and is usually completed through binary (merogony) or multiple (schizogony) fission. Once the sporoplasm has multiplied, it directs the development of new spores (sporogony) by first encapsulating meronts or schizonts (products of asexual multiplication) within a thick spore coating. New spores increase until the host cell membrane ruptures, releasing the progeny.

Microsporidia infections can result in a wide variety of conditions. These include hepatitis, pneumonia, skin lesions, diarrhea, weight loss, and wasting syndrome. Diagnosis is based on clinical manifestations and identification of microsporidia in fecal smears. Where possible, identification can be made based on the characteristic polar tubule coiled within a spore visualized by electron microscopy. Microsporidiosis is difficult to treat; albendazole (which inhibits tubulin and ATP synthesis in helminths) is often used for gastrointestinal or disseminated infections while fumagillin (an amebicide) is used for eye infections.

Pneumocystis Pneumonia

Pneumocystis jiroveci is a fungus found in the lungs of a wide variety of mammals. The life cycle of *P. jiroveci* is presented in **figure 40.26**. The disease that this fungus causes was originally termed *Pneumocystis carinii* pneumonia (PCP). The name of the fungus was changed to *Pneumocystis jiroveci* in honor of the

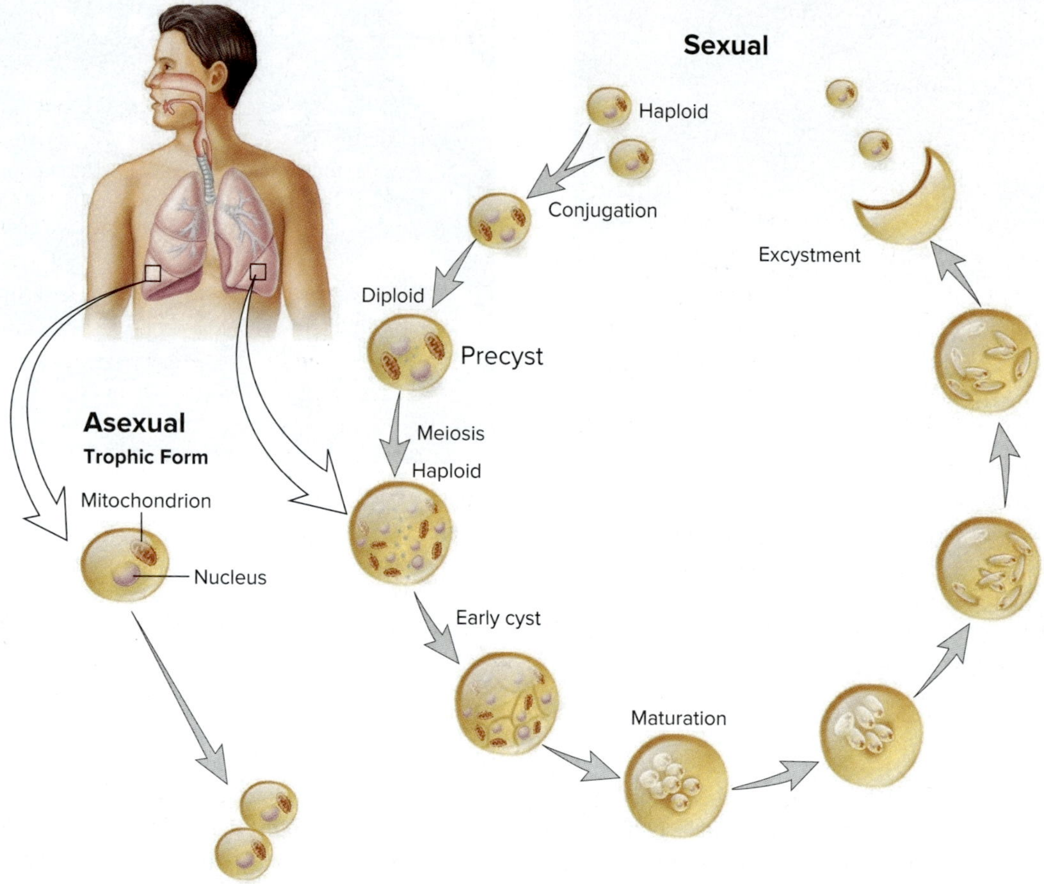

Figure 40.26 **Life Cycle of *Pneumocystis jiroveci*.** This fungus exhibits both sexual and asexual reproductive stages. Sexual reproduction occurs when haploid cells undergo conjugation followed by meiosis. Asexual reproduction occurs by binary fission. Both forms of reproduction can occur within the human host.

Czech microbiologist, Otto Jirovec, who first described this pathogen in humans. The acronym PCP for the disease has been retained despite the loss of the old species epithet (*carinii*). PCP now stands for *Pneumocystis* pneumonia.

Serological data indicate that most humans are exposed to *P. jiroveci* by age three or four. However, PCP occurs almost exclusively in immunocompromised hosts. Extensive use of immunosuppressive drugs and irradiation for the treatment of cancers and following organ transplants accounts for the formidable prevalence rates for PCP. This pneumonia also occurs in premature, malnourished infants and in more than 80% of HIV+ patients who have progressed to AIDS. Both the organism and the disease remain localized in the lungs—even in fatal cases. Within the lungs, *P. jiroveci* causes the alveoli to fill with a frothy exudate.

Laboratory diagnosis of *Pneumocystis* pneumonia can be made definitively only by microscopically demonstrating the presence of the microorganisms in infected lung material or by PCR analysis. Treatment is by means of oxygen therapy and a combination of trimethoprim and sulfamethoxazole (TMP-SMX). Prevention and control are through prophylaxis with TMP-SMX in susceptible persons.

Comprehension Check

1. Compare and contrast infections caused by *Entamoeba histolytica* and *Giardia intestinalis*.
2. How is *Toxoplasma gondii* transmitted?
3. Why are some mycotic diseases of humans called opportunistic mycoses?
4. What parts of the human body can be affected by *Candida* infections?
5. Note that antifungal agents are not used to treat infections caused by microsporidia and *P. jiroveci*. These microorganisms have been difficult to classify taxonomically. How do you think this may relate to their unusual antimicrobial sensitivities?

Key Concepts

40.1 Relatively Few Fungi and Protists Are Human Pathogens

- Human fungal diseases, or mycoses, can be divided into five groups according to the level and mode of entry into the host (**figure 40.1**). These are the superficial, cutaneous, subcutaneous, systemic, and opportunistic mycoses (**table 40.1**).
- Protozoa are responsible for some of the most serious human diseases that affect hundreds of millions of people worldwide (**table 40.2**) and can also be grouped by route of disease transmission (**table 40.3**).

40.2 Fungi Can Be Transmitted by Airborne Routes

- Most systemic mycoses that occur in humans are acquired by inhaling spores from soil where the free-living fungi are found.
- Three types can occur in humans: blastomycosis (**figure 40.2**), coccidioidomycosis (**figure 40.3**), and histoplasmosis (**figure 40.4**). All of these diseases begin as lung infections and then may spread to other regions of the body.

40.3 Arthropods Can Transmit Protozoal Disease

- The most important human protozoal pathogen is *Plasmodium*, the causative agent of malaria (**figures 40.5–40.8**). Human malaria is caused by five species of *Plasmodium: P. falciparum, P. vivax, P. malariae, P. knowlesi,* and *P. ovale*.
- Great strides have been made in the twenty-first century in reducing the global burden of malaria, with its incidence in Africa reduced by half.
- Flagellated protozoa are transmitted by arthropods and infect the blood and tissues of humans. Two major groups occur: the leishmanias, which cause the diseases collectively termed leishmaniasis (**figures 40.9** and **40.10**), and the trypanosomes (**figures 40.11** and **40.12**), which cause African trypanosomiasis and Chagas disease (American trypanosomiasis).

40.4 Direct Contact Diseases Can Be Caused by Fungi and Protists

- Superficial mycoses of the hair shaft are collectively called piedras. Two major types are black piedra and white piedra. Tinea versicolor is a third common superficial mycosis.
- The cutaneous fungi that parasitize the hair, nails, and outer layer of the skin are called dermatophytes, and their infections are termed dermatophytoses, ringworms, or tineas. Many types can occur in humans: tinea barbae (ringworm of the beard), tinea capitis (ringworm of the scalp; **figure 40.13**), tinea corporis (ringworm of the body), tinea pedis (ringworm of the feet), and tinea unguium (ringworm of the nails; **figure 40.14**).
- The dermatophytes that cause the subcutaneous mycoses are normal saprophytic inhabitants of soil and decaying vegetation. Three types of subcutaneous mycoses can occur in humans: chromoblastomycosis, maduromycosis, and sporotrichosis (**figures 40.15** and **40.16**).
- Trichomoniasis is a sexually transmitted disease caused by the protozoan flagellate *Trichomonas vaginalis*. *T. vaginalis* causes a foul-smelling, mucopurulent vaginal discharge (**figure 40.17**).

40.5 Food and Water Are Vehicles of Protozoal Diseases

- *Entamoeba histolytica* is the amoeboid protozoan responsible for amebiasis. This is a very common disease in warm climates throughout the world. It is acquired when cysts are ingested with contaminated food or water (**figure 40.18**).
- Freshwater protozoa in the genera *Naegleria* and *Acanthamoeba* can cause primary amebic meningoencephalitis and granulomatous amoebic encephalitis and keratitis, respectively.
- *Cryptosporidium parvum* is a common coccidial apicomplexan protozoan that causes severe diarrheal disease. It is acquired from contaminated food or water.
- *Cyclospora cayetanensis* is shed in the feces of a current host and infects the gastrointestinal tract of a new host after it has developed at 22° to 32°C for several days or weeks. It can be acquired from contaminated water used to rinse fruit or vegetables (**figure 40.19**).
- *Giardia intestinalis* is a flagellated protozoan that causes the common intestinal disease giardiasis (**figure 40.20**). This disease is distributed throughout the world, and in the United States, it is the most common cause of waterborne diarrheal disease.
- Toxoplasmosis is a disease caused by the protozoan *Toxoplasma gondii*. It can be acquired by handling cats or their litter boxes, by consuming undercooked or raw meat, or by maternal-fetal transmission (**figure 40.21**).

40.6 Opportunistic Diseases Can Be Caused by Fungi and Protists

- An opportunistic organism is one that is generally harmless in its normal environment but can become pathogenic in a compromised host.

- The most important opportunistic mycoses affecting humans include systemic aspergillosis (**figure 40.22**), candidiasis (**figure 40.23**), cryptococcosis (**figure 40.24**), and *Pneumocystis* pneumonia (**figure 40.26**).
- An emerging opportunistic fungal disease, especially of the immunocompromised, is caused by fungi known as microsporidia. They live as a spore form and have a unique organelle used for infecting new host cells (**figure 40.25**).

Active Learning

1. *Toxoplasma gondii* manipulates host-cell gene expression by modulating the level of key transcription factors; however, the mechanisms by which this is accomplished are unclear. Proteins delivered from the protozoan rhoptry and dense granules are known to be secreted into the host environment (*see figure 24.16*). Discuss how you might determine the function of one such protein.
2. When entering the host's skin, most *Leishmania* spp. are killed by professional phagocytes (neutrophils, macrophages, and dendritic cells). Promastigotes that lead to disease form a parasitophorous vacuole (PV) in macrophages that enables their growth. It is within PVs that promastigotes transform into amastigotes (figure 40.10). Discuss the molecular cues that might trigger the formation of PVs and differentiation of promastigotes into amastigotes. How would you test your predictions?
3. Trypanosomes are notorious for their ability to change their surface antigens frequently. Given the kinetics of a primary immune response (primary antibody production; *see figure 33.15*), how often would surface antigens need to change to stay "ahead" of host antibody specificity?
4. We are currently seeing an unprecedented reduction in the rate of malaria infection across Africa. To help sustain and expand the efforts that have given rise to this success in a cost-efficient manner, a group of British epidemiologists developed a dynamic model to examine the impact of preventative measures, medical interventions, as well as population, geographic, and environmental factors. If you were to develop such a model, what specific data would you want to consider?

   Read the original paper: Walker, P. G. T., et al. 2017. Estimating the most efficient allocation of interventions to achieve reductions in *Plasmodium falciparum* malaria burden and transmission in Africa: a modeling study. *Lancet Global Health* 4:e474–84. doi:10.1016/S2214-109X(16)30073-0.
5. *Cryptosporidium parvum* is a significant cause of waterborne diarrheal diseases worldwide, yet very little is known about its interaction with the human host. To address this, cultured human cells were mutated and screened for the capacity of *C. parvum* to bind and enter these cells. Discuss how you might design a similar experiment to ask what factors are needed for host-protozoan interaction. Be sure to consider the use of positive and negative controls and how you would screen the mutant *C. parvum* and/or host cells.

   Read the original paper: Yu, X., et al. 2017. Characterization of host cell mutants significantly resistant to *Cryptosporidium parvum* infection. *J. Eukaryot. Microbiol.* 64:843–9. doi:10.1111/jeu.12419.

41

Microbiology of Food

©Image Source

The Art, Science, and Genetics of Brewing Beer

Beer has a long history. Prehistoric nomads probably learned to make beer from grains and water before they figured out how to bake bread. Beer was on Noah's list of provisions. Six-thousand-year-old Babylonian tablets describe beer recipes. Ancient Chinese, Egyptians, and Romans brewed beer. It is even rumored that the Pilgrims made landfall in Plymouth in 1620 because their beer supply was running low. Apart from water, beer is considered the most frequently consumed liquid. With a global population over 7 billion, that's a lot of beer.

There are several essential steps to making beer. First, barley is soaked in warm water to trigger germination. Once germinated, the grain produces the enzymes gluconase, which breaks down plant cell walls, and amylase to degrade the plant starch to a sugar fermentable by yeast. Next the sprouted grain is heated and dried, denaturing the enzymes. This mixture is called malt, and it is flushed with warm water so yeast can begin fermenting the starch to ethanol and CO_2.

Predictably, many modifications have been made to the art and science of beer brewing. Since the mid-1990s, this has included the development of genetically modified (GM) ingredients. Most breweries find it more cost effective to add gluconases and amylases collected from bacteria that have been genetically engineered to produce these enzymes. Rice and corn genetically modified to be herbicide-resistant have been used for a number of years by many larger breweries. And there are a number of GM barley varieties. For instance, some are designed to be resistant to pathogenic fungi (*Rhizoctonia solani* and *R. oryzae*). Another uses nitrogen more efficiently, and then there is the GM barley that produces heat-stable gluconase.

The use of GM yeast, however, seems to be entirely off limits. Yeast geneticists developing new strains for brewers do it the old-fashioned way, by mating strains and selecting for new traits. Of course, this is the twenty-first century, so this is accomplished with high-throughput robotics. However, some labs are using CRISPR to introduce variation. Because CRISPR technology deletes or changes existing genes (rather than introducing new genes from another organism), the industry is not quite sure if yeast strains built by CRISPR represent genetic modification. Perhaps the bigger unanswered question is, Why is GM yeast taboo but GM grain is not?

As the history of beer illustrates, humankind long ago figured out how to harness microorganisms to transform raw foods into gastronomic delights, including chocolate, cheeses, pickles, sausages, soy sauce, and wines. It was not until the late nineteenth century, however, that the role microorganisms play in food spoilage was appreciated. This illustrates the two opposing roles of microorganisms in food production and preservation. This chapter begins with an overview of microbial spoilage of food; the second half discusses the importance of other microbes in food and beverage production.

Readiness Check:

Based on what you have learned previously, you should be able to:

- ✓ Compare and contrast the growth and physiology of obligate aerobes, facultative anaerobes, and strict anaerobes (section 7.5)
- ✓ Review physical and chemical agents commonly used to control microbial growth (sections 8.4 and 8.5)
- ✓ Describe the capacity of a bacterial endospore to survive in nature, and list example bacteria that form endospores (sections 3.9, 23.2, and 23.3)
- ✓ Summarize the possible fates of pyruvate in fermentation (section 11.8)
- ✓ Explain the principles of a toxin-mediated disease (section 35.4)
- ✓ Predict methods used to test for food-borne microorganisms (sections 19.3, 29.2, 37.2, and 37.3)
- ✓ Discuss the typical bacterial, viral, fungal, and protist diseases transmitted by food (sections 38.4, 39.4, and 40.5)

41.1 Microbial Growth Can Cause Food Spoilage

After reading this section, you should be able to:

a. Differentiate between intrinsic and extrinsic factors that influence food spoilage
b. Explain how the composition of a food item helps determine the kind of microbial growth that might occur
c. List two physical and two biological features of a food that influence spoilage

Table 41.1 Differences in Spoilage Processes in Relation to Food Characteristics

Substrate	Food Example	Chemical Reactions or Processes[1]	Typical Products (and Effects)
Pectin	Fruits	Pectinolysis	Methanol, uronic acids (loss of fruit structure, soft rots)
Proteins	Meat	Proteolysis, deamination	Amino acids, peptides, amines, H_2S, ammonia, indole (bitterness, souring, bad odor, sliminess)
Carbohydrates	Starchy foods	Hydrolysis, fermentations	Organic acids, CO_2, mixed alcohols (souring, acidification)
Lipids	Butter	Hydrolysis, fatty acid degradation	Glycerol and mixed fatty acids (rancidity, bitterness)

1 Other reactions also occur during the spoilage of these substrates.

Just as foods nourish us, so too can they support the growth of microorganisms that can cause spoilage and make us sick. Microbial growth is controlled by factors related to the food itself, called intrinsic factors; and extrinsic factors, which include the environment in which the food is stored. Here we review the most important intrinsic and extrinsic factors involved in food spoilage.

Intrinsic Factors

Food composition is a critical intrinsic factor that influences microbial growth (**table 41.1**). If a food consists primarily of carbohydrates, microbial spoilage will most likely occur at some point in time. Carbohydrate-rich foods with high water activity including some fruits and vegetables first show spoilage by fungi. For instance, even slight bruising of the tomato skin that exposes exposing the interior results in rapid fungal growth. Molds also produce enzymes that weaken and penetrate the protective outer skin of fruits and vegetables. Once molds degrade carbohydrates within the skin, bacteria can colonize the food. Colonizing bacteria include those that cause soft rots, such as *Erwinia carotovora,* which produce hydrolytic enzymes such as pectinase. ◀ *Solutes affect osmosis and water activity (section 7.5)*

Molds also rapidly grow on other, denser carbohydrates such as grains and corn, particularly when these products are stored in moist conditions (**figure 41.1**). Contamination of grains by the ascomycete *Claviceps purpurea* causes ergotism, a toxic condition (*see figure 25.12*). Hallucinogenic alkaloids produced by this fungus can lead to altered behavior, abortion, and death if infected grains are eaten.

In contrast to carbohydrates, meat and dairy products are rich in proteins and fats that support bacterial growth. The production of short-chained fatty acids from fats renders butter rancid. The anaerobic breakdown of proteins is called **putrefaction.** It yields foul-smelling amine compounds such as cadaverine (imagine the origin of that name) and putrescine (**figure 41.2**). Unpasteurized milk is a good example of a complex food containing fats, protein, and carbohydrates. This leads to a succession of microbes during spoilage: Acid production by *Lactococcus lactis* subspecies *lactis* is followed by additional acid production associated with the growth of more acid-tolerant organisms such as *Lactobacillus.* Next, yeasts and molds become dominant and degrade the accumulated lactic acid,

Figure 41.1 Food Spoilage. Typical examples of fungal spoilage include (a) bread, most likely supporting the growth of *Penicillium* sp. (green hyphae) and other fungi, and (b) ear rot in corn, which can result in major economic losses.
(a) ©Imagebroker/Alamy Stock Photos; (b) ©Daleen Loest/iStock/Getty Images

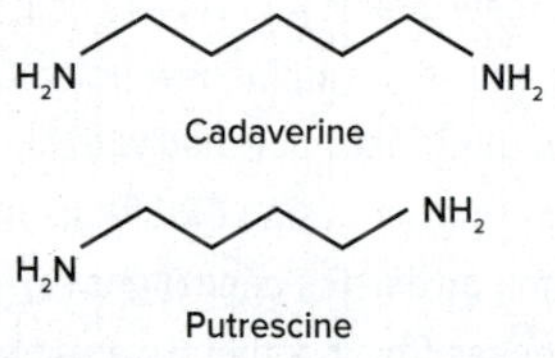

Figure 41.2 Cadaverine and Putrescine. These foul-smelling amines are the products of protein degradation.

Figure 41.3 Milk Spoilage. Fresh (left) and curdled (right) milk are shown. The spoilage process produces acids that denature and precipitate the milk protein casein to yield separated curds and whey.
©Donald Klein

and the acidity gradually decreases. Eventually protein-digesting bacteria become active, resulting in a putrid odor and bitter flavor. The milk, originally opaque, eventually becomes clear because the proteins and fats have coagulated (**figure 41.3**).

The pH and oxidation-reduction (redox) potential of a food also are critical intrinsic factors. A low pH favors the growth of yeasts and molds. In neutral or alkaline pH foods, such as meats, bacteria are more dominant in spoilage and putrefaction. Furthermore, when meat products are cooked, they present a reducing environment for microbial growth. These products, with their readily available amino acids, peptides, and growth factors, are ideal media for the growth of anaerobes, including clostridia.

The physical structure of a food also can affect the course and extent of spoilage. Grinding and mixing of foods such as sausage and hamburger increase the food surface area and distribute contaminating microorganisms throughout the food. This can result in rapid spoilage if such foods are stored improperly.

Many foods contain natural antimicrobial substances, including complex chemical inhibitors and enzymes. Coumarins found in fruits and vegetables have antimicrobial activity. Cow's milk and eggs also contain antimicrobial substances. Eggs are rich in the enzyme lysozyme, which lyses cell walls of contaminating Gram-positive bacteria (*see figure 32.4*). Herbs and spices, which were once the mainstay of food preservation, often possess significant antimicrobial substances; generally fungi are more sensitive than bacteria to these compounds. Sage and rosemary are two of the most antimicrobial spices. Aldehydic and phenolic compounds that inhibit microbial growth are found in cinnamon, mustard, and oregano. Other important inhibitors are garlic, which contains allicin; cloves, which have eugenol; and basil, which contains rosmarinic acid. Nonetheless, spices can sometimes contain pathogenic and spoilage organisms. Enteric bacteria, *Bacillus cereus, Clostridium perfringens,* and *Salmonella enterica* have been detected in spices.

Extrinsic Factors

Extrinsic factors are more easily controlled than intrinsic factors and are therefore frequently the basis for food preservation. Temperature is an important extrinsic factor in determining whether a food will spoil; indeed it is the basis of modern refrigeration. However, modulation of temperature during food processing can have a significant impact on shelf life. For instance, spoilage problems can occur with minimally processed, concentrated frozen citrus products. These were once prepared with little or no heat treatment, and Gram-positive bacteria and yeasts could grow upon thawing. Today most frozen concentrates are pasteurized (a heating process, discussed in section 41.2), filtered, and then concentrated by heating under vacuum. Vitamin C and flavors are added back to the product before packaging.

Humidity is also a key factor in controlling food spoilage. At higher relative humidities, microbial growth is initiated more rapidly, even at lower temperatures. When drier foods are placed in moist environments, moisture absorption occurs on the food surface, promoting microbial growth.

The atmosphere in which food is stored also is important. This is especially true with shrink-wrapped foods because many plastic films allow oxygen diffusion, which results in increased growth of surface-associated microorganisms. Excess CO_2 can decrease the solution pH, inhibiting microbial growth. Storing meat in a high CO_2 atmosphere inhibits Gram-negative bacteria, resulting in a population dominated by the lactobacilli.

Comprehension Check

1. What are some intrinsic factors that influence food spoilage and how do they exert their effects?
2. How would the spoilage of bread differ from that of a high-fat food like butter?
3. Why do sausage and other ground meat products provide a better environment for the growth of food spoilage organisms than solid cuts of meats?
4. What extrinsic factors can determine whether food spoilage will occur?

41.2 Various Methods Are Used to Control Food Spoilage

After reading this section, you should be able to:

a. Review the limitations of refrigeration
b. Explain why canning is not always a fail-safe means of preserving food
c. Compare and contrast two pasteurization methods
d. Describe how the principle of water availability can be used to preserve foods
e. Explain how commonly used chemicals preserve food
f. Defend an informed opinion on the safety and efficacy of food irradiation
g. Describe two microbe-based food preservatives
h. Examine the packaging of food you frequently consume and assess its capacity to prevent spoilage

Foods can be preserved by a variety of methods. The goal of each method is to eliminate or reduce the populations of spoilage and disease-causing microorganisms while maintaining food quality. We now briefly discuss some of these techniques.

Filtration

Microorganisms can be removed from water, wine, beer, juices, soft drinks, and other liquids by filtration. This can keep bacterial populations low or eliminate them entirely. Removal of large particulates by prefiltration and centrifugation maximizes filter life and effectiveness. Several major brands of beer are filtered rather than pasteurized to better preserve flavor and aroma.

Low Temperature

Refrigeration at 5°C retards microbial growth, although with extended storage, microorganisms eventually grow and produce spoilage. Of particular concern is the growth of *Listeria monocytogenes,* which can grow at temperatures used for refrigeration. It should be kept in mind that refrigeration slows the metabolic activity of most microbes, but it does not lead to significant decreases in existing microbial populations.

High Temperature

Controlling microbial populations in foods by means of high temperatures can significantly limit disease transmission and spoilage. Heating processes, first used by Nicholas Appert in 1809, provide a safe means of preserving foods, particularly when carried out in commercial canning operations (**figure 41.4**). Canned food is heated in special containers called retorts at about 115°C for intervals ranging from 25 to over 100 minutes. The precise time and temperature depend on the nature of the food. Sometimes canning does not kill all microorganisms but only those that will spoil the food, because growth of remaining bacteria is inhibited by an intrinsic property of the food, such as pH. After heat treatment, cans are cooled as rapidly as possible, usually with cold water. ◀ *Physical control methods alter microorganisms to make them nonviable (section 8.4)*

Despite efforts to eliminate spoilage microorganisms during canning, canned foods may become spoiled. This may be due to spoilage before canning, underprocessing during canning, and leakage of contaminated water through seams during cooling. Spoiled canned food is often recognized by changes in color, texture, odor, and taste (although sampling is not advisable).

Figure 41.4 Food Preparation for Canning. Microbial control is important in processing and preserving many foods. Here a worker is pouring peas into a large, clean vat during the preparation of vegetable soup. After preparation, the soup is transferred to cans. Each can is heated for a short period, sealed, processed at temperatures from 110° to 121°C in a canning retort to destroy spoilage microorganisms, and finally cooled.

MICRO INQUIRY *Canned products such as vegetable soup often contain salt. Other than serving as a flavoring, what is the purpose of salt in the canned product?*

Organic acids, sulfides, and gases (particularly CO_2 and H_2S) may be produced. If spoilage microorganisms produce gas, both ends of the can will bulge outward. Sometimes the swollen ends can be moved by thumb pressure (soft swells); in other cases, the gas pressure is so great that the ends cannot be dented by hand (hard swells). However, swelling is not always due to microbial spoilage; acid in low pH foods may react with iron in the can to release hydrogen and generate a hydrogen swell.

Pasteurization involves heating food to a temperature that kills disease-causing microorganisms and substantially reduces the levels of spoilage organisms. There are three types of pasteurization, which differ by temperature and length of time the product is heated (**table 41.2**). Shorter-term processing results in improved flavor and extended product shelf life. These require stainless steel heat exchangers specifically designed to rapidly heat and cool the liquid being pasteurized. The duration of pasteurization is based on the statistical probability that the number

Table 41.2 Methods of Pasteurization

Type of Pasteurization	Temperature and Duration of Heating	Products Pasteurized
Low-temperature holding (LTH)	68.2°C, 30 minutes	Beer, fruit juice, smaller volumes of milk
High-temperature short-time (HTST)	72°C, 15 seconds	Industry standard for milk and other dairy products in the United States
Ultra-high temperature (UHT)	138°C, 3 seconds	Milk that does not require refrigeration until opened

of remaining viable microorganisms will be below a certain level after a particular heating time at a specific temperature. These calculations are discussed in detail in section 8.4. *Food Pathogens and Temperature*

Water Availability

The presence and availability of water also affect the ability of microorganisms to colonize foods. Simply by drying a food, one can slow or prevent spoilage processes; this is commonly seen in dried fruit, which does not require refrigeration. Water can be made less available by adding solutes such as sugar and salt. Water availability is measured in terms of **water activity** (a_w). This represents the ratio of relative humidity of the air over a test solution (or food, in this case) compared with that of distilled water, which has an a_w of 1. When large quantities of salt or sugar are added to food, most microorganisms are dehydrated by the hypertonic conditions and cannot grow. However, even under these adverse conditions, osmophilic and xerophilic microorganisms may spoil food. In contrast to most spoilage bacteria that thrive at an a_w of about 0.9, these microbes grow best at an a_w of 0.6 to 0.7. Osmophilic (Greek *osmus,* impulse; *philein,* to love) microbes grow best in or on media with a high osmotic concentration (e.g., jams and jellies), whereas xerophilic (Greek *xerosis,* dry; *philein,* to love) microorganisms thrive in low a_w environments (e.g., cereals) and may not grow under high a_w conditions. Dehydration, such as lyophilization to produce freeze-dried foods, is a common means of preventing microbial growth. Freeze-drying increases the osmolarity of the product while at the same time reducing the a_w so microbial growth is restricted.

Chemical-Based Preservation

Various chemical agents can be used to preserve foods; these substances are closely regulated by the U.S. Food and Drug Administration (FDA) and are listed as being "*g*enerally *r*ecognized *a*s *s*afe" or **GRAS** (**table 41.3**). They include simple organic acids, sulfite, ethylene oxide as a gas sterilant, sodium nitrite, and ethyl formate. These chemical agents damage the microbial plasma membrane or denature various cell proteins. Other compounds interfere with the functioning of nucleic acids, thereby inhibiting cell reproduction.

Sodium nitrite is an important chemical used to help preserve ham, sausage, bacon, and other cured meats by inhibiting the growth of *Clostridium botulinum* and the germination of its endospores. This protects against botulism and reduces the rate of spoilage. Besides increasing meat safety, nitrite decomposes to nitric acid, which reacts with heme pigments to keep the meat red in color. However, there is concern about nitrite because it can react with amines to form carcinogenic nitrosamines. *Botulism (section 39.4)*

Low pH can also be used to hinder microbial spoilage. For example, acetic and lactic acids inhibit growth of *Listeria monocytogenes.* Organic acids (1–3%) can be used to treat meat carcasses, and poultry can be cleansed with 10% lactic acid/sodium lactate buffer (pH 3) prior to packaging. In addition, low pH can increase the activity of other chemical preservatives. Sodium propionate is most effective at lower pH values, where it is primarily undissociated. Breads, with their low pH values, often contain sodium propionate as a preservative. The use of citric acid is also common. *Organic acids (section 42.1)*

Table 41.3 Major Groups of Chemicals Used in Food Preservation

Preservatives	Approximate Maximum Use	Organisms Affected	Foods
Benzoic acid/benzoates	0.1%	Yeasts and molds	Margarine, pickle relishes, apple cider, soft drinks, tomato ketchup, salad dressings
Dehydroacetic acid	65 ppm	Insects	Pesticide on strawberries, squash
Ethylene/propylene oxides	700 ppm	Yeasts, molds, vermin	Fumigant for spices, nuts
Ethyl formate	15–200 ppm	Yeasts and molds	Dried fruits, nuts
Parabens[1]	0.1%	Yeasts and molds	Bakery products, soft drinks, pickles, salad dressings
Propionic acid/propionates	0.32%	Molds	Bread, cakes, some cheeses; inhibitor of ropy bread dough
Sodium diacetate	0.32%	Molds	Bread
Sodium nitrite	120 ppm	Clostridia	Cold cuts, hot dogs, sausages
Sorbic acid/sorbates	0.2%	Molds	Hard cheeses, figs, syrups, salad dressings, jellies, cakes
SO_2/sulfites	200–300 ppm	Insects and microorganisms	Molasses, dried fruits, wine, lemon juice (not used in meats or other foods recognized as sources of thiamine)

Source: Jay, James M. *Modern Food Microbiology*, 6th ed. New York: Springer, 2000.

1 *Methyl, propyl, and heptyl esters of p-hydroxybenzoic acid.*

High Hydrostatic Pressure

In the last decade, the use of high hydrostatic pressure (HHP), also called pascalization, has grown as an alternative way to preserve food. HHP is the application of pressures from 100 to 800 millipascals (mPa) without significant changes in temperature. Pressures typically range from 350 to 600 mPa for 5 to 15 minutes at temperatures from 5° to 35°C. HHP is most detrimental to cell membranes; thus eukaryotic microbes are more susceptible. Predictably, Gram-positive bacteria, particularly endospores, are more resistant to HHP than Gram-negative bacteria. At present, there is no industry standard for HHP conditions considered effective for various foods and their microbial contaminants.

Radiation

Although the public is wary of food radiation, it is an effective way to prevent food spoilage. Unfortunately, there is the common misperception that food becomes radioactive upon irradiation. This would be similar to thinking that you become radioactive following a chest or dental X ray. ◀ *Radiation (section 8.4)*

Gamma irradiation of food most often uses a cobalt-60 source; however, cesium-137 is employed in some facilities. Gamma radiation has excellent penetrating power but must be used with moist foods because radiation is effective only if it generates reactive oxygen species from water, resulting in oxidation of nucleic acids, lipids, and proteins. This process of **radappertization,** named after Nicholas Appert, can extend the shelf life of meats, seafoods, fruits, and vegetables. Unlike canning, radappertization does not heat the food product.

Microbial Product–Based Inhibition

Interest is increasing in the use of bacteriocins for the preservation of foods. **Bacteriocins** are proteins secreted by certain bacteria that kill closely related microbes by binding to specific sites on the target cell. They often affect plasma membrane integrity and function. Nisin, a small amphiphilic peptide produced by some strains of *Lactococcus lactis,* is the only widely used bacteriocin currently employed in food preservation. It is nontoxic to humans and affects Gram-positive bacteria by binding to lipid II during peptidoglycan synthesis and forming pores in the plasma membrane (*see figure 12.10*). Nisin was added to the GRAS list of food additives in the United States about 40 years ago. This means that it can be added directly to foods such as packaged meats, cheese, eggs, and canned vegetables. Nisin is effective in low-acid foods, where it improves inactivation of *Clostridium botulinum* during the canning process and inhibits germination of any surviving endospores. ◀ *Bacteriocins (section 32.3)*

Interestingly, a preparation of strains of lytic **bacteriophages** that specifically infect *Listeria monocytogenes* is used to spray ready-to-eat meats such as hot dogs and lunch meats. The spray is applied to the surface of the meats prior to packaging. A mixture of different phages are present; using multiple phage types significantly reduces the development of phage-resistant strains.

Packaging

Gases present in the atmosphere within a food storage package can have a significant affect on microbial growth. This has led to the development of **modified atmosphere packaging** (**MAP**). Modern shrink-wrap materials and vacuum technology make it possible to package foods with controlled atmospheres. These materials are largely impermeable to oxygen. This prolongs shelf life by a factor of two to five times compared to the same product packaged in air. With a CO_2 content of 60% or greater in the atmosphere surrounding a food, spoilage fungi will not grow, even if low levels of oxygen are present. High-oxygen MAP is also effective because it triggers the formation of superoxide anions within microbial cells. Some products currently packaged using MAP technology include delicatessen meats and cheeses, pizza, grated cheese, some bakery items, and dried products such as coffee.

Comprehension Check

1. Construct a table that lists the chemical, physical, and biological approaches to food preservation.
2. What types of chemicals can be used to preserve foods?
3. What cellular component does HHP target? What is radappertization?
4. How does nisin function? What bacterial genus produces this important peptide?
5. Consider the fact that *Listeria* phages must be propagated in the pathogen prior to producing an anti-*Listeria* spray. What safeguards do you think must be instituted to ensure the safety of the spray?
6. What are the major gases involved in MAP? How are their concentrations varied to inhibit microbial growth?

41.3 Food-Borne Disease Outbreaks

After reading this section, you should be able to:

a. Summarize a major food-borne infection that had substantial health and economic consequences
b. Describe the origin and consequences of a major food intoxication event in the United States
c. Explain why these cases of food-borne disease outbreaks serve as cautionary tales for government regulatory agencies, food producers, and consumers
d. Contrast and compare aflatoxins and fumonisins

Most food-borne illnesses are neither recognized nor reported. The source and the identity of a food pathogen are only discovered if it causes an illness dire enough that the victim visits a physician or hospital. That said, the Centers for Disease

Control and Prevention (CDC) estimates there are about 48 million cases of food-related illness annually in the United States, 128,000 hospitalizations, and about 3,000 deaths. Of these, only 14 million can be attributed to known pathogens. *Salmonella enterica* serovars, noroviruses, *Campylobacter jejuni,* strains of enterohemorrhagic *Escherichia coli* (e.g., strain O157:H7), and *Listeria monocytogenes* are major causes of food-borne diseases.

All food-borne illnesses are transmitted by the fecal-oral route. The "5-Fs"—feces, fingers, food, fomites, and flies—are critical for fecal-oral transmission. In this context, fomites often include faucets, drinking cups, and cutting boards. Individual food-borne diseases are covered in sections 38.4, 39.4, and 40.5; here we review several major outbreaks of food-borne diseases that illustrate the personal, economic, and societal impacts of tainted food. We divide our coverage into the two primary types of food-related diseases: food-borne infections and food intoxications.

Food-Borne Infection

©Noam Armonn/Alamy Stock Photos

A **food-borne infection** involves ingestion of the pathogen, followed by its growth in the host, including tissue invasion or the release of toxins. The most common microbes involved are summarized in **table 41.4** (*also see table 39.3*).

Listeriosis, caused by *Listeria monocytogenes,* was responsible for one of the largest meat recalls in U.S. history. In 2002 a seven-state listeriosis outbreak was linked to deli meats and hot dogs produced at a single meat-processing plant in Pennsylvania. Pregnant women, the young and the old, and immunocompromised individuals are especially vulnerable to *L. monocytogenes* infections. In this outbreak, 7 deaths, 3 stillbirths, and 46 illnesses were caused by consumption of contaminated meats. Microbiologists matched the strain of *L. monocytogenes*

Table 41.4 Some Food-Borne Bacteria That Cause Acute Vomiting and Diarrhea

Organism	Incubation Period (Hours)	Vomiting	Diarrhea	Fever	Food Source
Staphylococcus aureus	1–6	+++[1]	+	Rare	Meats, cold cuts, and salads (e.g., tuna, egg, pasta), cream-filled bakery goods
Bacillus cereus	1–6 (intoxication); 8–16 (infection)	+++	++	–[2]	Reheated fried rice, sprouts, cucumber
Clostridium perfringens	8–16	±[3]	+++	–	Rewarmed meat dishes
Clostridium botulinum	18–24	±	Rare	–	Canned goods contaminated during processing or packaging, raw honey
Escherichia coli (enterohemorrhagic)	3–5 days	±	++	±	Undercooked hamburger, unpasteurized fruit juices, raw vegetables
Escherichia coli (enterotoxigenic strain)	16–72	±	++	–	Contaminated drinking water; major cause of traveler's diarrhea.
Vibrio parahaemolyticus	6–96	+	++	±	Shellfish, particularly clams and oysters
Vibrio cholerae	24–72	+	+++	–	Contaminated water; shellfish
Shigella spp.	24–72	±	++	+	Salads, sandwiches, produce, contaminated water
Salmonella spp. (gastroenteritis)	8–48	±	++	+	Poultry, eggs, meats, dairy, produce, peanut butter
Salmonella enterica serovar Typhi (typhoid fever)	10–14 days	±	±	++	Spread from a healthy carrier to food via fecal-oral transmission.
Campylobacter jejuni	2–10 days	–	+++	++	Poultry, produce, raw milk
Yersinia enterocolitica	3–14 days	±	+	+	Pork, milk

Source: Brooks, Geo. F. et al., *Medical Microbiology,* 21st ed. Norwalk, CT: Appleton & Lange, 1998.

1 + *indicates condition is present, number of symbols indicates severity.*

2 – *indicates condition is absent.*

3 ± *indicates condition sometimes occurs.*

found in the contaminated food products with samples obtained from floor drains in a Wampler, Inc., packaging plant. This prompted the recall of 27.4 million pounds of meats that had been distributed over a 5-month period to stores, restaurants, and school lunch programs. Following the outbreak, the plant closed for a month and the Wampler brand name was phased out. This episode prompted the U.S. Department of Agriculture (USDA) to step up its environmental testing program for *L. monocytogenes* so that it now tests plants that do not regularly submit data to the USDA. It also performs surprise inspections of those that do. Physicians now routinely advise people at risk of contracting listeriosis to avoid eating soft cheeses (e.g., feta, Brie, Camembert), refrigerated smoked meats such as lox, as well as deli meats and undercooked hot dogs. In 2006 the Wampler plant in Pennsylvania was closed, having never recovered from the $100 million cost and the damage to its reputation caused by the 2002 outbreak. Thus a single outbreak resulted in death, illness, bankruptcy, changes to food safety inspection policy, and routine preventative medicine to protect at-risk patients from listeriosis.

Outbreaks involving the enterohemorrhagic microbe *E. coli* O157:H7 (EHEC) also illustrate the impact food infections can have. *E. coli* O157:H7 produces a toxin known as Shiga-like toxin, which is responsible for the potentially life-threatening complication hemolytic uremic syndrome, most commonly seen in children and the elderly. Although EHEC produces a toxin, the disease it causes is not an intoxication because the bacterium must be present to produce the toxin in the host. The first U.S. outbreaks of *E. coli* O157:H7 (in the 1980s) occurred in individuals who had consumed undercooked hamburgers. This microbe inhabits the intestines of a variety of mammals, and ground beef contaminated with cattle feces during slaughter and processing continues to be a common cause of these outbreaks. ◀ Escherichia coli *gastroenteritis (section 39.4)*

More recently other types of foods have been found to be contaminated with *E. coli* O157:H7. One example is minimally processed, ready-to-eat fresh salad greens, which have become a popular convenience and represent a multibillion-dollar business in the United States and Europe. The produce is repeatedly washed, then dried and packed into plastic bags. When transported, the produce must be kept cool. Unfortunately, this healthy approach to consuming vegetables is not always without risk. In the past decade, over a dozen outbreaks of enterohemorrhagic *E. coli* O157:H7 were traced to minimally processed produce. One of the largest and most well publicized was a spinach-associated outbreak in 2006, which sickened more than 200 people. This outbreak prompted intense investigation, and it was determined that contamination occurred while the spinach was still in the field. Surveillance studies determined that the most likely source was feral swine that burrowed under fences, traveled across the spinach fields, and contaminated irrigation water with their feces. This finding raises more questions than it answers, including: How can thousands of agricultural fields be protected from wild animals? Why didn't the postharvest washing procedure cleanse the spinach of *E. coli*? What factors allowed the initial bacterial inoculum on the plants to grow by at least a factor of 10? This single event prompted efforts by both government and industry to ensure that minimally processed produce remains a convenience, not a sporadic source of disease.

Food Intoxication

Microbial growth in food products can result in **food intoxication** (table 41.4). Intoxication produces symptoms shortly after the food is consumed because growth of the disease-causing microorganism is not required. The classic example is picnickers vomiting 4 to 6 hours after consuming salad contaminated with *Staphylococcus aureus* enterotoxin. The consumption of *S. aureus* cells is not necessary, as toxins produced in the food are sufficient to cause illness.

In addition to *S. aureus,* three Gram-positive, endospore-forming rods are also important sources of food intoxications and infections: *Clostridium botulinum, C. perfringens,* and *Bacillus cereus*. *C. botulinum* intoxication is discussed in chapter 39. There are typically 30 or fewer cases of botulism reported each year in the United States, and most are associated with home canning. *C. perfringens* typically causes a 24-hour infection. It has a relatively high infectious dose (10^8 cells) and results in abdominal cramps and diarrhea 8 to 22 hours after consumption. *B. cereus* causes two types of intoxication, each the result of a different toxin. Food poisoning following ingestion of *B. cereus*–contaminated meats, fish, or vegetables can cause a diarrheal illness that is very similar to that of *C. perfringens*. *B. cereus* intoxication, characterized by acute (less than 24 hours) vomiting, is generally associated with consumption of contaminated starchy foods such as rice or pastries.

Some fungus-derived toxins have more insidious health effects. The most potent are the fungal carcinogens aflatoxins and fumonisins. **Aflatoxins** are produced by *Aspergillus flavus* and *A. parasiticus*. These fungi grow and release toxin in moist grains and nuts. Aflatoxin consumption is most common in tropical countries where these products are stored in suboptimal conditions and are consumed with little or no post-harvest processing. Aflatoxins are planar, ringed compounds that wedge between strands of DNA (intercalate), thereby causing frameshift mutations that can cause cancer. This occurs primarily in the liver, where they are converted to unstable derivatives. About 20 different aflatoxins are known; the most important are shown in **figure 41.5**. Of these, aflatoxin B_1 is the most common and the most potent carcinogen. After ingestion by lactating animals (e.g., dairy cows), aflatoxins B_1 and B_2 are modified in the liver to yield the aflatoxins M_1 and M_2. If cattle consume aflatoxin-contaminated feeds, aflatoxins can contaminate milk and dairy products. Besides their importance in grains, aflatoxins have also been found in beer, cocoa, raisins, peanut butter, and soybean meal.

Diet appears to be related to aflatoxin exposure: The average aflatoxin intake in the typical European-style diet is 19 ng/day, whereas for some Asian diets it is estimated to be 103 ng/day. Aflatoxin sensitivity also can be influenced by prior disease exposure. Individuals who have had hepatitis B have a 30-fold higher risk of liver cancer upon exposure to aflatoxins than individuals who have never had this disease. It has been observed

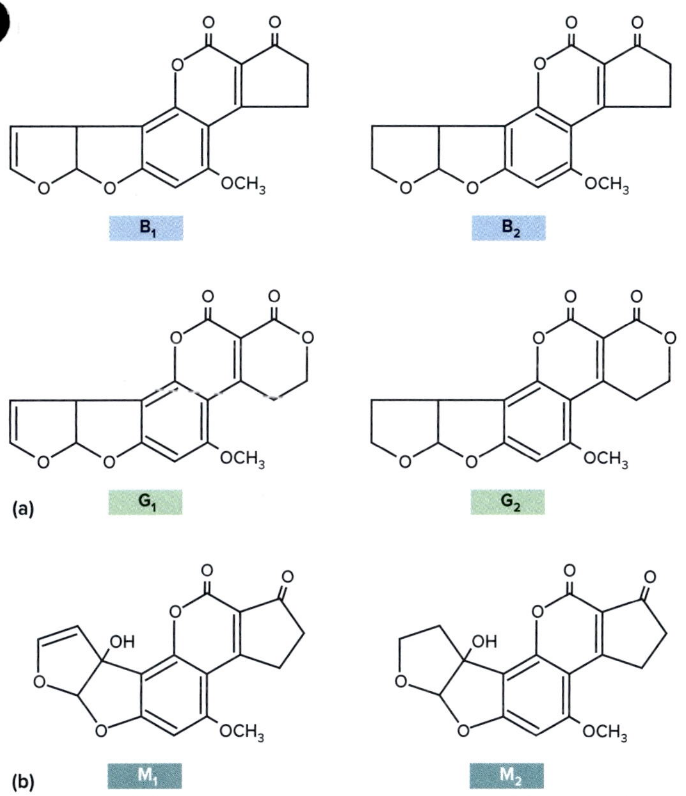

Figure 41.5 Aflatoxins. When *Aspergillus flavus* and related fungi grow on foods, carcinogenic aflatoxins can be formed. These have four basic structures. (a) The letter designations refer to the color of the compounds under ultraviolet light after extraction from the grain and separation by chromatography. The B_1 and B_2 compounds fluoresce with a blue color, and the G_1 and G_2 appear green. (b) The two type M aflatoxins are found in the milk of lactating animals that have ingested type B aflatoxins.

MICRO INQUIRY *What foods are most prone to aflatoxin contamination?*

that prevention of hepatitis B infections by vaccination and reduction of carrier populations help control the potential effects of aflatoxins.

The **fumonisins** are fungal contaminants of corn that were first isolated in 1988. They are produced by *Fusarium moniliforme* and cause leukoencephalomalacia in horses (also called blind staggers, which is fatal within 2 to 3 days), pulmonary edema in pigs, and esophageal cancer in humans. Fumonisins inhibit ceramide synthase, a key enzyme for the proper use of fatty molecules in the cell. This disrupts synthesis and metabolism of sphingolipids, important compounds that influence a wide variety of cell functions. At least 10 different fumonisins exist; the basic structures of fumonisins FB1 and FB2 are shown in **figure 41.6**. Corn and corn-based feeds and foods, including cornmeal and corn grits, can be contaminated. Thus it is extremely important to store corn

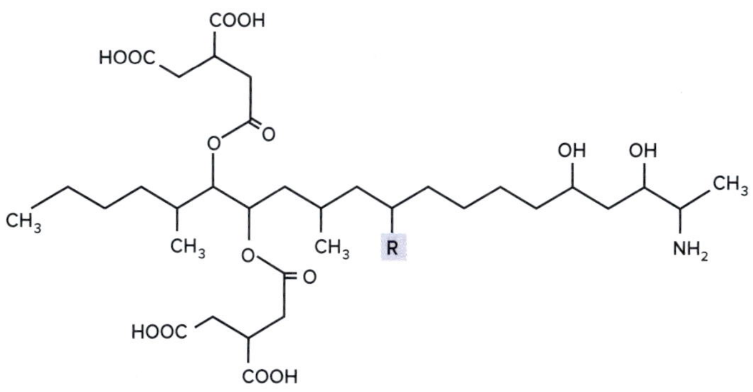

Figure 41.6 Fumonisin Structure. The basic structure of fumonisins FB1 and FB2 produced by *Fusarium moniliforme,* a fungal contaminant that can grow in improperly stored corn. A total of at least 10 different fumonisins have been isolated. These are strongly polar compounds that cause diseases in domestic animals and humans. FB1, R = OH; FB2, R = H.

and corn products under dry conditions, where these fungi cannot develop.

Other eukaryotic microorganisms also synthesize potent toxins. For example, algal toxins contaminate fish and thereby affect the health of marine animals higher in the food chain; they also contaminate shellfish and finfish, which are later consumed by humans. These toxins are produced during harmful algal blooms, as discussed in section 30.2.

Comprehension Check

1. What is the difference between a food-borne infection and a food intoxication? Which do you think is harder to prevent? Explain your answer.
2. What practical recommendations regarding food preparation could you make to someone concerned about *E. coli* O157:H7 contamination?
3. Aflatoxins are produced by members of which fungal genus? How do they damage animals that eat the contaminated food?
4. Members of what microbial genus produce fumonisins and why are these compounds of concern? If improperly stored, what are the major foods and feeds in which these chemicals might be found?

41.4 Detection of Food-Borne Pathogens Requires Government-Industry Cooperation

After reading this section, you should be able to:

a. List the principles that guide the development of food testing strategies
b. Describe how a food-borne pathogen consumed by an individual might be traced back to the food source

In 1905 Upton Sinclair's novel *The Jungle* exposed the horrific state of the meat packaging industry. The following year, the first in a series of federal and state laws was enacted in the United States aimed at improving food safety. However, between 1938 and 2001, when the FDA Food Safety Modernization Act was passed, there was no major legislation addressing this issue. Importantly, the goal of the 2001 act was to change the focus of federal regulators from responding to food contamination to preventing it. But while we may look to the government to keep food safe, food safety is a shared responsibility, as precautions must be defined and in place during production, transport, sale, preparation, and consumption of any food item. The length and complexity of the chain from producer to consumer have increased dramatically in the last 30 years with the rise in food imports and exports, making food safety efforts even more complicated. Here we provide a broad overview of food pathogen detection.

It is impossible to test each food item for every potential pathogen. Apart from the logistical implications, such an approach is economically untenable. For example, if the U.S. dairy industry were to test all dairy products daily at each farm with a $5-per-sample assay, it would cost about $150 million annually. Checking each milk tanker truck as it enters a processing plant reduces the cost to about $21 million. So the challenge in developing a testing strategy involves identifying which pathogens to detect, as well as when and where to detect them.

Several principles guide the development of the technologies, protocols, and policies used to keep foods safe. These include (1) specificity and sensitivity of the method for the given pathogen—any given test should rarely yield false negatives or positives (specificity) and the pathogen in question should be detected even in low concentrations (sensitivity); (2) speed, which is particularly important for fresh produce awaiting distribution; and (3) simplicity—that is, foods should be tested with limited sample preparation. Furthermore, the goal of each detection strategy should be clearly defined. Ideally "testing to prevent," that is, to confirm food is safe before it leaves the farm or processing plant, is the goal. When this is unreasonable due to logistics and cost, the next best approach is "testing to protect," which involves analysis before the food is accessible for consumption. These strategies are designed to avoid "testing to recover," when an outbreak has occurred and the origin of the contaminated food must be identified.

Growth of contaminating food-borne pathogens on selective media has long been the bedrock of the food safety lab. Once an organism has been isolated, a variety of methods can be used to verify its identity. For example, cultured microbes are commonly subject to biochemical tests, which may take up to a week to complete. These approaches are now being replaced by newer, molecular methods with excellent sensitivity, specificity, and turnaround time. While some molecular techniques still rely on culturing the pathogen first, many others are culture independent.

In general, three types of molecular methods are used in testing food: nucleic acid–based methods, immunological methods, and biosensors. Nucleic acid–based methods include end-point collection PCR, real-time PCR, multiplex PCR, and whole-genome sequencing (WGS). For example, multiplex PCR can be used to screen a food for several pathogens at once using primers specific for each microorganism of interest, or to amplify multiple genes from a single microbe for the application of multilocus sequence analysis or single nucleotide polymorphism analysis (*see figure 19.3*). WGS is now used in routine surveillance of certain food-borne pathogens. WGS does more than simply acknowledge the presence or absence of a pathogen; it enables analysis of virulence and antibiotic resistance evolution. Immunoassays include enzyme-linked immunosorbant assays (ELISA; *see figure 37.11*) and lateral flow assays that are simple, inexpensive, and importantly can be done in the field. The use of immunomagnetic beads is a more rapid approach than enrichment culture to separate a suspected pathogen from a food item. Immunomagnetic beads consist of a pathogen-specific antibody bound to a magnetic particle. Once the antibody has bound its target microbe, a magnetic field is applied, thereby separating the magnet-bound pathogen from other materials. Biosensors couple a biological material (e.g., cell receptor, enzyme, synthetic peptide) that detects a pathogen to a transducer that provides an optical, electrical, magnetic, or mechanical readout. Biosensors are valuable in food safety testing because they provide real-time pathogen detection and are typically portable so they can be used in the field. ◀ *Polymerase chain reaction amplifies targeted DNA (section 17.2); Next-generation genomic sequencing (section 18.2); Microbial taxonomy and phylogeny are largely based on molecular characterization (section 19.3); Culture-based methods (section 37.2)*

The CDC oversees a program called **PulseNet** for the early detection of food-borne pathogens, particularly during an outbreak. The basis for this program is the use of pulsed-field gel electrophoresis (PFGE) under carefully controlled and duplicated conditions. PFGE generates a strain-specific DNA electrophoresis pattern, or fingerprint, for each bacterial pathogen (**figure 41.7**). PFGE is similar to standard DNA gel electrophoresis but is capable of separating very large fragments of DNA. Because PFGE can resolve large (Mb) pieces of DNA, whole genomes are cut with restriction enzymes that cleave in only a few places to generate a DNA fingerprint. The patterns obtained for each strain of food-borne pathogen are found in a national database maintained by the CDC. This enables strain-specific identification of new isolates through comparison with the database. ◀ *Restriction enzymes (section 17.1); Gel electrophoresis (Techniques & Applications 17.1)*

Although the PulseNet system is still in use, the advent of inexpensive WGS through next-generation sequencing techniques is poised to replace it. The goal is to use WGS to provide precise and timely food safety information. In an outbreak,

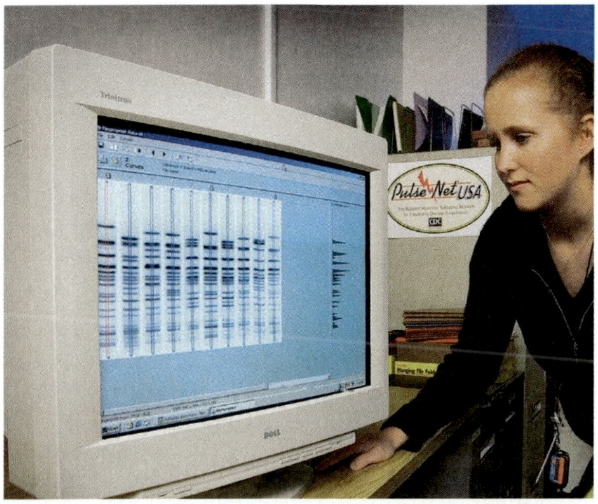

Figure 41.7 PulseNet. This network operates throughout the United States to identify microbes causing food-borne disease using pulsed-field gel electrophoresis of genomes extracted from contaminating microbes. Shown here, a food microbiologist examines the electrophoretic pattern, or fingerprint, generated by a bacterial pathogen.

MICRO INQUIRY *Why is it important that PulseNet coordinate the identities of food-borne pathogens on a national level?*
Source: CDC

WGS outpaces PulseNet in high-resolution subtyping of the causative organism and the evolution of antibiotic resistance genes, which can help guide treatment of the affected population. The 2011 outbreak of a novel *E. coli* pathogen in Europe that sickened more than 4,000 people and killed 50 individuals in 16 countries provides an excellent example of the application of WGS to a food-borne outbreak. In less than a month, the entire genomes of several *E. coli* O104:H4 strains recovered from patients were sequenced. It was determined that this was a novel strain and with an unusual constellation of virulence factors from both enterohemorrhagic and enteroaggregative *E. coli* strains. This enabled the origin of the outbreak to be traced to contaminated bean sprouts. ◀ *Next-generation genomic sequencing (section 18.2);* Escherichia coli *gastroenteritis (section 39.4)*

Comprehension Check

1. Consider the food supply chain for an imported vegetable. When and where do you think the vegetable should be screened for pathogens? Explain.
2. What are the three molecular approaches now used in food safety testing? Which can be used in the field?
3. You are on a team investigating an outbreak of *Salmonella enterica* food poisoning. You have PulseNet and WGS information from several patients and several contaminated foods. What questions can be answered with these data?

41.5 Microbiology of Fermented Foods: Beer, Cheese, and Much More

After reading this section, you should be able to:

a. Compare and contrast mesophilic and thermophilic fermentations
b. Differentiate between a yeast-lactic and a mold-lactic fermentation
c. Outline the process by which cheese is made
d. List some fermented meats
e. Provide an overview of the production of wine, champagne, beer, and distilled spirits
f. Describe a fermented food item not commonly eaten in the United States

Fermentation has been used in food production for thousands of years; the earliest records date as far back as 6,000 BCE. Microbial growth, either of natural or inoculated populations, causes chemical or textural changes to form a product that can be stored for extended periods. The fermentation process also is used to create pleasing food flavors and odors—such as cheese, yogurt, and chocolate (**Techniques & Applications 41.1**). Here we discuss the microbiology of the most common fermented foods. ◀ *Fermentation does not involve an electron transport chain (section 11.8)*

Fermented Milks

Although we are most familiar with yogurt, at least 400 different fermented milks are produced around the world. The majority of fermented milk products rely on **lactic acid bacteria** (**LAB**), which include species belonging to the genera *Lactobacillus, Lactococcus, Leuconostoc,* and *Streptococcus* (**figure 41.8**). These firmicutes tolerate acidic conditions, are nonsporing, and are aerotolerant with a strictly fermentative metabolism. ◀ *Order Lactobacillales (section 23.2)*

Regardless of the type of milk product, **starter cultures** of LAB are added to begin fermentation. For example, *Lactococcus lactis* and *Streptococcus cremoris* are used in cheese production and various *Leuconostoc* species are used to start buttermilk fermentation. A vexing problem for the dairy industry is the presence of bacteriophages that destroy starter cultures. Lactic acid production from a heavily phage-infected starter culture can cease within 30 minutes. Overcoming this problem has revealed that many reliable starter cultures consist of bacteria that prevent phage infection by a variety of mechanisms, such as preventing adsorption or phage DNA entry, the activity of restriction modification, or the use of a CRISPR/Cas system. Nonetheless, various phage control measures are available, and still more are in development. ◀ Responses to viral infection (section 14.6); *Restriction enzymes (section 17.1)*

Mesophilic Fermentations

Mesophilic milk fermentation results in the production of buttermilk and sour cream. This type of fermentation relies on the

TECHNIQUES & APPLICATIONS

41.1 Chocolate: The Sweet Side of Fermentation

Chocolate has been called the "world's favorite food," but few people realize that fermentation is an essential part of chocolate production. The Aztecs were the first to develop chocolate fermentation, serving a chocolate drink made from the seeds of the chocolate tree, *Theobroma cacao* (Greek *theos,* god; *broma,* food, or "food of the gods"). Chocolate trees now grow in Africa as well as South America.

The process of chocolate fermentation has changed very little over the past 500 years. Each tree produces large pods that each hold 30 to 40 seeds in a sticky pulp (**box figure**). Ripe pods are harvested and slashed open to release the pulp and seeds. The sooner the fermentation begins, the better the product, so fermentation occurs on the farm where the trees are grown. The seeds and pulp are placed in "sweat boxes" or in heaps in the ground and covered, usually with banana leaves.

Like most fermentations, this process involves a succession of microbes. First, a community of yeasts, including *Candida rugosa* and *Kluyveromyces marxianus,* hydrolyzes the pectin that covers the seeds and ferments sugars to release ethyl alcohol and CO_2. As the temperature and the alcohol concentration increase, the yeasts are inhibited and lactic acid bacteria increase in number. Lactic acid production drives the pH down, encouraging the growth of bacteria that produce acetic acid as a fermentation end product. Acetic acid is critical to the production of fine chocolate because it kills the sprout inside the seed and releases enzymes that further degrade proteins and carbohydrates, contributing to the overall taste of the chocolate. In addition, acetate esters, derived from acetic acid, are important for the development of good flavor. Fermentation takes 5 to 7 days. An experienced cocoa grower knows when fermentation is complete: If it is stopped too soon, the chocolate will be bitter and astringent; if it lasts too long, microbes start growing on the seeds instead of in the pulp. "Off tastes" arise when Gram-positive bacteria in the genus *Bacillus* and filamentous fungi belonging to the genera *Aspergillus, Penicillium,* and *Mucor* hydrolyze lipids in the seeds to release short-chain fatty acids. As the pH begins to rise, bacteria in the genera *Pseudomonas, Enterobacter,* and *Escherichia* also contribute to bad tastes and odor.

After fermentation, the seeds, now called beans, are spread out to dry. The dried beans are brown and lack pulp. They are bagged and sold to chocolate manufacturers, who first roast the beans to further reduce the bitter taste and kill most of the microbes. The beans are then ground and the nibs—the inner part of each bean—are removed. The nibs are crushed into a thick paste called a chocolate liquor, which contains cocoa solids and cocoa butter but no alcohol. Cocoa solids are brown and have a rich flavor, and cocoa butter has a high fat content and is off-white in color. The two components are separated, and the cocoa solids can be sold as cocoa, while the cocoa butter is used to make white chocolate or sold to cosmetics companies for use in lipsticks and lotions. However, the bulk of these two components will be used to make chocolate. The cocoa solids and butter are reunited in controlled ratios, and sugar, vanilla, and other flavors are added. The better the fermentation, the less sugar needs to be added (and the more expensive the chocolate).

The final product, delicious chocolate, is a combination of over 300 different chemical compounds. This mixture is so complex that no one has yet been able to make synthetic chocolate that can compete with the natural fermented plant. Microbiologists and food scientists are studying the fermentation process to determine the role of each microbe. But like chemists, they have had little luck in replicating the complex, imprecise fermentation that occurs on cocoa farms. In fact, the finest, most expensive chocolate starts as cocoa on farms where the details of fermentation have been handed down through generations. Chocolate production is truly an art as well as a science, while eating it is simply divine.

(a)

(b)

Cocoa Fermentation. (a) Cocoa pods growing on the cocoa tree. Each pod is 13 to 15 cm in length and contains 30 to 40 seeds in a sticky white pulp. (b) Seeds and pulp are fermented in large heaps for 5 to 7 days and then dried in the sun. Chocolate cannot be produced without fermentation.

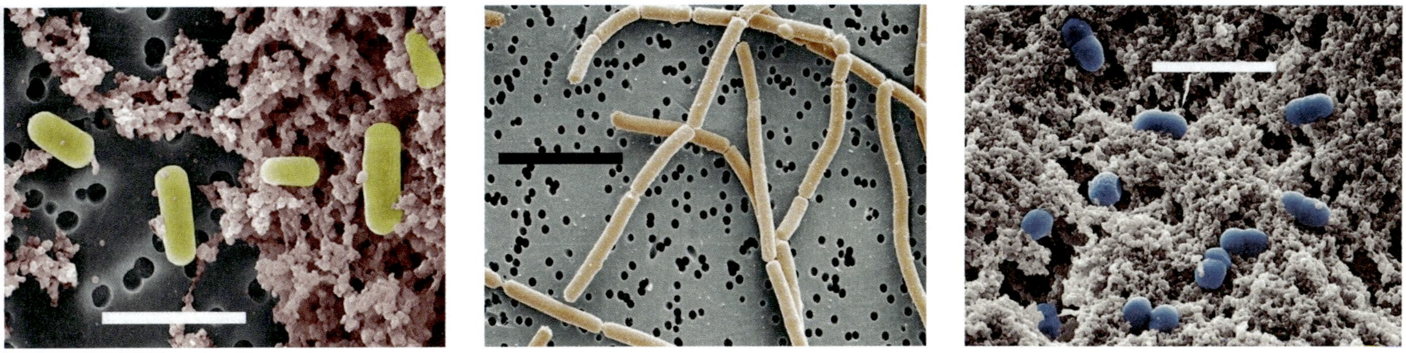

(a) (b) (c)

Figure 41.8 Lactic Acid Bacteria (LAB). Colorized scanning electron micrographs of LAB used as starter cultures. (a) *Lactobacillus helveticus*. (b) *Lactobacillus delbrueckii* subspecies *bulgaricus*. (c) *Lactococcus lactis*. The bacteria are supported by filters, seen as holes in the background. Scale bar = 5 μm.

MICRO INQUIRY *Can you name at least two features that make these bacteria well suited for milk fermentations?*

acidic products of fermentation to denature milk protein. To carry out the process, milk is typically inoculated with the desired starter culture and is then incubated at an optimum growth temperature (approximately 20° to 30°C). Microbial growth is stopped by cooling, and *Lactobacillus* spp. and *Lactococcus lactis* cultures are used for aroma and acid production. *Lactococcus lactis* subspecies *diacetylactis* converts CO_2 and milk citrate to diacetyl, which gives a richer, buttery flavor to the finished product.

Thermophilic Fermentations

Despite the name, thermophilic fermentations are not carried out at extremely hot temperatures, rather they are conducted around 45°C. An important example is yogurt production. In commercial production, milk is pasteurized, cooled to 43°C or lower, and inoculated with a 1:1 ratio of *Streptococcus salivarius* subspecies *thermophilus* (*S. thermophilus*) and *Lactobacillus delbrueckii* subspecies *bulgaricus* (*L. bulgaricus*). *S. thermophilus* grows more rapidly at first and renders the milk anoxic and weakly acidic. *L. bulgaricus* then acidifies the milk even more. Acting together, the two species ferment almost all of the lactose to lactic acid and flavor the yogurt with diacetyl (*S. thermophilus*) and acetaldehyde (*L. bulgaricus*). Many yogurts now contain probiotic bacterial strains, but these do not necessarily contribute to the fermentation process (*see figure 34.11*). Fruits or fruit flavors are pasteurized separately and combined with the yogurt. Freshly prepared yogurt contains about 10^9 bacteria per gram.

Yeast-Lactic Fermentation

Yeast-lactic fermentations include kefir, a product that can contain up to 2% ethanol. This fermented milk originated in the Caucasus Mountains. Kefir products tend to be foamy due to microbial carbon dioxide production. This process is based on the use of kefir "grains" as a starter culture. These are coagulated lumps of casein (milk protein) that contain yeasts, lactic acid bacteria, and acetic acid bacteria, which are recovered at the end of the fermentation. Originally kefir was produced in leather sacks hung by the front door during the day, and passers-by were expected to push and knead the sack to mix and stimulate the fermentation. Kefir has gained popularity and is now mass-produced and sold in supermarkets worldwide.

Mold-Lactic Fermentation

Mold-lactic fermentation results in a unique Finnish fermented milk called *viili*. Milk is inoculated with a mixture of the fungus *Geotrichum candidum* and lactic acid bacteria. The cream rises to the surface, and after incubation at 18° to 20°C for 24 hours, lactic acid reaches a concentration of 0.9%. The fungus forms a velvety layer across the top of the final product, which also can be made with a bottom fruit layer.

Cheese Production

Cheese is one of the oldest foods, probably developed roughly 8,000 years ago. About 2,000 distinct varieties of cheese are produced throughout the world, representing approximately 20 general types (**table 41.5**).

The production of cheese varies with the type of cheese, but a general outline includes the following steps. First, the milk may be pasteurized to remove spoilage microorganisms. It is then cooled to 32°C so that starter cultures can be added. The mixture is held at this temperature for 30 minutes. This ripening step allows fermentation to begin and the pH declines as acidic fermentation products accumulate. Next, rennet is added; rennet is a mixture of enzymes produced by mammals to aid in the digestion of milk. Rennin is the key enzyme in rennet because it

Table 41.5 Major Types of Cheese and Microorganisms Used in Their Production

	CONTRIBUTING MICROORGANISMS[1]	
Cheese (Country of Origin)	**Earlier Stages of Production**	**Later Stages of Production**
Soft, unripened		
Cottage	*Lactococcus lactis*	*Leuconostoc mesenteroides* subspecies *cremoris*
Cream	*Lactococcus cremoris, L. diacetylactis, Streptococcus thermophilus*	
Mozzarella (Italy)	*S. thermophilus, L. bulgaricus*	
Soft, ripened		
Brie (France)	*Lactococcus lactis, Lactococcus cremoris*	*Penicillium camemberti, P. candidum, Brevibacterium linens*
Camembert (France)	*L. lactis, Lactococcus cremoris*	*Penicillium camemberti, B. linens*
Semisoft		
Blue, Roquefort (France)	*L. lactis, Lactococcus cremoris*	*P. roqueforti*
Brick, Muenster (United States)	*L. lactis, Lactococcus cremoris*	*B. linens*
Limburger (Belgium)	*L. lactis, Lactococcus cremoris*	*B. linens*
Hard, ripened		
Cheddar, Colby (Britain)	*L. lactis, Lactococcus cremoris*	*Lactobacillus casei, L. plantarum*
Swiss (Switzerland)	*L. lactis, L. helveticus, S. thermophilus*	*Propionibacterium shermanii, P. freudenreichii*
Very hard, ripened		
Parmesan (Italy)	*L. lactis, Lactococcus cremoris, S. thermophilus*	*L. delbrueckii* subspecies *bulgaricus*

1 *Lactococcus lactis* stands for *L. lactis* subspecies *lactis*. *Lactococcus cremoris* is *L. lactis* subspecies *cremoris*, *Lactococcus diacetylactis* is *L. lactis* subspecies *diacetylactis*, and *Streptococcus thermophilus* is *Streptococcus salivarius* subspecies *thermophilus*.

catalyzes a complex reaction that results in hydrolysis of milk protein (casein), converting the milk to solid curds. The curds set for about 30 minutes to allow the formation of a firm coagulum. Further fermentation of curds continues until the pH reaches 6.4, at which point the cheese is cut into small pieces and heated to 38°C, which helps separate remaining liquid (whey) from the curd. After the whey is completely drained from the curd, the process of cheddaring begins (**figure 41.9**). Here curd is stacked and occasionally flipped over; this process helps expel residual whey while fermentation continues. When the curd mats, now knitted together, reach a pH of 5.1 to 5.5, they are cut into smaller pieces. Depending on what kind of cheese is being made, salt will either be added (e.g., cheddar cheese) or the cheese will be rolled in loaves and placed in a salt solution called brine (e.g., mozzarella). The next step is to take the salted cheese and form it into blocks so it can age for weeks to years. This will, in part, determine the final hardness and texture of the cheese. Finally, it will be packaged or waxed (e.g., Jarlsberg cheese).

Lactococcus lactis is used as a starter culture for a number of cheeses (table 41.5). Starter culture density is often over 10^9

Figure 41.9 Cheddar Cheese Production. Cheddar, a village in England, has given its name to a cheese made in many parts of the world. "Cheddaring" is the process of turning and piling the curd to expel whey and develop the desired cheese texture. It is used in the production of all cheeses.
©Joe Munroe/Science Source

colony-forming units (CFU) per gram of cheese curd before ripening. However, the high salt, low pH, and temperatures that characterize the cheese microenvironment quickly reduce these numbers. This enables other bacteria, sometimes called nonstarter lactic acid bacteria, to grow; their numbers can reach 10^7 to 10^9 CFU/g after several months of aging. Thus both starter and nonstarter LAB contribute to the final taste, texture, aroma, and appearance of the cheese.

In some cases, molds are used to further enhance the cheese. Obvious examples are Roquefort and blue cheese. For these cheeses, *Penicillium roqueforti* spores are added to the curds just before final cheese processing. Sometimes the surface of an already formed cheese is inoculated at the start of ripening; for example, Camembert cheese is inoculated with spores of *Penicillium camemberti*.

Meat and Fish

A variety of meat products can be fermented: sausage, country-cured hams, salami, cervelat, Lebanon bologna, fish sauces (processed by halophilic *Bacillus* spp.), *izushi,* and *katsuobushi*. *Pediococcus acidilactici* and *Lactobacillus plantarum* are most often involved in sausage fermentations. *Izushi* is based on the fermentation of fresh fish, rice, and vegetables by *Lactobacillus* spp.; *katsuobushi* results from the fermentation of tuna by *Aspergillus glaucus*. These fermentations originated in Japan.

Wines and Champagnes

Wine production, the focus of **enology** (Greek *oinos,* wine; *ology,* the science of), starts with the collection of grapes, which are then crushed so the liquid, called **must,** can be fermented. Wine production concludes with a variety of storage and aging steps (**figure 41.10**). All grapes have white juices. To make a red wine, skins of red grapes remain in contact with the must before fermentation. Wines can be produced by using the natural grape skin microorganisms, but this mixture of bacteria and yeasts gives unpredictable fermentation results. To avoid this, fresh must is treated with a sulfur dioxide fumigant and a desired strain of the yeast *Saccharomyces cerevisiae* or *S. ellipsoideus* is added. After inoculation, the juice is fermented for 3 to 5 days at temperatures between 20° and 28°C. Depending on the alcohol tolerance of the yeast strain (the alcohol eventually kills the yeast that produces it), the final product may contain 10 to 14% alcohol. Clearing and development of flavor occur during the aging process. Grape juice contains high levels of organic acids, including malic and tartaric acids. If the levels of these acids do not decrease during the fermentation process, the wine will be too acidic and have poor stability and "mouth feel." Malic acid levels are decreased when fermented by the bacteria *Leuconostoc* species *oenos, plantarum, hilgardii, brevis*, and *casei*. These microbes transform malic acid (a four-carbon dicarboxylic acid) to lactic acid (a three-carbon monocarboxylic acid) and carbon dioxide. This malolactic fermentation results in deacidification, improvement of flavor stability, and, in some cases, the possible

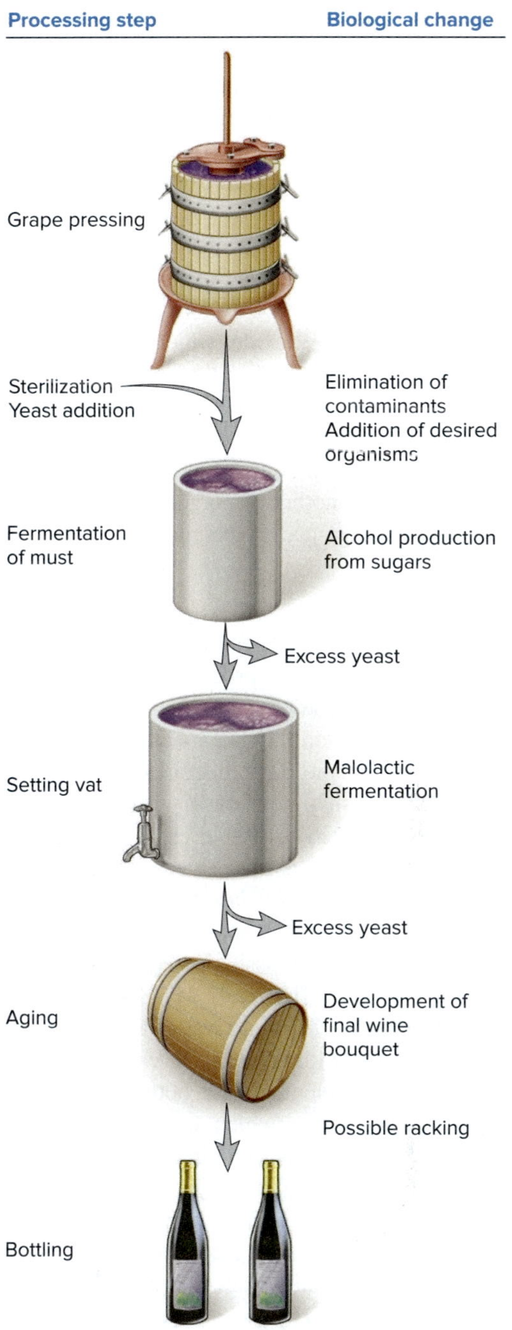

Figure 41.10 Wine Making. Once grapes are pressed, the sugars in the juice (the must) can be immediately fermented to produce wine. Must preparation, fermentation, and aging are critical steps.

accumulation of bacteriocins in the wines. ◀ Ascomycota *includes yeasts and molds (section 25.5); Bacteriocins (section 32.3)*

A critical part of wine making involves the choice of whether to produce a dry (no remaining free sugar) or a sweeter (varying amounts of free sugar) wine. This can be controlled by regulating the initial must sugar concentration. With higher levels of sugar, alcohol will accumulate and inhibit the fermentation before the sugar is completely used, producing a sweeter wine. During final fermentation in the aging process, flavoring compounds accumulate and influence the bouquet of the wine.

Microbial growth during fermentation produces sediments, which are removed during **racking.** Racking is simply storing the wine while it ages. It can be carried out at the time the fermented wine is transferred to bottles or casks for aging or after the wine is placed in bottles.

Champagnes and sparkling wines are essentially the same product, but a champagne must be made in the Champagne region of France. They result when the fermentation is continued in bottles to produce carbon dioxide, which forms bubbles. Sediments that remain are collected in the necks of inverted champagne bottles after the bottles have been carefully turned. The necks of the bottles are then frozen and the corks removed to disgorge accumulated sediments. The bottles are refilled with clear champagne from another disgorged bottle, and the product is ready for final packaging and labeling.

Beers and Ales

Beer and ale production uses cereal grains such as barley, wheat, and rice. The first step in beer production, known as **mashing,** involves germination of the barley grains and activation of their enzymes to produce a **malt** (**figure 41.11**). The malt is then mixed with water and the desired grains, and the mixture is transferred to the mash tun or cask where the active barley enzymes hydrolyze starch to usable carbohydrates. Once this process is completed, the **mash** is heated with **hops** (dried flowers of the female vine *Humulus lupulus*), which were originally added to the mash to inhibit spoilage microorganisms. The hops also provide flavor and assist in clarification of the liquid portion called the wort (**figure 41.12**). In this heating step, the hydrolytic enzymes are inactivated and the wort can be **pitched**—inoculated—with the desired yeast.

Most beers are fermented with bottom yeasts, related to *Saccharomyces pastorianus,* which settle at the bottom of the fermentation vat. Bottom yeasts require 7 to 12 days of fermentation to produce beer with a pH of 4.1 to 4.2. With a top yeast, such as *Saccharomyces cerevisiae,* the pH is lowered to 3.8 to produce ales. Freshly fermented (green) beers are aged, and when they are bottled, CO_2 is usually added. Beer can be pasteurized at 40°C or higher or sterilized by passage through membrane filters to minimize flavor changes.

Distilled Spirits

Distilled spirits are produced by an extension of beer production processes. Rather than starting with fresh grain, a **sour mash** is used. This is the liquid fermented grain collected from a previously prepared batch of alcohol. The mash is inoculated with a homolactic bacterium such as *Lactobacillus delbrueckii* subspecies *bulgaricus* (figure 41.8*b*), which can lower the mash pH to around 3.8 in 6 to 10 hours. This limits the development of undesirable organisms. It is boiled, and the volatile components are condensed to yield a product with a higher alcohol content than beer. Rye and bourbon are examples of whiskeys.

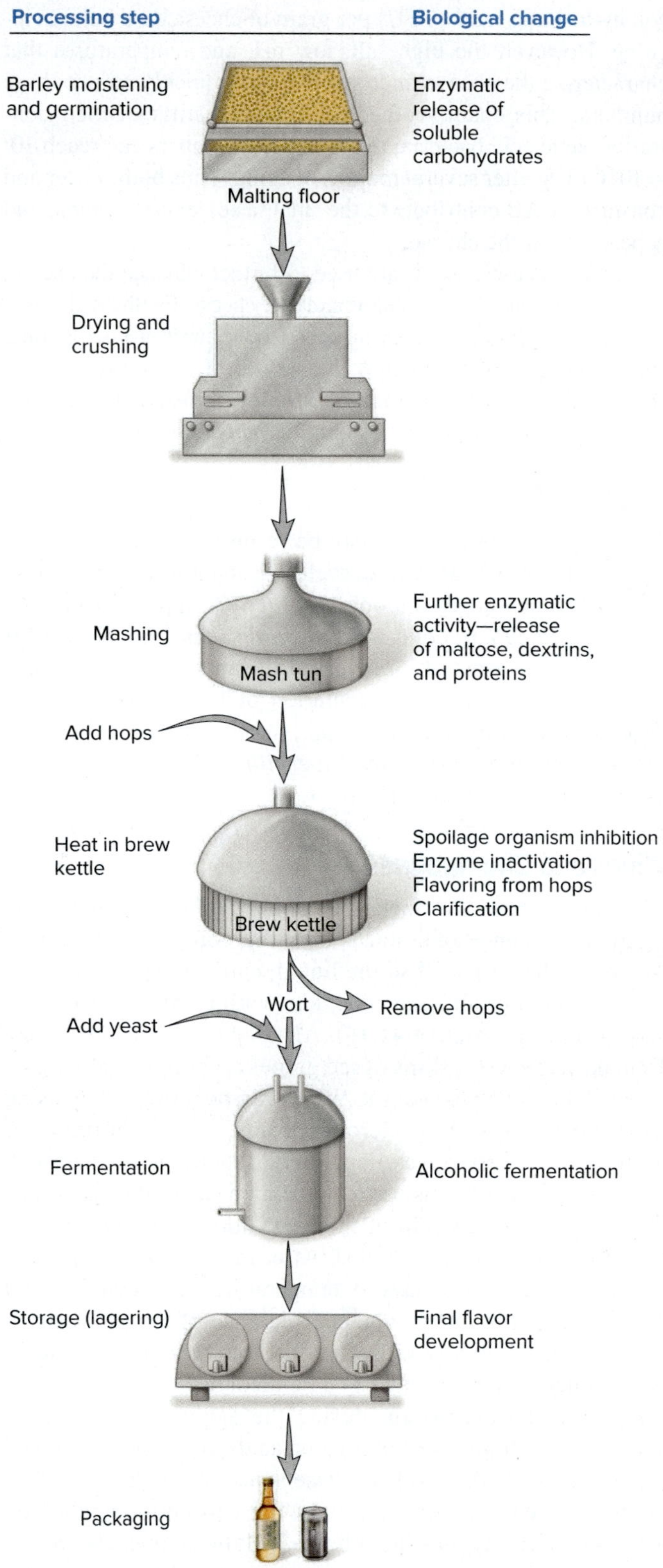

Figure 41.11 Producing Beer. To make beer, the complex carbohydrates in the grain must first be transformed into a fermentable substrate. Beer production requires mashing, the use of hops, boiling for clarification, flavor development, and inactivation of malting enzymes to produce the wort.

Figure 41.12 Brew Kettles Used for Preparation of Wort. In large-scale processes, copper brew kettles can be used for wort preparation, as shown here.

Rye whiskey must contain at least 51% rye grain, and bourbon must contain at least 51% corn. Scotch whiskey is made primarily of barley. Vodka and grain alcohols are also produced by distillation. Gin is vodka to which resinous flavoring agents—often juniper berries—have been added to provide a unique aroma and flavor.

Production of Breads

Bread is one of the most ancient of human foods. The use of yeasts to leaven bread is carefully depicted in paintings from ancient Egypt, and a bakery at the Giza Pyramid area, from the year 2575 BCE, has been excavated. In bread making, yeast growth is carried out under oxic conditions. This results in increased CO_2 production and minimum alcohol accumulation. The fermentation of bread involves several steps: α- and β-amylases present in the moistened dough release maltose and glucose from starch. The baker's strain of the yeast *Saccharomyces cerevisiae,* which produces maltase, invertase, and zymase enzymes, is added. The CO_2 produced by the yeast causes the bread to rise and results in the light texture of many breads. Traces of fermentation products contribute to the final flavor. Usually bakers add sufficient yeast to allow the bread to rise within 2 hours—the longer the rising time, the more additional growth by contaminating bacteria and fungi can occur, making the product less desirable. For instance, bread products can be spoiled by *Bacillus* species that produce ropiness.

Other Fermented Foods

Many other plant products can be fermented, as summarized in **table 41.6**. These include *sufu,* which is produced by the fermentation of tofu (a chemically coagulated soybean milk product).

Table 41.6 Fermented Foods Produced from Fruits, Vegetables, Beans, and Related Substrates

Foods	Raw Ingredients	Fermenting Microorganisms	Location
Bacteria-Fermented Foods			
Coffee	Coffee beans	*Erwinia dissolvens, Saccharomyces* spp.	Brazil, Congo, Hawaii, India
Gari	Cassava	*Corynebacterium manihot, Geotrichum* spp.	West Africa
Kimchi	Cabbage and other vegetables	Lactic acid bacteria	Korea
Ogi	Corn	*Lactobacillus plantarum, Lactococcus lactis, Z. rouxii*	Nigeria
Olives	Green olives	*Leuconostoc mesenteroides, Lactobacillus plantarum*	Worldwide
Pickles	Cucumbers	*Pediococcus cerevisiae, L. plantarum, L. brevis*	Worldwide
Poi	Taro roots	Lactic acid bacteria	Hawaii
Sauerkraut	Cabbage	*L. mesenteroides, L. plantarum, Lactobacillus brevis*	Worldwide
Mold-Fermented Foods			
Kenkey	Corn	*Aspergillus* spp., *Penicillium* spp., lactobacilli, yeasts	Ghana, Nigeria
Miso	Soybeans	*Aspergillus oryzae, Zygosaccharomyces rouxii*	Japan
Ontjom	Peanut presscake	*Neurospora sitophila*	Indonesia
Peujeum	Cassava	Molds	Indonesia
Soy sauce	Soybeans	*A. oryzae* or *A. soyae, Z. rouxii, Lactobacillus delbrueckii*	Japan
Sufu	Soybeans	*Actinimucor elegans, Mucor* spp.	China
Tao-si	Soybeans	*A. oryzae*	Philippines
Tempeh	Soybeans	*Rhizopus oligosporus, R. oryzae*	Indonesia, New Guinea, Surinam

Source: Jay, James M. *Modern Food Microbiology*, 6th ed. New York: Springer, 2000.

To carry out the fermentation, the tofu curd is cut into small chunks and dipped into a solution of salt and citric acid. After the cubes are heated to pasteurize their surfaces, the fungi *Actinomucor elegans* and some *Mucor* species are added. When a white mycelium develops, the cubes, now called *pehtze,* are aged in salted rice wine. Another popular product is tempeh, a soybean mash fermented by *Rhizopus* spp.

Sauerkraut or sour cabbage is produced from shredded cabbage. A concentration of 2.2 to 2.8% sodium chloride restricts the growth of Gram-negative bacteria while favoring the development of lactic acid bacteria. The primary microorganisms contributing to this product are *Leuconostoc mesenteroides* and *Lactobacillus plantarum.* A predictable microbial succession occurs in sauerkraut's development. The activities of the lactic acid–producing cocci usually cease when the acid content reaches 0.7 to 1%. At this point, *Lactobacillus plantarum* and *Lactobacillus brevis* continue to function. The final acidity is generally about pH 1.7, with lactic acid comprising 1 to 1.3% of the total acid in a satisfactory product.

Pickles are produced by placing cucumbers and components such as dill seeds in casks filled with a brine. The sodium chloride concentration begins at 5% and more is added to reach about 16% in 6 to 9 weeks. The salt not only inhibits the growth of undesirable bacteria but also extracts water and water-soluble constituents from the cucumbers. These soluble carbohydrates are fermented to lactic acid. The fermentation, which can require 10 to 12 days, involves the growth of the Gram-positive bacteria *L. mesenteroides, Enterococcus faecalis, Pediococcus acidilactici, Lactobacillus brevis,* and *L. plantarum. L. plantarum* plays the dominant role in this fermentation process. Sometimes, to achieve more uniform pickle quality, natural microorganisms are first destroyed and the cucumbers are fermented using pure cultures of *P. acidilactici* and *L. plantarum.*

When grass, chopped corn, and other fresh animal feeds are stored under moist anoxic conditions, they undergo a mixed acid fermentation that produces pleasant-smelling **silage.** Silage is used to feed livestock during the winter. Trenches or more traditional vertical steel or concrete silos are used to store silage.

Comprehension Check

1. What are the major types of milk fermentations?
2. Briefly compare how cultured buttermilk and yogurt are made.
3. What major steps are used to produce cheese? How is the cheese curd formed in this process?
4. Which fungal genus is often used in cheese making?
5. Describe and contrast the processes of wine and beer production.
6. Describe the fermentation of one food described here that is unfamiliar to you.

Key Concepts

41.1 Microbial Growth Can Cause Food Spoilage

- Most foods, especially when raw, provide an excellent environment for microbial growth. This growth can lead to spoilage or preservation, depending on the microorganisms present and environmental conditions.
- The course of microbial development in a food is influenced by the intrinsic characteristics of the food itself—pH, salt content, substrates present, water presence, and availability—and extrinsic factors, including temperature, relative humidity, and atmospheric composition (**table 41.1**).
- Microorganisms can spoil meat, dairy products, fruits, vegetables, and canned goods in several ways. Spices, with their antimicrobial compounds, sometimes protect foods.

41.2 Various Methods Are Used to Control Food Spoilage

- Foods can be preserved in a variety of physical and chemical ways, including filtration, alteration of temperature (cooling, pasteurization, sterilization), drying, addition of chemicals, radiation, and fermentation (**tables 41.2** and **41.3**).
- Modified atmosphere packaging (MAP) is used to control microbial growth in foods and to extend product shelf life. This process involves decreased oxygen and increased carbon dioxide levels in the space between the food surface and the wrapping material.
- Interest is increasing in using bacteriocins for food preservation. Nisin, a product of *Lactococcus lactis,* is a major substance approved for use in foods.

41.3 Food-Borne Disease Outbreaks

- Foods can be contaminated by pathogens at any point in the food production, storage, or preparation processes. Pathogens such as *Salmonella enterica, Campylobacter jejuni, Listeria monocytogenes,* and *E. coli* can be transmitted to susceptible consumers. Because such microbes are ingested and grow in the host, they are said to cause food-borne infections (**table 41.4**).
- If the pathogen grows in the food before consumption and forms toxins that affect food consumers without further microbial growth, the disease is a food intoxication. Examples are intoxications caused by *Staphylococcus, Clostridium,* and *Bacillus* spp. Fungi that grow in foods, especially cereals and grains, can produce important disease-causing chemicals, including the carcinogens aflatoxins (**figure 41.5**) and fumonisins (**figure 41.6**).

41.4 Detection of Food-Borne Pathogens Requires Government-Industry Cooperation

- Detection of food-borne pathogens is a major part of food microbiology. Many assays rely on culture-based approaches.
- Molecular techniques are replacing many culture-based approaches. Such approaches include the use of nucleic acid-based techniques, immunological methods, and biosensors. The PulseNet network is used to respond to food-borne illness outbreaks. This technology is being replaced by whole-genome sequencing (**figure 41.7**).

41.5 Microbiology of Fermented Foods: Beer, Cheese, and Much More

- Dairy products are fermented to yield a wide variety of cultured milk products. These include mesophilic, therapeutic probiotic, thermophilic, yeast-lactic, and mold-lactic products.
- Growth of lactic acid–forming bacteria, often with the additional use of rennin, coagulate milk solids. These solids are processed to yield a wide variety of cheeses, including soft unripened, soft ripened, semisoft, hard, and very hard cheeses (**table 41.5**). Both bacteria and fungi are used in these cheese production processes.
- Wines are produced from pressed grapes and are either dry or sweet, depending on the level of free sugar that remains at the end of the alcoholic fermentation (**figure 41.10**). Champagne is produced when fermentation resulting in CO_2 formation continues in the bottle.
- Beer and ale are produced from cereals and grains. The starches in these substrates are hydrolyzed, in the processes of malting and mashing, to produce a fermentable wort. *Saccharomyces cerevisiae* and *S. pastorianus* are yeasts commonly used in the production of beer and ale (**figure 41.11**).
- Many plant products can be fermented with bacteria, yeasts, and molds. Important products are breads, soy sauce, *sufu,* and tempeh (**table 41.6**). Sauerkraut and pickles are produced in a fermentation process in which natural populations of lactobacilli play a major role.

Active Learning

1. Compare the "sell-by" date of a package of hot dogs with that of a fresh cut of meat. What are the intrinsic and extrinsic factors that contribute to the difference?
2. You are going through a salad line in a cafeteria at the end of the day. Which types of foods would you tend to avoid? Or would you skip the salad altogether? Explain.
3. Keep a record of what you eat for a day or two. Determine if any of the food and beverages you ate were produced with the aid of microorganisms. Indicate at what levels microorganisms were deliberately used. Be sure to consider ingredients such as citric acid, which is produced at the industrial level by several species of fungi.
4. During cheese production, LAB convert lactose to lactate and casein (milk protein) to amino acids. Lactate and amino acids then become the substrates for further microbial growth, which results in aroma production and deacidification of the cheese. The yeast *Yarrowia lipolytica* grows on the surface of many cheeses; it is capable of both lactate and amino acid catabolism. When grown on a lactate plus amino acid medium, *Y. lipolytica* preferentially consumes amino acids. Amino acid degradation results in the release of ammonia, which increases the pH. Draw a flow chart that shows the LAB fermentation of milk, followed by the growth of *Y. lipolytica*. Indicate which substrates are consumed first and what happens to the pH. Based on this simplified scenario, why do you think most cheeses involve the activity of more than one yeast species?
5. In 2014-2015, an outbreak of listeriosis in 12 states resulted in the hospitalization of at least 34 people and 7 deaths. Scientists at the CDC traced the causative *Listeria monocytogenes* strain to commercially prepared, prepackaged caramel apples. Discuss the techniques likely used to determine the origin of contamination. Be sure to consider all the ingredients and processes needed to make this snack as well as the food supply chain.

Read the original paper: Angelo, K. M., et al. 2017. Multistate outbreak of *Listeria monocytogenes* infections linked to whole apples used in commercially produced, prepackaged caramel apples: United States, 2014-2015. *Epidemiol. Infect.* 145:848–56. doi:10.1017/S0950268816003083.

42

Biotechnology and Industrial Microbiology

©Rocketclips/123RF

Where Are the New Antibiotics?

Bill's dad takes a statin drug to lower his cholesterol, and his grandmother takes a beta-blocker for her heart. His mom has a steroid inhaler to keep her asthma under control. When Bill comes down with strep throat, his physician prescribes a week-long course of amoxicillin and he gets better.

Bill's relatives are benefiting from drugs pharmaceutical companies love to develop. They aren't so wild about helping Bill, though. It's not that they don't want to; they believe they can't afford it. Consider this: It costs over 2 billion U.S. dollars and at least a decade to bring a drug from discovery to market. Then the patent lasts for about another 10 years. Once the patent expires, the cost of the drug (and the profit therein) drops precipitously. Whereas Bill's relatives will very likely take their medicines for the rest of their lives, Bill takes his antibiotic for a week. When he's done: no more strep throat, no more profit for the drug manufacturer.

But in the meantime, amoxicillin (like all antibiotics) is losing its effectiveness as more and more bacteria develop resistance. So when new antibiotics are developed, they cannot be widely prescribed, lest resistance quickly develop. An antibiotic will never be a "blockbuster" drug.

Despite these disincentives, the need for new antibiotics grows more urgent with each passing year. Recall that two-thirds of antibiotics are natural microbial products. However, because only about 1% of all bacteria and archaea—and even fewer fungi—have been cultured, the environment holds great untapped resources. But history tells us we need new strategies to access this potential antibiotic reservoir. Between 1935 and 1962, the seven classes of antibiotics[1] we still use today were developed. But from 1962 to 2000, no new drug classes were discovered. Instead, while men went to the moon and the internet was born, existing antibiotics were chemically modified, generating multiple generations of existing drugs. It was not until the first decade of the twenty-first century that three new classes were introduced to clinical practice: oxazolidinones, lipopeptides, and mutilins. Currently, antibiotic discovery is becoming synonymous with microbial discovery: looking for novel genes in new habitats, such as deep-sea trenches and the symbionts of other animals, or inventing new ways to coax microbes from common environments (like soil) to grow in the lab.

Industrial microbiology harnesses the capacity of microbes to synthesize compounds that have important applications in medicine, agriculture, food preparation, and other industrial processes. These compounds are commonly called **natural products.** In this chapter, we discuss these products and how microorganisms (or their genes) are optimized for industrial manufacturing.

[1] In chronological order of discovery, these seven drug classes include sulfa drugs, β-lactams, chloramphenicol/tetracyclines, aminoglycosides, macrolides, glycopeptides, quinolones, and streptogramins.

Readiness Check:

Based on what you have learned previously, you should be able to:

- ✓ Compare and contrast at least five different classes of antibiotics (table 9.1)
- ✓ Describe the difference between batch and continuous culture techniques (sections 7.7 and 7.9)
- ✓ Describe the means by which DNA mutations can be induced (section 16.1)
- ✓ Explain the mechanisms of horizontal gene transfer (sections 16.4–16.9)
- ✓ Summarize the process and importance of PCR (section 17.2)
- ✓ Define metagenomics and functional genomics and discuss their applications (sections 18.3 and 18.5)
- ✓ Discuss the principles by which vaccines confer immunity (section 36.6)

42.1 Microbes Are the Source of Many Products of Industrial Importance

After reading this section, you should be able to:

a. Describe at least five compounds of industrial importance made by microbes
b. Define biocatalysis and explain its advantages over chemical synthesis

Table 42.1 Major Microbial Products and Processes of Industrial Importance

Substances	Microorganisms
Industrial Products	
Ethanol (from glucose)	*Saccharomyces cerevisiae*
Ethanol (from lactose)	*Kluyveromyces fragilis*
Acetone and butanol	*Clostridium acetobutylicum*
2,3-butanediol	*Enterobacter, Serratia*
Enzymes	*Aspergillus, Bacillus, Mucor, Trichoderma*
Agricultural Products	
Gibberellins	*Gibberella fujikuroi*
Food Additives	
Amino acids (e.g., lysine)	*Corynebacterium glutamicum*
Organic acids (citric acid)	*Aspergillus niger*
Nucleotides	*Corynebacterium glutamicum*
Vitamins	*Ashbya, Eremothecium, Blakeslea*
Polysaccharides	*Xanthomonas*
Medical Products	
Antibiotics	*Penicillium, Streptomyces, Bacillus*
Alkaloids	*Claviceps purpurea*
Steroid transformations	*Rhizopus, Arthrobacter*
Insulin, human growth hormone, somatostatin, interferons	*Escherichia coli, Saccharomyces cerevisiae*, and others (recombinant DNA technology)
Biofuels	
Hydrogen	Photosynthetic microorganisms
Methane	*Methanothermobacter*
Ethanol	*Zymomonas, Thermoanaerobacter*

Most people are not aware that many antibiotics and other medicines, food additives, enzymes with industrial applications, and biofuels are examples of microbial products (**table 42.1**). Here we introduce a few of the more common microbial products that have become an integral part of modern life.

Antibiotics

More than 65% of all antibiotics are produced by microorganisms. Currently most antibiotics are produced by members of the genus *Streptomyces* and by filamentous fungi (*see table 23.3*). Antibiotics are secreted by these microbes and either used in the native state or chemically modified to produce semisynthetic derivatives. Semisynthetic derivatives were developed to treat a broader spectrum of bacteria or bacteria that had become resistant to the parent drug.

The synthesis of penicillin and its derivatives illustrates how an antibiotic is commercially produced. As with any industrially important natural product, the manufacturer will determine how to maximize product yield. In this case, the growth of the fungus *Penicillium chrysogenum* requires careful adjustment of the medium, so that the slowly hydrolyzed disaccharide lactose is used in combination with a limited supply of nitrogen to stimulate a greater accumulation of penicillin after growth has stopped (**figure 42.1**). If a particular penicillin is needed,

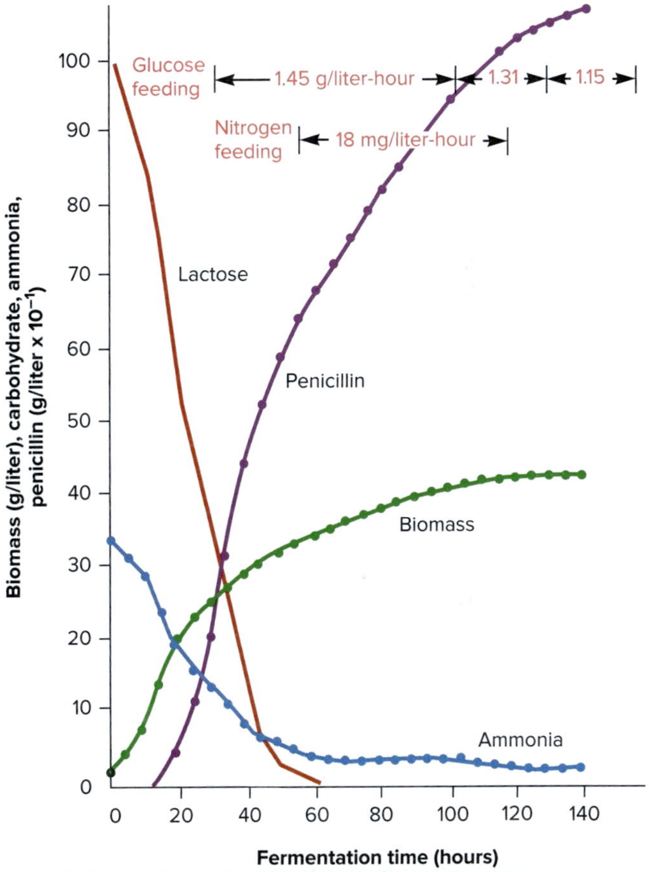

Figure 42.1 Penicillin Fermentation Involves Precise Control of Nutrients. The synthesis of penicillin begins when nitrogen from ammonia becomes limiting. After most of the lactose has been degraded, glucose is added along with a low level of nitrogen. This stimulates maximum transformation of the carbon sources to penicillin. The scale factor uses the convention recommended by the American Society for Microbiology; that is, a number on the axis should be multiplied by 0.10 to obtain the true value.

MICRO INQUIRY *Why are the cells first fed lactose and then glucose?*

a specific precursor is added to the medium. For example, phenylacetic acid is added to maximize production of penicillin G, which has a benzyl side chain (*see figure 9.5*). After about a week, the broth is separated from the fungal hyphae and processed by absorption, precipitation, and crystallization to yield the final product. This basic product can then be modified by chemical procedures to yield a variety of semisynthetic penicillins. Manufacture of semisynthetic β-lactam antibiotics, such as ampicillin and amoxicillin, begins with natural penicillin. The β-lactam ring is preserved, but the side chain is modified, in this case to provide a broader spectrum of activity. ◀ *Penicillins (section 9.4)*

Amino Acids

Amino acids such as lysine and glutamic acid are used in the food industry as nutritional supplements in bread products and as flavor-enhancing compounds such as monosodium glutamate (MSG). Amino acid production is typically carried out by regulatory mutants that overproduce a specific amino acid or a key intermediate. Production of glutamate and several other amino acids is carried out using mutants of *Corynebacterium glutamicum* that lack or have only a limited ability to process the TCA cycle intermediate α-ketoglutarate to succinyl-CoA (*see figure 11.8*). By keeping the biotin level low and adding fatty acid derivatives, membrane permeability is increased, allowing the secretion of high concentrations of glutamate. The impaired bacteria use the glyoxylate cycle (*see figure 12.22*) to meet their needs for essential biochemical intermediates, especially during the growth phase. By the time nutrients are depleted and growth stops, an almost complete molar conversion of isocitrate to glutamate occurs. ◀ *Synthesis of amino acids consumes many precursor metabolites (section 12.5)*

Organic Acids

Simply reading the ingredient list on most processed foods will illustrate the widespread use of organic acids such as citric, acetic, and lactic acids. These acids are principally used as preservatives. Organic acid production illustrates how the concentration of trace elements can influence product yield.

Citric acid is commonly added to food as a preservative or to impart a tart flavor. Its production involves limiting trace metals such as manganese and iron to stop *Aspergillus niger* growth at a specific point. Generally, high sugar concentrations (15 to 18%) are used. This reflects the regulation of glycolysis and the tricarboxylic acid cycle (recall that citric acid is a constituent of the TCA cycle). If the sugar levels are high after the active growth phase, citrate synthase activity increases and the activities of aconitase and isocitrate dehydrogenase decrease. This results in citric acid accumulation and excretion by the stressed microorganism.

Biopolymers

Biopolymers are microbially produced polymers, primarily polysaccharides, used to modify the flow characteristics of liquids and to serve as gelling agents. They have many uses in the pharmaceutical and food industries.

Biopolymers include (1) dextrans, which are used as blood expanders and absorbents; (2) *Erwinia* polysaccharides used in paints; (3) polyesters, derived from *Pseudomonas oleovorans,* which are used for specialty plastics; (4) cellulose microfibrils, produced by *Acetobacter* spp., that serve as a food thickener; (5) polysaccharides such as scleroglucan used by the oil industry as drilling mud additives; and (6) xanthan polymers, including xanthan gum produced by *Xanthomonas campestris,* used for a variety of applications as food additives. For example, xanthan polymers are commonly found in many condiments such as salad dressings and gluten-free foods, where they help ingredients stick together. They are also used to enhance oil recovery by thickening drilling mud.

Biosurfactants

Biosurfactants are amphiphilic molecules; that is, they possess both hydrophobic and hydrophilic regions. Thus they partition at the interface between fluids that differ in polarity, such as oil and water. For this reason, they are used for emulsification, increasing detergency, wetting, and phase dispersion, as well as solubilization. These properties are especially important in bioremediation, oil spill dispersion, and enhanced oil recovery. The most widely used microbially produced biosurfactants are glycolipids. These are carbohydrates that bear long-chain fatty acids. They can be isolated as extracellular products from a variety of microorganisms, including pseudomonads and yeasts.

Biocatalysts

Biocatalysts are enzymes used in the industrial production of chemicals and pharmaceuticals. The vast majority of such enzymes are of microbial origin and include dehydrogenases, oxygenases, hydrolases, transferases, and lyases. These enzymes may introduce minor changes in molecules, such as the insertion of a hydroxyl or keto function, or the saturation or desaturation of a complex cyclic structure. A typical bioconversion is the hydroxylation of a steroid as shown in **figure 42.2**. In this example, the water-insoluble steroid is dissolved in acetone and then added to the reaction system. The course of the modification is monitored, and the final, soluble product is extracted and purified.

Biocatalysis generally has the advantage over chemical synthesis in that natural enzymes have stereospecificity so that the product is not a mixture of isomers and is biologically active. Further, microbial enzymes catalyze reactions under mild conditions; some of these reactions would require harsh conditions (e.g., extreme pH, temperature) if performed by chemical means.

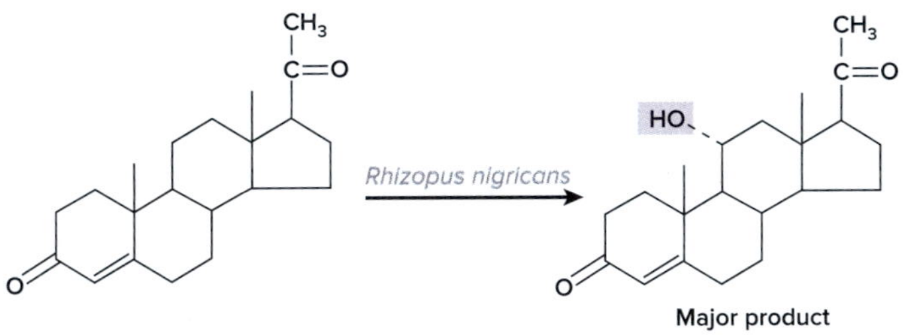

Figure 42.2 Biotransformation to Modify a Steroid. Hydroxylation of progesterone in the 11α position by *Rhizopus nigricans*. The steroid is dissolved in acetone before addition to the pregrown fungal culture.

Vaccines

Vaccines are the single most important public health weapon. The most common vaccines are live attenuated viruses (e.g., measles vaccine) or killed pathogens (influenza vaccine). However, different vaccine strategies must be used if the pathogen cannot be grown in vitro (e.g., hepatitis B virus), if there is too much genetic diversity among pathogenic strains, or if the pathogen produces molecules that are too similar to host molecules, thereby triggering autoimmunity. This is the case for *Neisseria meningitidis* serovar B (MenB), one of five strains of this bacterium to cause meningitis. A meningitis vaccine made of capsular polysaccharide of the other four serovars has been available for many years. However, the polysaccharide capsule of serovar B closely resembles a polysaccharide found on the surface of many human cells. To develop a vaccine against serovar B, a **reverse vaccinology** approach was used. Reverse vaccinology mines the genomic sequence of the pathogen to assemble a list of antigens that may be good vaccine targets. ◀ *Vaccines immunize susceptible populations (section 36.6)*

To understand reverse vaccinology, it is helpful to recall that an antigen is a foreign molecule targeted by the host immune system. Purified antigens, rather than whole organisms, can be used in vaccines. Such antigens must (1) be expressed by the pathogen during infection; (2) be secreted or found on the surface of the pathogen; (3) be found in all strains of the pathogen; (4) elicit a host immune response; and (5) be essential for the survival of the pathogen, at least while it is in the host.

Considering the thousands of genes revealed in each annotated genome, finding genes whose products meet all of these criteria might seem next to impossible. By using the complete genome sequence of serogroup B *N. meningitidis*, 600 surface proteins were identified and about 300 were tested in mice. Of these about 25 were worthy of further study, which led to the discovery of several antigens that in combination induce a robust immune response. In 2014 the first MenB vaccine was approved for use in the United States. Reverse vaccinology is currently being applied in the development of vaccines to protect against a variety of other pathogens, including *Staphylococcus aureus*, group B streptococcus (*S. agalactiae*), *Chlamydophila pneumoniae*, and *Bacillus anthracis*.

Comprehension Check

1. What critical limiting factors are used in the production of penicillin? Why do you think the microbe must be stressed to produce high levels of drug?
2. Why are regulatory mutants used to increase the production of glutamic acid by *Corynebacterium glutamicum*?
3. What kinds of molecular changes result from biocatalysis?
4. How does reverse vaccinology differ from conventional vaccine development?

42.2 Biofuel Production Is a Dynamic Field

After reading this section, you should be able to:

a. Compare the value of biofuel use with the challenges presented in its manufacture and transport
b. Discuss at least two types of biofuels

Although it is difficult to project future global energy consumption, by 2050 the world's population is expected to increase by at least 1.8 billion people, reaching more than 9 billion. As population and economic growth increases energy demands to new levels, most nations have pledged to take steps to limit atmospheric CO_2 to 450 ppm by 2100 (*see figure 28.11*). A key step to meeting this goal is to develop economically competitive alternatives to fossil fuel. **Microbial energy conversion** includes the microbial transformation of organic materials into **biofuels,** such as ethanol and hydrogen, that can fuel cars or other machines. In this section, we introduce the microbial production of these biofuels. ◀ *Global climate change (section 28.2)*

The U.S. Department of Energy has set a goal of substituting 30% of gasoline with biofuels by 2030. Currently ethanol is used as an additive to gasoline because it is easy to store and can be burned in cars designed to use pure gasoline. Corn is the substrate most commonly used for generating ethanol in the United States, while sugar and palm oil are used in other countries. Ethanol production involves the microbial degradation of plant starch using amylases and amyloglucosidases. The resulting sugars are then fermented to ethanol. However, the use of corn and other foods to make bioethanol has unintended consequences including increased pesticide and fertilizer use, and inflating worldwide food prices. Cellulose (or cellulosic) biofuel production using crop residues is beginning to replace corn-based biofuel production. Crop residues are plant materials left in the field after harvest as well as plants like switchgrass that grow on land not fertile enough for crops. This material consists of cellulose and hemicellulose—polymers of five different hexoses and pentoses: glucose, xylose, mannose, galactose, and arabinose. Degradation of cellulose and hemicellulose to release these monomers is commonly done by heating the plant material and treating it with acid, which is energy-intensive and corrosive. This is an expensive and potentially hazardous process. This has sparked interest in replacing chemical pretreatment with microbial degradation of these polymers. This has led to the development of a variety of newly isolated microbes and genetically engineered organisms that can degrade cellulose at temperatures in excess of 100°C. These are excellent examples of biocatalysis.

Even with cheaper, more sustainable plant materials for ethanol production, ethanol as a biofuel has several disadvantages. One is that it absorbs water and cannot be shipped through existing pipelines because they almost always contain some water. So ethanol must be distilled twice to remove the water: once when it is generated and then again after shipping. Furthermore, ethanol contains far less energy than other fuels, including gasoline and higher molecular weight alcohols such as butanol, which are easier to generate. This has motivated a number of biotechnology and energy companies to engineer microbes to produce other alcohols, isoprenoids, and hydrocarbons. These products more closely resemble gasoline (petrol) with low hygroscopy (i.e., they won't absorb water like ethanol), high energy density, and good fluidity even in the cold.

In many ways, hydrogen (H_2) as a biofuel compares very favorably to other fuels. With about three times more potential energy per unit weight than gasoline, H_2 is the highest energy-content fuel available. However, unlike ethanol, hydrogen cannot simply be mixed with gasoline and used in today's vehicles. This has led to the development of fuel cell electric vehicles (FCEVs), which emit only water vapor and warm air. California is leading the United States in providing H_2 fueling stations and FCEVs.

©David Carillet/Getty Images

Because most of the commercial H_2 is currently used to refine petroleum and produce fertilizer, there is much interest in developing new sources of H_2. A diverse group of microbes produces H_2, making its manufacture quite flexible. Like ethanol, H_2 is a product of fermentation. It is also produced by oxygenic photosynthetic microbes (cyanobacteria and algae) and anoxygenic photoheterotrophs. Two different enzymes (thus processes) generate H_2: hydrogenase and nitrogenase. Both are keenly sensitive to oxygen, so H_2 production is an anaerobic process.

Oxygenic photosynthetic microbes produce H_2 at night, when they oxidize cellular storage products such as poly-β-hydroxybutyrate. The use of aquatic phototrophs to generate H_2 holds promise because there is plenty of substrate (light and CO_2). However, hydrogenase is strongly inhibited by its product, so H_2 must be removed from the system as soon as it is created. By contrast, H_2 production by anoxygenic photoheterotrophic bacteria can occur via cellular hydrogenases or nitrogenases. This is of interest to industrial microbiologists because in an atmosphere artificially purged of all N_2, nitrogenase will form only H_2. This process requires a tremendous amount of ATP and reductant, but unlike hydrogenase, nitrogenase is not inhibited by the accumulation of H_2, and electron conversion to H_2 reaches efficiencies near 100%.

Although the future of specific biofuels remains to be determined, several things are clear. The growing desire for energy will eventually make fossil fuels rare and prohibitively expensive. At the same time, the need to increase food production will make it untenable to grow feedstocks for biofuel production on scarce arable land. It may be that synthetic biology will save the day, and instead of drilling for oil, we will be harvesting it from bioreactors. Or it may turn out that H_2 will become an important resource, especially in countries that have not yet built an extensive infrastructure to support a hydrocarbon-based lifestyle. But no matter what eventually transpires, these are exciting times for the microbial energy conversion field.

Comprehension Check

1. Why is there interest in developing alternatives to ethanol as the biofuel of choice in the United States?
2. What types of microbes are good candidates for the industrial production of H_2?

42.3 Growing Microbes in Industrial Settings Presents Challenges

After reading this section, you should be able to:

a. Explain the use of the term *fermenter* as it is used in an industrial setting
b. Describe two types of fermenters

As our review of natural products and processes points out, the development of appropriate culture media and conditions is critical to the manufacture of the desired product. Culturing microbes in industrial settings is much different than their growth in either nature or the laboratory. However, before describing microbial growth on an industrial scale, we must clarify the term **fermentation.** When used in a physiological sense, fermentation refers to the reduction of pyruvate or other organic compound to reoxidize NADH and conserve energy (*see figures 11.19 and 11.20*). However, to industrial microbiologists, fermentation means the mass culture of microorganisms, plant, and animal cells. Industrial fermentations require the development of appropriate culture media and the transfer of small-scale technologies to a much larger scale. In fact, the success of an industrially important microbial product rests on the process of **scale-up.** This is when a procedure developed in a small flask is modified for use in a large fermenter. The microenvironment of the small culture must be maintained despite increases in culture volume. Although the process of transitioning a culture originally developed in a 250-milliliter Erlenmeyer flask to a 100,000-liter reactor seems like it should be relatively straightforward, it is often very difficult to obtain the same yield of the desired product, and a series of troubleshooting experiments must be devised.

Microorganisms are often grown in stirred fermenters or other mass culture systems. Stirred **fermenters** can range in size from 3 liters to over 100,000 liters, depending on production requirements. A typical industrial stirred fermentation unit is illustrated in **figure 42.3**. Upon inoculation of microbes into sterile growth medium, aeration, pH and temperature adjustments, and sampling are carried out under rigorously controlled conditions. When required, foam control agents are added, especially with high-protein media. Computers are used to monitor outputs from probes that determine biomass, levels of critical metabolic products, pH, input and exhaust gas composition, and other parameters. Environmental conditions can be changed or held constant over time, depending on the requirements of the particular process.

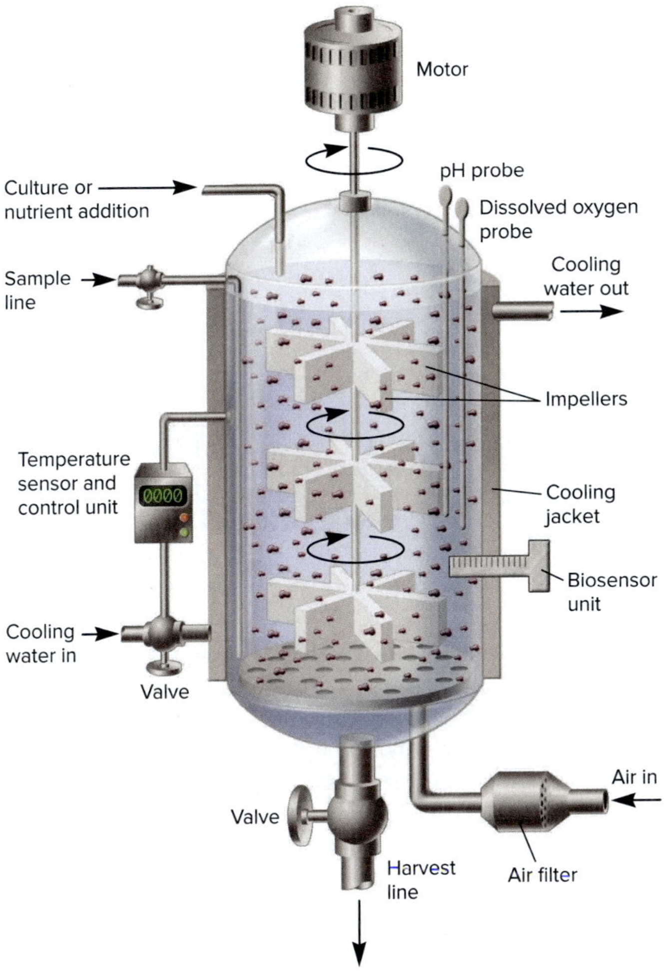

Figure 42.3 An Industrial Fermenter. This unit can be run under oxic or anoxic conditions, with nutrient additions, sampling, and fermentation monitoring carried out under aseptic conditions. Biosensors and infrared monitoring provide real-time information on the course of the fermentation. Specific substrates, metabolic intermediates, and final products can be detected.

Frequently a critical component in the medium is added continuously—a process called **continuous feed**—so that the microorganisms will not have excess substrate available at any given time. This is particularly important with glucose and other carbohydrates. If excess glucose is present at the beginning of a fermentation, it can be catabolized and lost as volatile end products, reducing the final yield of the desired product.

Other culturing approaches exist, among them continuous culture techniques using chemostats. These can sometimes markedly improve rates of substrate use and product yield because microorganisms can be maintained in logarithmic phase. However, for some industrial processes continuous active growth is undesirable, because the product of interest is not made during exponential growth. Microbial products can be classified as primary and secondary metabolites. **Primary metabolites** consist of compounds related to the synthesis of microbial cells during exponential growth. They include amino acids, nucleotides, and fermentation end products such as ethanol and organic acids. In addition, industrially useful enzymes, either associated with the microbial cells or exoenzymes, often are synthesized by microorganisms during log phase growth. **Secondary metabolites** usually accumulate when nutrients become limited or waste products accumulate following the active growth phase, and thus are not produced in chemostats. These compounds are sometimes considered part of a stress response as nutrients become scarce. Most antibiotics fall into this category. ◀ *Growth curves consist of five phases (section 7.4); Chemostats and turbidostats are used for continuous culture of microorganisms (section 7.9)*

Comprehension Check

1. What is the objective of the scale-up process and why is it so critical?
2. What parameters can be monitored in a modern, large-scale industrial fermentation?
3. Which of the following are secondary metabolites: pyruvate, a secreted amylase, an antibiotic, leucine? What kind of system (fermenter vs. chemostat) would most likely be used for the production of the streptomycete antibiotic streptomycin? Explain your answer.

42.4 Production Strains Maximize Output of Industrially Important Compounds

After reading this section, you should be able to:

a. List three techniques used to optimize microbial output of industrial products
b. Compare and contrast three directed evolution technologies
c. Describe how metagenomics increases the pool of microbial products available for screening
d. Describe the goal of synthetic biology and its advantages over more conventional techniques

Most microbes used in industrial processes were originally isolated from natural materials (e.g., soil) and are difficult to grow under standard laboratory conditions. Furthermore, even after appropriate growth conditions are determined, the original strains often produce very low amounts of the desired product. Starting with these microbes, microbiologists use a variety of strategies to create strains, called **production strains,** optimized for industrial purposes. Here we describe some of these approaches.

Chemical and Physical Mutagenesis

Once a microbe is found to produce a promising compound, a variety of mutagenic techniques can be used to increase product yield. For many years, biotechnologists were limited to chemical, ultraviolet (UV) light, and X-ray mutagenesis. For example, penicillin was first produced in cultures of *Penicillium notatum* grown in stationary vessels, but only low concentrations of penicillin were made. In 1943, strain NRRL 1951 of *Penicillium chrysogenum* was isolated and further improved using these methodologies (**figure 42.4**). Today most penicillin is produced with *Penicillium chrysogenum* grown in aerobic, stirred fermenters, yielding 55 times more penicillin than the original static cultures. ◀ *Mutations: heritable changes in a genome (section 16.1)*

Protoplast Fusion

Protoplast fusion is another technique that promotes genetic variability in microorganisms and plant cells. In this method, protoplasts—cells lacking a cell wall—are prepared by growing microbes in an isotonic solution in the presence of enzymes that degrade the cell wall (e.g., lysozyme for bacteria, chitinase for yeasts). The protoplasts of cells of the same or even different species physically fuse during coincubation. The cell wall is then regenerated during growth on osmotic stabilizers such as sucrose. Protoplast fusion allows for recombination between chromosomes of the two organisms that are now contained in a single cell. It is also inherently mutagenic. An example of protoplast fusion is the creation of new, useful *Penicillium* strains following the fusion of *P. roqueforti* and *P. chrysogenum* protoplasts.

Heterologous Gene Expression

In contrast to chemical and physical mutagenesis and protoplast fusion, in vitro genetic transfer of specific genes to a microbe that is easier to manipulate and grow enables a more targeted approach to strain construction. When genes from one organism are cloned into another and then transcribed and translated into protein, it is called **heterologous gene expression.** An important and early example of heterologous gene expression is the cloning and expression of the human insulin genes in *E. coli*. Heterologous gene expression enables the production of specific proteins and peptides without contamination by other products that might be synthesized in the original organism. This approach can decrease the time and cost of

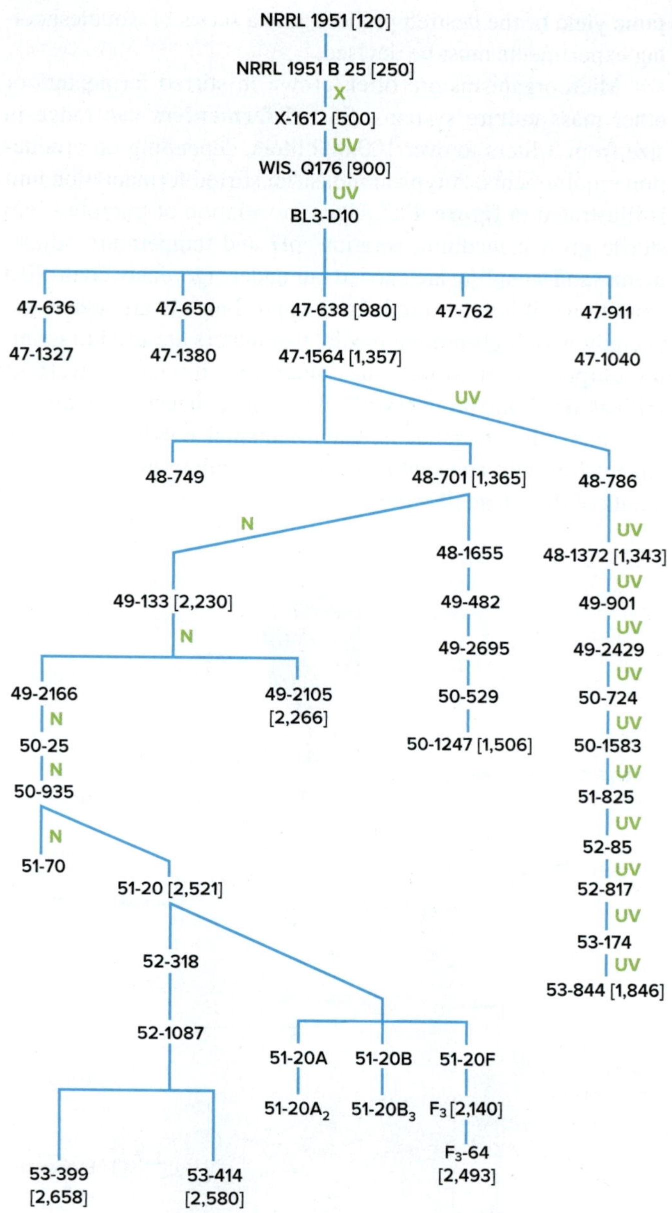

Figure 42.4 Mutation Makes It Possible to Increase Fermentation Yields. The genealogy of the mutational processes used to increase penicillin yields with *Penicillium chrysogenum*. These included X-ray treatment (X), UV treatment (UV), and mustard gas (N). Unmarked transfers were used for mutant growth and isolation. Yields in International Units/mL in brackets.

recovering and purifying a product. Another major advantage of engineered protein production is that only biologically active stereoisomers are produced. This specificity is required to avoid the possible harmful side effects of inactive stereoisomers. ◀ *Key discoveries led to the development of DNA cloning technology (section 17.1)*

Despite these advantages, production of nonnative proteins often disrupts the host cell's highly regulated balance of redox power (e.g., NADH/NAD^+) and ATP. Therefore it is frequently necessary to optimize the cell's regulatory and metabolic activities so that production is maximized. This is called **metabolic engineering.** The application of "omic" data (e.g., genomic, transcriptomic, proteomic, and metabolomic data) to guide genetic manipulation of the host strain for this purpose is called **systems metabolic engineering.**

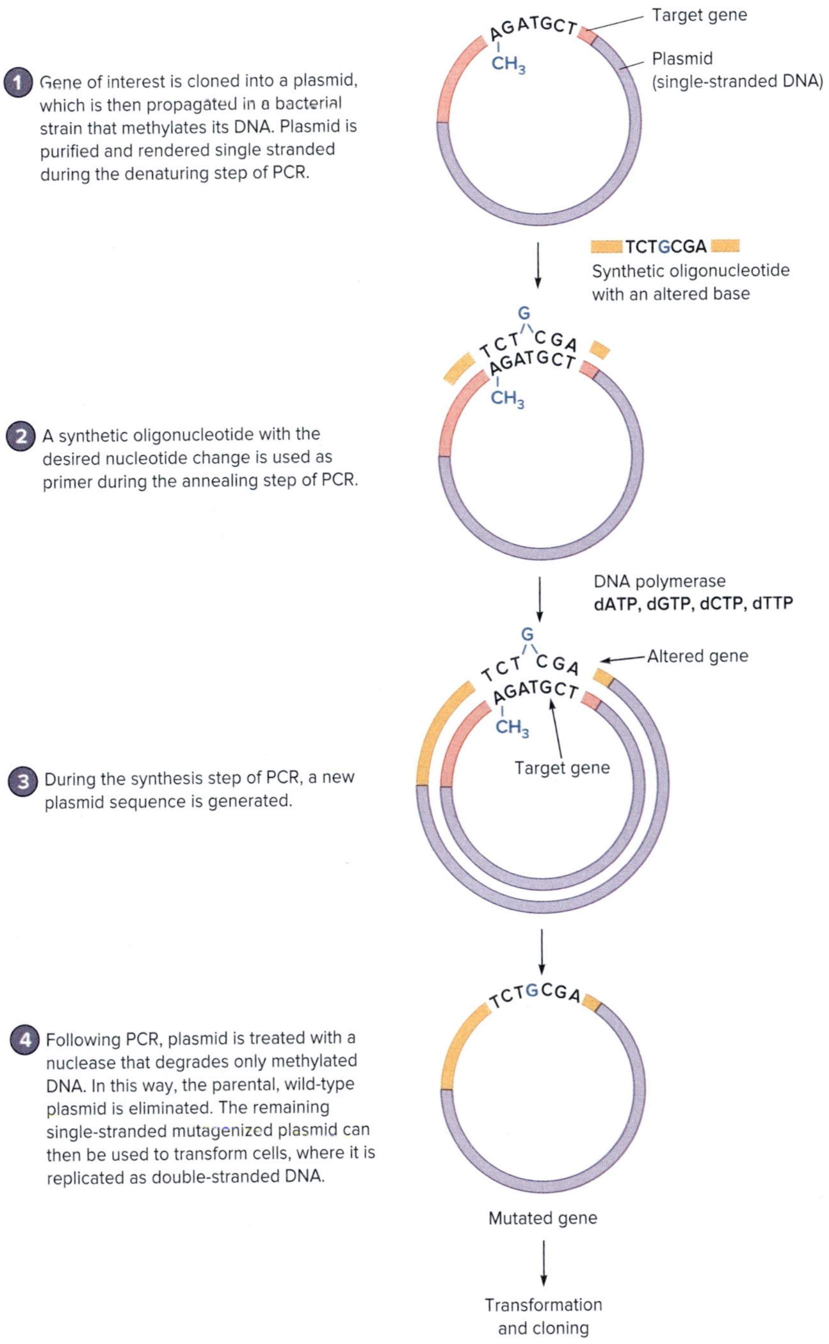

Figure 42.5 Site-Directed Mutagenesis. A synthetic oligonucleotide is used to add a specific mutation to a gene. In some microorganisms, the CRISPR/Cas system can be used to delete a gene or change its coding sequence.

Directed Evolution

As demonstrated in figure 42.4, mutants that overproduce a product or had some other desirable trait were traditionally obtained by "brute force" mutation. That is, mutations that conferred the trait were allowed to accumulate in a microbial strain after repeated rounds of induced mutagenesis and mutations that occur during transfer from culture to culture. Production strains created in this way have a genotype quite different from the parental strain. Indeed, until the advent of genomic sequencing, the number and sites of most mutations remained a mystery. **Directed evolution** takes a more rational approach to the construction of production strains because only genes of interest are targeted for mutagenesis. In most cases, this requires that the gene sequences be altered in vitro, then replaced in the original strain or a heterologous host that may be better suited to produce the compound at high levels. Biotechnologists have discovered that making even minor amino acid substitutions can lead to unexpected changes in protein function and stability, resulting in higher yields of existing products or products with new, desirable features. Regulatory sequences can also be altered, resulting in the overproduction of the desired product.

Several technical advancements were key to the development of direct evolution methodologies. The first is **site-directed mutagenesis.** In this technique, the nucleotide sequence of a specific gene is altered. The steps required are outlined in **figure 42.5**. More recently, CRISPR/Cas technology has become the principal method by which eukaryotic genomes are edited. Optimizing CRISPR/Cas systems for bacterial and archaeal genome editing has lagged behind, but is poised to become an important method in strain development. ◀ *Cas9 nuclease is a precise tool for genome editing (section 17.5)*

A variety of other directed evolution techniques have been developed that can make larger changes to the gene of interest. Some of these are listed in **table 42.2**, which also reveals that many natural products are not proteins and therefore cannot be modified by simply mutating a single, structural gene. Instead, altering these nonprotein products requires an understanding of the biosynthetic pathway by which they are catalytically assembled.

The biosynthetic pathways of certain types of natural products makes them particularly amenable to bioengineering. For instance, polyketide and nonribosomal peptide antibiotics have become models of bioengineering because the biosynthetic enzymes are grouped into large modular complexes,

Table 42.2 Some Directed Evolution Technologies

Technology	Overview of Approach	Examples of Products Generated
Genome-based strain reconstruction	A new strain is constructed based on the genotype of a production strain previously generated by "brute force." The new strain possesses only the mutations that are required for overproduction of the product.	Lysine
Metabolic pathway engineering	Rational approach to product improvement by either modification of specific biochemical reactions in a pathway, such as mutagenesis of biosynthetic enzyme genes or the use of recombinant DNA technology to introduce new genes into the production strain. The goal is to optimize the flux of metabolites so the pathway operates most efficiently and the highest product yields possible are obtained.	Antibiotics: cephamycin C, neomycin, spiramycin, erythromycin L-lysine, aromatic amino acids, ethanol, vitamins
Assembly of designed oligonucleotides (ADO)	If the gene of interest contains a conserved sequence flanking the region to be mutated, oligonucleotides can be designed that include the conserved region and a mixture of point mutations in the region to be mutated. This combination of different (called degenerate) oligonucleotides is used to create variants by in vitro homologous recombination.	Lipases
Error-prone PCR	Variants of a gene are produced by using a DNA polymerase that is particularly error-prone during PCR.	Lycopene
DNA shuffling	Similar genes from different species are randomly fragmented and pooled, and new DNA fragments are generated by in vitro homologous recombination. Progeny sequences encoding desirable traits are identified; these new genes are then shuffled (bred) over and over again, creating new progeny that contain multiple desirable mutations.	Cephalosporinase
Whole genome shuffling	Similar to DNA shuffling described above, except entire genomes are recombined. Protoplast fusion may be used to obtain a single cell with the genomes of two microbial species so that recombination can occur.	Lactic acid

Source: Adrio, J. L., and Demain, A. L. "Genetic improvement of processes yielding microbial products." *FEMS Microbiol. Rev.* 30, 2016, 187–214.

which function in an assembly-line fashion (**figure 42.6**). Each module within the complex catalyzes a specific type of reaction. For instance, the anticancer agents epothilones C and D are synthesized by a combination of polyketide synthase (PKS) and nonribosomal peptide synthetase (NRPS) enzymes. The first module is responsible for binding a starter unit and transferring it to the first extension module. Each extension module modifies the intermediate and hands it off to the next module in line until it reaches the last module and is released. The mixing and matching of domains encoding extension modules from hundreds of such modular multienzyme complexes can be used to generate new products. Many soil bacteria, such as members of the genera *Streptomyces* and *Bacillus*, possess PKS and NRPS genes. Biotechnologists are mining other microbes and metagenomes for novel PKS and NRPS genes.

The recognition that RNA is a versatile molecule that can have catalytic activity and control gene expression has given rise to a new class of therapeutic agents. Noting that RNA can fold into seemingly an infinite number of complex three-dimensional structures, molecular biologists use in vitro evolution to generate a library of up to 10^{15} different RNA molecules of interest. This procedure is called "systematic evolution of ligands by exponential enrichment" (SELEX) and is outlined in **figure 42.7**. Recall that a molecule that binds to a specific target or receptor is called a ligand, so in this case, the goal is to design an RNA ligand that has biological activity because it fits very tightly into a cleft or groove in a specific target molecule. This class of engineered RNAs have been given the name **aptamers** (Latin *aptus*, "fitting"). In 2005 the first aptamer-based therapy, for the progressive vision disease macular degeneration, began clinical use. ◀ *Ribozymes: catalytic RNA molecules (section 10.6)*

All directed evolution methodologies are designed to produce new genes, proteins, or RNAs that will create hundreds, if not thousands, of variants. Sorting through so many new candidate molecules requires a specific assay that assesses the desired trait, such as stability under reducing conditions or increased ability to bind a specific molecule. Selecting the best variant could represent a tremendous workflow bottleneck; however, the development of **high-throughput screening (HTS)** enables the rapid selection of a subset of desirable molecules from tens of thousands of candidates. HTS employs a combination of robotics and computer analysis to screen sample molecules and even whole cells for a specific trait, usually in 96-well microtiter plates. The combination of

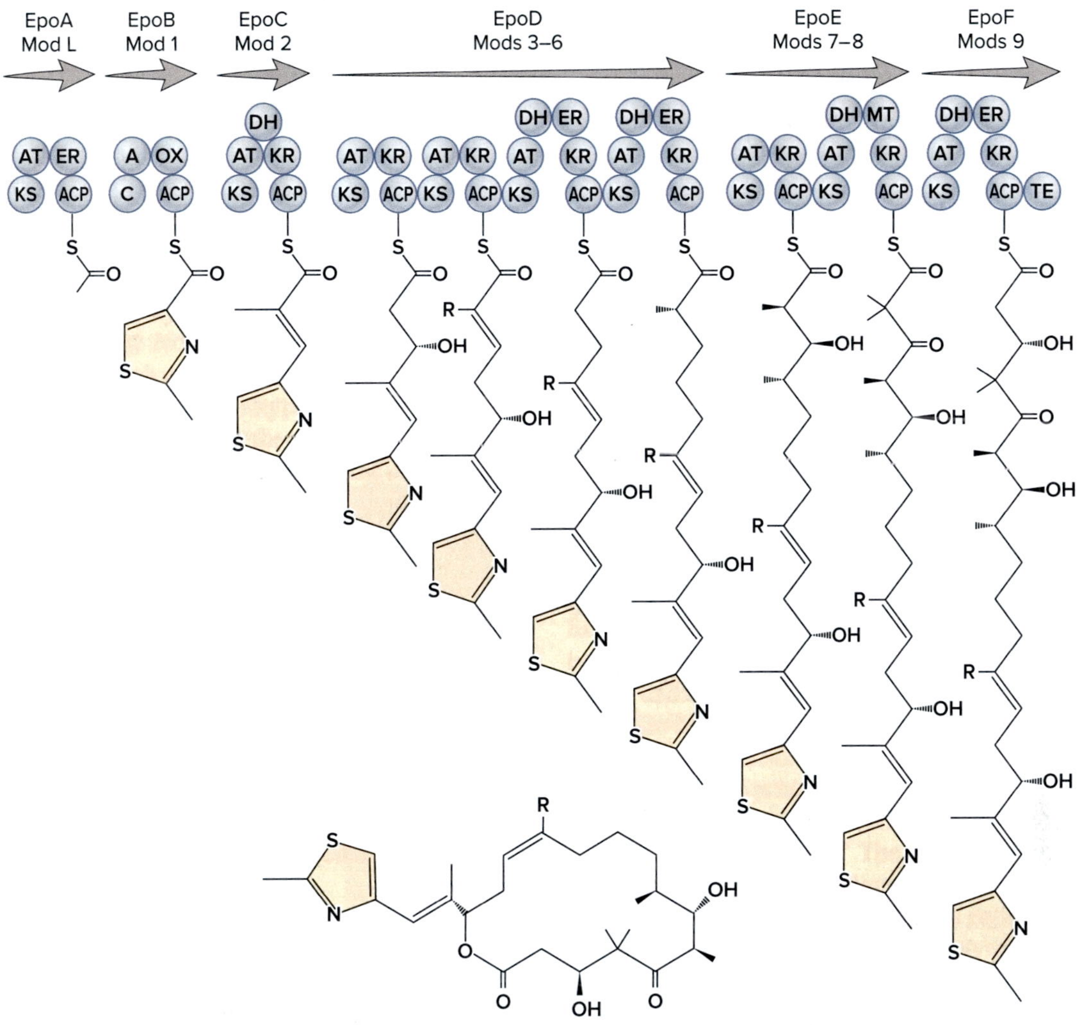

Figure 42.6 Modular Assembly of the Anticancer Agents Epothilone C and D. The genes encoding the enzymes that assemble this polyketide are found on the genome of the myxobacterium *Sorangium cellulosum.* They are arranged on the genome in the order of synthesis. The catalytic domains in the corresponding enzyme are shown as circles. These domains include keto-acyl synthase (KS), acyltransferase (AT), enoyl reductase (ER), acyl carrier protein (ACP), adenylation (A), condensation (C), oxidation (O), dehydratase (DH), ketoreductase (KR), and thioesterase (TE). The progression of synthesis of the molecule is shown left to right, and the mature structure is shown, bottom center.

MICRO INQUIRY *Why are modular genes particularly suitable as platforms for the development of new natural products of industrial interest?*

directed evolution approaches and HTS has propelled industrial microbiology to a new level of exploration and efficiency not previously envisioned.

Metagenomics

As we have learned, techniques for growing most microbes found in nature have not yet been developed. While having a pure culture of the microorganism that produces an interesting compound is always the best approach, industrial microbiologists would be at a great disadvantage if they limited their search for products to only those found in microbes that can be grown in the laboratory. The example of antibiotics made by actinobacteria is instructive. Worldwide, the top 10 centimeters of many soils are estimated to contain so many actinobacteria that have never been screened for antibiotic production that at the current rate, it would take another 128 years to screen them all, assuming they can be easily cultured. Industrial microbiologists have therefore turned to metagenomics to sample the genetic diversity from a number of environments. ◀ *Metagenomics provides access to uncultured microbes (section 18.3)*

Metagenomic libraries from natural samples are generally screened in two ways. First, all the cloned DNA from the environment is sequenced and compared to genes that have

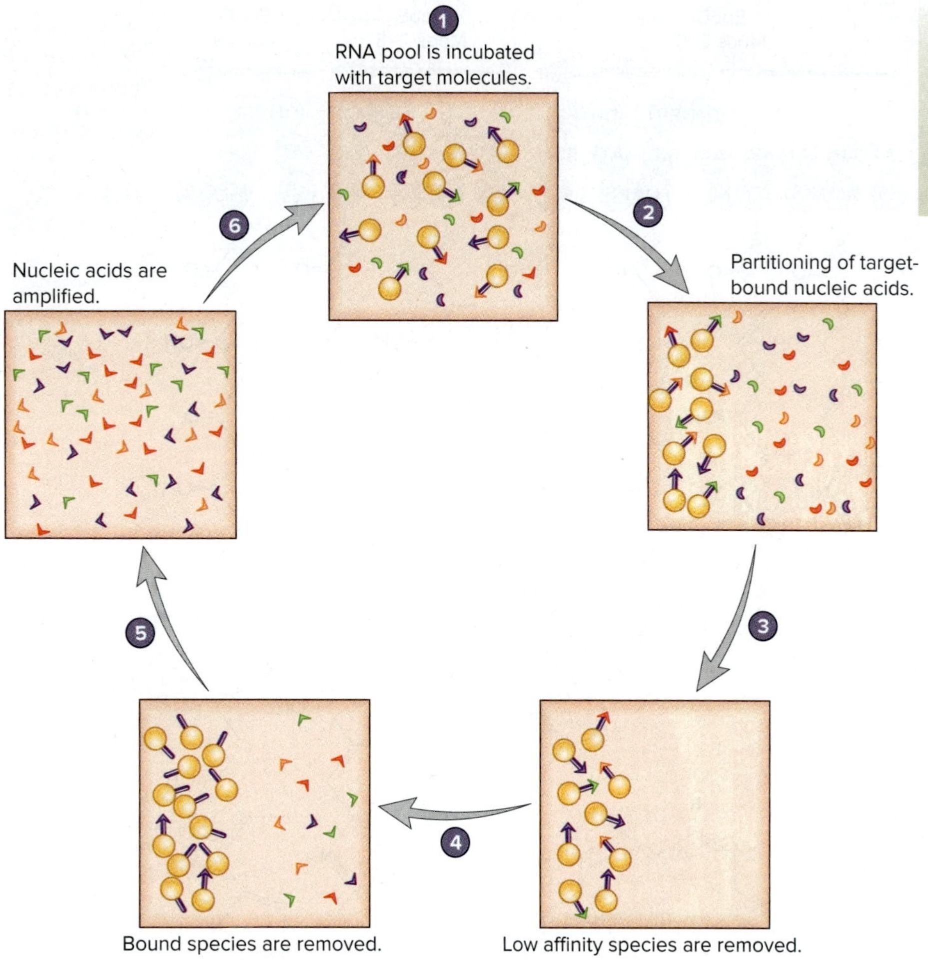

Figure 42.7 Directed Evolution of RNA Molecules. Systematic evolution of ligands by exponential enrichment (SELEX).

MICRO INQUIRY *Why is SELEX considered an "enrichment" technique?*

already been characterized (*see figure 18.8*). In this way, new versions of known genes (alleles) can be identified based upon similarity to genes that are already known. This means that genes for which no function can be deduced will be ignored. But what if one of these novel genes encodes a previously undiscovered protein with a highly desirable activity? A functional screen of the metagenomic library can be used to explore this possibility (**figure 42.8**). In a functional screen, the genes are cloned into vectors and expressed in microbes such as *E. coli* or *Bacillus subtilis*, which are then screened for the acquisition of a new, specific function. Of course, these two lines of inquiry are not mutually exclusive. Keep in mind that once interesting genes are identified, they can undergo directed evolution techniques to optimize products for industrial use. ◀ *Genomic libraries: cloning genomes in pieces (section 17.3)*

Synthetic Biology

So far we have discussed engineering existing regulatory genetic networks and biochemical pathways. **Synthetic biology** looks beyond the constraints of what is already present in an organism and constructs or repurposes entirely new capabilities. Simply put, the field of synthetic biology involves the forward engineering of biological systems (plants, microbes) to perform completely novel functions. Examples include the design of microorganisms that seek out and kill cancerous tumor cells and the redesign of photosynthesis to produce excess energy that can be fed into existing electrical grids.

Synthetic biology offers three chief advantages over the techniques discussed so far. It is more efficient to assemble large arrays of genes encoding enzymes that perform complex tasks than to try and find such an assembly already existing in nature. For example, the assembly of multiple genes from plants, fungi, and bacteria into a single host microorganism has resulted in a range of new biofuels. Next, the need for metabolic engineering can be limited or reduced entirely when new genes are synthesized in vitro for a specific host. This is because the genes can be modified in vitro to meet host specifications. Lastly, synthetic biology can mix-and-match regulatory mechanisms with the expression of genes normally governed by other means. An interesting example is the development of a bacterial strain that seeks out tumor cells and releases its anticancer drug cargo when triggered by a quorum sensing regulator (*see figure 14.23*).

There is, however, no set of instructions for something so complex as designing microbes to perform completely novel functions. Each project requires careful planning followed by the iterative process of design, build, test, redesign, rebuild, retest, and so on. For instance, to engineer a bacterium to produce a biofuel, the first step would be to review literature to determine what genes are needed (planning). The next step would be to figure out how these genes are going to be cloned and what bacterial species will host them (design). Once the genes are inserted into the host (build), the amount of biofuel made would be quantified (test). It is safe to say that perfection is never attained with the first try. However, the results inform the synthetic biologist of what steps need revision (or if the entire design should be scrapped and fresh approach taken). Synthetic biology has the potential to generate a vast diversity of products, and researchers at the J. Craig Venter Institute have prepared for such versatility by constructing and expressing a completely

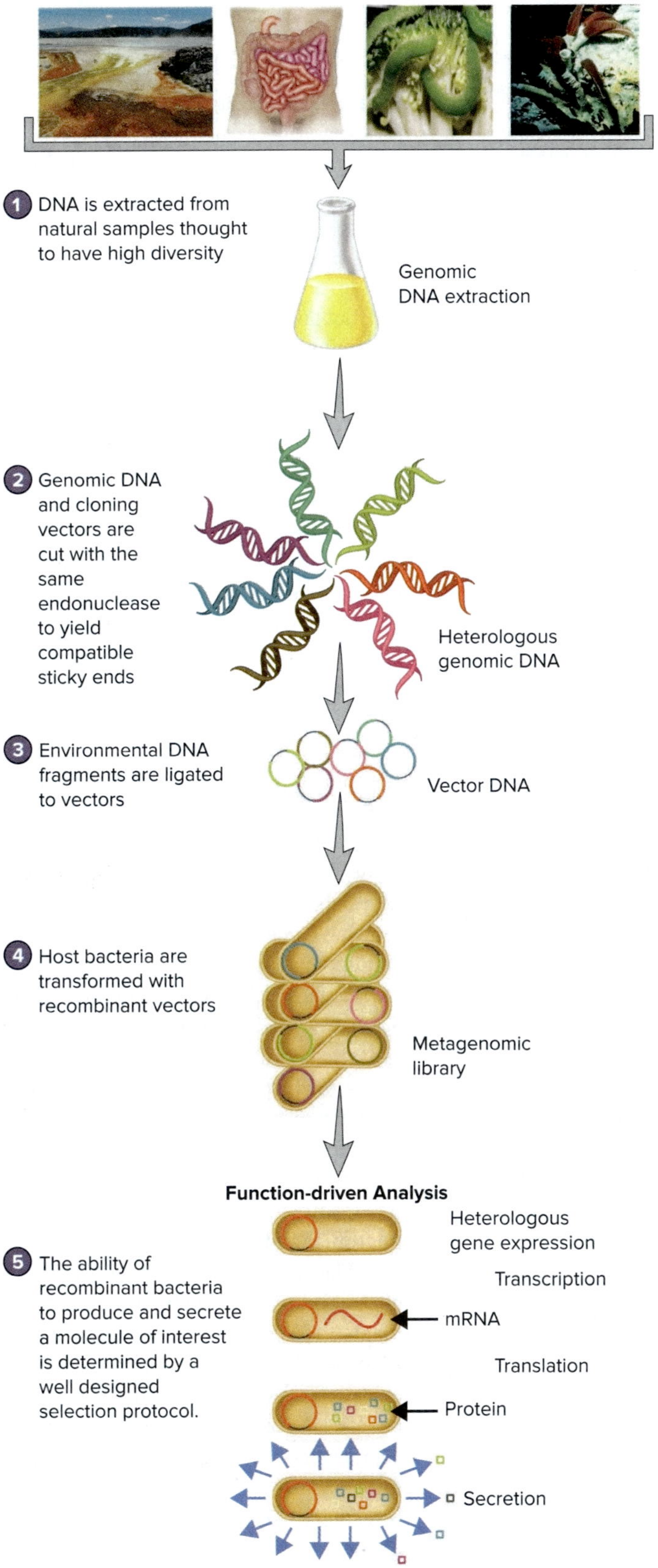

Figure 42.8 Functional Screening of Metagenomic Libraries. Rather than trying to culture every organism in a natural sample to look for the production of an important, novel compound, a metagenomic library can be constructed and expressed in host bacteria, which are screened for the desired protein function. This is usually performed in a high-throughput fashion.

(1a) ©Yi Xiang Yeng/iStock/Getty Images; (1b) ©McGraw-Hill Education; (1c) ©Nigel Cattlin/Science Source; (1d) Source: OAR/National Undersea Research Program (NURP)/College of William & Mary/NOAA

synthetic genome in mycoplasma cells as described in chapters 17 and 18. This genome serves as the platform into which new genes can be added for the production of novel compounds.

Comprehension Check

1. What might be the disadvantages of using production strains with undefined induced mutations?
2. Define protoplast fusion and list types of microorganisms used in this process.
3. What is metabolic engineering? Besides antibiotics, what other types of biologically active molecules do you think might be created using combinatorial biology?
4. What is high-throughput screening and why has it become so important?
5. Compare and contrast the development of a new product using synthetic biology and functional screening of metagenomic libraries.

42.5 Agricultural Biotechnology Relies on a Plant Pathogen

After reading this section, you should be able to:

a. Explain how the natural infection process used by *Agrobacterium tumefaciens* has been modified for the genetic modification of plants
b. Defend the use of Bt as a pesticide

A plasmid from the plant pathogenic bacterium *Agrobacterium tumefaciens* is chiefly responsible for the successful genetic engineering of plants. In nature, infection of dicotyledonous (dicot) plant cells by this bacterium transforms the cells into tumor cells, and crown gall disease develops. Only strains of *A. tumefaciens* possessing a large conjugative plasmid called the Ti (*t*umor *i*nducing) plasmid are pathogenic (*see figures 31.12 and 31.13*). This is because the Ti plasmid transfers a region known as T-DNA to the plant's chromosomes at sites where it is stably maintained.

The Ti plasmid and its T-DNA have been genetically modified for use as a vector for the insertion of recombinant DNA into plant chromosomes. The tumor-inducing genes have been deleted, while elements found in other vectors (e.g., selectable marker and multicloning site; *see figure 17.6*) have been added. The gene or genes to be expressed by the plant are inserted into the T-DNA region between the direct repeats needed for integration into the plant genome (**figure 42.9**). Then the plasmid is returned to *A. tumefaciens,* plant cells are infected with the bacterium, and transformants are selected by screening for a trait encoded by T-DNA. Finally, whole plants are regenerated from the transformed cells.

While it is relatively easy to use the *A. tumefaciens* Ti plasmid to modify dicots such as potato, tomato, celery, lettuce, and alfalfa, its use in monocots has only recently been optimized. The creation of new procedures for inserting DNA into monocot plant cells is leading to the use of recombinant DNA techniques with many important crop plants such as corn, wheat, and other grains.

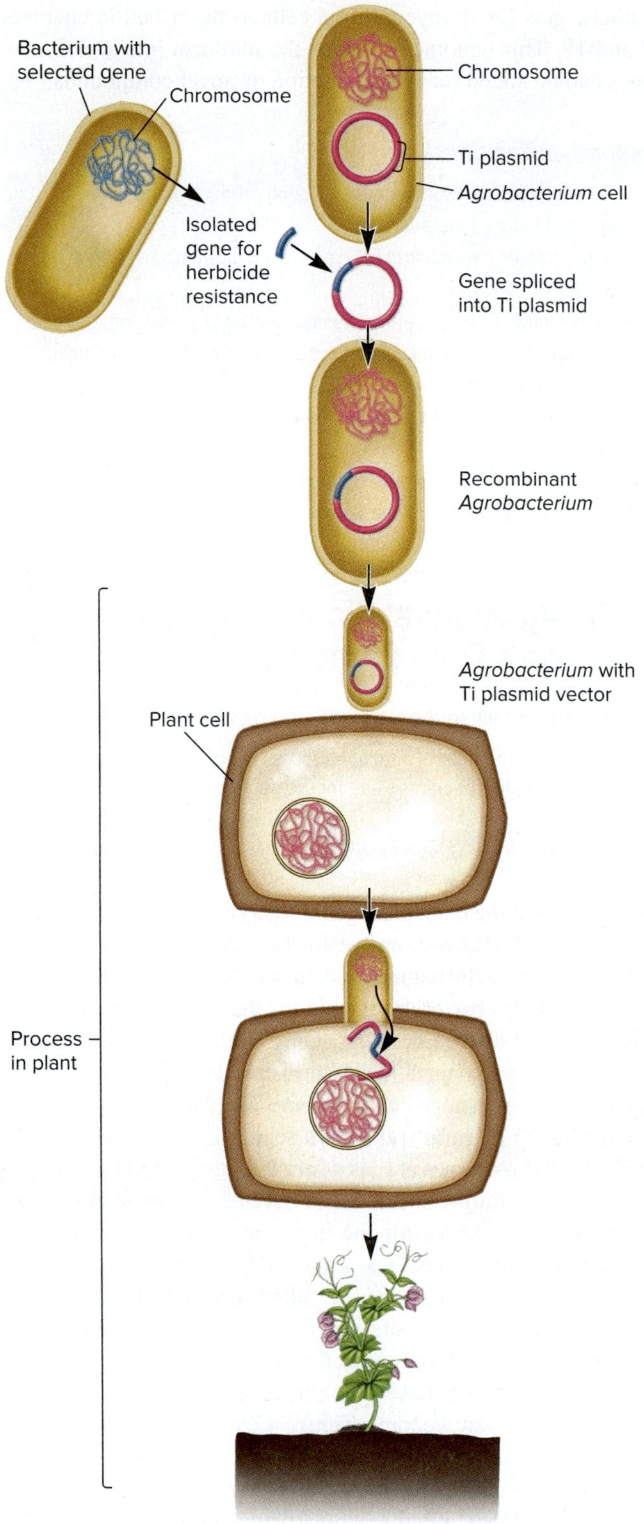

(a) The Ti plasmid of *Agrobacterium tumefaciens* is genetically modified to remove virulence genes and genes to be expressed by plant cells are inserted.

(b) The recombinant plasmids are taken up by the *Agrobacterium* cells, which multiply and copy the foreign gene.

(c) Genetically engineered *Agrobacterium* is inoculated into a culture of target plant cells and infects the cells.

(d) Fusion of the bacterium with the plant cell wall permits entrance of the Ti plasmid and incorporation of the herbicide gene into the plant chromosome. Mature plants are grown from single cells, and these transgenic plants express the new gene.

(e) Because the gene is part of the plant's genome, it is transmitted to offspring in seeds.

Figure 42.9 Bioengineering of Plants. Most techniques employ a genetically modified strain of the natural tumor-producing bacterium called *Agrobacterium tumefaciens*.

Another microbial application to agriculture involves the use of bacteria, fungi, and viruses as **bioinsecticides** and **biopesticides.** These are defined as biological agents; that is, microbes or their components that kill susceptible insects. Bacterial agents include a variety of *Bacillus* species; however, *B. thuringiensis* is most widely used. This bacterium is only weakly toxic to insects as a vegetative cell, but during sporulation, it produces an intracellular protein toxin crystal, the parasporal body, that kills insects (*see figure 23.19*). The parasporal crystal, after exposure to alkaline conditions in the insect hindgut, breaks apart to release protoxin. After protoxin reacts with a protease enzyme in the target insect's gut, active toxin is generated. Six of the active toxin units integrate into the plasma membrane to form a hexagonal-shaped pore through the midgut cell, as shown in **figure 42.10**. This leads to the loss of osmotic balance and ATP, and finally to cell lysis. ◀ *Order* Bacillales *(section 23.2)*

B. thuringiensis has become an important industrial microbe. The toxin is widely marketed as the insecticide **Bt,** and the toxin genes can be used to genetically modify plants. The manufacture of Bt begins by growing *B. thuringiensis* in fermenters. The endospores and crystals are released into the medium when cells lyse (*see figure 3.47*). The medium is then centrifuged and converted to a powder for application to plants. Bt has been used on a worldwide basis for over 40 years. Unlike chemical insecticides, Bt does not accumulate in the soil or in nontarget animals. Rather, it is readily lost from the environment by microbial and abiotic degradation.

The genes encoding the Bt toxin (*cry* genes, for *cry*stal) can be introduced into plants. Unlike many other genetically modified organisms, transgenic plants expressing the *cry* gene have generally been well accepted. The widespread acceptance of these plants reflects the history of safe application of Bt as an insecticide without adverse environmental or health impacts. In addition, it is well understood that the Cry protein can only be activated in target insects. Long-term studies have shown that Bt is nontoxic to mammals and is not an allergen in humans. One potential problem, the horizontal gene flow of the *cry* gene to weeds and other plants, has not been reliably demonstrated. On the contrary, the decreased need for insecticides when cultivating Bt-modified crops has resulted in the recovery of nontarget insects

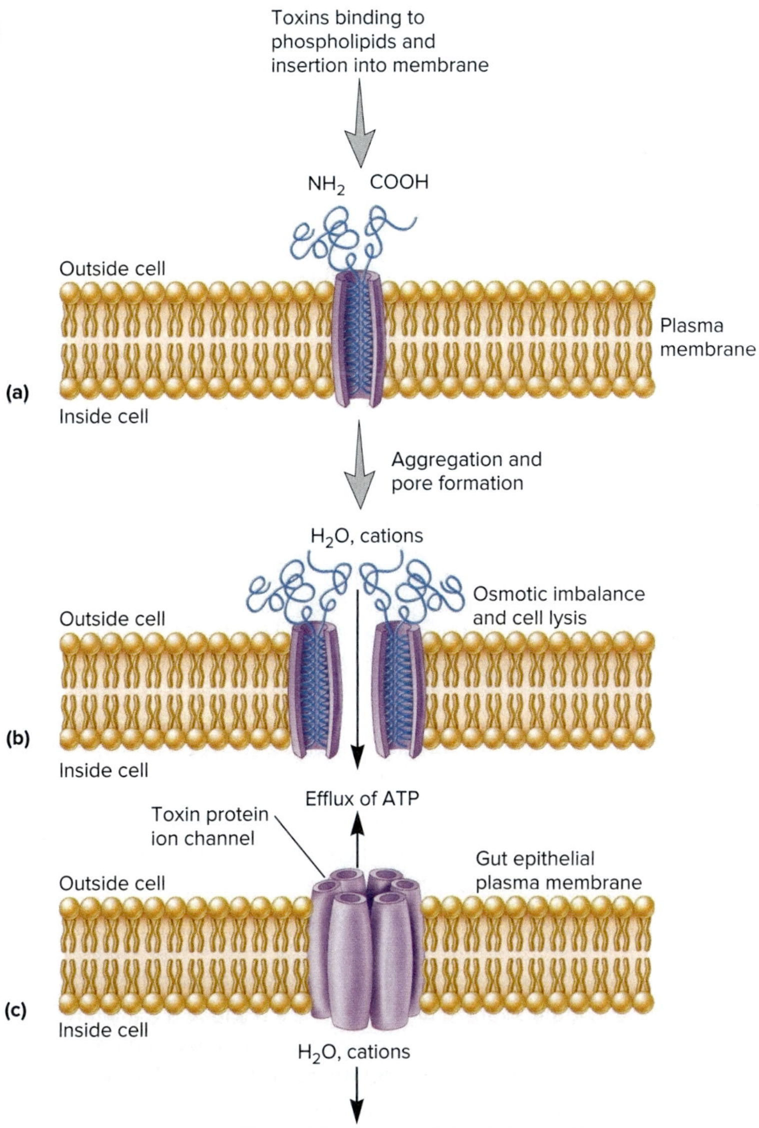

Figure 42.10 The Mode of Action of the *Bacillus thuringiensis* Toxin in the Target Insect. (a) Insertion of the 68 kDa active toxin molecules into the membrane. (b) Aggregation and pore formation, showing a cross section of the pore. (c) Final hexagonal pore, which causes an influx of water and cations as well as a loss of ATP, resulting in cell imbalance and lysis.

MICRO INQUIRY *How is Bt toxin activated within the insect gut?*

that were also eliminated by insecticide application. Many of these insects eliminate undesirable pest insects, resulting in enhanced levels of biocontrol where Bt crops have been sown.

Comprehension Check

1. What is the Ti plasmid and how is it modified for the genetic modification of plants?
2. How is Bt toxin produced and why is it so widely accepted?
3. Trace the steps needed to construct a transgenic plant that produces the Cry protein.

42.6 Some Microbes Are Products

After reading this section, you should be able to:

a. Describe the use of microbes in nanotechnology
b. Explain the utility of biosensors

So far, we have discussed the use of microbial products to meet defined goals. However, microbial cells can be marketed as valuable products. Perhaps the most common example is the inoculation of legume seeds with rhizobia to ensure efficient nodulation and nitrogen fixation. Here we introduce several other microbes and microbial structures that are of industrial or agricultural relevance. ◀ *Nitrogen-fixing bacteria are vital to agriculture (section 31.3)*

Diatoms have aroused the interest of nanotechnologists. These photosynthetic protists produce intricate silica shells that differ according to species (**figure 42.11**, *also see figure 24.19*). Nanotechnologists are interested in diatoms because they create precise structures at the micrometer scale that can be used for making optical, catalytic, or electrical materials. Three-dimensional structures in nanotechnology are currently built plane by plane, and meticulous care must be taken to etch each individual structure to its final, exact shape. Diatoms, on the other hand, build directly in three dimensions and do so while growing exponentially. The silica in these shells can be chemically converted to silicon to generate nanoelectronics. Another application involves the use of diatom shells to coat solar energy cells. This is attractive because diatom shells capture three times more electrons than the standard coatings. ◀ Stramenopila *(section 24.4)*

Figure 42.11 Marine Diatom Surface Features Are Attractive Platforms for Nanotechnology. The precise patterns on diatom silica shells have nanotechnology applications, including biosensing, drug delivery, and the development of electronic, photonic, and structural materials. A falsely colorized scanning electron micrograph of *Actinoptychus* sp., which ranges in diameter from 20 to 150 μm, is shown here.

Magnetotactic bacteria are also of interest to nanotechnologists. Magnetosomes are formed by certain bacteria that convert iron to magnetite. The sizes and shapes of the magnetosomes differ among species, but like diatom shells, they are perfectly formed despite the fact that they are only tens of nanometers in diameter. Although tiny, magnetic beads can be chemically synthesized, they are not as precisely formed and lack the membrane that surrounds magnetosomes. This membrane enables the attachment of useful biological molecules such as enzymes and antibodies. Potential applications for magnetosomes currently under investigation include their use as a contrast medium to improve magnetic resonance tomography (commonly called MRI) and as biological probes to detect cancer at early stages. ◀ *Inclusions (section 3.6);* Magnetospirillum *displays cytoplasmic magnetosomes (section 22.1)*

Another rapidly developing area of biotechnology is that of **biosensor** production. In this field of bioelectronics, living microorganisms (or their enzymes or organelles, and most recently diatom shells) are linked with electrodes, and biological reactions are converted into electrical currents. Biosensors are used to measure specific components in beer, monitor pollutants, detect flavor compounds in food, and study environmental processes such as changes in biofilm concentration gradients. Using biosensors, it is possible to measure the concentration of substances from many different environments. Biosensors have been developed using immunochemical-based detection systems. These biosensors detect pathogens, herbicides, toxins, proteins, and DNA.

Comprehension Check

1. Why are diatoms and magnetotactic bacteria of interest to nanotechnologists? What do you think might be special challenges when growing these microbes for industrial applications?
2. In what areas are biosensors being used to assist in chemical and biological monitoring efforts?

Key Concepts

42.1 Microbes Are the Source of Many Products of Industrial Importance

- A wide variety of compounds are produced in industrial microbiology that impact our lives in many ways. These include antibiotics, amino acids, organic acids, biopolymers, and biosurfactants (**table 42.1**).
- Microorganisms also can be used as biocatalysts to carry out specific chemical reactions.
- Vaccine development continues to challenge biotechnologists; reverse vaccinology identifies potential targets and uses their structure and function to inform vaccine production.

42.2 Biofuel Production Is a Dynamic Field

- The field of microbial energy conversion includes the production of biofuels and the development of microbial sources of H_2.
- Although corn is currently the most common substrate used for the microbial production of ethanol, the use of both corn as substrate and ethanol as product is being challenged.

42.3 Growing Microbes in Industrial Settings Presents Challenges

- Microorganisms can be grown in controlled environments of various types using fermenters and other culture systems (**figure 42.3**).
- If defined constituents are used, growth parameters can be chosen and varied in the course of growing a microorganism. This approach is used particularly for the production of amino acids, organic acids, and antibiotics.

42.4 Production Strains Maximize Output of Industrially Important Compounds

- Historically, microorganisms used in industry were isolated from nature and modified by the use of classic mutational techniques.
- Site-directed mutagenesis and protein engineering are used to modify genes and their expression. These approaches are leading to new and often different products with novel properties (**figures 42.5** and **42.6**).
- Directed evolution includes a suite of techniques designed to target specific molecules for change with the goal of optimizing their function in a particularly useful way (**table 42.2**).
- Metagenomics is an important tool in the search for new products, which are detected in a metagenomic functional screen (**figure 42.8**).
- Synthetic biology is the use of molecular methods to design and construct new cellular behaviors and products.

42.5 Agricultural Biotechnology Relies on a Plant Pathogen

- Many plants are genetically modified through the use of the Ti plasmid of *Agrobacterium tumefaciens*. The plasmid can be engineered to introduce genes whose products confer new traits to plants (**figure 42.9**).
- *Bacillus thuringiensis* is an important biopesticide, and the Bt gene has been incorporated into several important crop plants (**figure 42.10**).

42.6 Some Microbes Are Products

- Microorganisms are being used in a wide range of biotechnological applications such as nanotechnology and biosensors.

Active Learning

1. You have discovered a secondary metabolite made by an unusual fungus that has great industrial potential. Unfortunately, the fungus makes very little of this product, so you must develop a scheme to increase the yield. Discuss the pros and cons of each of the following strategies: (a) mutagenesis of the fungus followed by repeated passaging until a high-producing strain evolves, (b) heterologous expression of the genes and regulators responsible for producing this product in a bacterial host, (c) heterologous expression of the genes responsible for producing this product in a bacterial host but using a bacterial promoter that will drive constitutive expression of the product.
2. Brewer's yeast, *Saccharomyces cerevisiae,* does not metabolize lactose, but an engineered strain of *S. cerevisiae* able to consume lactose is of industrial interest. This is because such a strain can convert whey (an unusable by-product of cheese production) to ethanol. Such a strain has been engineered by cloning the genes that encode β-galactosidase and lactose permease from another yeast, *Kluyveromyces lactis,* into *S. cerevisiae.* However, the recombinant *S. cerevisiae* expresses the *K. lactis* genes only at a low level, so very little lactose is consumed. Discuss a directed evolution technique that could be applied to enhance β-galactosidase production by engineered *S. cerevisiae.*
3. The development of insect resistance to the Cry protein of *B. thuringiensis* may limit the use of Bt as an effective pesticide. A directed evolution approach can be taken to develop Cry proteins with high affinity to specific insect protein targets. Review table 42.2 and suggest a technique by which the different variations of the *cry* gene could be modified and Cry proteins with new capabilities selected. Be sure to consider how you will screen your new Cry proteins and consider if you will use continuous (iterative) evolution or if you have a specific end point in mind.

   Read the original paper: Badran, A. H., et al. 2016. Continuous evolution of *Bacillus thuringiensis* overcomes insect resistance. *Nature.* 533:58–63. doi:10.1038/nature17938.

43 Applied Environmental Microbiology

Source: U.S. Coast Guard

Deepwater Horizon Oil Consumed by Microbes

One of the planet's largest oil disasters occurred on April 20, 2010, when an explosion aboard BP's Deepwater Horizon oil rig resulted in the death of 11 people and the release of at least 200,000 tons of oil and natural gas (methane) over 87 days into the Gulf of Mexico. Unlike other high-profile oil spills, this one featured dramatic images of oil spewing from an uncapped well on the seafloor. About half the spill rose to the surface where it was physically removed by booms or formed droplets and either sank or washed ashore. But about 40% of the spill included compounds with fewer than 10 carbons (e.g., methane, ethane, propane, and butane). These formed lateral plumes trapped 800 to 1,200 meters below the less dense gas-free seawater.

Several years later, mass balance calculations revealed that about a quarter of the spilled hydrocarbons could not be accounted for. Could microbes be responsible for some (or all) of the "missing" hydrocarbons? Could they have mineralized the compounds to CO_2 or immobilized them as biomass? Arriving at a satisfactory answer to these questions required a combination of methods, including sequencing 16S rRNA genes, metagenomic and single-cell genome sequencing, and measuring the flux of ^{14}C-labeled test hydrocarbons, just to name a few. In the end, the scientific community agreed that microbes present in the sediments degraded a large (but unknown) fraction of the spill. Interestingly, the microbial community changed during this process. Early in the spill, when concentrations of saturated hydrocarbons were high, single-cell genome sequencing revealed genes for alkane degradation, chemotaxis, and motility. However, as alkane hydrocarbons declined, so did evidence of genes whose products degrade them. The loss of alkanes meant that polyaromatic hydrocarbons (PAHs) represented a larger fraction of the undegraded oil. Many large PAHs persisted until groups of microbes with related capabilities, called microbial consortia, grew in numbers and collectively degraded these compounds. However, the most complex PAHs remain trapped in sediments where they are toxic to eukaryotes. Thus, although microbial oxidation of many contaminants has been helpful, the persistence of PAHs in the environment a decade later has significant implications for future bioremediation efforts after the next catastrophe.

The Deepwater Horizon experience vividly illustrates that the natural decontamination of soils, waters, and wastewaters is largely a microbial process. Water cleanliness—whether one is referring to tap water or wastewater management—is a multidisciplinary endeavor, involving geology, biochemistry, and physics, among other fields. In this chapter, we address these seemingly opposing aspects of water purity. We also discuss the means by which microbes can contribute to meeting the world's insatiable demand for energy and their role in degrading environmental contaminants—bioremediation.

Readiness Check:

Based on what you have learned previously, you should be able to:

- ✓ Describe several waterborne diseases (sections 39.4 and 40.5)
- ✓ List environmental factors that influence microbial growth (section 7.5)
- ✓ Explain how the term *redox potential* is used to describe microbial habitats and how this influences the elements found in these habitats (section 28.1)
- ✓ Distinguish between mineralization and immobilization of an element (section 28.1)
- ✓ Compare and contrast carbon substrates that are readily degraded with those that are not (section 28.1)
- ✓ Sketch generalized carbon and nitrogen iron cycles (section 28.1)

43.1 Purification and Sanitary Analysis Ensure Safe Drinking Water

After reading this section, you should be able to:

a. Trace water purification steps from source to tap
b. Describe how the U.S. Environmental Protection Agency regulates water purity
c. Discuss the use of indicator organisms and how they are enumerated

Access to clean, drinkable water is a fundamental human right, yet over a billion people worldwide do not enjoy a reliable source of potable water. Not only must water be purified before drinking, it must be monitored for purity. The U.S. Environmental Protection Agency (EPA) has designated

Table 43.1 Microbial Water Contaminant Candidates Listed by the EPA

Organism	Concern
Adenoviruses	Viruses that cause respiratory illness, sometimes gastrointestinal disease
Caliciviruses	Viruses that cause self-limiting nausea, vomiting, diarrhea; includes norovirus
Campylobacter jejuni	Causes nausea, vomiting, and diarrhea; usually self-limiting
Enteroviruses	Includes poliovirus, coxsackie viruses, and echoviruses; on rare occasions cause central nervous system disease
E. coli O157:H7	Enterohemorrhagic *E. coli* that causes severe GI illness; can also cause kidney failure
Helicobacter pylori	Bacterium that causes gastric ulcers and cancer
Hepatitis A virus	Causes liver disease leading to jaundice
Legionella pneumophila	Found in hot water systems; causes bacterial pneumonia when inhaled. Usually limited to elderly patients.
Mycobacterium avium	Respiratory illness or disseminated disease in those with underlying disease or immunosuppression
Naegleria fowleri	Protozoan found in warm surface and groundwater; causes primary amoebic meningoencephalitis
Salmonella enterica	Causes GI illness; severity depends on subspecies
Shigella sonnei	Causes GI illness, including bloody diarrhea

certain bacterial, viral, and protozoan pathogens that can survive (or thrive) in water as "microbial containment candidates" because they represent particularly severe health risks (**table 43.1**).

Water purification is therefore a critical link in controlling disease transmission. The choice of purification method depends on the volume, source, and initial quality of the water. Many municipalities draw their water from surface sources, such as reservoirs. Surface water is usually purified by a process that consists of at least three steps (**figure 43.1**). First, the water may be held for a specified duration to allow larger particles to settle out in a **sedimentation basin.** Sometimes coagulants are also added in a procedure called **coagulation** or flocculation, which removes some microorganisms, organic matter, toxic contaminants, and suspended fine particles. The coagulated particles are called flocs. The partially clarified water is then moved to a **settling basin** and mixed with chemicals such as alum (aluminum sulfate) and lime to trigger additional precipitation. The water is further purified by passing it through **rapid sand filters.** Here sand grains roughly 1 mm in diameter physically trap fine particles and flocs to remove up to 99% of bacteria. After filtration, the water is disinfected. This step usually involves chlorination or ozonation. In the United States and some European countries, the chlorine dose must be large enough to leave residual free chlorine at a concentration of 0.2 to 2.0 milligrams per liter. The creation of **disinfection by-products (DBPs)** such as trihalomethanes (THMs), formed when chlorine reacts with organic matter, must be monitored because some DBPs are carcinogenic.

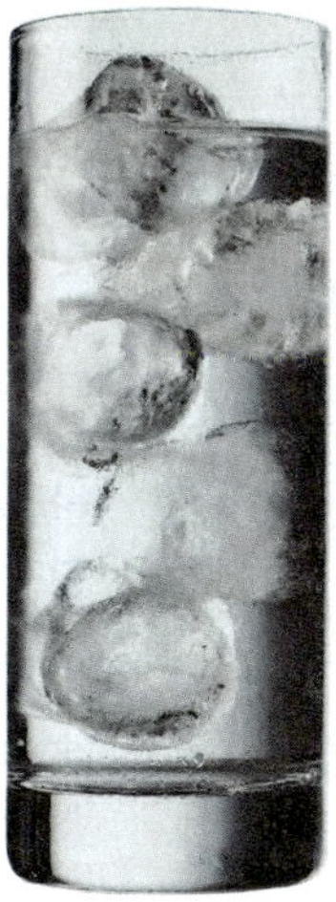

These purification processes remove or inactivate most bacteria. Unfortunately, these steps often do not remove *Giardia* cysts, *Cryptosporidium* oocysts, *Cyclospora* spp., and viruses. *Giardia intestinalis,* a cause of human diarrhea, is now recognized as the most commonly identified waterborne pathogen in the United States (*see figure 24.4*). *Cryptosporidium* spp. are also a significant problem. Both organisms are resistant to chlorine and most other disinfectants. A major source of *G. intestinalis* and *Cryptosporidium* spp. contamination is the Canada goose, which has an ever-expanding population. Water supplies that receive terrestrial runoff are therefore at particular risk of contamination by these protists. ◀ *Food and water are vehicles of fungal and protozoal diseases (section 40.5)*

In the United States, the EPA has developed regulations called the Long Term 2 Enhanced Surface Water Treatment (LT2) Rule as a means of ensuring protection against pathogens. The LT2 Rule applies to all public water systems that draw from surface waters or groundwater fed by surface waters. It sets a **maximum containment level goal (MCLG)** for specific pathogens. The EPA defines MCLGs as "health goals set at a level at which no known or anticipated adverse effects on health of persons occur and which allows an adequate margin of safety." MCLGs are set at zero for *Giardia, Cryptosporidium,* and *Legionella*

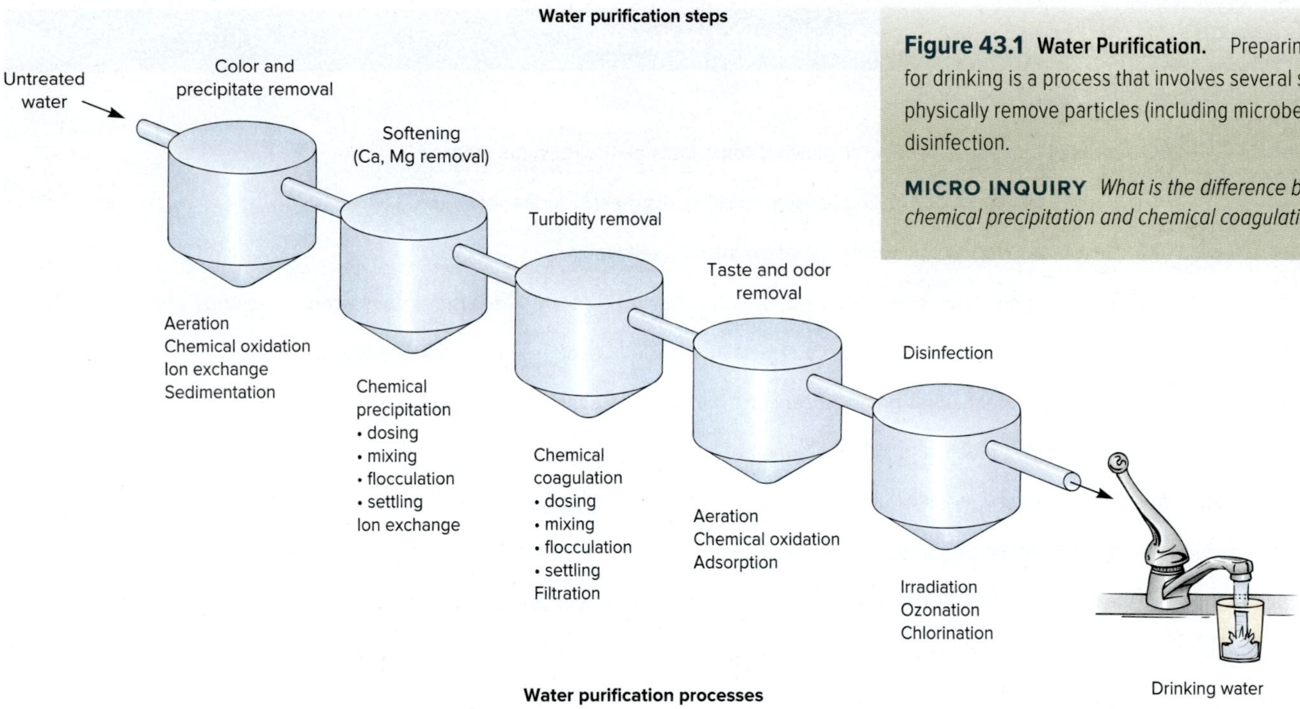

Figure 43.1 Water Purification. Preparing water for drinking is a process that involves several steps to physically remove particles (including microbes) and disinfection.

MICRO INQUIRY *What is the difference between chemical precipitation and chemical coagulation?*

spp., and certain viruses. In addition, the Total Coliform Rule, which applies to all public water systems regardless of supply source, sets an MCLG of zero for total and fecal coliform bacteria. Water systems found to be contaminated must develop additional water treatment strategies to provide 99.9% inactivation of the microbes. The treatment approach depends on the level of contamination but generally involves filtration techniques such as **slow sand filters.** This treatment involves the slow passage of water through a bed of sand in which a biofilm covers the surface of each sand grain. Waterborne microorganisms are removed by adhesion to the gelatinous surface microbial layer. Coagulation and filtration reduce virus levels about 90 to 99%. Further inactivation of viruses can be achieved by chemical oxidants, high pH, and photooxidation. ◀ *Biofilms are common in nature (section 7.6)*

Sanitary Analysis of Waters

Once water has been purified, it must be deemed potable. Although a wide range of viral, bacterial, and protozoan diseases result from the contamination of water with human and other animal fecal wastes (table 43.1), the detection of **indicator organisms** as an index of possible water contamination by human pathogens has long been the standard approach to monitoring drinking water safety. Researchers are still searching for the ideal indicator organism to use in sanitary microbiology. Among the suggested criteria for such a microbe are:

1. The indicator bacterium should be suitable for analysis of all types of water: tap, river, ground, reservoir, recreational, estuary, sea, and waste.
2. It should be present whenever enteric pathogens are present.
3. It should survive longer than the hardiest enteric pathogen.
4. It should not reproduce in the contaminated water, as this would produce an inflated value.
5. It should be harmless to humans.
6. Its level in contaminated water should have some direct relationship to the degree of fecal pollution.
7. The assay procedure for the indicator should have great specificity; in other words, the presence of other, nonpathogenic bacteria should not generate positive results. In addition, the procedure should have high sensitivity and detect low levels of the indicator.
8. The testing method should be easy to perform.

Coliforms, including *Escherichia coli,* are members of the family *Enterobacteriaceae.* These bacteria are commonly found in the intestines of humans and other animals, and are used widely as indicator organisms. They lose viability in freshwater at slower rates than most of the major intestinal bacterial pathogens. When such "foreign" enteric indicator bacteria are not detectable in a specific volume (generally 100 milliliters) of water, the water is considered potable (Latin *potabilis,* fit to drink). ◀ *Order* Enterobacteriales *includes pathogens and mutualists (section 22.3)*

However, coliforms include a wide range of bacteria whose primary source may not be the intestinal tract. To address this, tests have been developed that assay for the presence of **fecal coliforms.** These are coliforms from the intestines of warm-blooded animals, which grow at the more restrictive temperature of 44.5°C. Other indicator microorganisms include **fecal enterococci.** These microbes are used as an indicator of fecal contamination in brackish and marine water. In saltwater, these bacteria die more slowly than fecal coliforms, providing a more reliable indicator of possible recent pollution. ◀ *Order* Lactobacillales *(section 23.2)*

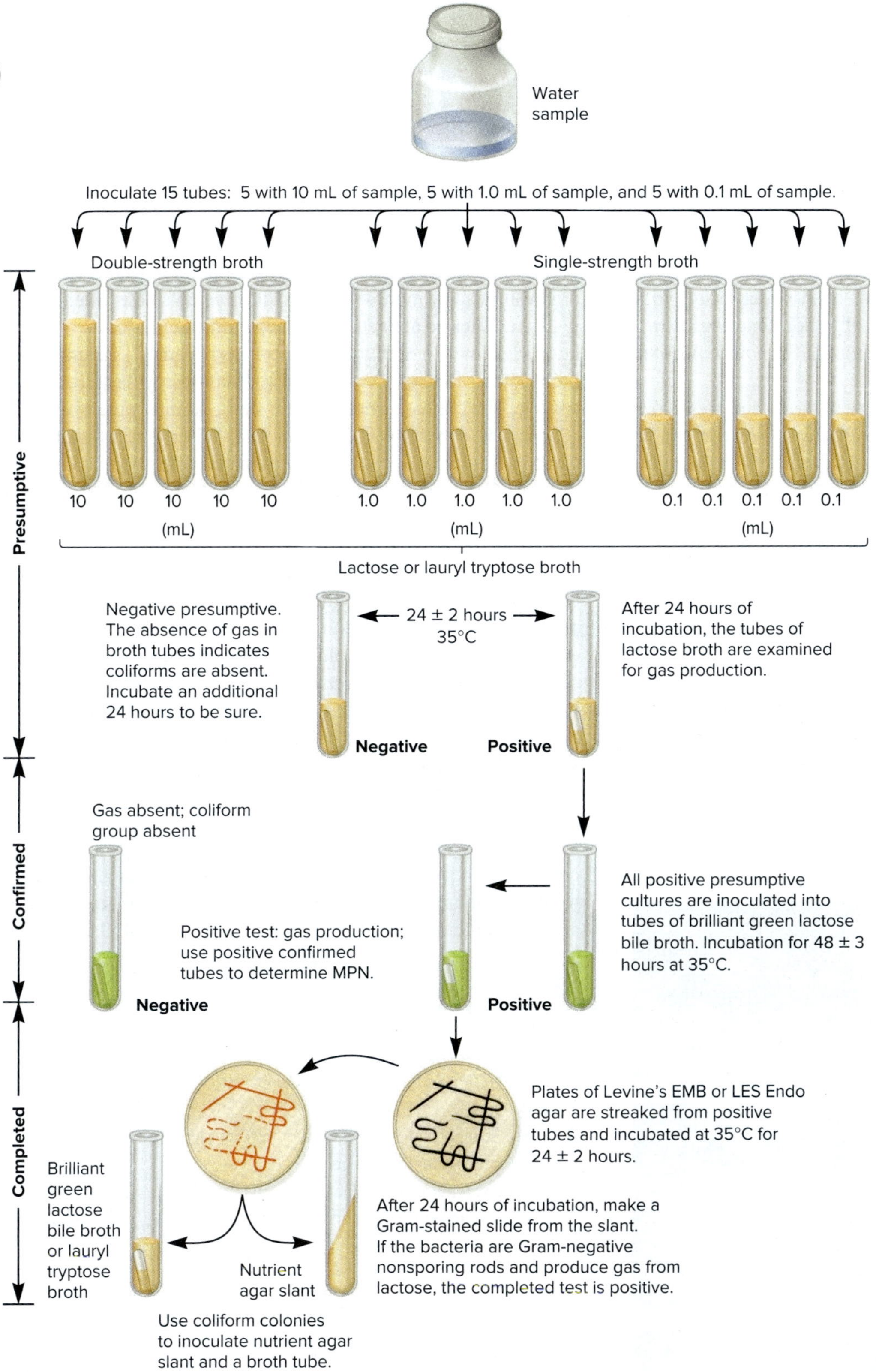

Figure 43.2 The Multiple-Tube Fermentation Test. The multiple-tube fermentation technique has been used for many years for the sanitary analysis of water. Lactose broth tubes are inoculated with different water volumes in the presumptive test. Microbes in tubes positive for gas production are inoculated into brilliant green lactose bile broth in the confirmed test, and positive tubes are used to calculate the most probable number (MPN) value. The completed test is used to establish that coliform bacteria are present.

The original test for coliforms involves presumptive, confirmed, and completed tests, as shown in figure 43.2. The presumptive step gives an estimate of the ***most probable number*** **(MPN)** of coliforms in the water (*see figure 29.3*). The complete process, including the confirmed and completed tests, requires at least 4 days of incubations and transfers.

The **membrane filter technique** is another common method of evaluating the microbiological characteristics of water. The water sample is passed through a membrane filter. The filter with its trapped bacteria is transferred to the surface of a solid medium or to an absorptive pad containing the desired liquid medium. Specific media enable the rapid detection of total coliforms, fecal coliforms, or fecal enterococci by the growth of their characteristic colonies. Samples may first be placed on a less selective resuscitation medium or incubated at a less stressful temperature prior to growth under the final set of selective conditions. An example of such a resuscitation step is the use of a 2-hour incubation on a pad soaked with lauryl sulfate broth, as is carried out in the LES Endo procedure. A resuscitation step often is needed with chlorinated samples, where the microorganisms are especially stressed. Membrane filters have been widely used with water that does not contain high levels of background organisms, sediment, or heavy metals. |◀ *Mechanical removal methods rely on barriers (section 8.3)*

More simplified tests for detecting coliforms and fecal coliforms are now available. The ***presence-absence* (P-A) test** can be used for coliforms. This is a modification of the most probable number procedure in which a larger water sample (100 milliliters) is incubated in a single culture bottle with a triple-strength broth containing lactose, lauryl tryptose, and bromocresol purple indicator. The P-A test is based on the assumption that no coliforms should be present in 100 milliliters of drinking water. A positive test results in the production of acid (a yellow color), which requires confirmation.

To test for both coliforms and *E. coli*, the related **Colilert defined substrate test** can be used. A water sample of 100 milliliters is added to a specialized medium containing *o*-nitrophenyl-β-D-galactopyranoside (ONPG) and 4-methylumbelliferyl-β-D-glucuronide (MUG) as the only nutrients. If coliforms are present, the medium will turn yellow within 24 hours at 35°C due to hydrolysis of ONPG, which releases *o*-nitrophenol (**figure 43.3**). To check for *E. coli,* the medium is observed under ultraviolet (UV) light for fluorescence. When *E. coli* is present, MUG is hydrolyzed to yield a fluorescent product. If the test is negative for the presence of coliforms, the water is considered acceptable for human consumption. If coliforms are present, tests for fecal coliforms or *E. coli* must be performed.

Figure 43.3 The Defined Substrate Test. This test is used to detect coliforms and fecal coliforms in single 100-milliliter water samples. The medium uses ONPG and MUG as defined substrates. (a) Sterile control. (b) Yellow color due to the presence of coliforms. (c) Fluorescence resulting from the presence of fecal coliforms.

(a, b) ©Donald Klein; (c) ©Ethan Shelkey and Joanne M. Willey, Ph.D.

The use of indicator organisms as a means of detecting all pathogenic microbes is limited. Some pathogens simply do not co-occur with indicator organisms. Another problem is that indicator organisms can be more sensitive to disinfectants; thus when drinking water is tested after such treatment, indicator-based assays may be misleading. In addition, culture-based methods generally require at least 18 hours to complete, and some assays require almost a week. The consequences of these shortcomings were evident in 1993, when about 400,000 Milwaukee, Wisconsin, residents became ill from drinking water contaminated with *Cryptosporidium parvum.* Currently the Centers for Disease Control and Prevention estimates that in the United States, 1 million people are sickened from drinking impure water and another 1,000 die annually.

It is clear that effective, reliable, rapid, and quantitative methods for determining the presence or absence of specific pathogens are needed. Molecular approaches can be all these things—with the ability to quickly detect multiple, specific pathogens in a single water sample. Some of the methodologies that have been applied to water testing include flow cytometry (*sections 29.2 and 37.3*), fluorescent in situ hybridization (FISH; *section 29.2*), quantitative PCR (*section 17.2*), and microarrays (*section 29.3*). One example is *Legionella pneumophila* contamination, which is routinely detected by PCR amplification. Many communities routinely use 16S rRNA amplification to detect coliforms. Unlike older methods, this technique can also be used to differentiate between nonpathogenic and enterotoxigenic strains, including Shiga toxin–producing *E. coli* O157:H7.

Comprehension Check

1. Why are three different steps used to precipitate material during drinking water purification (see figure 43.1)?
2. How can *Giardia* and *Cryptosporidium* cysts be removed from a water supply?
3. Why is it important that an indicator microorganism survive but not grow in the water to be tested?
4. Why do you think the age of pipes used in municipal water systems is a factor in introducing waterborne pathogens into the water supply?

43.2 Wastewater Treatment Maintains Human and Environmental Health

After reading this section, you should be able to:

a. Compare and contrast the processes that occur during primary, secondary, and tertiary wastewater treatment
b. Explain how, in general, septic systems operate
c. Identify at least two challenges for the future of wastewater treatment

Table 43.2 Common Contaminants of Wastewater

Contaminant	Problems Caused by Contaminant
Suspended solids	Can cause deposition of sludge leading to anoxic conditions when discharged into aquatic environments
Biodegradable organic compounds	Includes proteins, fats, and carbohydrates that when discharged into streams and rivers cause increased heterotrophic growth leading to anoxic conditions
Pathogenic microorganisms	Cause infectious disease
Priority pollutants	Include organic and inorganic compounds that may be toxic, carcinogenic, mutagenic, or teratogenic
Refractory organic compounds	Organic compounds such as phenols, surfactants, agricultural pesticides, and other toxic compounds that resist conventional wastewater treatment
Heavy metals	Usually discharged by industry; discharge into rivers and streams is toxic to all trophic levels
Dissolved inorganic constituents	Include calcium, sodium, and sulfate that may be added to domestic water supplies and may have to be removed prior to wastewater reuse

Source: Metcalf and Eddy, Inc., Tchobanoglous, G., and Stensel, H. David. *Wastewater Engineering*, 4th ed. New York: McGraw-Hill, 2002.

The flip side of ensuring safe drinking water is wastewater management. In fact, the combination of wastewater management, safe drinking water, vaccines, and antibiotics accounts for most of the gain in longevity accomplished in the last century. Citizens of Western nations take this for granted, yet about 40% of the Earth's population lacks basic sanitation.

Wastewater includes sewage as well as industrial and agricultural effluent and street runoff collected in storm sewers. These waters often contain high levels of organic matter, heavy metals, nutrients, and particulates (**table 43.2**). We begin our discussion of wastewater treatment with an overview of large-scale wastewater treatment processes, followed by the means by which water quality is monitored. We then review home treatment systems.

Table 43.3 Major Steps in Treatment of Wastes

Treatment Step	Processes
Primary	Removal of insoluble particulate materials by settling, screening, addition of alum and other coagulation agents, and other physical procedures
Secondary	Biological removal of dissolved organic matter Trickling filters Activated sludge Lagoons Extended aeration systems Anaerobic digesters
Tertiary	Biological removal of inorganic nutrients Chemical removal of inorganic nutrients Virus removal/inactivation Trace chemical removal

Wastewater Treatment Processes

Wastewater treatment involves at least three spatially segregated steps: primary, secondary, and tertiary treatment (**table 43.3** and **figure 43.4**). At the end of the process, water is usually chlorinated or treated with ozone before it is released into the environment.

Primary treatment prepares wastewater for treatment by physically removing much of the solid material. This can be accomplished in several ways, including screening, precipitation of small particulates, and settling in basins or tanks (**table 43.4**).

Figure 43.4 Aerial View of a Conventional Sewage Treatment Plant. Sewage treatment plants allow natural processes of self-purification that occur in rivers and lakes to be carried out under more intense, managed conditions in large concrete vessels.

©Mariusz Szczygiel/Shutterstock

MICRO INQUIRY *This plant includes anaerobic digesters. What are the advantages of this additional step?*

Table 43.4 Wastewater Treatment Methods

Nature of Operation	Process	Brief Explanation
Physical operations	Screening	Mechanical removal of large suspended materials with parallel bars, rods, wires, grating, wire mesh, or perforated plates
	Comminution	Comminutors are cutting instruments used to pulverize large floating material before wastewater enters primary settling tank.
	Flow equalization	A basin or basins used to adjust the rate of wastewater entry into secondary and tertiary treatment facilities
	Sedimentation	Also called clarification, this widely used procedure is based on the gravitational separation of large particles. Used in primary settling basins and to separate biological floc in activated sludge settling basins. Also used to precipitate chemically treated particulates when chemical coagulation is employed.
	Flotation	The removal of particles by bubbling air or other gas into the water. Particles float to the surface and are collected by skimming; removes fine particles more efficiently than sedimentation. Chemical additives can be added to improve particle removal.
	Granular-medium filtration	Wastewater is passed through a filter bed composed of granular material such as sand; used to remove suspended solids and chemically precipitated phosphorus.
Chemical operations	Chemical precipitation	Addition of coagulants such as alum, ferric chloride, ferric sulfate, or lime to wastewater prior to sedimentation; facilitates the flocculation of finely divided solids into flocs that will settle
	Adsorption	Soluble material is collected by adsorption of activated carbon; usually performed after biological treatment to remove remaining dissolved organic carbon
	Disinfection	Elimination of pathogens (and other microbes) by a variety of methods including UV light, heat, gamma radiation, and chemical agents such as chlorine, bromine, ozone, phenolic compounds, and others
	Dechlorination	Removal of all chlorine residues before discharge into receiving waters. These compounds are reactive and potentially dangerous.
Biological operations	Activated sludge process	An aerobic, continuous flow system that uses microbes to aerobically degrade organic matter; occurs after primary settling. Material is mechanically mixed to maintain aeration and prevent settling (figure 43.6*a*).
	Aerated lagoon	A basin 1 to 4 m in depth where wastewater is treated in a process similar to activated sludge treatment. Turbulence created by aeration keeps the contents of the lagoon mixed.
	Trickling filters	Most commonly used aerobic attached-growth biological treatment. A large basin filled with rock or plastic packing material provides a substrate for microbial biofilm growth and increased contact area for wastewater (figure 43.6*b*).
	Rotating biological contractors	Attached growth process that features large cylinders partially submerged in wastewater. Biofilms form on the cylinders, which are continuously rotated.
	Stabilization ponds	Shallow basin that is mixed naturally or mechanically. Aerobic ponds treat soluble organic wastes and effluents from wastewater treatment facilities; anaerobic ponds stabilize organic wastes. Wastewater retention times from 1 to 3 months
	Anaerobic digestion	Used for treatment of sludge and wastewater with high organic content (figure 43.7)
	Biological nutrient removal	Removal of nitrogen by nitrification-denitrification and anammox reaction (see *figure 21.13*) and phosphorus, often by proprietary processes (e.g., A/O, PhoStrip, and SBR process)

Source: United Nations. *Water-Water Treatment Technologies: A General Review.* 2003.

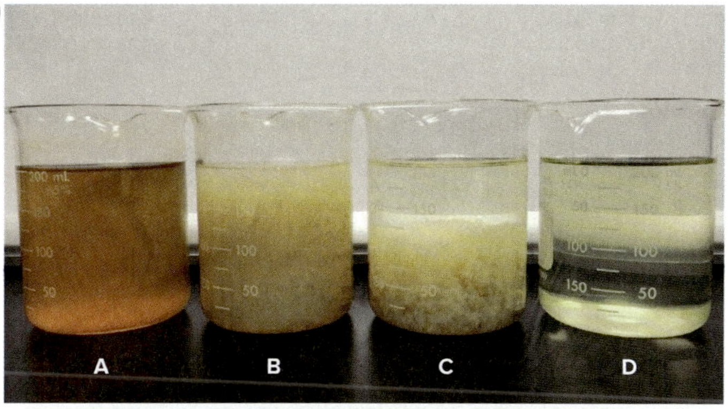

Figure 43.5 **Proper Floc Formation in Activated Sludge.** Microorganisms play a critical role in the functioning of activated sludge systems. Wastewater effluent in beaker A contains bacteria and other organic materials that resist settling. In this case, a commercial coagulant containing ferric chloride is added (beaker B). This accelerates the formation of stable flocs, which settle at the bottom of beaker C. Beaker D shows the effluent after the settled material has been removed.

©Ecologix Environmental Systems, LLC.

MICRO INQUIRY *Why is proper floc formation critical to the function of a treatment facility?*

The resulting solid material is called **sludge.** In general, the goal of primary treatment is to generate water (effluent) that can be further purified by biological treatment and produce sludge that can be treated before disposal.

Secondary treatment of effluent relies on microbial growth to convert dissolved organic matter (DOM) into microbial biomass and carbon dioxide. This can reduce degradable DOM by about 90 to 95%. When microbial growth is completed, under ideal conditions microorganisms will aggregate and form stable flocs, also called **granular sludge,** that settle. As shown in **figure 43.5**, commercial coagulants are often added to wastewater effluent to increase rate of flocculation and the amount of material that settles. Alternatively, bacteria that have been genetically modified to create a stable, granular sludge may be added. Efficient flocculation is important because a sewage treatment facility that generates stable granular sludge requires up to 75% less space and reduces cost by as much as 25%.

When these processes occur with lower O_2 levels or with a microbial community that is too young or too old, unsatisfactory floc formation and settling can occur. The result is a **bulking sludge,** caused by the massive development of filamentous bacteria such as those in the genera *Sphaerotilus* and *Thiothrix,* together with many poorly characterized filamentous microbes. These filamentous microorganisms form flocs that do not settle well and are thus problematic. ◀ *Order* Thiotrichales *includes large filamentous bacteria (section 22.3)*

An aerobic **activated sludge** system (**figure 43.6*a***) involves the horizontal flow of materials with recycling of sludge. Activated sludge systems can be designed with variations in mixing. In addition, the ratio of organic matter added to the active microbial biomass can be varied. A low-rate system (low nutrient input per unit of microbial biomass), with slower growing microorganisms, will produce an effluent with low residual levels of dissolved organic matter. A high-rate system (high nutrient input per unit of microbial biomass), with faster growing microorganisms, will remove more dissolved organic carbon per unit time but produce a poorer quality effluent.

Aerobic secondary treatment is often carried out with a **trickling filter** (figure 43.6*b*). The waste effluent is passed over rocks or other solid materials coated with microbial biofilms, and the microbial community degrades the organic waste. A sewage treatment plant can produce less sludge by employing the **extended aeration** process (figure 43.6*c*). Microorganisms grow on the dissolved organic matter, and the newly formed microbial biomass is eventually consumed to meet maintenance energy requirements. This requires extremely large aeration basins and long aeration times.

All aerobic processes produce excess microbial biomass, or sewage sludge, which contains many recalcitrant organics. Often the sludge from aerobic sewage treatment, together with the materials that settled in primary treatment, are further treated by anaerobic digestion. **Anaerobic digestion** occurs in large tanks designed to operate with continuous input of untreated sludge and removal of the final, stabilized sludge product (**figure 43.7**). Methane, the major component of natural gas, is vented and often burned for heat and electricity production. Within the anaerobic digester, at least three processes occur: (1) the fermentation of the sludge components to form organic acids, including acetate; (2) production of the other methanogenic substrates CO_2 and hydrogen; and (3) methanogenesis by archaea. These methanogenic processes involve critical balances between electron acceptors and donors (**table 43.5**). To function most efficiently, the hydrogen concentration must be maintained at a low level. If hydrogen and organic acids accumulate, the drop in pH will inhibit methane production, resulting in a stuck digester. Often lime is added to neutralize the pH and restore methanogen function. ◀ *Methanogens and methanotrophs (section 20.3)*

Anaerobic digestion has many advantages. Most of the microbial biomass produced in aerobic growth is degraded in the anaerobic digester. Also, because the process of methanogenesis is energetically very inefficient, methanogenic archaea consume about twice as many nutrients to produce an equivalent biomass as that of aerobic systems. Consequently less sludge is produced and it can be easily dried. Dried sludge removed from well-operated anaerobic systems can even be sold as organic garden fertilizer. However, sludge from poorly managed systems may contain high levels of heavy metals and other contaminants, which require removal prior to release into the environment.

Tertiary treatment further purifies wastewaters by removing nitrogen and phosphorus compounds that promote

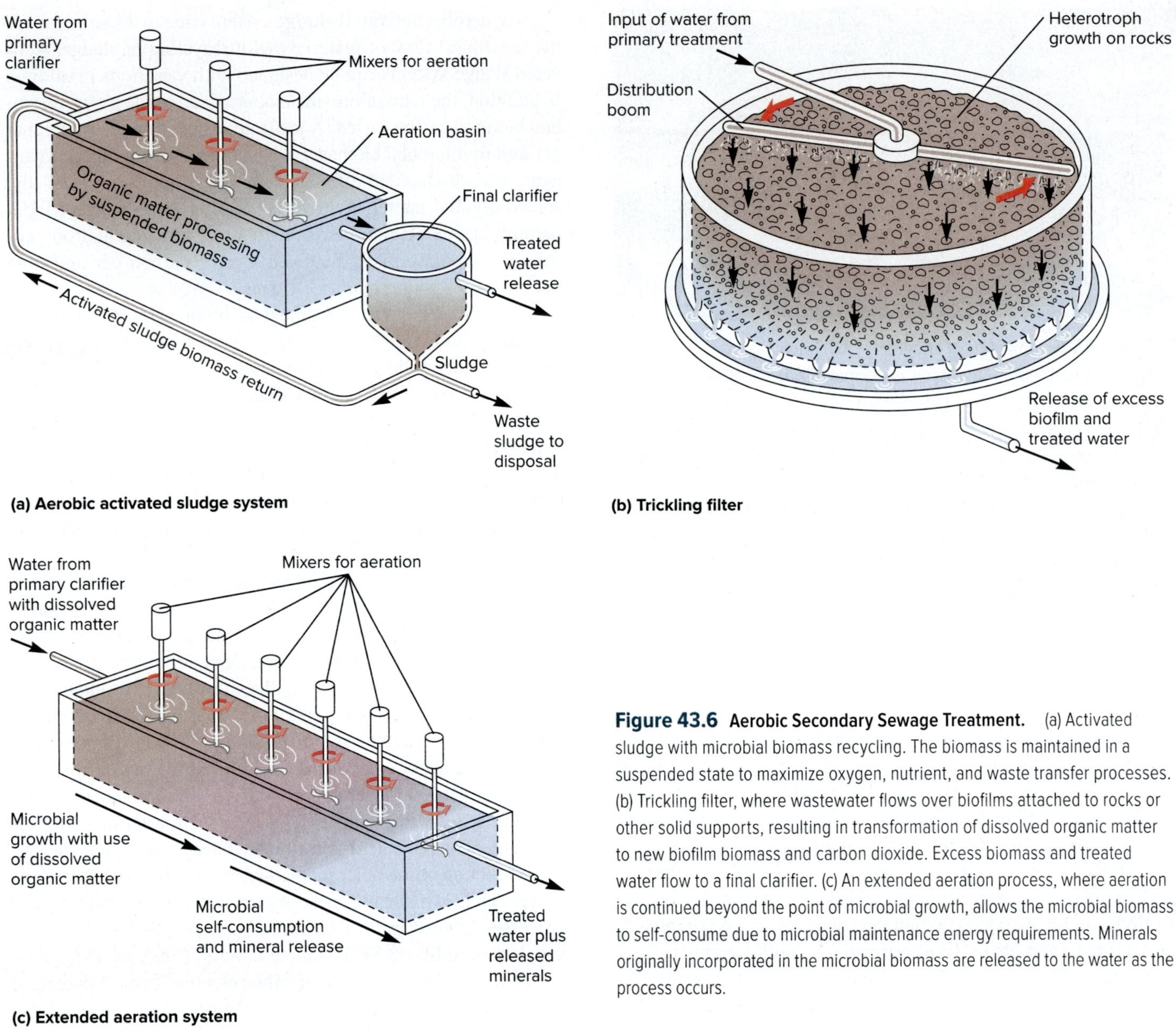

Figure 43.6 Aerobic Secondary Sewage Treatment. (a) Activated sludge with microbial biomass recycling. The biomass is maintained in a suspended state to maximize oxygen, nutrient, and waste transfer processes. (b) Trickling filter, where wastewater flows over biofilms attached to rocks or other solid supports, resulting in transformation of dissolved organic matter to new biofilm biomass and carbon dioxide. Excess biomass and treated water flow to a final clarifier. (c) An extended aeration process, where aeration is continued beyond the point of microbial growth, allows the microbial biomass to self-consume due to microbial maintenance energy requirements. Minerals originally incorporated in the microbial biomass are released to the water as the process occurs.

eutrophication. It also removes heavy metals, biodegradable organics, and many remaining microbes, including viruses. Organic pollutants can be removed with activated carbon filters. Phosphate usually is precipitated as calcium or iron phosphate (e.g., by the addition of lime). To remove phosphorus, oxic and anoxic conditions are used alternately in a series of treatments, and phosphorus accumulates in microbial biomass as polyphosphate. Excess nitrogen may be removed by "stripping," volatilization of NH_3 at high pH. Ammonia itself can be chlorinated to form dichloramine, which is then converted to molecular nitrogen. In some cases, microbial processes can be used to remove nitrogen and phosphorus. A widely used process for nitrogen removal is denitrification, whereby nitrate produced by microbes under aerobic conditions serves as an electron acceptor during anaerobic respiration. Nitrate reduction yields nitrogen gas (N_2) and nitrous oxide (N_2O) as the major products. In addition to denitrification, the anammox process is also important. In this reaction, ammonium ion (used as the electron donor) reacts with nitrite (the electron acceptor) to yield N_2. The anammox process can convert up to 80% of the initial ammonium ion to N_2 gas. Tertiary treatment is expensive and is usually not employed except where necessary to prevent obvious ecological disruption. ◀ *Phylum* Planctomycetes: *anammox bacteria (section 21.5); Nitrogen cycle (section 28.1)*

Measuring Treated Wastewater Quality

The process of wastewater treatment must be monitored to ensure that waters released into the environment do not pose

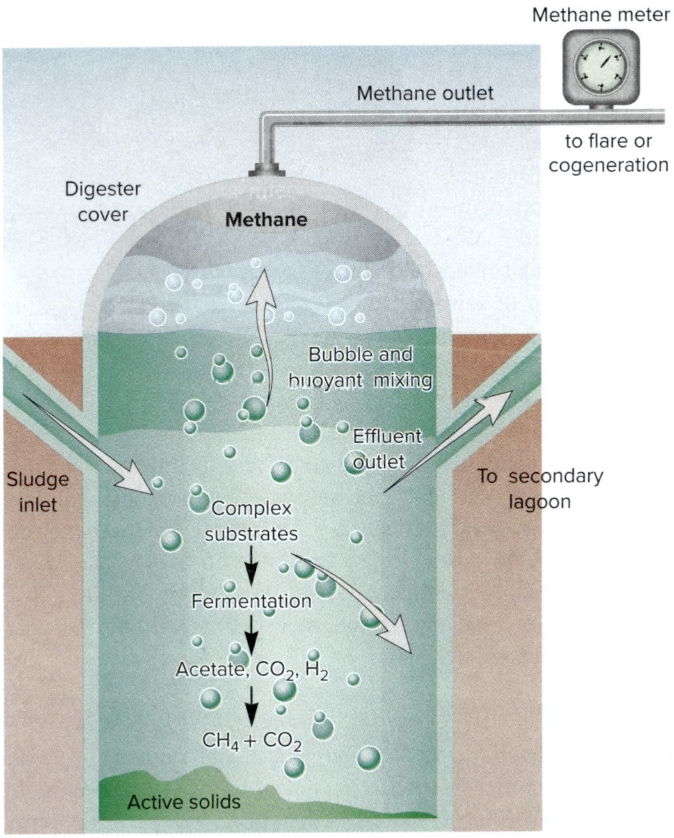

Figure 43.7 **Anaerobic Digesters Generate Methane for Local or Municipal Use.** The basic design of an anaerobic digester includes an inlet for the delivery of waste. Fermentation of complex substrates yields acetate, CO_2, and H_2, which archaea convert to CH_4. The effluent outlet functions to optimize the retention time (usually from 22 to 28 days) to maximize CH_4 capture.

environmental and health risks. Successfully treated water should have very little organic carbon. The amount of carbon during and after wastewater treatment can be measured as **total organic carbon** (**TOC**), as chemically oxidizable carbon by the **chemical oxygen demand** (**COD**) test, or as biologically usable carbon by the **biochemical oxygen demand** (**BOD**) test. The TOC includes all carbon, whether or not it can be used by microorganisms. It is assayed by oxidizing all organic matter in a sample to CO_2 at high temperature in an oxygen stream. The CO_2 produced is measured by infrared or potentiometric techniques. The COD gives a similar measurement, except that lignin (*see figure 31.2*) often does not react with the oxidizing chemical used in this procedure.

The BOD test measures only that portion of total carbon oxidized (i.e., consumed) by microorganisms in a 5-day period at 20°C. It measures the amount of dissolved O_2 needed for microbial oxidation of biodegradable organic material. Ammonia released during organic matter oxidation can also consume O_2 in the BOD test. To prevent this "bottle effect," nitrification is often inhibited by the addition of 2-chloro-6-(trichloromethyl) pyridine (nitrapyrin). Except when treated effluents are analyzed, nitrification is not a major concern.

It is critical to note that TOC, COD, and BOD measurements are concerned only with carbon removal. They do not directly address the removal of minerals such as nitrate, phosphate, and sulfate. These impact cyanobacterial and algal growth in lakes, rivers, and oceans by contributing to the process of eutrophication. ◀ *Biogeochemical cycling sustains life on Earth (section 28.1); Microorganisms in coastal ecosystems are adapted to a changeable environment (section 30.2)*

Comprehension Check

1. Explain the goal of primary, secondary, and tertiary wastewater treatment.
2. Why is granular sludge more desirable than bulking sludge?
3. What are the steps of organic matter processing that occur in anaerobic digestion?
4. After anaerobic digestion is completed, why is sludge disposal of concern?
5. Compare TOC, COD, and BOD. How are these similar and different?

Table 43.5 Sequential Reactions in the Anaerobic Digester

Process Step	Substrates	Products	Major Bacterial and Archaeal Genera	
Fermentation	Organic polymers	Butyrate, propionate, lactate, succinate, ethanol, acetate,[1] H_2,[1] CO_2[1]	*Clostridium* *Bacteroides* *Peptostreptococcus*	*Peptococcus* *Eubacterium* *Lactobacillus*
Acetogenic reactions	Butyrate, propionate, lactate, succinate, ethanol	Acetate, H_2, CO_2	*Syntrophomonas* *Syntrophobacter*	
Methanogenic reactions	Acetate H_2 and HCO_3^-	$CH_4 + CO_2$ CH_4	*Methanosarcina* *Methanosaeta* *Methanobrevibacter* *Methanomicrobium*	*Methanogenium* *Methanothermobacter* *Methanococcus* *Methanospirillum*

1 Methanogenic substrates produced in the initial fermentation step.

Home Treatment Systems

In the absence of municipal sewage systems, **septic tanks** are most commonly used. A conventional septic tank functions as a simple anaerobic digester, with an anaerobic liquefaction and digestion step. This is followed by organic matter adsorption and entrapment of microorganisms in an aerobic leach field, where biological oxidation occurs. Pathogenic microorganisms and dissolved organic matter are removed from the wastewater in home septic systems during subsurface passage through adsorption and trapping by fine sandy materials, clays, and organic matter (**figure 43.8**). Microorganisms associated with these materials—including predators such as protozoa—use the trapped pathogens as food. This results in purified water with a lower microbial population.

It is imperative that home septic systems be located far above groundwater, often an important source of potable water. Thus not all sites are suitable for septic systems. Prior to septic tank installation, a percolation test (or "perc test") must be performed. Perc tests determine if the surrounding soil can absorb the liquid produced by the septic tank, thereby preventing it from contaminating the groundwater beneath. Local governments develop specific protocols for perc tests, which are often witnessed when performed as the outcome of a perc test can determine if a piece of real estate can be developed.

A septic tank may not operate correctly for several reasons. If the retention time of the waste in the septic tank is too short, undigested solids move into the leach field, plugging the system. If the leach field floods and becomes anoxic, biological oxidation does not occur, and effective treatment ceases. Other problems can occur, especially when a suitable soil is not present and the septic tank outflow from a conventional system drains too rapidly to the deeper subsurface. Fractured rocks and coarse gravel materials do not effectively adsorb or filter wastewater. In addition, nitrogen and phosphorus from waste can pollute groundwater. This leads to nutrient enrichment of ponds, lakes, rivers, and estuaries as the subsurface water enters these environmentally sensitive ecosystems.

Domestic and commercial septic systems can be designed with nitrogen and phosphorus removal steps. Nitrogen is usually removed by nitrification and denitrification, with organic matter provided by sawdust or a similar material. A reductive iron dissolution process can be used for phosphorus removal.

Subsurface zones also can become contaminated with pollutants from other sources. Land disposal of sewage sludge, illegal dumping of septic tank sludge, improper toxic waste disposal, and runoff from agricultural operations all contribute to groundwater contamination with chemicals and microorganisms. Many pollutants that reach the subsurface will persist and may affect the quality of groundwater for extended periods. Much research is being conducted to find ways to treat groundwater in place—in situ treatment. As discussed in section 43.4, microorganisms and microbial processes are critical in many of these remediation efforts.

Future Challenges

Our planet now has over 7 billion people, and the need for clean drinking water and wastewater treatment have become ever more pressing. In an effort to address both at once, water reuse or recycling has become a reality for many. Water reuse is the purification of wastewater so that it can serve as drinking water. Currently a southern California municipality operates one of the largest water reclamation plants in the world.

Not surprisingly, water reclamation is complex. Not only must the obvious contaminants be removed (i.e., pathogens, heavy metals, nutrients that cause eutrophication), but twenty-first-century sewage also carries alarming levels of pharmaceuticals, introduced by both industry and the households. This has stimulated the development of methods to detect and remove the vast array of drugs common in sewage. In addition to recycling water, recent technologies enable the reclamation of biopolymers, plastics, and cellulose fibers in an economically viable fashion. The current challenge is the widespread application of these technologies.

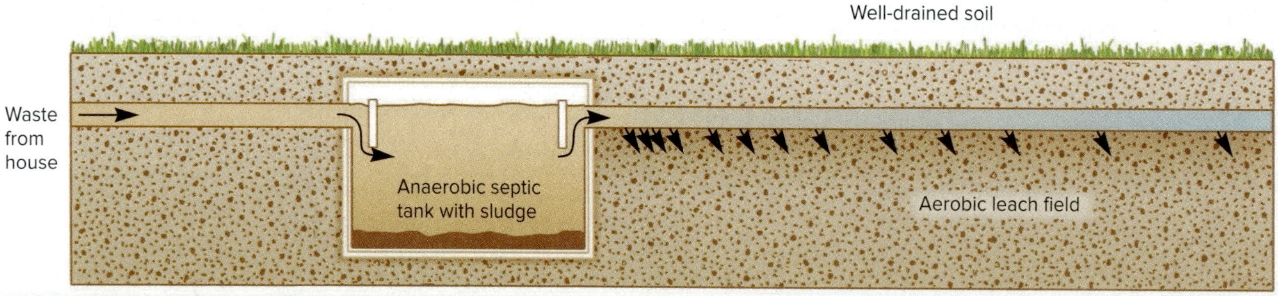

Figure 43.8 The Conventional Septic Tank Home Treatment System. This system combines an anaerobic waste liquefaction unit (the septic tank) with an aerobic leach field. Biological oxidation of the liquefied waste takes place in the leach field, unless the soil becomes flooded.

MICRO INQUIRY *Why would flooding halt biological oxidation of waste in the leach field?*

Population growth also stresses the capacity of municipal systems to carry and treat the ever-increasing volume of wastewater. This leads to sewage overflows during storms. The frequency of sewage overflow and the associated public health risks have resulted in reexamination of current infrastructure and, in some cases, construction of overflow basins.

Comprehension Check

1. How, in principle, does a conventional septic tank and leach field system work? What factors can reduce the effectiveness of this system?
2. What type of wastewater treatment system does your hometown use?
3. List two areas of active research in the wastewater treatment field. Which seem more pressing in your community? Explain your answer.

43.3 Microbial Fuel Cells: Batteries Powered by Microbes

After reading this section, you should be able to:

a. Diagram the basic design of a microbial fuel cell
b. Explain potential applications of microbial fuels cells

An exciting and emerging field in microbiology is the use of bacteria to generate electricity, not through the production of methane (as in anaerobic digesters, figure 43.7) but by directly capturing electrons from the microbe's electron transport chain (ETC). This is possible because when some heterotrophic microorganisms oxidize organic material, they can pass the electrons directly to an electrode. A **microbial fuel cell** (**MFC**) captures these electrons to generate electricity. To accomplish this, the microbes are continuously fed a rich diet of organic substrates to minimize de novo biosynthesis (recall that anabolic reactions require electrons carried by NADPH). In this way, much of the organic substrate is oxidized during catabolism and electrons are donated to the ETC. The basic design of a MFC is relatively simple (**figure 43.9**). It consists of two chambers, one oxic and the other anoxic, divided by a membrane through which protons can pass. Biomass is delivered to the bacteria in the anoxic side. The oxidation of this organic matter yields protons and electrons. The protons diffuse across the membrane to the oxic chamber. The electrons are deposited on an anode that connects the two chambers. Electrons flow toward the cathode but can be used to generate electricity while in transit. A catalyst in the oxic chamber combines the protons, electrons, and oxygen to generate water.

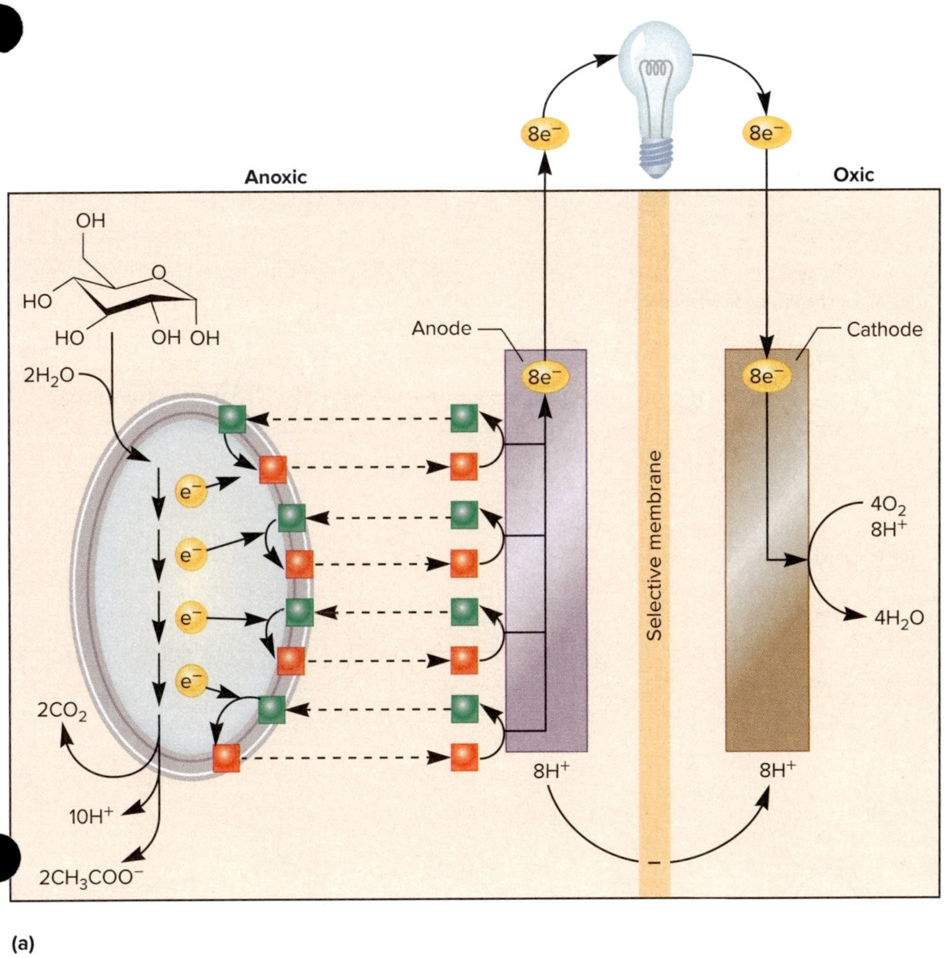

(a)

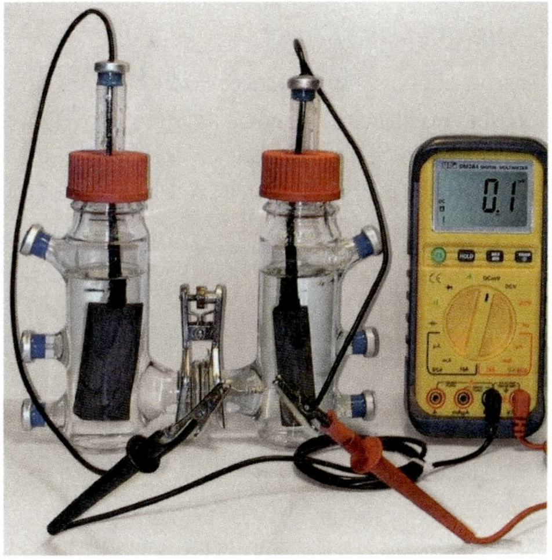

(b)

Figure 43.9 Microbial Fuel Cells.
(a) Biomass is catabolized anaerobically by microbes in the anoxic chamber. The resulting protons flow to the oxic chamber, and the electrons are shuttled from the cell membrane to an anode, where they travel to the oxic chamber. During electron transit, they can be used to generate electricity. Green boxes represent oxidized electron shuttles and red boxes represent reduced shuttles. (b) A prototypical two-chambered microbial fuel cell.
(b) ©Kelly P. Nevin

MICRO INQUIRY *Why must the membrane between the two chambers be selectively permeable?*

Many different types of heterotrophic microbes can produce power in MFCs. The only requirement is that they oxidize organic matter anaerobically. In fact, communities of microbes can be employed. The largest constraint to obtaining maximum efficiency appears to be delivery of electrons to the electrode. There are several ways this can be accomplished. Often a chemical mediator is used to shuttle electrons from inside the cell, across the cellular membrane, to the anode (figure 43.9). Some microbes, such as the γ-proteobacterium *Shewanella oneidensis,* produce "nanowires" to transmit electrons to the anode (in nature, these nanowires shuttle electrons to external electron acceptors such as Fe^{3+}; *see figure 22.25*).

MFC development is ongoing, and great strides have been made to maximize power output, which has increased orders of magnitude in just a decade. Applications of MFCs include drinking water purification and the treatment of landfill leachate: In both cases, unwanted organic material is used to feed the bacteria in the MFC. MFCs can also be used when energy requirements are small but batteries cannot be used. Environmental monitoring instruments that need to be left in the field for years at a time are good examples. In addition, because MFCs are self-contained, they are attractive for countries that lack well-developed power grids.

Comprehension Check

1. How does a microbial fuel cell work?
2. What would be the outcome if microbes in a fuel cell were carbon limited?
3. How might microbial fuel cells be involved in wastewater management?

43.4 Biodegradation and Bioremediation Harness Microbes to Clean the Environment

After reading this section, you should be able to:

a. Describe the importance of reductive dehalogenation in bioremediation
b. Discuss factors that influence the rate of biodegradation and bioremediation of a compound
c. Explain how endogenous microbes can be stimulated to degrade environmental pollutants
d. Summarize the rationale that supports the addition of microbes and microbial products to contaminated soils and waters

The metabolic activities of microbes can also be exploited in complex natural environments for beneficial outcomes. Examples are the use of resident microbial communities to carry out biodegradation, bioremediation, and environmental maintenance processes (see chapter opener), and the addition of microorganisms to soils or plants for the improvement of crop production. We discuss both of these applications, with an emphasis on bioremediation in this section.

Biodegradation Processes

Before discussing biodegradation, it is important to consider its definition. **Biodegradation** has at least three outcomes (**figure 43.10**): (a) a minor change in an organic molecule, leaving the main structure still intact; (b) fragmentation of a complex organic molecule in such a way that the fragments could be reassembled to yield the

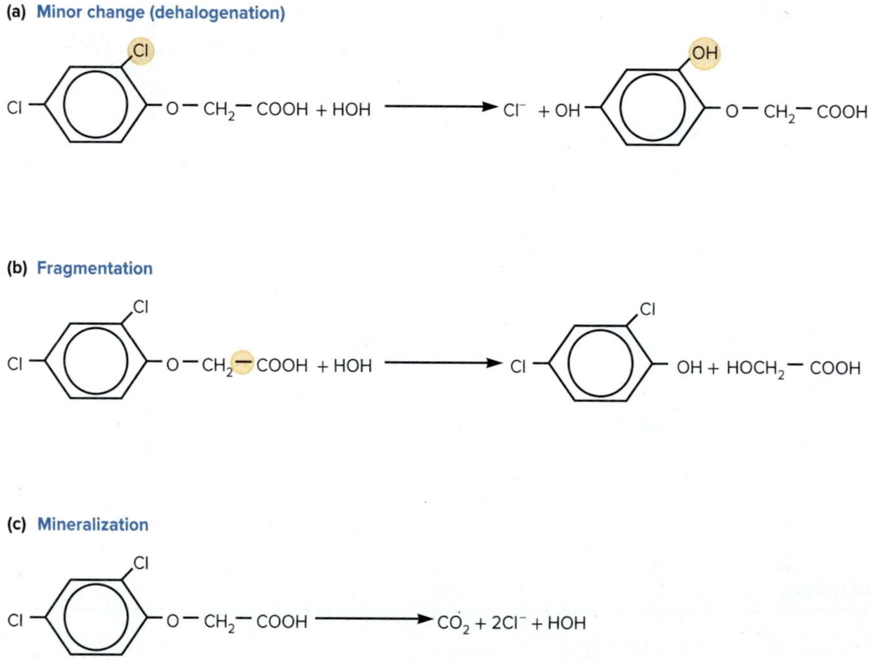

Figure 43.10 Biodegradation Has Several Meanings. Biodegradation is a term that can be used to describe three types of changes in a molecule: (a) a minor change in the functional groups attached to an organic compound, such as the substitution of a hydroxyl group for a chlorine group; (b) an actual breaking of the organic compound into organic fragments in such a way that the original molecule could be reconstructed; and (c) the complete degradation of an organic compound to minerals, which usually involves many organisms and the input of additional carbon substrates (note equation is not balanced for simplicity).

MICRO INQUIRY *Why does dehalogenation increase the likelihood that the carbon compound will be degraded?*

original structure; and (c) complete mineralization, which is the transformation of organic molecules to inorganic forms.

The removal of toxic industrial products in soils and aquatic environments is a daunting and necessary task. Compounds such as perchloroethylene (PCE), trichloroethylene (TCE), and polychlorinated biphenyls (PCBs) are common contaminants. These compounds adsorb onto organic matter in the environment, making decontamination using traditional approaches difficult or ineffective. The use of microbes to transform these contaminants to nontoxic degradation products is called **bioremediation.** To understand how bioremediation takes place, we first must consider the biochemistry of biodegradation.

Degradation of complex compounds requires several discrete stages, usually performed by different microbes. Initially contaminants are converted to less toxic compounds that are more readily degraded. The first step for many contaminants, including organochloride pesticides, alkyl solvents, and aryl halides, is **reductive dehalogenation.** This is the removal of a halogen substituent (e.g., chlorine, bromine, fluorine) while at the same time adding electrons to the molecule. This can occur in two ways. In hydrogenolysis, the halogen substituent is replaced by a hydrogen atom (**figure 43.11*a***). Alternatively dihaloelimination removes two halogen substituents from adjacent carbons while inserting an additional bond between the carbons (figure 43.11*b*). Both processes require an electron donor. The dehalogenation of PCBs uses electrons derived from water; alternatively hydrogen can be the electron donor for the dehalogenation of different chlorinated compounds. Major genera with species that carry out this process include *Desulfitobacterium, Dehalospirillum,* and *Desulfomonile.*

Reductive dehalogenation usually occurs under anoxic conditions. In fact, humic acids (polymeric residues of lignin decomposition that accumulate in soils and waters) can serve as terminal electron acceptors under what are called humic acid–reducing conditions. The use of humic acids as electron acceptors has been observed with the anaerobic dechlorination of vinyl chloride and dichloroethylene. Once the anaerobic dehalogenation steps are completed, degradation of the main structure of many pesticides and other xenobiotics (foreign compounds) often proceeds more rapidly in the presence of O_2. Thus the degradation of halogenated toxic compounds generally requires the action of several microbial genera, called a consortium.

Structure and stereochemistry are critical in predicting the fate of a specific chemical in nature. When a constituent is in the *meta* as opposed to the *ortho* position, the compound will be degraded at a much slower rate. This stereochemical difference is the reason that the common lawn herbicide 2,4-dichlorophenoxyacetic acid (2,4-D), with a chlorine in the *ortho* position, will be largely degraded in a single summer. In contrast, 2,4,5-trichlorophenoxyacetic acid, with a constituent in the *meta* position, will persist in the soils for several years and thus is used for long-term brush control.

Another important factor that influences biodegradation is a compound's chirality or handedness (*see figure A1.8*). Microorganisms often degrade one isomer of a substance but not the other. At least 25% of herbicides are chiral. Thus it is critical to add the herbicide isomer that is effective and also degradable. However, populations that make up a microbial community can change in response to the addition of inorganic or organic substrates. If a particular compound, such as a herbicide, is added repeatedly, the soil microbial community changes and faster rates of degradation occur. Eventually a microbial community can become so efficient at herbicide degradation that herbicide effectiveness is diminished.

Degradation processes that occur in soils also can be used in large-scale degradation of hydrocarbons or wastes from agricultural operations. The waste material is incorporated into soil or allowed to flow across the soil surface, where degradation occurs. While this is often very effective, sometimes such degradation processes do not reduce toxicity. An example of this problem is the microbial metabolism of 1,1,1-trichloro-2,2-bis-(p-chlorophenyl) ethane (DDT), a pesticide once commonly used in the United States. Degradation removes a chlorine function to give 1,1-dichloro-2,2-bis-(p-chlorophenyl) ethylene (DDE), which is also an environmental toxin (**figure 43.12*a***). Another important example is the degradation of trichloroethylene (TCE), a widely used solvent.

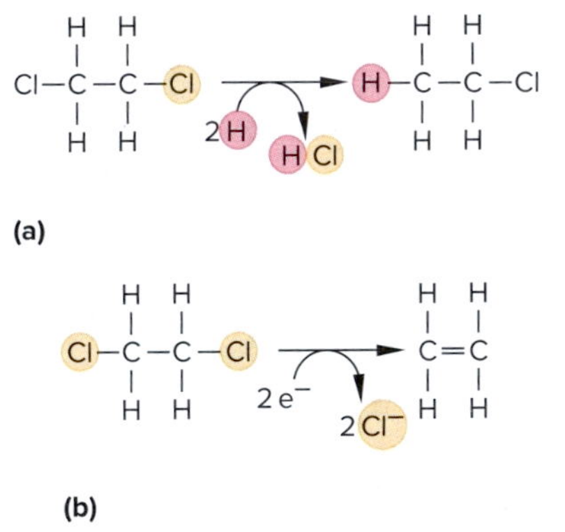

Figure 43.11 Reductive Dehalogenation. In these two examples, 1,2 dichlorobenzoate is dehalogenated by (a) alkyl hydrogenolysis and (b) dihaloelimination. Note that although the substrate is the same, these strategies result in different products.

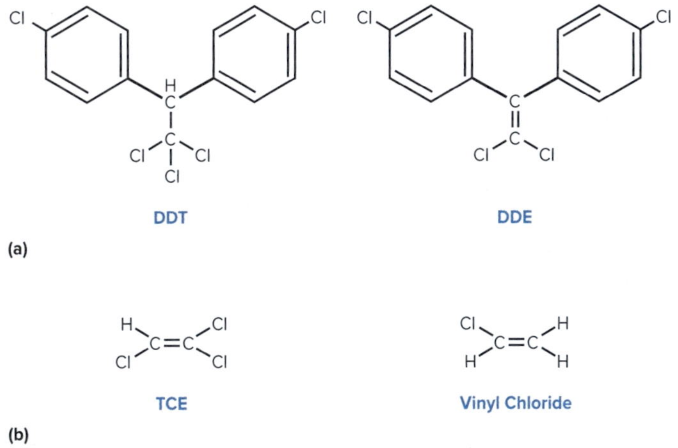

Figure 43.12 Toxic Degradation Products. The parent compounds DDT and TCE are metabolized to the toxins DDE and vinyl chloride, respectively.

Figure 43.13 **Bioremediation of the *Exxon Valdez* Oil Spill.** Workers sprayed the coastline with nitrogen and phosphorus-containing fertilizer to provide limiting nutrients to hydrocarbon-degrading microorganisms. This increased the rate at which oil was degraded.

When degraded under anoxic conditions, the dangerous carcinogen vinyl chloride can be formed (figure 43.12*b*).

An emerging approach to the bioremediation of contaminants in soils is the use of microbial fuel cells (figure 43.9). One of the major challenges to the microbial degradation of subsurface organic compounds and heavy metals is the provision of an electron acceptor that will participate in the oxidation of the contaminant. For instance, the biodegradation of petroleum in soils is frequently limited because it is very difficult and expensive to introduce oxygen into sediments. Microbial fuel cells can be inserted in sediments such that the anode can serve as the terminal electron acceptor by bacteria in the community during biodegradation. Several innovative applications of microbial fuel cells have been demonstrated, including the oxidation of toluene by *Geobacter metallireducens.* |◀ *Order* Desulfuromonales *includes metabolically flexible anaerobes (section 22.4)*

Bioremediation

Bioremediation usually involves stimulating the degradative activities of microorganisms already present in contaminated waters or soils. These existing microbial communities typically cannot carry out biodegradation processes at the desired rate due to limiting physical or nutritional factors. For example, biodegradation may be limited by low levels of oxygen, nitrogen, phosphorus, or other nutrients. In these cases, it is necessary to determine the limiting factors and supply the needed materials or modify the environment. Often the addition of easily metabolized organic matter such as glucose increases biodegradation of recalcitrant compounds that otherwise would not be used as carbon and energy sources by microorganisms. This process is called **cometabolism.**

Stimulating Hydrocarbon Degradation in Waters and Soils

Experience with oil spills in marine environments illustrates how effective stimulating natural microbial communities can be. Hydrocarbons lack phosphorus and nitrogen, so these elements limit their degradation. One of the first demonstrations that simple fertilization accelerates hydrocarbon degradation by natural populations occurred in 1989 following the *Exxon Valdez* oil spill in Prince William Sound, Alaska. As seen in **figure 43.13**, workers added nitrogen and phosphorus-containing fertilizer, which significantly diminished the oil in coastal sediments. Since that time, it has been recognized that another key variable in promoting bioremediation is the degree of contact between microorganisms and the hydrocarbon substrate. To increase hydrocarbon accessibility by microbes, pellets containing nutrients and an oleophilic (hydrocarbon soluble) preparation are often used. This technique accelerates the degradation of crude oil slicks by 30 to 40%, in comparison with control oil slicks where additional nutrients are not available. More recent research also focuses on the entire microbial community involved in the bioremediation process. Oil, which is a complex mixture of organic molecules, is degraded by a variety of microbes, but their capacity to do so is influenced by rates of predation by viruses and protists, the rate of nutrient recycling, and even the release of biosurfactants by other microbes (**figure 43.14**). Surfactants serve to disperse the oil; as described in the chapter

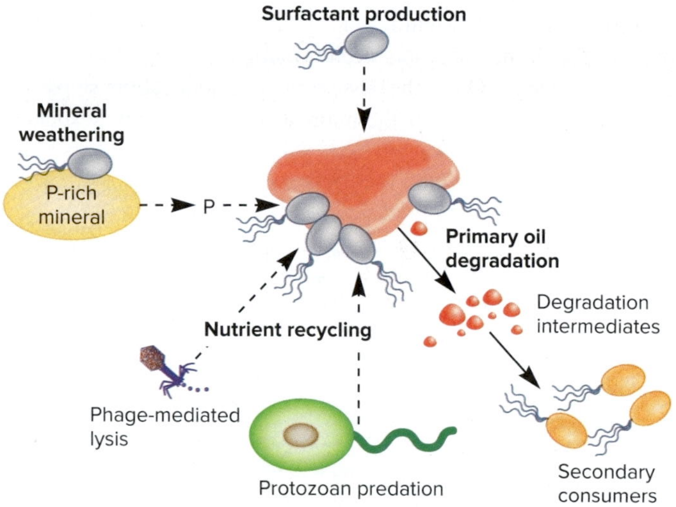

Figure 43.14 **Oil Degradation Requires a Microbial Community.** Primary oil degradation involves bacteria that begin oil degradation (blue microbes). These bacteria release a variety of degradation products that are further broken down by secondary consumers (orange microbes). Dashed lines represent indirect interactions that influence the rate of bioremediation, such as phage lysis and protozoan predation. P-rich material, phosphorus-rich material.

opener, so-called dispersants used during the BP oil spill in the Gulf of Mexico in 2010 became almost as controversial as the oil itself. This has prompted the development of microbial-based oil dispersants, described next.

Bioaugmentation

The acceleration of microbiological processes by the addition of known active microorganisms to soils, waters, or other complex systems is called **bioaugmentation.** This is not a new concept; commercial culture preparations have long been available to facilitate silage formation and to improve septic tank performance. However, the magnitude of contamination coupled with an increased awareness of some of the dangers cleanup can introduce has stimulated the growth of the bioaugmentation industry. For instance, an oil dispersant consisting of several nonpathogenic *Bacillus* spp., wetting agents, and nutrients is now commercially available.

Microorganisms can be added directly to natural communities, where they create their own microhabitats. For example, microorganisms in the water column overlying PCB-contaminated sand-clay soils have been observed to create their own clay hutches by binding clays to their outer surfaces with exopolysaccharides. However, in most cases, microbes are added with other components to ensure their viability. A blend of microorganisms can be purchased from commercial bioremediation engineering companies to treat petrochemical contamination of soils and water. These microbes are most often added in a slurry along with nutrients, enzymes, surfactants, and chemicals that release oxygen to promote aerobiosis. In other cases, microbes are introduced with protective inert microhabitats. This helps the foreign microbes establish an effective population in the face of competition from the natural, endogenous microbial community. Thus the application of principles of microbial physiology and ecology can facilitate the successful management of microbial communities in nature.

Metal Bioleaching

Bioleaching is the use of microorganisms that produce acids from reduced sulfur compounds to create acidic environments that solubilize desired metals for recovery. This approach is used to recover metals from ores and mining tailings with metal levels too low for smelting. Bioleaching carried out by natural populations of *Leptospirillum*-like species and thiobacilli, for example, allow recovery of up to 70% of the copper in low-grade ores. This involves the biological oxidation of copper present in these ores to produce soluble copper sulfate. In fact, this process is now used for copper mining.

Comprehension Check

1. List alternative definitions for the term *biodegradation.*
2. Explain the two mechanisms of reductive dehalogenation. What role do humic acids play in anaerobic degradation processes?
3. Discuss chirality and isomerization and their importance for understanding biodegradation.
4. What components are commonly added to microbes used to treat petrochemical contamination? Why are these added?
5. How is bioleaching carried out, and what microbes are involved?

Key Concepts

43.1 Purification and Sanitary Analysis Ensure Safe Drinking Water

- Water purification typically involves sedimentation, coagulation, chlorination, and rapid and slow sand filtration. Chlorination may lead to the formation of organic disinfection by-products, including trihalomethane compounds, which are potential carcinogens (**figure 43.1**).
- *Cryptosporidium, Cyclospora, Giardia* spp., and viruses are of concern, as conventional water purification and chlorination do not always ensure their removal and inactivation to acceptable limits.
- Indicator organisms are used to assess the presence of pathogenic microorganisms. Most probable number (MPN) and membrane filtration procedures are employed to estimate the number of indicator organisms present. The presence-absence (P-A) test is used for coliforms and defined substrate tests are used for coliforms and *E. coli* (**figures 43.2** and **43.3**).

43.2 Wastewater Treatment Maintains Human and Environmental Health

- Conventional sewage treatment is a controlled intensification of natural self-purification processes, and can involve primary, secondary, and tertiary treatment steps (**figures 43.4** and **43.5; tables 43.3** and **43.4**).
- The biochemical oxygen demand (BOD) test is an indirect measure of the organic matter that can be oxidized by the aerobic microbial community. The chemical oxygen demand (COD) and total organic carbon (TOC) tests provide information on carbon that is not biodegraded in the 5-day BOD test.
- Home treatment systems operate on general self-purification principles. The conventional septic tank provides anaerobic liquefaction and digestion, whereas the aerobic leach field allows oxidation of the soluble effluent. These systems are now designed to provide nitrogen and phosphorus removal to lessen impacts of on-site sewage treatment systems on vulnerable marine and freshwaters (**figure 43.8**).

43.3 Microbial Fuel Cells: Batteries Powered by Microbes

- Microbial fuel cells capture electrons donated to an anode during heterotrophic bacterial respiration. MFCs have a variety of applications, including the purification of waters while generating power and providing energy for monitoring environmental processes (**figure 43.9**).

43.4 Biodegradation and Bioremediation Harness Microbes to Clean the Environment

- Biodegradation is a critical part of natural systems mediated largely by microorganisms. This can involve minor changes in a molecule, fragmentation, or mineralization (**figure 43.10**).
- Biodegradation can be influenced by many factors, including oxygen levels, humic acids, and the presence of readily usable organic matter. Reductive dehalogenation occurs best under anoxic conditions (**figure 43.11**), and the presence of organic matter can facilitate modification of recalcitrant compounds in the process of cometabolism.
- Degradation management can be carried out in place, whether this be large marine oil spills, soils, or the subsurface. Such large-scale efforts often involve the use of natural microbial communities (**figures 43.13** and **43.14**).
- Bioremediation can also be accomplished by adding microbes directly to contaminated environments. These microbes are usually introduced with other components such as nutrients and dispersants to ensure their success despite competition from endogenous microbial communities.

Active Learning

1. You wish to build a house in a rural community where you will need to install a septic system and drill a well for drinking water. Where will you site the well in relation to the septic system? Following the perc test, you discover that the town will allow you to build your house only if you install a sand filter in addition to your septic tank. Why do you think this might be the case?
2. You are responsible for the bioremediation of soil contaminated by jet fuel. Your coworker suggests you fertilize the soil with nitrogen and phosphorus to promote the growth of naturally occurring hydrocarbon-degrading microbes. However, you just read about a new mixture of microbes sold in a slurry with nutrients and a surfactant. What data do you need to collect to decide which approach would be best? How will you obtain these data?
3. Biofilms form on surfaces in drinking water distribution networks and in some cases have been shown to be reservoirs of protists in the genera *Giardia* and *Cryptosporidium,* as well as a strain of *Poliovirus*. Discuss how you would assess the attachment to, growth within, and detachment from model biofilms in tap water in a model water distribution network. What specific variables would you consider and what microbial, chemical, and physical parameters would you measure?
4. Microbiologists studying the bioremediation of hydrocarbons in diesel-contaminated microcosm systems discovered that even when defined consortia of bacteria known to degrade oil were added, a bloom of bacteria in the *Alkanindiges* genus occurred. Prior to the addition of diesel, this microbe was present in very low numbers. Discuss the meaning and implications of their conclusion: "Rare community members presumably act as a reservoir of ecological functions in high-diversity environments, such as soils. This rare-to-dominant bacterial shift illustrates the potential role of a biosphere facing drastic environmental disturbances. Additionally, it supports the concept of 'conditionally rare taxa,' in which rareness is a temporary state conditioned by environmental constraints."

   Read the original paper: Fuentes, S., et al. 2017. From rare to dominant: A fine-tuned bacterial bloom during petroleum hydrocarbon bioremediation. *Appl. Environ. Microbiol.* 82:888–96. doi:10.1128/AEM.02625-15.

Appendix One

A Review of the Chemistry of Biological Molecules

Appendix I provides a brief summary of the chemistry of organic molecules with particular emphasis on the molecules present in microbial cells. Only basic concepts and terminology are presented; introductory textbooks in biology and chemistry should be consulted for a more extensive treatment of these topics.

Atoms and Molecules

Matter is made of elements that are composed of atoms. An element contains only one kind of atom and cannot be broken down to simpler components by chemical reactions. An atom is the smallest unit characteristic of an element and can exist alone or in combination with other atoms. When atoms combine, they form molecules. Molecules are the smallest particles of a substance. They have all the properties of the substance and are composed of two or more atoms.

Although atoms contain many subatomic particles, three directly influence their chemical behavior: protons, neutrons, and electrons. The atom's nucleus is located at its center and contains varying numbers of protons and neutrons (**figure AI.1**). Protons have a positive charge, and neutrons are uncharged. The mass of these particles and the atoms that they compose is given in terms of the atomic mass unit (AMU), which is 1/12 the mass of the most abundant carbon isotope. Often the term *dalton* (Da) is used to express the mass of molecules. It also is 1/12 the mass of an atom of ^{12}C or 1.661×10^{-24} grams. Both protons and neutrons have a mass of about 1 dalton. The atomic weight is the actual measured weight of an element and is almost identical to the mass number for the element, the total number of protons and neutrons in its nucleus. The mass number is indicated by a superscripted number preceding the element's symbol (e.g., ^{12}C, ^{16}O, and ^{14}N).

Negatively charged particles called electrons circle the atomic nucleus (figure AI.1). The number of electrons in a neutral atom equals the number of its protons and is given by the atomic number, the number of protons in an atomic nucleus. The atomic number is characteristic of a particular type of atom. For example, carbon has an atomic number of six, hydrogen's number is one, and oxygen's is eight (**table AI.1**).

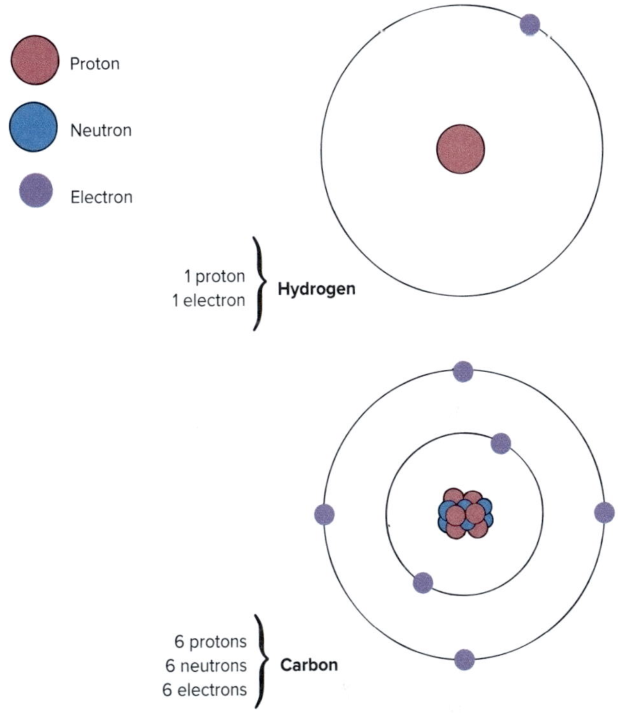

Figure AI.1 Diagrams of Hydrogen and Carbon Atoms. The electron orbitals are represented as concentric circles.

Table AI.1 Atoms Commonly Present in Organic Molecules

Atom	Symbol	Atomic Number	Atomic Weight	Number of Chemical Bonds
Hydrogen	H	1	1.01	1
Carbon	C	6	12.01	4
Nitrogen	N	7	14.01	3
Oxygen	O	8	16.00	2
Phosphorus	P	15	30.97	5
Sulfur	S	16	32.06	2

The electrons move constantly within a volume of space surrounding the nucleus, even though their precise location in this volume cannot be determined accurately. This volume of space in which an electron is located is called its orbital. Each orbital can contain two electrons. Orbitals are grouped into shells of different energy that surround the nucleus. The first shell is closest to the nucleus and has the lowest energy; it contains only one orbital. The second shell contains four orbitals, one circular and three shaped like dumbbells (**figure AI.2*a***). It can contain up to eight electrons. The third shell has even higher energy and holds more than eight electrons. Shells are filled beginning with the innermost and moving outward. For example, carbon has six electrons, two in its first shell and four in the second (figures AI.1 and AI.2*b*). The electrons in the outermost shell are the ones that participate in chemical reactions. The most stable condition is achieved when the outer shell is filled with electrons. Thus the number of bonds an element can form depends on the number of electrons required to fill the outer shell. Since carbon has four electrons in its outer shell and the shell is filled when it contains eight electrons, it can form four covalent bonds (table AI.1).

Chemical Bonds

Molecules are formed when two or more atoms associate through chemical bonding. Chemical bonds are attractive forces that hold together atoms, ions, or groups of atoms in a molecule or other substance. Many types of chemical bonds are present in organic molecules; three of the most important are covalent bonds, ionic bonds, and hydrogen bonds.

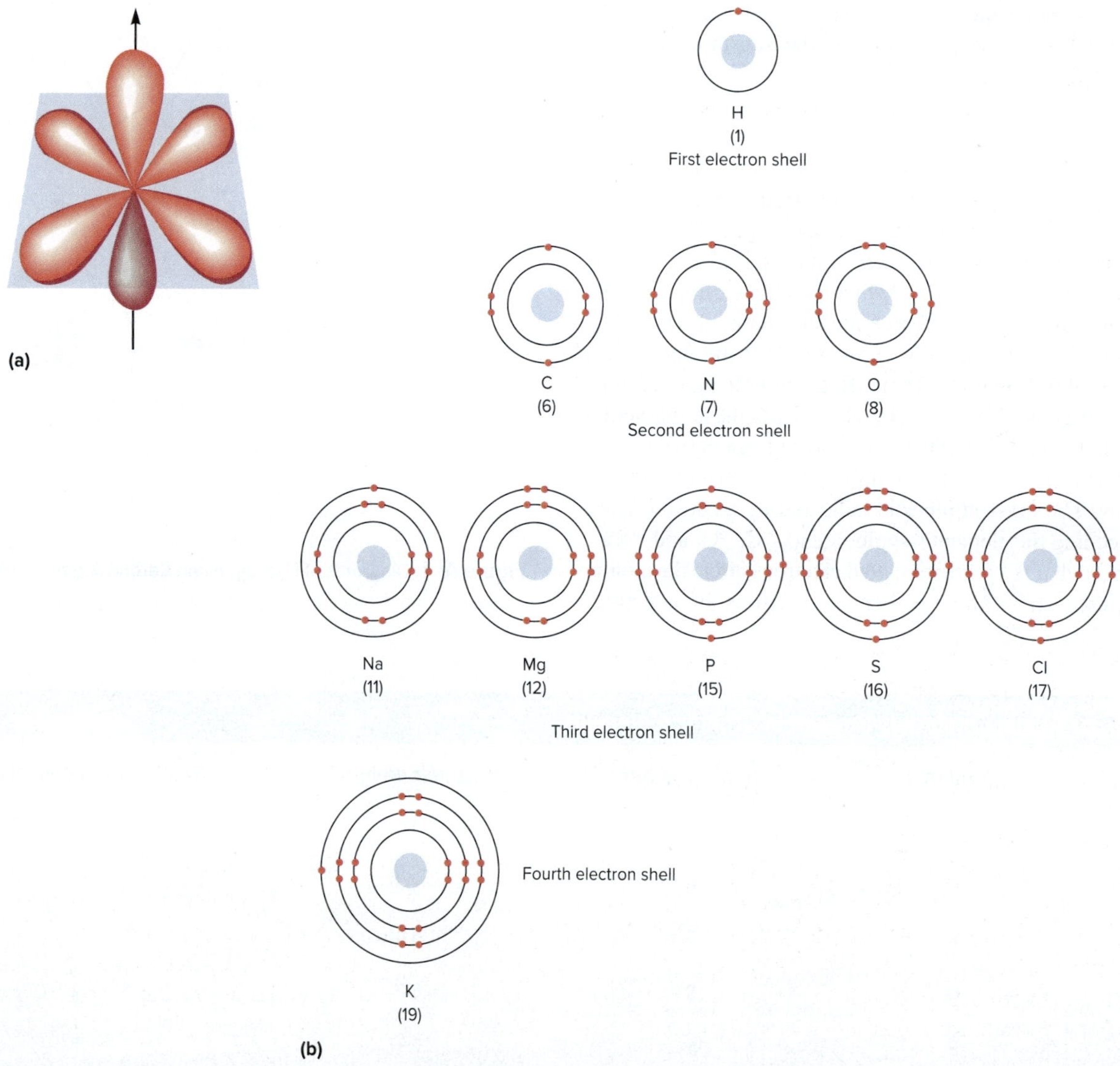

Figure AI.2 Electron Orbitals. (a) The three dumbbell-shaped orbitals of the second shell. The orbitals lie at right angles to each other. (b) The distribution of electrons in some common elements. Atomic numbers are given in parentheses.

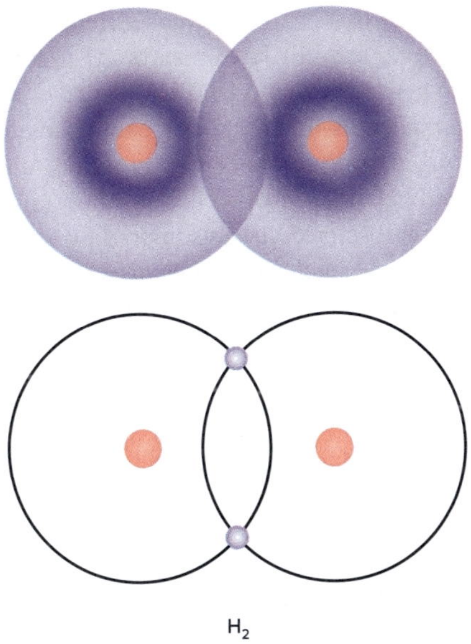

Figure AI.3 The Covalent Bond. A hydrogen molecule is formed when two hydrogen atoms share electrons.

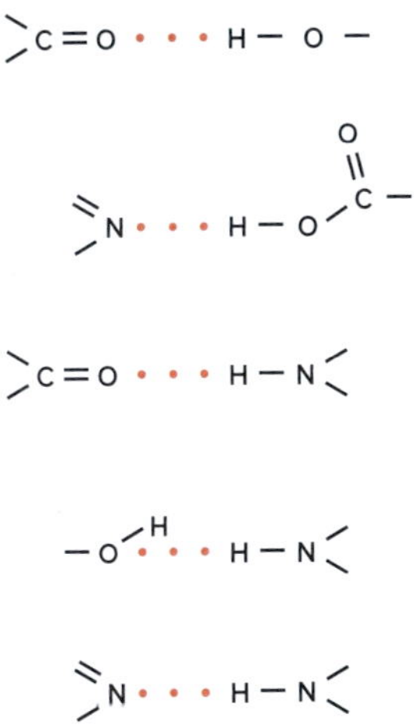

Figure AI.4 Hydrogen Bonds. Representative examples of hydrogen bonds present in biological molecules.

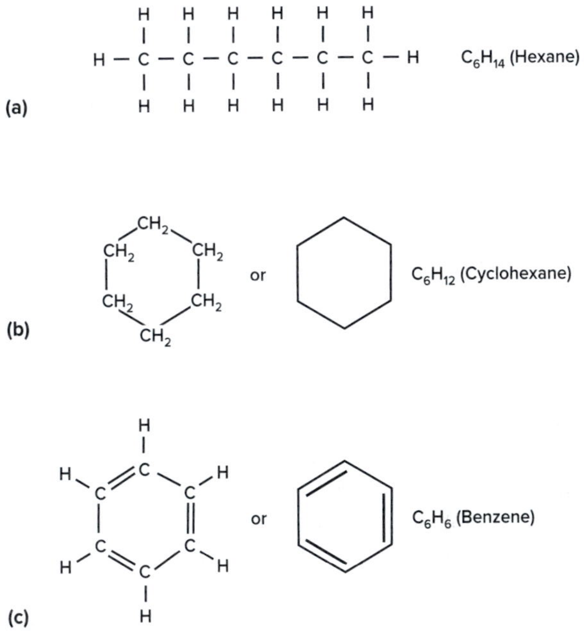

Figure AI.5 Hydrocarbons. Examples of hydrocarbons that are (a) linear, (b) cyclic, and (c) aromatic.

In covalent bonds, atoms are joined together by sharing pairs of electrons (**figure AI.3**). If the electrons are equally shared between identical atoms (e.g., in a carbon-carbon bond), the covalent bond is strong and nonpolar. When two different atoms such as carbon and oxygen share electrons, the covalent bond formed is polar because the electrons are pulled toward the more electronegative atom, the atom that more strongly attracts electrons (the oxygen atom). A single pair of electrons is shared in a single bond; a double bond is formed when two pairs of electrons are shared.

Atoms often contain either more or fewer electrons than the number of protons in their nuclei. When this is the case, they carry a net negative or positive charge and are called ions. Cations carry positive charges and anions have a net negative charge. When a cation and an anion approach each other, they are attracted by their opposite charges. This ionic attraction that holds two groups together is called an ionic bond. While it is difficult to make a direct comparison between the strength of ionic and covalent bonds, in an aqueous environment ionic bonds are easily disrupted, making them an important component of biological molecules.

When a hydrogen atom is covalently bonded to a more electronegative atom such as oxygen or nitrogen, the electrons are unequally shared and the hydrogen atom carries a partial positive charge. It will be attracted to an electronegative atom such as oxygen or nitrogen, which carries an unshared pair of electrons; this attraction is called a hydrogen bond (**figure AI.4**). Although an individual hydrogen bond is weak, there are so many hydrogen bonds in proteins and nucleic acids that they play a major role in determining protein and nucleic acid structure.

Organic Molecules

Most molecules in cells are organic molecules, molecules that contain carbon. Since carbon has four electrons in its outer shell, it forms four covalent bonds in order to fill its outer shell with eight electrons. This property makes it possible to form chains and rings of carbon atoms that also can bond with hydrogen and other atoms (**figure AI.5**). Although adjacent carbons usually are connected by single bonds, they may be joined by double or triple bonds. Rings that have alternating single and double bonds, such as the benzene ring, are called aromatic rings. The hydrocarbon chain or ring provides a chemically inactive skeleton to which more reactive groups of atoms may be attached. These reactive groups with specific properties are known as functional groups. They usually contain atoms of oxygen, nitrogen, phosphorus, or sulfur (**figure AI.6**) and are largely responsible for most characteristic chemical properties of organic molecules.

Organic molecules are often divided into classes based on the nature of their functional groups. Ketones have a carbonyl group within the carbon chain, whereas alcohols have a hydroxyl on the chain. Organic acids have a carboxyl group, and amines have an amino group (**figure AI.7**).

Organic molecules may have the same chemical composition and yet differ in their molecular structure and properties. Such molecules are called isomers. One important class of isomers is the stereoisomers. Stereoisomers have the same atoms arranged in the same nucleus-to-nucleus sequence but differ in the spatial arrangement of their atoms. For example, an amino acid such as alanine can form stereoisomers (**figure AI.8**). L-Alanine and other L-amino acids are the stereoisomer forms normally present in proteins.

Carbohydrates

Carbohydrates are aldehyde or ketone derivatives of polyhydroxy alcohols. The smallest and least complex carbohydrates are the simple sugars or monosaccharides. The most common sugars have five or six carbons. A sugar in its ring form has two isomeric structures, the α and β forms, that differ in the orientation of the hydroxyl on the aldehyde or ketone carbon, which is

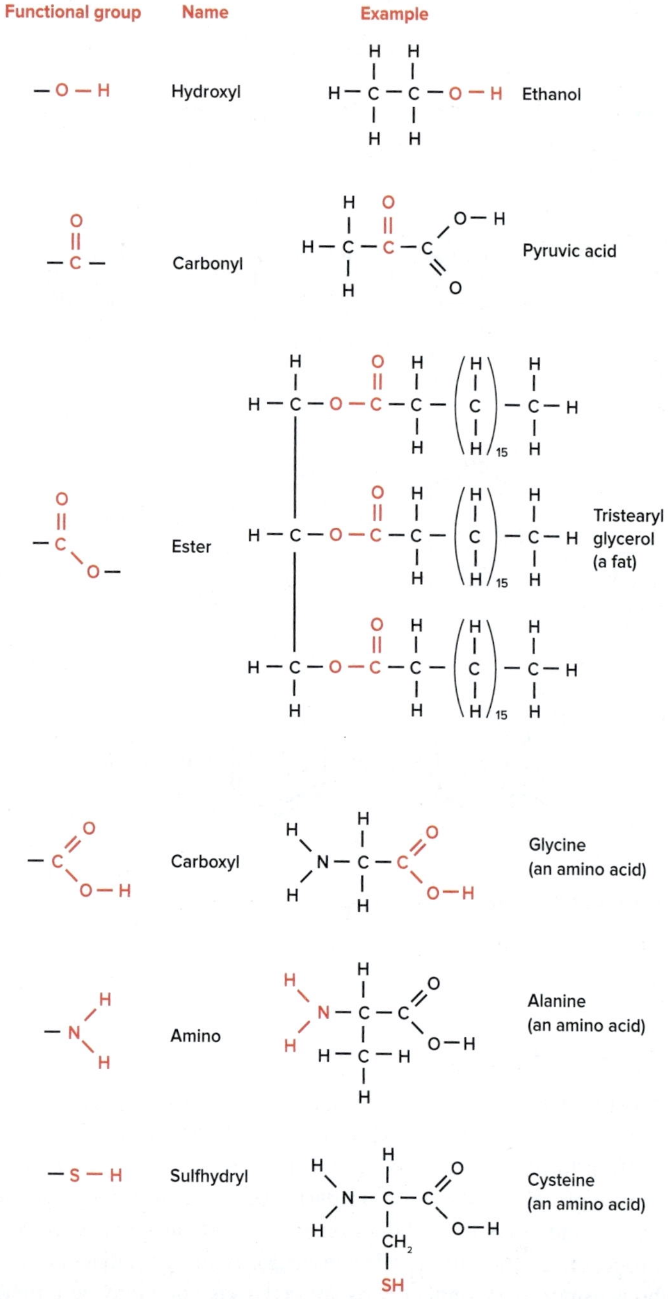

Figure AI.6 Functional Groups. Some common functional groups in organic molecules. The groups are shown in red.

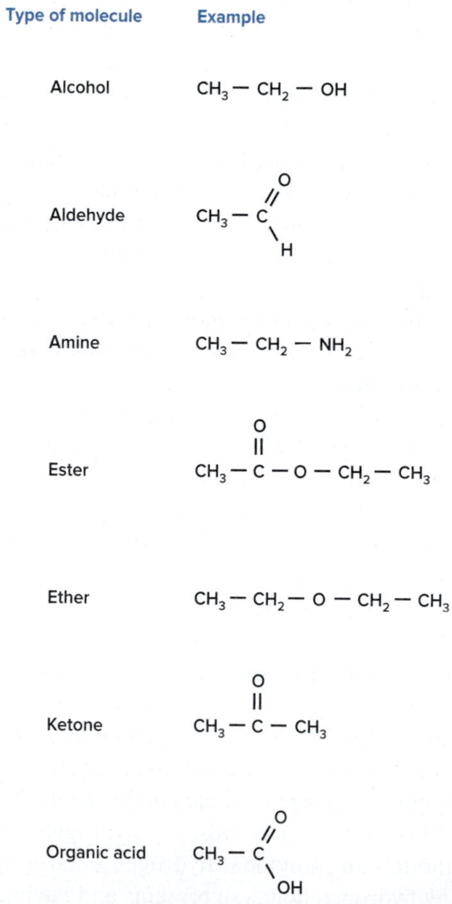

Figure AI.7 Types of Organic Molecules. These are classified on the basis of their functional groups.

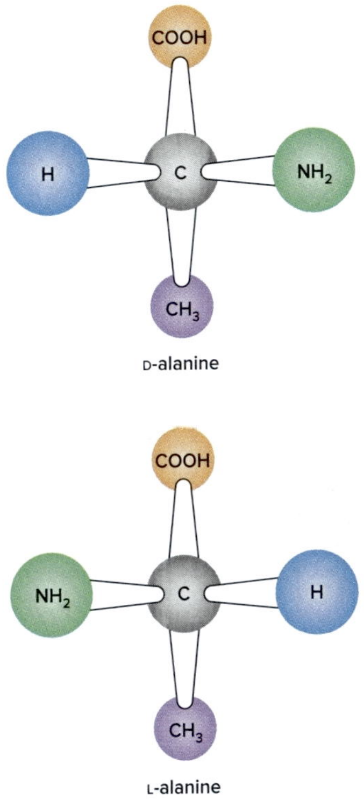

Figure AI.8 The Stereoisomers of Alanine. The α-carbon is in gray. L-alanine is the form usually present in proteins.

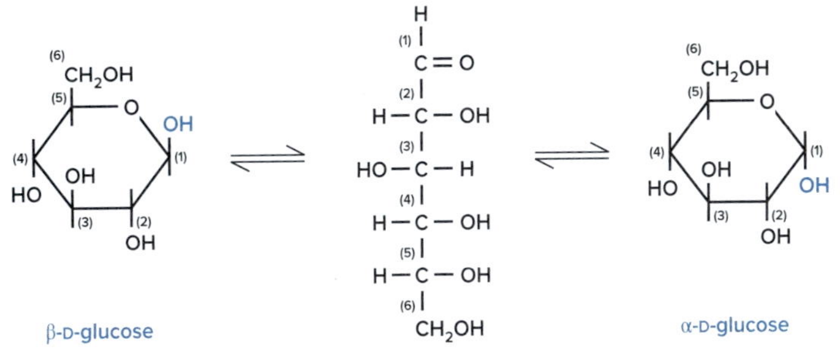

Figure AI.9 The Interconversion of Monosaccharide Structures. The open chain form of glucose and other sugars is in equilibrium with closed ring structures (depicted here with Haworth projections). Aldehyde sugars form cyclic hemiacetals, and keto sugars produce cyclic hemiketals. When the hydroxyl on carbon one of cyclic hemiacetals projects above the ring, the form is known as a β form. The α form has a hydroxyl that lies below the plane of the ring. The same convention is used in showing the α and β forms of hemiketals such as those formed by fructose.

called the anomeric or glycosidic carbon (**figure AI.9**). Microorganisms have many sugar derivatives in which a hydroxyl is replaced by an amino group or some other functional group (e.g., glucosamine).

Two monosaccharides can be joined by a bond between the anomeric carbon of one sugar and a hydroxyl or the anomeric carbon of the second (**figure AI.10**). The bond joining sugars is a glycosidic bond and may be either α or β depending on the orientation of the anomeric carbon. Two sugars linked in this way constitute a disaccharide. Some common disaccharides are maltose (two glucose molecules), lactose (glucose and galactose), and sucrose (glucose and fructose). If 10 or more sugars are linked together by glycosidic bonds, a polysaccharide is formed. For example, starch and glycogen are common polymers of glucose that are used as sources of carbon and energy (**figure AI.11**).

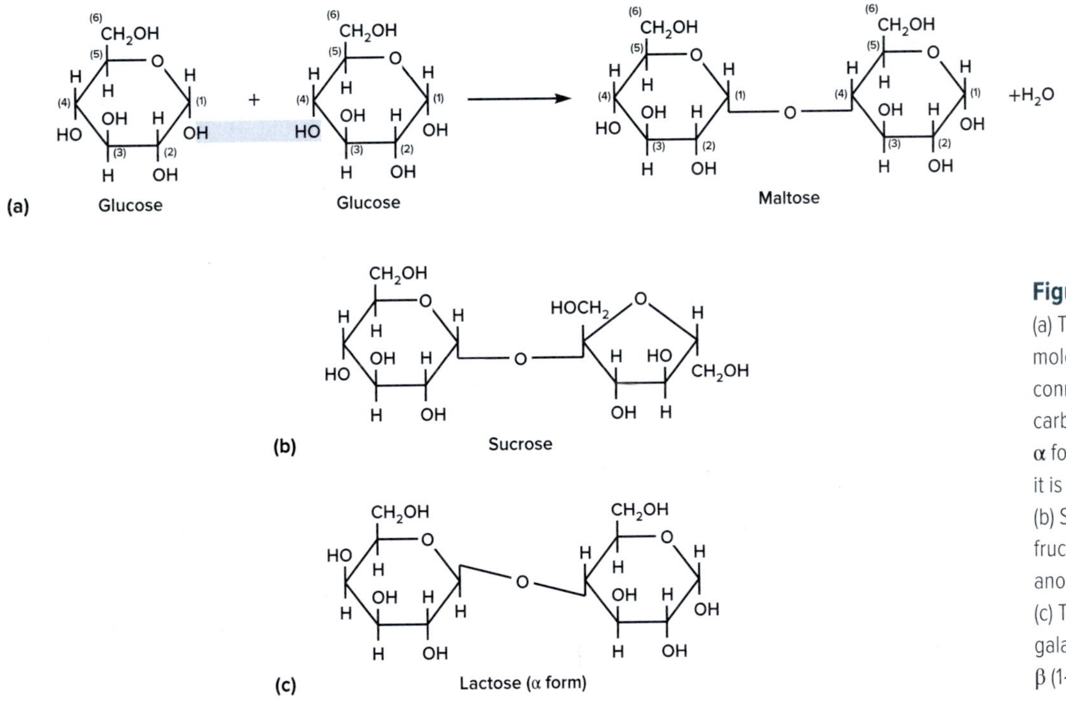

Figure AI.10 Common Disaccharides. (a) The formation of maltose from two molecules of α-glucose. The bond connecting the glucose extends between carbons one and four, and involves the α form of the anomeric carbon. Therefore, it is called an α (1→4) glycosidic bond. (b) Sucrose is composed of a glucose and a fructose joined to each other through their anomeric carbons, and αβ (1→2) bond. (c) The milk sugar lactose contains galactose and glucose joined by a β (1→4) glycosidic bond.

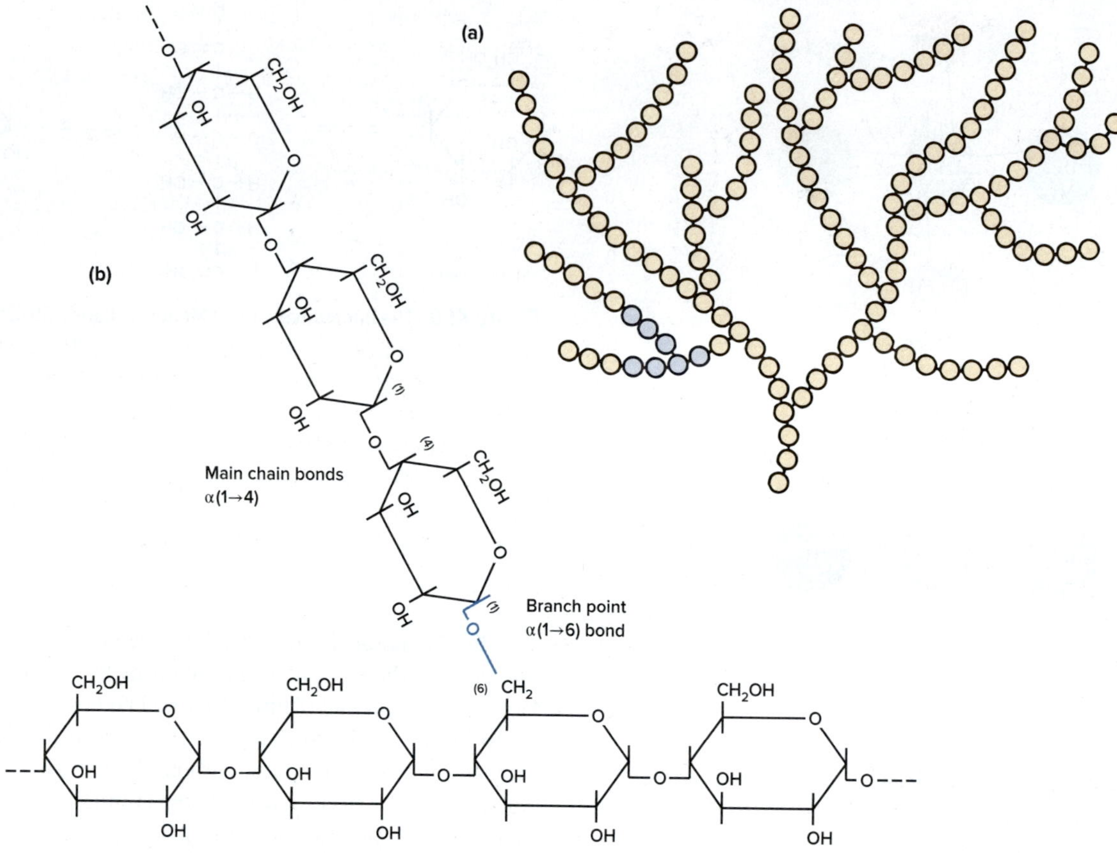

Figure AI.11 Glycogen and Starch Structure. (a) An overall view of the highly branched structure characteristic of glycogen and most starch. The circles represent glucose residues. (b) A close-up of a small part of the chain (shown in blue in part *a*) revealing a branch point with its α (1→6) glycosidic bond, which is colored blue.

Lipids

All cells contain a heterogeneous mixture of organic molecules that are relatively insoluble in water but very soluble in nonpolar solvents such as chloroform, ether, and benzene. These molecules are called lipids. Lipids vary greatly in structure and include triacylglycerols, phospholipids, steroids, carotenoids, and many other types. Among other functions, they serve as membrane components, storage forms for carbon and energy, precursors of other cell constituents, and protective barriers against water loss.

Most lipids contain fatty acids, which are monocarboxylic acids that often are straight chained but may be branched. Saturated fatty acids lack double bonds in their carbon chains, whereas unsaturated fatty acids have double bonds. The most common fatty acids are 16 or 18 carbons long.

Two good examples of common lipids are triacylglycerols and phospholipids. Triacylglycerols are composed of glycerol esterified to three fatty acids (**figure AI.12*a***). They are used to store carbon and energy. Phospholipids are lipids that contain at least one phosphate group and often have a nitrogenous constituent as well. Phosphatidylethanolamine is an important phospholipid frequently present in bacterial membranes (figure AI.12*b*). It is composed of two fatty acids esterified to glycerol. The third glycerol hydroxyl is joined with a phosphate group, and ethanolamine is attached to the phosphate. The resulting lipid is very asymmetric with a hydrophobic nonpolar end contributed by the fatty acids and a polar, hydrophilic end. In cell membranes, the hydrophobic end is buried in the interior of the membrane, while the polar-charged end is at the membrane surface and exposed to water.

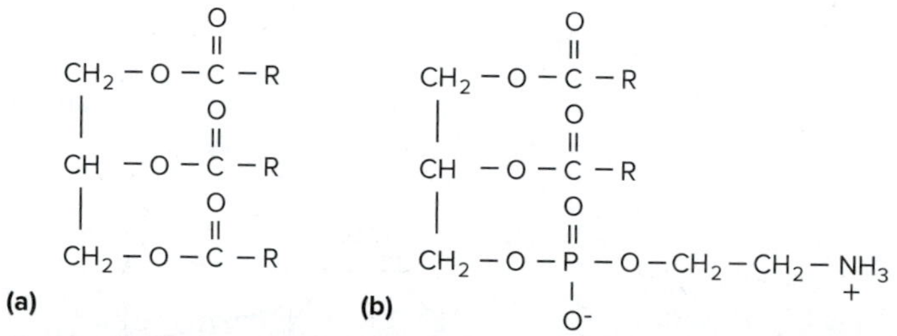

Figure AI.12 Examples of Common Lipids. (a) A triacylglycerol or neutral fat. (b) The phospholipid phosphatidylethanolamine. The R groups represent fatty acid side chains.

Proteins

The basic building blocks of proteins are amino acids. An amino acid contains a carboxyl group and an amino group on its alpha carbon (**figure AI.13**). About 20 amino acids are normally found in proteins; they differ from each other with respect to their side chains. In proteins, amino acids are linked together by peptide bonds between their carboxyls and α-amino groups to form linear polymers called peptides. If a peptide contains more than 50 amino acids, it usually is called a polypeptide. Each protein is composed of one or more polypeptide chains and has a molecular weight greater than about 6,000 to 7,000.

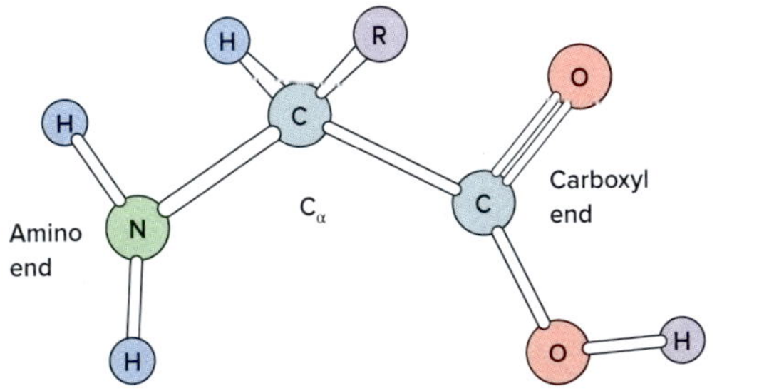

Figure AI.13 L-Amino Acid Structure. The uncharged form is shown.

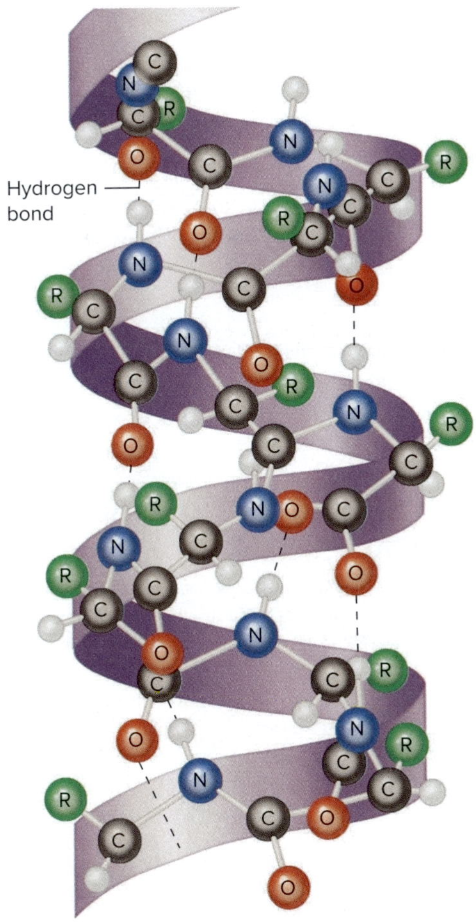

Figure AI.14 The α-Helix. A polypeptide twisted into one type of secondary structure, the α-helix. The helix is stabilized by hydrogen bonds joining peptide bonds that are separated by three amino acids.

Proteins have three or four levels of structural organization and complexity. The primary structure of a protein is the sequence of the amino acids in its polypeptide chain or chains. The structure of the polypeptide chain backbone is also considered part of the primary structure. Each different polypeptide has its own amino acid sequence that is a reflection of the nucleotide sequence in the gene that codes for its synthesis. The polypeptide chain can coil along one axis in space into various shapes such as the α-helix (**figure AI.14**). This arrangement of the polypeptide in space around a single axis is called the secondary structure. Secondary structure is formed and stabilized by the interactions of amino acids that are fairly close to one another on the polypeptide chain. The polypeptide with its primary and secondary structure can be coiled or organized in space along three axes to form a more complex, three-dimensional shape (**figure AI.15**). This level of organization is the tertiary structure (**figure AI.16**). Amino acids more distant from one another on the polypeptide chain contribute to tertiary structure. Secondary and tertiary structures are examples of conformation, molecular shape that can be changed by bond rotation and without breaking covalent bonds. When a protein contains more than one polypeptide chain, each chain with its own primary, secondary, and tertiary structure associates with the other chains to form the final molecule. The way in which polypeptides associate with each other

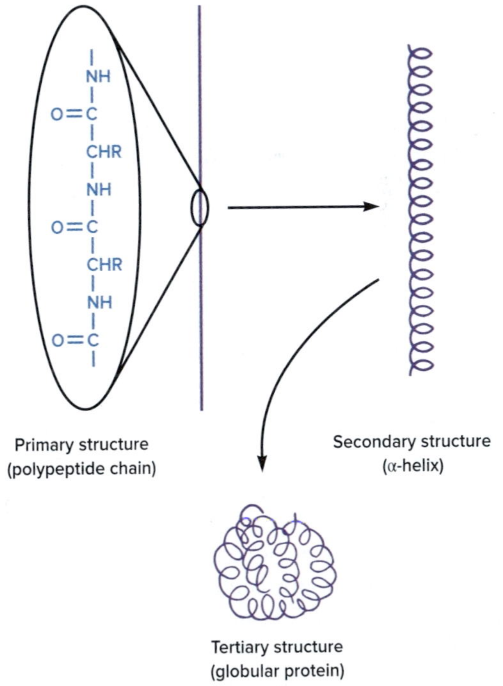

Figure AI.15 Secondary and Tertiary Protein Structures. The formation of secondary and tertiary protein structures by folding a polypeptide chain with its primary structure.

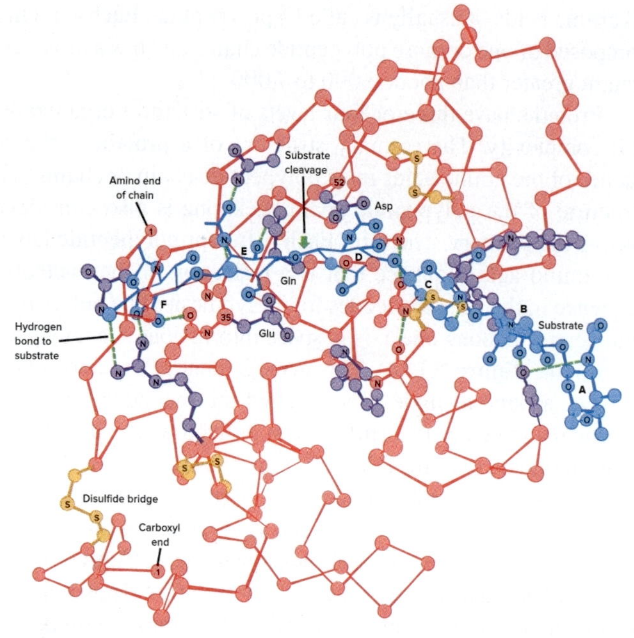

Figure AI.16 Lysozyme. The tertiary structure of the enzyme lysozyme. (a) A diagram of the protein's polypeptide backbone with the substrate hexasaccharide shown in blue. The point of substrate cleavage is indicated.

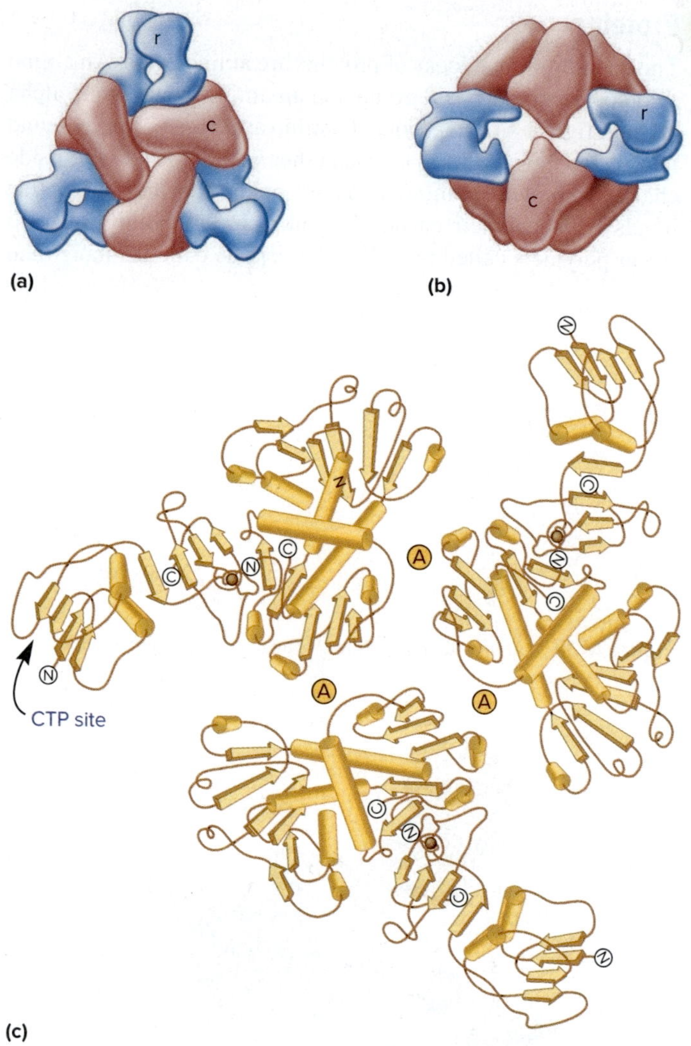

Figure AI.17 An Example of Quaternary Structure. The enzyme aspartate carbamoyltransferase from *Escherichia coli* has two types of subunits, catalytic and regulatory. The association between the two types of subunits is shown: (a) a top view, and (b) a side view of the enzyme. The catalytic (C) and regulator (r) subunits are shown in different colors. (c) The peptide chains shown when viewed from the top as in (a). The active sites of the enzyme are located at the positions indicated by A.

(a and b) Source: Krause, et al., in *Proceedings of the National Academy of Sciences,* V. 82, 1985, as appeared in *Biochemistry,* 3d edition, by Lubert Stryer; (c) Source: Kantrowitz, et al., in *Trends in Biochemical Science,* V. 5, 1980, as appeared in *Biochemistry,* 3d edition, by Lubert Stryer.

in space to form the final protein is called the protein's quaternary structure (**figure AI.17**). The final conformation of a protein is ultimately determined by the amino acid sequence of its polypeptide chains.

Protein secondary, tertiary, and quaternary structure is largely determined and stabilized by many weak noncovalent forces such as hydrogen bonds and ionic bonds. Because of this, protein shape often is very flexible and easily changed. This flexibility is very important in protein function and in the regulation of enzyme activity. Because of their flexibility, however, proteins readily lose their proper shape and activity when exposed to harsh conditions. The only covalent bond commonly involved in the secondary and tertiary structure of proteins is the disulfide bond. The disulfide bond is formed when two cysteines are linked through their sulfhydryl groups. Disulfide bonds generally strengthen or stabilize protein structure.

Nucleic Acids

The nucleic acids, deoxyribonucleic acid (DNA) and ribonucleic acid (RNA), are polymers of deoxyribonucleosides and ribonucleosides joined by phosphate groups. The nucleosides in DNA contain the purines adenine and guanine, and the pyrimidine bases thymine and cytosine. In RNA the pyrimidine uracil is substituted for thymine. Because of their importance for genetics and molecular biology, the chemistry of nucleic acids is introduced earlier in the text. The structure and synthesis of purines and pyrimidines are discussed in chapter 12 (*section 12.6*). The structures of DNA and RNA are described in chapter 13 (*section 13.2*).

Appendix Two

Common Metabolic Pathways

Appendix II contains a few of the more important pathways discussed in the text, particularly those involved in carbohydrate catabolism. Enzyme names and final end products are given in color. Consult the text for a description of each pathway and its physiologic role.

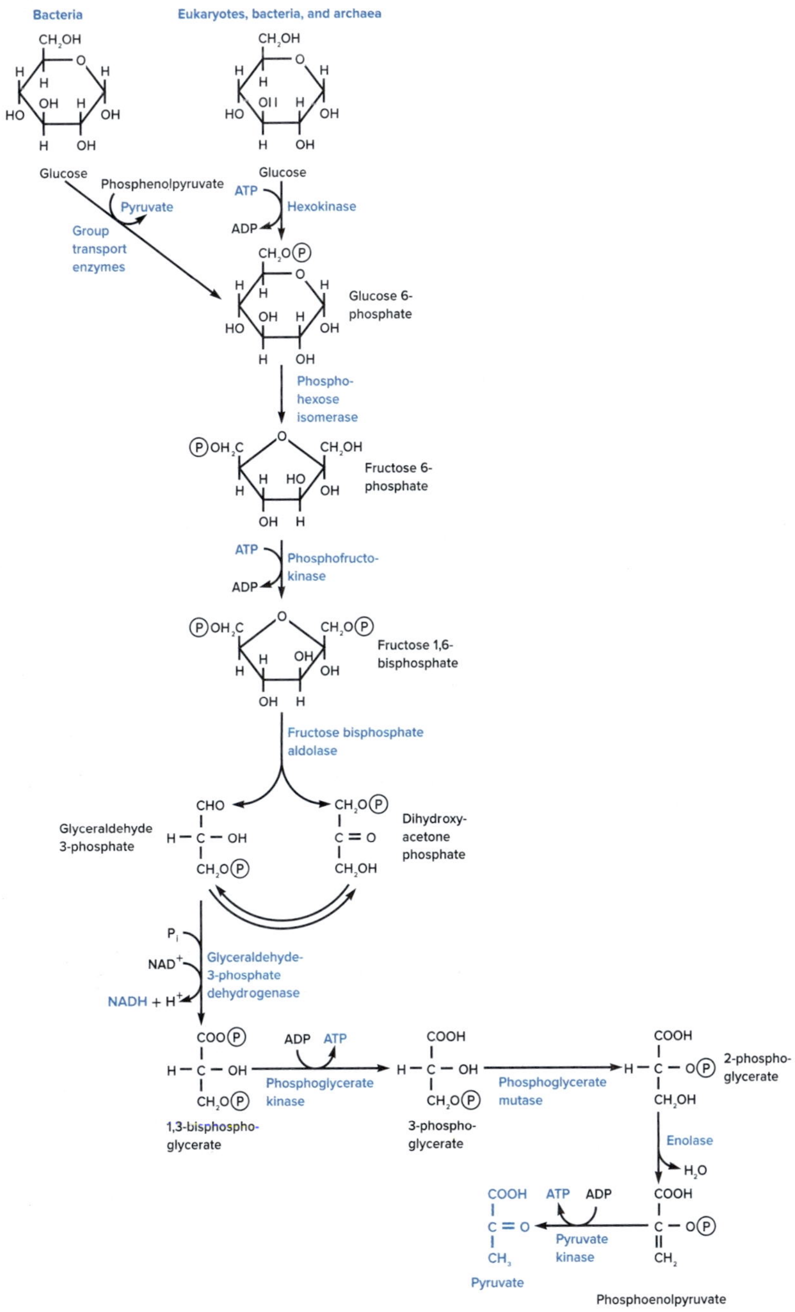

Figure AII.1 The Embden-Meyerhof Pathway. This pathway converts glucose and other sugars to pyruvate and generates NADH and ATP. In some bacteria, glucose is phosphorylated to glucose 6-phosphate during group translocation transport across the plasma membrane.

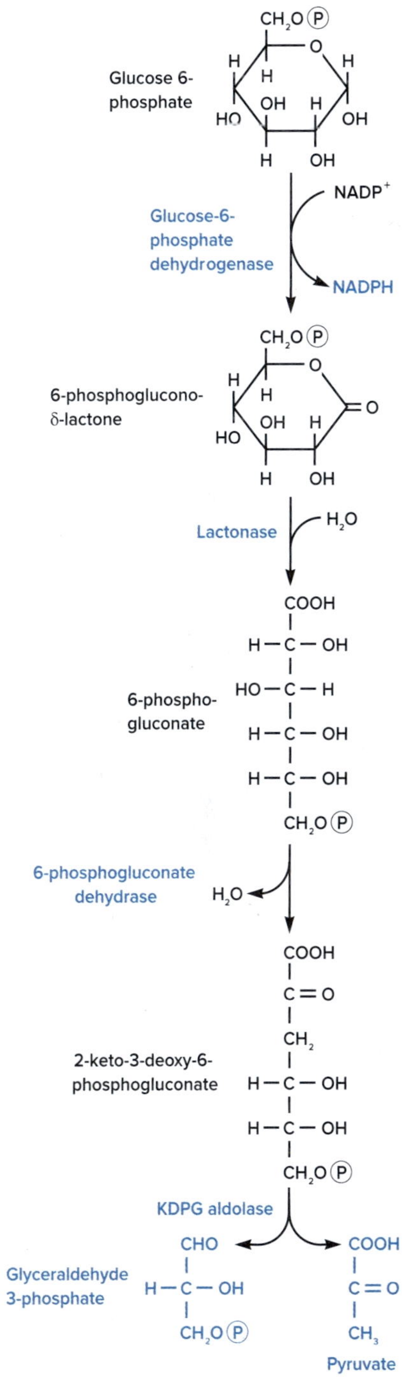

Figure AII.2 The Entner-Doudoroff Pathway.

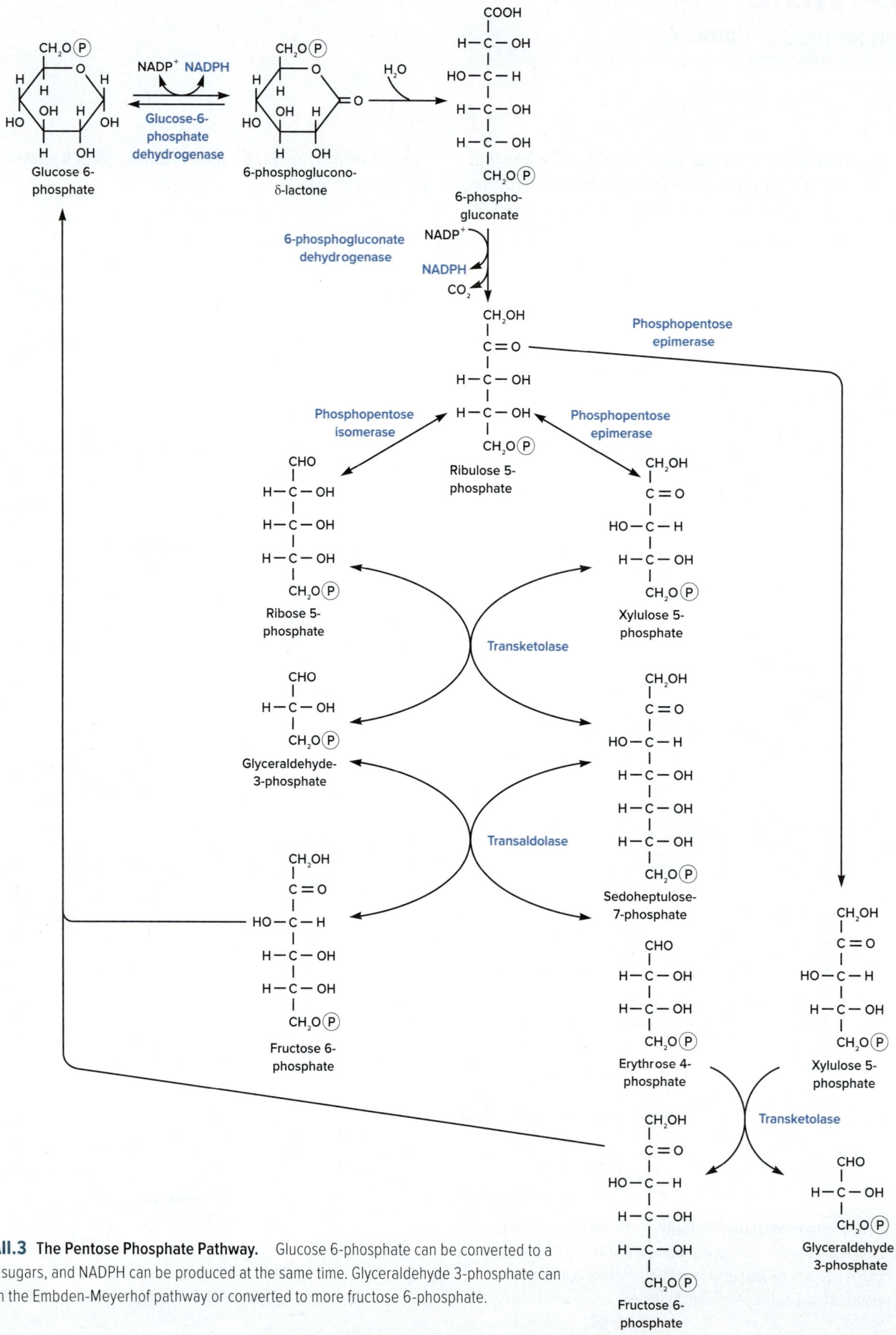

Figure AII.3 The Pentose Phosphate Pathway. Glucose 6-phosphate can be converted to a variety of sugars, and NADPH can be produced at the same time. Glyceraldehyde 3-phosphate can be used in the Embden-Meyerhof pathway or converted to more fructose 6-phosphate.

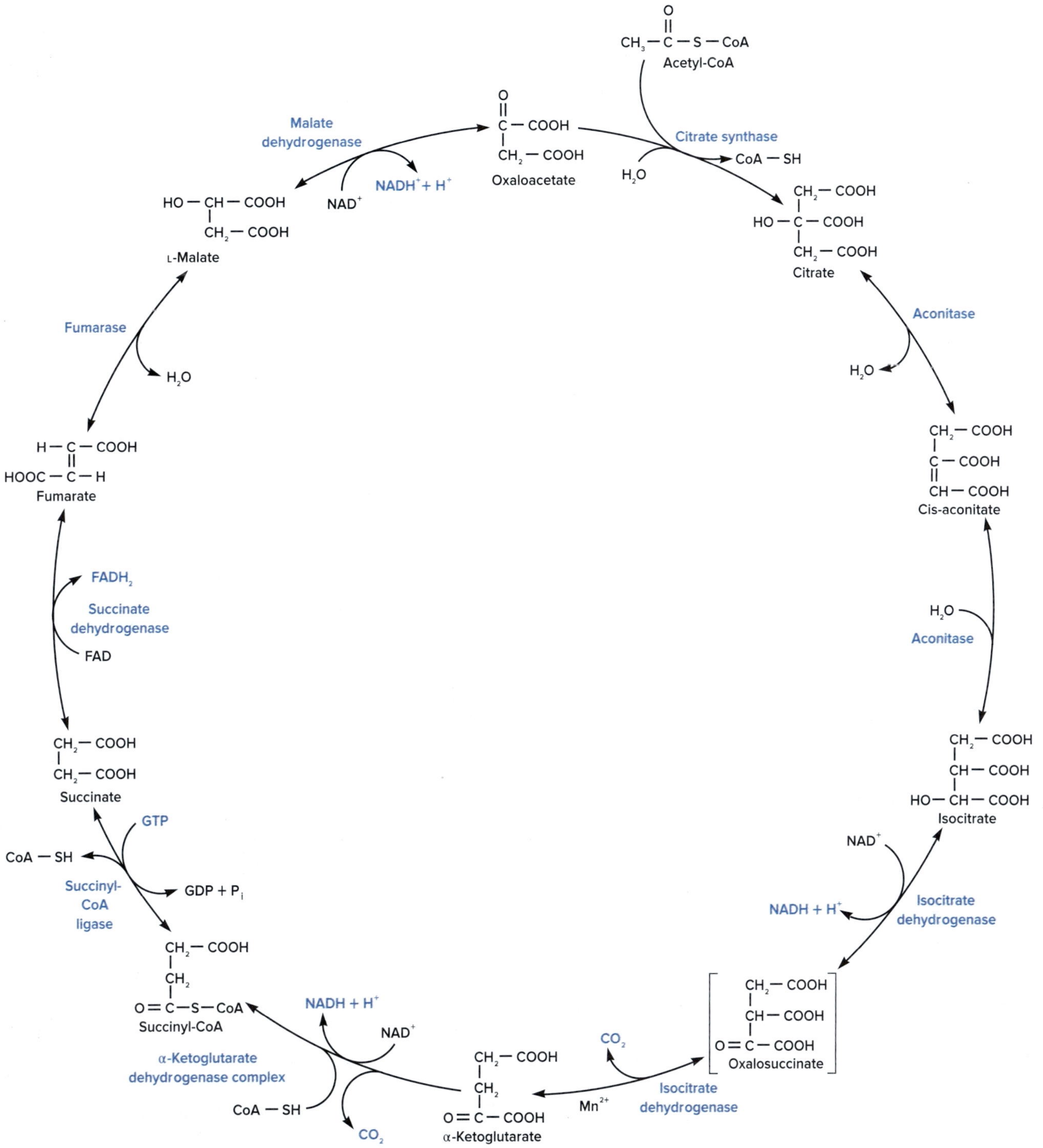

Figure AII.4 The Tricarboxylic Acid Cycle. Cis-aconitate and oxalosuccinate remain bound to aconitase and isocitrate dehydrogenase. Oxalosuccinate has been placed in brackets because it is so unstable.

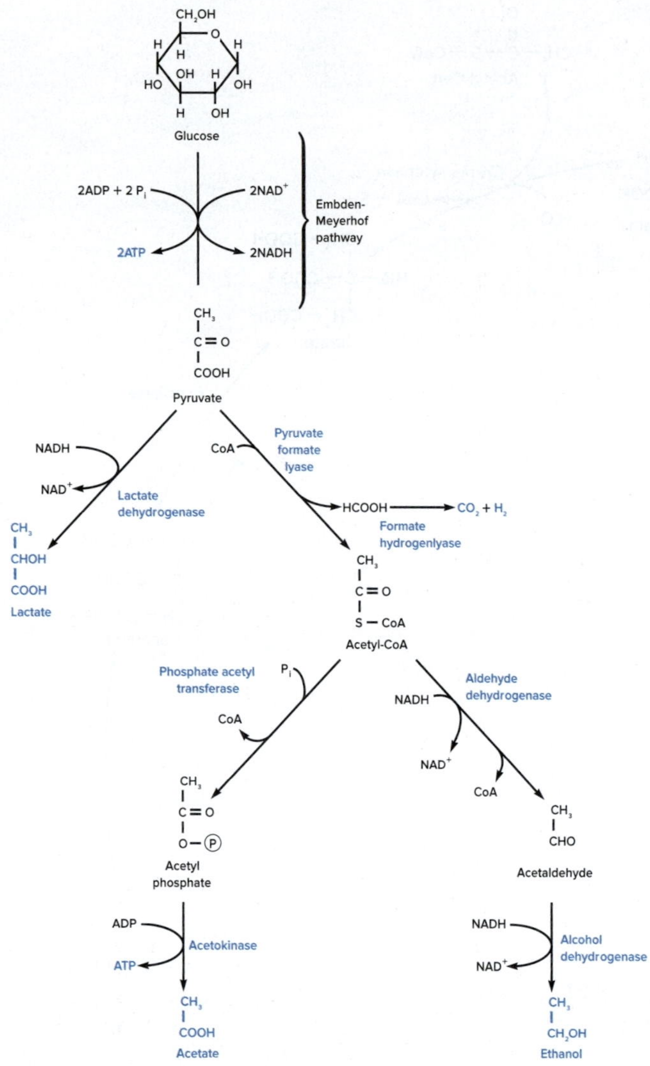

Figure AII.5 The Mixed Acid Fermentation Pathway. This pathway is characteristic of many members of the *Enterobacteriaceae* such as *Escherichia coli.*

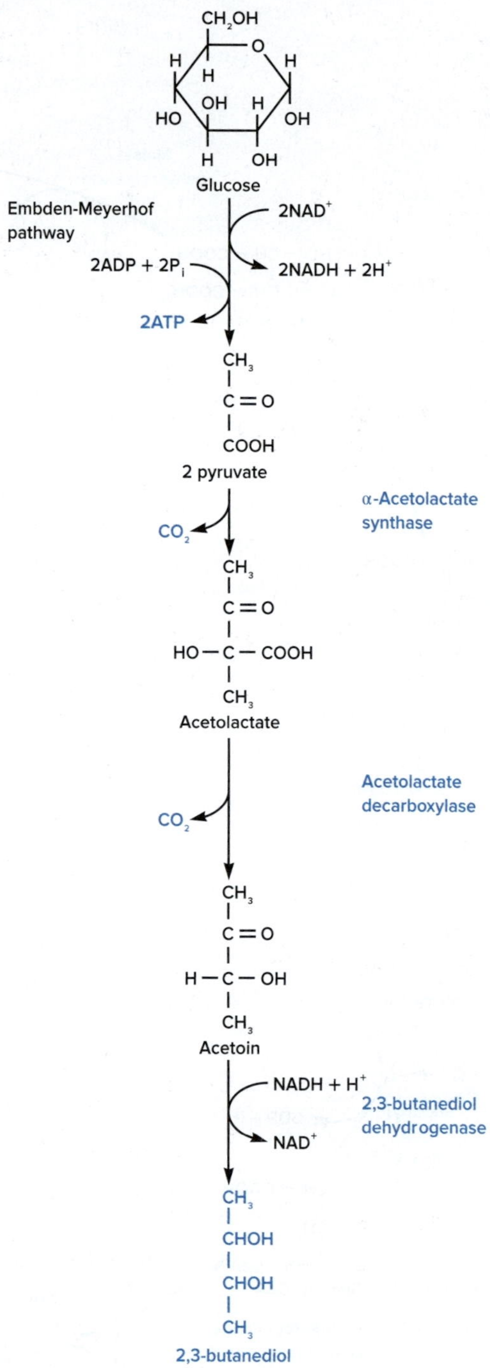

Figure AII.6 The Butanediol Fermentation Pathway. This pathway is characteristic of members of the *Enterobacteriaceae* such as *Enterobacter.* Other products may also be formed during butanediol fermentation.

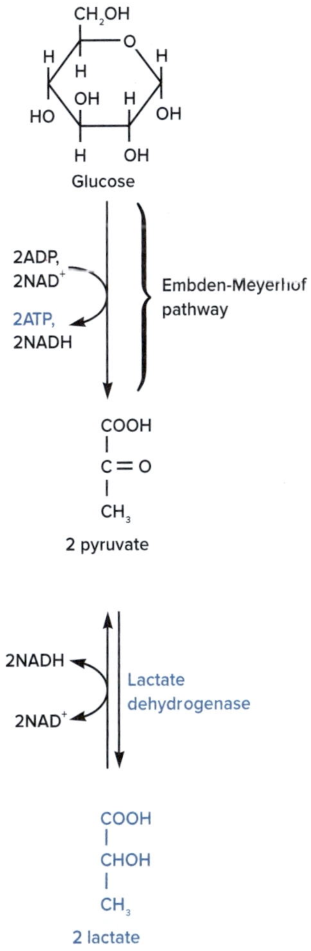

(a)

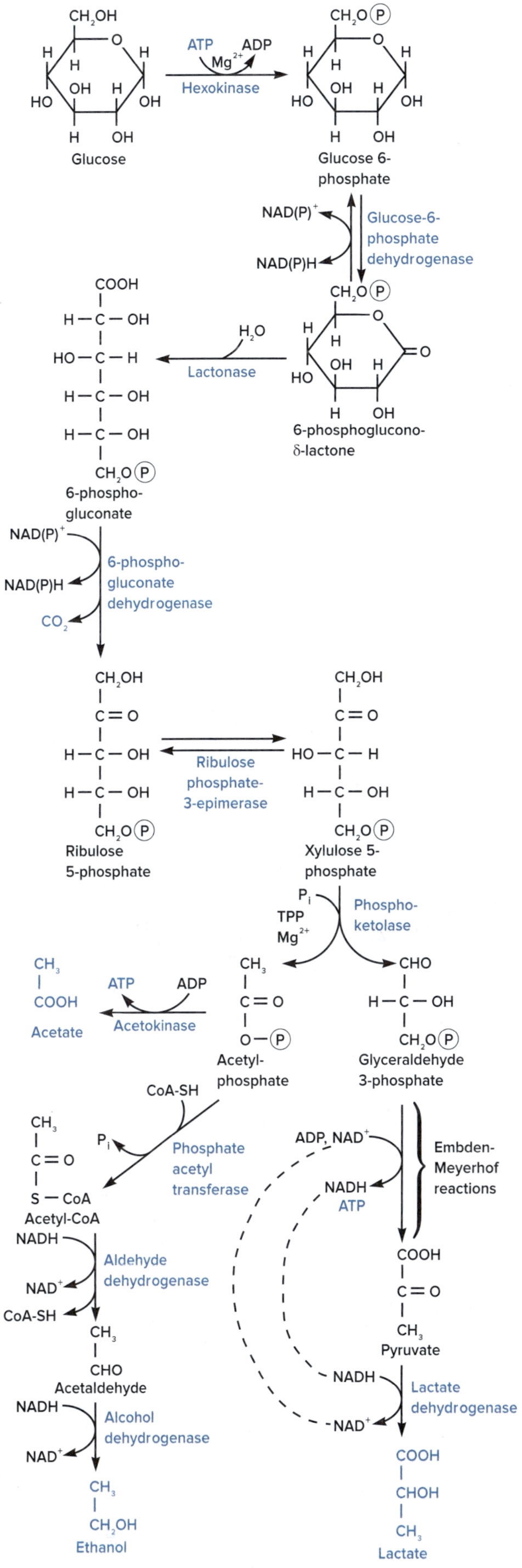

(b)

Figure AII.7 Lactic Acid Fermentations. (a) Homolactic fermentation pathway. (b) Heterolactic fermentation pathways.

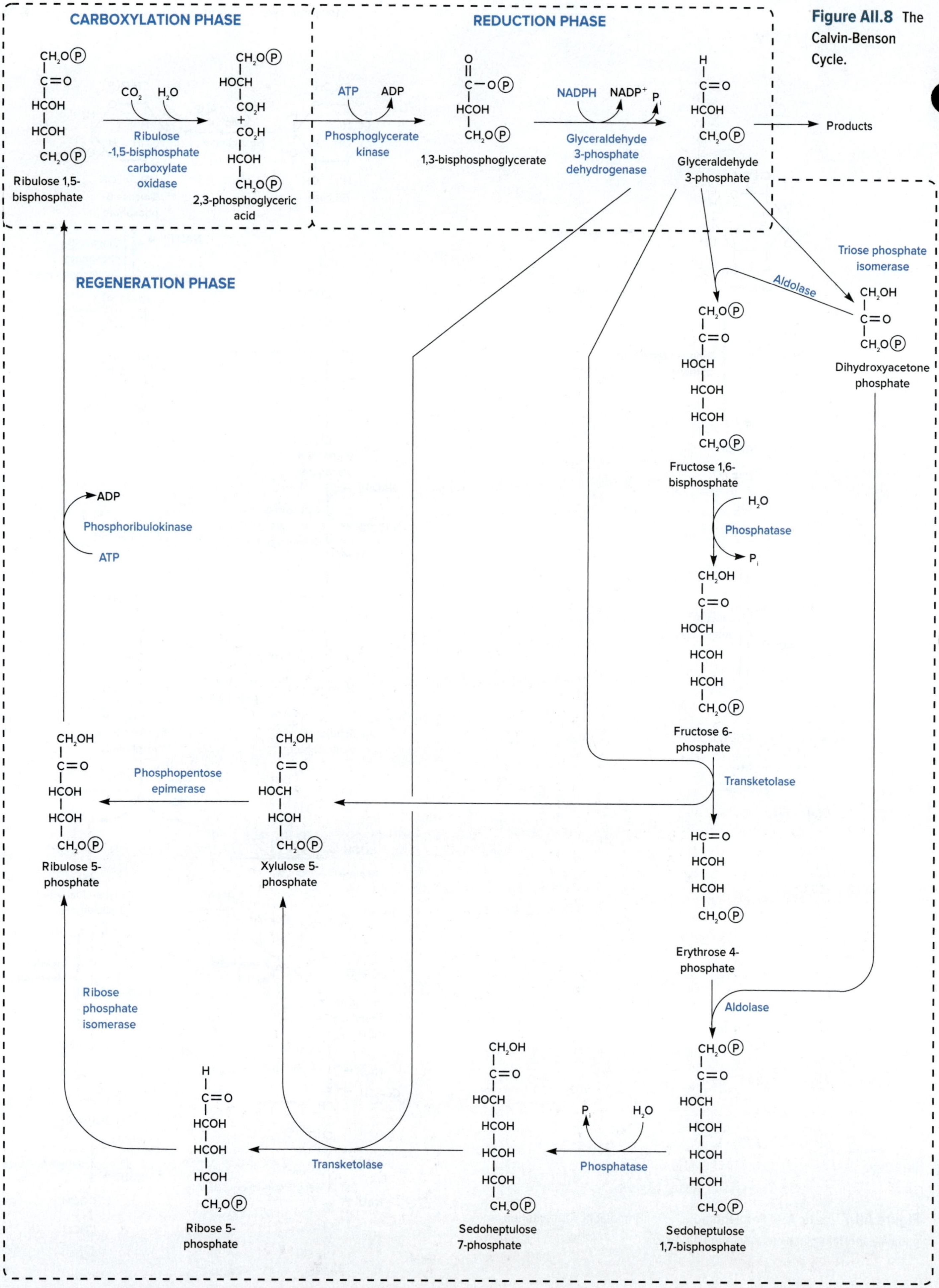

Figure AII.8 The Calvin-Benson Cycle.

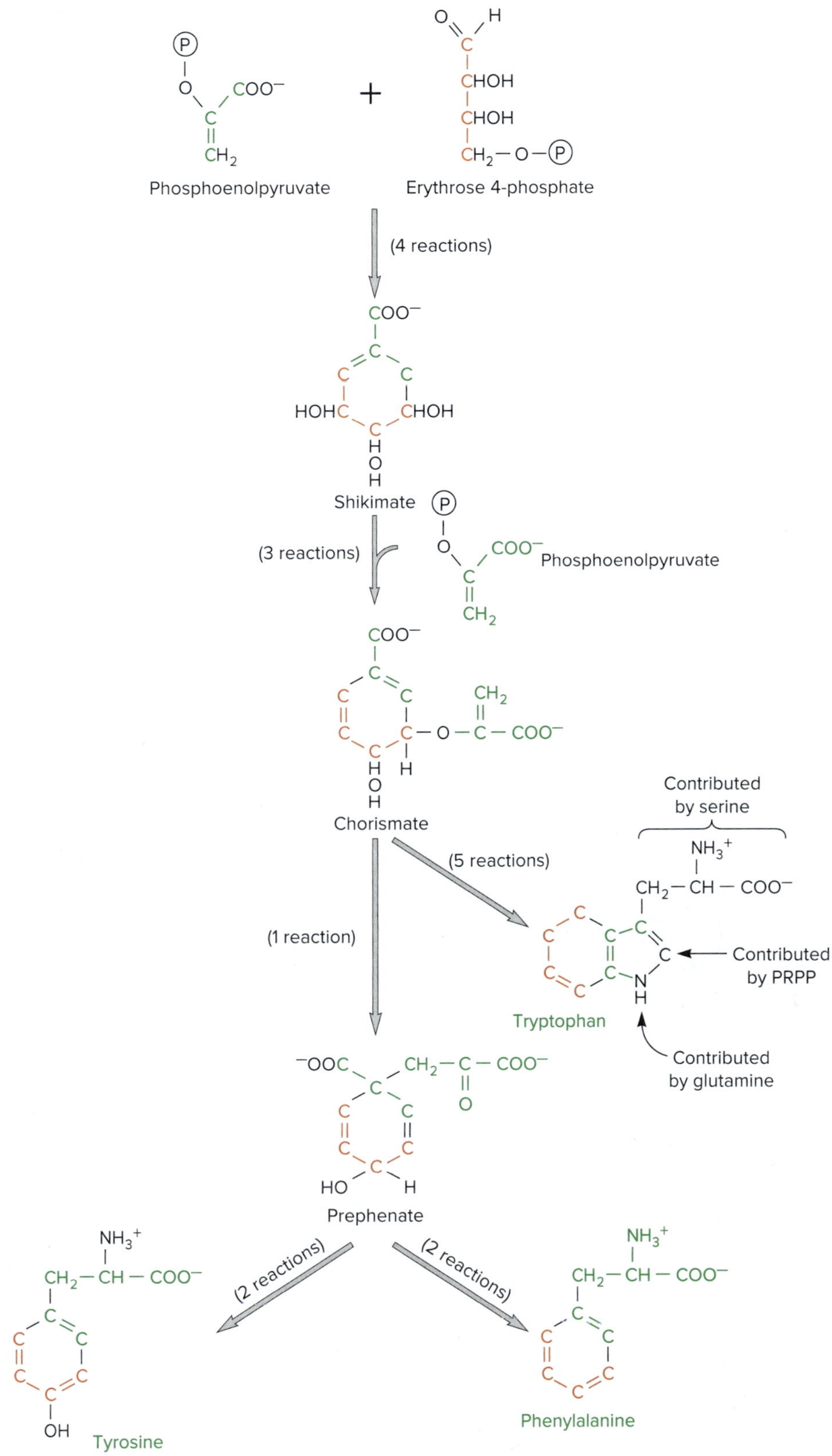

Figure AII.9 Aromatic Amino Acid Synthesis. The carbons arising from phosphoenolpyruvate (green) and erythrose 4-phosphate (red) are shown. The remaining carbons present in tryptophan are provided by 5-phosphoribosyl-1-pyrophosphate (PRPP) and the amino acid serine. PRPP is also important in purine biosynthesis (figure AII.10).

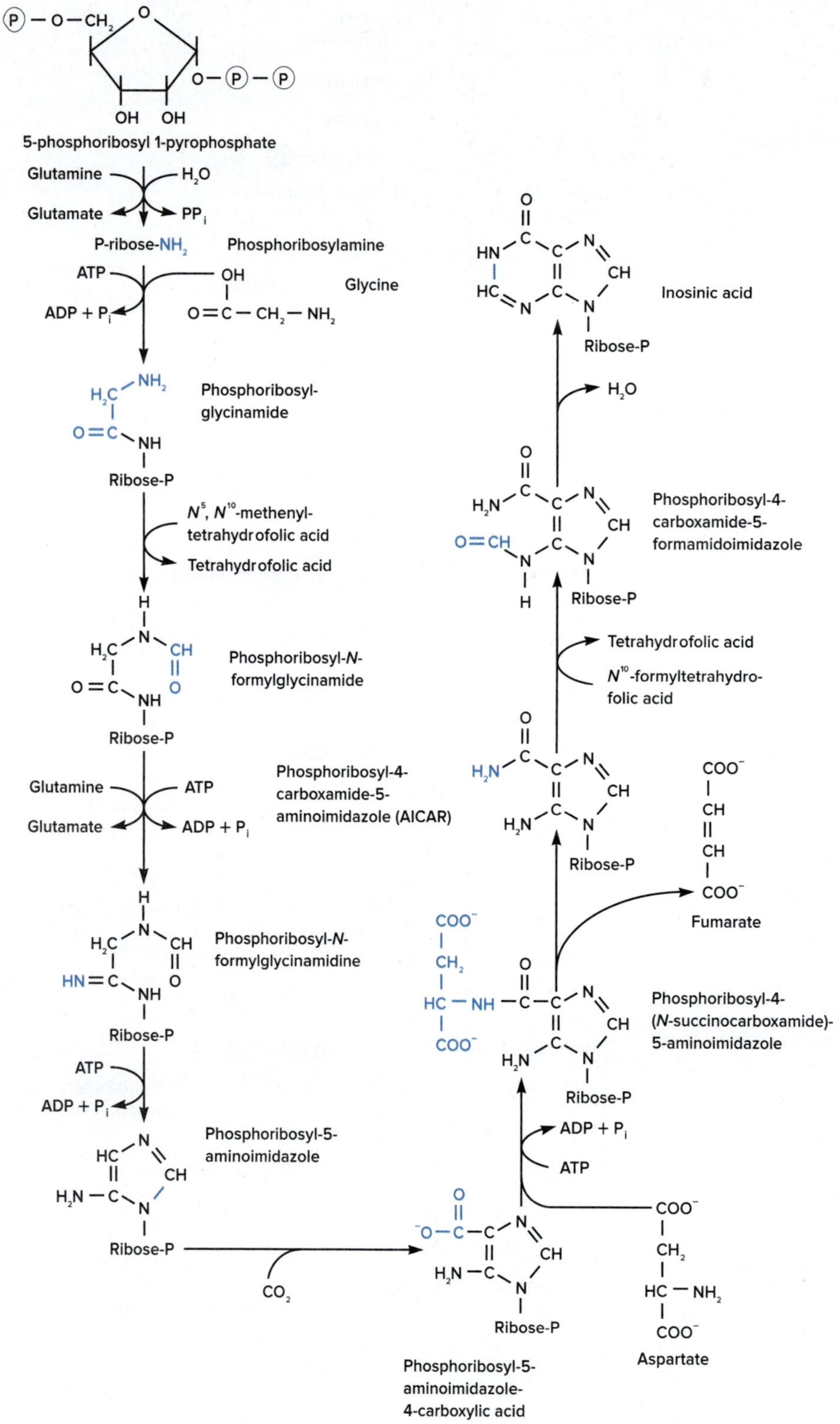

Figure AII.10 The Pathway for Purine Biosynthesis. Inosinic acid is the first purine end product. The purine skeleton is constructed while attached to a ribose phosphate.

Appendix Three
Microorganism Pronunciation Guide

The pronunciation of each name is given in parentheses. The phonetic spelling system is explained at the beginning of the glossary.

Bacteria

Acetobacter (ah-se″to bak′ter)
Acinetobacter (as″ĭ-net″o-bak′ter)
Actinomyces (ak″tĭ-no-mi′sēz)
Agrobacterium (ag″ro-bak-te′re-um)
Alcaligenes (al″kah-lij′ĕ-nēz)
Anabaena (ah-nab′ē-nah)
Arthrobacter (ar″thro-bak′ter)
Bacillus (bah-sil′lus)
Bacteroides (bak″tĕ-roi′dēz)
Bdellovibrio (del″o-vib′re-o)
Beggiatoa (bej″je-ah-to′ah)
Beijerinckia (bi″jer-ink′e-ah)
Bifidobacterium (bi″fid-o-bak-te′re-um)
Bordetella (bor″dĕ-tel′lah)
Borrelia (bŏ-rel′e ah)
Brucella (broo-sel′lah)
Campylobacter (kam″pī-lo-bak′ter)
Caulobacter (kaw″lo-bak′ter)
Chlamydia (klah-mid′e-ah)
Chlorobium (klo-ro′be-um)
Chromatium (kro-ma′te-um)
Citrobacter (sit″ro-bak′ter)
Clostridioides (klo-strid′e-oi″dez)
Clostridium (klo-strid′e-um)
Corynebacterium (ko-ri″ne-bak-te′re-um)
Coxiella (kok″se-el′lah)
Cytophaga (si-tof′ah-gah)
Desulfovibrio (de-sul″fo-vib′re-o)
Enterobacter (en″ter-o-bak′ter)
Erwinia (er-win′e-ah)
Escherichia (esh″er-i′ke-ah)
Flexibacter (flek″sĭ-bak′ter)
Francisella (fran-sĭ-sel′ah)
Frankia (frank′e-ah)
Gallionella (gal″le-o-nel′ah)
Haemophilus (he-mof′ĭ-lus)
Halobacterium (hal″o-bak-te′re-um)
Hydrogenomonas (hi-dro″jĕ-no-mo′nas)
Hyphomicrobium (hi″fo-mi-kro′be-um)
Klebsiella (kleb″se-el′lah)
Lactobacillus (lak″to-bah-sil′lus)
Legionella (le″jun-el′ah)
Leptospira (lep″to-spi′rah)
Leptothrix (lep′to-thriks)
Leuconostoc (loo″ko-nos′tok)
Listeria (lis-te′re-ah)
Methanothermobacter (meth″ah-no-ther″mo-bak′ter)
Methylococcus (meth″il-o-kok′-us)
Methylomonas (meth″il-o-mo′nas)
Micrococcus (mi″kro-kok′us)
Mycobacterium (mi″ko-bak-te′re-um)
Mycoplasma (mi″ko-plaz′mah)
Neisseria (nīs-se′re-ah)
Nitrobacter (ni″tro-bak′ter)
Nitrosomonas (ni-tro″so-mo′nas)
Nocardia (no-kar′de-ah)
Pasteurella (pas″tĕ-rel′ah)
Photobacterium (fo″to-bak-te′re-um)
Propionibacterium (pro″pe-on″e-bak-te′re-um)
Proteus (pro′te-us)
Pseudomonas (soo″do-mo′nas)
Rhizobium (ri-zo′be-um)
Rhodopseudomonas (ro″do-soo″do-mo′nas)
Rhodospirillum (ro″do-spi-ril′um)
Rickettsia (rĭ-ket′se-ah)
Salmonella (sal″mo-nel′ah)
Sarcina (sar′sĭ-nah)
Serratia (sĕ-ra′she-ah)
Shigella (shĭ-gel′ah)
Sphaerotilus (sfe-ro′tĭ-lus)
Spirillum (spi-ril′um)
Spirochaeta (spi″ro-ke′tah)
Spiroplasma (spi″ro-plaz′mah)
Staphylococcus (staf″ĭ-lo-kok′us)
Streptococcus (strep″to-kok′us)
Streptomyces (strep″to-mi′sēz)
Sulfolobus (sul″fo-lo′bus)
Thermoactinomyces (ther″mo-ak″tĭ-no-mi′sēz)
Thermoplasma (ther″mo-plaz′mah)
Thiobacillus (thi″o-bah-sil′us)
Thiothrix (thi′o-thriks)
Treponema (trep″o-ne′mah)
Ureaplasma (u-re′ah-plaz″ma)
Veillonella (va″yon-el′ah)
Vibrio (vib′re-o)
Xanthomonas (zan″tho-mo′nas)
Yersinia (yer-sin′e-ah)
Zoogloea (zo″o-gle′ah)

Viruses

Vernacular, nonscientific virus names are not written in italics, and thus the following names arc not italicized.

adenovirus (ad″ĕ-no-vi′rus)
arbovirus (ar″bo-vi′rus)
baculovirus (bak″u-lo-vi′rus)
coronavirus (kor″o-nah-vi′rus)
cytomegalovirus (si″to-meg″ah-lo-vi′rus)
Epstein-Barr virus (ep′stīn-bar′)
hepadnavirus (hep-ad″nə-vi′rus)
hepatitis virus (hep″ah-ti′tis)
herpesvirus (her″pēz-vi′rus)
influenza virus (in″flu-en′zah)
measles virus (me′zelz)
mumps virus (mumps)
orthomyxovirus (or″tho-mik″so-vi′rus)
papillomavirus (pap″ĭ-lo″mah-vi′rus)
paramyxovirus (par″ah-mik″so-vi′rus)
parvovirus (par″vo-vi′rus)
picornavirus (pi-kor″nah-vi′rus)
poliovirus (po″le-o-vi′rus)
polyomavirus (pol″e-o-mah-vi′rus)
poxvirus (poks-vi′rus)
rabies virus (ra′bēz)
reovirus (re″o-vi′rus)
retrovirus (re″tro-vi′rus)
rhabdovirus (rab″do-vi′rus)
rhinovirus (ri″no-vi′rus)
rotavirus (ro′tah-vi″rus)
rubella virus (roo-bel′ah)
togavirus (to″gah-vi′rus)
varicella-zoster virus (var″ĭ-sel′ah zos′ter)
variola virus (vah-re-o′-lah)

Fungi

Agaricus (ah-gar′ĭ-kus)
Amanita (am″ah-ni′tah)
Arthrobotrys (ar″thro-bo′tris)
Aspergillus (as″per-jil′us)
Blastomyces (blas″to-mi′sēz)
Candida (kan′dĭ-dah)
Cephalosporium (sef″ah-lo-spo′re-um)
Claviceps (klav′ĭ-seps)
Coccidioides (kok-sid″e-oi′dēz)
Cryptococcus (krip″to-kok′us)
Epidermophyton (ep″ĭ-der-mof′ĭ-ton)
Fusarium (fu-sa′re-um)
Histoplasma (his″to-plaz′mah)
Microsporum (mi-kros′po-rum)
Mucor (mu′kor)
Neurospora (nu-ros′po-rah)
Penicillium (pen″ĭ-sil′e-um)
Phytophthora (fi-tof′tho-rah)
Pneumocystis (noo″mo-sis′tis)
Rhizopus (ri-zo′pus)
Saccharomyces (sak″ah-ro-mi′sēz)
Saprolegnia (sap″ro-leg′ne-ah)
Sporothrix (spo′ro-thriks)
Trichoderma (trik-o-der′mah)
Trichophyton (tri-kof′ĭ-ton)

Protists

Acanthamoeba (ah-kan″thah-me′bah)
Acetabularia (as″ĕ-tab″u-la′re-ah)
Amoeba (ah-me′bah)
Balantidium (bal″an-tid′e-um)
Chlamydomonas (klah-mid″do-mo′nas)
Chlorella (klo-rel′ah)
Cryptosporidium (krip″to-spo-rid′e-um)
Entamoeba (en″tah-me′bah)
Euglena (u-gle′nah)
Giardia (je-ar′de-ah)
Gonyaulax (gon″e-aw′laks)
Leishmania (lēsh-ma′ne-ah)
Naegleria (na-gle′re-ah)
Paramecium (par″ah-me′se-um)
Plasmodium (plaz-mo′de-um)
Prototheca (pro″to-the′kah)
Spirogyra (spi″ro-ji′rah)
Tetrahymena (tet″rah-hi′mē-nah)
Toxoplasma (toks″o-plaz′mah)
Trichomonas (trik″o-mo′nas)
Trypanosoma (tri″pan-o-so′mah)
Volvox (vol′voks)

Glossary

Pronunciation Guide

Some of the boldface terms in this glossary are followed by a phonetic spelling in parentheses. These pronunciation aids usually come from *Dorland's Illustrated Medical Dictionary*. The following rules are taken from this dictionary and will help in using its phonetic spelling system.

1. An unmarked vowel ending a syllable (an open syllable) is long; thus, *ma* represents the pronunciation of *may; ne,* that of *knee; ri,* of *wry; so,* of *sew; too,* of *two;* and *vu,* of *view.*
2. An unmarked vowel in a syllable ending with a consonant (a closed syllable) is short; thus, *kat* represents *cat; bed, bed; hit, hit; not, knot; foot, foot;* and *kusp, cusp.*
3. A long vowel in a closed syllable is indicated by a macron; thus, *māt* stands for *mate; sēd,* for *seed; bīl,* for *bile; mōl,* for *mole; fūm,* for *fume;* and *f o͞o l,* for *fool.*
4. A short vowel that ends or itself constitutes a syllable is indicated by a breve; thus, *ĕ-fekt'* for *effect; ĭ-mūn'* for *immune;* and *ŭ-kl o͞o d'* for *occlude.*

Primary (′) and secondary (″) accents are shown in polysyllabic words. Unstressed syllables are followed by hyphens.

Some common vowels are pronounced as indicated here.

ə	sofa	ĕ	met	ŏ	got
ā	mate	ī	bite	ū	fuel
ă	bat	ĭ	bit	ŭ	but
ē	beam	ō	home		

Source: Dorland, W. A. Newman, *Dorland's Illustrated Medical Dictionary.* Philadelphia: Saunders, 1998.

A

AB toxins Exotoxins composed of two parts (A and B). The B portion is responsible for toxin binding to a cell but does not directly harm it; the A portion enters the cell and disrupts its function.

ABC protein secretion pathway *See* ATP-binding cassette transporters.

accessory pigments Photosynthetic pigments such as carotenoids and phycobiliproteins that aid chlorophyll in trapping light energy.

acellular slime mold *See* plasmodial slime mold.

acetyl-coenzyme A (acetyl-CoA) A molecule made of acetic acid and coenzyme A that is energy rich; it is produced by many catabolic pathways and is the substrate for the tricarboxylic acid cycle, fatty acid biosynthesis, and other pathways.

acid-fast Refers to bacteria (e.g., mycobacteria) that cannot be easily decolorized with acid alcohol after being stained with dyes such as basic fuchsin.

acid-fast staining A staining procedure that differentiates between bacteria based on their ability to retain a dye when washed with an acid alcohol solution.

acidophile (a-sid′o-fīl) A microorganism that has its growth optimum between about pH 0 and 5.5.

acquired immune tolerance The ability to produce antibodies against nonself antigens while "tolerating" (not producing antibodies against) self antigens.

acquired immunity *See* adaptive immunity.

actin filaments One of the proteinaceous components of the eukaryotic cytoskeleton; also called microfilaments.

actinobacteria (ak″tĭ-no-bak-tēr-e-ah) A group of Gram-positive bacteria containing actinomycetes and their high G + C relatives.

actinomycete (ak″tĭ-no-mi′sēt) An aerobic, Gram-positive bacterium that forms branching filaments (hyphae) and asexual spores.

actinorhizae Associations between actinomycetes and plant roots.

activated sludge Solid matter or sediment composed of actively growing microorganisms that participate in the aerobic portion of a biological sewage treatment process.

activation energy The energy required to bring reacting molecules together to reach the transition state in a chemical reaction.

activator-binding site *See* activator protein.

activator protein A transcriptional regulatory protein that binds to a specific site on DNA (activator-binding site) and enhances transcription initiation.

active (catalytic) site The part of an enzyme that binds the substrate to form an enzyme-substrate complex and catalyze the reaction. Also called the catalytic site.

active transport The transport of solute molecules across a membrane against a gradient; it requires a carrier protein and the input of energy. Three major types are primary active transport, which uses hydrolysis of ATP to power transport; secondary active transport, which uses ion gradients across a membrane to power active transport; and group translocation. *See also* group translocation.

acute-phase proteins Liver proteins that assist in the prevention of blood loss and ready the host for microbial invasion.

***N*-acylhomoserine lactone** *See* quorum sensing.

adaptive immunity Immunity that develops in response to a specific antigen; involves T and B lymphocyte differentiation and antibody production. Immunologic memory is formed to each antigen.

adaptive mutation The notion that mutations will arise when organisms are subjected to prolonged stress that will help the organism survive the stress.

adenine (ad′e-nēn) A purine derivative, 6-aminopurine, found in nucleosides, nucleotides, coenzymes, and nucleic acids.

adenosine diphosphate (ADP) (ah-den′o-sēn) The nucleoside diphosphate usually formed upon the breakdown of ATP when it provides energy for work.

adenosine 5′-triphosphate (ATP) A high energy molecule that serves as the cell's major form of energy currency.

adhesin (ad-he′zin) A molecule on the surface of a microorganism involved in adhesion to a substratum (e.g., host tissue).

adjuvant Material added to an antigen in a vaccine preparation that increases its immunogenicity.

aerial mycelium The mat of hyphae formed by actinomycetes or fungi that grows above the substrate, imparting a fuzzy appearance to colonies.

aerobic anoxygenic phototrophy (AAnP) Phototrophic process by which electron donors such as organic matter or sulfide are used under aerobic conditions.

aerobic respiration A metabolic process in which molecules, often organic, are oxidized with oxygen as the final electron acceptor.

aerotolerant anaerobes Microbes that grow equally well whether or not oxygen is present.

aflatoxin (af″lah-tok′sin) A polyketide secondary fungal metabolite that can cause cancer.

agar (ahg′ar) A complex sulfated polysaccharide, usually from red algae, that is used to solidify culture media.

agglutinates The visible aggregates or clumps formed by an agglutination reaction.

agglutination reaction The formation of an insoluble immune complex by the cross-linking of cells or particles.

airborne transmission The spread of a pathogen as it travels suspended in air over 1 meter or more from the source to the host.

akinetes (ā-kin′ēts) Specialized, nonmotile, dormant, thick-walled resting cells formed by some cyanobacteria.

alcoholic fermentation A fermentation process that produces ethanol and CO_2 from sugars.

alga (al′gah) A common term for several unrelated groups of photosynthetic eukaryotic microorganisms. Most are now considered protists.

alignment The comparison of gene, RNA, and protein sequences in a base-by-base (or amino acid by amino acid) fashion; used to calculate the degree of similarity between sequences.

alkaliphile A microorganism that grows best at pHs from about 8.5 to 11.5.

allele An alternative form of a gene.

allergen An antigen that induces an allergic response.

allergy *See* type I hypersensitivity.

allochthonous (ăl′ək-thə-nəs) Substances not native to a given environment (e.g., nutrient influx into freshwater ecosystems).

allograft A transplant between genetically different individuals of the same species.

allosteric effector *See* allosteric enzyme.

allosteric enzyme An enzyme whose activity is altered by the noncovalent binding of a small molecule (allosteric effector) at a regulatory site separate from the catalytic site; effector binding causes a conformational change in the enzyme's catalytic site, causing enzyme activation or inhibition.

αβ T cells CD4 or CD8 T cells that possess alpha/beta T-cell receptors.

α-hemolysis *See* hemolysis.

alphaproteobacteria Members of the bacterial class *Alphaproteobacteria.*

alternative complement pathway An antibody-independent pathway of complement activation.

alternate sigma factors Sigma is the subunit of bacterial RNA polymerase required for transcription initiation; alternate sigma factors direct the core RNA polymerase to transcribe distinct sets of inducible genes or operons, and is thereby a mechanism for global regulation.

alternative splicing The use of different exons during RNA splicing to generate different polypeptides from the same gene.

alveolar macrophage A vigorously phagocytic macrophage located on the epithelial surface of the lung alveoli, where it ingests inhaled particulate matter and microorganisms.

amensalism A relationship in which the product of one organism has a negative effect on another organism.

amino acid activation The preparatory step of protein synthesis in which amino acids are attached to transfer RNA molecules. The reaction is catalyzed by aminoacyl-tRNA synthetases.

aminoacyl or **acceptor site (A site)** The ribosomal site that contains an aminoacyl-tRNA at the beginning of the elongation cycle during protein synthesis.

aminoacyl-tRNA synthetases *See* amino acid activation.

aminoglycoside antibiotics A group of antibiotics that contain a cyclohexane ring and amino sugars; all aminoglycoside antibiotics inhibit protein synthesis by binding to the small ribosomal subunit.

amphibolic pathways Metabolic pathways that function both catabolically and anabolically.

amphipathic Term describing a molecule that has both hydrophilic and hydrophobic regions (e.g., phospholipids).

amphitrichous (am-fit′rĕ-kus) A cell with a single flagellum at each end.

anabolism The synthesis of complex molecules from simpler molecules with the input of energy and reducing power.

anaerobic digestion (an″a-er-o′bik) The microbiological treatment of sewage wastes under anoxic conditions to produce methane.

anaerobic respiration An energy-conserving process in which the terminal electron transport chain acceptor is a molecule other than oxygen.

anagenesis Changes in gene frequencies and distribution among species; the accumulation of small genetic changes within a population that introduces genetic variability but is not enough to result in either speciation or extinction.

anammoxosome The compartment within a planctomycete cell in which the anammox reaction occurs.

anammox reaction The coupled use of nitrite as an electron acceptor and ammonium ion as an electron donor under anoxic conditions to yield nitrogen gas.

anaphylaxis (an″ah-fĭ-lak′sis) A severe, immediate type I hypersensitivity reaction following exposure of a sensitized individual to the appropriate antigen.

anaplasia The reversion of an animal cell to a more primitive, undifferentiated state.

anaplerotic reactions (an′ah-plĕ-rot′ik) Reactions that replenish depleted tricarboxylic acid cycle intermediates.

anergy (an′ər-je) A state of immune cell unresponsiveness to antigens.

anoxic Without oxygen present.

anoxygenic photosynthesis Photosynthesis that does not oxidize water to produce oxygen.

antibiotic A microbial product or its derivative that kills susceptible microorganisms or inhibits their growth.

antibody (immunoglobulin) A glycoprotein made by plasma cells in response to the introduction of an antigen.

antibody affinity The strength of binding between an antigen and an antibody.

antibody-dependent cell-mediated cytotoxicity (ADCC) The killing of antibody-coated target cells by immune cells with Fc receptors that recognize the Fc region of the bound antibody.

antibody-dependent enhancement A phenomenon whereby the production of neutralizing antibodies to one serotype of dengue virus serves to opsonize rather than neutralize other serotypes.

antibody-mediated immunity *See* humoral immunity.

antibody class switching The process by which plasma cells change their expression from one isotype of immunoglobulin to a second isotype.

antibody titer An approximation of the antibody concentration.

anticodon The three bases on a tRNA that are complementary to the codon on mRNA.

antigen A substance (such as a protein, nucleoprotein, polysaccharide, or sometimes a glycolipid) to which lymphocytes respond.

antigen-binding fragment (Fab) "Fragment antigen binding." A monovalent antigen-binding fragment of an immunoglobulin molecule that consists of one light chain and part of one heavy chain, linked by interchain disulfide bonds.

antigen-presenting cells (APCs) Cells that take in protein antigens, process them, and present peptide fragments bound to MHC molecules to T cells; includes macrophages, dendritic cells, and B cells.

antigen processing The hydrolytic digestion of antigens to produce peptide fragments, which are collected and presented by class I or class II MHC molecules on the surface of an antigen-presenting cell.

antigenic determinant site *See* epitope.

antigenic drift A small change due to mutation in the antigenic character of an organism.

antigenic shift A major change in the antigenic character of an organism that makes it unrecognized by host immune mechanisms. This term is most often used to describe an antigenically distinct virus that results when two different strains of virus recombine genetic material.

antigenic variation The capacity of some microbes to change surface proteins when in a host organism, thereby evading the host immune system.

antimetabolite A compound that blocks metabolic pathway function by competitively inhibiting a key enzyme because it closely resembles the normal enzyme substrate.

antimicrobial agent An agent that kills microorganisms or inhibits their growth.

antiport Coupled transport of two molecules in which one molecule enters the cell as the other leaves the cell.

antisense RNA A single-stranded RNA with a base sequence complementary to a segment of a target RNA molecule; when bound to the target RNA, the target's activity is altered.

antisepsis The prevention of infection or sepsis.

antiseptic Chemical agent applied to tissue to prevent infection by killing or inhibiting pathogens.

antitoxin An antibody to a microbial toxin, usually a bacterial exotoxin, that binds and neutralizes the toxin.

apical complex A set of organelles characteristic of members of the protist subdivision *Apicomplexa*: includes polar rings, subpellicular microtubules, conoid, rhoptries, and micronemes.

apicomplexan (a'pĭ-kom-plek'san) A protist that has an apical complex and a spore-forming stage. It is either an intra- or extracellular parasite of animals; a member of the taxon *Apicomplexa*.

apicoplast (a'pĭ-ko-plast) Organelles in apicomplexan protists derived from cyanobacterial endosymbionts. Apicoplasts, however, are not involved in phototrophy; rather, they are the site of fatty acid, isoprenoid, and heme biosythesis.

apoenzyme (ap″o-en'zīm) The protein part of an enzyme that also has a nonprotein component.

apoptosis (ap″o-to'sis) Programmed cell death. A physiological suicide mechanism that results in fragmentation of a cell into membrane-bound particles that are eliminated by phagocytosis.

aporepressor The inactive form of a repressor protein; the repressor becomes active when the corepressor binds to it.

appressorium A flattened region of hypha found in some plant-infecting fungi that aids in penetrating the host plant cell wall. Occurs in both pathogenic fungi and nonpathogenic mycorrhizal fungi.

aptamer (ap'tăm-er) A single-stranded DNA or RNA molecule engineered to bind to a target protein to modify the activity of the protein.

arbuscular mycorrhizae (AM) The mycorrhizal fungi in a fungus-root association that penetrate the outer layer of the root, grow intracellularly, and form characteristic much-branched hyphal structures called arbuscules.

arbuscules *See* arbuscular mycorrhizal fungi.

archaea (ar'ke-ah) Members of the domain *Archaea*.

archaellum (ar'ke-el″um) Appendage on archaeal cells that functions in motility; functional equivalent of the bacteria flagellum.

archaerhodopsin A transmembranous protein to which retinal is bound; it functions as a light-driven proton pump resulting in photophosphorylation without chlorophyll or bacteriochlorophyll. Found in the purple membrane of halophilic archaea.

arthroconidia (arthrospores) Fungal spores formed by fragmentation.

artificially acquired active immunity The type of immunity that results from immunizing an animal with a vaccine. The immunized animal produces its own antibodies and activated lymphocytes.

artificially acquired passive immunity The temporary immunity that results from introducing into an animal antibodies that have been produced either in another animal or by in vitro methods.

ascocarp A multicellular structure in ascomycete fungi lined with specialized cells called asci.

ascogenous hypha A specialized hypha that gives rise to one or more asci.

ascomycetes (as″ko-mi-se'tēz) A group of fungi that form ascospores.

ascospore (as'ko-spor) A spore contained or produced in an ascus.

ascus (pl., **asci**) A specialized cell, characteristic of the ascomycetes, in which two haploid nuclei fuse to produce a zygote, which often immediately divides by meiosis; at maturity, an ascus contains ascospores.

aseptate (coenocytic) hypha A multinucleate hypha (filament) that lacks cross walls.

assimilatory reduction (e.g., **assimilatory nitrate reduction** and **assimilatory sulfate reduction**) The reduction of an inorganic molecule to incorporate it into organic material. No energy is conserved during this process.

atomic force microscope *See* scanning probe microscope.

atopic reaction *See* localized anaphylaxis.

ATP *See* adenosine 5′-triphosphate.

ATP-binding cassette transporters (ABC transporters) Transport systems that use ATP hydrolysis to drive translocation across the plasma membrane; can be used for nutrient uptake (ABC importer) or export of substances (ABC exporter), including protein secretion (ABC protein secretion pathway).

ATP synthase An enzyme that catalyzes synthesis of ATP from ADP and P_i, using energy derived from the proton motive force.

attack rate The proportional number of cases that develop in a population exposed to an infectious agent.

attenuated vaccine Live, nonpathogenic organisms used to activate adaptive immunity.

attenuation 1. A mechanism for the regulation of transcription termination of some bacterial operons by aminoacyl-tRNAs. 2. A procedure that reduces or abolishes the virulence of a pathogen without altering its immunogenicity.

attenuator A factor-independent transcription termination site in the leader sequence that is involved in attenuation.

autochthonous (ô-tŏk-thə-nəs) Substances (nutrients) that originate in a given environment.

autoclave An apparatus for sterilizing objects by the use of steam under pressure.

autoimmune disease A disease produced by the immune system attacking self antigens.

autoimmunity A condition characterized by the presence of serum autoantibodies and self-reactive lymphocytes that may be benign or pathogenic.

autoinducer With respect to quorum sensing, a small molecule that induces synthesis of the enzyme responsible for the autoinducer's synthesis. *See also* quorum sensing.

autoinduction In general terms, any molecule that increases its own level of production.

autolysins (aw-tol'ĭ-sins or aw″to-lye'sins) Enzymes that partially digest peptidoglycan in growing bacteria so that the cell wall can be enlarged.

autophagocytosis Ancient means of homeostasis whereby intracellular components are recycled.

autophagosome *See* macroautophagy.

autophagy *See* autophagocytosis and macroautophagy.

autoradiography A procedure that detects radioactively labeled materials using a photographic process.

autotroph (aw'to-trōf) An organism that uses CO_2 as its sole or principal source of carbon.

auxotroph (awk'so-trōf) An organism with a mutation that causes it to lose the ability to synthesize an essential nutrient; because of the mutation, the organism must obtain the nutrient or a precursor from its surroundings.

average nucleotide identity (ANI) A pairwise comparison between the whole or partial genome sequences of two microorganisms; expressed as percent identity.

avidity The combined strength of binding between an antigen and all the antibody-binding sites.

axenic (a-zen'ik) Not contaminated by any foreign organisms; the term is used in reference to pure microbial cultures. *See also* pure culture.

axopodium A thin, needlelike type of pseudopodium with a central core of microtubules.

B

B7 (CD80, CD86) proteins Glycoproteins on the surface of antigen-presenting cells (B cells, macrophages, dendritic cells, as well as T cells); it binds CD28 on T cells, resulting in second signal activation.

bacillus (bah-sil'lus) A rod-shaped bacterium or archaeon.

bacteremia The presence of viable bacteria in the blood.

bacteria Members of the domain *Bacteria*.

bacterial artificial chromosome (BAC) A cloning vector constructed from the *Escherichia coli* F-factor plasmid.

bactericide An agent that kills bacteria.

bacteriochlorophyll (bak-te″re-o-klo'ro-fil) A modified chlorophyll that serves as the primary light-trapping pigment in purple and green photosynthetic bacteria and heliobacteria.

bacteriocin (bak-te′re-o-sin) A secreted bacterial protein that kills other closely related bacteria.

bacteriophage (bak-te′re-o-fāj″) A virus that uses bacteria as its host; often called a phage.

bacteriostatic Inhibiting the growth and reproduction of bacteria.

bacteroid A modified, often pleomorphic, bacterial cell within the root nodule cells of legumes; it carries out nitrogen fixation.

baeocytes Small, spherical, reproductive cells produced by some cyanobacteria through multiple fission.

barophilic (bar″o-fil′ik) or **barophile** *See* piezophilic.

barotolerant Organisms that can grow and reproduce at high pressures but do not require them.

basal body *See* flagellum.

base analogues Molecules with minor chemical differences from normal DNA nucleotides that can substitute for them during DNA replication, leading to mutations. Also used as antiviral agents.

base excision repair *See* excision repair.

basic dyes Dyes that are cationic, or have positively charged groups, and bind to negatively charged cell structures.

basidiocarp (bah-sid′e-o-karp″) The fruiting body of a basidiomycete that contains the basidia.

basidiomycetes (bah-sid″e-o-mi-se′tēz) A group of fungi in which the spores are borne on club-shaped organs called basidia.

basidiospore (bah-sid′e-ō-spōr) A spore borne on the outside of a basidium.

basidium (bah-sid′e-um; pl., **basidia**) A structure formed by basidiomycetes; it bears on its surface basidiospores (typically four) that are formed following karyogamy and meiosis.

basophil A weakly phagocytic white blood cell in the granulocyte lineage. It synthesizes and stores vasoactive molecules (e.g., histamine) that are released in response to external triggers.

batch culture Growth of microorganisms in a closed culture vessel without adding fresh or removing old (spent) medium.

B cell (**B lymphocyte**) A type of lymphocyte that, following interaction with antigen, becomes a plasma cell, which synthesizes and secretes antibody molecules involved in humoral immunity.

B-cell receptor (**BCR**) A transmembrane immunoglobulin complex on the surface of a B cell that binds an antigen and stimulates the B cell. It is composed of a membrane-bound immunoglobulin, usually IgD or a modified IgM, complexed with the Ig-α/Ig-β heterodimer.

benthic Pertaining to the bottom of the sea or other body of water.

β-hemolysis *See* hemolysis.

β-lactam Referring to antibiotics containing a β-lactam ring (e.g., penicillins and cephalosporins).

β-lactam ring The cyclic chemical structure composed of one nitrogen and three carbon atoms. It has antibacterial activity, interfering with bacterial cell wall synthesis.

β-lactamase An enzyme that hydrolyzes the β-lactam ring, rendering the antibiotic inactive. Sometimes called penicillinase.

β-oxidation pathway The major pathway of fatty acid oxidation to produce NADH, $FADH_2$, and acetyl coenzyme A.

betaproteobacteria Members of the bacterial class *Betaproteobacteria.*

binal symmetry The symmetry of some virus capsids (e.g., those of complex phages) that is a combination of icosahedral and helical symmetry.

binary fission Asexual reproduction in which a cell separates into two identical daughter cells.

binomial system The nomenclature system in which an organism is given two names; the first is the capitalized generic name, and the second is the uncapitalized specific epithet.

bioaugmentation Addition of pregrown microbial cultures to an environment to perform a specific task.

biochemical oxygen demand (**BOD**) The amount of oxygen used by organisms in water under standard conditions; it provides an index of the amount of microbially oxidizable organic matter present.

biochemical pathways Sets of chemical reactions performed by organisms that convert a starting substrate into one or more products.

biocide Chemical or physical agent, usually with a broad spectrum of biological activity, that inactivates microorganisms.

biodegradation The breakdown of a complex chemical through biological processes that results in minor loss of functional groups, fragmentation into smaller constituents, or complete breakdown to CO_2 and minerals.

biofilms Organized microbial communities encased in extracellular polymeric substances and associated with surfaces, often with complex structural and functional characteristics.

biofuels Energy-dense compounds (i.e., fuels) produced by microbes, usually by the degradation of plant substrates.

biogeochemical cycling The oxidation and reduction of substances carried out by living organisms and abiotic processes that results in the cycling of elements within and between different parts of the ecosystem and the atmosphere.

bioinformatics The interdisciplinary field that manages and analyzes large biological data sets, including genome and protein sequences.

bioinsecticide A microbe or microbial product used to kill or disable unwanted insect pests.

biological safety cabinets Cabinets that use HEPA filters to project a curtain of sterile air across its opening, preventing microbes from entering or exiting.

biomagnification The increase in concentration of a substance in higher-level consumer organisms.

bioremediation The use of biologically mediated processes to remove or degrade pollutants from specific environments.

biosensor A device for the detection of a particular substance (an analyte) that combines a biological receptor with a physicochemical detector.

biotechnology The process in which living organisms are manipulated, particularly at the molecular genetic level, to form useful products.

bioterrorism The intentional or threatened use of microorganisms or organic toxins to produce death or disease in humans, animals, and plants.

biovar Variant strains of microbes characterized by biological or physiological differences.

BLAST (**basic local alignment search tool**) A publicly available computer program that identifies regions of similarity between nucleotide or amino acid sequences. It compares input or query sequences with sequence databases and determines the statistical significance of matches. Homologies can be used to infer gene or protein function and evolutionary relationships.

blastospores Fungal spores produced from a vegetative mother cell by budding.

bloom The growth of a single microbial species or a limited number of species in an aquatic environment, usually in response to a sudden pulse of nutrients. *See also* harmful algal bloom.

B lymphocyte *See* B cell.

bright-field microscope A microscope that illuminates the specimen directly with bright light and forms a dark image on a brighter background.

broad-spectrum drugs Chemotherapeutic agents that are effective against many different kinds of pathogens.

bronchial-associated lymphoid tissue (**BALT**) *See* mucosal-associated lymphoid tissue.

Bt An insecticide that consists of a toxin produced by *Bacillus thuringiensis.*

bubo (bu′bo) A tender, inflamed, enlarged lymph node that results from a variety of infections.

budding A vegetative outgrowth of yeast and some bacteria as a means of asexual reproduction; the daughter cell is smaller than the parent.

bulking sludge Sludge produced in sewage treatment that does not settle properly, usually due to the development of filamentous microorganisms.

butanediol fermentation A type of fermentation most often found in members of the family *Enterobacteriaceae* in which 2,3-butanediol is a major product; acetoin is an intermediate in the pathway and may be detected by the Voges-Proskauer test.

C

calorie The amount of heat needed to raise one gram of water from 14.5° to 15.5°C.

Calvin-Benson cycle The main pathway for the fixation (reduction and incorporation) of CO_2

into organic material by photoautotrophs and chemolithoautotrophs.

cancer A malignant tumor that expands locally by invasion of surrounding tissues and systemically by metastasis.

cannulae Tubelike structures extending from the surface of some archaeal cells. They are thought to function in attachment to surfaces.

5′ cap *See* pre-mRNA.

capsid The protein coat or shell that surrounds a virion's nucleic acid.

capsomer The ring-shaped morphological unit of icosahedral capsids.

capsule A layer of well-organized material, not easily washed off, lying outside the cell wall.

carbapenems A group of broad-spectrum β-lactam antibiotics typically resistant to β-lactamases.

carbonate equilibrium system The interchange among CO_2, HCO_3^-, and CO_3^{2-} that keeps oceans buffered between pH 7.6 and 8.2.

carbonosome A bacterial compartment for storage of carbon in the form of polyhydroxyalkanoate.

carbon to nitrogen ratio The ratio of carbon to nitrogen in a soil; it can be used to predict rates of organic decomposition and thus the cycling of nutrients in the system.

carboxysomes Polyhedral inclusions where CO_2 fixation occurs; contain the CO_2-fixation enzyme ribulose 1,5-bisphosphate carboxylase/oxygenase. They are a type of microcompartment.

cardinal temperatures The minimal, maximal, and optimum temperatures for growth.

catabolism That part of metabolism in which larger, more complex molecules are broken down into smaller, simpler molecules with the release of energy.

catabolite activator protein (CAP) A protein that regulates the expression of genes encoding catabolite repressible enzymes; also called cyclic AMP receptor protein (CRP).

catabolite repression Inhibition of the synthesis of several catabolic enzymes by a preferred carbon and energy source (e.g., glucose).

catalase An enzyme that catalyzes the destruction of hydrogen peroxide.

catalyst A substance that accelerates a reaction without being permanently changed itself.

catalytic site *See* active site.

catalyzed reporter deposition-fluorescent in situ hybridization (CARD-FISH) A FISH technique in which the signal is amplified so that low levels of fluorescence can be detected. *See also* fluorescent in situ hybridization (FISH).

catenanes (kăt′ə-nāns′) Circular, covalently closed nucleic acid molecules that are locked together like the links of a chain.

cathelicidins Antimicrobial cationic peptides that are produced by a variety of cells (e.g., neutrophils, respiratory epithelial cells, and alveolar macrophages).

caveolin-dependent endocytosis *See* endocytosis.

CD4⁺ cell *See* T-helper cell.

CD8⁺ cell *See* cytotoxic T lymphyocyte.

cell cycle The sequence of events in a cell's growth-division cycle between the end of one division and the end of the next.

cell envelope The plasma membrane plus all other external layers.

cellular (cell-mediated) immunity The type of immunity mediated by T cells.

cellular slime molds Protists with a vegetative phase consisting of amoeboid cells that aggregate to form a multicellular pseudoplasmodium. They belong to the taxon *Dictyostelium;* formerly considered fungi.

cell wall The strong structure that lies outside the plasma membrane; it supports and protects the membrane and helps maintain cell shape.

Centers for Disease Control and Prevention (CDC) Public health agency of the United States, headquartered in Atlanta, Georgia; serves as the national agency for developing and carrying out disease prevention and control, environmental health, and health promotion and education.

central metabolic pathways Those pathways central to the metabolism of an organism because they function catabolically and anabolically (e.g., glycolytic pathways and tricarboxylic acid cycle).

central tolerance The process by which self-reactive B cells are eliminated in the bone marrow and self-reactive T cells are eliminated in the thymus.

chain of infection *See* infectious disease cycle.

chain termination DNA sequencing method A DNA-sequencing method that uses dideoxynucleotides to cause termination of DNA replication at random sites.

chancre (shang′ker) The primary lesion of syphilis occurring at the site of entry of the pathogen.

chaperone proteins Proteins that assist in the folding and stabilization of other proteins. Some are also involved in directing newly synthesized proteins to protein secretion systems or to other locations in the cell.

chemical fixation *See* fixation.

chemical oxygen demand (COD) The amount of chemical oxidation required to convert organic matter in water and wastewater to CO_2.

chemiosmotic hypothesis (kem″e-os-mot′ik) The hypothesis that proton and electrochemical gradients are generated by electron transport and then used to drive ATP synthesis by oxidative phosphorylation or photophosphorylation.

chemoheterotroph (ke″mo-het′er-o-trōf″) *See* chemoorganoheterotroph.

chemokine A type of cytokine that stimulates chemotaxis and chemokinesis of immune cells.

chemolithoautotroph A microorganism that oxidizes reduced inorganic compounds to derive both energy and electrons; CO_2 is the carbon source.

chemolithoheterotroph A microorganism that uses reduced inorganic compounds to derive both energy and electrons; organic molecules are used as the carbon source.

chemolithotroph A microorganism that uses reduced inorganic compounds as a source of energy and electrons.

chemoorganoheterotroph An organism that uses organic compounds as sources of energy, electrons, and carbon for biosynthesis. Also called chemoheterotroph and chemoorganotrophic heterotroph.

chemoreceptors Proteins in the plasma membrane or periplasmic space that bind chemicals and trigger the appropriate chemotactic response.

chemostat A continuous culture apparatus that feeds medium into the culture vessel at the same rate as medium containing microorganisms is removed; the medium in a chemostat contains one essential nutrient in a limiting quantity.

chemotaxins Agents that cause white blood cells to migrate up its concentration gradient.

chemotaxis The pattern of cellular behavior in which the cell moves toward chemical attractants and away from repellents.

chemotherapy The use of chemical agents to treat disease.

chemotrophs Organisms that obtain energy from the oxidation of chemical compounds.

chlamydiae (klə-mid′e-e) Members of the genera *Chlamydia* and *Chlamydophila*: Gram-negative, obligate intracellular pathogens.

chlorophyll The green photosynthetic pigment that consists of a large tetrapyrrole ring with a magnesium atom in the center.

chloroplast A eukaryotic plastid that contains chlorophyll and is the site of photosynthesis.

chlorosomes Elongated, intramembranous vesicles found in the green sulfur and nonsulfur bacteria; they contain light-harvesting pigments. Sometimes called chlorobium vesicles.

chromatic adaptation The ability of cyanobacteria to change the relative concentrations of light-harvesting pigments in response to changes in wavelengths of incident light.

chromatin The complex of DNA and proteins, including histones, from which eukaryotic and some archaeal chromosomes are made.

chromatin immunoprecipitation (ChIP) A technique that enables the identification of protein-DNA interactions.

chromatin modification The addition of acetyl, methyl, phosphoryl, and other groups to a histone protein. These modifications change the level of transcription of nearby genes.

chromatin remodeling The alteration of chromatin structure by inserting or removing variant histone proteins from a nucleosome.

chromogen (kro′me-jen) A colorless substrate that is acted on by an enzyme to produce a colored end product.

chromophore group A chemical group with double bonds that absorbs visible light and gives a dye its color.

chromosomes The bodies that have most or all of the cell's DNA and contain most of its genetic information (mitochondria and chloroplasts also contain DNA).

chronic inflammation Inflammation that is characterized by a dense tissue infiltration of lymphocytes and macrophages into the affected site and the formation of new connective tissue, which usually causes permanent tissue damage.

chytrids A term used to describe members of *Chytridiomycota,* fungi that produce motile zoospores with single, posterior, whiplash flagella.

cidal Causing death.

cilia Threadlike appendages extending from the surface of some protists that beat rhythmically to propel them; cilia are membrane-bound cylinders with a complex internal array of microtubules, usually in a 9 + 2 pattern.

ciliates (*Ciliophora*) Protists that move by rapidly beating cilia and belong to *Alveolata.*

citric acid cycle *See* tricarboxylic acid cycle.

clade A group of organisms derived from a common ancestor.

class I MHC molecule *See* major histocompatibility complex.

class II MHC molecule *See* major histocompatibility complex.

classical complement pathway The antibody-dependent pathway of complement activation; it stimulates lysis of pathogens, phagocytosis, and other host defenses.

clathrin-dependent endocytosis *See* endocytosis.

clonal selection The process by which an antigen binds to the best-fitting B-cell receptor, activating that B cell, resulting in the synthesis of that specific antibody and clonal expansion.

clone (1) A group of genetically identical cells or organisms derived by asexual reproduction from a single parent. (2) A DNA sequence that has been isolated and replicated using a cloning vector.

cloning vector A DNA molecule that can replicate independently of the host chromosome and maintain a piece of inserted foreign DNA, such as a gene, into a recipient cell. It may be a plasmid, phage, cosmid, or artificial chromosome.

club fungi Common name for members of *Basidiomycota.*

cluster of differentiation molecules (CDs) Functional cell surface proteins that can be used to identify leukocyte subpopulations (e.g., interleukin-2 receptor, CD4, and CD8).

coagulation The process of adding chemicals to precipitate impurities from water during purification; also called flocculation.

coccolithophores Photosynthetic protists belonging to the group *Stramenopila;* characterized by coccoliths—intricate cell walls made of calcite.

coccus (kok′us, pl., **cocci**, kok′si) A roughly spherical bacterial or archaeal cell.

code degeneracy The presence of more than one codon for each amino acid.

coding sequences (CDS) In genomic analysis, open reading frames presumed to encode proteins but not tRNA or rRNA.

codon (ko′don) A sequence of three nucleotides in mRNA that directs the incorporation of an amino acid during protein synthesis or signals the stop of translation.

coenocytic (se″no-sit′ik) *See* aseptate hypha.

coenzyme A loosely bound cofactor that often dissociates from the enzyme active site after product has been formed.

coenzyme Q (CoQ, ubiquinone) *See* electron transport chain.

coevolution The evolution of two interacting organisms such that they optimize their survival with one another; that is, they develop reciprocal adaptations.

cofactor The nonprotein component of an enzyme required for catalytic activity.

colicin (kol′ĭ-sin) A plasmid-encoded protein produced by enteric bacteria; binds to receptors on the cell envelope of sensitive target bacteria, where it may cause lysis or attack specific intracellular sites such as ribosomes.

coliform A Gram-negative, nonsporing, facultative rod that ferments lactose with gas formation within 48 hours at 35°C.

Colilert defined substrate test An assay designed to detect the presence of coliforms in water.

colonization The establishment of a site of microbial reproduction on an inanimate surface or organism without necessarily resulting in tissue invasion or damage.

colonization resistance The capacity of microbes within the microbiome to consume nutrients, produce metabolites, and use other mechanisms to limit the growth of pathogens.

colony An assemblage of microorganisms growing on a solid surface; the assemblage often is directly visible.

colony forming units (CFU) The number of microorganisms that form colonies when cultured using spread plates or pour plates, an indication of the number of viable microorganisms in a sample.

colony stimulating factor (CSF) A protein that stimulates the growth and development of specific cell populations (e.g., granulocyte-CSF stimulates granulocytes to be made from precursor stem cells).

colorless sulfur bacteria A diverse group of nonphotosynthetic proteobacteria that can oxidize reduced sulfur compounds such as hydrogen sulfide. Many are chemolithotrophs and derive energy from sulfur oxidation.

comammox The capacity of some *Nitrospira* bacteria to perform both steps in the nitrification process, thereby performing *com*plete *amm*onia *ox*idation.

combinatorial joining The genomic recombination of V, D, and J regions during lymphocyte development that gives rise to diverse T-cell receptors, B-cell receptors, and antibodies.

cometabolism The modification of a compound not used for growth by a microorganism, which occurs in the presence of another organic material that serves as a carbon and energy source.

commensal Living on or within another organism without injuring or benefiting the other organism.

commensalism A type of symbiosis in which one individual gains from the association (the commensal) and the other is neither harmed nor benefited.

common-source epidemic An epidemic characterized by a sharp rise to a peak and then a rapid but not as pronounced decline in the number of individuals infected; it usually involves a single contaminated source that infects individuals.

communicable disease A disease associated with a pathogen that can be transmitted from one host to another.

comparative genomics Comparison of genomes from different organisms to discern differences and similarities.

compartmentation The differential distribution of enzymes and metabolites to separate cell structures or organelles.

compatible solute A low molecular weight molecule used to protect cells against changes in solute concentrations (osmolarity) in their habitat; it can exist at high concentrations within the cell without hindering metabolism or growth.

competent cell A cell that can take up free DNA fragments and incorporate them into its genome during transformation.

competition An interaction between two organisms attempting to use the same resource (nutrients, space, etc.).

competitive exclusion principle Two competing organisms overlap in resource use, which leads to the exclusion of one of the organisms.

competitive inhibitor A molecule that inhibits enzyme activity by binding the enzyme's active site.

complementary Refers to the matching of two strands of DNA or RNA based on the base-pairing rules.

complementary DNA (cDNA) A DNA copy of an RNA molecule (e.g., a DNA copy of an mRNA).

complement system A group of plasma proteins that plays a major role in innate immunity.

complex medium Culture medium that contains some ingredients of unknown chemical composition.

compromised host A host with lowered resistance to infection and disease.

concatemer A long DNA molecule consisting of several genomes linked together in a row.

conditional mutations Mutations with phenotypes that are expressed only under certain environmental parameters (e.g., temperature).

confocal microscope A light microscope in which monochromatic laser derived light scans across the specimen at a specific level. Stray light from above and below the plane of focus is blocked out to give an image with excellent contrast and resolution.

conidiospore (ko-nid′e-o-spōr) An asexual, thin-walled spore borne on hyphae and not contained within a sporangium; it may be produced singly or in chains.

conidium (ko-nid′e-um; pl., **conidia**) *See* conidiospore.

conjugate redox pair The electron acceptor and electron donor of a redox half reaction.

conjugation (1) The form of gene transfer and recombination in bacteria and archaea that requires direct cell-to-cell contact. (2) A complex form of sexual reproduction commonly employed by protists.

conjugative plasmid A plasmid that carries the genes that enable its transfer to other bacteria during conjugation (e.g., F plasmid).

conjugative transposon A mobile genetic element that is able to transfer itself from one bacterium to another by conjugation.

consensus sequence A commonly occurring sequence of nucleotides within a genetic element.

conserved hypothetical proteins In genomic analysis, proteins of unknown function for which at least one match exists in a database.

consortium A physical association of two or more different organisms, usually beneficial to all of the organisms.

constant region (**C_L** and **C_H**) The part of an antibody molecule that does not vary greatly in amino acid sequence among molecules of the same class, subclass, or type.

constitutive gene A gene that is expressed at nearly the same level at all times.

contact transmission Spread of a pathogen by contact of the source or reservoir of the pathogen.

contigs In genomic sequencing, fragments of DNA with identical sequences at the ends that have been assembled into a longer, continuous sequence of DNA.

continuous culture system A culture system with constant environmental conditions maintained through continual provision of nutrients and removal of wastes. *See also* chemostat and turbidostat.

continuous feed The constant addition of a carbon source during industrial fermentation.

contractile vacuole (vak′u-ōl) In protists and some animals, a clear fluid-filled vacuole that takes up water from within the cell and then contracts, releasing it to the outside through a pore in a cyclical manner. It functions primarily in osmoregulation and excretion.

cooperation A positive but not obligatory interaction between two different organisms.

coral bleaching The loss of photosynthetic pigments by either physiological inhibition or expulsion of the coral photosynthetic endosymbiont, zooxanthellae, a dinoflagellate.

core genome The common set of genes found in all genomes in a species or other taxon.

core polysaccharide *See* lipopolysaccharide.

corepressor A small molecule that binds a transcription repressor protein, thereby activating the repressor and inhibiting the synthesis of a repressible enzyme.

cosmid A plasmid vector with lambda phage *cos* sites that can be packaged in a phage capsid; it is useful for cloning large DNA fragments.

coverage The number of times a nucleotide is sequenced in any given sequencing project. Synonymous with depth of coverage, so "depth" and "coverage" are used interchangeably. Breadth of coverage refers to the percentage of genome sequenced.

CRISPR-Cas system An array of DNA sequences and a set of enzymes that recognize invading DNA of plasmid or bacteriophage origin and target them for destruction; found in many bacteria and most archaea.

cristae (kris′te) Infoldings of the inner mitochondrial membrane.

cryophile *See* psychrophile.

cryptidins Antimicrobial peptides produced by Paneth cells in the intestines.

crystallizable fragment (Fc) The stem of the Y portion of an antibody molecule. Cells such as macrophages bind to the Fc region; it also is involved in complement activation.

culturomics An approach to isolating and identifying microbes that involves cultivation under many different conditions followed by identification by mass spectrometry and SSU rRNA gene sequencing.

curing The loss of a plasmid from a cell.

cut-and-paste transposition *See* simple transposition.

cyanobacteria A large group of Gram-negative, morphologically diverse bacteria that carry out oxygenic photosynthesis.

cyanophycin Inclusions that store nitrogen; found in cyanobacteria.

3′, 5′-cyclic adenosine monophosphate (cAMP) A small nucleotide that is involved in regulating a number of cellular processes in bacteria and other organisms. In some bacteria, it functions in catabolite repression.

cyclic dimeric GMP (c-di-GMP) A small nucleotide consisting of two GMP molecules linked together by their phosphates. It functions as a second messenger in many bacteria and some eukaryotes.

cyclic photophosphorylation (fo″to-fos″for-ĭ-la′shun) The formation of ATP when light energy is used to move electrons cyclically through an electron transport chain during photosynthesis.

cyst A general term used for a specialized microbial cell enclosed in a wall. Cysts are formed by protists and a few bacteria. They may be dormant, resistant structures formed in response to adverse conditions or reproductive cysts that are a normal stage in the life cycle.

cystitis *See* urinary tract infection.

cytochromes (si′to-krōms) *See* electron transport chain.

cytokine (si′to-kīn) A general term for proteins released by immune cells in response to inducing stimuli; they mediate the activity of other cells.

cytokinesis Processes that apportion the cytoplasm and organelles, synthesize a septum, and divide a cell into two daughter cells during cell division.

cytopathic effect The observable change that occurs in cells as a result of viral replication.

cytoplasm All material in the cell enclosed by the plasma membrane, with the exception of the nucleus in eukaryotic cells.

cytoproct A specific site in certain protists (e.g., ciliates) where digested material is expelled.

cytosine (si′to-sēn) A pyrimidine 2-oxy-4-aminopyrimidine found in nucleosides, nucleotides, and nucleic acids.

cytoskeleton A network of structures made from filamentous proteins (e.g., actin and tubulin) and other components in the cytoplasm of cells.

cytosol Liquid component of the cytoplasm.

cytostome A permanent site in a ciliate protist in which food is ingested.

cytotoxic T lymphocyte (CTL) Effector T cell that can lyse target cells bearing class I MHC molecules complexed with antigenic peptides. CTLs arise from antigen-activated Tc cells.

D

Dane particle A 42-nm spherical particle that is one of three seen in hepatitis B virus infections. The Dane particle is the complete virion.

dark-field microscopy Microscopy in which the specimen is brightly illuminated while the background is dark.

dark reaction *See* photosynthesis.

deamination The removal of amino groups from amino acids.

decimal reduction time (***D*** or ***D* value**) The time required to kill 90% of the microorganisms or endospores in a sample at a specified temperature.

defensins A large family of antimicrobial peptides produced by neutrophils, macrophages, lymphocytes, and a variety of other host cells.

defined (synthetic) medium Culture medium made with components that can be described with chemical formulae.

delayed-type hypersensitivity *See* type IV hypersensitivity.

deltaproteobacteria Members of the bacterial class *Deltaproteobacteria.*

denaturation A loss of the three-dimensional protein shape or nucleic-acid structure.

denaturing gradient gel electrophoresis (DGGE) A technique by which DNA is rendered single stranded (denatured) while undergoing electrophoresis so that DNA fragments of the same size can be separated according to nucleotide sequence.

dendritic cell An antigen-presenting cell that has long membrane extensions resembling the dendrites of neurons.

denitrification The reduction of nitrate to gaseous products, primarily nitrogen gas, during anaerobic respiration.

deoxyribonucleic acid (DNA) (de-ok″se-ri″bo-nu-kle′ik) A polynucleotide that constitutes the genetic material of all cellular organisms. It is composed of deoxyribonucleotides connected by phosphodiester bonds.

depth filters Filters composed of fibrous or granular materials that are used to decrease microbial load and sometimes to sterilize solutions.

desensitization The process by which a sensitized or hypersensitive individual is made insensitive or nonreactive to a sensitizing agent (e.g., an allergen).

detergent An organic molecule, other than a soap, that serves as a wetting agent and emulsifier; it is normally used as cleanser, but some may be used as antimicrobial agents.

diatoms (*Bacillariophyta*) Photosynthetic protists with siliceous cell walls called frustules belonging to the group *Stramenopila*.

diauxic growth (di-awk′sik) Biphasic growth response in which a microorganism, when exposed to two nutrients, initially uses one of them for growth and then alters its metabolism to make use of the second.

differential interference contrast (DIC) microscope A light microscope that employs two beams of plane polarized light. The beams are combined after passing through the specimen and their interference is used to create the image.

differential media Culture media that distinguish between groups of microorganisms based on differences in their growth and metabolic products.

differential staining Staining procedures that divide bacteria into groups based on staining properties.

diglycerol tetraether lipids Archaeal lipids formed when isoprenoid hydrocarbons are linked to two glycerols by ether bonds.

dikaryotic stage (di-kar-e-ot′ik) In fungi, having two separate haploid nuclei, one from each parent.

dilution susceptibility tests A method by which antibiotics are evaluated for their ability to inhibit bacterial growth in vitro. A standardized concentration of bacteria is added to serially diluted antibiotics and incubated. Tubes lacking additional bacterial growth suggest antibiotic concentrations that are bacteriocidal or bacteriostatic.

dinoflagellate (di″no-flaj′e-lāt) A photosynthetic protist characterized by two flagella used in swimming in a spinning pattern; belong to the group *Alveolata*.

diplococcus (dip″lo-kok′us) A pair of cocci.

direct counts Methods for determining population size by examining a sample of the population and counting the cells within the sample.

direct immunofluorescence Technique for imaging of microscopic agents using fluorescently labeled antibodies to bind to the agent.

direct repair A type of DNA repair mechanism in which a damaged nitrogenous base is returned to its normal form (e.g., conversion of a thymine dimer back to two normal thymine bases).

directed evolution A variety of approaches used to mutate a specific gene or genes with the aim of increasing product yield or generating new molecules.

disease syndrome A set of signs and symptoms that are characteristic of a disease.

disinfectant An agent, usually chemical, that disinfects; normally, it is employed only with inanimate objects.

disinfection The killing, inhibition, or removal of microorganisms that may cause disease. It usually refers to the treatment of inanimate objects with chemicals.

disinfection by-products (DBPs) Chlorinated organic compounds such as trihalomethanes formed during chlorine use for water disinfection. Many are carcinogens.

dissimilatory reduction (e.g., **dissimilatory nitrate reduction** and **dissimilatory sulfate reduction**) The use of a substance as an electron acceptor for an electron transport chain. The acceptor (e.g., sulfate or nitrate) is reduced but not incorporated into organic matter.

dissolved organic matter (DOM) Nutrients that are available in the soluble, or dissolved, state.

divisome A collection of proteins that aggregate at the region in a dividing microbial cell where a septum will form.

DNA amplification The process of increasing the number of DNA molecules, usually by the polymerase chain reaction.

DNA-dependent DNA polymerase An enzyme that catalyzes the synthesis of DNA using a DNA molecule as a template.

DNA-dependent RNA polymerase An enzyme that catalyzes the synthesis of RNA using a DNA molecule as a template.

DNA-DNA hybridization (DDH) An assay in which the single-stranded DNA of one organism is hybridized to that of another, and the percent that hybridizes is calculated. There must be at least 70% similarity between two genomes as determined by DDH for microbes to be considered the same species; however, this is only one criterion upon which species assignments are made.

DNA gyrase A topoisomerase enzyme that relieves tension generated by the rapid unwinding of DNA during DNA replication or transcription.

DNA ligase An enzyme that joins two DNA fragments through the formation of a new phosphodiester bond.

DNA microarrays Solid supports that have DNA attached in organized arrays and are used to evaluate gene expression.

DNA polymerase (pol-im′er-ās) An enzyme that synthesizes new DNA using a parental nucleic acid strand (usually DNA) as a template.

double diffusion agar assay (Öuchterlony technique) An immunodiffusion reaction in which both antibody and antigen diffuse through agar to form stable immune complexes, which can be observed visually.

doubling time *See* generation time.

droplet nuclei Small particles (0 to 4 μm in diameter) that represent what is left from the evaporation of larger particles (10 μm or more in diameter) called droplets.

drug inactivation Physical or chemical alteration of chemotherapeutic agent resulting in its loss of biological activity.

D value *See* decimal reduction time.

dysbiosis An alteration in the microbiome that has a negative impact on the health of the host.

E

ecotype A population of an organism that is genetically very similar but ecologically distinct from others of the same species.

ectomycorrhizae A mutualistic association between fungi and plant roots in which the fungus surrounds the root with a sheath.

ectoplasm In some protists, the cytoplasm directly under the cell membrane (plasmalemma) is divided into an outer gelatinous region, the ectoplasm, and an inner fluid region, the endoplasm.

effector proteins Proteins secreted by pathogens that are targeted to host cells via a secretion system.

efflux pumps Membrane-associated protein assemblies that transport amphiphilic molecules from cytoplasm to the extracellular environment.

electron acceptor A compound that accepts electrons in an oxidation-reduction reaction. Often called an oxidizing agent or oxidant.

electron cryotomography A specialized electron microscopy procedure that involves rapid freezing of intact specimens, maintenance of the specimen in a frozen state while being examined, and imaging from different angles. The information from each angle is used to create a three-dimensional reconstruction.

electron donor An electron donor in an oxidation-reduction reaction. Often called a reducing agent or reductant.

electron shuttle A substance or structure along which electrons are passed from the electron transport chain to an external terminal electron acceptor.

electron transport chain (ETC) A series of electron carriers that operate together to transfer electrons from donors to acceptors such as oxygen. Molecules involved in electron transport include nicotinamide adenine dinucleotide (NAD^+), NAD phosphate ($NADP^+$), cytochromes, heme proteins, nonheme proteins (e.g., iron-sulfur proteins and ferredoxin), coenzyme Q, flavin adenine dinucleotide (FAD), and flavin mononucleotide (FMN). Also called an electron transport system.

electroporation The application of an electric field to create temporary pores in the plasma membrane to render a cell temporarily transformation competent.

elementary body (EB) A small, dormant body that serves as the agent of transmission between host cells in the chlamydial life cycle.

elongation factors Proteins that function in the elongation cycle of protein synthesis.

Embden-Meyerhof pathway A pathway that degrades glucose to pyruvate (i.e., a glycolytic pathway).

encystment The formation of a cyst.

endemic disease A disease that exists at a steady, low-level frequency at a moderate, regular interval.

endergonic reaction (end″er-gon′ik) A reaction that does not spontaneously go to completion as written; the standard free energy change is positive, and the equilibrium constant is less than one.

endocytic pathways *See* endocytosis.

endocytosis The process in which a cell takes up solutes or particles by enclosing them in vesicles pinched off from its plasma membrane. It often occurs at regions of the plasma membrane coated by proteins such as clathrin and caveolin. Endocytosis involving these proteins is called clathrin-dependent endocytosis and caveolin-dependent endocytosis, respectively. When the substance endocytosed first binds to a receptor, the process is called receptor-mediated endocytosis.

endogenous antigen processing Antigen degradation by cytoplasmic proteasomes and subsequent complexing with class I MHC molecules for presentation on the processing cell's surface.

endogenous pyrogen A host-derived chemical mediator (e.g., interleukin-1) that acts on the hypothalamus, stimulating a rise in core body temperature (i.e., it stimulates the fever response).

endomycorrhizae Referring to a mutualistic association of fungi and plant roots in which the fungus penetrates into the root cells.

endophyte A microorganism living within a plant but not necessarily parasitic on it.

endoplasm *See* ectoplasm.

endoplasmic reticulum (ER) A system of membranous tubules and flattened sacs (cisternae) in eukaryotic cells. Rough endoplasmic reticulum (RER) bears ribosomes on its surface; smooth endoplasmic reticulum (SER) lacks them.

endosome A membranous vesicle formed by endocytosis. It undergoes a maturational process that starts with early endosomes and proceeds to late endosomes and finally lysosomes.

endospore An extremely heat- and chemical-resistant, dormant, thick-walled spore that develops within some members of the bacterial phylum *Firmicutes*. It has a complex structure that includes (from outermost to innermost) exosporium, spore coat, cortex, spore cell wall, and spore core.

endosymbiosis A type of symbiosis in which one organism is found within another organism.

endosymbiotic hypothesis The hypothesis that mitochondria, and related organelles (e.g., hydrogenosomes), and chloroplasts arose from bacterial endosymbionts of ancestral eukaryotic cells.

endotoxin The lipid A component of lipopolysaccharide (LPS) from Gram-negative bacterial cell walls. Nanogram quantities can induce fever, activate complement and coagulation cascades, act as a mitogen to B cells, and stimulate cytokine release from a variety of cells. Systemic effects of endotoxin are referred to as septic shock.

end product inhibition *See* feedback inhibition.

energy The capacity to do work or cause particular changes.

enhancer A site in DNA to which a eukaryotic activator protein binds.

enology The science of wine making.

enriched media Media that contain nutrients to encourage growth of fastidious microbes.

enrichment culture The growth of specific microbes from natural samples by including media components or manipulating environmental conditions to promote growth of the desired microbes while suppressing the growth of other microbes.

enteric bacteria (enterobacteria) Members of the family *Enterobacteriaceae*; also can refer to bacteria that live in the intestinal tract.

enterotoxin A toxin specifically affecting the cells of the intestinal mucosa, causing vomiting and diarrhea.

enthalpy Heat content of a system.

Entner-Doudoroff pathway A pathway that converts glucose to pyruvate and glyceraldehyde 3-phosphate by producing 6-phosphogluconate and then dehydrating it (i.e., a glycolytic pathway).

entropy A measure of the randomness or disorder of a system; a measure of that part of the total energy in a system that is unavailable for useful work.

envelope In virology, an outer membranous layer that surrounds the nucleocapsid of some viruses.

enveloped virus A virus having virions with a nucleocapsid enclosed within a lipid bilayer.

environmental microbiology The study of microbial processes that occur in soil, water, and other natural habitats. Specific microorganisms are not necessarily investigated; rather, the cumulative impact of the microbial community on the specific biome, as well as global implications, are assessed.

enzyme A protein catalyst with specificity for the reaction catalyzed and its substrates.

enzyme-linked immunosorbent assay (ELISA) A serological assay in which bound antigen or antibody is detected by another antibody that is conjugated to an enzyme. The enzyme converts a colorless substrate to a colored product reporting the antibody capture of the antigen.

eosinophil (e″o-sin′o-fil) A leukocyte that has a two-lobed nucleus and cytoplasmic granules that release chemical mediators in response to parasite infections.

epidemic A disease that suddenly increases in occurrence above the normal level in a given population.

epidemiologist A person who specializes in epidemiology.

epidemiology The study of the factors determining and influencing the frequency and distribution of disease, injury, and other health-related events and their causes in defined human populations.

epifluorescence microscopy The illumination of a microscope sample with a high-intensity light from above. The light is passed through an exciter filter so that when light of a specific wavelength is focused on the specimen, it emits light, thereby making the sample visible as a specific color on a black background.

epilimnion The upper, warmer layer of water in a stratified lake.

epiphyte An organism that grows on the surface of plants.

episome A plasmid that can either exist independently of the host cell's chromosome or be integrated into it.

epitope (ep′i-tōp) An area of an antigen that stimulates the production of, and combines with, specific antibodies; also known as the antigenic determinant site.

epsilonproteobacteria Members of the bacterial class *Epsilonproteobacteria*.

equilibrium The state of a system in which no net change is occurring and free energy is at a minimum; in a chemical reaction at equilibrium, the rates in the forward and reverse directions exactly balance each other out.

equilibrium constant (K_{eq}) A value that relates the concentration of reactants and products to each other when a reaction is at equilibrium.

ergot The dried sclerotium of *Claviceps purpurea*. Also, an ascomycete that parasitizes rye and other plants; production of an alkaloid causes an animal disease called ergotism.

erythroblastosis fetalis Hemolytic disease of the newborn whereby red blood cells are destroyed by maternal antibody.

eukaryotic cells Cells that have a membrane-delimited nucleus; all plants, fungi, protists and animals are eukaryotic.

eutrophic Nutrient enriched.

eutrophication The enrichment of an aquatic environment with nutrients.

excision repair A type of DNA repair mechanism in which a section of a strand of damaged DNA is excised and replaced, using the complementary strand as a template. Two types are recognized: base excision repair and nucleotide excision repair.

excystment The escape of one or more cells or organisms from a cyst.

exergonic reaction (ek″ser-gon′ik) A reaction that spontaneously goes to completion as written; the standard free energy change is negative, and the equilibrium constant is greater than one.

green nonsulfur bacteria Anoxygenic photosynthetic bacteria that contain bacteriochlorophylls *a* and *c;* usually photoheterotrophic and display gliding motility. Include members of the phylum *Chloroflexi.*

green sulfur bacteria Anoxygenic photosynthetic bacteria that contain bacteriochlorophylls *a,* plus *c, d,* or *e;* photolithoautotrophic; use H_2, H_2S, or S as electron donor. Include members of the phylum *Chlorobi.*

group A streptococcus (GAS) A Gram-positive, coccus-shaped bacterium often found in the throat and on the skin of humans, having the Lancefield group A of surface carbohydrate.

group B streptococcus (GBS) A Gram-positive, coccus-shaped bacterium found occasionally on mucous membranes of humans, having the Lancefield group B surface carbohydrate.

group translocation An active transport process in which a molecule is moved across a membrane by carrier proteins while being chemically altered at the same time (e.g., phosphoenolpyruvate: sugar phosphotransferase system).

growth arrest A phenomenon in which bacterial cells are neither actively dividing nor dead. They are stressed by lack of nutrients or by deviation from optimal conditions.

growth factors Organic compounds that must be supplied in the diet for growth because they are essential cell components or precursors of such components and cannot be synthesized by the organism.

growth rate constant (*k*) The rate of microbial population growth expressed in terms of the number of generations per unit time.

guanine (gwan′in) A purine derivative, 2-amino-6-oxypurine found in nucleosides, nucleotides, and nucleic acids.

guanosine tetraphosphate (ppGpp) A small nucleotide that functions as a second messenger for the stringent response. *See also* stringent response.

guild A group of microbes defined by the same physiological activity.

gumma (gum′ah) A soft, gummy, benign tumor occurring in tertiary syphilis.

gut-associated lymphoid tissue (GALT) *See* mucosal-associated lymphoid tissue.

H

haloarchaea (extreme halophiles) A group of archaea that depend on high NaCl concentrations for growth and do not survive at a concentration below about 1.5 M NaCl.

halophile A microorganism that requires high levels of sodium chloride for growth.

halotolerant The ability to withstand large changes in salt concentration.

hami Structures extending from the surface of some archaeal cells that terminate in a grappling hooklike apparatus. They are thought to function in attachment to surfaces.

hapten A molecule not immunogenic by itself that, when coupled to a macromolecular carrier, can elicit antibodies directed against itself.

harmful algal bloom (HAB) In aquatic ecosystems, the growth of a single population of phototroph, either a protist (e.g., diatom, dinoflagellate) or a cyanobacterium, that produces a toxin poisonous to other organisms, sometimes including humans. Alternatively, in the absence of toxin production, the concentration of bloom microbe reaches levels that are inherently harmful to other organisms, such as filter-feeding bivalves.

Hartig net The area of nutrient exchange between ectomycorrhizal fungal hyphae and plant host cells.

health-care–associated infection An infection that is acquired during a patient's stay in a hospital or other type of clinical care facility.

heat fixation *See* fixation.

heat-shock proteins Proteins produced when cells are exposed to high temperatures or other stressful conditions. They protect the cells from damage and often aid in the proper folding of proteins.

helical capsid A viral capsid in the form of a helix.

helicases Enzymes that use ATP energy to unwind DNA ahead of the replication fork.

hemagglutination (hem″ah-gloo″tĭ-na′shun) The clumping of red blood cells by antibodies or components of virus capsids. *See also* viral hemagglutination.

hemagglutination assay A testing procedure based on a hemagglutination reaction.

hemagglutinin (HA) (1) Proteins that cause red blood cells to clump together (hemagglutination). (2) One of the envelope spikes of influenza and some other viruses. They are the basis for identifying different influenza virus strains.

hematopoiesis The process by which blood cells develop into specific lineages from stem cells. Red and white blood cells and platelets develop from this process.

hemolysis The disruption of red blood cells and release of their hemoglobin. In α-hemolysis, a greenish zone of incomplete hemolysis forms around a colony grown on blood agar. A clear zone of complete hemolysis without any obvious color change is formed during β-hemolysis.

herd immunity The resistance of a population to infection and spread of an infectious agent due to the immunity of a high percentage of the population.

heterocysts Specialized cells of cyanobacteria that are the sites of nitrogen fixation.

heterokont flagella A pattern of flagellation found in the members of the protist taxon *Stramenopila,* featuring two flagella, one extending anteriorly and the other posteriorly.

heterolactic fermenters Microorganisms that ferment sugars to form lactate and other products such as ethanol and CO_2.

heterologous gene expression The cloning, transcription, and translation of a gene that has been introduced (cloned) into an organism that normally does not possess the gene.

heterotroph An organism that uses reduced, preformed organic molecules as its principal carbon source.

hexose monophosphate pathway *See* pentose phosphate pathway.

Hfr conjugation Conjugation involving an Hfr strain and an F^- strain.

Hfr strain A strain of *Escherichia coli* that donates its chromosomal genes with high frequency to a recipient cell during conjugation because the F factor is integrated into the donor's chromosome.

hierarchical cluster analysis The organization of gene expression data such that induced and repressed genes, or genes of similar function, are grouped separately.

high-efficiency particulate (HEPA) filter Thick, fibrous filter constructed to remove 99.97% of particles that are 0.3 μm or larger.

high phosphate transfer potential A characteristic of a phosphorylated compound such that it readily transfers a phosphoryl group to another molecule concomitant with a large energy release.

high-throughput screening (HTS) A system that combines liquid handling devices, robotics, computers, data processing, and a sensitive detection system to screen thousands of compounds for a single capability.

histatin An antimicrobial peptide composed of 24 to 38 amino acids, heavily enriched with histidine, that targets fungal mitochondria.

histone A small basic protein with large amounts of lysine and arginine that is associated with eukaryotic DNA in chromatin. Related proteins are observed in many archaeal species, where they form archaeal nucleosomes.

holdfast A structure produced by some bacteria (e.g., *Caulobacter*) that attaches them to a solid object.

holoenzyme A complete enzyme consisting of the apoenzyme plus a cofactor. Also refers to a complete enzyme consisting of all protein subunits (e.g., RNA polymerase holoenzyme).

holozoic nutrition Acquisition of nutrients (e.g., bacteria) by endocytosis and the subsequent formation of a food vacuole or phagosome.

homolactic fermenters Organisms that ferment sugars almost completely to lactic acid.

homologous recombination Recombination involving two DNA molecules that are very similar in nucleotide sequence.

hopanoids Lipids found in bacterial membranes that are similar in structure and function to the sterols found in eukaryotic membranes.

hops Dried flowers of the plant *Humulus lupulus* used in the preparation of beer and lager.

horizontal (lateral) gene transfer (HGT/LGT) The process by which genes are transferred from one mature, independent organism to another. In bacteria and archaea, transformation, conjugation, and transduction are the major mechanisms by which HGT can occur.

hormogonia Small motile fragments produced by fragmentation of filamentous cyanobacteria; used for asexual reproduction and dispersal.

hospital-acquired infections *See* health-care–associated infections.

host An organism that harbors another organism.

housekeeping genes Genes that encode proteins that function throughout most of the life cycle of an organism (e.g., enzymes of the Embden-Meyerhof pathway).

human leukocyte antigen complex (HLA) A collection of genes on chromosome 6 that encodes proteins involved in host defenses; also called major histocompatibility complex.

human microbiome The sum of all the microorganisms that live on and in the human body.

humoral (antibody-mediated) immunity The type of immunity that results from the presence of soluble antibodies in blood and lymph.

hydrogen hypothesis A hypothesis that considers the origin of the eukaryotes through the development of the hydrogenosome. It suggests the organelle arose as the result of an endosymbiotic anaerobic bacterium that produced CO_2 and H_2 as the products of fermentation.

hydrogenosome One member of the mitochondrial family of organelles, it is found in some anaerobic protists that produce ATP by fermentation.

hydrophilic A polar substance that has a strong affinity for water (or is readily soluble in water).

hydrophobic A nonpolar substance lacking affinity for water (or which is not readily soluble in water).

3-hydroxypropionate bi-cycle A pathway used by green nonsulfur bacteria to fix CO_2.

hyperthermophile (hi″per-ther′mo-fīl) A microbe that has its growth optimum between 85° and about 120°C. Hyperthermophiles usually do not grow well below 55°C.

hypha (hi′fah; pl., **hyphae**) The unit of structure of most fungi and some bacteria; a tubular filament.

hypolimnion The colder, bottom layer of water in a stratified lake.

I

icosahedral capsid A viral capsid that has the shape of a regular polyhedron having 20 equilateral triangular faces and 12 corners.

IgA Immunoglobulin A; the class of dimeric immunoglobulins that is present in mucous membranes and in many body secretions.

IgD Immunoglobulin D; the class of immunoglobulins found on the surface of many B lymphocytes, where it serves as an antigen receptor in the stimulation of antibody synthesis.

IgE Immunoglobulin E; the immunoglobulin class that binds to mast cells and basophils, and is responsible for type I or anaphylactic hypersensitivity reactions such as hay fever and asthma. IgE is also involved in resistance to helminth parasites.

IgG Immunoglobulin G; the predominant immunoglobulin class in serum. It serves to neutralize toxins, opsonize bacteria, activate complement, and cross the placenta to protect the fetus and neonate.

IgM Immunoglobulin M; the class of serum antibody produced first upon infection. It is a large, pentameric molecule that agglutinates pathogens and activates complement. The monomeric form is present on the surface of some B lymphocytes.

immobilization The incorporation of a simple, soluble substance into the body of an organism, making it unavailable for use by other organisms.

immune complex The product of antibody binding to antigen; may also contain components of the complement system.

immune synapse The binding of a T cell, via its T-cell receptor and coreceptor (CD4 or CD8), to a major histocompatibility complex (MHC)–peptide on the surface of an antigen-presenting cell; it is the first step in T-cell activation.

immune system The defensive system in an animal consisting of the innate and adaptive immune responses. It is composed of widely distributed cells, tissues, and organs that recognize foreign substances and microorganisms and acts to neutralize or destroy them.

immunity Refers to the overall general ability of a host to resist a particular disease; the condition of being immune.

immunization The deliberate introduction of foreign materials into a host to stimulate an adaptive immune response. *See also* vaccine.

immunoblotting The electrophoretic transfer of proteins from polyacrylamide gels to filters to demonstrate the presence of specific proteins through reaction with labeled antibodies.

immunodeficiency The inability to produce a normal innate and/or adaptive immune response.

immunodiffusion A technique involving the diffusion of antigen or antibody within a semisolid gel to produce a precipitin reaction.

immunoelectrophoresis (pl., **immuno-electrophoreses**) The electrophoretic separation of protein antigens followed by diffusion and precipitation in gels using antibodies against the separated proteins.

immunofluorescence A technique used to identify particular antigens microscopically in cells or tissues by binding a fluorescent antibody conjugate.

immunoglobulin (Ig) *See* antibody.

immunology The study of host defenses against invading foreign materials, including pathogenic microorganisms, transformed or cancerous cells, and tissue transplants from other sources.

immunoprecipitation A reaction involving soluble antigens reacting with antibodies to form an aggregate that precipitates out of solution.

inactivated vaccines Nonviable infectious agents prepared to stimulate immunity but not cause infectious disease; also known as killed vaccines.

incidence The number of new cases of a disease in a population at risk.

inclusions (1) Granules, crystals, or globules of organic or inorganic material in the cytosol of cells. (2) Clusters of viral proteins or virions within the nucleus or cytoplasm of virus-infected cells.

incubation period The period after pathogen entry into a host and before signs and symptoms appear.

indel A blending of the words *insertion* and *deletion*; these genetic mutations can be taxonomically useful.

index case The first disease case in an outbreak or epidemic.

indicator organism An organism whose presence indicates the condition of a substance or environment, for example, the potential presence of pathogens. Coliforms are used as indicators of fecal pollution.

indirect immunofluorescence Technique for imaging microscopic agents using antibodies to target the agent and a second, fluorochrome-labeled reagent to bind to the antibodies.

induced mutations Mutations caused by exposure to a mutagen.

inducer A small molecule that stimulates the synthesis of an inducible enzyme.

inducible gene A gene whose expression level can be increased by a regulatory molecule.

induction (1) In virology, the events that trigger a virus to switch from a lysogenic mode to a lytic pathway. (2) In genetics, an increase in gene expression.

infection The invasion of a host by a microorganism with subsequent establishment and multiplication of the agent. An infection may or may not lead to overt disease.

infection thread A tubular structure formed during the infection of a root by nitrogen-fixing bacteria. The bacteria enter the root by way of the infection thread and stimulate the formation of the root nodule.

infectious disease Any change from a state of health in which part or all of the host's body cannot carry on its normal functions because of the presence of an infectious agent or its products.

infectious disease cycle (chain of infection) The linked events that must occur for an infectious disease to be expressed in an individual.

infectious dose 50 (ID_{50}) Refers to the number of organisms that infect 50% of an experimental group of hosts within a specified time period.

infectivity Infectiousness; the state or quality of being infectious or communicable.

inflammasomes A complex formed within the cytoplasm of certain immune cells that promotes inflammation; composed of oligomerized NOD-like receptors and formed in response to the capture of microbe-associated molecular patterns and danger-associated molecular patterns.

inflammation A localized protective response to tissue injury or destruction. Acute inflammation is characterized by pain, heat, swelling, and redness in the injured area.

initiation factors Proteins that function during initiation of protein synthesis.

initiator tRNA The tRNA that binds to the start codon during initiation of translation.

innate (natural) immunity Refers to nonspecific defenses that are mounted within minutes to hours after challenge by a microbial or other foreign antigen. Sometimes also called nonspecific immunity.

innate lymphoid cell (ILC) A lymphocyte that functions as a bridge between the innate and adaptive arms of immunity, amplifying and relaying information about infection and tissue damage.

innate resistance mechanisms *See* innate (natural) immunity.

insertion sequence A transposable element that contains genes only for those enzymes, such as transposase, that are required for transposition.

in silico analysis The study of biology through the examination of nucleic acid and amino acid sequence. *See also* bioinformatics.

in situ reverse transcriptase (ISRT)-FISH A method that uses fluorescently labeled probes to identify mRNA transcripts in natural samples. *See also* fluorescent in situ hybridization (FISH).

insulators Sites in eukaryotic DNA that limit the effects of regulatory proteins to only those genes for which they are specified.

integral membrane protein *See* plasma membrane.

integrase An enzyme observed in some viruses; it catalyzes the integration of provirus DNA into the host chromosome.

integrative conjugative elements (ICEs) *See* conjugative transposon.

integrins (in′tə-grinz) Cellular adhesion receptors that mediate cell-cell and cell-substratum interactions, usually by recognizing linear amino acid sequences on protein ligands.

integron A genetic element with an attachment site for site-specific recombination and an integrase gene. It can capture genes and gene cassettes.

intercalating agents Molecules that can insert between the stacked bases of a DNA double helix, thereby distorting the DNA and inducing frameshift mutations.

interferon (IFN) A cytokine that stimulates cells to produce antiviral proteins (type 1 IFNs), and regulates growth, differentiation, or function of immune system cells (type 2 IFNs).

interleukin (in″tər-loo′kin) A cytokine that regulates growth and differentiation, particularly of lymphocytes.

intermediate filaments One of the proteinaceous components of the eukaryotic cytoskeleton.

internal transcribed spacer region (ITSR) The region between small subunit rRNA genes found in many microbial genomes that is conserved and transcribed.

interspecies hydrogen transfer The linkage of H_2 production by fermentative microorganisms to the use of H_2 by archaea during methanogenesis.

intoxication A disease that results from the entrance of a specific toxin into the body of a host. The toxin can induce the disease in the absence of the toxin-producing organism.

intraepidermal lymphocytes T cells found in the epidermis of the skin.

intranuclear inclusion body A structure found within cells infected with cytomegalovirus.

intron A noncoding intervening sequence in a gene that codes for pre-mRNA and is removed (spliced) from the final RNA product.

invasiveness The ability of a microorganism to enter a host, grow and reproduce within the host, and spread throughout its body.

invertible DNA A DNA fragment flanked by inverted repeats that may alternate in orientation by recombination between those repeats. If one flanking sequence has a promoter, the inversion constitutes an on/off switch for the inverted DNA.

iodophor An antimicrobial agent consisting of an organic compound complexed with iodine.

ionizing radiation Radiation of very short wavelength and high energy that causes atoms to lose electrons (i.e., ionize).

iron-sulfur (Fe-S) protein *See* electron transport chain.

isoelectric focusing An electrophoretic technique in which proteins are separated based on their isoelectric point.

isoelectric point The pH value at which a protein or other molecule no longer carries a net charge.

isoenzyme (isozyme) An enzyme that carries out the same catalytic function but differs in terms of its amino acid sequence, regulatory properties, or other characteristics.

isolated lymphoid follicles Highly organized lymphoid tissue distributed throughout the small intestine.

isotope fractionation The preferential use of one stable isotope over another by microbes carrying out metabolic processes.

J

joule The SI (International System of Units) unit of measure for energy or work. One calorie is equivalent to 4.1840 joules.

K

kallikrein An enzyme that acts on kininogen, releasing the active bradykinin protein.

keratinocytes (kĕ-rat′ĭ-no-sīt) Cells of the skin that express keratin.

Kirby-Bauer method A disk diffusion test to determine the susceptibility of a microorganism to chemotherapeutic agents.

Koch's postulates (koks) A set of rules for proving that a microorganism causes a particular disease.

Koplik's spots White lesions that appear on mucous membranes in the mouth (buccal mucosa) characteristic of measles infection.

Krebs cycle *See* tricarboxylic acid (TCA) cycle.

L

labyrinthulids (lab-ah-rin-thu′lids) A subgroup of stramenopiles characterized by having heterokont flagellated zoospores.

lactic acid bacteria (LAB) Strictly fermentative, Gram-positive bacteria that produce lactic acid as the primary fermentation end product; used in the fermentation of dairy products.

lactic acid fermentation A fermentation that produces lactic acid as the sole or primary product.

lactoferrin An iron-sequestering protein released from macrophages and neutrophils into plasma.

lagging strand The strand of DNA synthesized discontinuously during DNA replication.

lag phase A period following the introduction of microorganisms into fresh culture medium when there is no increase in cell numbers or mass during batch culture.

Lancefield system (group) One of the serologically distinguishable groups (e.g., group A and group B) into which streptococci can be divided.

Langerhans cell Cell found in the skin that internalizes antigen and moves in the lymph to lymph nodes, where it differentiates into a dendritic cell.

lateral flow assay A simple, inexpensive diagnostic test in which patient sample is applied to a filter paper. The sample is wicked along the filter paper (lateral flow) where an antigen-antibody interaction is detected.

lateral gene transfer *See* horizontal gene transfer.

leader A sequence in a gene that lies between the promoter and the start codon. It is transcribed, becoming a nontranslated sequence at the 5′ end of mRNA. It often aids in initiation and regulation of transcription and translation.

leading strand The strand of DNA that is synthesized continuously during DNA replication.

lectin complement pathway An antibody independent pathway of complement activation that is initiated by microbial lectins (proteins that bind carbohydrates).

leghemoglobin A heme-containing pigment produced in leguminous plants. It functions to protect the nitrogenase enzyme from oxygen in nodule-forming, nitrogen-fixing bacteria.

leishmanias (lēsh″ma′ne-ăs) Trypanosomal protists of the genus *Leishmania.*

lentic An aquatic system that features slow-moving or still waters.

lethal dose 50 (LD_{50}) Refers to the number of organisms that kills 50% of an experimental group of hosts within a specified time period.

leukocyte (loo′ko-sīt) Any white blood cell.

ligand *See* receptor.

light reactions *See* photosynthesis.

lignin An irregularly branched molecule made of phenylpropene units; an important structural component of woody plants.

limnology The study of the biological, chemical, and physical aspects of freshwater systems.

lipid A The lipid component of a lipopolysaccharide; also called endotoxin.

lipidomics Study of all the lipids produced by an organism under specific circumstances.

lipopolysaccharide (LPS) (lip″o-pol″e-sak′ah-rīd) A molecule containing both lipid and polysaccharide, found in the outer membrane of the typical Gram-negative cell wall. In many bacteria, it consists of three components: lipid A, core polysaccharide, and O antigen.

lithoheterotrophy The metabolic strategy characterized by the use of inorganic chemicals as an electron and energy source and organic carbon as a source of carbon. *See also* chemolithoheterotroph.

lithotroph An organism that uses reduced inorganic compounds as its electron source.

littoral zone The coastal region of a lake, stream, or ocean.

lobopodia Rounded pseudopodia found in some amoeboid protists.

localized anaphylaxis (atopic reaction) Type I hypersensitivity that remains contained in one tissue space (e.g., hay fever).

long-term stationary phase The phase following the death phase of a growth curve in which the population size remains at a more or less constant low level for an extended period.

lophotrichous (lo-fot′rĭ-kus) A cell with a cluster of flagella at one or both ends.

lotic An aquatic system that features fast-moving, free-running waters.

lymph node A small secondary lymphoid organ that contains lymphocytes, macrophages, and dendritic cells. It serves as a site for (1) filtration and removal of foreign antigens and (2) activation and proliferation of lymphocytes.

lymphocyte A nonphagocytic, mononuclear leukocyte that is an immunologically competent cell, or its precursor. Lymphocytes are present in the blood, lymph, and lymphoid tissues. *See also* B cell and T cell.

lymphoid tissue inducer (LTi) cells Innate lymphoid cells that stimulate the production of secondary lymphoid tissue including lymph nodes and Peyer's patches.

lysis (li′sis) The rupture or physical disintegration of a cell.

lysogenic (li-so-jen′ik) *See* lysogens.

lysogenic conversion A change in the phenotype of a bacterium due to the presence of a prophage.

lysogenic cycle The phase of a temperate virus's life cycle in which it establishes and maintains lysogeny.

lysogens (li′so-jens) Bacterial and archaeal cells that carry a provirus and can produce viruses under the proper conditions.

lysogeny (li-soj′e-ne) The state in which a viral genome remains within a bacterial or archaeal cell after infection and reproduces along with it, rather than taking control of the host cell and destroying it.

lysosome A spherical membranous eukaryotic organelle that contains hydrolytic enzymes and is responsible for the intracellular digestion of substances.

lysozyme An enzyme that degrades peptidoglycan by hydrolyzing the $\beta(1 \rightarrow 4)$ bond that joins *N*-acetylmuramic acid and *N*-acetylglucosamine.

lytic cycle (lit′ik) A viral life cycle that results in the lysis of the host cell.

M

macroautophagy Digestion of cytoplasmic components that involves enclosing the material (e.g., an organelle) in a double-membrane structure called an autophagosome. The autophagosome delivers the material to a lysosome for digestion.

macroelement A nutrient that is required in relatively large amounts (e.g., carbon and nitrogen).

macrolide antibiotic An antibiotic containing a macrolide ring—a large lactone ring with multiple keto and hydroxyl groups—linked to one or more sugars.

macromolecular crowding A phenomenon that occurs in the cytoplasm from the high concentration of solutes, resulting in reduced solvent.

macromolecule A large molecule that is a polymer of smaller units joined together.

macronucleus The larger of the two nuclei in ciliate protists. It is normally polyploid and directs the routine activities of the cell.

macrophage The name for a large, mononuclear phagocytic antigen-presenting cell, present in blood, lymph, and other tissues.

madurose The sugar derivative 3-*O*-methyl-D-galactose, which is characteristic of several actinomycete genera that are collectively called maduromycetes.

magnetosomes Magnetic greigite or magnetite particles in magnetotactic bacteria that allow the bacteria to orient themselves in magnetic fields.

maintenance energy The energy a cell requires simply to maintain itself or remain alive and functioning properly. It does not include the energy needed for either growth or reproduction.

major histocompatibility complex (MHC) A chromosome locus encoding the histocompatibility antigens and other components of the immune system. Class I MHC molecules are cell surface glycoproteins present on all nucleated cells; class II MHC glycoproteins are on antigen-presenting cells.

malt Grain soaked in water to soften it, induce germination, and activate its enzymes. The malt is then used in brewing and distilling.

mannose-binding protein (MBP) Also known as mannose-binding lectin, a serum protein that binds to mannose residues in microbial cell walls, thereby initiating complement activation.

marine snow Organic matter that sinks out of the photic zone of the ocean. These flocculent particles are composed of fecal pellets, diatom frustules, and other materials that are not rapidly degraded.

mash The soluble materials released from germinated grains and prepared as a microbial growth medium.

mashing The process in which cereals are mixed with water and incubated to degrade their complex carbohydrates (e.g., starch) to more readily usable forms such as simple sugars.

massively parallel nucleotide sequencing *See* next-generation DNA sequencing.

mass spectrometry A type of spectrometry that determines mass-to-charge ratio of ions formed from the molecule being analyzed. The ratio can be used to identify structures and determine sequences of proteins.

mast cell A white blood cell that produces vasoactive molecules (e.g., histamine) and stores them in vacuoles near the cell membrane where they are released upon cell stimulation by external triggers.

maximum containment level goal The number of pathogens permitted in a water supply to prevent adverse health effects and provide a margin of safety.

M cell Specialized cell of the intestinal mucosa and other sites (e.g., urogenital tract) that delivers antigen from its apical face to lymphocytes clustered within the pocket in its basolateral face.

Mediator A complex of proteins in eukaryotic cells that is involved in transcription initiation and regulation.

meiosis (mi-o′sis) The type of cell division by which a diploid cell divides and forms four haploid cells.

melting temperature (T_m) The temperature at which double-stranded DNA separates into individual strands; used to compare genetic similarity in microbial taxonomy.

membrane attack complex (MAC) The complex of complement proteins (C5b–C9) that creates a pore in the plasma membrane of a target cell and leads to cell lysis.

membrane filter Porous material that retains microorganisms as the suspending liquid passes through the pores.

membrane filter technique The use of a thin porous filter to collect microorganisms from water, air, and food.

memory cell An inactive lymphocyte derived from a sensitized B or T cell capable of a rapid response to subsequent antigen exposure.

mesophile A microorganism with a growth optimum around 20° to 45°C, a minimum of 15° to 20°C, and a maximum about 45°C or lower.

messenger RNA (mRNA) Single-stranded RNA synthesized from a nucleic acid template (DNA in cellular organisms, RNA in some viruses) during transcription; mRNA binds to ribosomes and directs the synthesis of protein.

metabolic channeling The localization of metabolites and enzymes in different parts of a cell.

metabolic endotoxemia Increased permeability of the gut wall to lipopolysaccharide (LPS) observed in obese individuals that leads to a state of chronic inflammation, fatty liver disease, and atherosclerosis.

metabolic engineering The use of molecular techniques to develop novel biochemical pathways that either promote product yield or generate new products.

metabolism The total of all chemical reactions in the cell; almost all are enzyme catalyzed.

metabolite flux The rate of turnover of a metabolite.

metabolites Chemical compounds produced by the metabolic activities of organisms.

metabolome All the small molecules produced by a cell, including cytoplasmic and secreted molecules.

metabolomics Analysis of all the small metabolites present in a cell under specific circumstances.

metagenomic library A genomic library constructed of DNA (or cDNA derived from mRNA) extracted directly from the environment.

metagenomics The study of genomes recovered from environmental samples (including the human body) without first isolating members of the microbial community and growing them in cultures.

metalimnion In lakes, the region of the water column between the hypolimnion and the epilimnion.

metaproteomics The emerging field of examining all the proteins produced in a given microbial habitat at a specific time.

metatranscriptomics Large-scale sequencing of mRNAs collected from natural environments as a means to assess microbial activities.

methane hydrate Pools of trapped methane that accumulates in latticelike cages of crystalline water 500 m or more below the sediment surface in many regions of the world's oceans.

methanogenesis The production of methane by certain members of the archaeal phylum *Euryarchaeota*.

methanogens (meth′ə-no-jens″) Strictly anaerobic archaea that derive energy by converting CO_2, H_2, formate, acetate, and other compounds to either methane or methane and CO_2.

methanotroph A microbe that has the ability to grow on methane as its sole carbon and energy source.

methylotroph A bacterium that uses reduced one-carbon compounds such as methane and methanol as its sole source of carbon and energy.

metronidazole Antimicrobial chemotherapeutic agent of the nitromidazole family having biological activity on anaerobic bacteria and protozoa.

Michaelis constant (K_m) A value that represents the substrate concentration required for an enzyme to operate at half maximal velocity.

microaerophile (mi″kro-a′er-o-fīl) A microorganism that requires low levels of oxygen for growth, around 2 to 10%, but is damaged by normal atmospheric oxygen levels.

microautoradiography A technique whereby the uptake of a radioactive substrate by single cells is determined.

microbe-associated molecular pattern (MAMP) Conserved molecular structures that occur in patterns on microbial surfaces. The structures and their patterns are unique to particular types of microorganisms and invariant among members of a given microbial group.

microbial ecology The study of microbial populations and communities in natural habitats, such as soils, water, food, etc.

microbial energy conversion The processes by which microbes are used to generate useful forms of energy, such as biofuels (ethanol, H_2, methanol), or electricity through microbial fuel cells.

microbial flora (microbiota) *See* microbiota.

microbial fuel cell An apparatus designed to capture electrons produced during microbial respiration and convert these electrons to electricity.

microbial loop The cycling of organic matter synthesized by photosynthetic microorganisms among other microbes, such as bacteria and protozoa. This process "loops" organic nutrients, minerals, and carbon dioxide back for reuse by the primary producers and makes the organic matter unavailable to higher consumers.

microbial mat A firm structure of layered microorganisms with complementary physiological activities that can develop on surfaces in aquatic environments.

microbial transformation *See* biotransformation.

microbiology The study of organisms that are usually too small to be seen with the naked eye.

microbiome The totality of microorganisms and microbial genomes that constitute a host's normal microbiota.

microbiota The microbes found in a habitat.

microcosms Incubation chambers with conditions that mimic those in a natural setting.

microdomain Regions of the plasma membrane demarcated by phospholipids, farnesol, hopanoids, and carotenoids; these serve as platforms for the assembly of large protein complexes.

microdroplet culture A technique designed to obtain an axenic culture whereby cells from a microbial community are encapsulated in a gel matrix, which when emulsified generates a porous microdroplet that houses a single cell. The microdroplets can be incubated together, thereby allowing the exchange of small molecules. Following incubation, microdroplets containing colonial growth can be separated by flow cytometry.

microelectrode An electrode with a tip that is sufficiently small to perform nondestructive measurements at intervals of less than 1 mm.

micronucleus The smaller of the two nuclei in ciliate protists. Micronuclei are diploid and involved only in genetic recombination and the regeneration of macronuclei.

micronutrients Nutrients required in very small quantities for growth and reproduction. Also called trace elements.

microorganism An organism that is usually too small to be seen clearly with the naked eye and is often unicellular, or if multicellular, does not exhibit a high degree of differentiation.

micro RNAs (miRNAs) Small RNAs that inhibit translation of a target mRNA.

microsporidia Primitive obligate intracellular fungi; primarily infect vertebrates.

microtubules (mi″kro-tu′buls) One of the proteinaceous components of the eukaryotic cytoskeleton, flagella, and cilia; composed of tubulin proteins.

mineralization The conversion of organic nutrients into inorganic material during microbial growth and metabolism.

mineral soil Soil that contains less than 20% organic carbon.

minimal inhibitory concentration (MIC) The lowest concentration of a drug that will prevent the growth of a particular microorganism.

minimal lethal concentration (MLC) The lowest concentration of a drug that will kill a particular microorganism.

minus (negative) strand A viral single-stranded nucleic acid that is complementary to the viral mRNA.

mismatch repair A type of DNA repair in which a portion of a newly synthesized strand of DNA containing mismatched base pairs is removed and replaced, using the parental strand as a template.

missense mutation A single base substitution in DNA that changes a codon for one amino acid into a codon for another.

mitochondrion (mi″to-kon′dre-on) The eukaryotic organelle that is the site of electron transport, oxidative phosphorylation, and pathways such as the Krebs cycle; it provides most of a nonphotosynthetic cell's energy under aerobic conditions.

mitosis A process that takes place in the nucleus of a eukaryotic cell and results in the formation of two new nuclei, each with the same number of chromosomes as the parent.

mitosome A member of the mitochondria family of organelles found in certain protists that lack mitochondria or hydrogenosomes.

mixed acid fermentation A type of fermentation carried out by members of the family *Enterobacteriaceae* in which ethanol and a complex mixture of organic acids are produced.

mixotrophy A nutritional strategy that includes different types of metabolism (e.g., phototrophy and chemoorganotrophy).

mobilizable genomic island A DNA sequence transferred horizontally that encodes its own transfer functions in addition to recombination functions and often antibiotic resistance genes and genes promoting pathogenesis; these often comprise dozens of kilobases in a chromosome.

modified atmosphere packaging (MAP) Addition of gases such as nitrogen and carbon dioxide to packaged foods to inhibit the growth of spoilage organisms.

mold Any of a large group of fungi that exist as multicellular filamentous colonies; typically lack macroscopic fruiting bodies.

molecular chaperones *See* chaperone proteins.

monocistronic mRNA An mRNA containing a single coding region.

monocyte A mononuclear phagocytic leukocyte; precursor of macrophages and dendritic cells.

monotrichous (mon-ot′rĭ-kus) Having a single flagellum.

morbidity rate Measures the number of individuals who become ill as a result of a particular disease within a susceptible population during a specific time period.

mordant A substance that helps fix dye on or in a cell.

morphovar A variant strain of a microbe characterized by morphological differences.

mortality rate The ratio of the number of deaths from a given disease to the total number of cases of the disease.

most probable number (MPN) Statistical estimation of the size of a population in a liquid by diluting and determining end points for microbial growth.

mucociliary blanket The layer of cilia and mucus that lines certain portions of the respiratory system; it traps microorganisms and then transports them by ciliary action away from the lungs.

mucosal-associated lymphoid tissue (MALT) Organized and diffuse immune tissues found as part of the mucosal epithelium. It can be specialized to the gut (GALT) or the bronchial system or (BALT).

mucosal tolerance The ability of the immune system to not respond to, or tolerate, antigens (including microbes) present in and on mucosal tissues.

multicloning site (MCS) A region of DNA on a cloning vector that has a number of restriction enzyme recognition sequences to facilitate the cloning of a gene.

multilocus sequence typing (MLST) A method for genotypic classification of bacteria or archaea within a single genus using nucleotide differences among five to seven housekeeping genes.

multiple strand displacement A technique used to amplify very low concentrations of DNA; it can yield enough DNA from a single cell to perform DNA sequencing.

multiplex PCR A polymerase chain reaction that contains multiple primer sets to amplify several targets all in one tube.

murein *See* peptidoglycan.

must The juices of fruits, including grapes, that can be fermented for the production of alcohol.

mutagen (mu′tah-jen) A chemical or physical agent that causes mutations.

mutation A heritable change in the genetic material.

mutualism A type of symbiosis in which both partners gain from the association and are dependent on each other.

mutualist An organism associated with another in an obligatory relationship that is beneficial to both.

mycelium (mi-se′le-um) A mass of branching hyphae found in fungi and some bacteria.

Myc factor (mik) A lipochitooligosaccharide produced by some mycorrhizal fungi that prompts their uptake by the roots of host plants.

mycobiont The fungal partner in a lichen.

mycolic acids Complex 60 to 90 carbon fatty acids with a hydroxyl on the β-carbon and an aliphatic chain on the α-carbon; found in the cell walls of mycobacteria.

mycologist A person specializing in mycology; a student of mycology.

mycology The science and study of fungi.

mycoplasmas Common name for bacteria that are members of the class *Mollicutes,* phylum *Tenericutes.* They lack cell walls and cannot synthesize peptidoglycan precursors.

mycorrhizal fungi Fungi that form stable, mutualistic relationships on (ectomycorrhizal) or in (endomycorrhizal) the roots of vascular plants.

mycosis (mi-ko′sis; pl., **mycoses**) Any disease caused by a fungus.

mycotoxin A toxin produced by fungi.

myositis (mi″o-si′tis) Inflammation of a striated or voluntary muscle.

myxobacteria A group of Gram-negative, aerobic soil bacteria characterized by gliding motility, a complex life cycle with the production of fruiting bodies, and the formation of myxospores.

myxospores (mik′so-spōrs) Special dormant spores formed by myxobacteria.

N

naked amoeba An amoeboid protist that lacks a cell wall or other supporting structures.

naked viruses *See* nonenveloped virus.

nanowires Threadlike extensions of plasma membrane that transfer electrons from the terminal point in the electron transport chain to an external metal surface.

narrow-spectrum drugs Chemotherapeutic agents effective only against a limited variety of microorganisms.

natural classification A classification system that arranges organisms into groups whose members share many characteristics and reflect as much as possible the biological nature of organisms.

natural immunity *See* innate immunity.

natural killer (NK) cell A type of white blood cell that destroys host cells infected with intracellular pathogens and tumor cells.

naturally acquired active immunity The type of active immunity that develops when an individual's immune system encounters an appropriate antigen during the course of normal activities; it often arises as the result of recovering from an infection and lasts a long time.

naturally acquired passive immunity The type of temporary immunity that involves the transfer of antibodies from one individual to another.

natural products Industrially important compounds, such as antibiotics, produced by microbes.

negative selection In immunology, the process by which lymphocytes that recognize host (self) antigens undergo apoptosis or become anergic (inactive).

negative staining A staining procedure in which a dye is used to make the background dark while the specimen is unstained.

negative-strand DNA *See* minus strand.

negative-strand RNA *See* minus strand.

negative transcriptional control Regulation of transcription by a repressor protein. When bound to the repressor-binding site, transcription is inhibited.

neoplasia Abnormal cell growth and reproduction due to a loss of regulation of the cell cycle; produces a tumor in solid tissues.

neuraminidase An enzyme that cleaves the chemical bond linking neuraminic acids to the sugars present on the surface of animal cells; in virology, one type of envelope spike on influenza viruses has neuraminidase activity and is used to identify different strains.

neutralization Binding of specific immunoglobulins to toxins or viruses, inhibiting their biological activity.

neutral mutation A missense mutation that does not affect the function of the protein (e.g., a polar amino acid is substituted for a polar amino acid).

neutrophil A mature white blood cell in the granulocyte lineage. It has a nucleus with three to five lobes and is very phagocytic.

neutrophile A microorganism that grows best at a neutral pH range between pH 5.5 and 8.0.

next-generation nucleotide sequencing Methodologies that allow rapid genome sequencing without the prior construction of a genomic library. Instead, fragments of genomic DNA are tethered to a solid substrate, and each fragment is sequenced simultaneously or in a massively parallel fashion.

nicotinamide adenine dinucleotide (NAD^+) (nik″o-tin′ah-mīd) An electron-carrying coenzyme; it is particularly important in catabolic processes and usually transfers electrons from an electron source to an electron transport chain.

nicotinamide adenine dinucleotide phosphate ($NADP^+$) An electron-carrying coenzyme that most often participates as an electron carrier in biosynthetic metabolism.

nitrification The oxidation of ammonia to nitrate.

nitrifying archaea Mesophilic archaea (thaumarchaeotes) capable of oxidizing ammonium to nitrite.

nitrifying bacteria Chemolithotrophic, Gram-negative bacteria that are members of several families within the phylum *Proteobacteria* that either oxidize ammonia to nitrite or nitrite to nitrate.

nitrogenase (ni′tro-jen-ās) The enzyme that catalyzes biological nitrogen fixation.

nitrogen fixation The metabolic process in which atmospheric molecular nitrogen (N_2) is reduced to ammonia; carried out by nitrogen-fixing bacteria and archaea.

nocardioforms Bacteria that resemble members of the genus *Nocardia;* they develop a substrate mycelium that readily breaks up into rods and coccoid elements (a quality called fugacity).

Nod factors Signaling compounds that alter gene expression in the plant host of a rhizobium.

NOD-like receptors (NLRs) Receptors similar to nucleotide-binding and oligomerization domain (NOD) receptors that sense endogenous metabolites and regulate the production of inflammatory cytokines in response to microbe-associated molecular patterns and danger-associated molecular patterns.

nomenclature The branch of taxonomy concerned with the assignment of names to taxonomic groups in agreement with published rules.

noncompetitive inhibitor A chemical that inhibits enzyme activity by a mechanism that does not involve binding the active site of the enzyme.

noncyclic photophosphorylation The process in which light energy is used to make ATP when electrons are moved from water to $NADP^+$ during oxygenic photosynthesis; both photosystem I and photosystem II are involved.

nonenveloped (naked) virus A virus having virions composed only of a nucleocapsid (i.e., lacking an envelope).

nonheme iron protein *See* electron transport chain.

nonhomologous end joining A genetic repair phenomenon in which two linear chromosomal fragments are rejoined despite no sequence homology.

nonsense (stop) codon A codon that does not code for an amino acid but is a signal to terminate protein synthesis.

nonsense mutation A mutation that converts a sense codon to a nonsense (stop) codon.

normal microbiota *See* microbiota.

nosocomial infection (nos′o-ko″me-al) *See* health-care–associated infection.

nuclear envelope The complex double-membrane structure forming the outer boundary of the eukaryotic nucleus. It is covered by nuclear pores through which substances enter and leave the nucleus.

nuclear pore complex The nuclear pore plus about 30 proteins that form the pore and are involved in moving materials across the nuclear envelope.

nucleocapsid (nu″kle-o-kap′sid) The viral nucleic acid and its surrounding capsid; the basic unit of virion structure.

nucleoid An irregularly shaped region in a bacterial or archaeal cell containing genetic material.

nucleoid-associated protein (NAP) Proteins that serve an architectural role in the nucleoid and contribute to chromosome structure.

nucleoid occlusion A mechanism used by many bacteria to prevent premature formation of a septum during binary fission.

nucleolus (nu-kle′o-lus) The organelle, located within the nucleus and not bounded by a membrane, that is the location of ribosomal RNA synthesis and the assembly of ribosomal subunits.

nucleoside (nu′kle-o-sīd″) A combination of ribose or deoxyribose with a purine or pyrimidine base but lacking phosphate.

nucleosome (nu′kle-o-sōm″) A complex of histones and DNA found in eukaryotic chromatin and some archaea; the DNA is wrapped around the surface of the beadlike histone complex.

nucleotide (nu′kle-o-tīd) A combination of ribose or deoxyribose with phosphate and a purine or pyrimidine base; a nucleoside plus one or more phosphates.

nucleotide excision repair *See* excision repair.

nucleus The eukaryotic organelle enclosed by a double-membrane envelope that contains the cell's chromosomes.

numerical aperture The property of a microscope lens that determines how much light can enter and how great a resolution the lens can provide.

O

O antigen The terminal polysaccharide of the lipopolysaccharide found on some Gram-negative bacterial cell walls.

obligate aerobes Organisms that grow only in the presence of oxygen.

obligate anaerobes Microorganisms that cannot tolerate oxygen and die when exposed to it.

ocean acidification The decline in pH in ocean waters as a result of changes to the carbonate equilibrium system.

oceanography The study of the biological, chemical, and physical aspects of the oceans.

Okazaki fragments Short stretches of polynucleotides produced during discontinuous DNA replication.

oligonucleotide A short fragment of DNA or RNA, usually artificially synthesized, used in a number of molecular genetic techniques such as DNA sequencing and polymerase chain reaction.

oligonucleotide signature sequence Short, conserved nucleotide sequences that are specific for a phylogenetically defined group of organisms. The signature sequences found in small subunit rRNA molecules are most commonly used.

oligotrophic environment (ol″ĭ-go-trof′ik) An environment containing low levels of nutrients that support microbial growth.

oncogene (ong″ko-jēn) A gene whose activity is associated with the conversion of normal cells to cancer cells.

oncovirus A virus known to be associated with the development of cancer.

oocyst (o′o-sist) Cyst formed around a zygote of certain protozoa.

öomycetes (o″o-mi-se′tēz) A collective name for protists also known as water molds. Formerly thought to be fungi.

open reading frame (ORF) A sequence of DNA thought to encode a protein because it is not interrupted by a stop codon and has an apparent promoter and ribosome-binding site at the 5′ end and a terminator at the 3′ end.

operational taxonomic unit (OTU) A term that includes all sources of data used to construct phylogenetic trees; that is, it includes species, strains, and genomic sequences from organisms not yet grown in culture.

operator The segment of DNA in an operon to which the repressor protein binds; it controls the expression of the genes adjacent to it.

operon In bacteria, the sequence of bases in DNA that contains a promoter and one or more structural genes and often an operator or activator-binding site that controls their expression.

opportunistic microorganism or pathogen A microorganism that is usually free living or a part of the host's normal microbiota but may become pathogenic under certain circumstances, such as when the immune system is compromised.

opsonization (op″so-ni-za′shun) The coating of foreign substances by antibody, complement proteins, or fibronectin to make the substances more readily recognized by phagocytic cells.

optical tweezer A focused laser beam that can drag and isolate an individual microorganism from a complex microbial mixture.

organelle A structure within or on a cell that performs specific functions and is related to the cell in a way similar to that of an organ to the body of a multicellular organism.

organic soil A soil that contains at least 20% organic carbon.

organotrophs Organisms that use reduced organic compounds as their electron source.

origin of replication A site on a chromosome or plasmid where DNA replication is initiated.

origin recognition complex (ORC) A complex of six proteins that "marks" the origins of replication in eukaryotic chromosomes.

orthologue A gene found in the genomes of two or more different organisms that share a common ancestry. The products of orthologous genes are presumed to have similar functions.

osmophiles Microorganisms that grow best in or on media of high solute concentration.

osmotolerant Organisms that grow over a fairly wide range of water activity or solute concentration.

osmotrophy A form of nutrition in which soluble nutrients are absorbed through the cytoplasmic membrane; found in bacteria, archaea, fungi, and some protists.

Öuchterlony technique *See* double diffusion agar assay.

outbreak The sudden, unexpected occurrence of a disease in a given population.

outer membrane A membrane located outside the peptidoglycan layer in the cell envelope of typical Gram-negative bacteria.

outer membrane vesicle (OMV) A spherical portion of the Gram-negative outer membrane that pinches off to leave the cell. OMVs may carry proteins, toxins, and nucleic acids.

oxazolidinones Synthetic drugs, characterized by a heterocyclic five-membered ring, that inhibit protein synthesis.

oxidation-reduction (redox) reactions Reactions involving electron transfers; the electron donor (reductant) gives electrons to an electron acceptor (oxidant).

oxidative burst *See* respiratory burst.

oxidative phosphorylation (fos″for-ĭ-la′-shun) The synthesis of ATP from ADP using energy made available during electron transport initiated by the oxidation of a chemical energy source.

oxygenic photosynthesis Photosynthesis that oxidizes water to form oxygen; the form of photosynthesis characteristic of plants, protists, and cyanobacteria.

P

palisade arrangement Angular arrangements of bacteria whereby cells remain partially attached after snapping division.

pandemic An increase in the occurrence of a disease within a large and geographically widespread population (often refers to a worldwide epidemic).

Paneth cell (pah′net) A specialized epithelial cell of the intestine that secretes hydrolytic enzymes and antimicrobial proteins and peptides.

pangenome The collection of genes found in all strains that belong to a single species or other taxonomic group.

paralogue Two or more genes in the genome of a single organism that arose through duplication of a common ancestral gene. The products of paralogous genes may have slightly different functions.

parasite An organism that lives on or within another organism (the host) and benefits from the association while harming its host. Often the parasite obtains nutrients from the host.

parasitism A type of symbiosis in which one organism benefits from the other and the host is usually harmed.

parasporal body An intracellular, solid protein crystal made by the bacterium *Bacillus thuringiensis*. It is the basis of the bacterial insecticide Bt.

parfocal A microscope that retains proper focus when the objectives are changed.

particulate organic matter (POM) Nutrients that are not dissolved or soluble, generally referring to aquatic ecosystems. This includes microorganisms and their cellular debris after senescence or viral lysis.

passive diffusion The movement of molecules from a region of higher concentration to one of lower concentration as a result of random thermal agitation.

pasteurization The process of heating liquids to destroy microorganisms that can cause spoilage or disease.

pathogen (path′o-jən) Any organism that causes disease.

pathogenicity (path″o-je-nis′ĭ-te) The condition or quality of being pathogenic, or the ability to cause disease.

pathogenicity island A segment of DNA in some pathogens that contains genes responsible for virulence; often codes for a type III secretion system that allows the pathogen to secrete virulence proteins and damage host cells.

pattern recognition receptor (PRR) A receptor found on macrophages and other phagocytic cells that binds to microbe-associated molecular patterns on microbial surfaces.

PCR bias Artifactual or misleading results obtained when microbial populations are assessed by amplifying genes or genomes by the polymerase chain reaction (PCR); occurs because not all DNA or RNA is amplified at the same rate.

pelagic zone The central region of a lake or ocean not influenced by events or nutrients found in the coastal regions.

pellicle (pel′ĭ-k′l) (1) A relatively rigid layer of proteinaceous elements just beneath the plasma membrane in many protists. The plasma membrane is sometimes considered part of the pellicle. (2) A biofilm that forms at an air-liquid interface

penicillin A β-lactam antibiotic. The first true antibiotic to be discovered and used clinically.

penicillinase A β-lactamase enzyme that digests the active moiety of penicillin.

pentose phosphate (hexose monophosphate) pathway A glycolytic pathway that oxidizes glucose 6-phosphate to ribulose 5-phosphate and then converts it to a variety of three to seven carbon sugars.

peplomer (spike) (pep′lo-mer) A protein or protein complex that extends from the viral envelope and often is important in virion attachment to the host cell surface.

peptide interbridge A short peptide chain that connects the tetrapeptide chains in the peptidoglycan of some bacteria.

peptidoglycan (pep″tĭ-do-gli′kan) A large polymer composed of long chains of alternating *N*-acetylglucosamine and *N*-acetylmuramic acid residues. The polysaccharide chains are linked to each other through connections between tetrapeptide chains attached to the *N*-acetylmuramic acids. It provides much of the strength and rigidity possessed by bacterial cell walls; also called murein.

peptidyl or donor site (P site) The site on the ribosome that contains the peptidyl-tRNA at the beginning of the elongation cycle during protein synthesis.

peptidyl transferase The ribozyme that catalyzes the transpeptidation reaction in protein synthesis; in this reaction, an amino acid is added to the growing peptide chain.

perforin Pore-forming proteins secreted along with granzymes by cytotoxic T lymphocytes and NK cells.

peripheral membrane protein *See* plasma membrane.

peripheral tolerance The process by which self-reactive B and T cells are inhibited in body tissues other than bone marrow and thymus, respectively.

periplasm (per′ĭ-plaz-əm) The substance that fills the periplasmic space.

periplasmic flagella The flagella that lie under the outer sheath and extend from both ends of a spirochete cell to overlap in the middle and form the axial filament. Also called axial fibrils, axial filaments, and endoflagella.

periplasmic space The space between the plasma membrane and the outer membrane in typical Gram-negative bacteria, and between the plasma membrane and the cell wall in typical Gram-positive bacteria. A similar space is sometimes observed between the plasma membrane and the cell wall of some archaea.

peristalsis The muscular contractions of the gut that propel digested foods and waste through the intestinal tract.

peritrichous (pĕ-rit′rĭ-kus) A cell with flagella distributed over its surface.

peronosporomycetes Name for öomycetes, a subgroup of *Stramenopila,* distinguished by the formation of large egg cells when they reproduce.

persisters Variants of microbial cells in a population that are resistant to antibiotics.

Peyer's patches Highly organized lymphoid tissue in the small intestine.

phage (fāj) *See* bacteriophage.

phagocytic vacuole A membrane-delimited vacuole produced by cells carrying out phagocytosis. It is formed by the invagination of the plasma membrane and contains solid material.

phagocytosis The endocytotic process in which a cell encloses large particles in a phagocytic vacuole (phagosome) and engulfs them.

phagolysosome The vacuole that results from the fusion of a phagosome with a lysosome.

phagosome A membrane-enclosed vacuole formed by the invagination of the cell membrane during endocytosis.

phase-contrast microscope A microscope that converts slight differences in refractive index and cell density into easily observed differences in light intensity.

phenetic system A classification system that groups organisms together based on the similarity of their observable characteristics.

phenol coefficient test A measurement of the effectiveness of disinfectants that compares their activity against test bacteria with that of phenol.

phenotype Observable characteristics of an organism.

phenotypic rescue The identification of a cloned gene based on its ability to complement a genetic deficiency in the host cell.

phosphatase (fos′fah-tās″) An enzyme that catalyzes the hydrolytic removal of phosphate from molecules.

phosphoenolpyruvate:sugar phosphotransferase system (PTS) A group translocation system used by many bacteria. As a sugar is transported into the cell, the hydrolysis of a high-energy phosphate bond fuels its import, and the sugar is modified by the covalent attachment of the phosphoryl group.

phosphorelay system A set of proteins involved in the transfer of phosphate from one protein in the set to another. It is often used to regulate protein activity or transcription.

photic zone The illuminated area of an aquatic habitat from the surface to the depth to which the rate of photosynthesis equals that of respiration.

photoautotroph *See* photolithoautotroph.

photolithoautotroph An organism that uses light energy, an inorganic electron source (e.g., H_2O, H_2, H_2S), and CO_2 as its carbon source. Also called photolithotrophic autotroph or a photoautotroph.

photoorganoheterotroph A microorganism that uses light energy, organic electron sources, and organic molecules as a carbon source. Also called photoorganotrophic heterotroph.

photophosphorylation The synthesis of ATP from ADP using energy made available by the absorption of light.

photoreactivation The process in which blue light is used by an enzyme to repair thymine dimers in DNA.

photosynthate Nutrient material released from phototrophic organisms and contributes to the pool of dissolved organic matter.

photosynthesis The trapping of light energy and its conversion to chemical energy (light reactions), which is then used to reduce CO_2 and incorporate it into organic molecules (dark reactions).

photosystem I The photosystem in eukaryotic cells and cyanobacteria that absorbs longer wavelength light, usually greater than about 680 nm, and transfers the energy to chlorophyll P700 during photosynthesis; it is involved in both cyclic and noncyclic photophosphorylation.

photosystem II The photosystem in eukaryotic cells and cyanobacteria that absorbs shorter wavelength light, usually less than 680 nm, and transfers the energy to chlorophyll P680 during photosynthesis; it participates in noncyclic photophosphorylation.

phototaxis The ability of certain phototrophic organisms to move in response to a light source.

phototrophs Organisms that use light as their energy source.

phycobiliproteins Photosynthetic pigments found in cyanobacteria that are composed of proteins with attached tetrapyrroles.

phycobilisomes Particles on the membranes of cyanobacteria that contain photosynthetic pigments.

phycobiont (fi″ko-bi′ont) The photosynthetic protist or cyanobacterial partner in a lichen.

phycocyanin (fi″ko-si′an-in) A blue phycobiliprotein pigment used to trap light energy during photosynthesis.

phycoerythrin (fi″ko-er′i-thrin) A red photosynthetic phycobiliprotein pigment used to trap light energy.

phycology (fi-kol′o-je) The study of algae.

phyletic classification system *See* phylogenetic classification system.

phyllosphere The surface of plant leaves.

phylogenetic classification system A classification system based on evolutionary relationships, rather than the general similarity of characteristics.

phylogenetic tree A graph made of nodes and branches, much like a tree in shape, that shows phylogenetic relationships between groups of organisms and sometimes also indicates the evolutionary development of groups.

phylogeny The evolutionary development of a species.

phylotype A taxon that is characterized only by its nucleic acid sequence; generally discovered during metagenomic analysis.

physical map In genomics, this refers to a diagram of a chromosome, gene, or other genetic element that shows restriction endonuclease recognition sites. Specific genes and other genetic elements (e.g., origin of replication) may also be noted.

phytoplankton A community of floating photosynthetic organisms, composed of photosynthetic protists and cyanobacteria.

picoplankton Planktonic microbes between 0.2 and 2.0 μm in size, including the cyanobacterial genera *Prochlorococcus* and *Synechococcus.*

piezophilic (pi-e-zo-fil′ic) Describing organisms that prefer or require high pressures for growth and reproduction.

pili (s., **pilus**) A fine, hairlike protein appendage on many bacteria, archaea, and some fungi. They attach cells to surfaces, and some are involved in twitching motility.

pilin The protein monomer that comprises a pilus.

pitching Pertaining to inoculation of a nutrient medium with yeast, for example, in beer brewing.

plaque A clear area in a lawn of host cells that results from their lysis by viruses.

plaque assay A method used to determine the number of infectious virions.

plaque-forming unit The unit of measure of a plaque assay. It usually represents a single infectious virion.

plasma cell A mature, differentiated B lymphocyte that synthesizes and secretes antibody.

plasmalemma The plasma membrane in protists.

plasma membrane The selectively permeable membrane surrounding the cell's cytoplasm; also called the cell membrane, plasmalemma, or cytoplasmic membrane. For most cells it is a lipid bilayer (some archaea have a lipid monolayer) with proteins embedded in it (integral membrane proteins) and associated with the surface (peripheral membrane proteins).

plasmid A double-stranded DNA molecule that can exist and replicate independently of the chromosome. A plasmid is stably inherited but is not required for the host cell's growth and reproduction.

plasmodial (acellular) slime mold A member of the protist division *Amoebozoa* (*Myxogastria*) that exists as a thin, streaming, multinucleate mass of protoplasm (plasmodium) for some portion of its life cycle.

plasmolysis (plaz-mol′ĭ-sis) The process in which water osmotically leaves a cell, which causes the cytoplasm to shrivel up and pull the plasma membrane away from the cell wall.

plastid A cytoplasmic organelle of algae and higher plants that contains pigments such as chlorophyll, stores food reserves, and often carries out processes such as photosynthesis.

pleomorphic (ple″o-mor′fik) Refers to cells or viruses that are variable in shape and lack a single, characteristic form.

plus (positive) strand A viral nucleic acid strand that is equivalent in base sequence to the viral mRNA.

point mutation A mutation that affects only a single base pair.

polar flagellum A flagellum located at one end of an elongated cell.

3′ poly-A tail *See* pre-mRNA.

poly-β-hydroxybutyrate (PHB) A linear polymer of β-hydroxybutyrate used as a reserve of carbon and energy by many bacteria.

polycistronic mRNA An mRNA that has more than one coding region; formed when an operon is transcribed.

polyhistidine tagging (His-tagging) A method by which a protein is fused to six histidine residues, thereby facilitating the protein's purification through affinity chromatography.

polymerase chain reaction (PCR) An in vitro technique used to synthesize large quantities of specific nucleotide sequences from small amounts of DNA. It employs oligonucleotide primers complementary to specific sequences in the target gene and special heat-stable DNA polymerases (e.g., Taq polymerase).

polymorphonuclear leukocyte (pol″e-mor″fo-noo′kle-ər) A leukocyte that has a variety of nuclear forms.

polyphasic taxonomy An approach in which taxonomic schemes are developed using a wide range of phenotypic and genotypic information.

polyprotein A large polypeptide that contains several proteins, which are released when the polyprotein is cut by proteases. Polyproteins are often produced by viruses having positive-strand RNA genomes.

polyribosome (pol″e-ri′bo-sōm) A complex of several ribosomes translating a single messenger RNA.

population An assemblage of organisms of the same type.

porin proteins Proteins that form channels across the outer membrane of typical Gram-negative bacterial cell walls through which small molecules enter the periplasm.

portal of entry Body surface that is the initial site of pathogen access to the host.

positive transcriptional control Control of transcription by an activator protein. When the activator is bound to the activator-binding site, the level of transcription increases.

postherpetic neuralgia The severe pain after a herpes infection.

posttranslational regulation Alteration of the activity of a protein after it has been synthesized by a structural modification, such as phosphorylation or a conformational change. *See also* allosteric enzyme and reversible covalent modification.

pour plate A Petri dish of solid culture medium with isolated microbial colonies growing both on its surface and within the medium that has been prepared by mixing microorganisms with cooled, still-liquid medium and then allowing the medium to harden.

precipitation (precipitin) reaction The reaction of an antibody with a soluble antigen to form an insoluble precipitate.

precipitin The antibody responsible for a precipitation reaction.

precursor metabolites Intermediates of glycolytic pathways, TCA cycle, and other pathways that serve as starting molecules for biosynthetic pathways that generate monomers and other building blocks needed for synthesis of macromolecules.

pre-mRNA In eukaryotes, the RNA transcript of DNA made by RNA polymerase II prior to processing to form mRNA by the addition of the 5′ cap and 3′ poly-A tail, and the removal of introns.

presence-absence (P-A) test An assay designed to detect the presence of coliforms in water; a positive result must be confirmed by a different assay.

prevalence rate Refers to the total number of individuals infected at any one time in a given population regardless of when the disease began.

Pribnow box A base sequence in bacterial promoters that is recognized by RNA polymerase and is the site of initial polymerase binding.

primary active transport *See* active transport.

primary metabolites Microbial metabolites produced during active growth of an organism.

primary mRNA *See* pre-mRNA.

primary producer Photoautotrophic and chemoautotrophic organisms that incorporate carbon dioxide into organic carbon and thus provide new biomass for the ecosystem.

primary production The incorporation of carbon dioxide into organic matter by photosynthetic organisms and chemoautotrophic organisms.

primary treatment The first step of sewage treatment, in which physical settling and screening are used to remove particulate materials.

primase *See* primosome.

primers *See* polymerase chain reaction and primosome.

primosome In bacteria, a complex of proteins that includes the enzyme primase, responsible for synthesizing the RNA primers needed for DNA replication.

prion (pre′on) An infectious agent consisting only of protein; prions cause a variety of spongiform encephalopathies such as scrapie in sheep.

probiotic A living microorganism that may provide health benefits beyond its nutritional value when ingested.

prochlorophytes A group of cyanobacteria having both chlorophyll *a* and *b* but lacking phycobilins.

prodromal stage (pro-dro′məl) The period during the course of a disease when signs and symptoms are present, but they are not yet distinctive enough for an accurate diagnosis.

production strains Microbial strains used in industrial processes that have been engineered to generate a high yield of the desired product.

professional phagocyte Any one of the three types of immune cells that phagocytose microbes. These include neutrophils, macrophages, and dendritic cells.

programmed cell death (1) In eukaryotes, apoptosis. (2) In some bacteria, a mechanism proposed to account for the decline in cell numbers during the death phase of a growth curve. *See also* apoptosis.

promoter The region on DNA at the start of a gene that RNA polymerase binds to before beginning transcription.

proofreading The ability of enzymes to check their products to ensure that the correct product is made. For instance, DNA polymerase checks newly synthesized DNA and replaces an incorrect nucleotide with the correct one prior to further synthesis.

propagated epidemic An epidemic that is characterized by a relatively slow and prolonged rise and then a gradual decline in the number of individuals infected. It usually results from the introduction of an infected individual into a susceptible population and transmission of the pathogen from person to person.

prophage (pro′fāj) **(provirus)** (1) The latent form of a temperate bacterial or archaeal virus that remains within the lysogen, usually integrated into the host chromosome. (2) The form of a eukaryotic virus that remains within the host cell during a latent infection. Also refers to a retroviral genome after it has been integrated into the host's chromosome.

prostheca (pros-the′kah) An extension of a bacterial cell, including the plasma membrane and cell wall, that is narrower than the mature cell.

prosthetic group A tightly bound cofactor that remains at the active site of an enzyme during its catalytic activity.

protease (pro′te-ās) An enzyme that hydrolyzes proteins. Also called a proteinase.

proteasome A large, cylindrical protein complex observed in eukaryotic cells that degrades ubiquitin-labeled proteins to peptides in an ATP-dependent process. Also called 26S proteasome. Similar protein-degrading machinery has been observed in some archaea and bacteria.

protein disulfide isomerase A periplasmic enzyme that catalyzes the formation of disulfide bonds as a polypeptide folds into its functional form.

protein modeling The process by which the amino acid sequence of a protein is analyzed using software designed to predict the protein's three-dimensional structure.

proteome The complete collection of proteins that an organism produces.

proteomics The study of the structure and function of cellular proteins.

proteorhodopsin A rhodopsin molecule first identified in marine proteobacteria but since found in a variety of microbes. It enables rhodopsin-based phototrophy. *See also* archaerhodopsin.

protist Unicellular and sometimes colonial eukaryotic organisms that lack cellular differentiation into tissues. Many chemoorganotrophic protists are referred to as protozoa; many phototrophic protists are referred to as algae.

protistology The study of protists.

protomer An individual protein of a viral capsid; a capsomer is made of protomers.

proton motive force (PMF) The potential energy arising from a gradient of protons and a membrane potential that powers ATP synthesis and other processes.

proto-oncogene Normal cellular gene that when mutated or overexpressed contributes to malignant transformation of a cell.

protoplast (1) The plasma membrane and everything within it. (2) A bacterial, archaeal, or fungal cell with its cell wall completely removed. It is spherical in shape and osmotically sensitive.

prototroph A microorganism that requires the same nutrients as the majority of naturally occurring members of its species.

protozoa (pro″to-zo′a) A common term for a group of unrelated unicellular, chemoorganotrophic protists.

protozoology The study of protozoa.

provirus *See* prophage.

pseudogene A degraded, nonfunctional gene.

pseudomurein A complex polysaccharide observed in the cell walls of some archaea; it resembles peptidoglycan (murein) in structure and chemical make-up; also called pseudopeptidoglycan.

pseudopodium or pseudopod (soo″do-po′de-um) A nonpermanent cytoplasmic extension of the cell by which amoeboid protists move and feed.

psychrophile (si′kro-fīl) A microorganism that has an optimum growth temperature of 15°C or lower and a temperature maximum around 20°C.

psychrotolerant (psychrotroph) A microorganism that has a growth optimum between 20° and 30°C and a maximum of about 35°C.

PulseNet A program developed and overseen by the Centers for Disease Control and Prevention in which pulse-field gel electrophoresis is used to characterize pathogens found to contaminate food in an effort to detect and track these microbes.

punctuated equilibria The observation based on the fossil record that evolution does not proceed at a slow and linear pace but rather is periodically interrupted by rapid bursts of speciation and extinction driven by abrupt changes in environmental conditions.

pure (axenic) culture A population of cells that are identical because they arise from a single cell.

purine (pu′rēn) A heterocyclic, nitrogen-containing molecule with two joined rings that occurs in nucleic acids and other cell constituents (e.g., adenine and guanine).

purple membrane An area of the plasma membrane of *Halobacterium* that contains archaerhodopsin and is active in trapping light energy.

purple nonsulfur bacteria Anoxygenic photosynthetic α- and β-proteobacteria.

purple sulfur bacteria Anoxygenic photosynthetic γ-proteobacteria.

putrefaction The microbial decomposition of organic matter, especially the anaerobic breakdown of proteins, with the production of foul-smelling compounds such as hydrogen sulfide and amines.

pyelonephritis An infection of one or both kidneys, often as a result of an untreated urinary tract infection.

pyogenic Pus-producing.

pyrenoid (pi′rĕ-noid) The differentiated region of the chloroplast that is a center of starch formation in some photosynthetic protists.

pyrimidine (pi-rim′i-dēn) A nitrogen-containing molecule with one ring that occurs in nucleic acids and other cell constituents; the most important pyrimidines are cytosine, thymine, and uracil.

pyruvate dehydrogenase complex (PDH) A multienzyme complex that catalyzes pyruvate oxidation and decarboxylation to acetyl-CoA, which then enters the tricarboxylic acid cycle.

pyruvate formate-lyase (PFL) An enzyme system that cleaves pyruvate into formic acid and acetyl-CoA. It is used by enteric bacteria that perform mixed acid fermentation.

Q

Q cycle The process associated with proton movement across the membrane at complex III of an electron transport chain.

quinine drugs Antimicrobial chemotherapeutic agents derived from the bark of the cinchona tree and used to treat malaria.

quorum sensing The process in which bacteria monitor their own population density or the presence of other species of bacteria by sensing the levels of signal molecules (e.g., *N*-acylhomoserine lactone) released by the microorganisms. When these signal molecules reach a threshold concentration, quorum-dependent genes are expressed.

R

racking The removal of sediments from wine bottles.

radappertization The use of gamma rays from a cobalt source for control of microorganisms in foods.

radioimmunoassay (RIA) (ra″de-o-im″u-no-as′a) A sensitive assay in which a purified radioisotope-labeled antigen or antibody competes for antibody or antigen with unlabeled standard antigen or test antigen in experimental samples to determine the concentration of a substance in the samples.

rapid plasmid reagin A rapid diagnostic test for syphilis.

rapid sand filters A step in drinking water purification that removes particulates by passage over sand.

reaction-center chlorophyll pair The two chlorophyll molecules in a photosystem that are energized by the absorption of light and release electrons to an associated electron transport chain, thus initiating energy conservation by photophosphorylation.

reactive nitrogen intermediate (RNI) Charged nitrogen radicals.

reactive oxygen intermediate (ROI) *See* reactive oxygen species (ROS).

reactive oxygen species (ROS) Superoxide radical (O_2^-), hydrogen peroxide (H_2O_2), singlet oxygen ($^1O_2$), and hydroxyl radical (OH^-) derivatives of oxygen.

reading frame The way in which nucleotides in DNA and mRNA are grouped into codons for reading the message contained in the nucleotide sequence.

readthrough A phenomenon in which a ribosome ignores a stop codon and continues protein synthesis. This generates a different protein than that produced if the ribosome had stopped at the stop codon.

real-time PCR A type of polymerase chain reaction (PCR) that quantitatively measures the amount of template in a sample as the amount of fluorescently labeled amplified product.

RecA A protein that functions in DNA repair and recombination processes. Homologues of RecA have been found in all three domains of life.

receptor-mediated endocytosis *See* endocytosis.

receptors Proteins that bind signaling molecules (ligands), thereby initiating cellular responses.

recognition sites Specific nucleotide sequences to which DNA-binding proteins (e.g., regulatory proteins, sigma factors, restriction enzymes) bind.

recombinant DNA technology The techniques used in carrying out genetic engineering.

recombinants Organisms produced following a recombination event.

recombinant-vector vaccine A vaccine produced by the introduction of one or more of a pathogen's genes into attenuated viruses or bacteria, which serves as a vector, replicating within the vertebrate host and expressing the gene(s) of the pathogen. The pathogen's antigens induce an immune response.

recombinases Enzymes that catalyze integration and excision of DNA segments during site-specific recombination.

recombination The process in which a new recombinant chromosome is formed by combining genetic material from two different nucleic acids. In some cases, the nucleic acids are from two different organisms.

recombinational repair A DNA repair process that repairs damaged DNA when there is no remaining template; a piece of DNA from a sister molecule is used.

red tides Population blooms of dinoflagellates that release red pigments and toxins, which can lead to paralytic shellfish poisoning. *See also* harmful algal blooms.

redox potential The tendency of a chemical compound or system to accept electrons; more formally called oxidation-reduction potential.

reducing power Molecules such as NADH and NADPH that temporarily store electrons. The stored electrons are used in anabolic reactions such as CO_2 fixation and the synthesis of monomers (e.g., amino acids).

reductive acetyl-CoA pathway A biochemical pathway used by methanogens to fix CO_2 and by acetogens to generate acetic acid. Also known as the Wood-Ljungdahl pathway.

reductive amination pathway A mechanism used by many microbes to incorporate ammonia.

reductive dehalogenation The cleavage of carbon-halogen bonds by anaerobic bacteria that creates a strong electron-donating environment.

reductive TCA cycle A pathway used by some chemolithotrophs and some anoxygenic phototrophs to fix CO_2. It involves the TCA cycle running in the reverse direction; also called reverse citric acid cycle.

refractive index A measure of how much a substance deflects a light ray from a straight path as it passes from one medium (e.g., glass) to another (e.g., air). It is calculated as the ratio of the velocity of light passing through the first medium to that of light passing through the second medium.

regulatory site *See* allosteric enzyme.

regulatory T cells (Tregs) T cells that control the development of effector T cells and B cells.

regulon A collection of genes or operons that is controlled by a common regulatory protein.

remote sensing The gathering of data and transformation of data into maps.

replica plating A technique for isolating mutants by transferring cells from an agar plate with nonselective medium to a piece of velvet and then onto plates with selective media or environmental conditions.

replicase An RNA-dependent RNA polymerase used to replicate the genome of an RNA virus.

replication The process in which an exact copy of parental DNA (or viral RNA) is made with the parental molecule serving as a template.

replication fork The Y-shaped structure where DNA is replicated.

replicative form (RF) A double-stranded nucleic acid that is formed from a single-stranded viral genome and used to synthesize mRNA and new copies of the genome.

replicative transposition A transposition event in which a replicate of the transposon remains in its original site.

replicon A unit of the genome that contains an origin of replication and in which DNA is replicated.

replisome A large protein complex that copies the DNA double helix to form two daughter chromosomes.

repressible gene A gene that encodes a protein whose level drops in the presence of a small molecule. If the gene product is a biosynthetic enzyme, the small molecule is often an end product of the metabolic pathway in which it functions.

repressor protein A protein that can bind to a repressor-binding site and inhibit transcription.

reservoir A site, alternate host, or carrier that normally harbors pathogenic organisms and serves as a source from which other individuals can be infected.

reservoir host An organism other than a human infected with a pathogen that can also infect humans.

residual body A lysosome after digestion of its contents has occurred. It contains undigested material.

resolution The ability of a microscope to distinguish between small objects that are close together.

respiration An energy-yielding process in which the energy substrate is oxidized using an exogenous or externally derived electron acceptor.

respiratory burst An increase in oxygen consumption by an activated phagocytic cell to generate reactive oxygen species within the phagolysosome.

response regulator *See* two-component signal transduction system.

restriction A process used by bacteria to defend the cell against viral infection; accomplished by enzymes that cleave the viral DNA; also called host restriction.

restriction enzymes Enzymes produced by bacterial cells that cleave DNA at specific nucleotide sequences. They evolved to protect bacteria from viral infection and are used in genetic engineering.

restriction fragment length polymorphism (RFLP) Fragments of DNA of different sizes following endonuclease cleavage. The differences can be used to determine genetic similarity among organisms.

reticulate body (RB) The cellular form in the chlamydial life cycle that grows and reproduces within the host cell.

reticulopodia Netlike pseudopodia found in certain amoeboid protists.

retroviruses A group of viruses with RNA genomes that carry the enzyme reverse transcriptase and form a DNA copy of their genome during their life cycle.

reverse electron flow An energy-consuming process used by some chemolithotrophs and phototrophs to generate reducing power.

reverse transcriptase (RT) A multifunctional enzyme used by retroviruses and reverse transcribing DNA viruses during their life cycles. It has RNA-dependent DNA polymerase, DNA-dependent DNA polymerase, and RNAase activity. It synthesizes double-stranded DNA from single-stranded RNA.

reverse vaccinology The use of genomics to identify candidate target molecules on a pathogen for vaccine development.

reversible chain termination sequencing A next-generation DNA sequencing technique that incorporates fluorescently labeled nucleotides that bear a small molecule on the 3′ OH to block addition of the next nucleotide triphosphate. Once the incorporation of the modified base is recorded, the blocking molecule is removed, leaving an unmodified 3′ OH so the next labeled, modified base can be incorporated and recorded; the process continues in an iterative fashion.

reversible covalent modification A mechanism of enzyme regulation in which the enzyme's activity is either increased or decreased by the reversible covalent addition of a group such as phosphate or AMP to the protein.

reversion mutation A mutation in a mutant organism that changes its phenotype back to the wild type; the second mutation is at the same site as the initial mutation.

rhizobia Any one of a number of α- and β-proteobacteria that form symbiotic nitrogen-fixing nodules on the roots of leguminous plants.

rhizomorph A macroscopic, densely packed thread consisting of individual cells formed by some fungi. A rhizomorph can remain dormant or serve as a means of fungal dissemination.

rhizoplane The surface of a plant root.

rhizosphere A region around the plant root where materials released from the root increase the microbial population and its activities.

ribonucleic acid (RNA) A polynucleotide composed of ribonucleotides joined by phosphodiester bonds.

ribosomal frameshifting A phenomenon in which a ribosome shifts the reading frame of the message, thus generating a different protein than it would have if it had stopped at the stop codon.

ribosomal RNA (rRNA) The RNA present in ribosomes; contributes to ribosome structure and also directly involved in the mechanism of protein synthesis.

ribosome The organelle where protein synthesis occurs; the message encoded in mRNA is translated here.

ribosome-binding site (RBS) Nucleotide sequences on mRNA recognized by a ribosome, so that it orients properly on the mRNA and begins translation at the appropriate codon.

riboswitch A site in the leader of an mRNA molecule that interacts with a metabolite or other small molecule, causing the leader to change its folding pattern. In some riboswitches, this change can alter transcription; in others, it affects translation.

ribotyping The use of conserved rRNA sequences to probe chromosomal DNA for classifying bacterial strains.

ribozyme An RNA molecule with catalytic activity.

ribulose 1,5-bisphosphate carboxylase/oxygenase (RubisCO) The enzyme that catalyzes incorporation of CO_2 in the Calvin-Benson cycle.

rifamycins Semisynthetic antibacterial drugs that block transcription by binding to bacterial RNA polymerase.

RNA-dependent DNA polymerase An enzyme that synthesizes DNA using an RNA template; also called reverse transcriptase.

RNA-dependent RNA polymerase An enzyme that synthesizes RNA using an RNA template.

RNA polymerase An enzyme that catalyzes the synthesis of RNA.

RNA polymerase core enzyme The portion of bacterial RNA polymerase that synthesizes RNA. It consists of all subunits except sigma factor.

RNA polymerase holoenzyme Bacterial RNA polymerase core enzyme plus sigma factor.

RNA-seq Direct sequencing of total cellular mRNA.

RNA silencing A response made by eukaryotic cells that protects them from infections caused by double-stranded RNA viruses.

RNA world The theory that posits that the first self-replicating molecule was RNA and this led to the evolution of the first primitive cell.

rolling-circle replication A mode of DNA replication in which the replication fork moves around a circular DNA molecule, displacing a strand to give a 5′ tail that is also copied to produce new double-stranded DNA.

root nodule Gall-like structures on roots that contain endosymbiotic nitrogen-fixing bacteria.

rough endoplasmic reticulum (RER) *See* endoplasmic reticulum.

R plasmids (R factors) Plasmids bearing one or more drug-resistance genes.

S

sac fungi Common name for the *Ascomycota*.

sanitization Reduction of the microbial population on an inanimate object to levels judged safe by public health standards.

saprophyte (sap′ro-fīt) An organism that takes up nonliving organic nutrients in dissolved form and usually grows on decomposing organic matter.

SAR11 The most abundant group of microbes on Earth, this lineage of α-proteobacteria has been found in almost all marine ecosystems studied.

satellites Subviral infectious agents composed of DNA or RNA encapsidated with the aid of an unrelated helper virus.

scale-up The process of increasing the volume of a culture from a small, experimental system to a large-volume system for the production of an industrially important compound.

scanning electron microscope (SEM) An electron microscope that passes a beam of electrons over the surface of a specimen; an image is formed from the electrons that are emitted by the specimen.

scanning probe microscope A microscope used to study surface features by moving a sharp probe over the object's surface (e.g., atomic force and scanning tunneling microscopes).

scanning tunneling microscope *See* scanning probe microscope.

sclerotia (sklĕ-ro′she-ə) Compact masses of hyphae produced by some filamentous fungi that permit overwintering. In the spring, they germinate to produce either additional hyphae or conidia.

Sec system A system found in all domains of life that transports a protein through or inserts it into a membrane.

secondary active transport *See* active transport.

secondary metabolites Products of metabolism that are synthesized after growth has been completed. Antibiotics are considered secondary metabolites.

secondary treatment The biological degradation of dissolved organic matter in the process of sewage treatment; the organic material is either mineralized or changed to settleable solids.

second law of thermodynamics Physical and chemical processes proceed in such a way that the entropy of the universe (the system and its surroundings) increases to the maximum possible.

second messengers Small molecules that are made in response to an extracellular signal (the first messenger). They "pass" the message to effector molecules that bring about the appropriate response to the signal.

secretory IgA (sIgA) The class of immunoglobulin associated with mucous membranes. *See also* IgA.

secretory pathway The process used by eukaryotic cells to synthesize proteins and lipids, followed by secretion or delivery to organelles or the plasma membrane.

secretion The process by which proteins are transported through the plasma membrane of a bacterial cell.

sedimentation basin A containment vessel used during surface water purification. Following the addition of material to facilitate the formation of coagulated impurities, water is left until such material has precipitated, forming sediment at the bottom of the basin.

segmented genome A viral genome that is divided into several parts or fragments, each usually coding for a single polypeptide.

selectable marker A gene conferring a wild-type or mutant phenotype that can be determined by growth on specific media.

select agent Biological entities (bacteria, viruses, toxins, etc.) designated by the United States as having the potential to pose great harm to the public; as such laboratory use is highly regulated.

selectins A family of cell adhesion molecules displayed on activated endothelial cells; mediate leukocyte binding to the vascular endothelium.

selective media Culture media that favor the growth of specific microorganisms; this may be accomplished by inhibiting the growth of undesired microorganisms.

selective toxicity The ability of a chemotherapeutic agent to kill or inhibit a microbial pathogen while damaging the host as little as possible.

self-assembly The spontaneous formation of a complex structure from its component molecules without the aid of special enzymes or factors.

sense codon A codon that specifies an amino acid.

sense strand The strand of DNA that is not used as the transcription template.

sensor kinase *See* two-component signal transduction system.

sensory rhodopsin Microbial rhodopsin that senses the spectral quality of light. *See also* archaerhodopsin.

sepsis Systemic response to infection manifested by two or more of the following conditions: temperature >38°C or <36°C; heart rate >90 beats per min; respiratory rate >20 breaths per min, or pCO_2 <32 mm Hg; leukocyte count >12,000 cells per ml^3 or >10% immature (band) forms. Sepsis also has been defined as the presence of pathogens or their toxins in blood and other tissues.

septate hyphae Fungal hyphae (filaments) having cross walls.

septation The process of forming a cross wall between two daughter cells during cell division.

septicemia (sep″tĭ-se′me-ah) A disease associated with the presence in the blood of pathogens or bacterial toxins.

septic shock Sepsis associated with severe hypotension despite adequate fluid resuscitation, along with the presence of perfusion abnormalities that may include but are not limited to lactic acidosis, oliguria, or an acute alteration in mental status.

septic tank A tank used to process domestic sewage. Solid material settles out and is partially degraded by anaerobic bacteria as sewage slowly flows through the tank. The outflow is further treated or dispersed in oxic soil.

septum (pl., **septa**) A partition that is used to separate cellular material. Septa are formed during cell division to divide the two daughter cells. They also separate cells in filamentous organisms and a developing endospore from its mother cell.

sequencing by synthesis Any DNA sequencing technology that determines the nucleotide sequence as labeled nucleotides are incorporated.

serology The branch of immunology that is concerned with in vitro reactions involving one or more serum constituents (e.g., antibodies and complement).

serotyping A technique or serological procedure used to differentiate between strains (serovars or serotypes) of microorganisms that have differences in the antigenic composition of a structure or product.

serovar A variant strain of a microbe that has distinctive antigenic properties.

settling basin A basin used during water purification to chemically precipitate fine particles, microorganisms, and organic material by coagulation or flocculation.

sex pilus (pĭ′lus) A thin protein appendage required for bacterial conjugation. The cell with sex pili donates DNA to recipient cells.

sheath A hollow, tubelike structure surrounding a cell; present in some genera of bacteria.

Shine-Dalgarno sequence A segment in the leader of bacterial mRNA and some archaeal mRNA that binds to a sequence on the 16S rRNA of the small ribosomal subunit. This helps orient the mRNA on the ribosome.

shuttle vector A DNA vector (e.g., plasmid, cosmid) that has two origins of replication, each recognized by a different microorganism. Thus the vector can replicate in both microbes.

siderophore (sid′er-o-for″) A small molecule that complexes with ferric iron and supplies it to a cell by aiding in its transport across the plasma membrane.

sigma factor A protein that helps bacterial RNA polymerase core enzyme recognize the promoter at the start of a gene. Thus it is a transcription factor.

sign An objective change in a diseased body that can be directly observed (e.g., a fever or rash).

signal 1 Primary lymphocyte activation that occurs when antigen is presented to the lymphocyte by its appropriate MHC molecule on the presenting cell.

signal 2 Confirmatory event causing lymphocytes stimulated by signal 1 to complete activation and thus become effector cells.

signal peptidase The membrane-bound enzyme that cleaves the signal peptide from a polypeptide as it is transported through the plasma membrane.

signal peptide The amino terminal sequence on a protein destined for transport through or into a membrane. It delays protein folding and is recognized by the Sec system.

signal recognition particle A complex of a small RNA and a protein that directs a polypeptide and its translating ribosome to the Sec system during cotranslational translocation.

silage Fermented plant material with increased palatability and nutritional value for animals, which can be stored for extended periods.

silencers Sites on eukaryotic chromosomes to which repressor proteins bind.

silent mutation A mutation that does not change an organism's proteins or phenotype even though the DNA base sequence has been changed.

simple (cut-and-paste) transposition A transposition event in which the transposable element is cut out of one site and ligated into a new site.

single-cell genomic sequencing A method whereby the genome of a single microbial cell is copied many times by the multiple displacement amplification (MDA) technique. The genome is then sequenced by a next-generation technique.

single nucleotide polymorphism (SNP) Difference of one base in genes or genetic elements that are otherwise highly conserved among members of a given taxon.

single radial immunodiffusion (RID) assay An immunodiffusion technique that quantitates antigens by following their diffusion through a gel containing antibodies directed against the test antigens.

site-directed mutagenesis The in vitro process by which a specific nucleotide change is made to a particular gene.

site-specific recombination Recombination between two DNA molecules that does not involve extensive areas of homology between the two molecules. It is catalyzed by enzymes called recombinases distinct from RecA, and it is the mechanism by which transposition occurs.

skin-associated lymphoid tissue (SALT) The lymphoid tissue in the skin that forms a first-line defense as a part of innate immunity.

S-layer A regularly structured layer composed of protein or glycoprotein that lies on the surface of many bacteria and archaea.

slime The viscous extracellular glycoproteins or glycolipids produced by some bacteria. *See also* slime layer.

slime layer A layer of diffuse, unorganized, easily removed material lying outside an archaeal or bacterial cell wall.

slime mold A common term for members of the protist taxa *Myxogastria* and *Dictyostelia*.

slow sand filter A bed of sand through which water slowly flows; the gelatinous microbial layer on the sand grain surface removes waterborne microorganisms, particularly *Giardia*.

sludge A general term for the precipitated solid matter produced during water and sewage treatment; solid particles composed of organic matter and microorganisms that are involved in aerobic sewage treatment (activated sludge).

small interfering RNAs (siRNAs) Small RNA molecules that affect either transcription of target genes or translation of target mRNAs.

small RNAs (sRNAs) Small regulatory RNA molecules that do not function as messenger, ribosomal, or transfer RNAs.

small subunit rRNA (SSU rRNA) The rRNA associated with the small ribosomal subunit: 16S rRNA in bacterial and archaeal cells and 18S rRNA in eukaryotes.

smooth endoplasmic reticulum *See* endoplasmic reticulum.

snapping division A distinctive type of binary fission resulting in an angular or a palisade arrangement of cells; characteristic of members of the genera *Arthrobacter* and *Corynebacterium*.

somatic hypermutation A high rate of mutation within antibody-encoding genes that occurs in B cells following T-cell activation. This mechanism generates antibodies that are then screened for high avidity and affinity.

soil organic matter (SOM) The organic material within a soil; organic material is important because it helps to retain nutrients, maintain soil structure, and hold water for plant use.

sortase An enzyme that catalyzes the attachment of a secreted protein to the surface peptidoglycan of a Gram-positive bacterium.

sorting signal A sequence at the C-terminus of some Gram-positive secreted proteins that directs the protein to sortase enzymes for anchoring to the peptidoglycan.

SOS response A complex, inducible process that allows bacterial cells with extensive DNA damage to survive, although often in a mutated form; it involves cessation of cell division, upregulation of several DNA repair systems, and induction of translesion DNA synthesis.

sour mash Liquid mixture of fermented grain that is reserved from one batch of distilled spirits to start the fermentation of a new batch; used in the production of whiskey and bourbon.

specialized transduction A transduction process in which only a specific set of bacterial or archaeal genes is carried to a recipient cell by a temperate virus; the cell's genes are acquired because of a mistake in the excision of a provirus during the lysogenic life cycle.

species Species of higher organisms are groups of interbreeding or potentially interbreeding natural populations that are reproductively isolated. Archaeal and bacterial species are often defined as collections of strains that have many stable properties in common and differ significantly from other strains. However, there is currently considerable debate about the best definition of archaeal and bacterial species.

spheroplast (sfēr′o-plast) A spherical cell formed by the weakening or partial removal of the rigid cell wall component.

spike *See* peplomer.

spirillum (spi-ril′um) A rigid, spiral-shaped bacterium.

spirochete (spi′ro-kēt) A flexible, spiral-shaped bacterium with periplasmic flagella.

spleen A secondary lymphoid organ where old erythrocytes are destroyed and blood-borne antigens are trapped and presented to lymphocytes.

spliceosome A complex of proteins and small RNAs that carries out RNA splicing in eukaryotes.

spontaneous generation An early belief that living organisms could develop from nonliving matter.

spontaneous mutations Mutations that occur due to errors in replication or the occurrence of lesions in the DNA (e.g., apurinic sites).

the uptake of free DNA by a cell. (3) The conversion of a cell into a malignant, cancerous cell.

transition mutations Mutations that involve the substitution of a different purine base for the purine present at the site of the mutation or the substitution of a different pyrimidine for the normal pyrimidine.

translation Protein synthesis; the process by which the genetic message carried by mRNA directs the synthesis of polypeptides with the aid of ribosomes and other cell constituents.

translesion DNA synthesis A type of DNA synthesis that occurs during the SOS response. DNA synthesis occurs without an intact template and generates mutations as a result.

translocation (1) During the elongation cycle of protein synthesis, the movement of the ribosome relative to the mRNA that repositions the A site of the ribosome over the next codon to be translated; (2) the movement of a protein across or into a membrane.

transmissible spongiform encephalopathies Degenerative central nervous system diseases in which the brain has a spongy appearance; due to prions.

transmission electron microscope (TEM) A microscope in which an image is formed by passing an electron beam through a specimen and focusing the scattered electrons with magnetic lenses.

transpeptidation (1) The reaction that forms the peptide cross-links during peptidoglycan synthesis. (2) The reaction that forms a peptide bond during the elongation cycle of protein synthesis.

transposable element A small DNA molecule that carries the genes for transposition and thus can move around the genome.

transposase An enzyme that catalyzes the movement of a mobile genetic element.

transposition The movement of a piece of DNA around the chromosome.

transposon (tranz-po′zon) A mobile genetic element that encodes recombinase, which is needed for transposition, and contains other genes that are not required for transposition.

transversion mutations Mutations that result from the substitution of a purine base for the normal pyrimidine or vice versa.

tricarboxylic acid (TCA) cycle The cycle that oxidizes acetyl coenzyme A to CO_2 and generates NADH and $FADH_2$ for oxidation in an electron transport chain; also supplies precursor metabolites for biosynthesis. Also called citric acid cycle and Krebs cycle.

trichocyst (trĭk′o-sist) A small organelle beneath the pellicle of some protists. It can be extruded to function as either an anchoring device or as a form of defense.

trichome (tri′kōm) A filament of microbial cells that are in close contact with one another over a large area.

trickling filter A bed of rocks covered with a microbial film that aerobically degrades organic waste during secondary sewage treatment.

trigger factor A protein associated with the ribosome exit tunnel that facilitates proper folding of a nascent polypeptide.

trimethoprim A synthetic antibiotic that inhibits production of folic acid by binding to dihydrofolate reductase. Trimethoprim has a wide spectrum of activity and is bacteriostatic.

trophozoite (trof″o-zo′īt) The active, motile feeding stage of a protozoan organism.

tropism (1) The movement of living organisms toward or away from a focus of heat, light, or other stimulus. (2) The selective infection of certain organisms or host tissues by a virus due to the presence of the specific receptor to which the virus binds.

trypanosome (tri-pan′o-sōm) A flagellated protozoan of the genus *Trypanosoma;* often live in the blood of humans and other vertebrates and are transmitted by insect bites.

tubercle (too′ber-k′l) A small, rounded nodular lesion produced by *Mycobacterium tuberculosis.*

tuberculous cavity An air-filled cavity that results from a tubercle lesion caused by *M. tuberculosis.*

tumble Random turning or tumbling movements made by flagellated bacteria when they stop swimming in a straight line (a run).

tumor suppressor genes Genes encoding tumor suppressor proteins.

tumor suppressor proteins Proteins that regulate cell cycling or repair DNA. Their inactivation can contribute to the development of cancer.

turbidostat A continuous culture system equipped with a photocell that adjusts the flow of medium through the culture vessel to maintain a constant cell density or turbidity.

twitching motility A type of bacterial motility characterized by short, intermittent jerky motions due to extension and retraction of type IV pili.

two-component signal transduction system A regulatory system that uses the transfer of phosphoryl groups to control gene transcription and protein activity. It has two major components: a sensor kinase and a response regulator protein.

two-dimensional gel electrophoresis An electrophoretic procedure that separates proteins using two procedures: isoelectric focusing and SDS-PAGE.

tyndallization The process of repeated heating and incubation to destroy bacterial spores.

type I hypersensitivity A form of immediate hypersensitivity arising from the binding of antigen to IgE attached to mast cells, which then release anaphylaxis mediators such as histamine. Examples: hay fever, asthma, and food allergies.

type II hypersensitivity A form of immediate hypersensitivity involving the binding of antibodies to antigens on cell surfaces followed by destruction of the target cells (e.g., through complement attack, phagocytosis, or agglutination).

type III hypersensitivity A form of immediate hypersensitivity resulting from the exposure to excessive amounts of antigens to which antibodies bind. These antibody-antigen complexes activate complement and trigger an acute inflammatory response with subsequent tissue damage.

type IV hypersensitivity A delayed hypersensitivity response that results from the binding of antigen to activated T lymphocytes, which then release cytokines to trigger inflammation that damages tissue. Type IV hypersensitivity is seen in contact dermatitis from poison ivy, leprosy, and tertiary syphilis.

type I secretion system *See* ABC protein secretion pathway type II secretion system. A structure found in proteobacteria to transport proteins from the periplasm through the outer membrane.

type III secretion system A system in Gram-negative bacteria that secretes virulence factors and injects them into host cells.

type IV pilus Fine, hairlike appendage of some bacterial cells that can mediate twitching motility or DNA uptake.

type IV secretion system A system in Gram-negative bacteria that secretes proteins; some Gram-negative and Gram-positive bacteria use this secretion system to transfer DNA during bacterial conjugation.

type V secretion system A pathway in Gram-negative bacteria that translocates proteins across the outer membrane after they have been translocated across the plasma membrane by the Sec system.

type VI secretion system A protein secretion system in Gram-negative bacteria that is similar in structure to the machinery used by some bacteriophages to insert their DNA into host cells.

type VII secretion system A protein secretion system in Gram-positive bacteria that translocates proteins across the plasma membrane.

type IX secretion system A two-step protein secretion system in the *Bacteroidetes* that guides proteins to the outer surface of the cell.

type strain The microbial strain that is the nomenclatural type or holder of the species name. A type strain will remain within that species should nomenclature changes occur.

U

ubiquinone *See* electron transport chain.

ultraviolet (UV) radiation High-energy radiation of short wavelength, about 10 to 400 nm.

uniporters Carrier proteins that move a single solute across a membrane.

universal phylogenetic tree A phylogenetic tree that considers the evolutionary relationships among organisms from all three domains of life: *Bacteria, Archaea,* and *Eukarya.*

upstream activating sequences (UASs) Sequences near eukaryotic genes where activator proteins bind.

uracil (u′rah-sil) The pyrimidine 2,4-dioxypyrimidine, which is found in nucleosides, nucleotides, and RNA.

urinary tract infection Infection of the bladder most commonly caused by uropathogenic *E. coli*.

use dilution test Method of evaluating disinfectant effectiveness at a recommended dilution.

V

vaccine A preparation of antigens and adjuvant administered to induce development of an adaptive immune response that will protect the individual against a specific pathogen or a toxin.

vaccinomics The application of genomics to vaccine development.

valence The number of antigenic determinant sites on the surface of an antigen or the number of antigen-binding sites possessed by an antibody molecule.

vancomycin A glycopeptide antibiotic obtained from *Amycolatopsis orientalis* that is effective only against Gram-positive bacteria and inhibits peptidoglycan synthesis.

variable region (V_L and V_H) The region at the N-terminal end of immunoglobulin heavy and light chains whose amino acid sequence varies between antibodies of different specificity. Variable regions form the antigen-binding site.

vector (1) In genetic engineering, a DNA molecule that can replicate and transport a piece of inserted foreign DNA, such as a gene, into a recipient cell. It may be a plasmid, phage, cosmid or artificial chromosome. (2) In epidemiology, it is a living organism, usually an arthropod or other animal, that transfers an infective agent between hosts.

vector-borne transmission The transmission of an infectious pathogen between hosts by means of a vector.

vehicle An inanimate substance or medium that transmits a pathogen.

vertical gene transfer The transfer of genes from a parent organism to its offspring.

vertical transmission Pathogen transfer from mother to unborn child (in utero).

vesicular transport The movement of materials in small membrane-bound vesicles between the endoplasmic reticulum, Golgi apparatus, lysosomes, plasma membrane, and other organelles of the endocytic and secretory pathways in eukaryotic cells.

viable but nonculturable (VBNC) microorganisms Microbes known to be growing in a specific environment but that cannot be cultured under standard laboratory conditions.

viable counting methods Methods for determining population size that depend on a microbe's ability to reproduce when cultured. Also called standard plate counts.

vibrio (vib′re-o) A rod-shaped bacterial cell that is curved to form a commalike shape.

viral hemagglutination The clumping of red blood cells caused by the attachment of some viruses.

viral neutralization An antibody-mediated process in which IgG, IgM, and IgA antibodies bind to some viruses during their extracellular phase and inactivate or neutralize them.

viricide (vir′i-sīd) An agent that inactivates viruses so that they cannot reproduce within host cells.

virion A complete virus particle; at the simplest, it consists of a protein capsid surrounding a single nucleic acid molecule.

virioplankton Viruses that occur in waters; high levels are found in marine and freshwater environments.

viroid An infectious agent that is a single-stranded RNA not associated with any protein; the RNA does not code for any proteins and is not translated.

virology The branch of microbiology that is concerned with viruses and viral diseases.

virulence The degree or intensity of pathogenicity of an organism as indicated by case fatality rates and/or ability to invade host tissues and cause disease.

virulence factor A product, usually a protein or carbohydrate, that contributes to virulence or pathogenicity.

virulent (bacteriophages) viruses Viruses that lyse their host cells at the end of the viral life cycle.

virus An infectious agent having a simple acellular organization with a protein coat and a nucleic acid genome, lacking independent metabolism, and multiplying only within living host cells.

viruslike particles (VLP) Presumptive viruses detected by electron microscopy or fluorescence microscopy; because these approaches do not demonstrate that these entities can infect host cells, they cannot be considered viruses.

W

wastewater treatment The use of physical and biological processes to remove particulate and dissolved material from sewage and to control pathogens.

water activity (a_w) A quantitative measure of water availability in the habitat; the water activity of a solution is one-hundredth its relative humidity.

water molds *See* öomycetes.

white blood cell (WBC) Blood cell having innate or acquired immune function. These cells are named for the white or buffy layer in which they are found when blood is centrifuged.

whole-cell vaccine A vaccine made from complete pathogens, which can be either killed microorganisms or live, attenuated microbes.

whole-genome shotgun sequencing Genome sequencing in which random fragments of a complete genome are individually sequenced. The nucelotide sequences of the fragments are placed in the proper order based on overlapping identical sequences.

wild type Prevalent form of a gene or phenotype.

Winogradsky column A glass column with an anoxic lower zone and an oxic upper zone, which allows growth of microorganisms under conditions similar to those found in a nutrient rich lake.

wobble The loose base pairing between an anticodon and a codon at the third position of the codon.

World Health Organization An international health agency located in Geneva, Switzerland.

wort The filtrate of malted grains used as the substrate for the production of beer and ale by fermentation.

X

xenograft (zen″o-graft) A tissue graft between animals of different species.

xerotolerant Microorganisms that grow best under low water activity (a_w) conditions, and may not be able to grow at high a_w values.

Y

yeast A unicellular, uninuclear fungus that reproduces either asexually by budding or fission, or sexually through spore formation.

yeast artificial chromosome (YAC) Engineered DNA that contains all the elements required to propagate a chromosome in yeast and is used to clone foreign DNA fragments in yeast cells.

Y-M shift The change from yeast (Y) morphology in the host to mold (M) morphology in the environment.

Z

zoonosis (zo″o-no′sis; pl., **zoonoses**) A disease that can be transmitted from animals to humans.

zooxanthella (zo″o-zan-thel′ah) A dinoflagellate found living symbiotically within cnidarians (corals) and other invertebrates.

zygomycetes (zi″go-mi-se′tez) A group of fungi that usually has aspetate hyphae. Sexual reproduction normally involves the formation of zygospores.

zygospore A thick-walled, sexual, resting spore characteristic of the zygomycetous fungi.

Index

Page numbers followed by t and f designate tables and figures, respectively; page numbers in *italics* refer to definitions; page numbers in **boldface** refer to major discussions; page numbers with A refer to appendix.

A

B

C

D

E

G

H

J

K

L

N

O

P

Q

R

T

U

V

Creating America

A History of the United States

McDOUGAL LITTELL

Zitkala-Ša
Native American author

Benjamin Franklin
Author, printer, scientist, diplomat

Abigail Adams
Supporter of women's rights; married to President John Adams

Maya Lin
Designer of the Vietnam Veterans Memorial

Juan Seguín
A hero of the Texas Revolution

Amelia Earhart
Aviator; first woman to fly solo across the Atlantic Ocean

Ida B. Wells
Journalist and reformer who fought against discrimination

Abraham Lincoln
Sixteenth president of the United States

Dr. Martin Luther King, Jr.
Civil rights leader

Harry S. Truman
Thirty-third president of the United States

Creating America

A History of the United States

Jesus Garcia

Donna M. Ogle

C. Frederick Risinger

Joyce Stevos

Winthrop D. Jordan

McDougal Littell

Evanston, Illinois • Boston • Dallas

Senior Consultants

Jesus Garcia is Professor of Curriculum and Instruction at the University of Kentucky. A former elementary and secondary social studies teacher, Dr. Garcia has co-authored many books and articles on subjects that range from teaching social studies in elementary school to seeking diversity in education. Dr. Garcia has also worked for both the Chicago and Washington, D.C., public schools as a consultant on social studies standards.

Donna M. Ogle is Professor of Reading and Language Arts at National-Louis University in Evanston, Illinois, and is a specialist in reading in the content areas with an interest in social studies. She was recently elected president of the International Reading Association. A former social studies teacher, Dr. Ogle is also Director of a Goals 2000 grant for four Chicago high schools. She developed the K-W-L strategy that is so widely used in schools.

C. Frederick Risinger is Associate Director of the Social Studies Development Center at Indiana University. He is a past president of the National Council for the Social Studies. Dr. Risinger also served on the coordinating committee for the National History Standards Project. Dr. Risinger writes a monthly column on technology in the social studies classroom for *Social Education.*

Joyce Stevos is the Director of Strategic Planning and Professional Development for the Providence, Rhode Island, Public Schools. She is also an adjunct faculty member at Rhode Island College. Previously, she served 15 years as the social studies area supervisor in Providence, in which position she developed programs on Holocaust studies, the Armenian genocide, character education, voter education, and government and law.

Winthrop D. Jordan is the William F. Winter Professor of History and Professor of Afro-American Studies at the University of Mississippi, where he has taught since 1982. Since 1998, he has been the Frederick A. P. Barnard Distinguished Professor at that university. Dr. Jordan holds a Ph.D. from Brown University. He is the author of several books, including the highly acclaimed *White Over Black.* The book won four national awards, one of which was the National Book Award.

Acknowledgments begin on page R97.

ISBN 0-395-92899-0

Printed in the United States of America
1 2 3 4 5 6 7 8 9 – DWO – 06 05 04 03 02 01 00

Consultants and Reviewers

Content Consultants

The content consultants reviewed the manuscript for historical depth and accuracy and for clarity of presentation.

Roger Beck
Department of History
Eastern Illinois University
Charleston, Illinois

David Farber
Department of History
University of New Mexico
Albuquerque, New Mexico

Cheryl Johnson-Odim
Department of History
Loyola University
Chicago, Illinois

Joseph Kett
Department of History
University of Virginia
Charlottesville, Virginia

Jack N. Rakove
Department of History
Stanford University
Stanford, California

Virginia Stewart
Department of History
University of North Carolina, Wilmington
Wilmington, North Carolina

Christopher Waldrep
Department of History
Eastern Illinois University
Charleston, Illinois

Nancy Woloch
Department of History
Barnard College
New York, New York

Multicultural Advisory Board

The multicultural advisors reviewed the manuscript for appropriate historical content.

Betty Dean
Social Studies Consultant
Pearland, Texas

Tyrone C. Howard
College of Education
The Ohio State University
Columbus, Ohio

Jose C. Moya
Department of History
University of California at Los Angeles
Los Angeles, California

Pat Payne
Office of Multicultural Education
Indianapolis Public Schools
Indianapolis, Indiana

Betto Ramirez
Former Teacher, La Joya, Texas
Social Studies Consultant
Mission, Texas

Jon Reyhner
Department of Education
Northern Arizona University
Flagstaff, Arizona

Ronald Young
Department of History
Georgia Southern University
Statesboro, Georgia

Consultants and Reviewers

Teacher Consultants

The following educators contributed activity options for the Pupil's Edition and teaching ideas and activities for the Teacher's Edition.

Paul C. Beavers
J. T. Moore Middle School
Nashville, Tennessee

Holly West Brewer
Buena Vista Paideia Magnet School
Nashville, Tennessee

Ron Campana
Social Studies Consultant
New York, New York

Patricia B. Carlson
Swanson Middle School
Arlington, Virginia

Ann Cotton
Ft. Worth Independent School District
Ft. Worth, Texas

Kelly Ellis
Hamilton Junior High School
Cypress, Texas

James Grimes
Middlesex County Vocational–Technical High School
Woodbridge, New Jersey

Brent Heath
De Anza Middle School
Ontario, California

Suzanne Hidalgo
Serrano Middle School
Highland, California

Barbara Kennedy
Sylvan Middle School
Citrus Heights, California

Pamela Kniffin
Navasota Junior High School
Navasoto, Texas

Tammy Leiber
Navasota Junior High School
Navasoto, Texas

Lori Lesslie
Cedar Bluff Middle School
Knoxville, Tennessee

Brian McKenzie
Dr. Charles R. Drew Science Magnet School
Buffalo, New York

W. W. Bear Mills
Goddard Junior High School
Midland, Texas

Lindy Poling
Millbrook High School
Raleigh, North Carolina

Jean Price
T. H. Rogers Middle School
Houston, Texas

Meg Robbins
Wilbraham Middle School
Wilbraham, Massachusetts

Philip Rodriguez
McNair Middle School
San Antonio, Texas

Leslie Schubert
Parkland School
McHenry, Illinois

Robert Sisko
Carteret Middle School
Carteret, New Jersey

Marci Smith
Hurst-Euless-Bedford Independent School District
Bedford, Texas

James Sorenson
Chippewa Middle School
Des Plaines, Illinois

Nicholas G. Sysock
Carteret Middle School
Carteret, New Jersey

Lisa Williams
Lamberton Middle School
Carlisle, Pennsylvania

Michael Yell
Hudson Middle School
Hudson, Wisconsin

Teacher Panels

The following educators provided ongoing review during the development of prototypes, the table of contents, and key components of the program.

Bill Albright
Wilson Southern Junior High School
Sinking Spring, Pennsylvania

Henry Assetto
Gordon Middle School
Coatesville, Pennsylvania

James Berry
Kennedy Middle School
Grand Prairie, Texas

Ralph Burnley
Roosevelt Middle School
Philadelphia, Pennsylvania

Mary Ann Canamar
Garner Middle School
San Antonio, Texas

Zoe Carter
San Jacinto Junior High School
Midland, Texas

Stephen Cicero
Butler Area Junior High School
Butler, Pennsylvania

Charles Crescenzi
Dover Intermediate School
Dover, Pennsylvania

Sharon McDonald
Cook Junior High School
Houston, Texas

Phil Mifsud
Roosevelt Middle School
Erie, Pennsylvania

Joel Mumma
Centerville Middle School
Lancaster, Pennsylvania

September Olson
Richardson Middle School
El Paso, Texas

Donald Roberts
Frick International Studies Academy
Pittsburgh, Pennsylvania

Mary Rogers
Brookside Intermediate School
Friendswood, Texas

Lucy Sanchez
John F. Kennedy High School
San Antonio, Texas

Steve Seale
Hamilton Middle School
Houston, Texas

Yolanda Villalobos
Dallas Independent School District
Dallas, Texas

Diane Williams
Hill-Freedman Middle School
Philadelphia, Pennsylvania

Lisa Williams
Lamberton Middle School
Carlisle, Pennsylvania

Student Board

The following students reviewed pages for the textbook.

Adam Backhaus
Parkland School
McHenry, Illinois

Ben Barney
Hamilton Junior High School
Cypress, Texas

Amanda Berrier
Lamberton Middle School
Carlisle, Pennsylvania

Debra Hurwitz
T. H. Rogers Middle School
Houston, Texas

Reginald Jones
Dr. Charles R. Drew Science Magnet School
Buffalo, New York

Brendon Keinath
Wilbraham Middle School
Wilbraham, Massachusetts

Daniel MacDonald
Carteret Middle School
Carteret, New Jersey

Cameron Mote
Serrano Middle School
Highland, California

Kim Nguyen
Sylvan Middle School
Citrus Heights, California

Arianna G. Noriega
De Anza Middle School
Ontario, California

Nicholas Tofilon
Burlington Middle School
Burlington, Wisconsin

Emmanuel Zepeda
Haven Middle School
Evanston, Illinois

UNIT 1 Three Worlds Meet BEGINNINGS TO 1763

Pontiac

UNIT 2

Creating a New Nation 1763 – 1791

George Washington

The Early Republic 1789 – 1844

UNI 3

UNIT 4 A Changing Nation 1810 – 1860

California gold miner

Frances Ellen Watkins Harper

The Nation Divided and Rebuilt 1846 – 1877

Abraham Lincoln

UNIT 6 America Transformed 1860 – 1914

Immigrants approaching Ellis Island

Modern America Emerges 1880 – 1920

UNIT 7

Queen Liliuokalani

UNIT 8 Depression, War, and Recovery 1919 – 1960

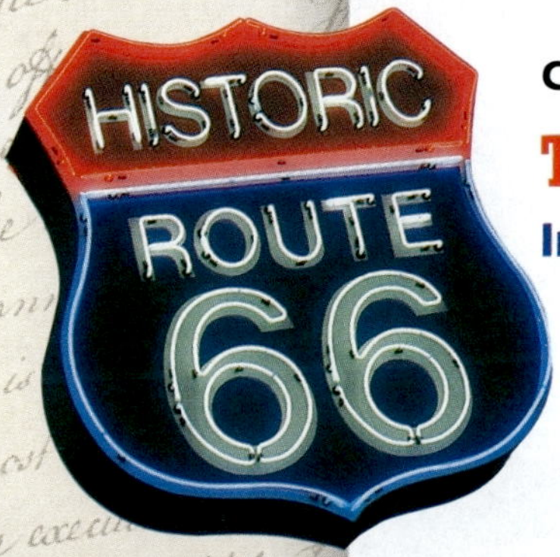

Tensions at Home and Abroad 1954 – PRESENT

UNIT 9

President John F. Kennedy

Buzz Aldrin

Features

Interdisciplinary CHALLENGE

GEOGRAPHY in HISTORY

Technology OF THE Time

INTERACTIVE PRIMARY SOURCES

Literature Connections

HISTORY WORKSHOP

CITIZENSHIP TODAY

Economics in History

America's HERITAGE

AMERICA'S HISTORY MAKERS

daily life

Features

HISTORY *through* ART

Now *and* then

Connections TO

STRANGE *but* True

Voices from the Past

A VOICE FROM THE PAST

These, with the pictures, busts [sculptures of the head and shoulders], and prints (of which copies upon copies are spread everywhere), have made your father's face as well known as that of the moon.

Benjamin Franklin,
letter to his daughter Sally

Voices from the Past

A VOICE FROM THE PAST

Eliza made her desperate retreat across the river just in the dusk of twilight. The gray mist of evening, rising slowly from the river, enveloped her as she disappeared up the bank, and the swollen current and floundering masses of ice presented a hopeless barrier between her and her pursuer.

Harriet Beecher Stowe,
Uncle Tom's Cabin

A VOICE FROM THE PAST

Is it possible . . . that nine millions of men can make effective progress in economic lines if they are deprived of political rights? . . . If history and reason give any distinct answer to these questions, it is an emphatic *No.*

W. E. B. Du Bois,
The Souls of Black Folk

Voices from the Past

Visual Primary Sources for Assessment

Historical Maps

Charts and Graphs

Charts

Causes of the War of 1812

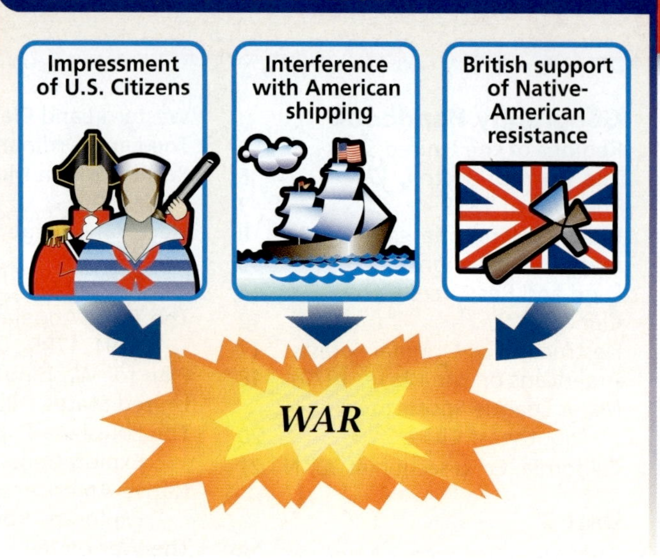

Graphs

CONNECTIONS TO MATH

Military Deaths in the American Revolution

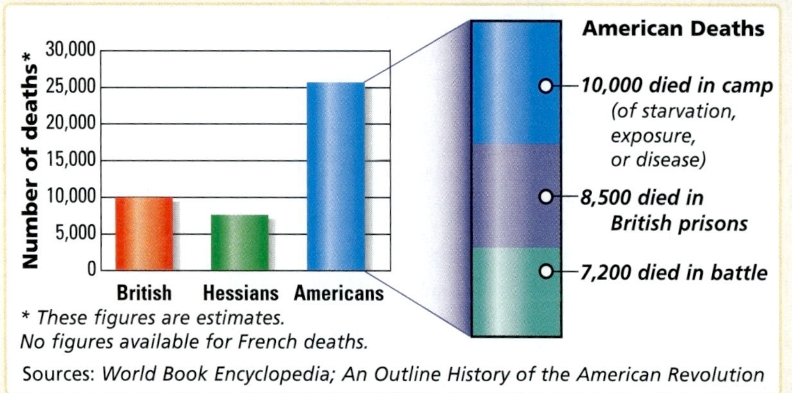

* These figures are estimates.
No figures available for French deaths.

Sources: *World Book Encyclopedia; An Outline History of the American Revolution*

Time Lines and Infographics

Time Lines

Infographics

The Rise and Decline of Feudalism

THEMES of AMERICAN HISTORY

Imagine life in Jamestown, America's first permanent English settlement. The nation we inhabit now is a much different place than it was then, more than three centuries ago. Yet there are repeating themes—ideas and issues—in American history that tie the past and present together. This book focuses on nine significant themes in U.S. history. Understanding these themes will help you to make sense of American history.

Democratic Ideals

From the day they declared themselves citizens of a new nation, Americans have built their society around the principles of democracy. In a democracy, power lies with the people, and every individual enjoys basic rights that cannot be taken away. Throughout the nation's history, however, some Americans—mainly women and minorities—have had to struggle to gain their full rights. Still, the ideals of democracy remain the guiding principles of this land.

What right or freedom do you consider the most important? Why?

Citizenship

The citizens of the United States enjoy rights and freedoms found in very few other places in the world. Yet Americans know that with such freedoms come responsibilities and duties. Whether they stand in line to vote or spend a weekend to clean up a local river, Americans recognize that citizen participation is what keeps a democracy strong.

How do citizens that you know contribute to your community?

Dr. Martin Luther King, Jr.

Impact of the Individual

The history of the United States is the story not only of governments and laws but of individuals. Indeed, individuals have made the United States what it is today through their extraordinary achievements. American history provides a variety of examples of the impact of the individual on society in both the United States and the world.

Name several individuals who have an impact on American society today. What impact do they have?

Diversity and Unity

The United States has been a land of many peoples, cultures, and faiths. Throughout the nation's history, this blend of ethnic, racial, and religious groups has helped to create a rich and uniquely American culture. The nation's many different peoples are united in their belief in American values and ideals.

What things do you enjoy that came to the United States from other cultures?

Immigration and Migration

The movement of people has played a vital role in American history. This country was settled by and has remained a magnet for immigrants. Even within the United States, large numbers of people have migrated to different regions of the country. However, movements to and within the United States have not always been voluntary. Africans were brought against their will to this country. Native Americans were forced from their homelands in order to make room for European settlers.

Why do you think people continue to immigrate to the United States?

A young Asian immigrant

Expansion

When the United States declared its independence from Great Britain, it was only a collection of states along the Atlantic Ocean. But the new country would not remain that way for long. Many Americans shared a sense of curiosity, adventure, and a strong belief that their destiny was to expand all the way to the Pacific Ocean. Driven by this belief, they pushed westward. Americans' efforts to increase the size of their nation is a recurring theme in early U.S. history.

Where do you predict that the exploration of space—the final frontier—will lead?

Poster for Buffalo Bill's Wild West show

America and the World

As the power and prestige of the United States have grown, the nation has played a much more active role in world affairs. Indeed, throughout the 20th century, the United States focused much of its energy on events beyond its borders. The nation fought in two world wars and tried to promote democracy, peace, and economic growth around the globe. As one of the world's political and economic leaders, the United States will continue to be a key player in world affairs throughout the new century.

What do you think the role of the United States in the world should be today?

Science and Technology

Americans have always been quick to embrace inventions and new ways of doing things. After all, this country was settled by people who turned away from old ways and tried new ones. In the past two centuries, new inventions, new technologies, and scientific breakthroughs have transformed the United States—and will continue to do so in the new century.

What recent inventions or innovations affect your life?

Economics in History

Economics has had a powerful impact on the course of U.S. history. For example, the desire for wealth led thousands to join the California Gold Rush in 1849. The nation as a whole has grown wealthy, thanks to its abundant resources and the hard work of its citizens. An important economic issue, however, has been how to make sure that all people have opportunities to share fully in the nation's wealth. This issue will continue to be important in the 21st century.

What do you think are the most exciting economic opportunities for Americans today?

Thomas Edison's first light bulb

READING to REMEMBER

Reading is a little like playing a sport; you'll improve if you practice and if you learn the right techniques. This book is designed to help you read better and to remember more of what you read. That, in turn, will help you to learn history.

Think about what it's like learning to play a sport, such as basketball. There are certain steps you can follow. You can use those same steps in reading to remember.

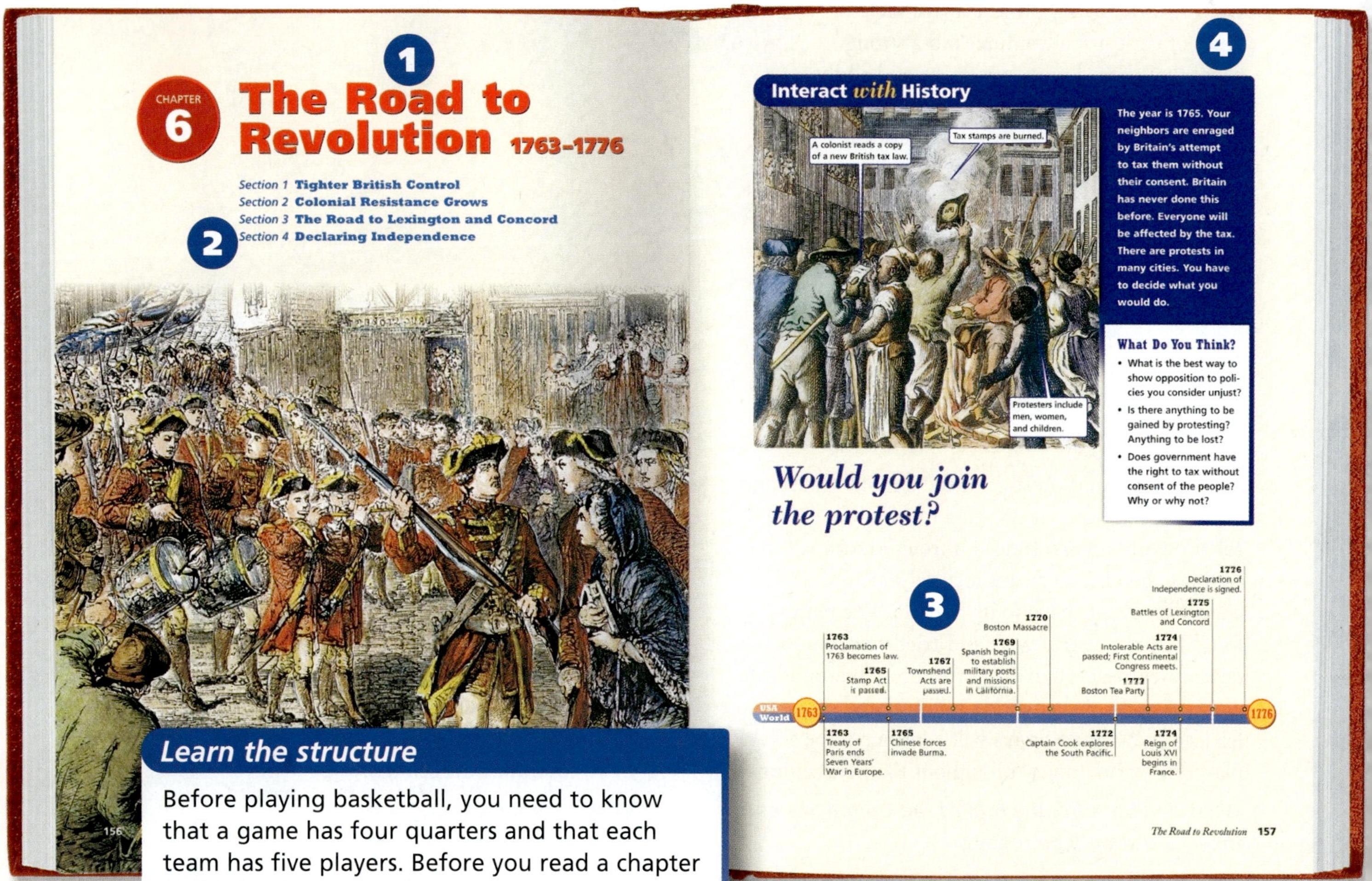

Learn the structure

Before playing basketball, you need to know that a game has four quarters and that each team has five players. Before you read a chapter in this book, preview its structure.

1. See what clues are in the chapter's title.
2. Notice how many sections there are and the title of each one.
3. Look at the time line on the chapter opener. What years does the chapter cover? What happened during those years?

Scan the chapter and look at the pictures, maps, and graphs. They will give you a hint of what the chapter is about.

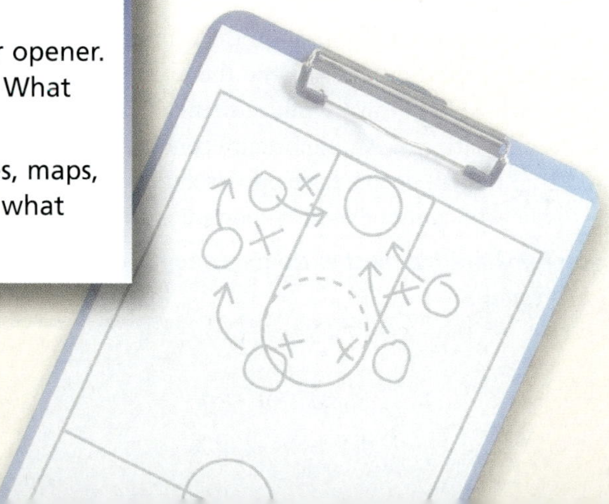

Do warm-up exercises

Before a game, an athlete does stretching exercises and takes practice shots. This book also has warm-up exercises for each chapter.

4. Interact with History (at left) will help you imagine what it was like to live in the past. Look at the pictures and decide how you would respond to the question.
5. Setting the Stage (below) will help you start thinking about the chapter's theme and time period. It also teaches a reading skill and gives you a graphic organizer to help you take notes.

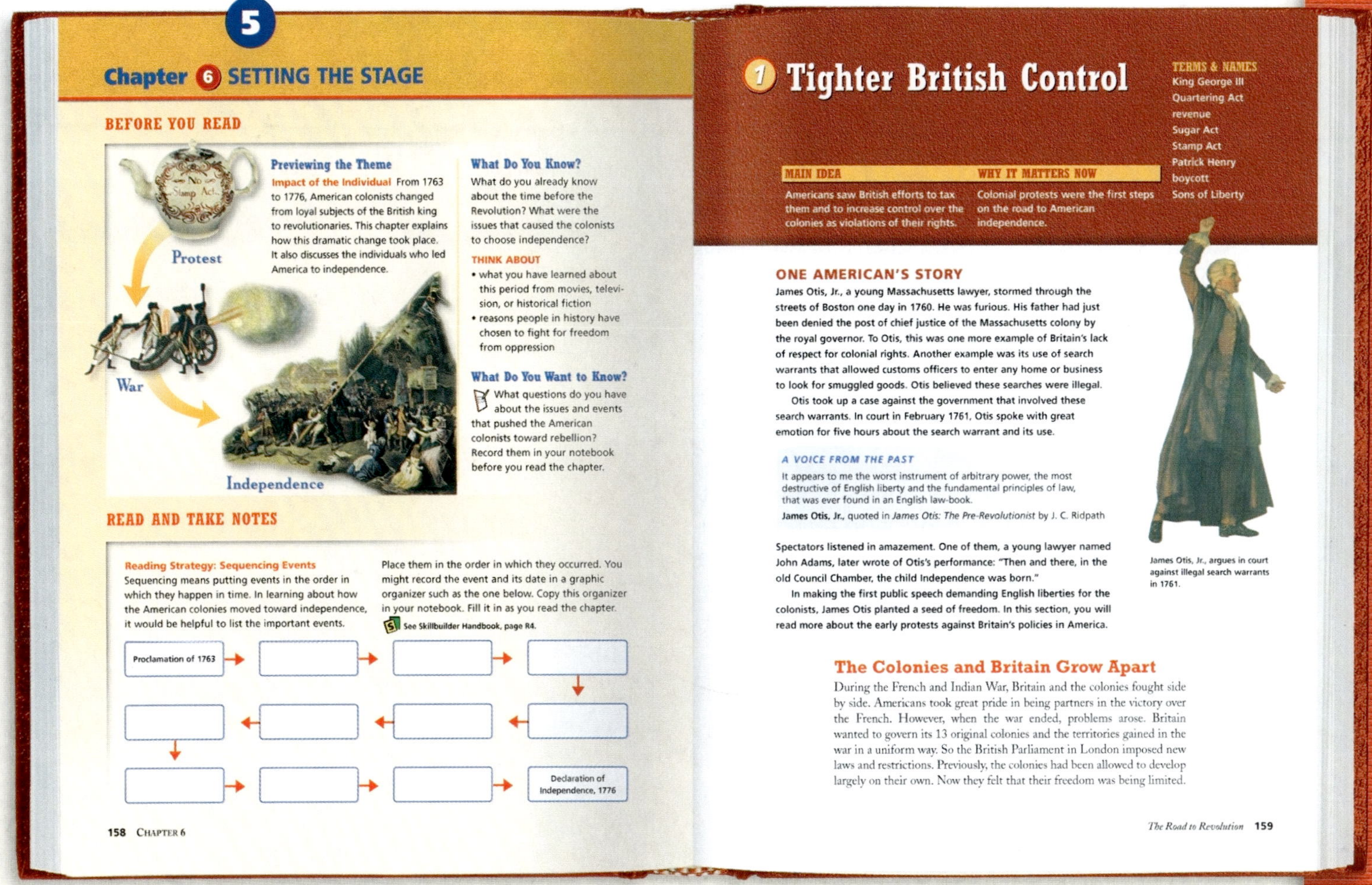

5

Chapter 6 SETTING THE STAGE

BEFORE YOU READ

Protest

War

Independence

Previewing the Theme

Impact of the Individual From 1763 to 1776, American colonists changed from loyal subjects of the British king to revolutionaries. This chapter explains how this dramatic change took place. It also discusses the individuals who led America to independence.

What Do You Know?

What do you already know about the time before the Revolution? What were the issues that caused the colonists to choose independence?

THINK ABOUT

- what you have learned about this period from movies, television, or historical fiction
- reasons people in history have chosen to fight for freedom from oppression

What Do You Want to Know?

What questions do you have about the issues and events that pushed the American colonists toward rebellion? Record them in your notebook before you read the chapter.

READ AND TAKE NOTES

Reading Strategy: Sequencing Events
Sequencing means putting events in the order in which they happen in time. In learning about how the American colonies moved toward independence, it would be helpful to list the important events. Place them in the order in which they occurred. You might record the event and its date in a graphic organizer such as the one below. Copy this organizer in your notebook. Fill it in as you read the chapter.

See Skillbuilder Handbook, page R4.

Proclamation of 1763 → … → Declaration of Independence, 1776

158 CHAPTER 6

1 Tighter British Control

TERMS & NAMES
King George III
Quartering Act
revenue
Sugar Act
Stamp Act
Patrick Henry
boycott
Sons of Liberty

MAIN IDEA
Americans saw British efforts to tax them and to increase control over the colonies as violations of their rights.

WHY IT MATTERS NOW
Colonial protests were the first steps on the road to American independence.

ONE AMERICAN'S STORY

James Otis, Jr., a young Massachusetts lawyer, stormed through the streets of Boston one day in 1760. He was furious. His father had just been denied the post of chief justice of the Massachusetts colony by the royal governor. To Otis, this was one more example of Britain's lack of respect for colonial rights. Another example was its use of search warrants that allowed customs officers to enter any home or business to look for smuggled goods. Otis believed these searches were illegal.

Otis took up a case against the government that involved these search warrants. In court in February 1761, Otis spoke with great emotion for five hours about the search warrant and its use.

A VOICE FROM THE PAST

It appears to me the worst instrument of arbitrary power, the most destructive of English liberty and the fundamental principles of law, that was ever found in an English law-book.

James Otis, Jr., quoted in *James Otis: The Pre-Revolutionist* by J. C. Ridpath

Spectators listened in amazement. One of them, a young lawyer named John Adams, later wrote of Otis's performance: "Then and there, in the old Council Chamber, the child Independence was born."

In making the first public speech demanding English liberties for the colonists, James Otis planted a seed of freedom. In this section, you will read more about the early protests against Britain's policies in America.

James Otis, Jr., argues in court against illegal search warrants in 1761.

The Colonies and Britain Grow Apart

During the French and Indian War, Britain and the colonies fought side by side. Americans took great pride in being partners in the victory over the French. However, when the war ended, problems arose. Britain wanted to govern its 13 original colonies and the territories gained in the war in a uniform way. So the British Parliament in London imposed new laws and restrictions. Previously, the colonies had been allowed to develop largely on their own. Now they felt that their freedom was being limited.

The Road to Revolution 159

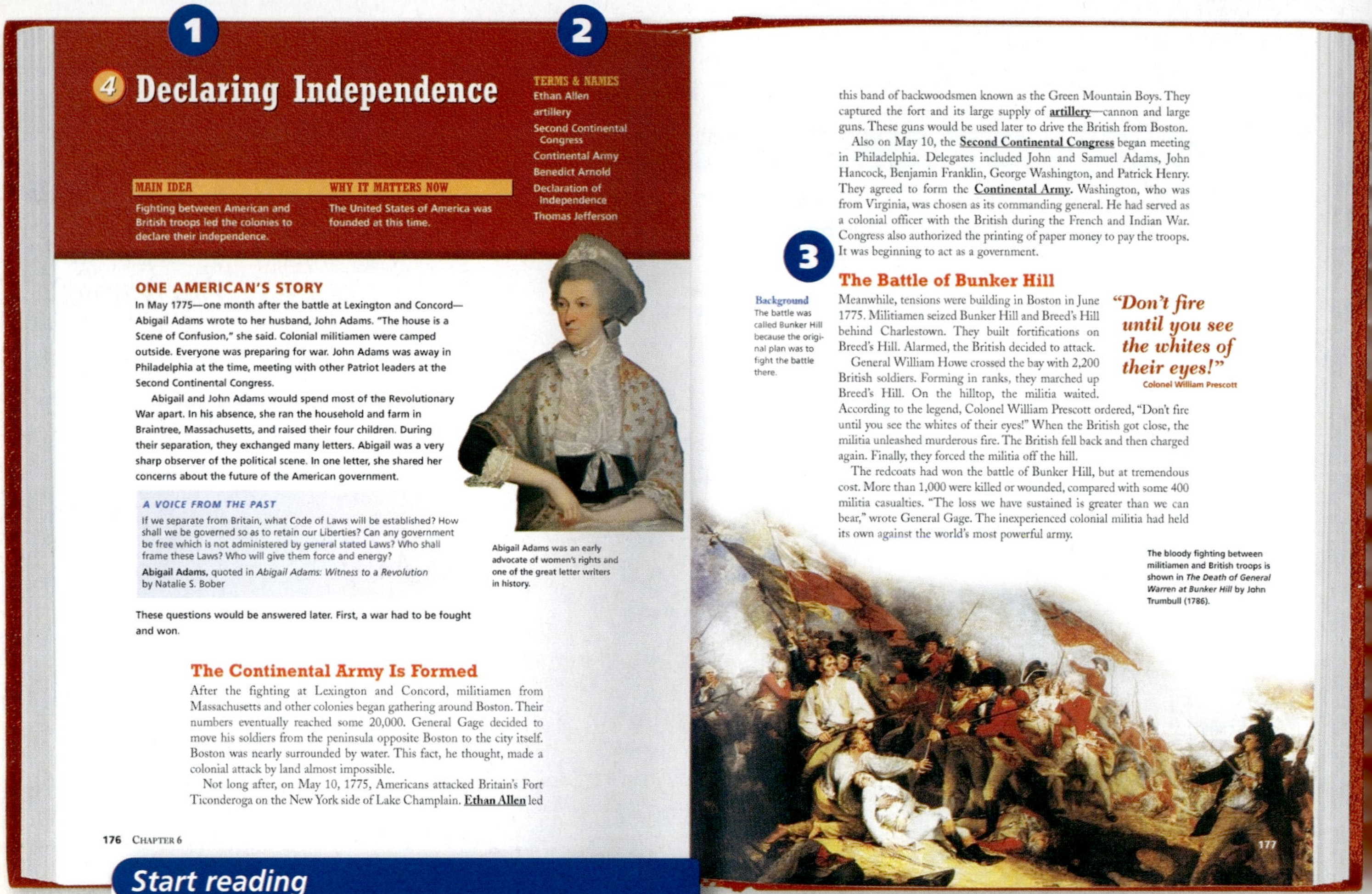

1 2

4 Declaring Independence

TERMS & NAMES
Ethan Allen
artillery
Second Continental Congress
Continental Army
Benedict Arnold
Declaration of Independence
Thomas Jefferson

MAIN IDEA
Fighting between American and British troops led the colonies to declare their independence.

WHY IT MATTERS NOW
The United States of America was founded at this time.

ONE AMERICAN'S STORY

In May 1775—one month after the battle at Lexington and Concord—Abigail Adams wrote to her husband, John Adams. "The house is a Scene of Confusion," she said. Colonial militiamen were camped outside. Everyone was preparing for war. John Adams was away in Philadelphia at the time, meeting with other Patriot leaders at the Second Continental Congress.

Abigail and John Adams would spend most of the Revolutionary War apart. In his absence, she ran the household and farm in Braintree, Massachusetts, and raised their four children. During their separation, they exchanged many letters. Abigail was a very sharp observer of the political scene. In one letter, she shared her concerns about the future of the American government.

A VOICE FROM THE PAST
If we separate from Britain, what Code of Laws will be established? How shall we be governed so as to retain our Liberties? Can any government be free which is not administered by general stated Laws? Who shall frame these Laws? Who will give them force and energy?
Abigail Adams, quoted in *Abigail Adams: Witness to a Revolution* by Natalie S. Bober

Abigail Adams was an early advocate of women's rights and one of the great letter writers in history.

These questions would be answered later. First, a war had to be fought and won.

The Continental Army Is Formed

After the fighting at Lexington and Concord, militiamen from Massachusetts and other colonies began gathering around Boston. Their numbers eventually reached some 20,000. General Gage decided to move his soldiers from the peninsula opposite Boston to the city itself. Boston was nearly surrounded by water. This fact, he thought, made a colonial attack by land almost impossible.

Not long after, on May 10, 1775, Americans attacked Britain's Fort Ticonderoga on the New York side of Lake Champlain. Ethan Allen led this band of backwoodsmen known as the Green Mountain Boys. They captured the fort and its large supply of artillery—cannon and large guns. These guns would be used later to drive the British from Boston.

Also on May 10, the Second Continental Congress began meeting in Philadelphia. Delegates included John and Samuel Adams, John Hancock, Benjamin Franklin, George Washington, and Patrick Henry. They agreed to form the Continental Army. Washington, who was from Virginia, was chosen as its commanding general. He had served as a colonial officer with the British during the French and Indian War. Congress also authorized the printing of paper money to pay the troops. It was beginning to act as a government.

3

The Battle of Bunker Hill

Background The battle was called Bunker Hill because the original plan was to fight the battle there.

Meanwhile, tensions were building in Boston in June 1775. Militiamen seized Bunker Hill and Breed's Hill behind Charlestown. They built fortifications on Breed's Hill. Alarmed, the British decided to attack.

"Don't fire until you see the whites of their eyes!" Colonel William Prescott

General William Howe crossed the bay with 2,200 British soldiers. Forming in ranks, they marched up Breed's Hill. On the hilltop, the militia waited. According to the legend, Colonel William Prescott ordered, "Don't fire until you see the whites of their eyes!" When the British got close, the militia unleashed murderous fire. The British fell back and then charged again. Finally, they forced the militia off the hill.

The redcoats had won the battle of Bunker Hill, but at tremendous cost. More than 1,000 were killed or wounded, compared with some 400 militia casualties. "The loss we have sustained is greater than we can bear," wrote General Gage. The inexperienced colonial militia had held its own against the world's most powerful army.

The bloody fighting between militiamen and British troops is shown in *The Death of General Warren at Bunker Hill* by John Trumbull (1786).

176 CHAPTER 6

177

Start reading

Once the preparation is over, an athlete starts playing the game. After preparation, you'll be ready to read the chapter.

1. Read one section at a time. Each chapter is divided into sections that are only three to six pages long.
2. First, read the Terms & Names in the bar at the top of the first page. Also read the sentences labeled Main Idea and Why It Matters Now. They tell you what's important in the material you're about to read.
3. Notice how the section is divided into smaller chunks, each with a red headline. Read each headline and ask yourself what the section will be about.

Then read the section from beginning to end.

Get helpful tips

This book is designed to offer you help as you read. On the inside margin of every page are three types of help.

4 Reading History questions allow you to stop and make sure you understand what you just read. You can also use them to review for a test.

5 Background notes explain some of the history behind an event.

6 Vocabulary notes give the definition of words that you may not know.

The first of Parliament's laws was the Proclamation of 1763. (See Chapter 5.) It said that colonists could not settle west of the Appalachian Mountains. Britain wanted this land to remain in the hands of its Native American allies to prevent another revolt like Pontiac's Uprising.

The proclamation angered colonists who had hoped to move to the fertile Ohio Valley. Many of these colonists had no land of their own. It also upset colonists who had bought land as an investment. As a result, many ignored the law.

British Troops and Taxes

King George III, the British monarch, wanted to enforce the proclamation and also keep peace with Britain's Native American allies. To do this, he decided to keep 10,000 soldiers in the colonies. In 1765, Parliament passed the Quartering Act. This was a cost-saving measure that required the colonies to quarter, or house, British soldiers and provide them with supplies. General Thomas Gage, commander of these forces, put most of the troops in New York.

Britain owed a large debt from the French and Indian War. Keeping troops in the colonies would raise that debt even higher. Britain needed more revenue, or income, to meet its expenses. So it attempted to have the colonies pay part of the war debt. It also wanted them to contribute toward the costs of frontier defense and colonial government.

In the past, the king had asked the colonial assemblies to pass taxes to support military actions that took place in the colonies. This time, however, Parliament voted to tax the Americans directly.

In 1764, Parliament passed the Sugar Act. This law placed a tax on sugar, molasses, and other products shipped to the colonies. It also called for strict enforcement of the act and harsh punishment of smugglers.

Colonial merchants, who often traded in smuggled goods, reacted with anger.

Colonial leaders such as James Otis claimed that Parliament had no right to tax the colonies, since the colonists were not represented in Parliament. As Otis exclaimed, "Taxation without representation is tyranny!" British finance minister George Grenville disagreed. The colonists were subjects of Britain, he said, and enjoyed the protection of its laws. For that reason, they were subject to taxation.

Britain Passes the Stamp Act

The Sugar Act was just the first in a series of acts that increased tension between the mother country and the colonies. In 1765, Parliament passed the Stamp Act. This law required all legal and commercial documents to carry an official stamp showing that a tax had been paid. All diplomas, contracts, and wills had to carry a stamp.

Reading History
A. Summarizing Who was upset by the Proclamation of 1763?

Background
Fought between 1754 and 1763, this war was part of a global conflict between Britain and France.

Vocabulary
tyranny: absolute power in the hands of a single ruler

The colonial view of the hated stamp tax is shown by the skull and crossbones on this emblem (above); a royal stamp is pictured at right.

160

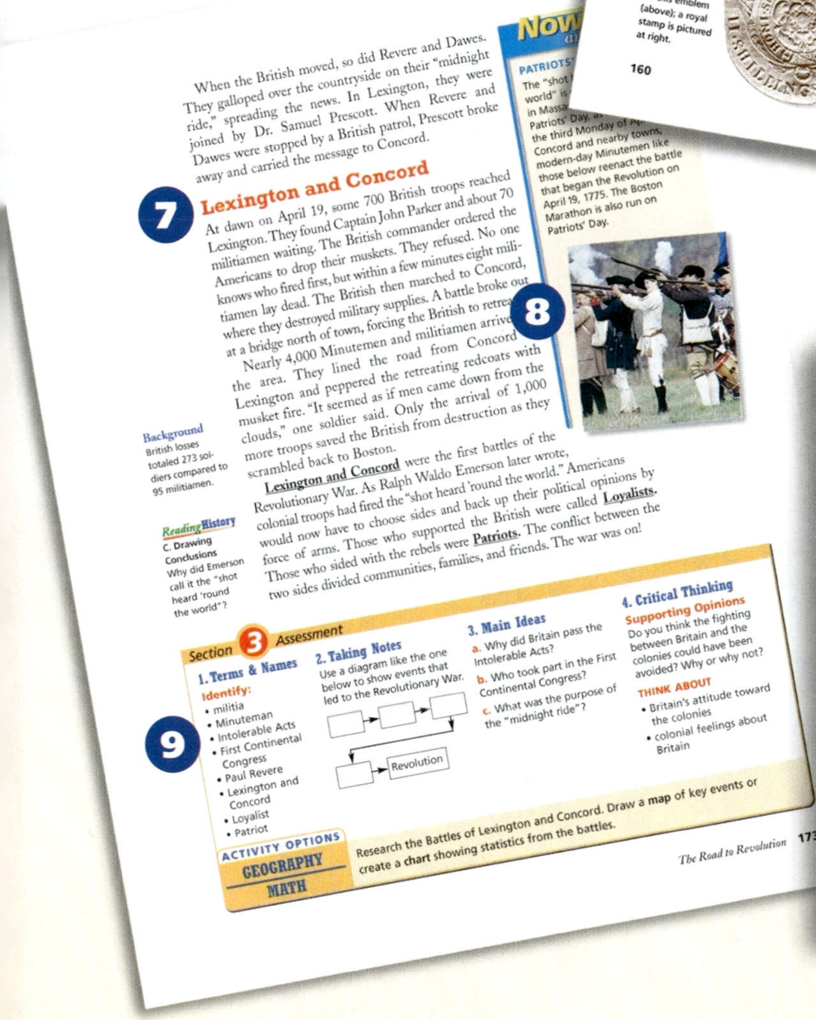

When the British moved, so did Revere and Dawes. They galloped over the countryside on their "midnight ride," spreading the news. In Lexington, they were joined by Dr. Samuel Prescott. When Revere and Dawes were stopped by a British patrol, Prescott broke away and carried the message to Concord.

Lexington and Concord

At dawn on April 19, some 700 British troops reached Lexington. They found Captain John Parker and about 70 militiamen waiting. The British commander ordered the Americans to drop their muskets. They refused. No one knows who fired first, but within a few minutes eight militiamen lay dead. The British then marched to Concord, where they destroyed military supplies. A battle broke out at a bridge north of town, forcing the British to retreat.

Nearly 4,000 Minutemen and militiamen arrived in the area. They lined the road from Concord to Lexington and peppered the retreating redcoats with musket fire. "It seemed as if men came down from the clouds," one soldier said. Only the arrival of 1,000 more troops saved the British from destruction as they scrambled back to Boston.

Lexington and Concord were the first battles of the Revolutionary War. As Ralph Waldo Emerson later wrote, colonial troops had fired the "shot heard 'round the world." Americans would now have to choose sides and back up their political opinions by force of arms. Those who supported the British were called Loyalists. Those who sided with the rebels were Patriots. The conflict between the two sides divided communities, families, and friends. The war was on!

Background
British losses totaled 273 soldiers compared to 95 militiamen.

Reading History
C. Drawing Conclusions Why did Emerson call it the "shot heard 'round the world"?

Now

PATRIOTS' DAY
The "shot heard 'round the world" is ... in Massachusetts ... Patriots' Day, ... the third Monday of April. ... Concord and nearby towns, modern-day Minutemen like those below reenact the battle that began the Revolution on April 19, 1775. The Boston Marathon is also run on Patriots' Day.

Section 3 Assessment

1. Terms & Names
Identify:
• militia
• Minuteman
• Intolerable Acts
• First Continental Congress
• Paul Revere
• Lexington and Concord
• Loyalist
• Patriot

2. Taking Notes
Use a diagram like the one below to show events that led to the Revolutionary War.
Revolution

3. Main Ideas
a. Why did Britain pass the Intolerable Acts?
b. Who took part in the First Continental Congress?
c. What was the purpose of the "midnight ride"?

4. Critical Thinking
Supporting Opinions Do you think the fighting between Britain and the colonies could have been avoided? Why or why not?
THINK ABOUT
• Britain's attitude toward the colonies
• colonial feelings about Britain

ACTIVITY OPTIONS
GEOGRAPHY
MATH
Research the Battles of Lexington and Concord. Draw a map of key events or create a chart showing statistics from the battles.

The Road to Revolution 173

Review what you read

When you finish reading a section, review and remember what you read. Don't worry if you have to go back and read something a second time. You can also use the following tips to review.

7 Reread the red headlines. Do you remember what was discussed in the paragraphs that follow?

8 Look at the pictures, maps, and graphs. Do you understand how they relate to what you just read?

9 Complete all the parts of the Section Assessment. They will help you review the section.

The Landscape of America

The best place to begin your study of American history is with the geography of America. Geography is more than the study of the land and people. It also involves the relationship between people and their environment.

The United States is part of the North American continent. The United States ranks third in both total area and population in the world. It is filled with an incredible variety of physical features, natural resources, climatic conditions, and people. This handbook will help you to learn about these factors and to understand how they affected the development of the United States.

A combine harvests wheat in the Midwest.

A timber company collects logs in the Northwest.

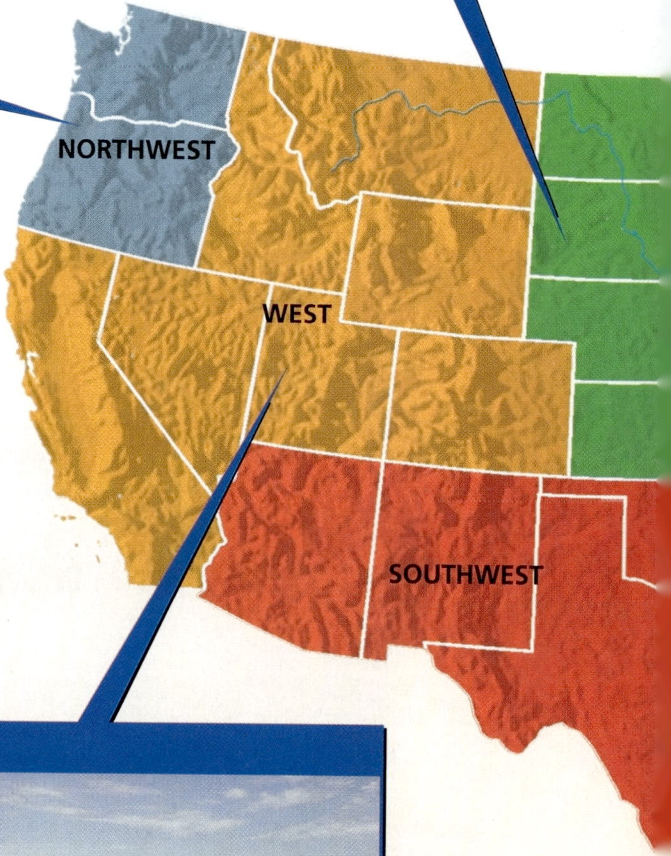

A cowboy drives cattle in the West.

Fishers haul their catch toward shore in the Northeast.

TABLE OF CONTENTS

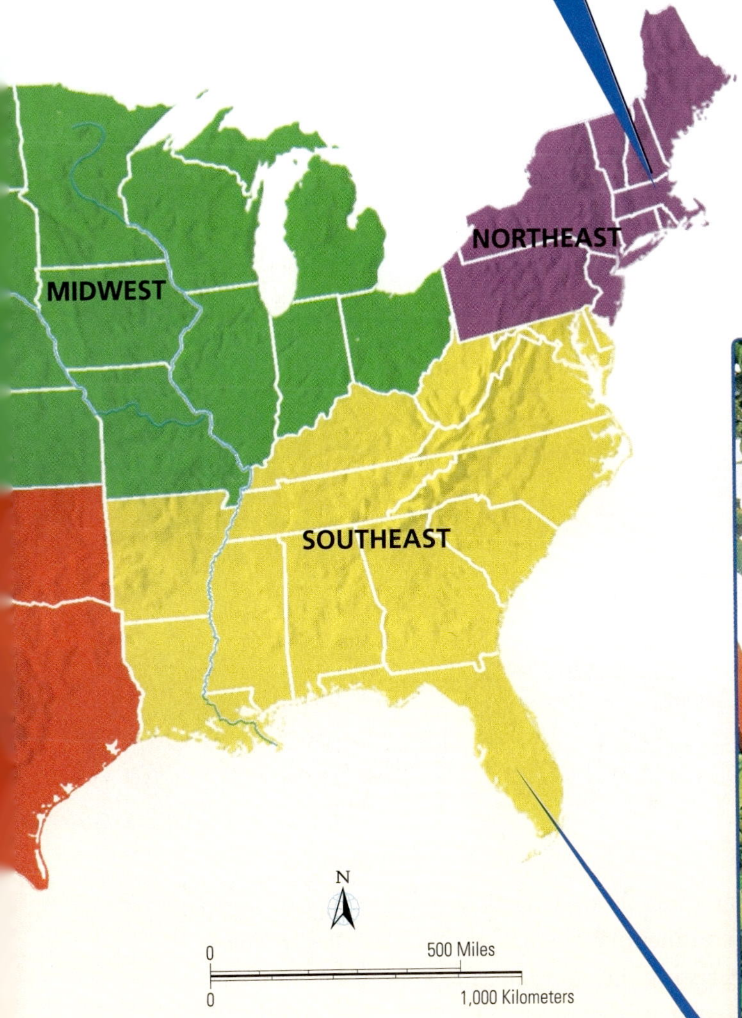

Oranges are big business in the Southeast.

Themes of Geography

One useful way to think about geography is in terms of major themes or ideas. These pages examine the five major themes of geography and show how they apply to Boston, Massachusetts. Recognizing and understanding these themes will help you to understand all the different aspects of geography.

Boston is located on the shores of the Atlantic Ocean.

Location

"Where am I?" Your answer to this question is your *location.* One way to answer it is to use *absolute* location. That means you'll use the coordinates of longitude and latitude to give your answer (see page 8). For example, if you're in Boston, its absolute location is approximately 42° north latitude and 71° west longitude.

Like most people, however, you'll probably use *relative* location to answer the question. Relative location describes where a certain area is in relation to another area. For example, Boston lies in the northeast corner of the United States, next to the Atlantic Ocean.

THINKING ABOUT GEOGRAPHY What is the relative location of your school?

Place

"What is Boston like?" *Place* can help you answer this question. Place refers to the physical and human factors that make one area different from another. Physical characteristics are natural features, such as physical setting, plants, animals, and weather. For example, Boston sits on a hilly peninsula.

Human characteristics include cultural diversity and the things people have made—including language, the arts, and architecture. For instance, Boston includes African Americans, as well as people of Irish, Italian, Chinese, and Hispanic ancestry.

THINKING ABOUT GEOGRAPHY What physical and human characteristics make where you live unique?

Boston has grown and changed since this 1722 map.

Region

Geographers can't easily study the whole world at one time. So they break the world into regions. A *region* can be as large as a continent or as small as a neighborhood. A region has certain shared characteristics that set it apart. These characteristics might include political division, climate, language, or religion. Boston is part of the northeast region. It shares a climate—continental temperate—with the cities of New York and Philadelphia.

THINKING ABOUT GEOGRAPHY What characteristics does your city or town share with nearby cities or towns?

Airplanes from Boston's Logan International Airport move people and ideas around the globe.

People shop, eat, and interact at Boston's famous Quincy Market.

Movement

Movement refers to the shifting of people, goods, and ideas from one place to another. People constantly move in search of better places to live, and they trade goods with one another over great distances. Movement also causes ideas to travel from place to place. In recent years, technology has quickened the movement of ideas and goods.

Boston became known as the *Cradle of Liberty* because of the movement of ideas. The concepts of freedom and self-government that developed in Boston spread to the other colonies and helped to start the American Revolution.

THINKING ABOUT GEOGRAPHY What are some of the different ways you spread information and ideas?

Human-Environment Interaction

Human-environment interaction refers to ways people interact with their environment, such as building a dam, cutting down a tree, or even sitting in the sun.

In Boston, human-environment interaction occurred when officials filled in swampy areas to make the city larger. In other ways, the environment has forced people to act. For example, people have had to invent ways to protect themselves from extreme weather and natural disasters.

THINKING ABOUT GEOGRAPHY What are ways that people in your city or town have changed their environment?

Themes of Geography Assessment

1. Main Ideas

a. What is the relative location of your home?

b. What are three characteristics of the region in which you live?

c. What are at least three ways in which you have recently interacted with the environment?

2. Critical Thinking

Forming and Supporting Opinions Which aspect of geography described in these themes do you think has most affected your life? Explain.

THINK ABOUT

- ways that you interact with your environment
- how you travel from place to place

Map Basics

Geographers use many different types of maps, and these maps all have a variety of features. The map on the next page gives you information on a historical event—the War of 1812. But you can use it to learn about different parts of a map, too.

Types of Maps

Physical maps Physical maps show mountains, hills, plains, rivers, lakes, oceans, and other physical features of an area.

Political maps Political maps show political units, such as countries, states, provinces, counties, districts, and towns. Each unit is normally shaded a different color, represented by a symbol, or shown with a different typeface.

Historical maps Historical maps illustrate such things as economic activity, migrations, battles, and changing national boundaries.

Tools of Geography

The ancient Greeks developed some of the first ways to study geography. Today, geographers and map makers use advanced technology to study geography.

GPS

A Global Positioning System (GPS) is a navigational system that uses at least three satellites to identify a person's absolute location. It is also used to study other aspects of geography.

Surveyors

An American surveys the land in the 19th century.

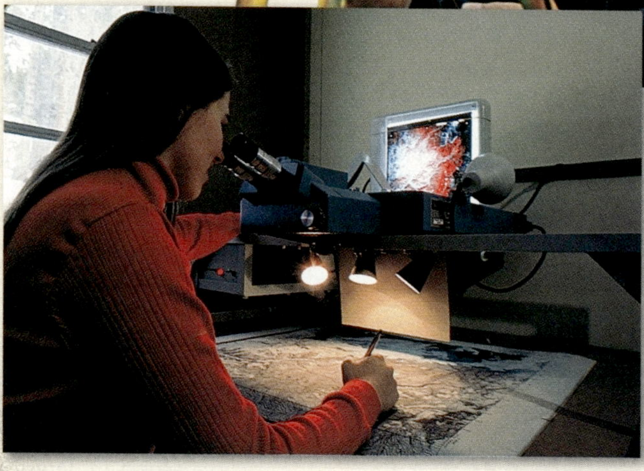

Computers

Computers can create electronic maps in which geographers can quickly add or remove features that keep the map current. Computers can also be used to monitor environmental problems such as deforestation and global warming.

Reading a Map

A **Lines** Lines indicate political boundaries, roads and highways, human movement, and rivers and other waterways.

B **Symbols** Symbols represent such items as capital cities, battle sites, or economic activities.

C **Labels** Labels are words or phrases that explain various items or activities on a map.

D **Compass Rose** A compass rose shows which way the directions north (N), south (S), east (E), and west (W) point on the map.

E **Scale** A scale shows the ratio between a unit of length on the map and a unit of distance on the earth. A typical one-inch scale indicates the number of miles and kilometers that length represents on the map.

F **Colors** Colors show a variety of information on a map, such as population density or the physical growth of a country.

G **Legend or Key** A legend or key lists and explains the symbols, lines, and colors on a map.

H **Lines of Longitude** These are imaginary, north-south lines that run around the globe.
Lines of Latitude These are imaginary, east-west lines that run around the globe. Together, latitude and longitude lines form a grid on a map or globe to indicate an area's absolute location.

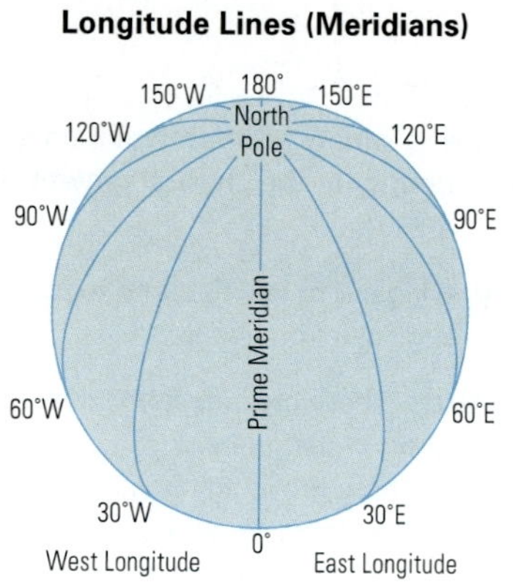

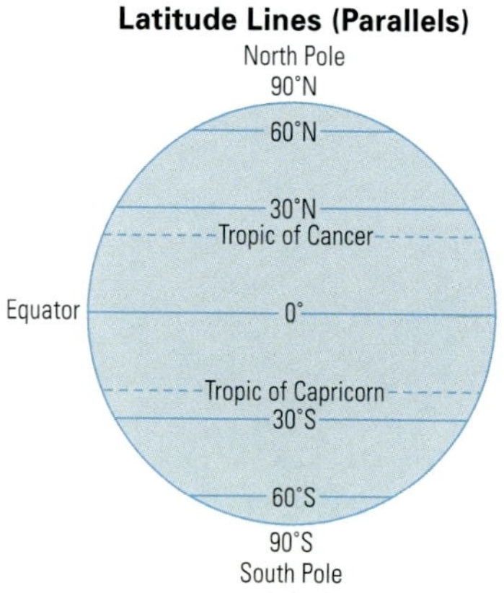

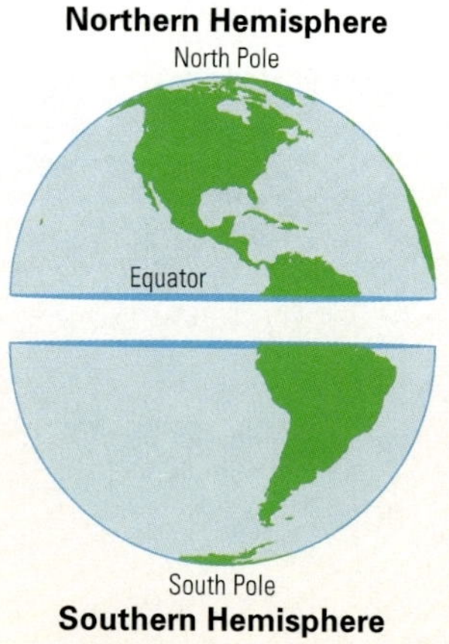

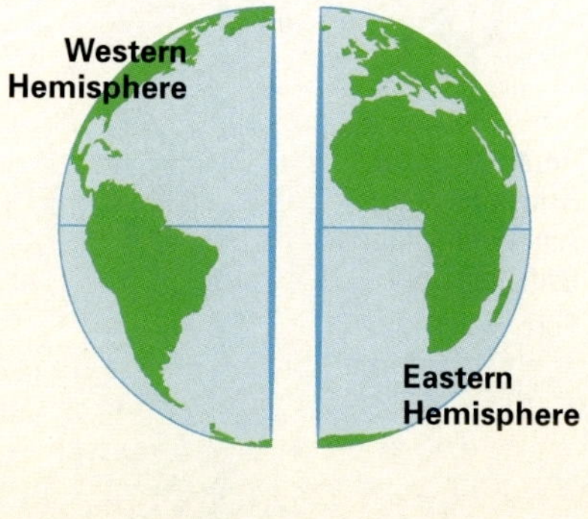

Longitude lines

- are imaginary lines that run north to south around the globe and are known as meridians
- show the distance in degrees east or west of the prime meridian

The prime meridian is a longitude line that runs from the North Pole to the South Pole. It passes through Greenwich, England, and measures 0° longitude.

Latitude lines

- are imaginary lines that run east to west around the globe and are known as parallels
- show distance in degrees north or south of the equator

The equator is a latitude line that circles the earth halfway between the North and South poles. It measures 0° latitude.

The tropics of Cancer and Capricorn are parallels that form the boundaries of the Tropics, a region that stays warm all year.

Latitude and longitude lines appear together on a map and allow you to pinpoint the absolute location of cities and other geographic features. You express this location through coordinates of intersecting lines. These are measured in degrees.

Hemisphere

Hemisphere is a term for half the globe. The globe can be divided into Northern and Southern hemispheres (separated by the equator) or into Eastern and Western hemispheres. The United States is located in the Northern and Western hemispheres.

Projections

A projection is a way of showing the curved surface of the earth on a flat map. Flat maps cannot show the size, shape, and direction of a globe all at once with total accuracy. As a result, all projections distort some aspect of the earth's surface. Some maps distort distances, while other maps distort angles. On the next page are four projections.

Mercator Projection

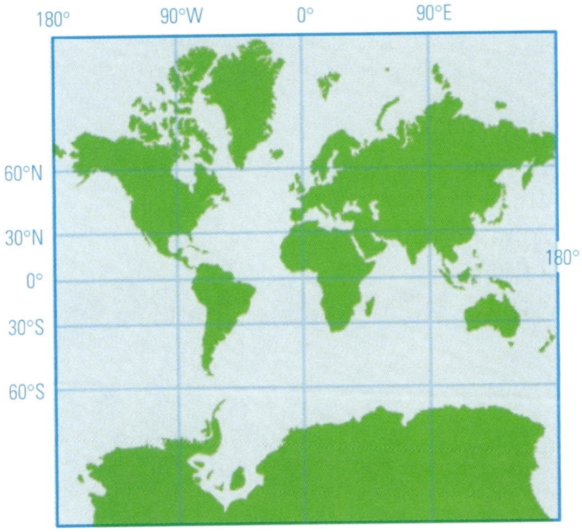

The Mercator projection shows most of the continents as they look on a globe. However, the projection stretches out the lands near the North and South poles. The Mercator is used for all kinds of navigation.

Azimuthal Projection

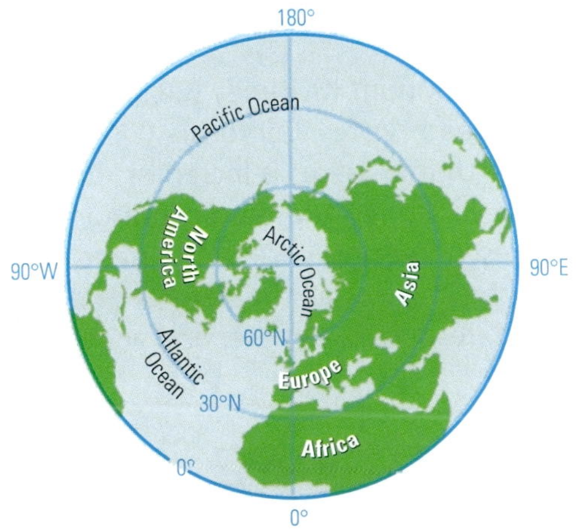

An azimuthal projection shows the earth so that a straight line from the central point to any other point on the map gives the shortest distance between the two points. Size and shape of the continents are also distorted.

Homolosine Projection

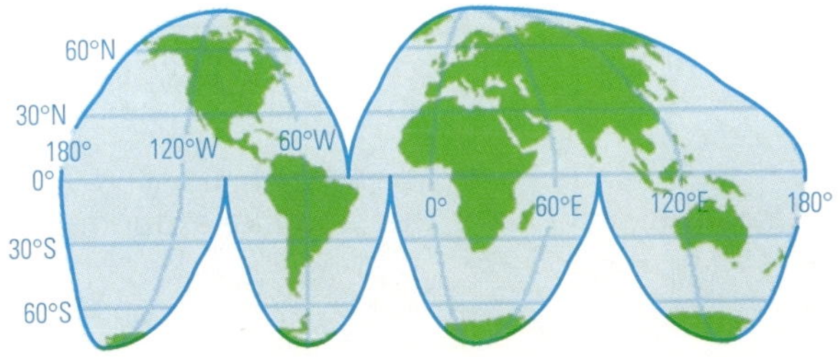

This projection shows the accurate shapes and sizes of the landmasses, but distances on the map are not correct.

Robinson Projection

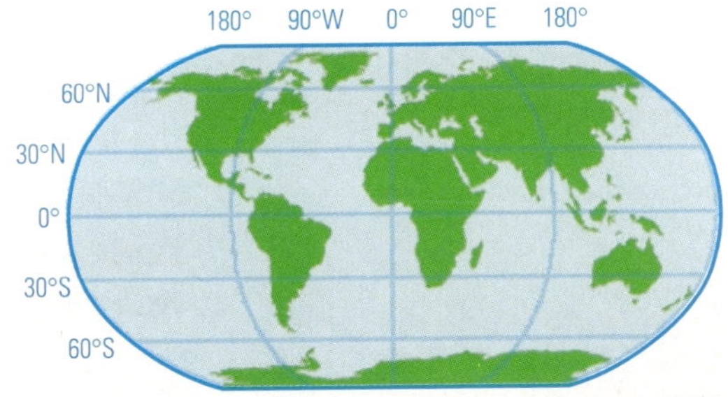

Textbook maps commonly use the Robinson projection. It shows the entire earth with nearly the true sizes and shapes of the continents and oceans. However, the shapes of the landforms near the poles appear flat.

Map Basics Assessment

1. Main Ideas

a. What is the longitude and latitude of your city or town?

b. What information is provided by the legend on the map on page 7?

c. What is a projection? Compare and contrast Antarctica on the Mercator and the Robinson projections.

2. Critical Thinking

Making Inferences Why do you think latitude and longitude are so important to sailors?

THINK ABOUT

- the landmarks you use to find your way around
- the landmarks available to sailors on the ocean

Physical Geography of the United States

From the heights of Mount McKinley (20,320 feet above sea level) in Alaska to the depths of Death Valley, California (282 feet below sea level), the geography of the United States is incredibly diverse. In between these extremes lie such varied features and conditions as scorching Arizona deserts, lush Oregon forests, freezing Vermont winters, and sunny Florida beaches. Physical geography involves all the natural features on the earth. This includes the land, resources, climate, and vegetation.

Flowers and brush cover the Coral Pink Sand Dunes in southern Utah.

Land

Separated from much of the world by two oceans, the United States covers 3,717,796 square miles and spans the entire width of North America. To the west, Hawaii stretches the United States into the Pacific Ocean. To the north, Alaska extends the United States to the Arctic Circle. On the U.S. mainland, a huge central plain separates large mountains in the West and low mountains in the East. Plains make up almost half of the country, while mountains and plateaus make up a quarter each.

An abundance of lakes—Alaska alone has three million—and rivers also dot the landscape. Twenty percent of the United States is farmed, providing the country with a steady food supply. Urban areas cover only about two percent of the nation. Refer to the map on the next page for a complete look at the U.S. landscape.

THINKING ABOUT GEOGRAPHY What is the land like around your city or state?

Oil drilled from Alaska helps power the nation's planes, trains, cars, and factories.

Resources

The United States has a variety of natural resources. Vast amounts of coal, oil, and natural gas lie underneath American soil. Valuable deposits of lead, zinc, uranium, gold, and silver also exist. These resources have helped the United States become the world's leading industrial nation—producing nearly 21 percent of the world's goods and services.

These resources have also helped the United States become both the world's largest producer of energy (natural gas, oil, coal, nuclear power, and electricity) and the world's largest consumer of it. Other natural resources include the Great Lakes, which are shared with Canada. They contain about 20 percent of the world's total supply of fresh surface water. Refer to the map on the next page to examine the nation's natural resources.

THINKING ABOUT GEOGRAPHY What are the different natural resources that you and your family use in your daily lives?

Miners extract such minerals as gold, silver, and copper from the Rocky Mountains.

The Appalachians are among the earth's oldest mountains.

Climate

People brave the harsh winters in the continental temperate climate of the upper Midwest.

Mount Washington in New Hampshire has experienced one of the world's lowest recorded wind chills of over –100° F, one of the world's highest wind speeds at 231 miles per hour, and a three-day snow fall of 8 feet.

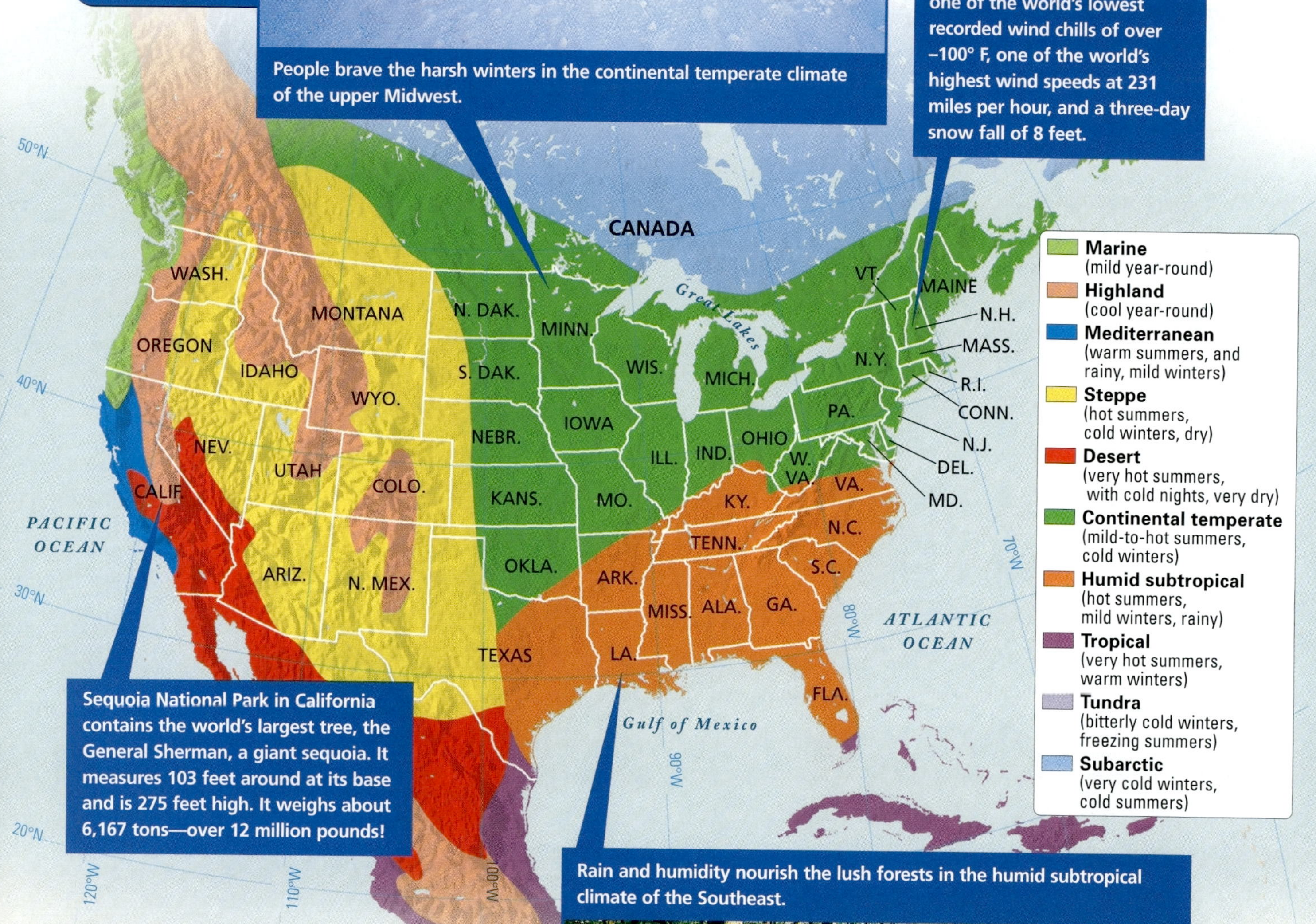

Sequoia National Park in California contains the world's largest tree, the General Sherman, a giant sequoia. It measures 103 feet around at its base and is 275 feet high. It weighs about 6,167 tons—over 12 million pounds!

Rain and humidity nourish the lush forests in the humid subtropical climate of the Southeast.

Tourists and residents bask in the sunshine of Waikiki Beach in Hawaii.

Climate

The United States contains a variety of climates. For example, the mean temperature in January in Miami, Florida, is 67° F, while it is 11° F in Minneapolis, Minnesota. Most of the United States experiences a continental climate, or distinct change of seasons. Some regional climatic differences include hot and humid summers in the Southeast versus hot and dry summers in the Southwest. Harsh winters and heavy snow can blanket parts of the Midwest, the Northeast, and the higher elevations of the West and Northwest. Refer to the map on the previous page to see the nation's climatic regions.

Human activities have affected the climate, too. For example, pollution from cars and factories can affect local weather conditions and may be contributing to a dangerous rise in the earth's temperature.

THINKING ABOUT GEOGRAPHY How would you describe the climate where you live?

This mountainside burns with the autumn foliage of New Hampshire.

Vegetation

Between 20,000 and 25,000 species and subspecies of plants and vegetation grow in the United States—including over 1,000 different kinds of trees. Climate often dictates the type of vegetation found in a region. For instance, cold autumns in the Northeast contribute to the brilliantly colored autumn leaves. Rain nourishes the forests in the Northwest and Southeast. The central plains, where rainfall is less heavy, are covered by grass. Cactus plants thrive in the dry southwestern deserts.

Along with natural vegetation, climate dictates the nation's variety of planted crops. For example, temperate weather in the Midwest helps wheat to grow, while warm weather nourishes citrus fruit in Florida and California.

THINKING ABOUT GEOGRAPHY What kinds of trees or plants grow in your region?

Physical Geography Assessment

1. Main Ideas

a. What are the different aspects of physical geography?

b. Which state contains the largest variety of climates?

c. What two states contain most of the country's oil resources?

2. Critical Thinking

Drawing Conclusions What do you think are the advantages of living in a country with diverse physical geography?

THINK ABOUT

- the different resources available in your region
- the variety of recreational activities in your region

Geographic Dictionary

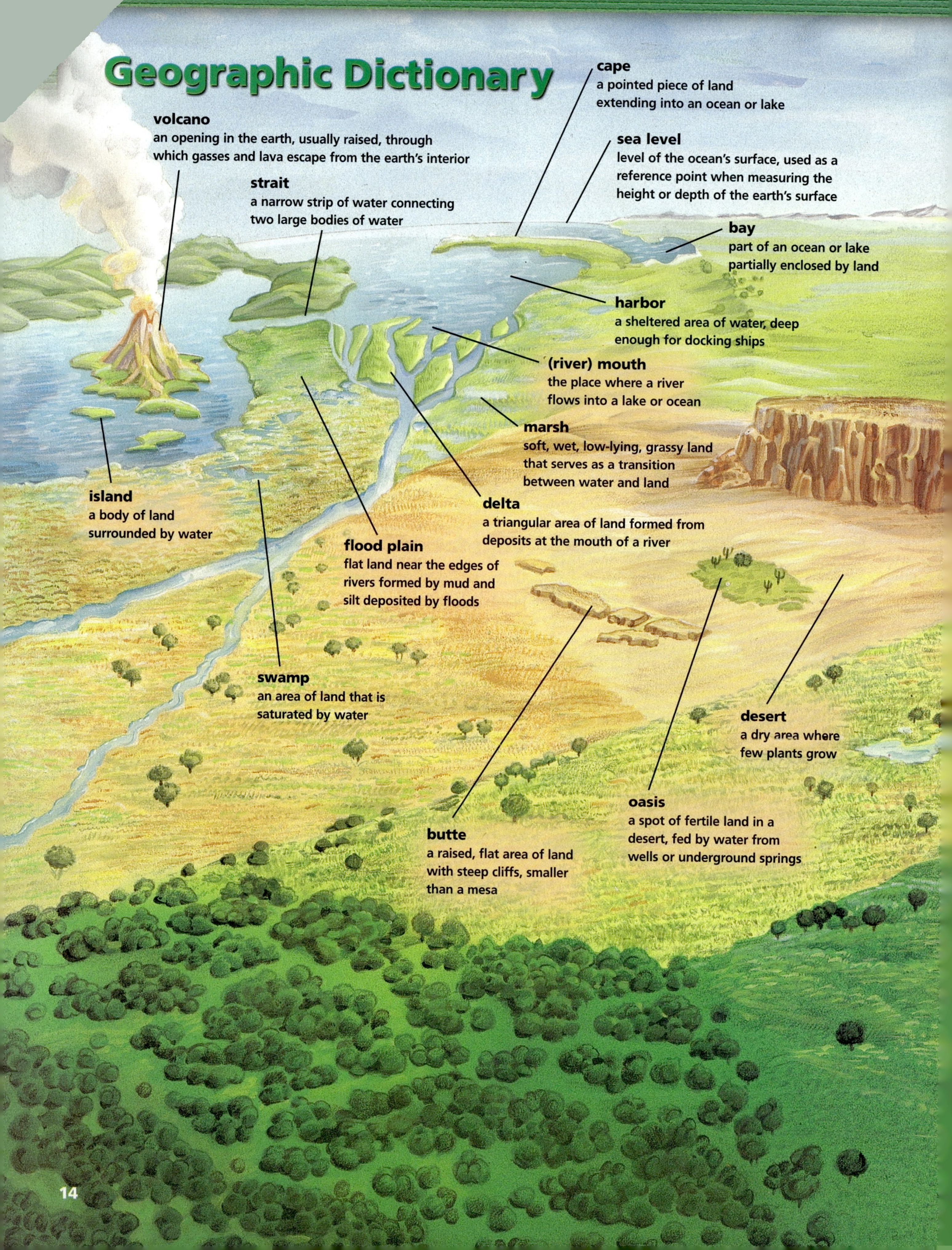

prairie
a large, level area of grassland with few or no trees
mountain
natural elevation of the earth's surface with steep sides and greater height than a hill
glacier
a large ice mass that moves slowly down a mountain or over land
steppe
a wide, treeless plain
valley
low land between hills or mountains
mesa
a wide, flat-topped mountain with steep sides, larger than a butte
cataract
a large, powerful waterfall
canyon
a narrow, deep valley with steep sides
cliff
the steep, almost vertical edge of a hill, mountain, or plain
plateau
a broad, flat area of land higher than the surrounding land

Human Geography of the United States

Human geography focuses on people's relationships with each other and the surrounding environment. It includes two main themes of geography: human-environment interaction and movement. The following pages will help you to better understand the link between people and geography.

Humans Adapt to Their Surroundings

Humans have always adapted to their environment. For example, in North America, many Native American tribes burned forest patches to create grazing area to attract animals and to clear area for farmland. In addition, Americans have adapted to their environment by building numerous dams, bridges, and tunnels. More recently, scientists and engineers have been developing building materials that will better withstand the earthquakes that occasionally strike California.

THINKING ABOUT GEOGRAPHY What are some of the ways in which you interact with your environment on a daily basis?

Early Americans of the Southwest protected themselves from the weather by building cliff dwellings.

The Hoover Dam, located on the Colorado River between Arizona and Nevada, provides electricity for Arizona, Nevada, and Southern California.

An oil spill from the ship *Exxon Valdez* harmed wildlife, such as this bird, in Prince William Sound, Alaska, in 1989.

Humans Affect the Environment

When humans interact with the environment, sometimes nature suffers. In the United States, for example, major oil leaks or spills occur each year—fouling shorelines and harming wildlife. Building suburbs and strip malls has also destroyed forests, farmland, and valuable wetlands.

THINKING ABOUT GEOGRAPHY What are some of the environmental problems in your city or town?

Destruction of Original Forests

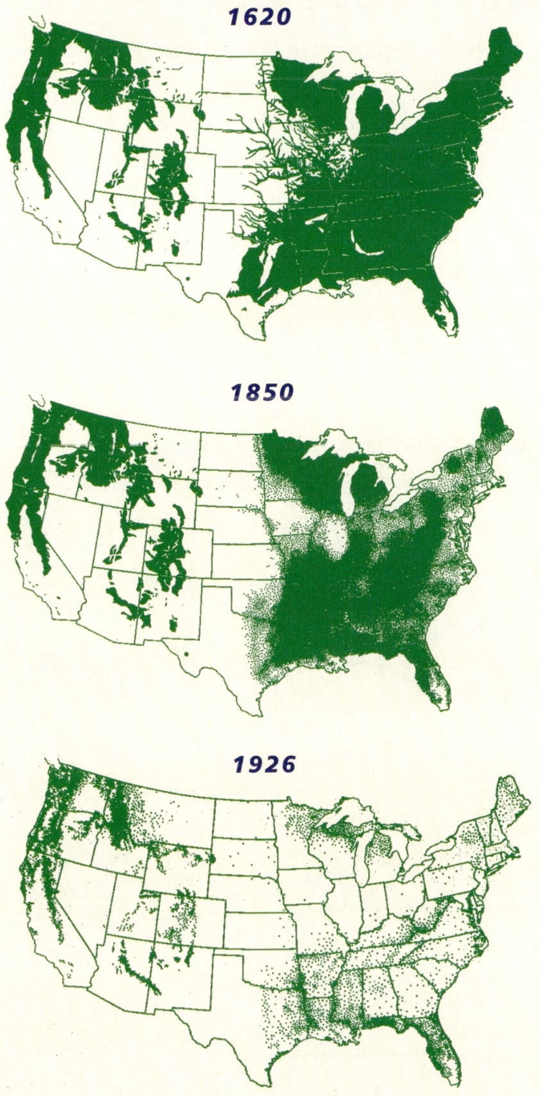

These maps show that, over the years, human beings have nearly cut down all the original forests in the United States. Each dot represents 25,000 acres.

Children plant trees along a Chicago expressway.

Preserving and Restoring

Americans—as well as people all over the world—have been working hard to balance economic progress with conservation. For example, car companies in the United States and around the world are working to develop pollution-free vehicles. In 1994, the average American family of four recycled around 1,100 pounds of waste. And, in the 1990s, Americans have planted more than two million acres of new trees each year.

THINKING ABOUT GEOGRAPHY What are some of the ways in which you help the environment?

Human Movement

In prehistoric times, people roamed the earth in search of food. Today in the United States, people move from place to place for many different reasons. Among them are cost of living, job availability, and climate. Since the 1970s, many Americans—as well as many new immigrants—moved to the Sunbelt. This region runs through the southern United States from Virginia to California. Between 1950 and 1990, that region's population soared from 52 million to 118 million.

THINKING ABOUT GEOGRAPHY Has your family ever moved? If so, what were some of the reasons?

This map shows human movement in the 1970s. The information below explains some of the results of this movement in the 1990s.

Americans on the Move, 1970s

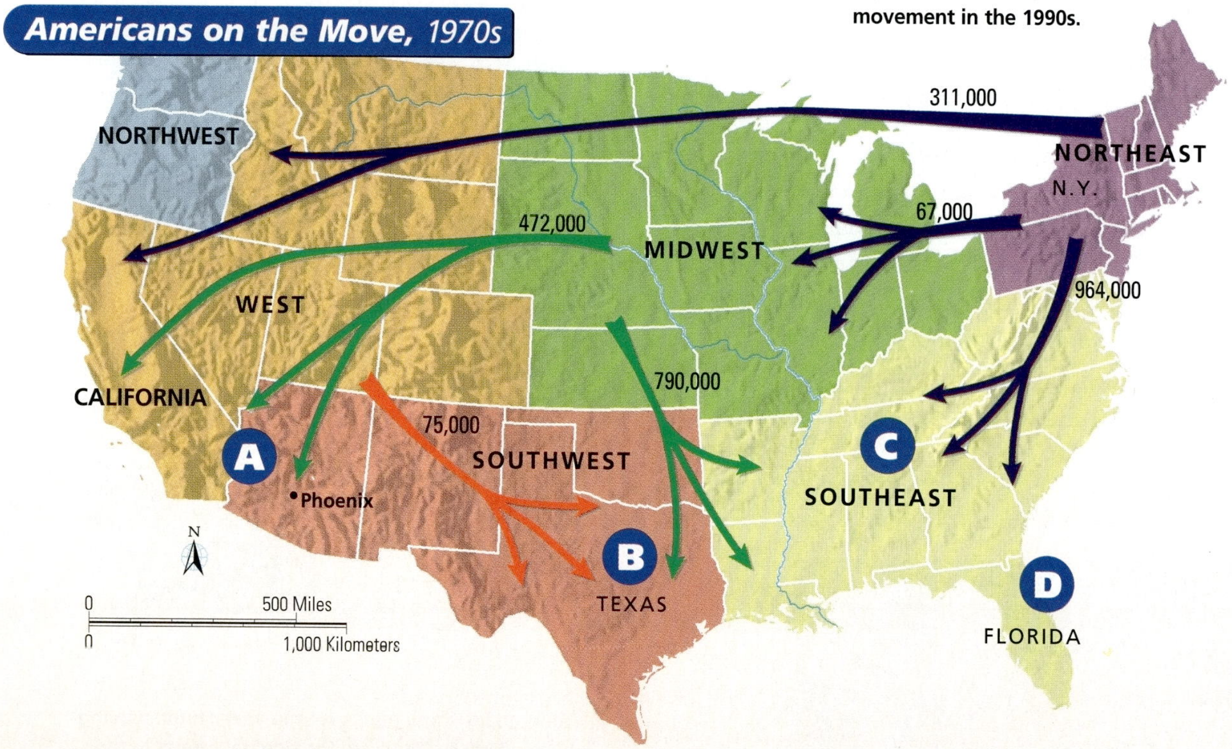

A By 1996, the Phoenix-Mesa metropolitan area reached a population of 2.75 million, more than the number of people living in the entire state in 1980.

B Between 1990 and 1994, Texas overtook New York as the nation's second most populous state, behind California.

C One of the nation's fastest growing areas was the Southeast, where population growth ranged from six to nine percent between 1990 and 1994. Jobs grew in the area by 14 percent.

D Florida's population is growing so much that it could become as populous as New York state by around 2020.

New home developments cover the desert in Las Vegas, Nevada.

In the late 19th century, millions of immigrants arrived on the shores of the United States.

The Sears Tower overlooks Chinatown in Chicago.

Humans Spread Ideas and Information

Throughout U.S. history, people from all over the world have come to the United States. They have brought with them food, music, language, technology, and other aspects of their culture. As a result, the United States is one of the most culturally rich and diverse nations in the world. Look around your town or city. You'll probably notice different people, languages, and foods.

Today, the spreading of ideas and customs does not rely solely on human movement. Technology—from the Internet to television to satellites—spreads ideas and information throughout the world faster than ever. This has created an ever-growing, interconnected world. As the 21st century opens, human geography will continue to play a key role in shaping the United States and the world.

THINKING ABOUT GEOGRAPHY How have computers and the Internet affected your life?

Human Geography Assessment

1. Main Ideas

a. What are some of the ways that people have helped to restore the environment?

b. What are some of the ways that residents of your region have successfully modified their landscape?

c. What are some of the reasons that people move from place to place?

2. Critical Thinking

Recognizing Effects In what ways has technology helped bring people in the world together?

THINK ABOUT

- the different ways in which people communicate today
- the speed in which people today can communicate over long distances

EOGRAPHY HANDBOOK ASSESSMENT

TERMS

Briefly explain and identify each of the following.

1. physical map
2. political map
3. longitude
4. latitude
5. hemisphere
6. projection
7. flood plain
8. sea level
9. human geography
10. human movement

REVIEW QUESTIONS

Themes of Geography (pages 4–5)

1. What is the difference between *absolute* location and *relative* location?
2. What is meant by the theme of place?
3. What are the themes of movement and human-environment interaction?

Map Basics (pages 6–9)

4. What do you think are some of the benefits of using technology to study geography?
5. What are the three major kinds of maps?
6. What are latitude and longitude lines?

Physical Geography (pages 10–13)

7. How have the natural resources in the United States helped its economic development?
8. What are the different climates within the United States?

Human Geography (pages 16–19)

9. How is human geography different from physical geography?
10. What aspects of human geography might cause people to move?

CRITICAL THINKING

1. **Forming and Supporting Opinions** Which of the five themes of geography do you think has had the most impact on history? Why?
2. **Analyzing Causes** How do the climate and natural resources of an area affect its economy?
3. **Drawing Conclusions** How have computers helped geographers make more accurate maps?
4. **Making Inferences** Why do you think the Mercator projection is used for all types of navigation?
5. **Recognizing Effects** How does a diverse landscape help or hurt the economy of an area?

GEOGRAPHY SKILLS

1. INTERPRETING MAPS: Movement

Basic Map Elements

a. What region of the United States is shown?

b. Compare the number of teams on the 1987 map and the 2000 map. How many more teams are on the 2000 map?

Interpreting the Map

c. What geographic theme(s) is most responsible for the increase in sports teams in this region?

d. According to the map, which sport enjoyed the biggest surge in popularity in this region?

Major League Sports in Southeast Cities

GEOGRAPHY SKILLS

2. INTERPRETING MAPS: Region

Study the map and then answer the questions.

California: Cross section at 38° N

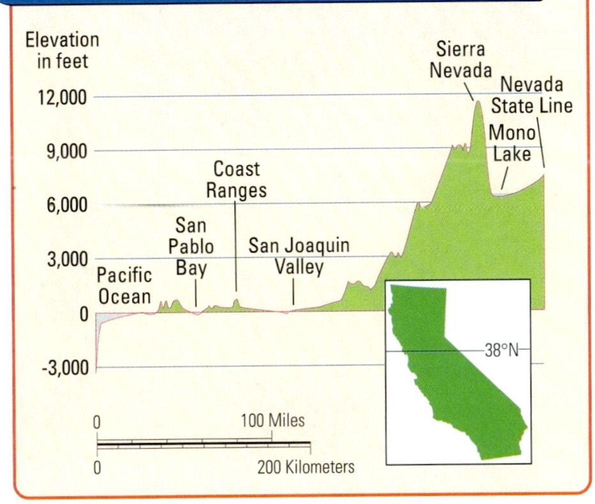

Basic Map Elements

a. What are the different landforms on the map?

Interpreting the Map

b. What is the level of the San Joaquin Valley? How many miles does it take to get from there to the highest point in California at the 38th parallel?

3. INTERPRETING PRIMARY SOURCES

In 1803, President Thomas Jefferson appointed Meriwether Lewis to explore the lands of the Louisiana Purchase. Jefferson gave him these instructions:

> The object of your mission is to explore the Missouri river . . . by its course & communication with the waters of the Pacific Ocean, may offer the most direct & practicable water communication across this continent, for the purposes of commerce. . . .
>
> Other object worthy of notice will be the soil & face of the country, its growth & vegetable productions . . . the mineral productions of every kind. . . . climate as characterized by the thermometer . . . the dates at which particular plants put forth or lose their flowers, or leaf, times of appearance of particular birds, reptiles, or insects.
>
> **Thomas Jefferson,** quoted in *The Journals of Lewis and Clark*

a. What was Jefferson expecting to find in the West?

b. Why might the president want to know about the land's soil and vegetable production?

c. What aspect of human geography might be of interest to the president?

ALTERNATIVE ASSESSMENT

1. INTERDISCIPLINARY ACTIVITY: Math

Plotting Latitude and Longitude On a piece of graph paper, sketch a map of the United States. Be sure to draw in state boundaries, too. Then, using an atlas as a reference, draw and mark the latitude and longitude lines that cross the nation at five degree intervals. Plot the estimated longitude and latitude location of your city or town. Determine at which degrees the lines intersect where you live. Repeat this exercise for at least five different places you have visited or would like to visit within the United States.

2. COOPERATIVE LEARNING ACTIVITY

Making a Map How well do you know the neighborhood around your school? Form groups of three to four students. Then work together to draw a map of the neighborhood around your school. Include:

- streets
- residences
- stores
- geographic features
- important landmarks

The map should be accurate but not too cluttered with unnecessary details. Compare your group's map with those of the other groups in the class.

3. TECHNOLOGY ACTIVITY

Writing Directions Several Internet sites provide detailed maps of the United States. They also provide driving directions to most places in the country.

- Locate one of these map sites on the Internet.
- Think of a place in the United States that you would like to visit.
- Work with the computer to find the best route for reaching it.

Write out clear directions as well as the total mileage of your trip. Also, note the type of map it is and the features it highlights.

Visit www.mcdougallittell.com for important geography sites.

4. HISTORY PORTFOLIO

Review your alternative assessment activities. Select the one that you think was your best work. Choose one of the options below.

Option 1 Use comments made by your teacher or classmates to improve your work.

Option 2 Depict your work visually, in a graph or chart, for example.

Add your work to your History portfolio.

UNIT 1

BEGINNINGS TO 1763

Three Worlds Meet

The ruins of the Anasazi Cliff Palace, in modern-day Colorado, stand as a reminder of the Native American societies that existed before the arrival of Europeans.

"The story of my people and the story of this place are one single story."

—Taos Pueblo man

CHAPTER **1**

The World in 1500 Beginnings–1500

Imagine how these Native Americans might feel at their first sight of strange ships.

Interact *with* History

The year is 1492, and you live on an island in the Caribbean. One day you see a giant boat topped by strange white cloths. Men climb into smaller boats and row toward you. You have never seen men like this. They have pale skin and wear heavy, colorful clothing. You wonder what will happen when they land.

What Do You Think?

- What can different societies learn from each other?
- What might they want to gain from each other?
- What positive and negative things might happen when they meet?

What happens when different societies meet?

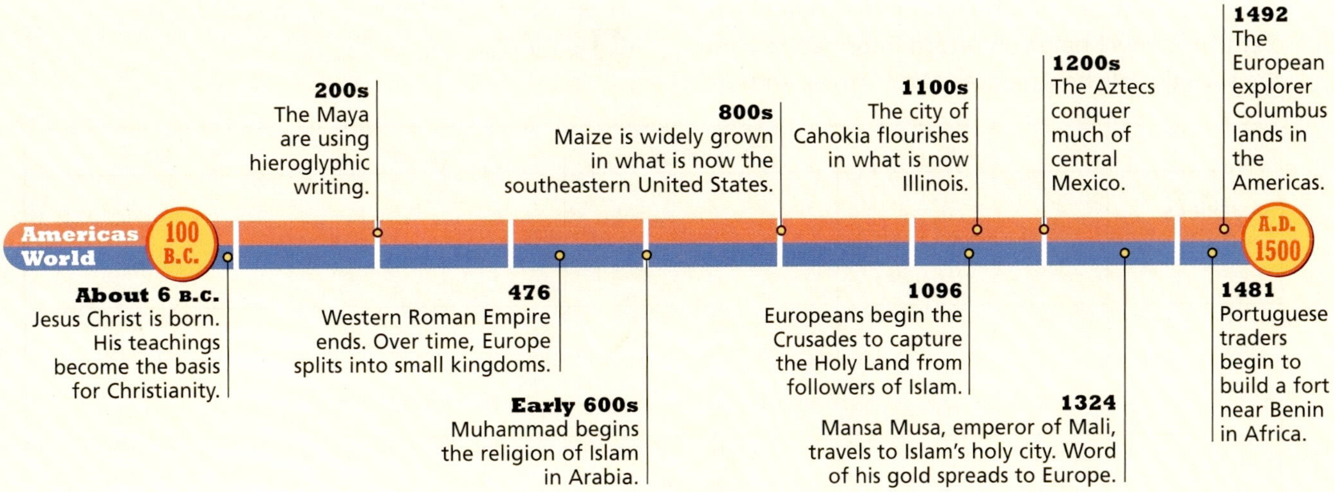

hapter 1 SETTING THE STAGE

BEFORE YOU READ

Previewing the Theme

Diversity and Unity Chapter 1 explains that by 1500, diverse societies had developed in North America, West Africa, and Europe. After 1500, economic, political, social, and religious forces brought West Africans and Europeans to North America. Those people and Native Americans helped create the United States.

CENTRAL AMERICA This pyramid at Chíchen Itzá, in what is now Mexico, was built between the 900s and 1200s.

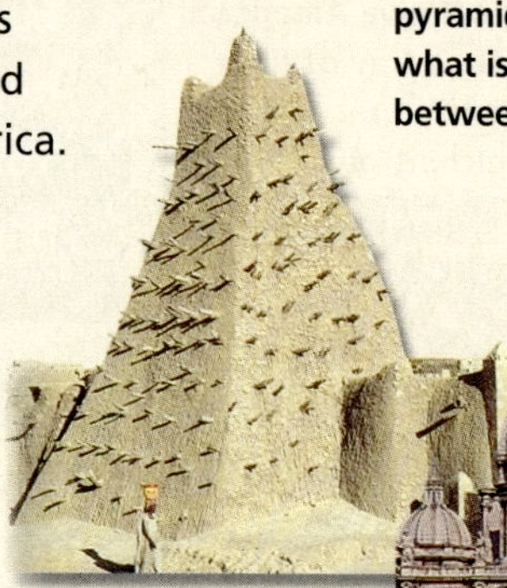

AFRICA This Muslim mosque in Timbuktu, Mali, was built in the 1300s and 1400s.

EUROPE St. Peter's Basilica (a Christian church) in Rome, Italy, was built in the 1500s and early 1600s.

What Do You Know?

What do you know about the history of North America, West Africa, and Europe? How advanced must a society be to build large structures like the ones at the left?

THINK ABOUT

- what you know about other societies, such as Egypt, that built large structures
- what you've read in books

What Do You Want to Know?

What questions do you have about the past societies of North America, West Africa, and Europe? What do you want to know about how they met? Record those questions in your notebook before you read the chapter.

READ AND TAKE NOTES

Reading Strategy: Categorizing One way to make better sense of what you read is to categorize. To categorize is to sort information into groups. The chart below will help you record information about the societies of the Americas, West Africa, and Europe. As you read, look for information relating to the categories of trade, technology, religion, and art. Record that information on your chart.

See Skillbuilder Handbook, page R6.

	Trade	Technology	Religion	Art
AMERICAS				
WEST AFRICA				
EUROPE				

1 Crossing to the Americas

TERMS & NAMES
archaeologist
artifact
migrate
culture
domestication
civilization
irrigation
Mound Builders

MAIN IDEA

Ancient peoples came from Asia to the Americas and over time developed complex civilizations.

WHY IT MATTERS NOW

Archaeologists and other scientists continue to make new discoveries about these ancient people.

ONE AMERICAN'S STORY

To do her work, Solveig Turpin must climb rugged cliffs, step over rattlesnakes, and dodge sharp cactus spines. For more than 20 years, she has searched the caves and cliffs of Texas for paintings that ancient people left on rock walls. Turpin is an archaeologist. That is a scientist who studies the human past by examining the things people left behind. One painting that Turpin found shows a red, 9-foot-long panther. She believes it shows a religious leader who turned himself into an animal.

Archaeologist Solveig Turpin wears a shirt displaying the rock art of ancient peoples as she discusses her work.

A VOICE FROM THE PAST

This is the Shaman [religious leader] who transforms into the largest and most powerful animal here. . . . I like to call [the shamans] supramen because they were over everything.

Solveig Turpin, quoted in *In Search of Ancient North America*

Archaeologists make theories about the past based on what they learn from bones and artifacts. Artifacts are tools and other objects that humans made. They give clues about who ancient people were and how they lived. This section discusses some theories about early Americans.

The First People in America

As many societies do, many Native Americans have stories explaining the origin of their people. Some believe the gods created their ancestors. Others believe their ancestors were born of Mother Earth. In contrast, scientists think that the first Americans **migrated**, or moved, here from Asia. But scientists disagree about how and when this move took place.

Some ancient people may have crossed a land bridge that joined Asia and North America during the last Ice Age. The Ice Age was a time of extreme cold that lasted for thousands of years. Glaciers trapped so much water that ocean levels dropped. A bridge of land, now called Beringia, appeared where the Bering Strait is now. (See map, page 28.) When the earth grew warm again, the glaciers melted and flooded Beringia. Some scientists who hold this theory believe the earliest Americans arrived

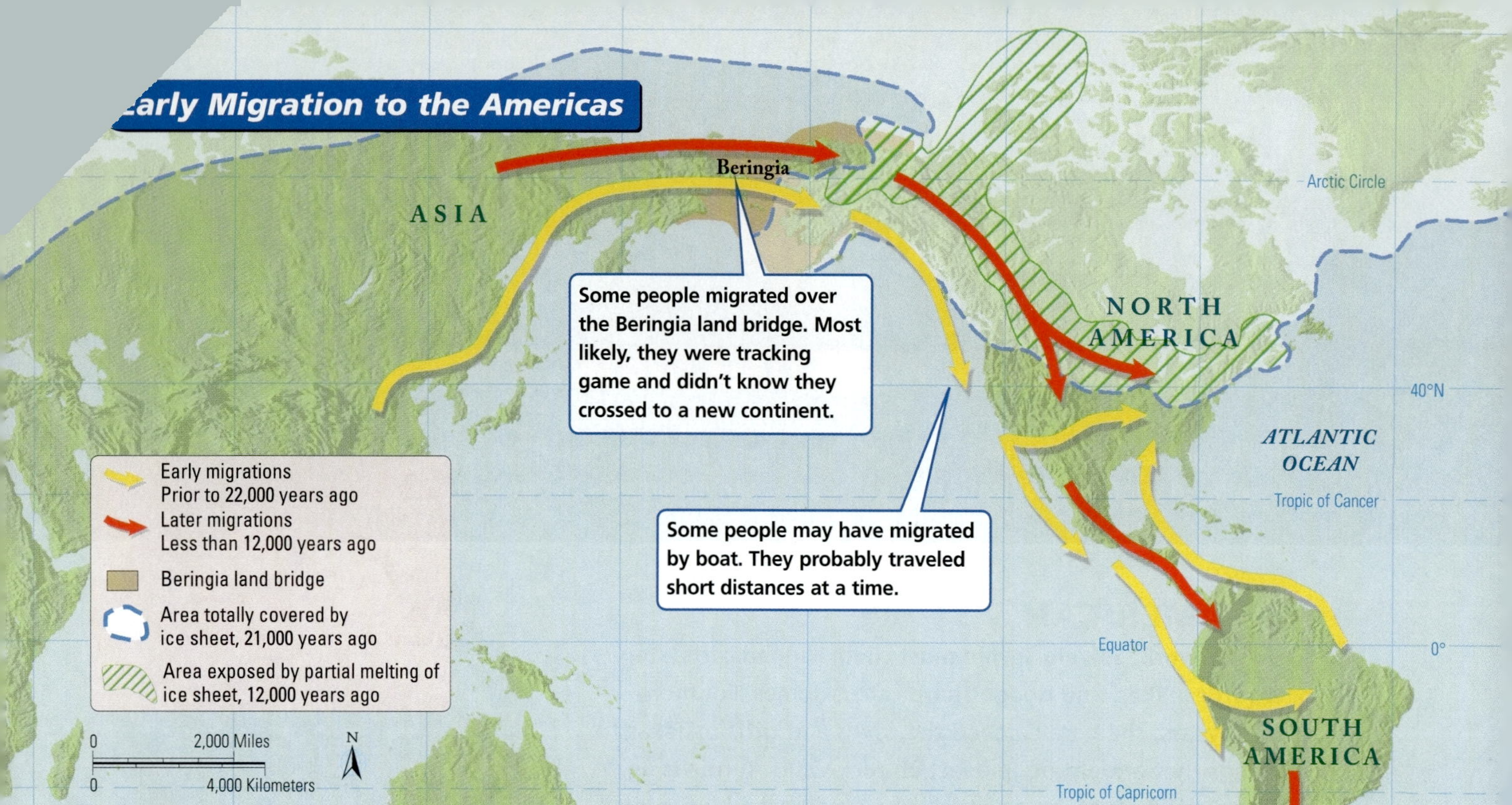

GEOGRAPHY SKILLBUILDER Interpreting Maps

1. **Movement** *The oldest artifacts have been found in Chile, in western South America. Which set of migrations probably led to human settlement there?*
2. **Human-Environment Interaction** *What geographic feature, shown on the map, do you think motivated ancient people to keep moving south?*

Ancient hunters used spear points such as this one, which is about 10,000 years old.

12,000 years ago. Other scientists believe humans came to the Americas much earlier. They have found artifacts in South America that tests show to be 30,000 years old. These scientists believe that people came to the Americas by many routes, over thousands of years. Some came by boat, sailing short distances from island to island. This theory may also change as scientists find more evidence of ancient Americans.

Agriculture Leads to Civilization

A **culture** is a way of life shared by people with similar arts, beliefs, and customs. The first Americans lived in hunting and gathering cultures. They hunted small animals, such as rabbits, and large animals, such as the woolly mammoth. They gathered wild seeds, nuts, and berries.

Vocabulary
woolly mammoth: a hairy ancestor of the elephant, now extinct

In time, people started to plant the seeds they found. This was the beginning of agriculture. About 5,000 years ago, humans began domestication. **Domestication** is the practice of breeding plants or taming animals to meet human needs. By trial and error, people in central Mexico learned which seeds grew the best crops. By selecting the right seeds, they improved the quality of maize, or corn, until its ears were large. Dried and stored for future use, corn became a main food source.

Knowledge of agriculture spread throughout the Americas. Having a stable food supply changed the way people lived. Once they no longer had to travel to find food, they built permanent villages. Farmers were able to produce large harvests, so that fewer people needed to farm.

Reading History

A. Drawing Conclusions Why would a culture need to learn agriculture before it could develop a civilization?

Some people began to practice other crafts, such as weaving or making pots. A few people became religious leaders.

Slowly, some cultures grew complex and became civilizations. A **civilization** has five features: (1) cities that are centers of trade, (2) specialized jobs for different people, (3) organized forms of government and religion, (4) a system of record keeping, and (5) advanced tools.

Early Mesoamerican Civilizations

About 1200 B.C., an advanced civilization arose in Mesoamerica, a region that stretches from central Mexico to present-day Nicaragua. For 800 years, a people called the Olmec thrived along the Gulf of Mexico. The Olmec set up a network of trade routes and constructed earthen mounds shaped like pyramids. They built large, busy cities like La Venta.

A VOICE FROM THE PAST

La Venta was not just an empty ceremonial spot visited by Olmec priests and nobles but a prosperous community of fishers, farmers, traders, and specialists, such as the artisans and the sculptors.

Rebecca González, quoted in "New Light on the Olmec," *National Geographic*

Around 400 B.C., the Olmec abandoned La Venta and other cities. Scientists don't know why. By then, Olmec culture had spread along trade routes and influenced others. Later people in Mesoamerica adapted Olmec religious practices and carved designs inspired by Olmec art.

By A.D. 250, about 650 years after the Olmec vanished, the Maya had developed a great civilization. Their cities were in southern Mexico and Guatemala, where they built pyramid mounds topped by temples. From artifacts, archaeologists know that the Maya had an accurate yearly calendar. They were the first people in the Americas to create a number system using zero. Their written language used picture symbols.

Background

Mathematicians in India also developed the idea of using a symbol for zero. Traders later carried the idea from Asia to Europe.

By 900, the Maya had abandoned many of their cities. Scientists think that revolts, disease, or crop failures may have caused their society to fail.

The Hohokam and the Anasazi

During the Mayan period, an agricultural people inhabited the American Southwest. The Hohokam lived in what is now Arizona from about 300 B.C. to A.D. 1400. That desert region has little rain, so farming is difficult. But the Hohokam altered their dry environment. They dug hundreds of miles of canals to carry river water to their crops. The practice of bringing water to crops is called **irrigation.**

The Hohokam raised corn, beans, and squash. They also gathered wild plants and hunted animals. They traded widely—with people in Mexico, the Southwest, and California. Hohokam pottery and religious practices show the influence of Mesoamerican cultures, which they learned about through trade.

Ancient peoples of the American Southwest used images like this to communicate with each other. Such images are called petroglyphs.

The Mound Builders

During the 1700s, Europeans discovered several mysterious earthen mounds in what is now the American Southeast and Midwest. They believed a lost civilization had built the mounds. Historians now know that different Native American groups, known as the Mound Builders, built these structures. The builders may have used the mounds for burial tombs, as a tribute to their gods, or for some other religious purpose. One famous mound is the Great Serpent Mound in present-day Ohio. The Adena, Fort Ancient, or Hopewell cultures possibly built it. An aerial photograph of the Great Serpent Mound is shown below.

The mound is a quarter mile long, averages 20 feet wide, and is about 5 feet high.

CONNECT TO HISTORY

1. **Drawing Conclusions** Review the five characteristics of a civilization on page 29. Which of these characteristics would a culture need to be able to build something like the Great Serpent Mound?

See Skillbuilder Handbook, page R12.

CONNECT TO TODAY

2. **Researching** How do modern monuments to the dead differ from those constructed by the Mound Builders?

Visit www.mcdougallittell.com to learn more about the Mound Builders.

Beginning about A.D. 100, the Anasazi lived in the area where Utah, Arizona, Colorado, and New Mexico now meet. Scientists don't know their origin. Like the Hohokam, the Anasazi were mainly farmers who also traded widely.

The Anasazi built houses with hundreds of rooms and many stories. For protection, they placed some buildings against overhanging canyon walls. The 800-room Pueblo Bonito in Chaco Canyon, New Mexico, housed perhaps 1,000 people. In the 1500s, when Spanish explorers first saw these houses, they called them *pueblos,* meaning villages. Around 1300, drought or warfare caused the Anasazi to leave their homes.

The Mound Builders

In the eastern part of what is now the United States lived several groups of people called Mound Builders. The **Mound Builders** were early Native Americans who built large earthen structures.

The two oldest Mound Builder societies were the Adena and the Hopewell. Archaeologists know little about the Adena. The Hopewell, located in what is now the Midwest, lived from 400 B.C. to A.D. 400. Like the Hohokam, they grew corn. Artifacts show that they had a large trade network. It stretched from the Atlantic to the Rocky Mountains, and from the Great Lakes to Florida. Hopewell mounds served as burial sites. Their tombs contained jewelry and other gifts for the dead.

Reading **History**
B. Comparing What other ancient American cultures built pyramid-shaped mounds?

The last group of Mound Builders, the Mississippians, lived from A.D. 800 to 1700. They built some of the first cities in North America. For example, Cahokia in Illinois has more than 100 mounds. One of them, Monks Mound, rises 100 feet and covers 16 acres. In some cities, the Mississippians built flat-topped, pyramid-shaped temple mounds.

By the 1700s, most of the Mississippians had died from diseases they caught from Europeans. But many Native American groups continued to thrive throughout the United States, as you will read in Section 2.

Section 1 Assessment

1. Terms & Names

Identify:
- archaeologist
- artifact
- migrate
- culture
- domestication
- civilization
- irrigation
- Mound Builders

2. Taking Notes

Use a chart like the one below to list ancient cultures of Mesoamerica and North America and their locations.

Ancient Culture	Location

Which of these cultures was closest to where you live?

3. Main Ideas

a. By what land bridge did some ancient people migrate to North America, and how was it created?

b. How did the development of farming lead to the growth of civilization?

c. How did trade help to spread culture?

4. Critical Thinking

Comparing How did the Hohokam and the Anasazi adapt to living in their environment?

THINK ABOUT
- Hohokam agriculture
- Anasazi dwellings

ACTIVITY OPTIONS

SCIENCE
LANGUAGE ARTS

Research the growing of corn. Draw a **diagram** of a corn plant with its parts labeled or write a **description** of how corn grows.

2 Societies of North America

TERMS & NAMES
technology
tundra
kayak
matrilineal
slash-and-burn agriculture
Deganawida
Iroquois League

MAIN IDEA

By 1500, a variety of Native American groups—each with a distinct culture—lived in North America.

WHY IT MATTERS NOW

Many Americans today claim one or more of these cultures as part of their heritage.

ONE AMERICAN'S STORY

Many Native Americans today work to save their culture, so it won't vanish as some ancient cultures did. Haida artist Bill Reid took part in this effort. When he was a teenager in the 1930s, totem poles and other Haida crafts existed mostly in museums. Few Haida carvers were making new ones. Reid himself did not learn about Haida arts until he was 23 and visited his grandparents' island village off the west coast of Canada. Reid's grandfather showed him the tools of a great Haida artist of the 1800s.

Reid began to study Northwest Coast native arts by reading books and visiting museums. After taking a class in jewelry making, he created gold jewelry with Haida designs. He also carved totem poles and sculptures. When Reid died in 1998, his work was praised.

A VOICE FROM THE PAST

Canada has lost one of its greatest artists. A descendant of a lineage of great Haida artists . . . , Bill Reid revived an artistic tradition that had survived only in museum collections.

Dr. George MacDonald, at Bill Reid's memorial service, March 24, 1998

Creating sculptures of traditional designs was one way Bill Reid kept Haida culture alive.

Written records and people like Reid have preserved knowledge of the cultures that flourished in the Americas when Europeans arrived. This section explains the diversity of Native American groups in 1500.

Native American Diversity

By 1500, Native Americans had divided into hundreds of cultural groups, speaking perhaps 2,000 languages. One reason Native Americans were so diverse was that each group adapted to its own environment—whether subzero ice fields, scorching deserts, or dense forests.

Environment shaped each group's economy, technology, and religion. **Technology** is the use of tools and knowledge to meet human needs. In some regions, Native Americans based their economy on farming. In others, they relied on hunting or fishing. Different environments caused technology to vary. In coastal areas, farmers made tools from shells. In

GEOGRAPHY SKILLBUILDER
Interpreting Maps

1. **Movement** *Which Native American culture was able to trade directly with the Aztecs?*
2. **Location** *Which culture lived around the Great Lakes?*

deserts, they used irrigation. Environment affected religion, too. Native Americans strongly believed that certain places were sacred—and that animals, plants, and natural forces had spiritual importance.

Although Native American groups had many differences, they all felt closely connected to nature, as shown in the following chant.

A VOICE FROM THE PAST

Earth's body has become my body
by means of these I shall live on.
Earth's mind has become my mind
by means of these I shall live on.

Navajo Blessing Way, quoted in *America in 1492*

In addition, trade linked Native Americans. Trading centers developed across North America, especially at points where two cultures met.

Peoples of the North and Northwest Coast

Background
The Inuit are also called the Eskimo.

The Aleut (uh•LOOT) and the Inuit (IHN•yoo•iht) were peoples of the far North. The Aleut lived on islands off Alaska, and the Inuit lived near the coast on tundra. **Tundra** is a treeless plain that remains frozen under its top layer of soil. Ice and snow cover the ground most of the year.

Because their climate was too cold for farming, the Inuit and Aleut were hunters. They paddled **kayaks**, small boats made of animal skins, into icy seas to spear whales, seals, and walruses. They hunted these mammals for food, and they made seal and walrus skins into clothes. Some Inuit religious ceremonies honored the spirits of the whales and seals they caught. The Inuit also hunted such land animals as caribou. They made arrowheads and spear points from bones and antlers.

Farther south, Northwest Coast people also hunted sea mammals. But they mostly fished for salmon. Living by forests, Northwest Coast people used wood for houses, boats, and carved objects. They traded such coastal products as shells for items from the inland, such as furs.

Reading History
A. Making Inferences Why would inland people trade for seashells?

"The term [potlatch] comes from Chinook . . . and means 'to give.'"
Gloria Cranmer Webster, U'mista Cultural Centre

Northwest Coast groups such as the Kwakiutl (KWAH•kee•OOT•uhl) and Haida had a special ceremony—the potlatch. Individuals would give away most or all of their goods as a way to claim status and benefit their community. They held potlatches to mark life events, such as naming a child or mourning the dead.

Peoples of the West

Unlike the Native Americans of the Northwest Coast, those of the West did not rely mainly on the sea. The peoples of the West included tribes in California, the Columbia Plateau, and the Great Basin. Much of the West is desert or is not suitable for farming. The people who lived there existed mainly by hunting and gathering.

Reading History
B. Reading a Map On the map on page 33, locate the cultures of California, the Plateau, and the Great Basin. Notice why these three together are called the peoples of the West.

The men hunted deer, elk, antelope, rabbits, and birds. They also fished, especially for salmon that swam up the western rivers. Women gathered such wild foods as nuts, seeds, and berries. Many western groups moved with the seasons to collect food.

The Aztec Calendar

This stone is the Aztec calender wheel. In the center is the sun god. Around it are symbols for the 20 days of the Aztec month. Three are enlarged below.

The women of some western tribes became expert weavers. Pomo women wove beautiful baskets that they used to gather and store food. They wove some baskets tightly enough to be watertight.

The peoples of the West had strong spiritual beliefs, often linked to nature. Some held ceremonies to ensure a large food supply. Others held dances to ask for rain, for plant growth, and for good hunting. Still others believed that their religious leaders could contact the spirit world.

Peoples of Mexico

Far to the south, the Aztecs ruled a great civilization in what is now central Mexico. The origin of the Aztecs is unclear. They may have been hunters and gatherers like the Native Americans of the West. Sometime during the 1100s, they migrated into the Valley of Mexico.

In 1325, they began to build their capital city, Tenochtitlán (teh•NAWCH•tee•TLAHN), on islands in Lake Texcoco. Two things helped the Aztecs become a strong empire. First, they drained swamps and built an irrigation system. This enabled them to grow plenty of food. Second, they were a warlike people who conquered most of their neighbors. The defeated people then had to send the Aztecs food and resources.

The Aztecs had a complex society. Rulers were the highest class. Priests and government workers ranked next. Slaves and servants were at the bottom. The Aztecs had elaborate religious ceremonies linked to their calendar and their study of the sun, moon, and stars. Many of their beliefs came from earlier Mesoamerican cultures.

The Aztecs' most important ritual involved feeding their sun god human blood. To do this, the Aztecs sacrificed prisoners of war by cutting out the person's heart while he was still alive. One reason the Aztecs fought so many wars was to capture prisoners to sacrifice.

daily life

KACHINA DANCES

Every year in summer the Hopi, Zuni, and other Pueblo Indians held a religious celebration. The ceremony called on the kachinas, or spirits of the ancestors. The Pueblo believed the kachinas had the power to bring a plentiful harvest. At the festival, masked dancers played the role of different kachinas. They danced and sang songs to bring rain in the year ahead. Today, the Pueblo also carve kachina dolls, shown below, as well as hold dances.

Peoples of the Southwest

North of the Aztec, in what is now the American Southwest, lived the Pueblo people. Their ancestors were the ancient Hohokam and Anasazi. Like their ancestors, the Pueblo used irrigation to alter their desert region for farming. They lived in many-storied houses of adobe—dried mud bricks. These large buildings sometimes held an entire village.

Pueblo Indian farmers raised corn, beans, and squash. For meat, they hunted game and raised turkeys. Men did most of the farming, hunting, weaving, and building. Women ground the corn and cooked the food, repaired the adobe houses, and crafted pottery.

The Navajo and the Apache were nomadic, or wandering, hunter-gatherers who came to the region later than the Pueblo. For food, they relied mainly on game and on cactus, roots, and piñon nuts. Often, they traded these wild products for crops that the Pueblo had grown. Over time, the Navajo adopted farming and other Pueblo practices.

Reading **History**

C. Drawing Conclusions Why do you think the Navajo adopted farming?

Peoples of the Great Plains

Farther north, the Great Plains is a flat grassland region stretching from the Mississippi River west to the Rocky Mountains. Today, most people think of Plains Indians on horseback, but originally they had no horses. The Spanish first brought horses to the Americas in the 1500s.

Background

Horses spread across North America through trade and by escaping from humans and wandering on their own.

Some Plains groups were nomads. Others lived in villages by rivers, where land was easier to farm. In summer, entire villages set out to track bison. Hunting bison on foot was difficult, but Plains tribes used their environment to help them. Working together, the villagers stampeded the herd over a cliff, so the fall would kill or disable the animals. Plains Indians not only ate the bison's meat. They also made its hide into clothes and its bones into tools.

In winter such northern Plains groups as the Mandans and Pawnee lived in large circular lodges. Wooden beams held up the earthen walls. A hole at the top provided air, light, and an outlet for smoke from the fire. Buried partly underground, the earth lodge protected the people from the extreme cold and wind of the Plains climate.

Vocabulary
lodge: house

The spiritual beliefs of Plains tribes varied. Some felt a close tie to regional animals such as the bison or plants such as corn. Some honored sacred places, such as the Black Hills of South Dakota and Wyoming. Many Plains tribes held a ceremony called the Sun Dance, which involved making a vow and asking the Creator for aid.

Peoples of the Southeast

The Southeast, which stretches from east Texas to the Atlantic Ocean, has mild winters and warm summers with plentiful rainfall. The long growing season led the Choctaw (CHAHK•taw), Chickasaw (CHIHK•uh•SAW), and other southeastern groups to become farmers. As many other Native Americans did, they grew corn, beans, squash, and pumpkins.

***Reading* History**
D. Analyzing Causes Why would a long growing season lead people to become farmers?

Women did most of the farming, while men hunted, fished, and cleared land. The men spent months in the forest tracking deer. In the Southeast, people traced their family ties through the women. Societies in which ancestry is traced through the mother are called **matrilineal.**

In southeastern villages, people gathered at a central square for public meetings and such religious ceremonies as the Green Corn Festival. Held once a year, this festival offered thanks for the corn harvest and also served as a kind of New Year's celebration. People cleaned their houses, threw away old pots, and settled quarrels as a sign of a fresh start for the year.

Peoples of the Eastern Woodlands

Like the Southeast, the Northeast had plenty of fish, game, and rain. But the climate was colder with snowy winters. Forests covered much of the region, so it is called the Eastern Woodlands. Most of the people living there spoke either an Iroquoian or Algonquian language.

Many Native Americans in the Southeast and Eastern Woodlands played lacrosse using sticks like these. Modern Americans have adopted the game.

Like all Native Americans, the Iroquois learned to live in their environment. They hunted wild game. They adapted the forest for farming by using slash-and-burn agriculture. In **slash-and-burn agriculture,** farmers chopped down and then burned trees on a plot of land. The ashes from the fire enriched the soil. When a field's soil became worn out, the farmer abandoned it and cleared a new field. The Iroquois lived in longhouses, bark-covered shelters as long as 300 feet. One longhouse held eight to ten families.

Background
Although it is a quick way to clear fields, slash-and-burn agriculture does cause environmental damage by destroying forests.

The Algonquin lived in wigwams, domelike houses covered with deerskin and slabs of bark. For protection, both the Iroquois and Algonquin surrounded their villages with high fences made of poles. Iroquois villagers often needed protection not only from the enemies of the Iroquois, but from each other. The Iroquois often raided neighboring villages for food and captives.

In the late 1500s, five northern Iroquois nations took the advice of a peace-seeking man named **Deganawida.** They stopped warring with each other and formed an alliance. This alliance of the Cayuga, Mohawk, Oneida, Onondaga, and Seneca was the **Iroquois League.** The League brought a long period of peace to the Iroquois. A council of leaders from each nation governed the League. They followed rules called the Great Law of Peace. The Iroquois were also a matrilineal society. If a leader did something wrong, the women of his clan could vote him out of office.

Across the Atlantic, the peoples of West Africa also adapted to their environment and engaged in trade. West Africa was the region from which most Africans were brought to the Americas. You will read about it in the next section.

AMERICA'S HISTORY MAKERS

DEGANAWIDA (THE PEACEMAKER)

Iroquois tradition honors Deganawida as the Peacemaker. Seeing how destructive warfare was for the Iroquois, Deganawida went from tribe to tribe and described his dream of peace. A poor speaker, he persuaded few warriors. Finally, an Iroquois chief named Hiawatha spoke for him. After long negotiations, the leaders of the warring nations made peace. However, Deganawida's own tribe, the Huron, did not join the League.

How did both Deganawida and Hiawatha lead the Iroquois toward peace?

Section 2 Assessment

1. Terms & Names

Identify:
- technology
- tundra
- kayak
- matrilineal
- slash-and-burn agriculture
- Deganawida
- Iroquois League

2. Taking Notes

Use a cluster diagram to record how Native Americans from each region adapted to their environment.

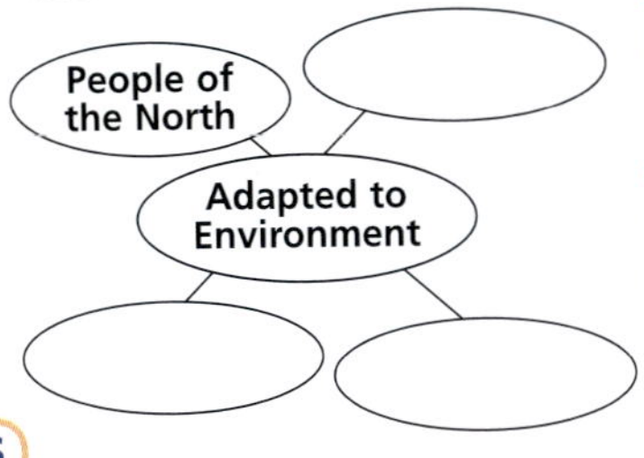

3. Main Ideas

a. What were some of the religious ceremonies of Native Americans?

b. How were the Pueblo like their ancestors, the Hohokam?

c. How did the formation of the Iroquois League benefit its member nations?

4. Critical Thinking

Drawing Conclusions
How did trade benefit both groups that took part in it?

THINK ABOUT
- who the Northwest Coast people traded with and what they exchanged
- what the Pueblo exchanged with nomadic groups

ACTIVITY OPTIONS

ART

LANGUAGE ARTS

Reread the Navajo chant on page 33. Draw an **illustration** to go with the chant or write additional **verses.**

The Iroquois Great Law of Peace

Setting the Stage The five nations of the Iroquois League created a constitution, called the Great Law of Peace, that had 117 laws and customs. These laws governed all aspects of life and war. In this excerpt, Deganawida introduces the Great Law by describing a tree that symbolizes the permanence and stability of the league. **See Primary Source Explorer** CD-ROM

1 I am Deganawida and with the Five Nations' Confederate **Lords**[1] I plant the Tree of Great Peace. I plant it in your territory, **Adodarhoh,**[2] and the Onondaga Nation, in the territory of you who are Firekeepers.

I name the tree the Tree of the Great Long Leaves. Under the shade of this Tree of the Great Peace we spread the soft white feathery down of the globe thistle as seats for you, Adodarhoh, and your cousin Lords.

We place you upon those seats, spread soft with the feathery down of the globe thistle, there beneath the shade of the spreading branches of the Tree of Peace. There shall you sit and watch the Council Fire of the **Confederacy of the Five Nations,**[3] and all the affairs of the Five Nations shall be transacted at this place before you, Adodarhoh, and your cousin Lords, by the Confederate Lords of the Five Nations.

2 Roots have spread out from the Tree of the Great Peace, one to the north, one to the east, one to the south, and one to the west. The name of these roots is The Great White Roots and their nature is Peace and Strength.

If any man or any nation outside the Five Nations shall obey the laws of the Great Peace and make known their disposition to the Lords of the Confederacy, they may trace the Roots to the Tree and if their minds are clean and they are obedient and promise to obey the wishes of the Confederate Council, they shall be welcomed to take shelter beneath the Tree of the Long Leaves.

A CLOSER LOOK

THE COUNCIL FIRE

The council fire of the Iroquois League was kept burning for about 200 years.

1. What do you think it would mean if the council fire were allowed to die?

A CLOSER LOOK

THE GREAT WHITE ROOTS

The roots of a tree help to anchor it in the ground, and they draw water and food from the soil.

2. Why might Deganawida say the nature of the roots is "Peace and Strength"?

1. **Lords:** chiefs.
2. **Adodarhoh:** the name of the office of the Onondaga chief.
3. **Confederacy of the Five Nations:** the Iroquois League.

Interactive Primary Source Assessment

1. Main Ideas

a. In what territory was the Tree of the Great Peace planted?

b. Where will the affairs of the Five Nations be conducted?

c. Where have the Tree's roots spread?

2. Critical Thinking

Making Inferences Were outsiders welcome to join the Iroquois League? Explain.

THINK ABOUT
- the phrase *they may trace the Roots to the Tree*
- the phrase *take shelter beneath the Tree*

3 Societies of West Africa

TERMS & NAMES
- Ghana
- Muslims
- Islam
- Mali
- Songhai
- Hausa
- Yoruba
- Benin

MAIN IDEA

The peoples of West Africa developed sophisticated kingdoms, trade networks, and artistic achievements.

WHY IT MATTERS NOW

It was from this region that many Africans were brought to the Americas.

ONE AFRICAN'S STORY

King Tenkaminen (TEHN•kah•MEE•nehn) of the West African empire of <u>Ghana</u> was a powerful ruler. He grew rich by taxing gold traders who traveled through his land. Travelers who visited Tenkaminen were impressed by his wealth. In 1067, a geographer wrote a description of the royal court.

A VOICE FROM THE PAST

The king adorns himself . . . wearing necklaces round his neck and bracelets on his forearms. . . . Behind the king stand ten pages holding shields and swords decorated with gold and on his right are the sons of the vassal [lower] kings of his country wearing splendid garments and their hair plaited [braided] with gold.

al-Bakri, quoted in *The Horizon History of Africa*

Kumasi, a modern West African chief, wears gold to show his status, just as the ancient king of Ghana did.

West Africa had several other kingdoms and empires that grew powerful through trade. It was also the homeland of many of the enslaved Africans who were brought to the Americas after 1500. You will read about West Africa in this section.

African Geography and World Trade

Africa is the world's second largest continent after Asia. (See the map on page 40.) Although Africa has a variety of land forms and climates, almost three quarters of it lies within the tropics. The equator runs east-west across the center of Africa. Dense rain forests stretch along the equator in central and western Africa. North and south of the rain forests are broad savannas, which are grassy plains with thorny bushes and scattered trees. Beyond the savanna in the North lies the Sahara, the world's largest desert. Beyond the savanna in the South lies the smaller Kalahari Desert.

By A.D. 1500, coastal ports had linked Africa with the rest of the world for many centuries. Ships from ports on the Mediterranean and the Red Sea carried goods to Arabia and Persia. On Africa's east coast, city-states carried on a brisk trade with ports across the Indian Ocean.

Like other parts of Africa, West Africa has rain forest along the equator and savanna to the north. The Niger River arcs across those grasslands and forests and then empties into the Atlantic Ocean. Along its northern edge, West Africa borders the Sahara.

Ghana Grows Wealthy

On a map, the Sahara appears to be a barrier between West Africa and the ports on the Mediterranean coast. But by A.D. 500, camel caravans led by eager merchants made regular journeys across the great desert. This connected West Africa to the wider world.

Background
The camel is used in the desert because it can travel up to 10 days without water.

Ghana became the first West African kingdom to grow rich through trade. From the 700s to the mid 1000s, Ghana prospered by controlling the busy trade in gold and salt. Located on the southern edge of the Sahara, Ghana became a marketplace for traders going north and south in search of salt and gold. (Ancient Ghana was northwest of modern Ghana.) Salt was important because it helps the human body retain water in hot weather. Traders carried salt from the Saharan salt mines in the north. In Ghana's markets, they met other traders offering gold from the forests of West Africa.

Reading History
A. Analyzing Causes Why did the king want taxes to be paid in gold and all gold nuggets to be given to him?

Ghana's king benefited from this trade. He imposed taxes on all gold and salt passing through his kingdom. The taxes had to be paid in gold. The king also claimed all gold nuggets found in his kingdom. Ghana's king used the resulting wealth to pay for an army and build an empire.

West African Empires, 800–1500

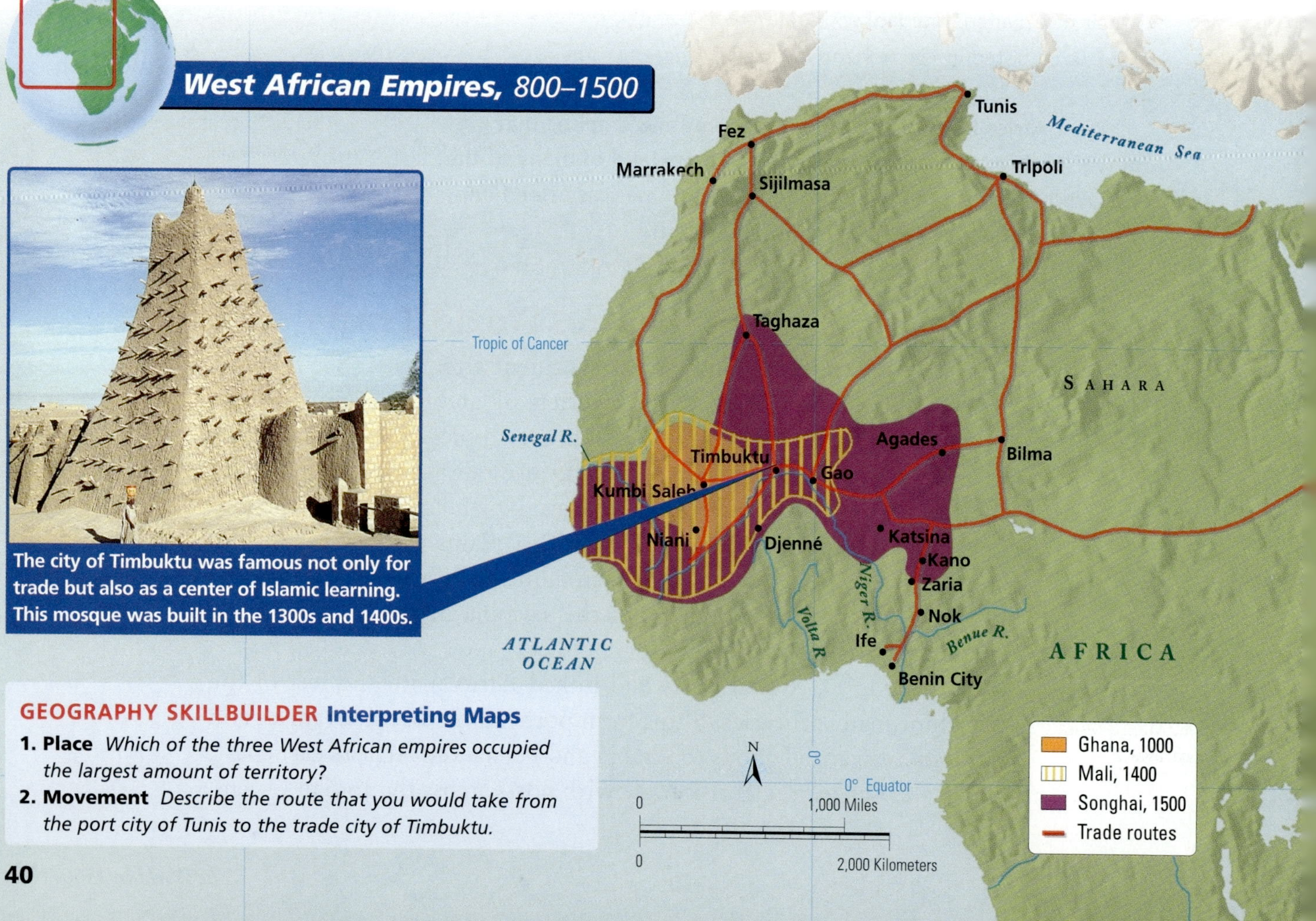

The city of Timbuktu was famous not only for trade but also as a center of Islamic learning. This mosque was built in the 1300s and 1400s.

GEOGRAPHY SKILLBUILDER Interpreting Maps

1. **Place** *Which of the three West African empires occupied the largest amount of territory?*
2. **Movement** *Describe the route that you would take from the port city of Tunis to the trade city of Timbuktu.*

Islam Enters Ghana

Vocabulary
pilgrimage: a trip to a holy place

Many of the traders who came to Ghana from North Africa were Muslims. **Muslims** are followers of the religion of Islam. Founded by the prophet Muhammad in the 600s, **Islam** teaches that there is one God, named Allah. Muslims must perform such duties as praying five times a day and making a pilgrimage to the holy city of Mecca in Arabia. Muslim traders crossing the Sahara brought Islam from North Africa to West Africa. Ghana's rulers allowed those Muslims to build mosques, or houses of worship, in Ghana's capital, Kumbi Saleh. In time, Ghana's rulers employed Muslims as advisers.

The Muslim empires of North Africa wanted to convert Ghana's people to Islam and to control Ghana's gold trade. In 1076, a Muslim army conquered Kumbi Saleh. This lessened Ghana's power. A number of local leaders took advantage of Ghana's weakness. They built up their own small states on the edges of the once mighty empire. Ghana never regained its former strength.

Over the next several centuries, more and more West Africans converted to Islam. In fact, many of the enslaved Africans who were brought to the Americas were Muslims.

Mali Replaces Ghana

By the 1200s, another West African kingdom had taken over most of Ghana's territory. This kingdom, called **Mali,** became West Africa's most powerful state. Its wealth also came from control of the gold-salt trade. But because it was located farther south than Ghana, Mali was better able to control the trade on the upper Niger River. (Ancient Mali stretched farther west than modern Mali and not as far north.)

A European mapmaker placed this picture of Mansa Musa on a map in 1375. It was drawn from his imagination.

Mali's first great ruler, Sundiata (sun•JAHT•ah), reigned from about 1230 to 1255. He came to power by crushing a cruel, unpopular leader. Sundiata's armies conquered many important trading cities. This made Mali's hold on trade stronger and made Mali more prosperous. Sundiata was a Muslim, but he did not force his people to accept Islam. Most of the people of Mali retained their traditional African beliefs.

Vocabulary
devout: very religious

Reading **History**
B. Making Inferences How do you think the Egyptians reacted to Mansa Musa's caravan?

Mali's other great leader was Mansa Musa (MAHN•sah moo•SAH), who was a devout Muslim. Mansa Musa came to the throne in 1312. Under his leadership, the empire became one of the largest in the world.

Mansa Musa is best remembered for making the Muslim pilgrimage to Mecca in 1324 and 1325. On his way to Mecca, he stopped in Cairo, Egypt. According to some stories, Mansa Musa entered the city leading a huge caravan that included 500 servants who waved staffs decorated with gold. Each of the 80 camels in his caravan struggled under the weight of a 300-pound sack of gold. The legend of Mali's wealth spread all the way to Europe. This was one reason that Europeans began to trade with Africa about 150 years later.

On his return to Mali, Mansa Musa brought back many Muslim scholars, artists, and architects. They helped spread Islamic culture and learning throughout the empire. The city of Timbuktu (TIHM•buhk•TOO) in eastern Mali became a leading center of trade and Islamic learning. After Mansa Musa's death in 1337, Mali slowly grew weaker.

The Empire of Songhai

As Mali's power decreased, the **Songhai** (SAWNG•HY) people living at the Great Bend in the Niger River broke away from its control. In 1464, under the leader Sunni Ali, they began their own empire. Sunni Ali was a Muslim, but he also practiced the traditional Songhai religion.

Under Sunni Ali, the Songhai captured the great city of Timbuktu. Then they put the important trading city of Djenné (jeh•NAY) under siege and captured it after seven years. In addition to conquering territory, Sunni Ali set up an organized system of government.

Vocabulary
siege: surrounding a castle or city with an army until it surrenders

After Sunni Ali died in 1492, conflicts arose. Some Muslims began a rebellion because they wanted Islam to be the only religion of Songhai. The leader of the revolt was Askia Muhammad, a devoted Muslim.

Askia Muhammad won his fight and became Songhai's second great ruler. For 35 years, he ably governed the empire. He chose capable officials who made the government run smoothly. He also expanded trade and set up an efficient tax system. Askia Muhammad used his wealth to build mosques and support Muslim scholars.

***Reading* History**
C. Recognizing Effects How would Askia Muhammad's actions promote Islam in Songhai?

After Askia Muhammad's reign, several weak rulers succeeded him. Even when a strong ruler took the throne again, the empire faced problems. In spite of Songhai's wealth and learning, it lacked modern weapons. In 1591, a Moroccan fighting force from North Africa invaded Songhai with gunpowder and cannon. They easily defeated Songhai's soldiers, who were defending their empire with swords and spears.

Now and then

AFRICAN HERITAGE

One way many African Americans show pride in their heritage is by wearing kente cloth. Kente cloth, shown below, is a colorful fabric woven by the Akan and Ewe people of Ghana.

Some African Americans celebrate the holiday of Kwanzaa in December. Based on traditional African harvest festivals, Kwanzaa lasts a week. Each day honors a value held by Africans: unity, self-determination, collective responsibility, cooperative economics, purpose, creativity, and faith.

Other West African Kingdoms

As empires rose and fell in some parts of West Africa, small city-states arose in other parts of the region. The **Hausa** (HOW•suh) states emerged after A.D. 1000 in what is now northern Nigeria. Hausa city-states, such as Katsina and Kano, thrived on trade. Although the Hausa people shared a language, their city-states were independent of each other.

The **Yoruba** (YAWR•uh•buh) lived in the forests southwest of the Niger River. Ife and Oyo, the largest Yoruba states, had kings who were considered to be partly divine. The Yoruba were mostly farmers, but they also had gifted artists, who carved wood and ivory and cast metal sculptures. Yoruba statues are still considered great art.

Another kingdom famous for its art was Benin. **Benin,** located in the delta of the Niger River, lay on main trade routes and prospered because of that. The capital, Benin City, was large and surrounded by thick, earthen walls. About 1600, a Dutch visitor compared Benin City to his home city of Amsterdam in Europe.

Reading **History**

D. Analyzing Points of View Do you think Dapper's view of Benin City is positive or negative? Explain.

A VOICE FROM THE PAST

The houses in this town stand in good order, each one close and evenly placed with its neighbor, just as the houses in Holland stand. . . . The king's court is very great. It is built around many square-shaped yards.

Olfert Dapper, quoted in *Centuries of Greatness*

Benin artists produced sophisticated bronze statues such as this figure of a horn-blower.

In the late 1400s, Europeans reached Benin. Portuguese ships arrived, and the Portuguese set up a trade center near Benin City. Benin traders sold the Portuguese pepper, ivory, and leopard skins in exchange for copper and guns. In time, the Portuguese and other Europeans also began to trade for enslaved Africans. The Europeans who came to West Africa were not seeking information about its rich history or culture. They wanted a supply of laborers to work on large farms, called plantations. Chapter 2 explains more about plantations and slavery.

Trade was just one reason Europeans were sailing far beyond their lands. Social changes were also spurring them to explore the world. Those changes are discussed in Section 4.

Section 3 Assessment

1. Terms & Names

Identify:

- Ghana
- Muslims
- Islam
- Mali
- Songhai
- Hausa
- Yoruba
- Benin

2. Taking Notes

Compare the Ghana Empire and the Mali Empire using a Venn diagram like the one shown.

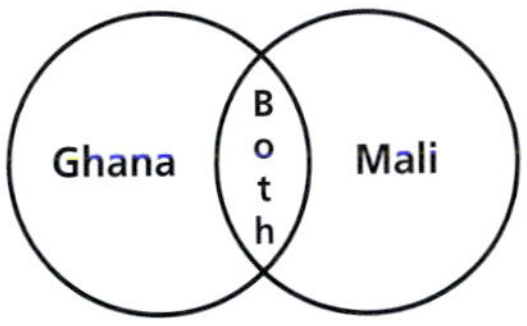

How was the influence of Islam different in each?

3. Main Ideas

a. How did Ghana's ruler benefit from controlling the gold-salt trade?

b. How did Islam spread within West Africa?

c. For what artistic achievements are the Yoruba and the people of Benin known?

4. Critical Thinking

Identifying Facts and Opinions Is the description of Benin City above mainly a statement of fact or opinion?

THINK ABOUT

- whether Dapper's statement can be proven by measurement or observation
- whether his statement expresses his own bias

ACTIVITY OPTIONS

SPEECH

GEOGRAPHY

Retell Mansa Musa's famous journey as an **oral history** or create a **map** that shows the route you think he took from Mali to Mecca.

Societies of Europe

TERMS & NAMES
European Middle Ages
feudalism
manor system
Crusades
Renaissance
printing press
Reformation
profit

MAIN IDEA

By 1500, Europe was going through a period of social change that sparked interest in learning and exploration.

WHY IT MATTERS NOW

The changes taking place in Europe led to the exploration of the Americas.

ONE EUROPEAN'S STORY

The peasants in this 11th-century drawing are probably plowing land owned by someone else. European peasants like Ermentrude hardly ever left the manor where they lived.

After telling her son to watch the baby, Ermentrude left her tiny hut. She gathered up a chicken and five eggs and went to see the steward. He was the man who managed the land where she lived for its owners. Ermentrude and her husband, Bodo, were farmers who worked on a small piece of a large estate owned by someone else.

When Ermentrude arrived at the big house on the estate, she gave the chicken and eggs to the steward as part of her rent. For a while, she visited some women. Then she hurried home to tend her grapevines, weave cloth, and cook supper before Bodo came home from plowing.

Ermentrude lived in the early 800s, but her life was typical of the way many Europeans lived for centuries. This section explains that way of life and how it had changed by 1500. The changes led Europeans to make voyages of exploration that took them to America.

Feudalism in Europe

Ermentrude lived in the **European Middle Ages,** which lasted from the late 400s, when the Western Roman Empire ended, to about the 1300s. (In some parts of Europe, the Middle Ages lasted to the 1400s.) The Romans used written laws and a mighty army to keep order. But over time, the empire grew weak. Germanic tribes from the east and north invaded the empire and contributed to its fall. The rough, uneducated Germanic tribes destroyed the strong Roman government and trade networks, and the tribes set up small kingdoms. With no trade, people stopped using money. They paid in goods, such as chickens and eggs.

Other groups also disrupted Europe. During the 800s to 1000s, Vikings swept down from the north. From their warships, they carried out lightning raids, looting villages and then racing back out to sea. To survive such difficult times, Europeans turned to feudalism. **Feudalism** is a political system in which a king allows nobles, or lords, to use lands that belong to him. In return, the lords owe the king military service and protection for the people living on the land.

Reading **History**
A. Identifying Problems What problems were Europeans trying to solve with feudalism and the manor system?

Along with feudalism, Europeans developed the **manor system.** In this system, lords divided their lands into manors, or large estates, that were farmed mostly by serfs. Serfs were landless peasants who weren't allowed to leave the manor. In return for the serfs' work, the lord promised to protect them. The lords built heavily walled castles where people could go in times of danger.

The Roman Catholic Church also gained power during these uncertain times. Taking on the roles once filled by government officials, the Church collected taxes, aided the sick, and punished criminals. It became a powerful, unifying force throughout Europe.

Revival of Trade and Towns

By the 1000s, feudalism had brought more stability to society. As strong lords gained more control over their lands, long periods of peace and security followed. Merchants once again felt safe to travel. New farming methods, such as better ways to plant and plow fields, led to a food surplus. With more to eat, the population increased. More people meant more demand for goods, which spurred trade. Old towns near busy trade routes revived, and new towns grew up near manor houses and churches. Money came back into use.

Vocabulary
craftspeople: those who work in skilled trades

As the economy grew, many serfs ran away to towns. Some became craftspeople who practiced such trades as shoemaking. Others became merchants who sold the goods that craftspeople made. Merchants and craftspeople formed a new social class, the middle class. They had fewer riches, rights, and privileges than lords, but far more freedom than they had known as serfs.

STRANGE *but* True

PEPPER MILLIONAIRES

Europeans were desperate to get spices, such as pepper and cloves. Before refrigeration, meat often spoiled. Spices helped disguise the taste of rotten meat.

In the 1500s, just one shipload of spices could make a merchant wealthy for life. The average working person would have to work at least 1,000 years to earn as much as a merchant could earn from one load of pepper!

Trade with the East

Trade increased, not only within Europe, but also with places outside Europe. Located on the Mediterranean, Italy had an advantage in this trade. Italian cities such as Venice traded with other port cities, such as Constantinople, located in what is now Turkey.

War also spurred trade. Many European Christians were angry that Muslims held the Holy Land, where Jesus had lived. In 1096, European Christians launched the **Crusades,** a series of wars to capture the Holy Land. They ultimately failed to take the Holy Land, but the Crusades changed European life. Italians supplied the ships that carried Crusaders to the Middle East. On the return trip, the ships brought Asian goods to Europe. These goods had traveled across the Indian Ocean and then overland to the Mediterranean.

Reading **History**
B. Recognizing Effects What effect did the Crusades have on Italian trade?

The Rise and Decline of Feudalism

After the Crusades, Italians continued to trade with Muslims in other Mediterranean cities.

An Italian merchant named Marco Polo also stirred European curiosity about distant lands. Polo had spent 24 years traveling in China and central Asia. A book written about Polo's travels described China's riches and wonders. It increased European interest in Asia.

The Decline of Feudalism

The growth of trade and towns weakened feudalism because so many serfs left the manors for town life. The power of the lords shrank because they had fewer people under their control. Beginning in 1347, a deadly disease also weakened feudalism. The bubonic plague swept across Europe, killing about one-fourth of the population and reducing the number of workers. Lords competed for the laborers who survived, so they began to pay wages to peasants, such as John of Cayworth.

> ***A VOICE FROM THE PAST***
>
> John of Cayworth . . . ought to carry in autumn beans or oats for 2 days with a cart and 3 animals of his own, the value of the work being 12 denarii [about a penny]. And he shall receive from the lord each day 3 meals.
>
> **Contract of John of Cayworth,** from *Readings in European History*

As feudal lords lost power, kings grew stronger. They won the support of townspeople because they could raise large armies to enforce order. In return, townspeople agreed to support their kings by paying taxes. The armies enforced order and imposed the king's authority over lesser lords. As countries became safer, trade flourished even more.

Reading **History**

C. Analyzing Causes What three causes led to the decline of feudalism?

The Renaissance and Reformation

Italy, which was thriving because of trade, became the birthplace of the **Renaissance**—a time of increased interest in art and learning. *Renaissance* is a French term meaning "rebirth." Lasting from the 1300s to 1600, the Renaissance spread from Italy throughout Europe.

Several forces led to this rebirth of learning. As feudalism weakened and the plague brought great suffering, Europeans began to question what life meant. In their search for new answers, some people turned to old sources. They read the writings and studied the art of the Greeks and Romans. The classical Greeks and Romans lived from about 750 B.C. to A.D. 476. As a result of these studies, European ideas changed.

1. The Greeks had praised human achievement. European scholars began humanism, the study of human worth, ideas, and potential.
2. Classical education stressed such subjects as history, philosophy, and literature. Europeans spent more time studying those subjects.
3. From classical art, European artists learned to make art more realistic. They created some of the world's finest paintings and statues.
4. Muslim scholars had saved classical manuscripts about science. Also, Muslim mathematicians had invented algebra. Contact with Muslim societies influenced European science and mathematics.

Vocabulary
philosophy: the study of the meaning of life

A new invention helped spread Renaissance ideas. In about 1455, a German named Johannes Gutenberg invented the **printing press,** a machine that mechanically prints pages. People no longer had to copy books by hand. Printers could make hundreds of copies of a book cheaply and accurately. More people read, and ideas spread quickly.

Background
A printing press uses movable type—blocks of metal or wood that have raised characters. The Chinese invented movable type in about 1045. Gutenberg re-invented it.

By the early 1500s, Renaissance ideas and other forces weakened the Catholic Church. Many church leaders were corrupt. Some claimed to grant God's forgiveness for money. Martin Luther, a German monk, publicly posted 95 statements that criticized such practices. This began the **Reformation,** a movement to correct problems in the Church.

The Reformation split the Church into two groups—Catholics and Protestants. In time, Protestants divided into many different churches.

HISTORY through ART

This painting, *School of Athens* by Raphael, shows many aspects of Renaissance art and culture.

- Like much Renaissance art, it looks more realistic than the art of the Middle Ages. (See page 44 for comparison.)
- It honors the Greek thinkers Aristotle and Plato, who are the two men in the center arch.
- It also honors Renaissance artists. Raphael himself is in the group to the right.

Why might Raphael have wanted to include himself in a painting with famous Greeks?

SKILLBUILDER Interpreting Charts

What economic activity was the primary cause of exploration?

When European colonists came to America, they carried their religious disagreements and hopes for religious freedom with them.

Changes in Trade

The Renaissance period saw not only changes in learning and religion, but also in trade. As trade grew, Italian merchants needed to improve the way they did business. They began to use more exact ways of keeping track of a business's income and its costs. By subtracting the costs from the income, the merchants determined the **profit.**

Italian merchants made huge profits by trading in Asian goods. Because Italians had done business with Muslims for centuries, they had a special relationship. In addition to that, the Italians used military strength to control the trade on the Mediterranean—and didn't allow other Europeans to take part in it.

Merchants in other European countries envied the profits made by Italian merchants. As a result, other Europeans began to want a share of the rich trade in Asian goods. They had to find different routes to Asia from the ones controlled by the Italians and Muslims. Other European countries began to search for a non-Mediterranean water route to Asia, as you will read in Section 5.

Reading History

D. Making Generalizations If a country tries to completely dominate trade in a certain area, how will other countries respond?

Section 4 Assessment

1. Terms & Names

Identify:

- European Middle Ages
- feudalism
- manor system
- Crusades
- Renaissance
- printing press
- Reformation
- profit

2. Taking Notes

On a chart like this one, list how the Renaissance changed art and learning.

Changes to Art and Learning
•
•
•
•

3. Main Ideas

a. What caused feudalism to develop?

b. What led to the revival of trade and towns?

c. How did Italy come to control European trade with Asia?

4. Critical Thinking

Contrasting How did the Renaissance differ from the European Middle Ages?

THINK ABOUT

- the economy
- how power was distributed
- the authority of the church

ACTIVITY OPTIONS

TECHNOLOGY

MUSIC

Design a **Web site** or compose a **song** advertising the great new Renaissance invention—the printing press.

5 Early European Explorers

TERMS & NAMES
navigator
caravel
Christopher Columbus

MAIN IDEA

As Europeans searched for sea routes to Asia, Christopher Columbus reached the Americas.

WHY IT MATTERS NOW

Columbus's journey permanently linked the Americas to the rest of the world.

ONE EUROPEAN'S STORY

Sailors seeking a route to Asia depended on the skill of their navigator. A **navigator** plans the course of a ship by using instruments to find its position. In the 1400s, Portugal had a famous prince called Henry the Navigator. Yet, Henry wasn't a navigator. He never sailed on any of the ships trying to find Asia. So how did he earn his name?

Henry lived at Sagres, on the southwestern tip of Portugal. It was a site that overlooked the Atlantic Ocean. He invited astronomers, mathematicians, mapmakers, and navigators to Sagres. There he began a school of navigation.

Henry decided to organize and pay for sailing expeditions to explore the Atlantic and the west coast of Africa. He was hoping to find African gold, to learn more about geography, and to spread Christianity. His ships traveled farther down the African coast than Europeans had ever gone. Because Henry sponsored the voyages, the English named him "the navigator." Those voyages began Europe's age of discovery. As you will read in this section, this age of discovery eventually led Europeans to the Americas.

Henry the Navigator sponsored voyages that helped Portugal find a water route to Asia.

A Water Route to Asia

Under Prince Henry, the Portuguese developed an improved ship called the **caravel**. The caravel had triangular sails as well as square sails. Square sails carried the ship forward when the wind was at its back. Triangular sails allowed the caravel to sail into the wind. The caravel was better than other European ships of the time at sailing into the wind.

In January 1488, the Portuguese explorer Bartolomeu Dias (DEE•uhs) reached the southern tip of Africa. After sailing around it, he returned to Portugal at the urging of his crew. Portugal's king named the tip the Cape of Good Hope because he hoped they had found a route to Asia.

Ten years later, another Portuguese explorer, Vasco da Gama, followed Dias's route around the cape. He continued north along the eastern coast of Africa. Then he sailed east across the Indian Ocean to India. At last, someone had found an all-water route to Asia.

AMERICA'S HISTORY MAKERS

CHRISTOPHER COLUMBUS

1451–1506

Christopher Columbus's son Ferdinand wrote that his father "took to the sea at the age of 14 and followed it ever after."

Columbus's early voyages nearly cost him his life. When he was 25, pirates off the coast of Portugal sank his ship. Columbus survived by grabbing a floating oar and swimming to shore.

But he also learned a lot from sailing on Portuguese ships. The sailors taught Columbus about Atlantic wind patterns. This knowledge later helped him on his history-making voyage.

What character traits, shown in Columbus's early life, might have made him a good leader?

That route meant that the Portuguese could now trade with Asia without dealing with the Muslims or Italians. Portugal took control of the valuable spice trade. The merchants of Lisbon, Portugal's capital, grew rich. Spain and other European rivals wanted to take part in this profitable trade. They began to look for their own water routes to Asia.

Reading **History**

A. Comparing Compare what happened after Portugal began to control the spice trade to what happened when Italy controlled it.

Columbus's Plan

By the time of da Gama's voyage, an Italian sailor named **Christopher Columbus** thought he knew a faster way to reach Asia. Europeans had known for centuries that the earth is round. Columbus decided that instead of sailing around Africa and then east, he would sail west across the Atlantic. He calculated that it would be a short journey.

But Columbus made several mistakes. First, he relied on the writings of two people—Marco Polo and a geographer named Paolo Toscanelli—who were wrong about the size of Asia. They claimed that Asia stretched farther from west to east than it really did.

Background

As you read in Section 4, Marco Polo's book about his travels had increased European interest in Asia.

Second, Columbus underestimated the distance around the globe. He thought the earth was only two thirds as large as it actually is! Because of Polo and Toscanelli, Columbus thought that Asia took up most of that distance. Therefore, he believed that the Atlantic Ocean must be small. And a voyage west to Asia would be short.

In 1483, Columbus asked the king of Portugal to finance a voyage across the Atlantic. The king's advisers opposed the plan. They argued that Columbus had miscalculated the distance to Asia. They also reminded the king of the progress that Portuguese explorers had made sailing down the coast of Africa looking for a route to Asia. The advisers persuaded the king not to finance the voyage. So in 1486, Columbus turned to Portugal's rival, Spain.

Help from Spain's Rulers

Spain's rulers, King Ferdinand and Queen Isabella, liked Columbus's plan because they wanted a share of the rich Asian trade. As a strong Catholic, the Queen also welcomed a chance to spread Christianity. But there were also reasons not to support Columbus. First, a royal council had doubts about Columbus's calculations and advised Ferdinand and Isabella not to finance him. Second, the Spanish monarchs were in the middle of a costly war to drive the Muslims out of Spain. Third, Columbus was asking a high payment for his services.

Vocabulary

monarch: a king or queen

The years of waiting had made Columbus determined to profit from his explorations. As a reward for his efforts, he demanded the high title

Admiral of the Ocean Sea and a percentage of any wealth he brought from Asia. He also expected to be made the ruler of the lands he found.

Finally in January of 1492, the Spanish conquered the last Muslim stronghold in Spain. The Spanish monarchs could now afford to finance Columbus but still had doubts about doing so. Columbus left the palace to return home. But after listening to a trusted adviser, the king and queen changed their minds and sent a rider on horseback to bring Columbus back. He and the rulers finally reached an agreement.

Reading **History**

B. Drawing Conclusions Did this agreement give Columbus what he was asking for? Explain.

A VOICE FROM THE PAST

Your Highnesses . . . accorded me great rewards and ennobled me so that from that time henceforth I might . . . be high admiral of the Ocean Sea and perpetual Governor of the islands and continent which I should discover.

Christopher Columbus, letter to King Ferdinand and Queen Isabella

Preparing to sail, Columbus assembled his ships—the *Niña,* the *Pinta,* and the *Santa María*—at the port of Palos de la Frontera in southern Spain.

Setting Sail

At first, Columbus had trouble finding a crew. Then a respected local shipowner agreed to sign on as captain of the *Pinta.* Other crew members soon followed. About 90 men loaded the ships with enough food for one year, casks of fresh water, firewood, and other necessities.

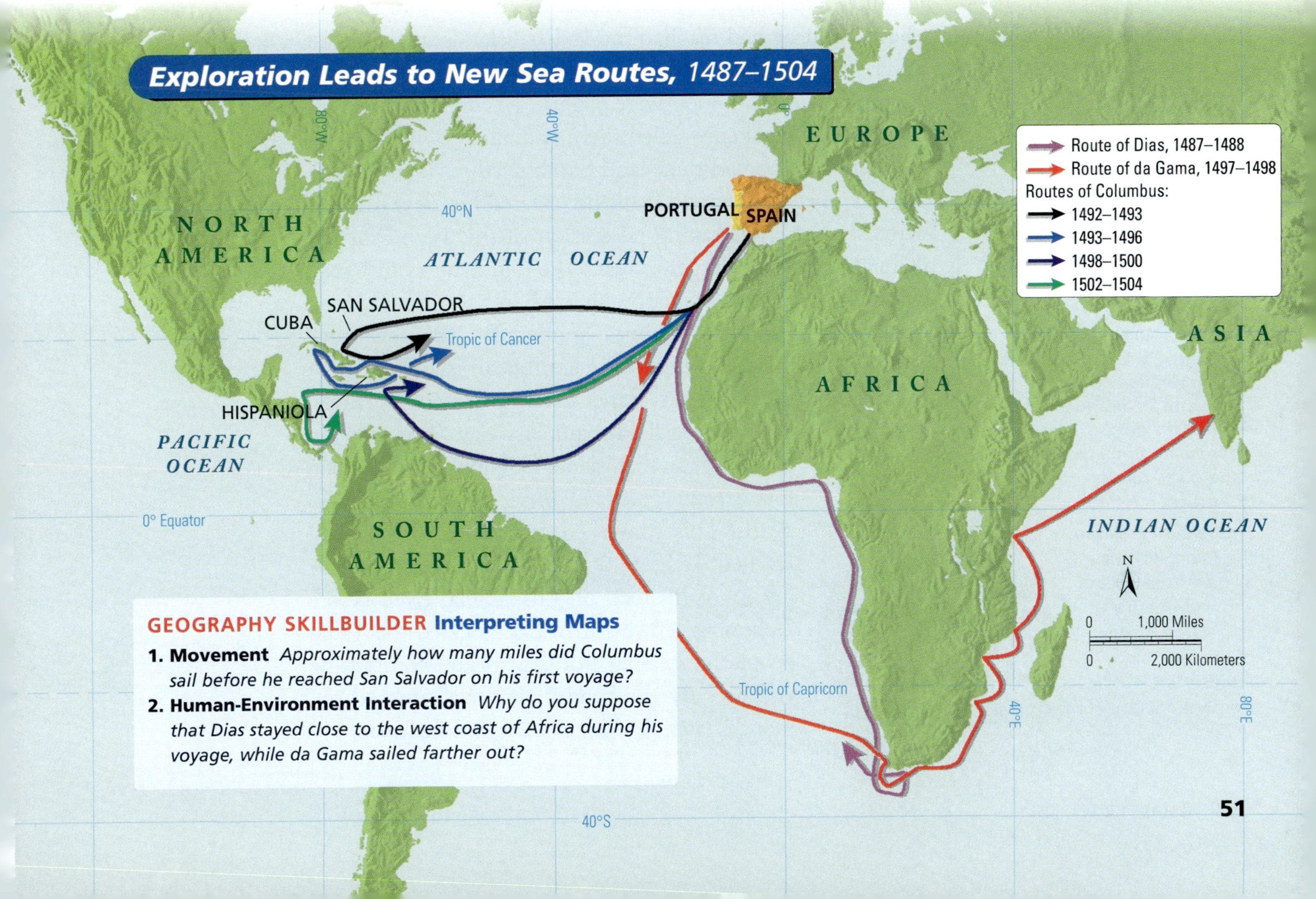

GEOGRAPHY SKILLBUILDER Interpreting Maps

1. **Movement** *Approximately how many miles did Columbus sail before he reached San Salvador on his first voyage?*
2. **Human-Environment Interaction** *Why do you suppose that Dias stayed close to the west coast of Africa during his voyage, while da Gama sailed farther out?*

The tiny fleet of wooden ships glided out of the harbor on August 3, 1492. First they sailed southwest toward the Canary Islands off the northwest coast of Africa. From there, Columbus was relying on trade winds that blew toward the west to speed his ships across the ocean.

Once aboard ship, Columbus kept a log, or daily record of each day's sailing. In fact, he kept two logs. One he showed to his men and one he kept secret. Columbus's secret log recorded the truth about the journey.

A VOICE FROM THE PAST

[We] made 15 leagues [this] day and . . . [I] decided to report less than those actually traveled so in case the voyage were long the men would not be frightened and lose courage.

Christopher Columbus, quoted in *Columbus and the Age of Discovery*

***Reading* History**

C. Analyzing Causes What caused Columbus to decide to keep two logs?

By October 10, the men had lost both courage and confidence in their leader. They had been at sea for almost ten weeks and had not seen land for over a month. Afraid that they would starve if the trip went on longer, they talked of returning home. To avoid mutiny, Columbus and the crew struck a bargain. The men agreed to sail on for three more days, and Columbus promised to turn back if they had not sighted land by then. Two days later in the early morning hours of October 12, a sailor on the *Pinta* called out "Tierra, tierra" [Land, land].

Reaching the Americas

By noon, the ships had landed on an island in the Caribbean Sea. Columbus believed that he had reached the Indies, islands in Southeast Asia where spices grew. The islanders who greeted Columbus and his men were Taino (TY•noh) people, but Columbus mistakenly called them Indians.

Background

Today, the Indies are called the East Indies. The islands of the Caribbean are called the West Indies.

Columbus named the island San Salvador. After unfurling the royal banner and flags, he ordered his crew to "bear witness that I was taking possession of this island for the King and Queen." Eager to reach the rich country of Japan, which he believed was nearby, he left San Salvador. He took six or seven Taino with him as guides. For the next three months, he visited several of the Caribbean islands.

Finally, he reached an island that he named Española, which we call Hispaniola today. (See map on page 51.) On that island, Columbus and his men found some gold and precious objects such as pearls. This convinced Columbus that he had reached Asia. He decided to return home, leaving 39 of his men on Hispaniola. Even before Columbus left, his men had angered the Taino people by stealing from them and committing violence. By the time Columbus returned ten months later, the Taino had killed the men.

Now and then

NATIVE AMERICAN VIEW OF COLUMBUS

In 1992, many Native Americans protested the 500th anniversary of Columbus's voyage. Suzan Shown Harjo, who is Cheyenne and Creek, explained why.

As Native American peoples in this red quarter of Mother Earth, we have no reason to celebrate an invasion that caused the demise [death] of so many of our people and is still causing destruction today.

The Spanish enslaved the Taino, who nearly all died from disease and bad treatment. This statue is one of the few Taino artifacts left from the 1500s.

In January 1493, he sailed back to Spain. Firmly believing that he had found a new water route to Asia, he wrote to Ferdinand and Isabella. The Spanish rulers called him to the royal court to report on his voyage. Neither Columbus nor the king and queen suspected that he had landed near continents entirely unknown to Europeans.

A French mapmaker uses an instrument to learn his exact position on the globe.

An Expanding Horizon

Columbus made three more voyages to the Americas, but never brought back the treasures he had promised Spain's rulers. He also failed to meet Queen Isabella's other goal. She wanted Christianity brought to new people. When she learned that Columbus had mistreated and enslaved the people of Hispaniola, she became angry.

After the fourth voyage, Spain's rulers refused to give Columbus any more help. He died in 1506, still believing he had reached Asia and bitter that he had not received the fame or fortune that he deserved.

In time, the geographic knowledge Columbus brought back changed European views of the world. People soon realized that Columbus had reached continents that had been unknown to them previously. And Europeans were eager to see if these continents could make them rich.

Reading **History**

D. Making Inferences How did the Atlantic become a bridge connecting Europe, Africa, and the Americas?

For centuries, Europeans had seen the ocean as a barrier. With one voyage, Columbus changed that. Instead of a barrier, the Atlantic Ocean became a bridge that connected Europe, Africa, and the Americas. As you will learn in Chapter 2, Columbus's explorations began an era of great wealth and power for Spain. As Spain grew rich, England, France, and other European countries also began to send ships to the Americas.

Section 5 Assessment

1. Terms & Names

Identify:

- navigator
- caravel
- Christopher Columbus

2. Taking Notes

On a diagram like the one shown, list the effects of Columbus's voyages.

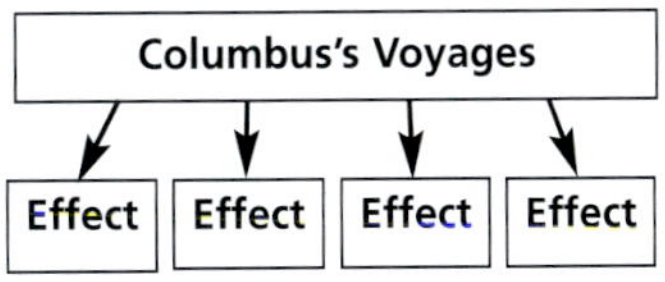

Which effects were negative and which were positive?

3. Main Ideas

a. Why was Prince Henry eager to find an all-water route to Asia?

b. Why did Spain's king and queen decide to support Columbus's first voyage?

c. Why was Columbus disappointed by the outcome of his four voyages to the Americas?

4. Critical Thinking

Analyzing Points of View

Explain how each of the following people might have viewed Columbus's first voyage. Give reasons for their points of view.

THINK ABOUT

- Columbus
- Queen Isabella
- a Taino chief

ACTIVITY OPTIONS

GEOGRAPHY

MATH

Use the map on page 31. Create an enlarged **map** of Columbus's first voyage, or measure the distance of each voyage to list on a **table**.

Chapter 1 ASSESSMENT

VISUAL SUMMARY

The World in 1500

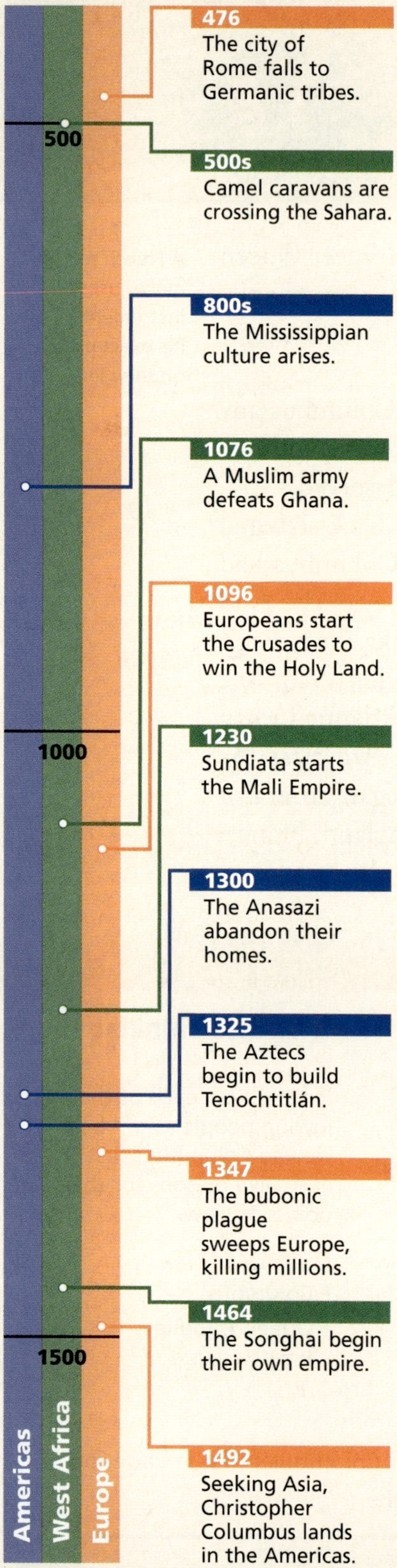

TERMS & NAMES

Briefly explain the importance of each of the following.

1. migrate
2. civilization
3. technology
4. Iroquois League
5. Islam
6. feudalism
7. Crusades
8. Renaissance
9. navigator
10. Christopher Columbus

REVIEW QUESTIONS

Crossing to the Americas (pages 27–31)

1. What are two theories about migration to the Americas?
2. For what purposes did the Mound Builders construct earthen mounds?

Societies of North America (pages 32–38)

3. What enabled the Aztecs to become a strong empire?
4. How did the Iroquois League come about?

Societies of West Africa (pages 39–43)

5. What enabled Ghana, Mali, and Songhai all to grow rich?
6. Did Islam become more or less influential in West Africa from the 700s to the 1400s? Explain.

Societies of Europe (pages 44–48)

7. How did the manor system work during the Middle Ages?
8. How did the Crusades increase European interest in trade?

Early European Explorers (pages 49–53)

9. Why did non-Italian Europeans seek new trade routes to Asia?
10. How did Columbus miscalculate the distance to Asia?

CRITICAL THINKING

1. USING YOUR NOTES

Using your completed chart, answer the questions below.

	Trade	Technology	Religion	Art
AMERICAS (Sections 1 and 2)				
WEST AFRICA (Section 3)				
EUROPE (Sections 4 and 5)				

a. What was one instance in which trade spread knowledge?
b. Which of the technologies that you listed are still used today?
c. What religions were practiced in each of the three regions?

2. ANALYZING LEADERSHIP

Do you think Columbus was a good leader or a bad one? Use details from the chapter to explain your answer.

3. THEME: DIVERSITY AND UNITY

How have Native Americans, Africans, and Europeans all influenced American culture? Give examples from your own experience.

4. MAKING GENERALIZATIONS

What types of goods are people most likely to seek through trade? Think about the trade goods mentioned in the chapter and why people wanted them.

5. APPLYING CITIZENSHIP SKILLS

Compare the Iroquois League to what you know of the U.S. government. How are they similar?

Interact *with* History

Think about the various encounters between societies mentioned in the chapter. What do you think happened when more Europeans came to the Americas and met Native Americans?

HISTORY SKILLS

1. INTERPRETING MAPS: Human-Environment Interaction

Study the map. Answer the questions.

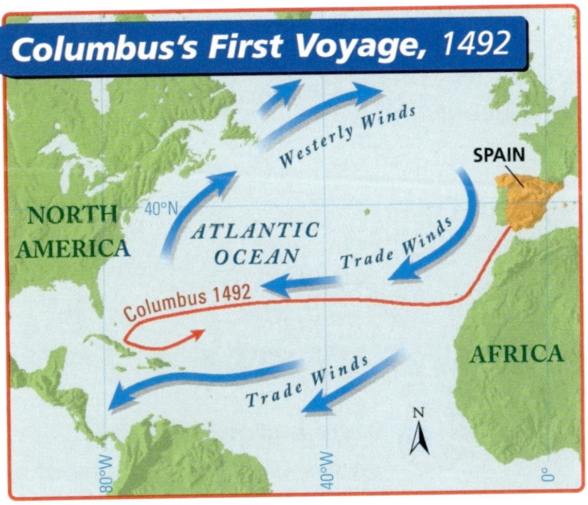

Basic Map Elements

a. What is the name of the winds blowing west across the Atlantic? blowing east?

b. In what direction did Columbus sail on his first journey from Europe?

Interpreting the Map

c. How did the winds affect Columbus's journey?

d. If Columbus's route had been farther north, would his voyage have taken more or less time? Explain.

2. INTERPRETING PRIMARY SOURCES

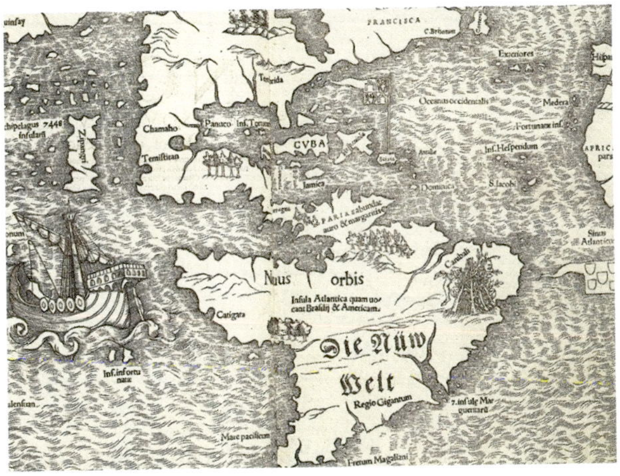

Drawn in 1570, this is the earliest map to show North America and South America as separate continents.

a. What are the names of any places that you recognize on this map?

b. Compare this map to the map of the world on page 51. Which continent on this map do you think looks more accurate? Explain why.

ALTERNATIVE ASSESSMENT

1. INTERDISCIPLINARY ACTIVITY: Literature

Retelling a Folk Tale Many North American Indian tales give insight into the relationship between Native Americans and their environment. Select a tale from a collection of Native American literature and retell the tale for the class.

2. COOPERATIVE LEARNING ACTIVITY

Creating a Sound Collage Columbus's log of his first voyage can be found in many public libraries. Have one or two members of your group select portions of the log for audiotaping. Other members of the group can decide on sound effects to be heard in the background. Consider using some of the following:

- sounds of the ocean or of storms
- shipboard sounds, such as bells, clanking chains, and flapping sails
- shouted commands to the crew
- seafaring songs or other music

You may want to use a sound effects tape, available in many libraries, or create your own sounds.

3. CD-ROM PRIMARY SOURCE EXPLORER

Choosing a Symbol Countries and alliances often use a symbol to represent who they are. The Iroquois League chose a tree. Think of a group you belong to and choose a symbol to represent it. Before you choose, use the CD-ROM, library, or Internet to find out more about why the Iroquois chose a tree.

- Think about what your group stands for. Write a list of adjectives to describe it.
- Brainstorm animals, plants, or objects that share those qualities. Choose one to be your symbol.
- Write an explanation of your symbol and why you chose it.

4. HISTORY PORTFOLIO

PORTFOLIO **Option 1** Review your section and chapter assessment activities. Select one that you think is your best work. Then use comments made by your teacher or classmates to improve your work and add it to your portfolio.

PORTFOLIO **Option 2** Review the questions that you wrote for What Do You Want to Know? on page 26. Then write a short report in which you explain the answers to your questions. If any questions were not answered, do research to answer them. Add your answers to your portfolio.

Create and Decode a Pictograph

Native Americans of the Southwest created thousands of images to communicate with each other. These images, known as pictographs, helped people recall certain events, ideas, or information. Even if the people who created them were no longer present, others could read the messages. Most images were painted or carved on the surfaces of rock. There are three types of pictographs: petroglyphs, petrograms, and geoglyphs. (See HELP DESK on the next page.)

ACTIVITY Create a pictograph that other students will decode, or figure out. Then, acting as an anthropologist, interview students in one other group about their pictograph.

TOOLBOX

Each group will need:

- drawing paper or poster board
- markers
- regular and colored pencils
- watercolor paints and brushes (optional)
- an envelope

The Fremont culture carved this petroglyph. It is currently located in Dinosaur National Monument—most of which sits in northwestern Colorado.

STEP BY STEP

1 Form a group of 4 or 5 students. Together, think of a message to tell someone living in the future. What might you want future generations to know about your culture, or way of life? If you're having trouble coming up with a message, copy the chart below into your notebook. Write information for each category that you think would be interesting to future generations. Then choose one of these categories for your message.

Sports	
Politics	
Fashion	
Music	
Entertainment	
Weather	
Daily Life	

2 Examine reference materials. In the library or on the Internet, research Native American pictographs. Use the information you find to help start your project. (See HELP DESK on the next page.)

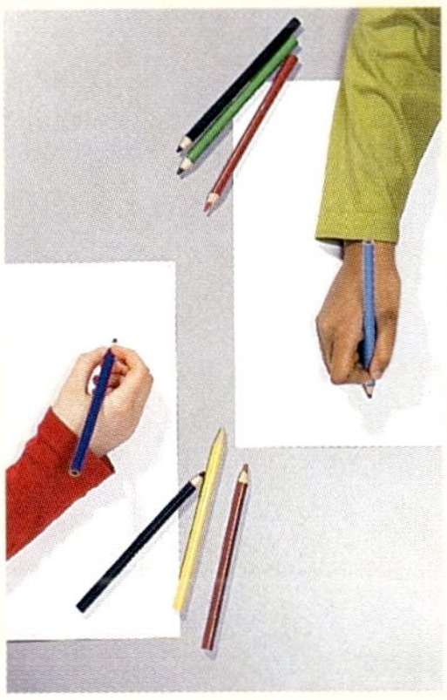

3 Create your pictograph. Communicate your message with symbols like the ones that you have researched. Sketch your pictograph on the drawing paper or poster board with a pencil first. Make the pictograph simple so that the decoders will understand your message. Remember to use symbols—not letters.

4 Decorate your pictograph. Use markers or watercolor paints to finish your pictograph. Also, record the translation of your pictograph in your notebook.

5 Exchange your pictograph with another group of students. Try to decode the message in the pictograph that the other group of students has given you. Write your translation and place it in your envelope. Give the envelope to the group whose pictograph you decoded.

6 Compare the other students' translation with your actual message. Did the other students understand your message? Let them know how accurate they were.

WRITE AND SPEAK

Using the information in the pictograph that you decoded, write a description of the people who created the message. Use the symbols as well as the message itself to help you in your description. Explain to the class how you came to your conclusions.

HELP DESK

For related information see Chapter 1, p. 29.

Researching Your Project

- *On the Trail of Spider Woman: Petroglyphs, Pictographs, and Myths of the Southwest* by Carol Patterson-Rudolph. Shows variety of actual pictographs.
- *21 Kinds of American Folk Art and How to Make Each One* by Jean and Cle Kinney. Explains process of making pictographs.

Visit www.mcdougallittell.com to learn more about pictographs.

Did You Know?

Petroglyphs are images carved into a rock using stone tools. **Petrograms** are images painted on a rock. **Geoglyphs** are images formed on the ground by scraping away soil or by arranging stones to form an image.

REFLECT & ASSESS

- **Which symbols in your pictograph were clear to the decoders? Which were not clear?**
- **What methods did you use to decode the messages of others?**
- **What did you learn about language and communication from doing this pictograph decoding activity?**

CHAPTER 2

European Exploration of the Americas

1492–1700

Ships in a harbor in Lisbon, Portugal, are preparing for a voyage of exploration.

Interact *with* History

The year is 1510. You live in a European port town and have heard exciting tales about mysterious lands across the sea. You decide to join a voyage of exploration in search of fortune.

What Do You Think?

- What do you think led people like Balboa to explore distant lands?
- What reasons would make you want to join a voyage of exploration?
- What reasons would keep you from joining such a voyage?

Would you join a voyage of exploration?

N. America

1497 Cabot searches for Northwest Passage.

1521 Cortés conquers Mexico.

1535 Cartier leads expedition up St. Lawrence River.

1539–1542 Coronado, de Soto, and Cabrillo explore different parts of North America.

1565 Spanish found St. Augustine.

1609 Hudson searches for Northwest Passage.

1626 Dutch buy Manhattan Island.

1680 Popé leads Pueblo Revolt and forces Spanish from New Mexico.

1492 — **1700**

World

1494 Spain and Portugal agree to Treaty of Tordesillas.

1542 King of Spain issues the New Laws for better treatment of Native Americans.

1588 English navy defeats Spanish Armada.

1644 Manchus establish Qing Dynasty in China.

1651 English Parliament passes Navigation Act.

Chapter 2 SETTING THE STAGE

BEFORE YOU READ

Previewing the Theme

Immigration and Migration From the 15th century to the 18th century, millions of people came to the Americas from other continents. Chapter 2 discusses why people came to the Americas and the effect that these migrations had on the people who already lived there.

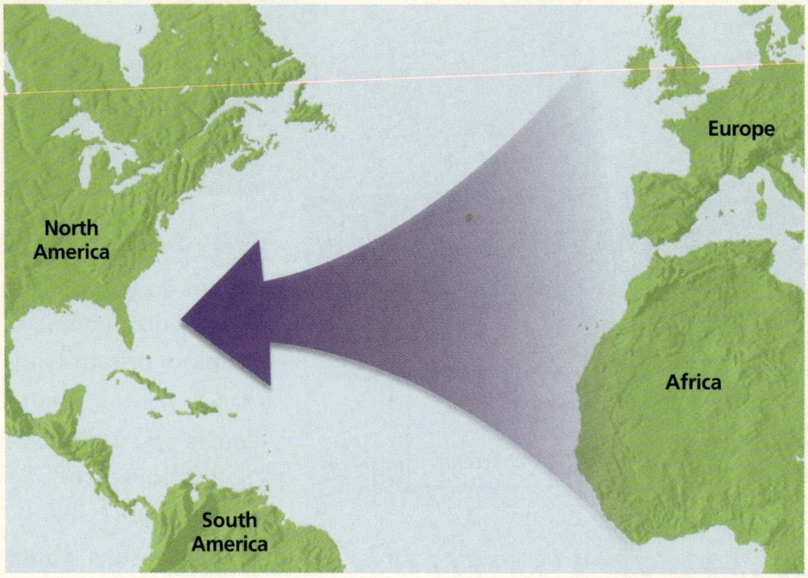

What Do You Know?

What comes to mind when someone uses the word *explorer*? Why do you think people explored different territories?

THINK ABOUT

- what you've learned about explorers from movies, school, or your parents
- reasons that people travel throughout the world today

What Do You Want to Know?

What questions do you have about exploration or the early colonization of the Americas? Write those questions in your notebook before you read the chapter.

READ AND TAKE NOTES

Reading Strategy: Taking Notes To help you remember what you read, take notes about the events and ideas discussed in the chapter. Taking notes means writing down important information.

The chart below lists the major events and ideas covered in the chapter. Use the chart to take notes about these important events and ideas.

See Skillbuilder Handbook, page R3.

Event/Idea	Notes
Exploration	
Establishing Colonies	
European Competition	
Columbian Exchange	
Origins of Slavery	

1 Spain Claims an Empire

TERMS & NAMES
Treaty of Tordesillas
missionary
mercantilism
Amerigo Vespucci
conquistador
Hernando Cortés
Montezuma
Francisco Pizarro

MAIN IDEA	WHY IT MATTERS NOW
Spain claimed a large empire in the Americas.	The influence of Spanish culture remains strong in modern America.

ONE EUROPEAN'S STORY

Pope Alexander VI had an important decision to make. In 1493, the rulers of Spain and Portugal wanted him to decide who would control the lands that European sailors were exploring. Ferdinand and Isabella of Spain expected Alexander VI to give Spain the rights over many of these lands. But King John II of Portugal claimed territories, too. What would the new pope do?

In May 1493, Alexander VI issued his ruling. He drew an imaginary line around the world. It was called the Line of Demarcation. Portugal could claim all non-Christian lands to the east of the line. Spain could claim the non-Christian lands to the west. In this section, you will learn how Spain and Portugal led Europe in the race to gain colonies in the Americas.

Pope Alexander VI

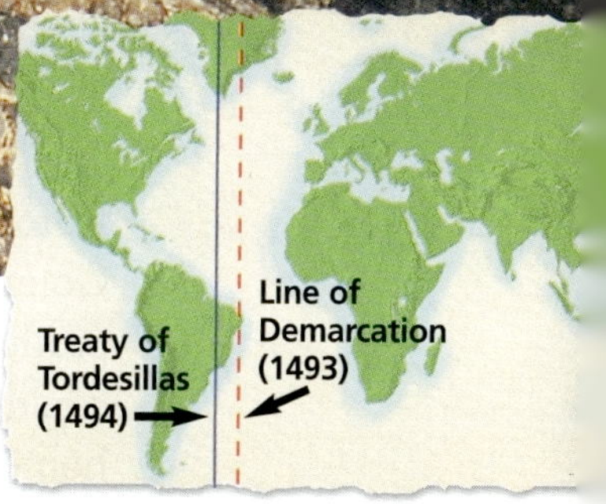

Spain and Portugal Compete

King John II was unhappy with the pope's placement of the line. He believed that it favored Spain. So he demanded that the Spanish rulers meet with him to change the pope's decision. In June 1494, the two countries agreed to the **Treaty of Tordesillas** (TAWR•day•SEEL•yahs). This treaty moved the Line of Demarcation more than 800 miles farther west.

The change eventually allowed Portugal to claim much of eastern South America, which later became the Portuguese colony of Brazil. After making this agreement, Spain and Portugal increased their voyages of exploration in search of wealth, power, and glory.

European countries had three main goals during this age of exploration. First, they wanted to spread Christianity beyond Europe. Each expedition included **missionaries,** or people sent to convert the native peoples to Christianity. Second, they wanted to expand their empires. Third, they wanted to become rich.

By increasing their wealth, European countries could gain power and security. An economic system called **mercantilism** describes how

Mercantilism

The main goal of mercantilism was to increase the money in a country's treasury by creating a favorable balance of trade. A country had a favorable balance of trade if it had more exports than imports. Colonies helped a country have the goods to maintain a favorable balance of trade.

For example, say Spain sold $500 in sugar to France, and France sold $300 in cloth to Spain. France would also have to pay Spain $200 worth of precious metals to pay for all the sugar. Spain would then have a favorable balance of trade because the value of its exports (sugar) was greater than the value of its imports (cloth). Spain would become richer because of the precious metals it received from France.

CONNECT TO HISTORY

1. **Finding Main Ideas** Under mercantilism, what did a country need to do to become rich? Discuss the way colonies enriched a country according to mercantilism.

See Skillbuilder Handbook, page R5.

CONNECT TO TODAY

2. **Making Inferences** Think about your own family budget. What do you think would happen if your family collected less money than it paid for goods for several years? Do you think this situation would be the same for a nation as it would for a family?

Visit www.mcdougallittell.com to learn more about mercantilism.

Europeans enriched their treasuries. (See *Economics in History,* above.) Colonies helped nations do this in several ways. They provided mines that produced gold and silver. They also produced goods such as crops that could be traded for gold and silver. Finally, they served as a market for the home country. The search for riches spurred European exploration.

Vocabulary
colony: a region or people that is politically and economically controlled by another country

Europeans Explore Foreign Lands

After Columbus's first voyage, many explorers went to sea. **Amerigo Vespucci** (vehs•POO•chee) was one of the first. He was an Italian sailor who set out in 1501 to find a sea route to Asia. Vespucci realized that the land he saw on this voyage was not Asia. A German mapmaker was impressed by Vespucci's account of the lands, so he named the continent "America" after him.

Another famous explorer was the Spaniard Vasco Núñez de Balboa. Balboa heard Native American reports of another ocean. In 1513, he led an expedition through the jungles of Panama and reached the Pacific Ocean. Raising his sword, Balboa stepped into the surf and claimed the ocean and all the lands around it for Spain. (See page 59.)

Perhaps no explorer was more capable than the Portuguese sailor Ferdinand Magellan. He proposed to reach Asia by sailing west around South America. The Spanish king agreed to fund Magellan's voyage.

In 1519, Magellan set out from Spain with five ships and about 240 men. After a stormy passage around South America, Magellan entered the Pacific Ocean. For several months his crew crossed the Pacific, suffering great hardship. A member of the crew described what they ate.

A VOICE FROM THE PAST

We were three months and twenty days without . . . fresh food. We ate biscuit, which was no longer biscuit, but powder of biscuits swarming with worms. . . . We drank . . . water that had been putrid for many days.

Antonio Pigafetta, quoted in *The Discoverers*

***Reading* History**

A. Finding Main Ideas What were the main contributions of Vespucci, Balboa, and Magellan as explorers?

Eventually, Magellan reached the Philippines, where he became involved in a local war and was killed. But his crew traveled on. In 1522, the one remaining ship arrived back in Spain. The sailors in Magellan's crew became the first people to sail around the world.

The Invasion of Mexico

While Magellan's crew was sailing around the world, the Spanish began their conquest of the Americas. Soldiers called ***conquistadors*** (kahn•KWIHS•tuh•DAWRZ), or conquerors, explored the Americas and claimed them for Spain. **Hernando Cortés** was one of these *conquistadors.* He landed on the Central American coast with 508 men in 1519.

The Spanish arrival shook the Aztec Empire, which dominated most of Mexico. The Aztec emperor **Montezuma** feared that Cortés had been

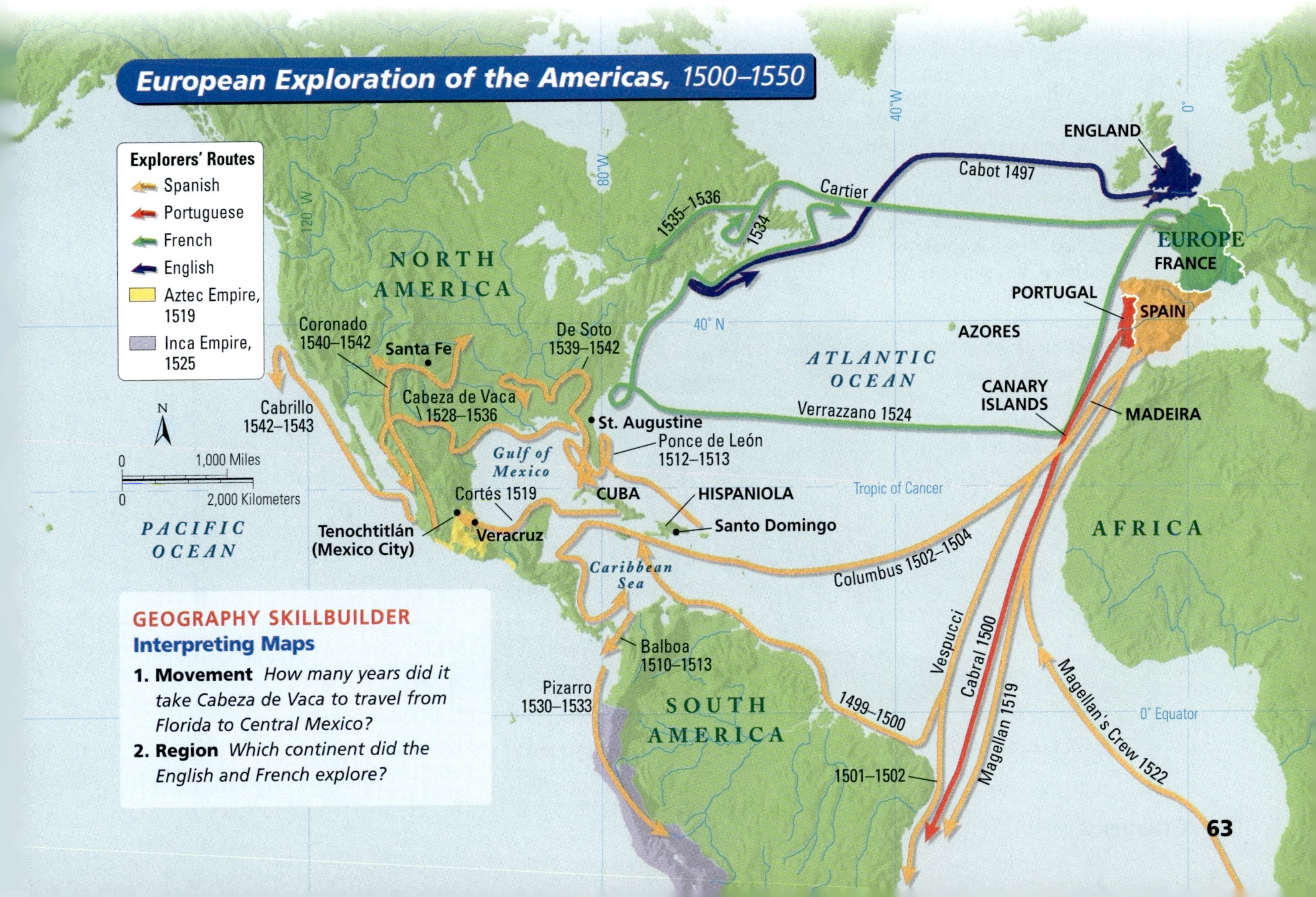

GEOGRAPHY SKILLBUILDER
Interpreting Maps

1. **Movement** *How many years did it take Cabeza de Vaca to travel from Florida to Central Mexico?*
2. **Region** *Which continent did the English and French explore?*

sent by an Aztec god to rule Mexico. Montezuma sent Cortés gifts—including two disks of solid gold and silver—to get him to leave. But the gifts only excited Spanish dreams of riches.

The Spaniards marched inland and formed alliances (agreements with friendly peoples) with the native peoples who hated Aztec rule. After a few months, Cortés reached the Aztec capital, Tenochtitlán (teh•NAWCH•tee•TLAHN). Montezuma received Cortés with great ceremony and housed the *conquistadors* in a magnificent palace. But Cortés took Montezuma captive and tried to rule the Aztec Empire by giving commands through Montezuma. The Aztecs rebelled.

Background
One of the people whom Cortés brought into his group was Malintzin. She was the daughter of a local chief and served as an interpreter for Cortés.

The Aztecs surrounded the Spaniards and their allies in their headquarters in Tenochtitlán. On the night of June 30, 1520, the Spaniards tried to sneak out of the city, but the Aztecs discovered them and vicious fighting broke out. About 800 Spaniards and more than 1,000 of their allies were killed that night. The Spaniards later called the event *La Noche Triste* (lah NAW•cheh TREES•teh)—the Sad Night.

Despite this defeat, the Spaniards and their allies regrouped. In May 1521, Cortés led his forces back to Tenochtitlán. At this point, the Spaniards got help from an invisible ally. Many Aztecs fell victim to an outbreak of smallpox, which severely weakened their ranks. The germs

AMERICA'S HISTORY MAKERS

HERNANDO CORTÉS

1485–1547

Hernando Cortés was born in Spain to a noble but poor family. In 1504, at the age of 19, he sailed to the Americas to seek his fortune. Although he became a wealthy landowner in Cuba, he was not satisfied. "I have come to win gold," he said, "not to plow the fields like a peasant." His great chance came when he was picked to lead the expedition to Mexico. Strong-willed, shrewd, and cruel, Cortés succeeded against great odds.

MONTEZUMA

1466–1520

Montezuma, ruler of the Aztec Empire, rose to the throne in 1502. Eloquent and wise, his words carried weight with his subjects. According to Juan de Tovar, a Jesuit, "When he spoke, he drew the sympathy of others by his subtle phrases and . . . by his profound reasoning." Montezuma lived in great luxury, receiving officials and commoners alike at his lavish palace. His subjects treated him almost as a god and were not allowed to look at him. Though brutal at times, he was said to be a just and effective ruler.

Which leader would you prefer to have lead your country? Why?

that caused this disease had been brought to America by the Europeans.

Vocabulary
siege: surrounding of a city

Cortés placed Tenochtitlán under siege for three months. When Tenochtitlán finally fell, the Aztec Empire lay in ruins. An Aztec poet described the scene.

A VOICE FROM THE PAST

Broken spears lie in the roads; we have torn our hair in our grief. The houses are roofless now, and their walls are red with blood. . . . We have pounded our hands in despair against the adobe walls, for our inheritance, our city, is lost and dead.

Aztec poet, quoted in *Seeds of Change*

This Aztec mask represents Quetzalcoatl, the god that Montezuma feared had sent Cortés.

On the rubble of the Aztec capital, the Spanish built Mexico City. Over time, the populations and cultures of Spain and Mexico merged and produced a new society, that of the present-day nation of Mexico.

The Conquest of the Incan Empire

Reading **History**
B. Reading a Map Use the map on page 63 to find the Incan Empire.

Despite the fall of the Aztecs, a people called the Inca still had a powerful empire centered in the Cuzco Valley in what is now Peru. By 1525, the Inca ruled a 2,000-mile-long territory in the Andes Mountains along the western coast of South America. The Inca also possessed much gold and silver.

Native American stories of Incan wealth reached the Spanish. In 1531, a *conquistador* named **Francisco Pizarro** led an expedition of 180 men into Peru. Like the Aztecs, the Incas feared that the Spanish might be gods. The Incan emperor Atahualpa (AH•tuh•WAHL•puh) ordered his troops not to fight. Then he went to meet the *conquistadors.* The Spanish attacked quickly. They killed thousands of Incas and took Atahualpa captive. In an attempt to free himself, the Incan emperor gave the Spanish a treasure of gold. The Spaniards strangled him anyway.

Reading **History**
C. Drawing Conclusions Why did the Incan Empire fall to the Spanish?

With Atahualpa dead, the Incan Empire collapsed. Having been ordered by Atahualpa not to fight, the Incas refused to defend themselves even after his death. Then Pizarro took control of this area for Spain. The Spanish called the area Peru.

Reasons for Spanish Victories

People have long been amazed that the great Aztec and Incan empires fell to such small groups of Spanish *conquistadors.* But Spanish success can be explained by four major reasons.

1. The spread of European diseases killed millions of Native Americans and weakened their resistance to conquest.
2. The Spanish were excellent soldiers and sailors. They also had superior weapons, such as guns, that helped them defeat much larger Native American armies.

3. Spain made alliances with Native Americans who were enemies of the Aztecs and Incas.
4. The Spanish *conquistadors* acted brutally toward the Native Americans under their control.

Having conquered the major Native American empires in Central and South America, the Spaniards began to explore other parts of North and South America.

Reading History

D. Drawing Conclusions What was the most important reason for the Spanish success in conquering territory in the Americas?

Other Spanish Explorers

The Spaniards hoped to collect treasures from North America as they had from Mexico and Peru. Rumors of golden cities kept Spanish hopes high. For example, a few men, including the Spaniard Álvar Núñez Cabeza de Vaca and Estevanico, a slave of North African descent, survived a shipwreck off the North American mainland. As the men wandered across the continent, they heard Native American stories about cities of gold. When they reached Mexico, Cabeza de Vaca and Estevanico thrilled the Spaniards with these rumors.

Estevanico was a slave who helped the Spanish explore parts of North America. He was killed during Coronado's search for golden cities.

Between 1539 and 1542, three expeditions set out to find these cities. Francisco Vázquez de Coronado traveled through present-day Arizona and New Mexico. Hernando de Soto set out from Florida to explore the southeast. Juan Rodríguez Cabrillo sailed up the California coast. But all three failed to find the fabled cities of gold.

For a while, it seemed that the Spaniards would explore the Americas all by themselves. As you will read in the next section, however, the Spanish would soon face competition from other Europeans.

Section 1 Assessment

1. Terms & Names

Identify:
- Treaty of Tordesillas
- missionary
- mercantilism
- Amerigo Vespucci
- *conquistador*
- Hernando Cortés
- Montezuma
- Francisco Pizarro

2. Taking Notes

Review the section and find four events to place on a time line that shows how Spain built its empire.

Spain Builds an Empire

1492 —|—|—|—|— 1542

Which event do you think is the most important? Why?

3. Main Ideas

a. Why did Europeans explore different territories?

b. Why did Spain succeed in conquering so much of the Americas?

c. What was significant about the Magellan expedition?

4. Critical Thinking

Comparing What was similar about the conquests of Mexico and Peru?

THINK ABOUT
- the *conquistadors*
- the Incan and Aztec leaders

ACTIVITY OPTIONS

ART

LANGUAGE ARTS

Use the library or the Internet to find a photograph of an Aztec or Incan artifact. Create a **replica** or write a **description** of the object.

2 European Competition in North America

MAIN IDEA

Other European countries competed with Spain for control over territory in the Americas.

WHY IT MATTERS NOW

European culture has strongly influenced American culture.

TERMS & NAMES

- Henry Hudson
- John Cabot
- Giovanni da Verrazzano
- Jacques Cartier
- Spanish Armada
- Samuel de Champlain
- New France

ONE EUROPEAN'S STORY

In 1609, an Englishman named **Henry Hudson** set sail from Europe. He sailed under the Dutch flag and hoped to find a route to China. Arriving at the coast of present-day New York, he sailed up the river that now bears his name. In his journal, Hudson described what he saw.

A VOICE FROM THE PAST

The land is the finest for cultivation that I ever in my life set foot upon, and it also abounds in trees of every description. The natives are a very good people; for, when they saw that I would not remain, they supposed that I was afraid of their bows, and taking the arrows, they broke them in pieces and threw them into the fire.

Henry Hudson, quoted in *Discoverers of America*

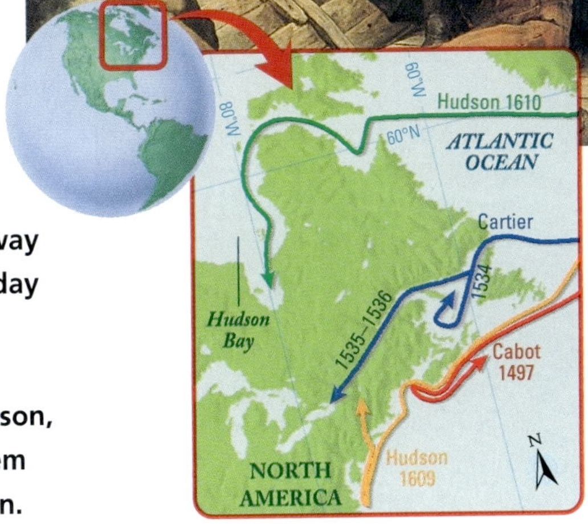

Hudson did not find a passage to Asia, but he led another expedition in 1610, this time sailing for the English. He made his way through ice-clogged waters in Canada and entered a large bay, today called Hudson Bay. There he sailed for months, but still found no westward passage.

After enduring a harsh winter, his crew rebelled. They put Hudson, his young son, and several loyal sailors in a small boat and set them adrift (shown at right). Hudson's party was never heard from again.

The Search for the Northwest Passage

Hudson's voyages showed that some European countries hoped to find a westward route to Asia as late as the 1600s. While Spain was taking control of the Americas, other Europeans were sending out expeditions to find the Northwest Passage, a water route through North America to Asia.

One of the first explorers to chart a northern route across the Atlantic in search of Asia was the Italian sailor **John Cabot**. In 1497, Cabot crossed the Atlantic Ocean to explore for the English. He landed in the area of Newfoundland, Canada. He was certain that he had reached Asia and claimed the land for England. The next year he set sail once more, hoping to reach Japan. He was never seen again. Even so, his voyages were the basis for future English colonies along North America's Atlantic shore.

In 1524, another Italian, **Giovanni da Verrazzano,** set out under the French flag to find the Northwest Passage. He explored the Atlantic coastline of North America, but there was no passage to be found.

France tried again between 1534 and 1536 with the voyages of **Jacques Cartier** (ZHAHK kahr•TYAY). Cartier traveled up the St. Lawrence River to the site of present-day Montreal. At that point, rapids blocked the way and ended his search for the Northwest Passage. It would be almost 75 years before the French would return to colonize the region.

Reading **History**

A. Reading a Map Use the maps on pages 63 and 67 to see the areas Cabot, Hudson, Verrazzano, and Cartier visited.

Spain Responds to Competition

French and English claims to North America angered Spain, which had claimed the land under the Treaty of Tordesillas. The tensions between Spain, England, and France stemmed from religious conflicts in Europe, such as the Reformation, which you read about in Chapter 1. These conflicts also led to fighting in the Americas.

Florida was one of the battlegrounds between the Spanish and the French. In 1564, a group of French Protestants, called Huguenots (HYOO•guh•NAHTS), founded a colony called Fort Caroline. Before long, Spanish troops under the command of Pedro Menéndez de Avilés arrived in that area. "This is the armada of the King of Spain," he announced, "who has sent me [here] to burn and hang the Lutheran [Protestant] French." Menéndez built a fort, St. Augustine, a short distance away. Then he brutally massacred the French.

Vocabulary

armada: a fleet of warships

America's **HERITAGE**

ST. AUGUSTINE

The thick stone walls of the fort at St. Augustine (shown below) still stand guard over the Florida coast today. Founded in 1565, St. Augustine is the oldest permanent European settlement in the United States. For more than two centuries, St. Augustine was an important outpost of Spain's empire in the Americas. Many Spanish colonial buildings remain at the site. The fort is now a national monument.

Spain and England Clash

Religious differences and the quest for national power also led to conflict between Spain and England. In 1558, Queen Elizabeth I, a Protestant, came to the English throne. Spain, which was Catholic, plotted to remove the Protestant queen. But Elizabeth fought to defend England and challenge Spain's power at sea.

Although England's navy was not as powerful as Spain's, the English fleet had many speedy ships with skillful sailors. Daring sailors, known as sea dogs, used these ships to attack the bulky Spanish sailing ships—called galleons—that brought gold and silver from the Americas.

Sir Francis Drake became the most famous of the sea dogs because of his bold adventures and attacks against the Spanish. In 1577, Drake began a three-year voyage that took him around the world. During this voyage,

The English navy used its smaller, quicker ships to defeat the larger, slower galleons of the Spanish Armada.

he raided Spanish ports and ships in South America. He stole great amounts of treasure from them. When he arrived home in 1580, he was a national hero. Not only had Drake and his men hounded the Spanish, but they were also the first Englishmen to sail around the world.

The Defeat of the Spanish Armada

The attacks of Drake and other sea dogs enraged Philip II, the Spanish king. Determined to teach the English a lesson, Philip sent the **Spanish Armada** to conquer England and restore Catholicism to that nation. This fleet, made up of 130 ships, set out for England in the summer of 1588.

The English and Spanish navies met in the English Channel, which separates England from the European continent. In their smaller but faster craft, the English darted among the Spanish warships, firing deadly rounds with their cannons. Confused and crippled, the armada was retreating when it was hit by a severe storm. With half of its ships destroyed, the armada barely made it home.

Spain was still quite strong after the defeat of the armada. It quickly rebuilt its navy and maintained its large colonial possessions. But Spain would never again be as powerful as it was in 1588.

Reading **History**

B. Drawing Conclusions Why was the defeat of the Spanish Armada important?

The English victory over Spain had two important effects. First, England remained independent and Protestant. Although England was less powerful than Spain, it had shown that it could defend itself. Second, Spain's image suffered. The world saw that Spain could be beaten. Other nations joined England in challenging Spain.

English adventurers like Drake continued to attack Spanish interests abroad. In addition, England challenged Spanish claims to lands in North America, such as California and Newfoundland. Even so, England took a cautious approach to overseas expansion. The English government refused to provide money to start colonies. Instead, private citizens had to provide the money for colonization. As a result, England did not establish a successful colony in America until after 1600.

The French and Dutch Seek Trade

France and the Netherlands were also looking for ways to gain wealth through exploration and colonization. At first, their goal in the Americas was to find the Northwest Passage to Asia. When that search failed, they began to focus on North America itself.

The Frenchman **Samuel de Champlain** (sham•PLAYN) explored the St. Lawrence River. In 1608, he founded a fur-trading post at Quebec. This post became the first permanent French settlement in North America. Champlain's activities opened a rich fur trade with local Native Americans. After a couple of decades, **New France,** as the colony was called, began to thrive.

At the same time, the Dutch were building a colony called New Netherland. It was located along the Hudson River in present-day New York. After Hudson's voyage up the river in 1609, the Dutch built Fort Nassau in 1614, near the site of the modern city of Albany.

In 1626, the Dutch bought Manhattan Island from Native Americans. The Dutch then founded the town of New Amsterdam on that site, where New York City is currently located. New Netherland was soon thriving from the fur trade with Native Americans.

These early French and Dutch colonies, however, were small compared to the large empire Spain was building in the Americas. You will read about the growth of Spain's American empire in the next section.

Reading **History**

C. Making Inferences Why do you think it took France and the Netherlands so long to set up colonies in the Americas?

Section 2 Assessment

1. Terms & Names

Identify:

- Henry Hudson
- John Cabot
- Giovanni da Verrazzano
- Jacques Cartier
- Spanish Armada
- Samuel de Champlain
- New France

2. Taking Notes

Use a chart like the one below to show how European nations competed for power.

England	
France	
Netherlands	
Spain	

3. Main Ideas

a. What were the English, French, and Dutch searching for in their early voyages of exploration?

b. How did England defeat the Spanish Armada?

c. Where did the French and Dutch set up their first American colonies?

4. Critical Thinking

Making Inferences Why do you think England founded colonies later than Spain did?

THINK ABOUT

- conditions in Spain and England
- the lands each country discovered

ACTIVITY OPTIONS

MUSIC

TECHNOLOGY

Research the life of one of the explorers discussed in this section. Compose a **song** or design a **Web page** about that person.

3 The Spanish and Native Americans

TERMS & NAMES
viceroyalty
encomienda
hacienda
mission
Popé
plantation
Bartolomé de Las Casas
Columbian Exchange

MAIN IDEA

Spanish rule in the Americas had terrible consequences for Native Americans.

WHY IT MATTERS NOW

The destruction of Native American cultures created social problems that continue today.

ONE AMERICAN'S STORY

Huamán Poma, a Peruvian Native American, was angry about the abuse the Spanish heaped upon Native Americans. He wrote to King Philip III of Spain to complain about the bad treatment.

A VOICE FROM THE PAST

It is their [the Spanish] practice to collect Indians into groups and send them to forced labor without wages, while they themselves receive the payment for the work. . . . The royal administrators and the other Spaniards lord it over the Indians with absolute power.

Huamán Poma, *Letter to a King*

In his letter, Poma asked the king to help the Native Americans and uphold the rule of law in Peru. If the king actually read the letter, it made no difference. Spanish colonists continued to mistreat Native Americans as the Spanish Empire expanded in the Americas.

A Spanish priest forces a Native American woman to work at a loom.

Spanish Colonies in the Americas

The Spanish Empire grew rapidly, despite efforts by other European countries to compete with Spain. By 1700, it controlled much of the Americas. Spain took several steps to establish an effective colonial government. First, it divided its American empire into two provinces called New Spain and Peru. Each province was called a **viceroyalty.** The top official of each viceroyalty was called the viceroy. He ruled in the king's name.

The Spanish also built new roads to transport people and goods across the empire. These roads stretched outward from the capitals at Mexico City and Lima. The roads helped Spain to control the colonies by allowing soldiers to move quickly from place to place. Roads also improved the Spanish economy because materials, such as gold and silver, could be transported efficiently to the coast and then to Spain.

GEOGRAPHY SKILLBUILDER Interpreting Maps

1. **Location** *Which viceroyalty included the West Indies?*
2. **Region** *Which viceroyalty covered more territory?*

Life in Spanish America

Spanish colonists received *encomiendas* to help them make the colonies productive. An ***encomienda*** was a grant of Native American labor. Hernando Cortés received a grant of more than 100,000 Native Americans to work his estate.

The Spanish rulers also created large estates, called ***haciendas,*** to provide food for the colony. *Haciendas* usually became large farms where Native Americans worked to grow cash crops, such as coffee and cotton. The *encomienda* and *hacienda* systems put much of the power and land in the hands of a few people.

Background
The problem of unequal wealth, especially in land, continues to trouble Latin American societies today.

The Spaniards made sure that people with Spanish backgrounds held power in the colonies. Spanish-born colonists such as Cortés made up the top layer of colonial society. Just below the Spanish were the Creoles—people of Spanish descent who were born in the colonies. The next step down the social order were the *mestizos. Mestizos* are people of mixed Spanish and Native American ancestry. The people with the least power and fewest rights were Native Americans and enslaved Africans.

The Role of the Church

The Catholic Church played an important role in Spanish colonial society. In places like New Mexico and California, the church built **missions,** settlements that included a church, a town, and farmlands. The goal of the missions was to convert Native Americans to Christianity. The missions also increased Spanish control over the land.

ReadingHistory
A. Summarizing How did the Spanish missions change the lives of Native Americans?

Missionaries helped the Native Americans to create a better supply of food. They also offered Native Americans protection against enemies. Many Native Americans learned how to read and write in the missions. Others developed skills such as carpentry and metalworking.

Over time, however, many Native Americans grew increasingly unhappy. The missionaries often worked them as if they were slaves. The missionaries also tried to replace Native American religions and traditions. As a result, some Native Americans ran away, while others rebelled. Some destroyed churches and killed missionaries.

In 1680, a man named **Popé** led the Pueblo Indians in a rebellion against the Spanish. His forces surrounded the Spanish settlement at

Santa Fe, in present-day New Mexico, and forced the colonists to flee. Popé ordered the churches and other Spanish buildings to be destroyed. He then tried to revive native customs that had been lost under Spanish rule. But before long, attacks from neighboring tribes weakened Pueblo control. In 1692, the Spanish regained control of Santa Fe.

Sugar Plantations Develop

The Spanish also forced Native Americans to work on **plantations,** large farms that raised cash crops. These crops were usually exported to Europe. The most important crop was sugar.

Although sugar was in great demand in Europe, there was not much land there to grow it. The resulting demand led to the development of sugar plantations in the Americas. On his second voyage to the Americas, in 1493, Columbus brought sugar cane to Hispaniola, one of the Caribbean islands he had landed on in 1492. He found ideal conditions for sugar production there. Spanish planters soon expanded operations to the nearby islands that Spain colonized.

Background
Other plantation crops included tobacco, cotton, cochineal (a dye), and cacao.

Sugar plantations required many workers, so the Spanish planters turned to native peoples, such as the Taino. Through *encomiendas,* the Spaniards forced thousands of Taino to work in the fields. The plantations thrived, but many of the Taino suffered and died.

The Abuse of Native Americans

Most Spaniards treated the Native Americans as little more than beasts of burden. According to Fray Toribio de Benavente, a Catholic missionary, the Spanish "do nothing but command. They are the drones who suck the honey which is made by the poor bees, the Indians."

Not all Spaniards approved of this treatment. One man in particular fought for better treatment of Native

HISTORY through ART

Theodore de Bry created this picture, *Sugar: the greatest gift of the Old World to the New,* in the 1600s. It shows workers processing sugar in the Americas. Europeans brought sugar production to the Americas from the Mediterranean.

How does the picture help explain why the Europeans used slaves to make sugar?

Americans. His name was **Bartolomé de Las Casas.** Las Casas had come to Hispaniola in 1502 and taken part in the conquest of Cuba a decade later. For his part in the conquest, he received an *encomienda.* Las Casas was also a Catholic priest, however, and he soon faced a moral dilemma: How can a person serve God and enslave Native Americans at the same time?

In 1514, Las Casas gave up his claim to the Native Americans who worked for him. For the next 50 years, he fought against the abuse of Native Americans, earning the title "Protector of the Indians."

Because of his efforts, the Spanish king issued the New Laws in 1542. These laws ordered the gradual freeing of all enslaved Native Americans. Holders of *encomiendas* who were found guilty of mistreating Native Americans had their *encomiendas* taken away. However, Spanish colonists strongly protested against the New Laws, and the king eventually reversed many of them.

The Columbian Exchange

The arrival of the Spanish in the Americas brought more than a clash of peoples and cultures. It also brought a movement of plants, animals, and diseases between the Eastern and Western hemispheres. This movement of living things between hemispheres is called the **Columbian Exchange.**

One result of the Columbian Exchange was the transfer of germs from Europe to the Americas. When Europeans came to America, they brought with them germs that caused diseases such as smallpox, measles, and influenza. Native Americans had no immunity to them.

Although exact numbers are unknown, historians estimate that diseases brought by Europeans killed more than 20 million Native Americans in Mexico in the first century after conquest. Many scholars agree that the population of Native Americans in Central America decreased by 90 to 95 percent between the years 1519 and 1619. The

Reading **History**

B. Making Inferences What might have happened if Native Americans had been immune to European diseases?

result was similar in Peru and other parts of the Americas. A Spanish missionary in Mexico described the effects of smallpox on the Aztecs.

A VOICE FROM THE PAST

There was a great havoc. Very many died of it. They could not walk. . . . They could not move; they could not stir; they could not change position, nor lie on one side; nor face down, nor on their backs. And if they stirred, much did they cry out. Great was its destruction.

Bernardino de Sahagún, quoted in *Seeds of Change*

Other effects of the Columbian Exchange were more positive. The Spanish brought many plants and animals to the Americas. European livestock—cattle, pigs, and horses—all thrived in the Americas. Crops from the Eastern Hemisphere, such as grapes, onions, and wheat, also thrived in the Western Hemisphere.

The Columbian Exchange benefited Europe, too. Many American crops became part of the European diet. Two that had a huge impact were potatoes and corn, which are highly nutritious. They helped feed European populations that might otherwise have gone hungry. Potatoes, for example, became an important food in Ireland, Russia, and other parts of northern Europe. Without potatoes, Europe's population might not have grown as rapidly as it did.

By mixing the products of two hemispheres, the Columbian Exchange brought the world closer together. Of course, people were also moving from one hemisphere to the other, blending their cultures in the process. The next section focuses on one important aspect of the movement of peoples: the forced migration of enslaved Africans to the Americas.

Background
In Ireland, the population increased from 3.2 million in 1754 to more than 8 million in 1845, largely because of the nutrition of potatoes.

Now and then

KILLER BEES

Even today, plant and animal species continue to move from one hemisphere to the other. A recent example of this is the killer bee (shown below).

Killer bees were first brought to Brazil from Africa to help make honey in the 1950s. Killer bees are aggressive, however, and can kill large animals when they swarm. After some of these bees escaped from a Brazilian laboratory in 1957, they began to migrate. In recent years, they have been responsible for the deaths of a number of pets in the American Southwest.

Section 3 Assessment

1. Terms & Names

Identify:
- viceroyalty
- *encomienda*
- *hacienda*
- mission
- Popé
- plantation
- Bartolomé de Las Casas
- Columbian Exchange

2. Taking Notes

Use a cluster diagram like the one below to show how Spain organized its colonies.

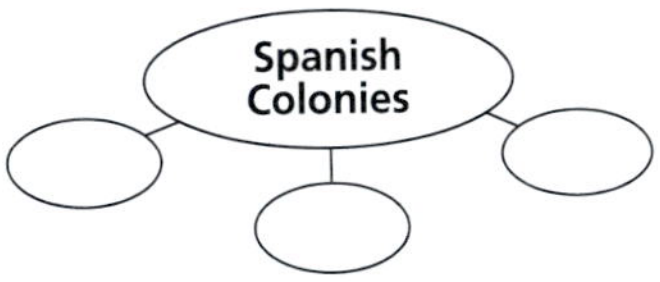

How did these actions help the Spanish control the Americas?

3. Main Ideas

a. What were the four levels of Spanish colonial society?

b. What was the main crop grown on colonial plantations?

c. How were Native Americans abused in the colonies?

4. Critical Thinking

Recognizing Effects
What were the positive and negative effects of the Columbian Exchange?

THINK ABOUT
- disease
- food
- livestock

ACTIVITY OPTIONS

ART
LANGUAGE ARTS

Make a **collage** that shows the plants and animals involved in the Columbian Exchange, or write a **story** that tells how Native Americans reacted to the animals.

4 Beginnings of Slavery in the Americas

TERMS & NAMES
slavery
African Diaspora
middle passage
slave codes
racism

MAIN IDEA

Slavery in the Americas began in order to provide cheap labor for the colonies.

WHY IT MATTERS NOW

The effects of slavery, including racism, helped shape attitudes and social conditions in the United States.

ONE AMERICAN'S STORY

In 1546, Diego de Campo may have been the most powerful man on the island of Hispaniola. He was the leader of 7,000 maroons, or runaway slaves. By contrast, there were only about 1,000 European men on the island.

The Spanish planters greatly feared de Campo. They did not dare to order their slaves around too harshly because the slaves might rebel with the help of the maroons. When the Spanish attacked the maroons, de Campo and his followers defeated the Spanish. Then the maroons burned the Spaniards' sugar mills.

Eventually the Spaniards captured de Campo. He pleaded to be spared and offered to lead the fight against the maroons. The Spanish accepted the offer. With de Campo's help, the Spanish brought the maroons under control, and slavery in Hispaniola grew. In this section, you will read how slave labor expanded and threatened the freedom of Native Americans and Africans.

Colonial troops searched for communities of maroons to destroy them.

The Origins of American Slavery

By the 1600s, **slavery,** the practice of holding a person in bondage for labor, was firmly established in the Americas. But slavery was not new. Its roots went back to the world's ancient civilizations.

Slavery took many different forms throughout history. In some societies, slaves were mainly domestic servants in wealthy households. Some slaves also labored in mines and fields.

People were often enslaved when they were captured in battle or sold to pay off debts. Some slaves were treated with respect. Some were allowed to marry and own property. The children of many slaves were allowed to go free.

Slavery began to change, however, with the rise of sugar plantations. Europeans had used slaves to grow sugar in the eastern Mediterranean since the 1100s. Then, in the 1400s and 1500s, Portugal and Spain set up sugar plantations on islands in the eastern Atlantic. To work these plantations, they used African slaves bought from traders in Africa.

When the Spanish and Portuguese founded their colonies in the Americas, they brought the plantation system with them. At first they tried to enslave Native Americans to work in the fields and mines. But the Native Americans quickly died from overwork and disease. In some cases, they rebelled with the help of local allies.

The Spaniards then looked to other sources of slave labor, including Spanish slaves, black Christian slaves, and Asian slaves. But there was not enough of any of these groups to meet demand.

*Reading*History

A. Drawing Conclusions Why did colonists decide that African slaves were more useful than Native American slaves?

Finally, the Spanish and Portuguese enslaved Africans to provide labor. They enslaved Africans for four basic reasons. First, Africans were immune to most European diseases. Second, Africans had no friends or family in the Americas to help them resist or escape enslavement. Third, enslaved Africans provided a permanent source of cheap labor. Even their children could be held in bondage. Fourth, many Africans had worked on farms in their native lands.

The Slave Trade

The slave trade grew slowly at first. In 1509, the Spanish governor of Hispaniola, Diego Colón—Columbus's son—wrote to King Ferdinand to complain about a labor shortage on the island. In response, the king sent 50 African slaves to Hispaniola. The slave trade increased with the demand for slaves to work in the colonies. Eventually the colonies came to depend on slave labor. As one Spanish official in Peru wrote, "The black slave is the basis of the *hacienda* and the source of all wealth which this realm produces."

European slave traders carried out the shipment of Africans to the Americas. The rulers of West African kingdoms participated in the trade, too. On the coast of Africa, local kings gathered captives from inland. The local kings then traded these captives for European goods, such as textiles, ironware, wine, and guns.

This trade made the coastal kingdoms rich while weakening inland African societies. In 1526, King Afonso, a West African ruler, protested against the slave trade in a letter to Portugal's king. Afonso wrote, "Everyday these [slave] merchants take our people. . . . So great is this corruption and evil that our country is becoming completely depopulated."

Vocabulary

depopulated: to lose population

CONNECTIONS TO MATH

Slaves Imported to the Americas, *1493–1810*

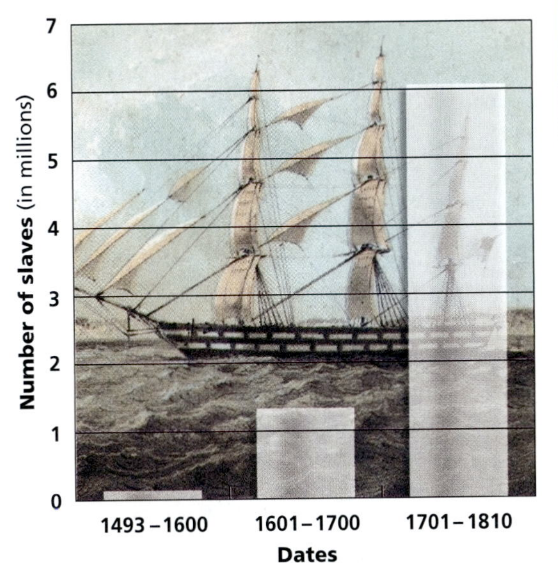

Source: Philip D. Curtin, *The Atlantic Slave Trade*

SKILLBUILDER Interpreting Graphs

1. *About how many slaves were imported to the Americas between 1493 and 1810?*
2. *Why do you think the numbers increased?*

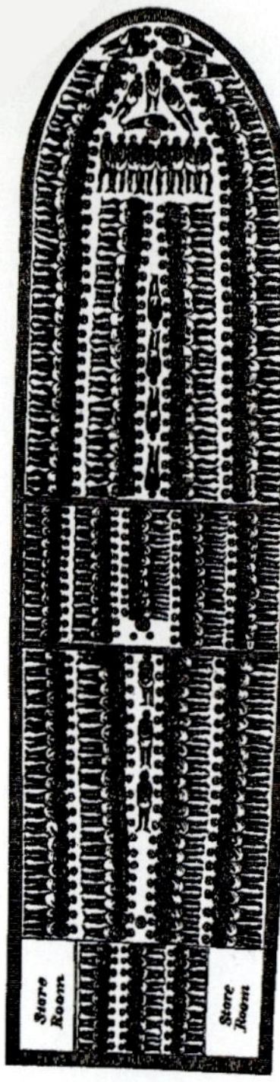

The diagram above shows how slave traders packed enslaved Africans onto slave ships for the middle passage. A British naval officer painted the picture on the right, which also shows the crowded conditions on slave ships.

The Middle Passage

Afonso's protest did not stop the forced removal of people from Africa. This removal has become known as the **African Diaspora.** Before the slave trade ended in the late 1800s, approximately 12 million Africans had been enslaved and shipped to the Western Hemisphere. Of these, perhaps two million died during the voyage.

Vocabulary
diaspora: the scattering of people outside their homeland

The voyage from Africa to the Americas was called the **middle passage.** The voyage was given this name because it was the middle leg of the triangular trade. The triangular trade refers to the movement of trade ships between Europe, Africa, and the Americas. You will learn more about the triangular trade in Chapter 4.

Olaudah Equiano (oh•LOW•duh EHK•wee•AHN•oh) was one of these kidnapped Africans. He made this journey in the 1700s. He was about 11 years old when he was taken from his home and sold into slavery. Later, after he bought his freedom, he wrote his life story and told what the middle passage was like.

A VOICE FROM THE PAST

The first object which saluted my eyes when I arrived on the coast, was the sea, and a slave ship . . . waiting for its cargo. These filled me with astonishment, which was soon converted into terror, when I was carried on board.

Olaudah Equiano, quoted in *Great Slave Narratives*

Equiano saw a row of men shackled together in chains. He also saw a large boiling kettle. He feared that he was going to be cooked and eaten "by those white men with horrible looks, red faces, and long hair."

The scene on the slave deck below was even worse. Several hundred slaves were crammed into a space so small that there was not even enough room to stand up. Foul smells and disease, along with the shrieks and groans of the dying, made the middle passage a terrifying experience. The captives who did not die faced new horrors in the Americas.

Reading History
B. Making Inferences Why would slave traders pack so many captives onto slave ships?

Slavery in the Americas

Once the enslaved Africans arrived in the colonies, they were sold at auction. Some were taken to large homes where they worked as servants. Most were forced to do hard labor in *haciendas* or mines. They were also fed and housed poorly.

Many slaves resisted slavery by running away. Across Peru and New Spain, maroons formed communities, often with Native Americans. Sometimes enslaved Africans rebelled. To prevent rebellion, the Spanish government passed **slave codes,** laws to regulate the treatment of slaves. Some of these laws tried to soften the harsh conditions of slavery, but most were designed to punish slaves and keep them in bondage.

*Reading*History

C. Analyzing Causes What could have caused slave traders to treat other humans with such cruelty?

Over time, Europeans came to associate slavery with black Africans. To many Europeans, dark skin color became a sign of inferiority. Slavery, which developed to provide a labor force, led to racism. **Racism** is the belief that some people are inferior because of their race.

The slave trade lasted for nearly 400 years, from the early 1500s to the mid-1800s. This contact between Africa and the Americas also formed part of the Columbian Exchange that you read about in Section 3. Africans brought to the Americas a vast knowledge about farming and animals. At the same time, American crops such as sweet potatoes, peanuts, and chilies made their way to Africa.

Enslaved Africans also brought with them a strong artistic heritage of dance, music, and storytelling. The slave trade brought together people from different parts of Africa with different cultural traditions. The experience of slavery helped create a common African-based culture in the Americas. By the 1700s, all the American colonies of European countries had African slaves. As you will read in the next chapter, African culture would be one of the forces that shaped life in the American colonies.

Section 4 Assessment

1. Terms & Names

Identify:
- slavery
- African Diaspora
- middle passage
- slave codes
- racism

2. Taking Notes

Use a diagram like the one below to compare the experience of Native Americans and Africans under slavery.

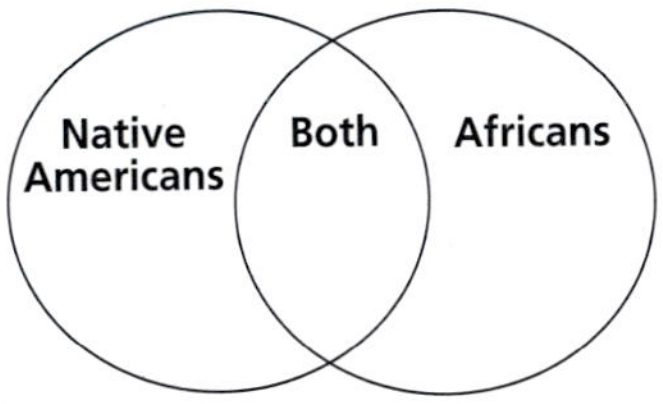

3. Main Ideas

a. When did slavery begin?

b. Why did Europeans bring Africans to the Americas?

c. What are three examples of bad conditions faced by enslaved Africans?

4. Critical Thinking

Recognizing Effects What were the long-term effects of slavery in the Americas?

THINK ABOUT
- the economy in the Americas
- the African Diaspora
- cultural diversity in the Americas

ACTIVITY OPTIONS

ART

MATH

Research some aspect of the slave trade, such as the middle passage or the number of people enslaved. Paint a **picture** or draw a **graph** to show what you learned.

Chapter 2 ASSESSMENT

TERMS & NAMES

Briefly explain the importance of each of the following.

1. mercantilism
2. Hernando Cortés
3. Montezuma
4. Spanish Armada
5. New France
6. *encomienda*
7. Columbian Exchange
8. slavery
9. African Diaspora
10. middle passage

REVIEW QUESTIONS

Spain Claims an Empire (pages 61–66)

1. What were three reasons for the European voyages of exploration in the 1400s and 1500s?
2. Who conquered the Aztecs and Incas?
3. What three reasons explain Spain's success in building an empire in the Americas?

European Competition in North America (pages 67–70)

4. What was the Northwest Passage?
5. Why did the Spanish Armada attack England?
6. What did the French and Dutch colonists trade?

The Spanish and Native Americans (pages 71–75)

7. How did Spanish rule affect Native Americans?
8. How did the Columbian Exchange affect Europe?

Beginnings of Slavery in the Americas (pages 76–79)

9. Why did the Spanish and Portuguese use slave labor in their colonies?
10. How did the slave trade work?

CRITICAL THINKING

1. USING YOUR NOTES

Event/Concept	Notes
Exploration	
Establishing Colonies	
European Competition	
Columbian Exchange	
Origins of Slavery	

Using your completed chart, answer the questions below.

a. What causes did European competition and exploration have in common?

b. How did the establishment of colonies in the Americas lead to slavery?

c. Which concept in the chart contributed most to the Columbian Exchange?

2. ANALYZING LEADERSHIP

Think about the explorers and *conquistadors* discussed in this chapter. What qualities did they possess that made them successful in their efforts?

3. THEME: IMMIGRATION AND MIGRATION

What were the causes and effects of the migration of Europeans and Africans to the Americas?

4. APPLYING CITIZENSHIP SKILLS

What kind of values did Bartolomé de Las Casas demonstrate in his actions? How effective was he in improving his society?

Interact *with* History

Have your answers about whether or not you would join a voyage of exploration changed after reading the chapter? Explain.

VISUAL SUMMARY

European Exploration of the Americas

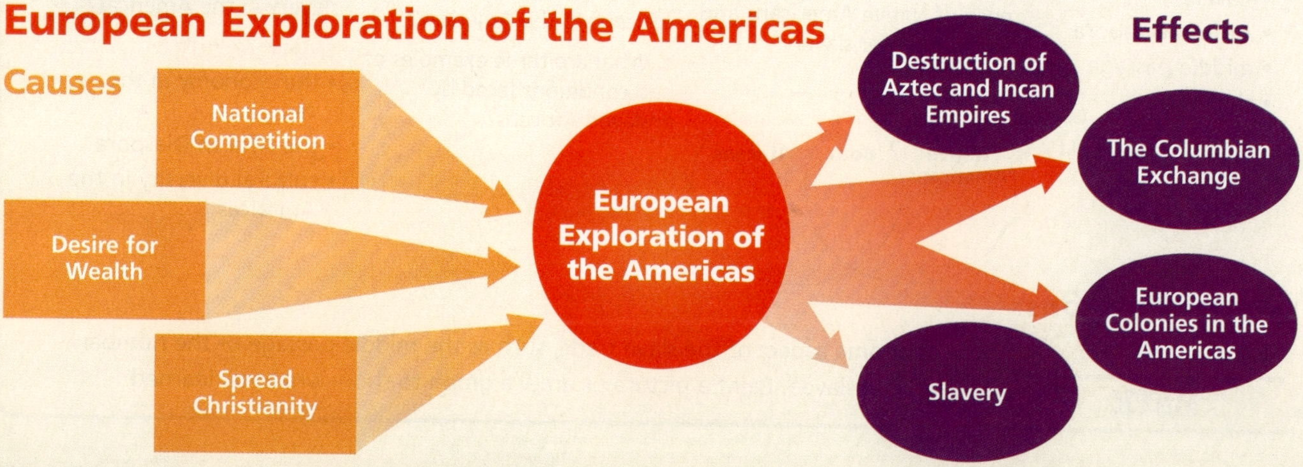

HISTORY SKILLS

1. INTERPRETING CHARTS

1601–1810

Slaves Imported to the Americas *(in thousands)*

REGION/COUNTRY	1601–1700	1701–1810
British N. America	*	348
British Caribbean	263.7	1,401.3
French Caribbean	155.8	1,348.4
Spanish America	292.5	578.6
Dutch Caribbean	40	460
Danish Caribbean	4	24
Brazil (Portugal)	560	1,891.4

*=less than 1,000

Source: Philip D. Curtin, *The Atlantic Slave Trade*

Basic Chart Elements

a. What is the subject of the chart?

b. How many years are covered in each column?

Interpreting the Chart

c. Which European nation imported the most slaves to the Americas?

d. Which region imported less than 1,000 slaves before 1700?

2. INTERPRETING PRIMARY SOURCES

Bernal Díaz del Castillo was a *conquistador* who accompanied Cortés during the conquest of Mexico. Díaz described what he saw when Cortés and Montezuma met. Read the passage. Then answer the questions.

> On our arrival we entered the large court, where the great Montezuma was awaiting our Captain [Cortés]. Taking him by the hand, the prince led him to his apartment in the hall where he was to lodge, which was very richly furnished. . . . Montezuma had ready for him a very rich necklace, made of golden crabs, a marvelous piece of work, which he hung round Cortés's neck. His captains were greatly astonished at this sign of honour.
>
> **Bernal Díaz del Castillo,** from *The Conquest of New Spain*

a. What can you tell about Montezuma from this passage?

b. How would you describe the relationship between Montezuma and Cortés?

ALTERNATIVE ASSESSMENT

1. INTERDISCIPLINARY ACTIVITY: Language Arts

Writing a News Report Research an event in the conquest of the Americas, such as *La Noche Triste* or the death of Atahualpa. Write a news report about the event that explains what happened and who was involved. You should also explain when, where, why, and how the event occurred.

2. COOPERATIVE LEARNING ACTIVITY

Creating a Diorama With a group of classmates, research the communities of Spanish America in the 1600s. Then create a diorama of one of those settlements, including a mission, a *hacienda,* roads, and mines or sugar mills. Use elements such as drawings, maps, and written text to show significant features of life there. Display your diorama and discuss it in class.

3. TECHNOLOGY ACTIVITY

Making a Class Presentation The Columbian Exchange refers to the movement of plants and animals around the world as a result of exploration. Today, species of plants and animals, such as killer bees, continue to move across the planet. Using the library or the Internet, find diaries and news articles about the effects of this continued movement of plants and animals.

Visit www.mcdougallittell.com to learn more about the migration of plants and animals.

Create a class presentation about the movement of plants and animals around the world, using the suggestions below.

- Choose a specific species of plant or animal. Use a map or globe to show where that species has moved in recent years.
- Choose one specific place, and list any species that are new to the area as well as their effects on the ecology.

4. HISTORY PORTFOLIO

Option 1 Review your section and chapter assessment activities. Select one that you think was your best work. Then revise your work based on the comments of your teacher or classmates and add your work to your portfolio.

Option 2 Review the questions that you wrote for What Do You Want to Know? on page 60. Then write a short report that explains the answers to your questions. If any questions were not answered, do research to answer them. Add your answers to your history portfolio.

CHAPTER 3 The English Establish 13 Colonies 1585–1732

Section 1 **Early Colonies Have Mixed Success**
Section 2 **New England Colonies**
Section 3 **Founding the Middle and Southern Colonies**

Early settlers build the fort at Jamestown.

Interact *with* History

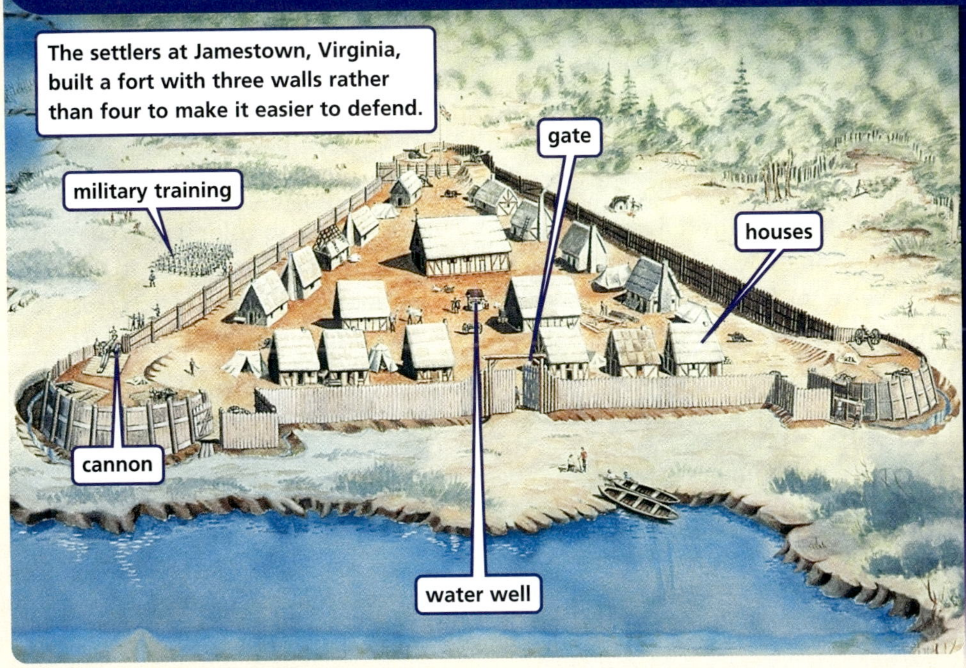

The year is 1607. You have just sailed across the ocean and arrived in a strange land. Your family has traveled to the eastern coast of North America in search of freedom and prosperity. Your first task in the new land is to decide what you need to do to survive.

What Do You Think?

- What do you need to survive in the wilderness?
- This settlement is actually a fort, with an armed force and high fences. What reasons might there be for building a fort?
- What kind of settlement would you build?

What dangers would you face as a settler?

N. America

1585 First English colony established at Roanoke.

1607 John Smith and other English settlers establish Jamestown.

1620 Pilgrims land at Plymouth.

1630 Puritans found Massachusetts Bay Colony.

1664 England takes New Amsterdam from Dutch.

1675 King Philip's War erupts.

1681 William Penn receives charter for Pennsylvania.

1692 Salem witchcraft trials are held.

1732 Colony of Georgia is founded by James Oglethorpe.

World

1587 Foreign missionaries are banished from Japan.

1588 England defeats Spanish Armada.

1605 Akbar, Mughal emperor of India, dies.

1649 Charles I of England is beheaded.

1660 English monarchy is restored.

1688 William and Mary take power in Britain's Glorious Revolution.

1585 — 1732

Chapter 3 SETTING THE STAGE

BEFORE YOU READ

Previewing the Theme Impact of the Individual

Beginning in 1585, English settlers started colonies along the eastern coast of North America. This chapter explains how the determination of a few leaders led to new colonies. It also explains how the colonies survived, gained more diverse settlers, and began to drive Native Americans (shown in the map below) off the land.

What Do You Know?

What do you already know about the American colonies? What sort of person might choose to leave his or her native country and cross the ocean to settle in a new land?

THINK ABOUT

- what you've learned about American settlers from movies, television, historical fiction, or science fiction about space travel
- opportunities and challenges offered in a new land

What Do You Want to Know?

What questions do you have about the Europeans who settled in North America? about those who were already here? Record your questions in your notebook before you read this chapter.

READ AND TAKE NOTES

Reading Strategy: Sequencing Events

Sequencing means putting events in order. In learning about the early colonies, for example, it will be useful to you to list the 13 original colonies and an important early date mentioned for each in the chapter. You might record the name and a date for each colony in a graphic organizer such as the one below. Copy this organizer in your notebook. Fill it in as you read the chapter.

See Skillbuilder Handbook, page R4.

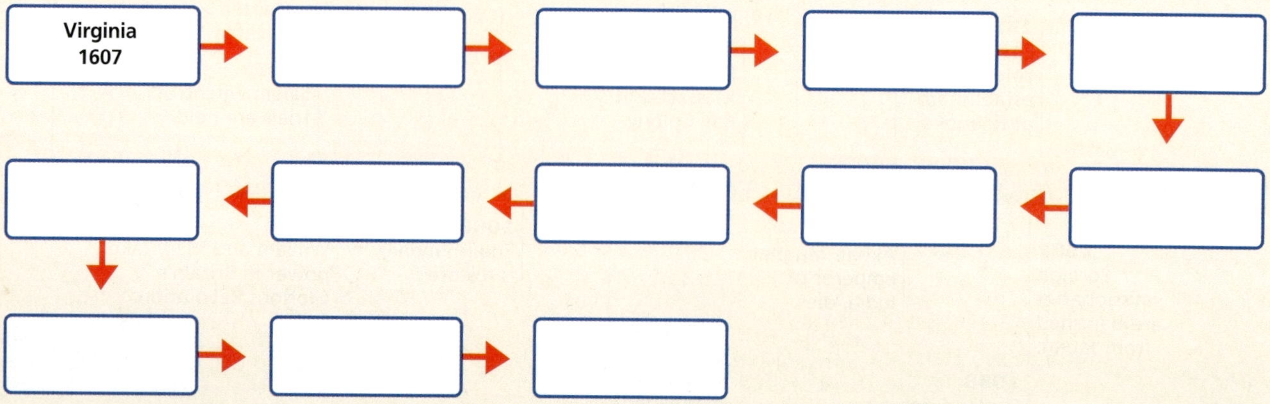

1 Early Colonies Have Mixed Success

TERMS & NAMES
joint-stock company
charter
Jamestown
John Smith
indentured servant
House of Burgesses
Bacon's Rebellion

MAIN IDEA

Two early English colonies failed, but Jamestown survived—partly through individual effort and hard work.

WHY IT MATTERS NOW

Jamestown's survival led to more English colonies and a lasting English influence in the United States.

ONE AMERICAN'S STORY

John White was a talented artist. He traveled with the first English expedition to Roanoke, an island off North Carolina, in 1585. While there, he painted scenes of Native American villages. White sailed back to England in 1586 and then returned to Roanoke as governor the next year, bringing with him more than 100 settlers. White's daughter Elinor gave birth to a baby girl, Virginia Dare, during their stay. John White described the event.

A VOICE FROM THE PAST

On August 18 a daughter was born to Elinor, . . . wife of Ananias Dare. . . . The child was christened on the following Sunday and was named Virginia because she was the first Christian born in Virginia.

John White, *The New World*

Drawing by John White of an old man of the Pomeiock tribe.

In 1587, White was forced to sail back to England a second time to get needed supplies. He left the colonists, including his granddaughter, Virginia, in Roanoke. Delayed by the Spanish Armada (a fleet of ships that attempted to invade England in 1588), White did not return to Roanoke until 1590.

To his shock and grief, he found no trace of the colonists or his granddaughter, all of whom had disappeared. The only clues to their whereabouts were the letters *CRO* carved in a tree and the word *Croatoan* carved in a doorpost. White never discovered the fate of his family and the other colonists. In this section, you will learn why English settlers such as White came to America despite such hardships. You'll also learn how they lived and what they believed.

The English Plan Colonies

As you read in Chapter 2, religious and political rivalries increased between England and Spain in the late 1500s. Spain had many colonies in the Americas, but England had none. England began directing its resources toward establishing colonies after its defeat of the Spanish Armada in 1588.

Richard Hakluyt (HAK•LOOT), an English geographer, urged England to start a colony. Hakluyt thought that colonies would provide a market for English exports. They also would serve as a source of raw materials. By having colonies, England hoped to increase its trade and build up its gold supply. This is the economic theory of mercantilism (see page 62). In mercantilism, the state controls trade and attempts to transfer wealth from colonies to the parent country. Hakluyt also thought that English colonies would help to plant the Protestant faith in the Americas.

The earliest English colonists had many reasons for going to America. The lack of economic opportunity in England forced many to seek their fortunes abroad. Stories of gold mines lured some to leave England. Others left to escape religious persecution.

Reading **History**
A. Summarizing Why did English colonists settle in America?

Two Early Colonies Fail

Sir Walter Raleigh was a soldier, statesman, and adventurer who served under Queen Elizabeth I of England. She gave him permission to sponsor the colony at Roanoke. He named England's first colony Virginia after the unmarried, or virgin, queen. Financed by Raleigh, the colony began in 1585 on Roanoke Island. The colonists relied on the Native Americans for food. But when the Native Americans realized that the settlers wanted their land, they cut off the colonists' food supply. Those who survived returned to England in 1586.

Vocabulary
financed: paid for; raised funds for

In 1587, artist John White convinced Raleigh to try again to establish the Roanoke colony, with the disastrous results described in One American's Story (page 85). To this day, no one knows for sure what happened. Some historians think that the colonists mingled with the neighboring Native Americans. Others believe that they moved to Chesapeake Bay and were killed by Native Americans defending their land.

In 1607, the Plymouth Company sponsored the Sagadahoc colony at the mouth of the Kennebec River in Maine. Some of the settlers were English convicts. One colonist wrote of George Popham, the governor, "He stocked or planted [the colony] out of all the jails of England." Within the first year, arguments among colonists, a harsh winter, fights with Native Americans, and food shortages forced most of the colonists to return to England.

Now and then

THE LUMBEE AND THE LOST COLONISTS

The Lumbee tribe lives mainly in North Carolina. Some of the Lumbee believe they are descendants of the lost colonists of Roanoke. Among the evidence cited is the fact that 41 of the 95 last names of the Lumbee were last names of the colonists.

Other Lumbee don't believe that they are descended from English ancestors. The Lumbee are trying to win federal recognition as a Native American tribe. English ancestry might weaken their claim for federal financial support.

Financing a Colony

Raleigh had financed the colony at Roanoke. When the colony failed, he lost his investment. The English learned from Raleigh's financial loss at Roanoke that one person could not finance a colony. To raise money, they turned to the **joint-stock company**. Joint-stock companies were backed by investors, people who put money into a project to earn profits. Each investor received pieces of ownership of the company called shares of stock. In this way, the investors split any profits and divided any losses.

Merchants organized the Virginia Company of London and the Virginia Company of Plymouth. King James I of England granted charters to both companies in 1606. A **charter** was a written contract, issued by a government, giving the holder the right to establish a colony.

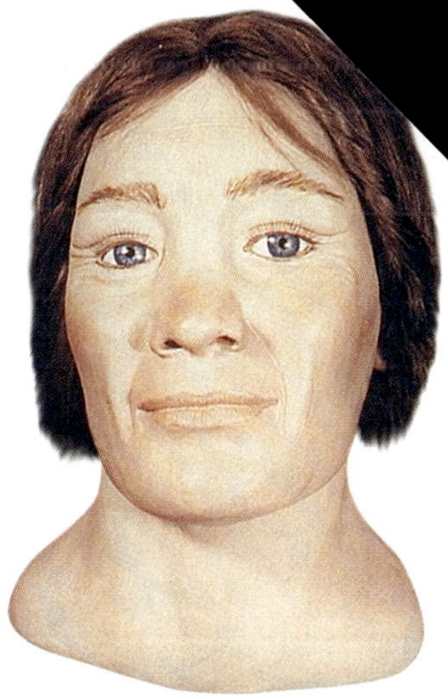

This is a computer reconstruction of the face of Mistress Forrest, believed to be the first woman to come to Jamestown.

Jamestown Is Founded in 1607

In 1607, the Virginia Company of London financed an expedition to Chesapeake Bay that included more than 100 colonists. They sailed up the James River until they found a spot to settle. They named the first permanent English settlement **Jamestown** in honor of King James.

Background
Historians used to believe that the colony's original site had been flooded by the James River. Recent archaeological digs, however, have discovered the site on higher ground.

From the start, the Jamestown colonists endured terrible hardships. The site of the colony was swampy and full of malaria-carrying mosquitoes. This disease made the colonists sick with fever. Many also became ill from drinking the river water. To make matters worse, the London Company had incorrectly told the settlers that the colony would be rich in gold. They spent their days searching for gold rather than building houses and growing food.

The climate was also a hardship. The colonists soon learned that the summers were hot and humid and the winters bitter cold. As one colonist recalled, "There were never Englishmen left in a foreign country in such misery as we were in this newly discovered Virginia."

Jamestown Grows

By January 1608, only 38 colonists remained alive. Later that year, **John Smith,** a soldier and adventurer, took control. To make sure the colonists worked, Smith announced, "He that will not work shall not eat." Smith's methods worked. He ordered an existing wall extended around Jamestown. He also persuaded the Powhatan tribe to trade their corn to the colonists. In 1609, Smith was injured in a gunpowder explosion and returned to England. That same year, 800 more English settlers arrived in Jamestown.

Reading **History**
B. Solving Problems If you had been John Smith, how would you have forced the colonists to work?

GEOGRAPHY SKILLBUILDER Interpreting Maps
1. **Location** *Which colony was located northeast of Jamestown? How many miles northeast was it?*
2. **Human-Environment Interaction** *Why did the colonists settle near the coast?*

Because of growing tensions between the settlers and Native Americans, the Powhatan stopped trading food and attacked the settlers. The settlers did not dare leave the fort. During the "starving time," the colonists ate rats, mice, and snakes. Only 60 of the colonists were still alive when two ships arrived in 1610. Lord De La Warr, the new governor, imposed discipline, and the "starving time" ended.

In 1612, John Rolfe developed a high-grade tobacco that the colonists learned to grow. It quickly became very popular in England. The success of tobacco growing changed Jamestown in many ways. The Virginia Company thought of the colonists as employees. The colonists, however, wanted a share of the profits.

The company responded by letting settlers own land. Settlers worked harder when the land was their own. The company offered a 50-acre land grant for each man, woman, or child who could pay his or her way to the colony. In 1619, the first African Americans arrived in Jamestown. The population of Virginia jumped from about 600 in 1619 to more than 2,000 in 1621.

Reading **History**

C. Analyzing Causes What was the main reason for the various arrangements the Virginia Company came up with to bring people to America?

Even more laborers were needed. Those who could not afford passage to America were encouraged to become **indentured servants.** These men and women sold their labor to the person who paid their passage to the colony. After working for a number of years, they were free to farm or take up a trade of their own.

The colonists soon became annoyed at the strict rule of the governor, who represented the Virginia Company's interests back in London. To provide for more local control, the company decided that burgesses, or elected representatives, of the colonists would meet once a year in an assembly. The **House of Burgesses,** created in 1619, became the first representative assembly in the American colonies.

AMERICA'S HISTORY MAKERS

POCAHONTAS

1595?–1617

Pocahontas met John Smith when she was about 12 years old. Smith taught her English and admired her spirit. She admired Smith's bravery and saved his life twice. After Smith returned to England, she married the colonist John Rolfe in 1614. Shown below is a portrait of Pocahontas, done in 1616.

Two years later, the Rolfes went to England to raise money for the Jamestown colony. While getting ready to sail home, Pocahontas died of smallpox.

How did Pocahontas show that Native Americans and white settlers might live in peace?

Conflicts with the Powhatan

Cultural differences put the Powhatan and the English on a collision course. At first, the Powhatan traded food with the colonists. Then, as more colonists arrived and wanted land, relations grew worse. In an effort to improve relations between the English colonists and the Powhatan, John Rolfe married Chief Powhatan's daughter, Pocahontas, in 1614.

For a time, there was an uneasy peace. The colonists learned from the Powhatan how to grow corn, catch fish, and capture wild fowl. However, the expanding tobacco plantations took over more and more Powhatan land. In 1622, in response to land grabs by the colonists, the Powhatan killed hundreds of Jamestown's residents.

Reading **History**

D. Finding Main Ideas What was the central dispute between the Powhatan and the settlers?

Bacon's Rebellion in 1676

As you have seen, many of the English colonists who came to Virginia during the 1600s fought with the Native Americans. They also battled one another. By the 1670s, one-fourth of the free white men were former indentured servants. These colonists, who did not own land, resented the wealthy eastern landowners. The poor settlers lived mostly on Virginia's western frontier, where they battled the Native Americans for land.

Nathaniel Bacon and a group of landless frontier settlers opposed Governor William Berkeley. They complained about high taxes and Governor Berkeley's favoritism toward large plantation owners. Bacon demanded that Berkeley approve a war against the Native Americans to seize their land for tobacco plantations. Governor Berkeley's refusal of Nathaniel Bacon's demand sparked **Bacon's Rebellion** in 1676.

Nathaniel Bacon (right) confronts Virginia governor William Berkeley at Jamestown in 1676.

Bacon marched into Jamestown, took control of the House of Burgesses, and burned Jamestown to the ground. Bacon's sudden illness and death ended the rebellion. Berkeley hanged Bacon's followers. Angered by Berkeley's actions, King Charles II recalled the governor to England. After that incident, the House of Burgesses passed laws to prevent a royal governor from assuming such power again. The burgesses had taken an important step against tyranny. In the next section, you will read about the New England colonies and their steps toward independence.

Vocabulary
tyranny: a government in which a single ruler has absolute power

Section 1 Assessment

1. Terms & Names

Identify:
- joint-stock company
- charter
- Jamestown
- John Smith
- indentured servant
- House of Burgesses
- Bacon's Rebellion

2. Taking Notes

Use a series-of-events chain to review events that led to the founding of Jamestown.

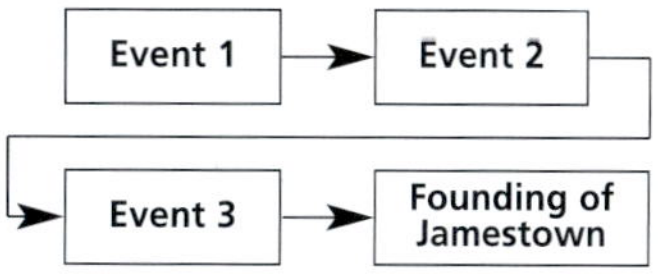

What were reasons England wanted colonies in America?

3. Main Ideas

a. Why did the first English settlement at Roanoke fail?

b. How did the English finance their colonies after 1606?

c. What was the outcome of Bacon's Rebellion?

4. Critical Thinking

Drawing Conclusions
What were the main reasons that Jamestown survived and prospered?

THINK ABOUT
- how, after the "starving time," Lord De La Warr took control
- John Rolfe's development of a high-grade tobacco plant

ACTIVITY OPTIONS

ART
LANGUAGE ARTS

You need indentured servants to work on your plantation. Draw a **poster** or write an **advertisement** that will attract people to your plantation.

Report from the New World

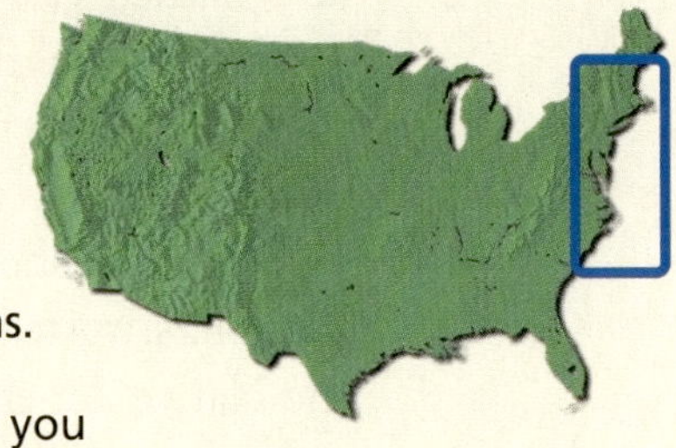

You are a settler who has landed on the wild eastern shore of North America. You and your 93 fellow colonists survived a frightening nine-week Atlantic voyage. Now you are struggling to build a new home in the wilderness. There are no roads, inns, or towns in this land. The game, berries, and fish here taste strange, sometimes unpleasant. Your only neighbors are small groups of Native Americans.

COOPERATIVE LEARNING On these pages are challenges you face as you put down roots in America. Working with a small group, decide how to deal with each challenge. Choose an option, assign a task to each group member, and do the activity. You will find useful information in the Data File. Be prepared to present your solutions to the class as part of a report to your sponsors back in England.

CIVICS CHALLENGE

"They had little or no care of any other thing, but to pamper their bellies."

As your colony takes root, most members work hard to farm, cook, wash, mend, trade, and defend the colony. But a few colonists think only of their own comfort. You call a meeting to set some rules about work. Present your solution to this problem using one of these options:

- **Make a poster for the meeting hall that states the new work rules and punishments.**
- **Write a report describing the problem and how the colony solved it.**

ECONOMICS CHALLENGE

"A bright tin dish most pleased him."

By the time spring arrives, your stores of English foods are running low. You and your friends decide to try trading with the neighboring Native Americans. They could provide a steady supply of meat, fish, and vegetables until your harvest comes in. Develop a plan for opening trade. Present your plan using one of these options:

- **As a group, role-play the meeting in which you create your trading plan.**
- **Write instructions for the team of colonists who will open trade with the Native Americans.**

ACTIVITY WRAP-UP

Present to the class **As a group, review your solution to each challenge. Consider the following:**

- **How well each solution meets its particular challenge**
- **Which solution shows the most creativity**

Once you have made your decision, present your solutions to the class. Each group member should take part in the presentation.

DATA FILE

The Journey

Distance: more than 5,000 nautical miles from Europe to the east coast of North America

Length: 6–14 weeks

Dangers: storms, scurvy, dysentery, malnutrition, seasickness, overcrowding

Food and Livestock Taken

barrels of salted beef, oatmeal, dried grains, cheese, oil, vinegar, and salt; seeds for peas, barley, herbs, and other crops; cows, horses, goats, pigs, sheep, and chickens

Equipment Taken

axes, hoes, nails, hooks for doors, hammers, chisels, hatchets, spades, pickaxes, iron pots, copper kettles, skillets, platters, dishes, wooden spoons, rugs

Weapons Taken

swords, muskets, daggers, gunpowder, light armor, cannon

Clothes Taken

shirts, several pairs of shoes, leather for mending, waistcoats, caps, skirts, jackets, trousers

Dangers in America

Biggest killers: typhoid, dysentery, famine

Other dangers: pneumonia, malaria, and other diseases; exposure to harsh weather; fire; wild animals; attacks by Native Americans

Benefits in America

religious and political freedom; opportunity to own land; abundant timber for shelters, forts, heat, ships, and trade; rich food resources

INTERNET **Visit www.mcdougallittell.com for more on the American colonies.**

2 New England Colonies

TERMS & NAMES
Pilgrims
Mayflower Compact
Puritans
Great Migration
Fundamental Orders of Connecticut
Roger Williams
Anne Hutchinson
King Philip's War

MAIN IDEA

Religion influenced the settlement and government of the New England colonies.

WHY IT MATTERS NOW

The Puritan work ethic and religious beliefs influence American culture today.

ONE AMERICAN'S STORY

In 1605, English fishermen captured and enslaved a Native American named Squanto and took him to England. While there, he learned to speak English. After a series of misadventures, including serving as a slave in Spain, Squanto returned to America in 1619. There he discovered that his Pawtuxet tribe had been wiped out by disease in the years 1616–1618. In 1621, Squanto set about helping the English plant corn, beans, and pumpkins on tribal lands. Colonist William Bradford made the following comment about Squanto.

A VOICE FROM THE PAST

Squanto . . . was a special instrument sent of God for their [the colonists'] good beyond their expectation. . . . He directed them how to set their corn, where to take fish, and to procure other commodities, and was also their pilot to bring them to unknown places.

William Bradford, quoted in *The Pilgrim Reader*

Thanks to Squanto, the first settlers in New England prospered and lived in peace with the Native Americans. In this section, you will learn about the Pilgrims and Puritans, their relations with the Native Americans, and their settlement of the New England colonies.

Squanto teaches the Pilgrims how to grow corn.

The Voyage of the *Mayflower*

In the early 1500s, King Henry VIII of England broke that country's ties with the Catholic Church and established the Church of England, an official state church under his control. In the early 1600s, a religious group called the Separatists called for a total break with the Church of England. They thought it was too much like the Catholic Church.

The **Pilgrims** were a Separatist group. King James attacked them for rejecting England's official church. To escape this harsh treatment, the Pilgrims fled to Holland, a country known for its acceptance of different opinions. Eventually, the Pilgrims became dissatisfied with life in Holland. They approached the Virginia Company and asked if they could settle in America "as a distinct body by themselves." The Virginia

The *Mayflower* brings the Pilgrims to Plymouth in 1620.

Company arranged for them to settle on land within its boundaries on the eastern coast of North America.

On a cold, raw November day in 1620, a ship called the *Mayflower* arrived off Cape Cod on the Massachusetts coast. Blown north of its course, the *Mayflower* landed in an area that John Smith had mapped and called New England. They landed at a site that had been named Plymouth.

Because the Pilgrims landed outside the limits of the Virginia Company, their charter did not apply. For the sake of order, the men aboard the *Mayflower* signed an agreement called the **Mayflower Compact.** In it, they vowed to obey laws agreed upon for the good of the colony. The Mayflower Compact helped establish the idea of self-government and majority rule. (See Interactive Primary Sources, page 98.)

The Pilgrims Found Plymouth

Like the early settlers at Jamestown, the Pilgrims at Plymouth endured a starving time. That first winter, disease and death struck with such fury that "the living were scarce able to bury the dead." Half the group had died by spring.

However, energy, hope, and help returned. One day a Native American walked up to a group of colonists. To their astonishment, he called out, "Welcome, Englishmen." This was Samoset, a Pemaquid who had learned to speak English from European fishermen. Samoset introduced the settlers to another Native American named Squanto, a Pawtuxet, who also spoke English.

> ***"Welcome, Englishmen."***
> **Samoset**

The Pilgrims had angered the Native Americans by taking their corn. Squanto acted as an interpreter between the Pilgrims and Chief Massasoit. He helped them to negotiate a peace treaty and showed them how to plant, hunt, and fish. While their crops grew, the colonists began trading with the Native Americans for furs and preparing lumber to ship back to England in order to make a profit.

Reading **History**

A. Making Inferences
Why do you think Squanto was so helpful to the Pilgrims?

Sometime in the fall—no one knows exactly when—the Plymouth settlement celebrated the blessings of a good harvest by holding a three-day feast. It was the first Thanksgiving. This Thanksgiving came to represent the peace that existed at that time between the Native Americans and Pilgrims.

America's HERITAGE

THE FIRST THANKSGIVING
It is hard to believe, but turkey was not on the menu at the first Thanksgiving. The Pilgrims and Native Americans ate venison (deer), roast duck, roast goose, clams and other shellfish, and eel (shown below). Other treats were white bread and corn bread, leeks, watercress, and salad herbs. The guests topped off their meal with wild plums and dried berries for dessert.

Thanks to the help of Squanto and other Native Americans, the Pilgrims learned to survive in their new environment. Soon more people would sail to New England seeking religious freedom.

The Puritans Come to Massachusetts Bay

Between about 1630 and 1640, a religious group called the **Puritans** left England to escape bad treatment by King James I. Unlike the Separatists, who wanted to break away from the Church of England, the Puritans wanted to reform, or "purify," its practices. By the thousands, Puritan families left for the Americas. Their leaving is known as the **Great Migration.** Many thousands of Puritans left their homeland to found new settlements around the world. Of these settlers, about 20,000 crossed the Atlantic Ocean to New England.

Background
During the Great Migration, the Puritans also went to Ireland, the Netherlands, the Rhineland, and the West Indies.

Many Puritan merchants had invested in the Massachusetts Bay Company. In 1629, the company received a royal charter to settle land in New England. In 1630, 11 well-supplied ships carried about 1,000 passengers to the Massachusetts Bay Colony. Unlike earlier colonists, the Puritans were well prepared and did not suffer through a starving time. John Winthrop was the colony's Puritan governor. He stated that the new colony would be a commonwealth, a community in which people work together for the good of the whole.

***Reading* History**
B. Making Inferences After Winthrop, politicians sometimes spoke of America as "a city upon a hill." What does this phrase suggest about America's role in the world?

A VOICE FROM THE PAST

So shall we keep the unity of the spirit, in the bond of peace. . . . Ten of us will be able to resist a thousand of our enemies. . . . For we must consider that we shall be as a City upon a Hill, the eyes of all people are on us.

John Winthrop, *"Model of Christian Charity"*

The New England Way

The basic unit of the commonwealth was the congregation—a group of people who belong to the same church. Each Puritan congregation set up its own town. The meetinghouse was the most important building in each town. There people gathered for town meetings, a form of self-government in which people made laws and other decisions for the community. In the Massachusetts Bay Colony, only male church members could vote or hold office. They elected representatives to a lawmaking body called the General Court, which in turn chose the governor.

By law, everyone in town had to attend church services held in the meetinghouse. The sermon, the most important part of the church service, provided instruction in the "New England Way." This was a term

Vocabulary
godliness: piety, reverence

used by the Puritans to describe both their beliefs and their society, which emphasized duty, godliness, hard work, and honesty. The Puritans thought that amusements such as dancing and playing games would lead to laziness. They believed that God required them to work long and hard at their vocation.

The Puritan work ethic helped contribute to the rapid growth and success of the New England colonies. The New England Way also depended on education. Because the Puritans wanted everyone to be able to read the Bible, laws required that all children learn to read.

*Reading*History
C. Summarizing What were some important elements of the New England Way?

Some Puritan congregations set up new colonies. In 1636, Thomas Hooker moved his congregation to the Connecticut Valley. There they wrote and adopted the **Fundamental Orders of Connecticut** in 1639 (see page 98). In effect, these laws were a constitution. The Fundamental Orders extended voting rights to non-church members and limited the power of the governor. They expanded the idea of representative government.

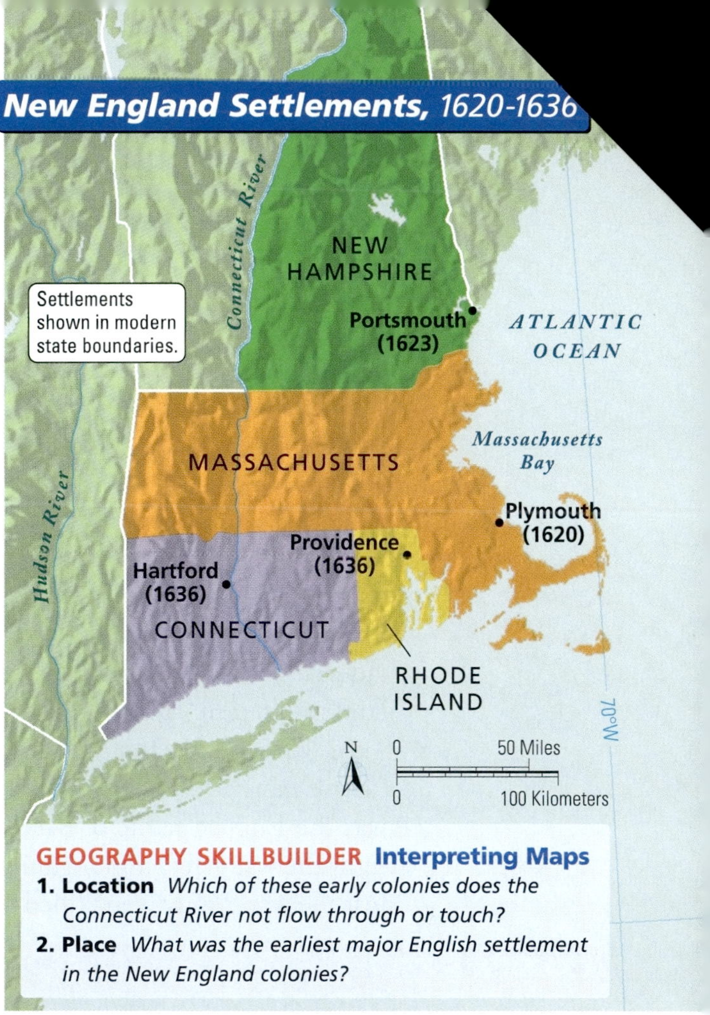

GEOGRAPHY SKILLBUILDER Interpreting Maps
1. **Location** *Which of these early colonies does the Connecticut River not flow through or touch?*
2. **Place** *What was the earliest major English settlement in the New England colonies?*

The first European settlement in New Hampshire was a village near Portsmouth in 1623. In 1638, John Wheelwright established the town of Exeter. The town's founders drew up the Exeter Compact, which was based on the Mayflower Compact.

Challenges to Puritan Leaders

Not everyone agreed with the New England Way. **Roger Williams** was a minister in Salem, Massachusetts, who founded the first Baptist church in America. He opposed forced attendance at church. He also opposed the English colonists' taking of Native American lands by force. Because of his beliefs, the General Court forced Williams to leave the colony. In 1636, he fled southward and founded the colony of Rhode Island, which guaranteed religious freedom and the separation of church and state.

Anne Hutchinson preaches in her home in Boston.

Anne Hutchinson believed that a person could worship God without the help of a church, minister, or Bible. She conducted discussions in her home that challenged church authority. Hutchinson was brought to trial and forced to leave Massachusetts. In 1638, she fled to Rhode Island.

Another religious group was the Quakers. Their name came from an early leader's statement that they should "tremble [quake] at the word of the Lord." Opponents coined the name as an insult. Quakers challenged the Massachusetts commonwealth. They believed that each person could know God directly through "an inner light." Neither ministers nor the Bible were needed. Quakers also believed in treating Native Americans fairly, which set them apart from other colonists. For such beliefs, Quakers were whipped, imprisoned, and hanged. Many left for Rhode Island.

Reading **History**

D. Forming Opinions Why is it odd that the Puritans persecuted certain individuals and groups for their religious beliefs?

King Philip's War

The growing population of colonists began to force the Native Americans from their land. Europeans and Native Americans defined land ownership differently. To Europeans, land could be owned by individuals. To Native Americans, land belonged to everyone. Conflict over land resulted in warfare.

In 1675–1676, the Puritan colonies fought a brutal war with the Native Americans. This was known as **King Philip's War.** "King Philip" was the English name of Metacom, leader of the Wampanoag. To help fight the war, Metacom organized an alliance of tribes. The Wampanoag lost the war. Many were killed, while others were sold into slavery in the West Indies. Those who remained lost their land and were forced to become laborers. English settlers expanded even further into Native American land.

Background

Metacom was the son of Massasoit, friend of the Pilgrims.

CAUSE & EFFECT: *King Philip's War, 1675–1676*

SKILLBUILDER Interpreting Charts

1. *Was there a greater loss of life among the settlers or the Native Americans?*
2. *How might the growing population of Europeans have created more conflict with the Native Americans?*

The Salem Witchcraft Trials

Puritan New England was originally a society centered on the church. By the late 1600s, however, this had begun to change. The younger generations did not share the strict religious views of their parents. Several Salem village girls were told frightening stories about witches by Tituba, a slave from the West Indies. Pretending to be bewitched, the girls falsely accused others of witchcraft. The witch-hunts began in 1692. The clergy viewed the Salem witch-hunts and trials as a sign from God for the village to return to a strict Puritan lifestyle.

HISTORY through ART

This mid-nineteenth-century oil painting, *The Trial of George Jacobs, August 5, 1692,* was painted by T. H. Matteson in 1855. It captures the horrors of the Salem witch trials. As the young women cry out, the accused tries to defend himself against charges that he bewitched them.

Jacobs's own granddaughter testified against him. He was tried and convicted on August 5, 1692, and executed two weeks later along with four neighbors.

How accurately do you think the painting shows the strong emotions in the courtroom?

Hysteria spread through Salem. Those accused were forced to name others as witches. More than 100 people were arrested and tried. Of those, 20 were found guilty and put to death. Nineteen persons were hanged, and another was pressed to death by heavy stones when he refused to enter a plea in response to the charge of witchcraft. The panic was short-lived, and Salem came to its senses. The experience showed, however, how a society can create scapegoats for its problems.

Vocabulary
scapegoat: one that is made to bear the blame of others

In the next section, you will read about the Middle and Southern colonies, how they were founded, and how they provided the new settlers with economic opportunities.

Section 2 Assessment

1. Terms & Names

Identify:
- Pilgrims
- Mayflower Compact
- Puritans
- Great Migration
- Fundamental Orders of Connecticut
- Roger Williams
- Anne Hutchinson
- King Philip's War

2. Taking Notes

Use a cluster diagram to review details about the New England Way.

Which parts would you find easy to accept? Which difficult?

3. Main Ideas

a. What is the Mayflower Compact?

b. What is the meaning of the term the "Great Migration"?

c. What were some of the causes of King Philip's War?

4. Critical Thinking

Recognizing Effects What impact did the arrival of the English in New England have on the Native Americans?

THINK ABOUT
- Squanto
- Chief Massasoit
- King Philip's War

ACTIVITY OPTIONS

LANGUAGE ARTS

SPEECH

Choose one of the Puritan dissenters from this section and retell his or her story. Either write a **newspaper article** about the person or give an **oral history.**

The Mayflower Compact

Setting the Stage In 1620, 41 of the colonists aboard the *Mayflower* drew up the Mayflower Compact. This document refers to the area where they landed as "Virginia" because the land grants of the Virginia Company extended into New England. The colonists provided for self-government under majority rule of the male voters. **See Primary Source Explorer** CD-ROM

A CLOSER LOOK

REASONS FOR VOYAGE

The three reasons the colonists give for their voyage to the eastern seaboard of North America are the glory of God, the advancement of Christianity, and the honor of the king.

1. Might there have been other, more practical reasons for the voyage?

A CLOSER LOOK

GUIDING PURPOSE

The general good of the colony is the guiding purpose of the colonists in signing the compact.

2. What does this suggest about the relationship between the individual and the community?

We, whose names are underwritten, . . . having undertaken for the glory of God, and advancement of the Christian faith, and the honor of our King and country, a voyage to plant the first colony in the northern parts of Virginia, do by these presents, solemnly and mutually in the presence of God and one another **covenant**[1] and combine ourselves together into a civil **body politic,**[2] for our better ordering and preservation; and furtherance of the ends aforesaid . . . do enact, constitute, and frame such just and equal laws, ordinances, acts, constitutions, and offices from time to time as shall be thought most [proper] and convenient for the general good of the colony unto which we promise all due submission and obedience. In witness whereof we have hereunto subscribed our names at Cape Cod the eleventh of November, in the year of our **sovereign**[3] lord King James of England . . . Anno Domini 1620.

From B. P. Poore, ed., *The Federal and State Constitutions,* Part I, p. 931.

1. **covenant:** promise in a binding agreement.
2. **body politic:** the people of a politically organized group.
3. **sovereign:** supreme.

The Fundamental Orders of Connecticut

Setting the Stage In January 1639, male citizens of three townships in Connecticut (Hartford, Windsor, and Wethersfield) assembled and drew up the Fundamental Orders of Connecticut. This document is often called the first written constitution in America. It contains a preamble, or introduction, and a set of laws. **See Primary Source Explorer** CD-ROM

Preamble

Forasmuch as it has pleased the Almighty God by the wise disposition of His Divine Providence so to order and dispose of things that we, the inhabitants and residents of Windsor, Hartford, and Wethersfield are now cohabiting and dwelling in and upon the river of Conectecotte [Connecticut] and the lands thereunto adjoining; and well knowing where a people are gathered together the Word of God requires that, to maintain the peace and union of such a people, there should be an orderly and decent government established according to God, to order and dispose of the affairs of the people at all seasons as occasion shall require; do therefore associate and **conjoin**[1] ourselves to be as one public state or commonwealth. . . . As also in our civil affairs to be guided and governed according to such laws, rules, orders, and decrees as shall be made, ordered and decreed, as follows:

Laws, Rules, and Orders

1. It is ordered, sentenced, and decreed that there shall be yearly two general assemblies or courts. . . . The first shall be called the Court of Election, wherein shall be yearly chosen . . . so many magistrates and other public officers as shall be found **requisite.**[2] . . .

4. It is ordered . . . that no person be chosen governor above once in two years, and that the governor be always a member of some approved congregation. . . .

5. It is ordered . . . that to the aforesaid Court of Election the several towns shall send their deputies. . . . Also, the other General Court in September shall be for making of laws, and any other public occasion which concerns the good of the Commonwealth. . . .

7. It is ordered . . . that after there are warrants given out for any of the said General Courts, the constable or constables of each town shall forthwith give notice distinctly to the inhabitants of the same . . . that at a place and time . . . they meet and assemble themselves together to elect and choose certain deputies to be at the General Court then following to [manage] the affairs of the Commonwealth. . . .

A CLOSER LOOK

GOOD GOVERNMENT

Good government is pleasing to God in the eyes of the colonists. An orderly and decent government helps to maintain peace and order within a community and between people.

3. How would you define good government today?

A CLOSER LOOK

THE GOVERNOR'S ROLE

The person serving as governor can serve only once every two years and must be a member of an approved congregation.

4. Why might the colonists have wished to limit the power of the chief executive?

A CLOSER LOOK

THE COURTS

The Court of Election chooses officials to serve; the General Court makes laws.

5. Why might it be a good idea to separate these two functions?

1. **conjoin:** unite.
2. **requisite:** required.

Interactive Primary Sources Assessment

1. Main Ideas

a. Whose rights did the Mayflower Compact protect?

b. Why are written documents useful in setting up a government?

c. How were the Fundamental Orders based on religion?

2. Critical Thinking

Comparing
In what way are these two documents alike?

THINK ABOUT
- self-government
- majority rule

Founding the Middle and Southern Colonies

TERMS & NAMES
Peter Stuyvesant
patroon
Duke of York
proprietary colony
William Penn
Quaker
royal colony
James Oglethorpe

MAIN IDEA

The founding of the Middle and Southern colonies provided settlers with many economic opportunities.

WHY IT MATTERS NOW

America is still a place where immigrants seek freedom and economic opportunity.

ONE AMERICAN'S STORY

The Dutch had founded the colony of New Netherland (later New York) on the eastern coast of North America in 1624. Peter Stuyvesant, the new governor, arrived in the city of New Amsterdam in May 1647. Because of his harsh personality and rough manner, he soon lost the support of the Dutch colonists. In 1664, a British fleet ordered the city of New Amsterdam to surrender itself to British control. Stuyvesant was unable to gain the support of the Dutch colonists against the British. He surrendered and then defended his decision to his superiors back in the Netherlands.

A VOICE FROM THE PAST

Powder and provisions failing, and no relief or reinforcements being expected, we were necessitated [forced] to come to terms with the enemy, not through neglect of duty or cowardice . . . but in consequence of an absolute impossibility to defend the fort, much less the city of New Amsterdam, and still less the country.

Peter Stuyvesant, quoted in *Peter Stuyvesant and His New York*

Peter Stuyvesant, governor of the Dutch colony of New Netherland, lost his leg in 1644 during a military action against the island of St. Martin in the Caribbean.

After the surrender, Stuyvesant retired to his farm. This land later became part of New York City. In this section, you will read about the founding of the Middle Colonies (such as New York) and the Southern Colonies. You will learn who settled there and why they came.

The Middle Colonies

The Middle Colonies were New York, New Jersey, Pennsylvania, and Delaware. They were located between New England to the north and the Chesapeake region to the south. (See the map on page 102.) Swedes, Dutch, English, Germans, and Africans were among the groups who came to these colonies.

Religious freedom attracted many groups, including Protestants, Catholics, Quakers, and Jews. The Hudson and Delaware rivers supported shipping and commerce. The river valleys had rich soil and mild winters. These conditions were favorable for farming and raising livestock.

New Netherland Becomes New York

In 1624, Dutch settlers financed by the Dutch West India Company founded the colony of New Netherland. New Netherland included the Hudson River valley, Long Island, and the land along the Delaware River.

To attract more settlers, the Dutch West India Company employed the patroon system. A **patroon** was a person who brought 50 settlers to New Netherland. As a reward, a patroon received a large land grant. He also received special privileges in hunting, fishing, and fur trading on his land.

In the early years, many different kinds of people settled in New Netherland. Twenty-three Jewish settlers arrived in 1654, and others soon followed. Later, Africans were brought to the colony as slaves and indentured servants. Many Puritans also came.

Peter Stuyvesant, the colony's governor, wanted to add land to New Netherland. He attacked the nearby charter colony of New Sweden in 1655. This colony was located along the Delaware River. The main settlement was Fort Christina (later named Wilmington, Delaware). It had been settled by Swedes in 1638. After an attack by the Dutch, the Swedes surrendered Fort Christina.

America's **HERITAGE**

THE LOG CABIN

Swedish colonists living in Delaware built the first log cabin in America in 1638. The log cabin was the perfect house to build where there were many trees. Settlers needed few tools to build such cabins, which were made of round logs with curved notches at the ends. After the ends were placed in the notches, the logs were secured. After 1780, the log cabin became the typical frontier home.

Background
The Duke of York became King James II in 1685.

England's King Charles II decided that his brother, the **Duke of York,** should drive the Dutch out of New Netherland. The Dutch colony was a threat to England because of its trade. It was also a threat because of its expanding settlements and its location. There were English colonies in New England to the north and Virginia to the south. As you have seen, when the duke's ships appeared off New Amsterdam in August 1664, the colony surrendered. New Netherland became the **proprietary colony** of New York. The Duke of York was now the proprietor, or owner, of the colony.

Reading **History**
A. Forming Opinions Why might the promise of religious freedom encourage a diverse population in a colony?

New Jersey, Pennsylvania, and Delaware

The Duke of York had become the largest single landowner in America. He gave part of his claim, the province of New Jersey, to his friends Sir George Carteret and Lord John Berkeley in 1664. They encouraged settlers to come by promising freedom of religion. They also promised large grants of land and a representative assembly.

William Penn became another large landowner in America. Born into a wealthy English family, Penn joined the **Quakers,** to his father's disapproval. The young Penn was attacked for his Quaker beliefs. King Charles II owed the Penn family money. In repayment, in 1681 he gave Penn a large piece of land in America that came to be called Pennsylvania. The name means "Penn's woods."

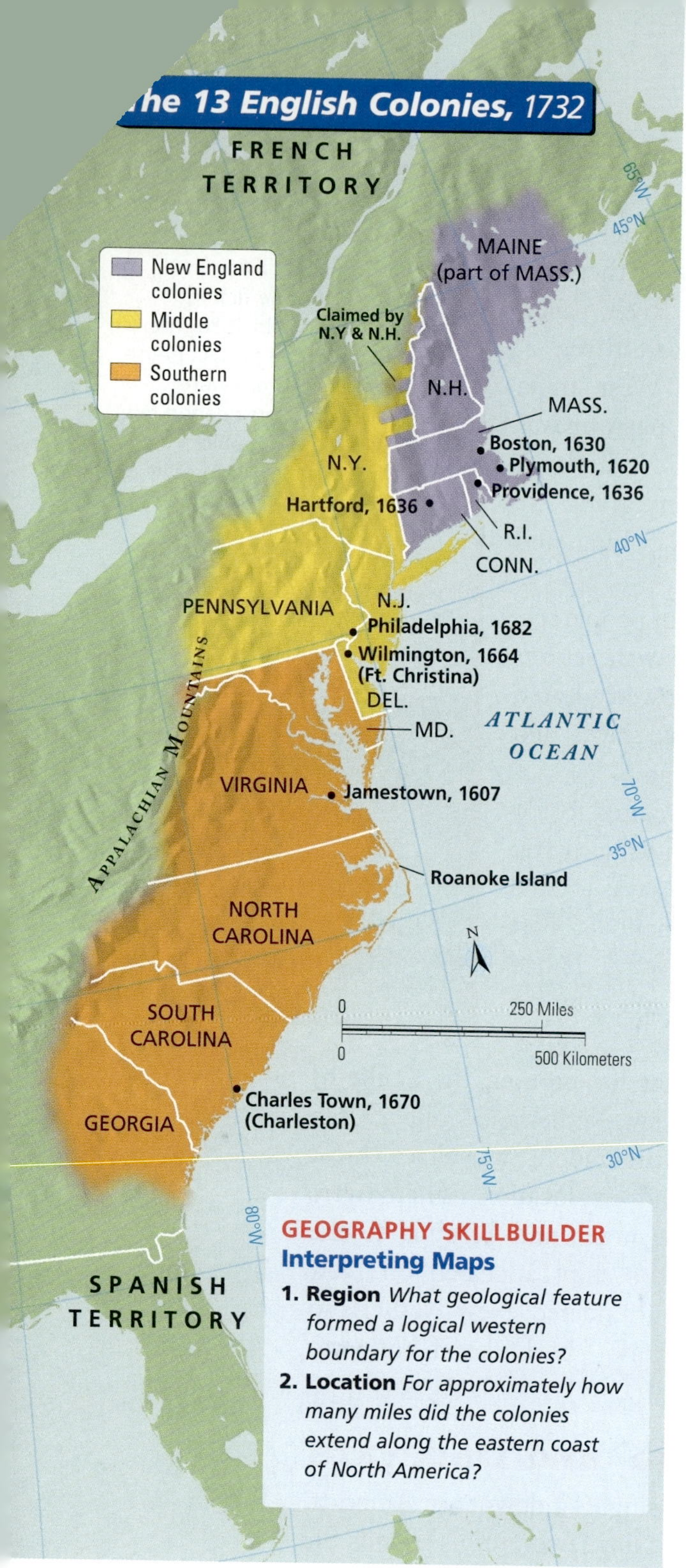

GEOGRAPHY SKILLBUILDER
Interpreting Maps
1. **Region** *What geological feature formed a logical western boundary for the colonies?*
2. **Location** *For approximately how many miles did the colonies extend along the eastern coast of North America?*

Penn used this land to create a colony where Quakers could live according to their beliefs. Among other things, the Quakers believed that all people should live in peace and harmony. They welcomed different religions and ethnic groups. In Pennsylvania, Penn extended religious freedom and equality to all. He especially wanted the Native Americans to be treated fairly. In a letter to them in 1681, Penn said, "May [we] always live together as neighbors and friends."

Reading **History**
B. Comparing and Contrasting How did Penn's policies toward Native Americans compare with those of other colonies you have read about?

Penn's policies helped make Pennsylvania one of the wealthiest of the American colonies. Many settlers came to Pennsylvania seeking religious freedom and a better life. In 1704, Penn granted the three lower counties of Delaware their own assembly. The counties later broke away to form the colony of Delaware.

The Southern Colonies

The new Southern Colonies were Maryland, the Carolinas, and Georgia. The Appalachian Mountains bordered parts of these colonies in the west. In the east, the colonies bordered the Atlantic Ocean. The soil and climate of this region were suitable for warm-weather crops such as tobacco, rice, and indigo.

Maryland and the Carolinas

Lord Baltimore established Maryland in 1632 for Roman Catholics fleeing persecution in England. To attract other settlers besides Catholics, Lord Baltimore promised religious freedom. In 1649, Maryland passed the Toleration Act.

Maryland based its economy on tobacco, which required backbreaking work. Every three or four years, the tobacco crop used up the soil, and workers had to clear new land. Most laborers came as either servants or slaves. Maryland attracted few women as settlers.

In 1663, Carolina was founded as a colony. English settlers from Barbados built Charles Town, later called Charleston, in 1670. They busied themselves cutting timber, raising cattle, and trading with the Native Americans. After 1685, Charleston became a refuge for Huguenots, French Protestants seeking religious freedom.

Vocabulary
Carolina: The name of the colony is based on a Latin form of "Charles," in honor of King Charles II.

Carolina's colonists needed laborers to grow rice and indigo. The English settlers encouraged the use of enslaved Africans. They also sold local Native Americans into slavery. As a result, wars broke out between the settlers and the Tuscarora and Yamasee tribes. The settlers' taking of tribal lands also fueled the wars.

Carolina's proprietors, or owners, refused to send help to stop a threatened Spanish attack on Charleston. Because of this, the colonists overthrew the colony's proprietary rule in 1719. In 1729, Carolina became a **royal colony**. Then it was ruled by governors appointed by the king. The colony was divided into North Carolina and South Carolina.

Georgia

In 1732, **James Oglethorpe** founded Georgia as a refuge for debtors. The English government wanted to use the colony as a military outpost against Spanish Florida to the south and French Louisiana to the west. In 1739, during a war between England and Spain, the Spanish tried to force the English colonists out of Georgia but were unsuccessful. English, German, Swiss, and Scottish colonists settled in Georgia. All religions were welcome. As the colony's leader, Oglethorpe set strict rules that upset the colonists. The king, in response to unrest, made Georgia a royal colony in 1752.

Reading**History**
C. Reading a Map Use the map on page 102 to check the location of Georgia in relation to the Spanish territory of Florida.

James Oglethorpe was the founder of Georgia.

By the early 1700s, there were 13 English colonies along the eastern coast of North America. In the next chapter, you will read about how these colonies developed.

Section 3 Assessment

1. Terms & Names
Identify:
- Peter Stuyvesant
- patroon
- Duke of York
- proprietary colony
- William Penn
- Quaker
- royal colony
- James Oglethorpe

2. Taking Notes
Identify an effect for each cause listed in the chart below.

Cause	Effect
New Netherland threat to English	
English attacked Quakers	
Laborers needed in Carolinas	
Oglethorpe too strict in Georgia	

3. Main Ideas
a. What were the goals of the patroon system?

b. What three Middle Colonies offered religious freedom?

c. What were three crops grown in the Southern Colonies?

4. Critical Thinking
Analyzing Causes Why did colonists in Maryland and the Carolinas enslave Native Americans and use African slaves?

THINK ABOUT
- the crops being grown
- the nature of farm work

ACTIVITY OPTIONS
LANGUAGE ARTS
SCIENCE

What are the health effects of tobacco? Write a **news article** or give a **television report** for a science show about the effects of tobacco on the body.

hapter 3 ASSESSMENT

TERMS & NAMES

Briefly explain the importance of each of the following.

1. joint-stock company
2. Jamestown
3. John Smith
4. House of Burgesses
5. Pilgrims
6. Mayflower Compact
7. Great Migration
8. Fundamental Orders of Connecticut
9. proprietary colony
10. William Penn

REVIEW QUESTIONS

Early Colonies Have Mixed Success (pages 85–91)

1. What were the reasons given by Richard Hakluyt that England should start a colony?
2. Why were Jamestown and Plymouth financed by joint-stock companies?
3. How did John Rolfe change the Virginia colony?

New England Colonies (pages 92–99)

4. What was John Winthrop's vision for Massachusetts Bay?
5. What was the system of government in the Massachusetts Bay Colony?
6. What were some of the effects of King Philip's War?

Founding the Middle and Southern Colonies (pages 100–103)

7. Why did Charles II want New Netherland?
8. What were relations like between Native Americans and settlers in Pennsylvania?
9. What was the Toleration Act of 1649?
10. What sorts of crops were particularly well suited to the soil and climate in the Southern colonies?

CRITICAL THINKING

1. USING YOUR NOTES

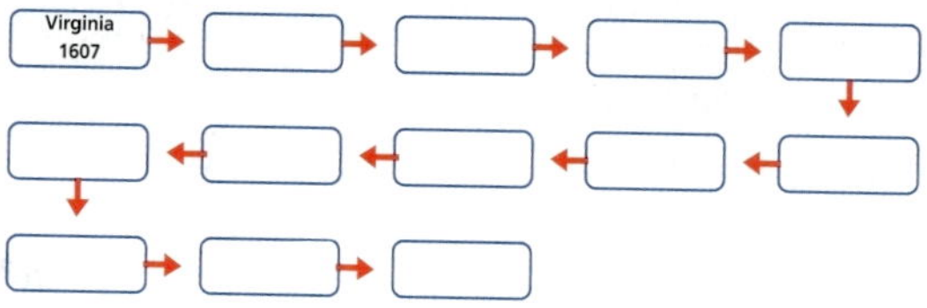

Using your completed chart, answer these questions:

a. Which was the earliest successful settlement in Virginia?

b. Which colony was founded last?

2. ANALYZING LEADERSHIP

Why do you think William Penn was a more successful leader than Peter Stuyvesant?

3. APPLYING CITIZENSHIP SKILLS

What were some of the common ideals that link the Mayflower Compact, the establishment of the House of Burgesses, and town meetings?

4. THEME: IMPACT OF THE INDIVIDUAL

How did individual effort help ensure the success of England's colonies in America?

5. ANALYZING CAUSES

Why might the Puritans have been intolerant of religious dissenters?

Interact *with* History

How do the dangers you discussed before you read this chapter compare with the dangers people actually faced?

VISUAL SUMMARY

The 13 Colonies

		Important Early Dates	*Founder(s)*
New England Colonies	Massachusetts	Plymouth,1620; Mass. Bay, 1630	Pilgrims; Puritans
	New Hampshire	Portsmouth, 1623	Proprietors
	Rhode Island	Providence, 1636	Roger Williams
	Connecticut	Hartford, 1636	Thomas Hooker
Middle Colonies	New York (New Netherland)	Dutch settlers arrive, 1624	Dutch West India Company
	Delaware	Fort Christina, 1638	Swedes
	New Jersey	Duke of York establishes, 1664	George Carteret, John Berkeley
	Pennsylvania	Charles II bestows land, 1681	William Penn
Southern Colonies	Virginia	Jamestown, 1607	Virginia Company of London
	Maryland	Founded as religious haven, 1632	Lord Baltimore
	North Carolina	Founded, 1663	Proprietors
	South Carolina	Founded, 1663	Proprietors
	Georgia	Founded as debtors' refuge, 1732	James Oglethorpe

HISTORY SKILLS

1. INTERPRETING GRAPHS

Study the graph and then answer the questions.

Population of the Colonies

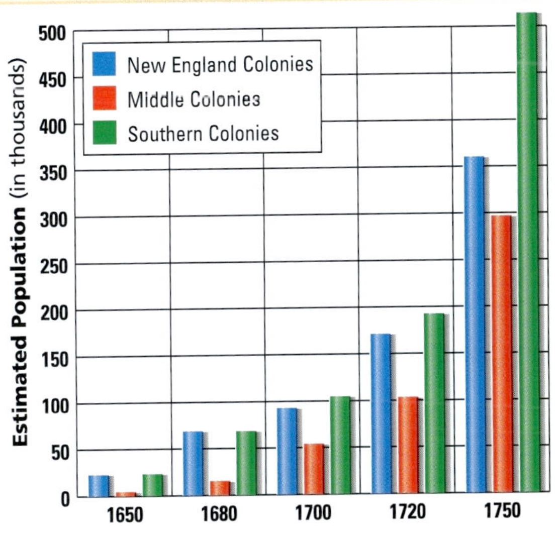

Source: *Historical Statistics of the United States,* series, Z 1–20.

a. How much did the population of the Southern colonies increase between 1720 and 1750?

b. What was the increase in the population of the New England colonies between 1700 and 1720?

c. Which region had the largest population over the years, and which region had the smallest?

2. INTERPRETING PRIMARY SOURCES

The early colonists used a surveyor's compass (below) to divide up the land they had come to settle.

a. What attitudes about ownership of the land are revealed by the use of a surveyor's compass?

b. How might the use of a surveyor's compass reflect differences in attitudes toward the land between European settlers and Native Americans?

ALTERNATIVE ASSESSMENT

1. INTERDISCIPLINARY ACTIVITY: Geography

Drawing a Map Draw a map of the New England, Middle, and Southern colonies. Place on the map the major cities and rivers of each colony. Explain how each region's geographic location contributed to the colonies' economic activities. Share your map with the class.

2. COOPERATIVE LEARNING ACTIVITY

Performing a Scene from a Play The "lost colonists" of Roanoke disappeared sometime between 1588 and 1590. Write and perform a play depicting a meeting of the colonists in which they try to decide what to do. John White has not returned with the supplies he promised to bring. How are the colonists to deal with food shortages, illness, and relations with the Native Americans?

Take the roles of Elinor Dare and her husband, Ananias; their child, Virginia; and other colonists. Come up with different solutions to their problems. Then vote on your preferred course of action.

3. CD-ROM PRIMARY SOURCE EXPLORER

Planning a Government As with any group of people living in a community, some sort of government was needed in Plymouth. The Pilgrims devised the Mayflower Compact. Using the CD-ROM, library, and Internet, find out more about the Mayflower Compact.

Create your own plan for a government using the following suggestions:

- Draw up a plan for a government that will apply to your class.
- Adapt ideas from the Mayflower Compact that you think will work for the class.
- Decide what rules are needed in your government. Decide who will hold office, how they will be appointed or selected, and how long they will serve.
- Decide whether there should be limits on majority rule in your government.

4. HISTORY PORTFOLIO

PORTFOLIO **Option 1** Review your section and chapter assessment activities. Select the one that you think is your best work. Then use comments made by your teacher or classmates to improve your work, and add it to your portfolio.

PORTFOLIO **Option 2** Review the questions that you wrote for What Do You Want to Know? on page 84. Then write a short report in which you explain the answers to your questions. Add your answers to your portfolio.

CHAPTER 4

The Colonies Develop 1700–1753

In 1702, a vast countryside surrounded Philadelphia. Most colonists earned their living in the country. Fewer than one in ten lived in cities.

Interact *with* History

It is the early 1700s when you arrive in one of America's larger port cities. After nearly a month of ocean travel, you are thrilled to see land. As you leave the ship, you wonder where you will live and how you will earn a living.

What Do You Think?

- Will you choose to live where other people from your homeland live? Or will you try somewhere new?
- How did you make a living in your old country? Will this influence your choice?

Would you settle on a farm or in a town?

USA

c. 1700 Colonial population reaches 257,000.

1712 Slave uprising occurs in New York City.

1718 French found city of New Orleans at mouth of Mississippi River.

Spanish priests build Alamo in Texas.

1739 Enslaved Africans revolt in Stono Rebellion.

1742 First European settlement west of Allegheny Mountains is established.

c. 1750 Population of the English colonies passes the one million mark.

1700 — **1753**

World

1701 War of the Spanish Succession begins in Europe.

1707 Act of Union unites England with Scotland and creates Great Britain.

1727 George II becomes King of Great Britain.

1747 Slave trading African kingdom of Dahomey is defeated by Oyo.

1752 China suppresses Tibetan rebellion and forces Dalai Lama to accept its authority.

BEFORE YOU READ

Previewing the Theme

Economics in History When immigrants came to the Americas, they settled in places with different climates and resources. These conditions affected the economic choices made by colonists. As Chapter 4 explains, those choices contributed to the formation of four different colonial regions.

What Do You Know?

What ideas and pictures come to mind when you hear people talk about "the South" or "the North"? Why do you think these distinct regions developed?

THINK ABOUT

- what you have learned about these regions from books or movies
- the way geography affects people's choices

What Do You Want to Know?

What questions do you have about how the four colonial regions developed? Record these questions in your notebook before you read the chapter.

READ AND TAKE NOTES

Reading Strategy: Analyzing Causes and Recognizing Effects As you read about history, it is important to understand not only what happened in the past, but also the reasons why it happened. Clue words that indicate cause—such as *because* and *since*—can help you look for causes of historical events. Use the chart below to list causes that contributed to the different economic developments in each of the colonial regions.

See Skillbuilder Handbook, page R10.

		NEW ENGLAND COLONIES	MIDDLE COLONIES	SOUTHERN COLONIES	BACKCOUNTRY
CAUSES	Climate				
	Resources				
	People				
EFFECT	Economic Development				

1 New England: Commerce and Religion

TERMS & NAMES
Backcountry
subsistence farming
triangular trade
Navigation Acts
smuggling

MAIN IDEA

Fishing and trade contributed to the growth and prosperity of the New England Colonies.

WHY IT MATTERS NOW

Coastal cities in New England continue to engage in trade.

Whales hunted by New Englanders, such as Peleg Folger, might weigh as much as 50 tons and be over 60 feet in length.

ONE AMERICAN'S STORY

Peleg Folger, a New England sailor, was only 18 years old when he began whaling. Folger kept a journal that describes what whaling was like in the 1750s. In one journal entry, Folger explained what happened after whales were sighted and small boats were launched to pursue them.

A VOICE FROM THE PAST

So we row'd about a mile and a Half from the [ship], and then a whale come up under us, & [smashed in] our boat . . . and threw us every man overboard [except] one. And we all came up and Got Hold of the Boat & Held to her until the other boat (which was a mile and half off) came up and took us in, all Safe, and not one man Hurt, which was remarkable, the boat being threshed to pieces very much.

Peleg Folger, quoted in *The Sea-Hunters*

When Folger and his mates did manage to kill a whale, they cut a hole in its head. Then "a man got in up to his armpits and Dipt out [barrels] of clear oil." When the ship returned to port, this oil was sold to colonists, who used it as fuel in their lamps.

Many settlers in the New England Colonies—Massachusetts, New Hampshire, Connecticut, and Rhode Island—turned to the Atlantic Ocean to make a living. The majority of New Englanders, however, were farmers.

Distinct Colonial Regions Develop

Between 1700 and 1750, the population of England's colonies in North America doubled and then doubled again. At the start of the century, the colonial population stood at about 257,000. By 1750, more than 1,170,000 settlers called the English colonies home.

By the 1700s, the colonies formed three distinct regions: the New England Colonies, the Middle Colonies, and the Southern Colonies. Another area was the **Backcountry**. It ran along the Appalachian Mountains through the far western part of the other regions.

Several factors made each colonial region distinct. Some of the most important were each region's climate, resources, and people.

1. New England had long winters and rocky soil. English settlers made up the largest group in the region's population.
2. The Middle Colonies had shorter winters and fertile soil. The region attracted immigrants from all over Europe.
3. The Southern Colonies had a warm climate and good soil. There, some settlers used enslaved Africans to work their plantations.
4. The Backcountry's climate and resources varied, depending on the latitude. Many Scots-Irish immigrants settled there.

Vocabulary
latitude: the distance north or south of the equator, measured in degrees

During the colonial era, the majority of people made their living by farming. However, the type of agriculture they practiced depended on the climate and resources in the region where they settled.

The Farms and Towns of New England

Life in New England was not easy. The growing season was short, and the soil was rocky. Most farmers practiced **subsistence farming.** That is, they produced just enough food for themselves and sometimes a little extra to trade in town.

Most New England farmers lived near a town. This was because colonial officials usually did not sell scattered plots of land to individual

The New England Colonies, 1750

GEOGRAPHY SKILLBUILDER Interpreting Maps
1. **Location** *All of the New England Colonies are to the east of what major river?*
2. **Region** *What ocean-related products was colonial New England known for?*

This New England meetinghouse is located in Sturbridge Village, Massachusetts.

farmers. Instead, they sold larger plots of land to groups of people—often to the congregation of a Puritan church. A congregation then settled the town and divided the land among the members of its church.

Reading **History**

A. Recognizing Effects How did the way land was sold in New England affect the way people lived?

This pattern of settlement led New England towns to develop in a unique way. Usually, a cluster of farmhouses surrounded a green—a central square where a meetinghouse was located and where public activities took place. Because people lived together in small towns, shopkeepers had enough customers to make a living. Also, if the townspeople needed a blacksmith or a carpenter, they could pool their money and hire one.

Harvesting the Sea

New England's rocky soil made farming difficult. In contrast, the Atlantic Ocean offered many economic opportunities. In one story, a group of settlers was standing on a hill overlooking the Atlantic. One of them pointed out to sea and exclaimed, "There is a great pasture where our children's grandchildren will go for bread!"

The settler's prediction came true. Not far off New England's coast were some of the world's best fishing grounds. The Atlantic was filled with mackerel, halibut, cod, and many other types of fish.

New England's forests provided everything needed to harvest these great "pastures" of fish. The wood cut from iron-hard oak trees made excellent ship hulls. Hundred-foot-tall white pines were ideal for masts. Shipbuilders used about 2,500 trees to produce just one ship!

New England's fish and timber were among its most valuable articles of trade. Coastal cities like Boston, Salem, New Haven, and Newport grew rich as a result of shipbuilding, fishing, and trade.

Background

In 1742, over 16,000 people lived in Boston.

Atlantic Trade

New England settlers engaged in three types of trade. First was the trade with other colonies. Second was the direct exchange of goods with Europe. The third type was the triangular trade.

Triangular trade was the name given to a trading route with three stops. For example, a ship might leave New England with a cargo of rum

Triangular Trade, *1750*

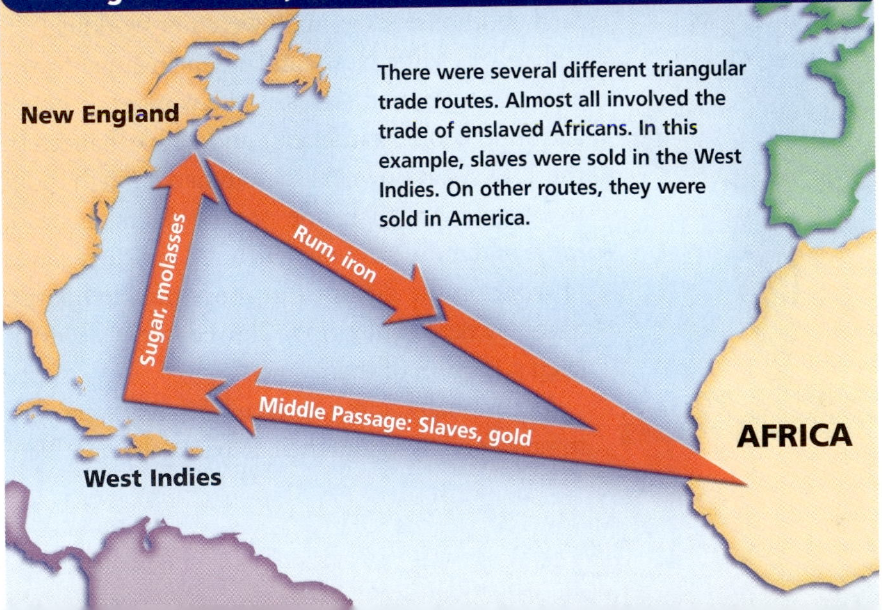

and iron. In Africa, the captain would trade his cargo for slaves. Slaves then endured the horrible Middle Passage to the West Indies, where they were exchanged for sugar and molasses. Traders then took the sugar and molasses back to New England. There, colonists used the molasses to make rum, and the pattern started over.

Background
See Olaudah Equiano's descriptions of the Middle Passage on page 78.

New England won enormous profits from trade. England wanted to make sure that it received part of those profits. So the English government began to pass the **Navigation Acts** in 1651. The Navigation Acts had four major provisions designed to ensure that England made money from its colonies' trade.

1. All shipping had to be done in English ships or ships made in the English colonies.
2. Products such as tobacco, wood, and sugar could be sold only to England or its colonies.
3. European imports to the colonies had to pass through English ports.
4. English officials were to tax any colonial goods not shipped to England.

But even after the passage of the Navigation Acts, England had trouble controlling colonial shipping. Merchants ignored the acts whenever possible. **Smuggling**—importing or exporting goods illegally—was common. England also had great difficulty preventing pirates—like the legendary Blackbeard—from interfering with colonial shipping.

STRANGE *but* True

BLACKBEARD THE PIRATE

Of all the pirates who attacked colonial ships, Blackbeard (shown below) was the most famous. He was a fearsome man known to stick matches in his hair to light up his face during battle.

Blackbeard's pirate career finally came to an end in 1718, when Virginia's governor sent an expedition against him. Nearly half the expedition's men died in the key battle. Blackbeard himself did not fall until he had suffered nearly 25 wounds. Before sailing back to port, sailors cut off his head and put it on the front of their ship.

African Americans in New England

There were few slaves in New England. Slavery simply was not economical in this region of small farms. Also, because the growing season was short, there was little work for slaves during the long winter months. Farmers could not afford to feed and house slaves who were not working.

***Reading* History**
B. Analyzing Causes Why were there relatively few enslaved workers in New England?

Even so, some New Englanders in larger towns and cities did own slaves. They worked as house servants, cooks, gardeners, and stable-hands. In the 1700s, slave owners seldom had enough room to house more than one or two slaves. Instead, more and more slave owners hired out their slaves to work on the docks or in shops or warehouses. Slave owners sometimes allowed their slaves to keep a portion of their wages.

Occasionally, some enslaved persons were able to save enough to buy their freedom. In fact, New

England was home to more free blacks than any other region. A free black man might become a merchant, sailor, printer, carpenter, or landowner. Still, white colonists did not treat free blacks as equals.

Changes in Puritan Society

The early 1700s saw many changes in New England society. One of the most important was the gradual decline of the Puritan religion. There were a number of reasons for this decline.

Reading **History**

C. Making Inferences Why might an interest in material things compete with the Puritan religion?

One reason was that the drive for economic success competed with Puritan ideas. Many colonists, especially those who lived along the coast, seemed to care as much about business and material things as they did about religion. One observer had this complaint.

"[Boston] is so conveniently Situated for Trade."

An observer in 1713

A VOICE FROM THE PAST

[Boston] is so conveniently Situated for Trade and the Genius of the people are so inclined to merchandise, that they seek no other Education for their children than writing and Arithmetick.

An observer in 1713, quoted in *A History of American Life*

Another reason for the decline of the Puritan religion was the increasing competition from other religious groups. Baptists and Anglicans established churches in Massachusetts and Connecticut, where Puritans had once been the most powerful group.

Political changes also weakened the Puritan community. In 1691, a new royal charter for Massachusetts guaranteed religious freedom for all Protestants, not just Puritans. The new charter also granted the vote based on property ownership instead of church membership. This change put an end to the Puritan churches' ability to control elections.

To the south of New England were the Middle Colonies, which developed in quite different ways—as the next section shows.

Section 1 Assessment

1. Terms & Names

Identify:

- Backcountry
- subsistence farming
- triangular trade
- Navigation Acts
- smuggling

2. Taking Notes

Use a chart like the one shown to record how New Englanders prospered from the Atlantic Ocean.

Economic Activity	Benefits to Colonists

How did some profit illegally from the ocean?

3. Main Ideas

a. How did most people in New England earn a living?

b. Why did England pass the Navigation Acts?

c. What factors led to the decline of the Puritan religion in New England?

4. Critical Thinking

Making Inferences What advantages might there be in living near other people in small towns, such as those in New England?

THINK ABOUT

- the transportation options available to colonists
- why shopkeepers chose to open businesses in towns

ACTIVITY OPTIONS

ART

TECHNOLOGY

Read more about whaling. Make a **mobile** that shows different kinds of whales or plan a **multimedia presentation** on whaling today.

The Middle Colonies: Farms and Cities

TERMS & NAMES
cash crop
gristmill
diversity
artisan
Conestoga wagon

MAIN IDEA

The people who settled in the Middle Colonies made a society of great diversity.

WHY IT MATTERS NOW

States in this region still boast some of the most diverse communities in the world.

ONE AMERICAN'S STORY

Elizabeth Ashbridge was only 19 years old when she arrived in America from England in the 1730s. Even though she was young, she had already been married and widowed. And although she was an indentured servant, she hoped to earn her freedom and find a way to express her strong religious feelings.

After several years, Elizabeth did gain freedom. She started to search for a religion that she could devote her life to. Finally, in the colony of Pennsylvania, she found what she was looking for—the Society of Friends, or Quakers. The new Quaker longed to share her beliefs openly.

A woman speaks out at a Quaker meeting. The Society of Friends allowed women a more active role than other religions.

A VOICE FROM THE PAST

I was permitted to see that all I had gone through was to prepare me for this day; and that the time was near, when it would be required of me, to go and declare to others what the God of mercy had done for my soul.

Elizabeth Ashbridge, *Some Account . . . of the Life of Elizabeth Ashbridge*

The Quakers believed that people of different beliefs could live together in harmony. They helped to create a climate of tolerance and acceptance in the Middle Colonies of New York, New Jersey, Pennsylvania, and Delaware. These colonies began to attract a wide variety of immigrants, as you will read in this section.

A Wealth of Resources

The Middle Colonies had much to offer in addition to a climate of tolerance. A Frenchman named Michel Guillaume Jean de Crèvecoeur (krehv•KUR) praised the region's "fair cities, substantial villages, extensive fields . . . decent houses, good roads, orchards, meadows, and bridges, where an hundred years ago all was wild, woody, and uncultivated."

The prosperity that Crèvecoeur described was typical of the Middle Colonies. Immigrants from all over Europe came to take advantage of this region's productive land. Their settlements soon crowded out Native Americans, who had lived in the region for thousands of years.

Among the immigrants who came to the Middle Colonies were Dutch and German farmers. They brought the advanced agricultural methods of their countries with them. Their skills, knowledge, and hard work would soon result in an abundance of foods.

The Middle Colonies boasted a longer growing season than New England and a soil rich enough to grow **cash crops**. These were crops raised to be sold for money. Common cash crops included fruits, vegetables, and, above all, grain. The Middle Colonies produced so much grain that people began calling them the "breadbasket" colonies.

The Importance of Mills

Vocabulary
grist: another name for grain, the one-seeded fruit of cereal grasses like wheat and rye

After harvesting their crops of corn, wheat, rye, or other grains, farmers took them to a **gristmill**. There, millers crushed the grain between heavy stones to produce flour or meal. Human or animal power fueled some of these mills. But water wheels built along the region's plentiful rivers powered most of the mills.

The bread that colonists baked with these products was crucial to their diet. Colonists ate about a pound of grain in some form each day—nearly three times more than Americans eat today. Even though colonists ate a great deal of grain, they had plenty left over to send to the region's coastal markets for sale.

Philipsburg Manor, in Sleepy Hollow, New York, has a working 18th-century farm and a water-powered gristmill.

GEOGRAPHY SKILLBUILDER Interpreting Maps

1. **Place** *What are the three major rivers in the Middle Colonies?*
2. **Movement** *Why might the Middle Colonies' rivers that empty into the ocean be important for farmers?*

The Cities Prosper

The excellent harbors along the coasts of the Middle Colonies were ideal sites for cities. New York City grew up at the mouth of the Hudson River, and Philadelphia was founded on the Delaware River. The merchants who lived in these growing port cities exported cash crops, especially grain, and imported manufactured goods.

Reading History

A. Reading a Map Locate New York and Philadelphia on the map on page 115. Note the rivers next to which they were built.

Because of its enormous trade, Philadelphia was the fastest growing city in the colonies. The city owed its expansion to a thriving trade in wheat and other cash crops. By 1720, it was home to a dozen large shipyards—places where ships are built or repaired.

The city's wealth also brought many public improvements. Large and graceful buildings, such as Philadelphia's statehouse—which was later renamed Independence Hall—graced the city's streets. Streetlights showed the way along paved roads. In 1748, a Swedish visitor named Peter Kalm exclaimed that Philadelphia had grown up overnight.

daily life

NAMES AND OCCUPATIONS

Many English colonists had names like Miller and Smith—names that reflected how their families had made a living in England. For example, a colonist named Miller probably had an ancestor who had operated a mill. Similarly, Smith probably had an ancestor who had been a blacksmith.

Sometimes colonists continued in the same occupations as their ancestors. But as time went on, colonists turned to other occupations, and their names no longer reflected how they earned a living. Yet names like Smith and Miller remain common in the United States, reflecting the country's past as English colonies.

A VOICE FROM THE PAST

And yet its natural advantages, trade, riches and power, are by no means inferior to any, even of the most ancient towns in Europe.

Peter Kalm, quoted in *America at 1750*

New York could also thank trade for its rapid growth. This bustling port handled flour, bread, furs, and whale oil. At midcentury, an English naval officer admired the city's elegant brick houses, paved streets, and roomy warehouses. "Such is this city," he said, "that very few in England can rival it in its show."

Background

In 1742, New York City's population was about 11,000, and nearly 13,000 people lived in Philadelphia.

A Diverse Region

Many different immigrant groups arrived in the port cities of the Middle Colonies. Soon, the region's population showed a remarkable

diversity, or variety, in its people. One of the largest immigrant groups in the region, after the English, was the Germans.

Many of the Germans arrived between 1710 and 1740. Most came as indentured servants fleeing religious intolerance. Known for their skillful farming, these immigrants soon made a mark on the Middle Colonies. "German communities," wrote one historian, "could be identified by the huge barns, the sleek cattle, and the stout workhorses."

Background
By the second half of the 1700s, more than one in three colonists in Pennsylvania claimed German ancestry.

The Middle Colonies, *1750*

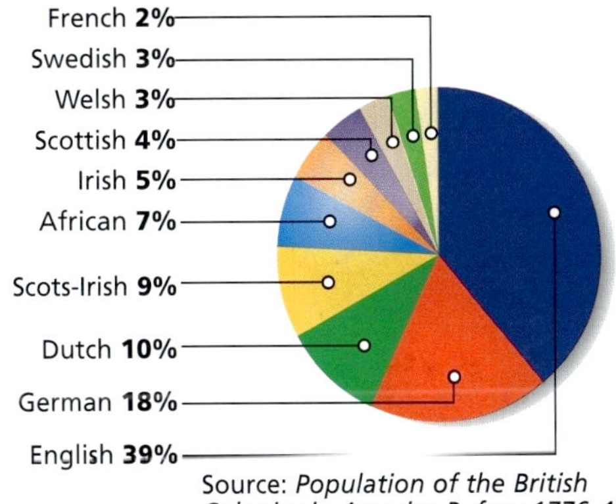

Source: *Population of the British Colonies in America Before 1776,* 1975

SKILLBUILDER Interpreting Graphs

1. *What group made up nearly one-fifth of the population in the Middle Colonies?*
2. *What were the two main languages spoken in the Middle Colonies?*

Germans also brought a strong tradition of craftsmanship to the Middle Colonies. For example, German gunsmiths first developed the long rifle. Other German **artisans,** or craftspeople, became ironworkers and makers of glass, furniture, and kitchenware.

Germans built **Conestoga wagons** to carry their produce to town. These wagons used wide wheels suitable for dirt roads, and the wagons' curved beds prevented spilling when climbing up and down hills. The wagons' canvas covers offered protection from rain. Conestoga wagons would later be important in settling the West.

Reading **History**
B. Summarizing How would you describe the population of the Middle Colonies?

The Middle Colonies became home to many people besides the Germans. There were also the English, Dutch, Scots-Irish, African, Irish, Scottish, Welsh, Swedish, and French. Because of the diversity in the Middle Colonies, different groups had to learn to accept, or at least tolerate, one another.

A Climate of Tolerance

While the English Puritans shaped life in the New England Colonies, many different groups contributed to the culture of the Middle Colonies. Because of the greater number of different groups, it was difficult for any single group to dominate the others. Thus, the region's diversity helped to create a climate of tolerance. Some of the region's religious groups also helped to promote tolerance.

The Middle Colonies' earliest settlers, the Dutch in New York and the Quakers in Pennsylvania, both practiced religious tolerance. That is, they honored the right of religious groups to follow their own beliefs without interference. Quakers also insisted on the equality of men and women. As a result, Quaker women served as preachers, and female missionaries traveled the world spreading the Quaker message.

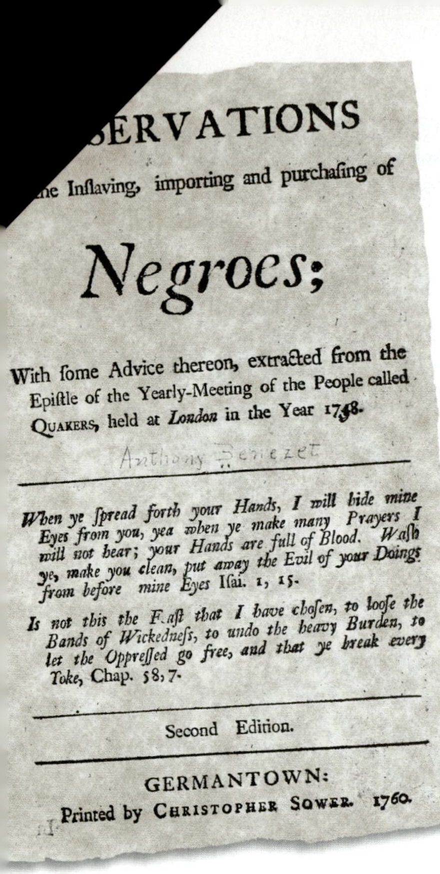
ERVATIONS
e Inſlaving, importing and purchaſing of

Negroes;

With ſome Advice thereon, extracted from the Epiſtle of the Yearly-Meeting of the People called QUAKERS, held at *London* in the Year 1758.

Anthony Benezet

When ye ſpread forth your Hands, I will hide mine Eyes from you, yea when ye make many Prayers I will not hear; your Hands are full of Blood. Waſh ye, make you clean, put away the Evil of your Doings from before mine Eyes Iſai. 1, 15.

Is not this the Faſt that I have choſen, to looſe the Bands of Wickedneſs, to undo the heavy Burden, to let the Oppreſſed go free, and that ye break every Yoke, Chap. 58, 7.

Second Edition.

GERMANTOWN:
Printed by CHRISTOPHER SOWER. 1760.

Most Quakers were opposed to slavery. Shown here is a Quaker antislavery pamphlet printed in the Middle Colonies.

Quakers were also the first to raise their voices against slavery. Quaker ideals influenced immigrants in the Middle Colonies—and eventually the whole nation.

African Americans in the Middle Colonies

The tolerant attitude of many settlers in the Middle Colonies did not prevent slavery in the region. In 1750, about 7 percent of the Middle Colonies' population was enslaved. As in New England, many people of African descent lived and worked in cities.

New York City had a larger number of people of African descent than any other city in the Northern colonies. In New York City, enslaved persons worked as manual laborers, servants, drivers, and as assistants to artisans and craftspeople. Free African-American men and women also made their way to the city, where they worked as laborers, servants, or sailors.

Reading **History**
C. Forming Opinions Why do you think that force was needed to keep Africans enslaved?

Tensions existed between the races in New York City, sometimes leading to violence. In 1712, for example, about 24 rebellious slaves set fire to a building. They then killed nine whites and wounded several others who came to put out the fire. Armed colonists caught the suspects, who were punished horribly. Such punishments showed that whites would resort to force and violence to control slaves. Even so, the use of violence did little to prevent the outbreak of other slave rebellions.

Force would also be used in the South, which had far more enslaved Africans than the North. In the next section, you will learn how the South's plantation economy came to depend on the labor of enslaved Africans.

Section 2 Assessment

1. Terms & Names

Identify:
- cash crop
- gristmill
- diversity
- artisan
- Conestoga wagon

2. Taking Notes

Use a cluster diagram like the one shown to indicate where different immigrants in the Middle Colonies came from.

What was the third largest group in the region?

3. Main Ideas

a. What attracted settlers to the Middle Colonies?

b. What service was performed at gristmills?

c. Why might enslaved Africans be able to join in rebellion more easily in the city than the country?

4. Critical Thinking

Analyzing Causes What factors allowed large coastal cities to develop in the Middle Colonies?

THINK ABOUT
- geography
- people
- trade

ACTIVITY OPTIONS

MATH
GEOGRAPHY

Read more about Philadelphia. Create a **database** of the city's population growth in the 1700s or draw a **map** that shows its physical growth.

3 The Southern Colonies: Plantations and Slavery

TERMS & NAMES
indigo
Eliza Lucas
William Byrd II
overseer
Stono Rebellion

MAIN IDEA

The economy of the Southern Colonies relied heavily on slave labor.

WHY IT MATTERS NOW

The existence of slavery deeply affected the South and the nation.

George Mason was active in local affairs in Virginia. He would later play a role in the drafting of the United States Constitution.

ONE AMERICAN'S STORY

George Mason was born to a wealthy Virginia family in 1725. Mason—who later described the slave trade as "disgraceful to mankind"—wrote about the contributions of enslaved persons on his family's plantation.

A VOICE FROM THE PAST

My father had among his slaves carpenters, coopers [barrel makers], sawyers, blacksmiths, tanners, curriers, shoemakers, spinners, weavers and knitters, and even a distiller. . . . His woods furnished timber and plank for the carpenters and coopers, and charcoal for the blacksmith; his cattle killed for his own consumption and for sale supplied skins for the tanners, curriers, and shoemakers, and his sheep gave wool and his fields produced cotton and flax for the weavers and spinners, and his orchards fruit for the distiller.

George Mason, quoted in *Common Landscape of America*

Because the Masons and other wealthy landowners produced all that they needed on their own plantations, they appeared to be independent. But their independence usually depended on the labor of enslaved Africans. Although planters were only a small part of the Southern population, the plantation economy and slavery shaped life in the Southern Colonies: Maryland, Virginia, the Carolinas, and Georgia.

The Plantation Economy

The South's soil and almost year-round growing season were ideal for plantation crops like rice and tobacco. These valuable plants required much labor to produce, but with enough workers they could be grown as cash crops. Planters had no trouble transporting their crops because the region's many waterways made it easy for oceangoing ships to tie up at plantation docks.

Like George Mason's boyhood home, most plantations were largely self-sufficient. That is, nearly everything that planters, their families, and their workers needed was produced on the plantation. Because plantations were so self-sufficient, large cities like those in the North were rare

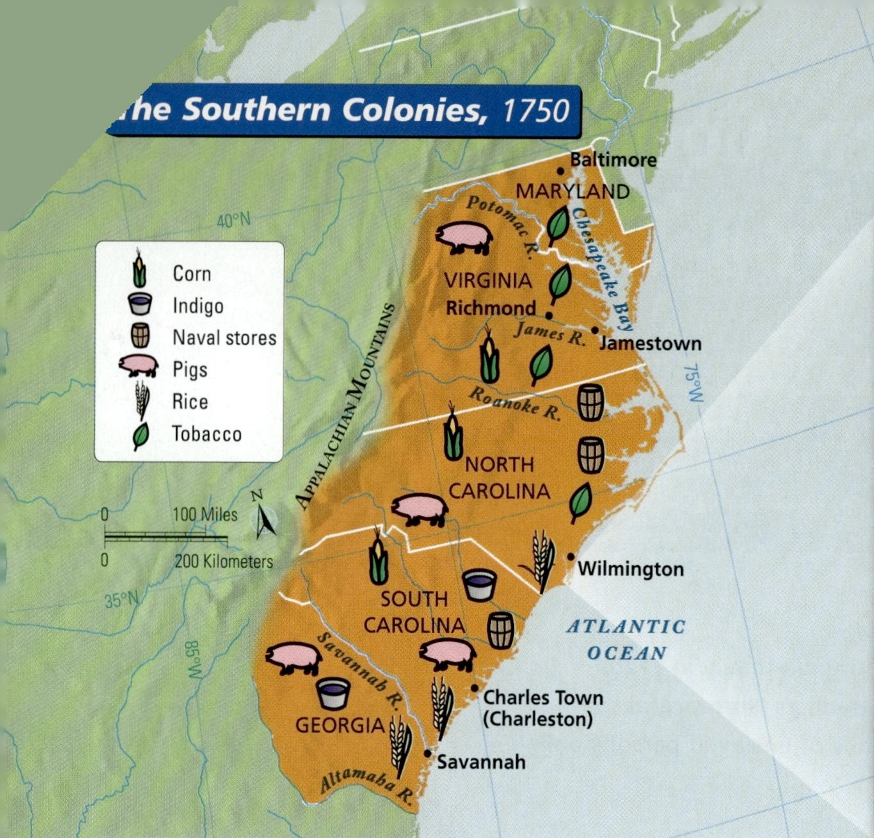

GEOGRAPHY SKILLBUILDER Interpreting Maps

1. Location *The Southern Colonies were south of what latitude?*

2. Place *Which Southern Colonies grew crops of both rice and indigo?*

The Orton plantation, south of Wilmington, North Carolina, was founded around 1725. Such plantations were representative of the economic and political power held by Southern planters.

in the Southern Colonies. The port city of Charles Town (later called Charleston) in South Carolina was an early exception.

As the plantation economy continued to grow, planters began to have difficulty finding enough laborers to work their plantations. Toward the end of the 1600s, the planters began to turn to enslaved Africans for labor.

Background
In 1742, Charles Town's population was 6,800.

The Turn to Slavery

For the first half of the 1600s, there were few Africans in Virginia, whether enslaved or free. In 1665, fewer than 500 Africans had been brought into the colony. At that time, African and European indentured servants worked in the fields together.

U.S. Slave Population

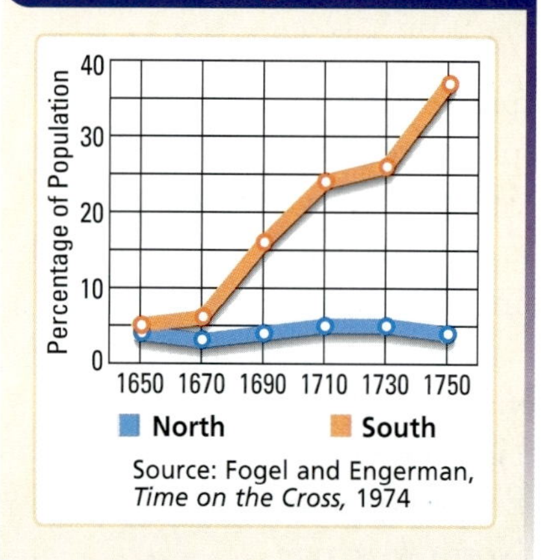

Starting in the 1660s, the labor system began to change as indentured white servants started to leave the plantations. One reason they left was the large amount of land available in the Americas. It was fairly easy for white men to save enough money to buy land and start their own farms. White servants could not be kept on the plantations permanently. As Bacon's rebellion showed, it was also politically dangerous for planters to try to keep them there (see page 89). As a result, the landowners had to find another source of labor.

Reading **History**
A. Drawing Conclusions Why did white workers choose not to remain on the plantations as laborers?

Planters tried to force Native Americans to work for them. But European diseases caused many Native Americans to die. Those who survived usually knew the country well enough to run away.

Reading **History**

B. Analyzing Causes What factors led to the importation of enslaved Africans into the South?

To meet their labor needs, the planters turned to enslaved Africans. As a result, the population of people of African descent began to grow rapidly. By 1750, there were over 235,000 enslaved Africans in America. About 85 percent lived in the Southern Colonies. Enslaved Africans made up about 40 percent of the South's population.

Plantations Expand

The growth of slavery allowed plantation farming to expand in South Carolina and Georgia. Without slave labor, there probably would have been no rice plantations in the region's swampy lowlands.

Enslaved workers drained swamps, raked fields, burned stubble, and broke ground before planting. They also had to flood, drain, dry, hoe, and weed the same fields several times before the harvest.

The cultivation of rice required not only back-breaking labor but also considerable skill. Because West Africans had these skills, planters sought out slaves who came from Africa's rice-growing regions.

On higher ground, planters grew **indigo,** a plant that yields a deep blue dye. A young woman named **Eliza Lucas** had introduced indigo as a successful plantation crop after her father sent her to supervise his South Carolina plantations when she was 17.

AMERICA'S HISTORY MAKERS

WILLIAM BYRD II
1674–1744

William Byrd II was one of the best known of the Southern planters. His family owned a large estate in Virginia. After his father died, Byrd took on his father's responsibilities, including membership in the House of Burgesses.

But Byrd is best remembered for his writing. His most famous work is *History of the Dividing Line betwixt Virginia and North Carolina.* In it, Byrd celebrates the land and climate of the South. At times, however, he is critical of its people. Even today, the book creates a vivid picture of life in the Southern Colonies.

How did William Byrd II demonstrate his leadership abilities?

The Planter Class

Slave labor allowed planters, such as the Byrd family of Virginia, to become even wealthier. These families formed an elite planter class. They had money or credit to buy the most slaves. And because they had more slaves, they could grow more tobacco, rice, or indigo to sell.

Small landowners with just one or two slaves simply could not compete. Many gave up their land and moved westward. As a result, the powerful planter class gained control of the rich land along the coast. The planter class was relatively small compared to the rest of the population. However, this upper class soon took control of political and economic power in the South. A foreign traveler in the South commented that the planters "think and act precisely as do the nobility in other countries."

Reading **History**

C. Recognizing Effects How did the growth of slavery affect political power in the South?

Some planters, following the traditions of nobility, did feel responsible for the welfare of their enslaved

HISTORY *through* ART

Benjamin Henry Latrobe's watercolor sketch, *An Overseer Doing His Duty,* shows enslaved African women on a Virginia plantation. An overseer looks on as the two women work to remove tree stumps.

What opinion do you think Latrobe had of the conditions on plantations?

workers. Power, they believed, brought with it the responsibility to do good. Many planters, though, were tyrants. They held complete authority over everyone in their households. Planters frequently used violence against slaves to enforce their will.

Vocabulary
tyrant: harsh ruler

Life Under Slavery

On large Southern plantations, slaves toiled in groups of about 20 to 25 under the supervision of **overseers.** Overseers were men hired by planters to watch over and direct the work of slaves. Enslaved persons performed strenuous and exhausting work, often for 15 hours a day at the peak of the harvest season. If slaves did not appear to be doing their full share of work, they were often whipped by the overseer.

Enslaved people usually lived in small, one-room cabins that were furnished only with sleeping cots. For a week's food, a slave might receive only around a quarter bushel of corn and a pound of pork. Some planters allowed their slaves to add to this meager ration by letting them raise their own potatoes, greens, fruit, or chicken.

In spite of the brutal living conditions, Africans preserved many customs and beliefs from their homelands. These included music, dances, stories, and, for a time, African religions—including Islam. African kinship customs became the basis of African-American family culture. A network of kin was a source of strength even when families were separated.

***Reading* History**
D. Finding Main Ideas What customs and beliefs from their homelands provided strength for enslaved Africans?

Resistance to Slavery

At the same time that enslaved Africans struggled to maintain their own culture, they fought against their enslavement. They sometimes worked slowly, damaged goods, or purposely carried out orders the wrong way. A British traveler in 1746 noted that many slaves pretended not to understand tasks they often had performed as farmers in West Africa.

A VOICE FROM THE PAST

You would really be surpriz'd at their Perseverance; let an hundred Men shew him how to hoe, or drive a wheelbarrow, he'll still take the one by the Bottom, and the other by the Wheel; and they often die before they can be conquer'd.

Edward Kimber, quoted in *White over Black*

At times, slaves became so angry and frustrated by their loss of freedom that they rose up in rebellion. One of the most famous incidents was the **Stono Rebellion.** In September 1739, about 20 slaves gathered at the Stono River just south of Charles Town. Wielding guns and other weapons, they killed several planter families and marched south, beating drums and loudly inviting other slaves to join them in their plan to seek freedom in Spanish-held Florida. By late that afternoon, however, a white militia had surrounded the group of escaping slaves. The two sides clashed, and many slaves died in the fighting. Those captured were executed.

Background
Slave codes were laws designed to control slaves and keep them in bondage.

Stono and similar revolts led planters to make slave codes even stricter. Slaves were now forbidden from leaving plantations without permission. The laws also made it illegal for slaves to meet with free blacks. Such laws made the conditions of slavery even more inhumane.

The Southern Colonies' plantation economy and widespread use of slaves set the region on a very different path from that of the New England and Middle Colonies. In the next section, you will learn how settlers used the unique resources of the Backcountry to create settlements there.

Section 3 Assessment

1. Terms & Names

Identify:
- indigo
- Eliza Lucas
- William Byrd II
- overseer
- Stono Rebellion

2. Taking Notes

Use a diagram like the one shown to review the factors that led to the use of slaves in the South.

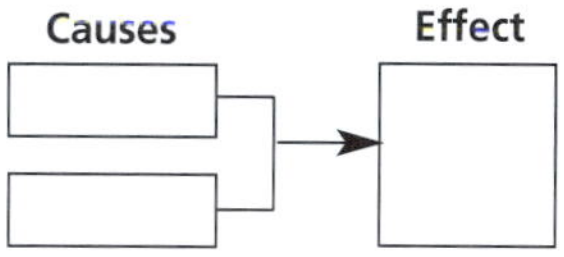

Why didn't planters use Native American workers?

3. Main Ideas

a. What percentage of the South's population was enslaved in 1750?

b. What crops did plantations in Georgia and South Carolina grow?

c. How did enslaved persons resist their slavery?

4. Critical Thinking

Contrasting How did geographic differences between the Southern Colonies and the New England Colonies affect their labor systems?

THINK ABOUT
- the climate of the regions
- the nature of the soil

ACTIVITY OPTIONS

ART

SCIENCE

Do more research on rice plantations. Draw a **diagram** of a typical plantation or write a **report** on how rice is cultivated today.

REGION AND HUMAN-ENVIRONMENT INTERACTION

Differences Among the Colonies

Many factors shape a region's economy and the way its settlers make a living. One of the most important is its physical geography—the climate, soil, and natural resources of the region. The geography of the American colonies varied from one colony to another. For example, in some areas, farmers could dig into rich, fertile soil. In others, they could not stick their shovels in the ground without hitting rocks.

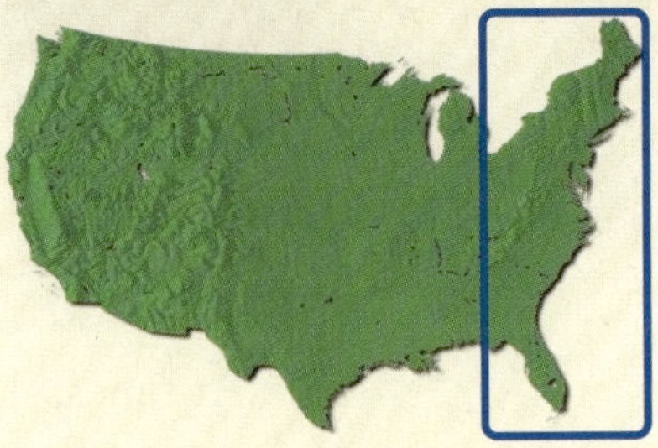

Major Regional Exports (by export value*)

NEW ENGLAND COLONIES

New England had a short growing season and rocky soil. Colonists took advantage of other opportunities in the region, especially fishing and whaling.

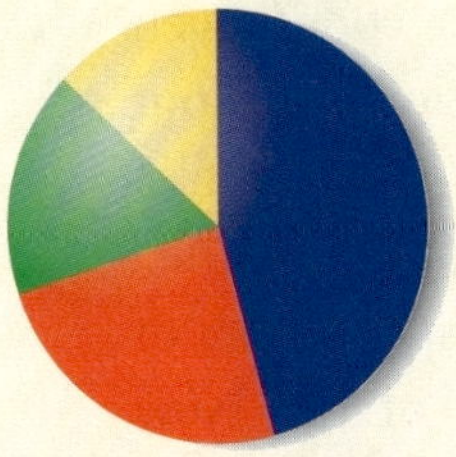

Dried Fish and Whale Oil	44%
Livestock	17%
Wood Products	13%
Other	26%

MIDDLE COLONIES

The longer growing season of the Middle Colonies—the "breadbasket colonies"—allowed farmers to grow cash crops of grain.

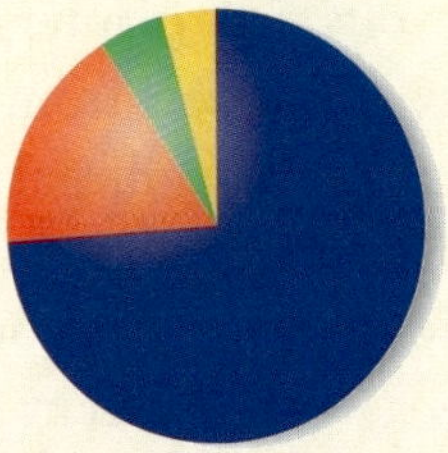

Grain	73%
Iron	5%
Wood Products	5%
Other	17%

SOUTHERN COLONIES

The South had a nearly year-round growing season. The use of enslaved Africans allowed Southern planters to produce cash crops of tobacco and rice.

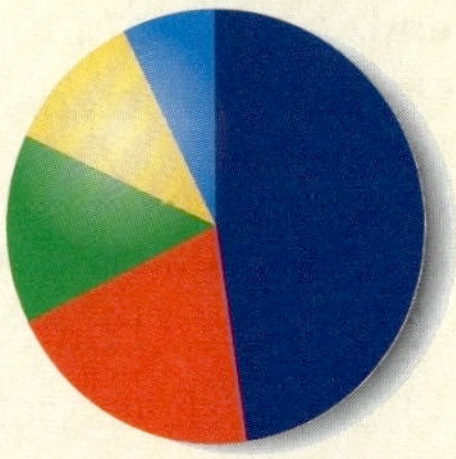

Tobacco	48%
Rice	20%
Bread, Flour, Grain (not rice)	13%
Indigo	7%
Other	12%

*Export Value in Pounds Sterling (Five-Year Average, 1768–1772)

Source: James F. Shepherd and Gary M. Walton, *Shipping, Maritime Trade, and the Economic Development of Colonial North America* (Cambridge: Cambridge University Press, 1972.)

ARTIFACT FILE

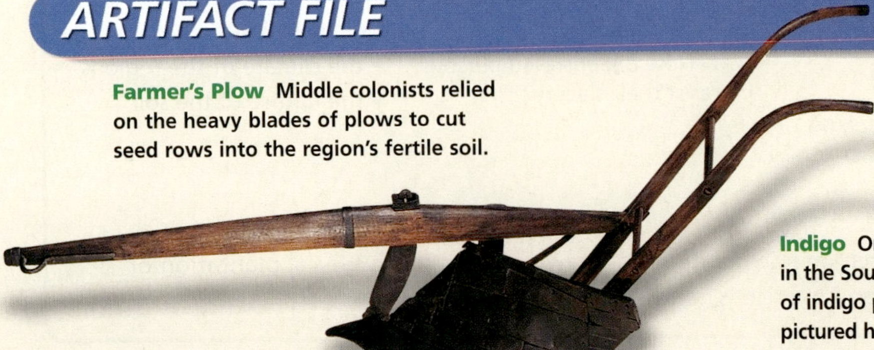

Farmer's Plow Middle colonists relied on the heavy blades of plows to cut seed rows into the region's fertile soil.

Indigo On some plantations in the South, planters grew crops of indigo plants—like the one pictured here—to produce the rich blue dyes used to color this yarn.

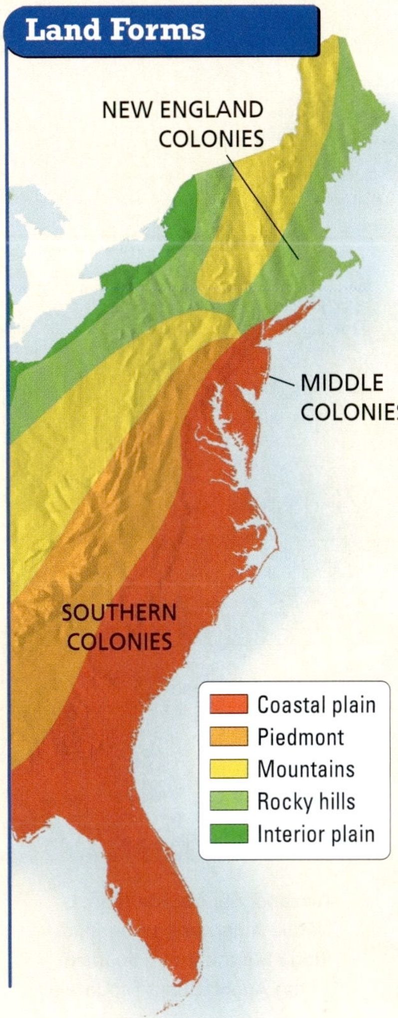

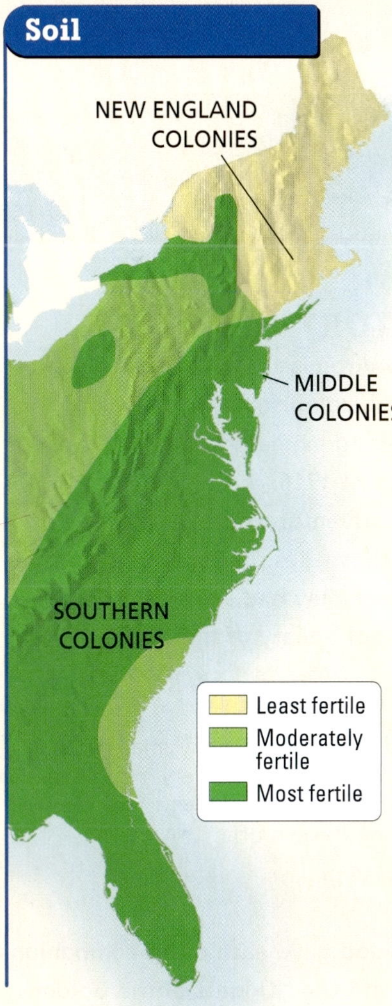

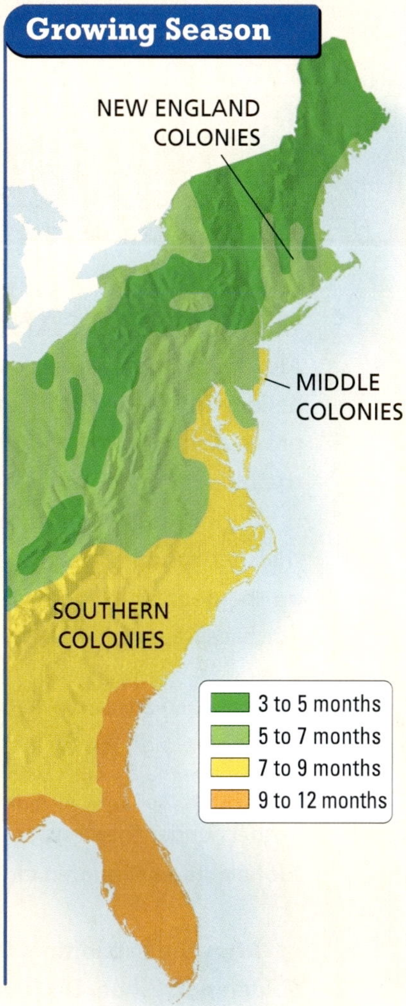

Physical Geography The maps above show the different types of land forms, soil, and growing seasons that were found in the different colonial regions. These factors helped to shape the economies of each of the regions, which were quite different, as the pie graphs show on the previous page.

CONNECT TO GEOGRAPHY

1. **Region** How long was the growing season in most of the Southern Colonies?
2. **Human-Environment Interaction** How might the soil quality in the Middle Colonies have influenced the region's population?

See Geography Handbook, pages 10–13.

CONNECT TO HISTORY

3. **Analyzing Causes** Why did the land forms and soil of New England cause many to turn to the Atlantic Ocean for a living?

On-Line Field Trip

The New Bedford Whaling Museum in Massachusetts has many objects related to whaling, including bone or ivory objects called scrimshaws. A sailor carved this whale's tooth with a jackknife or sail needle and colored the design with ink.

Visit www.mcdougallittell.com for more information.

4 The Backcountry

TERMS & NAMES
Appalachian Mountains
fall line
piedmont
clan

MAIN IDEA
Settlers moved to the Backcountry because land was cheap and plentiful.

WHY IT MATTERS NOW
Backcountry settlers established a rural way of life that still exists in certain parts of the country.

ONE AMERICAN'S STORY

Alexander Spotswood governed Virginia from 1710 to 1722. He believed that the future of English colonists lay to the west. To prove his point, he led a month-long expedition over the crest of the Blue Ridge Mountains in August 1716.

During the 400-mile journey, adventurers braved dense thickets, muddy streams, and rattlesnakes. At night, they feasted on the deer, wild turkeys, and bear they had shot. John Fontaine, who accompanied Spotswood, kept a diary of the trip.

A VOICE FROM THE PAST

We had a rugged way; we passed over a great many small runs of water, some of which were very deep, and others very miry. Several of our company were dismounted, some were down with their horses, others under their horses, and some thrown off.

John Fontaine, quoted in *Colonial Virginia*

After the expedition, Spotswood gave each of his companions a golden horseshoe. His journey with the "Knights of the Golden Horseshoe" is considered a symbol of Virginia's westward expansion.

Alexander Spotswood meets Native Americans in the Blue Ridge Mountains—a segment of the Appalachians Mountains.

Geography of the Backcountry

Just as Spotswood predicted, settlers soon began to move into the Backcountry. This was a region of dense forests and rushing streams in or near the **Appalachian Mountains.** The Appalachians stretch from eastern Canada south to Alabama.

In the South, the Backcountry began at the **fall line.** The fall line is where waterfalls prevent large boats from moving farther upriver. Beyond the fall line is the **piedmont.** Piedmont means "foot of the mountains." It is the broad plateau that leads to the Blue Ridge Mountains of the Appalachian range.

The Backcountry's resources made it relatively easy for a family to start a small farm. The region's many springs and streams provided water, and forests furnished wood that settlers could use for log cabins and fences.

Backcountry Settlers

The first Europeans in the Backcountry made a living by trading with the Native Americans. Backcountry settlers paid for goods with deerskins. A unit of value was one buckskin or, for short, a "buck."

Vocabulary
buck: an adult male deer; the adult female is called a *doe*

Farmers soon followed the traders into the region, but they had to be cautious. As the number of settlements grew, the farmers often clashed with the Native Americans whose land they were taking.

Farmers sheltered their families in log cabins. They filled holes between the logs with mud, moss, and clay. Then they sawed out doors and windows. Lacking glass, settlers used paper smeared with animal fat to cover their windows.

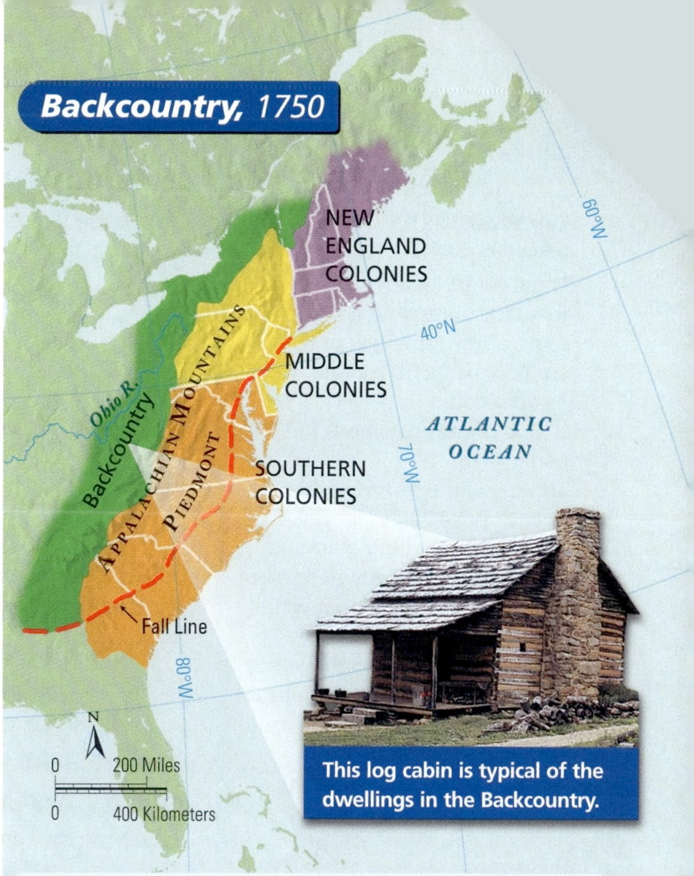

This log cabin is typical of the dwellings in the Backcountry.

GEOGRAPHY SKILLBUILDER Interpreting Maps
Region *What geographical feature did the northern and southern areas of the Backcountry have in common?*

Reading History
A. Analyzing Points of View What was William Byrd's attitude toward Backcountry settlers?

William Byrd—on his expedition to establish the southern border of Virginia—described a long night that he spent in one such cabin. He complained that he and at least ten other people were "forct to pig together in a Room . . . troubled with the Squalling of peevish, dirty children into the Bargain."

Backcountry life may have been harsh, but by the late 1600s many families had chosen to move there. Some of them went to escape the plantation system, which had crowded out many small farmers closer to the seacoast. Then, in the 1700s, a new group of emigrants—the Scots-Irish—began to move into the Backcountry.

The Scots-Irish

The Scots-Irish came from the borderland between Scotland and England. Most of them had lived for a time in northern Ireland. In 1707, England and Scotland merged and formed Great Britain. The merger caused many hardships for the Scots-Irish. Poverty and crop failures made this bad situation even worse.

As a result, Scots-Irish headed to America by the thousands. After they arrived, they quickly moved into the Backcountry. The Scots-Irish brought their clan system with them to the Backcountry. **Clans** are large groups of families—sometimes in the thousands—that claim a common ancestor. Clan members were suspicious of outsiders and banded together when danger threatened. These clans helped families to deal with the dangers and problems of the Backcountry.

Vocabulary
clan: comes from an Old Irish word that means offspring, or descendants

Now and then

BACKCOUNTRY SPORTS TODAY

Three centuries ago, crowds in the Backcountry were thrilled by some of the same games that are now part of track and field competitions.

One of these games is the hammer throw. In this event, an athlete swings around a 16-pound metal ball on a wire-rope handle. After whirling around several times, the athlete lets go of the hammer, hoping it will travel the farthest distance.

The Scots-Irish brought other games to America, including the shotput, high jump, and long jump.

Backcountry Life

Life in the Backcountry was very different from life along the seaboard. Settlers along the coast carried on a lively trade with England. But in the Backcountry, rough roads and rivers made it almost impossible to move goods.

As a result, Backcountry farmers learned quickly to depend on themselves. They built log cabins and furnished them with cornhusk mattresses and homemade benches and tables. They fed their families with the hogs and cattle they raised and with the fish and game they killed. They grew yellow corn to feed their livestock and white corn to eat. Popcorn was probably their only snack food. To protect their precious corn from pests, daytime patrols of women, children, and the elderly served as human scarecrows.

Reading **History**

B. Making Inferences How would you describe the way people in the Backcountry lived?

Women in the Backcountry worked in the cabin and fields, but they also learned to use guns and axes. An explorer who traveled in the region described one of these hardy Backcountry women.

A VOICE FROM THE PAST

She is a very civil woman and shows nothing of ruggedness or Immodesty in her carriage, yett she will carry a gunn in the woods and kill deer, turkeys, etc., shoot doun wild cattle, catch and tye hoggs, knock down [cattle] with an ax and perform the most manfull Exercises.

A visitor to the Backcountry, quoted in *A History of American Life*

Settlers in the Backcountry often acted as if there were no other people in the region, but this was not so. In the woods and meadows that surrounded their cabins, settlers often encountered Native Americans and other groups that had made America their home.

Other Peoples in North America

The Backcountry settlers started a westward movement that would play a critical role in American history. Most settlers' motivation to move west was simple—the desire for land.

Reading **History**

C. Summarizing As England's colonies expanded westward, what groups did they encounter?

Yet the push to the west brought settlers into contact with other peoples of North America. Native Americans had made their homes there for thousands of years. In addition, France and Spain claimed considerable territory in North America.

Sometimes this contact led to changes in people's cultures. For instance, North America had no horses until the Spanish colonists brought them into Mexico in the 1500s. Horses migrated north, and Native Americans caught them and made them an important part of their culture.

Contact also led to conflict. As English settlers pushed into the Backcountry, they put pressure on Native American tribes. Some tribes reacted by raiding isolated homesteads and small settlements. White settlers struck back, leading to more bloodshed.

This painting shows Native Americans catching wild horses. Many would later use the horses to hunt buffalo on the Great Plains.

The English colonists also came into conflict with the French. The French had colonized eastern Canada and had moved into the territories, rich with fur, along the Mississippi River. French fur traders wanted to prevent English settlers from moving west and taking away part of the trade. One Native American told an Englishman, "You and the French are like two edges of a pair of shears, and we are the cloth that is cut to pieces between them."

Vocabulary
shears: scissors

Spain also controlled large areas of North America—including territories that today form part or all of the states of Arizona, California, Colorado, Florida, Nevada, New Mexico, Texas, Utah, and Wyoming. Spanish settlers were farmers, ranchers, and priests. Priests, who established missions to convert Native Americans, built forts near the missions for protection. In 1718, Spaniards built Fort San Antonio de Bexar to guard the mission of San Antonio de Valero, later renamed the Alamo.

These different groups continued to compete—and sometimes fight—with one another. Frequently, England's colonies had to unite against these other groups. As a result, a common American identity began to take shape, as you will read in Chapter 5.

Section 4 Assessment

1. Terms & Names

Identify:
- Appalachian Mountains
- fall line
- piedmont
- clan

2. Taking Notes

Use a chart like the one shown to list some of the geographic characteristics of the Backcountry.

Backcountry Geography
1.
2.
3.
4.

3. Main Ideas

a. Which settlers migrated to the Backcountry?

b. How did clans help the Scots-Irish survive?

c. What economic activities did women carry out in the region?

4. Critical Thinking

Identifying Problems As England's colonies expanded farther west, what problems would they face?

THINK ABOUT
- other inhabitants of the Americas
- the resources desired by the colonists

ACTIVITY OPTIONS

LANGUAGE ARTS
ART

Read an account of the Backcountry written in the 1700s. Write a **newspaper article** or draw a series of **cartoons** that describe what you have read.

Chapter 4 ASSESSMENT

VISUAL SUMMARY

The Colonies Develop

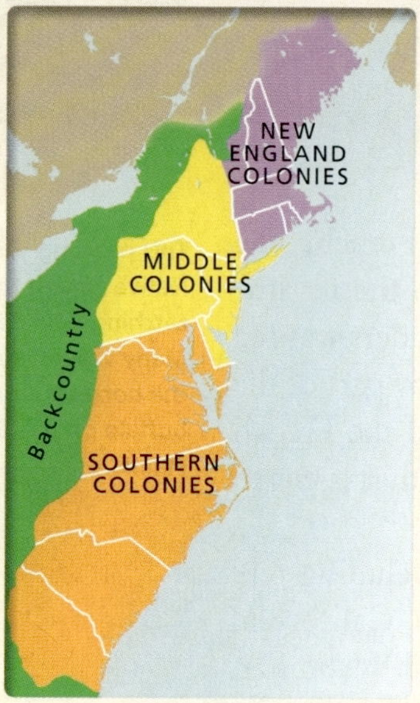

New England: Commerce and Religion

New England was distinguished by its small farming towns and profitable fishing and trade.

The Middle Colonies: Farms and Cities

Middle Colony farms produced large cash crops that fueled trade in its coastal cities.

The Southern Colonies: Plantations and Slavery

The South's plantation economy and large number of enslaved Africans made it different from the other regions.

The Backcountry

The Backcountry was distant from the denser coastal populations, so settlers there developed an independent and rugged way of life.

TERMS & NAMES

Briefly explain the importance of the following.

1. Backcountry
2. subsistence farming
3. triangular trade
4. Navigation Acts
5. cash crop
6. gristmill
7. Conestoga wagon
8. overseer
9. Stono Rebellion
10. Appalachian Mountains

REVIEW QUESTIONS

New England: Commerce and Religion (pages 109–113)

1. How would you describe the life of a New England farmer?
2. In what ways did settlers in the region take advantage of the Atlantic Ocean?
3. How were New England towns settled?

The Middle Colonies: Farms and Cities (pages 114–118)

4. How were farms in the Middle Colonies different than those in New England?
5. What characterized the population of the Middle Colonies?

The Southern Colonies: Plantations and Slavery (pages 119–125)

6. Why did Southern planters infrequently travel to towns to sell their crops or to buy food and supplies?
7. Why did planters turn to enslaved Africans for labor?
8. In what ways did slaves resist?

The Backcountry (pages 126–129)

9. Where was the Backcountry located in the 1700s?
10. How was life in the Back-country different from that along the coast?

CRITICAL THINKING

1. USING YOUR NOTES

Using your completed chart, answer the questions below.

		NEW ENGLAND COLONIES	MIDDLE COLONIES	SOUTHERN COLONIES	BACKCOUNTRY
CAUSES	Climate				
	Resources				
	People				
EFFECT	Economic Development				

a. How was the Middle Colonies' climate different from the Backcountry's?

b. How did the South's labor system differ from the North's?

c. How did the resources of New England affect its economy?

2. ANALYZING LEADERSHIP

How did the South's plantation economy influence who became leaders in the region?

3. THEME: ECONOMICS IN HISTORY

What factors influenced the economic development of each of the four colonial regions?

4. APPLYING CITIZENSHIP SKILLS

How did the Quaker influence in the Middle Colonies contribute to the behavior of citizens of the region?

5. SEQUENCING EVENTS

What changes took place in the population and treatment of African Americans between 1650 and 1750?

Interact *with* History

How would the choice that you made at the beginning of the chapter have varied according to the region in which you lived? Would you still make the same choice?

HISTORY SKILLS

1. INTERPRETING MAPS: Human-Environment Interaction

Study the map. Answer the questions.

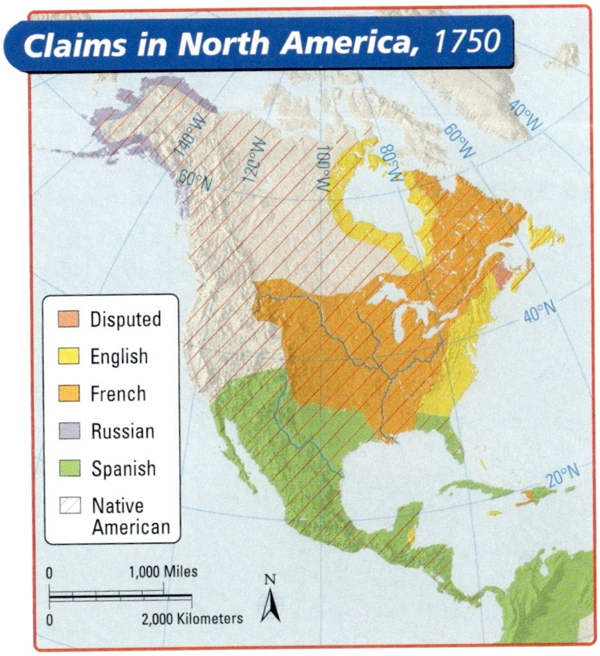

Basic Map Elements

a. What is the subject of the map?

Interpreting the Map

b. Which of the groups shown inhabited the largest area of North America?

c. Which groups claimed the northernmost territory?

2. INTERPRETING PRIMARY SOURCES

In the backwoods of North Carolina, William Byrd met a family he suspected of being escaped slaves. Read the selection below and answer the questions.

> [They] called themselves free, though by the shyness of the master of the house, who took care to keep least in sight, their freedom seemed a little doubtful. . . . Many slaves [hide] in this obscure part of the world, nor will any of their righteous neighbors discover them. On the contrary, [their neighbors profit by] settling such fugitives on some out-of-the-way corner of their land to raise stocks for a mean and inconsiderable share, well knowing their condition makes it necessary for them to [accept any pay they are offered].
>
> **William Byrd,** from *Secret History of the Dividing Line*

a. Why does Byrd suspect the family members are escaped slaves?

b. Why don't their neighbors turn them in?

ALTERNATIVE ASSESSMENT

1. INTERDISCIPLINARY ACTIVITY: Geography

Making a Map Using the library or the Internet, read more about the history of the slave trade in the 1700s. Create a map of Africa that shows countries that were major sources of enslaved persons.

INTERNET **Visit www.mcdougallittell.com to learn more about the history of Africa.**

2. COOPERATIVE LEARNING ACTIVITY

Building a Log Cabin Do some more research on the Backcountry and the history of log cabins. Work with others to record details about the location where you will build your cabin. Then design and construct a model of a log cabin that could be compared to the cabins in which Backcountry settlers lived. Be sure to build a setting for your cabin that shows the landscape where you chose to make your home.

3. TECHNOLOGY ACTIVITY

Making a Class Presentation Life on a farm in colonial New England was a real challenge. Using the library or the Internet, find accounts of how New England farmers and their families lived. Then design a multimedia presentation that focuses on a typical New England farmer and his family.

INTERNET **Visit www.mcdougallittell.com to learn more about the way people lived during colonial times.**

- Create a map of the town in which the farmer and his family lived.
- Dress up like a farmer to discuss the challenges of New England agriculture.
- Create a chart that lists the differences between your family's lifestyle and the colonial family's.

4. HISTORY PORTFOLIO

PORTFOLIO **Option 1** Review your section and chapter assessment activities. Select one that you think is your best work. Then use comments made by your teacher or classmates to improve your work and add it to your portfolio.

PORTFOLIO **Option 2** Review the questions that you wrote for What Do You Want to Know? on page 108. Then write a short report in which you explain the answers to your questions. If any questions were not answered, do research to answer them. Add your answers to your portfolio.

CHAPTER 5

Beginnings of an American Identity 1689–1763

Conflicts with the British government helped shape a separate identity for British colonists. Here, Patrick Henry argues against the king in 1763.

Interact *with* History

Conflicts with Native Americans and French colonists also helped shape the identity of British colonists. In response to attacks on the Pennsylvania frontier, Benjamin Franklin published this cartoon in 1754 urging the colonies to unite for defense.

The initials stand for the different British colonies.

A colonial superstition was that a snake cut into pieces would come back to life if the pieces were joined.

What do you have in common with other British colonists?

You have seen this cartoon in the *Pennsylvania Gazette*. You are outraged by the attacks on British traders and settlers. You wonder whether it is wise to join with other colonies, though. Will it mean that Virginians or New Englanders will be able to make laws for Pennsylvania?

What Do You Think?

- What are some good reasons to join with the other British colonies?
- How great are the differences between the British colonies?
- What separates British colonists from French colonists?

USA 1680–1763

1689 Massachusetts colonists overthrow royal governor Andros.

1704 *Boston Newsletter* is founded.

1735 Decision in the Zenger trial supports freedom of the press.

1738 Minister George Whitefield arrives in Georgia.

1754 French and Indian War begins.

1759 Quebec falls to the British.

1763 French and Indian War ends.

World

1689 William and Mary replace James II as rulers of England.

1707 England and Scotland join to form Great Britain.

1709 About 13,500 people leave the German states and emigrate to England.

1756 Seven Years' War between France and Britain is declared.

Chapter 5 SETTING THE STAGE

BEFORE YOU READ

Previewing the Theme

Democratic Ideals In the 1700s and even earlier, American colonists demanded protection of their "rights as Englishmen." Among these was the right to elect representatives to government. Chapter 5 explains how a heritage of English rights was one of many forces drawing the different British colonies together.

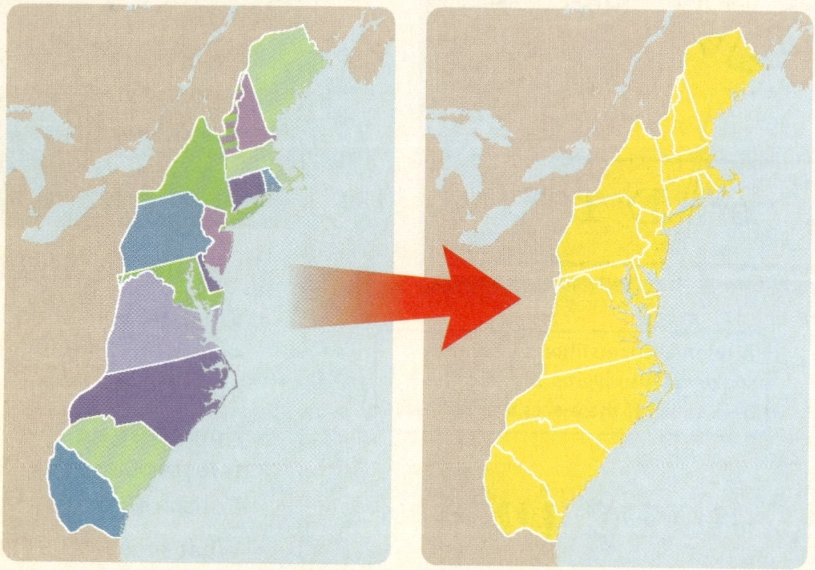

What Do You Know?

What beliefs do you consider American? How do you think people in Britain's American colonies saw themselves?

THINK ABOUT

- your own beliefs as an American
- what you know about the regions where the colonies were established
- what you know about the backgrounds and beliefs of colonists in different regions

What Do You Want to Know?

What questions do you have about colonial America in the early and middle 1700s? Write them down in your notebook before you read this chapter.

READ AND TAKE NOTES

Reading Strategy: Finding Main Ideas To recognize a main idea, you must notice how smaller details are connected. In your notebook, copy a web like the one shown here. Write brief notes about the main things people in the British colonies had in common—the beliefs and experiences that formed an American identity.

- Read and remember the Main Idea at the beginning of each section.
- At the end of each group of paragraphs under a heading, ask yourself, "Have I learned about something that united the colonists?"

See Skillbuilder Handbook, page R5.

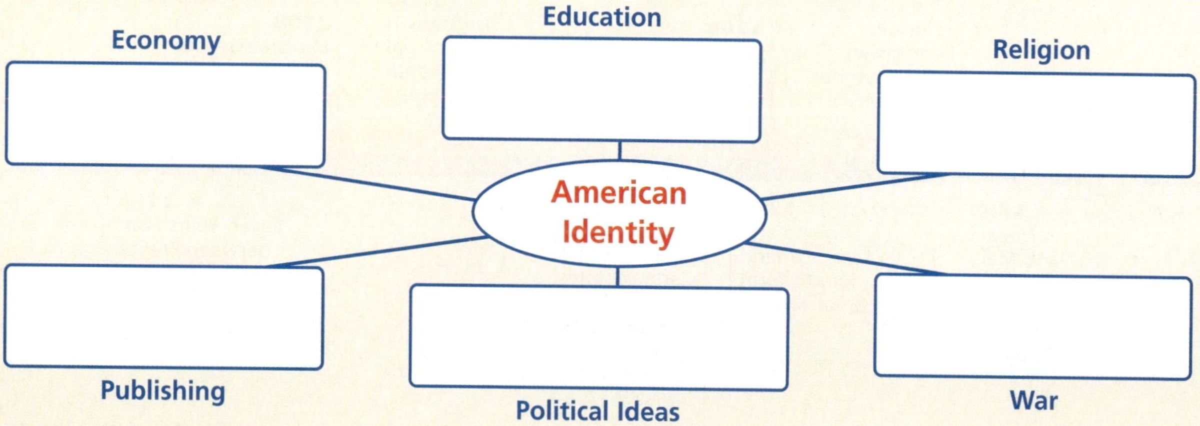

1 Early American Culture

TERMS & NAMES
apprentice
Great Awakening
Jonathan Edwards
George Whitefield
Enlightenment
Benjamin Franklin
John Locke

MAIN IDEA

The British colonies were shaped by prosperity, literacy, and new movements in religion and thought.

WHY IT MATTERS NOW

These forces began to create an American identity that is still developing today.

ONE AMERICAN'S STORY

On October 2, 1704, Sarah Kemble Knight set out on horseback from her home in Boston. She was riding all the way to New Haven, Connecticut. Today the ride is two hours by car, but then it took five days.

In her journal, Madam Knight described her travel hardships and commented on people she met—country girls, tobacco-chewing farmers, and rude housewives.

A VOICE FROM THE PAST

We hoped to reach the french town and Lodg there that night, but unhapily lost our way about four miles short. . . . A surly old shee Creature, not worthy the name of woman, . . . would hardly let us go into her Door, though the weather was so stormy none but shee would have turnd out a Dogg.

Sarah Kemble Knight, *The Journal of Madam Knight*

This reprint of *The Journal of Madam Knight* is a 1920 edition.

Her attitude toward people from other colonies was typical. In the early 1700s, people of the different British colonies did not think of themselves as living in one country. They were separated by distance and customs. In this section, you will learn what began to draw the colonies together.

Land, Rights, and Wealth

At the time of Madam Knight's journey, the colonies were thriving. Cheap farmland and plentiful natural resources gave colonists a chance to prosper. They would have had less opportunity in Europe. In England, fewer than 5 percent of the people owned land. In fact, land rarely went up for sale. By contrast, in the colonies, land was plentiful—once Native American groups were forced to give up their claims. Colonists who owned land were free to use or sell whatever it produced.

Land ownership gave colonists political rights as well as prosperity. Generally, only white male landowners or property owners could vote. There were some exceptions. City dwellers could vote by paying a fee.

Colonial Social Ranks
HIGH
• large landowners
• church officials
• government officials
• wealthy merchants
UPPER MIDDLE
• small farmers
• tradespeople
LOWER MIDDLE
• renters
• unskilled workers
LOW
• indentured servants
• slaves

Land ownership also helped determine colonists' social position. Unlike England, America had no class of nobles whose titles passed from parent to child. But people were still divided into high, middle, and low ranks, as they were in England. Large landholders were high in rank. Small farmers who owned their land were in the middle rank. Most colonists fit this category. People who did not own land, such as servants, slaves, or hired workers, were low in rank. Colonial women held the same rank as their husbands or fathers.

Reading **History**
A. Finding Main Ideas What did colonists gain by owning land?

Colonists showed respect to their "betters" by curtsying or tipping a hat, for example. Seats in church were assigned by rank, with wealthy families in the front pews and poor people in the back. Despite such divisions, the wealthy were expected to aid the poor.

Women and the Economy

Although women were not landholders, their work was essential to the colonial economy. As you learned in Chapter 4, enslaved African women helped raise cash crops such as tobacco and indigo. Most white women were farm wives who performed tasks and made products their families needed. They cooked, churned butter, made soap and candles, spun fibers, wove cloth, sewed and knitted clothes, and did many other chores. They usually tended a garden and looked after farm animals. At harvest time, they often worked in the fields alongside men and older children.

Because cash was scarce, farm wives bartered, or traded, with their neighbors for goods and services. For example, a woman who nursed a sick neighbor or helped deliver a baby might be paid in sugar or cloth.

Women in towns and cities usually did the same types of housework that rural women did. In addition, some urban women ran inns or other businesses. Madam Knight, whose journey was described in One American's Story on page 135, sold writing paper, taught handwriting, and rented rooms to guests. A few women, usually the wives or widows of tradesmen, practiced trades themselves.

Reading **History**
B. Finding Main Ideas In what ways was women's work essential to the economy?

Although women contributed to the colonial economy, they did not have many rights. Women could not vote. In most churches, they could not preach or hold office. (Quaker meetings were an exception.) A married woman could not own property without her husband's permission. By law, even the money a woman earned belonged to her husband.

Young People at Work

Children's work also supported the colonial economy. Families were large. New England families, for example, had an average of six to eight children. More children meant more workers. Children as young as three or four were expected to be useful. They might help look after farm animals, gather berries, and watch younger children.

Around age six, boys were "breeched." This meant that they no longer wore the skirts or smocks of all young children but were given a pair of pants. They then began to help their fathers at work. Sons of farmers worked all day clearing land and learning to farm. Sons of craftsmen tended their fathers' shops and learned their fathers' trades.

Around age 11, many boys left their fathers to become apprentices. An **apprentice** learned a trade from an experienced craftsman. The apprentice received food, clothing, lodging, and a general education, as well as training in the specific craft or business. He worked for free, usually for four to seven years, until his contract was fulfilled. Then he could work for wages or start his own business.

Girls rarely were apprenticed. They learned sewing and other household skills from their mothers. In New England, girls of 13 or 14 often were sent away to other households to learn specialized skills such as weaving or cheese making. Orphaned girls and boys worked as servants for families who housed and fed them until adulthood.

Reading **History**

C. Contrasting
How did the training of boys and girls differ?

Colonial Schooling

If land, wealth, and hard work were valued across the colonies, so was education. Most children were taught to read so that they could understand the Bible. Only children from wealthy families went beyond reading to learn writing and arithmetic. These children learned either from private tutors or in private schools. Poorer children sometimes learned to read from their mothers.

HISTORY *through* **ART**

This drawing shows the inside of an 18th-century one-room schoolhouse.

What does the picture suggest to you about colonial schooling?

daily life

THE SCHOOL OF MANNERS

Colonial children learned proper behavior from *The School of Manners,* a book published in 1701. Here are examples of rules and an illustration from the book.

"Spit not in the Room, but in a corner, and rub it out with thy Foot, or rather go out and do it abroad."

"If thou meetest the scholars of any other School jeer not nor affront them, but show them love and respect and quietly let them pass along."

Or they attended "dame schools," where women taught the alphabet and used the Bible to teach reading. Most children finished their formal education at age seven.

Children's textbooks emphasized religion. The widely used *New England Primer* paired the letter *A* with the verse "In *Adam's* fall / We Sinned all." Beside the letter *B* was a picture of the Bible. The primer contained the Lord's Prayer and *The Shorter Catechism,* more than 100 questions and answers about religion.

Colonial America had a high literacy rate, as measured by the number of people who could sign their names. In New England, 85 percent of white men were literate, compared with 60 percent of men in England. In the Middle Colonies, 65 percent of white men were literate, and in the South, about 50 percent were. In each region of the colonies, roughly half as many white women as men were literate. Most colonists thought schooling was more important for males. Educated African Americans were rare. If they were enslaved, teaching them to read was illegal. If they were free, they were often kept out of schools.

Reading History
D. Contrasting How was colonial education different from education today?

Newspapers and Books

Colonial readers supported a publishing industry that also drew the colonies together. In the early 1700s, the colonies had only one local newspaper, the *Boston News-letter.* But over the next 70 years, almost 80 different newspapers appeared in America. Many were published for decades.

Most books in the colonies were imported from England, but colonists slowly began to publish their own books. Almanacs were very popular. A typical almanac included a calendar, weather predictions, star charts, farming advice, home remedies, recipes, jokes, and proverbs. In 1732, Benjamin Franklin began to publish *Poor Richard's Almanack.* It contained sayings that are still repeated today, such as "Haste makes waste."

Colonists also published poetry, regional histories, and autobiographies. Most personal stories told of struggles to maintain religious faith during hard times. A form of literature unique to the Americas was the captivity narrative. In it, a colonist captured by Native Americans described living among them.

Reading History
E. Categorizing What were some types of colonial literature?

Mary Rowlandson's 1682 captivity narrative, *The Sovereignty and Goodness of God,* was one of the first colonial bestsellers. Native Americans attacked Rowlandson's Massachusetts village in 1676, during King Philip's War. They held her hostage for 11 weeks. During that time, she was a servant to a Narragansett chieftain, knitting stockings and making shirts for his family and others. "I told them it was Sabbath day," she recalled, "and desired them to let me rest, and told them I

Vocabulary
ransom: to pay for a captive's release

would do as much more tomorrow. To which they answered me, they would break my face." After townspeople raised money to ransom Rowlandson, she was released. Although she mourned a young daughter who had died in captivity, she praised God for returning her safely.

The Great Awakening

Mary Rowlandson's religious faith was central to her life. But in the early 1700s, many colonists feared they had lost the religious passion that had driven their ancestors to found the colonies. Religion seemed dry, dull, and distant, even to regular churchgoers.

In the 1730s and 1740s, a religious movement called the **Great Awakening** swept through the colonies. The traveling ministers of this movement preached that inner religious emotion was more important than outward religious behavior. Their sermons appealed to the heart and drew large crowds. **Jonathan Edwards,** one of the best-known preachers, terrified listeners with images of God's anger but promised they could be saved.

Background
Religious meetings with large, intensely emotional crowds remain part of American religious tradition.

A VOICE FROM THE PAST

And now you have an extraordinary opportunity, a day wherein Christ has thrown the door of mercy wide open, and stands in calling and crying with a loud voice to poor sinners. . . . How awful it is to be left behind at such a day!

Jonathan Edwards, "Sinners in the Hands of an Angry God"

The Great Awakening lasted for years and changed colonial culture. Congregations argued over religious practices and often split apart. People left their old churches and joined other Protestant groups such as Baptists. Some of these groups welcomed women, African Americans, and Native Americans. Overall, churches gained 20,000 to 50,000 new members. To train ministers, religious groups founded colleges such as Princeton and Brown.

The Great Awakening inspired colonists to help others. **George Whitefield** (HWIT•feeld) drew thousands of people with his sermons and raised funds to start a home for orphans. Other ministers taught religion and reading to Native Americans and African Americans. The Great

George Whitefield preaching to a crowd

AMERICA'S HISTORY MAKERS

BENJAMIN FRANKLIN
1706–1790

As an Enlightenment thinker, Benjamin Franklin used reason to improve society. At 42, he retired from business to devote his life to science and public service. He proved that lightning was a form of electricity. Then he invented the lightning rod to protect buildings. The Franklin stove and bifocal eyeglasses were also his inventions. He organized a fire department, a lending library, and a society to discuss philosophy. Later he helped draft the Declaration of Independence.

How did Franklin help improve colonial society?

Awakening encouraged ideas of individual worth, equality, and the right to challenge authority. In this way, the movement helped prepare colonists to declare independence from England years later.

*Reading*History
F. Recognizing Effects What were five effects of the Great Awakening?

The Enlightenment

Unlike the Great Awakening, which stressed religious emotion, the **Enlightenment** emphasized reason and science as the paths to knowledge. **Benjamin Franklin** was a famous American Enlightenment figure. This intellectual movement appealed mostly to wealthy, educated men. But it, too, had far-reaching effects on the colonies.

The Enlightenment began in Europe, as scientists discovered natural laws governing the universe. Isaac Newton, for example, explained the law of gravity.

Other Enlightenment thinkers applied the idea of natural law to human societies. The English philosopher **John Locke** argued that people have natural rights. These are rights to life, liberty, and property. People create governments to protect their natural rights, he claimed. If a government fails in this duty, people have the right to change it. Locke challenged the belief that kings had a God-given right to rule.

Enlightenment ideas of natural rights and government by agreement influenced leaders across Europe and the colonies. As you will see in Section 2, colonists began to wonder whether the British government protected their rights and freedoms. Eventually, they would rebel and form a new government.

Section 1 Assessment

1. Terms & Names

Identify:
- apprentice
- Great Awakening
- Jonathan Edwards
- George Whitefield
- Enlightenment
- Benjamin Franklin
- John Locke

2. Taking Notes

Describe the parts of colonial culture in a chart.

Economy	
Education	
Publishing	
Religion	

Why was each important in colonial culture?

3. Main Ideas

a. Why was land ownership so important to the colonists?

b. How did women and young people contribute to the colonial economy?

c. How did the Great Awakening affect the colonies?

4. Critical Thinking

Contrasting How were the Great Awakening and the Enlightenment different?

THINK ABOUT
- the ideas each movement promoted
- the people to whom each movement appealed

ACTIVITY OPTIONS

LANGUAGE ARTS
ART

Make up a **saying** that reflects some part of colonial culture, or draw an **illustration** of a saying from colonial times.

2 Roots of Representative Government

TERMS & NAMES
Magna Carta
Parliament
Edmund Andros
Glorious Revolution
English Bill of Rights
salutary neglect
John Peter Zenger

MAIN IDEA

Colonists expected their government to preserve their basic rights as English subjects.

WHY IT MATTERS NOW

U.S. citizens expect these same rights, such as the right to a trial by jury.

ONE AMERICAN'S STORY

On April 7, 1688, the famous Puritan minister Increase Mather set sail for England. He was to speak to King James II to get relief for the Massachusetts colony. The English government had canceled the charter of Massachusetts and sent a royal governor to rule.

The colonists thought that the governor trampled their rights as English subjects. Mather stayed in England for four years. During this time, he saw the king driven out and replaced by new rulers. In the end, he came home with a new charter that he hoped would satisfy the New England colonists.

This is a detail of *Increase Mather* by Jan van der Spriett.

A VOICE FROM THE PAST

For all English liberties are restored to them: No Persons shall have a Penny of their Estates taken from them; nor any Laws imposed on them, without their own Consent by Representatives chosen by themselves.

Increase Mather, quoted in *The Last American Puritan*

Mather called the new charter "a Magna Carta for New England." In this section, you will learn about the rights of English people set forth in the Magna Carta and later documents. These rights are the basis for the rights Americans enjoy today, such as the right not to be jailed without cause and the right to a jury trial.

The Rights of Englishmen

English colonists expected certain rights that came from living under an English government. These "rights of Englishmen" had developed over centuries.

The first step toward guaranteeing these rights came in 1215. That year, a group of English noblemen forced King John to accept the **Magna Carta** (Great Charter). The king needed the nobles' money to finance a war. This document guaranteed important rights to noblemen and freemen—those not bound to a master. They could not have their property seized by the king or his officials. They could not be taxed, in most cases, unless a council of prominent men agreed. They could not be put to trial based only on an official's word, without witnesses. They could be punished only by a jury of their peers, people of the same social rank.

A VOICE FROM THE PAST

No freeman shall be seized, imprisoned, dispossessed, outlawed, or exiled, . . . nor will we proceed against or prosecute him except by the lawful judgment of his peers, or by the law of the land.

Magna Carta, translated in *A Documentary History of England*

Reading **History**

A. Comparing What rights from the Magna Carta remain rights in America today?

The Magna Carta limited the powers of the king. Over time, the rights it listed were granted to all English people, not just noblemen and freemen.

Parliament and Colonial Government

One of the most important English rights was the right to elect representatives to government. **Parliament,** England's chief lawmaking body, was the colonists' model for representative government. Parliament was made up of two houses. Members of the House of Commons were elected by the people. Members of the House of Lords were nonelected nobles, judges, and church officials.

The king and Parliament were too far away to manage every detail of the colonies. Also, like the citizens of England, English colonists in America wanted to have a say in the laws governing them. So they formed

CITIZENSHIP TODAY

The Importance of Juries

The right to a trial by jury, established in the Magna Carta, is an important legal right. When you become an adult, you will likely be asked to serve on a jury.

Many young people in Knox County, Illinois, have already served as jurors on a teen court (shown below, with an advisor). They decide the best punishment for other teenagers who have admitted breaking a law. For example, shoplifters might be sentenced to write an apology to the store. Knox County is one of more than 500 U.S. communities that have teen courts.

Knox County Teen Court volunteers

How Can You Serve on a Teen Court?

1. Search the library or Internet to learn more about teen courts.
2. Ask the police department whether your town has a teen court. If it does, volunteer.
3. If you want to start a teen court, seek advice from a community that has one.
4. Invite a lawyer to your class to talk about a juror's role.
5. Find a group to sponsor your court, and get support from youth officers and judges.

See the Citizenship Handbook, page 280.

Visit www.mcdougallittell.com to learn more about courts and juries.

their own elected assemblies, similar to the House of Commons. Virginia's House of Burgesses was the first of these. In Pennsylvania, William Penn allowed colonists to have their own General Assembly. These Virginia and Pennsylvania assemblies imposed taxes and managed the colonies.

Reading **History**
B. Making Inferences Why did the colonists dislike laws passed by Parliament?

Although the colonists governed themselves in some ways, England still had authority over them. The king appointed royal governors to rule some colonies on his behalf. Parliament had no representatives from the colonies. Even so, it passed laws that affected the colonies. The colonists disliked these laws, and they began to clash with royal governors over how much power England should have in America. These conflicts became more intense in the late 1600s.

A Royal Governor's Rule

"You have no more privileges left you."
a Boston court official

The reign of James II threatened the colonies' tradition of self-government. James became king in 1685. He wanted to rule England and its colonies with total authority. One of his first orders changed the way the Northern colonies were governed. These colonies, especially Massachusetts, had been smuggling goods and ignoring the Navigation Acts (see Chapter 4). When challenged, the people of Massachusetts had claimed that England had no right to make laws for them. The previous king, Charles II, had then canceled their charter.

The colonists hated Governor Andros.

Reading **History**
C. Recognizing Effects How did James II weaken self-government in the colonies?

King James combined Massachusetts and the other Northern colonies into one Dominion of New England, ruled by royal governor **Edmund Andros**. Andros angered the colonists by ending their representative assemblies and allowing town meetings to be held only once a year.

With their assemblies outlawed, some colonists refused to pay taxes. They said that being taxed without having a voice in government violated their rights. Andros jailed the loudest complainers. At their trial, they were told, "You have no more privileges left you than not to be Sould [sold] for Slaves."

The colonists sent Increase Mather to England to plead with King James (see One American's Story on page 141). However, a revolution in England swept King James and Governor Andros from power.

Background
England had become Protestant in the 16th century. Catholics were kept out of high office.

England's Glorious Revolution

The English Parliament had decided to overthrow King James for not respecting its rights. Events came to a head in 1688. King James, a Catholic, had been trying to pack his next Parliament with officials who would overturn anti-Catholic laws. He had dismissed the last Parliament in 1685. The Protestant leaders of Parliament were outraged. They offered

The French and Indian War

TERMS & NAMES
French and Indian War
Albany Plan of Union
Battle of Quebec
Treaty of Paris
Pontiac's Rebellion
Proclamation of 1763

MAIN IDEA

Britain's victory in the French and Indian War forced France to give up its North American colonies.

WHY IT MATTERS NOW

British influence spread over North America, though French populations and place names still exist here.

ONE AMERICAN'S STORY

Charles de Langlade, born in 1729, was the son of a French fur trader and his Ottawa wife. His family controlled the fur trade around what is now Green Bay, Wisconsin.

In 1752, Charles commanded 250 Ottawa and Chippewa warriors in an attack on the village of Pickawillany, in present-day Ohio. His reason: the Miami people who lived there had stopped trading with the French and were now trading with the British. Charles and his men destroyed the village's British trading post and killed the Miami chief. This attack helped lead to the French and Indian War.

This section describes the war, in which French forces fought British forces in North America. Each side had Native American allies. Charles de Langlade led several successful attacks against the British. But in the end, he saw the British drive French armies from the continent.

This 1903 painting by Edward Deming shows Charles de Langlade attacking British forces in 1755.

France Claims Western Lands

As you learned in Chapters 2 and 4, the French were exploring the North American interior while English colonists were settling the eastern coast. By the late 1600s, French explorers had claimed the Ohio River valley, the Mississippi River valley, and the entire Great Lakes region. The French territory of Louisiana, claimed by the explorer La Salle in 1682, stretched from the Appalachian Mountains to the Rocky Mountains.

The French built their main settlements, Quebec and Montreal, along the St. Lawrence River in Canada. (See the map on page 148.) They also built forts along the Great Lakes and along rivers draining into the Mississippi. By 1760, the French colony, New France, had a European population of about 80,000. By contrast, the British colonies had more than a million settlers.

Some Europeans in New France were Jesuit priests. They wanted to convert Native Americans to Christianity. Other Europeans in New France worked as fur traders. Native Americans brought furs to French forts and

Background
Often French traders lived among and married Native Americans.

exchanged them for goods such as iron pots and steel knives. Many French traders carried goods by canoe into remote parts of New France.

Native American Alliances

The English competed with the French for furs. Also, different Native American groups competed to supply furs to the Europeans. The fur trade created economic and military alliances between the Europeans and their Native American trading partners. The Huron and Algonquin peoples of the Great Lakes region were allied with the French. The Iroquois of upper New York often were allied with the Dutch and, later, the English.

Background
The Iroquois were a union of six nations.

Alliances between Europeans and Native Americans led to their involvement in each other's wars. For example, by the mid-1600s, the Iroquois had trapped all the beavers in their own lands. To get more furs, they made war on their Huron and Algonquin neighbors, driving them west. Eventually the Iroquois controlled an area ranging from Maine west to the Ohio Valley and north to Lake Michigan. Iroquois expansion threatened the French fur trade. In response, the French armed the Huron and Algonquin peoples to fight the Iroquois. The Iroquois were armed by the English.

Reading **History**
A. Recognizing Effects How did the fur trade lead to wars?

A French trader visits a Native American family.

When France and England declared war on each other in Europe in 1689, French and English colonists in America also began to fight. With their Native American allies, they attacked each other's settlements and forts. During the 1700s, two more wars between France and England fueled wars in their colonies. Neither side won a clear victory in these wars. A final war, the **French and Indian War** (1754–1763), decided which nation would control the northern and eastern parts of North America.

Conflict in the Ohio River Valley

The seeds for the French and Indian War were planted when British fur traders began moving into the Ohio River valley in the 1750s. British land companies were also planning to settle colonists there. The French and their Native American allies became alarmed. To keep the British out of the valley, Charles de Langlade destroyed the village of Pickawillany and its British trading post (see One American's Story on page 146).

Reading **History**
B. Making Inferences Why was the Ohio River Valley important to the French and British governments?

The British traders left, and the French built forts to protect the region linking their Canadian and Louisiana settlements. This upset the Virginia colony, which claimed title to the land. In 1753, the lieutenant governor of Virginia sent a small group of soldiers to tell the French to leave. Their leader was a 21-year-old major named George Washington. Washington reported the French commander's reply.

A VOICE FROM THE PAST

He told me the Country belong'd to them, that no English Man had a right to trade upon them Waters; & that he had Orders to make every Person Prisoner that attempted it on the Ohio or the Waters of it.

George Washington, *"Journey to the French Commandant"*

Virginia's lieutenant governor sent about 40 men to build a fort at the head of the Ohio River, where Pittsburgh stands today. French and Native American troops seized the partially built fort in April 1754 and completed it themselves. The French named it Fort Duquesne (du•KAYN).

War Begins and Spreads

George Washington was on his way to defend Fort Duquesne when he learned of its surrender. He and his men pushed on and built another small fort, Fort Necessity. Following Washington's surprise attack on a French force, the French and their allies attacked Fort Necessity on July 3, 1754. After Washington surrendered, the French let him march back to Virginia. The French and Indian War had begun. This war became part of the Seven Years' War (1756–1763), a worldwide struggle for empire between France and Great Britain.

Background
The Seven Years' War was fought not only in North America but also in the Caribbean, throughout Europe, and in India and Africa.

GEOGRAPHY SKILLBUILDER
Interpreting Maps

1. **Place** *Which nation controlled territory along the St. Lawrence and Ohio rivers?*
2. **Place** *Which forts were the sites of British victories?*

While Washington was surrendering Fort Necessity, representatives from the British colonies and the Iroquois nations were meeting at Albany, New York. The colonists wanted the Iroquois to fight with them against the French. The Iroquois would not commit to this alliance.

Benjamin Franklin, who admired the union of the six Iroquois nations, suggested that the colonies band together for defense. His **Albany Plan of Union** was the first formal proposal to unite the colonies. The plan called for each colony to send representatives to a Grand Council. This council would be able to collect taxes, raise armies, make treaties, and start new settlements. The leaders in Albany supported Franklin's plan, but the colonial legislatures later defeated it because they did not want to give up control of their own affairs.

Braddock's Defeat

Britain realized that to win the war, it could not rely solely on the colonists for funding or for troops. Therefore, the British sent General Edward Braddock and two regiments to Virginia. In 1755, Braddock marched toward the French at Fort Duquesne. George Washington was at his side. Their red-coated army of 2,100 moved slowly over the mountains, weighed down by a huge cannon.

On July 9, on a narrow trail eight miles from Fort Duquesne, fewer than 900 French and Indian troops surprised Braddock's forces. Washington suggested that his men break formation and fight from behind the trees, but Braddock would not listen. The general held his position and had four horses shot out from under him. Washington lost two horses. Four bullets went through Washington's coat, but, miraculously, none hit him. In the end, nearly 1,000 men were killed or wounded. General Braddock died from his wounds. American colonists were stunned by Braddock's defeat and by many other British losses over the next two years.

Reading **History**

C. Drawing Conclusions Why was Braddock defeated by a smaller enemy force?

The British Take Quebec

In 1757, Britain had a new secretary of state, William Pitt, who was determined to win the war in the colonies. He sent the nation's best generals to America and borrowed money to pay colonial troops for fighting. The British controlled six French forts by August 1759, including Fort Duquesne (rebuilt as Fort Pitt). In late summer, the British began to attack New France at its capital, Quebec.

Background

Because the British seemed likely to win the war, some Iroquois had joined them as allies.

America's **HERITAGE**

ACADIANS TO CAJUNS

Braddock's defeat and other early losses in the war increased British concern about the loyalty of the French people in Acadia (now Nova Scotia). The British had won Acadia from France in 1713.

In 1755, British officers forced out 6,000 Acadians who would not take a loyalty oath. The British burned Acadian villages and spread the people to various British colonies, as shown. Eventually, some Acadians made their way to the French territory of Louisiana. There they became known as Cajuns.

Quebec sat on cliffs 300 feet above the St. Lawrence River. Cannon and thousands of soldiers guarded its thick walls. British general James Wolfe sailed around the fort for two months, unable to capture it. Then, in September, a scout found a steep, unguarded path up the cliffs to the plains just west of Quebec. At night, Wolfe and 4,000 of his men floated to the path and secretly climbed the cliffs.

When the French awoke, the British were lined up on the plains, ready to attack. In the short, fierce battle that followed, Wolfe was killed. The French commander, Montcalm, died of his wounds the next day. Quebec surrendered to the British. The **Battle of Quebec** was the turning point of the war. When Montreal fell the next year, all of Canada was in British hands.

Reading **History**
D. Finding Main Ideas How were the British able to capture Quebec?

The Treaty of Paris

Britain and France battled in other parts of the world for almost three more years. Spain made a pact in 1761 to aid France, but its help came too late. When the Seven Years' War ended in 1763, Britain had won.

By the **Treaty of Paris,** Britain claimed all of North America east of the Mississippi River. To reward Spain for its help, France gave it New Orleans and Louisiana, the French territory west of the Mississippi. Britain, which had seized Cuba and the Philippines from Spain, gave them back in exchange for Florida. The treaty ended French power in North America.

Background
France kept only a few islands near Newfoundland and in the West Indies.

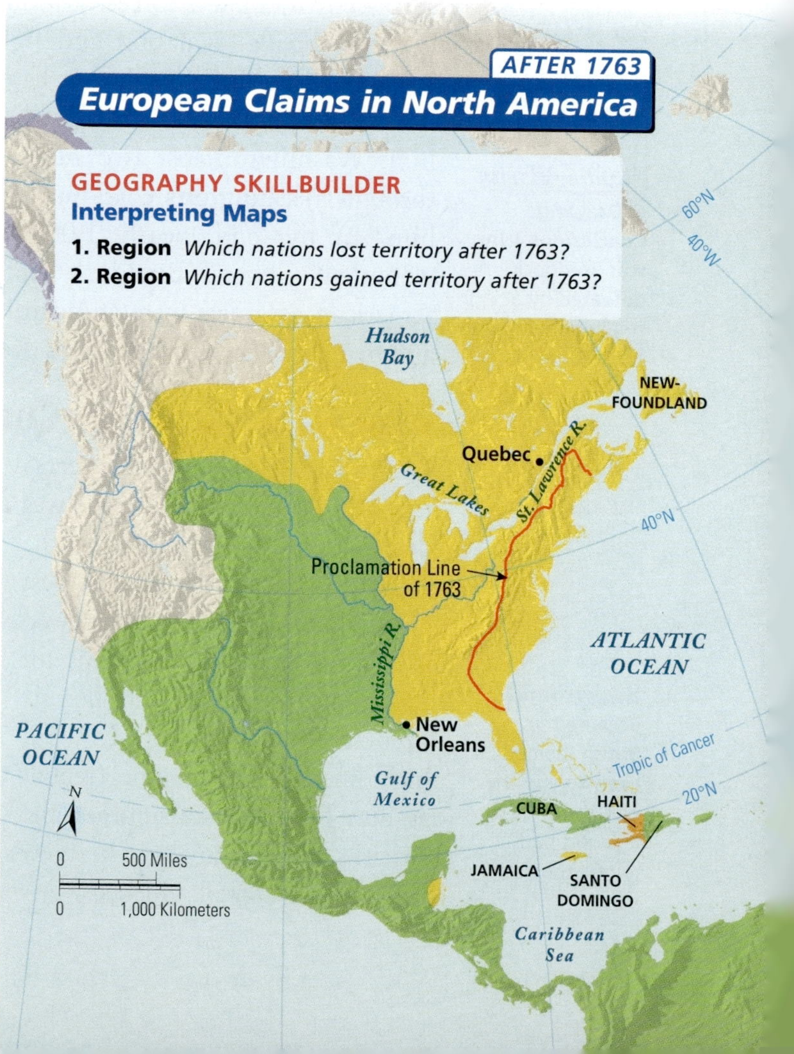

Pontiac's Rebellion

After French forces withdrew, the British took over their forts. They refused to give supplies to the Native Americans, as the French had. British settlers also moved across the mountains onto Native American land. In the spring and summer of 1763, Native American groups responded by attacking settlers and destroying almost every British fort west of the Appalachians. They surrounded the three remaining forts. This revolt was called **Pontiac's Rebellion,** although the Ottawa war leader Pontiac was only one of many organizers.

Reading **History**
E. Analyzing Points of View
Why did the Native Americans attack the British?

British settlers reacted with equal viciousness, killing even Indians who had not attacked them. British officers came up with a brutal plan to end the Delaware siege at Fort Pitt.

Pontiac

A VOICE FROM THE PAST

Could it not be contrived to send the Small Pox among those disaffected [angry] tribes of Indians? We must on this occasion use every stratagem in our power to reduce them.

Major General Jeffrey Amherst, quoted in *The Conspiracy of Pontiac*

The officers invited Delaware war leaders in to talk and then gave them smallpox-infected blankets as gifts. This started a deadly outbreak.

By the fall, the Native Americans had retreated. Even so, the uprising made the British government see that defending Western lands would be costly. Therefore, they issued the **Proclamation of 1763,** which forbade colonists to settle west of the Appalachians.

Reading **History**
F. Reading a Map
Find the Proclamation Line of 1763 on the map on page 150.

The colonists were angry. They thought they had won the right to settle the Ohio River Valley. The British government was angry at the colonists, who did not want to pay for their own defense. This hostility helped cause the war for American independence, as you will read.

Section 3 Assessment

1. Terms & Names

Identify:
- French and Indian War
- Albany Plan of Union
- Battle of Quebec
- Treaty of Paris
- Pontiac's Rebellion
- Proclamation of 1763

2. Taking Notes

Write the month and year each battle occurred. Classify each as a French or British victory.

Date	Incident	Victor
	Seizure of Fort Duquesne	
	Surrender of Fort Necessity	
	Braddock's defeat	
	Battle of Quebec	

Which was most important?

3. Main Ideas

a. How did the fur trade contribute to the French and Indian War?

b. Why did the British begin to win the war after 1758?

c. What were some causes and effects of Pontiac's Rebellion?

4. Critical Thinking

Analyzing Points of View
Why did the French, British, and Native Americans fight over the Ohio River Valley?

THINK ABOUT
- how the British viewed the valley
- how the French viewed it
- how the Native Americans viewed it

ACTIVITY OPTIONS

GEOGRAPHY
MUSIC

Learn more about the Battle of Quebec and its setting. Make a three-dimensional **model** of the battle or write a **song** about it.

VISUAL SUMMARY

Beginnings of an American Identity

Early American Culture

English colonists shared certain values, such as land ownership and hard work. The Great Awakening and the Enlightenment also drew colonists together.

Roots of Representative Government

English colonists expected the right to elect representatives to government and other political rights that had developed in England over centuries.

The French and Indian War

English colonists were also drawn together as they fought against common enemies—the French and their Native American allies.

TERMS & NAMES

Briefly explain the importance of each of the following.

1. Great Awakening
2. Enlightenment
3. John Peter Zenger
4. Magna Carta
5. Parliament
6. Glorious Revolution
7. Edmund Andros
8. French and Indian War
9. Treaty of Paris
10. Proclamation of 1763

REVIEW QUESTIONS

Early American Culture (pages 135–140)

1. Why did colonists want to own land?
2. What was women's role in the colonial economy?
3. What were three effects of the Great Awakening on colonial culture?

Roots of Representative Government (pages 141–145)

4. Why did colonies have representative assemblies?
5. What was one important right granted in the Magna Carta?
6. How did the Zenger trial help lead to freedom of the press?
7. How was the English Bill of Rights related to the Glorious Revolution?

The French and Indian War (pages 146–151)

8. What was George Washington's role in the French and Indian War?
9. What did England gain as a result of the French and Indian War?
10. What was one reason for Pontiac's Rebellion?

CRITICAL THINKING

1. USING YOUR NOTES

Using your completed chart, answer the questions below.

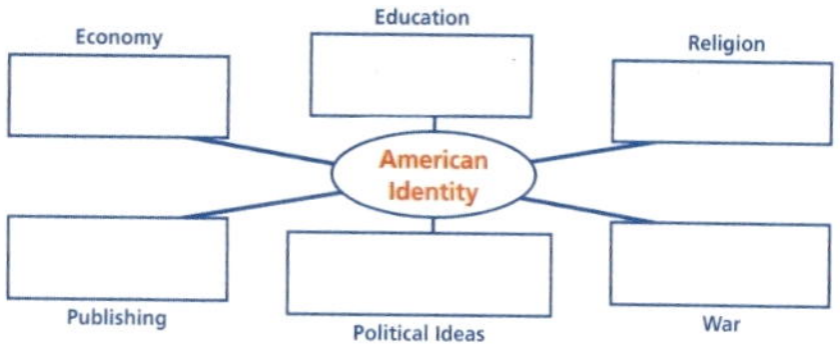

a. What were some political ideas shared by people in the American colonies?

b. How was religion important to American identity?

c. How did publishing help build an American identity?

2. THEME: DEMOCRATIC IDEALS

What democratic ideals did Americans inherit from England?

3. APPLYING CITIZENSHIP SKILLS

Why is jury duty an example of responsible citizenship?

4. CONTRASTING

How did colonial government differ from present-day government in the United States?

5. ANALYZING CAUSES

What do you think was the most important cause of the French and Indian War?

6. ANALYZING LEADERSHIP

Give an example of bad military or political leadership from the chapter. What mistake was made?

Interact *with* History

Now that you have read the chapter, what would you say British colonists in America had in common?

HISTORY SKILLS

1. INTERPRETING MAPS: Movement

Study the map. Answer the questions.

French Explorers on the Mississippi

Basic Map Elements

a. What do the colors indicate?

Interpreting the Map

b. Who traveled earlier?

c. Who reached the Gulf of Mexico?

d. Along whose route were Fort Detroit and New Orleans later founded?

2. INTERPRETING PRIMARY SOURCES

An unnamed Frenchman who knew Pontiac quotes a speech Pontiac gave to support an attack on British soldiers. Read the quote carefully. Answer the questions.

> When I go to see the English commander and say to him that some of our comrades are dead, instead of bewailing their death, as our French brothers do, he laughs at me and at you. If I ask anything for our sick, he refuses with the reply that he has no use for us. From all this you can well see that they are seeking our ruin. Therefore, my brothers, we must all swear their destruction and wait no longer.
>
> **Journal of Pontiac's Conspiracy,** 1763

a. Why does Pontiac want to attack the English?

b. What is Pontiac's attitude toward the French?

ALTERNATIVE ASSESSMENT

1. INTERDISCIPLINARY ACTIVITY: Language Arts

Making a Speech Do research to learn more about freedom of the press. Then make a speech to convince people that it is important.

2. COOPERATIVE LEARNING ACTIVITY

Holding a Diplomatic Council Working in a small group, review the causes of the French and Indian War, and do further research. Then hold a diplomatic council to try to prevent the war. Role-play representatives from the following groups.

- officials of the English government
- officials of the French government
- English settlers
- French fur traders
- English-allied Iroquois
- French-allied Huron or Algonquin

3. TECHNOLOGY ACTIVITY

Making a Class Presentation Colonial American culture was not like modern American culture. Using the library or the Internet, find images, literature, and informative articles that tell you about daily life in the early and middle 1700s.

INTERNET **Visit www.mcdougallittell.com to learn more about colonial America.**

Create a class presentation about colonial culture using the suggestions below.

- Stage a fashion show that illustrates what people of different ranks and ethnic backgrounds wore.
- Act out an interview with an interesting figure from the chapter, such as Madam Sarah Knight, Benjamin Franklin, or Pontiac.
- Illustrate sayings from *Poor Richard's Almanack.*
- Give an oral book report on a colonial captivity narrative. For example, find out what happened to Mary Rowlandson among the Narragansett.

4. HISTORY PORTFOLIO

PORTFOLIO **Option 1** Review your section and chapter assessment activities. Select one that you think is your best work. Then use comments made by your teacher or classmates to improve your work and add it to your portfolio.

PORTFOLIO **Option 2** Review the questions that you wrote for What Do You Want to Know? on page 134. Then write a short report in which you explain the answers to your questions. If any questions were not answered, do research to answer them. Add your answers to your portfolio.

UNIT 2

1763 – 1791

Creating a New Nation

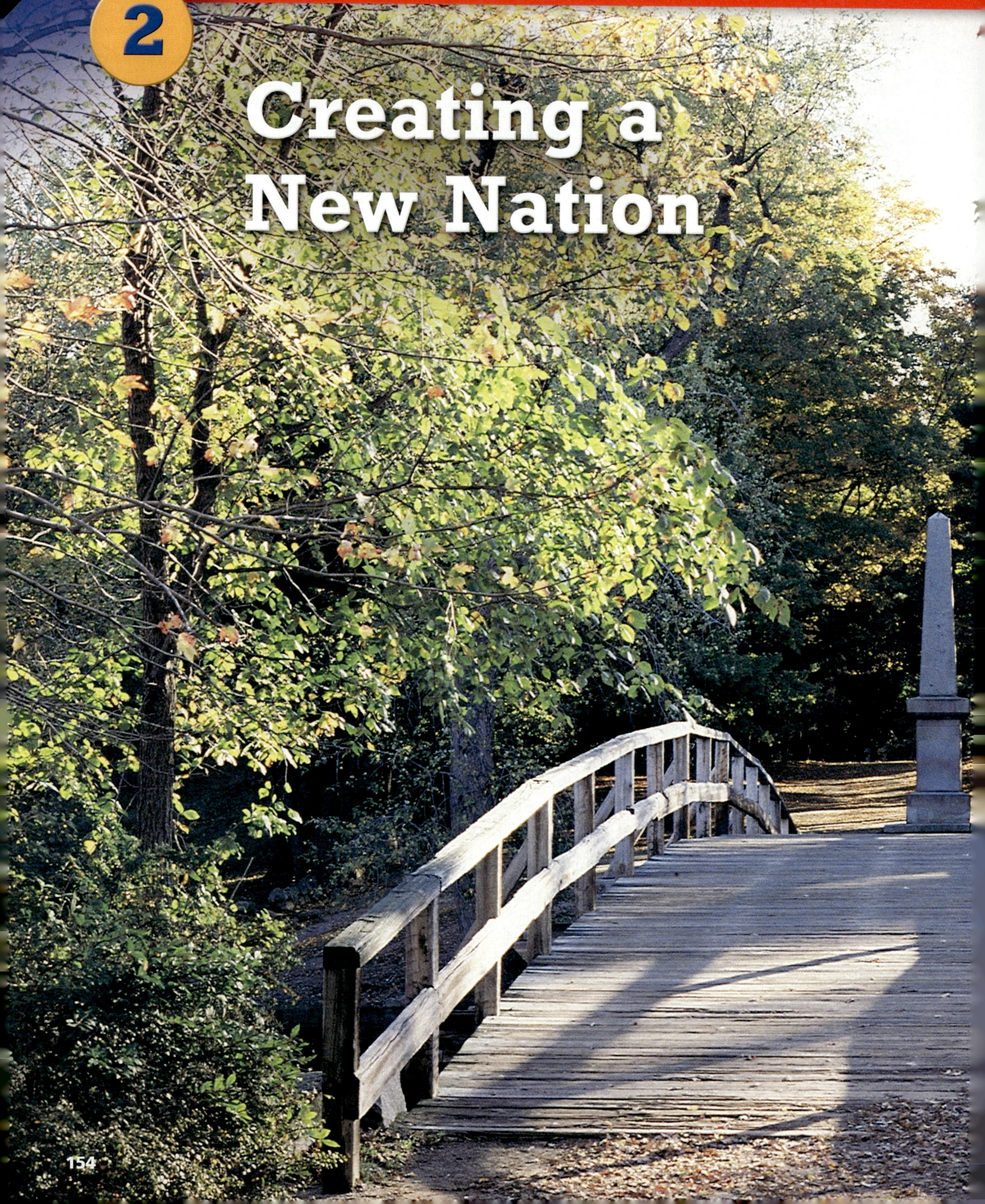

*"By the rude bridge . . .
the embattled farmers
stood and fired the shot
heard round the world."*

—Ralph Waldo Emerson, "Concord Hymn"

Americans battled British troops at Old North Bridge, Concord, Massachusetts, in April 1775.

CHAPTER 6

The Road to Revolution 1763–1776

Angry colonists watch the arrival of British troops in Boston.

Interact *with* History

The year is 1765. Your neighbors are enraged by Britain's attempt to tax them without their consent. Britain has never done this before. Everyone will be affected by the tax. There are protests in many cities. You have to decide what you would do.

What Do You Think?

- What is the best way to show opposition to policies you consider unjust?
- Is there anything to be gained by protesting? Anything to be lost?
- Does government have the right to tax without consent of the people? Why or why not?

Would you join the protest?

USA

- **1763** Proclamation of 1763 becomes law.
- **1765** Stamp Act is passed.
- **1767** Townshend Acts are passed.
- **1769** Spanish begin to establish military posts and missions in California.
- **1770** Boston Massacre
- **1773** Boston Tea Party
- **1774** Intolerable Acts are passed; First Continental Congress meets.
- **1775** Battles of Lexington and Concord
- **1776** Declaration of Independence is signed.

World

- **1763** Treaty of Paris ends Seven Years' War in Europe.
- **1765** Chinese forces invade Burma.
- **1772** Captain Cook explores the South Pacific.
- **1774** Reign of Louis XVI begins in France.

1763 — 1776

Chapter 6 SETTING THE STAGE

BEFORE YOU READ

Previewing the Theme

Impact of the Individual From 1763 to 1776, American colonists changed from loyal subjects of the British king to revolutionaries. This chapter explains how this dramatic change took place. It also discusses the individuals who led America to independence.

What Do You Know?

What do you already know about the time before the Revolution? What were the issues that caused the colonists to choose independence?

THINK ABOUT

- what you have learned about this period from movies, television, or historical fiction
- reasons people in history have chosen to fight for freedom from oppression

What Do You Want to Know?

What questions do you have about the issues and events that pushed the American colonists toward rebellion? Record them in your notebook before you read the chapter.

READ AND TAKE NOTES

Reading Strategy: Sequencing Events
Sequencing means putting events in the order in which they happen in time. In learning about how the American colonies moved toward independence, it would be helpful to list the important events. Place them in the order in which they occurred. You might record the event and its date in a graphic organizer such as the one below. Copy this organizer in your notebook. Fill it in as you read the chapter.

See Skillbuilder Handbook, page R4.

Proclamation of 1763 → [ ] → [ ] → [ ] → [ ] → [ ] → [ ] → [ ] → [ ] → [ ] → [ ] → Declaration of Independence, 1776

1 Tighter British Control

TERMS & NAMES
King George III
Quartering Act
revenue
Sugar Act
Stamp Act
Patrick Henry
boycott
Sons of Liberty

MAIN IDEA

Americans saw British efforts to tax them and to increase control over the colonies as violations of their rights.

WHY IT MATTERS NOW

Colonial protests were the first steps on the road to American independence.

James Otis, Jr., argues in court against illegal search warrants in 1761.

ONE AMERICAN'S STORY

James Otis, Jr., a young Massachusetts lawyer, stormed through the streets of Boston one day in 1760. He was furious. His father had just been denied the post of chief justice of the Massachusetts colony by the royal governor. To Otis, this was one more example of Britain's lack of respect for colonial rights. Another example was its use of search warrants that allowed customs officers to enter any home or business to look for smuggled goods. Otis believed these searches were illegal.

Otis took up a case against the government that involved these search warrants. In court in February 1761, Otis spoke with great emotion for five hours about the search warrant and its use.

A VOICE FROM THE PAST

It appears to me the worst instrument of arbitrary power, the most destructive of English liberty and the fundamental principles of law, that was ever found in an English law-book.

James Otis, Jr., quoted in *James Otis: The Pre-Revolutionist* by J. C. Ridpath

Spectators listened in amazement. One of them, a young lawyer named John Adams, later wrote of Otis's performance: "Then and there, in the old Council Chamber, the child Independence was born."

In making the first public speech demanding English liberties for the colonists, James Otis planted a seed of freedom. In this section, you will read more about the early protests against Britain's policies in America.

The Colonies and Britain Grow Apart

During the French and Indian War, Britain and the colonies fought side by side. Americans took great pride in being partners in the victory over the French. However, when the war ended, problems arose. Britain wanted to govern its 13 original colonies and the territories gained in the war in a uniform way. So the British Parliament in London imposed new laws and restrictions. Previously, the colonies had been allowed to develop largely on their own. Now they felt that their freedom was being limited.

The first of Parliament's laws was the Proclamation of 1763. (See Chapter 5.) It said that colonists could not settle west of the Appalachian Mountains. Britain wanted this land to remain in the hands of its Native American allies to prevent another revolt like Pontiac's Rebellion.

The proclamation angered colonists who had hoped to move to the fertile Ohio Valley. Many of these colonists had no land of their own. It also upset colonists who had bought land as an investment. As a result, many ignored the law.

Reading **History**

A. Summarizing Who was upset by the Proclamation of 1763?

British Troops and Taxes

King George III, the British monarch, wanted to enforce the proclamation and also keep peace with Britain's Native American allies. To do this, he decided to keep 10,000 soldiers in the colonies. In 1765, Parliament passed the **Quartering Act.** This was a cost-saving measure that required the colonies to quarter, or house, British soldiers and provide them with supplies. General Thomas Gage, commander of these forces, put most of the troops in New York.

Britain owed a large debt from the French and Indian War. Keeping troops in the colonies would raise that debt even higher. Britain needed more **revenue,** or income, to meet its expenses. So it attempted to have the colonies pay part of the war debt. It also wanted them to contribute toward the costs of frontier defense and colonial government.

In the past, the king had asked the colonial assemblies to pass taxes to support military actions that took place in the colonies. This time, however, Parliament voted to tax the Americans directly.

In 1764, Parliament passed the **Sugar Act.** This law placed a tax on sugar, molasses, and other products shipped to the colonies. It also called for strict enforcement of the act and harsh punishment of smugglers. Colonial merchants, who often traded in smuggled goods, reacted with anger.

Colonial leaders such as James Otis claimed that Parliament had no right to tax the colonies, since the colonists were not represented in Parliament. As Otis exclaimed, "Taxation without representation is tyranny!" British finance minister George Grenville disagreed. The colonists were subjects of Britain, he said, and enjoyed the protection of its laws. For that reason, they were subject to taxation.

Vocabulary

tyranny: absolute power in the hands of a single ruler

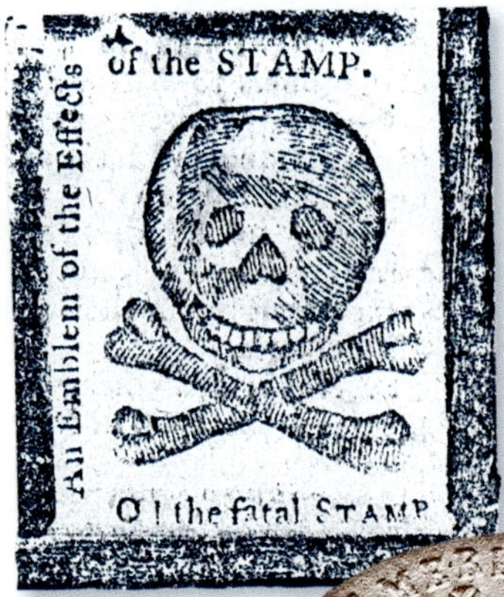

The colonial view of the hated stamp tax is shown by the skull and crossbones on this emblem (above); a royal stamp is pictured at right.

Britain Passes the Stamp Act

The Sugar Act was just the first in a series of acts that increased tension between the mother country and the colonies. In 1765, Parliament passed the **Stamp Act.** This law required all legal and commercial documents to carry an official stamp showing that a tax had been paid. All diplomas, contracts, and wills had to carry a stamp.

Even published materials such as newspapers had to be written on special stamped paper.

The Stamp Act was a new kind of tax for the colonies. The Sugar Act had been a tax on imported goods. It mainly affected merchants. In contrast, the Stamp Act was a tax applied within the colonies. It fell directly on all colonists. Even more, the colonists had to pay for stamps in silver coin—a scarce item in the colonies.

Colonial leaders vigorously protested. For them, the issue was clear. They were being taxed without their consent by a Parliament in which they had no voice. If Britain could pass the Stamp Act, what other taxes might it pass in the future? Samuel Adams, a leader in the Massachusetts legislature, asked, "Why not our lands? Why not the produce of our lands and, in short, everything we possess and make use of?" **Patrick Henry,** a member of Virginia's House of Burgesses, called for resistance to the tax. When another member shouted that resistance was treason, Henry replied, "If this be treason, make the most of it!"

The Colonies Protest the Stamp Act

Colonial assemblies and newspapers took up the cry—"No taxation without representation!" In October 1765, nine colonies sent delegates to the Stamp Act Congress in New York City. This was the first time the colonies met to consider acting together in protest. Delegates drew up a petition to the king protesting the Stamp Act. The petition declared that the right to tax the colonies belonged to the colonial assemblies, not to Parliament. Later, colonial merchants organized a **boycott** of British goods. A boycott is a refusal to buy.

Reading **History**
B. Making Inferences Why did the colonists boycott goods?

Meanwhile, some colonists formed secret societies to oppose British policies. The most famous of these groups was the **Sons of Liberty**. Many Sons of Liberty were lawyers, merchants, and craftspeople—the colonists most affected by the Stamp Act. These groups staged protests against the act.

Background
To voice their protests, the Sons of Liberty in Boston met under a huge, 120-year-old elm tree that they called the Liberty Tree.

Not all of their protests were peaceful. The Sons of Liberty burned the stamped paper whenever they could find it. They also attacked customs officials, whom they covered with hot tar and feathers and paraded in public. Fearing for their safety, many customs officials quit their jobs.

The protests in the colonies had an effect in Britain. Merchants thought that their trade with America would be hurt. Some British political leaders, including

Colonists protest the Stamp Act.

Bostonians Paying the Taxman

In this British political cartoon, Americans are depicted as barbarians who would tar and feather a customs official, or tax collector, and pour hot tea down his throat.

- **A** Liberty Tree as a gallows
- **B** Stamp Act posted upside down
- **C** Protesters in Boston
- **D** Customs official tarred and feathered

the popular parliamentary leader William Pitt, agreed with American thinking about taxing the colonies. Pitt spoke out against the Stamp Act.

A VOICE FROM THE PAST

> The Americans have not acted in all things with prudence and [good] temper. They have been driven to madness by injustice. Will you punish them for the madness you have [caused]? . . . My opinion . . . is that the Stamp Act be repealed absolutely, totally and immediately.
>
> **William Pitt,** quoted in *Patriots* by A. J. Langguth

Parliament finally saw that the Stamp Act was a mistake and repealed it in 1766. But at the same time, Parliament passed another law—the Declaratory Act. This law said that Parliament had supreme authority to govern the colonies. The Americans celebrated the repeal of the Stamp Act and tried to ignore the Declaratory Act. A great tug of war between Parliament and the colonies had begun. The central issue was control of the colonies, as you will learn in the next section.

Reading **History**

C. Drawing Conclusions Why was it important for Parliament to pass the Declaratory Act?

Section 1 Assessment

1. Terms & Names

Identify:

- King George III
- Quartering Act
- revenue
- Sugar Act
- Stamp Act
- Patrick Henry
- boycott
- Sons of Liberty

2. Taking Notes

Use a cluster diagram like the one below to review points of conflict between Britain and the colonies.

Which do you think was the most serious? Explain.

3. Main Ideas

a. Why did the Proclamation of 1763 anger colonists?

b. How did colonists react to the Stamp Act?

c. What was the goal of secret societies such as the Sons of Liberty?

4. Critical Thinking

Analyzing Points of View
What were the two sides in the debate over British taxation of the colonies?

THINK ABOUT

- how Parliament viewed the colonies
- what concerned the colonists about taxes

ACTIVITY OPTIONS

ART

MUSIC

Imagine that you are a colonial leader who wants to get your fellow colonists to protest British policy. Design a **poster** or write a **song of protest.**

2 Colonial Resistance Grows

MAIN IDEA

Many Americans began to organize to oppose British policies.

WHY IT MATTERS NOW

Americans continue to protest what they view as wrongs and injustices.

TERMS & NAMES

Crispus Attucks
Townshend Acts
writs of assistance
Samuel Adams
Boston Massacre
John Adams
committee of correspondence
Boston Tea Party

ONE AMERICAN'S STORY

Crispus Attucks knew about the struggle for freedom. The son of an African-American father and a Native American mother, Attucks was born into slavery in Framingham, Massachusetts, around 1723. As a young man, Attucks escaped by running away to sea. He spent the next 20 years as a sailor, working on whaling boats. To avoid recapture, he used a false name, calling himself Michael Johnson.

In March 1770, Attucks found himself in Boston, where feelings against British rule were reaching a fever pitch. The words *freedom* and *liberty* seemed to be on everyone's lips. One night Attucks heard about a disturbance involving colonists and British troops and decided to investigate. He had no idea that he was about to play a key role in American history—losing his life to a British bullet in a protest that came to be known as the Boston Massacre. In this section, you will read how the tension between Britain and its colonies led to violence.

Crispus Attucks, a sailor of African-American and Native American ancestry, was an early hero of America's struggle for freedom.

The Townshend Acts Are Passed

After the uproar over the Stamp Act, Britain hoped to avoid further conflict. Even so, it still needed to raise money to pay for troops and other expenses in America. The Quartering Act was not working. Most of the British army was in New York, and New York saw that as an unfair burden. Its assembly refused to pay to house the troops.

The king's finance minister, Charles Townshend, told Parliament that he had a way to raise revenue in the colonies. So in 1767, Parliament passed his plan, known as the **Townshend Acts.**

The first of the Townshend Acts suspended New York's assembly until New Yorkers agreed to provide housing for the troops. The other acts placed duties, or import taxes, on various goods brought into the colonies, such as glass, paper, paint, lead, and tea. Townshend thought that duties, which were collected before the goods entered the colonies, would anger the colonists less than the direct taxes of the Stamp Act. The money raised would be used to pay the salaries of British governors and other officials in the colonies. To enforce the acts, British officers would use **writs of assistance,** or search warrants, to enter homes or businesses to search for smuggled goods.

The Reasons for Protest

Protests immediately broke out at news of the Townshend Acts. New Yorkers were angry that their elected assembly had been suspended. People throughout the colonies were upset that Britain was placing new taxes on them. "The issue," said John Dickinson, an important Pennsylvania lawyer, was "whether Parliament can legally take money out of our pockets without our consent." He explained his opposition to the Townshend Acts in essays called *Letters from a Farmer in Pennsylvania,* published in 1767.

daily life

WOMEN AND PROTEST

Women were not allowed to participate in political life in the colonies. So their role in protesting British actions was not as prominent as that of men. However, women made their beliefs known by taking part in demonstrations.

Also, some women formed the Daughters of Liberty. This was a patriotic organization that joined in the boycott of British tea and other goods. The refusal of these colonial women to use British imports caused them personal hardship. They were forced to make many of the boycotted items, such as clothing, themselves.

A VOICE FROM THE PAST

Let these truths be . . . impressed on our minds—that we cannot be happy without being free—that we cannot be free without being secure in our property—that we cannot be secure in our property if without our consent others may . . . take it away—that taxes imposed on us by Parliament do thus take it away—that duties laid for the sole purpose of raising money are taxes—that attempts to lay such duties should be instantly and firmly opposed.

John Dickinson, quoted in *A New Age Now Begins* by Page Smith

Reading History

A. Making Inferences Why does Dickinson believe that taxes interfere with happiness?

The colonists were also angry about the writs of assistance. Many believed, as James Otis had argued (see page 159), that the writs went against their natural rights. These rights had been described by English philosopher John Locke during the Enlightenment. The law of nature, said Locke, teaches that "no one ought to harm another in his life, health, liberty, or possessions." The colonists felt that the Townshend Acts were a serious threat to their rights and freedoms.

Tools of Protest

To protest the Townshend Acts, colonists in Boston announced another boycott of British goods in October 1767. The driving force behind this protest was **Samuel Adams,** a leader of the Boston Sons of Liberty. Adams urged colonists to continue to resist British controls.

The boycott spread throughout the colonies. The Sons of Liberty pressured shopkeepers not to sell imported goods. The Daughters of Liberty called on colonists to weave their own cloth and use American products. As a result, trade with Britain fell sharply.

Colonial leaders asked for peaceful protests. Articles in the *Boston Gazette* asked the people to remain calm—

This engraving, *The Bloody Massacre Perpetrated in King Street* by Boston silversmith Paul Revere, appeared in the *Boston Gazette.*

"no mobs. . . . Constitutional methods are best." However, tempers were running high. When customs officers in Boston tried to seize the American merchant ship *Liberty,* which was carrying smuggled wine, a riot broke out. The rioters forced the customs officers to flee.

Fearing a loss of control, officials called for more British troops. A defiant Samuel Adams replied, "We will destroy every soldier that dares put his foot on shore. . . . I look upon them as foreign enemies."

The Boston Massacre

In the fall of 1768, 1,000 British soldiers (known as redcoats for their bright red jackets) arrived in Boston under the command of General Thomas Gage. With their arrival, tension filled the streets of Boston.

Since the soldiers were poorly paid, they hired themselves out as workers, usually at rates lower than those of American workers. Resentment against the redcoats grew. Soldiers and street youths often yelled insults at each other. "Lobsters for sale!" the youths would yell, referring to the soldiers' red coats. "Yankees!" the soldiers jeered. *Yankee* was supposed to be an insult, but the colonists soon took pride in the name.

On March 5, 1770, tensions finally exploded into violence. A group of youths and dockworkers—among them Crispus Attucks—started trading insults in front of the Custom House. A fight broke out, and the soldiers began firing. Attucks and four laborers were killed.

Reading **History**

B. Recognizing Propaganda How did the use of the word *massacre* show an anti-British view?

The Sons of Liberty called the shooting the **Boston Massacre.** They said that Attucks and the four others had given their lives for freedom. The incident became a tool for anti-British propaganda in newspaper articles, pamphlets, and posters. The people of Boston were outraged.

Meanwhile, the redcoats who had fired the shots were arrested for murder. **John Adams,** a lawyer and cousin of Samuel Adams, defended them in court. Adams was criticized for taking the case. He replied that the law should be "deaf . . . to the clamors of the populace." He supported

AMERICA'S HISTORY MAKERS

SAMUEL ADAMS
1722–1803

Samuel Adams was a Harvard graduate. But unlike his cousin John, also a Harvard graduate, he showed little skill for the law. Later, when he took control of the family business, he lost his father's fortune. Yet he succeeded in one important undertaking—moving America toward independence.

Adams's true talent lay in rousing people to action in support of a cause. A fiery orator and a master of propaganda, he used words as a weapon. One British official said that "every dip of his pen stings."

JOHN ADAMS
1735–1826

John Adams, unlike Samuel, was considered a moderate in the struggle against Britain. He was an important voice of reason and at first opposed resisting by force.

Adams believed in the rule of law. He called his defense of the soldiers in the Boston Massacre "one of the best pieces of service I ever rendered my country."

Eventually, Adams became convinced that only outright resistance would gain liberty for America. He said, "Britain has at last driven America, to the last Step, a compleat Seperation from her."

How did the cousins John and Samuel Adams differ in the way they protested British actions?

the colonial cause but wanted to show that the colonists followed the rule of law. Adams argued that the soldiers had acted in self-defense. The jury agreed. To many colonists, however, the Boston Massacre would stand as a symbol of British tyranny.

The Tea Act

The colonists were unaware that on the day of the Boston Massacre, Parliament proposed the repeal of the Townshend Acts. One month later, all the acts except the tax on tea were repealed. The colonial boycott had been effective—British trade had been hurt. But Parliament kept the tea tax to show that it still had the right to tax the colonists. For most Americans, the crisis was over.

Samuel Adams, however, wanted to make sure people did not forget the cause of liberty. He started a drive to form **committees of correspondence** in various towns in Massachusetts. These groups exchanged letters on colonial affairs. Before long, committees throughout Massachusetts were corresponding with one another and with committees in other colonies.

Reading **History**

C. Drawing Conclusions Why did Samuel Adams think that the colonists might forget the cause of liberty?

Then, in 1773, Parliament opened up old wounds when it passed the Tea Act. Tea was very popular in the colonies, but much of it was smuggled in from Holland. The Tea Act gave the British East India Company control over the American tea trade. The tea would arrive in the colonies only in the trading company's ships and be sold there by its merchants. Colonists who had not been paying any tax on smuggled tea would now have to pay a tax on this regulated tea. This enraged colonial shippers and merchants. The colonists wondered what Parliament would do next.

Colonists dumped hundreds of chests of tea into Boston Harbor in 1773 to protest the Tea Act.

The Boston Tea Party

Protests against the Tea Act took place all over the colonies. In Charleston, South Carolina, colonists unloaded tea and let it rot on the docks. In New York City and Philadelphia, colonists blocked tea ships from landing. In Boston, the Sons of Liberty organized what came to be known as the **Boston Tea Party.**

Reading **History**

D. Reading a Map Find Boston Harbor on the map on page 172.

On the evening of December 16, 1773, a group of men disguised as Native Americans boarded three tea ships docked in Boston Harbor. One of the men, George Hewes, a Boston shoemaker, later recalled the events.

> **A VOICE FROM THE PAST**
>
> We then were ordered by our commander to open the hatches and take out all the chests of tea and throw them overboard. . . . In about three hours from the time we went on board, we had thus broken and thrown overboard every tea chest to be found on the ship, while those in the other ships were disposing of the tea in the same way, at the same time.
>
> **George Hewes,** quoted in *A Retrospect of the Boston Tea-Party*

That night, Hewes and the others destroyed 342 chests of tea. Many colonists rejoiced at the news. They believed that Britain would now see how strongly colonists opposed taxation without representation.

Others doubted that destroying property was the best way to settle the tax debate. Some colonial leaders offered to pay for the tea if Parliament would repeal the Tea Act. Britain rejected the offer. It not only wanted repayment, but it also wanted the men who destroyed the tea to be brought to trial. The British reaction to the Boston Tea Party would fan the flames of rebellion in the 13 colonies, as you will read in the next section.

Reading **History**

E. Recognizing Effects How did Britain react to the Tea Party?

Section 2 Assessment

1. Terms & Names

Identify:

- Crispus Attucks
- Townshend Acts
- writs of assistance
- Samuel Adams
- Boston Massacre
- John Adams
- committee of correspondence
- Boston Tea Party

2. Taking Notes

Create a time line like the one below to show important events described in this section.

1767 — 1773

Which event do you think was the most important? Explain.

3. Main Ideas

a. Why did colonists oppose the Townshend Acts?

b. Why were British troops sent to Boston?

c. What prompted the Boston Tea Party?

4. Critical Thinking

Drawing Conclusions

Do you think colonial outrage over the Boston Massacre was justified? Explain.

THINK ABOUT

- how the British troops were taunted
- whether troops have the right to fire on citizens

ACTIVITY OPTIONS

SPEECH
TECHNOLOGY

Read more about the Boston Massacre or the Boston Tea Party. Present an **oral report** or plan a **multimedia presentation** about the event.

Fight for Representative Government!

You are a colonist living in Boston on the eve of the American Revolution. Nearly a decade of protest against British policies has failed to secure American rights. Redcoats continue to be quartered in the city. The Tea Act still stands. Now the dumping of tea in Boston Harbor by some Patriots has charged the atmosphere with tension. Trouble lies ahead, but you are determined to fight for a government that will protect your rights.

COOPERATIVE LEARNING On this page are two challenges that you face as the conflict with Britain unfolds. Working with a small group, decide how to deal with each challenge. Choose an option, assign a task to each group member, and do the activity. You will find useful information in the Data File. Present your solutions to the class.

ART CHALLENGE

"Do not . . . sip the accursed, dutied STUFF"

People all over Boston are worried about the Tea Act. It taxes tea, but it also lets the British East India Company sell tea through its own agents. In time, the plan could drive American tea sellers out of business. How can you protest these threats to American commerce and liberty? Present your viewpoint using one of these options:

- **Design an anti–Tea Act poster.**
- **Draw a political cartoon showing the dangers of the Tea Act.**

MATH CHALLENGE

"Wear none but your own country linen"

Years of struggle have taken their toll on Boston. People are tired of soldiers and of boycotting British goods, such as clothing. But the Tea Act presents a huge threat. The Boston Tea Party took care of only one shipment. How can you help encourage the boycott of other British goods, such as clothing? Look at the Data File for help. Present your appeal using one of these options:

- Make a graph showing the effect of colonial boycotts on imports of British goods to America.
- Write an editorial using statistics to show how American boycotts have hurt the British.

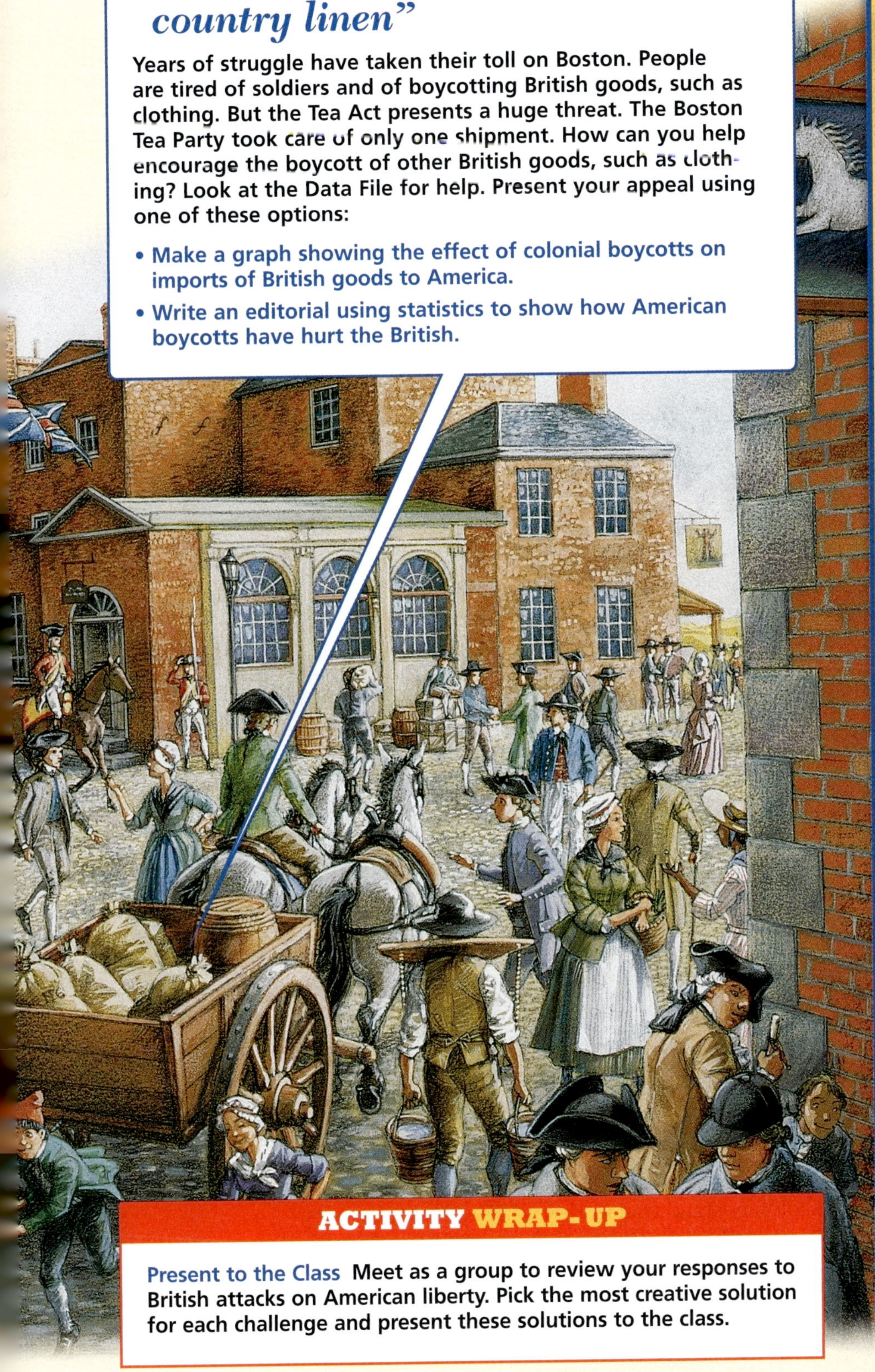

ACTIVITY WRAP-UP

Present to the Class Meet as a group to review your responses to British attacks on American liberty. Pick the most creative solution for each challenge and present these solutions to the class.

DATA FILE

Population in 1774–1775

Britain: 7,860,000
London: 700,000

The 13 colonies: 2,350,000
Philadelphia: 33,000
New York: 22,000
Boston: 16,000

North American Imports from Britain

(in millions of pounds sterling)

1763	1.6	1770	1.9
1764	2.3	1771	4.2
1765	1.9	1772	3.0
1766	1.8	1773	2.1
1767	1.9	1774	2.6
1768	2.2	1775	0.2
1769	1.3	1776	0.1

North American Exports to Britain

(in millions of pounds sterling)

1763	1.1	1770	1.0
1764	1.1	1771	1.3
1765	1.2	1772	1.3
1766	1.0	1773	1.4
1767	1.1	1774	1.4
1768	1.3	1775	1.9
1769	1.1	1776	0.1

Key Boycott Dates

1764	Boycott after passage of Sugar Act
1765	Boycott after passage of Stamp Act
1766	Boycott relaxed after Stamp Act repealed
1767	Boycott after passage of Townshend Acts
1770	Townshend Acts repealed
1774	Boycott after passage of Intolerable Acts

Sales and Consumption of Tea at the Time of the Boston Tea Party

British sales: fourth most important product shipped to America
American consumption: 1.2 million pounds per year

INTERNET Visit www.mcdougallittell.com for more on Revolutionary America.

The Road to Lexington and Concord

TERMS & NAMES
- militia
- Minuteman
- Intolerable Acts
- First Continental Congress
- Paul Revere
- Lexington and Concord
- Loyalist
- Patriot

MAIN IDEA

The tensions between Britain and the colonies led to armed conflict in Massachusetts.

WHY IT MATTERS NOW

Americans at times still find themselves called upon to fight for their principles.

ONE AMERICAN'S STORY

At dawn on April 19, 1775, some 70 militiamen gathered on the grassy common at the center of Lexington, Massachusetts, a small town near Boston. Captain John Parker, a veteran of the French and Indian War, was their commander. The **militia** was a force of armed civilians pledged to defend their community. About one-third of the Lexington militia were **Minutemen**, trained to be "ready to act at a minute's warning." Everyone had heard the news—the British were coming!

Each militiaman was equipped with a musket, a bayonet, and ammunition. Parker had spent months drilling his troops, but they had never faced British soldiers. Soon they would meet the British on Lexington Green in the first battle of the Revolutionary War. According to tradition, Parker told his men, "Stand your ground; don't fire unless fired upon, but if they mean to have war, let it begin here."

In this section, you will read how colonial protests eventually turned into violent revolution.

This statue of Captain John Parker stands in Lexington, Massachusetts.

The Intolerable Acts

The Boston Tea Party had aroused fury in Britain. One British official said that the people of Boston "ought to be knocked about their ears." King George III declared, "We must master them or totally leave them to themselves and treat them as aliens." Britain chose to "master" the colonies.

In 1774, Parliament passed a series of laws to punish the Massachusetts colony and to serve as a warning to other colonies. The British called these laws the Coercive Acts, but they were so harsh that the colonists called them the **Intolerable Acts**.

One of the acts would close the port of Boston until colonists paid for the destroyed tea. Others banned committees of correspondence, allowed Britain to house troops wherever necessary, and let British officials accused of crimes in the colonies stand trial in Britain. To enforce the acts, Parliament appointed General Thomas Gage governor of Massachusetts.

In 1773, Sam Adams had written, "I wish we could arouse the continent." The Intolerable Acts answered his wish. Other colonies immediately offered Massachusetts their support. They sent food and money to Boston. The committees of correspondence also called for a meeting of colonial delegates to discuss what to do next.

The First Continental Congress Meets

In September 1774, delegates from all the colonies except Georgia met in Philadelphia. At this meeting, called the **First Continental Congress,** delegates voted to ban all trade with Britain until the Intolerable Acts were repealed. They also called on each colony to begin training troops. Georgia agreed to be a part of the actions of the Congress even though it had voted not to send delegates.

Reading **History**
A. Evaluating Why do you think the First Continental Congress was important?

The First Continental Congress marked a key step in American history. Although most delegates were not ready to call for independence, they were determined to uphold colonial rights. This meeting planted the seeds of a future independent government. John Adams called it "a nursery of American statesmen." The delegates agreed to meet in seven months, if necessary. By that time, however, fighting with Britain had begun.

Between War and Peace

The colonists hoped that the trade boycott would force a repeal of the Intolerable Acts. After all, past boycotts had led to the repeal of the Stamp Act and the Townshend Acts. This time, however, Parliament stood firm. It even increased restrictions on colonial trade and sent more troops.

By the end of 1774, some colonists were preparing to fight. In Massachusetts, John Hancock headed the Committee of Safety, which had the power to call out the militia. The colonial troops continued to train.

CAUSE AND EFFECT: *Growing Conflict Between Britain and America*

DATE	BRITISH ACTION	COLONIAL REACTION
1763	Proclamation of 1763 issued	Proclamation leads to anger
1765	Stamp Act passed	Boycott of British goods; Stamp Act Resolves passed
1766	Stamp Act repealed; Declaration Act passed	Boycott ended
1767	Townshend Acts passed	New boycotts; Boston Massacre (March 1770)
1770	Townshend Acts repealed (April)	Tension between colonies and Britain reduced
1773	Tea Act passed	Boston Tea Party
1774	Intolerable Acts passed	First Continental Congress bans trade; militias organized
1775	Troops ordered to Lexington and Concord, Massachusetts	Militia fights British troops; Second Continental Congress; Continental Army established

SKILLBUILDER Interpreting Charts

1. *What British action caused the first violence in the growing conflict between Britain and America?*
2. *How might the Intolerable Acts be seen as a reaction as well as an action?*

Most colonial leaders believed that any fight with Britain would be short. They thought that a show of force would make Britain change its policies. Few expected a war. One who did was Patrick Henry.

> ***A VOICE FROM THE PAST***
>
> Gentlemen may cry peace, peace—but there is no peace. The war is actually begun! The next gale that sweeps from the north will bring to our ears the clash of resounding arms! Our brethren are already in the field! Why should we idle here? . . . I know not what course others may take. But as for me, give me liberty or give me death.
>
> **Patrick Henry,** quoted in *Patriots* by A. J. Langguth

Henry delivered what became his most famous speech in the Virginia House of Burgesses in March 1775.

The Midnight Ride

Meanwhile, spies were busy on both sides. Sam Adams had built a spy network to keep watch over British activities. The British had their spies too. They were Americans who were loyal to Britain. From them, General Gage learned that the Massachusetts militia was storing arms and ammunition in Concord, about 20 miles northwest of Boston. He also heard that Sam Adams and John Hancock were in Lexington. On the night of April 18, 1775, Gage ordered his troops to arrest Adams and Hancock in Lexington and to destroy the supplies in Concord.

*Reading*History

B. Recognizing Effects What effect might spying have had on the people of Boston?

The Sons of Liberty had prepared for this moment. **Paul Revere,** a Boston silversmith, and a second messenger, William Dawes, were charged with spreading the news about British troop movements. Revere had arranged a system of signals to alert colonists in Charlestown, on the shore opposite Boston. If one lantern burned in the North Church steeple, the British troops were coming by land; if two, they were coming by water. Revere would go across the water from Boston to Charlestown and ride to Lexington and Concord from there. Dawes would take the land route.

Background

The signals were a backup system in case Revere was captured.

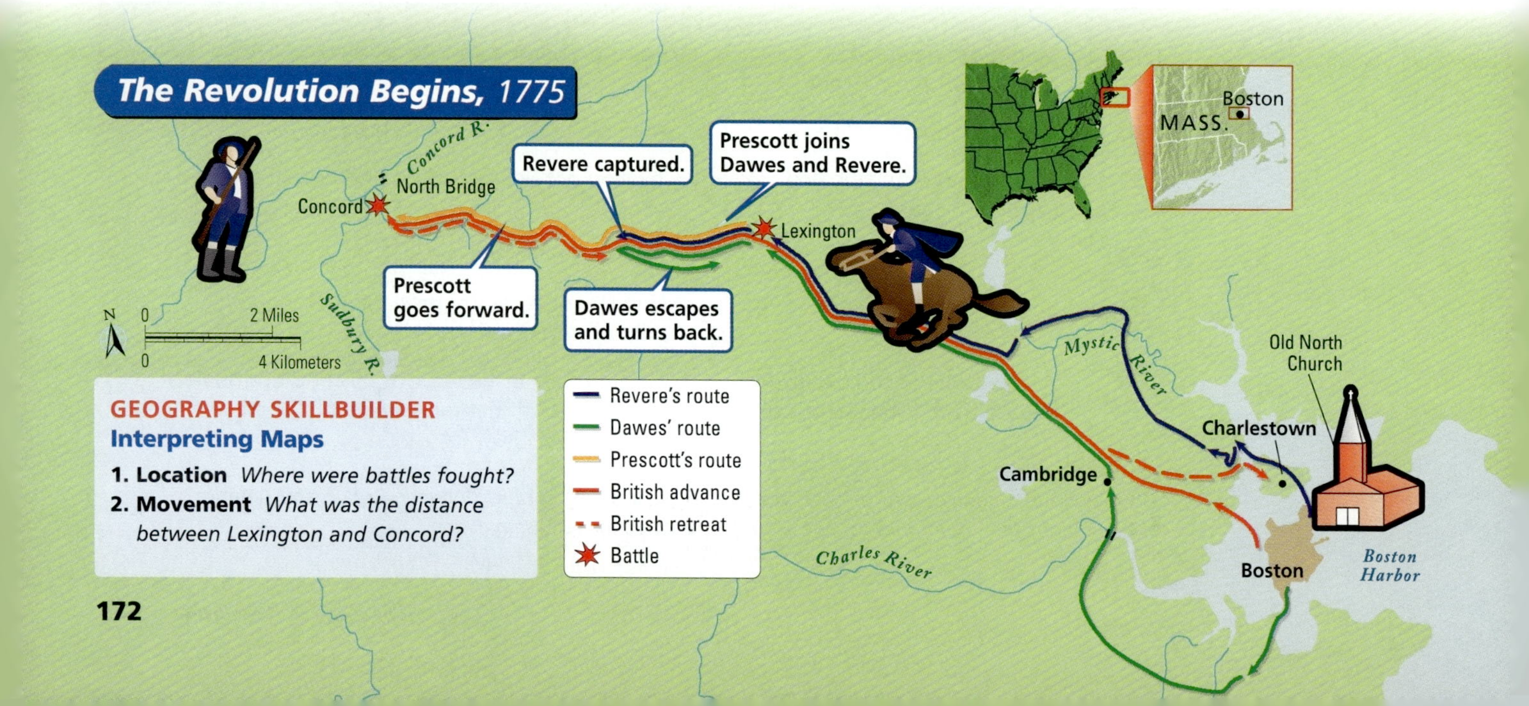

GEOGRAPHY SKILLBUILDER
Interpreting Maps

1. **Location** *Where were battles fought?*
2. **Movement** *What was the distance between Lexington and Concord?*

When the British moved, so did Revere and Dawes. They galloped over the countryside on their "midnight ride," spreading the news. In Lexington, they were joined by Dr. Samuel Prescott. When Revere and Dawes were stopped by a British patrol, Prescott broke away and carried the message to Concord.

Lexington and Concord

At dawn on April 19, some 700 British troops reached Lexington. They found Captain John Parker and about 70 militiamen waiting. The British commander ordered the Americans to drop their muskets. They refused. No one knows who fired first, but within a few minutes eight militiamen lay dead. The British then marched to Concord, where they destroyed military supplies. A battle broke out at a bridge north of town, forcing the British to retreat.

Nearly 4,000 Minutemen and militiamen arrived in the area. They lined the road from Concord to Lexington and peppered the retreating redcoats with musket fire. "It seemed as if men came down from the clouds," one soldier said. Only the arrival of 1,000 more troops saved the British from destruction as they scrambled back to Boston.

Lexington and Concord were the first battles of the Revolutionary War. As Ralph Waldo Emerson later wrote, colonial troops had fired the "shot heard 'round the world." Americans would now have to choose sides and back up their political opinions by force of arms. Those who supported the British were called **Loyalists.** Those who sided with the rebels were **Patriots.** The conflict between the two sides divided communities, families, and friends. The war was on!

Background
British losses totaled 273 soldiers compared to 95 militiamen.

***Reading* History**
C. Drawing Conclusions
Why did Emerson call it the "shot heard 'round the world"?

Now and then

PATRIOTS' DAY

The "shot heard 'round the world" is celebrated every year in Massachusetts and Maine. Patriots' Day, as it is called, is the third Monday of April. In Concord and nearby towns, modern-day Minutemen like those below reenact the battle that began the Revolution on April 19, 1775. The Boston Marathon is also run on Patriots' Day.

Section 3 Assessment

1. Terms & Names

Identify:
- militia
- Minuteman
- Intolerable Acts
- First Continental Congress
- Paul Revere
- Lexington and Concord
- Loyalist
- Patriot

2. Taking Notes

Use a diagram like the one below to show events that led to the Revolutionary War.

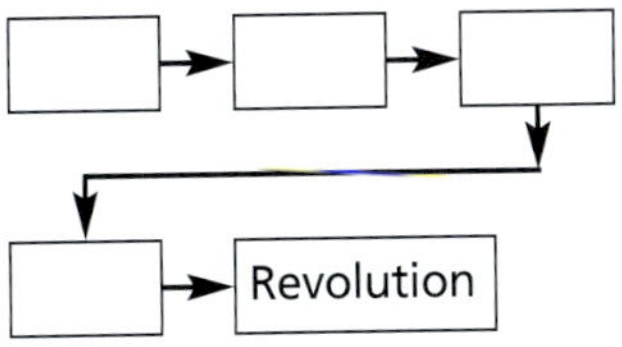

3. Main Ideas

a. Why did Britain pass the Intolerable Acts?

b. Who took part in the First Continental Congress?

c. What was the purpose of the "midnight ride"?

4. Critical Thinking

Supporting Opinions
Do you think the fighting between Britain and the colonies could have been avoided? Why or why not?

THINK ABOUT
- Britain's attitude toward the colonies
- colonial feelings about Britain

ACTIVITY OPTIONS

GEOGRAPHY
MATH

Research the Battles of Lexington and Concord. Draw a **map** of key events or create a **chart** showing statistics from the battles.

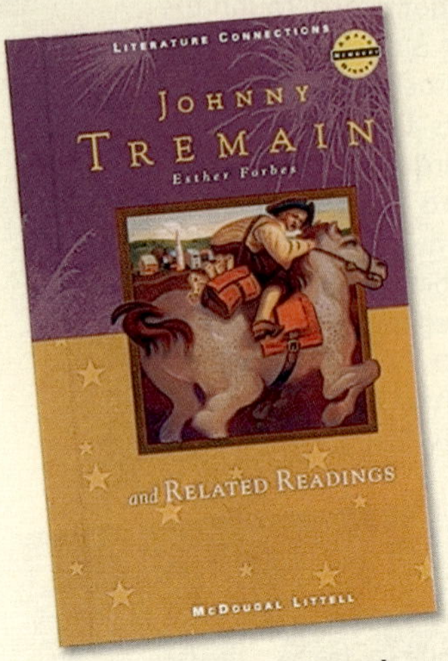

From Johnny Tremain

by Esther Forbes

In 1775, 16-year-old Johnny Tremain lives in Boston and works as a delivery boy for a newspaper. Because he travels so much around the city, he is able to help the Patriots gather information about what the British are doing.

On the night of April 18, Johnny learns that British troops will be leaving on an expedition to seize the gunpowder at Lexington and Concord. He rushes to tell this news to Dr. Joseph Warren, who is a Patriot. Then Johnny goes to bed, wondering if the war has started and worried about his friend Rab, who has gone to join the Minutemen at Lexington.

So Johnny slept. It was daylight when he woke with Warren's hand upon his shoulder. Outside on Tremont Street he could hear the clumping of army boots. A sergeant was swearing at his men. The soldiers were paraded so close to the house, which stood **flush**[1] with the sidewalkless street, that Johnny at first thought they must be in the room.

Doctor Warren dared speak no louder than a whisper.

"I'm going now."

"Something's happened?"

"Yes." He motioned Johnny to follow him into the kitchen. This room was on the back of the house. They could talk without danger of being overheard by the troops in the street.

Doctor Warren had on the same clothes as the day before. He had not been to bed. But now his hat was on his head. His black bag of instruments and medicines was packed and on the table. Silently he put milk, bread, herrings beside it, and gestured to Johnny to join him.

"Where did it begin?" asked Johnny.

"Lexington."

"Who won?"

"They did. Seven hundred against seventy. It wasn't a battle. It was . . . just target practice . . . for them. Some of our men were killed and the British **huzzaed**[2] and took the road to Concord."

"And did they get our supplies there?"

"I don't know. Paul Revere sent for me just after the firing on Lexington Green."

The young man's usually fresh-colored face was **haggard**[3]. He knew the seriousness of this day for himself and for his country.

"But everywhere the alarm is spreading. Men are grabbing their guns—marching for Concord. Paul Revere did get through in time last night. Billy Dawes a little later. Hundreds—maybe thousands—of Minute Men are on the march. Before the day's over, there'll be real fighting—not target practice. But Gage doesn't know that it's begun. You see, long before Colonel Smith got to Lexington—just as soon as he heard that Revere had warned the country—he sent back for reinforcements. For Earl Percy. You and I, Johnny, are just about the only people in Boston who know that blood has already been shed."

"Were many killed—at Lexington?"

"No, not many. They stood up—just a handful. The British fired on them. It was dawn."

Johnny licked his lips. "Did they tell you the names of those killed?"

1. **flush:** in a line with.

2. **huzzaed:** cheered.

3. **haggard:** tired.

"No. Did Rab get out in time?"

"Yes. Last Sunday."

The Doctor's clear blue eyes darkened. He knew what was in Johnny's mind. He picked up his bag. "I've got to get to them. They'll need surgeons. Then, too, I'd rather die fighting than on a gallows. Gage won't be so **lenient**[4] now—soon as he learns war has begun."

"Wait until I get my shoes on."

"No, Johnny, you are to stay here today. Pick up for me any information. For instance, out of my bedroom window I can see soldiers standing the length of the street 'way over to the Common. You find out what regiments are being sent—and all that. And today go about and listen to what folk are saying. And the names of any the British arrest. We know Gage expects to move his men back here tonight. If so, there'll be a lot of confusion getting them into town. You watch your chance and slip out to me."

"Where'll I find you?"

" . . . Ask about."

"I will do so."

"They've begun it. We'll end it, but this war . . . it may last quite a long time."

They shook hands silently. Johnny knew that Warren was always conscious of the fact that he had a crippled hand. Everybody else had accepted and forgotten it. The back door closed softly. Warren was gone.

Johnny went to the **surgery,**[5] put on his boots and jacket. The wall clock said eight o'clock. It was time to be about. There was no leaving by the front door. The soldiers were leaning against it. Through the curtains of the windows he could see the muskets. He noticed the facings on their uniforms. The Twenty-Third Regiment. The narrow course of Tremont Street was filled to the brim and overflowing with the waiting scarlet-coated men. Like a river of blood. He left by the kitchen.

4. **lenient:** not strict.

5. **surgery:** operating room.

CONNECT TO HISTORY

1. **Recognizing Effects** What was Johnny's reaction to the news about Lexington? Discuss what roles Johnny and Dr. Warren were to play in the early days of the Revolutionary War.

See Skillbuilder Handbook, page R10.

CONNECT TO TODAY

2. **Researching** Where are there revolutions in the world today?

Visit www.mcdougallittell.com to learn more about revolutions.

British troops fire on the Lexington militia on April 19, 1775. The war begins here!

Declaring Independence

TERMS & NAMES
- Ethan Allen
- artillery
- Second Continental Congress
- Continental Army
- Benedict Arnold
- Declaration of Independence
- Thomas Jefferson

MAIN IDEA

Fighting between American and British troops led the colonies to declare their independence.

WHY IT MATTERS NOW

The United States of America was founded at this time.

ONE AMERICAN'S STORY

In May 1775—one month after the battle at Lexington and Concord—Abigail Adams wrote to her husband, John Adams. "The house is a Scene of Confusion," she said. Colonial militiamen were camped outside. Everyone was preparing for war. John Adams was away in Philadelphia at the time, meeting with other Patriot leaders at the Second Continental Congress.

Abigail and John Adams would spend most of the Revolutionary War apart. In his absence, she ran the household and farm in Braintree, Massachusetts, and raised their four children. During their separation, they exchanged many letters. Abigail was a very sharp observer of the political scene. In one letter, she shared her concerns about the future of the American government.

A VOICE FROM THE PAST

If we separate from Britain, what Code of Laws will be established? How shall we be governed so as to retain our Liberties? Can any government be free which is not administered by general stated Laws? Who shall frame these Laws? Who will give them force and energy?

Abigail Adams, quoted in *Abigail Adams: Witness to a Revolution* by Natalie S. Bober

Abigail Adams was an early advocate of women's rights and one of the great letter writers in history.

These questions would be answered later. First, a war had to be fought and won.

The Continental Army Is Formed

After the fighting at Lexington and Concord, militiamen from Massachusetts and other colonies began gathering around Boston. Their numbers eventually reached some 20,000. General Gage decided to move his soldiers from the peninsula opposite Boston to the city itself. Boston was nearly surrounded by water. This fact, he thought, made a colonial attack by land almost impossible.

Not long after, on May 10, 1775, Americans attacked Britain's Fort Ticonderoga on the New York side of Lake Champlain. **Ethan Allen** led this band of backwoodsmen known as the Green Mountain Boys. They captured the fort and its large supply of **artillery**—cannon and large guns. These guns would be used later to drive the British from Boston.

Also on May 10, the **Second Continental Congress** began meeting in Philadelphia. Delegates included John and Samuel Adams, John Hancock, Benjamin Franklin, George Washington, and Patrick Henry. They agreed to form the **Continental Army.** Washington, who was from Virginia, was chosen as its commanding general. He had served as a colonial officer with the British during the French and Indian War. Congress also authorized the printing of paper money to pay the troops. It was beginning to act as a government.

The Battle of Bunker Hill

Background
The battle was called Bunker Hill because the original plan was to fight the battle there.

Meanwhile, tensions were building in Boston in June 1775. Militiamen seized Bunker Hill and Breed's Hill behind Charlestown. They built fortifications on Breed's Hill. Alarmed, the British decided to attack.

"Don't fire until you see the whites of their eyes!"
Colonel William Prescott

General William Howe crossed the bay with 2,200 British soldiers. Forming in ranks, they marched up Breed's Hill. On the hilltop, the militia waited. According to the legend, Colonel William Prescott ordered, "Don't fire until you see the whites of their eyes!" When the British got close, the militia unleashed murderous fire. The British fell back and then charged again. Finally, they forced the militia off the hill.

The redcoats had won the battle of Bunker Hill, but at tremendous cost. More than 1,000 were killed or wounded, compared with some 400 militia casualties. "The loss we have sustained is greater than we can bear," wrote General Gage. The inexperienced colonial militia had held its own against the world's most powerful army.

The bloody fighting between militiamen and British troops is shown in *The Death of General Warren at Bunker Hill* by John Trumbull (1786).

Connections TO LITERATURE

PHILLIS WHEATLEY

Phillis Wheatley was America's first important African-American poet. She was born in Africa about 1753 and sold into slavery as a child. She was a household servant for the Wheatley family of Boston but was raised and educated as a family member.

Some of Wheatley's poems were about the Patriot cause. Of George Washington, she wrote:

Proceed, great chief, with
virtue on thy side,
Thy ev'ry action let the
goddess guide.
A crown, a mansion, and
a throne that shine,
With gold unfading,
Washington! be thine.

In other poems, Wheatley connected America's fight against British oppression with the struggle for freedom for enslaved African Americans.

A Last Attempt at Peace

Despite this deepening conflict, most colonists still hoped for peace. Even some Patriot leaders considered themselves loyal subjects of the king. They blamed Parliament for the terrible events taking place.

In July 1775, moderates in Congress drafted the Olive Branch Petition and sent it to London. This document asked the king to restore harmony between Britain and the colonies. Some members opposed the petition but signed it anyway as a last hope.

Background
The olive branch is considered a symbol of peace.

The king rejected the petition, however, and announced new measures to punish the colonies. He would use the British navy to block American ships from leaving their ports. He also would send thousands of hired German soldiers, called Hessians, to fight in America. "When once these rebels have felt a smart blow, they will submit," he declared.

***Reading* History**
A. Analyzing Points of View
Why did King George reject the petition?

The colonial forces were not going to back down, though. They thought they were equal to the British troops. George Washington knew otherwise. The British soldiers were professionals, while the colonial troops had little training and were poorly equipped. The Massachusetts militia barely had enough gunpowder to fight one battle.

During the summer of 1775, Washington arrived at the militia camp near Boston. He immediately began to gather supplies and train the army. In the fall, Washington approved a bold plan. Continental Army troops would invade Quebec, in eastern Canada. They hoped to defeat British forces there and draw Canadians into the Patriot camp. One of the leaders of this expedition was **Benedict Arnold.** He was an officer who had played a role in the victory at Fort Ticonderoga.

After a grueling march across Maine, Arnold arrived at Quebec in November 1775. By that time, however, winter had set in. Under harsh conditions, the Americans launched their attack but failed. After several months, they limped home in defeat.

The British Retreat from Boston

In Massachusetts, the Continental Army had surrounded British forces in Boston. Neither side was able or willing to break the standoff. However, help for Washington was on the way. Cannons were being hauled from Fort Ticonderoga. This was a rough job, since there were no roads across the snow-covered mountains. It took soldiers two months to drag the 59 heavy weapons to Boston, where they arrived in January 1776.

Armed with these cannons, Washington moved his troops to Dorchester Heights, overlooking Boston. The Americans threatened to bombard the city. General Howe, who was now in charge of the British forces, decided to withdraw his troops. On March 17, about 9,000 British soldiers departed Boston in more than 100 ships. Boston Patriots joyfully reclaimed their city. Although the British had damaged homes and destroyed possessions, Boston was still standing.

More than 1,000 Loyalist supporters left along with the British troops. Anti-British feeling in Boston was so strong that the Loyalists feared for their safety. Some Patriots even called for Loyalists to be hanged as traitors. This did not happen, but Loyalists' homes and property were seized.

Reading **History**

B. Forming Opinions Did the Loyalists deserve punishment? Explain.

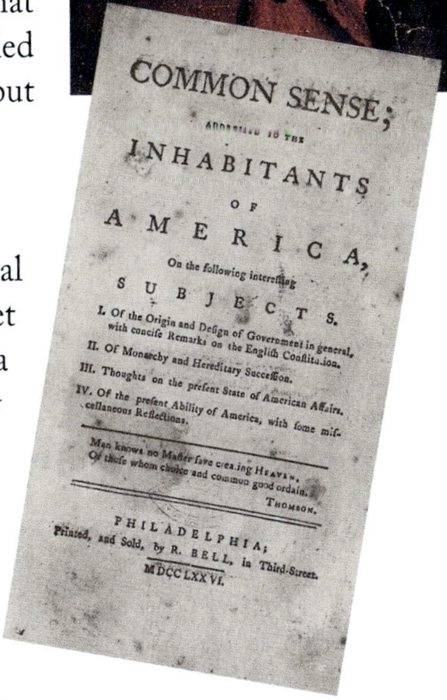

COMMON SENSE;
ADDRESSED TO THE
INHABITANTS
OF
AMERICA,
On the following interesting
SUBJECTS.
I. Of the Origin and Design of Government in general, with concise Remarks on the English Constitution.
II. Of Monarchy and Hereditary Succession.
III. Thoughts on the present State of American Affairs.
IV. Of the present Ability of America, with some miscellaneous Reflections.

Man knows no Master save creating HEAVEN,
Or those whom choice and common good ordain.
THOMSON.

PHILADELPHIA;
Printed, and Sold, by R. BELL, in Third-Street.
MDCCLXXVI.

This is the front page of *Common Sense* by Thomas Paine (above). It was one of the most influential political documents in history.

Common Sense Is Published

In early 1776, most Americans still wanted to avoid a final break with Britain. However, the publication of a pamphlet titled *Common Sense* helped convince many Americans that a complete break with Britain was necessary. Written by Thomas Paine, a recent immigrant from England, this pamphlet made a strong case for American independence.

Paine ridiculed the idea that kings ruled by the will of God. Calling George III "the Royal Brute," Paine argued that all monarchies were corrupt. He also disagreed with the economic arguments for remaining with Britain. "Our corn," he said, "will fetch its price in any market in Europe." He believed that America should follow its own destiny.

A VOICE FROM THE PAST

Everything that is right or natural pleads for separation. The blood of the slain, the weeping voice of nature cries, "'Tis time to part." Even the distance at which the Almighty has placed England and America is a strong and natural proof that the authority of the one over the other was never the design of heaven.

Thomas Paine, *Common Sense*

Common Sense was an instant success. Published in January, it sold more than 100,000 copies in three months. The call for independence had become a roar.

A Time of Decision

The Continental Congress remained undecided. A majority of the delegates still did not support independence. Even so, in May 1776, Congress adopted a resolution authorizing each of the 13 colonies to establish its own government.

On June 7, Richard Henry Lee of Virginia introduced a key resolution. It called the colonies "free and independent states" and declared

HISTORY *through* ART

The Declaration of Independence is presented for adoption to the Continental Congress by John Adams, Roger Sherman, Robert Livingston, Thomas Jefferson, and Benjamin Franklin (left to right). John Trumbull painted this work many years after the adoption of the Declaration on July 4, 1776.

What is the artist trying to show about the mood of the American leaders as they declare independence?

that "all political connection between them and the state of Great Britain is . . . totally dissolved."

Congress debated the resolution, but not all the delegates were ready to vote on it. They did, however, appoint a committee to draft a **Declaration of Independence**. The committee included Benjamin Franklin, John Adams, Roger Sherman, Robert Livingston, and **Thomas Jefferson**.

The group chose Jefferson to compose the Declaration. Two reasons for selecting Jefferson were that he was an excellent writer and that he came from Virginia. The members knew that no independence movement could succeed without Virginia's support. Jefferson immediately went to work. In two weeks, he had prepared most of the Declaration. (See pages 182–185.) On July 2, 1776, Congress considered Lee's resolution again. Despite some strong opposition, the measure passed. From this point forward, the colonies considered themselves independent.

The Liberty Bell was rung to announce the first public reading of the Declaration of Independence, in Philadelphia on July 8, 1776.

The Declaration Is Adopted

Two days later, on July 4, 1776, Congress adopted the document that proclaimed independence—the Declaration of Independence. John Hancock, the president of the Congress, was the first to sign the Declaration. According to tradition, he wrote in large letters and commented, "There, I guess King George will be able to read that." The core idea of the Declaration is based on the philosophy of John Locke. This idea is that people have unalienable rights, or rights that government cannot take away. Jefferson stated this belief in what was to become the Declaration's best-known passage.

A VOICE FROM THE PAST

We hold these truths to be self-evident, that all men are created equal, that they are endowed by their Creator with certain unalienable Rights, that among these are Life, Liberty and the pursuit of Happiness.

Thomas Jefferson, The Declaration of Independence

Reading **History**

C. Summarizing When does the Declaration say it is right to overthrow an established government?

If a government disregards these rights, Jefferson explained, it loses its right to govern. The people then have the right to abolish that government, by force if necessary. They can form a new government that will protect their rights. When Jefferson spoke of "the people," however, he meant only free white men. Women and enslaved persons were left out of the Declaration.

The Declaration also explained the reasons for breaking with Britain. It then declared the colonies to be free and independent states. This was a very serious action—treason from the British point of view—and the delegates knew it. John Hancock urged the delegates to stand together in mutual defense. Each realized that if the war were to be lost, they would most likely be hanged.

The Declaration closed with this pledge: "And for the support of this Declaration, with a firm reliance on the protection of divine Providence, we mutually pledge to each other our Lives, our Fortunes, and our sacred Honor."

Americans had declared independence. Now they had to win their freedom on the battlefield.

AMERICA'S HISTORY MAKERS

THOMAS JEFFERSON
1743–1826

Jefferson was just 33 when chosen to write the Declaration of Independence. He was already a brilliant thinker and writer and a highly respected political leader. Jefferson came from a wealthy Virginia family. As a child, he was interested in everything, and he became an inventor, scientist, and architect, among other things. In 1769, he began his political career in the House of Burgesses.

Jefferson felt that writing the Declaration was a major achievement of his life. He had that fact carved on his tombstone.

Why do you think Jefferson felt the Declaration was one of his greatest achievements?

Section 4 Assessment

1. Terms & Names

Identify:

- Ethan Allen
- artillery
- Second Continental Congress
- Continental Army
- Benedict Arnold
- Declaration of Independence
- Thomas Jefferson

2. Taking Notes

Use the chart below to explain colonial views for and against independence.

Views About Independence

For	
Against	

What is the strongest reason for independence? against independence?

3. Main Ideas

a. What challenges did George Washington face in forming the army?

b. What forced the British to leave Boston?

c. What is *Common Sense*?

4. Critical Thinking

Drawing Conclusions
Why did it take colonists so long to declare independence?

THINK ABOUT

- the colonists' British traditions
- the risk of revolution

ACTIVITY OPTIONS

ART
LANGUAGE ARTS

Find out more about a person discussed in this section. Create a **trading card** or write a **biography** of that person.

The Declaration of Independence

Setting the Stage On July 4, 1776, the Second Continental Congress adopted what became one of America's most cherished documents. Written by Thomas Jefferson, the Declaration of Independence voiced the reasons for separating from Britain and provided the principles of government upon which the United States would be built. **See Primary Source Explorer**

[Preamble]

When in the Course of human events, it becomes necessary for one people to dissolve the political bands which have connected them with another, and to assume among the powers of the earth, the separate and equal station to which the Laws of Nature and of Nature's God entitle them, a decent respect to the opinions of mankind requires that they should declare the causes which impel them to the separation.

[The Right of the People to Control Their Government]

We hold these truths to be self-evident, that all men are created equal, that they are **endowed**[1] by their Creator with certain **unalienable**[2] Rights, that among these are Life, Liberty and the pursuit of Happiness; that, to secure these rights, Governments are instituted among Men, deriving their just powers from the consent of the governed; that whenever any Form of Government becomes destructive of these ends, it is the Right of the People to alter or to abolish it, and to institute new Government, laying its foundation on such principles and organizing its powers in such form, as to them shall seem most likely to effect their Safety and Happiness. Prudence, indeed, will dictate that Governments long established should not be changed for light and transient causes; and accordingly all experience hath shewn that mankind are more disposed to suffer, while evils are sufferable, than to right themselves by abolishing the forms to which they are accustomed. But when a long train of abuses and **usurpations,**[3] pursuing invariably the same Object, evinces a design to reduce them under absolute **Despotism,**[4] it is their right, it is their duty, to throw off such Government, and to provide new Guards for their future security.

Such has been the patient sufferance of these Colonies; and such is now the necessity which constrains them to alter their former Systems of Government. The history of the present King of Great Britain is a history of repeated injuries and usurpations, all having in direct object the establishment of an absolute Tyranny over these States. To prove this, let facts be submitted to a **candid**[5] world.

A CLOSER LOOK

RIGHTS OF THE PEOPLE

The ideas in this passage reflect the views of John Locke, an English philosopher. His belief that a government's power comes from the consent of the governed is the foundation of modern democracy.

1. In what way can American voters bring about changes in their government?

---

1. **endowed:** provided.
2. **unalienable:** unable to be taken away.
3. **usurpations:** unjust seizures of power.
4. **Despotism:** rule by a tyrant with absolute power.
5. **candid:** fair, impartial.

[Tyrannical Acts of the British King]

He has refused his Assent to Laws, the most wholesome and necessary for the public good.

He has forbidden his Governors to pass Laws of immediate and pressing importance, unless suspended in their operation till his assent should be obtained; and, when so suspended, he has utterly neglected to attend to them.

He has refused to pass other Laws for the accommodation of large districts of people, unless those people would **relinquish**[6] the right of Representation in the Legislature, a right inestimable to them, and formidable to tyrants only.

He has called together legislative bodies at places unusual, uncomfortable, and distant from the depository of their public Records, for the sole purpose of fatiguing them into compliance with his measures.

He has dissolved Representative Houses repeatedly, for opposing with manly firmness his invasions on the rights of the people.

He has refused for a long time, after such dissolutions, to cause others to be elected; whereby the Legislative powers, incapable of Annihilation, have returned to the people at large for their exercise; the State remaining in the mean time exposed to all the dangers of invasions from without, and **convulsions**[7] within.

He has endeavoured to prevent the population of these States; for that purpose obstructing the Laws for **Naturalization**[8] of Foreigners; refusing to pass others to encourage their migration hither, and raising the conditions of new Appropriations of Lands.

He has obstructed the Administration of Justice, by refusing his Assent to Laws for establishing Judiciary powers.

He has made Judges dependent on his Will alone, for the **tenure**[9] of their offices, and the amount and payment of their salaries.

He has erected a multitude of New Offices, and sent hither swarms of Officers to harass our people and **eat out their substance.**[10]

He has kept among us, in times of peace, Standing Armies, without the Consent of our legislatures.

He has affected to render the Military independent of and superior to the Civil power. He has combined with others to subject us to a jurisdiction foreign to our constitution and unacknowledged by our laws; giving his Assent to their Acts of pretended Legislation:

For **quartering**[11] large bodies of armed troops among us;

For protecting them, by a mock Trial, from punishment for any Murders which they should commit on the Inhabitants of these States;

For cutting off our Trade with all parts of the world;

6. **relinquish:** give up.
7. **convulsions:** violent disturbances.
8. **Naturalization:** process of becoming a citizen.
9. **tenure:** term.
10. **eat out their substance:** drain their resources.
11. **quartering:** housing or giving lodging to.

A CLOSER LOOK

GRIEVANCES AGAINST BRITAIN

The list contains 27 offenses by the British king and others against the colonies. It helps explain why it became necessary to seek independence.

2. Which offense do you think was the worst? Why?

A CLOSER LOOK

LOSS OF REPRESENTATIVE GOVERNMENT

One of the Intolerable Acts of 1774 stripped the Massachusetts Legislature of many powers and gave them to the colony's British governor.

3. Why was this action so "intolerable"?

A CLOSER LOOK

QUARTERING TROOPS WITHOUT CONSENT

The Quartering Act of 1765 required colonists to provide housing and supplies for British troops in America.

4. Why did colonists object to this act?

A CLOSER LOOK

TAXATION WITHOUT REPRESENTATION

The colonists believed in the long-standing British tradition that Parliament could tax only those citizens it represented—and the colonists claimed to have no representation in Parliament.

5. How do persons today give consent to taxation?

For imposing Taxes on us without our Consent;

For depriving us, in many cases, of the benefits of Trial by Jury;

For transporting us beyond Seas to be tried for pretended offenses;

For abolishing the free System of English Laws in a neighboring Province, establishing therein an **Arbitrary**[12] government, and enlarging its Boundaries so as to render it at once an example and fit instrument for introducing the same absolute rule into these Colonies;

For taking away our Charters, abolishing our most valuable laws, and altering fundamentally the Forms of our Governments;

For suspending our own Legislatures, and declaring themselves invested with power to legislate for us in all cases whatsoever.

He has **abdicated**[13] Government here, by declaring us out of his Protection and waging War against us.

He has plundered our seas, ravaged our Coasts, burnt our towns, and destroyed the lives of our people.

He is at this time transporting large Armies of **foreign Mercenaries**[14] to compleat the works of death, desolation, and tyranny, already begun with circumstances of Cruelty & **perfidy**[15] scarcely paralleled in the most barbarous ages, and totally unworthy the Head of a civilized nation.

He has constrained our fellow Citizens, taken Captive on the high Seas, to bear Arms against their Country, to become the executioners of their friends and Brethren, or to fall themselves by their Hands.

He has excited **domestic insurrections**[16] amongst us, and has endeavoured to bring on the inhabitants of our frontiers the merciless Indian Savages, whose known rule of warfare is an undistinguished destruction of all ages, sexes and conditions.

A CLOSER LOOK

PETITIONING THE KING

The colonists sent many petitions to King George III. In the Olive Branch Petition of 1775, the colonists expressed their desire to achieve "a happy and permanent reconciliation." The king rejected the petition.

6. Why did the colonists at first attempt to solve the dispute and remain loyal?

[Efforts of the Colonies to Avoid Separation]

In every stage of these Oppressions We have **Petitioned for Redress**[17] in the most humble terms; Our repeated Petitions have been answered only by repeated injury. A Prince, whose character is thus marked by every act which may define a Tyrant, is unfit to be the ruler of a free people.

Nor have We been wanting in attentions to our British brethren. We have warned them from time to time of attempts by their legislature to extend an unwarrantable jurisdiction over us. We have reminded them of the circumstances of our emigration and settlement here. We have appealed to their native justice and **magnanimity,**[18] and we have conjured them by the ties of our common kindred, to disavow these usurpations, which would inevitably interrupt our connections and correspondence. They too have been deaf to

12. **Arbitrary:** not limited by law.
13. **abdicated:** given up.
14. **foreign Mercenaries:** professional soldiers hired to serve in a foreign army.
15. **perfidy:** dishonesty, disloyalty.
16. **domestic insurrections:** rebellions at home.
17. **Petitioned for Redress:** asked for the correction of wrongs.
18. **magnanimity:** generosity, forgiveness.

the voice of justice and of **consanguinity**.[19] We must, therefore, **acquiesce**[20] in the necessity, which denounces our Separation, and hold them, as we hold the rest of mankind, Enemies in War, in Peace Friends.

[The Colonies Are Declared Free and Independent]

We, therefore, the Representatives of the United States of America, in General Congress, Assembled, appealing to the Supreme Judge of the world for the **rectitude**[21] of our intentions, do, in the name, and by the Authority of the good People of these Colonies solemnly publish and declare, That these United Colonies are, and of Right ought to be, Free and Independent States; that they are Absolved from all Allegiance to the British Crown, and that all political connection between them and the State of Great Britain is, and ought to be, totally dissolved; and that as Free and Independent States, they have full Power to levy War, conclude Peace, contract Alliances, establish Commerce, and do all other Acts and Things which Independent States may of right do.

And for the support of this Declaration, with a firm reliance on the protection of divine Providence, we mutually pledge to each other our Lives, our Fortunes, and our sacred Honor. [Signed by]

John Hancock *President, from Massachusetts*
[Georgia] Button Gwinnett; Lyman Hall; George Walton
[Rhode Island] Stephen Hopkins; William Ellery
[Connecticut] Roger Sherman; Samuel Huntington; William Williams; Oliver Wolcott
[North Carolina] William Hooper; Joseph Hewes; John Penn
[South Carolina] Edward Rutledge; Thomas Heyward, Jr.; Thomas Lynch, Jr.; Arthur Middleton
[Maryland] Samuel Chase; William Paca; Thomas Stone; Charles Carroll
[Virginia] George Wythe; Richard Henry Lee; Thomas Jefferson; Benjamin Harrison; Thomas Nelson, Jr.; Francis Lightfoot Lee; Carter Braxton
[Pennsylvania] Robert Morris; Benjamin Rush; Benjamin Franklin; John Morton; George Clymer; James Smith; George Taylor; James Wilson; George Ross
[Delaware] Caesar Rodney; George Read; Thomas McKean
[New York] William Floyd; Philip Livingston; Francis Lewis; Lewis Morris
[New Jersey] Richard Stockton; John Witherspoon; Francis Hopkinson; John Hart; Abraham Clark
[New Hampshire] Josiah Bartlett; William Whipple; Matthew Thornton
[Massachusetts] Samuel Adams; John Adams; Robert Treat Paine; Elbridge Gerry

19. **consanguinity:** relationship by a common ancestor; close connection.

20. **acquiesce:** accept without protest.

21. **rectitude:** moral uprightness.

A CLOSER LOOK

POWERS OF AN INDEPENDENT GOVERNMENT

The colonists identified the ability to wage war and agree to peace; to make alliances with other nations; and to set up an economic system as powers of a free and independent government.

7. What other powers are held by an independent government?

A CLOSER LOOK

DECLARATION SIGNERS

The Declaration was signed by 56 representatives from the 13 original states.

8. Which signers do you recognize? Write one line about each of those signers.

Interactive Primary Source Assessment

1. Main Ideas

a. What is the purpose of the Declaration of Independence as stated in the Preamble?

b. What are the five main parts of the Declaration?

c. What are three rights that all people have?

2. Critical Thinking

Drawing Conclusions Why did the colonies feel that they had to declare their independence?

THINK ABOUT

- colonial grievances against Britain
- Britain's response to these grievances

Chapter 6 ASSESSMENT

VISUAL SUMMARY

The Road to Revolution

1763 Proclamation of 1763

1764 Sugar Act

1765 Quartering Act; Stamp Act; Sons of Liberty; Stamp Act Congress

1766 Repeal of Stamp Act; Declaratory Act

1767 Townshend Acts; Suspension of New York Assembly

1768 Occupation of Boston by British troops

1769 Daughters of Liberty

1770 Boston Massacre; Repeal of all Townshend Acts except tea tax

1772 Committees of Correspondence

1773 Tea Act; Boston Tea Party

1774 Intolerable Acts; First Continental Congress; Boycott of British goods

1775 Battles of Lexington and Concord; Second Continental Congress; Appointment of Washington as commander of Continental Army; Battle of Bunker Hill; Olive Branch Petition

1776 *Common Sense;* Declaration of Independence

TERMS & NAMES

Briefly explain the importance of each of the following.

1. Stamp Act
2. Sons of Liberty
3. writs of assistance
4. Samuel Adams
5. Boston Tea Party
6. militia
7. Lexington and Concord
8. Loyalist
9. Declaration of Independence
10. Thomas Jefferson

REVIEW QUESTIONS

Tighter British Control (pages 159–162)

1. How did relations between Britain and the colonies change after the Seven Years' War?
2. Why did Britain try to tax the colonies?
3. Why did the colonists cry, "No taxation without representation"?

Colonial Resistance Grows (pages 163–169)

4. How did the colonists protest the Townshend Acts?
5. How was the Boston Massacre used for propaganda purposes?
6. How did the committees of correspondence help keep people informed?

The Road to Lexington and Concord (pages 170–175)

7. Why was the First Continental Congress held?
8. What was the Midnight Ride?

Declaring Independence (pages 176–185)

9. What was the Battle of Bunker Hill?
10. What was the core idea of the Declaration of Independence?

CRITICAL THINKING

1. USING YOUR NOTES

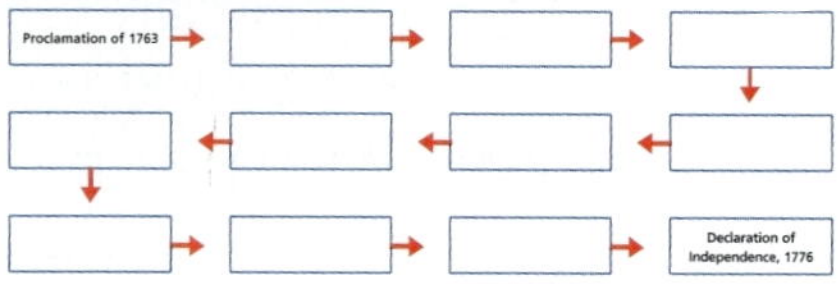

Using your completed chart, answer the questions below.

a. What city was the site of early protest activity?

b. What event happened after the Tea Act?

2. ANALYZING LEADERSHIP

How did colonial leaders differ in their methods of defending and securing basic rights for the colonies?

3. APPLYING CITIZENSHIP SKILLS

Did colonial leaders have a responsibility to include women, African Americans, and other groups in the Declaration of Independence? Explain.

4. THEME: IMPACT OF THE INDIVIDUAL

How did John Adams's role as lawyer for the British soldiers involved in the Boston Massacre help set a tone for the Revolutionary cause?

5. DRAWING CONCLUSIONS

What factors and events led the colonies to seek independence?

6. SUPPORTING OPINIONS

Do you think the American Revolution would have occurred if Britain had not taxed the colonies? Why or why not?

Interact *with* History

Now that you have read about the road to revolution, do you consider your decision made at the beginning of the chapter to join or not join the protest a wise choice or a poor choice? Explain.

HISTORY SKILLS

1. INTERPRETING MAPS: Location

Study the map and then answer the questions.

Basic Map Elements

a. What is the subject of the map?

b. What do the colors yellow and orange represent?

Interpreting the Map

c. What is the approximate distance between Great Britain and the United States?

d. Who do you think had the advantage in a war fought on the North American mainland?

2. INTERPRETING PRIMARY SOURCES

This British political cartoon is titled *Poor Old England Endeavoring to Reclaim His Wicked Children*. Study the cartoon carefully. Answer the questions.

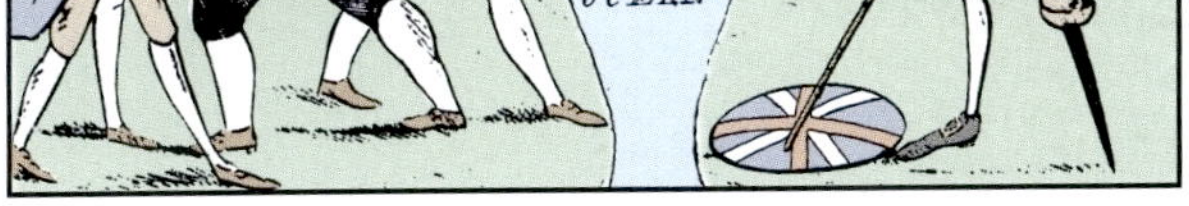

a. How is England pictured? America?

b. Does the cartoon suggest England will be successful? Explain.

ALTERNATIVE ASSESSMENT

1. INTERDISCIPLINARY ACTIVITY: World History

Giving a Report Do research about another country, colony, or region of the world during the 1760s and 1770s. Give a brief oral report, comparing and contrasting events and conditions there with those in the British colonies.

2. COOPERATIVE LEARNING ACTIVITY

Debating During discussion of the Declaration of Independence, members of the Continental Congress held heated debates on the issue of independence and whether to abolish the slave trade. Imagine that you and your classmates are delegates to the Congress. Hold a debate on these issues, choosing one of the following roles to play:

- a strong supporter of independence and of an end to the slave trade (such as Samuel Adams or John Adams)
- a supporter of independence but an opponent of ending the slave trade (such as George Washington)
- an independence opponent (such as John Dickinson)

3. CD-ROM PRIMARY SOURCE EXPLORER

Writing a Play The Declaration of Independence announced to the world that the colonies had chosen independence and why. Using the CD-ROM, your textbook, and library research, find out more about the Declaration and the people and events surrounding it. Then write a one-act play using the suggestions below:

- Set the play after the Declaration has been drafted.
- Have Thomas Jefferson, as the author, answer questions about what he included and perhaps did not include in the Declaration.
- Have Jefferson questioned by delegates to the Congress, or have him appear before a town meeting of ordinary citizens.
- Prepare introductory material to describe the setting and write dialogue for the characters.

4. HISTORY PORTFOLIO

PORTFOLIO **Option 1** Review your section and chapter assessment activities. Select one that you think was your best work. Then give consideration to suggestions made by your teacher or classmates and revise your work. Add it to your portfolio.

PORTFOLIO **Option 2** Review the questions that you wrote for What Do You Want to Know? on page 158. Then write a short report explaining your answers. Add the answers to your portfolio.

Raise the Liberty Pole

In 1765, the Sons of Liberty gathered around a huge elm tree in Boston that they named the Liberty Tree. It became a meeting place where people voiced their protests against British policies. Replicas of the Liberty Tree—giant poles sometimes decorated with the flags of the colonies—were raised throughout the colonies. These liberty poles represented the unity of the American colonies as they struggled to break away from British rule.

ACTIVITY Like the American Patriots, each group of students will raise its own liberty pole. Each group also will write and deliver a persuasive speech supporting the cause of the American colonies.

TOOLBOX

Each group will need:

- scissors
- poster board
- pencil
- markers
- masking tape
- 3 cardboard tubes from wrapping paper
- construction paper
- twine
- stapler

STEP BY STEP

1 Form groups. Each group should consist of four or five students. The members of your group will do the following jobs:

- research each colony
- design and create flags
- construct a pole
- write and deliver a speech

2 Do research on the 13 colonies. For each colony, your group should find a person, place, or object that represents that colony. For example, a Pilgrim's hat might represent Massachusetts. The 13 colonies are listed below.

New England Colonies	Middle Colonies	Southern Colonies
Massachusetts (including Maine)	New York	North Carolina
New Hampshire	Delaware	Virginia
Connecticut	New Jersey	Maryland
Rhode Island	Pennsylvania	South Carolina
		Georgia

Members of the Sons of Liberty raise a liberty pole in July 1776 to celebrate America's independence.

3 Design and create 13 flags for the colonies. Decide what person, place, or object you will use on your flag for each colony. Cut each flag out of the poster board. Sketch your design on the flag with a pencil. Then use markers to decorate it. On the back of each flag, explain how your design portrays the characteristics of that colony.

4 Construct the pole. Using masking tape, fasten the three cardboard tubes together to form one long tube. Then reinforce the tube by taping construction paper around it.

5 String the flags on the pole. Feed a piece of twine through the open ends of the long tube. Tie the ends of the twine together to form a tight loop. Now staple all 13 flags to the twine.

6 Raise your liberty pole. Lean your liberty pole next to a small table or desk. Take turns with members of your group and visit other liberty poles. As students visit your station, explain the significance of your flag designs.

WRITE AND SPEAK

Write a persuasive speech to recruit others to join the cause of liberty. In your speech, explain what is wrong with British policies. Give reasons why the colonies should become independent. Then read your speech to the other groups as part of the recruitment process.

HELP DESK

For related information on the Liberty Tree, see pages 161–162 in Chapter 6.

Researching Your Project

- *The Revolutionary War* by Bart McDowell
- *The American Revolutionaries* edited by Milton Meltzer

Visit www.mcdougallittell.com for more on the Revolution.

Did You Know?

The numbers 45 and 92 played an important part in the history of these liberty poles. The 45th issue of a British newspaper openly criticized the king in 1763 and was reprinted in the colonies. In 1768, 92 members of the Massachusetts General Assembly voted against canceling a letter to the other 12 colonies that called for action against Britain. To represent the numbers, 92 members of the Sons of Liberty would often raise liberty poles to a height of about 45 feet.

REFLECT & ASSESS

- **How well do your flags represent the colonies?**
- **How clearly does your speech explain grievances against the British?**
- **Why do you think the practice of raising liberty poles spread to many of the colonies?**

CHAPTER 7

The American Revolution 1776–1783

This painting shows General Washington and his ragged troops traveling to their winter camp at Valley Forge.

Interact *with* History

It is 1777. Your brother is an American soldier. In his last letter to you, he wrote that the army has no shoes or bullets and little food. But he plans to keep fighting.

Now, a British army is coming toward your farm. You hear that they are stealing crops to feed themselves and their horses.

What Do You Think?

- What sacrifices do civilians like Mrs. Schuyler make during wartime?
- What sacrifices do soldiers make?
- Is it worth such sacrifices to win independence for your country? Why or why not?

What would you sacrifice to win freedom?

USA

1776
British capture New York, but Americans win a battle in New Jersey.

1777
Battles of Saratoga convince Europeans that America might win the war.

1778
France enters the war on the American side.

1779
American frontier soldiers take a British fort in what is now Indiana.

1780
Moving south, the British capture Charles Town (Charleston).

1781
In the war's last big battle, the Americans and French defeat the British at Yorktown.

1783
Treaty of Paris ends the war.

World

1776
Scotsman Adam Smith publishes a book saying that government shouldn't control the economy.

1778
Voltaire, who wrote about the rights of people, dies in Paris.

1781
Emperor Joseph II of Austria ends serfdom, which forced peasants to farm someone else's land for life.

1783
Simón Bolívar, who will become a revolutionary leader in South America, is born in what is now Venezuela.

Chapter 7 SETTING THE STAGE

BEFORE YOU READ

This Patriot has climbed a flagpole to tear down the British flag and replace it with an American one.

Previewing the Theme

Citizenship During the Revolution, many Americans saw themselves as citizens of a new country—the United States. They made great sacrifices and risked their lives and fortunes to win independence. This chapter explains how those Patriots overcame enormous odds to defeat the powerful British Empire.

What Do You Know?

What stories do you know about the people or events of the Revolution? How do people display courage and self-sacrifice during wartime?

THINK ABOUT

- what you've learned about the American Revolution from books, movies, and other classes
- news stories you've heard about revolutions or civil wars in other countries today

What Do You Want to Know?

What would you like to learn about the steps that people took to win the American Revolution? In your notebook, record what you hope to learn from this chapter.

READ AND TAKE NOTES

Reading Strategy: Sequencing Events To sequence is to put events in the order in which they happened. You learned this skill in Chapter 6 by sequencing the events that led to the American Revolution. Now as you read Chapter 7, practice sequencing again. Put the major battles and events of the war in order by recording them on a time line. Copy the time line below in your notebook. You may want to make it bigger.

See Skillbuilder Handbook, page R4.

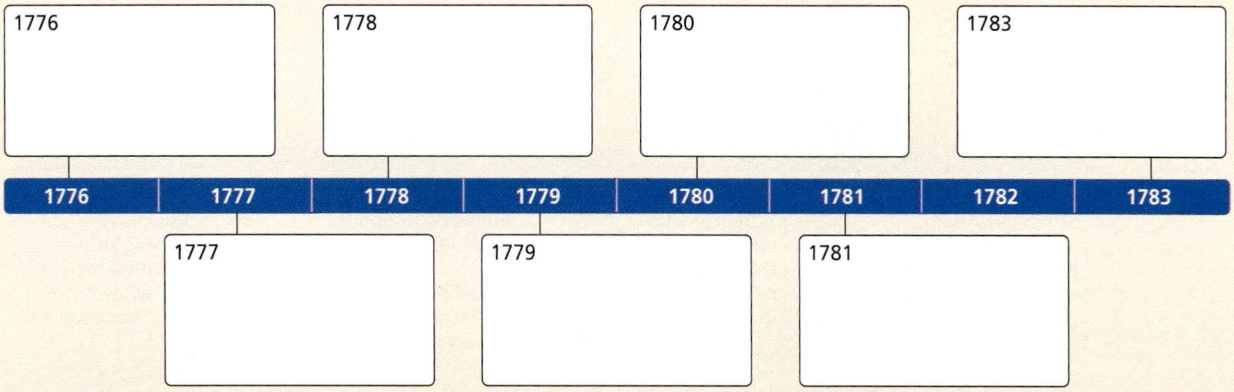

1 The Early Years of the War

TERMS & NAMES
George Washington
mercenary
strategy
rendezvous
Battles of Saratoga

MAIN IDEA

The American desire to gain rights and liberties led them to fight for independence from Britain.

WHY IT MATTERS NOW

Today those same rights and liberties are protected by the U.S. Constitution.

ONE AMERICAN'S STORY

In search of liberty, Haym Salomon moved from eastern Europe to America sometime between 1764 and 1775. He was a Jew from Poland. Arriving in New York City, Salomon soon became a successful merchant and banker. After the war broke out, Salomon supported the Patriot cause.

When the British captured New York in 1776, many Patriots fled but Salomon stayed. The British arrested him as a spy. Salomon spoke many languages. The British thought he could help their supply officers deal with foreign merchants, so they let him out of prison. Salomon used this opportunity to help other prisoners escape.

In 1778, the British wanted to arrest Salomon again, so he fled to Philadelphia. His earlier time in the cold, damp prison had permanently damaged his health. Even so, he continued to aid the Patriots. He loaned the new government more than $600,000, which was never repaid.

Like Salomon, many people made hard choices about which side to support during the Revolutionary War. This section discusses those choices and the obstacles Americans faced in the war's early years.

Haym Salomon sacrificed his health and his fortune to help his new country.

Americans Divided

The issue of separating from Great Britain divided American society. Opinion polls did not exist in the 1700s, so we don't know exactly how many people were on each side. But historians estimate that roughly 20 to 30 percent of Americans were Loyalists, roughly 40 to 45 percent were Patriots, and the rest remained neutral. Most Americans did not support the Revolution.

Both Patriots and Loyalists came from all walks of life and all parts of America. In general, New England and Virginia had high numbers of Patriots. Loyalists were numerous in cities, in New York State, and in the

Choosing Sides

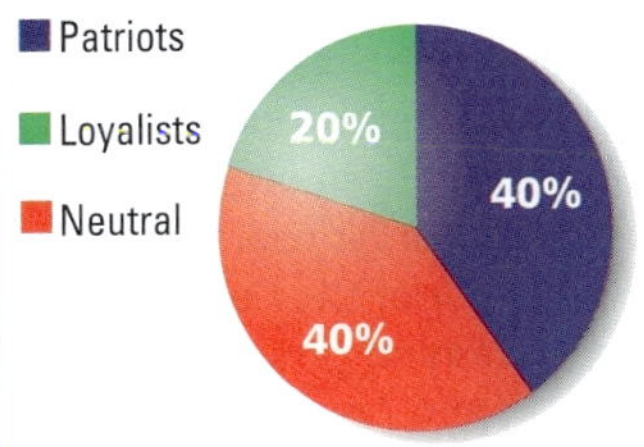

Source: *Blackwell Encyclopedia of the American Revolution*

South. Many Loyalists worked for the British government or were clergy in the Church of England. Some Quakers were Loyalists, although many wanted peace. (Their faith taught that war was wrong.)

The war divided Native Americans, too. For instance, some Iroquois nations fought with the British and others with the Americans. Those Native Americans who joined the British feared that if the Americans won, they would take Native American land. Some Native Americans who lived near colonists and interacted with them sided with the Americans.

Background
The Iroquois League had generally kept peace among the Iroquois nations for about 200 years—until the American Revolution.

African Americans also fought on both sides. At first, slave owners feared that African Americans who had guns might lead slave revolts. Therefore, few states allowed African Americans to enlist, or sign up with the army. Then a British governor offered freedom to any enslaved person who joined the British army. Many slaves ran away to fight for the British. In response, most states began to accept African-American soldiers. In all, about 5,000 African Americans served in the Continental Army. Many African Americans who did so hoped that American independence would bring greater equality.

Differences over the war split families, too. For example, Benjamin Franklin's son William took Britain's side. The father and son stopped speaking.

AMERICA'S HISTORY MAKERS

GEORGE WASHINGTON
1732–1799

At the age of 16, George Washington worked as a surveyor, setting land boundaries in the wilderness. He learned to handle hardship by hunting for food and sleeping outdoors.

In the French and Indian War, Washington had many brushes with death. Yet he wrote, "I heard the bullets whistle, and, believe me, there is something charming in the sound."

That war made him the most famous American officer. People loved him for his courage. As commander of the Continental Army, Washington's popularity helped unite Americans.

How did Washington's early life prepare him to lead the army?

Creating an Army

Because not everyone supported the war, raising an army was difficult. The army also faced other problems. In June 1775, **George Washington** became the commander of the Continental Army. At first, this new national army was formed from state militias, made up of untrained and undisciplined volunteers.

Reading History
A. Analyzing Causes Why would the lack of agreement about the war make it hard to raise an army?

After Congress created the Continental Army, men began to enlist, but most of them didn't stay long. At the start of the war, Congress asked men to enlist only for one year. Later Congress did lengthen the term of service. When the soldiers' time was up, they went home. As a result, Washington's army never numbered more than 17,000 men.

Congress's inability to supply the army also frustrated Washington. The soldiers needed everything—blankets, shoes, food, and even guns and ammunition. Angrily, Washington wrote, "Could I have foreseen what I have, and am likely to experience, no consideration upon earth should have induced [persuaded] me to accept this command."

Many women tried to help the army. Martha Washington and other wives followed their husbands to army camps. The women cooked, did laundry, and nursed sick or wounded soldiers. A few women even helped fight. Mary Hays earned the nickname "Molly Pitcher" by carrying water to tired soldiers during a battle. Deborah Sampson dressed as a man, enlisted, and fought in several engagements.

Reading **History**

B. Solving Problems How did Washington try to solve the problem of leading a small, inexperienced force against a large professional army?

Building an army was crucial to Washington's plan. To the British, the Americans were disorganized, inexperienced rebels. The British thought that if they won a decisive battle, the Americans would give up. By contrast, Washington's main goal was to survive. To do so, he needed to keep an army in the field, win some battles—no matter how small—and avoid a crushing defeat. He knew he could not hope to win a major battle until he had a large, well-equipped army.

These Hessian boots weighed about 12 pounds a pair.

Struggle for the Middle States

As Chapter 6 explains, Washington had forced the British to retreat from Boston in March 1776. He then hurried his army to New York City, where he expected the British to go next. One British goal was to occupy coastal cities so that their navy could land troops and supplies in those cities. From there, they could launch their military campaigns.

Washington's hunch was correct. In July 1776, Britain's General William Howe arrived in New York with a large army. Then in August, more soldiers arrived, including about 9,000 Hessian mercenaries. A **mercenary** is a professional soldier hired to fight for a foreign country. British soldiers usually signed up for life—which discouraged enlistment. So Britain needed mercenaries, whom it hired from the German states.

Background

The British-hired mercenaries came from a part of Germany called Hesse, which is the origin of the term *Hessian.*

For several months, the British and American armies fought for New York State. Finally, the British forced Washington to retreat through New Jersey. By December, when the American army crossed the Delaware River into Pennsylvania, it was in terrible condition. Charles Willson Peale, a Philadelphia painter who watched the crossing, saw one muddy soldier who "had lost all his clothes. He was in an old, dirty blanket jacket, his beard long, and his face so full of sores he could not clean it." To Peale's shock, the soldier called his name. He was Peale's brother!

Political writer Thomas Paine also witnessed the hard conditions and the soldiers' low spirits on the retreat. To

GEOGRAPHY SKILLBUILDER Interpreting Maps

1. **Place** *In what state did the American victories take place?*
2. **Movement** *How did the British general Howe travel from New York to Brandywine?*

This famous painting of Washington crossing the Delaware River is inaccurate. Instead of the boats shown here, Washington used flat-bottomed boats that were 40- to 60-feet long.

urge them to keep fighting, Paine published the first in a series of pamphlets called *The American Crisis.*

A VOICE FROM THE PAST

These are the times that try men's souls. The summer soldier and the sunshine patriot will, in this crisis, shrink from the service of their country; but he that stands it *now,* deserves the love and thanks of man and woman.

Thomas Paine, *The American Crisis*

Reading **History**

C. Recognizing Propaganda How does this passage promote the American cause?

Washington hoped a victory would encourage his weary men. He also knew that he must attack the British quickly because most of his soldiers would leave once their enlistments ended on December 31.

Late on December 25, 1776, Washington's troops rowed across the icy Delaware River to New Jersey. From there, they marched in bitter, early-morning cold to Trenton to surprise the Hessians, some of whom were sleeping after their Christmas celebration. The Americans captured or killed more than 900 Hessians and gained needed supplies. Washington's army won another victory at Princeton eight days later. These victories proved that the American general was better than the British had thought. The American army began to attract new recruits.

Britain's Strategy

Meanwhile, the British were pursuing a **strategy**—an overall plan of action—to seize the Hudson River Valley. If successful, they would cut off New England from the other states. The strategy called for three armies to meet at Albany, New York. General John Burgoyne would lead a force south from Canada. Lieutenant Colonel Barry St. Leger would lead his army from Lake Ontario down the Mohawk Valley. Burgoyne expected General Howe to follow the Hudson north from New York City.

Burgoyne left Canada in June 1777 with an army that included British, Hessians, and Iroquois. In July, they captured Fort Ticonderoga.

Called "Gentleman Johnny" by his soldiers, Burgoyne enjoyed traveling slowly and throwing parties to celebrate victories. After Ticonderoga, his delays gave the Americans time to cut down trees to block his route. They also burned crops and drove off cattle, leaving the countryside bare of supplies for the British troops.

Things grew rougher during the last 25 miles of Burgoyne's march to Albany. On a map, the route looked easy, but it really crossed a swampy wilderness. The army had to build bridges and roads. Burgoyne took four weeks to reach the Hudson. Still confident, he looked forward to the **rendezvous,** or meeting, with St. Leger and Howe in Albany.

GEOGRAPHY SKILLBUILDER Interpreting Maps

1. **Place** *From which two cities did the British invade the United States?*
2. **Movement** *What did St. Leger want to capture by taking the longer route by way of Lake Ontario?*

On August 4, Burgoyne received a message from Howe. He would not be coming north, Howe wrote, because he had decided to invade Pennsylvania to try to capture General Washington and Philadelphia—where the Continental Congress met. "Success be ever with you," wrote Howe. Yet Burgoyne needed Howe's soldiers, not his good wishes.

Reading **History**
D. Evaluating Review Howe's two goals for his invasion of Pennsylvania. Was he successful?

Howe did invade Pennsylvania. In September 1777, he defeated but did not capture Washington at the Battle of Brandywine. Howe then occupied Philadelphia. In October, Washington attacked Howe at Germantown. Washington lost the battle, however, and retreated.

Battles Along the Mohawk

As Burgoyne received Howe's message, St. Leger faced his own obstacle in reaching Albany. In the summer of 1777, he was trying to defeat a small American force at Fort Stanwix in the Mohawk River valley of New York. St. Leger's forces included Iroquois led by Mohawk chief Joseph Brant, also called Thayendanegea (THĪ•ehn•DAHG•ee).

The Iroquois chief Joseph Brant was a British ally.

Brant and his sister, Molly, had strong ties to the British. Molly was a British official's wife, and Joseph was a convert to the Church of England. Both Joseph and Molly tried to convince the Iroquois to fight for the British, who upheld Iroquois rights to their land.

Vocabulary
convert: a person who changes religions

During August 1777, American general Benedict Arnold led a small army up the Mohawk River. He wanted to chase the British away from

Fort Stanwix. Arnold sent a captured Loyalist and some Iroquois who were American allies to spread the rumor that he had a large army.

The trick worked. St. Leger's troops were afraid they were about to be outnumbered. The army retreated so fast that it left behind tents, cannon, and supplies. Because of St. Leger's flight and Howe's refusal to follow the strategy, no one was left to rendezvous with Burgoyne.

Saratoga: A Turning Point

By this time, Burgoyne's army was running out of supplies, and it needed horses. The general sent a raiding party into Vermont to see what it could find. The raiding party encountered New England troops, who badly defeated it at the Battle of Bennington on August 16, 1777.

Despite this setback, Burgoyne's army headed slowly toward Albany. On the way, it met a powerful Continental Army force led by General Horatio Gates. They were waiting on a ridge called Bemis Heights, near Saratoga, New York. There the Americans had created fortifications, or built-up earthen walls, behind which to fight. The Polish engineer Tadeusz Kosciuszko (TAH•deh•oosh KAWSH•choosh•kaw) had helped the Americans do this.

ReadingHistory
E. Reading a Map Find Saratoga on the map on page 197. Notice how close it is to Burgoyne's goal of Albany.

Burgoyne would have to break through the fortifications to proceed to Albany. On September 19, he attacked. While Gates commanded the Americans on the ridge, Benedict Arnold led an attack on nearby

CITIZENSHIP TODAY

Exercising Free Speech

The British could have charged Thomas Paine with a crime for writing *The American Crisis.* The crime was sedition, or stirring up rebellion. By saying what he thought, Paine risked going to prison. Today U.S. citizens have the right to speak freely without fear of jail.

Like Thomas Paine, some students have used free speech to urge people to take action. For example, the Sidney Lanier Middle School in Houston, Texas, has published its school newspaper on the Internet. In October 1996, one writer urged other students to get involved in that year's election, saying, "Even though you will not be able to vote yet, you can still influence your parents to do so."

These students are working together to produce a school newspaper.

How Do You Exercise Free Speech?

1. Working in a small group, choose an issue that you care about. Look through newsmagazines for ideas.
2. Use a cluster diagram to record your feelings and opinions about the issue.
3. As a group, decide what action you think people should take on the issue.
4. Write an article expressing the group's opinion. Each member should read the article and suggest changes. Revise the article.
5. Send the revised article to the editorial page of your school or local newspaper.

See Citizenship Handbook, page 282.

Visit www.mcdougallittell.com to learn more about free speech.

Freeman's Farm. His men repeatedly charged the British and inflicted heavy casualties. Still, the British held their position.

On October 7, another battle broke out. Again Arnold led daring charges against the British. Although hundreds of muskets were firing at him, he galloped through the battlefield "like a madman," a sergeant later said. Frightened, Burgoyne's Hessian mercenaries began to fall back. Eventually, a bullet tore into Arnold's leg and stopped him. Even so, the Americans forced Burgoyne to retreat.

Burgoyne's army moved slowly through heavy rain to a former army camp at Saratoga. By the time they arrived, the men were exhausted. Some fell in the mud and slept in their wet uniforms. The Continental Army then surrounded Burgoyne's army and fired on it day and night without stopping. Burgoyne decided to surrender. The series of conflicts that led to this surrender is known as the **Battles of Saratoga**.

The Battles of Saratoga had two very different consequences. As Benedict Arnold was recovering from his wound, he married a woman who was a Loyalist. Over time, Arnold came to feel that Congress had not rewarded him enough for his heroic actions at Saratoga and other battles. Influenced by his bitterness and his wife, he betrayed his army. In 1780, he agreed to turn over an American fort to the British. Although his plot was discovered before he could carry it out, he escaped. Even today, the name *Benedict Arnold* is used to mean traitor.

On the positive side, the victory at Saratoga was a turning point in the Revolution. It caused European nations to think that the Americans might win their war for independence. As you will read in Section 2, several European nations decided to help America in its struggle.

America's HERITAGE

THE FIRST FLAG

June 14 is Flag Day in the United States. On June 14, 1777, the Continental Congress adopted the stars and stripes design for the U.S. flag. According to legend, a Philadelphia seamstress named Betsy Ross designed the first flag, illustrated below. Historians have found no evidence to support this legend. However, Ross did make flags for the Pennsylvania navy.

***Reading* History**

F. Forming and Supporting Opinions Which of the two consequences of the Battles of Saratoga was more significant? Why?

Section 1 Assessment

1. Terms & Names

Identify:
- George Washington
- mercenary
- strategy
- rendezvous
- Battles of Saratoga

2. Taking Notes

Use a cluster diagram like the one shown to list the difficulties Americans faced in the early years of the war.

Which difficulty do you think was hardest to overcome?

3. Main Ideas

a. How were Americans divided over the issue of separating from Great Britain?

b. Why was it difficult for George Washington to form and keep a large army?

c. How did the Battles of Saratoga mark a turning point in the war?

4. Critical Thinking

Contrasting How did the British and American strategies differ during the early years of the war?

THINK ABOUT
- what the British expected from the Americans
- Washington's main goals for the American army
- why Burgoyne invaded from Canada

ACTIVITY OPTIONS

LANGUAGE ARTS
ART

Learn more about a Revolutionary War leader. Write a brief **biography** or create a **trading card** with a picture on one side and important facts on the other.

2 The War Expands

TERMS & NAMES
ally
Marquis de Lafayette
bayonet
desert
privateer
James Forten
John Paul Jones

MAIN IDEA
Some Europeans decided to help America. As the war continued, it spread to the sea and the frontier.

WHY IT MATTERS NOW
This was the beginning of the United States' formal relationships with other nations.

ONE AMERICAN'S STORY

To defeat the mighty British Empire, the United States needed a foreign ally. An **ally** is a country that agrees to help another country achieve a common goal. The ideal ally would share America's goal of defeating Britain. It also had to be able to provide money, troops, and ships. So the United States turned to France—Britain's long-time enemy.

In the fall of 1776, Congress sent Benjamin Franklin to the French capital, Paris. His job was to persuade France to be the ally of the United States. Franklin was already famous for his experiments with electricity. When he reached Paris, he became a celebrity. He wrote to his daughter, saying that medallions with his likeness were popular there.

Franklin's simple Quaker coat and fur hat amused the French. The clothes fit the image they had of him—a wise, noble man from a wild country.

A VOICE FROM THE PAST

These, with the pictures, busts [sculptures of the head and shoulders], and prints (of which copies upon copies are spread everywhere), have made your father's face as well known as that of the moon.

Benjamin Franklin, letter to his daughter Sally

In spite of his popularity, Franklin couldn't convince the French to become America's formal ally until after the victory at Saratoga. Then the French agreed to an alliance. This section explains how the war expanded after foreign allies joined the American side.

Help from Abroad

France was still bitter over its defeat by Britain in the French and Indian War, in which France lost its North American colonies. The French hoped to take revenge on the British by helping Britain's American colonies break free. In 1776, France began to give secret aid to the Americans. However, the French didn't want to lose to Britain a second time. That is why they didn't publicly ally themselves with the United States until after the Americans had proved they could win battles.

After hearing of the American victory at Saratoga, King Louis XVI of France recognized U.S. independence. In 1778, France signed two treaties of alliance with the United States. By doing so, France went to war with Britain. As part of its new alliance, France sent badly needed funds, supplies, and troops to America.

In 1779, France persuaded its ally Spain to help the Americans. Spain was also Britain's rival. The Spanish governor of Louisiana, General Bernardo de Gálvez, acted quickly. He captured the British strongholds of Natchez and Baton Rouge in the lower Mississippi Valley.

Background
Galveston, Texas, is named for Gálvez.

From there, his small army went on to take Mobile, and in 1781 Pensacola in West Florida. These victories prevented the British from attacking the United States from the southwest. In addition, Britain had to keep thousands of troops fighting Gálvez—instead of fighting the Americans. However, like France, Spain's motives were not simply to help the United States. Gálvez's victories helped extend Spain's empire in North America.

Reading **History**
A. Recognizing Effects How did America's allies prevent Britain from focusing all its might on the Americans?

By entering the war on America's side, France and Spain forced the British to fight a number of enemies on land and sea. The British had to spread their military resources over many fronts. For example, they were afraid they might have to fight the French in the West Indies, so they sent troops there. This prevented the British from concentrating their strength to defeat the inexperienced Americans.

Europeans Help Washington

The Americans gained some of the military experience they needed from Europe. Several European military officers came to Washington's aid, including men from France, Poland, and the German states.

Background
Many of these European officers were professional soldiers looking for an army that would hire them. Some, like Lafayette, also believed in the American cause.

The **Marquis de Lafayette** (LAF•ee•EHT) was a 19-year-old French nobleman who volunteered to serve in Washington's army. He wanted a military career, and he believed in the American cause. He quickly gained Washington's confidence and was given the command of an army division. Lafayette won respect and love from his men by sharing their hardships. Called "the soldier's friend," he used his own money to buy warm clothing for his ragged troops. Washington regarded him almost as a son.

Lafayette fought in many battles and also persuaded the French king to send a 6,000-man army to America. He became a hero in both France and the United States. Later he took part in France's own revolution.

Along with Lafayette came the Baron de Kalb, a German officer who had served in the French army. He became one of Washington's generals and earned a reputation for bravery. In 1780, he received 11 wounds in the Battle of Camden and died.

Another German, Baron von Steuben, helped turn the inexperienced Americans

Lafayette stands with the slave James Armistead, whose owner allowed him to spy for Lafayette. After the war, the state of Virginia set Armistead free. Armistead then took Lafayette's last name as his own.

into a skilled fighting force. Washington asked him to train the army. In 1778, Steuben began by forming a model company of 100 men. Then he taught them how to move in lines and columns and how to handle weapons properly. Under Steuben's direction, the soldiers practiced making charges with **bayonets**—long steel knives attached to the ends of guns. Within a month, the troops were executing drills with speed and precision. Once the model company succeeded, the rest of the army adopted Steuben's methods.

Winter at Valley Forge

Help from Europeans came at a time when the Americans desperately needed it. In late 1777, Britain's General Howe forced Washington to retreat from Philadelphia. Beginning in the winter of 1777–1778, Washington and his army camped at Valley Forge in southeast Pennsylvania.

On the march to Valley Forge, Washington's army was so short on supplies that many soldiers had only blankets to cover themselves. They also lacked shoes. The barefoot men left tracks of blood on the frozen ground as they marched. The soldiers' condition did not improve at camp. The Marquis de Lafayette described what he saw.

A VOICE FROM THE PAST

The unfortunate soldiers were in want of everything; they had neither coats, nor hats, nor shirts, nor shoes; their feet and their legs froze till they grew black and it was often necessary to amputate them. . . . The Army frequently passed whole days without food.

Marquis de Lafayette, quoted in *Valley Forge: Pinnacle of Courage*

***Reading* History**

B. Identifying Facts and Opinions Is Lafayette mainly recounting facts or expressing opinions? Explain.

Because of this, the name *Valley Forge* came to stand for the great hardships that Americans endured in the Revolutionary War. Over the winter, the soldiers at Valley Forge grew weak from not having enough food or warm clothing. Roughly a quarter of them died from malnutrition, exposure to the cold, or diseases such as smallpox and typhoid fever.

daily life

CAMP LIFE IN WINTER

At Valley Forge, soldiers slept in small huts, 12 men to a hut. They slept in shifts so they could take turns using the scarce blankets. The men also shared clothing. If one went on guard duty, the others lent him their clothes and stayed by the fire in the hut until he came back. Guards had to stand in old hats to keep their shoeless feet warm.

The soldiers cooked on hot stones, in iron kettles, or on portable iron braziers. Often the only food they had was fire cakes—a bread made of flour and water paste.

These iron kettles were so heavy that soldiers often threw them away on a march.

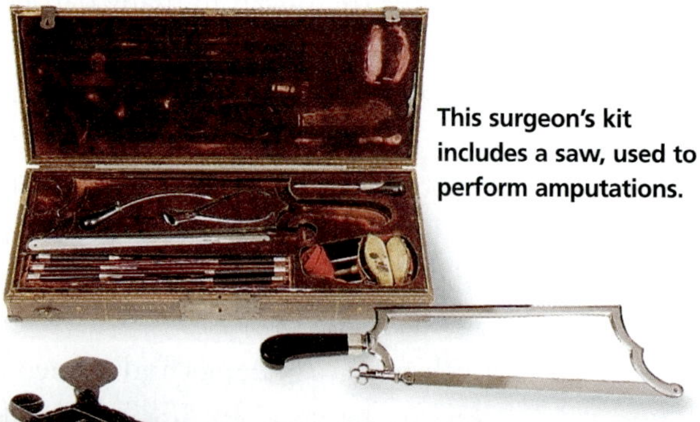

This surgeon's kit includes a saw, used to perform amputations.

Soldiers would place burning coals in braziers like this. Braziers were used to cook food and heat huts.

Washington appealed to Congress to send the soldiers supplies, but it was slow in responding. Luckily, private citizens sometimes came to the soldiers' aid. According to one story, on New Year's Day 1778, a group of Philadelphia women drove ten teams of oxen into camp. The oxen were pulling wagons loaded with supplies and 2,000 shirts. The women had the oxen killed to provide food for the men.

Despite the hardships, Washington and his soldiers showed amazing endurance. Under such circumstances, soldiers often **desert,** or leave military duty without intending to return. Some soldiers did desert, but Lieutenant Colonel John Brooks wrote that the army stayed together because of "Love of our Country." The men also stayed because of Washington. Private Samuel Downing declared that the soldiers "loved him. They'd sell their lives for him."

Reading **History**
C. Analyzing Points of View
What are two different explanations for why American soldiers did not desert?

War on the Frontier

Elsewhere, other Americans also took on difficult challenges. In 1777, a 24-year-old frontiersman named George Rogers Clark walked into the office of Virginia's governor, Patrick Henry. Clark said he had come to take part in defending the Western frontier. He lived in Kentucky, which was claimed by Virginia. Clark wanted Virginia to defend that region against British soldiers and their Native American allies in what is now Indiana and Illinois. "If a country is not worth protecting," he said, "it is not worth claiming."

"If a country is not worth protecting, it is not worth claiming."
George Rogers Clark

Background
In the late 1700s, the Western frontier was the region between the Appalachian Mountains and the Mississippi River.

Clark was difficult to ignore. He stood six feet tall, had red hair, and displayed a dramatic personality. He persuaded Governor Henry that he was right. The governor told Clark to raise an army to capture British posts on the Western frontier.

In May of 1778, Clark and a group of frontiersmen began to travel down the Ohio River. He recruited others on the way, until he had a force of 175 to 200. They went by boat and later on foot to Kaskaskia, a British post on the Mississippi River. They captured Kaskaskia without a fight.

Then they moved east to take Fort Sackville at Vincennes, in present-day Indiana. Earlier, a small force sent by Clark had taken Vincennes, but British forces under Henry Hamilton had recaptured it. Settlers called Hamilton the "Hair Buyer" because he supposedly paid rewards for American scalps.

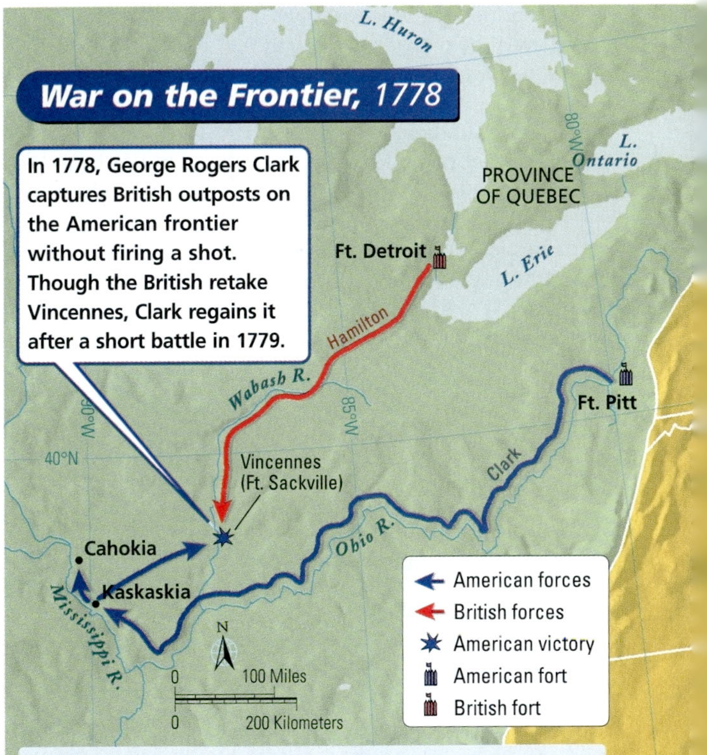

GEOGRAPHY SKILLBUILDER Interpreting Maps
1. **Movement** *From what fort did British general Hamilton travel to Vincennes?*
2. **Region** *What rivers form the boundaries of the region captured by Clark and his men?*

Determined to retake Fort Sackville, Clark and his men set out for Vincennes from Kaskaskia in February 1779. Hamilton wasn't expecting an attack because the rivers were overflowing their banks and the woods were flooded. Clark's men slogged through miles of icy swamps and waded through chest-deep water. They caught the British at Vincennes by surprise.

When Hamilton and his troops tried to remain in the fort, Clark pretended to have a larger force than he really had. He also found a way to frighten the British into leaving. Clark and his men had captured several Native Americans, who were allies of the British and had American scalps on their belts. Clark executed some of them in plain view of the fort. He promised to do the same to Hamilton and his men if they didn't surrender immediately. The British gave up.

Reading **History**
D. Making Decisions What factors do you think influenced Clark's decision to execute his prisoners?

Clark's victory gave the Americans a hold on the vast region between the Great Lakes and the Ohio River. This area was more than half the total size of the original 13 states. However, Fort Detroit on Lake Erie remained in the hands of the British.

STRANGE *but* True

THE FIRST COMBAT SUBMARINE

During the Revolution, the Americans built the first combat submarine—the *Turtle,* shown below. It held only a pilot, who steered with one hand and cranked a propeller with the other. To submerge, the pilot used a foot pump to let in water.

In 1776, the *Turtle* failed on its mission to attach a bomb to a British warship in New York harbor. It reached the ship but couldn't drill through its copper-clad hull. The *Turtle* failed at later missions, too.

War at Sea

The war expanded not only to the frontier but also to the sea. By 1777, Britain had about 100 warships off the American coast. This allowed Britain to control the Atlantic trade routes. There was no way the Americans could defeat the powerful British navy.

Vocabulary
merchant ship: a ship used in trade

But American privateers attacked British merchant ships. A **privateer** is a privately owned ship that a wartime government gives permission to attack an enemy's merchant ships. After capturing a British merchant ship, the crew of a privateer sold its cargo and shared the money. As a result, a desire for profit as well as patriotism motivated privateers. The states and Congress commissioned more than 1,000 privateers to prey on the British. During the war, they captured hundreds of British ships. This disrupted trade, causing British merchants to call for the war to end.

Many men answered the privateers' call for volunteers. Among them was 14-year-old **James Forten,** who was the son of a free African-American sail maker. In 1780, Forten signed up to sail on the *Royal Louis* to earn money for his family after his father died. When a British ship captured the *Royal Louis* in 1781, the British offered Forten a free trip to England. Reportedly, Forten refused, saying he would never betray his country. Released from a British prison after the war, Forten walked barefoot from New York to his home in Philadelphia. He later became famous for his efforts to end slavery.

A Naval Hero

Though outnumbered, the Continental Navy scored several victories against the British. An officer named **John Paul Jones** won the most famous sea battle.

In 1779, Jones became the commander of a ship named *Bonhomme Richard.* With four other ships, he patrolled the English coast. In September, Jones's vessels approached a convoy in which two British warships were guarding a number of supply ships.

Vocabulary
convoy: a group of ships traveling together for safety

Jones closed in on the *Serapis,* the larger of the two warships. At one point, the *Bonhomme Richard* rammed the better-armed British vessel. As the two ships locked together, the confident British captain demanded that Jones surrender. In words that have become a famous U.S. Navy slogan, Jones replied, "I have not yet begun to fight!"

The two warships were so close together that the muzzles of their guns almost touched. They blasted away, each seriously damaging the other. On the shore, crowds of Britons gathered under a full moon to watch the fighting. After a fierce three-and-a-half-hour battle, the main mast of the *Serapis* cracked and fell. The ship's captain then surrendered. The *Bonhomme Richard* was so full of holes that it eventually sank, so Jones and his crew had to sail away in the *Serapis*!

Jones's success against the best navy in the world angered the British and inspired the Americans. Even so, the Americans knew that the war had to be won on land. The next section discusses the major land battles in the closing years of the war.

AMERICA'S HISTORY MAKERS

JOHN PAUL JONES
1747–1792

The most famous naval officer of the Revolution is known by a fake name. He was born John Paul in Scotland and first went to sea as a 12-year-old. By age 21, he had command of a merchant ship.

In 1773, Paul killed the leader of a mutiny on his ship. To avoid a murder trial, he fled to America and added Jones to his name.

Bold and daring, Jones scored many victories against the British. But his battle with the *Serapis* is what earned his place in history.

What are the two words that you think best describe Jones's character? Explain.

Section 2 Assessment

1. Terms & Names

Identify:

- ally
- Marquis de Lafayette
- bayonet
- desert
- privateer
- James Forten
- John Paul Jones

2. Taking Notes

Use this diagram to list the effects of the entry of France and Spain into the war.

Cause: France and Spain enter the war.

Effect | Effect | Effect

3. Main Ideas

a. What role did Benjamin Franklin play in helping America win the Revolution?

b. How did European officers such as Lafayette aid America in the Revolutionary War?

c. What was John Paul Jones's major contribution during the war, and why was it important?

4. Critical Thinking

Analyzing Points of View
Why do you think George Rogers Clark thought the frontier was important to defend?

THINK ABOUT

- why General Hamilton was called "Hair Buyer"
- why America might have wanted the frontier region after the war

ACTIVITY OPTIONS

ART
SPEECH

Imagine yourself at Valley Forge in the winter of 1777–1778. Create a **comic strip** or give a **talk** describing your response to the harsh conditions.

The Path to Victory

TERMS & NAMES
Lord Cornwallis
guerrillas
pacifist
Battle of Yorktown

MAIN IDEA
Seeking Loyalist support, the British invaded the South—but ultimately lost the war there.

WHY IT MATTERS NOW
For more than two centuries, the American Revolution has inspired other people to fight tyranny.

ONE AMERICAN'S STORY

Patriot Nancy Hart glared at the five armed Loyalists who burst into her Georgia cabin. Tradition says that the men had shot her last turkey and ordered her to cook it for them. Raids like this were common in the South, where feuding neighbors used the war as an excuse to fight each other. Both Patriots and Loyalists took part in the raids. Many women and children had moved out of Georgia, but the six-foot-tall, freckled Hart chose to stay and fight. She could shoot a gun as accurately as any man.

As she prepared the food, Hart planned her attack. When dinner was ready, the men sat down to eat. Seizing one of their muskets, Hart quickly shot and killed one man and wounded another. She kept the gun aimed on the others as her daughter ran for help. A group of nearby Patriots arrived and hanged the Loyalists.

As Nancy Hart's story demonstrates, the fighting between Patriots and Loyalists in the South was vicious. In this section, you will learn why the British war effort shifted to the South and why it failed.

The state of Georgia named a county after Nancy Hart, who is shown here holding Loyalists prisoner.

Savannah and Charles Town

The British believed that most Southerners were Loyalists. Because of this, in 1778 the British decided to move the war to the South. After three years of fighting in the North, the British were no closer to victory. Although they had captured Northern cities, they couldn't control the countryside because they did not have enough troops to occupy it. The British believed that if they gained territory in the South, Southern Loyalists would hold it for them.

The British also expected large numbers of Southern slaves to join them because they had promised to grant the slaves freedom. Although thousands of African Americans did run away to join the British, not all of them were set free. Instead, some British officers sold African Americans into slavery in the West Indies.

Reading **History**

A. Drawing Conclusions Why was it an advantage to be able to move troops between the West Indies and the South?

Britain's West Indian colonies were a third reason the British invaded the South. Southern seaports were closer to the West Indies, where British troops were stationed. If the British captured Southern ports, they could move troops back and forth between the two regions.

In December 1778, the British captured the port of Savannah, Georgia. Using Savannah as a base, they then conquered most of Georgia. In 1780, a British army led by General Henry Clinton landed in South Carolina. They trapped American forces in Charles Town (now Charleston), which was the largest Southern city. When the city's 5,000 defenders surrendered, the Americans lost almost their entire Southern army. It was the worst American defeat of the war.

The Swamp Fox and Guerrilla Fighting

After that loss, Congress assigned General Horatio Gates—the victor at Saratoga—to form a new Southern army. Continental soldiers led by Baron de Kalb formed the army's core. Gates added about 2,000 new and untrained militia. He then headed for Camden, South Carolina, to challenge the army led by the British general **Lord Cornwallis.**

On the way, a band of Patriots from South Carolina approached Gates. "Their number did not exceed 20 men and boys, some white, some black, and all mounted, but most of them miserably equipped," wrote an officer. Their leader was Francis Marion, called the "Swamp Fox." He provided Gates with helpful knowledge of South Carolina's coastal swamplands. Gates sent Marion to destroy boats on the Santee River behind Camden. (See the map on page 209.) This would cut off British communications with Charles Town.

In August 1780, Gates's army ran into British troops outside Camden. The Americans were in no condition to fight. They were out of supplies and half-starved. Even worse, Gates put the inexperienced militia along part of the frontline instead of behind the veterans. When the British attacked, the militia panicked and ran. Gates also fled, but Kalb remained with his soldiers and received fatal wounds. This second defeat in the South ended Gates's term as head of an army and caused American spirits to fall to a new low.

Reading **History**

B. Evaluating What do you think was most responsible for the American loss at Camden?

After Camden, a small British force set out for Charles Town with a column of American prisoners. Marion's band overwhelmed the British and freed the prisoners. Fighting from a base in the swamps, Marion's men cut the British supply line that led inland and north from Charles Town. Marion used the methods of a guerrilla. **Guerrillas** are small bands of fighters who weaken the enemy with surprise raids and hit-and-run attacks. Both Patriots and Loyalists formed guerrilla bands in the South. They carried out vicious raids.

Now *and* then

BATTLE TACTICS

A difference in battle tactics affected future warfare. British soldiers marched shoulder to shoulder in three rows. When they neared the enemy, the first row knelt, the second crouched, and the third remained standing. They all fired without aiming.

The Americans were better shots. They often marched in rows, but sometimes they hid in woods or behind walls to take aim. Guerrillas attacked swiftly, then fled into the countryside. Those tactics succeeded and are still used today. This photograph shows modern U.S. soldiers learning guerrilla tactics.

Artillery of the Revolution

Artillery—large guns and cannon—played a key role in the American Revolution. The ability of these guns to kill and destroy from a distance made them essential in war. One witness of a battle described the destruction: "Many men were badly injured and mortally wounded by the fragments of bombs, . . . their arms and legs severed or themselves struck dead." Most cannon used in the Revolution were made of cast bronze. During the 1700s, artillery design did not change significantly. However, artillery became more mobile (more easily moved).

After each shot, a soldier sponged the inside of the barrel. This put out sparks and cleaned away any dirt left from the last shot.

A soldier loaded the cannon with gunpowder and a cannonball. He did so by ramming them down the barrel.

Soldiers aimed the gun by turning the entire carriage. An instrument called a quadrant told them how high to raise the barrel to reach their target.

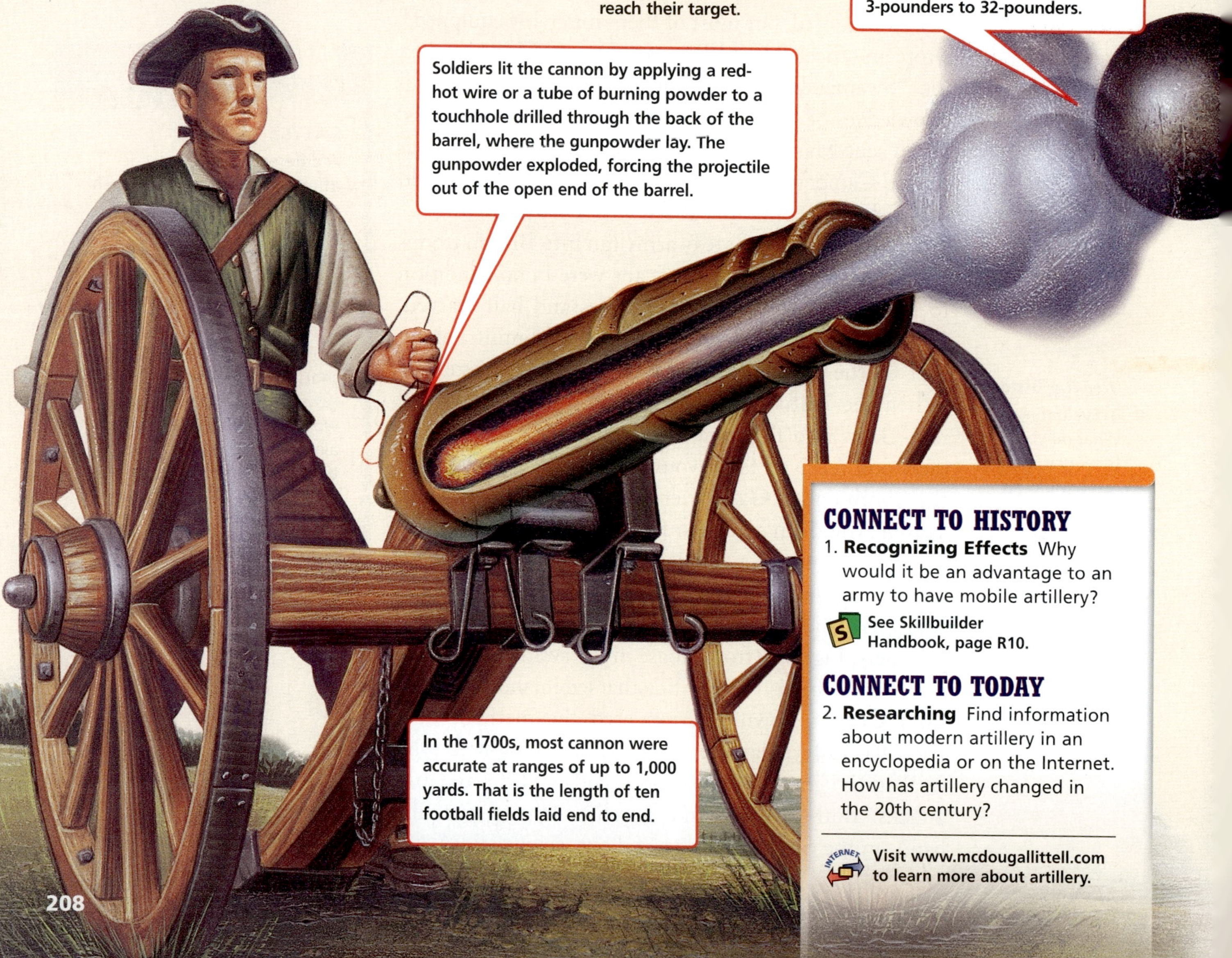

Cannon were classified by the weight of the iron ball they fired. American artillery ranged from 3-pounders to 32-pounders.

Soldiers lit the cannon by applying a red-hot wire or a tube of burning powder to a touchhole drilled through the back of the barrel, where the gunpowder lay. The gunpowder exploded, forcing the projectile out of the open end of the barrel.

In the 1700s, most cannon were accurate at ranges of up to 1,000 yards. That is the length of ten football fields laid end to end.

CONNECT TO HISTORY

1. **Recognizing Effects** Why would it be an advantage to an army to have mobile artillery?

See Skillbuilder Handbook, page R10.

CONNECT TO TODAY

2. **Researching** Find information about modern artillery in an encyclopedia or on the Internet. How has artillery changed in the 20th century?

Visit www.mcdougallittell.com to learn more about artillery.

The Tide Turns

Even battles in the South sometimes turned vicious. One example was the Battle of Kings Mountain, fought on the border of North and South Carolina in October 1780. After surrounding a force of about 1,000 Loyalist militia and British soldiers, the Americans slaughtered most of them. James P. Collins, a 16-year-old American, described the scene.

A VOICE FROM THE PAST

The dead lay in heaps on all sides, while the groans of the wounded were heard in every direction. I could not help turning away from the scene before me with horror and, though exulting in victory, could not refrain from shedding tears.

James P. Collins, quoted in *The Spirit of Seventy-Six*

Many of the dead had been shot or hanged after they surrendered. The Americans killed them in revenge for Loyalist raids and an earlier incident in which the British had butchered Americans. Kings Mountain was one of Britain's first losses in the South. It soon suffered more.

After Gates's defeat at Camden, Washington put a new general, Nathanael Greene, in charge of the Southern army. Greene was one of Washington's most able officers. He had been a Quaker, but his church had cast him out because of his belief in the armed struggle against the British. Most Quakers are **pacifist,** or opposed to war.

Under Greene's command, the American army avoided full-scale battles, in which the British had the edge because of superior firepower. So the American forces let the British chase them around the countryside and wear themselves out. When the Americans did fight, they did their best to make sure the British suffered heavy losses.

Reading **History**

C. Contrasting How did Greene's strategy as a general differ from that of Gates at Camden?

As the fighting dragged on into its sixth year, opposition to the war grew in Britain. As a result, some British leaders began to think that American independence would not be so bad.

The End of the War

In 1781, most of the fighting took place in Virginia. In July of that year, the British general Cornwallis set up his base at Yorktown, located on a peninsula in Chesapeake Bay. From there, his army could receive supplies by ship from New York.

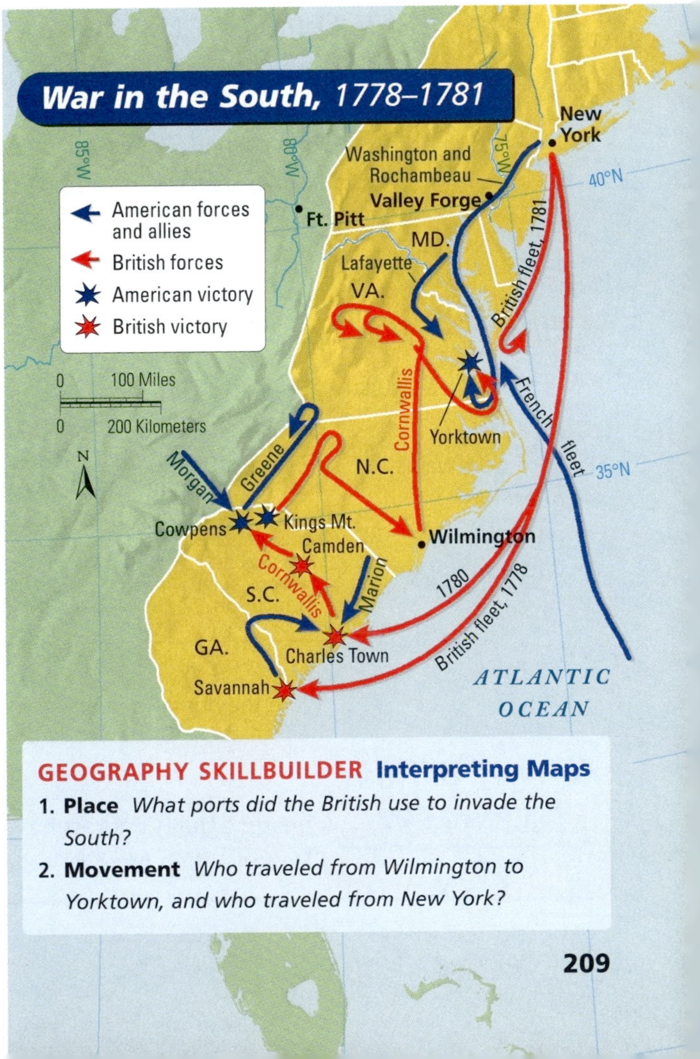

War in the South, *1778–1781*

GEOGRAPHY SKILLBUILDER Interpreting Maps

1. **Place** *What ports did the British use to invade the South?*
2. **Movement** *Who traveled from Wilmington to Yorktown, and who traveled from New York?*

The victorious American forces accept the British surrender at Yorktown. George Washington is to the left of the American flag.

Washington saw Cornwallis's decision as a golden opportunity. In August 1781, a large French fleet arrived from the West Indies and blocked Chesapeake Bay. These ships prevented the British from receiving supplies—and from escaping. They also allowed Washington to come from the North and trap Cornwallis on the peninsula. Washington had enough men to do this because a large French force led by General Jean Rochambeau had joined his army.

Washington and Rochambeau moved south. When British ships tried to reach Cornwallis, French ships drove them back. In the **Battle of Yorktown,** the American and French troops bombarded Yorktown with cannon fire, turning its buildings to rubble. Cornwallis had no way out. On October 19, 1781, he surrendered his force of about 8,000.

Vocabulary
bombard: to attack with artillery

Although some fighting continued, Yorktown was the last major battle of the war. When the British prime minister, Lord North, heard the news, he gasped, "It is all over!" Indeed, he and other British leaders were soon forced to resign. Britain's new leaders began to negotiate a peace treaty, which is discussed in the next section.

Section 3 Assessment

1. Terms & Names

Identify:
- Lord Cornwallis
- guerrillas
- pacifist
- Battle of Yorktown

2. Taking Notes

Use a chart like the one below to list the geographic factors that made the British move their war effort to the South.

Physical factors, such as location	Human factors, such as who lived there

Were the human factors as helpful as the British hoped?

3. Main Ideas

a. Why did the fighting between Patriots and Loyalists in the South turn vicious?

b. What type of warfare did Francis Marion and his men employ?

c. How did Gates's errors in leadership contribute to the American loss at Camden?

4. Critical Thinking

Analyzing Causes How did each of the following help bring about the British defeat at Yorktown?

THINK ABOUT
- the location chosen by Cornwallis
- the French fleet
- the French troops under Rochambeau
- Washington's planning

ACTIVITY OPTIONS

TECHNOLOGY
MUSIC

Imagine that Congress has asked you to commemorate the Battle of Yorktown. Design a **Web page** or write a **song** celebrating the U.S. victory.

4 The Legacy of the War

TERMS & NAMES
Treaty of Paris of 1783
republicanism
Elizabeth Freeman
Richard Allen

MAIN IDEA

After the war, the new nation faced issues such as a high national debt and calls for equality.

WHY IT MATTERS NOW

To promote liberty, some states passed laws outlawing slavery and protecting religious freedom.

Although this painting is not of Joseph Plumb Martin himself, he may have dressed like this American soldier.

ONE AMERICAN'S STORY

In 1776, 15-year-old Joseph Plumb Martin of Connecticut signed up to fight for the Americans. He stayed with the army until the war ended and rose in rank from private to sergeant. Among his experiences were the terrible winter at Valley Forge and the winning battle at Yorktown.

One of the hardest things Martin faced was leaving the army after the war was over. Many years later, he wrote about that day.

A VOICE FROM THE PAST

There was as much sorrow as joy. . . . We had lived together as a family of brothers for several years, setting aside some little family squabbles, like most other families, had shared with each other the hardships, dangers, and sufferings incident to a soldier's life; had sympathized with each other in trouble and sickness; had assisted in bearing each other's burdens. . . . And now we were to be . . . parted forever.

Joseph Plumb Martin, quoted in *The Revolutionaries*

At war's end, Martin and his country faced an uncertain future. How would the United States recover from the war? What issues would confront the new nation? Section 4 discusses those questions.

Why the Americans Won

In November 1783, the last British ships and troops left New York City, and American troops marched in. As Washington said good-bye to his officers in a New York tavern, he hugged each one. Tears ran down his face. He became so upset that he had to leave the room.

Earlier in the fall, Washington had written a farewell letter to his armies. In it, he praised them by saying that their endurance "through almost every possible suffering and discouragement for the space of eight long years, was little short of a standing miracle."

By their persistence, the Americans won independence even though they faced many obstacles. As you have read, they lacked training and experience. They were often short of supplies and weapons. By contrast, the British forces ranked among the best trained in the world. They were experienced and well-supplied professional soldiers. Yet the Americans had certain advantages that enabled them to win.

1. **Better leadership.** British generals were overconfident and made poor decisions. By contrast, Washington learned from his mistakes. After early defeats, he developed the strategy of dragging out the war to wear down the British. Despite difficulties, he never gave up.
2. **Foreign aid.** Britain's rivals, especially France, helped America. Foreign loans and military aid were essential to America's victory.
3. **Knowledge of the land.** The Americans knew the land where the war took place and used that knowledge well. The British could control coastal cities but could not extend their control to the interior.
4. **Motivation.** The Americans had more reason to fight. At stake were not only their lives but also their property and their dream of liberty.

Reading **History**

A. Evaluating What do you think was Washington's best characteristic as a leader?

The Treaty of Paris

As the winners, the Americans won favorable terms in the **Treaty of Paris of 1783,** which ended the Revolutionary War. The treaty included the following six conditions:

1. The United States was independent.
2. Its boundaries would be the Mississippi River on the west, Canada on the north, and Spanish Florida on the south.
3. The United States would receive the right to fish off Canada's Atlantic Coast, near Newfoundland and Nova Scotia.
4. Each side would repay debts it owed the other.
5. The British would return any enslaved persons they had captured.
6. Congress would recommend that the states return any property they had seized from Loyalists.

Neither Britain nor the United States fully lived up to the treaty's terms. Americans did not repay the prewar debts they owed British merchants or return Loyalist property. For their part, the British did not return

HISTORY *through* ART

The American painter Benjamin West began a portrait of the men who negotiated the Treaty of Paris. But the British officials refused to pose, so West never finished the painting. From left to right are the American officials John Jay, John Adams, Benjamin Franklin, and two others.

What does this painting reveal about the British response to losing the war?

runaway slaves. They also refused to give up military outposts in the Great Lakes area, such as Fort Detroit.

Background
Even after George Rogers Clark's Western victories, the British stayed at Fort Detroit.

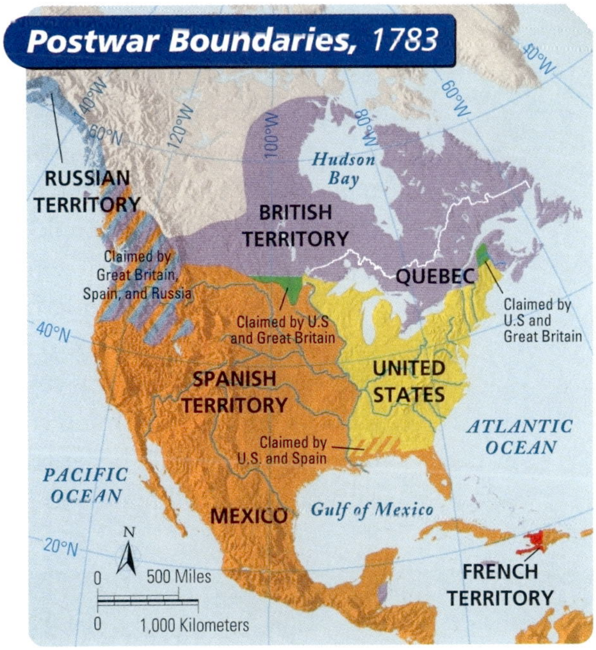

Costs of the War

No one knows exactly how many people died in the war, but eight years of fighting took a terrible toll. An estimated 25,700 Americans died in the war, and 1,400 remained missing. About 8,200 Americans were wounded. Some were left with permanent disabilities, such as amputated limbs. The British suffered about 10,000 military deaths.

Many soldiers who survived the war left the army with no money. They had received little or no pay for their service. Instead of back pay, the government gave some soldiers certificates for land in the West. Many men sold that land to get money for food and other basic needs.

Both the Congress and the states had borrowed money to finance the conflict. The war left the nation with a debt of about $27 million—a debt that would prove difficult to pay off.

The losers of the war also suffered. Thousands of Loyalists lost their property. Between 60,000 and 100,000 Loyalists left the United States during and after the war. Among them were several thousand African Americans and Native Americans, including Joseph Brant. Most of the Loyalists went to Canada. There they settled new towns and provinces. They also brought English traditions to areas that the French had settled. Even today, Canada has both French and English as official languages.

*Reading***History**
B. Analyzing Causes Why do you think the Loyalists left the United States?

CONNECTIONS TO MATH

Military Deaths in the American Revolution

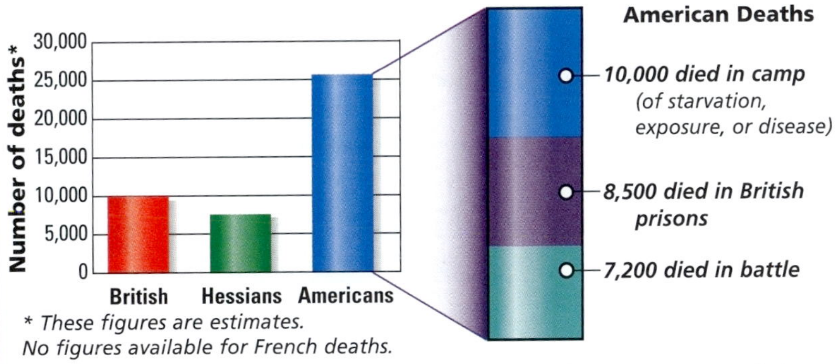

*These figures are estimates.
No figures available for French deaths.

Sources: *World Book Encyclopedia; An Outline History of the American Revolution*

SKILLBUILDER Interpreting Charts
1. *How many more deaths did the Americans suffer than the British?*
2. *What percentage of American deaths occurred in battle?*

Issues After the War

The American Revolution was not just a war, but a change in ideas about government. Before the war, Americans had demanded their rights as English citizens. But after declaring their independence, they replaced that goal with the idea of **republicanism.** This idea stated that instead of a king, the people would rule. The government would obtain its authority from the citizens and be responsible to them.

For this system to work, individuals would have to place the good of the country above their own interests. At first, only men were allowed to take part in governing by voting or holding public office—and not even all men. However, women could help the nation by teaching their children the virtues that benefited public life. Such virtues included honesty, duty, and the willingness to make sacrifices.

Economics *in* History

Free Enterprise

One cause of the Revolution was the colonists' resentment of British mercantilism. Parliament passed laws to discourage the colonists from developing their own manufacturing and to force them to buy British goods. During the war, British economic control weakened. British exports of woolens to the colonies dropped from £645,900 in 1774 to only £2,540 in 1776. As a result, the colonists were able to make more economic choices—for example, they could choose to manufacture wool clothing.

The end of Britain's mercantilist control allowed free enterprise to begin to develop in the United States. In a free-enterprise system, business can be conducted freely based on the choices of individuals. The government does not control the system, but only protects and regulates it.

CONNECT TO HISTORY

1. **Analyzing Causes** Why do you think the colonists were able to manufacture their own wool clothing during the war?

See Skillbuilder Handbook, page R10.

CONNECT TO TODAY

2. **Comparing** Think about a mall where you shop. Name examples of businesses that compete with each other. Compare the methods they use to attract customers.

Visit www.mcdougallittell.com to learn more about free enterprise.

- **A** Competition encourages businesses to improve goods and services and to keep prices down.
- **B** Property is owned by individuals and businesses.
- **C** The desire to make a profit motivates businesspeople.
- **D** Individuals, not the government, decide what to buy and what to manufacture and sell.
- **E** The government protects private property and makes sure businesses operate fairly.

As part of their liberty, Americans called for more religious freedom. Before the war, some laws discriminated against certain religions. Some states had not allowed Jews or Catholics to hold public office. After the war, states began to abolish those laws. They also ended the practice of using tax money to support churches.

Background
Only Northern states ended slavery after the war. In the North, slavery was not as important a part of the economy as in the South.

Many people began to see a conflict between slavery and the ideal of liberty. Vermont outlawed slavery, and Pennsylvania passed a law to free slaves gradually. Individual African Americans also tried to end slavery. For example, **Elizabeth Freeman** sued for her freedom in a Massachusetts court and won. Her victory in 1781 and other similar cases ended slavery in that state. Freeman later described her desire for freedom.

Elizabeth Freeman fought a court case that helped end slavery in Massachusetts.

A VOICE FROM THE PAST

Anytime while I was a slave, if one minute's freedom had been offered to me, and I had been told I must die at the end of that minute, I would have taken it—just to stand one minute on God's earth a free woman.

Elizabeth Freeman, quoted in *Notable Black American Women*

Reading **History**
C. Solving Problems How did free African Americans take on the responsibility of trying to improve their lives?

With freedom, African Americans began to form their own institutions. For example, the preacher **Richard Allen** helped start the Free African Society. That society encouraged African Americans to help each other. Allen also founded the African Methodist Episcopal Church, the first African-American church in the United States.

Perhaps the main issue facing Americans after the war was how to shape their national government. American anger over British taxes, violation of rights, and control of trade had caused the war. Now the United States needed a government that would protect citizens' rights and economic freedom. In Chapter 8, you will read how U.S. leaders worked to create such a government.

Section 4 Assessment

1. Terms & Names

Identify:
- Treaty of Paris of 1783
- republicanism
- Elizabeth Freeman
- Richard Allen

2. Taking Notes

Use a chart like the one below to classify the terms of the Treaty of Paris according to which side they favored. (Do not list terms that don't favor either side.)

Terms of the Treaty of Paris	
Favorable to America	Favorable to Britain

3. Main Ideas

a. What advantages helped the Americans win the Revolutionary War?

b. How did the end of the war affect Loyalists?

c. What were the economic costs of the war to individuals and to the government?

4. Critical Thinking

Recognizing Effects How did republicanism shape the United States after the war?

THINK ABOUT
- American ideas about government
- the roles men and women could play in public life
- religious freedom
- the antislavery movement

ACTIVITY OPTIONS
SPEECH
MATH

Look up the U.S. population in 1780. Calculate what percentage of American people died in the war. Report your findings in a **speech** or a **pie graph.**

Chapter 7 ASSESSMENT

TERMS & NAMES

Briefly explain the importance of each of the following.

1. George Washington
2. mercenary
3. Battles of Saratoga
4. ally
5. Marquis de Lafayette
6. John Paul Jones
7. Lord Cornwallis
8. Battle of Yorktown
9. Treaty of Paris of 1783
10. republicanism

REVIEW QUESTIONS

The Early Years of the War (pages 193–199)

1. What motives led African Americans to fight for the British? The Americans?
2. How did women help the American war effort?
3. What events led to the British defeat at Saratoga?

The War Expands (pages 200–205)

4. What foreign countries helped America? How?
5. What were conditions like at Valley Forge?

The Path to Victory (pages 206–210)

6. What two Southern ports did the British capture?
7. How did America's ally France contribute to the victory at Yorktown?

The Legacy of the War (pages 211–215)

8. For what did Washington praise his army in his farewell letter?
9. What land did the United States acquire from Britain as a result of the Treaty of Paris?
10. What three states outlawed slavery after the war?

CRITICAL THINKING

1. USING YOUR NOTES

Using your completed time line, answer the questions below.

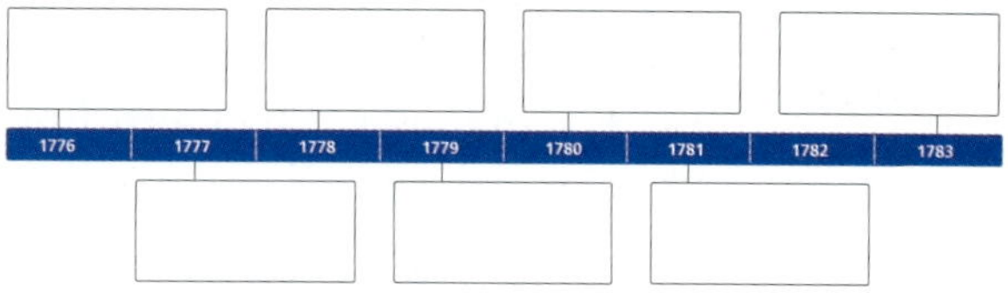

a. What were the main events of 1776 and 1777?
b. While George Rogers Clark was capturing Kaskaskia, what was happening in the South?

2. ANALYZING LEADERSHIP

George Washington was the most beloved American leader of his time. What qualities do you think made him such a respected leader?

3. THEME: CITIZENSHIP

What Revolutionary leaders displayed civic virtue by putting the good of the nation ahead of their own interests? Explain your answer.

4. RECOGNIZING EFFECTS

How did Britain's loss in the war allow free enterprise to develop in the United States?

5. APPLYING CITIZENSHIP SKILLS

How was the writing of *The American Crisis* an example of good U.S. citizenship?

Interact *with* History

How did the sacrifices you discussed before you read the chapter compare with what Patriots really did?

VISUAL SUMMARY

The American Revolution

People and Events of the Revolution

Military		Civilian	
George Washington	commanded the Continental Army.	Haym Salomon	helped finance the war for America.
Marquis de Lafayette	fought for the Americans.	Molly Pitcher	aided soldiers by bringing them water in battle.
John Burgoyne	surrendered to the Americans at Saratoga.	Thomas Paine	wrote *The American Crisis* to inspire Americans.
John Paul Jones	won a major naval victory for America.	Benjamin Franklin	was a diplomat to France and Britain.
George Rogers Clark	helped hold the Western frontier for America.	James Forten	was captured by the British but would not betray America.
Lord Cornwallis	surrendered at Yorktown, ending the war.	Nancy Hart	defended her Georgia home against Loyalist raiders.

HISTORY SKILLS

1. INTERPRETING MAPS: Movement

Study the map, then answer the questions.

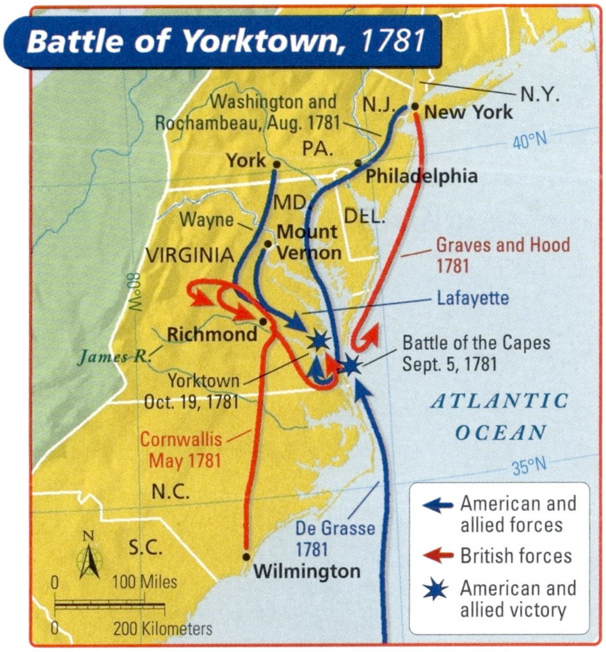

Basic Map Elements

a. In what state was the Battle of Yorktown fought?

b. In what battle did De Grasse's fleet defeat the British fleet led by Graves and Hood?

Interpreting the Map

c. Describe the route Washington took to reach Yorktown.

d. What British force tried to come to Cornwallis's aid?

2. INTERPRETING PRIMARY SOURCES

Read the passage below, which is a longer version of the quotation you read on page 196.

> These are the times that try men's souls. The summer soldier and the sunshine patriot will, in this crisis, shrink from the service of their country; but he that stands it *now,* deserves the love and thanks of man and woman. Tyranny, like hell, is not easily conquered; yet we have this consolation with us, that the harder the conflict, the more glorious the triumph. What we obtain too cheap, we esteem too lightly: it is dearness only that gives everything its value.
>
> **Thomas Paine,** *The American Crisis*

a. In your own words, explain who "the summer soldier and the sunshine patriot" are.

b. What does Paine promise will be the reward of a hard conflict?

ALTERNATIVE ASSESSMENT

1. INTERDISCIPLINARY ACTIVITY: Science

Creating a Poster Two of the diseases that killed American soldiers in camp were typhoid fever and smallpox. Do research to learn the following about one of these diseases: its cause, its symptoms, its treatment, and how common it is today. Use the information to create a public health poster to inform people about the disease.

2. COOPERATIVE LEARNING ACTIVITY

Performing a Talk Show In a small group, prepare to hold a talk show in which the guests discuss which side to take in the Revolutionary War. One member of your group should be the talk show host. The others should play the roles of various Americans. Each person should explain the reasons for his or her position. Perform the talk show before the class. Students might choose from the following roles:

a. the wife of an American soldier

b. an enslaved African American

c. an Iroquois chief

d. a Quaker minister

e. an employee of the British government

3. TECHNOLOGY ACTIVITY

Creating a Multimedia Presentation Use the Internet, books, and other reference materials to create a multimedia presentation on one of the major battles of the Revolution. Consider including the following:

- paintings of the conflict and photographs of artifacts
- quotations from participants and witnesses
- music of the time period
- recorded sound effects
- graphs showing battle statistics

Show your presentation to the class.

Visit www.mcdougallittell.com to learn more about Revolutionary battles.

4. HISTORY PORTFOLIO

Option 1 Review your section and chapter assessment activities. Select one that you think is your best work. Then use comments made by your teacher or classmates to improve your work. Add the completed work to your portfolio.

Option 2 Review the questions that you wrote for What Do You Want to Know? on page 192. Then write a short report in which you explain the answers to your questions. If any questions were not answered, do research to answer them. Add your report to your portfolio.

CHAPTER 8

Confederation to Constitution

1776–1791

Delegates to the Constitutional Convention in 1787 gathered in Philadelphia. They held their meetings in the Pennsylvania State House, now known as Independence Hall.

Interact *with* History

The year is 1787, and your young country needs to reform its government. Now everyone is wondering what the new government will be like. You have been called to a convention to decide how the new government should be organized.

What Do You Think?

- What will be your main goal in creating a new government?
- How will you get the people at the convention to agree on important issues?

How do you form a government?

USA

1776

1777 Patriots win Battles of Saratoga.

Continental Congress passes the Articles of Confederation.

1781 Articles of Confederation go into effect.

British surrender at Yorktown.

1783 Treaty of Paris formally ends the Revolutionary War.

1786–1787 Daniel Shays leads a rebellion of Massachusetts farmers.

1787 Constitutional Convention is held in Philadelphia.

1788 U.S. Constitution is ratified.

1789 George Washington becomes the first president of the United States.

1791 Bill of Rights is ratified.

1791

World

1779 Spain declares war on Britain.

1781 Joseph II allows religious freedom for Christians in Austria.

1787 Freetown, Sierra Leone, is made a home for freed slaves.

1789 French Revolution begins.

BEFORE YOU READ

Previewing the Theme

Democratic Ideals Between 1776 and 1791, the United States struggled to set up a national government. The Articles of Confederation established the first federal government. Chapter 8 explains how the weaknesses of the Articles led Americans to write a new constitution for the United States.

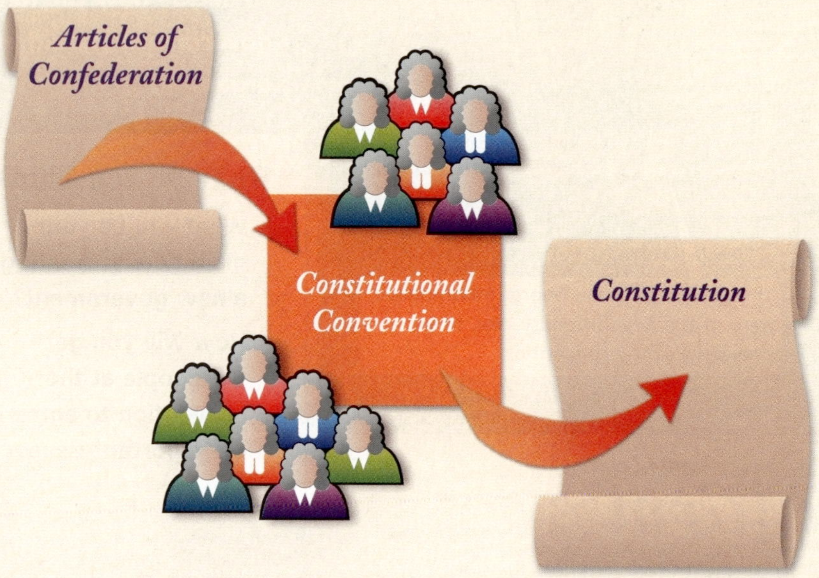

What Do You Know?

What do you think of when people talk about the U.S. government? Why do nations have governments? What does the U.S. government do?

THINK ABOUT

- what you've learned about the U.S. government from the news or your teachers
- what the purpose of a government is
- how the government affects your everyday life

What Do You Want to Know?

What questions do you have about the creation of the U.S. Constitution? Write those questions in your notebook before you read the chapter.

READ AND TAKE NOTES

Reading Strategy: Solving Problems When you read history, look for how people solved problems they faced in the past. Copy the chart below in your notebook. Use it to identify the methods that Americans used to solve the problems faced by the nation after declaring its independence.

See Skillbuilder Handbook, page R17.

Problems	Solutions
Western lands	
Postwar depression	
Representation in the new government	
Slavery	

1 The Confederation Era

TERMS & NAMES
- Wilderness Road
- republic
- Articles of Confederation
- Land Ordinance of 1785
- Northwest Territory
- Northwest Ordinance
- Shays's Rebellion

MAIN IDEA

The Articles of Confederation were too weak to govern the nation after the war ended.

WHY IT MATTERS NOW

The weakness of the Articles of Confederation led to the writing of the U.S. Constitution.

ONE AMERICAN'S STORY

In 1775, Daniel Boone and 30 woodsmen cut a road over the Appalachian Mountains into Kentucky. They hacked through brush, chopped down trees, and bridged creeks. They labored like this for about 250 miles. Eventually, they arrived in a grassy meadow along the banks of the Kentucky River. Felix Walker, a member of Boone's party, described what they saw.

A VOICE FROM THE PAST

On entering the plain we were permitted to view a very interesting and romantic sight. A number of buffaloes . . . supposed to be between two and three hundred, made off . . . in every direction. . . . Such a sight some of us never saw before, nor perhaps ever may again.

Felix Walker, quoted in *The Life and Adventures of Daniel Boone*

Early travel to Kentucky is shown in this detail of *Daniel Boone Escorting Settlers Through the Cumberland Gap* (1851–1852) by George Caleb Bingham.

Boone was one of the earliest American settlers in Kentucky. In the late 1700s, most Americans thought of Kentucky as the wild frontier. Some, like Boone, looked at the frontier and saw a world of opportunity. Exploring and governing these lands was only one of the many challenges that faced the new government of the United States.

Moving West

The trail into Kentucky that Daniel Boone helped build was called the **Wilderness Road.** This road was not easy to travel. It was too narrow for carts or wagons, but it became the main road into Kentucky. The settlers came on foot or on horseback. Settlers were drawn to Kentucky's rich river valleys, where few Native Americans lived. But some Native Americans, such as the Shawnee, did live, hunt, and fish in the area.

Tensions between Native Americans and settlers led to violent confrontations. But the settlers did not stop coming. By the early 1790s, about 100,000 Americans lived there. While settlers headed into the Western territories, the people in the East began to create new state governments.

New State Governments

Once the American colonies declared independence, each of the states set out to create its own government. The framers, or creators, of the state constitutions did not want to destroy the political systems that they had had as colonies. They simply wanted to make those systems more democratic. Some states experimented with creating separate branches of government, giving different powers to different branches. By creating separate branches, Americans hoped to prevent the government from becoming too powerful.

Background
Two states, Connecticut and Rhode Island, kept their old colonial charters as their constitutions. The other 11 states wrote new constitutions.

Some states included a bill of rights in their constitutions as a way to keep the government under control. The idea of a bill of rights came from the English Bill of Rights of 1689. This was a list of rights that the government guaranteed to English citizens.

Although not all the states had a bill of rights, all of them did have a republican form of government. In a **republic,** the people choose representatives to govern them.

The Articles of Confederation

While the states were setting up their governments, Americans also discussed the form of their national government. During the Revolutionary War, Americans realized that they needed to unite to win the war against Britain. As Silas Deane, a diplomat from Connecticut, wrote, "United we stand, divided we fall."

"United we stand, divided we fall."
Silas Deane

In 1776, the Continental Congress began to develop a plan for a national government. Congress agreed that the government should be a republic. But the delegates disagreed about whether each state should have one vote or voting should be based on population. They also disagreed about whether the national government or the individual states should control the lands west of the Appalachians.

The Continental Congress eventually arrived at a final plan, called the **Articles of Confederation.** In the Articles, the national government had few powers, because many Americans were afraid that a strong government would lead to tyranny, or oppressive rule. The national government was run by a Confederation Congress. Each state had only one vote in the Congress. The national government had the power to wage war, make peace, sign treaties, and issue money.

But the Articles left most important powers to the states. These powers included the authority to set taxes and enforce national laws. The Articles proposed to leave the states in control of the lands west of the Appalachian Mountains.

Reading **History**
A. Reading a Map Look at the map on page 223 to see which states claimed territories in the West.

The Continental Congress passed the Articles of Confederation in November 1777. It then sent the Articles to the states for ratification, or approval. By July 1778, eight states had ratified the Articles. But some of the small states that did not have Western land claims refused to sign.

These states felt that unless the Western lands were placed under the control of the national government, they would be at a disadvantage. The states with Western lands could sell them to pay off debts left from the Revolution. But states without lands would have difficulty paying off the high war debts.

Over the next three years, all the states gave up their claims to Western lands. This led the small states to ratify the Articles. In 1781, Maryland became the 13th state to accept the Articles. As a result, the United States finally had an official government.

Reading **History**

B. Finding Main Ideas Why did the states without Western land claims want the other states to give up their claims?

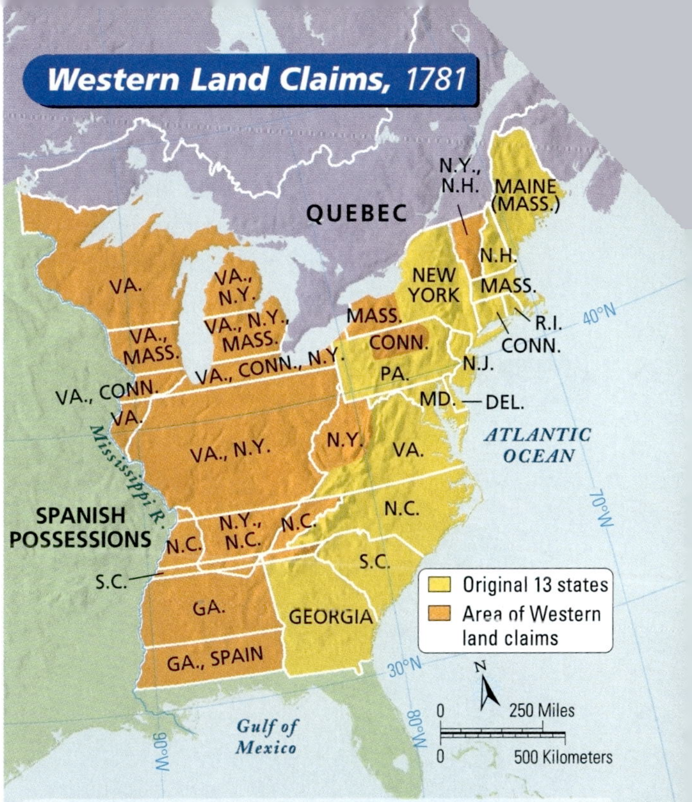

GEOGRAPHY SKILLBUILDER Interpreting Maps

1. **Location** *Which of the original 13 states had Western land claims?*
2. **Location** *To what geographic feature did the Western land claims extend?*

The Northwest Ordinance

One of the most important questions that the Confederation Congress faced was what to do with the Western lands that it now controlled. Congress passed important laws on how to divide and govern these lands—the Land Ordinance of 1785 and the Northwest Ordinance (1787). (See Geography in History on pages 226–227.)

The **Land Ordinance of 1785** called for surveyors to stake out six-mile-square plots, called townships, in the Western lands. These lands later became known as the **Northwest Territory.** The Northwest Territory included land that formed the states of Ohio, Indiana, Michigan, Illinois, and Wisconsin and part of Minnesota.

The **Northwest Ordinance** (1787) described how the Northwest Territory was to be governed. As the territory grew in population, it would gain rights to self-government. When there were 5,000 free males in an area, men who owned at least 50 acres of land could elect an assembly. When there were 60,000 people, they could apply to become a new state.

The Northwest Ordinance also set conditions for settlement in the Northwest Territory and outlined the settlers' rights. Slavery was outlawed, and the rivers were to be open to navigation by all. Freedom of religion and trial by jury were guaranteed.

The Northwest Ordinance was important because it set a pattern for the orderly growth of the United States. As the nation grew, it followed this pattern in territories added after the Northwest Territory.

Background

According to the Northwest Ordinance, Native Americans were to be treated fairly, and their lands were not to be taken from them.

Weaknesses of the Articles

Aside from its handling of land issues, however, the Confederation Congress had few successes. By the end of the Revolutionary War, the United States faced serious problems, and the Confederation Congress did not have enough power to solve them.

U.S. Government, 1776–1787

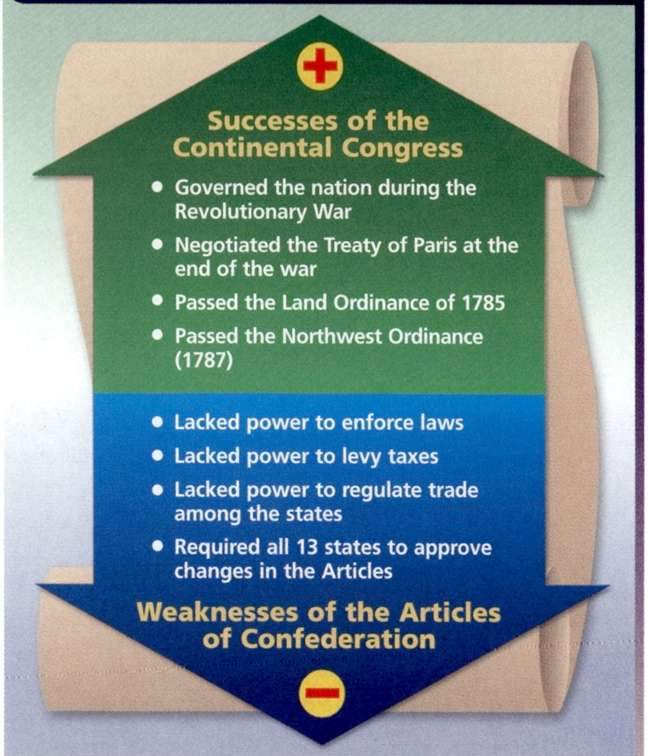

SKILLBUILDER Interpreting Graphs

1. *What do you think was the greatest success of the Continental Congress?*
2. *What do you think was the greatest weakness of the Articles of Confederation?*

Debt was a critical problem for the government. Congress had borrowed large sums to pay for the Revolutionary War. Much of that money was owed to soldiers of its own army. Upset at not being paid, several hundred soldiers surrounded the Pennsylvania State House where Congress was meeting in June 1783. The soldiers threatened the legislators, thrusting their bayonets through the windows. The delegates were forced to flee the city. The event was a clear sign of Congress's weakness.

***Reading*History**
C. Analyzing Causes How did debt cause problems for the U.S. government under the Articles of Confederation?

Even if Congress wanted to pay the soldiers, it did not have the power to levy taxes. The national government depended on the states to send money to Congress. But the states sent very little money.

Congress was not alone in facing economic crises. People throughout the nation faced hard times. In Massachusetts, the economy was so bad that people rose up in arms against the government.

Shays's Rebellion

In the mid-1780s, Massachusetts faced economic problems, as did other states. People had little money, but the state continued to levy high taxes. The average family owed $200 in taxes per year—more money than most farmers made. Many Massachusetts farmers fell deeply into debt. Debt laws at the time were strict. Anyone who could not repay his debts would have his property auctioned off. If the auction didn't raise enough money to settle the debts, the debtor could be put in jail. In western Massachusetts, many jails were packed with debtors.

Farmers asked the Massachusetts legislature to provide debt relief. But the legislature refused—and the farmers rebelled. One of the leaders of the rebellion was a Revolutionary War veteran named Daniel Shays. He commanded a group of about 1,500 men.

Shays's rebels take over a Massachusetts courthouse. A stone marker rests on the spot of the rebellion.

In January 1787, Shays and his men marched on a federal arsenal, a place to store weapons. The arsenal was defended by 900 soldiers from the state militia. The militia quickly defeated Shays's men. But even though the militia put down **Shays's Rebellion,** as the uprising came to be known, the farmers won the sympathy of many people. America's leaders realized that an armed uprising of common farmers spelled danger for the nation.

Background
In 1788, Daniel Shays was pardoned for his actions.

Some leaders hoped that the nation's ills could be solved by strengthening the national government. In the next section, you'll read how Americans held a convention to change the Articles of Confederation.

Section 1 Assessment

1. Terms & Names

Identify:
- Wilderness Road
- republic
- Articles of Confederation
- Land Ordinance of 1785
- Northwest Territory
- Northwest Ordinance
- Shays's Rebellion

2. Taking Notes

Use a diagram like the one below to list some of the challenges Americans faced in shaping a new government.

Which challenge do you think was the toughest? Why?

3. Main Ideas

a. What issues affected the Western territories between 1775 and 1787?

b. What were three successes of the Continental Congress?

c. What were three weaknesses of the Articles of Confederation?

4. Critical Thinking

Forming and Supporting Opinions Which side would you have supported during Shays's Rebellion—the farmers or the officials who called out the militia? Why?

THINK ABOUT
- the farmers' problems
- the farmers' march on the arsenal
- the job of the government

ACTIVITY OPTIONS

CIVICS
GEOGRAPHY

Write an **opinion article** about how the United States should govern the Western territories or draw a **map** showing how you would have divided the lands.

The Northwest Territory

The Northwest Territory was officially known as "the Territory Northwest of the River Ohio." In the mid-1780s, Congress decided to sell the land in the territory to settlers. The sale of land solved two problems. First, it provided cash for the government. Second, it increased American control over the land.

The Land Ordinance of 1785 outlined how the land in the Northwest Territory would be divided. Congress split the land into grids with clearly defined boundaries. It created townships that could be divided into sections, as shown on the map below. Each township was six miles by six miles. This was an improvement over earlier methods of setting boundaries. Previously, people had used rocks, trees, or other landmarks to set boundaries. There had been constant fights over disputed claims.

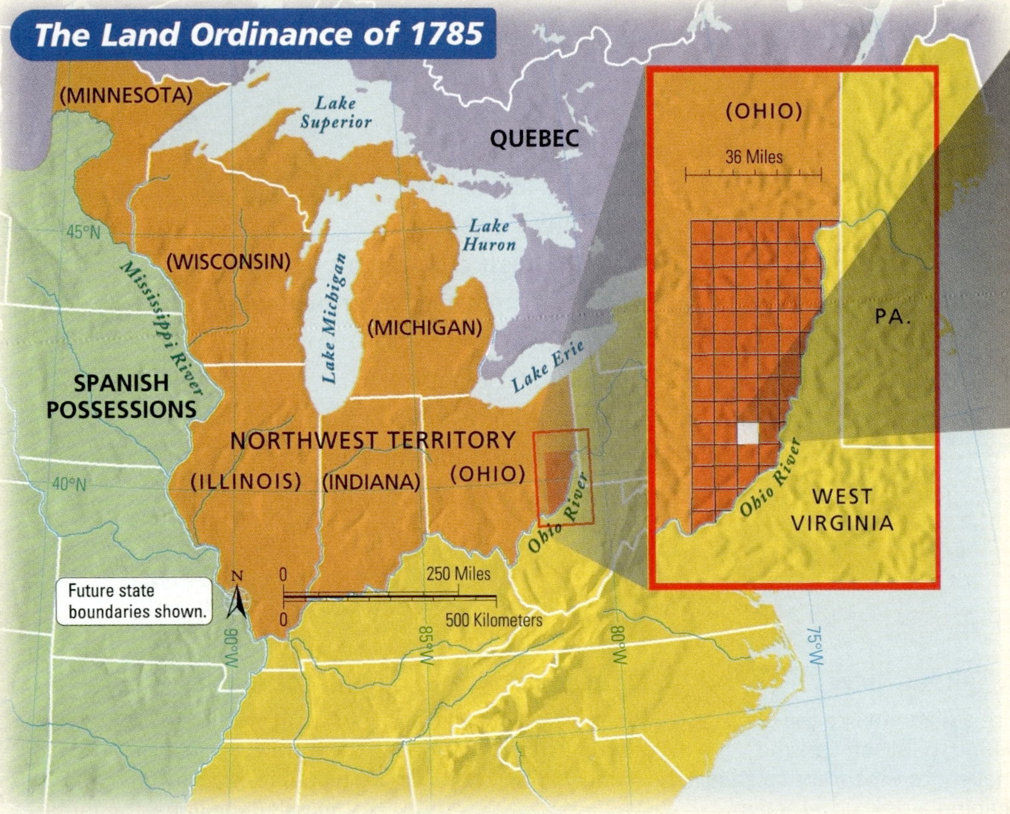

TOWNSHIP, 1785

36	30	24	18	12	6
35	29	23	17	11	5
34	28	22	16	10	4
33	27	21	15	9	3
32	26	20	14	8	2
31	25	19	13	7	1

Each township contained 36 sections. Each section was one square mile.

ARTIFACT FILE

The Theodolite The theodolite is a surveying tool. It consists of a telescope that can be moved from side to side and up and down. A theodolite measures angles and determines alignment. These functions are necessary for land surveyors to establish accurate boundaries for land claims.

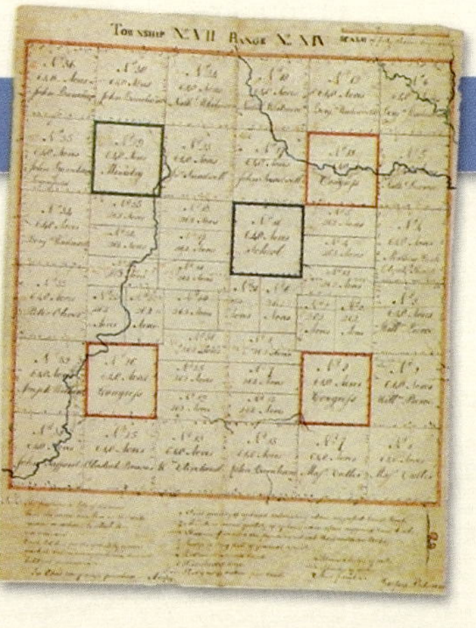

Township Map Congress reserved several plots (outlined on map) for special purposes. A few were set aside for later sale to raise money for the government. One plot was reserved to support a local school.

1 The first things settlers needed were food and shelter. Cutting trees provided fields for crops and wood for log cabins. The first crop most farmers planted was corn. Even if the land was not fully cleared of trees, farmers planted corn between the stumps.

2 A shortage of labor meant that a farmer working alone was doing well if he cleared several acres a year. As a result, few farms were completely fenced in, and forest covered most of the property. Hogs were allowed to find food in the woods. Farmers collected apples from trees and used sap to make syrup.

3 Over time, families planted fruits and vegetables. Cattle raising also became more common. Beef cattle supplied families with meat. Dairy cattle provided milk. Families could sell extra fruits, vegetables, and dairy products, such as butter and cheese.

On-Line Field Trip

The Shelby County Historical Society is located in Sidney, Ohio. Its collections include documents on the early settlement of Ohio, such as this picture of settlers clearing land.

Visit www.mcdougallittell.com for more information.

CONNECT TO GEOGRAPHY

1. **Region** What was the land in the Northwest Territory like before Americans settled there?
2. **Human-Environment Interaction** How did American settlers affect the landscape in the territory?

See Geography Handbook, pages 4–5.

CONNECT TO HISTORY

3. **Making Inferences** Why did so many people buy land in the new territory?

2 Creating the Constitution

TERMS & NAMES
- Constitutional Convention
- James Madison
- Virginia Plan
- New Jersey Plan
- Great Compromise
- Three-Fifths Compromise

MAIN IDEA

The states sent delegates to a convention to solve the problems of the Articles of Confederation.

WHY IT MATTERS NOW

The Constitutional Convention formed the plan of government that the United States still has today.

ONE AMERICAN'S STORY

On the afternoon of May 15, 1787, Edmund Randolph, the young governor of Virginia, arrived in Philadelphia for the Constitutional Convention. The young nation faced violence and lawlessness, as Shays's Rebellion had shown. And now delegates from throughout the states were coming to Philadelphia to discuss reforming the government.

Randolph knew the serious task he and the other delegates were about to undertake. Early in the convention, Randolph rose to speak. He looked squarely at the delegates and reminded them of their grave responsibility.

A VOICE FROM THE PAST

Let us not be afraid to view with a steady eye the [dangers] with which we are surrounded. . . . Are we not on the eve of [a civil] war, which is only to be prevented by the hopes from this convention?

Edmund Randolph, quoted in *Edmund Randolph: A Biography*

Over the next four months, the delegates debated how best to keep the United States from falling apart. In this section, you will read about the Convention of 1787 and the creation of the U.S. Constitution.

Edmund Randolph (left) and the other delegates gathered in the Pennsylvania State House (above) to discuss creating a new government for the United States.

A Constitutional Convention Is Called

In 1786, a series of events began that would eventually lead to a new form of government for the United States. In September of that year, delegates from five states met in Annapolis, Maryland, to discuss ways to promote trade among their states. At the time, most states placed high taxes on goods from other states. The delegates believed that creating national trade laws would help the economies of all the states.

Making such changes required amending the Articles of Confederation, because the national government had been granted no power to regulate trade among the states. The Annapolis delegates, led by Alexander Hamilton of New York, called for the states to send representatives to

Philadelphia the following May to discuss such changes.

At first, many Americans doubted that the national government needed strengthening. But news of Shays's Rebellion in late 1786 and early 1787 quickly changed many people's minds. Fearing that rebellion might spread, 12 states sent delegates to the meeting in Philadelphia in the summer of 1787. Only Rhode Island refused to participate.

Background
Rhode Island did not send delegates because it feared that a strong national government would force people to repay the war debts on difficult terms.

The Convention's Delegates

The 55 delegates to the **Constitutional Convention,** as the Philadelphia meeting became known, were a very impressive group. About half were lawyers. Others were planters, merchants, and doctors. Three-fourths of them had been representatives in the Continental Congress. Many had been members of their state legislatures and had helped write their state constitutions. Along with other leaders of the time, these delegates are called the Founders, or Founding Fathers, of the United States.

America's most famous men were at the Constitutional Convention. George Washington, the hero of the Revolution, came out of retirement for the meeting. Benjamin Franklin, the famous scientist and statesman, lent his wit and wisdom to the convention. One of the ablest delegates was **James Madison.** Madison had read more than a hundred books on government in preparation for the meeting. When Thomas Jefferson, serving as ambassador to France, read the list of delegates, he wrote, "It is really an assembly of demigods [highly honored people]."

Reading **History**
A. Evaluating
How well do the characteristics of the Founders serve as models of civic virtue?

Not everyone was at the Constitutional Convention. Thomas Jefferson and John Adams were overseas at their diplomatic posts. But they wrote home to encourage the delegates. Others had a less positive outlook on the convention. For example, Patrick Henry, who had been elected as a delegate from Virginia, refused to attend. He said he "smelled a rat in Philadelphia, tending toward monarchy."

Also, the convention did not reflect the diverse U.S. population of the 1780s. There were no Native Americans, African Americans, or women among the delegates. The nation's early leaders did not consider these groups of people to be citizens and did not invite any of them to attend. However, the framework of government the Founders established is the very one that would eventually provide full rights and responsibilities to all Americans.

The Delegates Assemble

Most of the delegates arrived at the Constitutional Convention without a clear idea of what to expect. Some thought they would only draft

America's **HERITAGE**

INDEPENDENCE HALL

The Pennsylvania State House, where the Constitutional Convention took place, is now called Independence Hall. It is protected as part of a national park in Philadelphia.

The State House itself was the site where George Washington received his commission to lead the Continental Army and where the Declaration of Independence was signed. The Liberty Bell is nearby. Many visitors come to Philadelphia to stand in the building where much of America's early history as a nation was made.

AMERICA'S
HISTORY MAKERS

JAMES MADISON

1751–1836

James Madison was a short, soft-spoken man, but he may have made the greatest contribution of any of the Founders at the Constitutional Convention. He took thorough notes of the convention's proceedings. His notes are the most detailed picture we have of the debates and drama of the convention.

But Madison did not just observe the convention. He was perhaps the most important participant. One of the other delegates called him "the best informed Man of any point in debate." Madison was so important that he earned the title "Father of the Constitution."

How did Madison contribute to the Constitutional Convention?

amendments to the Articles of Confederation. Others thought they would design an entirely new plan for the government. But they all agreed that the government should protect people's rights.

Back in 1776, many Americans thought that government was the main threat to people's rights. But by 1787, many realized that the people often came into conflict and needed a government that could maintain order. As a result, the government had to be strong enough to protect people's rights but not too strong to be controlled. Madison later wrote about this problem.

A VOICE FROM THE PAST

If men were angels, no government would be necessary. If angels were to govern men, neither external nor internal controls on government would be necessary. In framing a government which is to be administered by men over men, the great difficulty lies in this: you must first enable the government to control the governed; and in the next place oblige it [the government] to control itself.

James Madison, *The Federalist* "Number 51"

Reading **History**

B. Using Primary Sources According to Madison, what is the central problem in framing a government?

This was the challenge that faced the delegates: how to set up a strong but limited federal government. By May 25, 1787, at least two delegates from each of seven states had arrived in Philadelphia. With 29 delegates in attendance, the convention was officially under way.

The Convention Begins

The first order of business was to elect a president for the convention. Robert Morris of Pennsylvania nominated George Washington. No American was more respected or admired than Washington. Every delegate voted for him. Washington's quiet and dignified leadership set a solemn and serious tone for the convention.

At their next meeting, the delegates decided on the rules for the convention. They wanted to be able to consider all ideas and to be able to change sides in any debate. They did not want to be pressured by the politics of the day. For these reasons, they decided that their discussions would remain secret. To ensure privacy, the windows in their meeting room were kept shut even though it was summer. Guards were posted outside the door. Whenever the door was opened, the delegates stopped talking. With the secrecy rule approved, they got down to business.

Reading **History**

C. Making Decisions Do you agree with the Founders' decision to keep the convention secret? Why or why not?

The Virginia Plan

On May 29, the delegates began the real work of designing a new national government. Presiding over the convention, George Washington

The delegates at the Constitutional Convention debated the Constitution intensely.

recognized Edmund Randolph as the first speaker. Randolph offered a plan for a whole new government. The plan became known as the **Virginia Plan.** Madison, Randolph, and the other Virginia delegates had drawn up the plan while they waited for the convention to open.

The Virginia Plan proposed a government that would have three branches. The first branch of government was the legislature, which made the laws. The second branch was the executive, which enforced the laws. The third branch was the judiciary, which interpreted the laws.

"If men were angels, no government would be necessary."
James Madison

The Virginia Plan proposed a legislature with two houses. In both houses, the number of representatives from each state would be based on the state's population or its wealth. The legislature would have the power to levy taxes, regulate commerce, and make laws "in all cases where the separate states are incompetent [unable]."

Reading **History**
D. Summarizing What was the Virginia Plan?

The Virginia Plan led to weeks of debate. Because they had larger populations, larger states supported the plan. It would give them greater representation in the legislature. The smaller states opposed this plan. They worried that the larger states would end up ruling the others. Delaware delegate John Dickinson voiced the concerns of the small states.

A VOICE FROM THE PAST

Some of the members from the small states wish for two branches in the general legislature and are friends to a good [strong] national government; but we would sooner submit [give in] to a foreign power than submit to be deprived, in both branches of the legislature, of an equal suffrage [vote], and thereby be thrown under the domination of the larger states.

John Dickinson, quoted in *Mr. Madison's Constitution*

The Great Compromise

In response to the Virginia Plan, New Jersey delegate William Paterson presented an alternative on June 15. The **New Jersey Plan** called for a legislature with only one house. In it, each state would have one vote. In providing equal representation to each state, the New Jersey Plan was similar to the Articles of Confederation.

Even though the New Jersey Plan gave the legislature the power to regulate trade and to raise money by taxing foreign goods, it did not offer the broad powers proposed by the Virginia Plan. The delegates

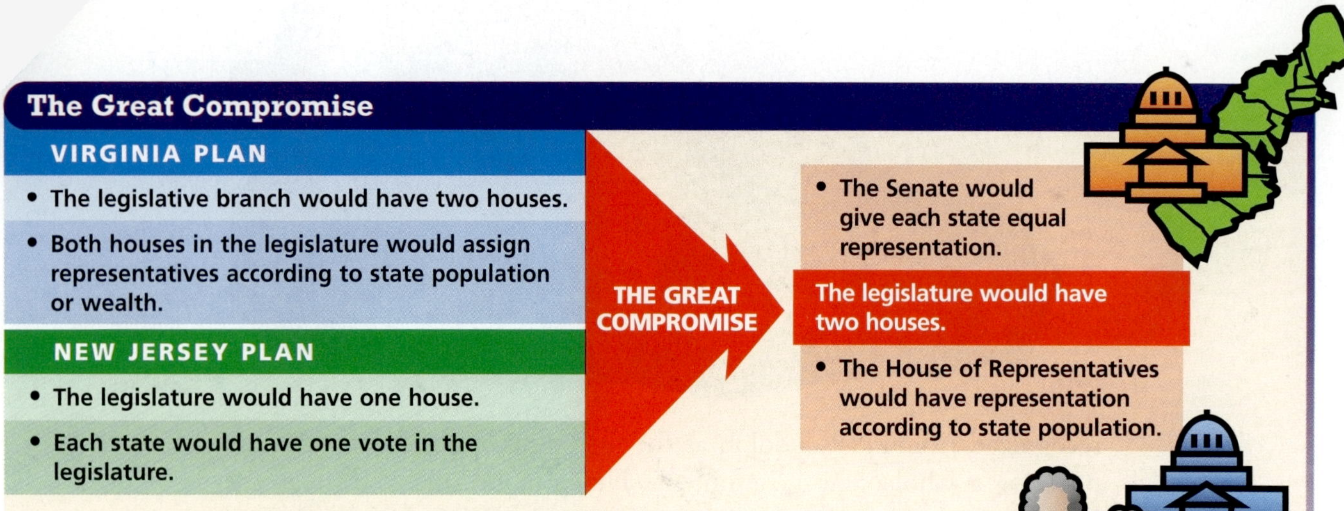

SKILLBUILDER Interpreting Charts

1. *Which plan appealed more to the small states?*
2. *Did the Great Compromise include more of what the large states wanted or more of what the small states wanted?*

voted on these two plans on June 19. The Virginia Plan won and became the framework for drafting the Constitution.

During the rest of June, the delegates argued over representation in the legislature. Emotions ran high as the delegates struggled for a solution. In desperation, the delegates selected a committee to work out a compromise in early July. The committee offered the **Great Compromise.** (Some people also refer to it as the Connecticut Compromise.)

To satisfy the smaller states, each state would have an equal number of votes in the Senate. To satisfy the larger states, the committee set representation in the House of Representatives according to state populations. More than a week of arguing followed the introduction of the plan, but on July 16, 1787, the convention passed it.

Slavery and the Constitution

Because representation in the House of Representatives would be based on the population of each state, the delegates had to decide who would be counted in that population. The Southern states had many more slaves than the Northern states. Southerners wanted the slaves to be counted as part of the general population for representation but not for taxation. Northerners argued that slaves were not citizens and should not be counted for representation but should be counted for taxation.

On this issue, the delegates reached another compromise, known as the **Three-Fifths Compromise.** Under this compromise, three-fifths of the slave population would be counted when setting direct taxes on the states. This three-fifths ratio also would be used to determine representation in the legislature.

The delegates had another heated debate about the slave trade. Slavery had already been outlawed in several Northern states. All of the Northern states and several of the Southern states had banned the

ReadingHistory
E. Forming and Supporting Opinions Did the delegates do the right thing in agreeing to the Three-Fifths Compromise? Explain.

importation of slaves. Many Northerners wanted to see this ban extended to the rest of the nation. But Southern slaveholders strongly disagreed. The delegates from South Carolina and Georgia stated that they would never accept any plan "unless their right to import slaves be untouched." Again, the delegates settled on a compromise. On August 29, they agreed that Congress could not ban the slave trade until 1808.

Now and then

PRESERVING THE CONSTITUTION

The National Archives is responsible for preserving the 200-year-old sheets of parchment on which the original Constitution was first written.

The Archives stores the document in an airtight glass case enclosed in a 55-ton vault of steel and concrete. Every few years, scientists examine the pages with the latest technology. For the last examination in 1995, they used fiber-optic light sources and computer-guided electronic cameras designed for space exploration.

Regulating Trade

In the compromise over slavery, the North gave in to the South by delaying the end of the African slave trade until 1808. But the South gave in to the North on a few issues as well. For example, the South agreed to a tax on the slave trade.

Southerners also agreed to allow the national government to pass laws on how goods could be exported. Since the Southern economy was based on exports, such laws would affect them greatly. The delegates from most states, however, were glad to grant the national government the power to regulate trade. After all, this was one of the issues that had led to the Annapolis Convention the previous year.

The Constitutional Convention continued to meet into September. On Saturday, September 15, 1787, the delegates voted their support for the Constitution in its final form. On Sunday, it was written out on four sheets of thick parchment. On Monday, all but three delegates signed the Constitution. It was then sent, with a letter signed by George Washington, to the Confederation Congress, which sent it to the states for ratification, or approval. In the next section, you will read about the debate over ratification.

Section 2 Assessment

1. Terms & Names

- Constitutional Convention
- James Madison
- Virginia Plan
- New Jersey Plan
- Great Compromise
- Three-Fifths Compromise

2. Taking Notes

Use a chart like the one below to take notes on the contributions made by the leading delegates at the Constitutional Convention.

Delegate	Contribution

3. Main Ideas

a. What was the relationship between the Annapolis Convention and the Constitutional Convention?

b. What were the delegates to the Constitutional Convention like as a group?

c. How did the Constitutional Convention address the issue of slavery?

4. Critical Thinking

Analyzing Points of View
How did the delegates at the convention differ on the issue of representation in the new government?

THINK ABOUT

- the large states and the small states
- the Virginia Plan
- the New Jersey Plan
- the Great Compromise

ACTIVITY OPTIONS

TECHNOLOGY
ART

Think about the Three-Fifths Compromise. Make an **audio recording** of a speech or draw a **political cartoon** that expresses your views on the issue.

Ratifying the Constitution

TERMS & NAMES
federalism
Federalists
Antifederalists
The Federalist papers
George Mason
Bill of Rights

MAIN IDEA

Americans across the nation debated whether the Constitution would produce the best government.

WHY IT MATTERS NOW

American liberties today are protected by the U.S. Constitution, including the Bill of Rights.

ONE AMERICAN'S STORY

For a week in early January 1788, a church in Hartford, Connecticut, was filled to capacity. Inside, 168 delegates were meeting to decide whether their state should ratify the U.S. Constitution. Samuel Huntington, Connecticut's governor, addressed the assembly.

A VOICE FROM THE PAST

This is a new event in the history of mankind. Heretofore, most governments have been formed by tyrants and imposed on mankind by force. Never before did a people, in time of peace and tranquillity, meet together by their representatives and, with calm deliberation, frame for themselves a system of government.

Samuel Huntington, quoted in *Original Meanings*

The governor supported the new Constitution and wanted to see it ratified. Not everyone agreed with him. In this section, you will learn about the debates that led to the ratification of the Constitution.

Samuel Huntington

Federalists and Antifederalists

By the time the convention in Connecticut opened, Americans had already been debating the new Constitution for months. The document had been printed in newspapers and handed out in pamphlets across the United States. The framers of the Constitution knew that the document would cause controversy. They immediately began to campaign for ratification, or approval, of the Constitution.

The framers suspected that people might be afraid the Constitution would take too much power away from the states. To address this fear, the framers explained that the Constitution was based on federalism. **Federalism** is a system of government in which power is shared between the central (or federal) government and the states. Linking themselves to the idea of federalism, the people who supported the Constitution took the name **Federalists.**

People who opposed the Constitution were called **Antifederalists.** They thought the Constitution took too much power away from the states and did not guarantee rights for the people. Some were afraid that a strong president might be declared king. Others thought the Senate might turn into a powerful aristocracy. In either case, the liberties won at great cost during the Revolution might be lost.

Vocabulary
aristocracy: a group or class considered superior to others

Antifederalists published their views about the Constitution in newspapers and pamphlets. They used logical arguments to convince people to oppose the Constitution. But they also tried to stir people's emotions by charging that it would destroy American liberties. As one Antifederalist wrote, "After so recent a triumph over British despots [oppressive rulers], . . . it is truly astonishing that a set of men among ourselves should have had the effrontery [nerve] to attempt the destruction of our liberties."

The Federalist Papers

The Federalists did not sit still while the Antifederalists attacked the Constitution. They wrote essays to answer the Antifederalists' attacks. The best known of the Federalist essays are ***The Federalist* papers.** These essays first appeared as letters in New York newspapers. They were later published together in a book called *The Federalist.*

Three well-known politicians wrote *The Federalist* papers—James Madison, Alexander Hamilton, and John Jay, the secretary of foreign affairs for the Confederation Congress. Like the Antifederalists, the Federalists appealed to reason and emotion. In *The Federalist* papers, Hamilton described why people should support ratification.

***Reading* History**
A. Making Inferences What does Hamilton think will happen if the Constitution is not ratified?

A VOICE FROM THE PAST

Yes, my countrymen, . . . I am clearly of opinion it is in your interest to adopt it [the Constitution]. I am convinced that this is the safest course for your liberty, your dignity, and your happiness.

Alexander Hamilton, *The Federalist* "Number 1"

Federalists and Antifederalists

FEDERALISTS	ANTIFEDERALISTS
• Supported removing some powers from the states and giving more powers to the national government	• Wanted important political powers to remain with the states
• Favored dividing powers among different branches of government	• Wanted the legislative branch to have more power than the executive
• Proposed a single person to lead the executive branch	• Feared that a strong executive might become a king or tyrant
	• Believed a bill of rights needed to be added to the Constitution to protect people's rights

SKILLBUILDER Interpreting Charts
1. *Which group wanted a stronger central government?*
2. *If you had been alive in 1787, would you have been a Federalist or an Antifederalist?*

John Jay

George Mason

HISTORY *through* ART

Supporters of the Constitution turned out in parades like this one in New York in 1788. The "Ship of State" float has Alexander Hamilton's name on it to celebrate his role in creating the Constitution.

What does the picture indicate about the importance of the Constitution in people's lives?

The Federalists had an important advantage over the Antifederalists. Most of the newspapers supported the Constitution, giving the Federalists more publicity than the Antifederalists. Even so, there was strong opposition to ratification in Massachusetts, North Carolina, Rhode Island, New York, and Virginia. If some of these states failed to ratify the Constitution, the United States might not survive.

The Battle for Ratification

The first four state conventions to ratify the Constitution were held in December 1787. It was a good month for the Federalists. Delaware, New Jersey, and Pennsylvania voted for ratification. In January 1788, Georgia and Connecticut ratified the Constitution. Massachusetts joined these states in early February.

By late June, nine states had voted to ratify the Constitution. That meant that the document was now officially ratified. But New York and Virginia had not yet cast their votes. There were many powerful Antifederalists in both of those states. Without Virginia, the new government would lack the support of the largest state. Without New York, the nation would be separated into two parts geographically.

Virginia's convention opened the first week in June. The patriot Patrick Henry fought against ratification. **George Mason,** perhaps the most influential Virginian aside from Washington, also was opposed to it. Mason had been a delegate to the Constitutional Convention in Philadelphia, but he had refused to sign the final document. Both Henry and Mason would not consider voting for the Constitution until a bill of rights was added. A bill of rights is a set of rules that defines people's rights.

Reading **History**

B. Drawing Conclusions How did the lack of a bill of rights endanger the Constitution?

James Madison was also at Virginia's convention. He suggested that Virginia follow Massachusetts's lead and ratify the Constitution, and he recommended the addition of a bill of rights. With the addition of a bill of rights likely, Virginia ratified the Constitution at the end of June.

The news of Virginia's vote arrived while the New York convention was in debate. The Antifederalists had outnumbered the Federalists when the convention had begun. But with the news of Virginia's ratification, New Yorkers decided to join the Union. New York also called for a bill of rights.

It was another year before North Carolina ratified the Constitution. In 1790, Rhode Island became the last state to ratify it. By then, the new Congress had already written a bill of rights and submitted it to the states for approval.

The Bill of Rights

Background
The seven states that asked for a bill of rights were Massachusetts, South Carolina, New Hampshire, Virginia, New York, North Carolina, and Rhode Island.

At the same time that seven of the states ratified the Constitution, they asked that it be amended to include a bill of rights. Supporters of a bill of rights hoped that it would set forth the rights of all Americans. They believed it was needed to protect people against the power of the national government.

Madison, who was elected to the new Congress in the winter of 1789, took up the cause. He proposed a set of changes to the Constitution. Congress edited Madison's list and proposed placing the amendments at the end of the Constitution in a separate section.

The amendments went to the states for ratification. As with the Constitution, three-quarters of the states had to ratify the amendments for them to take effect. With Virginia's vote in 1791, ten of the amendments were ratified and became law. These ten amendments to the U.S. Constitution became known as the **Bill of Rights.** (See the Constitution Handbook, page 242.)

The passage of the Bill of Rights was one of the first acts of the new government. In the next chapter, you will read about other issues that faced the new government.

America's HERITAGE

RELIGIOUS FREEDOM

Freedom of religion was an important part of the First Amendment. Jefferson and Madison believed that government enforcement of religious laws was the source of much social conflict. They supported freedom of religion as a way to prevent such conflict.

Even before Madison wrote the Bill of Rights, he worked to ensure religious liberty in Virginia. In 1786, he helped pass the Virginia Statute for Religious Freedom, originally written by Jefferson in 1777.

Section 3 Assessment

1. Terms & Names

- federalism
- Federalists
- Antifederalists
- *The Federalist* papers
- George Mason
- Bill of Rights

2. Taking Notes

Use a diagram like the one below to compare and contrast the Federalists and the Antifederalists.

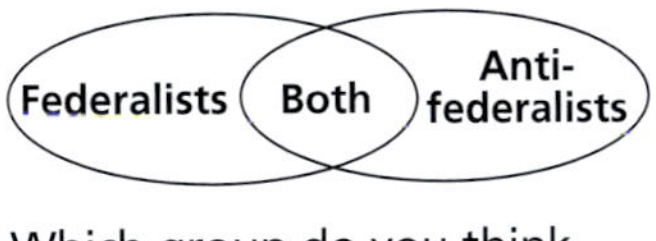

Which group do you think made the stronger argument about ratification? Why?

3. Main Ideas

a. How was the Constitution ratified?

b. How did the Federalists and the Antifederalists try to convince people to take their sides in the debate over the Constitution?

c. What was the significance of the Bill of Rights?

4. Critical Thinking

Recognizing Propaganda
Reread the quotation by Hamilton on page 235. Is it an example of propaganda? Why or why not?

THINK ABOUT
- Hamilton's use of the word *countrymen*
- Hamilton's reference to liberty, dignity, and happiness

ACTIVITY OPTIONS

SPEECH
LANGUAGE ARTS

Review the political fight over ratification of the Constitution. Hold a **press conference** or write a **news report** on the ratification debate.

The Federalist "Number 51"

Setting the Stage James Madison wrote 29 essays in *The Federalist* papers to argue in favor of ratifying the Constitution. In *The Federalist* "Number 51," Madison explains how the government set up by the Constitution will protect the rights of the people by weakening the power of any interest, or group, to dominate the government. **See Primary Source Explorer** CD-ROM

It is of great importance in a republic not only to guard the society against the oppression of its rulers, but to guard one part of the society against the injustice of the other part. Different interests necessarily exist in different classes of citizens. If a majority be united by a common interest, the rights of the minority will be insecure. There are but two methods of providing against this evil: the one by creating a will in the community independent of the majority—that is, of the society itself; the other, by **comprehending**[1] in the society so many separate descriptions of citizens as will render an unjust combination of a majority of the whole very improbable, if not **impracticable**.[2] . . .

Whilst[3] all authority in it will be derived from and dependent on the society, the society itself will be broken into so many parts, interests and classes of citizens, that the rights of individuals, or of the minority, will be in little danger from interested combinations of the majority. In a free government the security for civil rights must be the same as that for religious rights. It consists in the one case in the multiplicity of interests, and in the other in the **multiplicity of sects**.[4] . . .

In the extended republic of the United States, and among the great variety of interests, parties, and sects which it embraces, a **coalition**[5] of a majority of the whole society could seldom take place on any other principles than those of justice and the general good. . . .

It is no less certain than it is important . . . that the larger the society, provided it lie within a practicable sphere, the more duly capable it will be of self-government. And happily for the republican cause, the practicable sphere may be carried to a very great extent by a **judicious modification**[6] and mixture of the *federal principle.*

—James Madison

A CLOSER LOOK

MINORITY RIGHTS

In the 1700s, people feared that democratic majorities could turn into mobs that would violate other people's rights. Madison had to explain how the Constitution would prevent this.

1. What two methods does Madison suggest a society can use to protect minority rights?

A CLOSER LOOK

REPUBLICS IN LARGE SOCIETIES

For centuries, people believed that only small societies could be republics. But Madison argues that large societies are more likely to remain republics.

2. Why does Madison believe that a large republic is likely to protect justice?

1. **comprehending:** understanding.
2. **impracticable:** not practical or realistic.
3. **whilst:** while.
4. **multiplicity of sects:** large number of groups.
5. **coalition:** alliance of groups.
6. **judicious modification:** careful change.

Objections to the Constitution

Setting the Stage George Mason was one of the leading Antifederalists. In "Objections to the Constitution of Government Formed by the Convention," he listed his reasons for opposing ratification. Above all, he feared that the Constitution created a government that would destroy democracy in the young nation. **See Primary Source Explorer**

There is no Declaration of Rights; and the Laws of the general Government being **paramount**[1] to the Laws and Constitutions of several States, the Declaration of Rights in the separate States are no Security. Nor are the people secured even in the Enjoyment of the Benefits of the common-Law; can never produce proper Information in the Legislature, or inspire Confidence in the People; the Laws will therefore be generally made by Men little concern'd in, and **unacquainted**[2] with their Effects and Consequences.

The Senate have the Power of altering all Money-Bills, and of originating Appropriations of Money and the **Sallerys**[3] of the Officers of their own Appointment in **Conjunction**[4] with the President of the United States; altho' they are not the Representatives of the People, or **amenable**[5] to them. . . .

The President of the United States has the unrestrained Power of granting Pardon for Treason; which may be sometimes exercised to screen from Punishment those whom he had secretly **instigated**[6] to commit the Crime, and thereby prevent a Discovery of his own Guilt.

This Government will **commence**[7] in a moderate **Aristocracy;**[8] it is at present impossible to foresee whether it will, in it's Operation, produce a **Monarchy,**[9] or a corrupt oppressive Aristocracy; it will most probably vibrate some Years between the two, and then terminate in the one or the other.

—*George Mason*

1. **paramount:** most important.
2. **unacquainted:** unfamiliar.
3. **sallerys:** salaries.
4. **conjunction:** joining.
5. **amenable:** agreeable.
6. **instigated:** caused.
7. **commence:** begin.
8. **aristocracy:** rule by a few, usually nobles.
9. **monarchy:** rule by one, usually a king.

A CLOSER LOOK

DECLARATION OF RIGHTS

At the time of the ratification debate, Americans across the nation complained that the Constitution did not include a bill of rights.

3. What arguments does Mason make about the lack of a Declaration of Rights?

A CLOSER LOOK

ABUSE OF POWER

Mason believed that presidents might abuse the power to grant pardons for treason in order to protect the guilty.

4. Can you think of any presidents who have granted pardons?

Interactive Primary Sources Assessment

1. Main Ideas

a. Why does Madison believe that a society broken into many parts will not endanger minority rights?

b. What does Mason argue might happen if the president had the power to pardon people?

c. For each writer, what is one example of a fact and one example of an opinion?

2. Critical Thinking

Drawing Conclusions Who do you think makes the stronger argument? Explain your reasons.

THINK ABOUT

- what you know about the history of the United States
- the evidence used by each writer

Chapter 8 ASSESSMENT

VISUAL SUMMARY

Confederation to Constitution

1777
Continental Congress passes the Articles of Confederation.

1777–1781
States debate ratification of the Articles of Confederation.

1781
Articles of Confederation go into effect.

1786
Annapolis Convention is held.

1786–1787
Shays's Rebellion occurs.

1787
Constitutional Convention is held in Philadelphia.

1788
U.S. Constitution is ratified.

1789
Government created by the new Constitution takes power.

1791
Bill of Rights is added to the Constitution.

TERMS & NAMES

Briefly explain the importance of each of the following.

1. republic
2. Articles of Confederation
3. Northwest Ordinance
4. Shays's Rebellion
5. Constitutional Convention
6. James Madison
7. Great Compromise
8. Federalists
9. George Mason
10. Bill of Rights

REVIEW QUESTIONS

The Confederation Era (pages 221–227)

1. What is the Wilderness Road, and where did it lead?
2. What problems did the Continental Congress successfully address?
3. What powers did the government have under the Articles of Confederation?
4. How did Shays's Rebellion affect people's views on the Articles of Confederation?

Creating the Constitution (pages 228–233)

5. What groups of people were not represented at the Constitutional Convention?
6. What were some things the delegates agreed on at the convention?
7. What compromises did the delegates make during the convention?

Ratifying the Constitution (pages 234–239)

8. What is federalism?
9. Why were Virginia and New York important in the battle for ratification of the Constitution?
10. Why did some states think that it was necessary to add a bill of rights to the Constitution?

CRITICAL THINKING

1. USING YOUR NOTES

Problems	Solutions
Western lands	
Postwar depression	
Representation in the new government	
Slavery	

Using your completed chart, answer the questions below.

a. What do you think was the most serious problem the young nation faced?

b. Was the nation able to solve this problem? How?

2. ANALYZING LEADERSHIP

Think about the leaders discussed in this chapter. Based on their actions, which leader do you think made the greatest contribution to the Constitutional Convention? Why?

3. THEME: DEMOCRATIC IDEALS

How do the Articles of Confederation and the Constitution each carry out democratic ideals?

4. APPLYING CITIZENSHIP SKILLS

Do you think the Founders were right to make the compromises they did in the Constitution on the issues of representation and slavery? What might have happened if they had not compromised?

5. RECOGNIZING EFFECTS

How might U.S. history be different if Virginia had refused to ratify the Constitution? If New York had refused? If both had refused?

Interact *with* History

How did your ideas about how you would form a government change after reading this chapter?

HISTORY SKILLS

1. INTERPRETING MAPS: Region

Study the map and then answer the questions.

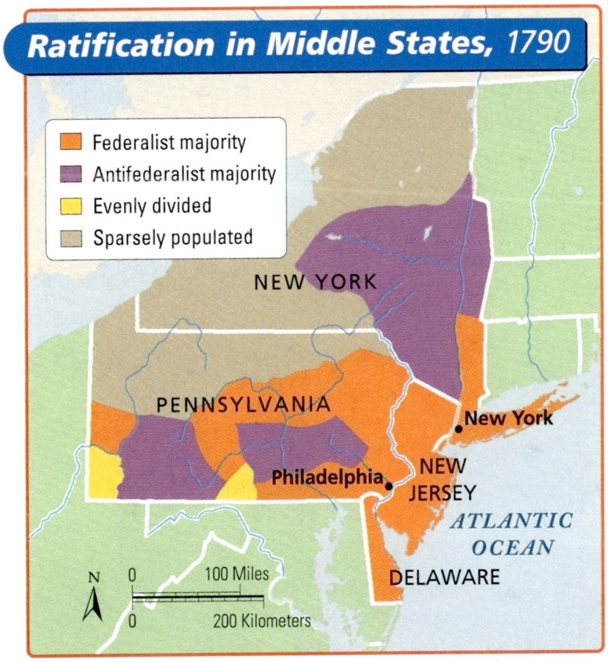

Source: *American Heritage Pictorial Atlas of United States History*

Basic Map Elements

a. Which states are identified on the map?

b. In which states did the Federalists have statewide majorities?

Interpreting the Map

c. Why do you think the two cities on the map were strong Federalist supporters?

2. INTERPRETING PRIMARY SOURCES

The following law was put into effect in Virginia in 1786. Read the law and answer the questions.

> *Be it enacted by the General Assembly,* That no man shall be compelled to frequent or support any religious worship, place, or ministry, whatsoever . . . but that all men shall be free to profess, and by argument maintain, their opinion in matters of religion, and that the same shall in no way diminish, enlarge, or affect their civil capacities.
>
> *The Statute of Virginia for Religious Freedom, 1786*

a. How would you summarize this law?

b. Which right in the Bill of Rights was based on this law?

c. Based on what you know about colonial history, how had American society changed between the early 1600s and the late 1700s?

ALTERNATIVE ASSESSMENT

1. INTERDISCIPLINARY ACTIVITY: Government

Making a Chart Do research to learn how the U.S. Constitution has been used as a model by other nations. Make a chart to summarize the information you find about one specific nation. Include the country, the date the country's constitution was ratified, and two ways in which that nation's constitution is similar to and different from the U.S. Constitution.

2. COOPERATIVE LEARNING ACTIVITY

Staging a Debate Stage a debate between a Federalist and an Antifederalist. Work in small groups to read some of the different arguments each side used. Look for discussions of one or more of the issues listed below. Pick one of the issues and stage a debate, using the strongest arguments from each side. Let the class determine who won the debate and why.

a. the representation of people in Congress

b. the strength of the president and Senate

c. the need for a bill of rights

3. CD-ROM PRIMARY SOURCE EXPLORER

Creating a Museum Exhibit The creation of the U.S. Constitution was one of the most important events in the nation's history. There is a great amount of information about the Constitution. Using the Primary Source Explorer CD-ROM and your local library, collect information on different topics relating to the Constitution.

Create a museum exhibit about the Constitution using the suggestions below.

- Include information on the historical background of the Constitutional Convention, such as Shays's Rebellion and Enlightenment ideas about government.
- Find biographies about the delegates to the convention, including portraits.
- Collect important primary sources, such as Madison's notes and *The Federalist* papers.
- Include photographs or facsimiles of the documents.
- Draw a diagram that shows a layout for the exhibit.

4. HISTORY PORTFOLIO

Option 1 Review your section and chapter assessment activities. Select one that you think shows your best work. Then use comments made by your teacher or classmates to improve your work and add it to your portfolio.

Option 2 Review the questions that you wrote for What Do You Want to Know? on page 220. Then write a report that explains the answers to your questions. If any questions were not answered, do research to answer them. Add your answers to your portfolio.

The Living Constitution

The Framers of the Constitution created a flexible plan for governing the United States far into the future. They also described ways to allow changes in the Constitution. For over 200 years, the Constitution has guided the American people. It remains a "living document." The Constitution still thrives, in part, because it echoes the principles the delegates valued. Each generation of Americans renews the meaning of the Constitution's timeless ideas. These two pages show you some ways in which the Constitution has shaped events in American history. **See Primary Source Explorer** CD-ROM

"In framing a system which we wish to last for ages, we should not lose sight of the changes which ages will produce."

—JAMES MADISON, CONSTITUTIONAL CONVENTION

1787

Delegates in Philadelphia sign the Constitution.

1965

Civil rights leaders protest to end the violation of their constitutional rights. Dr. Martin Luther King, Jr., Coretta Scott King, and others march from Selma toward Montgomery, Alabama, to gain voting rights.

1971

The 26th Amendment to the Constitution gives young people "18 years of age or older" the right to vote.

1981

A Supreme Court decision rules that Congress can exclude women from the draft. Still, many women who have joined the armed forces have served in combat.

TABLE OF CONTENTS

HOW TO READ THE CONSTITUTION

The complete text of the Constitution of the United States begins on page 248. The main column has the actual text. Some of the spellings and punctuation have been updated for easier reading. Headings and subheadings have been added to the Constitution to help you find specific topics. Those parts of the Constitution that are no longer in use have been crossed out. "A Closer Look" notes and charts will help you understand issues related to the Constitution.

1999

The Senate tries President Bill Clinton for the impeachment charges brought against him by the House of Representatives. As required by the Constitution, the Senate needs a two-thirds majority vote to convict him. This rule saves his presidency.

Seven Principles of the Constitution

The Framers of the Constitution constructed a new system of government. Seven principles supported their efforts. To picture how these principles work, imagine seven building blocks. Together they form the foundation of the United States Constitution. In the pages that follow, you will find the definitions and main ideas of the principles shown in the graphic below.

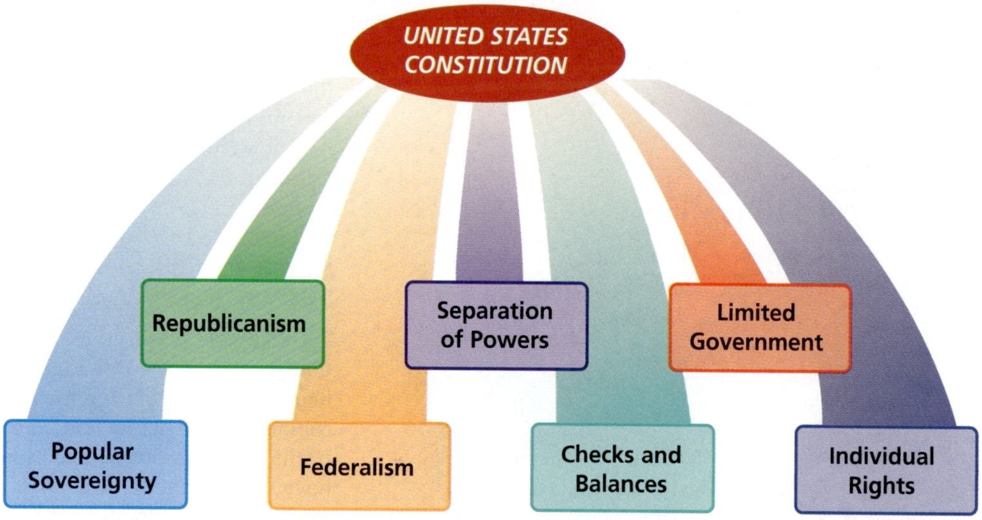

1 Popular Sovereignty
Who Gives the Government Its Power?

"We the people of the United States . . . establish this Constitution for the United States of America." These words from the Preamble, or introduction, to the Constitution clearly spell out the source of the government's power. The Constitution rests on the idea of **popular sovereignty**—a government in which the people rule. As the nation changed and grew, popular sovereignty took on new meaning. A broader range of Americans shared in the power to govern themselves.

In 1987, Americans gathered in Washington, D.C., to celebrate the 200th anniversary of the Constitution. The banner proudly displays that the power to govern belongs to the people.

In a republican government, voting citizens make their voices heard at the polls. The power of the ballot prompts candidates to listen to people's concerns.

2 Republicanism
How Are People's Views Represented in Government?

The Framers of the Constitution wanted the people to have a voice in government. Yet the Framers also feared that public opinion might stand in the way of sound decision making. To solve this problem, they looked to republicanism as a model of government.

Republicanism is based on this belief: The people exercise their power by voting for their political representatives. According to the Framers, these lawmakers played the key role in making a republican government work. Article 4, Section 4, of the Constitution also calls for every state to have a "republican form of government."

3 Federalism
How Is Power Shared?

The Framers wanted the states and the nation to become partners in governing. To build cooperation, the Framers turned to federalism. **Federalism** is a system of government in which the states and national government share powers.

The Framers used federalism to structure the Constitution. The Constitution assigns certain powers to the national government. These are *delegated powers*. Powers kept by the states are *reserved powers*. Powers shared or exercised by national and state governments are known as *concurrent powers*.

Federalism

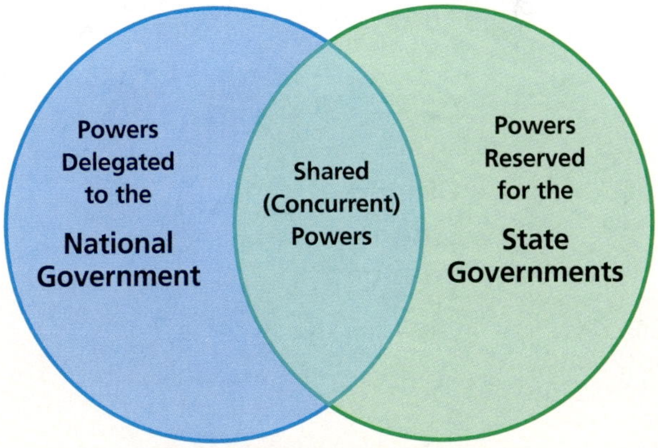

The overlapping spheres of power bind the American people together.

4 Separation of Powers
How Is Power Divided?

The Framers were concerned that too much power might fall into the hands of a single group. To avoid this problem, they built the idea of **separation of powers** into the Constitution. This principle means the division of basic government roles into branches. No one branch is given all the power. Articles 1, 2, and 3 of the Constitution detail how powers are split among the three branches.

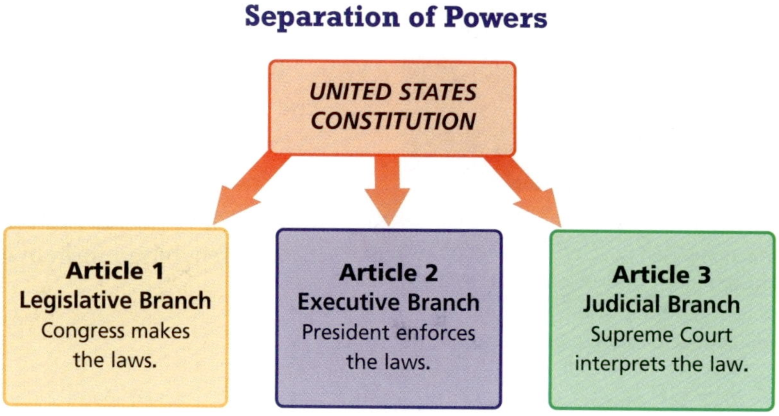

5 Checks and Balances
How Is Power Evenly Distributed?

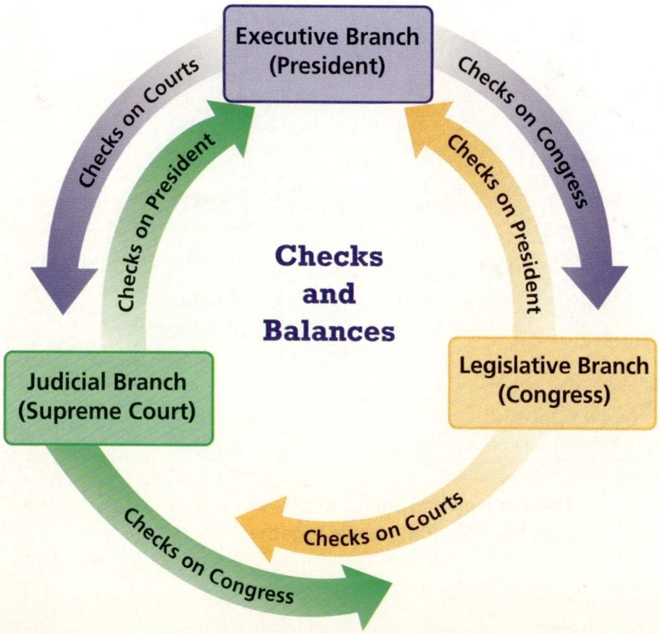

Baron de Montesquieu, an 18th-century French thinker, wrote, "Power should be a check to power." His comment refers to the principle of **checks and balances.** Each branch of government can exercise checks, or controls, over the other branches. Though the branches of government are separate, they rely on one another to perform the work of government.

The Framers included a system of checks and balances in the Constitution to help make sure that the branches work together fairly. For example, only Congress can pass laws. Yet the president can check this power by refusing to sign a law into action. In turn, the Supreme Court can declare that a law, passed by Congress and signed by the president, violates the Constitution.

6 Limited Government
How Is Abuse of Power Prevented?

The Framers restricted the power of government. Article 1, Section 9, of the Constitution lists the powers denied to the Congress. Article 1, Section 10, forbids the states to take certain actions.

The principle of **limited government** is also closely related to the "rule of law": In the American government everyone, citizens and powerful leaders alike, must obey the law. Individuals or groups cannot twist or bypass the law to serve their own interests.

In this political cartoon, President Richard Nixon shakes his fist as he defies the "rule of law." Faced with charges of violating the Constitution, Nixon resigned as president in 1974.

Students exercise their right to protest. They urge the community to protect the environment.

7 Individual Rights
How Are Personal Freedoms Protected?

The first ten amendments to the Constitution shield people from an overly powerful government. These amendments are called the Bill of Rights. The Bill of Rights guarantees certain **individual rights,** or personal liberties and privileges. For example, government cannot control what people write or say. People also have the right to meet peacefully and to ask the government to correct a problem. Later amendments to the Constitution also advanced the cause of individual rights.

Assessment: Principles of the Constitution

1. Main Ideas

a. What are the seven principles of government?

b. How does the Constitution reflect the principle of separation of powers?

c. Why did the Framers include a system of checks and balances in the Constitution?

2. Critical Thinking

Recognizing Effects What might happen to the U.S. republic if most Americans did not exercise their right to vote?

THINK ABOUT

- the rights and responsibilities of citizens
- the views of elected representatives

The Constitution of the United States

See Primary Source Explorer CD-ROM

"In 1787, I was not included in that 'We, the people.' . . . But through the process of amendment, interpretation, and court decision, I have finally been included in 'We, the people.'"

—BARBARA JORDAN, 1974
The first African-American congresswoman from the deep South (Texas)

Preamble. *Purpose of the Constitution*

We the people of the United States, in order to form a more perfect Union, establish justice, insure domestic tranquility, provide for the common defense, promote the general welfare, and secure the blessings of liberty to ourselves and our posterity, do ordain and establish this Constitution for the United States of America.

A CLOSER LOOK Goals of the Preamble

PREAMBLE	EXPLANATION	EXAMPLES
"Form a more perfect Union"	Create a nation in which states work together	• Interstate road network • U.S. coins, paper money
"Establish justice"	Make laws and set up courts that are fair	• Court system • Jury system
"Insure domestic tranquility"	Keep peace within the country	• National Guard • Federal marshals
"Provide for the common defense"	Safeguard the country against attack	• Army • Navy
"Promote the general welfare"	Contribute to the happiness and well-being of all the people	• Safety in the workplace • Aid to the poor
"Secure the blessings of liberty to ourselves and our posterity"	Make sure future citizens remain free	• Commission on Civil Rights • Federal Election Commission

SKILLBUILDER Interpreting Charts

1. *Which goal of the Preamble do you think is most important? Why?*
2. *How does the Preamble reflect the principle of popular sovereignty?*

Article 1. *The Legislature*

MAIN IDEA The main role of Congress, the legislative branch, is to make laws. Congress is made up of two houses—the Senate and the House of Representatives. Candidates for each house must meet certain requirements. Congress performs specific duties, also called delegated powers.

WHY IT MATTERS NOW Representatives in Congress still voice the views and concerns of the people.

Section 1. Congress All legislative powers herein granted shall be vested in a Congress of the United States, which shall consist of a Senate and House of Representatives.

Section 2. The House of Representatives

1. Elections The House of Representatives shall be composed of members chosen every second year by the people of the several states, and the **electors** in each state shall have the qualifications requisite for electors of the most numerous branch of the state legislature.

2. Qualifications No person shall be a Representative who shall not have attained to the age of twenty-five years, and been seven years a citizen of the United States, and who shall not, when elected, be an inhabitant of that state in which he shall be chosen.

3. Number of Representatives Representatives ~~and direct taxes~~ shall be apportioned among the several states which may be included within this Union, according to their respective numbers, ~~which shall be determined by adding to the whole number of free persons, including those bound to service for a term of years, and excluding Indians not taxed, three-fifths of all other Persons.~~ The actual **enumeration** shall be made within three years after the first meeting of the Congress of the United States, and within every subsequent term of ten years, in such manner as they shall by law direct. The number of Representatives shall not exceed one for every thirty thousand, but each state shall have at least one Representative; ~~and until such enumeration shall be made, the state of New Hampshire shall be entitled to choose three, Massachusetts eight, Rhode Island and Providence Plantations one, Connecticut five, New York six, New Jersey four, Pennsylvania eight, Delaware one, Maryland six, Virginia ten, North Carolina five, South Carolina five, and Georgia three.~~

4. Vacancies When vacancies happen in the representation from any state, the executive authority thereof shall issue writs of election to fill such vacancies.

5. Officers and Impeachment The House of Representatives shall choose their Speaker and other officers; and shall have the sole power of **impeachment.**

VOCABULARY

electors voters

enumeration an official count, such as a census

impeachment the process of accusing a public official of wrongdoing

A CLOSER LOOK

ELECTIONS

Representatives are elected every two years. There are no limits on the number of terms a person can serve.

1. What do you think are the advantages of holding frequent elections of representatives?

A CLOSER LOOK

REPRESENTATION

The 435 members of the House are divided according to state population as determined by a census every ten years.

2. How many representatives does your state have?

Section 3. The Senate

1. Numbers The Senate of the United States shall be composed of two Senators from each state, ~~chosen by the legislature thereof~~, for six years; and each Senator shall have one vote.

2. Classifying Terms Immediately after they shall be assembled in consequence of the first election, they shall be divided as equally as may be into three classes. The seats of the Senators of the first class shall be vacated at the expiration of the second year, of the second class at the expiration of the fourth year, and of the third class at the expiration of the sixth year, so that one-third may be chosen every second year; ~~and if vacancies happen by resignation, or otherwise, during the recess of the legislature of any state, the executive thereof may make temporary appointments until the next meeting of the legislature, which shall then fill such vacancies.~~

3. Qualifications No person shall be a Senator who shall not have attained to the age of thirty years, and been nine years a citizen of the United States, and who shall not, when elected, be an inhabitant of that state for which he shall be chosen.

VOCABULARY

pro tempore for the time being

indictment a written statement issued by a grand jury charging a person with a crime

quorum the minimum number of members that must be present for official business to take place

***A CLOSER LOOK* Federal Office Terms and Requirements**

POSITION	TERM	MINIMUM AGE	RESIDENCY	CITIZENSHIP
Representative	2 years	25	state in which elected	7 years
Senator	6 years	30	state in which elected	9 years
President	4 years	35	14 years in the U.S.	natural-born
Supreme Court Justice	unlimited	none	none	none

SKILLBUILDER Interpreting Charts

Why do you think the term and qualifications for a senator are more demanding than for a representative?

4. Role of Vice-President The Vice-President of the United States shall be President of the Senate, but shall have no vote, unless they be equally divided.

5. Officers The Senate shall choose their other officers, and also a President **pro tempore**, in the absence of the Vice-President, or when he shall exercise the office of President of the United States.

6. Impeachment Trials The Senate shall have the sole power to try all impeachments. When sitting for that purpose, they shall be on oath or affirmation. When the President of the United States is tried, the Chief Justice shall preside: and no person shall be convicted without the concurrence of two-thirds of the members present.

7. Punishment for Impeachment Judgment in cases of impeachment shall not extend further than to removal from office, and disqualification to hold and enjoy any office of honor, trust or profit under the United States; but the party convicted shall nevertheless be liable and subject to **indictment,** trial, judgment and punishment, according to law.

A CLOSER LOOK

IMPEACHMENT

The House brings charges against the president. The Senate acts as the jury. The Chief Justice of the Supreme Court presides over the hearings.

3. How many presidents have been impeached?

Section 4. Congressional Elections

1. Regulations The times, places and manner of holding elections for Senators and Representatives shall be prescribed in each state by the legislature thereof; but the Congress may at any time by law make or alter such regulations, except as to the places of choosing Senators.

2. Sessions The Congress shall assemble at least once in every year, ~~and such meeting shall be on the first Monday in December, unless they shall by law appoint a different day.~~

Section 5. Rules and Procedures

1. Quorum Each house shall be the judge of the elections, returns and qualifications of its own members, and a majority of each shall constitute a **quorum** to do business; but a smaller number may adjourn from day to day, and may be authorized to compel the attendance of absent members, in such manner, and under such penalties as each house may provide.

2. Rules and Conduct Each house may determine the rules of its proceedings, punish its members for disorderly behavior, and, with the concurrence of two-thirds, expel a member.

3. Congressional Records Each house shall keep a journal of its proceedings, and from time to time publish the same, excepting such parts as may in their judgment require secrecy; and the yeas and nays of the members of either house on any question shall, at the desire of one-fifth of those present, be entered on the journal.

4. Adjournment Neither house, during the session of Congress, shall, without the consent of the other, adjourn for more than three days, nor to any other place than that in which the two houses shall be sitting.

A CLOSER LOOK

SENATE RULES

Senate rules allow for debate on the floor. Using a tactic called filibustering, senators give long speeches to block the passage of a bill. Senator Strom Thurmond holds the filibustering record—24 hours, 18 minutes.

4. Why might a senator choose filibustering as a tactic to block a bill?

Section 6. Payment and Privileges

1. Salary The Senators and Representatives shall receive a compensation for their services, to be ascertained by law, and paid out of the treasury of the United States. They shall in all cases, except treason, felony and breach of the peace, be privileged from arrest during their attendance at the session of their respective houses, and in going to and returning from the same; and for any speech or debate in either house, they shall not be questioned in any other place.

2. Restrictions No Senator or Representative shall, during the time for which he was elected, be appointed to any civil office under the authority of the United States, which shall have been created, or the emoluments whereof shall have been increased during such time; and no person holding any office under the United States, shall be a member of either house during his continuance in office.

A CLOSER LOOK

SALARIES

Senators and representatives are paid $136,700 a year. The Speaker of the House is paid $175,400—the same as the vice-president.

5. How do the salaries of members of Congress compare to those of adults you know?

VOCABULARY

revenue income a government collects to cover expenses

naturalization a way to give full citizenship to a person of foreign birth

tribunals courts

felonies serious crimes

appropriation public funds set aside for a specific purpose

Section 7. How a Bill Becomes a Law

1. Tax Bills All bills for raising **revenue** shall originate in the House of Representatives; but the Senate may propose or concur with amendments as on other Bills.

2. Lawmaking Process Every bill which shall have passed the House of Representatives and the Senate, shall, before it become a law, be presented to the President of the United States; if he approves he shall sign it, but if not he shall return it, with his objections to that house in which it shall have originated, who shall enter the objections at large on their journal, and proceed to reconsider it. If after such reconsideration two-thirds of that house shall agree to pass the bill, it shall be sent, together with the objections, to the other house, by which it shall likewise be reconsidered, and if approved by two-thirds of that house, it shall become a law. But in all such cases the votes of both houses shall be determined by yeas and nays, and the names of the persons voting for and against the bill shall be entered on the journal of each house respectively. If any bill shall not be returned by the President within ten days (Sundays excepted) after it shall have been presented to him, the same shall be a law, in like manner as if he had signed it, unless the Congress by their adjournment prevent its return, in which case it shall not be a law.

3. Role of the President Every order, resolution, or vote to which the concurrence of the Senate and House of Representatives may be necessary (except on a question of adjournment) shall be presented to the President of the United States; and before the same shall take effect, shall be approved by him, or being disapproved by him, shall be repassed by two-thirds of the Senate and House of Representatives, according to the rules and limitations prescribed in the case of a bill.

A CLOSER LOOK How a Bill Becomes a Law

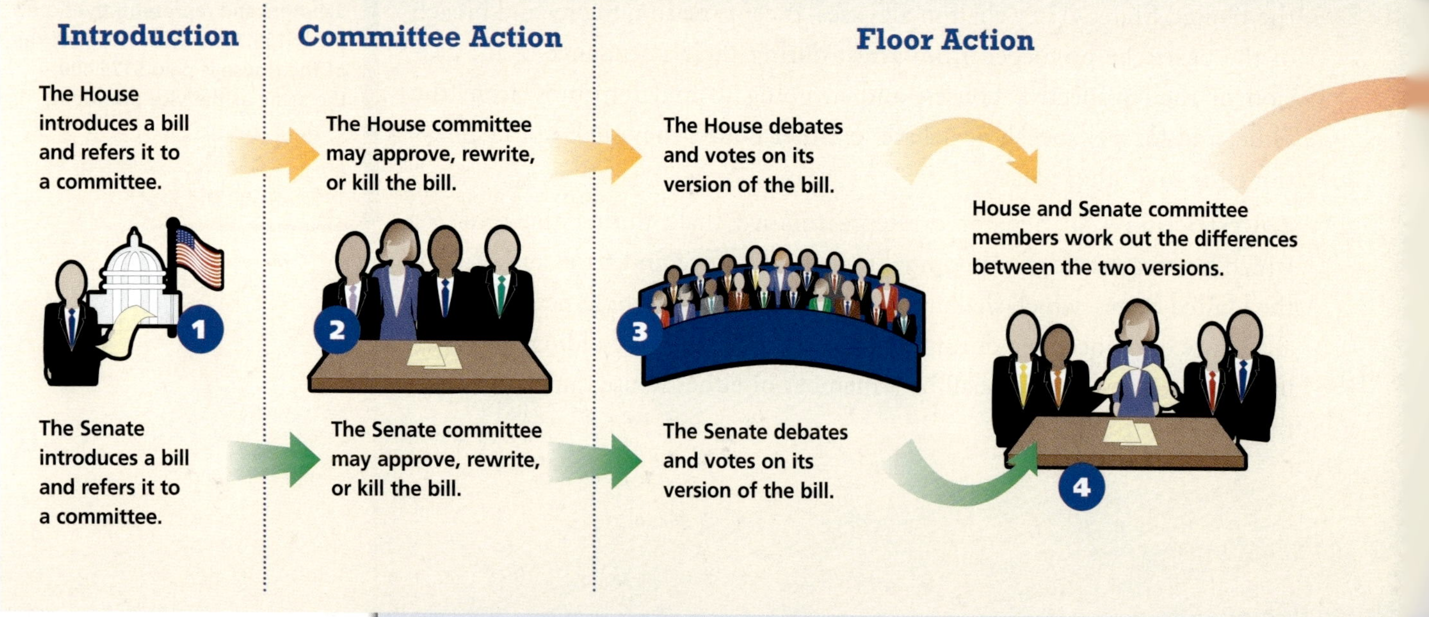

Section 8. Powers Granted to Congress

1. Taxation The Congress shall have power to lay and collect taxes, duties, imposts and excises, to pay the debts and provide for the common defense and general welfare of the United States; but all duties, imposts and excises shall be uniform throughout the United States;

2. Credit To borrow money on the credit of the United States;

3. Commerce To regulate commerce with foreign nations, and among the several states, and with the Indian tribes;

4. Naturalization, Bankruptcy To establish a uniform rule of **naturalization,** and uniform laws on the subject of bankruptcies throughout the United States;

5. Money To coin money, regulate the value thereof, and of foreign coin, and fix the standard of weights and measures;

6. Counterfeiting To provide for the punishment of counterfeiting the securities and current coin of the United States;

7. Post Office To establish post offices and post roads;

8. Patents, Copyrights To promote the progress of science and useful arts, by securing for limited times to authors and inventors the exclusive right to their respective writings and discoveries;

9. Federal Courts To constitute **tribunals** inferior to the Supreme Court;

10. International Law To define and punish piracies and **felonies** committed on the high seas, and offenses against the law of nations;

11. War To declare war, grant letters of marque and reprisal, and make rules concerning captures on land and water;

12. Army To raise and support armies, but no **appropriation** of money to that use shall be for a longer term than two years;

13. Navy To provide and maintain a navy;

A CLOSER LOOK

REGULATING COMMERCE

Commerce can also apply to travelers crossing state lines. Congress's power to regulate the movement of people from state to state paved the way for the Civil Rights Act of 1964. This act included fair treatment of interstate travelers. People of all races can use public places, such as hotels and bus stations.

6. To what other areas might the commerce clause apply?

A CLOSER LOOK

DECLARING WAR

Only Congress can declare war. Yet in the following "undeclared" wars, Congress bowed to the president's power to take military action and send troops overseas: Korean War (1950–1953), Vietnam War (1957–1975), Persian Gulf War (1991), and Kosovo crisis (1999).

7. Why do you think the Constitution sets limits on the president's war-making powers?

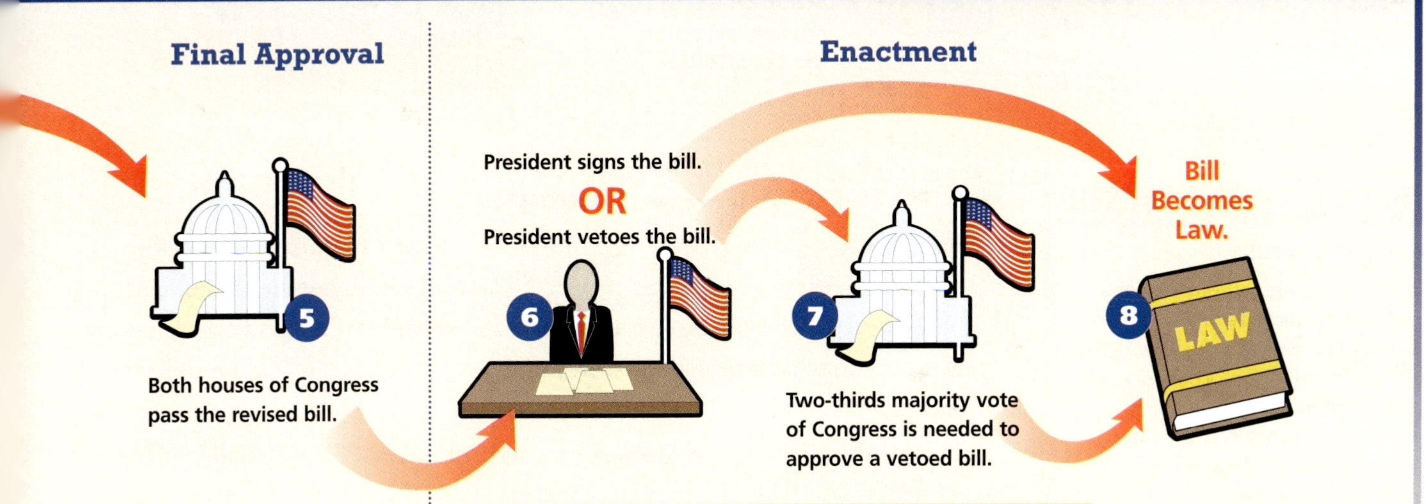

SKILLBUILDER Interpreting Charts

1. *How can a president block a bill?*
2. *What examples of checks and balances are shown in the chart?*

VOCABULARY

militia an emergency military force, such as the National Guard, that is not part of the regular army

bill of attainder a law that condemns a person without a trial in court

ex post facto law a law that would make an act a criminal offense after it was committed

tender money

14. Regulation of Armed Forces To make rules for the government and regulation of the land and naval forces;

15. Militia To provide for calling forth the **militia** to execute the laws of the Union, suppress insurrections and repel invasions;

16. Regulations for Militia To provide for organizing, arming, and disciplining the militia, and for governing such part of them as may be employed in the service of the United States, reserving to the states respectively the appointment of the officers, and the authority of training the militia according to the discipline prescribed by Congress;

17. District of Columbia To exercise exclusive legislation in all cases whatsoever, over such district (not exceeding ten miles square) as may, by cession of particular states, and the acceptance of Congress, become the seat of the government of the United States, and to exercise like authority over all places purchased by the consent of the legislature of the state in which the same shall be, for the erection of forts, magazines, arsenals, dockyards, and other needful buildings;—and

18. Elastic Clause To make all laws which shall be necessary and proper for carrying into execution the foregoing powers, and all other powers vested by this Constitution in the government of the United States, or in any department or officer thereof.

A CLOSER LOOK The Elastic Clause

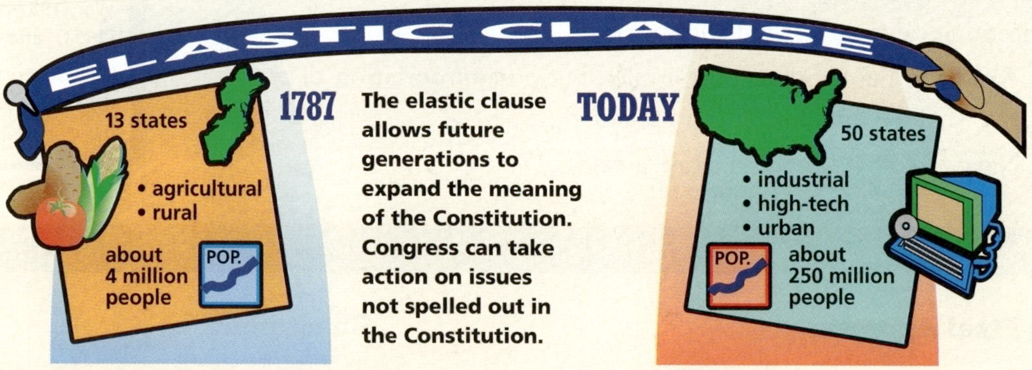

A CLOSER LOOK

HABEAS CORPUS

A writ of habeas corpus is a legal order. It protects people from being held in prison or jail without formal charges of a crime. In 1992, the Supreme Court recognized that "habeas corpus is the [basic] instrument for safeguarding individual freedom."

8. How does habeas corpus help ensure fairness and justice?

Section 9. Powers Denied Congress

~~**1. Slave Trade** The migration or importation of such persons as any of the states now existing shall think proper to admit, shall not be prohibited by the Congress prior to the year one thousand eight hundred and eight, but a tax or duty may be imposed on such importation, not exceeding ten dollars for each person.~~

2. Habeas Corpus The privilege of the writ of habeas corpus shall not be suspended, unless when in cases of rebellion or invasion the public safety may require it.

3. Illegal Punishment No **bill of attainder** or **ex post facto law** shall be passed.

4. Direct Taxes No capitation, ~~or other direct,~~ tax shall be laid, ~~unless in proportion to the census or enumeration herein before directed to be taken.~~

5. Export Taxes No tax or duty shall be laid on articles exported from any state.

6. No Favorites No preference shall be given by any regulation of commerce or revenue to the ports of one state over those of another: nor shall vessels bound to, or from, one state be obliged to enter, clear, or pay duties in another.

7. Public Money No money shall be drawn from the treasury, but in consequence of appropriations made by law; and a regular statement and account of the receipts and expenditures of all public money shall be published from time to time.

8. Titles of Nobility No title of nobility shall be granted by the United States: and no person holding any office of profit or trust under them shall, without the consent of the Congress, accept of any present, emolument, office, or title, of any kind whatever, from any king, prince, or foreign state.

A CLOSER LOOK

DIRECT TAX

In 1913, the 16th Amendment allowed Congress to collect an income tax—a direct tax on the amount of money a person earns. Americans today pay much more in taxes than their ancestors would have imagined.

9. Why do you think the issue of taxes is so important to people?

A CLOSER LOOK

TITLES OF NOBILITY

The Framers disapproved of titles of nobility. The Declaration of Independence, signed in 1776, listed King George III's abuses of power. One symbol of this abuse was English titles of nobility, such as "king," "queen," and "duke." The Framers said clearly that there would be no such titles in the new republic.

10. How do TV news reporters address members of Congress and the president?

Section 10. Powers Denied the States

1. Restrictions No state shall enter into any treaty, alliance, or confederation; grant letters of marque and reprisal; coin money; emit bills of credit; make anything but gold and silver coin a **tender** in payment of debts; pass any bill of attainder, ex post facto law, or law impairing the obligation of contracts, or grant any title of nobility.

2. Import and Export Taxes No state shall, without the consent of the Congress, lay any imposts or duties on imports or exports, except what may be absolutely necessary for executing its inspection laws; and the net produce of all duties and imposts, laid by any state on imports or exports, shall be for the use of the treasury of the United States; and all such laws shall be subject to the revision and control of the Congress.

3. Peacetime and War Restraints No state shall, without the consent of Congress, lay any duty of tonnage, keep troops or ships of war in time of peace, enter into any agreement or compact with another state, or with a foreign power, or engage in war, unless actually invaded, or in such imminent danger as will not admit of delay.

Article 1 Assessment

1. Main Ideas

a. What is the main job of the legislative branch?

b. What role does the vice-president of the United States play in the Senate?

c. Why are there more members in the House of Representatives than the Senate?

d. What is one of the powers denied to Congress?

2. Critical Thinking

Drawing Conclusions How does Article 1 show that the Constitution is a clearly defined yet flexible document?

THINK ABOUT

- the powers of Congress
- the "elastic clause"

VOCABULARY

natural-born citizen a citizen born in the United States or a U.S. commonwealth, or to parents who are U.S. citizens living outside the country

affirmation a statement declaring that something is true

Article 2. *The Executive*

MAIN IDEA **The president and vice-president are the leaders of the executive branch. Their main role is to enforce the laws. The president commands the military and makes foreign treaties with the Senate's approval.**

WHY IT MATTERS NOW **As the United States has become a world power, the authority of the president has also expanded.**

Section 1. The Presidency

1. Terms of Office The executive power shall be vested in a President of the United States of America. He shall hold his office during the term of four years, and, together with the Vice-President, chosen for the same term, be elected, as follows:

2. Electoral College Each state shall appoint, in such manner as the Legislature thereof may direct, a number of electors, equal to the whole number of Senators and Representatives to which the State may be entitled in the Congress; but no Senator or Representative, or person holding an office of trust or profit under the United States, shall be appointed an elector.

A CLOSER LOOK Electoral College *(1990 Census)*

American voters do not choose their president directly. Members of a group called the electoral college actually elect the president. Each state has electors. Together they form the electoral college. In most states, the winner takes all. Except for Maine and Nebraska, all the electoral votes of a state go to one set of candidates.

number of electors for each state = total number of its senators and representatives

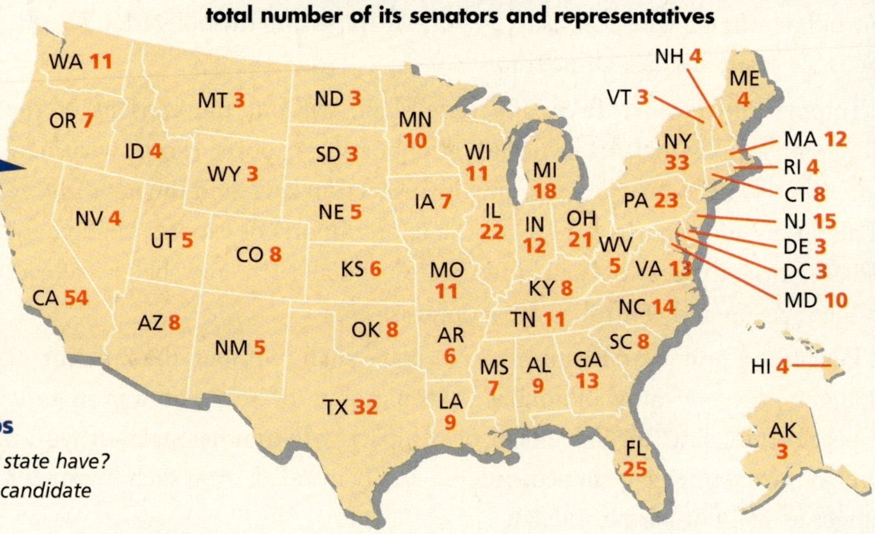

SKILLBUILDER Interpreting Maps

1. *How many electoral votes does your state have?*
2. *In which states would a presidential candidate campaign most heavily? Why?*

~~**3. Former Method of Electing President** The electors shall meet in their respective states, and vote by ballot for two persons, of whom one at least shall not be an inhabitant of the same state with themselves. And they shall make a list of all the persons voted for, and of the number of votes for each; which list they shall sign and certify, and transmit sealed to the seat of the government of the United States, directed to the President of the Senate. The President of the Senate shall, in the presence of the Senate and House of Representatives, open all the certificates, and the votes shall then be counted. The person having the greatest number of votes shall be the~~

President, if such number be a majority of the whole number of electors appointed; and if there be more than one who have such majority, and have an equal number of votes, then the House of Representatives shall immediately choose by ballot one of them for President; and if no person have a majority, then from the five highest on the list the said House shall in like manner choose the President. But in choosing the President, the votes shall be taken by States, the representation from each state having one vote; a quorum for this purpose shall consist of a member or members from two-thirds of the states, and a majority of all the states shall be necessary to a choice. In every case, after the choice of the President, the person having the greatest number of votes of the electors shall be the Vice-President. But if there should remain two or more who have equal votes, the Senate shall choose from them by ballot the Vice-President.

A CLOSER LOOK

Vice-President Lyndon Johnson, next in line of succession, takes the oath of office after the assassination of President John F. Kennedy in 1963. Johnson, like every U.S. president, promises to uphold the Constitution. The 25th Amendment sets up clearer procedures for presidential succession.

4. Election Day The Congress may determine the time of choosing the electors, and the day on which they shall give their votes, which day shall be the same throughout the United States.

5. Qualifications No person except a **natural-born citizen,** or a citizen of the United States at the time of the adoption of this Constitution, shall be eligible to the office of President; neither shall any person be eligible to that office who shall not have attained to the age of thirty-five years, and been fourteen years a resident within the United States.

6. Succession In case of the removal of the President from office, or of his death, resignation, or inability to discharge the powers and duties of the said office, the same shall devolve on the Vice-President, and the Congress may by law provide for the case of removal, death, resignation or inability, both of the President and Vice-President, declaring what officer shall then act as President, and such officer shall act accordingly, until the disability be removed, or a President shall be elected.

7. Salary The President shall, at stated times, receive for his services, a compensation, which shall neither be increased nor diminished during the period for which he shall have been elected, and he shall not receive within that period any other emolument from the United States, or any of them.

8. Oath of Office Before he enter on the execution of his office, he shall take the following oath or **affirmation:**—"I do solemnly swear (or affirm) that I will faithfully execute the office of President of the United States, and will to the best of my ability, preserve, protect and defend the Constitution of the United States."

A CLOSER LOOK

PRESIDENT'S SALARY

The president's yearly salary is $200,000. The president also gets special allowances, such as $100,000 for travel expenses. Here are some other benefits:

- living in a mansion, the White House
- vacationing at Camp David, an estate in Maryland
- using *Air Force One*, a personal jet plane

11. Why do you think the president needs to have a plane and a vacation spot?

A CLOSER LOOK Roles of the President

Commander in Chief

As a military leader, President Abraham Lincoln meets with his generals during the Civil War.

Chief Executive

Like a business executive, the president solves problems and makes key decisions. President John F. Kennedy is shown in the oval office in 1962.

Chief Diplomat and Chief of State

As a foreign policy maker, President Richard M. Nixon visits the People's Republic of China in 1972.

Legislative Leader

President Lyndon Johnson signs the Civil Rights Act of 1964. All modern presidents have legislative programs they want Congress to pass.

Head of a Political Party

President Ronald Reagan rallies support at the 1984 Republican Convention. By this time, Reagan had put together a strong bloc of voters who supported the Republican Party's policies. During his presidency (1981–1989), Reagan helped build new unity among party members.

Section 2. Powers of the President

1. Military Powers The President shall be commander in chief of the Army and Navy of the United States, and of the militia of the several states, when called into the actual service of the United States; he may require the opinion, in writing, of the principal officer in each of the executive departments, upon any subject relating to the duties of their respective offices, and he shall have power to grant **reprieves** and pardons for offenses against the United States, except in cases of impeachment.

2. Treaties, Appointments He shall have power, by and with the advice and consent of the Senate, to make treaties, provided two-thirds of the Senators present concur; and he shall nominate, and by and with the advice and consent of the Senate, shall appoint ambassadors, other public ministers and consuls, judges of the Supreme Court, and all other officers of the United States, whose appointments are not herein otherwise provided for, and which shall be established by law; but the Congress may by law vest the appointment of such inferior officers, as they think proper, in the President alone, in the courts of law, or in the heads of departments.

3. Vacancies The President shall have power to fill up all vacancies that may happen during the recess of the Senate, by granting commissions which shall expire at the end of their next session.

Section 3. Presidential Duties He shall from time to time give to the Congress information of the State of the Union, and recommend to their consideration such measures as he shall judge necessary and expedient; he may, on extraordinary occasions, **convene** both houses, or either of them, and in case of disagreement between them, with respect to the time of adjournment, he may adjourn them to such time as he shall think proper; he shall receive ambassadors and other public ministers; he shall take care that the laws be faithfully executed, and shall commission all the officers of the United States.

Section 4. Impeachment The President, Vice-President and all civil officers of the United States shall be removed from office on impeachment for, and conviction of, treason, bribery, or other high crimes and **misdemeanors**.

VOCABULARY

reprieves delays or cancellations of punishment

convene call together

misdemeanors violations of the law

A CLOSER LOOK

SUPREME COURT APPOINTMENTS

Recent presidents have used their power of appointment to add minorities and women to the Supreme Court. In 1967, President Lyndon Johnson appointed the first African-American justice, Thurgood Marshall. In 1981, President Ronald Reagan appointed the first woman, Sandra Day O'Connor.

12. What do you think influences a president's choice for a Supreme Court justice?

A CLOSER LOOK

STATE OF THE UNION

Major TV networks broadcast the State of the Union address to the whole nation. In this yearly message, the president urges Congress to achieve certain lawmaking goals. The president's speech also must gain the attention of TV viewers.

13. Why is the president's power to persuade an important political skill?

Article 2 Assessment

1. Main Ideas

a. What is the chief purpose of the executive branch?

b. What are the requirements for becoming president?

c. How does the Constitution limit the president's power to make appointments and treaties?

2. Critical Thinking

Analyzing Issues Why do you think the Constitution states that the president must seek approval from the Senate for most political appointments and treaties?

THINK ABOUT
- the abuse of power
- the will of the voters

VOCABULARY

inferior courts courts with less authority than the Supreme Court

appellate having power to review court decisions

A CLOSER LOOK

ORGANIZING FEDERAL COURTS

The Judiciary Act of 1789, passed by the First Congress, included establishing a Supreme Court with a chief justice and five associate justices and other lower federal courts.

14. How many Supreme Court justices are there today?

A CLOSER LOOK

JUDICIAL POWER

Judicial power gives the Supreme Court and other federal courts the authority to hear certain kinds of cases. These courts have the power to rule in cases involving the Constitution, national laws, treaties, and states' conflicts.

15. What federal cases have you seen reported on TV?

Article 3. *The Judiciary*

MAIN IDEA **The judicial branch interprets the laws. This branch includes the Supreme Court, the highest court in the nation, and other federal courts.**

WHY IT MATTERS NOW **Supreme Court rulings can shape government policies on hotly debated issues.**

Section 1. Federal Courts and Judges The judicial power of the United States shall be vested in one Supreme Court, and in such **inferior courts** as the Congress may from time to time ordain and establish. The judges, both of the Supreme and inferior courts, shall hold their offices during good behavior, and shall, at stated times, receive for their services a compensation, which shall not be diminished during their continuance in office.

Section 2. The Courts' Authority

1. General Authority The judicial power shall extend to all cases, in law and equity, arising under this Constitution, the laws of the United States, and treaties made, or which shall be made, under their authority;—to all cases affecting ambassadors, other public ministers and consuls;—to all cases of admiralty and maritime jurisdiction;—to controversies to which the United States shall be a party;—to controversies between two or more states;—~~between a state and citizens of another state;—~~between citizens of different states;—between citizens of the same state claiming lands under grants of different states, ~~and between a state, or the citizens thereof, and foreign states, citizens or subjects.~~

A CLOSER LOOK Judicial Review

Judicial review allows the Supreme Court and other federal courts to play a key role in lawmaking. The judges examine a law or government activity. They then decide whether it violates the Constitution. The Supreme Court established this important right in the case of *Marbury* v. *Madison* (1803). (See Chapter 10.)

2. Supreme Court In all cases affecting ambassadors, other public ministers and consuls, and those in which a state shall be party, the Supreme Court shall have original jurisdiction. In all the other cases before mentioned, the Supreme Court shall have **appellate** jurisdiction, both as to law and fact, with such exceptions, and under such regulations, as the Congress shall make.

A CLOSER LOOK Checks and Balances

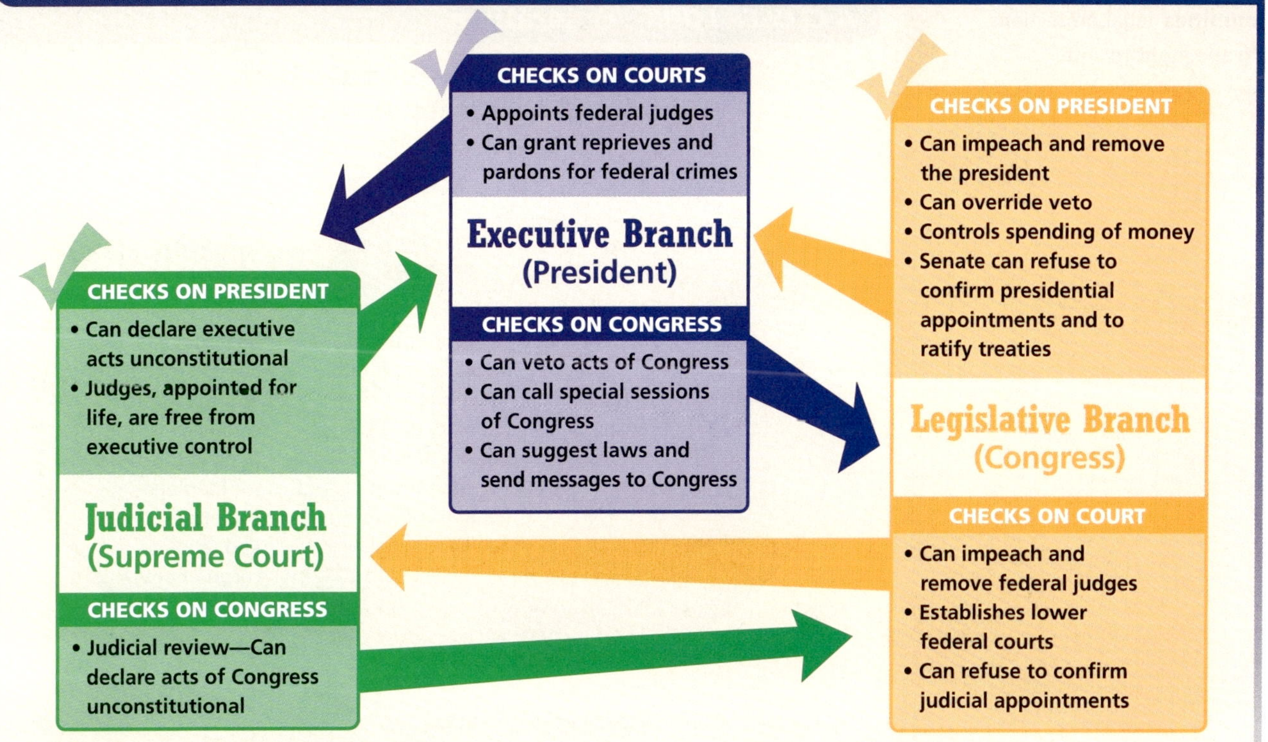

SKILLBUILDER Interpreting Charts

1. *Why is judicial review an important action of the Supreme Court?*
2. *Which check do you think is most powerful? Why?*

3. Trial by Jury The trial of all crimes, except in cases of impeachment, shall be by jury; and such trial shall be held in the state where the said crimes shall have been committed; but when not committed within any state, the trial shall be at such place or places as the Congress may by law have directed.

Section 3. Treason

1. Definition Treason against the United States shall consist only in levying war against them, or in adhering to their enemies, giving them aid and comfort. No person shall be convicted of treason unless on the testimony of two witnesses to the same overt act, or on confession in open court.

2. Punishment The Congress shall have power to declare the punishment of treason, but no attainder of treason shall work corruption of blood, or forfeiture except during the life of the person attained.

Article 3 Assessment

1. Main Ideas

a. What is the main purpose of the judicial branch?

b. What is judicial review?

c. What are two kinds of cases that can begin in the Supreme Court?

2. Critical Thinking

Drawing Conclusions Why might the Supreme Court feel less political pressure than Congress in making judgments about the Constitution?

THINK ABOUT

- the appointment of Supreme Court justices
- Congress members' obligation to voters

VOCABULARY

immunities legal protections

suffrage right to vote

A CLOSER LOOK Federalism

Americans live under both national and state governments.

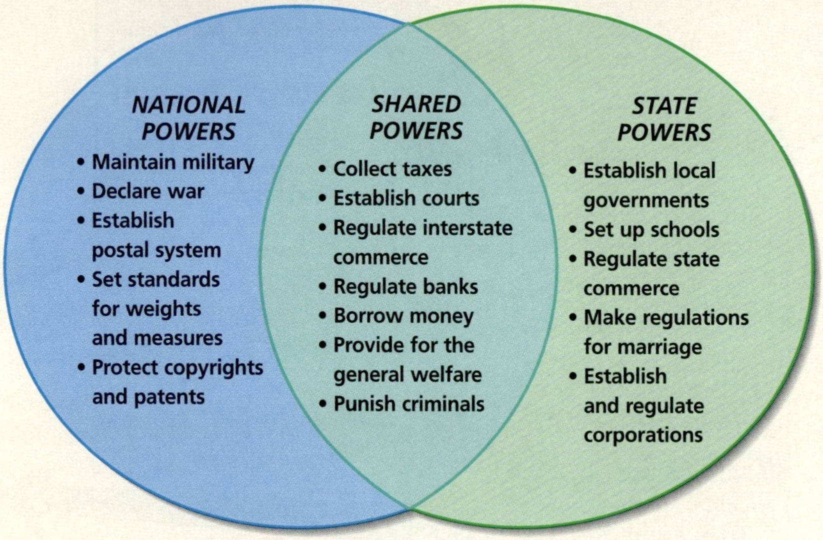

SKILLBUILDER Interpreting Charts

What do you think is the purpose of dividing the powers between national and state governments?

Article 4. *Relations Among States*

MAIN IDEA **States must honor one another's laws, records, and court rulings.**

WHY IT MATTERS NOW **Article 4 promotes cooperation, equality, and fair treatment of citizens from all the states.**

Section 1. State Acts and Records Full faith and credit shall be given in each state to the public acts, records, and judicial proceedings of every other state. And the Congress may by general laws prescribe the manner in which such acts, records and proceedings shall be proved, and the effect thereof.

Section 2. Rights of Citizens

1. Citizenship The citizens of each state shall be entitled to all privileges and **immunities** of citizens in the several states.

2. Extradition A person charged in any state with treason, felony, or other crime, who shall flee from justice, and be found in another state, shall on demand of the executive authority of the state from which he fled, be delivered up, to be removed to the state having jurisdiction of the crime.

~~**3. Fugitive Slaves** No person held to service or labor in one state, under the laws thereof, escaping into another, shall, in consequence of any law or regulation therein, be discharged from such service or labor, but shall be delivered up on claim of the party to whom such service or labor may be due.~~

A CLOSER LOOK

EXTRADITION

Persons charged with serious crimes cannot escape punishment by fleeing to another state. They must be returned to the first state and stand trial there.

16. Why do you think the Framers included the power of extradition?

Section 3. New States

1. Admission New states may be admitted by the Congress into this Union; but no new state shall be formed or erected within the jurisdiction of any other state; nor any state be formed by the junction of two or more states, or parts of states, without the consent of the legislatures of the states concerned as well as of the Congress.

2. Congressional Authority The Congress shall have power to dispose of and make all needful rules and regulations respecting the territory or other property belonging to the United States; and nothing in this Constitution shall be so construed as to prejudice any claims of the United States, or of any particular state.

Section 4. Guarantees to the States

The United States shall guarantee to every state in this Union a republican form of government, and shall protect each of them against invasion; and on application of the legislature, or of the executive (when the legislature cannot be convened) against domestic violence.

A CLOSER LOOK

ADMISSION TO STATEHOOD

In 1998, Puerto Ricans voted against their island becoming the 51st state. A lawyer in Puerto Rico summed up a main reason: "Puerto Ricans want to have ties to the U. S., but they want to protect their language and culture." Also, as a U.S. commonwealth, Puerto Rico makes its own laws and handles its own finances.

17. Do you think Puerto Rico should become a state? Why or why not?

Article 5. *Amending the Constitution*

MAIN IDEA **The Constitution can be amended, or formally changed.**

WHY IT MATTERS NOW **The amendment process allows the Constitution to adapt to modern times.**

The Congress, whenever two-thirds of both houses shall deem it necessary, shall propose amendments to this Constitution, or, on the application of the legislatures of two-thirds of the several states, shall call a convention for proposing amendments, which, in either case, shall be valid to all intents and purposes, as part of this Constitution, when ratified by the legislatures of three-fourths of the several states, or by conventions in three-fourths thereof, as the one or the other mode of ratification may be proposed by the Congress; ~~provided that no amendment which may be made prior to the year one thousand eight hundred and eight shall in any manner affect the first and fourth clauses in the ninth section of the first article~~; and that no state, without its consent, shall be deprived of its equal **suffrage** in the Senate.

A CLOSER LOOK **Process for Amending the Constitution**

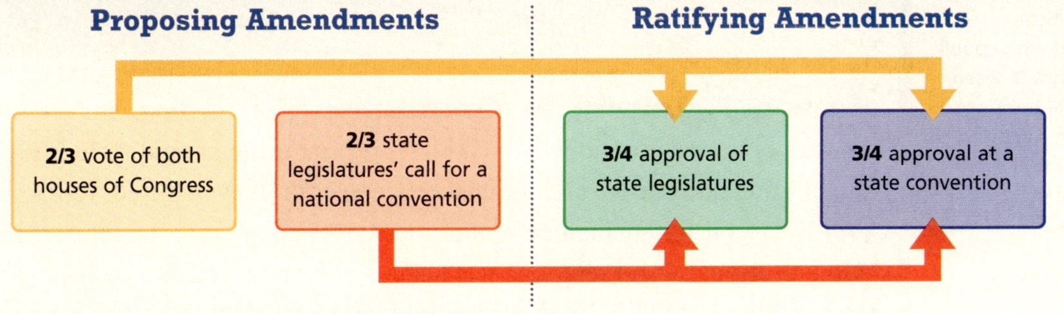

SKILLBUILDER **Interpreting Charts**

Why do you think more votes are needed to ratify an amendment than to propose one?

VOCABULARY

ratification official approval

unanimous consent complete agreement

A CLOSER LOOK

PAYING DEBTS

The U.S. government agreed to pay all debts held under the Articles of Confederation. For example, the United States still owed money from the costs of the Revolutionary War.

18. What problems might arise in a country that has a huge national debt?

Article 6. *Supremacy of the National Government*

MAIN IDEA The Constitution, national laws, and treaties are the supreme, or highest, law of the land. All government officials must promise to support the Constitution.

WHY IT MATTERS NOW The authority of federal laws over state laws helps keep the nation unified.

Section 1. Valid Debts All debts contracted and engagements entered into, before the adoption of this Constitution, shall be as valid against the United States under this Constitution, as under the Confederation.

Section 2. Supreme Law This Constitution, and the laws of the United States which shall be made in pursuance thereof; and all treaties made, or which shall be made, under the authority of the United States, shall be the supreme law of the land; and the judges in every state shall be bound thereby, anything in the constitution or laws of any state to the contrary notwithstanding.

A CLOSER LOOK

In 1957, the "supreme law of the land" was put to a test. The governor of Arkansas defied a Supreme Court order. The Court ruled that African-American students could go to all-white public schools. President Dwight D. Eisenhower then sent federal troops to protect the first African-American students to enroll in Central High School in Little Rock, Arkansas.

Section 3. Loyalty to Constitution The Senators and Representatives before mentioned, and the members of the several state legislatures, and all executive and judicial officers, both of the United States and of the several states, shall be bound by oath or affirmation to support this Constitution; but no religious test shall ever be required as a qualification to any office or public trust under the United States.

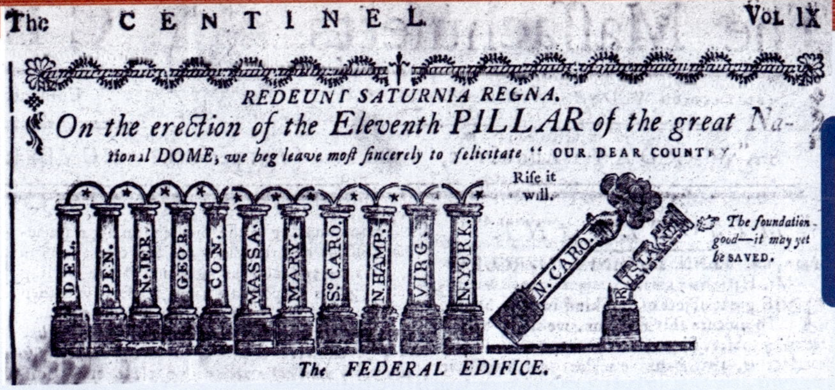

A CLOSER LOOK

This political cartoon shows that New York was the 11th state to ratify the Constitution. Each of the 13 states is represented by a pillar.

Article 7. *Ratification*

MAIN IDEA Nine of the 13 states had to ratify, or approve, the Constitution before it could go into effect.

WHY IT MATTERS NOW The approval of the Constitution launched a new plan of government still followed today.

The **ratification** of the conventions of nine states shall be sufficient for the establishment of this Constitution between the states so ratifying the same. Done in convention by the **unanimous consent** of the states present, the seventeenth day of September in the year of our Lord one thousand seven hundred and eighty-seven and of the independence of the United States of America the twelfth. In witness whereof we have hereunto subscribed our names.

George Washington—President and deputy from Virginia
New Hampshire: *John Langdon, Nicholas Gilman*
Massachusetts: *Nathaniel Gorham, Rufus King*
Connecticut: *William Samuel Johnson, Roger Sherman*
New York: *Alexander Hamilton*
New Jersey: *William Livingston, David Brearley, William Paterson, Jonathan Dayton*
Pennsylvania: *Benjamin Franklin, Thomas Mifflin, Robert Morris, George Clymer, Thomas FitzSimons, Jared Ingersoll, James Wilson, Gouverneur Morris*
Delaware: *George Read, Gunning Bedford, Jr., John Dickinson, Richard Bassett, Jacob Broom*
Maryland: *James McHenry, Dan of St. Thomas Jenifer, Daniel Carroll*
Virginia: *John Blair, James Madison, Jr.*
North Carolina: *William Blount, Richard Dobbs Spaight, Hugh Williamson*
South Carolina: *John Rutledge, Charles Cotesworth Pinckney, Charles Pinckney, Pierce Butler*
Georgia: *William Few, Abraham Baldwin*

A CLOSER LOOK

THE SIGNERS

The 39 men who signed the Constitution were wealthy and well-educated. About half of them were trained in law. Others were doctors, merchants, bankers, and slaveholding planters. Missing from the list of signatures are the names of African Americans, Native Americans, and women. These groups reflected the varied population of the United States in the 1780s.

19. How do you think the absence of these groups affected the decisions made in creating the Constitution?

Articles 4–7 Assessment

1. Main Ideas

a. What rights does Article 4 guarantee to citizens if they go to other states in the nation?

b. What are two ways of proposing an amendment to the Constitution?

c. What makes up "the supreme law of the land"?

2. Critical Thinking

Forming and Supporting Opinions Should the Framers of the Constitution have allowed the people to vote directly for ratification of the Constitution? Why or why not?

THINK ABOUT

- the idea that the government belongs to the people
- the general public's ability to make sound political decisions

The Bill of Rights and Amendments 11–27

In 1787, Thomas Jefferson sent James Madison a letter about the Constitution. Jefferson wrote, "I will now add what I do not like . . . [there is no] bill of rights." He explained his reasons: "A bill of rights is what the people are entitled to against every government on earth . . . and what no just government should refuse." Jefferson's disapproval is not surprising. In writing the Declaration of Independence, he spelled out basic individual rights that cannot be taken way. These are "life, liberty, and the pursuit of happiness." The Declaration states that governments are formed to protect these rights.

Several states approved the Constitution only if a list of guaranteed freedoms was added. While serving in the nation's first Congress, James Madison helped draft the Bill of Rights. In 1791, these first ten amendments became part of the Constitution.

AMENDMENTS 1–10. *The Bill of Rights*

MAIN IDEA The Bill of Rights protects citizens from government interference.

WHY IT MATTERS NOW Issues related to the Bill of Rights are still being applied, tested, and interpreted.

AMENDMENT 1. Religious and Political Freedom (1791)

Congress shall make no law respecting an establishment of religion, or prohibiting the free exercise thereof; or **abridging** the freedom of speech, or of the press; or the right of the people peaceably to assemble, and to petition the Government for a redress of grievances.

A CLOSER LOOK The Five Freedoms

SKILLBUILDER Interpreting Charts

1. *Why is freedom of speech and the press important in a democratic society?*
2. *What impact has religious freedom had on the American way of life?*

AMENDMENT 2. Right to Bear Arms (1791) A well-regulated militia, being necessary to the security of a free state, the right of the people to keep and bear arms, shall not be infringed.

AMENDMENT 3. Quartering Troops (1791) No soldier shall, in time of peace be **quartered** in any house, without the consent of the owner, nor in time of war, but in a manner to be prescribed by law.

AMENDMENT 4. Search and Seizure (1791) The right of the people to be secure in their persons, houses, papers, and effects, against unreasonable searches and seizures, shall not be violated, and no warrants shall issue, but upon probable cause, supported by oath or affirmation, and particularly describing the place to be searched, and the persons or things to be seized.

AMENDMENT 5. Rights of Accused Persons (1791) No person shall be held to answer for a capital, or otherwise infamous crime, unless on a presentment or indictment of a Grand Jury, except in cases arising in the land or naval forces, or in the militia, when in actual service in time of war or public danger; nor shall any person be subject for the same offense to be twice put in jeopardy of life or limb; nor shall be compelled in any criminal case to be a witness against himself, nor be deprived of life, liberty, or property, without **due process of law;** nor shall private property be taken for public use, without just compensation.

AMENDMENT 6. Right to a Speedy, Public Trial (1791) In all criminal prosecutions, the accused shall enjoy the right to a speedy and public trial, by an impartial jury of the State and district wherein the crime shall have been committed, which district shall have been previously ascertained by law, and to be informed of the nature and cause of the accusation; to be confronted with the witnesses against him; to have **compulsory process** for obtaining witnesses in his favor, and to have the assistance of **counsel** for his defense.

VOCABULARY

abridging reducing

quartered given a place to stay

due process of law fair treatment under the law

compulsory process required procedure

counsel a lawyer

A CLOSER LOOK

SEARCHES

Metal detectors at airports search passengers. Airline workers search all carry-on luggage. Do these actions violate the 4th Amendment? The courts say no. They have cited many situations that allow for searches without a warrant, or written order. A person's right to privacy is balanced against the government's need to prevent crime.

20. What does the right to privacy mean to you at home and at school?

A CLOSER LOOK

In 1966, the Supreme Court made a decision based on the 5th and 6th Amendments. The warnings outlined in this ruling are often called "Miranda rights." Miranda rights protect suspects from giving forced confessions. Police must read these rights to a suspect they are questioning. For example:

- "You have the right to remain silent."
- "Anything that you say can and will be used against you in a court of law."
- "You have the right to an attorney."

A CLOSER LOOK

Protesters claim that the death penalty violates the 8th Amendment, which protects people against "cruel and unusual punishments." Supporters believe that the death penalty is a justly deserved punishment.

AMENDMENT 7. Trial by Jury in Civil Cases (1791) In suits at **common law,** where the value in controversy shall exceed twenty dollars, the right of trial by jury shall be preserved, and no fact tried by a jury, shall be otherwise reexamined in any court of the United States, than according to the rules of the common law.

AMENDMENT 8. Limits of Fines and Punishments (1791) Excessive **bail** shall not be required, nor excessive fines imposed, nor cruel and unusual punishments inflicted.

AMENDMENT 9. Rights of People (1791) The enumeration in the Constitution of certain rights shall not be construed to deny or disparage others retained by the people.

AMENDMENT 10. Powers of States and People (1791) The powers not delegated to the United States by the Constitution, nor prohibited by it to the States, are reserved to the States respectively, or to the people.

A CLOSER LOOK

STATES' POWERS

The 10th Amendment gives the states reserved powers. Any powers not clearly given to the national government by the U.S. Constitution belong to the states. State constitutions sometimes assume authority in unexpected areas. For example, California's constitution sets rules for governing the use of fishing nets.

21. What are some common areas in which states have authority?

Bill of Rights Assessment

1. Main Ideas

a. Which amendment protects your privacy?

b. Which amendments guarantee fair legal treatment?

c. Which amendment prevents the federal government from taking powers away from the states and the people?

2. Critical Thinking

Forming and Supporting Opinions The 4th, 5th, 6th, 7th, and 8th Amendments protect innocent people accused of crimes. Do you think these five amendments also favor the rights of actual criminals? Explain.

THINK ABOUT

- criminals who go free if valuable evidence is found after their trials
- criminals released on bail

Amendments 11–27

MAIN IDEA The Constitution has adapted to social changes and historical trends.

WHY IT MATTERS NOW Amendments 11–27 show that the Constitution is a living document.

AMENDMENT 11. Lawsuits Against States (1798)

Passed by Congress March 4, 1794. Ratified February 7, 1795. Proclaimed 1798.
Note: Article 3, Section 2, of the Constitution was modified by Amendment 11.

The Judicial power of the United States shall not be construed to extend to any suit in law or **equity**, commenced or prosecuted against one of the United States by citizens of another state, or by citizens or subjects of any foreign state.

AMENDMENT 12. Election of Executives (1804)

Passed by Congress December 9, 1803. Ratified June 15, 1804.
Note: Part of Article 2, Section 1, of the Constitution was replaced by the 12th Amendment.

The electors shall meet in their respective states and vote by ballot for President and Vice-President, one of whom, at least, shall not be an inhabitant of the same state with themselves; they shall name in their ballots the person voted for as President, and in distinct ballots the person voted for as Vice-President, and they shall make distinct lists of all persons voted for as President, and of all persons voted for as Vice-President, and of the number of votes for each, which lists they shall sign and certify, and transmit sealed to the seat of the government of the United States, directed to the President of the Senate;—the President of the Senate shall, in the presence of the Senate and House of Representatives, open all the certificates and the votes shall then be counted;—the person having the greatest number of votes for President, shall be the President, if such number be a majority of the whole number of electors appointed; and if no person have such majority, then from the persons having the highest numbers not exceeding three on the list of those voted for as President, the House of Representatives shall choose immediately, by ballot, the President. But in choosing the President, the votes shall be taken by states, the representation from each state having one vote; a quorum for this purpose shall consist of a member or members from two-thirds of the states, and a majority of all the states shall be necessary to a choice. And if the House of Representatives shall not choose a President whenever the right of choice shall devolve upon them, ~~before the fourth day of March next following,~~ then the Vice-President shall act as President, as in the case of the death or other constitutional disability of the President. The person having the greatest number of votes as Vice-President, shall be the Vice-President, if such number be a majority of the whole number of Electors appointed, and if no person have a majority, then from the two highest numbers on the list, the Senate shall choose the Vice-President; a quorum for the purpose shall consist of two-thirds of the whole number of Senators, and a majority of the whole number shall be necessary to a choice. But no person constitutionally ineligible to the office of President shall be eligible to that of Vice-President of the United States.

VOCABULARY

common law a system of law developed in England, based on customs and previous court decisions

bail money paid by arrested persons to guarantee they will return for trial

equity a system of justice not covered under common law

A CLOSER LOOK

SEPARATE BALLOTS

The presidential election of 1800 ended in a tie between Thomas Jefferson and Aaron Burr. At this time, the candidate with the most votes became president. The runner-up became vice-president. The 12th Amendment calls for separate ballots for the president and vice-president. The vice-president is specifically elected to the office, rather than being the presidential candidate with the second-most votes.

22. Why do you think it's important for a presidential election to result in a clear-cut winner?

AMENDMENT 13. Slavery Abolished (1865)

Passed by Congress January 31, 1865. Ratified December 6, 1865.

Note: A portion of Article 4, Section 2, of the Constitution was superseded by the 13th Amendment.

Section 1. Neither slavery nor involuntary **servitude,** except as a punishment for crime whereof the party shall have been duly convicted, shall exist within the United States, or any place subject to their jurisdiction.

Section 2. Congress shall have power to enforce this article by appropriate legislation.

AMENDMENT 14. Civil Rights (1868)

Passed by Congress June 13, 1866. Ratified July 9, 1868.

Note: Article 1, Section 2, of the Constitution was modified by Section 2 of the 14th Amendment.

Section 1. All persons born or **naturalized** in the United States, and subject to the jurisdiction thereof, are citizens of the United States and of the state wherein they reside. No state shall make or enforce any law which shall abridge the privileges or immunities of citizens of the United States; nor shall any state deprive any person of life, liberty, or property, without due process of law; nor deny to any person within its jurisdiction the equal protection of the laws.

Section 2. Representatives shall be apportioned among the several states according to their respective numbers, counting the whole number of persons in each state, excluding Indians not taxed. But when the right to vote at any election for the choice of electors for President and Vice-President of the United States, Representatives in Congress, the executive and judicial officers of a state, or the members of the legislature thereof, is denied to any of the male inhabitants of such state, being twenty-one years of age, and citizens of the United States, or in any way abridged, except for participation in rebellion, or other crime, the basis of representation therein shall be reduced in the proportion which the number of such male citizens shall bear to the whole number of male citizens twenty-one years of age in such state.

VOCABULARY

servitude being under the authority of an owner or master

naturalized granted nationality

insurrection revolt against authority

bounties rewards

A CLOSER LOOK

The 14th Amendment laid the groundwork for many civil rights laws, such as the Americans with Disabilities Act (1990). This act gave people with mental or physical disabilities "equal protection of the laws." For example, public places had to be designed for wheelchair use. Wider doors and ramps allow disabled people to go in and out of buildings.

Section 3. No person shall be a Senator or Representative in Congress, or elector of President and Vice-President, or hold any office, civil or military, under the United States, or under any state, who, having previously taken an oath, as a member of Congress, or as an officer of the United States, or as a member of any state legislature, or as an executive or judicial officer of any state, to support the Constitution of the United States, shall have engaged in **insurrection** or rebellion against the same, or given aid or comfort to the enemies thereof. But Congress may, by a vote of two-thirds of each house, remove such disability.

Section 4. The validity of the public debt of the United States, authorized by law, including debts incurred for payment of pensions and **bounties** for services in suppressing insurrection or rebellion, shall not be questioned. But neither the United States nor any state shall assume or pay any debt or obligation incurred in aid of insurrection or rebellion against the United States, or any claim for the loss or emancipation of any slave; but all such debts, obligations and claims shall be held illegal and void.

Section 5. The Congress shall have power to enforce, by appropriate legislation, the provisions of this article.

AMENDMENT 15. Right to Vote (1870)

Passed by Congress February 26, 1869. Ratified February 3, 1870.

Section 1. The right of citizens of the United States to vote shall not be denied or abridged by the United States or by any state on account of race, color, or previous condition of servitude.

Section 2. The Congress shall have power to enforce this article by appropriate legislation.

A CLOSER LOOK Reconstruction Amendments

The 13th, 14th, and 15th Amendments are often called the Reconstruction Amendments. They were passed after the Civil War during the government's attempt to rebuild the Union and to grant rights to recently freed African Americans.

Amendment 13

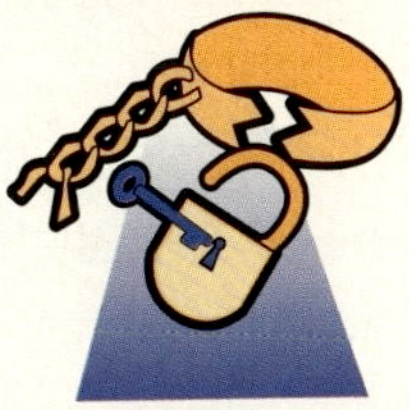

1865

- Ended slavery in the United States

Amendment 14

1868

- Defined national and state citizenship
- Protected citizens' rights
- Promised "equal protection of the laws"

Amendment 15

1870

- Designed to protect African Americans' voting rights

SKILLBUILDER Interpreting Charts

What problems did these amendments try to solve?

A CLOSER LOOK

VOTING RIGHTS

The Voting Rights Act of 1965 extended the 15th Amendment. To qualify as voters, African Americans were no longer required to take tests proving that they could read and write. Also, federal examiners could help register voters. As a result, the number of African-American voters rose sharply.

23. What effect do you think the Voting Rights Act had on candidates running for office?

CLOSER LOOK

INCOME TAX

People below the poverty level, as defined by the federal government, do not have to pay income tax. In 1997, the poverty level for a family of four was $16,400 per year. About 13.3 percent of all Americans were considered poor in 1997.

24. Why do you think people below the poverty level do not pay any income tax?

AMENDMENT 16. Income Tax (1913)

Passed by Congress July 12, 1909. Ratified February 3, 1913.

Note: Article 1, Section 9, of the Constitution was modified by the 16th Amendment.

The Congress shall have power to lay and collect taxes on incomes, from whatever source derived, without apportionment among the several states, and without regard to any census or enumeration.

AMENDMENT 17. Direct Election of Senators (1913)

Passed by Congress May 13, 1912. Ratified April 8, 1913.

Note: Article 1, Section 3, of the Constitution was modified by the 17th Amendment.

Section 1. The Senate of the United States shall be composed of two Senators from each state, elected by the people thereof, for six years; and each Senator shall have one vote. The electors in each state shall have the qualifications requisite for electors of the most numerous branch of the state legislatures.

Section 2. When vacancies happen in the representation of any state in the Senate, the executive authority of such state shall issue writs of election to fill such vacancies: Provided, that the legislature of any state may empower the executive thereof to make temporary appointments until the people fill the vacancies by election as the legislature may direct.

Section 3. This amendment shall not be so construed as to affect the election or term of any Senator chosen before it becomes valid as part of the Constitution.

A CLOSER LOOK

Under Prohibition, people broke the law if they made, sold, or shipped alcoholic beverages. Powerful crime gangs turned selling illegal liquor into a big business. This photo shows federal agents getting ready to smash containers of illegal whiskey. The 21st Amendment ended Prohibition.

AMENDMENT 18. Prohibition (1919)

Passed by Congress December 18, 1917. Ratified January 16, 1919. Repealed by the 21st Amendment.

~~**Section 1.** After one year from the ratification of this article the manufacture, sale, or transportation of intoxicating liquors within, the importation thereof into, or the exportation thereof from the United States and all territory subject to the jurisdiction thereof for beverage purposes is hereby prohibited.~~

~~**Section 2.** The Congress and the several states shall have concurrent power to enforce this article by appropriate legislation.~~

~~**Section 3.** This article shall be inoperative unless it shall have been ratified as an amendment to the Constitution by the legislatures of the several states, as provided in the Constitution, within seven years from the date of the submission hereof to the states by the Congress.~~

A CLOSER LOOK

At left, marchers campaign for the 19th Amendment—woman suffrage. Since winning the right to vote in 1920, women have slowly gained political power. Pictured below are Congress members who belong to the Congressional Caucus for Women's Issues.

AMENDMENT 19. Woman Suffrage (1920)

Passed by Congress June 4, 1919. Ratified August 18, 1920.

Section 1. The right of citizens of the United States to vote shall not be denied or abridged by the United States or by any state on account of sex.

Section 2. Congress shall have power to enforce this article by appropriate legislation.

AMENDMENT 20. "Lame Duck" Sessions (1933)

Passed by Congress March 2, 1932. Ratified January 23, 1933.

Note: Article 1, Section 4, of the Constitution was modified by Section 2 of this amendment. In addition, a portion of the 12th Amendment was superseded by Section 3.

Section 1. The terms of the President and Vice-President shall end at noon on the 20th day of January, and the terms of Senators and Representatives at noon on the 3rd day of January, of the years in which such terms would have ended if this article had not been ratified; and the terms of their successors shall then begin.

Section 2. The Congress shall assemble at least once in every year, and such meeting shall begin at noon on the 3rd day of January, unless they shall by law appoint a different day.

Section 3. If, at the time fixed for the beginning of the term of the President, the President elect shall have died, the Vice-President elect shall become President. If a President shall not have been chosen before the time fixed for the beginning of his term, or if the President elect shall have failed to qualify, then the Vice-President elect shall act as President until a President shall have qualified; and the Congress may by law provide for the case wherein neither a President elect nor a Vice-President elect shall have

A CLOSER LOOK

LAME DUCKS

Before the 20th Amendment, an official defeated in a November election stayed in office until March. These politicians were compared to "lame ducks" whose wings were clipped. The 20th Amendment shortens the time that a newly elected president, vice-president, or Congress member must wait to take office.

25. What are the benefits of having a shorter "lame duck" period?

VOCABULARY

inoperative no longer in force

primary an election in which registered members of a political party nominate candidates for office

qualified, declaring who shall then act as President, or the manner in which one who is to act shall be selected, and such person shall act accordingly until a President or Vice-President shall have qualified.

Section 4. The Congress may by law provide for the case of the death of any of the persons from whom the House of Representatives may choose a President whenever the right of choice shall have devolved upon them, and for the case of the death of any of the persons from whom the Senate may choose a Vice-President whenever the right of choice shall have devolved upon them.

Section 5. Sections 1 and 2 shall take effect on the 15th day of October following the ratification of this article.

Section 6. This article shall be **inoperative** unless it shall have been ratified as an amendment to the Constitution by the legislatures of three-fourths of the several states within seven years from the date of its submission.

AMENDMENT 21. Repeal of Prohibition (1933)

Passed by Congress February 20, 1933. Ratified December 5, 1933.

Section 1. The eighteenth article of amendment to the Constitution of the United States is hereby repealed.

Section 2. The transportation or importation into any state, territory, or possession of the United States for delivery or use therein of intoxicating liquors, in violation of the laws thereof, is hereby prohibited.

Section 3. This article shall be inoperative unless it shall have been ratified as an amendment to the Constitution by conventions in the several states, as provided in the Constitution, within seven years from the date of the submission hereof to the states by the Congress.

A CLOSER LOOK

George Washington set the tradition of limiting the presidency to two terms. Franklin Roosevelt broke this custom when he was elected president four terms in a row—1932, 1936, 1940, and 1944. His record-long presidency led to the 22nd Amendment. A two-term limit, written into the Constitution, checks the president's power.

AMENDMENT 22. Limit on Presidential Terms (1951)

Passed by Congress March 21, 1947. Ratified February 27, 1951.

Section 1. No person shall be elected to the office of the President more than twice, and no person who has held the office of President, or acted as President, for more than two years of a term to which some other person was elected President shall be elected to the office of the President more than once. ~~But this article shall not apply to any person holding the office of President when this article was proposed by the Congress, and shall not prevent any person who may be holding the office of President, or acting as President, during the term within which this article becomes operative from holding the office of President or acting as President during the remainder of such term.~~

Section 2. This article shall be inoperative unless it shall have been ratified as an amendment to the Constitution by the legislatures of three-fourths of the several states within seven years from the date of its submission to the states by the Congress.

AMENDMENT 23. Voting in District of Columbia (1961)

Passed by Congress June 17, 1960. Ratified March 29, 1961.

Section 1. The district constituting the seat of government of the United States shall appoint in such manner as Congress may direct: a number of electors of President and Vice-President equal to the whole number of Senators and Representatives in Congress to which the district would be entitled if it were a state, but in no event more than the least populous state; they shall be in addition to those appointed by the states, but they shall be considered, for the purposes of the election of President and Vice-President, to be electors appointed by a state; and they shall meet in the district and perform such duties as provided by the twelfth article of amendment.

Section 2. The Congress shall have power to enforce this article by appropriate legislation.

AMENDMENT 24. Abolition of Poll Taxes (1964)

Passed by Congress August 27, 1962. Ratified January 23, 1964.

Section 1. The right of citizens of the United States to vote in any **primary** or other election for President or Vice-President, for electors for President or Vice-President, or for Senator or Representative in Congress, shall not be denied or abridged by the United States or any state by reason of failure to pay any poll tax or other tax.

Section 2. The Congress shall have power to enforce this article by appropriate legislation.

A CLOSER LOOK

POLL TAX

The poll tax was aimed at preventing African Americans from exercising their rights. Many could not afford to pay this fee required for voting.

26. How do you think the 24th Amendment affected elections?

AMENDMENT 25. Presidential Disability, Succession (1967)

Passed by Congress July 6, 1965. Ratified February 10, 1967.

Note: Article 2, Section 1, of the Constitution was affected by the 25th Amendment.

Section 1. In case of the removal of the President from office or of his death or resignation, the Vice-President shall become President.

Section 2. Whenever there is a vacancy in the office of the Vice-President, the President shall nominate a Vice-President who shall take office upon confirmation by a majority vote of both houses of Congress.

Section 3. Whenever the President transmits to the President pro tempore of the Senate and the Speaker of the House of Representatives his written declaration that he is unable to discharge the powers and duties of his office, and until he transmits to them a written declaration to the contrary, such powers and duties shall be discharged by the Vice-President as Acting President.

A CLOSER LOOK

PRESIDENTIAL DISABILITY

President John F. Kennedy's death in 1963 signaled the need for the 25th Amendment. The Constitution did not explain what to do in the case of a disabled president. James Reston, a writer for *The New York Times,* summed up the problem: Suppose Kennedy was "strong enough to survive [the bullet wounds], but too weak to govern." The 25th Amendment provides for an orderly transfer of power.

27. What do you think can happen in a country where the rules for succession are not clear?

A CLOSER LOOK

SUCCESSION

Who takes over if a president dies in office or is unable to serve? The top five in the line of succession follow:

- **vice-president**
- **speaker of the house**
- **president pro tempore of the Senate**
- **secretary of state**
- **secretary of the treasury**

28. Why should voters know the views of the vice-president?

Section 4. Whenever the Vice-President and a majority of either the principal officers of the executive departments or of such other body as Congress may by law provide, transmit to the President pro tempore of the Senate and the Speaker of the House of Representatives their written declaration that the President is unable to discharge the powers and duties of his office, the Vice-President shall immediately assume the powers and duties of the office as Acting President. Thereafter, when the President transmits to the President pro tempore of the Senate and the Speaker of the House of Representatives his written declaration that no inability exists, he shall resume the powers and duties of his office unless the Vice-President and a majority of either the principal officers of the executive department[s] or of such other body as Congress may by law provide, transmit within four days to the President pro tempore of the Senate and the Speaker of the House of Representatives their written declaration that the President is unable to discharge the powers and duties of his office. Thereupon Congress shall decide the issue, assembling within forty-eight hours for that purpose if not in session. If the Congress, within twenty-one days after receipt of the latter written declaration, or, if Congress is not in session, within twenty-one days after Congress is required to assemble, determines by two thirds vote of both houses that the President is unable to discharge the powers and duties of his office, the Vice-President shall continue to discharge the same as Acting President; otherwise, the President shall resume the powers and duties of his office.

A CLOSER LOOK Amendments Time Line *1791–1992*

Use the key below to help you categorize the amendments.

- **Voting Rights**
- **Social Changes**
- **Overturned Supreme Court Decisions**
- **Election Procedures and Conditions of Office**

1790

Bill of Rights Amendments 1–10
1791

Amendment 11
1798
Protects state from lawsuits filed by citizens of other states or countries.

Amendment 12
1804
Requires separate electoral ballots for president and vice-president.

Amendment 13
1865
Bans slavery.

Amendment 14
1868
Defines American citizenship and citizens' rights.

Amendment 15
1870
Stops national and state governments from denying the vote based on race.

AMENDMENT 26. 18-year-old Vote (1971)

Passed by Congress March 23, 1971. Ratified July 1, 1971.

Note: Amendment 14, Section 2, of the Constitution was modified by Section 1 of the 26th Amendment.

Section 1. The right of citizens of the United States, who are eighteen years of age or older, to vote shall not be denied or abridged by the United States or by any state on account of age.

Section 2. The Congress shall have power to enforce this article by appropriate legislation.

AMENDMENT 27. Congressional Pay (1992)

Passed by Congress September 25, 1789. Ratified May 7, 1992.

No law, varying the compensation for the services of the Senators and Representatives, shall take effect, until an election of Representatives shall have intervened.

ROCK THE VOTE

A CLOSER LOOK

Members of the recording industry founded Rock the Vote. They urge young people to vote in elections.

Amendments 11–27 Assessment

1. Main Ideas

a. Which amendments affected the office of president?

b. Which pair of amendments shows the failure of laws to solve a social problem?

c. Which amendments corrected unfair treatment toward African Americans and women?

2. Critical Thinking

Supporting Opinions Which of the amendments has had the greatest effect on you?

THINK ABOUT

- your rights as a citizen
- your responsibilities as a citizen
- your role in government

Amendment 16
1913
Allows Congress to tax incomes.

Amendment 17
1913
Establishes direct election of U.S. senators.

Amendment 18
1919
Prohibits making, selling, and shipping alcoholic beverages.

Amendment 19
1920
Extends the vote to women.

Amendment 20
1933
Changes date for starting new Congress and inaugurating new president.

Amendment 21
1933
Repeals Amendment 18.

Amendment 22
1951
Limits terms president can serve to two.

Amendment 23
1961
Gives citizens of Washington, D.C., right to vote in presidential elections.

Amendment 24
1964
Bans poll taxes.

Amendment 25
1967
Sets procedures for presidential succession.

Amendment 26
1971
Gives 18-year-olds right to vote in federal and state elections.

Amendment 27
1992
Limits ability of Congress to increase its pay.

2000

VISUAL SUMMARY

The Constitution of the United States

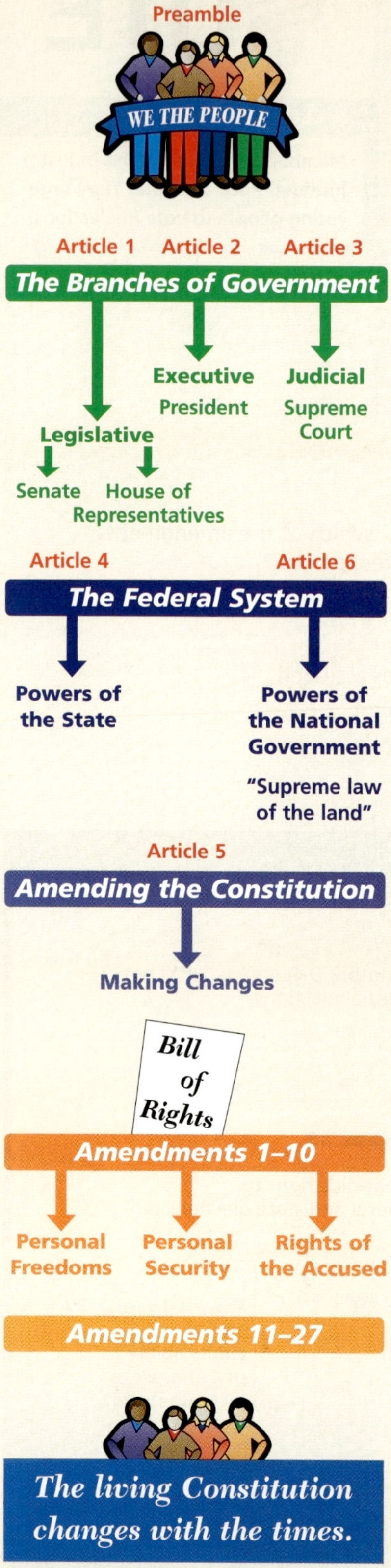

Constitution ASSESSMENT

VOCABULARY

Briefly explain the importance of each of the following.

1. electors
2. impeachment
3. naturalization
4. felonies
5. bill of attainder
6. ex post facto law
7. suffrage
8. due process of law
9. servitude
10. primary

SEVEN PRINCIPLES OF THE CONSTITUTION

Make a chart like the one shown. Then fill it in with a definition of each principle and an example from the Constitution.

Principle	Definition	Example
1. popular sovereignty		
2. republicanism		
3. federalism		
4. separation of powers		
5. checks and balances		
6. limited government		
7. individual rights		

REVIEW QUESTIONS

Article 1 (pages 249–255)

1. What are the requirements for becoming a member of the House of Representatives and the Senate?
2. What are two military powers granted to Congress?

Article 2 (pages 256–259)

3. How does the electoral college choose the president?
4. What are three powers of the president?

Article 3 (pages 260–261)

5. What are the two most important powers of the federal courts?

Articles 4–7 (pages 262–265)

6. How can the Constitution be changed?
7. If a state law and a federal law conflict, which law must be obeyed? Why?
8. How can the Constitution be ratified, or approved?

Bill of Rights and Amendments 11–27 (pages 266–277)

9. What five freedoms are guaranteed in the First Amendment?
10. Which amendments extend voting rights to a broader range of Americans?

CRITICAL THINKING

1. DRAWING CONCLUSIONS

In a two-column chart, summarize the processes for changing the Constitution. Then use your completed chart to answer the questions below.

Proposing Amendments	Ratifying Amendments
1.	1.
2.	2.

a. What role can citizens play in proposing amendments?

b. What do you think are the main reasons for changing the Constitution?

2. MAKING INFERENCES

Explain how the "elastic clause" in Article 1 gives Congress the authority to take action on other issues unknown to the Framers of the Constitution.

3. ANALYZING LEADERSHIP

Think about the president's roles described in the Constitution. What qualities does a president need to succeed as a leader in so many different areas?

4. RECOGNIZING EFFECTS

How would you describe the impact of the 14th, 15th, and 16th Amendments on life in the United States?

5. APPLYING CITIZENSHIP

Suppose you and your family go on a road trip across several states. According to Article 4 of the Constitution, what citizens' rights do you have in the states you are visiting?

HISTORY SKILLS

INTERPRETING PRIMARY SOURCES

In 1937, President Franklin D. Roosevelt gave a speech over the radio. He used interesting comparisons to explain how the government works.

> I described the American form of government as a three-horse team provided by the Constitution to the American people so that their field might be plowed. The three horses are, of course, the three branches of government—the Congress, the Executive, and the Courts. . . . It is the American people themselves who are in the driver's seat. It is the American people themselves who want the furrow plowed.
>
> **Franklin D. Roosevelt,** Radio Address

- How does Roosevelt describe the separation of powers?
- How does Roosevelt explain popular sovereignty?

ALTERNATIVE ASSESSMENT

1. INTERDISCIPLINARY ACTIVITY: Math

Creating a Data Display Do research on methods that political candidates use to raise money for their campaigns. Examples include public money, political action committees, direct mail, and so on. Summarize your findings in a graph or chart.

2. COOPERATIVE LEARNING ACTIVITY

Drafting a Constitution Imagine you are asked to write a constitution for a newly formed country. Working with a group, make an outline like the one shown to help you organize your draft.

I. Purpose of the Constitution (Preamble)
II. Making Laws (Legislative Branch)
III. Carrying Out the Laws (Executive Branch)
IV. Making Laws Fair (Judicial Branch)
V. Choosing Leaders
VI. Citizens' Rights (Bill of Rights)

3. CD-ROM PRIMARY SOURCE EXPLORER

Making a Learning Center Creating the U.S. Constitution was one of the most important events in the nation's history. Use the CD-ROM and the library to collect information on different topics related to the Constitution.

Create a learning center featuring the suggestions below.

- Find biographies and portraits of the Framers.
- Collect important primary sources such as James Madison's notes and *The Federalist* papers.
- Gather recent pictures and news articles about the Congress, the president, the Supreme Court, and the Bill of Rights.

4. HISTORY PORTFOLIO

Review your draft of the constitution you wrote for the assessment activity. Choose one of these options below.

PORTFOLIO **Option 1** Use comments made by your teacher or classmates to improve your work.

PORTFOLIO **Option 2** Illustrate your constitution. Add your work to your history portfolio.

The Role of the Citizen

Citizens of the United States enjoy many basic rights and freedoms. Freedom of speech and religion are examples. These rights are guaranteed by the Constitution, the Bill of Rights, and other amendments to the Constitution. Along with these rights, however, come responsibilities. Obeying rules and laws, voting, and serving on juries are some examples.

Active citizenship is not limited to adults. Younger citizens can help their communities become better places. The following pages will help you to learn about your rights and responsibilities. Knowing them will help you to become an active and involved citizen of your community, state, and nation.

In this book you will find examples of active citizenship by young people like yourself. **Look for the Citizenship Today features.**

President John F. Kennedy urged all Americans to become active citizens and work to improve their communities.

The weather was sunny but cold on January 20, 1961—the day that John F. Kennedy became the 35th president of the United States. In his first speech as president, he urged all Americans to serve their country. Since then, Kennedy's words have inspired millions of Americans to become more active citizens.

> ***"Ask not what your country can do for you—ask what you can do for your country!"***
>
> —JOHN F. KENNEDY

What Is a Citizen?

A citizen is a legal member of a nation and pledges loyalty to that nation. A citizen has certain guaranteed rights, protections, and responsibilities. A citizen is a member of a community and wants to make it a good place to live.

Today in the United States there are a number of ways to become a citizen. The most familiar are citizenship by birth and citizenship by naturalization. All citizens have the right to equal protection under the law.

CITIZENSHIP BY BIRTH A child born in the United States is a citizen by birth. Children born to U.S. citizens traveling or living outside the country, such as military personnel, are citizens. Even children born in the United States to parents who are not citizens of the United States are considered U.S. citizens. These children have dual citizenship. This means they are citizens of two countries—both the United States and the country of their parents' citizenship. At the age of 18, the child may choose one of the countries for permanent citizenship.

CITIZENSHIP BY NATURALIZATION A person who is not a citizen of the United States may become one through a process called naturalization. The steps in this process are shown below. To become a naturalized citizen, a person must meet certain requirements.

- Be at least 18 years old. Children under the age of 18 automatically become naturalized citizens when their parents do.
- Enter the United States legally.
- Live in the United States for at least five years immediately prior to application.
- Read, write, and speak English.
- Show knowledge of American history and government.

See Citizenship Today: Becoming a Citizen, p. 467

Steps in the Naturalization Process

1. **File an application.**
2. **Take an examination.**
3. **File a legal petition for naturalization.**
4. **Appear at a court hearing.**
5. **Take an oath of allegiance.**

Hundreds of people become new citizens at a single ceremony in San Antonio, Texas.

What Are Your Rights?

Citizens of the United States are guaranteed rights by the U.S. Constitution, state constitutions, and state and federal laws. All citizens have three kinds of rights: basic freedoms, protection from unfair government actions, and equal treatment under the law.

Citizens' basic rights and freedoms are sometimes called **civil rights**. Some of these rights are personal, and others are political.

The U.S. Constitution grants these five basic freedoms.

BASIC FREEDOMS

- Freedom of religion
- Freedom of speech
- Freedom of the press
- Freedom of peaceful assembly
- Freedom to petition the government for change

The second category of rights is intended to protect citizens from unfair government actions.

PERSONAL PROTECTIONS

- The right to bear arms
- Freedom from being forced to house soldiers
- Protection from unreasonable search and seizure
- The right to a speedy public trial by an impartial jury
- No excessive bail or fines
- Protection from cruel and unusual punishment

Other parts of the Bill of Rights grant these rights.

The third category is the right to equal treatment under the law. The government cannot treat one individual or group differently from another.

RIghts of citizenship have expanded over the years.

EQUAL PROTECTION UNDER THE LAW

- No slavery
- The right to vote to all male citizens over 21 years old
- The right to vote to women
- The right to vote to 18 year olds
- The Civil Rights Acts of 1964 protects voting rights and prevents discrimination.
- The Americans with Disabilities Act of 1990 protects the rights of disabled citizens.

LIMITS TO RIGHTS The rights guaranteed to citizens have sensible limits. For example, the right to free speech does not allow a person to falsely shout, "Fire!" at a crowded concert. The government may place limits on certain rights to protect national security or to provide equal opportunities for all citizens. And rights come with responsibilities.

What Are Your Responsibilities?

For American democracy to work, citizens must carry out important responsibilities. There are two kinds of responsibilities—personal and civic. Personal responsibilities include taking care of yourself, helping your family, knowing right from wrong, and behaving in a respectful way.

Civic responsibilities are those that involve your government and community. They include obeying rules and laws, serving on juries, paying taxes, and defending your country when called upon. One of the most important responsibilities is voting. When you turn 18, you will have that right

As a young person, you can be a good citizen in a number of ways. You might work with other people in your community to make it a fair and just place to live. Working for a political party or writing to your elected officials about issues that concern you are some other examples.

The chart below shows how responsibilities change with a citizen's age. Notice that all citizens share the responsibility to obey the laws of their communities.

See Citizenship Today: Obeying Rules and Laws, p. 300

Responsibilities of a Citizen

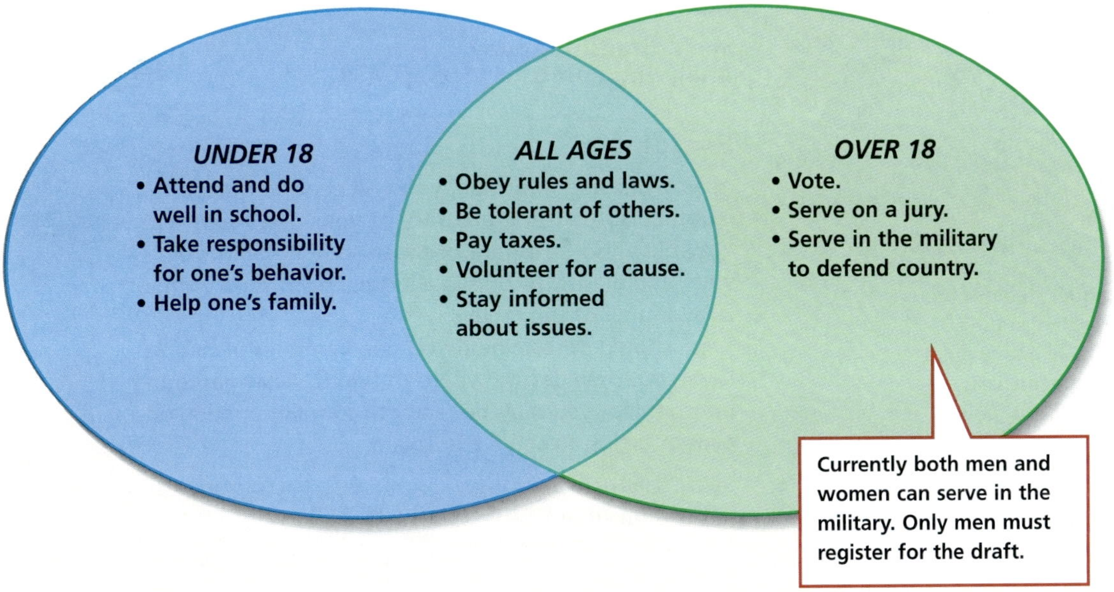

CITIZENSHIP ACTIVITIES

1. Interview a recently naturalized citizen. Ask about the test he or she took to become a U.S. citizen. Write a report of your findings.

2. Using newspapers or magazines, find examples of citizens using their unalienable rights or practicing responsible citizenship. Cut out five articles to illustrate the points. Mount them and write a one-sentence explanation of each article.

Building Citizenship Skills

Good citizenship skills include **staying informed, solving problems** or **making decisions,** and **taking action.** Every citizen can find ways to build citizenship skills. By showing respect for the law and for the rights of others in your daily life, you promote democracy. You can also work to change conditions in your community to make sure all citizens experience freedom and justice.

How Do You Stay Informed?

Americans can sometimes feel that they have access to too much information. It may seem overwhelming. Even so, you should stay informed on issues that affect your life. Staying informed gives you the information you need to make wise decisions and helps you find ways to solve problems.

These Texas middle school students are staying informed by talking to their Texas State Representative. Many public officials enjoy having students visit and ask them questions about their jobs, and about issues students think are important.

Watch, Listen, and Read

The first step in practicing good citizenship is to know how to find information that you need.

Sources of information include broadcast and print media and the Internet. Public officials and civic organizations are also good sources for additional information. Remember as you are reading to evaluate your sources.

See Citizenship Today:
Analyzing Points of View, p. 467

Evaluate

As you become informed, you will need to make judgments about the accuracy of your news sources. You must also be aware of those sources' points of view and biases. (A bias is a judgment formed without knowing all of the facts.)

You should determine if you need more information. If you do, then decide where to find it. After gathering information, you may be ready to form an opinion or a plan of action to solve a problem.

See Citizenship Today:
Detecting Bias in the Media, p. 661

Communicate

To bring about change in their communities, active citizens may need to contact public officials. In today's world, making contact is easy.

You can reach most public officials by telephone, voice mail, fax, or letter. Many public officials also have Internet pages or e-mail that encourage input from the public.

See Citizenship Today:
Writing Government Officials, p. 774

How Do You Make Wise Decisions?

Civic life involves making important decisions. As a voter, whom should you vote for? As a juror, should you find the defendant guilty or not guilty? As an informed citizen, should you support or oppose a proposed government action? Unlike decisions about which video to rent, civic decisions cannot be made by a process as easy as tossing a coin. Instead, you should use a problem-solving approach like the one shown in the chart below. Decision-making won't always proceed directly from step to step. Sometimes it's necessary to backtrack a little. For example, you may get to the "Analyze Information" step and realize that you don't have enough information to analyze. Then you can go back a step and gather more information.

Problem-Solving and Decision-Making Process

Problem-solving and decision-making involves many steps. This diagram shows you how to take those steps. Notice that you may have to repeat some steps depending on the information you gather.

IDENTIFY THE PROBLEM
Decide what the main issues are and what your goal is.

GATHER INFORMATION
Get to know the basics of the problem. Find out as much as possible about the issues.

ANALYZE THE INFORMATION
Look at the information and determine what it reveals about solving the problem.

CONSIDER OPTIONS
Think of as many ways as possible to solve the problem. Don't be afraid to include ideas that others might think are unacceptable.

CHOOSE A SOLUTION
Choose the solution you believe will best solve the problem and help you reach your goal.

IMPLEMENT THE SOLUTION
Take action or plan to take action on a chosen solution.

EVALUATE THE SOLUTION
Review the results of putting your solution into action. Did the solution work? Do you need to adjust the solution in some way?

Students working on an environmental project are gathering and analyzing information to help them make decisions.

How Do You Participate in Your Community?

Across the country many young people have come up with ways to make their communities better places to live. Thirteen-year-old Aubyn Burnside of Hickory, North Carolina, is just one example. Aubyn felt sorry for foster children she saw moving their belongings in plastic trash bags. She founded Suitcases for Kids. This program provides used luggage for foster children who are moving from one home to another. Her program has been adopted by other young people in several states. Below are some ways in which you can participate in your community.

See Citizenship Today: Community Service, p. 610

Students participate in a rally to promote safety in their school.

Find a Cause

How can you become involved in your community? First, select a community problem or issue that interests you. Some ideas from other young people include starting a support group for children with cancer, publishing a neighborhood newspaper with children's stories and art, and putting on performances to entertain people in shelters and hospitals.

Develop Solutions

Once you have found a cause on which you want to work, develop a plan for solving the problem. Use the decision-making or problem-solving skills you have learned to find ways to approach the problem. You may want to involve other people in your activities.

Follow Through

Solving problems takes time. You'll need to be patient in developing a plan. You can show leadership in working with your group by following through on meetings you set up and plans you make. When you finally solve the problem, you will feel proud of your accomplishments.

CITIZENSHIP ACTIVITIES

1. Use the telephone directory to make a list of names, addresses, and phone numbers of public officials or organizations that could provide information about solving problems in your community.
2. Copy the steps in the problem-solving and decision-making diagram and show how you followed them to solve a problem or make a decision. Be sure to clearly state the problem and the final decision.

Practicing Citizenship Skills

You have learned that good citizenship involves three skills: staying informed, solving problems, and taking action. Below are some activities to help you improve your citizenship skills. By practicing these skills you can work to make a difference in your own life and in the lives of those in your community.

CITIZENSHIP ACTIVITIES

Stay Informed

CREATE A PAMPHLET OR RECRUITING COMMERCIAL

Ask your school counselors or write to your state department of education to get information on state-run colleges, universities, or technical schools. Use this information to create a brochure or recruiting commercial showing these schools and the different programs and degrees they offer.

KEEP IN MIND

What's there for me? It may help you think about what areas students are interested in and may want to pursue in later life.

Where is it? You may want to have a map showing where the schools are located in your state.

How can I afford it? Students might want to know if financial aid is available to attend the schools you have featured.

Make Wise Decisions

CREATE A GAME BOARD OR SKIT

Study the decision-making diagram on page 285. With a small group, develop a skit that explains the steps in problem solving. Present your skit to younger students in your school. As an alternative, create a game board that would help younger students understand the steps in making a decision.

KEEP IN MIND

What do children this age understand? Be sure to create a skit or game at an age-appropriate level.

What kinds of decisions do younger students make? Think about the kinds of decisions that the viewers of your skit or players of the game might make.

How can I make it interesting? Use visual aids to help students understand the steps in decision making.

Take Action

CREATE A BULLETIN BOARD FOR YOUR CLASS

Do some research on the Internet or consult the yellow pages under "Social Services" to find the names of organizations that have volunteer opportunities for young people. Call or write for more information. Then create a bulletin board for your class showing groups that would like volunteer help.

KEEP IN MIND

What kinds of jobs are they? You may want to list the types of skills or jobs volunteer groups are looking for.

How old do I have to be? Some groups may be looking for younger volunteers; others may need older persons.

How do I get there? How easy is it to get to the volunteer group's location?

UNIT 3

1789 – 1844

The Early Republic

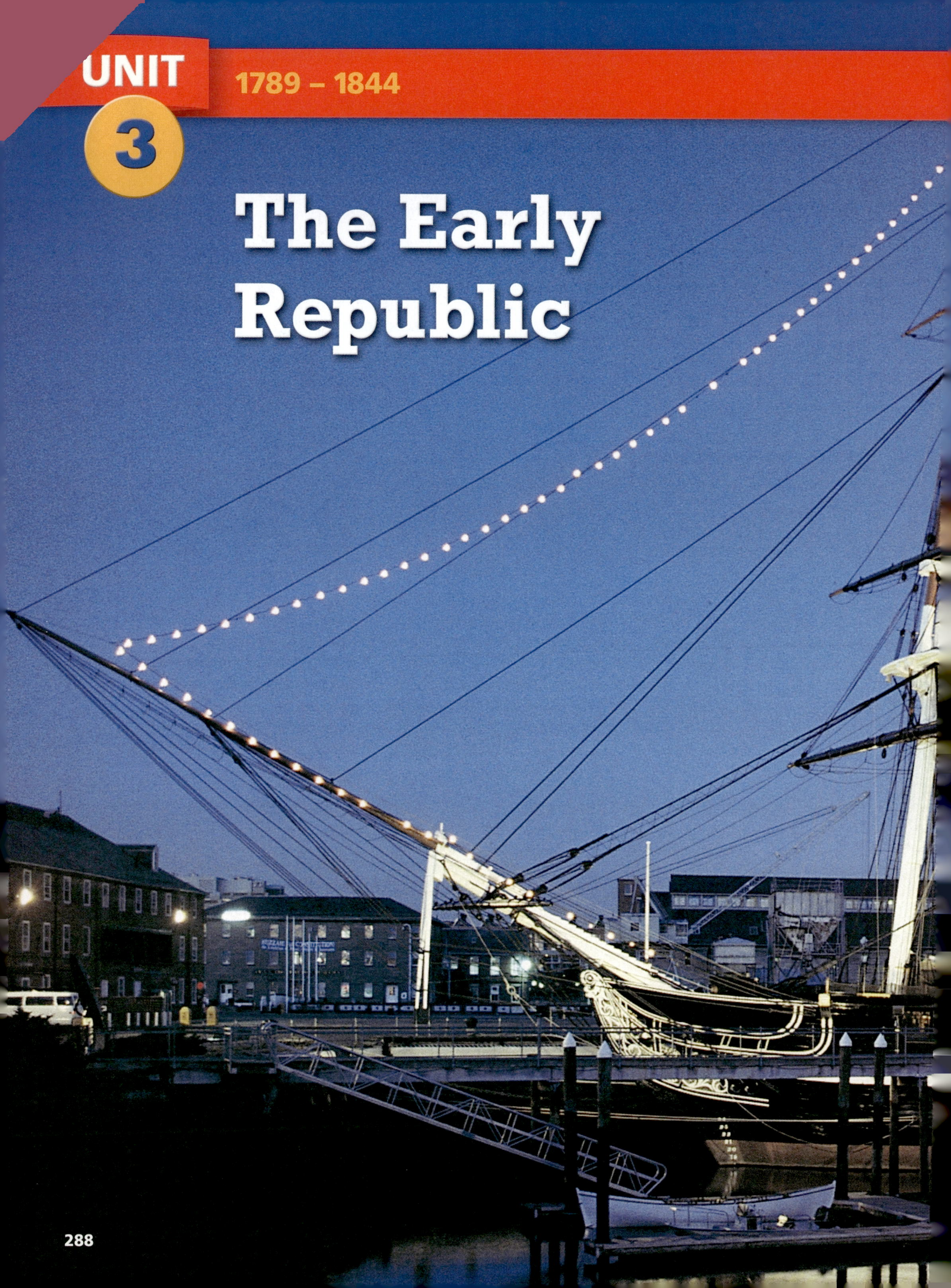

"Our country! May she always be in the right; but our country, right or wrong."

—Stephen Decatur

Known as "Old Ironsides," the U.S.S. *Constitution* won more battles than any other early American warship. Today it rests in Boston Harbor.

CHAPTER 9

Launching a New Republic 1789–1800

Section 1 **Washington's Presidency**
Section 2 **Challenges to the New Government**
Section 3 **The Federalists in Charge**

George Washington arrives by boat in New York on April 23, 1789, for his presidential inauguration.

Interact *with* History

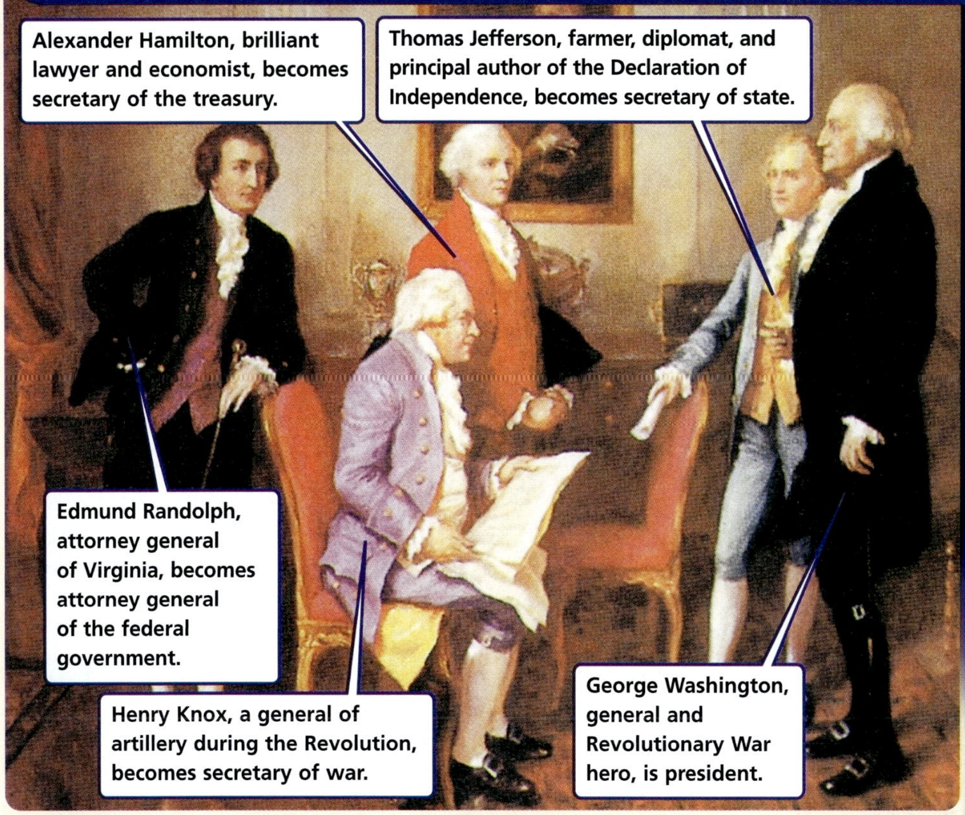

The year is 1789, and George Washington has been inaugurated as the first president of the United States. It quickly becomes obvious to you and to others that the president will need help. He chooses people with different talents and experience to help him govern.

What Do You Think?

- Why might you want people with different viewpoints in your government?
- How would you go about setting up a government?
- What do you think your biggest challenges would be?

What kind of person would you choose to help you govern?

USA

1789 George Washington inaugurated as president.

1791 The first Bank of the United States is established.

1792 Washington re-elected president.

1794 Whiskey Rebellion occurs.

1796 John Adams elected president.

1798 Alien and Sedition Acts

1800 Thomas Jefferson elected president.

World

1789 French Revolution begins.

1791 Slaves revolt in Santo Domingo.

1793 French king Louis XVI executed; Reign of Terror begins in France.

1797 Britain appoints Richard Wellesley Governor-General of India.

1798 French Expedition to Egypt

1800 Napoleon becomes First Consul of France.

Chapter 9 SETTING THE STAGE

BEFORE YOU READ

Inaugural coat buttons, 1789, proclaim the beginning of the Washington presidency.

Previewing the Theme

Democratic Ideals During the Federalist era (1789–1801), the leaders of the United States faced many challenges and difficulties. In this chapter, you will see how the way in which they responded to those obstacles and opportunities established a democratic foundation.

What Do You Know?

What do you think of when you hear the words *democracy* and *republic?* Why do you think the citizens and leaders of the new country wanted to establish a republic governed by laws?

THINK ABOUT

- the experience of the colonists under British rule
- the effect of the Revolutionary War and the period immediately after the war

What Do You Want to Know?

What questions do you have about the people who created the U.S. government? Record your questions in your notebook before you read the chapter.

READ AND TAKE NOTES

Reading Strategy: Identifying and Solving Problems As you read history, try to identify problems in past times and the solutions that people came up with to solve their problems. A graphic organizer such as the chart below can help you to keep track of problems and their solutions. In the middle of the chart, four headings categorize the major issues faced by the young nation. Copy the chart into your notebook and then record problems and the proposed solutions in each category.

See Skillbuilder Handbook, page R17.

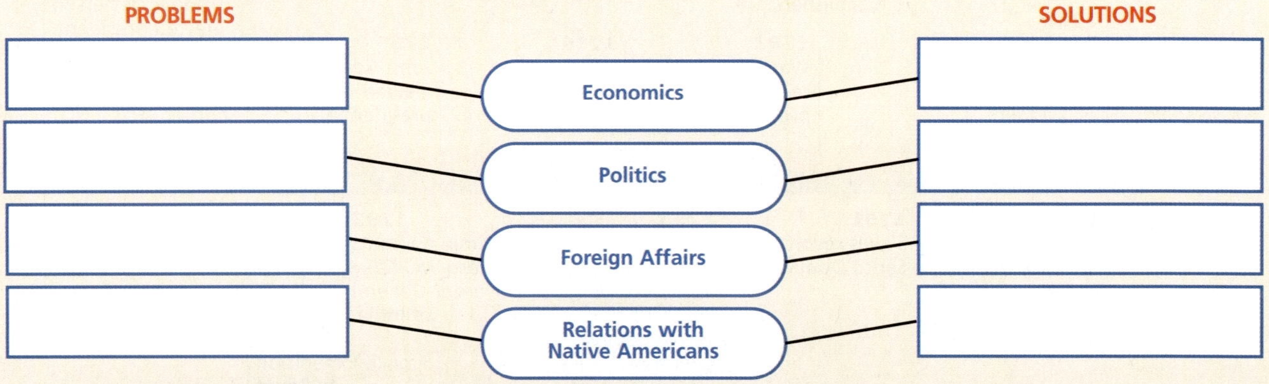

1 Washington's Presidency

TERMS & NAMES
inaugurate
Federal Judiciary Act
cabinet
tariff

MAIN IDEA	WHY IT MATTERS NOW
The president and the Congress began to set up the new government.	The strength of the U.S. today is due to the decisions of the Founders about how to organize the government.

ONE AMERICAN'S STORY

Charles Thomson had known George Washington for many years. Thomson had served as secretary of the Continental Congress when delegates from the colonies first met in Philadelphia in 1774.

Now, 15 years later, on April 14, 1789, he had a very different job to do. He had come to Mount Vernon in Virginia with a letter for George Washington. Washington knew the reason for the visit. Thomson's letter was to tell him that he had been elected the nation's first president. Before giving Washington the letter, Thomson made a short speech.

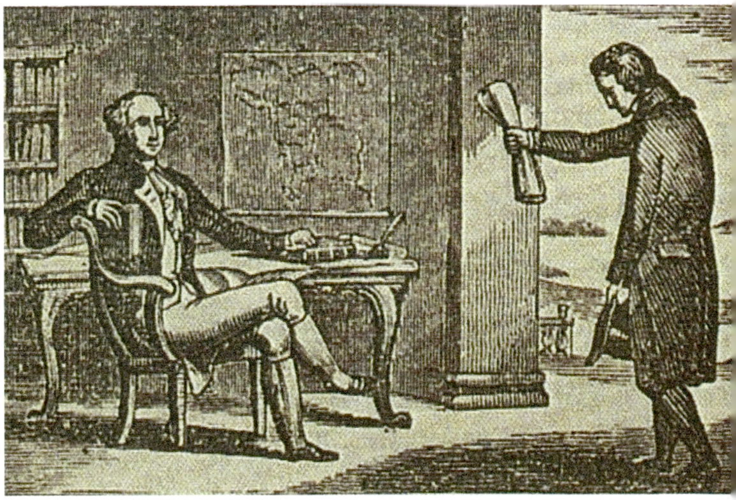

Charles Thomson delivers the letter to Washington announcing his election as president.

A VOICE FROM THE PAST

I have now Sir to inform you that . . . your patriotism and your readiness to sacrifice . . . private enjoyments to preserve the liberty and promote the happiness of your Country [convinced the Congress that you would accept] this important Office to which you are called not only by the unanimous votes of the Electors but by the voice of America.

Charles Thomson, quoted in Washington's Papers, Library of Congress

As you will read in this section, Washington accepted the honor and the burden of his new office. He guided the nation through its early years.

Washington Takes Office

Washington had been elected only a few months before. Each member of the electoral college had written down two names. The top vote-getter, Washington, became president. The runner-up, John Adams, became vice-president. Washington left Mount Vernon on April 16, 1789. He traveled north through Baltimore and Philadelphia to New York City, the nation's capital. On April 30 at Federal Hall, Washington was **inaugurated,** or sworn in, as president. John Adams of Massachusetts was his vice-president.

As the nation's first president, Washington faced a difficult task. He knew that all eyes would be on him. His every action as president would set a precedent—an example that would become standard practice. People argued over what to call him. Some, including John Adams, suggested "His Excellency" or "His Highness." Others argued that such titles would suggest that he was a king. The debate tied up Congress for a month. Finally, "Mr. President" was agreed upon. Congress had to settle other differences about how the new government should be run.

*Reading*History

A. Making Inferences Why were people so concerned about how to address the president?

Setting Up the Courts

The writers of the Constitution had left many matters to be decided by Congress. For example, the Constitution created a Supreme Court but left it to Congress to decide the number of justices. Leaders also argued about how much power the Supreme Court should have. One reason for disagreement was that the states already had their own courts. How would authority be divided between the state and federal courts?

To create a court system, Congress passed the **Federal Judiciary Act** of 1789. This act gave the Supreme Court six members: a chief justice, or judge, and five associate justices. Over time, that number has grown to nine. The act also provided for other lower, less powerful federal courts. Washington appointed John Jay, the prominent New York lawyer and diplomat, as chief justice.

Vocabulary

judiciary: system of courts and judges

Washington's Cabinet

The Constitution also gave Congress the task of creating departments to help the president lead the nation. The president had the power to appoint the heads of these departments, who were to assist the president with the many issues and problems he had to face. These heads of departments became his **cabinet.**

The Congress created three departments. In his first major task as president, Washington chose talented people to run them. For secretary of war, he picked Henry Knox, a trusted general during the Revolution. It was Knox's job to oversee the nation's defenses. For secretary of state, Washington chose Thomas Jefferson. He had been serving as U.S. minister to France. The State Department oversaw relations between the United States and other countries. Washington turned to the brilliant Alexander Hamilton to be the secretary of the treasury. Hamilton had to manage the government's money. The secretary's ties to the president began during the war when he had served as one of Washington's aides. To advise the

Now and then

THE PRESIDENT'S CABINET

The president's cabinet has more than tripled in size since it began with the secretaries of state, war, and treasury, and the attorney general. As the nation has faced new challenges, the government has added new departments. In 1977, concerns about oil shortages led to the creation of the Department of Energy. The Department of Veterans' Affairs was added in 1989. Today the cabinet (shown below) includes the heads of 14 departments.

government on legal matters, Washington picked Edmund Randolph as attorney general.

These department heads and the attorney general made up Washington's cabinet. The Constitution made no mention of a cabinet. However, Washington began the practice of calling his department heads together to advise him.

Economic Problems

As secretary of the treasury, Alexander Hamilton faced the task of straightening out the nation's finances. First of all, the new government needed to pay its war debts. During the Revolution, the United States had borrowed millions of dollars from France, the Netherlands, and Spain. Within the United States, merchants and other private citizens had loaned money to the government. State governments also had wartime debts to pay back. By 1789, the national debt totaled more than $52 million.

Most government leaders agreed that the nation must repay its debts to win the respect of both foreign nations and its own citizens. Hamilton saw that the new nation must assure other countries that it was responsible about money. These nations would do business with the United States if they saw that the country would pay its debts. If the nation failed to do so, no country would lend it money in the future.

Hamilton came up with a financial plan that reflected his belief in a strong central government. He thought the power of the national government should be stronger than that of the state governments. Hamilton also believed that government should encourage business and industry. He sought the support of the nation's wealthy merchants and manufacturers. He thought that the nation's prosperity depended on them. The government owed money to many of these rich men. By paying them back, Hamilton hoped to win their support for the new government.

AMERICA'S HISTORY MAKERS

ALEXANDER HAMILTON
1755?–1804

Alexander Hamilton was born into poverty in the British West Indies. When he was ten years old, the young Alexander went to work as a clerk. He so impressed his employers that they helped to send him to school at King's College (now Columbia University) in New York.

During the Revolutionary War, he became an aide to General Washington. Hamilton moved up quickly in the army and later in political life. Although of humble origins, Hamilton had little faith in the common people and put his trust in the wealthy and educated to govern.

Why is it odd that Hamilton distrusted the common people to govern?

Reading **History**

B. Making Inferences Why might merchants and manufacturers support a strong central government?

Hamilton's Financial Plan

In 1790, Hamilton presented his plan to Congress. He proposed three steps to improve the nation's finances.

1. paying off all war debts
2. raising government revenues
3. creating a national bank

Hamilton wanted the federal government to pay off the war debts of the states. However, sectional differences arose over repayment of state debts. Virginia, Georgia, and many other Southern states had already repaid their debts and did not like being asked to help Northern states pay theirs.

Hamilton asked Thomas Jefferson of Virginia to help him gain Southern support. They reached a compromise. In exchange for Southern support of the plan, Northerners agreed to place the new nation's capital in the South. The location chosen was on the Potomac River between Virginia and Maryland.

The secretary of the treasury favored tariffs. A **tariff** is a tax on imported goods. It serves two purposes: raising money for the government and encouraging the growth of American industry. The government placed the highest tariffs on foreign goods—such as shoes and textiles—that Americans bought in great quantities. This ensured a steady flow of income to the government. In addition, since tariffs made foreign goods more expensive, they encouraged people to buy American goods.

Hamilton also called for the creation of a national bank. Such a bank would meet many needs. It would give the government a safe place to keep

Economics *in* History

How Banks Work

Why did Hamilton want to create a national bank? He believed that such a bank could help the economy of the new nation. It would create a partnership between the federal government and American business.

Let's say you deposited money into a bank account. Then you went back another day to withdraw some of the money. What happened in the meantime? Did the money just sit in the bank until you wanted it back? No—the bank used your money, and in doing so, helped fuel economic growth. In this way, money flows in a circular path from people like you into the general economy and back to you again. In the process, money can create goods and services, jobs, and profits, as the diagram explains.

CONNECT TO HISTORY

1. **Analyzing Points of View** Do you think that the people who feared a strong central government supported Hamilton's idea of a national bank? Why or why not?

See Skillbuilder Handbook, page R8.

CONNECT TO TODAY

2. **Making Inferences** How do banks make money?

Visit www.mcdougallittell.com to learn more about banking.

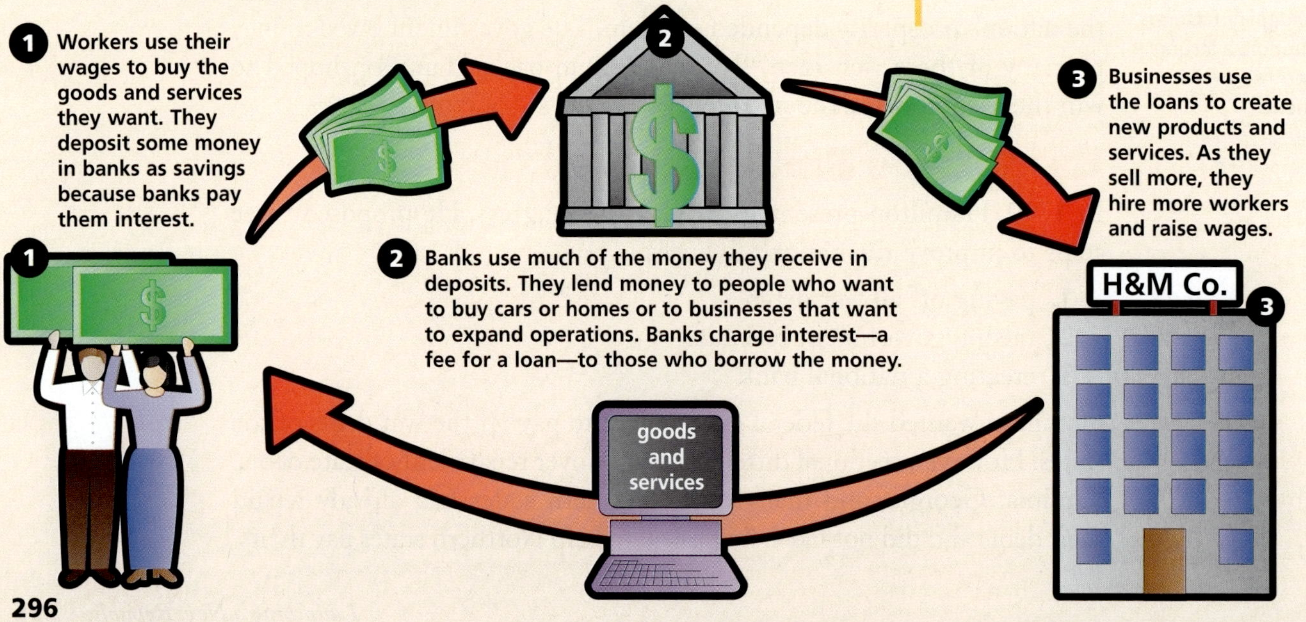

Vocabulary
currency: money

money. It would also make loans to businesses and government. Most important, it would issue bank notes—paper money that could be used as currency. Overall, Hamilton's plan would strengthen the central government. However, this worried Jefferson and Madison.

Two of the first U.S. coins, 1792

Interpreting the Constitution

Jefferson and Madison believed that the Constitution discouraged the concentration of power in the federal government. The Constitution's writers had tried to make the document general enough so that it would be flexible. As a result, disagreements sometimes arose over the document's meaning.

The debate over Hamilton's plan for a national bank exposed differences about how to interpret the Constitution. Madison and Jefferson argued that the Constitution did not give the government the power to set up a bank. They believed in the strict construction—narrow or strict interpretation—of the Constitution. They stated that the government has only those powers that the Constitution clearly says it has. Therefore, since the Constitution does not mention a national bank, the government cannot create one.

Hamilton disagreed. He favored a loose construction—broad or flexible interpretation—of the Constitution. Pointing to the elastic clause in the document, he argued that the bank was "necessary and proper" to carry out the government's duties. (See The Living Constitution, page 254.) According to this view, when the Constitution grants a power to Congress, it also grants Congress the "necessary and proper" means to carry out that power. Jefferson and Hamilton argued their positions to Washington. Hamilton won, and the Bank of the United States was set up in 1791. The president, meanwhile, was dealing with other challenges at home and abroad, which you will read about in Section 2.

Reading **History**
C. Contrasting What is the main difference between strict and loose interpretations of the Constitution?

Section 1 Assessment

1. Terms & Names

Identify:
- inaugurate
- Federal Judiciary Act
- cabinet
- tariff

2. Taking Notes

In a chart, list members of Washington's cabinet and their responsibilities.

Cabinet member	Responsibilities

Which cabinet member had the greatest responsibilities? Explain.

3. Main Ideas

a. What was the purpose of Washington's cabinet?

b. What economic problems did the new government face?

c. How did Hamilton's financial plan attempt to solve the nation's economic problems?

4. Critical Thinking

Contrasting How did Hamilton and Jefferson differ in their interpretation of the Constitution?

THINK ABOUT
- views on the national bank
- views on the role of government

ACTIVITY OPTIONS

LANGUAGE ARTS
ART

Imagine you oppose or support Hamilton's plan for the nation's finances. Write a **letter to the editor** or draw a **political cartoon** expressing your opinion.

2 Challenges to the New Government

TERMS & NAMES
- Battle of Fallen Timbers
- Treaty of Greenville
- Whiskey Rebellion
- French Revolution
- neutral
- Jay's Treaty
- Pinckney's Treaty

MAIN IDEA

Washington established central authority at home and avoided war with European powers.

WHY IT MATTERS NOW

Washington's policies at home and abroad set an example for later presidents.

ONE AMERICAN'S STORY

Pioneers had been moving west since before the Revolution. However, the settlers met fierce resistance from Native Americans. One of their most respected military leaders was Chief Little Turtle of the Miami tribe of Ohio. In 1790 and 1791, he had won decisive victories against U.S. troops.

Now, two years later, the Miami and their allies again faced attack by American forces. At a council meeting, Little Turtle gave a warning to his people about the troops led by General Anthony Wayne.

General Anthony Wayne negotiates with a Miami war chief.

A VOICE FROM THE PAST

We have beaten the enemy twice under different commanders. . . . The Americans are now led by a chief [Wayne] who never sleeps. . . . During all the time he has been marching on our villages . . . we have not been able to surprise him. Think well of it. . . . It would be prudent [wise] to listen to his offers of peace.

Little Turtle, quoted in *The Life and Times of Little Turtle*

While the council members weighed Little Turtle's warning, President Washington was making plans to secure—guard or protect—the western borders of the new nation.

Securing the Northwest Territory

As a general, Washington had skillfully waged war. As the nation's president, however, he saw that the country needed peace in order to prosper. But in spite of his desire for peace, he considered military action as trouble brewed in the Trans-Appalachian West, the land between the Appalachian Mountains and the Mississippi River. The 1783 Treaty of Paris had attempted to resolve the claims. The source of the trouble was competing claims for these lands. Some years later, however, Spain, Britain, the United States, and Native Americans claimed parts of the area as their own.

Spain held much of North America west of the Mississippi. It also claimed Florida and the port of New Orleans at the mouth of the

Mississippi. For American settlers in the West, this port was key to trade. They carried their goods to market by flatboat down the Mississippi to New Orleans. They took Spanish threats to close the port very seriously. The Spanish also stirred up trouble between the white settlers and the Creeks, Choctaws, and other Native American groups in the Southeast.

Reading **History**

A. Making Inferences What expectations might the Native Americans have had of the British as the tribes came into conflict with white settlers?

The strongest resistance to white settlement came from Native Americans in the Northwest Territory. This territory was bordered by the Ohio River to the south and Canada to the north. Native Americans in that territory hoped to join together to form an independent Native American nation. In violation of the Treaty of Paris, the British still held forts north of the Ohio River. The British supported Native Americans in order to maintain their access to fur in these territories. Eventually, Native Americans and white settlers clashed over the Northwest Territory.

Battle of Fallen Timbers

Believing the Northwest Territory was critical to the security and growth of the new nation, Washington sent troops to the Ohio Valley. As you read in One American's Story, this first federal army took a beating from warriors led by Little Turtle in 1790. The chief's force came from many tribes, including the Shawnee, Ottawa, and Chippewa, who joined in a confederation to defeat the federal army.

After a second defeat in 1791 of an army headed by General Arthur St. Clair, Washington ordered another army west. This time Anthony Wayne, known as "Mad Anthony" for his reckless courage, was at its head.

The other chiefs ignored Little Turtle's advice to negotiate. They replaced him with a less able leader. Expecting British help, Native American warriors gathered at British-held Fort Miami. On August 20, 1794, a fighting force of around 2,000 Native Americans clashed with Wayne's troops. The site was covered with trees that had been struck down by a storm. The Native Americans were defeated in what became known as the **Battle of Fallen Timbers**.

The Battle of Fallen Timbers memorial sculpture below shows two American soldiers and a Native American.

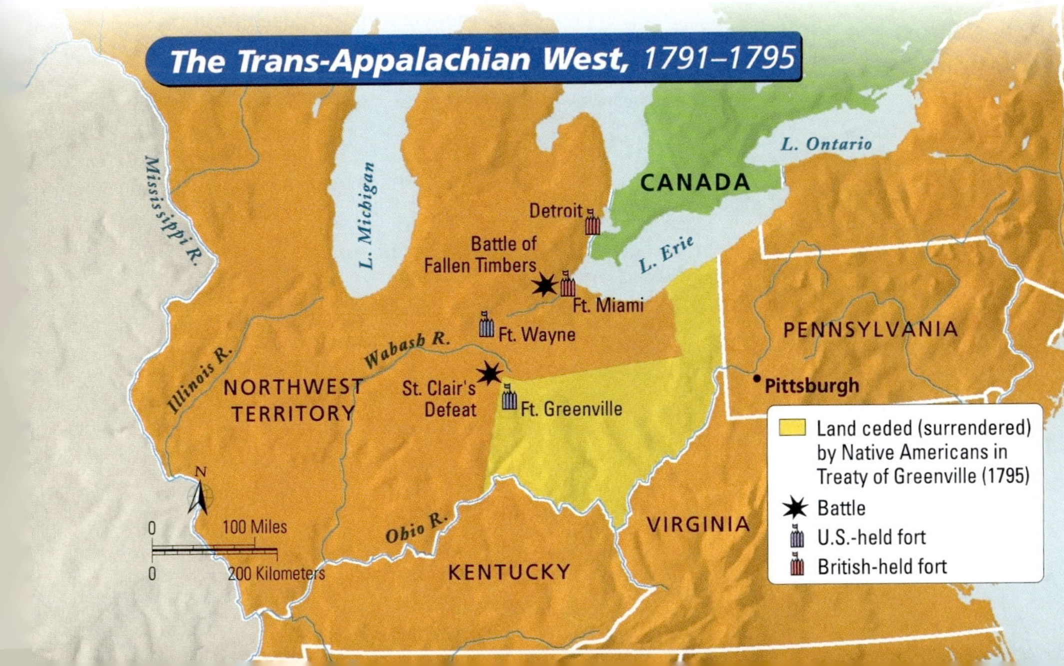

GEOGRAPHY SKILLBUILDER
Interpreting Maps

1. **Region** *What does the yellow area of the map represent?*
2. **Location** *Why might the British forts be located near water?*

The Native Americans retreated to Fort Miami. The British, not wanting war with the United States, refused to help them. The Battle of Fallen Timbers crushed Native American hopes of keeping their land in the Northwest Territory. Twelve tribes signed the **Treaty of Greenville** in 1795. They agreed to cede, or surrender, much of present-day Ohio and Indiana to the U.S. government.

*Reading*History

B. Reading a Map Use the map on page 299 to see which two states to the south bordered the land ceded by Native Americans.

The Whiskey Rebellion

Not long after the Battle of Fallen Timbers, Washington put another army into the field. The conflict arose over the government's tax on whiskey, part of Hamilton's financial plan. From Pennsylvania to Georgia, outraged farmers resisted the tax. For them, whiskey—and the grain it was made from—were important products.

Because of poor roads, backcountry farmers had trouble getting their grain to market. Crops such as wheat and rye were more easily carried to market in liquid form, so farmers made their grain into whiskey. A farmer's horse could haul only two bushels of rye but could carry two barrels of rye whiskey. This was an amount equal to 24 bushels of the grain. In addition, their customers paid more for whiskey than grain. With little cash to buy goods, let alone pay the tax, farmers often traded whiskey for salt, sugar, and other goods. The farmers used whiskey as money to get whatever supplies they needed.

CITIZENSHIP TODAY

Obeying Rules and Laws

As the Whiskey Rebellion shows, since the earliest days of the republic our government has made laws and punished those who broke them. These laws affect not only adult citizens, but young people as well.

Today, for example, communities across the country are trying to control the problem of juvenile crime by imposing curfews on young people. These laws require minors to be off the streets after a certain time, often ten or eleven at night. Penalties can be harsh. In certain communities, minors who break curfew laws can be detained, and their parents can be fined.

People who favor curfews believe such laws cut crime. Those who oppose curfews think such limits are the responsibility of parents and not the government.

Why Should You Obey Rules and Laws?

1. What are some arguments in favor of curfew laws? What are arguments against them? Make a list of each.
2. Poll your classmates to see how many agree with each position.
3. Write an essay expressing your opinion on this issue.
4. Brainstorm changes or adaptations to curfew laws that you think would make them more flexible.

See Citizenship Handbook, page 283.

Visit www.mcdougallittell.com to learn more about young people and the law.

In the summer of 1794, a group of farmers in western Pennsylvania staged the **Whiskey Rebellion** against the tax. One armed group beat up a tax collector, coated him with tar and feathers, and stole his horse. Others threatened an armed attack on Pittsburgh.

Washington, urged on by Hamilton, was prepared to enforce the tax and crush the Whiskey Rebellion. They feared that not to act might undermine the new government and weaken its authority. Hamilton condemned the rebels for resisting the law.

"Such a resistance is treason."
Alexander Hamilton

A VOICE FROM THE PAST

Such a resistance is treason against society, against liberty, against everything that ought to be dear to a free, enlightened, and prudent people. To tolerate it were to abandon your most precious interests. Not to subdue it were to tolerate it.

Alexander Hamilton, *The Works of Alexander Hamilton*

In October 1794, General Henry Lee, with Hamilton at his side, led an army of 13,000 soldiers into western Pennsylvania to put down the uprising. As news of the army's approach spread, the rebels fled. After much effort, federal troops rounded up 20 barefoot, ragged prisoners. Washington had proved his point. He had shown that the government had the power and the will to enforce its laws. Meanwhile, events in Europe gave Washington a different kind of challenge.

The French Revolution

In 1789, the French launched a revolution for liberty and equality. At first, the **French Revolution** won strong support from the United States. However, by 1793 the revolution had turned violent. The king of France, Louis XVI, was beheaded, along with thousands of French citizens accused of being enemies of the state. In the same year, France declared war on Britain, Spain, and Holland. The monarchs of these nations had long feared that the revolution would put their own thrones at risk. Britain led the fight to defeat revolutionary France.

The war between France and Britain put the United States in an awkward position. France had been America's ally in the Revolution against the British. A 1778 treaty still bound

Connections TO WORLD HISTORY

EYEWITNESS TO REVOLUTION

In 1789, an American citizen with a strange first name, Gouverneur Morris, went to Paris as a private business agent. Three years later, President Washington appointed him U.S. minister to France. An eyewitness to the French Revolution, Morris kept a detailed record of what he saw, including the execution of the king and queen by guillotine, as shown below.

Here is part of a letter he wrote on October 18, 1793:

"Terror is the order of the Day. . . . The Queen was executed the Day before yesterday. Insulted during her Trial and reviled in her last Moments, she behav'd with Dignity throughout."

the two nations together. In addition, many saw France's revolution as proof that the American cause had been just. Jefferson felt that a move to crush the French Revolution was an attack on liberty everywhere. Hamilton, though, pointed out that Britain was the United States' most important trading partner, and British trade was too important to risk war.

In April 1793, Washington declared that the United States would remain **neutral,** not siding with one country or the other. He stated that the nation would be "friendly and impartial" to both sides. Congress then passed a law forbidding the United States to help either side.

Reading **History**

C. Drawing Conclusions What sort of U.S. obligation to France did the wartime alliance and treaty of 1778 create?

Remaining Neutral

Britain made it hard for the United States to remain neutral. Late in 1792, the British began seizing the cargoes of American ships carrying goods from the French West Indies.

Washington sent Chief Justice John Jay to England for talks about the seizure of U.S. ships. Jay also hoped to persuade the British to give up their forts on the Northwest frontier. During the talks in 1794, news came of the U.S. victory at the Battle of Fallen Timbers. Fearing another entanglement, the British agreed to leave the Ohio Valley by 1796. In **Jay's Treaty,** the British also agreed to pay damages for U.S. vessels they had seized. Jay failed, however, to open up the profitable British West Indies trade to Americans. Because of this, Jay's Treaty was unpopular.

Like Jay, Thomas Pinckney helped the United States reduce tensions along the frontier. In 1795, **Pinckney's Treaty** with Spain gave Americans the right to travel freely on the Mississippi River. It also gave them the right to store goods at the port of New Orleans without paying customs duties. In addition, Spain accepted the 31st parallel as the northern boundary of Florida and the southern boundary of the United States.

Meanwhile, more American settlers moved west. As you will read in the next section, change was coming back east as Washington stepped down.

Reading **History**

D. Evaluating What were some of the advantages to the new nation of remaining neutral?

Section 2 Assessment

1. Terms & Names

Identify:
- Battle of Fallen Timbers
- Treaty of Greenville
- Whiskey Rebellion
- French Revolution
- neutral
- Jay's Treaty
- Pinckney's Treaty

2. Taking Notes

Use a chart to record U.S. responses to various challenges.

Challenge	Response
From Spain	
From Britain	
From France	

Which challenge seemed greatest? Why?

3. Main Ideas

a. What military and other actions secured the West for the United States?

b. Why did Washington consider it important to put down the Whiskey Rebellion?

c. How did the French Revolution create problems for the United States?

4. Critical Thinking

Drawing Conclusions Why was neutrality a difficult policy for the United States to maintain?

THINK ABOUT
- ties with France
- ties with Britain
- restrictions on trade

ACTIVITY OPTIONS

GEOGRAPHY | **ART**

Make a **map** that describes the Battle of Fallen Timbers, or draw a **scene** from that battle.

3 The Federalists in Charge

TERMS & NAMES
foreign policy
political party
XYZ Affair
Alien and Sedition Acts
states' rights

MAIN IDEA
The split between Hamilton and Jefferson led to the growth of political parties.

WHY IT MATTERS NOW
The two-party system is still a major feature of politics in the United States.

ONE AMERICAN'S STORY

In 1796, President George Washington decided that two terms in office was enough. The president was fed up with political quarreling. He wanted to return to Mount Vernon, his estate in Virginia. But as he left office, he feared the development of political parties would split the nation into enemy camps. With Hamilton's help, in 1796 he wrote a final address to the nation.

This painting portrays Mount Vernon in 1792.

A VOICE FROM THE PAST

Let me now . . . warn you . . . against the [harmful] effects of the spirit of party. . . . This spirit, unfortunately . . . exists in different shapes in all governments . . . but in those of the popular form, it is seen in its greatest rankness and is truly their worst enemy.

George Washington, Farewell Address

In his address, Washington warned of the dangers of political division, or what he termed "the spirit of party." As you will see in this section, few people took his advice.

Washington Retires

Washington had come to the presidency greatly admired by the American people. Throughout his eight years in office (1789–1797), he had tried to serve as a symbol of national unity. In large part, he succeeded. During his second term, however, opponents of Jay's Treaty led attacks on the president. Thomas Paine called Washington "treacherous in private friendship . . . and a hypocrite in public life" because he failed to support the French Revolution.

Washington saw such attacks as the outcome of political disagreements. In his farewell address, he warned that such differences could weaken the nation. Despite his advice, political parties became a part of American politics.

Americans listened more closely to Washington's parting advice on **foreign policy**—relations with the governments of other countries. He urged the nation's leaders to remain neutral and "steer clear of permanent alliances with any portion of the foreign world." He warned that agreements with foreign nations might work against U.S. interests. His advice served to guide U.S. foreign policy into the twentieth century.

Growth of Political Parties

Despite Washington's warning against political parties, Americans were deeply divided over how the nation should be run. During Washington's first term (1789–1792), Hamilton and Jefferson had hotly debated the direction the new nation should take. Then Jefferson returned to Virginia in 1793. During Washington's second term, Madison took Jefferson's place in the debates with Hamilton.

Both sides disagreed on how to interpret the Constitution and on economic policy. Hamilton favored the British government and opposed the French Revolution. Jefferson and Madison were the opposite. Hamilton fought for a strong central government. Jefferson and Madison feared such a government might lead to tyranny. They had different visions of what the nation should become. Hamilton wanted a United States in which trade, manufacturing, and cities grew. Jefferson and Madison pictured a rural nation of planters and farmers.

These differences on foreign and domestic policy led to the nation's first political parties. A **political party** is a group of people that tries to promote its ideas and influence government. It also backs candidates for office. Together, Jefferson and Madison founded the Democratic-Republican Party. The party name reflected their strong belief in democracy and the republican system. Their ideas drew farmers and workers to the new party. Hamilton and his friends formed the Federalist Party. Many Northern merchants and manufacturers became Federalists.

*Reading*History

A. Summarizing What were the major beliefs of each party?

The First Political Parties

FEDERALISTS	DEMOCRATIC-REPUBLICANS
Strong national government	Limited national government
Fear of mob rule	Fear of rule by one person or a powerful few
Loose construction (interpretation) of the Constitution	Strict construction (interpretation) of the Constitution
Favored national bank	Opposed national bank
Economy based on manufacturing and shipping	Economy based on farming
Supporters: lawyers, merchants, manufacturers, clergy	Supporters: farmers, tradespeople

SKILLBUILDER Interpreting Charts

1. *Which economic interests were served by the Federalists?*
2. *Which party favored a ruling elite? Which put more trust in the common people?*

America's HERITAGE

WASHINGTON, D.C., AND BENJAMIN BANNEKER

Benjamin Banneker was a free African-American farmer. He was a self-taught mathematician and astronomer. He also wrote an almanac (see below). He was named to the survey commission appointed to lay out the boundaries of the nation's new capital. Working with chief planner Pierre L'Enfant, Banneker helped to decide where the White House and Capitol would be located. Their final design is shown at the left.

John Adams Takes Office

In 1796, the United States held its first elections in which political parties competed. The Federalists picked Washington's vice-president, John Adams, as their candidate for president. An experienced public servant, Adams had been a leader during the Revolution and at the Continental Congress. He had also been a diplomat in France, the Netherlands, and Britain before serving with Washington. The Democratic-Republicans chose Jefferson.

In the electoral college, Adams received 71 votes and Jefferson 68. The Constitution stated that the runner-up should become vice-president. Therefore, the country had a Federalist president and a Democratic-Republican vice-president. Adams became president in 1797. His chief rival, Jefferson, entered office as his vice-president. In 1800, Adams became the first president to govern from the nation's new capital city, Washington, D.C.

Problems with France

When Washington left office in 1797, relations between France and the United States were tense. With Britain and France still at war, the French began seizing U.S. ships to prevent them from trading with the British. Within the year, the French had looted more than 300 U.S. ships.

Although some Federalists called for war with France, Adams hoped talks would restore calm. To this end, he sent Charles Pinckney, Elbridge Gerry, and John Marshall to Paris. Arriving there, they requested a meeting with the French minister of foreign affairs. For weeks, they were

STORY *through* ART

American newspapers fueled public anger over the XYZ Affair by publishing editorials and cartoons like this one. Here the five-man group ruling France demands money at dagger point from the three Americans. The American diplomats respond, "Cease bawling, monster! We will not give you sixpence!"

What attitude does the cartoonist have toward France's role in this affair? How can you tell?

ignored. Then three French agents—later referred to as X, Y, and Z—took the Americans aside to tell them the minister would hold talks. However, the talks would occur only if the Americans agreed to loan France $10 million and to pay the minister a bribe of $250,000. The Americans refused. "No, no, not a sixpence," Pinckney shot back.

Adams received a full report of what became known as the **XYZ Affair.** After Congress and an outraged public learned of it, the press turned Pinckney's words into a popular slogan: "Millions for defense, not one cent for tribute!" In 1798, Congress canceled its treaties with France and allowed U.S. ships to seize French vessels. Congress also set aside money to expand the navy and the army.

Reading **History**

B. Drawing Conclusions How did the XYZ Affair show the young nation's growing confidence?

The Alien and Sedition Acts

The conflict with France made Adams and the Federalists popular with the public. Many Democratic-Republicans, however, were sympathetic to France. One Democratic-Republican newspaper called Adams "the blasted tyrant of America." In turn, Federalists labeled Democratic-Republicans "democrats, mobcrats, and other kinds of rats."

Angered by criticism in a time of crisis, Adams blamed the Democratic-Republican newspapers and new immigrants. Many of the immigrants were Democratic-Republicans. To silence their critics, the Federalist Congress passed the **Alien and Sedition Acts** in 1798. These acts targeted aliens—immigrants who were not yet citizens. One act increased the waiting period for becoming a U.S. citizen from 5 to 14 years. Other acts gave the president the power to arrest disloyal aliens or order them out of the country during wartime. A fourth act outlawed sedition, saying or writing anything false or harmful about the government.

With these acts, the Federalists clamped down on freedom of speech and the press. About 25 Democratic-Republican newspaper editors were charged under this act, and 10 were convicted of expressing opinions damaging to the government. A Vermont congressman, Matthew Lyon, was also locked up for saying that the president should be sent "to a mad house." The voters re-elected Lyon while he was in jail.

Reading **History**

C. Making Inferences How might the theory of states' rights undermine the federal government?

The Democratic-Republicans, led by Jefferson and Madison, searched for a way to fight the Alien and Sedition Acts. They found it in a theory called **states' rights**. According to this theory, states had the right to judge when the federal government had passed an unconstitutional law.

Jefferson and Madison wrote resolutions (or statements) passed by the Kentucky and Virginia legislatures in 1798 and 1799. These Kentucky and Virginia Resolutions declared that the Alien and Sedition Acts violated the Constitution. No other states supported Kentucky and Virginia. However, within two years the Democratic-Republicans won control of Congress, and they either repealed the Alien and Sedition Acts or let them expire between 1800 and 1802.

Peace with France

While Federalists and Democratic-Republicans continued to battle at home, the United States made peace with France. Although war fever was high at home, Adams reopened talks with France. This time the two sides quickly signed an agreement to stop all naval attacks. Called the Convention of 1800, this treaty cleared the way for U.S. and French ships to sail the ocean in peace.

Adams's actions made him enemies among the Federalists. Despite this, he spoke proudly of having saved the nation from bloodshed. "I desire no other inscription over my gravestone than: 'Here lies John Adams, who took upon himself the responsibility of the peace with France in the year 1800.'"

Adams lost the presidential election of 1800 to Thomas Jefferson. You will read more about Jefferson in the next chapter.

Section 3 Assessment

1. Terms & Names

Identify:

- foreign policy
- political party
- XYZ Affair
- Alien and Sedition Acts
- states' rights

2. Taking Notes

Use a cluster diagram to review details about the Alien and Sedition Acts.

What was the worst effect of the Alien and Sedition Acts? Why?

3. Main Ideas

a. What two pieces of advice did Washington give in his Farewell Address?

b. What led to the rise of political parties?

c. Why did Congress pass the Alien and Sedition Acts? How did Kentucky and Virginia respond?

4. Critical Thinking

Evaluating Do you think Washington's warning about political parties was good advice? Explain.

THINK ABOUT

- roles of political parties
- advantages of parties
- disadvantages of parties

ACTIVITY OPTIONS

TECHNOLOGY

SPEECH

Read more about Benjamin Banneker. Plan part of a **video presentation** on him or present **dramatic readings** of excerpts from the almanac he wrote.

Chapter 9 ASSESSMENT

TERMS & NAMES

Briefly explain the importance of each of the following.

1. inaugurate
2. cabinet
3. tariff
4. Battle of Fallen Timbers
5. Whiskey Rebellion
6. neutral
7. foreign policy
8. political party
9. Alien and Sedition Acts
10. states' rights

REVIEW QUESTIONS

Washington's Presidency (pages 293–297)

1. What questions about the judiciary were left open by the Constitution? How were they answered?
2. What financial problems did the new nation face?
3. How did Hamilton and Jefferson interpret the Constitution differently?

Challenges to the New Government (pages 298–302)

4. What did Washington do to secure the West?
5. Why did Washington send troops to put down the Whiskey Rebellion?
6. Why did Washington favor neutrality in the conflict between France and Britain?
7. What problems did the Jay and Pinckney treaties address?

The Federalists in Charge (pages 303–307)

8. Why did Washington oppose political parties?
9. What was the XYZ Affair?
10. Why did Federalists pass the Alien and Sedition Acts? How did Republicans respond?

CRITICAL THINKING

1. USING YOUR NOTES

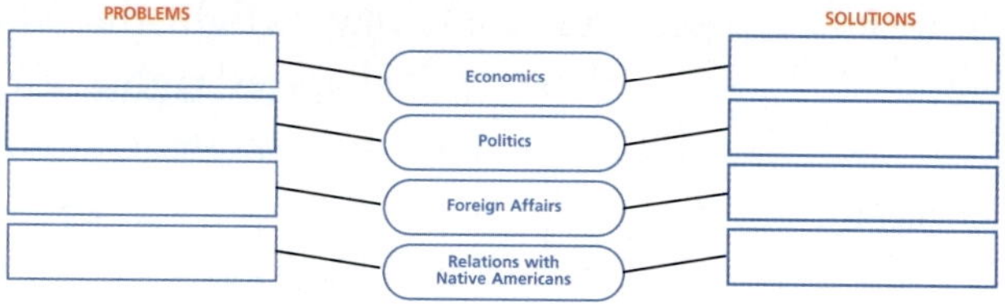

Using your completed chart, answer the questions.

a. Which of the problems listed on the chart were solved peacefully?
b. How might an economic problem require a political solution?

2. ANALYZING LEADERSHIP

How did Washington's efforts to serve as a symbol of national unity help the new nation?

3. APPLYING CITIZENSHIP SKILLS

How might the farmers in the Whiskey Rebellion have expressed their disapproval of the whiskey tax while staying within the law?

4. THEME: DEMOCRATIC IDEALS

Did the formation of political parties make the nation more or less democratic?

Interact *with* History

How did the challenges of setting up a government that you discussed before you read the chapter compare with the actual challenges you read about?

VISUAL SUMMARY

The First Presidents

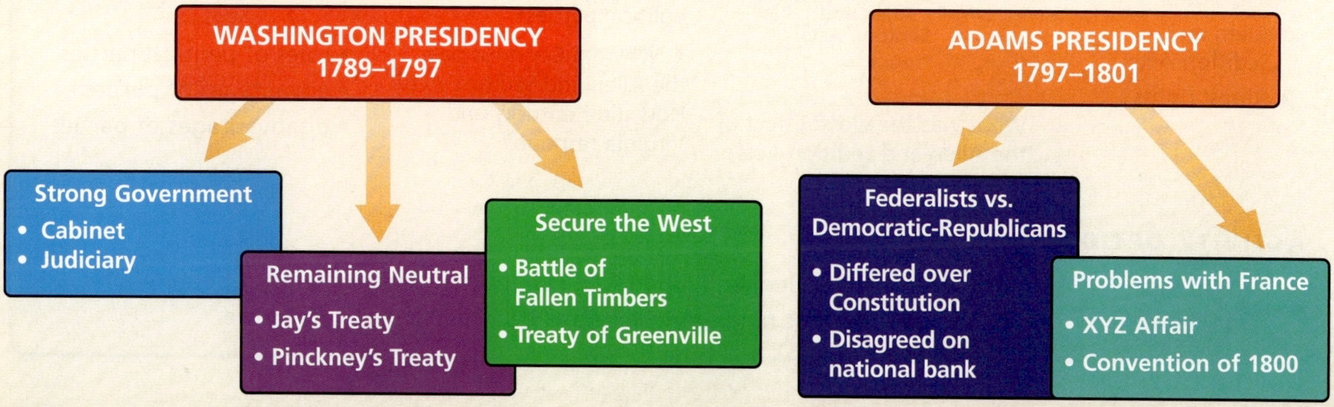

HISTORY SKILLS

1. INTERPRETING CHARTS

The following chart shows the money problems of the new nation between 1789 and 1791. The numbers have been rounded off. Study the chart and then answer the questions.

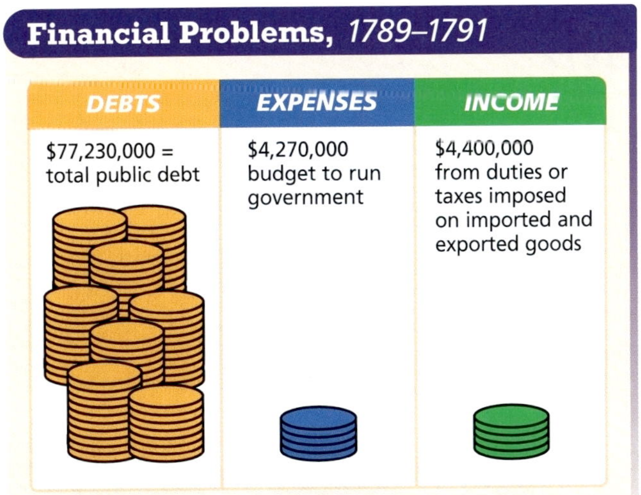

Source: *Historical Statistics of the United States*

a. What was the government's total income in these years?

b. How much money did the government owe in these same years?

c. How might the government try to raise more money?

2. INTERPRETING PRIMARY SOURCES

In a letter to her sister, John Adams's wife, Abigail, had the following to say about President Washington. Read the quotation and answer the questions.

> He is polite with dignity, affable without familiarity, distant without haughtiness, . . . modest, wise, and good.
>
> **Abigail Adams,** letter of January 5, 1790

a. What qualities seemed to set Washington apart from other political leaders?

b. Is Mrs. Adams's overall impression of Washington positive or negative? Explain your answer.

c. What qualities or characteristics did the people of Washington's time seem to expect in a leader?

ALTERNATIVE ASSESSMENT

1. INTERDISCIPLINARY ACTIVITY: World History

Writing a Letter Imagine that you are a U.S. citizen during the French Revolution. Write a letter to the secretary of state recommending a policy you think the U.S. government should follow.

2. COOPERATIVE LEARNING ACTIVITY

Holding a Debate The controversy over the Alien and Sedition Acts deeply divided Federalists and Democratic Republicans. Were the acts constitutional or an abuse of basic rights? Should criticism of the government be allowed in a time of near war?

Working in three groups, do research on the Alien and Sedition Acts and the positions taken by both political parties on these acts. Then have Federalist and Democratic-Republican groups pick representatives to debate the questions posed above, while the remaining group acts as audience and judge for the debate.

3. TECHNOLOGY ACTIVITY

Creating a Television Commercial People of the eighteenth century had different expectations of their political leaders than do people today. The quotation from Abigail Adams on this page is an example of the different perspective people held in Washington's time. Using the library or the Internet, find pictures of, and quotations about, political leaders of the time.

INTERNET **Visit www.mcdougallittell.com to learn more about the first presidents.**

Create a 30-second television commercial that advertises an American history theme park, using the suggestions below.

- Try to find quotations that suggest the most important qualities of leaders such as Washington, Adams, Hamilton, and Jefferson.
- Try to find more than one image of each leader; find images that suggest different qualities in the person.
- Use quotations from and images of leaders as part of your commercial for a theme park.
- Show your commercial to the class.

4. HISTORY PORTFOLIO

PORTFOLIO **Option 1** Review your section and chapter assessment activities. Select one that you think is your best work. Then use comments made by your teacher or classmates to improve your work and add it to your portfolio.

PORTFOLIO **Option 2** Review the questions that you wrote for What Do You Want to Know? on page 292. Then write a short report in which you explain the answers to your questions. If any questions were not answered, do research to answer them. Add your answers to your portfolio.

CHAPTER 10

The Jefferson Era 1800–1816

Albert Bierstadt's painting, *The Rocky Mountains* (1863), celebrates the beauty of the American West.

Interact *with* History

You have been chosen to participate in an expedition to the West in the early 1800s. You are excited and curious, but also a little scared. You know that you will see and experience many new things. But you know there are risks involved, too.

What Do You Think?

- Notice the land features in these scenes. What problems might they hold for an explorer?
- What other people might you meet on the expedition?
- How will you find food during the long winter?

What dangers will you face on an expedition west?

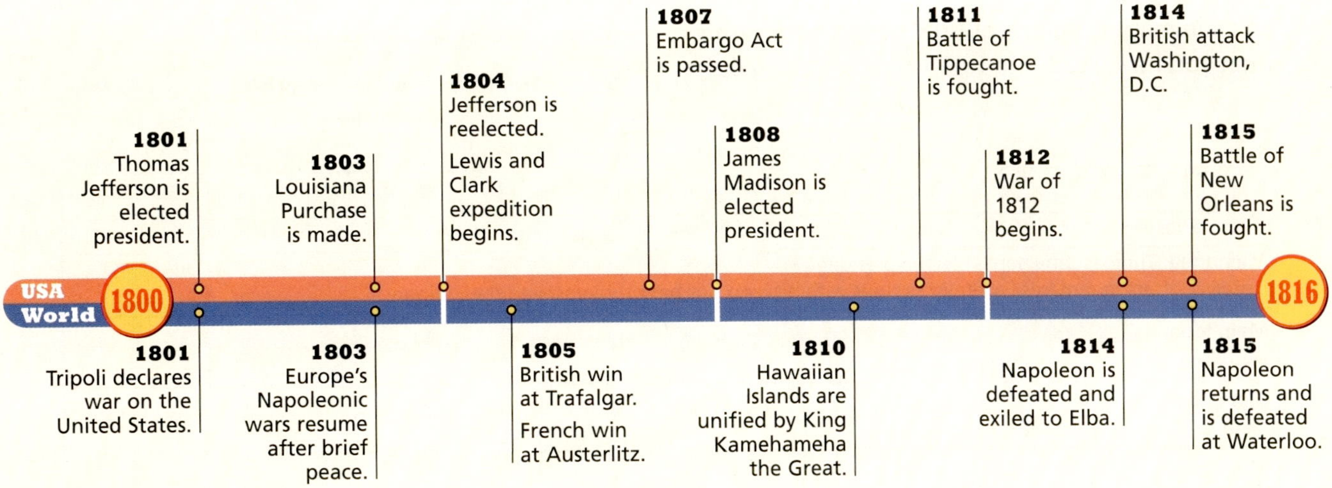

Chapter 10 SETTING THE STAGE

BEFORE YOU READ

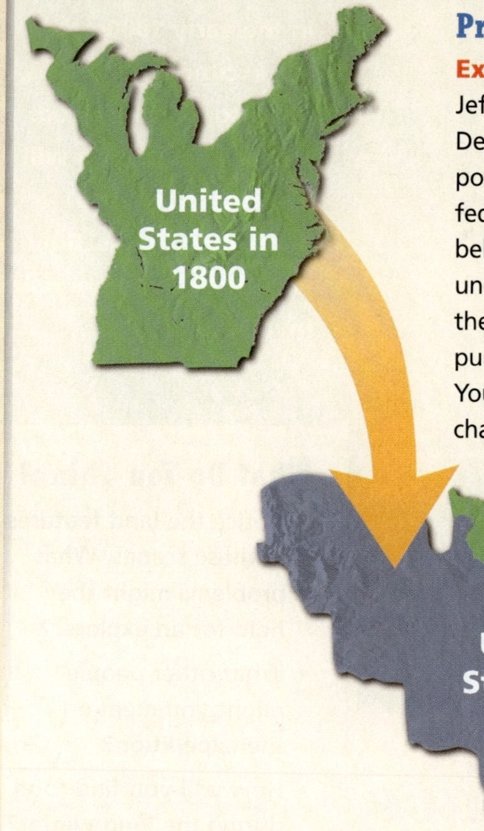

Previewing the Theme

Expansion In 1801, Thomas Jefferson and his party—the Democratic-Republicans—took power. They reduced the size of the federal government, which they believed had become too powerful under the Federalists. They doubled the size of the country, however, by purchasing the Louisiana Territory. You will read about these dramatic changes in Chapter 10.

What Do You Know?

What parts of the United States today were not part of the country in 1800? What should leaders consider before they buy land for their countries?

THINK ABOUT

- where the money for the purchase will come from
- what should be done if people already live on the land

What Do You Want to Know?

What details do you need to help you understand the nation's expansion in the early 1800s? Make a list of information you need in your notebook before you read the chapter.

READ AND TAKE NOTES

Reading Strategy: Summarizing When you study history, it is important to clearly understand what you read. One way to achieve a clear understanding is to summarize. When you summarize, you restate what you have read into fewer words, stating only the main ideas and essential details. It is important to use your own words in a summary. Use the chart below to record your summaries of the main ideas and essential details in Chapter 10.

See Skillbuilder Handbook, page R2.

The Jefferson Era
Summaries
Main Idea: Thomas Jefferson is elected president. **Details:** Jefferson replaces Federalist policies with his own but has problems with the judiciary.
Main Idea: **Details:**
Main Idea: **Details:**
Main Idea: **Details:**

1 Jefferson Takes Office

TERMS & NAMES
radical
Judiciary Act of 1801
John Marshall
Marbury v. Madison
unconstitutional
judicial review

MAIN IDEA	WHY IT MATTERS NOW
When Jefferson became president in 1801, his party replaced Federalist programs with its own.	Today's Democratic party traces its roots to the party of Jefferson, the Democratic-Republicans.

ONE AMERICAN'S STORY

Supporters of John Adams and Thomas Jefferson—competitors in the presidential election of 1800—fought for their candidates with nasty personal attacks. Scottish immigrant James Callender, a Jefferson supporter, wrote some of the harshest criticisms. During the campaign, he warned voters not to reelect President John Adams.

A VOICE FROM THE PAST

In the fall of 1796 . . . the country fell into a more dangerous juncture than almost any the old confederation ever endured. The tardiness and timidity of Mr. Washington were succeeded by the rancour [bitterness] and insolence [arrogance] of Mr. Adams. . . . Think what you have been, what you are, and what, under [Adams], you are likely to become.

James Callender, quoted in *American Aurora*

In the presidential election of 1800, Thomas Jefferson was the candidate of the Democratic-Republican party. John Adams represented the Federalists.

Adams's defenders were just as vicious. One went so far as to claim that if Jefferson won, "the soil will be soaked with blood, and the nation black with crimes." In spite of the campaign's nastiness, the election ended with a peaceful transfer of power from one party to another. The 1800 election was more than a personal battle, though. It was a contest between two parties with different ideas about the role of government.

The Election of 1800

The two parties contesting the election of 1800 were the Federalists, led by President John Adams, and the Democratic-Republicans, represented by Thomas Jefferson. Each party believed that the other was endangering the Constitution and the American republic.

The Democratic-Republicans thought they were saving the nation from monarchy and oppression. They argued, again and again, that the Alien and Sedition Acts supported by the Federalists violated the Bill of Rights. (See pages 306–307.) The Federalists thought that the nation was about to be ruined by **radicals**—people who take extreme political positions. They remembered the violence of the French Revolution, in which radicals executed thousands in the name of liberty.

When election day came, the Democratic-Republicans won the presidency. Jefferson received 73 votes in the electoral college, and Adams earned 65. But there was a problem. Aaron Burr, whom the Democratic-Republicans wanted as vice president, also received 73 votes.

Background
In 1804, the Twelfth Amendment solved this problem by creating separate ballots for president and vice president.

STRANGE but True

HAMILTON-BURR DUEL

In 1804, the Democratic-Republicans replaced Aaron Burr as their candidate for vice president. Burr then decided to run for governor of New York.

Alexander Hamilton questioned Burr's fitness for public office. He wrote that Burr was a "dangerous man . . . who ought not to be trusted with the reins of government."

Burr lost the election. Furious, he challenged Hamilton to a duel. Hamilton went to the duel but resolved not to fire. Burr, however, shot Hamilton, who died the next day.

Breaking the Tie

According to the Constitution, the House of Representatives had to choose between Burr and Jefferson. The Democratic-Republicans clearly intended for Jefferson to be president. However, the new House of Representatives, dominated by Jefferson's party, would not take office for some months. Federalists still held a majority in the House, and their votes would decide the winner.

The Federalists were divided. Some feared Jefferson so much that they decided to back Burr. Others, such as Alexander Hamilton, considered Burr an unreliable man and urged the election of Jefferson. Hamilton did not like Jefferson, but he believed that Jefferson would do more for the good of the nation than Burr. "If there be a man in the world I ought to hate," he said, "it is Jefferson. . . . But the public good must be [more important than] every private consideration."

Reading History
A. Analyzing Points of View Why did Hamilton think that Jefferson was the better choice for president?

Over a period of seven days, the House voted 35 times without determining a winner. Finally, two weeks before the inauguration, Alexander Hamilton's friend James A. Bayard persuaded several Federalists not to vote for Burr. On the thirty-sixth ballot, Jefferson was elected president. Aaron Burr, who became vice president, would never forget Hamilton's insults.

People were overjoyed by Jefferson's election. A Philadelphia newspaper reported that bells rang, guns fired, dogs barked, cats meowed, and children cried over the news of Jefferson's victory.

The Talented Jefferson

In over 200 years, the United States has had more than 40 presidents. Many of them were great leaders. But no president has ever matched Thomas Jefferson in the variety of his achievements.

Jefferson's talents went beyond politics. He was still a young lawyer when he became interested in the architecture of classical Greece and Rome. The look of our nation's capital today reflects that interest. When Washington, D.C., was being built during the 1790s, Jefferson advised its architects and designers.

Jefferson's passion for classical styles can also be seen in his plan of Monticello, his Virginia home. For this elegant mansion, Jefferson designed storm windows, a seven-day clock, and a dumbwaiter—a small elevator that brought bottles of wine from the cellar.

Jefferson was a skilled violinist, horseman, amateur scientist, and a devoted reader, too. His book collection later became the core of the Library of Congress. After his election, Jefferson applied his many talents and ideas to the government of the United States.

Jefferson's Philosophy

The new president had strong opinions about what kind of country the United States ought to be. But his first order of business was to calm the nation's political quarrels.

> ***A VOICE FROM THE PAST***
>
> Let us, then, fellow-citizens, unite with one heart and one mind. . . . Every difference of opinion is not a difference of principle. . . . We are all Republicans, we are all Federalists.
>
> **Thomas Jefferson,** First Inaugural Address

Reading **History**

B. Summarizing How did Thomas Jefferson try to unite the nation after he was elected?

One way Jefferson tried to unite Americans was by promoting a common way of life. He wanted the United States to remain a nation of small independent farmers. Such a nation, he believed, would uphold the strong morals and democratic values that he associated with country living. He hoped that the enormous amount of available land would prevent Americans from crowding into cities, as people had in Europe.

As president, Jefferson behaved more like a gentleman farmer than a privileged politician. Instead of riding in a fancy carriage to his inauguration, Jefferson walked the two blocks from his boarding house to the Capitol. Though his chef served elegant meals, the president's guests ate at round tables so that no one could sit at the head of the table.

To the end, Jefferson refused to elevate himself because of his office. For his tombstone, he chose this simple epitaph: "Here was buried

The Talented Jefferson

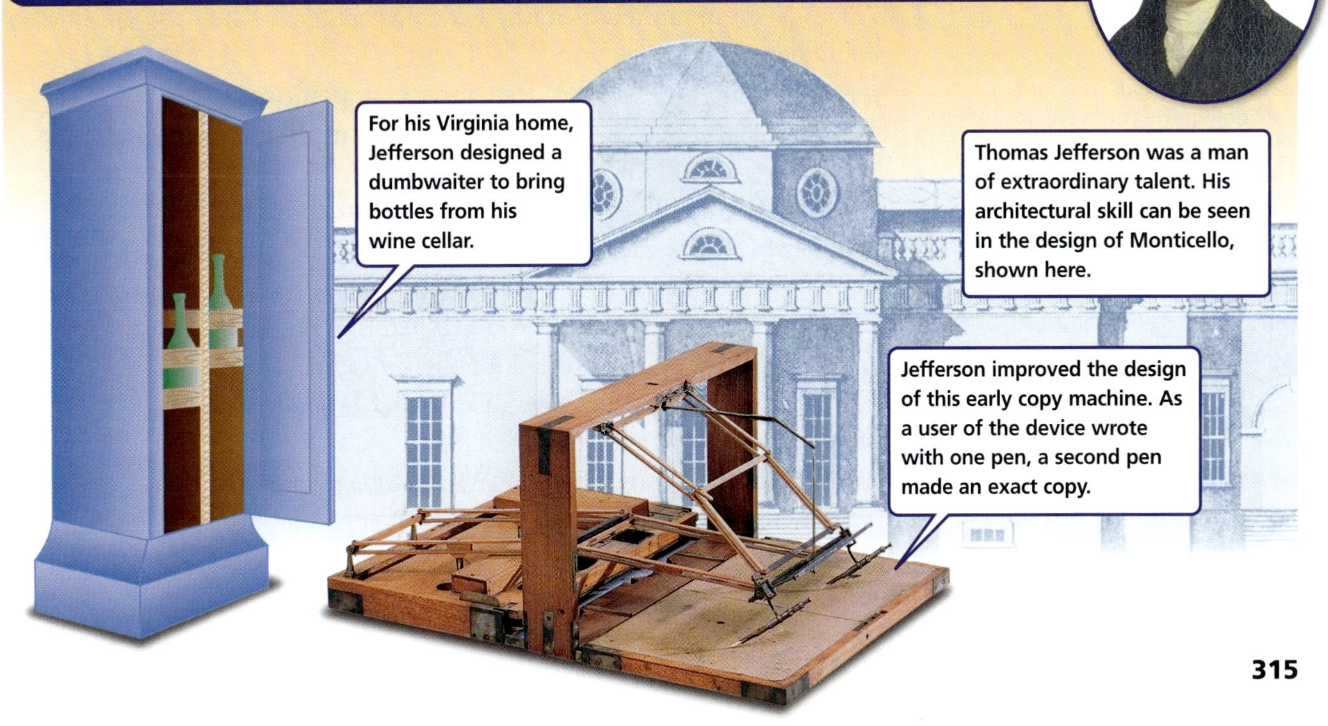

Thomas Jefferson, author of the Declaration of American Independence, of the statute of Virginia for religious freedom, and father of the University of Virginia." Jefferson chose not to list his presidency. His belief in a modest role for the central government is reflected in the changes he made during his presidency.

Background
In addition to founding the University of Virginia in 1819, Jefferson designed its buildings and supervised their construction.

Undoing Federalist Programs

Jefferson believed that the federal government should have less power than it had had under the Federalists. During his term of office, he sought to end many Federalist programs.

At the president's urging, Congress—now controlled by Democratic-Republicans—allowed the Alien and Sedition Acts to end. Jefferson then released prisoners convicted under the acts—among them, James Callender. Congress also ended many taxes, including the unpopular whiskey tax. Because the loss of tax revenue lowered the government's income, Jefferson reduced the number of federal employees to cut costs. He also reduced the size of the military.

Jefferson next made changes to the Federalists' financial policies. Alexander Hamilton had created a system that depended on a certain amount of public debt. He believed that people who were owed money by their government would make sure the government was run properly. But Jefferson opposed public debt. He used revenues from tariffs and land sales to reduce the amount of money owed by the government.

AMERICA'S HISTORY MAKERS

JOHN MARSHALL
1755–1835

John Marshall was born, the first of 15 children, in Virginia's back-country. He had little formal schooling. He received most of his education from his parents and a minister who lived with the family one year.

Even so, the lasting strength of the U.S. Constitution is partly due to Marshall's brilliant legal mind. In his long tenure as Chief Justice, John Marshall participated in more than 1,000 decisions and wrote 519 of them himself.

How does Marshall's record as Chief Justice demonstrate his decision-making abilities?

Marshall and the Judiciary

Though Jefferson ended many Federalist programs, he had little power over the courts. John Adams had seen to that with the **Judiciary Act of 1801**. Under this act, Adams had appointed as many Federalist judges as he could between the election of 1800 and Jefferson's inauguration in 1801. These last-minute appointments meant that the new Democratic-Republican president would face a firmly Federalist judiciary.

Reading History
C. Analyzing Causes Why did the Federalists retain a great deal of power even after they were defeated by the Democratic-Republicans?

Jefferson often felt frustrated by Federalist control of the courts. Yet because judges received their appointments for life, the president could do little.

Before he left office in 1801, President Adams also appointed a new Chief Justice of the Supreme Court. He chose a 45-year-old Federalist, **John Marshall**. He guessed that Marshall would be around for a long time to check the power of the Democratic-Republicans. He was right. Marshall served as Chief Justice for over three decades. Under Marshall, the Supreme Court upheld federal authority and strengthened federal courts. One of the most important decisions of the Marshall Court was ***Marbury* v. *Madison*** (1803).

Marbury v. *Madison*

Vocabulary
justice of the peace: a low-level official with limited authority, including the power to perform marriages

William Marbury was one of Adams's last-minute appointments. Adams had named him as a justice of the peace for the District of Columbia.

Marbury was supposed to be installed in his position by Secretary of State James Madison. When Madison refused to give him the job, Marbury sued. The case went to the Supreme Court, which ruled that the law under which Marbury sued was **unconstitutional**—that is, it contradicted the law of the Constitution.

Although the Court denied Marbury's claim, it did establish the principle of **judicial review**. This principle states that the Supreme Court has the final say in interpreting the Constitution. In his decision, Marshall declared, "It is emphatically the province and duty of the Judicial Department to say what the law is." If the Supreme Court decides that a law violates the Constitution, then that law cannot be put into effect.

Jefferson and Madison were angry when Marshall seized this new power for the Supreme Court, but they could hardly fight his decision. After all, he had decided *Marbury* v. *Madison* in their favor.

By establishing judicial review, Marshall helped to create a lasting balance among the three branches of government. The strength of this balance would be tested as the United States grew. In the next section, you will read about a period of great national growth.

Now and then

THE SUPREME COURT TODAY

The principle of judicial review is still a major force in American society. In June 1999, the Supreme Court used this power to restrict the ability of the federal government to enforce its laws in the 50 states.

In one case, *Alden* v. *Maine,* the Court ruled that employees of a state government cannot sue their state even when the state violates federal labor laws—such as those that set guidelines for overtime wages.

Section 1 Assessment

1. Terms & Names

Identify:
- radical
- Judiciary Act of 1801
- John Marshall
- *Marbury* v. *Madison*
- unconstitutional
- judicial review

2. Taking Notes

Use a chart like the one below to list some of the changes made by Jefferson and his party.

Changes made by Democratic-Republicans
1.

What branch of government gave Jefferson trouble?

3. Main Ideas

a. How was the tie between Jefferson and Burr settled after the election of 1800?

b. In what ways did Jefferson's talents reach beyond politics?

c. How did the opinions of Jefferson and Hamilton regarding the public debt differ?

4. Critical Thinking

Making Generalizations
How was Thomas Jefferson's philosophy reflected in his personal life?

THINK ABOUT
- how he behaved after being elected
- how he felt about his presidency later in life

ACTIVITY OPTIONS

TECHNOLOGY
ART

Read more about Thomas Jefferson. Design Jefferson's **Internet home page** showing his inventions or create a **model** of a building he designed.

2 The Louisiana Purchase and Exploration

MAIN IDEA

Jefferson purchased the Louisiana Territory in 1803 and doubled the size of the United States.

WHY IT MATTERS NOW

Thirteen more states were eventually organized on the land acquired by the Louisiana Purchase.

TERMS & NAMES

Louisiana Purchase
Meriwether Lewis
William Clark
Lewis and Clark expedition
Sacagawea
Zebulon Pike

ONE AMERICAN'S STORY

In 1790, Captain Robert Gray became the first American to sail around the world. Two years later, Gray explored a harbor in what is now Washington state. This harbor was later named Gray's Harbor, and Washington's largest river was named after Gray's ship, the *Columbia.*

New England merchants like Captain Gray had to sail all the way around South America to reach the profitable trading regions of the Oregon Country. (See the map on page 320.) In spite of the long trip, merchants from Boston soon began to appear there frequently. They appeared so often that the Native Americans they traded with began calling all white people "Bostons."

Gray's explorations helped to establish U.S. claims to the Pacific Northwest. In this section, you will learn how a lucky land purchase and a daring expedition further hastened westward expansion.

Robert Gray sailed his ship *Columbia* on trading voyages to the Northwest and China.

The West in 1800

In 1800, when Americans talked about the "West," they meant the area between the Appalachian Mountains and the Mississippi River. Thousands of settlers were moving westward across the Appalachians to settle in this region. Many moved onto land long-inhabited by Native Americans. Even so, several U.S. territories soon declared statehood. Kentucky and Tennessee had become states by 1800, and Ohio entered the union in 1803.

Although the Mississippi River was the western border of the United States, there was a great deal of activity further west. In 1800, France and Spain were negotiating for ownership of the Louisiana Territory—the vast region between the Mississippi River and the Rocky Mountains.

The Pacific coast region and the Oregon Country, as you read in One American's Story, also attracted increasing attention. In California, Spain had a chain of 21 missions stretching from San Diego to San Francisco. Starting just north of San Francisco, Russian settlements dotted the Pacific coast all the way to Alaska. Great Britain also claimed land in the region.

Reading **History**

A. Reading a Map Use the map on page 320 to find the location of New Orleans.

As the number of westerners grew, so did their political influence. A vital issue for many settlers was the use of the Mississippi River. Farmers and merchants used the river to move their products to the port of New Orleans, and from there to east coast markets. Threats to the free navigation of the Mississippi and the use of the port at New Orleans brought America to the brink of war.

Napoleon and New Orleans

"There is on the globe one single spot the possessor of which is our natural and habitual enemy," President Jefferson wrote. That spot was New Orleans. This strategic port was originally claimed by France. After losing the French and Indian War, France turned over the Louisiana Territory—including New Orleans—to Spain. But in a secret treaty in 1800, Spain returned Louisiana and the port to France's powerful leader, Napoleon. Now Napoleon planned to colonize the American territory.

Background

In 1799, Napoleon was made the top official of the French Republic. In 1804, he became emperor.

In 1802, these developments nearly resulted in war. Just before turning Louisiana over to France, Spain closed New Orleans to American shipping. Angry westerners called for war against both Spain and France. To avoid hostilities, Jefferson offered to buy New Orleans from France. He received a surprising offer back. The French asked if the United States wanted to buy all of the Louisiana Territory—a tract of land even larger than the United States at that time.

The Louisiana Purchase

A number of factors may have led Napoleon to make his surprising offer. He was probably alarmed by America's fierce determination to keep the port of New Orleans open. Also, his enthusiasm for a colony in America may have been lessened by events in a French colony in the West Indies. There, a revolt led by Toussaint L'Ouverture (too•SAN loo•vehr•TOOR) had resulted in disastrous losses for the French. Another factor was France's costly war against Britain. America's money may have been more valuable to Napoleon than its land.

Connections TO WORLD HISTORY

TOUSSAINT L'OUVERTURE

Toussaint L'Ouverture was born in Hispaniola, an island in the West Indies once colonized by both France and Spain. In 1791, L'Ouverture helped to lead a slave revolt against the French-controlled part of Hispaniola. A natural leader, L'Ouverture won admiration when he preached harmony between former slaves and planters.

In 1801, L'Ouverture overran the Spanish part of the island. He then freed all the slaves and put himself in charge of the entire island.

Hoping to regain their territory, the French invaded in 1802. They arrested L'Ouverture but failed to end the rebellion.

Jefferson was thrilled by Napoleon's offer. However, the Constitution said nothing about the president's right to buy land. This troubled Jefferson, who believed in the strict interpretation of the Constitution. But he also believed in a republic of small farmers, and that required land. So, on April 30, 1803, the **Louisiana Purchase** was approved for $15 million—about three cents per acre. The purchase doubled the size of the United States. At the time, Americans knew little about the territory. But that would soon change.

Reading **History**

B. Making Inferences Why did Jefferson purchase Louisiana even though the Constitution said nothing about the president's right to buy land?

Lewis and Clark Explore

Since 1802, Thomas Jefferson had planned an expedition to explore the Louisiana country. Now that the Louisiana Purchase had been made, learning about the territory became even more important.

Jefferson chose a young officer, Captain **Meriwether Lewis,** to lead the expedition. In Jefferson's map-lined study, the two men eagerly planned the trip. Lewis turned to his old friend, Lieutenant **William Clark,** to select and oversee a volunteer force, which they called the Corps of Discovery. Clark was a skilled mapmaker and outdoorsman and proved to be a natural leader. The Corps of Discovery soon became known as the **Lewis and Clark expedition.**

Vocabulary
corps (kor): a number of people acting together for a similar purpose

Clark was accompanied by York, his African-American slave. York's hunting skills won him many admirers among the Native Americans met by the explorers. The first black man that many Indians had ever seen, York became something of a celebrity among them.

Lewis and Clark set out in the summer of 1803. By winter, they reached St. Louis. Located on the western bank of the Mississippi River, St. Louis would soon become the gateway to the West. But in 1803, the city was a sleepy town with just 180 houses. Lewis and Clark spent the winter at St. Louis and waited for the ceremony that would mark the transfer of Louisiana to the United States. In March 1804, the American flag flew over St. Louis for the first time.

The Rocky Mountain summit of Pikes Peak is 14,110 feet high.

GEOGRAPHY SKILLBUILDER Interpreting Maps

1. **Location** *What two rivers met at the starting point of the Lewis and Clark expedition?*
2. **Movement** *How were Lewis and Clark's return routes different from each other?*

AMERICA'S HISTORY MAKERS

MERIWETHER LEWIS

1774–1809

Meriwether Lewis was well qualified for the first overland expedition to the Pacific Northwest. In Virginia, he had become an expert hunter. From 1801 to 1803, he worked for President Jefferson, who had him trained in geography, mineralogy, and astronomy.

The journals Lewis kept tell what the West was like in the early 1800s and are still exciting to read. In one entry, dated September 17, 1804, Lewis describes the "immense herds of Buffaloe, deer Elk and Antelopes which we saw in every direction feeding on the hills and plains."

WILLIAM CLARK

1770–1838

William Clark was an army friend of Meriwether Lewis. Lewis personally chose him to be co-captain of the Corps of Discovery.

Clark's experience in his state militia and the U.S. Army had taught him how to build forts, draw maps, and lead expeditions through enemy territory.

He had less formal training than Lewis, but with his six feet of height and muscular build, he was a more rugged explorer.

Clark's leadership skills smoothed disputes. Also, his artistic skills made the expedition's maps and drawings both accurate and beautiful.

What were the different skills of Lewis and Clark that qualified them as co-leaders of the expedition?

Up the Missouri River

Reading **History**

C. Making Inferences Why did Lewis and Clark travel on the Missouri River?

The explorers, who numbered about 40, set out from St. Louis in May of 1804. They headed up the Missouri River in one shallow-bottomed riverboat and two pirogues—canoes made from hollowed-out tree trunks. They had instructions from President Jefferson to explore the river and hoped to find a water route across the continent. Lewis and Clark were also told to establish good relations with Native Americans and describe the landscape, plants, and animals they saw.

The explorers inched up the Missouri. The first afternoon, they traveled only about three miles. Sometimes the men had to pull, rather than row or sail, their boats against the current. In late October, they reached the Mandan Indian villages in what is now North Dakota.

The explorers built a small fort and spent the winter with the friendly Mandan. There, they also met British and French-Canadian trappers and traders. They were not happy to see the Americans. They suspected that the Americans would soon compete with them for the rich trade in beaver furs—and they were right.

In the spring of 1805, the expedition set out again. A French trapper, his 17-year-old-wife, **Sacagawea** (SAK•uh•juh•WEE•uh), and their baby went with them. Sacagawea was a Shoshone woman whose language skills and knowledge of geography would be of great value to Lewis and Clark—especially when they reached the area where she was born.

On to the Pacific Ocean

On their way west, the expedition had to stop at the Great Falls of the Missouri. Lewis called this ten-mile-long series of waterfalls "the grandest sight I ever beheld." He described his approach to the falls.

A VOICE FROM THE PAST

I had proceeded on this course about two miles . . . whin my ears were saluted with the agreeable sound of a fall of water and advancing a little further I saw the spray arrise above the plain like a collumn of smoke. . . . (It) soon began to make a roaring too tremendious to be mistaken for any cause short of the great falls of the Missouri.

Meriwether Lewis, quoted in *Undaunted Courage*

Lewis and Clark kept beautiful journals that provided priceless information about the West.

To get around the Great Falls, the explorers had to carry their boats and heavy supplies for 18 miles. They built wheels from cottonwood trees to move the boats. Even with wheels, the trek took nearly two weeks. Rattlesnakes, bears, and even a hailstorm slowed their steps.

As they approached the Rocky Mountains, Sacagawea excitedly pointed out Shoshone lands. Eager to make contact with the tribe, Lewis and a small party made their way overland. Lewis soon found the Shoshone, whose chief recognized Sacagawea as his sister. The chief traded horses to Lewis and Clark, and the Shoshone helped them cross the Rocky Mountains.

The explorers then journeyed to the mighty Columbia River, which leads to the Pacific Ocean. In November 1805, Clark wrote in his journal, "Ocian in view! O! The joy." They soon arrived at the Pacific Coast. There, they spent a rain-soaked winter before returning to St. Louis the following year.

The Lewis and Clark expedition brought back a wealth of scientific and geographic information. Though they learned that an all-water route across the continent did not exist, Americans received an exciting report of what lay to the west.

Reading **History**

D. Finding Main Ideas Why was the Lewis and Clark expedition valuable?

Pike's Expedition

Lewis and Clark explored the northern part of the Louisiana Purchase. In 1806, an expedition led by **Zebulon Pike** left St. Louis on a southerly route. (Refer to the map on page 320.) Pike's mission was to find the sources of the Arkansas and Red rivers. The Red River formed a boundary between Spanish territory and Louisiana.

Pike's party of two dozen men headed westward across the Great Plains. When they reached the Arkansas River, they followed it toward the Rocky Mountains. From 150 miles away, Pike spied the Rocky Mountain peak that would later bear his name—Pikes Peak. However, he failed in his attempt to climb it. Then they turned south, hoping that they would eventually run into the Red River. Instead, they ran into the

Background

The previous year, Pike had led a 5,000-mile expedition to search for the source of the Mississippi River.

EFFECTS: Exploration of the West, 1804–1807

1. **Accurate maps**
   Lewis and Clark and Pike produced the first good maps of the Louisiana Territory. Later travelers would use these maps to make their way west.
2. **Growth of fur trade**
   Exploration boosted interest in the fur trade. Hunters and trappers would add to the knowledge of the West.
3. **Mistaken view of Great Plains**
   Pike inaccurately described the treeless Great Plains as a desert. This led many Americans to believe that the Plains were useless for farming.

SKILLBUILDER Interpreting Charts
Why might Pike's description of the Great Plains have led to the idea that Native Americans east of the Mississippi should be moved there?

Compass used by Lewis and Clark

Vocabulary
Rio Grande: Spanish for *big river*

Rio Grande, which was in Spanish territory. There, they were arrested by Spanish troops.

The explorers returned to the United States after being released by Spanish officials in 1807. Though Pike and his men never explored the Red River, they did bring back valuable descriptions of the Great Plains and the Rio Grande River Valley.

The Effects of Exploration

The first American explorers of the West brought back tales of adventure as well as scientific and geographical information. As the chart above shows, this information would have long-lasting effects.

Early in Jefferson's presidency, events at home occupied much of the new president's time. In the next section, you will learn about foreign affairs during the same time period.

Section 2 Assessment

1. Terms & Names

Identify:
- Louisiana Purchase
- Meriwether Lewis
- William Clark
- Lewis and Clark expedition
- Sacagawea
- Zebulon Pike

2. Taking Notes

Use a chart like the one below to record the factors that might have led Napoleon to sell the Louisiana Territory.

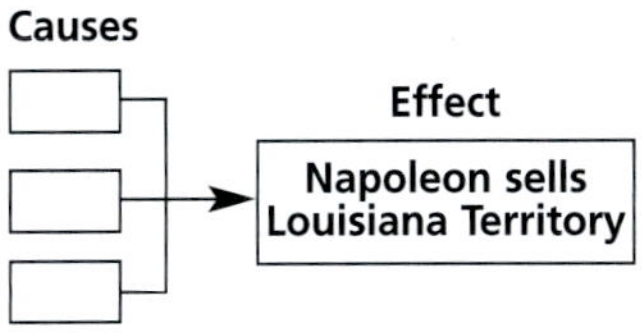

3. Main Ideas

a. What groups might dispute European land claims in the West?

b. Why was New Orleans important to Americans?

c. How did Sacagewea help Lewis and Clark?

4. Critical Thinking

Recognizing Effects What were some of the effects of the explorations of the West in the 1800s?

THINK ABOUT
- how other people might use the information brought back by the explorers
- the economic effects of the expedition

ACTIVITY OPTIONS

WORLD HISTORY
GEOGRAPHY

Read more about New Orleans. Make an illustrated **time line** of the French, Spanish, and U.S. ownership of the city or create a **map** of its port.

Native Americans on the Explorers' Route

When Thomas Jefferson bought the Louisiana Territory, Native Americans had already been living in that area for thousands of years. Before Lewis and Clark began their trip, Jefferson instructed them to deal with Native Americans in a peaceful manner and to make it clear that the United States wished to be "friendly and useful to them." On their journey, Lewis and Clark met almost 50 different tribes.

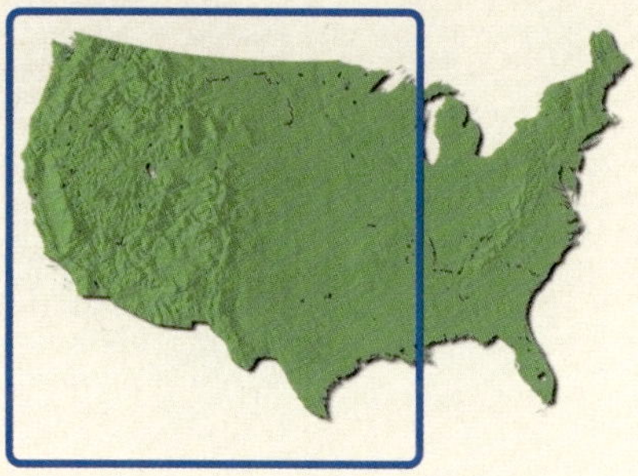

Sacagawea
In 1805, the explorers arrived in Shoshone territory near the Rocky Mountains. A Shoshone chief, Cameahwait, confirmed that there was no all-water route to the Pacific. Later, when Cameahwait recognized Sacagawea as his sister, he agreed to sell the explorers the horses they needed to cross the mountains.

1 Oto
In 1804, Lewis and Clark met the Oto, a buffalo-hunting people. This was the first formal meeting of U.S. representatives with western Native Americans. Lewis told the Oto that they were "children" of a new great "father"—President Thomas Jefferson.

ARTIFACT FILE

Buffalo Robe Pictured to the right is a section of a Mandan buffalo robe. On it, a Mandan painted a battle scene between the Mandan and the Sioux.

2 Mandan

Neighboring tribes as well as exploring Europeans relied on the mainstay of Mandan culture—corn. The Mandan also crafted beautiful leatherwork and pottery. Lewis and Clark spent an entire winter with the Mandan.

3 Teton Sioux

Upon meeting the Teton Sioux, Lewis and Clark showed off an air gun. Known for their aggressiveness, the Teton already viewed the Americans as competitors for trade in this region. As a result, confrontation marked Lewis and Clark's visit.

On-Line Field Trip

The Peabody Museum in Cambridge, Massachusetts, holds an important collection of Native American artifacts. This rain hat was worn by Chinook whalers of the Northwest. The Chinook made these water-repellent hats out of cedar bark and bear grass.

Visit www.mcdougallittell.com for more information.

CONNECT TO GEOGRAPHY

1. **Place** What fort was built where the Columbia River empties into the Pacific Ocean?
2. **Location** In what mountain range did the Shoshone tribe live?

See Geography Handbook, page 4.

CONNECT TO HISTORY

3. **Forming Opinions** What do you think the Native Americans that Lewis and Clark met thought about the explorers?

3 Problems with Foreign Powers

TERMS & NAMES
impressment
Embargo Act of 1807
Tecumseh
War Hawk

MAIN IDEA
Jefferson tried to avoid involvement in the problems of other nations.

WHY IT MATTERS NOW
British interference with the affairs of the United States led to the War of 1812.

ONE AMERICAN'S STORY

In 1804, U.S. Navy Lieutenant Stephen Decatur was on a daring mission overseas. The United States was at war with Tripoli, a state on the North African coast. The war, which began in 1801, was the result of repeated attacks on American merchant ships by African pirates. Decatur's mission was to destroy the U.S. warship *Philadelphia*—which had been captured by Tripoli—so that it could not be used by the enemy.

Decatur bravely sailed into Tripoli's harbor and set fire to the *Philadelphia.* He then managed to escape under enemy fire with only one man wounded. Decatur later issued this rallying cry for all Americans.

A VOICE FROM THE PAST

Our country! In her [relationships] with foreign nations may she always be in the right; but our country, right or wrong.

Stephen Decatur, 1816

Stephen Decatur struggles in hand-to-hand combat with African pirates.

Decatur's attack was one of the most celebrated events of the war, which ended in 1805. The conflict showed how hard it was for the United States to stay out of foreign affairs while its citizens participated so heavily in overseas trade. In this section, you will learn how President Jefferson handled problems with other nations.

Jefferson's Foreign Policy

When Thomas Jefferson took office in 1801, he expected to concentrate on domestic concerns. In his inaugural address, he happily noted that America was "kindly separated by nature and a wide ocean from the exterminating havoc [wars] of one quarter of the globe." Jefferson advised the United States to seek the friendship of all nations, but to enter into "entangling alliances with none."

However, the president's desire to keep the United States separated from other nations and their problems was doomed to fail. For one thing, American merchants were busily engaged in trade all over the world. For another, the Louisiana Purchase and the Lewis and Clark expedition were about to open the country to westward expansion. Expansion would bring Americans into closer contact with people from other nations who had already established settlements in the West.

Reading **History**
A. Analyzing Causes Why was it hard for the United States to avoid other nations' problems?

Finally, the United States had little control over the actions of foreign nations—as North African interference with U.S. shipping had shown. Staying out of the ongoing conflict between France and England would be just as difficult.

Problems with France and England

For a long time, the United States managed not to get involved in the European wars that followed the French Revolution. At times, the nation even benefited from the conflict. Busy with affairs in Europe, France sold the Louisiana Territory to the United States. And American shippers eagerly took over the trade interrupted by the war.

By 1805, however, the British began to clamp down on U.S. shipping. They did not want Americans to provide their enemies with food and supplies. After the United States threatened to take action, the British decided to set up a partial blockade. This would only allow some American ships to bring provisions to Europe.

This partial blockade angered France, which enacted its own laws to control foreign shipping. These changes put American merchants in a difficult position. If they obeyed the British rules, their ships could be seized by the French. If they obeyed the French rules, their ships could be seized by the British.

Britain also interfered with U.S. trade by the **impressment,** or kidnapping, of American sailors to work on British ships. Between 1803 and 1812, the British impressed about 6,000 American sailors.

One of the most famous incidents occurred in 1807. The British ship *Leopard* attacked an American naval ship, the *Chesapeake,* off the coast of Virginia. Three Americans lost their lives in the battle. The attack aroused widespread anger. Had Congress been in session, America might have declared war. But Jefferson, who had been re-elected in 1804, decided against it. One critic, furious at the president's caution, called Jefferson a "dish of skim milk curdling at the head of our nation."

British officers seize an American sailor at gunpoint.

Trade as a Weapon

Vocabulary
coercion: the practice of forcing someone to act in a certain way by use of pressure or threats

Instead of declaring war, Jefferson asked Congress to pass legislation that would stop all foreign trade. "Peaceable coercion," as the president described his policy, would prevent further bloodshed.

In December, Congress passed the **Embargo Act of 1807**. Now American ships were no longer allowed to sail to foreign ports. The act also closed American ports to British ships.

Jefferson's policy was a disaster. It was more harmful to the United States than to the British and French. American farmers and merchants were especially hard hit. Southern and Western farmers, for example, lost important markets for their grain, cotton, and tobacco. Shippers lost income, and many chose to violate the embargo by making false claims about where they were going. One New Englander said the embargo was like "cutting one's throat to cure the nosebleed."

Reading **History**

B. Recognizing Effects What were the results of the Embargo Act?

The embargo became a major issue in the election of 1808. Jefferson's old friend James Madison won the election. By the time he took office, Congress had already repealed the embargo.

Madison's solution to the problem was a law that allowed merchants to trade with any country except France and Britain. Trade with these countries would start again when they agreed to respect U.S. ships. But this law proved no more effective than the embargo.

Tecumseh and Native American Unity

British interference with American shipping and impressment of U.S. citizens made Americans furious. They also were angered by Britain's actions in the Northwest. Many settlers believed that the British were stirring up Native American resistance to frontier settlements.

The Shawnee chief Tecumseh led Native American resistance to white rule in the Ohio River Valley.

Since the Battle of Fallen Timbers in 1794 (see page 299), Native Americans continued to lose their land. Thousands of white settlers had swarmed into Ohio and then into Indiana.

Tecumseh, a Shawnee chief, vowed to stop the loss of Native American land. He believed that the reason Native Americans continued to lose their land was because they were separated into many different tribes. He concluded that Native Americans had to do what white Americans had done: unite. Events in 1809 proved him right.

That September, William Henry Harrison, governor of the Indiana Territory, signed the Treaty of Fort Wayne with chiefs of the Miami, Delaware, and Potawatomi tribes. They agreed to sell over three million acres of land. But Tecumseh declared the treaty meaningless.

Reading **History**

C. Forming Opinions Why did Tecumseh declare the Treaty of Fort Wayne meaningless?

A VOICE FROM THE PAST

[Whites] have taken upon themselves to say this [land] belongs to the Miamis, this to the Delawares and so on. But the Great Spirit intended [Native American land] to be the common property of all the tribes, [and it cannot] be sold without the consent of all.

Tecumseh, quoted in *Tecumseh and the Quest for Indian Leadership*

After the Treaty of Fort Wayne, many Native Americans began to answer Tecumseh's call for unity. But his efforts ultimately failed. In November 1811, while Tecumseh was away recruiting tribes for his alliance, the Shawnee were defeated by Harrison's forces at the Battle of

Tippecanoe. It was a severe setback for Tecumseh's movement.

War Hawks

After the battle of Tippecanoe, Tecumseh and his warriors found a warm welcome with the British in Canada. At that point, the Native Americans and the British became allies. Tecumseh's welcome in Canada raised even higher the anti-British feelings in the West.

Leaders such as Congressman Henry Clay of Kentucky angrily demanded war against Britain. Westerners who called for war were known as **War Hawks**. They wanted British aid to Native Americans stopped, and they wanted the British out of Canada. Conquering Canada would open up a vast new empire for Americans.

> **Vocabulary**
> **hawk:** a person who favors the use of military force to carry out foreign policy

Causes of the War of 1812

SKILLBUILDER Interpreting Charts
Which cause of the War of 1812 was not related to activities on the sea?

Other Americans sought war because of the British violations of American rights at sea. Future president Andrew Jackson said hostilities were necessary "for the protection of our maritime citizens impressed on board British ships of war," and to "open a market for the productions of our soil."

Urged on by Jackson and the War Hawks, Congress declared war on Britain on June 18, 1812. In the next section, you will read about the second—and final—war between the United States and Great Britain.

Section 3 Assessment

1. Terms & Names

Identify:
- impressment
- Embargo Act of 1807
- Tecumseh
- War Hawk

2. Taking Notes

Use a chart like the one below to record the effects of Jefferson's Embargo Act.

Causes: Embargo Act

Effects: [ ] [ ]

Why didn't the act work?

3. Main Ideas

a. How did the British and French interfere with American shipping?

b. How did Jefferson respond to the interference?

c. Why did the War Hawks favor war?

4. Critical Thinking

Analyzing Points of View
Why did Tecumseh think it was important for Native Americans to unite?

THINK ABOUT
- what he learned about white men
- what Native Americans would lose if they did not act together

ACTIVITY OPTIONS

ART / **SPEECH**

Do research on the Battle of Tippecanoe. Draw a **comic strip story** of the battle or hold a **press conference** to describe the battle's outcome.

The War of 1812

TERMS & NAMES
Oliver Hazard Perry
Battle of the Thames
Francis Scott Key
Treaty of Ghent

MAIN IDEA	WHY IT MATTERS NOW
Angered by Britain's interference in the nation's affairs, the United States went to war.	The War of 1812 showed that the United States was willing and able to protect its national interests.

ONE AMERICAN'S STORY

The war between the United States and Britain had begun in 1812. Two years later, British troops were marching toward Washington, D.C. Dolley Madison, the president's wife, stayed behind until the last minute. With bombs bursting in the distance, she hurried to save important historical objects from the White House.

A VOICE FROM THE PAST

I have had [a wagon] filled with . . . the most valuable portable articles belonging to the house. . . . I insist on waiting until the large picture of General Washington is secured . . . It is done! and the precious portrait placed in the hands of two gentlemen of New York, for safe keeping.

Dolley Madison, from a letter sent to her sister

When the British troops arrived in the city, they set fire to many public buildings, including the White House and the Capitol. The next day, a violent storm caused even more damage. Fortunately, the heavy rains that accompanied the storm helped put out the fires. You will learn about other events of the War of 1812 in this section.

Before British troops set fire to the president's mansion, Dolley Madison saved priceless historical objects.

The War Begins

Britain did not really want a war with the United States because it was already involved in another war with France. To try to avoid war, the British announced that they would no longer interfere with American shipping. But the slow mails of the day prevented this news from reaching the United States until weeks after June 18th, when Congress approved Madison's request for a declaration of war.

The War of 1812 had two main phases. From 1812 to 1814, Britain concentrated on its war against France. It devoted little energy to the conflict in North America, although it did send ships to blockade the American coast. The second phase of the war began after the British defeated France in April 1814. With their European war nearly at an end, the British could turn their complete attention to the United States.

The United States military was weak when the war was declared. Democratic-Republicans had reduced the size of the armed forces. When the war began, the Navy had only about 16 ships. The army had fewer than 7,000 men. These men were poorly trained and equipped, and were often led by inexperienced officers. A young Virginia army officer complained that the older officers were victims of "sloth, ignorance, or habits of [excessive] drinking."

The First Phase of the War

In spite of its small size, the United States Navy rose to the challenge. Its warships were the fastest afloat. American naval officers had gained valuable experience fighting pirates in the Mediterranean Sea. Early in the war, before the British blockaded the coast, ships such as the *Constitution* and the *United States* won stirring victories. These victories on the high seas boosted American confidence.

The most important U.S. naval victory took place on Lake Erie. In the winter of 1812-1813, the Americans had begun to build a fleet on the shores of Lake Erie. **Oliver Hazard Perry,** an experienced officer, took charge of this infant fleet. In September 1813, the small British force on the lake set out to attack the American ships. Commodore Perry, who had predicted that this would be "the most important day of my life," sailed out to meet the enemy. Perry's ship, the *Lawrence,* flew a banner declaring, "Don't give up the ship."

BATTLES OF THE WAR

1. **Atlantic Ocean.** British navy blockades American coast, 1813
2. **Lake Erie.** Perry's fleet defeats a British fleet, 1813
3. **Thames River.** Harrison defeats British in Canada, killing Tecumseh, 1813
4. **Washington, D.C.** British burn the capital but later fail to capture nearby Baltimore, 1814
5. **Lake Champlain.** American ships defeat British, who retreat to Canada, 1814
6. **New Orleans.** Jackson's army defeats British in Battle of New Orleans, 1815

GEOGRAPHY SKILLBUILDER

Interpreting Maps

1. **Location** *Where was Fort McHenry located?*
2. **Movement** *Which battle required American troops to march into Canada?*

For two hours, the British and Americans exchanged cannon shots. Perry's ship was demolished and the guns put out of action. He grabbed his ship's banner and leaped into a rowboat. Under British fire, he and four companions rowed to another ship. In command of the second ship, Perry destroyed two of the enemy's ships and soon forced the British to surrender. After the battle, Perry sent a message to General Harrison: "We have met the enemy and they are ours."

"Don't give up the ship."
Banner on Perry's ship, the *Lawrence*

When General Harrison received Perry's note, he set out to attack the British. But when Harrison transported his army across Lake Erie to Detroit, he discovered that the British had retreated into Canada. Harrison pursued the British forces and defeated them at the **Battle of the Thames** in October. This victory put an end to the British threat to the Northwest—and also claimed the life of Tecumseh, who died in the battle fighting for the British.

Reading **History**
A. Drawing Conclusions What was the overall result of the Battle of the Thames?

The Second Phase of the War

After defeating Napoleon in April 1814, Britain turned its full attention to the United States. As you read in One American's Story, British forces burned the Capitol building and the president's mansion in August. The British then attacked Fort McHenry at Baltimore.

The commander of Fort McHenry had earlier requested a flag "so large that the British will have no difficulty in seeing it." Detained on a British ship, a Washington lawyer named **Francis Scott Key** watched the all-night battle. At dawn, Key discovered that the flag was still flying. He expressed his pride in what became the U.S. national anthem.

A VOICE FROM THE PAST

Oh say can you see by the dawn's early light
What so proudly we hail'd at the twilight's last gleaming,
Whose broad stripes and bright stars through the perilous fight
O'er the ramparts we watch'd were so gallantly streaming?
And the rockets' red glare, the bombs bursting in air,
Gave proof through the night that our flag was still there.
Oh, say does that star-spangled banner yet wave
O'er the land of the free and the home of the brave?

Francis Scott Key

America's HERITAGE

THE STAR-SPANGLED BANNER

The "Star Spangled Banner," inspired by the flag that flew over Fort McHenry (see below), continues to move Americans. On hearing this national anthem, patriotic listeners stand, take off their hats, and put their hands over their hearts. These actions pay respect to the American flag and the song that celebrates it.

Francis Scott Key's song enjoyed widespread popularity for more than 100 years before an act of Congress made it the national anthem in 1931.

Meanwhile, in the north, the British sent a force from Canada across Lake Champlain. Its goal was to push south and cut off New England. The plan failed when the American fleet defeated the British in the Battle of Lake Champlain in September 1814.

In the south, the British moved against the strategic port of New Orleans. In December 1814, dozens of ships carrying 7,500 British troops approached Louisiana. To fight them, the Americans patched together an army under the command of General Andrew Jackson.

Reading **History**
B. Reading a Map Locate the battles of the second phase of the war on the map on page 331. Note how far apart the sites were.

*Reading*History

C. Making Inferences Why did Jackson fight the British at New Orleans after a peace treaty was signed?

The British attacked Jackson's forces on January 8, 1815. Protected by earthworks, American riflemen mowed down the advancing redcoats. It was a great victory for Jackson. American casualties totaled 71, compared to Britain's 2,000. Though the Battle of New Orleans made Jackson a hero, it was unnecessary. Slow mails from Europe had delayed news of the **Treaty of Ghent,** which ended the War of 1812. It had been signed two weeks earlier, on December 24, 1814.

The Legacy of the War

The treaty showed that the war had no clear winner. No territory changed hands, and trade disputes were left unresolved. Still, the war had important consequences. First, the heroic exploits of men such as Andrew Jackson and Oliver Perry increased American patriotism. Second, the war broke the strength of Native Americans, who had sided with the British. Finally, when war interrupted trade, the Americans were forced to make many of the goods they had previously imported. This encouraged the growth of U.S. manufactures.

The United States had also proved that it could defend itself against the mightiest military power of the era. For perhaps the first time, Americans believed that the young nation would survive and prosper. You will learn about the country's growing prosperity in Chapter 11.

The Effects of the War

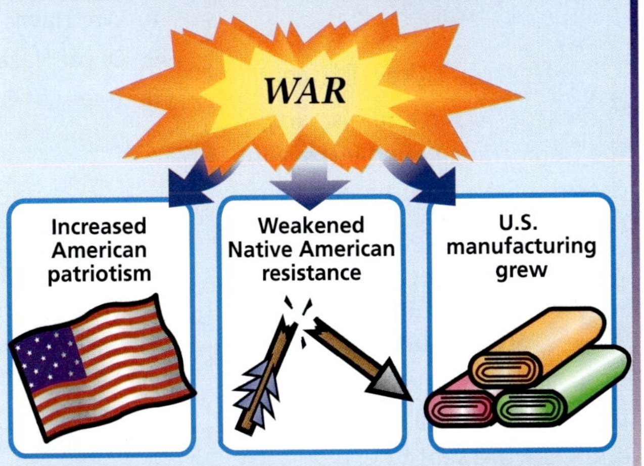

SKILLBUILDER Interpreting Charts
Which effect do you think resulted from the war's interruption of U.S. trade?

Section 4 Assessment

1. Terms & Names

Identify:

- Oliver Hazard Perry
- Battle of the Thames
- Francis Scott Key
- Treaty of Ghent

2. Taking Notes

Use a chart like the one shown to record military events of the War of 1812.

First Phase of War	Second Phase of War
1.	2.

Why was the war divided into two phases?

3. Main Ideas

a. What was the state of the U.S. military when the war began?

b. What were the results of General Harrison's victory at the Battle of the Thames?

c. Where did the British focus their attacks during the second phase of the war?

4. Critical Thinking

Recognizing Effects What was the legacy of the War of 1812?

THINK ABOUT

- Americans' feelings toward their country
- U.S. relations with Native Americans
- possible economic effects

ACTIVITY OPTIONS

LANGUAGE ARTS

ART

Research the U.S.S. *Constitution.* Write a **poem** to commemorate one of its victories or design a **model** to show its parts.

Chapter 10 ASSESSMENT

VISUAL SUMMARY

The Jefferson Era

Jefferson Takes Office

Thomas Jefferson and his party, the Democratic-Republicans, win control of the government from the Federalists.

The Louisiana Purchase and Exploration

After Jefferson purchases Louisiana from France, Lewis and Clark are sent to explore the new American territory.

Problems With Foreign Powers

Other countries' interference makes it difficult for Jefferson to stay out of foreign affairs.

The War of 1812

When Britain continues to interfere in American affairs, the two nations battle in the War of 1812.

TERMS & NAMES

Briefly explain the importance of each of the following:

1. *Marbury* v. *Madison*
2. judicial review
3. Louisiana Purchase
4. Lewis and Clark expedition
5. impressment
6. Embargo Act of 1807
7. Tecumseh
8. War Hawk
9. Oliver Hazard Perry
10. Treaty of Ghent

REVIEW QUESTIONS

Jefferson Takes Office (pages 313–317)

1. What were the main parties in the election of 1800, and how did their views differ?
2. How did Jefferson envision the future of America?

The Louisiana Purchase and Exploration (pages 318–325)

3. What was the extent of U.S. territory after the Louisiana Purchase?
4. What difficulties did Lewis and Clark face on their expedition?
5. What troubles did Zebulon Pike have on his 1806-1807 trip?

Problems With Foreign Powers (pages 326–329)

6. Why did Jefferson have difficulty staying out of foreign affairs?
7. How did Tecumseh intend to prevent the loss of Native American land?
8. What were some of the causes of the War of 1812?

The War of 1812 (pages 330–333)

9. Which battle ended the British threat to the U.S. Northwest?
10. What event preceded the second phase of the war?

CRITICAL THINKING

1. USING YOUR NOTES

Using your completed chart, answer the questions below.

The Jefferson Era
Summaries
Main Idea: Thomas Jefferson is elected president. **Details:** Jefferson replaces Federalist policies with his own but has problems with the judiciary.
Main Idea: **Details:**
Main Idea: **Details:**
Main Idea: **Details:**

a. What were the threats to U.S. trade and sailors?

b. How did Jefferson respond to foreign interference with U.S. trade?

2. ANALYZING LEADERSHIP

How do you think Thomas Jefferson's behavior as president might have affected the way future presidents viewed the office?

3. THEME: EXPANSION

How did the expansion of the United States affect its foreign policy?

4. RECOGNIZING PROPAGANDA

Before elections, supporters of different candidates sometimes make outrageous claims. How was the election of 1800 an example of this?

5. APPLYING CITIZENSHIP SKILLS

In what ways did Jefferson's behavior as president reflect his idea of good citizenship?

Interact *with* History

How did the dangers you predicted before you read the chapter compare to those experienced by people on expeditions west?

HISTORY SKILLS

1. INTERPRETING GRAPHS

Study the graph. Answer the questions.

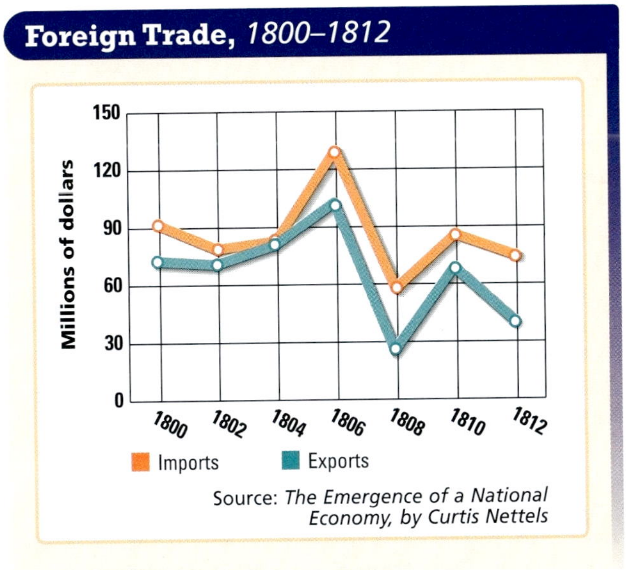

Source: *The Emergence of a National Economy*, by Curtis Nettels

Basic Graph Elements

a. What do the numbers on the left side of the chart indicate?

Interpreting the Graph

b. In what years did the value of U.S. trade decrease dramatically?

c. What was the value of exports in 1806? In 1808?

2. INTERPRETING PRIMARY SOURCES

Examine the cartoon below, which comments on the United States trade policies you read about in Section 3.

a. What does *Ograbme* spell when written backwards?

b. What other clues tell you that this cartoon is about the embargo against Great Britain?

ALTERNATIVE ASSESSMENT

1. INTERDISCIPLINARY ACTIVITY: World History

Making a Time Line The history of the Jefferson Era is closely related to events in Europe that were occurring at the same time. Do research to learn about the Napoleonic Wars. Then make a time line with events from U.S. history on one side and events from the Napoleonic Wars on the other.

2. COOPERATIVE LEARNING ACTIVITY

Designing a Plan for Economic Action Form a group that will develop a plan to end British interference with U.S. shipping. Select a leader for your group. Then pick one member of your group to represent farmers, another to represent shippers, and another to speak for those who are demanding war. Discuss the interests of all these groups and then write down your policy and present it to the class. Be sure to consider:

- the superior power of the British Navy
- where the farmers will sell their products
- how the shippers will earn their money

3. TECHNOLOGY ACTIVITY

Creating a Multimedia Presentation Use the Internet, books, and other reference materials to create a multimedia presentation on one of the major battles of the War of 1812. Consider including the following:

- paintings or written descriptions of the battle
- pictures of the weapons that were used
- music from the time period
- recorded sound effects
- graphs showing battle statistics

INTERNET **Visit www.mcdougallittell.com to learn more about the War of 1812.**

4. HISTORY PORTFOLIO

PORTFOLIO **Option 1** Review your section and chapter assessment activities. Select one that you think is your best work. Then use comments made by your teacher or classmates to improve your work and add it to your portfolio.

PORTFOLIO **Option 2** Review the questions that you wrote for What Do You Want to Know? on page 312. Then write a short report in which you explain the answers to your questions. If any questions were not answered, do research to answer them. Add your answers to your portfolio.

Making Explorers' Field Notes

On their expedition in the early 1800s, Lewis and Clark filled their journals with field notes—detailed observations and scientific illustrations of the land, plants, and wildlife they saw. Lewis made drawings of plants and animals. Clark drew detailed maps. For many years, their journals were the main source of information about the West.

ACTIVITY Create a journal of field notes that includes illustrations of plants, animals, and terrain found in your neighborhood. Then write a comparison article between your field notes and those of Lewis and Clark.

TOOLBOX

Each group will need:

- drawing paper
- poster board for covers
- pencil and pen
- scissors
- ruler
- string
- hole punch and hole reinforcers

The scientific and artistic skills of Lewis and Clark made their journals both accurate and beautiful.

STEP BY STEP

1 **Form groups.** Each group should consist of 3 to 4 students. The members of your group will do the following tasks:

- design and create a handmade journal
- take a walk in your neighborhood and record observations as field notes
- compare the field notes you have created with those of Lewis and Clark

2 **Make your journal.** Your group will need a journal to make field notes on the nature walk. Each page of the journal should be six inches wide and roughly eight inches long, approximately the size of the one used by Lewis and Clark. Cut 10–15 sheets of that size out of the drawing paper. Create a front and back cover for the journal using the poster board.

3 **Put your journal together.** Punch three holes in the left side of the pages, including the covers. Place hole reinforcers around the holes to ensure that the pages won't tear. Bind the pages and the covers together with string.

4 **Explore your neighborhood.** Take a walk in your neighborhood, in the area around your school, or in a nearby woods. You might want to divide the tasks of observing, drawing, describing, and mapping among the members of the group.

5 **Model your field notes on those of Lewis and Clark.** In their journals, Lewis and Clark included drawings of animals and plants, as well as detailed observations about them. Remember to draw the plants, insects, and animals as if you have never seen them before.

6 **Make a map of your route.** In addition to the drawings, create a map of your walk. Include any interesting landmarks as well as a detailed description of the terrain. Remember to sketch the route as if it's unexplored territory.

WRITE AND SPEAK

Write and present a paper that compares and contrasts Lewis and Clark's journal with the one your group has completed. Also, explain how your journal might help someone who has just moved to your school or neighborhood.

HELP DESK

For related information, see pages 320–325 in Chapter 10.

Researching Your Project

- You can find copies of Lewis and Clark's journals in many libraries.
- *Undaunted Courage* by Stephen Ambrose gives a fascinating account of the expedition.

Visit www.mcdougallittell.com to learn more about Lewis and Clark.

Did You Know?

- The Corps of Discovery sent Jefferson six live specimens of animals, including a prairie dog.
- At one point, the men on the expedition consumed nine pounds of buffalo meat a day.
- Swarms of mosquitoes plagued the expedition. The explorers often found it impossible to eat without inhaling some of the pesky insects.
- Clark estimated that the expedition traveled 4,162 miles. His guess was only 40 miles off the actual distance.

REFLECT & ASSESS

- **What process did your group use to observe, draw, and describe your route?**
- **How did you decide what information to include in your journal?**
- **How important are the illustrations to understanding the area in which your group took its walk?**

CHAPTER 11

National and Regional Growth

1800–1844

The Erie Canal, which opened in 1825, increased western trade and migration.

Interact *with* History

This Currier and Ives print, *Progress of the Century,* shows some inventions of the 1800s.

From 1790 to 1840, you have seen an explosion of new inventions. These include the cotton gin, the steamboat, the steel plow, and the telegraph. You have also seen neighbors leave their farms to run machines in new factories. You sense that the country is changing.

What Do You Think?

- What would it mean to be able to grow more grain and cotton?
- What would it mean to communicate and travel more quickly?
- How might it feel to do factory work instead of farm work?

How will new inventions change your country?

USA

1800

1807
Robert Fulton launches a steamboat on the Hudson River.

1808
Congress bans the African slave trade.

1812
War of 1812 disrupts U.S. shipping.

1813
Weaving factory built in Waltham, Massachusetts.

1820
Missouri Compromise balances number of slave and free states.

1823
Monroe Doctrine issued.

1825
Erie Canal completed.

1831
Nat Turner leads slave rebellion in Virginia.

1844
Telegraph line connects Washington, D.C., and Baltimore.

World

1804
Haiti wins independence from France.

1815
Napoleon defeated at Waterloo.

1821
Peru and Mexico gain independence from Spain.

1825
First public railroad operates in England.

1833
Slavery is abolished in British Empire.

1839
Louis Daguerre is recognized for his photographic process.

1844

Chapter 11 SETTING THE STAGE

BEFORE YOU READ

Previewing the Theme

Science and Technology

In this chapter, you will learn how new inventions helped regions of the country grow in the 1800s and also helped pull regions together as a nation. For example, the power loom made the Northeast a cloth manufacturing center. The cotton gin turned the South into a Cotton Kingdom. The telegraph let people communicate instantly across regions.

What Do You Know?

What connects you to someone who lives in the same region? When have you felt a bond with someone from a different region?

THINK ABOUT

- the activities of people in different regions
- the things that unite people as a nation

What Do You Want to Know?

What would you ask people from different regions—a factory worker, wheat farmer, plantation owner, or field slave—about their lives in the 1800s? Write these questions in your notebook. Read to see if they are answered in Chapter 11.

READ AND TAKE NOTES

Reading Strategy: Analyzing Causes and Recognizing Effects To help you understand the development of regional growth, sectional tensions, and national unity in Chapter 11, pay attention to causes and effects. Growth, tensions, and unity each had more than one cause. As you read, identify different causes and note them on the chart below. Often a topic sentence at the beginning of a paragraph will state a cause and effect. Be alert for such clue words as "led to," "as a result," or "changed."

See Skillbuilder Handbook, page R10.

Causes		Effects
	→	REGIONAL GROWTH
	→	SECTIONAL TENSIONS
	→	NATIONAL UNITY

1 Early Industry and Inventions

TERMS & NAMES
- Samuel Slater
- Industrial Revolution
- factory system
- Lowell mills
- interchangeable parts
- Robert Fulton
- Samuel F. B. Morse

MAIN IDEA

New machines and factories changed the way people lived and worked in the late 1700s and early 1800s.

WHY IT MATTERS NOW

The industrial development that began more than 200 years ago continues today.

ONE AMERICAN'S STORY

In 1789, the Englishman <u>Samuel Slater</u> sailed to the United States under a false name. It was illegal for textile workers like him to leave the country. Britain wanted no other nation to copy its new machines for making thread and cloth. But Slater was going to bring the secret to America. When he got to New York, he wrote a letter to Rhode Island investor Moses Brown.

A VOICE FROM THE PAST

A few days ago I was informed that you wanted a manager of *cotton spinning* . . . in which business I flatter myself that I can give the greatest satisfaction, in making machinery, making good yarn, either for *stockings* or *twist,* as any that is made in England.

Samuel Slater, quoted in *Samuel Slater: Father of American Manufactures*

Samuel Slater's mill was located in Pawtucket, Rhode Island.

With Brown's backing, Slater built the first successful water-powered textile mill in America. You will learn in Section 1 how the development of industries changed the ways Americans lived and worked.

The Industrial Revolution Begins

The War of 1812 brought great economic changes to the United States. It sowed the seeds for an Industrial Revolution like the one begun in Britain during the late 18th century. During the **<u>Industrial Revolution,</u>** factory machines replaced hand tools, and large-scale manufacturing replaced farming as the main form of work. For example, before the Industrial Revolution, women spun thread and wove cloth at home using spinning wheels and hand looms. The invention of such machines as the spinning jenny and the power loom made it possible for unskilled workers to produce cloth. These workers, who were often children, could produce more cloth, more quickly.

The **<u>factory system</u>** brought many workers and machines together under one roof. Most factories were built near a source of water to power the machines. People left their farms and crowded into cities where the factories were. They worked for wages, on a set schedule. Their way of life changed, and not always for the better.

Many Americans, such as Thomas Jefferson, did not want the United States to industrialize. But the War of 1812 led the country in that direction. Because the British naval blockade kept imported goods from reaching U.S. shores, Americans had to start manufacturing their own goods. The blockade also stopped investors from spending money on shipping and trade. Instead, they invested in new American industries. Taking advantage of the country's free enterprise system, American businessmen built their own factories, starting in New England. These businessmen and their region grew wealthier.

Vocabulary
industrialize: to develop factories

ReadingHistory
A. Recognizing Effects How did the War of 1812 cause economic changes in America?

Factories Come to New England

New England was a good place to set up factories for several reasons. Factories needed water power, and New England had many fast-moving rivers. For transportation, it also had ships and access to the ocean. In addition, New England had a willing labor force. The area's first factory workers were families who were tired of scraping a living from their stony fields.

Samuel Slater built his first spinning mill in Pawtucket, Rhode Island, in 1790. He hired eight children between the ages of 7 and 12, paying them a low wage. Later, he built a larger mill and employed whole families. As Slater influenced others to start mills, his family system of employment spread through Rhode Island, Connecticut, and southern Massachusetts.

The Lowell Mills Hire Women

In 1813, the American textile industry leaped forward when Francis Cabot Lowell built a factory in Waltham, in eastern Massachusetts. This factory not only spun raw cotton into yarn, but wove it into cloth on power looms. Lowell had seen power looms in English mills and had figured out how to build them. Like Samuel Slater, he had brought secrets to America.

The Waltham factory was so successful that Lowell and his partners built a new factory town, Lowell, near the Merrimack and Concord rivers. The **Lowell mills,** textile mills in the village, employed farm girls who lived in company-owned boardinghouses. "Lowell girls" worked $12\frac{1}{2}$-hour days in deafening noise.

Background
Founded in 1826, the town was named for Lowell, who died in 1817.

Lowell girls published a literary magazine.

A VOICE FROM THE PAST

At first the hours seemed very long . . . and when I went out at night the sound of the mill was in my ears . . . . You know that people learn to sleep with the thunder of Niagara [falls] in their ears, and a cotton mill is no worse, though you wonder that we do not have to hold our breath in such a noise.

"Letters from Susan," quoted in the *Lowell Offering*

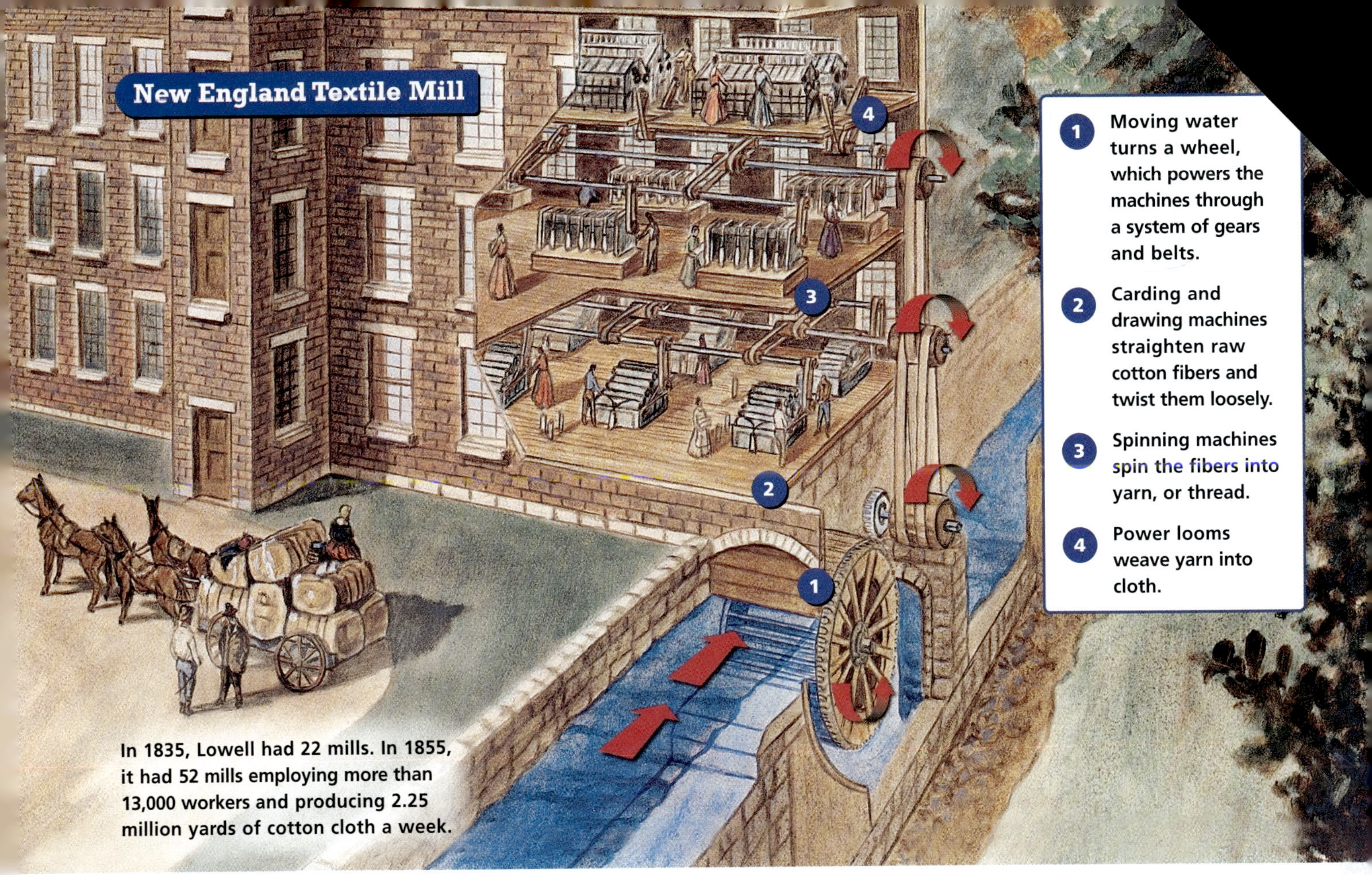

Young women came to Lowell in spite of the noise. In the early years, wages were high—between two and four dollars a week. Older women supervised the girls, making them follow strict rules and attend church. Girls read books, went to lectures, and even published a literary magazine—the *Lowell Offering*. Usually they worked for only a few years, until they married. By the 1830s, however, falling profits meant that wages dropped and working conditions worsened for the Lowell girls.

Reading **History**
B. Contrasting
How did the Lowell mills differ from Slater's mill?

The Lowell mills and other early factories ran on water power. Factories built after the 1830s were run by more powerful steam engines. Because steam engines used coal and wood, not fast-moving water, factories could be built away from rivers and beyond New England.

A New Way to Manufacture

New manufacturing methods changed the style of work in other industries besides the textile industry. In 1797, the U.S. government hired the inventor Eli Whitney to make 10,000 muskets for the army. He was to have the guns ready in two years. Before this time, guns were made one at a time by gunsmiths, from start to finish. Each gun differed slightly. If a part broke, a new part had to be created to match the broken one.

Whitney sought a better way to make guns. In 1801, he went to Washington with a box containing piles of musket parts. He took a part from each pile and assembled a musket in seconds. He had just demonstrated the use of **interchangeable parts,** parts that are exactly alike.

Robert Fulton invented the *Clermont,* a steamboat.

Machines that produced exactly matching parts soon became standard in industries. Interchangeable parts speeded up production, made repairs easy, and allowed the use of lower-paid, less-skilled workers. But the new system also required a new style of management, with inspectors to make sure each piece was uniform. Workers who were used to more independence disliked such close supervision.

Reading **History**

C. Recognizing Effects What were the effects of using interchangeable parts?

Moving People, Goods, and Messages

New inventions increased factory production. They also improved transportation and communication. Steamboats carried people and goods farther and faster and led to the growth of cities like New Orleans and St. Louis. **Robert Fulton** invented a steamboat that could move against the current or a strong wind. He launched the *Clermont* on the Hudson River in 1807. Its steam engine turned two side paddle wheels, which pulled the boat through the water.

The *Clermont* was dubbed "Fulton's Folly" and described as "looking precisely like a backwoods saw-mill mounted on a scow [boat] and set on fire." But it made the 300-mile trip from New York to Albany and back in a record 62 hours. Even Fulton had not expected to travel so quickly.

A VOICE FROM THE PAST

I overtook many sloops and schooners, beating to the windward, and parted with them as if they had been at anchor. The power of propelling boats by steam is now fully proved.

Robert Fulton, quoted in *Robert Fulton and the "Clermont"*

In 1811, the first steamship traveled down the Mississippi and Ohio rivers. But its engine was not powerful enough to return upriver, against the current. Henry Miller Shreve, a trader on the Mississippi, designed a more powerful engine. He installed it on a double-decker boat with a paddle wheel in the back. In 1816, he sailed this boat up the Mississippi and launched a new era of trade and transportation on the river.

Reading **History**
D. Recognizing Effects What made the steamboat and telegraph such important inventions?

In 1837, **Samuel F. B. Morse** first demonstrated his telegraph. This machine sent long and short pulses of electricity along a wire. These pulses could be translated into letters of a message. With the telegraph, it took only seconds to communicate with someone in another city. In 1844, the first long-distance telegraph line carried news from Baltimore to Washington, D.C., about who had been nominated for president. Telegraph lines spanned the country by 1861, bringing people closer as a nation. Both the telegraph and the steamboat brought more national unity.

Technology Improves Farming

Other new inventions increased farm production. In 1836, the blacksmith John Deere invented a lightweight plow with a steel cutting edge. Older cast-iron plows were designed for the light, sandy soil of New England. But rich, heavy Midwestern soil clung to the bottom of these plows and slowed farmers down. Deere's new plow made preparing ground much less work. As a result, more farmers began to move to the Midwest.

Vocabulary
patented: protected by a patent, which gives an inventor the sole right to make, use, or sell an invention

The mechanical reaper and the threshing machine were other inventions that improved agriculture. Cyrus McCormick's reaper, patented in 1834, cut ripe grain. The threshing machine separated kernels of wheat from husks.

John Deere invented the steel plow.

New technologies linked regions and contributed to national unity. With new farm equipment, Midwestern farmers grew food to feed Northeastern factory workers. In turn, Midwestern farmers became a market for Northeastern manufactured goods. The growth of Northeastern textile mills increased demand for Southern cotton. This led to the expansion of slavery in the South, as you will learn in Section 2.

Section 1 Assessment

1. Terms & Names

Identify:
- Samuel Slater
- Industrial Revolution
- factory system
- Lowell mills
- interchangeable parts
- Robert Fulton
- Samuel F. B. Morse

2. Taking Notes

On a chart like the one below, note new inventions, their dates, and their effects on the United States.

Invention	Date	Effects

Which inventions did most to link the nation? Explain.

3. Main Ideas

a. Why was New England a good place to build early factories?

b. What were working conditions like in Lowell mills?

c. How were different U.S. regions linked economically?

4. Critical Thinking

Evaluating How would you judge Samuel Slater and Francis Lowell, who brought secrets to the United States illegally?

THINK ABOUT
- what they gained
- how they affected the United States and England
- what you believe about keeping technology secret

ACTIVITY OPTIONS

SCIENCE
SPEECH

Explain how an invention from this chapter works, either in an **oral report** or a **labeled diagram.**

Interdisciplinary CHALLENGE

Run a Mill Town

You are the owner of a new water-powered textile mill that will soon open in New England. Mills have been around for more than 20 years, and you have studied their operations closely. Even so, you face many problems as you start your business. Machinery failures, labor problems, demanding investors—all will be part of your life from now on.

COOPERATIVE LEARNING On this page are two challenges you face as the owner of a textile mill. Working with a small group, decide how to deal with each challenge. Choose an option, assign a task to each group member, and do the activity. You will find useful information in the Data File. Be prepared to present your solutions to the class.

HEALTH CHALLENGE

"Anna Tripp lost three fingers today."

A neighboring mill owner has just left after sharing some bad news. Today 12-year-old Anna Tripp lost three fingers in one of his machines. Last week, one of his workers was hit and nearly killed by the flying end of a broken belt. Several girls went home because they had trouble breathing. They blamed the closed, damp machine rooms with lint-filled air. How will you reduce the number of costly health problems like these in your mill? Present your plan using one of these options:

- **Write a speech to workers outlining the company's safety measures.**
- **Design a sign for each floor of the mill stating the company's safety rules.**

MATH CHALLENGE

"So many applicants for employment"

To make a profit, you must operate at top capacity for the lowest cost. Your mill generates 2 mill power. This dictates how many spinners you can hire. You need about two-and-a-half times that many weavers. What will be your weekly payroll for spinners and weavers? How many men will you hire? Women? Look at the Data File for help. Present your hiring plan for spinners and weavers using one of these options:

- **Write a report telling investors whom you plan to hire.**
- **Design want-ad posters aimed at the workers you are looking for.**

ACTIVITY WRAP-UP

Present to the Class **Meet as a group to review your responses to running a mill town. Pick the most creative solution for each challenge and present these solutions to the class.**

DATA FILE

Water Power

Potential energy: energy released when water falls from a height.

Kinetic energy: energy provided by fast-moving water.

1 mill power: power produced by 25 cubic feet of water per second dropping over a 30-foot fall; about 60 horsepower.

1 mill power: runs 3,584 spindles.

Waterwheels

Overshot

Undershot

Mid-wheel

Wage Rates

Men: $.85–$2.09 per day, depending on skill

Women: $.52–$.78 per day, depending on skill

Positions

Pickers: clean raw cotton.

Carders: feed cotton into machine that makes a thick strand of fibers.

Spinners: operate a machine that twists thick fibers into yarn and winds it on bobbins fastened to moving spindles. One worker operates 128 spindles.

Dressers: treat finished yarn with a starch paste.

Drawing-in hands: attach dressed yarn to the mechanical loom for weavers.

Weavers: weave dressed yarn into finished cloth. One worker operates two looms.

Work Hours

12 hours per day, 6 days per week, 309 days per year with holidays on Fast Day (spring), the Fourth of July, and Thanksgiving

Visit www.mcdougallittell.com to learn more about textile mills.

2 Plantations and Slavery Spread

TERMS & NAMES
Eli Whitney
cotton gin
spirituals
Nat Turner

MAIN IDEA
The invention of the cotton gin and the demand for cotton caused slavery to spread in the South.

WHY IT MATTERS NOW
The spread of slavery created lasting racial and sectional tensions.

ONE AMERICAN'S STORY

Catherine Beale was born into slavery in 1838. At the age of 91, in 1929, she recalled her childhood on a Virginia plantation. When asked what games she had played, Catherine replied that enslaved children never played games—they were too busy with chores. Among the tasks were picking and cleaning cotton.

A VOICE FROM THE PAST

We had to work in the field in the day and at night we had to pick out the seed before we went to bed. And we had to clean the wool, we had to pick the burrs and sticks out so it would be clean and could be carded and spun and wove.

Catherine Beale, quoted in *Slave Testimony*

Enslaved workers labor in the cotton fields.

Catherine had to clean cotton by hand because the plantation didn't have a cotton gin. This machine made it easier for enslaved workers to clean cotton. But it also made cotton growing and slave owning more profitable. In this section, you will learn how slavery expanded in the South and how it affected the lives of people living under it.

The Cotton Boom

Eli Whitney invented a machine for cleaning cotton in 1793, after visiting the Georgia plantation of Catherine Greene, the widow of a Revolutionary War general. Mrs. Greene was struggling to make her plantation profitable. English textile mills had created a huge demand for cotton, but the short-fibered cotton that grew in most parts of the South was hard to clean by hand. A worker could clean just one pound of this cotton in a day.

Whitney's **cotton gin** (short for "engine") made the cotton-cleaning process far more efficient. With the new machine, one worker could now clean as much as 50 pounds of cotton a day. The cotton gin helped set the South on a different course of development from the North. It made

The Cotton Gin

short-fibered cotton a commercial product and changed Southern life in four important ways.

Reading History

A. Reading a Map Use the map on page 350 to find cotton-growing areas in 1840.

1. It triggered a vast move westward. Cotton farming moved beyond the Atlantic coastal states, where long-fibered, easy-to-clean cotton grew. Cotton plantations began to spread into northern Florida, Alabama, and Mississippi. Then they crossed into Louisiana and Arkansas. After 1840, they reached Texas.
2. Because cotton was so valuable, planters put most of their efforts into growing it rather than other crops. To Southerners, cotton was king.
3. More Native American groups were driven off Southern land as it was taken over for cotton plantations.
4. Growing cotton required a large work force, and slavery continued to be important as a source of labor. Many slaves from the east were sold south and west to new cotton plantations.

Reading History

B. Recognizing Effects What impact did the cotton gin have on the South?

Slavery Expands

From 1790 to 1860, cotton production rose greatly. So did the number of enslaved people in the South. Using slave labor, the South raised millions of bales of cotton each year for the textile mills of England and the American Northeast. (See the graph on page 350.) In 1820, the South earned $22 million from cotton exports. By the late 1830s, earnings from cotton exports were nearly ten times greater, close to $200 million.

As cotton earnings rose, so did the price of slaves. A male field hand sold for $300 in the 1790s. By the late 1830s, the price had jumped to

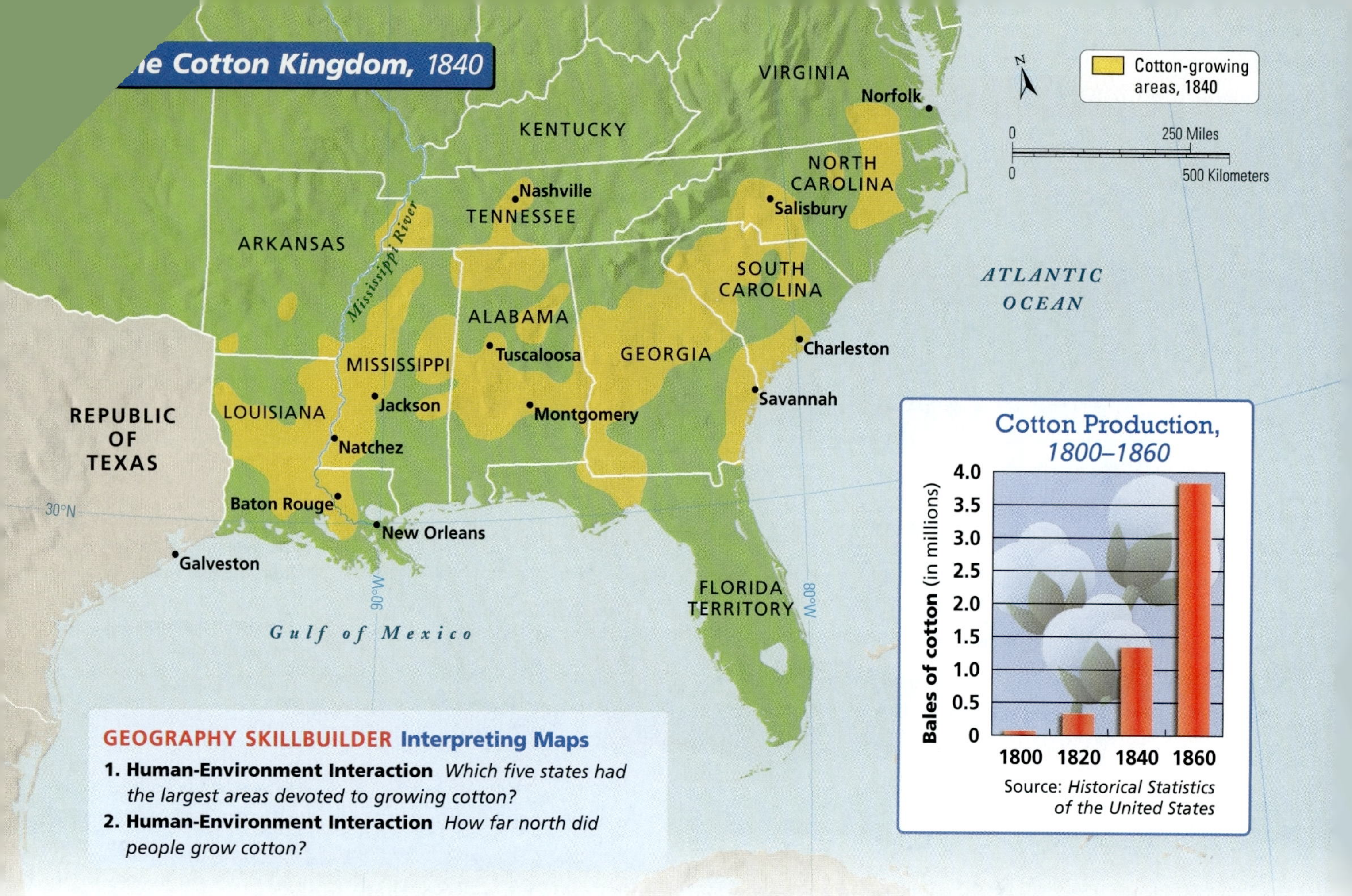

GEOGRAPHY SKILLBUILDER Interpreting Maps

1. **Human-Environment Interaction** *Which five states had the largest areas devoted to growing cotton?*
2. **Human-Environment Interaction** *How far north did people grow cotton?*

$1,000. After 1808, when it became illegal to import Africans for use as slaves, the trading of slaves already in the country increased.

The expansion of slavery had a major impact on the South's economy. But its effect on the people living there was even greater.

Slavery Divides the South

Slavery divided white Southerners into those who held slaves and those who did not. Slaveholders with large plantations were the wealthiest and most powerful people in the South, but they were relatively few in number. Only about one-third of white families owned slaves in 1840. Of these slave-owning families, only about one-tenth had large plantations with 20 or more slaves.

Most white Southern farmers owned few or no slaves. Still, many supported slavery anyway. They worked their small farms themselves and hoped to buy slaves someday, which would allow them to raise more cotton and earn more money. For both small farmers and large planters, slavery had become necessary for increasing profits.

Reading **History**
C. Analyzing Points of View Why did many white farmers without slaves still support slavery?

African Americans in the South

Slavery also divided black Southerners into those who were enslaved and those who were free. Enslaved African Americans formed about one-third of the South's population in 1840. About half of them worked on large plantations with white overseers. A former slave described the routine.

A VOICE FROM THE PAST

The overseer was 'straddle his big horse at three o'clock in the mornin', roustin' the hands off to the field. . . . The rows was a mile long and no matter how much grass [weeds] was in them, if you [left] one sprig on your row they [beat] you nearly to death.

Wes Brady, quoted in *Remembering Slavery*

Reading **History**

D. Contrasting How was plantation slavery different from slavery in cities?

Not all slaves faced the back-breaking conditions of plantations. In cities, enslaved persons worked as domestic servants, skilled craftsmen, factory hands, and day laborers. Sometimes they were hired out and allowed to keep part of their earnings. Frederick Douglass, an African-American speaker and publisher, once commented, "A city slave is almost a freeman, compared with a slave on the plantation." But they were still enslaved.

In 1840, about 8 percent of African Americans in the South were free. They had either been born free, been freed by an owner, or bought their own freedom. Many free African Americans in the South lived in cities such as Baltimore and Washington, D.C.

Though not enslaved, free blacks faced many problems. Some states made them leave once they gained their freedom. Most states did not permit them to vote or receive an education. Many employers refused to hire them. But their biggest threat was the possibility of being captured and sold into slavery.

daily *life*

SPIRITUALS

Singing spirituals offered comfort for pain, bound people together at religious meetings, and eased the boredom of daily tasks. This verse came from a spiritual sung by slaves in Missouri.

Dear Lord, dear lord,
when slavery'll cease
Then we poor souls
will have our peace;—
There's a better day a coming,
Will you go along with me?
There's a better day a coming,
Go sound the jubilee!

Finding Strength in Religion

An African-American culture had emerged on plantations by the early 1800s. Slaves relied on that culture—with its strong religious convictions, close personal bonds, and abundance of music—to help them endure the brutal conditions of plantation life.

Reading **History**

E. Making Inferences Why would enslaved African Americans be inspired by the biblical story of Moses?

Some slaveholders tried to use religion to make slaves accept their treatment. White ministers stressed such Bible passages as "Servants, obey your masters." But enslaved people took their own messages from the Bible. They were particularly inspired by the story of Moses leading the Hebrews out of bondage in Egypt.

Enslaved people expressed their religious beliefs in **spirituals,** religious folk songs. Spirituals often contained coded messages about a planned escape or an owner's unexpected return. African-American spirituals later influenced blues, jazz, and other forms of American music.

Detail of *Plantation Burial,* (1860), John Antrobus.

Families Under Slavery

Perhaps the cruelest part of slavery was the sale of family members away from one another. Although some slaveholders would not part mothers from children, many did, causing unforgettable grief. When enslaved people ran away, it was often to escape separation or to see family again.

When slave families could manage to be together, they took comfort in their family life. They married, though their marriages were not legally recognized. They tried to raise children, despite interference from owners. Most slave children lived with their mothers, who tried to protect them from punishment. Parents who lived on other plantations often stole away to visit their children, even at the cost of a whipping. Frederick Douglass recalled visits from his mother, who lived 12 miles away.

Reading **History**

F. Recognizing Effects How did slavery harm family life?

A VOICE FROM THE PAST

I do not recollect of [remember] ever seeing my mother by the light of day. She was with me in the night. She would lie down with me, and get me to sleep, but long before I waked she was gone.

Frederick Douglass, *Narrative of the Life of Frederick Douglass*

Douglass's mother resisted slavery by the simple act of visiting her child. Douglass later rebelled by escaping to the North. Other enslaved people rebelled in more violent ways.

A slave auction threatens to split a family apart.

Slave Rebellions

Armed rebellion was an extreme form of resistance to slavery. Gabriel Prosser planned an attack on Richmond, Virginia, in 1800. In 1822, Denmark Vesey planned a revolt in Charleston, South Carolina. Both plots were betrayed, and the leaders were hanged.

The most famous rebellion was led by **Nat Turner** in Virginia in 1831. On August 21, Turner and 70 followers killed 55 white men, women, and children. Later, witnesses claimed that he spoke these words.

> ***A VOICE FROM THE PAST***
>
> We do not go forth for the sake of blood and carnage; . . . Remember that ours is not a war for robbery, . . . it is a struggle for freedom.
>
> **Nat Turner,** quoted in *Nat Turner,* by Terry Bisson

Most of Turner's men were captured when their ammunition ran out, and 16 were killed. When Turner was caught, he was tried and hanged.

Turner's rebellion spread fear in the South. Whites killed more than 200 African Americans in revenge. State legislatures passed harsh laws that kept free blacks and slaves from having weapons or buying liquor. Slaves could not hold religious services unless whites were present. Postmasters stopped delivering antislavery publications.

Reading **History**

G. Recognizing Effects How did Nat Turner's rebellion affect white Southerners?

After Turner's rebellion, the grip of slavery grew even tighter in the South. Tension over slavery increased between the South and the North, as you will see in the next section.

AMERICA'S HISTORY MAKERS

NAT TURNER

1800–1831

Nat Turner was born on a plantation in Virginia. As a child, Turner learned to read and write. He became an enthusiastic reader of the Bible. Slaves gathered in forest clearings to listen to his powerful sermons. Turner believed that God wanted him to free the slaves, even if by armed rebellion. He defended the justice of his cause in what came to be known as *Confessions of Nat Turner,* which he dictated to a white lawyer before his execution.

How did Turner justify his rebellion?

Section 2 Assessment

1. Terms & Names

Identify:

- Eli Whitney
- cotton gin
- spirituals
- Nat Turner

2. Taking Notes

In a chart like the one below, note facts about each group of Southerners.

Group	Facts
slaveholding whites	
nonslaveholding whites	
enslaved blacks	
free blacks	

Why do you think many free blacks lived in cities?

3. Main Ideas

a. How did the cotton gin lead to the spread of slavery?

b. How was life different for plantation slaves, city slaves, and free blacks in the South?

c. What were three ways that enslaved people resisted slavery?

4. Critical Thinking

Drawing Conclusions Why do you think Southern whites reacted as they did to Nat Turner's rebellion?

THINK ABOUT

- Turner's trial and hanging
- the killings that followed the rebellion
- the new laws that were passed

ACTIVITY OPTIONS

LANGUAGE ARTS

SPEECH

Write a **book report** on a slave narrative, or perform an **oral interpretation** of a passage from one.

3 Nationalism and Sectionalism

TERMS & NAMES
nationalism
Henry Clay
American System
Erie Canal
James Monroe
sectionalism
Missouri Compromise
Monroe Doctrine

MAIN IDEA

Patriotic pride united the states, but tension between the North and South emerged.

WHY IT MATTERS NOW

The tension led to the Civil War, and regional differences can still be found in the United States today.

ONE AMERICAN'S STORY

In the early 1800s, as you have read, the North began to industrialize and the South relied more heavily on growing cotton. At the same time, a rising sense of nationalism pulled people from different regions together. Nationalism is a feeling of pride, loyalty, and protectiveness toward your country. The War of 1812 sent a wave of nationalist feeling through the United States.

Representative Henry Clay, from Kentucky, was a strong nationalist. After the war, President James Madison supported Clay's plan to strengthen the country and unify its different regions.

A VOICE FROM THE PAST

Every nation should anxiously endeavor to establish its absolute independence, and consequently be able to feed and clothe and defend itself. If it rely upon a foreign supply that may be cut off . . . it cannot be independent.

Henry Clay, quoted in *The Annals of America*

Henry Clay

In this section, you will learn how nationalism affected U.S. economic growth and foreign policy. You'll also see how Americans were beginning to be torn between the interests of their own regions and those of the country as a whole.

Nationalism Unites the Country

In 1815, President Madison presented a plan to Congress for making the United States economically self-sufficient. In other words, the country would prosper and grow by itself, without foreign products or foreign markets.

The plan—which Henry Clay promoted as the **American System**—included three main actions.

1. **Establish a protective tariff,** a tax on imported goods that protects a nation's businesses from foreign competition. Congress passed a tariff in 1816. It made European goods more expensive and encouraged Americans to buy cheaper American-made products.

A. Recognizing Effects How would the three parts of the American System help to make the country self-sufficient?

2. **Establish a national bank** that would promote a single currency, making trade easier. (Most regional banks issued their own money.) In 1816, Congress set up the second Bank of the United States.
3. **Improve the country's transportation systems,** which were important for a strong economy. Poor roads made transportation slow and costly.

Roads and Canals Link Cities

Representative John C. Calhoun of South Carolina also called for better transportation systems. "Let us bind the Republic together with a perfect system of roads and canals," he declared in 1817. Earlier, in 1806, Congress had funded a road from Cumberland, Maryland, to Wheeling, Virginia. By 1841, the National Road, designed as the country's main east-west route, had been extended to Vandalia, Illinois.

Water transportation improved, too, with the building of canals. In fact, the period from 1825 to 1850 is often called the Age of Canals. Completed in 1825, the massive **Erie Canal** created a water route between New York City and Buffalo, New York. The canal opened the upper Ohio Valley and the Great Lakes region to settlement and trade. It also fueled nationalism by unifying these two sections of the country.

The Erie Canal allowed farm products from the Great Lakes region to flow east and people and manufactured goods from the East to flow

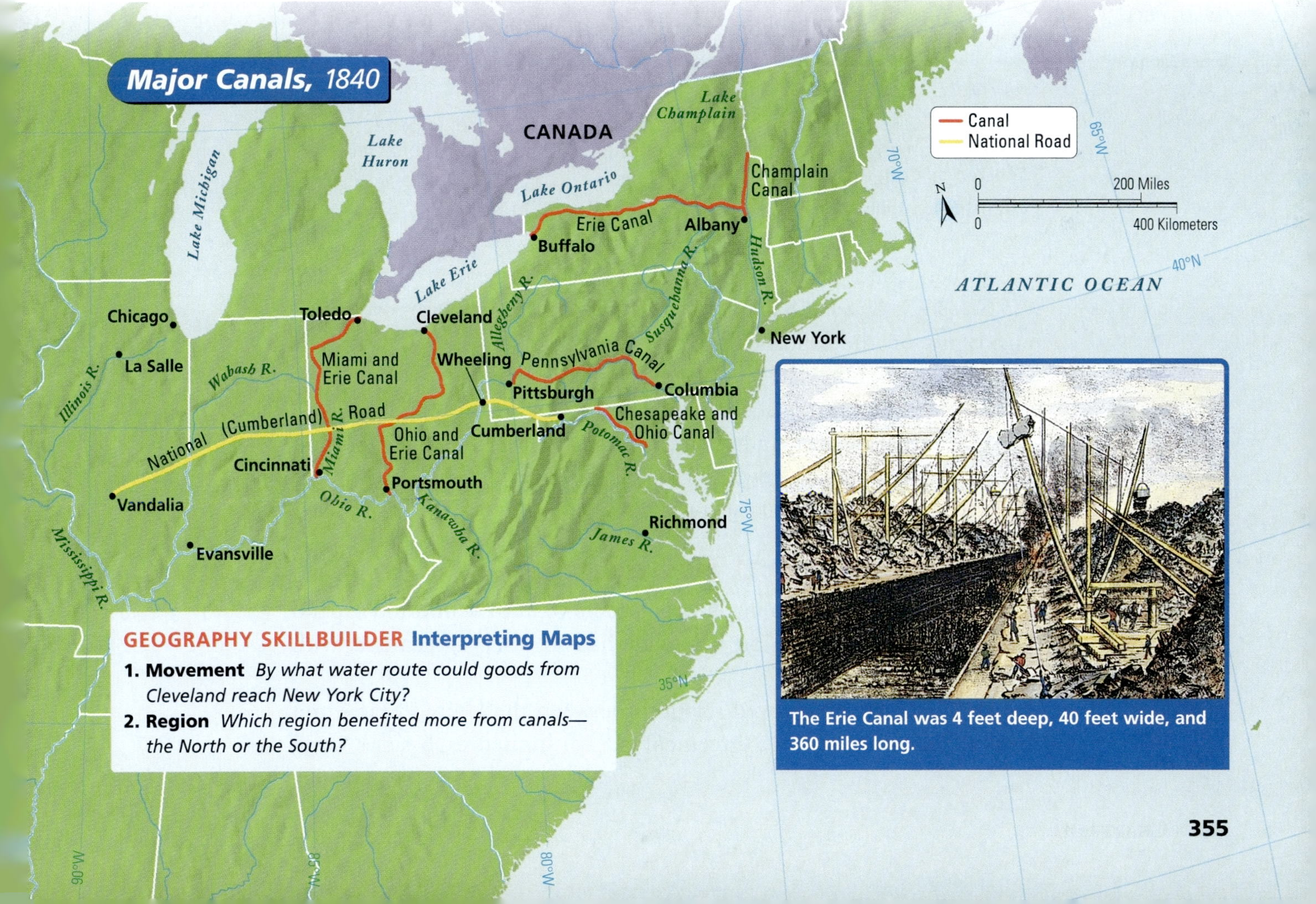

GEOGRAPHY SKILLBUILDER Interpreting Maps

1. **Movement** *By what water route could goods from Cleveland reach New York City?*
2. **Region** *Which region benefited more from canals—the North or the South?*

The Erie Canal was 4 feet deep, 40 feet wide, and 360 miles long.

west. Trade stimulated by the canal helped New York City become the nation's largest city. Between 1820 and 1830, its population swelled from less than 125,000 to more than 200,000.

Around the 1830s, the nation began to use steam-powered trains for transportation. In 1830, only about 30 miles of track existed in the United States. But by 1850, the number had climbed to 9,000 miles. Improvements in rail travel led to a decline in the use of canals.

The Era of Good Feelings

James Monroe

As nationalist feelings spread, people slowly shifted their loyalty away from state governments and more toward the federal government. Democratic-Republican **James Monroe** won the presidency in 1816 with a large majority of electoral votes. The Federalist Party provided little opposition to Monroe and soon disappeared. Political differences gave way to what one Boston newspaper called the Era of Good Feelings.

During the Monroe administration, several landmark Supreme Court decisions promoted national unity by strengthening the federal government. For example, in *McCulloch* v. *Maryland* (1819), the state of Maryland wanted to tax its branch of the national bank. If this tax were allowed, the states could claim to have power over the federal government. The Court upheld federal authority by ruling that a state could not tax a national bank.

Background
Maryland also argued that Congress had no power to create the bank, but the court ruled that it did have such power.

A VOICE FROM THE PAST

The States have no power, by taxation or otherwise, to retard, impede, burden, or in any manner control the operations of the constitutional laws enacted by Congress.

Chief Justice John Marshall, *McCulloch* v. *Maryland* (1819)

Another Court decision that strengthened the federal government was *Gibbons* v. *Ogden* (1824). Two steamship operators fought over shipping rights on the Hudson River in New York and New Jersey. The Court ruled that interstate commerce could be regulated only by the federal government, not the state governments.

The Supreme Court under John Marshall clearly stated important powers of the federal government. A stronger federal government reflected a growing nationalist spirit.

Reading History
B. Finding Main Ideas How did the Supreme Court strengthen the federal government?

Settling National Boundaries

This nationalist spirit also made U.S. leaders want to define and expand the country's borders. To do this, they had to reach agreements with Britain and Spain.

Two agreements improved relations between the United States and Britain. The Rush-Bagot Agreement (1817) limited each side's naval

U.S. Boundary Settlements, 1818 and 1819

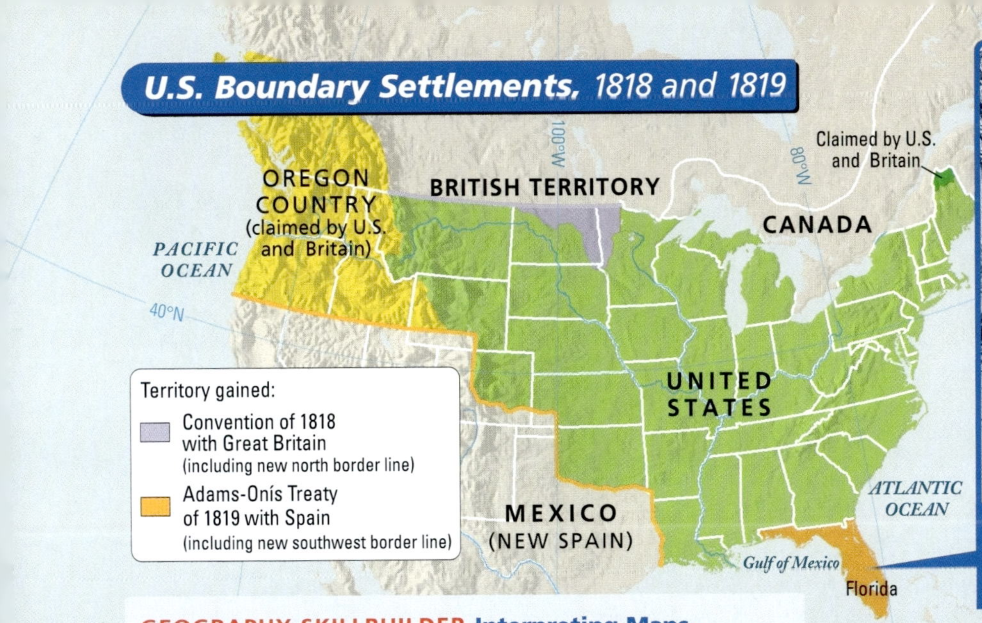

Spain gave up Florida to the United States after Andrew Jackson invaded the territory to capture raiding Seminoles.

GEOGRAPHY SKILLBUILDER Interpreting Maps

1. **Location** *How far west did the Adams-Onís Treaty Line extend?*
2. **Region** *Who claimed the Oregon Country?*

forces on the Great Lakes. In the Convention of 1818, the two countries set the 49th parallel as the U.S.-Canadian border as far west as the Rocky Mountains.

But U.S. relations with Spain were tense. The two nations disagreed on the boundaries of the Louisiana Purchase and the ownership of West Florida. Meanwhile, pirates and runaway slaves used Spanish-held East Florida as a refuge. In addition, the Seminoles of East Florida raided white settlements in Georgia to reclaim lost lands.

Reading **History**
C. Analyzing Causes Why did Andrew Jackson invade East Florida?

In 1817, President Monroe ordered General Andrew Jackson to stop the Seminole raids, but not to confront the Spanish. Jackson followed the Seminoles into Spanish territory and then claimed the Floridas for the United States.

Monroe ordered Jackson to withdraw but gave Spain a choice. It could either police the Floridas or turn them over to the United States. In the Adams-Onís Treaty of 1819, Spain handed Florida to the United States and gave up claims to the Oregon Country. The map above shows boundaries drawn and territories gained in 1818 and 1819.

Sectional Tensions Increase

At the same time nationalism was unifying the country, sectionalism was threatening to drive it apart. **Sectionalism** is loyalty to the interests of your own region or section of the country, rather than to the nation as a whole. Economic changes had created some divisions within the United States. As you have seen, white Southerners were relying more on cotton and slavery. In the Northeast, wealth was based on manufacturing and trade. In the West, settlers wanted cheap land and good transportation. The interests of these sections were often in conflict.

Sectionalism became a major issue when Missouri applied for statehood in 1817. People living in Missouri wanted to allow slavery in their state. At the time, the United States consisted of 11 slave states and 11

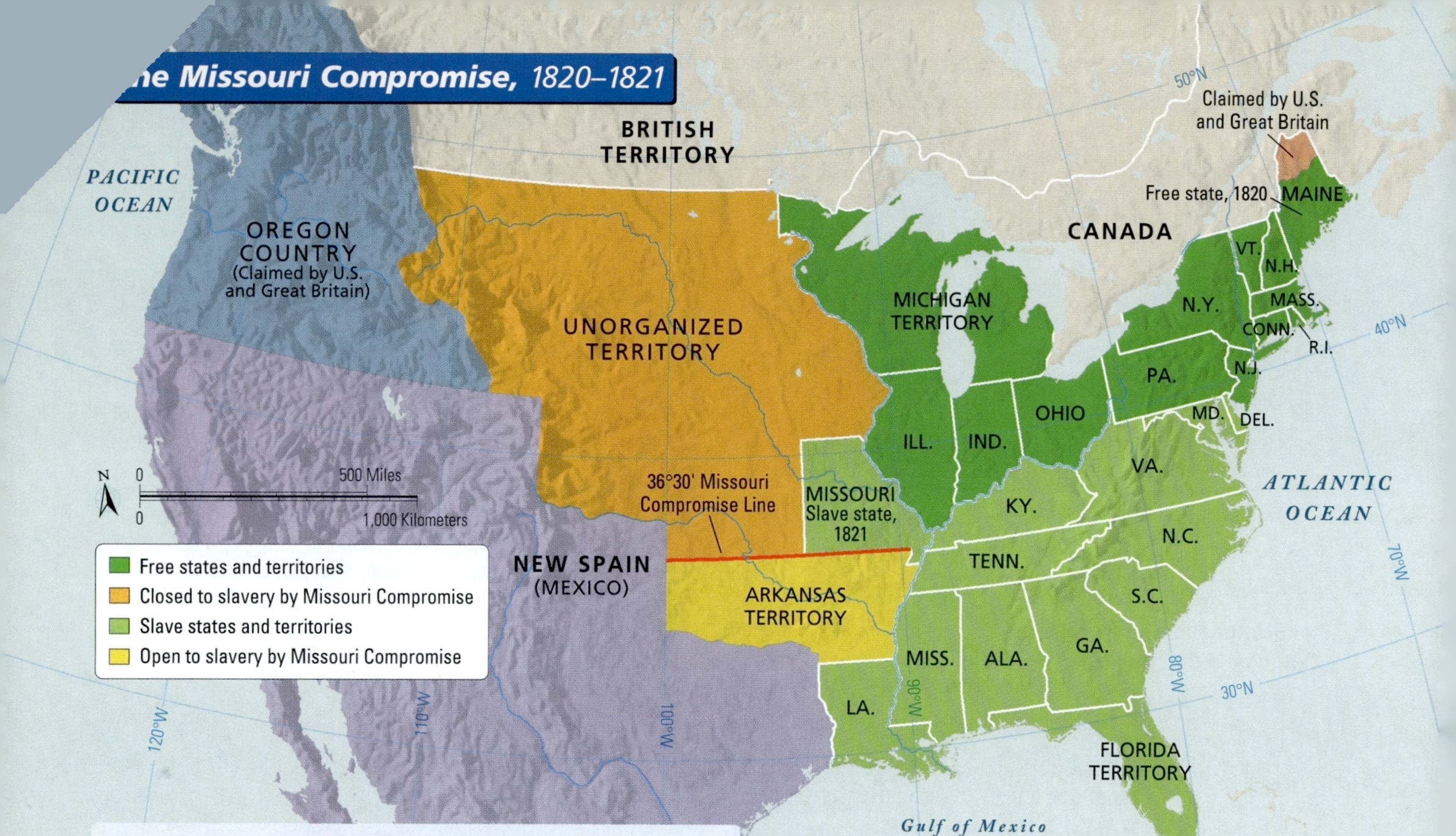

GEOGRAPHY SKILLBUILDER Interpreting Maps

1. **Location** *At what latitude was the Missouri Compromise Line?*
2. **Region** *What territory was opened to slavery by the Missouri Compromise?*

free states. Adding Missouri as a slave state would upset the balance of power in Congress. The question of Missouri soon divided the nation.

The Missouri Compromise

For months, the nation argued over admitting Missouri as a slave state or a free state. Debate raged in Congress over a proposal made by James Tallmadge of New York to ban slavery in Missouri. Angry Southerners claimed that the Constitution did not give Congress the power to ban slavery. They worried that free states could form a majority in Congress and ban slavery altogether. Representative Thomas Cobb of Georgia expressed the Southerners' point of view when he responded to Tallmadge.

ReadingHistory
D. Analyzing Points of View
Why was it so important to Southerners to admit Missouri as a slave state?

A VOICE FROM THE PAST

If you persist, the Union will be dissolved. You have kindled a fire which all the waters of the ocean cannot put out, which seas of blood can only extinguish.

Thomas Cobb, quoted in *Henry Clay: Statesman for the Union*

Meanwhile, Maine, which had been part of Massachusetts, also wanted statehood. Henry Clay, the Speaker of the House, saw a chance for compromise. He suggested that Missouri be admitted as a slave state and Maine as a free state. Congress passed Clay's plan, known as the **Missouri Compromise,** in 1820. It kept the balance of power in the Senate between the slave states and free states. It also called for slavery to be banned from the Louisiana Territory north of the parallel 36° 30', Missouri's southern border.

> **"If you persist, the Union will be dissolved."**
> Thomas Cobb

Thomas Jefferson, nearing 80 years old and living quietly in Virginia, was troubled by the Missouri Compromise. Worried that sectionalism would destroy the country, Jefferson wrote: "In the gloomiest moment of the Revolutionary War I never had any apprehension equal to what I feel from this source."

The Monroe Doctrine

Background
Latin America refers to the nations of the Western Hemisphere south of the United States.

The nation felt threatened not only by sectionalism, but by events elsewhere in the Americas. In Latin America, several countries had successfully fought for their independence from Spain and Portugal. Some European monarchies planned to help Spain and Portugal regain their colonies, hoping to keep the urge to revolt from reaching Europe. U.S. leaders feared that if this happened, their own government would be in danger.

Russian colonies in the Pacific Northwest also concerned Americans. The Russians entered Alaska in 1784. By 1812, their trading posts reached almost to San Francisco.

In December 1823, President Monroe issued a statement that became known as the **Monroe Doctrine.** (See Interactive Primary Source, page 360.) Monroe said that the Americas were closed to further colonization. He also warned that European efforts to reestablish colonies would be considered "dangerous to our peace and safety." Finally, he promised that the United States would stay out of European affairs. The Monroe Doctrine showed that the United States saw itself as a world power and protector of Latin America.

In Chapter 12, you will learn how a new democratic spirit grew—and how Native Americans suffered—during Andrew Jackson's presidency.

Section 3 Assessment

1. Terms & Names

Identify:
- nationalism
- Henry Clay
- American System
- Erie Canal
- James Monroe
- sectionalism
- Missouri Compromise
- Monroe Doctrine

2. Taking Notes

On a diagram like the one below, name things that contributed to national unity in the early 1800s.

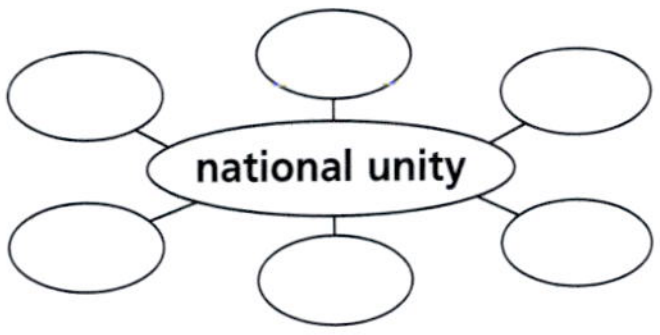

Which of these are still important for national unity?

3. Main Ideas

a. How did the Erie Canal help the nation grow?

b. How did the Missouri Compromise resolve a conflict between the North and South?

c. What was the main message of the Monroe Doctrine, and who was it directed toward?

4. Critical Thinking

Recognizing Effects If the Supreme Court had decided differently in *Gibbons* v. *Ogden* or *McCulloch* v. *Maryland,* what might be one result today?

THINK ABOUT
- if states could interfere with federal laws
- if states controlled interstate commerce

ACTIVITY OPTIONS

LANGUAGE ARTS
ART

In an **editorial** or a **political cartoon,** give your opinion of either the Missouri Compromise or the Monroe Doctrine.

The Monroe Doctrine

Setting the Stage On December 6, 1823, President James Monroe gave a State of the Union address. Part of the speech became known as the Monroe Doctrine. The "allied powers" Monroe refers to are Russia, Prussia, Austria, and France. Earlier in the year, these European monarchies had crushed a revolution in Spain and restored the Spanish king to his throne. They were threatening to help Spain regain its Latin American colonies. See **Primary Source Explorer** CD-ROM

A CLOSER LOOK

NO FUTURE COLONIES

Monroe declares that European countries may not start any new colonies in the Americas.

1. Why might it threaten the United States to have new European colonies near them?

A CLOSER LOOK

NEUTRALITY TOWARD EUROPE

Monroe says that the United States will not take sides in European wars.

2. Why might the United States want to remain neutral toward conflicts in Europe?

[T]he occasion has been judged proper for asserting, as a principle in which the rights and interests of the United States are involved, that the American continents, by the free and independent condition which they have assumed and maintain, are henceforth not to be considered as subjects for future colonization by any European powers. . . .

It was stated at the commencement of the last session that great effort was then making in Spain and Portugal to improve the condition of the people of those countries and that it appeared to be conducted with extraordinary moderation. It need scarcely be remarked that the result has been so far very different from what was then anticipated. . . . The citizens of the United States cherish sentiments the most friendly in favor of the liberty and happiness of their fellowmen on that side of the Atlantic. In the wars of the European powers in matters relating to themselves we have never taken any part, nor does it **comport**[1] with our policy so to do. It is only when our rights are invaded or seriously **menaced**[2] that we resent injuries or make preparation for our defense.

With the movements in this hemisphere we are of necessity more immediately connected, and by causes which must be obvious to all enlightened and impartial observers. The political system of the allied powers is essentially different in this respect from that of America. This difference proceeds from that which exists in their respective governments; and to the defense of our own, which has been achieved by the loss of so much blood and treasure, and matured by the wisdom of their most enlightened citizens, and under which we have enjoyed **unexampled felicity,**[3] this whole nation is devoted. We owe it, therefore, to **candor**[4] and to the **amicable**[5] relations existing between the

1. **comport:** agree with.
2. **menaced:** threatened.
3. **unexampled felicity:** the greatest happiness.
4. **candor:** honesty.
5. **amicable:** friendly.

United States and those powers to declare that we should consider any attempt on their part to extend their system to any portion of this hemisphere as dangerous to our peace and safety.

With the existing colonies or dependencies of any European power we have not interfered and shall not interfere. But with the governments who have declared their independence and maintained it, and whose independence we have, on great consideration and on just principles, acknowledged, we could not view any **interposition**[6] for the purpose of oppressing them, or controlling in any other manner their destiny, by any European power in any other light than as the manifestation of an unfriendly disposition toward the United States. In the war between those new governments and Spain we declared our neutrality at the time of their recognition, and to this we have adhered and shall continue to adhere, provided no change shall occur which, in the judgment of the competent authorities of this government, shall make a corresponding change on the part of the United States indispensable to their security.

The late events in Spain and Portugal show that Europe is still unsettled. Of this important fact no stronger proof can be adduced than that the allied powers should have thought it proper, on any principle satisfactory to themselves, to have interposed by force in the internal concerns of Spain. To what extent such interposition may be carried, on the same principle, is a question in which all independent powers whose governments differ from theirs are interested, even those most remote, and surely none more so than the United States.

—*James Monroe*

6. **interposition:** interference.

A CLOSER LOOK

A DIFFERENT SYSTEM

Monroe states that the United States will defend its republican form of government and would be threatened if Europeans set up monarchies in the Americas.

3. Why would U.S. citizens want their government to be a republic and not an absolute monarchy?

A CLOSER LOOK

NO INTERFERENCE

Monroe warns that if Europeans invade the newly independent republics in Latin America, this would be considered hostile to the United States as well.

4. What would the United States have to fear if these republics were overthrown?

Interactive Primary Source Assessment

1. Main Ideas

a. Why might the United States want no more European colonies in the Americas, particularly in Latin America?

b. How would staying neutral in European wars protect the United States?

c. How might the U.S. system of government be threatened if Europeans regained control of former colonies in the Americas?

2. Critical Thinking

Making Inferences For decades, the United States lacked the military power to enforce the Monroe Doctrine and depended on the British navy to keep other European powers out of Latin America. Why, then, did the United States proclaim the Monroe Doctrine?

THINK ABOUT

- what the doctrine shows about the values and wishes of the United States
- what it shows about how the country saw itself or wanted to be seen

VISUAL SUMMARY

Early Industry and Inventions

New machines allowed the Northeast to industrialize and the Midwest to increase farm production.

Plantations and Slavery Spread

The cotton gin led to the expansion of plantations and slavery in the South.

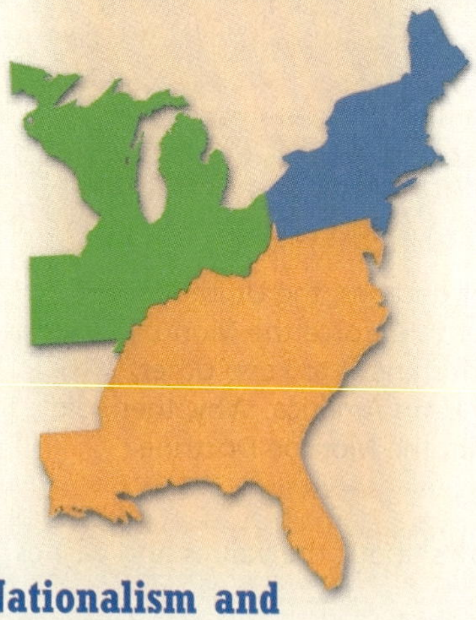

Nationalism and Sectionalism

Nationalism drew regions together. At the same time, economic differences created tension between regions.

TERMS & NAMES

Briefly explain the importance of each of the following.

1. Samuel Slater
2. Industrial Revolution
3. Robert Fulton
4. Eli Whitney
5. cotton gin
6. Nat Turner
7. nationalism
8. sectionalism
9. Missouri Compromise
10. Monroe Doctrine

REVIEW QUESTIONS

Early Industry and Inventions (pages 341–347)

1. How did the War of 1812 push the United States to build factories?
2. Why did its many rivers make the Northeast a good place to build early factories?
3. What was one effect of the steamboat?
4. How did interchangeable parts transform the manufacturing process?

Plantations and Slavery Spread (pages 348–353)

5. Why did slavery spread in the South?
6. What were three hardships faced by enslaved people on plantations?
7. How did religion help people endure or resist slavery?

Nationalism and Sectionalism (pages 354–361)

8. How did the Supreme Court's ruling in *McCulloch v. Maryland* strengthen the federal government?
9. How did the United States gain the territory of Florida?
10. What were the terms of the Missouri Compromise?

CRITICAL THINKING

1. USING YOUR NOTES

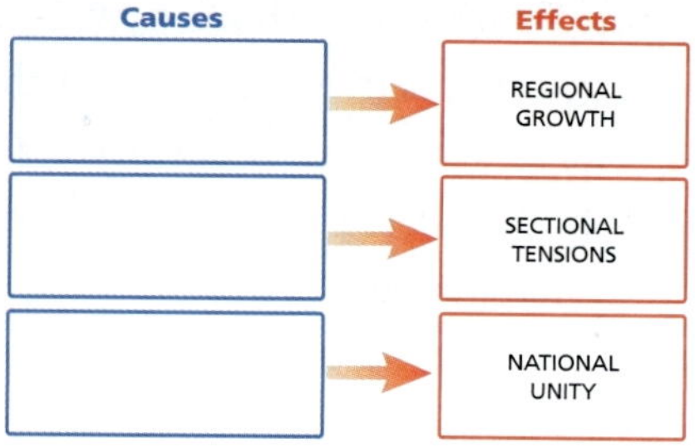

Using your completed chart, answer the questions.

a. What were three causes leading to national unity?

b. What was one cause of sectional tension?

2. THEME: SCIENCE AND TECHNOLOGY

Of all the new inventions mentioned in the chapter, which do you think was most important and why?

3. ANALYZING CAUSES

How did geographic differences between regions lead to economic differences between them?

4. APPLYING CITIZENSHIP SKILLS

Do you think the Missouri Compromise was a wise decision? Consider what might have happened without it, and also why it made Jefferson so uneasy.

5. ANALYZING LEADERSHIP

Think about the Monroe Doctrine and the boundary settlements achieved during the Monroe administration. How would you judge Monroe's foreign policy?

Interact *with* History

Did you predict the ways that new inventions would change the country? What surprised you?

HISTORY SKILLS

1. INTERPRETING MAPS: Region

Study the map. Answer the questions.

Independence in Latin America, 1830

Basic Map Elements

a. What region is the subject of the map?

b. What do the dates on the map mean?

Interpreting the Map

c. What countries were independent by 1823, when the Monroe Doctrine was issued?

d. When did Mexico become independent?

2. INTERPRETING PRIMARY SOURCES

The following verse is from a well-known folk song. Read the verse and answer the questions.

Low Bridge, Everybody Down

I've got a mule and her name is Sal,
Fifteen miles on the Erie Canal.
She's a good old worker and a good old pal,
Fifteen miles on the Erie Canal.
We've hauled some barges in our day,
Filled with lumber, coal, and hay,
And we know every inch of the way
From Albany to Buffalo.

a. In the song, how is the Erie Canal used?

b. What feeling does the song give you about working on the Erie Canal?

ALTERNATIVE ASSESSMENT

1. INTERDISCIPLINARY ACTIVITY: Science

Making a Presentation Do research to learn how inventions of the early 1800s have been improved upon today. For example, learn what kind of engine powers modern boats, and why it works better than a steam engine. Or find out what modern farmers use instead of the McCormick reaper. Share your findings in an oral presentation with visual aids.

2. COOPERATIVE LEARNING ACTIVITY

Planning an Exhibit As a class, plan a museum exhibit to show what slavery was like on cotton plantations. Break into small groups to research different topics—for example, what enslaved people wore, what their houses were like, what rules they lived under, and what stories they told. Bring back your research and decide how you can best share what you learned with an audience. Part of your exhibit might be a model of a plantation or dramatic readings from slave narratives.

3. PRIMARY SOURCE EXPLORER

Planning Foreign Policy The Monroe Doctrine was President Monroe's outline for U.S. foreign policy early in the 19th century. Using the Primary Source Explorer CD-ROM, library, and Internet, find out more about the Monroe Doctrine.

Imagine that you are president of the United States. Come up with four main principles of foreign policy that this country should follow in the 21st century.

- With classmates, talk about broad principles from the Monroe Doctrine. Keeping out of European conflicts would be one example. Protecting free republics would be another.
- Decide whether you agree or disagree with these principles. Think of current U.S. policies that follow or reject them.
- As president, decide how you will communicate U.S. foreign policy for the 21st century to the public. If you make a televised speech, what facts and visual aids would be most persuasive?

4. HISTORY PORTFOLIO

Option 1 Review your section and chapter assessment activities. Choose one that you think is your best work. Try to improve your work, using suggestions from your teacher or classmates. Add the new version to your portfolio.

Option 2 Review the questions you listed for What Do You Want to Know? on page 340. In a brief report, write what you learned about the lives of people from different regions in the early 1800s.

UNIT 4

1810 – 1860

A Changing Nation

Scotts Bluff in Nebraska became a landmark for settlers migrating west on the Oregon Trail.

"Our manifest destiny is to overspread the continent."

—John L. O'Sullivan

CHAPTER 12

The Age of Jackson 1824–1840

The people came by the thousands to the White House on Inauguration Day to see their president—Andrew Jackson.

Interact *with* History

General Andrew Jackson commands his troops in battle.

The year is 1828. You will vote for president for the first time. Important economic, social, and political issues face the country. The favored candidate is Andrew Jackson, a military hero. Before you vote, you should decide what qualities make a strong leader.

What Do You Think?

- What qualities are suggested by this image?
- Which earlier presidents would you consider strong leaders and which not?
- Would qualities that make a military leader also make a good president? Why or why not?

What qualities do you think make a strong leader?

USA

- **1824** John Quincy Adams is elected president.
- **1828** Tariff of Abominations signed into law. Andrew Jackson is elected president.
- **1830** Indian Removal Act is passed.
- **1832** Jackson vetoes charter of Bank of the United States. South Carolina nullifies tariffs. Jackson is reelected.
- **1836** Martin Van Buren is elected president.
- **1838** Cherokees begin to travel the Trail of Tears.
- **1840** William Henry Harrison is elected president.

World

- **1824** Simón Bolívar becomes president of Peru.
- **1830** Revolutions occur in Belgium, France, and Poland.
- **1832** Reform Act increases number of voters in Britain.
- **1837** Victoria becomes queen of Great Britain.
- **1838** Zulu clash with Boer settlers in South Africa.

Chapter 12 SETTING THE STAGE

BEFORE YOU READ

Previewing the Theme

Economics in History Americans elected Andrew Jackson president in 1828. Many believed that he would bring sweeping changes to the government. This chapter explains how President Jackson made decisions that had far-reaching effects on the American economy and on political life. In fact, because he so dominated the life of the nation, his time in office has been called the Age of Jackson.

What Do You Know?

What do you already know about the issues that faced the nation in the first half of the 19th century? How did presidents before Jackson deal with problems?

THINK ABOUT

- what you have learned about Andrew Jackson from books and movies
- how American life is affected by the actions of a president, by conflicts among different parts of the country, and by the will of the people

What Do You Want to Know?

What questions do you have about Jackson and his presidency? Record them in your notebook before reading the chapter.

READ AND TAKE NOTES

Reading Strategy: Finding Main Ideas To make it easier for you to understand what you read, learn to find the main idea of each paragraph, topic heading, and section. Remember that the supporting details help explain the main ideas. On the chart below, write down the main ideas about the political, economic, and social changes during Jackson's presidency.

See Skillbuilder Handbook, page R5.

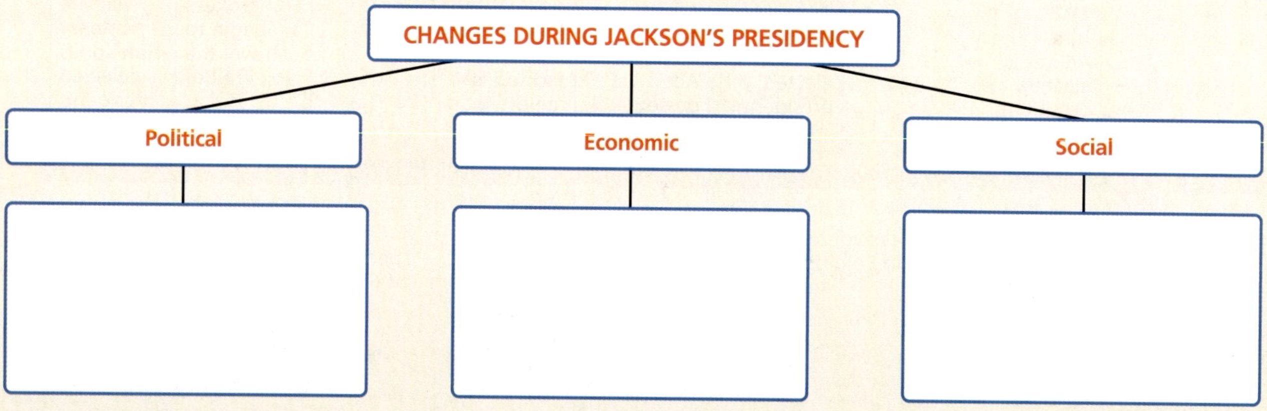

1 Politics of the People

TERMS & NAMES
John Quincy Adams
Andrew Jackson
Jacksonian democracy
spoils system

MAIN IDEA

Andrew Jackson's election to the presidency in 1828 brought a new era of popular democracy.

WHY IT MATTERS NOW

Jackson's use of presidential powers laid the foundation of the modern presidency.

ONE AMERICAN'S STORY

Margaret Bayard Smith was 22 years old when she married and moved to Washington, D.C., in 1800. For the next 40 years, she and her husband, a government official, were central figures in the political and social life of Washington. They entertained presidents from Jefferson to Jackson.

Smith wrote magazine articles and numerous letters describing life in Washington. In 1824, she described how John Quincy Adams reacted to his election as president.

A VOICE FROM THE PAST

When the news of his election was communicated to Mr. Adams by the Committee . . . the sweat rolled down his face—he shook from head to foot and was so agitated that he could scarcely stand or speak.

Margaret Bayard Smith, *The First Forty Years of Washington Society*

Margaret Bayard Smith wrote about life in the nation's capital in the first half of the 19th century.

Adams had reason to be shaken by his election. It had been hotly contested, and he knew that he would face much opposition as he tried to govern. In this section, you will learn how Adams defeated Andrew Jackson in 1824, only to lose to him four years later.

The Election of 1824

In 1824, regional differences led to a fierce fight over the presidency. The Democratic-Republican Party split apart, with four men hoping to replace James Monroe as president. **John Quincy Adams,** Monroe's secretary of state, was New England's choice. The South backed William Crawford of Georgia. Westerners supported Henry Clay, the "Great Compromiser," and **Andrew Jackson,** a former military hero from Tennessee.

Jackson won the most popular votes. But he did not receive a majority of electoral votes. According to the Constitution, if no person wins a majority of electoral votes, the House of Representatives must choose the president. The selection was made from the top three vote getters.

Clay had come in fourth and was out of the running. In the House vote, he threw his support to Adams, who then won. Because Adams

AMERICA'S HISTORY MAKERS

JOHN QUINCY ADAMS

1767–1848

John Quincy Adams was born into wealth and social position. He was the son of President John Adams. Like his father, he had a sharp mind, spoke eloquently, worked tirelessly in public service, and had high principles. But he was sometimes vain, and unwilling to compromise. This made him unpopular with many people and often ineffective. After his presidency, he served with distinction in Congress.

ANDREW JACKSON

1767–1845

Andrew Jackson was the son of a poor farm couple from South Carolina. Orphaned by age 14, he was a wild and reckless youth.

Jackson moved on to become a successful lawyer and plantation owner in Tennessee. But his quick temper still got him into brawls and duels. Bullets in his body from two duels frequently caused him pain.

Jackson's humble background and reputation for toughness endeared him to voters. They considered him one of their own.

Why do you think Jackson was popular but Adams was not?

later named Clay as his secretary of state, Jackson's supporters claimed that Adams gained the presidency by making a deal with Clay. Charges of a "corrupt bargain" followed Adams throughout his term.

Adams had many plans for his presidency. He wanted to build roads and canals, aid education and science, and regulate the use of natural resources. But Congress, led by Jackson supporters, defeated his proposals.

Reading **History**
A. Analyzing Causes What was the main reason John Quincy Adams was not effective as president?

Jacksonian Democracy

Jackson felt that the 1824 election had been stolen from him—that the will of the people had been ignored. Jackson and his supporters were outraged. He immediately set to work to gain the presidency in 1828.

For the next four years, the split in the Democratic-Republican Party between the supporters of Jackson and of Adams grew wider. Jackson claimed to represent the "common man." He said Adams represented a group of privileged, wealthy Easterners. This division eventually created two parties. The Democrats came from among the Jackson supporters, while the National Republicans grew out of the Adams camp.

The election of 1828 again matched Jackson against Adams. It was a bitter campaign—both sides made vicious personal attacks. Even Jackson's wife, Rachel, became a target. During the campaign, Jackson crusaded against control of the government by the wealthy. He promised to look out for the interests of common people. He also promoted the concept of majority rule. The idea of spreading political power to all the people and ensuring majority rule became known as **Jacksonian democracy.**

Actually, the process of spreading political power had begun before Jackson ran for office. When Jefferson was president in the early 1800s, additional people had gained the right to vote as states reduced restrictions on who could vote. Before, for example, only those who owned property or paid taxes could vote in many states. This easing of voting restrictions increased the number of voters. But voting was still limited to adult white males.

The expansion of voting rights helped Jackson achieve an overwhelming win in the 1828 presidential election. Jackson's triumph was hailed as a victory for common people. Large numbers of Western farmers as well as workers in the nation's cities supported him. Their vote put an end to the idea that the government should be controlled by an educated elite. Now, the common people would be governed by one of their own. (See chart "Changes in Ideas About Democracy," page 373.)

Reading **History**

B. Recognizing Effects What factor made Jackson's appeal to the "common man" especially important in the election of 1828?

STRANGE *but* True

ADAMS AND JEFFERSON

John Adams and Thomas Jefferson died on the same day—the Fourth of July, 1826, the 50th anniversary of the adoption of the Declaration of Independence.

Both Adams and Jefferson were founders of the nation, signers of the Declaration, and presidents. They were also political enemies who had become friends late in life.

Adams was 90; Jefferson, 83. Adams's last words were "Jefferson still survives." He was unaware that Jefferson had died hours earlier.

The People's President

Jackson's humble background, and his reputation as a war hero, helped make him president. Many saw his rise above hardship as a real American success story. He was the first president not from an aristocratic Massachusetts or Virginia family, and the first from the West.

Jackson indeed had had a hard life. His father died shortly before his birth, and Jackson grew up on a frontier farm in South Carolina. At 13, he joined the militia with his older brother to fight in the Revolutionary War. In 1781, they were taken prisoner by the British. While captive, he allegedly refused when commanded to shine an officer's boots. The officer struck Jackson with a sword, leaving scars on his hand and head. Later, Jackson's mother obtained her sons' release from a military prison, where they had become ill with smallpox. Jackson's brother died, but his mother nursed Jackson back to health. A short time later, she also died. Jackson's experiences during the Revolution left him with a lifelong hatred of the British.

After the war, Jackson moved to the Tennessee frontier. In 1784, he began to study law. He built a successful legal practice and also bought and sold land. Jackson then purchased a plantation near Nashville and ran successfully for Congress. After the War of 1812 broke out, he was appointed a general in the army. At the Battle of New Orleans in 1815, Jackson soundly defeated the British even though his troops were greatly outnumbered. He became a national war hero. He earned the nickname "Old Hickory," after a soldier claimed that he was "tough as hickory."

Jackson usually wore this miniature oil portrait of his beloved wife, Rachel, around his neck.

Jackson Takes Office

Jackson's success in the presidential election of 1828 came at a high price. Shortly after he won, his wife, Rachel, died of a heart attack. Jackson believed that the campaign attacks on her reputation had killed her. She was a private woman who preferred a quiet life. In fact, she had said that she would "rather be a doorkeeper in the house of God than . . . live in that palace at Washington." Margaret Bayard Smith described Rachel's importance to Jackson, saying she "not only made him a happier, but a better man."

Jackson looked thin, pale, and sad at his inauguration on March 4, 1829. But the capital was full of joy and excitement. Thousands of people were there. Senator Daniel Webster wrote about the inauguration.

A VOICE FROM THE PAST

I have never seen such a crowd before. Persons have come five hundred miles to see General Jackson, and they really seem to think that the country has been rescued from some dreadful danger.

Daniel Webster, *Correspondence*

Reading **History**

C. Drawing Conclusions Why did Jackson's supporters react with such enthusiasm at his inauguration?

At the inauguration ceremony, the crowd shouted, waved, applauded, and saluted its hero. He bowed low to the people in turn. A throng followed Jackson to the White House reception. One person described the crowd as containing "all sorts of people, from the highest and most polished, down to the most vulgar and gross in the nation."

The crowd grew rowdy. People broke china and glasses as they grabbed for the food and drinks. The pushing and shoving finally drove the new president to flee the White House. As Supreme Court Justice Joseph Story observed, "The reign of King Mob seemed triumphant."

CITIZENSHIP TODAY

Exercising the Vote

During the Age of Jackson, rules on who could vote were eased. This increased the number of voters. But voting was still limited to adult white males. Over the years, other groups gained the right to vote, including African Americans, women, and Native Americans. Today's elections are open to all citizens aged 18 and over.

Future voters can practice casting their votes in mock, or pretend, elections. The National Student/Parent Mock Election teaches students to be informed voters. Mock presidential elections attract coverage by the media. Television stations may even broadcast live from schools, interviewing student voters.

One high school student, Charlie Tran from San Jose, California, said, "Students seem to catch the important political events surrounding them. Some students are taking their views . . . to a new level by campaigning for the candidate they support."

Students register to vote in a mock election.

How Do You Set Up a Mock Election?

1. Choose issues and candidates and then set up a mock election in your classroom. (You could focus on the national, state, or local level.)
2. Create the materials of an election, such as the polling place, ballots, and posters.
3. Campaign for the candidates or the issues you support.
4. Conduct the voting.
5. Prepare mock media reports on the election's outcome. You may want to interview voters.

See Citizenship Handbook, page 283.

Visit www.mcdougallittell.com to learn more about citizenship and voting.

Changes in Ideas About Democracy

JEFFERSONIAN DEMOCRACY	JACKSONIAN DEMOCRACY
government for the people by capable, well-educated leaders	government by the people
democracy in political life	democracy in social, economic, and political life
championed the cause of the farmer in a mainly agricultural society	championed the cause of the farmer and the laborer in an agricultural and industrial society
limited government	limited government, but with a strong president

SKILLBUILDER Interpreting Charts

1. *What do you think was the most important change in democracy?*
2. *Did Jefferson or Jackson exercise more power?*

A New Political Era Begins

Jackson's inauguration began a new political era. In his campaign, he had promised to reform government. He started by replacing many government officials with his supporters. This practice of giving government jobs to political backers became known as the **spoils system.** The name comes from a statement that "to the victor belong the spoils [possessions] of the enemy." Jackson's opponents charged that the practice was corrupt. But he defended it, noting that it broke up one group's hold on government.

As president, Jackson would face three major issues—the status of Native Americans, the rights of the states, and the role of the Bank of the United States. In the next section, you will learn how Jackson's policies affected Native Americans.

Section 1 Assessment

1. Terms & Names

Identify:

- John Quincy Adams
- Andrew Jackson
- Jacksonian democracy
- spoils system

2. Taking Notes

Use a chart to identify important biographical information about Andrew Jackson.

Life of Andrew Jackson	
Youthful life	
Road to Congress	
War hero	
Appeal to voters	

3. Main Ideas

a. How did Andrew Jackson react to the election of 1824? Why?

b. What factors helped Jackson win the presidency in 1828?

c. What was the effect of expanding voting rights?

4. Critical Thinking

Analyzing Points of View
What are reasons for and against the spoils system?

THINK ABOUT

- the effects of giving government workers lifetime jobs
- the effects of rewarding political supporters

ACTIVITY OPTIONS

GEOGRAPHY
MATH

Find out which states Jackson and Adams won in the 1828 election. Show the results on a **map** or **chart** that includes vote totals and percentages.

2 Jackson's Policy Toward Native Americans

TERMS & NAMES
Sequoya
Indian Removal Act
Indian Territory
Trail of Tears
Osceola

MAIN IDEA

During Jackson's presidency, Native Americans were forced to move west of the Mississippi River.

WHY IT MATTERS NOW

This forced removal forever changed the lives of Native Americans in the United States.

ONE AMERICAN'S STORY

For 12 years, a brilliant Cherokee named **Sequoya** (sih KWOY uh) tried to find a way to "teach the Cherokees to talk on paper like the white man." In 1821, he reached his goal. Sequoya invented a writing system for the Cherokee language without ever having learned to read or write in any other language. Helped by his young daughter, he identified all the sounds in Cherokee and created 86 characters to stand for syllables.

Using this simple system, the Cherokees soon learned to read and write. They even published a newspaper and books in their own language. A traveler in 1828 marveled at how many Cherokees had learned to read and write without schools or even paper and pens.

Sequoya invented a writing system of 86 characters, shown here, for the Cherokee language.

A VOICE FROM THE PAST

I frequently saw as I rode from place to place, Cherokee letters painted or cut on the trees by the roadside, on fences, houses, and often on pieces of bark or board, lying about the houses.

Anonymous traveler, quoted in the *Advocate*

Sequoya hoped that by gaining literacy—the ability to read and write—his people could share the power of whites and keep their independence. But even Sequoya's invention could not save the Cherokees from the upheaval to come. In this section, you will learn about President Jackson's policy toward Native Americans and its effects.

Native Americans in the Southeast

Since the 1600s, white settlers had pushed Native Americans westward as they took more and more of their land. However, there were still many Native Americans in the East in the early 1800s. Some whites hoped that the Native Americans could adapt to the white people's way of life. Others wanted the Native Americans to move. They believed this was the only way to avoid conflict over land. Also, many whites felt that Native Americans were "uncivilized" and did not want to live near them.

By the 1820s, about 100,000 Native Americans remained east of the Mississippi River. The majority were in the Southeast. The major tribes

Reading **History**

A. Reading a Map Use the map on page 376 to locate Native American lands in the Southeast.

were the Cherokee, Chickasaw, Choctaw, Creek, and Seminole. Whites called them the Five Civilized Tribes because they had adopted many aspects of white culture. They held large areas of land in Georgia, the Carolinas, Alabama, Mississippi, and Tennessee.

The Cherokee Nation

More than any other Southeastern tribe, the Cherokee had adopted white customs, including their way of dressing. Cherokees owned prosperous farms and cattle ranches. Some even had slaves. From Sequoya, they acquired a written language, and they published their own newspaper, the *Cherokee Phoenix*. Some of their children attended missionary schools. In 1827, the Cherokees drew up a constitution based on the U.S. Constitution and founded the Cherokee Nation.

A year after the Cherokees adopted their constitution, gold was discovered on their land in Georgia. Now, not only settlers but also miners wanted these lands. The discovery of gold increased demands by whites to move the Cherokees. The federal government responded with a plan to remove all Native Americans from the Southeast.

Now and then

CHEROKEE PEOPLE TODAY

Today, there are more than 300,000 Cherokees. They are part of three main groups—the Cherokee Nation of Oklahoma, the United Keetoowah Band in Oklahoma, and the Eastern Band of Cherokee Indians of North Carolina.

Wilma Mankiller, shown below, was the first woman elected principal chief of the Cherokee Nation of Oklahoma. She has said that the "Cherokee people possess an extraordinary ability to face down adversity and continue moving forward. We are able to do that because our culture, though certainly diminished, has sustained us since time immemorial."

Jackson's Removal Policy

Andrew Jackson had long supported a policy of moving Native Americans west of the Mississippi. He first dealt with the Southeastern tribes after the War of 1812. The federal government ordered Jackson, then acting as Indian treaty commissioner, to make treaties with the Native Americans of the region. Through these treaties forced on the tribes, the government gained large tracts of land.

Jackson believed that the government had the right to regulate where Native Americans could live. He viewed them as conquered subjects who lived within the borders of the United States. He thought that Native Americans had one of two choices. They could adopt white culture and become citizens of the United States. Or they could move into the Western territories. They could not, however, have their own governments within the nation's borders.

After the discovery of gold, whites began to move onto Cherokee land. Georgia and other Southern states passed laws that gave them the right to take over Native American lands. When the Cherokee and other tribes protested, Jackson supported the states.

To solve the problem, Jackson asked Congress to pass a law that would require Native Americans to either move west or submit to state laws. Many Americans objected to Jackson's proposal. Massachusetts congressman Edward Everett opposed removing Native Americans against their will to a distant land. There, he said, they would face "the

Removal of Native Americans, 1820–1840

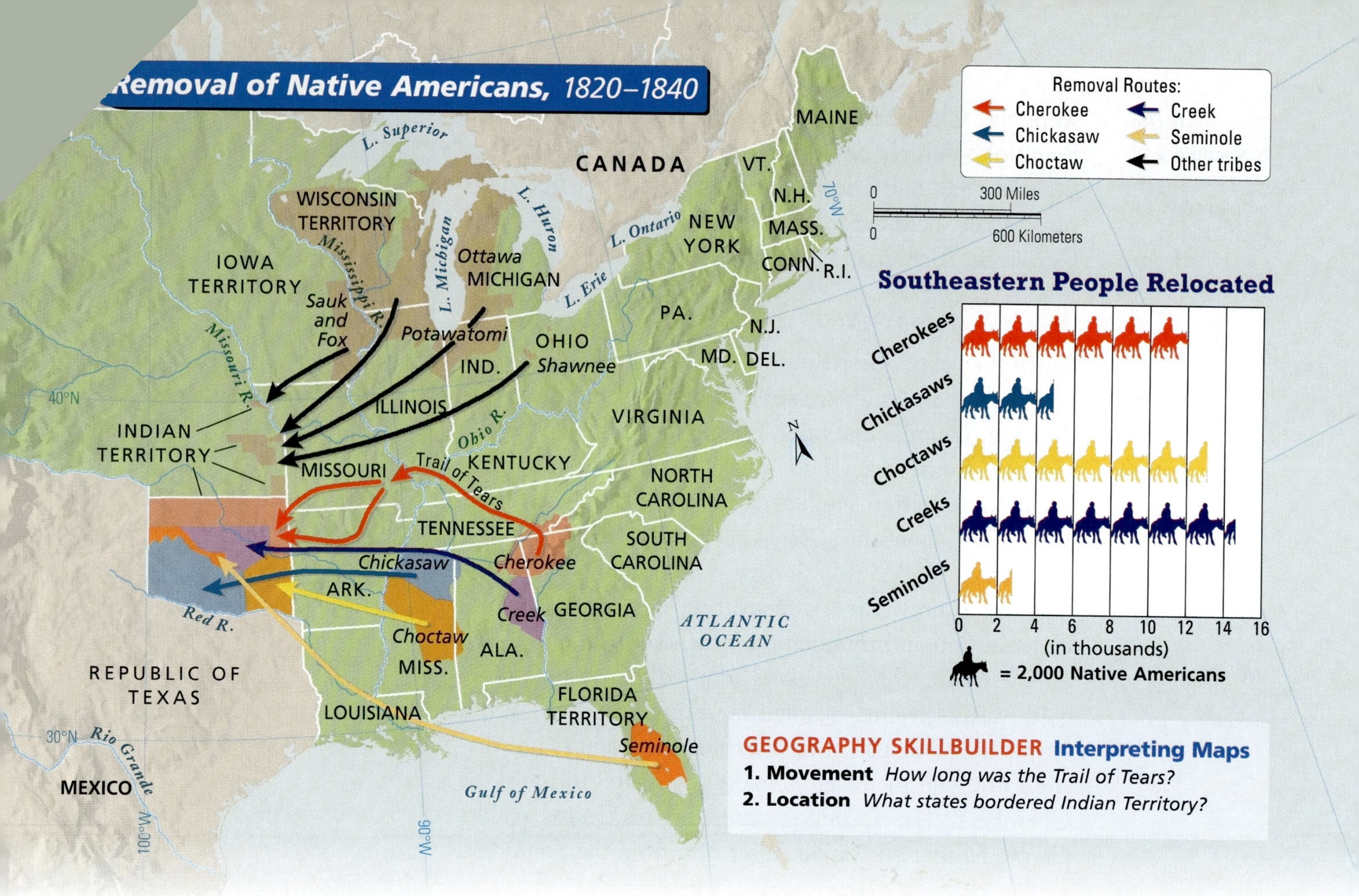

GEOGRAPHY SKILLBUILDER Interpreting Maps
1. **Movement** *How long was the Trail of Tears?*
2. **Location** *What states bordered Indian Territory?*

perils and hardships of a wilderness." Religious groups such as the Quakers also opposed forced removal of Native Americans. After heated debate, Congress passed the **Indian Removal Act** in 1830. The act called for the government to negotiate treaties that would require Native Americans to relocate west.

Jackson immediately set out to enforce the law. He thought his policy was "just and liberal" and would allow Native Americans to keep their way of life. Instead, his policy caused much hardship and forever changed relations between whites and Native Americans.

*Reading*History
B. Drawing Conclusions What were reasons for and against the Indian Removal Act?

The Trail of Tears

As whites invaded their homelands, many Native Americans saw no other choice but to sign treaties exchanging their land for land in the West. Under the treaties, Native Americans would be moved to an area that covered what is now Oklahoma and parts of Kansas and Nebraska. This area came to be called **Indian Territory.**

Beginning in the fall of 1831, the Choctaw and other Southeast tribes were removed from their lands and relocated to Indian Territory. The Cherokees, however, first appealed to the U.S. Supreme Court to protect their land from being seized by Georgia. In 1832, the court, led by Chief Justice John Marshall, ruled that only the federal government, not the states, could make laws governing the Cherokees. This ruling meant that the Georgia laws did not apply to the Cherokee Nation. However, both Georgia and President Jackson ignored the Supreme Court. Jackson said, "John Marshall has made his decision. . . . Now let him enforce it."

A small group of Cherokees gave up and signed a treaty to move west. But the majority of the Cherokees, led by John Ross, opposed the treaty. Jackson refused to negotiate with these Cherokees.

Reading **History**

C. Recognizing Effects What happened to the Cherokees as a result of the Indian Removal Act?

In 1838, federal troops commanded by General Winfield Scott rounded up about 16,000 Cherokees and forced them into camps. Soldiers took people from their homes with nothing but the clothes on their backs. Over the fall and winter of 1838–1839, these Cherokees set out on the long journey west. Forced to march in the cold, rain, and snow without adequate clothing, many grew weak and ill. One-fourth died. The dead included John Ross's wife. One soldier never forgot what he witnessed on the trail.

A VOICE FROM THE PAST

Murder is murder and somebody must answer, somebody must explain the streams of blood that flowed in the Indian country in . . . 1838. Somebody must explain the four-thousand silent graves that mark the trail of the Cherokees to their exile. I wish I could forget it all, but the picture of six-hundred and forty-five wagons lumbering over the frozen ground with their Cargo of suffering humanity still lingers in my memory.

John G. Burnett, quoted in *The Native Americans,* edited by Betty and Ian Ballantine

This harsh journey of the Cherokee from their homeland to Indian Territory became known as the **Trail of Tears.**

HISTORY *through* ART

In 1838, the Cherokees left their homeland by wagon, horse, donkey, and foot, forced to travel hundreds of miles along the Trail of Tears. This painting is by Robert Lindneux, a 20th-century artist.

How does the artist show the suffering on the Trail of Tears?

Osceola led the Seminoles in their fight against removal.

Native American Resistance

Not all the Cherokees moved west in 1838. That fall, soldiers had rounded up an old Cherokee farmer named Tsali and his family, including his grown sons. On the way to the stockade, they fought the soldiers. A soldier was killed before Tsali fled with his family to the Great Smoky Mountains in North Carolina. There they found other Cherokees. The U.S. Army sent a message to Tsali. If he and his sons would give themselves up, the others could remain. They surrendered, and all except the youngest son were shot. Their sacrifice allowed some Cherokees to stay in their homeland.

Other Southeast tribes also resisted relocation. In 1835, the Seminoles refused to leave Florida. This refusal led to the Second Seminole War. One elderly Seminole explained why he could not leave: "If suddenly we tear our hearts from the homes around which they are twined [wrapped around], our heart strings will snap."

Background
The Seminoles fought three wars against the U.S. government between 1817 and 1858, when their resistance ended.

One of the most important leaders in the war was **Osceola** (AHS ee OH luh). Hiding in the Everglades, Osceola and his band used surprise attacks to defeat the U.S. Army in many battles. In 1837, Osceola was tricked into capture when he came to peace talks during a truce. He later died in prison. But the Seminoles continued to fight. Some went deeper into the Everglades, where their descendants live today. Others moved west. The Second Seminole War ended in 1842.

Some tribes north of the Ohio River also resisted relocation. The Shawnee, Ottawa, Potawatomi, Sauk, and Fox were removed to Indian Territory. But in 1832, a Sauk chief named Black Hawk led a band of Sauk and Fox back to their lands in Illinois. In the Black Hawk War, the Illinois militia and the U.S. Army crushed the uprising.

In the next section, you will learn about other issues Jackson faced, especially increasing tensions between various sections of the country.

Section 2 Assessment

1. Terms & Names

Identify:
- Sequoya
- Indian Removal Act
- Indian Territory
- Trail of Tears
- Osceola

2. Taking Notes

Use a chart to list the reasons for Jackson's Native American removal policy.

Reasons Native Americans Were Forced West		
Economic	Political	Social

What do you think was the main reason?

3. Main Ideas

a. How did President Jackson justify the Indian Removal Act?

b. In what ways did Native Americans resist the Indian Removal Act?

c. What were the consequences of the Indian Removal Act?

4. Critical Thinking

Recognizing Effects
What were some economic effects of the Indian Removal Act on Native Americans? On whites?

THINK ABOUT
- what the Native Americans lost
- what the white settlers gained

ACTIVITY OPTIONS

GEOGRAPHY
MATH

Use the map on page 376 to estimate the distance traveled by each of the five Southeastern tribes. Show your calculation on a **map** or **chart.**

3 Conflicts Over States' Rights

MAIN IDEA

Jackson struggled to keep Southern states from breaking away from the Union over the issue of tariffs.

WHY IT MATTERS NOW

Disputes about states' rights and federal power remain important in national politics.

TERMS & NAMES

John C. Calhoun
Tariff of Abominations
doctrine of nullification
Webster-Hayne debate
Daniel Webster
secession

ONE AMERICAN'S STORY

Early in his political career, **John C. Calhoun** was hailed as "one of the master-spirits who stamp their name upon the age in which they live." This was praise indeed for someone from the backwoods of South Carolina who had little formal education before age 18. Elected to the U.S. Congress at 28, Calhoun soon was one of its leaders. He supported the need for a strong central government and became something of a hero to the nation's young people. He spoke out against sectionalism.

A VOICE FROM THE PAST

What is necessary for the common good may apparently be opposed to the interest of particular sections. It must be submitted to [accepted] as the condition of our [nation's] greatness.

John C. Calhoun, quoted in *John C. Calhoun: American Portrait* by Margaret L. Coit

But Calhoun's concern for the economic and political well-being of his home state of South Carolina, and the South in general, later caused him to change his beliefs. He became the foremost champion of states' rights, rigid in his views and increasingly bitter.

In this section, you will learn how two strong-willed men—Calhoun and Jackson—came in conflict over the issue of states' rights.

The bitter debate over state's rights took a physical toll on John C. Calhoun. He is shown here in about 1825 and in 1849.

Rising Sectional Differences

Andrew Jackson had taken office in 1829. At the time, the country was being pulled apart by conflicts among its three main sections—the Northeast, the South, and the West. Legislators from these regions were arguing over three major economic issues: the sale of public lands, internal improvements, and tariffs.

The federal government had acquired vast areas of land through conquests, treaties, and purchases. It raised money partly by selling these public lands. However, Northeasterners did not want public lands in the West to be sold at low prices. The cheap land would attract workers who were needed in the factories of the Northeast. But Westerners wanted low land prices to encourage settlement. The more people who moved West, the more political power the section would have.

The issue of internal improvements also pulled the sections apart. Business leaders in the Northeast and West backed government spending on internal improvements, such as new roads and canals. Good transportation would help bring food and raw materials to the Northeast and take manufactured goods to Western markets. Southerners opposed more federal spending on internal improvements because the government financed these projects through tariffs, which were taxes on imported goods. The South did not want any increase in tariffs.

Since 1816, tariffs had risen steadily. They had become the government's main source of income. Northerners supported high tariffs because they made imported goods more expensive than American-made goods. The Northeast had most of the nation's manufacturing. Tariffs helped

Background
During the Jackson era, the West included states that are now considered part of the Midwest.

Economics *in* History

How Tariffs Work

Tariffs are taxes added to the cost of goods imported from another country. There are two kinds of tariffs—revenue tariffs and protective tariffs. **Revenue tariffs** are used to raise money, like the sales taxes that states add to purchases today. These tariffs tend to be fairly low. **Protective tariffs** usually are much higher. They have another goal: to persuade consumers to buy goods made in their own country instead of purchasing foreign-made products. Congress passed a protective tariff in 1828 to help American companies.

The illustration shows how a protective tariff works. A British-made teapot sells for $3.50, and a similar teapot made in the United States sells for $4.00. Most shoppers will buy the British teapot and save 50 cents. But when the government adds a 40 percent tariff to British goods, the price of the British teapot soars to $4.90. The result: consumers buy the now-cheaper American teapots.

CONNECT TO HISTORY

1. **Recognizing Effects** Do consumers benefit from high tariffs? Why or why not?

See Skillbuilder Handbook, page R10.

CONNECT TO TODAY

2. **Making Inferences** Today, many leaders around the world promote the idea of "free trade." What do you think "free trade" means?

Visit www.mcdougallittell.com to learn more about tariffs.

American manufacturers sell their products at a lower price than imported goods.

Reading **History**
A. Analyzing Causes Why did the three sections of the country differ on the sale of public lands, internal improvements, and tariffs?

The South opposed rising tariffs because its economy depended on foreign trade. Southern planters sold most of their cotton to foreign buyers. They were not paid in money but were given credit. They then used the credit to buy foreign manufactured goods. Because of higher tariffs, these foreign goods cost more. Eventually, the tariff issue would lead to conflict between North and South.

Tariff of Abominations

In 1828, in the last months of John Quincy Adams's presidency, Congress passed a bill that significantly raised the tariffs on raw materials and manufactured goods. Southerners were outraged. They had to sell their cotton at low prices to be competitive. Yet tariffs forced them to pay high prices for manufactured goods. Southerners felt that the economic interests of the Northeast were determining national policy. They hated the tariff and called it the **Tariff of Abominations** (an abomination is a hateful thing).

Differences over the tariff helped Jackson win the election of 1828. Southerners blamed Adams for the tariff, since it was passed during his administration. So they voted against him.

A South Carolina woman sews a palmetto emblem (inset) to her hat to show her support for nullification. The palmetto is a South Carolina symbol.

Crisis over Nullification

The Tariff of Abominations hit South Carolinians especially hard because their economy was in a slump. Some leaders in the state even spoke of leaving the Union over the issue of tariffs. John C. Calhoun, then Jackson's vice-president, understood the problems of South Carolina's farmers because he was one himself. But he wanted to find a way to keep South Carolina from leaving the Union. The answer he arrived at was the **doctrine of nullification.** A state, Calhoun said, had the right to nullify, or reject, a federal law that it considers unconstitutional.

Reading **History**
B. Summarizing How did the issue of tariffs threaten to tear the Union apart?

Calhoun was not the first person to propose the doctrine of nullification. It had been stated earlier by Thomas Jefferson. Jefferson argued that the Union was a league of sovereign, or self-governing, states that had the right to limit the federal government. Calhoun developed the doctrine further. He said that any state could nullify, or refuse to recognize, a federal law within its borders. He believed that Congress had no right to impose a tariff that favored one section of the country. Therefore, South Carolina had the right to nullify the tariff. Calhoun's doctrine was an extreme form of states' rights—the theory that states have the right to judge whether a law of Congress is unconstitutional.

In the summer of 1828, Calhoun wrote a document called the "South Carolina Exposition and Protest." It stated his theory. Calhoun allowed the document to be published, but he did not sign his name. He knew his ideas would cause controversy.

Vocabulary
controversy: a public dispute

Daniel Webster (standing) and Robert Y. Hayne (seated, with hands extended) debated nullification in the U.S. Senate in 1830.

The States' Rights Debate

Calhoun was right. His ideas added fuel to the debate over the nature of the federal union. This debate had been going on since independence from Britain. More and more people took sides. Some supported a strong federal government. Others defended the rights of the states. This question would be a major political issue from this time until the Civil War was fought to resolve it some 30 years later.

One of the great debates in American history took place in the U.S. Senate over the doctrine of nullification—the **Webster-Hayne debate** of 1830. On one side was **Daniel Webster,** a senator from Massachusetts and the most powerful speaker of his time. On the other was Robert Y. Hayne, a senator from South Carolina. Hayne defended nullification. He argued that it gave the states a lawful way to protest and to maintain their freedom. He also said that the real enemies of the Union were those "who are constantly stealing power from the States, and adding strength to the Federal Government."

Webster argued that it was the people and not the states that made the Union. In words that were printed and spread across the country, Webster declared that freedom and the Union go together.

A VOICE FROM THE PAST

When my eyes shall be turned to behold for the last time the sun in heaven, may I not see him shining on the broken and dishonored fragments of a once glorious Union. . . . Liberty and Union, now and forever, one and inseparable!

Daniel Webster, a speech in the U.S. Senate, January 26, 1830

Jackson had not yet stated his position on the issue of states' rights, even though Calhoun was his vice-president. He got his chance in April at a dinner in honor of the birthday of Thomas Jefferson. Calhoun and other

Reading History

C. Analyzing Points of View What do you think Calhoun meant by "the benefits and burdens of the Union" should be equally distributed?

supporters of nullification planned to use the event to win support for their position. Jackson learned of their plans and went to the dinner prepared.

After dinner, Jackson was invited to make a toast. He stood up, looked directly at Calhoun, and stated bluntly, "Our Federal Union—it must be preserved." As Calhoun raised his glass, his hand trembled. Called on to make the next toast, Calhoun stood slowly and said, "The Union—next to our liberty, the most dear; may we all remember that it can only be preserved by respecting the rights of the states and distributing equally the benefits and burdens of the Union." From that time, the two men were political enemies.

South Carolina Threatens to Secede

Even though Jackson made it clear that he opposed the doctrine of nullification, he did not want to drive the South out of the Union. He asked Congress to reduce the tariff, and Congress did so in 1832. But Southerners thought the reduced rates were still too high. South Carolina nullified the tariff acts of 1828 and 1832 and voted to build its own army. South Carolina's leaders threatened **secession,** or withdrawal from the Union, if the federal government tried to collect tariffs.

Jackson was enraged. He told a South Carolina congressman that if the state's leaders defied federal laws, he would "hang the first man of them I can get my hands on." Jackson ran for reelection in 1832, this time without Calhoun as his running mate. After he won, he made it clear that he would use force to see that federal laws were obeyed and the Union preserved.

In the Senate, Henry Clay came forward with a compromise tariff in 1833. He hoped that it would settle the issue and prevent bloodshed. Congress quickly passed the bill, and the crisis ended. South Carolina stayed in the Union. In the next section, you will read about another issue of Jackson's presidency—his war on the national bank.

Section 3 Assessment

1. Terms & Names

Identify:

- John C. Calhoun
- Tariff of Abominations
- doctrine of nullification
- Webster-Hayne debate
- Daniel Webster
- secession

2. Taking Notes

Use a chart to indicate how each section stood on these issues.

	North-east	West	South
Sale of public lands			
Internal improvements			
High tariffs			

3. Main Ideas

a. Why did the South oppose high tariffs?

b. What were Calhoun's reasons for proposing the doctrine of nullification?

c. Why did South Carolina threaten secession, and how was the crisis resolved?

4. Critical Thinking

Recognizing Effects In what ways would the doctrine of nullification have made it difficult for the federal government to operate?

THINK ABOUT

- its effect on the enforcement of laws
- its effect on the power of the federal government

ACTIVITY OPTIONS

SPEECH

TECHNOLOGY

Research Daniel Webster's speech; a part of it appears on page 382. Deliver a **speech** for or against nullification to the class, or record it on an **audiocassette.**

Prosperity and Panic

TERMS & NAMES
inflation
Martin Van Buren
Panic of 1837
depression
Whig Party
William Henry Harrison
John Tyler

MAIN IDEA

Jackson's policies caused the economy to collapse after he left office and affected the next election.

WHY IT MATTERS NOW

The condition of the economy continues to affect the outcomes of presidential elections.

ONE AMERICAN'S STORY

Nicholas Biddle was the kind of person that Andrew Jackson neither liked nor trusted. Biddle was wealthy, well educated, and came from a socially prominent Philadelphia family. He was also the influential president of the powerful Second Bank of the United States—the bank that Jackson believed to be a monster of corruption. Jackson declared war on Biddle and the bank during his 1832 reelection campaign. But Biddle felt sure of his political power.

A VOICE FROM THE PAST

I have always deplored making the Bank a [political] question, but since the President will have it so, he must pay the penalty of his own rashness. . . . [m]y hope is that it will contribute to relieve the country of the domination of these miserable [Jackson] people.

Nicholas Biddle, from a letter to Henry Clay dated August 1, 1832

For his part, Jackson vowed to "kill" the bank. In this section, you will read about his war on the bank and its effect on the economy.

Nicholas Biddle was the president of the powerful Second Bank of the United States, located in Philadelphia.

Mr. Biddle's Bank

The Second Bank of the United States was the most powerful bank in the country. It held government funds and issued money. As its president, Nicholas Biddle set policies that controlled the nation's money supply.

Although the bank was run efficiently, Jackson had many reasons to dislike it. For one thing, he had come to distrust banks after losing money in financial deals early in his career. He also thought the bank had too much power. The bank made loans to members of Congress, and Biddle openly boasted that he could influence Congress. In addition, Jackson felt the bank's lending policies favored wealthy clients and hurt the average person.

To operate, the bank had to have a charter, or a written grant, from the federal government. In 1832, Biddle asked Congress to renew the bank's charter, even though it would not expire until 1836. Because 1832 was an election year, he thought Jackson would agree to renewal rather than risk angering its supporters. But Jackson took the risk.

Jackson's War on the Bank

When Congress voted to renew the bank's charter, Jackson vetoed the renewal. In a strongly worded message to Congress, Jackson claimed the bank was unconstitutional. He said the bank was a monopoly that favored the few at the expense of the many. The Supreme Court earlier had ruled that the bank was constitutional. But Jackson claimed elected officials had to judge the constitutionality of a law for themselves. They did not need to rely on the Supreme Court. His veto message also contained this attack on the bank.

Vocabulary
monopoly: a company or group with complete control over a product or service

A VOICE FROM THE PAST

It is to be regretted that the rich and powerful too often bend the acts of government to their selfish purposes. . . . Distinctions in society will always exist under every just government. . . . [B]ut when the laws undertake to . . . make the rich richer and the potent more powerful, the humble members of society . . . have a right to complain of the injustice of their Government.

Andrew Jackson, veto message, July 10, 1832

Reading **History**
A. Analyzing Points of View What reasons did Jackson have for wanting to destroy the Second Bank of the United States?

Jackson's war on the bank became the main issue in the presidential campaign of 1832. The National Republican Party and its candidate, Henry Clay, called Jackson a tyrant. They said he wanted too much power as president. The Democrats portrayed Jackson as a defender of the people. When he won reelection, Jackson took it as a sign that the public approved his war on the bank.

In his second term, Jackson set out to destroy the bank before its charter ended in 1836. He had government funds deposited in state banks, which opponents called Jackson's "pet banks." Biddle fought back by making it harder for people to borrow money. He hoped the resulting economic troubles would force Jackson to return government deposits to the bank. Instead, the people rallied to Jackson's position. Eventually, the bank went out of business. Jackson had won the war, but the economy would be a victim.

Jackson Fights the Second Bank

In this political cartoon, Jackson fights the many-headed monster—the Second Bank of the United States and its branches—with a cane labeled "VETO."

- **A** President Jackson
- **B** Cane labeled "VETO"
- **C** Nicholas Biddle
- **D** Vice-President Van Buren

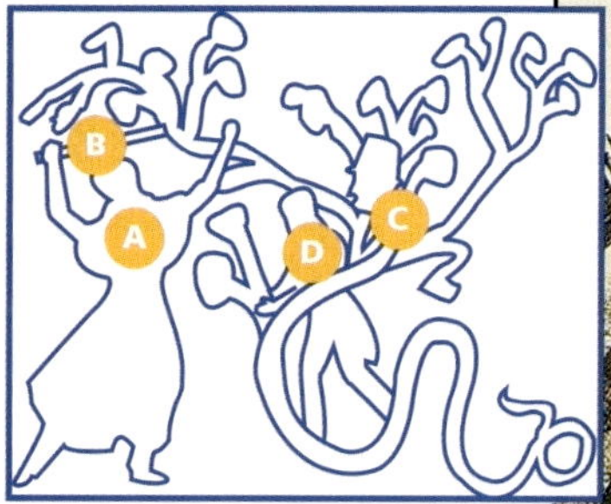

Prosperity Becomes Panic

Most of the nation prospered during Jackson's last years in office. Because it was easier to borrow money, people took out loans to buy public lands, and the economy boomed. But the "pet banks" issued too much paper money. The rise in the money supply made each dollar worth less. As a result, prices rose. **Inflation,** which is an increase in prices and decrease in the value of money, was the outcome. To fight inflation, Jackson issued an order that required people to pay in gold or silver for public lands.

Jackson left office proud of the nation's prosperity. But it was a puffed-up prosperity. Like a balloon, it had little substance. Because of Jackson's popularity, his vice-president, **Martin Van Buren,** was elected president in 1836. Within a few months after Van Buren took office, a panic—a widespread fear about the state of the economy—spread throughout the country. It became known as the **Panic of 1837**.

People took their paper money to the banks and demanded gold or silver in exchange. The banks quickly ran out of gold and silver. When the government tried to get its money from the state banks, the banks could not pay. The banks defaulted, or went out of business. A **depression,** or severe economic slump, followed.

The depression caused much hardship. Because people had little money, manufacturers no longer had customers for their goods. Almost 90 percent of factories in the East closed in 1837. Jobless workers had no way of buying food or paying rent. People went hungry. They lived in shelters or on the streets, where many froze in the winter. Every section of the country suffered, but the depression hit hardest in the cities. Farmers were hurt less because they could at least grow their own food. The depression affected politics, too.

Reading History

B. Recognizing Effects What were the short-term and long-term effects of Jackson's war on the bank?

America's HERITAGE

POLITICAL PARTIES

Today's Democratic and Republican parties were born more than a century ago. Andrew Jackson's supporters first called themselves Democratic-Republicans. But in the 1830s, they became known simply as Democrats. They stood for states' rights and saw themselves as defenders of the common people. The modern Republican Party was formed in 1854 as the successor to the Whig Party, founded in 1834.

In the Jackson era, political parties campaigned for their candidates by staging parades and rallies. Participants often carried banners like the one below with a log cabin, the symbol for the campaign of William H. Harrison in 1840.

The Rise of the Whig Party

In the depths of the depression, Senators Henry Clay and Daniel Webster argued that the government needed to help the economy. Van Buren disagreed. He believed that the economy would improve if left alone. He argued that "the less government interferes with private pursuits the better for the general prosperity." Many Americans blamed Van Buren for the Panic, though he had taken office only weeks before it started. The continuing depression made it almost impossible for him to win reelection in 1840.

Van Buren faced a new political party in that election. During Jackson's war on the national bank, Clay, Webster, and other Jackson opponents had formed the **Whig Party.** It was named after a British party that opposed royal power. The Whigs opposed the concentration of power in the chief executive—whom they mockingly called "King Andrew" Jackson. In 1840, the Whigs chose **William Henry Harrison** of Ohio to run for president and **John Tyler** of Virginia to run for vice-president.

Reading **History**

C. Making Inferences Why did the Whigs want to nominate a candidate like Harrison, who did not have strong political views?

The Whigs nominated Harrison largely because of his military record and his lack of strong political views. Harrison had led the army that defeated the Shawnees in 1811 at the Battle of Tippecanoe. He also had been a hero during the War of 1812. The Whigs made the most of Harrison's military record and his nickname, "Old Tippecanoe." The phrase "Tippecanoe and Tyler too" became the Whig election slogan.

The Election of 1840

During the 1840 election campaign, the Whigs emphasized personalities more than issues. They tried to appeal to the common people, as Andrew Jackson had done. Harrison was the son of a Virginia plantation owner. However, because he had settled on a farm in Ohio, the Whigs said Harrison was a true Westerner. They used symbols of the frontier, such as a log cabin, to represent Harrison. The Whigs contrasted Harrison with the wealthy Van Buren. Harrison won in a close election.

At his inauguration, the 68-year-old president spoke for nearly two hours in cold March weather with no hat or coat. Later, he was caught in the rain. He came down with a cold that developed into pneumonia. On April 4, 1841, one month after being inaugurated, Harrison died—the first president to die in office. Vice-President Tyler became president.

The election of 1840 showed the importance of the West in American politics. In the next chapter, you'll learn more about the lure of the West and the westward expansion of the United States.

Section 4 Assessment

1. Terms & Names

Identify:

- inflation
- Martin Van Buren
- Panic of 1837
- depression
- Whig Party
- William Henry Harrison
- John Tyler

2. Taking Notes

Use a diagram to list the events that led to the closing of the Second Bank of the United States.

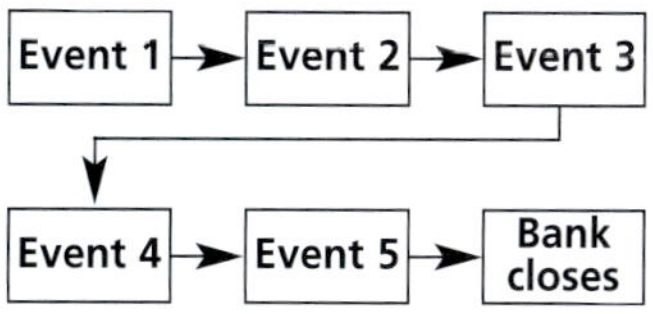

What was the most significant event?

3. Main Ideas

a. Why did Jackson declare war on the Second Bank of the United States?

b. How did Jackson kill the bank?

c. What role did Jackson's popularity play in the elections of 1836 and 1840?

4. Critical Thinking

Comparing What strategy did the Whig Party use in the 1840 election?

THINK ABOUT

- how Harrison was portrayed
- what group of voters it was trying to attract

ACTIVITY OPTIONS

LANGUAGE ARTS

ART

Imagine yourself as a presidential candidate in 1840. Focusing on the economy as an issue, write a campaign **slogan** or create a **banner** to rally support.

Chapter 12 ASSESSMENT

TERMS & NAMES

Briefly explain the importance of each of the following.

1. John Quincy Adams
2. Jacksonian democracy
3. spoils system
4. Sequoya
5. Indian Removal Act
6. Trail of Tears
7. secession
8. inflation
9. depression
10. Whig Party

REVIEW QUESTIONS

Politics of the People (pages 369–373)

1. How was Jackson different from earlier presidents?
2. How did Jackson appeal to voters in his election campaign of 1828?

Jackson's Policy Toward Native Americans (pages 374–378)

3. Why did white settlers want Native American land?
4. What was Jackson's position on Native Americans in the United States?
5. How did the Indian Removal Act affect Native Americans?

Conflicts over States' Rights (pages 379–383)

6. How did the issue of tariffs divide the country?
7. Why did nullification threaten the nation?
8. How was the nullification crisis resolved?

Prosperity and Panic (pages 384–387)

9. Why did Jackson oppose the Second Bank of the United States?
10. What were the effects of Jackson's war on the bank?

CRITICAL THINKING

1. USING YOUR NOTES

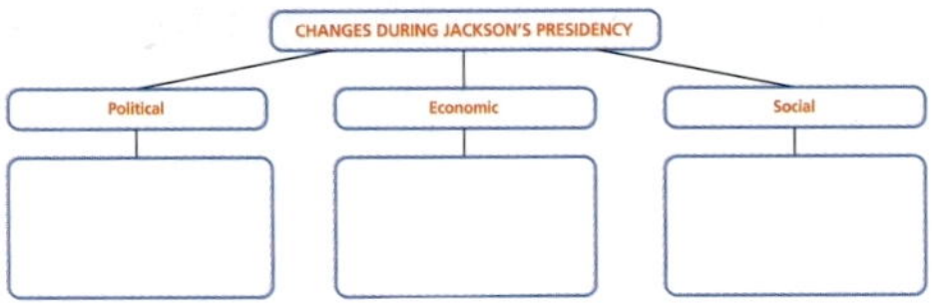

Use your completed chart to answer the questions.

a. What do you think was the most positive change of the Jackson era? Explain.
b. What was the most negative change? Explain.
c. Which change had the most far-reaching effects? Give evidence to support your answer.

2. ANALYZING LEADERSHIP

What was the basis of Andrew Jackson's power as president?

3. APPLYING CITIZENSHIP SKILLS

How did the majority of voters in the presidential elections of 1828 and 1840 exercise their vote in a similar way?

4. THEME: ECONOMICS IN HISTORY

Based on its economic effects, was Jackson's decision to end the national bank a good one? Explain.

5. RECOGNIZING BIAS

In what ways did Andrew Jackson's policy toward Native Americans reflect bias?

Interact *with* History

Now that you have read the chapter, do you think the qualities that made Jackson a strong military leader made him a good president? Explain your answer.

VISUAL SUMMARY

Major Issues of Jackson's Presidency

POLICY TOWARD NATIVE AMERICANS	CONFLICT OVER STATES' RIGHTS	WAR ON BANK OF THE UNITED STATES
White settlers wanted Native American lands.	Sectional differences developed.	Second Bank of the United States had economic and political power.
Jackson proposed Indian Removal Act of 1830.	Jackson supported strong central government.	Jackson opposed bank and vetoed renewal of its charter.
Thousands of Native Americans removed to Indian Territory.	South Carolina threatened to secede over tariff issue, but compromise reached.	Bank driven out of business, but Jackson's policies eventually led to inflation and depression.

HISTORY SKILLS

1. INTERPRETING GRAPHS

Study the graph. Answer the questions.

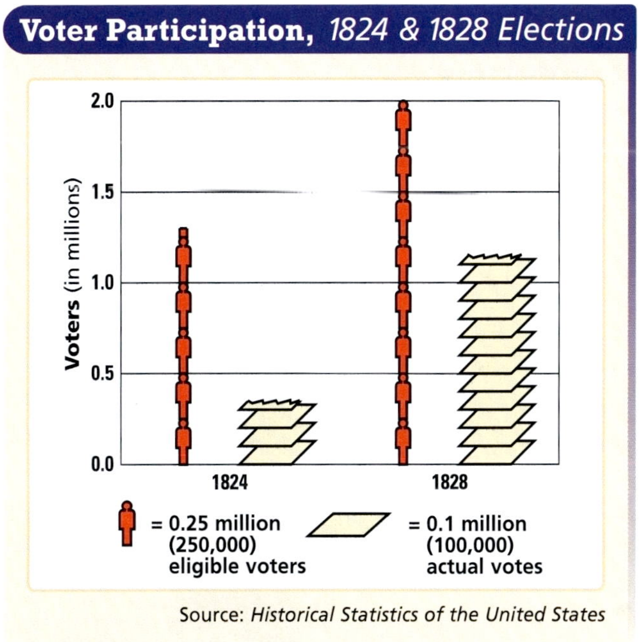

Basic Graph Elements

a. What is the subject of the graph?

b. What do the symbols represent?

Interpreting the Graph

c. How many more eligible voters were there in 1828 than in 1824?

d. What percentage of eligible voters cast their ballots in 1824? in 1828?

2. INTERPRETING PRIMARY SOURCES

This is a political cartoon from the 1832 election campaign. It is entitled "King Andrew the First."

a. Is the comparison of Jackson to a king meant as praise or criticism? Explain.

b. What does it mean that he is standing on torn papers entitled "U.S. Constitution" and "U.S. Bank"?

ALTERNATIVE ASSESSMENT

1. INTERDISCIPLINARY ACTIVITY: Speech

Presenting a Debate Working with a partner, take sides on the doctrine of nullification and prepare a debate for the class.

2. COOPERATIVE LEARNING ACTIVITY

Preparing a Problem-Solving Proposal Plan and write a proposal outlining a solution to the problems between white settlers and Native Americans in the Southeast in the early 1800s. Working in a small group, first brainstorm ideas and make a list. Then identify the positives and negatives of each idea. Agree on an outline for a final plan. Then write, revise, and make a final copy of your proposal. Present your proposal to the class and defend it.

3. TECHNOLOGY ACTIVITY

Designing a Political Campaign Web Site Andrew Jackson was one of the more colorful figures in American history. His election as president brought significant changes to the nation. Plan a Web site for candidate Jackson for the 1828 presidential campaign. Use the Internet or search the library for information on Jackson's life before the presidency and on the 1828 election.

INTERNET **Visit www.mcdougallittell.com to learn more about Andrew Jackson.**

Design the Web site using the suggestions below.

- Include biographical facts.
- Select images that tell Jackson's story.
- Present his views on the major issues of the day using quotations from speeches and other documents.
- Locate appropriate links for visitors to your Web site.

4. HISTORY PORTFOLIO

PORTFOLIO **Option 1** Review your section and chapter assessment activities. Select one that you think was your best work. Then use comments made by your teacher or classmates to improve your work and add it to your portfolio.

PORTFOLIO **Option 2** Review the questions that you wrote for What Do You Want to Know? on page 368. Then write a short report in which you explain the answers to your questions. Add your work to your history portfolio.

CHAPTER 13

Manifest Destiny 1810–1853

Weary from their trip west, this pioneer family stops for a rest.

Interact *with* History

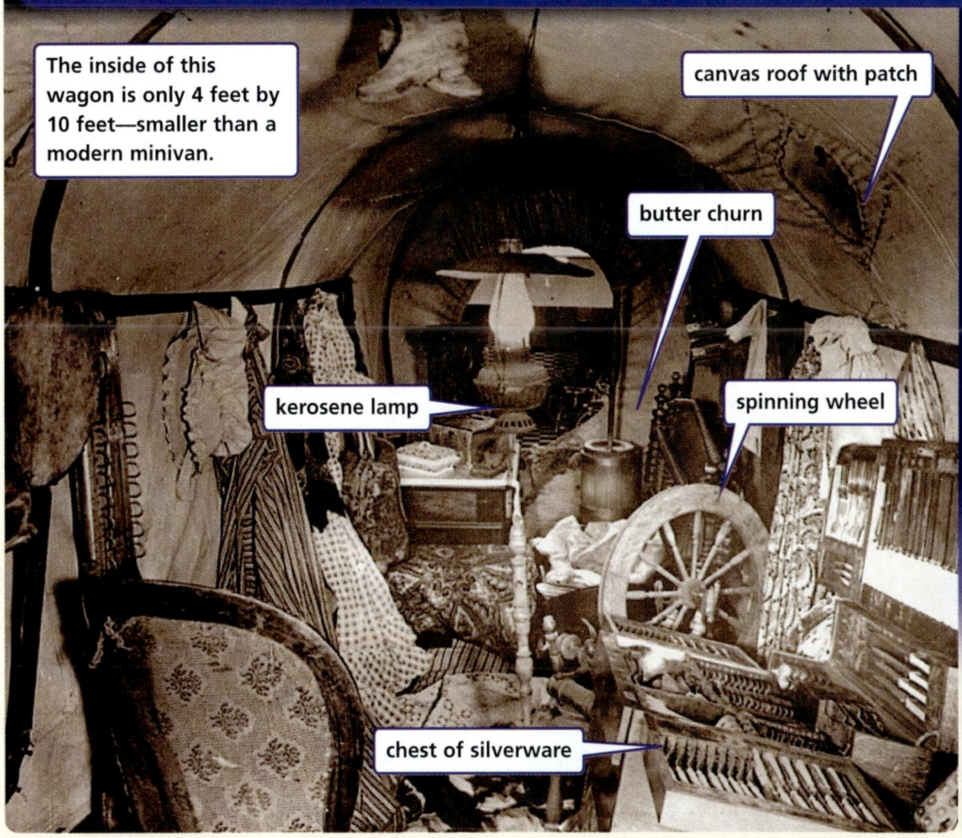

The year is 1844, and you live on a small rocky farm in Massachusetts. Your family has decided to move to Oregon to gain cheap, fertile land. Your father says this move will make your family better off—and give you a brighter future.

What Do You Think?

- What do you think daily life on the trail west might be like?
- What might be the greatest obstacles that you face?
- Notice the necessities packed in this crowded wagon. What might have been left behind?

What might you gain and lose by going west?

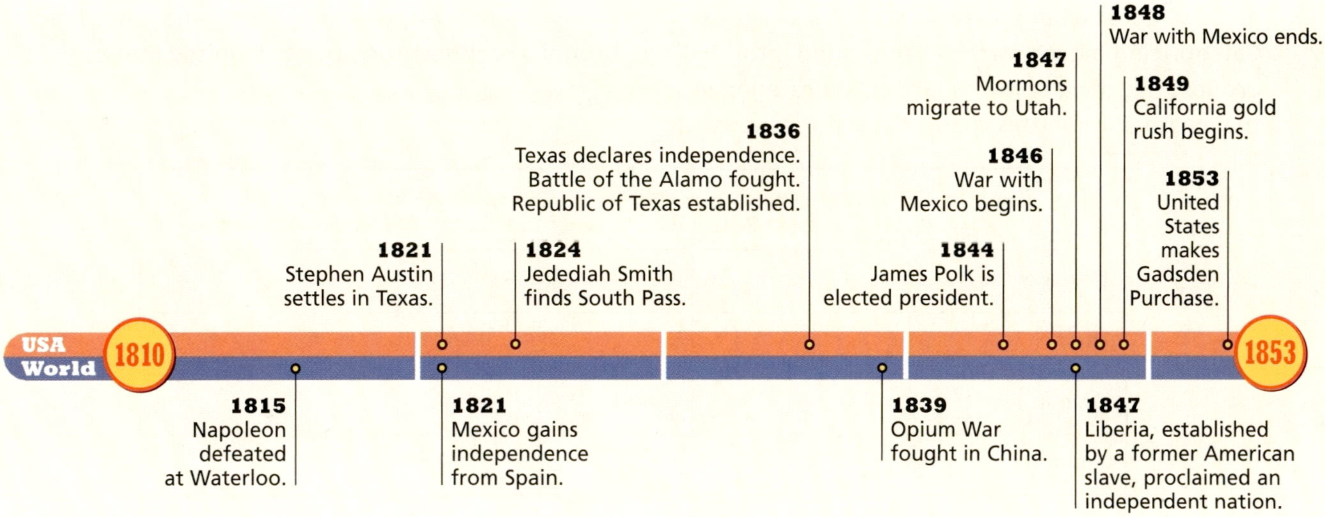

Chapter 13 SETTING THE STAGE

BEFORE YOU READ

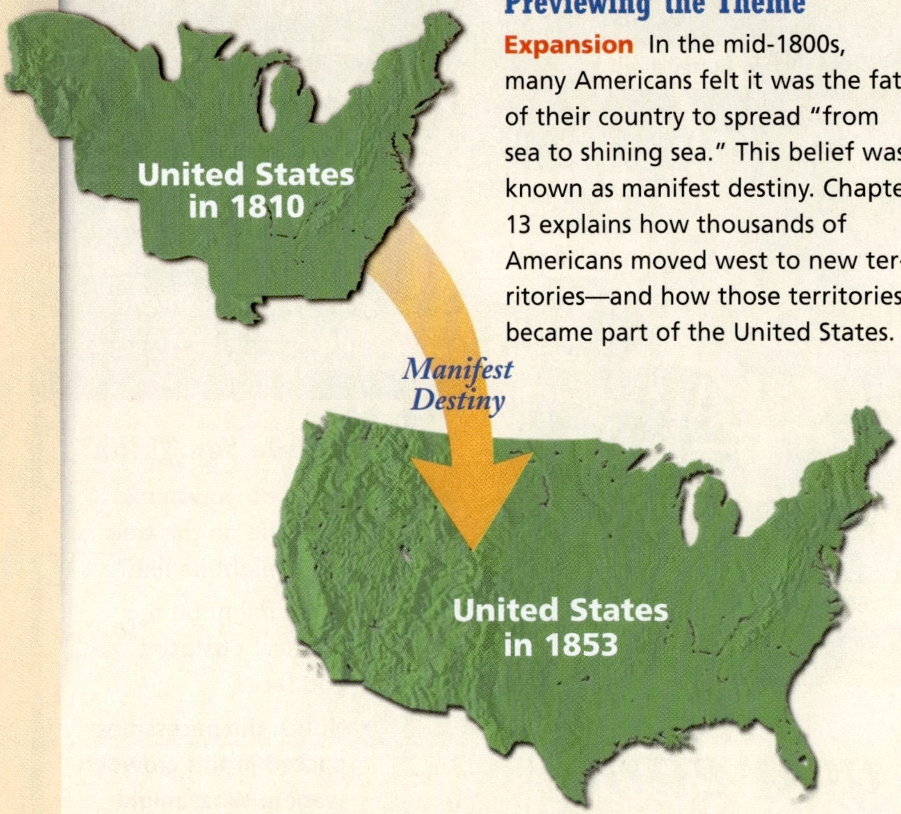

Previewing the Theme

Expansion In the mid-1800s, many Americans felt it was the fate of their country to spread "from sea to shining sea." This belief was known as manifest destiny. Chapter 13 explains how thousands of Americans moved west to new territories—and how those territories became part of the United States.

What Do You Know?

What do you think of when you hear the phrase "the West"? Who do you think moved west in the early 1800s? What do you think drew them to the West?

THINK ABOUT

- what you've learned about the West from movies or travel
- reasons that people move to new places today

What Do You Want to Know?

What questions do you have about the westward movement of the 1800s? Record those questions in your notebook before you read the chapter.

READ AND TAKE NOTES

Reading Strategy: Categorizing To help you make sense of what you read, learn to categorize. Categorizing means sorting information into groups. The chart below will help you categorize the information in this chapter about the westward movement. Use the chart to take notes on what groups went west, why they went, and what events brought each territory into the United States.

See Skillbuilder Handbook, page R6.

	Types of people who traveled there	Why they went there	Key events that brought the territory into the United States
New Mexico			
Utah			
Oregon			
Texas			
California			

1 Trails West

TERMS & NAMES
Jedediah Smith
mountain man
Jim Beckwourth
land speculator
Santa Fe Trail
Oregon Trail
Mormon
Brigham Young

MAIN IDEA

Thousands of settlers followed trails through the West to gain land and a chance to make a fortune.

WHY IT MATTERS NOW

This migration brought Americans to the territories that became New Mexico, Oregon, and Utah.

ONE AMERICAN'S STORY

The mountain man Jedediah Smith was leading an expedition to find a route through the Rocky Mountains when a grizzly bear attacked. The bear seized Smith's head in its mouth, shredded his face, and partially tore off one ear. Smith's men chased the bear away. Jim Clyman recalled the scene.

A VOICE FROM THE PAST

I asked [Smith] what was best. He said, "One or two go for water and if you have a needle and thread get it out and sew up my wounds around my head." . . . I told him I could do nothing for his ear. "Oh, you must try to stitch it up some way or other," said he. Then I put in my needle and stitched it through and through.

Jim Clyman, quoted in *The West,* by Geoffrey C. Ward

This likeness of Jedediah Smith shows the ruggedness and spirit of the mountain man.

Ten days after this attack, Smith was ready to continue exploring. The following spring, he found what he was looking for—a pass through the Rocky Mountains.

Jedediah Smith was one of the daring fur trappers and explorers known as mountain men. The mountain men opened up the West by discovering the best trails through the Rockies. Later, thousands of pioneers followed these trails. In this section, you will learn about the trails—and why people followed them west.

Mountain Men and the Rendezvous

Mountain men survived by being tough and resourceful. They spent most of the year alone, trapping small animals such as beavers. Easterners wanted beaver furs to make the men's hats that were in fashion at the time. To obtain furs, mountain men roamed the Great Plains and the Far West, the regions between the Mississippi River and the Pacific Ocean, and set traps in icy mountain streams.

Because of their adventures, mountain men such as Jedediah Smith and **Jim Beckwourth** became famous as rugged loners. However, they were not as independent as the legends have portrayed them. Instead, they were connected economically to the businessmen who bought their furs.

AMERICA'S HISTORY MAKERS

JIM BECKWOURTH

1798–1867

Jim Beckwourth was born in slavery and set free by his owner. At the age of 25, Beckwourth joined a group of fur traders going west and in time became a daring mountain man.

For several years, Beckwourth lived with a Crow tribe. Later, he worked as an army scout and gold prospector. In 1850, he discovered a mountain pass that became the route into present-day northern California. This pass is still called Beckwourth Pass.

What was Beckwourth's most important contribution to the westward movement?

One businessman, William Henry Ashley, created a trading arrangement called the rendezvous system. Under this system, individual trappers came to a prearranged site for a rendezvous with traders from the east. The trappers bought supplies from those traders and paid them in furs. The rendezvous took place every summer from 1825 to 1840. In that year, silk hats replaced beaver hats as the fashion, and the fur trade died out.

Vocabulary
rendezvous (RAHN•day•voo): meeting; from a French word meaning "present yourselves"

Mountain Men Open the West

During the height of the fur trade, mountain men worked some streams so heavily that they killed off the animals. This forced the trappers to search for new streams where beaver lived. The mountain men's explorations provided Americans with some of the earliest firsthand knowledge of the Far West. This knowledge, and the trails the mountain men blazed, made it possible for later pioneers to move west.

For example, thousands of pioneers used South Pass, the wide valley through the Rockies that Jedediah Smith had publicized. Smith learned of this pass, in present-day Wyoming, from Native Americans. Unlike the high northern passes used by Lewis and Clark, South Pass was low, so snow did not block it as often as it blocked higher passes. Also, because South Pass was wide and less steep, wagon trails could run through it.

Reading History
A. Reading a Map Find South Pass on the map on page 395. Notice which two trails used that pass.

Smith wrote to his brother that he wanted to help people in need: "It is for this that I go for days without eating, and am pretty well satisfied if I can gather a few roots, a few snails, . . . a piece of horseflesh, or a fine roasted dog."

The Lure of the West

Few of the people who went west shared Smith's noble motive. To many, the West with its vast stretches of land offered a golden chance to make money. The Louisiana Purchase had doubled the size of the United States, and some Americans wanted to take the land away from Native Americans who inhabited this territory.

People called **land speculators** bought huge areas of land. To speculate means to buy something in the hope that it will increase in value. If land value did go up, speculators divided their land holdings into smaller sections. They made great profits by selling those sections to the thousands of settlers who dreamed of owning their own farms.

Manufacturers and merchants soon followed the settlers west. They hoped to earn money by making and selling items that farmers needed. Other people made the trip to find jobs or to escape people to whom they owed money.

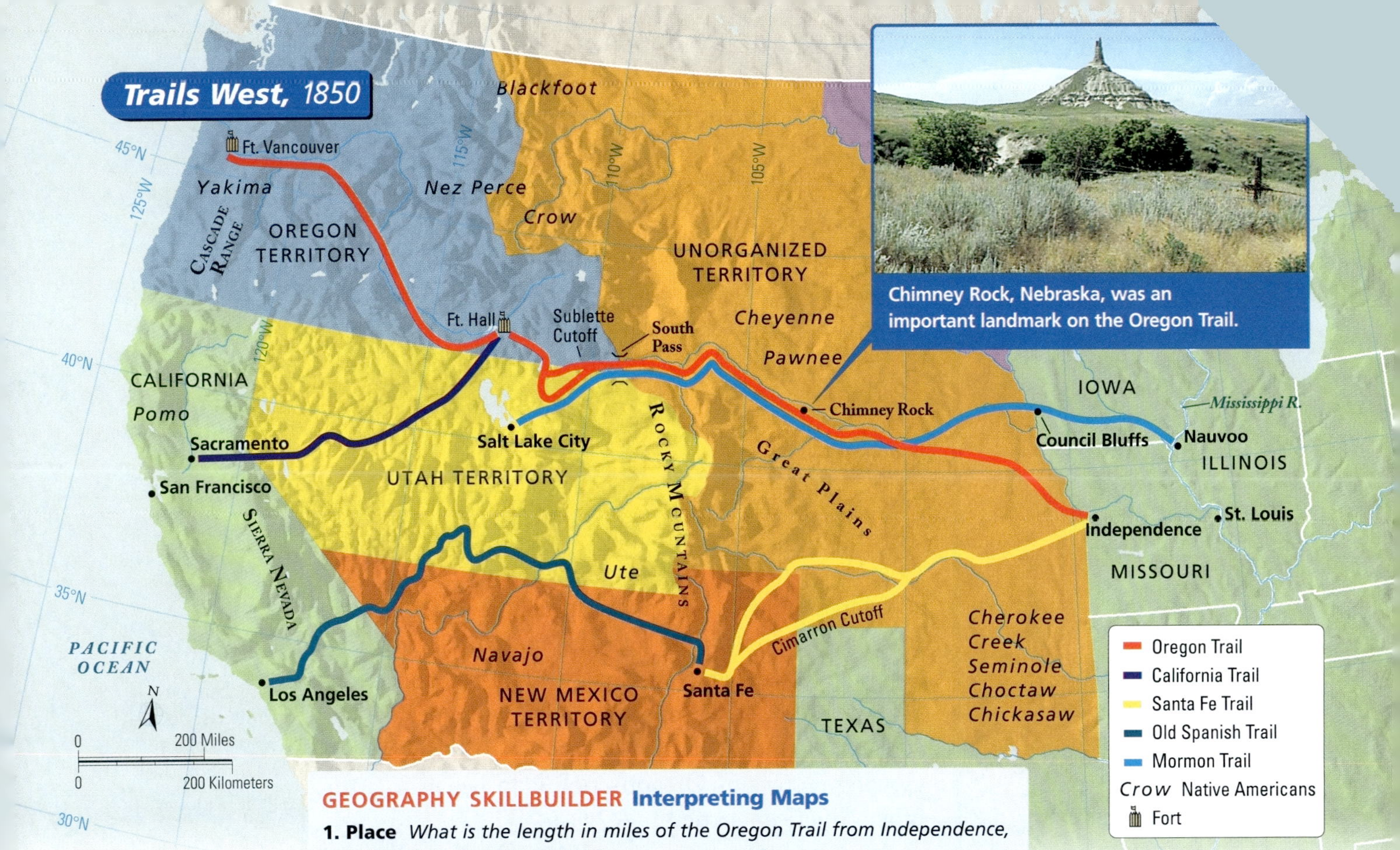

Chimney Rock, Nebraska, was an important landmark on the Oregon Trail.

GEOGRAPHY SKILLBUILDER Interpreting Maps

1. **Place** *What is the length in miles of the Oregon Trail from Independence, Missouri, to Fort Vancouver?*
2. **Human-Environment Interaction** *Why do you think some of the trails ran along rivers?*

The Trail to Santa Fe

Traders also traveled west in search of markets. After Mexico gained independence from Spain in 1821, it opened its borders to American traders, whom Spain had kept out. In response, the Missouri trader William Becknell set out with hardware, cloth, and china for Santa Fe, capital of the Mexican province of New Mexico. By doing so, he opened the **Santa Fe Trail,** which led from Missouri to Santa Fe. Once in Santa Fe, he made a large profit because the New Mexicans were eager for new merchandise.

When Becknell returned to Missouri weeks later, a curious crowd met him. One man picked up one of Becknell's bags and slit it open with a knife. As gold and silver coins spilled onto the street, the onlookers gasped. The news spread that New Mexico was a place where traders could become rich.

Reading History

B. Making Inferences What do you think other Missourians might decide to do after seeing Becknell's wealth?

The following spring, Becknell headed to Santa Fe again. This time he loaded his trade goods into covered wagons, which Westerners called prairie schooners. Their billowing white canvas tops made them look like schooners, or sailing ships.

Becknell could not haul wagons over the mountain pass he had used on his first trip to Santa Fe. Instead, he found a cutoff, a shortcut that avoided steep slopes but passed through a deadly desert to the south. As his traders crossed the burning sands, they ran out of water. Crazed by

daily life

DINNER ON THE TRAIL

To add to their limited supplies, pioneers on the trail gathered berries and wild onions. They also hunted buffalo and small game. Below is a recipe that many might have used.

Fricasseed Squirrel
1 squirrel, skinned
3 slices of bacon, chopped
1 tablespoon chopped onions
2 teaspoons lemon juice
⅓ cup water
salt, pepper, & flour
Cut squirrel in pieces. Rub pieces with salt, pepper, and flour. Fry with bacon for 30 minutes. Add onion, lemon juice, and water. Cover tightly. Cook for 1½ hours.

Nebraska Centennial First Ladies Cookbook

thirst, they lopped off mules' ears and killed their dogs to drink the animals' blood. Finally, the men found a stream. The water saved them from death, and they reached Santa Fe.

Becknell returned home with another huge profit. Before long, hundreds of traders and prairie schooners braved the cutoff to make the 800-mile journey from Missouri to New Mexico each year.

Oregon Fever

Hundreds of settlers also began migrating west on the **Oregon Trail,** which ran from Independence, Missouri, to the Oregon Territory. The first whites to cross the continent to Oregon were missionaries, such as Marcus and Narcissa Whitman in 1836. At that time, the United States and Britain were locked in an argument about which country owned Oregon. To the Whitmans' great disappointment, they made few converts among the Native Americans. However, their glowing reports of Oregon's rich land began to attract other American settlers.

Vocabulary
converts: people who accept a new religious belief

Amazing stories spread about Oregon. The sun always shone there. Wheat grew as tall as a man. One tale claimed that pigs were "running about, . . . round and fat, and already cooked, with knives and forks sticking in them so you can cut off a slice whenever you are hungry."

Such stories tempted many people to make the 2,000-mile journey to Oregon. In 1843, nearly 1,000 people traveled from Missouri to Oregon. The next year, twice as many came. "The Oregon Fever has broken out," observed a Boston newspaper, "and is now raging."

One Family Heads West

The experiences of the Sager family show how difficult the trail could be. In 1844, Henry Sager, his wife, and six children left Missouri to find cheap, fertile land in Oregon. They had already moved four times in the past four years. Henry's daughter Catherine explained her family's moves.

A VOICE FROM THE PAST

Father was one of those restless men who are not content to remain in one place long at a time. . . . [He] had been talking of going to Texas. But mother, hearing much said about the healthfulness of Oregon, preferred to go there.

Catherine Sager, quoted in *The West,* by Geoffrey C. Ward

The Oregon Trail was dangerous, so pioneers joined wagon trains. They knew their survival would depend on cooperation. Before setting out, the wagon train members agreed on rules and elected leaders to enforce them.

Even so, life on the trail was full of hardship. The Sagers had barely begun the trip when Mrs. Sager gave birth to her seventh child. Two months later, nine-year-old Catherine fell under a moving wagon, which crushed her left leg. Later, "camp fever" killed both of the Sager parents.

Reading **History**
C. Finding Main Ideas What difficulties did families like the Sagers face?

Even though the Sager parents had died, the other families in the train cooperated to help the Sager orphans make it to Oregon. There, the Whitmans agreed to adopt them. When Narcissa met them, Catherine recalled, "We thought as we shyly looked at her that she was the prettiest woman we had ever seen."

The Mormon Trail

While most pioneers went west in search of wealth, one group migrated for religious reasons. The **Mormons,** who settled Utah, were members of the Church of Jesus Christ of Latter-Day Saints. Joseph Smith had founded this church in upstate New York in 1830. The Mormons lived in close communities, worked hard, shared their goods, and prospered.

The Mormons, though, also made enemies. Some people reacted angrily to the Mormons' teachings. They saw the Mormon practice of polygamy—allowing a man to have more than one wife at a time—as immoral. Others objected to their holding property in common.

Reading **History**
D. Analyzing Causes Why did Brigham Young lead the Mormons to Utah?

In 1844, an anti-Mormon mob in Illinois killed Smith. **Brigham Young,** the next Mormon leader, moved his people out of the United States. His destination was Utah, then part of Mexico. In this desolate region, he hoped his people would be left to follow their faith in peace.

In 1847, about 1,600 Mormons followed part of the Oregon Trail to Utah. There they built a new settlement by the Great Salt Lake. Because Utah has little rainfall, the Mormons had to work together to build dams and canals. These structures captured water in the hills and carried it to the farms in the valleys below. Through teamwork, they made their desert homeland bloom.

In the meantime, changes were taking place in Texas. As you will read in Section 2, Americans had been moving into that Mexican territory, too.

Section 1 Assessment

1. Terms & Names

Identify:
- Jedediah Smith
- mountain man
- Jim Beckwourth
- land speculator
- Santa Fe Trail
- Oregon Trail
- Mormon
- Brigham Young

2. Taking Notes

Use a cluster diagram like the one shown to review details about the trails west.

Which trail would you have wanted to travel? Why?

3. Main Ideas

a. How did the mountain men open up the West for later settlement?

b. What are two examples of pioneer groups who used cooperation to overcome hardship?

c. What were the different motives that drew people west?

4. Critical Thinking

Drawing Conclusions Of all the hardships faced by people who went west, what do you think was the worst? Explain.

THINK ABOUT
- the mountain men
- William Becknell
- the Sagers
- the Mormons

ACTIVITY OPTIONS

LANGUAGE ARTS
ART

Research a pioneer from this section and either write a **letter** from his or her point of view to a friend or illustrate a **journal entry** with sights from your journey.

Survive the Oregon Trail!

You are part of a wagon train heading west on the Oregon Trail. During your journey, you will cross endless flat prairies and mountains that climb more steeply than a staircase. You will suffer through blazing heat and icy snowstorms. Food is scarce in the land you travel through—and human settlements are even more scarce.

COOPERATIVE LEARNING On this page are three challenges you will face on your journey. Working with a small group, create a solution for each challenge. To help your group work together, assign a task to each group member. You will find helpful information in the Data File. Be prepared to present your solutions to the class.

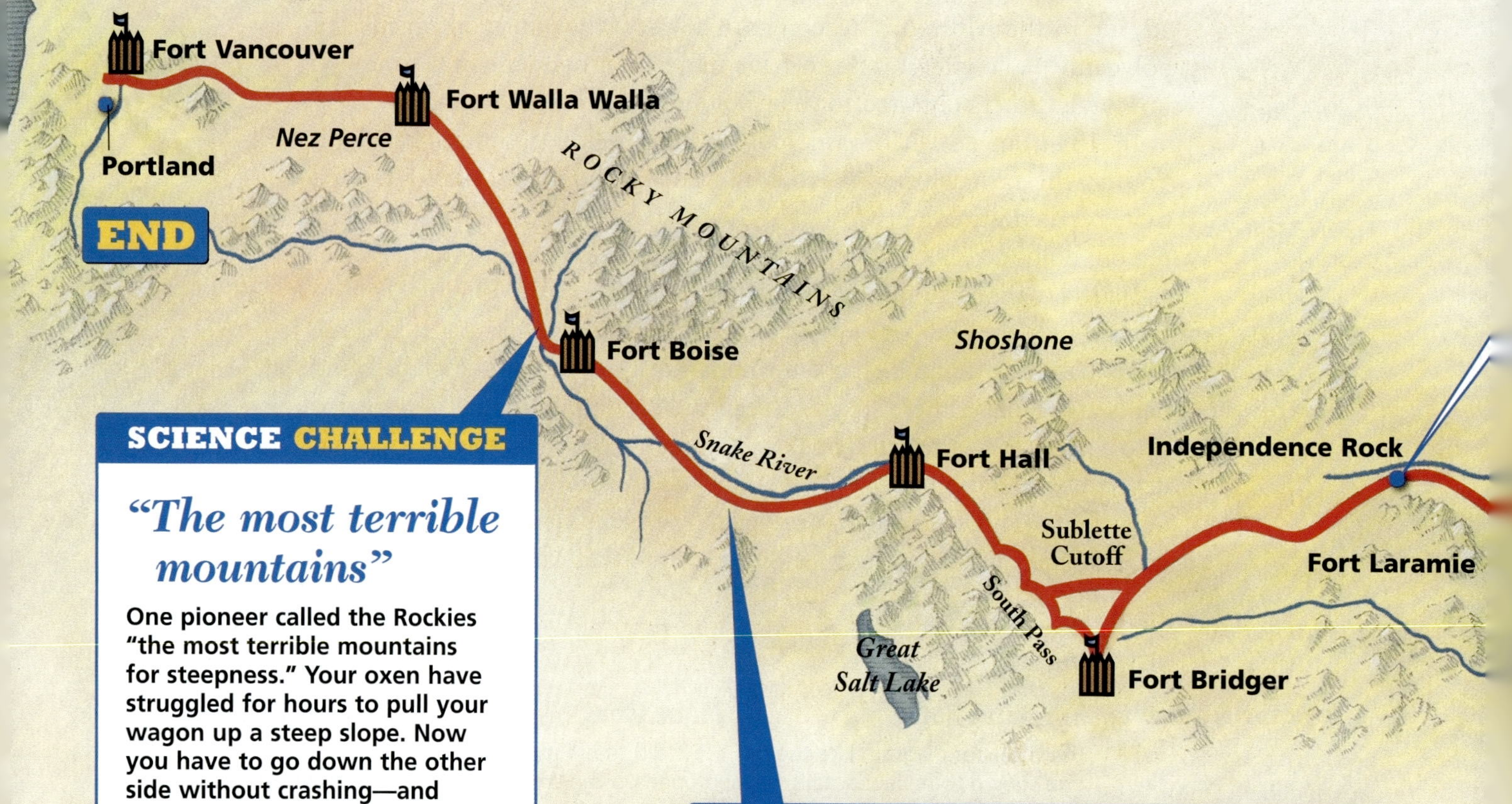

SCIENCE CHALLENGE

"The most terrible mountains"

One pioneer called the Rockies "the most terrible mountains for steepness." Your oxen have struggled for hours to pull your wagon up a steep slope. Now you have to go down the other side without crashing—and wagon wheels have no brakes. How will you slow your descent? Use the Data File for help. Present your ideas using one of these options:

- **Write instructions for climbing and descending the mountain.**
- **Illustrate your solution in a how-to diagram.**

CIVICS CHALLENGE

"A thieving scoundrel"

You wake one morning to the sound of shouting. One of the men in the wagon train has been caught stealing another family's ox—to replace an ox that died from drinking bad water. The wagon train leader asks you to help decide how to punish the thief. Present your decision using one of these options:

- **As a group, role-play a discussion of what the punishment should be.**
- **Write an explanation of a punishment to the wagon train leader.**

HEALTH CHALLENGE

"A grand blow-out"

On July 4, your wagon train stops near Independence Rock for "a grand blow-out" to celebrate your progress. Each family will bring a dish to the party. What will you cook that is tasty and nutritious? You might use supplies from your wagon and berries or animals from the area. Look at the Data File for help. Present your choice using one of these options:

- Draw a picture of your dish and describe it.
- Write an original recipe.

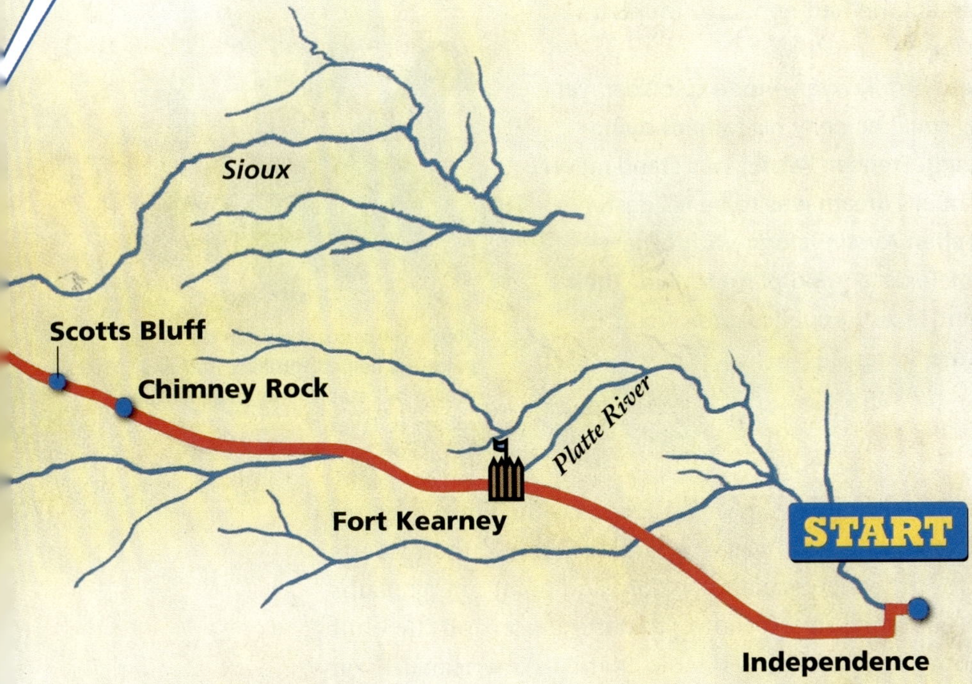

ACTIVITY WRAP-UP

Present to the class As a group, review your responses to the challenges of the Oregon Trail. Choose the most creative solution for each challenge and present these solutions to the class.

DATA FILE

The Journey

Distance: 2,000 miles

Length: 4–6 months

Average daily distance: 12–15 miles

Best time to travel: April–September

The Wagon

Size of box: 4 x 10 feet

Load: 1,600–2,500 pounds

Oxen needed per wagon: 4–8

Animals Taken

oxen, horses, dairy cows, cattle, chickens, mules, pigs, dogs, cats

Food Supplies

flour, corn meal, salt, baking soda, sugar, crackers, dried beans, rice, dried fruit, bacon, coffee, vinegar

Cooking Equipment

Dutch oven, large kettle, frying pan, bread pan, coffee grinder, rolling pin, tin cups and plates, water kegs, knives, spoons

Other Equipment

bedding, spare wagon parts, tar, rope, chains, pulleys, tools, fishing poles, guns, ammunition, matches, soap, medicines

Trail Hazards

Worst discomforts: heat, cold, wind, rain, dust, mud, mosquitoes, hunger, thirst

Biggest killer: disease

Most common accidents: shooting, drowning, crushing by wagon wheels, injuries from animals

Visit www.mcdougallittell.com for more information on the Oregon Trail.

2 The Texas Revolution

TERMS & NAMES
Stephen Austin
Tejano
Antonio López de Santa Anna
Sam Houston
William Travis
Juan Seguín
Battle of the Alamo
Lone Star Republic

MAIN IDEA

American and *Tejano* citizens led Texas to independence from Mexico.

WHY IT MATTERS NOW

The diverse culture of Texas has developed from the contributions of many different groups.

ONE AMERICAN'S STORY

Son of a bankrupt Missouri mine owner, Stephen Austin read his mother's letter, written in 1821, in stunned silence. His father, Moses Austin, was dead. In his last moments, she told her son, "He called me to his bedside, . . . he begged me to tell you to take his place . . . to go on . . . in the same way he would have done."

Stephen knew what that meant. Moses Austin had spent the last years of his life chasing a crazy dream. He had hoped to found a colony for Americans in Spanish Texas.

Stephen's dream, though, was to be a lawyer—not a colonizer. Yet as a loving and obedient son, how could he deny his father's dying wish? A week after his father's death, Stephen Austin was standing on Texas soil. From that day on, his father's dream was to be his destiny.

This section explains how Stephen Austin, along with others, worked hard to make the lands of Texas a good place to live. Their spirit would create an independent Texas Republic. Later, Texas would become a state in the United States.

Stephen Austin, shown in this painting, helped fulfill his father's dream by establishing an American colony in Texas.

Spanish Texas

The Spanish land called *Tejas* (Tay•HAHS) bordered the United States territory called Louisiana. The land was rich and desirable. It had forests in the east, rich soil for growing corn and cotton, and great grassy plains for grazing animals. It also had rivers leading to natural ports on the Gulf of Mexico. It was home to Plains and Pueblo Native Americans. Even though *Tejas* was a state in the Spanish colony of New Spain, it had few Spanish settlers. Around 1819, Spanish soldiers drove off Americans trying to claim those lands as a part of the Louisiana Purchase.

In 1821, only about 4,000 *Tejanos* (Tay•HAH•nohs) lived in Texas. ***Tejanos*** are people of Spanish heritage who consider Texas their home. The Comanche, Apache, and other tribes fought fiercely against Spanish settlement of Texas. The Spanish officials wanted many more settlers to move to Texas. They hoped that new colonists would help to defend against Native Americans and Americans who illegally sneaked into Texas.

Vocabulary
empresarios: individuals who agreed to recruit settlers for the land

To attract more people to Texas, the Spanish government offered huge tracts of land to *empresarios.* But they were unable to attract Spanish settlers. So, when Moses Austin asked for permission to start a colony in Texas, Spain agreed. Austin was promised a large section of land. He had to agree that settlers on his land had to follow Spanish laws.

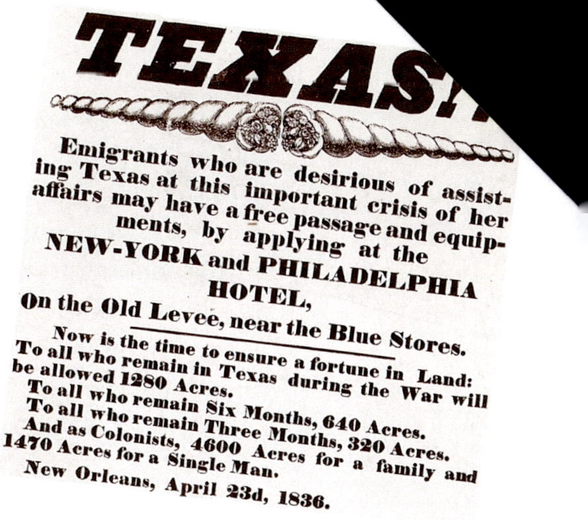

Mexican Independence Changes Texas

Shortly after Stephen Austin arrived in Texas in 1821, Mexico successfully gained its independence from Spain. *Tejas* was now a part of Mexico. With the change in government, the Spanish land grant given to Austin's father was worthless. Stephen Austin traveled to Mexico City to persuade the new Mexican government to let him start his colony. It took him almost a year to get permission. And the Mexican government would consent only if the new settlers agreed to become Mexican citizens and members of the Roman Catholic Church.

Posters such as the one above encouraged Americans from the East to settle in Texas. Some people scrawled G.T.T. on their doors to indicate they had "gone to Texas."

Between 1821 and 1827, Austin attracted 297 families to his new settlement. These original Texas settler families are known as the "Old Three Hundred." He demanded evidence that each family head was moral, worked hard, and did not drink. So law-abiding were his colonists that Austin could write to a new settler, "You will be astonished to see all our houses with no other fastening than a wooden pin or door latch."

The success of Austin's colony attracted more land speculators and settlers to Texas from the United States. Some were looking for a new life, some were escaping from the law, and others were looking for a chance to grow rich. By 1830, the population had swelled to about 30,000, with Americans outnumbering the *Tejanos* six to one.

Rising Tensions in Texas

Reading **History**
A. Analyzing Causes Why was there growing tension between Americans and *Tejanos?*

As more and more Americans settled in Texas, tensions between them and the *Tejanos* increased. Used to governing themselves, Americans resented following Mexican laws. Since few Americans spoke Spanish, they were unhappy that all official documents had to be in that language. Slave owners were especially upset when Mexico outlawed slavery in 1829. They wanted to maintain slavery so they could grow cotton. Austin persuaded the government to allow slave owners to keep their slaves.

On the other hand, the *Tejanos* found the Americans difficult to live with, too. *Tejanos* thought that the Americans believed they were superior and deserved special privileges. The Americans seemed unwilling to adapt to Mexican laws.

The Mexican government sent an official to Texas to investigate the tensions. He was not happy with what he found. In 1829, he reported to his government, "I am warning you to take timely measures . . . Texas could throw this whole nation into revolution." His advice turned out to be right.

Responding to the warnings, the Mexican government cracked down on Texas. First, it closed the state to further American immigration. Next, it required Texans to pay taxes for the first time. Finally, to enforce these new laws, the government sent more Mexican troops to Texas.

Reading **History**

B. Summarizing What three actions did the Mexican government take to control Texas?

Texans Revolt Against Mexico

These actions caused angry protests. Some Texans even talked of breaking away from Mexico. Most, however, listened to Austin, who remained loyal to Mexico. In 1833, Austin set off for Mexico City with a petition. This document listed reforms supported by both Americans and *Tejanos*. The most important request was that Texas become a self-governing state within Mexico.

In Mexico City, Austin met General **Antonio López de Santa Anna,** the Mexican president. At first, the general agreed to most of the reforms in Austin's petition. But then Santa Anna learned of a letter Austin had written. The letter said that if the changes weren't approved Austin would support breaking away from Mexican rule. This was rebellion! The general had Austin jailed for an entire year. The Texans were furious and ready to rebel.

Santa Anna's answer to talk of rebellion was to send more troops to Texas. In late September 1835, Mexican soldiers marched to the town of Gonzales. They had orders to seize a cannon used by the Texans for protection against Native Americans. Texas volunteers had hung a flag on the big gun that said, "Come and Take It."

The Mexican troops failed to capture the cannon. Two months later, Texans drove Mexican troops out of an old mission in San Antonio that was used as a fortress. It was called the Alamo. Among the Texas volunteers were free African Americans such as Hendrick Arnold and Greenbury Logan. Angered by these insults, Santa Anna and 6,000 troops headed for Texas.

Reading **History**

C. Analyzing Causes What Texan actions moved Santa Anna to head toward Texas?

The Fight for the Alamo

On March 1–2, 1836, Texans met at a settlement called Washington-on-the-Brazos to decide what to do about Santa Anna's troops. They believed they could do only one thing: to declare Texas a free and independent republic. **Sam Houston,** the only man at the meeting with military experience, was placed in command of the Texas army.

America's HERITAGE

REMEMBER THE ALAMO!

Today the Alamo, shown below, is again under siege. Moisture seeps into the limestone walls and causes them to crumble. Many people view the mission as a memorial to Americans' willingness to fight for freedom, so a Texas group has begun attempts to preserve the Alamo from further damage. The Alamo looks quite different from the battle site of 1836. The famous bell-shaped front was added in the 1850s.

The Texas army hardly existed. At that moment, there were two small forces ready to stand up to Santa Anna's army. One was a company of 420 men, led by James Fannin, stationed at Goliad, a fort in southeast Texas. The second was a company of 183 volunteers at the Alamo. Headed by **William Travis,** this small force included such famous frontiersmen as Davy Crockett and Jim Bowie. In addition, **Juan Seguín** (wahn seh•GEEN) led a band of 25 *Tejanos* in support of revolt.

"Remember the Alamo!"
a Texan soldier

On February 23, 1836, Santa Anna's troops surrounded San Antonio. The next day, Mexicans began their siege of the Alamo. Two nights later, Travis scrawled a message to the world.

Reading **History**

D. Making Inferences Why would William Travis address his message to all Americans?

A VOICE FROM THE PAST

The enemy has demanded a surrender. . . . I have answered the demand with a cannon shot, and our flag still waves proudly from the walls. I shall never surrender or retreat.

William Travis, "To the People of Texas and all the Americans in the World."

Because Juan Seguín spoke Spanish, he was chosen to carry the plea through enemy lines. Seguín got the message through to other Texas defenders. But when he returned, he saw the Alamo in flames.

The Alamo's defenders held off the Mexican attack for 12 violent days. Travis and the defenders stubbornly refused to surrender. On the 13th day, Santa Anna ordered more than 1,800 men to storm the fortress. The Texans met the attackers with a hailstorm of cannon and gun fire. Then suddenly it became strangely quiet. The Texans had run out of ammunition. At day's end, all but five Texans were dead. The **Battle of the Alamo** was over.

HISTORY *through* ART

The Battle of the Alamo was so intense that Davy Crockett did not have time to reload his gun, which he called "Betsy." He used it as a club. This print is by a 20th-century illustrator, Frederick Yohn.

What does the print reveal about the battle?

AMERICA'S HISTORY MAKERS

JUAN SEGUÍN

1806–1890

Juan Seguín was a *Tejano* hero of the Texas Revolution. It was Seguín who dashed through enemy lines at the Alamo with a last desperate attempt for aid.

And after the war, it was Seguín who arranged for the remains of the Alamo defenders to be buried with full military honors.

Newcomers to Texas who disliked all *Tejanos* falsely accused Sequín of planning rebellion. Fearing for his life, he fled to Mexico in 1842, there "to seek a refuge amongst my enemies."

SAM HOUSTON

1793–1863

Raised by a widowed mother, Sam Houston grew up in Tennessee. He lived with the Cherokee for about three years. Later, he served in the U.S. Army, in Congress, and as the governor of Tennessee.

"I was a General without an army," wrote Houston, after taking command of the Texas forces in 1836. Yet by the time the war was over, he and his troops had defeated Santa Anna's larger army.

Houston was elected the first president of the Republic of Texas. When Texas became a state, he served as a U.S. senator.

In what ways did the experiences of Seguín and Houston differ?

Those men who had not died in the fighting were executed at Santa Anna's command. A total of 183 Alamo defenders died. A few women and children were not killed. Susanna Dickinson, one of the survivors, was ordered by Santa Anna to tell the story of the Alamo to other Texans. He hoped the story would discourage more rebellion. The slaughter at the Alamo shocked Texans—and showed them how hard they would have to fight for their freedom from Mexico.

Victory at San Jacinto

With Santa Anna on the attack, Texans—both soldiers and settlers—fled eastward. Houston sent a message to the men at Goliad, ordering them to retreat. They were captured by Mexican forces, who executed more than 300. The Texans would not soon forget the massacre at Goliad. But even in retreat and defeat, Houston's army doubled. Now it was a fighting force of 800 angry men. It included *Tejanos,* American settlers, volunteers from the United States, and many free and enslaved African Americans.

Reading **History**

E. Reading a Map Use the map on page 405 to see where battles were taking place.

In late April, Santa Anna caught up with Houston near the San Jacinto (san juh•SIN•toh) River. Late in the afternoon of April 21, 1836, the Texans advanced on the Mexican army "with the stillness of death." When close to Santa Anna's camp, they raced forward, rifles ready, screaming "Remember the Alamo!" "Remember Goliad!"

In just 18 minutes, the Texans killed more than half of the Mexican army. Santa Anna was forced to sign a treaty giving Texas its freedom. With the Battle of San Jacinto, Texas was now independent.

Lone Star Republic

In September 1836, Texans raised a flag with a single star. They adopted a nickname—**Lone Star Republic**—and proclaimed Texas an independent nation. The new nation set up its own army and navy. Sam Houston was elected president of the Lone Star Republic by a landslide.

Vocabulary
annex: to join territory to an existing country

Many Texans did not want Texas to remain independent for long. They considered themselves Americans and wanted to be a part of the United States. In 1836, the Texas government asked Congress to annex Texas to the Union.

Many Northerners objected. They argued that Texas would become a slave state, and they opposed any expansion of slavery. If Texas joined the Union, slave states would outnumber free states and have a voting advantage in Congress. Other people feared that annexing Texas would lead to war with Mexico.

GEOGRAPHY SKILLBUILDER Interpreting Maps

1. **Movement** *About how many total miles did Santa Anna travel from Mexico to San Jacinto?*
2. **Movement** *Look at the distances traveled by Mexican forces and those traveled by the Texans. Which side do you think had an advantage? Explain.*

In response, Congress voted against annexation. Texas remained an independent republic for almost ten years. In the next section, you will learn that the question of annexing Texas did lead to a war between the United States and Mexico.

ReadingHistory
F. Analyzing Causes Why didn't Congress annex Texas?

Section 2 Assessment

1. Terms & Names

Identify:

- Stephen Austin
- *Tejano*
- Antonio López de Santa Anna
- Sam Houston
- William Travis
- Juan Seguín
- Battle of the Alamo
- Lone Star Republic

2. Taking Notes

Use a diagram like the one shown to review events that led to Texan independence and put them in order.

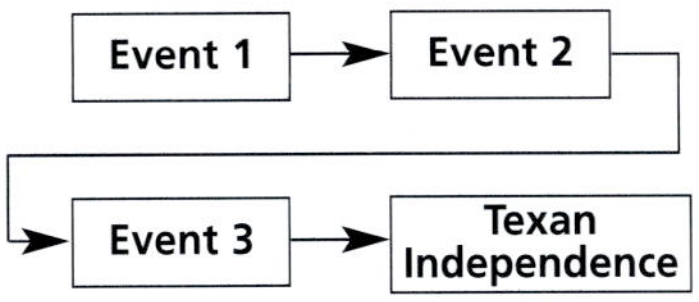

3. Main Ideas

a. Why did Americans want to move to Texas?

b. How did the Mexican government respond to the Texas rebellion?

c. Why did Congress refuse to annex Texas?

4. Critical Thinking

Recognizing Effects How did losing the Battle of the Alamo help the Texans win their independence?

THINK ABOUT

- the Texans' and Americans' shock over the loss of the battle to the Mexicans
- the need to recruit more forces to fight with the Texas army

ACTIVITY OPTIONS

ART
TECHNOLOGY

Research a figure from the Texas Revolution. Create a **trading card** or design that person's **Web page** for the Internet.

3 The War with Mexico

TERMS & NAMES
- James K. Polk
- manifest destiny
- Zachary Taylor
- Bear Flag Revolt
- Winfield Scott
- Treaty of Guadalupe Hidalgo
- Mexican Cession

MAIN IDEA

The United States expanded its territory westward to stretch from the Atlantic to the Pacific coast.

WHY IT MATTERS NOW

Today, one-third of all Americans live in the areas added to the United States in 1848.

ONE AMERICAN'S STORY

Henry Clay sneered, "Who is **James K. Polk**?" Clay had just learned the name of the man nominated by Democrats to run against him for president in 1844. "A mistake!" answered Washington insiders.

News of Polk's nomination was flashed to the capital by the newly invented telegraph machine. People were convinced that the machine didn't work. How could the Democrats choose Polk? A joke!

Polk was America's first "dark horse," a candidate who received unexpected support. The Democrats had nominated this little-known man only when they could not agree on anyone else.

Still, Polk wasn't a complete nobody. He had been governor of Tennessee and served seven terms in Congress. Polk was committed to national expansion. He vowed to annex Texas and take over Oregon. Americans listened and voted.

When those votes were counted, Clay had his answer. James Knox Polk was the eleventh president of the United States.

During his campaign, Polk's ideas about expanding the country captured the attention of Americans. As you will read in this section, after his election Polk looked for ways to expand the nation.

James Polk's presidential campaign emphasized expansion of the United States.

Americans Support Manifest Destiny

The abundance of land in the West seemed to hold great promise for Americans. Although populated with Native Americans and Mexicans, those lands were viewed by white settlers as unoccupied. Many Americans wanted to settle those lands themselves, and they worried about competition from other nations. Mexico occupied the southwest lands, and Britain shared the northwest Oregon Territory with the United States. Many Americans believed that the United States was destined to stretch across the continent from the Atlantic Ocean to the Pacific Ocean. In 1845, a newspaper editor named John O'Sullivan gave a name to that belief.

A VOICE FROM THE PAST

Our manifest destiny [is] to overspread and possess the whole of the continent which Providence [God] has given us for the development of the great experiment of liberty and . . . self-government.

John O'Sullivan, *United States Magazine and Democratic Review*

Reading History
A. Drawing Conclusions What were the positives and negatives of the idea of manifest destiny?

John O'Sullivan used the word *manifest* to mean clear or obvious. The word *destiny* means events sure to happen. Therefore, **manifest destiny** suggested that expansion was not only good but bound to happen—even if it meant pushing Mexicans and Native Americans out of the way. After Polk's election in 1844, manifest destiny became government policy.

The term "manifest destiny" was new, but the idea was not. By the 1840s, thousands of Americans had moved into the Oregon Territory. Since 1818, Oregon had been occupied jointly by the United States and Britain. In his campaign, Polk had talked of taking over all of Oregon. "Fifty-four forty or fight!" screamed one of his slogans. The parallel of 54° 40' N latitude was the northern boundary of the shared Oregon Territory.

"Our manifest destiny [is] to . . . possess the whole of the continent."
John O'Sullivan

Rather than fight for all of Oregon, however, Polk settled for half. In 1846, the United States and Great Britain agreed to divide Oregon at the 49th parallel. This agreement extended the boundary line already drawn between Canada and the United States. Today this line still serves as the border between much of the United States and Canada.

Troubles with Mexico

Polk had good reason for avoiding war with Britain over Oregon. By 1846, he had much bigger troubles brewing with Mexico over Texas.

In 1845, Congress admitted Texas as a slave state, in spite of Northern objections to the spread of slavery. However, Mexico still claimed Texas as its own. Mexico angrily viewed this annexation as an act of war. To make matters worse, Texas and Mexico could not agree on the official border between them. Texas claimed the Rio Grande, a river south of San Antonio, as its southern boundary. Mexico insisted on the Nueces (noo•AY•sis) River as the border of Texas. The difference in the distance between the two rivers was more than 100 miles at some points. Many thousands of miles of territory were at stake.

Reading History
B. Reading a Map Use the map on page 408 to find the locations of the disputed border between Texas and Mexico.

Mexico said it would fight to defend its claim. Hoping to settle the dispute peacefully, Polk sent John Slidell, a Spanish-speaking ambassador, to offer Mexico $25 million for Texas, California, and New Mexico. But Slidell's diplomacy failed.

Believing that the American people supported his expansion plans, Polk wanted to force the issue with Mexico. He purposely ordered General **Zachary Taylor** to station troops on the northern bank of the Rio Grande. This river bank was part of the disputed territory. Viewing this as an act of war, Mexico moved an army into place on the southern bank. On April 25, 1846, a Mexican cavalry unit crossed the Rio Grande. They ambushed an American patrol and killed or wounded 16 American soldiers.

When news of the attack reached Washington, Polk sent a rousing war message to Congress, saying, "Mexico has invaded our territory and shed American blood upon American soil." Two days later, Congress declared war. The War with Mexico had begun. Thousands of volunteers, mostly from western states, rushed to enlist in the army. Santa Anna, who was president of Mexico, built up the Mexican army.

However, Americans had mixed reactions to Polk's call for war. Illinois representative Abraham Lincoln questioned the truthfulness of the president's message and the need to declare war. Northeasterners questioned the justice of men dying in such a war. Slavery became an issue in the debates over the war. Southerners saw expansion into Texas as an opportunity to extend slavery and to increase their power in Congress. To

Reading **History**

C. Analyzing Causes How did the War with Mexico start?

The War with Mexico, 1846–1847

A symbol of the Bear Flag Revolt, this flag has a grizzly bear and a star made with blackberry juice. The revolt declared California independent from Mexico.

GEOGRAPHY SKILLBUILDER Interpreting Maps

1. **Movement** *About how far did General Kearny's troops move to reach San Diego?*
2. **Location** *Why would defending Texas be much easier than defending Mexico?*

prevent this from happening, antislavery representatives introduced a bill to prohibit slavery in any lands taken from Mexico. Frederick Douglass, the abolitionist, summarized the arguments.

A VOICE FROM THE PAST

The determination of our slaveholding President to prosecute the war, and the probability of his success in wringing from the people men and money to carry it on, is made evident, . . . None seem willing to take their stand for peace at all risks; and all seem willing that the war should be carried on in some form or other.

Frederick Douglass in *The North Star,* January 21, 1848

Despite opposition, the United States plunged into war. In May 1846, General Taylor led troops into Mexico. Many Americans thought it would be easy to defeat the Mexicans, and the war would end quickly.

Capturing New Mexico and California

Not long after the war began, General Stephen Kearny (KAHR•nee)—a U.S. army officer—and his men left Fort Leavenworth, Kansas, with orders to occupy New Mexico. Then they were to continue west to California. As his troops marched along the Santa Fe Trail, they sang songs like this one.

ReadingHistory

D. Making Inferences
How does this song support the idea of manifest destiny?

A VOICE FROM THE PAST

Old Colonel Kearny, you can bet,
Will keep the boys in motion,
Till Yankee Land includes the sand
On the Pacific Ocean.

Six weeks and 650 hot and rugged miles later, Kearny's army entered New Mexico. Using persuasion instead of force, he convinced the Mexican troops that he meant to withdraw. This allowed him to take New Mexico without firing a shot. Then Kearny and a small force of soldiers marched on toward California, which had only 6,000 Mexican residents. The remainder of the force moved south toward Mexico.

In California, Americans led by the explorer John C. Frémont rebelled against Mexican rule in the **Bear Flag Revolt.** They arrested the Mexican commander of Northern California and raised a crude flag showing a grizzly bear sketched in blackberry juice. The rebels declared California independent of Mexico and named it the Republic of California. In the fall, U.S. troops reached California and joined forces with the rebels. Within weeks, Americans controlled all of California.

STRANGE but True

SANTA ANNA'S LOST LEG

Santa Anna lost his left leg in a battle with the French. In 1842, he held a funeral for his severed limb. On that day, church and political officials followed the dictator's leg through the streets of Mexico City to its final resting place—an urn placed on a column.

Two years later, an angry mob broke the urn and threw the leg away. The leg was rescued by a loyal soldier who took it home and hid it.

Thirty years later, that soldier visited Santa Anna and returned the bones of his long-lost leg.

The Invasion of Mexico

The defeat of Mexico proved far more difficult. The Mexican army was much larger, but the U.S. troops were led by well-trained officers. American forces invaded Mexico from two directions.

General Taylor battled his way south from Texas toward the city of Monterrey in northern Mexico. On February 22, 1847, his 4,800 troops met General Santa Anna's 15,000 Mexican soldiers near a ranch called Buena Vista. After the first day of fighting, Santa Anna sent Taylor a note offering him a chance to surrender. Taylor declined. At the end of the second bloody day of fighting, Santa Anna reported that "both armies have been cut to pieces." However, it was Santa Anna who retreated after the Battle of Buena Vista. The war in the north of Mexico was over.

In southern Mexico, fighting continued. A second force led by General **Winfield Scott** landed at Veracruz on the Gulf of Mexico and battled inland toward Mexico City. Outside the capital, Scott met fierce resistance at the castle of Chapultepec (chuh•POOL•tuh•pek). About 1,000 soldiers and 100 young military cadets bravely defended the fortress. Despite their determined resistance, Mexico City fell to Scott in September 1847. As he watched, a Mexican officer sighed and said, "God is a Yankee."

Background
General Winfield Scott had become a national hero during the War of 1812.

The Mexican Cession

On February 2, 1848, the war officially ended with the **Treaty of Guadalupe Hidalgo** (gwah•duh•LOOP•ay hih•DAHL•go). In this treaty, Mexico recognized that Texas was part of the United States, and the

Growth of the United States, *1783–1853*

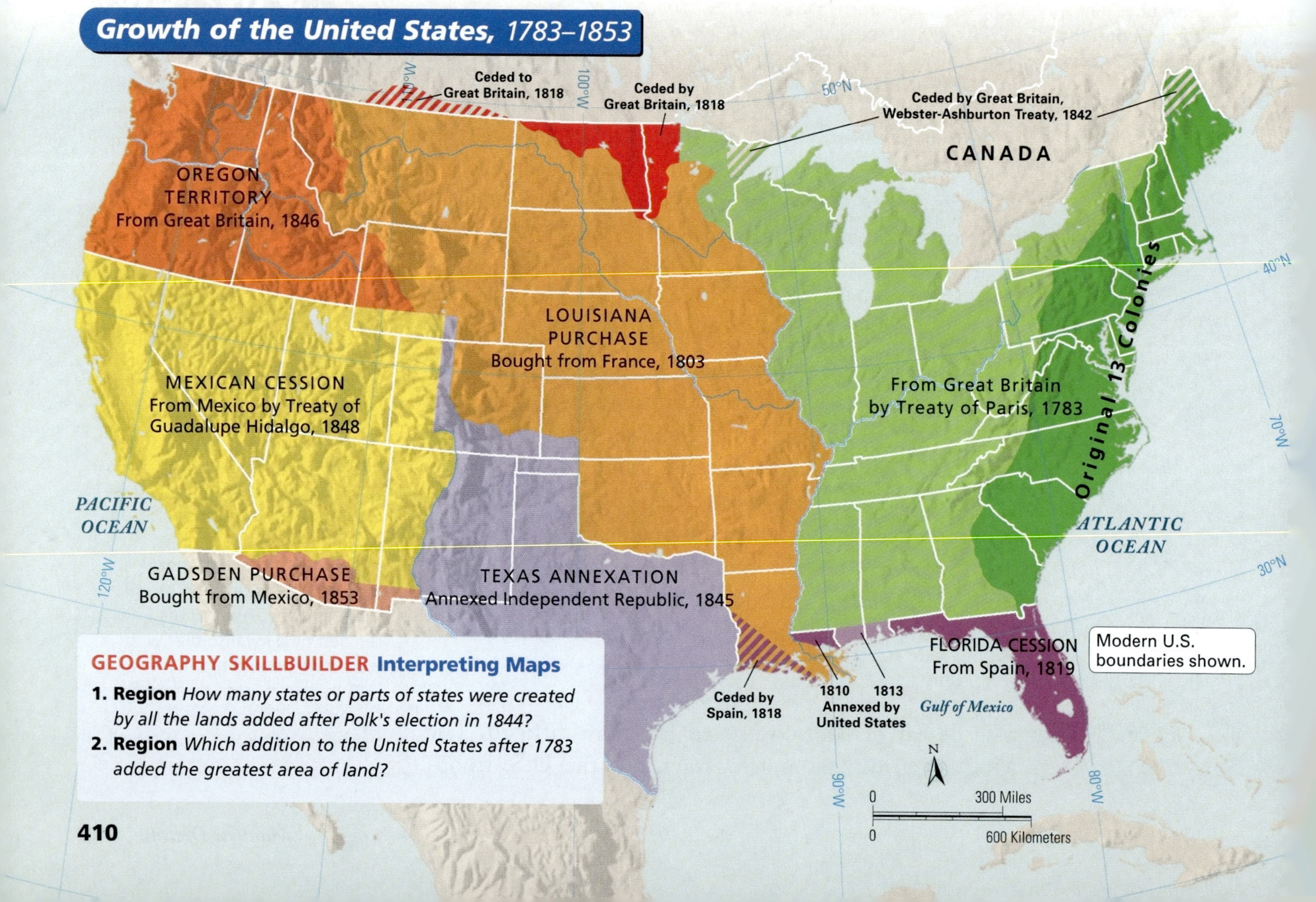

GEOGRAPHY SKILLBUILDER Interpreting Maps

1. **Region** *How many states or parts of states were created by all the lands added after Polk's election in 1844?*
2. **Region** *Which addition to the United States after 1783 added the greatest area of land?*

Reading History
E. Finding Main Ideas What were the three main parts of the Treaty of Guadalupe Hidalgo?

Rio Grande was the border between the nations. Mexico also ceded, or gave up, a vast region known as the **Mexican Cession.** This area included the present-day states of California, Nevada, Utah, most of Arizona, and parts of New Mexico, Colorado, and Wyoming. Together with Texas, this land amounted to almost one-half of Mexico. The loss was a bitter defeat for Mexico, particularly because many Mexicans felt that the United States had provoked the war in the hope of gaining Mexican territory.

In return, the United States agreed to pay Mexico $15 million. The United States would also pay the $3.25 million of claims U.S. citizens had against Mexico. Finally, it also promised to protect the 80,000 Mexicans living in Texas and the Mexican Cession.

Mexicans living in the United States saw the conquest of their land differently. Suddenly they were a minority in a nation with a strange language, culture, and legal system. At the same time, they would make important contributions to their new country. They taught new settlers how to develop the land for farming, ranching, and mining. A rich new culture resulted from the blend of many cultures in the Mexican Cession.

Reading History
F. Making Inferences Why did the United States pay a large price for the Gadsden Purchase?

"From Sea to Shining Sea"

The last bit of territory added to the continental United States was a strip of land across what is now southern New Mexico and Arizona. The government wanted the land as a location for a southern transcontinental railroad. In 1853, Mexico sold the land—called the Gadsden Purchase—to the United States for $10 million.

On July 4, 1848, in Washington, President Polk laid the cornerstone of a monument to honor George Washington. In Washington's day, the western border of the United States was the Mississippi River. The United States in 1848 now stretched "from sea to shining sea." In August, Polk learned that gold had been found in California. In the next section, you will read about the California gold rush.

Section 3 Assessment

1. Terms & Names

Identify:
- James K. Polk
- manifest destiny
- Zachary Taylor
- Bear Flag Revolt
- Winfield Scott
- Treaty of Guadalupe Hidalgo
- Mexican Cession

2. Taking Notes

Review the chapter and find five key events to place on a time line as shown.

War with Mexico

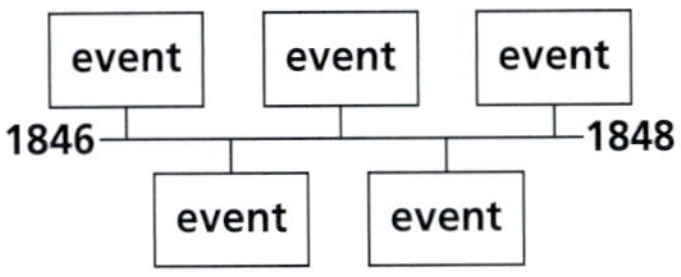

3. Main Ideas

a. How did the acquisitions of Oregon and the Mexican Cession relate to the idea of manifest destiny?

b. Why were some people opposed to the War with Mexico?

c. What does the phrase "sea to shining sea" mean?

4. Critical Thinking

Comparing Compare the different ways land was acquired by the United States in the period of manifest destiny from 1844 to 1853.

THINK ABOUT
- the acquisition of the Oregon Territory
- lands in the Southwest

ACTIVITY OPTIONS

MATH
GEOGRAPHY

In an almanac, find the current population of the states formed from the Mexican Cession. Create a **graph** or a **map** to display the information.

4 The California Gold Rush

TERMS & NAMES
forty-niner
Californio
Mariano Vallejo
John Sutter
James Marshall
California gold rush

MAIN IDEA

Gold was found in California, and thousands rushed to that territory. California quickly became a state.

WHY IT MATTERS NOW

The gold rush made California grow rapidly and helped bring about California's cultural diversity.

ONE AMERICAN'S STORY

Luzena Wilson said of the year 1849, "The gold excitement spread like wildfire." The year before, James Marshall had discovered gold in California. Luzena's husband decided to become a **forty-niner**—someone who went to California to find gold, starting in 1849.

Most forty-niners left their families behind, but Luzena traveled to California with her husband. She later said, "I thought where he could go I could, and where I went I could take my two little toddling babies."

Luzena discovered that women—and their homemaking skills—were rare in California. Shortly after she arrived, a miner offered her five dollars for the biscuits she was baking. Shocked, she just stared at him. He quickly doubled his offer and paid in gold. Luzena realized she could make money by feeding miners, so she opened a hotel.

Like the Wilsons, thousands of people from around the world became forty-niners. In this section, you will read about the forty-niners and what their mining experiences were like. You will learn how the rapid growth of California's population caused problems for the people who lived there before 1849. You will also discover how the gold rush boosted California's economy and changed the nation's history.

This woman is carrying food to miners, just as Luzena Wilson did.

California Before the Rush

Before the forty-niners came, California was populated by as many as 150,000 Native Americans and 6,000 ***Californios***—settlers of Spanish or Mexican descent. Most *Californios* lived on huge cattle ranches. They had acquired their estates when the Mexican government took away the land that once belonged to the California missions.

One important *Californio* was **Mariano Vallejo** (mah•RYAH•noh vah•YEH•hoh). A member of one of the oldest Spanish families in America, he owned 250,000 acres of land. Proudly describing the accomplishments of the *Californios*, Vallejo wrote, "We were the pioneers of the Pacific coast . . . while General Washington was carrying on the war of the Revolution." Vallejo himself had been the commander of Northern California when it belonged to Mexico.

Vocabulary
immigration: the movement of people into a country or region where they were not born

When Mexico owned California, its government feared American immigration and rarely gave land to foreigners. But **John Sutter,** a Swiss immigrant, was one exception. Dressed in a secondhand French army uniform, Sutter had visited the Mexican governor in 1839. A charming man, Sutter persuaded the governor to grant him 50,000 acres in the unsettled Sacramento Valley. Sutter built a fort on his land and dreamed of creating his own personal empire based on agriculture.

In 1848, Sutter sent a carpenter named **James Marshall** to build a sawmill on the nearby American River. One day Marshall inspected the canal that brought water to Sutter's Mill. He later said, "My eye was caught by a glimpse of something shining. . . . I reached my hand down and picked it up; it made my heart thump for I felt certain it was gold."

Rush for Gold

News of Marshall's thrilling discovery spread rapidly. From all over California, people raced to the American River—starting the **California gold rush.** A gold rush occurs when large numbers of people move to a site where gold has been found. Throughout history, people have valued gold because it is scarce, beautiful, easy to shape, and resistant to tarnish.

Miners soon found gold in other streams flowing out of the Sierra Nevada Mountains. Colonel R. B. Mason, the military governor of California, estimated that the region held enough gold to "pay the cost of the present war with Mexico a hundred times over." He sent this news to Washington with a box of gold dust as proof.

The following year thousands of gold seekers set out to make their fortunes. A forty-niner who wished to reach California from the East had a choice of three routes, all of them dangerous:

1. Sail 18,000 miles around South America and up the Pacific coast—suffering from storms, seasickness, and spoiled food.
2. Sail to the narrow Isthmus of Panama, cross overland (and risk catching a deadly tropical disease), and then sail to California.
3. Travel the trails across North America—braving rivers, prairies, mountains, and all the hardships of the trail.

Because the adventure was so difficult, most gold seekers were young men. "A gray beard is almost as rare as a petticoat," observed one miner. Luzena Wilson said that during the six months she lived in the mining city of Sacramento, she saw only two other women.

HISTORY *through* **ART**

Clipper ship companies used advertising cards such as this one to convince Easterners that their line provided the fastest, most pleasant voyages.

How has the artist tried to project a positive image for sailing west?

Reading History
A. Categorizing What were the three different types of transportation that people took to get to California?

Now and then

LEVI'S BLUE JEANS

Nearly everyone in the United States owns at least one pair of faded, comfortable blue jeans. The first jeans were invented for California miners.

In 1873, a man named Levi Strauss wanted to sell sturdy pants to miners. Strauss made his pants out of the strongest fabric he could buy—cotton denim. He reinforced the pockets with copper rivets so that they could hold heavy tools without ripping.

For more than 125 years, jeans have remained popular. Levi Strauss's pants have proved to be durable in more ways than one.

Life in the Mining Camps

The mining camps had colorful names like Mad Mule Gulch, Hangtown, and Coyote Diggings. They began as rows of tents along the streams flowing out of the Sierra Nevada. Gradually, the tents gave way to rough wooden buildings that housed stores and saloons.

Mining camps could be dangerous. One woman who lived in the region wrote about camp violence.

A VOICE FROM THE PAST

In the short space of twenty-four days, we have had murders, fearful accidents, bloody deaths, a mob, whippings, a hanging, . . . and a fatal duel.

Louise Clappe, quoted in *Frontier Women*

The mining life was hard for other reasons. Camp gossip told of miners who grew rich overnight by finding eight-pound nuggets, but in reality, such easy pickings were rare. Miners spent their days standing knee-deep in icy streams, where they sifted through tons of mud and sand to find small amounts of gold. Exhaustion, poor food, and disease all damaged the miners' health.

Not only was acquiring gold brutally difficult, but the miners had to pay outrageously high prices for basic supplies. In addition, gamblers and con artists swarmed into the camps to swindle the miners of their money. As a result, few miners grew rich.

Reading **History**
B. Making Inferences Why do you think life in the mining camps was so rough?

Miners from Around the World

About two-thirds of the forty-niners were Americans. Most of these were white men—many from New England. However, Native Americans, free blacks, and enslaved African Americans also worked the mines.

Thousands of experienced miners came from Sonora in Mexico. Other foreign miners came from Europe, South America, Australia, and China. Most of the Chinese miners were peasant farmers who fled from a region that had suffered several crop failures. By the end of 1851, one of every ten immigrants was Chinese.

Used to backbreaking labor in their homeland, the Chinese proved to be patient miners. They would take over sites that American miners had abandoned because the easy gold was gone. Through steady, hard work, the Chinese made these "played-out" sites yield profits. American miners resented the success of the Chinese and were suspicious of their different foods, dress, and customs. As the numbers of Chinese miners grew, American anger toward them also increased.

Reading **History**
C. Analyzing Causes Why did some Americans resent Chinese miners?

Surface Mining

Gold is found in cracks, called veins, in the earth's rocky crust. As mountains and other outcrops of rock erode, the gold veins come to the surface. The gold breaks apart into nuggets, flakes, and dust. Flood waters then wash it downhill into stream beds. To mine this surface gold, forty-niners had to use tools designed to separate it from the mud and sand around it. American miners learned some technology from Mexicans who came from the mining region of Sonora.

CONNECT TO HISTORY

1. **Drawing Conclusions** Which mining method could be used by an individual miner and which needed a group of miners? Explain your answer.

See Skillbuilder Handbook, page R12.

CONNECT TO TODAY

2. **Researching** How is gold mined today?

Visit www.mcdougallittell.com to learn more about the California gold rush.

Conflicts Among Miners

A mixture of greed, anger, and prejudice caused some miners to cheat others. For example, I. B. Gilman promised to free an enslaved African American named Tom if he saved enough gold. For more than a year, Tom mined for himself after each day's work was done. When he finally had $1,000, Gilman gave him a paper saying he was free. The next day, the paper suspiciously disappeared. Even though Tom was certain he had been robbed, he couldn't prove it. He had to work for another year before Gilman would free him.

Once the easy-to-find gold was gone, American miners began to force Native Americans and foreigners such as Mexicans and Chinese out of the gold fields to reduce competition. This practice increased after California became a state in 1850. One of the first acts of the California state legislature was to pass the Foreign Miners Tax, which imposed a tax of $20 a month on miners from other countries. That was more than most could afford to pay. As the tax collectors arrived in the camps, most foreigners left.

*Reading*History

D. Analyzing Causes How did the state government make mining harder for foreigners?

Driven from the mines, the Chinese opened shops, restaurants, and laundries. So many Chinese owned businesses in San Francisco that their neighborhood was called Chinatown, a name it still goes by today.

The Impact of the Gold Rush

By 1852, the gold rush was over. While it lasted, about 250,000 people flooded into California. This huge migration caused economic growth that changed California permanently. The port city of San Francisco grew to become a center of banking, manufacturing, shipping, and trade. Its population exploded from around 400 in 1845 to 35,000 in 1850. Sacramento became the center of a productive farming region.

However, the gold rush ruined many *Californios.* The newcomers did not respect *Californios,* their customs, or their legal rights. In many cases,

CAUSE AND EFFECT: *U.S. Expansion, 1846–1853*

CAUSE	EFFECT
Westward trails move thousands to new territories.	Oregon Territory acquired by the United States.
Austin and others colonize Texas.	Texas Revolution
United States annexes Texas.	War with Mexico
Mexican Cession acquired by the United States.	United States expands "sea to sea."
Transcontinental railroad route needed.	Gadsden Purchase
Thousands of gold seekers rush to California.	California becomes a state.

SKILLBUILDER Interpreting Charts

1. *Which two causes are related to transportation?*
2. *Which cause fulfilled the nation's "manifest destiny"?*

Reading **History**

E. Recognizing Effects What impact did the gold rush have on the people who lived in California before the forty-niners came?

Americans seized their property. For example, Mariano Vallejo lost all but 300 acres of his huge estate. Even so, their Spanish heritage became an important part of California culture.

Native Americans suffered even more. Thousands of them died from diseases brought by the newcomers. The miners hunted down and killed thousands more. The reason was the Anglo-American belief that Native Americans stood in the way of progress. By 1870, California's Native American population had fallen from 150,000 to only about 58,000.

A final effect of the gold rush was that by 1849 California had enough people to apply for statehood. Skipping the territorial stage, California applied to Congress for admission to the Union and was admitted as a free state in 1850. Although its constitution outlawed slavery, it did not grant African Americans the vote.

Mariano Vallejo, unhappy that *Californio* culture was ignored in the new American California, named his home "Tear of the Mountain."

For some people, California's statehood proved to be the opportunity of a lifetime. The enslaved woman Nancy Gooch gained her freedom because of the law against slavery. She then worked as a cook and washerwoman until she saved enough money to buy the freedom of her son and daughter-in-law in Missouri. Nancy Gooch's family moved to California to join her. Eventually, they became so prosperous that they bought Sutter's sawmill, where the gold rush first started.

On a national level, California's statehood created turmoil. Before 1850, there was an equal number of free states and slave states. Southerners feared that because the statehood of California made free states outnumber slave states, Northerners might use their majority to abolish slavery. As Chapter 18 explains, conflict over this issue threatened the survival of the Union.

Section 4 Assessment

1. Terms & Names

Identify:
- forty-niner
- *Californio*
- Mariano Vallejo
- John Sutter
- James Marshall
- California gold rush

2. Taking Notes

Use a chart like the one shown to review and record hardships faced by the forty-niners.

HARDSHIPS	
In the camps	
At work mining	

Which hardships would you have found most difficult?

3. Main Ideas

a. How did the California gold rush get started?

b. Why didn't many forty-niners become rich?

c. How did California's statehood affect the nation?

4. Critical Thinking

Recognizing Effects What were some of the effects of the California gold rush?

THINK ABOUT
- changes in San Francisco
- California's bid for statehood

ACTIVITY OPTIONS

SCIENCE
TECHNOLOGY

Research the hazards of mining gold and either plan a **science exhibit** or give an **electronic presentation.**

Chapter 13 ASSESSMENT

VISUAL SUMMARY

Manifest Destiny

Trails West

Mountain men and traders opened trails in the Far West. Pioneers then went west to gain land, wealth, or religious freedom.

The Texas Revolution

Americans moved into the Mexican territory of Texas. Conflicts led those Americans to revolt, and Texas gained independence.

The War with Mexico

President Polk wanted to expand the nation. He negotiated to gain Oregon. The United States fought Mexico to gain much of the Southwest.

The California Gold Rush

The discovery of gold lured thousands of people to California. California's economy and population grew, resulting in statehood.

TERMS & NAMES

Briefly explain the importance of each of the following.

1. mountain man
2. Oregon Trail
3. Stephen Austin
4. *Tejano*
5. Antonio López de Santa Anna
6. manifest destiny
7. Bear Flag Revolt
8. Mexican Cession
9. forty-niner
10. California gold rush

REVIEW QUESTIONS

Trails West (pages 393–399)

1. What were three reasons why people moved west?
2. What were the three main trails that led to the West?
3. How did the Mormons make the land in Utah productive?

The Texas Revolution (pages 400–405)

4. Why were Texans unhappy with Mexican rule?
5. Why were the battles of the Alamo and San Jacinto important to the Texas Revolution?

The War with Mexico (pages 406–411)

6. What areas did the United States gain as a result of Americans' belief in manifest destiny?
7. How is the Bear Flag Revolt related to the War with Mexico?
8. What lands did the United States acquire as a result of the Treaty of Guadalupe Hidalgo?

The California Gold Rush (pages 412–417)

9. Who were four groups of people who became forty-niners?
10. What were three ways California changed because of the gold rush?

CRITICAL THINKING

1. USING YOUR NOTES

	Types of people who traveled there	Why they went there	Key events that allowed the U.S. to take ownership of the territory
New Mexico			
Utah			
Oregon			
Texas			
California			

Using your completed chart, answer the questions below.

a. In what ways were the reasons people went west similar?

b. Which of the five regions listed on your chart entered the United States peacefully?

c. Which event added the most territory to the United States?

2. ANALYZING LEADERSHIP

Think about the leaders discussed in this chapter. What characteristics did they have that made them good leaders?

3. THEME: EXPANSION

How did the idea of manifest destiny help bring about the expansion of the United States?

4. DRAWING CONCLUSIONS

How did the War with Mexico and the California gold rush contribute to the cultural diversity of the United States?

5. APPLYING CITIZENSHIP SKILLS

What were the different viewpoints that people held about the War with Mexico?

Interact *with* History

Based on this chapter, what do you think you would have gained or lost by going west?

HISTORY SKILLS

1. INTERPRETING MAPS: Movement

Study the map. Answer the questions.

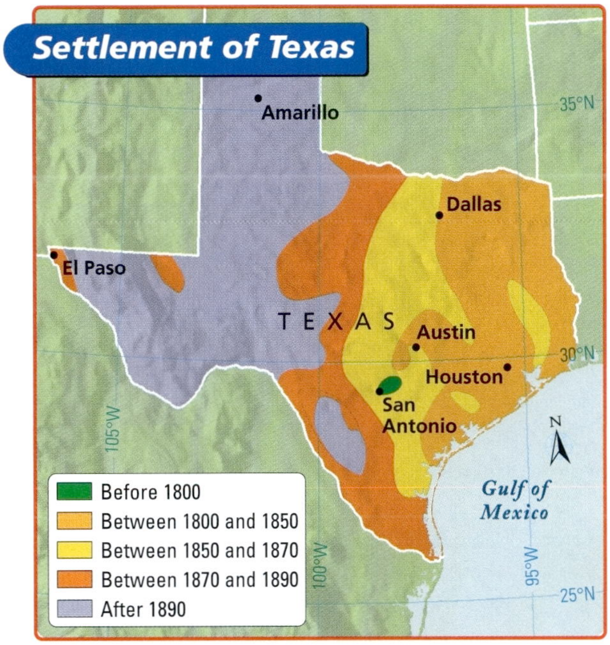

Basic Map Elements

a. What is the subject of the map?

b. What years are covered by the map?

c. What do the colors indicate?

Interpreting the Map

d. Which area of Texas was settled first?

e. In what general direction was Texas settled?

2. INTERPRETING PRIMARY SOURCES

This photograph was taken of a man who planned to go to California to find gold. Study the photo carefully. Answer the questions.

a. What does the photo reveal about the man's expectations of danger?

b. What does the photo suggest about how successful he hopes to be?

ALTERNATIVE ASSESSMENT

1. INTERDISCIPLINARY ACTIVITY: Science

Creating a Diagram Do research to learn how gold is deposited into veins in the earth and how erosion later exposes the gold. Draw diagrams showing the processes of gold vein formation and erosion. Share your diagram with the class.

2. COOPERATIVE LEARNING ACTIVITY

Creating a News Magazine Show With the support of President Polk, Congress declared war on Mexico in 1846. Though many Americans supported the decision, some groups felt that war with Mexico was unnecessary and unjust. Working with a small group, create a news magazine show that explores the different viewpoints surrounding the Mexican War. Research these opinions. Then write and perform the news magazine for the class. One student should be the moderator, while the other students in the group should choose one of the following groups to represent.

a. President Polk and his supporters

b. Northerners, including Abraham Lincoln and Frederick Douglass

c. Southerners

3. TECHNOLOGY ACTIVITY

Making a Class Presentation Life in the mining camps was not like life "back east." Information about the camps comes from primary sources, like diaries and newspaper articles. Using the Internet and library, find sources about life in the mining camps.

Visit www.mcdougallittell.com to learn more about gold mining.

Your sources might include:

- images of mines, miners, or miners' shacks
- images of items that the general store sold to miners
- tales of the gold fields
- information about the diversity of cultures in the camps

4. HISTORY PORTFOLIO

Option 1 Review your section and chapter assessment activities. Select one that you think is your best work. Then use comments made by your teacher or classmates to improve your work and add it to your portfolio.

Option 2 Review the questions that you wrote for What Do You Want to Know? on page 392. Then write a short report in which you explain the answers to your questions. If any questions were not answered, do research to answer them. Add your answers to your portfolio.

CHAPTER 14

A New Spirit of Change 1820–1860

Mathew Brady took this picture of New York City in 1860. It is one of the earliest photographs of a poor city neighborhood.

Interact *with* History

You are a writer who moves to New York in the mid-1800s. A newspaper hires you to write about reform. One day, you hear a speaker call for the end of slavery. Another day you talk to a factory worker whose pay has been cut. In the city, you see great poverty and suffering.

What Do You Think?

- How might you persuade Americans to change life in the city?
- What are the problems that you see in these two pictures?
- Should reform come about through new laws or through individual actions?

What reforms do you think will most benefit American society?

USA

1828
Noah Webster publishes the *American Dictionary of the English Language*.

1829
David Walker prints *Appeal*, a pamphlet urging slaves to revolt.

1836
The Lowell Mill girls go on strike to demand better conditions.

1843
Dorothea Dix asks the Massachusetts legislature to improve the care of the mentally ill.

1848
The Seneca Falls Convention demands women's rights.

1851
Maine passes a law banning the sale of alcohol.

1820 — **1860**

World

1824
The British Parliament makes trade unions illegal.

1829
Louis Braille invents a raised type that allows blind people to read.

1845
Ireland's potato crop fails, causing famine. Thousands flee to America.

1848
A revolution in Germany fails. Some Germans move to America.

1854
Brazil's first railway opens.

BEFORE YOU READ

Previewing the Theme

Impact of the Individual In the mid-1800s, millions of Europeans moved to the United States and changed its culture. Writers and artists also shaped American culture. In addition, many individuals worked to reform society in such areas as education and the antislavery movement. Chapter 14 describes how individuals changed America.

This engraving shows a woman teaching a young slave to read, even though it was illegal to do so. The boy grew up, escaped, and became a great reformer—Frederick Douglass.

What Do You Know?

What do you think were the worst problems in the United States in the mid-1800s? How do you think people tried to solve them?

THINK ABOUT

- stories or films that are set in this period
- problems that exist now
- the actions people take to solve today's problems

What Do You Want to Know?

What would you like to learn about the way individuals changed the United States in the mid-1800s? Record your questions in your notebook before you read the chapter.

READ AND TAKE NOTES

Reading Strategy: Comparing To understand the many influences on U.S. culture, learn to compare. Comparing means examining the similarities between people, actions, or ideas. The chart below will help you compare the influences that various people had upon America in the middle of the 19th century. Use the chart to take notes about how people changed America. Also take notes about people who tried to have an influence but failed.

See Skillbuilder Handbook, page R8.

	How People Influenced America in the Mid-1800s
Immigrants	
Writers	
Reformers	
Abolitionists	
Women	

1 The Hopes of Immigrants

TERMS & NAMES
emigrant
immigrant
steerage
push-pull factor
famine
prejudice
nativist

MAIN IDEA

In the mid-1800s, millions of Europeans came to the United States hoping to build a better life.

WHY IT MATTERS NOW

These Germans, Irish, and Scandinavians had a strong influence on American culture.

ONE AMERICAN'S STORY

In June 1831, Gjert Hovland (YEHRT HAHV•LIHND) and his family left Norway for America. After a few years, Hovland wrote to a friend in Norway. He boasted that in the United States a poor man's vote counted as much as a rich man's vote. Americans could travel and work freely. The United States had so much opportunity that Hovland wondered why anyone would choose to stay hungry in Norway.

A VOICE FROM THE PAST

It would greatly please me to learn that all of you who are in need and have little chance of supporting yourselves and your families have decided to leave Norway and come to America; for, even if many more come, there will still be room here for all. Those who are willing to work will not lack employment or business here.

Gjert Hovland, letter to Torjuls Maeland, April 22, 1835

Advertisements for land attracted immigrants, who came to the United States with only what could fit in trunks like the one shown above.

Millions of people like Hovland decided to become <u>emigrants</u>, or people who leave a country. Arriving in the United States, they became <u>immigrants</u>, or people who settle in a new country. This section explains how immigrants enriched the United States with their work and their cultures.

Why People Migrated

Most immigrants endured hardships to come to America. Although some, like Hovland, brought their families, many immigrant men came alone and suffered loneliness. Nearly all immigrants made the ocean voyage in **<u>steerage</u>,** the cheapest deck on a ship. In steerage, hundreds of people lived jammed together for ten days to a month. Conditions were filthy. Many passengers became ill or died on the journey.

Despite the hard passage, immigrants flocked to the United States during the mid-1800s. They came from Britain, Ireland, Germany, Scandinavia (Sweden, Denmark, and Norway), and China. Most came from Europe. What made them come to America? Historians talk about

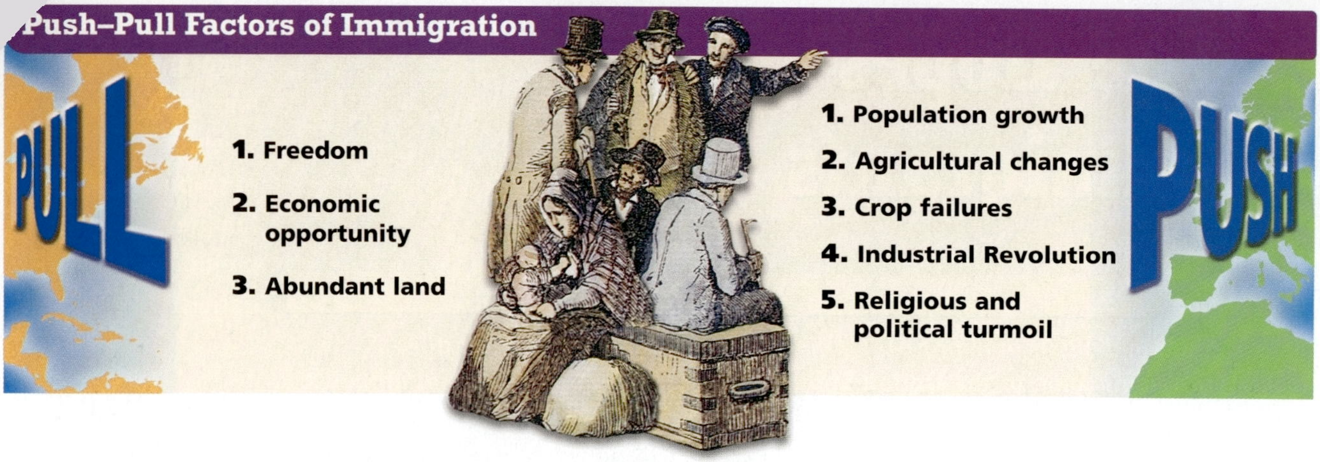

push-pull factors. These forces push people out of their native lands and pull them toward a new place. **Push factors** included the following:

1. **Population growth.** Better food and sanitation caused Europe's population to boom after 1750, and the land became overcrowded.
2. **Agricultural changes.** As Europe's population grew, so did cities. Landowners wanted to make money selling food to those cities. New methods made it more efficient to farm large areas of land than to rent small plots to tenants. So landlords forced tenants off the land.
3. **Crop failures.** Poor harvests made it difficult for small farmers to pay their debts. Some of these farmers chose to start over in America. Crop failures also led to hunger, causing people to emigrate.
4. **Industrial Revolution.** Goods produced in factories became cheaper than goods produced by artisans. Suddenly out of work, some artisans took factory jobs. Others emigrated.
5. **Religious and political turmoil.** To escape religious persecution, Quakers fled Norway and Jews left Germany. Also, many Germans came to America after a revolution in Germany failed in 1848.

Vocabulary
tenant: renter

Vocabulary
artisan: skilled worker

Immigrants chose the United States because of three main **pull factors**:

1. **Freedom.** As Gjert Hovland wrote, "Everyone has the freedom to practice the teaching and religion he prefers."
2. **Economic opportunity.** Immigrants sought a land where they could support their families and have a better future. Immigration often rose during times of U.S. prosperity and fell during hard times.
3. **Abundant land.** The acquisition of the Louisiana Purchase and the Mexican Cession gave the United States millions more acres of land. To land-starved Europeans, America was a land of opportunity.

***Reading* History**
A. Solving Problems Many of the push factors were problems. Which pull factors were solutions to which problems?

Scandinavians Seek Land

Public land in America was sold for $1.25 an acre, which lured thousands of Scandinavians. At first, their governments tried to keep them at home. A Swedish law of 1768 restricted the right to emigrate. But growing poverty in Scandinavia caused officials to cancel this law in 1840.

Scandinavian clergymen also tried to halt the emigration. At first, they warned their church members against leaving the homeland. Eventually, though, the preachers realized their words had little effect. Some of them even went to America themselves.

> ***Reading* History**
> **B. Making Inferences** Why do you think Scandinavians moved to places that felt familiar?

In the United States, Scandinavians chose regions that felt familiar. Many settled in the Midwest, especially Minnesota and Wisconsin. These states had lakes, forests, and cold winters like their homelands. A high proportion of Scandinavian immigrants became farmers.

Germans Pursue Economic Opportunity

Like the Scandinavians, many Germans moved to the Midwest. Germans especially liked Wisconsin because the climate allowed them to grow their traditional crop of oats. Some moved to Milwaukee, Wisconsin, because the Catholic bishop there was German. (In the 1800s, German Christian immigrants included both Catholics and Protestants.)

Germans also settled in Texas. In New Braunfels, a group of German nobles bought land and sold it in parcels to German immigrants. The town had to survive poor harvests and conflicts with Native Americans, but it eventually prospered. Germans also founded Fredericksburg, Texas, which still retains its German culture today.

Immigrants from Germany settled in cities as well as on farms or the frontier. German artisans opened businesses as bakers, butchers, carpenters, printers, shoemakers, and tailors. Many German immigrants achieved great success. For instance, in 1853 John Jacob Bausch and Henry Lomb started a firm to make eyeglasses and other lenses. Their company became the world's largest lens maker.

Immigration and Settlement, 1820–1860

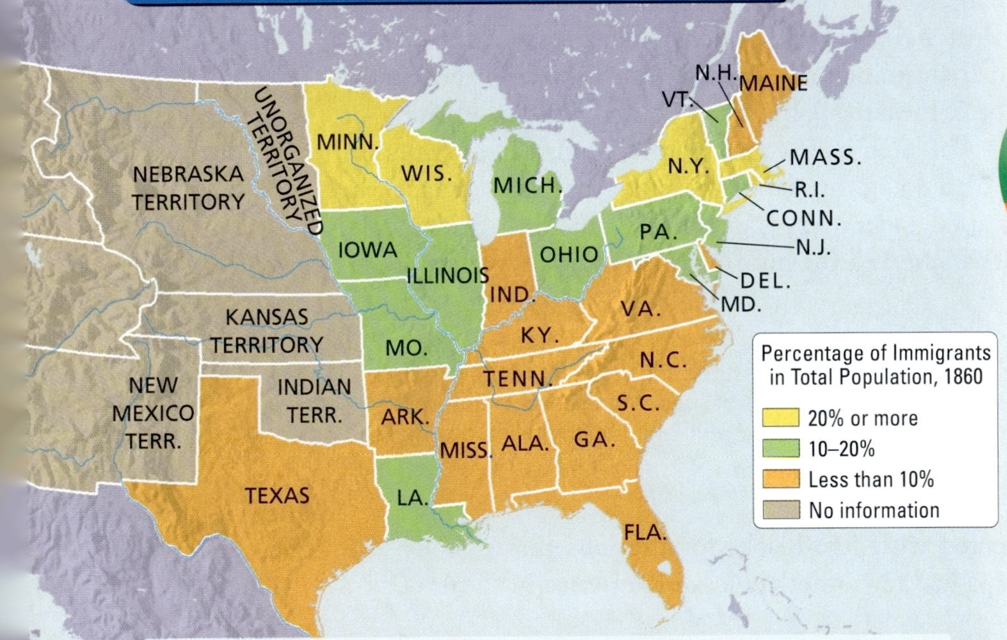

Sources of Immigration, *1820–1860*

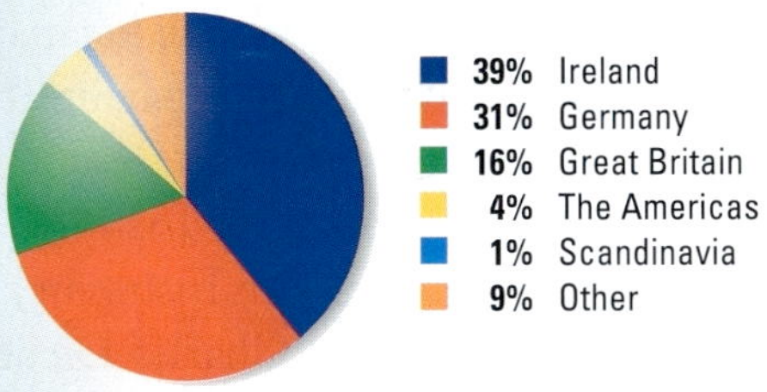

Immigration to the United States
(by decade)

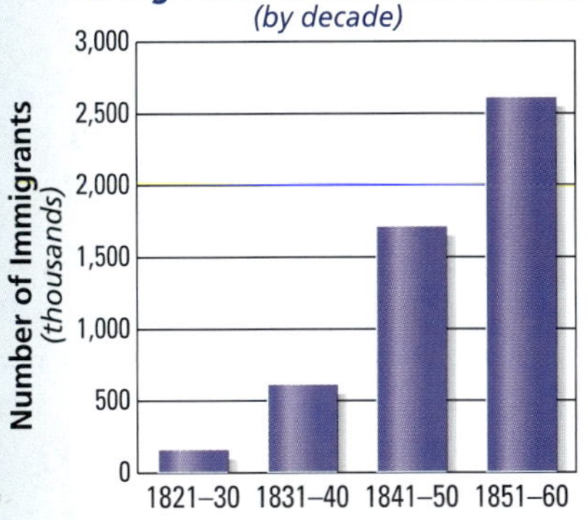

Source: *Historical Statistics of the United States*

GEOGRAPHY SKILLBUILDER Interpreting Maps

1. **Place** *Which states had the highest percentage of immigrants in 1860?*
2. **Region** *Compare the settlement of immigrants in the North and the South. What do you know about the South that might account for this difference?*

Some German immigrants were Jews. Many of them worked as traveling salespeople. They brought pins, needles, pots—and news—to frontier homes and mining camps. In time, some opened their own general stores. Other Jews settled in cities, where many found success. For example, Alexander Rothschild worked as a grocer upon arriving in Hartford, Connecticut, in the 1840s. By 1851, he ran a popular hotel.

daily life

IMMIGRANT CULTURE

To maintain their culture, immigrants continued many of their traditional activities in the United States. For example, German culture is rich in music. German immigrants put together marching bands, symphony orchestras, and choruses.

In Ireland, many of the Irish had poured their energy into defying the British. This gave them experience with political organization. As a result, Irish immigrants became active in U.S. politics, especially in the cities.

The Germans were the largest immigrant group of the 1800s and strongly influenced American culture. Many things we think of as originating in America came from Germany—the Christmas tree, gymnasiums, kindergartens, and the hamburger and frankfurter.

Background

The hamburger and frankfurter are named after the German cities Hamburg and Frankfurt am Main.

The Irish Flee Hunger

Most Irish immigrants were Catholic. Protestant Britain had ruled Ireland for centuries—and controlled the Catholic majority by denying them rights. Irish Catholics could not vote, hold office, own land, or go to school. Because of the poverty produced by Britain's rule, some Irish came to America in the early 1800s.

Then, in 1845, a disease attacked Ireland's main food crop, the potato, causing a severe food shortage called a **famine.** The Irish Potato Famine killed 1 million people and forced many to emigrate. By 1854, between 1.5 and 2 million Irish had fled their homeland.

In America, Irish farmers became city-dwellers. Arriving with little or no savings, many of these immigrants had to settle in the port cities where their ships had docked. By 1850, the Irish made up one-fourth of the population of Boston, New York, Philadelphia, and Baltimore.

The uneducated Irish immigrants arrived with few skills and had to take low-paying, back-breaking jobs. Irish women took in washing or worked as servants. The men built canals and railroads across America. So many Irishmen died doing this dangerous work that people said there was "an Irishman buried under every [railroad] tie." In 1841, British novelist Charles Dickens observed the huts in which railroad workers lived.

Reading History

C. Drawing Conclusions How did the effects of British rule make it hard for Irish immigrants to America to find good jobs?

A VOICE FROM THE PAST

The best were poor protection from the weather; the worst let in the wind and rain through the wide breaches in the roofs of sodden grass and in the walls of mud; some had neither door nor window; some had nearly fallen down.

Charles Dickens, quoted in *To Seek America*

The Irish competed with free blacks for the jobs that nobody else wanted. Both groups had few other choices in America in the 1800s.

U.S. Cities Face Overcrowding

Immigrants like the Irish and Germans flocked to American cities. So did native-born Americans, who hoped for the chance to make a better living. Between 1800 and 1830, New York's population jumped from 60,489 to 202,589. St. Louis doubled its population every nine years. Cincinnati grew even faster, doubling every seven years.

Rapid urban growth brought problems. Not enough housing existed for all the newcomers. Greedy landlords profited from the housing shortage by squeezing large apartment buildings onto small lots. Using every inch of space for rooms, these cramped living quarters lacked sunlight and fresh air. Their outdoor toilets overflowed, spreading disease. In such depressing urban neighborhoods, crime flourished.

American cities were unprepared to tackle these problems. In fact, before 1845, New York City had no public police force. Until the 1860s, it had only a volunteer fire department. And in 1857, the rapidly growing city had only 138 miles of sewers for 500 miles of streets.

Reading **History**
D. Identifying Problems What problems were politicians trying to solve by offering to help new immigrants?

Most immigrant groups set up aid societies to help newcomers from their country. Many city politicians also offered to assist immigrants in exchange for votes. The politicians set up organizations to help new arrivals find housing and work.

Some Americans Oppose Immigration

Some native-born Americans feared that immigrants were too foreign to learn American ways. Others feared that immigrants might come to outnumber natives. As a result, immigrants faced anger and prejudice. **Prejudice** is a negative opinion that is not based on facts. For example,

CITIZENSHIP TODAY

Becoming a Citizen

Most immigrants who came to America in the 1800s shared one thing: an appreciation for the nation's values and laws. As a result, many chose to become U.S. citizens.

This trend continues today. In recent decades, more than half a million Vietnamese have immigrated to the United States. Many became citizens of their new country. One of them was Lam Ton, who is a successful restaurant owner in Chicago. Ton viewed U.S. citizenship as both a privilege and a duty. "We have to stick to this country and help it do better," he said.

Each year, immigrants from around the world are sworn in as U.S. citizens on Citizenship Day, September 17. But first they must pass a test on English, the U.S. political system, and the rights and duties of citizenship.

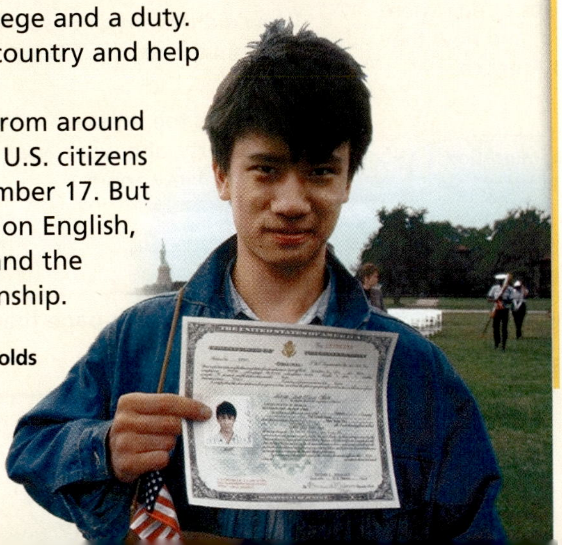

This young immigrant proudly holds up his certificate of citizenship.

How Does Someone Become a Citizen?

1. In a small group, discuss what questions you would ask those seeking to become U.S. citizens.
2. Create a citizenship test using your questions.
3. Have another group take the test and record their scores.
4. Use the McDougal Littell Internet site to link to the actual U.S. citizenship test. Compare it to your test.

See Citizenship Handbook, page 281.

Visit www.mcdougallittell.com to learn more about becoming a U.S. citizen.

In 1844, a riot took place between Catholics and non-Catholics in Philadelphia. Several people were killed.

some Protestants in the 1800s believed that Catholics threatened democracy. Those Protestants feared that the Pope, the head of the Roman Catholic Church, was plotting to overthrow democracy in America.

Background
Protestants feared the Pope because in many European countries, the Catholic Church worked closely with the ruling monarchs.

Native-born Americans who wanted to eliminate foreign influence called themselves **nativists**. Some nativists refused to hire immigrants and put up signs like "No Irish need apply." In cities like New York and Boston, nativists formed a secret society. Members promised not to vote for any Catholics or immigrants running for political office. If asked about their secret group, they said, "I know nothing about it."

In the 1850s, nativists started a political party. Because of the members' answers to questions about their party, it was called the Know-Nothing Party. It wanted to ban Catholics and the foreign-born from holding office. It also called for a cut in immigration and a 21-year wait to become an American citizen. The Know-Nothings did elect six governors. But they disappeared quickly as a national party. Their northern and southern branches couldn't agree on the issue of slavery.

In spite of such barriers as prejudice, the immigrants of the 1800s had a strong impact on American culture. Writers and artists of the 1800s also shaped American culture. Section 2 discusses their influence.

Section 1 Assessment

1. Terms & Names

Identify:
- emigrant
- immigrant
- steerage
- push-pull factor
- famine
- prejudice
- nativist

2. Taking Notes

Use a cluster diagram like the one below to record details about immigration, such as which groups came, where they settled, and how they influenced America.

3. Main Ideas

a. What were the push-pull factors that led to immigration?

b. How did the arrival of so many immigrants affect U.S. cities?

c. What was the Know-Nothing Party, and what was its point of view about immigration?

4. Critical Thinking

Analyzing Causes How did the rapid increase in immigration cause conflict?

THINK ABOUT
- why Irish immigrants and free blacks competed for jobs
- the growth of cities and the problems it created
- the prejudices of nativists
- religious differences

ACTIVITY OPTIONS

TECHNOLOGY
ART

Plan a **multimedia presentation** or design a **Web page** that shows immigrants the advantages of settling in the United States.

2 American Literature and Art

TERMS & NAMES
romanticism
Hudson River school
transcendentalism
civil disobedience

MAIN IDEA

Inspired by nature and democratic ideals, writers and artists produced some of America's greatest works.

WHY IT MATTERS NOW

Nineteenth-century writers such as Hawthorne and Thoreau laid the foundation for American literature.

ONE AMERICAN'S STORY

As a young man, Washington Irving published articles that made fun of society in the early 1800s. Although he studied to be a lawyer, he eventually made writing his full-time career.

Irving wrote some of the first stories to describe America. For example, "Rip Van Winkle" tells of a man in New York State. Rip wakes up after a 20-year nap to find everything changed. He goes to the inn, which once had a picture of King George on its sign.

A VOICE FROM THE PAST

The red coat was changed for one of blue and buff, a sword was held in the hand instead of a sceptre [staff of authority], the head was decorated with a cocked hat, and underneath was painted in large characters, GENERAL WASHINGTON.

Washington Irving, "Rip Van Winkle"

While Rip slept, the Americans had fought and won their revolution!

Irving's work helped to win European respect for American writing for the first time. This section discusses other individuals of the 1800s who created uniquely American literature and art.

In another Irving tale, "The Legend of Sleepy Hollow," a spooky creature—perhaps a ghost from the Revolution—chases a teacher.

Writing About America

Irving and other writers were influenced by a style of European art called **romanticism.** It stressed the individual, imagination, creativity, and emotion. It drew inspiration from nature. American writers turned their interest in nature into a celebration of the American wilderness.

Many books featured the wilderness. James Fenimore Cooper wrote five novels about the dramatic adventures of wilderness scout Natty Bumppo. One that remains popular is *The Last of the Mohicans.* Francis Parkman wrote a travel book, *The Oregon Trail,* about the frontier trail.

HISTORY through ART

Asher Durand was a founder of the Hudson River school of painting. His best-known work, *Kindred Spirits,* was painted in 1849. This romantic work shows two artists inspired by a beautiful landscape. The figures in the painting are Durand's friends, the poet William Cullen Bryant and the painter Thomas Cole.

What aspects of the painting make it romantic?

In addition, writers began to use a more American style. A teacher and lawyer named Noah Webster gave guidelines to that style in his *American Dictionary of the English Language.* Webster first published his dictionary in 1828. He later revised it in 1840. The dictionary gave American, not British, spellings and included American slang.

Other writers besides Irving celebrated America's past. Henry Wadsworth Longfellow wrote many poems that retold stories from history. For example, "Paul Revere's Ride" depicted the Revolutionary hero's ride to warn of a British attack. Generations of students memorized lines from the poem, such as, "One if by land, and two if by sea; / And I on the opposite shore will be."

Creating American Art

Influenced by romanticism, American painters also focused on nature. One group of painters worked in the Hudson River Valley in New York State. **Hudson River school** artists painted peaceful landscapes of mountains, forests, and rivers. Several members of the Hudson River school traveled to find more exciting scenery. For example, Albert Bierstadt took several trips to America's mountainous West. He produced huge paintings that convey the majesty of the American landscape. (See page 310.)

Background
The National Audubon Society, whose goal is the protection of wildlife today, is named for John James Audubon.

Other artists also went west. John James Audubon came to the United States from France at age 18. Traveling across the continent, Audubon sketched the birds and animals of his adopted country.

Enslaved African Americans also contributed to American art. They made beautiful baskets, quilts, and pottery. Most of these slaves remained anonymous, but one did not. David Drake worked in a South Carolina pottery factory and signed the pottery he created. He was the only factory worker to do so.

Following One's Conscience

By the 1840s, Americans took new pride in their emerging culture. Ralph Waldo Emerson, a New England writer, encouraged this pride. He urged Americans to cast off European influence and develop their own beliefs. His advice was to learn about life from self-examination and from nature as well as books.

"No law can be sacred to me but that of my nature."
Ralph Waldo Emerson

Emerson's student, Henry David Thoreau, followed that advice. In 1845, Thoreau moved to a simple cabin he had built by Walden Pond near the town of Concord, Massachusetts. Thoreau furnished it with only a bed, a table, a desk, and three chairs. He wrote about his life in the woods in *Walden*. Thoreau said that people should live by their own individual standards.

***Reading*History**
A. Making Inferences What do you think it means to "hear a different drummer"?

A VOICE FROM THE PAST

If a man does not keep pace with his companions, perhaps it is because he hears a different drummer. Let him step to the music which he hears, however measured or far away.

Henry David Thoreau, *Walden*

Emerson and Thoreau belonged to a group of thinkers with a new philosophy called **transcendentalism.** It taught that the spiritual world is more important than the physical world. It also taught that people can find the truth within themselves—through feeling and intuition.

Because Thoreau believed in the importance of individual conscience, he urged people not to obey laws they considered unjust. Instead of protesting with violence, they should peacefully refuse to obey those laws. This form of protest is called **civil disobedience.** For example, Thoreau did not want to support the U.S. government, which allowed slavery and fought the War with Mexico. Instead of paying taxes that helped to finance the war, Thoreau went to jail.

Another New England transcendentalist, Margaret Fuller, also called for change. In her magazine, *The Dial*, and in her book, *Woman in the Nineteenth Century*, Fuller argued for women's rights.

Connections TO LITERATURE

"CIVIL DISOBEDIENCE"

In his essay "Civil Disobedience," Thoreau wrote that "Under a government which imprisons any unjustly the true place for a just man is also a prison."

Thoreau did land in prison when he refused to pay his taxes. According to legend, Emerson visited Thoreau in jail and asked, "Why are you here?" Thoreau replied, "Why are you not here?"

In the 20th century, Mohandas K. Gandhi of India and Martin Luther King, Jr., of the United States both used civil disobedience to fight injustice.

STRANGE *but* True

GIFTS ON POE'S GRAVE

Every year a mysterious figure dressed in black celebrates Edgar Allan Poe's birthday. He leaves three roses on the author's Baltimore grave at 3:00 A.M.

The puzzling tradition began in 1949, exactly 100 years after Poe's death. In 1993, a new black-coated visitor took over the tradition. The person who began the ritual was ill—and later died in 1999.

Although many witnesses watch the ritual each year, none ask the visitor his name. Poe's fans have always liked mysteries.

Exploring the Human Heart

Like Thoreau, other writers broke with tradition. In 1855, poet Walt Whitman published *Leaves of Grass,* a book that changed American poetry. His bold, unrhymed poems praised ordinary people. Emily Dickinson lived in her family's home almost her entire life. She wrote poems on small pieces of paper that she sewed into booklets. Her subjects include God, nature, love, and death. Most of her 1,775 poems were published only after her death. Both Whitman and Dickinson shaped modern poetry by experimenting with language.

Fiction writers of the 1800s also shaped American literature. Edgar Allan Poe wrote terrifying tales that influence today's horror story writers. He also wrote the first detective story, "The Murders in the Rue Morgue."

Reading **History**

B. Recognizing Effects How did Poe influence the fiction that people read today?

Nathaniel Hawthorne depicted love, guilt, and revenge during Puritan times in *The Scarlet Letter.* The novel shows that harsh judgment without mercy can lead to tragedy. Hawthorne may have learned that lesson from his family history. One of his ancestors condemned people at the Salem witchcraft trials.

Herman Melville won fame by writing thrilling novels about his experiences as a sailor. In 1851, Melville published his masterpiece, *Moby Dick.* This novel tells about a man's destructive desire to kill a white whale. Although the novel was not popular when it was published, it is widely read now. Several movie versions exist.

These fiction writers portrayed the harmful effects of cruel actions. Other people thought that individuals could alter society for good. Section 3 describes those reformers.

Section 2 Assessment

1. Terms & Names

Identify:
- romanticism
- Hudson River school
- transcendentalism
- civil disobedience

2. Taking Notes

Use a chart like the one below to list important individual writers and artists. For each one, name or describe one of his or her works.

Writer or artist	His or her work

Which one would you like to learn more about? Why?

3. Main Ideas

a. What was romanticism and how did Americans adapt it?

b. What is civil disobedience and what did Thoreau do that is an example of it?

c. How did the writers of the mid-1800s shape modern literature?

4. Critical Thinking

Evaluating Why do you think the literature and art of the mid-1800s are still valued?

THINK ABOUT
- the way they feature U.S. history and culture
- their universal themes—themes that relate to all people in all time periods
- the way they reflect changes happening at that time

ACTIVITY OPTIONS

ART / TECHNOLOGY

Choose an American painting, sketch it, and make it into a **jigsaw puzzle;** or make an **audio recording** of a museum guide's description of it.

3 Reforming American Society

TERMS & NAMES
revival
Second Great Awakening
temperance movement
labor union
strike
Horace Mann
Dorothea Dix

MAIN IDEA

In the mid-1800s, several reform movements worked to improve American education and society.

WHY IT MATTERS NOW

Several laws and institutions, such as public schools, date back to this period.

ONE AMERICAN'S STORY

Anne Newport Royall was a travel writer. In her 1830 book *Letters from Alabama,* Royall recorded America's growing interest in religion. She also described hearing a preacher at a Tennessee revival, or meeting to reawaken religious faith.

A VOICE FROM THE PAST

His text was, "He that hath ears to hear, let him hear." The people must have been deaf indeed that could not have heard him. . . . He began low but soon bawled to deafening. He spit in his hands, rubbed them against each other, and then would smite them together, till he made the woods ring.

Anne Newport Royall, *Letters from Alabama*

This revival meeting took place during the Second Great Awakening—a rebirth of religious faith named after the Great Awakening of the 1700s.

Some preachers, like the one Royall saw, were circuit riders. A circuit rider rode from town to town, often holding his meetings in a tent. The preacher gave a sermon urging individuals to give up their sins. This section explains how, in the mid-1800s, many individuals called on Americans to reform, or to improve themselves and their society.

A Spirit of Revival

The renewal of religious faith in the 1790s and early 1800s is called the **Second Great Awakening.** Revivalist preachers said that anyone could choose salvation. This appealed to equality-loving Americans. Revivals spread quickly across the frontier. Settlers eagerly awaited the visits of preachers like Peter Cartwright. At the age of 16, Cartwright had given up a life of gambling and joined a Methodist Church. He became a minister and spent more than 60 years preaching on the frontier.

The revival also traveled to Eastern cities. There, former lawyer Charles Grandison Finney held large revival meetings. He preached that "all sin consists in selfishness" and that religious faith led people to help others. Such teaching helped awaken a spirit of reform. Americans began to believe that they could act to make things better.

Temperance pledges often displayed inspiring pictures and mottoes.

Temperance Societies

Led by churches, some Americans began the **temperance movement,** which is a campaign to stop the drinking of alcohol. Heavy drinking was common in the early 1800s. Some workers spent most of their wages on alcohol—leaving their families without enough money to live on. As a result, many women joined the temperance movement. "There is no reform in which women can act better or more appropriately than temperance," said Mary C. Vaughan.

Temperance workers handed out pamphlets urging people to stop drinking. One pamphlet read, "Drink is the . . . source . . . of nearly all the ills that afflict the human family." Temperance speakers traveled across America asking people to sign a paper pledging to give up alcohol. By 1838, a million people had signed.

Reading **History**
A. Evaluating How did the temperance movement encourage people to take responsibility for their families?

Temperance also won the support of business owners. Industry needed workers who could keep schedules and run machines. Alcohol made it hard for workers to do either. New England businessman Neal Dow led the fight to make it illegal to sell alcohol. In 1851, Maine banned the sale of liquor. By 1855, 13 other states passed similar laws. But many people opposed these laws, and most were repealed. Still, the movement to ban alcohol remained strong, even into the 20th century.

Vocabulary
repeal: to cancel

Fighting for Workers' Rights

As business owners tried to improve workers' habits, workers called for improvements in working conditions. Factory work was noisy, boring, and unsafe. In the 1830s, American workers began to organize.

The young women mill workers in Lowell, Massachusetts, started a labor union. A **labor union** is a group of workers who band together to seek better working conditions. In 1836, the mill owners raised the rent of the company-owned boarding houses where the women lived. About 1,500 women went on **strike,** stopping work to demand better conditions. Eleven-year-old Harriet Hanson helped lead the strikers.

A VOICE FROM THE PAST

I . . . started on ahead, saying, . . . "I don't care what you do, I am going to turn out, whether anyone else does or not," and I marched out, and was followed by the others. As I looked back at the long line that followed me, I was more proud than I have ever been since.

Harriet Hanson, quoted in *A People's History of the United States*

Other workers called for shorter hours and higher wages. In 1835 and 1836, 140 strikes took place in the eastern United States. Then the Panic of 1837 brought hard times. Jobs were scarce, and workers were afraid to cause trouble. The young labor movement fell apart. Even so, workers achieved a few goals. For example, in 1840 President Martin Van Buren ordered a ten-hour workday for government workers.

Background
President Van Buren's order reduced the workweek from 70 to 60 hours.

Improving Education

In the 1830s, Americans also began to demand better schools. In 1837, Massachusetts set up the first state board of education in the United States. Its head was **Horace Mann.** Mann called public education "the great equalizer." He also argued that "education creates or develops new treasures—treasures never before possessed or dreamed of by any one." By 1850, many Northern states had opened public elementary schools.

Boston opened the first public high school in 1821. A few other Northern cities followed suit. In addition, churches and other groups founded hundreds of private colleges in the following decades. Many were located in states carved from the Northwest Territory. These included Antioch and Oberlin Colleges in Ohio, the University of Notre Dame in Indiana, and Northwestern University in Illinois.

Women could not attend most colleges. One exception was Oberlin. It was the first college to accept women as well as men. In 1849, English immigrant Elizabeth Blackwell became the first woman to earn a medical degree in the United States. Despite such individual efforts, it was rare for a woman to attend college until the late 1800s.

Mary Jane Patterson was the first African-American woman to earn a college degree. She graduated from Oberlin in 1862 and went on to work as a teacher.

African Americans also faced obstacles to getting an education. This was especially true in the South. There, teaching an enslaved person to read had been illegal since the Nat Turner Rebellion in 1831. Enslaved African Americans who tried to learn were brutally punished. Even in the North, most public schools barred African-American children.

Reading History
B. Making Inferences Why do you think women and African Americans had a hard time getting an education?

Few colleges accepted African Americans. Those that did often took only one or two blacks at a time. The first African American to receive a college degree was Alexander Twilight in 1823. John Russwurm received one in 1826 and later began the first African-American newspaper.

Caring for the Needy

As some people promoted education, others tried to improve society's care for its weakest members. In 1841, **Dorothea Dix,** a reformer from Boston, was teaching Sunday school at a women's jail. She discovered some women who were locked in cold, filthy cells

AMERICA'S HISTORY MAKERS

HORACE MANN
1796–1859

Horace Mann once said in a speech to students, "Be ashamed to die until you have won some victory for humanity." Mann had no reason to be ashamed. As a child, he knew poverty and hardship. He educated himself and later fought for public education for other people.

Toward the end of his life, Mann became president of Antioch College. It committed itself to education for both men and women and equal rights for African Americans.

DOROTHEA DIX
1802–1887

At the age of 12, Dorothea Dix left an unhappy home to go live with her grandparents in Boston. Just two years later, she began teaching little children.

In 1841, Dix saw the harsh treatment of mentally ill women. Society frowned upon women traveling alone, but Dix defied custom. She went by train to several places where the mentally ill were housed.

Dix wrote a report about her research. (See page 438.) That report changed the care of the mentally ill.

How might their backgrounds have motivated Dorothea Dix and Horace Mann to become leaders in reform movements?

simply because they were mentally ill. Visiting other jails, Dix learned that the mentally ill often received no treatment. Instead, they were chained and beaten. Dix pleaded with the Massachusetts Legislature to improve the care of the mentally ill. Later, she traveled all over the United States on behalf of the mentally ill. Her efforts led to the building of 32 new hospitals.

Some reformers worked to improve life for people with other disabilities. Thomas H. Gallaudet started the first American school for deaf children in 1817. Samuel G. Howe founded the Perkins School for the Blind in Boston in the 1830s.

Reformers also tried to improve prisons. In the early 1800s, debtors, lifelong criminals, and child offenders were put in the same cells. Reformers demanded that children go to special jails. They also called for the rehabilitation of adult prisoners. Rehabilitation means preparing people to live useful lives after their release from prison.

Reading **History**

C. Recognizing Effects How did reformers change the treatment of the mentally ill, the disabled, and prisoners?

Spreading Ideas Through Print

During this period of reform, Americans began to receive more information about how they should lead their lives. In the 1830s, cheaper newsprint and the invention of the steam-driven press lowered the price of a newspaper to a penny. Average Americans could afford to buy the "penny papers." Penny papers were also popular because, in addition to serious news, they published gripping stories of fires and crimes.

Hundreds of new magazines also appeared. One was the *Ladies' Magazine.* Its editor was Sarah Hale, a widow who used writing to support her family. The magazine advocated education for women. It also suggested that men and women were responsible for different, but equally important, areas of life. The magazine taught that a woman's area was the home and the world of "human ties." A man's area was politics and the business of earning a living for his family. Later, Hale edited *Godey's Lady's Book,* which published poems and stories as well as articles.

Creating Ideal Communities

While magazines sought to tell people how to live and reform movements tried to change society, some individuals decided to start over. They aimed to build an ideal society, called a utopia.

Two attempts at utopias were New Harmony, Indiana, and Brook Farm, Massachusetts. In both, residents received food and other necessities of life in exchange for work. However, both utopias experienced conflicts and financial difficulties. They ended after only a few years.

Reading **History**
D. Forming and Supporting Opinions Why do you think it was hard for utopias to succeed? Give reasons.

Religious belief led to some utopias. For example, the Shakers followed the beliefs of Ann Lee. She preached that people should lead holy lives in communities that demonstrate God's love to the world. When a person became a Shaker, he or she vowed not to marry or have children. Shakers shared their goods with each other, believed that men and women are equal, and refused to fight for any reason. Shakers set up communities in New York, New England, and on the frontier.

People called them *Shakers* because they shook with emotion during church services. Otherwise, Shaker life was calm. Shakers farmed and built simple furniture in styles that remain popular today. The childless Shakers depended on converts and adopting children to keep their communities going. In the 1840s, the Shakers had 6,000 members—their highest number. In 1999, only seven Shakers remained.

In the 1840s and 1850s, reform found a new direction. Many individuals began to try to win rights for two oppressed groups—women and enslaved persons. Section 4 discusses these efforts.

Section 3 Assessment

1. Terms & Names

Identify:
- revival
- Second Great Awakening
- temperance movement
- labor union
- strike
- Horace Mann
- Dorothea Dix

2. Taking Notes

Create a chart like the one below. Use it to list problems identified by reformers and their solutions to them.

Problem	Reformer's Solution

3. Main Ideas

a. How did the Second Great Awakening influence the reform movement?

b. How did labor unions try to force business owners to improve working conditions?

c. What were women's contributions to the reform movement?

4. Critical Thinking

Recognizing Effects What was the long-term impact of the reform movement that took place in the mid-1800s?

THINK ABOUT
- the changes reformers made in education, temperance, prisons, and the care of the disabled
- which of those changes are still in effect today

ACTIVITY OPTIONS

SPEECH
CIVICS

Think of a modern problem that is similar to an issue discussed in this section. Give a **speech** or write a **letter** to a government official suggesting a reform.

Report to the Massachusetts Legislature

Setting the Stage After traveling to several places where the mentally ill were kept, Dorothea Dix wrote a report describing the conditions she had discovered. In 1843, she presented her report to lawmakers to alert them to the horrible treatment of the mentally ill. This report has been called the "first piece of social research ever conducted in America." An excerpt from Dorothea Dix's report follows. **See Primary Source Explorer** CD-ROM

Report to the Massachusetts Legislature

Gentlemen: . . . I come to present the strong claims of suffering humanity. I come to place before the Legislature of Massachusetts the condition of the miserable, the desolate, the outcast. I come as the **advocate**[1] of helpless, forgotten, insane, and idiotic men and women; of beings sunk to a condition from which the most unconcerned would start with real horror; of beings wretched in our prisons, and more wretched in our **almshouses.**[2]

I must confine myself to a few examples, but am ready to furnish other and more complete details, if required.

I proceed, gentlemen, briefly to call your attention to the *present* state of insane persons confined within this **Commonwealth,**[3] in *cages, closets, cellars, stalls, pens! Chained, naked, beaten with rods,* and *lashed* into obedience.

I offer the following extracts from my notebook and journal.

Springfield: In the jail, one lunatic woman, furiously mad, a state **pauper,**[4] improperly situated, both in regard to the prisoners, the keepers, and herself. It is a case of extreme self-forgetfulness and oblivion to all the decencies of life, to describe which would be to repeat only the grossest scenes. She is much worse since leaving Worcester. In the almshouse of the same town is a woman apparently only needing **judicious**[5] care and some well-chosen employment to make it unnecessary to confine her in solitude in a dreary unfurnished room. Her appeals for employment and companionship are most touching, but the mistress replied "she had no time to attend to her."

Lincoln: A woman in a cage. *Medford:* One idiotic subject chained, and one in a close stall for seventeen years. *Pepperell:* One often doubly chained, hand and foot; another violent; several peaceable now. *Brookfield:* One man caged, comfortable. *Granville:* One often closely confined, now losing the use of his

A CLOSER LOOK

ADVOCATE OF THE HELPLESS

In earlier times, the term *idiotic* did not mean stupid. It was used to describe someone who was mentally retarded.

1. For what groups of people is Dix pleading for help?

A CLOSER LOOK

JUDICIOUS CARE

Dix describes a woman who needs only some care and a useful task to do.

2. What did the woman's keeper say when Dix pointed that out?

1. **advocate:** a person who pleads another person's cause.
2. **almshouses:** homes for poor people.
3. **Commonwealth:** one of four U.S. states whose constitution uses this term to describe their form of self-government; in this case, Massachusetts.
4. **pauper:** a person who lives on the state's charity.
5. **judicious:** wise and careful.

limbs from want of exercise. *Charlemont:* One man caged. *Savoy:* One man caged. *Lenox:* Two in the jail, against whose unfit condition there the jailer protests.

Dedham: The insane **disadvantageously**[6] placed in the jail. In the almshouse, two females in stalls, situated in the main building, lie in wooden bunks filled with straw; always shut up. One of these subjects is supposed curable. The overseers of the poor have declined giving her a trial at the hospital, as I was informed, on account of expense.

Besides the above, I have seen many who, part of the year, are chained or caged. The use of cages is all but universal. Hardly a town but can refer to some not distant period of using them; chains are less common; **negligences**[7] frequent; willful abuse less frequent than sufferings proceeding from ignorance, or want of consideration. I encountered during the last three months many poor creatures wandering reckless and unprotected through the country. . . . But I cannot **particularize.**[8] In traversing the state, I have found hundreds of insane persons in every variety of circumstance and condition, many whose situation could not and need not be improved; a less number, but that very large, whose lives are the saddest pictures of human suffering and degradation.

I give a few illustrations; but description fades before reality. . . .

Men of Massachusetts, I beg, I implore, I demand pity and protection for these of my suffering, outraged sex. . . . Become the benefactors of your race, the just guardians of the solemn rights you hold in trust. Raise up the fallen, **succor**[9] the desolate, restore the outcast, defend the helpless, and for your eternal and great reward receive the benediction, "Well done, good and faithful servants, become rulers over many things!"

A CLOSER LOOK

I HAVE SEEN MANY

Notice that Dix cites evidence from many different towns.

3. Why do you think she includes so many specific details in her report?

A CLOSER LOOK

MEN OF MASSACHUSETTS

When Dix says "Men of Massachusetts," she is still speaking to the members of the state legislature.

4. What does Dix want the Massachusetts Legislature to do?

6. **disadvantageously:** harmfully.
7. **negligences:** careless actions.
8. **particularize:** to name in detail.
9. **succor:** to give help during a time of need.

Interactive Primary Source Assessment

1. Main Ideas

a. On what evidence did Dorothea Dix base her report about "suffering humanity"?

b. How were the mentally ill treated in Massachusetts?

c. Who did Dorothea Dix ask to help to improve the care of the mentally ill?

2. Critical Thinking

Evaluating Dix succeeded in convincing the legislature to provide funds for new hospitals. What do you think made her report so persuasive?

THINK ABOUT

- the details included in the report
- how Dix got the information to write her report
- the techniques you would use to persuade someone

4 Abolition and Women's Rights

TERMS & NAMES
abolition
Frederick Douglass
Sojourner Truth
Underground Railroad
Harriet Tubman
Elizabeth Cady Stanton
Seneca Falls Convention
suffrage

MAIN IDEA

The spread of democracy led to calls for freedom for slaves and more rights for women.

WHY IT MATTERS NOW

The abolitionists and women reformers of this time inspired 20th-century reformers.

ONE AMERICAN'S STORY

African-American poet Frances Ellen Watkins Harper was free but grew up in the slave state of Maryland. Harper often wrote about the suffering of enslaved persons. For example, the poem excerpt below describes a woman whose child has been taken from her and sold.

A VOICE FROM THE PAST

They tear him from her circling arms,
Her last and fond embrace.
Oh! never more may her sad eyes
Gaze on his mournful face.

No marvel, then, these bitter shrieks
Disturb the listening air:
She is a mother, and her heart
Is breaking in despair.

Frances Ellen Watkins Harper, "The Slave Mother"

Frances Ellen Watkins Harper impressed audiences with her speaking ability as she called for reform.

In the 1850s, Harper lectured against slavery throughout the North. Later in her life, she called for other reforms, such as the right to vote for women. As this section explains, many individuals in the mid-1800s demanded equal rights for African Americans and women.

Abolitionists Call for Ending Slavery

Abolition, the movement to end slavery, began in the late 1700s. By 1804, most Northern states had outlawed slavery. In 1807, Congress banned the importation of African slaves into the United States. Abolitionists then began to demand a law ending slavery in the South.

David Walker, a free African American in Boston, printed a pamphlet in 1829 urging slaves to revolt. Copies of this pamphlet, *Appeal . . . to the Colored Citizens of the World,* made their way into the South. This angered slaveholders. When Walker heard that his life was in danger, he refused to run away. Shortly afterward, he died mysteriously.

Northern whites also fought slavery. In 1831, William Lloyd Garrison began to publish an abolitionist newspaper, *The Liberator,* in Boston. Of his antislavery stand, he wrote, "I will not retreat a single inch—AND I WILL BE HEARD." Many people hated his views. In 1834, a furious mob in Boston grabbed Garrison and dragged him toward a park to hang him. The mayor stepped in and saved his life.

Reading History
A. Drawing Conclusions How would the Grimké sisters' background help them as abolitionist speakers?

Two famous abolitionists were Southerners who had grown up on a plantation. Sisters Sarah and Angelina Grimké believed that slavery was morally wrong. Moving north, they became Quakers and joined the American Anti-Slavery Society. Women were not supposed to lecture in public in the mid-1800s. However, the Grimké sisters spoke out for abolition even though they received criticism for it.

In 1840, abolitionists closely watched the trial of a group of Africans who had rebelled on a slave ship, *Amistad*. Would the rebellious Africans be sent to slavery in Cuba or set free? To win proslavery votes, President Martin Van Buren tried to send them to Cuba. But former president John Quincy Adams helped win their freedom before the Supreme Court in 1841. The Africans returned home in 1842.

Eyewitnesses to Slavery

Two moving abolitionist speakers, **Frederick Douglass** and **Sojourner Truth,** spoke from their own experience of slavery. In 1842, Douglass announced in a speech that he was a thief: "I stole this head, these limbs, this body from my master, and ran off with them."

Douglass's courage and talent at public speaking won him a career as a lecturer for the Massachusetts Anti-Slavery Society. Poet James Russell Lowell said of him, "The very look and bearing of Douglass are an irresistible logic against the oppression of his race."

People who opposed abolition spread rumors that the brilliant speaker could never have been a slave. To prove them wrong, in 1845 Douglass published an autobiography that vividly narrated his slave experiences. Afterwards, he feared recapture by his owner, so he left America for a two-year speaking tour of Great Britain and Ireland. When Douglass returned, he bought his freedom. He began to publish an antislavery newspaper.

Sojourner Truth also began life enslaved. Originally named Isabella, Sojourner Truth was born in New York State. In 1827, she fled her owners and went to live with Quakers, who set her free. They also helped her win a court battle to recover her young son. He had been sold illegally into slavery in the South.

Reading History
B. Comparing How were Frederick Douglass and Sojourner Truth similar as abolitionists?

A devout Christian, Truth changed her name in 1843 to reflect her life's work: to sojourn (or stay temporarily in a place) and "declare the truth to the people." Speaking for abolition, she drew huge crowds throughout the North.

AMERICA'S HISTORY MAKERS

FREDERICK DOUGLASS
1817–1895

Douglass, born Frederick Bailey, was the son of a black mother and a white father. When he was eight, his owner sent him to be a servant for the Auld family. Mrs. Auld defied state law and taught young Frederick to read.

At the age of 16, Douglass returned to the plantation as a field hand. He endured so many whippings he later wrote, "I was seldom free from a sore back."

In 1838, he escaped to the North by hopping a train with a borrowed pass. To avoid recapture, he changed his last name.

How did Mrs. Auld unknowingly help Douglass become an abolitionist leader? Explain.

The Underground Railroad

Some abolitionists wanted to do more than campaign for laws ending slavery. Some brave people helped slaves escape to freedom along the Underground Railroad. Neither underground nor a railroad, the **Underground Railroad** was actually an aboveground series of escape routes from the South to the North. On these routes, runaway slaves traveled on foot. They also took wagons, boats, and trains.

Some enslaved persons found more unusual routes to freedom. For example, Henry Brown persuaded a white carpenter named Samuel A. Smith to pack him in a wooden box and ship him to Philadelphia. The box was only two and one half feet deep, two feet wide, and three feet long. It bore the label "This side up with care." Despite the label, Brown spent several miserable hours traveling head down. At the end of about 24 hours, Henry "Box" Brown climbed out of his box a free man in Philadelphia. Brown eventually made his way to Boston and worked on the Underground Railroad.

On the Underground Railroad, the runaways usually traveled by night and hid by day in places called stations. Stables, attics, and cellars all served as stations. At his home in Rochester, New York, Frederick Douglass hid up to 11 runaways at a time.

Reading **History**

C. Reading a Map The map on page 447 shows the routes of the Underground Railroad. Notice that most of these routes led to Canada.

Now and then

THE UNDERGROUND RAILROAD

In 1996, historian Anthony Cohen took six weeks to travel from Maryland to Canada. Cohen followed the paths runaway slaves had taken 150 years earlier. He is shown below arriving in Canada.

Cohen walked, sometimes as much as 37 miles in a day. He also hitched rides on trains and canal boats.

About those long-ago slaves fleeing toward the hope of freedom, Cohen said, "They had no choice. . . . Nobody would do this if they didn't have to."

Harriet Tubman

The people who led the runaways to freedom were called conductors. One of the most famous conductors was **Harriet Tubman.** Born into slavery in Maryland, the 13-year-old Tubman once tried to save another slave from punishment. The angry overseer fractured Tubman's skull with a two-pound weight. She suffered fainting spells for the rest of her life but did not let that stop her from working for freedom. In 1849, Tubman learned that her owner was about to sell her. Instead, she escaped. She later described her feelings as she crossed into the free state of Pennsylvania: "I looked at my hands to see if I was the same person now that I was free. There was such a glory over everything."

After her escape, Harriet Tubman made 19 dangerous journeys to free enslaved persons. The tiny woman carried a pistol to frighten off slave hunters and medicine to quiet crying babies. Her enemies offered $40,000 for her capture, but no one caught her. "I never run my train off the track and I never lost a passenger," she proudly declared. Among the people she saved were her parents.

Reading **History**

D. Forming and Supporting Opinions Why do you think escaped slaves such as Brown, Douglass, and Tubman risked their lives to help free others?

Reformers' Hall of Fame

William Lloyd Garrison

Even after being threatened with hanging, Garrison continued to publish his antislavery newspaper, *The Liberator.*

Sojourner Truth and Harriet Tubman

Truth spoke out for both abolition and women's rights. Tubman risked her life leading people to freedom on the Underground Railroad.

Lucretia Mott and Susan B. Anthony

An abolitionist, Mott also helped lead the movement for women's rights. Anthony fought for women's suffrage into the 20th century.

Women Reformers Face Barriers

Vocabulary
delegation: a group that represents a larger group

Other women besides the Grimké sisters and Sojourner Truth were abolitionists. Two of these were Lucretia Mott and **Elizabeth Cady Stanton.** Mott and Stanton were part of an American delegation that attended the World Anti-Slavery Convention in London in 1840. These women had much to say about their work. Yet when they tried to enter the convention, they were not allowed to do so. Men angrily claimed that it was not a woman's place to speak in public. Instead, the women had to sit silent behind a heavy curtain.

To show his support, William Lloyd Garrison joined them. He said, "After battling so many long years for the liberties of African slaves, I can take no part in a convention that strikes down the most sacred rights of all women."

Stanton applauded Garrison for giving up his chance to speak on abolition, the cause for which he had fought so long. "It was a great act of self-sacrifice that should never be forgotten by women."

However, most people agreed with the men who said that women should stay out of public life. Women in the 1800s enjoyed few legal or political rights. They could not vote, sit on juries, or hold public office. Many laws treated women—especially married women—as children. Single women enjoyed some freedoms, such as being able to manage their own property. But in most states, a husband controlled any property his wife inherited and any wages she might earn.

As the convention ended, Stanton and Mott decided it was time to demand equality for women. They made up their minds to hold a convention for women's rights when they returned home.

AMERICA'S HISTORY MAKERS

ELIZABETH CADY STANTON

1815–1902

Elizabeth Cady Stanton's first memory was the birth of a sister when she was four. So many people said, "What a pity it is she's a girl!" that Stanton felt sorry for the new baby. She later wrote, "I did not understand at that time that girls were considered an inferior order of beings."

When Stanton was 11, her only brother died. Her father said, "Oh, my daughter, I wish you were a boy!" That sealed Stanton's determination to prove that girls were just as important as boys.

How did Stanton's childhood experiences motivate her to help other people besides herself?

The Seneca Falls Convention

Stanton and Mott held the **Seneca Falls Convention** for women's rights in Seneca Falls, New York, on July 19 and 20, 1848. The convention attracted between 100 and 300 women and men, including Frederick Douglass.

Before the meeting opened, a small group of planners debated how to present their complaints. One woman read aloud the Declaration of Independence. This inspired the planners to write a document modeled on it. The women called their document the Declaration of Sentiments and Resolutions. Just as the Declaration of Independence said that "All men are created equal," the Declaration of Sentiments stated that "All men and women are created equal." It went on to list several complaints or resolutions. Then it concluded with a demand for rights.

A VOICE FROM THE PAST

Now, in view of this entire disenfranchisement [denying the right to vote] of one-half the people of this country, their social and religious degradation—in view of the unjust laws above mentioned, and because women do feel themselves aggrieved, oppressed, and fraudulently deprived of their most sacred rights, we insist that they have immediate admission to all the rights and privileges which belong to them as citizens of the United States.

Seneca Falls Declaration of Sentiments and Resolutions, 1848

Reading **History**

E. Using Primary Sources Why did the women at the Seneca Falls Convention believe they deserved rights and privileges?

Every resolution won unanimous approval from the group except **suffrage,** or the right to vote. Some argued that the public would laugh at women if they asked for the vote. But Elizabeth Cady Stanton and Frederick Douglass fought for the resolution. They argued that the right to vote would give women political power that would help them win other rights. The resolution for suffrage won by a slim margin.

The women's rights movement was ridiculed. In 1852, the *New York Herald* poked fun at women who wanted "to vote, and to hustle with the rowdies at the polls" and to be men's equals. The editorial questioned what would happen if a pregnant woman gave birth "on the floor of Congress, in a storm at sea, or in the raging tempest of battle."

Continued Calls for Women's Rights

In the mid-1800s, three women lent powerful voices to the growing women's movement. Sojourner Truth, Maria Mitchell, and Susan B. Anthony each offered a special talent.

In 1851, Sojourner Truth rose to speak at a convention for women's rights in Ohio. Some participants hissed their disapproval. Because Truth supported the controversial cause of abolition, they feared her appearance would make their own cause less popular. But Truth won applause with her speech that urged men to grant women their rights.

A VOICE FROM THE PAST

I have heard much about the sexes being equal. I can carry as much as any man, and can eat as much too, if I can get it. I am as strong as any man. . . . If you have woman's rights give it to her and you will feel better. You will have your own rights, and they won't be so much trouble.

Sojourner Truth, quoted by Marius Robinson, convention secretary

The scientist Maria Mitchell fought for women's equality by helping to found the Association for the Advancement of Women. Mitchell was an astronomer who discovered a comet in 1847. She became the first woman elected to the American Academy of Arts and Sciences.

Susan B. Anthony was a skilled organizer who worked in the temperance and antislavery movements. She built the women's movement into a national organization. Anthony argued that a woman must "have a purse [money] of her own." To this end, she supported laws that would give married women rights to their own property and wages. Mississippi passed the first such law in 1839. New York passed a property law in 1848 and a wages law in 1860. By 1865, 29 states had similar laws. (Anthony also fought for suffrage. See Chapter 22.)

But women's suffrage stayed out of reach until the 1900s, and the U.S. government did not fully abolish slavery until 1865. As you will read in the next chapter, the issue of slavery began to tear the nation apart in the mid-1800s.

This drawing shows a husband and wife fighting over who will "wear the pants in the family"—that is, who will rule the household.

Section 4 Assessment

1. Terms & Names

Identify:
- abolition
- Frederick Douglass
- Sojourner Truth
- Underground Railroad
- Harriet Tubman
- Elizabeth Cady Stanton
- Seneca Falls Convention
- suffrage

2. Taking Notes

On a time line like the one below, record important events in the historical development of the abolition movement.

1807 — 1865

Why does the time line end in 1865?

3. Main Ideas

a. Why were freedom of speech and freedom of the press important to the abolitionist movement?

b. What were Frederick Douglass's contributions to the abolitionist movement?

c. What were Elizabeth Cady Stanton's contributions to the women's rights movement?

4. Critical Thinking

Drawing Conclusions

Why do you think that many of the people who fought for abolition also fought for women's rights?

THINK ABOUT
- why they opposed slavery
- the social and economic position of women
- what the two causes had in common

ACTIVITY OPTIONS

TECHNOLOGY

DRAMA

With a partner, act out a meeting between a reformer from Section 3 and one from Section 4. **Videotape** their conversation or **perform** it for the class.

GEOGRAPHY in HISTORY

PLACE AND MOVEMENT

The Underground Railroad

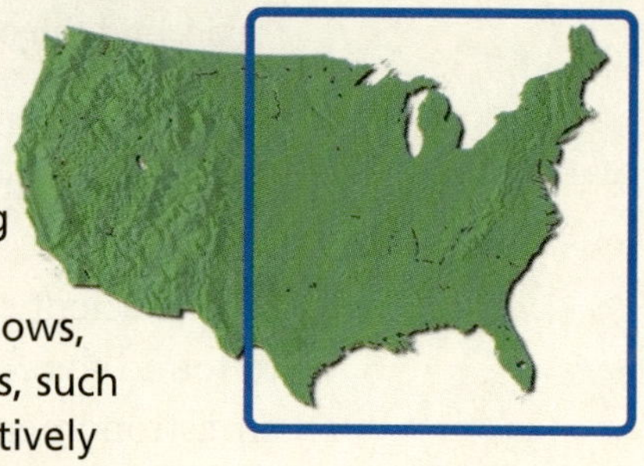

The Underground Railroad was a network of people and places that hid escaping slaves and helped them reach safety in the North or in Canada. One reason slaves often went to Canada is that a U.S. federal law required people to return runaway slaves to their owners. Defying this law, both whites and blacks helped slaves to escape.

The map on page 447 shows the main escape routes. As the map shows, most of the slaves who escaped came from states bordering free states, such as Kentucky and Virginia. Distances from there to the North were relatively short, increasing the chances of reaching freedom. However, the number of slaves who escaped from the Deep South, such as Georgia and South Carolina, was very small, because of the long distances that had to be traveled. While no one knows the exact number, historians estimate that 40,000 to 100,000 people may have used the Underground Railroad on their journey from slavery to freedom.

Among the many people who helped slaves to freedom was former slave Harriet Tubman (far left). She became a well-known guide on the Underground Railroad. She is pictured with her husband (third from left), along with other formerly enslaved people.

ARTIFACT FILE

Identity Tag Enslaved persons were forced to wear tags that identified to whom they belonged.

Freedom Marker The "P" on the rock shown here told slaves that they were in Pennsylvania, a free state.

CONNECT TO GEOGRAPHY

1. **Place** What geographic feature made it more likely that a slave in Missouri would escape to Michigan than to New York?
2. **Movement** In what way did the Underground Railroad differ from other migrations?

See Geography Handbook, pages 4–5.

CONNECT TO HISTORY

3. **Drawing Conclusions** How did the Underground Railroad reflect the American people's division over slavery?

On-Line Field Trip

The National Underground Railroad Freedom Center is being built in Cincinnati, Ohio. Its collections will include artifacts and primary sources like this poster, which shows that substantial rewards were offered for the recapture of slaves.

Visit www.mcdougallittell.com for more information.

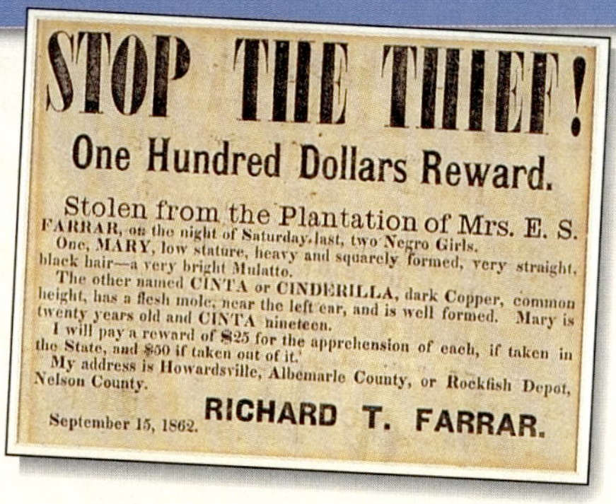

STOP THE THIEF!

One Hundred Dollars Reward.

Stolen from the Plantation of Mrs. E. S. FARRAR, on the night of Saturday last, two Negro Girls.

One, MARY, low stature, heavy and squarely formed, very straight, black hair—a very bright Mulatto.

The other named CINTA or CINDERILLA, dark Copper, common height, has a flesh mole, near the left ear, and is well formed. Mary is twenty years old and CINTA nineteen.

I will pay a reward of $25 for the apprehension of each, if taken in the State, and $50 if taken out of it.

My address is Howardsville, Albemarle County, or Rockfish Depot, Nelson County.

September 15, 1862. RICHARD T. FARRAR.

Chapter 14 ASSESSMENT

TERMS & NAMES

Briefly explain the importance of each of the following.

1. immigrant
2. push-pull factors
3. civil disobedience
4. revival
5. Second Great Awakening
6. labor union
7. abolition
8. Underground Railroad
9. Seneca Falls Convention
10. suffrage

REVIEW QUESTIONS

The Hopes of Immigrants (pages 423–428)

1. What factors influenced so many immigrants to come to America in the 1800s?
2. What did Germans contribute to U.S. culture?
3. How did the potato famine affect Irish emigration?

American Literature and Art (pages 429–432)

4. How did American artists display the love of nature in their paintings?
5. What did the Transcendentalists believe?

Reforming American Society (pages 433–439)

6. Why did many business owners support the temperance movement?
7. Why was it hard for African Americans to receive an education?

Abolition and Women's Rights (pages 440–447)

8. Who published antislavery writings?
9. How did the Underground Railroad work?
10. What was the Seneca Falls Declaration of Sentiments and Resolutions?

CRITICAL THINKING

1. USING YOUR NOTES

	How People Influenced America in the mid-1800s
Immigrants	
Writers	
Reformers	
Abolitionists	
Women	

Using your chart, answer the questions below.

a. Who influenced America to make reforms?

b. Compare the goals of abolitionists and women. How are they alike?

2. ANALYZING LEADERSHIP

Who is someone from this chapter who exercised leadership by standing up for an unpopular position?

3. THEME: IMPACT OF THE INDIVIDUAL

Judging from what you read in this chapter, what methods can individuals use to influence their society?

4. APPLYING CITIZENSHIP SKILLS

Who in this chapter displayed good citizenship by taking responsibility for their own behavior or by providing for their families? Give examples.

5. FORMING AND SUPPORTING OPINIONS

If someone asked you what was the most important reform of this period, what would you say? Why?

Interact *with* History

Think about the laws you proposed before you read the chapter. Has your opinion changed since you read the chapter?

VISUAL SUMMARY

A New Spirit of Change

HISTORY SKILLS

1. INTERPRETING GRAPHS

School Enrollment, *1840–1870*

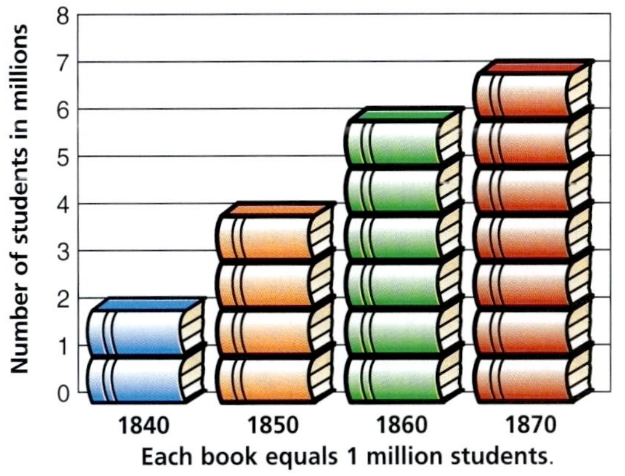

Source: *Census of the United States*

Basic Graph Elements

a. On the graph, what does each book stand for?

Interpreting the Graph

b. How did school enrollment change during the period from 1840 to 1870?

c. What is the difference in school enrollment between 1840 and 1870?

2. INTERPRETING PRIMARY SOURCES

Philip Younger escaped from slavery and eventually went to Canada. He dictated this narrative to Benjamin Drew, who included it in his *A North-Side View of Slavery.*

> I served in slavery fifty-five years, and am now nearly seventy-two years old. . . . I got off by skill. I have children and grandchildren in slavery. I had rather starve to death here, being a free man, than to have plenty in slavery. I cannot be a slave any more—nobody could hold me as a slave now, except in irons. Old as I am, I would rather face the Russian fire, or die at the point of the sword, than go into slavery.
>
> **Philip Younger,** quoted in *A North-Side View of Slavery*

a. What do you think Younger means by saying, "I got off by skill"?

b. How important is his freedom to him?

ALTERNATIVE ASSESSMENT

1. INTERDISCIPLINARY ACTIVITY: Literature

Reading a Poem Aloud Choose a poem by Henry Wadsworth Longfellow that includes references to American history. Research the actual event to which the poem refers. Write an introduction that explains whether the poem portrays history accurately. Practice reading the poem until you are comfortable. Then present the poem to the class.

2. COOPERATIVE LEARNING ACTIVITY

Making an Annotated Map Working in a small group, create an annotated map of the Underground Railroad. Make an enlargement of the map of the Underground Railroad on page 447. Then annotate it, using some of the following suggestions.

- Do research to find the stories of slaves who escaped along the Underground Railroad.
- Put brief, typed summaries of a few stories on the map, with arrows connecting them to the right route.
- Draw illustrations of escape narratives and put them on the map.
- Research to discover appropriate songs and create a recording that can play as people view the map.

3. CD-ROM PRIMARY SOURCE EXPLORER

Preparing a Report Dorothea Dix did not just stand before the legislature and talk about her own opinions. She gathered evidence and then wrote a persuasive report. Using the CD-ROM, library, and Internet, find out more about her report. Plan your own report that you could present to the student government.

- Choose a problem in your school that you think needs addressing.
- Find evidence of the problem. This may involve interviewing other students or observing events.
- Decide what you want the student government or school board to do about the problem.
- Write your report. Make sure to refer to the evidence that you gathered.

4. HISTORY PORTFOLIO

PORTFOLIO **Option 1** Review your section and chapter assessment activities. Select one that you think is your best work. Then use comments made by your teacher or classmates to improve your work and add it to your portfolio.

PORTFOLIO **Option 2** Review the questions that you wrote for What Do You Want to Know? on page 422. Write a short report in which you explain the answers to your questions. If any questions were not answered, do research to answer them. Add your report to your portfolio.

Pack Your Trunk

For immigrants, packing up to go to a new land required making hard decisions. Wealthy people could ship belongings ahead. Most immigrants, though, carried their belongings in burlap bags, knotted sheets, large baskets, or small trunks. Even children carried small bundles. Only the very basic items or very precious ones could be taken to the United States. Baggage contained practical items such as tools and household items. But some had personal items such as portraits of loved ones. A few people even carried bags of dirt from their home country!

ACTIVITY Pack a trunk with items needed for a new life in the United States. Explain why you chose the items that you have packed. Finally, write a letter to a friend or relative back in Europe about your journey to the United States.

TOOLBOX

Each group will need:

- a shoebox
- assorted magazines (optional)
- craft sticks
- drawing paper or posterboard
- markers or colored pencils
- 3 x 5 note cards
- masking tape
- styrofoam (optional)

STEP BY STEP

1 Form groups. Each group should consist of about four or five students. Assign group members the following tasks:

- Do research on what people brought with them when immigrating to the United States.
- Choose ten items that you will need for your new life.
- Present your items in class and give reasons for selecting them.

2 Research what immigrants brought with them. In the library or on the Internet (see Researching Your Project on the next page), research what immigrants brought with them to the United States. Make a list of everything you think you'll need for a new life in the United States. Some basic items included

- books
- favorite or special clothing
- toys
- important documents

Real immigrants brought these items to the United States: a mortar and pestle (used to grind spices or medicines), a shoe brush, a coffee grinder, and a paisley shawl.

3 Create your items. From your list, choose ten items that you think will be most important to starting a new life. Then draw pictures of the selected items or cut pictures of them out of a magazine. Attach a craft stick to the back of the picture with masking tape.

4 Write reasons. Think of why you selected each of the ten items. Write the reasons for each item on a separate 3 x 5 note card. Attach each note card to the back of the corresponding picture.

5 Decorate the shoebox to look like a trunk. Using the masking tape, affix your pictures to the rim of the shoebox, or use styrofoam in the bottom of the box to insert the pictures.

6 Examine other groups' trunks. Walk around the room and examine the contents of the other groups' trunks. Compare your trunk with that of your classmates. Share your reasons for selecting certain items.

WRITE AND SPEAK

Write a descriptive letter. Use the point of view of an immigrant. Write a letter to someone in your homeland describing your journey to the United States. Your letter might also describe what you miss most (personal belongings or people, for example) since the move. Read the letters to others in your class.

HELP DESK

For related information, see pages 423–428 in Chapter 14.

Researching Your Project

- *They Sought a New World* by William Kurdek and Margaret S. Englehart
- *American Immigration* by Edward G. Hartmann

Visit www.mcdougallittell.com for more information about the immigrant experience.

Did You Know?

Most immigrants traveled in **steerage** or third class. It was the lowest area of the ship, where a steering mechanism was located. A family "berth," or space allotted, in steerage was about six feet square.

Before World War I, fares in steerage to the United States from Europe were never more than $35 and by 1900 were as low as $10.

Shipping companies often fed herring (a kind of fish) to the immigrants. Herring was cheap and nourishing. It was also thought to help prevent seasickness.

REFLECT & ASSESS

- What priorities did you use in selecting items for your trunk?
- Which items would have to be left behind if you only had a small bag for your belongings?
- Why do you think other groups selected items different from yours?

UNIT 5

1846 – 1877

A Nation Divided and Rebuilt

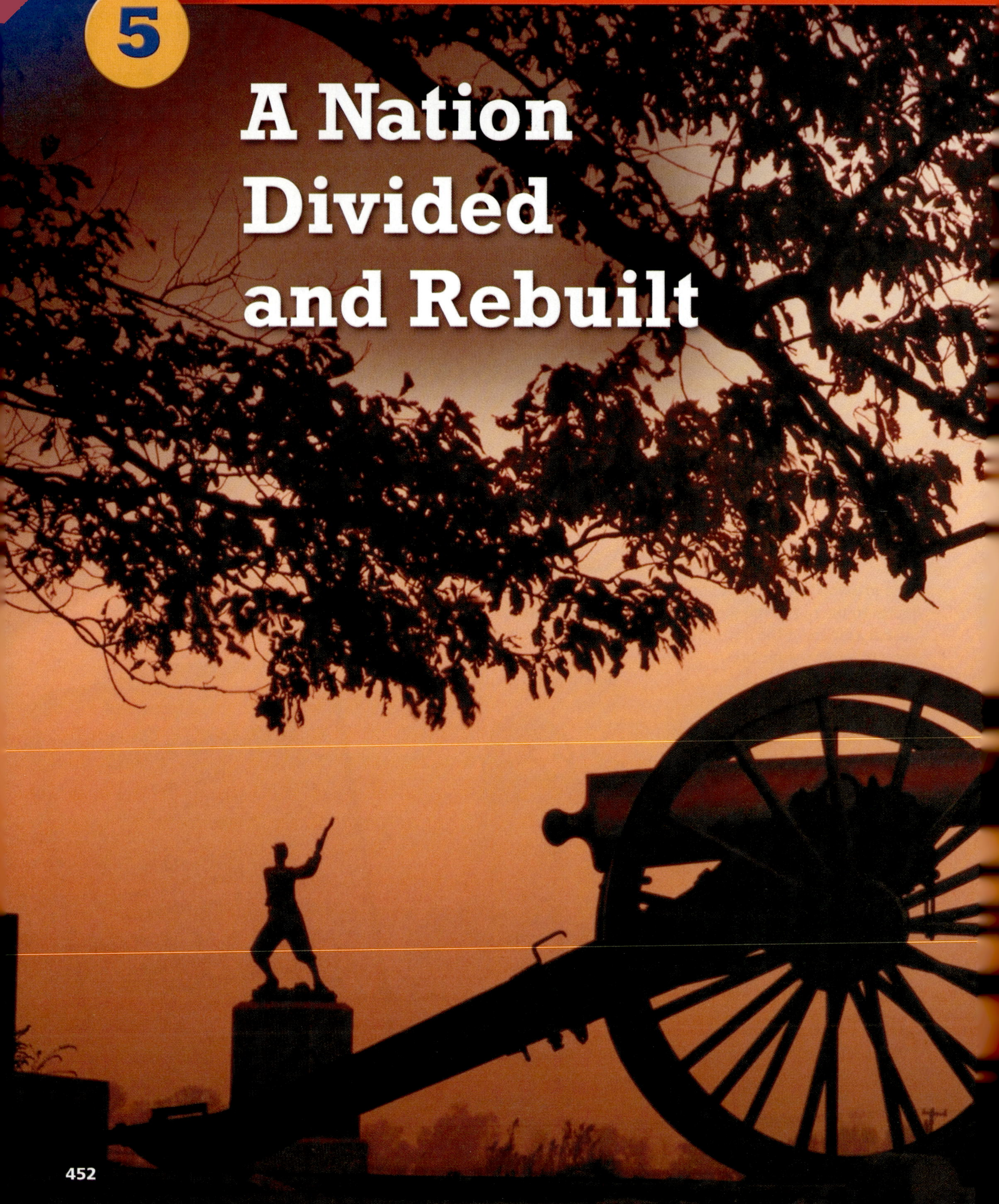

"Government of the people, by the people, for the people, shall not perish from the earth."

—Abraham Lincoln, Gettysburg Address

Gettysburg, Pennsylvania, was the site of the most decisive battle of the Civil War.

CHAPTER 15

The Nation Breaking Apart

1846–1861

Abraham Lincoln and Stephen A. Douglas debate the issue of slavery in the 1858 Senate campaign in Illinois.

Interact *with* History

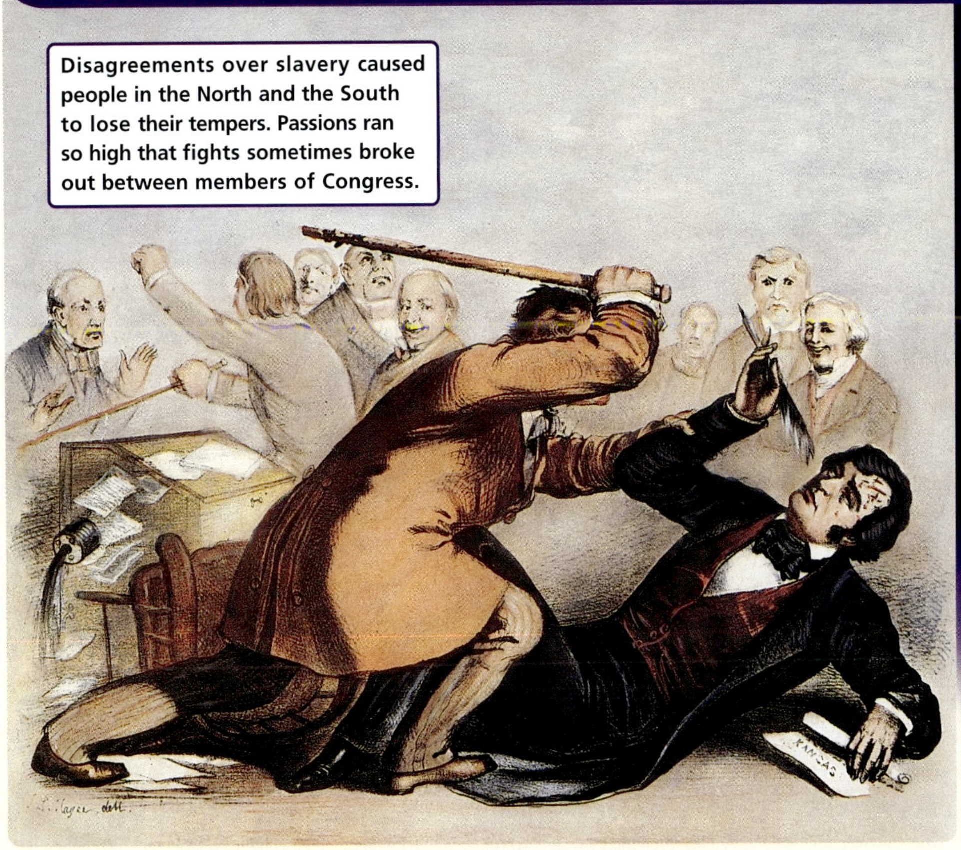

Disagreements over slavery caused people in the North and the South to lose their tempers. Passions ran so high that fights sometimes broke out between members of Congress.

You are a representative in Congress in the 1850s. The issue of slavery is causing heated debates. Tensions over slavery have risen so high that respectable men have turned to violence to settle their differences. You worry that soon this violence may affect the entire nation.

What Do You Think?

- Why do you think people feel so strongly about slavery?
- Do you think debates, such as those between Lincoln and Douglas, could settle emotional issues without leading to violence?

How would you keep the nation together?

USA

1846
War with Mexico begins.
Wilmot Proviso is introduced.

1848
Zachary Taylor is elected president.

1850
President Taylor dies. Millard Fillmore becomes president.
Congress passes the Compromise of 1850.

1852
Uncle Tom's Cabin is published.
Franklin Pierce is elected president.

1854
Congress passes the Kansas–Nebraska Act.

1856
James Buchanan is elected president.

1857
Chief Justice Roger B. Taney delivers his opinion in the *Dred Scott* case.

1860
Abraham Lincoln is elected president.

1861
The Confederate States of America is formed.

World

1848
Rebellions erupt across Europe.

1850
Taiping Rebellion begins in China.

1856
War breaks out between Britain and Persia.

1861
Czar Alexander II frees the serfs in Russia.

BEFORE YOU READ

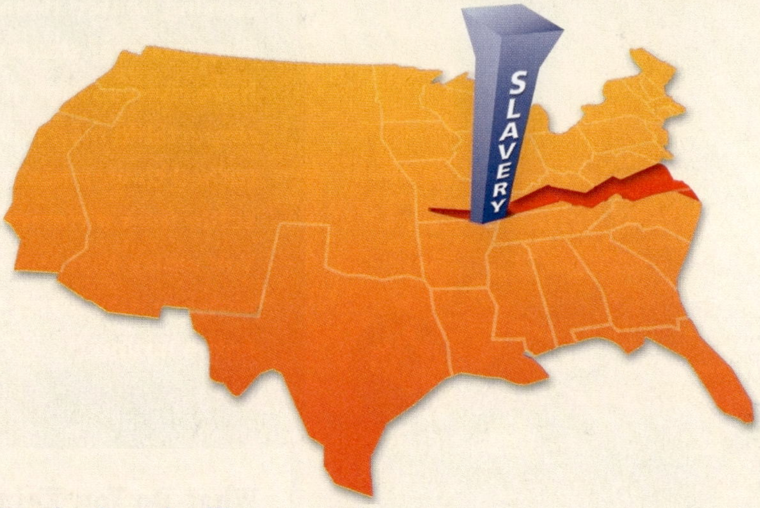

Previewing the Theme

Diversity and Unity As you read in Chapter 14, slavery heightened tensions between Northerners and Southerners. Many people feared that the issue might tear the nation apart. But most Americans hoped that they would be able to solve their problems peacefully. Chapter 15 describes how these tensions led to crisis. It also explains how Americans tried to keep their nation united and how those efforts failed.

What Do You Know?

What do you think about when you hear the terms *slavery* and *abolition?* Why do you think the issue of slavery caused so much anger and resentment?

THINK ABOUT

- what you've learned about differences between the North and the South from books, travel, television, or movies
- reasons people have violent conflicts today

What Do You Want to Know?

What questions do you have about the sectional crisis that led to the Civil War? Record them in your notebook before you read this chapter.

READ AND TAKE NOTES

Reading Strategy: Analyzing Causes Analyzing causes means looking closely at events and describing why they happened. The diagram below will help you analyze some of the causes of secession.

Use the diagram to take notes on how each issue drove the North and the South farther apart.

See Skillbuilder Handbook, page R10.

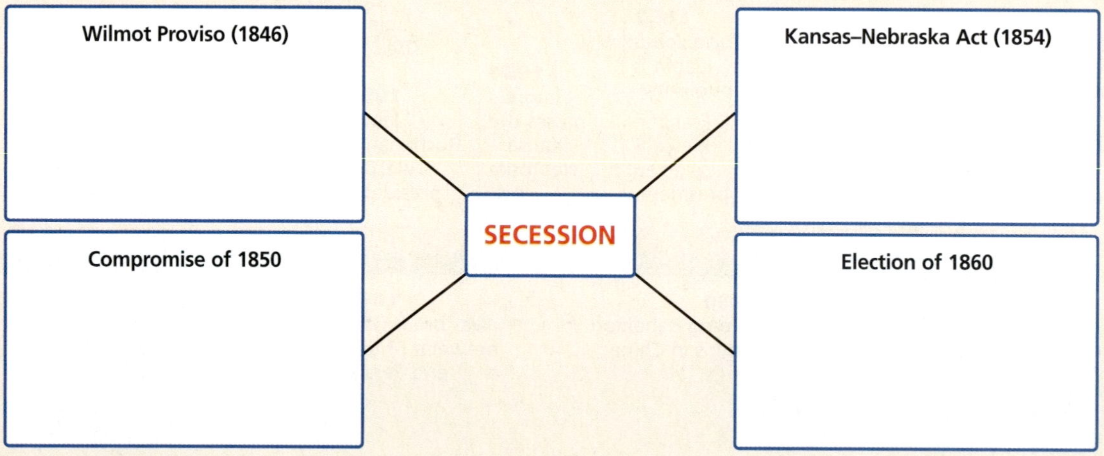

1 Growing Tensions Between North and South

TERMS & NAMES
Wilmot Proviso
Free-Soil Party
Henry Clay
Daniel Webster
Stephen A. Douglas
Compromise of 1850

MAIN IDEA

Disagreements between the North and the South, especially over the issue of slavery, led to political conflict.

WHY IT MATTERS NOW

Regional differences can make national problems difficult to resolve.

ONE EUROPEAN'S STORY

Alexis de Tocqueville [TOHK•vihl] was a young French government official from a wealthy family. In 1831, he sailed across the Atlantic Ocean to study American prisons and politics.

At one point, Tocqueville traveled in a steamship down the Ohio River. The river was the border between Ohio, a free state, and Kentucky, a slave state. Tocqueville noted what he saw on both sides of the river.

Alexis de Tocqueville

A VOICE FROM THE PAST

The State of Ohio is separated from Kentucky just by one river; on either side of it the soil is equally fertile, and the situation equally favourable, and yet everything is different. Here [on the Ohio side] a population devoured by feverish activity, trying every means to make its fortune. . . . There [on the Kentucky side] are people who make others work for them and show little compassion, a people without energy, mettle or the spirit of enterprise. . . . These differences cannot be attributed to any other cause but slavery. It degrades the black population and enervates [saps the energy of] the white.

Alexis de Tocqueville, *Journey to America*

In this section, you will read about the differences between the North and the South. You will also read about how these differences caused political tensions that threatened to tear the nation apart.

North and South Take Different Paths

As you read in Chapter 11, the economies of the North and the South developed differently in the early 1800s. Although both economies were mostly agricultural, the North began to develop more industry and commerce. By contrast, the Southern economy relied on plantation farming.

The growth of industry in the North helped lead to the rapid growth of Northern cities. Much of this population growth came from immigration. In addition, immigrants and Easterners moved west and built farms in the new states formed from the Northwest Territory. Most canals and railroads ran east and west, helping the Eastern and Midwestern states develop strong ties with each other.

onomics *in* History

Trade

Trade is based on a simple idea. If you have something someone else needs or wants, and that person has something you need or want, you exchange, or trade, those two things. After the trade, you should both be better off than before.

The concept of trade works similarly for groups of people. For example, in the early 1800s, the South had few factories. Planters who wanted manufactured goods usually had to buy them from manufacturers in the North or in Europe. To have the cash to buy those goods, Southerners sold other goods, such as cotton, to the North and other countries. Each sold the goods they could produce in order to get money to buy the goods they could not make.

CONNECT TO HISTORY

1. **Solving Problems** What problem does trade help a country solve? How else could a country solve this problem?

See Skillbuilder Handbook, page R17.

CONNECT TO TODAY

2. **Comparing** What goods do Americans sell to other countries today? What goods do Americans buy from other countries?

Visit www.mcdougallittell.com for more about trade.

The South developed differently than the North. A few wealthy planters controlled Southern society. They made great profits from the labor of their slaves. Much of this profit came from trade. Planters relied on exports, especially cotton. Because these plantations were so profitable, planters invested in slaves instead of industry. As a result, the South developed little industry.

Most Southern whites were poor farmers who owned no slaves. Many of these people resented the powerful slaveholders. But poor whites accepted slavery because it kept them off the bottom of society.

Reading **History**
A. Making Generalizations How did the economies of the North and the South differ?

Antislavery and Racism

The issue of slavery caused tension between the North and the South. In the North, the antislavery movement had slowly been gaining strength since the 1830s. Abolitionists believed that slavery was unjust and should be abolished immediately. Many Northerners who opposed slavery took a less extreme position. Some Northern workers and immigrants opposed slavery because it was an economic threat to them. Because slaves did not work for pay, free workers feared that managers would employ slaves rather than them. Some workers were even afraid that the expansion of slavery might force workers into slavery to find jobs.

Despite their opposition to slavery, most Northerners, even abolitionists, were racist by modern standards. Many whites refused to go to school with, work with, or live near African Americans. In most states, African Americans could not vote.

Vocabulary
racist: having prejudice based on race

When Northern attacks on slavery increased, slaveholders defended slavery. Most offered the openly racist argument that white people were superior to blacks. Many also claimed that slavery helped slaves by introducing them to Christianity, as well as providing them with food, clothing, and shelter throughout their lives. Slaveholders were determined to defend slavery and their way of life. In this way, the different ideas about slavery brought the North and the South into conflict.

The Wilmot Proviso

After the Missouri Compromise in 1820, political disagreements over slavery seemed to go away. But new disagreements arose with the outbreak of the War with Mexico in 1846. Many Northerners believed that Southerners wanted to take territory from Mexico in order to extend slavery. To prevent that, Representative David Wilmot of Pennsylvania proposed a bill, known as the **Wilmot Proviso,** to outlaw slavery in any territory the United States might acquire from the War with Mexico.

But slaveholders believed that Congress had no right to prevent them from bringing slaves into any of the territories. They viewed slaves as property. The Constitution, they claimed, gave equal protection to the property rights of all U.S. citizens. The Wilmot Proviso removed the right of slaveholders to take their slaves, which they regarded as property, anywhere in the United States or its territories. Southerners claimed that the bill was unconstitutional.

The Wilmot Proviso divided Congress along regional lines. The bill passed the House of Representatives. But Southerners prevented it from passing the Senate.

Even though the Wilmot Proviso never became law, it had important effects. It led to the creation of the **Free-Soil Party,** a political party dedicated to stopping the expansion of slavery. The party's slogan expressed its ideals—"Free Soil, Free Speech, Free Labor, and Free Men." The Free-Soil Party won more than ten seats in Congress in the election of 1848. More important, the party made slavery a key issue in national politics. Politicians could ignore slavery no longer.

Connections TO WORLD HISTORY

EXPANDING SLAVERY

William Walker, a Tennessee-born adventurer, wanted to take over land in Central America. In 1855, he joined an army of Nicaraguan rebels and seized power. Walker declared himself president of Nicaragua in 1856. As president, he legalized slavery there.

Troops from nearby countries drove him from power in 1857. The actions of men like Walker helped to convince Northerners that slaveholders were intent on expanding slavery beyond the U.S. South.

***Reading*History**
B. Recognizing Effects What were the effects of the Wilmot Proviso?

Controversy over Territories

By 1848, the nation's leaders had begun to debate how to deal with slavery in the lands gained from the War with Mexico. The proposed addition of new states threatened the balance in Congress between North and South. The discovery of gold in California brought thousands of people into that territory. There would soon be enough people in California for it to apply for statehood. Most California residents wanted their state to be a free state. But this would tip the balance of power clearly in favor of the North. Southerners wanted to divide California in half, making the northern half a free state and the southern half a slave state.

In 1849, President Zachary Taylor proposed that California submit a plan for statehood that year, without going through the territorial stage. By skipping this stage, Taylor's plan gave Southern slaveholders little time to move to California with their slaves.

In March 1850, California applied to be admitted as a free state. With California as a free state, slave states would become a minority in the Senate just as they were in the House. Jefferson Davis, a senator from Mississippi, warned, "For the first time, we are about permanently to destroy the balance of power between the sections."

Background
U.S. land gains from the War with Mexico included all or parts of the future states of California, Nevada, Utah, Arizona, New Mexico, and Colorado.

The Compromise of 1850

California could not gain statehood, however, without the approval of Congress. And Congress was divided over the issue. Behind the scenes, statesmen sought compromise. Taking the lead was Senator **Henry Clay**

This engraving dramatically portrays the Senate debate over the Compromise of 1850.

Henry Clay led the Congress in creating compromises on several important issues during his long career.

John C. Calhoun of South Carolina opposed the Compromise of 1850. He believed the South had no reason to compromise on the issue of slavery.

Daniel Webster spoke eloquently in favor of the compromise.

of Kentucky. Clay had helped create the Missouri Compromise in 1820. Now Clay crafted a plan to settle the California problem.

Reading **History**

C. Reading a Map Look at the map on page 464 to see how the Compromise of 1850 affected the territories open to slavery.

1. To please the North, California would be admitted as a free state, and the slave trade would be abolished in Washington, D.C.
2. To please the South, Congress would not pass laws regarding slavery for the rest of the territories won from Mexico, and Congress would pass a stronger law to help slaveholders recapture runaway slaves.

Many people on both sides felt they had to give up too much in this plan. But others were tired of the regional bickering. They wanted to hold the Union together. **Daniel Webster,** senator from Massachusetts, supported the compromise for the sake of the Union.

A VOICE FROM THE PAST

I wish to speak today, not as a Massachusetts man, nor as a Northern man, but as an American. . . . I speak today for the preservation of the Union. Hear me for my cause.

Daniel Webster, quoted in *The Annals of America*

The job of winning passage of the plan fell to Senator **Stephen A. Douglas** of Illinois. By the end of September, Douglas succeeded, and the plan, now known as the **Compromise of 1850,** became law.

Some people celebrated the compromise, believing that it had saved the Union. But the compromise would not bring peace. In the next section, you will learn how sectional tensions continued to rise.

AMERICA'S HISTORY MAKERS

STEPHEN A. DOUGLAS

1813–1861

Stephen A. Douglas was one of the most powerful members of Congress in the mid-1800s. In fact, he was called the "Little Giant" because he commanded great respect even though he was only five feet four inches tall.

Perhaps the most important issue that Douglas faced during his career was the expansion of slavery into the territories. Douglas privately hated slavery. But he did not believe a debate on morality would do any good. He suggested that the people of each territory should decide whether or not to allow slavery.

What groups of Americans agreed with Douglas's position on slavery?

Section 1 Assessment

1. Terms & Names

Identify:

- Wilmot Proviso
- Free-Soil Party
- Henry Clay
- Daniel Webster
- Stephen A. Douglas
- Compromise of 1850

2. Taking Notes

Use a chart like the one below to explain the effects of each cause.

Causes	Effects
Abolitionism	
Wilmot Proviso	
California's application for statehood	

Which issue do you think most threatened national unity?

3. Main Ideas

a. What were two ways that the North and the South differed by the mid-1800s?

b. In what ways was racism common in both the North and the South?

c. How did the War with Mexico lead to conflict between the North and the South?

4. Critical Thinking

Comparing and Contrasting How was the Compromise of 1850 similar to and different from the Missouri Compromise?

THINK ABOUT

- the regional tensions at the time the compromises were proposed
- who proposed each bill
- the provisions of the bills

ACTIVITY OPTIONS

TECHNOLOGY

SPEECH

Imagine you are a television news director. Plan a five-minute **documentary** or organize a **panel discussion** on the Compromise of 1850.

2 The Crisis Deepens

TERMS & NAMES
- Harriet Beecher Stowe
- *Uncle Tom's Cabin*
- Fugitive Slave Act
- popular sovereignty
- Kansas–Nebraska Act
- John Brown

MAIN IDEA

Turmoil over slavery led to acts of violence.

WHY IT MATTERS NOW

Violence can make compromise more difficult.

ONE AMERICAN'S STORY

Harriet Beecher Stowe was outraged when she heard about the part of the Compromise of 1850 that would help slaveholders recapture runaway slaves. She described her feelings about the law.

A VOICE FROM THE PAST

Since the legislative act of 1850, when [I] heard . . . Christian and humane people actually recommending the remanding [returning of] escaped fugitives into slavery, as a duty binding on good citizens, . . . [I] could only think, These men and Christians cannot know what slavery is.

Harriet Beecher Stowe, *Uncle Tom's Cabin*

Stowe's anger motivated her to write *Uncle Tom's Cabin,* a novel that portrayed slavery as brutal and immoral. In this section, you will learn how the Compromise of 1850 deepened the division between the North and the South.

Harriet Beecher Stowe

The Fugitive Slave Act

The 1850 law to help slaveholders recapture runaway slaves was called the **Fugitive Slave Act**. People accused of being fugitives under this law could be held without an arrest warrant. In addition, they had no right to a jury trial. Instead, a federal commissioner ruled on each case. The commissioner received five dollars for releasing the defendant and ten dollars for turning the defendant over to a slaveholder.

Southerners felt that the Fugitive Slave Act was justified because they considered slaves to be property. But Northerners resented the Fugitive Slave Act. It required Northerners to help recapture runaway slaves. It placed fines on people who would not cooperate and jail terms on people who helped the fugitives escape. In addition, Southern slave catchers roamed the North, sometimes capturing free African Americans.

The presence of slave catchers throughout the North brought home the issue of slavery to Northerners. They could no longer ignore the fact that, by supporting the Fugitive Slave Act, they played an important role in supporting slavery. They faced a moral choice. Should they obey the law and support slavery, or should they break the law and oppose slavery?

Uncle Tom's Cabin

Reading History
A. Analyzing Points of View How does Harriet Beecher Stowe present slavery in *Uncle Tom's Cabin*?

The novel *Uncle Tom's Cabin,* published by Harriet Beecher Stowe in 1852, expressed the moral issues of slavery in a highly dramatic way. The main character of Stowe's novel was Uncle Tom, a respected older slave. The central plot tells of Tom's life under three owners. Two of the owners were kind, but the third—the evil Simon Legree—had Tom beaten. The novel includes dramatic episodes, such as the dangerous journey of an escaped slave named Eliza and her baby across the Ohio River.

A VOICE FROM THE PAST

Eliza made her desperate retreat across the river just in the dusk of twilight. The gray mist of evening, rising slowly from the river, enveloped her as she disappeared up the bank, and the swollen current and floundering masses of ice presented a hopeless barrier between her and her pursuer.

Harriet Beecher Stowe, *Uncle Tom's Cabin*

Stowe's book was wildly popular in the North. But white Southerners believed the book falsely criticized the South and slavery.

The Kansas–Nebraska Act

While the Fugitive Slave Act and *Uncle Tom's Cabin* heightened the conflicts between the North and the South, the issue of slavery in the territories brought bloodshed to the West. In 1854, Senator Stephen A. Douglas of Illinois drafted a bill to organize territorial governments for the Nebraska Territory. He proposed that it be divided into two territories—Nebraska and Kansas.

Background
The Nebraska Territory was part of the Louisiana Purchase. It lay north of the 36° 30′ line, so the Missouri Compromise banned slavery there.

To get Southern support for the bill, he suggested that the decision about whether to allow slavery in each of these territories be settled by popular sovereignty. **Popular sovereignty** is a system where the residents vote to decide an issue. If this bill passed, it would result in getting rid of the Missouri Compromise by

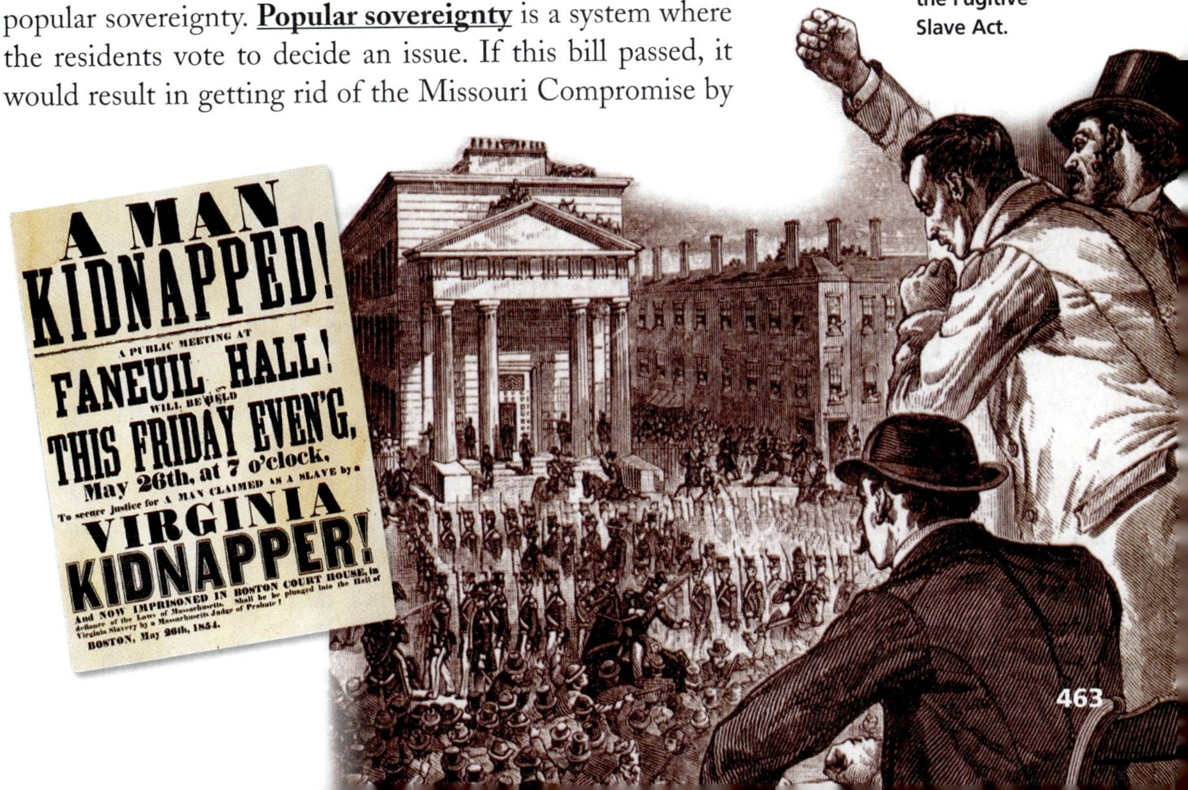

In 1854, Bostonians protested the capture of an African American by federal marshals under the Fugitive Slave Act.

...ee and Slave States and Territories, 1820–1854

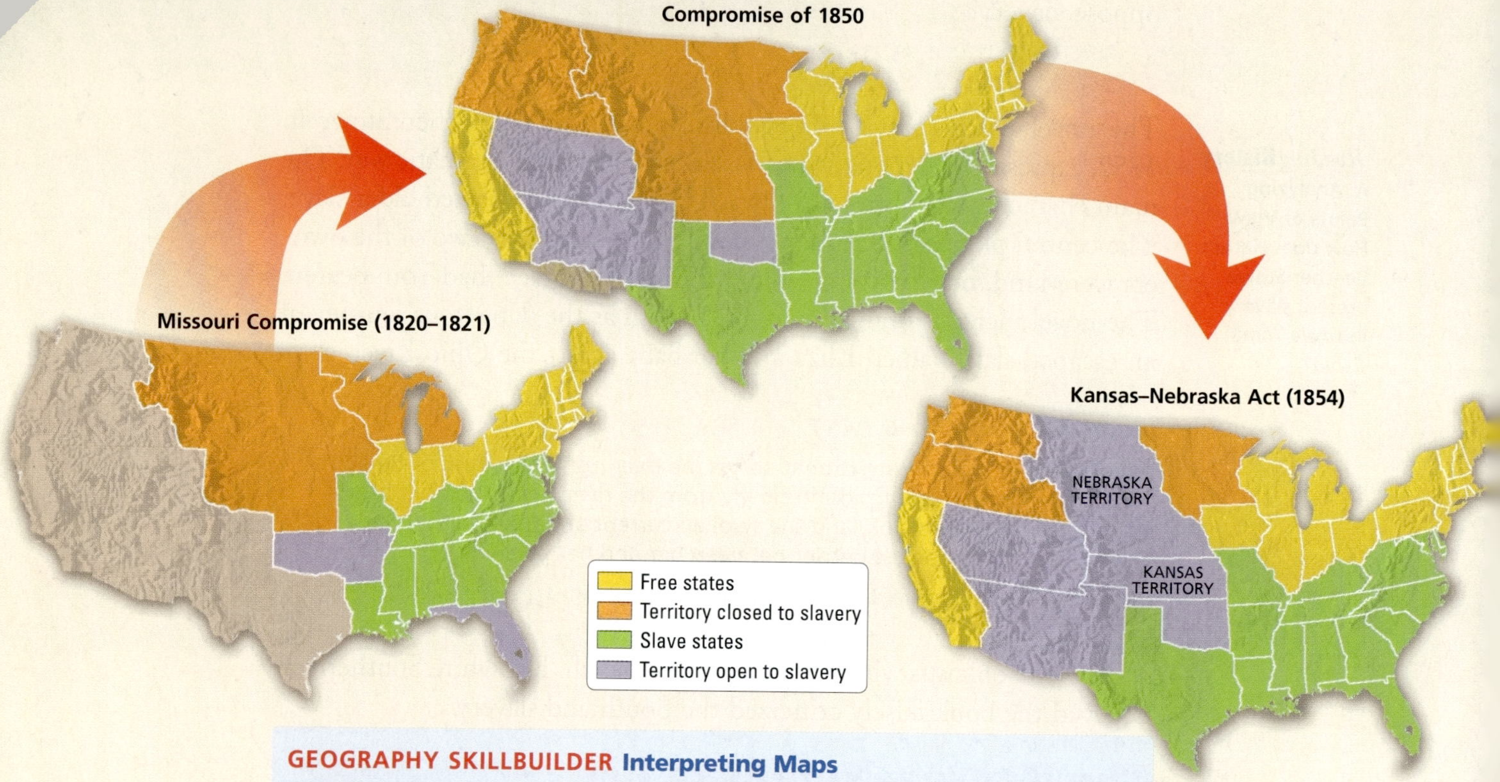

GEOGRAPHY SKILLBUILDER Interpreting Maps

1. **Region** *What new free states and slave states entered the Union between 1820 and 1854?*
2. **Region** *How did the Kansas–Nebraska Act change the amount of territory open to slavery?*

allowing people to vote for slavery in territories where the Missouri Compromise had banned it.

As Douglas hoped, Southerners applauded the repeal of the Missouri Compromise and supported the bill. Even though the bill angered opponents of slavery, it passed. It became known as the **Kansas–Nebraska Act.** Few people realized that the act would soon turn Kansas into a battleground over slavery.

"Bleeding Kansas"

Proslavery and antislavery settlers rushed into the Kansas Territory, just west of Missouri, to vote for the territorial legislature. At the time of the election in March 1855, there were more proslavery settlers than antislavery settlers in the territory. But the proslavery forces did not want to risk losing the election. Five thousand Missourians came and voted in the election illegally. As a result, the official Kansas legislature was packed with proslavery representatives.

Antislavery settlers boycotted the official government and formed a government of their own. With political authority in dispute, settlers on both sides armed themselves. In May, a proslavery mob attacked the town of Lawrence, Kansas. The attackers destroyed offices and the

Vocabulary
boycott: refuse to participate in

house of the governor of the antislavery government. This attack came to be known as the Sack of Lawrence.

Onto this explosive scene came **John Brown,** an extreme abolitionist. To avenge the Sack of Lawrence, Brown and seven other men went to the cabins of several of his proslavery neighbors and murdered five people. This attack is known as the Pottawatomie Massacre, after the creek near where the victims were found. As news of the violence spread, civil war broke out in Kansas. It continued for three years, and the territory came to be called "Bleeding Kansas."

Reading **History**

B. Sequencing Events What events in Kansas preceded the Pottawatomie Massacre?

Violence in Congress

While violence was spreading in Kansas in the spring of 1856, blood was also being shed in the nation's capital. In late May, Senator Charles Sumner of Massachusetts delivered a speech attacking the proslavery forces in Kansas. His speech was packed with insults. Sumner even made fun of A. P. Butler, a senator from South Carolina.

Preston Brooks, a relative of Butler, heard about Sumner's speech. To defend Butler and the South, he attacked Sumner, who was sitting at his desk. Brooks hit Sumner over the head with his cane. Sumner tried to defend himself, but his legs were trapped. Brooks hit him 30 times or more, breaking his cane in the assault. (The painting on page 455 shows this event.)

Many Southerners cheered Brooks's defense of the South. But most Northerners were shocked at the violence in the Senate. "Bleeding Kansas" and "Bleeding Sumner" became rallying cries for antislavery Northerners and slogans for a new political party. In the next section, you will learn about the creation of the Republican Party.

STRANGE *but* True

PRESTON BROOKS'S CANE

Many Americans, Northerners and Southerners alike, were ashamed of the behavior of Sumner and Brooks. But sectional tensions were so high at the time that a large number of Southerners cheered Brooks for his actions.

A number of Brooks's supporters sent him new canes to replace the one he had broken while hitting Sumner on the head. Some of the canes were inscribed with mottoes such as "Hit Him Again."

Section 2 Assessment

1. Terms & Names

Identify:

- Harriet Beecher Stowe
- *Uncle Tom's Cabin*
- Fugitive Slave Act
- popular sovereignty
- Kansas–Nebraska Act
- John Brown

2. Taking Notes

Use a chart like the one below to compare Northern and Southern views of the issues listed.

Northern View	Issue	Southern View
	Fugitive Slave Act	
	Kansas–Nebraska Act	
	"Bleeding Kansas"	

3. Main Ideas

a. How did the book *Uncle Tom's Cabin* influence national politics?

b. Why was the Kansas–Nebraska Act so controversial?

c. What was the cause of "Bleeding Kansas"?

4. Critical Thinking

Solving Problems What would you have done to prevent the violence in Kansas?

THINK ABOUT

- the repeal of the Missouri Compromise
- popular sovereignty
- the actions of John Brown

ACTIVITY OPTIONS

LITERATURE

ART

Read a chapter of *Uncle Tom's Cabin.* Write a **book review** or make a series of **drawings** illustrating the story.

3 Slavery Dominates Politics

TERMS & NAMES
Republican Party
John C. Frémont
James Buchanan
Dred Scott v. *Sandford*
Roger B. Taney
Abraham Lincoln
Harpers Ferry

MAIN IDEA

Disagreement over slavery led to the formation of the Republican Party and heightened sectional tensions.

WHY IT MATTERS NOW

The Democrats and the Republicans are the major political parties of today.

This medal shows an early motto of the Republican Party. It also makes clear the Republican connection to the antislavery position of the Free-Soil Party.

ONE AMERICAN'S STORY

Joseph Warren, editor of the Detroit *Tribune,* wanted the antislavery parties of Michigan to join forces. In 1854, his newspaper pushed them to unite.

A VOICE FROM THE PAST

[A convention should be called] irrespective of the old party organizations, for the purpose of agreeing upon some plan of action that shall combine the whole anti-Nebraska, anti-slavery sentiment of the State, upon one ticket [set of candidates endorsed by a political party].

Detroit *Tribune,* quoted in *The Origins of the Republican Party*

By July, antislavery politicians from various parties, including the Whigs, Free-Soilers, and some Democrats, had settled their differences. On July 6, they met to form a new party "to concentrate the popular sentiment of this state against the aggression of the slave power." In memory of Thomas Jefferson, they called themselves Republicans. In this section, you will learn why the Republican Party was formed and how it changed American politics in the 1850s.

The Republican Party Forms

The creation of the Republican Party grew out of the problems caused by the Kansas–Nebraska Act of 1854. The law immediately caused a political crisis for the Whig Party. Southern Whigs had supported the bill for the same reason that Northern Whigs had opposed it: the bill proposed to open new territories to slavery. There was no room for compromise, so the party split.

The Southern Whigs were destroyed by the split. A few joined the Democratic Party. But most searched for leaders who supported slavery and the Union. The Northern Whigs, however, joined with other opponents of slavery and formed the **Republican Party**.

The Republicans quickly gained strength in the North. "Bleeding Kansas" was the key to the Republican rise. Many people blamed the violence on the Democrats. With the 1856 elections nearing, the

Republicans believed that they had an excellent opportunity to gain seats in Congress and win the presidency.

The Republicans needed a strong presidential candidate in 1856 to strengthen their young party. They nominated **John C. Frémont.** Young and handsome, Frémont was a national hero for his explorations in the West, which earned him the nickname the "Pathfinder."

*Reading***History**
A. Summarizing Why did the Republicans nominate Frémont for president in 1856?

Republicans liked Frémont for a couple of reasons. He had spoken in favor of admitting both California and Kansas as free states. Also, he had little political experience and did not have a controversial record to defend. Even so, the Republican position on slavery was so unpopular in the South that Frémont's name did not appear on the ballot there.

The Election of 1856

The Democrats nominated **James Buchanan** to run for the presidency in 1856. As minister to Great Britain, he had been in England since 1853 and had spoken neither for nor against the Kansas–Nebraska Act.

Buchanan took advantage of his absence from the country. He said little about slavery and claimed that his goal was to maintain the Union. Buchanan appealed to Southerners, to many people in the upper South and the border states, and to Northerners who were afraid that Frémont's election could tear the nation apart.

The American, or Know-Nothing, Party also nominated a presidential candidate in 1856. They chose Millard Fillmore, who had been president, following the death of Zachary Taylor, from 1850 until 1853. But the Know-Nothings were divided over slavery and had little strength.

The 1856 presidential election broke down into two separate races. In the North, it was Buchanan against Frémont. In the South, it was Buchanan against Fillmore. Buchanan won. He carried all the slave states except Maryland, where Fillmore claimed his only victory. Buchanan also won several Northern states.

Although he lost the election, Frémont won 11 Northern states. These results showed two things. First, the Republican Party was a major force in the North. Second, the nation was sharply split over slavery.

The Case of Dred Scott

The split in the country was made worse by the Supreme Court decision in the case of Dred Scott. Scott had been a slave in Missouri. His owner took him to live in territories where slavery was illegal. Then they returned to Missouri. After his owner's death, Scott sued for his freedom. He argued that he was a free man because he had lived in territories where slavery was illegal. His case, ***Dred Scott v. Sandford***, reached the Supreme Court in 1856.

Now *and* then

THIRD-PARTY CANDIDATES

American politics has usually been dominated by two parties. Most third-party candidates, such as the Republican Frémont in 1856, lose elections. In 1998, Minnesotans broke that pattern when they elected for governor Jesse Ventura of the Reform Party (below).

One reason for Ventura's victory was his celebrity status. He had been a professional wrestler and the host of a popular radio show. His plain speaking appealed to many Minnesotans who felt the major parties were out of touch with the people.

Ventura was the first Reform Party candidate to gain a major political office.

Dred Scott (above) first sued for his freedom in 1846. The Supreme Court, led by Chief Justice Roger B. Taney (right), did not rule on the case until 1857.

In 1857, the Court ruled against Scott. Chief Justice **Roger B. Taney** [TAW•nee] delivered his opinion in the case. In it, he said that Dred Scott was not a U.S. citizen. As a result, he could not sue in U.S. courts. Taney also ruled that Scott was bound by Missouri's slave code because he lived in Missouri. As a result, Scott's time in free territory did not matter in his case.

In addition, Taney argued that Congress could not ban slavery in the territories. To do so would violate the slaveholders' property rights, protected by the Fifth Amendment. In effect, Taney declared legislation such as the Missouri Compromise unconstitutional.

Reading History

B. Recognizing Effects How did Taney's opinion affect the Missouri Compromise?

Southerners cheered the Court's decision. Many Northerners were outraged and looked to the Republican Party to halt the growing power of Southern slaveholders.

Lincoln and Douglas Debate

After the *Dred Scott* decision, the Republicans charged that the Democrats wanted to legalize slavery not only in all U.S. territories but also in all the states. They used this charge to attack individual Democrats. Stephen A. Douglas, sponsor of the Kansas–Nebraska Act, was one of their main targets in 1858. That year, Illinois Republicans nominated **Abraham Lincoln** to challenge Douglas for his U.S. Senate seat. In his first campaign speech, Lincoln expressed the Northern fear that Southerners wanted to expand slavery to the entire nation. He set the stage for his argument by using a metaphor from the Bible.

A VOICE FROM THE PAST

"A house divided against itself cannot stand." I believe this government cannot endure, permanently half slave and half free. I do not expect the Union to be dissolved—I do not expect the house to fall—but I do expect it will cease to be divided. It will become all one thing, or all the other.

Abraham Lincoln, Springfield, Illinois, June 16, 1858

Later in the year, the two men held formal debates across Illinois. The Lincoln–Douglas debates are now seen as models of political debate. At the time, the debates allowed people to compare the short, stocky, well-dressed Douglas with the tall, thin, gawky Lincoln.

The two men squarely addressed the nation's most pressing issue: the expansion of slavery. For Lincoln, slavery was "a moral, a social and a political wrong." But he did not suggest abolishing slavery where it already existed. He argued only that slavery should not be expanded.

Douglas did not share Lincoln's belief that it was the national government's role to prevent the expansion of slavery. Instead, he argued

CITIZENSHIP TODAY

Debating Points of View

Debate has long been an important method of exploring public issues. The Lincoln–Douglas debates drew crowds from all over Illinois to hear Lincoln and Douglas discuss the issues of the day. Debates such as these can help people find out about candidates' views.

Today, the National Forensic League (NFL) sponsors Lincoln–Douglas Debates, competitions for high school students. Many judges, actors, news commentators, and talk show hosts began to develop their debating skills in such competitions.

High school students can benefit from learning to defend their positions in debates. One student explained what she learned from NFL debates. "I learned about how to think really fast and how to respond."

Jessica Bailey of Apple Valley High School in Minnesota won second place in the national Lincoln–Douglas Debates competition in 1998.

How Do You Debate an Issue?

1. Choose a debate opponent and an issue to debate. (One NFL topic for national competition was whether the federal government should establish an educational policy to increase academic achievement in secondary schools in the United States.)
2. Research the topic you chose.
3. Agree on a format for your debate—how many minutes for presentation, rebuttal, and closing.
4. Debate your opponent in front of the class.
5. Find out how many students in the audience agree with each side, then ask for their reasons.

See Citizenship Handbook, page 248.

Visit www.mcdougallittell.com for more on debates.

that popular sovereignty was the best way to address the issue because it was the most democratic method to do so.

Reading **History**
C. Making Inferences Why were popular sovereignty and the opinion in the *Dred Scott* case inconsistent?

But popular sovereignty was a problem for Douglas. The Supreme Court decision in the *Dred Scott* case had made popular sovereignty unconstitutional. Why? It said that people could not vote to ban slavery, because doing so would take away slaveholders' property rights. In the debates, Lincoln asked Douglas if he thought people in a territory who were against slavery could legally prohibit it—despite the *Dred Scott* decision.

Douglas replied that it did not matter what the Supreme Court might decide about slavery because "the people have the lawful means to introduce it or exclude it as they please." Douglas won reelection. Lincoln, despite his loss, became a national figure and strengthened his standing in the Republican Party.

John Brown Attacks Harpers Ferry

In 1859, John Brown, who had murdered proslavery Kansans three years before, added to the sectional tensions. Brown had a plan. He wanted to inspire slaves to fight for their freedom. To do this, he planned to capture the weapons in the U.S. arsenal at **Harpers Ferry,** Virginia.

Vocabulary
arsenal: stock of weapons

On October 16, 1859, Brown and 18 followers—13 whites and 5 blacks—captured the Harpers Ferry arsenal. They killed four people in the raid. Brown then sent out the word to rally and arm local slaves.

HISTORY *through* ART

John Steuart Curry painted *The Tragic Prelude* between 1937 and 1942. He shows a wild-eyed John Brown standing on the bodies of Civil War soldiers.

What do you think Curry's views were on John Brown's role in U.S. history?

But no slaves joined the fight. The U.S. Marines attacked Brown at Harpers Ferry. Some of his men escaped. But Brown and six others were captured, and ten men were killed.

Brown was then tried for murder and treason. He was convicted and sentenced to hang. On the day he was hanged, abolitionists tolled bells and fired guns in salute. Southerners were enraged by Brown's actions and horrified by Northern reactions to his death.

As the nation headed toward the election of 1860, the issue of slavery had raised sectional tensions to the breaking point. In the next section, you will read about the election of 1860 and its effect on the nation.

Section 3 Assessment

1. Terms & Names

Identify:

- Republican Party
- John C. Frémont
- James Buchanan
- *Dred Scott* v. *Sandford*
- Roger B. Taney
- Abraham Lincoln
- Harpers Ferry

2. Taking Notes

Use a chart like the one below to take notes on the major events discussed in this section.

Election of 1856	
Dred Scott v. Sandford	
Lincoln–Douglas debates	
Harpers Ferry	

3. Main Ideas

a. What issues led to the creation of the Republican Party?

b. What was Chief Justice Taney's opinion in the *Dred Scott* case?

c. How did John Brown's attack on Harpers Ferry increase tensions between the North and the South?

4. Critical Thinking

Identifying Facts and Opinions How did Lincoln and Douglas disagree about slavery? Which of their views were facts, and which were opinions?

THINK ABOUT

- Lincoln's speech at Springfield in 1858
- Douglas's support of popular sovereignty

ACTIVITY OPTIONS

MATH

GEOGRAPHY

Do research to find election returns from the 1856 presidential election. Make **graphs** or draw a **map** to illustrate the results.

4 Lincoln's Election and Southern Secession

TERMS & NAMES
platform
secede
Confederate States of America
Jefferson Davis
Crittenden Plan

MAIN IDEA
The election of Lincoln led the Southern states to secede from the Union.

WHY IT MATTERS NOW
This was the only time in U.S. history that states seceded from the Union.

In 1860, the Republican delegates met in Chicago at a convention hall known as the Wigwam.

ONE AMERICAN'S STORY

In May 1860, Murat Halstead, a reporter for the Cincinnati *Commercial,* traveled to Chicago to cover the Republican convention. Most people assumed that William Seward of New York would win the party's presidential nomination. But there were other candidates, too.

Throughout the convention, candidates tried to win away Seward's delegates. As the other candidates hoped, Seward failed to win the nomination on the first ballot. Throughout the tense voting, Abraham Lincoln, a lesser-known candidate from Illinois, gained strength. Halstead described the scene as Lincoln received the winning votes.

A VOICE FROM THE PAST

There was a moment's silence. The nerves of the thousands, which through the hours of suspense had been subjected to terrible tension, relaxed, and as deep breaths of relief were taken, there was a noise in the Wigwam [convention hall] like the rush of a great wind [just before] a storm—and in another breath, the storm was there. There were thousands cheering with the energy of insanity.

Murat Halstead, *Caucuses of 1860*

Having won the nomination, Lincoln could turn his attention to winning the general election. In this section, you will learn about the election of 1860 and its role in pushing the nation toward civil war.

Political Parties Splinter

A few weeks before the Republicans nominated Abraham Lincoln, the Democrats held their convention in Charleston, South Carolina. Northern and Southern Democrats disagreed over what to say about slavery in the party's **platform,** or statement of beliefs.

The Southerners wanted the party to defend slavery in the platform.

But Northerners wanted the platform to support popular sovereignty as a way of deciding whether a territory became a free state or a slave state. The Northerners won the platform vote, causing 50 Southern delegates to walk out of the convention.

The remaining delegates tried to nominate a presidential candidate. Stephen A. Douglas was the leading contender, but the Southerners who stayed refused to back him because of his support for popular sovereignty. Douglas could not win enough votes to gain the nomination.

Finally, the Democrats gave up and decided to meet again six weeks later in Baltimore to choose a candidate. But as the Baltimore convention opened, Northerners and Southerners remained at odds. This time, almost all the Southerners left the meeting.

With the Southerners gone, the Northern Democrats nominated Douglas. Meanwhile, the Southern Democrats decided to nominate their own candidate. They chose John Breckinridge of Kentucky, the current vice-president and a supporter of slavery.

As you read in One American's Story on page 471, the Republicans had already nominated Abraham Lincoln. In addition to Lincoln, Douglas, and Breckinridge, a candidate from a fourth party entered the race. This party was called the Constitutional Union Party, and its members had one aim—to preserve the Union. They nominated John Bell of Tennessee to run for president.

Reading **History**

A. Recognizing Effects How did slavery affect U.S. political parties in 1860?

The Election of 1860

The election of 1860 turned into two different races for the presidency, one in the North and one in the South. Lincoln and Douglas were the only candidates with much support in the North. Breckinridge and Bell competed for Southern votes.

Lincoln and Breckinridge were considered to have the most extreme views on slavery. Lincoln opposed the expansion of slavery into the territories. Breckinridge insisted that the federal government be required to protect slavery in any territory. Douglas and Bell were considered moderates because neither wanted the federal government to pass new laws on slavery.

This cartoon of the long-legged Abe Lincoln shows him to be the fittest candidate in the 1860 presidential election.

The outcome of the election made it clear that the nation was tired of compromise. Lincoln defeated Douglas in the North. Breckinridge carried most of the South. Douglas and Bell managed to win only in the states between the North and the Deep South. Because the North had more people in it than the South, Lincoln won the election.

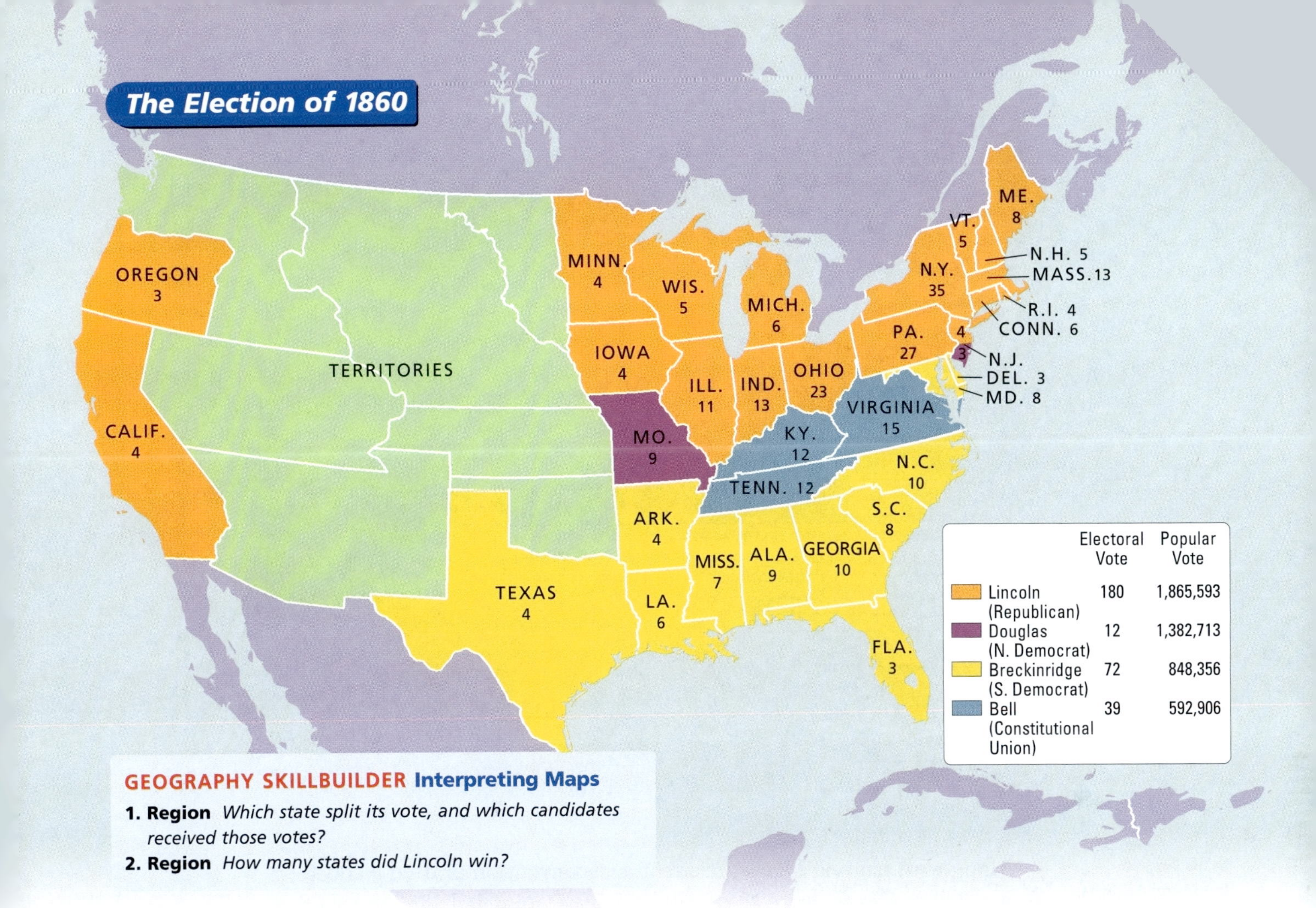

GEOGRAPHY SKILLBUILDER Interpreting Maps

1. **Region** *Which state split its vote, and which candidates received those votes?*
2. **Region** *How many states did Lincoln win?*

Despite Lincoln's statements that he would do nothing to abolish slavery in the South, white Southerners did not trust him. Many were sure that he and the other Republicans would move to ban slavery. As a result, white Southerners saw the Republican victory as a threat to the Southern way of life.

Southern States Secede

Before the 1860 presidential election, many Southerners had warned that if Lincoln won, the Southern states would **secede,** or withdraw from the Union. Supporters of secession based their arguments on the idea of states' rights. They argued that the states had voluntarily joined the Union. Consequently, they claimed that the states also had the right to leave the Union.

Background
Before becoming president of the Confederacy, Davis had been a hero during the War with Mexico and a U.S. senator.

On December 20, 1860, South Carolina became the first state to secede. Other states in the Deep South, where slave labor and cotton production were most common, also considered secession. During the next six weeks, Mississippi, Florida, Alabama, Georgia, Louisiana, and Texas joined South Carolina in secession.

In early February 1861, the states that had seceded met in Montgomery, Alabama. They formed the **Confederate States of America.** The convention named **Jefferson Davis** president of the Confederacy.

In his First Inaugural Address, Lincoln argued passionately for the North and the South to preserve the Union.

Along with naming Davis president, the convention drafted a constitution. The Confederate Constitution was modeled on the U.S. Constitution. But there were a few important differences. For example, the Confederate Constitution supported states' rights. It also protected slavery in the Confederacy, including any territories it might acquire.

Having formed its government, the Confederate states made plans to defend their separation from the Union. Some believed that war between the states could not be avoided. But everyone waited to see what the Union government would do in response.

The Union Responds to Secession

Northerners considered the secession of the Southern states to be unconstitutional. During his last months in office, President James Buchanan argued against secession. He believed that the states did not have the right to withdraw from the Union because the federal government, not the state governments, was sovereign. If secession were permitted, the Union would become weak, like a "rope of sand." He believed that the U.S. Constitution was framed to prevent such a thing from happening.

Vocabulary
sovereign: supreme, self-governing authority

In addition to these issues, secession raised the issue of majority rule. Southerners complained that Northerners intended to use their majority to force the South to abolish slavery. But Northerners responded that Southerners simply did not want to live by the rules of democracy. They complained that Southerners were not willing to live with the election results. As Northern writer James Russell Lowell wrote, "[The Southerners'] quarrel is not with the Republican Party, but with the theory of Democracy."

Efforts to Compromise Fail

With the states in the lower South forming a new government in Montgomery, Alabama, some people continued to seek compromise. Senator John J. Crittenden of Kentucky developed a compromise plan. The **Crittenden Plan** was presented to Congress in late February 1861, but it did not pass.

"We must not be enemies."
Abraham Lincoln

With the hopes for compromise fading, Americans waited for Lincoln's inauguration. What would the new president do about the crisis? On March 4, Lincoln took the oath of office and gave his First Inaugural Address. He assured the South that he had no intention of abolishing slavery there. But he spoke forcefully against secession. Then he ended his speech with an appeal to friendship.

Reading History
B. Making Inferences What do you think Lincoln meant by "mystic chords of memory"?

A VOICE FROM THE PAST

We are not enemies, but friends. We must not be enemies. Though passion may have strained, it must not break our bonds of affection. The mystic chords of memory, stretching from every battle-field and patriot grave, to every living heart and hearthstone, all over this broad land, will yet swell the chorus of the Union, when again touched, as surely they will be, by the better angels of our nature.

Abraham Lincoln, *First Inaugural Address*

Lincoln would not press the South. He wanted no invasion. But he would not abandon the government's property there. Several forts in the South, including Fort Sumter in South Carolina, were still in Union hands. These forts would soon need to be resupplied. Throughout March and into April, Northerners and Southerners waited anxiously to see what would happen next. You will find out in the next chapter.

Section 4 Assessment

1. Terms & Names

Identify:
- platform
- secede
- Confederate States of America
- Jefferson Davis
- Crittenden Plan

2. Taking Notes

Use a time line to fill in the main events that occurred between April 1860 and March 1861.

April 1860 — May 1860 — June 1860 — Nov. 1860 — Feb. 1861 — March 1861

Do you think secession could have been avoided? Why?

3. Main Ideas

a. Who were the candidates in the 1860 presidential election, and what policies did each candidate stand for?

b. Which states seceded right after Lincoln's election? How did they justify this action?

c. What attempts did the North and the South make to compromise? What were the results?

4. Critical Thinking

Analyzing Points of View Do you think the Southern states seceded to protect slavery or states' rights?

THINK ABOUT
- the Southern view of the Fugitive Slave Act
- the Confederate Constitution
- slaveholders' views of the Republican Party

ACTIVITY OPTIONS

SPEECH
TECHNOLOGY

Read Lincoln's First Inaugural Address. Deliver a section of the **speech** before the class or plan an **electronic presentation** about that day and Lincoln's message.

Chapter 15 ASSESSMENT

VISUAL SUMMARY

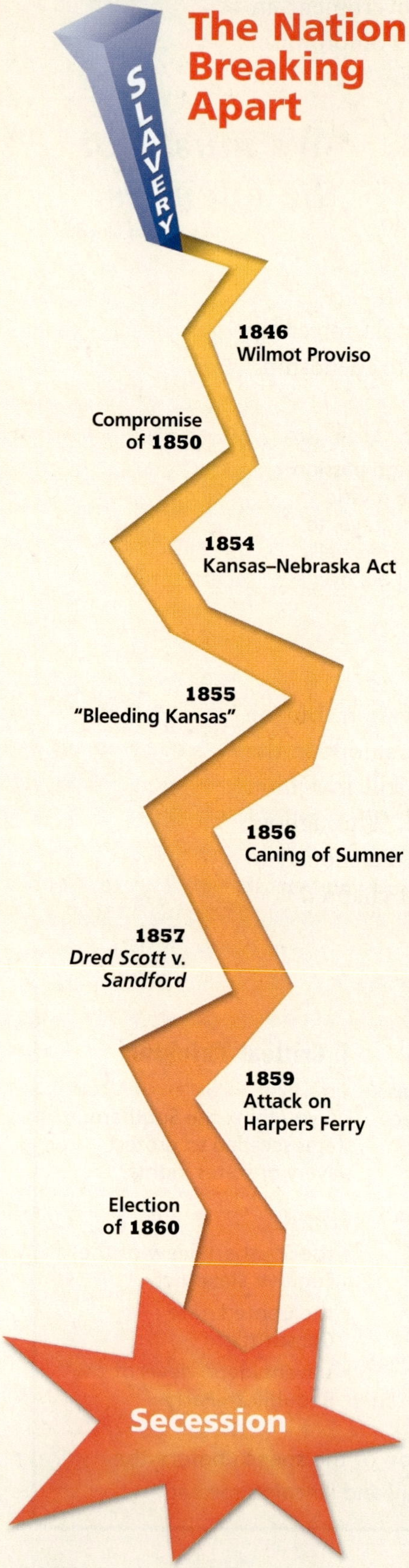

TERMS & NAMES

Briefly explain the importance of each of the following.

1. Wilmot Proviso
2. Compromise of 1850
3. *Uncle Tom's Cabin*
4. popular sovereignty
5. Kansas–Nebraska Act
6. John Brown
7. John C. Frémont
8. *Dred Scott* v. *Sandford*
9. secede
10. Confederate States of America

REVIEW QUESTIONS

Growing Tensions Between North and South (pages 457–461)

1. How did the North and the South differ in the 1840s?
2. How did Southerners react to the Wilmot Proviso?
3. What was Stephen A. Douglas's role in passing the Compromise of 1850?

The Crisis Deepens (pages 462–465)

4. How did Northerners react to the Fugitive Slave Act?
5. Why did most Northerners and Southerners disagree about the Kansas–Nebraska Act?
6. How did "Bleeding Kansas" cause problems for Democrats?

Slavery Dominates Politics (pages 466–470)

7. What positions did Lincoln and Douglas take in their debates?
8. What was the result of John Brown's raid on Harpers Ferry?

Lincoln's Election and Southern Secession (pages 471–475)

9. What were the results of the election of 1860, and what did these results show?
10. How did Southerners justify secession?

CRITICAL THINKING

1. USING YOUR NOTES

Using your completed diagram, answer the questions below.

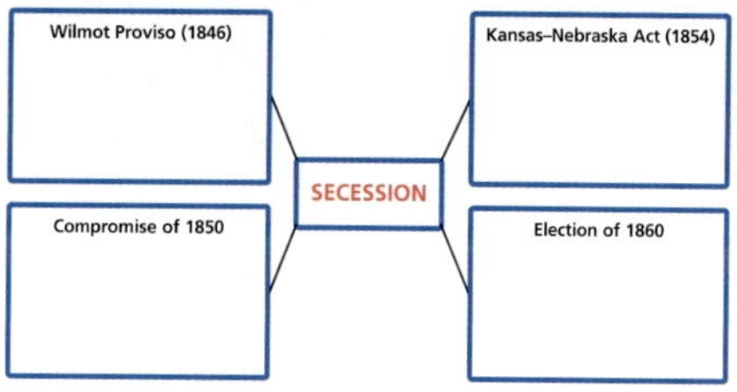

a. What did the Compromise of 1850 and the Kansas–Nebraska Act have in common?

b. Which event do you think caused the most damage to the relationship between the North and the South? Explain.

2. ANALYZING LEADERSHIP

Why were the nation's leaders in 1860 unable to compromise like the leaders in 1820 and 1850? Does their failure to compromise in 1860 mean that they were not as capable as earlier leaders?

3. APPLYING CITIZENSHIP SKILLS

What alternatives did the states in the lower South have to secession? Which of these alternatives do you think would have been the best choice?

4. SOLVING PROBLEMS

How did slavery divide Americans in the 1850s?

5. THEME: DIVERSITY AND UNITY

What could have been done in the 1850s to prevent the Southern states from seceding? What did Americans have in common that could have overcome their differences over slavery?

Interact *with* History

Now that you have read about the sectional crisis of the 1850s, do you think the solution you came up with at the start of the chapter would have helped keep the Union together? Explain.

HISTORY SKILLS

1. INTERPRETING MAPS: Region

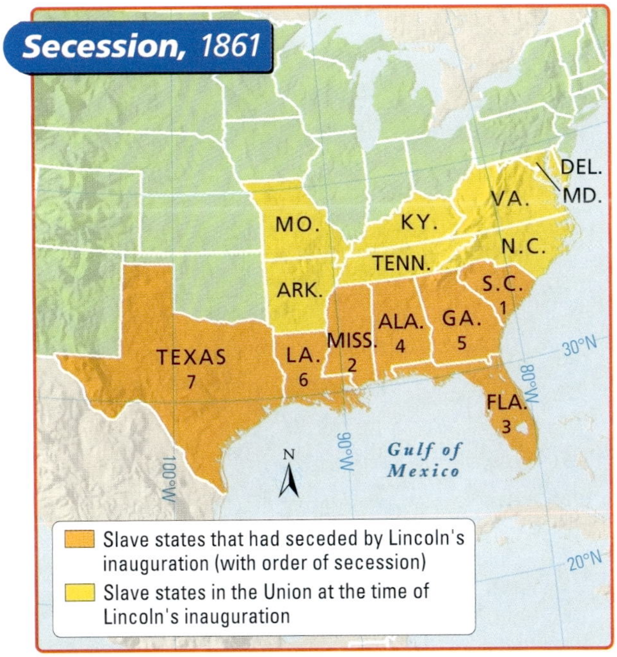

Basic Map Elements

a. Which states seceded before Lincoln's inauguration in March, 1861?

Interpreting the Map

b. What geographical characteristics did these states have in common?

c. Why do you think these states were the first to secede?

2. INTERPRETING PRIMARY SOURCES

A detail of Thomas Hovenden's painting, *The Last Moments of John Brown* (1884), shows Brown being led to the gallows.

a. How does Brown appear in this image?

b. Based on Hovenden's presentation of Brown, what do you think were the artist's views on Brown?

ALTERNATIVE ASSESSMENT

1. INTERDISCIPLINARY ACTIVITY: Language Arts

Conducting an Interview Pick one of the important people in this chapter, such as Stephen A. Douglas or Harriet Beecher Stowe. Think about what questions you would ask that person in an interview. Then research that person's biographies, speeches, or writings to see how he or she might answer your questions. Write the interview in question-and-answer format. Use direct quotes as often as possible.

2. COOPERATIVE LEARNING ACTIVITY

Presenting a Mock Trial The case of Dred Scott went to the Supreme Court. The Court, led by Chief Justice Roger B. Taney, decided against Scott. Did the court reach the right verdict?

Working in a small group, do research on the Dred Scott case. Present a mock trial of the case. Look for information about the roles of Taney, the other justices, Scott, and the other major participants. Have the class render a verdict at the end of the trial. Compare your verdict with the actual historical results.

3. TECHNOLOGY ACTIVITY

Making a Class Presentation American political parties went through extreme changes in the 1840s and 1850s. Using the Internet or the library, find the election returns of the presidential elections from 1848 to 1860.

INTERNET **Visit www.mcdougallittell.com to learn more about presidential elections.**

Use presentation software to teach your class about U.S. presidential elections. Here are some suggestions.

- Use pie charts to show the percentage of votes that went to each party in each election.
- Use bar graphs to show the growth in total popular vote (nationwide) for each party for each election.

4. HISTORY PORTFOLIO

PORTFOLIO **Option 1** Review your section and chapter assessment activities. Select one that you think was your best work. Then use the comments made by your teacher or classmates to improve your work and add it to your portfolio.

PORTFOLIO **Option 2** Review the questions that you wrote for What Do You Want to Know? on page 456. Then write a short report in which you explain the answers to your questions. If any questions were not answered, do research to answer them. Add your answers to your portfolio.

CHAPTER 16

The Civil War Begins 1861–1862

In this vivid engraving, South Carolina shore guns fire on Fort Sumter in Charleston's harbor.

Interact *with* History

The date is April 12, 1861. You and other residents of Charleston, South Carolina, watch the bombardment of Fort Sumter by Confederate forces. This event signals the beginning of the Civil War—a war between factions or regions of the same country.

What Do You Think?

- What sort of physical destruction might take place in a civil war?
- What social, political, and economic trouble might be likely to occur in a civil war?
- What might happen when a civil war breaks out?

How might a civil war be worse than other wars?

USA 1861–1862

March 4, 1861
Abraham Lincoln inaugurated as president.

April 12, 1861
Confederate forces fire on Fort Sumter.

July 21, 1861
First Battle of Bull Run (Manassas) occurs.

March 9, 1862
The warships *Monitor* and *Merrimack* (or *Virginia)* clash.

April 6, 1862
Battle of Shiloh takes place.

April 25, 1862
New Orleans falls to Union forces.

September 17, 1862
Battle of Antietam (Sharpsburg) occurs.

World 1861–1862

March, 1861
Italy unified under King Victor Emmanuel II.

May 13, 1861
Britain declares neutrality in American Civil War.

April 13, 1862
France annexes Cochin China (southern Vietnam).

May 5, 1862
French troops are defeated at Puebla, Mexico.

June 25, 1862
Imperial decree expels foreigners from Japan.

September, 1862
Bismarck becomes prime minister of Prussia.

Chapter 16 SETTING THE STAGE

BEFORE YOU READ

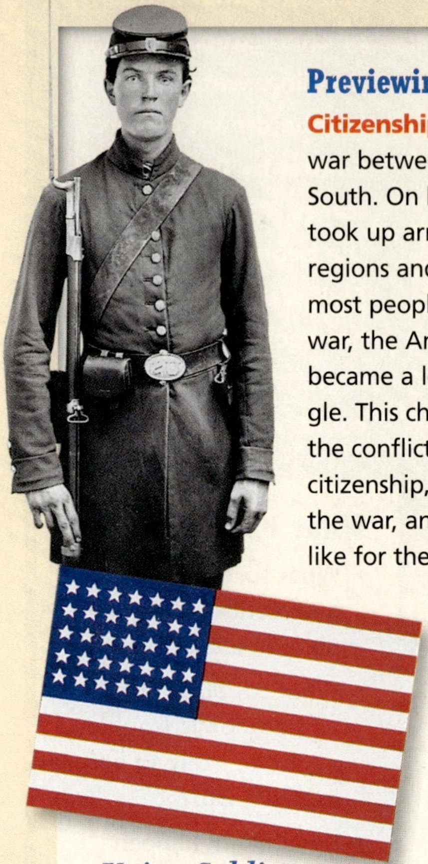

Union Soldier

Previewing the Theme

Citizenship Secession led to war between the North and South. On both sides, citizens took up arms to defend their regions and ideals. Although most people expected a short war, the American Civil War became a long, bloody struggle. This chapter describes the conflicting demands of citizenship, the early years of the war, and what life was like for the soldiers.

Confederate Soldier

What Do You Know?

What do you think of when you hear the phrase *civil war?* What would it be like to fight in a war of brother against brother? Where and how did the Civil War begin?

THINK ABOUT

- what a civil war is
- what you've learned about the Civil War from movies, television, and books
- reasons that countries threaten to break apart in today's world

What Do You Want to Know?

What details do you need to help you understand the outbreak of the Civil War? Make a list of those details in your notebook before you read the chapter.

READ AND TAKE NOTES

Reading Strategy: Comparing and Contrasting When you compare, you look for similarities between two or more objects, ideas, events, or people. When you contrast, you look for differences. Comparing and contrasting can be a useful strategy for studying the two sides in a war. Use the chart shown here to compare and contrast the North and the South in the early years of the Civil War.

See Skillbuilder Handbook, page R9.

	North	South
Reasons for fighting		
Advantages		
Disadvantages		
Military strategy		
Battle victories		

1 War Erupts

TERMS & NAMES
Fort Sumter
Robert E. Lee
border state
King Cotton
Anaconda Plan
blockade
First Battle of Bull Run

MAIN IDEA
The secession of the Southern states quickly led to armed conflict between the North and the South.

WHY IT MATTERS NOW
The nation's identity was in part forged by the Civil War.

ONE AMERICAN'S STORY

Two months before the Civil War broke out, 22-year-old Emma Holmes of Charleston began keeping a detailed diary. Like other South Carolinians, Holmes got caught up in the passions that led her state to secede. From a rooftop, she witnessed the event that started the war. She wrote about South Carolina's attack on Fort Sumter, a federal fort in Charleston's harbor, in her diary.

A VOICE FROM THE PAST

[A]t half past four this morning, the heavy booming of cannons woke the city from its slumbers. . . . Every body seems relieved that what has been so long dreaded has come at last and so confident of victory that they seem not to think of the danger of their friends. . . . I had a splendid view of the harbor with the naked eye. We could distinctly see flames amidst the smoke. All the barracks were on fire. . . . With the telescope I saw the shots as they struck the fort and [saw] the masonry crumbling.

Emma Holmes, *The Diary of Emma Holmes 1861–1866*

This photograph of Emma Holmes was taken in 1900.

Many Southerners expected a short war that they would easily win. Northerners expected the same. In this section, you will learn how the war started, how the states divided, and how each side planned to win.

First Shots at Fort Sumter

As they seceded from the Union (the states loyal to the United States of America during the Civil War), the Southern states took over most of the federal forts inside their borders. President Abraham Lincoln had to decide what to do about the forts that remained under federal control. Major Robert Anderson and his garrison held on to **Fort Sumter** in the harbor of Charleston, South Carolina, but they were running out of supplies.

If Lincoln supplied the garrison, he risked war. If he ordered the troops to leave the fort, he would be giving in to the rebels. Lincoln informed South Carolina that he was sending supply ships to Fort Sumter. Leaders of the Confederacy (the nation formed by Southern states in 1861) decided to prevent the federal government from holding onto the fort by attacking before the supply ships arrived.

At 4:30 A.M. on April 12, 1861, shore guns opened fire on the island fort. For 34 hours, the Confederates fired shells into the fort until Anderson was forced to surrender. No one was killed, but the South's attack on Fort Sumter was the beginning of the Civil War.

Lincoln Calls Out the Militia

Two days after the surrender of Fort Sumter, President Lincoln asked the Union states to provide 75,000 militiamen for 90 days to put down the uprising in the South. Citizens of the North responded with enthusiasm to the call to arms. A New York woman wrote, "It seems as if we never were alive till now; never had a country till now."

Background
The state militias were armies of ordinary citizens rather than professional soldiers.

In the upper South, however, state leaders responded with anger. The governor of Kentucky said that the state would "furnish no troops for the wicked purpose of subduing her sister Southern States." In the weeks that followed, Virginia, North Carolina, Tennessee, and Arkansas voted to join the Confederacy.

As each state seceded, volunteers rushed to enlist, just as citizens did in the North. A young Arkansas enlistee wrote, "So impatient did I become for starting that I felt like ten thousand pins were pricking me in every part of the body, and started off a week in advance of my brothers." Some feared the war would be over before they got the chance to fight.

With Virginia on its side, the Confederacy had a much better chance for victory. Virginia was wealthy and populous, and the Confederacy in May of 1861 moved its capital to Richmond. Virginia also was the home of **Robert E. Lee,** a talented military leader. When Virginia seceded, Lee resigned from the United States Army and joined the Confederacy. Although Lee opposed slavery and secession, he explained, "I cannot raise my hand against my birthplace, my home, my children." He eventually became the commanding general of the Army of Northern Virginia.

***Reading* History**
A. Comparing
Why might citizens in both the North and the South have been eager to fight in the Civil War?

AMERICA'S HISTORY MAKERS

ABRAHAM LINCOLN
1809–1865

Today, Abraham Lincoln is considered one of the great men of all time. Yet early in his presidency, he was widely criticized and ridiculed. Critics labeled him ignorant, incompetent, and socially crude. As Lincoln grew into his job, however, he gained the respect and affection of many Northerners.

Even as a youth, Lincoln had displayed a gift for public speaking. During the Civil War, through his speeches and writings, Lincoln inspired fellow Americans to "dare to do our duty as we understand it."

Why would the ability to inspire people be important in a wartime leader?

Choosing Sides

After Virginia seceded, both sides knew that the border states would play a key role in the war's outcome. The **border states**—Delaware, Maryland, Kentucky, and Missouri—were slave states that bordered states in which slavery was illegal. Because of their location and resources, the border states could tip the scales toward one side.

Keeping Maryland in the Union was important for the North. If Maryland seceded, then Washington, D.C., would be cut off from the Union. To hold on to the state, Lincoln considered arresting Maryland lawmakers who backed the South, but he decided against it.

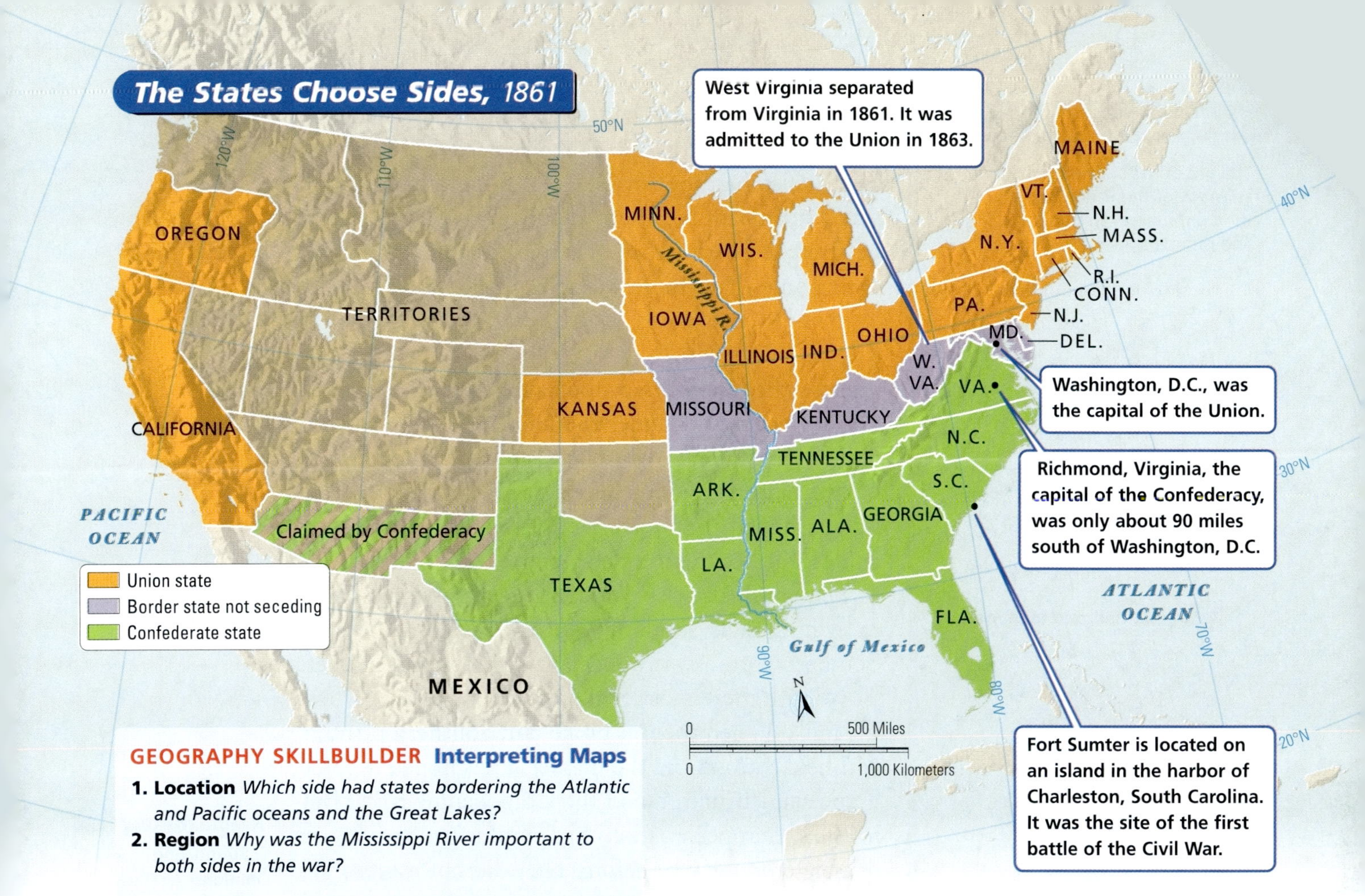

GEOGRAPHY SKILLBUILDER Interpreting Maps

1. **Location** *Which side had states bordering the Atlantic and Pacific oceans and the Great Lakes?*
2. **Region** *Why was the Mississippi River important to both sides in the war?*

Pro-Union leaders eventually gained control of the Maryland legislature, and the state stayed in the Union.

Kentucky was also important to both sides because of its rivers. For the Union, the rivers could provide an invasion route into the South. For the South, the rivers could provide a barrier. Kentuckians were deeply divided over secession. However, a Confederate invasion in 1861 prompted the state to stay in the Union.

Reading **History**
B. Summarizing Why were the border states critical to the war's outcome?

Both Missouri and Delaware also stayed in the Union. In Virginia, federal troops helped a group of western counties break away. These counties formed the state of West Virginia and returned to the Union. In the end, 24 states made up the Union and 11 joined the Confederacy.

Strengths and Weaknesses

The Union had huge advantages in manpower and resources. The North had about 22 million people. The Confederacy had roughly 9 million, of whom about 3.5 million were slaves. About 85 percent of the nation's factories were located in the North. The North had more than double the railroad mileage of the South. Almost all the naval power and shipyards belonged to the North.

The Union's greatest asset, however, was President Abraham Lincoln. He developed into a remarkable leader. Lincoln convinced Northerners that democracy depended on preserving the Union.

Resources, *1860*

The pie charts show the relative strength of the Union and the Confederacy in population and industry.

Total U.S. Population

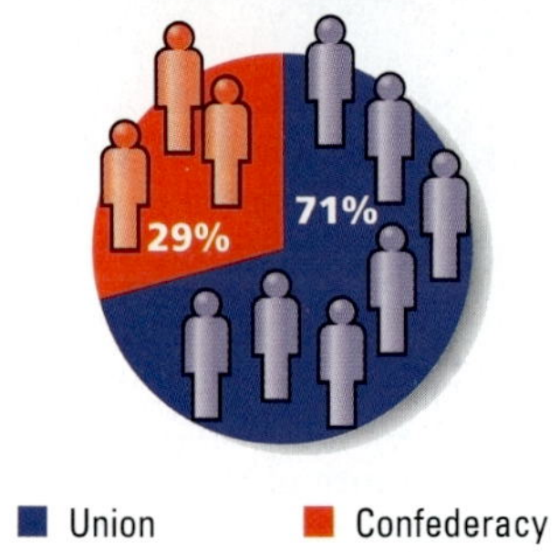

Total U.S. Railroad Mileage

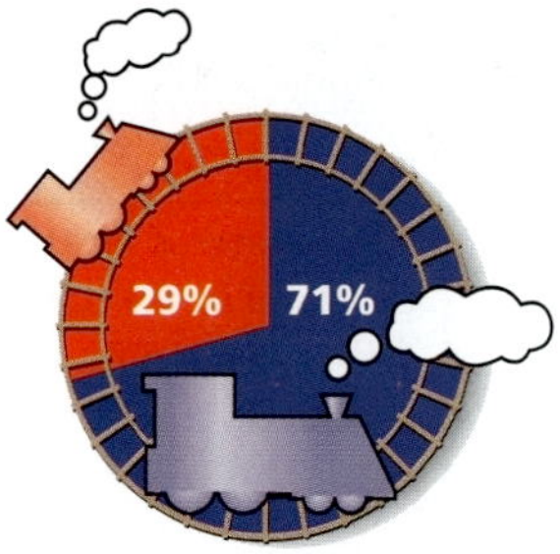

Total U.S. Manufacturing Plants

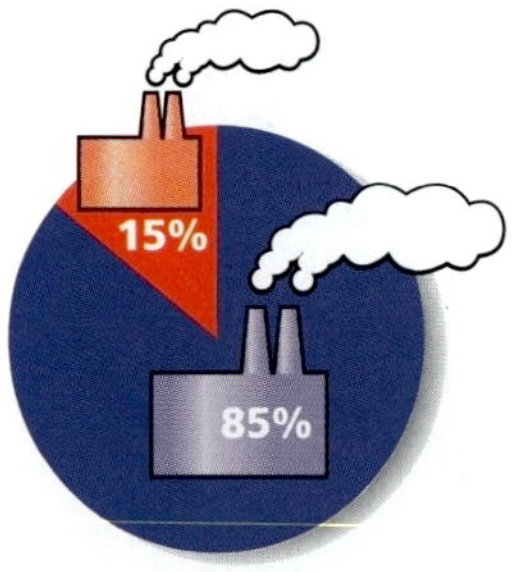

Total U.S. Industrial Workers

Source: Encyclopedia Americana

SKILLBUILDER
Interpreting Charts

1. *Which side had more resources?*
2. *How might the North's railways and factories have helped its armies?*

The Confederacy had some advantages, too. It began the war with able generals, such as Robert E. Lee. It also had the advantage of fighting a defensive war. This meant Northern supply lines would have to be stretched very far. In addition, soldiers defending their homes have more will to fight than invaders do.

The Confederate Strategy

At first, the Confederacy took a defensive position. It did not want to conquer the North—it only wanted to be independent. "All we ask is to be let alone," said Confederate President Jefferson Davis. Confederate leaders hoped the North would soon tire of the war and accept Southern independence.

The South also depended on **King Cotton** as a way to win foreign support. Cotton was king because Southern cotton was important in the world market. The South grew most of the cotton for Europe's textile mills. When the war broke out, Southern planters withheld cotton from the market. They hoped to force France and Britain to aid the Confederate cause. But in 1861, European nations had surplus cotton because of a big crop the year before. They did not want to get involved in the American war.

Reading **History**

C. Supporting Opinions At the beginning of the Civil War, which side would you have predicted to win? Why?

As the war heated up, the South soon moved away from its cautious plans. It began to take the offensive and try for big victories to wreck Northern morale.

The Union Strategy

The North wanted to bring the Southern states back into the Union. To do this, the North developed an offensive strategy based on General Winfield Scott's **Anaconda Plan.** This plan was designed to smother the South's economy like a giant anaconda snake squeezing its prey.

The plan called for a naval blockade of the South's coastline. In a **blockade,** armed forces prevent the transportation of goods or people into or out of an area. The plan also called for the Union to gain control of the Mississippi River. This would split the Confederacy in two.

One of the drawbacks of Scott's plan was that it would take time to work. But many people, eager for action, were calling for an immediate attack on Richmond, the Confederate capital. Lincoln ordered an invasion of Virginia in the summer of 1861.

Battle of Bull Run

To take Richmond, the Union army would first have to defeat the Confederate troops stationed at the town of Manassas, Virginia. This was a railway center southwest of Washington, D.C.

Background
In the South, the battle was called the First Battle of Manassas. In most cases, the South named a battle after a nearby town. The North used a landmark near the fighting, usually a stream.

On July 21, 1861, Union forces commanded by General Irvin McDowell clashed with Confederate forces headed by General Pierre Beauregard near a little creek called Bull Run north of Manassas. In the North, this battle came to be known as the **First Battle of Bull Run.**

The Confederate Army passes in review before General Pierre Beauregard.

At one point in the battle, a Confederate officer rallied his troops by pointing his sword toward Southern General Thomas J. Jackson. The officer cried, "There is Jackson standing like a stone wall! Rally behind the Virginians!" From this incident, Jackson won the nickname "Stonewall" Jackson. His men held fast against the Union assault.

As fresh troops arrived, the Confederates equaled the Union forces in number and launched a countercharge. Attacking the Union line, they let out a blood-curdling scream. This scream, later called the "rebel yell," caused the Union troops to panic. They broke ranks and scattered.

The Confederate victory in the First Battle of Bull Run thrilled the South and shocked the North. Many in the South thought the war was won. The North realized it had underestimated its opponent. Lincoln sent the 90-day militias home and called for a real army of 500,000 volunteers for three years. In the next section, you will learn what army life was like.

Section 1 Assessment

1. Terms & Names

Identify:
- Fort Sumter
- Robert E. Lee
- border state
- King Cotton
- Anaconda Plan
- blockade
- First Battle of Bull Run

2. Taking Notes

Use a Venn diagram to compare and contrast the strengths of the North and the South.

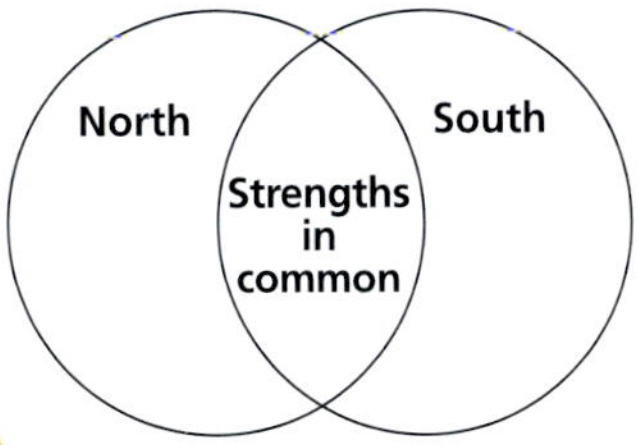

3. Main Ideas

a. How did citizens in the North and the South respond to the outbreak of the Civil War?

b. Why were the border states important to both sides in the Civil War?

c. What kind of military strategy did each side develop?

4. Critical Thinking

Comparing How was the South's situation in the Civil War similar to the situation of the Patriots in the Revolutionary War?

THINK ABOUT
- their reasons for fighting
- their opponents' strengths

ACTIVITY OPTIONS

LANGUAGE ARTS
TECHNOLOGY

Read an account of the First Battle of Bull Run. Use the information to write a **news article** or plan the battle's **home page** for the Internet.

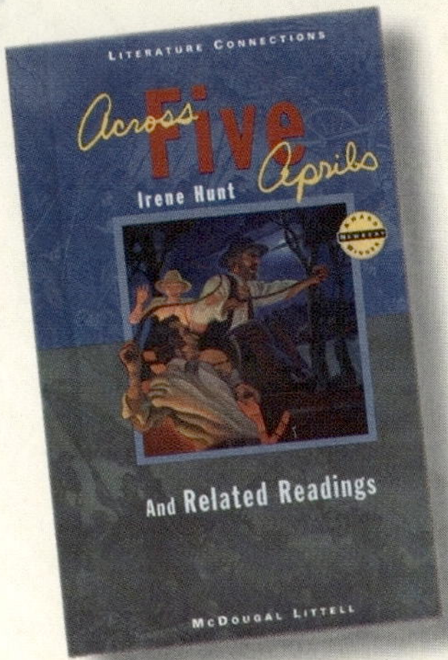

from

Across Five Aprils

by Irene Hunt

At the beginning of the Civil War, nine-year-old Jethro Creighton and his family live on a farm in southern Illinois. Although Illinois is a free state that remains in the Union, many people in the area have ties to states where slavery is legal. For example, the Creightons have relatives in "Kaintuck," or Kentucky. When this excerpt begins, Jethro has already seen his brother Tom and his cousin Eb go off to fight for the Union. His other brothers, John and Bill, are trying to decide which side to fight for.

A line of wild geese flew southward far overhead, and Jethro stood motionless as he watched them disappear from sight. So **engrossed**[1] he was with the flight of the geese that he did not hear Bill's footsteps until his brother was quite near. He caught his breath at sight of Bill's face, which was swollen and beginning to grow discolored from a deep cut and many bruises.

"What's hurt you, Bill?" he asked, his voice barely **audible**,[2] for he was pretty sure he knew.

"We had a fight, Jeth, about an hour ago. We fit like two madmen, I guess."

"You and John?"

Bill's sigh was almost a moan. "Yes, me and John. Me and my brother John."

Jethro could not answer. He stared at the cut above Bill's right eye, from which blood still trickled down his cheek. Somewhere, far off in another field, a man shouted to his horses, and the shout died away in a cry that ran frightened over the brown water of the creek and into the darkening woods.

He had heard cries often that autumn, all through the countryside. They came at night, wakened him, and then lapsed into silence, leaving him in fear and **perplexity**.[3] Sounds once familiar were no longer as they had seemed in other days—his father calling cattle in from the pasture, the sheep dog's bark coming through the fog, the distant creak of the pulley as Ellen drew water for her chickens—all these once familiar sounds had taken on overtones of wailing, and he seemed to hear an echo of that wailing now. He shivered and looked away from his brother's face.

Bill sat down on the ground beside him. "Did ever Ma tell you, Jeth, about when John and me was little and was goin' to school fer the first time? At night I'd git a book and I'd say to Pa, 'What air that word, Pa?' and when he would tell me, I'd turn to John, jest a scant year older, and I'd say, 'Did Pa call it right, Johnny?' Ma and Pa used to laugh at that, but they was pleased to talk about it. They was always set up at John and me bein' so close."

"I know it." Jethro's words came from a tight throat. "What made you fight, Bill?"

"Hard feelin's that have been buildin' up fer weeks, hard feelin's that fin'ly come out in hard words." He held his hand across his eyes for a minute and then spoke quickly. "I'm leavin', Jeth; it ain't that I want to, but it's that I must. The day is comin' when I've got to fight, and I won't fight fer **arrogance**[4] and big money aginst the southern farmer. I won't do it. You tell Pa that. Tell him, too, that I'm takin' my brown mare—she's mine, and I hev the right. Still, it will leave him short, so you tell him that I'm leavin' money I

---

1. **engrossed:** absorbed; occupied.
2. **audible:** able to be heard.
3. **perplexity:** confusion.
4. **arrogance:** overbearing pride.

made at the sawmill and at corn shuckin'; it's inside the cover of his Bible. You tell him to take it and buy another horse."

Jethro was crying unashamedly in the face of his grief. "Don't go, Bill. Don't do it," he begged.

"Jeth . . ."

"I don't want you to go, Bill. I don't think I kin stand it."

"Listen to me, Jeth; you're gittin to be a sizable boy. There's goin' to be a lot of things in the years ahead that you'll have to stand. There'll be things that tear you apart, but you'll have to stand 'em. You can't count on cryin' to make 'em right."

The colors were beginning to fade on Walnut Hill. A light wind bent the dried grass and weeds. Jethro felt choked with grief, but he drew a sleeve across his eyes and tried to look at his brother without further weeping.

"Where will you go, Bill?"

"To Kaintuck. I'll go to Wilse's place first. From there—I don't know."

"Will you fight fer the Rebs?"

Bill hesitated a few seconds. "I've studied this thing, Jeth, and I've hurt over it. My heart ain't in this war; I've told you that. And while I say that the right ain't all on the side of the North, I know jest as well that it ain't all on the side of the South either. But if I hev to fight, I reckon it will be fer the South."

Jethro nodded. There were things you had to endure. After a while he asked, "Air you goin' tonight, Bill?"

"Right away. I've had things packed in that holler tree fer a couple days. I've knowed that this was comin' on, but I couldn't make myself leave. Now I'm goin'. The little mare is saddled and tied down at the molasses press. I'll go as fur as Newton tonight; in the morning I'll take out early."

He got to his feet. "There's lots of things I want to say, but I reckon I best not talk." Without looking at Jethro he laid his hand on the boy's shoulder. "Git all the larnin' you kin—and take keer of yoreself, Jeth," he said and turned abruptly away.

"Take keer of yoreself, Bill," Jethro called after him.

Across the prairies, through the woods, over the brown water of the creek, there was a sound of crying. Jethro ran to a tree and hid his face. He had heard his mother say that if you watch a loved one as he leaves you for a long journey, it's like as not to be the last look at him that you'll ever have.

Like the Creighton brothers, the Terrill brothers had different loyalties. William Rufus Terrill (left) fought for the Union as an artillery officer at Shiloh. His brother, James Terrill (right), was the commanding officer of a Virginia infantry regiment. Both men died in the war. William was killed at Perryville, Kentucky, in 1862; James was killed in battle in 1864. After the war, the family built a memorial to the brothers, which contained the words, "God Alone Knows Which Was Right."

CONNECT TO HISTORY

1. **Recognizing Effects** How has the Civil War affected the Creighton family? Discuss how the effect reflects what is happening in the country.

See Skillbuilder Handbook, page R10.

CONNECT TO TODAY

2. **Researching** Where have civil wars or internal rebellions taken place in the recent past?

Visit www.mcdougallittell.com to learn more about other civil wars.

2 Life in the Army

TERMS & NAMES
hygiene
rifle
minié ball
ironclad

MAIN IDEA

Both Union and Confederate soldiers endured many hardships serving in the army during the Civil War.

WHY IT MATTERS NOW

The hardships endured led to long-lasting bitterness on both sides.

ONE AMERICAN'S STORY

When the Civil War began, Peter Vredenburgh, Jr., the son of a well-known judge, was working as a lawyer in Eatontown, New Jersey. In 1862, he answered President Lincoln's call for an additional 300,000 soldiers. Nearly 26 years old, Vredenburgh became a major in the 14th Regiment New Jersey Volunteer Infantry. Less than two months after joining the regiment, he wrote a letter urging his parents to keep his 18-year-old brother from enlisting.

A VOICE FROM THE PAST

I am glad that Jim has not joined any Regt. [regiment] and I hope he never will. I would not have him go for all my pay; it would be very improbable that we could both go through this war and come out unharmed. Let him come here and see the thousands with their arms and legs off, or if that won't do, let him go as I did the other day through the Frederick hospitals and see how little account a man's life and limbs are held in by others and what little return he gets in reputation or money for the risk and privations of enlisting and his ideas of the fun of the thing will vanish in thin air.

Major Peter Vredenburgh, Jr., quoted in *Upon the Tented Field*

Major Peter Vredenburgh, Jr., was an officer in the Union army.

On September 19, 1864, Vredenburgh was killed in battle. In this section, you will learn more about other soldiers and what their experiences were like.

Those Who Fought

Like Peter Vredenburgh, the majority of soldiers in the Civil War were between 18 and 30 years of age. But both the Confederate and Union armies had younger and older soldiers. Charles Carter Hay was just 11 years old when he joined an Alabama regiment. William Wilkins was 83 when he became one of the Pennsylvania Home Guards.

Farmers made up the largest group among Civil War soldiers. About half the soldiers on both sides came from farms. Having rarely traveled far from their fields, many viewed going off to war as an exciting adventure. Some rode a train for the first time.

Although the majority of soldiers in the war were born in the United States, immigrants from other countries also served. German and Irish immigrants made up the largest ethnic groups. One regiment from New York had soldiers who were born in 15 foreign countries. The commanding officer gave orders in seven languages.

At the beginning of the war, African Americans wanted to fight. They saw the war as a way to end slavery. However, neither the North nor the South accepted African Americans into their armies. As the war dragged on, the North finally took African Americans into its ranks. Native Americans served on both sides.

Reading **History**

A. Summarizing How did most men in the North and the South feel about going off to war?

In all, about 2 million American soldiers served the Union, and fewer than 1 million served the Confederacy. The vast majority were volunteers. Why did so many Americans volunteer to fight? Many sought adventure and glory. Some sought an escape from the boredom of farm and factory work. Some signed up because their friends and neighbors were doing it. Others signed up for the recruitment money offered by both sides. Soldiers also fought because they were loyal to their country or state.

Turning Civilians into Soldiers

After enlisting, a volunteer was sent to a nearby army camp for training. A typical camp looked like a sea of canvas tents. The tents were grouped by company, and each tent held from two to twenty men. In winter, the soldiers lived in log huts or in heavy tents positioned on a log base. In the Civil War, recruits in training elected their company officers. Both the Union and Confederate armies followed this practice.

A soldier in training followed a set schedule. A bugle or drum awakened the soldier at dawn. After roll call and breakfast, the soldier had the first of several drill sessions. In between drills and meals, soldiers performed guard duty, cut wood for the campfires, dug trenches for latrines (outdoor toilets), and cleaned up the camp.

Shortly after they came to camp, new recruits were given uniforms and equipment. Union soldiers wore blue uniforms, and Confederate soldiers wore gray or

daily *life*

DRILL SESSIONS

"The first thing in the morning is drill. Then drill, then drill again. Then drill, drill, a little more drill. Then drill, and lastly drill." That is the way one soldier described his day in camp.

A soldier in training might have as many as five drill sessions a day, each lasting up to two hours. The soldiers learned to stand straight and march in formation. They also learned to load and fire their guns. Shown drilling below are soldiers of the 22nd New York State Militia near Harpers Ferry, Virginia, in 1862.

yellowish-brown uniforms. Getting a uniform of the right size was a problem, however. On both sides, soldiers traded items to get clothing that fit properly.

Background
Before uniforms became standardized, soldiers dressed in outfits supplied from home. This caused confusion on the battlefield.

Early in the war, Northern soldiers received clothing of very poor quality. Contractors took advantage of the government's need and supplied shoddy goods. Shoes made of imitation leather, for example, fell apart when they got wet. In the Confederacy, some states had trouble providing uniforms at all, while others had surpluses. Because the states did not always cooperate and share supplies, Confederate soldiers sometimes lacked shoes. Like soldiers in the Revolutionary War, they marched over frozen ground in bare feet. After battles, needy soldiers took coats, boots, and other clothing from the dead.

At the beginning of the war, most soldiers in army camps received plenty of food. Their rations included beef or salt pork, flour, vegetables, and coffee. But when they were in the field, the soldiers' diet became more limited. Some soldiers went hungry because supply trains could not reach them.

STRANGE *but* True

DEADLIER THAN BULLETS

"Look at our company—21 have died of disease, 18 have become so unhealthy as to be discharged, and only four have been killed in battle." So a Louisiana officer explained the high death rate in the Civil War.

More than twice as many men died of disease as died of battle wounds. Intestinal disorders, including typhoid fever, diarrhea, and dysentery, killed the most. Pneumonia, tuberculosis, and malaria killed many others. Bad water and food, poor diet, exposure to cold and rain, unsanitary conditions, and disease-carrying insects all contributed to the high rate of disease.

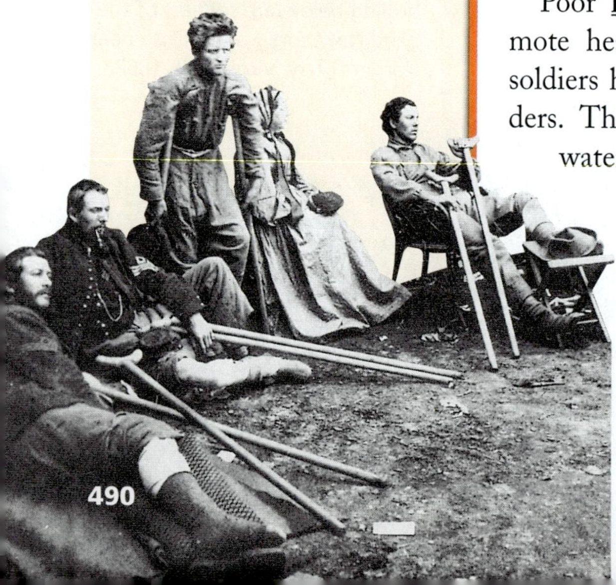

Hardships of Army Life

Civil War soldiers in the field were often wet, muddy, or cold from marching outdoors and living in crude shelters. Many camps were unsanitary and smelled from the odors of garbage and latrines. One Union soldier described a camp near Washington. In the camp, cattle were killed to provide the troops with meat.

A VOICE FROM THE PAST

The hides and [waste parts] of the [cattle] for miles upon miles around, under a sweltering sun and sultry showers, would gender such swarms of flies, armies of worms, blasts of stench and oceans of filth as to make life miserable.

William Keesy, quoted in *The Civil War Infantryman*

Not only were the camps filthy, but so were the soldiers. They often went weeks without bathing or washing their clothes. Their bodies, clothing, and bedding became infested with lice and fleas.

Poor **hygiene**—conditions and practices that promote health—resulted in widespread sickness. Most soldiers had chronic diarrhea or other intestinal disorders. These disorders were caused by contaminated water or food or by germ-carrying insects. People did not know that germs cause diseases. Doctors failed to wash their hands or their instruments. An observer described how surgeons "armed with long, bloody knives and saws, cut and sawed away with frightful rapidity, throwing the mangled limbs on a pile nearby as soon as removed."

Reading **History**
B. Making Inferences What changes could have helped lower the spread of disease among soldiers?

The naval duel between the Union *Monitor* and the Confederate *Merrimack* (or *Virginia)* took place on March 9, 1862.

Changes in Military Technology

While camp life remained rough, military technology advanced. Improvements in the weapons of war had far-reaching effects. Battle tactics changed, and casualties soared.

Vocabulary
casualties: number of people killed or injured

Rifles that used minié balls contributed to the high casualty rate in the Civil War. A **rifle** is a gun with a grooved barrel that causes a bullet to spin through the air. This spin gives the bullet more distance and accuracy. The **minié ball** is a bullet with a hollow base. The bullet expands upon firing to fit the grooves in the barrel. Rifles with minié balls could shoot farther and more accurately than old-fashioned muskets. As a result, mounted charges and infantry assaults did not work as well. Defenders using rifles could shoot more of the attackers before they got close.

***Reading*History**
C. Drawing Conclusions
Which changes in military technology had an effect on the average soldier? Why?

Ironclads, warships covered with iron, proved to be a vast improvement over wooden ships. In the first ironclad battle, the Confederate *Virginia* (originally named the *Merrimack*) battled the Union *Monitor* off the coast of Virginia in 1862. After hammering away for about four hours, the battle ended in a draw. (See page 492 for more information on ironclads.)

Despite new technology and tactics, neither side gained a decisive victory in the first two years of the war, as you will see in the next section.

Section 2 Assessment

1. Terms & Names

Identify:
- hygiene
- rifle
- minié ball
- ironclad

2. Taking Notes

Complete the chart below.

The Typical Civil War Soldier	
Age	
Occupation	
Training	
Hardships	

Which hardship do you think would have been most difficult to endure? Why?

3. Main Ideas

a. How were the wartime experiences of Northern and Southern soldiers alike?

b. What factors contributed to the spread of disease among soldiers?

c. How did the use of the rifle and minié ball change combat tactics in the Civil War?

4. Critical Thinking

Forming and Supporting Opinions What were the motives that led individual soldiers to fight in the Civil War?

THINK ABOUT
- the multiple reasons that people had for enlisting
- what you consider valid reasons for fighting

ACTIVITY OPTIONS

LANGUAGE ARTS
ART

Imagine you are a soldier in the Civil War. Write a **letter** home to your parents about your experience or draw an **illustrated map** of your training camp.

echnology OF THE Time

Ironclads

They moved through the water, as one observer put it, "like a huge, half-submerged crocodile." To crew members of traditional wooden ships, the ironclads indeed may have seemed like horrible mechanical monsters.

With a powerful iron hull almost entirely under water and a rotating gun turret, or short tower, an ironclad easily destroyed the older vessels it met. When the *Monitor* and the *Merrimack* (or *Virginia*) clashed during the Civil War in the first battle ever waged between ironclads, a new era of naval warfare had begun. Below is a closer look at the Union's *Monitor*.

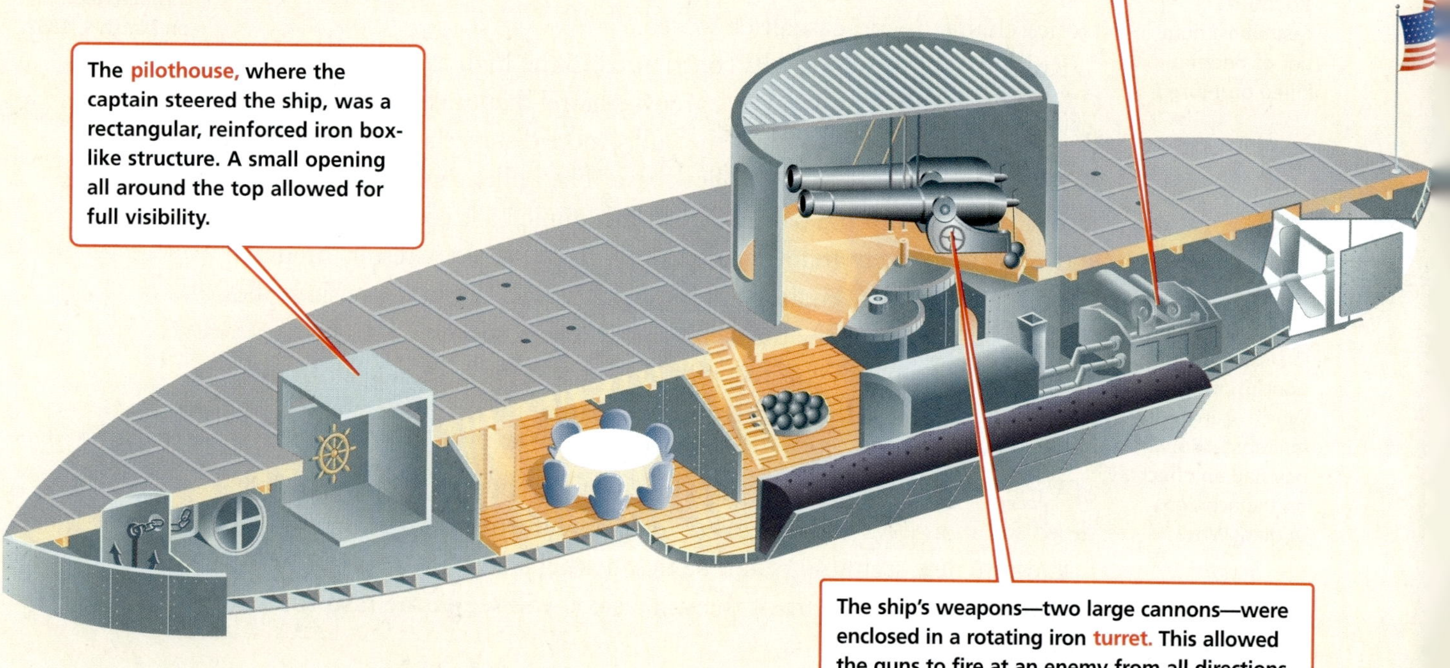

SCALE AND SIZE OF WARSHIPS

Class, Name, Launch Date	Length	Weight	Number of Crew	Weaponry
Ironclad (USS *Monitor*) 1862	172 ft.	987 tons	49	two 11-inch smoothbore cannons
Battleship (USS *Maine*) 1889	319 ft.	6,682 tons	374	four 10-inch guns, six 6-inch guns, seven rapid-fire 6-pounders, four torpedo tubes
Nuclear Submarine (USS *Nautilus*) 1954	324 ft.	4,092 tons	105	six torpedo tubes
Aircraft Carrier (USS *Nimitz*) 1972	1,092 ft.	95,000 tons	6,000	24 F-14A Tomcat warplanes, 16 radar guided missiles, six-barrel, 20-millimeter Gatling gun

CONNECT TO HISTORY

1. **Drawing Conclusions** Why would a rotating gun be an advantage in a naval battle?

See Skillbuilder Handbook, page R12.

CONNECT TO TODAY

2. **Researching** Find out more about modern battleships or aircraft carriers, and write a brief report about their capabilities in battle.

Visit www.mcdougallittell.com for more on military ships.

3 No End in Sight

TERMS & NAMES
Ulysses S. Grant
Battle of Shiloh
cavalry
Seven Days' Battles
Battle of Antietam

MAIN IDEA

In the first two years of the war, neither side gained a decisive victory over the other.

WHY IT MATTERS NOW

A long war can cause much death and destruction and leave a bitter legacy.

ONE AMERICAN'S STORY

In the summer of 1861, President Lincoln gave George McClellan command of the Union army in the East. The army had recently been defeated at Bull Run. McClellan faced the task of restoring the soldiers' confidence while organizing and training an army that could defeat the Confederates.

Within months, McClellan had accomplished the task and won the devotion of his troops. The entire nation expected great things. In November 1861, Lincoln made McClellan general in chief of the entire Union army. But while Lincoln kept urging him to attack Richmond, McClellan kept drilling his troops.

A VOICE FROM THE PAST

[S]oon as I feel that my army is well organized and well disciplined and strong enough, I will advance and force the Rebels to a battle on a field of my own selection. A long time must elapse before I can do that.

General George McClellan, quoted in *Civil War Journal: The Leaders*

Lincoln said McClellan had "the slows." While McClellan was stalling in the East, another general was winning victories in the West.

President Lincoln (right) meets with General McClellan (left) on the Antietam battlefield in 1862.

Union Victories in the West

That victorious Union general in the West was **Ulysses S. Grant**. In civilian life, he had failed at many things. But Grant had a simple strategy of war: "Find out where your enemy is, get at him as soon as you can, strike at him as hard as you can, and keep moving on."

In February 1862, Grant made a bold move to take Tennessee. Using ironclad gunboats, Grant's forces captured two Confederate river forts. These were Fort Henry on the Tennessee and Fort Donelson on the nearby Cumberland. (See map on next page.) The seizure of Fort Henry opened up a river highway into the heart of the South. Union gunboats could now travel on the river as far as northern Alabama. When the people of Nashville, Tennessee, heard the forts were lost, they fled the city in panic. A week later, Union troops marched into Nashville.

The Battle of Shiloh

After Grant's river victories, Albert S. Johnston, Confederate commander on the Western front, ordered a retreat to Corinth, Mississippi. Grant followed. By early April, Grant's troops had reached Pittsburg Landing on the Tennessee River. There he waited for more troops from Nashville. Johnston, however, decided to attack before Grant gained reinforcements. Marching his troops north from Corinth on April 6, 1862, Johnston surprised the Union forces near Shiloh Church. The **Battle of Shiloh** in Tennessee turned into the fiercest fighting the Civil War had yet seen.

Commanders on each side rode into the thick of battle to rally their troops. One Union general, William Tecumseh Sherman, had three horses shot out from under him. General Johnston was killed, and the command passed to General Pierre Beauregard. By the end of the day, each side believed that dawn would bring victory.

That night, there was a terrible thunderstorm. Lightning lit up the battlefield, where dead and dying soldiers lay in water and mud. During the night, Union boats ran upriver to ferry fresh troops to Grant's camp. Grant then led an attack at dawn and forced the exhausted Southern troops to retreat.

Reading **History**

A. Contrasting
How did Grant differ from McClellan as a military leader?

The cost of the Union victory was staggering. Union casualties at Shiloh numbered over 13,000, about one-fourth of those who had fought. The Confederates lost nearly 11,000 out of 41,000 soldiers. Describing

the piles of mangled bodies, General Sherman wrote home, "The scenes on this field would have cured anybody of war." Congressmen criticized Grant for the high casualties and urged Lincoln to replace him. But Lincoln replied, "I can't spare this man—he fights."

> **"I can't spare this man–he fights."**
> Abraham Lincoln, describing General Grant

The Fall of New Orleans

The spring of 1862 brought other bad news for the Confederacy. On April 25, a Union fleet led by David Farragut captured New Orleans, the largest city in the South. Rebel gunboats tried to ram the Union warships and succeeded in sinking one. Farragut's ships had to run through cannon fire and then dodge burning rafts in order to reach the city. Residents stood on the docks and cursed the Yankee invaders, but they were powerless to stop them.

The fall of New Orleans was a heavy blow to the South. Mary Chesnut of South Carolina, the wife of an aide to President Davis, wrote in her diary, "New Orleans gone—and with it the Confederacy. Are we not cut in two?" Indeed, after the victories of General Grant and Admiral Farragut, only a 150-mile stretch of the Mississippi remained in Southern hands. The Union was well on its way to achieving its goal of cutting the Confederacy in two. But guarding the remaining stretch of the river was the heavily armed Confederate fort at Vicksburg, Mississippi.

Reading **History**
B. Recognizing Effects Why was the fall of New Orleans significant?

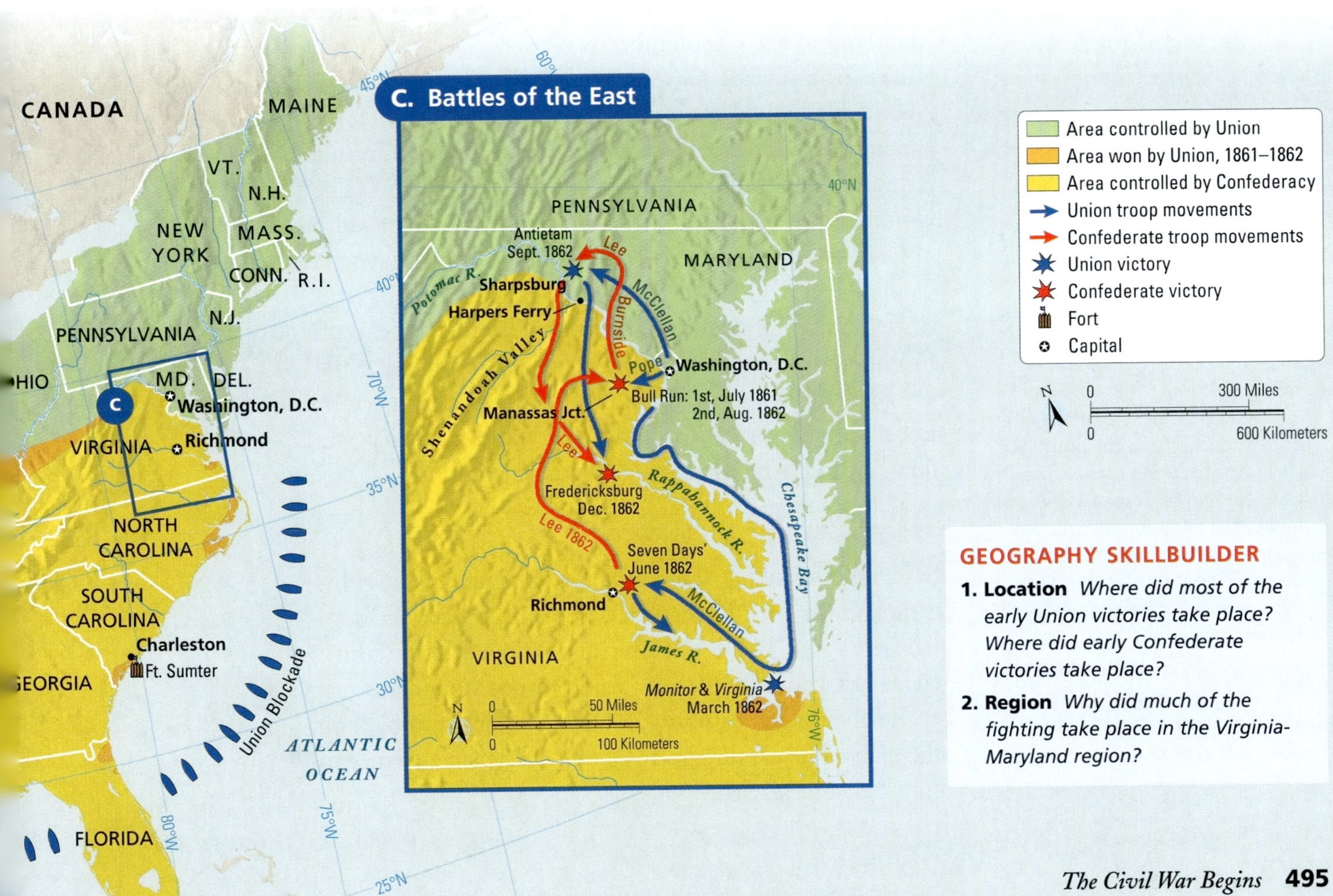

GEOGRAPHY SKILLBUILDER

1. **Location** *Where did most of the early Union victories take place? Where did early Confederate victories take place?*
2. **Region** *Why did much of the fighting take place in the Virginia-Maryland region?*

Lee Claims Victories in the East

Meanwhile, also in the spring of 1862, McClellan finally made his move to try to capture Richmond. He planned to attack the Confederate capital by way of a stretch of land between the York and James rivers. McClellan succeeded in bringing his troops within a few miles of Richmond.

But in June 1862, Robert E. Lee took charge of the Army of Northern Virginia and proceeded to turn the situation around. Lee sent Jeb Stuart and his **cavalry**—soldiers on horseback—to spy on McClellan. With about 1,000 men, Stuart rode around the whole Union army in a few days and reported its size back to Lee. Lee then attacked McClellan's army. The two sides clashed for a week, from June 25 to July 1, 1862, in what became known as the **Seven Days' Battles.** The Army of Northern Virginia suffered heavier losses, but it forced McClellan's army to retreat.

Reading **History**

C. Making Inferences How was Lee's appointment fortunate for the South?

In late August, the Confederates won a second victory at Bull Run, and Union troops withdrew back to Washington. Within just a few months, Lee had ended the Union threat in Virginia.

AMERICA'S HISTORY MAKERS

JEFFERSON DAVIS

1808–1889

Jefferson Davis expected to be given a military command when the Confederacy was formed in 1861. But Davis was chosen President of the Confederacy instead, which stunned and saddened him.

Because of his strong sense of duty and loyalty to the South, Davis accepted the unwelcome post. He had to immediately form a national government and prepare for war at the same time. Davis found it hard to compromise or accept disagreement with his opinions.

How do the qualities required in a military leader differ from those required in a political leader?

Lee Invades the North

Riding a wave of victories, General Lee decided to invade the Union. He wrote to tell President Davis of his plan. Lee thought it was a crucial time, with the North at a low point. Without waiting for Davis's response, Lee crossed the Potomac with his army and invaded Maryland in early September 1862.

Reading **History**

D. Reading a Map Use the map on page 495 to follow Lee's movements into the North.

Lee had several reasons for taking the war to the North. He hoped a victory in the North might force Lincoln to talk peace. The invasion would give Virginia farmers a rest from war during the harvest season. The Confederates could plunder Northern farms for food.

Lee hoped the invasion would show that the Confederacy could indeed win the war, which might convince Europe to side with the South. By this time, both Britain and France were leaning toward recognizing the Confederacy as a separate nation. They were impressed by Lee's military successes, and their textile industry was now hurting from the lack of Southern cotton.

Bloody Antietam

Soon after invading Maryland, Lee drew up a plan for his campaign in the North. A Confederate officer accidentally left a copy of Lee's battle plans wrapped around three cigars at a campsite. When Union troops stopped to rest at the abandoned campsite, a Union soldier stumbled on the plans. The captured plans gave McClellan a chance to stop Lee and his army.

Confederate artillery soldiers lie dead after the Battle of Antietam.

McClellan went on the attack, though he moved slowly as always. On September 17, 1862, at Antietam Creek near Sharpsburg, Maryland, McClellan's army clashed with Lee's. The resulting **Battle of Antietam** was the bloodiest day in all of American history. A Confederate officer later described the battle.

A VOICE FROM THE PAST

Again and again . . . by charges and counter-charges, this portion of the field was lost and recovered, until the green corn that grew upon it looked as if it had been struck by a storm of bloody hail. . . . From sheer exhaustion, both sides, like battered and bleeding athletes, seemed willing to rest.

John B. Gordon, quoted in *Voices of the Civil War*

After fighting all day, neither side had gained any ground by nightfall. The only difference was that about 25,000 men were dead or wounded. Lee, who lost as much as one-third of his fighting force, withdrew to Virginia. The cautious McClellan did not follow, missing a chance to finish off the crippled Southern army. Lincoln was so fed up that he fired McClellan in November, 1862. In the next chapter, you will learn about the historic action Lincoln took after the Battle of Antietam.

Section 3 Assessment

1. Terms & Names

Identify:

- Ulysses S. Grant
- Battle of Shiloh
- cavalry
- Seven Days' Battles
- Battle of Antietam

2. Taking Notes

Review the section and find five key events to place on a time line as shown.

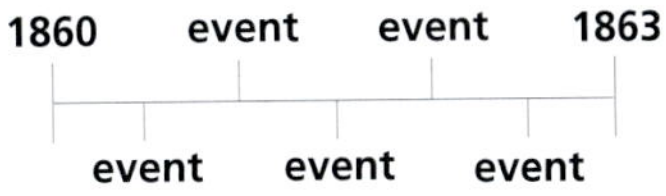

Which of these events do you think was most important?

3. Main Ideas

a. Why were Union victories in the West and the fall of New Orleans significant to the Union cause?

b. Why did Lee go on the offensive against the North?

c. How did the South's fortunes change after Lee took command of the Army of Northern Virginia?

4. Critical Thinking

Making Inferences What does Lee's invasion of the North suggest about his qualities as a general and a leader?

THINK ABOUT

- Lee's military skills and style
- the North's resources

ACTIVITY OPTIONS

GEOGRAPHY

ART

Develop a new military strategy for either the North or the South. Show your strategy on a **map** or in a **diagram** of troop movements.

Chapter 16 ASSESSMENT

TERMS & NAMES

Briefly explain the importance of each of the following.

1. Fort Sumter
2. Robert E. Lee
3. border state
4. blockade
5. hygiene
6. rifle
7. ironclad
8. Ulysses S. Grant
9. Battle of Shiloh
10. Battle of Antietam

REVIEW QUESTIONS

War Erupts (pages 481–487)

1. How and when did the Civil War start?
2. What advantages did the North have at the beginning of the war?
3. What were the war strategies of the two sides?

Life in the Army (pages 488–492)

4. What was the typical Civil War soldier like?
5. Why did so many people volunteer to fight in the Civil War?
6. Why was the incidence of disease so high among Civil War soldiers?
7. How did the use of rifles and minié balls change war tactics?

No End in Sight (pages 493–497)

8. What goal of the Union strategy did Grant further, and how did he do it?
9. Why did the North have such a hard time capturing Richmond, Virginia?
10. How did Lee's appointment to head the Army of Northern Virginia affect the course of the war?

CRITICAL THINKING

1. USING YOUR NOTES

	North	South
Reasons for fighting		
Advantages		
Disadvantages		
Military strategy		
Battle victories		

Using your completed chart, answer the questions.

a. Which side seemed likelier to win the war? Why?

b. Which side followed more closely its original strategy in the first two years of the war?

2. ANALYZING LEADERSHIP

Think about the leaders discussed in this chapter. Choose one. What character traits helped make him an effective leader?

3. APPLYING CITIZENSHIP SKILLS

Which individuals or groups of people demonstrated good and poor citizenship during the war? Explain your choices.

4. THEME: CITIZENSHIP

How could people on both sides of the Civil War believe that they were being good citizens by fighting?

5. MAKING DECISIONS

In your opinion, was Lincoln correct in deciding to go to war to save the Union? Explain your answer.

Interact *with* History

How did the consequences and effects of civil war that you predicted before you read the chapter compare with the actual conditions you read about?

VISUAL SUMMARY

The Civil War, 1861–1862

1861

March 4, 1861
Lincoln inaugurated.

April 12, 1861
Fort Sumter fired upon by rebel forces.

April 15, 1861
Lincoln calls on states to provide 75,000 militiamen.

May 1861
Confederate Congress votes to set up capital in Richmond.

July 21, 1861
First Battle of Bull Run (Manassas)

February 1862
Union forces capture Fort Henry and Fort Donelson.

March 9, 1862
Monitor and *Merrimack* (*Virginia*) clash.

April 6, 1862
Battle of Shiloh

April 25, 1862
Fall of New Orleans

June 25 to July 1, 1862
Seven Days' Battles

September 17, 1862
Battle of Antietam (Sharpsburg)

November 1862
Lincoln relieves McClellan of command.

1862

HISTORY SKILLS

1. INTERPRETING MAPS: Movement

Study the map and answer the questions.

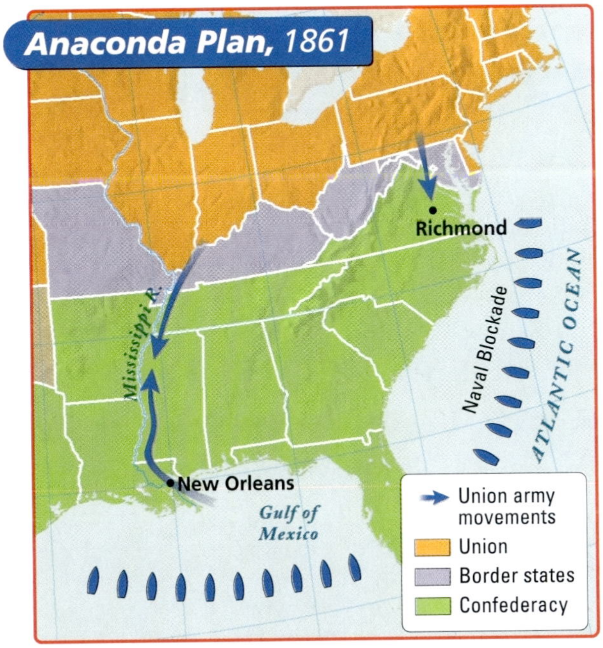

Basic Map Elements

a. What is the subject of the map?

b. What do the colors of the states indicate?

Interpreting the Map

c. What does the arrow in the East indicate?

d. What two bodies of water did the blockade cover?

2. INTERPRETING PRIMARY SOURCES

The following quotation comes from a letter that Robert E. Lee wrote to his sister after Virginia had seceded from the Union and he had resigned from the U.S. Army. Read the quotation and then answer the questions.

> With all my devotion to the Union and the feeling of loyalty and duty of an American citizen, I have not been able to make up my mind to raise my hand against my relatives, my children, my home. I have, therefore, resigned my commission in the Army, and, save in defense of my native state, with the sincere hope that my poor services may never be needed, I hope I may never be called on to draw my sword.
>
> **Robert E. Lee,** quoted in *The Annals of America*

a. What inner conflict does Lee express?

b. What obligation or loyalty does Lee consider greater than his duty to his country?

ALTERNATIVE ASSESSMENT

1. INTERDISCIPLINARY ACTIVITY: Literature

Reading Letters Using the library or the Internet, find firsthand accounts of the Civil War and read some letters written by soldiers. Choose one letter to read aloud to the class. Then, pretend you are the person receiving the letter, and describe your reaction to it.

2. COOPERATIVE LEARNING ACTIVITY

Developing a Peace Proposal The Civil War broke out because the two sides could not reach a compromise. Work in a small group to develop a compromise that would have resolved the conflict between the North and the South. Have one person in the group take the position of a Northerner, another a Southerner, and the third a mediator between the two. Present your proposal to the class for their comments.

3. TECHNOLOGY ACTIVITY

Making a Class Presentation Life in the army training camps during the Civil War was hard. Information about life in the camps comes from primary sources. Using the library and the Internet, find diaries, letters, photographs, and news articles about daily life.

INTERNET **Visit www.mcdougallittell.com to learn more about army camps in the Civil War.**

Plan an electronic presentation on army life for your class. Use the list of suggested topics below to begin brainstorming.

- Drawings, maps, and photos that show the design of a typical camp
- Items used for cooking, lodging, sanitation, military drilling, and recreation
- Drills and maneuvers soldiers learned
- Personal belongings of the Yankees and rebels
- Firsthand accounts of camp life from letters and diaries
- Changes in camp life as the war dragged on

4. HISTORY PORTFOLIO

PORTFOLIO **Option 1** Review your section and chapter assessment activities. Select one that you think was your best work. Then use comments made by your teacher or classmates to improve your work and add it to your portfolio.

PORTFOLIO **Option 2** Review the questions that you wrote for What Do You Want to Know? on page 480. Then write a short report in which you explain the answers to your questions. If any questions were not answered, do research to answer them. Add your answers to your portfolio.

CHAPTER 17

The Tide of War Turns 1863–1865

Confederate and Union cavalry clash at Yellow Tavern, Virginia, on May 11, 1864.

Interact *with* History

In 1863, you have been a Civil War soldier for two years. The life of a soldier is a hard one. The food is awful. Disease is common. Worst of all is the horrible violence and death. Often you feel the urge to run away and go home.

What Do You Think?

- What would you be willing to sacrifice for your country? What if your country fought for something you did not believe in?
- How would the attitudes of fellow soldiers influence your decision?

What would inspire you to keep fighting?

USA

January 1863 Emancipation Proclamation is issued.

July 1863 Battle of Gettysburg takes place.
Union takes Vicksburg.

March 1864 Grant is put in charge of all Union armies.

November 1864 Lincoln is reelected.

April 1865 Union takes Richmond.
Lee surrenders at Appomattox Court House.
Lincoln is assassinated.

1863 — **1866**

World

January 1863 Polish nationalists revolt against Russian rule.

July 1863 Source of Nile River is found at Lake Victoria in present-day Uganda.

June 1864 Archduke Maximilian of Austria becomes emperor of Mexico.

September 1864 First International Workingmen's Association is established, and Karl Marx becomes its leader.

September 1865 English officials arrest *Fenian* leaders of planned uprising in Ireland.

Chapter 17 SETTING THE STAGE

BEFORE YOU READ

Previewing the Theme

Impact of the Individual During the Civil War, thousands of citizens and soldiers acted with great dignity and courage. Leading these men and women were some of the most extraordinary leaders the nation has ever had. You will read more about these citizens and soldiers in Chapter 17.

Robert E. Lee was the military genius at the head of the Confederate armies.

Abraham Lincoln is one of the greatest presidents the United States has ever had.

Jefferson Davis, the Confederate president, had the difficult task of keeping the South united.

Ulysses S. Grant took charge of the Union armies in March 1864.

What Do You Know?

What advantages and disadvantages did the North and the South have? Did particular individuals give either side an advantage during the Civil War?

THINK ABOUT

- what qualities contribute to the success of military leaders
- the importance of obeying orders for soldiers even if it might mean death

What Do You Want to Know?

What questions do you have about the later part of the Civil War and how it ended? Make a list of those questions before you read the chapter.

READ AND TAKE NOTES

Reading Strategy: Comparing and Contrasting When you study historical events, it is important to compare and contrast the effects that events had on different individuals and groups. A single event might affect two groups of people in completely different ways. Use the chart below to compare and contrast the impact of events on the Union and the Confederacy in the later years of the Civil War.

See Skillbuilder Handbook, page R9.

	North	South
Emancipation Proclamation		
War's Impact		
Northern Victories in Battle		
Union Wins Civil War		

1 The Emancipation Proclamation

TERMS & NAMES
Emancipation Proclamation
54th Massachusetts Regiment

MAIN IDEA
In 1863, President Lincoln issued the Emancipation Proclamation, which helped to change the war's course.

WHY IT MATTERS NOW
The Emancipation Proclamation was an important step in ending slavery in the United States.

ONE AMERICAN'S STORY

During the Civil War, abolitionists like Frederick Douglass continued their fight against slavery. Douglass urged President Lincoln to emancipate, or free, enslaved Americans. "Sound policy . . . demands the instant liberation of every slave in the rebel states," he declared.

A VOICE FROM THE PAST

To fight against slaveholders, without fighting against slavery, is but a half-hearted business, and paralyzes the hands engaged in it. . . . Fire must be met with water. . . . War for the destruction of liberty [by the South] must be met with war for the destruction of slavery.

Frederick Douglass, quoted in *Battle Cry of Freedom*

Douglass pointed out that the Confederate war effort depended on slave labor. Enslaved Americans worked in Southern mines, fields, and factories. They also built forts and hauled supplies for rebel armies. For both practical and moral reasons, he said, Lincoln should free the slaves. In this section, you will learn how ending slavery became an important goal of the Civil War.

During the Civil War, Frederick Douglass offered advice to President Lincoln. He urged the president to make the conflict a war against slavery.

Calls for Emancipation

Throughout the war, abolitionists such as Frederick Douglass had been urging Lincoln to emancipate enslaved persons. Many criticized the president for being too cautious. Some even charged that Lincoln's lack of action aided the Confederate cause.

Still, Lincoln hesitated. He did not believe he had the power under the Constitution to abolish slavery where it already existed. Nor did he want to anger the four slave states that remained in the Union. He also knew that most Northern Democrats, and many Republicans, opposed emancipation.

Lincoln did not want the issue of slavery to divide the nation further than it already had. Although he disliked slavery, the president's first priority was to preserve the Union. "If I could save the Union without freeing *any* slave I would do it," he declared. "If I could save it by freeing *all* the slaves I would do it; and if I could save it by freeing some and leaving others alone, I would also do that."

By the summer of 1862, however, Lincoln had decided in favor of emancipation. The war was taking a terrible toll. If freeing the slaves helped weaken the South, then he would do it. Lincoln waited, however, for a moment when he was in a position of strength. After General Lee's forces were stopped at Antietam, Lincoln decided to act.

The Emancipation Proclamation

On January 1, 1863, Lincoln issued the **Emancipation Proclamation,** which freed all slaves in Confederate territory. The proclamation had a tremendous impact on the public. However, it freed very few slaves. Most of the slaves that Lincoln intended to liberate lived in areas distant from the Union troops that could enforce his proclamation.

Background
In September 1862, Lincoln issued an early proclamation that gave rebellious states a chance to preserve slavery by rejoining the Union.

Lincoln presents the Emancipation Proclamation to his cabinet.

A VOICE FROM THE PAST

On the first day of January, in the year of our Lord one thousand eight hundred and sixty-three, all persons held as slaves within any State or designated part of a State, the people whereof shall then be in rebellion against the United States, shall be then, [thenceforth], and forever free.

Abraham Lincoln, from the *Emancipation Proclamation*

Why, critics charged, did Lincoln free slaves only in the South? The answer was in the Constitution. Because freeing Southern slaves weakened the Confederacy, the proclamation could be seen as a military action. As commander-in-chief, Lincoln had this authority. Yet the Constitution did not give the president the power to free slaves within the Union. But Lincoln did ask Congress to abolish slavery gradually throughout the land.

Reading History
A. Drawing Conclusions Why did Lincoln choose to limit his proclamation mostly to rebellious states?

Although the Emancipation Proclamation did not free many enslaved people at the time it was issued, it was important as a symbolic measure. For the North, the Civil War was no longer a limited war whose main goal was to preserve the Union. It was a war of liberation.

Response to the Proclamation

Abolitionists were thrilled that Lincoln had finally issued the Emancipation Proclamation. "We shout for joy that we live to record this righteous decree," wrote Frederick Douglass. Still, many believed the law should have gone further. They were upset that Lincoln had not freed *all* enslaved persons, including those in the border states.

B. Summarizing Why did Northern Democrats oppose the Emancipation Proclamation?

Other people in the North, especially Democrats, were angered by the president's decision. Northern Democrats, the majority of whom were against emancipating even Southern slaves, claimed that the proclamation would only make the war longer by continuing to anger the South. A newspaperman in Ohio called Lincoln's proclamation "monstrous, impudent, and heinous . . . insulting to God as to man."

Most Union soldiers, though, welcomed emancipation. One officer noted that, although few soldiers were abolitionists, most were happy "to destroy everything that . . . gives the rebels strength."

White Southerners reacted to the proclamation with rage. Although it had limited impact in areas outside the reach of Northern armies, many slaves began to run away to Union lines. At the same time that these slaves deprived the Confederacy of labor, they also began to provide the Union with soldiers.

African-American Soldiers

In addition to freeing slaves, the Emancipation Proclamation declared that African-American men willing to fight "will be received into the armed service of the United States."

Frederick Douglass had argued for the recruitment of African-American soldiers since the start of the war. He declared, "Once [you] let the black man get upon his person the brass letters, U.S. . . . there is no power on earth which can deny that he has earned the right to citizenship."

Before the proclamation, the federal government had discouraged the enlistment of African Americans, and only a few regiments were formed. After emancipation, African Americans rushed to join the army. By war's end, about 180,000 black soldiers wore the blue uniform of the Union army.

African-American soldiers were organized in all-black regiments, usually led by white officers. They were often given the worst jobs

Thousands of African Americans, such as these men of the 4th U.S. Colored Troops, fought for the Union during the Civil War.

Now and then

AFRICAN AMERICANS IN THE MILITARY

During the Civil War, no African-American soldier was promoted above the rank of captain. But times have changed. In 1989, General Colin Powell (shown below) was made a four-star general and named chairman of the Joint Chiefs of Staff—the highest position in the military.

General Powell's appointment was the climax of a long struggle to fully integrate American armed forces. From the Civil War through World War II, African-American soldiers were kept apart from white soldiers and denied equal rights. However, in 1948, President Harry Truman ended segregation in the armed forces. Today the American military is fully integrated.

to do and were paid less than white soldiers. Despite these obstacles, African-American soldiers showed great courage on the battlefield and wore their uniforms with pride. More than one regiment insisted on fighting without pay rather than accepting lower pay than the white soldiers.

Reading **History**

C. Identifying Facts How did many black soldiers protest when they were offered lower pay than white soldiers?

The 54th Massachusetts

One unit that insisted on fighting without pay was the **54th Massachusetts Regiment,** one of the first African-American regiments organized in the North. The soldiers of the 54th—among whom were two sons of Frederick Douglass—soon made the regiment the most famous of the Civil War.

The 54th Massachusetts earned its greatest fame in July 1863, when it led a heroic attack on Fort Wagner in South Carolina. The soldiers' bravery at Fort Wagner made the 54th a household name in the North and increased African-American enlistment.

The soldiers of the 54th Massachusetts and other African-American regiments faced grave dangers if captured. Rather than take African Americans as prisoners, Confederate soldiers often shot them or returned them to slavery.

The war demanded great sacrifices, not only from soldiers and prisoners, but also from people back home. In the next section, you will read about the hardships that the Civil War placed on the civilian populations in both the North and the South.

Section 1 Assessment

1. Terms & Names

Identify:
- Emancipation Proclamation
- 54th Massachusetts Regiment

2. Taking Notes

Use a chart to record responses to the Emancipation Proclamation.

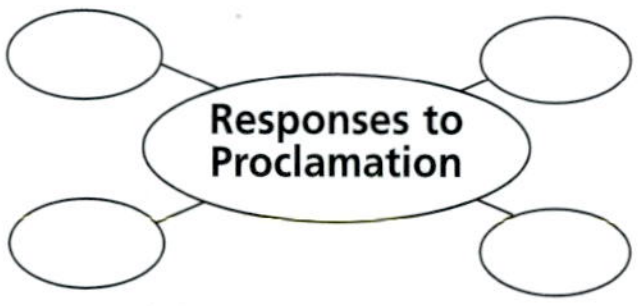

How did the proclamation change Northerners' views of the war?

3. Main Ideas

a. What was Lincoln's reason for not emancipating slaves when the war began?

b. Why was the immediate impact of the Emancipation Proclamation limited?

c. Why did black soldiers often face greater hardships than white soldiers?

4. Critical Thinking

Recognizing Effects How did the Emancipation Proclamation change the role of African Americans in the war?

THINK ABOUT
- how the proclamation changed military policy
- the response of many Southern slaves to the proclamation

ACTIVITY OPTIONS

TECHNOLOGY
MUSIC

Do research on the 54th Massachusetts Regiment. Create a **Web site** for the regiment or write a **song** about the soldiers' heroism at Fort Wagner.

2 War Affects Society

TERMS & NAMES
Copperhead
conscription
bounty
income tax
greenback
Clara Barton

MAIN IDEA

The Civil War caused social, economic, and political changes in the North and the South.

WHY IT MATTERS NOW

Some changes, like the growth of industry, affected Americans long after the end of Civil War.

ONE AMERICAN'S STORY

As the Civil War moved into its third year, the constant demand for men and resources began to take its toll back home. Sometimes, the hardships endured by civilians resulted in angry scenes like that witnessed by Agnes, a resident of Richmond, Virginia.

On April 3, 1863, Agnes went for her morning walk and soon came upon a group of hungry women and children, who had gathered in front of the capitol. She described the scene as these women and children were joined by other people who were upset by the shortage of food.

Food became scarce in many places during the Civil War. Here, women demand milk for their hungry families.

A VOICE FROM THE PAST

The crowd now rapidly increased, and numbered, I am sure, more than a thousand women and children. It grew and grew until it reached the dignity of a mob—a bread riot.

Agnes, quoted in *Reminiscences of Peace and War*

The mob then went out of control. It broke into shops and stole food, clothing, and other goods. Only the arrival of Confederate president Jefferson Davis and the threat of force ended the riot.

In this section, you will read more about hardships that the Civil War caused on the home front. These hardships caused changes in civilian society in both the North and the South.

Disagreement About the War

In the spring of 1863, riots like the one in Richmond broke out in a number of Southern towns. Southerners were growing weary of the war and the constant sacrifices it demanded.

Confederate soldiers began to leave the army in increasing numbers. By the end of the year, the Confederate army had lost nearly 40 percent of its men. Some of these men were on leave, but many others were deserters.

In this political cartoon, the Union defends itself against "Copperheads." This was the name given to Northerners who sympathized with the South.

Faced with the difficulties of waging war, the Confederate states fell into disagreement. The same principle of states' rights that led them to break with the Union kept them from coordinating their war effort. As one Southern governor put it, "I am *still* a rebel . . . no matter who may be in power."

Disagreements over the conduct of the war also arose in the North. Lincoln's main opponents were the **Copperheads,** Northern Democrats who favored peace with the South. (A copperhead is a poisonous snake that strikes without warning.) Lincoln had protesters arrested. He also suspended the writ of habeas corpus, which prevents the government from holding citizens without a trial.

Vocabulary
writ: a written order issued by a court of law

The Draft Laws

As the war dragged on, both the North and the South needed more soldiers. As a result, both sides passed laws of **conscription,** also known as the draft. These laws required men to serve in the military.

The Confederates had been drafting soldiers since the spring of 1862. By 1863, all able-bodied white men between the ages of 18 and 45 were required to join the army. However, there were a number of exceptions. Planters who owned 20 or more slaves could avoid military service. In addition, wealthy men could hire substitutes to serve in their place. By 1863, substitutes might cost as much as $6,000. The fact that wealthy men could avoid service caused poor Southerners to complain that it was a "rich man's war but a poor man's fight."

Reading History
A. Drawing Conclusions Why were many soldiers dissatisfied with the draft laws?

The Union draft law was passed in March 1863. Like the Confederacy, the Union allowed draftees to hire substitutes. However, the North also offered $300 **bounties,** or cash payments, to men who volunteered to serve. As a result, only a small percentage of men in the North were drafted. Most men volunteered and received the bounty.

Even so, the draft was extremely unpopular. In July 1863, anger over the draft and simmering racial tensions led to the New York City draft riots. For four days, rioters destroyed property and attacked people on the streets. Over 100 people were killed—many of them African Americans.

Economic Effects of the War

Many people suffered economic hardship during the war. The suffering was severe in the South, where most battles were fought, but the North also experienced difficulties.

Food shortages were very common in the South, partly because so many farmers were fighting in the Confederate army. Moreover, food sometimes could not get to market because trains were now being used to carry war materials. The Confederate army also seized food and other supplies for its own needs.

Reading **History**

B. Analyzing Causes Why were economic problems particularly bad in the South?

Another problem, especially in the South, was inflation—an increase in price and decrease in the value of money. The average family food bill in the South increased from $6.65 a month in 1861 to $68 by mid–1863. Over the course of the war, prices rose 9,000 percent in the South.

Inflation in the North was much lower, but prices still rose faster than wages, making life harder for working people. Some people took advantage of wartime demand and sold goods for high prices.

Overall, though, war production boosted Northern industry and fueled the economy. In the short term, this gave the North an economic advantage over the South. In the long term, industry would begin to replace farming as the basis of the national economy.

During the war, the federal government passed two important economic measures. In 1861, it established the first **income tax**—a tax on earnings. The following year, the government issued a new paper currency, known as **greenbacks** because of their color. The new currency helped the Northern economy by ensuring that people had money to spend. It also helped the Union to pay for the war.

Some Southerners in the border states took advantage of the stronger Union economy by selling cotton to Northern traders, in violation of Confederate law. "Yankee gold," wrote one Confederate officer, "is fast accomplishing what Yankee arms could never achieve—the subjugation of our people."

Vocabulary

subjugate: to bring under control or to conquer

daily life

INFLATION IN THE SOUTH

During the Civil War, inflation caused hardship in the North and the South. But inflation was especially severe in the Confederacy, where prices could become outrageously high.

The food prices shown below are from 1864. Consider how many days it took a Confederate soldier to earn enough money to buy each of these foods.

Resistance by Slaves

Another factor that affected the South was the growing resistance from slaves. To hurt the Southern economy, slaves slowed their pace of work or stopped working altogether. Some carried out sabotage, destroying crops and farm equipment to hurt the plantation economy. When white planters fled advancing Union armies, slaves often refused to go along. They stayed behind, waiting for Union soldiers to free them.

Some enslaved people even rose up in rebellion against their overseers. More commonly, though, slaves ran away from plantations to join the Union forces as they pushed farther into Confederate territory. One Union officer described a common sight.

A VOICE FROM THE PAST

It was very touching to see the vast numbers of colored [African-American] women following after us with babies in their arms, and little ones like our Anna clinging to their tattered skirts. One poor creature, while nobody was looking, hid two boys, five years old, in a wagon, intending, I suppose that they should see the land of freedom if she couldn't.

Union officer, quoted in *The Civil War*

After Lincoln issued the Emancipation Proclamation, the number of slaves fleeing Southern plantations greatly increased. By the end of the war, as many as half a million had fled to Union lines.

AMERICA'S HISTORY MAKERS

CLARA BARTON

1821–1912

Trained as a schoolteacher, Clara Barton was working for the government when the Civil War began. She organized a relief agency to help with the war effort. "While our soldiers stand and fight," she said, "I can stand and feed and nurse them."

She also made food for soldiers in camp and tended to the wounded and dying on the battlefield. At Antietam, she held a doctor's operating table steady as cannon shells burst all around them. The doctor called her "the angel of the battlefield." After the war, Barton founded the American Red Cross.

How did Clara Barton demonstrate her leadership abilities?

Women Aid the War Effort

With so many men away at war, women in both the North and the South assumed increased responsibilities. Women plowed fields and ran farms and plantations. They also took over jobs in offices and factories that had previously been done only by men.

Other social changes came about because of the thousands of women who served on the front lines as volunteer workers and nurses. Susie King Taylor was an African-American woman who wrote an account of her experiences as a volunteer with an African-American regiment. She asked her readers to remember that "many lives were lost,—not men alone but noble women as well."

Relief agencies put women to work washing clothes, gathering supplies, and cooking food for soldiers. Also, nursing became a respectable profession for many women. By the end of the war, around 3,000 nurses had worked under the leadership of Dorothea Dix in Union hospitals. Southern women were also active as nurses and as volunteers on the front.

Reading History

C. Summarizing How did women participate in the Civil War?

Women also played a key role as spies in both the North and the South. Harriet Tubman served as a spy for Union forces along the coast of South Carolina. The most famous Confederate spy was Belle Boyd. Although she was arrested six times, she continued her work through much of the war. At one point, she even sent messages from her jail cell by putting them in little rubber balls and tossing them out the window.

Civil War Prison Camps

Women caught spying were thrown into jail, but soldiers captured in battle suffered far more. At prison camps in both the North and the South, prisoners of war faced terrible conditions.

One of the worst prison camps in the North was in Elmira, New York. Perhaps the harshest feature of a prisoner's life at the camp was the New York winter. One prisoner called Elmira "an excellent summer prison for southern soldiers, but an excellent place for them to find their graves in the winter." In just one year, more than 24 percent of Elmira's 12,121 prisoners died of sickness and exposure to severe weather.

Reading **History**

D. Making Inferences Why were death rates so high at many Civil War prison camps?

Conditions were also horrible in the South. The camp with the worst reputation was at Andersonville, Georgia. Built to hold 10,000 prisoners, at one point it housed 33,000. Inmates had little shelter from the heat or cold. Most slept in holes scratched in the dirt. Drinking water came from one tiny creek that also served as a sewer. As many as 100 men per day died at Andersonville from starvation, disease, and exposure.

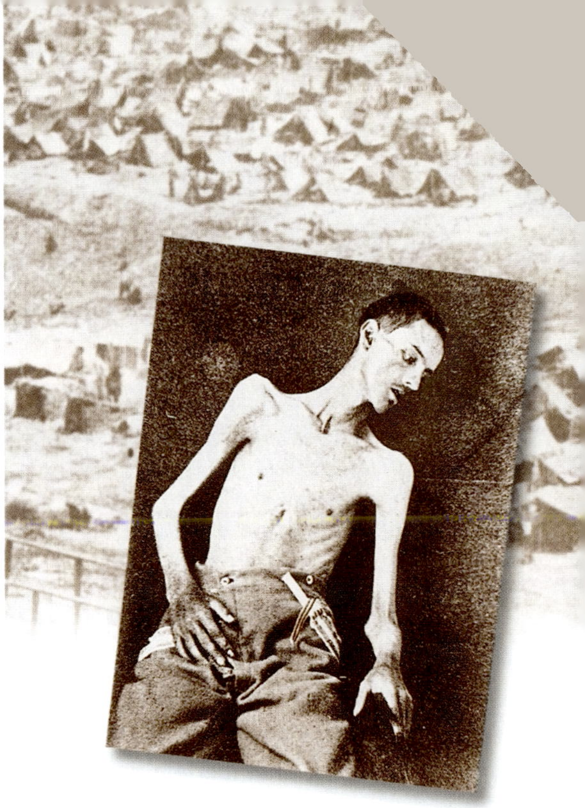

The terrible conditions at Civil War prison camps caused much suffering and death.

People who saw the camps were shocked by the condition of the soldiers. The poet Walt Whitman—who served as a Union nurse—described a group of soldiers who returned from a prison camp. He exclaimed, "Can those be *men?* . . . are they not really mummied, dwindled corpses?"

Around 50,000 men died in Civil War prison camps. But this number was dwarfed by the number of dead on the battlefronts and even more from disease in army camps. In the next section, you will read about the bloody battles that led to the end of the Civil War.

Section 2 Assessment

1. Terms & Names

Identify:

- Copperhead
- conscription
- bounty
- income tax
- greenback
- Clara Barton

2. Taking Notes

Use a diagram like the one below to compare conditions in the North and South during the later years of war.

Conditions During the War

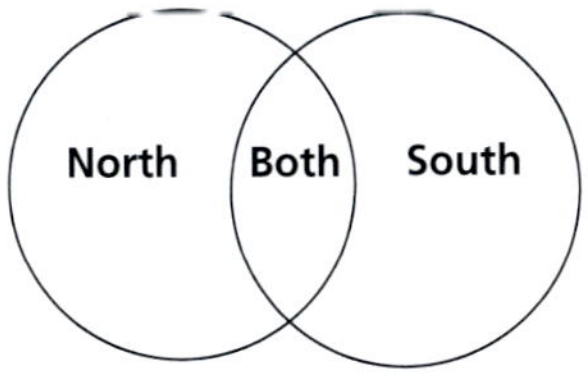

3. Main Ideas

a. How did the South's principle of states' rights undermine the Confederate war effort?

b. How did the draft laws in the North and South differ?

c. What conditions at prison camps caused so many to suffer behind enemy lines?

4. Critical Thinking

Making Generalizations What economic changes took place during the Civil War?

THINK ABOUT

- the war's effect on prices
- industry and agriculture
- new economic measures begun by the government

ACTIVITY OPTIONS

GEOGRAPHY

SPEECH

Study Civil War prison camps. Make a **map** showing where they were located or give a **speech** explaining why prisoners should be treated better.

3 The North Wins

TERMS & NAMES
Battle of Gettysburg
Pickett's Charge
Ulysses S. Grant
Robert E. Lee
Siege of Vicksburg
William Tecumseh Sherman
Appomattox Court House

MAIN IDEA

Thanks to victories, beginning with Gettysburg and ending with Richmond, the Union survived.

WHY IT MATTERS NOW

If the Union had lost the war, the United States might look very different now.

ONE AMERICAN'S STORY

Joshua Lawrence Chamberlain was a 32-year-old college professor when the war began. Determined to fight for the Union, he left his job and took command of troops from his home state of Maine. Like most soldiers, Chamberlain had to get accustomed to the carnage of the Civil War. His description of the aftermath of one battle shows how soldiers got used to the war's violence.

A VOICE FROM THE PAST

It seemed best to [put] myself between two dead men among the many left there by earlier assaults, and to draw another crosswise for a pillow out of the trampled, blood-soaked sod, pulling the flap of his coat over my face to fend off the chilling winds, and still more chilling, the deep, many voiced moan [of the wounded] that overspread the field.

Joshua Lawrence Chamberlain, quoted in *The Civil War*

During the war, Chamberlain fought in 24 battles. He was wounded six times and had six horses shot out from under him. He is best remembered for his actions at the Battle of Gettysburg, where he courageously held off a fierce rebel attack. In this section, you will read about that battle and others that led to the end of the Civil War.

In 1862, Joshua Chamberlain was offered a year's travel with pay to study languages in Europe. He chose to fight for the Union instead.

The Road to Gettysburg

In September 1862, General McClellan stopped General Lee's Northern attack at the Battle of Antietam. But the cautious McClellan failed to finish off Lee's army, which retreated safely to Virginia.

President Lincoln, who was frustrated by McClellan, replaced him with Ambrose Burnside. But Burnside also proved to be a disappointment. At the Battle of Fredericksburg, Virginia, in December 1862, Burnside attacked Confederate troops who had dug trenches. The bloody result was 12,600 Union casualties. This disastrous attack led General Lee to remark, "It is well that war is so terrible—we should grow too fond of it!"

Lincoln replaced Burnside with General Joseph Hooker, who faced Lee the following May at Chancellorsville, Virginia. The result was yet another Union disaster. With half as many men as Hooker, Lee still managed to cut the Union forces to pieces. However, the South paid a high price for its victory. As General "Stonewall" Jackson returned from a patrol on May 2, Confederate guards thought he was a Union soldier and shot him in the arm. Shortly after a surgeon amputated the arm, Jackson caught pneumonia. On May 10, Lee's prized general was dead.

In spite of Jackson's tragic death, Lee decided to head North once again. He hoped that a Confederate victory in Union territory would fuel Northern discontent with the war and bring calls for peace. He also hoped a Southern victory would lead European nations to give diplomatic recognition and aid to the Confederacy.

The Battle of Gettysburg

In late June 1863, Lee crossed into southern Pennsylvania. The Confederates learned of a supply of shoes in the town of Gettysburg and went to investigate. There, on July 1, they ran into Union troops. Both sides called for reinforcements, and the **Battle of Gettysburg** was on.

Reading **History**

A. Reading a Map Use the map and illustration on pages 514–515 to study Gettysburg's geography.

The fighting raged for three days. On the rocky hills and fields around Gettysburg, 90,000 Union troops, under the command of General George Meade, clashed with 75,000 Confederates.

During the struggle, Union forces tried to hold their ground on Cemetery Ridge, just south of town, while rebel soldiers tried to dislodge them. At times, the air seemed full of bullets. "The balls were whizzing so thick," said one Texan, "that it looked like a man could hold out a hat and catch it full."

The turning point came on July 3, when Lee ordered General George Pickett to mount a direct attack on the middle of the Union line. It was a deadly mistake. Some 13,000 rebel troops charged up the ridge into heavy Union fire. One soldier recalled "bayonet thrusts, sabre strokes, pistol shots . . . men going down on their hands and knees, spinning round like tops . . . ghastly heaps of dead men."

Reading **History**

B. Making Inferences Why might Lincoln have been disappointed after the Union victory at Gettysburg?

Pickett's Charge, as this attack came to be known, was torn to pieces. The Confederates retreated and waited for a Union counterattack. But once again, Lincoln's generals failed to finish off Lee's army. The furious Lincoln wondered when he would find a general who would defeat Lee once and for all.

Even so, the Union rejoiced over the victory at Gettysburg. Lee's hopes for a Confederate victory in the North were crushed. The North had lost 23,000 men, but Southern losses were even greater. Over one-third of Lee's army, 28,000 men, lay dead or wounded. Sick at heart, Lee led his army back to Virginia.

America's HERITAGE

THE GETTYSBURG ADDRESS

On November 19, 1863, President Lincoln spoke at the dedication of a cemetery in Gettysburg for the 3,500 soldiers buried there. His speech was short, and few who heard it were impressed. Lincoln himself called it "a flat failure."

Even so, the Gettysburg Address has since been recognized as one of the greatest speeches of all time. In it, Lincoln declared that the nation was founded on "the proposition that all men are created equal." He ended with a plea to continue the fight for democracy so that "government of the people, by the people, for the people shall not perish from the earth."

See page 524 for the full text of the Gettysburg Address.

PLACE AND HUMAN-ENVIRONMENT INTERACTION

Battle of Gettysburg

A monument stands today near a ridge at the Gettysburg battlefield. Labeled the "High Water Mark of the Rebellion," it shows how far Confederate troops advanced against Union lines. There, on July 3, 1863, the South came closest to winning the Civil War.

The fighting began on July 1. When a Confederate force captured Gettysburg, Union defenders took up new positions in the hills south of town. The next day, Confederate troops attacked across a wheat field and peach orchard in an attempt to seize the hill called Little Round Top. But Union forces held their ground.

July 3 was the decisive day. Lee, having failed to crack the side of General Meade's Union line, attacked its center. In an assault that came to be known as Pickett's Charge, some 13,000 men charged uphill across an open field toward the Union lines along Cemetery Ridge. Union soldiers covered the field with rifle and cannon fire. "Pickett's Charge" was a Confederate disaster.

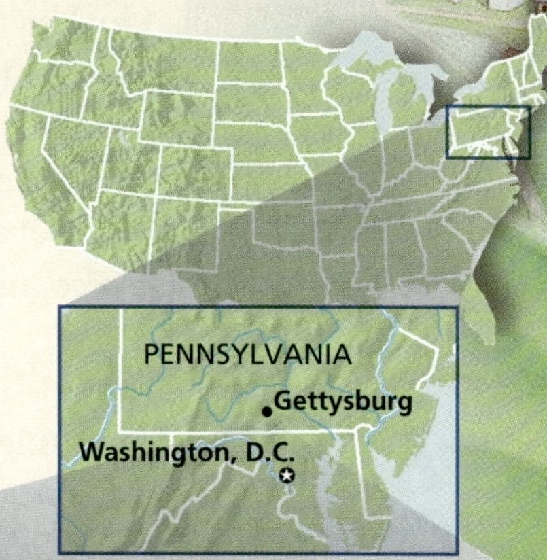

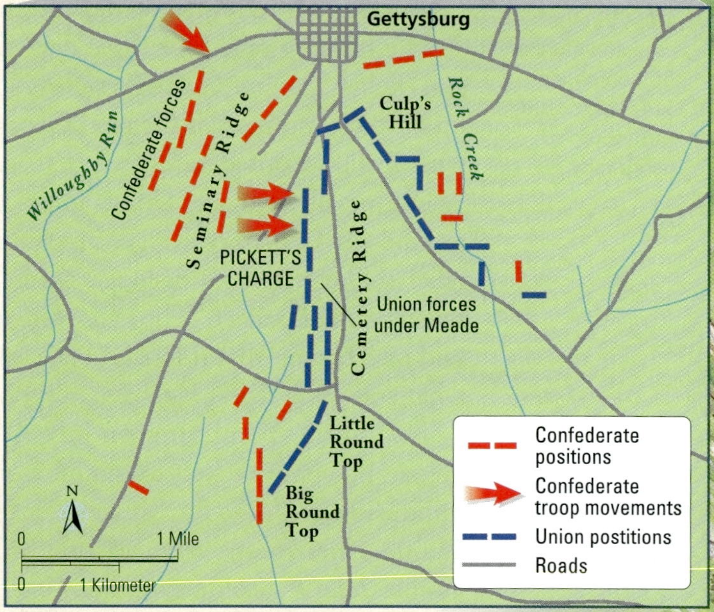

Before beginning the charge named for him, Major General Pickett wrote to his fiancée, "My brave Virginians are to attack in front. Oh, may God in mercy help me."

ARTIFACT FILE

Regimental Flag Flags helped soldiers to identify the different sides during battle. Often, a regiment's flag would show the names of battles it had fought. This flag, which belonged to the 28th North Carolina, was captured at Pickett's Charge.

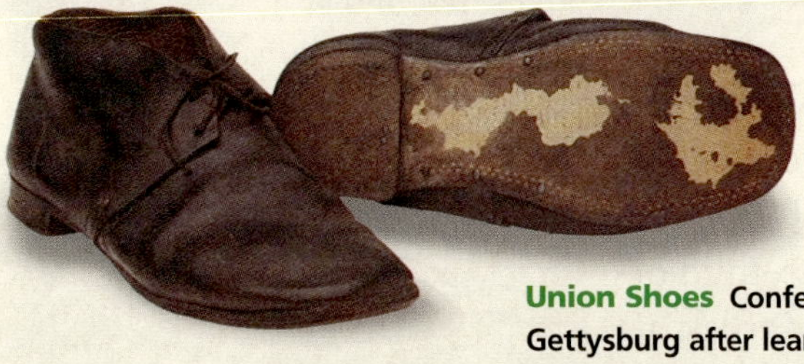

Union Shoes Confederate troops first went to Gettysburg after learning of a supply of shoes in the town. The shoes pictured here were cut down by a Union soldier to make them more comfortable.

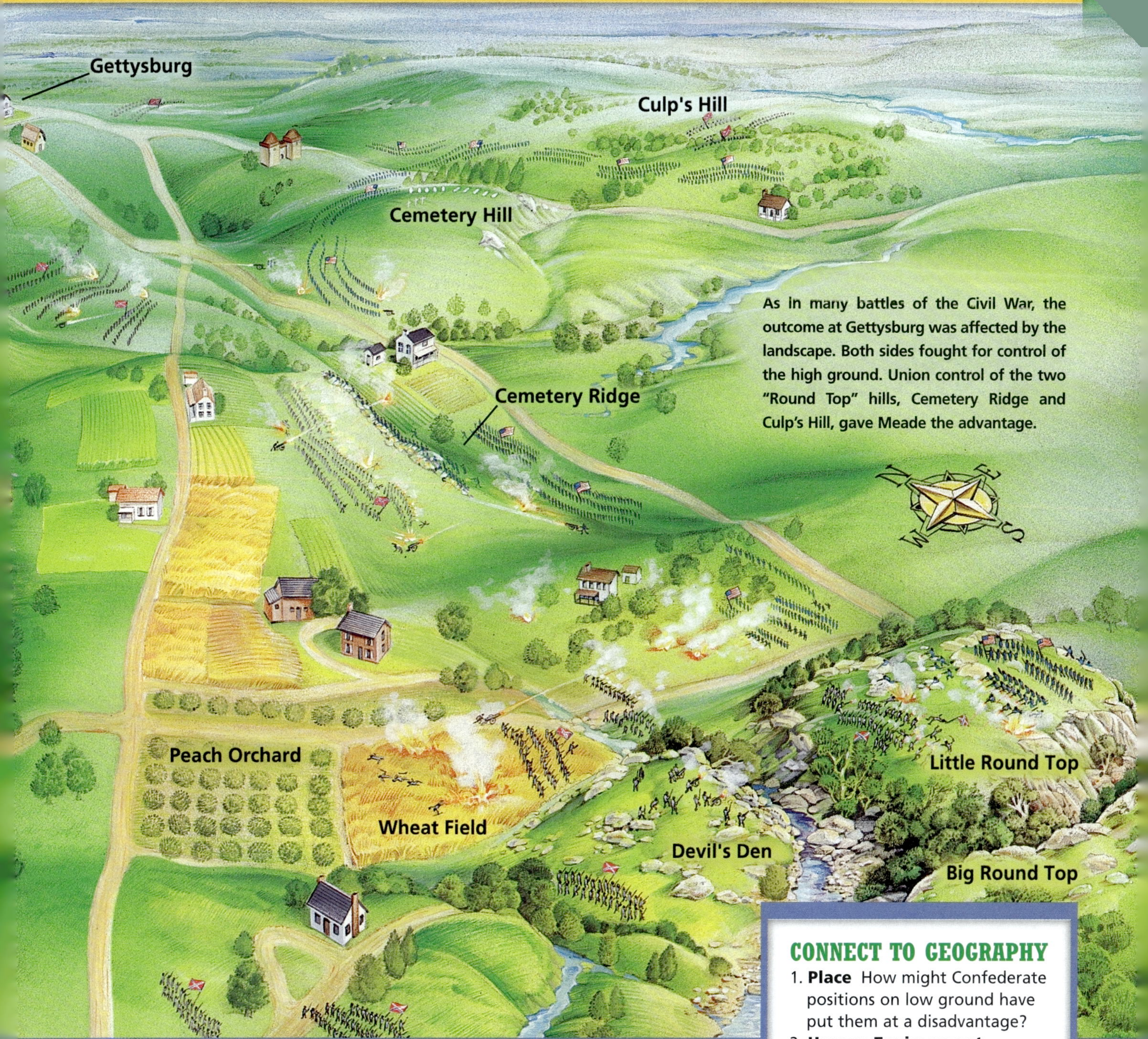

As in many battles of the Civil War, the outcome at Gettysburg was affected by the landscape. Both sides fought for control of the high ground. Union control of the two "Round Top" hills, Cemetery Ridge and Culp's Hill, gave Meade the advantage.

On-Line Field Trip

The Gettysburg National Military Park Museum contains many objects relating to the Battle of Gettysburg, including this federal bass drum. This heavy drum—two feet in diameter—was harnessed to the neck of a soldier, who beat time with leather-covered wooden mallets.

Visit www.mcdougallittell.com for more information.

CONNECT TO GEOGRAPHY

1. **Place** How might Confederate positions on low ground have put them at a disadvantage?
2. **Human-Environment Interaction** How might the attitudes of Union soldiers have been affected by fighting in their own territory?

See Geography Handbook, pages 4–5.

CONNECT TO HISTORY

3. **Making Inferences** Why do you think a Southern victory on Northern soil would have been so significant?

AMERICA'S HISTORY MAKERS

ULYSSES S. GRANT
1822–1885

<u>General Ulysses S. Grant</u> was an unlikely war hero. Although educated at West Point Military Academy, he was a poor student and showed little interest in an army career. With his quiet manner and rumpled uniform, he often failed to impress his fellow officers.

Yet on the battlefield, Grant proved to be a brilliant general. Highly focused and cool under fire, he won the first major Union victories of the war.

Grant was willing to fight Lee—even if the costs were high. He told his generals, "Wherever Lee goes, there you will go also."

ROBERT E. LEE
1807–1870

<u>Robert E. Lee</u> seemed destined for greatness. In his crisp uniform and trim, white beard, Lee was a dashing figure on the battlefield.

Born to a leading Virginia family, Lee was a top student at West Point and won praise for his actions in the Mexican War. General Winfield Scott called him "the very best soldier I have ever seen in the field."

Lee did not want to fight the Union, but he felt he had to stand by Virginia. "I did only what my duty demanded," Lee said. "I could have taken no other course without dishonor."

How did the tough decisions made by Grant and Lee affect the Civil War?

The Siege of Vicksburg

On July 4, 1863, the day after Pickett's Charge, the Union received more good news. In Mississippi, General Ulysses S. Grant had defeated Confederate troops at the **<u>Siege of Vicksburg</u>**.

Vocabulary
siege: the surrounding of a city, town, or fortress by an army trying to capture it

The previous year, Grant had won important victories in the West that opened up the Mississippi River for travel deep into the South. Vicksburg was the last major Confederate stronghold on the river. Grant had begun his attack on Vicksburg in May 1863. But when his direct attacks failed, he settled in for a long siege. Grant's troops surrounded the city and prevented the delivery of food and supplies. Eventually, the Confederates ran out of food. In desperation, they ate mules, dogs, and even rats. Finally, after nearly a month and a half, they surrendered.

The Union victory fulfilled a major part of the Anaconda Plan. The North had taken New Orleans the previous spring. Now, with complete control over the Mississippi River, the South was split in two.

Background
The Anaconda Plan called for blockading Southern ports, taking control of the Mississippi, and capturing Richmond.

With the victories at Vicksburg and Gettysburg, the tide of war turned in favor of the North. Britain gave up all thought of supporting the South. And, in General Grant, President Lincoln found a man who was willing to fight General Lee.

Sherman's Total War

In March 1864, President Lincoln named General Grant commander of all the Union armies. Grant then developed a plan to defeat the Confederacy. He would pursue Lee's army in Virginia, while Union forces under General **<u>William Tecumseh Sherman</u>** pushed through the Deep South to Atlanta and the Atlantic coast.

Battling southward from Tennessee, Sherman took Atlanta in September 1864. He then set out on a march to the sea, cutting a path of destruction up to 60 miles wide and 300 miles long through Georgia.

Sherman waged total war: a war not only against enemy troops, but against everything that supports the enemy. His troops tore up rail lines, destroyed crops, and burned and looted towns.

Reading **History**

C. Drawing Conclusions How might the political situation in the North have been different if Sherman had not taken Atlanta?

Sherman's triumph in Atlanta was important for Lincoln. In 1864, the president was running for reelection, but his prospects were not good. Northerners were tired of war, and Democrats—who had nominated George McClellan—stood a good chance of winning on an antiwar platform.

Sherman's success changed all that. Suddenly, Northerners could sense victory. Lincoln took 55 percent of the popular vote and won re-election. In his second inaugural speech, Lincoln hoped for a speedy end to the war: "With malice towards none; with charity for all; . . . let us strive on to finish the work we are in; to bind up the nation's wounds; . . . to do all which may achieve and cherish a just, and a lasting peace." (See page 525 for more of Lincoln's Second Inaugural Address.)

"Let us strive . . . to bind up the nation's wounds."

Abraham Lincoln

In December, Sherman took Savannah, Georgia. He then sent a telegram to Lincoln: "I beg to present you, as a Christmas gift, the city of Savannah, with 150 heavy guns and . . . about 25,000 bales of cotton."

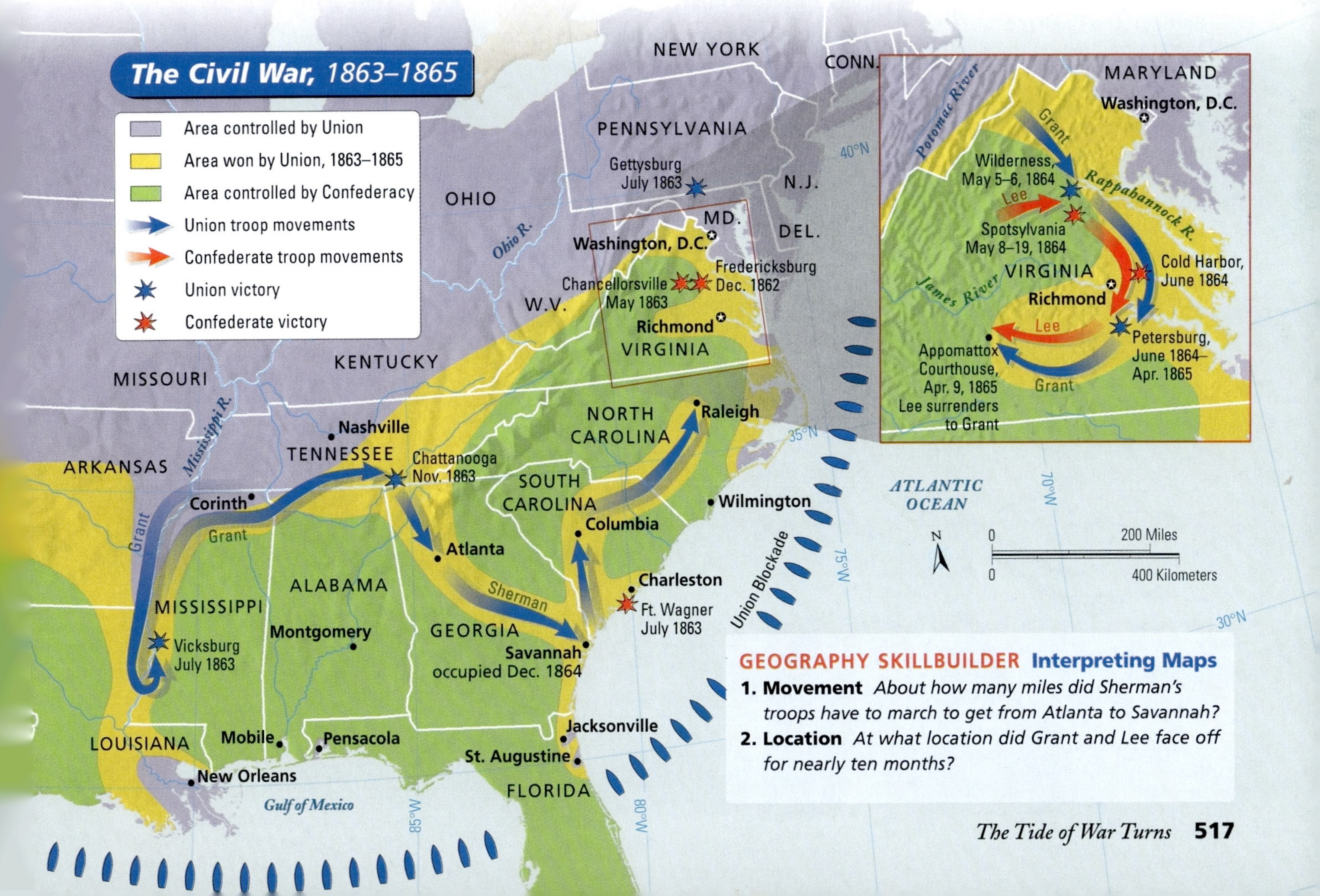

GEOGRAPHY SKILLBUILDER **Interpreting Maps**

1. **Movement** *About how many miles did Sherman's troops have to march to get from Atlanta to Savannah?*
2. **Location** *At what location did Grant and Lee face off for nearly ten months?*

In 1861, Congress created the Medal of Honor to reward individual bravery in combat.

Grant's Virginia Campaign

After taking Savannah, Sherman moved north through the Carolinas seeking to meet up with Grant's troops in Virginia. Since May 1864, Grant and his generals had been fighting savage battles against Lee's forces. In battle after battle, Grant would attack, rest, then attack again, all the while moving south toward Richmond.

*Reading*History

D. Reading a Map Use the map on page 517 to find the locations of the major battles of Grant's Virginia campaign.

At the Battle of the Wilderness in May 1864, Union and Confederate forces fought in a tangle of trees and brush so thick that they could barely see each other. Grant lost over 17,000 men, but he pushed on. "Whatever happens," he told Lincoln, "we will not retreat."

At Spotsylvania and Cold Harbor, the fighting continued. Again, the losses were staggering. Grant's attack in June, at Cold Harbor, cost him 7,000 men, most in the first few minutes of battle. Some Union troops were so sure they would die in battle that they pinned their names and addresses to their jackets so their bodies could be identified later.

In June 1864, Grant's armies arrived at Petersburg, just south of Richmond. Unable to break through the Confederate defenses, the Union forces dug trenches and settled in for a long siege. The two sides faced off for ten months.

In the end, though, Lee could not hold out. Grant was drawing a noose around Richmond. So Lee pulled out, leaving the Confederate capital undefended. The Union army marched into Richmond on April 3. One Richmond woman recalled, "Exactly at eight o'clock the Confederate flag that fluttered above the Capitol came

HISTORY *through* ART

This photograph shows Union officers before the Battle of the Wilderness. Next to the tree on the right is the photographer Mathew Brady. Photography was still a new art when the Civil War began. Brady's Civil War photos represent one of the first examples of photojournalism.

How might people's attitudes toward war be affected when they can see pictures from the front lines?

down and the Stars and Stripes were run up. . . . We covered our faces and cried aloud."

Surrender at Appomattox

From Richmond and Petersburg, Lee fled west, while Grant followed in pursuit. Lee wanted to continue fighting, but he knew that his situation was hopeless. He sent a message to General Grant that he was ready to surrender.

On April 9, 1865, Lee and Grant met in the small Virginia town of **Appomattox Court House** to arrange the surrender. Grant later wrote that his joy at that moment was mixed with sadness.

A VOICE FROM THE PAST

I felt like anything rather than rejoicing at the downfall of a foe who had fought so long and valiantly, and had suffered so much for a cause, though that cause was, I believe, one of the worst for which a people ever fought, and one for which there was the least excuse. I do not question, however, the sincerity of the great mass of those who were opposed to us.

Ulysses S. Grant, *Personal Memoirs*

STRANGE *but* True

WILMER MCLEAN

The first major battle of the Civil War was fought on the property of Wilmer McLean. McLean lived in Manassas, Virginia, the site of the Battle of Bull Run. After the battle, McLean decided to move to a more peaceful place. He chose the village of Appomattox Court House (see map on page 517).

When Lee made the decision to surrender in April 1865, he sent Colonel Charles Marshall to find a location for a meeting with Grant. Marshall stopped the first man he saw in the deserted streets of Appomattox Court House. It was Wilmer McLean.

McLean reluctantly offered his home. Thus, the war that began in McLean's back yard ended in his parlor.

Grant offered generous terms of surrender. After laying down their arms, the Confederates could return home in peace, taking their private possessions and horses with them. Grant also gave food to the hungry Confederate soldiers.

After four long years, the Civil War was coming to a close. Its effects would continue, however, changing the country forever. In the next section, you will learn about the long-term consequences of the Civil War.

Section 3 Assessment

1. Terms & Names

Identify:

- Battle of Gettysburg
- Pickett's Charge
- Ulysses S. Grant
- Robert E. Lee
- Siege of Vicksburg
- William Tecumseh Sherman
- Appomattox Court House

2. Taking Notes

Use a time line like the one below to record key events from Section 3.

Which event is considered the turning point of the war?

3. Main Ideas

a. Why was the Battle of Gettysburg important?

b. Why was Northern success in the Siege of Vicksburg important?

c. How did Grant treat Confederate soldiers after the surrender at Appomattox Court House?

4. Critical Thinking

Contrasting How was the Civil War different from wars that Americans had previously fought?

THINK ABOUT

- the role of civilians
- Sherman's military strategy

ACTIVITY OPTIONS

GEOGRAPHY
LANGUAGE ARTS

Research the Siege of Vicksburg. Make a **topographic map** of the area or write an **article** describing the soldiers' hardships during the siege.

4 The Legacy of the War

TERMS & NAMES
Thirteenth Amendment
John Wilkes Booth

MAIN IDEA	WHY IT MATTERS NOW
The Civil War brought great changes and new challenges to the United States.	The most important change was the liberation of 4 million enslaved persons.

ONE AMERICAN'S STORY

In the spring of 1864, a year before the end of the Civil War, the Union army was running out of cemetery space to bury its war dead. The secretary of war ordered Quartermaster General Montgomery Meigs to find a new site for a cemetery. Without hesitation, Meigs chose Robert E. Lee's plantation in Arlington, Virginia, just across the Potomac River from Washington, D.C. "The grounds about the mansion are admirably adapted to such a use," wrote Meigs in June 1864.

Meigs was from Georgia and had served under Lee in the U.S. Army before the war. Unlike Lee, however, Meigs remained loyal to the Union and disagreed strongly with Lee's decision to join the Confederacy. His decision to turn Lee's plantation into a Union cemetery was highly symbolic. The Union soldiers who died fighting Lee's army would be buried in Lee's front yard. That site became Arlington National Cemetery.

During the Civil War, the government turned Robert E. Lee's Virginia plantation into a graveyard. That graveyard eventually became Arlington National Cemetery.

Costs of the War

Many Northerners shared Montgomery Meigs's bitter feelings toward the South. At the same time, many Southerners felt great resentment toward the North. After the war, President Lincoln hoped to heal the nation and bring North and South together again. The generous terms of surrender offered to Lee were part of that effort. Hard feelings remained, however, in part because the costs of the war were so great.

The Civil War was the deadliest war in American history. In four years of fighting, approximately 620,000 soldiers died—360,000 for the Union and 260,000 for the Confederacy. Another 275,000 Union soldiers and 260,000 Confederate soldiers were wounded. Many suffered from their wounds for the rest of their lives.

Altogether, some 3,000,000 men served in the armies of the North and South—around 10 percent of the population. Along with the soldiers, many other Americans had their lives disrupted by the war.

Reading **History**

A. Contrasting How did government spending during the Civil War compare to that during previous years?

The war also had great economic costs. Together, the North and South spent more than five times the amount spent by the government in the previous eight decades. Many years after the fighting was over, the federal government was still paying interest on loans taken out during the war.

The Thirteenth Amendment

One of the greatest effects of the war was the freeing of millions of enslaved persons. As the Union army moved through the South during and after the war, Union soldiers released African Americans from bondage. One of those released was Booker T. Washington, who later became a famous educator and reformer. He recalled the day a Union officer came to his plantation to read the Emancipation Proclamation.

A VOICE FROM THE PAST

> After the reading we were told that we were all free, and could go when and where we pleased. My mother, who was standing by my side, leaned over and kissed her children, while tears of joy ran down her cheeks. She explained to us what it all meant, that this was the day for which she had been so long praying, but fearing that she would never live to see.
>
> **Booker T. Washington,** quoted in his autobiography, *Up from Slavery*

Reading **History**

B. Making Inferences Why was an amendment needed to free enslaved persons even after the Emancipation Proclamation?

The Emancipation Proclamation applied primarily to slaves in the Confederacy, however. Many African Americans in the border states were still enslaved. In 1864, with the war still under way, President Lincoln had approved of a constitutional amendment to end slavery entirely, but it failed to pass Congress.

In January 1865, Lincoln urged Congress to try again to end slavery. This time, the measure—known as the **Thirteenth Amendment**—passed. By year's end, 27 states, including eight in the South, had ratified the amendment. From that point on, slavery was banned in the United States.

CONNECTIONS TO MAT

Costs of the Civil War

CONFEDERATE CASUALTIES

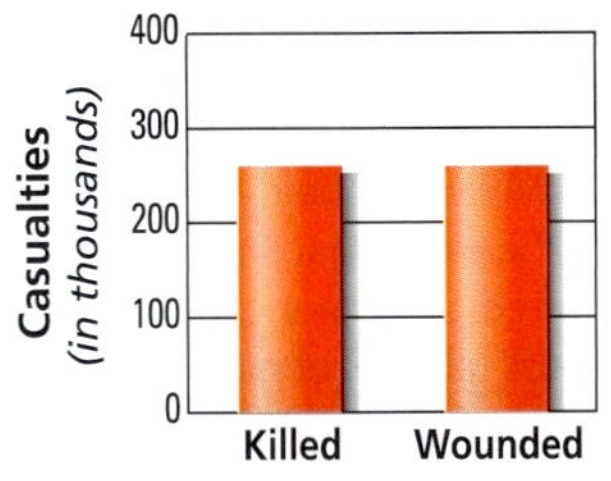

UNION CASUALTIES

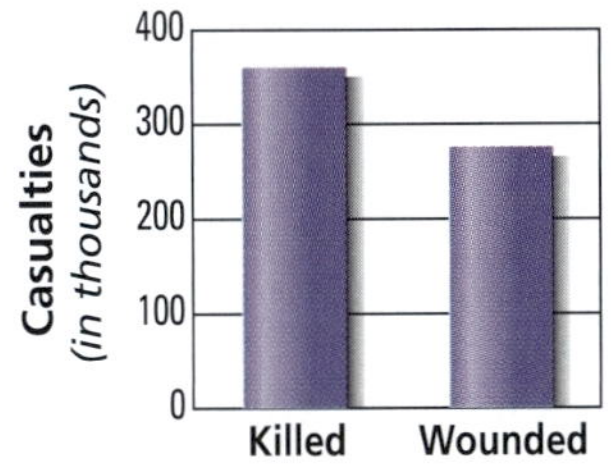

Source: *World Book; Historical Statistics of the United States; The United States Civil War Center*

ECONOMIC COSTS

- **Federal loans and taxes to finance the war totaled $2.6 billion.**
- **Federal debt on June 30, 1865, rose to $2.7 billion.**
- **Confederate debt ran over $700 million.**
- **Union inflation reached 182% in 1864 and 179% in 1865.**
- **Confederate inflation rose to 9,000% by the end of the war.**

SKILLBUILDER

Interpreting Graphs

1. *About how many Confederate soldiers were killed in the Civil War?*
2. *Approximately how many soldiers were wounded in the war?*

Lincoln's Assassination

Lincoln did not live to see the end of slavery, however. Five days after Lee's surrender at Appomattox, the president and his wife went to see a play at Ford's Theatre in Washington, D.C. During the play, a Confederate supporter, **John Wilkes Booth,** crept into the balcony where the president sat and shot him in the back of the head. Booth then jumped over the railing and landed on the stage. Although he broke his leg in the leap, he managed to escape the theater.

Connections TO LITERATURE

WALT WHITMAN

1819–1892

One of the greatest American poets, Walt Whitman (below) was a large, bearded man whose poetry captured the American spirit. His most famous book of poems, *Leaves of Grass,* praised the values of freedom and democracy.

Whitman was 41 when the Civil War began. Too old for the army, he offered his services as a nurse when his younger brother was wounded at Fredericksburg. He stayed on after that to help at hospitals in Washington, D.C.

Whitman wrote a book of poetry about war. Later editions of the book, which appeared after Lincoln's assassination, included several poems about the president.

That same evening, an accomplice of Booth stabbed Secretary of State William Seward, who later recovered. Another man was supposed to assassinate Vice-President Johnson, but he failed to carry out the attack.

Vocabulary
accomplice: someone who aids a lawbreaker

Although Booth had managed to escape after shooting the president, Union troops found and killed him several days later. Soldiers also hunted down Booth's accomplices, whom they either hanged or imprisoned.

After Lincoln was shot, he was carried to a house across the street from the theater. The bullet in his brain could not be removed, however. The next morning, April 15, 1865, the president died. He was the first American president to be assassinated.

Lincoln's murder stunned the nation and caused intense grief. In Washington, D.C., people wept in the streets. One man who mourned the nation's loss was the poet Walt Whitman. In one poem, Whitman considered the president's legacy.

ReadingHistory
C. Summarizing How did Americans react to the assassination of Lincoln?

A VOICE FROM THE PAST

This dust was once the man,
Gentle, plain, just and resolute, under whose cautious hand,
Against the foulest crime in history known in any land or age,
Was saved the Union of these States.

Walt Whitman, *This Dust Was Once the Man*

The loss of Lincoln's vast experience and great political skills was a terrible setback for a people faced by the challenge of rebuilding their nation. In both the North and the South, life would never be the same after the Civil War.

Consequences of the War

In the North, the war changed the way people thought about the country. In fighting to defend the Union, people came to see the United States as a single nation rather than a collection of states. After 1865, people no longer said "the United States *are*" but "the United States *is.*"

The war also caused the national government to expand. Before the war, the government was relatively small and had limited powers. With the demands of war, however, the government grew larger and more powerful. Along with a new paper currency and income tax, the government established a new federal banking system. It also funded railroads, gave western land to settlers, and provided for state colleges. This growth of federal power continued long after the war was over.

Background
In the 1850s, an improved way of making steel—the Bessemer process—had been perfected, allowing for the mass production of steel.

The war also changed the Northern economy. New industries such as steel, petroleum, food processing, and manufacturing grew rapidly. By

CAUSE AND EFFECT: *The Civil War, 1861–1865*

CAUSES	IMMEDIATE EFFECTS
Conflict over slavery in territories	Abolition of slavery
Economic differences between North and South	Devastation of South
Failure of Congress to compromise	Reconstruction of South
Election of Lincoln as president	**LONG-TERM EFFECTS**
Secession of Southern states	Growth of industry
Firing on Fort Sumter	Government more powerful
	Nation reunited

SKILLBUILDER Interpreting Charts

1. *What military event is among the causes of the Civil War?*

2. *What effect did the Civil War have on the federal government?*

the late 1800s, industry had begun to replace farming as the basis of the national economy.

For the South, however, the war brought economic disaster. Farms and plantations were destroyed. About 40 percent of the South's livestock was killed. Fifty percent of its farm machinery was wrecked. Factories were also demolished, and thousands of miles of railroad tracks were torn up. Also gone was the labor system that the South had used—slavery.

Before the war, the South accounted for 30 percent of the nation's wealth. After the war it accounted for only 12 percent. These economic differences between the North and the South would last for decades.

The country faced difficult challenges after the war. How would the South be brought back into the Union, and how would four million former slaves be integrated into national life? You will read more about these challenges in the next chapter.

Section 4 Assessment

1. Terms & Names

Identify:

- Thirteenth Amendment
- John Wilkes Booth

2. Taking Notes

Use a chart like the one below to record the social, economic, and political legacy of the Civil War.

Legacy of the Civil War

Society	Economy	Politics

Is the legacy of the Civil War still apparent today? How?

3. Main Ideas

a. What were some of the human costs of the Civil War?

b. What did the Thirteenth Amendment achieve?

c. What was the state of the Southern economy after the Civil War?

4. Critical Thinking

Making Inferences How do you think the assassination of President Lincoln affected the nation?

THINK ABOUT

- the reaction of ordinary citizens
- its impact on government

ACTIVITY OPTIONS

MATH

TECHNOLOGY

Read about the postwar economy. Create a **database** on industry in the North or make a **storyboard** for a video on the problems in the South.

The Gettysburg Address (1863)

Setting the Stage On November 19, 1863, officials gathered in Gettysburg, Pennsylvania. They were there to dedicate a national cemetery on the ground where the decisive Battle of Gettysburg had taken place nearly five months earlier. Following the ceremony's main address, which lasted nearly two hours, President Lincoln delivered his Gettysburg Address in just over two minutes. In this famous speech, Lincoln expressed his hopes for the nation. **See Primary Source Explorer** CD-ROM

Four **score**[1] and seven years ago our fathers brought forth on this continent a new nation, conceived in liberty, and dedicated to the proposition that all men are created equal.

Now we are engaged in a great civil war, testing whether that nation or any nation so conceived and so dedicated, can long endure. We are met on a great battlefield of that war. We have come to dedicate a portion of that field, as a final resting place for those who here gave their lives that that nation might live. It is altogether fitting and proper that we should do this.

But, in a larger sense, we cannot dedicate—we can not **consecrate**[2]—we can not hallow—this ground. The brave men, living and dead, who struggled here, have consecrated it, far above our poor power to add or **detract.**[3] The world will little note, nor long remember what we say here, but it can never forget what they did here. It is for us the living, rather, to be dedicated here to the unfinished work which they who fought here have thus far so nobly advanced. It is rather for us to be here dedicated to the great task remaining before us—that from these honored dead we **take increased devotion to**[4] that cause for which they **gave the last full measure of devotion**[5]—that we here highly resolve that these dead shall not have died **in vain**[6]—that this nation, under God, shall have a new birth of freedom—and that government of the people, by the people, for the people, shall not perish from the earth.

1. **score:** a group of 20.
2. **consecrate:** to declare as sacred.
3. **detract:** to take away from.
4. **take increased devotion to:** work harder for.
5. **gave the last full measure of devotion:** sacrificed their lives.
6. **in vain:** for nothing.

A CLOSER LOOK

LINCOLN'S MODESTY

Lincoln claimed that what he said at Gettysburg would not be long remembered. However, the address soon came to be recognized as one of the best speeches of all time.

1. What features of Lincoln's address make it so memorable?

A CLOSER LOOK

FIGHTING FOR A CAUSE

Different people fought for different causes during the Civil War. Sometimes, the causes for which people fought changed over the course of the war.

2. What cause is Lincoln referring to in the Gettysburg Address?

Second Inaugural Address (1865)

Setting the Stage President Lincoln delivered his Second Inaugural Address just before the end of the Civil War. In this excerpt, he recalled the major cause of the war and vowed to fight for the restoration of peace and unity. **See Primary Source Explorer** CD-ROM

One-eighth of the whole population were colored slaves. . . . These slaves constituted a peculiar and powerful interest. All knew that this interest was somehow the cause of the war. To strengthen, perpetuate, and extend this interest was the object for which the **insurgents**[1] would rend the Union even by war, while the Government claimed no right to do more than to restrict the territorial enlargement of it. Neither party expected for the war the magnitude or the duration which it has already attained. Neither anticipated that the cause of the conflict might cease with or even before the conflict itself should cease. Each looked for an easier triumph, and a result less fundamental and astounding. Both read the same Bible and pray to the same God, and each invokes His aid against the other. . . . Fondly do we hope, fervently do we pray, that this mighty **scourge**[2] of war may speedily pass away. Yet, if God wills that it continue until all the wealth piled by the **bondsman's**[3] two hundred and fifty years of **unrequited**[4] toil shall be sunk, and until every drop of blood drawn with the lash shall be paid by another drawn with the sword, as was said three thousand years ago, so still it must be said "the judgments of the Lord are true and righteous altogether."

With malice toward none, with charity for all, with firmness in the right as God gives us to see the right, let us strive on to finish the work we are in, to bind up the nation's wounds, to care for him who shall have borne the battle and for his widow and his orphan, to do all which may achieve and cherish a just and lasting peace among ourselves and with all nations.

1. **insurgent:** one that revolts against civil authority.
2. **scourge:** a source of suffering and devastation.
3. **bondsman:** enslaved person.
4. **unrequited:** not paid for.

A CLOSER LOOK

SLAVERY IN TERRITORIES

Before the Civil War, Northern states wanted to prohibit slavery in territories that would eventually become new states. Southern states fought to expand slavery, fearing the prohibition would threaten slavery where it already existed.

1. Why did the Southerners fear that prohibiting slavery in new territories might threaten slavery where it already existed?

A CLOSER LOOK

MALICE TOWARD NONE

As Northerners became more confident in victory, many looked forward to punishing Southerners, whom they blamed for the war. Lincoln, however, urged citizens to care for one another and work for a just and lasting peace.

2. Why do you think that Lincoln believed it would be wiser for Americans not to place blame or seek revenge on one another?

Interactive Primary Sources Assessment

1. Main Ideas

a. Why might President Lincoln have begun the Gettysburg Address by noting that the country was "dedicated to the proposition that all men are created equal"?

b. According to Lincoln's Second Inaugural Address, why did the Confederacy go to war?

c. To what did Lincoln refer with the phrase "the bondsman's two hundred and fifty years of unrequited toil?"

2. Critical Thinking

Making Inferences In 1865, if the South had asked to rejoin the Union without ending slavery, do you think Lincoln would have agreed?

THINK ABOUT

- what Lincoln identifies as the cause of the war
- what might happen if the war ended but slavery did not

Chapter 17 ASSESSMENT

TERMS & NAMES

Briefly explain the importance of each of the following.

1. Emancipation Proclamation
2. 54th Massachusetts Regiment
3. conscription
4. Battle of Gettysburg
5. Ulysses S. Grant
6. Robert E. Lee
7. Siege of Vicksburg
8. William Tecumseh Sherman
9. Appomattox Court House
10. Thirteenth Amendment

REVIEW QUESTIONS

The Emancipation Proclamation (pages 503–506)

1. Why did Lincoln issue the Emancipation Proclamation?
2. How did black soldiers aid the war effort?

War Affects Society (pages 507–511)

3. How did events on the home front show the toll that war was taking there?
4. Why did some people say the Civil War was a "rich man's war but a poor man's fight"?
5. How did enslaved persons help the Union?

The North Wins (pages 512–519)

6. Why was the Battle of Gettysburg so important?
7. How did Sherman's march help the Union?
8. How did Grant defeat Lee?

The Legacy of the War (pages 520–523)

9. How was the Thirteenth Amendment different from the Emancipation Proclamation?
10. How did the war change the national economy?

CRITICAL THINKING

1. USING YOUR NOTES

	North	South
Emancipation Proclamation		
War's Impact		
Northern Victories in Battle		
Union Wins Civil War		

a. How did white and black Southerners react to the Emancipation Proclamation?

b. How did inflation affect the North and the South?

2. ANALYZING LEADERSHIP

What qualities made Lincoln an effective leader?

3. THEME: IMPACT OF THE INDIVIDUAL

How did General Grant's actions in the war make a crucial difference to the outcome?

4. FORMING AND SUPPORTING OPINIONS

One Union relief worker said, "The suffering of men in battle is nothing next to the agony that women feel sending forth their loved ones to war." Do you agree with this statement? Explain why or why not.

5. APPLYING CITIZENSHIP SKILLS

How might the behavior of General Grant and President Lincoln toward the Confederacy have helped to begin healing the war-torn nation?

Interact *with* History

Having read about the ferocity of battle during the Civil War, do you still believe that you would be inspired to continue the fighting? Why or why not?

VISUAL SUMMARY

The Civil War, 1863–1865

1863

January 1863 Emancipation Proclamation is issued.

March 1863 Union passes the draft law.

July 1863 Battle of Gettysburg takes place. Union takes Vicksburg.

March 1864 Grant is put in charge of all Union armies.

May 1864 Battle of the Wilderness and Spotsylvania occur.

June 1864 Battle of Cold Harbor occurs.

September 1864 Sherman takes Atlanta, Georgia.

November 1864 Lincoln is reelected.

December 1864 Sherman takes Savannah, Georgia.

January 1865 Thirteenth Amendment is passed by Congress.

April 1865 Lee surrenders at Appomattox Court House. Lincoln is assassinated.

1866

HISTORY SKILLS

1. INTERPRETING MAPS: Location

Study the map. Answer the questions.

Basic Map Elements

a. Which color represents Confederate states?

Interpreting the Map

b. Next to what river was Vicksburg built?

c. Why might Vicksburg be important for Southerners supplying the Confederacy from Texas?

2. INTERPRETING SECONDARY SOURCES

Historian Bruce Catton describes a brief meeting between two brothers during the Siege of Vicksburg. Read the passage and answer the questions.

> Both boys [were] from Missouri, one of them in Confederate gray and the other in Federal blue. The Confederate had a roll of bills in his hand and gave them to his brother to send to their mother. . . . He couldn't get things out from Vicksburg through the Union lines, Vicksburg being completely surrounded, so he asked his brother to send them to her, and the brother did. There was no shooting while these arrangements were made, then the brothers shook hands and retired to their individual lines, and the shooting started up again.
>
> **Bruce Catton,** from *Reflections on the Civil War*

a. What does this account say about the nature of war?

b. Why might the fact that the brothers came from Missouri pertain to their fighting for different sides?

ALTERNATIVE ASSESSMENT

1. INTERDISCIPLINARY ACTIVITY: Science

Writing a Report Do research on the advances in military technology during the Civil War. Write a report that describes the new or improved weapons that were used during the war. Present the report to your class with charts that illustrate the changes and innovations that you describe.

2. COOPERATIVE LEARNING ACTIVITY

Creating a Newspaper Work with a group of classmates to produce a simple newspaper that covers a specific period of the Civil War. With your group, choose the period you want to cover and decide whether you will take a Northern or Southern viewpoint. Research events and conditions during that period, then select the topics you will cover in your newspaper. Examples include:

- important battles and military strategies
- political news, including government actions and foreign involvement
- social and economic conditions
- profiles of important figures

3. CD-ROM PRIMARY SOURCE EXPLORER

Use the Internet, books, and other resources to do research for a multimedia presentation on a Civil War battle. Using presentation software, consider including the following content:

- paintings or written descriptions of the battle
- images of the destruction caused by the conflict
- music from the time period
- statistics on casualties
- diary and journal entries

4. HISTORY PORTFOLIO

PORTFOLIO **Option 1** Review your section and chapter assessment activities. Select one that you think is your best work. Then use comments made by your teacher or classmates to improve your work and add it to your portfolio.

PORTFOLIO **Option 2** Review the questions that you wrote for What Do You Want to Know? on page 502. Then write a short report in which you explain the answers to your questions. If any questions were not answered, do research to answer them. Add your answers to your portfolio.

Create a Medal of Honor

In 1782, George Washington established the nation's first award to recognize the bravery of American soldiers. It was a purple heart made of cloth and was called the Badge of Military Merit. But the award was not used much after the Revolutionary War. Then, in 1861, Congress created a new award—the Medal of Honor. In the Civil War, 1,520 Union soldiers, including 20 African Americans, received the medal. Today, that medal is the highest United States military award for individual bravery and is commonly referred to as the Congressional Medal of Honor.

ACTIVITY Create a medal of honor for a hero or heroine from either side of the Civil War. Write a letter recommending your hero for the medal. Then read a speech in class as you award your medal to the deserving individual.

TOOLBOX

Each group will need:

- tops of juice cans or cardboard
- aluminum foil
- pieces of ribbon
- safety pins
- glue
- markers
- scissors
- writing paper

STEP BY STEP

1 **Form groups.** Each group should consist of three or four students. Each group will:

- identify a Civil War hero or heroine
- create a medal
- write a letter explaining why your group considers that person a hero
- award the medal and read a speech in class

2 **Research a hero.** First, brainstorm characteristics that you think a hero should have. Then, using this chapter, books on the Civil War, or the Internet, select an individual who you think was a hero for either side of the Civil War. Take notes on the actions of the person you selected. The actions should show how that person meets your standards for being a hero.

The Congressional Medal of Honor rewards military personnel who risked their lives, "above and beyond the call of duty." In the Civil War, 1,520 Union soldiers, including 20 African Americans, received the Medal of Honor.

3 Design your medal. Think about the shape, images, and words that you will use to make your medal of honor. You may want to look at the example from Chapter 17 and in Civil War books. Sketch a design of your medal in pencil.

4 Construct your medal. Using art supplies, construct a medal based on your design. First, construct the medal itself by using a juice can lid or a cardboard pattern. Decorate the medal by using foil and markers. Then add a ribbon and pin.

5 Write a letter. The letter should be addressed to the Congress of the United States of America or the Congress of the Confederate States of America. Your letter should give reasons for recommending your hero based on the person's actions. Carefully recopy your letter after you have made any corrections.

6 Pin up the medal. Create a Medal of Honor board with other groups in your class. Pin up all the medals you have created. Discuss with other groups the standards your group set for calling a person a hero.

WRITE AND SPEAK

Write a Speech As a Civil War military leader, write a speech praising the courage and bravery of the individual who is receiving the medal. Choose another member of your group to act as the recipient of the award. Read your speech to the class as you award the medal to that person.

HELP DESK

For related information, see page 518 in Chapter 17.

Researching Your Project

Visit the library to learn more about the history of military awards or go to your local historical museum to examine actual medals.

Visit www.mcdougallittell.com to learn more about military awards.

Did You Know?

Since the Congressional Medal of Honor was created in 1861, 3,408 persons have received the award. Nineteen persons have received two such awards. Only one woman has received the award.

The Confederate States of America also awarded 43 of its finest men with the Confederate Medal of Honor.

The practice of awarding medals began in Europe during the Middle Ages. Kings realized that giving land to valued knights was too costly, so they gave medals instead.

REFLECT & ASSESS

- What requirements did you set for awarding a medal of honor?
- What reasons did your letter and speech give for awarding the medal?
- What symbols did you use for your medal? Why did you select them?

CHAPTER 18

Reconstruction

1865–1877

Freed African Americans stand in the ruins of a Southern city after the Civil War.

Interact *with* History

The Civil War has just ended, and the Southern states are back in the Union. But the Southern economy is in ruins. Slavery has been abolished. Northerners and Southerners feel deep anger toward one another. As a member of Congress, you must help rebuild the nation.

What Do You Think?

- What problems would you face in rebuilding the nation?
- How would you ease tensions between North and South?
- How would you help freed African Americans?

How would you rebuild the Union?

USA

- **1865** Andrew Johnson becomes president after Lincoln's assassination. Thirteenth Amendment abolishes slavery.
- **1866** Civil Rights Act is passed.
- **1867** Reconstruction Acts are passed.
- **1868** Fourteenth Amendment extends full citizenship to African Americans. Ulysses S. Grant is elected president.
- **1870** Fifteenth Amendment guarantees voting rights to African Americans.
- **1872** Grant is reelected president.
- **1873** Financial panic leads to an economic depression.
- **1877** Rutherford B. Hayes is inaugurated as president and ends Reconstruction.

World

- **1867** Emperor Maximilian is executed in Mexico.
- **1870** Unification of Italy is completed.
- **1871** Unification of Germany is completed.
- **1876** Korea becomes an independent nation.

1865 — 1877

Chapter 18 SETTING THE STAGE

BEFORE YOU READ

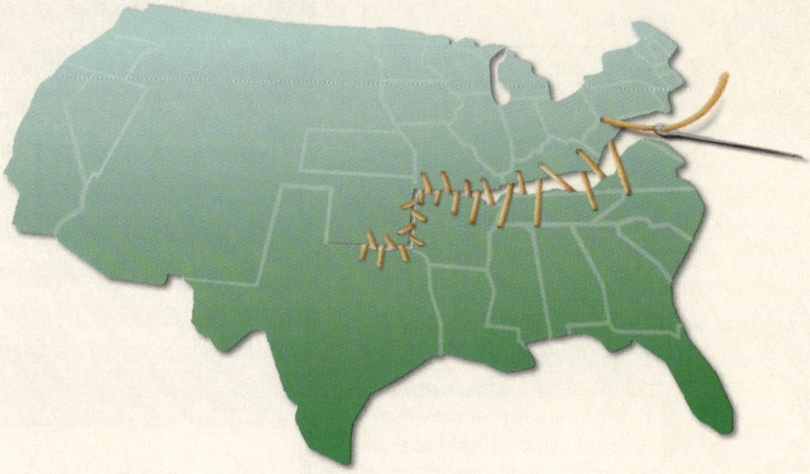

Previewing the Theme

Democratic Ideals After the Union won the war, the nation faced the question of how a democratic government should treat the people who rebelled against it. The Union also faced the question of how to include the millions of freed African Americans in the political process. In this chapter, you will learn how the country struggled to move toward the ideal of equality during Reconstruction.

What Do You Know?

What do you think it means to reconstruct something? What kinds of things did the U.S. government need to reconstruct after the Civil War? Which of these issues do you think was most important?

THINK ABOUT

- what you learned about the Civil War in the last two chapters
- what you've learned about civil rights in the United States from television and movies

What Do You Want to Know?

What questions do you have about Reconstruction? Record those questions in your notebook before you read this chapter.

READ AND TAKE NOTES

Reading Strategy: Identifying and Solving Problems Sometimes, to understand what you read, you must learn to identify problems and solutions. As you read through this chapter, use a diagram like the one below to take notes on the problems the United States faced during Reconstruction and the actions the nation took to solve them.

See Skillbuilder Handbook, page R17.

Problems	Solutions
Black codes →	
President Johnson →	
Education →	
Economy →	
Ku Klux Klan →	
Voting →	

1 Rebuilding the Union

TERMS & NAMES
Radical Republicans
Reconstruction
Freedmen's Bureau
Andrew Johnson
black codes
civil rights
Fourteenth Amendment

MAIN IDEA

During Reconstruction, the president and Congress fought over how to rebuild the South.

WHY IT MATTERS NOW

Reconstruction was an important step in the African-American struggle for civil rights.

ONE AMERICAN'S STORY

After the Civil War, Pennsylvania congressman Thaddeus Stevens became a leader of the Radical Republicans. This group of congressmen favored using federal power to create a new order in the South and to promote full citizenship for freed African Americans.

A VOICE FROM THE PAST

The whole fabric of southern society *must* be changed. . . . If the South is ever to be made a safe Republic let her lands be cultivated by the toil of the owners, or the free labor of intelligent citizens.

Thaddeus Stevens, quoted in *The Era of Reconstruction* by Kenneth Stampp

In this section, you will learn how political leaders battled over how to bring the Southern states back into the Union.

Thaddeus Stevens addresses Congress.

Reconstruction Begins

After the Civil War ended in 1865, the South faced the challenge of building a new society not based on slavery. The process the federal government used to readmit the Confederate states to the Union is known as **Reconstruction.** Reconstruction lasted from 1865 to 1877.

In his Second Inaugural Address, in March 1865, Lincoln promised to reunify the nation "with malice [harm] toward none, with charity for all." Lincoln's plan included pardoning Confederate officials. It also called for allowing the Confederate states to quickly form new governments and send representatives to Congress.

To assist former slaves, the president established the **Freedmen's Bureau.** This federal agency set up schools and hospitals for African Americans and distributed clothes, food, and fuel throughout the South.

When Lincoln was killed in April 1865, Vice-President **Andrew Johnson** became president. Johnson was a Democrat. The Republicans

AMERICA'S HISTORY MAKERS

ANDREW JOHNSON
1808–1875

Andrew Johnson was a self-educated man whose strong will led to trouble with Congress.

As a former slaveholder from Tennessee, Johnson called for a mild program for bringing the South back into the Union. In particular, he let states decide whether to give voting rights to freed African Americans.

Johnson's policies led to a break with the Radical Republicans in Congress and, finally, to his impeachment trial (see page 537).

Why might Johnson have chosen not to punish the South?

had put him on the ticket in 1864 to help win support in the nation's border states. Johnson was a former slaveholder and, unlike Lincoln, a stubborn, unyielding man.

Johnson believed that Reconstruction was the job of the president, not Congress. His policies were based on Lincoln's goals. He insisted that the new state governments ratify the Thirteenth Amendment, which prohibited slavery. He also insisted that they accept the supreme power of the federal government.

Johnson offered amnesty, or official pardon, to most white Southerners. He promised to return their property. In return, they had to pledge loyalty to the United States. At first, the large plantation owners, top military officers, and ex-Confederate leaders were not included in this offer. But they, too, eventually won amnesty.

Background
Not all Confederate leaders were pardoned. Former Confederate president Jefferson Davis, for example, was imprisoned for treason for two years.

Rebuilding Brings Conflict

As the Southern states rebuilt, they set up new state governments that seemed very much like the old ones. Some states flatly refused to ratify the Thirteenth Amendment. "This is a white man's government," said the governor of South Carolina, "and intended for white men only."

The Southern states passed laws, known as **black codes,** which limited the freedom of former slaves. In Mississippi, for instance, one law said that African Americans had to have written proof of employment. Anyone without such proof could be put to work on a plantation. African Americans were forbidden to meet in unsupervised groups or carry guns. Because of such laws, many people in the North suspected that white Southerners were trying to bring back the "old South."

Reading **History**
A. Analyzing Causes What was the main reason Southern states passed black codes?

When Congress met in December 1865, its members refused to seat representatives from the South. Many of these Southern representatives had been Confederate leaders only months before.

Under the Constitution, Congress has the right to decide whether its members are qualified to hold office. So instead of admitting the Southerners, Congress set up a committee to study conditions in the South and decide whether the Southern states should be represented. By taking such action, Congress let the president know that it planned to play a role in Reconstruction.

Republicans outnumbered Democrats in both houses of Congress. Most Republicans were moderates who believed that the federal government should stay out of the affairs of individuals and the states.

Vocabulary
moderates: people opposed to extreme views

The Radical Republicans, however, wanted the federal government to play an active role in remaking Southern politics and society. Led by Thaddeus Stevens and Massachusetts senator Charles Sumner, the

Radical Republicans pose for a formal portrait. Standing (left to right): James F. Wilson, George S. Boutwell, and John A. Logan. Seated: Benjamin F. Butler, Thaddeus Stevens, Thomas Williams, and John A. Bingham.

group also demanded full and equal citizenship for African Americans. Their aim was to destroy the South's old ruling class and turn the region into a place of small farms, free schools, respect for labor, and political equality for all citizens.

The Civil Rights Act

Urged on by the Radicals, Congress passed a bill promoting **civil rights**—those rights granted to all citizens. The Civil Rights Act of 1866 declared that all persons born in the United States (except Native Americans) were citizens. It also stated that all citizens were entitled to equal rights regardless of their race.

Republicans were shocked when President Johnson vetoed the bill. Johnson argued that federal protection of civil rights would lead "towards centralization" of the national government. He also insisted that making African Americans full citizens would "operate against the white race." Congress voted to override Johnson's veto. That is, two-thirds of the House and two-thirds of the Senate voted for the bill after the president's veto, and the bill became law.

The Fourteenth Amendment

Reading **History**

B. Making Inferences How did the Fourteenth Amendment encourage states to give African Americans the vote?

Republicans were not satisfied with passing laws that ensured equal rights. They wanted equality to be protected by the Constitution itself. To achieve this goal, Congress proposed the **Fourteenth Amendment** in 1866. It stated that all people born in the United States were citizens and had the same rights. All citizens were to be granted "equal protection of the laws." However, the amendment did not establish black suffrage. Instead, it declared that any state that kept African Americans from voting would lose representatives in Congress. This meant that the Southern states would have less power if they did not grant black men the vote.

Johnson refused to support the amendment. So did every former Confederate state except Tennessee. This rejection outraged both moderate and Radical Republicans. As a result, the two groups agreed to join forces and passed the Reconstruction Acts of 1867. The passage of these acts began a period known as Radical Reconstruction. From this point on, Congress controlled Reconstruction.

One of the Reconstruction Acts of 1867 divided the South into five military districts, each run by an army commander. Members of the ruling class before the war lost their voting rights. The law also stated that before the Southern states could reenter the Union, they would have to do two things:

1. They must approve new state constitutions that gave the vote to all adult men, including African Americans.
2. They must ratify the Fourteenth Amendment.

Background Attempts to secure voting rights for African Americans applied only to men. Women were not allowed to vote until 1919.

The New Southern Governments

In 1867, Southern voters chose delegates to draft their new state constitutions. About three-fourths of the delegates were Republicans. About half of the Republicans were poor white farmers. Angry at planters for starting what they called the "rich man's war," these delegates were called scalawags (scoundrels) for going along with Radical Reconstruction.

Another one-fourth of the Republican delegates were known as carpetbaggers—white Northerners who had rushed to the South after the war. Many Southerners accused them, often unfairly, of seeking only to get rich or gain political power.

Background Carpetbaggers were said to have headed south carrying only a cheap suitcase, known as a carpetbag.

African Americans made up the rest of the Republican delegates. Of these, half had been free before the war. Most were ministers, teachers, or skilled workers. About 80 percent of them could read.

HISTORY through ART

His First Vote, an 1868 oil painting by Thomas Waterman Wood, shows a new African-American voter.

How do you think the man felt about voting?

The new constitutions written by these delegates set up public schools and gave the vote to all adult males. By 1870, voters in all the Southern states had approved their new constitutions. As a result, the former Confederate states were let back into the Union and allowed to send representatives to Congress.

During Reconstruction, more than 600 African Americans served in state legislatures throughout the South, and 14 of the new U.S. congressmen from the South were African Americans. Two African Americans served as U.S. senators during this time. One was Hiram Revels of Mississippi, a minister in the African Methodist Episcopal Church. He had recruited African Americans to fight for the Union during the Civil War.

Reading **History**

C. Finding Main Ideas What political gains did African Americans make during Reconstruction?

Johnson Is Impeached

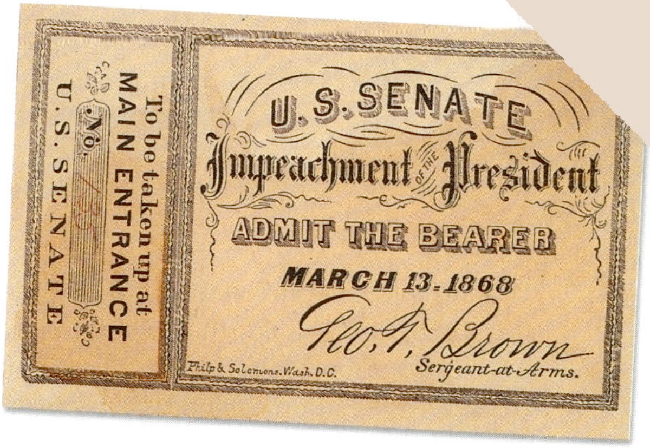

This is a ticket to the 1868 impeachment trial of President Johnson.

President Johnson fought against many of Congress's reform efforts during Radical Reconstruction. For instance, he chose people friendly to ex-Confederates to serve as military commanders in the South. The conflict between Johnson and Congress soon brought a showdown.

In 1867, Congress passed the Tenure of Office Act, which prohibited the president from firing government officials without the Senate's approval. In February 1868, Johnson fired his secretary of war, Edwin Stanton, over disagreements about Reconstruction. Three days later, the House of Representatives voted to impeach the president. This means that the House formally accused him of improper conduct while in office. By removing Johnson from office, they hoped to strengthen Congress's role in Reconstruction.

Reading **History**

D. Drawing Conclusions Why did Congress decide to impeach President Johnson?

The case moved to the Senate for a trial. After several weeks of testimony, the senators prepared to vote. George Julian, a 20-year congressman from Indiana, recalled the tension in the air.

A VOICE FROM THE PAST

The galleries were packed, and an indescribable anxiety was written on every face. Some of the members of the House near me grew pale and sick under the burden of suspense. Such stillness prevailed that the breathing in the galleries could be heard at the announcement of each Senator's voice.

George Julian, quoted in *Grand Inquests*

Vocabulary

acquitted: cleared of a charge

In the end, President Johnson was acquitted by a single vote. But much work remained to be done in rebuilding the South. In the next section, you will learn how African Americans in the South worked to improve their lives.

Section 1 Assessment

1. Terms & Names

Identify:

- Radical Republicans
- Reconstruction
- Freedmen's Bureau
- Andrew Johnson
- black codes
- civil rights
- Fourteenth Amendment

2. Taking Notes

Use a diagram to review the events that led to Johnson's impeachment.

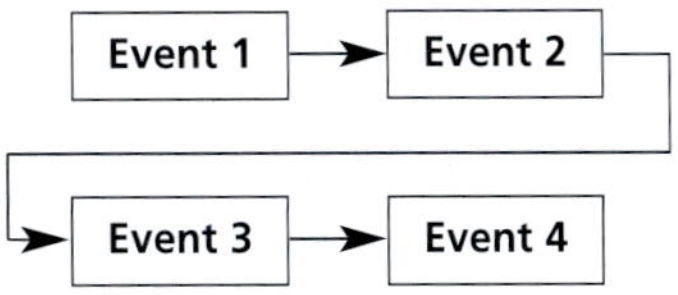

Which event seems most important and why?

3. Main Ideas

a. What was Lincoln's Reconstruction plan?

b. How did white Southerners plan to restore the "old South"?

c. What impact did the Reconstruction Acts of 1867 have on the South?

4. Critical Thinking

Evaluating Do you think the House was justified in impeaching President Johnson? Why or why not?

THINK ABOUT

- the clash over Reconstruction policies
- Congress's motives for impeaching Johnson

ACTIVITY OPTIONS

TECHNOLOGY

SPEECH

Research an African American who served in Congress during Reconstruction. Design his Internet **home page,** or make a **speech** about his accomplishments.

Rebuilding Richmond

You live in Richmond, Virginia, the capital of the Confederacy. It is 1865, and the South faces defeat in the Civil War. On April 2, Confederate officials set fire to supplies in Richmond to prevent the approaching Union army from using them. The fire spreads out of control and destroys downtown Richmond. The next day, Union troops march into the city and take command. You must now help rebuild the city.

COOPERATIVE LEARNING On this page are two challenges you face as a resident of Richmond. Working with a small group, decide how to solve one of these problems. Divide the work among the group members. You will find useful information in the Data File. Be prepared to present your solutions to the class.

This picture shows a street in Richmond before the fire.

ARTS CHALLENGE

"We want the burnt district of Richmond to . . . sit proudly again."

The smell of charred wood still floats in the breeze. However, spirit and determination fill the air. Warehouses are opening. Newly cleared streets bustle with activity. The rebuilding of Richmond has begun. How would you design one block of Richmond's new downtown business district? Use the Data File for help. Then present your plan using one of these options:

- **Make a model of your new city block.**
- **Create a written proposal with a rough sketch to illustrate.**

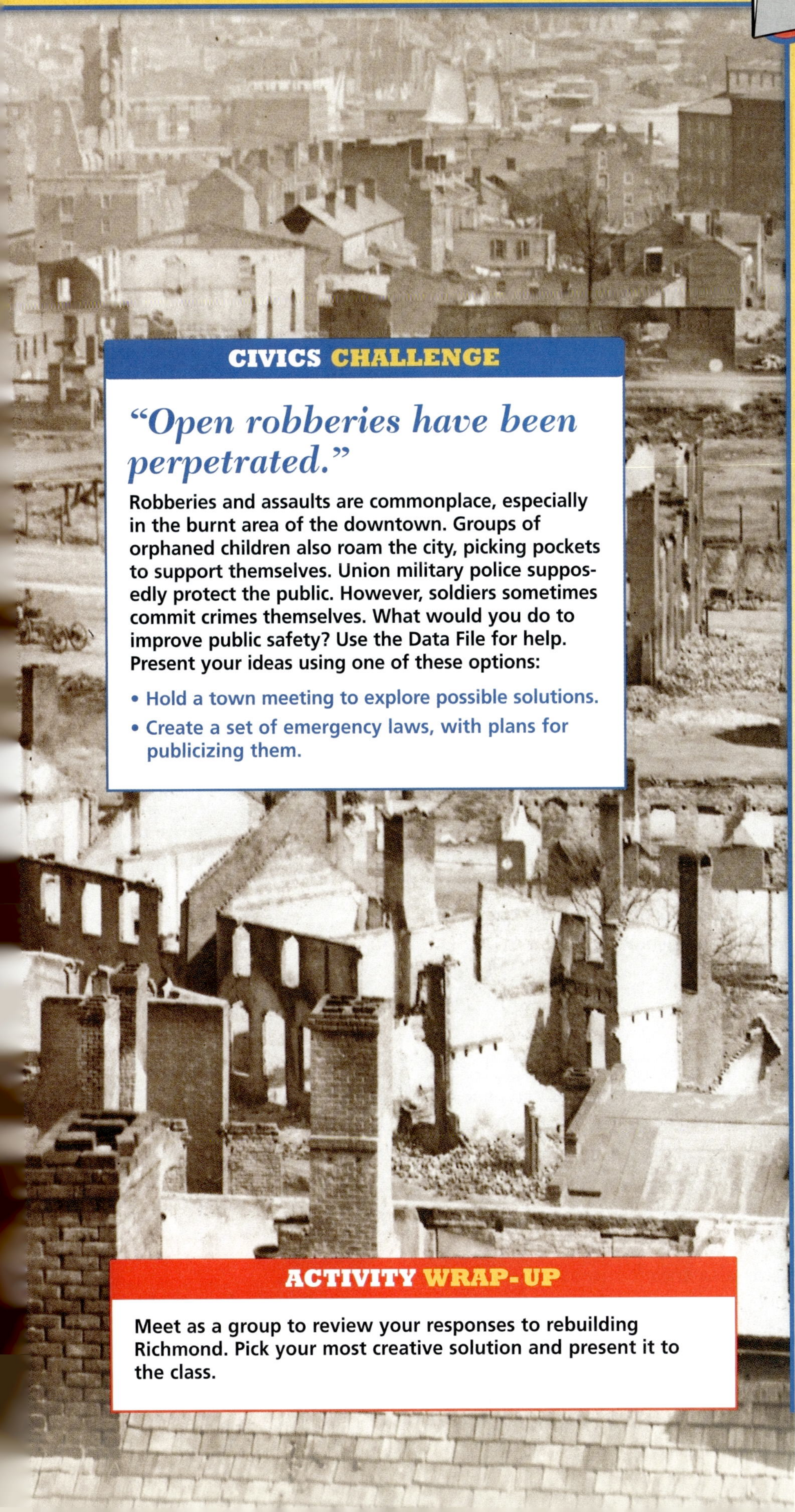

CIVICS CHALLENGE

"Open robberies have been perpetrated."

Robberies and assaults are commonplace, especially in the burnt area of the downtown. Groups of orphaned children also roam the city, picking pockets to support themselves. Union military police supposedly protect the public. However, soldiers sometimes commit crimes themselves. What would you do to improve public safety? Use the Data File for help. Present your ideas using one of these options:

- Hold a town meeting to explore possible solutions.
- Create a set of emergency laws, with plans for publicizing them.

ACTIVITY WRAP-UP

Meet as a group to review your responses to rebuilding Richmond. Pick your most creative solution and present it to the class.

DATA FILE

THE BURNT DISTRICT

- about $30 million damage
- 20 city blocks destroyed, including 900 buildings

Destroyed Property
all banks, 20 law offices, 24 grocery stores, 36 merchant shops, 2 carriage factories, 2 paper mills, 7 book and stationery stores, 2 train depots, 3 bridges, a church, a machine shop, a tin shop, a pottery factory, several flour mills and printing offices

Surviving Property
capitol and city hall, residential areas, ironworks

EMERGENCY SERVICES

Union Army
- distributes 13,000 food rations
- provides medical help
- guards homes; patrols streets

American Union Commission
- hands out food tickets
- distributes 80,000 pounds of flour; feeds soup to 800 people a day
- provides garden seeds and sells shovels at cost to farmers

REBUILDING

April 1865
- rubble is cleared
- markets sell meat, fish, produce
- hotels and bakeries open
- one bridge is rebuilt

May 1865
- two banks open
- gas and telegraph service is restored
- river opens to steamboat traffic

Summer 1865
- horse-drawn buses operate
- city government is reinstated

Fall 1865
- ironworks reopens
- 100 buildings are now under construction

Visit www.mcdougallittell.com to learn more about Reconstruction.

2 Reconstruction and Daily Life

TERMS & NAMES
freedmen's school
sharecropping
Ku Klux Klan
lynch

MAIN IDEA
As the South rebuilt, millions of newly freed African Americans worked to improve their lives.

WHY IT MATTERS NOW
Many important African-American institutions, including colleges, began during Reconstruction.

ONE AMERICAN'S STORY

One day, as the Civil War came to a close, two enslaved women named Mill and Jule saw a fleet of Union gunboats coming up the Mississippi River. Yankee soldiers came ashore and offered them and other slaves passage aboard their boats. On that day, Mill and Jule left the plantation where they had toiled for so long.

A VOICE FROM THE PAST

An' we all got on the boat in a hurry . . . we all give three times three cheers for the gunboat boys, and three times three cheers for big Yankee [soldiers], an' three times three cheers for gov'ment; an' I tell you every one of us, big and little, cheered loud and long and strong, an' made the old river just ring ag'in.

Mill and Jule, quoted in *We Are Your Sisters*

Freed people and a federal soldier pose on a South Carolina plantation.

The Union's victory in the Civil War spelled the end of slavery in America and a new beginning for the nation's millions of newly freed African Americans. In this section, you will learn about the gains and setbacks of former slaves during Reconstruction.

Responding to Freedom

African Americans' first reaction to freedom was to leave the plantations. No longer needing passes to travel, they journeyed throughout the region. "Right off colored folks started on the move," recalled one freedman. "They seemed to want to get closer to freedom, so they'd know what it was—like it was a place or a city." Some former slaves returned to the places where they were born. Others went looking for more economic opportunity. Still others traveled just because they could.

African Americans also traveled in search of family members separated from them during slavery. One man walked 600 miles from Georgia to North Carolina to find his family. To locate relatives, people placed advertisements in newspapers. The Freedmen's Bureau helped many families reunite. A Union officer wrote in 1865, "Men are taking

Reading History
A. Analyzing Causes For what reasons did former slaves move?

their wives and children, families which had been for a long time broken up are united and oh! such happiness."

Freedom allowed African Americans to strengthen their family ties. Former slaves could marry legally. They could raise families without fearing that their children might be sold. Many families adopted children of dead relatives and friends to keep family ties strong.

Starting Schools

With freedom, African Americans no longer had to work for an owner's benefit. They could now work to provide for their families. To reach their goal of economic independence, however, most had to learn to read and write. As a result, children and adults flocked to **freedmen's schools** set up to educate newly freed African Americans. Such schools were started by the Freedmen's Bureau, Northern missionary groups, and African-American organizations. Freed people in cities held classes in warehouses, billiard rooms, and former slave markets. In rural areas, classes were held in churches and houses. Children who went to school often taught their parents to read at home.

Reading History
B. Finding Main Ideas Why did freed people desire an education?

In the years after the war, African-American groups raised more than $1 million for education. However, the federal government and private groups in the North paid most of the cost of building schools and hiring teachers. Between 1865 and 1870, the Freedmen's Bureau spent $5 million for this purpose.

More than 150,000 African-American students were attending 3,000 schools by 1869. About 10 percent of the South's African-American adults could read. A number of them became teachers. Northern teachers, black and white, also went South to teach freed people. Many white Southerners, however, worked against these teachers' efforts. White racists even killed teachers and burned freedmen's schools in some parts of the South. Despite these setbacks, African Americans kept working toward an education.

Background
Most African Americans were illiterate because teaching slaves to read and write had been illegal.

America's HERITAGE

BLACK COLLEGES
Some of today's African-American colleges and universities date back to Reconstruction. The Freedmen's Bureau and other societies raised funds to build many of the schools. Howard University, shown in this photograph, opened in 1867. It was named for General Oliver Otis Howard, head of the Freedmen's Bureau. During Reconstruction, these colleges offered courses ranging from basic reading and writing to medicine and law. They also trained much-needed teachers.

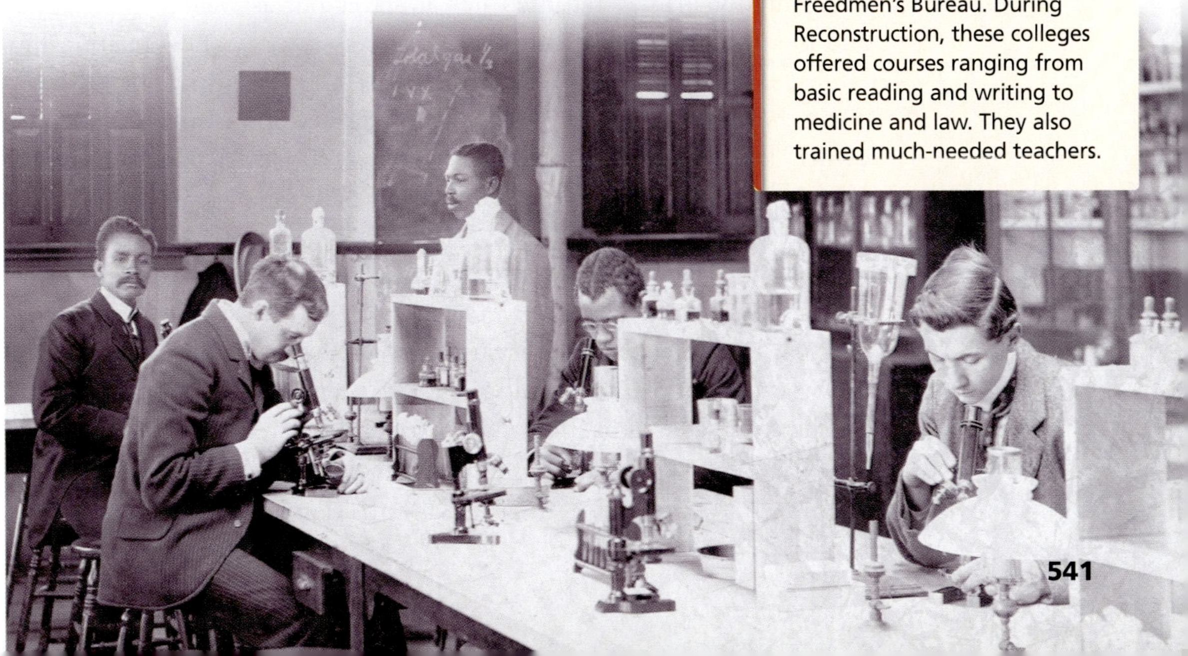

African–American families hoped to own land but were often disappointed.

40 Acres and a Mule

More than anything else, freed people wanted to own land. As one freedman said, "Give us our own land and we take care of ourselves, but without land, the old masters can hire us or starve us, as they please."

As the Civil War ended, General William T. Sherman suggested that abandoned land in coastal South Carolina be split into 40-acre parcels and given to freedmen. The rumor then spread that all freedmen would get 40 acres and a mule. Most African Americans thought they deserved at least that much. In the end, however, most freedmen never received land. Those who did often had to return it to its former owners after the owners were pardoned by President Johnson. One freedman, Bayley Wyat, protested.

A VOICE FROM THE PAST

Our wives, our children, our husbands, [have] been sold over and over again to purchase the lands we now [locate] upon; for that reason we have a divine right to the land. . . . And then didn't we clear the land, and raise the crops of corn, of cotton, of tobacco, of rice, of sugar, of everything.

Bayley Wyat, quoted in *Reconstruction: America's Unfinished Revolution*

Radical Republican leaders Thaddeus Stevens and Charles Sumner pushed to make land reform part of the Reconstruction Acts of 1867. Stevens proposed a plan to Congress that would have taken land from plantation owners and given it to freed people.

Many moderate Republicans and even some Radicals were against the plan. They believed that new civil and voting rights were enough to give African Americans a better life.

***Reading*History**

C. Analyzing Points of View
What were some arguments in favor of giving land to freed people?

Supporters of the plan argued that civil rights meant little without economic independence. Land could provide that independence, they claimed. However, Congress did not pass the land-reform plan.

The Contract System

Without their own property, many African Americans returned to work on plantations. They returned not as slaves but as wage earners. They and the planters both had trouble getting used to this new relationship. "It seems humiliating to be compelled to bargain and haggle with our own servants about wages," wrote the daughter of a Georgia plantation owner. For their part, many freed workers assumed that wages were extra. They thought that the planters still had to house and feed them.

After the Civil War, planters desperately needed workers to raise cotton, still the South's main cash crop. African Americans reacted to this demand for labor by choosing the best contract offers. The contract system was far better than slavery. African Americans could decide whom to work for, and planters could not abuse them or split up families.

Background
Civil War deaths and the departure of slaves from plantations created a labor shortage in the South.

The contract system still had drawbacks, however. Even the best contracts paid very low wages. Workers often could not leave the plantations

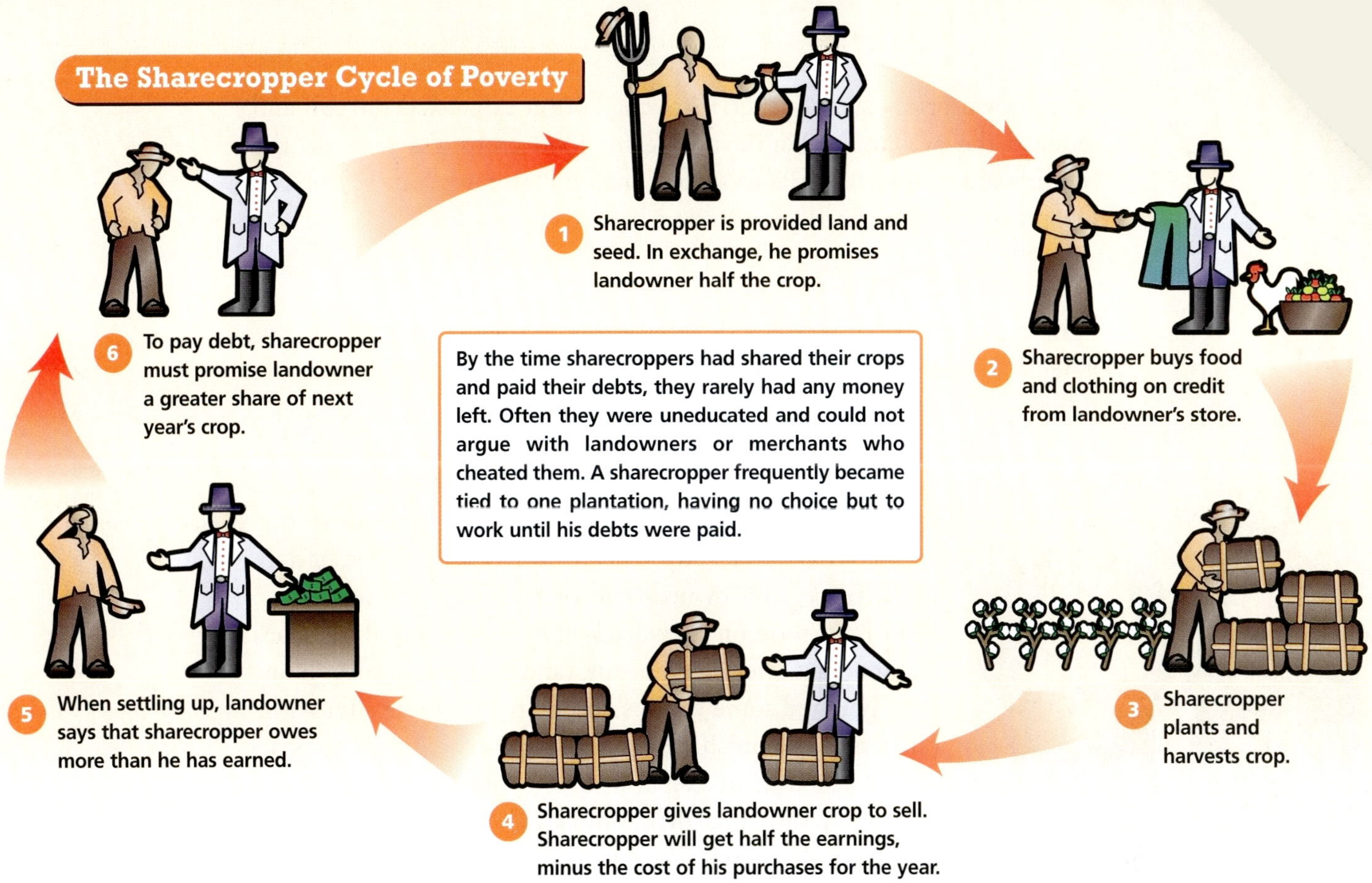

without permission. Many owners cheated workers out of wages and other benefits. Worse yet, laws punished workers for breaking their contracts, even if the plantation owners were abusing or cheating them. These drawbacks made many African Americans turn to sharecropping.

Sharecropping and Debt

Under the **sharecropping** system, a worker rented a plot of land to farm. The landowner provided the tools, seed, and housing. When harvest time came, the sharecropper gave the landowner a share of the crop. This system gave families without land a place to farm and gave landowners cheap labor.

But problems soon arose with the sharecropping system. One cause of these problems was that farmers and landowners had opposite goals. Farmers wanted to grow food to feed their families, but landowners forced them to grow cash crops, such as cotton. As a result, farmers had to buy food from the local store—which was usually owned by the landlord. Most farmers did not have the money to pay for goods. As a result, many were caught in a cycle of debt, as shown in the diagram above. Often farmers had to use one year's harvest to pay the previous year's bills.

Reading **History**
D. Recognizing Effects What were some problems with the sharecropping system?

White farmers also became sharecroppers. Many had lost their land in the war. Others had lost it to taxes. By 1880, one-third of the white farmers in the Deep South worked someone else's land.

No matter who worked the plantations, much of what they grew was cotton. After the war, the value of cotton dropped. Southern planters responded by trying to produce more of the cash crop—a move that drove down prices even further. Growing cotton exhausted the soil and reduced the amount of land available for food crops. As a result, the South had to import half its food. Relying on cotton was one reason the Deep South experienced years of rural poverty.

The Ku Klux Klan

African Americans in the South faced other problems besides poverty. They also faced violent racism. Many planters and former Confederate soldiers did not want African Americans to have more rights. In 1866, such feelings spurred the rise of a secret group called the **Ku Klux Klan.** The Klan's goals were to restore Democratic control of the South and keep former slaves powerless.

Reading History
E. Finding Main Ideas What were the goals of the Ku Klux Klan?

These Ku Klux Klan members were arrested after an 1868 riot in Alabama.

The Klan attacked African Americans. Often it targeted those who owned land or had become prosperous. Klansmen rode on horseback and dressed in white robes and hoods. They beat people and burned homes. They even **lynched** some victims, killing them on the spot without a trial as punishment for a supposed crime. The Klan also attacked white Republicans.

Klan victims had little protection. Military authorities in the South often ignored the violence. President Johnson had appointed most of these authorities, and they were against Reconstruction.

The Klan's terrorism served the Democratic Party. As gun-toting Klansmen kept Republicans away from the polls, the Democrats increased their power.

In the next section you will see how planters took back control of the South. You also will learn how they blocked African Americans' attempts to win more rights.

Section 2 Assessment

1. Terms & Names

Identify:
- freedmen's school
- sharecropping
- Ku Klux Klan
- lynch

2. Taking Notes

Use a cluster diagram like the one below to review details about sharecropping.

For farmers, what were the advantages and disadvantages of sharecropping?

3. Main Ideas

a. How did freedom help strengthen African-American families?

b. How were African Americans educated during Reconstruction?

c. What were the main reasons African Americans wanted their own land?

4. Critical Thinking

Analyzing Causes
Despite greater civil rights, why did African Americans still face difficulty in improving their lives?

THINK ABOUT
- the defeat of the land-reform bill
- the Ku Klux Klan's rise
- the attitude of military authorities in the South

ACTIVITY OPTIONS

SPEECH
ART

Make a **speech** to President Johnson or design a **mural** explaining why land should be given to newly freed African Americans.

3 End of Reconstruction

TERMS & NAMES
Fifteenth Amendment
Panic of 1873
Compromise of 1877

MAIN IDEA
As white Southerners regained power, Reconstruction ended, as did black advances toward equality.

WHY IT MATTERS NOW
Reforms made during Reconstruction made later civil rights gains possible.

ONE AMERICAN'S STORY

Robert B. Elliott was a U.S. congressman from South Carolina during Reconstruction. In 1874, he made a stirring speech supporting a civil rights bill that would outlaw racial discrimination in public services. (See Interact with History, page 531.)

A VOICE FROM THE PAST

The passage of this bill will determine the civil status, not only of the negro but of any other class of citizens who may feel themselves discriminated against. It will form the capstone of that temple of liberty begun on this continent.

Robert B. Elliott, quoted in *The Glorious Failure*

Robert B. Elliott lost his political office when Reconstruction ended.

Elliott was elected South Carolina's attorney general in 1876. He began his term in 1877, just as Reconstruction was ending. That year, federal troops left the South. White Southerners took back control of the region. Quickly, they forced African Americans, including Elliott, out of office.

In this section, you will learn about the events that ended Reconstruction. You will also see how Reconstruction's end meant setbacks in the fight for civil rights and equality.

The Election of Grant

The Republican Party seemed stronger than ever in 1868. That year, its candidate, General Ulysses S. Grant, won the presidency. During the campaign, the Democrats attacked the Republicans' Reconstruction policies. They blamed the party for granting rights to African Americans.

On Election Day, however, the Republicans won. Grant received 214 electoral votes. His Democratic opponent received only 80. The popular count was much closer. Grant had a majority of only 306,000 votes.

Grant would not have had such a majority without the freedmen's vote. Despite attacks by the Ku Klux Klan, about 500,000 African Americans voted in the South. Most cast their ballots for Grant.

Now and then

AFRICAN AMERICANS IN CONGRESS

Between 1870 and 1877, 16 African Americans served in Congress. Seven are shown in the picture below. Two were senators: Hiram R. Revels and Blanche K. Bruce, both of whom were from Mississippi.

In 1999, there were 38 African Americans in Congress. The longest-serving member was John Conyers, a representative from Michigan elected in 1964. Only two African-American senators were elected in the 20th century. Massachusetts senator Edward W. Brooke served from 1967 to 1979. Illinois senator Carol Moseley-Braun served from 1993 to 1999.

The Fifteenth Amendment

After Grant's victory, Radical Republicans worried that the Southern states might try to keep African Americans from voting in future elections. To prevent this, Radical leaders proposed a new constitutional amendment.

The **Fifteenth Amendment** stated that citizens could not be stopped from voting "on account of race, color, or previous condition of servitude." (This amendment, like the Fourteenth Amendment, did not apply to Native Americans on tribal lands.) The amendment was ratified in 1870.

Reading **History**

A. Comparing How was the Fifteenth Amendment a step beyond the Fourteenth Amendment?

The Fifteenth Amendment was not aimed only at the South. African-American men could not vote in 16 states. "We have no moral right to impose an obligation on one part of the land which the rest will not accept," one Radical wrote. With the Fifteenth Amendment, the nation again turned toward democracy.

The Fifteenth Amendment did not apply to women. This made many white women angry. Why couldn't they vote when black men—former slaves—could? Suffragist Elizabeth Cady Stanton protested the idea of uneducated immigrants and freedmen "who never read the Declaration of Independence" making laws for educated white women. Most African-American women were not as angry. To Frances E. W. Harper, a black suffragist and writer, it was important for African Americans to gain voting rights, even if that meant only men at first.

Vocabulary

suffragist: someone who favors equal voting rights, especially for women

Grant Fights the Klan

Despite gaining the vote, African Americans in the South continued to be terrorized by the Ku Klux Klan. In 1871, to stop the terror, President Grant asked Congress to pass a tough law against the Klan. Joseph Rainey, a black congressman from South Carolina, had received death threats from the Klan. He urged his fellow lawmakers to support the bill.

A VOICE FROM THE PAST

When myself and colleagues shall leave these Halls and turn our footsteps toward our southern home we know not but that the assassin may await our coming. Be it as it may we have resolved to be loyal and firm, and if we perish, we perish! I earnestly hope the bill will pass.

Joseph Rainey, quoted in *The Trouble They Seen*

Congress approved the anti-Klan bill. Federal marshals then arrested thousands of Klansmen. Klan attacks on African-American voters declined. As a result, the 1872 presidential election was both fair and peaceful in the South. Grant won a second term.

Scandal and Panic Weaken Republicans

Under the Grant administration, support for the Republicans and Reconstruction weakened. Scandals hurt the administration and caused divisions in the Republican Party. A financial panic further hurt the Republicans and turned the country's attention away from Reconstruction.

President Grant did not choose his advisers well. He put his former army friends and his wife's relatives in government positions. Many of these people were unqualified. Some Grant appointees took bribes. Grant's private secretary, for instance, took money from whiskey distillers who wanted to avoid paying taxes. Grant's secretary of war, General William Belknap, left office after people accused him of taking bribes.

Such scandals deeply outraged many Republicans. In 1872, some Republican officials broke away and formed the new Liberal Republican Party. The Republicans, no longer unified, became less willing to impose tough Reconstruction policies on the South.

Reading History
B. Making Inferences How did Republican scandals hurt Reconstruction?

In 1873, political corruption and Republican quarreling gave way to a more serious problem. When several powerful Eastern banks ran out of money after making bad loans, a financial panic swept the country. In the **Panic of 1873,** banks across the land closed. The stock market temporarily collapsed. The panic caused an economic depression, a time of low business activity and high unemployment. The railroad industry, which relied on banks for loans, suffered. Within a year, 89 of the country's 364 railroads went broke. Railroad failures left Midwestern farmers with no way to move their crops, and many farmers were ruined.

This cartoon from *Puck* magazine shows President Grant weighed down by corruption in his administration.

The depression, which lasted about five years, touched nearly all parts of the economy. By 1875, more than 18,000 companies had folded. Hundreds of workers had lost their jobs. Many Americans blamed the crisis on the Republicans—the party in power. As a result, Democrats won victories in the 1874 congressional and state elections. In the middle of the depression, Americans grew tired of hearing about the South's problems. The nation was losing interest in Reconstruction.

Reading History
C. Recognizing Effects What resulted from the Panic of 1873?

Supreme Court Reversals

To make matters worse for the Republicans, the Supreme Court began to undo some of the changes that had been made in the South. In an 1876 case, *U.S.* v. *Cruikshank,* the Court ruled that the federal government could not punish individuals who violated the civil rights of African Americans. Only the states had that power, the Court declared. Southern state officials often would not punish those who attacked African Americans. As a result, violence against them increased.

In the 1876 case *U.S.* v. *Reese,* the Court ruled in favor of white Southerners who barred African Americans from voting. The Court stated that the Fifteenth Amendment did not give everyone the right to vote—it merely listed the grounds on which states could not deny the vote. In other words, states could prevent African Americans from voting for other reasons. States later imposed poll taxes and literacy tests to restrict the vote. These Court decisions weakened Reconstruction and blocked African-American efforts to gain full equality.

Reading **History**

D. Recognizing Effects How did the Reese and Cruikshank rulings affect African Americans' efforts to gain civil rights?

Reconstruction Ends

The final blow to Reconstruction came with the 1876 presidential election. The Democrats nominated Samuel J. Tilden, governor of New York. The Republicans chose Rutherford B. Hayes, governor of Ohio. The race was very close. Victory depended on the electoral votes of South Carolina, Louisiana, and Florida. The votes in those states were so close that both the Democrats and the Republicans claimed victory. A special commission of eight Republicans and seven Democrats made a deal. Under the **Compromise of 1877,** Hayes became president. In return, the Republicans compromised with the Southern Democrats on several issues.

This cartoon from *Harper's Weekly* shows a federal soldier as the freedman's only defense against white Southerners.

1. The government would remove federal troops from the South.
2. The government would provide land grants and loans for the construction of railroads linking the South to the West Coast.
3. Southern officials would receive federal funds for construction and improvement projects.
4. Hayes would appoint a Democrat to his cabinet.
5. The Democrats promised to respect African Americans' civil and political rights.

Reading **History**

E. Summarizing What events led to a weakening of support for Reconstruction?

Abolitionist Wendell Phillips was against the compromise. He doubted that the South would respect black rights. "The whole soil of the South is hidden by successive layers of broken promises," he said. "To trust a Southern promise would be fair evidence of insanity."

After the 1876 presidential election, the Reconstruction governments in the South collapsed. The Democrats returned to power, believing that they were the redeemers, or rescuers, of the South.

Reconstruction: Civil Rights Amendments and Laws	
Civil Rights Act of 1866	• Granted citizenship and equal rights to all persons born in the United States (except Native Americans)
Fourteenth Amendment (1868)	• Granted citizenship and equal protection of the laws to all persons born in the United States (except Native Americans)
Fifteenth Amendment (1870)	• Protected the voting rights of African Americans
Civil Rights Act of 1875	• Outlawed racial segregation in public services • Ensured the right of African Americans to serve as jurors

SKILLBUILDER Interpreting Charts

1. *Which amendment and law are most similar?*
2. *Which amendment specifically protects voting rights?*

The Legacy of Reconstruction

Historians still argue about the success of Reconstruction. The nation did rebuild and reunite. However, Reconstruction did not achieve equality for African Americans.

After Reconstruction, most African Americans still lived in poverty. Legally, they could vote and hold public office. But few took part in politics. They continued to face widespread violence and prejudice.

During this period, however, African Americans did make lasting gains. Protection of civil rights became part of the U.S. Constitution. The Fourteenth and Fifteenth amendments would provide a legal basis for civil rights laws of the 20th century. Black schools and churches begun during Reconstruction also endured. Reconstruction changed society, putting African Americans on the path toward full equality. In the next unit, you will learn about other changes in American society after the Civil War.

Section 3 Assessment

1. Terms & Names

Identify:

- Fifteenth Amendment
- Panic of 1873
- Compromise of 1877

2. Taking Notes

Review the chapter and find five key events to place on a time line as shown.

1865 — event — event — 1877

event — event — event

Which event was most important and why?

3. Main Ideas

a. What did the Fifteenth Amendment declare?

b. What effect did scandals in the Grant administration have on the Republican Party?

c. What demands did Southern Democrats make in the Compromise of 1877?

4. Critical Thinking

Drawing Conclusions

Why do you think the Republicans were willing to agree to the Compromise of 1877 and end Reconstruction?

THINK ABOUT

- the election of 1876
- the Panic of 1873
- the Supreme Court rulings

ACTIVITY OPTIONS

LANGUAGE ARTS

CIVICS

Research Ku Klux Klan activities barring African Americans from voting. Then write a protest **letter to the editor** or propose a **law** to protect voting rights.

Chapter 18 ASSESSMENT

TERMS & NAMES

Briefly explain the importance of each of the following.

1. Reconstruction
2. Andrew Johnson
3. black codes
4. civil rights
5. Fourteenth Amendment
6. sharecropping
7. lynch
8. Fifteenth Amendment
9. Panic of 1873
10. Compromise of 1877

REVIEW QUESTIONS

Rebuilding the Union (pages 533–539)

1. What was the Freedmen's Bureau?
2. What were the main parts of President Johnson's Reconstruction plan?
3. Who were scalawags and carpetbaggers?
4. What reason did the House give for impeaching President Johnson?

Reconstruction and Daily Life (pages 540–544)

5. Why did Congress not pass a land-reform plan?
6. What new systems of labor developed in the South after the Civil War?
7. How did the Ku Klux Klan serve the Democratic Party?

End of Reconstruction (pages 545–549)

8. Why did the Fifteenth Amendment arouse anger in many women?
9. What caused an economic depression in the 1870s?
10. How did Supreme Court rulings during Reconstruction help weaken African Americans' civil rights?

CRITICAL THINKING

1. USING YOUR NOTES

Problems	Solutions
Black codes	
President Johnson	
Education	
Economy	
Ku Klux Klan	
Voting	

Using your diagram, answer the following questions.

a. What was the solution to the problem of educating African Americans?

b. What was the solution to the problem of Ku Klux Klan violence?

2. ANALYZING LEADERSHIP

Why might Reconstruction be considered a time in which the presidency was weak?

3. THEME: DEMOCRATIC IDEALS

How did the Fourteenth and Fifteenth amendments promote greater equality for African Americans? How were the amendments limited?

4. APPLYING CITIZENSHIP SKILLS

What were the different viewpoints of Elizabeth Cady Stanton and Frances E. W. Harper regarding the Fifteenth Amendment's failure to give women an important right of citizenship—the right to vote?

5. ANALYZING CAUSES

What aspect of the Compromise of 1877 likely played the greatest role in ending Reconstruction?

Interact *with* History

How did your solutions to rebuilding the nation compare with the actual solutions carried out?

VISUAL SUMMARY

Reconstruction

Rebuilding the Union

During Reconstruction, Congress decided how the Southern states would be readmitted to the Union and passed laws to improve conditions for freed people.

Reconstruction and Daily Life

After slavery ended, freed African Americans reunited their families, attended school, and began working for pay. Racist violence and lack of land slowed their progress.

End of Reconstruction

In the 1870s, hostile Supreme Court decisions, the Southern Democrats' return to power, and the withdrawal of federal troops from the South ended Reconstruction.

HISTORY SKILLS

1. INTERPRETING MAPS: Region

Study the map and then answer the questions.

Election of 1876

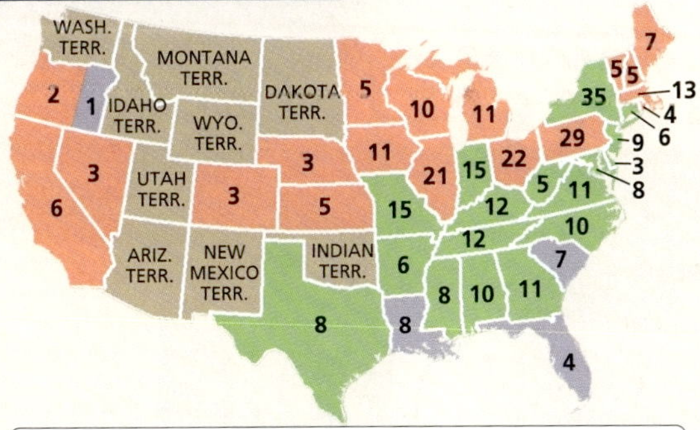

		Uncontested Electoral Vote	Final Electoral Vote	Popular Vote	Percentage of Popular Vote
Republican Hayes	(orange)	165	185	4,036,298	48
Democratic Tilden	(green)	184	184	4,300,590	52
	(purple)	Disputed			

Basic Map Elements

a. What is the subject of the map?

Interpreting the Map

b. In what region of the country were most of the disputed votes located?

c. What regions voted mostly Republican?

2. INTERPRETING PRIMARY SOURCES

This political cartoon shows the effect of the Ku Klux Klan on African-American families in the South. Study the cartoon carefully and then answer the questions.

a. Based on the cartoon, what were the goals of the Ku Klux Klan?

b. What does the cartoon suggest about how the Klan intended to achieve its goals?

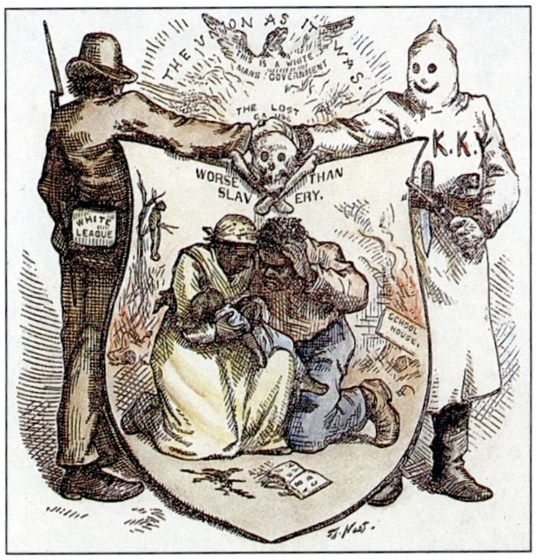

ALTERNATIVE ASSESSMENT

1. INTERDISCIPLINARY ACTIVITY: Language Arts

Writing Letters Write letters that you imagine the following three people would write about Reconstruction: 1) a member of the old Southern upper class; 2) a newly freed African American; and 3) a white Northern carpetbagger.

2. COOPERATIVE LEARNING ACTIVITY

Conducting an Impeachment Trial The impeachment trial of Andrew Johnson was a dramatic and colorful event. Many officials pleaded their case either for or against the president.

Working in a small group, research the trial using resources such as diaries, journals, autobiographies, letters, and books. Each group member should choose an official who spoke at the trial and collect some of his quotes. The group should then perform its own trial in front of the class, with each member portraying the official he or she chose. Some possible officials include Thaddeus Stevens, Edmund G. Ross, James W. Grimes, Benjamin Butler, and Chief Justice Samuel Chase.

3. TECHNOLOGY ACTIVITY

Making an Electronic Presentation Life under the sharecropping system was not easy. Information about sharecropping comes from a variety of sources. Using the library or the Internet, find diaries, memoirs, images, and news articles about life as a sharecropper.

Visit www.mcdougallittell.com to learn more about sharecropping.

Use presentation software to share your information about sharecropping. Consider the suggestions below to get started.

- images or descriptions of a sharecropper's shack
- examples of the crops that were grown by a sharecropper
- facts and quotations about a sharecropper's life

4. HISTORY PORTFOLIO

Option 1 Review your section and chapter assessment activities. Select one that you think is your best work. Then use comments made by your teacher or classmates to improve your work and add it to your portfolio.

Option 2 Review the questions that you wrote for What Do You Want to Know? on page 532. Then write a short report in which you explain the answers to your questions. If any questions were not answered, do research to answer them. Add your answers to your portfolio.

UNIT 6

1860 – 1914

America Transformed

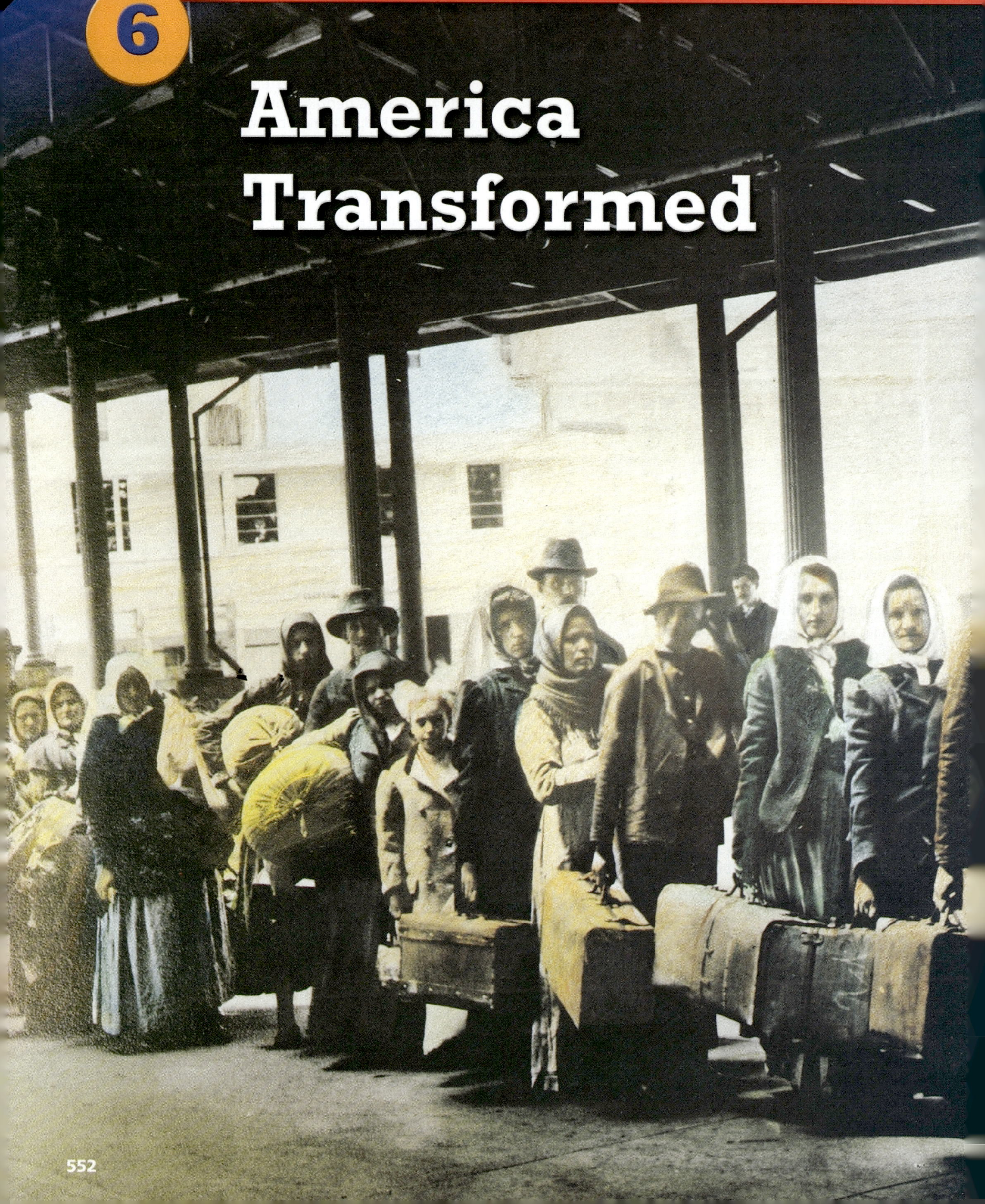

European immigrants such as those shown in this photograph (taken around 1900) streamed into Ellis Island at the turn of the century.

CHAPTER 19 Growth in the West 1860–1900

Life in the West was hard and dangerous, as this Charles M. Russell painting of cowhands shows.

Interact *with* History

A Nebraska "sodbuster" family takes time away from their chores to pose in front of their sod house.

It is 1865, and the Civil War has just ended. You are drawn to the West by stories of gold, silver, fertile soil, and free land, and by tales of adventure and new opportunities. Yet you know there would be hardships and unknown dangers. Your life would never be the same.

What Do You Think?

- What might be some of the ways to make a living in the West?
- What do you think your daily life would be like in the West?
- What would be the biggest difference in your life?

How might your life change in the West?

USA / **World** 1860–1900

USA:

- **1862** Congress passes the Homestead Act.
- **1864** Sand Creek Massacre
- **1867** The Grange is founded.
- **1876** Battle of the Little Bighorn is fought.
- **1880** James Garfield is elected president.
- **1881** Chester Arthur becomes president after Garfield is assassinated.
- **1884** Grover Cleveland is elected president.
- **1889** Oklahoma land rush begins.
- **1890** Wounded Knee Massacre
- **1891** Farmers organize the Populist Party.
- **1896** William McKinley is elected president.

World:

- **1861** Serfs are freed in Russia.
- **1871** Franco-Prussian War ends.
- **1885** Berlin Conference on African affairs divides Africa among European nations.
- **1889** First Pan-American Conference is held.
- **1900** Boxer Rebellion takes place in China.

Chapter 19 SETTING THE STAGE

BEFORE YOU READ

Preview the Theme

Diversity and Unity As Chapter 19 explains, hundreds of thousands of men, women, and children packed up their belongings and went to the West after the Civil War. Most were looking for new opportunities and land of their own; some were seeking freedom or adventure. Their arrival led to conflict with the Native Americans, who were the first occupants of the area.

What Do You Know?

What do you think about when you hear terms like *cowboy* and *Wild West?* What do you already know about the people, places, and events in the West in the last half of the 19th century?

THINK ABOUT

- what you have learned about the West from books, movies, and television
- what happens when different cultures clash

What Do You Want to Know?

What details do you need to help you understand the settling of the West? Make a list of those details in your notebook before you read the chapter.

READ AND TAKE NOTES

Reading Strategy: Finding Main Ideas To make it easier for you to understand what you read, learn to find the main idea of each paragraph, topic heading, and section. Remember that the supporting details help to explain the main ideas. On the chart below, write down the main ideas about the many diverse people who settled the West.

See Skillbuilder Handbook, page R5.

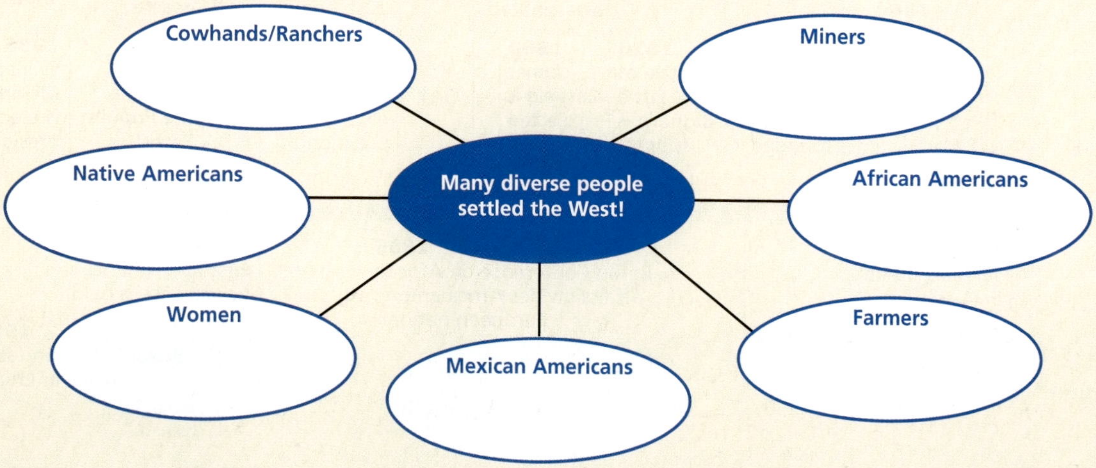

1 Miners, Ranchers, and Cowhands

TERMS & NAMES
frontier
Great Plains
boomtown
long drive
vaquero
vigilante

MAIN IDEA

Miners, ranchers, and cowhands settled in the West seeking economic opportunities.

WHY IT MATTERS NOW

The mining and cattle industries that developed then still contribute to American economic growth.

ONE AMERICAN'S STORY

Nat Love was born a slave in Tennessee in 1854. After the Civil War, he was one of thousands of African Americans who left the South and went west. In 1869, Love headed for Dodge City, Kansas. He was 15 and now free.

Love's horse taming skills landed him a job as a cowhand. For 20 years, he took part in the cattle drives that brought Texas cattle to Kansas stockyards. He became well known for his expert horsemanship and his rodeo riding and roping. In his 1907 autobiography, Love offered a lively but exaggerated account of his life. He told how he braved hailstorms, fought wild animals, and held off human attackers.

A VOICE FROM THE PAST

I carry the marks of fourteen bullet wounds on different parts of my body, most any one of which would be sufficient to kill an ordinary man. . . . Horses were shot from under me, men killed around me, but always I escaped with a trifling wound at the worst.

Nat Love, *The Life and Adventures of Nat Love*

As you will read in this section, few cowhands led lives as exciting as that described by Nat Love, but they all helped to open a new chapter in the history of the American West.

Nat Love was an African-American cowhand who became a rodeo star.

Geography and Population of the West

In the mid-1800s, towns such as St. Joseph and Independence, Missouri, were jumping-off places for settlers going west. They were the last cities and towns before the frontier. The **frontier** was the unsettled or sparsely settled area of the country occupied largely by Native Americans.

Many white settlers thought of the **Great Plains**—the area from the Missouri River to the Rocky Mountains—as empty. (See map on page 558.) Few had been attracted to its rolling plains, dry plateaus, and deserts. However, west of the Rockies, on the Pacific Coast, settlers had followed miners streaming into California after the 1849 gold rush. By 1850, California had gained statehood. Oregon followed in 1859.

The Western Frontier, 1850–1890

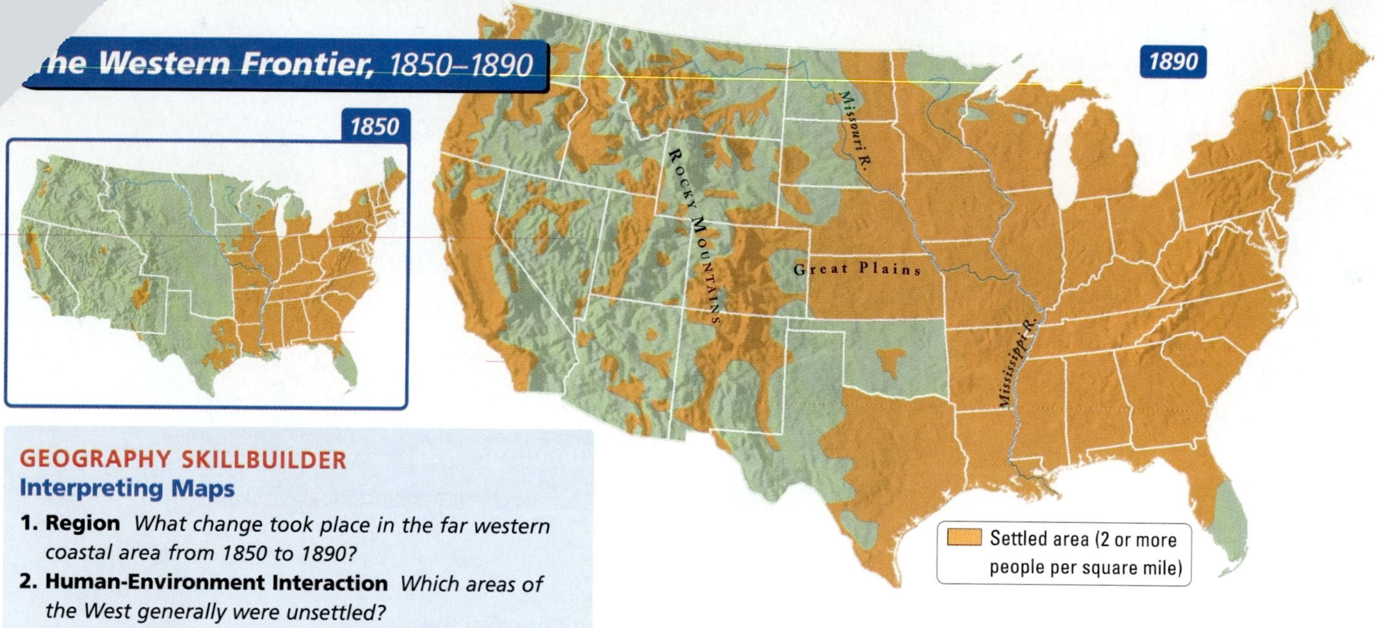

GEOGRAPHY SKILLBUILDER
Interpreting Maps

1. **Region** *What change took place in the far western coastal area from 1850 to 1890?*
2. **Human-Environment Interaction** *Which areas of the West generally were unsettled?*

The Great Plains had few trees, but its grasslands were home to about 300,000 Native Americans in the mid-1800s. Most followed the buffalo herds that rumbled across the open plains. Despite the presence of these peoples, the United States claimed ownership of the area.

Railroads played a key role in settling the western United States. Trains carried the natural resources of the West—minerals, timber, crops, and cattle—to eastern markets. In turn, trains brought miners, ranchers, and farmers west to develop these resources further. As the railroads opened new areas to white settlement, they also helped to bring an end to the way of life of the West's first settlers—the Native Americans.

Reading **History**
A. Making Inferences Why were railroads so important to the West?

Mining in the West

In 1859, gold and silver strikes drew fortune seekers to Colorado and Nevada. As many as 100,000 miners raced to the Rocky Mountains in Colorado after gold was discovered near Pikes Peak. Also in 1859, prospectors hit "pay dirt" at the Comstock Lode in western Nevada. (A lode is a deposit of a valuable mineral buried in layers of rock.) From 1859 to 1880, the Comstock mine produced some $300 million in silver and gold.

Vocabulary
strike: valuable discovery of a precious mineral

Nearby Virginia City, Nevada, became a **boomtown,** a town that has a sudden burst of economic or population growth. Population jumped from 3,000 in the 1860s to over 20,000 in the 1870s. The writer Samuel Clemens, better known as Mark Twain, captured the excitement of life there.

A VOICE FROM THE PAST

The sidewalks swarmed with people. . . . Money was as plenty as dust; [everyone] considered himself wealthy. . . . There were . . . fire companies, brass bands, banks, hotels, theaters . . . gambling palaces . . . street-fights, murders, . . . riots, . . . and a half dozen jails . . . in full operation.

Mark Twain, *Roughing It*

Other major strikes took place in the Black Hills of South Dakota in 1874 and at Cripple Creek, Colorado, in 1891. In 1896, gold was discovered in

Miners brought ore, like the gold nugget shown, from underground mines that dotted the hillsides.

Canada's Yukon Territory. News of the strike set off a fresh epidemic of gold fever. Prospectors rushed to the Yukon's Klondike region.

The chance to strike it rich drew Americans from both east and west coasts. Gold fever also attracted miners from other parts of the world, including Europe, South America, Mexico, and China. Unfortunately, few prospectors became rich. Most left, disappointed and broke.

Early miners used panning and sluicing to wash sand and gravel from a stream to separate out any bits of precious metal, as you read in Chapter 13. Large mining companies moved in after surface mines no longer yielded gold or silver. Only they could buy the costly, heavy equipment needed to take the precious metals from underground. Water cannons blasted away hillsides to expose gold deposits. In other places, workers sank shafts thousands of feet into the ground to create underground mines. These new methods recovered more precious metals, but in the process stripped hillsides of vegetation and left rivers polluted.

Reading **History**

B. Recognizing Effects Why did large mining companies replace individual prospectors?

Paid workers in company mines replaced independent prospectors. The work was hard and dangerous. Dust caused lung problems, and deadly cave-ins could trap miners hundreds of feet below the surface.

By the 1890s, the mining boom was over. Many mines closed because the costs had become too high, and the quality of the ore had dropped. Jobless workers moved elsewhere. Once-thriving communities became ghost towns. Still, the mining boom had lasting effects. Nevada, Colorado, and South Dakota all grew so rapidly that they soon gained statehood.

The Rise of the Cattle Industry

The cattle trade had existed in the Southwest since the Spanish arrived there in the 1500s. But cattle herds remained small until the Civil War. There were few buyers for western beef because there was no efficient way to get the beef to markets in the more heavily populated cities of the East. The ranchers mostly sold their cattle locally.

The extension of railroad lines from Chicago and St. Louis into Kansas by the 1860s brought changes. An Illinois livestock dealer named Joseph McCoy realized that railroads could bring cattle from Texas ranches to meat-hungry eastern cities. Cowhands had only to drive cattle herds north from Texas to his stockyards in Abilene, Kansas. From there, the beef could be shipped to Chicago and points east by rail car.

McCoy's plan turned cattle ranching into a very profitable business. Cattle fed on the open range for a year or two and cost the rancher nothing. Ranchers then hired cowhands to round up the cattle and take them to Abilene. There they were sold for as much as ten times their original price. The success of the Abilene stockyards spurred the growth of other Kansas cow towns, including Wichita and Dodge City. The cattle drives to cow towns along the railways were called the **long drives.**

Over time, cowhands followed specific trails across the plains. The first was the Chisholm Trail, which stretched from San Antonio, Texas, to Abilene, Kansas. It was named for Jesse Chisholm, a trader who marked the northern part of the route. From 1867 to 1884, about four million cattle were driven to market on this trail. As cattle raising became more profitable, ranching spread north across the plains from Texas to Montana.

ReadingHistory
C. Reading a Map Use the map on page 581 to locate the Chisholm Trail.

Vaqueros and Cowhands

The first cowhands, or ***vaqueros,*** as they were known in Spanish, came from Mexico with the Spaniards in the 1500s. They settled in the Southwest. The *vaqueros* helped Spanish, and later Mexican, ranchers manage their herds. From the *vaquero,* the American cowhand learned to rope and ride. Cowhands also adapted the saddle, spurs, lariat (which they used to rope a calf or steer), and chaps of the *vaqueros.*

Background
Chaps, from the Spanish word *chaperejos,* were seatless leather pants worn over trousers to protect legs from scrub brush, snakes, and cactus.

About one in three cowhands in the West were either Mexican or African American. Many Mexican cowhands were descendants of the *vaqueros.* Some African-American cowhands were former slaves. They came west at the end of Reconstruction because the enactment of Black Codes in the South put restrictions on their freedom. Also among the cowhands were a large number of former Confederate and Union soldiers.

The "Wild West"

At first, the rapidly growing cow towns had no local governments. There were no law officers to handle the fights that broke out as cowhands drank and gambled after a long drive. A more serious threat to law and order came from "con men." These swindlers saw new towns as places to get rich quick by cheating others.

Vocabulary
con man: a person who cheats victims by first gaining their confidence

daily *life*

LIFE OF A COWHAND: THE ROUNDUP

During some parts of the year, the cowhand's life was downright dull. While cattle grazed on the open range, cowhands sat around the ranch, repairing their gear and doing odd jobs. The pace quickened at roundup time in the spring and fall.

For several weeks, 150 to 250 cowhands from nearby ranches rode hundreds of miles locating cattle. Cowhands from each ranch collected their cattle, removed sick or weak animals, and branded new calves. Then the cowhands were ready for the long drive. A roundup by *vaqueros* is shown in this painting by James Walker.

Some Union and Confederate veterans were led to crime by hard feelings left over from the Civil War. Outlaws like John Wesley Hardin, "Billy the Kid," and Jesse and Frank James made crime a way of life. Some women became outlaws, too. Belle Starr, better known as the Bandit Queen, was a legendary horse thief.

For protection, citizens formed vigilante groups. **Vigilantes** were people who took the law into their own hands. They caught suspected criminals and punished them without a trial. Vigilante justice often consisted of hanging suspects from the nearest tree or shooting them on the spot. As towns became more settled, citizens elected a local sheriff or asked the federal government for a marshal. These law officers would arrest lawbreakers and hold them in jail until the time of trial.

Bandit Queen Belle Starr sits atop a horse she just might have stolen.

End of the Long Drives

For about 20 years, the cattle industry boomed. As the railroads extended farther west and south into Texas, the long drives grew shorter. The future looked bright. But by 1886, several developments had brought the cattle boom to an end. First, the price of beef dropped sharply as the supply increased in the early 1880s. It fell from more than $30 a head to $7. Then came the newly invented barbed wire. As more settlers moved to the Great Plains to farm or raise sheep, they fenced in their lands with barbed wire. The open range disappeared, and cattle could no longer pass freely over the trails. Finally, in the harsh winter of 1886–1887, thousands of cattle on the northern Plains froze to death. Many ranchers were put out of business.

Meanwhile, as the mining and cattle industries were developing, the Native Americans of the Great Plains were being pushed off their land, as you will read in the next section.

*Reading***History**
D. Analyzing Causes What caused the decline of cattle ranching on the open range?

Section 1 Assessment

1. Terms & Names

Identify:
- frontier
- Great Plains
- boomtown
- long drive
- *vaquero*
- vigilante

2. Taking Notes

Use a diagram to review the rise and fall of the cattle industry.

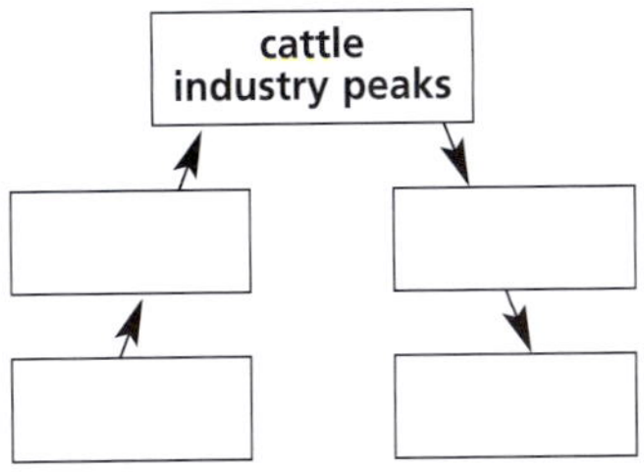

3. Main Ideas

a. What economic opportunities drew large numbers of people to the West beginning in the 1860s?

b. How did the transcontinental railroad spur western settlement?

c. What did cowhands learn from the *vaqueros?*

4. Critical Thinking

Evaluating Could cattle ranchers have stopped the decline of the cattle industry that occurred in the late 1880s?

THINK ABOUT
- economic causes
- impact of weather
- changing settlement patterns

ACTIVITY OPTIONS

LANGUAGE ARTS
MUSIC

Do research on a legendary figure of the West such as Wyatt Earp, "Calamity Jane," or Nat Love. Then write a **biographical sketch** or **song** about the person.

2 Native Americans Fight to Survive

MAIN IDEA

The Native Americans of the Great Plains fought to maintain their way of life as settlers poured onto their lands.

WHY IT MATTERS NOW

The taking of their lands led to social and economic problems for Native Americans that continue to this day.

TERMS & NAMES

reservation
Sand Creek Massacre
Sitting Bull
George A. Custer
Battle of the Little Bighorn
Wounded Knee Massacre
Dawes Act

ONE AMERICAN'S STORY

Buffalo Bird Woman was a Hidatsa who lived almost 100 years. She was born in 1840. As a child, she and her family made their home along the Missouri River. Later the federal government forced her family onto a reservation. A <u>reservation</u> is land set aside for Native American tribes.

The federal government attempted to "Americanize" Native American children, including Buffalo Bird Woman, by sending them away to boarding schools. But Buffalo Bird Woman struggled to hold on to Hidatsa customs. She spoke only her native language and wore traditional dress. As an old woman, she looked back on her early years.

A VOICE FROM THE PAST

Sometimes at evening I sit, looking out on the . . . Missouri [river]. . . . In the shadows I seem . . . to see our Indian village, with smoke curling upward from the earth lodges; and in the river's roar I hear the yells of the warriors, the laughter of . . . children as of old. It is but an old woman's dream. . . . Our Indian life, I know, is gone forever.

Buffalo Bird Woman, quoted in *Native American Testimony,* edited by Peter Nabokov

Buffalo Bird Woman's life spanned the years when Native Americans waged their final fight to keep lands guaranteed to them by treaties. As white settlers claimed Native American hunting grounds, Plains peoples fought a losing battle to save not only their land but their way of life.

Buffalo Bird Woman saw the Native American way of life change forever during her almost 100 years of life.

Native American Life on the Plains

Before the arrival of Europeans in the 1500s, most Plains tribes lived in villages along rivers and streams. The women tended crops of beans, corn, and squash. The men hunted deer and elk and in the summer stalked the vast buffalo herds that inhabited the Plains.

In the early 1540s, the Spanish brought the first horses to the Great Plains. The arrival of horses changed the way of life of the Plains people. They quickly became expert riders. By the late 1700s, most Plains tribes kept their own herds of horses. Mounted on horseback, hunters traveled far from their villages seeking buffalo.

The buffalo was central to the life of Plains tribes. Its meat became the chief food in their diet, while its skins served as portable shelters called tepees. Plains women turned buffalo hides into clothing, shoes, and blankets and used buffalo chips (dried manure) as cooking fuel. Bones and horns became tools and bowls. Over time, many Plains tribes developed a nomadic way of life tied to buffalo hunting.

Vocabulary
nomadic: wandering from place to place

A Clash of Cultures

When the federal government first forced Native American tribes of the Southeast to move west of the Mississippi in the 1830s, it settled them in Indian Territory. This territory was a huge area that included almost all of the land between the Missouri River and Oregon Territory. Most treaties made by the government with Native Americans promised that this land would remain theirs "as long as Grass grows or water runs."

Reading History
A. Analyzing Causes What was the major source of conflict between white settlers and Native Americans?

Unfortunately, these treaty promises would be broken. Government policy was based on the belief that white settlers were not interested in the Plains. The land was considered too dry for farming. However, as wagon trains bound for Oregon and California crossed the Great Plains in the 1850s, some pioneers saw possibilities for farming and ranching on its grasslands. Soon white settlers moved onto the prairies.

These settlers pressured the federal government for more land. They also wanted protection from Native Americans in the area. In 1851, the government responded by calling the Sioux, Cheyenne, Arapaho, and

GEOGRAPHY SKILLBUILDER
Interpreting Maps
1. **Location** *In which area of the West did most of the major battles take place?*
2. **Movement** *What was the major change that took place in the West between 1850 and 1890?*

other Plains tribes together near Fort Laramie in present-day Wyoming. Government officials tried to buy back some Native American land and also set boundaries for tribal lands. Many Plains tribes signed the First Treaty of Fort Laramie (1851)—they saw no other choice.

But some Cheyennes and Sioux resisted. They preferred conflict with settlers and soldiers to the restrictions of reservation life. In southeastern Colorado, bands of Cheyenne warriors attacked miners and soldiers. In response, about 1,200 Colorado militia led by Col. John Chivington opened fire on a peaceful Cheyenne village along Sand Creek in 1864. More than 150 Cheyenne men, women, and children were killed in what came to be known as the **Sand Creek Massacre.**

Reading **History**

B. Reading a Map Locate the site of the Sand Creek Massacre on the map on page 563.

The Plains tribes reacted to such attacks by raiding white settlements. One of the fiercest battles took place in Montana. There the government had begun to build a road called the Bozeman Trail across Sioux hunting grounds. To stop construction, the Sioux attacked construction workers. In 1866, Captain W. J. Fetterman and 80 troopers stumbled into a deadly ambush set by the Sioux. All the soldiers were killed in what was called the Fetterman Massacre.

Such incidents finally forced the government to try to find a way to end the fighting. In 1868, U.S. officials signed the Second Treaty of Fort Laramie with the Sioux, Northern Cheyenne, and Arapaho. The treaty gave these tribes a large reservation in the Black Hills of South Dakota.

Battle of the Little Bighorn

The Second Treaty of Fort Laramie did not end the trouble between the Sioux and white settlers, though. In 1874, white prospectors discovered gold in the Black Hills. Paying no attention to the Fort Laramie treaty, thousands of miners rushed onto Sioux land. Tribal leaders angrily rejected a government offer to buy back the land. Many Sioux warriors fled the reservation during the

HISTORY *through* ART

Artist Edgar S. Paxson researched the Battle of the Little Bighorn for 20 years to try to accurately re-create the last moments of the fighting. Custer is at the center clutching at a bullet wound in his chest.

What do you think this painting shows about the fighting at the Little Bighorn?

winter of 1875–1876. They united under the leadership of two Sioux chiefs—**Sitting Bull** and Crazy Horse—to push back the intruders.

The Seventh Cavalry set out to return the Sioux to the reservations. It was commanded by Lieutenant Colonel **George A. Custer,** a hero of the Civil War and of other campaigns against Plains tribes. On June 25, his forces met several thousand Sioux and Cheyennes near the Little Bighorn River in Montana in the **Battle of the Little Bighorn.** In less than two hours, Custer and his men—211 in all—were wiped out.

News of Custer's defeat shocked the nation. The government responded by stepping up military action. As a result, Little Bighorn was the last major Native American victory. In 1877, Crazy Horse surrendered and Sitting Bull and his followers fled to Canada. In 1881, Sitting Bull's starving band surrendered to U.S. troops and were returned to the reservation.

Reading **History**

C. Recognizing Effects What were the results of the Battle of the Little Bighorn?

Resistance in the Northwest and Southwest

Background

Nez Perce means "pierced nose" in French. French-Canadian trappers gave this name to these Native American people because some of them wore jewelry in their noses.

The Nez Perce (nehz PURS) was a Northwest tribe that lived in eastern Oregon and Idaho. Until the 1860s, the Nez Perce lived peacefully on land guaranteed to them by an 1855 treaty. However, as white settlement increased, the government forced them to sell most of their land and move to a narrow strip of territory in Idaho. Most reluctantly agreed, but a group of Nez Perce led by Chief Joseph refused.

In 1877, Chief Joseph and his followers fled north to seek refuge in Canada. For four months, the Nez Perce traveled across 1,000 miles of rugged terrain with army troops in pursuit. About 40 miles from the Canadian border, the army caught up with them. Greatly outnumbered, the Nez Perce surrendered. Chief Joseph spoke for his people when he said, "I will fight no more, forever."

In the Southwest, both the Navajos and Apaches fought against being removed to reservations. U.S. troops ended Navajo resistance in Arizona in 1863 by burning Navajo homes and crops. Most Navajos surrendered. Nearly 8,000 took what they called the "Long Walk," a brutal journey of 300 miles to a reservation in eastern New Mexico. Hundreds died on the way. Their new home was a parched strip of land near the Pecos River. After four years, the government allowed the Navajos to return to Arizona, where many live today.

Native American Leaders

Sitting Bull ▶
(c. 1831–1890)
Sioux chief: "We did not give our country to you; you stole it."

◀ Chief Joseph
(1840–1904)
Nez Perce chief: "It makes my heart sick when I remember all the good words and all the broken promises."

Geronimo ▶
(1829–1909)
Apache leader: "Once I moved about like the wind. Now I surrender."

"Once I moved about like the wind. Now I surrender."

Geronimo

In the early 1870s, the government forced many Apaches to settle on a barren reservation in eastern Arizona. But a group led by Geronimo refused to remain. Escaping the reservation, these Apaches survived by raiding settlers' homes. Geronimo was captured many times but always managed to escape. In 1886, however, he finally surrendered and was sent to prison.

A Way of Life Ends

As the Native Americans of the Plains battled to remain free, the buffalo herds that they depended upon for survival dwindled. At one time, 30 million buffalo roamed the Plains. However, hired hunters killed the animals to feed crews building railroads. Others shot buffalo as a sport or to supply Eastern factories with leather for robes, shoes, and belts. From 1872 to 1882, hunters killed more than one million buffalo each year.

Background
In 1889, fewer than 100 buffalo remained.

By the 1880s, most Plains tribes had been forced onto reservations. With their hunting grounds fast disappearing, some turned in despair to a Paiute prophet named Wovoka. He preached a vision of a new age in which whites would be removed and Native Americans would once again freely hunt the buffalo. To prepare for this time, Wovoka urged Native Americans to perform the chants and movements of the Ghost Dance. Wovoka's hopeful vision quickly spread among the Plains peoples.

Many of Wovoka's followers, especially among the Sioux, fled their reservations and gathered at the Pine Ridge Reservation in South Dakota. White settlers and government officials began to fear that they were preparing for war. The army was sent to track down the Ghost Dancers. They rounded them up, and a temporary camp was made along Wounded Knee Creek in South Dakota, on December 28, 1890. The next day, as the Sioux were giving up their weapons, someone fired a shot. The troopers responded to the gunfire, killing about 300 men, women, and children. The **Wounded Knee Massacre,** as it was called, ended armed resistance in the West.

***Reading* History**
D. Drawing Conclusions What was the most important factor in the defeat of the Native Americans?

These Native American students in Oklahoma Territory in 1901 posed for a class picture at the school where they were sent to learn the culture of white people.

The Dawes Act Fails

Some white Americans had been calling for better treatment of Native Americans for years. In 1881, Helen Hunt Jackson published *A Century of Dishonor*, which listed the failures of the federal government's policies toward Native Americans. About the same time, Sarah Winnemucca, a Paiute reformer, lectured in the East about the injustices of reservation life.

Reading History

E. Analyzing Points of View Why did reformers support assimilation?

Many well-meaning reformers felt that assimilation was the only way for Native Americans to survive. Assimilation meant adopting the culture of the people around them. Reformers wanted to make Native Americans like whites—to "Americanize" them.

The **Dawes Act,** passed in 1887, was intended to encourage Native Americans to give up their traditional ways and become farmers. The act divided reservations into individual plots of land for each family. The government sold leftover land to white settlers. The government also sent many Native American children to special boarding schools where they were taught white culture. In "One American's Story," you read about the effort to Americanize Buffalo Bird Woman. But these attempts to Americanize the children still did not make them part of white society.

In the end, the Dawes Act did little to benefit Native Americans. Not all of them wanted to be farmers. Those who did lacked the tools, training, and money to be successful. Over time, many sold their land for a fraction of its real value to white land promoters or settlers.

The situation of Native Americans at the end of the 1800s was tragic. Their lands had been taken and their culture treated with contempt. Not until decades later would the federal government recognize the importance of their way of life. In the next section, you will read about some of the people who settled on Native American lands.

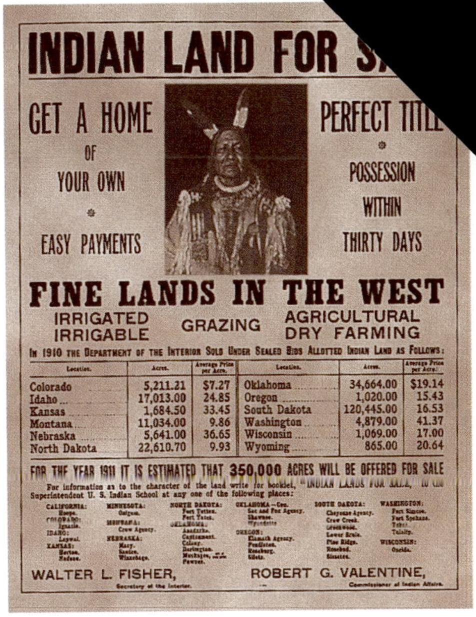

This poster advertised the sale of Native American lands to white settlers.

Section 2 Assessment

1. Terms & Names

Identify:
- reservation
- Sand Creek Massacre
- Sitting Bull
- George A. Custer
- Battle of the Little Bighorn
- Wounded Knee Massacre
- Dawes Act

2. Taking Notes

Use a chart to compare the life of Plains people before and after the arrival of white settlers.

	Before	After
Meeting survival needs		
Customs		
Land use		

3. Main Ideas

a. How did federal government policy toward Native Americans change as white settlers moved to the West?

b. How did the destruction of the buffalo affect Plains peoples?

c. Why was Wounded Knee a turning point in relations between Native Americans and the government?

4. Critical Thinking

Recognizing Effects How were the effects of the Dawes Act different from what was intended?

THINK ABOUT
- goals of the act
- impact on the land use, culture, and independence of the Plains peoples

ACTIVITY OPTIONS

TECHNOLOGY

SPEECH

Research the life of a Native American leader discussed in this section. Create that person's **Web page** or give a **speech** from this person's perspective.

3 Life in the West

TERMS & NAMES
homestead
Mexicano
William "Buffalo Bill" Cody
buffalo soldier

MAIN IDEA
Diverse groups of people helped to shape both the reality and the myth of the West.

WHY IT MATTERS NOW
The myth of the West continues to be a part of our culture.

ONE AMERICAN'S STORY

Abigail Scott was born in Illinois in 1834. She was told that her mother remarked at the time, "Poor baby! She'll be a woman some day! . . . A woman's lot is so hard!" At 17, Abigail moved to Oregon by wagon train with her family. Her mother died on the journey. In Oregon, Abigail taught school until she married a farmer named Benjamin Duniway in 1853. When he was disabled in an accident, Abigail assumed the support of her family. She wrote about a day on a pioneer farm with its endless chores.

A VOICE FROM THE PAST

[W]ashing, scrubbing, churning . . . preparing . . . meals in our lean-to kitchen . . . [having] to bake and clean and stew and fry; to be in short, a general pioneer drudge, with never a penny of my own, was not pleasant business.

Abigail Scott Duniway, in her autobiography, *Path Breaking*

Like Abigail Scott Duniway, this pioneer woman worked long and hard. She is shown here with buffalo chips she has collected on the treeless prairie to use as fuel.

Later, Duniway grew committed to the cause of women's rights. In 1871, she started a weekly pro-suffrage newspaper. Two years later, she founded the Oregon Equal Suffrage Association.

Oregon honored Duniway for her part in the suffrage struggle by registering her as the state's first woman voter. As you will read in this section, women like Duniway helped to shape the West.

Women in the West

Women often were not given recognition for their efforts to turn scattered western farms and ranches into settled communities. In their letters and diaries, many women recorded the harshness of pioneer life. Others talked about the loneliness. While men went to town for supplies or did farm chores with other men, women rarely saw their neighbors. Mari Sandoz lived in Nebraska on a **homestead,** a piece of land and the house on it. She wrote that women "had only the wind and the cold and the problems of clothing [and] shelter." Living miles from others, women were their family's doctors—setting broken bones and delivering babies—as well as cooks.

Despite its challenges, western life provided opportunities for women. Most who worked held traditional jobs. They were teachers or servants or gave their families financial support by taking in sewing or laundry. However, a few became sheriffs, gamblers, and even outlaws. In mining camps and cow towns, some even ran dance halls and boarding houses.

Western lawmakers recognized the contributions women made to western settlement by giving them more legal rights than women had in the East. In most territories, women could own property and control their own money. In 1869, Wyoming Territory led the nation in giving women the vote. Esther Morris, who headed the suffrage fight there, convinced lawmakers women would bring law and order to the territory.

When Wyoming sought statehood in 1890, many in Congress demanded that the state repeal its woman suffrage law. But Wyoming lawmakers stood firm. They told Washington, "We may stay out of the Union for 100 years, but we will come in with our women." Congress backed down. By 1900, women had also won the right to vote in Colorado, Utah, and Idaho.

Reading **History**

A. Making Inferences Why were women in the West the first to win the right to vote?

The Rise of Western Cities

Cities seemed to grow overnight in the West. Gold and silver strikes made instant cities of places like Denver in Colorado Territory and brought new life to sleepy towns like San Francisco in California. These cities prospered, while much of the area around them remained barely settled. San Francisco grew from a small town to a city of about 25,000 in just one year after the 1849 gold rush.

Miners who flocked to the "Pikes Peak" gold rush of 1859 stopped first in Denver to buy supplies. Not even a town in 1857, Denver was the capital of Colorado Territory by 1867. A decade later, it became the state capital when Colorado was admitted into the Union. The decision by Denver citizens to build a railroad to link their city with the transcontinental railroad sent population soaring. In 1870, it had about 4,800 residents. In 1890, it had nearly 107,000.

The railroads also brought rapid growth to other towns in the West. Omaha, Nebraska, flourished as a meat-processing center for cattle ranches in the area. Portland, Oregon, became a regional

Population of Western Cities

CITY	1860	1890
Denver	2,603*	106,713
Omaha	1,883	140,452
Portland	2,874	46,385
San Francisco	56,802	298,997

SKILLBUILDER Interpreting Charts
Which city had the largest increase in numbers of people, 1860–1890?

*1861 Territorial Census
Sources: *Population Abstract of the United States; Colorado Republic*

San Francisco, 1847

San Francisco, 1850

market for fish, grain, and lumber. While these cities were growing on the Great Plains and Pacific coast, the Southwest was also developing.

Mexicanos in the Southwest

The Southwest included what are now New Mexico, Texas, Arizona, and California. For centuries, it had been home to people of Spanish descent whose ancestors had come from Mexico. These Spanish-speaking southwesterners called themselves **Mexicanos.**

Reading **History**
B. Summarizing Who were the Mexicanos?

In the 1840s, the annexation of Texas and Mexico's defeat in the Mexican War brought much of the Southwest under the control of the United States. Soon after, English-speaking white settlers—called Anglos by the Mexicanos—began arriving. These pioneers were attracted to the Southwest by opportunities in ranching, farming, and mining. Their numbers grew in the 1880s and 1890s, as railroads connected the region with the rest of the country.

As American settlers crowded into the Southwest, the Mexicanos lost economic and political power. Many also lost land. They claimed their land through grants from Spain and Mexico. But American courts did not usually recognize these grants. One Mexicano remarked that "the North Americans . . . consider us unworthy to form with them one nation and one society." Only in New Mexico Territory did Hispanic society survive despite Anglo-American settlement.

The Myth of the Old West

America's love affair with the West began just as the cowboy way of life was vanishing in the late 1800s. To most Americans, the West had become a larger-than-life place where brave men and women tested themselves against hazards of all kinds and won. Easterners eagerly bought "dime novels" filled with tales of daring adventures. Sometimes the hero was a real person like Wyatt Earp or "Calamity Jane." But the plots were fiction or exaggerated accounts of real-life incidents.

Also adding to the myth were more serious works of fiction, like Owen Wister's bestselling novel about Wyoming cowhands, *The Virginian* (1902). Such works showed little of the drabness of daily life in the West. White settlers—miners, ranchers, farmers, cowhands, and law officers—played heroic roles not only in novels but also in plays and, later, in movies.

Legends of the Old West

"Billy the Kid" ▶
(Henry McCarty, 1859–1881)
He was an outlaw and hired gun who called himself William Bonney and was killed at 21.

▲ Wyatt Earp
(1848–1929)
He was a frontier peace officer, gunfighter, and gambler in towns such as Dodge City, Kansas, and Tombstone, Arizona.

◀ "Calamity Jane"
(Martha Jane Canary, c. 1852–1903)
She had a legendary career as a wagon driver, scout, and Wild West Show performer.

Native Americans generally appeared as villains. African Americans were not even mentioned.

William "Buffalo Bill" Cody, a buffalo hunter turned showman, brought the West to the rest of the world through his Wild West show. Cody recognized people's fascination with the West. His show, with its reenactments of frontier life, played before enthusiastic audiences across the country and in Europe.

The Real West

Background
From 1866 to 1898, some 12,500 African Americans served in the West in the 9th and 10th Cavalry and the 24th and 25th Infantry regiments.

The myth of the Old West overlooked the contributions of many peoples. The first cowhands, as you read earlier, were the Mexican *vaqueros*. Native Americans and African Americans played a role in cattle ranching, too. Many African Americans also served in the U.S. Army in the West, where Native Americans nicknamed them "**buffalo soldiers**." And the railroads would not have been built without the labor of Chinese immigrants.

Western legends often highlighted the attacks by Native Americans on soldiers or settlers. But the misunderstandings and broken treaties that led to the conflicts were usually overlooked.

Reading History
C. Comparing How does the real West compare to the myth of the West?

Historians also say that the image of the self-reliant westerner who tames the wild frontier ignores the important role played by the government in western settlement. Settlers needed the help of the army to remove Native Americans. The government also aided in the building of the railroads and gave the free land that drew homesteaders to the West. You will read about these homesteaders and the problems that they faced in the next section.

Now and then

THE WEST IN POPULAR CULTURE

The distorted picture of life in the West that had been part of popular culture for decades eventually changed. Starting in the late 1970s, the view became more realistic, especially in motion pictures.

For example, the hardships of the cattle drives were shown in *Lonesome Dove* (1989). The sufferings of Native Americans were portrayed in *Dances With Wolves* (1990). The role of women in the West was dramatized in *Sarah, Plain and Tall* (1991). And the contributions of African Americans were noted in *Buffalo Soldiers* (1997), pictured below.

Section 3 Assessment

1. Terms & Names

Identify
- homestead
- Mexicano
- William "Buffalo Bill" Cody
- buffalo soldier

2. Taking Notes

Use a chart to compare Wild West myths with the realities of western life.

Myth	Real life

What do you think is the most well-known myth about the West?

3. Main Ideas

a. How were women's contributions to the West recognized by western lawmakers?

b. Which factors led to the growth of such western cities as Denver, Omaha, and San Francisco?

c. How did the arrival of Anglo-Americans change life for Spanish-speaking residents of the Southwest?

4. Critical Thinking

Making Inferences What changed the attitudes of western lawmakers about giving women voting rights?

THINK ABOUT
- new roles for women
- need for stability in western communities
- part played by women in settlement

ACTIVITY OPTIONS

MATH
SPEECH

Pick a western city mentioned in this section. Create a **database** of information about the city or give a **short speech** describing its growth.

Stage a Wild West Show!

You are the manager of Buffalo Bill's Wild West, the biggest and most famous of the 50 outdoor circuses and shows performing in the United States in the 1880s. Your job is to keep the show running smoothly. You oversee publicity and keep performers and livestock housed, fed, and supplied with the gear they need. You also manage finances.

COOPERATIVE LEARNING On this page are three challenges you face as the manager of Buffalo Bill's Wild West show. Working with a small group, decide how to deal with each challenge. Choose an option, assign a task to each group member, and do the activity. You will find useful information in the Data File.

MATH CHALLENGE

"over $280,000"

What a season you had from May to September of 1886! The show took in more than $280,000. But you had big expenses, too. You want to compare income and expenses so you can make changes to increase profits. Use the Data File for more statistics. Then present your summary using one of these options:

- Make a line graph comparing income and expenses.
- Make a pie chart showing how you spent the income—that is, what your expenses were. Include one sector for profit (the difference between total income and total expenses). Label each sector with the percentage.

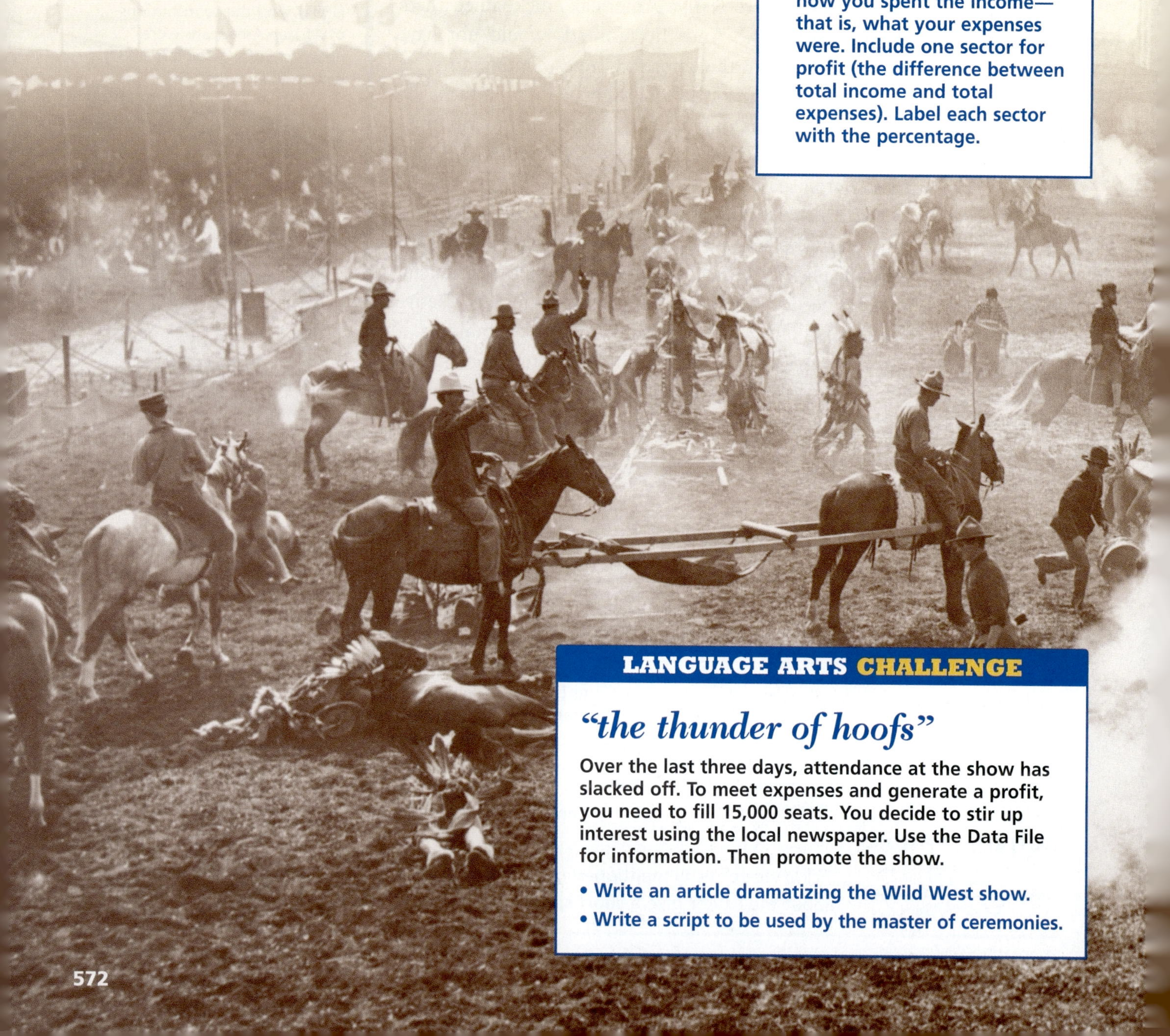

LANGUAGE ARTS CHALLENGE

"the thunder of hoofs"

Over the last three days, attendance at the show has slacked off. To meet expenses and generate a profit, you need to fill 15,000 seats. You decide to stir up interest using the local newspaper. Use the Data File for information. Then promote the show.

- Write an article dramatizing the Wild West show.
- Write a script to be used by the master of ceremonies.

ART CHALLENGE

"History . . . in Living Legends"

You need more posters right away because splashy posters are your best advertising. They dazzle people with thrilling Western scenes. They also give show times and admission fees. Use the Data File for help. Then present your ideas using one of these options:

- Design a 2' x 3' poster showing the excitement of the Wild West show.
- Sketch two small action posters, each showing one event, for shop windows.

ACTIVITY WRAP-UP

Present to the class Meet as a group to review your methods of promoting and charting the success of your Wild West show. Evaluate which of your solutions is the best for each challenge. Once you have chosen one solution for each, make a class presentation. Each group member should take part.

DATA FILE

WILD WEST SHOW

- 3-hour show; admission, 50¢

Major Show Acts

- "Star-Spangled Banner" Overture
- Grand Review of Buffalo Bill Cody and cast on horseback
- Annie Oakley shoots card targets, apple off her poodle's head
- Reenactments of covered wagons crossing the prairie; a Native American buffalo hunt; outlaws attacking a mail coach; a battle between army and Native Americans
- Sharpshooting by Buffalo Bill

Show Crew and Gear

- crew of 700—including Buffalo Bill, Annie Oakley, Sitting Bull, cowhands, Native Americans, 36-piece band, cooks, blacksmiths, teams to set up grandstand and tents
- animals—including 500 horses, 10 mules, 5 steers, 18 buffalo
- gear—including covered wagons and tepees, guns and ammunition, living and dressing room tents, dining tents, booths selling popcorn, souvenirs, and canvas scenery backdrops, 26-car train for transport

INCOME AND EXPENSES

May 9–September 25, 1886

Total income: $287,000 (from attendance)

Total expenses: $285,000

Owners', managers', key performers' salaries: $55,000

Rents: $58,000

Advertising: $17,000

Printing: $4,000

Groceries/ammunition: $7,000

Miscellaneous expenses: $144,000 (other wages, livestock feed, electricity, medical, security, etc.)

Visit www.mcdougallittell.com to learn more about the Old West.

4 Farming and Populism

TERMS & NAMES
- Homestead Act
- Exoduster
- sodbuster
- Grange
- cooperative
- Populist Party
- gold standard
- William Jennings Bryan

MAIN IDEA	WHY IT MATTERS NOW
A wave of farmers moved to the Plains in the 1800s and faced many economic problems.	Farmers are facing similar economic problems today.

ONE AMERICAN'S STORY

From 1865 to 1900, about 800,000 Swedes left their homeland in northern Europe. Sweden's population was soaring, and good farmland was becoming scarce. Most Swedes were farmers. They were drawn to the United States by the promise of more and better land.

For Olaf Olsson, the acres of free land offered to settlers by the U.S. government was an unbelievable opportunity. Shortly after he arrived in 1869, Olsson wrote home to tell friends and family what awaited them in America.

A VOICE FROM THE PAST

We do not dig gold with pocket knives, we do not expect to become . . . rich in a few days or in a few years, but what we aim at is to own our own homes. . . . The advantage which America offers is not to make everyone rich at once, without toil or trouble, but . . . that the poor . . . [can] secure a large piece of good land almost without cost, that they can work up little by little.

Olaf Olsson, quoted in *The Swedish Americans,* by Allyson McGill

As you will read in this section, many Americans as well as immigrants from all parts of Europe shared Olsson's optimism. They uprooted their families to start a new life on the Plains.

Railroad posters advertise lands in the West.

U.S. Government Encourages Settlement

For years, people had been calling on the federal government to sell western land at low prices. Before the Civil War, Southern states fought such a policy. They feared that a big westward migration would result in more nonslave states. Once the South left the Union, however, the way was clear for a new land policy. To interest both American and immigrant families like the Olssons in going west, the federal government passed the **Homestead Act** in 1862. This law offered 160 acres of land free to anyone who agreed to live on and improve the land for five years.

After Reconstruction ended in 1877, African Americans in the South faced harsh new forms of discrimination. (See Chapter 18.) By 1879, leaders like Benjamin "Pap" Singleton of Tennessee had convinced thousands to migrate to new homes in Kansas. They compared themselves to the biblical Hebrews led out of slavery in Egypt and called themselves **Exodusters.** One of them, John Solomon Lewis, remarked, "When I landed on the soil of Kansas, I looked on the ground and I says this is free ground." In all, some 50,000 African Americans settled in Kansas, Missouri, Indiana, and Illinois.

Thousands of European immigrants also sought a new start in the West. Swedes, like Olaf Olsson, joined Germans, Norwegians, Ukrainians, and Russians on the Great Plains. They often first learned about the West from land agents for American railroad companies. These salesmen traveled throughout Europe with pamphlets proclaiming "Land for the Landless! Homes for the Homeless."

Reading **History**

A. Analyzing Causes How did the railroads help to settle the West?

From 1850 to 1870, the government gave millions of acres of public land to the railroads to promote railroad expansion. The railroads resold much of the land to settlers. This not only made the railroad companies rich, but it also supplied new customers for railroad services. The railroads' sales pitch worked. In the 1860s, so many Swedes and Norwegians settled in Minnesota that a local editor wrote, "It seems as if the Scandinavian Kingdoms were being emptied into this state."

Life on the Farming Frontier

Once pioneers reached their new homes on the plains, they faced many challenges never mentioned by the land agents. The plains were nearly treeless. So farmers were forced to build their first homes from blocks of sod. Sod is the top layer of prairie soil that is thickly matted with grass roots.

For fuel, the **sodbusters,** as the farmers were called, burned corn cobs or "cow chips" (dried manure). In many places, sodbusters had to dig wells more than 280 feet deep to reach the only water. Blizzards, prairie fires, hailstorms, tornadoes, grasshoppers, and drought added to the misery of life on the plains. Many settlers, such as Katherine Kirk of South Dakota, wondered whether they had the courage "to stick it out."

Connections TO SCIENCE

SOD HOUSES

To build their dwellings, Plains farmers, or sodbusters, like the Nebraska family pictured here, cut the tough buffalo grass of the prairie into two-or three-foot strips. Then they laid chunks of sod into two rows as walls. The walls were often 36 inches thick.

Prairie grass was thick. Its roots grew outward under the soil, often connecting with one another. This held the sod together. The roots also provided a layer of insulation. So sod houses, or soddies, stayed warm in the winter and cool in the summer. But their roofs leaked rain and dirt, and the walls housed mice, snakes, and insects.

New inventions helped farmers to meet some of these challenges. A steel plow invented by John Deere in 1838 and improved upon by James Oliver in 1868 sliced through the tough sod. Windmills adapted to the plains pumped water from deep wells to the surface. Barbed wire allowed farmers to fence in land and livestock. Reapers made the harvesting of crops much easier, and threshers helped farmers to separate grain or seed from straw. These inventions also made farm work more efficient. From 1860 to 1890, farmers doubled their production of wheat.

The Problems of Farmers

As farmers became more efficient, they grew more and more food. The result was that farmers in the West and South watched with alarm as prices for farm crops began to drop lower and lower in the 1870s.

Economics *in* History

Supply and Demand

Farmers in the West were having economic problems in the 1880s. The supply of food was increasing rapidly, but consumer demand was growing slowly. To attract more consumers, farmers had to drop the prices of their products.

The farmers were experiencing the **law of supply and demand.** The amount of economic goods available for sale is the **supply.** The willingness and ability of consumers to spend money for goods and services is **demand.** The price of goods is set by the supply of that good and the demand for that good.

At a lower price, businesses produce less of a good because they will make less money. As the price rises, they produce more. Consumer demand works in the opposite way. Consumers want to buy more of the good when the price is lower—after all, it costs them less. They buy less when the price is higher. The actual price of a good results from a compromise—how much consumers are willing to pay and how little businesses are willing to take for the good.

CONNECT TO HISTORY

1. **Recognizing Effects** Suppose farmers found a new market for their wheat—the people in another country, for instance. What effect would that have on price? Why?

See Skillbuilder Handbook, page R10.

CONNECT TO TODAY

2. **Comparing** How does the price of blue jeans show the law of supply and demand?

Visit www.mcdougallittell.com to learn more about supply and demand.

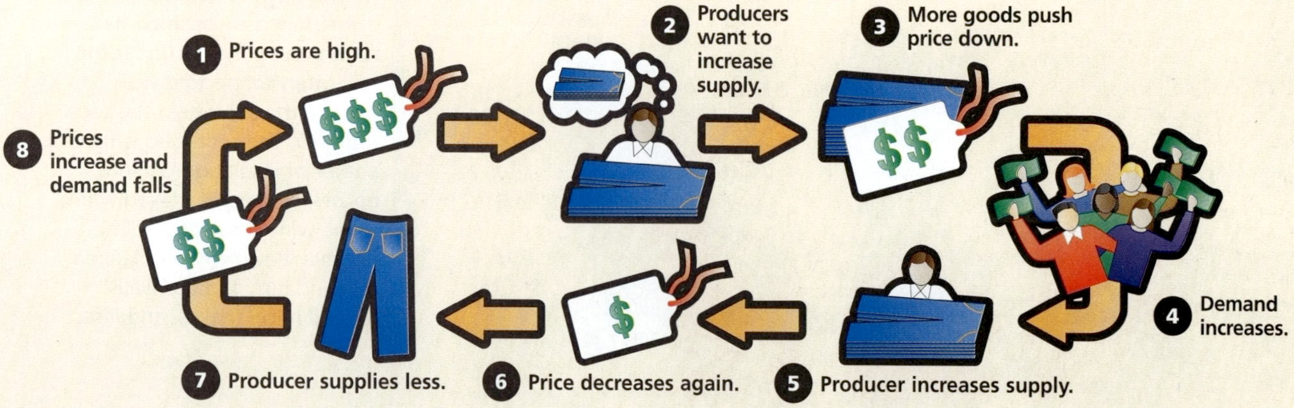

(See "Economics in History" on page 576.) Wheat that sold for $1.45 a bushel after the Civil War was 49 cents 30 years later. One reason for lower prices was overproduction. Farmers were growing more food because additional farmland had been opened up and farming methods and machines had improved.

"The Grange Awakening the Sleepers" (1873) shows a farmer trying to warn the country about the menace of the railroads.

Receiving less money for their crops was bad enough. But at the same time farmers had to spend more to run a farm. New farm machinery and railroad rates were especially costly. Railroads charged the farmers high fees to carry their crops to market. The railroads also usually owned the grain elevators where crops were stored until shipment. Farmers had no choice but to pay the high costs of storage that railroads charged.

Farmers were angry. They began to work together to seek solutions to their problems. In 1867, farmers had formed the **Grange,** officially known as the Patrons of Husbandry. The group's main purpose at first had been to meet the social needs of farm families who lived great distances from one another. However, as economic conditions got worse, Grange members took action. They formed **cooperatives.** These are organizations owned and run by their members. The cooperatives bought grain elevators and sold crops directly to merchants. This allowed farmers to keep more of their profits.

Farmers also began to demand action from the government to change their circumstances. For example, Grangers asked states to regulate railroad freight rates and storage charges. Illinois, Minnesota, Wisconsin, and Iowa did so. In 1877, the Supreme Court backed the farmers in their fight against the railroads. In *Munn* v. *Illinois,* the Court ruled that states and the federal government could regulate the railroads because they were businesses that served the public interest.

The Rise of Populism

In 1890, several farm groups joined together to try to gain political power. They formed the **Populist Party,** or People's Party. The Populists wanted the government to adopt a free silver policy, that is, the unlimited coining of silver. Since silver was plentiful, more money would be put in circulation. They believed that increasing the money supply would cause inflation. Inflation, in turn, would result in rising prices. Higher prices for crops would help farmers pay back the money that they had borrowed to improve their farms.

Reading **History**

B. Summarizing What steps did farmers take to seek solutions to their problems?

Opponents of free silver wanted to keep the gold standard. Under the **gold standard,** the government backs every dollar with a certain amount

AMERICA'S HISTORY MAKERS

WILLIAM JENNINGS BRYAN

1860–1925

William Jennings Bryan was known as the "silver-tongued orator from Nebraska." His powerful voice and his strong belief in the "common people" won him two terms in Congress and a presidential nomination at age 36. He made and lost two more bids for president in 1900 and 1908.

Although he never again held elective office, Bryan remained influential in the Democratic Party. Many reforms that he fought for, such as an eight-hour workday and woman suffrage, later became law.

How do you think Bryan was able to influence reform without being elected president?

of gold. Since the gold supply is limited, fewer dollars are in circulation. Inflation is less likely. This protects the value of money by keeping prices down.

In 1892, the Populist Party platform called for free silver to expand the money supply, government ownership of railroads, shorter working hours, and other political reforms. The Populist presidential candidate, James B. Weaver, lost to Grover Cleveland. But he won more than a million votes—a good showing for a third-party candidate.

The Election of 1896

By the next presidential campaign, money issues mattered much more to voters. The nation had suffered through a serious depression, the Panic of 1893. The Republican candidate, William McKinley, favored the gold standard. He warned that "free silver" would mean higher prices for food and other goods.

The Populists joined the Democratic Party in supporting **William Jennings Bryan** of Nebraska. Bryan urged the Democratic convention to support free silver in his stirring "Cross of Gold" speech.

A VOICE FROM THE PAST

Burn down your cities and leave our farms, and your cities will spring up again as if by magic; but destroy our farms and the grass will grow in the street of every city in the country. . . . [We] . . . answer . . . their demand for a gold standard by saying . . . : You shall not press down upon the brow of labor this crown of thorns. You shall not crucify mankind upon a cross of gold.

William Jennings Bryan, Democratic Convention speech, July 8, 1896

Reading **History**

C. Analyzing Points of View What point was William Jennings Bryan making about the importance of farms?

Farmers in the South and the West voted overwhelmingly for Bryan. But McKinley, who was backed by industrialists, bankers, and other business leaders, won the East and the election by about half a million votes. This election was the beginning of the end for the Populist Party.

The Closing of the Frontier

By the late 1880s, fenced-in fields had replaced open plains. The last remaining open land was in Indian Territory. The Oklahoma land rush of 1889 symbolized the closing of the frontier. At the blast of the starting gun on April 22, thousands of white settlers rushed to claim two million acres of land that had once belonged to Native Americans. In May 1890, this part of Indian Territory officially became Oklahoma Territory. In 1890, 17 million people lived between the Mississippi and the Pacific. That year the Census Bureau declared that the country no longer had a continuous frontier line—the frontier no longer existed.

Thousands of settlers rushed into Oklahoma Territory in 1889 to claim the last open land on the frontier.

To many, the frontier was what had made America unique. In 1893, historian Frederick Jackson Turner wrote an influential essay on the frontier. Turner said that the frontier was a promise to all Americans, no matter how poor, that they could advance as far as their abilities allowed. To Turner the frontier meant opportunity, and its closing marked the end of an era.

Reading **History**

D. Using Secondary Sources What did Frederick Jackson Turner believe about the frontier?

A VOICE FROM THE PAST

Up to our own day American history has been in a large degree the history of the colonization of the Great West. The existence of an area of free land, its continuous recession, and the advance of American settlement westward, explain American development.

Frederick Jackson Turner, "The Significance of the American Frontier"

Today many historians question Turner's view. They think he gave too much importance to the frontier in the nation's development and in shaping a special American character. These historians point out that the United States remains a land of opportunity long after the frontier's closing.

In the next chapter, you will learn how an industrial society developed in the East during the same period that the West was settled.

Section 4 Assessment

1. Terms & Names

Identify:

- Homestead Act
- Exoduster
- sodbuster
- Grange
- cooperative
- Populist Party
- gold standard
- William Jennings Bryan

2. Taking Notes

Review the chapter and find five key events to place on a time line as shown.

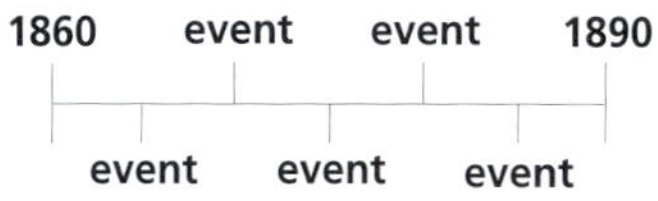

What do you think was the most important event?

3. Main Ideas

a. How did the federal government encourage and support settlement of the Plains?

b. Which groups of people moved onto the Plains in the late 1800s? Why did they come?

c. Which problems in the 1890s led farmers to take political action?

4. Critical Thinking

Drawing Conclusions

Why did the Grange favor government regulation of the railroads?

THINK ABOUT

- powers of monopolies
- importance of railroads to farmers
- railroad freight and storage rates

ACTIVITY OPTIONS

TECHNOLOGY

SPEECH

Pick one invention that helped farmers on the Plains. Plan an **electronic presentation** on the invention or deliver a **sales pitch** to potential buyers.

Chapter 19 ASSESSMENT

VISUAL SUMMARY

Growth of the West

Miners

Miners were attracted to the West by gold and silver strikes. Mining contributed to the population growth of many western territories.

Cowhands & Ranchers

Ranchers and cowhands established a thriving cattle industry. New settlement, barbed wire, and bad weather ended the cattle boom.

Native Americans

Government policies, wars, and the destruction of the buffalo led to the defeat of the Plains peoples and to their placement on reservations.

Homesteaders

Hundreds of thousands of homesteaders settled on the Plains. Their life was hard, but they used new technologies to increase their output.

TERMS & NAMES

Briefly explain the importance of each of the following.

1. frontier
2. long drive
3. reservation
4. Battle of the Little Bighorn
5. Dawes Act
6. homestead
7. Mexicano
8. Homestead Act
9. sodbuster
10. Populist Party

REVIEW QUESTIONS

Miners, Ranchers, and Cowhands (pages 557–561)

1. What role did miners play in the settlement of the West?
2. What made cattle ranching so profitable in the late 1800s?
3. What ended the boom in the cattle business?

Native Americans Fight to Survive (pages 562–567)

4. What caused conflict between Native Americans and white settlers on the Great Plains?
5. How did Native Americans resist white settlement?

Life in the West (pages 568–573)

6. What rights did women in the West gain before women in Eastern states?
7. How has the myth of the "Wild West" been revised?

Farming and Populism (pages 574–579)

8. How did the federal government encourage people to settle on the Great Plains?
9. What were the goals of the Grange?
10. What marked the closing of the frontier?

CRITICAL THINKING

1. USING YOUR NOTES

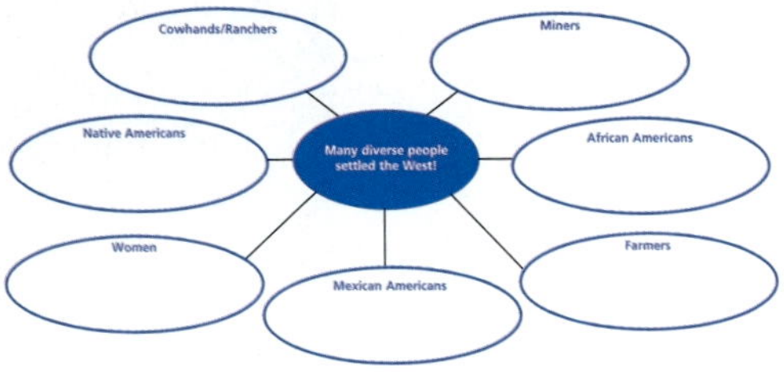

Using your completed chart, answer the questions below.

a. What were the main reasons that drew people to the West?

b. Which groups do you think benefited from being in the West and which groups did not? Explain.

2. APPLYING CITIZENSHIP SKILLS

What are the dangers of vigilante justice?

3. THEME: DIVERSITY AND UNITY

Why might the contributions of women and Native Americans, African Americans, and other ethnic groups have been overlooked in early books and films on the West?

4. ANALYZING LEADERSHIP

Why did the Nez Perce Chief Joseph decide to surrender? What other choices might he have made?

5. CONTRASTING

How did ranchers and sodbusters differ over land use? Why did these differences lead to conflict?

6. FORMING AND SUPPORTING OPINIONS

What do you think would be the most difficult challenge in starting a new life on the Great Plains? Give reasons for your answer.

Interact *with* History

Now that you have read the chapter, would you still make the same statements about how your life would change in the West? Explain.

HISTORY SKILLS

1. INTERPRETING MAPS: Movement

Study the map. Answer the questions.

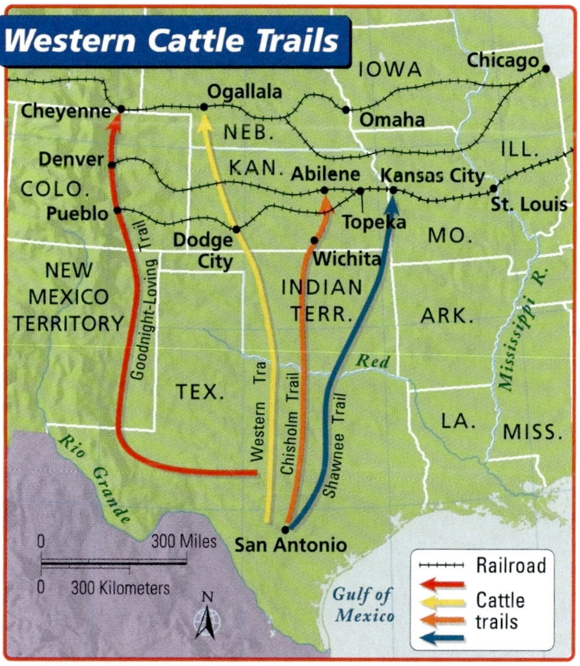

Basic Map Elements

a. What is the subject of the map?

b. What do the arrows indicate?

Interpreting the Map

c. What is the longest cattle trail?

d. How many miles did it cover?

2. INTERPRETING PRIMARY SOURCES

This quotation by Charles Goodnight, a Texas rancher and founder of the Goodnight-Loving Trail (see above map), appeared in *The West* by Geoffrey Ward.

> All in all, my years on the trail were the happiest I ever lived. There were many hardships and dangers, of course, that called on all a man had of endurance and bravery; but when all went well there was no other life so pleasant. Most of the time we were solitary adventurers in a great land as fresh and new as a spring morning, and we were free and full of the zest of darers.

a. What does Goodnight think about life on a cattle trail?

b. How does Goodnight describe the West?

ALTERNATIVE ASSESSMENT

1. INTERDISCIPLINARY ACTIVITIES: Art

Analyzing Visual Images of the West Look through magazines and books to find images that portray the West today. Create a display of pictures with captions that explains whether each picture portrays the real or legendary Wild West.

2. COOPERATIVE LEARNING ACTIVITY

Performing Dramatic Readings Drawing on the many journals, diaries, and other primary sources left by sodbusters, create a Voices of the Plains play. Working in small groups, have students research such topics as "the journey west," "first impressions," "dugouts and soddies," "farming the plains," "women's work and worries." Within groups, members can create presentations using these suggestions.

- Choose and research a topic.
- Select quotations related to the topic.
- Pick writers to compose lines to introduce and make transitions between quotations, and choose readers to perform readings.

3. TECHNOLOGY ACTIVITY

Designing a "Legends of the West" Web Site The "Wild West" of the late 1800s was a land of myth and legend. It was peopled by a cast of memorable characters whose exploits, real and imagined, have been portrayed in books, motion pictures, and television. Use the library or search the Internet for information on the "Wild West."

INTERNET **Visit www.mcdougallittell.com to learn more about the West.**

Design a "Wild West" Web site following the suggestions below.

- Select the legendary personalities to be featured.
- Include words and images that capture the flavor of the West.
- Choose musical selections to add background.
- Find Web sites that would be good links for visitors to your page.

4. PORTFOLIO ACTIVITY

PORTFOLIO **Option 1** Review your section and chapter assessment activities. Select one that you think was your best work. Use comments made by your teacher or classmates to improve your work. Add this activity to your history portfolio

PORTFOLIO **Option 2** Review the list of details that you wanted to know about the West on page 556. Then write a short report in which you explain the details you have learned. Add this report to your portfolio.

CHAPTER 20

An Industrial Society 1860–1914

These laborers are working in a foundry, a place where metal is cast.

Interact *with* History

The year is 1894. You work in a factory that is unheated and badly lit. The machine that you operate is dangerous. The economy is doing poorly, so the factory has cut your wages. Some of your coworkers have gone out on strike. They want better pay and working conditions.

What Do You Think?

- What are some risks you would be taking if you join the strike?
- What might you gain if you take part in the strike?
- What other methods might you use to persuade your employer to meet your demands?

Would you join the strike? Why or why not?

USA / **World** — 1860–1914

USA:

- **1863** Two companies begin to build a transcontinental railroad across the United States.
- **1876** Alexander Graham Bell patents the telephone.
- **1882** Thomas Edison installs electric lights in New York City.
- **1888** Benjamin Harrison is elected president.
- **1892** Grover Cleveland is elected president for the second time.
- **1894** Pullman Strike halts rail traffic across the nation.
- **1901** Oil drillers discover a huge oil field in Texas.
- **1905** Supreme Court overturns a New York law establishing a 60-hour workweek for bakers.

World:

- **1869** Suez Canal opens in Egypt.
- **1889** The world's tallest structure to date, the Eiffel Tower in Paris, is built of iron.
- **1896** The Italian engineer Guglielmo Marconi patents the radio.
- **1904** Russia finishes building the first Trans-Siberian railway across Asia.

BEFORE YOU READ

Business leader John D. Rockefeller is shown as a wealthy king. Notice which industries are the "jewels in his crown."

Previewing the Theme

Economics in History As Chapter 20 explains, natural resources and new inventions caused industry to boom in the late 1800s. Some businesses grew very large and made great profits by wiping out competitors and paying low wages. Workers banded together to demand higher pay and better working conditions.

What Do You Know?

Do you know of any businesses that started back in the 1800s? How do businesses grow?

THINK ABOUT

- businesses that you see in your community
- businesses that are advertised on television, in magazines or newspapers, or on the Internet

What Do You Want to Know?

What facts and details would help you understand how a nation of small businesses became a nation of giant corporations? In your notebook, list the facts and details you hope to learn from this chapter.

READ AND TAKE NOTES

Reading Strategy: Analyzing Causes and Recognizing Effects The conditions or actions that lead to a historical event are its causes. The consequences of an event are its effects. As you read the chapter, look for the causes and effects of industrial and railroad growth. Causes include geographical factors and actions by individuals and the government. Effects include both benefits and problems. Use the diagram below to record both causes and effects.

See Skillbuilder Handbook, page R10.

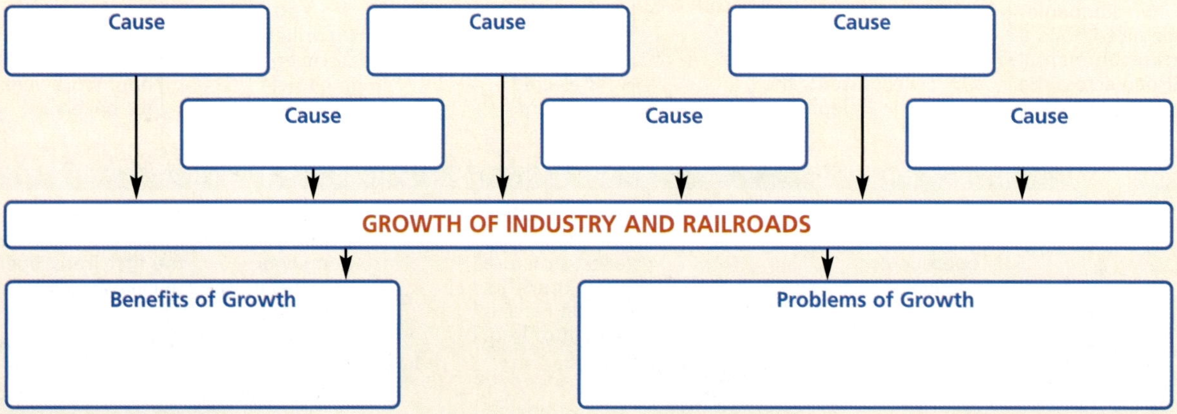

1 The Growth of Industry

MAIN IDEA

The growth of industry during the years 1860 to 1914 transformed life in America.

WHY IT MATTERS NOW

Modern businesses rely on many of the inventions and products developed during that time.

TERMS & NAMES

- petroleum
- patent
- business cycle
- Bessemer steel process
- generator
- Thomas Edison
- Alexander Graham Bell
- Centennial Exhibition

ONE AMERICAN'S STORY

In the 1850s, most Americans lit their homes with oil lamps. They could have used kerosene, an oil made from coal, but it was expensive. Then, in 1855, a chemist released a report saying that kerosene could be made more cheaply from an oily flammable liquid called petroleum.

However, people didn't know how to obtain petroleum from its source deep in the earth. They just gathered the petroleum that had seeped to the surface, which was a time-consuming process. So several men decided to try drilling for oil. One of these men was Edwin Drake, who in 1857 visited a site in Pennsylvania where petroleum oozed to the surface.

A VOICE FROM THE PAST

Within ten minutes after my arrival . . . I had made up my mind that [petroleum] could be obtained in large quantities by Boreing as for Salt Water.

Edwin Drake, quoted in *The Americans: The Democratic Experience*

Drake began drilling through bedrock in 1859. In August, he struck oil, an event that launched the oil industry. It was one of many new industries that developed after the mid-1800s, as this section explains.

The wooden structure is Drake's first oil well.

The Industrial Revolution Continues

In the late 1700s, the establishment of textile mills in New England was an early sign of industrialization, or the growth of industry, in the United States. By the mid-1800s, the use of factories had spread to other industries and other regions. The pace of industrialization became more rapid after the Civil War. Several factors helped U.S. industries grow.

1. **Plentiful natural resources.** America had immense forests and large supplies of water. It also had vast mineral wealth, including coal, iron, copper, silver, and gold. Industry used these resources to manufacture a variety of goods.

2. **Improved transportation.** In the early 1800s, steamboats, canals, and railroads made it possible to ship items long distances. Railroad building boomed after the Civil War. As shipping raw materials and finished goods to markets became even easier, industry grew.
3. **Growing population.** From 1860 to 1900, the U.S. population grew from 31.5 million to 76 million. This led to a growing need for goods. The demand for goods spurred the growth of industry.
4. **High immigration.** Between 1860 and 1900, about 14 million people immigrated to the United States. Many of them knew specialized trades, such as metalworking. Such knowledge was valuable to industries. In addition, unskilled immigrants supplied the labor that growing industry needed.
5. **New inventions.** New machines and improved processes helped industry produce goods more efficiently. Inventors applied for patents for the machines or processes they invented. A **patent** is a government document giving an inventor the exclusive right to make and sell his or her invention for a specific number of years.
6. **Investment capital.** When the economy was thriving, many businesses made large profits. Hoping to share in those profits, banks and wealthy people lent businesses money. The businesses used this capital to build factories and buy equipment.

Background
Some of these canals were built because of Henry Clay's American System. (See Chapter 11.)

Vocabulary
capital: money and property used in a business

The Business Cycle

American industry did not grow at a steady pace; it experienced ups and downs. This pattern of good and bad times is called the **business cycle.** During good times, called booms, people buy more, and some invest in business. As a result, industries and businesses grow. During bad times, called busts, spending and investing decrease. Industries lay off workers and make fewer goods. Businesses may shrink—or even close. Such a period of low economic activity is a depression.

America experienced depressions in 1837 and 1857. Both were eventually followed by periods of strong economic growth. In the late 1800s, there were two harsh depressions, also called panics. The depression of 1873 lasted five years. At its height, three million people were out of work. During the depression that began in 1893, thousands of businesses failed, including more than 300 railroads.

The Business Cycle

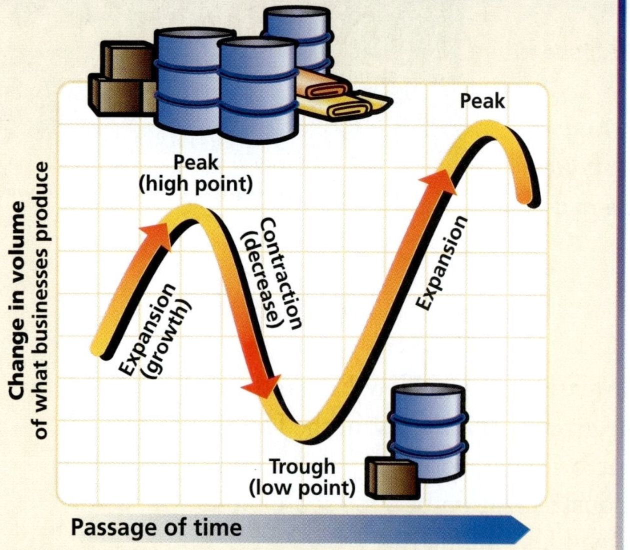

SKILLBUILDER Interpreting Charts
1. *How does the amount of goods produced at the peak compare to the amount at the trough?*
2. *Are all peaks equally prosperous? Explain.*

Reading History
A. Recognizing Effects How do you think depressions affect ordinary people?

Even with these economic highs and lows, industries in the United States grew tremendously between 1860 and 1900. Overall, the amount of manufactured goods increased six times during these years.

Steel: The Backbone of Industry

The steel industry contributed to America's industrial growth. Before the mid-1800s, steel was very expensive to manufacture because the steel-making process used huge amounts of coal. In the 1850s, William Kelly in the United States and Henry Bessemer in England independently developed a new process for making steel. It used less than one-seventh of the coal that the older process used. This new manufacturing technique was called the **Bessemer steel process.**

Because the Bessemer process cut the cost of steel, the nation's steel output increased 500 times between 1867 and 1900. Industry began to make many products out of steel instead of iron. These products included plows, barbed wire, nails, and beams for buildings. But the main use of steel throughout the late 1800s was for rails for the expanding railroads. (See Section 2.)

Reading **History**

B. Drawing Conclusions Which industries benefited from the steel products mentioned here?

Connections TO SCIENCE

IRON VS. STEEL

Why is the comic book hero Superman also called the man of steel? People often use the word *steel* as a synonym for strength.

Steel is an iron alloy—a blend of iron and other materials such as carbon. But steel is stronger and more durable than iron. That is why steel replaced iron in many industries in the late 1800s. A giant ladle (bucket) used to pour melted steel is pictured below.

Stainless steel, invented in the early 1900s, has an additional benefit: it doesn't rust. Stainless steel is used in tools, machines, and many household items, such as pots, pans, and utensils.

Edison and Electricity

Another industry that grew during the late 1800s was the electric-power industry. By the 1870s, inventors had designed efficient generators. A **generator** is a machine that produces electric current. As a result, people grew eager to tap the power of electricity.

The inventor who found the most ways to use electricity was **Thomas Edison.** In 1876, he opened a laboratory in Menlo Park, New Jersey. He employed many assistants, whom he organized into teams to do research. Edison's laboratory invented so many things that Edison received more than 1,000 U.S. patents, more than any other individual inventor.

Edison would start with an idea for a possible invention. Then he would work hard to make that idea a reality—even if problems arose.

A VOICE FROM THE PAST

It has been just so in all my inventions. The first step is an intuition—and comes with a burst, *then* difficulties arise. . . . "Bugs"—as such little faults and difficulties are called—show themselves and months of anxious watching, study and labor are requisite [needed] before commercial success—or failure—is certainly reached.

Thomas Edison, quoted in *Edison* by Matthew Josephson

Reading **History**

C. Finding Main Ideas According to Edison, is inventing easy?

Edison's most famous invention was practical electric lighting. Other inventors had already created electric lights, but they were too bright and

CONNECTIONS TO SCIENCE

American Inventors, *1870–1900*

Thomas A. Edison

Imagine life without being able to burn lights 24 hours a day. Or without movies and recorded music. Edison invented not only the light bulb but also the phonograph and a moving-picture viewer.

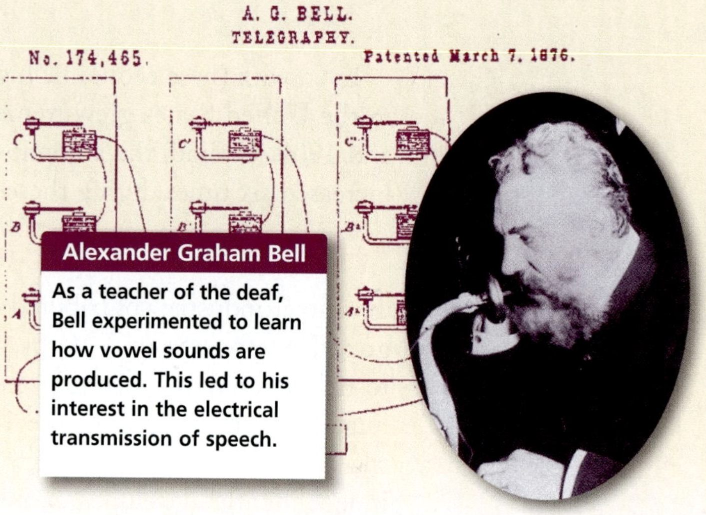

Alexander Graham Bell

As a teacher of the deaf, Bell experimented to learn how vowel sounds are produced. This led to his interest in the electrical transmission of speech.

flickery for home use. Edison figured out how to make a safe, steady light bulb. He also invented a system to deliver electricity to buildings.

By 1882, he had installed electric lighting in a half-mile-square area of New York City. Electric lighting quickly replaced gaslights. By the late 1880s, Edison's factory produced about a million light bulbs a year.

Bell and the Telephone

Electricity played a role in communications devices invented during the 1800s. In 1835, Samuel Morse developed the telegraph. It allowed people to use electrical impulses to send messages over long distances.

The next step in communications was the telephone, invented by **Alexander Graham Bell.** He was a Scottish immigrant who taught deaf students in Boston. At night, Bell and his assistant, Thomas Watson, tried to invent a device to transmit human speech using electricity.

After years of experiments, Bell succeeded. One day in March 1876, he was adjusting the transmitter in the laboratory in his apartment. Watson was in another room with the receiver. The two doors between the rooms were shut. According to Watson's memoirs, Bell accidentally spilled acid on himself and said, "Mr. Watson, come here. I want you." Watson rushed down the hall. He burst into the laboratory, exclaiming that he had heard and understood Bell's words through the receiver.

Reading **History**

D. Analyzing Points of View
Why do you think the scientists said this about the telephone?

Bell showed his telephone at the **Centennial Exhibition** in June 1876. That was an exhibition in Philadelphia to celebrate America's 100th birthday. There, several of the world's leading scientists and the emperor of Brazil saw his demonstration. Afterward, they declared, "Here is the greatest marvel ever achieved in electrical science."

Inventions Change Industry

The telephone industry grew rapidly. By 1880, more than 50,000 telephones had been sold. The invention of the switchboard allowed more and more people to connect into a telephone network. Women commonly worked in the new job of switchboard operator.

The typewriter also opened jobs for women. Christopher Latham Sholes helped invent the first practical typewriter in 1867. He also

Jan Matzeliger

An immigrant from Dutch Guiana, Matzeliger worked in a shoe factory. To reduce the time needed to fasten shoe leather to the sole by hand, he invented a machine to do the job. It increased production by 1,400 percent!

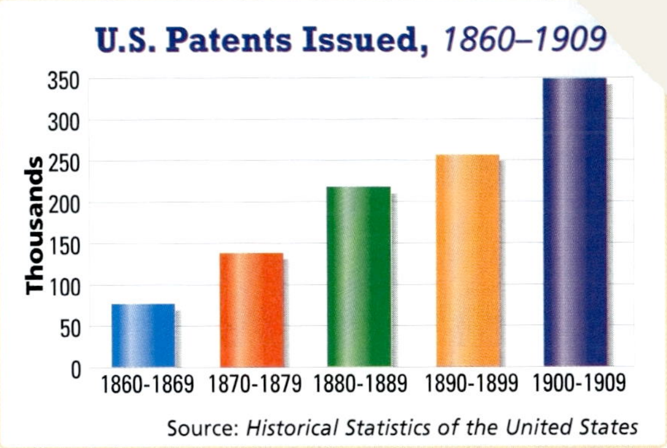

SKILLBUILDER Interpreting Graphs

1. *How many more patents were issued from 1900 to 1909 than from 1860 to 1869?*
2. *Was this a time of increasing or decreasing inventiveness?*

improved the machine and sold his rights to it to a manufacturer who began to make typewriters in the 1870s.

The sewing machine also changed American life. Elias Howe first patented it in 1846. In the next few years, the sewing machine received many design improvements. Isaac Singer patented a sewing machine in 1851 and continued to improve it. It became a bestseller and led to a new industry. In factories, people produced ready-made clothes. Instead of being fitted to each buyer, clothes came in standard sizes and popular styles. Increasingly, people bought clothes instead of making their own.

Other inventors helped industry advance. African-American inventor Granville T. Woods patented devices to improve telephone and telegraph systems. Margaret Knight invented machines for the packaging and shoemaking industries and also improved motors and engines.

Of all the up-and-coming industries of the middle 1800s, one would have a larger impact on American life than any other. That was the railroad industry. You will read about railroads in Section 2.

Section 1 Assessment

1. Terms & Names

Identify:
- petroleum
- patent
- business cycle
- Bessemer steel process
- generator
- Thomas Edison
- Alexander Graham Bell
- Centennial Exhibition

2. Taking Notes

Use a cluster diagram like the one below to list some of the inventions of the late 1800s.

How has one of these inventions recently been improved?

3. Main Ideas

a. What factors contributed to industrial growth in the United States?

b. What is the business cycle?

c. What caused the steel-making industry to boom and why?

4. Critical Thinking

Recognizing Effects How did the inventions of the late 1800s make it easier to do business?

THINK ABOUT
- electric generators and light bulbs
- the telephone
- the typewriter

ACTIVITY OPTIONS

SCIENCE

TECHNOLOGY

Choose an invention and learn more about it. Create a **display** explaining how it works or design a **Web page** linking to sites with more information.

2 Railroads Transform the Nation

TERMS & NAMES
transcontinental railroad
standard time

MAIN IDEA	WHY IT MATTERS NOW
The railroads tied the nation together, speeded industrial growth, and changed U.S. life.	The railroad first made possible our modern system of shipping goods across the country.

ONE AMERICAN'S STORY

Ah Goong was one of thousands of Chinese workers on the Western railroads in the late 1800s. Building a railroad across steep mountains was dangerous. In some places, the workers had to blast rock from a cliff wall to build bridges and tunnels. The lightest Chinese climbed into wicker baskets at the top of the cliff. Others lowered the baskets hundreds of feet to the blasting site. Years later, Ah Goong's granddaughter described her grandfather's job.

A VOICE FROM THE PAST

Swinging near the cliff, Ah Goong . . . dug holes, then inserted gunpowder and fuses. He worked neither too fast nor too slow, keeping even with the others. The basketmen signaled one another to light the fuses. He struck match after match and dropped the burnt matches over the sides. At last his fuse caught; he waved, and the men above pulled hand over hand hauling him up, pulleys creaking.

Maxine Hong Kingston, *China Men*

Chinese immigrants—like the one at the lower left—helped build several railroads in the West.

Irishmen and Americans also worked on the railroads. This section discusses the building of the railroads and how they changed America.

Deciding to Span the Continent

Americans had talked about building a **transcontinental railroad**—one that spanned the entire continent—for years. Such a railroad would encourage people to settle the West and develop its economy. In 1862, Congress passed a bill that called for two companies to build a transcontinental railroad across the center of the United States.

The Central Pacific was to start in Sacramento, California, and build east. The Union Pacific was to start in Omaha, Nebraska, and build west. To build the railroad, these two companies had to raise large sums of money. The government lent them millions of dollars. It also gave them 20 square miles of public land for every mile of track they laid. The railroad companies could then sell the land to raise money.

With the guarantees of loans and land, the railroads attracted many investors. The Central Pacific began to lay its first track in 1863. The

Union Pacific laid its first rail in July 1865 (after the Civil War had ended).

Building the Railroad

The Central Pacific faced a labor shortage because most men preferred to try to strike it rich as miners. Desperate for workers, the Central Pacific's managers overcame the widespread prejudice against the Chinese and hired several dozen of them. The Chinese were small and weighed, on average, no more than 110 pounds. But they were efficient, fearless, and hard working.

They also followed their own customs, which led to an unexpected benefit for the railroad company. The Chinese drank tea instead of unboiled water, so they were sick less often than other workers. Pleased with the Chinese workers, the company brought more men over from China. At the peak of construction, more than 10,000 Chinese worked on the Central Pacific.

Background
Boiling water kills germs.

The Union Pacific hired workers from a variety of backgrounds. After the Civil War ended in 1865, former soldiers from both North and South flocked to work on the railroad. Freed slaves came, too. But one of the largest groups of Union Pacific workers was immigrants, many from Ireland.

Reading History
A. Drawing Conclusions Why did the Union Pacific have a larger supply of workers?

Both railroads occasionally hired Native Americans. Washos, Shoshones, and Paiutes all assisted the race of the rails across the deserts of Nevada and Utah.

daily life

RAILROAD CAMPS

Union Pacific workers often worked 12-hour days. Graders had the job of leveling the roadbed. After a day of hard labor, they slept in small dirt shanties like the one below.

Track layers lived together in groups of 100 to 135, in railroad cars with three layers of bunk beds. The cars were parked at the end of the just-finished track. Workers ate in a dining car with their plates nailed to the table. They gobbled a quick meal of beef, beans, and bread. As soon as one group of 125 workers was done, the next group filed in.

Railroads Tie the Nation Together

Only short, undergrown trees dotted the vast open space. To the south shimmered the Great Salt Lake. In the east rose the bluish shapes of the Rocky Mountains. Across that space, from opposite directions, the workers of the Central Pacific and the Union Pacific toiled. By May 10, 1869, Central Pacific workers had laid 690 miles of track. Union Pacific workers had laid 1,086 miles. Only one span of track separated the two lines at their meeting point at Promontory, Utah.

Reading History
B. Reading a Map Using the map on page 592, find the Union Pacific and Central Pacific Railroads. Notice how they connect Omaha to Sacramento.

Hundreds of railroad workers, managers, spectators, and journalists gathered on that cool, windy day to see the transcontinental railroad completed. Millions of Americans waited to hear the news by telegraph. A band played as a Chinese crew and an Irish crew laid the last rails. The last spike, a golden one, was set in place. First, the president of the Central Pacific raised a hammer to drive in the spike. After he swung the hammer down, the crowd roared with laughter. He had missed. The vice-president of the Union Pacific took a turn and also missed. But the telegraph operator couldn't see and had already sent the message: "done." People across the nation celebrated.

This golden spike united the Central Pacific and Union Pacific Railroads.

The Union Pacific-Central Pacific line was the first transcontinental railroad. By 1895, four more U.S. lines had been built across the continent. Between 1869 and 1890, the amount of money railroads earned carrying freight grew from $300 million to $734 million per year.

Background
Canada had also built a transcontinental railroad, so there were six altogether.

Railroad Time

The railroads changed America in a surprising way: they altered time. Before the railroads, each community determined its own time, based on calculations about the sun's travels. This system was called "solar time." Solar time caused problems for people who scheduled trains crossing several time zones and for travelers.

A VOICE FROM THE PAST

I have been annoyed and perplexed by the changes in the time schedules of connecting railroads. My watch could give me no information as to the arrival and departure of trains, nor of the time for meals.

John Rodgers, quoted in *Passage to Union*

To solve this problem, the railroad companies set up **standard time.** It was a system that divided the United States into four time zones. Although the plan went into effect on November 18, 1883, Congress did not adopt standard time until 1918. By then, most Americans saw its benefit because following schedules had become part of daily life.

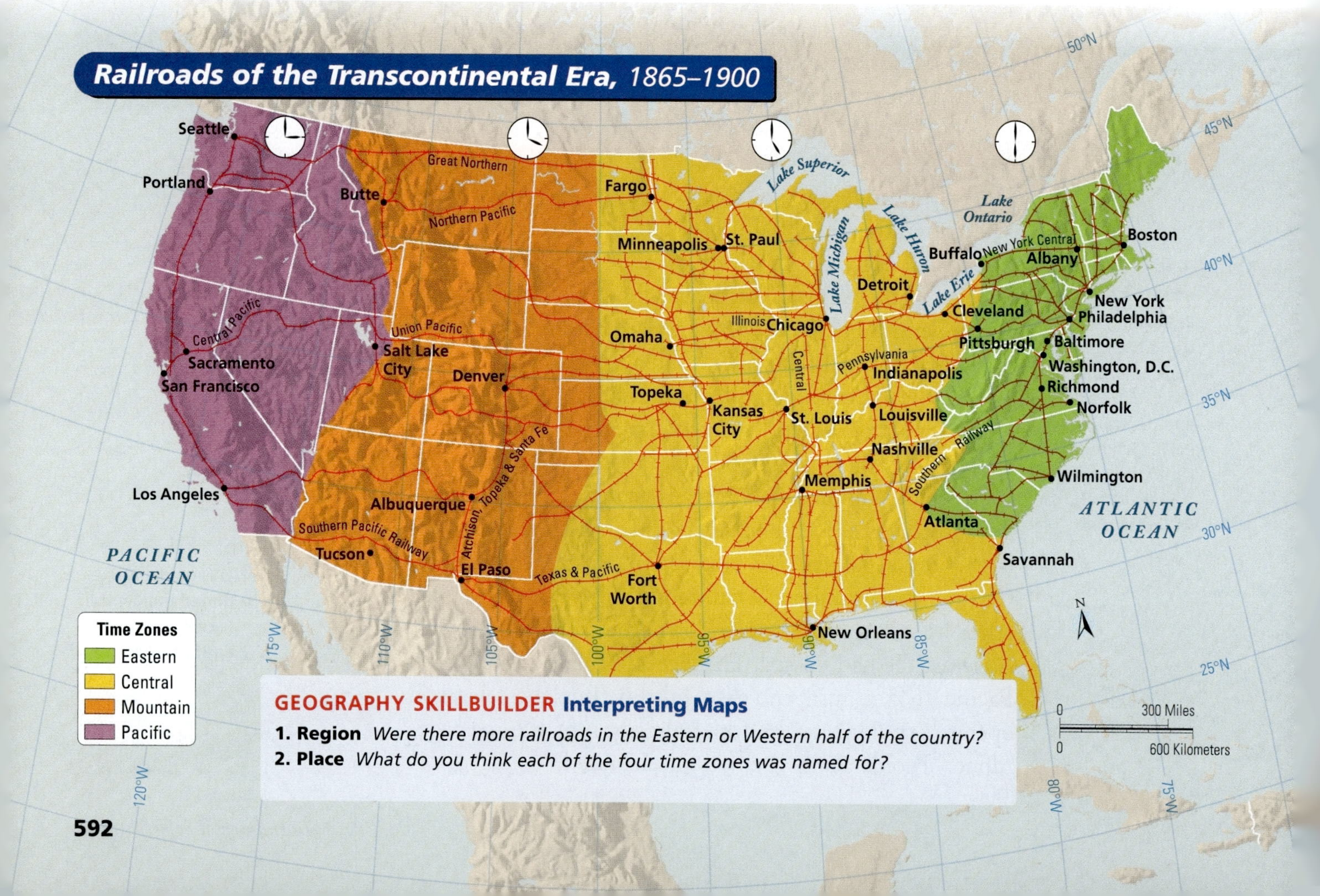

Railroads of the Transcontinental Era, *1865–1900*

GEOGRAPHY SKILLBUILDER Interpreting Maps

1. **Region** *Were there more railroads in the Eastern or Western half of the country?*
2. **Place** *What do you think each of the four time zones was named for?*

Economic and Social Changes

The railroads changed people's lives in many other important ways. They helped create modern America.

1. **Linked the economies of the West and East.** From the West, the railroads carried eastward raw materials such as lumber, livestock, and grain. Materials like these were processed in Midwestern cities such as Chicago and Cleveland. (See Geography in History on pages 598–599.) From Eastern cities, in turn, came manufactured goods, which were sold to Westerners.
2. **Helped people settle the West.** Railroads were lifelines for settlers. Trains brought them lumber, farm equipment, food, and other necessities and hauled their crops to market.
3. **Weakened the Native American hold on the West.** As Chapter 19 explained, the railroads carried hunters who killed off the herds of buffalo. They also brought settlers and miners who laid claim to Native American land.
4. **Gave people more control of the environment.** Before railroads, people lived mainly where there were waterways, such as rivers. Roads were primitive. Railroads made possible cities such as Denver, Colorado, which had no usable waterways.

Just as railroads changed life for many Americans in the late 1800s, so did big business. You will read about big business in Section 3.

Reading **History**

C. Evaluating Which of these four changes do you think were positive, and which were negative?

Connections TO ART & MUSIC

RAILROAD HEROES

Several American songs celebrate railroad heroes. One tells of Casey Jones, an engineer who saved lives. He slammed on the brakes as his train rounded a bend and plowed into a stalled freight train. He died but slowed the train enough to save his passengers.

Another song tells of a mythical worker named John Henry, shown below. This ballad celebrates an African American's strength in a track-laying race against a steam-driven machine.

Section 2 Assessment

1. Terms & Names

Identify:
- transcontinental railroad
- standard time

2. Taking Notes

Using a chart like the one below, record which groups of people helped build the transcontinental railroad.

Central Pacific	Union Pacific

Which group worked on both railroads?

3. Main Ideas

a. Why did the federal government want a transcontinental railroad built?

b. How did the government encourage the building of the railroad?

c. Why was standard time created?

4. Critical Thinking

Recognizing Effects
Which of the trends started by railroads are still part of the modern business world?

THINK ABOUT
- railroads' effect on time
- the way they linked the economy
- the way they changed where people settled

ACTIVITY OPTIONS

ART
TECHNOLOGY

You have been asked to honor those who built the transcontinental railroad. Design a **memorial** or create the opening screen of a **multimedia presentation.**

3 The Rise of Big Business

TERMS & NAMES
robber baron
corporation
John D. Rockefeller
Andrew Carnegie
monopoly
trust
philanthropist
Gilded Age

MAIN IDEA	WHY IT MATTERS NOW
Business leaders guided industrial expansion and created new ways of doing business.	These leaders developed the modern corporation, which dominates business today.

ONE AMERICAN'S STORY

In 1853, when Jay Gould was 17, he visited New York. Big-city wealth impressed Gould. After returning to his small hometown, he told a friend, "Crosby, I'm going to be rich. I've seen enough to realize what can be accomplished by means of riches, and I tell you I'm going to be rich."

"What's your plan?" his friend asked.

"I have no immediate plan," Gould replied. "I only see the goal. Plans must be formed along the way."

Gould achieved his goal. By the time he died in 1892, he was worth $77 million. He made a lot of his money using methods that are illegal today—such as bribing officials and selling fake stock. Most of his deals involved railroads, including the Union Pacific.

Jay Gould was a robber baron. A <u>robber baron</u> was a business leader who became wealthy through dishonest methods. This section discusses other business leaders and their companies.

Jay Gould used methods such as trickery and false reports to "bowl over" his competition.

The Growth of Corporations

Until the late 1800s, most businesses were owned directly by one person or by a few partners. Then advances in technology made many business owners want to buy new equipment. One way to raise money to do so was to turn their businesses into corporations. A **corporation** is a business owned by investors who buy part of the company through shares of stock. A corporation has advantages over a privately owned business:

1. By selling stock, a corporation can raise large amounts of money.
2. A corporation has a special legal status and continues to exist after its founders die. Banks are more likely to lend a corporation money.
3. A corporation limits the risks to its investors, who do not have to pay off the corporation's debts.

In the late 1800s, few laws regulated corporations. This led to the growth of a few giant corporations that dominated American industry. The oil and steel industries are examples of this process.

The Oil and Steel Industries

As Section 1 explained, the oil and steel industries began to grow in the late 1800s. Two men dominated these industries. **John D. Rockefeller** led the oil industry, and **Andrew Carnegie** controlled the steel industry.

Vocabulary
refinery: a plant that purifies oil

John D. Rockefeller built his first refinery in 1863. He decided that the best way to make money was to put his competitors out of business. A company that wipes out its competitors and controls an industry is a **monopoly**. Rockefeller bought other refineries. He made secret deals with railroads to carry his oil at a lower rate than his competitors' oil. He also built and purchased his own pipelines to carry oil.

Rockefeller's most famous move to end competition was to develop the trust in 1882. A **trust** is a legal body created to hold stock in many companies, often in the same industry. Rockefeller persuaded other oil companies to join his Standard Oil Trust. By 1880, the trust controlled 95 percent of all oil refining in the United States—and was able to set a high price for oil. The public had to pay that price because they couldn't buy oil from anyone else. As head of Standard Oil, Rockefeller earned millions of dollars. He also gained a reputation as a ruthless robber baron.

Reading **History**
A. Analyzing Points of View Why do you think people thought monopolies were unfair?

Businessmen in other industries began to follow Rockefeller's example. Trusts were formed in the sugar, cottonseed oil, and lead-mining industries. Many people felt that these monopolies were unfair and hurt the economy. But the government was slow to regulate them.

Rockefeller tried to control all the companies in his industry. By contrast, Andrew Carnegie tried to beat his competition in the steel industry

AMERICA'S HISTORY MAKERS

JOHN D. ROCKEFELLER
1839–1937

John D. Rockefeller was born to a poor family in upstate New York. From his mother, he learned the habit of frugality—he avoided unnecessary spending. "Willful waste makes woeful want" was a saying that Rockefeller's mother passed down to him.

By 1897, he had made millions and millions of dollars. Instead of keeping all that vast fortune for himself and his family, he spent the rest of his life donating money to several worthy causes.

ANDREW CARNEGIE
1835–1919

When Andrew Carnegie was 12, he and his family moved from Scotland to Pennsylvania. Carnegie's first job was in a cotton mill.

Later he worked in a telegraph office. There he was noticed by a railroad superintendent, who hired Carnegie as his assistant. Carnegie learned not only about running a big business but also about investing money. Eventually, he quit to start his own business.

Despite his fortune, Carnegie once wrote that none of his earnings gave him as much happiness as his first week's pay.

Compare the characters of Rockefeller and Carnegie. What do you think made each of them successful?

by making the best and cheapest product. To do so, he sought to control all the processes related to the manufacture of steel. He bought the mines that supplied his iron ore, and the ships and railroads that carried that ore to his mills. Carnegie's company dominated the U.S. steel industry from 1889 to 1901, when he sold it.

Reading **History**

B. Contrasting How did the methods that Carnegie and Rockefeller used to eliminate competition differ?

Rockefeller and Carnegie were multimillionaires. They also were both **philanthropists,** people who give large sums of money to charities. Rockefeller donated money to the University of Chicago and Rockefeller University in New York. Carnegie also gave money to universities, and he built hundreds of public libraries. During his life, Rockefeller gave away more than $500 million. Carnegie gave away more than $350 million.

The Gilded Age

The rags-to-riches stories of people such as Rockefeller and Carnegie inspired many Americans to believe that they too could grow rich. Stories like theirs also inspired writer Horatio Alger. He wrote popular stories about poor boys who worked hard and became quite successful.

Inspiring as these stories were, they hid an important truth. Most people who made millions of dollars had not been raised in poverty. Many belonged to the upper classes and had attended college. Most began their careers with the advantage of money or family connections.

For the rich, the late 1800s was a time of fabulous wealth. Writers Mark Twain and Charles Warner named the era the **Gilded Age.** To

HISTORY *through* ART

Artist Seymour Guy painted this portrait of the Vanderbilts, one of the wealthiest families in America. It shows them in their New York mansion. Notice the gilded picture frames and the expensive furnishings.

Photographer Jessie Tarbox Beals shot this photograph of a poor family in a tenement. A tenement is an apartment house that is usually rundown and very overcrowded. This family probably had only this tiny space.

What do these two images tell you about the differences between the way rich and poor people lived during the Gilded Age?

Vocabulary
gold leaf: gold that has been pounded into thin sheets

gild is to coat an object with gold leaf. Gilded decorations were popular during the era. But the name has a deeper meaning. Just as gold leaf can disguise an object of lesser value, so did the wealth of a few people mask society's problems, including corrupt politics and widespread poverty.

The South Remains Agricultural

One region that knew great poverty was the South. The Civil War had left the South in ruins. Industry did grow in some Southern areas, such as Birmingham, Alabama. Founded in 1871, Birmingham developed as an iron- and steel-producing town. In addition, cotton mills opened from southern Virginia to Alabama. Compared with the Northern economy, however, the Southern economy grew very slowly after the war.

Most of the South remained agricultural. As you have read, many Southern landowners rented their land to sharecroppers who paid a large portion of their crops as rent. Often sharecroppers had to buy their seed and tools on credit. The price of cotton, the South's main crop, was very low. Sharecroppers made little money from selling cotton and had difficulty paying what they owed. And because most sharecroppers had little education, merchants cheated them, increasing their debt.

***Reading*History**
C. Making Inferences What is Fortune implying about the storekeeper?

A VOICE FROM THE PAST

My father once kept an account . . . of the things he "took up" at the store as well as the storekeeper. When the accounts were footed [added] up at the end of the year the thing became serious. The storekeeper had $150 more against my father than appeared on the latter's book. . . . It is by this means that [sharecroppers] are swindled and kept forever in debt.

T. Thomas Fortune, testimony to a Senate committee, 1883

At the same time that sharecroppers struggled to break free of debt, workers in the industrial North also faced injustices. In the next section, you will learn how labor unions tried to fight back.

Section 3 Assessment

1. Terms & Names

Identify:
- robber baron
- corporation
- John D. Rockefeller
- Andrew Carnegie
- monopoly
- trust
- philanthropist
- Gilded Age

2. Taking Notes

Use a Venn diagram like the one shown to compare and contrast Rockefeller and Carnegie.

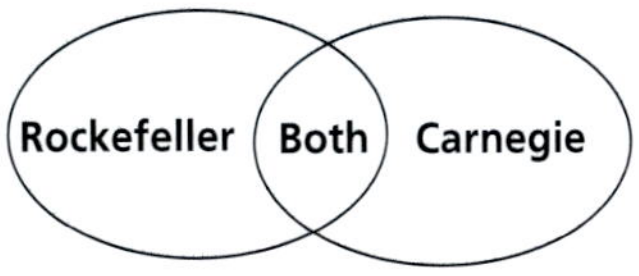

Whose business methods do you agree with more?

3. Main Ideas

a. Why did the number of corporations grow in the late 1800s?

b. Who is an example of a robber baron? Why?

c. Why was the South so much less industrial than the North?

4. Critical Thinking

Forming and Supporting Opinions Do you think that wealthy people have a duty to become philanthropists? Explain your opinion.

THINK ABOUT
- Carnegie and Rockefeller
- how most wealthy people gain their money
- the differences between the rich and the poor

ACTIVITY OPTIONS

MATH
SPEECH

In your local or school library, look up the business cycle. Create a **graph** of the cycle for the last century or prepare an **oral report** to Congress on the trends.

GEOGRAPHY in HISTORY

REGION AND HUMAN-ENVIRONMENT INTERACTION

Industry in the Midwest

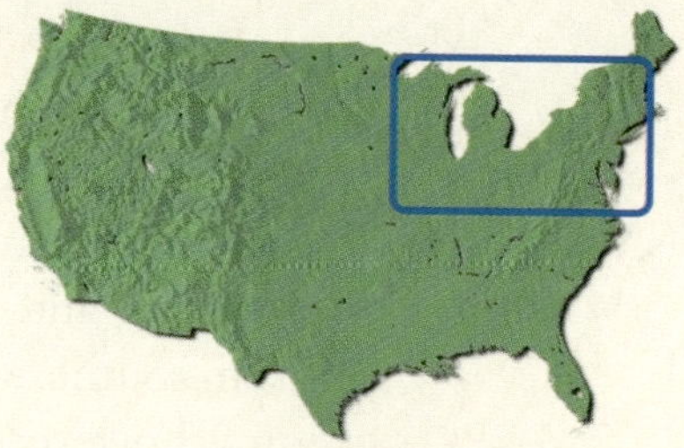

The Midwest is the region around the Great Lakes and the Upper Mississippi Valley. The region saw explosive growth during the 1800s. The first wave came after 1825, when the Erie Canal linked the East with the Great Lakes region. The second wave, caused by investments in products related to the Civil War (1861–1865), saw a boom in mining, farming, forestry, and meat-packing. By 1890, 29 percent of the country's manufacturing employment was in the Midwest, and the next big wave of growth was just beginning. New industries included steel and steel products, such as train rails and skyscraper beams.

Transportation and resources spurred the region's growth. Coal, oil, iron ore, limestone, and lumber were abundant, and the land was fertile. Trains, rivers, and lakes connected the Midwest to markets in the East and South and brought in raw materials from the West. The map on page 599 shows the resources of the lower Great Lakes and how transportation by rail and water joined regions.

The industries of the Midwest used raw materials that came both from their own region and from other regions of the country. For example, the cattle in this photograph of the Chicago stockyards came by rail from the ranches of the West. In contrast, the logs being floated down the river came from the pine forests of Michigan and Wisconsin.

ARTIFACT FILE

A Quick Dinner **Midwestern meat-packing companies advertised canned meats as a way to save time feeding a hungry family.**

Affordable Housing **People began to build with wooden siding over a frame of wooden two-by-fours. These homes were cheap and quick to construct.**

Iron ore from the Lake Superior region and coal from southern Illinois were used to manufacture steel.

CONNECT TO GEOGRAPHY

1. **Region** What advantages did the Midwest have that helped it become highly industrialized?
2. **Human-Environment Interaction** How did the development of railroads add to the region's advantages?

See Geography Handbook, pages 4–5.

CONNECT TO HISTORY

3. **Analyzing Causes** Chicago was a big meatpacking center. Why do you think that industry chose to locate there?

On-Line Field Trip

The Chicago Historical Society in Chicago, Illinois, contains photographs, documents, and artifacts such as this Western Electric typewriter, made in 1900. Typewriters enabled office workers to produce neat, clean documents quickly.

Visit www.mcdougallittell.com for more information.

4 Workers Organize

TERMS & NAMES
sweatshop
Knights of Labor
socialism
Haymarket affair
Pullman Strike
Eugene V. Debs
Samuel Gompers
American Federation of Labor (AFL)

MAIN IDEA

To increase their ability to bargain with management, workers formed labor unions.

WHY IT MATTERS NOW

Many of the modern benefits that workers take for granted were won by early unions.

ONE AMERICAN'S STORY

In 1867, Mary Harris Jones lost her husband and four children during a yellow fever epidemic in Memphis. For the rest of her life, she dressed in black as a sign of mourning. Moving from Memphis to Chicago, Jones started a dressmaking business. But the great Chicago fire of 1871 destroyed everything she owned. Instead of giving up in despair, Jones found a cause to fight for.

A VOICE FROM THE PAST

From the time of the Chicago fire I became more and more engrossed [interested] in the labor struggle and I decided to take an active part in the efforts of the working people to better the conditions under which they worked and lived.

Mary Harris Jones, *Autobiography of Mother Jones*

Jones became an effective labor leader who organized meetings, gave speeches, and helped strikers. Workers loved her so much that they called her Mother Jones. In this section, you will learn why workers went on strike in the late 1800s and the results of those strikes.

Mother Jones won the love of working people by fighting for their rights.

Workers Face Hardships

Business owners of the late 1800s wanted to keep their profits high, so they ran their factories as cheaply as possible. Some cut costs by requiring workers to buy their own tools or to bring coal to heat the factories. Others refused to buy safety equipment. For example, railroads would not buy air brakes or automatic train-car couplers. Because of this, 30,000 railroad workers were injured and 2,000 killed every year.

If a factory became too crowded, the owner rarely built a larger one. Instead, the owner sent part of the work to be done by smaller businesses that critics called sweatshops. **Sweatshops** were places where workers labored long hours under poor conditions for low wages. Often both children and adults worked there.

Factory and sweatshop workers did the same jobs, such as sewing collars or making buttonholes, all day long. They grew bored and did not experience the satisfaction that came from making an entire product themselves. Further, both factory and sweatshop owners kept wages low. In the 1880s, the average weekly wage was less than $10. This barely paid a family's expenses. If a worker missed work due to illness or had any unexpected bills, the family went into debt. Workers began to feel that only other working people could understand their troubles.

Child labor was common in the late 1800s, and as this boy's bare feet demonstrate, safety practices were rare.

A VOICE FROM THE PAST

They know what it is to bring up a family on ninety cents a day, to live on beans and corn meal week in and week out, to run in debt at the stores until you cannot get trusted [credit] any longer, to see the wife breaking down . . . , and the children growing sharp and fierce like wolves day after day because they don't get enough to eat.

A railroad worker, quoted in the *Philadelphia Inquirer,* July 23, 1877

Vocabulary
negotiate: to discuss something in order to reach an agreement

So discontented workers joined together to try to improve their lives. They formed labor unions—groups of workers that negotiated with business owners to obtain better wages and working conditions.

Early Unions

As you read in Chapter 14, the first labor unions began in the mid-1800s but were unable to win many improvements for workers. After the Civil War, some unions started to form national organizations. One of these was the **Knights of Labor.** This was a loose federation of workers from all different trades. Unlike many labor organizations, the Knights allowed women and, after 1878, African-American workers to join their union. They inspired many people to support their cause.

Then, beginning in 1873, the United States fell into a serious economic depression. Over the next four years, millions of workers took pay cuts, and about one-fifth lost their jobs. In July 1877, the Baltimore and Ohio (B & O) Railroad declared a wage cut of 10 percent. The day the pay cut was to go into effect, B & O workers in Martinsburg, West Virginia, refused to run the trains. No labor union had called the strike. The workers themselves had stopped working on their own.

"[Working people] know what it is to bring up a family on ninety cents a day."
A railroad worker, 1877

Reading History
A. Analyzing Causes Why do you think the president acted to stop the strike?

This work stoppage was the Railroad Strike of 1877. As the news spread, workers in many cities and in other industries joined in. This threw the country into turmoil. In several cities, state militias battled angry mobs. President Rutherford B. Hayes called out federal troops. Before the two-week strike ended, dozens of people were killed.

The strike did not prevent the railroad pay cut, but it showed how angry American workers had become. In 1884–1885, railroaders again went out on strike. This time they went on strike against the Union

Pacific and two other railroads. The strikers, who were members of the Knights of Labor, gained nationwide attention when they won their strike. Hundreds of thousands of new workers joined the union.

Union Setbacks

The growth of labor unions scared many business leaders. They blamed the labor movement on socialists and anarchists. Socialists believe in **socialism.** In that economic system, all members of a society are equal owners of all businesses—they share the work and the profits. Anarchists are far more extreme. They want to abolish all governments.

Business and government leaders feared that unions might spread such ideas, so they tried to break union power. In Chicago in 1886, the McCormick Harvester Company locked out striking union members and hired strikebreakers to replace them. On May 3, union members, strikebreakers, and police clashed. One union member was killed.

*Reading*History
B. Recognizing Effects Did the action of the bomber make it seem more or less likely that anarchists were behind union activity? Explain.

The next day, union leaders called a protest meeting at Haymarket Square. Held on a rainy evening, the rally was small. As police moved in to end the meeting, an unknown person threw a bomb. It killed 7 police and wounded about 60. The police then opened fire on the crowd, killing several people and wounding about a hundred. This conflict was called the **Haymarket affair.**

Afterward, the Chicago police arrested hundreds of union leaders, socialists, and anarchists. Opposition to unions increased. The membership in the Knights of Labor dropped rapidly—even though that wasn't the union that had called the meeting at Haymarket Square.

The Homestead and Pullman Strikes

Labor conflicts grew more bitter. In 1892, Andrew Carnegie reduced wages at his steel mills in Homestead, Pennsylvania, but the union refused to accept the cut. The company responded by locking out union workers from the mills and announcing that it would hire nonunion labor. The company also hired 300 armed guards. In response, the locked-out workers gathered weapons. The guards arrived on July 6, and a battle broke out that left ten people dead. The Pennsylvania state militia began to escort the nonunion workers to the mills. After four months, the strike collapsed, breaking the union.

One night during the Pullman Strike, some 600 freight cars were burned.

*Reading*History
C. Analyzing Causes Why was it so difficult for early unions to win against big business?

Workers lost another dispute in 1894. In that depression year, many railroad companies went bankrupt. To stay in business, the Pullman Palace Car Company, which made railroad cars, cut workers' pay 25 percent. But Pullman did not lower the rent it charged workers to live in company housing. After their rent was deducted from the lower pay, many Pullman workers took home almost nothing.

The Pullman workers began the **Pullman Strike,** a strike which spread throughout the rail industry in 1894. When the Pullman Company refused to negotiate, American Railway Union president **Eugene V. Debs** called on all U.S. railroad workers to refuse to handle Pullman cars. Rail traffic in much of the country came to a halt. President Grover Cleveland called out federal troops, which ended the strike. Debs was put in jail.

Now and then

MODERN BENEFITS WON BY UNIONS

Many people today begin work at 9:00 A.M. and end at 5:00 P.M. Contrast this 8-hour workday with the 10- to 12-hour day of most 19th-century workers. The 8-hour day was one benefit won by labor unions. Others include workers' compensation (insurance that pays for injuries received on the job), pensions, laws against child labor, two-day weekends, sick pay, and vacation pay. By winning these benefits, unions improved the work lives of many Americans.

Gompers Founds the AFL

Not all companies treated workers as harshly as Carnegie and Pullman did. For instance, in the 1880s, the soap company Procter & Gamble began to give its employees an extra half day off a week. It also began a profit-sharing plan, in which a company gives part of its profits to workers.

However, workers at most companies received low wages and few benefits. So in spite of the opposition to unions, the labor movement did not die. In 1886, labor leader **Samuel Gompers** helped found a new national organization of unions called the **American Federation of Labor (AFL).** Gompers served as AFL president for 37 years.

Reading **History**
D. Identifying Problems What problems did the AFL try to solve?

The AFL focused on improving working conditions. By using strikes, boycotts, and negotiation, the AFL won shorter working hours and better pay for workers. By 1904, it had about 1.7 million members.

In the next few decades, labor unions helped change the way all Americans worked. At the same time, city growth and immigration transformed America. You will read about that in Chapter 21.

Section 4 Assessment

1. Terms & Names

Identify:
- sweatshop
- Knights of Labor
- socialism
- Haymarket affair
- Pullman Strike
- Eugene V. Debs
- Samuel Gompers
- American Federation of Labor (AFL)

2. Taking Notes

Review this section and find five key events to place on a time line like the one below.

1870 — event — event — event — event — event — 1910

3. Main Ideas

a. What hardships did workers face in the late 1800s?

b. What happened to unions after the protest at Haymarket Square?

c. How did Carnegie's company break the union at the Homestead mills?

4. Critical Thinking

Drawing Conclusions In your opinion, was the government more supportive of unions or business in the late 1800s? Explain.

THINK ABOUT
- the Railroad Strike of 1877
- the Homestead Strike
- the Pullman Strike

ACTIVITY OPTIONS

LANGUAGE ARTS
ART

Decide whether unions should be encouraged. Write an **editorial** or create a **public message poster** expressing your opinion.

Chapter 20 ASSESSMENT

VISUAL SUMMARY

An Industrial Society

TERMS & NAMES

Briefly explain the importance of each of the following.

1. patent
2. business cycle
3. transcontinental railroad
4. standard time
5. corporation
6. John D. Rockefeller
7. Andrew Carnegie
8. Haymarket affair
9. Pullman Strike
10. American Federation of Labor (AFL)

REVIEW QUESTIONS

The Growth of Industry (pages 585–589)

1. How do inventors protect their rights to what they invent?
2. What did Thomas Edison and Alexander Graham Bell invent?

Railroads Transform the Nation (pages 590–593)

3. What geographic feature made building the Central Pacific difficult?
4. What took place when workers connected the Central Pacific and the Union Pacific?

The Rise of Big Business (pages 594–599)

5. What is a monopoly?
6. What are trusts, and why did some people think they were bad for the country?
7. Why did writers Mark Twain and Charles Warner name the late 1800s the Gilded Age?

Workers Organize (pages 600–603)

8. What ideas did business leaders fear that unions would spread?
9. How did the Pullman Strike begin and end?
10. Which unions were led by Eugene V. Debs and Samuel Gompers?

CRITICAL THINKING

1. USING YOUR NOTES

Using your completed chart, answer the questions below.

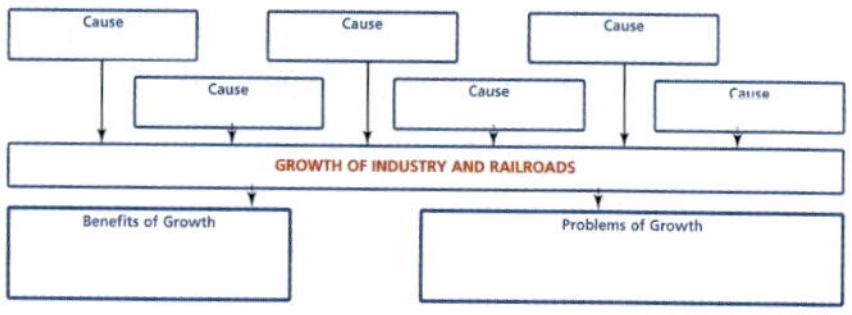

a. How did the growth of railroads act as a cause of industrial growth?

b. Who do you think benefited most from the growth of industry? Explain.

2. THEME: ECONOMICS IN HISTORY

Who is someone from this chapter that might view the United States as a land of economic opportunity? Explain your answer.

3. ANALYZING LEADERSHIP

What characteristics of a good leader did Mother Jones possess?

4. APPLYING CITIZENSHIP SKILLS

Were John D. Rockefeller and Andrew Carnegie good citizens? Support your answer with details from this chapter.

5. COMPARING

How were the problems of sharecroppers, described on page 597, similar to those of Pullman workers, described on pages 602–603?

6. DRAWING CONCLUSIONS

Why do you think unions were more successful at attracting members in the late 1800s than in the early 1800s?

Interact *with* History

Now that you have read the chapter, would you change your mind about joining the strike? Explain.

HISTORY SKILLS

1. INTERPRETING GRAPHS

U.S. Rails Produced, *1860–1909*

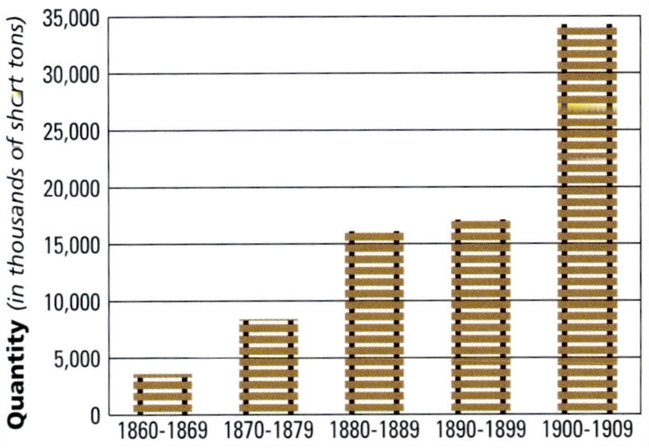

Source: *Historical Statistics of the United States*

Basic Graph Elements

a. What period of time does this graph cover?

b. What was the general trend in rail production?

Interpreting the Graph

c. Which decade saw the biggest change in rail production?

d. What does this graph indicate about what was happening in the railroad industry?

2. INTERPRETING PRIMARY SOURCES

This memoir was written by James Davis, who was secretary of labor during the 1920s. In 1886, when he was 12, he worked in a mill making wrought iron.

> In this mill there is a constant din by day and night. Patches of white heat glare from the opened furnace doors like the teeth of some great, dark dingy devil grinning across the smoky vapors of the Pit. Half-naked, soot-smeared fellows fight the furnace hearths with hooks, rabbles, and paddles. Their scowling faces are lit with fire like sailors manning their guns in a night fight when a blazing fire ship is bearing down upon them. The sweat runs down their backs and arms and glistens in the changing lights. Brilliant blues and rays of green and bronze come from the coruscating [sparkling] metal, molten yet crystallizing into white-hot frost within the furnace puddle.
>
> **James Davis,** *The Iron Puddler*

a. What were working conditions like in the mill where Davis worked?

b. Do you think Davis liked his work? Use details from the passage to support your answer.

ALTERNATIVE ASSESSMENT

1. INTERDISCIPLINARY ACTIVITY: Science

Diagramming a Process Do research to find out how iron is converted to steel in the Bessemer process. Make a diagram that shows the process. Use terms from science to describe what happens to different elements during the heating process. Share your diagram with the class.

2. COOPERATIVE LEARNING ACTIVITY

Performing a Monologue Working in a small group, research the life of a business or labor leader from this chapter. Write and perform a monologue in which the leader discusses his or her life. After the presentation, ask the class to discuss what they learned from your monologue. Assign group members one or more of the following tasks:

a. researching the person's life
b. writing the monologue
c. editing the monologue
d. finding costumes and props
e. creating sound effects (if necessary)
f. directing the rehearsals
g. acting out the monologue

3. TECHNOLOGY ACTIVITY

Creating a Database Using the Internet, reference materials such as *Historical Statistics of the United States,* or other printed material, compile statistics that could be used to create a database of information about industrialization in the United States. Consider researching the following topics:

- production figures for various goods
- number of strikes or work stoppages
- income of various industries
- value of the plants and machinery owned by various industries

Visit www.mcdougallittell.com to learn more about industrialization in the United States.

4. HISTORY PORTFOLIO

Option 1 Review your section and chapter assessment activities. Select one that you think is your best work. Then use comments made by your teacher or classmates to improve your work and add it to your portfolio.

Option 2 Review the questions that you wrote for What Do You Want to Know? on page 584. Then write a short report in which you explain the answers to your questions. If any questions were not answered, do research to answer them. Add your report to your portfolio.

CHAPTER 21

Changes in American Life

1880–1914

Crowds of people walk, work, and shop on Mulberry Street in New York's Lower East Side.

Interact *with* History

It is 1900, and you have decided to leave your native country. After a long and difficult voyage, you arrive in the United States. Now you need to find a new home and a job. You have to create a new life in a strange land.

What Do You Think?

- What caused you to leave your native country?
- What problems did you face on your voyage?
- What do you hope to find in the United States?

How will you make a home in your new country?

USA

1882 Congress passes the Chinese Exclusion Act.

1888 Electric trolleys are set up in Richmond, Virginia.

1889 The first electric elevator is installed.

Jane Addams founds Hull House.

1892 Immigration center opens on Ellis Island.

1893 Columbian Exposition opens in Chicago.

1896 Supreme Court establishes "separate-but-equal" doctrine in *Plessy* v. *Ferguson* case.

1906 Earthquake and fire devastate San Francisco.

1909 National Association for the Advancement of Colored People (NAACP) is founded.

World

1880 — **1914**

1889 Brazil becomes a republic.

1893 Karl Benz invents the modern automobile.

1896 First modern Olympic Games are held in Athens, Greece.

1910 Mexican Revolution begins.

Chapter 21 SETTING THE STAGE

BEFORE YOU READ

Previewing the Theme

Diversity and Unity Chapter 21 discusses how American society went through deep changes at the end of the 19th century. The economy became more industrial. Millions of immigrants came to the United States. Cities grew rapidly. These changes caused tensions in the nation. Immigrants and racial minorities faced discrimination.

What Do You Know?

What do you think about when you hear the term *immigration*? Why do people move to different countries? What kinds of challenges might immigrants face in their new country?

THINK ABOUT

- what you know about immigration from your own experience or the experience of your family
- what would make you want to move away from your home

What Do You Want to Know?

What questions do you have about American life around 1900? Write them in your notebook before you read the chapter.

READ AND TAKE NOTES

Reading Strategy: Categorizing To help you make sense of what you read, learn to categorize. Categorizing means sorting information into groups. The chart below will help you take notes and categorize the changes in American life that occurred during the late 19th and early 20th centuries.

See Skillbuilder Handbook, page R6.

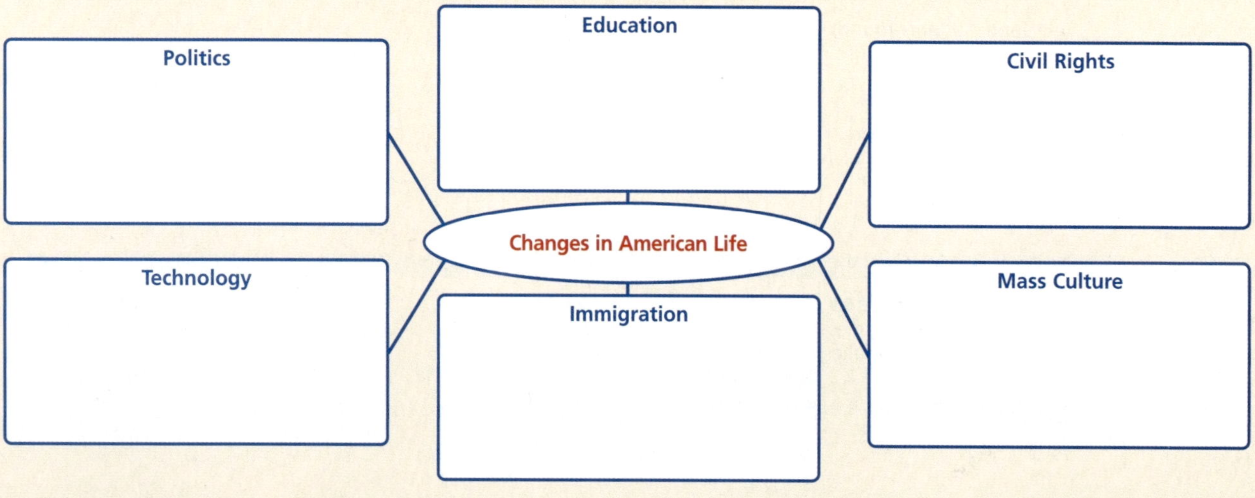

1 Cities Grow and Change

TERMS & NAMES
urbanization
tenement
slum
social gospel
Jane Addams
Hull House
political machine
Tammany Hall

MAIN IDEA

Industrialization and immigration caused American cities to grow rapidly.

WHY IT MATTERS NOW

Modern American city life first emerged during this period.

ONE AMERICAN'S STORY

Carl Jensen came to the United States from Denmark in 1906. Like most of the millions of immigrants who came to America around the turn of the century, he immediately began to look for work. He described the crush of people, including himself, who were searching for jobs in New York.

A VOICE FROM THE PAST

Along thirty miles of water front I wandered in search of work . . . waiting through rain and sleet and snow with gangs of longshoremen [dockworkers] to reach the boss before he finished picking the men he wanted. . . . Strong men crushed each other to the ground in their passion for work.

Carl Jensen, quoted in *A Sunday between Wars*

Shipyards in growing cities provided jobs for many Americans.

Jensen eventually found work in New York's garment, or clothing, district. At the turn of the century, the promise of work drew millions of people like Carl Jensen from around the world to American cities. In this section, you will read about the rapid growth of American cities.

Industrialization Expands Cities

The Industrial Revolution, which had been changing how people worked, also changed *where* people worked. Since colonial days, most Americans had lived and worked in rural areas. But in the late 1800s, that began to change as more and more people moved to cities to find jobs.

Industries were drawn to cities because cities offered good transportation and plentiful workers. Increasing numbers of factory jobs appeared in America's cities, followed by more workers to fill those jobs. The growth of cities that resulted from these changes is called **urbanization**.

Many of the people who moved to American cities were immigrants like Carl Jensen. People also migrated from America's farms to the cities. Once there, even workers with few skills could usually find steady work.

CONNECTIONS TO SCIENCE
Building a Skyscraper

Modern cities depend on skyscrapers to increase the space for people to live and work. Steel, electricity, and elevators make skyscrapers possible.

1 **STEEL FRAMES** Steel beams can carry much more weight than brick or stone walls. The strength of the steel allows architects to design extremely tall buildings.

2 **WINDOWS** In skyscrapers, the outer walls do not support the weight of the building; the steel beams do. As a result, many skyscrapers have outer walls made of glass to allow sunlight inside.

3 **ELEVATOR** Tall buildings would be useless if people could not reach all of the floors. Elevators powered by electricity make such tall buildings practical.

Technology Changes City Life

New technologies helped cities absorb the millions of people who flocked there. For example, new technologies made possible the construction of skyscrapers, buildings that looked tall enough to scrape the sky. Skyscrapers helped cities grow and made modern city life possible.

The elevator was a key invention for constructing tall buildings that could hold greater numbers of people. Before the 1860s, buildings rarely rose higher than four stories because it was hard for people to climb to the top. In 1889, the Otis Elevator Company installed the first electric elevator. Now buildings could be more than a few stories tall because people no longer had to walk up to the higher floors. As a result, buildings could hold more people.

The use of steel also helped to increase the height of buildings. In 1885, the Home Insurance Building in Chicago boasted an iron and steel skeleton that could hold the immense weight of the skyscraper's floors and walls. The building climbed to ten stories. Skyscrapers changed city skylines forever.

Reading **History**
A. Recognizing Effects How did industry and technology help cities grow?

The Streetcar City

As electricity helped change the way people traveled inside buildings, it also changed how people traveled around cities. Before industrialization, people walked or used horse-drawn vehicles to travel over land. But by 1900, electric streetcars in American cities were carrying more than 5 billion passengers a year. Streetcars and trains changed the walking city into the streetcar city.

Background
Streetcars are also called trolleys.

In 1888, Richmond, Virginia, became the first American city to have a transportation system powered by electricity. Other cities soon installed their own electric streetcars. The streetcars could quickly carry people to work and play all over the city. Some cities, such as Chicago, moved their electric streetcars above the street level, creating elevated, or "el," lines. Other cities, like New York, placed their city rail lines in underground tunnels, making subways.

The streetcar city spread outward from the city's center in ways the walking city never could. The ability to live farther away from work helped new suburbs to develop around cities. Some people in the suburbs wanted to become part of the city they bordered. That way they also could be served by the city's transportation system. Largely due to public transportation, cities expanded. For example, in 1889, Chicago annexed several suburbs and more than doubled its population as well as its area.

Vocabulary
annex: to add

Urban Disasters and Slums

The concentration of people in cities increased the danger of disasters because people and buildings were packed closely together. For example, in 1906, a powerful earthquake rocked San Francisco. The tremors caused large fires to tear through the city. The central business district was destroyed. About 700 people died, and nearly $400 million in property was damaged. But natural disasters were not the only source of danger for the people of the cities. Poverty and disease also threatened their lives.

As people flocked to cities, overcrowding became a serious problem. It was especially serious for families who could not afford to buy a house. Such families usually lived in rented apartments or tenements. A **tenement** is an apartment house that is usually run-down and overcrowded.

Old buildings, landlord neglect, poor design, and little government control led to dangerous conditions in many tenements. Poor families who could not afford to rent a place of their own often needed to move in with other families. This resulted in severely overcrowded tenements. Inadequate garbage pick-up also caused problems. Tenants sometimes dumped their garbage into the narrow air shafts between tenements. There was little fresh air, and the smell was awful.

Reading History
B. Summarizing What was it like to live in a turn-of-the-century tenement?

Many tenements had no running water. Residents had to collect water

HISTORY through ART

This photograph by Lewis Hine shows a family of Italian immigrants in their cramped, decaying tenement in New York City in 1912. Often photographers, such as Hine, had their subjects pose for their pictures to create the strongest effect.

What effect do you think Hine wanted this photograph to have?

from a faucet on the street. The water could be heated for bathing. But it was often unsafe for drinking. Sewage flowed in open gutters and threatened to spread disease among tenement dwellers.

A neighborhood with such overcrowded, dangerous housing was called a **slum.** The most famous example was New York City's Lower East Side. But every city had slums. After visiting Chicago's slums, the British writer Rudyard Kipling wrote in disgust, "Having seen it [Chicago], I urgently desire never to see it again."

Reformers Attack Urban Problems

Many Americans were also disgusted by poverty and slums. Some people fought to reform, or create changes, that would solve these problems. They were known as urban reformers.

The social gospel movement provided one basis for these beliefs. The **social gospel** movement aimed to improve the lives of the poor. Led by Protestant ministers, the ideas of the movement were based on Christian values. The most important concerns of the social gospel movement were labor reforms, such as abolishing child labor. Some reformers inspired by the movement opened settlement houses. They helped the poor and immigrants improve their lives. Settlement houses offered services such as daycare, education, and health care to needy people in slum neighborhoods.

Reading **History**
C. Making Inferences How did Christian values support the social gospel movement?

CITIZENSHIP TODAY

Community Service

Since the United States began, citizens have shared concerns about their communities. Many citizens, such as Jane Addams in 1889, have identified problems and proposed solutions to them.

In 1993, sixth-grader David Levitt asked his principal if the leftover food from the school cafeteria could be sent to a program to feed needy people. David was told that many restrictions prevented giving away the food.

Determined to get food to people who needed it, David talked to the school board, the state health department, and private companies to convince them to back his program. Today, more than 500,000 pounds of food from schools has been given to hungry people in the Seminole, Florida, area.

David Levitt carries supplies for his food pantry program.

How Do You Participate in Your Community?

1. In a small group, think about problems within your community. Make a list of those problems.
2. Choose one problem to work on.
3. Gather information about the problem. Keep a log of your sources to use again.
4. After you gather information, brainstorm solutions to the problem. Create a plan to carry out one solution.
5. Present the problem and your plan to the class.

See the Citizenship Handbook, page 286.

Visit www.mcdougallittell.com to learn more about community service.

Many settlement house founders were educated middle-class women. **Jane Addams** founded Chicago's **Hull House** in 1889 with Ellen Gates Starr. Hull House soon became a model for other settlement houses, including New York's Henry Street Settlement House, which Lillian D. Wald established in 1889.

Political Machines Run Cities

Political machines were another type of organization that addressed the problems of the city. A **political machine** is an organization that influences enough votes to control a local government.

Political machines gained support by trading favors for votes. For example, machine bosses gave jobs or food to supporters. In return, supporters worked and voted for the machine. Political machines also did many illegal things. They broke rules to win elections. They accepted bribes to affect government actions.

The most famous political machine was **Tammany Hall** in New York City. It was led by William Marcy Tweed. Along with his greedy friends, "Boss" Tweed stole enormous amounts of money from the city.

Despite such corruption, political machines did a number of good things for cities. They built parks, sewers, schools, roads, and orphanages in many cities. In addition, machine politicians often helped immigrants get settled in the United States by helping them find jobs or homes. Many immigrants gratefully supported the political machine after this kind of help. In the next section, you will learn more about immigration.

Reading **History**
D. Comparing and Contrasting How were settlement houses and political machines similar? How were they different?

AMERICA'S HISTORY MAKERS

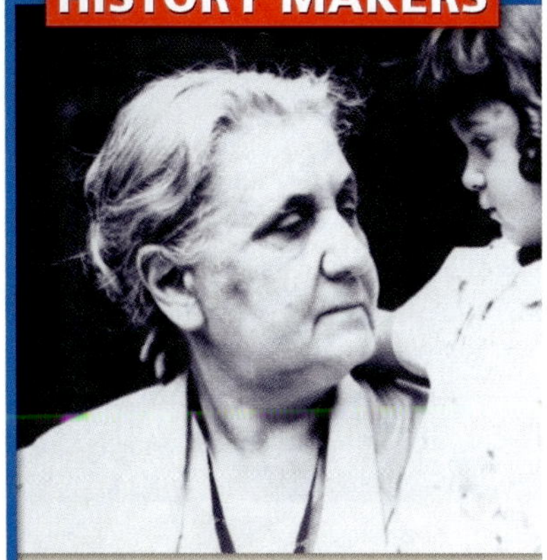

JANE ADDAMS
1860–1935

Jane Addams founded Hull House as an "effort to aid in the solution of the social and industrial problems which are [caused] by the modern conditions of life in a great city."

In addition to Hull House, Addams was active in many other areas. She fought for the passage of laws to protect women workers and outlaw child labor. She also worked to improve housing and public health. In 1931, she was awarded a share of the Nobel Peace Prize for her efforts.

Why did Jane Addams found Hull House?

Section 1 Assessment

1. Terms & Names

Identify:
- urbanization
- tenement
- slum
- social gospel
- Jane Addams
- Hull House
- political machine
- Tammany Hall

2. Taking Notes

Use a chart like the one below to show the causes and effects of urban growth.

Cause	Effect
Steel	
Elevators	
Streetcars	
Immigration	

3. Main Ideas

a. Why did immigrants and farmers settle in big cities at the end of the 19th century?

b. What are two inventions that made modern city life possible?

c. What urban problems did reformers try to solve?

4. Critical Thinking

Evaluating What were some of the advantages and disadvantages of machine politics?

THINK ABOUT
- the problems faced by immigrants and cities
- Tammany Hall and "Boss" Tweed

ACTIVITY OPTIONS

LANGUAGE ARTS
ART

It is 1900, and you have just moved to an American city. Write a **letter** to friends back home or draw a **picture** that describes your new home.

2 The New Immigrants

TERMS & NAMES
new immigrants
Ellis Island
Angel Island
melting pot
assimilation
Chinese Exclusion Act

MAIN IDEA

Millions of immigrants—mostly from southern and eastern Europe—moved to the United States.

WHY IT MATTERS NOW

The new immigrants had an important role in shaping American culture in the 20th century.

ONE AMERICAN'S STORY

In 1907, 10-year-old Edward Corsi left Italy to come to America. After two weeks at sea, he caught his first sight of the Statue of Liberty. He described the reaction of the people on the ship.

A VOICE FROM THE PAST

This symbol of America . . . inspired awe in the hopeful immigrants. Many older persons among us, burdened with a thousand memories of what they were leaving behind, had been openly weeping. . . . Now somehow steadied, I suppose, by the concreteness of the symbol of America's freedom, they dried their tears.

Edward Corsi, *In the Shadow of Liberty*

Every day, thousands of immigrants like Corsi streamed through Ellis Island, the nation's immigration station. In this section, you will learn about the immigrants who came to the United States around 1900 and how they affected the nation.

The Statue of Liberty and Ellis Island were two of the first things many immigrants saw of the United States.

The New Immigrants

Until the 1890s, most immigrants to the United States had come from northern and western Europe. But after 1900, fewer northern Europeans immigrated, and more southern and eastern Europeans did. This later group of immigrants came to be known as the **new immigrants.** Southern Italy sent large numbers of immigrants. Many Jews from eastern Europe and Slavic peoples, such as Poles and Russians, also immigrated.

Ellis Island was the first stop for most immigrants from Europe. There, they were processed before they could enter the United States. First, they had to pass a physical examination. Those with serious health problems or diseases were sent home. Next, they were asked a series of questions: Name? Occupation? How much money do you have?

Louis Adamic came to America from Slovenia, in southeastern Europe, in 1913. Adamic described the night he spent on Ellis Island. He and many other immigrants slept on bunk beds in a huge hall. Lacking a warm blanket, the young man "shivered, sleepless, all night, listening to snores" and dreams "in perhaps a dozen different languages."

U.S. Immigration, *1841–1900*

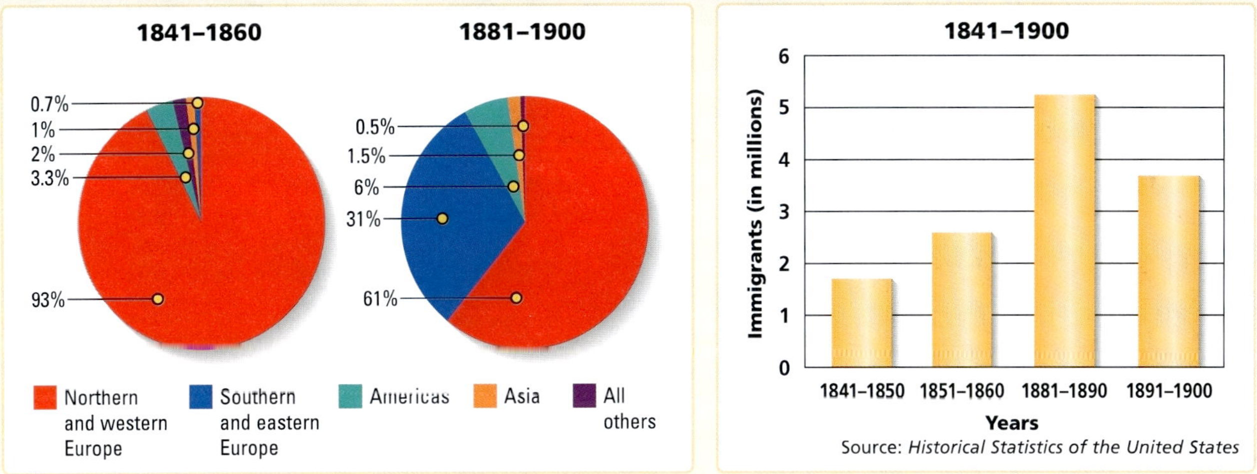

SKILLBUILDER Interpreting Graphs

1. *About how many immigrants came to the United States from 1841 to 1860?*
2. *About how many southern and eastern European immigrants came to the United States from 1881 to 1900?*

While European immigrants passed through Ellis Island on the East Coast, Asians landed at **Angel Island** in San Francisco Bay. In Angel Island's filthy buildings, most Chinese immigrants were held for several weeks. One unhappy prisoner carved in the wall, "For what reason must I sit in jail? It is only because my country is weak and my family poor."

Many Mexican immigrants entered the United States through Texas. Jesús [heh•SOOS] Garza recalled how simple his journey was. "I paid my $8, passed my examination, then changed my Mexican coins for American money and went to San Antonio, Texas."

Settling in America

Immigrants settled where they could find jobs. Many found work in American factories. The immigrants contributed to the growth of cities such as New York, Boston, Philadelphia, Pittsburgh, and Chicago. About half of the new immigrants settled in four industrial states: Massachusetts, New York, Pennsylvania, and Illinois.

Once in America, newer immigrants looked for people from the same village in the old country to help them find jobs and housing. People with similar ethnic backgrounds often moved to the same neighborhoods. As a result, ethnic neighborhoods with names like "Little Italy" and "Chinatown" became common in American cities.

The immigrants living in these communities pooled money to build places of worship for their neighborhoods. They published newspapers in their native languages. They commonly supported political machines, often led by politicians who had also come from their country of origin. Such politicians could speak the native language and help new arrivals feel comfortable. Most importantly, politicians could help immigrants find jobs.

Reading **History**
A. Identifying Problems How did immigrants show creativity in solving problems?

"I paid my $8, passed my examination, . . . and went to San Antonio."
Jesús Garza

Labor unions helped immigrants fit into American life. The various languages on the signs at this rally show the ethnic diversity in the labor movement.

Immigrants Take Tough Jobs

Immigrants took whatever jobs they could get. Many immigrants worked in Northern factories. As you read in Chapter 20, most factories offered low wages, long hours, and unsafe conditions. Many European immigrants who had settled in the East found jobs in sweatshops for about $10 a week. One observer of textile sweatshops noted, "The faces, hands, and arms to the elbows of everyone in the room are black with the color of the cloth on which they are working."

While European immigrants settled mostly in the East and Midwest, Asian immigrants settled mostly in the West. Many Chinese immigrants worked on the railroad. Others settled in Western cities where they set up businesses such as restaurants and stores. Large numbers of Japanese immigrants first came to Hawaii in 1885 to work on sugar plantations. Others settled on the mainland, where they fished, farmed, and worked in mines.

Immigrants from Mexico came to the Southwest. Mexican immigration increased after 1910 when revolution in that country forced people to flee. Growers and ranchers in California and Texas used the cheap labor Mexican immigrants offered. Owners of copper mines in Arizona hired Mexicans as well.

Reading **History**

B. Making Inferences
What did all the immigrants seem to have in common?

Becoming Americans

Some Americans have described the United States as a **melting pot,** or a place where cultures blend. The new immigrants blended into American society as earlier immigrants had. This process of blending into society is called **assimilation.** Most new immigrants were eager to assimilate. To do so, they studied English and how to be American citizens.

Many workers began to assimilate at work. Employers and labor unions both tried to "Americanize" immigrant workers by offering classes in citizenship and English. A Lithuanian worker explained that his labor union helped him learn to "read and speak and enjoy life like an American." He then became an interpreter for the union to help other Lithuanians become Americans.

At the same time the immigrants were learning about America, they were also *changing* America. Immigrants did not give up their cultures right away. Bits and pieces of immigrant languages, foods, and music worked their way into the rest of American culture.

Despite their efforts to assimilate, immigrants faced prejudice from native-born Americans. Many Protestants feared the arrival of Catholics and Jews. Other native-born Americans thought immigrants would not fit into democratic society because they would be controlled by political machines. Such prejudices led some native-born Americans to push for restrictions to reduce the numbers of new immigrants coming to America.

Now and then

LATE 20TH-CENTURY IMMIGRATION

Historians refer to the people who came to the United States around 1900 as the "new immigrants." But an even newer wave of immigrants has been coming to the United States since the 1980s.

From 1981 to 1996, nearly 13.5 million people immigrated to the United States. About 6.5 million came from other nations in the Western Hemisphere. In the same period, 4.8 million people came from Asia.

Restrictions on Immigration

Many native-born Americans also feared they would have to compete with immigrants for jobs. Immigrants were desperate for jobs and would often take work for lower wages in worse conditions than other Americans. Some Americans worried that there would not be enough jobs for everyone.

In 1882, Congress began to pass laws to restrict immigration. They placed taxes on new immigrants and banned specific groups, such as beggars and people with diseases. Nonwhites faced deeper prejudice than European immigrants, and Asians faced some of the worst. In 1882, Congress passed the **Chinese Exclusion Act**. It banned Chinese immigration for ten years.

Background
The Chinese Exclusion Act was renewed in 1892. In 1902, the ban was made permanent. It was not repealed until 1943.

The Chinese Exclusion Act was not the only example of prejudice in America around 1900. As you will read in the next section, racial discrimination was common throughout the United States.

Section 2 Assessment

1. Terms & Names

Identify:
- new immigrants
- Ellis Island
- Angel Island
- melting pot
- assimilation
- Chinese Exclusion Act

2. Taking Notes

Use a chart to take notes on immigrant experiences in the United States.

Voyage → Housing → Work → Assimilation

Which part of the immigrant experience was the most difficult?

3. Main Ideas

a. How were the new immigrants different from earlier immigrants?

b. How did immigrants support one another?

c. Why did nonwhite immigrants have a harder time assimilating than European immigrants did?

4. Critical Thinking

Making Generalizations
How well does the idea of the melting pot reflect U.S. immigration around 1900?

THINK ABOUT
- assimilation
- immigrant languages and cultures
- ethnic neighborhoods

ACTIVITY OPTIONS

MATH
GEOGRAPHY

Research immigration to your city or state. Create a **spreadsheet** of this information or draw a **map** showing immigration routes.

from

Dragonwings

by Laurence Yep

In 1903, eight-year-old Moon Shadow makes the long journey from China to join his father in San Francisco. After living for a time in Chinatown, the two move into a white neighborhood where Moon Shadow's father takes a job as a handyman working for Miss Whitlaw, who runs a boarding house for elderly people. Moon Shadow makes friends with Miss Whitlaw's niece Robin. In April of 1906, their world is turned upside down. As Moon Shadow goes outside to fetch water from the pump, an earthquake hits San Francisco, endangering rich and poor, young and old, American and immigrant alike.

The morning was filled with that soft, gentle twilight of spring, when everything is filled with soft, dreamy colors and shapes; so when the earthquake hit, I did not believe it at first. It seemed like a nightmare where everything you take to be the rock-hard, solid basis for reality becomes unreal.

Wood and stone and brick and the very earth became fluidlike. The pail beneath the pump jumped and rattled like a spider dancing on a hot stove. The ground deliberately seemed to slide right out from under me. I landed on my back hard enough to drive the wind from my lungs. The whole world had become unglued. Our stable and Miss Whitlaw's house and the tenements to either side heaved and bobbed up and down, riding the ground like ships on a heavy sea. Down the alley mouth, I could see the cobblestone street **undulate**[1] and twist like a red-backed snake.

From inside our stable, I could hear the cups and plates begin to rattle on their shelves, and the equipment on Father's work table clattered and rumbled **ominously.**[2]

Suddenly the door banged open and Father stumbled out with his clothes all in a bundle. "It's an earthquake, I think," he shouted. He had washed his hair the night before and had not had time to twist it into a **queue,**[3] so it hung down his back long and black.

He looked around in the back yard. It was such a wide, open space that we were fairly safe there. Certainly more safe than in the frame doorway of our stable. He got into his pants and shirt and then his socks and boots.

"Do you think one of the mean dragons is doing all this?" I asked him.

"Maybe. Maybe not." Father had sat down to stuff his feet into his boots. "Time to wonder about that later. Now you wait here."

He started to get to his feet when the second tremor shook and he fell forward flat on his face. I heard the city bells ringing. They were rung by no human hand—the earthquake had just shaken them in their steeples. The second tremor was worse than the first. From all over came an immense wall of noise: of metal tearing, of bricks crashing, of wood breaking free from wood nails, and all. Everywhere, what man had built came undone. I was looking at a tenement house to our right and it just seemed to shudder and then collapse. One moment there were solid wooden walls and the next moment it had fallen with the cracking of wood and the tinkling of glass and the screams of people inside.

1. **undulate:** to move like a wave.
2. **ominously:** threateningly.
3. **queue:** a long braid of hair hanging down the back.

Mercifully, for a moment, it was lost to view in the cloud of dust that rose up. The debris surged against Miss Whitlaw's fence and toppled it over with a creak and a groan and a crash. I saw an arm sticking up from the mound of rubble and the hand was twisted at an impossible angle from the wrist. Coughing, Father pulled at my arm. "Stay here now," he ordered and started for Miss Whitlaw's.

I turned. Her house was still standing, but the tenement house to the left had partially collapsed; the wall on our side and part of the front and back had just fallen down, revealing the apartments within: the laundry hanging from lines, the old brass beds, and a few lucky if astonished people just looking out dazedly on what had once been walls. I could see Jack sitting up in bed with his two brothers. His mother and father were standing by the bed holding on to Maisie. Their whole family crowded into a tiny two-room apartment. Then they were gone, disappearing in a cloud of dust and debris as the walls and floor collapsed. Father held me as I cried.

Miss Whitlaw came out onto her porch in her nightdress and a shawl. She pulled the shawl tighter about her shoulders. *"Are you all right?"*

"Yes," Father said, patting me on the back. "Aren't we, Moon Shadow?"

"Yes." I wiped my eyes on my sleeves.

"Is everyone okay inside?" Father asked Miss Whitlaw.

She nodded. We joined her on the porch and walked with her into her house. Robin was sitting on the stairs that led up to the second floor. She huddled up, looking no longer like the noisy, **boisterous**[4] girl I knew. The front door was open before her. She must have gone outside to look. *"Just about the whole street's gone."*

From up the stairs we could hear the **querulous**[5] old voices of the boarders demanding to know what had happened. Miss Whitlaw shouted up the stairs, "Everything's all right."

"Are you sure?" Father asked quietly.

Miss Whitlaw laughed. *"From top to bottom. Papa always built well. He said he wanted a house that could hold a herd of thundering elephants—that was what he always called Mama's folks. He never liked them much."*

"It's gone," Robin repeated. *"Just about the whole street's gone."*

"Oh, really now." Miss Whitlaw walked past Robin. We followed her out the front door to the front porch. Robin was right.

---

4. **boisterous:** loud, noisy.

5. **querulous:** complaining.

San Franciscans watch the destruction caused by the 1906 earthquake and the resulting fire.

CONNECT TO HISTORY

1. **Finding Main Ideas** How does the earthquake affect the neighborhood that Moon Shadow lives in?

See Skillbuilder Handbook, page R5.

CONNECT TO TODAY

2. **Researching** What happened in San Francisco after the quake? Did it affect the immigrants differently than others?

Visit www.mcdougallittell.com to learn more about the San Francisco earthquake.

3 Segregation and Discrimination

MAIN IDEA

Racial discrimination ran through American society in the late 19th and early 20th centuries.

WHY IT MATTERS NOW

Modern American society continues to face the problems caused by racism and discrimination.

TERMS & NAMES

racial discrimination
Jim Crow
segregation
Plessy v. *Ferguson*
Booker T. Washington
W. E. B. Du Bois
NAACP
Ida B. Wells

ONE AMERICAN'S STORY

African-American sisters Bessie and Sadie Delany grew up in Raleigh, North Carolina, at the turn of the century. Almost 100 years later, they still remembered their first taste of racial discrimination, different treatment on the basis of race.

A VOICE FROM THE PAST

We were about five and seven years old at the time. Mama and Papa used to take us to Pullen Park in Raleigh for picnics, and that particular day, the trolley driver told us to go to the back. We children objected loudly, because we always liked to sit in the front, where the breeze would blow your hair. That had been part of the fun for us. But Mama and Papa just gently told us to hush and took us to the back without making a fuss.

Sarah L. Delany and A. Elizabeth Delany, *Having Our Say*

Bessie (left) and Sadie Delany

Millions of Americans were familiar with the racial discrimination described by the Delanys. As you will read in this section, it was common throughout the United States.

Racism Causes Discrimination

As you read in earlier chapters, racist attitudes had been developing in America since the introduction of slavery. The low social rank held by slaves led many whites to believe that whites were superior to blacks. Whites held similar attitudes towards Asians, Native Americans, and Latin Americans. Even most scientists of the day believed that whites were superior to nonwhites. However, no scientists believe this today.

Such attitudes led whites to discriminate against nonwhites across the country. The most obvious example of racial discrimination was in the South. Southern blacks had their first taste of political power during Reconstruction. (See Chapter 18.) But when Reconstruction ended in 1877, Southern states began to restrict African Americans' rights.

Segregation Expands in the South

One way for whites to weaken African-American political power was to restrict their voting rights. For example, Southern states passed laws that set up literacy, or reading, tests and poll taxes to prevent African Americans from voting. White officials made sure that blacks failed literacy tests by giving unfair exams. For example, white officials sometimes gave blacks tests written in Latin. Poll taxes kept many blacks from voting because they didn't have enough cash to pay the tax.

Reading **History**

A. Recognizing Effects What was the purpose behind literacy tests, poll taxes, and grandfather clauses?

Such laws threatened to prevent poor whites from voting, too. To keep them from losing the vote, several Southern states added grandfather clauses to their constitutions. Grandfather clauses stated that a man could vote if he or an ancestor, such as a grandfather, had been eligible to vote before 1867. Before that date, most African Americans, free or enslaved, did not have the right to vote. Whites could use the grandfather clause to protect their voting rights. Blacks could not.

In addition to voting restrictions, African Americans faced Jim Crow laws. **Jim Crow** laws were meant to enforce **segregation,** or separation, of white and black people in public places. As a result, separate schools, trolley seats, and restrooms were common throughout the South.

Plessy v. Ferguson

African Americans resisted segregation, but they had little power to stop it. In 1892, Homer Plessy, an African American, sued a railroad company, arguing that segregated seating violated his Fourteenth Amendment right to "equal protection of the laws."

In 1896, the case of ***Plessy v. Ferguson*** reached the Supreme Court. The Court ruled against Plessy. It argued that "separate but equal" facilities did not violate the Fourteenth Amendment. This decision allowed Southern states to maintain segregated institutions.

Reading **History**

B. Identifying Problems Why was a policy of "separate but equal" unfair?

But the separate facilities were not equal. White-controlled governments and companies allowed the facilities for African Americans to decay. African Americans would have to organize to fight for equality.

Segregation forced African Americans to use separate entrances from whites and to attend separate, usually inferior, schools like the one shown below.

African Americans Organize

Booker T. Washington was an early leader in the effort to achieve equality. He had been born into slavery. But after the Civil War, he became a teacher. In 1881, he founded the Tuskegee Institute in Alabama to help African Americans learn trades and gain economic strength. Washington hired talented teachers and scholars, such as George Washington Carver.

Background
Carver made important discoveries to improve farming.

To gain white support for Tuskegee, Washington did not openly challenge segregation. As he said in an 1895 speech in Atlanta, in "purely social matters" whites and blacks "can be as separate as the fingers, yet one as the hand in all things essential to mutual progress."

However, some blacks disagreed with Washington's views. **W. E. B. Du Bois** (doo•BOYS) encouraged African Americans to reject segregation.

A VOICE FROM THE PAST

Is it possible . . . that nine millions of men can make effective progress in economic lines if they are deprived of political rights? . . . If history and reason give any distinct answer to these questions, it is an emphatic *No.*

W. E. B. Du Bois, *The Souls of Black Folk*

Reading History
C. Making Inferences In what way did Washington and Du Bois disagree about how to achieve African-American progress?

In 1909, Du Bois and other reformers founded the National Association for the Advancement of Colored People, or the **NAACP**. The NAACP played a major role in ending segregation in the 20th century.

AMERICA'S HISTORY MAKERS

W. E. B. DU BOIS
1868–1963

W. E. B. Du Bois grew up in a middle-class home. He went to college and earned his doctorate at Harvard. Du Bois became one of the most distinguished scholars of the 20th century.

Du Bois fought against segregation. He believed that the best way to end it would be to have educated African Americans lead the fight. He referred to this group of educated African Americans as the "Talented Tenth"—the most educated 10 percent of African Americans.

Why do you think Du Bois believed the Talented Tenth should lead the fight against segregation?

Violence in the South and North

Besides discrimination, African Americans in the South also faced violence. The Ku Klux Klan, which first appeared during Reconstruction, used violence to keep blacks from challenging segregation. More than 2,500 African Americans were lynched between 1885 and 1900.

Ida B. Wells, an African-American journalist from Memphis, led the fight against lynching. After three of her friends were lynched in 1892, she mounted an antilynching campaign in her newspaper. When whites called for Wells herself to be lynched, she moved to Chicago. But she continued her work against lynching. (See Interactive Primary Sources, page 624.)

Like Wells, many blacks moved north to escape discrimination. Public facilities there were not segregated by law. But Northern whites still discriminated against blacks. Blacks could not get housing in white neighborhoods and usually were denied good jobs. Anti-black feelings among whites sometimes led to violence. In 1908, whites in Springfield, Illinois, attacked blacks who had moved there. The whites lynched two blacks within a half mile of Abraham Lincoln's home.

Racism in the West

Chinese immigrants who came to the West in the 1800s also faced severe discrimination. Chinese laborers received lower wages than whites for the same work. Sometimes, Chinese workers faced violence. In 1885, white workers in Rock Springs, Wyoming, refused to work in the same mine as Chinese workers. The whites stormed through the Chinese part of town, shooting Chinese people and burning buildings. During the attack, 28 Chinese people were killed and 15 were wounded.

The Workingmen's Party of California produced this anti-Chinese poster during the 1880s.

At the same time, Mexicans and African Americans who came to the American Southwest were forced into peonage (PEE•uh•nihj). In this system of labor, people are forced to work until they have paid off debts. Congress outlawed peonage in 1867, but some workers were still forced to work to repay debts. In 1911, the U.S. Supreme Court declared such labor to be the same as peonage. As a result, the Court struck down such forms of labor as a violation of the Thirteenth Amendment.

Background
The Thirteenth Amendment banned "involuntary servitude"—another term for slavery.

Despite the problems caused by racism, many Americans had new opportunities to enjoy their lives at the turn of the century. In the next section, you will learn about changes in people's daily lives.

Section 3 Assessment

1. Terms & Names

Identify:
- racial discrimination
- Jim Crow
- segregation
- *Plessy* v. *Ferguson*
- Booker T. Washington
- W. E. B. Du Bois
- NAACP
- Ida B. Wells

2. Taking Notes

Use a chart to identify people and events related to racial discrimination at the turn of the century.

People	
Events	

Which person do you think did the most to end racial discrimination?

3. Main Ideas

a. What were Jim Crow laws?

b. How did discrimination against African Americans in the North differ from discrimination in the South?

c. What did Chinese immigrants and Mexican immigrants have in common?

4. Critical Thinking

Solving Problems What could have been done to end racial discrimination against nonwhites in the United States at the turn of the century?

THINK ABOUT
- attitudes of whites about nonwhites
- the efforts of nonwhites to find jobs and security
- competition for jobs

ACTIVITY OPTIONS

LANGUAGE ARTS
TECHNOLOGY

Research a civil rights leader from the turn of the century. Write a short **biography** of that person or design a **Web site** devoted to the work of that person.

from *Crusade for Justice*

Setting the Stage Ida B. Wells was the editor of the *Free Speech and Headlight,* a small Baptist newspaper in Memphis, Tennessee. She used the paper to attack the evils of Jim Crow, especially lynching. In her autobiography, *Crusade for Justice,* she described the events that led to the lynching of three of her friends. **See Primary Source Explorer**

A CLOSER LOOK

ECONOMIC COMPETITION

Moss, McDowell, and Stewart were African Americans who opened a grocery store near a white-owned store in a black neighborhood.

1. Why might the opening of the black-owned grocery store lead to problems?

A CLOSER LOOK

LYNCHINGS

There was a sharp increase in the number of lynchings in the United States in the 1890s. From 1891 to 1900, more than 1,100 African Americans were lynched.

2. Why do you think the number of lynchings increased in this period?

A CLOSER LOOK

THE GREAT MIGRATION

Between 1890 and 1920, hundreds of thousands of African Americans left the South to escape racism. This movement is called the Great Migration.

3. Why does Wells's newspaper advise African Americans to move away in the wake of the lynching?

While I was thus carrying on the work of my newspaper . . . there came the lynching in Memphis which changed the whole course of my life. . . .

Thomas Moss, Calvin McDowell, and Henry Stewart owned and operated a grocery store. . . . There was already a grocery owned and operated by a white man who **hitherto**[1] had had a **monopoly**[2] on the trade of this thickly populated colored suburb. Thomas's grocery changed all that, and he and his **associates**[3] were made to feel that they were not welcome by the white grocer. . . .

About ten o'clock that [one Saturday] night, . . . shots rang out in the back room of the store. The men stationed there had seen several white men stealing through the rear door and fired on them without a moment's pause. Three of these men were wounded, and others fled and gave the alarm.

Sunday morning's paper came out with **lurid**[4] headlines telling how officers of the law had been wounded while in the **discharge**[5] of their duties. . . . The same newspaper told of the arrest and jailing of the **proprietor**[6] of the store and many of the colored people. . . .

On Tuesday following, . . . a body of picked [white] men was admitted to the jail. . . . This mob took out of their cells Thomas Moss, Calvin McDowell, and Henry Stewart. . . . They were loaded on a switch engine of the railroad which ran back of the jail, carried a mile north of the city limits, and horribly shot to death.

Although stunned by the events of that hectic week, the *Free Speech* [Wells's newspaper] felt that it must carry on. Its [lead article] for that week said:

> The city of Memphis has demonstrated that neither character nor standing **avails**[7] the Negro if he dares to protect himself against the white man or become his rival. There is nothing we can do about the lynching now, as we are out-numbered and without arms. The white mob could help itself to ammunition without pay, but the order was rigidly enforced against the selling of guns to Negroes. There is therefore only one thing left that we can do; save our money and leave a town which will neither protect our lives and property, nor give us a fair trial in the courts, but takes us out and murders us in cold blood when accused by white persons.

1. **hitherto:** until this time.
2. **monopoly:** exclusive control by one person or group.
3. **associates:** friends or partners.
4. **lurid:** causing shock or horror.
5. **discharge:** performance of duty.
6. **proprietor:** owner.
7. **avails:** helps.

Like Country Pretty Much

Setting the Stage Kee Low was a Chinese immigrant. He had come to the United States in 1876. He was interviewed in 1924 as part of a project by scholars to create a "Survey of Race Relations." This is an excerpt from that interview. In it, Kee Low tells his story. Despite the racism, he still "like country pretty much." **See Primary Source Explorer** CD-ROM

I arrived in San Francisco in 1876, 49 years ago. Come to San Francisco when country one hundred years old. People treat Chinese rotten then. Don't blame people much at that time. Chinese and European not educated as much then as today. More civilized today. People drive Chinese out of country. . . .

I was living on the waterfront, and they told me to get out one day. Sunday morning, they come together and drive Chinese out. . . . They want to get us out to San Francisco, to go on steamer, and we stayed on the **wharf**[1] all night, and they bring us little black coffee and little bread in morning. We pretty hungry. The last day, some of the citizens, Judge Greene, Judge Hanford, United States Attorney, nice fellow want to help us. . . . Judge Greene told the Chinese that those who wanted to stay and make good citizens could stay, and those who wanted to go could go. One half wanted to go, and one half wanted to stay. . . .

There were so many around the streets that they had to have somebody to protect these people. Some of the **hobos**[2] tried to make them go back to the wharf, but volunteers tried to keep these fellows away. They **commenced**[3] shooting and kill one of them. So Chinese people get excited when gun begin to sound, so they throw shoes, blankets and everything and run. I was uptown myself. I didn't intend to go. I ran outside to see what happened because I was so excited. . . . Call up one or two friends of mine and tell them get killed, and we better get out of the way. We run out in woods. Build fire. Pretty cold. I told friends, we got to protect ourselves. We got to get out of here.

1. **wharf:** landing place for ships.
2. **hobos:** homeless people.
3. **commenced:** began.

A CLOSER LOOK

RACIST ATTITUDES

Some people believe that racism is caused by ignorance.

4. Why does Kee Low believe that discrimination against the Chinese was worse in the 1870s than in the 1920s?

A CLOSER LOOK

REASONS TO STAY

Despite the violence that they faced for having Asian ancestry, half of the Chinese with Kee Low wanted to stay in the United States.

5. Why do you think Asian-Americans stayed in the United States despite discrimination?

Interactive Primary Sources Assessment

1. Main Ideas

a. What do the accounts of Wells and Low have in common?

b. How did the officers of the law behave differently in the report by Low than in the one by Wells?

c. What conclusions do Wells and Low come to about how someone should respond to discrimination?

2. Critical Thinking

Forming and Supporting Opinions Do you think Wells and Low were right to flee racism? Why?

THINK ABOUT

- the causes of racism
- the threat of violence to Wells and Low

4

Society and Mass Culture

MAIN IDEA

Industrialization and new technologies created a mass culture in the United States.

WHY IT MATTERS NOW

Modern American mass culture had its beginnings during this period.

TERMS & NAMES

- mass culture
- Joseph Pulitzer
- William Randolph Hearst
- department store
- mail-order catalog
- leisure
- vaudeville
- ragtime

ONE AMERICAN'S STORY

Mary Ellen Chase dreaded her first day of teaching at a new school. Many of the boys were larger than she and were known to be troublemakers. But she would do her best to control the class.

A VOICE FROM THE PAST

I stormed up and down. . . . This pathetic pretense of courage, aided by the mad flourishing of my razor strop, brought forth . . . the expression of respectful fear on the faces of the young giants.

Mary Ellen Chase, quoted in *The Good Old Days—They Were Terrible!*

Students work on their lessons in this New York City classroom in 1906.

Discipline was a key goal of education in the 1800s. Education also helped create a common culture for the millions of Americans who went to school. In this section, you will learn how education helped create an American **mass culture**—a common culture experienced by large numbers of people.

Education and Publishing Grow

Immigration caused enormous growth in American schools. To teach citizenship and English to immigrants, new city and state laws required children to attend school. Between 1880 and 1920, the number of children attending school more than doubled. To serve the growing number of students, the number of public high schools increased from 2,526 in 1890 to 14,326 in 1920.

The growth of education increased American literacy. Reading became more popular. Americans read large numbers of novels. Dime novels were especially popular. They sold for ten cents each and told exciting tales of romance and adventure, often set in the West or on the high seas.

Americans also read more newspapers. Tough competition pushed newspaper publishers to try all sorts of gimmicks to outsell their rivals. For example, **Joseph Pulitzer,** owner of the *New York World,* and

Background
By 1898, each of these newspapers sold more than a million copies a day.

William Randolph Hearst, owner of the *New York Morning Journal,* were fierce competitors. They filled the pages of their papers with spectacular stories. They also added special features, such as comics and sports.

Modern Advertising and New Products

Newspapers had a wide influence on American life, including the rise of modern advertising. Advertisers used images of celebrities in newspapers and magazines to tempt people to buy products. They advertised everything from cereal to jewelry to soap. Some ads played on people's fears. For example, advertisers might scare a young woman concerned about her appearance into buying a particular brand of face cream. Advertising was effective in turning brand names into household words.

Advertisements also helped people learn about new products. At the turn of the century, new inventions, such as the electric washing machine, promised to help people do their household chores more easily. Because women did most of these chores as well as most of the shopping, manufacturers marketed these new devices to women.

One of the places people could buy these—and many other—goods was in department stores. **Department stores** sold everything from clothing to furniture to hardware. The Chicago businessman Marshall Field discovered as a sales clerk that he could increase his sales by paying close attention to each woman customer. Field opened his own department store in downtown Chicago with the motto, "Give the lady what she wants."

People who did not live near a department store could order goods through the mail. Companies like Montgomery Ward and Sears Roebuck sent catalogs to customers. These **mail-order catalogs** included pictures and descriptions of merchandise. People could place their orders by mail, and the company would deliver the product. Richard Sears claimed that he sold 10,000 items a minute.

Reading **History**
A. Summarizing What developments changed American methods of selling at the turn of the century?

In 1896, the post office made it easier for people to receive goods through the mail by establishing a new delivery system. Rural free delivery brought packages directly to homes in rural areas. Now people in these areas could get the same goods as people in the cities.

STRANGE *but* True

BICYCLES TO AIRPLANES

At the turn of the century, two bicycle mechanics invented a machine that would help advertisers and businessmen reach new customers. In 1892, Orville and Wilbur Wright opened a bicycle shop in Ohio. They used the profits to fund experiments in aeronautics, the construction of aircraft.

In 1903, the Wright brothers took a gasoline-powered airplane that they had designed to a sandy hill outside Kitty Hawk, North Carolina. On December 17 of that year, Orville made the first successful flight of a powered aircraft in history. By 1918, the U.S. Postal Service began airmail service that made it faster and easier for people to get goods.

Urban Parks and World's Fairs

Advertising and shopping were not the only daily activities changing at this time. **Leisure,** or free time, activities also changed. In cities, new parks provided people with entertainment. The increasing number of people working in factories and offices liked going to parks to get some sunshine and fresh air. Parks helped bring grass and trees back into city landscapes.

Leisure Activities

▼ **Coney Island**
Visitors to New York's Coney Island cool off in the Steeple-chase Pool.

▼ **World's Fair**
Visitors to the 1893 world's fair in Chicago saw exotic sights, such as elephants.

▼ **Football**
Excited fans watch the 1881 Harvard–Yale football game at the Polo Grounds in New York.

Central Park in New York City is the nation's best-known urban park. Opened in 1876, Central Park looked like the country. Trees and shrubs dotted its gently rolling landscape. Winding walkways let city dwellers imagine they were strolling in the woods. People could also ride bicycles and play sports in the park.

In addition to urban parks, amusement parks provided a place people could go for fun. The most famous amusement park was Coney Island in New York City. Completed in 1904, Coney Island had shops, food vendors, and exciting rides like roller coasters. One immigrant woman said Coney Island "is just like what I see when I dream of heaven!"

World's fairs provided another wildly popular form of entertainment for Americans. Between 1876 and 1916, several U.S. cities, including Philadelphia, Chicago, St. Louis, and San Francisco, hosted world's fairs. The fairs were designed to show off American technology. The 1876 fair in Philadelphia displayed Alexander Graham Bell's newly invented telephone. Millions of people attended these fairs. Nearly 10 million attended the Philadelphia fair alone. Visitors were drawn to foods, shows, and amusements. The historian Thomas Schlereth described the giant wheel built by George Ferris at the 1893 Chicago fair.

Reading **History**
B. Comparing and Contrasting What did urban parks and world's fairs have in common?

A VOICE FROM THE PAST

Chicago's answer to Paris's 1889 Eiffel Tower, Ferris's 264-foot bicycle wheel in the sky dominated the landscape. With thirty-six cars, each larger than a Pullman coach and capable of holding 60 people, the wheel, when fully loaded, rotated 2,160 people in the air.

Thomas Schlereth, *Victorian America*

Spectator Sports

During this time, spectator sports also became popular entertainment. Baseball, football, boxing, and many other sports drew thousands of people to fields and gyms around the country.

Baseball was the most popular sport. Summer games drew crowds of enthusiastic fans. By the 1890s, baseball had standardized rules and a published schedule of games. Racial discrimination kept African-American baseball players out of baseball's American and National Leagues. In order to compete, African Americans formed their own teams in the Negro American League and the Negro National League. (See Geography in History, pages 722–723.)

Going to the Show

In addition to sports, other forms of live entertainment attracted large audiences. **Vaudeville,** for example, featured a mixture of song, dance, and comedy. A show would have a series of acts leading up to an exciting end, which advertisers billed as the "wow finish."

New types of music also began to be heard. **Ragtime,** a blend of African-American songs and European musical forms, was an important new musical form. African-American composer Scott Joplin heard ragtime while he traveled through black communities from New Orleans to Chicago. Joplin's "Maple Leaf Rag," published in 1899, became a hit in the first decade of the 20th century.

Early in the 20th century, movies began to compete with live entertainment. The first movies were silent and were added as the final feature of a vaudeville show. Soon storefront theaters appeared that showed only movies. After 1905, these movie theaters were called nickelodeons because they charged just a nickel for admission.

Movies, music, sports, and advertising contributed to shaping modern American mass culture. People across the nation experienced many of these things. In the next chapter, you will learn about different nationwide changes—the reform movements of the Progressive era.

Reading **History**

C. Making Inferences How do you think movies contributed to mass culture?

America's HERITAGE

RAGTIME

Tired of slow waltz music, young people eagerly embraced ragtime at the turn of the century. The name probably came from a description of the rhythm of black dance music as "ragged time." Ragtime's exciting beat inspired the names of such songs as "Irresistible Fox Trot Rag," "That Fascinating Rag," and "That Nifty Rag."

Ragtime had an enormous influence on American music. Throughout the 20th century, American musical styles such as jazz, blues, rock-and-roll, rap, and rhythm-and-blues built on the style of ragtime.

Scott Joplin

Section 4 Assessment

1. Terms & Names

Identify:

- mass culture
- Joseph Pulitzer
- William Randolph Hearst
- department store
- mail-order catalog
- leisure
- vaudeville
- ragtime

2. Taking Notes

Use a diagram like the one below to note the changes that created a mass culture at the turn of the century.

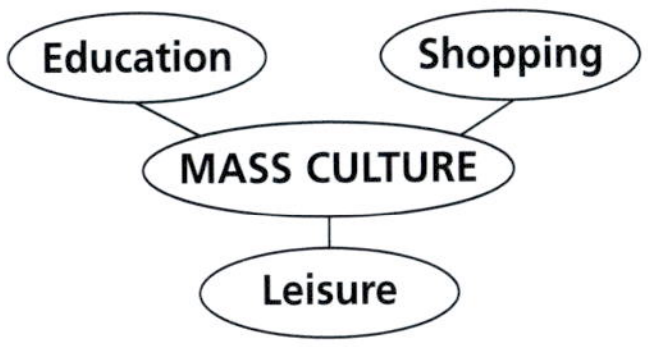

3. Main Ideas

a. What did dime novels and newspapers have in common?

b. How did mail-order catalogs affect the way people shopped?

c. What did visitors see at world's fairs?

4. Critical Thinking

Making Inferences Why did mass culture emerge during this period?

THINK ABOUT

- the impact of newspapers
- advertising and catalogs
- the development of leisure time

ACTIVITY OPTIONS

ART

LANGUAGE ARTS

Research a world's fair from the turn of the century. Then make a **poster** or write a **newspaper advertisement** that will attract people to the fair.

TERMS & NAMES

Briefly explain the importance of each of the following.

1. urbanization
2. Jane Addams
3. political machine
4. Ellis Island
5. assimilation
6. Jim Crow
7. *Plessy* v. *Ferguson*
8. W. E. B. Du Bois
9. Joseph Pulitzer
10. leisure

REVIEW QUESTIONS

Cities Grow and Change (pages 609–613)

1. How did public transportation change city life?
2. What dangers did urban overcrowding pose to tenement dwellers?
3. How did big-city political machines keep their power?

The New Immigrants (pages 614–619)

4. Where did most American immigrants come from around 1900?
5. How did immigrants enter the United States?
6. Why have some people described the United States as a melting pot?

Segregation and Discrimination (pages 620–625)

7. Why was *Plessy* v. *Ferguson* an important Supreme Court decision?
8. What did African-American leaders do to fight discrimination?

Society and Mass Culture (pages 626–629)

9. What is mass culture?
10. How did city parks improve city life?

CRITICAL THINKING

1. USING YOUR NOTES

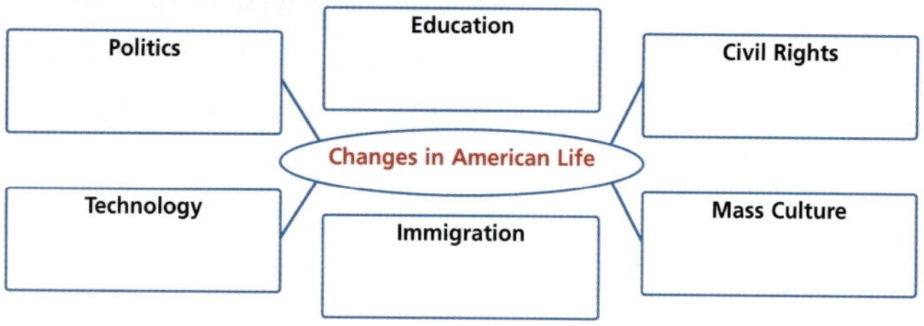

Using your completed chart, answer the questions below:

a. What changes did increased immigration cause?
b. How did the growing popularity of spectator sports and movies help bring about mass culture?
c. Which changes helped immigrants assimilate into American life?

2. ANALYZING LEADERSHIP

Think about the actions of Booker T. Washington and W. E. B. Du Bois. What approach did each take against discrimination? Whose approach do you think was the most likely to be effective?

3. THEME: DIVERSITY AND UNITY

How do you think the emergence of mass culture around 1900 affected immigrants and nonwhites?

4. APPLYING CITIZENSHIP SKILLS

What kinds of things prevented African Americans and immigrants from having full citizenship? How did they attempt to participate in American politics?

Interact *with* History

Have your ideas about how you'll make a home in the United States changed after reading the chapter? Explain.

VISUAL SUMMARY

Changes in American Life

Cities Grow and Change
Industrialization caused American cities to grow.

American Life Around 1900

Segregation and Discrimination
Racial and ethnic minorities faced discrimination across the country.

The New Immigrants
Large numbers of immigrants, especially from southern and eastern Europe, came to the United States.

Society and Mass Culture
New leisure activities and mass culture emerged at this time.

HISTORY SKILLS

1. INTERPRETING GRAPHS

Study the graph and answer the questions.

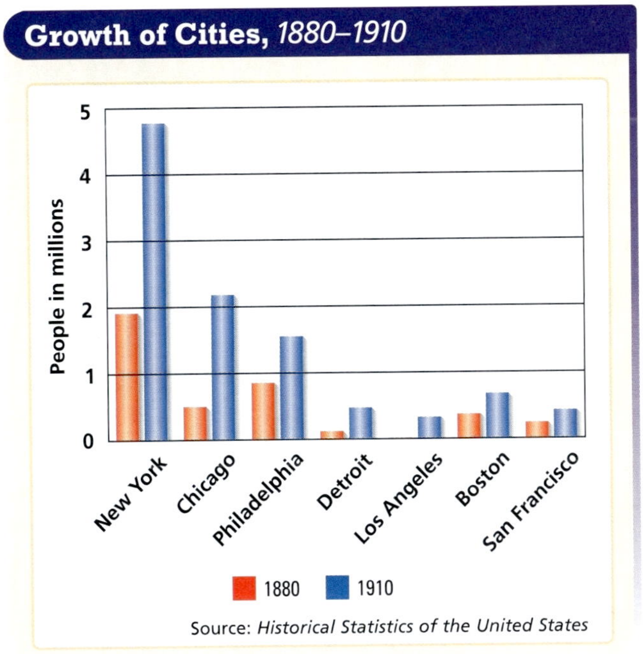

Basic Graph Elements

a. What cities are shown on the graph?

b. What were the three largest U.S. cities in 1910?

Interpreting the Graph

c. Which city's population increased by the greatest amount between 1880 and 1910?

d. Which city's population increased by the greatest percentage between 1880 and 1910?

2. INTERPRETING PRIMARY SOURCES

Advertisements urged Americans to buy all sorts of goods, such as bicycles. Study the ad carefully. Answer the questions.

a. How does the picture portray the activity of cycling?

b. How do you think this advertisement might convince people to buy bicycles?

ALTERNATIVE ASSESSMENT

1. INTERDISCIPLINARY ACTIVITY: Math

Making a Graph Do research to determine the number of immigrants who came from various countries between 1880 and 1914. Find out where immigrants come from today. Make a graph that displays your findings. Explain your graph to the class.

2. COOPERATIVE LEARNING ACTIVITY

Designing a Park Cities across the United States built parks at the turn of the century. They wanted to provide city dwellers with places to enjoy grass, trees, and fresh air as well as all sorts of leisure activities. What kinds of things would you want to be able to do in a park?

Working in small groups, research city parks built in the late 1800s. Design a park that could please as many urban residents as possible. Make sure to take into account the following.

a. sports fields

b. areas for rest and relaxation

c. places for nature and wildlife

d. buildings for food and restrooms

3. CD-ROM PRIMARY SOURCE EXPLORER

Creating a Museum Exhibit Racial discrimination has been a sad feature of American history. There is a large amount of information about racism and the ways that racial and ethnic minorities have tried to overcome it. Using the CD-ROM or your local library, collect information on discrimination and civil rights.

Create a museum exhibit about discrimination and civil rights in your hometown using the ideas below.

- Include information on groups that faced discrimination.
- Find biographies about any important civil rights leaders in your town.
- Collect primary sources, such as newspaper articles or autobiographies, that discuss important events.
- Draw a diagram that shows a floor plan for your exhibit.

4. HISTORY PORTFOLIO

PORTFOLIO **Option 1** Review your section and chapter assessment activities. Select one that you think was your best work. Then use comments made by your teacher or classmates to improve your work, and add it to your portfolio.

PORTFOLIO **Option 2** Review the questions that you wrote for What Do You Want to Know? on page 608. Then write a short report in which you explain the answers to your questions. If any questions were not answered, do research to answer them. Add your answers to your history portfolio.

Create an Exhibit

People stroll down The Great Pike at the St. Louis World's Fair.

In 1904 the Louisiana Purchase Exposition, better known as the St. Louis World's Fair, opened to great fanfare. The event celebrated the 100th anniversary of the U.S. purchase of the Louisiana Territory from France in 1803. Taking five years to plan and opening a year late, the fair focused on education and American technology. The automobile was among the most notable attractions at the fair. People from 63 countries and 43 states gathered in St. Louis.

ACTIVITY Create an exhibit that reflects some aspect of technology at the end of the 19th century. Then make a classroom fair. Write an article about it and give a speech describing your favorite exhibit.

TOOLBOX

Each group will need:

- bifold (type of poster board that folds open)
- poster board
- drawing paper
- markers
- scissors
- pencils
- glue
- cardboard

The New York-to-St. Louis Automobile Parade arrives at the St. Louis World's Fair.

STEP BY STEP

1 Form groups. Each group should consist of three or four students. During the workshop, each group will be expected to:

- research technology and inventions just prior to 1904
- design and create an exhibit for a classroom fair
- write a news report about the fair
- give a speech in praise of a favorite exhibit at the fair

2 Research the fair. Using this chapter, books on the St. Louis fair, or the Internet, find out what kinds of exhibits were displayed. Also research the technology and inventions of the time. Some themes of the fair's massive exhibit halls are listed below. Pick one theme on which to focus. Then brainstorm ideas for your exhibit and choose the best one.

World's Fair Themes	
transportation	education
technology	the arts

3 Design your exhibit.

Think about what your group wants to create. Using drawing paper, sketch a design of the exhibit in pencil. Next to your sketch, list all the items you'll need for the exhibit. Assign each person in your group certain items to bring for the next class period.

4 Lay out your display.

Use the images and text you found to visually organize the three-panel display. Vary the size of the images, type-size of the text, and include color to make your layout clear and interesting. Remember to create a title.

5 Create a mini St. Louis fair.

Along with the other groups, arrange the exhibits around the classroom. Walk around the room and look at the other groups' exhibits. Discuss with other groups how you created your exhibit.

WRITE AND SPEAK

Write a newspaper article. Cover the fair as a journalist from another city. Write an article about the classroom fair, describing the atmosphere as well as the exhibits. Then give a speech in praise of the outstanding exhibit of the fair.

HELP DESK

For related information, see pages 627–628 in Chapter 21.

Researching Your Project

- *The Song of the Molimo* by Jane Cutler

Visit www.mcdougallittell.com for links to sites about world's fairs.

Did You Know?

One vender at the fair had difficulty selling tea in the hot St. Louis summer. As a result, he began putting ice cubes in the tea, and sales of his "iced tea" soared.

Though there has been some controversy over who invented the ice-cream cone, the St. Louis World's Fair was the place it became popular. One story states that a vendor at the fair rolled a waffle into a cone-shaped holder when another vendor ran out of dishes. However, Italo Marchiony of New York City claimed to have been selling ice-cream cones since 1896.

REFLECT & ASSESS

- **How did your group come up with its idea?**
- **How does your design fit into the theme of the St. Louis fair?**
- **What criteria did you use when judging the exhibits?**

UNIT 7

1880 – 1920

Modern America Emerges

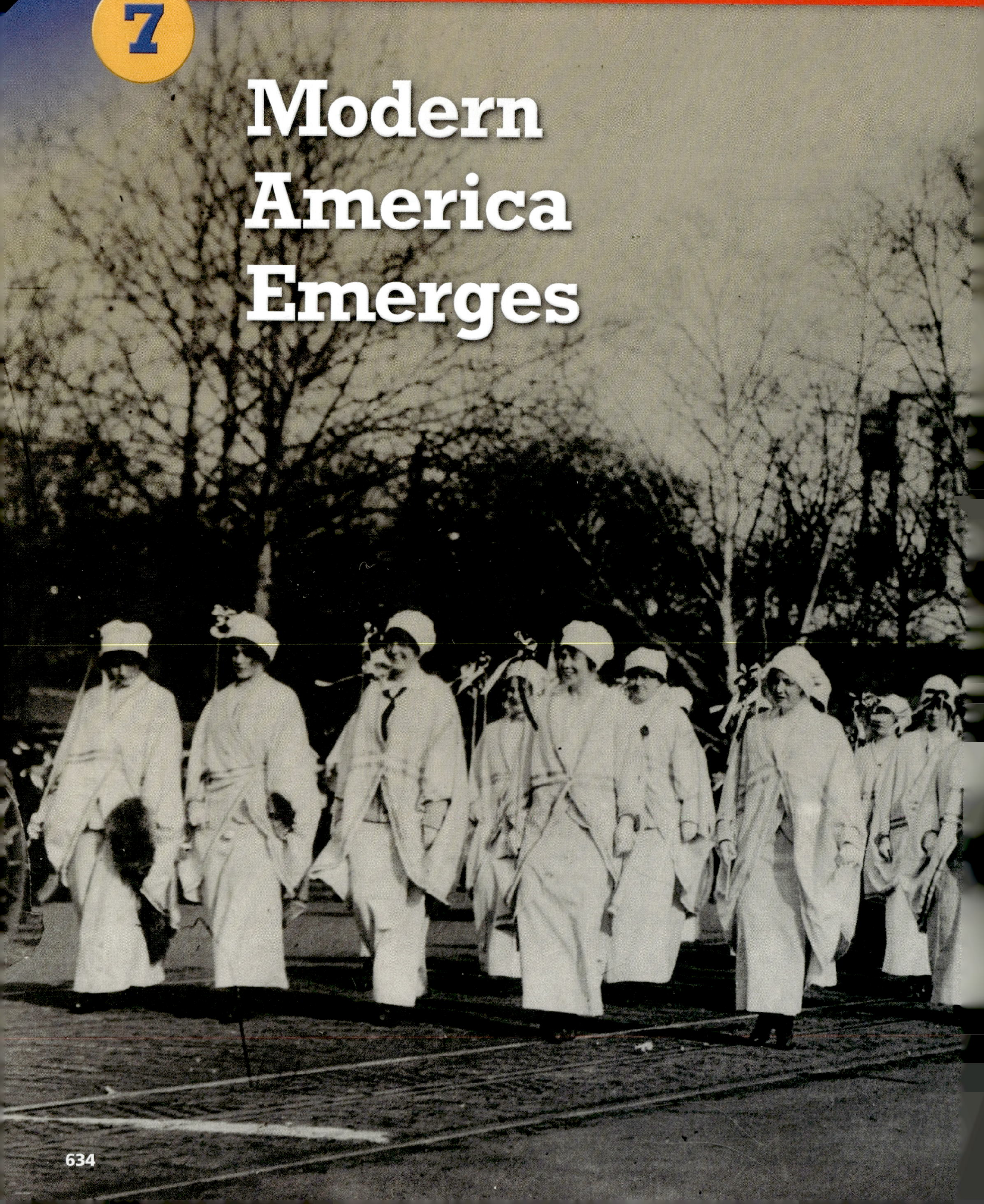

"Woman must not depend upon the protection of man, but must be taught to protect herself."

—*Susan B. Anthony*

Suffragists marched in Washington, D.C., in 1914 in support of a constitutional amendment giving women the right to vote.

CHAPTER 22

The Progressive Era 1890–1920

Homelessness—Children sleep in the street.

Poor sanitation—Dead horse rots in city street while children play in the gutter.

Poverty—Family earns money by making artificial flowers in its tenement.

Child labor—Children work in a Pennsylvania coal mine.

Interact *with* History

Theodore Roosevelt gives a speech. He faces difficult social problems, such as those shown to the left.

It is 1901, and Theodore Roosevelt has suddenly become president. You and all Americans are counting on him to help end child labor, poverty, business abuses, and political corruption. You're anxious to see what actions the new president will take to solve these problems.

What Do You Think?

- What different problems do the photographs show?
- What qualities would a leader need to tackle such problems?
- What might be the cause of these different problems?

How would you solve one of these problems?

USA

1890 Congress passes Sherman Antitrust Act.

1896 William McKinley is elected president.

1901 McKinley is assassinated, and Theodore Roosevelt becomes president.

1904 Roosevelt is re-elected president.

1908 William Howard Taft is elected president.

1912 Woodrow Wilson is elected president.

1913 17th Amendment provides for direct election of senators.

1919 18th Amendment outlaws alcohol.

1920 19th Amendment grants women the right to vote.

World

1890 German leader Bismarck is dismissed by Kaiser Wilhelm II.

1892 Gladstone becomes prime minister of Great Britain.

1894 Uganda becomes a British protectorate.

1900 Boxer uprising against foreigners begins in China.

1910 Union of South Africa is established.

1913 Gandhi, leader of Indian resistance movement, is arrested.

Chapter 22 SETTING THE STAGE

BEFORE YOU READ

Previewing the Theme

Impact of the Individual The problems caused by the growth of industries and cities sparked a variety of reform movements. This chapter shows that individual reformers and reform groups shared a strong desire to solve problems in American society.

A social worker pays a visit to a poor family.

What Do You Know?

What do you know about life in American cities in the early 1900s? What problems plagued the cities? How have people living in cities overcome obstacles?

THINK ABOUT

- what you've learned in previous chapters
- what you know about urban problems today

What Do You Want to Know?

What questions do you have about the reform movements of the early 1900s? Record your questions in your notebook before you read the chapter.

READ AND TAKE NOTES

Reading Strategy: Identifying and Solving Problems This chapter focuses on the problems that Americans faced at the turn of the century and how they worked to solve those problems. A graphic organizer can help you keep track of problems and solutions. Major problems faced by the nation at the turn of the century are listed in the first column of the chart below. As you read, record solutions for these problems in the second column of the chart.

See Skillbuilder Handbook, page R17.

PROBLEM	SOLUTION
Political: patronage; limited suffrage and democracy	
Social: poverty; alcohol abuse	
Economic: power of big corporations; unemployment	
Environmental: impure food and water; diminishing natural resources	

1 Roosevelt and Progressivism

TERMS & NAMES
progressivism
muckrakers
direct primary
initiative
referendum
recall
Sherman Antitrust Act
Theodore Roosevelt

MAIN IDEA

Reformers tried to solve the problems of the cities. They gained a champion in Theodore Roosevelt.

WHY IT MATTERS NOW

Many of the reforms of the Progressive Era have had an effect on life in America today.

ONE AMERICAN'S STORY

Newspaper journalist Nellie Bly worked for *The New York World.* In 1887, Bly wanted to investigate the Women's Lunatic Asylum in New York City. An asylum is a place where people with mental illness can get help. She faked mental illness and fooled doctors so that she could become a patient there. After spending ten days in the asylum, Bly wrote a newspaper article about what she had witnessed. She described being forced to take ice cold baths.

A VOICE FROM THE PAST

My teeth chattered and my limbs were goose-fleshed and blue with cold. Suddenly I got, one after the other, three buckets of water over my head—ice-cold water, too—into my eyes, my ears, my nose and my mouth.

Nellie Bly, quoted in *Nellie Bly: Daredevil, Reporter, Feminist*

She reported that nurses choked and beat patients. Shortly after Bly's stories appeared, conditions at the asylum improved.

Bly also wrote about poor conditions in slums, factories, prisons, and nursing homes. Like other reformers, she wanted to correct the wrongs in American society. All of these reformers made up the Progressive movement around the turn of the century.

Nellie Bly

The Rise of Progressivism

As you saw in Chapter 21, the rapid growth of cities and industries in the United States at the turn of the century brought many problems. Among them were poverty, the spread of slums, and poor conditions in factories. A depression in the 1890s made problems worse. In addition, corrupt political machines had won control of many city and state governments. Big corporations had gained power over the economy and government.

To attack these problems, individuals organized a number of reform movements. These reformers believed in the basic goodness of people. They also believed in democracy. The reformers were mostly native born and middle-class. They could be found in either political party. Their reform movements came to be grouped under the label **progressivism.**

Now *and* then

BIG BUSINESS AND COMPETITION

In the late 1800s, John D. Rockefeller made a fortune as he gained control of most of the nation's oil refineries, oil fields, and pipelines. In 1906, the government filed an antitrust suit against Rockefeller's Standard Oil. This resulted in its breakup in 1911. The cartoon below shows Standard Oil as an octopus.

In the 1990s, Bill Gates became the richest man in the world as he built Seattle-based Microsoft into a computer software giant. In 1998, the government filed an antitrust suit against Microsoft. It charged the company with using illegal tactics to gain a monopoly with its computer operating system and Web browser.

About 1900, a new group of writers began to expose corruption in American society. They were called **muckrakers.** The muckrakers created a public demand for reform. Muckraker Ida Tarbell, for example, accused Standard Oil of using unfair tactics to force small companies out of business.

The progressive reformers shared at least one of three basic goals: first, to reform government and expand democracy; second, to promote social welfare; third, to create economic reform.

Reforming Government and Expanding Democracy

In the 1870s and 1880s, elected officials often handed out government jobs and contracts. In return, they won political support. This practice was called patronage. It became a hot political issue during the presidencies of Rutherford B. Hayes, James Garfield, and Chester Arthur. Finally, Congress passed the Pendleton Civil Service Act in 1883. This law required people to take civil service exams for certain government jobs. It also prevented elected officials from firing civil service workers for political reasons.

In the 1890s and early 1900s, progressive leaders in a number of states sought to expand democracy. They wanted to give voters more control over their government. In 1903, under progressive governor Robert M. La Follette, Wisconsin became the first state to establish a direct primary. In a **direct primary,** voters, rather than party conventions, choose candidates to run for public office.

Reading **History**

A. Finding Main Ideas What was the main goal behind the progressive reforms of government?

In Oregon, newspaper editor William S. U'Ren promoted three reforms besides the direct primary.

1. **Initiative**—This reform allowed voters to propose a law directly.
2. **Referendum**—In this reform, a proposed law was submitted to the vote of the people.
3. **Recall**—This reform allowed people to vote an official out of office.

In the years that followed, many other states adopted one or more of these progressive reforms.

Promoting Social Welfare

This goal addressed such problems as poverty, unemployment, and poor working conditions. You read about the social gospel and settlement house movements in Chapter 21. Leaders in these movements promoted many social-welfare reforms. For example, Jane Addams provided social services to the poor at Hull House. She also worked to help the unemployed. Florence Kelley, also from Hull House, pushed for minimum wage laws and limits on women's working hours.

Another group of reformers who wanted to improve social welfare were the prohibitionists. They worked to prevent alcohol from ruining people's lives. The prohibitionists built on the temperance movement of the 1800s.

This photograph shows an immigrant family in a crowded tenement at the turn of the century.

Creating Economic Reform

The third progressive goal was to create economic reform. This meant limiting the power of big business and regulating its activities. By the late 1800s, business leaders in some major industries had formed trusts. These were combinations of businesses. The business firms in a trust worked together to cut prices and squeeze out competitors. Then the trust would raise prices and make larger profits.

Reading **History**
B. Summarizing How did progressives pursue their three basic goals?

The **Sherman Antitrust Act** of 1890 made it illegal for corporations to gain control of industries by forming trusts. However, the government did not enforce the law at first. Enforcement required a strong president.

Roosevelt and the Square Deal

Theodore Roosevelt—the first progressive president—provided this strength and leadership. He came to the presidency by accident, however. In 1898, Roosevelt won fame fighting in the Spanish-American War in Cuba. He returned from Cuba a war hero and was elected governor of New York. In 1900, Roosevelt ran on the Republican ticket as President McKinley's vice president.

Then an assassin shot McKinley, just six months after his inauguration. Roosevelt became president when McKinley died on September 14, 1901. At age 42, Roosevelt was the youngest person ever to become president. He brought his boundless energy to the office. The president often joined his six children in playing in the White House. Americans admired Roosevelt's zest for living. He gained the public's support for reform.

"I believe in a square deal."
Theodore Roosevelt

Roosevelt began his reforms with an effort to break up the corporate trusts. He thought industries should be regulated for the public interest.

A VOICE FROM THE PAST

When I say I believe in a square deal I do not mean, and nobody who speaks the truth can mean, that he believes it possible to give every man the best hand. If the cards do not come to any man, or if they do come, and he has not got the power to play them, that is his affair. All I mean is that there shall not be any crookedness in the dealing.

Theodore Roosevelt, speech on April 5, 1905

Roosevelt saw government as an umpire. Its purpose was to ensure fairness, or a "square deal," for workers, consumers, and big business.

To root out "crookedness," Roosevelt used the Sherman Antitrust Act. Since its passage in 1890, many corporations had ignored the law, which was intended to regulate the trusts. No one had enforced it—no one, that is, until Roosevelt became president in 1901.

At the end of 1901, the nation's railroads were run by a handful of companies. The power of railroads continued to grow. It was not surprising, therefore, that one of Roosevelt's first targets was the railroads. He used the Sherman Antitrust Act to bust up a railroad trust.

Roosevelt was not against big business as such. However, he opposed any trust he thought worked against the national interest. In addition to the railroad trust, Roosevelt broke up the Standard Oil Company and a tobacco trust. In all, the government filed suit against 44 corporations during Roosevelt's presidency.

Reading History

C. Making Inferences How do you think big business leaders regarded President Roosevelt? Why?

Roosevelt Leads Progressive Reforms

As president, Roosevelt had a great deal of power to push progressive ideas. To make such ideas into law, however, he needed help. Roosevelt got it as voters began pressuring their senators and representatives. As a result, Congress passed laws that helped change American society.

Roosevelt acted to regulate the meat-packing industry after reading Upton Sinclair's *The Jungle.* The novel describes a packing plant in which dead rats end up in the sausage. Sinclair focused attention on the poor sanitary conditions under which the meat-packers worked. "I aimed at the public's heart, and by accident I hit it in the stomach," he noted.

Roosevelt launched an investigation of the meat-packing industry. In 1906, he signed the Meat Inspection Act. This act created a government meat inspection program. Roosevelt also signed the Pure Food and Drug Act. This law banned the sale of impure foods and medicines.

While Roosevelt tried to win a square deal for most Americans, he did not push for civil rights for African Americans. He believed that discrimination was morally wrong. However, he did not take the political risk of leading a fight for civil rights.

Shown at the left is the cover of Upton Sinclair's novel, *The Jungle.* The photograph shows immigrant workers stuffing sausages in a Chicago meat-packing house.

Conservation

Roosevelt was a strong crusader for conservation—controlling how America's natural resources were used. As an outdoorsman and hunter, he had observed the gradual loss of natural resources. He camped with naturalist John Muir for four days in Yosemite, California. Because he loved the Yosemite Valley so much, he set out to preserve Yosemite and other areas for people's "children and their children's children."

Roosevelt preserved more than 200 million acres of public lands. He established the nation's first wildlife refuge at Pelican Island, Florida. He doubled the number of national parks in the United States. At one point, Congress refused to establish any more national parks. Roosevelt used the Antiquities Act to create national monuments instead. In this way, he preserved the Grand Canyon and the Petrified Forest in Arizona. Roosevelt spoke of the glories of the Grand Canyon while visiting the site in 1903.

Reading **History**

D. Finding Main Ideas Why did President Roosevelt bypass Congress?

A VOICE FROM THE PAST

Leave it as it is. You cannot improve on it. The ages have been at work on it, and man can only mar it.

Theodore Roosevelt, quoted in *Yellowstone*

Both the Grand Canyon and the Petrified Forest have since become national parks. America's next president, William Howard Taft, was not as interested in conservation. However, he did continue Roosevelt's progressive reforms, as you will read in the next section.

AMERICA'S HISTORY MAKERS

THEODORE ROOSEVELT

1858–1919

From his youth on, Theodore Roosevelt lived what he called the "strenuous life." He rode horses, hiked, boxed, wrestled, and played tennis. In winter, he swam in the icy Potomac River. He hunted rhinoceros in Africa, harpooned devilfish in Florida, and boated down the Amazon.

Americans loved reading of his exploits and affectionately referred to him as "Teddy" or "T.R." Once, on a hunting trip, he refused to shoot a bear cub. News of the event resulted in a new toy—the teddy bear.

How did Roosevelt's active style of living carry over into his presidency?

Section 1 Assessment

1. Terms & Names

Identify:

- progressivism
- muckrakers
- direct primary
- initiative
- referendum
- recall
- Sherman Antitrust Act
- Theodore Roosevelt

2. Taking Notes

Use a chart to list examples of progressive reforms.

Goals	Reforms
To expand democracy	
To protect social welfare	
To create economic reform	

Which reform was most important? Explain.

3. Main Ideas

a. What kinds of problems did progressives attempt to solve?

b. What did President Roosevelt mean by a "square deal," and how did he try to achieve it?

c. What were Roosevelt's achievements in the field of conservation?

4. Critical Thinking

Recognizing Effects In what ways do the reforms that President Roosevelt promoted affect your life today?

THINK ABOUT

- the quality of the food you eat
- natural resources that have been preserved

ACTIVITY OPTIONS

ART

GEOGRAPHY

Do research on one of the natural areas that President Roosevelt preserved. Create a **travel brochure** or an illustrated **map** of the area.

GEOGRAPHY in HISTORY

REGION AND HUMAN-ENVIRONMENT INTERACTION

The National Parks Movement

As the United States expanded westward, two things became evident. First, this was a land of astonishing beauty. Second, this unspoiled beauty would not last if it wasn't protected.

President Theodore Roosevelt may have given the conservation movement its most significant boost. An outdoorsman, naturalist, and visionary, he established the U.S. Forest Service and set aside more than 200 million acres of public lands as national parks, forests, monuments, and wildlife refuges.

Creating parks was just the first step in protecting these lands. Problems arose that had not been foreseen. These problems included a lack of funds and growing numbers of tourists and researchers. In 1916, the National Park Service was established with the goal of saving the parks for future generations.

In 1903, Teddy Roosevelt (left) joined conservationist John Muir (right) for a camping trip. Their trip took them from the "big trees" of the Sequoia forest to the wonders of the Yosemite Valley. This photo of Roosevelt and Muir was taken at Glacier Point in Yosemite. Both men wanted to protect the magnificent beauty of America's most spectacular regions.

ARTIFACT FILE

Sequoia National Park is a land of giants. In a forest where many trees are more than 250 feet high, it is difficult to get a sense of scale when looking at the biggest of these giants. The General Sherman Tree, shown here, is the largest tree by volume in the world. A number of trees in Sequoia National Park are named for Civil War generals.

Everglades National Park in Florida is part of the approximately 1,500,000-acre Everglades region. This wetland habitat is home to birds, especially waders such as herons, egrets, and ibis, and is famous for its alligators.

The National Parks Today

National parks are identified on the map above. The National Park System includes many areas, all of which are under the management of the National Park Service (NPS).

On-Line Field Trip

Yellowstone National Park was designated the world's first national park in 1872. Covering about 2,200,000 acres, Yellowstone is still the largest national park in the United States. There are many geysers in Yellowstone, including Old Faithful (at right).

Visit www.mcdougallittell.com for more information.

CONNECT TO GEOGRAPHY

1. **Region** What might be two reasons the national parks are concentrated where they are?
2. **Human-Environment Interaction** What effects might visits from many tourists have on a national park?

See Geography Handbook, page 17.

CONNECT TO HISTORY

3. **Analyzing Causes** What general mood of the era made the late 1800s a likely time for successfully starting a national park?

2 Taft and Wilson as Progressives

TERMS & NAMES
William Howard Taft
Sixteenth Amendment
Seventeenth Amendment
Clayton Antitrust Act
Federal Reserve Act

MAIN IDEA

Progressive reforms continued under William Howard Taft and Woodrow Wilson.

WHY IT MATTERS NOW

Constitutional amendments passed during this time affect Americans today.

ONE AMERICAN'S STORY

During the Progressive Era, many Americans became disturbed by the problems caused by capitalism. Some even turned to socialism. This is a system in which business and industry are totally controlled by the state. Labor leader Eugene V. Debs became a socialist while serving time in prison for his role in a labor strike. In 1894, as head of the American Railway Union, Debs supported a strike by the workers who made railroad cars. When the federal government broke up the strike, Debs defied the courts. He was sent to prison for six months.

In the 1908 presidential election, Debs ran as the Socialist Party candidate. In his campaign, he urged American workers to consider what competition was like in a capitalist system.

The forceful speeches of Eugene V. Debs attracted large audiences.

A VOICE FROM THE PAST

Competition was natural enough at one time, but do you think you are competing today? . . . Against whom? Against Rockefeller? About as I would if I had a wheelbarrow and competed with the Santa Fe [railroad] from here to Kansas City.

Eugene V. Debs, quoted in *The Annals of America*

Debs made a decent showing in the election, winning more than 420,000 votes. However, the Republican candidate Taft did better and was elected.

Taft and Progressivism

In the 1908 election, Debs ran against Republican **William Howard Taft** and Democrat William Jennings Bryan. Neither Debs nor Bryan stood much of a chance against Taft. He was Roosevelt's handpicked successor. Roosevelt's popularity swayed many people to vote for Taft, who promised to follow Roosevelt's progressive policies.

Taft continued Roosevelt's attack on trusts. During his four years in office, Taft pursued almost twice as many antitrust suits as Roosevelt had in nearly eight years in office. But Taft received less credit for his progressivism because he became allied with conservative Republicans rather than Roosevelt's progressive Republicans. Nevertheless, Taft did move forward with progressive reforms. His reforms addressed the progressive goals of democracy, social welfare, and economic reform. Two of the major progressive achievements under President Taft were constitutional amendments.

Two Progressive Amendments

The **Sixteenth Amendment** was passed in 1909 and ratified in 1913. It gave Congress the power to create income taxes. The Constitution previously did not allow direct taxes on an individual's income. This amendment was intended to provide a means of spreading the cost of running the government among more people. The income tax soon became the main source of federal revenue.

Economics *in* History

Types of Taxes

The Sixteenth Amendment, ratified in 1913, made it constitutional for the federal government to have an income tax. Congress quickly passed an income tax law the same year. The income tax provides revenue to the federal government by taxing profits and earnings. In a graduated income tax, larger incomes are taxed at higher rates than smaller incomes. The income tax is only one of several taxes that governments use to raise money.

TYPES OF TAXES

INCOME TAXES

1. **Individual:** You pay a percentage of what you earn at work or from investments. Under the payroll deduction plan, income taxes are deducted (taken out) from your wages or salary before you get your paycheck.
2. **Corporate:** Corporations pay a percentage of their profits in income tax.

PROPERTY TAXES

People pay taxes on property they own, such as land or a house. Property taxes are often used to support public services such as schools.

SALES TAXES

Sales tax is imposed on the retail price of merchandise and collected by the retailer. For example, when you buy a pair of jeans, you pay sales tax, which will be listed on your receipt.

ESTATE TAXES

This tax is charged against the value of the property of a person who has died. It is also called the "death tax" because it is collected from the dead person's estate before the estate is passed on to the heirs.

CONNECT TO HISTORY

1. **Making Inferences** How might a corporate income tax fit the goals of the Progressive Era?

See Skillbuilder Handbook, page R11.

CONNECT TO TODAY

2. **Drawing Conclusions** Some states that have a sales tax do not charge that tax on the purchase of goods like food or clothing. Why do you think they make an exception for these purchases?

INTERNET Visit www.mcdougallittell.com to learn more about taxes.

STRANGE *but* True

FROM PRESIDENT TO CHIEF JUSTICE

William Howard Taft was the only man in American history to serve first as president and then as chief justice of the U.S. Supreme Court. He had always wanted to be a Supreme Court justice. Even his mother said, "I do not want my son to be President. His is a judicial mind and he loves the law."

Taft was unhappy as president. When he left office, he said: "I'm glad to be going. This is the lonesomest place in the world." Eight years later, in 1921, President Warren G. Harding appointed Taft to the Supreme Court. Taft is shown here in his judicial robes.

The **Seventeenth Amendment** was ratified in 1913. It provided for the direct election of U.S. senators by voters in each state. Formerly, state legislatures had chosen U.S. senators. Under this system, many senators obtained their positions through corrupt bargains. Because of this, the Senate was called the "Millionaires' Club." The Seventeenth Amendment gave people a more direct voice in the government.

Reading **History**
A. Drawing Conclusions Why are the Sixteenth and Seventeenth amendments considered progressive?

The Election of 1912

Taft achieved a number of progressive reforms. However, a deep split developed between him and progressive leaders in the Republican Party. Still, with the support of conservative Republicans, Taft won the party's nomination as its presidential candidate in 1912.

However, many progressive Republicans supported Theodore Roosevelt. He had entered the race and formed the Progressive Party, also known as the Bull Moose Party.

The Democrats chose Governor Woodrow Wilson of New Jersey as their presidential candidate. Eugene Debs again entered the race as the Socialist candidate. With the Republicans deeply divided, Wilson won the election.

The Wilson Presidency

As president, Wilson established a progressive record. Wilson believed that "bigness" itself was dangerous. He wanted the government to use its powers to break up monopolies—groups that sought complete control over an industry. He also wanted the government to help workers in their struggles against business owners.

At Wilson's urging, Congress passed the **Clayton Antitrust Act** of 1914. The new law laid down rules forbidding business practices that lessened competition. A business, for example, could no longer buy the stock of a competitor. The Clayton Act gave the government more power to regulate trusts. In addition, the Clayton Act was also prolabor:

1. It said labor unions and farm organizations could merge and expand.
2. It limited the ability of the courts to force workers to end strikes.
3. It legalized such labor tactics as strikes, picketing, and boycotts.

Vocabulary
boycott: an attempt to pressure a business by refusing to buy a product or use a service

During Wilson's two terms, reforms to the nation's financial system occurred. In 1913, the **Federal Reserve Act** was passed. This improved the nation's monetary and banking system. The law created the modern banking system, which resembles a pyramid. At the top is the Federal Reserve Board, which is appointed by the president. Next are 12 Federal Reserve Banks for different regions of the country. These are "bankers' banks." They serve the bottom level—the member banks.

The Federal Reserve Act created a more flexible currency system by allowing banks to control the money supply. To raise money, for example, the Federal Reserve Board, or "Fed," lowers the interest rate that it charges member banks. These banks then borrow more from the Fed and thus have more money to lend to people and businesses.

President Wilson throws out a baseball at the opening game of the 1916 season.

ReadingHistory
B. Summarizing What were some of Wilson's achievements as a progressive president?

Wilson did no more to advance civil rights for African Americans than Roosevelt did. In fact, Wilson approved the segregation, or separation, of African-American and white employees in the federal government. Throughout the Progressive Era, presidents Roosevelt, Taft, and Wilson did not actively promote civil rights for African Americans.

The Eighteenth Amendment

Another amendment passed during the Progressive Era was the Eighteenth Amendment. This is also called the Prohibition Amendment. During Wilson's administration, supporters of prohibition gained strength. Reformers thought an alcohol ban would reduce poverty. They argued that liquor added to unemployment and violence. Business leaders saw that alcohol made workers less efficient. Finally, in 1917, Congress passed a constitutional amendment. The Eighteenth Amendment prohibited the manufacture and sale of alcoholic beverages. The states ratified the amendment in 1919.

In the next section, you will read about the most important amendment of the era—the Nineteenth Amendment, which gave women the vote.

Section 2 Assessment

1. Terms & Names

Identify:

- William Howard Taft
- Sixteenth Amendment
- Seventeenth Amendment
- Clayton Antitrust Act
- Federal Reserve Act

2. Taking Notes

Complete the chart to review some of the major reforms of both the Taft and Wilson administrations.

Law	Description
Sixteenth Amendment	
Clayton Antitrust Act	
Federal Reserve Act	

3. Main Ideas

a. What caused the Republican Party to split in 1912?

b. What were the major progressive accomplishments of Wilson's presidency?

c. What did the Federal Reserve Act do?

4. Critical Thinking

Making Inferences Why did progressive presidents do little to advance civil rights for African Americans?

THINK ABOUT

- the goals of progressivism
- the groups of people that progressivism aimed to help

ACTIVITY OPTIONS

LANGUAGE ARTS
TECHNOLOGY

Research one of the people mentioned in this section. Then write the **script** for the first 10 minutes of his documentary or design his **Web page**.

Women Win New Rights

TERMS & NAMES
Susan B. Anthony
Carrie Chapman Catt
Nineteenth Amendment

MAIN IDEA

Women became leaders in social reform movements and won the right to vote during the Progressive Era.

WHY IT MATTERS NOW

Today, American women enjoy the right to vote because of women reformers in the Progressive Era.

ONE AMERICAN'S STORY

After graduating from nursing school in 1891, Lillian Wald briefly studied medicine. Then a friend asked her to teach a home nursing class at a school for immigrants in New York City. One day a child asked Wald to help her sick mother, who was absent from the class. Following the child home, Wald was shocked by what she saw.

A VOICE FROM THE PAST

Over broken asphalt, over dirty mattresses and heaps of refuse we went. The tall houses reeked with rubbish. . . . There were two rooms and a family of seven not only lived here but shared their quarters with boarders.

Lillian Wald, quoted in *Always a Sister*

The experience gave Wald a new mission. Inspired to help such poor immigrants, she founded the Nurses' Settlement. This was later called the Henry Street Settlement. It was the first visiting nurse program in the country not run by a religious group. The program mainly helped poor women and children.

In this section, you will read about others like Wald who worked to make life better for all women.

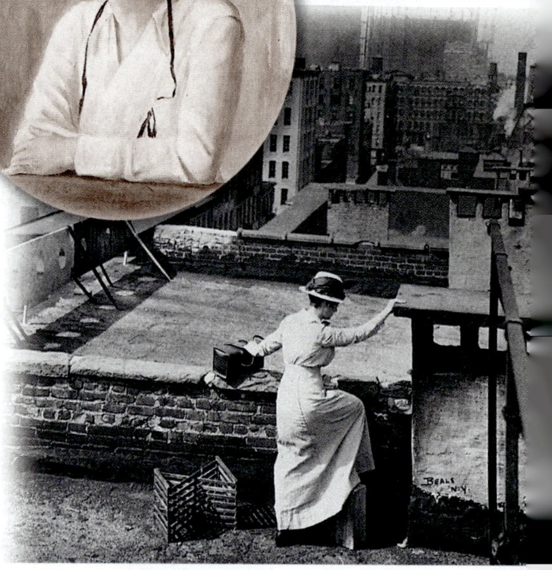

(Above left) Lillian Wald. (Above) A visiting nurse takes a shortcut between two tenements.

New Roles for Women

The social reform movements of the Progressive Era were led by educated, middle-class women. At the turn of the century, women like Wald were looking for new roles outside the home. The growth of industry had changed many urban, middle-class homes. These homes now had indoor running water and electric power for lamps and vacuum cleaners.

In addition, factories produced the products that women once made in the home, such as soap, clothing, and canned goods. Such technological advances reduced some of the unpleasant work of homemaking. At the same time, families were becoming smaller as women had fewer children.

As a result, the homemaker's role began to change. High schools, colleges, and women's clubs offered courses in home economics and domestic science. In these courses, women were encouraged to apply the latest methods to running their homes.

Other women responded to changes in the home by taking jobs in factories, offices, and stores. Women worked as telephone operators, store clerks, and typists. Those who gained a college education could pursue a profession. The choices were limited to such fields as teaching and nursing. Women who could afford to were expected to quit their jobs when they married. In 1890, approximately 30 percent of women between the ages of 20 and 24 worked outside the home. However, only about 15 percent between the ages of 25 and 44 did so.

ReadingHistory

A. Finding Main Ideas How and why did women's roles begin to change around the turn of the century?

Women Progressives

The social reform movements that many middle-class, college-educated women took part in were focused on helping people. These included the settlement house and prohibition movements. A settlement house is a community center providing assistance to residents—particularly immigrants—in a slum neighborhood.

Jane Addams was a good example of the progressive female leader. After graduating from college, Addams sought a meaningful way to participate in society. She was financially independent. A visit to a settlement house in a London slum inspired her to start a similar program in Chicago. She was helped by her friend Ellen Starr.

With donations from wealthy Chicagoans, Addams and Starr rented an old mansion. Hull House was located in a poor, immigrant neighborhood. Within just a few years, they organized a full program of services, classes, and clubs. These were run by a group of young women residents and over 90 volunteers. Hull House served as an information bureau for new immigrants. It also helped the unemployed find jobs. It offered a kindergarten, a day nursery, after-school youth clubs, nutrition classes, and a concert program. Workers also pressured politicians for improved city services for the neighborhood.

Connections TO LITERATURE

WOMEN OUTSIDE THE HOME

Charlotte Perkins Gilman (shown below) was an influential writer on women's rights. She wanted to free women from housework to pursue careers. In *Women and Economics* (1898), Gilman argued that a wife's dependency on her husband limited her personal development.

In *Concerning Children* (1900) and *The Home* (1903), she proposed that families live in large apartments. These would have centralized nurseries and a staff devoted to cooking, cleaning, and child-care. This support would free women to work outside the home.

A VOICE FROM THE PAST

One function of the settlement to its neighborhood somewhat resembled that of the big brother whose mere presence in the playground protected the little ones from bullies.

Jane Addams, quoted in *Women and the American Experience*

The young women residents of Hull House received no salary and had to pay for their room and board. This meant that they had to be financially independent. For some, Hull House provided training for other public service. Florence Kelley, for example, worked at Hull House from 1891 to 1898. She later became secretary of the National Consumers' League. This group promoted better working conditions in factories and stores.

Another prominent but controversial progressive leader was Carry Nation. She campaigned for prohibition. Nation had once been married to an alcoholic. Tall and strong, she adopted dramatic methods in her opposition to alcoholic beverages. In the 1890s, she smashed saloons with a hatchet. This caused law enforcement officials to arrest her for disturbing the peace. Although some people criticized Nation, her efforts helped bring about passage of the Eighteenth Amendment in 1919.

Suffrage for Women

Many women progressives were active in the struggle for woman suffrage, or the right to vote. American women fought longer for the right to vote than they did for any other reform. Some leaders in the fight died before realizing their goal.

In 1890, two separate woman suffrage groups merged to form the National American Woman Suffrage Association (NAWSA). Elizabeth Cady Stanton served as its first president. Two years later, in 1892, **Susan B. Anthony** became president. She held the position until 1900. Expressing their frustration over the difficulty of gaining suffrage, Elizabeth Cady Stanton and Susan B. Anthony wrote, "Words can not describe the indignation, the humiliation a proud woman feels for her sex in disfranchisement [being deprived of the right to vote]."

Background
Because she helped organize the woman suffrage movement, Susan B. Anthony became the first woman to be pictured on a U.S. coin.

NAWSA at first focused on state campaigns to win the right to vote, since earlier efforts at passing a federal amendment had failed. But by 1896, only four states allowed women to vote. These were Wyoming, Utah, Idaho, and Colorado. Between 1896 and 1910, women did not gain the right to vote in a single state. Then, between 1910 and 1914, seven more Western states approved full suffrage for women.

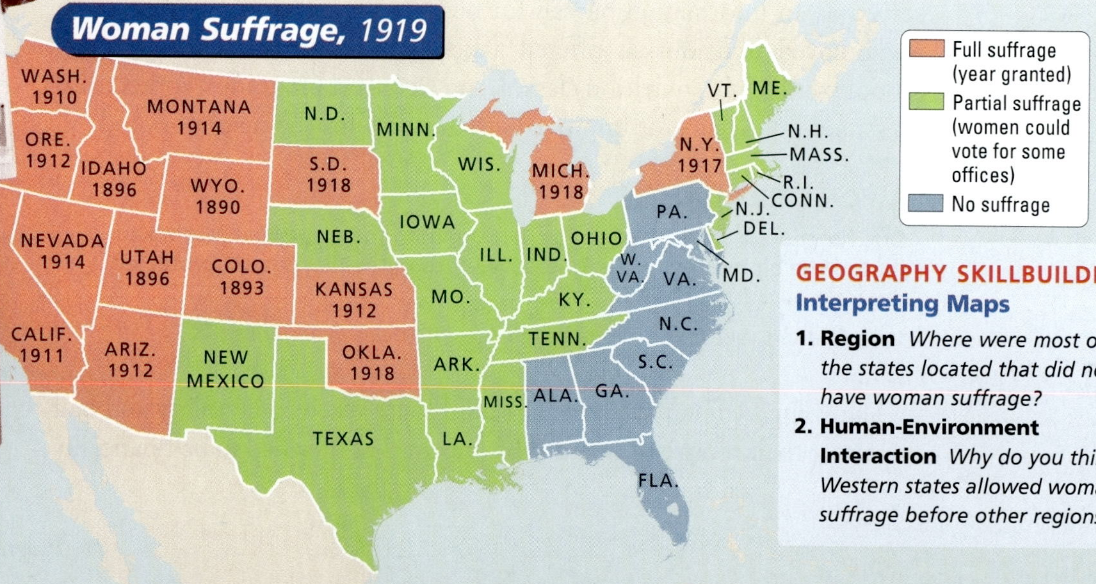

GEOGRAPHY SKILLBUILDER
Interpreting Maps

1. **Region** *Where were most of the states located that did not have woman suffrage?*
2. **Human-Environment Interaction** *Why do you think Western states allowed woman suffrage before other regions?*

The Progressive Amendments, *1909–1920*

Number	Description	Passed by Congress	Ratified by States
16th	Federal income tax	1909	1913
17th	Senators elected by people rather than state legislatures	1912	1913
18th	Manufacture, sale, or transport of alcohol prohibited	1917	1919
19th	Woman suffrage	1919	1920

SKILLBUILDER Interpreting Charts

1. *For which amendment was there the longest gap between passage by Congress and ratification by states?*
2. *What do the dates 1909 and 1920 represent in this chart?*

The Nineteenth Amendment

The Western successes turned the tide in favor of woman suffrage. The United States' entry into World War I in 1917 made the final difference. During the war, membership in NAWSA reached 2 million. **Carrie Chapman Catt,** president of NAWSA, argued that the nation could no longer deny the right to vote to women, who were supporting the war effort by selling war bonds and organizing benefits. President Wilson urged the Senate to pass a women's suffrage amendment. He called passage "vital to the winning of the war."

In 1918, the House passed the **Nineteenth Amendment,** which gave women full voting rights. The Senate approved the amendment in 1919. In 1920, the states ratified it. In the final state campaigns, women staged marches, parades, and rallies around the country. Charlotte Woodard had attended the first women's rights convention in 1848 at Seneca Falls as a teenager. In 1920, the 91-year-old Woodard voted in a presidential election for the first time.

Reading **History**

B. Recognizing Effects What factors helped women gain the right to vote?

Section 3 Assessment

1. Terms & Names

Identify:
- Susan B. Anthony
- Carrie Chapman Catt
- Nineteenth Amendment

2. Taking Notes

Use a chart to record the achievements of some women leaders of the era.

Progressive Achievements	
Lillian Wald	
Jane Addams	
Florence Kelley	

Which achievement seems greatest and why?

3. Main Ideas

a. How did women's roles expand near the turn of the century?

b. What was the background of many women who became leaders in social reform movements?

c. How did World War I influence the passage of the Nineteenth Amendment?

4. Critical Thinking

Comparing and Contrasting In what ways was the struggle for woman suffrage similar to and different from African Americans' struggle for equal rights?

THINK ABOUT
- the restrictions that both groups faced
- how long they struggled for basic rights

ACTIVITY OPTIONS

LANGUAGE ARTS
TECHNOLOGY

Research one of the women reformers discussed in this chapter. Then write the **script** for the first 10 minutes of her documentary or design her **Web page.**

CHAPTER 23

Becoming a World Power

1880–1917

Section 1 **The United States Continues to Expand**

Section 2 **The Spanish-American War**

Section 3 **U.S. Involvement Overseas**

The United States "Great White Fleet" symbolized the nation's presence as a global power at the beginning of the 20th century.

Interact *with* History

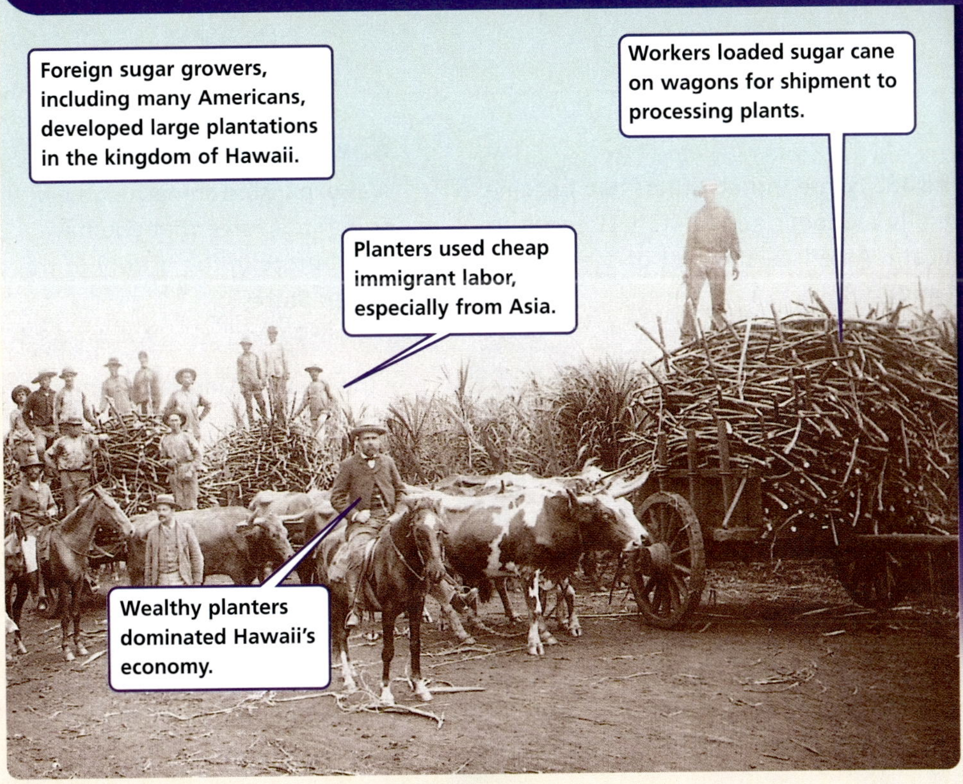

In 1893, American sugar planters in the kingdom of Hawaii think that they could make more money if Hawaii were an American state. So they stage a revolt and take control of the government. Now they want you, the U.S. president, to take control of Hawaii.

What Do You Think?

- What interests does the United States have in other countries?
- How important is protecting those interests?
- How important are the opinions of another country's citizens?

When should you get involved in the affairs of another country?

USA

1880

1887 Hawaii grants United States exclusive use of Pearl Harbor.

1892 Grover Cleveland is elected president.

1893 Planters overthrow the Hawaiian queen, Liliuokalani.

1896 William McKinley is elected president.

1898 Spanish-American War takes place.

1901 Theodore Roosevelt becomes president after McKinley is assassinated.

1904 Roosevelt is elected president.

1908 William Howard Taft is elected president.

1912 Woodrow Wilson is elected president.

1914 Panama Canal opens.

1917

World

1885 European nations divide Central and West Africa at the Berlin West Africa Conference.

1895 Sino-Japanese War ends with Japanese victory over China.

1900 Boxer Rebellion occurs in China.

1903 Republic of Panama established.

1904 Russo-Japanese War begins.

1910 Mexican Revolution begins.

1916 Pancho Villa makes a raid in the United States.

Chapter 23 SETTING THE STAGE

BEFORE YOU READ

Previewing the Theme

Expansion By the end of the 1800s, the United States had become a world power. The nation used its economic and military strength to expand its influence in both Latin America—the part of the Americas south of the United States—and Asia. Some Americans worried that U.S. expansion overseas might weaken the nation's democratic principles. These are the issues and events explored in Chapter 23.

The American eagle spreads its wings over Asia and Latin America in this political cartoon from 1904.

What Do You Know?

Was expansion something new, or was it a force that you have seen before in the history of the United States?

THINK ABOUT

- the idea of Manifest Destiny
- the Louisiana Purchase
- the War with Mexico

What Do You Want to Know?

What do you think might have caused the United States to expand overseas at the end of the 1800s? Make a list of the possible reasons before you read the chapter.

READ AND TAKE NOTES

Reading Strategy: Finding Main Ideas An important skill for reading history is the ability to find main ideas. Identifying main ideas helps you to organize and understand the variety of details and examples that support those ideas. Use a chart like the one below to write main ideas about U.S. expansion overseas.

See Skillbuilder Handbook, page R5.

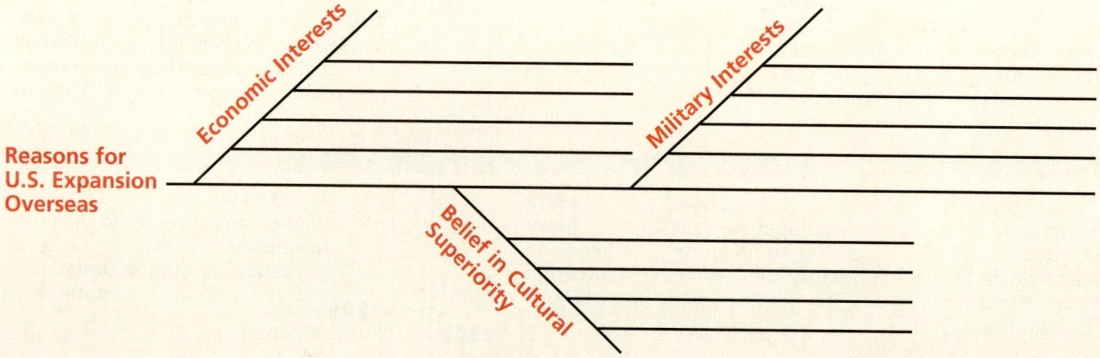

1 The United States Continues to Expand

TERMS & NAMES
imperialism
William Seward
Queen Liliuokalani

MAIN IDEA
The United States expanded its interest in world affairs and acquired new territories.

WHY IT MATTERS NOW
During this period, the United States acquired Alaska and Hawaii as territories.

ONE AMERICAN'S STORY

Alfred T. Mahan joined the U.S. Navy in the 1850s and served for nearly 40 years. In the 1890s, Mahan made use of his decades of experience to write several books on the historical importance of sea power. In one passage, he discussed the economic importance of trading stations and colonies.

A VOICE FROM THE PAST

The trading-station . . . [was] the same as the . . . colony. In both cases the mother-country had won a foothold in a foreign land, seeking a new outlet for what it had to sell, a new sphere for its shipping, more employment for its people, and more comfort and wealth for itself.

A. T. Mahan, *The Influence of Sea Power upon History, 1660–1805*

Naval historian Alfred Thayer Mahan at the turn of the century

Mahan encouraged government officials to build up American naval forces so that the United States could compete with other powerful nations. In this section you will learn how the United States began to extend its influence beyond the national boundaries.

Reasons for U.S. Expansion

Americans had always sought to expand the size of their nation. Throughout the 19th century, they extended their control toward the Pacific Coast. By the 1880s, however, many leaders became convinced that the United States should join the imperialist powers of Europe and establish colonies overseas. **Imperialism**—the policy by which stronger nations extend their economic, political, or military control over weaker territories—was a trend around the world.

European nations had been establishing colonies for centuries. In the late 19th century, Africa became a major area of European expansion. By the early 20th century, only two countries in Africa—Ethiopia and Liberia—remained independent.

Imperialist countries also competed for territory in Asia, especially in China. There, European nations had to compete with Japan, which had also become a world power by the end of the 1800s.

Most Americans gradually came to approve of the idea of expansion overseas. Three factors helped to fuel the development of American imperialism.

1. **Economic Interests.** Economic leaders argued that expansion would increase U.S. financial prosperity. Industry had greatly expanded after the Civil War. Many industrialists saw new colonies as a potential source of cheap raw materials. Agriculture had also expanded. Farmers pointed out that colonies would mean new markets for their products.
2. **Military Interests.** In his books, Alfred T. Mahan had argued that economic interests went hand-in-hand with military interests. Foreign policy experts agreed. They urged U.S. leaders to follow the European example and establish a military presence overseas.
3. **Belief in Cultural Superiority.** Many Americans believed that their government, religion, and even race were superior to those of other societies. Some people hoped to spread democratic ideas overseas. Others saw a chance to advance Christianity. Racist ideas about the inferiority of the nonwhite populations in many foreign countries were also used to justify American imperialism.

Reading **History**

A. Making Inferences Why might economic and military interests go hand in hand?

Each of these developments—economic interests, military interests, and a belief in cultural superiority—led the United States to a larger role on the world stage.

GEOGRAPHY SKILLBUILDER
Interpreting Maps

1. **Location** *What country lies to the west of Alaska?*
2. **Location** *On which Hawaiian island is Pearl Harbor?*

Seward and Alaska

A strong backer of expansion was **William Seward,** Secretary of State under presidents Abraham Lincoln and Andrew Johnson. Seward made his biggest move in 1867, when he arranged the purchase of Alaska from Russia.

Not everyone was pleased by Seward's move, though. At the time, the $7.2-million deal was widely criticized. Newspapers called Alaska a "Polar Bear Garden" and "Seward's Icebox." Even so, the purchase of the resource-rich territory turned out to be a great bargain for the United States.

Throughout his career, Seward continued to pursue new territory. Before he retired in 1869, he considered acquiring the Hawaiian Islands, a group of volcanic and coral islands in the central Pacific Ocean. That would not happen, however, for almost 30 more years.

Background

In the late 1800s, large gold fields were discovered in Alaska. The territory was also rich in fur-bearing animals, timber, copper, coal, and oil.

The Annexation of Hawaii

Reading History
B. Reading a Map Locate the Hawaiian Islands on the map on page 660.

In the early 1800s, Christian missionaries from the United States had moved to the Kingdom of Hawaii to convert the local population. Some of the missionaries' descendants started sugar plantations. By the late 1800s, wealthy planters dominated Hawaii's economy.

In 1891, **Queen Liliuokalani** (lee•LEE•oo•oh•kah•LAH•nee) became the leader of Hawaii. Believing that planters had too much influence, she wanted to limit their power. Around the same time, U.S. trade laws changed to favor sugar grown exclusively in American states.

American planters in Hawaii were upset by these threats to their political and economic interests. In January 1893, they staged a revolt. With the help of U.S. Marines, they overthrew the queen and set up their own government. They then asked to be annexed by the United States.

Vocabulary
annex: to add

U.S. leaders already understood the value of the islands. In 1887, they had pressured Hawaii to allow a U.S. naval base at Pearl Harbor, the kingdom's best port. The base became an important refueling station for American merchant and military ships bound for Asia.

Thus, when President Benjamin Harrison received the planters' request in 1893, he gave his approval and sent a treaty to the Senate. But before the Senate could act, Grover Cleveland became president. He did not approve of the planters' actions and withdrew the treaty. Hawaii would not be annexed until 1898, during the Spanish-American War. In the next section, you will read about the events that led to that war.

AMERICA'S HISTORY MAKERS

QUEEN LILIUOKALANI
1838–1917

As a young princess, Liliuokalani received a western education and toured the world. Although she learned about many cultures, she remained committed to Hawaii. An excellent musician, she wrote the famous Hawaiian song "Aloha Oe [Farewell to Thee]."

She was the first queen of Hawaii and proved to be a good leader. She resisted the foreign takeover of Hawaii and inspired a revolt against the planters. Only in 1895, when the safety of her supporters was threatened, did she agree to give up her throne.

How did Queen Liliuokalani protect her followers after planters seized power?

Section 1 Assessment

1. Terms & Names

Identify:
- imperialism
- William Seward
- Queen Liliuokalani

2. Taking Notes

Use a chart like the one shown to record causes of U.S. expansion overseas in the late 1800s.

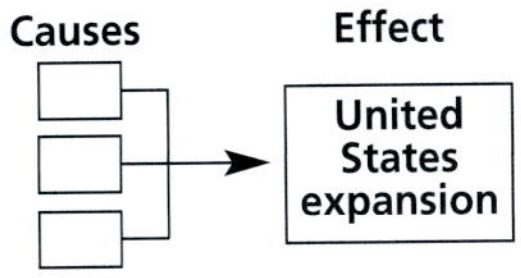

Which was the most important cause?

3. Main Ideas

a. Where was the focus of U.S. expansion before the late 1800s?

b. How did William Seward contribute to U.S. expansion?

c. Why did the American planters' request for the annexation of Hawaii fail in the early 1890s?

4. Critical Thinking

Making Inferences What benefits were American planters looking for when they staged a revolt in 1893?

THINK ABOUT
- the new policies of Queen Liliuokalani
- changes in U.S. trade laws

ACTIVITY OPTIONS

TECHNOLOGY
ART

Read more about Hawaii's Queen Liliuokalani. Outline a **video presentation** on the overthrow of the queen or plan a **mural** that depicts the event.

2 The Spanish-American War

MAIN IDEA

Independence movements in Spanish colonies led to the Spanish-American War in 1898.

WHY IT MATTERS NOW

U.S. involvement in Latin America and Asia expanded greatly after the Spanish-American War.

TERMS & NAMES

- yellow journalism
- U.S.S. *Maine*
- Spanish-American War
- Rough Riders
- Platt Amendment
- Anti-Imperialist League
- Luis Muñoz Rivera

ONE AMERICAN'S STORY

José Martí was forced to leave Cuba in the 1870s, when he was still a teenager. In those years, the Caribbean island was a Spanish colony, and he had spoken out for independence. Martí later described the terrible conditions that existed under Spanish rule.

A VOICE FROM THE PAST

Cuba's children . . . suffer in indescribable bitterness as they see their fertile nation enchained and also their human dignity stifled . . . all for the necessities and vices of the [Spanish] monarchy.

José Martí, quoted in *José Martí, Mentor of the Cuban Nation*

After being forced out of Cuba, Martí spent much of his life in the United States. In 1892, he was elected to lead the Cuban Revolutionary Party. At the Party's headquarters in New York City, Martí began to plan a revolt against Spain that began in 1895.

Martí's lifelong struggle for Cuban independence made him a symbol of liberty throughout Latin America. In this section, you will read how U.S. disapproval of Spain's treatment of Cubans led to the Spanish-American War.

José Martí dedicated his life to the Cuban struggle for independence from Spain.

Rebellion Against Spain

The Spanish empire was crumbling at the end of the 19th century. Spain had once controlled most of the Americas, including land that became part of the United States. By the 1890s, however, it owned only a few colonies. Among them were the Philippine Islands in the Pacific and the Caribbean islands of Cuba and Puerto Rico. (See the maps on page 665.) Many of the inhabitants of these colonies had begun to demand independence.

Cubans had revolted against Spain several times in the second half of the nineteenth century. Each time, Spanish soldiers defeated the rebels. In 1895, an ongoing economic depression had increased Cubans' anger over Spanish rule, and they rebelled again. José Martí, who had helped to organize the rebellion from New York, returned to Cuba. He was killed in a skirmish with Spanish troops shortly after, but the revolt continued.

Spain sent General Valeriano "the Butcher" Weyler to crush the rebels. Weyler's methods were harsh. He forced many Cubans from their homes and placed them in camps guarded by Spanish troops. Thousands died of starvation and disease in the camps.

The revolt in Cuba caused alarm in the United States. Business leaders were concerned because the fighting disrupted U.S. trade with Cuba. Most Americans, however, became outraged when the press began to describe the brutality of Spanish officials. Two New York City newspapers, in particular, stirred up people's emotions.

*Reading*History

A. Forming Opinions How can newspapers affect public opinion?

The *World,* owned by Joseph Pulitzer, and the *New York Journal,* owned by William Randolph Hearst, were battling for customers. Both owners were able to attract readers by printing stories that described—and often exaggerated—news about Spanish cruelty. This sensational style of writing was known as **yellow journalism.** It was named after "The Yellow Kid," a popular comic strip that ran in the two New York papers.

The United States Goes to War

William McKinley, the U.S. president in 1898, did not want war. "I have been through [the Civil War]," he told a friend. "I have seen the dead piled up, and I do not want to see another."

Even so, public opinion—stirred up by sensational newspaper reports—forced McKinley to take action. He demanded that Spain halt its harsh treatment of Cubans. Spain did bring General Weyler home, but conditions remained severe.

In January 1898, McKinley sent the **U.S.S. *Maine*** to Cuba. Riots had broken out in the capital, Havana, and the battleship was dispatched to protect U.S. citizens. Then, the following month, the *Maine* exploded and sank in Havana's harbor, killing 260 sailors.

No one knows what caused the explosion. Most historians today believe that it was an accident. For example, a spark might have set off an explosion in the ship's coal bunker. Even so, Americans blamed Spain.

The explosion of the *Maine* and accounts of the event by yellow journalists led many Americans to favor war against Spain.

Detecting Bias in the Media

Professional journalists try to report the news without bias—that is, without letting their personal opinions or those of their employer influence what they write. Unbiased reporting allows citizens to weigh the facts, come to their own understanding of issues and events, and act accordingly.

As you have read, journalists and their employers do not always avoid bias. Before the United States declared war on Spain in 1898, "yellow journalists" deliberately exaggerated stories to help sell newspapers. These stories helped to turn public opinion in favor of the Spanish-American War. They reflected a bias—that the United States should declare war on Spain.

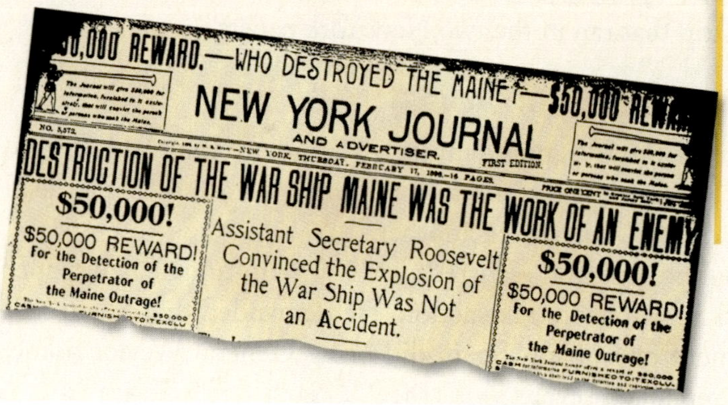
,000 REWARD.—WHO DESTROYED THE MAINE?—$50,000 REWA

NEW YORK JOURNAL AND ADVERTISER.

FIRST EDITION.

DESTRUCTION OF THE WAR SHIP MAINE WAS THE WORK OF AN ENEMY

$50,000!

$50,000 REWARD! For the Detection of the Perpetrator of the Maine Outrage!

Assistant Secretary Roosevelt Convinced the Explosion of the War Ship Was Not an Accident.

$50,000!

$50,000 REWARD! For the Detection of the Perpetrator of the Maine Outrage!

William Randolph Hearst ran this headline in his *New York Journal* before authorities had a chance to determine the cause of the *Maine's* explosion.

Can You Find Bias in the Media?

1. With a small group, collect news stories from different sources that cover the same issue or event.
2. Record any differences in the way a specific issue or event is covered by the sources you have selected.
3. Review the differences and decide whether any of the sources showed bias in their coverage.
4. Write a report that describes any bias you might detect. Explain why the biased source might have reported the story the way it did.

See Citizenship Handbook, page 284.

Visit www.mcdougallittell.com to learn more about the news media.

"Remember the Maine!" became a call to arms. On April 20, 1898, President McKinley signed a congressional resolution that called for Cuba's independence and demanded a withdrawal of Spanish forces. He gave Spain three days to respond. Spain refused, and the **Spanish-American War** began.

The War in the Philippines

The United States went to war to fight for Cuban freedom. But the first major battle of the Spanish-American War took place in a Spanish colony on the other side of the world—the Philippine Islands. Many Filipinos, as the inhabitants of the islands were called, had also revolted against Spanish rule in the 1890s.

Before the war began, the Filipino independence movement had attracted the attention of Theodore Roosevelt. At that time, Roosevelt was assistant secretary of the navy. He put a fleet of American ships in Hong Kong on alert. Their leader, Commodore George Dewey, prepared his forces and made contact with the head of the Filipino rebel forces, Emilio Aguinaldo (eh•MEE•lyoh AH•gee•NAHL•doh).

Reading **History**

B. Making Inferences Why did Theodore Roosevelt put the U.S. fleet in Hong Kong on alert?

When the war began, Dewey set out for Manila, the Philippine capital, where part of the Spanish fleet was located. The battle in Manila Bay began early on the morning of May 1, 1898. By a little past noon,

Dewey's forces had destroyed the Spanish fleet. About 380 Spanish sailors were dead or wounded. No Americans died. U.S. troops, aided by Filipino rebels, took control of Manila in August.

Dewey became an instant hero in the United States. Thousands of babies born at the time of the victory in Manila Bay were named for him, and a chewing gum called "Dewey's Chewies" became popular.

The War in the Caribbean

When the Spanish-American War began, the U.S. Army had only 28,000 men. Within four months, over 200,000 more joined up. Among the new recruits was Theodore Roosevelt, who had resigned from the Navy Department to volunteer.

Roosevelt helped to organize the First United States Volunteer Cavalry. This unit was nicknamed the **Rough Riders.** Its recruits included cowboys, miners, college students, New York policemen, athletes, and Native Americans.

In June, the Rough Riders and about 16,000 other soldiers—nearly a quarter of them African American—gathered in Tampa, Florida. They then set out for Santiago, a Spanish stronghold in southern Cuba. When the Rough Riders arrived, their dark-blue wool uniforms were too hot for the Cuban climate. Also, many of the soldiers came down with tropical diseases. Even so, they fought their way toward Santiago.

In order to gain control of Santiago's port, American troops had to capture San Juan Hill. They attacked the Spanish on July 1.

GEOGRAPHY SKILLBUILDER Interpreting Maps

1. **Movement** *Where was Dewey's fleet before it steamed toward the Philippines?*
2. **Location** *About how far is Havana from the tip of southern Florida?*

HISTORY through ART

On July 1, 1898, U.S. troops, including the Rough Riders, attacked San Juan Hill outside of Santiago, Cuba. This painting shows Theodore Roosevelt leading a cavalry charge up the hill—even though the regiment's horses had been left behind in Florida.

Why did the artist show the Rough Riders on horses when they made their charge on foot?

African-American soldiers from the Tenth Cavalry began to drive the Spanish back. Roosevelt and the Rough Riders joined them as they rushed forward and captured the hill.

Two days later, American ships destroyed Spain's fleet as it tried to escape Santiago Harbor. On July 17, the city surrendered. A week later, U.S. forces took Puerto Rico. Finally, on August 12, 1898, Spain signed a truce. To U.S. Secretary of State John Hay, it had been "a splendid little war." For Spain, four centuries of glory had come to an end.

Results of the War

Although the war had been fought over Cuba, U.S. leaders demanded that Spain give up other colonies after the war—including Puerto Rico, the island of Guam, and the Philippines. Spain had no choice but to agree. The final peace treaty was signed in Paris in December 1898.

Reading **History**

C. Making Inferences Why did the United States demand that Spain give up territories in addition to Cuba?

One of the most difficult questions for U.S. leaders after the war was what to do with the Philippines. Filipinos had fought alongside Americans during the war and believed that Spain's defeat would bring them independence. But President McKinley eventually decided that the Philippines should become an American colony.

Filipinos were bitterly disappointed. Led by Emilio Aguinaldo, they began to fight against their new colonial rulers. American troops sent to put down the resistance were not able to restore order until 1902.

The United States was also reluctant to grant Cuba complete independence. First, Cuba had to add the **Platt Amendment** to its constitution. This gave the United States the right to intervene in Cuban affairs anytime there was a threat to "life, property, and individual liberty." Cuba also had to allow a U.S. naval base at Guantánamo Bay.

Puerto Rico became an American territory. The United States set up a government and appointed the top officials. Puerto Ricans had little to say in their own affairs. Only in 1917 would the United States agree to make Puerto Rico a self-governing territory and grant U.S. citizenship to all Puerto Ricans.

The Anti-Imperialist League

U.S. treatment of Spain's former colonies after the Spanish-American War disappointed many people in the United States.

Several well-known Americans, including businessman Andrew Carnegie, reformer Jane Addams, and writer Mark Twain, joined with others to form the **Anti-Imperialist League.** Members of the League believed that Americans should not deny other people the right to govern themselves.

Reading **History**

C. Finding Main Ideas Why did the Anti-Imperialist League oppose U.S. efforts to collect colonies?

A VOICE FROM THE PAST

We hold that the policy known as imperialism is hostile to liberty. . . . We regret that it has become necessary in the land of Washington and Lincoln to reaffirm that all men, of whatever race or color, are entitled to life, liberty, and the pursuit of happiness.

From the *Platform of the American Anti-Imperialist League*

The voice of the Anti-Imperialist League was lost, however, in the roar of popular approval of the Spanish-American War.

Many Americans hoped that their nation would surpass the glory of the old Spanish empire. In the next section, you will read more about how the United States continued its involvement overseas.

AMERICA'S HISTORY MAKERS

LUIS MUÑOZ RIVERA

1859–1916

Luis Muñoz Rivera devoted his life to obtaining self-government for Puerto Rico—first from Spain and then from the United States.

After Spain granted Puerto Rico self-rule in 1897, Muñoz Rivera joined the government. He resigned and renewed his struggle when Puerto Rico became a U.S. territory.

Muñoz Rivera died just before the United States granted Puerto Ricans U.S. citizenship and a large measure of self-government.

In what ways did Muñoz Rivera use his leadership skills to help his country?

Section 2 Assessment

1. Terms & Names

Identify:

- yellow journalism
- U.S.S. *Maine*
- Spanish-American War
- Rough Riders
- Platt Amendment
- Anti-Imperialist League
- Luis Muñoz Rivera

2. Taking Notes

Use a time line to record the major events of the Spanish-American War.

Spanish-American War, 1898

About how long did the Spanish-American War last?

3. Main Ideas

a. What led to the Cuban rebellion against Spain in 1895?

b. What was the first major military event of the Spanish-American War?

c. What happened in the Philippines after the war?

4. Critical Thinking

Forming Opinions Did the United States betray its democratic principles when it made the Philippines a colony?

THINK ABOUT

- the public's response to yellow journalists and U.S. military victories
- the work of the Anti-Imperialist League

ACTIVITY OPTIONS

LANGUAGE ARTS

MATH

Research the Spanish-American War. Write a **television news script** covering a major battle or create a **database** of wartime casualties.

3 U.S. Involvement Overseas

TERMS & NAMES
sphere of influence
Open Door Policy
Boxer Rebellion
Panama Canal
Roosevelt Corollary

MAIN IDEA
In the early 1900s, the United States expanded its involvement in Asia and Latin America.

WHY IT MATTERS NOW
The United States still trades extensively with Asian and Latin American countries.

A Japanese artist portrayed Commodore Matthew Perry's meeting with Japanese officials in 1853.

ONE AMERICAN'S STORY

In 1852, President Millard Fillmore sent Commodore Matthew Perry on a mission to open Japan to U.S. trade. For over two centuries, Japan's rulers had kept the country closed to most foreigners. Perry wanted to break Japan's traditional policy with a demonstration of American power.

A VOICE FROM THE PAST

[I was determined] to adopt an entirely contrary plan of proceedings from that of all others who had . . . visited Japan on the same errand [to open up trade]: to demand as a right and not to [ask] as a favor those acts of courtesy which are due from one civilized nation to another.

Commodore Matthew Perry, *Personal Journal*

Perry arrived in Japan in July 1853 but was not able to win a trade agreement. He departed but returned the next year with more warships. Under the threat of force, the Japanese gave in. In March 1854, Japan signed a treaty giving American ships access to its ports. In this section, you will read more about U.S. involvement in Asia, as well as in Latin America.

A Power in the Pacific

Throughout the 1800s, the United States continued to expand its involvement in Asia. Toward the end of the century, the United States acquired a chain of islands—including Hawaii and Guam—that stretched across the Pacific Ocean to Asia.

During the Spanish-American War, Americans fought in the Philippine Islands, a Spanish colony in eastern Asia. After the war, the United States annexed the islands and put down the Filipino independence movement.

Some Americans objected to the annexation of the Philippines. However, supporters of imperialism, such as Indiana senator Albert Beveridge, applauded U.S. actions. Beveridge boasted, "The Philippines are ours forever. And just beyond the Philippines are China's [unlimited] markets. We will not retreat from either. . . . The power that rules the Pacific is the power that rules the world."

Many Americans looked forward to the profits promised by Asian markets and resources. Others saw a chance to extend U.S. democracy and culture in the region. The Philippines would provide a base for these activities.

> *"The power that rules the Pacific . . . rules the world."*
> **Albert Beveridge**

The United States in China

As Senator Beveridge noted, control of the Philippines gave Americans greater access to China. However, by the time the United States acquired the islands, other imperialist nations, including Japan, were already deeply involved in China.

When Commodore Perry opened Japan to U.S. trade in the 1850s, he also opened the nation to western ideas. After Perry's voyages, Japan began to modernize and soon emerged as a world power. In the 1890s, Japan demonstrated its strength in a successful war against China.

After the war, both Japan and the major European powers expanded their **spheres of influence** in China. These were areas where foreign nations claimed special rights and economic privileges. By the late 1890s, France, Germany, Britain, Japan, and Russia had established prosperous settlements along the coast of China. They also claimed exclusive rights to railroad construction and mining development in the nation's interior.

The competition for spheres of influence worried U.S. leaders who wanted access to China's markets and resources. In 1899, Secretary of State John Hay asked nations involved in the region to follow an **Open Door Policy**. This meant that no single country should have a monopoly on trade with China. Eventually, most of the nations accepted Hay's proposal.

Reading **History**
A. Recognizing Causes Why did John Hay propose the Open Door Policy?

Many Chinese people were not pleased by the presence of foreigners. One group, called the "Boxers," was angered by the privileges given to foreigners and the disrespect they showed toward Chinese traditions. In 1900, Chinese resentment toward foreigners' attitude of cultural superiority led to a violent uprising known as the **Boxer Rebellion**. Many foreigners were killed before the uprising was put down by an international force.

GEOGRAPHY SKILLBUILDER Interpreting Maps

1. **Place** *What country controlled the port of Macao?*
2. **Region** *What country had the largest sphere of influence in the coastal region of China?*

The Panama Canal

As American interests in the Pacific expanded, easy access to the region became vital. For that reason, U.S. leaders proposed a canal to connect the Atlantic and Pacific oceans. A canal would mean that U.S. ships would not have to travel around South America. The Spanish-American War, fought in both oceans, also made clear the need for such a shortcut.

The South American nation of Colombia controlled the best spot for the canal—the Isthmus of Panama. But Colombia was unwilling to give up this land. Ignoring Colombia's right to control its territory, President Roosevelt sent the U.S. Navy to support a revolution on the isthmus. Out of this revolution, the new nation of Panama was created in 1903.

Vocabulary
isthmus: a narrow strip of land connecting two larger masses of land

The new Panamanian leaders gave the U.S. government rights to a ten-mile-wide strip of land called the Canal Zone. There, the United States would build the **Panama Canal,** the shortcut that would connect the Atlantic and Pacific oceans.

Some people in Latin America and the United States opposed Roosevelt's actions. They believed that he had interfered in Colombia's affairs in order to cheat it out of land. In 1921, the United States finally paid Colombia $25 million for the loss of Panama.

***Reading*History**
B. Summarizing What political difficulty faced U.S. leaders who wanted to build the Panama Canal?

Building the Canal

Building the canal was extremely difficult. The land was swampy and full of mosquitoes that carried the organism that causes malaria. In spite of the difficulties, the project moved forward. When Roosevelt visited Panama in 1906, he wrote a letter describing the work.

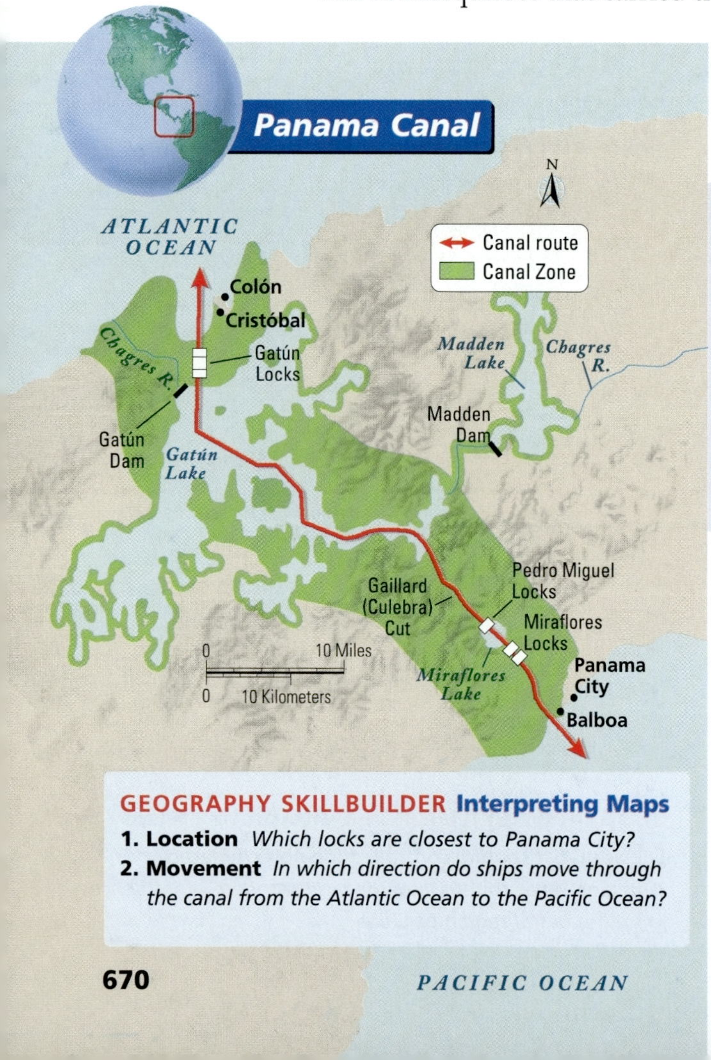

GEOGRAPHY SKILLBUILDER Interpreting Maps
1. **Location** *Which locks are closest to Panama City?*
2. **Movement** *In which direction do ships move through the canal from the Atlantic Ocean to the Pacific Ocean?*

A VOICE FROM THE PAST

Steam shovels are hard at it; scooping huge masses of rock and gravel and dirt previously loosened by the drillers and dynamite blasters, loading it on trains which take it away. . . . They are eating steadily into the mountain cutting it down and down. . . . It is an epic feat.

Theodore Roosevelt, from a letter sent to his son

More than 45,000 workers, including many black West Indians, labored for years on the canal. They did not finish the work until 1914. The canal cost $352 million, the most expensive project up to that time. It was expensive in human terms, too. More than 5,000 workers died from diseases or accidents.

Background
In 1977, the United States signed a treaty that transfers ownership of the canal to Panama on December 31, 1999.

Technology OF THE Time

How the Panama Canal Works

Engineers faced a problem in building the Panama Canal. Because of the region's different landscape elevations, no waterway would remain level. They solved this dilemma by building three sets of *locks*—water-filled chambers that raise or lower ships to match a canal's different water levels.

1. The lock gates open on one end to allow the ship to enter.
2. The gates close, and water is pumped in or out depending on whether the ship is moving up or down.
3. Once the water in the chamber and the canal ahead is level, the second gate opens and the ship moves on.

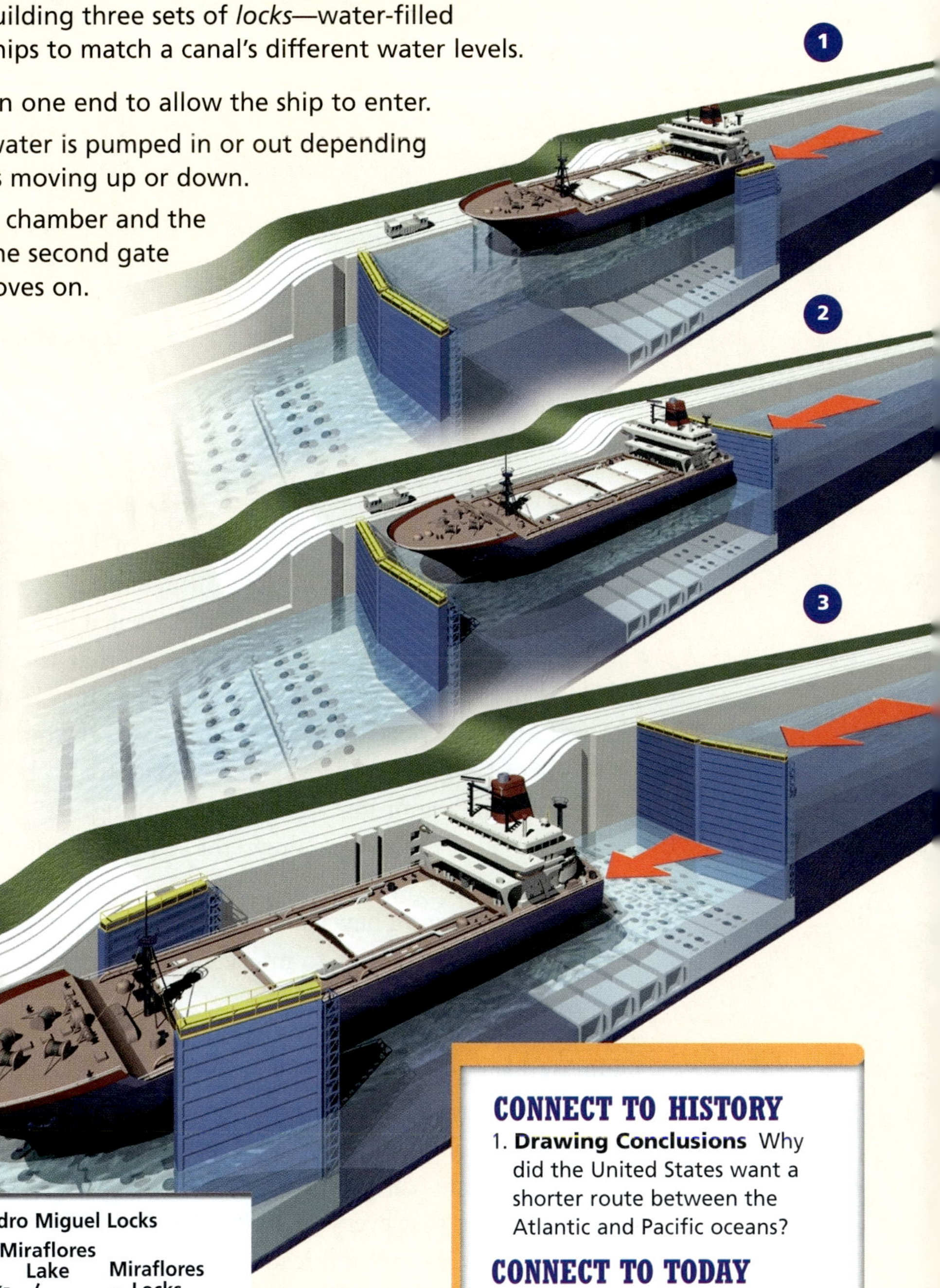

The locks, whose steel gates rise six stories high, can hold as much as 26 million gallons of water—enough to supply a major U.S. city for one day.

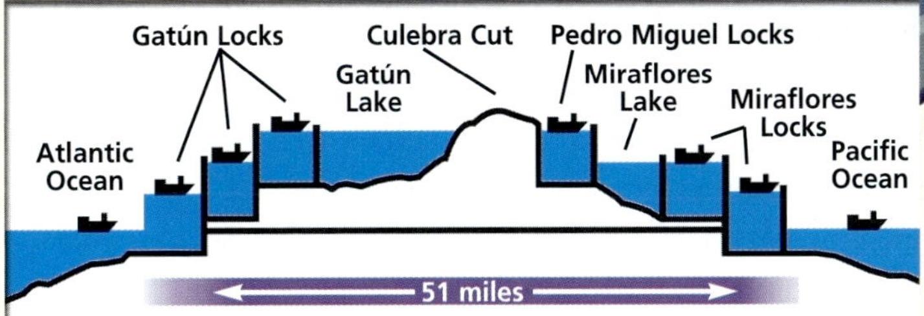

This cross-section shows the different elevations and locks that a ship moves through on the 8–9 hour trip through the canal. Before the canal was built, a trip around South America could take two months.

CONNECT TO HISTORY

1. **Drawing Conclusions** Why did the United States want a shorter route between the Atlantic and Pacific oceans?

CONNECT TO TODAY

2. **Researching** What is the economic and political status of the Panama Canal today?

Visit www.mcdougallittell.com to learn more about the Panama Canal.

U.S. in Latin America, 1898–1917

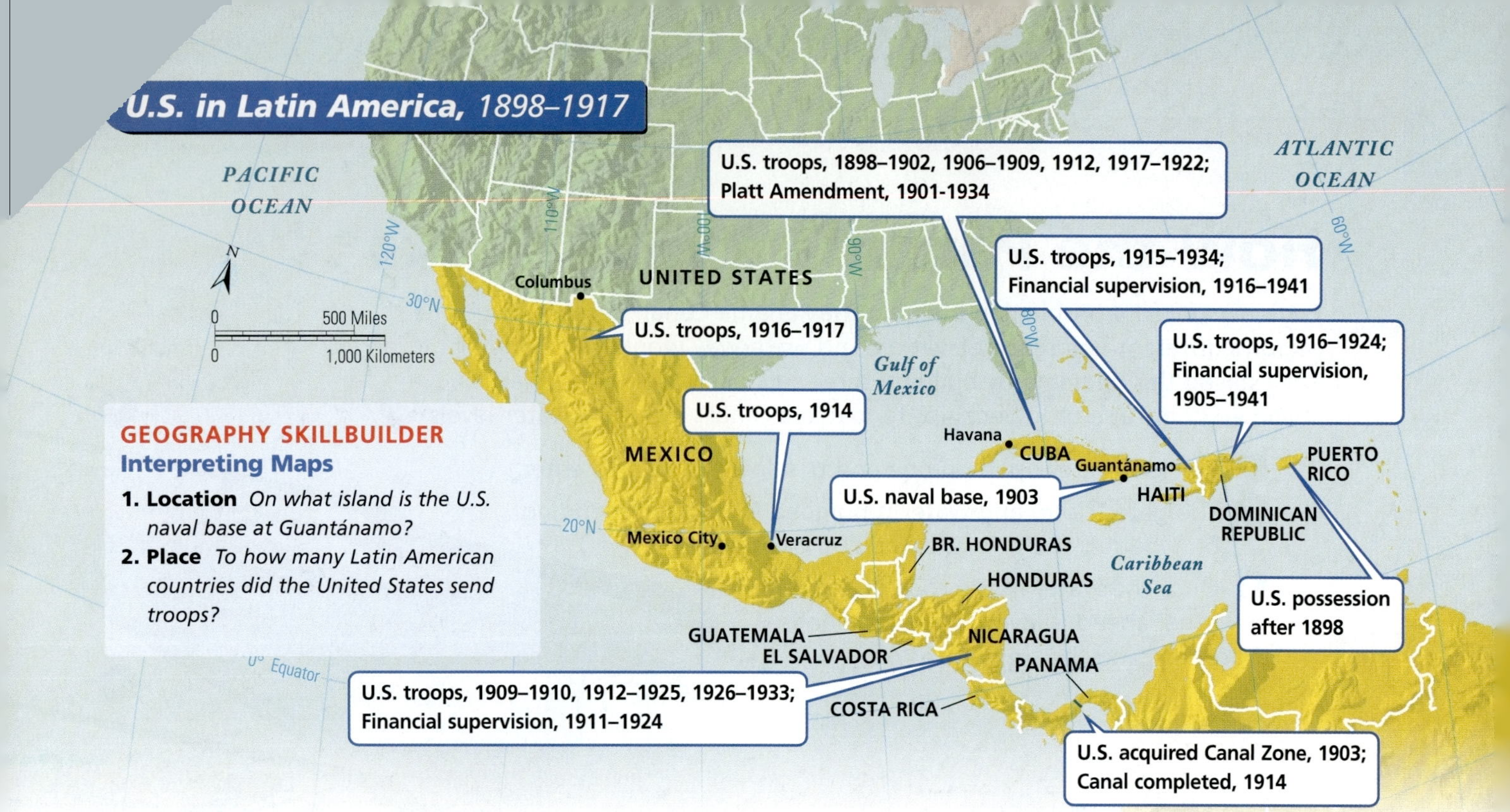

GEOGRAPHY SKILLBUILDER
Interpreting Maps

1. **Location** *On what island is the U.S. naval base at Guantánamo?*
2. **Place** *To how many Latin American countries did the United States send troops?*

U.S. Involvement in Latin America

The Panama Canal was only one sign of U.S. involvement in Latin America. As the U.S. economy continued to grow, so did Americans' interest in the resources of their southern neighbors.

Businesses in the United States found that they could cheaply buy food and raw materials—for example, bananas, coffee, and copper—from Latin America. They shipped these goods to the United States and sold them for higher prices. U.S. companies also bought large amounts of land in the region for farming and mining.

As economic interests drew the United States deeper into Latin American affairs, U.S. leaders became concerned about political stability in the region. They were especially worried that instability might tempt European nations to intervene in the region.

Reading **History**
C. Making Inferences Why was the United States interested in the political stability of Latin America?

Policing the Hemisphere

During his presidency, Theodore Roosevelt made it clear that the United States would remain the dominant power in the Western Hemisphere. He summed up his foreign policy toward the region with an African saying: "Speak softly, but carry a big stick." Roosevelt, however, rarely spoke softly. He made sure that everyone knew the United States would use military force if its interests were threatened.

Roosevelt reminded European powers of the Monroe Doctrine—the policy that prevented other nations from intervening in Latin America. In 1904, he added the **Roosevelt Corollary**. Now, the doctrine would not only prevent European intervention in Latin America; it also authorized the United States to act as a "policeman" in the region. That is, U.S. leaders would now intervene in Latin America's domestic affairs

Vocabulary
corollary: a statement that follows logically from an earlier statement

when they believed that such action was necessary to maintain stability.

In 1905, the United States used the Roosevelt Corollary to take control of the Dominican Republic's finances after the country failed to pay its foreign debts. A year later, when a revolt threatened Cuba's government, the policy was used to send troops there.

Later presidents expanded on Roosevelt's "big stick diplomacy." William Howard Taft urged American businesses to invest in Latin America, promising military action if anything threatened these investments. He kept his word. In 1912, Taft sent marines to Nicaragua to restore order.

Background
Taft's policy was known as "dollar diplomacy."

President Taft's successor, Woodrow Wilson, also intervened in Latin America. When a revolution in Mexico began to threaten U.S. interests, Wilson took action. In 1914, he sent a fleet to Veracruz after U.S. sailors were arrested. Two years later, he sent troops into Mexico after a Mexican revolutionary named Pancho Villa (PAHN•choh VEE•yah) raided New Mexico and killed 19 Americans in the town of Columbus.

Americans rarely questioned U.S. actions in Latin America. They saw their nation as a good police officer, maintaining peace and preventing disorder. But many Latin Americans saw the United States as an imperial power that cared only about its own interests. This mistrust continues to trouble U.S. relations with its neighbors. In the next chapter, you will read about U.S. involvement in another part of the world—Europe.

Now and then

GLOBO COP?

In the early 1900s, the United States used its "police powers" in the Western Hemisphere. Today, U.S. forces participate in police actions all over the globe. This fact has led some journalists to call the United States the "Globo Cop."

In the 1990s, U.S. forces helped lead international police actions in Somalia (see photo below), Yugoslavia, and other areas in crisis. The United States also led the Gulf War forces that liberated Kuwait after it was seized by Iraq.

The United States continues to patrol its own hemisphere, too. In 1989, U.S. troops invaded Panama to overthrow dictator Manuel Noriega.

Section 3 Assessment

1. Terms & Names

Identify:
- sphere of influence
- Open Door Policy
- Boxer Rebellion
- Panama Canal
- Roosevelt Corollary

2. Taking Notes

Use a chart like the one below to record details about U.S. involvement in Asia and Latin America.

Asia	Latin America

How was U.S. involvement in Asia different from that in Latin America?

3. Main Ideas

a. Why was the United States interested in the Philippines?

b. Why was the nation of Panama created in 1903?

c. How did the Roosevelt Corollary change U.S. foreign policy?

4. Critical Thinking

Drawing Conclusions
Why did the United States become so heavily involved in Asia and Latin America?

THINK ABOUT
- U.S. economic growth
- American military interests

ACTIVITY OPTIONS

SCIENCE
MATH

Research the Panama Canal. Build a simple **model** of the canal or create a **graph** that shows how many ships use the canal each year.

VISUAL SUMMARY

Becoming a World Power

The United States Continues to Expand

In the late 1800s, the United States began to expand overseas.

- Alaska was purchased from Russia.
- Planters took over Hawaii's government.

The Spanish-American War

Events in Cuba, a Spanish colony in the Caribbean, led to the Spanish-American War.

- U.S. forces won victories in the Caribbean and in Asia.
- Spain gave up its colonies in Cuba, Puerto Rico, the Philippines, and Guam.

U.S. Involvement Overseas

In both Asia and Latin America, the United States began to play a larger role.

- U.S. leaders insisted on an Open Door Policy in China.
- The United States built the Panama Canal.

TERMS & NAMES

1. imperialism
2. Queen Liliuokalani
3. yellow journalism
4. Spanish-American War
5. Rough Riders
6. Anti-Imperialist League
7. Open Door Policy
8. Boxer Rebellion
9. Panama Canal
10. Roosevelt Corollary

REVIEW QUESTIONS

The United States Continues to Expand (pages 659–661)

1. Why did Americans become interested in overseas expansion in the late 1800s?
2. How did the public react when William Seward negotiated the purchase of Alaska in 1867?
3. Why did the United States take an interest in Hawaii?
4. Why might President Cleveland have wanted to restore Liliuokalani to the Hawaiian throne?

The Spanish-American War (pages 662–667)

5. How did the Spanish-American War begin?
6. What were the most important battles of the war?
7. What territory did the United States take control of as a result of its victory over the Spanish?

U.S. Involvement Overseas (pages 668–673)

8. Why did U.S. leaders want access to China's markets after the Spanish-American War?
9. Why was there an interest in building a canal across Latin America?
10. How were the Latin American policies of Roosevelt, Taft, and Wilson similar?

CRITICAL THINKING

1. USING YOUR NOTES

Using your completed chart, answer the questions below.

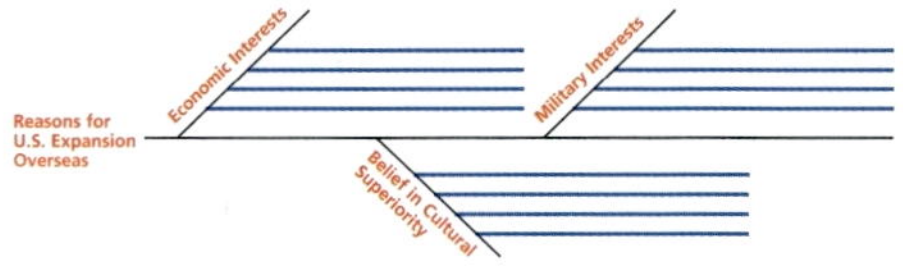

a. How did U.S. economic interests in Latin America influence the foreign policy of the United States?

b. In what ways was the Boxer Rebellion a reaction to the attitude of foreigners in China?

2. ANALYZING LEADERSHIP

What qualities made Theodore Roosevelt an effective leader?

3. THEME: EXPANSION

How did U.S. expansion at the end of the 19th century compare with expansion that occurred earlier? Discuss both similarities and differences.

4. APPLYING CITIZENSHIP SKILLS

How might the activities of the Anti-Imperialist League have helped to remind citizens of their democratic responsibilities?

5. FORMING OPINIONS

The "yellow journalism" of major newspapers influenced U.S. foreign policy at the turn of the century. How does modern media, such as television, shape public opinion today?

Interact *with* History

How has your study of U.S. involvement overseas at the turn of the century influenced your opinion about getting involved in the affairs of another country?

HISTORY SKILLS

1. INTERPRETING GRAPHS

Study the graph and then answer the questions.

U.S. Trade Expansion, *1865–1915*

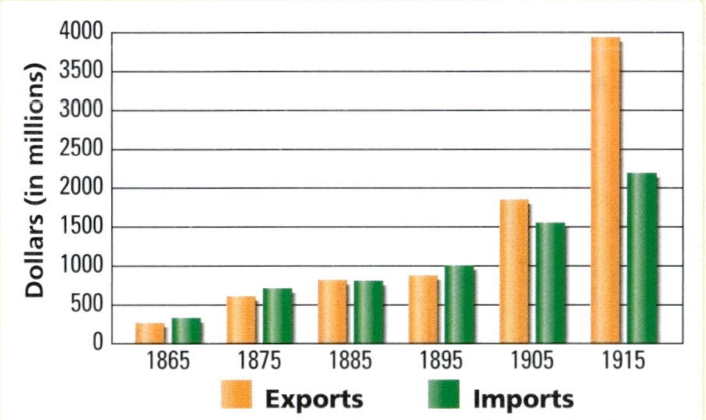

Source: *Historical Statistics of the United States*

Basic Graph Elements

a. What does this graph represent?

Interpreting the Graph

b. What were the values of U.S. exports in 1865 and 1915?

c. What were the values of U.S. imports in 1865 and 1915?

2. INTERPRETING PRIMARY SOURCES

W. A. Rogers's political cartoon from 1904 shows Theodore Roosevelt tugging the U.S. Navy around the Caribbean.

a. What do you think this cartoon is about?

b. Why did the cartoonist portray Roosevelt with a big stick?

ALTERNATIVE ASSESSMENT

1. INTERDISCIPLINARY ACTIVITY: Geography

Making a Map During the 19th century, the United States acquired several islands in the Pacific Ocean. Do more research on U.S. possessions in the Pacific and make a map that shows the islands' locations.

2. COOPERATIVE LEARNING ACTIVITY

Creating a News Story Imagine that you are a journalist in the late 1800s. The publisher of the *New York Journal,* William Randolph Hearst, has asked you to put together a team of reporters and artists. With your team, plan and write an illustrated news story for the *Journal* that features an important event from the Spanish-American War.

3. TECHNOLOGY ACTIVITY

Creating a Multimedia Presentation When the United States annexed the Philippines after the Spanish-American War, Filipinos rose in rebellion. Use the Internet, books, and other resources for a multimedia presentation on the Philippine-American war that resulted from the rebellion.

Visit www.mcdougallittell.com to learn more about U.S. expansion overseas.

Using presentation software, consider including the following content:

- descriptions or images of battles
- the views of Filipino and American soldiers, including African-American troops
- public opinion in the two countries
- statistics of casualties suffered by both sides during the conflict

4. HISTORY PORTFOLIO

Option 1 Review your section and chapter assessment activities. Select one that you think is your best work. Then use comments made by your teacher or classmates to improve your work and add it to your portfolio.

Option 2 Review the questions that you wrote for What Do You Want to Know? on page 658. Then write a short report in which you explain the answers to your questions. If any questions were not answered, do research to answer them. Add your answers to your portfolio.

CHAPTER 24

World War I

1914–1920

This photograph shows a battlefield view of trench warfare during World War I.

Interact *with* History

This 1918 poster urged Americans not to waste food during the war.

Kaiser Wilhelm II of Germany was blamed for the war.

People with war gardens bought less food, so more could be sent to Europe.

The year is 1918, and the United States has been drawn into World War I. Each citizen is called upon to help the war effort. Some will join the American armed forces and go to fight in Europe. Others will work in factories at home, producing weapons and supplies. Even children will do their part.

What Do You Think?

- How can Americans at home help win the war?
- What might U.S. soldiers experience in Europe?
- How might being at war affect the country?

How will you support the war effort?

USA

1914

August 15, 1914
U.S.-built Panama Canal officially opens.

May 7, 1915
Many Americans die as German U-boat sinks *Lusitania.*

November 7, 1916
Woodrow Wilson is reelected president.

April 2, 1917
Wilson asks Congress to declare war on Germany.

January 8, 1918
President Wilson proposes League of Nations.

November 2, 1920
Warren G. Harding is elected president.

1920

World

June 28, 1914
Austria-Hungary's Archduke Franz Ferdinand is assassinated, starting World War I.

February–December, 1915
Allies and Central Powers clash at Gallipoli in the Ottoman Empire.

July–November, 1916
French, British, and Germans suffer huge losses at the Battle of the Somme.

March 3, 1918
Russia withdraws from the war.

November 11, 1918
The Allies defeat the Central Powers, ending World War I.

June 28, 1919
The Allies and Germany sign the Treaty of Versailles.

Chapter 24 SETTING THE STAGE

BEFORE YOU READ

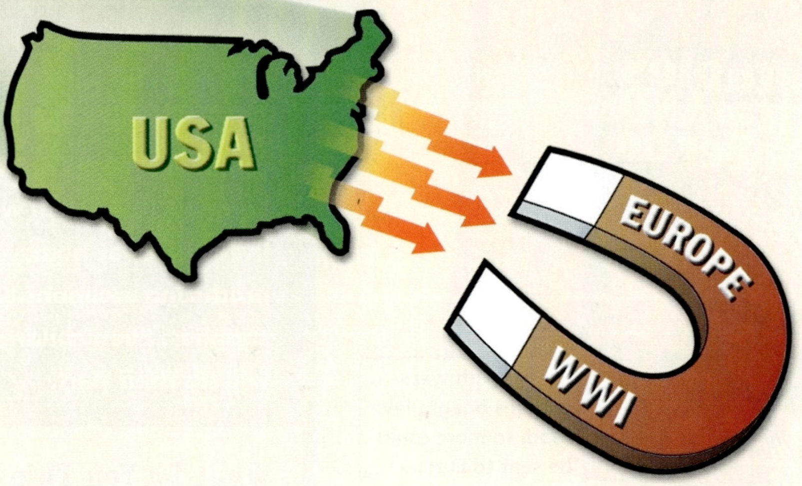

What Do You Know?

What do you think of when you hear the phrase "world war"? How many countries might have participated in the war? Where did most of the fighting take place?

THINK ABOUT

- what you've learned about World War I from movies or television
- reasons that millions of people might choose to risk their lives in a global conflict

Previewing the Theme

America in the World Although the United States was reluctant to join World War I, American involvement helped the Allies win the war. The war also brought about permanent changes in American society. Chapter 24 explains how the United States became involved in the war, how it mobilized for war, and how it tried to establish a lasting peace.

What Do You Want to Know?

What details do you need to help you understand what is involved in waging a world war? Make a list of these details in your notebook before you read the chapter.

READ AND TAKE NOTES

Reading Strategy: Recognizing Effects To help you make sense of what you read, learn to analyze the effects of important historical events. The chart below will help you analyze some of the effects of World War I, both on the world and on the United States. In each box, fill in a different effect. Add more boxes if you need to.

See Skillbuilder Handbook, page R10.

EFFECTS ON THE WORLD

EFFECTS ON THE UNITED STATES

1 War Breaks Out in Europe

MAIN IDEA

After World War I broke out, the United States eventually joined the Allied side.

WHY IT MATTERS NOW

This was the first time that the United States was involved in a European conflict.

TERMS & NAMES

militarism
Central Powers
Allies
trench warfare
U-boat
Woodrow Wilson
neutrality
Zimmermann telegram

ONE AMERICAN'S STORY

While the United States was forming its own empire, European nations were competing to expand their empires. Rivalry poisoned relationships among these nations. In the spring of 1914, tensions were running high in Europe. President Woodrow Wilson sent Colonel Edward M. House, his trusted advisor, to study the situation.

House gave the president a troubling report. He compared Europe to an open keg of gunpowder. "It only requires a spark," he said, "to set the whole thing off." Soon the spark ignited. On June 28, 1914, a young Serbian man shot and killed Archduke Franz Ferdinand. The archduke was the heir to the throne of Austria-Hungary. One month later, Austria declared war on Serbia. One by one, the nations of Europe chose sides and the Great War, later called World War I, began.

Archduke Franz Ferdinand and his wife are murdered at Sarajevo on June 28, 1914.

Causes of World War I

A single action, the assassination of the archduke, started World War I. But the conflict had many underlying causes.

1. **Imperialism.** Britain, France, Germany, and Italy competed for colonies in Africa and Asia. Because it had fewer colonies than Britain and France, Germany felt it deserved more colonies to provide it with resources and buy its goods.
2. **Nationalism.** Europeans were very nationalistic, meaning that they had strong feelings of pride, loyalty, and protectiveness toward their own countries. They wanted to prove their nations were the best. They placed their countries' interests above all other concerns. In addition, some ethnic groups hoped to form their own separate nations and were willing to fight for such a cause.
3. **Militarism.** The belief that a nation needs a large military force is **militarism**. In the decades before the war, the major powers built up their armies and navies.

A Divided Europe, Summer 1914

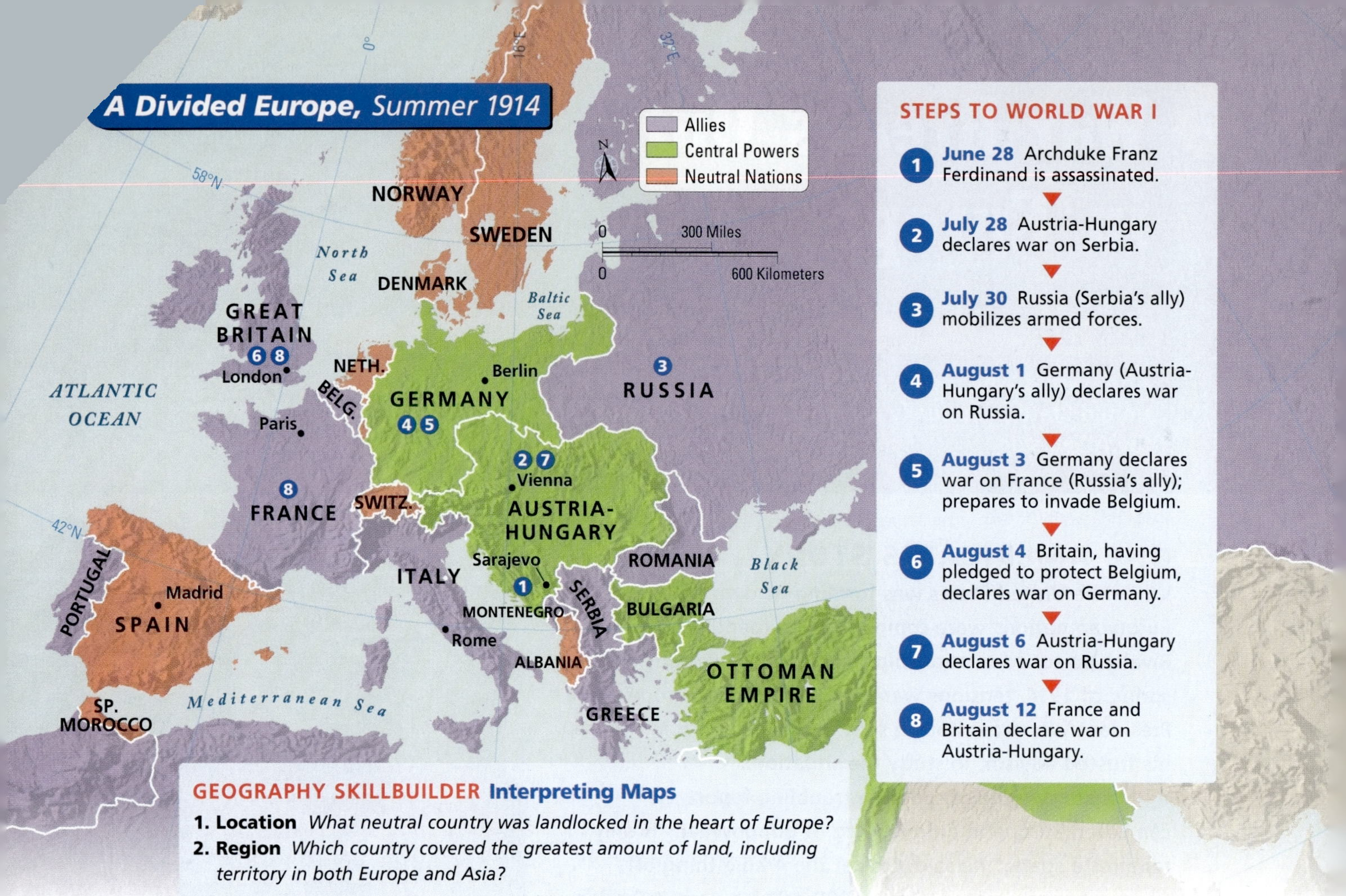

STEPS TO WORLD WAR I

1. **June 28** Archduke Franz Ferdinand is assassinated.
2. **July 28** Austria-Hungary declares war on Serbia.
3. **July 30** Russia (Serbia's ally) mobilizes armed forces.
4. **August 1** Germany (Austria-Hungary's ally) declares war on Russia.
5. **August 3** Germany declares war on France (Russia's ally); prepares to invade Belgium.
6. **August 4** Britain, having pledged to protect Belgium, declares war on Germany.
7. **August 6** Austria-Hungary declares war on Russia.
8. **August 12** France and Britain declare war on Austria-Hungary.

GEOGRAPHY SKILLBUILDER Interpreting Maps

1. **Location** *What neutral country was landlocked in the heart of Europe?*
2. **Region** *Which country covered the greatest amount of land, including territory in both Europe and Asia?*

4. **Alliances** In 1914, a tangled network of competing alliances bound European nations together. An attack on one nation forced all its allies to come to its aid. Any small conflict could become a larger war.

European nations had divided into two opposing alliances. The **Central Powers** were made up of Austria-Hungary, Germany, the Ottoman Empire, and Bulgaria. They faced the Allied Powers, or **Allies,** consisting of Serbia, Russia, France, Great Britain, Italy, and seven other countries.

Background
The Ottoman Empire included modern-day Turkey and Syria.

Stalemate in the Trenches

When the war began in August, most people on both sides assumed it would be over within a few months. With France as its goal, the German army invaded Belgium on August 4, 1914. Despite stiff resistance, the Germans fought their way west into France. They reached the Marne River about 40 miles from Paris. There the French, supported by the British, rallied and prepared to fight back. The First Battle of the Marne, in September 1914, stopped the German advance.

Reading History
A. Reading a Map On the map on page 688, find the site of the first Battle of the Marne.

Instead of one side quickly defeating the other, the two sides stayed stuck in the mud for more than three years. The soldiers were fighting a new kind of battle, **trench warfare.** Troops huddled at the bottom of rat-infested trenches. They fired artillery and machine guns at each other. Lines of trenches stretched across France from the English Channel to the border with Switzerland. (See pages 684–685 for an

Vocabulary
trench: a long, deep ditch dug for protection

illustration of the trenches.) For more than three years, the battle lines remained almost unchanged. Neither side could win a clear victory.

In the trenches, soldiers faced the constant threat of sniper fire. Artillery shelling turned the area between the two opposing armies into a "no man's land" too dangerous to occupy. When soldiers left their trenches to attack enemy lines, they rushed into a hail of bullets and clouds of poison gas.

*Reading*History
B. Reading a Map Find the site of the Battle of the Somme on the map on page 688.

When battles did take place, they cost many thousands of lives, often without gaining an inch for either side. The Battle of the Somme (SAHM), between July and November 1916, resulted in more than 1.2 million casualties. British dead or wounded numbered over 400,000. German losses totaled over 600,000, and French nearly 200,000. Despite this, the Allies gained only about seven miles.

A War of New Technology

New technology raised the death toll. The tank, a British invention, smashed through barbed wire, crossed trenches, and cleared paths through no man's land. Soldiers also had machine guns that fired 600 bullets a minute. Poison gas, used by both sides, burned and blinded soldiers.

World War I was the first major conflict in which airplanes were used in combat. By 1917, fighter planes fought each other far above the clouds. Manfred von Richthofen, known as the Red Baron, was Germany's top ace. An ace was an aviator who had downed five or more enemy aircraft. Von Richthofen shot down over 80 enemy planes.

Background
U-boat was short for "undersea boat."

At sea, the Germans used submarines, which they called **U-boats,** to block trade. They were equipped with both guns and torpedoes. German U-boats sank over 11 million tons of Allied shipping.

New Technology of War

First used effectively during World War I, these new weapons caused high casualties.

NARD

EUROPE VIA LIVERPOOL

LUSITANIA

Fastest and Largest Steamer now in Atlantic Service Sails SATURDAY, MAY 1, 10 A. M.

Transylvania, Fri., May 7, 5 P.M.
Orduna, - - Tues., May 18, 10 A.M.
Tuscania, - - Fri., May 21, 5 P.M.
LUSITANIA, Sat., May 29, 10 A.M.
Transylvania, Fri., June 4, 5 P.M.

Gibraltar–Genoa–Naples–Piraeus
S.S. Carpathia, Thur., May 13, Noon

ROUND THE WORLD TOURS
Through bookings to all principal Ports of the World.
Company's Office, 21-24 State St., N. Y.

NOTICE!

TRAVELLERS intending to embark on the Atlantic voyage are reminded that a state of war exists between Germany and her allies and Great Britain and her allies; that the zone of war includes the waters adjacent to the British Isles; that, in accordance with formal notice given by the Imperial German Government, vessels flying the flag of Great Britain, or of any of her allies, are liable to destruction in those waters and that travellers sailing in the war zone on ships of Great Britain or her allies do so at their own risk.

IMPERIAL GERMAN EMBASSY

WASHINGTON, D. C., APRIL 22, 1915.

The British liner *Lusitania* is sunk off the Irish coast by a German submarine on May 7, 1915.

America's Path to War

When the war started in 1914, President **Woodrow Wilson** announced a policy of **neutrality,** refusing to take sides in the war. A popular song, "I Didn't Raise My Boy to Be a Soldier," expressed the antiwar sentiment of many Americans.

Over time, however, German attacks shifted public opinion to the Allied cause. In the fall of 1914, Britain set up a naval blockade of German ports, seizing all goods bound for Germany. In response, German submarines sank all Allied merchant ships they found off the British coast. In May 1915, a German U-boat torpedoed the British passenger ship *Lusitania,* killing 1,198 people, including 128 Americans. The sinking turned many Americans against Germany.

Reading **History**

C. Making Inferences Why did the sinking of the *Lusitania* turn Americans against Germany?

But President Wilson kept the United States neutral. He demanded that the German government halt unrestricted submarine warfare, and it agreed. In the election of 1916, the Democratic party's campaign slogan, "He kept us out of war," appealed to voters. Wilson won reelection.

Desperate to defeat Britain, Germany resumed unrestricted submarine warfare at the end of January 1917. Its military leaders knew this action would bring the United States into the war. However, they hoped to win the war before the Americans arrived.

The next month, another blow to German-American relations came from the **Zimmermann telegram.** The telegram was discovered by the British, who passed it on to the Americans. In it, Arthur Zimmermann, the German foreign minister, told the German ambassador in Mexico to propose that Mexico join the Germans. In exchange, Germany would help Mexico get back its "lost" territories of Texas, New Mexico, and Arizona. Americans were furious.

In March, German submarines sank three American ships. President Wilson asked for a declaration of war.

> **A VOICE FROM THE PAST**
>
> The world must be made safe for democracy. . . . We desire no conquest. . . . We are but one of the champions of the rights of mankind. We shall be satisfied when those rights have been made . . . secure.
>
> **Woodrow Wilson,** message to Congress, April 2, 1917

"The world must be made safe for democracy."
Woodrow Wilson

Six senators and 50 representatives, including the first woman in Congress, Jeanette Rankin of Montana, voted against going to war. But the majority shared the president's commitment to join the Allies.

Revolution in Russia

Events in Russia made U.S. entry into the war more urgent for the Allies. By early 1915, the huge Russian army had been outfought by a smaller German army led by better-trained officers. In August 1915, Czar Nicholas II insisted on taking control of the troops himself. His poor leadership was blamed for more deaths. By 1917, food shortages led to riots, and soaring inflation led to strikes by angry workers in Russia.

In March 1917, Czar Nicholas II was forced to step down. A temporary government continued the unpopular war until November. In that month the Bolsheviks, a communist group led by Vladimir Ilich Lenin, took power. Communism is a political system in which the government owns key parts of the economy, and there is no private property.

Because the war had devastated Russia, Lenin at once began peace talks with Germany. In March 1918, Russia withdrew from the war by signing the Treaty of Brest-Litovsk. German troops could now turn from Russia to the Western front. The Allies urged American troops to come quickly, as you will read in the next section.

***Reading*History**
D. Analyzing Causes What led Russia to pull out of the war?

Section 1 Assessment

1. Terms & Names

Identify:
- militarism
- Central Powers
- Allies
- trench warfare
- U-boat
- Woodrow Wilson
- neutrality
- Zimmermann telegram

2. Taking Notes

Write at least four events that brought the United States into World War I.

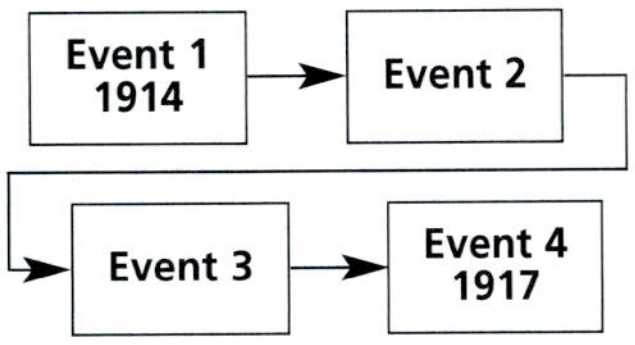

Which of these events was most important? Why?

3. Main Ideas

a. What were the long-term causes of World War I?

b. Why were Americans divided over the issue of remaining neutral?

c. Why was Russia's withdrawal from the war in 1917 a blow to Allies?

4. Critical Thinking

Analyzing Causes How did imperialism, nationalism, and militarism work to reinforce each other?

THINK ABOUT
- the goals of each
- how nationalism might encourage military buildup
- how nationalism contributed to the race for colonies

ACTIVITY OPTIONS

SCIENCE
ART

Research one of the new weapons of World War I. Explain how it works using a **model,** or draw an illustrated **diagram** of a defense against the weapon.

Survive Trench Warfare

You are a platoon leader assigned to a section of the front in central France. You have 60 men under your command. Day and night, through constant rain, earthen trenches full of sticky mud serve as your only protection. Sometimes, you think, the cold, rain, mud, rats, and fatigue are tougher to endure than a German bombardment.

COOPERATIVE LEARNING On this page are two challenges you face as a soldier during World War I. Working with a small group, decide how to deal with each challenge. Choose an option, assign a task to each group member, and do the activity. You will find useful information in the Data File. Be prepared to present your solutions to the class.

PHYSICAL EDUCATION CHALLENGE

"They must have 20 or 30 pounds of mud on them."

Until now, no one thought the trenches would be a permanent part of this war. So you, like the other soldiers along the front, weren't trained to cope with heavy, thick mud, 70-pound backpacks, and the other demands of living and fighting in these conditions. You learned on the job. Now you've been ordered to contribute ideas for a training program that will prepare recruits for the trenches. Look at the Data File for help. Then present your ideas using one of these options:

- **Design an exercise regimen to strengthen troops for the trenches.**
- **Write a booklet of survival tips based on your platoon's experiences.**

HEALTH CHALLENGE

"Your feet swell to two or three times . . . normal size."

You're worried about your men getting trench foot. You've heard horror stories about men whose feet swelled so much they couldn't pull off their boots. Some of these men developed gangrene and had their feet amputated. The key to preventing trench foot is staying dry. What will you do? Use the Data File for help. Then present your solution using one of these options:

- Come up with a way to keep the men's feet dry.
- Role-play a conversation with veteran soldiers about preventing trench foot.

ACTIVITY WRAP-UP

Present to the Class As a group, review your methods of surviving the trenches. Pick the most creative solution for each challenge, and present these solutions to the class.

DATA FILE

THE TRENCHES

- Trenches covered about 450 miles between the North Sea and the Swiss border.
- In France, ten-foot-deep trenches were dug into the ground and topped with sandbag parapets.
- Inside was a fire step, a ledge two or three feet up from bottom of the trench, used by sentries or troops firing.
- The sides were held up by sandbags and timber.

A SOLDIER'S GEAR

60–75 pounds of gear, including blankets, waterproof groundsheet, extra boots and occasionally waterproof gum boots, quilted coat, shovel for digging trenches, helmet, wire clippers, pail for rations, 2 quarts of water, 4 days' food, 200 cartridges, 6 hand grenades, gas mask, 3 pairs of socks, soap, toothbrush, bottle of whale oil, towel, rifle, bayonet

TRENCH FOOD

beef stew, corned beef, bread, hard biscuits, pork and beans, tins of jam, butter, sugar, tea

PROBLEMS

- **trench foot:** condition caused by feet staying wet 24 hours a day; feet swell, turn numb and blue; if not treated, gangrene sets in and feet must be amputated; helped by rubbing whale oil on feet, changing to dry socks three times daily
- **mud:** mud traps the wounded until some drown, clogs rifles and gear, weighs men down, causes trench walls to fall in
- **rats:** huge rats, big as rabbits, infest the trenches

INTERNET Visit www.mcdougallittell.com for more on trench warfare.

2 America Joins the Fight

TERMS & NAMES
- John J. Pershing
- American Expeditionary Force
- convoy system
- Second Battle of the Marne
- Alvin York
- armistice

MAIN IDEA

U.S. forces helped the Allies win World War I.

WHY IT MATTERS NOW

For the first time, the United States asserted itself as a world power.

ONE AMERICAN'S STORY

Eddie Rickenbacker was America's most famous flying ace. He was one of the first Americans to get a look at the trenches from the cockpit of an airplane. The date was March 6, 1918. It was, recalled Rickenbacker, the first flyover of the battlefield by a "made-in-America Squadron." What Rickenbacker saw shocked him.

A VOICE FROM THE PAST

[T]here appeared to be nothing below but these old battered ditches . . . and billions of shell holes. . . . [N]ot a tree, a fence . . . nothing but . . . ruin and desolation. The whole scene was appalling.

Eddie Rickenbacker, *Fighting the Flying Circus*

Rickenbacker went on to distinguished service as a wartime aviator. As you will read in this section, he and other U.S. soldiers helped the Allies win the war.

An American gun crew advances against German positions in 1918. The ruin that Rickenbacker described is apparent.

Raising an Army and a Navy

The U.S. Army was not ready for war. American fighting forces consisted of fewer than 200,000 soldiers, many of them recent recruits. To meet its need for troops, the government began a draft. This system of choosing people for forced military service was first used during the Civil War. In May 1917, Congress passed the Selective Service Act. This act required all males between the ages of 21 and 30 to sign up for military service. By the end of 1918, nearly 3 million men had been drafted.

About 2 million American soldiers went to France. They served under General **John J. Pershing** as the **American Expeditionary Force,** or AEF. British commanders asked the U.S. government to have AEF troops join existing French and British combat units. Wilson refused. He believed that having "distinct and separate" American combat units would guarantee the United States a major role in the peace talks at war's end. Most U.S. troops fought separately, but some fought under Allied command.

Close to 50,000 American women also served in World War I. Some volunteered for overseas duty with the American Red Cross. However, for the first time in American history, women also served in the military. The Navy, desperate for clerical workers, took about 12,000 female volunteers. The Marine Corps accepted 305 female recruits, known as Marinettes. Over 1,000 women went overseas for the Army. Nurses made up the largest group of females in the armed forces. However, women also acted as interpreters, operated switchboards, entertained troops, and drove ambulances for the AEF.

Reading **History**
A. Finding Main Ideas How did women serve in the U.S. armed forces?

Around 400,000 African Americans served in the armed forces. More than half of them served in France. As they had at home, African-American troops overseas faced discrimination. However, it came from white American soldiers rather than from their European allies. At first, the Army refused to take black draftees. However, responding to pressure from African-American groups, the military eventually created two African-American combat divisions.

American Ships Make a Difference

In the first years of the war, German U-boat attacks on supply ships were a serious threat to the Allied war effort. American Rear Admiral William S. Sims convinced the Allies to adopt a system of protection. In a **convoy system,** a heavy guard of destroyers escorted merchant ships across the Atlantic in groups. Begun in May 1917, this strategy quickly reduced the loss rate.

Another American tactic gave the Allies added protection from the U-boat menace. Beginning in June 1918, the Allies laid a barrier of 70,000 mines in the North Sea. The 180-mile-long minefield made U-boat access to the North Atlantic almost impossible. Admiral Sims called the North Sea minefield "one of the wonders of the war."

Vocabulary
mines: hidden explosive devices

Convoy System

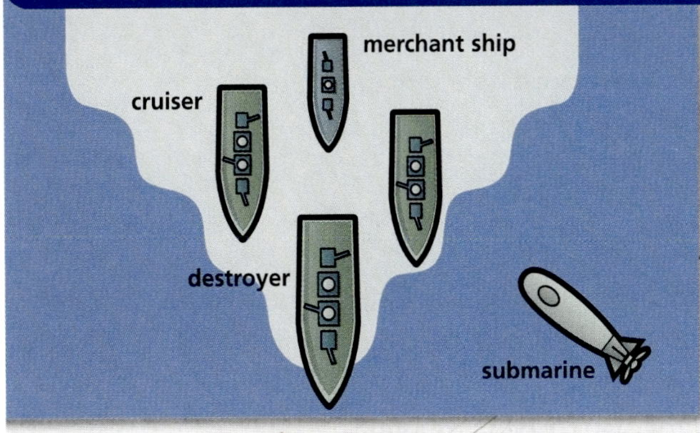

The battleship *New Jersey* is pictured in camouflage, or disguise, around 1918.

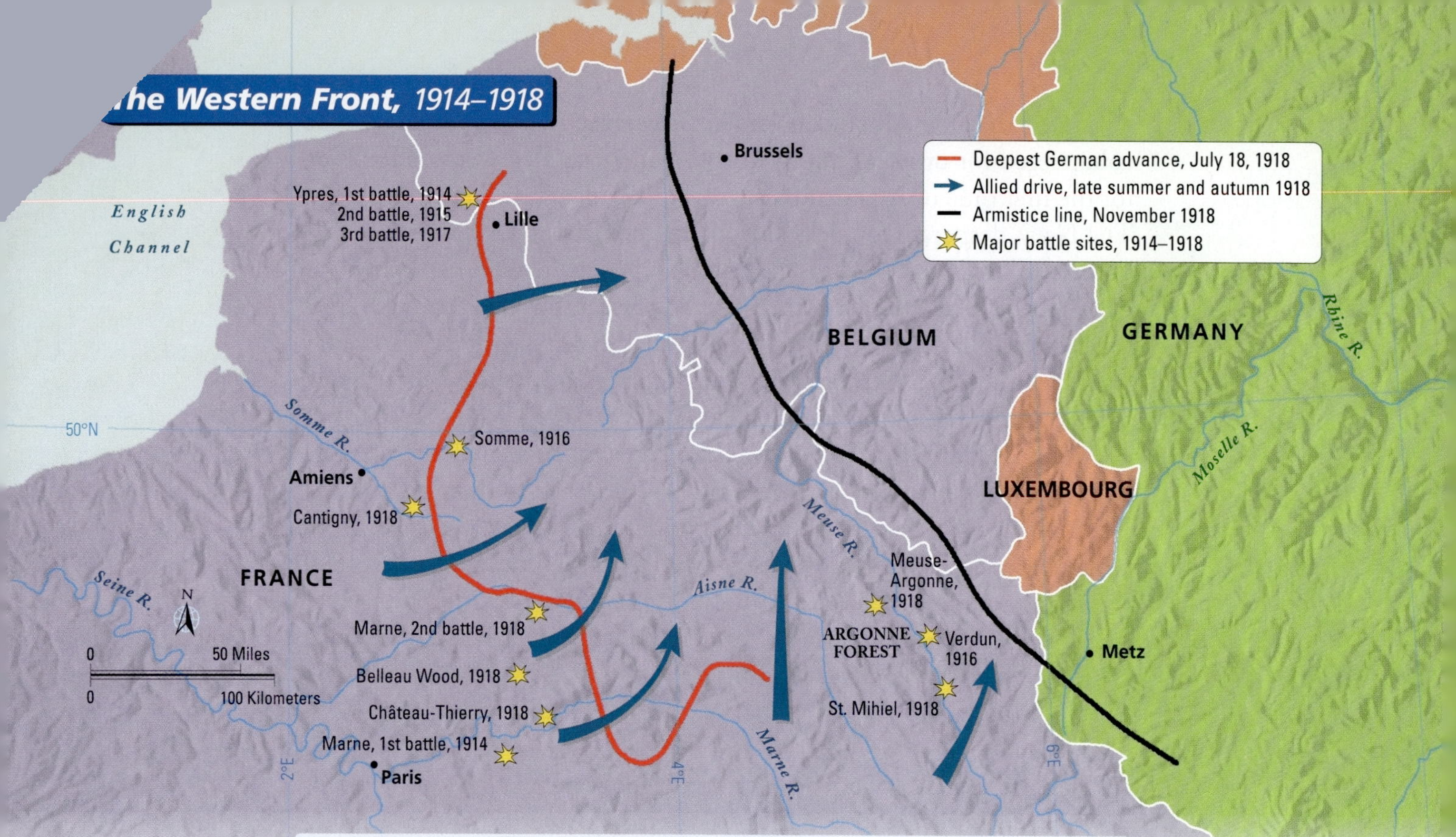

GEOGRAPHY SKILLBUILDER Interpreting Maps

1. **Location** *What three battles occurred closest to the Armistice line?*
2. **Movement** *In what two directions did the Allied drives move?*

American Troops Enter the War

By the time the first American troops arrived in France in June 1917, the Allies had been at war for almost three years. The small force of 14,000 Yanks boosted the morale of the battle-weary Allies. However, almost a year would pass before the bulk of the American troops landed in Europe.

Background
American soldiers were also called *doughboys.* This term was used even during the Civil War.

After their Russian opponents withdrew from the war, the Germans and the other Central Powers prepared to finish the fight in France. In March 1918, the Germans launched an offensive to end the war before the Americans arrived in force. Within two months, they had smashed through the French lines, reaching the Marne River only 50 miles from Paris. Just in time, in May 1918, one million fresh American troops arrived ready for action.

On May 28, American soldiers attacked the French town of Cantigny (kahn•tee•NYEE), which was occupied by the Germans. The soldiers advanced into the town, blasting enemy soldiers out of trenches and dragging them from cellars. Within two hours, the Yanks had taken control of Cantigny. The American victory lifted Allied morale.

When the Germans moved against the town of Château-Thierry (shah•toh•tyeh•REE), the Americans held their ground. They helped the French stop the German advance. Encouraged by these successes, French General Ferdinand Foch, commander of the Allied forces, ordered General Pershing's American forces to retake Belleau (beh•LOH) Wood.

This was a forest near the Marne River well defended by German troops. American soldiers succeeded, but at a fearful cost. One unit lost 380 of its 400 men. However, the Americans had proved themselves in combat.

Pushing the Germans Back

The **Second Battle of the Marne** in the summer of 1918 was the turning point of the war. It began with a German drive against the French line. During three days of heavy fighting, about 85,000 Americans helped the Allies halt the German advance. The Allies then took the initiative. They cut the enemy off from its supply lines and forced the Germans back.

For the rest of the war, the Allies advanced steadily. By early September, the Germans had lost all the territory they had gained since the spring. September 26, 1918, marked the beginning of the final Meuse-Argonne (myooz•ahr•GAHN) offensive. Around 1.2 million U.S. soldiers took part in a massive drive to push back the German line between the Argonne Forest and the Meuse River. The war's final battle left 26,000 Americans dead. But by November, the Germans were retreating.

Reading **History**
B. Recognizing Effects What was the effect of the Meuse-Argonne offensive?

The Meuse-Argonne offensive made a hero of American soldier **Alvin York.** At first, Tennessee-born Sergeant York seemed an unlikely candidate for military fame. Because of his religious beliefs, he tried unsuccessfully to avoid the draft. He refused to bear arms on religious grounds. An army captain convinced him to change his mind. In October 1918, in the Argonne Forest, York attacked German machine gunners, killing 25 of them. Other German soldiers surrendered, and York returned to the American lines with 132 captives.

Reading **History**
C. Evaluating What was heroic about Sergeant York?

Another American hero was pilot Eddie Rickenbacker. He won fame as the U.S. "ace of aces" for shooting down a total of 26 enemy planes. Just before the Meuse-Argonne offensive, he attacked seven German planes, sending two of them crashing to the ground. This action won him the Medal of Honor.

Four African-American combat units also received recognition for their battlefield valor. Fighting under French commanders, the 369th, 371st, and 372nd regiments (and part of the 370th) were awarded France's highest honor, the Croix de Guerre. The 369th spent more continuous time on the front lines than any other American unit. Although under intense fire for 191 days, it never lost a foot of ground.

Connections TO LITERATURE

LITERATURE OF WORLD WAR I

Several notable American writers served in World War I. They included Ernest Hemingway, the poet E. E. Cummings, and John Dos Passos. Hemingway drove an ambulance for the Italian army. He put this experience into his war novel *A Farewell to Arms.* Cummings wrote of his time in France in *The Enormous Room.*

Dos Passos, who also worked as an ambulance driver, once explained what attracted him to the battlefront: "What was war like, we wanted to see with our own eyes. I wanted to see the show."

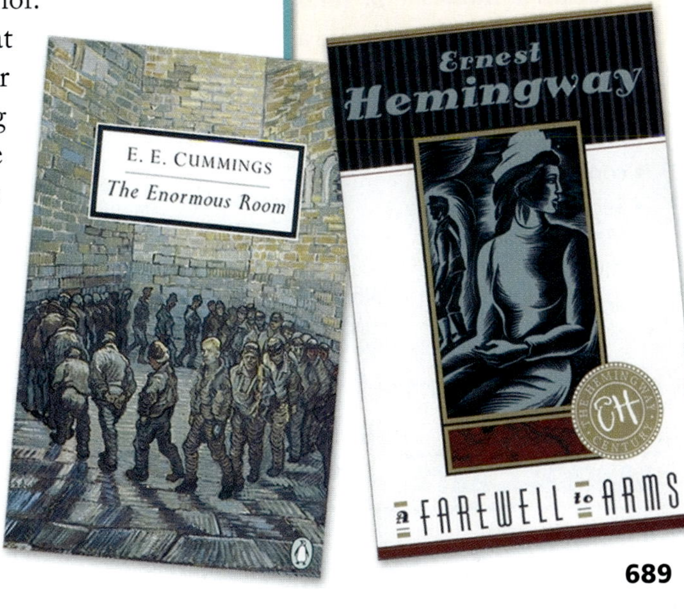

Americans were proud of the contribution their troops made to the war effort. They helped shift the balance in favor of the Allies.

Germany Stops Fighting

After the defeat of the Meuse-Argonne, General Erich Ludendorff advised the German government to seek peace. In early November, Germany's navy mutinied and its allies dropped out. On November 9, the Kaiser stepped down. Two days later Germany agreed to an **armistice,** an end to fighting. On November 11, 1918, at 11:00 A.M.—the 11th hour of the 11th day of the 11th month—all fighting ceased.

Background
For many years after the war, Americans celebrated Armistice Day as a national holiday.

About 8.5 million soldiers died in the war, and about 21 million were wounded. Before he was killed in battle, one British soldier summed up the war's tragic costs.

A VOICE FROM THE PAST

The sufferings of the men at the Front, of the wounded whose flesh and bodies are torn in a way you cannot conceive; the sorrow of those at home. . . . What a cruel and mad diversion of human activity!

William John Mason, quoted in *The Lost Generation of 1914*

Millions of civilians in Europe, Asia, and Africa also died in the war—from starvation and disease. In the next section, you will learn how the war affected U.S. civilians.

CONNECTIONS TO MATH

Military Deaths in World War I*

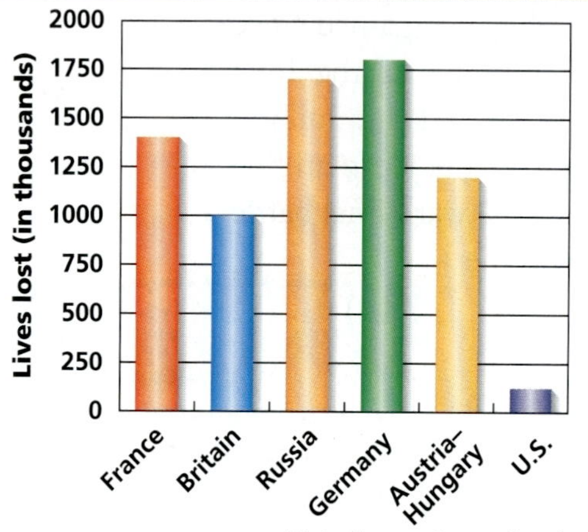

*Not all countries are listed.
Source: *Over There,* by Byron Farwell

SKILLBUILDER
Interpreting Charts

1. *Which two nations on the chart suffered the most deaths?*
2. *U.S. deaths were about what percentage of combined French and British deaths?*

Section 2 Assessment

1. Terms & Names

Identify:
- John J. Pershing
- American Expeditionary Force
- convoy system
- Second Battle of the Marne
- Alvin York
- armistice

2. Taking Notes

Create a web to show how American groups or individuals helped fight the war.

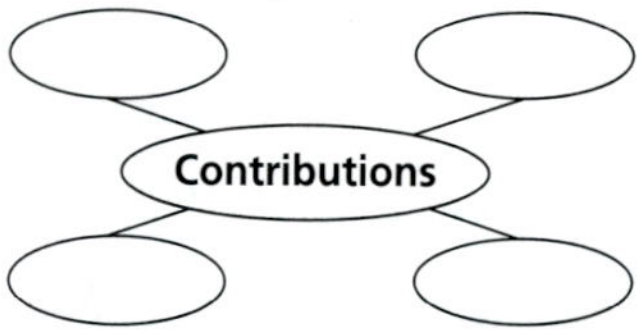

Whose contribution was most surprising?

3. Main Ideas

a. Why did Wilson want U.S. forces to fight as a separate American combat unit?

b. What were two ways the U.S. Navy countered the U-boat threat?

c. Why was the Meuse-Argonne offensive a turning point in the war?

4. Critical Thinking

Recognizing Effects How important was America's entry into the war to the Allied cause?

THINK ABOUT
- the morale of Allied troops
- troop strength
- performance in battle

ACTIVITY OPTIONS

MUSIC
LANGUAGE ARTS

Make an **audiotape** of music or sounds that suggest the stages of the war, or write a **letter** in the voice of a soldier in the war.

3 Life on the Home Front

TERMS & NAMES
war bonds
propaganda
Espionage Act
Sedition Act
Oliver Wendell Holmes
Great Migration

MAIN IDEA

The war required sacrifice for Americans at home and changed life in other ways.

WHY IT MATTERS NOW

Some wartime changes were permanent, such as black migration to Northern cities.

ONE AMERICAN'S STORY

On the home front, the war opened up new jobs for women. Most of the women who took these jobs were already in the work force. Carrie Fearing worked for the Railroad Administration. When the war ended, female workers were laid off. Fearing wrote to the railroad director, hoping to keep her job.

A VOICE FROM THE PAST

We never took a soldier's place, a soldier would not do the work we did . . . such as sweeping, picking up waste and paper and hauling steel shavings. . . . We . . . were liked and respected by all who knew us. . . . We like our job very much and I hope you will . . . place us back at the shop.

Carrie Fearing, quoted in *Women, War, and Work*

In May 1918, these women worked in the Union Pacific Railroad freight yard in Cheyenne, Wyoming.

Like Fearing, most women who helped the country get ready for war were pleased to have wartime jobs. They were proud of the part they had played in getting the country ready for war.

Mobilizing for War

To prepare for war, the government needed money. World War I cost the United States $35.5 billion. Americans helped pay almost two-thirds of that amount by buying government war bonds. **War bonds** were low-interest loans by civilians to the government, meant to be repaid in a number of years. To sell the bonds, officials held Liberty Loan drives. Posters urged citizens to "Come Across or the Kaiser Will." Hollywood actors like Charlie Chaplin toured the country selling bonds to starstruck audiences.

Schoolchildren rolled bandages and collected tin cans, paper, toothpaste tubes, and apricot pits. The pits were burned and made into charcoal for gas mask filters. Some Boy Scout troops even sold war bonds. So that more food could be sent to soldiers, people planted "victory gardens" in backyards and vacant lots. Women's groups came together in homes and churches to knit socks and sweaters and sew hospital gowns.

To persuade women to buy war bonds, this poster appealed to their love of family.

Patriotic citizens also saved food by observing wheatless Mondays and Wednesdays, when they ate no bread, and meatless Tuesdays. To save gas, they stopped their Sunday pleasure drives. The government limited civilian use of steel and other metals. Women donated their corsets with metal stays to scrap drives. Manufacturers stopped making tin toys for children and removed metal from caskets.

Reading **History**

A. Finding Main Ideas What were civilians asked to do for the war effort?

The war brought more government control of the economy. To produce needed war supplies, in 1917 President Wilson set up the War Industries Board. The board had great power. It managed the buying and distributing of war materials. It also set production goals and ordered construction of new factories. With the president's approval, the board also set prices. Another government agency, the National War Labor Board, settled conflicts between workers and factory owners.

To rally citizen support, Wilson created the Committee on Public Information. The committee's writers, artists, photographers, and filmmakers produced **propaganda,** opinions expressed for the purpose of influencing the actions of others. They sold the war through posters, pamphlets, and movies. One popular pamphlet, "How the War Came to America," came out in Polish, German, Swedish, Bohemian, and Spanish. In movie houses, audiences watched such patriotic films as *Under Four Flags* and *Pershing's Crusaders.*

Intolerance and Suspicion

Patriotic propaganda did much to win support for the war. But its anti-German, anti-foreign focus also fueled prejudice. Suddenly people distrusted anything German. A number of towns with German names changed their names. Berlin, Maryland, became Brunswick. People called sauerkraut "liberty cabbage," and hamburger became "Salisbury steak." Owners of German shepherds took to calling their pets "police dogs."

Reading **History**

B. Recognizing Effects How did war propaganda fuel prejudice?

On June 15, 1917, Congress passed the **Espionage Act**. The **Sedition Act** followed in May 1918. These laws set heavy fines and long prison terms for such anti-war activities as encouraging draft resisters. The laws made it illegal to criticize the war. U.S. courts tried more than 1,500 pacifists, socialists, and other war critics. Hundreds went to jail. Socialist party leader Eugene Debs gave a speech arguing that the war was fought by poor workingmen for the profit of wealthy business owners. For this talk, a judge sentenced him to ten years in prison.

The government ignored complaints that the rights of Americans were being trampled. In the 1919 decision in *Schenck* v. *United States,* the Supreme Court upheld the Espionage Act. Schenck, convicted of distributing pamphlets against the draft, had argued that the Espionage Act violated his right to free speech. Justice **Oliver Wendell Holmes** wrote the court's opinion.

A VOICE FROM THE PAST

The most stringent [strict] protection of free speech would not protect a man in falsely shouting fire in a theater and causing a panic. . . . The question in every case is whether the words used . . . are of such a nature as to create a clear and present danger that they will bring about . . . evils that Congress has a right to prevent.

Oliver Wendell Holmes, *Schenck* v. *United States, 1919*

Reading **History**
C. Analyzing Points of View Why did Justice Holmes believe that free speech could be limited?

Justice Holmes argued that free speech, guaranteed by the First Amendment, could be limited, especially in wartime.

New Jobs and the Great Migration

As soldiers went off to battle, the United States faced a labor shortage. Northern factories gearing up for war were suddenly willing to hire workers they had once rejected. Throughout the South, African Americans heeded the call. Between 1910 and 1920, about 500,000 African Americans moved north to such cities as New York, Chicago, Detroit, Cleveland, and St. Louis. This movement became known as the **Great Migration.** African Americans left to escape the bigotry, poverty, and racial violence of the South. They hoped for a better life in the North.

HISTORY *through* ART

The Migration of the Negro, Panel No. 1 (1940–41), by Jacob Lawrence, shows three of the most common destinations for African Americans leaving the South.

What does the painting suggest about the departure of African Americans from the South?

New jobs were opening up in the American Southwest. These jobs were fueled by the growth of railroads and irrigated farming. A revolution was under way in Mexico, and the chaos led many Mexicans to flee across the border after 1910. Many immigrants settled in Texas, Arizona, Colorado, and California. Most became farm workers. During the war years, some went to Northern cities to take better-paying factory jobs.

The wartime labor shortage also meant new job choices for women. Women replaced male workers in steel mills, ammunition factories, and assembly lines. Women served as streetcar conductors and elevator operators. The war created few permanent openings for women, but their presence in these jobs gave the public a wider view of their abilities. Women's contributions during the war helped them win the vote.

Reading **History**

D. Recognizing Effects What groups gained new jobs as a result of the war?

Now and then

THE FLU EPIDEMIC

In 1918, flu victims often came down with pneumonia and died within a week. Today, bacterial infections such as pneumonia resulting from the flu can be controlled with antibiotics.

The 1998 discovery of the frozen remains of a 1918 flu victim in an Alaskan cemetery may one day lead to a better understanding of the virus. Scientists have found a genetic link between the 1918 flu virus and swine flu, a virus first found in pigs. The Alaskan find may help scientists develop vaccines to protect against future flu outbreaks.

The Flu Epidemic of 1918

Another result of the war was a deadly flu epidemic that swept the globe in 1918. It killed more than 20 million people on six continents by the time it disappeared in 1919. It had no known cure. Spread around the world by soldiers, the virus took some 500,000 American lives. People tried desperately to protect themselves. Everywhere, schools and other public places shut down to limit the flu's spread.

In the army, more than a quarter of the soldiers caught the disease. In some AEF units, one-third of the troops died. Germans fell victim in even larger numbers than the Allies. World War I brought death and disease to millions. It would also have longer-term effects, as you will read in Section 4.

Section 3 Assessment

1. Terms & Names

Identify:

- war bonds
- propaganda
- Espionage Act
- Sedition Act
- Oliver Wendell Holmes
- Great Migration

2. Taking Notes

Make a chart like the one below to show reasons for wartime shifts in population.

	Shift	Reason(s)
African Americans		
Mexicans		

How similar were the two groups' reasons for moving?

3. Main Ideas

a. What were three ways American families could contribute to the war effort?

b. What was the purpose of the Espionage and Sedition Acts? What groups were most affected by them?

c. What kinds of new job opportunities did the war create for women and minorities?

4. Critical Thinking

Making Inferences What were the positive and the negative consequences of American wartime propaganda?

THINK ABOUT

- contributions to war effort
- effect on opponents of war and on German-Americans

ACTIVITY OPTIONS

SPEECH

MATH

Deliver a **radio broadcast** on the importance of conserving food, or make a **calculation** of the amount of food your class wastes monthly.

4 The Legacy of World War I

TERMS & NAMES
League of Nations
Fourteen Points
Treaty of Versailles
reparations
Red Scare
Palmer raids

MAIN IDEA

After the war, Americans were divided over foreign policy and domestic issues.

WHY IT MATTERS NOW

The war affected the role the United States played in the world during the rest of the century.

ONE AMERICAN'S STORY

Massachusetts Senator Henry Cabot Lodge had favored U.S. entry into the war. However, he opposed President Wilson's idea that the United States join an international organization. The **League of Nations** was such an organization set up to settle conflicts through negotiation. Lodge warned against joining an alliance that would require the United States to guarantee the freedom of other nations.

A VOICE FROM THE PAST

If we guarantee any country . . . its independence . . . we must [keep] at any cost . . . our word. . . . I wish [the American people] carefully to consider . . . whether they are willing to have the youth of America ordered to war by other nations.

Henry Cabot Lodge, speech to the Senate, February 28, 1919

Senator Henry Cabot Lodge (1850–1924) opposed U.S. entry into the League of Nations.

Speeches like Lodge's helped turn the public against the League and gave President Wilson the most crushing defeat of his political career. In this section, you will learn how the United States and Europe adjusted to the end of the war.

Wilson's Fourteen Points

In January 1918, President Wilson spoke to Congress about his goals for peace. This was ten months before the end of the war. His statement became known as the **Fourteen Points** (see page 699). The speech called for smaller military forces, an end to secret treaties, freedom of the seas, and free trade. It also called for changes in national boundaries. Most of these changes gave independence to peoples formerly ruled by Austria-Hungary or the Ottoman Empire.

For Wilson, the fourteenth point mattered most. He called for an association of nations to peacefully settle disputes. This association was to become the League of Nations, which Republicans like Lodge opposed. Wilson firmly believed that acceptance of his Fourteen Points by the warring parties would bring about what he called a "peace without victory."

Treaty of Versailles

Wilson led the U.S. delegation to the peace conference in France. Though many Europeans considered him a hero, conference leaders did not. The leaders of Britain, France, and Italy did not share Wilson's vision of "peace without victory." They wanted Germany to pay heavily for its part in the war.

The **Treaty of Versailles** (vuhr•SY) forced Germany to accept full blame for the war. Germany was stripped of its colonies and most of its armed forces. It was also burdened with $33 billion in **reparations**—money that a defeated nation pays for the destruction caused by a war. The treaty divided up the empires of Austria-Hungary and the Ottomans. It created Yugoslavia and Czechoslovakia and recognized Poland's independence.

Reading **History**

A. Recognizing Effects How were the Central Powers punished by the Treaty of Versailles?

Wilson managed to include the League of Nations in the treaty. He firmly believed the League would help to keep the peace. He returned home to seek Senate approval for the treaty. But the Republican-run Senate was dead set against it. Senator Henry Cabot Lodge kept delaying a vote on the treaty.

After weeks of delay, Wilson decided to make his case to the public. In September of 1919, he began a cross-country speaking trip to build support for the League. In about 21 days, he traveled almost 10,000 miles and gave over 30 speeches.

A VOICE FROM THE PAST

In the covenant [agreement] of the League of Nations, the moral forces of the world are mobilized . . . . They consent . . . to submit every matter of difference between them to the judgment of mankind, and just so certainly as they do that, . . . war will be pushed out of the foreground of terror in which it has kept the world.

Woodrow Wilson, speech in Pueblo, Colorado, on September 25, 1919

Shortly after giving this speech, Wilson collapsed from strain. Later, he suffered a stroke from which he never fully recovered.

Negotiations to get the treaty through Congress continued, but Americans were not eager for more foreign commitments. Lodge and his supporters offered to accept the treaty if major changes were made in the League. Wilson refused to compromise. As a result, the United States did not ratify the treaty. The League of Nations was formed without the United States.

Reading **History**

B. Analyzing Causes Why didn't the United States ratify the Treaty of Versailles?

The war and the Treaty of Versailles failed to make Europe "safe for democracy." In the next decades, Germany's resentment of the treaty grew. The treaty planted the seeds of World War II, an even more deadly conflict to come.

AMERICA'S HISTORY MAKERS

WOODROW WILSON

1856–1924

A gifted speaker, Woodrow Wilson had a strong sense of duty, and he inspired great loyalty. Yet he could be a harsh judge of others, stiff and unbending in his relations with people. Sculptor Jo Davidson remarked that "He invoked fear and respect . . . but not affection." Though not America's best-loved president, he still commands respect. When historians list the nation's best presidents, Wilson often ranks in the top ten.

How might Wilson's character have worked against approval of the Treaty of Versailles?

Postwar Europe, 1919

GEOGRAPHY SKILLBUILDER Interpreting Maps

1. **Region** *What new nations were created after the war?*
2. **Region** *In what part of Europe were most of the new nations located?*

EFFECTS OF WORLD WAR I ON EUROPE

IMMEDIATE EFFECTS	LONG-TERM EFFECTS
Revolution in Russia	Breakup of empires
Allied victory	Formation of League of Nations
Destruction in Europe	Resentments leading to World War II

Strikes and the Red Scare

Background
In 1919, police, steelworkers, and coal miners also went on strike.

The Treaty of Versailles was not the only issue that divided Americans after the war. Shortly after the war ended, the United States experienced a number of labor strikes. For example, in Seattle, Washington, in February 1919, more than 55,000 workers took part in a peaceful general strike. The shutdown paralyzed the city.

Some Americans saw efforts to organize labor unions as the work of radicals, people who favor extreme measures to bring about change. The strikes sparked fears of a communist revolution like the one that toppled the Russian czar. In 1919–1920, this fear created a wave of panic called the **Red Scare** (communists were called *reds*). Public fear was heightened by the discovery of mail bombs sent to government officials. Many believed the bombs were the work of anarchists. Anarchists are radicals who do not believe in any form of government.

ReadingHistory
C. Recognizing Effects What resulted from the Red Scare?

In January 1920, Attorney General A. Mitchell Palmer took action. He ordered federal agents and local police to raid the homes and headquarters of suspected radicals. His agents arrested at least 6,000 people in the **Palmer raids.** Without search warrants, agents burst into homes and offices and dragged citizens off to jail.

The Red Scare was not only antiradical but also antiforeign. During the Red Scare, two Italian-born anarchists, Nicola Sacco and Bartolomeo Vanzetti, were arrested for killing two men in an armed robbery in

Massachusetts. They claimed they were innocent, but both were found guilty and executed. Their trial attracted worldwide attention.

Racial Tensions Increase

Americans also saw a rise in racial tensions after the war. Between 1910 and 1920, the Great Migration brought a half million African Americans to northern cities. In the cities where African Americans had settled in large numbers, whites and blacks competed for factory jobs and housing.

On July 2, 1917, tensions erupted into a race riot in East St. Louis, Illinois. The trouble began when blacks were brought in to take the jobs of white union members who had gone on strike. A shooting incident touched off a full-scale riot.

Reading **History**
D. Analyzing Causes How did the war contribute to racial tensions?

Two years later, African-American soldiers returning from the war found their social plight unchanged. They had fought to make the world "safe for democracy." At home, though, they were still second-class citizens.

Simmering resentments over housing, job competition, and segregation exploded during the summer of 1919. In 25 cities around the country, race riots flared. In Chicago, a black man swimming in Lake Michigan drifted into the white section of a beach. Whites stoned him until he drowned. Thirteen days of rioting followed. Before it ended, 38 people were dead.

Longing for "Normalcy"

By the time campaigning began for the 1920 election, Americans felt drained. Labor strikes, race riots, the Red Scare, and the fight over the Versailles Treaty and the League of Nations had worn them out. Voters were ready for a break. Republican candidate Warren G. Harding of Ohio offered them one. His promise to "return to normalcy" appealed to voters. Harding won a landslide victory. In the next chapter, you will learn about American life after his election.

Section 4 Assessment

1. Terms & Names

Identify:

- League of Nations
- Fourteen Points
- Treaty of Versailles
- reparations
- Red Scare
- Palmer raids

2. Taking Notes

Create a diagram to examine the war's effects on Europe and America.

Effects of World War I	
Europe	United States

Which effects were positive and which were negative?

3. Main Ideas

a. Why did Germany resent the Treaty of Versailles?

b. Why did Lodge and other Republicans oppose joining the League of Nations?

c. What caused the Red Scare? Who was most affected by it?

4. Critical Thinking

Analyzing Points of View
Why was Wilson unable to get other powers to accept his goals for the peace conference?

THINK ABOUT

- conflicting goals
- practicality of Wilson's aims
- attitudes of other nations towards U.S. contributions during the war

ACTIVITY OPTIONS

LANGUAGE ARTS
ART

Imagine that you work for a newspaper. Write an **editorial** about the Palmer raids, or draw a political **cartoon** about the raids.

INTERACTIVE PRIMARY SOURCE

The Fourteen Points

Setting the Stage Nine months after the United States entered World War I, President Wilson delivered to Congress a statement of war aims. This statement became known as the "Fourteen Points." In the speech, President Wilson set forth 14 proposals for reducing the risk of war in the future. Numbers have been inserted to help identify the main points, as well as those omitted. **See Primary Source Explorer** CD-ROM

All the peoples of the world are in effect partners . . . , and for our own part we see very clearly that unless justice be done to others it will not be done to us. The program of the world's peace, therefore, is our program; and that program, . . . as we see it, is this:

[1] Open **covenants**[1] of peace, openly arrived at, after which there shall be no private international understandings of any kind but diplomacy shall proceed always frankly and in the public view.

[2] Absolute freedom of navigation upon the seas . . . in peace and in war. . . .

[3] The removal, so far as possible, of all economic barriers and the establishment of an equality of trade conditions among all the nations. . . .

[4] Adequate guarantees given and taken that national **armaments**[2] will be reduced. . . .

[5] A free, open-minded, and absolutely impartial adjustment of all colonial claims, based upon . . . the principle that . . . the interests of the populations concerned must have equal weight with the . . . claims of the government whose title is to be determined.

[6–13: These eight points deal with specific boundary changes.]

[14] A general association of nations must be formed under specific covenants for the purpose of affording mutual guarantees of political independence and territorial **integrity**[3] to great and small states alike.

—*Woodrow Wilson*

1. **covenants:** binding agreements.

2. **armaments:** weapons and supplies of war.

3. **integrity:** the condition of being whole or undivided; completeness.

A CLOSER LOOK

THE VALUE OF OPENNESS

The first of Wilson's points attempts to solve one of the problems that caused the outbreak of World War I—agreements between nations arrived at in secret.

1. How might agreements arrived at in public prevent another world war?

A CLOSER LOOK

BALANCING CLAIMS

Wilson frequently appeals to fairness, balance, and impartiality in settling competing claims.

2. What might be unusual about a leader such as Wilson calling for an impartial adjustment of colonial claims?

A CLOSER LOOK

LEAGUE OF NATIONS

Wilson proposes that nations join a formal organization to protect one another.

3. Why did Wilson believe that such an organization would benefit the world?

Interactive Primary Source Assessment

1. Main Ideas

a. Why should diplomacy avoid private dealings and proceed in public view?

b. How might equality of trade be important to keeping the peace?

c. What must nations join together to guarantee?

2. Critical Thinking

Evaluating The first five points address issues that Wilson believed had caused the war. How successful do you think Wilson's ideas have been in the rest of the 20th century?

THINK ABOUT

- other conflicts since World War I
- peacekeeping efforts around the world

Chapter 24 ASSESSMENT

TERMS & NAMES

Briefly explain the importance of each of the following.

1. militarism
2. Allies
3. trench warfare
4. Zimmermann telegram
5. American Expeditionary Force
6. convoy system
7. propaganda
8. Great Migration
9. Treaty of Versailles
10. Red Scare

REVIEW QUESTIONS

War Breaks Out in Europe (pages 679–685)

1. What were the sources of tension between the European powers that led to war?
2. Why did the United States at first remain neutral in the war between the Allies and the Central Powers?
3. What brought the United States into the war on the Allied side?

America Joins the Fight (pages 686–690)

4. How did the Allies fight the German U-boat threat?
5. How did U.S. entry into the war affect the Allies?
6. What led Germany to agree to an armistice?

Life on the Home Front (pages 691–694)

7. How did U.S. civilians aid the war effort?
8. How did Congress contribute to increased prejudice and intolerance on the home front?

The Legacy of World War I (pages 695–699)

9. How did Wilson's goals for the peace conference differ from those of his European allies?
10. Why did the Senate reject the Treaty of Versailles?

CRITICAL THINKING

1. USING YOUR NOTES

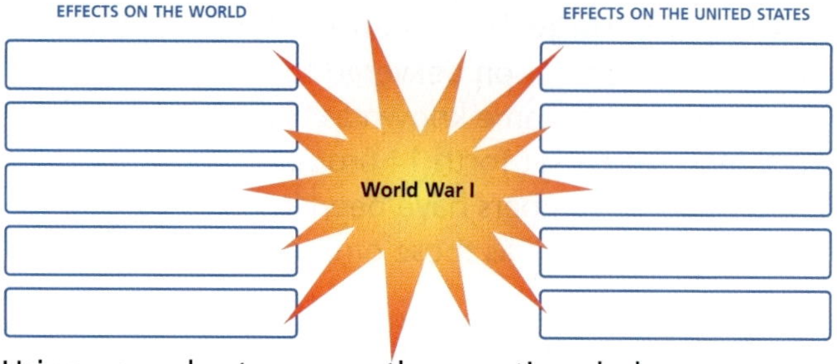

Using your chart, answer the questions below.

a. Were the effects of the war greater in Europe or the United States?
b. What political effects did the war have on the United States?
c. How did the war affect African-American civilians?

2. APPLYING CITIZENSHIP SKILLS

Are limitations on freedom of speech justified by war? Explain your opinion.

3. THEME: AMERICA IN THE WORLD

How did Wilson's view of the role the United States should play in world affairs compare with Theodore Roosevelt's view of America's role?

4. ANALYZING LEADERSHIP

Do you think Wilson's refusal to compromise to get the Treaty of Versailles through Congress was a good decision? Why?

Interact *with* History

How accurately did you predict the ways in which American citizens might support the war effort?

VISUAL SUMMARY

World War I

War Breaks Out in Europe

When the Allies and the Central Powers went to war in Europe, the United States reluctantly joined the Allies.

America Joins the Fight

Millions of U.S. soldiers and civilian volunteers went abroad and helped the Allies win the war.

Life on the Home Front

The war required Americans to sacrifice many things, even political freedoms. The war also brought new jobs.

The Legacy of World War I

The war broke up European empires and left lasting social changes in the United States.

HISTORY SKILLS

1. INTERPRETING MAPS: Movement

Study the map. Answer the questions.

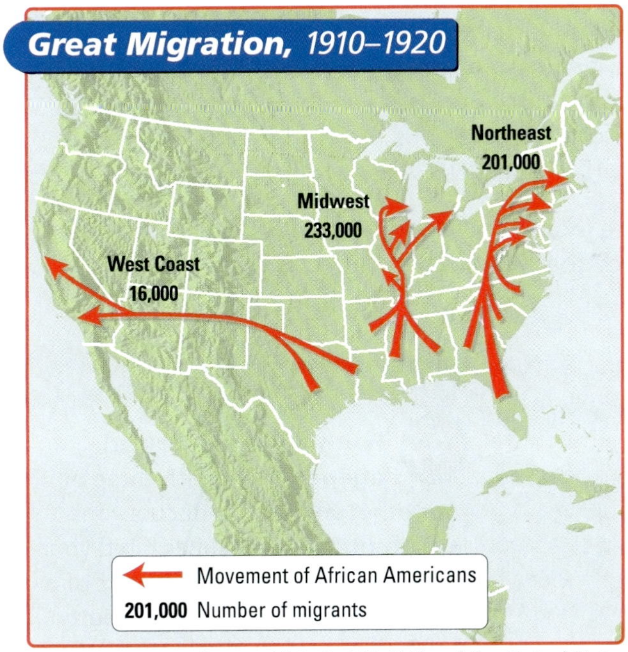

Source: *Historical Statistics of the United States*

Basic Map Elements

a. What is the subject of the map?

b. What do the arrows indicate?

Interpreting the Map

c. In what general directions did migration take place?

d. Of these directions, in which direction did the fewest African Americans move?

2. INTERPRETING PRIMARY SOURCES

This famous World War I poster shows Uncle Sam, a national symbol. Study the poster and answer the questions.

a. How does Uncle Sam's clothing show that he stands for the United States?

b. What was the purpose of this propaganda poster?

ALTERNATIVE ASSESSMENT

1. INTERDISCIPLINARY ACTIVITY: Music

Analyzing Music of the War Years During World War I, songs such as "Johnny I Hardly Knew You" protested the war, while "Over There" and "Pack Up Your Troubles" cheered the troops. Analyze the lyrics of a wartime song, examining its attitude toward the war.

2. COOPERATIVE LEARNING ACTIVITY

Negotiating a Treaty Working in seven groups, do research on the Paris peace conference that created the Treaty of Versailles. Have each group represent one of these nations: Germany, France, Britain, the United States, Italy, Japan, Poland. Hold a mock conference in which delegates present their goals and debate resolutions to be included in the treaty. Compare your treaty with the actual historical results.

- Identify the national interests of your country for the postwar settlement.
- Make a list of goals and issues you want discussed at the conference.
- Choose a chief spokesperson for your group.

3. CD-ROM PRIMARY SOURCE EXPLORER

Planning Peace Shortly after America's entry into World War I, President Wilson was planning for peace. He wanted to reduce the risk of war in the future. For this purpose, he came up with his Fourteen Points. Using the CD-ROM, the library, and the Internet, find out more about the Fourteen Points.

Create your own plan for peace using the suggestions below.

- Draw up a plan for peace that might apply to your classroom or community.
- Adapt ideas from the Fourteen Points that you think will work for your classroom or community.
- Decide how peace can be enforced in the classroom and in wider communities.
- Discuss whether working to establish peace might involve setting restrictions on personal liberty.

4. HISTORY PORTFOLIO

Option 1 Review your section and chapter assessment activities. Select one that you think is your best work. Then use comments made by your teacher or classmates to improve your work and add it to your portfolio.

Option 2 Review the details that you listed for What Do You Want to Know? on page 678. Then write a short report in which you explain the detailed information that you acquired. Add your answers to your portfolio.

Campaign for Liberty Bonds

To rally Americans to support World War I, the government set up the Committee on Public Information (CPI). This agency called on creative individuals to join "the world's greatest adventure in advertising." Speakers gave patriotic speeches in theaters, hotels, and restaurants. Artists designed posters persuading Americans to buy Liberty Bonds. These loans to the government helped fund the war effort. Liberty Bonds were actually sold through four Liberty Loan drives in 1917 and 1918.

ACTIVITY Create a poster to help the government raise money for World War I. In addition, write and present a patriotic speech that wins public support of the war.

TOOLBOX

Each group will need:

- poster board
- colored markers
- pencils
- drawing paper
- glue
- scissors

Posters such as this one appealed to patriotism and love of family to sell Liberty Bonds.

STEP BY STEP

1 Form an imaginary ad agency. Meet with three or four other students to discuss your latest contract: The CPI has hired your agency to create a poster as part of a nationwide campaign to sell Liberty Bonds and promote World War I. Your group will:

- do research on Liberty Bonds
- design and create a poster advertising Liberty Bonds
- write and deliver a "pep talk" persuading people to buy Liberty Bonds

2 Research Liberty Bonds. Look on the Internet, in this chapter, or in books about World War I to find out more about Liberty Bonds and to see actual posters. As you look over the posters, think about the feelings the posters bring out. What images and words seem most powerful or persuasive?

3. **Choose a theme for your poster.** Persuading people to buy Liberty Bonds means that you need to show that winning World War I is important. One way is to appeal to people's emotions. For example, the poster can appeal to their sense of fear, pride, or love of family.

4. **Sketch out your idea.** Write out a slogan and choose images based on the theme of your poster. Make sure your words and pictures communicate the same feeling and message. Draw an outline of the images and the letters. Then cut both the letters and images out. Be sure that they're large enough to be seen from several yards away.

5. **Create the poster.** Decide where the art and writing will appear. Experiment with the arrangement of the art and the writing. Move them around. Do not overwhelm your viewers with too many images or too many words. Use vivid, patriotic colors for your poster.

6. **Create a bulletin board display.** Pin or tape your poster on the wall, along with the posters of the other groups. As you examine the other posters, compare and contrast your poster with the others.

WRITE AND SPEAK

Write a patriotic speech. As a group, write a two-minute "pep talk" persuading people to buy Liberty Bonds and to support the soldiers fighting overseas. Include reasons why the war is worth fighting. Each group member should be prepared to deliver the speech, using the poster you made as a visual aid.

HELP DESK

For related information, see pages 691–692 in Chapter 24.

Researching Your Project

- *World War I* by Gail Stewart
- *Causes and Consequences of World War I* by Stewart Ross

Visit www.mcdougallittell.com for links to sites about World War I.

Did You Know?

The CPI used about 75,000 lecturers. They gave around 755,190 speeches to about 300 million people in 5,000 towns.

Even children were moved by advertising slogans to help fund the war: "Lick a stamp and lick the kaiser." Children filled books with war stamps, each worth 25 cents. These stamps were then converted into government bonds.

Even President Wilson helped to raise money for the war effort. He sold wool from sheep raised on the White House lawn.

REFLECT & ASSESS

- **What aspects of your poster do you think will inspire people to buy Liberty Bonds?**
- **How well does your speech inspire patriotic feeling about the war?**
- **Which do you think is a more powerful means of persuasion—your poster or your speech?**

UNIT 8

1919 – 1960

Depression, War, and Recovery

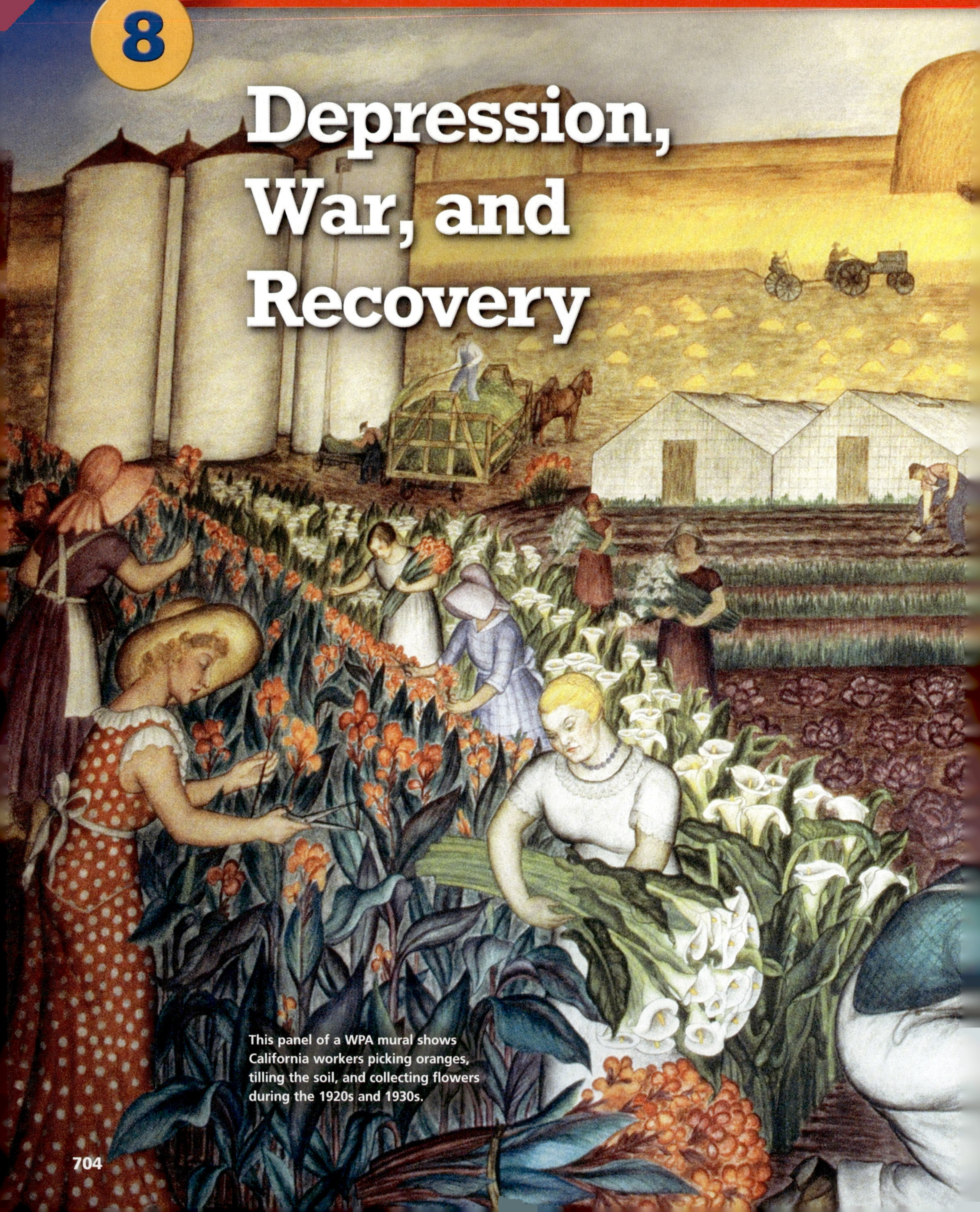

This panel of a WPA mural shows California workers picking oranges, tilling the soil, and collecting flowers during the 1920s and 1930s.

"The test of our progress . . . is whether we provide enough for those who have too little."

—Franklin Delano Roosevelt

CHAPTER 25

The Roaring Twenties 1919–1929

The carefree spirit of the Roaring Twenties is captured on this magazine cover from 1926.

Interact *with* History

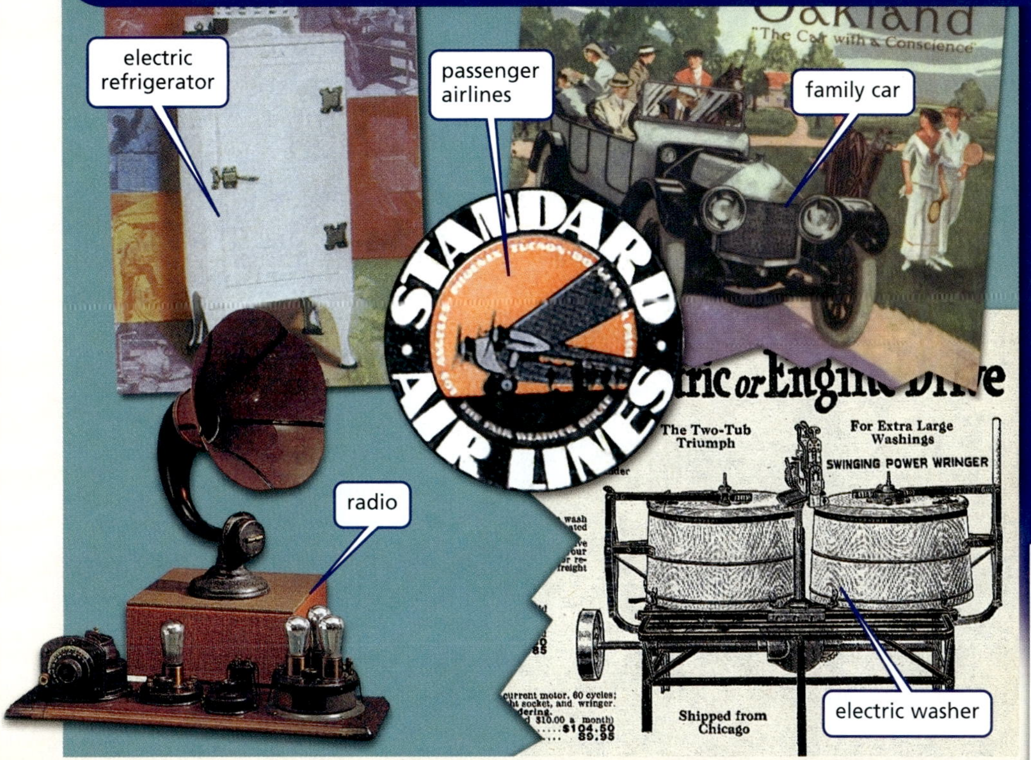

World War I is over, and a new decade has begun. There is peace in the world and prosperity at home. It is a time of exciting social, cultural, economic, and technological change. You see new products and new ideas coming into your life.

What Do You Think?

- How will these new ideas and products change your life?
- Will these changes make life better and easier? How?
- Which class of people will be affected most by these changes?

Which changes in culture or technology will affect your life the most?

USA

1919

1920 Warren G. Harding is elected president.

1923 Calvin Coolidge becomes president.

1924 Coolidge is elected president.

Nellie Tayloe Ross is first woman elected governor.

1925 Scopes Trial is held.

Harlem Renaissance flourishes.

1927 Lindbergh makes first transatlantic solo flight.

First movie with sound, *The Jazz Singer,* released.

1928 Herbert Hoover is elected president.

1929

World

1921 Chinese Communist Party is founded.

1922 Benito Mussolini is named Italy's prime minister.

1923 Adolf Hitler tries but fails to gain power in southern Germany.

1926 Hirohito becomes emperor of Japan.

1928 Kellogg-Briand Pact signed.

1929 National Revolutionary Party organized in Mexico.

BEFORE YOU READ

Previewing the Theme

Science and Technology As Chapter 25 explains, the 1920s were a time of peace and economic prosperity for many Americans. Consumer buying, the growth of the automobile industry, and the development of new technologies helped business to expand and changed the way people lived. The Roaring Twenties also brought new ideas, new attitudes, and new forms of entertainment.

More cars meant traffic jams—like this one in St. Louis in 1920.

What Do You Know?

What do you already know about the Roaring Twenties? What were the issues and who were the personalities that made this decade "roar"?

THINK ABOUT

- how the 1920s have been portrayed in movies, television, and historical fiction
- what happens to a country when rapid changes take place

What Do You Want to Know?

What additional information do you want about the issues and personalities of the 1920s? Record questions you may have in your notebook before you read the chapter.

READ AND TAKE NOTES

Reading Strategy: Finding Main Ideas To understand what you read, learn to find the main idea of each paragraph, topic heading, and section. Remember that the supporting details help to explain the main idea. On the chart below, write down the main idea in this chapter for each category of American life.

See Skillbuilder Handbook, page R5.

Categories	Main Ideas
Government	
Business	
Agriculture	
Technology	
Society	
Popular Culture	

1 The Business of America

TERMS & NAMES
- Warren G. Harding
- Teapot Dome Scandal
- Calvin Coolidge
- laissez faire
- isolationist
- Kellogg-Briand Pact
- assembly line
- installment buying

MAIN IDEA

The government supported business and kept a hands-off policy in other matters.

WHY IT MATTERS NOW

How involved the government should be in the economy remains an issue today.

ONE AMERICAN'S STORY

Warren G. Harding was a pleasant man of whom it was said he "looked like a president." He was happiest relaxing or playing cards with his closest friends. But urged on by his wealthy and ambitious wife, Florence Kling Harding, he rose from small-town newspaper publisher, to U.S. senator from Ohio, to Republican presidential candidate.

The advice from Republican Party leaders in 1920 was "Keep Warren at home. Don't let him make any speeches." So Harding spent most of the 1920 election race campaigning from his front porch in Marion, Ohio. But Harding was what the voters wanted. He promised them prosperity at home and peace abroad, and they elected him president. Mrs. Harding supposedly said, "Well, Warren Harding, I have got you the Presidency; what are you going to do with it?"

In this section, you will read about Presidents Warren G. Harding and Calvin Coolidge, the booming economy of the Roaring Twenties, and the new technologies that helped businesses to grow.

Warren G. Harding and his wife, Florence Kling Harding, at their home in Marion, Ohio.

Harding and the "Return to Normalcy"

After some 20 years of reform and war, Americans were ready for the "normalcy" promised by Harding in the election and at his inauguration.

A VOICE FROM THE PAST

Our supreme task is the resumption of our onward, normal way. Reconstruction, readjustment, restoration all these must follow. I would like to hasten them.

Warren G. Harding, Inaugural Address, March 4, 1921

As president, Harding wanted to lift the burden of taxes and regulations from the shoulders of Americans. To do this, he proposed lower taxes and "less government in business and more business in government." He also sought higher tariffs on foreign goods to help American companies.

Harding chose a pro-business cabinet. The secretary of the treasury was Andrew W. Mellon, one of the wealthiest men in the United States.

STRANGE *but* True

PRESIDENT SWORN IN BY FATHER

On August 2, 1923, Vice-President Calvin Coolidge went to bed early at his family home in Plymouth Notch, Vermont. During the night, Coolidge's father received news by telegram that Harding had died. He then woke his son. The U.S. attorney general urged Coolidge to take the oath of office as soon as possible. At 2:47 A.M. on August 3, 1923, John Coolidge, a justice of the peace, administered the oath to his son.

The New York Times. EXTRA

PRESIDENT HARDING DIES SUDDENLY; STROKE OF APOPLEXY AT 7:30 P.M.; CALVIN COOLIDGE IS PRESIDENT

COOLIDGE TAKES THE OATH OF OFFICE

His Father, Who Is a Notary Public, Administers It After Form Is Found By Him in His Library.

DEATH STROKE CAME WITHOUT WARNING

Mrs. Harding Was Reading to Her Husband When First Sign Appeared —She Ran for Doctor

Mellon persuaded Congress to lower taxes and balance the budget. Herbert Hoover, an engineer who organized aid to Europe in World War I, was secretary of commerce. He worked to cut federal government waste.

While some of Harding's cabinet choices, like Mellon and Hoover, were excellent, a number were unqualified, and even corrupt. These men had been Harding's friends back in Ohio and were known as the "Ohio Gang." They used their government positions to make money illegally. Their actions helped to wreck the Harding presidency. The worst scandal involved Secretary of the Interior Albert Fall. It was called the **Teapot Dome Scandal.** Fall took bribes and made illegal deals with oil executives to drill on oil-rich government land in Teapot Dome, Wyoming.

Reading History
A. Drawing Conclusions How did members of the Ohio Gang take advantage of their friendship with Harding?

Rumors of corruption in the Harding administration began to be heard in 1923. Harding, who was politically and personally honest, was alarmed. He had once said, "I knew that this job would be too much for me." Tired and depressed, Harding went on a speaking tour in the summer of 1923. It was then that he learned the full extent of the corruption. He died suddenly while on the trip, on August 2, 1923. The American people mourned his death, but they were shocked when the scandals became public.

Coolidge Takes Over

Vice-President **Calvin Coolidge** became president when Harding died. He moved quickly to try to clean up the scandals. His efforts limited the political damage to the Republican Party, and Coolidge was elected president in his own right in 1924. He defeated Democrat John W. Davis and Robert M. La Follette, the Progressive Party nominee.

Coolidge and those who voted for him felt that prosperity would be the reward of those who worked hard. As a friend of business, Coolidge agreed with the economic theory of **laissez faire.** It stated that business, if left unregulated by the government, would act in a way that would benefit the nation. In 1925, Coolidge stated his belief that "the chief business of the American people is business." He said that Americans were concerned with "prospering in the world." Under the Coolidge administration, business prospered and so did many Americans.

Vocabulary
laissez faire: to allow to do (French)

"The chief business of the American people is business."
Calvin Coolidge

Coolidge also believed that it was not the government's job to help people with social and economic problems. Farmers were one group that Coolidge refused to help. Because new machinery had been introduced, farmers were producing more food than the nation needed. So food prices were dropping.

Congress passed a bill that required the government to buy the extra food. This would have raised prices. But Coolidge vetoed the bill.

Like Harding, Coolidge was an **isolationist.** Both believed that the United States should stay out of other nations' affairs except in matters of self-defense. Both supported efforts to avoid war.

Coolidge's major peace effort was the **Kellogg-Briand Pact** of 1928. This pact, or treaty, was signed by 15 nations who pledged not to make war against one another except in self-defense. Most Americans supported the treaty. They hoped that if war were outlawed, it would disappear. Then they could concentrate on their own lives.

Technology Changes American Life

The economy was booming in the 1920s. Both Harding and Coolidge kept government regulation to a minimum, and business flourished. Part of the "roar" in the Roaring Twenties was the growth in the nation's wealth. The average annual income per person rose more than 35 percent during the period—from $522 to $716. This increase in income gave Americans more money to buy goods and to spend on leisure activities.

Automobiles had the greatest impact on life during the 1920s. Henry Ford, who built his first successful automobile in 1896, was determined to make a car that most people could afford. At the Ford Motor Company in Detroit, his dream came true with a car called the Model T. In 1920, Ford produced more than a million automobiles, at a rate of one per minute. Each car cost the consumer $335.

Reading **History**

B. Recognizing Effects What effect did the assembly line have on the price of cars?

To speed up production and lower costs and prices, Ford used an **assembly line.** In an assembly line, the product moves along a conveyor belt across the factory. Workers at various stations add parts as the belt moves past them. By the mid-1920s, a Model T came off a Ford assembly line every ten seconds.

1923 Model T Ford

How an Automobile Assembly Line Works

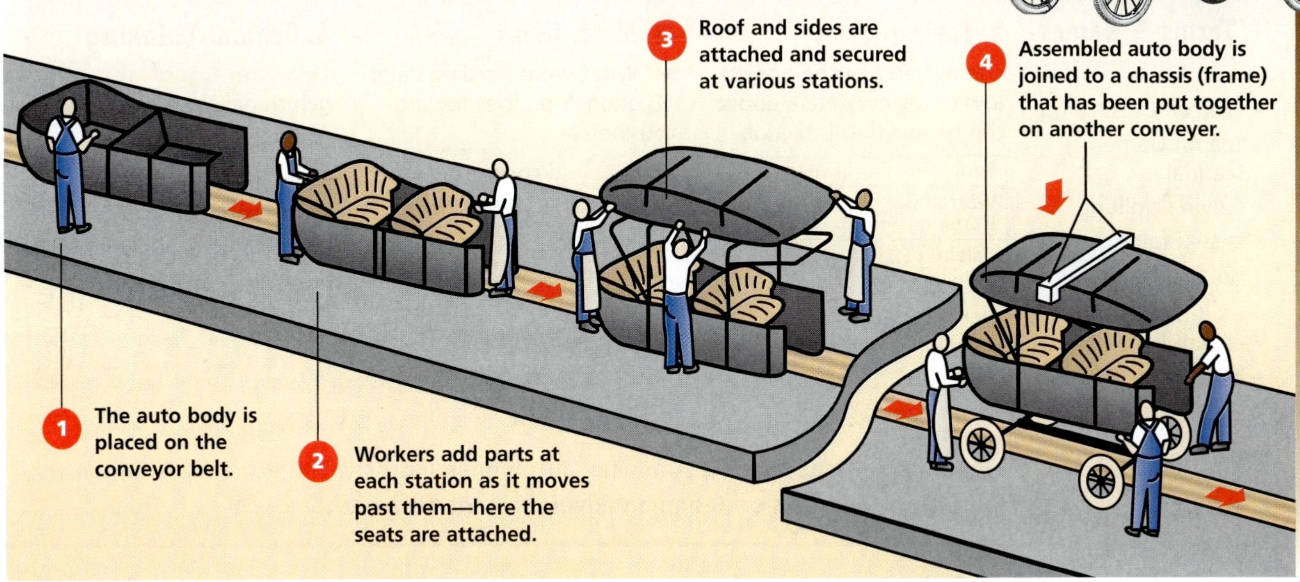

Credit allowed consumers to buy the latest products—$50 down and small monthly payments bought this new oil heater.

Other advances in technology improved life. New machines turned out products faster and cheaper. Once-costly items were now available to many consumers. Some consumers used credit and paid for their purchases through **installment buying.** This allowed repaying the amount borrowed in small monthly payments. National advertising also got its start at this time, as a way of helping to promote new products.

Cheap fuel powered the new prosperity. Petroleum and electricity became widely available. These power sources made possible new inventions and advances in technology that made life easier, such as electric vacuum cleaners, washers, sewing machines, toasters, and fans. However, it was mostly only the white middle class that could afford these new products.

Reading **History**

C. Summarizing How did advances in technology change the lives of Americans?

The Air Age Begins

The 1920s also marked the beginning of the air age. After World War I, many former military pilots bought old war planes and worked as crop-dusters, stunt fliers, and flight instructors. In 1918, the Post Office Department began air mail service. Airplanes had found new uses.

Transatlantic flights by Charles A. Lindbergh in 1927 and Amelia Earhart in 1928 and 1932 helped to promote the idea of commercial air transportation. Pan American Airways, founded in 1927, became the nation's first passenger airline. By the end of the decade, its operations were drawing distant cities closer together both in North and South America.

In the next section, you will read about more changes in life in the United States and the conflicts these changes caused.

Section 1 Assessment

1. Terms & Names

Identify:
- Warren G. Harding
- Teapot Dome Scandal
- Calvin Coolidge
- laissez faire
- isolationist
- Kellogg-Briand Pact
- assembly line
- installment buying

2. Taking Notes

Use a chart like the one below to review details about the people in this section.

People	Details
Warren G. Harding	
Calvin Coolidge	
Henry Ford	

3. Main Ideas

a. What were Harding's and Coolidge's policies toward business?

b. How did corruption affect the Harding administration?

c. How did new technology help business to grow during the 1920s?

4. Critical Thinking

Drawing Conclusions Which developments in the 1920s added to prosperity?

THINK ABOUT
- government's role in the economy
- advances made in technology

ACTIVITY OPTIONS

TECHNOLOGY

ART

Research an aspect of the American automobile industry. Either draw a **diagram** of how a car works or design an **advertisement** for an automobile.

2 Changes in Society

TERMS & NAMES
flapper
Prohibition
Al Capone
NAACP
Marcus Garvey
fundamentalism
Ku Klux Klan

MAIN IDEA

Changes in society in the 1920s brought new attitudes and lifestyles but also caused divisions and conflict.

WHY IT MATTERS NOW

Many of the social issues of the 1920s continue to challenge American society today.

ONE AMERICAN'S STORY

Poet Edna St. Vincent Millay was one of many young people who rebelled against traditional values in the 1920s. She had left her home in Maine to study poetry at Vassar College in New York. She graduated in 1917 as World War I neared its end. Then she moved to the Greenwich Village section of New York City. There Millay lived among artists and writers whose ideas were different from those traditionally held by society. She wrote poems about love and the carefree lifestyle of the 1920s.

A VOICE FROM THE PAST

My candle burns at both ends;
It will not last the night;
But ah, my foes, and oh, my friends—
It gives a lovely light!

Edna St. Vincent Millay, "First Fig," from *A Few Figs from Thistles*

Edna St. Vincent Millay became a bestselling poet and a symbol of her time.

Millay was a symbol of the 1920s woman. In this section, you will read about the changing roles of young people and women, problems facing African Americans, and conflicts that came to divide society.

Youth in the Roaring Twenties

The 1920s were called the Roaring Twenties. According to author F. Scott Fitzgerald, "The uncertainties of 1919 were over. America was going on the greatest, gaudiest spree in history." During the decade, youth and its culture were celebrated. For the first time, young people as a group rebelled against the values of the past and the authority of their elders. The under-25 generation wanted fun and freedom. Many of them experimented with new fashions, attitudes, and ways of behavior.

Young people stayed in school longer, and more went to college. School became a place for socializing as well as learning. Young people expressed their new freedom in daring new clothes, lively songs and dances, and silly fads. Men wore extra-wide floppy pants and sported hair slicked down close to the head. Women wore a shorter hairstyle called a *bob* to match the shorter dresses of the period.

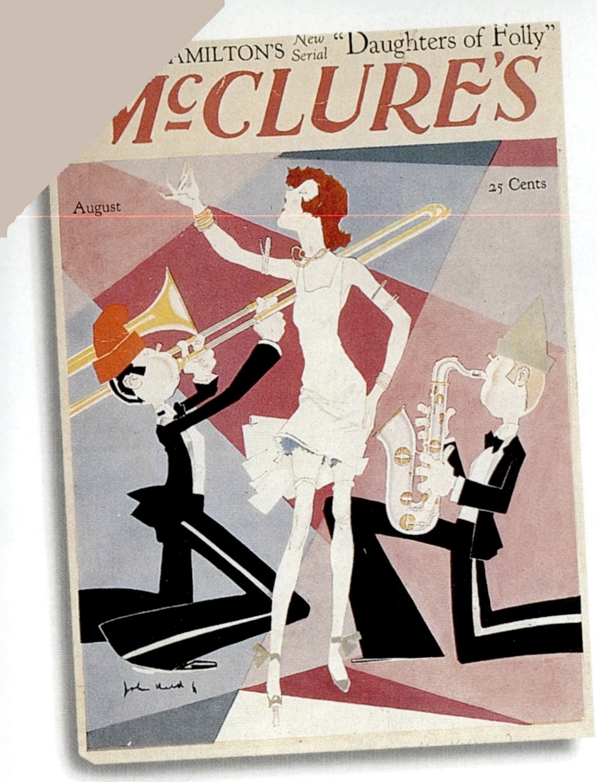

The flapper appeared on many magazine covers during the 1920s.

The Charleston was a favorite dance. It involved wild, flailing movements of the arms and legs. Dance marathons became the rage. In these contests, couples would dance nonstop for days. Songs also captured the high spirits of the decade. Among the most popular tunes were "Runnin' Wild" and "Ain't We Got Fun." Many young people imitated the behavior of favorite stars from Hollywood movies. Other fads included crossword puzzles, mah-jongg, and flagpole sitting (sitting on a platform on top of a flagpole for days).

Background
Mah-jongg was a game from China played with small painted tiles.

The spirited behavior of young women during the decade was just one way women's lives changed.

New Roles for Women

The symbol of the 1920s American woman was the **flapper.** The flapper was the creation of John Held, Jr., a magazine illustrator. Flappers often wore bobbed hair, makeup, and dresses that fell to just below the knee. They were always eager to try something new, whether it was a new fashion, behavior, dance, or fad.

During the 1920s, women took more active roles in their life than ever before. They had more personal freedom. They drove cars, played sports, went to college, and took jobs. Margaret Sanger, a reformer who focused on women's health issues, described these women.

A VOICE FROM THE PAST

Today women are on the whole much more individual. They possess as strong likes and dislikes as men. They live more and more on the plane of social equality with men . . . [and] there is more enjoyable companionship and real friendship between men and women.

Margaret Sanger, quoted in *A More Perfect Union*

The prosperity of the 1920s opened new job opportunities for women in business offices, retail stores, factories, and various professions. College graduates most often became teachers and nurses, but also librarians, social workers, and bankers. Women with less education worked in factories or in offices as typists and secretaries or in stores as clerks and cashiers. Attitudes toward marriage also changed. Men and women came to view marriage as more of an equal partnership. Women still had the responsibility of housework and child rearing. But labor-saving appliances and timesaving convenience foods made life easier.

Reading **History**
A. Recognizing Effects What were some of the effects of women's greater opportunities?

Because of the 19th Amendment, women were able to vote for the first time in 1920. Some even ran for political office. In 1924, two were elected governor—Nellie Tayloe Ross in Wyoming and Miriam "Ma" Ferguson in Texas. But political gains for women were slow. By 1928, only 145 women held seats in state legislatures across the country.

Prohibition and Lawlessness

Another change in American society came on January 16, 1920. That was the date when the 18th Amendment went into effect. The amendment was commonly called **Prohibition,** the ban on the manufacture and sale of alcohol. Many people saw Prohibition as a victory of small-town, Protestant Americans over city dwellers. Supporters felt that Prohibition would promote morality and good health. To enforce the ban, Congress had passed the Volstead Act in 1919.

Saloons were forced to close their doors. But many Americans did not consider drinking harmful or sinful. They resented government interference. People who wanted alcohol found endless ways to get it. For instance, illegal nightclubs known as speakeasies sold liquor. People called bootleggers made their living by transporting and selling liquor illegally. Others simply brewed their own homemade liquor.

Background
Bootlegger came from the old smugglers' practice of carrying liquor in the legs of boots.

One unfortunate result of Prohibition was the growth of organized crime. In nearly every major city, criminal gangs battled for control of bootlegging operations. The most ruthless crime boss of the era was **Al Capone** in Chicago. With a private army of 700 criminals, he violently seized control of the city's 10,000 speakeasies. By the late 1920s, most Americans had come to see Prohibition as a failure. It was repealed by the 21st Amendment in 1933. Prohibition ended, but organized crime did not end with it.

Changes for African Americans

*Reading*History
B. Reading a Map
Locate cities with significant African-American populations on the map on page 723.

The 1920s also brought major changes to the lives of many African Americans. To find better jobs, African Americans had begun moving north in the early 1900s. As you read in Chapter 24, this movement was called the Great Migration. The jobs that they held in industries during World War I raised their expectations for a better life.

In the North, African Americans gained some economic and political power. But they still faced discrimination in jobs and housing. Rising tensions between African Americans and whites in Northern cities led to over 25 race riots in 1919 alone. The movement of an additional 1.5 million African Americans to these cities during the 1920s increased tensions even more.

The National Association for the Advancement of Colored People **(NAACP)** tried to protect the constitutional rights of African Americans. The NAACP worked to make people aware of crimes against African Americans. But it was unable to get Congress to pass legislation to help African Americans fight against discrimination.

*Reading*History
C. Analyzing Points of View
What action did Marcus Garvey believe would improve the lives of African Americans?

Daily threats and discrimination made some African Americans lose faith in America. **Marcus Garvey,** the founder of the Universal Negro Improvement Association, called for a return to Africa and the formation of a separate nation there. He said, "If Europe is for the Europeans, then Africa shall be for the black peoples of the world." Few African Americans migrated to Africa. But Garvey set an example for future black political movements.

Marcus Garvey led a Back-to-Africa movement in the 1920s.

More than 40,000 Ku Klux Klan members march in Washington, D.C., in 1925, to show their growing political power.

A Divided Society

Some groups felt threatened by the changes in society in the 1920s. Conflicts developed over ideas and values. Divisions between groups resulted—between African Americans and whites, the native-born and immigrants, and the urban and rural communities. Science and religion also were in conflict.

In religion, a movement called **fundamentalism** gained both recognition and political power. Fundamentalists believed in a literal, or word-for-word, interpretation of the Bible. They did not want the theory of evolution taught in public schools because it opposed their belief in the biblical story of creation. Evolution is the scientific theory that living things developed over millions of years from earlier and simpler forms of life.

Fundamentalists succeeded in banning the teaching of evolution in Tennessee and 12 other states. In 1925, in Dayton, Tennessee, biology teacher John Scopes broke this law. He took this action to test whether the law could be enforced. Scopes's trial attracted national attention. The jury found Scopes guilty, but the Tennessee Supreme Court reversed the decision. Controversy over the teaching of evolution continues today.

*Reading*History

D. Analyzing Causes What action taken by fundamentalists caused John Scopes to break the law in Tennessee?

Another reaction to changes in society was the rebirth of the **Ku Klux Klan.** The Klan called for a "racially and morally pure" America. It became strong in several states, including some outside the South. By 1924, the Klan claimed as many as five million members. It tried to influence national, state, and local politics by using violence against African Americans and other groups. Its power began to decrease by the end of the decade because of personal and financial scandals in the organization.

In this section, you read about divisions in society. In the next, you will learn how mass media and popular culture brought Americans together.

Section 2 Assessment

1. Terms & Names

- flapper
- Prohibition
- Al Capone
- NAACP
- Marcus Garvey
- fundamentalism
- Ku Klux Klan

2. Taking Notes

Use a cluster diagram to review the fads of the Roaring Twenties.

Which fads of the 1920s had lasting influence?

3. Main Ideas

a. How did the Roaring Twenties change the lives of young people?

b. What factors were responsible for the changes in women's lives?

c. What were the conflicts that divided society?

4. Critical Thinking

Recognizing Effects How was American society transformed in the 1920s?

THINK ABOUT

- roles of young people and women
- migration of African Americans
- conflicts between groups

ACTIVITY OPTIONS

ART / **MUSIC**

Draw a **poster** with an image that represents the Roaring Twenties or write a **song** capturing the spirit of the times.

3 The Jazz Age and the Harlem Renaissance

TERMS & NAMES
- jazz
- mass media
- popular culture
- Harlem Renaissance
- Lost Generation
- expatriate

MAIN IDEA

Popular culture was influenced by the mass media, sports, and the contributions of African Americans.

WHY IT MATTERS NOW

Much of today's popular culture had its origins in this period.

ONE AMERICAN'S STORY

The decade known as the Roaring Twenties was also called the Jazz Age, because the lively, loose beat of **jazz** captured the carefree spirit of the times. Jazz was developed by African-American musicians in New Orleans. That city was the home of Louis Armstrong, who became one of the world's great jazz musicians.

As a child, Armstrong often listened to jazz played at funeral processions and dance halls. He was raised by a poor, single mother and started working at age seven. His job was collecting junk in a horse-drawn wagon. While in the wagon, Armstrong often played a small tin horn.

Louis Armstrong brought New Orleans jazz to the North in the 1920s.

A VOICE FROM THE PAST

I had a little tin horn, the kind the people celebrate with. I would blow this long tin horn without the top on it. Just hold my fingers close together. Blow it as a call for old rags, bones, bottles or anything that people had to sell. . . . The kids loved the sounds of my tin horn!

Louis Armstrong, quoted in *Louis Armstrong* by Sandford Brown

Later, Armstrong learned to play the trumpet. With other jazz musicians, he spread this new music to other parts of the country—from Chicago to New York's Harlem—and then to Europe.

In this section, you will read more about the spread of popular culture, the Harlem Renaissance, and the artists of the Lost Generation.

More Leisure Time for Americans

Laborsaving appliances and shorter working hours gave Americans more leisure time. Higher wages also gave them money to spend on leisure activities. People wanted more fun, and they were willing to spend money to have it. Americans paid 25 cents or more to see a movie—an increase of at least 5 times the price in the previous decade. By the end of the 1920s, there were more than 100 million weekly moviegoers.

In addition to attending movies, some Americans went to museums and public libraries. Others bought books and magazines. Sales rose by

Celebrities of the 1920s

◀ **Charlie Chaplin** was the most popular male film star during the 1920s. He was known as the "great comedian."

Babe Ruth ▶ helped to popularize baseball. In 1927, he became the first player to hit 60 home runs in one year.

◀ **Helen Wills** dominated women's tennis in the 1920s.

▼ **Duke Ellington** was a world-famous jazz pianist and composer. His band played at the Cotton Club in Harlem in the 1920s.

50 percent. Americans also spent time listening to the radio, talking on the telephone, playing games, and driving their cars. In 1929, Americans spent about $4 billion on entertainment—a 100 percent jump in a decade.

But not all Americans were able to take part equally in leisure-time activities or in the consumer culture of the 1920s. Some, like African Americans and Hispanic Americans, had their time and choices limited by factors such as income and race.

Reading **History**

A. Recognizing Effects What was the main effect that laborsaving devices and reduced working hours had on Americans' lives?

Mass Media and Popular Culture

New types of **mass media**—communications that reach a large audience—began to take hold in the 1920s. Radio and movies provided entertainment and spread the latest ideas about fashions and lifestyles.

The first commercial radio broadcast took place in Pittsburgh at station KDKA in 1920. Other radio stations soon emerged. The number of households with radios jumped from about 60,000 in 1922 to 10 million in 1929. Radio stations broadcast news, sports, music, comedy, and commercials. Not only were Americans better informed than before, but listening to the same radio programs united the nation.

Of all the powerful new influences of the 1920s, none shaped the ideas and dreams of Americans more than motion pictures. The moviemaking industry was centered in Hollywood, California.

Movies gave people an escape into worlds of glamour and excitement they could never enter. Audiences flocked to movie theaters to see their favorite actors and actresses. These included Charlie Chaplin, Mary Pickford, Douglas Fairbanks, Clara Bow, and Rudolph Valentino. Movies also spread American popular culture to Europe. **Popular culture** included songs, dances, fashions, and even slang expressions like scram (leave in a hurry) and ritzy (elegant).

Moviemakers like Samuel Goldwyn, the Warner brothers, and Louis B. Mayer made fortunes overnight. For most of the 1920s, films were silent. In 1927, *The Jazz Singer* introduced sound. Another *talkie* caused a sensation in 1928—Walt Disney's cartoon *Steamboat Willie,* featuring Mickey Mouse. Within a few years, all movies were talkies.

A Search for Heroes

Another leisure activity was watching sporting events and listening to them on the radio. Sporting events of all types—baseball, football, hockey, boxing, golf, and tennis—enjoyed rising attendance. Boxing became very popular. Fans who could not attend the fights listened to matches on the radio or saw them on newsreels shown at movie theaters. The Jack Dempsey–Gene Tunney boxing match of 1926 drew 120,000 fans.

Reading **History**

B. Summarizing What were some of the changes that came about in popular entertainment in the 1920s?

In the 1920s, professional baseball gained many new fans because games were broadcast on radio. As a result, fans flocked to major league ballparks. In New York City, fans went to Yankee Stadium, which opened in 1923, to watch the "Bronx Bombers"—the nickname for the New York Yankees. Even college football and basketball attracted huge crowds.

Sports figures captured the imagination of the American public. They became heroes because they restored Americans' belief in the power of the individual to improve his or her life. Babe Ruth of the Yankees was baseball's top home-run hitter. Someone once asked Ruth why his $80,000 salary was higher than the president's. Ruth supposedly replied, "Well, I had a better year."

Baseball players weren't the only sports heroes. Golfers idolized Bobby Jones. People cheered Helen Wills and Bill Tilden on the tennis courts. In 1926, New York City threw a huge homecoming parade for Gertrude Ederle, the first woman to swim the English Channel. Americans also made national heroes of two daring young fliers—Charles A. Lindbergh and Amelia Earhart.

AMERICA'S HISTORY MAKERS

CHARLES A. LINDBERGH

1902–1974

Charles A. Lindbergh took flying lessons in 1922 and bought his first airplane in 1923. Four years later, in May 1927, he became the first person to fly nonstop alone across the Atlantic Ocean.

Lindbergh had heard about an offer of $25,000 to anyone who could fly nonstop from New York to Paris. Piloting his single engine monoplane, the *Spirit of St. Louis,* without radio or parachute, Lindbergh flew some 3,600 miles in 33½ hours. "Lucky Lindy" became an instant hero.

AMELIA EARHART

1897–1937

Amelia Earhart was often called "Lady Lindy" because of both her physical resemblance to Charles Lindbergh and her similar accomplishments as a pilot.

Earhart took flying lessons in 1921 and bought her first plane in 1922. Noted for her courage and independence, she flew where no women had gone before. She was the first woman to cross the Atlantic in a plane (as a passenger) in 1928 and the first to fly solo across the Atlantic in 1932. She disappeared on a round-the-world flight in 1937. What happened remains a mystery to this day.

Why do you think Lindbergh and Earhart became American heroes?

HISTORY *through* ART

Artists of the Harlem Renaissance celebrated the cultural traditions and the life experiences of African Americans. This painting by Lois Mailou Jones is entitled *The Ascent of Ethiopia.*

How does the artist show the link between African and American culture?

The Harlem Renaissance

Wartime military service and work in war industries had given African Americans a new sense of freedom. They migrated to many cities across the country, but it was New York City that turned into the unofficial capital of black America. In the 1920s, Harlem, a neighborhood on New York's West Side, was the world's largest black urban community.

The migrants from the South brought with them new ideas and a new kind of music called jazz. Soon Harlem produced a burst of African-American cultural activity known as the **Harlem Renaissance,** which began in the 1920s and lasted into the 1930s. It was called a renaissance because it symbolized a rebirth of hope for African Americans.

Vocabulary
renaissance: rebirth (French)

Harlem became home to writers, musicians, singers, painters, sculptors, and scholars. There they were able to exchange ideas and develop their creativity. Among Harlem's residents were poets Langston Hughes, James Weldon Johnson, and Countee Cullen and novelists Claude McKay and Zora Neale Hurston. Hughes was perhaps Harlem's most famous writer. He wrote about the difficult conditions under which African Americans lived.

Jazz became widely popular in the 1920s. It was a form of music that combined African rhythms, blues, and ragtime to produce a unique sound. Jazz spread from its birthplace in New Orleans to other parts of the country and made its way into the nightclubs of Harlem. These nightclubs featured popular jazz musicians such as Louis Armstrong and Duke Ellington, and singers such as the jazz and blues great, Bessie Smith. Harlem's most famous nightclub was the Cotton Club. It made stars of many African-American performers, but only white customers were allowed in the club.

***Reading* History**
C. Recognizing Effects What changes to popular culture resulted from African Americans migrating to the North?

The Lost Generation

For some artists and writers, the decade after the war was not a time of celebration but a time of deep despair. They had seen the ideas of the Progressives end in a senseless war. They were filled with resentment and they saw little hope for the future. They were called the **Lost Generation.**

For many of them, only one place offered freedom and tolerance. That was Paris. The French capital became a gathering place for American **expatriates,** people who choose to live in a country other than their own. Among the American expatriates living in Paris was the young novelist Ernest Hemingway. As an ambulance driver in Europe during World War I, he had seen the war's worst. His early novels, *The Sun Also Rises* and *A Farewell to Arms,* reflected the mood of despair that followed the war.

Reading **History**
D. Analyzing Causes Why did many American writers become expatriates and live in Paris?

Novelists F. Scott Fitzgerald and Sinclair Lewis were two other members of the Lost Generation. Fitzgerald and his wife, Zelda, lived the whirlwind life of the Jazz Age—fast cars, nightclubs, wild parties, and trips to Paris. His masterpiece, *The Great Gatsby,* is a tragic story of wealthy New Yorkers whose lives spin out of control. The novel is a portrait of the dark side of the Roaring Twenties.

Lewis wrote *Babbitt,* a novel that satirized, or made fun of, the American middle class and its concern for material possessions.

F. Scott Fitzgerald is pictured here in France with his wife, Zelda. He published his masterpiece, *The Great Gatsby,* while living there.

A VOICE FROM THE PAST

It's the fellow with four to ten thousand a year . . . and an automobile and a nice little family in a bungalow . . . that makes the wheels of progress go round! . . . That's the type of fellow that's ruling America today; in fact, it's the ideal type to which the entire world must tend, if there's to be a decent, well-balanced . . . future for this little old planet!

Sinclair Lewis, *Babbitt*

The social values and materialistic lifestyles criticized by Lewis soon came to an end. As you will read in the next chapter, the soaring economy that brought prosperity in the 1920s came to a crashing halt. It was followed by a worldwide economic depression in the 1930s.

Section 3 Assessment

1. Terms & Names

Identify
- jazz
- mass media
- popular culture
- Harlem Renaissance
- Lost Generation
- expatriate

2. Taking Notes

Use the chart to review facts about mass media.

Radio	Movies

How did mass media change the lives of Americans?

3. Main Ideas

a. Which two factors gave Americans more leisure time?

b. What effect did radio have on sports?

c. Why was Harlem called the unofficial capital of black America?

4. Critical Thinking

Evaluating What contributions to popular culture occurred in the 1920s?

THINK ABOUT
- the impact of World War I
- the power of mass media
- new social values

ACTIVITY OPTIONS

ART
TECHNOLOGY

Find an image and important facts about a noted person in this section. Draw a **trading card** or plan that person's **home page** for the Internet.

African-American Baseball Leagues

More than a million African Americans left the South from 1917 to 1929. They were lured to large cities by the offer of higher wages and the increased demand for labor. It was during this period of growing urbanization that the Negro baseball leagues were formed. The map on the next page shows cities with notable teams in the 1920s and 1930s. Each city had an African-American population large enough and wealthy enough to support a team.

Most teams were owned by African Americans. To raise money for expenses, teams needed to play as many games as possible. Thus, they traveled constantly. Stopping in big cities and small towns, they played other African-American teams or white amateur and professional teams. Eventually, teams of African-American all-stars played exhibitions against professional white all-stars.

In 1920, Andrew "Rube" Foster persuaded owners of seven other teams to join him in forming the first Negro baseball league—the Negro National League. Foster (pictured in suit at right) was a former player and then owner of the Chicago American Giants. The Negro American League got started in 1937. After major league baseball was integrated by Jackie Robinson in 1947, the Negro leagues began to decline.

ARTIFACT FILE

Pittsburgh Crawfords

The Pittsburgh Crawfords, shown here with the team's bus, was one of the best teams in the Negro leagues in the 1930s. The success of traveling teams helped to boost revenues of African-American-owned hotels and restaurants in every city that they played.

Memorabilia

The baseball jersey and shoes pictured here are part of the uniform worn by a player from one of the traveling teams of the period.

Cities with Notable African-American Baseball Teams, *1920s–1930s*

On-Line Field Trip

National Baseball Hall of Fame and Museum

The Negro leagues have been widely honored. This poster is from the National Baseball Hall of Fame and Museum in Cooperstown, New York. In Kansas City, Missouri, the Negro Leagues Baseball Museum also keeps the memory of these teams alive.

Visit www.mcdougallittell.com for more information.

CONNECT TO GEOGRAPHY

1. **Location** Which cities had notable teams in the 1920s and 1930s?
2. **Region** Why do you think that African-American teams were located mainly in cities in the East?

See Geography Handbook, pp. 4–5.

CONNECT TO HISTORY

3. **Evaluating** How did the migration of African Americans from the South to Northern cities lead to the rise of Negro baseball leagues?

Chapter 25 ASSESSMENT

VISUAL SUMMARY

The Roaring Twenties

Politics

Republican presidents Warren G. Harding and Calvin Coolidge supported business in the United States and isolationism in foreign relations.

Economics

Business prospered in the 1920s, helped by government support and the development of new technologies. But some groups, notably farmers, faced hardships.

Technology

Technological developments, such as the assembly line, and cheap, available sources of power, such as electricity and petroleum, powered the new prosperity.

Society and Culture

Changes in society brought new attitudes and lifestyles, especially for young people and women. Movies, radio, jazz, and sports became popular forms of entertainment.

TERMS & NAMES

Briefly explain the importance of each of the following.

1. Warren G. Harding
2. Calvin Coolidge
3. isolationist
4. Kellogg-Briand Pact
5. NAACP
6. Marcus Garvey
7. fundamentalism
8. mass media
9. Harlem Renaissance
10. Lost Generation

REVIEW QUESTIONS

The Business of America (pages 709–712)

1. How was the foreign policy of Harding and Coolidge isolationist?
2. What was the economic theory of laissez faire?
3. Which factors contributed to the nation's growing wealth during the 1920s?

Changes in Society (pages 713–716)

4. What changes took place in the behavior and values of young people during the 1920s?
5. What was the image of the *flapper?*
6. How did the 19th Amendment change women's lives?
7. What were some of the divisions in society in the 1920s?

The Jazz Age and the Harlem Renaissance (pages 717–723)

8. What were three examples of American popular culture?
9. Which factors contributed to the popularity of sports?
10. Why are the 1920s also called the Jazz Age?

CRITICAL THINKING

1. USING YOUR NOTES

Categories	Main Ideas
Government	
Business	
Agriculture	
Technology	
Society	
Popular Culture	

Using your completed chart, answer the questions below.

a. What role did government choose to play in the economy?
b. How did technology affect life during the 1920s?
c. How did popular culture change the habits of society?

2. ANALYZING LEADERSHIP

Think about the political leaders discussed in this chapter. Which of their characteristics made them good or poor leaders?

3. APPLYING CITIZENSHIP SKILLS

In what other ways could people have protested Prohibition besides disregarding the law?

4. THEME: SCIENCE AND TECHNOLOGY

How did advances in technology contribute to the prosperity of the United States during the 1920s?

5. DRAWING CONCLUSIONS

Explain how the African-American migration to the North and the spread of jazz contributed to the cultural diversity of the United States.

Interact *with* History

In your opinion, which changes in American life discussed in this chapter would have affected you the most?

HISTORY SKILLS

1. INTERPRETING GRAPHS

Study the graph and then answer the questions.

Urbanization of America, *1910–1930*

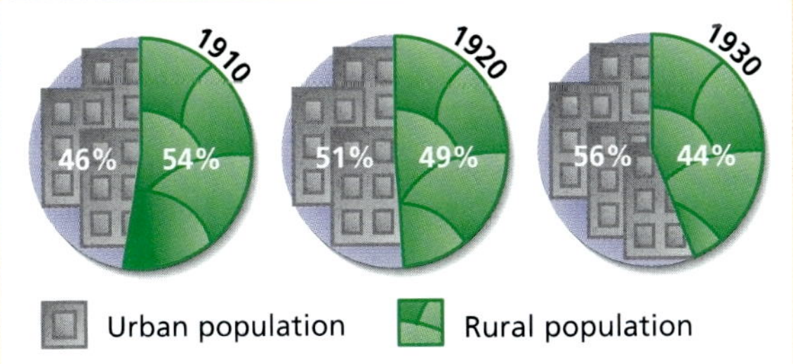

Source: *Historical Statistics of the United States*

a. What was the total percentage increase in the urban population from 1910 to 1930?

b. During which 10-year period did the United States become more urban than rural?

2. INTERPRETING PRIMARY SOURCES

Read the poem below, which was written by Langston Hughes, an African American. Then answer the questions that follow.

I, too, sing America.

I am the darker brother.
They send me to eat in the kitchen
When company comes,
But I laugh,
And eat well,
And grow strong.

Tomorrow,
I'll be at the table
When company comes.
Nobody'll dare
Say to me,
"Eat in the kitchen,"
Then.

Besides,
They'll see how beautiful I am
And be ashamed—

I, too, am America.

Langston Hughes, "I, Too"

a. What is the subject of the poem?

b. What change does Hughes see taking place?

ALTERNATIVE ASSESSMENT

1. INTERDISCIPLINARY ACTIVITY: Language Arts

Writing a Report Do research to learn how cars are made on an assembly line today. Describe the procedure for manufacturing a car from start to finish. Include the method of production, materials used, and the time it takes to make a car. Share your report with the class.

2. COOPERATIVE LEARNING ACTIVITY

Writing or Drawing an Advertisement National advertising became an important way to promote new products to the public during the 1920s. Working in a small group, write a radio ad or draw a magazine ad for a product that made life easier during the 1920s, such as a car, refrigerator, toaster, vacuum cleaner, sewing machine, fan, or washer. Make sure your advertisement covers the following ideas:

a. Do research on the product. Describe the product in the 1920s.

b. How will the product improve life for Americans?

c. How much does the product cost?

3. TECHNOLOGY ACTIVITY

Planning a Web Page on the Roaring Twenties The 1920s were years of great creativity. American popular culture was being spread throughout the nation and abroad by radio and movies. Use the library or search the Internet for information about the music, fashions, fads, and celebrities of the period.

Visit www.mcdougallittell.com to learn more about the Roaring Twenties.

Create a Roaring Twenties Web page by following the suggestions below.

- Select appropriate images of personalities, fashions, and fads.
- Include biographical information and quotations.
- Choose music that captures the spirit of the era.
- Decide which Web sites would be good links for visitors to your page.

4. PORTFOLIO ACTIVITY

Option 1 Review your section and chapter assessment activities. Select one that you think was your best work. Use comments made by your teacher or classmates to improve your work, and then add it to your portfolio.

Option 2 Review the questions that you wrote for "What Do You Want to Know?" on page 708. Then write a brief report explaining the answers to your questions. Add your work to your history portfolio.

CHAPTER 26

The Great Depression and New Deal 1929–1940

Homeless people used scrap materials to build shacks in a New York City alley during the Great Depression.

Interact *with* History

It's 1932. The economy is bad, and millions of people are out of work. Some are starving.

Two men are running for president. One says the government should give money to the poor. The other says this will make people stop looking for jobs. He wants charity groups to help people in need.

What Do You Think?

- Is the government responsible for everyone's well-being?
- What responsibility do individuals have to help others?
- What is the best way to help people out of poverty?

Who do you think should help the poor?

USA

1929 U.S. stock market crashes. Great Depression begins.

1931 President Hoover declares that the country will work itself out of the Depression.

1932 Americans turn against Hoover. Franklin Delano Roosevelt is elected president.

1933 Roosevelt initiates government programs to help the economy.

1935 Congress passes the Social Security Act.

1936 Roosevelt is reelected.

1937 Roosevelt tries but fails to add justices to the Supreme Court.

1939 John Steinbeck publishes *The Grapes of Wrath* about migrant workers.

World

1931 Affected by the Depression, Japan invades Manchuria, partially to expand its economy.

1933 Adolf Hitler becomes dictator of Germany.

1936 Léon Blum, socialist premier of France, introduces reforms such as the 40-hour workweek.

1939 Germany invades Poland, starting World War II.

1929 — 1940

Chapter 26 SETTING THE STAGE

BEFORE YOU READ

Previewing the Theme

Economics in History Chapter 26 explains how the prosperity of the 1920s ended suddenly. The years of unemployment and hard times that followed are known as the Great Depression. To improve economic conditions, President Roosevelt and Congress enacted programs that permanently changed the U.S. government.

What Do You Know?

What do people do to get by when they are unable to find work? Where can they turn for help besides the government?

THINK ABOUT

- stories you may have heard about the Great Depression
- movies and books about people with economic struggles

What Do You Want to Know?

What questions do you have you want to know about how the Great Depression affected Americans? What facts and details would you like to learn about government actions during that period? In your notebook, list the things you hope to learn from this chapter.

READ AND TAKE NOTES

Reading Strategy: Evaluating To evaluate is to make a judgment about something. As you read this chapter, look for details about how the following people responded to the Great Depression: President Herbert Hoover, President Franklin Delano Roosevelt, and ordinary citizens (civilians). Record those details in a chart like the one below. Then evaluate how effective those responses were at making the situation better.

See Skillbuilder Handbook, page R18.

GREAT DEPRESSION

Hoover's Responses	FDR's Responses	Citizens' Responses
Effectiveness	Effectiveness	Effectiveness

1 Hoover and the Crash

TERMS & NAMES
- Herbert Hoover
- speculation
- buying on margin
- Black Tuesday
- Crash of 1929
- Great Depression
- public works projects
- Bonus Army

MAIN IDEA

After the stock market crash of 1929, the U.S. economy sank into the worst depression in its history.

WHY IT MATTERS NOW

Today the government regulates banking and the stock exchange to prevent such severe depressions.

ONE AMERICAN'S STORY

Not everyone prospered in the 1920s. For example, many farmers suffered poverty because farm prices stayed low. Republican senator George Norris from Nebraska joined a group of lawmakers called the farm bloc, who sought to pass new laws to aid farmers. Norris also criticized bankers for not caring about farmers' problems.

A VOICE FROM THE PAST

When the great leaders of banking and industry can see no further than the artificial prosperity that comes to Big Business while those who toil on farms are getting no return for their labor, then indeed we have a right to question the wisdom of our financial leaders.

George Norris, "The Farmers' Situation, a National Danger"

Instead of holding to pure Republican ideas, Norris said he "would rather be right than regular."

Norris also urged the building of government-owned power plants in rural areas. He was unusual because most Republicans believed that government should not interfere in the economy. Section 1 explains how Republican policies came under attack when the economy failed.

Problems in the Economy

Secretary of Commerce **Herbert Hoover** became the Republican candidate for president in 1928. In a speech, he stated, "We shall soon . . . be in sight of the day when poverty will be banished from this nation." But the overall prosperity of the 1920s hid the fact that some industries were in trouble. These included agriculture, railroads, textile mills, and mines.

The growing wealth of the richest Americans also hid the struggle of the majority. By 1929, 71 percent of American families earned less than $2,500 per year, the minimum needed to live decently. Some people had no jobs. African Americans in particular had high jobless rates.

During the 1920s, industries had improved efficiency and begun to produce more goods. But the income of middle-class and poor people didn't rise enough for them to purchase the extra goods. Products piled up in warehouses, causing a problem. Unless businesses sold their products, they couldn't pay for materials, salaries, equipment, or shipping.

CONNECTIONS TO MATH

Economic Problems, *1920–1929*

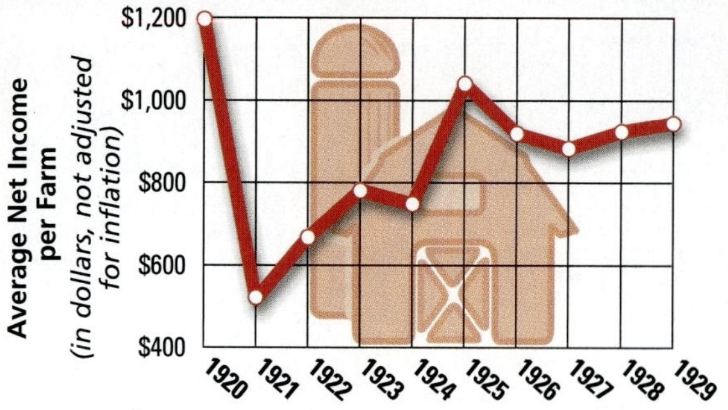

Average Net Income per Farm Notice the big drop in average farm income between 1920 and 1921. Although farm income generally rose during the decade, it did not reach the 1920 level.

Value of Exports Notice the drop between 1920 and 1921. For the next eight years, exports remained generally the same. U.S. businesses barely increased the amount they sold overseas.

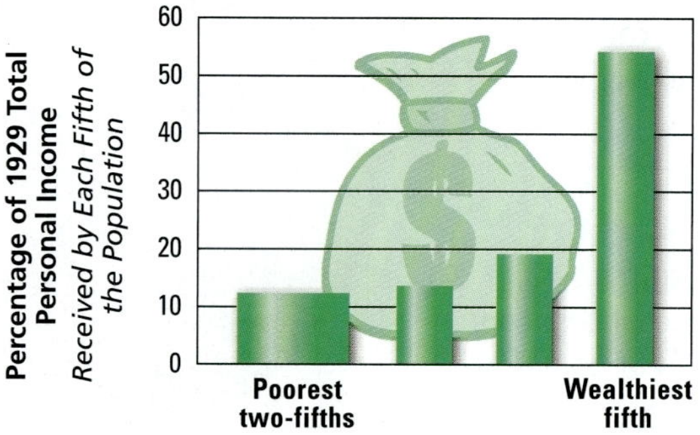

Income Distribution In 1929, the wealthiest fifth of the population had more income than the other four-fifths combined. Poor and middle-class people could not afford to buy many consumer goods. This hurt business.

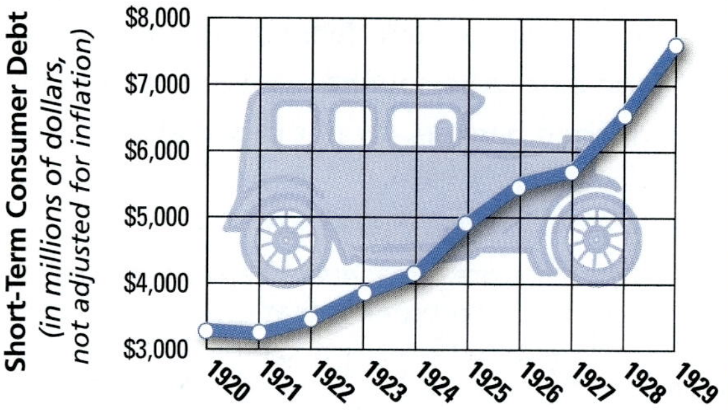

Short-Term Consumer Debt If people do not earn enough to buy what they want, they may take out loans or buy on credit. Short-term consumer debt skyrocketed from 1920 to 1929.

Sources: *Historical Statistics of the United States; A Study of Saving in the United States*

SKILLBUILDER Interpreting Graphs

1. *Which of these problems affected individuals, and which affected businesses?*
2. *By how much did short-term consumer debt rise from 1920 to 1929?*

Some consumers bought goods anyway—on credit. But when their debt grew higher than they could repay, they stopped buying new items. Unsold goods piled up even more.

Yet stock market prices kept climbing. Americans who could afford it rushed to buy stocks. Increasingly, investors bought on **speculation,** buying and selling stocks in the hope of making a quick profit.

Investors also began **buying on margin**—they paid a small part of a stock's price as a down payment and borrowed the rest. When they sold the stock, they repaid the loan and kept the profit. The system worked as long as prices rose. But if prices fell, borrowers couldn't repay their loans because they had to sell the stock for less than they paid for it.

Despite these problems, people believed Hoover when he predicted growing prosperity. In 1928, he won the presidency by a landslide.

Vocabulary
credit: an agreement to pay over time, instead of all at once

The Crash and the Great Depression

On September 3, 1929, the value of stocks on the New York Stock Exchange reached a high point. Then prices drifted downward. On October 23, prices dropped sharply. The next morning, people tried to sell thousands of shares before their value dropped further. Many of those who had bought on margin were forced to sell stocks to pay off their loans.

This heavy selling drove prices even lower. Because more people wanted to sell stocks than to buy them, prices had to go down to attract purchasers. But the quickly falling prices scared off buyers. Meanwhile, sellers hurried to unload their shares at the best price they could get.

On October 24, a record 12.9 million shares were traded. But the worst was yet to come. On October 29, **Black Tuesday,** investors sold 16.4 million shares of stocks at prices much lower than they had been selling for a month earlier. The plunge in stock market prices, called the **Crash of 1929,** was the first event of a terrible economic depression.

After the stock market crash, banks began to demand that people pay back the money they had borrowed to buy stocks. When people could not repay these loans, banks ran short of money. This frightening news sent people running to the banks to withdraw their savings. But banks typically do not keep enough cash on hand to pay all of their depositors at once. Unable to pay their depositors, many banks simply closed. By March 1933, about 9,000 banks had gone out of business.

Background
This pattern of selling stocks and withdrawing deposits because of fear is the reason the depressions of the 1800s are also called panics.

Businesses felt the impact next. Many already had warehouses filled with more goods than they could sell. Because the economy's problems scared people, they stopped buying new goods. As a result, businesses sold less and less. Tens of thousands of businesses went bankrupt.

To survive, many businesses fired workers. Unemployment grew to 25 percent by 1933. As more people lost jobs, they bought fewer products—and companies laid off even more workers. Unable to pay their bills, thousands of people lost their homes, and millions went hungry.

The United States had experienced economic depressions before. But no depression caused as much suffering or lasted as long as this one—from 1929 to World War II. Therefore, it is called the **Great Depression.**

The Great Depression also affected millions of people around the world. For example, many European countries had borrowed money from U.S. banks to rebuild after World War I. When the American economy failed, so did Europe's. Hard times spread around the globe.

Reading **History**
A. Analyzing Causes Why do you think people expected Hoover to end the Depression?

Hoover Acts Conservatively

Americans looked to the president to end the hard times. Along with most Republicans, Hoover feared that government interference might hurt the economy even more. Even so, he did try to fight the Depression, but his actions often backfired. For example, he tried to balance the

Recession and Depression

Many people hoped the economy would fix itself because they believed depressions were a natural part of the business cycle (shown below). Economies go through ups and downs. The period when an economy is at its worst is a trough. There are two kinds of troughs—recessions and depressions. A depression is more severe.

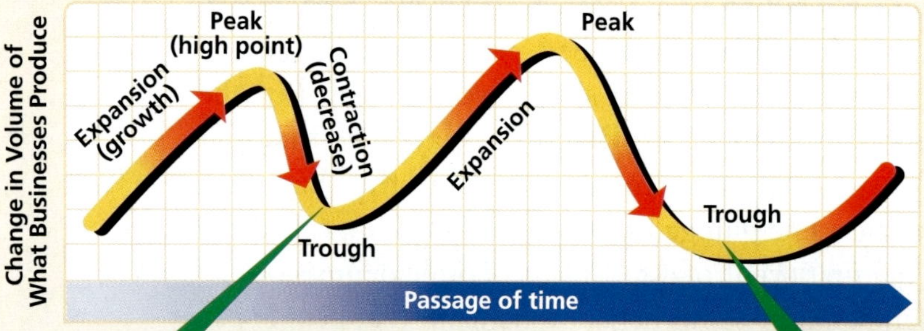

Recession

- The production of a nation's goods and services goes down each month for six months.
- Business owners produce less and invest less in new equipment and facilities. They also lay off workers.
- Consumers buy fewer goods.

Depression

- The production of goods and services drops lower than in a recession.
- The period of no economic growth is longer than in a recession. Unemployment is higher.
- The slowdown may spread to other countries; international trade declines dramatically.

CONNECT TO HISTORY

1. **Evaluating** Why were the hard times of the 1930s a depression, not a recession?

See Skillbuilder Handbook, page R18.

CONNECT TO TODAY

2. **Making Inferences** How would today's world be affected differently by a depression in the United States than by one in Holland? Explain.

Visit www.mcdougallittell.com to learn more about economics.

federal budget by cutting government spending and raising taxes. This pulled money out of the economy, which made the slump worse.

Some of Hoover's actions made him very unpopular. For example, he believed that federal relief—aid to the poor—would make people too dependent on government. So he would not support giving relief. In his view, one answer to the hard times was a quality he called "rugged individualism." In 1931, Hoover said, "We cannot legislate ourselves out of a world economic depression. We can and will work ourselves out."

Hoover also stressed the value of volunteer efforts. He encouraged churches and private charities such as the Salvation Army and the Red Cross to help needy Americans. In 1932, private giving did reach a record level—but it still wasn't enough to help everyone in need.

As unemployment, hunger, and homelessness grew, people became bitter toward the president. They blamed him for their suffering. They called empty pockets turned inside out "Hoover flags." And they called villages of wretched huts that housed homeless people "Hoovervilles."

Finally, Hoover softened his stand against government relief. States received some federal money to give to the needy. And in 1932, Hoover set up an agency to lend money to states, cities, and towns. This money would be used for **public works projects**—government-funded projects to build public resources such as roads and dams. These projects would create jobs. But Hoover's actions proved to be too little and too late.

Reading **History**

B. Solving Problems What were Hoover's solutions to the problems caused by the Depression, and did they succeed?

Hoover Loses to Roosevelt

U.S. troops used tanks and grenades to force the Bonus Army from their camps.

In the summer of 1932, a dramatic event made Hoover even more unpopular. Congress had promised World War I veterans a bonus for wartime service. The bonuses were not due to be paid until the 1940s. But many of the ex-soldiers were jobless and wanted early payment. Some decided to march to Washington, D.C., to ask Congress to pass such a law. Through May and June, the **Bonus Army** of 12,000 to 15,000 veterans poured into Washington and set up camps around the city. Some of them were accompanied by their families.

The Senate, backed by Hoover, voted down a bill that called for a bonus payment. Most veterans gave up and returned home, but a few thousand remained to protest.

At the end of July, General Douglas MacArthur decided on his own to drive the Bonus Army from Washington. MacArthur's troops threw tear gas and prodded the veterans and their children with bayonets. One veteran was shot to death. The American public reacted angrily.

Reading **History**
C. Making Inferences Why do you think Evalyn McLean became so angry?

A VOICE FROM THE PAST

I saw in a news reel the tanks, the cavalry, and the gas-bomb throwers running those wretched Americans out of our capital. I was so raging mad I could have torn the theater down.

Evalyn Walsh McLean, *Father Struck It Rich*

Because of the attack, Americans turned against Hoover more than ever. In the 1932 presidential election, Democratic candidate Franklin Delano Roosevelt carried all but six states. Section 2 explains how Roosevelt tried to end the Depression with new federal programs.

Section 1 Assessment

1. Terms & Names

Identify:
- Herbert Hoover
- speculation
- buying on margin
- Black Tuesday
- Crash of 1929
- Great Depression
- public works projects
- Bonus Army

2. Taking Notes

Use a diagram like the one below to show the sequence of events that led from the stock market crash to massive unemployment.

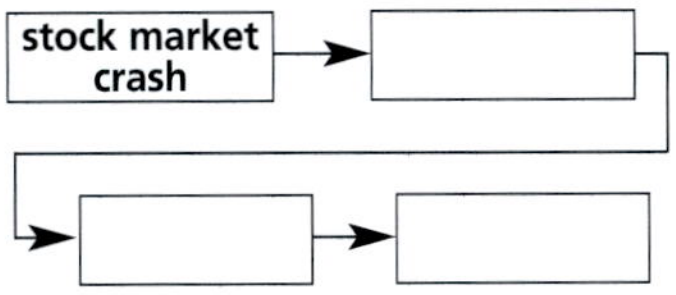

3. Main Ideas

a. What weaknesses existed in the economy during the 1920s?

b. What is buying on margin, and how was it a problem?

c. Why did Hoover become unpopular with many Americans?

4. Critical Thinking

Contrasting How did Hoover's view of the federal government and that of most Americans differ?

THINK ABOUT
- Hoover's attitude about federal relief
- why Americans blamed Hoover for their suffering
- what Americans might have expected from Hoover

ACTIVITY OPTIONS

MATH
SPEECH

Research the changes in stock market value from September through October 1929. Create a **graph** or give a series of **radio news bulletins** about the changes.

2 Roosevelt and the New Deal

TERMS & NAMES

- Franklin Delano Roosevelt
- fireside chat
- New Deal
- Hundred Days
- Social Security Act
- Second New Deal
- deficit spending

MAIN IDEA

After becoming president, Franklin D. Roosevelt took many actions to fight the Great Depression.

WHY IT MATTERS NOW

Roosevelt increased government's role in helping needy Americans and regulating the financial industry.

ONE AMERICAN'S STORY

Dynamite Garland's father had worked for the railroad. But when the Depression struck, Dynamite's father lost his job, and her family had to move into a garage that a landlord let them have rent-free.

A VOICE FROM THE PAST

We had a coal stove, and we had to each take turns, the three of us kids, to warm our legs. It was awfully cold when you opened those garage doors. . . . In the morning, we'd get out and get some snow and put it on the stove and melt it and wash around our faces. Never the neck or anything. Put on our two pairs of socks on each hand and two pairs of socks on our feet, and long underwear and lace it up with Goodwill shoes. Off we'd walk, three, four miles to school.

Dynamite Garland, quoted in *Hard Times*

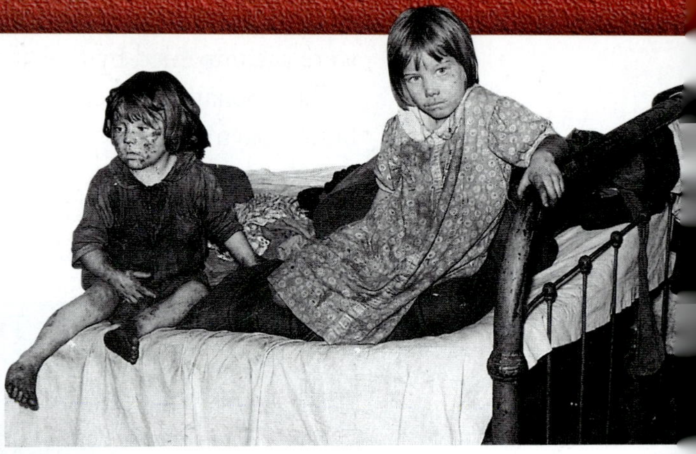

Children such as Dynamite Garland and the girls in this photograph suffered greatly from hunger and poverty during the Depression.

Starting in 1933, Franklin Delano Roosevelt, the new Democratic president, created a number of programs to help the economy and people like Dynamite. This section describes those programs.

Roosevelt Takes Charge

Millions of people lacked food and shelter. Yet the country had to endure a frustrating four-month wait from the November election to Roosevelt's inauguration. The Twentieth Amendment, which moved the inauguration date to January, was not ratified until 1933. President Roosevelt, nicknamed FDR, was finally inaugurated on March 4, 1933.

Roosevelt differed from Hoover in two important ways. First, he gave Americans hope, beginning with his inaugural address: "Let me assert my firm belief that the only thing we have to fear is fear itself." Second, he was willing to try new ideas and change the way government worked. Though he had no fixed plan to end the Depression, he set up a "brain trust" of advisers, including college professors and economists.

Roosevelt took three immediate steps that boosted public confidence. First, he declared a "bank holiday"—a temporary shutdown of all banks.

Second, he promised that only the banks that were in good shape would be allowed to reopen.

Third, the day before the banks reopened, FDR gave the first of many **fireside chats.** In these radio talks, he explained his policies in a warm, friendly style. He said it was safer to "keep your money in a reopened bank than under the mattress." The next day, people deposited more money into the banks than they withdrew.

Reading **History**

A. Drawing Conclusions Did Roosevelt's first fireside chat affect the public the way he wanted? Explain.

The Hundred Days

During the campaign, FDR had pledged a "new deal" for Americans. This snappy phrase, the **New Deal,** came to stand for FDR's programs to fight the Depression.

In the session of Congress lasting from March 9 to mid-June 1933, Roosevelt sent Congress a pile of new bills. Many of them passed with little debate in this famous session of Congress, called the **Hundred Days.**

The laws passed during the Hundred Days had three major goals, known as the "three Rs."

1. **relief** for the hungry and jobless
2. **recovery** for agriculture and industry
3. **reforms** to change the way the economy worked

FDR wanted not only to ease suffering but also to try to prevent such a severe depression from happening again. The major programs passed during the Hundred Days included relief, recovery, and reform plans that related to jobs, banking, wages, and agriculture. The chart on page 737 lists several major programs and explains what those programs accomplished.

AMERICA'S HISTORY MAKERS

FRANKLIN DELANO ROOSEVELT

1882–1945

A distant cousin of Theodore Roosevelt, Franklin D. Roosevelt became a New York state senator when he was 29. Later, he served as assistant secretary of the Navy.

At the age of 39, FDR caught polio. For the rest of his life, he walked with braces or rode in a wheelchair. Despite this, he continued in politics and was elected governor of New York in 1928.

The public rarely saw photos revealing FDR's disability. Even so, many Americans sensed that he was a man who understood trouble. This quality helped him as a leader during the Depression.

How would an understanding of trouble help Roosevelt to lead?

Responses to the New Deal

Some conservatives thought the New Deal went too far. They opposed the growth of the federal government and questioned how it would pay for all the new programs. They also feared that the New Deal was moving the country toward socialism.

Vocabulary

socialism: an economic system in which businesses are owned by the government, not individuals

Yet, other critics charged that the New Deal didn't go far enough. Louisiana senator Huey Long declared, "Unless we provide for redistribution of wealth in this country, the country is doomed." But Long's motives were far from noble. For years, he had ruled his state like a dictator. Attacking FDR was a way to increase his own power.

Father Charles Coughlin, a priest with a popular radio program, also argued for changing the economy to help the poor. He eventually began to blame Jews for the nation's problems. In the 1940s, the Catholic Church stopped his broadcasts. Another critic, Francis Townshend, proposed giving $200 a month to every American over age 60. He said a sales tax would pay for the pension, but economists disputed his figures.

CONNECTIONS TO SCIENCE

The Tennessee Valley Authority

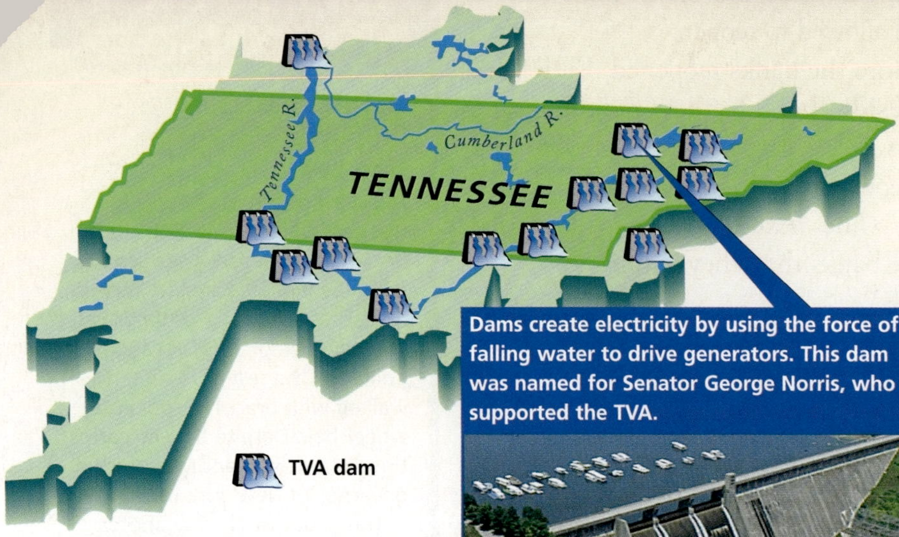

Dams create electricity by using the force of falling water to drive generators. This dam was named for Senator George Norris, who supported the TVA.

FACTS ABOUT THE TVA

- The TVA supplies power to an area of about 80,000 square miles, shown in dark and light green to the left. The region uses more than 100 billion kilowatt-hours of electricity. This is 65 times as much as in 1933.
- There are more than 40 TVA dams.
- Users of TVA power pay about a third less for their electricity than other Americans.
- The TVA dams also help control flooding on the Tennessee River.

In the 1934 congressional elections, voters had a chance to react to these criticisms. The party in power usually loses seats in a nonpresidential election. But, indicating their support for Roosevelt, voters in this election sent even more Democrats to Congress.

The Second New Deal

Although he rejected Townshend's plan, FDR did want to help the elderly. Bank failures and the stock market crash had stolen the savings of many old people. Some had lost their homes or had to beg for food.

Background
Townshend's plan had been very popular with the public. FDR knew his own popularity would increase if he proposed a pension.

In August 1935, Congress passed one of the most important bills of the century. Under the **Social Security Act,** workers and employers made payments into a special fund, from which they would draw a pension after they retired. The act also gave help to laid-off workers, disabled workers, and needy families with dependent children.

> ***A VOICE FROM THE PAST***
>
> We have tried to frame a law which will give some measure of protection to the average citizen and to his family against the loss of a job and against poverty-ridden old age.
>
> **Franklin D. Roosevelt,** quoted in *Promises to Keep*

Social Security was part of a set of programs passed in 1935. These became known as the **Second New Deal.** Other programs of the Second New Deal are listed in the chart on the next page.

In 1936, the Democrats nominated Roosevelt for a second term. Business leaders opposed his reelection because they feared higher taxes. They also thought he was increasing government power at their expense. But a widespread alliance of working-class Americans supported FDR.

Major Programs of the New Deal

	PROGRAMS	ACCOMPLISHMENTS
Hundred Days, 1933	**FERA** (Federal Emergency Relief Administration)	Provided federal money for relief projects to the roughly 13 million unemployed
Hundred Days, 1933	**PWA** (Public Works Administration)	Created jobs by having people build highways, bridges, and other public works
Hundred Days, 1933	**AAA** (Agricultural Adjustment Administration)	Regulated farm production and promoted soil conservation
Hundred Days, 1933	**TVA** (Tennessee Valley Authority)	Planned development of the Tennessee Valley region
Hundred Days, 1933	**CCC** (Civilian Conservation Corps)	Hired young men to plant trees, build dams, and work on other conservation projects
Hundred Days, 1933	**FDIC** (Federal Deposit Insurance Corporation)	Protected the money of depositors in insured banks
Hundred Days, 1933	**NRA** (National Recovery Administration)	Regulated industry and raised wages and prices
Second New Deal, 1935	**WPA** (Works Progress Administration)	Established large-scale national works programs to create jobs
Second New Deal, 1935	**REA** (Rural Electrification Administration)	Brought electricity to rural areas
Second New Deal, 1935	**NYA** (National Youth Administration)	Set up job programs for young people and helped them continue their education
Second New Deal, 1935	**Wagner Act**	Protected labor's right to form unions and set up a board to hear labor disputes
Second New Deal, 1935	**Social Security Act**	Provided workers with unemployment insurance and retirement benefits

SKILLBUILDER Interpreting Charts

1. *How did the PWA and the CCC help both those who were hired and the nation as a whole?*
2. *How did the Second New Deal help both young and old workers?*

Reading **History**

B. Analyzing Causes Why did many African Americans switch to the Democratic Party?

They included African Americans, who until then had remained loyal to the Republican Party—the party of Lincoln, who had issued the Emancipation Proclamation during the Civil War. However, FDR's programs to help the poor convinced many African Americans to vote Democrat. On Election Day, FDR won every state except Maine and Vermont.

Roosevelt Fights the Supreme Court

From the high point of his 1936 victory, Roosevelt's presidency took a downward turn. Most of the nine justices of the Supreme Court didn't support FDR's programs. Using the power of judicial review (see Chapter 10), in 1935 the Court struck down laws that it believed gave the federal government too much power. These actions threatened to destroy the New Deal.

In 1937, FDR asked Congress to pass a bill allowing him to add up to six justices to the Supreme Court. He planned to appoint justices who shared his ideas about government. This would give him the majority he

"Dr. Roosevelt" reassures the Constitution that the New Deal won't harm it. Instead, FDR blames "Nurse Supreme Court" for any problems.

needed to save his programs from being overturned. Both Republicans and Democrats harshly criticized FDR's "court-packing" bill. They said it interfered with the system of checks and balances that were set up by the U.S. Constitution. Congress agreed and voted it down.

In the end, Roosevelt did achieve his goal of a more sympathetic Court. Within the next two and a half years, retirements and deaths allowed Roosevelt to name five liberal justices to the bench. But the president may have lost more than he won. His clumsy attempt to pack the Court with allies damaged his image and gave ammunition to his critics.

Reading History

C. Recognizing Effects What was the outcome of Roosevelt's attempt to pack the Court?

The New Deal Slows Down

Opposition to Roosevelt grew after the "court-packing" attempt. Then in late 1937, the economy worsened again. The amount of goods produced by industry fell, and unemployment rose. Many blamed FDR for the downturn.

Critics also attacked Roosevelt's use of **deficit spending,** or using borrowed money to fund government programs. Roosevelt himself had doubts about it. Even though some economists said that huge amounts of deficit spending were needed to boost the economy, FDR hesitated to take that course. He proposed few new programs in his second term. Meanwhile, as Section 3 explains, Americans continued to suffer from harsh economic conditions.

Section 2 Assessment

1. Terms & Names

Identify:

- Franklin Delano Roosevelt
- fireside chat
- New Deal
- Hundred Days
- Social Security Act
- Second New Deal
- deficit spending

2. Taking Notes

Use a chart like the one below to list FDR's major programs and whether you think each program's goal was relief, recovery, or reform (or a combination of these).

Program	Goal

Which programs created jobs?

3. Main Ideas

a. How did Roosevelt give Americans hope?

b. What happened during the period known as the Hundred Days?

c. What were the consequences of FDR's attempt to increase the size of the Supreme Court?

4. Critical Thinking

Analyzing Points of View
What were some of the different reasons that people criticized FDR?

THINK ABOUT

- the conservatives
- Huey Long, Father Coughlin, and Francis Townshend
- those who opposed the "court-packing" bill

ACTIVITY OPTIONS

ART

TECHNOLOGY

Choose one aspect of the New Deal that you have an opinion about. Create a **political cartoon** or design a **Web page** expressing your opinion.

3 Life During the Depression

TERMS & NAMES
- Dust Bowl
- Eleanor Roosevelt
- Congress of Industrial Organizations (CIO)
- sit-down strike

MAIN IDEA
During the Depression, most Americans knew great hardship.

WHY IT MATTERS NOW
Because of this, a generation was scarred by suffering in ways that later generations were not.

ONE AMERICAN'S STORY

Born to former slaves, Mary McLeod Bethune refused to accept the racial stereotype that all she could be was a servant. She gained an education and went on to found Bethune-Cookman College. In 1936, FDR named her director of the Division of Negro Affairs at the National Youth Administration. Bethune ran training programs for 600,000 African-American young people. She also supervised her college from afar. Because of her health, her doctor told her to stop working so hard.

A VOICE FROM THE PAST

I promise to reform, but in an hour the promise is forgotten. For I am my mother's daughter, and the drums of Africa still beat in my heart. They will not let me rest while there is a single Negro boy or girl without a chance to prove his worth.

Mary McLeod Bethune, "Faith That Moved a Dump Heap"

Section 3 discusses the difficulty of living during the Depression—and the efforts of people like Bethune to make the hard times easier.

A friend of Mrs. Roosevelt, Mary McLeod Bethune became the first African-American woman to head a federal agency.

The Dust Bowl Destroys Lives

In the early 1930s, a drought hit the Great Plains and lasted for several years. Even before then, the overgrazing of livestock and the overplowing of fields had damaged the land by destroying the natural grasses whose roots anchored the soil. A journalist wrote, "You could hear the fields crack and dry, and the only movement in the down-driving heat was the dead withering of the dry blighted leaves on the twigs."

Winds picked up dirt from the dry, exposed fields. During dust storms, noon turned into night as walls of dust filled the air and hid the sun. Dust damaged farms across a 150,000-square-mile region called the **Dust Bowl,** which covered parts of Kansas, Oklahoma, Texas, Colorado, and New Mexico. Dust storms ripped through the plains for years until rain and improved farming methods finally brought relief.

With their crops buried under layers of dirt, ruined farmers loaded their belongings onto trucks and set off with their families to find work.

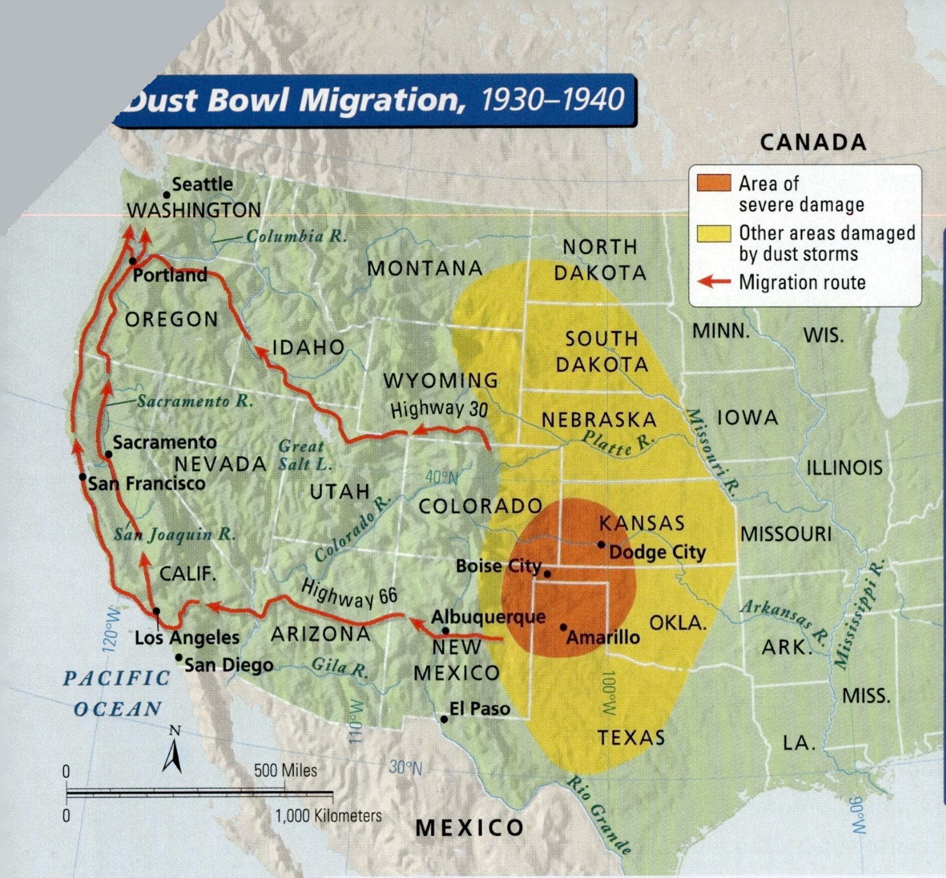

This photograph demonstrates why the dust storms were often called "black blizzards."

GEOGRAPHY SKILLBUILDER Interpreting Maps

1. **Place** *Which states suffered the most damage from the dust storms?*
2. **Movement** *What were the main highways that people took to leave the Dust Bowl, and where did they lead?*

Many drove west on Route 66, the main highway to California. They had heard that California's farms needed workers.

But as the newcomers poured in, California farm towns quickly became overcrowded. Families lived in tiny shacks. By 1940, about 2.5 million people fleeing the Dust Bowl had made their way to California and other Pacific coast states. Because many had come from Oklahoma, they were sometimes called "Okies."

Reading **History**

A. Reading a Map Look at the map above. Notice the three states that most Dust Bowl migrants went to.

Living Through Hard Times

Not just in the Dust Bowl, but all over the country, families suffered. Even after the recovery measures of the New Deal, unemployment remained high. In 1936, for example, 9 million people had no jobs.

Without work, families couldn't afford to buy food. Bread lines offering food to the hungry appeared across the country. In January 1931, New York's 82 bread lines served an average of 85,000 small meals a day: bread and soup or bread and stew. Men, women, and children waited in these lines for their daily food. Some fainted from hunger while they waited.

Many people also lost their homes. Thousands of homeless people sought shelter under bridges and overpasses. One woman remembered "people living in old, rusted out car bodies. . . . There were people living in shacks made of orange crates. One family with a whole lot of kids were living in a piano box."

Children had to grow up fast during the Depression. To add to their family's income, boys worked after school or even dropped out of school. Often girls had to stay home to look after younger children. Sometimes teenagers who failed to find jobs ran away from home to avoid burdening their families. By late 1932, perhaps a quarter million teens roamed the country. They sneaked onto freight trains, begged for food, and lived in squatter camps along the railroad tracks.

Reading **History**

B. Summarizing What were the various hardships suffered by families during the Depression?

Family life suffered as many unemployed men felt a loss of status. They sometimes became irritable and quarreled with their families. Working women came under pressure to give up their jobs to jobless men. In fact, some New Deal projects would hire a woman only if her husband had a job. Even so, poverty forced many women to work as servants or at other low-paying jobs that men didn't want.

Artists Portray the Struggle

Many books of the period described the hard times. *Let Us Now Praise Famous Men* (1941) by James Agee and Walker Evans reported on the harsh lives of tenant farmers. John Steinbeck's novel *The Grapes of Wrath* (1939) told of Okies who had been evicted from their farms.

Vocabulary

evicted: forced to leave property

A VOICE FROM THE PAST

Carloads, caravans, homeless and hungry; twenty thousand and fifty thousand and a hundred thousand and two hundred thousand. They streamed over the mountains, hungry and restless—restless as ants, scurrying to find work to do—to lift, to push, to pull, to pick, to cut—anything, any burden to bear, for food. The kids are hungry. We got no place to live. Like ants scurrying for work, for food, and most of all for land.

John Steinbeck, *The Grapes of Wrath*

The African-American writer Richard Wright was one of many writers who were hired by the Works Progress Administration. Freed from his economic worries, he also wrote creatively in his spare time and produced the novel *Native Son* (1940). It depicts one African American's anger about society's racism.

daily life

HAVING FUN DURING HARD TIMES

To forget life's troubles, people went to the movies. At first, the Depression caused attendance to decline. But audiences soon grew as hard times made people eager for entertainment.

Viewers flocked to escapist movies such as *The Wizard of Oz.* People also saw realistic movies about the times, such as *The Grapes of Wrath,* based on John Steinbeck's novel.

Reading comic books about superheroes also was a popular pastime. Comic books were first published in 1933. *Superman* was introduced in 1938.

▼ *The Grapes of Wrath*

▼ *Gone with the Wind*

◀ *The Wizard of Oz*

AMERICA'S HISTORY MAKERS

ELEANOR ROOSEVELT

1884–1962

Eleanor Roosevelt had an unhappy childhood. Her parents died when she was young, and the grandmother who raised her criticized Eleanor frequently.

In 1905, she married her distant cousin Franklin. When FDR entered politics, she was shy and disliked attending public events.

During World War I, Eleanor did volunteer work and discovered the joy of helping others. This gave her confidence. By the time FDR became governor of New York, Eleanor was actively working for social reform. She continued to do so as first lady.

How did Eleanor Roosevelt change from an insecure person to a leader for social reform?

Photographers also captured Depression-era suffering. Dorothea Lange was one of several photographers whom the government hired to document the times. Her pictures show the hard lives of poor people during the Depression. (See pages 745 and 751.)

Women in the New Deal

The first lady, **Eleanor Roosevelt,** worked to help poor Americans. Because her husband had a disability, Mrs. Roosevelt acted as his "eyes and ears." She toured the country, visiting coal mines, work camps, and hospitals to find out how programs were working. Then she told the president what she learned and made suggestions.

In March 1933, Eleanor Roosevelt began to hold regular press conferences for women reporters. At these, the first lady introduced the women who ran New Deal programs. During Roosevelt's presidency, more women held positions with the government than ever before.

Reading **History**

C. Making Inferences Why do you think Eleanor Roosevelt publicized the women who worked in government?

In 1933, the president named Frances Perkins secretary of labor, which made her the first female cabinet officer. Years earlier, Perkins had assisted Jane Addams at Chicago's Hull House. As secretary of labor, she supported laws granting a minimum wage, a limit on child employment, and unemployment compensation.

Minorities and the Depression

Mary McLeod Bethune was one of several African Americans who played a role in the government. They were called FDR's "Black Cabinet." This group included William Hastie and Robert C. Weaver. Hastie was a brilliant young lawyer who worked in the Department of the Interior. Weaver, an economist who had graduated from Harvard, became the president's adviser on racial issues.

Though he included more African Americans in government, FDR failed to back civil rights laws. For example, he did not support an antilynching bill. FDR opposed lynching but feared upsetting Southern white congressmen. Roosevelt said, "If I come out for the antilynching bill now, [the Southerners] will block every bill I ask Congress to pass to keep America from collapsing." In spite of this, African Americans remained loyal to the president because of his efforts to help the poor.

The Depression also greatly affected Mexican Americans. Many lived in rural areas, especially in the Southwest. Increasingly, migrants from other areas competed with them for jobs. Mexican Americans living in cities also had difficulty finding scarce jobs. While many Mexican Americans did benefit from New Deal programs, in general they received less aid than other groups.

Reading **History**

D. Contrasting How did African Americans and Mexican Americans benefit differently from the New Deal?

During the 1930s, immigration from Mexico declined. In addition, many immigrants returned to Mexico. Some left on their own; the federal government deported others. Some of those who were forced to leave were U.S. citizens whose rights were ignored. Because they feared deportation, many Mexican Americans stopped applying for aid.

Life improved somewhat for Native Americans. In 1934, Congress passed the Indian Reorganization Act, which restored some reservation lands to Indian ownership. It also created the Indian Arts and Crafts Board to promote native arts.

In early 1939, a women's group refused to rent a hall for opera star Marian Anderson's performance because of her race. So Mrs. Roosevelt asked her to sing at the Lincoln Memorial.

Unions Gain Strength

Some minorities joined a new labor organization. The country's largest labor organization was the American Federation of Labor (AFL). It was open only to skilled workers, such as plumbers and electricians. Labor leader John L. Lewis wanted industrywide unions that included both skilled and unskilled workers. He and other leaders founded the **Congress of Industrial Organizations (CIO),** which broke from the AFL in 1938. It was more open to women and minorities than the AFL.

Background
John L. Lewis had been president of the United Mine Workers of America since 1920.

In the 1930s, the labor movement used an effective bargaining tactic called the **sit-down strike.** Instead of walking off their jobs, striking workers remained idle inside the plant. As a result, factory owners could not hire strikebreakers to do the work.

The Wagner Act, passed in 1935, gave unions the ability to negotiate better working conditions. Union membership jumped from 2.7 million in 1933 to 7.0 million in 1937. The growing strength of labor unions was just one legacy of the New Deal. Section 4 discusses other legacies of the Great Depression and the New Deal.

Section 3 Assessment

1. Terms & Names

Identify:
- Dust Bowl
- Eleanor Roosevelt
- Congress of Industrial Organizations (CIO)
- sit-down strike

2. Taking Notes

Use a cluster diagram like the one below to record details about life during the Depression.

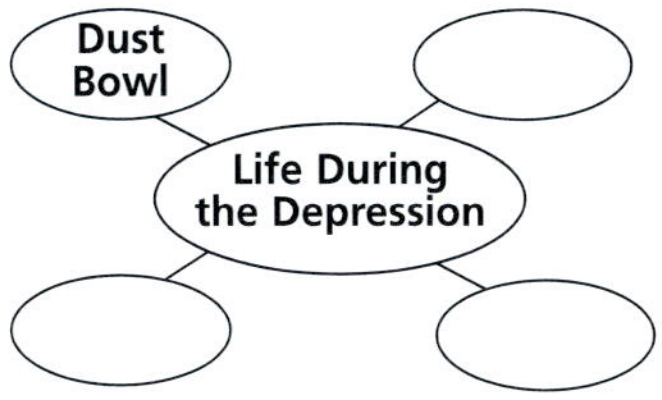

3. Main Ideas

a. How did storms in the Dust Bowl contribute to economic problems?

b. What effect did the Depression have on families?

c. How did Eleanor Roosevelt help her husband, the president?

4. Critical Thinking

Recognizing Effects What were some positive and negative results of the government's policies toward minorities during the Depression?

THINK ABOUT
- African Americans
- Mexican Americans
- Native Americans

ACTIVITY OPTIONS

SPEECH
ART

You have been asked to teach young children about life during the Depression. Write and perform a **monologue** or create a **comic strip** about it.

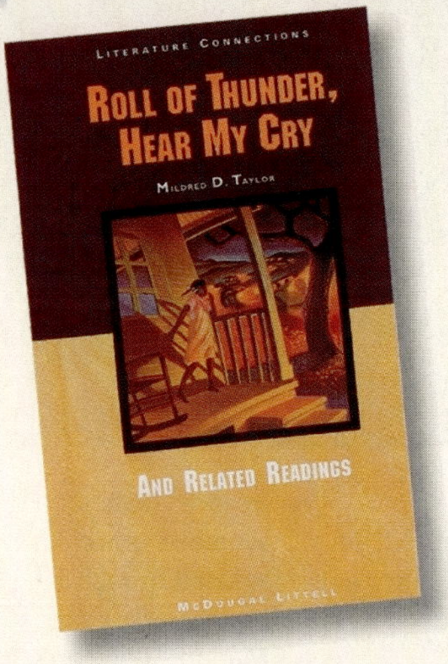

from

Roll of Thunder, Hear My Cry

by Mildred D. Taylor

Even before the Depression, many African Americans struggled with poverty. When the Depression began, hard times made it more difficult for them to get ahead—or to keep what they had. Because the Logan family owns land in Mississippi, they are better off than their sharecropper neighbors. But they also have more to lose.

"Little Man, would you come on? You keep it up and you're gonna make us late."

My youngest brother paid no attention to me. Grasping more firmly his newspaper-wrapped notebook and his tin-can lunch of cornbread and oil sausages, he continued to concentrate on the dusty road. He lagged several feet behind my other brothers, Stacey and Christopher-John, and me, attempting to keep the rusty Mississippi dust from swelling with each step and drifting back upon his shiny black shoes and the cuffs of his corduroy pants by lifting each foot high before setting it gently down again. Always meticulously neat, six-year-old Little Man never allowed dirt or tears or stains to mar anything he owned. Today was no exception.

"You keep it up and make us late for school, Mama's gonna wear you out," I threatened, pulling with exasperation at the high collar of the Sunday dress Mama had made me wear for the first day of school—as if that event were something special. It seemed to me that showing up at school at all on a bright August-like October morning made for running the cool forest trails and wading barefoot in the forest pond was concession enough; Sunday clothing was asking too much. Christopher-John and Stacey were not too pleased about the clothing or school either. Only Little Man, just beginning his school career, found the prospects of both intriguing.

"Y'all go ahead and get dirty if y'all wanna," he replied without even looking up from his studied steps. "Me, I'm gonna stay clean."

"I betcha Mama's gonna 'clean' you, you keep it up," I grumbled.

"Ah, Cassie, leave him be," Stacey admonished, frowning and kicking testily at the road.

"I ain't said nothing but—"

Stacey cut me a wicked look and I grew silent. His disposition had been irritatingly sour lately. If I hadn't known the cause of it, I could have forgotten very easily that he was, at twelve, bigger than I, and that I had promised Mama to arrive at school looking clean and ladylike. "Shoot," I mumbled finally, unable to restrain myself from further comment, "it ain't my fault you gotta be in Mama's class this year."

Stacey's frown deepened and he jammed his fists into his pockets, but said nothing.

Christopher-John, walking between Stacey and me, glanced uneasily at both of us but did not interfere. A short, round boy of seven, he took little interest in troublesome things, preferring to remain on good terms with everyone. Yet he was always sensitive to others and now, shifting the handle of his lunch can from his right hand to his right wrist and his smudged notebook from his left hand to his left armpit, he stuffed his free hands into his pockets and attempted to make his face as moody as Stacey's and as cranky as mine. But after a few moments he seemed to forget that he was supposed to be grouchy and began whistling cheerfully. There was little that could make Christopher-John unhappy for very long, not even the thought of school.

I tugged again at my collar and dragged my feet in the dust, allowing it to sift back onto my socks and shoes like gritty red snow. I hated the dress. And the shoes. There was little I could do in a dress, and as for shoes, they imprisoned freedom-loving feet accustomed to the feel of the warm earth.

"Cassie, stop that," Stacey snapped as the dust billowed in swirling clouds around my feet. I looked up sharply, ready to protest. Christopher-John's whistling increased to a raucous, nervous shrill, and grudgingly I let the matter drop and trudged along in moody silence, my brothers growing as pensively quiet as I.

Before us the narrow, sun-splotched road wound like a lazy red serpent dividing the high forest bank of quiet, old trees on the left from the cotton field, forested by giant green-and-purple stalks, on the right. A barbed-wire fence ran the length of the deep field, stretching eastward for over a quarter of a mile until it met the sloping green pasture that signaled the end of our family's four hundred acres. An ancient oak tree on the slope, visible even now, was the official dividing mark between Logan land and the beginning of a dense forest. Beyond the protective fencing of the forest, vast farming fields, worked by a multitude of sharecropping families, covered two thirds of a ten-square-mile plantation. That was Harlan Granger land.

Once our land had been Granger land too, but the Grangers had sold it during Reconstruction to a Yankee for tax money. In 1887, when the land was up for sell again, Grandpa had bought two hundred acres of it, and in 1918, after the first two hundred acres had been paid off, he had bought another two hundred. It was good rich land, much of it still virgin forest, and there was no debt on half of it. But there was a **mortgage**[1] on the two hundred acres bought in 1918 and there were taxes on the full four hundred, and for the past three years there had not been enough money from the cotton to pay both and live on too.

That was why Papa had gone to work on the railroad.

In 1930 the price of cotton dropped. And so, in the spring of 1931, Papa set out looking for work, going as far north as Memphis and as far south as the Delta country. He had gone west too, into Louisiana. It was there he found work laying track for the railroad. He worked the remainder of the year away from us, not returning until the deep winter when the ground was cold and barren. The following spring after the planting was finished, he did the same. Now it was 1933, and Papa was again in Louisiana laying track.

I asked him once why he had to go away, why the land was so important. He took my hand and said in his quiet way: "Look out there, Cassie girl. All that belongs to you. You ain't never had to live on nobody's place but your own and long as I live and the family survives, you'll never have to. That's important. You may not understand that now, but one day you will. Then you'll see."

I looked at Papa strangely when he said that, for I knew that all the land did not belong to me. Some of it belonged to Stacey, Christopher-John, and Little Man, not to mention the part that belonged to Big Ma, Mama, and Uncle Hammer, Paper's older brother who lived in Chicago. But Papa never divided the land in his mind; it was simply Logan land. For it he would work the long, hot summer pounding steel; Mama would teach and run the farm; Big Ma, in her sixties, would work like a woman of twenty in the fields and keep the house; and the boys and I would wear threadbare clothing washed to dishwater color; but always, the taxes and the mortgage would be paid.

1. mortgage: the transfer of a deed to property, usually in exchange for a loan.

This Dorothea Lange photograph shows sharecroppers like the Logans' neighbors.

CONNECT TO HISTORY

1. **Analyzing Causes** Which of the economic problems shown in the graphs on page 730 has been making life hard for the Logan family?

See Skillbuilder Handbook, page R10.

CONNECT TO TODAY

2. **Researching** What are current interest rates on mortgages?

Visit www.mcdougallittell.com to learn more about the Great Depression.

4 The Effects of the New Deal

TERMS & NAMES
Securities and Exchange Commission
liberal
conservative

MAIN IDEA
The Depression and the New Deal had many long-term effects on U.S. government and society.

WHY IT MATTERS NOW
Politicians still debate how large a role government should play in American life.

ONE AMERICAN'S STORY

Until 1935, Ward James worked as a writer for a New York publisher. Then he lost his job. His wife and son went to Ohio to live with her parents. James moved in with a friend and applied for government relief. Even after he got a steady writing job with the WPA, Ward James continued to worry about what would happen next.

A VOICE FROM THE PAST

Everyone was emotionally affected. We developed a fear of the future which was very difficult to overcome. Even though I eventually went into some fairly good jobs, there was still this constant dread: everything would be cut out from under you and you wouldn't know what to do. It would be even harder, because you were older.

Ward James, quoted in *Hard Times*

The fear and despair described by Ward James was felt by many unemployed people during the Depression.

Years after the Depression, James still had "a little fear . . . that it might happen again." As this section explains, both the Depression and the New Deal had lasting effects on Americans and their government.

Lasting Effects of the Depression

Americans like Ward James who lived through the Depression often saw themselves as the survivors of a terrible battle. For the rest of their lives, many feared losing their money and property again. One elderly government worker bought land whenever she could afford it so that if the Depression returned, she would "have something to live off."

Virginia Durr, who had worked for the federal government under FDR, said that the Depression affected people in two ways. "The great majority reacted by thinking money is the most important thing in the world. . . . And there was a small number of people who felt the whole system was lousy. You have to change it."

The New Deal did not end the Depression. Even with all the new programs, the government still wasn't spending enough money to jump-start a stalled economy. Then, in the 1940s, World War II changed the situation. To fight in that war, the government had to purchase guns, tanks, ships, airplanes, and other military equipment. The defense industry hired many people, who then had more money to spend. The U.S. economy started growing again.

Although the New Deal didn't end the Depression, it forever changed the U.S. government. As Supreme Court justice John Clarke told FDR, "You have put a new face upon the social and political life of our country."

> **"We developed a fear of the future."**
> Ward James

A Larger Role for Government

*Reading***History**
A. Recognizing Effects How did FDR increase the president's power?

President Roosevelt increased the president's power. Under FDR, the White House became the center of government. More than other early-20th-century presidents, Roosevelt proposed bills and programs for Congress to consider instead of waiting for Congress to act.

Other nations also saw the rise of strong leaders. But at the same time, those nations saw a loss of freedom. For example, during the Depression, Germany elected Adolf Hitler, who became a dictator. The United States did have some leaders who abused power—such as Huey Long—but they never became president. FDR's leadership and his concern for the poor helped Americans keep their faith in democracy.

As well as increasing the president's power, Roosevelt also expanded the federal government. Because of the New Deal, the federal government became directly responsible for people's well-being in a way it had not

CONNECTIONS TO MATH

Effects of the New Deal, *1929–1941*

Although Franklin Roosevelt's New Deal programs did not end the Depression, they did make some economic conditions better. Use these graphs to determine how the New Deal—begun in 1933—affected the unemployment rate, the number of bank closings, and the number of business failures.

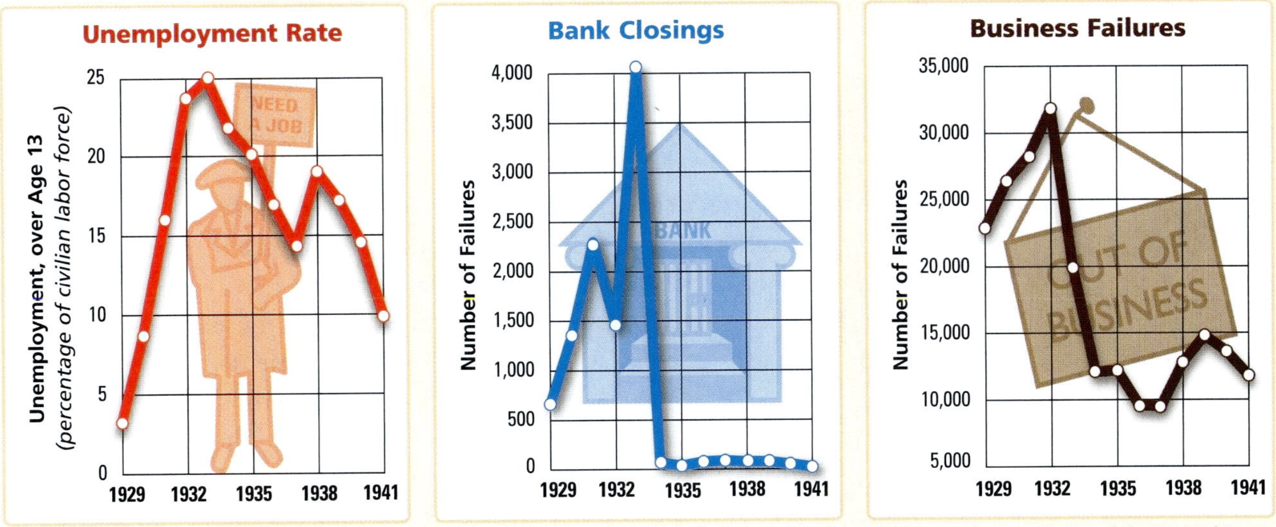

Sources: *Historical Statistics of the United States*

SKILLBUILDER Interpreting Graphs

1. *Judging from these graphs, did the Depression's negative effects on business improve after FDR took office in 1933? Explain.*
2. *In which area was there the biggest change for the better?*

HISTORY through ART

The Works Progress Administration (WPA) created many jobs. One of the WPA's most enduring legacies is the art that it commissioned. Much WPA art was used to decorate public places, such as post offices and government buildings. This mural by William Gropper shows the building of a dam.

What attitude did Gropper want to convey about laborers?

been before. It now made relief payments, served school lunches, and ran a program providing pensions. People came to see the federal government, not their state or local governments, as the protector of their welfare.

The federal government went into debt to provide this aid. FDR used deficit spending both to fund the New Deal and to pay for the war. Since then, deficit spending has often been part of the federal budget.

New Deal Programs Today

Several of FDR's New Deal programs continue to help Americans today. Some of the more important programs that still exist offer the following benefits and protections.

1. **A National Pension System.** The Social Security system pays out old-age pensions (and has been expanded to include aid to other groups). It is funded through taxes on employers and employees.
2. **Oversight of Labor Practices.** Created by the Wagner Act, the National Labor Relations Board (NLRB) oversees labor unions. It also investigates disputes between management and labor.
3. **Agricultural Price Supports.** This program pays farmers to raise crops for domestic use rather than export. To receive payments, farmers must agree to limit the space they devote to certain crops.
4. **Protection for Savings.** After the bank holiday of 1933, the Federal Deposit Insurance Corporation (FDIC) was created. The FDIC insures bank deposits up to $100,000. It replaces the deposits of individuals if banks close.
5. **Regulation of the Stock Market.** A federal agency called the **Securities and Exchange Commission** watches the stock market. It makes sure companies follow fair practices for trading stocks.

Reading **History**

B. Analyzing Causes Why do you think FDR wanted to create an agency to oversee the stock market?

An Ongoing Political Debate

The issues that came out of the New Deal continue to shape American politics. For example, Democrats and Republicans still argue about whether federal or local government should be responsible for various programs. In addition, Democrats are more likely to be liberal and Republicans are more likely to be conservative in their political beliefs. A **liberal** in politics favors government action to bring about social and economic reform. A **conservative** favors fewer government controls and more individual freedom in economic matters.

Despite these lingering disagreements, some New Deal programs are still so popular that everyone supports them. For example, neither party wants to end Social Security, even though the system is in trouble. The amounts that people pay in through payroll taxes today do not completely pay for pensions. The system may run out of money sometime in the future.

In early 1999, President Bill Clinton announced a plan to save Social Security by using extra money from the federal budget. The Republicans accepted his plan. They knew that saving Social Security is a priority for Americans and that voters might grow angry if they made a political fight of the issue.

FDR probably would have approved. "The great public," Roosevelt said, "is interested more in government than in politics." Roosevelt felt that party labels mattered little as long as politicians "did the big job that their times demanded to be done."

In the 1940s, President Roosevelt would face another big job. He had to lead the country in fighting a world war. Chapter 27 discusses World War II and America's role in it.

Now and then

SOCIAL SECURITY

Today, many young people worry that the Social Security trust fund won't have enough money for their pensions. For one thing, people live longer now than they did in the 1930s. Also, the percentage of people receiving benefits keeps increasing compared to the number paying into the system. Last, Social Security benefits have been expanded since the 1930s.

Reading **History**

C. Analyzing Points of View How would you summarize FDR's view of government's role?

Section 4 Assessment

1. Terms & Names

Identify:

- Securities and Exchange Commission
- liberal
- conservative

2. Taking Notes

Use a bulleted list like the one shown below to list the legacy of the Depression and New Deal.

Legacy of the Depression and New Deal

- ____________
- ____________
- ____________
- ____________

What part of the legacy affects politics today?

3. Main Ideas

a. What psychological impact did the Depression have on many Americans?

b. What finally pulled the United States out of its economic depression?

c. How do today's political differences date back to the Depression?

4. Critical Thinking

Drawing Conclusions Of the following New Deal programs, which one do you think affects your life the most? Explain.

THINK ABOUT

- Social Security
- Federal Deposit Insurance Corporation
- Securities and Exchange Commission

ACTIVITY OPTIONS

LANGUAGE ARTS

TECHNOLOGY

Ask your grandparents or other older relatives what they think the legacy of the Depression and New Deal is. Prepare a **written interview** or an **audio recording.**

VISUAL SUMMARY

The Great Depression and New Deal

CAUSES

- problems in agriculture and some industries
- unequal income distribution
- too much inventory
- too much debt
- stock market speculation

BROOKLYN DAILY EAGLE
And Complete Long Island News
LATE NEWS
WALL ST. IN PANIC AS STOCKS CRASH

DEPRESSION

- stock market crash
- bank and business failures
- high unemployment
- hunger and homelessness

NEW DEAL

- relief for the hungry and jobless
- recovery for agriculture and industry
- reforms to change the way the economy worked

LONG-TERM EFFECTS

- fear of future
- more influence by federal government
- more presidential power
- long-term government programs
- debate between conservatives and liberals

Chapter 26 ASSESSMENT

TERMS & NAMES

Briefly explain the importance of each of the following.

1. Herbert Hoover
2. Crash of 1929
3. Great Depression
4. Franklin Delano Roosevelt
5. New Deal
6. deficit spending
7. Dust Bowl
8. Eleanor Roosevelt
9. liberal
10. conservative

REVIEW QUESTIONS

Hoover and the Crash (pages 729–733)

1. Why did stock prices fall so quickly during the stock market crash?
2. Who did President Hoover think should help the needy?
3. How did MacArthur's attack on the Bonus Army affect the 1932 election?

Roosevelt and the New Deal (pages 734–738)

4. What was the "brain trust"?
5. What were fireside chats, and how did they affect the country?
6. Why didn't Roosevelt propose many new programs during his second term?

Life During the Depression (pages 739–745)

7. How did writers and filmmakers respond to the hard times?
8. What new bargaining tactic did labor unions use, and how did it work?

The Effects of the New Deal (pages 746–749)

9. How did the New Deal change the role of the federal government in American life?
10. What New Deal program remains popular even though it is in financial trouble?

CRITICAL THINKING

1. USING YOUR NOTES

Using your completed chart, answer the questions below.

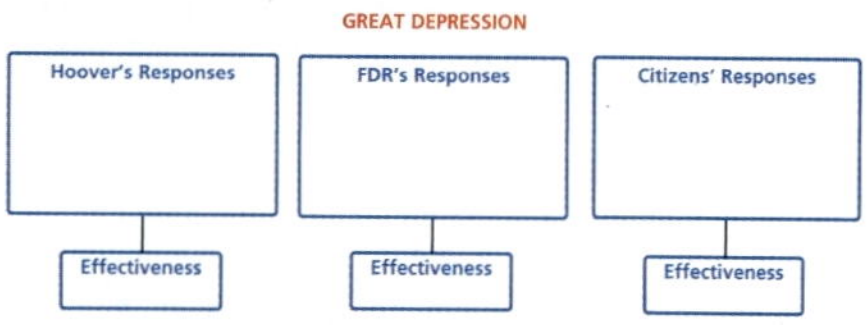

a. What do you think was Hoover's most successful response to the Depression?

b. Judging from his responses to the Depression, do you think that FDR was one of our greatest presidents? Explain.

2. ANALYZING LEADERSHIP

During the Depression, many countries turned to strong leaders. How did the Depression-era leadership of Germany differ from that of the United States?

3. ANALYZING CAUSES

Review the economic problems that led to the Depression. What similar problems exist in the economy today?

4. THEME: ECONOMICS IN HISTORY

What reforms did FDR make to ensure that the United States would never again experience such a severe depression?

5. APPLYING CITIZENSHIP SKILLS

How were each of the following citizenship skills important during the Depression: voting, staying informed about issues, and community service?

Interact *with* History

Now that you have read the chapter, would you give the same answer to the question, "Who do you think should help the poor?" Explain why or why not.

HISTORY SKILLS

1. INTERPRETING CHARTS

Study the chart. Then answer the questions.

Roosevelt's Presidential Elections

ELECTION YEAR	TOTAL VOTES	VOTES FOR FDR
1932	39,749,382	22,825,016
1936	45,642,303	27,747,636
1940	49,840,443	27,263,448
1944	47,974,819	25,611,936

Source: *New York Times Almanac*

Basic Chart Elements

a. What presidential elections does this chart cover?

b. Who won each of the elections?

Interpreting the Chart

c. What percentage of the vote did FDR win in each election?

d. Most of FDR's programs were proposed in 1933 and 1935. Based on the election numbers, did his programs make him more or less popular?

2. INTERPRETING PRIMARY SOURCES

This photograph, taken by Dorothea Lange, is one of the most famous images of the Great Depression. It shows a mother and her children in a camp for migrant workers in 1936.

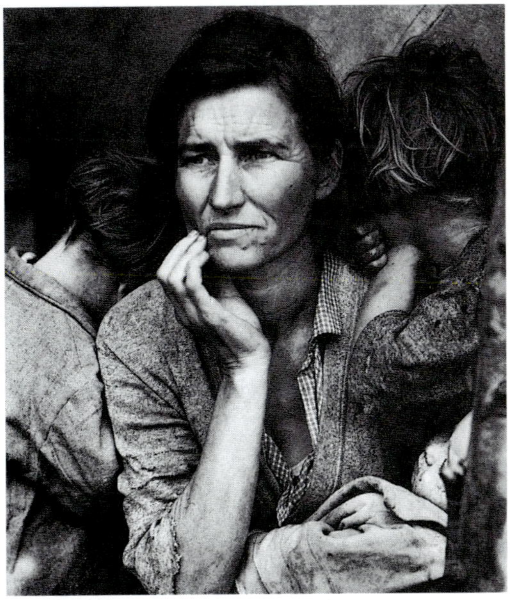

a. How would you describe this woman's expression?

b. Why do you think this photograph is often used to illustrate the effects of the Depression on individuals?

ALTERNATIVE ASSESSMENT

1. INTERDISCIPLINARY ACTIVITY: Science

Diagramming a Dam Find out how hydroelectric dams, such as those built by the TVA, turn the energy of flowing water into electricity. Draw a diagram or illustration explaining this process.

2. COOPERATIVE LEARNING ACTIVITY

Debating Government Policy Working in a small group, have each group member take on the role of a Depression-era leader and hold a panel discussion in which you debate the following question: Should the government go into debt to help the poor? Before holding the discussion, do further research on your leader's beliefs. Possible leaders to choose include the following.

- Herbert Hoover
- Franklin Delano Roosevelt
- Eleanor Roosevelt
- Huey Long
- Father Coughlin
- Francis Townshend
- a conservative Republican

3. TECHNOLOGY ACTIVITY

Preparing a Multimedia Presentation Choose an aspect of the Great Depression that you would like to learn more about. Using books, the Internet, and software, research your topic. Then prepare a multimedia presentation. Consider including oral histories, quotations from FDR's speeches, Depression-era photographs and art, excerpts from literature, and videotaped interviews with older relatives.

Visit www.mcdougallittell.com to learn more about the Great Depression.

4. HISTORY PORTFOLIO

Option 1 Review your section and chapter assessment activities. Select one that you think is your best work. Then use comments made by your teacher or classmates to improve your work and add it to your portfolio.

Option 2 Review the questions that you wrote for What Do You Want to Know? on page 728. Then write a short report in which you explain the answers to your questions. If any questions were not answered, do research to answer them. Add your answers to your portfolio.

Paint a WPA Mural

Franklin Roosevelt once remarked that "the very soundness of our democratic institutions depends on the determination of our government to give employment to idle [people]." Indeed, one of Roosevelt's New Deal programs, the Works Progress Administration's (WPA) Federal Art Project (FAP), provided creative work for artists during the Great Depression. These artists created more than 2,500 murals and 100,000 paintings. Many of the works celebrate American workers. The paintings mainly appeared in schools, post offices, and other public buildings.

A man shoes a horse in part of the mural shown above.

ACTIVITY Create your own WPA mural, as artists did during the Great Depression. Then write an exhibit note that tells viewers about your mural. Finally, give a speech dedicating your mural to your school.

TOOLBOX

Each group will need:

- pencils
- drawing paper
- newspapers to protect desks and floor
- yardstick
- poster-sized/oversized sheets of art paper
- water-based paints
- paintbrushes

Warren Hunter painted this mural in 1939 on the wall of the post office in Alice, Texas.

STEP BY STEP

1 Form artist groups. Meet with three or four other students to discuss your new art project. Imagine that you have all signed a work contract with the WPA. Your job is to create a mural to show Americans at work during the Great Depression.

2 Research the WPA and the FAP. Use the Internet, encyclopedias, or books about the Great Depression to brainstorm ideas for your mural. Jot down the types of work people were doing, the structures they were building, and the machinery they were using. With your group, discuss these ideas and answer these questions to help plan your mural:

- Who should be in the mural?
- What should they look like?
- What should they be doing?
- What colors would best represent the subject of your mural?

3 Sketch your mural. Using a pencil, accurately sketch your mural on drawing paper. This drawing should be a much smaller version of the mural you will paint later.

4 **Draw a grid over the sketch.** A grid consists of lines drawn across and down a drawing to form squares. Like individual puzzle pieces, each square contains a part of the picture. Use a yardstick to make your grid lines straight.

5 **Transfer the design to larger paper.** Draw another grid on oversized paper that has the same number of squares as the scale drawing. This time, though, the grid should be much larger—as large as the oversized paper. Make sure the grid lines are faint. Now copy the original drawing one square at a time.

6 **Paint the mural.** Each group member can paint one grid area at a time. This process is like filling in the outlines of a gigantic coloring book. When your mural is completed, hang it on a wall along with the other groups' paintings.

WRITE AND SPEAK

Write an exhibit note about how the mural reflects the type of work people did during the Great Depression. Attach the explanation to the back of the mural. Then, in a classroom dedication ceremony, deliver a speech dedicating your mural to your school.

HELP DESK

For related information, see page 748 in Chapter 26.

Researching Your Project

- *Life During the Great Depression* by Dennis Nishi
- *The New Deal* by Gail Stewart

Visit www.mcdougallittell.com to learn more about the Great Depression.

Did You Know?

Between 1935 and 1943, the WPA employed 8.5 million people and spent $11 billion. The average wage was $55 a month. The government poverty level at the time was $100 a month.

The agency built

- 651,087 miles of roads
- 125,110 public buildings
- 8,192 parks
- 853 airports

Among these projects were New York City's famed Lincoln Tunnel and the Fort Knox gold depository, located in Kentucky.

REFLECT & ASSESS

- Compare and contrast your image with those of the other groups.
- Why did you choose the particular image in your mural?
- How well does your mural represent the history, social issues, and culture of the time?

CHAPTER 27

The Rise of Dictators and World War II

1931–1945

The battleship U.S.S. *West Virginia* burns in Pearl Harbor in Hawaii after the Japanese attack on December 7, 1941.

Interact *with* History

The year is 1941, and the American naval base at Pearl Harbor has been bombed. Now the United States has joined the Allies in World War II. The Allies face dangerous opponents in leaders such as Hitler, Mussolini, and Tojo. You must do your part to help defeat them.

What Do You Think?

- What threat do dictators pose to the world?
- What would you be willing to sacrifice to defeat dictators?

Would you risk your life to fight against dictators?

USA

- **1932** Franklin Roosevelt is elected president.
- **1935** Congress passes first Neutrality Act.
- **1936** Franklin Roosevelt is elected to a second term.
- **1940** Roosevelt is elected to a third term.
- **1941** Japan bombs American naval base at Pearl Harbor, Hawaii.
- **1944** Roosevelt is elected to a fourth term.
- **1945** United States drops atomic bombs on Hiroshima and Nagasaki in Japan.

World

- **1931** Japan invades Manchuria.
- **1935** Italy invades Ethiopia.
- **1937** Japan invades China.
- **1939** Germany invades Poland.
- **1943** Soviets defeat Germans at Stalingrad.
- **1944** Allies invade Europe at Normandy.
- **1945** Germany and Japan surrender.

1931 — 1945

Chapter 27 SETTING THE STAGE

BEFORE YOU READ

Previewing the Theme

America in the World The United States tried to stay out of World War II. But after the attack on Pearl Harbor, the United States entered the war. Chapter 27 describes U.S. participation in the war. It also explains how the war changed the role of the United States in the world.

What Do You Know?

What do you think of when you hear the word *dictator* and the phrase *World War II*? Where did the fighting take place in this war?

THINK ABOUT

- what you've heard about dictators from the news
- what you have learned about World War II from books, movies, or television

What Do You Want to Know?

What questions do you have about World War II? Record these questions in your notebook before you read the chapter.

READ AND TAKE NOTES

Reading Strategy: Sequencing Events
Sequencing means putting events in the order in which they occurred. In learning about World War II, you will find it useful to list important events in order. For example, you might record important battles and their dates in a graphic organizer such as the one shown below. Copy this organizer in your notebook. Fill it in as you read the chapter.

See Skillbuilder Handbook, page R4.

IMPORTANT BATTLES IN EUROPE, AFRICA, AND THE PACIFIC

September 1, 1939—Germany invades Poland → [ ] → [ ] → [ ] → [ ] → [ ] → [ ] → [ ] → April 1945—U.S. Marines invade Okinawa

1 Steps to War

TERMS & NAMES
fascism
Adolf Hitler
Nazi Party
Joseph Stalin
Axis
appeasement
Lend-Lease Act
Pearl Harbor

MAIN IDEA
The rise of dictators in Europe and Asia led to World War II.

WHY IT MATTERS NOW
Aggressive rulers still threaten peace today.

ONE AMERICAN'S STORY

One of George Messersmith's duties as a U.S. diplomat in Austria in the 1930s was to watch events in Central Europe closely. What he saw happening in Germany worried him. Although Germany had been devastated after its defeat in World War I, it was again on the rise. In March 1936, Messersmith described what he saw.

Adolf Hitler greets a crowd of more than one million people at a Harvest Festival in 1937.

A VOICE FROM THE PAST

The National Socialist [Nazi] regime in Germany is based on a program of ruthless force, which program has for its aim, first, the enslavement of the German population to a National Socialist social and political program, and then to use the force of these 67 million people for the extension of German political and economic sovereignty over South-Eastern Europe—thus putting it into a position to dominate Europe completely.

George Messersmith, quoted in *The Making of the Second World War*

Messersmith's predictions would soon prove true. In the coming years, Germany and its allies threw the world into war, as you will read in this section.

The Rise of Dictators

By the mid-1930s, dictators, or absolute rulers, had seized control in several countries—Italy, Germany, Japan, and the Soviet Union. Their rise to power was due to economic and political factors that dated back to the end of World War I.

The treaties that ended the war had left many nations feeling betrayed. Japan and Italy, for example, had helped to win the war. However, both were dissatisfied by the peace treaties. Italy gained less territory than it wanted. Japan felt ignored by the European powers. Of the losing countries, Germany was treated the most severely. The winners stripped Germany of more than 10 percent of its territory and all of its overseas colonies. The winners also forced Germany to disarm. And they made Germany pay for war damages and accept responsibility for the war.

Connections TO SCIENCE

GERMAN SCIENTISTS

Many scientists left Germany or gave up their German citizenship after the Nazis took power. The most famous German scientist to do so was the physicist Albert Einstein (below).

Einstein, a German Jew, was visiting the United States when Hitler took control of Germany in 1933. Einstein announced he would not return home. "I shall live only in a country where civil liberty, tolerance, and equality of all citizens before the law prevail," he said. Einstein played a key role in convincing President Roosevelt to support research that would lead to the development of nuclear weapons.

Meanwhile, World War I had left the economies of Europe in ruins. Both sides emerged from the war heavily in debt. There was some economic growth in the 1920s. But the world economic situation collapsed with the Great Depression of the 1930s. Mass unemployment caused widespread unrest. Many Europeans turned to new leaders to solve these problems.

*Reading*History

A. Finding Main Ideas What factors led to the rise of dictators after World War I?

Mussolini, Hitler, and Stalin

One new leader was Benito Mussolini of Italy. Shortly after World War I, Mussolini began a political movement known as **fascism** (FASH•IZ•uhm). Fascists preached an extreme form of patriotism and nationalism that was often linked to racism. They oppressed people who did not share their views. In 1922, Mussolini became prime minister of Italy. In 1925, he established a dictatorship and took the title *Il Duce* (eel DOO•chay), or "the Leader."

In Germany, **Adolf Hitler** led the fascist National Socialist German Workers' Party, or **Nazi Party.** Throughout the early 1930s, the Nazis gained power by preaching German racial superiority. They also promised to avenge the nation's defeat in World War I. In 1933, the Nazis won control of the government. Hitler then overthrew the constitution. He called himself *der Führer* (duhr FYUR•uhr), or "the Leader."

Vocabulary

avenge: to get revenge

In the Soviet Union, the Communists tightened their grip on power during the 1920s and 1930s. V.I. Lenin, who led the Communist takeover of Russia in 1917, died in 1924. His successor was **Joseph Stalin.** Under Stalin, the government tried to control every aspect of life in the nation. It crushed any form of opposition.

Background

In theory, Communists and fascists have opposing ideas about government and society. Despite these differences, Stalin and Hitler were both brutal dictators.

Dictators Seek to Expand Territory

While dictators were gaining power in Europe, the military was gaining increasing power in Japan. By 1931, the Japanese military pushed the island nation to grab more land and resources. That year, the Japanese attacked Manchuria, a province in northern China rich in natural resources. They conquered the region within months.

Both Italy and Germany also sought new territory. In 1935, Italy attacked Ethiopia, one of the few independent African nations. Italian troops roared in with machine guns, tanks, and airplanes. By the spring of 1936, *Il Duce* had his first conquest.

That same year, Hitler moved troops into the Rhineland, a region of Germany along the French border. Under the Treaty of Versailles, the Rhineland was to remain free of German forces. The French government was outraged by the treaty violation. However, it took no action. Nor did the League of Nations.

HISTORY *through* ART

Spanish artist Pablo Picasso expresses the horrors of war in his painting *Guernica* (GUAHR•nih•keh), shown above. Picasso created this work after German planes destroyed much of the Spanish town of Guernica in April 1937, during the Spanish Civil War. Through Picasso's painting, the town became a symbol of the destructiveness of air warfare.

What characteristics of war does the painting bring out?

Background
Some Americans went to Spain to fight against the fascists. They were known as the Abraham Lincoln Battalion.

In 1936, Hitler and Mussolini formed an alliance known as the Rome-Berlin Axis. After this treaty, Germany, Italy, and their allies became known as the **Axis.** That year, a civil war erupted in Spain. The conflict pitted Spain's fascist-style military against the country's elected government. Hitler and Mussolini supplied the fascist forces with troops, weapons, and aircraft. In April 1939, Spain's army declared victory over the government and established a dictatorship.

In 1938, Hitler invaded Austria, home to mostly German-speaking peoples. He insisted that the Austrians wanted to be part of Germany. Many residents of Austria and Germany welcomed the unification.

Appeasement at Munich

After taking over Austria, Hitler set his sights on the Sudetenland. This was a region of Czechoslovakia where many people of German descent lived. Czechoslovakia, though, did not want to give up the region.

France and Russia pledged their support to Czechoslovakia if Germany attacked. Suddenly, Europe teetered on the brink of another war. Britain's prime minister, Neville Chamberlain, stepped in. He met with Hitler in an attempt to calm the situation. But their talks made little progress.

On September 29, 1938, Hitler and Chamberlain met in Munich, Germany. By the next day, the two sides had made a breakthrough and signed an agreement. Germany gained control of the Sudetenland. In return, Hitler promised to stop seeking any more territory.

Reading **History**
B. Analyzing Points of View What were the different points of view about the policy of appeasement?

The Munich Agreement was an example of the British and French policy known as **appeasement.** Under this policy, they met Germany's demands in order to avoid war. Chamberlain returned home from Munich and triumphantly announced that he had achieved "peace in our time."

Others, however, disagreed with appeasement. Winston Churchill reportedly wrote of the agreement: "[Britain and France] had to choose between war and shame. They chose shame. They will get war, too."

Germany Starts the War

Hitler soon broke the promise he had made in Munich. In March 1939, his troops moved in and conquered the rest of Czechoslovakia. The *Führer* then declared his intent to seize territory from Poland. Britain and France warned that an attack on Poland would mean war.

Britain and France assumed they had an ally in Stalin. After all, the Soviet Union and Germany were bitter enemies. However, in August 1939, Germany and the Soviet Union signed a nonaggression pact. In it, they agreed not to declare war on each other. On September 1, 1939, Germany invaded Poland. Great Britain and France declared war on Germany two days after the invasion of Poland. World War II had begun.

Reading History

C. Making Inferences Why do you think Stalin signed a nonaggression pact with Hitler?

The Germans introduced a new method of warfare known as *blitzkrieg* ("lightning war"). It stressed speed and surprise in the use of tanks, troops, and planes. German forces drove deep into Poland. As Germany conquered western Poland, the Soviet Union invaded from the east. In less than a month, Poland fell to the invading armies.

In April 1940, Hitler conquered Denmark and overran Norway. A month later, Germany launched a *blitzkrieg* against Belgium, Luxembourg, and the Netherlands. British and French troops could do little to stop the advancing Germans.

World War II in Europe, 1939–1941

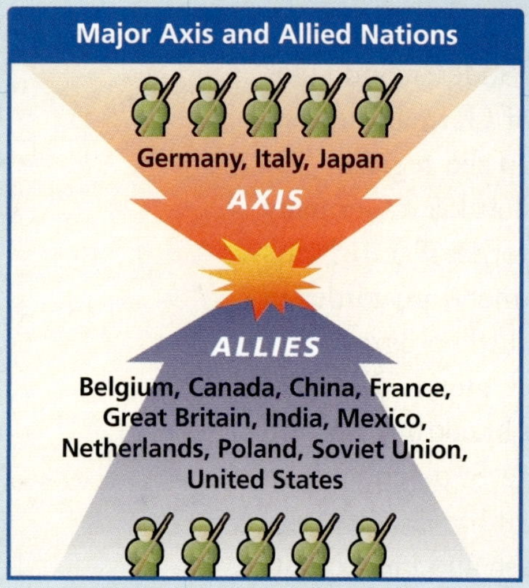

GEOGRAPHY SKILLBUILDER
Interpreting Maps

1. **Movement** *In which direction did the Germans attack in 1940? in 1941?*
2. **Place** *Which European nations were neutral in World War II?*

As each nation surrendered, British soldiers retreated to the French seaport of Dunkirk on the English Channel. Under heavy German bombardment, British vessels evacuated nearly 340,000 British, French, and Belgian troops.

In June 1940, the Germans launched a major offensive against France. In less than two weeks, they reached Paris. Days later, France surrendered. Hitler believed that Great Britain would seek peace after France fell.

Even though France had fallen, the British had no intention of quitting. Churchill, the new British prime minister, declared, "We shall defend every village, every town, and every city." Hitler soon made plans to invade Britain. To do so, however, he needed to destroy Britain's Royal Air Force, often called the RAF. In the summer of 1940, the German air force, or Luftwaffe (LUFT•VAHF•eh), and the RAF fought in the skies over Britain.

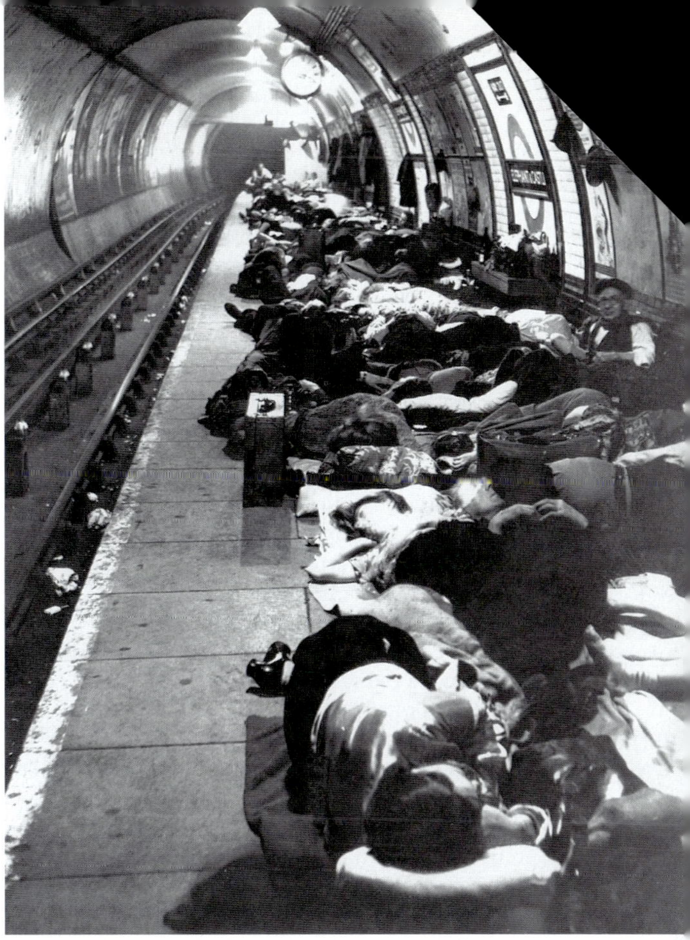

British civilians sleep in a subway station being used as an air raid shelter during the Battle of Britain in 1940.

German planes also unleashed massive bombing attacks on London and other civilian targets. By September, however, the Battle of Britain had left Hitler frustrated. The RAF was holding off the Luftwaffe. And despite constant bombing, the British people did not surrender.

Germany Attacks the Soviet Union

While Hitler's forces conquered Western Europe, Stalin's troops invaded Finland in November 1939. The Soviets then seized the countries of Estonia, Latvia, and Lithuania along the Baltic Sea. Despite their partnership, Hitler and Stalin distrusted each other. Hitler feared Soviet ambitions in Europe. He also wanted Soviet wheat and oil fields.

As a result, Hitler invaded the Soviet Union in June 1941. German forces moved easily through the giant country. They inflicted heavy casualties on Soviet troops. Then Hitler made a major mistake. He decided not to concentrate all his forces against Moscow. Instead, he reinforced his armies heading north toward Leningrad and south toward the Crimean Peninsula. The Germans tried to capture Leningrad from September 1941 to January 1944. About one million citizens died, many from starvation. But the city never fell to the Germans.

Reading **History**

D. Making Inferences Why did Hitler attack the Soviet Union just months after signing the nonaggression pact?

As German troops approached Moscow in December 1941, they ran into the harshest Russian winter in decades. Many German soldiers suffered frostbite. German tanks and weapons broke down in the cold. The Nazi advance had ground to a halt, and Soviet forces drove the Germans back.

The United States Aids the Allies

While the Nazis advanced, President Roosevelt tried to help the Allies by supplying them with arms and other materials. "We must be the great arsenal of democracy," he declared. He proposed the **Lend-Lease Act** to address this issue. This measure allowed the United States to lend or lease raw materials, equipment, and weapons to the Allied nations. Congress approved the act in 1941. Under Lend-Lease, the United States sent about $50 billion worth of war goods to the Allies.

America's HERITAGE

U.S.S. *ARIZONA* MEMORIAL

The U.S.S. *Arizona* suffered extensive damage during the attack on Pearl Harbor. The ship sank, and 1,177 of its crew died. The nation chose not to raise the ship. Instead, officials created a memorial (shown below) that sits above the sunken hull.

The names of all the crewmen who perished aboard the ship are carved on the memorial. To commemorate the 50th anniversary of the attack, President George Bush visited the site and dropped flowers in the water above the ship.

Japan Attacks Pearl Harbor

In 1940, Japan joined the alliance with Germany and Italy. In 1941, an even more warlike government came to power in Japan. Its leader was Hideki Tojo (HEE•deh•kee TOH•JOH), an army general. The Tojo government made plans to invade the Dutch East Indies—a source of oil—and Asian territories.

In the eyes of Japan's rulers, only one thing stood in their way—the United States Navy. On December 7, 1941, Japanese warplanes bombed the huge American naval base at **Pearl Harbor** in Hawaii. Before the day was over, about 2,400 Americans—both servicemen and civilians—died. Many of the American warplanes and ships were destroyed or damaged.

President Roosevelt asked Congress to declare war on Japan. He called December 7, 1941, "a date which will live in infamy." The nation quickly united behind him. On December 11, Germany and Italy declared war on the United States. In the next section, you will read about U.S. participation in the war in Europe.

Reading **History**

E. Analyzing Causes What was the main source of conflict between Japan and the United States?

Section 1 Assessment

1. Terms & Names

Identify:
- fascism
- Adolf Hitler
- Nazi Party
- Joseph Stalin
- Axis
- appeasement
- Lend-Lease Act
- Pearl Harbor

2. Taking Notes

Use a diagram to review events that led to American participation in World War II.

Event 1 → Event 2 → Event 3 → Event 4

Which of these events do you think was the most important? Why?

3. Main Ideas

a. Who were the main Axis powers? Who were the main Allied powers?

b. Why was Hitler unable to conquer Great Britain?

c. What event prompted U.S. entry into the war?

4. Critical Thinking

Analyzing Causes Why do you think dictators such as Hitler and Mussolini were able to gain such power in the years before World War II?

THINK ABOUT
- the peace treaties of World War I
- the worldwide depression

ACTIVITY OPTIONS

LANGUAGE ARTS
GEOGRAPHY

Imagine that you are a citizen of one of the countries invaded by Germany. Write a **journal entry** describing the invasion or draw a **map** of the invasion route.

2 War in Africa and Europe

TERMS & NAMES
- Dwight D. Eisenhower
- D-Day
- Battle of the Bulge
- Yalta Conference
- Holocaust

MAIN IDEA

The Allies defeated the Axis powers in Europe and Africa.

WHY IT MATTERS NOW

During World War II, the United States assumed a leading role in world affairs that continues today.

ONE AMERICAN'S STORY

Private First Class Richard Courtney could hardly believe it. Rumors had been circulating for weeks that his 26th Infantry Division was heading overseas to fight the Nazis in Europe. Now it was finally happening. His ship pulled out of New York harbor on a late summer morning in 1944. As the ship started down the river and headed out to sea, Courtney, a 19-year-old native of Altoona, Pennsylvania, described his feelings.

A VOICE FROM THE PAST

I was eager to see all the ships in the harbor and to look up at the *Statue of Liberty,* which I had seen two years before on a trip to New York with my father. . . . For a moment I considered missing Mass and staying on deck with the others to see the statue. Then my better sense took over, and I headed down the stairs to Mass. As soon as Mass ended, I hurried up the stairs and rushed out on deck to see water, nothing but water. Well, Old Girl [Statue of Liberty], I will just have to wait for the return trip to see you again.

Richard Courtney, *Normandy to the Bulge*

U.S. troops stand beside their train as they wait to travel overseas for duty in World War II.

Courtney was just one of millions of soldiers who left American shores to fight around the world. In this section, you will learn how American troops, along with those of its allies, defeated Germany and Italy and freed Europe.

Mobilizing for War

The Japanese attack on Pearl Harbor pulled the United States into World War II. Now, there was little time to waste. The nation quickly had to build up its armed forces. Millions of Americans volunteered for duty. Millions more were drafted, or selected for military service. Under the Selective Service Act, all men between the ages of 18 and 38 had to register for military service.

Those who served represented many of the nation's ethnic and racial groups. For example, more than 300,000 Mexican Americans fought in Europe as well as Asia. Nearly one million African Americans served in the armed forces. Native Americans and Asian Americans also took part in the struggle. African-American and some Japanese-American soldiers fought in segregated, or separate, units. For example, the 99th Fighter Squadron, known as the Tuskegee Airmen, consisted of African-American pilots. They served in North Africa and Italy.

More than 300,000 women also served in the U.S. armed forces. Many worked for the Women's Army Corps (WAC) as mechanics, drivers, and clerks. Others joined the Army and Navy Nurse Corps. Thousands of women also joined the U.S. Navy and Coast Guard, where they performed important noncombat duties.

Battles in Africa and Italy

The Allies began making plans to invade Europe. The Americans wanted to land in France as soon as possible. Stalin agreed. But Churchill believed the Allies were not prepared for such an invasion. He convinced the Americans that the Allies should first drive the Germans out of North Africa. This action would help the Allies gain control of the Mediterranean and open the way to invade Europe through Italy.

Reading **History**

A. Making Decisions Why did the Allies decide to attack the Nazis in North Africa before invading France?

Since the summer of 1940, Britain had been battling Axis forces for control of northern Africa—especially Egypt. Without Egypt, the British would lose access to the Suez Canal. The canal was the shortest sea route to Asia and the Middle Eastern oil fields.

World War II in Europe and Africa, 1942–1945

The Tuskegee Airmen were an all-black unit of pilots that fought in North Africa and Italy.

GEOGRAPHY SKILLBUILDER

Interpreting Maps

1. **Movement** *Which Allied power captured Berlin?*
2. **Place** *What was the last territory in North Africa held by the Axis?*

British troops in northern Africa faced a tough opponent in Germany's General Erwin Rommel. Rommel's skills had earned him the nickname "The Desert Fox." He commanded Germany's Afrika Korps, including two powerful tank divisions. In June 1942, Rommel's tanks pushed the British lines to the Egyptian town of El Alamein. The Desert Fox was just 200 miles from the Suez Canal.

He would go no further, however. The British stopped the German advance at El Alamein and then forced them to retreat. A wave of Allied troops, led by American General **Dwight D. Eisenhower,** landed in northern Africa in November 1942. They advanced toward Rommel's army in Tunisia. In February 1943, the two sides clashed. The inexperienced Americans fell to Rommel's forces. However, the Allies regrouped and continued attacking. In May, the Axis powers in northern Africa surrendered. The Allies now could establish bases from which to attack southern Europe.

The invasion of Italy got under way with an attack on the island of Sicily in July 1943. Allied and German forces engaged in a month of bitter fighting. American nurse June Wandrey recalled trying to help the wounded.

A VOICE FROM THE PAST

Many wounded soldiers' faces still haunt my memory. I recall one eighteen year old who had just been brought in from the ambulance to the shock ward. I went to him immediately. He looked up at me trustingly, sighed and asked, "How am I doing, Nurse?" . . . I put my hands around his face, kissed his forehead and said, "You are doing just fine, soldier." He smiled sweetly and said, "I was just checking up." Then he died. Many of us shed tears in private.

June Wandrey, quoted in *We're in This War, Too*

AMERICA'S HISTORY MAKERS

DWIGHT D. EISENHOWER

1890–1969

If ever there was a general who cared about his troops, it was General Dwight Eisenhower. As Allied forces battled in Italy, Eisenhower learned that he and another general were scheduled to stay in two large villas. He was not happy. He exploded.

> That's *not* my villa! And that's not General Spaatz' villa! None of those will belong to any general as long as I'm Boss around here. This is supposed to be a rest center—for combat men—not a playground for the Brass!

How might Eisenhower's concern for the common soldier have affected his standing with the troops?

The Allies forced the Germans out of Sicily and then swept into Italy. By this time, the Italians had turned on Mussolini. Officials had imprisoned their leader. However, he escaped. The new Italian government surrendered to the Allies in September 1943.

The Allied Advance and D-Day

Meanwhile, Germany's difficulties in the Soviet Union had grown worse. In September 1942, German forces attacked the Russian city of Stalingrad, an important industrial center. A brutal battle took place. The Soviet army fiercely defended the city. As winter approached, the German commander begged Hitler to let him retreat. The *Führer* refused.

Reading **History**

B. Summarizing What prevented the Germans from conquering the Soviet Union?

Fighting continued through the winter. The trapped Germans had no food or supplies. Many thousands of Nazi soldiers froze or starved to death. In February 1943, the remaining German troops surrendered.

American troops storm Omaha Beach in Normandy in northern France on June 6, 1944.

Each side had suffered staggering losses. With Germany's defeat at Stalingrad, its hopes of conquering the Soviet Union appeared gone.

Hitler soon had other things to worry about in the West. In June 1944, the Allies' plan to invade France got under way. On the morning of June 6, more than 5,000 ships and landing craft carried more than 130,000 soldiers across the English Channel to a region in northern France called Normandy. The attackers included American, British, and Canadian forces. The day of this historic assault became known as **D-Day.** It was the largest seaborne invasion in history.

The attack surprised the German forces positioned along the beach. Nonetheless, they defended the region fiercely. As Allied troops hit the shore, they endured a hail of gun and mortar fire. More than 10,000 Allied soldiers were killed or wounded as they attempted to move inland. By the end of the day, however, the Allies had secured the beaches.

Background
The Germans were surprised by the attack at Normandy because many, including Hitler, thought it would occur at Calais—150 miles away—where the English Channel is narrowest.

By the end of June 1944, 850,000 Allied troops had poured into France. They moved inland toward Paris, battling German troops along the way. On August 25, Allied forces liberated, or freed, the French capital. As they continued fighting to recapture the rest of France from the Germans, numerous American heroes emerged. One of them was Audie Murphy, the most decorated U.S. soldier of World War II. In January 1945, German troops attacked Murphy's unit in France. The 20-year-old Murphy climbed on a burning tank destroyer and used its machine gun to kill about 50 enemy troops. The U.S. government awarded him the Medal of Honor, the nation's highest military award.

As Allied forces advanced through Europe from the west, Soviet troops were beating back Hitler's army in the East. In December 1944, the German leader launched one final assault. In what became known as the **Battle of the Bulge,** German troops attacked Allied forces in the Ardennes region in Belgium and Luxembourg. The Nazi troops overwhelmed the Allies and pushed them back. U.S. forces regrouped and defeated the Germans. The Battle of the Bulge was costly. German casualties totalled 120,000. Meanwhile, nearly 80,000 Americans were killed, captured, or wounded.

…n Europe

…1945, the Germans were retreating everywhere. That …leaders met in the Soviet resort of Yalta. Attending the …**nce** were the "Big Three" as they were called—Roosevelt, … Stalin. During the conference, these leaders made plans …the war and the future of Europe.

…ised to declare war on Japan after Germany surrendered. …ers also agreed to establish a postwar international peace-…nization. In addition, …l the type of govern-…uld be set up in Eastern …he war.

Churchill, Roosevelt, and Stalin meet during the Yalta Conference in 1945.

… of the Yalta Conference, …sevelt was in poor health. …, just months after being …fourth term, the president …lt's vice-president, Harry …succeeded him. As the …ed Roosevelt's death, the … continued the war effort.

…il 1945, the Russians reached Berlin. Deep inside his air-…dolf Hitler sensed the end was near. On April 30, the man who had conquered much of Europe committed suicide.

On May 2, the Soviet Army captured Berlin. Five days later, German leaders officially signed an unconditional surrender at General Eisenhower's headquarters in France. The Allies declared the next day, May 8, as V-E Day, or Victory in Europe Day. The war in Europe was finally over.

The Horrors of the Holocaust

Vocabulary
concentration camp: place where Germans held persecuted groups during World War II

As the Allies fought toward Berlin, they made a shocking discovery. Scattered throughout German-occupied Europe were concentration camps where Jews and people of other persecuted groups had been murdered. The world would soon learn of the horrifying events that took place behind German lines during the war. In what has become known as the **Holocaust,** the Nazis killed about 6 million Jewish men, women, and children—more than two-thirds of the Jews in Europe. The Nazis also killed millions of people of other ethnic groups, including Gypsies, Russians, and Poles. An estimated 11 million people were killed in all.

The roots of the Holocaust lay in Adolf Hitler's intense racism. He preached that other groups, particularly the Jews, were inferior to Germans. As he rose to power in the 1930s, Hitler blamed the Jews for many of Germany's troubles. After becoming leader of Germany, Hitler enforced anti-Semitism, prejudice against Jews, in numerous ways. He denied Jews many of their rights and possessions.

Survivors of the concentration camp at Buchenwald in central Germany stand behind a fence in April 1945.

Soon after war broke out, Germany's anti-Semitic policies took an even darker turn. In a policy decision labeled "The Final Solution," Nazi leaders set out to murder every Jew under German rule. To accomplish this evil scheme, the Germans built huge facilities known as concentration camps. Officials crammed Jews into railroad boxcars and sent them to these camps. They forced able-bodied people to work. All others were slaughtered. The Germans carried out their killings with terrible efficiency. For example, they killed hundreds of people at a time in gas chambers disguised as showers. They then burned the bodies in large ovens or open pits. The largest concentration camp was Auschwitz in Poland. More than 1 million people are thought to have been murdered there.

On reaching the camps, the advancing Allies were outraged by what they saw. The Allies would battle this type of hate and bias by bringing German leaders to trial for what they had done. First, however, they had to defeat the Japanese. In the next section, you will read about the war in the Pacific.

Section 2 Assessment

1. Terms & Names

Identify:

- Dwight D. Eisenhower
- D-Day
- Battle of the Bulge
- Yalta Conference
- Holocaust

2. Taking Notes

Use a cluster diagram like the one shown below to identify the key battles and events that led to the Allies' victory in Europe.

3. Main Ideas

a. How did the United States build an army for the war?

b. Why did the Allies try to conquer North Africa before attacking southern Europe?

c. Why was the Battle of Stalingrad considered the turning point of the war in the east?

4. Critical Thinking

Supporting Opinions How might the war have been different if Hitler had decided to fight alongside the Soviet Union instead of against it?

THINK ABOUT

- the difficulties of fighting a two-front war
- the resources of Germany and the Soviet Union

ACTIVITY OPTIONS

GEOGRAPHY

TECHNOLOGY

Research the El Alamein battle. Draw a **map** of the battle or make a **database** showing the resources, such as the weapons and troops, of each side.

INTERACTIVE PRIMARY SOURCE

A Voice from the Holocaust

Setting the Stage Elie Wiesel (EHL•ee vee•ZEHL) was a Jewish boy from Romania. In 1944, when Wiesel was just 15, the Nazis sent the Jews of his town to Auschwitz in Poland. Wiesel's mother and one of his sisters died there. Wiesel and his father were sent to the Buchenwald concentration camp, where Wiesel's father died just a few months before the camp was liberated. In this excerpt from *Night,* Wiesel describes the terror he experienced on his way to Auschwitz. **See Primary Source Explorer** CD-ROM

The train stopped at Kaschau, a little town on the **Czechoslovak frontier.**[1] We realized then that we were not going to stay in Hungary. Our eyes were opened, but too late.

The door of the car slid open. A German officer, accompanied by a Hungarian lieutenant-interpreter, came up and introduced himself.

"From this moment, you come under the authority of the German army. Those of you who still have gold, silver, or watches in your possession must give them up now. Anyone who is later found to have kept anything will be shot on the spot. Secondly, anyone who feels ill may go to the hospital car. That's all."

The Hungarian lieutenant went among us with a basket and collected the last possessions from those who no longer wished to taste the bitterness of terror. "There are eighty of you in this wagon," added the German officer. "If anyone is missing, you'll all be shot, like dogs. . . ."

They disappeared. The doors were closed. We were caught in a trap, right up to our necks. The doors were nailed up; the way back was finally cut off. The world was a cattle wagon **hermetically**[2] sealed.

—Elie Wiesel

1. **Czechoslovak frontier:** the border of Czechoslovakia, a former European country occupied by Germany during World War II.

2. **hermetically:** thoroughly.

A CLOSER LOOK

A REIGN OF TERROR

The Germans attempt to rule their captives with a combination of brutality and terror.

1. What does the narrator mean when he describes "those who no longer wished to taste the bitterness of terror"?

A CLOSER LOOK

A CLOSED WORLD

The Jews as well as others were transported to the death camps in railway wagons.

2. What might be the effect of sealing people up in railway cars?

Interactive Primary Source Assessment

1. Main Ideas

a. What does the narrator mean when he says, "Our eyes were opened, but too late"?

b. What would be the effect on people of uprooting them from their homes?

c. This excerpt is from a book called *Night.* What might be the meaning of the title?

2. Critical Thinking

Analyzing Causes The horrors of the Holocaust followed from viewing and treating other people as less than human. What elements in this excerpt show the Germans treating the Jews this way?

THINK ABOUT

- the words and images used, both by the narrator and those he quotes
- the relations between those with power and those without

3 War in the Pacific

TERMS & NAMES
- Bataan Death March
- Battle of Midway
- island hopping
- Manhattan Project
- Hiroshima

MAIN IDEA

After early losses, the Allies defeated the Japanese in the Pacific.

WHY IT MATTERS NOW

Since the war, the United States has continued to play a major role in Asia.

ONE AMERICAN'S STORY

In April 1942, more than 70,000 Filipino and American troops surrendered to the Japanese on the Bataan Peninsula in the Philippines. From there, the Japanese marched the starving, exhausted soldiers about 60 miles to a prison camp. Along the way, about 10,000 prisoners lost their lives to shootings, beatings, and starvation. Sergeant Sidney Stewart was an American soldier in the Bataan Death March.

Thousands of American prisoners endure the Bataan Death March.

A VOICE FROM THE PAST

The sun beat down on my throbbing head. I thought only of bringing my feet up, putting them down, bringing them up. Along the road the jungle was a misty green haze, swimming before my sweat-filled eyes.

The hours dragged by, and a great many of the prisoners reached the end of their endurance. The drop-outs became more numerous. They fell by the hundreds in the road. . . .

There was a crack of a pistol and the shot rang out across the jungle. There was another shot, and more shots, and I knew that, straggling along behind us, was a clean-up squad of Japanese, killing their helpless victims on the white dusty road. . . . The shots continued, goading us on. I gritted my teeth. "Oh, God, I've got to keep going. I can't stop. I can't die like that."

Sidney Stewart, *Give Us This Day*

Allied and Japanese forces fought for more than three years in the Pacific. As you will read in this section, the fighting was brutal before the Allies emerged victorious.

Japan Expands Its Empire

At the same time as the attack on Pearl Harbor, Japanese forces launched attacks throughout the Pacific. By Christmas, Japan controlled Hong Kong, Thailand, and the U.S. islands of Guam and Wake.

The Japanese also pushed further into Southeast Asia, attacking Malaya and Burma. Great Britain, which ruled these lands and Hong Kong, fought back. But British forces proved to be no match for the Japanese invaders. Japan conquered the region within a few months.

But it took Japan longer to conquer the Philippines. They invaded the islands in December 1941 and pushed the Allied forces from the capital city of Manila onto the Bataan Peninsula. American and Filipino troops, led by U.S. General Douglas MacArthur, then fought the Japanese to a standstill for several months.

As fighting raged in the Philippines, the Allies feared that the Japanese might invade Australia. President Roosevelt ordered MacArthur to withdraw to Australia in March 1942. But MacArthur promised, on reaching Australia, "I shall return." Shortly after MacArthur left, the Japanese mounted an offensive. The U.S. troops on Bataan surrendered and endured the brutal Bataan Death March. The situation looked bleak for the Allies. But the momentum would soon turn.

"I shall return."
Gen. Douglas MacArthur

The Allies Turn the Tide at Midway

In the spring of 1942, the Allies began to turn the tide against the Japanese. The push began in April, with a daring air raid on Japanese cities, including Tokyo. Lieutenant Colonel James Doolittle led 16 bombers in the attack. Doolittle's raid caused little damage. But it shocked Japan's leaders and boosted the Allies' morale.

In May, the U.S. Navy clashed with Japanese forces in the Coral Sea off Australia. For the first time in naval history, enemy ships fought a battle without seeing each other. Instead, war planes launched from aircraft carriers fought the battle. Neither side won a clear victory in the Battle of the Coral Sea. However, the Americans had successfully blocked Japan's push toward Australia.

*Reading***History**
A. Evaluating What was the significance of the Battle of the Coral Sea?

The opposing navies clashed again in June off the island of Midway in the central Pacific. The U.S. Navy destroyed four Japanese carriers and at least 250 planes. America lost one carrier and about 150 planes. The **Battle of Midway,** in June 1942, was a turning point in the war.

GEOGRAPHY SKILLBUILDER
Interpreting Maps
1. **Location** *Which battle was fought closest to the Japanese mainland?*
2. **Location** *How many miles is Hawaii from Japan?*

The Allies Advance

After the Battle of Midway, the Allies went on the attack to liberate the lands Japan had conquered. Rather than attempt to retake every Japanese-held island, the Allies decided to invade islands that were not heavily defended by the Japanese. The Allies could then use the captured islands to stage further attacks. This strategy was known as **island hopping.**

*Reading***History**

B. Finding Main Ideas What was the Allies' strategy in the Pacific?

The two sides fought an important battle on the island of Guadalcanal. U.S. Marines marched ashore in August 1942. Six months of bitter fighting followed. In February 1943, the Allies finally won. They had gained their first major land victory against the Japanese.

Playing a role in this victory—and many others throughout the Pacific—was a group of Navajo Indians. To keep Japanese intelligence from breaking its codes, the U.S. military had begun using the Navajo language to transmit important messages. The marines recruited about 400 Navajos to serve as Code Talkers. They accompanied troops into battle and helped them communicate safely.

In October 1944, Allied forces invaded the Philippines. The effort included a massive naval battle off the Philippine island Leyte (LAY•tee). About 280 ships participated. The Allies won the three-day battle. Japan's navy was so badly damaged that it was no longer a threat. Allied forces came ashore. They liberated Manila in March 1945. General MacArthur, three years after leaving the Philippines, had returned.

Although they lost the fight in the Philippines, the Japanese increased their use of a new weapon—the *kamikaze* (KAH•mih• KAH•zee), or suicide pilot. *Kamikazes* filled their planes with explosives and crashed them into Allied warships. Japanese pilots volunteered for these suicide missions. But they couldn't stop Allied advances.

U.S. Marines raise a flag atop Mount Suribachi on Iwo Jima.

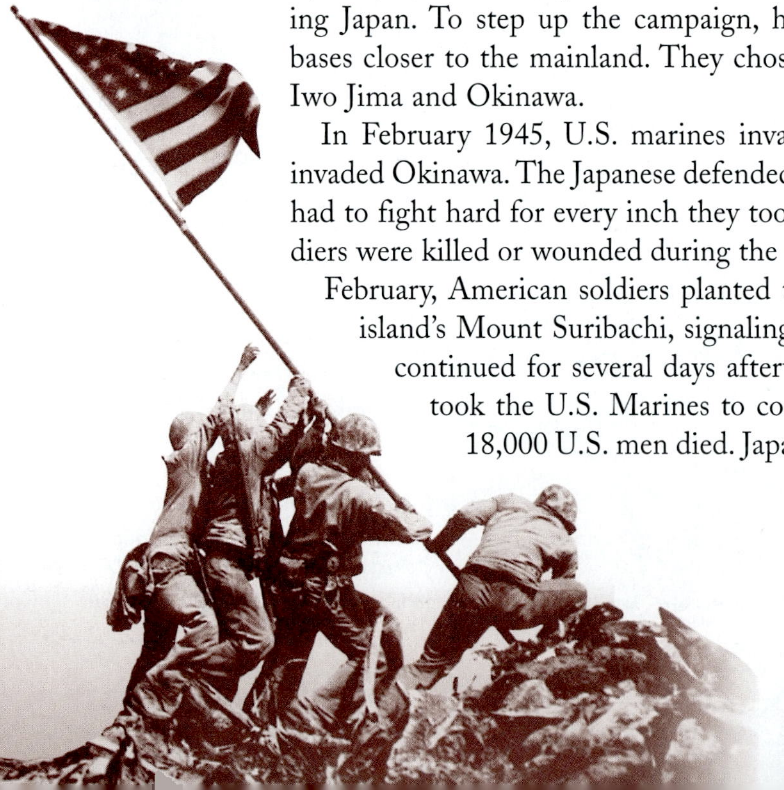

Iwo Jima and Okinawa

By early 1945, with Japan's defenses weakened, the Allies began bombing Japan. To step up the campaign, however, they had to establish bases closer to the mainland. They chose the Japanese-held islands of Iwo Jima and Okinawa.

In February 1945, U.S. marines invaded Iwo Jima. In April, they invaded Okinawa. The Japanese defended the islands fiercely. The Allies had to fight hard for every inch they took. More than 23,000 U.S. soldiers were killed or wounded during the campaign for Iwo Jima. In late February, American soldiers planted the U.S. flag at the top of the island's Mount Suribachi, signaling their victory, though fighting continued for several days afterward. In the several months it took the U.S. Marines to conquer both islands, more than 18,000 U.S. men died. Japanese deaths exceeded 120,000.

*Reading***History**

C. Summarizing What happened during the battles for Iwo Jima and Okinawa?

Atomic Weapons End the War

In the summer of 1945, Japan continued to fight. The Allies planned to invade Japan in November 1945. American military leaders feared that an invasion of mainland Japan might cost 200,000 American casualties. Therefore, American officials considered the use of an atomic bomb.

Shortly after entering the war, the United States set up the **Manhattan Project** in 1942. This was a top-secret program to build an atomic bomb. Led by American scientist J. Robert Oppenheimer, the project team worked for three years to construct the weapon.

The Japanese city of Hiroshima was leveled by the atomic bomb.

Soon after officials successfully tested the bomb, Truman told Japan that if it did not surrender, it faced destruction. The Japanese refused to give in. On August 6, 1945, the B-29 bomber *Enola Gay* dropped an atomic bomb on the city of **Hiroshima.** The explosion killed more than 70,000 people and turned five square miles into a wasteland. Still, the Japanese refused to surrender. On August 9, the United States dropped a second atomic bomb on Nagasaki, killing another 40,000. On August 14, Japan surrendered.

Background
By the end of 1945, another 70,000 people had died due to injuries and radiation caused by the atomic bomb dropped on Hiroshima.

On September 2, 1945, Japanese and Allied leaders met aboard the U.S. battleship *Missouri* in Tokyo Bay. There, Japanese officials signed an official letter of surrender.

The war changed forever the lives of the soldiers who fought in it. In the next section, you will learn about how the war affected Americans back home.

Section 3 Assessment

1. Terms & Names

Identify:
- Bataan Death March
- Battle of Midway
- island hopping
- Manhattan Project
- Hiroshima

2. Taking Notes

Use a diagram like the one shown to list events that led to the defeat of Japan.

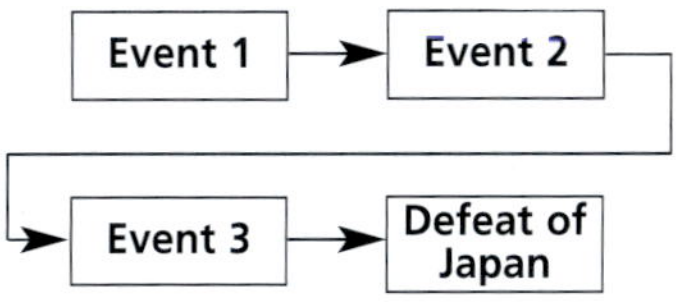

Which event do you think was most important, and why?

3. Main Ideas

a. Why was the Battle of Midway considered such an important victory for the Allies?

b. Why did the Allies want to conquer the islands of Iwo Jima and Okinawa?

c. What event finally prompted Japan to surrender?

4. Critical Thinking

Forming Opinions What might be the arguments for and against using the atomic bomb on Japan?

THINK ABOUT
- the consequences of invading Japan
- the bomb's destructive power

ACTIVITY OPTIONS

LANGUAGE ARTS
SCIENCE

Research the Manhattan Project. Write a **biography** of one of the scientists on the project or draw a **diagram** explaining how the atomic bomb worked.

4 The Home Front

TERMS & NAMES
War Production Board
rationing
Rosie the Riveter
A. Philip Randolph
bracero program
Japanese-American internment

MAIN IDEA
Americans at home made great contributions to the Allied victory.

WHY IT MATTERS NOW
World War II caused lasting changes in the lives of civilians.

ONE AMERICAN'S STORY

Margaret "Peggy" Hooper of San Pedro, California, was 17 years old when the United States entered the war. Her father went off to fight. Eventually, Hooper took a job as an "incoming inspector" at an aircraft plant. Her duties included keeping time sheets and inspecting materials. She often described her work in her letters to a friend serving with the Pacific fleet.

A VOICE FROM THE PAST

Gosh, we have been working hard at work lately. Just rushed to death and never getting through. Our production schedule has been doubled and still we work harder and put out more all the time. . . .

You had better be careful how you talk to me 'cause I have developed a big muscle in my right arm and a good strong one in my left arm, so take it easy, kid.

Margaret Hooper, quoted in *Since You Went Away*

Women factory workers rivet the interior of an airplane during World War II.

World War II created jobs for thousands of citizens such as Peggy Hooper. Americans on the home front worked together to help achieve an Allied victory, as you will read in this section.

Wartime Production

The effort to defeat the Axis powers took more than just soldiers. American forces needed planes, tanks, weapons, parachutes, and other supplies. Under the guidance of the **War Production Board** (WPB), factories churned out materials around the clock. By 1945, the country had built about 300,000 aircraft and 75,000 ships. The United States was producing 60 percent of all Allied ammunition.

With so many factories in need of workers, jobs became easy to find. In effect, the war ended the Great Depression. Shortly after the war began, the nation's unemployment rate fell. The country's yearly gross national product (GNP) rose to new heights during the war. The GNP is the total value of all the goods and services produced by a nation during a year. Between 1939 and 1945, the U.S. GNP soared from $90.5 billion to nearly $212 billion.

Because the armed forces needed so many materials, some of the items Americans took for granted became scarce. For example, American auto makers did not produce any cars between 1942 and 1945. Instead, they built tanks, jeeps, and airplanes. Items such as gasoline, tires, shoes, meat, and sugar were also in short supply. To divide these scarce goods among its citizens, the government established a system of **rationing.** Under this system, families received a fixed amount of a certain item.

Reading **History**
A. Recognizing Effects How did World War II affect the U.S. economy?

The war was expensive. To help pay the cost, the government raised income taxes and sold war bonds. These bonds were loans that the government promised to repay with interest. Movie stars urged people to buy war bonds. Americans bought billions of dollars worth of bonds.

Opportunities for Women and Minorities

With so many men fighting overseas, the demand for women workers rose sharply. In 1940, about 14 million women worked—about 25 percent of the nation's labor force. By 1945, that number had climbed to more than 19 million—roughly 30 percent of the work force. Women worked in munitions factories, shipyards, and offices.

Much of the nation welcomed the growing numbers of women into the workplace. The country promoted **"Rosie the Riveter"**—an image of a strong woman hard at work at an arms factory—as its cherished symbol for its new group of wage earners.

The war also created new job opportunities for minorities. More than 1 million African Americans worked in the defense industry during the war years. Many of these jobs were along the West Coast and in the North. As a result, more than 1 million African Americans migrated from the South during the war. Many traveled to California and such Northern cities as Detroit and Chicago. The inflow of African Americans often inflamed racial tensions. In 1943, a terrible race riot broke out in Detroit. Federal troops had to restore order after 34 people were killed.

On paper, at least, African Americans enjoyed equal rights in some workplaces. **A. Philip Randolph,** an African-American labor leader, had helped achieve these rights in 1941. Randolph had threatened to lead an African-American protest march for better jobs through Washington, D.C. President Roosevelt sought to avoid such a march. As a result, he issued Executive Order 8802. It outlawed job discrimination in defense industries working for the federal government.

Background
A. Philip Randolph was the leader of the Brotherhood of Sleeping Car Porters, a powerful African-American labor union.

A. Philip Randolph and Eleanor Roosevelt chat at a labor rally in 1946.

Other minorities lent their hand to the home-front effort. Some 46,000 Native Americans left their reservations to work in the nation's war industries. Tens of thousands of Hispanics—people with ancestors from Spanish-speaking lands—also joined the ranks of the country's war-related laborers. Included in this group were thousands of Mexicans who migrated to the United States at the government's request. During the war years, the nation faced a serious shortage of farm workers. The government responded by hiring Mexicans to perform the much-needed labor. This policy was known as the **_bracero_ program.** By mid-1945, more than 120,000 *braceros* worked on farms throughout the country.

Meanwhile, Mexican Americans struggled against prejudice and sometimes violence. In Los Angeles, for example, U.S. sailors often fought with "zoot suiters." These were young Mexican-American men who wore zoot suits—an outfit consisting of a broad-brimmed hat, a knee-length jacket, and baggy-legged pants. In what became known as the zoot-suit riots, groups of American servicemen attacked Mexican Americans. Beginning the night of June 3, 1943, the violence lasted 10 days before it was brought under control by police.

Reading **History**
B. Comparing and Contrasting What did the African-American and Hispanic-American experiences during World War II have in common?

The Internment of Japanese Americans

In the aftermath of Pearl Harbor, a growing number of Americans began to direct their anger toward people of Japanese ancestry. Many Americans saw Asian immigrants as a threat to their jobs. Many also believed that Asians could never fit into American society. As a result,

CITIZENSHIP TODAY

Writing to Government Officials

In the late 1970s, Japanese Americans asked the government to redress, or make up for, the injustice of the World War II internment. A letter-writing campaign by Japanese Americans helped secure passage of the Civil Liberties Act of 1988. This act included a formal apology and authorized payments of $20,000 each to Japanese Americans who were interned. The act also established a public-education program to prevent such discrimination in the future.

President Ronald Reagan signs the Civil Liberties Act of 1988.

How Do You Write to Government Officials?

1. Think about public issues that are important to you.
2. Choose one issue about which you would like the government to adopt a certain policy or take a certain action.
3. Gather information about the issue.
4. Refine your position on the issue in terms of a specific policy or action that you would like the government to follow.
5. Write a letter to your congressional representative or senator in which you urge him or her to take a particular stand on the issue.

See Citizenship Handbook, page 284.

Visit www.mcdougallittell.com for more information on contacting elected officials.

Congress banned practically all immigrants from Asia in 1924.

In the days and weeks after Pearl Harbor, several newspapers declared Japanese Americans to be a security threat. President Roosevelt eventually responded to the growing anti-Japanese hysteria. In February 1942, he signed an order that allowed for the removal of Japanese and Japanese Americans from the Pacific Coast. This action came to be known as the **Japanese-American internment.** More than 110,000 men, women, and children were rounded up. They had to sell their homes and possessions and leave their jobs.

Soldiers stand by as Japanese Americans in San Francisco board a bus to take them to an internment camp.

These citizens were placed in internment camps, areas where they were kept under guard. In these camps, families lived in single rooms with little privacy. About two-thirds of the people interned were *Nisei* (NEE•say), Japanese Americans born in the United States.

The nation's fear of disloyalty from Japanese Americans was unfounded. Many of the camp internees raised the American flag each morning. In addition, thousands of young men in the camps volunteered to fight for the United States. The all-*Nisei* units, the 442nd Infantry and the 100th Infantry, fought in Europe. They were among the most highly decorated units in the war. One member, Daniel Inouye, showed extreme courage. After being severely wounded, he continued to lead his platoon in an attack in Italy. He lost his right arm, but earned the Distinguished Service Cross. In the next section, you will learn about other effects that World War II had on the both the United States and the world.

Background
Daniel Inouye later became a U.S. senator from the state of Hawaii.

Section 4 Assessment

1. Terms & Names

Identify:

- War Production Board
- rationing
- Rosie the Riveter
- A. Philip Randolph
- *bracero* program
- Japanese-American internment

2. Taking Notes

Use a cluster diagram like the one shown to review the ways in which Americans at home contributed to the war effort.

3. Main Ideas

a. How did the war lift the nation out of the Great Depression?

b. How did the war spur an African-American migration at home?

c. What action did the U.S. government take against many Japanese Americans during the war?

4. Critical Thinking

Comparing and Contrasting How were the war years a time of both opportunity and struggle for American women and minorities?

THINK ABOUT

- Rosie the Riveter
- African-American migrants
- zoot-suit riots

ACTIVITY OPTIONS

LANGUAGE ARTS

ART

Research the wartime life of one of the groups mentioned in this section. Write a **report** or design a **mural** about its members' experiences during the war.

Build Morale on the Home Front

You are an American helping to fight the war from the home front. You have friends and family stationed in Europe and the Pacific. You worry about them every day. To help them and the rest of your family and friends through the war, you do everything you can to keep morale high.

COOPERATIVE LEARNING On this page are three challenges you face as an American on the home front during World War II. Working with a small group, decide how to meet each challenge. Choose an option, assign a task to each group member, and do the activity. You will find useful information in the Data File. Be prepared to present your solutions to the class.

MUSIC CHALLENGE

"Jukebox Saturday Night"

Everybody loves the Andrews Sisters and Glenn Miller's big band sound. People whistle tunes like "Jukebox Saturday Night," a favorite of jitterbugging bobbysoxers. You want to write a snappy song with catchy lyrics to brighten people's moods in these tough times. Present your song using one of these options:

- **Sing your song in an audition for the bandleader.**
- **Tape record your song to send to the bandleader.**

HOME ECONOMICS CHALLENGE

"Make it do, or do without"

It's up to you to help with the war effort. How will you do it? You know that factories need certain materials to manufacture supplies for the war effort. And the troops need food. What can you do to help the troops fighting overseas? Use the Data File for help. Then present your ideas using one of these options:

- **Create a list of activities that you and your friends can participate in to help the war effort, both by collecting materials and saving food.**
- **Write a letter to a relative serving in the armed forces explaining how the activities of you and your friends have helped him and his fellow soldiers.**

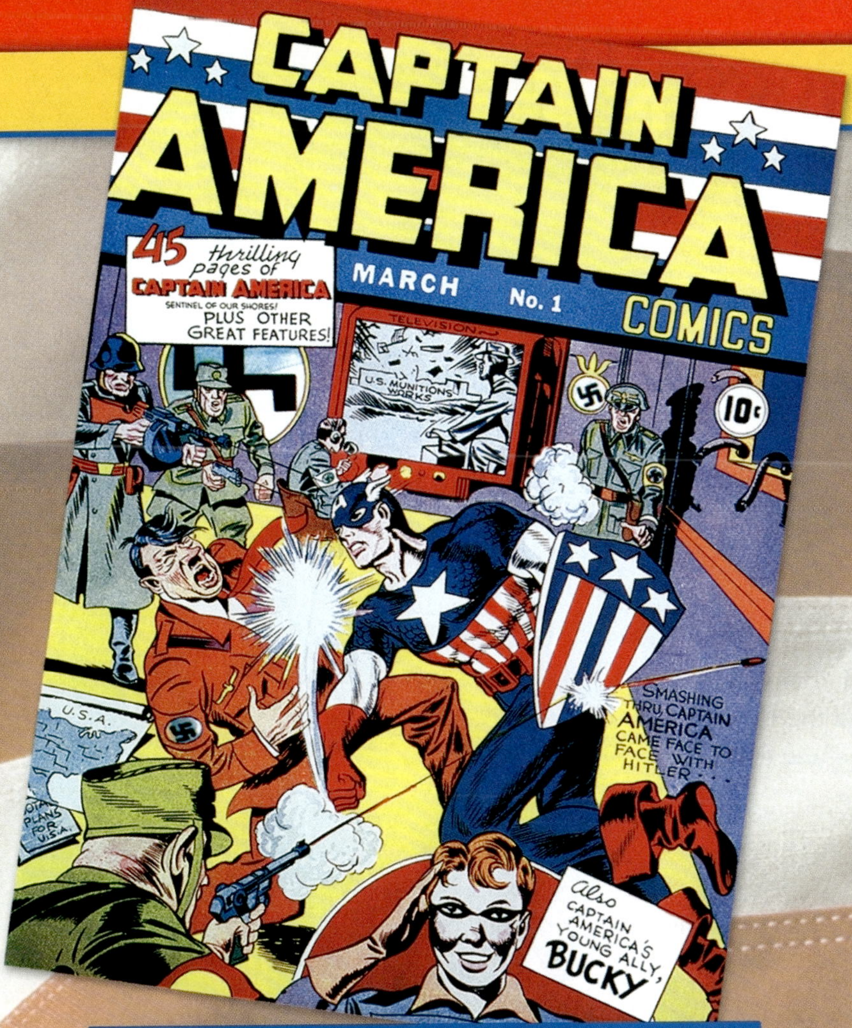

ART CHALLENGE

"POW!". . . "OW!!"

These days, battle stories crackle over the radio and fill the daily papers, bringing worry into every waking hour. The comic book adventures of Captain America help you cope with all the frightening news. This bold superpatriot and his pal Bucky find and defeat Nazis working in American factories, radio stations, and transportation systems. A recent Captain America tale gave you the idea for a super-spy character who helps the troops fighting overseas. You decide to send your idea to Joe Simon and Jack Kirby, creators of Captain America. Present your idea using one of these options:

- Create a short comic book of your hero's exploits.
- Design a cover illustrating your comic superhero.

ACTIVITY WRAP-UP

Present to the class As a group, review your methods of boosting wartime morale. Evaluate which of your solutions is the best.

Present your solutions to the class.

DATA FILE

Life on the Home Front

- People collect tinfoil, old tools, scrap metal, lard, and bacon grease for making arms and ammunition.
- Rationed gas and tires force workers to rise early and take buses, trolleys, and trains to work. People stay home at night.
- People share housing to cut costs and provide child care for working mothers. Neighbors use each other's appliances.
- Bobby pins, can openers, flash-light batteries, boxed candy, lawn mowers, alarm clocks, and other everyday items are scarce.
- People save fuel by keeping houses at 65 degrees.

Victory Gardens

- Americans plant gardens on rooftops, in backyards, and on vacant lots, producing beans, radishes, carrots, squash, corn, and tomatoes.
- Housewives preserve corn relish, stewed tomatoes, and fruit jams.

Popular Music

Big Bands The Glenn Miller Orchestra is one of the most popular bands. It creates a special sound, mixing a clarinet and saxophones. The band's hits include "Don't Sit Under the Apple Tree (With Anyone Else But Me)," "When Johnny Comes Marching Home," "My Prayer," "Moonlight Serenade," and "A String of Pearls."

Andrews Sisters These three sisters harmonize on songs, including "I'll Be With You in Apple Blossom Time" and "Boogie Woogie Bugle Boy of Company B."

Visit www.mcdougallittell.com to learn more about the home front.

The Legacy of the War

TERMS & NAMES
Marshall Plan
G.I. Bill of Rights
Nuremberg trials
United Nations

MAIN IDEA

World War II had deep and lasting effects on the United States and the world.

WHY IT MATTERS NOW

As a result of World War II, the United States became the dominant power in the world.

ONE AMERICAN'S STORY

When the end of the war came, Elliot Johnson was excited. He was finally going home. At an army dismissal parade, however, one of his captains told the troops that it might not be so easy to put the war behind them.

A VOICE FROM THE PAST

When it was over, we all threw our hats in the air and screamed and yelled and cheered. . . . I recall very well one of the captains standing and looking at us without cheering. "You guys are anxious to get home and put this all behind you," he said. "But you don't understand how big a part of your life this has been. "You'll put it all behind you for about ten years, and then someday you'll hear a marching band. You'll pick up the beat and it will all come back to you and you'll be right back here on the parade ground marching again." And he was right.

Elliot Johnson, quoted in *The Homefront*

Soldiers celebrate being discharged from the service on May 12, 1945, at Fort Dix, New Jersey.

From the soldiers who survived it to the families who lost a loved one, World War II affected millions of Americans. The great struggle also touched the United States and the world in many other ways, as you will read in this section.

The War's Human Cost

No war has claimed so many lives or caused so much destruction as World War II. The human cost on both sides was immense. About 20 million soldiers were killed, and millions more were wounded. The Soviet Union suffered the greatest losses, with at least 7.5 million military deaths and another 5 million people wounded. More than 400,000 American soldiers died and more than 600,000 were wounded.

Civilian casualties also numbered in the millions. Both the Allied and Axis powers had fought a war without boundaries. They bombed cities, destroyed villages, and brought destruction to civilian life. Again, the Soviet Union experienced the worst losses. All told, about 20 million

Soviet citizens died in the struggle. China, which also endured years of attack from Japan in the 1930s, lost about 10 million civilians.

The war also created an enormous wave of refugees. They included orphans, prisoners of war, survivors of Nazi concentration camps, and those who fled advancing armies. After the war, 21 million refugees, most starving and homeless, tried to put their lives back together amid the ruins of Europe and Asia.

World War II Military Casualties, *1939–1945*

NATION	DEAD	WOUNDED
Soviet Union	7,500,000	5,000,000
Germany	3,500,000	7,250,000
China	2,200,000	1,762,000
Japan	1,219,000	295,247
United States	405,399	671,278
Great Britain	329,208	348,403
France	210,671	390,000
Italy	77,494	120,000

Source: *World Book*

SKILLBUILDER Interpreting Charts

1. *Which two nations suffered the most casualties in World War II?*
2. *Which of the major combatants suffered the fewest casualties?*

Economic Winners and Losers

Reading **History**

A. Finding Main Ideas Why did the United States emerge from World War II so strong?

The war left many of the world's economies in ruins. Bombing campaigns had destroyed factories, transportation centers, and other important buildings. Only the United States—where no major battles were fought (except for Pearl Harbor)—came out of the war with a strong economy. The boom in industry during the war had pulled the nation out of the Great Depression. After the war, the U.S. economy continued to grow.

With the world's strongest economy, the United States set out to help rebuild the shattered economies of Europe and Japan. U.S. forces occupied Japan for several years after the war. During that time, they introduced programs that put Japan on the road to recovery. In 1948, Congress approved the **Marshall Plan** to help boost the economies of Europe. The plan was named after the man who came up with it, Secretary of State George C. Marshall. Under the plan, the United States gave more than $13 billion to help the nations of Europe get back on their feet.

Changes in American Society

The nation faced important social changes in the years following the war. For one thing, the country had to deal with the return of millions of soldiers. With so many servicemen suddenly back home, the competition for jobs and education was great. The government responded by passing a law that is commonly known as the **G.I. Bill of Rights** or G.I. Bill. This measure provided educational and economic help to veterans. The government paid for returning soldiers' schooling and provided them with a living allowance. More than 7.8 million World War II veterans attended school under the G.I. Bill.

The return of so many fighting men also created a great demand for housing. The Truman administration took steps to address the country's housing shortage. However, many Americans had to live in crowded urban slums or in country shacks.

The U.S. soldiers who returned home found an America that had changed. During the war, millions of Americans had moved to find war-related jobs in California and in the cities. Included in this group was a large number of African Americans. By war's end, hundreds of thousands of African Americans had moved from the South to various Northern cities and California. There, they lived in overcrowded ghettos and experienced prejudice. However, many also found economic opportunity.

Now and then

WAR CRIMES

More than 40 years after the Nuremberg trials (shown below), the world community once again brought army officials to trial for war crimes. These crimes were committed during brutal civil wars in the former Yugoslavia from 1991 to 1999. An international tribunal met in The Hague in 1996 to begin trying persons for their role in the conflicts.

These civil wars pitted Serbs, Croats, Bosnians, and Albanians against each other. Many people, especially Serbs, were accused of undertaking a policy of "ethnic cleansing"—the systematic attempt to rid a region of people from certain ethnic groups, often by killing them.

The Nuremberg Trials

As the United States dealt with important matters at home, the nation also joined the world in dealing with war crimes. The international community put together a court to try Nazi leaders for their role in World War II.

The trial opened in November 1945 in Nuremberg, Germany. The original 24 defendants included some of Hitler's top officials. The charges against them included crimes against humanity. These crimes referred to the Nazis' murder of millions of Jews and others. In his opening argument, the U.S. chief counsel at Nuremberg spelled out why a trial was necessary.

A VOICE FROM THE PAST

What makes this inquest significant is that these prisoners represent sinister influences that will lurk in the world long after their bodies have returned to dust. They are living symbols of racial hatreds, of terrorism and violence, and of the arrogance and cruelty of power.

Robert H. Jackson, *The Nürnberg Case*

After nearly a year-long trial, 19 of the defendants were found guilty. Twelve were sentenced to death. About 185 other Nazi leaders were found guilty in later trials. The **Nuremberg trials** upheld an important idea: People are responsible for their actions, even in wartime.

Creation of the United Nations

The war helped to establish another principle—nations must work together in order to secure world peace. The outbreak of World War II demonstrated the weakness of the League of Nations, the international peacekeeping body created after the First World War. The League was weak in large part because the United States had refused to join out of a strong desire to stay out of foreign affairs. Toward the end of World War II, President Roosevelt urged his fellow Americans not to turn their backs on the world again.

Reading **History**

B. Solving Problems Why did President Roosevelt support U.S. participation in the United Nations?

The country listened. In April 1945, delegates from 50 nations—including the United States—met in San Francisco to discuss creating a new international peace organization. In June, all 50 nations approved the charter creating the new peacekeeping body known as the **United Nations,** or UN.

D.R. Fitzpatrick drew this cartoon, entitled ". . . Shall Not Have Died in Vain." He hoped the memory of U.S. losses would push Americans to support the UN and preserve peace.

International Tensions

The horrors of World War II had caused many countries to work together toward lasting peace. However, tensions still arose among nations in the wake of the war. For example, in 1948 the United Nations helped found the nation of Israel to create a homeland for the Jews in Palestine. Fighting immediately broke out as neighboring Arab nations attacked Israel. In addition, colonies around the world began fighting for their independence.

Reading History

C. Reading a Map Look at the map on page R33 to find out where Israel is.

The United States, however, was more concerned with the rise of the Soviet Union. Despite suffering so much damage and loss of life, the Soviet Union emerged from World War II as a great power. It had conquered much of Eastern Europe.

During the war, the United States and the Soviet Union had been uneasy partners. After the war, Stalin angered the United States by breaking a wartime promise to promote democracy in the nations he had occupied in Eastern Europe. Instead, Stalin forced the countries to live under Communist regimes. The Soviet Union wanted to spread communism. The United States wanted to halt it. This led to future conflict.

Background

An important reason for U.S. leaders to drop the atomic bombs on Japan was to make the Soviets fear U.S. power.

Finally, the end of the war marked the beginning of the atomic age. The atomic bombs dropped on Japan showed the world a powerful new weapon. In the next chapter, you will learn how atomic weapons increased tensions between the United States and the Soviet Union.

Section 5 Assessment

1. Terms & Names

Identify:

- Marshall Plan
- G.I. Bill of Rights
- Nuremberg trials
- United Nations

2. Taking Notes

Use a cluster diagram like the one shown to review the effects of World War II.

Which effect seems the most important to you?

3. Main Ideas

a. What was the Marshall Plan?

b. How did the G.I. Bill of Rights help World War II veterans?

c. What principles did the Nuremberg trials establish?

4. Critical Thinking

Analyzing Causes Why did the United States emerge from the war so much better off than other nations?

THINK ABOUT

- the geographic location of the United States
- the role of American industry

ACTIVITY OPTIONS

LANGUAGE ARTS

SPEECH

As a reporter, research and write a **news article** on a defendant at the Nuremberg trials or, as a lawyer, deliver a closing **speech** against a defendant.

VISUAL SUMMARY

The Rise of Dictators and World War II

■ Asia and the Pacific ■ Europe and Africa

1931–1941

- 1931 Japan invades Manchuria.
- 1933 Hitler comes to power in Germany.

- 1936 Germany and Italy form Axis.
- 1938 Germany takes over Austria.
- 1939 Germany invades Poland. World War II begins.
- 1940 Germany invades France. Battle of Britain fought.
- 1941 Japanese attack Pearl Harbor.

1942

- Japanese stopped at Battle of Coral Sea.
- Japanese defeated at Battle of Midway.

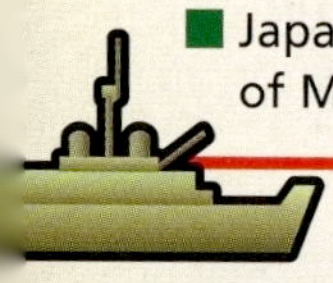

1943

- Soviets defeat Germans at Stalingrad.
- Allies stop Axis advance in North Africa.
- Allies invade Italy. Italy surrenders.

1944

- Allies invade Europe at Normandy.
- Allies invade the Philippines.

1945

- Allies invade Iwo Jima and Okinawa.
- Germany surrenders.
- United States drops atomic bombs on Hiroshima and Nagasaki.
- Japan surrenders.

TERMS & NAMES

Briefly explain the importance of each of the following.

1. Adolf Hitler
2. Pearl Harbor
3. D-Day
4. Holocaust
5. Battle of Midway
6. Hiroshima
7. rationing
8. Japanese-American internment
9. G.I. Bill of Rights
10. Nuremberg trials

REVIEW QUESTIONS

Steps to War (pages 757–762)

1. Why did Hitler attack the Soviet Union?
2. What was the Lend-Lease program?

War in Africa and Europe (pages 763–769)

3. What role did women play in the war?
4. What was D-Day and why was it significant?

War in the Pacific (pages 770–773)

5. What was the strategy of island hopping?
6. What was the Manhattan Project?

The Home Front (pages 774–779)

7. In what ways did Americans at home contribute to the war effort?
8. Why did the nation put thousands of Japanese Americans in internment camps during the war?

The Legacy of the War (pages 780–783)

9. Which nation lost the most soldiers and civilians in the war?
10. What international tensions arose after World War II?

CRITICAL THINKING

1. USING YOUR NOTES

IMPORTANT BATTLES IN EUROPE, AFRICA, AND THE PACIFIC

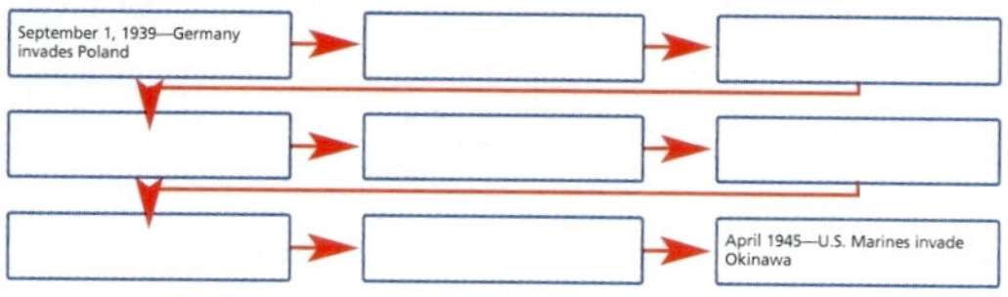

Use your chart to answer these questions.

a. Which battle that you listed occurred first?

b. Which battle was the most important?

2. ANALYZING LEADERSHIP

Do you agree or disagree with Neville Chamberlain's policy of appeasement? Explain.

3. APPLYING CITIZENSHIP SKILLS

Imagine you are a Japanese American in an internment camp. If you were to write a letter of protest to the government, what violations of your rights would you describe in the letter?

4. THEME: AMERICA IN THE WORLD

Why do think the United States joined the United Nations after World War II, when it had refused to join the League of Nations after World War I?

5. COMPARING AND CONTRASTING

What role did racism play in the Holocaust and the internment of Japanese Americans? How was the level of racism different?

Interact *with* History

After reading the chapter, would you make the same choice about whether to risk your life to fight against dictators that you made at the beginning of the chapter? Explain.

HISTORY SKILLS

1. INTERPRETING MAPS: Movement

Study the map and answer the questions.

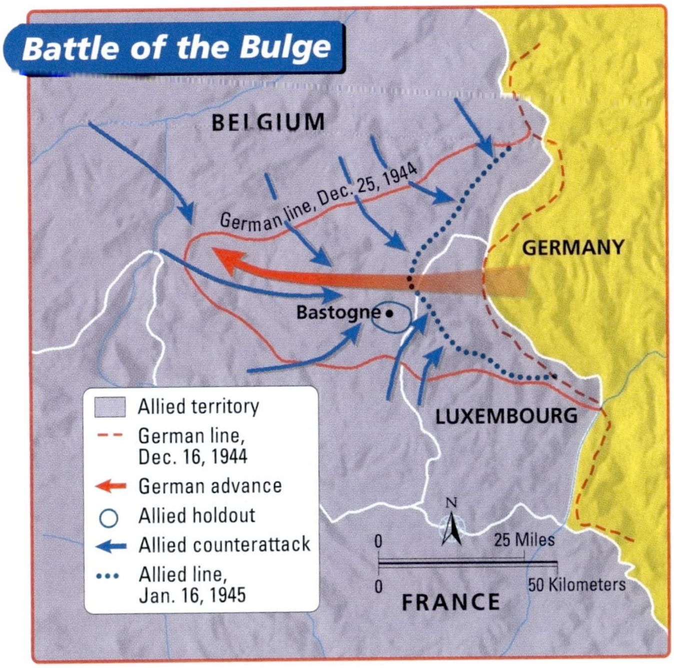

Basic Map Elements

a. In what countries was the Battle of the Bulge fought?

b. During what time period was the battle fought?

Interpreting the Map

c. About how many miles westward had the Germans advanced by Christmas 1944?

d. Why was this battle called the Battle of the Bulge?

2. INTERPRETING PRIMARY SOURCES

This poster shows a woman factory worker during World War II.

a. What does this poster reveal about women's roles in the war?

b. What does this image suggest about the abilities of women?

ALTERNATIVE ASSESSMENT

1. INTERDISCIPLINARY ACTIVITY: Language Arts

Writing a Diary Imagine you are a woman who has found a job as a welder in a shipyard during the war. Write a diary entry describing the changes the war has brought to your life.

2. COOPERATIVE LEARNING ACTIVITY

Creating a News Broadcast With a group of three to five students, create a news show about the experiences of Americans at home or abroad during World War II. Choose a specific year between 1941 and 1945 and a specific location, such as a hospital, factory, or battle front.

- Take on the roles of anchor, reporter, or interviewees.
- Write a script for your role.
- Conduct your "broadcast" in front of the class.

3. CD-ROM PRIMARY SOURCE EXPLORER

Making a Class Presentation Using the Internet, library, and the CD-ROM, do further research on the Holocaust and present your findings to the class. Some suggested topic ideas are given below.

- Examine in more depth one of the Nazi camps in Europe.
- Research individual experiences of the Holocaust, such as those of Elie Wiesel, Anne Frank, or Gerda Weissman Klein.
- Build a Holocaust collage, using photographs and other visuals along with corresponding captions.
- Create a map depicting the location of all of the Nazi concentration camps in Europe.
- Examine U.S. response to the Holocaust from 1942 to 1945.

4. HISTORY PORTFOLIO

PORTFOLIO **Option 1** Review your section and chapter assessment activities. Select one that you think is your best work. Then use comments made by your teacher or classmates to improve your work and add it to your portfolio.

PORTFOLIO **Option 2** Review the questions that you wrote for What Do You Want to Know? on page 756. Then write a short report in which you explain the answers to your questions. Add your answers to your portfolio.

CHAPTER 28

The Cold War and the American Dream 1945–1960

Section 1 **Peacetime Adjustments and the Cold War**
Section 2 **The Korean War and McCarthyism**
Section 3 **The Fifties**

The photograph above shows the window of an Edsel showroom. The new car was introduced in 1957. Shown at right is the Edsel.

Interact *with* History

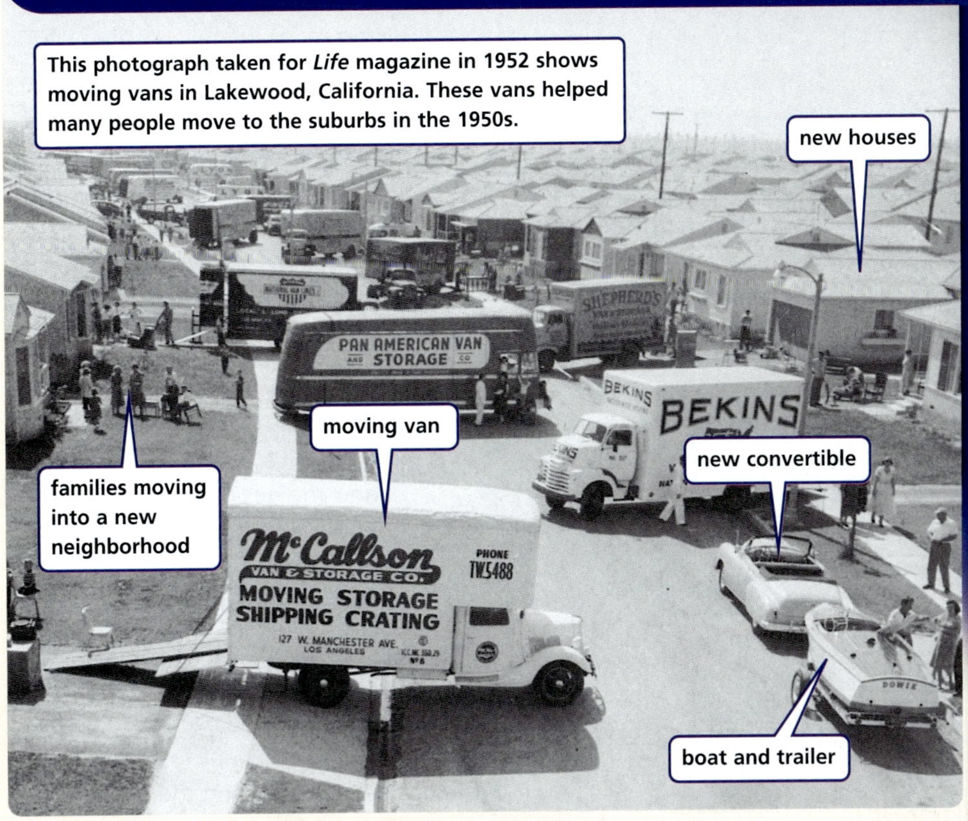

This photograph taken for *Life* magazine in 1952 shows moving vans in Lakewood, California. These vans helped many people move to the suburbs in the 1950s.

In the 1950s, American technology produced a flood of consumer goods. These included cars and houses in suburbs springing up across the country. You and your family have moved to a new house in a growing suburb—which some people think of as the American Dream.

What Do You Think?

- How might the American Dream be connected to prosperity?
- How might the American Dream involve helping others?
- How might the American Dream be connected to democracy, equality, and justice?

What is the American Dream to you?

USA

1945 United Nations is established.

1947 Truman Doctrine and Marshall Plan are established.

1948 Harry S. Truman is elected president.

1949 United States joins NATO.

1952 Dwight D. Eisenhower is elected president.

United States explodes first hydrogen bomb.

1953 Rosenbergs are executed as spies.

1954 Senator Joseph McCarthy claims Communist influence in U.S. Army.

1956 Highway Act is passed.

Eisenhower is reelected president.

1960 John F. Kennedy is elected president.

World

1946 Churchill gives "Iron Curtain" speech.

1948 Berlin Airlift begins.

1949 Germany is partitioned.

China becomes Communist.

1950 Korean War begins.

1952 Mau Mau revolt shakes Kenya.

1953 Korean War cease-fire is agreed to.

1956 Soviets suppress Hungarian Revolution.

1957 Soviets launch *Sputnik*.

Chapter 28 SETTING THE STAGE

BEFORE YOU READ

Previewing the Theme

Economics in History After more than 15 years of depression and war, the United States entered a time of economic prosperity and rapid change. This chapter explains how the United States came into conflict with the Soviet Union.

(Right) Eisenhower campaign glasses. (Top) Drive-in restaurant in the 1950s.

What Do You Know?

What do you think of when you hear the phrase *the American Dream*? What sorts of dreams might a people and a nation have?

THINK ABOUT

- what you have learned about the 1950s from television and movies
- hopes and dreams that might be achieved both through and outside of politics

What Do You Want to Know?

What additional information do you want to know about the Cold War and the American Dream? Record the sort of information you want in your notebook before you read the chapter.

READ AND TAKE NOTES

Reading Strategy: Categorizing The presidencies of Harry S. Truman and Dwight D. Eisenhower spanned the years between 1945 and 1960. There were a number of important issues, foreign and domestic, that both presidents had to deal with. Use the chart below to categorize each president's policy or action on the issues.

See Skillbuilder Handbook, page R6.

ISSUES	PRESIDENTS	
Domestic	**Truman**	**Eisenhower**
Labor unions and big business		
Communist threat at home		
Foreign		
Korea		
Communism in Europe		

1 Peacetime Adjustments and the Cold War

TERMS & NAMES
Harry S. Truman
Fair Deal
Cold War
containment
Truman Doctrine
NATO
Marshall Plan

MAIN IDEA

Americans looked for prosperity after World War II. They also fought communism in the Cold War.

WHY IT MATTERS NOW

The U.S. economy grew rapidly, and the nation's role in the world expanded after World War II.

ONE AMERICAN'S STORY

Harold Russell was a soldier, not an actor. Even so, in 1946 he won an Academy Award for best supporting actor in the Hollywood film *The Best Years of Our Lives.* Russell played an amputee struggling to adjust to civilian life after World War II. It was a role he knew well. As a paratrooper, Sergeant Russell had lost both hands in a grenade explosion. He was then fitted with hooks. Russell had to teach himself and others to accept his disability.

Sergeant Harold Russell demonstrates to other disabled veterans how to drink a cup of coffee.

A VOICE FROM THE PAST

I was all right. My problem was to make the people I met feel at ease. I just acted myself and didn't sulk in corners hiding the hooks. When my neighborhood friends saw I was okay and laughing they said to themselves, "Why should we feel sorry for him? He's getting along better than we are."

Harold Russell, quoted in *Life,* December 16, 1946

Russell won a second special Academy Award for "bringing hope and courage to his fellow veterans." As you will read in this section, millions of returning soldiers like Russell were preparing to restart their lives at the end of World War II.

Adjusting to Peace

The United States had spent the years 1941–1945 fighting World War II. Now, the country was at peace. The aircraft industry and other defense plants were changing over to making goods for peacetime. As part of this process, most industries reduced their work force. Factories shut down, and more than 10 million returning war veterans were looking for work.

Returning servicemen flooded the job market. Veterans won out over female workers in the competition for jobs in the first years after the war. In the aircraft industry, more than 800,000 workers—mostly women—were laid off. Several years after the war ended, as the economy boomed, employment rates for women began to return to wartime peaks. However, often these jobs were in traditional women's fields, such as office work and teaching.

CONNECTIONS TO MATH

A Changing Economy

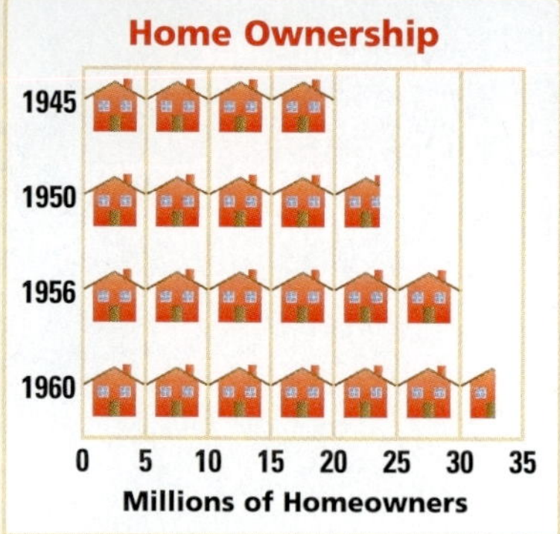

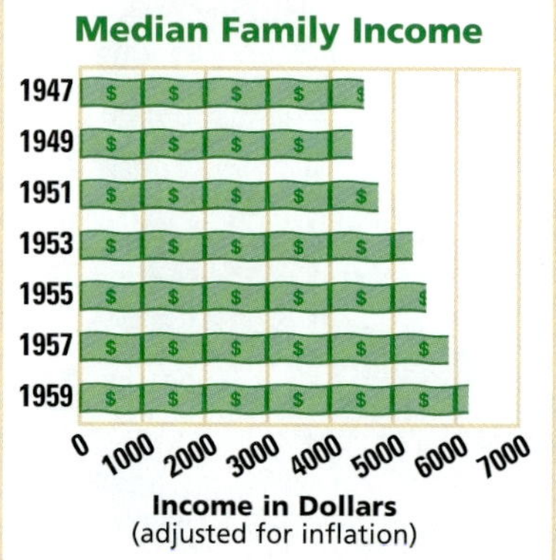

Source: *Historical Statistics of the United States*

SKILLBUILDER
Interpreting Charts

1. *What period showed the biggest increase in home ownership?*
2. *About how much did family income increase in the years 1947 to 1959?*

The Postwar Economy

Instead of slowing down, as many had feared, the nation's economy boomed. During the war years, few consumer goods had been produced. After the war, people were starting families and buying new homes. They wanted cars, washing machines, toasters, and all the other goods they had put off buying during wartime. American factories were fitted out with new machinery and tools to make different products.

The spending spree led to inflation, or a rise in prices. During the war, the government had put controls on prices and wages. After the war, in 1946, the controls on prices were lifted. Consumer goods were still in short supply. People had plenty of money to spend, but few goods to buy. As a result, the demand for goods increased, and prices skyrocketed.

After the war, the number of marriages increased. At first, a housing shortage forced many newlyweds to move in with relatives. Government-guaranteed housing loans for veterans under the G.I. Bill spurred the demand for new houses. Businessman William Levitt saw a way to meet this demand. He applied assembly-line techniques to home building. His mass-produced houses were so cheap that many people could afford them. He built Levittown in 1947 on Long Island, New York. Three years later, 17,500 homes had turned the farmland into an instant suburb. However, not everyone benefited from the postwar boom.

Reading **History**
A. Making Inferences Why did Levitt's houses become so popular?

Labor Unrest and Civil Rights

During the war, unions had agreed to give up pay raises. When the government put controls on wages, the unions agreed not to strike. But with the war over, workers faced with rising prices demanded better pay. In 1946, the United States entered one of the most strike-torn years in its history. More than one million workers joined strikes in the automobile, steel, meatpacking, and electrical industries.

Later that year, both miners and railway workers also went out on strike. Although President **Harry S. Truman** was a friend of labor, he feared these strikes would cripple the nation. During the railroad strike, he threatened to draft all railroad workers into the army. He said he would have the army run the trains. But the strike was settled before Truman could carry out his threat.

African Americans were still excluded from prosperity and full equality in the postwar world. World War II had raised the hopes of African Americans for greater equality. Many African-American veterans expected their wartime service to be recognized. Particularly in the

South, however, little had changed. In many Southern states, African Americans who attempted to vote were threatened, fired from their jobs, and even murdered.

To deal with these problems, President Truman created a commission on civil rights. He issued an executive order in July 1948 ending racial segregation and discrimination in the armed forces. Truman also asked Congress for an anti-lynching law and an end to the poll tax as a requirement for voting. However, when southern Democrats in Congress balked at the proposals, Truman backed off. Nonetheless, he was the first president to make equal rights a national issue. The action he took began the federal government's effort to deal with racial issues.

Background
In some southern states a poll tax had to be paid in order to vote.

The Fair Deal

In 1946, fears about the economy hurt the Democrats. Voters sent a Republican majority to Congress. The new Congress wanted to block Truman's programs. Congress turned down his plans to provide federal funds for housing, education, and health care. Congress also limited the power of unions by passing the 1947 Taft-Hartley Act. This act outlawed the closed shop, a workplace that hired only union members. The act also gave the president the power to require an 80-day cooling-off period before a strike.

As the 1948 presidential election campaign opened, few political experts believed President Truman would keep his job. Polls showed the Republican candidate, New York Governor Thomas E. Dewey, to be the clear favorite. Even within his own party, few of Truman's supporters thought that he could win.

Ignoring politicians and pollsters, Truman took his campaign to the people. Hiring a special train, he made a tour through hundreds of cities and small towns. Wherever his train stopped, Truman blasted the "do-nothing" Republican Congress because it passed little legislation. His strategy worked. When the votes were counted, he had won an upset victory over Dewey. The Democrats had also regained control of Congress.

After his victory, Truman presented Congress with a package of reforms he called the Fair Deal. He hoped to extend the social programs begun with FDR's New Deal. The **Fair Deal** called for new projects to create jobs, build public housing, and end racial discrimination in hiring. Many Republicans and southern Democrats worked together to block his plans. Congress passed few of his proposals. Only his low-cost public housing measure became law. In addition to problems at home, Truman faced major problems abroad.

AMERICA'S HISTORY MAKERS

HARRY S. TRUMAN
1884–1972

During Harry Truman's 1948 whistle-stop campaign, he traveled many thousands of miles in eight weeks. He was often accompanied by his daughter and wife.

Audiences enjoyed his straightforward manner and his spirited attacks on the Republicans. He blamed the Republicans for the nation's problems. When the votes were counted, Truman won an upset victory over his opponent, Thomas Dewey. A Chicago newspaper (above) mistakenly declared Dewey the winner in the close contest.

What might Truman's whistle-stop campaign suggest about his character?

Origins of the Cold War

After World War II, the capitalist Western democracies came into increasing conflict with the communist Soviet Union. Their differing economic and political systems resulted in misunderstandings. During the war, the Western democracies were allied with the Soviet Union. They were united in the struggle to defeat Nazi Germany. However, as victory grew nearer, the West and the Soviet Union distrusted each other.

The most difficult issue was the political future of Eastern Europe. In the final battles of the war, the Soviets freed Eastern European states from Nazi rule. Then Soviet forces occupied those states, including the eastern sector of Germany. The Soviet leader Joseph Stalin promised free elections to the nations of Eastern Europe. However, when the war ended, Stalin installed pro-Soviet governments throughout Eastern Europe.

Reading History

B. Comparing and Contrasting How did the Soviet Union and the West view Eastern Europe?

Stalin feared that free elections in Eastern Europe might result in the election of anti-Soviet governments on its borders. The Western democracies saw Stalin's occupation of Eastern Europe differently. President Truman believed that Stalin intended to spread communism worldwide.

Truman was determined to protect Western Europe from the threat of Soviet expansion. As the gap between the Soviet Union and the Western democracies widened, tensions developed. The resulting **Cold War** was a conflict that pitted the United States against the Soviet Union. The two nations never directly confronted each other on the battlefield. However, the threat of deadly conflict lasted for decades.

Europe after World War II, 1955

GEOGRAPHY SKILLBUILDER Interpreting Maps

1. **Location** *What westernmost country in Europe was a member of NATO?*
2. **Region** *Which NATO members directly bordered the Warsaw Pact countries?*

Containing Communism Abroad

As tensions between the United States and the Soviet Union were increasing, Britain's Winston Churchill visited the United States in 1946. He warned the world of Soviet aims.

A VOICE FROM THE PAST

From Stettin in the Baltic to Trieste in the Adriatic, an iron curtain has descended across the continent. Behind that line lie all the . . . states of Central and Eastern Europe. . . . All these . . . populations . . . lie in the Soviet sphere and all are subject . . . not only to Soviet influence but to . . . increasing . . . control from Moscow.

Winston Churchill, "Iron Curtain" speech, Fulton, Missouri

"An iron curtain has descended across the continent."
Winston Churchill

Background
Truman's containment policy was first announced in 1947 in response to Soviet pressure on Greece and Turkey.

The Truman Administration's main strategy in the Cold War was its containment policy. The goal of **containment** was to stop the spread of communism. This meant that the United States would work in military and nonmilitary ways to contain communism. Next, Truman announced the **Truman Doctrine,** which promised aid to people struggling to resist threats to democratic freedom.

In 1948, there was alarm over communist control of Eastern Europe. This led to formation of the North Atlantic Treaty Organization (NATO). The **NATO** alliance included the United States, Canada, and ten Western European nations. In response, the Soviet Union and Eastern European nations formed the Warsaw Pact (see map on page 792).

Marshall Plan and Berlin Airlift

Hoping to prevent the spread of communism, the United States came up with a plan to revive the war-torn economies of Europe. The plan was named for Truman's Secretary of State, George C. Marshall. The **Marshall Plan** offered $13 billion in aid to western and southern Europe. The plan helped the nations of Europe rebuild.

Reading History
C. Reading a Map Use the map on page 792 to see how Berlin was divided among the four nations.

The European nation in which the Cold War almost turned hot was Germany. In June of 1945, the Allies had agreed to a temporary division of Germany into four zones. These were controlled by the Soviet Union, France, Great Britain, and the United States. The Western powers merged their zones and made plans to unite them as West Germany. Stalin feared a united Germany might threaten the Soviet Union.

Berlin, Germany's former capital, lay within the eastern zone, still held by the Soviet Union. Like Germany, it too had been divided into East and West Berlin. In 1948, Stalin hoped to force the Western powers to abandon the city. His forces blocked access to Berlin.

Truman responded by approving a huge airlift of food, fuel, and equipment into the city. For nearly a year,

An American plane brings supplies to Berlin during the airlift.

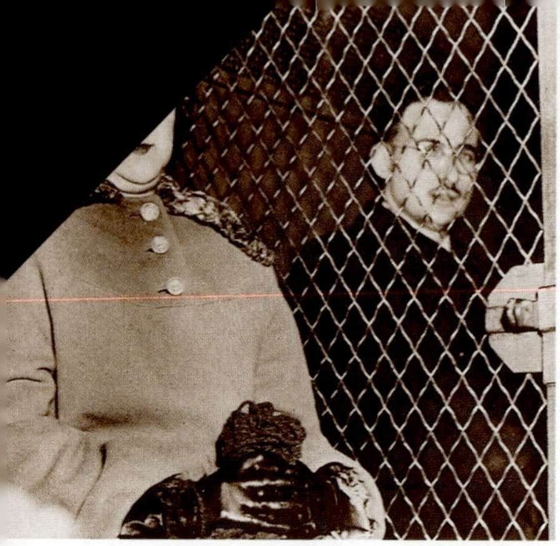

Ethel and Julius Rosenberg were executed despite numerous pleas to spare their lives.

U.S. and British cargo planes made 275,000 flights into Berlin. They carried supplies to the city's residents. In 1949, Stalin called off the blockade. By May 1949, Germany had been divided into communist East Germany and democratic West Germany.

Fear of Communism at Home

After World War II, a growing number of Americans feared that communism would gain strength within the United States. In part this was a response to the Soviet occupation of Eastern Europe. At first, attention focused on Americans who belonged to the U.S. Communist party who, it was feared, might spy for Russia.

Two famous spy trials made such fears believable. Alger Hiss was a former State Department official. He was accused of passing military information to the Soviet Union. Tried for lying under oath, he was sentenced to five years in prison in 1950. Ethel and Julius Rosenberg were members of the American Communist party. In 1951, they were convicted of passing atomic secrets to the Russians. They were executed in 1953.

Truman fought Republican charges that his administration was soft on communism. He issued an executive order requiring 3 million government workers to undergo loyalty checks. Federal workers who objected to signing loyalty oaths lost their jobs. Between 1947 and 1951, loyalty boards forced over 3,000 government workers to resign.

The anticommunist crusade gave new life to the House Un-American Activities Committee (HUAC). In 1947, HUAC began targeting actors, directors, and writers in the movie industry for suspected communist ties. Within the entertainment industry, lists of names circulated among the Hollywood movie studios. These were blacklists—unofficial lists of people thought to be communists. The careers of the people on these lists were ruined. As you will read in the next section, fear of communism dominated American life in the early 1950s.

Section 1 Assessment

1. Terms & Names

Identify:
- Harry S. Truman
- Fair Deal
- Cold War
- containment
- Truman Doctrine
- NATO
- Marshall Plan

2. Taking Notes

In a chart, explain the goals of these Cold War programs.

Program	Goal
Containment policy	
Truman Doctrine	
Marshall Plan	
NATO	

3. Main Ideas

a. Why was inflation a concern in the early postwar period?

b. What were the causes of the Cold War?

c. Why did the United States experience fear of communism after the war?

4. Critical Thinking

Forming Opinions Do you think an exaggerated fear of communism could occur again? Explain.

THINK ABOUT
- relations between the United States and Russia today
- American attitudes toward opposing views
- beliefs about communism

ACTIVITY OPTIONS

LANGUAGE ARTS
ART

Imagine that you were a child in Berlin during the airlift. Write a **letter** to a pen pal in the United States, or draw a **picture** describing your experiences.

2 The Korean War and McCarthyism

TERMS & NAMES
Mao Zedong
38th parallel
Korean War
Joseph McCarthy
brinksmanship
arms race
H-bomb
space race

MAIN IDEA
The Cold War and the Korean War produced a far-reaching form of anticommunism.

WHY IT MATTERS NOW
Reckless charges damaged personal lives and set up a climate of suspicion that affected Americans for years.

ONE AMERICAN'S STORY

John Stewart Service was one of thousands of Americans whose lives were turned upside down by the anticommunism of the postwar era. Service was born in China of missionary parents. He spent his childhood and teenage years there before coming to the United States for college. His firsthand knowledge of China made him a respected member of the Foreign Service. As a China expert, he warned the State Department of the weakness of China's anticommunist Nationalist party.

In 1949, the Communists took control of China. Angry Americans wanted someone to blame. Service's good advice was forgotten. He became one of the first State Department officials blamed for the loss of China to the Communists. Although a loyalty board cleared him of charges of disloyalty, he lost his job. The Supreme Court later ruled that he had been unfairly dismissed. As you will read in this section, many innocent Americans suffered a similar fate.

John Stewart Service defends himself before the Senate Foreign Relations Committee against charges that he worked with Communists.

Origins of the Korean War

In September 1949, the Communists defeated the anticommunist Nationalists in a civil war in China. The Nationalists were supported by the United States. **Mao Zedong** became head of the new Communist state. The Nationalist government, headed by Chiang Kai-shek, fled to the island of Taiwan, formerly Formosa, off the coast of the Chinese mainland. Many Americans were shocked by the fall of the Nationalist government. They viewed the takeover as part of a Communist plot to rule the world. They blamed the State Department for failing to stop the Communist revolution. American fear of communism grew. Events in Korea contributed to this fear.

Korea had been a Japanese colony for half a century when Japan surrendered to the Allies at the end of World War II. In 1945, Soviet troops occupied Korea north of the **38th parallel,** or line of latitude. American forces took control south of this line. Aided by the Soviets, a Communist government came to power in North Korea. In South Korea, a noncommunist leader supported by the United States governed.

Fighting Breaks Out in Korea

In June 1950, North Korean forces crossed the 38th parallel into South Korea. The conflict that followed became known as the **Korean War**. President Truman viewed Korea as a test case for his containment policy. He responded promptly. The United States appealed to the United Nations (UN) to stop the Communist move into South Korea. Sixteen nations provided soldiers for a UN force. However, U.S. troops made up most of the force and did most of the fighting. General Douglas MacArthur, former World War II hero in the Pacific, served as commander of all UN forces.

In early fighting, the North Koreans pushed the South Koreans back almost to Pusan. This city was on the southeastern tip of the Korean peninsula. MacArthur reversed the situation by landing his troops at Inchon. This was a port city behind the North Korean lines. It was a daring, dangerous plan, but it worked.

Squeezed between enemy troops coming at them from the north and south, the North Koreans soon retreated across the 38th parallel. General MacArthur requested permission of his superiors to pursue the enemy into North Korea. The UN and President Truman agreed. The president hoped the invasion might lead to a reunion of the two Koreas. The UN forces pushed northward beyond the 38th parallel (latitude) toward the Yalu

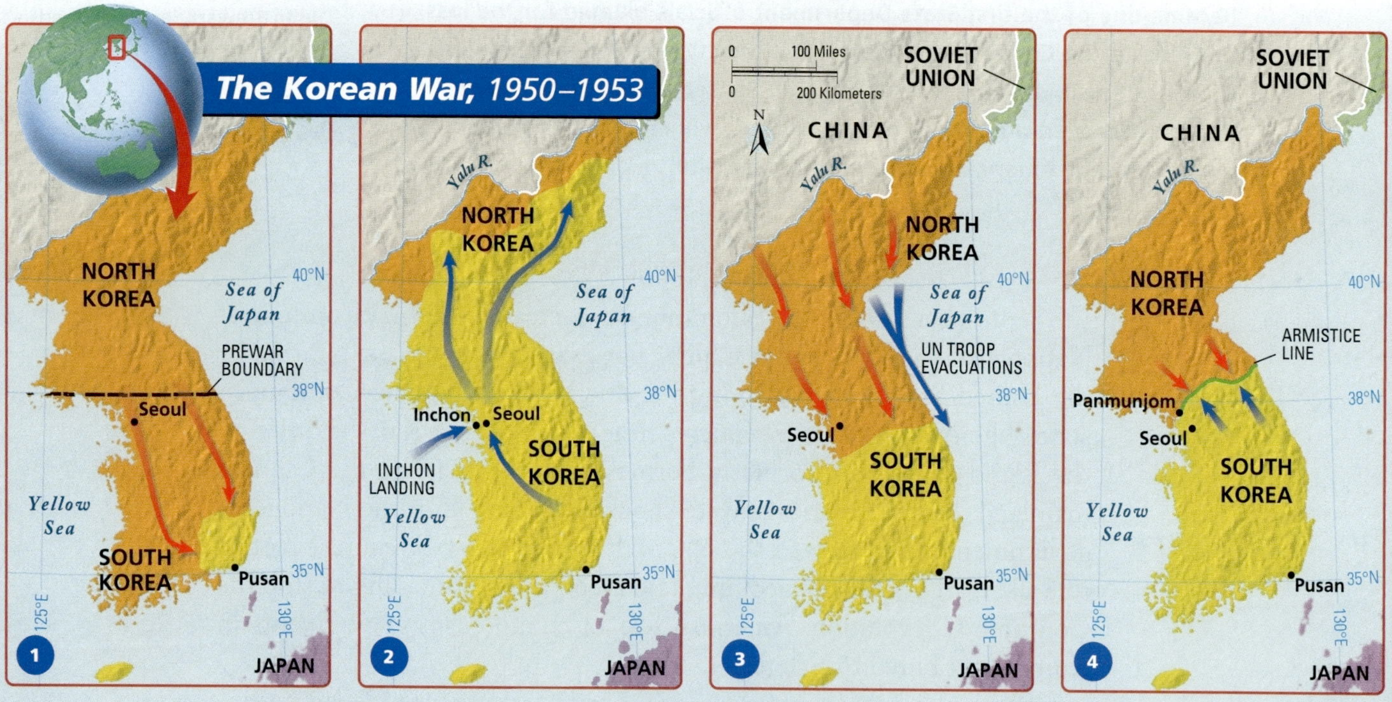

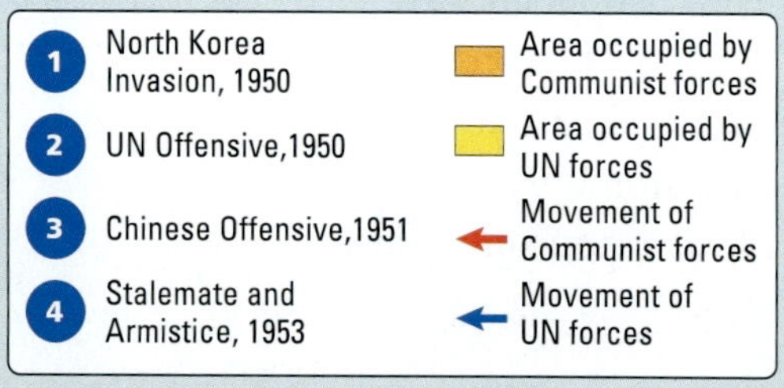

GEOGRAPHY SKILLBUILDER Interpreting Maps

1. **Movement** *In map 1, which forces moved south almost to Pusan?*
2. **Region** *Compare the prewar boundary in map 1 with the armistice line in map 4. Which side gained slightly more territory?*

River, the boundary separating China from North Korea. The Chinese warned them to stop.

American and South Korean troops search prisoners after the Inchon landing.

China Enters the Conflict

Communist China saw the movement of UN forces into North Korea as a threat to China's security. Chinese leaders warned that a further advance would force them to enter the war. Ignoring this warning, UN forces pushed on toward the Yalu River. On November 25, 1950, hundreds of thousands of Chinese Communist troops attacked in human waves across the Yalu River into North Korea. They drove UN troops back to South Korea. By early 1951, the two sides were deadlocked along the 38th parallel.

Reading **History**
A. Reading a Map Find the 38th parallel on the maps on page 796. Notice the movement of communist and UN forces back and forth across this parallel.

General MacArthur requested permission to blockade China's coastline and bomb China. Truman refused. The president feared that such action would draw the Soviet Union in and make the conflict a world war. The general went over the president's head to win support for his war aims. He spoke and wrote to newspaper and magazine publishers. He also wrote to Republican leaders.

As president, Truman was commander-in-chief of the armed forces. He viewed MacArthur's public criticism of his orders as undermining civilian control of the military. As a result, Truman fired MacArthur and ordered him home. When MacArthur returned to the United States, his admirers treated him as a hero. His farewell speech to Congress won the sympathy of many. "I now close my military career and just fade away—an old soldier who tried to do his duty as God gave him the light to see that duty. Good-bye." Despite support for MacArthur, Truman refused to back down. Most Americans came to agree with the president's actions.

Reading **History**
B. Making Inferences What effect might MacArthur's actions have had on the idea of civilian control of the military?

War Ends in Stalemate

As the war dragged on, it became more unpopular. In July 1951, Truman accepted a Soviet suggestion that truce talks begin. The talks dragged on for two years. They continued through the 1952 presidential campaign. When Truman decided not to run again, the Democrats chose Illinois governor Adlai Stevenson as their candidate. The Republicans picked World War II hero General Dwight D. Eisenhower. Ike, as voters liked to call him, criticized the unpopular war. He promised to go to Korea to seek a speedy end to the conflict.

Eisenhower made good on his promise when he won a landslide victory. During talks with the North Koreans and Chinese, he agreed to compromise to end the war. But he also warned privately that he was ready to use nuclear weapons and carry the war into China. A cease-fire ended the fighting in July 1953. The two Koreas were left more or less where they had been in 1950 with a border near the 38th parallel. Communism had been contained in Korea. However, Americans felt frustrated by the indecisive war. Some politicians selfishly made use of this frustration.

McCarthy and Communism

Senator Joseph McCarthy during the 1954 Army-McCarthy hearings

One such politician was **Joseph McCarthy,** a Republican senator from Wisconsin. He used the Korean War to fan Americans' fears of communism. In February 1950, McCarthy declared that he had a list of 205 State Department officials who belonged to the Communist Party. These charges were never proven. Nonetheless, McCarthy's claim launched a hunt for Communists that wrecked the careers of thousands of people. The term *McCarthyism* came to stand for reckless charges against innocent citizens.

In the spring of 1954, the Senate held hearings. During these nationally televised Army-McCarthy hearings, McCarthy accused the U.S. Army of "coddling Communists." Army spokesmen then charged McCarthy's staff with improper conduct. McCarthy responded with unsupported charges against a young lawyer helping to represent the Army. Joseph Welch, the Army counsel, spoke out against McCarthy.

A VOICE FROM THE PAST

Until this moment, Senator, I think I never really gauged your cruelty or your recklessness. . . . Senator. You have done enough. Have you no sense of decency, sir, at long last? Have you left no sense of decency?

Joseph Welch, Army-McCarthy hearings, April 22, 1954

Reading **History**

C. Summarizing What were some of McCarthy's charges?

Americans watching the exchange between McCarthy and Welch were shocked by McCarthy's conduct. After the Senate issued a statement censuring, or criticizing, his conduct, he faded from public view.

Eisenhower and the Cold War

Like Truman, President Eisenhower waged the Cold War. Eisenhower's Secretary of State was John Foster Dulles. Dulles rejected Truman's containment policy. He favored a more aggressive stand. He urged the overthrow of Communist governments. In 1956, Dulles announced that the United States would go to the brink of war to combat communism. This approach was known as **brinksmanship.**

In August 1949, Americans learned that the Soviet Union had produced an atomic bomb, in part by using information stolen by Soviet spies. The two superpowers were soon locked in an **arms race**, developing weapons with more destructive power. In 1952, the United States built a hydrogen bomb, or **H-bomb.** Three years later the Soviets tested their H-bomb. Fear led both sides to build up huge nuclear stockpiles.

In the 1950s, both the United States and the Soviet Union helped allies and weakened enemies around the world. In 1953 in Iran, the U.S. government's Central Intelligence Agency (CIA) helped topple a leader whom they thought might seek Soviet aid. In 1954, the CIA trained an army that succeeded in overthrowing Guatemala's President Jacobo Arbenz Guzmán. The United States believed he favored communism.

During Eisenhower's presidency, the Suez Canal in Egypt, which connected the Mediterranean Sea and the Red Sea, was at the center of another Cold War conflict. In 1955, Egypt's ties with the Soviet Union angered Britain and the United States. The two Western powers withdrew aid to Egypt. Gamal Abdel Nasser, Egypt's leader, reacted by seizing the canal, which was owned by France and Britain. France, Britain, and Israel jointly attacked Egypt. The Soviet Union threatened to support Egypt. The United States, along with the Soviets and the rest of the UN, pressured France, Britain, and Israel to withdraw from Egypt. The UN imposed a cease-fire.

In 1957, the superpowers began a **space race.** The Soviet Union stunned the world by launching the world's first space satellite. They sent *Sputnik* into orbit around the earth. This meant that the Soviet Union had a missile powerful enough to reach the United States. American scientists raced to launch a satellite. Congress set aside billions of dollars for space research.

Eisenhower suggested easing tensions through face-to-face talks. A setback to such efforts occurred in May 1960. The president was to meet in Paris with Soviet Premier Nikita Khrushchev. Two weeks before the meeting, the Soviets shot down an American U-2 plane. The spy plane had been flying over the Soviet Union. Eisenhower denied the aircraft was a spy plane until the pilot was captured. Khrushchev demanded an apology. When the president refused, the talks collapsed. Meanwhile, America was changing at home, as you will read in the next section.

Connections TO SCIENCE

SPUTNIK

The 184-pound *Sputnik 1* (shown below), whose name means "traveling companion," was the first man-made object to orbit the Earth. Circling every 96 minutes, it remained in orbit until early 1958. *Sputnik 2* carried a dog into space.

In 1961, the Soviet Union sent Yuri Gagarin into space to orbit the earth. The Americans lagged behind because the rockets that carried U.S. satellites were smaller and less powerful. The early Soviet lead disappeared, however, as American scientists and engineers found ways to improve rocket design, construction, and testing.

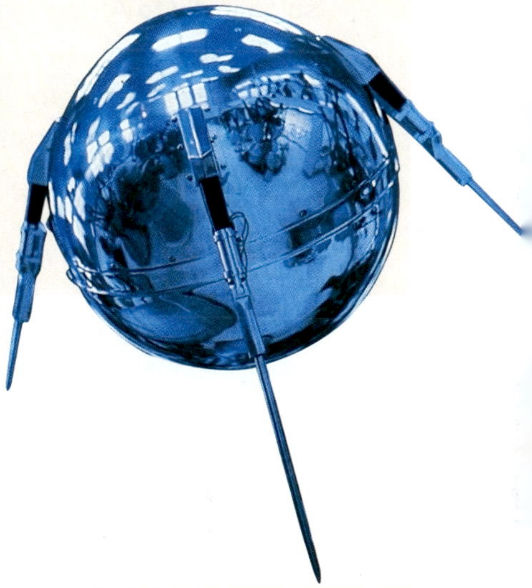

Section 2 Assessment

1. Terms & Names

Identify:
- Mao Zedong
- 38th parallel
- Korean War
- Joseph McCarthy
- brinksmanship
- arms race
- H-bomb
- space race

2. Taking Notes

Create a time line of up to five events that played a part in the Korean War from its beginning to end.

1950 — event — event — 1953

event — event — event

Which event do you think was most important? Why?

3. Main Ideas

a. Why did war break out in Korea? How did it end?

b. Why was McCarthy able to wield so much power during the 1950s?

c. How did Eisenhower's approach to the Cold War differ from Truman's?

4. Critical Thinking

Drawing Conclusions
How was U.S. involvement in Korea an example of the Truman Doctrine in action?

THINK ABOUT
- U.S. concerns about North Korean leadership
- U.S. goals for ending conflict
- the conflict's outcome

ACTIVITY OPTIONS

SCIENCE / **ART**

Research the problems of putting a satellite in orbit. Prepare a **report** explaining how these problems were solved, or draw a **design** of a rocket.

The Fifties

TERMS & NAMES
suburb
baby boom
sunbelt
rock 'n' roll

MAIN IDEA	WHY IT MATTERS NOW
While the United States was locked in a Cold War, social and economic changes took place in American life.	The American economy and popular culture continue to spread their influence around the globe.

This is the cover of a long-playing record album by LaVern Baker.

ONE AMERICAN'S STORY

LaVern Baker was one of many talented African-American artists to play a part in the popular music of the 1950s. Among her best known songs are "Tweedlee Dee," "Jim Dandy," and "See See Rider." As was the case with many other black rhythm-and-blues musicians, her records at first sold mostly to African-American teenagers. White singers covered, or copied, her songs. These remakes became hits. Disk jockeys played them over and over for their white radio audiences. Baker was annoyed that remakes of her songs by white singers outsold her originals. She made the following comment about one such singer.

A VOICE FROM THE PAST

When I went to Australia with Bill Haley, Big Joe Turner, the Platters, and Freddy Bell and the Bellboys, I left her my [flight] insurance policy. I sent it to her with a letter, "Since I'll be away and you won't have anything new to copy, you might as well take this."

LaVern Baker, quoted in *USA Today,* March 12, 1997

In 1990, Baker's importance as a recording artist was confirmed. The Rock and Roll Hall of Fame honored Baker by making her a member. The following section describes social, political, and cultural changes during the 1950s.

The Domestic Scene in the Fifties

Not everyone prospered in the 1950s. In 1957, nearly one out of every five Americans lived in poverty. Many of the nation's poorest were in cities. In his book *The Other America* (1962), Michael Harrington called attention to the forgotten poor. They were the people left behind as more well-to-do Americans headed for the **suburbs**—residential areas surrounding a city. Shops and businesses moved to suburbia as well.

Fewer people remained in the city to pay taxes for such services as garbage collection, firefighting, and road repair. Often those most affected by urban decay were African Americans and Latinos. Many could not afford homes in the suburbs.

Vocabulary
bracero: laborer, from the Spanish word *brazo,* which means "arm"

In the 1950s, immigration from Mexico increased greatly. Many people crossed the border illegally. Others came through the government-sponsored *bracero,* or temporary worker, program. The *braceros* found jobs on farms in the Southwest and Midwest. There they often earned low wages and endured difficult living and working conditions.

Despite this, when the program ended, many stayed on illegally. Employers took advantage of them. Fearful of being sent back to Mexico, workers were forced to work longer hours for lower pay.

When Eisenhower ran for president, he promised to steer a middle course. Once elected, Eisenhower pleased business leaders and conservatives without upsetting moderates and liberals. Although he disliked big government spending programs, Eisenhower kept most New Deal programs. He agreed to expand Social Security. He increased the minimum wage for workers. He also created the Department of Health, Education, and Welfare. He even backed some new spending. For example, Congress passed the Highway Act of 1956. This act provided $32 billion to build 41,000 miles of highway.

Changes Sweep America

In the postwar period, Americans began to feel more prosperous. The Depression and World War II had led many couples to put off marrying and starting a family. Now with the economy booming, Americans were getting married and having children. During the 1950s, the United States grew by almost 30 million people. This increase was mostly because of the **baby boom,** a sharp increase in the U.S. birthrate following World War II (from about 1946 through 1961). The number of families with three or four children increased dramatically.

The baby boom also spurred the growth of suburbs. Growing families left crowded city apartments for a house in the suburbs. Irving, Texas, a suburb of Dallas, was typical. In 1950, it had around 2,600 residents. Ten years later, 45,000 people lived there. To serve the suburbs, shopping centers, movie theaters, and restaurants sprouted up on what was once farmland. As suburbs grew, car sales exploded. In the suburbs, owning a car was a necessity. Few buses or other forms of public transportation existed.

In the 1950s, Americans not only moved from city to suburb. They also moved from the north and east to the south and west. The movement of people to the **sunbelt** increased the population of the warmer states of the South and Southwest. In the 1960s, California surpassed New York as the nation's most populous state.

STRANGE but True

FROM AUTOBAHN TO INTERSTATE

In the 1930s, Germany began to build a vast network of limited access, four-lane highways called *autobahns.* Germans believed these roads would have great military value.

During World War II, General Eisenhower saw the German road system firsthand. He was impressed by the way these highways enabled Germans to quickly move troops and supplies. President Eisenhower remembered Germany's *autobahns* when he called on Congress to pass the Highway Act of 1956. This act created the nation's first interstate highway system. A cloverleaf interchange is shown below.

Faces of the Fifties

◀ **Willie Mays,** the great center fielder, is shown at bat for the San Francisco Giants.

Elvis Presley ▶ dressed in a gold suit for the cover of his album ***50,000,000 Elvis Fans Can't Be Wrong*** (RCA).

◀ **Marilyn Monroe** was a popular film star in the 1950s.

Lucille Ball and Desi Arnaz starred in the popular television show *I Love Lucy.* ▼

The American Dream in the Fifties

For millions of mainly white Americans, life in the suburbs was the American dream. They were happy to live in affordable, single-family houses. People welcomed the chance to send their children to good schools. Americans shopped in malls where parking was free and easy to find. They didn't care if their houses looked alike. Parents wanted a safe place in which to raise their children.

Many critics worried that Americans were being forced to fit into suburban life. Some argued that in business offices and suburbs, people felt pressured to conform—that is, to agree with the beliefs and ideas of the majority. Yet most Americans seemed willing to conform for the rewards of a comfortable life.

Reading **History**
A. Drawing Conclusions What might be some of the advantages and disadvantages of fitting in?

In the 1950s, popular magazines, films, and television programs praised women for their roles as homemakers. *Time* magazine called women the "keeper[s] of the suburban dream." But not all women felt fulfilled in this role. Some felt bored or isolated. Those working outside the home had limited job choices. Openings were largely in nursing, teaching, and office work.

By the mid-1950s, American industry was churning out goods for consumers to buy. The economy was booming. Americans filled their houses with dishwashers, washing machines, clothes dryers, and vacuum cleaners. The suburban living room or den showed off the family's television, tape recorder, and high-fidelity record player. The garage held a lawn mower. Barbecue equipment and patio furniture filled the backyard. Owning the latest car or appliance came to be a symbol of social standing and success. The advertising industry encouraged consumers to join the spending spree. Television helped advertisers lure buyers to stores and car showrooms.

Pop Culture and Rock 'n' Roll

In the 1950s, Hollywood cranked out westerns, musicals, and romances. However, movie attendance plummeted as more and more people stayed home to watch TV. By 1960, nine out of ten households owned a set. One of the most popular shows of the decade was the situation comedy (sitcom) *I Love Lucy.* It starred Lucille Ball as the zany wife of bandleader Desi Arnaz. In *Father Knows Best* and many other Fifties sitcoms,

ReadingHistory
B. Making Inferences Do you think such television shows reflected life in the average American family? Why?

cheerful moms kept the house spotless. The dads worked to support the family. On *Father Knows Best,* Mr. Anderson exercised kindly but firm control over his children, who seldom rebelled.

Young children watched *Lassie, The Lone Ranger, The Howdy Doody Show,* and the *Mickey Mouse Club.* Their teenage sisters and brothers had fallen head over heels for another form of entertainment—**rock 'n' roll** music. In 1955, Bill Haley and His Comets hit it big with "Rock Around the Clock." By the mid-1950s, Chuck Berry, Little Richard, Fats Domino, and other black musicians held the spotlight with white rockers like Jerry Lee Lewis. But the largest fan club belonged to Elvis Presley. With such songs as "Heartbreak Hotel" and "Don't Be Cruel," he became the king of rock 'n' roll. His onstage bumps and shakes delighted teenagers.

In the mid-1950s, Allen Ginsberg and Jack Kerouac led a group of poets and writers. They protested what they saw as the shallowness and conformity of American society. Known as "beatniks," their followers filled coffeehouses to hear their heated attacks on "square" society. A Democratic presidential candidate, John F. Kennedy, also wanted to shake up the dullness of the Eisenhower years.

The Election of 1960

The 1960 presidential election was one of the closest in U.S. history. John Fitzgerald Kennedy, Democratic senator from Massachusetts, defeated Richard M. Nixon, Eisenhower's vice president. At age 43, Kennedy was the nation's youngest elected president. He was also the first Catholic president. Kennedy had campaigned to "get this country moving again" after the Eisenhower years. Kennedy and Nixon staged the first televised presidential debates. Kennedy's youthful energy and confidence helped him to win. In the next chapter, you will read about Kennedy's role in setting domestic policy, including civil rights.

Section 3 Assessment

1. Terms & Names

Identify:
- suburb
- baby boom
- sunbelt
- rock 'n' roll

2. Taking Notes

Create a web like the one below to examine the way life was changing the United States in the 1950s.

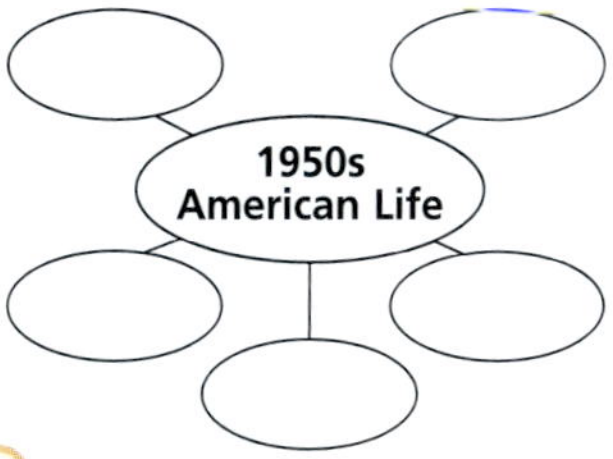

3. Main Ideas

a. How did the movement to the suburbs affect the urban poor?

b. What caused the 1950s baby boom? How did the baby boom contribute to suburban growth?

c. How did television affect American life in the 1950s?

4. Critical Thinking

Contrasting Do you think the American Dream for most Americans today would be the same as it was in the 1950s? Why?

THINK ABOUT
- expectations about suburban/urban living
- changes in transportation and workplace

ACTIVITY OPTIONS

TECHNOLOGY
MUSIC

Research one aspect of music in the 1950s, and either plan a **Web page** to share your information, or write your own **song** that fits this time period.

Route 66

America was changing, and few things contributed more to that change than U.S. Highway Route 66. Completed in the summer of 1926, this road connected small towns from Chicago to Los Angeles. Its course across the Heartland enabled farmers to move grain and produce to the big cities.

In the 1930s, farmers escaping the Great Plains' Dust Bowl fled westward along this highway. The first service stations—full-service gas stations—were built along Route 66. In the 1940s, it became an important route for the movement of troops and supplies. By the 1950s, a culture had developed along the highway. This roadside culture included the motor hotel (or motel), roadside diners, and tourist traps.

So many people were on the road that bigger, faster, wider highways were needed. These highways didn't go through the small towns connected by Route 66. With the new superhighways bypassing them, many of the well-known sights along Route 66 vanished.

ARTIFACT FILE

America's Main Street
Many of the towns, tourist traps, and beauty spots along Route 66 became popular destinations. Route 66 was often called "America's Main Street" because it ran through the centers of the small towns it connected.

Roadside Drive-In
Roadside food stands such as this one in Seligman, Arizona, were found all along Route 66.

On-Line Field Trip

The National Route 66 Museum is in Elk City, Oklahoma. Many such organizations preserve historic landmarks along Route 66. Each of the states along the road has its own Route 66 Association. There is also a National Historic Route 66 Federation.

Visit www.mcdougallittell.com for more information.

CONNECT TO GEOGRAPHY

1. **Movement** As people traveled along Route 66, what goods and services might have spread across the country in their wake?
2. **Human-Environment Interaction** In what ways did the movement of Americans over this highway change the environment?

See Geography Handbook, page 5.

CONNECT TO HISTORY

3. **Drawing Conclusions** In what ways did Route 66 contribute to the American Dream?

Chapter 28 ASSESSMENT

VISUAL SUMMARY

The Cold War and the American Dream

At Home

1940s:
- Truman faces labor unrest.
- Fear of communism spreads.
- Fair Deal is proposed.
- Equal rights for all remains a problem.

1950s:
- The economy booms under Eisenhower.
- McCarthy gains and loses power.
- Billions are spent on space research.
- Rock 'n' roll transforms popular culture.

Abroad

1940s:
- Truman Doctrine is announced.
- Marshall Plan offers aid to Europe.
- Berlin airlift takes place.
- NATO is formed.

1950s:
- Korean War ends in stalemate.
- Dulles practices brinksmanship.
- *Sputnik* is launched.
- Arms race takes place between superpowers.

TERMS & NAMES

Briefly explain the importance of each of the following.

1. Cold War
2. containment
3. Truman Doctrine
4. NATO
5. Korean War
6. brinksmanship
7. space race
8. baby boom
9. sunbelt
10. rock 'n' roll

REVIEW QUESTIONS

Peacetime Adjustments and the Cold War (pages 789–794)

1. How did the federal government help veterans?
2. Why was inflation a bigger problem than recession in the postwar period?
3. Why was the fate of Eastern Europe an issue that divided the Soviet Union from its former allies?
4. How did the Marshall Plan and the formation of NATO reflect Truman's containment policy?

The Korean War and McCarthyism (pages 795–799)

5. Why did the United States become involved in the Korean War?
6. Why were Americans frustrated by the outcome of the Korean War?
7. How was McCarthy able to gain such a powerful hold on the government and the American public?

The Fifties (pages 800–805)

8. What groups were left out of postwar prosperity?
9. What factors boosted the growth of suburbs?
10. Why did Americans become bigger consumers in the 1950s?

CRITICAL THINKING

1. USING YOUR NOTES

ISSUES	PRESIDENTS	
Domestic	Truman	Eisenhower
Labor unions and big business		
Communist threat at home		
Foreign		
Korea		
Communism in Europe		

Using your completed chart, answer the questions below.

a. Which policy or action might have increased the chances of war?

b. What policies or actions might have led to a stalemate?

c. In your opinion, could anything have been done to end the Korean War sooner?

2. APPLYING CITIZENSHIP SKILLS

Was McCarthyism or communism a greater threat to the American way of life? Explain (or support) your opinion.

3. THEME: ECONOMICS IN HISTORY

What factors contributed most strongly to the economic prosperity of the 1950s?

4. ANALYZING LEADERSHIP

Why did Truman consider it his duty as president to fire MacArthur? What might have been the consequences of allowing him to remain in Korea?

5. RECOGNIZING EFFECTS

Soviet and American leaders had different views of the events of war and the challenges of the post-war period. How did these different views contribute to the mistrust and fear of the Cold-War era?

Interact *with* History

How did the American Dream you discussed before you read the chapter compare with the dreams that people actually pursued?

HISTORY SKILLS

1. INTERPRETING GRAPHS

Gross National Product (GNP) means the total value of all goods and services produced by a nation. Study the graph and then answer the questions.

Gross National Product, *1945–1959*

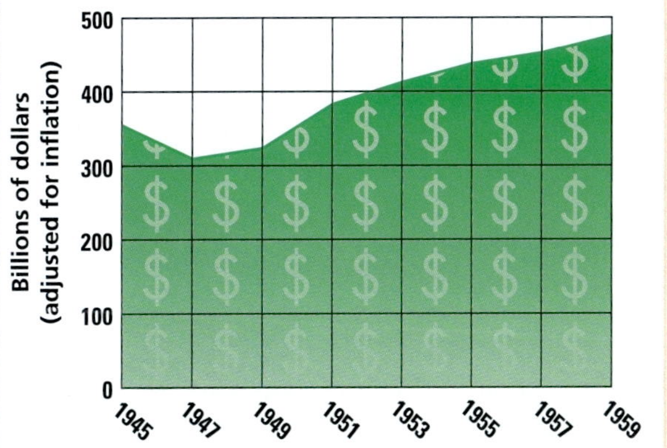

Source: *Historical Statistics of the United States*

a. In dollars, approximately how much did the GNP increase between 1945 and 1951?

b. About how much did the GNP increase between 1951 and 1959?

2. INTERPRETING PRIMARY SOURCES

The following cartoon shows a buyer or perhaps salesman discussing cars with Uncle Sam. Study the cartoon and then answer the questions that follow.

So Russia Launched a Satellite, but Has It Made a Car With Fins Yet?

a. Which is more important for a country's security—consumer goods or space technology?

b. In the opinion of the cartoonist, how might the American Dream be rooted in technology?

ALTERNATIVE ASSESSMENT

1. INTERDISCIPLINARY ACTIVITY: Speech

Debating NATO's Role Since the collapse of the Soviet Union, many former Warsaw Pact nations have wanted to become NATO members. Research and debate the question: What are some of the advantages and disadvantages of expanding NATO?

2. COOPERATIVE LEARNING ACTIVITY

Examining Images of the 1950s Examine images of the American family and family life in the media of the 1950s. Working in groups, look at advertisements in magazines and books of the era, and if possible view TV reruns of 1950s sitcoms. Present your findings as a short TV show about a 1950s family. Groups can prepare for the presentations using these suggestions.

- Make a list of media images you want to present.
- Assign group members roles as writers, actors, directors, and narrator.
- Identify any props or costumes needed.

3. TECHNOLOGY ACTIVITY

Making a Class Presentation Popular culture in the 1950s included music, television, and movies. Using the library and the Internet, find articles and pictures about popular shows, music, and celebrities of the period.

Visit www.mcdougallittell.com to learn more about the Fifties.

Prepare an electronic presentation about one aspect of popular culture of the Fifties. Use these suggestions or a topic of your own.

- Music: lyrics, bands, performers, record companies
- Movies: stars, popular films, drive-ins
- Fashion: clothing, dress, hairstyles, teen life
- Television: stars, shows, viewing habits

4. HISTORY PORTFOLIO

Option 1 Review your section and chapter assessment activities. Select one that you think was your best work. Then use comments made by your teacher or classmates to improve your work, and add it to your portfolio.

Option 2 Review the information that you hoped to acquire for What Do You Want to Know? on page 788. Then write a short report in which you explain the information. If any information seems incomplete, do research to expand it. Add your report to your portfolio.

UNIT 9

1954 to PRESENT

Tensions at Home and Abroad

Dr. Martin Luther King, Jr., waves to demonstrators at the March on Washington in 1963.

"I have a dream that my four little children . . . will not be judged by the color of their skin, but by the content of their character."

—Dr. Martin Luther King, Jr.

CHAPTER 29 The Civil Rights Era 1954–1975

Civil rights marchers sing at the March on Washington in 1963.

Interact *with* History

It is 1960, and you live in a Southern city. For decades, African Americans in the South have endured racial segregation. Now they are protesting against it—in spite of the risk of being attacked. You must decide whether or not you will participate in the protests and in what way.

What Do You Think?

- What policy do the students in the picture hope to change?
- How far would you be willing to go to help the protesters?
- In what ways, besides protesting, could you help to end segregation?

How would you stop injustice in society?

USA

- **1954** The Supreme Court decides *Brown v. Board of Education of Topeka.*
- **1955** Montgomery bus boycott begins.
- **1957** Federal troops are sent to desegregate Little Rock Central High School.
- **1963** The March on Washington takes place. Kennedy is assassinated, and Johnson becomes president.
- **1964** Congress passes Civil Rights Act of 1964.
- **1965** Congress passes the Voting Rights Act.
- **1968** Dr. Martin Luther King, Jr., is assassinated.
- **1970** *La Raza Unida* is founded.
- **1972** Members of AIM occupy the Bureau of Indian Affairs.

World

- **1957** African nation of Ghana wins independence.
- **1962** African National Congress leader Nelson Mandela is imprisoned.
- **1967** Civil war rages in Nigeria.
- **1971** India and Pakistan go to war.

1954 — 1975

BEFORE YOU READ

Previewing the Theme

Democratic Ideals For a century after the Civil War, the United States refused to provide equal rights for African Americans. Chapter 29 tells about the African-American struggle for equality. It also explains how African-American protests for equal rights spread to other groups, including Hispanics, Native Americans, and women.

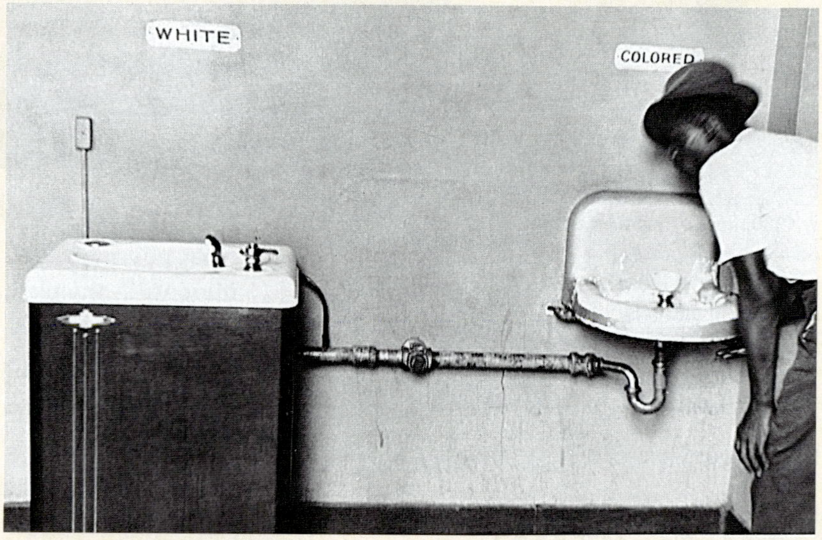

What Do You Know?

What do you think of when people talk about civil rights? Why have some people in the United States been denied their civil rights? Why are civil rights important in a democratic society?

THINK ABOUT

- what you have learned about racism, segregation, and discrimination in earlier chapters
- what you have learned about famous civil rights leaders from your parents or teachers

What Do You Want to Know?

What questions do you have about civil rights? Write those questions in your notebook before you read the chapter.

READ AND TAKE NOTES

Reading Strategy: Analyzing Causes Copy the chart below in your notebook. Use it to take notes on the cause of each event listed in the chart.

See Skillbuilder Handbook, page R10.

Causes	Events
	Brown v. *Board of Education of Topeka*
	Montgomery bus boycott
	Civil Rights Act of 1964
	Voting Rights Act of 1965
	La Raza Unida
	Publication of *The Feminine Mystique,* by Betty Friedan
	The federal government ends its "termination policy."

1 Origins of the Civil Rights Movement

TERMS & NAMES
Thurgood Marshall
Brown v. Board of Education of Topeka
Montgomery bus boycott
Dr. Martin Luther King, Jr.
SCLC
sit-in
SNCC

MAIN IDEA

Changes after World War II helped African Americans make progress in their struggle for equality.

WHY IT MATTERS NOW

The African-American struggle for equality became a model for modern protest movements.

Rosa Parks rides a bus in Montgomery, Alabama, in 1956.

ONE AMERICAN'S STORY

On December 1, 1955, Rosa Parks, an African-American woman from Montgomery, Alabama, boarded a bus to go home after work. Along the way, a group of white people climbed aboard. The bus driver told Parks and a few other African Americans to give up their seats for the whites and move to the back of the bus. All of them except Parks got up to move. She described what happened next.

A VOICE FROM THE PAST

The driver of the bus saw me still sitting there, and he asked was I going to stand up. I said, "No." He said, "Well, I'm going to have you arrested." Then I said, "You may do that."

Rosa Parks, *Rosa Parks: My Story*

The bus driver called the police, who arrested Parks. The arrest angered Montgomery's African-American community. They respected Parks for her long history of fighting for civil rights. She had been an officer in the local chapter of the NAACP. She had also been put off buses in the past for defying segregation. As Section 1 explains, her arrest would spark a movement that began to tear down segregation in America.

Postwar Changes Strengthen Protests

Since the Civil War, African Americans had fought for equality. Their goals included full political rights, better job opportunities, and an end to segregation. But before World War II, they had had little success. Several changes made their efforts more successful after the war.

First, more Americans began to see racism as evil. Racist attitudes had supported the discrimination that oppressed African Americans. Many white Americans saw that racist beliefs had contributed to the rise of Adolf Hitler and the Holocaust. As a result, they began to recognize that racism had no place in the United States.

In addition, the war made African Americans more determined than ever to win equality at home. Having fought for freedom in Europe, African Americans wanted a share of it in the United States, too.

African Americans also gained important resources to help them fight segregation. More blacks had moved into cities to work. They made more money and formed more contacts with one another at work, on the street, and in churches. These changes helped to make the civil rights protests successful.

Background
Churches played a key role in the movement. They offered spiritual and moral support, buildings for meetings, and ministers as leaders.

Brown Overturns *Plessy*

The NAACP, the oldest civil rights organization in the United States, benefited from these changes. Before the war, it had established a fund to pay for legal challenges to segregation. Even so, the "separate but equal" doctrine remained in effect well into the 1950s. This doctrine had been established by *Plessy* v. *Ferguson* in 1896.

In the early 1950s, African Americans in several states sued to end segregation in, or integrate, public schools. Up to this point, white-controlled school boards had provided white children with better schoolhouses and newer books and equipment than they provided to black children. **Thurgood Marshall,** the NAACP counsel, led the attorneys who challenged the segregation laws in the courts.

Vocabulary
integrate: to open to people of all races or ethnic groups; to desegregate

In the early 1950s, the Supreme Court heard these cases under the name ***Brown* v. *Board of Education of Topeka*.** On May 17, 1954, Chief Justice Earl Warren delivered the Court's historic opinion on these cases.

In 1954, Thurgood Marshall persuaded the Supreme Court that racial segregation in public schools was not constitutional.

A VOICE FROM THE PAST

We conclude that in the field of public education the doctrine of "separate but equal" has no place. Separate educational facilities are inherently unequal.

Chief Justice Earl Warren, *Brown* v. *Board of Education of Topeka*

The *Brown* decision was limited to public schools. But many people hoped that it would eventually end segregation in other public facilities. In the meantime, civil rights supporters hoped that black children would receive the same educational opportunities as white children. But the Supreme Court did not say how desegregation was to occur until a year later.

Reading History
A. Making Inferences Why might the *Brown* decision lead to the end of segregation in other public facilities?

At that time, the Court ordered public schools to desegregate "with all deliberate speed." This ruling, which was known as *Brown II,* actually gave segregated school districts *more* time to desegregate. A few places, such as Washington, D.C., desegregated quickly. But in most places, white-controlled schools resisted desegregation.

Montgomery Bus Boycott

In 1955, about six months after the *Brown II* decision, Rosa Parks was arrested, as you read in One American's Story on page 813. News of her arrest quickly reached the other members of her church. The church members issued a notice to other African-American churches and local groups. It said, "If Negroes did not ride the buses, they [the buses] could not operate. We are, therefore, asking every Negro to stay off the buses Monday in protest of the arrest and trial." This protest, called the **Montgomery bus boycott,** began that day.

Background
African Americans made up 70 percent of the riders on Montgomery's buses.

That evening, local NAACP leaders held a meeting to decide whether to continue the boycott. A 26-year-old Baptist minister from Atlanta, Georgia, named **Dr. Martin Luther King, Jr.,** spoke to the group.

A VOICE FROM THE PAST

There comes a time that people get tired. We are here this evening to say to those who have mistreated us so long that we are tired—tired of being segregated and humiliated; tired of being kicked about by the brutal feet of oppression.

Martin Luther King, Jr., quoted in *Stride Toward Freedom*

The church members vowed to continue the boycott. It went on for 13 months. Boycotters, including some whites, organized car pools, rode bikes, or walked to their jobs and schools. King and other leaders endured death threats, bombings, and jailings. The violent reactions of whites to the nonviolent boycott gained the attention of the national media.

Meanwhile, the Montgomery bus segregation law had been challenged in court. On November 13, 1956, the Supreme Court ruled that the law was unconstitutional. African Americans once again boarded the buses in Montgomery. This time they sat wherever they pleased.

The boycott had several important results. First, it ended segregation on Montgomery buses. Second, it led to the founding of the Southern Christian Leadership Conference (**SCLC**). SCLC coordinated civil rights protests across the South. Third, the boycott made Dr. King one of the best-known civil rights leaders in the nation.

***Reading* History**
B. Recognizing Effects What were the most important results of the Montgomery bus boycott?

AMERICA'S
HISTORY MAKERS

MARTIN LUTHER KING, JR.
1929–1968

Fresh out of school, King had been in Montgomery about a year when he became leader of the bus boycott. But his courage and brilliant speaking abilities made him the ideal leader for the civil rights movement.

King learned about nonviolence by studying writers and thinkers such as Mohandas Gandhi. He came to believe that only love could convert people to the side of justice. He described the power of nonviolent resisters: "We will wear you down by our capacity to suffer. And in winning our freedom . . . we will win you in the process."

Why do you think King was well-suited to lead a nonviolent protest?

Massive Resistance

Civil rights victories upset many Southern whites. Polls showed that more than 80 percent opposed school desegregation. Segregationists fought back against African Americans and civil rights organizations. The Ku Klux Klan used beatings, arson, and murder to threaten African Americans who pursued their civil rights.

Many whites, especially among the middle class, organized groups known as White Citizens Councils to prevent desegregation. The opposition of whites to desegregation became known as massive resistance. It was very effective in delaying desegregation.

Showdown in Little Rock

Massive resistance threatened the desegregation of schools in Little Rock, Arkansas, in 1957. Following the *Brown* case, the Little Rock school board made plans to integrate. It called for nine African-American students to enroll at Central High School in September 1957.

Background
The nine African-American students chosen to integrate Central High School became known as the Little Rock Nine.

As the start of the school year neared, segregationists tried to block the integration of the school. Arkansas governor Orval Faubus sided with the segregationists. On September 3, he ordered National Guard troops to prevent the African-American students from entering the school the next morning.

Eight of the students had received phone calls saying someone would drive them to the high school for their safety. When they arrived, the National Guard troops turned them away. The family of the ninth student, Elizabeth Eckford, had no telephone. She took a bus to school alone that morning. When she arrived, a mob of angry whites followed her toward the school's doors.

She saw a guard let some white students pass, so she went up to him. But he did not move out of the way. Later Eckford wrote, "When I tried to squeeze past him, he raised his bayonet. . . . Somebody started yelling, 'Lynch her! Lynch her!'" Finally, a white woman guided Eckford away from the mob and took her home.

Vocabulary
lynch: to execute illegally, especially by hanging

For three weeks, Faubus refused to allow the African-American students into the school—even after meeting with President Dwight Eisenhower. The president did not want to force the governor to obey the law, but he eventually realized that it was his only choice.

HISTORY *through* ART

Elizabeth Eckford faced an angry mob as she walked to Little Rock Central High School on September 4, 1957. She had made her black-and-white dress especially for her first day of school at Central High.

How does the photograph show the emotions stirred by desegregation?

On September 24, Eisenhower ordered the 101st Airborne Division into Little Rock. The Little Rock Nine rode to school, escorted by jeeps armed with machine guns. Paratroopers lined the streets and protected the students as they entered Central High.

Sit-Ins Energize the Movement

Victories like the one in Little Rock encouraged civil rights supporters to continue their fight. In February 1960, four African-American college students began a sit-in to desegregate a lunch counter at a store in Greensboro, North Carolina. A **sit-in** is a protest in which people sit in a place and refuse to move until their demands are met. The students sat down at the lunch counter and ordered coffee. The waitress refused to serve them because they were African Americans.

> ***"The doctrine of 'separate but equal' has no place."***
> From *Brown* v. *Board of Education*

That first day, the students stayed for 45 minutes. They came back each day that week with more protesters. By Thursday, there were more than 100 protesters, including some whites. Over the following weeks, thousands of protesters took part in sit-ins across the South.

Reading **History**
C. Solving Problems How did African Americans end discrimination at many lunch counters?

As the sit-ins spread, segregationists began to abuse the protesters. They covered the protesters with ammonia and itching powder. They yelled at them, beat them, and burned them with cigarettes. Some protesters went to jail. But other protesters replaced them at the counters. The sit-ins were an effective protest tactic. They forced many stores with lunch counters to serve African Americans.

Many civil rights leaders saw the success of the sit-ins and supported an organization for young people. Out of this movement, the Student Nonviolent Coordinating Committee (**SNCC**) was formed. Through SNCC, SCLC, and other groups, the civil rights movement increased the pressure for change in the 1960s, as you will read in the next section.

Section 1 Assessment

1. Terms & Names

Identify:
- Thurgood Marshall
- *Brown* v. *Board of Education of Topeka*
- Montgomery bus boycott
- Dr. Martin Luther King, Jr.
- SCLC
- sit-in
- SNCC

2. Taking Notes

Use a cluster diagram to record details about the early civil rights movement.

3. Main Ideas

a. How did World War II help lead to the civil rights movement?

b. What role did Thurgood Marshall play in challenging segregation?

c. How did Martin Luther King, Jr., become a well-known civil rights leader?

4. Critical Thinking

Contrasting How did the tactics used by civil rights protesters differ from the response of many Southern whites?

THINK ABOUT
- the Montgomery bus boycott
- the events in Little Rock
- the nature of sit-ins

ACTIVITY OPTIONS

ART
TECHNOLOGY

You have been asked to honor people in the civil rights movement. Create a **wall of fame,** or plan a **Web page** that pays tribute to several of them.

2

Kennedy, Johnson, and Civil Rights

TERMS & NAMES
Freedom Ride
CORE
March on Washington
Civil Rights Act of 1964
Freedom Summer
Voting Rights Act
Great Society
Malcolm X

MAIN IDEA

The civil rights movement led to the end of legal segregation.

WHY IT MATTERS NOW

African Americans still face discrimination but now have more opportunities than before.

ONE AMERICAN'S STORY

Jim Zwerg, a white student from Wisconsin, joined a Freedom Ride in May 1961. Freedom Rides were protests against segregation on interstate busing in the South. During the rides, whites would sit in the back of a bus. African Americans would sit in the front and refuse to move. At bus terminals along the route, black riders would try to use "whites only" facilities.

Zwerg and the other freedom riders expected trouble from segregationists on the journey. In Montgomery, Alabama, segregationists savagely attacked the freedom riders.

Zwerg was beaten unconscious. But he was not the only victim. Black or white, man or woman, few freedom riders were spared. In this section, you will read how the Freedom Rides and other protests helped African Americans win support for civil rights.

John Lewis (left) and Jim Zwerg of SNCC are covered with blood after being beaten in Montgomery.

Kennedy and Civil Rights

In 1960, Americans elected a new president. Although civil rights was not the main issue in the campaign, it played an important role. The Democrats nominated John F. Kennedy, a senator from Massachusetts. The Republicans nominated Vice-President Richard Nixon. During the campaign, the candidates had similar positions on most issues. But many Americans thought that Kennedy was more dynamic.

Late in the campaign, police arrested Martin Luther King, Jr., in Georgia. Kennedy called King's wife, and Robert Kennedy, the candidate's brother, arranged for King's release. That one act dramatically increased African-American support for Kennedy. The election was one of the closest in U.S. history, and Kennedy won.

As president, Kennedy had to work with a Congress that was reluctant to act on civil rights issues. In the early 1960s, Southern Democrats supported segregation. Kennedy did not want to anger Southern Democrats because they could weaken his presidency.

Even so, activists continued to pressure the federal government. In May 1961, the Congress of Racial Equality (CORE) planned Freedom Rides to desegregate interstate buses, or buses that travel between states.

Despite attacks on the freedom riders by segregationists along the route, the riders would not give up. Kennedy had to do something. Finally, he sent a group of federal marshals to protect the riders. Four months later, the federal government issued an order integrating interstate bus facilities. The riders had achieved their goal.

Protests in Birmingham

In the early 1960s, the civil rights movement gained strength across the South. African Americans in Birmingham, Alabama, wanted to integrate public facilities and gain better job and housing opportunities. Local civil rights leaders invited King and SCLC to join the protests.

Reading **History**

A. Drawing Conclusions Why did civil rights leaders choose to protest in cities where they were likely to face violence?

Birmingham was a great opportunity for protesters to expose the evils of segregation. They knew that Eugene "Bull" Connor, the city's Public Safety commissioner, was likely to use violence to stop the protests. They also knew that the sight of segregationists attacking nonviolent protesters would increase the pressure for change.

The protests began in April 1963. After about a week, the police arrested King, who had traveled to Birmingham for the protests. From jail, King wrote an eloquent defense of the protests.

A VOICE FROM THE PAST

I guess it is easy for those who have never felt the stinging darts of segregation to say, "Wait." . . . [But] there comes a time when the cup of endurance runs over, and men are no longer willing to be plunged into an abyss [a bottomless pit] of injustice.

Martin Luther King, Jr., "Letter from Birmingham Jail"

SCLC recruited children for the Birmingham marches. The police used dogs and firehoses on the marchers. People across the nation saw this on television and were horrified. Soon, Birmingham's white leaders agreed to

The firehoses used on Birmingham protesters had enough force to tear people's clothes and send small children skidding down the street.

desegregate lunch counters, remove segregation signs, and employ more African Americans in downtown stores.

The March on Washington

The events in Birmingham caused many Americans to support passage of new laws to protect the civil rights of all people. Civil rights organizations planned a huge demonstration in Washington, D.C., to build support for civil rights legislation.

On August 28, 1963, about 250,000 people took part in the **March on Washington,** as the demonstration became known. The march ended at the Lincoln Memorial. The high point of the march came when King delivered his "I Have a Dream" speech. (For a section from King's speech, see page 828.) During this speech, King spoke these famous words.

ReadingHistory

B. Making Inferences Why might marchers have chosen the Lincoln Memorial as the place to end their march?

A VOICE FROM THE PAST

I have a dream that my four little children will one day live in a nation where they will not be judged by the color of their skin, but by the content of their character.

Martin Luther King, Jr., from "I Have a Dream"

The March on Washington united many groups that called for passage of civil rights laws. President Kennedy promised support.

New Civil Rights Laws

Tragically, though, President Kennedy did not live long enough to fulfill this promise. On November 22, 1963, Kennedy and Vice-President Lyndon Baines Johnson went to Texas to campaign. As the presidential motorcade passed through Dallas, thousands of people greeted the president. Suddenly, shots rang out. Kennedy slumped forward; he'd been hit. The president died within an hour.

Background

The man accused of shooting Kennedy, Lee Harvey Oswald, was killed two days later by another assassin.

The tragedy deeply saddened the nation. Schools, factories, and businesses closed as citizens mourned their slain leader. The assassination of President Kennedy would remain a central event in the memories of many Americans in the decades to come.

Lyndon Johnson became president after Kennedy's death. He promised to continue Kennedy's policies. Johnson moved quickly on civil rights. He argued that "no memorial oration or eulogy could more eloquently honor President Kennedy's memory than the earliest possible passage of the civil rights bill." The nation's grief led to broad support for the bill.

In July, the **Civil Rights Act of 1964** was signed into law. The law banned segregation in public places, such as hotels, restaurants, and theaters. It also created the Equal Employment Opportunity Commission to prevent job discrimination. At long last, segregation was officially illegal throughout the United States.

Kennedy's flag-draped casket is drawn through the streets of Washington, D.C., during the slain leader's funeral procession.

Fighting for Voting Rights

White Southerners had long used literacy tests, poll taxes, and violence to keep African Americans from voting. The Civil Rights Act of 1964 barred states from using different voting standards for blacks and whites. In the same year, the states ratified the Twenty-Fourth Amendment. It outlawed poll taxes. Even so, African Americans in the South still found it difficult to vote. As a result, they lobbied Congress to pass a strong voting rights law.

In 1964, SNCC organized a voter-registration drive for Southern blacks. The program was called **Freedom Summer.** It brought Northern college students into Mississippi to work with SNCC organizers. The young volunteers endured bombings, beatings, arrests, and murder while performing their work. Even so, they managed to add about 1,200 African Americans to voter registration rolls.

Reading **History**

C. Drawing Conclusions How did television help to advance the civil rights movement?

Early in 1965, King and SCLC organized voter-registration drives in Selma, Alabama, including a protest march to Montgomery. On March 7, as the marchers crossed a bridge at the edge of Selma, state troopers on horseback attacked them. Americans watched as the violence was broadcast on national television. Pressure for federal action rose.

President Johnson told Alabama Governor George Wallace that he would not tolerate any more violence. When the march to Montgomery resumed, the president sent troops to protect it. He also used the public's anger at the incident to push for action on voting rights.

Reading **History**

D. Reading a Map Use the map below to see how the registration of African-American voters increased throughout the South.

Johnson used the considerable political skills he had acquired when he was a senator to push a voting rights bill through Congress. On August 6, 1965, he signed the **Voting Rights Act** into law. It banned literacy tests and other laws that kept blacks from registering to vote. It also sent federal officials to register voters. Within weeks, the percentage of African Americans in Selma who registered to vote increased from 10 percent to 60 percent.

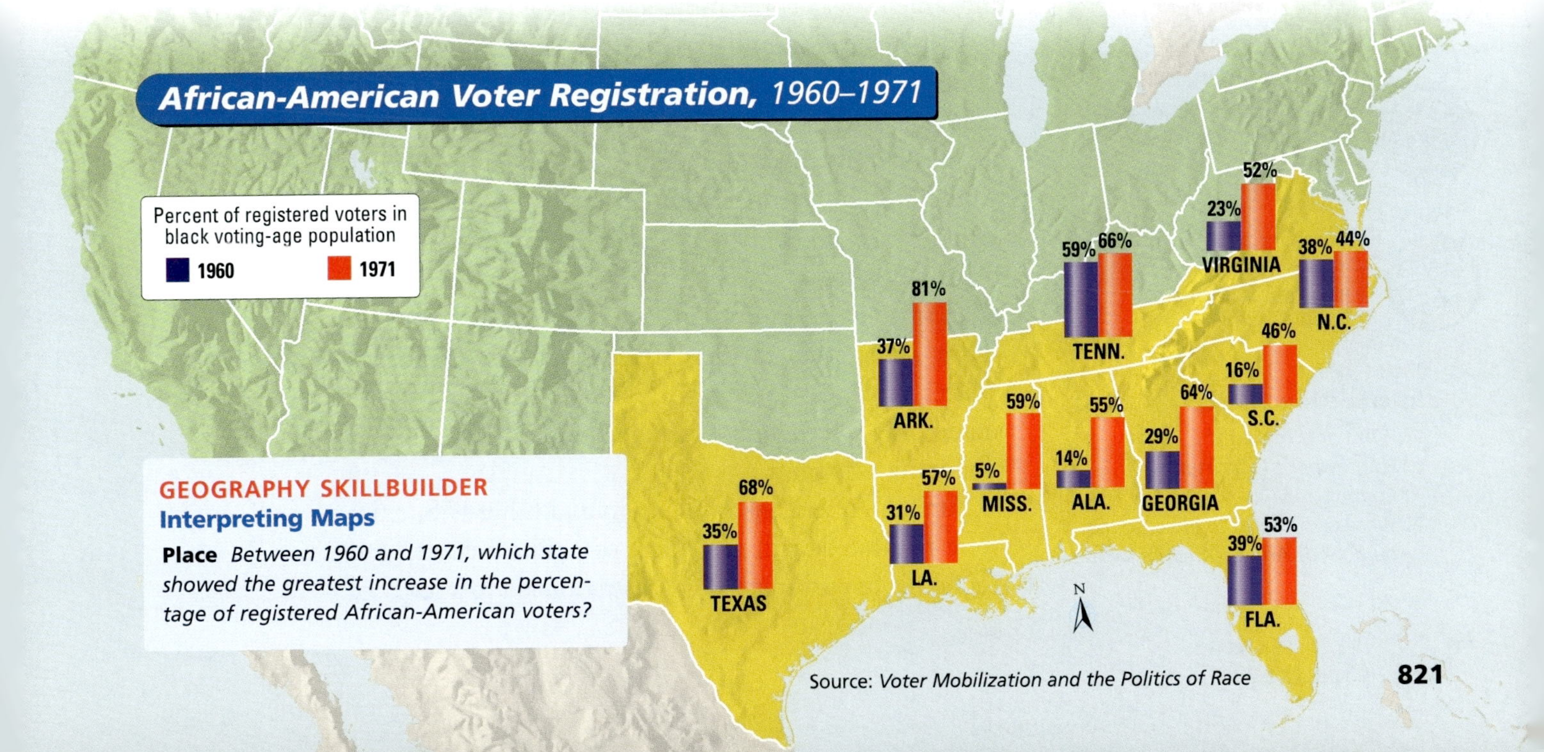

GEOGRAPHY SKILLBUILDER
Interpreting Maps
Place *Between 1960 and 1971, which state showed the greatest increase in the percentage of registered African-American voters?*

Source: *Voter Mobilization and the Politics of Race*

President Johnson signs the Medicare bill into law on July 30, 1965.

Johnson and the Great Society

The Civil Rights Act and the Voting Rights Act were important parts of the reform plan supported by President Johnson. Shortly after taking office, Johnson asked Americans to seek a "great society [that] demands an end to poverty and racial injustice." His program was called the **Great Society**. It provided a series of programs to help the disenfranchised, the poor, the elderly, and women. It also included legislation to promote education, end discrimination, and protect the environment.

Vocabulary
disenfranchised: people deprived of the rights of citizenship, especially the right to vote

Many of the programs were passed and still exist today, such as Medicare and Medicaid. Medicare provides health insurance for senior citizens, while Medicaid provides medical care for the poor.

In addition, Congress passed the Elementary and Secondary School Act, which provided new federal funds for education. Laws were also passed to protect the environment. Congress strengthened the 1960 Clean Water and 1963 Clean Air acts. And it passed legislation to protect endangered species and to preserve millions of acres of wilderness.

The Great Society

Civil Rights Act (1964)
Outlawed discrimination in public places, created the Equal Employment Opportunity Commission, and barred states from using different standards for voter registration for whites and blacks

Voting Rights Act (1965)
Banned literacy tests and used federal registrars to register voters

Medical Care Act (1965)
Established Medicare and Medicaid programs to assist the aged and the poor with medical care

Elementary and Secondary School Act (1965)
Provided federal aid to education

SKILLBUILDER
Interpreting Charts

1. *What act created the Equal Employment Opportunity Commission?*
2. *What acts helped to increase the number of African-American voters?*

Divisions in the Civil Rights Movement

In the late 1960s, with the civil rights laws of Johnson's Great Society already passed, civil rights leaders disagreed about what steps to take next. The SCLC and other organizations wanted to expand the nonviolent struggle. But some groups wanted the movement to become more aggressive.

In 1966, King and the SCLC joined protests in Chicago. In the North, there were no laws that denied African Americans their civil rights—white people simply discriminated against them. Whites would not sell property in certain areas to African Americans, and some white employers refused to hire black workers.

In spite of the protests, white Chicagoans were no more interested in desegregation than Southern whites had been. Chicago Mayor Richard J. Daley made only a few minor changes before SCLC abandoned its campaign.

African Americans in Chicago and other U.S. cities were frustrated with their lack of political power and economic opportunity. This frustration led to a series of riots in the late 1960s. Nationwide, 164 riots broke out in the first 9 months of 1967. Then, on April 4, 1968, Martin Luther King, Jr., was assassinated in Memphis, Tennessee. As the nation mourned the slain civil rights leader, African-American neighborhoods across the country exploded in anger. Over 45 people died in the rioting.

Reading **History**

E. Forming Opinions Why did race riots take place in Northern cities?

Some African Americans had begun to reject nonviolence and cooperation with whites. In 1966, SNCC's black members forced white members out of the organization. Stokely Carmichael, the new leader of SNCC, began to call for "black power." Carmichael and others wanted blacks to create their own organizations under their own control to fight white racism.

Malcolm X speaks at a rally in Harlem in 1963.

The Nation of Islam, a branch of Islam founded in the United States, also urged African Americans to separate from whites. In the 1960s, the Nation was led by Elijah Muhammad, but the group's most popular personality was **Malcolm X.**

By the mid-1960s, Malcolm X rejected the separatist ideas of the Nation of Islam and left the group. During a trip to Mecca, in Saudi Arabia, he had met Muslims of all races. He began to picture a world where all races could live together in peace. But he had little time to spread his new message. In 1965, he was gunned down by members of the Nation of Islam.

Background

Mecca is the holiest city for Muslims.

From the late 1960s on, civil rights progress came slowly for African Americans. But the African-American struggle for equality encouraged civil rights movements among other oppressed groups, as you will read in the next section.

Section 2 Assessment

1. Terms & Names

Identify:
- Freedom Ride
- CORE
- March on Washington
- Civil Rights Act of 1964
- Freedom Summer
- Voting Rights Act
- Great Society
- Malcolm X

2. Taking Notes

Use a time line like the one shown to record important events of the civil rights movement.

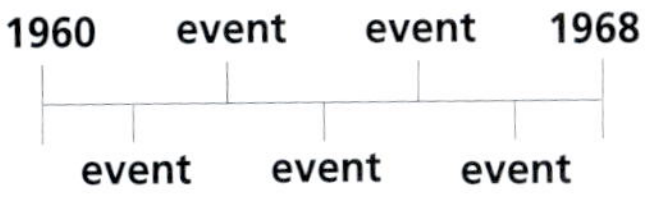

In what year was the voter registration drive in Selma?

3. Main Ideas

a. Why did civil rights workers believe that Birmingham was a good place to protest?

b. How did civil rights workers fight to improve African-American voting rights?

c. Why did the movement begin to break apart?

4. Critical Thinking

Making Inferences Why do you think African Americans placed so much importance on the right to vote?

THINK ABOUT
- who and what they might want to vote for
- what they were willing to endure to win voting rights
- how Southern whites kept them from voting

ACTIVITY OPTIONS

ART

SPEECH

Imagine that you are taking part in the March on Washington. Design a **poster** you could carry, or deliver a **speech** in favor of civil rights.

3

The Equal Rights Struggle Expands

TERMS & NAMES
Cesar Chavez
National Congress of American Indians
Betty Friedan
NOW
ERA

MAIN IDEA

The African-American struggle for equality inspired other groups to fight for equality.

WHY IT MATTERS NOW

Nonwhites and women continue to fight for equality today.

Cesar Chavez, head of the National Farm Workers Association, marches with striking grape pickers in the 1960s. (*Huelga* is the Spanish word for strike.)

ONE AMERICAN'S STORY

Cesar Chavez was born in Yuma, Arizona, in 1927. In the 1940s, he and his family worked as migrant laborers in the California fields. (Migrant workers travel from place to place in search of work.) One time, they found work picking peas. The whole family, parents and six children, worked. Chavez described the poor pay for such hard work.

A VOICE FROM THE PAST

They [the managers] would take only the peas they thought were good, and they only paid you for those. The pay was twenty cents a hamper, which had to weigh in at twenty-five pounds. So in about three hours, the whole family made only twenty cents.

Cesar Chavez, *Cesar Chavez: Autobiography of* La Causa

In 1962, Chavez decided to start a union for farm workers. But the owners refused to recognize the union. Chavez used the example set by Martin Luther King, Jr., to change their minds.

Responding to Chavez's call, workers went on strike. Then Chavez asked people not to buy produce harvested by nonunion workers. The tactics worked. In 1970, 26 major California growers signed a contract with the union. It gave the workers higher wages and new benefits. The victory of Chavez and his union showed how the fight for equal rights spread beyond African Americans, as you will read in this section.

Mexican Americans Organize

The farm workers' struggle inspired other Mexican Americans. By the 1960s, most Mexican Americans lived in cities in the Southwest and California. In 1970, Mexican Americans formed *La Raza Unida* (lah RAH•sah oo•NEE•dah)—"the united people." *La Raza* fought for better jobs, pay, education, and housing. It also worked to elect Mexican Americans to public office.

Mexican-American students also began to organize. They wanted reform in the school system. The students demanded such changes as

Background
Many Mexican Americans prefer to be called Chicanos, or Chicanas if they are female.

better facilities, more courses on the Mexican-American experience, and more Mexican-American teachers.

In 1968, students in Los Angeles walked out of classes to press their demands. At first, school authorities reacted harshly to the walkouts and arrested many of the protesters. Even so, they eventually admitted to the poor conditions of many schools and met with protesters to discuss solutions. By the early 1970s, many of the reforms that the students demanded had been made.

Hispanic Diversity

Hispanics, including Mexican Americans, trace their roots to Spanish-speaking countries and cultures. Because these countries are commonly known as Latin America, some people from these areas refer to themselves as Latinos.

Because Hispanic Americans come from many different countries, they sometimes have little in common. For example, among Mexican Americans, immigration and citizenship are important issues. Puerto Ricans, however, are already U.S. citizens and are not troubled by such issues.

Reading **History**
A. Summarizing What are the factors that unite and separate Hispanics?

Similarly, many Cubans came to the United States as political refugees after Communists took power in Cuba. They tend to be more politically conservative than other Hispanics. Such differences make it difficult for Hispanic Americans to achieve political unity.

Native Americans Unite

Native Americans, like Hispanics, often had difficulty uniting to address common problems. But in the 1950s, that began to change. In 1953, the federal government began a "termination policy" that ended federal protection of land and other assets held by Native American tribes. One result of this policy was the decline of traditional Native American cultures.

Native Americans protested against these policies. The **National Congress of American Indians** (NCAI)—founded in 1944 to promote the "common welfare" of Native Americans—led the protests. Under pressure, the federal government changed the policy in 1958.

The success of these protests inspired a new generation of Native American activists to fight for their rights. In 1961, more than 400 Native Americans from dozens of tribes met in Chicago. They issued a statement they titled the Declaration of Indian Purpose. In it, they demanded the "right to choose our own way of life" and the "responsibility of preserving precious heritage."

In 1968, a group of Native Americans founded the American Indian Movement (AIM). AIM was more aggressive than other organizations in demanding rights for people on reservations and greater recognition of

America's **HERITAGE**

CINCO DE MAYO

During the civil rights era, many Americans learned to appreciate the variety of cultural traditions that make their nation unique.

For example, many Americans now join in the celebration of the Mexican holiday Cinco de Mayo (5th of May). This holiday commemorates the 1862 victory of Mexican troops over the French Army in Puebla, Mexico.

The holiday reflects the national pride of Mexicans over the defeat of a superior fighting power. But the holiday has also taken on a broader meaning for all Americans as a celebration of the right of all people to self-determination. A Cinco de Mayo gathering is shown below.

Native Americans petition for the return of tribal artifacts at the state capital in Albany, New York, in 1970.

tribal laws. In 1972, members of AIM occupied the Bureau of Indian Affairs in Washington, D.C., for seven days. Russell Means, one of the group's leaders, declared, "We don't want civil rights in the white man's society—we want our own sovereign rights."

In the early 1970s, Native Americans protested to force the government to provide them with more federal aid. In addition, the Indian Self-Determination Act of 1975 gave tribal governments more control over social programs, law enforcement, and education. Other victories came through winning court cases. Native Americans won back some of their lands. They have also gone to court over rights to water, hunting, and fishing.

Reading **History**

B. Comparing Did any African-American organizations during the civil rights era have goals similar to AIM?

The Women's Movement

In the 1960s, women also demanded equal rights. Early in the decade, women were kept out of many jobs. They faced discrimination in many male-dominated businesses. For example, there were few female police officers. The military also limited the jobs open to women.

Women also had limited legal rights. Married women, for example, faced problems in signing contracts, selling property, and getting credit. A woman could lose her job if she became pregnant. In addition, society pressured women to quit their jobs when they married. Women who wanted to work at jobs outside their homes were seen as "unnatural." **Betty Friedan** described the problems women faced in her 1963 book, *The Feminine Mystique.*

A VOICE FROM THE PAST

We can no longer ignore that voice within women that says: "I want something more than my husband and my children and my home."

Betty Friedan, *The Feminine Mystique*

Friedan's words helped give direction to a movement for women's liberation. In 1966, Friedan helped to found the National Organization for Women (**NOW**). Some of NOW's major goals were to help women get good jobs and equal pay for their work.

In response to women's groups, Congress passed the Equal Rights Amendment (**ERA**) in 1972 and sent it to the states for ratification. The proposed amendment stated, "Equality of rights under the law shall not be denied or abridged by the United States or any State on account of sex." For the amendment to be added to the Constitution, 38 of the 50 states had to ratify it.

Background

An Equal Rights Amendment was initially proposed in 1923, after women gained the right to vote.

The amendment faced strong opposition, even from some women. Phyllis Schlafly, ERA's most famous opponent, argued that it would

Women's Rights Leaders, *1972*

In July 1972, women's rights leaders met in Washington, D.C., to demand that women play a larger role in the upcoming presidential conventions.

1. **Bella Abzug** was a U.S. representative from New York.
2. **Gloria Steinem** was the founding editor of *Ms.* magazine.
3. **Shirley Chisholm** was a U.S. representative from New York and ran for president in 1972.
4. **Betty Friedan** wrote *The Feminine Mystique* and was the cofounder and first president of NOW.

destroy American families and that the problems of feminists were not the government's business. She wrote, "It would be a tragic mistake for our nation to succumb to the tirades and demands of a few women who are seeking a constitutional cure for their personal problems."

Reading **History**

C. Analyzing Causes Why wasn't the ERA added to the Constitution even though it was passed in 1972?

In 1982, the extended deadline for ratification passed. The amendment died because it fell three states short of ratification. Other reforms, however, reduced the inequality between women and men. Sections of the Civil Rights Act of 1964 and the Higher Education Act of 1972 outlawed discrimination against women. These two laws helped to expand opportunities for women in education, sports, and the workplace.

The civil rights movements that followed World War II greatly changed life in the United States. As you will read in the next chapter, the war in Vietnam also changed the nation.

Section 3 Assessment

1. Terms & Names

Identify:

- Cesar Chavez
- National Congress of American Indians
- Betty Friedan
- NOW
- ERA

2. Taking Notes

Use a chart like the one shown to record important details about the struggle for equal rights.

Mexican Americans	Native Americans	Women

How do the positions women hold today reflect changes won in the civil rights era?

3. Main Ideas

a. What was *La Raza Unida*, and what did it do?

b. What was the Declaration of Indian Purpose?

c. How did Betty Friedan help to launch the women's liberation movement?

4. Critical Thinking

Analyzing Points of View
What were the different opinions about the ERA?

THINK ABOUT

- what NOW and other women's groups would have thought of it
- what Phyllis Schlafly thought of it

ACTIVITY OPTIONS

TECHNOLOGY
LANGUAGE ARTS

Plan part of a **multimedia presentation** focusing on one of the groups mentioned in this section, or write a **pamphlet** explaining that group's goals.

I Have a Dream

Setting the Stage On August 28, 1963, Martin Luther King, Jr., gave his most famous speech at the March on Washington. In it, he shared his dream of equality for all. **See Primary Source Explorer**

I am happy to join with you today in what will go down in history as the greatest demonstration for freedom in the history of our nation.

Five score years ago, a great American, in whose symbolic shadow we stand today, signed the Emancipation Proclamation. . . . But one hundred years later, the Negro still is not free. One hundred years later, the life of the Negro is still sadly crippled by the **manacle**[1] of segregation and the chains of discrimination.

So we've come here today to dramatize a shameful condition. . . .

I say to you today, my friends, that even though we face the difficulties of today and tomorrow, I still have a dream. It is a dream deeply rooted in the American dream.

I have a dream that one day this nation will rise up and live out the true meaning of its **creed**[2] —we hold these truths to be self-evident that all men are created equal.

I have a dream that my four little children will one day live in a nation where they will not be judged by the color of their skin but by the content of their character.

I have a dream today!

This is our hope. This is the faith that I will go back to the South with. . . . With this faith we will be able to work together, to pray together, to struggle together, to go to jail together, to stand up for freedom together, knowing that we will be free one day. This will be the day, this will be the day when all of God's children will be able to sing with new meaning "My country 'tis of thee, sweet land of liberty, of thee I sing. Land where my fathers died, land of the Pilgrim's pride, from every mountainside, let freedom ring!" And if America is to be a great nation, this must become true.

And when this happens, when we allow freedom to ring, when we let it ring from every tenement and every hamlet, from every state and every city, we will be able to speed up that day when all of God's children, black men and white men, Jews and **Gentiles,**[3] Protestants and Catholics, will be able to join hands and sing in the words of the old Negro spiritual, "Free at last, free at last. Thank God Almighty, we are free at last."

Martin Luther King, Jr.

1. **manacle:** handcuff.
2. **creed:** statement of belief.
3. **Gentile:** Non-Jewish person.

A CLOSER LOOK

ABRAHAM LINCOLN

In his speech, King made references to Abraham Lincoln. He specifically referred to the Emancipation Proclamation and used the phrase "Five score years ago" to remind his listeners of the opening of the Gettysburg Address.

1. Why do you think King would refer to Lincoln's speeches?

A CLOSER LOOK

THE AMERICAN DREAM

The American dream can mean different things to different people. Usually, however, it refers to the freedom and opportunity for Americans to lead their own lives.

2. How do civil rights fit into the American dream?

A CLOSER LOOK

GOING TO JAIL

Dr. King—who was arrested 30 times for civil rights activities—said that, with the faith that some day all men will be treated as equals, people will be ready to go to jail together.

3. Why do you think civil rights workers were willing to go to jail?

An Open Letter

Setting the Stage In 1969, Cesar Chavez wrote a letter in which he denied accusations that he had used violence to win decent wages and better benefits for farm workers. **See Primary Source Explorer** CD-ROM

Today . . . we remember the life and sacrifice of Martin Luther King, Jr., who gave himself totally to the nonviolent struggle for peace and justice. In his letter from Birmingham Jail, Dr. King describes better than I could our hopes for the strike and boycott: "Injustice must be exposed, with all the tension its exposure creates, to the light of human conscience and the air of national opinion before it can be cured." For our part, I admit that we have seized upon every tactic and strategy consistent with the morality of our cause to expose that injustice and thus to heighten the sensitivity of the American conscience so that farmworkers will have without bloodshed their own union and the dignity of bargaining with their **agribusiness**[1] employers. . . .

Our strikers here in **Delano**[2] and those who represent us throughout the world are well trained for this struggle. . . . They have been taught not to lie down and die or to flee in shame, but to resist with every ounce of human endurance and spirit. To resist not with retaliation in kind but to overcome with love and compassion, with **ingenuity**[3] and creativity, with hard work and longer hours, with stamina and patient **tenacity**[4], with truth and public appeal, with friends and allies, with mobility and discipline, with politics and law, and with prayer and fasting. They were not trained in a month or even a year; after all, this new harvest season will mark our fourth full year of strike and even now we continue to plan and prepare for the years to come. . . .

We shall overcome and change it not by retaliation or bloodshed but by a determined nonviolent struggle carried on by those masses of farmworkers who intend to be free and human.

Cesar E. Chavez

A CLOSER LOOK

AGRIBUSINESS

Farm workers were excluded from the National Labor Relations Act of 1935—the law that gives most Americans the right to organize a union. By 1975, pressure from Chavez had convinced lawmakers in California to allow farm workers in the state to organize unions.

1. Why do you think farm workers wanted to organize a union?

A CLOSER LOOK

TRAINING FOR PROTESTS

During the civil rights era, protesters often received extensive training before participating in marches, demonstrations, and sit-ins. Chavez explains that farm workers were also trained for the struggle.

2. Why do you think that training for nonviolent protest might be necessary?

1. **agribusiness:** farming as a large-scale business operation.
2. **Delano:** a farming city in California.
3. **ingenuity:** imagination or cleverness.
4. **tenacity:** persistence.

Interactive Primary Sources Assessment

1. Main Ideas

a. Why does King declare that the United States is not living up to its creed?

b. What does King say must happen before America can be considered a truly great nation?

c. Why do you think Cesar Chavez refers to King in his speech?

2. Critical Thinking

Comparing and Contrasting In what ways were the problems that King and Chavez wrote about similar and different?

THINK ABOUT

- the kinds of discrimination they faced
- their hopes for the future

Chapter 29 ASSESSMENT

VISUAL SUMMARY

The Civil Rights Era

1954

1954
Brown v. *Board of Education of Topeka*

1955
The Montgomery bus boycott begins.

1957
Federal troops are sent to desegregate Little Rock Central High School.

1960
John F. Kennedy is elected president.

1961
Native Americans issue the Declaration of Indian Purpose.

1962
Cesar Chavez starts a union for farm workers.

1963
March on Washington

Kennedy is assassinated, and Johnson becomes president.

1964
Civil Rights Act

1965
Voting Rights Act

Malcolm X is assassinated.

1968
Martin Luther King, Jr., is assassinated.

1970
La Raza Unida is founded.

1972
Congress passes the Equal Rights Amendment, but it is never ratified by the states.

1975

TERMS & NAMES

Briefly explain the importance of each of the following.

1. *Brown* v. *Board of Education of Topeka*
2. Montgomery bus boycott
3. Martin Luther King, Jr.
4. March on Washington
5. Civil Rights Act of 1964
6. Voting Rights Act
7. Great Society
8. Cesar Chavez
9. National Congress of American Indians
10. ERA

REVIEW QUESTIONS

Origins of the Civil Rights Movement (pages 813–817)

1. What factors helped to give strength to the demands of the civil rights movement?
2. What were the immediate and long-term effects of *Brown* v. *Board of Education of Topeka?*
3. How did white people react to civil rights protests?

Kennedy, Johnson, and Civil Rights (pages 818–823)

4. What factors made it difficult for Kennedy to act on civil rights?
5. Why did Congress eventually pass civil rights legislation?
6. What effects did Johnson's Great Society legislation have?

The Equal Rights Struggle Expands (pages 824–829)

7. How did farm workers participate in the equal rights movement?
8. What challenges did Hispanics face in their civil rights struggle?
9. Why did Native Americans protest U.S. government policy?
10. What kinds of discrimination did women challenge during the civil rights era?

CRITICAL THINKING

1. USING YOUR NOTES

Using your completed chart, answer the questions below.

Causes	Events
	Brown v. *Board of Education of Topeka*
	Montgomery bus boycott
	Civil Rights Act of 1964
	Voting Rights Act of 1965
	La Raza Unida
	Publication of *The Feminine Mystique*, by Betty Friedan
	The federal government ends its "termination policy."

a. What causes resulted in the Supreme Court decision *Brown* v. *Board of Education of Topeka?*

b. What forces led Betty Friedan to write *The Feminine Mystique?*

2. ANALYZING LEADERSHIP

What qualities do you think made Martin Luther King, Jr., an effective leader?

3. THEME: DEMOCRATIC IDEALS

How did participants in the civil rights movement advance the democratic ideals of the United States?

4. APPLYING CITIZENSHIP SKILLS

How did the nonviolent methods used by protesters during the civil rights movement demonstrate good citizenship?

5. MAKING GENERALIZATIONS

In what ways do you think the lives of Americans today might differ from the lives of Americans who lived before the civil rights movement?

Interact *with* History

How has your study of the civil rights era influenced your decision about the ways in which you would act to change injustices in society?

HISTORY SKILLS

1. INTERPRETING GRAPHS

Study the graph. Answer the questions.

Poverty in America, *1962–1974*

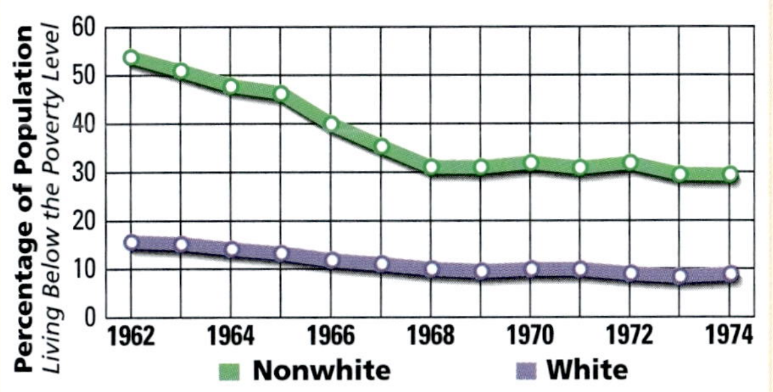

Source: *Statistical Abstract of the United States, 1968–1975*

Basic Graph Elements

a. What does the graph show?

Interpreting the Graph

b. Which group of people does the chart show as having a lower rate of poverty?

c. What was the difference between the poverty rates of whites and nonwhites in 1974?

2. INTERPRETING PRIMARY SOURCES

In 1976, Anthony Gauthier painted *Freedom*. Examine the painting and answer the questions.

a. Why do you think the artist showed a Native American with Martin Luther King, Jr.?

b. How did the artist change the U.S. flag to make a statement?

ALTERNATIVE ASSESSMENT

1. INTERDISCIPLINARY ACTIVITY: Language Arts

Writing Your Representative The House of Representatives is considering the Voting Rights Act. Your representative is against the act because it will reduce his or her chances for reelection. Write a letter trying to persuade your representative to support the bill.

2. COOPERATIVE LEARNING ACTIVITY

Organizing a March With a small group, think of a way in which you could make your community a better place for everyone to live. Then organize a march that will raise awareness of the group's plan. Remember to consider the following:

- How will you recruit and organize marchers?
- What would be the best route for the march?
- Is permission from local officials required to march?
- Will you need to consider food and sanitation?

3. CD-ROM PRIMARY SOURCE EXPLORER

Creating a Radio Special Many people say that the civil rights movement reached its climax on August 28, 1963. On that day, about 250,000 people arrived in Washington, D.C., to peacefully protest racial discrimination and show support for civil rights legislation. Use the CD-ROM, Internet, books, and other resources to create a special radio broadcast on the March on Washington. Consider including the following content:

- descriptions of and sound bytes from marchers
- the recollections of participants in the march
- recordings of speeches made at the gathering
- different newspaper accounts of the march

4. HISTORY PORTFOLIO

Option 1 Review your section and chapter assessment activities. Select one that you think is your best work. Then use comments made by your teacher or classmates to improve your work, and add it to your portfolio.

Option 2 Review the questions that you wrote for What Do You Want to Know? on page 812. Then write a short report in which you explain the answers to your questions. If any questions were not answered, do research to answer them. Add your answers to your portfolio.

CHAPTER 30

The Vietnam War Years 1954–1975

Visitors honor the dead at the Vietnam Veterans Memorial in Washington, D.C.

Interact *with* History

Prowar rally in New York in 1970

Antiwar rally in Washington in 1969

You are a young person in 1969. Your country is at war to stop Communists from taking over South Vietnam. College students have organized huge protests against the war and the draft. Many people think such protests are unpatriotic and an insult to the soldiers who are fighting.

What Do You Think?

- Should the United States try to stop the spread of communism in Vietnam?
- Why are so many people against the war?
- Is it unpatriotic to criticize the government? To refuse army service?

Would you support the war?

USA

1954 United States gives economic aid to South Vietnam.

1960 John F. Kennedy is elected president.

1963 Lyndon B. Johnson becomes president after Kennedy's assassination.

1964 Johnson is reelected president.

Tonkin Gulf Resolution is passed.

1965 First U.S. ground troops go to Vietnam.

1968 Richard M. Nixon is elected president.

1969 President Nixon begins troop withdrawals from Vietnam.

1970 Nixon orders invasion of Cambodia.

Four students are killed during an antiwar protest at Kent State University.

1973 U.S. involvement in war ends.

World

1954 Vietnam is divided into North and South.

1960 National Liberation Front (Viet Cong) organizes in South Vietnam.

1962 Soviet Union places nuclear missiles in Cuba.

1966 Communist leader Mao Zedong begins Cultural Revolution in China.

1968 Viet Cong launch Tet offensive.

Soviets invade Czechoslovakia to stop reforms.

1975 South Vietnam surrenders to Communists.

1954 — **1975**

Chapter 30 SETTING THE STAGE

BEFORE YOU READ

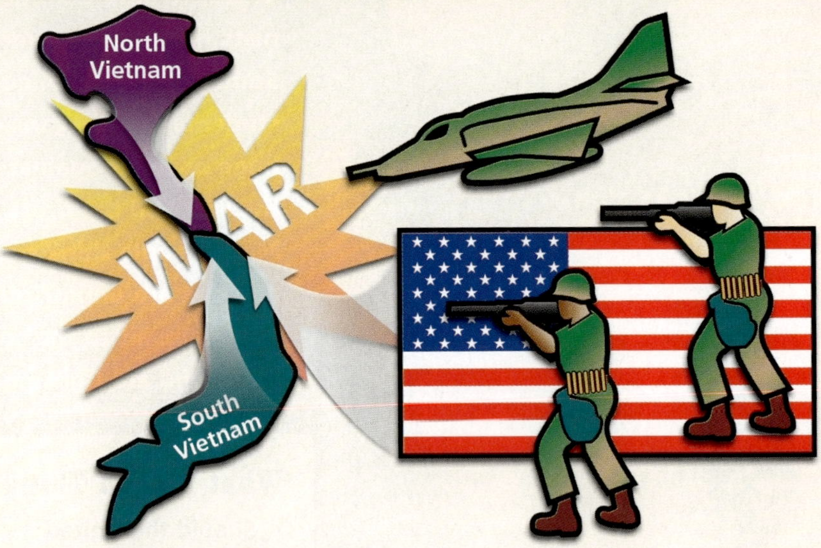

Previewing the Theme

America in the World The United States became involved in the Vietnam War to stop the spread of communism in Asia. As more U.S. troops were sent there, doubts grew about whether they could achieve their mission. Eventually, the United States withdrew its soldiers and became more cautious about fighting foreign wars. Chapter 30 explains the history and legacy of the war in Vietnam.

What Do You Know?

What images spring to mind when you hear the word *Vietnam?* What have you learned about the Vietnam War from movies, books, or relatives' stories? Have you seen the Vietnam Veterans Memorial or any other Vietnam memorials?

THINK ABOUT

- whether Americans seem proud of or ashamed of the war
- how this war differed from World War II or other wars you know about

What Do You Want to Know?

In your notebook, write down any questions you have about the war. Later, note the answers if you learn them in this chapter.

READ AND TAKE NOTES

Reading Strategy: Analyzing Points of View The Vietnam War bitterly divided Americans. Presidents, military strategists, ordinary soldiers, and college students, among others, all had their reasons for supporting or opposing the war. On a chart like the one below, note these reasons as you read. This will help you understand how the war split the country and why Americans still argue over it today.

See Skillbuilder Handbook, page R8.

Reasons for Involvement in Vietnam
French alliance
Domino theory
Nation-building
Cold War crises
U.S. weaponry

Reasons Against Involvement in Vietnam
U.S. interests
Draft
Social programs
Vietnamese civilians
Domestic unrest

1 Cold War Roots of the Conflict

TERMS & NAMES
Ngo Dinh Diem
French Indochina
Ho Chi Minh
domino theory
Viet Cong
Ho Chi Minh Trail
Cuban missile crisis

MAIN IDEA

The United States entered the Vietnam War to stop the spread of communism.

WHY IT MATTERS NOW

The United States still becomes involved in foreign struggles for political reasons.

ONE AMERICAN'S STORY

Edward Lansdale, a U.S. military officer, went to South Vietnam in June 1954. His mission: to stop the spread of communism in Vietnam. He would try to do this by helping the non-Communist government of South Vietnam resist being taken over by Communist North Vietnam. Lansdale believed that the United States could defeat communism by "exporting the American way" and "winning the hearts and minds of the people" with generous economic and military aid.

Lansdale became a trusted adviser to <u>Ngo Dinh Diem</u> (uhng•oh dihn zih•ehm), the leader of South Vietnam. At Lansdale's urging, the United States helped support Diem's unpopular government. Within a year, Lansdale reported, "The Free Vietnamese are now becoming unified and learning how to cope with the Communist enemy."

Lansdale was too optimistic. U.S. involvement in Vietnam grew into the longest war the United States ever fought—and one in which it failed. The war would deeply divide not only the Vietnamese but also Americans. In this section, you will learn how the United States first became involved in Vietnam.

Edward Lansdale was one of the earliest U.S. military advisers sent to Vietnam.

Vietnam After World War II

From the late 1800s until World War II, France ruled Vietnam as part of its colony of **<u>French Indochina</u>.** The colony also included neighboring Laos (LAH•ohs) and Cambodia. (See the map on page 837.) During this colonial period, France increased its wealth by exporting rice and rubber from Vietnam. But Vietnamese peasants lost their land and grew poor.

The Vietnamese never accepted French rule. Various groups of nationalists, who wanted Vietnam to become an independent nation, staged revolts against the French. In 1930, a revolutionary leader named **<u>Ho Chi Minh</u>** (hoh chee mihn) united three Communist groups to form the Indochinese Communist Party (ICP). This new party called for an independent Vietnam controlled by peasants and other workers.

Ho Chi Minh speaks to a French audience in 1946.

The Indochinese Communist Party organized protests by peasants against the French government. The French responded by arresting suspected Communists and executing a number of leaders. Ho Chi Minh, who was living in China, was sentenced to death without being present.

In 1940, during World War II, Japan took over Indochina. The next year, Ho Chi Minh secretly returned to Vietnam and hid in a jungle camp. Under his direction, the ICP joined with other nationalists to form an organization called the Viet Minh. The Viet Minh trained soldiers to fight to make Vietnam independent of all foreign rulers. Because Japan was an enemy of the United States in World War II, the U.S. government aided Ho Chi Minh and the Viet Minh in their fight against the Japanese.

After the Japanese surrendered to the Allies in August 1945, Ho Chi Minh declared Vietnam's independence before a cheering crowd in Hanoi. But France soon tried to regain control of Vietnam. Ho Chi Minh sought a peaceful solution to the conflict with France.

Reading **History**

A. Analyzing Causes What was the original source of the conflict in Vietnam?

A VOICE FROM THE PAST

If they force us into war, we will fight. The struggle will be atrocious [terrible], but the Vietnamese people will suffer anything rather than renounce [give up] their freedom.

Ho Chi Minh, quoted in *Vietnam: A History* by Stanley Karnow

Reading **History**

B. Reading a Map Use the map on page 837 to find Haiphong and Hanoi.

In 1946, war broke out between the Viet Minh and France. The French bombed Haiphong, and the Viet Minh attacked Hanoi.

Truman and Eisenhower Aid the French

As France fought to hold on to power in Vietnam, the United States struggled against the Soviet Union in the Cold War. President Truman followed a policy of containment, working to prevent the spread of communism in Western Europe.

In the fall of 1949, Communists gained control of China. This event made American leaders worry about the spread of communism in Asia. When France asked the United States for aid to help them fight the Viet Minh, the United States agreed. One reason was that U.S. leaders needed French support in opposing the Soviets in Europe. Another reason was that the United States did not want Vietnam to become Communist.

The United States entered the conflict in Vietnam in 1950, when President Truman offered $10 million in military aid to the French. After Dwight D. Eisenhower became president in 1953, he continued aiding the French war effort in Vietnam.

Reading **History**

C. Analyzing Points of View Why did Truman and Eisenhower support the French in Vietnam?

Both Truman and Eisenhower used the **domino theory** to explain the need to support anti-Communists in Vietnam. According to this theory, if a country fell to communism, nearby countries would also topple, like a row of dominoes standing on end. U.S. leaders feared that if Vietnam became Communist, the rest of Southeast Asia would follow.

Dividing North and South

Even with limited U.S. support, France could not defeat the Viet Minh. In 1954, the Viet Minh overran French forces at Dien Bien Phu, in northwestern Vietnam. In May 1954, France met with the Viet Minh for peace talks in Geneva, Switzerland. The two sides reached an agreement called the Geneva Accords. This agreement divided Vietnam into North and South along the 17th parallel, or at 17°N latitude. Surrounding this line was a demilitarized zone, or DMZ. The split was meant to be temporary, however. The two sides agreed to hold elections in 1956 for a single government that would reunify the country.

Until then, the Geneva Accords allowed for separate governments in the North and the South. Ho Chi Minh and the Communists controlled North Vietnam. Ngo Dinh Diem, an anti-Communist, became prime minister and, later, president of South Vietnam. Thousands of anti-Communists from the North fled to the South. The United States provided ships for their transportation.

North Vietnamese Communists carry supplies to South Vietnamese Communist rebels.

GEOGRAPHY SKILLBUILDER Interpreting Maps

1. **Location** *What country lies along the northern border of North Vietnam?*
2. **Location** *What cities on the map are in North Vietnam? In South Vietnam?*

Ho Chi Minh enjoyed great popularity in North Vietnam, while Diem had little support from the people of South Vietnam. As a result, Diem refused to hold national elections in 1956. President Eisenhower supported him, later saying, "If the elections had been held in 1956, Ho Chi Minh would have won 80% of the vote."

Instead, Eisenhower sent more aid and advisers to South Vietnam to help the Diem government. U.S. advisers described their mission as "nation-building."

The Viet Cong Oppose Diem

In spite of U.S. aid, Diem did not establish a democratic government in South Vietnam. Instead, his government was corrupt. In the countryside, for example, he let landlords take back land given to peasants. In addition, he jailed, tortured, and killed opponents.

Reading **History**
D. Making Inferences Why was the Diem government unpopular?

Diem's opponents included South Vietnamese Communists. In 1960, they joined with other dissatisfied South Vietnamese to form the National Liberation Front. Diem ridiculed the group by calling them the **Viet Cong,** for Vietnamese Communists. This name became the commonly used term for the group.

The Viet Cong fought to overthrow the Diem government and reunite the country under Communist rule. North Vietnam supported the Viet Cong, sending soldiers and supplies along a network of paths called the **Ho Chi Minh Trail.** This supply line wove through the jungles and mountains of neighboring Laos and Cambodia. By 1963, when John F. Kennedy was in the White House, the Viet Cong were close to victory.

Reading **History**
E. Reading a Map Find the Ho Chi Minh Trail on the map on page 837.

Kennedy Faces Communist Threats

President Kennedy continued to send military advisers and equipment to South Vietnam. By late 1963, the United States had more than 16,000 military personnel there. Kennedy faced a number of Cold War crises that influenced him to keep supporting the fight against communism in Vietnam.

The first was the Bay of Pigs invasion in April 1961. An army of Cuban exiles, trained by the United States, invaded Cuba. They planned to overthrow the country's Communist leader, Fidel Castro. Cuban troops easily crushed the invasion, humiliating the United States.

A woman in West Berlin talks across the Berlin Wall to her mother in East Berlin.

Then in June 1961, the Soviet Union threatened to close off Western access to West Berlin because so many East Germans were fleeing there to escape communism. Tensions rose when Kennedy insisted on West Berlin's independence. The Soviets and East Germans then built the Berlin Wall, a heavily guarded barrier dividing West Berlin from Communist East Berlin and East Germany. The wall, which made it harder for East Germans to flee, became a symbol of Communist oppression.

The **Cuban missile crisis** in October 1962 was Kennedy's most serious confrontation with the Soviets. Fidel Castro, believing the United States planned another attack on Cuba, had asked for more Soviet military aid. The United States learned that the Soviets had put nuclear missiles in Cuba. These missiles could reach U.S. cities within minutes. Kennedy weighed his choices. "The greatest danger of all," he told the country, "would be to do nothing." In a frightening showdown between the two superpowers, the Soviets agreed to remove the missiles, and the United States promised not to invade Cuba.

Reading **History**
F. Analyzing Causes What Cold War crises made Kennedy increase his commitment to fight communism in Asia?

These Cold War crises fed American fears that the Soviet Union might become the strongest world power. In this climate of fear and suspicion, the United States made a greater effort to contain communism in Asia by sending more money and military advisers to South Vietnam.

The Diem Government Falls

As U.S. aid increased, so did South Vietnamese opposition to Diem. American officials told Diem to make political, economic, and military reforms. But he refused.

The Kennedy administration lost faith in Diem. With U.S. support, a military coup overthrew Diem on November 1, 1963. Against Kennedy's wishes, the coup's leaders killed Diem. In a terrible and unrelated turn of events, President Kennedy was assassinated three weeks later. Vice-President Lyndon Johnson became president. He deepened U.S. involvement in the Vietnam War, as you will see in the next section.

Vocabulary
coup (koo): a sudden takeover by a small group

AMERICA'S HISTORY MAKERS

JOHN F. KENNEDY
1917–1963

In 1960, at age 43, John F. Kennedy became the youngest U.S. president ever elected. Handsome and energetic, he inspired belief in the country's capabilities.

Many Americans regard the Cuban missile crisis as Kennedy's finest moment of leadership. This conflict brought the United States to the brink of nuclear war. Kennedy considered bombing the Soviet missile sites in Cuba and invading the country before deciding it was safer to blockade Cuba and keep bargaining with Soviet leader Nikita Khrushchev.

How did the missile crisis show Kennedy's leadership?

Section 1 Assessment

1. Terms & Names

Identify:
- Ngo Dinh Diem
- French Indochina
- Ho Chi Minh
- domino theory
- Viet Cong
- Ho Chi Minh Trail
- Cuban missile crisis

2. Taking Notes

Review the section and identify a key event for each year on the time line.

1930 1945 1950 1960

1940 1946 1954 1963

What event brought the United States into the Vietnam conflict?

3. Main Ideas

a. What were Ho Chi Minh's goals for Vietnam?

b. How did the Cold War affect American decisions regarding Vietnam?

c. What level of involvement did the Truman, Eisenhower, and Kennedy administrations have in Vietnam?

4. Critical Thinking

Evaluating How did U.S. support of the Diem government involve a conflict of values?

THINK ABOUT
- American beliefs in democracy and individual rights
- the actions of the Diem government

ACTIVITY OPTIONS

SPEECH
ART

Record imaginary **radio interviews** with Ho Chi Minh and Ngo Dinh Diem about Vietnam, or construct **signs** that their supporters might carry in a demonstration.

2 War Expands in Vietnam

TERMS & NAMES
- Gulf of Tonkin Resolution
- escalation
- William Westmoreland
- guerrilla warfare
- napalm
- Agent Orange
- Tet offensive

MAIN IDEA

America sent ground troops to Vietnam expecting victory, but soldiers soon grew frustrated.

WHY IT MATTERS NOW

The Vietnam War taught Americans that superior military strength does not always ensure victory.

ONE AMERICAN'S STORY

Reginald Edwards landed in Vietnam in 1965, the first year American combat troops were sent there. Despite two years of training, he was unprepared for the experience. It was nothing like his dream of "landing on this beach like they did in World War II." On night patrols through the dark countryside, Edwards and his fellow marines shot at whatever moved. Their first large-scale attack on the enemy proved disastrous.

A VOICE FROM THE PAST

We had received fire. All of a sudden we could see people in front of us. Instead of waiting for air [support], we returned the fire, and you could see people fall. . . . Come to find out it was Bravo Company. What the VC [Viet Cong] had done was [lure] Bravo Company in front of us. . . . It was our own people. That's the bodies we saw falling. . . . I think we shot up maybe 40 guys in Bravo Company.

Private Reginald "Malik" Edwards, quoted in *Bloods: An Oral History of the Vietnam War by Black Veterans*

American soldiers take aim in the Vietnamese jungle.

In this section, you will learn why the Vietnam War created such confusion and why people questioned how the war was conducted.

U.S. Troops in Vietnam

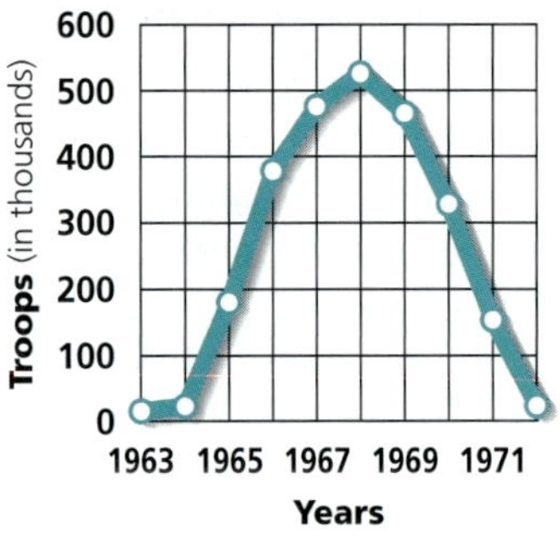

Source: Stanley I. Kutler, *Encyclopedia of the Vietnam War*

Johnson Sends Combat Troops

The assassination of President Diem in 1963 brought chaos to South Vietnam. One ineffective leader after another headed the government. Meanwhile, the North Vietnamese kept shipping more aid to the Viet Cong. By late 1964, combined Viet Cong and North Vietnamese forces controlled much of the South Vietnamese countryside.

Like earlier presidents, Lyndon Johnson did not want to lose Vietnam to communism. As a result, he increased U.S. efforts in Vietnam.

In the summer of 1964, Johnson's military advisers made plans to bomb North Vietnam. They wanted to pressure Ho Chi Minh to stop supporting the Viet Cong. But no bombing could start unless Congress approved the plan. A shooting incident off the coast of North Vietnam spurred Congress to give its approval.

The U.S. destroyer *Maddox* had been patrolling in the Gulf of Tonkin when North Vietnamese torpedo boats fired on it. Two days later, on August 4, the *Maddox* and another destroyer reported a second attack. However, no one could confirm this attack. There had been thunderstorms that night, and the weather could have affected the radar screens. U.S. jet pilots flying overhead said they had seen no North Vietnamese boats.

Reading History

A. Finding Main Ideas What did the Gulf of Tonkin Resolution do?

Despite doubts about the second attack, Johnson asked Congress to pass the **Gulf of Tonkin Resolution.** This gave the president the power to use military force in Vietnam. All but two senators voted for the resolution. The "yes" vote in the House was unanimous.

In March 1965, Johnson began bombing North Vietnam. At about the same time, he sent the first combat ground troops to Vietnam. Their numbers grew from 75,000 in the middle of 1965 to 184,000 by the end of 1965.

This policy of **escalation,** or increasing military involvement in Vietnam, continued over the next few years. General **William Westmoreland,** the commander of U.S. forces in South Vietnam, asked for more and more troops. By the end of 1968, there were more than 536,000 American military personnel in South Vietnam.

AMERICA'S HISTORY MAKERS

LYNDON JOHNSON
1908–1973

Lyndon Johnson wanted people to remember him as a social reformer. The Great Society was the name for his wide-ranging domestic programs, which included reforms in education, medical care for the elderly, aid to cities, and support for civil rights.

The Vietnam War overshadowed Johnson's achievements. He was tormented by the deaths and the social divisions the war brought. His goals for the country were blocked by the war, which still had not ended when he died.

How would you describe Johnson's presidency?

A Frustrating War

Many Americans thought that, with their superior weapons, U.S. ground forces would quickly defeat the Viet Cong and drive them out of the villages. Many conditions frustrated American soldiers, however. First of all, they could wage only a limited war, partly because the government feared drawing China into the conflict.

Background

The average World War II soldier was 26 and served for three years. Many soldiers in Vietnam had fathers who had fought in World War II.

Also, most U.S. soldiers in Vietnam were young and inexperienced. The average soldier was 19 and served a one-year tour of duty. Officers served even shorter tours, on average six months. The short tours meant that by the time soldiers and officers had gained enough experience, their tours of duty were over.

The Vietnam War differed from World War II in that there was no frontline. The Viet Cong mixed with the general population and operated everywhere, attacking U.S. troops in the countryside and in the cities. Even a shoeshine boy on a city street corner might toss a grenade into an army bus carrying American soldiers. Marine captain E. J. Banks described his frustration: "You never knew who was the enemy and who was the friend. . . . The enemy was all around you."

Vietnam Warfare

Booby Traps
The Viet Cong hid deadly booby traps made of sharpened sticks.

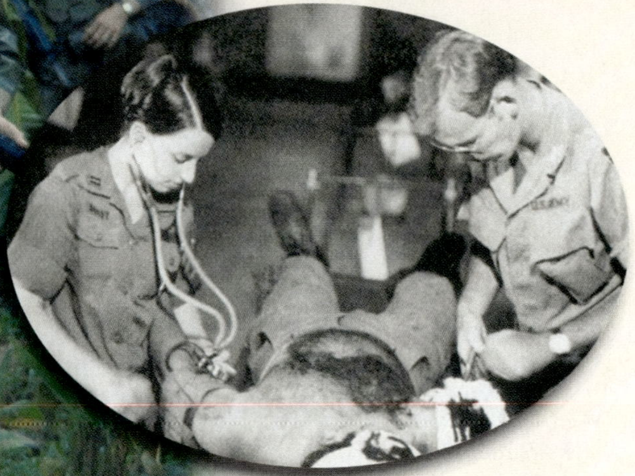

Guerrilla Warfare
In the jungles, surprise attacks could come at any moment. Helicopters quickly took the wounded to hospitals.

The style of fighting in Vietnam also differed from that in World War II. Because they could not match American firepower, the Viet Cong relied on **guerrilla warfare,** surprise attacks by small bands of fighters. Viet Cong guerrillas would suddenly emerge from networks of underground tunnels to fight. Then they would disappear back into the tunnels. They riddled the countryside and jungles with land mines and booby traps, such as bamboo stakes hidden in covered pits. They hung grenades from trees and hid them in bushes. Every day, U.S. Army and Navy nurses treated young soldiers with gruesome wounds.

Background
Nearly 7,500 women served in Vietnam as nurses.

Even the land and climate of Vietnam proved difficult. The heat was suffocating and the rain almost constant. Soldiers sweated through tangled jungles. After wading through flooded rice paddies, they had to pick leeches off their feet and legs. American soldier Warren Wooten said, "It seemed like the whole country was an enemy. The animals, the reptiles, the insects, the plants. And the people."

Finally, the Viet Cong were a very dedicated enemy. They took heavy losses, built up their ranks again, and kept on fighting year after year because they believed in their cause. An American who interviewed Viet Cong prisoners noted, "They see the war entirely as one of defense of their country against the invading Americans, who, in turn, are seen merely as successors to the French."

*Reading*History
B. Analyzing Causes Why was the war so hard for the United States to win?

Stripping the Jungle

One of the strengths of the Viet Cong was their ability to hide in the jungle and in underground tunnels. To reveal and destroy Viet Cong hideouts, American troops used chemicals that ruined the landscape.

Search-and-Destroy Missions
U.S. soldiers destroyed villages suspected of hiding Viet Cong.

Napalm
U.S. planes dropped fiery napalm bombs to wipe out Viet Cong bases. Napalm is jellied gasoline.

Over wide areas, U.S. planes dropped bombs of **napalm,** jellied gasoline that burns violently. Planes also sprayed **Agent Orange,** a chemical that kills plants, over the jungles.

Such chemicals helped destroy the hideouts and food supplies of the Viet Cong. But in the process, they also harmed innocent Vietnamese villagers. This undermined the villagers' support for the United States. Later, people learned that Agent Orange harmed U.S. soldiers as well. Veterans exposed to it have suffered from skin diseases and cancers.

Search-and-Destroy Missions

Search-and-destroy missions were another American war tactic that terrorized Vietnamese villagers. In such missions, soldiers hunted Viet Cong and burned or bombed villages thought to be sheltering them. Marine sergeant William Ehrhart described how search-and-destroy missions affected South Vietnamese peasants.

A VOICE FROM THE PAST

Their homes had been wrecked, their chickens killed, their rice confiscated [taken away]—and if they weren't pro-Vietcong before we got there, they sure . . . were by the time we left.

William Ehrhart, quoted in *Vietnam: A History* by Stanley Karnow

Reading **History**

C. Drawing Conclusions Why did Americans fail to "win the minds and hearts of the people" in Vietnam?

These destructive methods defeated the purpose of "winning the hearts and minds" of the villagers and turning them against communism. Furthermore, even if the tactics did clear a village of Viet Cong temporarily, the Viet Cong usually returned later.

The Surprise Tet Offensive

By the end of 1967, the war had caused great destruction, but neither side was close to victory. Still, U.S. military officials claimed that they would soon win. Then, on January 30 and 31, 1968, the Communists launched the **Tet offensive.** This was a surprise attack on U.S. military bases and more than 100 cities and towns in South Vietnam. It came during Tet, the Vietnamese celebration of the lunar New Year.

In preparation for the Tet offensive, the Viet Cong hid weapons in vegetable trucks, food trucks, peddlers' carts, and even coffins. They smuggled these weapons into South Vietnamese cities. Soldiers dressed in civilian clothes entered the cities on buses, on motorcycles, and on foot. No one could tell them apart from the war refugees who streamed into the cities from the countryside or from visitors coming for the holiday.

The Viet Cong fought to take over the cities during the offensive. They killed not only enemy soldiers but also government officials, schoolteachers, doctors, and priests.

Background
In the Tet offensive, enemy forces invaded the U.S. embassy compound in Saigon, killing five Americans.

The Tet offensive was a military defeat for the Communists. They gained no cities and lost 45,000 soldiers, while the South Vietnamese lost 2,300 soldiers and the United States 1,100.

But the attack stunned Americans. General Westmoreland had recently declared, "We have turned the corner," suggesting that victory was in sight. The Tet offensive raised doubts that this was true. Many government and business leaders began to think that the United States could not win the war, except at too high a price.

GEOGRAPHY SKILLBUILDER Interpreting Maps
1. **Location** *In what country were the major battles?*
2. **Location** *What did the number and location of attacks suggest about the country's security?*

The Tet offensive also made many Americans ask whether the U.S. mission in Vietnam was wise. To retake some cities, troops had to almost level them with bombing and shelling. Speaking of the city of Ben Tre, a U.S. major said, "It became necessary to destroy the town in order to save it." The quote became an example of what many considered the senselessness of the war.

Because of the doubts it raised, the Tet offensive became a turning point in the war. Afterward, President Johnson changed his war policy. When General Westmoreland asked for 206,000 more troops, to take advantage of the enemy's weakness, President Johnson said no. Then, on March 31, 1968, Johnson said that he would stop bombing most of North Vietnam and would seek to

***Reading*History**
D. Drawing Conclusions How was the Tet offensive both a failure and a success for the Viet Cong?

bargain for peace. In the same speech, he announced that he would not run for another term as president.

U.S. Morale Sinks

As the Vietnam War went on, it wore down American soldiers. They fought hard and bravely, but many were losing faith that the United States could win the war. The South Vietnamese government did not have the loyalty of the people. In addition, the South Vietnamese army often avoided fighting. American soldiers asked why they were fighting a war the Vietnamese did not want to fight themselves.

"I DON'T KNOW IF EITHER SIDE IS WINNING, BUT I KNOW WHO'S LOSING."

This 1968 cartoon by Herblock comments on the war's cost to civilians.

The low morale of American forces in Vietnam became clear when news of the My Lai (mee ly) massacre broke in 1969. The incident happened on March 16, 1968. A U.S. platoon led by Lieutenant William Calley, Jr., rounded up and shot between 175 and 500 unarmed civilians, mostly women, children, and old men. A U.S. helicopter pilot rescued some civilians by threatening to fire on the soldiers. To Americans, My Lai represented a horrifying breakdown in morality and discipline in the armed forces. In 1971, a colonel warned that U.S. forces were "in a state approaching collapse."

Background
Calley was jailed briefly, but many saw him as a scapegoat because charges were dropped against higher-ranking officers.

In the next section, you'll learn how the United States withdrew from the Vietnam War.

Section 2 Assessment

1. Terms & Names

Identify:
- Gulf of Tonkin Resolution
- escalation
- William Westmoreland
- guerrilla warfare
- napalm
- Agent Orange
- Tet offensive

2. Taking Notes

On a chart like the one below, note the war's effects on Vietnamese villagers and on U.S. soldiers.

Effects of War

Villagers	U.S. Soldiers

3. Main Ideas

a. How did President Johnson escalate U.S. involvement in the Vietnam War?

b. What made fighting the war so frustrating for American soldiers?

c. How was the Tet offensive a turning point in the war?

4. Critical Thinking

Analyzing Points of View
Were the Viet Cong right to see the Americans "merely as successors to the French"?

THINK ABOUT
- the goals of the French in Vietnam
- the goals of the Americans in Vietnam
- the actions of the French and the Americans

ACTIVITY OPTIONS

SCIENCE
HEALTH

Investigate the health effects of Agent Orange reported by Vietnam veterans. Design a science **exhibit** or create a **warning label** to share your findings.

3 The Vietnam War Ends

TERMS & NAMES
doves
hawks
Richard Nixon
Vietnamization
Cambodia
Twenty-sixth Amendment
War Powers Act

MAIN IDEA

The Vietnam War divided Americans and had lasting effects in the United States and Southeast Asia.

WHY IT MATTERS NOW

Lessons of the Vietnam War still influence the United States whenever it gets involved in a foreign conflict.

ONE AMERICAN'S STORY

Mary Ann Vecchio was just 14 years old when she suddenly became a symbol of the anguish that the Vietnam War caused Americans. A student journalist snapped Vecchio's picture as she knelt over a dead student at Kent State University in Ohio. The youth was Jeffrey Glenn Miller, one of four students killed by the National Guard during an antiwar demonstration on May 4, 1970. Looking back 25 years later, Vecchio described her feelings of disbelief.

Mary Ann Vecchio cries out in horror after the Kent State shootings in May 1970.

A VOICE FROM THE PAST

I couldn't believe that people would kill people over what they thought, just because he demonstrated against the Vietnam War—that they would shoot you over it. I couldn't believe . . . . There was nothing I could do for Jeffrey or any of the other students. And that's part of history and that'll remain with me for the rest of my life.

Mary Ann Vecchio Gillum, conference at Emerson College, April 23, 1995

In this section, you will learn how growing opposition to the war eventually led the United States to pull its troops out of Vietnam.

A Growing Antiwar Movement

As the war escalated in the mid-1960s, antiwar feeling grew among Americans at home. Religious leaders, civil rights leaders, teachers, students, journalists, and others protested the war for a variety of reasons. Some believed that the United States had no business involving itself in another country's civil war. Others believed that the methods of fighting the war were immoral. Still others thought that the costs to American society were too high.

College students formed a large and vocal group of protesters. They particularly opposed the draft, which required young men to serve in the military. In demonstrations around the country, young men burned their draft cards. About 50,000 people staged such a protest in front of the Pentagon on October 21, 1967.

Opponents of the draft pointed out its unfairness. Most draftees were poor. Middle- and upper-class youths could delay being drafted by enrolling in college. They also sought advice from draft counselors, doctors, and lawyers to help them avoid service. Certain medical conditions or religious beliefs, for example, could keep them out of the military.

Reading **History**

A. Analyzing Points of View Why did people think the draft was unfair?

Another unfair aspect of the draft was the high number of African Americans called to serve. African Americans made up about 20 percent of combat troops in Vietnam. In 1965, they accounted for 24 percent of U.S. Army combat deaths. Yet they were only 11 percent of the male population in the United States.

For this and other reasons, the antiwar movement became linked with the civil rights movement. In 1967, civil rights leader Martin Luther King, Jr., spoke out against the Vietnam War. Noting that the war took money away from antipoverty programs, he declared, "I was increasingly compelled to see the war as an enemy of the poor."

By 1967, it was clear that the Vietnam War was dividing Americans into two camps. Even within families, people took opposite sides. Those who opposed the war were called **doves.** Those who supported it were known as **hawks.** Supporters of the war staged marches of their own. Believing that antiwar protesters were unpatriotic, they popularized such slogans as "America—love it or leave it."

1968—A Turning Point

As you learned in Section 2, the Tet offensive in January 1968 made Americans doubt that they could win the war. Walter Cronkite, a respected TV news anchorman, visited Vietnam in February. After returning, he ended a special report with his own opinion of the war. He concluded that the United States was not winning but was in a deadlock.

A VOICE FROM THE PAST

[T]he only rational way out, then, will be to negotiate, not as victors, but as an honorable people who lived up to their pledge to defend democracy, and did the best they could.

Walter Cronkite, *A Reporter's Life*

President Johnson took Cronkite's words to heart. "If I've lost Cronkite, I've lost middle America," he reportedly said.

That summer, the Democratic National Convention in Chicago reflected the country's turmoil. Democrats chose Hubert Humphrey, Johnson's vice-president, as their nominee. Outside the convention hall, TV

daily *life*

THE "TELEVISION WAR"

The Vietnam War was the first "television war," broadcast each night on the evening news. Reports rarely showed actual battles, partly because much of the fighting occurred off and on and at night, between small units.

Networks also tried to avoid gruesome scenes because they did not want to offend viewers. In addition, the networks agreed not to show any American dead or wounded so that their families would not see them on the screen. Still, the images of war shocked TV audiences.

cameras showed police clubbing antiwar demonstrators and bystanders. The chaos helped Republican candidate **Richard Nixon** win the presidency in 1968. In his campaign, Nixon promised to "bring an honorable end to the war in Vietnam."

Nixon's Vietnam Strategy

In July 1969, Nixon announced his strategy of **Vietnamization.** It called for gradually withdrawing U.S. forces and turning the ground fighting over to the South Vietnamese. Nixon promised to withdraw 25,000 of the 543,000 U.S. ground troops in Vietnam by the end of the year. However, Nixon had already begun secret bombing raids of **Cambodia,** a country bordering Vietnam. This bombing was meant to stop North Vietnamese troops and supplies from moving along the Ho Chi Minh Trail. Many people grew angry when they learned that the government had widened the war and hidden its actions.

Background
In 1970, a U.S. invasion of Cambodia sparked protests at Kent State and many other colleges.

Public anger and distrust of the government grew after Daniel Ellsberg released the Pentagon Papers to the *New York Times* in 1971. Ellsberg had helped research and write these secret Defense Department papers. They showed that the four previous presidential administrations had not been honest with the public about U.S. involvement and goals in Vietnam.

Withdrawal from Vietnam

Promising that peace was at hand, Nixon was reelected by a landslide in 1972. On January 27, 1973, the United States and South Vietnam signed a peace agreement with North Vietnam and the Viet Cong. The United States agreed to withdraw all its troops, and North Vietnam agreed not to invade South Vietnam. On March 29, the last U.S. troops left Vietnam. For the United States, the war was over.

Thousands try to escape during the fall of Saigon.

But for the Vietnamese, the war continued. In 1975, North Vietnam launched a massive invasion of South Vietnam. On April 30 of that year, Communist forces captured Saigon, which they renamed Ho Chi Minh City. The war then ended.

Reading History
B. Making Inferences Why did fighting begin again after the peace agreement?

Legacy of the Vietnam War

The Vietnam War caused terrible destruction and suffering in Southeast Asia. More than 1.2 million North and South Vietnamese died in the conflict. American bombing and chemical spraying caused lasting damage to farmland and forests. The war ruined Vietnam's economy, leaving many in poverty. After the North Vietnamese set up Communist rule in the reunited country, many Vietnamese fled. By 1980, almost 173,000 had come to the United States.

The Vietnam War also took a heavy toll on American soldiers. About 58,000 died, and more than 300,000 were wounded. Many suffered permanent, disabling injuries. Returning soldiers often had recurring nightmares and other stress-related problems. To make things worse, they came home to a public that treated them coldly.

Background
Although there is no draft now, 18-year-old men must still register with the Selective Service in case Congress orders a draft in the future.

The Vietnam War had far-reaching political effects in the United States. The **Twenty-sixth Amendment,** passed in 1971, lowered the voting age from 21 to 18. Its supporters argued persuasively that anyone old enough to be drafted should be allowed to vote. The government ended the draft in 1973 because so many people opposed it. The nation now relies on an all-volunteer military.

Another legacy of Vietnam is that Americans have been less willing to get involved in overseas wars. In 1973, Congress passed the **War Powers Act,** which limits the president's war-making powers. The president must report to Congress within 48 hours if troops have been sent into a hostile situation without a declaration of war. They can remain for no more than 90 days unless Congress permits them to.

***Reading*History**
C. Making Inferences Why did the Vietnam War influence Congress to pass the War Powers Act?

Finally, the war made many Americans distrust government leaders, who sometimes misled the public about actions in Vietnam. In Chapter 31, you will read about the Watergate scandal, which further shook confidence in government and brought down Nixon's presidency.

America's HERITAGE

VIETNAM VETERANS MEMORIAL

The public finally honored those who served in Vietnam with the Vietnam Veterans Memorial, designed by Maya Lin and unveiled in Washington, D.C., on November 13, 1982. Etched into this V-shaped black granite wall are the names of all Americans killed in the war.

At the Wall, as it is known, visitors leave letters, flowers, and mementos. Many believe that the monument has helped heal the divisions created by the war. As the father of a veteran noted, "It doesn't say whether the war was right or wrong. . . . It just says, 'Here is the price we paid.'"

Section 3 Assessment

1. Terms & Names

Identify:
- doves
- hawks
- Richard Nixon
- Vietnamization
- Cambodia
- Twenty-sixth Amendment
- War Powers Act

2. Taking Notes

Use a chart like this one to review information about the antiwar movement.

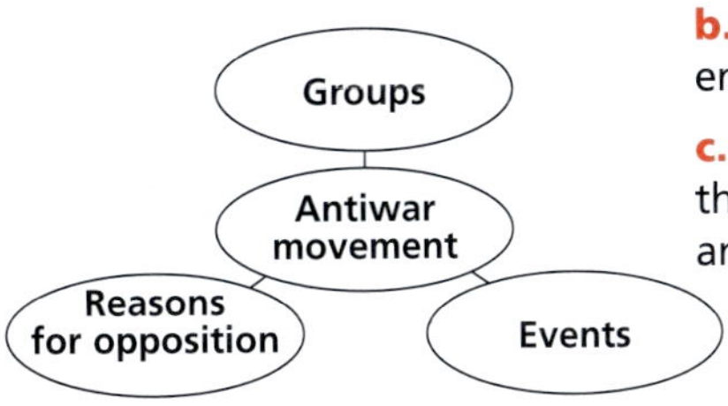

3. Main Ideas

a. Why did more and more Americans oppose the war after 1968?

b. How did the Vietnam War end?

c. In what major ways did the war affect Southeast Asia and the United States?

4. Critical Thinking

Forming and Supporting Opinions What is your opinion of the way the United States ended its involvement in the Vietnam War?

THINK ABOUT
- what happened to South Vietnam
- what options the United States had

ACTIVITY OPTIONS

LITERATURE
MATH

Compare the Vietnam War to World War II by writing an **essay** on two pieces of war literature or by presenting comparable **statistics.**

Chapter 30 ASSESSMENT

VISUAL SUMMARY

The Vietnam War Years

1946

1946
Communist-led Vietnamese nationalists struggle for independence from France.

1950
United States sends aid to the French.

1954
French are defeated. Vietnam divides into Communist North and non-Communist South.

1965
United States bombs North Vietnam and sends combat ground troops to South Vietnam.

1973
United States withdraws its troops.

1975

1975
South Vietnam surrenders to the Communists.

TERMS & NAMES

Briefly explain the importance of each of the following.

1. Ho Chi Minh
2. domino theory
3. Viet Cong
4. Ho Chi Minh Trail
5. guerrilla warfare
6. Agent Orange
7. Tet offensive
8. Richard Nixon
9. Vietnamization
10. Twenty-sixth Amendment

REVIEW QUESTIONS

Cold War Roots of the Conflict (pages 835–839)

1. Why did the United States decide to support France in its fight against the Viet Minh?
2. What decisions about Vietnam were laid out in the 1954 Geneva Accords?
3. What Cold War crises made President Kennedy continue to aid Ngo Dinh Diem's government in South Vietnam?

War Expands in Vietnam (pages 840–845)

4. How was President Johnson's Vietnam policy different from President Kennedy's?
5. What kept U.S. troops from defeating the Viet Cong quickly?
6. How did American war tactics affect Vietnamese villagers?
7. How was the Tet offensive both a defeat and a victory for the Communists?

The Vietnam War Ends (pages 846–849)

8. Why did many Americans protest against the war?
9. What policies did President Nixon pursue in Vietnam?
10. What long-term political effects did the Vietnam War have on the United States?

CRITICAL THINKING

1. USING YOUR NOTES

Reasons for involvement in Vietnam
French alliance
Domino theory
Nation building
Cold War crises
U.S. weaponry

Reasons against involvement in Vietnam
U.S. interests
Draft
Social programs
Vietnamese civilians
Domestic unrest

Using your completed chart from the beginning of this chapter, answer the questions.

a. Which do you consider the strongest reason in support of American involvement in the Vietnam War?

b. Which do you consider the strongest reason against American involvement in the Vietnam War?

c. Which side do you think is more persuasive? Why?

2. ANALYZING LEADERSHIP

How would you evaluate President Johnson's leadership during the Vietnam War and his decision not to seek a second term as president?

3. APPLYING CITIZENSHIP SKILLS

During the Vietnam War, many Americans had to choose between obeying laws and following their consciences. In your opinion, what is the right thing to do in such a situation?

4.THEME: AMERICA IN THE WORLD

How do you think involvement in the Vietnam War affected the reputation of the United States among other nations? Why?

5. DRAWING CONCLUSIONS

Would the United States have become involved in the Vietnam War if the Cold War had not been going on? Explain your opinion.

Interact *with* History

If you had lived during the Vietnam War, what would have determined whether you supported or opposed the war?

HISTORY SKILLS

1. INTERPRETING CHARTS

Read the chart and then answer the questions.

U.S. Deaths in Four Wars

WAR	BATTLE DEATHS	OTHER DEATHS*
World War I	53,513	63,195
World War II	292,131	115,185
Korean War	33,629	20,617
Vietnam War	47,244	10,446

*accidents, diseases, etc.

Source: Harry G. Summers, Jr., *Vietnam War Almanac*

Basic Chart Elements

a. What is the subject of the chart?

b. What is the difference between the second and third columns?

Interpreting the Chart

c. Which war or wars caused more battle deaths than the Vietnam War?

d. Which war had the lowest ratio of nonbattle deaths to battle deaths?

2. INTERPRETING PRIMARY SOURCES

Read this passage by a Vietnamese woman, addressed to an American veteran of the Vietnam War. Then answer the questions.

> You came to Vietnam, willingly or not, because your country demanded it. Most of you did not know, or fully understand, the different wars my people were fighting when you got here. For you, it was a simple thing: democracy against communism. For us, that was not our fight at all. How could it be? We knew little of democracy and even less about communism. For most of us it was a fight for independence—like the American Revolution.
>
> **Le Ly Hayslip,** *When Heaven and Earth Changed Places*

a. What did American soldiers misunderstand, according to Hayslip?

b. What seems to be Hayslip's attitude toward Americans?

ALTERNATIVE ASSESSMENT

1. INTERDISCIPLINARY ACTIVITY: Music

Analyzing Antiwar Songs Research antiwar songs of the late 1960s and early 1970s. Choose one song, play it for the class, and talk about the meaning of its lyrics. Explain what the song adds to your understanding of the Vietnam era.

2. COOPERATIVE ACTIVITY

Making a Video Documentary Work in a group to create and present a video documentary about the Vietnam War and its impact on the country or your town. As part of your research, talk to adults who lived through the Vietnam era and who have different perspectives on the war. You might try to locate a soldier or nurse who served, someone who faced the draft but was not called to serve, an antiwar demonstrator, or someone who lost a loved one in the war.

3. TECHNOLOGY ACTIVITY

Planning a Web Site Most Americans still know little about the country of Vietnam. Plan a Web site that will give future classes geographical, historical, and cultural background on Vietnam before they study the war. Use the library and the Internet, or contact colleges, museums, or Vietnamese organizations to get started.

INTERNET **Visit www.mcdougallittell.com for more information on Vietnam.**

Your site might include the following things.

- Detailed maps of Vietnam and its major cities
- Photographs of people, landscapes, and art objects
- A listing of products grown or made in Vietnam
- Biographies of important people in Vietnamese history
- Excerpts from stories, poetry, or memoirs by Vietnamese writers
- Internet links to other sites

4. HISTORY PORTFOLIO

PORTFOLIO **Option 1** Review your section and chapter assessment activities. Select one that you think is your best work. Use comments made by your teacher or classmates to improve your work, then add it to your portfolio.

PORTFOLIO **Option 2** Review the questions that you wrote for What Do You Want to Know? on page 834. Then write a short report in which you explain the answers to your questions. If any questions were not answered, do research to answer them.

CHAPTER 31

Years of Doubt
1969–1981

Section 1 **Nixon Confronts Problems**
Section 2 **Watergate Brings Down Nixon**
Section 3 **Issues of the Seventies**

Interact *with* History

It is the evening of August 8, 1974. You are watching television when your favorite program is interrupted by a speech from President Richard M. Nixon. Looking grim and tired, Nixon says that he will resign the next day. You wonder how the country will cope with the crisis.

What Do You Think?

- What happens when a president resigns from office? Has a president resigned before?
- How has the government functioned when a president has died?
- What other crises has the country faced?

How is America able to survive a crisis and move on?

USA 1969 – 1981

1969 Richard M. Nixon is inaugurated as president. U.S. astronauts land on moon.

1972 Watergate break-in occurs. Nixon is reelected.

1974 Nixon resigns. Gerald Ford becomes president.

1976 The U.S. celebrates its bicentennial. Jimmy Carter is elected president.

1980 Ronald Reagan is elected president.

World

1969 Golda Meir becomes prime minister of Israel.

1970 Anwar el-Sadat becomes president of Egypt.

1971 General Idi Amin seizes power in Uganda.

1973 U.S. involvement in Vietnam War ends. Military overthrows government of Chile.

1975 South Vietnam surrenders to North Vietnam.

1978 Camp David Accords are signed.

1979 Americans taken hostage in Iran. Soviet Union invades Afghanistan.

1981 American hostages are released by Iran.

Chapter 31 SETTING THE STAGE

BEFORE YOU READ

Previewing the Theme

Citizenship The 1970s were a period of conflict and change in the United States. Inflation, gas shortages, environmental disasters, and weak political leadership were only a few of the problems that confronted the nation. Chapter 31 presents the challenges America faced during the 1970s—and the tough decisions its citizens and leaders faced.

What Do You Know?

What do you already know about the problems of the 1970s, such as inflation and the energy crisis? Were they different from problems the country faced at other times in its history?

THINK ABOUT

- what you have learned about the 1970s from books, movies, television, and popular music
- how the nation has reacted to economic, political, and social problems in the past

What Do You Want to Know?

What would you like to know about Watergate and other crises of the 1970s? In your notebook, record what you hope to learn from this chapter.

READ AND TAKE NOTES

Reading Strategy: Taking Notes To help you remember what you read, take notes about the major issues and events discussed in this chapter. Taking notes means writing down important information. The chart below lists the three presidents of the 1970s—Richard Nixon, Gerald Ford, and Jimmy Carter. As you read, use a chart like the one below to take notes on the issues and events of their presidencies.

See Skillbuilder Handbook, page R3.

President	Issues and Events
Nixon	
Ford	
Carter	

1 Nixon Confronts Problems

TERMS & NAMES
Richard M. Nixon
Henry Kissinger
revenue sharing
détente
SALT

MAIN IDEA
President Richard M. Nixon faced the challenge of governing a deeply divided America.

WHY IT MATTERS NOW
Social, economic, and political divisions are still part of American life.

ONE AMERICAN'S STORY

Shortly after his election as president in November 1968, **Richard M. Nixon** began selecting his cabinet and closest advisers. Most of his choices were wealthy business executives or lawyers. But one who did not fit this description was **Henry Kissinger**. Kissinger was a professor at Harvard University and a Jewish refugee from Nazi Germany. He had become a well-known foreign-policy expert.

Nixon picked Kissinger to be his national security adviser because he felt that Kissinger had great intelligence, insight, and experience. Nixon also found that he and Kissinger had very similar ideas about politics. Both men were practical politicians who did not let their ideologies, or beliefs, get in the way of their actions.

A VOICE FROM THE PAST

The statesman manipulates [shrewdly manages] reality; his first goal is survival. . . . He is conscious of many great hopes which have failed, of many good intentions that could not be realized.

Henry A. Kissinger, *American Foreign Policy*

That statement might just as easily have been written by President Nixon. In this section, you will learn how Nixon used a practical approach to try to solve problems at home and, with Kissinger's help, abroad.

President Richard Nixon (left) walks with Henry Kissinger, his national security adviser.

A Divided America

As president, Richard M. Nixon would have liked to focus on foreign policy and leave domestic issues to his cabinet. But he became president at a time when America was being torn apart by inflation, racial problems, and conflict over the war in Vietnam. Also, he had won the presidency in 1968 by only a narrow margin. His major support came from conservatives. They wanted him to shrink the federal government and end Lyndon Johnson's Great Society programs.

Nixon hoped to cut the cost of running the federal government and turn some of its activities over to the states. But both houses of Congress were controlled by the Democrats. As a result, he took a more moderate

Now and then

DESTINATION MOON

On July 20, 1969, with 600 million people viewing on television, Apollo 11 astronaut Neil Armstrong stepped onto the moon and said, "That's one small step for a man, one giant leap for mankind." Edwin Aldrin (shown below) soon joined him. The excitement of the first moon landing gave Americans a sense of unity during a troubled time.

On July 31, 1999, an American space probe was sent crashing into the moon's surface. Its purpose was to see whether there was water on the moon. This flight took place almost 30 years to the day after the historic first landing.

approach at first. But he did veto many spending bills passed by the Democratic Congress. Included were cuts in funds for education and low-income housing—Great Society programs.

In 1969, President Nixon attempted to change the existing welfare system, which had been criticized by conservatives. But he could not get congressional agreement. He was more successful in starting **revenue sharing**. Under revenue sharing, the federal government gave back some tax money it collected to state and local governments. These governments could spend the money on any number of different programs. The plan was supported by both parties and passed Congress in 1972.

Reading History
A. Recognizing Effects How did Nixon try to shrink the size of the federal government?

Law-and-Order Politics

During the presidential campaign, Nixon had said that he would restore law and order. He promised an end to the social unrest and rioting of the late 1960s. Nixon also promised a return to traditional values.

One way to his goal, Nixon thought, was to appoint more conservative justices to the Supreme Court. He believed that these justices would rule against loose interpretations of the law. Such interpretations, Nixon and his supporters felt, were partly to blame for rising crime. During his first term, he appointed four new justices.

Nixon also used all the powers of the federal government to "crack down" on crime and protest. He directed the Central Intelligence Agency and the Federal Bureau of Investigation to investigate some of his political enemies. Nixon would later excuse any illegal acts by claiming they were needed for the nation's security.

A Troubled Economy

Civil unrest and crime were not the only problems President Nixon had to face. He inherited serious economic troubles. Under the Johnson administration, the government began spending huge amounts of money on programs to aid the poor *and* on fighting the Vietnam War. But taxes were not raised to cover these expenses. As a result, the government spent more money than it collected in taxes. This practice is called deficit spending.

Reading History
B. Analyzing Causes What were causes of the nation's troubled economy?

Deficit spending put a great deal of money into circulation in the late 1960s. At the same time, the economy began to slow down. By 1970, the economy had gone into a recession. With the recession, fewer goods were produced. The increase in the money supply and the decrease in products to buy caused inflation. (See Economics in History on the next page.)

Inflation caused problems for all Americans, but especially the poor. One mother said, "I used to be able to go to the store with $50 and come back with six or seven bags of groceries. Now I'm lucky if I come back with three." When Nixon took office in 1969, the inflation rate was about 5 percent. That was twice the rate it had been earlier in the decade. Increasing unemployment was another problem. It doubled from about 3 percent in 1969 to 6 percent in 1971.

Nixon tried different ways to help the economy. He cut spending. He also placed a temporary freeze on wage and price increases. Inflation dropped but only temporarily. Then, in 1973, the economy was jolted by the actions of the Organization of Petroleum Exporting Countries (OPEC).

OPEC raised its prices and cut its shipments of oil to the United States. OPEC, which was made up mainly of Arab nations, took these actions in part to protest America's support of Israel in the Yom Kippur War of 1973. That war was fought between Israel and its Arab neighbors. In a few months, gas prices quadrupled. Inflation and unemployment soared to new heights.

Vocabulary
quadrupled: multiplied by four

Besides these serious economic problems at home, Nixon also faced tough challenges in foreign policy.

Economics *in* History

Inflation

Inflation is a rise in prices across the economy. When the price of milk goes up, that's a price increase. When the prices of gasoline, clothing, food, housing, and health care all go up, that's inflation.

Sometimes prices rise because there's more demand for goods than there are goods available to buy. This happened after World War II. During the war years, factories had produced weapons and supplies for the war effort. After the war, it took a while for factories to produce enough peacetime goods to meet consumer demand.

Sometimes prices rise because businesses increase their prices to cover their rising costs. This was the case in the 1970s, when the rising price of oil sent up the prices of many different goods and services.

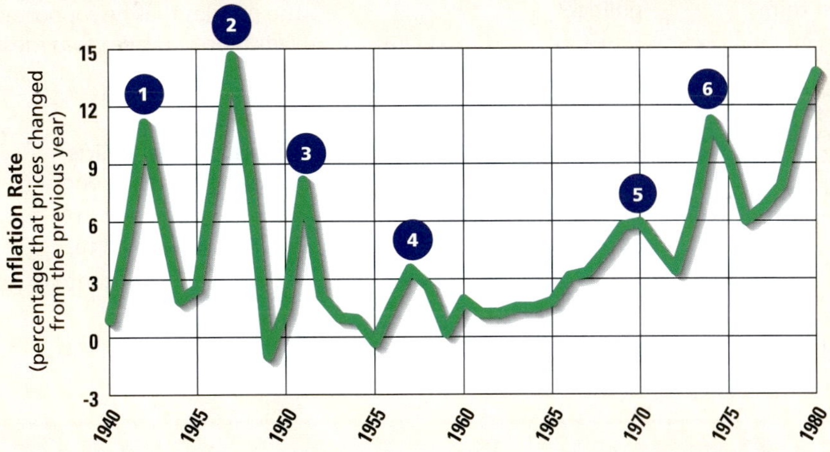

Source: *U.S. Bureau of the Census*

1. World War II
2. Postwar boom
3. Korean War
4. Suez crisis
5. Vietnam War
6. Arab oil embargo—oil prices increase

CONNECT TO HISTORY

1. **Making Inferences** Why do you think oil price increases had such a widespread effect on the economy?

See Skillbuilder Handbook, page R11.

CONNECT TO TODAY

2. **Researching** Find out what the rate of inflation has been in the most recent two years. How does today's inflation rate compare with that when Nixon came to office and when he left?

Visit www.mcdougallittell.com to learn more about inflation.

AMERICA'S HISTORY MAKERS

RICHARD M. NIXON

1913–1994

Richard Nixon entered politics in 1946 when he ran as a Republican for Congress. He was bright and ambitious. Some opponents, though, said that he would do anything to gain political power. Within seven years, he was vice-president of the United States.

In 1960, Nixon lost the presidential election to John Kennedy. In 1962, he was defeated for governor of California. Most thought his career over. But he battled back and was elected president in 1968. He is shown here in China in 1972 with his wife, Pat.

What characteristics did Nixon show as a leader?

Nixon Eases the Cold War

Nixon's main foreign-policy goal was world stability. During his 1968 presidential campaign, Nixon pledged to end the Vietnam War quickly and honorably. As you learned in Chapter 30, it took four years to negotiate a cease-fire with the North Vietnamese.

Nixon's most important triumph came in dealings with the People's Republic of China. He had long opposed the Communists, who took power in China in 1949. But Nixon believed a nation of a billion people could not be ignored. He asked Henry Kissinger to find a way to improve relations with China, even though they knew many Americans would be opposed. Kissinger arranged for Nixon to visit China in February 1972. This trip led to the opening of diplomacy and trade with the Chinese.

Nixon's China trip affected Amcrican relations with the Soviet Union, which was having conflicts with China. The Soviets feared closer relations between the United States and China. So they invited Nixon to Moscow in May 1972. As a result, Soviet-American relations improved. This easing of tensions between rivals is called **détente**—a French word. The policy of détente led the two nations to sign the Strategic Arms Limitation Treaty of 1972 (**SALT**). This pact limited the number of each country's nuclear weapons, easing fears of nuclear war.

Reading **History**

C. Making Inferences What effect did improved American-Chinese relations have on Soviet-American relations?

Nixon's triumphs in foreign policy helped make him look like a sure winner for reelection in 1972. But, as you will read in the next section, events that occurred in that campaign eventually destroyed his presidency.

Section 1 Assessment

1. Terms & Names

Identify:
- Richard M. Nixon
- Henry Kissinger
- revenue sharing
- détente
- SALT

2. Taking Notes

Use a spider diagram to describe problems Nixon faced in his first term.

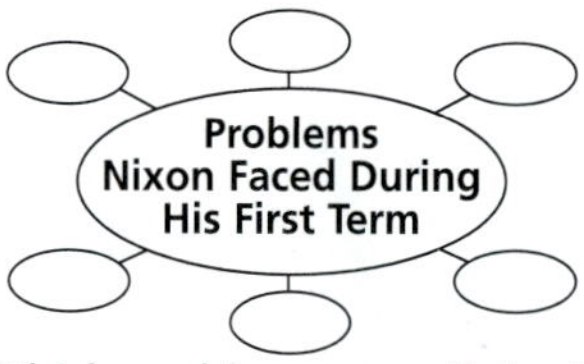

Which problems were linked?

3. Main Ideas

a. How did Nixon try to show support for law-and-order politics?

b. What economic problems developed during Nixon's first term as president?

c. How did Nixon change the country's relationship with China? with Russia?

4. Critical Thinking

Making Inferences Nixon often surprised Americans in the policies that he supported. Which do you think was most surprising? Why?

THINK ABOUT
- his economic policies, such as deficit spending
- his appointment of conservative justices
- the policy of détente

ACTIVITY OPTIONS

ART / **LANGUAGE ARTS**

Draw a **political cartoon** or write a **press release** describing one of Nixon's policies or achievements from his first term as president.

2 Watergate Brings Down Nixon

TERMS & NAMES

Committee to Reelect the President

Watergate scandal

MAIN IDEA

Nixon's involvement in the Watergate scandal caused a political crisis that forced him to resign.

WHY IT MATTERS NOW

Watergate led many Americans to have less confidence in government and politicians.

ONE AMERICAN'S STORY

Barbara Jordan grew up in a poor home in Houston, Texas, in the 1940s. Early on, she realized that her voice was one of her greatest assets. She had a strong, deep, beautiful voice and a gift for reciting dramatic poetry. Jordan was also highly intelligent. After graduating from Boston University Law School in 1959, she became interested in politics. She began her career as a volunteer worker for John Kennedy in his 1960 presidential campaign against Richard Nixon.

Jordan went on to become the first African-American woman from the South elected to the U.S. Congress. The 1972 election that sent her to the House of Representatives also saw Nixon's reelection as president. Two years later, at impeachment hearings in the House Judiciary Committee, Jordan warned that Nixon's actions during that campaign and his presidency were threats to the Constitution.

As a member of the House Judiciary Committee, Representative Barbara Jordan of Texas recommended the impeachment of President Nixon.

A VOICE FROM THE PAST

My faith in the Constitution is whole, it is complete, it is total, and I am not going to sit here and be an idle spectator to the diminution [lessening], the subversion [undermining], the destruction of the Constitution.

Barbara Jordan, speech in House Judiciary Committee, July 25, 1974

In this section, you will read about the events that led Nixon, Jordan, and the nation into the political nightmare that ended his presidency.

The 1972 Presidential Election

The 1972 presidential campaign did not appear to be much of a race for President Nixon. His diplomatic successes in China and the Soviet Union and Kissinger's negotiations to end the Vietnam War were triumphs for the president. The Republicans nominated him overwhelmingly for reelection.

Nixon's Democratic opponent was George McGovern. McGovern was a liberal senator from South Dakota who spoke out against the Vietnam War. McGovern had strong support from young people, African Americans, and members of the women's movement. But Nixon won with the largest victory of any Republican candidate to that time.

HISTORY *through* ART

The Watergate scandal was the subject of thousands of political cartoons. Here, cartoonist Paul Conrad shows Nixon attempting to tap the telephones at the Democratic Party headquarters in the Watergate building.

What do you think the cartoon suggests about Nixon's involvement in the Watergate scandal?

"HE SAYS HE'S FROM THE PHONE COMPANY..."

The Watergate Scandal

An almost certain victory in the 1972 presidential election had not been enough for Nixon. He had wanted to win big. By doing so, he would help Republicans take control of Congress, and he would gain more power. To ensure this landslide victory, many people working for him engaged in various illegal activities.

Reading **History**
A. Making Inferences Why did Nixon want a big win in the 1972 presidential election?

These activities started coming to light on June 17, 1972. Five men were caught breaking into Democratic Party headquarters in the Watergate office-apartment complex in Washington, D.C. The burglars had cameras and listening devices for the telephones. They were linked to Nixon's reelection campaign staff, called the **Committee to Reelect the President**. Nixon may not have known in advance about the break-in. But in less than a week, he was talking to his aides about covering up any White House connection to the Watergate burglary to avoid a scandal.

The cover-up involved lies told by President Nixon and his aides. It involved payments to the Watergate burglars and others to lie. It involved using the CIA to halt an FBI investigation of Watergate. These illegal actions by Nixon and his aides to cover up Watergate and other related crimes came to be called the **Watergate scandal.**

The Watergate break-in stayed on the back pages of most newspapers in the 1972 campaign. But some reporters kept investigating. They found more evidence tying Nixon to Watergate. In February 1973, the Senate began an investigation. The threat to Nixon's presidency was building.

Nixon Resigns

The Senate Watergate investigation began with the questioning of members of the reelection committee and the White House staff. Within six weeks, the investigation was closing in on the president's closest advisers—H. R. Haldeman, John Ehrlichman, former Attorney-General John Mitchell, and John Dean. On March 21, 1973, Dean, one of the president's attorneys, spoke to Nixon about the worsening situation.

Background
Eventually, 25 members of the Nixon administration were convicted and served prison terms for crimes connected to Watergate.

A VOICE FROM THE PAST

I think that there is no doubt about the seriousness of the problem we've got. We have a cancer within, close to the Presidency, that is growing. It is growing daily.

John Dean, quoted in *The White House Transcripts*

Dean was warning about money needed to keep the burglars quiet. Nixon agreed to pay the "hush money." He now clearly had committed a crime.

As more Watergate information was being uncovered, Dean decided to tell all to the Senate. He said that the president had been involved in the cover-up for months. But Nixon denied any knowledge. It was Dean's word against Nixon's until mid-July. Then a White House aide revealed that Nixon had been taping conversations in his office. A long battle over the tapes began in the courts.

Meanwhile, more bad news came for Nixon. It was revealed that Vice-President Spiro Agnew had accepted bribes as governor of Maryland and continued taking them as vice-president. Not wanting to face impeachment, Agnew resigned in October 1973. Nixon then nominated Congressman Gerald Ford of Michigan as the new vice-president.

President Nixon put on an upbeat face as he left Washington after resigning on August 9, 1974.

Evidence of the president's role in the cover-up continued to grow, but he told the country that he had done nothing wrong. He said, "I am not a crook." Then, in January 1974, the House Judiciary Committee began an impeachment investigation. It reviewed court testimony, Senate transcripts, and documents from special prosecutors. Nixon did not give them his tapes but released edited transcripts. The committee felt it had a strong case even without the tapes. In July, both Democrats and Republicans on the committee approved impeachment charges.

On August 5, Nixon was forced by a court order to release full transcripts of the tapes. The evidence that he had been involved in the cover-up from the beginning—what investigators called "a smoking gun"—had been recorded on the tapes. On August 9, 1974, Richard Nixon resigned. Vice-President Gerald Ford was then sworn in as the next president.

Reading **History**

B. Recognizing Effects How did the Watergate scandal change many Americans' view of their government?

Watergate was one of the worst political scandals in the nation's history. Many Americans lost faith in the government and its leaders. This lack of confidence weakened the government, especially the president. As you will read in the following section, the next two presidents—Gerald Ford and Jimmy Carter—worked hard to try to restore the presidency.

Section 2 Assessment

1. Terms & Names

Identify:

- Committee to Reelect the President
- Watergate scandal

2. Taking Notes

Use a time line like the one below to trace the events of the Watergate scandal.

June 1972 — event — event — event — event — August 1974

Which event made Nixon's downfall almost certain?

3. Main Ideas

a. Why did Nixon want a big win in the 1972 election?

b. What kinds of illegal activities was Nixon involved with in the Watergate scandal?

c. What was the outcome of the Watergate scandal?

4. Critical Thinking

Forming and Supporting Opinions What do you think would have happened if President Nixon had apologized for Watergate rather than trying to cover it up?

THINK ABOUT

- when the break-in occurred
- how the nation reacted to past scandals

ACTIVITY OPTIONS

TECHNOLOGY

LANGUAGE ARTS

Interview someone who remembers Watergate. Either create a **radio broadcast** of the interview or write it as a **question-and-answer magazine feature.**

3 Issues of the Seventies

TERMS & NAMES
Gerald Ford
Jimmy Carter
Camp David Accords
environmentalism
Rachel Carson
Iran hostage crisis

MAIN IDEA

Presidents Ford and Carter had a difficult time solving the nation's problems after Watergate.

WHY IT MATTERS NOW

These problems helped lead to a conservative mood in the United States.

ONE AMERICAN'S STORY

On August 9, 1974, Vice-President Gerald Ford became president after Richard Nixon resigned. A year earlier, Nixon had chosen Ford as his vice-president. He replaced Spiro Agnew, who had resigned in disgrace. During 25 years as a congressman from Michigan, Ford had gained a reputation for integrity and openness.

As president, Ford inherited a nation that had suffered through the years of Vietnam and Watergate. He tried to reassure Americans that the turmoil of the Nixon years was behind them in his first speech.

President Gerald Ford speaks to the nation from the White House.

A VOICE FROM THE PAST

My fellow Americans, our long national nightmare is over. Our Constitution works; our great Republic is a Government of laws and not of men. Here the people rule. . . . As we bind up the internal wounds of Watergate, more painful and more poisonous than those of foreign wars, let us restore the golden rule to our political process, and let brotherly love purge our hearts of suspicion and of hate.

Gerald Ford, speech on August 9, 1974

As part of the healing process, Ford decided to pardon former President Nixon. Ford's pardon of Nixon brought him much criticism and added to the divisions in the country.

In this section, you will learn more about the difficulties Presidents Ford and Jimmy Carter faced as they tried to govern in the 1970s.

Ford Takes Over

In his first weeks in office, Ford set out to restore confidence in the presidency. He soothed the nation with his plain speaking, openness, and willingness to talk to the press and to work with Congress.

However, within a month, Ford lost the support of many Americans when he pardoned Richard Nixon for any crimes he might have committed during the Watergate scandal. It had been Ford's hope to spare the country the spectacle of a former president being brought to trial. But

Reading History
A. Drawing Conclusions Why were many Americans upset with President Ford over Nixon's pardon?

many people felt strongly that Nixon should be charged with crimes because of the Watergate cover-up. Ford's popularity dropped sharply.

Making life even tougher for Ford was the fact that the economy was not in good shape. Inflation was spiraling higher while a recession was throwing more people out of work. Many Americans were having a hard time making ends meet. Ford proposed a voluntary campaign to "Whip Inflation Now." He asked Americans to cut spending and energy use. The WIN plan received much publicity but failed to help the economy.

In foreign affairs, Ford had mixed success. He asked Congress to help South Vietnam when the cease-fire in the Vietnam War broke down in 1974. But Congress refused. In 1975, he negotiated a treaty with European nations and Canada called the Helsinki Accords. This pact spelled out basic human rights for the citizens of the signer nations.

Vocabulary
human rights: basic rights and freedoms to which all human beings are entitled, such as freedom of speech

Ford's pardon of Nixon and his difficulty in improving the economy caused him problems in the 1976 presidential campaign. He only narrowly won his party's nomination over California governor Ronald Reagan.Then he lost the presidency in a close election to Democrat **Jimmy Carter**.

Carter was a former peanut farmer and governor of Georgia. He had run for president as a Washington outsider and one who would "never lie" to the American people. He promised honesty in government and support for human rights throughout the world.

America's HERITAGE

THE BICENTENNIAL

On July 4, 1976, the United States celebrated the 200th anniversary of the Declaration of Independence. The Bicentennial, as it was called, was a year-long birthday party. People in every corner of the nation celebrated with special events.

Americans also used the time to reflect on the nation's progress. Women, African Americans, Hispanic Americans, Native Americans, and others still faced discrimination. Yet Americans were proud of what had been achieved, and the celebration helped the nation move beyond the turmoil of the early 1970s.

Carter as President

Reading History
B. Making Inferences How did Jimmy Carter benefit from the Watergate scandal?

Many Americans were still suspicious of their government as Carter took office in 1977. Since Carter had never served in Washington, Americans hoped that he would bring fresh ideas to the presidency.

Carter immediately tried to show that he was one of the people. On inauguration day, he and his family walked from the Capitol to the White House rather than take the traditional limousine. However, being a Washington outsider would make political life difficult for Carter.

Carter and Congress often clashed. One point of conflict was the energy crisis. Early in 1977, shortages of oil and natural gas forced many schools and businesses to close. In response, Carter asked Americans to conserve energy. He also sent a national energy program to Congress. It would cut oil imports, increase production of oil and natural gas at home, and promote alternative energy sources like coal and nuclear and solar energy.

After months of debate, Congress passed some of the measures. But they were of little help when OPEC again sharply raised oil prices. Inflation surged beyond 10 percent. Unemployment rose. Like Nixon and Ford, Carter could not solve the nation's economic problems.

Connections TO WORLD HISTORY

MIDDLE EAST PEACE PROCESS

The signing of the Camp David Accords in 1978 by (left to right) Prime Minister Menachem Begin of Israel, President Carter, and President Anwar el-Sadat of Egypt was the first important step in the attempt to bring peace to the Middle East.

In 1998, another step was taken when President Bill Clinton brought Palestinian leader Yasir Arafat and Israeli Prime Minister Benjamin Netanyahu together to sign a peace agreement called the Wye River Accords.

Carter had more success in accomplishing his foreign-policy goals. He wanted to end the long-standing conflict with Panama over the Panama Canal. As you read in Chapter 23, the Panama Canal was built and controlled by the United States. Most Americans wanted it to stay that way. But Carter thought winning the good will of Latin America was worth losing control of the canal. Under treaties signed in 1977, the United States agreed to give the canal to Panama in 2000.

Carter also tried to reduce tensions in the Middle East. In 1978, he helped to negotiate the **Camp David Accords.** Under these agreements, Egypt and Israel signed the first peace treaty between Israel and an Arab nation, thus ending 30 years of conflict.

The Environmental Movement Begins

Protection of the environment was also a goal of Carter's. He supported a movement to save the environment that had gained momentum in the 1970s. Actually, the first laws to protect the nation's natural environment had been passed in the late 1800s. But **environmentalism,** or work toward protecting the environment, only began to attract wide public attention in the 1960s. In 1962, biologist **Rachel Carson** wrote of the dangers of heavy pesticide use in her bestseller, *Silent Spring.* She warned that some of these chemicals could kill animals, cause disease, and destroy the environment unless their use was limited or stopped.

Reading **History**

C. Finding Main Ideas What role did Rachel Carson play in the environmental movement?

In 1969, a huge oil spill near Santa Barbara, California, polluted miles of beaches and killed many marine animals. The cry for tougher laws to protect the environment grew. In the 1970s, Nixon, Ford, and Carter all proposed laws to restrict pesticide use, to regulate the cleanup of oil spills, and to curb air and water pollution. With these laws in effect, many polluted lakes began to recover, and high levels of some air pollutants began to drop. But environmental disasters continued to occur.

In 1979, an accident occurred at the Three Mile Island nuclear power plant in Pennsylvania. Radioactive water leaked out of the plant, causing fears that the nuclear reactor might explode. Within a week, the reactor was shut down. Disaster was averted. To assure people it was safe, Carter visited Three Mile Island. But not long after, he faced another disaster—a political one—in Iran.

Reagan and the Conservatives Win

*Reading*History

D. Reading a Map Locate Tehran, the capital of Iran, in the map on this page.

For decades, the United States had supported the Shah (king) of Iran. In 1979, Muslim leaders overthrew his government. When Carter allowed the Shah to come to the United States for medical treatment, Iranians struck back at the United States. On November 4, 1979, they overran the American embassy in Iran's capital of Tehran and took 52 Americans hostage. The **Iran hostage crisis** had begun.

Carter tried negotiating to get Iran's leaders to release the hostages but without success. He approved a secret military mission, but it failed. The continuing crisis affected the election in 1980. Americans blamed Carter for the plight of the hostages, for the nation's economic ills, and for making America look weak to the world.

Meanwhile, support for conservative ideas had been growing for more than a decade. The Republicans chose a conservative—Ronald Reagan, a former actor and California governor—to be their candidate in 1980. He vowed that, if elected president, he would not allow the United States to be pushed around. Reagan's get-tough talk appealed to many voters.

Carter eventually won release of the hostages. But the majority of voters had already decided that it was time for a change and elected Reagan president. The hostages left Iran on January 20, 1981, the day Reagan was inaugurated. In the next chapter, you will read how Ronald Reagan took the nation in a more conservative direction.

Section 3 Assessment

1. Terms & Names

Identify:
- Gerald Ford
- Jimmy Carter
- Camp David Accords
- environmentalism
- Rachel Carson
- Iran hostage crisis

2. Taking Notes

Use a chart to list the high and low points of both the Ford and Carter presidencies.

Presidency	High points	Low points
Gerald Ford		
Jimmy Carter		

Which do you think was the highest point of each? Why?

3. Main Ideas

a. Why did Carter win the 1976 election over Ford?

b. What were some problems related to energy use that occurred during Carter's term?

c. What progress did the environmental movement make during the 1960s and the 1970s?

4. Critical Thinking

Comparing How would you compare the strengths and weaknesses of Ford and Carter? Use events from their presidencies to support your answer.

THINK ABOUT
- actions each took to draw Americans together
- response to economic and foreign-policy issues

ACTIVITY OPTIONS

MATH
ART

Research inflation in the 1970s. Make a **graph** to show the annual rate change or create a **political cartoon** about inflation's effect on people's lives.

Save the Environment!

You are a citizen of the 1970s who has joined the fight against pollution, waste, and the destruction of the Earth's natural resources. Each year on Earth Day, you help out with community clean-ups. You are interested in finding ways to rally public opinion and help save the planet.

COOPERATIVE LEARNING On this page are three challenges you face as an American concerned about the environment. Working with a small group, decide how to deal with each challenge. Select an option, assign a task to each group member, and do the activity. You will find useful information in the Data File. Present your solutions to the class.

LANGUAGE ARTS CHALLENGE

"the community of living things"

You are worried that lack of concern about the environment may result in a permanently damaged world. The destruction of natural resources by individuals and industry continues. You decide to promote a campaign to save the environment. Present your ideas using one of these options:

- **Write an editorial describing environmental dangers and promoting conservation.**
- **Make a speech to rally people in support of conservation.**

Take Public Transportation

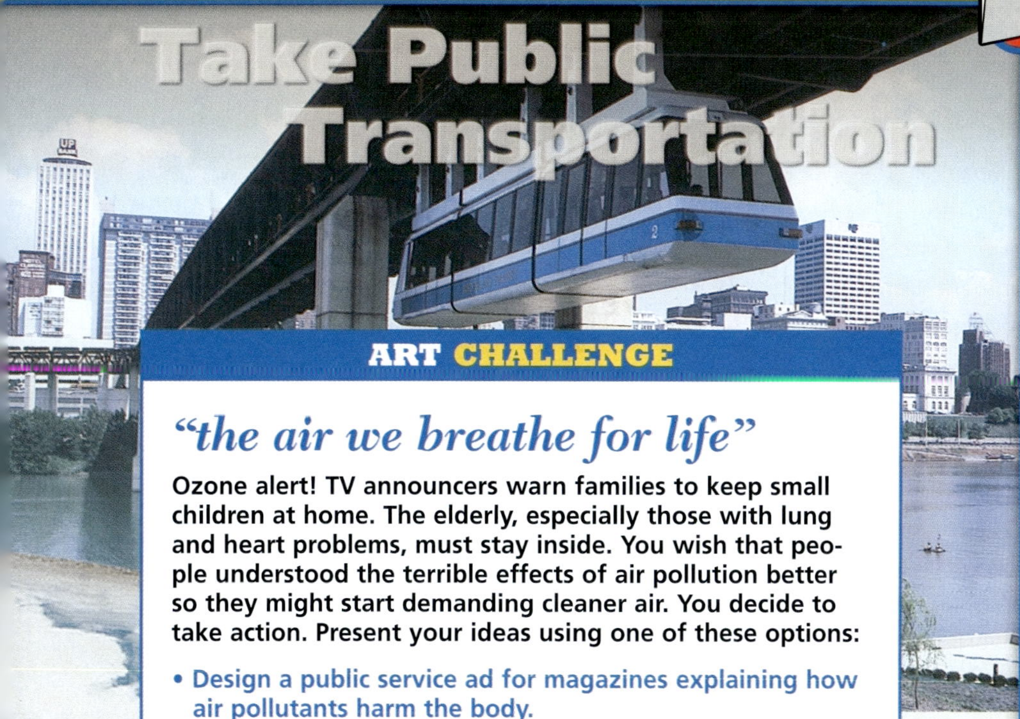

ART CHALLENGE

"the air we breathe for life"

Ozone alert! TV announcers warn families to keep small children at home. The elderly, especially those with lung and heart problems, must stay inside. You wish that people understood the terrible effects of air pollution better so they might start demanding cleaner air. You decide to take action. Present your ideas using one of these options:

- Design a public service ad for magazines explaining how air pollutants harm the body.
- Draw a cartoon asking people to use public transportation.

Prevent Pollution

SCIENCE CHALLENGE

"the lakes and rivers and oceans"

News of an oil spill off California's shores has sparked your concern for the environment. You are sickened by pictures of oil-soaked beaches and thousands of injured and dead animals. You want to help rid the oceans of this menace. Use the Data File for information. Then present your ideas, using one of these options:

- Create a labeled diagram for your local paper showing why oil pollution harms most sea life.
- Role-play a community meeting in which you explain how offshore oil pollution damages sea life.

ACTIVITY WRAP-UP

Present to the class Meet with the group to review your methods of helping to save the environment. Select a solution that best meets each challenge and then present them to the class.

DATA FILE

ENVIRONMENTAL PROBLEMS

Air and water pollution, acid rain, destruction of the ozone layer, toxic chemical disposal, extinction of wildlife, oil spills, destruction of forests, overpopulation

OCEAN LIFE ZONES

- Shorelines support crabs, oysters, shore birds, and other marine life.
- Open ocean over the continental shelf supports the largest amount of sea life, including algae, lobsters, thousands of fish species, turtles, seals, sharks, and whales.
- Deep ocean supports a smaller variety.

SOURCES OF OCEAN OIL POLLUTION

Offshore oil production, vessel accidents, natural seepage, nontanker and tanker shipping operations, waste car oil

AIR POLLUTANTS

- sources—fossil fuels, cars, power plants, industry, ozone
- effects—smog; eye irritation; impaired judgment; chest pains; lung, nerve, and kidney damage; cancer; birth defects

WAYS TO CLEAN WATER

- Install oil pollution prevention equipment on ships.
- Car owners, cities, and industry recycle used oil.

REDUCE AIR POLLUTION

- Burn fewer fossil fuels.
- Take public transportation.
- Improve car efficiency.
- Use sun and wind power.

INTERNET Visit www.mcdougallittell.com to learn more about the environment.

Chapter 31 ASSESSMENT

TERMS & NAMES

Briefly explain the importance of each of the following.

1. Richard M. Nixon
2. Henry Kissinger
3. revenue sharing
4. détente
5. Watergate scandal
6. Gerald Ford
7. Jimmy Carter
8. Camp David Accords
9. environmentalism
10. Iran hostage crisis

REVIEW QUESTIONS

Nixon Confronts Problems (pages 855–858)

1. What kinds of government programs did Nixon propose to deal with America's social problems?
2. How did Nixon try to promote "law and order"?
3. What were the economic problems Nixon faced?
4. How was Nixon's foreign policy toward China different from that of previous presidents?

Watergate Brings Down Nixon (pages 859–861)

5. What were the chief causes of the Watergate scandal?
6. What did reporters and others find when they looked more deeply into the Watergate burglary?
7. What were the events that caused Nixon to resign?

Issues of the Seventies (pages 862–867)

8. What were the main reasons people did not vote for Gerald Ford in the 1976 election?
9. What successes did Carter have with foreign policy and what problems did he have at home?
10. What appealed to voters about Ronald Reagan in the 1980 election?

CRITICAL THINKING

1. USING YOUR NOTES

President	Issues and Events
Nixon	
Ford	
Carter	

Use your completed chart to answer these questions.

a. What issues did all three presidents have to deal with during their terms? Which were the most frequently occurring issues of their presidencies?

b. Which of the presidents do you think faced the most difficult issues? Explain.

2. ANALYZING LEADERSHIP

What characteristics did Nixon, Ford, and Carter have that made them good leaders? Poor leaders?

3. THEME: CITIZENSHIP

What could Americans do to help their country during the 1970s?

4. FORMING AND SUPPORTING OPINIONS

Do you think Nixon's actions showed that his main goal was to survive during the events of Watergate? Explain your answer.

5. EVALUATING

Do you think that Ford made a good decision in pardoning Nixon? Explain why or why not.

Interact *with* History

Did your ideas about how the nation survives a crisis agree or disagree with what you read?

VISUAL SUMMARY

Years of Doubt

The Nixon Administration

- Revenue sharing
- Law-and-order politics
- Inflation, recession, and unemployment
- Opening to China
- Détente with the Soviet Union
- Watergate scandal
- Nixon resignation

The Ford Administration

- Unelected president
- Nixon pardon
- Whip Inflation Now program
- Helsinki Accords

The Carter Administration

- Energy crisis
- Worsening inflation
- Panama Canal Treaties
- Camp David Accords
- Environmental protection
- Iran hostage crisis

HISTORY SKILLS

1. INTERPRETING GRAPHS

Study the graph and then answer the questions.

Major Energy Sources, *1850–1980*

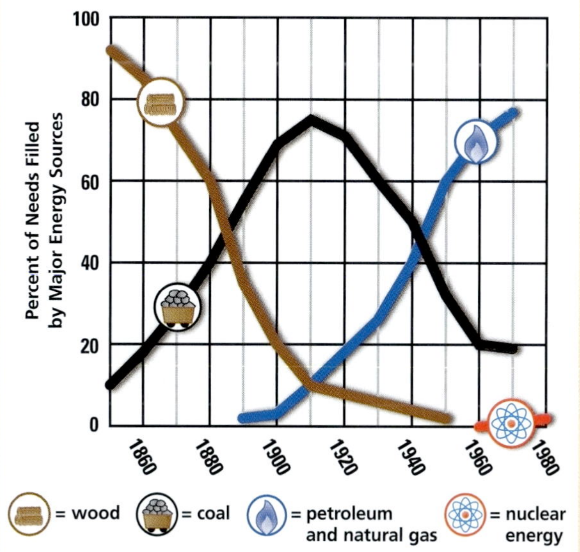

Source: *U.S. Department of Energy*

a. What fuel supplied most of the nation's energy between 1850 and 1880?

b. When did petroleum and natural gas begin being used as fuel?

c. What was the percentage increase in the use of coal from 1850 to 1910?

2. INTERPRETING PRIMARY SOURCES

The following excerpt was written by a citizens' group after a 1970 oil spill off the coast of Santa Barbara, California. Read the excerpt and then answer the questions.

> Centuries of careless neglect of the environment have brought [hu]mankind to a final crossroads. The quality of our lives is eroded and our very existence threatened by our abuse of the natural world. . . . We, therefore, resolve to act. We propose a revolution in conduct toward an environment which is rising in revolt against us. Granted that ideas and institutions long established are not easily changed; yet today is the first day of the rest of our life on this planet. We will begin anew.
>
> Santa Barbara Declaration of Environmental Rights, 1970

a. What kinds of problems does this excerpt from the *Santa Barbara Declaration* discuss?

b. What is the general solution that it proposes?

ALTERNATIVE ASSESSMENT

1. INTERDISCIPLINARY ACTIVITY: Geography

Making a Map Research the territory that was the subject of the negotiations at Camp David in 1979. Make a three-dimensional map that shows the land, including its location relative to Egypt, Israel, and the other countries of the Middle East. Use your map to identify the resources of the territory.

2. COOPERATIVE LEARNING ACTIVITY

Performing a Play The fall of President Richard Nixon was one of the most dramatic events of modern American history. Working in a small group, research Nixon's last two years in office. Pick one event from this time and write a one-act play about the incident. Use primary sources, such as diaries or tape transcripts, to create your dialogue. Present your play and ask for comments from your classmates.

3. TECHNOLOGY ACTIVITY

Designing a Bicentennial Web Site The 200th anniversary of independence was a cause for celebration for Americans everywhere. President Ford told the nation to "break out the flag, strike up the band, light up the sky." It was a time to reflect on how the nation began, where it was then, and what might happen next. Use the library and the Internet to find information about the celebration. Also make use of what you have learned about the nation's history from your textbook.

INTERNET **Visit www.mcdougallittell.com to learn more about the United States Bicentennial celebration.**

Design a Bicentennial Web site. Use the suggestions below to get started.

- Locate images showing Bicentennial highlights.
- Choose music that captures the spirit of the celebration.
- Highlight important events in the nation's history.
- Include the words and images of historic figures.
- Select Web sites that would be good links for visitors to your site.

4. PORTFOLIO ACTIVITY

PORTFOLIO **Option 1** Review your section and chapter assessment activities. Select one that you think is your best work. Then use comments made by your teacher or classmates to improve your work and add it to your portfolio.

PORTFOLIO **Option 2** Review the questions that you wrote in your notebook for What Do You Want to Know? on page 854. Then write a short report explaining your answers. Add your work to your history portfolio.

CHAPTER 32

Entering a New Millennium

1981–present

Section 1 **Conservatives Reshape Politics**

Section 2 **Technological and Economic Changes**

Section 3 **The New Americans**

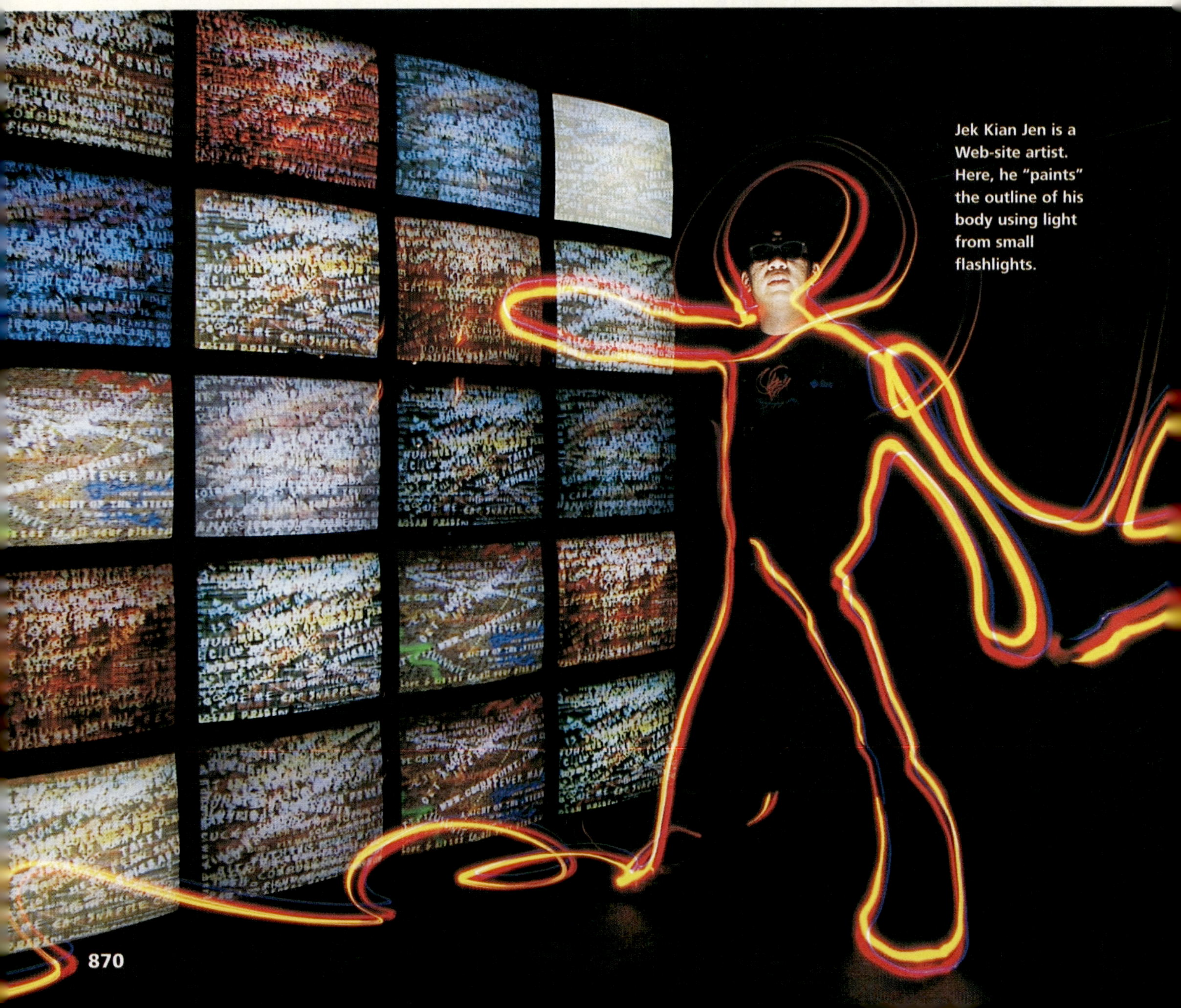

Jek Kian Jen is a Web-site artist. Here, he "paints" the outline of his body using light from small flashlights.

Interact *with* History

Sixteen-year-old Andre McGregor is already a high-tech businessman.

Shown holding a computer board, he designs Web pages. He also tutors adults in how to use technology.

The time is four years from now, and you're about to finish high school. Your friends ask about your plans for the future. At graduation, the speaker urges your class not just to focus on earning money but also to care about making the world a better place. How can you do that?

What Do You Think?

- What are your talents, and how could you use them to benefit both yourself and society?
- What things do you really enjoy doing?
- What do you think the United States and the world need most from your generation?

What can you contribute to the future?

USA

1981

1982 Barney Clark receives first artificial heart.

1984 President Reagan is reelected.

1986 The space shuttle *Challenger* explodes, killing all seven astronauts.

1988 Reagan's vice-president, George Bush, is elected president.

1990 The Hubble Space Telescope is launched.

1992 Democrat Bill Clinton is elected president.

1994 The Republicans gain control of both houses of Congress.

1996 Clinton is reelected.

1998 Clinton is impeached.

present

World

1982 Great Britain defeats Argentina in a war over the Falkland Islands.

1985 An earthquake in Mexico City kills thousands.

1989 The Berlin Wall is taken down.

1991 The Soviet Union breaks apart.

1994 In South Africa's first all-race election, Nelson Mandela is elected president.

1997 Scottish scientists clone a sheep.

BEFORE YOU READ

This magazine cover is symbolic of U.S. diversity. To create it, *Time* used a computer to blend the features of people of many different races.

Previewing the Theme

Diversity and Unity As this chapter explains, during the 1980s and 1990s, Republicans and Democrats fought for control of the national government. World events also affected U.S. politics. At the same time, technology changed daily life, and U.S. society became more diverse as immigrants from around the world continued to come to America.

What Do You Know?

How closely do you follow current events? What do you think are the major trends in the United States today?

THINK ABOUT

- political events in the news
- the latest changes in technology
- how immigration is affecting the United States

What Do You Want to Know?

What would you like to know about our most recent presidents? What questions do you have about technology? What do you want to learn about recent patterns of immigration? Record your questions in your notebook before you read this chapter.

READ AND TAKE NOTES

Reading Strategy: Recognizing Effects The consequences of an event are its effects. As you read the chapter, look for major events in the categories of politics, technological and economic change, and immigration. Notice how these events affected U.S. society. Record your information on a chart like the one below.

See Skillbuilder Handbook, page R10.

Influences on America, 1981–Today

Politics	Technological and Economic Change	Immigration
Effects	Effects	Effects

1 Conservatives Reshape Politics

MAIN IDEA

The country became more conservative, leading to Republican political victories.

WHY IT MATTERS NOW

In response, the Democratic Party became less liberal and even adopted some conservative ideas.

TERMS & NAMES

- Ronald Reagan
- supply-side economics
- Iran-Contra affair
- George Bush
- Persian Gulf War
- Bill Clinton
- NAFTA

ONE AMERICAN'S STORY

As the owner of a popular restaurant, Mike Savic knew the mood in his working-class Chicago neighborhood. In the fall of 1980, most of his customers were frightened. Unemployment and inflation were rising. President Jimmy Carter seemed to have no answers. Yet, Carter was asking the American people to reelect him. Most of Savic's neighbors had supported Carter in 1976, but now they turned to Carter's Republican opponent, former California governor <u>Ronald Reagan</u>.

This campaign button is from the 1980 debate between Jimmy Carter and Ronald Reagan.

A VOICE FROM THE PAST

People have been talking. . . . Some of them in this neighborhood have been laid off four, six months. Or their neighbor is out of work. People don't like to see their neighbors out of work. And they've been scared. . . . The ones who'd been talking that they were for Carter, well, they changed. They were talking different. They were going for Reagan.

Mike Savic, quoted in the *Chicago Tribune,* November 6, 1980

In 1980, millions of Democrats voted for Reagan in the hope that he could fix the economy. With their help, Reagan won the election. This section covers the presidencies of Reagan and his two successors.

Reagan's Conservative Goals

President Reagan was a conservative. In his 1981 inaugural address, he declared, "Government is not the answer to the problem; government is the problem." Reagan pursued the following conservative goals.

1. **Lower Taxes.** Reagan preached **<u>supply-side economics</u>.** This theory held that if people paid fewer taxes, they would save more money. Banks could loan that money to businesses, who could invest in ways to improve productivity. The supply of goods would increase, driving down prices. At Reagan's urging, Congress lowered income taxes by 25 percent over three years.
2. **Deregulation.** The president deregulated, or eased restrictions on, many industries. Reagan believed that business would grow more rapidly if government interfered with it less.

AMERICA'S HISTORY MAKERS

RONALD REAGAN

1911–

One reason for Ronald Reagan's popularity was his unfailing optimism at a time when many Americans felt uncertain. Not even an assassination attempt could dampen his spirits.

In 1981, a gunman shot the president. When Reagan's wife, Nancy, arrived at the hospital, Reagan told her, "Honey, I forgot to duck."

In addition, he remarked to the doctors who were about to operate on him, "Please tell me you're Republicans."

Why might it be important for a leader to convey a sense of optimism and high spirits?

3. **Fewer Government Programs.** Reagan fought to end or weaken many government programs, from affirmative action to environmental regulations.
4. **A Conservative Supreme Court.** Reagan named three conservative judges to the Supreme Court. One of them, Sandra Day O'Connor, was the first woman to sit on the nation's highest court.

At first, inflation rose, and unemployment stayed high. But by 1983, inflation decreased, and more people found jobs. Business boomed. Even so, Reagan's policies created a problem. Because of the tax cut, the federal government took in less money and had to resort to deficit spending. As a result, the national debt doubled from 1981 to 1986.

Vocabulary
deficit spending: using borrowed money to fund government programs

A Tough Anti-Communist Stand

Reagan opposed communism. To compete militarily with the Soviet Union, he began the most expensive arms buildup in history. It cost more than $2 trillion.

In 1985, Mikhail Gorbachev became the new leader of the Soviet Union. He and Reagan met four times to discuss improving U.S.-Soviet relations and easing the threat of nuclear war. They signed the Intermediate-Range Nuclear Force (INF) Treaty in 1987. Under that treaty, the two countries agreed to destroy all of their medium-range missiles.

The Reagan administration also decided to support the anti-Communist side in several conflicts, including two Central American civil wars. In El Salvador, the United States backed the government against Communist-led rebels. In Nicaragua, America provided aid to anti-Communist rebels known as Contras.

This aid resulted in a scandal, known as the **Iran-Contra affair**. In 1986, Americans learned that the U.S. government had sold weapons to Iran in return for help in freeing American hostages in the Middle East. The money from these sales went to the Contras. This action violated a law that barred the U.S. government from funding the rebels. President Reagan claimed he never knew about the deal. But investigators concluded that he should have kept track of what his administration was doing.

Reading History
A. Analyzing Causes Why did investigators hold Reagan responsible for the Iran-Contra affair?

Bush and a Changing World

Despite the scandal, Reagan and his administration remained popular. In 1988, Reagan's vice-president, **George Bush**, ran for president and won. During his presidency, dramatic foreign events took place.

In 1989, several Eastern European countries ended Communist rule. This angered old-time Communists in the Soviet Union. In August 1991, a group of them tried to take over the Soviet government. Boris Yeltsin, a Russian reform leader, fought the takeover attempt and won.

Unlike Gorbachev, who wanted to reform communism, Yeltsin and others wanted to get rid of it. One by one, the republics that made up the Soviet Union declared their independence from it. In December 1991, Yeltsin and the leaders of these nations joined in a loose alliance called the Commonwealth of Independent States (CIS). The Soviet Union, once a superpower, was gone. Its breakup marked the end of the Cold War.

A crisis also erupted in the Middle East. In August 1990, Iraq invaded its neighbor Kuwait—a major supplier of oil. The United States, led by Bush, and the United Nations (UN) organized a group of 39 nations to free Kuwait by fighting the **Persian Gulf War.** In mid-January 1991, they began bombing Iraqi military targets. A month later, ground forces moved into Kuwait and drove the Iraqis out of that country.

Reading **History**

B. Summarizing What were two major world events that happened during George Bush's presidency?

The war's success boosted George Bush's popularity, and he seemed certain to win reelection in 1992. Then the economy stalled. By the spring of 1992, the U.S. unemployment rate had climbed to around 7 percent—a six-year high. Americans began to think that Bush was good at foreign policy but ineffective with problems at home. In November, they elected his Democratic opponent, Arkansas governor **Bill Clinton.**

Connections TO WORLD HISTORY

THE BERLIN WALL FALLS

Communists built the Berlin Wall in 1961 to separate Communist East Berlin from West Berlin. In November 1989, as communism began to fall, East Germans tore down the wall. The photograph below shows Germans celebrating the opening of the wall. Andreas Ramos witnessed the event.

The final slab was moved away. A stream of East Germans began to pour through. . . . Looking around, I saw an indescribable joy in people's faces. It was the end of the government telling people what not to do, it was the end of the Wall, the war, the East, the West.

Clinton's Fights with Congress

One of President Clinton's first acts was to try to reform the health-care system. Millions of Americans had no insurance, and health-care costs were rising sharply. Clinton asked his wife, Hillary Rodham Clinton, to design a health-care plan—which she presented to Congress in 1993. Opponents criticized the plan as costing too much, and Congress chose not to vote on it. But it did pass a law allowing workers to keep their medical insurance when they change jobs.

Background

Most developed countries have nationalized health insurance, a government program that pays for the health care of most citizens. The United States does not.

Clinton did win passage of the North American Free Trade Agreement **(NAFTA)** in 1993. NAFTA would eventually eliminate tariffs in an effort to increase trade among Mexico, Canada, and the United States.

In the 1994 elections, Republicans took control of both houses of Congress. Clinton and the new Congress could not agree on a budget for 1995. Because funding was uncertain, government agencies shut down twice in late 1995. The American public grew angry with both sides. Finally, they reached a budget agreement.

The two sides did work together to pass welfare reform, a long-time conservative goal. In August 1996, Clinton signed a bill ending the federal government's guaranteed aid to needy families. The law gave states grants of money to spend on welfare. It also cut the length of time people could receive benefits.

Weather
Today: Variably cloudy, showers. High 53. Low 44.
Monday: Variably cloudy, warmer. High 62. Low 45.
Details, Page B2.

The Washington Post

Inside: Book World, TV Week, The Post Magazine, Comics
Today's Contents on Page A2

122nd Year No. 15 ... ... ABCFCK | Sunday, December 20, 1998 | $1.50

Prices may vary in areas outside metropolitan Washington. (See box on Page A2)

Clinton Impeached

House Approves Articles Charging Perjury, Obstruction

Although the House impeached President Clinton, the Senate did not find him guilty.

In 1996, Clinton was reelected. But his second term in office was marked by scandal. Independent Prosecutor Kenneth Starr was assigned to investigate a land deal that Clinton took part in during the 1970s. During that investigation, Starr learned damaging information. The president had had an improper relationship with a young White House intern. And he allegedly had lied under oath about it.

When the charges became public in January 1998, Clinton denied them. Later, he admitted to the relationship but insisted that he hadn't lied under oath. In December 1998, the House of Representatives voted to impeach him. The House approved two articles of impeachment, charging Clinton with perjury and obstruction of justice. In general, Republicans voted for impeachment while Democrats voted against it. This led many people to believe that the prosecution of Clinton was part of a political attack.

Background
Clinton was the second president to be impeached, or formally charged with wrongdoing. The first was Andrew Johnson. (See Chapter 18.)

In spite of the charges, Clinton remained popular. The Senate opened its trial of President Clinton in January 1999. Nearly a month later, the Senate acquitted him on both charges. Bill Clinton remained in office.

War in Kosovo

Despite his troubles at home, the president still had to act as a world leader. In 1999, Clinton helped to lead a group of nations dealing with a crisis in Europe. The Eastern European country of Yugoslavia had been created after World War I. It had six republics, the largest of which was Serbia. Many ethnic and religious groups lived in Yugoslavia. Some of them distrusted the other groups because of ancient grudges.

Vocabulary
ethnic: cultural

In the early 1990s, four Yugoslav republics declared independence. The Serbs tried to stop them, leading to a bloody war. In time, the UN negotiated peace in the region. The four republics remained independent—leaving only Serbia and Montenegro within Yugoslavia.

Serbia has a region called Kosovo (KAW•suh•VOH), inhabited mostly by people of Albanian descent. The Albanians speak a different language and practice a different religion from the Serbs. Serbia tried to wipe out Albanian culture in Kosovo, so Kosovo sought independence. In the late 1990s, a group called the Kosovo Liberation Army (KLA) began to carry out violent attacks against Serbian officials. The Serbian government bombed villages and fought the KLA. It also began trying to drive the Albanians out of their homes in Serbia by using violence and murder.

***Reading* History**
C. Analyzing Causes How did both sides contribute to the Kosovo conflict?

In 1999, the North Atlantic Treaty Organization (NATO) took action against Serbia. The United States took a lead role in this effort. In March, NATO began to drop bombs on Serbia. Several months of bombing and international negotiations followed. In June 1999, the Serbian leader agreed to withdraw his troops from Kosovo. After the troops pulled out, officials discovered horrifying evidence of torture and massacres carried out by Serbian forces.

Ongoing Issues

The conflict in Kosovo left many Americans feeling concerned about U.S. intervention overseas. Even after the Serbian troops withdrew, people wondered how a permanent peace could come to a region that had known such bitter conflicts.

The impeachment of Clinton also caused anxiety. Some people feared that the scandal would fuel the public's growing disgust with politics. Others wondered if the intense media coverage of politicians would discourage citizens from running for public office. Journalist Jonathan Alter saw both happening.

Reading **History**

D. Recognizing Effects What effect does Alter believe the impeachment will have on politics?

A VOICE FROM THE PAST

The country will grow even more contemptuous [scornful] of those who serve in what Teddy Roosevelt called "the arena." Getting good people to run for office will be hard in this meaner political culture.

Jonathan Alter, *Newsweek,* October 19, 1998

Despite these worries, many Americans felt excited about the future because of advances in technology. Section 2 discusses those changes.

Section 1 Assessment

1. Terms & Names

Identify:

- Ronald Reagan
- supply-side economics
- Iran-Contra affair
- George Bush
- Persian Gulf War
- Bill Clinton
- NAFTA

2. Taking Notes

Use a chart like the one shown to record important details about the terms of three presidents.

Reagan	
Bush	
Clinton	

What do you think was the greatest achievement by any of these presidents?

3. Main Ideas

a. What were the positive and negative effects of Reagan's economic policies?

b. Why was Bush defeated when he ran for reelection?

c. Why did Clinton and other NATO leaders decide to bomb Serbia?

4. Critical Thinking

Drawing Conclusions
What do you think was the strongest force shaping U.S. politics from 1981 to 2000—economics or foreign affairs?

THINK ABOUT

- Reagan's goals and actions
- the events of Bush's presidency
- the events of Clinton's presidency

ACTIVITY OPTIONS

ART

TECHNOLOGY

You have been asked to summarize the politics of the years 1981–2000. Create an **illustrated time line,** or plan part of a **multimedia presentation.**

2 Technological and Economic Changes

TERMS & NAMES
Internet
e-commerce
service economy
information revolution
downsizing

MAIN IDEA

Advances in science and technology have improved daily life and created a global economy.

WHY IT MATTERS NOW

Technology continues to change American homes, leisure activities, and workplaces.

ONE AMERICAN'S STORY

In the late 1990s, technology transformed business. One reason was the **Internet**, a worldwide computer network. Some enterprising people developed **e-commerce**, business that is conducted over the Internet.

Jeff Bezos was a pioneer of e-commerce. In 1994, he decided to open a bookstore on-line. The idea caught on quickly. From 1997 to 1999, the number of customers using his bookstore grew from 2 million to 11 million. Bezos doesn't define his job as selling books but as giving buyers more choice.

A VOICE FROM THE PAST

Our business is helping customers make purchasing decisions. . . . It all has to do with the balance of power shifting away from companies and toward consumers.

Jeff Bezos, quoted in "Companies Wired for the Bottom Line," *Newsweek*, September 20, 1999

Jeff Bezos revolutionized the bookselling industry by putting it on-line.

Although 1999 passed without his company's making a profit, Bezos wasn't worried. He was investing his earnings to make the business grow for the future. Other businesspeople seemed to agree with Bezos's optimism about e-commerce. The number of Internet businesses exploded as the year 2000 approached. Section 2 explains other ways that the Internet and technology changed American life.

Technology and Daily Life

As the 20th century drew to a close, thousands of institutions from hospitals to airports to banks relied on computers to perform essential tasks. Computer use also grew in homes and schools. By 1997, about 35 percent of all U.S. households had a personal computer. And from 1985 to 1998, the number of computers in classrooms leaped from 630,000 to more than 8 million.

Computers and the Internet revolutionized both communication and research. Using the Internet, a person can track down information on nearly any subject. Internet users can also send and receive electronic messages called e-mail. In addition, they can shop at on-line stores.

Other forms of new technology have transformed American life. One popular example is the battery-powered cellular telephone. People can carry these phones with them anywhere. Between 1990 and 1997, the number of cellular phone subscribers in the United States grew dramatically from 5.3 million to 55.3 million.

STRANGE *but* True

INTERNET 911

In April 1997, 12-year-old Sean Redden was using the Internet at his home in Texas when he saw a startling message: "Help me. I'm having trouble breathing."

The message was from 20-year-old Tarja Laitinen of Finland. While using the Internet, she'd had a severe attack of asthma.

Sean's mother called 911. This led to a series of phone calls that ended up with Tarja's being rushed to a Finnish hospital. Sean, with the help of the Internet, had saved Tarja's life!

A Changing Economy

For much of the 20th century, manufacturing made up a major share of the U.S. economy. By the year 2000, that had changed drastically. The computer age fueled the growth of the service economy. In a **service economy,** most jobs provide services instead of producing goods. By 1996, about 71 percent of all workers had jobs in the service industry. Many of these jobs—such as lawyers, teachers, engineers, and Web-site designers—focused on providing knowledge and information.

Reading **History**

A. Drawing Conclusions Why is the Internet an important part of the information revolution?

Some experts said the United States was going through an **information revolution.** This meant that technology had radically changed the way information was delivered—and gave people access to far more information than ever before. This improved the productivity of many industries.

Computers changed not just industries, but also the lives of most individual workers. For example, engineers began to use computers to test new car designs. In large companies, workers from many departments were able to share information through computer networks. Filmmakers used computers to create animated characters that would react on-screen with live actors.

Economic Change Affects Workers

The dramatic changes in the economy were both good news and bad news for U.S. workers. During the 1990s, high-tech industries created many new, high-paying jobs for skilled workers.

Yet other workers faced hard times. The decline of manufacturing meant the loss of factory jobs. The newly unemployed workers faced an uncertain future. They had earned good wages in the factory, although many of them were not highly skilled or educated. Most new jobs required specialized skills. As a result, job seekers had to obtain education and training. Or they had to settle for lower-paying jobs.

Background

Downsizing most often occurs during hard times but also can take place when the economy is strong.

Highly skilled workers also lost jobs. Many corporations engaged in **downsizing**—reducing the number of workers to increase company profits. During the 1990s, companies let go of hundreds of thousands of workers. In addition, many mergers—in which two companies join—took place. When two companies merge, it creates a situation in which people hold duplicate jobs, so mergers usually lead to layoffs.

Technology OF THE Time

The Computer Revolution

Would you be impressed if a scientist came to you and said, "I can take a machine, reduce it to one-thousandth of its present size, and yet increase its power"? That is exactly what happened with computers during the last half of the 20th century. Today, personal computers can perform more operations more quickly than the first giant computers did. In addition, the first computers were very expensive and difficult to maintain. The development of inexpensive personal computers made it possible for small businesses and ordinary families to use the latest technology.

1940s

ENIAC One of the first general-purpose electronic computers, the ENIAC used 18,000 vacuum tubes (which looked a bit like light bulbs). ENIAC took up 1,500 square feet, about one-third of a basketball court. It was used for mathematical calculations and could do 5,000 additions a second.

1970s

Commodore PET Computer chips made the personal computer possible. The Commodore PET was the first personal computer designed to be sold to the public. It cost only $595—but had just 12K of memory.

Computer Chip Vacuum tubes were replaced first by transistors and later by tiny computer chips. By the 1970s, all of a computer's operations were contained on a small number of chips, wired to a control board.

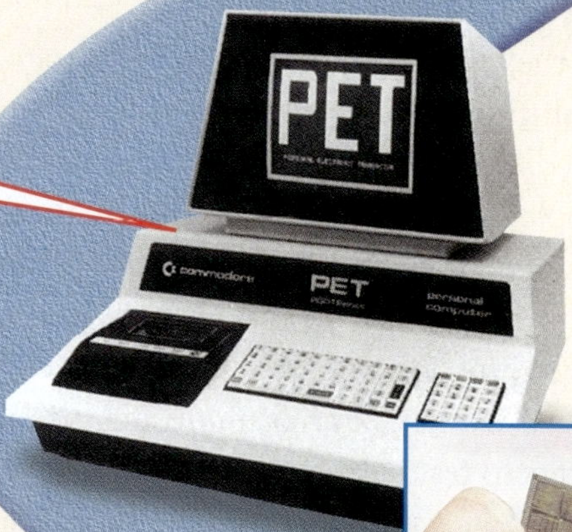

1990s

Virtual Reality The personal computers of the 1990s grew increasingly powerful and were able to perform sophisticated jobs. Here, an interior designer uses a virtual-environment program to recreate a room. Computers were also used in engineering and film animation.

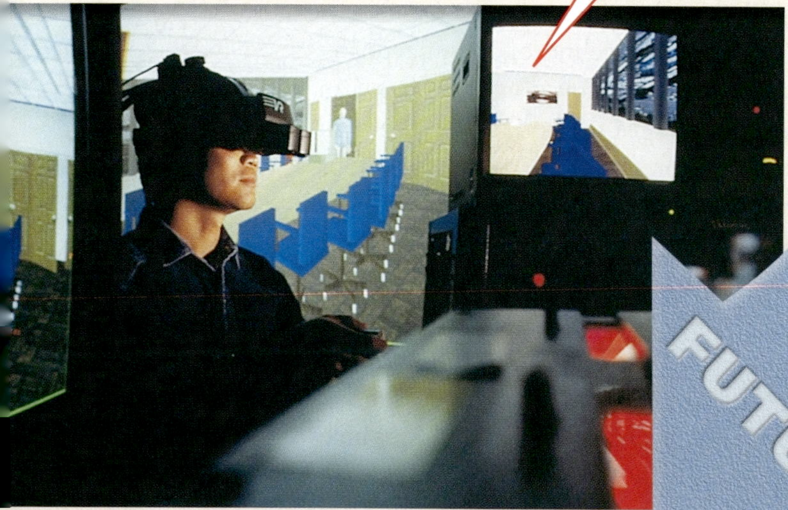

FUTURE

CONNECT TO HISTORY

1. **Recognizing Effects** Do you think the development of the personal computer had a positive or negative effect on the economy? Explain.

See Skillbuilder Handbook, page R10.

CONNECT TO TODAY

2. **Identifying Problems** What problem might the computer industry try to solve next?

Visit www.mcdougallittell.com to learn more about computers.

Eventually, most laid-off workers did find other jobs. But in an effort to cut costs, many companies began to hire people as part-time or temporary workers. Part-time and temporary jobs generally offered less pay and fewer benefits than full-time jobs.

Reading **History**

B. Analyzing Points of View Why do you think Golly believes working for a corporation is unstable?

Because of these trends, more people began to work for themselves. For example, Jeanne Golly grew frustrated after losing three different jobs with corporations. In 1995, she started her own consulting firm. "Anything has to be more stable than corporate life," she said. By 1997, the number of self-employed persons had risen to more than 10 million.

Although layoffs made headlines, more jobs were created than lost during the 1990s. The advances in technology helped to create a strong economy in which both unemployment and inflation were low.

A More Global Economy

Technology helped to build a more global economy—in which countries around the world are linked through business. Through the Internet, companies on different continents could do business as if they were in the same city. Trade and investment among nations expanded.

Corporations also built factories and offices in other countries. By 1995, for example, more than 3 million Japanese cars were made at U.S. plants. And businesses in different countries merged to form multinational corporations. For example, in May 1998, the German company Daimler-Benz AG bought the Chrysler Corporation. This created a worldwide auto giant, Daimler Chrysler.

President Clinton saw the growth of world trade as a chance for America to sell more goods and create more jobs. This was why he urged Congress to pass NAFTA.

Background

NAFTA is the North American Free Trade Agreement. See Section 1.

The global economy caused problems for some U.S. workers. To cut costs, some businesses moved their operations from the United States to countries where wages were lower. In the mid-1990s, for example, a large U.S. clothing maker moved many of its sewing operations to Mexico, the Caribbean, and Central America. Thousands of U.S. workers lost their jobs.

Because of global trade, the economies of various countries had become more closely linked. Nations were more likely to suffer from each other's financial woes. For example, in October 1997, Hong Kong's stock market fell, which caused stock markets from Europe to North America to drop. In the summer of 1998, Russia temporarily experienced an economic collapse, which also caused markets to fall.

Now *and* **then**

PREPARING FOR TOMORROW'S JOBS

In the past, many workers spent their entire careers at one company. By the 1990s, workers were changing jobs often to gain more pay or more challenging assignments.

Workers today also have more chances to work overseas. For example, 27-year-old Anne Larlarb has already worked in England and Thailand. In the photo above, she uses e-mail to keep in touch with friends.

If these trends continue, future workers will need to be flexible, quick learners, and open to other cultures.

Scientific Breakthroughs

In the last decades of the 20th century, the world of medicine saw many breakthroughs. Engineers developed smaller, more precise surgical

This firefighter has received an artificial arm with the ability to sense heat.

instruments. These and new technologies such as lasers allowed doctors to perform surgery through tiny incisions in the body, which heal more quickly than large cuts. New tests helped doctors to make better diagnoses.

Scientists developed new drugs that offer greater hope for a cure for cancer. New drugs and treatments also slowed the rate at which AIDS (acquired immune deficiency syndrome) kills infected people.

In 1997, Scottish scientists cloned the first mammal—a sheep. This set off a furious debate. Many feared that cloning human beings could be next. Some insisted that cloning would help to improve the human species. Others argued that the process is unethical.

Vocabulary
clone: to make a genetic duplicate of a living being

A VOICE FROM THE PAST

Creating life in the laboratory is totally inappropriate and so far removed from the process of marriage and parenting that . . . we must rebel against the very concept of human cloning. It is simply wrong to experiment with the creation of human life in this way.

U.S. Representative Vernon Ehlers, statement to U.S. Congress

***Reading* History**
C. Identifying Facts and Opinions Is Representative Ehlers expressing a fact or an opinion? Explain.

In nonmedical science, the United States and other nations began to build an international space station. The station was scheduled to be finished in the early 2000s. Nations will use it to research the stars, planets, and galaxies.

While technology and science shape the years to come, so too will people. Section 3 discusses the nation's diverse population and how these many groups are shaping American society.

Section 2 Assessment

1. Terms & Names

Identify:
- Internet
- e-commerce
- service economy
- information revolution
- downsizing

2. Taking Notes

Use a cluster diagram like the one shown to list changes in technology, the economy, and science.

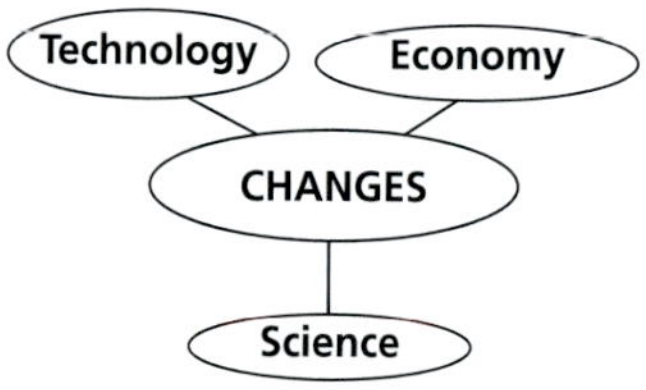

3. Main Ideas

a. How did the information revolution change jobs?

b. Why did the rise of the global economy cause some workers to worry?

c. What advances did scientists make in the field of medicine?

4. Critical Thinking

Forming and Supporting Opinions What job skills do you think you will need for the future? Why?

THINK ABOUT
- the changes in technology and science
- the changes in the economy

ACTIVITY OPTIONS

TECHNOLOGY
DRAMA

What technological marvel would you like to see invented? Design a **Web page** advertising the new technology, or perform a **skit** showing someone using it.

3 The New Americans

TERMS & NAMES
Immigration Reform and Control Act of 1986

MAIN IDEA
Due to immigration, the United States grew more diverse.

WHY IT MATTERS NOW
Americans of all backgrounds share common goals: the desire for equal rights and economic opportunity.

Junot Díaz won praise for his stories about being Dominican.

ONE AMERICAN'S STORY

Born in the Dominican Republic, Junot Díaz came to the United States with his family when he was seven. He started writing when he was just 13. In 1996, at the age of 28, he published his first book of stories. Díaz writes about being Dominican but believes his stories are universal; that is, they have meaning for everyone.

A VOICE FROM THE PAST

I am Dominican and that for me is important, but I also know that there is this whole idea that if you are a Dominican that's not universal, that's not American. But, I argue that [my stories] are universal and American.

Junot Díaz, quoted in *Frontera*

Díaz and millions of other recent immigrants have made important contributions to the nation's growth. This section discusses how these new Americans are making the United States a more diverse nation.

Immigrants Affect American Society

From 1981 to 1996, nearly 13.5 million people came to the United States. These new immigrants increased U.S. diversity. Most of the immigrants who arrived in America during earlier periods had come from Europe. Nearly 85 percent of the arrivals since 1981 came from either Latin America or Asia. The Census Bureau predicts that the U.S. Hispanic population will increase from 11 percent to 16 percent by 2020. The Asian population will climb from 3 percent to nearly 6 percent by then.

One cause of the recent surge in immigration is the Immigration and Nationality Act of 1965. It allowed people from a greater variety of countries to enter the United States. The lure of America also plays a role. As earlier immigrants did, many of the newcomers came to the United States seeking economic opportunity and, in some cases, political freedom.

U.S. citizens have mixed feelings about immigration. Some argue that immigrants take jobs from citizens. Many Americans also worry about the number of immigrants who enter the United States illegally. Officials

The American People

Ancestry of Americans, 1990	(descendants, in thousands)
1. German	58,000
2. Irish	39,000
3. English	33,000
4. African	24,000
5. Italian	15,000
6. Mexican	12,000
7. French	10,000
8. Native American	9,000
9. Polish	9,000
10. Dutch	6,000

Origins of Immigrants, 1981–1996	(immigrants, in thousands)
1. Mexico	3,300
2. Philippines	840
3. China*	730
4. Vietnam	720
5. Dominican Republic	510
6. India	500
7. Korea	450
8. Soviet Union†	420
9. El Salvador	360
10. Jamaica	320

* China includes Taiwan.

† The Soviet Union broke apart in 1991. This figure includes the former Soviet republics.

Source: *U.S. Bureau of the Census*

SKILLBUILDER Interpreting Charts

1. *Compare the origins of recent immigrants to the ancestry of Americans overall. Are any the same?*
2. *How might the list of top ten ancestry groups change in the future?*

estimate that in 1996, about 5 million illegal immigrants lived in the United States. Roughly 2.7 million were thought to be from Mexico.

Because illegal immigrants are here secretly, they do not pay income taxes. Yet they receive government services, such as police protection. As a result, some people feel that they are a drain on the U.S. economy. Congress passed the **Immigration Reform and Control Act of 1986** to strengthen immigration laws and enforcement measures. But illegal immigrants continued to cross America's borders.

Background
An illegal immigrant is someone who enters the United States secretly and without filling out the appropriate government forms.

Immigrant Contributions

Recent immigrants have brought, and continue to bring, many talents to the United States. The National Science Foundation estimates that 23 percent of all U.S. residents with doctorate degrees in engineering and science are foreign-born. High-tech industries, such as those located in Silicon Valley, California, have benefited from their skills.

In addition, immigrants are an important source of labor. Some studies indicate that without immigrants, the workforce might actually begin to shrink by 2015. In other words, U.S. businesses wouldn't be able to hire enough people to maintain their productivity.

Reading History
A. Drawing Conclusions What have immigrants contributed to the economy?

Immigrants also make sports much more exciting. In 1998 and 1999, Dominican-born Sammy Sosa thrilled baseball fans by battling Mark McGwire for the home run record. Immigrants have also starred in other sports such as basketball, football, soccer, and golf.

Many immigrants enrich American arts and culture. Latin music, for example, has become very popular. Chinese-born author Bette Bao Lord has written popular books for children and adults. In addition, immigrants, and their sons and daughters, are acting in a greater number of movies. "When immigrants come to America they bring their culture, and that culture becomes part of a new country," noted Cuban-born singer Gloria Estefan. "It makes everyone stronger."

What Americans Have In Common

While immigrants bring their culture to America, many of them also have embraced American ways. They wear American clothes, adopt American customs, and learn English.

They also share the American belief in certain ideals. The ideas of democracy and freedom that motivated people in 1776 still inspire Americans today. Ken Burns, a documentary filmmaker, explained that these beliefs make America unique in the world.

Reading History

B. Making Inferences Judging from this quotation, how does Ken Burns view the future of the United States?

A VOICE FROM THE PAST

There is no other country on Earth that is configured like ours. Every other nation is there because of race, religion, language, ethnicity, or geography. We are here only because we agreed to subscribe to the words on four pieces of paper—the U.S. Constitution. Unlike every other country, which sees itself as an end unto itself, we see ourselves as evolving. We're not satisfied. We're not willing to rest on our laurels. We think we can get better. We think we've got someplace to go.

Ken Burns, quoted in *America West*

Today the United States is a very different nation from the one founded in 1776. Democratic rights have expanded to include more and more people. As the United States moves into the future, it will no doubt continue to change. Tolerance and cooperation will be essential.

Citizens of all colors and backgrounds will play a vital role in shaping what America will be. So will today's students. You have a part to play in helping the United States embrace people from every culture and land. You are the generation that will create the America of the future.

Section 3 Assessment

1. Terms & Names

Identify:

- Immigration Reform and Control Act of 1986

2. Taking Notes

Use a diagram like the one shown to record the effects of recent immigration on the United States.

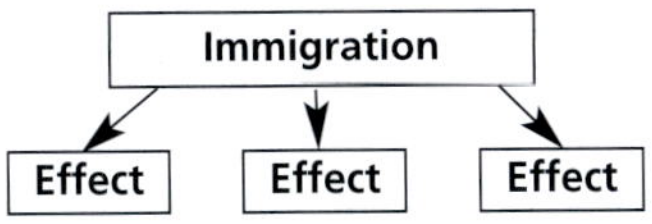

How has immigration affected your life?

3. Main Ideas

a. How did the Immigration and Nationality Act of 1965 help create a more diverse population?

b. In what areas of American society have immigrants made contributions?

c. What ideals do both immigrant and native-born Americans believe in?

4. Critical Thinking

Comparing and Contrasting How was the immigration that occurred in the years 1981–2000 similar to and different from earlier waves of immigration?

THINK ABOUT

- the immigration you read about in Chapter 14
- the immigration you read about in Chapter 21

ACTIVITY OPTIONS

MATH

SPEECH

Survey ten people outside your class to learn their ethnic background. Present your findings as a **table** like the one on page 884 or in a **speech.**

VISUAL SUMMARY

Entering a New Millennium

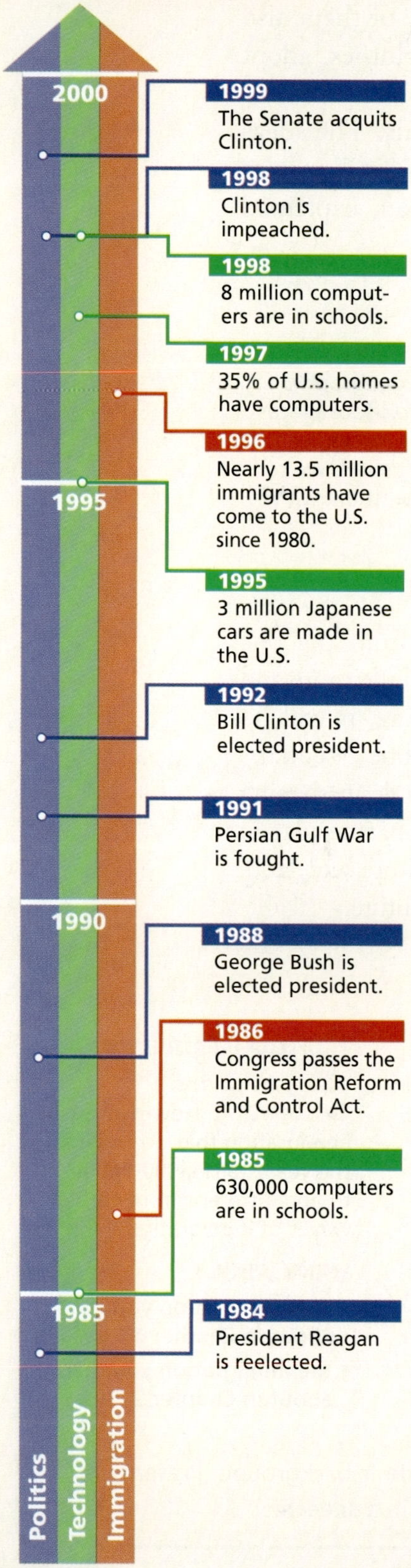

Chapter 32 ASSESSMENT

TERMS & NAMES

Briefly explain the importance of each of the following.

1. Ronald Reagan
2. supply-side economics
3. Iran-Contra affair
4. George Bush
5. Persian Gulf War
6. Bill Clinton
7. Internet
8. service economy
9. information revolution
10. Immigration Reform and Control Act of 1986

REVIEW QUESTIONS

Conservatives Reshape Politics (pages 873–877)

1. How did Reagan try to improve the economy?
2. What happened in the Iran-Contra affair?
3. What role did the United States play during the Persian Gulf War?
4. What happened during Clinton's impeachment and trial?

Technological and Economic Changes (pages 878–882)

5. What did the Internet allow its users to do?
6. How did corporate mergers affect individuals?
7. What were the different reactions to cloning?

The New Americans (pages 883–885)

8. Why did the recent wave of immigrants come to the United States?
9. What do immigrants contribute to the economy?
10. Why do experts think the United States will become more diverse?

CRITICAL THINKING

1. USING YOUR NOTES

Using your completed chart, answer the questions below.

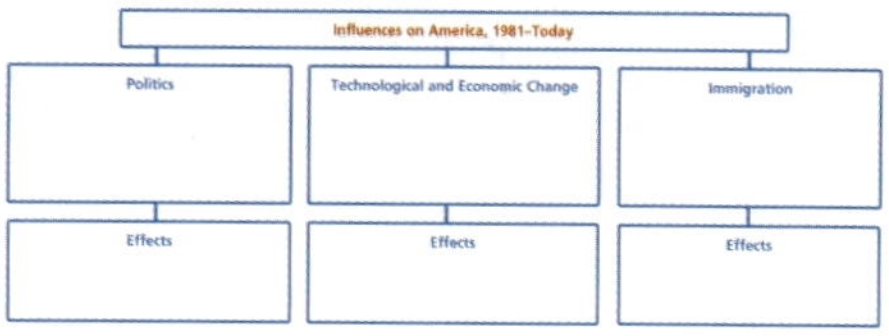

a. What were the effects of recent political events?

b. What were the main technological and economic changes?

2. ANALYZING LEADERSHIP

Judging from what you read in this chapter, what issues will a U.S. president face in the 21st century?

3. APPLYING CITIZENSHIP SKILLS

In what ways have recent immigrants demonstrated good citizenship here in the United States?

4. THEME: DIVERSITY AND UNITY

What ideas and goals help to unify Americans of different racial and ethnic backgrounds?

5. MAKING INFERENCES

Are U.S. citizens likely to be more or less welcoming to immigrants in good economic times? Explain.

Interact *with* History

Now that you have read about recent changes in the United States, what new ideas do you have about how you can contribute to the future?

HISTORY SKILLS

1. INTERPRETING MAPS: Movement

Study the map. Then answer the questions.

Immigrants by State, 1996

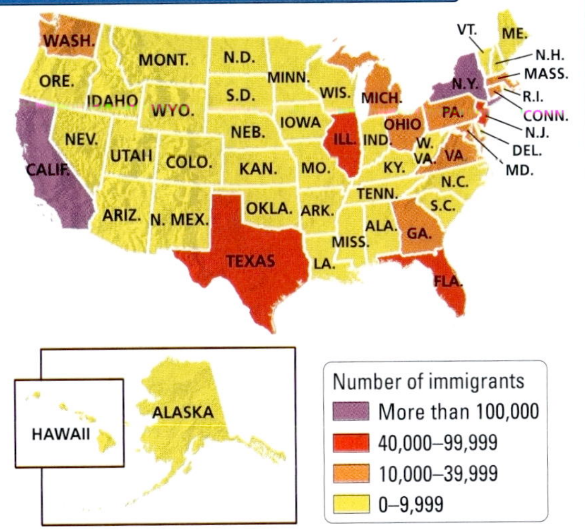

Source: *The World Almanac, 1999*

Basic Map Elements

a. What year's statistics does this map show?

b. What do the various colors of the states indicate?

Interpreting the Map

c. Which states have the highest numbers of immigrants?

d. How would you describe what those states have in common?

2. INTERPRETING PRIMARY SOURCES

This photograph shows a robot that was designed to play a keyboard instrument.

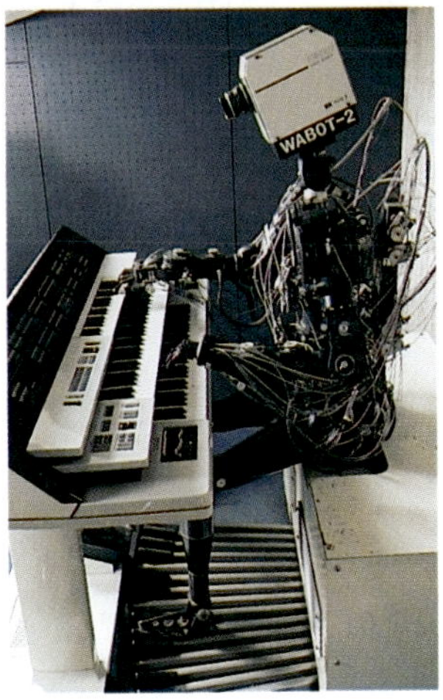

a. What are some of the separate skills that humans need to play keyboard instruments? Think of both physical and mental skills.

b. If a robot can play a keyboard instrument, what else might it be able to do? Consider how else it might use the separate skills involved in the task.

ALTERNATIVE ASSESSMENT

1. INTERDISCIPLINARY ACTIVITY: Science

Explaining a Medical Advance Research a recent advance in medical science, such as a new medicine, treatment, or diagnostic test. Give a brief speech explaining the advance to the class.

2. COOPERATIVE LEARNING ACTIVITY

Creating a Front Page With three other students, create a newspaper front page that reports on the major events of the late 20th century. Three of the students will each write an article. The fourth student will be responsible for finding images for each story. Possible story topics are

- the arrival of a new device, the computer
- the number of recent immigrants to America reaches 13 million
- the most recent presidential election

Type and print out the stories and arrange them with their images on a large piece of paper. Show the front page to the class and have each student read his or her story aloud.

3. TECHNOLOGY ACTIVITY

Designing a Web Page What do you think the United States will be like in ten years? Design a Web page that presents your vision. Your Web page may include pictures and descriptions that portray your ideas. It should also include links to other sites that discuss technology, politics, or social issues. Draw a design of what the page will look like. Also provide written descriptions and addresses of other Web sites that are linked to yours.

INTERNET **Visit www.mcdougallittell.com to see sample Web pages.**

4. HISTORY PORTFOLIO

PORTFOLIO **Option 1** Review your section and chapter assessment activities. Select one that you think is your best work. Then use comments by your teacher or classmates to improve your work, and add it to your portfolio.

PORTFOLIO **Option 2** Review the questions that you wrote for What Do you Want to Know? on page 872. Then write a short report in which you explain the answers to your questions. If any questions were not answered, do research to answer them. Add your answers to your portfolio.

Make a Time Capsule

A time capsule is a sealed container that preserves records and artifacts from the present for people in the future. Ancient Babylonians and Egyptians began this custom by carving messages inside their temples. These records give people today an idea of what life was like thousands of years ago. One of the first modern time capsules is in Oglethorpe, Georgia. Sealed in 1940, it is not due to be opened until 8113. It contains such things as a Donald Duck doll and 640,000 pages on microfilm.

ACTIVITY With a group, construct a time capsule. Write an explanation of the contents and your predictions or dreams for the future. Then record this explanation on a cassette tape.

A family from the 1960s poses in front of their house.

TOOLBOX

Each group will need:

posterboard	stapler
aluminum foil	markers
cellophane	3 x 5 notecards
scissors	blank cassette tape
duct or masking tape	cassette recorder

Hairstyles and fashions in the 1970s were very different from those in the 1950s.

STEP BY STEP

1. **Form groups.** Meet with three to four students to create a time capsule. To complete this project, groups will be expected to
   - do research on a particular decade
   - identify artifacts from that decade to place in a time capsule
   - construct a time capsule
   - write an explanation of the contents, and record it on an audio cassette

2. **Choose the 1950s, 1960s, 1970s, 1980s, or 1990s.** Use the Internet, encyclopedias, or books about that decade to brainstorm ideas for the contents of your time capsule. Each person should jot down items on a piece of paper. A sample is shown below. With your group, discuss these items and decide which ones you'd like to use. Be sure to include enough items so that people opening the capsule many years from now will have a complete picture of your decade.

1960s items
photo of Vietnam war
Beatles record
peace sign

3 Build the time capsule.

Build the time capsule. First, cut a small square out of the middle of the posterboard to make a window. Using tape or a stapler, place the cellophane over the hole. Then roll the posterboard so that it forms a large cylinder. Tape or staple the ends of the posterboard together. Place aluminum foil over one end of your time capsule—using tape to secure it.

4 Decorate the time capsule.

Decorate the time capsule. Use markers to decorate the outside of the time capsule. These decorations should reflect your decade. Also, include a message that is connected with your decade. Be sure to explain when the capsule should be opened.

5 Select your items.

Select your items. Use the actual item, or a picture. On each 3 x 5 card, write an explanation of the items you've selected and explain their purpose. Attach a card to each picture or item. Be sure to include the year the item was first used.

6 Present the contents to the class.

Present the contents to the class. Using your notecards, explain to the class why you chose these items. Place the notecards inside the time capsule. Wait until you make your audio cassette to seal the time capsule—and be sure to include it.

WRITE AND SPEAK

Write a paragraph describing your time capsule. Include your predictions and dreams for future generations. Then record your message on an audio cassette and place it inside the time capsule. After sealing the time capsule, put it on display with the other time capsules.

HELP DESK

For related information, see Chapters 29, 30, 31, and 32.

Researching Your Project

"Time Immemorial," *Popular Mechanics,* February 1999.

INTERNET **Visit www.mcdougallittell.com for links to sites about time capsules.**

Did You Know?

Westinghouse Electric Company coined the term *time capsule.* The company built one for the New York World's Fair of 1939–1940. The capsule is due to be opened in 6939. To preserve the capsule's contents, the company formed a new metal alloy called Cupaloy. It combined the durability of steel with the ability of copper to prevent corrosion.

The designers figured people living in the 70th century may not know the time capsule exists. So they created *The Book of Record.* It gives the time capsule's location and was placed in libraries all over the world.

REFLECT & ASSESS

- **Why did you choose the particular artifacts in your time capsule?**
- **How do you think future generations might react to the articles in your time capsule?**
- **How well do your items represent the history, social issues, and culture of your decade?**

Reference Section

Creating America

A History of the United States

Table of Contents

1.1 Summarizing

Defining the Skill

When you **summarize,** you restate a paragraph, passage, or chapter in fewer words. You include only the main ideas and most important details. It is important to use your own words when summarizing.

Applying the Skill

The passage below tells about Harriet Tubman, a prominent member of the Underground Railroad. She helped runaway slaves to freedom. Use the strategies listed below to help you summarize the passage.

How to Summarize

Strategy 1 Look for topic sentences stating the main idea. These are often at the beginning of a section or paragraph. Briefly restate each main idea—in your own words.

Strategy 2 Include key facts and any numbers, dates, amounts, or percentages from the text.

Strategy 3 After writing your summary, review it to see that you have included only the most important details.

HARRIET TUBMAN

1 One of the most famous conductors on the Underground Railroad was Harriet Tubman. 2 Born into slavery in Maryland, the 13-year-old Tubman once tried to save another slave from punishment. The angry overseer fractured Tubman's skull with a two-pound weight. She suffered fainting spells for the rest of her life but did not let that stop her from working for freedom. When she was 25, Tubman learned that her owner was about to sell her. Instead, 2 she escaped.

After her escape, 2 Harriet Tubman made 19 dangerous journeys to free enslaved persons. The tiny woman carried a pistol to frighten off slave hunters and medicine to quiet crying babies. Her enemies offered $40,000 for her capture, but 2 no one caught her. "I never run my train off the track and I never lost a passenger," she proudly declared. Among the people she saved were her parents.

Write a Summary

You can write your summary in a paragraph. The paragraph at right summarizes the passage you just read.

3 *Harriet Tubman was one of the most famous conductors on the Underground Railroad. She had been a slave, but she escaped. She later made 19 dangerous journeys to free other slaves. She was never captured.*

Practicing the Skill

Turn to Chapter 6, Section 2, "Colonial Resistance Grows." Read "The Boston Tea Party" and write a paragraph summarizing the passage.

1.2 Taking Notes

Defining the Skill

When you **take notes,** you write down the important ideas and details of a paragraph, passage, or chapter. A chart or an outline can help you organize your notes to use in the future.

Applying the Skill

The following passage describes President Washington's cabinet. Use the strategies listed below to help you take notes on the passage.

How to Take and Organize Notes

Strategy 1 Look at the title to find the main topic of the passage.

Strategy 2 Identify the main ideas and details of the passage. Then summarize the main idea and details in your notes.

Strategy 3 Identify key terms and define them. The term *cabinet* is shown in boldface type and underlined; both techniques signal that it is a key term.

Strategy 4 In your notes, use abbreviations to save time and space. You can abbreviate words such as *department (dept.), secretary (sec.), United States (U.S.),* and *president (pres.)* to save time and space.

1 WASHINGTON'S CABINET

2 The Constitution gave Congress the task of creating departments to help the president lead the nation. The 2 president had the power to appoint the heads of these departments, which became his 3 **cabinet.**

Congress created three departments. Washington chose talented people to run them. 2 For secretary of war, he picked Henry Knox, a trusted general during the Revolution. 2 For secretary of state, Washington chose Thomas Jefferson. He had been serving as ambassador to France. The State Department oversaw U.S. foreign relations. For secretary of the treasury, Washington turned to the brilliant 2 Alexander Hamilton.

Make a Chart

Making a chart can help you take notes on a passage. The chart below contains notes from the passage you just read.

2 *Item*	*Notes*
1. 3 cabinet	*heads of 4 depts.; 4 pres. appoints heads*
a. War Dept.	*Henry Knox; 4 sec. of war; former Revolutionary War general*
b. State Dept.	*Thomas Jefferson; sec. of state; oversees relations between 4 U.S. and other countries*
c. Treasury Dept.	*Alexander Hamilton; sec. of the treasury*

Practicing the Skill

Turn to Chapter 3, Section 3, "Founding the Middle and Southern Colonies." Read "Maryland and the Carolinas" and use a chart to take notes on the passage.

1.3 Sequencing Events

Defining the Skill

Sequence is the order in which events follow one another. By being able to follow the sequence of events through history, you can get an accurate sense of the relationship among events.

Applying the Skill

The following passage describes the sequence of events involved in Britain's plan to capture the Hudson River Valley during the American Revolution. Use the strategies listed below to help you follow the sequence of events.

How to Find the Sequence of Events

Strategy 1 Look for specific dates provided in the text. If several months within a year are included, the year is usually not repeated.

Strategy 2 Look for clues about time that allow you to order events according to sequence. Words such as *day, week, month,* or *year* may help to sequence the events.

BRITAIN'S STRATEGY

Burgoyne captured Fort Ticonderoga in 1 July 1777. From there, it was 25 miles to the Hudson River, which ran to Albany. 2 Burgoyne took three weeks to reach the Hudson. On 1 August 3, Burgoyne received a message from Howe. He would not be coming north, Howe wrote, because he had decided to invade Pennsylvania to try to capture Philadelphia and General Washington. "Success be ever with you," Howe's message said. But General Burgoyne needed Howe's soldiers, not his good wishes. Howe did invade Pennsylvania. In 1 September 1777, he defeated—but did not capture—Washington at the Battle of Brandywine.

Make a Time Line

Making a time line can help you sequence events. The time line below shows the sequence of events in the passage you just read.

Practicing the Skill

Turn to Chapter 2, Section 1, "Spain Claims an Empire." Read "Europeans Explore Foreign Lands" and make a time line showing the sequence of events in that passage.

1.4 Finding Main Ideas

Defining the Skill

The **main idea** is a statement that summarizes the main point of a speech, an article, a section of a book, or a paragraph. Main ideas can be stated or unstated. The main idea of a paragraph is often stated in the first or last sentence. If it is the first sentence, it is followed by sentences that support that main idea. If it is the last sentence, the details build up to the main idea. To find an unstated idea, you must use the details of the paragraph as clues.

Applying the Skill

The following paragraph describes the role of women in the American Revolution. Use the strategies listed below to help you identify the main idea.

How to Find the Main Idea

Strategy 1 Identify what you think may be the stated main idea. Check the first and last sentences of the paragraph to see if either could be the stated main idea.

Strategy 2 Identify details that support that idea. Some details explain the main idea. Others give examples of what is stated in the main idea.

WOMEN IN THE REVOLUTION

1 Many women tried to help the army. Martha Washington and other wives followed their husbands to army camps. 2 The wives cooked, did laundry, and nursed sick or wounded soldiers. 2 A few women even helped to fight. 2 Mary Hays earned the nickname "Molly Pitcher" by carrying water to tired soldiers during a battle. 2 Deborah Sampson dressed as a man, enlisted, and fought in several engagements.

Make a Chart

Making a chart can help you identify the main idea and details in a passage or paragraph. The chart below identifies the main idea and details in the paragraph you just read.

Main Idea: *Women helped the army during the Revolution.*

Detail: *They cooked and did laundry.*
Detail: *They nursed the wounded and sick soldiers.*
Detail: *They helped to fight.*
Detail: *One woman, Molly Pitcher, carried water to soldiers during battles.*

Practicing the Skill

Turn to Chapter 5, Section 1, "Early American Culture." Read "Women and the Economy" and create a chart that identifies the main idea and the supporting details.

1.5 Categorizing

Defining the Skill

To **categorize** is to sort people, objects, ideas, or other information into groups, called categories. Historians categorize information to help them identify and understand patterns in historical events.

Applying the Skill

The following passage contains information about the reasons people went west during the mid-1800s. Use the strategies listed below to help you categorize information.

How to Categorize

Strategy 1 First, decide what kind of information needs to be categorized. Decide what the passage is about and how that information can be sorted into categories.

For example, find the different motives people had for moving west.

Strategy 2 Then find out what the categories will be. To find why many different groups of people moved west, look for clue words such as *some, other,* and *another.*

Strategy 3 Once you have chosen the categories, sort information into them. Of the people who went west, which ones had which motives?

THE LURE OF THE WEST

1 People had many different motives for going west. 2 One motive was to make money. 2 *Some* people called speculators bought huge areas of land and made great profits by selling it to thousands of settlers. 2 *Other* settlers included farmers who dreamed of owning their own farms in the West because land was difficult to acquire in the East. 2 *Another* group to move west was merchants. They hoped to earn money by selling items that farmers needed. Finally, 2 *some* people went west for religious reasons. These people included 2 missionaries, who wanted to convert the Native Americans to Christianity, and Mormons, who wanted a place where they could practice their faith without interference.

Make a Chart

Making a chart can help you categorize information. You should have as many columns as you have categories. The chart below shows how the information from the passage you just read can be categorized.

3

Motives	Money	Land	Religion
Groups	• speculators • merchants	• farmers	• missionaries • Mormons

Practicing the Skill

Turn to Chapter 14, Section 3, "Reforming American Society." Read "Improving Education" and make a chart in which you categorize the changes happening in elementary, high school, and college education.

1.6 Making Public Speeches

Defining the Skill

A speech is a talk given in public to an audience. Some speeches are given to persuade the audience to think or act in a certain way, or to support a cause. You can learn how to **make public speeches** effectively by analyzing great speeches in history.

Applying the Skill

The following is an excerpt from the "I Have a Dream" speech delivered by Martin Luther King, Jr., in 1963 in Washington, D.C. Use the strategies listed below to help you analyze King's speech and prepare a speech of your own.

How to Analyze and Prepare a Speech

Strategy 1 Choose one central idea or theme and organize your speech to support it. King organized his speech around his dream of equality.

Strategy 2 Use words or images that will win over your audience. King referred to the Declaration of Independence when he used the words "all men are created equal."

Strategy 3 Repeat words or images to drive home your main point—as if it is the "hook" of a pop song. King repeats the phrase "I have a dream."

I HAVE A DREAM

1 I have a dream that one day this nation will rise up and live out the true meaning of its creed—we hold these truths to be 2 self-evident that all men are created equal.

3 I have a dream that one day on the red hills of Georgia the sons of former slaves and the sons of former slave owners will be able to sit down together at the table of brotherhood.

3 I have a dream that my four little children will one day live in a nation where they will not be judged by the color of their skin but by the content of their character.

3 I have a dream today!

Make an Outline

Making an outline like the one to the right will help you make an effective public speech.

Title: I Have a Dream

I. **Introduce Theme:** I have a dream
- A. This nation will live up to its creed
- B. Quote from the Declaration of Independence: that all men are created equal

II. **Repeat theme:** I have a dream
- A. Sons of former slaves and slave owners will sit together in brotherhood
- B. My four children will be judged by their character, not by their skin color

III. **Conclude:** I have a dream

Practicing the Skill

Turn to Chapter 12, Section 2, "Jackson's Policy Toward Native Americans." Read the section and choose a topic for a speech. First, make an outline like the one to the right to organize your ideas. Then write your speech. Next, practice giving your speech. Make it a three-minute speech.

2.1 Analyzing Points of View

Defining the Skill

Analyzing points of view means looking closely at a person's arguments to understand the reasons behind that person's beliefs. The goal of analyzing a point of view is to understand a historical figure's thoughts and opinions about a topic.

Applying the Skill

The following passage describes the Panic of 1837 and two politicians' points of view about it. Use the strategies listed below to help you analyze their points of view.

How to Analyze Points of View

Strategy 1 Look for statements that show you a person's view on an issue. For example, Van Buren said he believed the economy would improve if he took no action. Clay thought the government should do something to help the people.

Strategy 2 Think about why people might take the positions they took. Clay was in a different political party than Van Buren. Why do you think he would disagree with Van Buren's argument?

Strategy 3 Write a summary that explains why different people took different positions on the issue.

THE PANIC OF 1837

The Panic of 1837 caused severe hardship. People had little money, so manufacturers had few customers for their goods. Almost 90 percent of factories in the East closed. Jobless workers could not afford food or rent. Many people went hungry.

1 Whig senator Henry Clay wanted the government to do something to help the people. 1 President Van Buren, a Democrat, disagreed. He believed that the economy would improve if left alone. He argued that "the less government interferes with private pursuits the better for the general prosperity." Many Americans blamed Van Buren for the Panic, though he had taken office only weeks before it started. The continuing depression made it difficult for him to win reelection in 1840.

Make a Diagram

Using a diagram can help you analyze points of view. The diagram below analyzes the views of Clay and Van Buren in the passage you just read.

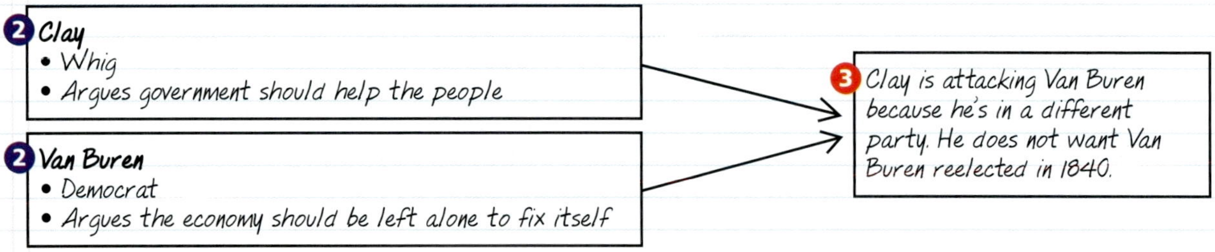

Practicing the Skill

Turn to the Interactive Primary Sources on pages 238 and 239. Read the selections by James Madison and George Mason. Make a chart to analyze their opposing points of view on the Constitution.

2.2 Comparing and Contrasting

Defining the Skill

Comparing means looking at the similarities and differences between two or more things. **Contrasting** means examining only the differences between them. Historians compare and contrast events, personalities, behaviors, beliefs, and situations in order to understand them.

Applying the Skill

The following paragraph describes the American and British troops during the Revolutionary War. Use the strategies listed below to help you compare and contrast these two armies.

How to Compare and Contrast

Strategy 1 Look for two aspects of the subject that may be compared and contrasted. This passage compares the British and American troops to show why the Americans won the war.

Strategy 2 To contrast, look for clue words that show how two things differ. Clue words include *by contrast, however, except,* and *yet.*

Strategy 3 To find similarities, look for clue words indicating that two things are alike. Clue words include *both, like, as,* and *similarly.*

WHY THE AMERICANS WON

1 By their persistence, the Americans defeated the British even though they faced many obstacles. The Americans lacked training and experience. They were often short of supplies and weapons. 2 *By contrast,* the British forces ranked among the best trained in the world. They were experienced and well-supplied professional soldiers. 2 *Yet,* the Americans also had advantages that enabled them to win. These advantages over the British were better leadership, foreign aid, a knowledge of the land, and motivation. Although 3 *both* the British and the Americans were fighting for their lives, 2 the Americans were also fighting for their property and their dream of liberty.

Make a Venn Diagram

Making a Venn diagram will help you identify similarities and differences between two things. In the overlapping area, list characteristics shared by both subjects. Then, in the separate ovals, list the characteristics of each subject not shared by the other. This Venn diagram compares and contrasts the British and American soldiers.

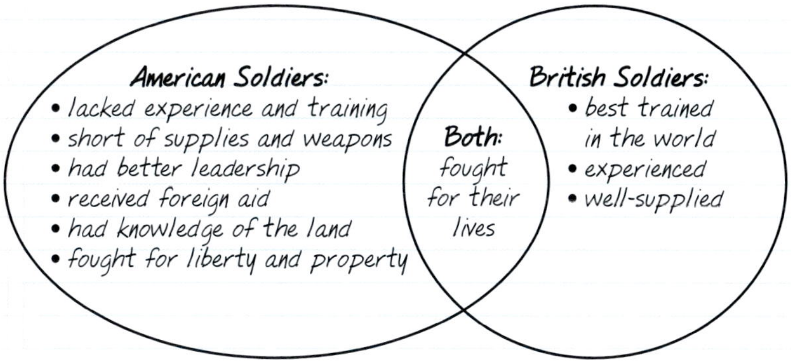

Practicing the Skill

Turn to Chapter 5, Section 1, "Early American Culture." Read "Young People at Work" and make a Venn diagram showing the similarities and differences between the roles of boys and girls in colonial America.

2.3 Analyzing Causes; Recognizing Effects

Defining the Skill

A **cause** is an action in history that makes something happen. An **effect** is the historical event that is the result of the cause. A single event may have several causes. It is also possible for one cause to result in several effects. Historians identify cause-and-effect relationships to help them understand why historical events took place.

Applying the Skill

The following paragraph describes events that caused changes in Puritan New England. Use the strategies listed below to help you identify the cause-and-effect relationships.

How to Analyze Causes and Recognize Effects

Strategy 1 Ask why an action took place. Ask yourself a question about the title and topic sentence, such as, "What caused changes in Puritan society?"

Strategy 2 Look for effects. Ask yourself, "What happened?" (the effect). Then ask, "Why did it happen?" (the cause). For example, What caused the decline of Puritan religion in New England?

Strategy 3 Look for clue words that signal causes, such as *cause* and *led to.*

1 CHANGES IN PURITAN SOCIETY

1 The early 1700s saw many changes in New England society. 2 One of the most important changes was the gradual decline of the Puritan religion in New England. There were a number of reasons for that decline.

3 One *cause* of this decline was the increasing competition from other religious groups. Baptists and Anglicans established churches in Massachusetts and Connecticut, where Puritans had once been the most powerful group. 3 Political changes also *led to* a weakening of the Puritan community. In 1691, a new royal charter for Massachusetts granted the vote based on property ownership instead of church membership.

Make a Diagram

Using a diagram can help you understand causes and effects. The diagram below shows two causes and an effect for the passage you just read.

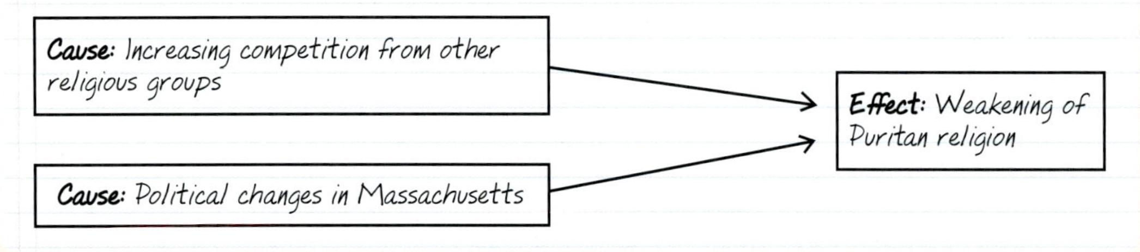

Practicing the Skill

Turn to Chapter 2, Section 1, "Spain Claims an Empire." Read "Reasons for Spanish Victories" and make a diagram about the causes and effects of the Spanish conquest of the Americas.

2.4 Making Inferences

Defining the Skill

Inferences are ideas that the author has not directly stated. **Making inferences** involves reading between the lines to interpret the information you read. You can make inferences by studying what is stated and using your common sense and previous knowledge.

Applying the Skill

The passage below describes the strengths and weaknesses of the North and the South as the Civil War began. Use the strategies listed below to help you make inferences from the passage.

How to Make Inferences

Strategy 1 Read to find statements of facts and ideas. Knowing the facts will give you a good basis for making inferences.

Strategy 2 Use your knowledge, logic, and common sense to make inferences that are based on facts. Ask yourself, "What does the author want me to understand?" For example, from the facts about population, you can make the inference that the North would have a larger army than the South. See other inferences in the chart below.

ADVANTAGES OF THE NORTH AND THE SOUTH

The North had more people and resources than the South. 1 The North had about 22 million people. 1 The South had roughly 9 million, of whom about 3.5 million were slaves. In addition, 1 the North had more than 80 percent of the nation's factories and almost all of the shipyards and naval power. The South had some advantages, too. 1 It had able generals, such as Robert E. Lee. 1 It also had the advantage of fighting a defensive war. Soldiers defending their homes have more will to fight than invaders do.

Make a Chart

Making a chart will help you organize information and make logical inferences. The chart below organizes information from the passage you just read.

1 Stated Facts and Ideas	2 Inferences
The North had about 22 million people. The Confederacy had about 9 million, less the 3.5 million slaves.	The North would have a larger army than the South.
The North had more factories, naval power, and shipyards.	The North could provide more weapons, ammunition, and ships for the war.
The Confederacy had excellent generals.	The Confederacy had better generals, which would help it overcome other disadvantages.
The Confederacy was fighting a defensive war.	Confederate soldiers would fight harder because they were defending their homes and families.

Practicing the Skill

Turn to Chapter 11, Section 1, "Early Industry and Inventions." Read "The Industrial Revolution Begins" and use a chart like the one above to make inferences about early industry.

2.5 Drawing Conclusions

Defining the Skill

Drawing conclusions means analyzing what you have read and forming an opinion about its meaning. To draw conclusions, look at the facts and then use your own common sense and experience to decide what the facts mean.

Applying the Skill

The following passage presents information about the Intolerable Acts and the colonists' reactions to them. Use the strategies listed below to help you draw conclusions about those acts.

How to Draw Conclusions

Strategy 1 Read carefully to identify and understand all the facts, or statements, that can be proven true.

Strategy 2 List the facts in a diagram and review them. Use your own experiences and common sense to understand how the facts relate to each other.

Strategy 3 After reviewing the facts, write down the conclusion you have drawn about them.

THE INTOLERABLE ACTS

1 In 1774, Parliament passed a series of laws to punish the Massachusetts colony and serve as a warning to other colonies. 1 These laws were so harsh that colonists called them the **Intolerable Acts.** One of the acts closed the port of Boston. Others banned committees of correspondence and allowed Britain to house troops wherever necessary.

In 1773, Sam Adams had written, "I wish we could arouse the continent." 1 The Intolerable Acts answered his wish. Other colonies immediately offered Massachusetts their support.

Make a Diagram

Making a diagram can help you draw conclusions. The diagram below shows how to organize facts and inferences to draw a conclusion about the passage you just read.

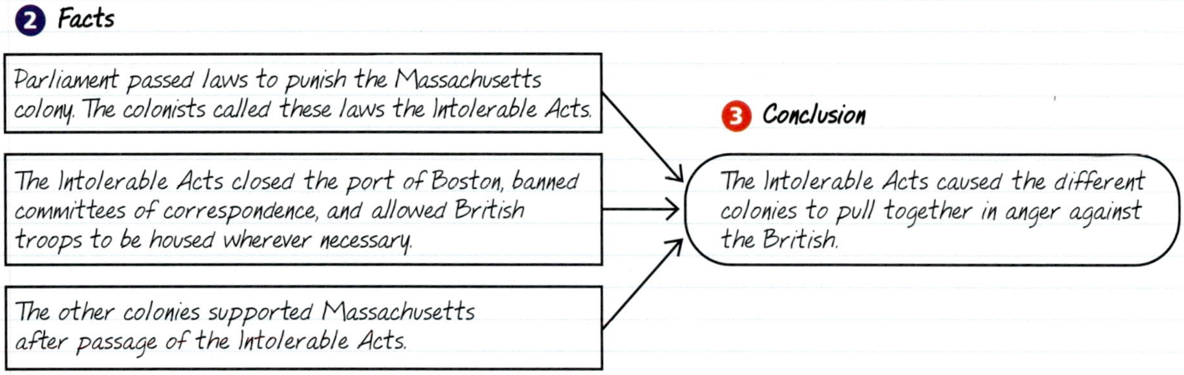

Practicing the Skill

Turn to Chapter 3, Section 2, "New England Colonies." Read "The Salem Witchcraft Trials" and use the diagram above as a model to draw conclusions about the trials.

2.6 Making Decisions

Defining the Skill

Making decisions involves choosing between two or more options, or courses of action. In most cases, decisions have consequences, or results. Sometimes decisions may lead to new problems. By understanding how historical figures made decisions, you can learn how to improve your decision-making skills.

Applying the Skill

The following passage describes Lincoln's decisions regarding federal forts after the Southern states seceded. Use the strategies listed below to help you analyze his decisions.

How to Make Decisions

Strategy 1 Identify a decision that needs to be made. Think about what factors make the decision difficult.

Strategy 2 Identify possible consequences of the decision. Remember that there can be more than one consequence to a decision.

Strategy 3 Identify the decision that was made.

Strategy 4 Identify actual consequences that resulted from the decision.

FIRST SHOTS AT FORT SUMTER

1 Lincoln had to decide what to do about the forts in the South that remained under federal control. A Union garrison still held **Fort Sumter**, but they were running out of supplies. 2 If Lincoln supplied the garrison, he risked war. 2 If he withdrew the garrison, he would be giving in to the rebels. 3 Lincoln informed South Carolina that he was sending supply ships to Fort Sumter. 4 Confederate leaders decided to prevent the federal government from holding on to the fort by attacking before the supply ships arrived. No one was killed, but 4 the South's attack on Fort Sumter signaled the beginning of the Civil War.

Make a Flow Chart

A flow chart can help you identify the process of making a decision. The flow chart below shows the decision-making process in the passage you just read.

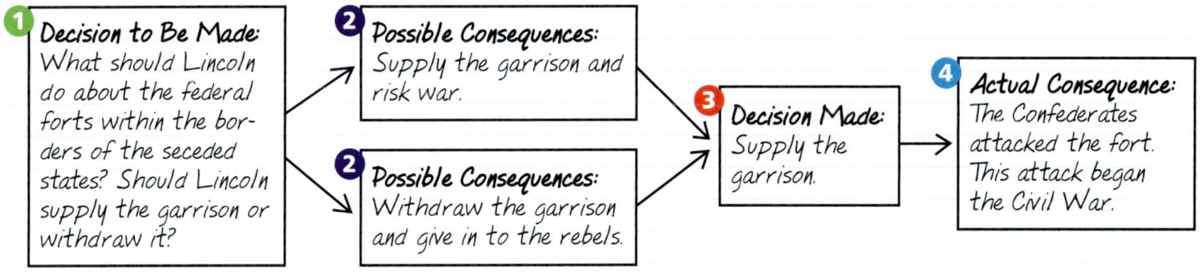

Practicing the Skill

Turn to Chapter 6, Section 1, "Tighter British Control." Read "The Colonies Protest the Stamp Act" and make a flow chart to identify a decision and its consequences described in that section.

2.7 Recognizing Propaganda

Defining the Skill

Propaganda is communication that aims to influence people's opinions, emotions, or actions. Propaganda is not always factual. Rather, it uses one-sided language or striking symbols to sway people's emotions. Modern advertising often uses propaganda. By thinking critically, you will avoid being swayed by propaganda.

Applying the Skill

The following political cartoon shows Andrew Jackson dressed as a king. Use the strategies listed below to help you understand how it works as propaganda.

How to Recognize Propaganda

Strategy 1 Identify the aim, or purpose, of the cartoon. Point out the subject and explain the point of view.

Strategy 2 Identify those images on the cartoon that viewers might respond to emotionally and identify the emotions.

Strategy 3 Think critically about the cartoon. What facts has the cartoon ignored?

Make a Chart

Making a chart will help you think critically about a piece of propaganda. The chart below summarizes the information from the anti-Jackson cartoon.

1 *Identify Purpose*	*The cartoon portrays Jackson negatively by showing him as a king.*
2 *Identify Emotions*	*The cartoonist knows that Americans like democracy. So he portrays Jackson as a king because kings are not usually supporters of democracy. He also shows Jackson standing on a torn U.S. Constitution—another thing that Americans love.*
3 *Think Critically*	*The cartoon shows Jackson vetoing laws. But it ignores the fact that those actions were not against the Constitution. The president has the power to veto legislation. In this case, Jackson was exercising the power of the presidency, not acting like a king.*

Practicing the Skill

Turn to Chapter 6, Section 2, "Colonial Resistance Grows," and look at the engraving *The Bloody Massacre* on page 165. Use a chart like the one above to think critically about the engraving as an example of propaganda.

2.8 Identifying Facts and Opinions

Defining the Skill

Facts are events, dates, statistics, or statements that can be proved to be true. **Opinions** are the judgments, beliefs, and feelings of a writer or speaker. By identifying facts and opinions, you will be able to think critically when a person is trying to influence your own opinion.

Applying the Skill

The following passage tells about the Virginia Plan for legislative representation offered at the Constitutional Convention of 1787. Use the strategies listed below to help you distinguish facts from opinions.

How to Recognize Facts and Opinions

Strategy 1 Look for specific information that can be proved or checked for accuracy.

Strategy 2 Look for assertions, claims, and judgments that express opinions. In this case, one speaker's opinion is expressed in a direct quote.

Strategy 3 Think about whether statements can be checked for accuracy. Then, identify the facts and opinions in a chart.

ANTIFEDERALIST VIEWS

1 Antifederalists published their views about the Constitution in newspapers and pamphlets. 1 They thought the Constitution took too much power away from the states and did not protect the rights of the people. They charged that the Constitution would destroy American liberties. As one Antifederalist wrote, 2 "it is truly astonishing that a set of men among ourselves should have had the [nerve] to attempt the destruction of our liberties."

Make a Chart

The chart below analyzes the facts and opinions from the passage above.

Statement	3 Can It Be Proved?	3 Fact or Opinion
Antifederalists published their views in newspapers and pamphlets.	Yes. Check newspapers and other historical documents.	Fact
They thought the Constitution took too much power away from the states.	Yes. Check newspapers and other historical documents.	Fact
It is astonishing that some Americans would try to destroy American liberties.	No. This cannot be proved. It is what one speaker believes.	Opinion

Practicing the Skill

Turn to Chapter 11, Section 3, and read the section entitled "The Missouri Compromise." Make a chart in which you analyze key statements to determine whether they are facts or opinions.

2.9 Forming and Supporting Opinions

Defining the Skill

When you **form opinions,** you interpret and judge the importance of events and people in history. You should always **support your opinions** with facts, examples, and quotes.

Applying the Skill

The following passage describes events that followed the gold rush. Use the strategies listed below to form and support your opinions about the events.

How to Form and Support Opinions

Strategy 1 Look for important information about the events. Information can include facts, quotations, and examples.

Strategy 2 Form an opinion about the event by asking yourself questions about the information. For example, How important was the event? What were its effects?

Strategy 3 Support your opinions with facts, quotations, and examples. If the facts do not support the opinion, then rewrite your opinion so it is supported by the facts.

THE IMPACT OF THE GOLD RUSH

By 1852, the gold rush was over. 1 While it lasted, about 250,000 people flooded into California. 1 This huge migration caused economic growth that changed California. 1 The port city San Francisco grew to become a center of banking, manufacturing, shipping, and trade. 1 However, the gold rush ruined many *Californios*. *Californios* are the Hispanic people of California. The newcomers did not respect *Californios,* their customs, or their legal rights. 1 In many cases, Americans seized their property.

Native Americans suffered even more. 1 Thousands died from diseases brought by the newcomers. 1 Miners hunted down and killed thousands more. 1 By 1870, California's Native American population had fallen from 150,000 to only about 30,000.

Make a Chart

Making a chart can help you organize your opinions and supporting facts. The following chart summarizes one possible opinion about the impact of the gold rush.

2 *Opinion*	*The effects of the gold rush were more negative than positive.*
3 *Facts*	*Californios were not respected, and their land was stolen.*
	Many Native Americans died from diseases, and others were killed by miners. Their population dropped from 150,000 to about 30,000.

Practicing the Skill

Turn to Chapter 11, Section 3, "Nationalism and Sectionalism." Read "The Missouri Compromise" and form your own opinion about the compromise and its impact. Make a chart like the one above to summarize your opinion and the supporting facts and examples.

2.10 Identifying and Solving Problems

Defining the Skill

Identifying problems means finding and understanding the difficulties faced by a particular group of people during a certain time. **Solving problems** means understanding how people tried to remedy those problems. By studying how people solved problems in the past, you can learn ways to solve problems today.

Applying the Skill

The following paragraph describes problems that the Constitutional Convention faced on the issues of taxation, representation, and slavery. Use the strategies listed below to help you see how the Founders tried to solve these problems.

How to Identify Problems and Solutions

Strategy 1 Look for the difficulties, or problems, people faced.

Strategy 2 Consider how the problem affected people with different points of view. For example, the main problem described here was how to count the population of each state.

Strategy 3 Look for solutions people tried to deal with each problem. Think about whether the solution was a good one for people with differing points of view.

SLAVERY AND THE CONSTITUTION

Because the House of Representatives would have members according to the population of each state, 1 the delegates had to decide who would be counted in the population of each state. The Southern states had many more slaves than the Northern states had. 2 Southerners wanted the slaves to be counted as part of the general population for representation but not for taxation. 2 Northerners argued that slaves were not citizens and should not be counted for representation, but that slaves should be counted for taxation. 3 The delegates decided that three-fifths of the slave population would be counted in the population to determine both representation and taxes.

Make a Chart

Making a chart will help you identify and organize information about problems and solutions. The chart below shows problems and solutions included in the passage you just read.

1 *Problem*	2 *Differing Points of View*	3 *Solution*
Northerners and Southerners couldn't agree on how to count population because of slavery in the South.	*Southerners wanted slaves counted for representation but not for taxation. Northerners wanted slaves counted for taxation but not for representation.*	*Delegates decided that three-fifths of the slave population should be counted.*

Practicing the Skill

Turn to Chapter 8, Section 2, "Creating the Constitution." Read "The Convention Assembles" and "The Virginia Plan." Then make a chart that summarizes the problems faced by the delegates at the Constitutional Convention and the solutions they agreed on.

2.11 Evaluating

Defining the Skill

To **evaluate** is to make a judgment about something. Historians evaluate the actions of people in history. One way to do this is to examine both the positives and negatives of a historical action, then decide which is stronger—the positive or the negative.

Applying the Skill

The following passage describes Susan B. Anthony's fight for women's rights. Use the strategies listed below to evaluate how successful she was.

How to Evaluate

Strategy 1 Before you evaluate a person's actions, first determine what that person was trying to do. In this case, think about what Anthony wanted to accomplish.

Strategy 2 Look for statements that show the positive, or successful, results of her actions. For example, Did she achieve her goals?

Strategy 3 Also look for statements that show the negative, or unsuccessful, results of her actions. Did she fail to achieve something she tried to do?

Strategy 4 Write an overall evaluation of the person's actions.

SUSAN B. ANTHONY

1 Susan B. Anthony was a skilled organizer who fought for women's rights. 2 She successfully built the women's movement into a national organization. Anthony believed that a woman must have money of her own. To this end, she supported laws that would give married women rights to control their own property and wages. 2 Mississippi passed the first such law in 1839. New York passed a property law in 1848 and a wages law in 1860. 3 Anthony also wanted to win the vote for women but failed to convince lawmakers to pass this reform in her lifetime. This reform did go through in 1920, 14 years after her death.

Make a Diagram for Evaluating

Using a diagram can help you evaluate. List the positives and negatives of the historical person's actions and decisions. Then make an overall judgment. The diagram below shows how the information from the passage you just read can be diagrammed.

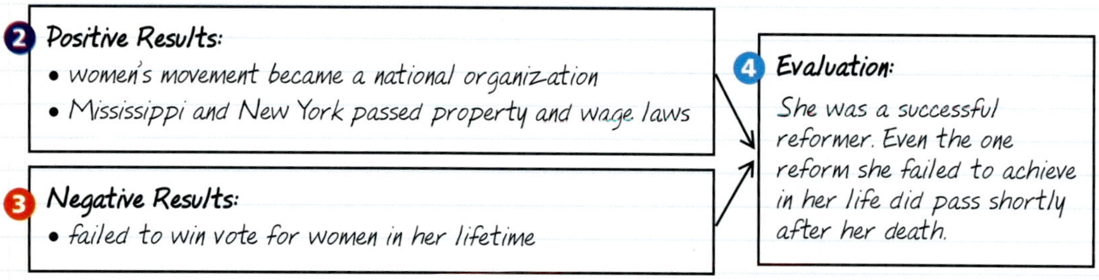

Practicing the Skill

Turn to Chapter 2, Section 3, "The Spanish and Native Americans." Read "The Columbian Exchange" and make a diagram in which you evaluate whether the Columbian Exchange had mainly a positive or negative impact on the world.

2.12 Making Generalizations

Defining the Skill

To **make generalizations** means to make broad judgments based on information. When you make generalizations, you should gather information from several sources.

Applying the Skill

The following three passages contain different views on George Washington. Use the strategies listed below to make a generalization about these views.

How to Make Generalizations

Strategy 1 Look for information that the sources have in common. These three sources all discuss George Washington's ability as a military leader.

Strategy 2 Form a generalization that describes Washington in a way that all three sources would agree with. State your generalization in a sentence.

WASHINGTON'S LEADERSHIP

1 Washington learned from his mistakes. After early defeats, he developed the strategy of dragging out the war to wear down the British. 1 Despite difficulties, he never gave up.

—*Creating America*

1 [Washington] was no military genius. . . . But he was a great war leader. Creating an army out of unpromising material, he kept it in being against great odds.

—*The Limits of Liberty*

1 [Washington] certainly deserves some merit as a general, that he . . . can keep General Howe dancing from one town to another for two years together, with such an army as he has.

—*The Journal of Nicholas Cresswell, July 13, 1777*

Make a Chart

Using a chart can help you make generalizations. The chart below shows how the information you just read can be used to generalize about people's views of Washington.

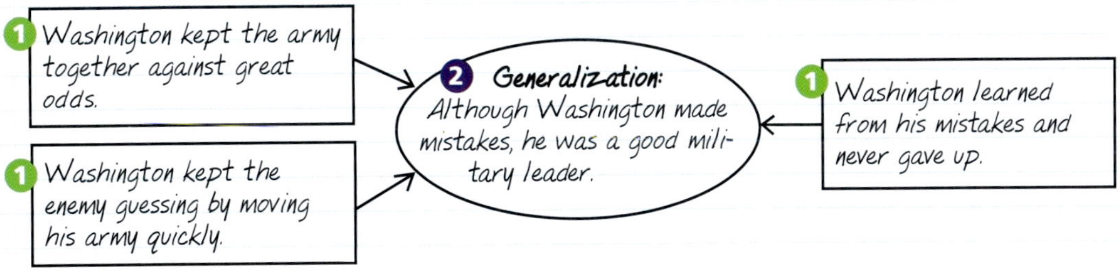

Practicing the Skill

Turn to Chapter 16, Section 1, "War Erupts." Read "Choosing Sides." Also read the excerpt from *Across Five Aprils* on pages 486–487. Then use a chart like the one above to make a generalization about how the outbreak of the Civil War affected Americans.

3.1 Using Primary and Secondary Sources

Defining the Skill

Primary sources are materials written or made by people who lived during historical events and witnessed them. Primary sources can be letters, journal entries, speeches, autobiographies, or artwork. Other kinds of primary sources are government documents, census surveys, and financial records. **Secondary sources** are materials written by people who did not participate in an event. History books are secondary sources.

Applying the Skill

The following passage contains both a primary source and a secondary source. Use the strategies listed below to help you read them.

How to Read Primary and Secondary Sources

Strategy 1 Distinguish secondary sources from primary sources. The first paragraph is a secondary source. The Declaration of Independence is a primary source. The secondary source explains something about the primary source.

Strategy 2 Analyze the primary source and consider why the author produced it. Consider what the document was supposed to achieve and who would read it.

Strategy 3 Identify the author of the primary source and note when and where it was written.

1 The core idea of the Declaration is based on the philosophy of John Locke. This idea is that people have unalienable rights, or rights that government cannot take away. Jefferson stated this belief in what was to become the Declaration's best-known passage.

> 2 We hold these truths to be self-evident, that all men are created equal, that they are endowed by their Creator with certain unalienable Rights, that among these are Life, Liberty and the pursuit of Happiness.
>
> 3 —Thomas Jefferson, *The Declaration of Independence*, 1776

Make a Chart

Making a chart will help you summarize information from primary sources and secondary sources. The chart below summarizes the information from the passage you just read.

Author	Thomas Jefferson
Document	The Declaration of Independence
Notes on Primary Source	The Declaration says that "all men are created equal." It also says that people have "unalienable rights." These rights include the right to life and the right to liberty, as well as a right to pursue happiness.
Notes on Secondary Source	Jefferson based his ideas on those of John Locke. Locke had written about rights that governments could not take away from the people.

Practicing the Skill

Turn to Chapter 6, Section 3, "The Road to Lexington and Concord." Read "Between War and Peace" and make a chart like the one above to summarize the information in the primary source and the secondary source.

3.2 Interpreting Graphs

Defining the Skill

Graphs use pictures and symbols, instead of words, to show information. There are many different kinds of graphs. Bar graphs, line graphs, and pie graphs are the most common. Bar graphs compare numbers or sets of numbers. The length of each bar shows a quantity. It is easy to see how different categories compare on a bar graph.

Applying the Skill

The bar graph below shows numbers of immigrants coming to the United States between 1821 and 1860. Use the strategies listed below to help you interpret the graph.

How to Interpret a Graph

Strategy 1 Read the title to identify the main idea of the graph.

Strategy 2 Read the vertical axis (the one that goes up and down) on the left side of the graph. This one shows the number of immigrants in thousands. Each bar represents the number of immigrants during a particular decade.

Strategy 3 Read the horizontal axis (the one that runs across the bottom of the graph). This one shows the four decades from 1821 to 1860.

Strategy 4 Summarize the information shown in each part of the graph. Use the title to help you focus on what information the graph is presenting.

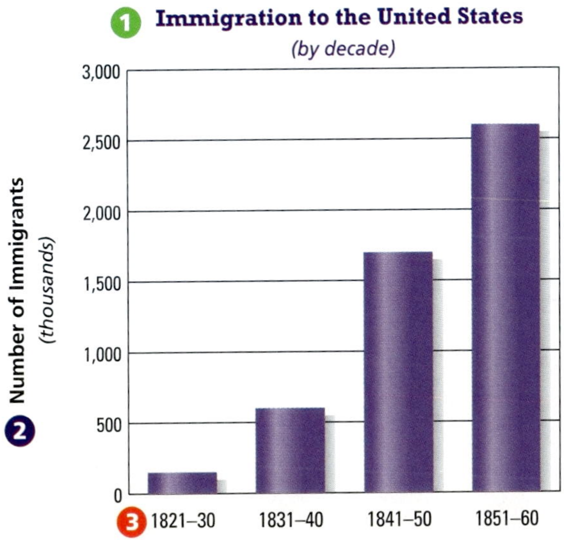

Write a Summary

Writing a summary will help you understand the information in the graph. The paragraph to the right summarizes the information from the bar graph.

4 *Immigration to the United States increased between 1821 and 1860. Between 1821 and 1830, fewer than 200,000 immigrants arrived. In the next decade, more than 500,000 immigrants came. During the 1840s, more than 1.5 million immigrants arrived, and that number increased to more than 2.5 million in the 1850s.*

Practicing the Skill

Turn to Chapter 2, Section 4, "Beginnings of Slavery in the Americas." Look at the graph entitled "Slaves Imported to the Americas, 1493–1810" and write a paragraph in which you summarize what you learned from it.

3.3 Interpreting Charts

Defining the Skill

Charts present information in a visual form. They organize, simplify, and summarize information in a way that makes that information easy to read and understand. Tables and diagrams are examples of commonly used charts.

Applying the Skill

The chart below shows the number of slaves who were imported to the Americas between 1601 and 1810. Use the strategies listed below to help you interpret the information in the chart.

How to Interpret a Chart

Strategy 1 Read the title. It will tell you what the chart is about.

Strategy 2 Read the labels to see how the information in the chart is organized. In this chart, it is organized by region and years.

Strategy 3 Study the data in the chart to understand the facts that the chart intends to show.

Strategy 4 Summarize the information shown in each part of the chart. Use the title to help you focus on what information the chart is presenting.

1601–1810

1 **Slaves Imported to the Americas** *(in thousands)*

2 REGION/COUNTRY	1601–1700	1701–1810
3 British N. America	*	348
British Caribbean	263.7	1,401.3
French Caribbean	155.8	1,348.4
Spanish America	292.5	578.6
Dutch Caribbean	40	460
Danish Caribbean	4	24
Brazil (Portugal)	560	1,891.4

*= less than 1,000

Source: Philip D. Curtin, *The Atlantic Slave Trade*

Write a Summary

Writing a summary can help you understand the information given in a chart. The paragraph to the right summarizes the information in the chart "Slaves Imported to the Americas, 1601–1810."

4 The chart shows how many slaves were imported to the Americas between 1601 and 1810. It divides the Americas into seven regions. It also divides the time period into two parts: 1601–1700 and 1701–1810. The number of slaves imported increased greatly from the 1600s to the 1700s. More slaves were imported to Brazil than to any other region.

Practicing the Skill

Turn to Chapter 27, Section 5, and look at the chart entitled "World War II Military Casualties, 1939–1945." Study the chart and write a paragraph in which you summarize what you learned from it.

3.4 Interpreting Time Lines

Defining the Skill

A **time line** is a visual list of events and dates shown in the order in which they occurred. Time lines can be horizontal or vertical. On horizontal time lines, the earliest date is on the left. On vertical time lines, the earliest date is often at the top.

Applying the Skill

The time line below lists dates and events during the presidencies of John Adams, Andrew Jackson, and Martin Van Buren. Use the strategies listed below to help you interpret the information.

How to Read a Time Line

Strategy 1 Read the dates at the beginning and end of the time line. These will show the period of history that is covered. The time line below is a dual time line. It includes items related to two topics. The labels show that the information covers U.S. events and world events.

Strategy 2 Read the dates and events in sequential order, beginning with the earliest one. Pay particular attention to how the entries relate to each other. Think about which events caused later events.

Strategy 3 Summarize the focus, or main idea, of the time line. Try to write a main idea sentence that describes the time line.

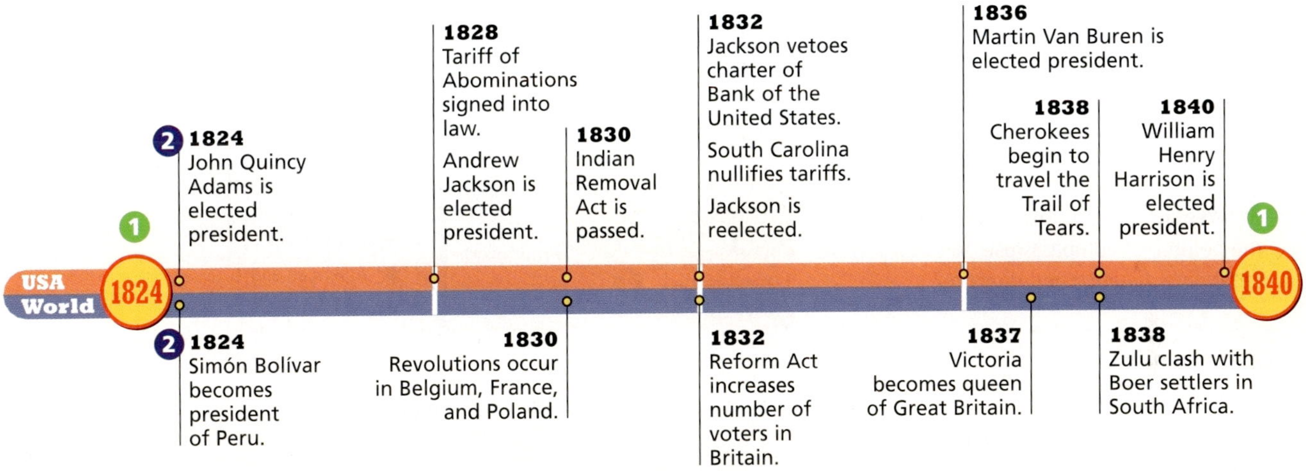

Write a Summary

Writing a summary can help you understand information shown on a time line. The summary to the right states the main idea of the time line and tells how the events are related.

3 *The time line covers the period between 1824, when John Quincy Adams was elected president, and 1840, when William Henry Harrison was elected president. During that period of time, Andrew Jackson and Martin Van Buren also served as president. The time line shows that the important issues in the United States were tariffs, banking, and relations with Native Americans.*

Practicing the Skill

Turn to Chapter 15, page 455, and write a summary of the information shown on the time line.

3.5 Reading a Map

Defining the Skill

Maps are representations of features on the earth's surface. Some maps show political features, such as national borders. Other maps show physical features, such as mountains and bodies of water. You can understand maps better once you know how to recognize these elements.

Applying the Skill

The following map shows the Battle of Yorktown during the Revolution. Use the strategies listed below to help you identify the elements common to most maps.

How to Read a Map

Strategy 1 Read the title. This identifies the main idea of the map.

Strategy 2 Look for the grid of lines that form a pattern of squares over the map. These numbered lines are the lines of latitude (horizontal) and longitude (vertical). They indicate the location of the area on the earth.

Strategy 3 Read the map key. It is usually in a box. This will give you the information you need to interpret the symbols or colors on the map.

Strategy 4 Use the scale and the pointer, or compass rose, to determine distance and direction.

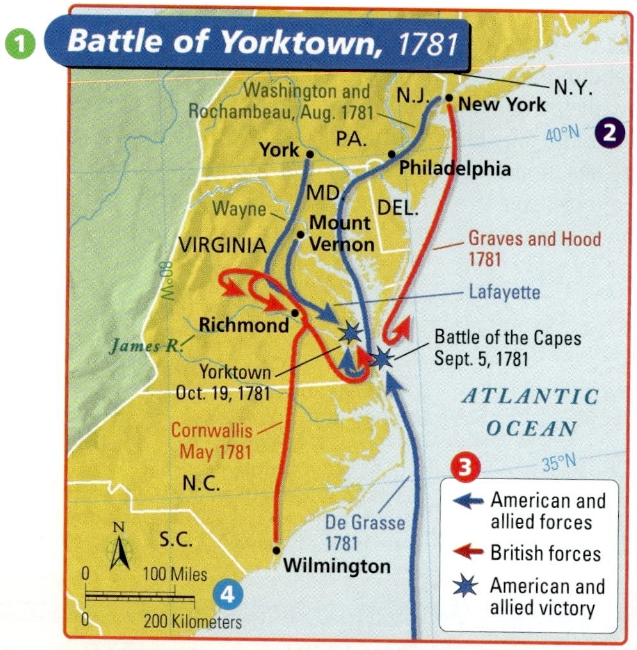

Make a Chart

A chart can help you organize information given on maps. The chart below summarizes information about the map you just studied.

Title	Battle of Yorktown, 1781
Location	between latitude 40° N and 35° N, just east of longitude 80° W
Map Key Information	blue = American and allied forces, red = British forces
Scale	7/16 in. = 100 miles, 9/16 in. = 200 km
Summary	British commanders Graves and Hood sailed south from New York. They were defeated by De Grasse at the Battle of the Capes. British commander Cornwallis marched north from Wilmington, North Carolina, to Virginia, where he was defeated by American forces.

Practicing the Skill

Turn to Chapter 1, Section 5, "Early European Explorers." Read the map entitled "Exploration Leads to New Sea Routes" and make a chart to identify information on the map.

3.6 Reading a Special-Purpose Map

Defining the Skill

Special-purpose maps help people focus on a particular aspect of a region, such as economic development in the South. These kinds of maps often use symbols to indicate information.

Applying the Skill

The following special-purpose map indicates the products of the Southern colonies. Use the strategies listed below to help you identify the information shown on the map.

How to Read a Special-Purpose Map

Strategy 1 Read the title. It tells you what the map is intended to show.

Strategy 2 Read the legend. This tells you what each symbol stands for. This legend shows the crops that were grown in various Southern colonies.

Strategy 3 Look for the places on the map where the symbol appears. These tell you the places where each crop was grown.

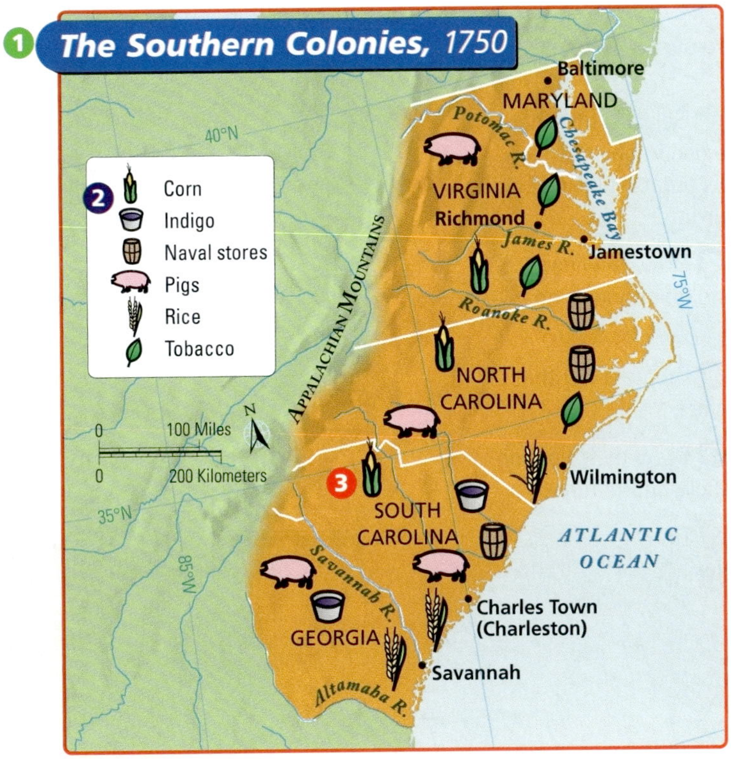

Make a Chart

A chart can help you understand special-purpose maps. The chart below shows information about the special-purpose map you just studied.

	Corn	Indigo	Naval stores	Pigs	Rice	Tobacco
Maryland						x
Virginia	x			x		x
North Carolina	x		x	x	x	x
South Carolina	x	x	x	x	x	
Georgia		x		x	x	

Practicing the Skill

Turn to Chapter 4, Section 1, "New England: Land and Sea." Look at the special-purpose map entitled "The New England Colonies" and make a chart that shows information about products from New England.

3.7 Creating a Map

Defining the Skill

Creating a map involves representing geographical information. When you draw a map, it is easiest to use an existing map as a guide. On the map you draw, you can show geographical information. You can also show other kinds of information, such as data on climates, population trends, resources, or routes. Often, this data comes from a graph or a chart.

Applying the Skill

Below is a map that a student created to show information about the number of slaves in 1750. Read the strategies listed below to see how the map was created.

How to Create a Map

Strategy 1 Select a title that identifies the geographical area and the map's purpose. Include a date in your title.

Strategy 2 Draw the lines of latitude and longitude using short dashes.

Strategy 3 Create a key that shows the colors.

Strategy 4 Draw the colors on the map to show information.

Strategy 5 Draw a compass rose and scale.

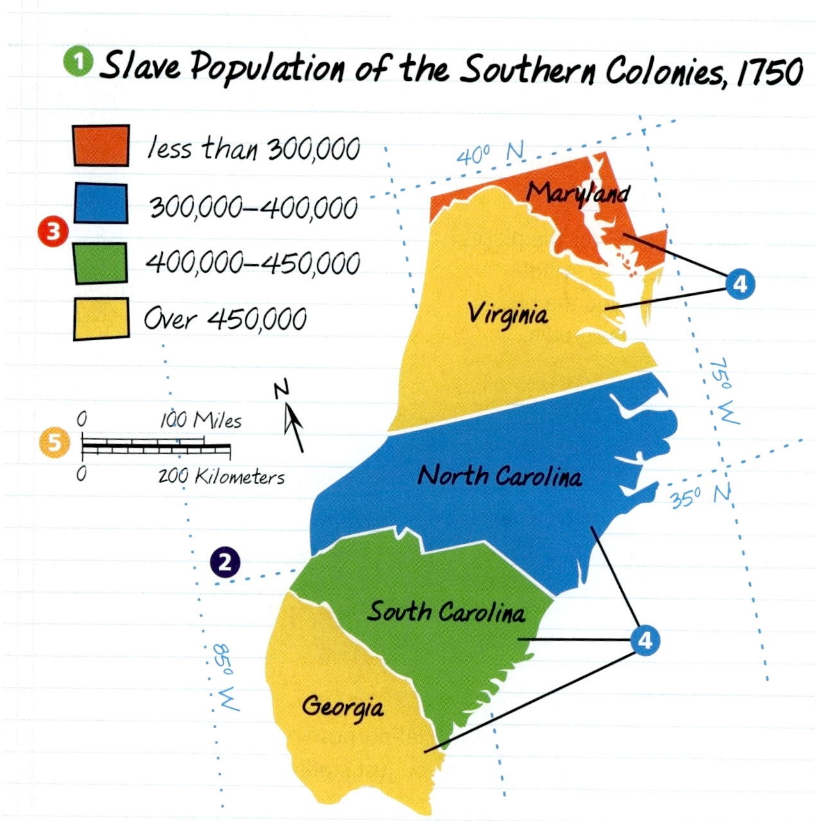

Practicing the Skill

Make your own map. Turn to page 104 in Chapter 3 and study the graph entitled "The 13 Colonies." Use the strategies described above to create a map that shows the 13 colonies and the dates that they were founded. You can use the map on page 102 of that chapter as a guide.

3.8 Interpreting Political Cartoons

Defining the Skill

Political cartoons are cartoons that use humor to make a serious point. Political cartoons often express a point of view on an issue better than words do. Understanding signs and symbols will help you to interpret political cartoons.

Applying the Skill

The cartoon below shows Abraham Lincoln and the other candidates running for the presidency in 1860. Use the strategies listed below to help you understand the cartoon.

How to Interpret a Political Cartoon

Strategy 1 Identify the subject by reading the title of the cartoon and looking at the cartoon as a whole.

Strategy 2 Identify important symbols and details. The cartoonist uses the image of a running race to discuss a political campaign. The White House is the finish line.

Strategy 3 Interpret the message. Why is Lincoln drawn so much taller than the other candidates? How does that make him the fittest candidate?

Make a Chart

Making a chart will help you summarize information from a political cartoon. The chart below summarizes the information from the cartoon above.

Subject	*"A Political Race" (The Election of 1860)*
Symbols and Details	*Running is a symbol for a political campaign. Lincoln is the tallest and fastest candidate.*
Message	3 *Lincoln is pulling ahead of the other candidates in the campaign for the presidency.*

Practicing the Skill

Turn to Chapter 18, Section 3, "End of Reconstruction." Look at the political cartoon on page 547. It shows a cartoonist's view of corruption in President Grant's administration. Use a chart like the one above and the strategies outlined to interpret the cartoon.

4.1 Using an Electronic Card Catalog

Defining the Skill

An **electronic card catalog** is a library's computerized search program that will help you find information about the books and other materials in the library. You can search the catalog by entering a book title, an author's name, or a subject of interest to you. The electronic card catalog will give you information about the materials in the library. This information is called bibliographic information. You can use an electronic card catalog to create a bibliography (a list of books) on any topic you are interested in.

Applying the Skill

The screen shown below is from an electronic search for information about Thomas Jefferson. Use the strategies listed below to help you use the information on the screen.

How to Use an Electronic Card Catalog

Strategy 1 Begin searching by choosing either subject, title, or author, depending on the topic of your search. For this search, the user chose "Subject" and typed in the words "Jefferson, Thomas."

Strategy 2 Once you have selected a book from the results of your search, identify the author, title, city, publisher, and date of publication.

Strategy 3 Look for any special features in the book. This book is illustrated, and it includes bibliographical references and an index.

Strategy 4 Locate the call number for the book. The call number indicates the section in the library where you will find the book. You can also find out if the book is available in the library you are using. If not, it may be in another library in the network.

```
Search Request:
1 Subject          Title          Author

Find  Options  Locations  Backup  Startover  Help

2 Miller, Douglas T. Thomas Jefferson and the
creation of America. New York: Facts on File,
1997.
        AUTHOR: Miller, Douglas T.
      2  TITLE: Thomas Jefferson and the
                creation of America/Douglas T.
                Miller.
   2 PUBLISHED: New York: Facts on File, ©1997.
        PAGING: vi, 122p. : ill ; 24 cm.
        SERIES: Makers of America.
       3 NOTES: Includes bibliographical
                references (p. 117-118) and index.
 4 CALL NUMBER: 1. 973.46 N61T 1997–Book Available–
```

Practicing the Skill

Turn to Chapter 10, "The Jefferson Era," and find a topic that interests you, such as the Federalists, the Louisiana Purchase, the Lewis and Clark expedition, or the War of 1812. Use the SUBJECT search on an electronic card catalog to find information about your topic. Make a bibliography of books about the subject. Be sure to include the author, title, city, publisher, and date of publication for all the books included.

4.2 Creating a Database

Defining the Skill

A **database** is a collection of data, or information, that is organized so that you can find and retrieve information on a specific topic quickly and easily. Once a computerized database is set up, you can search it to find specific information without going through the entire database. The database will provide a list of all information in the database related to your topic. Learning how to use a database will help you learn how to create one.

Applying the Skill

The chart below is a database for the significant battles of the Civil War. Use the strategies listed below to help you understand and use the database.

How to Create a Database

Strategy 1 Identify the topic of the database. The keywords, or most important words, in this title are "Civil War" and "Battles." These words were used to begin the research for this database.

Strategy 2 Identify the kind of data you need to enter in your database. These will be the column headings of your database. The key words "Battle," "Date," "Location," and "Significance" were chosen to focus the research.

Strategy 3 Identify the entries included under each heading.

Strategy 4 Use the database to help you find information quickly. For example, in this database you could search for "Union victories" to find a list of significant battles won by the North.

1 SIGNIFICANT CIVIL WAR BATTLES

2 BATTLE	DATE	LOCATION	SIGNIFICANCE
3 Fort Sumter	April 12, 1861	Charleston, SC	Beginning of the Civil War
First Battle of Bull Run (Manassas)	July 21, 1861	Virginia	Confederate victory
Shiloh	April 6–7, 1862	Tennessee (near Shiloh Church)	4 Union victory
Antietam	September 17, 1862	Sharpsburg, MD	No clear victory; considered bloodiest battle of war
Gettysburg	July 1–3, 1863	Gettysburg, PA	Retreat of Confederacy
Vicksburg	Three-month siege ending July 3, 1863	Vicksburg, MS	Union gained control of Mississippi River
Chattanooga	November 23–25, 1863	Chattanooga, TN	4 Union victory
Atlanta	September 2, 1864	Atlanta, GA	4 Union victory; helped convince Confederacy of defeat

Practicing the Skill

Create a database for U.S. presidents through the Civil War that shows each president's home state, political party, and years served as president. Use the informaton in "Presidents of the United States" on pages R40–R42 to provide the data. Use a format like the one above for your database.

4.3 Using the Internet

Defining the Skill

The Internet is a computer network that connects to universities, libraries, news organizations, government agencies, businesses, and private individuals throughout the world. Each location on the Internet has a home page with its own address, or URL (universal resource locator). With a computer connected to the Internet, you can reach the home pages of many organizations and services. The international collection of home pages, known as the World Wide Web, is a good source of up-to-date information about current events as well as research on subjects in history.

Applying the Skill

The Web page below shows the links for Chapter 6 of *Creating America.* Use the strategies listed below to help you understand how to use the Web page.

How to Use the Internet

Strategy ❶ Go directly to a Web page. For example, type http://www.mcdougallittell.com in the box at the top of the screen and press ENTER (or RETURN). The Web page will appear on your screen. Then click on ClassZone and find the link to *Creating America.*

Strategy ❷ Explore the *Creating America* links. Click on any one of the links to find out more about a specific subject. These links take you to other pages at this Web site. Some pages include links to related information that can be found at other places on the Internet.

Strategy ❸ When using the Internet for research, you should confirm the information you find. Web sites set up by universities, government agencies, and reputable news sources are more reliable than other sources. You can often find information about the creator of a site by looking for copyright information.

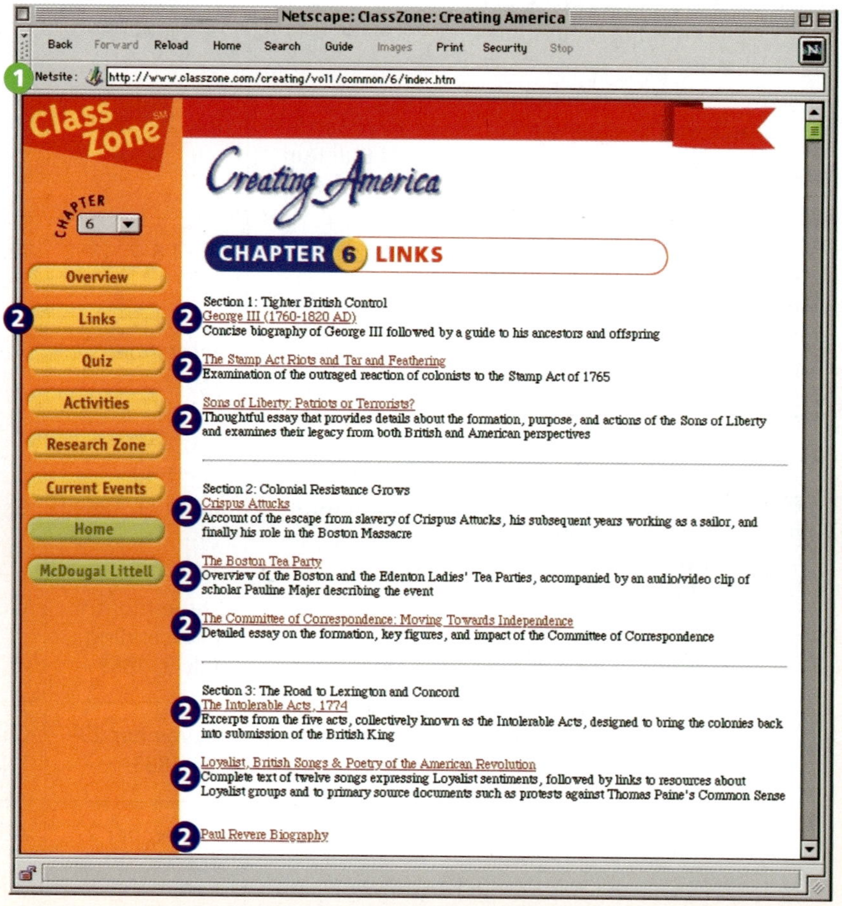

Practicing the Skill

Turn to Chapter 14, Section 2, "American Literature and Art." Read the section and make a list of topics you would like to research. If you have Internet access, go to the McDougal Littell home page at http://www.mcdougallittell.com and click on ClassZone. There you will find links that provide more information about the topics in the section.

4.4 Creating a Multimedia Presentation

Defining the Skill

Movies, CD-ROMs, television, and computer software are different kinds of media. To **create a multimedia presentation,** you need to collect information in different media and organize them into one presentation.

Applying the Skill

The scene below shows students using computers to create a multimedia presentation. Use the strategies listed below to help you create your own multimedia presentation.

How to Create a Multimedia Presentation

Strategy 1 Identify the topic of your presentation and decide which media are best for an effective presentation. For example, you may want to use slides or posters to show visual images of your topic. Or, you may want to use CDs or audiotapes to provide music or spoken words.

Strategy 2 Research the topic in a variety of sources. Images, text, props, and background music should reflect the historical period of the event you choose.

Strategy 3 Write the script for the presentation. You could use a narrator and characters' voices to tell the story. Primary sources are an excellent source for script material.

Strategy 4 Videotape the presentation. Videotaping the presentation will preserve it for future viewing and allow you to show it to different groups of people.

Practicing the Skill

Turn to Chapter 24, "World War I." Choose a topic from the chapter and use the strategies listed above to create a multimedia presentation about it.

World Political

ARCTIC OCEAN
Barents Sea
Norwegian Sea
ICELAND
NORWAY
FINLAND
RUSSIA
SWEDEN
UNITED KINGDOM
IRELAND
GERMANY
POLAND
BELARUS
UKRAINE
FRANCE
KAZAKHSTAN
MONGOLIA
Caspian Sea
Black Sea
UZBEKISTAN
KYRGYZSTAN
ITALY
PORTUGAL
SPAIN
TURKMENISTAN
TAJIKISTAN
TURKEY
NORTH KOREA
SOUTH KOREA
JAPAN
AZORES (PORT.)
Mediterranean Sea
SYRIA
CHINA
LEBANON
IRAQ
IRAN
AFGHANISTAN
CANARY IS. (SP.)
MOROCCO
ISRAEL
BHUTAN
ALGERIA
LIBYA
EGYPT
JORDAN
KUWAIT
PAKISTAN
NEPAL
WESTERN SAHARA
QATAR
TAIWAN
SAUDI ARABIA
U.A.E.
INDIA
PACIFIC OCEAN
MYANMAR
Red Sea
MAURITANIA
MALI
NIGER
ERITREA
OMAN
BANGLADESH
LAOS
NO. MARIANA IS. (U.S.)
VIETNAM
CHAD
YEMEN
Arabian Sea
THAILAND
South China Sea
CAPE VERDE
SUDAN
Bay of Bengal
PHILIPPINES
DJIBOUTI
CAMBODIA
FED. STATES OF MICRONESIA
NIGERIA
ETHIOPIA
CEN. AFR. REP.
BRUNEI
PALAU
CAMEROON
SOMALIA
MALDIVES
SRI LANKA
MALAYSIA
UGANDA
Gulf of Guinea
RWANDA
KENYA
Borneo
Sulawesi
Equator
PAPUA NEW GUINEA
GABON
DEM. REP. OF THE CONGO
Sumatra
PEOP. REP. OF THE CONGO
BURUNDI
SEYCHELLES
INDONESIA
TANZANIA
Java
New Guinea
CHAGOS ARCH. (U.K.)
COMOROS
MALAWI
ANGOLA
ZAMBIA
MOZAMBIQUE
MADAGASCAR
MAURITIUS
INDIAN OCEAN
ZIMBABWE
NAMIBIA
BOTSWANA
RÉUNION (FR.)
AUSTRALIA
SWAZILAND
ATLANTIC OCEAN
SOUTH AFRICA
LESOTHO
N
KERGUELEN IS. (FR.)
0
2,000 Miles
SO. SANDWICH ISLANDS (U.K.)
0
4,000 Kilometers
60°N
40°N
20°N
0°
20°S
60°S
20°W
20°E
40°E
60°E
80°E
100°E
120°E
140°E
80°N

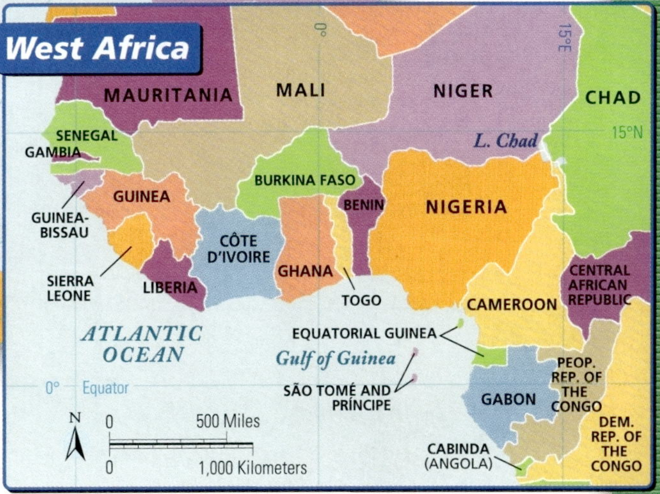
West Africa
MAURITANIA
MALI
NIGER
CHAD
SENEGAL
GAMBIA
L. Chad
BURKINA FASO
GUINEA
BENIN
NIGERIA
GUINEA-BISSAU
CÔTE D'IVOIRE
SIERRA LEONE
GHANA
LIBERIA
TOGO
CENTRAL AFRICAN REPUBLIC
CAMEROON
ATLANTIC OCEAN
EQUATORIAL GUINEA
Gulf of Guinea
SÃO TOMÉ AND PRÍNCIPE
PEOP. REP. OF THE CONGO
GABON
Equator
DEM. REP. OF THE CONGO
CABINDA (ANGOLA)
N
0
500 Miles
0
1,000 Kilometers
0°
15°E
15°N

Europe
FINLAND
NORWAY
North Sea
SWEDEN
ESTONIA
0
500 Miles
0
1,000 Kilometers
DENMARK
Baltic Sea
LATVIA
LITHUANIA
RUSSIA
IRELAND
UNITED KINGDOM
NETH.
BELARUS
POLAND
KAZAKHSTAN
GERMANY
ATLANTIC OCEAN
BELG.
LUX.
CZECH REP.
SLOVAKIA
UKRAINE
Caspian Sea
MOLDOVA
SWITZ.
AUSTRIA
HUNGARY
FRANCE
SLOVENIA
ROMANIA
CROATIA
PORTUGAL
GEORGIA
BOS.-HERZ.
YUGO.
Black Sea
BULGARIA
ARMENIA
ITALY
AZERBAIJAN
SPAIN
ALBANIA
MACEDONIA
TURKEY
Mediterranean Sea
GREECE
SYRIA
IRAQ
ALGERIA
TUNISIA
CYPRUS
50°N
40°N
0°
20°E
40°E
N

United States Political

CANADA
Lake Nipigon
Lake Superior
Gouin Reservoir
Lake St. John
Cabonga Reservoir
St. Lawrence R.
Ottawa R.
Lake Nipissing
Georgian Bay
Lake Simcoe
Lake Huron
Lake Ontario
Lake St. Clair
Lake Erie
Lake Michigan
Lake Champlain
Bay of Fundy
MAINE
Augusta
Montpelier
VERMONT
NEW HAMPSHIRE
Concord
Connecticut R.
Hudson R.
Boston
MASSACHUSETTS
Providence
RHODE ISLAND
CONNECTICUT
Hartford
Long Island
Albany
NEW YORK
New York
NEW JERSEY
Trenton
Philadelphia
Buffalo
Susquehanna R.
PENNSYLVANIA
Harrisburg
Pittsburgh
Dover
Delaware Bay
DELAWARE
Baltimore
Washington, D.C.
Annapolis
MARYLAND
Chesapeake Bay
WISCONSIN
Lake Winnebago
Madison
Milwaukee
MICHIGAN
Lansing
Detroit
Toledo
Cleveland
Mississippi River
Chicago
ILLINOIS
Springfield
INDIANA
Indianapolis
OHIO
Columbus
Cincinnati
WEST VIRGINIA
Charleston
Richmond
VIRGINIA
Ohio River
Frankfort
KENTUCKY
Jefferson City
St. Louis
MISSOURI
Cumberland River
Winston-Salem
Raleigh
NORTH CAROLINA
Pamlico Sound
Knoxville
Nashville
TENNESSEE
ARKANSAS
Memphis
Tennessee River
River
Little Rock
SOUTH CAROLINA
Columbia
Savannah River
Atlanta
GEORGIA
Birmingham
ALABAMA
MISSISSIPPI
Mississippi River
Montgomery
Alabama R.
LOUISIANA
Jackson
Jacksonville
Tallahassee
Baton Rouge
Lake Pontchartrain
New Orleans
Breton Sound
FLORIDA
Tampa
St. Petersburg
Lake Okeechobee
Miami
Gulf of Mexico
Straits of Florida
BAHAMAS
ATLANTIC OCEAN
National capital
State capital
N
0 500 Miles
0 1000 Kilometers
45°N
40°N
35°N
30°N
25°N
70°W
75°W
80°W
85°W
90°W
PUERTO RICO
ATLANTIC OCEAN
San Juan
PUERTO RICO
Caguas
Ponce
66°W
18°N
0 100 Miles
0 100 Kilometers
Caribbean Sea

United States Physical

HUDSON BAY LOWLANDS
ONTARIO
CANADA
Lake Nipigon
Lake Abitibi
Harricana River
QUEBEC
Gouin Reservoir
Saguenay R.
Lake St. John
LAURENTIDE SCARP
NOTRE DAME MTNS.
St. Maurice R.
Madeleine Is.
PRINCE EDWARD ISLAND
Charlottetown
NEW BRUNSWICK
Fredericton
Québec
Halifax
NOVA SCOTIA
Bay of Fundy
Cabonga Reservoir
Isle Royale
Lake Superior
LAURENTIAN SCARP
Ottawa R.
St. Lawrence R.
Montreal
MAINE
Augusta
Cape Sable
Lake Nipissing
Ottawa
Manitoulin I.
Montpelier
Mt. Washington
1917 m.
6288 ft.
Lake Champlain
Connecticut R.
VERMONT
NEW HAMPSHIRE
Concord
Lake Huron
Lake Simcoe
WISCONSIN
Wisconsin R.
Minneapolis
St. Paul
Mississippi River
MICHIGAN
Lake Michigan
Toronto
Lake Ontario
Syracuse
Albany
Hudson R.
Worcester
Boston
MASSACHUSETTS
Hamilton
Rochester
Springfield
Cape Cod
Providence
London
Buffalo
NEW YORK
Hartford
Nantucket Island
Martha's Vineyard
RHODE ISLAND
Madison
Milwaukee
Lansing
Flint
Lake St. Clair
New Haven
Long Island
CONNECTICUT
Grand Rapids
Detroit
Windsor
Lake Erie
New York
IOWA
Chicago
Toledo
Cleveland
PENNSYLVANIA
Harrisburg
Trenton
NEW JERSEY
Des Moines
Akron
Pittsburgh
Philadelphia
OHIO
Susquehanna R.
Wilmington
Columbus
Dover
Cape May
ATLANTIC OCEAN
ILLINOIS
Wabash River
Indianapolis
Baltimore
Annapolis
Delaware Bay
Springfield
Dayton
Ohio River
WEST VIRGINIA
Washington
DELAWARE
INDIANA
Cincinnati
MARYLAND
MISSOURI
Charleston
VIRGINIA
Chesapeake Bay
Jefferson City
St. Louis
Louisville
Frankfort
James R.
Richmond
Cape Charles
STATES
KENTUCKY
Norfolk
APPALACHIAN MOUNTAINS
Roanoke River
OZARK PLATEAU
CUMBERLAND PLATEAU
Cumberland River
Knoxville
BLUE RIDGE MTNS.
Hatteras Island
Pamlico Sound
Cape Hatteras
Raleigh
Nashville
TENNESSEE
Mt. Mitchell
2037 m.
6684 ft.
Pee Dee R.
NORTH CAROLINA
ARKANSAS
Charlotte
Arkansas River
Chattanooga
SOUTH CAROLINA
Memphis
Tennessee R.
Columbia
Savannah River
Little Rock
Birmingham
Atlanta
Charleston
COASTAL PLAIN
MISSISSIPPI
Tombigbee R.
ALABAMA
Chattahoochee R.
GEORGIA
Mississippi River
Alabama R.
Montgomery
Savannah
Jackson
LOUISIANA
Jacksonville
Mobile
Tallahassee
Baton Rouge
PLAIN
Lafayette
New Orleans
Lake Pontchartrain
Breton Sound
Cape San Blas
FLORIDA
Cape Canaveral
Gulf of Mexico
Tampa
St. Petersburg
Lake Okeechobee
BAHAMAS
Miami
Nassau
Cape Sable
Florida Keys
Straits of Florida
Land Elevation
Higher than 16,000 ft.
8,000 ft. to 16,000 ft.
4,000 ft. to 8,000 ft.
2,000 ft. to 4,000 ft.
1,000 ft. to 2,000 ft.
0 ft. to 1,000 ft.
Below sea level
Water Depth
0 ft. to 700 ft.
700 ft. to 9,800 ft.
9,800 ft. 19,700 ft.
Deeper than 19,700 ft.
National capital
State capital
Heights and depths
N
0
500 Miles
0
1000 Kilometers
PUERTO RICO
San Juan
Caguas
Ponce
18°N
100 Miles
100 Kilometers
66°W

ACTS About the STATES

Alabama
4,352,000 people
52,237 sq. mi.
Rank in area: 30
Entered Union in 1819

Alaska
614,000 people
615,230 sq. mi.
Rank in area: 1
Entered Union in 1959

Arizona
4,669,000 people
114,006 sq. mi.
Rank in area: 6
Entered Union in 1912

Arkansas
2,538,000 people
53,182 sq. mi.
Rank in area: 28
Entered Union in 1836

California
32,667,000 people
158,869 sq. mi.
Rank in area: 3
Entered Union in 1850

Colorado
3,971,000 people
104,100 sq. mi.
Rank in area: 8
Entered Union in 1876

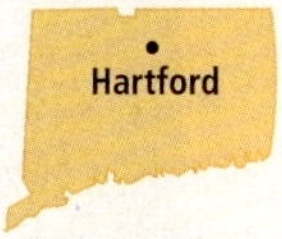

Connecticut
3,274,000 people
5,544 sq. mi.
Rank in area: 48
Entered Union in 1788

Delaware
744,000 people
2,397 sq. mi.
Rank in area: 49
Entered Union in 1787

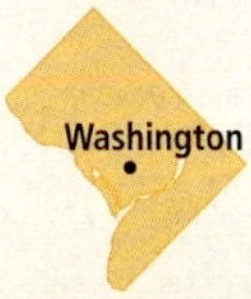

District of Columbia
523,000 people
68 sq. mi.

Florida
14,916,000 people
59,928 sq. mi.
Rank in area: 23
Entered Union in 1845

Georgia
7,642,000 people
58,977 sq. mi.
Rank in area: 24
Entered Union in 1788

Hawaii
1,193,000 people
6,459 sq. mi.
Rank in area: 47
Entered Union in 1959

Idaho
1,229,000 people
83,574 sq. mi.
Rank in area: 14
Entered Union in 1890

Illinois
12,045,000 people
57,918 sq. mi.
Rank in area: 25
Entered Union in 1818

Indiana
5,899,000 people
36,420 sq. mi.
Rank in area: 38
Entered Union in 1816

Iowa
2,862,000 people
56,276 sq. mi.
Rank in area: 26
Entered Union in 1846

Kansas
2,629,000 people
82,282 sq. mi.
Rank in area: 15
Entered Union in 1861

Kentucky
3,936,000 people
40,411 sq. mi.
Rank in area: 37
Entered Union in 1792

Louisiana
4,369,000 people
49,650 sq. mi.
Rank in area: 31
Entered Union in 1812

Maine
1,244,000 people
33,741 sq. mi.
Rank in area: 39
Entered Union in 1820

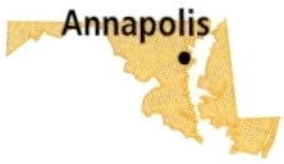

Maryland
5,135,000 people
12,297 sq. mi.
Rank in area: 42
Entered Union in 1788

Massachusetts
6,147,000 people
9,241 sq. mi.
Rank in area: 45
Entered Union in 1788

Michigan
9,817,000 people
96,705 sq. mi.
Rank in area: 11
Entered Union in 1837

Minnesota
4,725,000 people
86,943 sq. mi.
Rank in area: 12
Entered Union in 1858

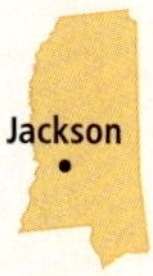

Mississippi
2,752,000 people
48,286 sq. mi.
Rank in area: 32
Entered Union in 1817

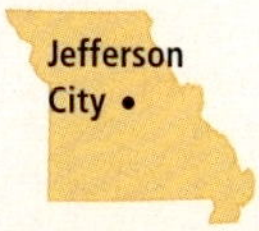

Missouri
5,439,000 people
69,709 sq. mi.
Rank in area: 21
Entered Union in 1821

Montana
880,000 people
147,046 sq. mi.
Rank in area: 4
Entered Union in 1889

Sources: U.S. Bureau of the Census, 1998 population estimates.
World Almanac and Book of Facts, 1999
Statistical Abstract of the United States, 1998

Nebraska
1,663,000 people
77,359 sq. mi.
Rank in area: 16
Entered Union in 1867

Ohio
11,209,000 people
44,828 sq. mi.
Rank in area: 34
Entered Union in 1803

Texas
19,760,000 people
267,277 sq. mi.
Rank in area: 2
Entered Union in 1845

Nevada
1,747,000 people
110,567 sq. mi.
Rank in area: 7
Entered Union in 1864

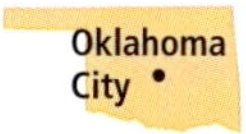

Oklahoma
3,347,000 people
69,903 sq. mi.
Rank in area: 20
Entered Union in 1907

Utah
2,100,000 people
84,904 sq. mi.
Rank in area: 13
Entered Union in 1896

New Hampshire
1,185,000 people
9,283 sq. mi.
Rank in area: 44
Entered Union in 1788

Oregon
3,282,000 people
97,132 sq. mi.
Rank in area: 10
Entered Union in 1859

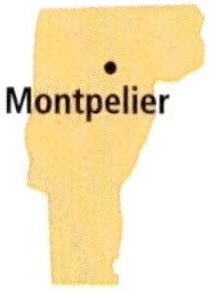

Vermont
591,000 people
9,615 sq. mi.
Rank in area: 43
Entered Union in 1791

New Jersey
8,115,000 people
8,215 sq. mi.
Rank in area: 46
Entered Union in 1787

Pennsylvania
12,001,000 people
46,058 sq. mi.
Rank in area: 33
Entered Union in 1787

Virginia
6,791,000 people
42,326 sq. mi.
Rank in area: 35
Entered Union in 1788

New Mexico
1,737,000 people
121,598 sq. mi.
Rank in area: 5
Entered Union in 1912

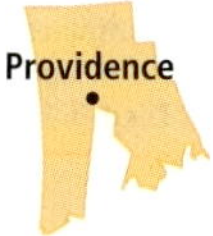

Rhode Island
988,000 people
1,231 sq. mi.
Rank in area: 50
Entered Union in 1790

Washington
5,689,000 people
70,637 sq. mi.
Rank in area: 19
Entered Union in 1889

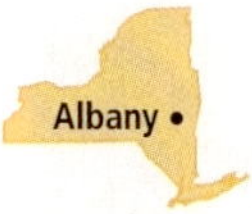

New York
18,175,000 people
53,989 sq. mi.
Rank in area: 27
Entered Union in 1788

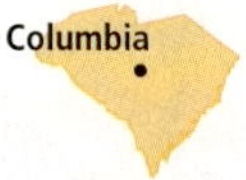

South Carolina
3,836,000 people
31,189 sq. mi.
Rank in area: 40
Entered Union in 1788

West Virginia
1,811,000 people
24,232 sq. mi.
Rank in area: 41
Entered Union in 1863

North Carolina
7,546,000 people
52,672 sq. mi.
Rank in area: 29
Entered Union in 1789

South Dakota
738,000 people
77,121 sq. mi.
Rank in area: 17
Entered Union in 1889

Wisconsin
5,224,000 people
65,500 sq. mi.
Rank in area: 22
Entered Union in 1848

North Dakota
638,000 people
70,704 sq. mi.
Rank in area: 18
Entered Union in 1889

Tennessee
5,431,000 people
42,145 sq. mi.
Rank in area: 36
Entered Union in 1796

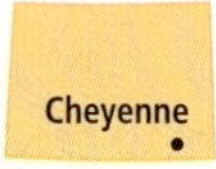

Wyoming
481,000 people
97,819 sq. mi.
Rank in area: 9
Entered Union in 1890

United States: *Major Dependencies*

- American Samoa—63,786 people; 90 sq. mi.
- Guam—151,716 people; 217 sq. mi.
- Commonwealth of Puerto Rico—3,887,652 people; 3,508 sq. mi.
- Virgin Islands of the United States—119,827 people; 171 sq. mi.
- Midway Islands—no indigenous inhabitants; 2 sq. mi.
- Wake Atoll—no indigenous inhabitants; 3 sq. mi.

RESIDENTS of the UNITED STATES

Here are some little-known facts about the presidents of the United States:

- Only former president to serve in Congress: John Quincy Adams
- First president born in the new United States: Martin Van Buren (eighth president)
- Only president who was a bachelor: James Buchanan
- First left-handed president: James A. Garfield
- Largest president: William H. Taft (6 feet 2 inches, 326 pounds)
- Youngest president: Theodore Roosevelt (42 years old)
- Oldest president: Ronald Reagan (77 years old when he left office in 1989)
- First president born west of the Mississippi River: Herbert Hoover (born in West Branch, Iowa)
- First president born in the 20th century: John F. Kennedy (born May 29, 1917)

1 **George Washington**
1789–1797
No Political Party
Birthplace: Virginia
Born: February 22, 1732
Died: December 14, 1799

2 **John Adams**
1797–1801
Federalist
Birthplace: Massachusetts
Born: October 30, 1735
Died: July 4, 1826

3 **Thomas Jefferson**
1801–1809
Democratic Republican
Birthplace: Virginia
Born: April 13, 1743
Died: July 4, 1826

4 **James Madison**
1809–1817
Democratic Republican
Birthplace: Virginia
Born: March 16, 1751
Died: June 28, 1836

5 **James Monroe**
1817–1825
Democratic Republican
Birthplace: Virginia
Born: April 28, 1758
Died: July 4, 1831

6 **John Quincy Adams**
1825–1829
Democratic Republican
Birthplace: Massachusetts
Born: July 11, 1767
Died: February 23, 1848

7 **Andrew Jackson**
1829–1837
Democrat
Birthplace: South Carolina
Born: March 15, 1767
Died: June 8, 1845

8 **Martin Van Buren**
1837–1841
Democrat
Birthplace: New York
Born: December 5, 1782
Died: July 24, 1862

9 **William H. Harrison**
1841
Whig
Birthplace: Virginia
Born: February 9, 1773
Died: April 4, 1841

10 **John Tyler**
1841–1845
Whig
Birthplace: Virginia
Born: March 29, 1790
Died: January 18, 1862

11 **James K. Polk**
1845–1849
Democrat
Birthplace: North Carolina
Born: November 2, 1795
Died: June 15, 1849

12 **Zachary Taylor**
1849–1850
Whig
Birthplace: Virginia
Born: November 24, 1784
Died: July 9, 1850

13 Millard Fillmore
1850–1853
Whig
Birthplace: New York
Born: January 7, 1800
Died: March 8, 1874

14 Franklin Pierce
1853–1857
Democrat
Birthplace: New Hampshire
Born: November 23, 1804
Died: October 8, 1869

15 James Buchanan
1857–1861
Democrat
Birthplace: Pennsylvania
Born: April 23, 1791
Died: June 1, 1868

16 Abraham Lincoln
1861–1865
Republican
Birthplace: Kentucky
Born: February 12, 1809
Died: April 15, 1865

17 Andrew Johnson
1865–1869
National Union
Birthplace: North Carolina
Born: December 29, 1808
Died: July 31, 1875

18 Ulysses S. Grant
1869–1877
Republican
Birthplace: Ohio
Born: April 27, 1822
Died: July 23, 1885

19 Rutherford B. Hayes
1877–1881
Republican
Birthplace: Ohio
Born: October 4, 1822
Died: January 17, 1893

20 James A. Garfield
1881
Republican
Birthplace: Ohio
Born: November 19, 1831
Died: September 19, 1881

21 Chester A. Arthur
1881–1885
Republican
Birthplace: Vermont
Born: October 5, 1829
Died: November 18, 1886

22 24 Grover Cleveland
1885–1889, 1893–1897
Democrat
Birthplace: New Jersey
Born: March 18, 1837
Died: June 24, 1908

23 Benjamin Harrison
1889–1893
Republican
Birthplace: Ohio
Born: August 20, 1833
Died: March 13, 1901

25 William McKinley
1897–1901
Republican
Birthplace: Ohio
Born: January 29, 1843
Died: September 14, 1901

26 Theodore Roosevelt
1901–1909
Republican
Birthplace: New York
Born: October 27, 1858
Died: January 6, 1919

27 William H. Taft
1909–1913
Republican
Birthplace: Ohio
Born: September 15, 1857
Died: March 8, 1930

28 Woodrow Wilson
1913–1921
Democrat
Birthplace: Virginia
Born: December 28, 1856
Died: February 3, 1924

29 Warren G. Harding
1921–1923
Republican
Birthplace: Ohio
Born: November 2, 1865
Died: August 2, 1923

RESIDENTS of the UNITED STATES

30 Calvin Coolidge
1923–1929
Republican
Birthplace: Vermont
Born: July 4, 1872
Died: January 5, 1933

31 Herbert C. Hoover
1929–1933
Republican
Birthplace: Iowa
Born: August 10, 1874
Died: October 20, 1964

32 Franklin D. Roosevelt
1933–1945
Democrat
Birthplace: New York
Born: January 30, 1882
Died: April 12, 1945

33 Harry S. Truman
1945–1953
Democrat
Birthplace: Missouri
Born: May 8, 1884
Died: December 26, 1972

34 Dwight D. Eisenhower
1953–1961
Republican
Birthplace: Texas
Born: October 14, 1890
Died: March 28, 1969

35 John F. Kennedy
1961–1963
Democrat
Birthplace: Massachusetts
Born: May 29, 1917
Died: November 22, 1963

36 Lyndon B. Johnson
1963–1969
Democrat
Birthplace: Texas
Born: August 27, 1908
Died: January 22, 1973

37 Richard M. Nixon
1969–1974
Republican
Birthplace: California
Born: January 9, 1913
Died: April 22, 1994

38 Gerald R. Ford
1974–1977
Republican
Birthplace: Nebraska
Born: July 14, 1913

39 James E. Carter, Jr.
1977–1981
Democrat
Birthplace: Georgia
Born: October 1, 1924

40 Ronald W. Reagan
1981–1989
Republican
Birthplace: Illinois
Born: February 6, 1911

41 George H. W. Bush
1989–1993
Republican
Birthplace: Massachusetts
Born: June 12, 1924

42 William J. Clinton
1993–
Democrat
Birthplace: Arkansas
Born: August 19, 1946

The Gazetteer identifies important places and geographical features in this book. Entries include a short description, often followed by two page numbers. The first number refers to a text page on which the entry is discussed, and the second, in italics, refers to a map where the place appears. (The reference *Atlas* is to the section of U.S. and world maps on pages R32–R37.) In addition, some entries include rounded-off geographical coordinates. There are entries for all U.S. states (with capital cities).

Africa world's second largest continent. *Atlas*

Alabama 22nd state. Capital: Montgomery. *Atlas*

Alamo Texas mission in San Antonio captured by Mexico in 1836. (29°N 98°W), 402, *m405*

Alaska 49th state. Capital: Juneau. *Atlas*

Antarctica continent at the South Pole. *Atlas*

Antietam Maryland creek; site of bloodiest day's fighting in the Civil War. (39°N 77°W), 497, *m495*

Appalachian Mountains mountain range running from Alabama into Canada. 126, *m127*

Appomattox Court House town near Appomattox, Virginia, where Lee surrendered to Grant on April 9, 1865. (37°N 79°W), 519, *m517*

Arizona 48th state. Capital: Phoenix. *Atlas*

Arkansas 25th state. Capital: Little Rock. *Atlas*

Asia world's largest continent. *Atlas*

Atlantic Ocean ocean forming east boundary of the United States. *Atlas*

Australia island country between Indian and Pacific oceans; also the world's smallest continent. *Atlas*

Austria-Hungary one of the Central Powers in World War I; after the war, divided into smaller countries. 680, *m680*

Aztec Empire former region of Mexico once under Aztec control. 63, *m63*

Backcountry identification for former undeveloped region beginning in the Appalachian Mountains and extending west. 126, *m127*

Baltimore Maryland city on Chesapeake Bay. (39°N 77°W), 332, *m331*

Bay of Pigs (Bahía de Cochinos) inlet on south coast of Cuba; site of 1961 ill-fated, U.S.-backed Cuban invasion attempt. (22°N 81°W), 838

Beringia former land bridge connecting Asia with North America and now under waters of Bering Strait. (66°N 169°W), 27, *m28*

Berlin capital of Germany; divided into East and West Berlin, 1948–1989. (53°N 13°E), 793, *m792*

Boston capital of Massachusetts; site of early colonial unrest and conflict. (42°N 71°W), 165, *m172*

Bull Run stream 30 miles southwest of Washington, D.C.; site of first land battle of Civil War. (39°N 78°W), 485, *m495*

Bunker Hill hill now part of Boston; its name misidentifies Revolutionary War battle fought at nearby Breed's Hill. (42°N 71°W), 177

Cahokia Illinois Mound Builders site; village taken from British by Clark in 1778. (39°N 90°W), 31, *m203*

California 31st state. Capital: Sacramento. *Atlas*

Canada nation sharing northern U.S. border. *Atlas*

Caribbean Sea expanse of the Atlantic Ocean between the Gulf of Mexico and South America. *Atlas*

Central America area of North America between Mexico and South America. *m72, Atlas*

Charleston as Charles Town, largest Southern colonial city; South Carolina site of first Civil War shots, at offshore Fort Sumter. (33°N 80°W), 481, *m483*

Charlestown former town, now part of Boston; site of both Bunker and Breed's hills. (42°N 71°W), 177, *m172*

Chicago large Illinois city on Lake Michigan. (42°N 88°W), 602, *Atlas*

China large nation in Asia. *Atlas*

Colorado 38th state. Capital: Denver. *Atlas*

Concord Massachusetts city and site of second battle of the Revolutionary War. (42°N 71°W), 172, *m172*

Confederate States of America nation formed by 11 Southern states during the Civil War. Capital: Richmond, Virginia. 473 and 482, *m483*

Connecticut 5th state. Capital: Hartford. *Atlas*

Cuba Caribbean island south of Florida. 662, *m665*

Delaware 1st state. Capital: Dover. *Atlas*

District of Columbia (D.C.) self-governing federal district between Virginia and Maryland, made up entirely of the city of Washington, the U.S. capital. (39°N 77°W), 305, *Atlas*

Dominican Republic nation sharing the island of Hispaniola with Haiti. 673, *m672*

England southern part of Great Britain. *Atlas*

English Channel narrow waterway separating Great Britain from France. 69, *m764*

Erie Canal all-water channel dug out to connect the Hudson River with Lake Erie. 355, *m355*

Europe second smallest continent, actually a peninsula of the Eurasian landmass. *Atlas*

Florida 27th state. Capital: Tallahassee. *Atlas*

Fort McHenry fort in Baltimore harbor where 1814 British attack inspired U.S. national anthem. (39°N 77°N), 332, *m331*

Fort Sumter fort in Charleston, South Carolina, harbor where 1861 attack by Confederates began the Civil War. (33°N 80°W), 481, *m483*

France nation in western Europe; it aided America in the Revolutionary War. *Atlas*

Gadsden Purchase last territory (from Mexico, 1853) added to continental United States. 411, *m410*

Georgia 4th state. Capital: Atlanta. *Atlas*

Germany nation in central Europe; once divided into West and East Germany, 1949–1990. *Atlas*

Gettysburg Pennsylvania town and site of 1863 Civil War victory for the North that is considered war's turning point. (40°N 77°W), 513, *m514*

Ghana first powerful West African trading empire. 40, *m40*

Great Britain European island nation across from France; it consists of England, Scotland, and Wales. *Atlas*

Great Lakes five connected lakes—Ontario, Erie, Huron, Michigan, and Superior—on the U.S. border with Canada. 355, *m355*

Great Plains vast grassland region in the central United States. 393, *m395*

Gulf of Mexico body of water forming southern U.S. boundary from east Texas to west Florida. *Atlas*

Haiti nation sharing the island of Hispaniola with Dominican Republic. *Atlas*

Harpers Ferry village today in extreme eastern West Virginia where John Brown raided stored U.S. weapons in 1859. (39°N 78°W), 469, *m495*

Hawaii 50th state. Capital: Honolulu. *Atlas*

Hiroshima Japanese city destroyed by U.S. atomic bomb dropped to end World War II. (34°N 132°E), 771, *m771*

Hispaniola West Indies island (shared today by Dominican Republic and Haiti) that Columbus mistook for Asia. 52, *m51*

Hudson River large river in eastern New York. 100, *m95*

Idaho 43rd state. Capital: Boise. *Atlas*

Illinois 21st state. Capital: Springfield. *Atlas*

Indiana 19th state. Capital: Indianapolis. *Atlas*

Indian Territory area, mainly of present-day Oklahoma, that in the 1800s became land for relocated Native Americans. 376, *m376*

Iowa 29th state. Capital: Des Moines. *Atlas*

Iran Middle East nation. 865, *m865*

Iraq Middle East nation whose 1990 invasion of Kuwait led to the Persian Gulf War. 875, *m865*

Ireland island country west of England whose mid-1800s famine caused more than one million people to emigrate to America. 426, *Atlas*

Israel Jewish nation in the Middle East. 864, *Atlas*

Italy nation in southern Europe. *Atlas*

Jamestown community in Virginia that was the first permanent English settlement in North America. 87, *m87*

Japan island nation in east Asia. *Atlas*

Kansas 34th state. Capital: Topeka. *Atlas*

Kentucky 15th state. Capital: Frankfort. *Atlas*

Kosovo province of the Yugoslavian republic of Serbia. 876, *m877*

Kuwait tiny, oil-rich Middle East nation. 875, *m865*

Latin America region made up of Mexico, Caribbean Islands, and Central and South America, where Latin-based languages of Spanish, French, or Portuguese are spoken. 359, *m672*

Lexington Massachusetts city and site of first Revolutionary War battle in 1775. (42°N 71°W), 173, *m172*

Little Bighorn River Montana site of Sioux and Cheyenne victory over Custer. (46°N 108°W), 565, *m563*

Little Rock capital of Arkansas and site of 1957 school-desegregation conflict. 816, *Atlas*

Los Angeles 2nd largest U.S. city, on California's coast. 824, *m804*

Louisiana 18th state. Capital: Baton Rouge. *Atlas*

Louisiana Purchase land west of the Mississippi River purchased from France in 1803. 319, *m320*

Lowell Massachusetts city built in early 1800s as planned factory town. (43°N 71°W), 342

Maine 23rd state. Capital: Augusta. *Atlas*

Mali early West African trading empire succeeding Ghana empire. 41, *m40*

Maryland 7th state. Capital: Annapolis. *Atlas*

Massachusetts 6th state. Capital: Boston. *Atlas*

Mexico nation sharing U.S. southern border. *Atlas*

Michigan 26th state. Capital: Lansing. *Atlas*

Middle East eastern Mediterranean region that includes countries such as Iran, Iraq, Syria, Kuwait, Jordan, Saudi Arabia, Israel, and Egypt. 864, *m764, m865*

Minnesota 32nd state. Capital: St. Paul. *Atlas*

Mississippi 20th state. Capital: Jackson. *Atlas*

Mississippi River second longest U.S. river, south from Minnesota to the Gulf of Mexico. 146, *m153*

Missouri 24th state. Capital: Jefferson City. *Atlas*

Missouri River longest U.S. river, east from the Rockies to the Mississippi River. 303, *m302*

Montana 41st state. Capital: Helena. *Atlas*

Montgomery Alabama capital and site of 1955 African-American bus boycott. (32°N 86°W), 813, *m350*

Nagasaki Japanese port city, one-third of which was destroyed by U.S. atomic bomb dropped to end World War II. (33°N 130°E), 773, *m771*

Nebraska 37th state. Capital: Lincoln. *Atlas*

Nevada 36th state. Capital: Carson City. *Atlas*

New England northeast U.S. region made up of Maine, New Hampshire, Vermont, Massachusetts, Rhode Island, and Connecticut. 109, *m110*

New France first permanent French colony in North America. 70, *m148*

New Hampshire 9th state. Capital: Concord. *Atlas*

New Jersey 3rd state. Capital: Trenton. *Atlas*

New Mexico 47th state. Capital: Santa Fe. *Atlas*

New Netherland early Dutch colony that became New York in 1664. 70

New Orleans Louisiana port city at mouth of the Mississippi River. *Atlas*

New Spain former North American province of the Spanish Empire, made up mostly of present-day Mexico and the southwest United States. 71, *m72*

New York 11th state. Capital: Albany. *Atlas*

New York City largest U.S. city, at the mouth of the Hudson River; temporary U.S. capital, 1785–1790. *Atlas*

Normandy region of northern France where Allied invasion in 1944 turned tide of World War II. 766, *m766*

North America continent of Western Hemisphere north of Panama-Colombia border. *Atlas*

North Carolina 12th state. Capital: Raleigh. *Atlas*

North Dakota 39th state. Capital: Bismarck. *Atlas*

North Korea Communist country in Asia, bordering eastern China. 795, *m796*

North Vietnam northern region of Vietnam, established in 1954; reunified with South Vietnam in 1975 after Vietnam War. 837, *m837*

Northwest Territory U.S. land north of the Ohio River to the Great Lakes and west to the Mississippi River; acquired in 1783. 223, *m226*

Ohio 17th state. Capital: Columbus. *Atlas*

Ohio River river that flows from western Pennsylvania to the Mississippi River. *Atlas*

Oklahoma 46th state. Capital: Oklahoma City. *Atlas*

Oregon 33rd state. Capital: Salem. *Atlas*

Oregon Country former region of northwest North America claimed jointly by Britain and the United States until 1846. 318, *m320*

Oregon Trail pioneer wagon route from Missouri to the Oregon Territory in the 1840s and 1850s. 394, *m393*

Pacific Ocean world's largest ocean, on the west coast of the United States. *Atlas*

Panama Canal ship passageway cut through Panama in Central America, linking Atlantic and Pacific oceans. (8°N 80°W), 670, *m670*

Pearl Harbor naval base in Hawaii; site of surprise Japanese aerial attack in 1941. (21°N 158°W), 661 *m660*

Pennsylvania 2nd state. Capital: Harrisburg. *Atlas*

Persian Gulf waterway between Saudi Arabia and Iran, leading to Kuwait and Iraq. 875, *m865*

Philadelphia large port city in Pennsylvania; U.S. capital, 1790–1800. (40°N 76°W), 229, *Atlas*

Philippine Islands Pacific island country off the southeast coast of China. 662, *m665*

Plymouth town on Massachusetts coast and site of Pilgrim landing and colony. (42°N 71°W), 93, *m95*

Portugal nation in southwestern Europe; leader in early oceanic explorations. 49, *m51*

Potomac River historic river separating Virginia from Maryland and Washington, D.C. 496, *m495*

Puerto Rico Caribbean island that has been U.S. territory since 1898. 667, *m665*

Quebec major early Canadian city; also a province of eastern Canada. 146, *m148*

Rhode Island 13th state. Capital: Providence. *Atlas*

Richmond Virginia capital that was also the capital of the Confederacy. (38°N 77°W), 482, *m483*

Rio Grande river that forms part of the border between the United States and Mexico. *Atlas*

Roanoke Island island off the coast of North Carolina; 1585 site of the first English colony in the Americas. (36°N 76°W), 85, *m87*

Rocky Mountains mountain range in the western United States and Canada. *Atlas*

Russia large Eurasian country, the major republic of the former Soviet Union (1922–1991). 680, *m680*

St. Augustine oldest permanent European settlement (1565) in the United States, on Florida's northeast coast. (30°N 81°W), 68, *m63*

St. Lawrence River Atlantic-to-Great Lakes waterway used by early explorers of mid-North America. 146, *m148*

St. Louis Missouri city at the junction of the Missouri and Mississippi rivers. (39°N 90°W), 320, *Atlas*

San Antonio Texas city and site of the Alamo. (29°N 99°W), 402, *m405*

San Francisco major port city in northern California. (38°N 123°W), 416, *m592*

San Salvador West Indies island near the Bahamas where Columbus first landed in the Americas. (24°N 74°W), 52, *m51*

Santa Fe Trail old wagon route from Missouri to Santa Fe in Mexican province of New Mexico. 395, *m395*

Songhai early West African trading empire succeeding Mali empire. 42, *m40*

South America continent of Western Hemisphere south of Panama-Colombia border. *Atlas*

South Carolina 8th state. Capital: Columbia. *Atlas*

South Dakota 40th state. Capital: Pierre. *Atlas*

South Korea East Asian country bordering North Korea. 795, *m796*

South Vietnam southern region of Vietnam, established in 1954; reunified with North Vietnam in 1975 after Vietnam War. 837, *m837*

Soviet Union country created in 1922 by joining Russia and other republics; in 1991, broken into independent states. 757, *m760*

Spain nation in southwestern Europe; early empire builder in the Americas. 50, *m51*

Tennessee 16th state. Capital: Nashville. *Atlas*

Tenochtitlán Aztec Empire capital; now site of Mexico City. 64, *m63*

Texas 28th state. Capital: Austin. *Atlas*

Utah 45th state. Capital: Salt Lake City. *Atlas*

Valley Forge village in southeast Pennsylvania and site of Washington's army camp during winter of 1777–1778. (40°N 75°W), 202, *m209*

Vermont 14th state. Capital: Montpelier. *Atlas*

Vicksburg Mississippi River site of major Union victory (1863) in Civil War. (32°N 91°W), 516, *m517*

Vietnam country in Southeast Asia; divided into two regions (1954–1975), North and South, until end of Vietnam War. 835, *m837*

Virginia 10th state. Capital: Richmond. *Atlas*

Washington 42nd state. Capital: Olympia. *Atlas*

Washington, D.C. capital of the United States since 1800; makes up whole of District of Columbia (D.C.). (39°N 77°W), 305, *Atlas*

West Africa region from which most Africans were brought to the Americas. 39, *m40*

Western Hemisphere the half of the world that includes the Americas. 75, *m74*

West Indies numerous islands in the Caribbean Sea, between Florida and South America. 111, *m111*

West Virginia 35th state. Capital: Charleston. *Atlas*

Wisconsin 30th state. Capital: Madison. *Atlas*

Wounded Knee South Dakota site that was scene of 1890 massacre of Sioux. (43°N 102°W), 566, *m563*

Wyoming 44th state. Capital: Cheyenne. *Atlas*

Yorktown Virginia village and site of American victory that sealed British defeat in Revolutionary War. (37°N 77°W), 209, *m209*

A

abolition (AB uh LIHSH uhn) *n.* the movement to end slavery. (p. 440)

abridge (uh BRIHJ) *v.* to reduce. (p. 266)

AEF *n.* the American Expeditionary Force, U.S. forces during World War I. (p. 686)

affirmation (AF uhr MAY shuhn) *n.* a statement declaring that something is true. (p. 257)

African Diaspora (AF rih kuhn dy AS puhr uh) *n.* the forced removal of Africans from their homelands to serve as slave labor in the Americas. (p. 78)

Agent Orange *n.* a chemical that kills plants. (p. 843)

Albany Plan of Union *n.* the first formal proposal to unite the American colonies, put forth by Benjamin Franklin. (p. 149)

Alien and Sedition (si DISH uhn) **Acts** *n.* a series of four laws enacted in 1798 to reduce the political power of recent immigrants to the United States. (p. 306)

allies (AL yz) *n.* an alliance of Serbia, Russia, France, Great Britain, Italy, and seven other countries during World War I. (p. 680)

ally (AL eye) *n.* a country that agrees to help another country achieve a common goal. (p. 200)

American Federation of Labor (AFL) *n.* a national organization of labor unions founded in 1886. (p. 603)

American System *n.* a plan introduced in 1815 to make the United States economically self-sufficient. (p. 354)

Anaconda (AN uh KAH duh) **Plan** *n.* a strategy by which the Union proposed to defeat the Confederacy in the Civil War. (p. 484)

Angel Island *n.* the first stop in the United States for most immigrants coming from Asia. (p. 615)

Antifederalist (AN tee FED uhr uh list) *n.* a person who opposed the ratification of the U.S. Constitution. (p. 234)

Anti-Imperialist (AN tee im PEER y uh LIZT) **League** *n.* a group of well-known Americans that believed the United States should not deny other people the right to govern themselves. (p. 667)

Appalachian (AP uh LAY chee uhn) **Mountains** *n.* a mountain range that stretches from eastern Canada south to Alabama. (p. 126)

appeasement (uh PEEZ muhnt) *n.* the granting of concessions to a hostile power in order to keep the peace. (p. 757)

appellate (uh PEL it) *adj.* having power to review court decisions. (p. 260)

Appomattox (AP uh MAT uhks) **Court House** *n.* the Virginia town where Robert E. Lee surrendered to Ulysses S. Grant in 1865, ending the Civil War. (p. 519)

apprentice (uh PREN tis) *n.* a beginner who learns a trade or a craft from an experienced master. (p. 137)

appropriation (uh PROH pree AY shuhn) *n.* public funds set aside for a specific purpose. (p. 253)

archaeologist (AHR kee AHL uh jist) *n.* a scientist who studies the human past by examining the things people left behind. (p. 27)

armistice (AHR mi stis) *n.* an end to fighting. (p. 690)

arms race *n.* from the late 1940s to the late 1980s, the United States and the Soviet Union tried to top each other by developing weapons with great destructive power. (p. 798)

Articles of Confederation *n.* a document, adopted by the Continental Congress in 1777 and finally approved by the states in 1781, that outlined the form of government of the new United States. (p. 222)

artifact (AHR tuh FAKT) *n.* a tool or other object made by humans. (p. 27)

artillery (ahr TIL uhr ee) *n.* a cannon or large gun. (p. 177)

artisan (AHR ti zuhn) *n.* a skilled worker, such as a weaver or a potter, who makes goods by hand; a craftsperson. (p. 117)

assimilation (uh SIM uh LAY shuhn) *n.* the process of blending into society. (p. 616)

Axis (AK sis) *n.* Germany, Italy, and their allies during World War II. (p. 757)

B

baby boom *n.* the term for the generation born between 1946 and 1961, when the U.S. birthrate sharply increased following World War II. (p. 801)

Backcountry *n.* a colonial region that ran along the Appalachian Mountains through the far western part of the New England, Middle, and Southern colonies. (p. 109)

Bacon's Rebellion *n.* a revolt against powerful colonial authority in Jamestown by Nathaniel Bacon and a group of landless frontier settlers that resulted in the burning of Jamestown in 1676. (p. 89)

bail (bayl) *n.* money paid as security by arrested persons to guarantee they will return for trial. (p. 268)

Bataan (buh TAN) **Death March** *n.* in 1942, the Japanese marched 70,000 Filipino and American soldiers 60 miles to a prison camp. (p. 768)

Battle of Antietam (an TEE tuhm) *n.* a Civil War battle in 1862 in which 25,000 men were killed or wounded. (p. 497)

Battle of Fallen Timbers *n.* in 1794, an American army defeated 2,000 Native Americans in a clash over control of the Northwest Territory. (p. 299)

Battle of Gettysburg (GET eez BURG) *n.* an 1863 battle in the Civil War in which the Union defeated the Confederacy, ending hopes for a Confederate victory in the North. (p. 513)

Battle of Midway *n.* a victory for the United States over the Japanese in a 1942 naval battle that was a turning point of World War II. (p. 770)

Battle of Quebec (kwi BEK) *n.* a battle won by the British over the French, and the turning point in the French and Indian War. (p. 150)

Battle of Shiloh (SHY loh) *n.* an 1862 battle in which the Union forced the Confederacy to retreat in some of the fiercest fighting in the Civil War. (p. 494)

Battle of Yorktown *n.* the last major battle of the Revolutionary War, which resulted in the surrender of British forces in 1781. (p. 210)

Battle of the Alamo (AL uh MOH) *n.* in 1836, Texans defended a church called the Alamo against the Mexican army; all but five Texans were killed. (p. 403)

Battle of the Bulge *n.* a month-long battle of World War II in which the Allies turned back the last major German offensive of the war. (p. 764)

Battle of the Little Bighorn *n.* an 1876 battle in which the Sioux and the Cheyenne wiped out an entire force of U.S. troops. (p. 565)

Battle of the Thames (temz) *n.* an American victory over the British in the War of 1812, which ended the British threat to the Northwest Territory. (p. 332)

Battles of Saratoga (SAR uh TOH guh) *n.* a series of conflicts between British soldiers and the Continental Army in 1777 that proved to be a turning point in the Revolutionary War. (p. 199)

bayonet (BAY uh net) *n.* a long steel knife attached to the end of a gun. (p. 202)

Bear Flag Revolt *n.* the 1846 rebellion by Americans against Mexican rule in California. (p. 409)

Benin (buh NIN) *n.* a West African kingdom that arose near the Niger River delta in the 1300s. (p. 43)

Bessemer (BES uh muhr) **steel process** *n.* a new way of making steel that was developed in the 1850s and caused steel production to soar. (p. 587)

bill of attainder (uh TAYN duhr) *n.* a law that condemns a person without a trial in court. (p. 255)

Bill of Rights *n.* the first ten amendments to the U.S. Constitution, added in 1791, and consisting of a formal list of citizens' rights and freedoms. (p. 237)

black code *n. a* law passed by Southern states that limited the freedom of former slaves. (p. 534)

Black Tuesday *n.* a name given to October 29, 1929, when stock prices fell sharply. (p. 731)

blockade *n.* when armed forces prevent the transportation of goods or people into or out of an area. (p. 484)

Bonus Army *n.* in 1932, thousands of veterans streamed into Washington demanding bonuses that they never received. (p. 733)

boomtown *n.* a town that has a sudden burst of economic or population growth. (p. 558)

border state *n.* a slave state that bordered states in which slavery was illegal. (p. 482)

Boston Massacre (MAS uh kuhr) *n.* a clash between British soldiers and Boston colonists in 1770, in which five of the colonists, including Crispus Attucks, were killed. (p. 165)

Boston Tea Party *n.* the dumping of 342 chests of tea into Boston Harbor by colonists in 1773 to protest the Tea Act. (p. 167)

bounty (BOWN tee) *n.* a reward or cash payment given by a government. (pp. 271, 508)

Boxer Rebellion *n.* in 1900, Chinese resentment toward foreigners' attitude of cultural superiority resulted in this violent uprising. (p. 669)

boycott (BOI KOT) *n.* a refusal to buy certain goods. (p. 161)

bracero (bruh SAIR oh) **program** *n.* the hiring of Mexicans to perform much-needed labor during World War II. (p. 774)

brinksmanship (BRINGKS muhn SHIP) *n.* in international politics, the act of pushing a dangerous situation to the limits; for example, the United States going to the brink of war to stop Communism. (p. 798)

Brown* v. *Board of Education of Topeka, Kansas *n.* a 1954 case in which the Supreme Court ruled that "separate but equal" education for black and white students was unconstitutional. (p. 814)

buck *n.* a buckskin from an adult male deer was a unit of money for settlers. (p. 127)

buffalo soldier *n.* a name given by Native Americans to African Americans serving in the U.S. army in the West. (p. 571)

business cycle *n.* the pattern of good times and bad times in the economy. (p. 586)

buy on margin *v.* to pay a small part of a stock's price and then borrow money to pay for the rest. (p. 730)

C

cabinet *n.* a group of department heads who serve as the president's chief advisers. (p. 294)

California gold rush *n.* in 1849, large numbers of people moved to California because gold had been discovered there. (p. 413)

Cambodia (kam BOW dee uh) *n.* a country bordering Vietnam. (p. 848)

Camp David Accords *n.* in 1979, under these agreements, Egypt and Israel signed a peace treaty that ended 30 years of conflict. (p. 864)

caravel (KAR uh VEL) *n.* a ship with triangular sails that allowed it to sail into the wind and with square sails that carried it forward when the wind was at its back. (p. 49)

cash crop *n.* a crop grown by a farmer to be sold for money rather than for personal use. (p. 115)

cavalry *n.* soldiers on horseback. (p. 496)

Centennial (sen TEN ee uhl) **Exhibition** *n.* an exhibition in Philadelphia in 1876 that celebrated America's 100th birthday. (p. 588)

Central Powers *n.* an alliance of Austria-Hungary, Germany, the Ottoman Empire, and Bulgaria during World War I. (p. 680)

charter *n.* a written contract issued by a government giving the holder the right to establish a colony. (p. 87)

checks and balances *n.* the ability of each branch of government to exercise checks, or controls, over the other branches. (p. 246)

Chinese Exclusion Act *n.* enacted in 1882, this law banned Chinese immigration for ten years. (p. 617)

civil disobedience (DIS uh BEE dee uhns) *n.* peacefully refusing to obey laws one considers unjust. (p. 431)

civilization (SIV uh li ZAY shuhn) *n.* a form of culture characterized by city trade centers, specialized workers, organized forms of government and religion, systems of record keeping, and advanced tools. (p. 29)

civil rights *n.* rights granted to all citizens. (p. 535)

Civil Rights Act of 1964 *n.* this act banned segregation in public places and created the Equal Employment Opportunity Commission. (p. 820)

clan *n.* a large group of families that claim a common ancestor. (p. 127)

Clayton Antitrust Act *n.* a law passed in 1914 that laid down rules forbidding business practices that lessened competition; it gave the government more power to regulate trusts. (p. 648)

Cold War *n.* the state of hostility, without direct military conflict, that developed between the United States and the Soviet Union after World War II. (p. 792)

Columbian (kuh LUM bee uhn) **Exchange** *n.* the transfer of plants, animals, and diseases between the Western and the Eastern hemispheres. (p. 74)

committee of correspondence *n.* a group of people in the colonies who exchanged letters on colonial affairs. (p. 166)

Committee to Reelect the President *n.* an organization linked to the break-in at the Democratic National Committee headquarters that set off the Watergate scandal. (p. 860)

common law *n.* a system of law developed in England, based on customs and previous court decisions. (p. 268)

Compromise of 1850 *n.* a series of Congressional laws intended to settle the major disagreements between free states and slave states. (p. 461)

Compromise of 1877 *n.* the agreement that resolved an 1876 election dispute: Rutherford B. Hayes became president and then removed the last federal troops from the South. (p. 548)

compulsory process *n.* a required procedure. (p. 267)

Conestoga (KON i STOW guh) **wagon** *n.* a vehicle with wide wheels, a curved bed, and a canvas cover used by American pioneers traveling west. (p. 117)

Confederate States of America *n.* the confederation formed in 1861 by the Southern states after their secession from the Union. (p. 473)

Congress of Industrial Organizations (CIO) *n.* a labor organization that broke away from the American Federation of Labor in 1938. (p. 743)

conquistador (kon KWIS tuh DAWR) *n.* a Spaniard who traveled to the Americas as an explorer and a conqueror in the 16th century. (p. 63)

conscription (kuhn SKRIP shuhn) *n.* a law that required men to serve in the military or be drafted. (p. 508)

conservative *n.* a person who favors fewer government controls and more individual freedom in economic matters. (p. 749)

Constitutional Convention *n.* a meeting held in 1787 to consider changes to the Articles of Confederation; resulted in the drafting of the Constitution. (p. 229)

containment (kuhn TAYN muhnt) *n.* the blocking by one nation of another nation's attempts to spread influence—especially the efforts of the United States to block the spread of Soviet Communism during the late 1940s and early 1950s. (p. 793)

Continental Army *n.* a colonial force authorized by the Second Continental Congress in 1775, with George Washington as its commanding general. (p. 177)

convene (kuhn VEEN) *v.* to call together. (p. 259)

convoy system *n.* a heavy guard of destroyers that escorts merchant ships during wartime. (p. 687)

cooperative (koh OP uhr uh tiv) *n.* an organization owned and run by its members. (p. 577)

Copperheads *n.* Abraham Lincoln's main political opponents; they favored peace with the South. (p. 508)

CORE *n.* the Congress of Racial Equality, a group that planned Freedom Rides to desegregate interstate buses. (p. 818)

corporation *n.* a business owned by investors who buy part of the company through shares of stock. (p. 594)

cotton gin *n.* a machine invented in 1793 that cleaned cotton much faster and far more efficiently than human workers. (p. 348)

counsel (KOWN suhl) *n.* a lawyer. (p. 267)

Crash of 1929 *n.* the plunge in stock market prices. (p. 731)

Crittenden (KRIT uhn duhn) **Plan** *n.* a compromise introduced in 1861 that might have prevented secession. (p. 475)

Crusades (kroo SAYDZ) *n.* a series of wars to capture the Holy Land, launched in 1096 by European Christians. (p. 45)

Cuban Missile Crisis *n.* in 1962, the United States and the Soviet Union almost went to war because the Soviets had placed nuclear missiles in Cuba. (p. 839)

culture (KUL chuhr) *n.* a way of life shared by people with similar arts, beliefs, and customs. (p. 28)

D

Dawes (dawz) **Act** *n.* a law, enacted in 1887, that distributed reservation land to individual owners. (p. 567)

D-Day *n.* June 6, 1944, the day the Allies invaded France during World War II. (p. 764)

Declaration of Independence *n.* the document, written in 1776, in which the colonies declared independence from Britain. (p. 180)

deficit (DEF i sit) **spend** *v.* to use borrowed money to fund government programs. (p. 738)

department store *n.* a store that sells everything from clothing to furniture to hardware. (p. 627)

depression *n.* a severe economic slump. (p. 386)

desert (di ZURT) *v.* to leave military duty without intending to return. (p. 203)

détente (day TAHNT) *n.* an easing of tensions between the United States and the Soviet Union during the Cold War. (p. 858)

direct primary *n.* voters, rather than party conventions, choose candidates to run for public office. (p. 640)

diversity (di VUR si tee) *n.* a variety of people. (p. 117)

doctrine of nullification (NUL uh fi KAY shuhn) *n.* a right of a state to reject a federal law that it considers unconstitutional. (p. 381)

domestication (doh MES ti KAY shuhn) *n.* the practice of breeding plants or taming animals to meet human needs. (p. 28)

domino (DOM uh NOH) **theory** *n.* a theory stating that if a country fell to communism, nearby countries would also fall to communism. (p. 837)

dove *n.* a person opposed to war. (p. 847)

downsize *v.* to reduce the number of workers in order to increase company profits. (p. 879)

Dred Scott* v. *Sandford *n.* an 1856 Supreme Court case in which a slave, Dred Scott, sued for his freedom because he had been taken to live in territories where slavery was illegal; the Court ruled against Scott. (p. 467)

due process of law *n.* fair treatment under the law. (p. 267)

dust bowl *n.* the area of dust-damaged farms across a 150,000-square-mile region during the early 1930s. (p. 739)

E

elector *n.* a voter. (p. 249)

Ellis Island *n.* the first stop in the United States for most immigrants coming from Europe. (p. 614)

Emancipation (i MAN suh PAY shuhn) **Proclamation** *n.* an executive order issued by Abraham Lincoln on January 1, 1863, freeing the slaves in all regions in rebellion against the Union. (p. 504)

Embargo (em BAHR goh) **Act of 1807** *n.* an act that stated that American ships were no longer allowed to sail to foreign ports, and it also closed American ports to British ships. (p. 328)

emigrant (EM i gruhnt) *n.* a person who leaves a country. (p. 423)

encomienda (en koh mee YEN duh) *n.* a grant of Native American labor. (p. 72)

English Bill of Rights *n.* an agreement signed by William and Mary to respect the rights of English citizens and of Parliament, including the right to free elections. (p. 144)

enlightenment (en LYT n muhnt) *n.* an 18th-century movement that emphasized the use of reason and the scientific method to obtain knowledge. (p. 140)

enumeration (i NOO muh RAY shuhn) *n.* an official count, such as a census. (p. 249)

environmentalism (en VY ruhn MEN tl IZ uhm) *n.* work toward protecting the environment. (p. 864)

equity (EK wi tee) *n.* a system of justice not covered under common law. (p. 269)

ERA *n.* the Equal Rights Amendment, a proposed amendment that would give equality of rights regardless of sex; the amendment died in 1982. (p. 826)

Erie (EER ee) **Canal** *n.* completed in 1825, this waterway connected New York City and Buffalo, New York. (p. 355)

escalation (ES kuh LAY shuhn) *n.* the policy of increasing military involvement, as in Vietnam. (p. 841)

Espionage (ES pee uh NAHZH) **Act** *n.* passed in 1917, this law set heavy fines and long prison terms for antiwar activities and for encouraging draft resisters. (p. 692)

European Middle Ages *n.* a period from the late 400s to about the 1300s, during which Europeans turned to feudalism and the manor system. (p. 44)

exoduster (EKS suh duhs tuhr) *n.* an African American who left the South for the West and compared himself or herself to Biblical Hebrews who left slavery in Egypt. (p. 575)

expatriate (ek SPAY tree it) *n.* a citizen of one country who takes up residence in another country. (p. 721)

ex post facto (EKS pohst FAK toh) **law** *n.* a law that would make an act a criminal offense after it was committed. (p. 255)

F

factory system *n.* a method of production that brought many workers and machines together into one building. (p. 341)

Fair Deal *n.* a program under Harry Truman that called for new projects to create jobs, new public housing, and an end to racial discrimination in hiring. (p. 791)

fall line *n.* the point at which a waterfall prevents large boats from moving farther upriver. (p. 126)

famine (FAM in) *n.* a severe food shortage. (p. 426)

fascism (FASH iz uhm) *n.* a political philosophy that advocates a strong, centralized, nationalistic government headed by a powerful dictator. (p. 756)

federalism *n.* a system of government where power is shared among the central (or federal) government and the states. (pp. 234, 245)

Federalists *n.* supporters of the Constitution. (p. 234)

Federalist Papers *n.* a series of essays defending and explaining the Constitution. (p. 235)

Federal Judiciary (joo DISH ee ER ee) **Act** *n.* it helped create a court system and gave the Supreme Court six members. (p. 294)

Federal Reserve Act *n.* a law passed in 1913 that "created" the nation's banking system and instituted a flexible currency system. (p. 648)

felony (FEL uh nee) *n.* a serious crime. (p. 253)

feudalism (FYOOD l IZ uhm) *n.* a political system in which the king allows nobles the use of his land in exchange for their military service and their protection of people living on the land. (p. 44)

Fifteenth Amendment *n.* passed in 1870, this amendment to the U.S. Constitution stated that citizens could not be stopped from voting "on account of race, color, or previous condition of servitude." (p. 546)

54th Massachusetts Regiment *n.* one of the first African-American regiments organized to fight for the Union in the Civil War. (p. 506)

fireside chat *n.* the name of Franklin Roosevelt's radio broadcasts in which he explained his policies. (p. 735)

First Battle of Bull Run *n.* an 1861 battle of the Civil War in which the South shocked the North with a victory. (p. 485)

flapper *n.* a young woman who embraced the fashions and urban attitudes of the 1920s. (p. 714)

foreign (FAWR in) **policy** *n.* relations with the governments of other countries. (p. 304)

Fort Sumter *n.* a federal fort located in the harbor of Charleston, South Carolina; the Southern attack on Fort Sumter marked the beginning of the Civil War. (p. 481)

forty-niner *n.* a person who went to California to find gold, starting in 1849. (p. 412)

Fourteen Points *n.* President Woodrow Wilson's goals for peace after World War I. (p. 695)

Fourteenth Amendment *n.* an amendment to the U.S. Constitution, passed in 1868, that made all persons born or naturalized in the United States—including former slaves—citizens of the country. (p. 535)

First Continental Congress *n.* a meeting of delegates in 1774 from all the colonies except Georgia to uphold colonial rights. (p. 171)

Freedmen's Bureau *n.* a federal agency set up to help former slaves after the Civil War. (p. 533)

freedmen's school *n.* a school set up to educate newly freed African Americans. (p. 541)

Freedom Ride *n.* a protest against segregation on interstate busing in the South. (p. 818)

Freedom Summer *n.* in 1964, the SNCC organized a voter-registration drive. (p. 821)

Free Soil Party *n.* a political party dedicated to stopping the expansion of slavery. (p. 459)

French and Indian War *n.* a conflict in North America from 1754 to 1763 that was part of a worldwide struggle between France and Britain; Britain defeated France and gained French Canada. (p. 147)

French Indochina (IN doh CHY nuh) *n.* a French colony that included present-day Vietnam, Laos, and Cambodia. (p. 835)

frontier (frun TEER) *n.* unsettled or sparsely settled area occupied largely by Native Americans. (p. 557)

Fugitive Slave Act *n.* an 1850 law to help slaveholders recapture runaway slaves. (p. 462)

French Revolution *n.* in 1789, the French launched a movement for liberty and equality. (p. 301)

fundamentalist *n.* a person who believes in a literal, or word-for-word, interpretation of the bible. (p. 716)

Fundamental Orders of Connecticut *n.* a set of laws that were established in 1639 by a Puritan congregation who had settled in the Connecticut Valley and that expanded the idea of representative government. (p. 95)

G

generator *n.* a machine that produces electric current. (p. 587)

Ghana (GAH nuh) *n.* a West African empire in the 8th–11th centuries A.D. (p. 39)

G.I. Bill of Rights *n.* passed in 1944, this bill provided educational and economic help to veterans. (p. 779)

Gilded (gil did) **Age** *n.* an era during the late 1800s of fabulous wealth. (p. 596)

"Glorious Revolution" *n.* the overthrow of English King James II in 1688 and his replacement by William and Mary. (p. 144)

gold standard *n.* a policy under which the government backs every dollar with a certain amount of gold. (p. 577)

Grange (graynj) *n.* formed in 1867, the Patrons of Husbandry tried to meet the social needs of farm families. (p. 577)

Great Awakening *n.* a revival of religious feeling in the American colonies during the 1730s and 1740s. (p. 139)

Great Compromise *n.* the Constitutional Convention's agreement to establish a two-house national legislature, with all states having equal representation in one house and each state having representation based on its population in the other house. (p. 232)

Great Depression *n.* a period, lasting from 1929 to 1941, in which the U.S. economy was in severe decline and millions of Americans were unemployed. (p. 731)

Great Migration *n.* the movement of Puritans from England to establish settlements around the world, including 20,000 who sailed for America (p. 94); the movement of African Americans between 1910 and 1920 to northern cities from the South. (p. 693)

Great Plains *n.* the area from the Missouri River to the Rocky Mountains. (p. 557)

Great Society *n.* a program started by President Lyndon Johnson that provided help to the poor, the elderly, and women, and also promoted education and outlawed discrimination. (p. 822)

greenback *n.* paper currency issued by the federal government during the Civil War. (p. 509)

gristmill (GRIST MIL) *n.* a mill in which grain is ground to produce flour or meal. (p. 115)

guerrilla (guh RIL uh) *n.* a soldier who weakens the enemy with surprise raids and hit-and-run attacks. (p. 207)

guerrilla warfare *n.* surprise attacks by small bands of fighters. (p. 842)

Gulf of Tonkin Resolution *n.* congressional resolution that gave the president power to use military force in Vietnam. (p. 841)

H

hacienda (HAH see EN duh) *n.* a large farm or estate. (p. 72)

Harlem Renaissance *n.* a flowering of African-American artistic creativity during the 1920s, centered in the Harlem community of New York City. (p. 720)

Harpers Ferry *n.* a federal arsenal in Virginia that was captured in 1859 during a slave revolt. (p. 469)

Hausa (HOW suh) *n.* a West African people who lived in what is now northern Nigeria after A.D. 1000. (p. 42)

hawk *n.* a person who supports war. (p. 847)

Haymarket affair *n.* in 1886, a union protest resulted in about 100 dead after an unknown person threw a bomb, and police opened fire on the crowd. (p. 602)

H-bomb *n.* a hydrogen bomb. (p. 798)

Hiroshima (HEER uh SHEE muh) *n.* the first city in Japan that was hit by an atomic bomb on August 6, 1945. (p. 771)

Ho Chi Minh (HOH CHEE MIN) **Trail** *n.* a network of paths that the Viet Cong used to move soldiers and supplies during the Vietnam War. (p. 838)

Holocaust (HOL uh KAWST) *n.* the systematic killing by Germany during World War II of about six million Jews as well as millions from other ethnic groups. (p. 765)

homestead *n.* land to settle on and farm. (p. 568)

Homestead Act *n.* passed in 1862, this law offered 160 acres of land free to anyone who agreed to live on and improve the land for five years. (p. 574)

House of Burgesses *n.* created in 1619, the first representative assembly in the American colonies. (p. 88)

Hudson River school *n.* a group of artists living in the Hudson River Valley in New York. (p. 430)

Hull House *n.* founded in 1889, a model for other settlement houses of the time. (p. 613)

Hundred Days *n.* in his first hundred days, from March 9 to mid-June 1933, Franklin Roosevelt sent Congress many new bills. (p. 735)

hygiene (HY JEEN) *n.* conditions and practices that promote health. (p. 490)

I

immigrant *n.* a person who settles in a new country. (p. 423)

Immigration Reform and Control Act of 1986 *n.* a law that is designed to strengthen immigration laws and enforcement measures. (p. 884)

immunity *n.* legal protection. (p. 262)

impeachment *n.* the process of accusing a public official of wrongdoing. (p. 249)

imperialism *n.* the policy by which stronger nations extend their economic, political, or military control over weaker nations or territories. (p. 659)

impressment *n.* the act of seizing by force. (p. 327)

inaugurate (in AW gyuh RAYT) *v.* to swear in or induct into office in a formal ceremony. (p. 293)

income tax *n.* a tax on earnings. (p. 509)

indentured servant *n.* a person who sold his or her labor in exchange for passage to America. (p. 88)

Indian Removal Act *n.* this 1830 act called for the government to negotiate treaties that would require Native Americans to relocate west. (p. 376)

Indian Territory *n.* present-day Oklahoma and parts of Kansas and Nebraska to which Native Americans were moved under the Indian Removal Act of 1830. (p. 376)

indictment (in DYT muhnt) *n.* a written statement issued by a grand jury charging a person with a crime. (p. 250)

indigo *n.* a plant grown in the Southern colonies that yields a deep blue dye. (p. 121)

individual right *n.* a personal liberty and privilege guaranteed to U.S. citizens by the Bill of Rights. (p. 247)

Industrial Revolution *n.* in late 18th-century Britain, factory machines began replacing hand tools and manufacturing replaced farming as the main form of work. (p. 341)

inferior court *n.* a court with less authority than the Supreme Court. (p. 260)

inflation *n.* an increase in the price of goods and services and a decrease in the value of money. (p. 386)

information revolution *n.* a time when technology has radically changed how much information and the way information is delivered. (p. 879)

initiative (i NISH uh tiv) *n.* the procedure that allows voters to propose a law directly. (p. 640)

inoperative *adj.* no longer in force. (p. 274)

installment buy *v.* to buy something by making small monthly payments. (p. 712)

insurrection (IN suh REK shuhn) *n.* open revolt against a government. (p. 271)

interchangeable part *n.* a part that is exactly like another part. (p. 343)

Internet *n.* a worldwide computer network. (p. 878)

Intolerable Acts *n.* a series of laws enacted by Parliament in 1774 to punish Massachusetts colonists for the Boston Tea Party. (p. 170)

Iran-Contra affair *n.* in 1986, the U.S. government sold weapons to Iran for help in freeing American hostages in the Middle East, and the money from the sale went to the Contra rebels in El Salvador. (p. 874)

Iran hostage crisis *n.* on November 4, 1979, a group of Iranians overran the American embassy in Iran's capital of Tehran and took 52 Americans hostage. (p. 865)

ironclad *n.* a warship covered with iron. (p. 491)

Iroquois (IR uh KWOH) **League** *n.* a 16th-century alliance of the Cayuga, Mohawk, Oneida, Onondaga, and Seneca Native American groups living in the eastern Great Lakes region. (p. 37)

irrigation *n.* the practice of bringing water to crops. (p. 29)

Islam (is LAHM) *n.* a religion founded by the prophet Muhammad in the 600s, which teaches that there is one God, named Allah. (p. 41)

island hopping *n.* a World War II strategy in which the Allies invaded islands that the Japanese weakly defended in order to stage further attacks. (p. 770)

isolationist *n.* a person who believed that the United States should stay out of other nations' affairs except in self-defense. (p. 711)

J

Jacksonian Democracy *n.* the idea of spreading political power to all the people, thereby ensuring majority rule. (p. 370)

Jamestown *n.* the first permanent English settlement in North America. (p. 87)

Jay's Treaty *n.* the agreement that ended dispute over American shipping during the French Revolution. (p. 302)

jazz *n.* a new kind of music in the 1920s that captured the carefree spirit of the times. (p. 717)

Jim Crow *n.* laws meant to enforce separation of white and black people in public places in the South. (p. 621)

joint-stock company *n.* a business in which investors pool their wealth in order to turn a profit. (p. 86)

judicial (joo DISH uhl) **review** *n.* the principle that the Supreme Court has the final say in interpreting the Constitution. (p. 317)

Judiciary (joo DISH ee ER ee) **Act of 1801** *n.* a law that increased the number of federal judges, allowing President John Adams to fill most of the new spots with Federalists. (p. 316)

K

Kansas-Nebraska Act *n.* an 1854 law that established the territories of Kansas and Nebraska and gave their residents the right to decide whether to allow slavery. (p. 464)

kayak (KY AK) *n.* a small boat made of animal skins. (p. 33)

Kellogg-Briand Pact *n.* in 1928, this pact was signed by many nations who pledged not to make war against each other except in self-defense. (p. 711)

King Cotton *n.* cotton was called king because cotton was important to the world market, and the South grew most of the cotton for Europe's mills. (p. 484)

King Philip's War *n.* a war between the Puritan colonies and Native Americans in 1675–1676. (p. 96)

Knights of Labor *n.* an organization of workers from all different trades formed after the Civil War. (p. 601)

Korean War *n.* a conflict between North Korea and South Korea, lasting from 1950 to 1953; the United States, along with other UN countries, fought on the side of the South Koreans, and China fought on the side of the North Koreans. (p. 796)

Ku Klux Klan *n.* a group formed in 1866 that wanted to restore Democratic control of the South and to keep former slaves powerless; the group called for a "racially and morally pure" America. (pp. 544, 716)

L

labor union *n.* a group of workers who band together to seek better working conditions. (p. 434)

laissez faire (LES ay FAIR) *n.* a theory that stated that business, if unregulated, would act in a way that would benefit the nation. (p. 710)

Land Ordinance of 1785 *n.* a law that established a plan for surveying and selling the federally owned lands west of the Appalachian Mountains. (p. 223)

land speculator *n.* a person who buys huge areas of land for a low price and then sells off small sections of it at high prices. (p. 394)

League of Nations *n.* an organization set up after World War I to settle international conflicts. (p. 695)

leisure (LEE zhuhr) *n.* free time. (p. 627)

Lend-Lease *n.* a 1941 law that allowed the United States to ship arms and supplies, without immediate payment, to nations fighting the Axis powers. (p. 760)

Lewis and Clark expedition *n.* a group led by Meriwether Lewis and William Clark who explored the lands of the Louisiana Purchase beginning in 1803. (p. 320)

Lexington and Concord *n.* sites in Massachusetts of the first battles of the American Revolution. (p. 173)

liberal *n.* a person who favors government action to bring about social and economic reform. (p. 749)

limited government *n.* the principle that requires all U.S. citizens, including government leaders, to obey the law. (p. 247)

lode *n.* a deposit of mineral buried in rock. (p. 558)

Lone Star Republic *n.* the nickname of the republic of Texas, given in 1836. (p. 405)

long drive *n.* taking cattle by foot to a railway. (p. 560)

Lost Generation *n.* the generation of the 1920s after World War I, when men and women saw little hope for the future. (p. 720)

Louisiana (loo EE zee AN uh) **Purchase** *n.* the 1803 purchase of the Louisiana Territory from France. (p. 319)

Lowell mills *n.* textile mills located in the factory town of Lowell, Massachusetts, founded in 1826. (p. 342)

Loyalist *n.* an American colonist who supported the British in the American Revolution. (p. 173)

M

mail-order catalog *n.* a publication that contains pictures and descriptions of items so that people can order by mail. (p. 627)

Magna Carta *n.* "Great Charter;" a document guaranteeing basic political rights in England, approved by King John in 1215. (p. 141)

Mali (MAH lee) *n.* a West African empire from the 13th–15th centuries that grew rich from trade. (p. 41)

Manhattan Project *n.* the top-secret program set up in 1942 to build an atomic bomb. (p. 771)

manifest destiny *n.* the belief that the United States was destined to stretch across the continent from the Atlantic Ocean to the Pacific Ocean. (p. 407)

manor system *n.* a system in which lords divided their lands into estates, which were farmed mostly by serfs who received protection from the lord in return. (p. 45)

Marbury* v. *Madison *n.* an 1803 case in which the Supreme Court ruled that it had the power to abolish laws by declaring them unconstitutional. (p. 317)

March on Washington *n.* a huge civil rights demonstration in Washington, D.C., in 1963. (p. 820)

Marshall Plan *n.* approved in 1948, the United States gave more than $13 billion to help the nations of Europe after World War II. (p. 779)

mass culture *n.* a common culture experienced by large numbers of people. (p. 626)

mass media *n.* communications that reach a large audience. (p. 718)

matrilineal (MAT ruh LIN ee uhl) *adj.* a society in which ancestry is traced through the mother. (p. 36)

Mayflower Compact *n.* an agreement established by the men who sailed to America on the *Mayflower,* which called for laws for the good of the colony and set forth the idea of self-government. (p. 93)

melting pot *n.* a place where cultures blend. (p. 616)

mercantilism (MUHR kuhn tee LIZ uhm) *n.* an economic system in which nations increase their wealth and power by obtaining gold and silver and by establishing a favorable balance of trade. (p. 61)

mercenary (MUR suh NER ee) *n.* a professional soldier hired to fight for a foreign country. (p. 195)

Mexican Cession (sesh uhn) *n.* a vast region given up by Mexico after the War with Mexico; it included the present-day states of California, Nevada, Utah, most of Arizona, and parts of New Mexico, Colorado, and Wyoming. (p. 411)

Mexicano (may hi KAH noh) *n.* a person of Spanish descent whose ancestors had come from Mexico and settled in the Southwest. (p. 570)

Middle Passage *n.* the middle leg of the triangular trade route—the voyage from Africa to the Americas—that brought captured Africans into slavery. (p. 78)

migrate *v.* to move from one location to another. (p. 27)

militarism *n.* the belief that a nation needs a large military force. (p. 679)

militia (muh LISH uh) *n.* a force of armed civilians pledged to defend their community during the American Revolution. (p. 170); an emergency military force that is not part of the regular army. (p. 254)

minié (MIN ee) **ball** *n.* a bullet with a hollow base. (p. 491)

Minuteman *n.* a member of the colonial militia who was trained to respond "at a minute's warning." (p. 170)

misdemeanor (mis di MEE nuhr) *n.* a violation of the law. (p. 259)

mission *n.* a settlement created by the Church in order to convert Native Americans to Christianity. (p. 72)

missionary *n.* a person sent by the Church to preach, teach, and convert native peoples to Christianity. (p. 61)

Missouri Compromise *n.* a series of laws enacted in 1820 to maintain the balance of power between slave states and free states. (p. 358)

monopoly *n.* a company that eliminates its competitors and controls an industry. (p. 595)

Monroe Doctrine *n.* a policy of U.S. opposition to any European interference in the Western Hemisphere, announced by President Monroe in 1823. (p. 359)

Montgomery bus boycott *n.* in 1955, African Americans boycotted the public buses in Montgomery, Alabama, in response to the arrest of Rosa Parks, who refused to give up her seat to a white person. (p. 815)

Mormon *n.* a member of a church founded by Joseph Smith in 1830. (p. 397)

Mound Builder *n.* an early Native American who built large earthen structures. (p. 31)

mountain man *n.* a fur trapper or explorer who opened up the West by finding the best trails through the Rocky Mountains. (p. 393)

muckraker *n.* around 1900, the term for a journalist who exposed corruption in American society. (p. 640)

Muslim (MUZ luhm) *n.* a follower of Islam. (p. 41)

N

NAACP *n.* formed in 1909, the National Association for the Advancement of Colored People. (pp. 622, 715)

NAFTA *n.* passed in 1993, the North American Free Trade Agreement created a free trade block among the United States, Mexico, and Canada. (p. 875)

napalm (NAY PAHM) *n.* a jellied gasoline that burns violently. (p. 843)

nationalism *n.* a feeling of pride, loyalty, and protectiveness toward one's country. (p. 354)

nativist *n.* a native-born American who wanted to eliminate foreign influence. (p. 428)

NATO *n.* the North Atlantic Treaty Organization is a military alliance formed in 1949 by ten Western European countries, the United States, and Canada. (p. 793)

natural-born citizen *n.* a citizen born in the United States or a commonwealth of the United States or to parents who are U.S. citizens living outside the country. (p. 257)

naturalization *n.* a way to give full citizenship to a person born in another country. (pp. 253, 270)

Navigation Acts *n.* a series of laws passed by Parliament, beginning in 1651, to ensure that England made money from its colonies' trade. (p. 112)

navigator *n.* a person who plans the course of a ship while at sea. (p. 49)

Nazi (NAHT see) **Party** *n.* the National Socialist German Workers' Party; came to power under Adolf Hitler in the 1930s. (p. 756)

NCAI *n.* the National Congress of American Indians was founded in 1944 and aimed to promote the "common welfare" of Native Americans. (p. 825)

neutral (NOO truhl) *adj.* not siding with one country or the other. (p. 302)

neutrality (noo TRAL i tee) *n.* refusing to take sides in a war. (p. 682)

New Deal *n.* President Franklin Roosevelt's programs to fight the Great Depression. (p. 735)

New France *n.* a fur-trading post established in 1608 that became the first permanent French settlement in North America. (p. 70)

new immigrant *n.* a person from southern or eastern Europe who entered the United States after 1900. (p. 614)

New Jersey Plan *n.* a plan of government proposed at the Constitutional Convention in 1787 that called for a one-house legislature in which each state would have one vote. (p. 231)

Nineteenth Amendment *n.* an amendment to the U.S. Constitution, ratified in 1920, which gave women full voting rights. (p. 653)

Nisei (NEE say) *n.* a Japanese American born in the United States. (p. 775)

Northwest Ordinance *n.* it described how the Northwest Territory was to be governed and set conditions for settlement and settlers' rights. (p. 223)

Northwest Territory *n.* territory covered by the Land Ordinance of 1785, which included land that formed the states of Ohio, Indiana, Michigan, Illinois, Wisconsin, and part of Minnesota. (p. 223)

NOW *n.* founded in 1966, the National Organization for Women pushed to get women good jobs at equal pay. (p. 826)

Nuremberg (NOOR uhm BURG) **Trials** *n.* the court proceedings held in Nuremberg, Germany, after World War II, in which Nazi leaders were tried for war crimes. (p. 780)

O

Open Door Policy *n.* in 1899, the United States asked nations involved in Asia to follow a policy in which no one country controlled trade with China. (p. 669)

Oregon Trail *n.* a trail that ran westward from Independence, Missouri, to the Oregon Territory. (p. 396)

overseer *n.* a worker hired by a planter to watch over and direct the work of slaves. (p. 122)

P

pacifist (PAS uh fist) *n.* a person morally opposed to war. (p. 209)

Palmer raids *n.* in 1920, federal agents and police raided the homes of suspected radicals. (p. 697)

Panama (PAN uh MAH) **Canal** *n.* a shortcut through Panama that connects the Atlantic and the Pacific oceans. (p. 670)

Panic of 1837 *n.* a financial crisis in which banks closed and the credit system collapsed. (p. 386)

Panic of 1873 *n.* a financial crisis in which banks closed and the stock market collapsed. (p. 547)

Parliament (PAHR luh muhnt) *n.* England's chief lawmaking body. (p. 142)

patent *n.* a government document giving an inventor the exclusive right to make or sell his or her invention for a specific number of years. (p. 586)

Patriot *n.* an American colonist who sided with the rebels in the American Revolution. (p. 173)

patroon (puh TROON) *n.* a person who brought 50 settlers to New Netherland and in return received a large land grant and other special privileges. (p. 101)

Pearl Harbor *n.* a naval base in Hawaii that was hit in a surprise attack by Japan on December 7, 1941. (p. 760)

Persian (PUR zhen) **Gulf War** *n.* in 1990–1991, the United States and the UN drove Iraq out of Kuwait, a country the Iraqis had invaded in 1990. (p. 875)

petroleum *n.* an oily, flammable liquid. (p. 585)

philanthropist (fil LAN thruh pist) *n.* a person who gives large sums of money to charities. (p. 596)

Pickett's Charge *n.* General George Pickett led a direct attack on Union troops during the 1863 Civil War battle at Gettysburg; the attack failed. (p. 513)

piedmont *n.* a broad plateau that leads to the foot of a mountain range. (p. 126)

Pilgrim *n.* a member of the group that rejected the Church of England, sailed to America, and founded the Plymouth Colony in 1620. (p. 92)

Pinckney's (PINGK neez) **Treaty** *n.* a 1795 treaty with Spain that allowed Americans to use the Mississippi River and to store goods in New Orleans; made the 31st parallel the southern U.S. border. (p. 302)

plantation *n.* a large farm that raises cash crops. (p. 73)

platform *n.* a statement of beliefs. (p. 471)

Platt Amendment *n.* a result of the Spanish-American War, which gave the United States the right to intervene in Cuban affairs when there was a threat to "life, property, and individual liberty." (p. 666)

Plessy* v. *Ferguson *n.* an 1896 case in which the Supreme Court ruled that separation of the races in public accommodations was legal. (p. 621)

political machine *n.* an organization that influences enough votes to control a local government. (p. 613)

political party *n.* a group of people that tries to promote its ideas and influence government, and also backs candidates for office. (p. 304)

Pontiac's (PON tee AKS) **Rebellion** *n.* a revolt against British forts and American settlers in 1763, led in part by Ottawa war leader Pontiac, in response to settlers' claims of Native American lands and to harsh treatment by British soldiers. (p. 151)

popular culture *n.* items such as music, fashion, and movies that are popular among a large number of people. (p. 718)

popular sovereignty (SOV uhr in tee) *n.* a government in which the people rule (p. 244); a system in which the residents vote to decide an issue. (p. 463)

Populist Party *n.* also known as the People's Party and formed in the 1890s, this group wanted a policy that would raise crop prices. (p. 577)

prejudice (PREJ uh dis) *n.* a negative opinion that is not based on facts. (p. 427)

printing press *n.* a machine invented about 1455 by Johannes Gutenberg. (p. 47)

privateer (PRY vuh TEER) *n.* a privately owned ship that has government permission during wartime to attack an enemy's merchant ships. (p. 204)

Proclamation (PRAHK luh MAY shuhn) **of 1763** *n.* an order in which Britain prohibited its American colonists from settling west of the Appalachian Mountains. (p. 151)

profit *n.* money a business makes, after subtracting the costs of doing business from the income. (p. 48)

progressivism (pruh GREHS ih VIHZ uhm) *n.* an early 20th-century reform movement seeking to return control of the government to the people, to restore economic opportunities, and to correct injustices in American life. (p. 639)

prohibition (PROH uh BIHSH uhn) *n.* the banning of the manufacture, sale, and possession of alcoholic beverages. (p. 715)

propaganda (PRAHP uh GAN duh) *n.* an opinion expressed for the purpose of influencing the actions of others. (p. 692)

proprietary (pruh PRY ih TEHR ee) **colony** *n.* a colony with a single owner. (p. 101)

pro tempore (proh TEHM puh ree) *adv.* Latin phrase meaning "for the time being." (p. 250)

public works project *n.* a government-funded project to build public resources such as roads and dams. (p. 732)

Pullman Strike *n.* a nationwide railway strike that spread throughout the rail industry in 1894. (p. 603)

Puritan *n.* a member of a group from England that settled the Massachusetts Bay Colony in 1630 and sought to reform the practices of the Church of England. (p. 94)

push-pull factor *n.* a factor that pushes people out of their native lands and pulls them toward a new place. (p. 424)

Q

Quaker (KWAY kuhr) *n.* a person who believed all people should live in peace and harmony; accepted different religions and ethnic groups. (p. 101)

quarter *v.* to give a place to stay. (p. 267)

Quartering Act *n.* a law passed by Parliament in 1765 that required the colonies to house and supply British soldiers. (p. 160)

quorum (KWAWR uhm) *n.* the minimum number of members that must be present for official business to take place. (p. 251)

R

racial (RAY shuhl) **discrimination** (dih SKRIHM uh NAY shuhn) *n.* different treatment based on a person's race. (p. 620)

racism (RAY SIHZ uhm) *n.* the belief that some people are inferior because of their race. (p. 79)

radical (RAD ih kuhl) *n.* a person who takes extreme political positions. (p. 313)

Radical Republican (rih PUHB lih kuhn) *n.* a congressman who, after the Civil War, favored using the government to create a new order in the South and to give African Americans full citizenship and the right to vote. (p. 533)

ragtime *n.* a blend of African-American songs and European musical forms. (p. 629)

ratification (RAT uh fih KAY shuhn) *n.* official approval. (p. 264)

ration (RASH uhn) *v.* to distribute a fixed amount of a certain item. (p. 773)

recall *v.* to vote an official out of office. (p. 640)

Reconstruction *n.* the process the U.S. government used to readmit the Confederate states to the Union after the Civil War. (p. 533)

Red Scare *n.* in 1919–1920, a wave of panic from fear of a Communist revolution. (p. 697)

referendum (REHF uh REHN duhm) *n.* when a proposed law is submitted to a vote of the people. (p. 640)

Reformation *n.* a 16th-century religious movement to correct problems in the Roman Catholic Church. (p. 47)

Renaissance (REHN ih SAHNS) *n.* a period of European history, lasting from the 1300s to 1600, that brought increased interest in art and learning. (p. 46)

rendezvous (RAHN day VOO) *n.* a meeting. (p. 197)

reprieve (rih PREEV) *n.* a delay or cancellation of punishment. (p. 259)

republic (rih PUHB lihk) *n.* a government in which people elect representatives to govern for them. (p. 222)

republicanism (rih PUHB lih keh NIHZ uhm) *n.* the belief that government should be based on the consent of the people; people exercise their power by voting for political representatives. (pp. 214, 245)

Republican Party *n.* the political party formed in 1854 by opponents of slavery in the territories. (p. 466)

reservation *n.* land set aside by the U.S. government for Native American tribes. (p. 562)

revenue (REHV uh noo) *n.* income a government collects to cover expenses. (pp. 160, 252)

revenue sharing *n.* the distribution of federal money to state and local governments with few or no restrictions on how it is spent. (p. 856)

revival (rih VY vuhl) *n.* a meeting designed to reawaken religious faith. (p. 433)

rifle *n.* a gun with a grooved barrel that causes a bullet to spin through the air. (p. 491)

robber baron *n.* a business leader who became wealthy through dishonest methods. (p. 594)

rock 'n' roll *n.* a form of popular music, characterized by heavy rhythms and simple melodies, that developed from rhythm and blues in the 1950s. (p. 803)

romanticism (roh MAN tih SIHZ uhm) *n.* a European artistic movement that stressed the individual, imagination, creativity, and emotion. (p. 429)

Roosevelt Corollary (KAWR uh lehr ee) *n.* a 1904 addition to the Monroe Doctrine allowing the United States to be the "policeman" in Latin America. (p. 672)

Rosie the Riveter (RIHV iht uhr) *n.* an image of a strong woman hard at work at an arms factory during World War II. (p. 773)

Rough Rider *n.* a member of the First United States Volunteer Cavalry, organized by Theodore Roosevelt during the Spanish-American War. (p. 665)

royal colony *n.* a colony ruled by governors appointed by a king. (p. 103)

S

salutary (SAL yuh TEHR ee) **neglect** *n.* a hands-off policy of England toward its American colonies during the first half of the 1700s. (p. 144)

SALT *n.* the Strategic Arms Limitation Treaty, a treaty signed in 1972 between the United States and the Soviet Union; it limited nuclear weapons. (p. 858)

Sand Creek Massacre (MAS uh kuhr) *n.* an 1864 attack in which more than 150 Cheyenne men, women, and children were killed by the Colorado militia. (p. 564)

Santa Fe (SAN tuh FAY) **Trail** *n.* a trail that began in Missouri and ended in Santa Fe, New Mexico. (p. 395)

SCLC *n.* the Southern Christian Leadership Conference, a group that coordinated civil rights protests across the South. (p. 815)

secede (sih SEED) *v.* to withdraw. (p. 473)

secession (sih SEHSH uhn) *n.* withdrawal. (p. 383)

Second Battle of the Marne (mahrn) *n.* a 1918 battle during World War I that marked the turning point in the war; allied troops along with Americans halted the German advance into France. (p. 689)

Second Continental Congress *n.* a governing body whose delegates agreed, in May 1775, to form the Continental Army and to approve the Declaration of Independence. (p. 177)

Second Great Awakening *n.* the renewal of religious faith in the 1790s and early 1800s. (p. 433)

Second New Deal *n.* a set of programs passed in 1935 to fight the Great Depression. (p. 736)

sectionalism (SEHK shuh nuh LIHZ uhm) *n.* the placing of the interests of one's own region ahead of the interests of the nation as a whole. (p. 357)

Securities and Exchange Commission *n.* an agency that watches the stock market and makes sure companies follow fair practices for trading stocks. (p. 748)

Sedition (sih DIHSH uhn) **Act** *n.* a 1918 law that made it illegal to criticize the war; it set heavy fines and long prison terms for those who engaged in antiwar activities. (p. 692)

segregation (SEHG rih GAY shuhn) *n.* separation, especially of races. (p. 621)

Seneca (SEHN ih kuh) **Falls Convention** *n.* a women's rights convention held in Seneca Falls, New York, in 1848. (p. 444)

separation of powers *n.* the division of basic government roles into branches. (p. 246)

service economy *n.* an economy in which most jobs provide services instead of producing goods. (p. 879)

servitude (SUR vih TOOD) *n.* a state of belonging to an owner or master. (p. 270)

Seven Days' Battles *n.* an 1862 Civil War battle in which the Confederacy forced the Union to retreat before it could capture the Southern capital of Richmond. (p. 496)

Seventeenth Amendment *n.* an amendment to the U.S. Constitution, ratified in 1913, that provided for the direct election of U.S. senators. (p. 648)

sharecropping *n.* a system in which landowners gave farm workers land, seed, and tools in return for a part of the crops they raised. (p. 543)

Shays's (SHAY zuhz) **Rebellion** *n.* an uprising of debt-ridden Massachusetts farmers in 1787. (p. 225)

Sherman Antitrust Act *n.* a law passed in 1890 that made it illegal for corporations to gain control of industries by forming trusts. (p. 641)

Siege (seej) **of Vicksburg** *n.* an 1863 Union victory in the Civil War that enabled the Union to control the entire Mississippi River. (p. 516)

sit-down strike *n.* a strike in which workers remain idle inside the plant or factory. (p. 743)

Sixteenth Amendment *n.* an amendment to the U.S. Constitution, ratified in 1913, that gave Congress the power to create income taxes. (p. 647)

slash-and-burn agriculture (ag rih kuhl chuhr) *n.* a farming method in which people clear fields by cutting and burning trees and grasses, the ashes of which fertilize the soil. (p. 37)

slave code *n.* a law passed to regulate the treatment of slaves. (p. 79)

slavery *n.* the practice of holding a person in bondage for labor. (p. 76)

slum *n.* a neighborhood with overcrowded and dangerous housing. (p. 612)

smuggle *v.* to illegally import or export goods. (p. 112)

SNCC *n.* formed in 1960, the Student Nonviolent Coordinating Committee was created to give young people a larger role in the civil rights movement. (p. 817)

social gospel (GAHS puhl) *n.* a movement aimed at improving the lives of the poor. (p. 612)

socialism *n.* an economic system in which all members of a society are equal owners of all businesses; members share the work and the profits. (p. 602)

Social Security Act *n.* a law, passed in 1935, that requires workers and employers to make payments into a fund, from which they draw a pension after they retired. (p. 736)

sodbuster *n.* a farmer on the frontier. (p. 575)

Songhai (SAWNG HY) *n.* a West African empire that succeeded Mali and controlled trade from the 1400s to 1591. (p. 42)

Sons of Liberty *n.* a group of colonists who formed a secret society to oppose British policies at the time of the American Revolution. (p. 161)

space race *n.* a competition, beginning in 1957, between the Soviet Union and the United States in the exploration of space. (p. 799)

Spanish-American War *n.* a war in 1898 that began when the United States demanded Cuba's independence from Spain. (p. 664)

Spanish Armada (ahr MAH duh) *n.* a fleet of ships sent in 1588 by Philip II, the Spanish king, to invade England and restore Roman Catholicism. (p. 69)

speculation (SPEHK yuh LAY shuhn) *n.* buying and selling of a stock in the hope of making a quick profit. (p. 730)

sphere of influence *n.* an area where foreign nations claim special rights and economic privileges. (p. 669)

spiritual *n.* a religious folk song. (p. 351)

spoils system *n.* the practice of winning candidates giving government jobs to political backers or supporters. (p. 373)

Stamp Act *n.* a 1765 law passed by Parliament that required all legal and commercial documents to carry an official stamp showing a tax had been paid. (p. 160)

standard time *n.* a system adopted in 1918 that divided the United States into four time zones. (p. 592)

states' rights *n.* theory that said that states had the right to judge when the federal government had passed an unconstitutional law. (p. 307)

steerage *n.* the cheapest deck or place on a ship. (p. 423)

Stono (STOH noh) **Rebellion** *n.* a 1739 uprising of slaves in South Carolina, leading to the tightening of already harsh slave laws. (p. 123)

strategy *n.* an overall plan of action. (p. 196)

strike *v.* to stop work to demand better working conditions. (p. 434)

subsistence farm *n.* a farm that produces enough food for the family with a small additional amount for trade. (p. 110)

suburb *n.* a residential area that surrounds a city. (p. 800)

suffrage *n.* the right to vote. (pp. 262, 444)

Sugar Act *n.* a law passed by Parliament in 1764 that placed a tax on sugar, molasses, and other products shipped to the colonies; also called for harsh punishment of smugglers. (p. 160)

sunbelt *n.* the warmer states of the South and Southwest. (p. 801)

supply-side economics *n.* the idea that lowering taxes will lead to increases in jobs, savings, investments, and so lead to an increase in government revenue. (p. 873)

sweatshop *n.* a place where workers labored long hours under poor conditions for low wages. (p. 600)

T

Tammany (TAM uh nee) **Hall** *n.* a famous political machine, located in New York City in the late 19th century. (p. 613)

tariff *n.* a tax on imported goods. (p. 296)

Tariff of Abominations *n.* an 1828 law that raised the tariffs on raw materials and manufactured goods; it upset Southerners who felt that economic interests of the Northeast were determining national economic policy. (p. 381)

Teapot Dome Scandal *n.* episode caused by Secretary of the Interior Albert B. Fall's leasing of oil-rich public land to private companies for money and land. (p. 710)

technology *n.* the use of tools and knowledge to meet human needs. (p. 32)

Tejano (tuh HAH noh) *n.* a person of Spanish heritage who considered Texas to be home. (p. 400)

temperance movement *n.* a campaign to stop the drinking of alcohol. (p. 434)

tender *n.* money. (p. 255)

tenement *n.* an apartment building that is usually run-down and overcrowded. (p. 611)

Tet (tet) **offensive** *n.* in 1968, a surprise attack by the Viet Cong on U.S. military bases and more than 100 cities and towns in South Vietnam during Tet, the Vietnamese celebration of the lunar New Year. (p. 844)

Thirteenth Amendment *n.* an amendment to the U.S. Constitution, adopted in 1865, banning slavery and involuntary servitude in the United States. (p. 521)

38th parallel *n.* the area north of this latitude in Korea occupied by Soviet troops in 1945. (p. 795)

Three-Fifths Compromise *n.* the Constitutional Convention's agreement to count three-fifths of a state's slaves as population for purposes of representation and taxation. (p. 232)

Townshend (TOWN zuhnd) **Acts** *n.* a series of laws passed by Parliament in 1767 that suspended New York's assembly and established taxes on goods brought into the British colonies. (p. 163)

Trail of Tears *n.* the tragic journey of the Cherokee people from their homeland to Indian Territory between 1838 and 1839; thousands of Cherokee died. (p. 377)

transcendentalism (TRAN sen DEN tl IZ uhm) *n.* a 19th-century philosophy that taught the spiritual world is more important than the physical world and that people can find truth within themselves through feeling and intuition. (p. 431)

transcontinental (TRANS kon tuh NEN tl) **railroad** *n.* a railroad that spanned the entire continent. (p. 590)

Treaty of Ghent (gent) *n.* treaty, signed in 1814, which ended the War of 1812; no territory exchanged hands and trade disputes were not resolved. (p. 333)

Treaty of Greenville *n.* a 1795 agreement in which 12 Native American tribes surrendered much of present-day Ohio and Indiana to the U.S. government. (p. 300)

Treaty of Guadalupe Hidalgo (GWAHD loop hi DAH goh) *n.* the 1848 treaty ending the U.S. war with Mexico; Mexico ceded nearly one-half of its land to the United States. (p. 410)

Treaty of Paris *n.* the 1763 treaty that ended the French and Indian War; Britain gave up all of North America east of the Mississippi River. (p. 150)

Treaty of Paris of 1783 *n.* the treaty that ended the Revolutionary War, confirming the independence of the United States and setting the boundaries of the new nation. (p. 212)

Treaty of Tordesillas (TAWR duh SEE uhs) *n.* the 1494 treaty in which Spain and Portugal agreed to divide the lands of the Western Hemisphere between them and moved the Line of Demarcation to the west. (p. 61)

Treaty of Versailles (vuhr SY) *n.* the 1919 treaty that ended World War I. (p. 696)

trench warfare *n.* a kind of warfare during World War I in which troops huddled at the bottom of trenches and fired artillery and machine guns at each other. (p. 680)

triangular trade *n.* the transatlantic system of trade in which goods, including slaves, were exchanged between Africa, England, Europe, the West Indies, and the colonies in North America. (p. 111)

tribunal (try BYOO nuhl) *n.* a court. (p. 253)

Truman Doctrine *n.* a policy that promised aid to people struggling to resist threats to democratic freedom. (p. 793)

trust *n.* a legal body created to hold stock in many companies, often in the same industry. (p. 595)

tundra (TUN druh) *n.* a treeless plain that remains frozen under its top layer of soil. (p. 33)

Twenty-sixth Amendment *n.* an amendment to the U.S. Constitution, adopted in 1971 and lowering the voting age from 21 to 18. (p. 849)

U

unanimous (yoo NAN uh muhs) **consent** *n.* complete agreement. (p. 264)

Uncle Tom's Cabin *n.* a novel published by Harriet Beecher Stowe in 1852 that portrayed slavery as brutal and immoral. (p. 462)

unconstitutional *n.* something that contradicts the law of the Constitution. (p. 317)

Underground Railroad *n.* a series of escape routes used by slaves escaping the South. (p. 442)

United Nations *n.* an international peacekeeping organization to which most nations in the world belong, founded in 1945 to promote world peace, security, and economic development. (p. 781)

urbanization *n.* growth of cities resulting from industrialization. (p. 609)

U.S.S. *Maine* *n.* a U.S. warship that mysteriously exploded and sank in the harbor of Havana, Cuba, on February 15, 1898. (p. 663)

V

vaquero (vah KAIR oh) *n.* a cowhand that came from Mexico with the Spaniards in the 1500s. (p. 560)

vaudeville (VAWD vil) *n.* a form of live stage entertainment with a mixture of songs, dance, and comedy. (p. 629)

viceroyalty (VYS ROI uhl tee) *n.* a province ruled by a viceroy, who ruled in the king's name. (p. 71)

Viet Cong *n.* a Vietnamese Communist. (p. 838)

Vietnamization (vee ET nuh mi ZAY shuhn) *n.* a strategy of gradually withdrawing U.S. forces and turning the ground fighting over to the South Vietnamese during the Vietnam War. (p. 848)

vigilante (vij uh LAN tee) *n.* a person willing to take the law into his or her own hands. (p. 561)

Virginia Plan *n.* a plan proposed by Edmund Randolph, a delegate to the Constitutional Convention in 1787, that proposed a government with three branches and a two-house legislature in which representation would be based on a state's population or wealth. (p. 231)

Voting Rights Act of 1965 *n.* this law banned literacy tests and other laws that kept African Americans from registering to vote. (p. 821)

W

war bond *n.* a low-interest loan by civilians to the government, meant to be repaid in a number of years. (p. 691)

War Hawk *n.* a westerner who supported the War of 1812. (p. 329)

War Powers Act *n.* passed in 1973, this limits the president's war-making powers without consulting Congress. (p. 849)

War Production Board *n.* an agency established during World War II to coordinate the production of military supplies by U.S. industries. (p. 772)

Watergate scandal *n.* a scandal resulting from the Nixon administration's attempt to cover up its involvement in the 1972 break-in at the Democratic National Committee headquarters in the Watergate apartment complex in Washington, D.C. (p. 860)

Webster-Hayne debate *n.* an 1830 debate between Daniel Webster and Robert Hayne over the doctrine of nullification. (p. 382)

Whig (hwig) **Party** *n.* a political party organized in 1834 to oppose the policies of Andrew Jackson. (p. 387)

Whiskey Rebellion *n.* a 1794 protest against the government's tax on whiskey, which was valuable to the livelihood of backcountry farmers. (p. 301)

Wilderness Road *n.* the trail into Kentucky that woodsman Daniel Boone helped to build. (p. 221)

Wilmot (WIL muht) **Proviso** (pruh VY zoh) *n.* an 1846 proposal that outlawed slavery in any territory gained from the War with Mexico. (p. 459)

Wounded Knee Massacre *n.* the massacre by U.S. soldiers of 300 unarmed Native Americans at Wounded Knee Creek, South Dakota, in 1890. (p. 566)

writ (rit) **of assistance** *n.* a search warrant that allowed British officers to enter colonial homes or businesses to search for smuggled goods. (p. 164)

X

XYZ Affair *n.* a 1797 incident in which French officials demanded a bribe from U.S. diplomats. (p. 306)

Y

Y2K *n.* a computer problem caused by computer programs using only the last two digits of a year and complicated by the arrival of the year 2000. (p. 878)

Yalta (YAWL tuh) **Conference** *n.* in 1945, Franklin Roosevelt, Winston Churchill, and Joseph Stalin discussed plans for the end of World War II and the future of Europe. (p. 765)

yellow journalism *n.* a style of journalism that exaggerates and sensationalizes the news. (p. 663)

Yoruba (YOH roo bah) *n.* a West African people who formed several states southwest of the Niger River. (p. 42)

Z

Zimmermann telegram *n.* a message sent in 1917 by the German foreign minister to the German ambassador in Mexico, proposing a German-Mexican alliance and promising to help Mexico regain Texas, New Mexico, and Arizona if the United States entered World War I. (p. 682)

A

abolition [abolición] *s.* movimiento para eliminar la esclavitud. (p. 440)

abridge [abreviar] *v.* reducir. (p. 266)

AEF *s.* Fuerza Expedicionaria Estadounidense, fuerzas de EE. UU. durante la primera guerra mundial. (p. 686)

affirmation [afirmación] *s.* declaración de que algo es cierto. (p. 257)

African Diaspora [diáspora africana] *s.* traslado forzado de los africanos, desde su patria a las Américas para trabajar allí como esclavos. (p. 78)

Agent Orange [agente naranja] *s.* herbicida que mata las plantas. (p. 843)

Albany Plan of Union [Plan de la Unión de Albany] *s.* primera propuesta formal para unir las colonias norteamericanas, presentado por Benjamín Franklin. (p. 149)

Alien and Sedition Acts [leyes de Extranjeros y Sedición] *s.* serie de cuatro leyes promulgadas en 1798 para reducir el poder político de inmigrantes recién llegados a Estados Unidos. (p. 306)

allies [aliados] *s.* alianza de Serbia, Rusia, Francia, Gran Bretaña, Italia y otros siete países durante la primera guerra mundial. (p. 680)

ally [aliado] *s.* país que acuerda ayudar a otro país a alcanzar un objetivo común. (p. 200)

American Federation of Labor (AFL) [Federación Norteamericana del Trabajo] *s.* organización nacional de sindicatos obreros fundada en 1886. (p. 603)

American System [Sistema Americano] *s.* plan presentado en 1815 para hacer autosuficiente a Estados Unidos. (p. 354)

Anaconda Plan [Plan Anaconda] *s.* estrategia de tres pasos mediante la cual la Unión se proponía derrotar a la Confederación durante la guerra civil estadounidense. (p. 484)

Angel Island [isla del Ángel] *s.* primera parada en Estados Unidos para la mayoría de los inmigrantes que venían de Asia. (p. 615)

Antifederalist [antifederalista] *s.* persona que se oponía a la ratificación de la Constitución de los Estados Unidos. (p. 234)

Anti-Imperialist League [Liga Antiimperialista] *s.* grupo de estadounidenses importantes que creían que Estados Unidos no debía negarle a otras personas el derecho de gobernarse a sí mismas. (p. 667)

Appalachian Mountains [montes Apalaches] *s.* cadena de montañas que se extiende desde el este de Canadá hacia el sur, hasta Alabama. (p. 126)

appeasement [apaciguamiento] *s.* otorgamiento de concesiones a una potencia hostil con el fin de mantener la paz. (p. 757)

appellate [de apelación] *adj.* que tiene el poder de reexaminar decisiones de las cortes. (p. 260)

Appomattox Court House [Appomattox] *s.* pueblo de Virginia donde Robert E. Lee se rindió a Ulysses s. Grant en 1865, finalizando así la guerra civil. (p. 519)

apprentice [aprendiz] *s.* joven que aprende un oficio o una artesanía de un maestro experto. (p. 137)

appropiation [apropiación] *s.* fondos públicos que se reservan para un propósito específico. (p. 253)

archaeologist [arqueólogo] *s.* científico que estudia el pasado humano examinando artículos que dejó la gente. (p. 27)

armistice [armisticio] *s.* suspención de la lucha en una guerra. (p. 690)

arms race [carrera de armamento] *s.* desde fines de los años cuarenta hasta fines de los años ochenta, Estados Unidos y la Unión Soviética trataron de superarse una a la otra desarrollando armas de mayor poder destructivo. (p. 798)

Articles of Confederation [Artículos de Confederación] *s.* documento, adoptado por el Congreso Continental en 1777 y finalmente aprobado por los estados en 1781, que delineaba la forma de gobierno de los nuevos Estados Unidos. (p. 222)

artifact [artefacto] *s.* herramienta u otro artículo hecho por seres humanos. (p. 27)

artillery [artillería] *s.* cañon o arma grande. (p. 177)

artisan [artesano] *s.* obrero especializado, como un tejedor a telar o un alfarero, que hace artículos a mano; artífice. (p. 117)

assimilation [asimilación] *s.* proceso de integrarse a una sociedad. (p. 616)

Axis [Eje] *s.* Alemania, Italia y sus aliados durante la segunda guerra mundial. (p. 757)

B

baby boom *s.* término para la generación que nació en Estados Unidos entre 1946 y 1961, cuando el índice de natalidad aumentó marcadamente después de la segunda guerra mundial. (p. 801)

Backcountry [tierras fronterizas] *s.* región colonial que se extendía a lo largo de los montes Apalaches a través de la sección oeste de Nueva Inglaterra y las colonias del centro y del sur. (p. 109)

Bacon´s Rebellion [Rebelión de Bacon] *s.* levantamiento contra la poderosa autoridad colonial de Jamestown por Nathaniel Bacon y un grupo de habitantes de la frontera que resultó en la quema de Jamestown en 1676. (p. 89)

bail [fianza] *s.* dinero que pagan como fianza las personas arrestadas para garantizar que van a regresar para el juicio. (p. 268)

Bataan Death March [Marcha de la Muerte de Bataan] *s.* en 1942 los japoneses forzaron a 70,000 soldados filipinos y estadounidenses a marchar 60 millas a un campo de prisioneros. (p. 768)

Battle of Antietam [batalla de Antietam] *s.* batalla de la guerra civil, en 1862, en que murieron o resultaron heridos 25,000 hombres. (p. 497)

Battle of Fallen Timbers *s.* en 1794 el ejército estadounidense derrotó a 2,000 amerindios en un enfrentamiento por el control del territorio del Noroeste. (p. 299)

Battle of Gettysburg [batalla de Gettysburg) *s.* batalla de 1863 de la guerra civil en que la Unión derrotó a la Confederación, poniendo fin a la esperanza de una victoria confederada en el Norte. (p. 513)

Battle of Midway *s.* victoria de Estados Unidos sobre los japoneses en una batalla naval de 1942 que señaló un cambio decisivo en la segunda guerra mundial. (p. 770)

Battle of Quebec [batalla de Quebec] *s.* batalla en la que los británicos derrotaron a los franceses y cambio decisivo en la guerra Francesa y Amerindia. (p. 150)

Battle of Shiloh [batalla de Shiloh] *s.* batalla de 1862 en que la Unión obligó a la Confederación a retroceder; fue una de las batallas más encarnizadas de la guerra civil. (p. 494)

Battle of Yorktown [batalla de Yorktown] *s.* última batalla importante de la guerra Revolucionaria que resultó en la capitulación de las fuerzas británicas en 1781. (p. 210)

Battle of the Alamo [batalla de El Álamo] *s.* en 1836 los texanos defendieron contra el ejército mexicano una misión llamada El Álamo; sobrevivieron sólo cinco texanos (p. 403)

Battle of the Bulge [batalla del Bolsón] *s.* batalla de la segunda guerra mundial de un mes de duración en que los aliados lograron rechazar la última gran ofensiva alemana de la guerra. (p. 764)

Battle of the Little Bighorn [batalla del Little Bighorn] *s.* batalla de 1876 en que los sioux y los cheyennes aniquilaron toda una partida militar estadounidense. (p. 565)

Battle of the Thames [batalla de Thames] *s.* victoria estadounidense sobre los británicos en la guerra de 1812 que puso fin a la amenaza británica en el Territorio del Noroeste. (p. 332)

Battles of Saratoga [batallas de Saratoga] *s.* serie de conflictos en 1777, entre soldados británicos y el Ejército Continental que resultó en un cambio decisivo en la guerra Revolucionaria. (p. 199)

bayonet [bayoneta] *s.* largo cuchillo de acero colocado en el extremo de un arma de fuego. (p. 202)

Bear Flag Revolt [revuelta de la Bandera del Oso] *s.* rebelión de 1846 por los estadounidenses contra el dominio mexicano en California. (p. 409)

Benin [Benín] *s.* reino de África Occidental que se estableció cerca del delta del río Níger en el siglo XIV y se transformó en un estado importante en el siglo XV. (p. 43)

Bessemer steel process [proceso siderúrgico Bessemer] *s.* manera nueva de producir acero desarrollada hacia 1850 que causó un gran incremento en la producción siderúrgica. (p. 587)

bill of attainder [decreto de proscripción] *s.* ley que condena a una persona sin juicio ante un tribunal. (p. 255)

Bill of Rights [Carta de Derechos] *s.* diez primeras enmiendas a la Constitución de Estados Unidos, adoptadas en 1791, que consisten en una lista formal de los derechos y libertades de los ciudadanos. (p. 237)

black code [código negro] *s.* ley pasada por los estados sureños que limitaba la libertad de los antiguos esclavos. (p. 534)

Black Tuesday [martes negro] *s.* nombre que se le da al 29 de octubre de 1929, cuando se desplomó el precio de las acciones. (p. 731)

blockade [bloqueo] *s.* acción de las fuerzas armadas que impide la entrada o salida de mercaderías o personas. (p. 484)

Bonus Army [Ejército de la Prima] *s.* en 1932 miles de veteranos marcharon a Washington demandando el pago de una prima que nunca habían recibido. (p. 733)

boomtown [pueblo en auge] *s.* pueblo que tiene una explosión repentina de crecimiento económico o demográfico. (p. 558)

border state [estado fronterizo] *s.* estados esclavistas fronterizos a estados en que la esclavitud era ilegal. (p. 482)

Boston Massacre [Matanza de Boston] *s.* choque en 1770 entre soldados británicos y colonos de Boston en que perecieron cinco de los colonistas, incluso Crispus Attucks. (p. 165)

Boston Tea Party [Motín del Té de Boston] *s.* como protesta contra el Acta del té, en 1773 los colonos arrojaron al puerto de Boston 342 cajones de té. (p. 167)

bounty [gratificación] *s.* recompensa o pago en dinero que da un gobierno. (pp. 271, 508)

Boxer Rebellion [Rebelión bóxer] *s.* en 1900 el resentimiento chino contra la actitud de superioridad cultural de los extranjeros resultó en este violento levantamiento. (p. 669)

boycott [boicot] *v.* negarse a comprar ciertos productos. (p. 161)

***bracero* program** [programa bracero] *s.* uso de trabajadores mexicanos en la época de escasez de labriegos durante la segunda guerra mundial. (p. 774)

brinksmanship *s.* política internacional, el acto de empujar al límite una situación peligrosa; por ejemplo: los Estados Unidos yendo al borde de la guerra para parar el comunismo. (p. 798)

Brown v. Board of Education of Topeka, Kansas [Brown contra el Consejo de Educación de Topeka, Kansas] *s.* caso de 1954 en que la Corte Suprema declaró que la doctrina educativa de "iguales pero separados" para los blancos y los afroamericanos no era constitucional. (p. 814)

buck [ciervo] gamuza obtenida de la piel de un ciervo adulto, o unidad de dinero de los colonos. (p. 127)

buffalo soldier [soldado búfalo] *s.* apodo que los amerindios les dieron a los afroamericanos que servían en el ejército estadounidense del oeste. (p. 571)

business cycle [ciclo económico] *s.* serie de períodos de la economía buenos y malos. (p. 586)

buy on margin [comprar valores a crédito] *s.* pagar una pequeña parte del precio de una acción y pagar el resto con un préstamo. (p. 730)

C

cabinet [gabinete ministerial] *s.* grupo de ministros que actúan como los asesores princpales del presidente. (p. 294)

California gold rush [fiebre del oro de California] *s.* en 1849 gran cantidad de gente se fue a California porque allí se había descubierto oro. (p. 413)

Cambodia [Camboya] *s.* país fronterizo de Vietnam. (p. 848)

Camp David Accords [acuerdos de Camp David] *s.* en 1979, basados en estos acuerdos, Egipto e Israel firmaron un tratado de paz que puso fin a 30 años de conflicto. (p. 864)

caravel [carabela] *s.* barco con velas triangulares que le permitían navegar hacia el viento y con velas cuadradas que lo llevaban hacia delante cuando soplaba viento en popa. (p. 49)

cash crop [cultivo comercial] *s.* cultivo que produce un agricultor para venderlo por dinero y no para su uso personal. (p. 115)

cavalry [caballería] *s.* soldados montados a caballo. (p. 496)

Centennial Exhibition [Exposición del Centenario] *s.* exposición de 1876 en Filadelfia que celebró el centésimo cumpleaños de Estados Unidos. (p. 588)

Central Powers [Potencias Centrales] *s.* alianza de Austria-Hungría, Alemania, el Imperio otomano y Bulgaria durante la primera guerra mundial. (p. 680)

charter [cédula] *s.* contrato escrito que concede un gobierno otorgando al que lo recibe el derecho a establecer una colonia. (p. 87)

checks and balances [frenos y cortapisas] *s.* capacidad de cada rama del gobierno de usar frenos o controles sobre las otras ramas. (p. 246)

Chinese Exclusion Act [ley de Exclusión para chinos] *s.* aprobada en 1882, esta ley prohibía la inmigración china por diez años. (p. 617)

civil disobedience [desobediencia civil] *s.* negarse pacíficamente a obedecer leyes que uno considera injustas. (p. 431)

civilization [civilización] *s.* forma de cultura caracterizada por ciudades con centros de comercio, trabajadores especializados, formas de gobierno y religión organizadas, sistemas de mantener registros, y herramientas avanzadas. (p. 29)

civil rights [derechos civiles] *s.* derecho otorgado a todos los ciudadanos. (p. 535)

Civil Rights Act of 1964 [ley de Derechos Civiles de 1964] *s.* esta ley prohibía la segregación racial en los lugares públicos y creó la Comisión para la Igualdad de Oportunidades de Empleo. (p. 820)

clan [clan] *s.* grupo grande de familias procedentes de un antepasado común. (p. 127)

Clayton Antitrust Act [ley Anti-trust Clayton] *s.* ley aprobada en 1914 que establecía reglas que prohibían prácticas comerciales que disminuyeran la competencia y le daba al gobierno más poder para reglamentar los trusts. (p. 648)

Cold War [guerra fría] *s.* estado de hostilidad, sin conflicto militar directo, que se desarrolló entre Estados Unidos y la Unión Soviética después de la segunda guerra mundial. (p. 792)

Columbian Exchange [transferencia colombina] *s.* transferencia de plantas, animales y enfermedades entre el hemisferio occidental y el oriental. (p. 74)

committee of correspondence [comité de correspondencia] *s.* grupo de personas de las colonias que se intercambiaban cartas sobre asuntos coloniales. (p. 166)

Committee to Reelect the President [Comité de Reelección del Presidente] *s.* organización cuya conexión con el allanamiento de la Sede Central del Partido Demócrata hizo estallar el escándalo Watergate. (p. 860)

common law [derecho consuetudinario] *s.* sistema de leyes desarrollado en Inglaterra, basado en costumbres y decisiones jurídicas anteriores. (p. 268)

Compromise of 1850 [Acuerdo de 1850] *s.* serie de medidas del Congreso para resolver los desacuerdos principales entre los estados libres y los esclavistas. (p. 461)

Compromise of 1877 [Acuerdo de 1877] *s.* acuerdo que resolvió la disputa sobre las elecciones de 1876: se declaró presidente a Rutherford B. Hayes, quien entonces retiró las tropas federales que quedaban en el Sur. (p. 548)

compulsory process [proceso obligatorio] *s.* procedimiento requerido. (p. 267)

Conestoga wagon [carreta conestoga] *s.* vehículo que tenía ruedas anchas, caja de carro curvada y capota de lona y se usaba para transportar gente y artículos. (p. 117)

Confederate States of America [Estados Confederados de América] *s.* confederación constituida en 1861 por los estados sureños después de separarse de la Unión. (p. 473)

Congress of Industrial Organizations (CIO) [Congreso de Organizaciones Industriales] *s.* organización sindical que en 1938 se separó de la Federación Norteamericana del Trabajo. (p. 743)

conquistador [conquistador] *s.* español que en el siglo XVI viajó a las Américas para explorar y conquistar. (p. 63)

conscription [conscripción] *s.* ley que requería que los hombres sirvieran en las fuerzas armadas o que fueran reclutados. (p. 508)

conservative [conservador] *s.* persona que está a favor de menos controles gubernamentales y más libertad individual en cuestiones de la economía. (p. 749)

Constitutional Convention [Convención Constitucional] *s.* reunión realizada en 1787 para considerar cambios a los Artículos de Confederación, que resultó en la redacción de la Constitución. (p. 229)

containment [contención] *s.* bloqueo de una nación en la expansión de la influencia de otras naciones, especialmente los esfuerzos de Estados Unidos por bloquear la expansión de la influencia soviética hacia fines de los años cuarenta y comienzos de los cincuenta. (p. 793)

Continental Army [Ejército Continental] *s.* fuerzas coloniales autorizadas en 1775 por el segundo Congreso Continental, con George Washington como su comandante en jefe. (p. 177)

convene [convocar] *v.* llamar a reunión. (p. 259)

convoy system [sistema de convoyes] *s.* fuerte flotilla de destructores que escolta a los barcos mercantes durante épocas de guerra. (p. 687)

cooperative [cooperativa] *s.* organización propiedad de los asociados que la dirigen. (p. 577)

Copperheads [víboras cobrizas] *s.* los principales adversarios políticos de Abraham Lincoln; abogaban por la paz con el Sur. (p. 508)

CORE [Congreso para la Igualdad Racial] *s.* grupo que planeó freedom rides o viajes en autobús por todo el Sur para eliminar la segregación racial en los autobuses interestatales. (p. 818)

corporation [corporación] *s.* empresa propiedad de inversionistas que compran parte de la compañía mediante acciones. (p. 594)

cotton gin [desmontadora de algodón] *s.* máquina inventada en 1793 que limpiaba el algodón con mucha más rapidez y eficiencia que los obreros humanos. (p. 348)

Crash of 1929 [Crack de 1929] *s.* el desplome de los precios de las acciones. (p. 731)

Crittenden Plan [Plan de Crittenden] *s.* acuerdo presentado en 1861 que podría haber evitado la secesión. (p. 475)

Crusades [cruzadas] *s.* serie de guerras para capturar la Tierra Santa, iniciada en 1096 por cristianos europeos. (p. 45)

Cuban Missile Crisis [crisis de los misiles cubanos] *s.* en 1962 casi estalló la guerra entre Estados Unidos y la Unión Soviética porque ésta había instalado misiles nucleares en Cuba. (p. 839)

culture [cultura] *s.* manera de vida compartida por gente que tiene artes, creencias y costumbres semejantes. (p. 28)

D

Dawes Act [ley Dawes] *s.* ley, aprobada en 1887, que distribuía la tierra de las reservas amerindias a dueños individuales. (p. 567)

D-Day [día D] *s.* 6 de junio de 1944, día en que los aliados invadieron a Francia durante la segunda guerra mundial. (p. 764)

Declaration of Independence [Declaración de Independencia] *s.* documento, escrito en 1776, en que las colonias declararon su independendia de Gran Bretaña. (p. 180)

deficit [déficit] spend *v.* usar dinero prestado para financiar programas del gobierno. (p. 738)

department store [almacén departamental] *s.* tienda que vende de todo, desde ropa a muebles a artículos de ferretería. (p. 627)

depression [depresión] *s.* aguda crisis económica. (p. 386)

desert [desertar] *v.* abandonar el servicio militar sin intenciones de regresar. (p. 203)

détente [distensión] *s.* disminución de las tensiones entre EE. UU. y la Unión Soviética durante la guerra fría. (p. 858)

direct primary [elecciones primarias directas] *s.* el electorado, y no las convenciones de partido, eligen a los candidatos para los cargos públicos. (p. 640)

diversity [diversidad] *s.* variedad de gente. (p. 117)

doctrine of nullification [doctrina de la invalidación] *s.* derecho de un estado a rechazar una ley federal que considerase inconstitucional. (p. 381)

domestication [domesticación] *s.* práctica de criar plantas o amansar animales para satisfacer las necesidades humanas. (p. 28)

domino theory [teoría del dominó] *s.* teoría que sostenía que si un país caía en la órbita comunista, los países vecinos también caerían en el comunismo. (p. 837)

dove [paloma] *s.* persona opuesta a la guerra. (p. 847)

downsize [reducir el tamaño] *v.* disminuir una empresa el número de sus empleados para incrementar las ganancias. (p. 879)

Dred Scott* v. *Sandford [Dred Scott contra Sandford] *s.* caso de 1865 de la Corte Suprema en que un esclavo, Dred Scott, entabló juicio por su libertad porque su amo lo había llevado a vivir en territorios donde la esclavitud era ilegal; la Corte dictaminó contra Scott. (p. 467)

due process of law [proceso legal debido] *s.* tratamiento justo bajo la ley. (p. 267)

dust bowl [cuenca de polvo] *s.* fincas arruinadas por el polvo a comienzos de los años treinta, en una región de unas 150,000 millas cuadradas. (p. 739)

E

elector [elector] *s.* votante. (p. 249)

Ellis Island [isla Ellis] *s.* para la mayoría de los inmigrantes que vienen de Europa, la primera parada en Estados Unidos. (p. 614)

Emancipation Proclamation [Proclama de Emancipación] *s.* orden ejecutiva dictada por Abraham Lincoln el 1.° de enero de 1863, que liberaba a los esclavos de todas las regiones insurgentes contra la Unión. (p. 504)

Embargo Act of 1807 [ley de Embargo de 1807] *s.* ley que dictaminaba que los barcos estadounidenses ya no estaban autorizados para ir a puertos extranjeros y que también cerraba los puertos de Estados Unidos a los barcos británicos. (p. 328)

emigrant [emigrante] *s.* persona que abandona un país. (p. 423)

encomienda [encomienda] *s.* concesión del trabajo de los amerindios. (p. 72)

English Bill of Rights [Carta de Derechos Ingleses] *s.* acuerdo firmado por Guillermo y María por el que prometían respetar los derechos del Parlamento y los ciudadanos ingleses, incluso el derecho a elecciones libres. (p. 144)

enlightenment [Ilustración] *s.* movimiento del siglo XVIII que enfatizaba el uso de la razón y el método científico para obtener conocimiento. (p. 140)

enumeration [enumeración] *s.* recuento oficial, como un censo. (p. 249)

environmentalism [ecologismo] *s.* trabajo dedicado a proteger el medio ambiente. (p. 864)

equity [equidad] *s.* sistema de justicia no cubierto bajo la ley común. (p. 269)

ERA *s.* Enmienda para la Igualdad de Derechos, enmienda constitucional propuesta para dar igualdad de derechos sin consideración de sexo; la propuesta murió en 1982. (p. 826)

Erie Canal [canal de Erie] *s.* completado en 1825, esta vía navegable conectaba a la ciudad de Nueva York con Bufalo, New York. (p. 355)

escalation [escalada] *s.* política de aumentar la intervención militar en Vietnam. (p. 841)

Espionage Act [ley sobre el Espionaje] *s.* aprobada en 1917, esta ley establecía multas severas y muchos años de prisión para quienes participaran en actividades contra la guerra o alentaran a los que resistían la conscripción. (p. 692)

European Middle Ages [Edad Media europea] *s.* período desde fines del siglo V hasta aproximadamente el siglo XIV, durante el cual los europeos adoptaron el feudalismo y el sistema señorial. (p. 44)

exodusters *s.* afroamericanos que abandonaron el Sur para irse al Oeste y se comparaban a los hebreos bíblicos que habían escapado la esclavitud de Egipto. (p. 575)

expatriate [expatriado] *s.* ciudadano de un país que establece su residencia en otro país. (p. 721)

ex post facto law [ley ex post facto] *s.* ley que hace que un acto sea una ofensa criminal aprobada después de cometido el acto. (p. 255)

F

factory system [sistema fabril] *s.* un método de producción que juntó a obreros y máquinas en el mismo edificio. (p. 341)

Fair Deal [Trato Justo] *s.* programa presentado por Harry Truman que proponía proyectos nuevos para crear trabajos, construir viviendas públicas y acabar con la discriminación racial en el empleo. (p. 791)

fall line *s.* punto a partir del cual una catarata impide que los barcos grandes continúen río arriba. (p. 126)

famine [hambruna] *s.* severa escasez de alimentos. (p. 426)

fascism [fascismo] *s.* filosofía política que propugna un fuerte gobierno nacionalista centralizado, con un dictador poderoso a la cabeza. (p. 756)

federalism [federalismo] *s.* sistema de gobierno en que el poder está dividido entre el gobierno central (o federal) y los estados. (pp. 234, 245)

Federalists [federalistas] *s.* partidarios de la Constitución. (p. 234)

Federalist Papers [*El federalista*] *s.* serie de ensayos que defienden y explican la Constitución, escritos por Alexander Hamilton, James Madison y John Jay. (p. 235)

Federal Judiciary Act [ley de la Judicatura Federal] *s.* ayudó a establecer un sistema de tribunales; le dio al Tribunal Supremo seis miembros. (p. 294)

Federal Reserve Act [ley de la Reserva Federal] *s.* ley aprobada en 1913 que creó el sistema bancario de la nación e instituyó un sistema monetario flexible. (p. 648)

felony [felonía] *s.* delito grave. (p. 253)

feudalism [feudalismo] *s.* sistema político en que el rey concedía a sus nobles el uso de sus tierras a cambio de su prestación militar y la protección de la gente que vivía en esas tierras. (p. 44)

Fifteenth Amendment [Enmienda Decimoquinta] *s.* aprobada en 1870, esta enmienda a la Constitución de Estados Unidos declaraba que a los ciudadanos no se les podía impedir que votaran "por motivo de raza, color ni condición anterior de esclavitud". (p. 546)

54th Massachusetts Regiment [54.° Regimiento de Massachusetts] *s.* regimientos afroamericano organizado para luchar por la Unión en la guerra civil. (p. 506)

fireside chats [charlas al calor de la lumbre] *s.* nombre dado a las radioemisiones de Franklin Roosevelt en las que explicaba sus medidas. (p. 735)

First Battle of Bull Run [primera batalla de Bull Run] *s.* una batalla de la guerra civil, de 1861, en que el Sur horrorizó al Norte con una victoria. (p. 485)

flapper *s.* jovencita librepensadora que abrazaba las modas y actitudes urbanas nuevas de los años veinte. (p. 714)

foreign policy [política exterior] *s.* relaciones con los gobiernos de otros países. (p. 304)

Fort Sumter [fuerte Sumter] *s.* fuerte federal ubicado en el puerto de Charleston, Carolina del Sur; el ataque sureño al fuerte Sumter marcó el comienzo de la guerra civil. (p. 481)

forty-niner [buscador de ventura del 49] *s.* persona que fue a California en búsqueda de oro, empezando en 1849. (p. 412)

Fourteen Points [Catorce puntos] *s.* los objetivos del presidente Woodrow Wilson para la paz que siguió a la primera guerra mundial. (p. 695)

Fourteenth Amendment [Enmienda Decimocuarta] *s.* enmienda a la Constitución de Estados Unidos, aprobada en 1868, que hizo ciudadanos del país a todas las personas nacidas en Estados Unidos o naturalizadas, incluso a los antiguos esclavos. (p. 535)

First Continental Congress [primer Congreso Continental] *s.* reunión en 1774 de delegados de todas las colonias, excepto Georgia, para defender los derechos coloniales. (p. 171)

Freedmen's Bureau [Agencia de Manumisos] *s.* agencia federal establecida para ayudar a los antiguos esclavos después de la guerra civil. (p. 533)

freedmen's school [escuela para los manumisos] *s.* escuela establecida por la Agencia de Manumisos para educar a los recientes libertos afroamericanos. (p. 541)

Freedom Ride [Viaje por la Libertad] *s.* protesta contra la segregación racial en los autobuses interestatales del Sur. (p. 818)

Freedom Summer [Verano de la Libertad] *s.* en 1964 el Comité de Estudiantes no Violentos organizó una campaña de registro de votantes. (p. 821)

Free Soil Party [Partido del Suelo Libre] *s.* partido político dedicado a parar la expansión de la esclavitud. (p. 459)

French and Indian War [guerra Francesa y Amerindia] *s.* conflicto en Norteamérica, entre 1754 y 1763, que fue parte de una lucha mundial entre Francia y Gran Bretaña y que terminó con la derrota de Francia y el traspaso del Canadá francés a Gran Bretaña. (p. 147)

French Indochina [Indochina Francesa] *s.* colonia francesa que incluía lo que es hoy Vietnam, Laos y Camboya. (p. 835)

frontier [frontera] *s.* región sin o con muy pocos asentamientos ocupada mayormente por amerindios. (p. 557)

Fugitive Slave Act [ley de los Esclavos Fugitivos] *s.* ley de 1850 para ayudar a los dueños de esclavos a recapturar los esclavos fugados. (p. 462)

French Revolution [Revolución francesa] *s.* en 1789 los franceses iniciaron un movimiento por la libertad y la igualdad. (p. 301)

fundamentalist [fundamentalista] *s.* persona que cree en la interpretación textual, o palabra por palabra, de la Biblia. (p. 716)

Fundamental Orders of Connecticut [Órdenes Fundamentales de Connecticut] *s.* conjunto de leyes establecidas en 1639 por una congregación puritana que se había asentado en el valle del río Connecticut y que ampliaban la idea de un gobierno representativo. (p. 95)

G

generator [generador] *s.* máquina que produce corriente eléctrica. (p. 587)

Ghana [Ghana] *s.* imperio del África Occidental entre los siglos VIII y XI d.de C. (p. 39)

G.I. Bill of Rights [Carta de Derechos del Soldado] *s.* aprobada en 1944, esta ley ofrecía ayuda educacional y económica a los veteranos. (p. 779)

Gilded Age [Edad Dorada] *s.* época de fines del siglo XIX de fabulosa riqueza. (p. 596)

"Glorious Revolution" [Revolución gloriosa] *s.* derrocamiento, en 1688, del rey inglés Jacobo II y su substitución por Guillermo y María. (p. 144)

gold standard [patrón oro] *s.* sistema en que el gobierno garantiza cada dólar con una cierta cantidad de oro. (p. 577)

Grange [La Quinta] *s.* creada en 1867 por un grupo de agricultores para tratar de satisfacer las necesidades sociales de las familias granjeras. (p. 577)

Great Awakening [Gran Despertar] *s.* renovación del sentimiento religioso en las colonias norteamericanas durante las décadas de 1730 a 1750. (p. 139)

Great Compromise [Gran Compromiso] *s.* acuerdo en la Convención Constitucional que estableció una legislatura nacional de dos cámaras; en una de estas cámaras, todos los estados tendrían representación igual, en la otra, cada estado tendría representación basada en su población. (p. 232)

Great Depression [gran depresión] *s.* período que duró desde 1929 hasta 1941, en que la economía de Estados Unidos declinó severamente y millones de estadounidenses estaban sin empleo. (p. 731)

Great Migration [Gran Emigración] *s.* movimiento de puritanos que salieron de Inglaterra para establecer asentamientos por todo el mundo, incluyendo a 20,000 que partieron para América (p. 94); el movimiento de afroamericanos entre 1910 y 1920 del Sur hacia las ciudades del Norte. (p. 693)

Great Plains [Grandes Llanuras] *s.* región desde el río Missouri hasta las montañas Rocosas. (p. 557)

Great Society [gran sociedad] *s.* programa iniciado por Lyndon Johnson para ayudan a los pobres, los ancianos y las mujeres y también promovía la educación, prohibía la discriminación racial y protegia el medio ambiente. (p. 822)

greenback [billete verde] *s.* papel moneda emitido por el gobierno federal durante la guerra civil. (p. 509)

gristmill [molino harinero] *s.* molino en que el grano se muele para producir cualquier tipo de harina. (p. 115)

guerrilla [guerrillero] *s.* soldado que debilita al enemigo con asaltos inesperados y ataques relámpagos. (p. 207)

guerrilla warfare [guerra de guerrillas] *s.* ataques inesperados por bandas pequeñas de guerrilleros. (p. 842)

Gulf of Tonkin Resolution [Resolución del golfo de Tonkín] *s.* resolución del Congreso que dio al presidente el poder de usar fuerza militar en Vietnam. (p. 841)

H

hacienda [hacienda] *s.* granja grande o finca. (p. 72)

Harlem Renaissance [renacimiento de Harlem] *s.* florecimiento de la creatividad artística afroamericana durante los años veinte, centrada en la comunidad de Harlem de la ciudad de Nueva York. (p. 720)

Harpers Ferry *s.* arsenal federal en Virginia, capturado en 1859 durante un levantamiento de esclavos. (p. 469)

Hausa [hausa] *s.* gente de África Occidental que después del año 1000 d.C. vivió en lo que ahora es la región norte de Nigeria. (p. 42)

hawk [halcón] *s.* persona que apoya la guerra. (p. 847)

Haymarket affair [asunto Haymarket] *s.* mitín de protesta sindicalista que resultó aproximadamente en un centenar de muertes después de que un desconocido tiró una bomba y la policía abrió fuego contra la multitud. (p. 602)

H-bomb [bomba H] *s.* bomba de hidrógeno. (p. 798)

Hiroshima [Hiroshima] *s.* primera ciudad del Japón contra la cual se lanzó la bomba atómica el 6 de agosto de 1945 durante la segunda guerra mundial. (p. 771)

Ho Chi Minh Trail [ruta de Ho Chi Minh] *s.* red de sendas que usó el Vietcong para mover soldados y provisiones durante la guerra de Vietnam. (p. 838)

Holocaust [Holocausto] *s.* matanza sistemática en Alemania, durante la segunda guerra mundial, de unos seis millones de judíos así como millones de otros grupos étnicos. (p. 765)

homestead [residencia] *s.* tierra para asentarse y construir una casa. (p. 568)

Homestead Act [ley de Residencia] *s.* aprobada en 1862, esta ley ofrecía 160 acres de tierra gratis a cualquiera que acordara ocuparla y trabajarla por cinco años. (p. 574)

House of Burgesses [Cámara de los Burgueses] *s.* creada en 1619, la primera asamblea representativa de las colonias norteamericanas. (p. 88)

Hudson River School [Escuela del río Hudson] *s.* grupo de artistas que vivían en el valle del río Hudson del estado de New York. (p. 430)

Hull House [Casa Hull] *s.* fundada en 1889, fue modelo para otras casas de acogida de la época. (p. 613)

Hundred Days [Cien Días] *s.* durante sus cien primeros días, del 9 de marzo a mediados de junio de 1933, Franklin Roosevelt mandó al Congreso muchos proyectos de ley nuevos. (p. 735)

hygiene [higiene] *s.* condiciones y prácticas que fomentan la buena salud. (p. 490)

I

immigrant [inmigrante] *s.* persona que se establece en un país nuevo. (p. 423)

Immigration Reform and Control Act of 1986 [ley de Reforma y Control de la Inmigración] *s.* ley pasada para reforzar las leyes de inmigración y las medidas para su cumplimiento. (p. 884)

immunity [inmunidad] *s.* protección legal. (p. 262)

impeachment [imputación] *s.* proceso de acusar a un funcionario público de un delito en el desempeño de sus funciones. (p. 249)

imperialism [imperialismo] *s.* política por la cual las naciones más poderosas extienden su control económico, político o militar sobre territorios o naciones más débiles. (p. 659)

impressment [secuestro] *s.* el acto de capturar personas a la fuerza. (p. 327)

inaugurate [investir] *v.* conferir un cargo oficial en una ceremonia formal. (p. 293)

income tax [impuesto sobre la renta] *s.* impuesto sobre los ingresos. (p. 509)

indentured servant [siervo escriturado] *s.* persona que vendía su trabajo a cambio de pasaje a Norteamérica. (p. 88)

Indian Removal Act [ley del Traslado de los Indígenas] *s.* esta ley de 1830 requería que el gobierno negociara tratados para el traslado de los amerindios al oeste. (p. 376)

Indian Territory [territorio Indio] *s.* lo que hoy son Oklahoma y partes de Kansas y Nebraska a la cual se trasladó a los amerindios bajo la ley del Traslado de los Indígenas de 1830. (p. 376)

indictment [acusación] *s.* declaración escrita dictada por un jurado de acusación que inculpa a una persona de un delito. (p. 250)

indigo [añil] *s.* planta que cultivaban las colonias sureñas de la cual se obtiene un colorante azul oscuro. (p. 121)

individual right [derecho individual] *s.* libertad y privilegio personal que garantiza a los ciudadanos estadounidenses la Carta de Derechos. (p. 247)

Industrial Revolution [revolución industrial] *s.* en la Inglaterra de fines del siglo XVIII, las maquinarias de fábrica empezaron a reemplazar las herramientas manuales, y la producción de bienes manufacturados reemplazó la agricultura como el principal modo de trabajo. (p. 341)

inferior court [tribunal inferior] *s.* corte con autoridad menor que la del Tribunal Supremo. (p. 260)

inflation [inflación] *s.* subida en el precio de los productos y los servicios y disminución del valor del dinero. (p. 386)

information revolution [revolución de la información] *s.* proceso tecnológico que ha cambiado radicalmente el volumen y el modo de transmitir información. (p. 879)

initiative [iniciativa] *s.* el procedimiento que permite a los votantes proponer una ley directamente. (p. 640)

inoperative [inoperante] *adj.* que no está vigente. (p. 274)

installment buy [comprar a cuotas] *v.* comprar algo haciendo pequeños pagos mensuales. (p. 712)

insurrection [insurrección] *s.* levantamiento abierto contra un gobierno. (p. 271)

interchangeable part [parte intercambiable] *s.* parte que es exactamente igual a otra parte. (p. 343)

Internet [Internet] *s.* interconexión mundial de redes informáticas. (p. 878)

Intolerable Acts [leyes Intolerables] *s.* serie de leyes aprobadas en 1774 por el Parlamento para castigar a Massachusetts por el Motín del Té de Boston. (p. 170)

Iran-Contra affair [asunto Irán-Contra] *s.* en 1986 el gobierno de Estados Unidos vendió armas a Irán a cambio de ayuda para liberar a los rehenes estadounidenses en el Oriente Medio y el dinero de la venta fue a los rebeldes Contra de El Salvador. (p. 874)

Iran hostage crisis [crisis de los rehenes de Irán] *s.* el 4 de noviembre de 1979 un grupo de iraníes invadieron la embajada de Estados Unidos en Teherán, capital de Irán, y tomaron de rehenes a 52 estadounidenses. (p. 865)

ironclad [acorazado] *s.* buque de guerra cubierto de hierro. (p. 491)

Iroquois League [Liga Iroquesa] *s.* alianza del siglo XVI entre los pueblos amerindios cayuga, mohawk, oneida, onondaga y seneca, que vivían en la región oriental de los Grandes lagos. (p. 37)

irrigation [irrigación] *s.* práctica de llevar agua a los cultivos. (p. 29)

Islam [islam] *s.* religión fundada por el profeta Mahoma en el siglo VII, que enseña que hay un solo Dios: Alá. (p. 41)

island hopping [brincar de isla a isla] *v.* estrategia usada durante la segunda guerra mundial según la cual los aliados invadían islas débilmente defendidas por los japoneses para así lanzar nuevos ataques. (p. 770)

isolationist [aislacionista] *s.* persona que creía que Estados Unidos debía mantenerse apartado de los asuntos de las otras naciones, excepto en defensa propia. (p. 711)

J

Jacksonian Democracy [democracia jacksoniana] *s.* idea de extender el poder político a toda la gente asegurando de ese modo el gobierno de la mayoría. (p. 370)

Jamestown. *s.* primer asentamiento inglés permanente en Norteamérica. (p. 87)

Jay's Treaty [Tratado de Jay] *s.* el acuerdo que puso fin a la disputa sobre los derechos marítimos estadounidenses durante la Revolución francesa. (p. 302)

jazz [jazz] *s.* tipo nuevo de música en los años veinte que capturó el despreocupado espíritu de la época. (p. 717)

Jim Crow [ley Jim Crow] *s.* ley que imponía la separación entre la gente blanca y la de piel negra en los lugares públicos del Sur. (p. 621)

joint-stock company [sociedad por acciones] *s.* empresa en que los inversionistas colocan su dinero en un fondo común con la intención de sacar ganancias. (p. 86)

judicial review [revisión judicial] *s.* principio de que el Tribunal Supremo tiene la última palabra en la interpretación de la Constitución. (p. 317)

Judiciary Act of 1801 [ley Judicial de 1801] *s.* ley que aumentó el número de jueces federales, permitiéndole al presidente John Adams cubrir la mayoría de los puestos nuevos con federalistas. (p. 316)

K

Kansas-Nebraska Act [ley de Kansas-Nebraska] *s.* ley de 1854 que estableció los territorios de Kansas y Nebraska y otorgó a sus habitantes el derecho a decidir si querían o no permitir la esclavitud. (p. 464)

kayak [kayak] *s.* embarcación pequeña hecha de piel de animales. (p. 33)

Kellogg-Briand Pact [Pacto Kellogg-Briand] *s.* en 1928 firmaron este pacto muchas naciones que prometieron no declararse en guerra entre ellas excepto en defensa propia. (p. 711)

King Cotton [rey Algodón] *s.* al algodón se lo llamaba rey porque era importante en el mercado mundial y el Sur cultivaba la mayor parte del algodón que usaban las fábricas textiles de Europa. (p. 484)

King Philip's War [guerra del rey Felipe] *s.* guerra entre las colonias puritanas y los amerindios que se libró entre 1675 y 1676. (p. 96)

Knights of Labor [Caballeros del Trabajo] *s.* organización de obreros de oficios diferentes formada después de la guerra civil. (p. 601)

Korean War [guerra de Corea] *s.* conflicto entre Corea del Norte y Corea del Sur que duró desde 1950 a 1953; Estados Unidos, junto con otros países de las Naciones Unidas, luchó del lado de los coreanos del Sur y China luchó del lado de los coreanos del Norte. (p. 796)

Ku Klux Klan *s.* grupo constituido en 1866 que quería restaurar el control del Sur a los demócratas y mantener sumisos a los antiguos esclavos. (p. 544)

L

labor union [sindicato laboral] *s.* obreros que se unen para tratar de conseguir mejores condiciones de trabajo. (p. 434)

laissez faire [dejad hacer] *s.* teoría que declaraba que el comercio no sujeto a regulaciones actuaría de manera beneficiosa a la nación. (p. 710)

Land Ordinance of 1785 [Ordenanza de Tierras de 1785] *s.* ley que establecía un plan para la agrimensura y venta de las tierras públicas al oeste de los montes Apalaches. (p. 223)

land speculator [especulador en tierras] *s.* persona que compra grandes extensiones de terreno a precio bajo y luego vende secciones pequeñas a precios altos. (p. 394)

League of Nations [Sociedad de Naciones] *s.* organización establecida después de la primera guerra mundial para resolver conflictos internacionales. (p. 695)

leisure *s.* tiempo libre. (p. 627)

Lend-Lease [ley de Préstamo y Arriendo] *s.* ley que le permitió a Estados Unidos mandar a las naciones que luchaban contra el Eje armas y suministros sin requerir pago inmediato. (p. 760)

Lewis and Clark expedition [expedición de Lewis y Clark] *s.* grupo dirigido por Meriwether Lewis y William Clark que exploró las tierras de la Compra de Luisiana empezando en 1803. (p. 320)

Lexington and Concord *s.* escenarios, en Massachusetts, de las primeras batallas de la Revolución norteamericana. (p. 173)

liberal [liberal] *s.* persona que favorece acción por parte del gobierno para lograr la reforma social y económica. (p. 749)

limited government [gobierno limitado] *s.* principio que requiere que todos los ciudadanos estadounidenses, incluso los líderes gubernamentales, obedezcan la ley. (p. 247)

lode [veta] *s.* mineral enterrado entre capas de roca. (p. 558)

Lone Star Republic [República de la Estrella Solitaria] *s.* apodo de la República de Texas, que se le dio en 1836. (p. 405)

long drive [largo arreo] *s.* arreo de ganado al ferrocarril. (p. 560)

Lost Generation [generación perdida] *s.* la generación de los años veinte, después de la primera guerra mundial cuando hombres y mujeres veían pocas esperanzas para el futuro. (p. 720)

Louisiana Purchase [Compra de Luisiana] *s.* en 1803, la compra a Francia del Territorio de Luisiana. (p. 319)

Lowell mills [fábricas de Lowell] *s.* fábricas textiles ubicadas en el pueblo manufacturero de Lowell, Massachusetts, fundado en 1826. (p. 342)

Loyalist [realista]. *s.* colono norteamericano que apoyaba a los británicos durante la Revolución norteamericana. (p. 173)

M

mail-order catalog [catálogo de venta por correo] *s.* publicación que contenía fotografías y la descripción de los artículos para que la gente encargara por correo. (p. 627)

Magna Carta [Carta Magna] *s.* "Gran Cédula Real"; documento que garantizaba los derechos políticos básicos en Inglaterra, aprobada por el rey Juan en el año 1215. (p. 141)

Mali [Malí] *s.* imperio del África Occidental desde el siglo XIII hasta el siglo XVI y que se enriqueció gracias al comercio. (p. 41)

Manhattan Project [Proyecto Manhattan] *s.* programa ultrasecreto establecido en 1942 para construir una bomba atómica. (p. 771)

manifest destiny [destino manifiesto] *s.* creencia de que era el destino de Estados Unidos extenderse por todo el continente, desde al océano Atlántico al océano Pacífico. (p. 407)

manor system [sistema señorial] *s.* sistema en que los nobles dividían sus tierras en propiedades, cultivadas mayormente por los siervos, quienes recibían la protección del noble. (p. 45)

Marbury v. Madison [Marbury contra Madison] *s.* caso de 1803 en que el Tribunal Supremo dictaminó que tenía el poder de invalidar leyes declarándolas inconstitucionales. (p. 317)

March on Washington [Marcha a Washington] *s.* enorme manifestación por los derechos civiles en Washington, D.C. en 1963. (p. 820)

Marshall Plan [Plan Marshall] *s.* aprobado en 1948, autorizó a Estados Unidos a dar más de trece mil millones de dólares para ayudar a la recuperación de las naciones de Europa después de la segunda guerra mundial. (p. 779)

mass culture [cultura de masas] *s.* cultura común compartida por grandes números de personas. (p. 626)

mass media [medios de comunicación de masas] *s.* comunicaciones que alcanzan a un público muy grande. (p. 718)

matrilineal [por línea materna] *adj.* sociedad en que la ascendencia se determina mediante la madre. (p. 36)

Mayflower Compact [Pacto del Mayflower] *s.* acuerdo firmado por los hombres que viajaron a América en el Mayflower, que requería leyes para el bien de la colonia y establecía el concepto de autogobierno. (p. 93)

melting pot [crisol de culturas] *s.* lugar donde las culturas se amalgaman. (p. 616)

mercantilism [mercantilismo] *s.* sistema económico en que las naciones tratan de aumentar su riqueza y poder obteniendo oro y plata y estableciendo una balanza comercial favorable. (p. 61)

mercenary [mercenario] *s.* soldado profesional contratado para luchar por un país extranjero. (p. 195)

Mexican Cession [Cesión mexicana] *s.* extensa región cedida por México después de la guerra con México; incluía los actuales estados de California, Nevada, Utah, la mayor parte de Arizona y partes de Nuevo México, Colorado y Wyoming. (p. 411)

Mexicano [mexicano] *s.* persona de ascendencia española cuyos antepasados habían venido de México y se habían establecido en el sudoeste. (p. 570)

middle passage [travesía intermedia] *s.* parte intermedia de la ruta del comercio triangular (el viaje de África a las Américas) que traía africanos capturados a la esclavitud. (p. 78)

migrate [migrar] *v.* mudarse de un lugar a otro. (p. 27)

militarism [militarismo] *s.* creencia de que una nación necesita una fuerza militar grande. (p. 679)

militia [milicia] *s.* fuerza de civiles armados comprometidos a defender su comunidad durante la Revolución norteamericana. (p. 170); fuerza militar de emergencia, que no es parte del ejército profesional. (p. 254)

minié ball [bala minié] *s.* bala con base hueca. (p. 491)

Minuteman [minutero] *s.* miembro de la milicia colonial entrenado para responder "con un minuto de aviso". (p. 170)

misdemeanor [fechoría] *s.* violación de la ley. (p. 259)

mission [misión] *s.* asentamiento creado por la Iglesia con el propósito de convertir a los amerindios al cristianismo. (p. 72)

missionary [misionero] *s.* persona enviada por la Iglesia para predicar, enseñar y convertir a los indígenas al cristianismo. (p. 61)

Missouri Compromise [Acuerdo de Missouri] *s.* serie de leyes aprobadas en 1820 para mantener el equilibrio del poder político entre los estados esclavistas y los libres. (p. 358)

monopoly [monopolio] *s.* compañía que elimina a sus competidores y controla una industria. (p. 595)

Monroe Doctrine [Doctrina Monroe] *s.* política de oposición estadounidense a cualquier interferencia europea en el hemisferio occidental, proclamada por el presidente Monroe en 1823. (p. 359)

Montgomery bus boycott [Boicoteo al transporte público de Montgomery] *s.* en 1955 los afroamericanos boicotearon el transporte público de Montgomery, Alabama, en respuesta al arresto de Rosa Parks, quien se había negado a dejar su asiento a una persona blanca. (p. 815)

Mormon [mormón] *s.* miembro de una iglesia fundada por Joseph Smith en 1830. (p. 397)

Mound Builder [constructor de túmulos] *s.* amerindio primitivo que construía grandes estructuras de tierra. (p. 31)

mountain man [hombre de las montañas] *s.* trampero o explorador que abrió el oeste hallando sendas a través de las montañas Rocosas. (p. 393)

muckraker [revuelve estiércol] *s.* hacia comienzos del siglo XX, periodista que exponía la corrupción dentro de la sociedad estadounidense. (p. 640)

Muslim [musulmán] *s.* adherente del islam. (p. 41)

N

NAACP *s.* constituida en 1909, la Asociación Nacional para el Progreso de la Gente de Color. (pp. 622, 715)

NAFTA [Tratado Norteamericano de Libre Comercio] *s.* aprobado en 1993, el Tratado Norteamericano de Libre Comercio o zona de libre comercio entre Estados Unidos, México y Canadá. (p. 875)

napalm [napalm] *s.* gasolina gelatinosa que arde violentamente. (p. 843)

nationalism [nacionalismo] *s.* sentido de orgullo, lealtad y protección hacia el país de uno. (p. 354)

nativist [nativista] *s.* estadounidense nativo que quería eliminar toda influencia extranjera. (p. 428)

NATO [OTAN] *s.* la Organización del Tratado del Atlántico Norte es una alianza militar constituida en 1949 por diez países de la Europa occidental, Estados Unidos y Canadá. (p. 793)

natural-born citizen [ciudadano nato] *s.* ciudadano nacido en Estados Unidos o en un estado asociado o dependencia de Estados Unidos, o a padres que son ciudadanos estadounidenses que viven fuera del país. (p. 257)

naturalization [naturalización] *s.* manera de darle ciudadanía completa a una persona nacida en otro país. (pp. 253, 270)

Navigation Acts [Actas de Navegación] *s.* serie de leyes aprobadas por el Parlamento, empezando en 1651, para asegurarse Inglaterra de que el comercio de sus colonias le rindiera ganancias económicas. (p. 112)

navigator [oficial de derrota] *s.* persona que planea el rumbo de un barco mientras está en el mar. (p. 49)

Nazi Party [Partido Nazi] *s.* partido Alemán Nacionalsocialista de los Trabajadores, llegó al poder bajo Adolfo Hitler en 1930. (p. 756)

NCAI *s.* Congreso Nacional de Amerindios, fundado en 1944 para promover el "bienestar común" de los amerindios. (p. 825)

neutral [neutral] *adj.* que no apoya ni a un país ni al otro. (p. 302)

neutrality [neutralidad] *s.* rechazo de la idea de apoyar a un país u otro durante una guerra. (p. 682)

New Deal [Nuevo Trato] *s.* programas de Franklin Roosevelt para luchar contra la depresión. (p. 735)

New France [Nueva Francia] *s.* puesto para el comercio de las pieles establecido en 1608 que se convirtió en el primer asentamiento francés permanente en Norteamérica. (p. 70)

new immigrant [inmigrante nuevo] *s.* persona del sur y el este de Europa que entró a Estados Unidos después de 1900. (p. 614)

New Jersey Plan [Plan de Nueva Jersey] *s.* plan de gobierno propuesto en 1787 en la Convención Constitucional que proponía una cámara legislativa única en que cada estado tendría un solo voto. (p. 231)

Nineteenth Amendment [Enmienda Decimonovena] *s.* enmienda a la Constitución de Estados Unidos ratificada en 1920 que dio a las mujeres el derecho absoluto a votar. (p. 653)

Nisei [nisei] *s.* estadounidense japonés nacido en Estados Unidos. (p. 775)

Northwest Ordinance [Ordenanza del Noroeste] *s.* describe cómo se iba a gobernar el territorio del Noroeste y establecía las condiciones para el asentamiento, así como los derechos de los colonos. (p. 223)

Northwest Territory [territorio del Noroeste] *s.* territorio organizado por la Ordenanza de Tierras de 1785, que incluía tierras que formaron los estados de Ohio, Indiana, Michigan, Illinois, Wisconsin y parte de Minnesota. (p. 223)

NOW *s.* fundada en 1966, la Organización Nacional de la Mujer adelantó una campaña para que las mujeres consiguieran empleos con paga igual a la de los hombres. (p. 826)

Nuremberg Trials [Juicios de Nuremberg] *s.* procesos judiciales que tuvieron lugar en Nuremberg, Alemania, después de la segunda guerra mundial, en que se enjuició a líderes nazis por sus crímenes de guerra. (p. 780)

O

Open Door Policy [política de puertas abiertas] *s.* en 1899 Estados Unidos instó a las naciones que tenían intereses en China a que siguieran una política según la cual ningún país controlaría el comercio con China. (p. 669)

Oregon Trail [Camino de Oregón] *s.* camino hacia el oeste que iba de Independence, Missouri, al territorio de Oregón. (p. 396)

overseer [capataz] *s.* persona contratada por el dueño de una plantación para vigilar a los esclavos y dirigir su trabajo. (p. 122)

P

pacifist [pacifista] *s.* persona moralmente opuesta a la guerra. (p. 209)

Palmer raids [allanamientos de Palmer] *s.* en 1920 agentes federales y la policía allanaron los hogares de personas que se sospechaba que eran radicales. (p. 697)

Panama Canal [canal de Panamá] *s.* atajo a través de Panamá que conecta los océanos Atlántico y Pacífico. (p. 670)

Panic of 1837 [pánico de 1837] *s.* crisis financiera con clausura de bancos y colapso del sistema crediticio, que resultó en muchas quiebras y serio desempleo. (p. 386)

Panic of 1873 [pánico de 1873) *s.* crisis financiera en que los bancos se clausuraron y la bolsa de comercio se derrumbó. (p. 547)

Parliament [Parlamento] *s.* el cuerpo legislativo principal de Inglaterra. (p. 142)

patent [patente] *s.* documento del gobierno que otorga a un inventor el derecho exclusivo a hacer o vender su invención durante un determinado número de años. (p. 586)

Patriot [patriota] *s.* colono norteamericano que durante la Revolución norteamericana estaba a favor de los rebeldes. (p. 173)

patroon [patrono] *s.* persona que traía 50 colonos a Nueva Holanda y a cambio recibía una gran concesión de tierras y otros privilegios especiales. (p. 101)

Pearl Harbor *s.* base naval en Hawai atacada de sorpresa por Japón el 7 de diciembre de 1941. (p. 760)

Persian Gulf War [guerra del Golfo Pérsico] *s.* en 1991 Estados Unidos y las Naciones Unidas echaron a Iraq de Kuwait, país que los iraquíes habían invadido en 1990. (p. 875)

petroleum [petróleo] *s.* líquido aceitoso e inflamable. (p. 585)

philanthropist [filántropo] *s.* persona que da grandes sumas de dinero a las organizaciones benéficas. (p. 596)

Pickett's Charge [carga de Pickett] *s.* en 1863 el general George Pickett dirigió una carga frontal contra las fuerzas de la Unión durante la batalla de la guerra civil en Gettysburg; el ataque fracasó. (p. 513)

piedmont [tierras bajas] *s.* ancha extensión de tierra llana al pie de una cadena de montañas. (p. 126)

Pilgrim [peregrino] *s.* miembros del grupo que rechazó la Iglesia de Inglaterra, viajó a América y fundó la colonia de Plymouth en 1620. (p. 92)

Pinckney's Treaty [Tratado de Pinckney] *s.* tratado de 1795 con España que otorgaba a Estados Unidos el uso del río Mississippi y el derecho a depositar bienes en Nueva Orleans; creó el paralelo 31 como el límite sur de Estados Unidos. (p. 302)

plantation [plantación] *s.* finca grande para cultivos comerciales. (p. 73)

platform [plataforma] *s.* declaración de creencias. (p. 471)

Platt Amendment [Enmienda Platt] *s.* resultado de la guerra entre Estados Unidos y España, dio a Estados Unidos el derecho a intervenir en los asuntos de Cuba cuando existiera amenaza a la "vida, propiedad y libertad individual". (p. 666)

Plessy v. Ferguson [Plessy contra Ferguson] *s.* caso de 1896 en que el Tribunal Supremo dictaminó que la separación de las razas en las instalaciones públicas era legal. (p. 621)

political machine [maquinaria política] *s.* organización que logra votos suficientes para controlar un gobierno local. (p. 613)

political party [partido político] *s.* grupo de personas que trata de promover sus ideas y ejercer su influencia sobre el gobierno, y que también apoya a candidatos a los cargos públicos. (p. 304)

Pontiac's Rebellion [rebelión de Pontiac] *s.* rebelión de 1763 contra los fuertes británicos y los colonos norteamericanos, dirigida en parte por el líder ottawa Pontiac, en reacción a los colonos que demandaban las tierras de los amerindios, así como a la severidad con que trataban a éstos los soldados británicos. (p. 151)

popular culture [cultura popular] *s.* cosas como la música, la moda y las películas que son populares dentro de una cantidad grande de personas. (p. 718)

popular sovereignty [soberanía popular] *s.* gobierno en que gobierna la gente (p. 244); sistema en que los ciudadanos votan para decidir un asunto. (p. 463)

Populist Party [Partido Populista] *s.* también conocido como el Partido del Pueblo y constituido en 1890, este grupo quería una política que aumentara el precio de lo que cultivaban los granjeros. (p. 577)

prejudice [prejuicio] *s.* opinión negativa no basada en los hechos. (p. 427)

printing press [imprenta] *s.* máquina inventada por Johannes Gutenberg alrededor de 1455. (p. 47)

privateer [corsario] *s.* barco de propiedad particular autorizado por un gobierno que está en guerra a atacar los barcos de la marina mercante enemiga. (p. 204)

Proclamation of 1763 [Proclama de 1763] *s.* orden por la cual Gran Bretaña les prohibía a los colonos norteamericanos establecer asentamientos al oeste de los montes Apalaches. (p. 151)

profit [ganancia] *s.* cantidad de dinero que saca una empresa depués de descontar a las entradas los gastos empresariales. (p. 48)

progressivism [progresismo] *s.* movimiento reformista de principios del siglo XX que buscaba devolver el control del gobierno al pueblo, restablecer oportunidades económicas y corregir las injusticias de la vida estadounidense. (p. 639)

prohibition [prohibición] *s.* proscripción de la fábrica, venta y posesión de bebidas alcohólicas. (p. 715)

propaganda [propaganda] *s.* opinión expresada con el propósito de influir en las acciones de otras personas. (p. 692)

proprietary colony [colonia de proprietario] *s.* colonia de un único dueño. (p. 101)

pro tempore *adv.* latín para "por el momento". (p. 250)

public works project [proyecto de obras públicas] *s.* proyecto patrocinado por el gobierno para construir recursos públicos, como caminos y diques. (p. 732)

Pullman Strike [huelga de Pullman] *s.* huelga nacional de ferrocarriles que se extendió por toda la industria ferroviaria en 1894. (p. 603)

Puritan [puritano] *s.* miembro de un grupo de Inglaterra que se asentó en la Colonia de la bahía de Massachusetts en 1630 y trató de reformar las prácticas de la Iglesia de Inglaterra. (p. 94)

push-pull factor *s.* factor que empuja a la gente a irse de su tierra natal y las atrae a un lugar nuevo. (p. 424)

Q

Quaker [cuáquero] *s.* persona que creía que todas las personas debieran vivir en paz y armonía; aceptaba a religiones y grupos étnicos diferentes. (p. 101)

quarter [acuartelar] *v.* dar alojamiento. (p. 267)

Quartering Act [ley de Acuartelamiento] *s.* ley aprobada por el Parlamento en 1765 que requería que las colonias alojaran a los soldados británicos y los aprovisionaran. (p. 160)

quorum [quórum] *s.* número mínimo de miembros que deben estar presentes para que pueda empezar a deliberar oficialmente una asamblea. (p. 251)

R

racial discrimination [discriminación racial] *s.* tratamiento diferente por motivos de raza. (p. 620)

racism [racismo] *s.* creencia de que alguna gente es inferior a causa de su raza. (p. 79)

radical [radical] *s.* persona que adopta posiciones políticas extremas. (p. 313)

Radical Republican [republicano radical] *s.* diputado que después de la guerra civil estaba a favor de usar el gobierno para crear un nuevo orden en el Sur y dar la ciudadanía total y derecho al voto a los afroamericanos. (p. 533)

ragtime *s.* mezcla de canciones afroamericanas y formas musicales europeas. (p. 629)

ratification [ratificación] *s.* aprobación oficial. (p. 264)

ration [racionar] *v.* distribuir una cantidad fija de cierto artículo. (p. 773)

recall [destituir] *v.* votar para sacar a un funcionario de su cargo. (p. 640)

Reconstruction [Reconstrucción] *s.* proceso que usó el gobierno de Estados Unidos para readmitir a los estados confederados a la Unión después de la guerra civil. (p. 533)

Red Scare [Terror Rojo] *s.* entre 1919 y 1920, ola de pánico sobre una posible revolución comunista. (p. 697)

referendum [referéndum] *s.* cuando una ley que se ha propuesto se somete al voto del pueblo. (p. 640)

Reformation [Reforma] *s.* movimiento religioso del siglo XVI para corregir los problemas de la Iglesia Católica Romana. (p. 47)

Renaissance [Renacimiento] *s.* período de la historia europea que duró desde el siglo XIV hasta comienzos del XVII y que acrecentó el interés por el arte y el saber. (p. 46)

rendezvous [encuentro] *s.* reunión. (p. 197)

reprieve [indultar] *n.* cancelación de un castigo. (p. 259)

republic [república] *s.* gobierno en que el pueblo elige representantes para que lo gobiernen. (p. 222)

republicanism [republicanismo] *s.* creencia de que el gobierno se debe basar en el consentimiento del pueblo; el pueblo ejercita su poder votando por representantes políticos. (pp. 214, 245)

Republican Party [Partido Republicano] *s.* el partido político constituído en 1854 por los que se oponían a la esclavitud en los territorios. (p. 466)

reservation [reserva] *s.* tierras destinadas por el gobierno de Estados Unidos para las tribus amerindias. (p. 562)

revenue [rentas públicas] *s.* entradas que recibe un gobierno para cubrir sus gastos. (pp. 160, 252)

revenue sharing [reparto de las rentas públicas] *s.* la distribución de dinero federal a los estados y gobiernos locales con pocas o no restricciones en la manera cómo se gasta. (p. 856)

revival [renacimiento religioso] *s.* reunión diseñada para revivir la fe religiosa. (p. 433)

rifle [rifle] *s.* arma de barril estriado que hace que la bala vaya girando por el aire. (p. 491)

robber baron [capitalista inescrupuloso] *s.* líder industrial que se hizo acaudalado usando medios deshonestos. (p. 594)

rock 'n' roll *s.* tipo de música popular caracterizada por ritmos pesados y melodías simples que se desarrolló del rhythm and blues en los años cincuenta. (p. 803)

romanticism [romanticismo] *s.* movimiento artístico europeo que acentuaba al individuo, la imaginación, la creatividad y la emoción. (p. 429)

Roosevelt Corollary [Corolario de Roosevelt] *s.* la adición en 1904 a la Doctrina Monroe permitiendo a los Estados Unidos a actuar de "policía" en Latinoamérica. (p. 672)

Rosie the Riveter [Rosita la Remachadora] *s.* imagen de una mujer fuerte trabajando duro en una fábrica de armas durante la segunda guerra mundial. (p. 773)

Rough Rider *s.* miembro del Primer Regimiento Estadounidense de Voluntarios de Caballería que organizó Theodore Roosevelt y que luchó en la guerra entre España y Estados Unidos. (p. 665)

royal colony [colonia real] *s.* colonia regida por gobernadores nombrados por un rey. (p. 103)

S

salutary neglect [indiferencia beneficiosa] *s.* política de no interferir Inglaterra en los asuntos de sus colonias norteamericanas durante la primera mitad del siglo XVIII. (p. 144)

SALT *s.* Tratado sobre Limitación de Armas Estratégicas firmado en 1972 entre Estados Unidos y la Unión Soviética; limitaba las armas nucleares. (p. 858)

Sand Creek Massacre [masacre de Sand Creek] *s.* un ataque en 1864, en el que más de 150 hombres, mujeres y niños del pueblo cheyene murieron a manos de la milicia de Colorado. (p. 564)

Santa Fe Trail [Camino de Santa Fe] *s.* camino hacia el oeste que iba de Missouri a Santa Fe, New Mexico. (p. 395)

SCLC *s.* Conferencia de Líderes Cristianos del Sur, grupo que coordinó por todo el sur las protestas para promover los derechos civiles. (p. 815)

secede [separarse] *v.* retirarse. (p. 473)

secession [secesión] *s.* separarse como parte de Estados Unidos. (p. 383)

Second Battle of the Marne [segunda batalla del Marne] *s.* en 1918 esta batalla de la primera guerra mundial marcó el cambio decisivo en el curso de la guerra; las tropas aliadas junto con las estadounidenses detuvieron el avance alemán hacia el interior de Francia. (p. 689)

Second Continental Congress [segundo Congreso Continental] *s.* organismo de gobierno cuyos delegados acordaron en mayo de 1775 organizar el Ejército Continental y aprobar la Declaración de Independencia. (p. 177)

Second Great Awakening [segundo Gran Despertar] *s.* renovación de la fe religiosa durante fines del siglo XVIII y comienzos del XIX. (p. 433)

Second New Deal [segundo Nuevo Trato] *s.* conjunto de programas que se pasaron en 1935 para luchar contra la depresión. (p. 736)

sectionalism [seccionalismo] *s.* colocar los intereses de la región propia por encima de los de la nación como unidad. (p. 357)

Securities and Exchange Commission *s.* [Comisión de Valores y Bolsa] *s.* agencia que vigila la bolsa de valores y hace que las empresas sigan procedimientos honestos en la venta de acciones. (p. 748)

Sedition Act [ley de Sedición] *s.* ley aprobada en 1918 que hacía ilegal la crítica de la guerra. Imponía fuertes multas y largos períodos de encarcelamiento para los que participaran en actividades contra la guerra. (p. 692)

segregation [segregación] *s.* separación. (p. 621)

Seneca Falls Convention [convención de Seneca Falls] *s.* convención sobre los derechos de la mujer llevada a cabo en Seneca Falls, New York, en 1848. (p. 444)

separation of powers [separación de poderes] *s.* división de las funciones básicas del gobierno en tres ramas. (p. 246)

service economy [economía de servicios] *s.* economía en que la mayoría de los empleos suministran servicios en lugar de producir mercancías. (p. 879)

servitude [servitud] *s.* práctica de pertenecer a un dueño o amo. (p. 270)

Seven Days' Battles [batallas de los Siete Días] *s.* batalla de 1862 de la guerra civil en que la Confederación forzó a la Unión a retroceder después de un intento fracasado de capturar la capital sureña de Richmond. (p. 496)

Seventeenth Amendment [Enmienda Decimoséptima] *s.* enmienda a la Constitución estadounidense, ratificada en 1913, que autorizaba la elección directa de los senadores estadounidenses por el electorado de cada estado. (p. 648)

sharecropping [aparcería] *s.* sistema en que los terratenientes daban a los agricultores tierra, semilla y herramientas a cambio de parte de la cosecha. (p. 543)

Shays's Rebellion [revuelta de Shay] *s.* sublevación de granjeros adeudados de Massachusetts en 1787. (p. 225)

Sherman Antitrust Act [ley Antitrust Sherman] *s.* ley aprobada en 1890 que declaró ilegal que las corporaciones obtuvieran control de las industrias formando trusts. (p. 641)

Siege of Vicksburg [sitio de Vicksburg] *s.* victoria unionista de 1863, durante la guerra civil que le permitió a la Unión controlar todo el río Misisipí. (p. 516)

sit-down strike *s.* [huelga de brazos caídos] huelga en que los obreros permanecen dentro de la fábrica pero se niegan a trabajar. (p. 743)

Sixteenth Amendment [Enmienda Decimosexta] *s.* enmienda a la Constitución de Estados Unidos, ratificada en 1913, que dio al Congreso el poder de crear impuestos a las rentas. (p. 647)

slash-and-burn agriculture [agricultura de corte y quema] *s.* método agrícola en que la gente preparaba los campos cortando y quemando árboles y pastos, cuyas cenizas fertilizaban la tierra. (p. 37)

slave code [código de los esclavos] *s.* ley pasada para reglamentar el tratamiento de los esclavos. (p. 79)

slavery [esclavitud] *s.* sistema de servidumbre humana involuntaria. (p. 76)

slum [barrio bajo] *s.* barrio de casas abarrotadas de gente y peligrosas. (p. 612)

smuggle [contrabandear] *v.* importar o exportar mercancías ilegalmente. (p. 112)

SNCC *s.* organizado en 1960, el Comité Coordinador No Violento de Estudiantes se creó para dar a los jóvenes un papel más importante en el movimiento por los derechos civiles. (p. 817)

social gospel [evangelio social] *s.* movimiento cuyo objetivo era mejorar la vida de los pobres. (p. 612)

socialism [socialismo] *s.* sistema económico en que todos los miembros de una sociedad son propietarios por igual de todas las empresas; los miembros comparten el trabajo y las ganancias (p. 602)

Social Security Act [ley de Seguridad Social] *s.* según esta ley aprobada en 1935, los empleados y los empresarios hacían pagos a un fondo especial, del cual podían recibir una jubilación después de retirarse. (p. 736)

sodbuster *s.* granjero de la frontera. (p. 575)

Songhai *s.* imperio de África Occidental que sucedió a Malí y controló el comercio desde el siglo XV hasta fines del XVI. (p. 42)

Sons of Liberty [Hijos de la Libertad] *s.* grupo de colonos que formaron una sociedad secreta para oponerse a las políticas británicas en los tiempos de la revolución norteamericana. (p. 161)

space race [carrera para conquistar el espacio] *s.* empezando en 1957, la Unión Soviética y Estados Unidos comenzaron a competir en la exploración del espacio. (p. 799)

Spanish-American War *s.* guerra de 1898 que comenzó cuando Estados Unidos demandó que España le concediera la independencia a Cuba. (p. 664)

Spanish Armada [Armada española] *s.* flota de buques enviada en 1588 por el rey español Felipe Segundo para invadir a Inglaterra y restaurar allí el catolicismo romano. (p. 69)

speculation [especulación] *s.* comprar y vender acciones con la esperanza de obtener una ganancia rápida. (p. 730)

sphere of influence [esfera de influencia] *s.* región donde las naciones extranjeras demandan derechos especiales y privilegios económicos. (p. 669)

spiritual [canción espiritual] *s.* canción folklórica religiosa. (p. 351)

spoils system [sistema de prebendas] *s.* práctica de otorgar los candidatos elegidos empleos gubernamentales a los simpatizantes políticos. (p. 373)

Stamp Act [ley del Timbre] *s.* ley aprobada por el Parlamento en 1765 que requería que todos los documentos comerciales y legales llevaran un timbre oficial que indicaba que se había pagado un impuesto. (p. 160)

standard time [hora oficial] *s.* sistema adoptado en 1918 que dividió a Estados Unidos en cuatro zonas horarias. (p. 592)

states' rights [derechos estatales] *s.* teoría que sostenía que los estados tenían el derecho a decidir cuándo el gobierno federal había pasado una ley inconstitucional. (p. 307)

steerage [tercera clase] *s.* nivel o lugar más barato de un barco. (p. 423)

Stono Rebellion [Rebelión de Stono] *s.* sublevación de esclavos de 1739, en Carolina del Sur, que resultó en que se hicieran aún más estrictas las leyes que controlaban a los esclavos. (p. 123)

strategy [estrategia] *s.* plan general de acción. (p. 196)

strike [declararse en huelga] *v.* suspender los obreros el trabajo para tratar de conseguir condiciones de trabajo mejores. (p. 434)

subsistence farm *v.* una granja que produce bastante alimento para la familia con sólo una pequeña cantidad para vender. (p. 110)

suburb [barrio residencial] *s.* área residencial que rodea a una ciudad. (p. 800)

suffrage [sufragio] *s.* derecho a votar. (pp. 262, 444)

Sugar Act [ley del Azúcar] *s.* ley aprobada por el Parlamento en 1764 que impuso impuestos al azúcar, la melaza y otros productos que llegaban a las colonias; también establecía severos castigos para los contrabandistas. (p. 160)

sunbelt *s.* estados más cálidos del sur y el suroeste. (p. 801)

supply-side economics [economía de la oferta] *s.* idea de que si se reducen los impuestos aumenta el número de empleos así como los ahorros y las inversiones, todo lo cual hace que aumenten las entradas del gobierno. (p. 873)

sweatshop *s.* lugar donde los obreros trabajaban largas horas en condiciones muy malas por salarios muy bajos. (p. 600)

T

Tammany Hall *s.* famosa maquinaria política de la ciudad de Nueva York de fines del siglo XIX. (p. 613)

tariff [arance aduanero] *s.* impuesto a las mercancías importadas. (p. 296)

Tariff of Abominations [arancel de las Abominaciones] *s.* ley de 1828 que subió los aranceles de las materias primas y las manufacturas; alteró a los sureños, quienes sentían que la política económica nacional la estaban determinando los intereses económicos del noreste. (p. 381)

Teapot Dome Scandal [escándalo de Teapot Dome] *s.* episodio causado por el ministro del Interior Albert B. Fall quien arrendó ricas reservas públicas de petróleo a compañías privadas a cambio de dinero y tierras. (p. 710)

technology [tecnología] *s.* uso de herramientas y conocimiento para satisfacer las necesidades humanas. (p. 32)

Tejano [tejano] *s.* persona de ascendencia española que consideraba a Texas su hogar. (p. 400)

temperance movement [movimiento de la templanza] *s.* campaña para acabar con el consumo de las bebidas alcohólicas. (p. 434)

tender *s.* dinero (p. 255)

tenement [casa de vecindad] *s.* edificio de apartamentos generalmente en muy malas condiciones y atestado. (p. 611)

Tet offensive [ofensiva del Tet] *s.* en 1968, ataque sorpresa por las tropas del Vietcong contra las bases militares estadounidenses y más de 100 ciudades y pueblos de Vietnam del Sur durante el Tet, la celebración del año nuevo lunar vietnamita. (p. 844)

Thirteenth Amendment [Enmienda Decimotercera] *s.* enmienda a la Constitución de Estados Unidos adoptada en 1865 que abolía la esclavitud y la servidumbre involuntaria en Estados Unidos. (p. 521)

38th parallel *s.* la región al norte de esta latitud ocupada por las tropas soviéticas en 1945. (p. 795)

Three-Fifths Compromise [Acuerdo de los Tres Quintos] *s.* acuerdo de la Convención Constitucional que establecía que, para efectos de la representación y del cobro de impuestos, se contarían como parte de la población tres quintos de los esclavos de un estado. (p. 232)

Townshend Acts [leyes de Townshend] *s.* serie de leyes aprobadas por el Parlamento en 1767 que suspendieron la Asamblea de Nueva York y establecieron impuestos a las mercancías importadas a las colonias británicas. (p. 163)

Trail of Tears [Sendero de las Lágrimas] *s.* trágica marcha del pueblo cherokee desde sus tierras hasta el Territorio Indio, entre 1838 y 1839; miles de cherokees murieron. (p. 377)

transcendentalism [trascendentalismo] *s.* filosofía del siglo XIX que enseñaba que el mundo espiritual es más importante que el mundo físico y que las personas pueden hallar la verdad dentro de sí mismas mediante los sentimientos y la intuición. (p. 431)

transcontinental railroad [ferrocarril transcontinental] *s.* ferrocarril que se extendía por todo el continente. (p. 590)

Treaty of Ghent [Tratado de Gante] *s.* tratado firmado en 1814 que puso fin a la guerra de 1812; no cambió de dueño ningún territorio ni se resolvieron los conflictos comerciales. (p. 333)

Treaty of Greenville [Tratado de Greenville] *s.* acuerdo de 1795 por el cual 12 tribus amerindias cedieron al gobierno de Estados Unidos gran parte de lo que hoy son los estados de Ohio e Indiana. (p. 300)

Treaty of Guadalupe Hidalgo [Tratado de Guadalupe Hidalgo] *s.* tratado de 1848 que puso fin a la guerra estadounidense con México; México cedió California y Nuevo México a los Estados Unidos. (p. 410)

Treaty of Paris [Tratado de París] *s.* tratado de 1763 que puso fin a la guerra Francesa y Amerindia; Inglaterra entregó toda norteamérica del este del río Mississippi. (p. 150)

Treaty of Paris of 1783 [Tratado de París de 1783] *s.* tratado que puso fin a la guerra Revolucionaria, confirmó la independencia de Estados Unidos y estableció los límites de la nueva nación. (p. 212)

Treaty of Tordesillas [Tratado de Tordesillas] *s.* tratado de 1494 por el cual España y Portugal acordaron repartirse las tierras del hemisferio occidental y movieron la línea de demarcación hacia el oeste. (p. 61)

Treaty of Versailles [Tratado de Versailles] *s.* tratado de 1919 que puso fin a la primera guerra mundial. (p. 696)

trench warfare [guerra de trincheras] *s.* clase de guerra durante la primera guerra mundial en que los combatientes se apiñaban en zanjas fortificadas y se disparaban artillería y fuego de ametralladora. (p. 680)

triangular trade [comercio triangular] *s.* sistema de comercio transatlántico en que se intercambiaban mercancías, incluso esclavos, entre África, Inglaterra, Europa, las Antillas y las colonias de Norte América. (p. 111)

tribunal [tribunal] *s.* corte. (p. 253)

Truman Doctrine [Doctrina Truman] *s.* política que prometía ayuda a la gente que luchaba por resistir las amenazas a la libertad democrática. (p. 793)

trust *s.* cuerpo legal creado para tener una cartera de acciones de varias empresas, con frecuencia en la misma industria. (p. 595)

tundra [tundra] *s.* pradera sin árboles que está permanentemente helada debajo de su capa de superior de tierra. (p. 33)

Twenty-sixth Amendment [Enmienda Vigésima Sexta] *s.* enmienda a la Constitución de Estados Unidos, adoptada en 1971, redujo la edad del derecho a votar de los 21 a los 18 años. (p. 849)

U

unanimous consent [consentimiento unánime] *s.* acuerdo completo. (p. 264)

Uncle Tom's Cabin [*La cabaña del tío Tom*] *s.* novela publicada por Harriet Beecher Stowe en 1852, que mostraba la esclavitud como brutal e inmoral. (p. 462)

unconstitutional [inconstitucional] *adj.* que contradice la ley de la constitución. (p. 317)

Underground Railroad [ferrocarril clandestino] *s.* serie de rutas de escape que usaban los esclavos para escaparse del Sur. (p. 442)

United Nations [Naciones Unidas] *s.* organización internacional para mantener la paz a la que pertenecen la mayoría de las naciones del mundo, creada en 1945 para promover la paz, seguridad y desarrollo económico del mundo. (p. 781)

urbanization [urbanización] *s.* crecimiento de las ciudades como resultado de la industrialización. (p. 609)

U.S.S. *Maine* *s.* barco de guerra estadounidense que explotó misteriosamente y se hundió en el puerto de La Habana, Cuba, el 15 de febrero de 1898. (p. 663)

V

vaquero [vaquero] *s.* peón de ganado que vino de México con los españoles en el siglo XVI. (p. 560)

vaudeville [vodevil] *s.* tipo de espectáculo teatral en vivo con mezcla de canciones, baile y comedia. (p. 629)

viceroyalty [virreinato] *s.* provincia regida por un virrey, que gobernaba en nombre del rey. (p. 71)

Viet Cong *s.* comunista vietnamita. (p. 838)

Vietnamization [vietnamización] *s.* estrategia de la guerra de Vietnam de retirar las fuerzas estadounidenses gradualmente y dejar la lucha terrestre en manos de los vietnamitas del Sur. (p. 848)

vigilante [vigilante] *s.* persona dispuesta a tomar la ley en sus propias manos. (p. 561)

Virginia Plan *s.* plan presentado por Edmund Randolph, delegado a la Convención Constitucional de 1787, que proponía un gobierno de tres ramas y una legislatura bicameral en la que la representación se basaría en la población o la riqueza de un estado. (p. 231)

Voting Rights Act of 1965 [ley de los Derechos al Voto] *s.* esta ley prohibía las pruebas de lectura y escritura y otras leyes que impedían que los afroamericanos se anotaran para votar. (p. 821)

W

war bond [bono de guerra] *s.* préstamo de interés bajo de la población civil al gobierno, que se pagaría dentro de un número de años. (p. 691)

War Hawk *s.* habitante de las regiones del oeste que apoyaba la guerra de 1812. (p. 329)

War Powers Act [ley de Poderes de Guerra] *s.* aprobada en 1973, esta ley limita los poderes del presidente para declarar la guerra sin consultar el Congreso. (p. 849)

War Production Board [Junta de Producción Bélica] *s.* agencia establecida durante la segunda guerra mundial para coordinar la producción de suministros militares por las empresas estadounidenses. (p. 772)

Watergate scandal [escándalo de Watergate] *s.* escándalo que resultó de los esfuerzos del gobierno de Nixon por encubrir su participación en el allanamiento de la Sede Central del Partido Demócrata en el edificio de apartamentos Watergate. (p. 860)

Webster-Hayne debate [debate Webster-Hayne] *s.* debate de 1830 entre Daniel Webster y Robert Hayne sobre la doctrina de la invalidación. (p. 382)

Whig Party [Partido Whig] *s.* partido político organizado en 1834 en oposición a las políticas de Andrew Jackson. (p. 387)

Whiskey Rebellion [rebelión del Whisky] *s.* protesta de 1794 contra el impuesto que impuso el gobierno al whisky, valioso medio económico de los granjeros de la frontera. (p. 301)

Wilderness Road [Camino al Desierto] *s.* sendero a Kentucky que ayudó a construir el pionero Daniel Boone. (p. 221)

Wilmot Proviso [Claúsula de Wilmot] *s.* propuesta de 1846 que excluía la esclavitud de cualquier territorio adquirido como resultado de la guerra con México. (p. 459)

Wounded Knee Massacre [masacre de Wounded Knee] *s.* masacre de 1890 por soldados estadounidenses de 300 amerindios desarmados, en Wounded Knee Creek, Dakota del Sur. (p. 566)

writ of assistance [mandato judicial de transferencia] *s.* orden de registro que permitía a los oficiales británicos entrar en los hogares o comercios coloniales en busca de contrabando. (p. 164)

X

XYZ Affair [asunto XYZ] *s.* incidente de 1797 en que funcionarios franceses demandaron que los diplomáticos estadounidenses les pagaran un soborno. (p. 306)

Y

Y2K [año 2000] *s.* problema de computadora causado por programas de computadoras que usan sólo los dos últimos dígitos de un año; complicado por la llegada del año 2000. (p. 878)

Yalta Conference [Conferencia de Yalta] *s.* en 1945 Franklin Roosevelt, Winston Churchill y Joseph Stalin discutieron planes para el fin de la segunda guerra mundial y el futuro de Europa. (p. 765)

yellow journalism [periodismo amarillo] *s.* estilo de periodismo que usa la exageración y el sensacionalismo para presentar las noticias. (p. 663)

Yoruba [Yoruba] *s.* gente de África Occidental que constituyó varios estados al sudoeste del río Níger. (p. 42)

Z

Zimmermann telegram [telegrama de Zimmerman] *s.* mensaje enviado en 1917 por el ministro de Relaciones Exteriores alemán al embajador alemán en México proponiendo una alianza entre México y Alemania y prometiendo ayudar a México a recuperar Texas, New Mexico y Arizona si Estados Unidos entraba en la guerra. (p. 682)

INDEX

An *i* preceding a page reference in italics indicates that there is an illustration, and usually text information as well, on that page. An *m* or a *c* preceding an italic page reference indicates a map or a chart, as well as text information on that page.

A

B

D

E

F

G

I

J

K

L

N

O

P

Q

R

S

U

V

W

X

Y

Z

ACKNOWLEDGMENTS

TEXT ACKNOWLEDGMENTS

Chapter 1, page 28: map, "New Paths to the Americas," *U.S. News & World Report,* October 12, 1998, page 62. Copyright © October 12, 1998, U.S. News and World Report. Visit us at our Web site at www.usnews.com for additional information. Adapted by permission of U.S. News & World Report.

page 33: Excerpt from "Navajo Blessing Way," from *Language and Art in the Navajo Universe* by Gary Witherspoon, page 26 (Ann Arbor: The University of Michigan Press, 1977). Reprinted by permission of The University of Michigan Press.

Chapter 6, pages 174–175: Excerpt from *Johnny Tremain* by Esther Forbes. Copyright © 1943 by Esther Forbes Hoskins, copyright renewed by Linwood M. Erskine, Jr., executor of the Estate of Esther Forbes Hoskins. Reprinted by permission of Houghton Mifflin Company. All rights reserved.

Chapter 9, page 300: "Obeying Rules and Laws," from *Government in America* by Richard J. Hardy. Copyright © 1995 by Houghton Mifflin Company. All rights reserved. Reprinted by permission.

Chapter 16, pages 486–487: Excerpt from *Across Five Aprils* by Irene Hunt. Copyright © 1964 by Irene Hunt. All rights reserved. Reprinted by permission of Shirley Beem on behalf of the author.

Chapter 17, pages 508, 509, 510, 512, 513, 516, 518, 519: Excerpts from *The Civil War* by Kenneth Burns, Ric Burns, and Geoffrey Ward. Copyright © 1990 by American Documentaries, Inc. Reprinted by permission of Alfred A. Knopf, a division of Random House, Inc.

Chapter 19, page 562: Excerpt from "Wahenee: An Indian Girl's Story Told by Herself to Gilbert L. Wilson," *North Dakota History,* Vol. 38, Nos. 1 & 2. Copyright © 1971 State Historical Society of North Dakota. Used by permission.

Chapter 21, pages 618–619: Excerpt from *Dragonwings* by Laurence Yep. Copyright © 1975 by Laurence Yep. Used by permission of HarperCollins Publishers.

pages 624–625: Excerpt from *Crusade for Justice: The Autobiography of Ida B. Wells,* edited by Alfreda M. Duster. Copyright © 1970 by The University of Chicago. All rights reserved. Reprinted by permission of The University of Chicago Press.

Chapter 25, page 713: "First Fig" by Edna St. Vincent Millay, from *Collected Poems,* HarperCollins Publishers. Copyright 1922, 1950 by Edna St. Vincent Millay. All rights reserved. Reprinted by permission of Elizabeth Barnett, literary executor.

page 725: "I, Too," from *Collected Poems* by Langston Hughes. Copyright © 1994 by the Estate of Langston Hughes. Reprinted by permission of Alfred A. Knopf, a division of Random House, Inc.

Chapter 26, page 744–745: Excerpt from *Roll of Thunder, Hear My Cry* by Mildred D. Taylor. Copyright © 1976 by Mildred D. Taylor. Used by permission of Dial Books for Young Readers, a division of Penguin Putnam Inc.

Chapter 27, page 767: Excerpt from *Night* by Elie Wiesel, translated by Stella Rodway. Copyright © 1960 by MacGibbon & Kee. Copyright renewed 1988 by The Collins Publishing Group. Reprinted by permission of Hill and Wang, a division of Farrar, Straus & Giroux, Inc.

Chapter 28, page 800: Excerpt from "Remembering LaVern Baker, a strong-willed R&B original" by Steve Jones, *USA Today,* March 12, 1997. Copyright © 1997 USA Today. Reprinted with permission.

Chapter 29, page 815: Excerpt from *Stride Toward Freedom: The Montgomery Story* by Martin Luther King, Jr. Copyright © 1958 by Martin Luther King, Jr., copyright renewed 1986 by Coretta Scott King. Reprinted by arrangement with The Heirs to the Estate of Martin Luther King, Jr., c/o Writers House, Inc., as agent for the proprietor.

page 819: Excerpt from "Letter from the Birmingham Jail" by Martin Luther King, Jr. Copyright © 1963 by Martin Luther King, Jr., copyright renewed 1991 by Coretta Scott King. Reprinted by arrangement with The Heirs to the Estate of Martin Luther King, Jr., c/o Writers House, Inc., as agent for the proprietor.

pages 820, 828: Excerpts from "I Have a Dream" by Martin Luther King, Jr. Copyright © 1963 by Martin Luther King, Jr., copyright renewed 1991 by Coretta Scott King. Reprinted by arrangement with The Heirs to the Estate of Martin Luther King, Jr., c/o Writers House, Inc., as agent for the proprietor.

Copyright © César E. Chávez Foundation. Reprinted by permission of the César E. Chávez Foundation.

Chapter 32, page 873: Excerpt from "Election no surprise to ordinary folk" by Anne Keegan, *Chicago Tribune,* November 6, 1980. Copyright © 1980 by the Chicago Tribune Company. All rights reserved. Used with permission.

page 883: Excerpt from "Dominican Dominion" by Maximo Zeledon, *Frontera Magazine,* Issue 5. Reprinted by permission of Frontera Magazine.

McDougal Littell Inc. has made every effort to locate the copyright holders for selections used in this book and to make full acknowledgment for their use. Omissions brought to our attention will be corrected in a subsequent edition.

ART CREDITS

Cover and Frontispiece

cover background Copyright © Bob Gelberg/SharpShooters.
frontispiece background The Granger Collection, New York.
Amelia Earhart: Copyright © Albert L. Bresnik.
Maya Lin: Copyright © 1999 Richard Howard/Black Star.
Juan Seguín: Detail of *Juan Seguin* (1838), Jefferson Wright. Texas State Library and Archives Commission.
Harry S. Truman: White House Collection. Copyright © White House Historical Association. Courtesy of the Harry S. Truman Library.
Ida Bell Wells: The Granger Collection, New York.
Abigail Adams: *Portrait traditionally said to be Abigail Adams* (about 1795), artist unknown. Oil on canvas, 30 1/4" x 26 1/2", N-150.55. Photo by Richard Walker. Copyright © New York State Historical Association, Cooperstown, New York.
Zitkala-Să: Negative no. Mss 299, William F. Hansen Collection, Photographic Archives, Harold B. Lee Library, Brigham Young University, Provo, Utah.
Benjamin Franklin: Copyright © Joseph-Siffrede Duplessis/Wood River Gallery/PNI.
Abraham Lincoln: The Library of Congress.
Martin Luther King, Jr: Photo by Howard Sochrer/Life Magazine, copyright © Time Inc.

Table of Contents

viii–xvii *background* The Granger Collection, New York; **viii** *top* Benin *Horn-Blower* (about 1550–1680) unknown African artist. Nigeria, Court of Benin, Bini Tribe. Bronze, 24 7/8" x 11 5/8" x 6 3/4" (63 cm x 29.4 cm x 17.2 cm). The Metropolitan Museum of Art, The Michael C. Rockefeller Memorial Collection, gift of Nelson A. Rockefeller, 1972 (1978.412.310). Photograph copyright © 1983 The Metropolitan Museum of Art; *bottom* The Granger Collection, New York; **ix** *top* Copyright © Michael Gadomski/Photo Researchers, Inc.; *center* The Granger Collection, New York; *bottom* Library of Congress; **x** *top* Copyright © R. Kord/H. Armstrong Roberts; *center* Detail of *George Washington in the Uniform of a British Colonial Colonel* (1772), Charles Willson Peale. Washington-Custis-Lee Collection. Washington and Lee University, Lexington, Virginia; *bottom* Corbis; **xi** *top* Copyright © 1998 North Wind Pictures; *center* National Museum of American History, Smithsonian Institution; *bottom* Illustration by Patrick Whelan; **xii** *top* From the collections of the Minnesota Historical Society; *bottom left* The Granger Collection, New York; *bottom right* Copyright © 1998 Louis Psihoyos/Matrix; **xiii** *top* Copyright © George Peter Alexandre Healy/Wood River Gallery/PNI; *center* Courtesy of Stamatelos Brothers Collection. Photo by Larry Sherer. Copyright © 1991 Time-Life Books Inc.; *bottom* The Granger Collection, New York; **xiv** *top, Northern Pacific Railroad. The Pioneer Route to Fargo Moorhead Town Bismark Dakota and Montana and the Famous Valley of the Yellowstone* (about 1885), Creator–Poole Brothers, Printers. Broadside. Chicago Historical Society; *bottom* The Granger Collection, New York; **xv** *top* Doubleday, Page and Company, New York, 1906, second issue; *center, Portrait of Queen Liliuokalani,* date and artist unknown. Bishop Museum; *bottom left* The Granger Collection, New York; *bottom right* Culver Pictures; **xvi** *top* Copyright © Stock Montage; *center* The Granger Collection, New York; *bottom* Copyright © Rob Boudreau/Tony Stone Images; **xvii** *top* Corbis-Bettmann; *center* NASA; *bottom* Copyright © Charles Feil/Stock Boston/PNI.

Geography Handbook

2 *top right* Copyright © H. Abernathy/H. Armstrong Roberts; *center left* Copyright © Warren Morgan/H. Armstrong Roberts; *bottom* Copyright © 1996 Denver A. Bryan; **3** *top* Copyright © Nathan Benn/Stock Boston; *bottom right* Copyright © Andy Sacks/Tony Stone Images; **4–5** The Granger Collection, New York;

Copyright © 1997 David Noble/FPG International; **5** *top right* © John Coletti/Stock Boston; *center* Copyright © Bill Horsman/Stock Boston; **6** *top* Copyright © George Mobley/NGS Image Collection; *center* Copyright © Lowell Georgia/NGS Image Collection; *bottom left* The Granger Collection, New York; **10** *top* Copyright © T. Algire/H. Armstrong Roberts; *bottom* Copyright © Ken Graham/Tony Stone Images; **11** *top* Copyright © SuperStock; *bottom* Copyright © Tom Dietrich/Tony Stone Images; **12** *top* Copyright © Zane Williams/Tony Stone Images; *bottom* Copyright © Eastcott/Momatiuk/Tony Stone Images; **13** *top* Copyright © SuperStock; *bottom* Copyright © F. Sieb/H. Armstrong Roberts; **14–15** Illustration by Ken Goldammer; **16** *top* Copyright © M. Schneiders/H. Armstrong Roberts; *bottom* Copyright © D. Frazier/H. Armstrong Roberts; **17** *top left* Copyright © Ron Levy/Liaison Agency; *top right, center right, bottom right* Copyright © Steve Adams/NGS Image Collection; *bottom left* Copyright © Cathlyn Melloan/Tony Stone Images; **18** *bottom* Copyright © J. Marshall/The Image Works; **19** *left* Culver Pictures; *right* Copyright © Santi Visalli/The Image Bank.

Unit 1

22–23 Copyright © R. Kord/H. Armstrong Roberts.

Chapter 1, 24, 25 The Granger Collection, New York; **26** *top* Werner Forman/Art Resource, New York; *center* Copyright © Aldona Sabalis/Photo Researchers, Inc.; *bottom* Copyright © Stuart Dee/The Image Bank; **27** Courtesy of Rock Art Foundation, San Antonio, Texas; **28** *center left* Courtesy Arizona State Museum, University of Arizona, Tucson, Arizona. Photo copyright © 1996 Jerry Jacka; **29** Copyright © Francois Gohier/Photo Researchers, Inc.; **30** Copyright © SuperStock; **32** *background* Photo of Bill Reid by R. Bettner. Collection Canadian Museum of Civilization, #K96-215 [or K96-216]. Reprinted with permission of Ms. Martine Reid; *foreground* Bear sculpture (1962), Bill Reid. 248 cm x 136 cm x 126 cm, U.B.C. Gift of Dr. Walter Koerner. Courtesy U.B.C. Museum of Anthropology, Vancouver, Canada; **34** National Museum of Anthropology, Mexico City, Mexico/Werner Forman Archive/Art Resource, New York; **35** Smithsonian Institution, Washington, D.C.; **36** *background* Copyright © Richard Saker/AllSport; *bottom left* Copyright © Salamander Picture Library; **39** Copyright © Thomas D. W. Friedmann/Photo Researchers, Inc.; **40** *left* Copyright © Aldona Sabalis/Photo Researchers, Inc.; **41** The Granger Collection, New York; **42** PNI; **43** The Metropolitan Museum of Art, The Michael C. Rockefeller Memorial Collection, gift of Nelson A. Rockefeller, 1972 (1978.412.310). Photograph copyright © 1983 The Metropolitan Museum of Art; **44** Copyright © Stock Montage; **45** Illustration by Alexander Verbitsky; **47** *The School of Athens* (1508), Raphael (Raffaello Santi). Stanza della Segnatura, Vatican City, Vatican. Photo copyright © Erich Lessing/Art Resource, New York; **49** Detail of *St. Vincent Polyptych* (date unknown), Nuno Gonçalves. Museu Nacional de Arte Antiga, Lisbon, Portugal. Copyright © Scala/Art Resource, New York; **50** *Portrait of a Man, Called Christopher Columbus* (1519), Sebastiano del Piombo. Oil on canvas, 42" x 34 3/4". The Metropolitan Museum of Art, gift of J. Pierpont Morgan, 1900; **52** Archivio Fotografico del Museo Preistorico Etnografico L. Pigorini, Rome. Photo by Lorenzo de Masi; **53** By courtesy of the Bibliothèque Nationale, Paris; **55** *bottom* Courtesy of John Goss; **56** Copyright © Francois Gohier/Photo Researchers, Inc.; **57** Photos by Sharon Hoogstraten.

Chapter 2, 58, 59 The Granger Collection, New York; **61** Detail of *Pope Alexander VI Borgia Kneeling in Prayer* (date unknown), Bernardino Pinturicchio. Sala die Misteri della Fede, Appartamento Borgia, Vatican Palace, Vatican State/Scala/Art Resource, New York; **64** Corbis; **65** Aztec mask (early 16th century), artist unknown. Turquoise, pearl shell. Werner Forman Archive/British Museum, London/Art Resource, New York; **66** The Granger Collection, New York; **67** *top, The Last Voyage of Henry Hudson* (1881), John Collier. Tate Gallery, London/Art Resource, New York; **68** PNI; **69** *Sea Battle between the Spanish Armada and English Naval Forces* (about 1600), Hendrik Corneliez Vroom. Oil on canvas, 91 cm x 153 cm. Landesmuseum Ferdinandeum, Innsbruck, Austria. Photo copyright © Erich Lessing/Art Resource, New York; **71** The Granger Collection, New York; **73** The British Library; **75** PNI; **76** Copyright © The Fotomas Index; **77** Corbis; **78** *left* The Newberry Library, Chicago; *right, Slaves Below Deck of Albanez* (date unknown), Francis Meynell. Copyright © National Maritime Museum Picture Library, London.

Chapter 3, 82 The Granger Collection, New York; **83** Colonial Williamsburg Foundation; **85** British Museum; **87** AP/Wide World Photos; **88** *Pocahontas* (1616), Simon van de Passe. Engraving, 6 7/8" x 4 3/4" (17.5 cm x 12 cm). Published in the *Baziliologia*, London, 1618. National Portrait Gallery, Smithsonian Institution/Art Resource, New York; **89** The Granger Collection, New York; **90–91** Illustration by Ivan Lapper. Copyright © The Reader's Digest Association; **92** The Granger Collection, New York; **93** *The Mayflower in Plymouth Harbor* (1882), William Halsall. Courtesy of the Pilgrim Society, Plymouth, Massachusetts; **94** *top* Copyright © SuperStock; *bottom* Copyright © Andrew J. Martinez/Photo Researchers, Inc.; **95** *bottom* The Granger Collection, New York; **96** *left* Courtesy of American Antiquarian Society; *right* The New York Public Library; **97** *Trial of George Jacobs for Witchcraft, 1692* (1855), T. H. Matteson. Oil on canvas, acc. #1246. Peabody Essex Museum, Salem, Massachusetts. Photo by Mark Sexton; **100** Detail of *Peter Stuyvesant and the Trumpeter (The Wrath of Peter Stuyvesant)* (1835), Asher B. Durand. Oil on canvas, 24 1/4" x 30 1/4", accession no. 1858.28. Copyright © Collection of The New-York Historical Society; **101** Copyright © Tom Dietrich/Tony Stone Images; **103** Brown Brothers; **105** Colonial Williamsburg Foundation.

Chapter 4, 106, 107 The Granger Collection, New York; **109** *Sperm Whaling: No. 2, The Capture* (1862), A. Van Best and R. S. Gifford. Lithograph by Endicott and Company, corrected by Benjamin Russell, 16 3/4 x 25 3/4" (43 cm x 65 cm). N.M.A.H., Harry Peters "America on Stone" Lithography Collection; **110** *right* Copyright © Thomas Neill; **112** North Carolina Collection, University of North Carolina Library at Chapel Hill; **114** Detail of *Quaker Meeting* (date unknown), Egbert Van Heemskerk. The Quaker Collection, Haverford (Pennsylvania) College Library; **115** *right* Reprinted with permission of Philipsburg Manor, Tarrytown, New York; **116** The Granger Collection, New York; **118** Library of Congress; **119** Detail of *George Mason* (1811), Dominic W. Boudet. Oil on canvas, 30" x 25" (76.2 cm x 63.5 cm). Virginia Museum of Fine Arts, Richmond. Gift of David K. E. Bruce. Photo by Ron Jennings. Copyright © Virginia Museum of Fine Arts; **120** *right* Copyright © Chip Henderson/Tony Stone Images; **121** Virginia Historical Society, Richmond, Virginia; **122** The Granger Collection, New York; **124** *bottom left* Smithsonian Institution, Washington, D.C.; *bottom right background* Photo copyright © Dorothy Miller; *bottom right foreground* Copyright © The New York Botanical Garden; **125** *bottom center* Copyright © Old Dartmouth Historical Society, New Bedford (Massachusetts) Whaling Museum; **126** The Granger Collection, New York; **127** *right* Copyright © Michael P. Gadomski/Photo Researchers, Inc.; **128** Copyright © Mark McGehearty/AllSport; **129** *Catching the Wild Horse* (date unknown), George Catlin. Oil on canvas, 12 1/4" x 16 1/4". The Thomas Gilcrease Institute of American History and Art, Tulsa, Oklahoma.

Chapter 5, 132 Virginia Historical Society, Richmond, Virginia; **133** Library of Congress; **137, 139** The Granger Collection, New York; **140** Corbis; **141** Detail of *Increase Mather* (1688), John Vander Spriett. Courtesy of the Massachusetts Historical Society; **142** Courtesy of Knox County, Illinois Teen Court; **143** The Granger Collection, New York; **145** *left* Corbis; *right* The Newberry Library, Chicago; **146** Detail of *Braddock's Defeat* (1903), Edward Deming. State Historical Society of Wisconsin Museum Collection; **147** Library of Congress; **149, 151** The Granger Collection, New York.

Unit 2

154–155 Copyright © William Johnson/Stock Boston.

Chapter 6, 156, 157 The Granger Collection, New York; **158** *top* Colonial Williamsburg Foundation; *center* Detail of *The Battle of Princeton* (date unknown), William Mercer. The Historical Society of Pennsylvania; *bottom* The Granger Collection, New York; **159** *James Otis Arguing Against the Writs of Assistance in the Old Towne House* (1901), Robert Reid. Courtesy Commonwealth of Massachusetts Art Commission; **160** *top* Rare Books Division, The New York Public Library. Astor, Lenox, and Tilden Foundations; *bottom* Emmet Collection, Manuscripts and Archives Division, The New York Public Library. Astor, Lenox, and Tilden Foundations; **161, 162** The Granger Collection, New York; **163** Copyright © Stock Montage; **164** Copyright © Collection of The New-York Historical Society; **165** The Granger Collection, New York; **166** *left* The Granger Collection, New York; *right, John Adams* (date unknown), John Trumbull. National Portrait Gallery, Smithsonian Institution/Art Resource, New York; **167** The Granger Collection, New York; **168–169** Copyright © McDougal Littell Inc.; **170** Copyright © Paul Mozell/Stock Boston; **173** Copyright © Charles Winters/Stock Boston; **174** Copyright © McDougal Littell Inc.; **175** The Granger Collection, New York; **176** *Portrait traditionally said to be*

Abigail Adams (about 1795), artist unknown. Oil on canvas, 30 1/4" x 26 1/2", N-150.55. Photo by Richard Walker. Copyright © New York State Historical Association, Cooperstown, New York; **177** Detail of *The Death of General Warren at the Battle of Bunker's Hill, 17 June 1775* (date unknown), John Trumbull. Oil on canvas. Yale University Art Gallery, Trumbull Collection; **178** Library of Congress; **179** *top* The Granger Collection, New York; *bottom* Library of Congress; **180** *top* Detail of *The Declaration of Independence, 4 July 1776* (date unknown), John Trumbull. Oil on canvas. Yale University Art Gallery, Trumbull Collection; *bottom* Copyright © R. Kord/H. Armstrong Roberts; **181, 182** The Granger Collection, New York; **187** *bottom* Clement Library, University of Michigan; **88** The Granger Collection, New York; **189** Photos by Sharon Hoogstraten.

Chapter 7, 190 *March to Valley Forge* (1883), William B. T. Trego. Courtesy of Stacey Swigart, The Valley Forge Historical Society; **191, 192** The Granger Collection, New York; **193** Haym Salomon Home for Nursing and Rehabilitation, Brooklyn, New York. Courtesy of American Jewish Historical Society, Waltham, Massachusetts, and New York, New York; **194** *George Washington in the Uniform of a British Colonial Colonel* (1772), Charles Willson Peale. Washington-Custis-Lee Collection. Washington and Lee University, Lexington, Virginia; **195** *top* From *American Story: The Revolutionaries.* Courtesy of the Jamestown-Yorktown Educational Trust. Photo by Katherine Wetzel. Copyright © 1996 Time-Life Books Inc.; **196** *Washington Crossing the Delaware* (1851), Emanuel Gottlieb Leutze. Oil on canvas, 149" x 255". The Metropolitan Museum of Art, gift of John S. Kennedy, 1897; **197** *bottom* The Granger Collection, New York; **198** Copyright © Michael Newman/PhotoEdit/PNI; **199** *Two American Flags Flown by John Paul Jones in 1779* (date unknown), unknown Dutch artist. Watercolor. Chicago Historical Society; **200, 201** The Granger Collection, New York; **202** From *American Story: The Revolutionaries.* Courtesy of the Jamestown-Yorktown Educational Trust. Photo by Katherine Wetzel. Copyright © 1996 Time-Life Books Inc.; **205** The Granger Collection, New York; **206** Library of Congress; **208** Illustration by Bill Cigliano; **210** *Surrender of Lord Cornwallis at Yorktown* (date unknown), John Trumbull. Yale University Art Gallery, Trumbull Collection; **211** Private collection; **212** The Granger Collection, New York; **215** Courtesy of the Massachusetts Historical Society; **216** Private collection.

Chapter 8, 218 The Granger Collection, New York; **219** *The Signing of the Constitution* (date unknown), H. C. Chandler. Art Resource, New York; **221** Detail of *Daniel Boone Escorting Settlers Through the Cumberland Gap* (1851–1852), George Caleb Bingham. Oil on canvas, 36 1/2" x 50 1/4". Washington University Gallery of Art, St. Louis (Missouri). Gift of Nathaniel Phillips, 1890; **222, 225** *top* The Granger Collection, New York; **225** *bottom* Copyright © Thad Samuels Abell II/NGS Image Collection; **226** *bottom left* Copyright © John E. Fletcher & Arlan R. Wiler/NGS Image Collection; *bottom right* William L. Clements Library, Map Division, University of Michigan; **227** *top* The Granger Collection, New York; *bottom* Courtesy of the Shelby County (Ohio) Historical Society; **228** Corbis; *inset* Library of Congress; **229** Copyright © Joseph Nettis/Tony Stone Images; **230** The Granger Collection, New York; **231** PNI; **234** The Granger Collection, New York; **235** *left, Portrait of John Jay* (c. 1783, 1804–1808), Gilbert Stuart, believed to have been begun by and finished by John Trumbull. National Portrait Gallery, Smithsonian Institution/Art Resource, New York; **235** *right,* **236** The Granger Collection, New York.

Constitution Handbook

242 *left* The Granger Collection, New York; *right* Copyright © Ivan Massar/Black Star; **243** *top* Copyright © Bob Daemmrich/The Image Works; *center* Copyright © Topham/The Image Works; *bottom* Courtesy of The New York Times; **244** *bottom* Copyright © J. L. Atlan/Sygma; **245** *top* Copyright © Robert E. Daemmrich/Tony Stone Images; **247** *top* Copyright © Engelhardt/St. Louis Post-Dispatch; *bottom* Copyright © Patrick Forden/Sygma; **248** *background* The Granger Collection, New York; **248** *inset,* **257** AP/Wide World Photos; **258** *top right* Copyright © Archive Photos; *top left* Culver Pictures; *center right* Copyright © 1972 Magnum Photos, Inc.; *center left* AP/Wide World Photos; *bottom right* Courtesy Ronald Reagan Library; **264** AP/Wide World Photos; **265** Copyright © Collection of The New-York Historical Society; **267** Copyright © Baron Wolman/Tony Stone Images; **268** *left* Copyright © Jean-Marc Giboux/Liaison Agency; *right* Copyright © Bob Dremmrich/Sygma; **270** Copyright © 1993 Ron Rovtar/FPG International; **272** The Granger Collection, New York; **273** *left* Copyright © 1998 FPG International; *right* Copyright © Cynthia Johnson/Liaison Agency; **274** Franklin D. Roosevelt Library; **277** Copyright © Rock the Vote Inc.

Citizenship Handbook

280 AP/Wide World Photos; **281** *background* Copyright © Bob Daemmrich; *foreground* Copyright © 1994 Mark Harmel/FPG International; **284** Copyright © Bob Daemmrich/The Image Works; **285** Copyright © David Young-Wolff/Tony Stone Images; **286** AP/Wide World Photos.

Unit 3

288–289 Copyright © Marvin E. Newman/The Image Bank.

Chapter 9, 290 The Granger Collection, New York; **291** *Washington's First Cabinet* (date unknown), Alonzo Chappel. Courtesy of CNA; **292** Washington inaugural buttons, 1789. National Museum of American History, Smithsonian Institution; **294** AP/Wide World Photos; **295** *Portrait of Alexander Hamilton* (about 1796), James Sharples (the Elder). National Portrait Gallery, Smithsonian Institution/Art Resource, New York; **297** Copyright © Larry Stevens/Nawrocki Stock Photo Inc.; **298** Chicago Historical Society; **299** *right* Ohio Historical Society; **301** *Execution of Louis XVI* (18th century), anonymous. Colored engraving. Musée de la Ville de Paris, Musée Carnavalet, Paris; **303** Mount Vernon Ladies Association; **305** *top* The Granger Collection, New York; *center right* Maryland Historical Society; **306** The Granger Collection, New York.

Chapter 10, 310 *The Rocky Mountains, Lander's Peak* (1863), Albert Bierstadt (1830–1902). Oil on canvas, 73 1/2" x 120 3/4". The Metropolitan Museum of Art, Rogers Fund, 1907 (07.123). Photograph copyright © 1979 The Metropolitan Museum of Art; **311** *Chasing Off Grizzlies.* Tab. 36 of atlas accompanying the English edition of Prince Maximilian of Wied's *Travels in the Interior of North America.* By Lucas Webber after Carl Bodmer, published by Ackermann and Company, London, December 1, 1842. Colored engraving, 12" x 17 1/2"; plate size, 15 1/2" x 21 1/4"; on trimmed stock, 17" x 23 1/2"; **313** *left, Thomas Jefferson* (date unknown), Rembrandt Peale. Copyright © Collection of the New-York Historical Society; *right* Boston Athenaeum; **314** The Granger Collection, New York; **315** *center right* Kirby Collection of Historical Paintings, Lafayette College; *bottom background* Monticello/Thomas Jefferson Memorial Foundation, Inc.; *bottom center* Polygraph, 1806. John Isaac Hawkins, Charles Willson Peale. University of Virginia. Photo courtesy of Monticello/Thomas Jefferson Memorial Foundation, Inc.; **316** Boston Athenaeum; **317** Copyright © James Blair/NGS Image Collection; **318, 319** The Granger Collection, New York; **320** *inset* Copyright © Addison Geary/Stock Boston/PNI; **321** *Lewis and Clark at Three Forks* (date unknown), E. S. Paxson. Mural in the Montana State Capitol. Courtesy of the Montana Historical Society. Photo by John Reddy; **322** Copyright © 1998 North Wind Pictures; **323** National Museum of American History, Smithsonian Institution; **324** *center left* Copyright © Stock Montage; *center right, The Surrounder, Chief of the Tribe* (1833), George Catlin. Oto tribe. National Museum of American Art, Smithsonian Institution/Art Resource, New York; *bottom center, bottom right* Peabody Museum, Harvard University. Copyright © President and Fellows of Harvard College. Photo by Hillel Burger; **325** *center left, Mink, a Pretty Girl* (1832), George Catlin. National Museum of American Art, Smithsonian Institution/Art Resource, New York; *center right, Steep Wing, a Brave of the Bad Arrow Points Band* (1832), George Catlin. Teton Dakota (Western Sioux). National Museum of American Art, Smithsonian Institution/Art Resource, New York; *bottom* Whaling Chief's Hat, Makah or Nootka. Cedar bark, bear grass, unidentified mammal hair, remnant feathers, leather thong. H 22 cm, D 27 cm. PM# 99-12-10/53080. Peabody Museum, Harvard University. Copyright © President and Fellows of Harvard College. Photo by Hillel Burger; **326** Detail of *Decatur Boarding the Tripolitan Gunboat* (date unknown), Dennis Malone Carter. Oil on canvas. Courtesy of the Naval Historical Foundation; **327** Library of Congress; **328** The Granger Collection, New York; **330** *left* White House Collection. Copyright © White House Historical Association; *right* The Granger Collection, New York; **332** National Museum of American History, Smithsonian Institution; **334** *top, Thomas Jefferson* (date unknown), Rembrandt Peale. Copyright © Collection of the New-York Historical Society; *bottom* National Museum of American History, Smithsonian Institution; **335** *Ograbme, or, The American Snapping Turtle* (1807), D. Longworth. Negative no. 7278. Copyright © Collection of The New-York Historical Society; **336** *top left* Elk skin bound journal (1805), William Clark. Ink on paper. Missouri Historical Society;

bottom Copyright © 1998 North Wind Pictures; **337** ...tos by Sharon Hoogstraten.

Chapter 11, 338, 339, 341 Corbis-Bettmann; **342** The Granger Collection, New York; **343** Illustration by Patrick Whelan; **344** Corbis-Bettmann; **345** Copyright © Stock Montage; **346–347** Illustrations by Randal Birkey; **348** The Granger Collection, New York; **349** Illustration by Patrick Whelan; **351** Detail of *Plantation Burial* (1860), John Antrobus. Williams Research Center, The Historic New Orleans Collection. Photo copyright © Jan White Brantley; **352** Corbis-Bettmann; **353** The Granger Collection, New York; **354** Corbis-Bettmann; **355** *right inset* The Granger Collection, New York; **356, 357** *right inset* Corbis-Bettmann.

Unit 4

364–365 Layne Kennedy/Corbis.

Chapter 12, 366 The Granger Collection, New York; **367** Chicago Historical Society; **368** Copyright © Collection of the New-York Historical Society; **369** Redwood Library and Athenaeum, Newport, Rhode Island; **370** *left, John Quincy Adams* (1858), George Healy. Oil on canvas, 62" x 47". White House Collection. Copyright © White House Historical Association; *right, Andrew Jackson* (1845), Thomas Sully. Oil on canvas, 20 3/8" x 17 1/4". Copyright © 1996 Board of Trustees, National Gallery of Art, Washington, D.C. Andrew W. Mellon Collection; **371** The Hermitage: Home of President Andrew Jackson, Nashville, Tennessee. Photo of ribbon by Sharon Hoogstraten; **372** Copyright © The News-Sentinel, Fort Wayne, Indiana; **373** *left* The Granger Collection, New York; *right* Culver Pictures; **374** *Sequoyah, Indian Statesman* (date unknown), anonymous, after Charles Bird King. Hand-colored lithograph, J. T. Bowen lithography company. National Portrait Gallery, Smithsonian Institution/Art Resource, New York; **375** Copyright © J. Pat Carter/Liaison Agency; **377** *The Trail of Tears* (date unknown), Robert Lindneux. Woolaroc Museum, Bartlesville, Oklahoma; **378** The Granger Collection, New York; **379** *top, John Caldwell Calhoun* (about 1818–1825), attributed to Charles Bird Kind. Oil on canvas, 76.2 cm x 63.5 cm. National Portrait Gallery, Smithsonian Institution/Art Resource, New York; **379** *bottom,* **381** *top* The Granger Collection, New York; **381** *bottom* The Museum of the Confederacy, Richmond, Virginia. Photo by Katherine Wetzel; **382** Detail of *Webster's Reply to Hayne* (date unknown), George Healy. Courtesy Boston Art Commission; **384** *top* The Granger Collection, New York; *bottom inset, Nicholas Biddle* (1839), Henry Inman. Oil on canvas [1978.2]. The Historical Society of Pennsylvania; **385** The Granger Collection, New York; **386** Copyright © Collection of The New-York Historical Society; **389** Library of Congress.

Chapter 13, 390 Denver Public Library, Western History Department; **391** National Archives; **393** *Jedediah Smith in the Badlands* (date unknown), Harvey Dunn. The South Dakota Art Museum Collection; **394** Copyright © United States Postal Service; **395** *inset* Copyright © Ric Ergenbright; **396** Skillet, negative no. OrHi 97119. Oregon Historical Society; **400** The Granger Collection, New York; **401** *top* Broadsides Collection, The Center for American History, The University of Texas at Austin; **402** Copyright © Bob Daemmrich/Stock Boston/PNI; **403** *The Battle of the Alamo* (about 1913), Frederick C. Yohn. Courtesy Continental Insurance; **404** *left* Texas State Library and Archives Commission; **404** *right,* **406** The Granger Collection, New York; **408** *inset,* **412** Courtesy of the California History Room, California State Library, Sacramento, California; **413** Copyright © Collection of The New-York Historical Society; **414** Courtesy of Levi Strauss Company; **415, 417** Courtesy of the California History Room, California State Library, Sacramento, California; **419** *bottom* From the collections of the Minnesota Historical Society.

Chapter 14, 420 Courtesy Meserve-Kunhardt Collection, Mount Kisco, New York; **421** The Granger Collection, New York; **422** The New York Public Library; **423** *background* From the collections of the Minnesota Historical Society; *foreground* Copyright © Collection of The New-York Historical Society; **424** *center* The Granger Collection, New York; **427** Michael S. Yamashita/Corbis; **428** The Granger Collection, New York; **429** From *The Headless Horseman* by Natalie Standiford, illustrated by Donald Cook. Illustration copyright © 1992 Donald Cook. Reprinted by permission of Random House, Inc.; **430** The Granger Collection, New York; **432** *background* Photo by Simon Marsden/The Marsden Archive; **433** The Granger Collection, New York; **434** Copyright © Archive Photos; **435** Department of Archives, Oberlin (Ohio) College; **436, 440** The Granger Collection, New York; **441** Chester County Historical Society, West Chester, Pennsylvania; **442** Copyright © 1996 Wayne Sorce; **443** *from left to right* The Granger Collection, New York; The Granger Collection, New York; The Granger Collection, New York; The Granger Collection, New York; Corbis-Bettmann; **444** The Granger Collection, New York; **445** *The Discord* (1885), F. Heppenheimer. Color lithograph, negative no. 51038. Copyright © Collection of The New-York Historical Society; **446** *center* Library of Congress; *bottom right, bottom left* Copyright © 1998 Louis Psihoyos/Matrix; **447** *bottom* From the Collections of the National Underground Railroad Freedom Center, Cincinnati, Ohio; **450** *paisley shawl* Photo by Sharon Hoogstraten; *all other items* From the collections of the Minnesota Historical Society; **451** Photos by Sharon Hoogstraten.

Unit 5

452–453 Corbis.

Chapter 15, 454, 455, 457, 460 The Granger Collection, New York; **461** Corbis; **462** Hulton-Deutsch Collection/Corbis; **463** *left* The Granger Collection, New York; *right* Corbis-Bettmann; **466** From the Collection of David J. and Janice L. Frent; **467** AFP/Corbis; **468, 471** The Granger Collection, New York; **472** Lloyd Ostendorf Collection; **474** *background* Hulton-Deutsch Collection/Corbis; *inset* Copyright © George Peter Alexandre Healy/Wood River Gallery/PNI; **477** *The Last Moments of John Brown* (1884), Thomas Hovenden. Oil on canvas, 77 3/8" x 63 1/4". The Metropolitan Museum of Art, gift of Mr. and Mrs. Carl Stoeckel, 1897. Photograph copyright © 1982 The Metropolitan Museum of Art.

Chapter 16, 478, 479 The Granger Collection, New York; **480** *top left, top right* Library of Congress; *bottom left, bottom right* The Granger Collection, New York; **482** The Lincoln Museum, Fort Wayne, Indiana. #0-43; **485** Copyright © H. Armstrong Roberts; **486** Copyright © McDougal Littell Inc.; **487** *left* Courtesy of Brian C. Pohanka; *right* Massachusetts Commandery Military Order of the Loyal Legion and the U.S. Army Military History Institute; **488** Bureau of Archives and History, New Jersey State Library; **489** Corbis-Bettmann; **490** Culver Pictures; **491** The Granger Collection, New York; **492** Illustration by Alexander Verbitsky; **493** Library of Congress; **496** U.S. Signal Corps photo no. 111-B-4146 (Brady Collection) in the National Archives; **497** Chicago Historical Society; **498** *left, center* Photos by Larry Sherer. Copyright © Time-Life Books Inc.; *bottom right* Fort Sumter National Park.

Chapter 17, 500 *Cavalry Charge at Yellow Tavern, Virginia, May 11, 1864* (1871), H. W. Chaloner. Oil on canvas, 31" x 47 7/8". Garbisch no. 61.1, purchased from Kenneth E. Snow by Garbisches (1961). Presented by the Estate of Edgar William and Bernice Chrysler Garbisch. Photo by West Point Museum Collections, United States Military Academy; **501** Corbis-Bettmann; **502** *top left, top right, bottom left* Library of Congress; *bottom right* The Granger Collection, New York; **503** Corbis; **504** The Granger Collection, New York; **505** *top* Copyright © Archive Photos; *bottom* Library of Congress; **507** Culver Pictures; **508** The Granger Collection, New York; **510** American Red Cross; **511** *background* Corbis; *foreground* Massachusetts Commandery Military Order of the Loyal Legion and the U.S. Army Military History Institute; **512** The Pejepscot Historical Society; **513** Copyright © 1994 Kunio Owaki/The Stock Market; **514–515** *background* Illustration by Ken Goldammer; **514** *center* Library of Congress; *bottom left* From *Echoes of Glory: Arms & Equipment of The Union.* Photo by Larry Sherer. Copyright © 1991 Time-Life Books Inc.; *bottom right* The Museum of the Confederacy, Richmond, Virginia. Photo by Katherine Wetzel; **515** *bottom* From *Echoes of Glory: Arms & Equipment of The Union.* Photo by Larry Sherer. Copyright © 1991 Time-Life Books Inc.; **516** Library of Congress; **518** *top left* Courtesy of Stamatelos Brothers Collection. Photo by Larry Sherer. Copyright © 1991 Time-Life Books Inc.; *bottom* Courtesy Meserve-Kunhardt Collection, Mount Kisco, New York; **520** Copyright © Robert M. Anderson/Uniphoto, Inc.; **522** National Archives; **526** *top right* Manassas National Battlefield Park, National Park Service. Photo by Larry Sherer; *bottom right* Collection of Old Capitol Museum of Mississippi History; **528** *center left, center* High Impact Photography/Copyright © Time-Life Books Inc.; *center right* Courtesy of Stamatelos Brothers Collection. Photo by Larry Sherer. Copyright © 1991 Time-Life Books Inc.; **529** Photos by Sharon Hoogstraten.

Chapter 18, 530 Corbis-Bettmann; **531** *The Shackle Broken—by the Genius of Freedom.* Color lithograph. Pub. by E. Sachs & Co., Baltimore, 1874. Chicago Historical Society; **533, 534** The Granger Collection, New York; **535** Corbis; **536** *His First Vote*

(1868), Thomas Waterman Wood. Oil on board. Cheekwood Museum of Art, Nashville, Tennessee; **537, 538–539** *background* The Granger Collection, New York; **538** *top right,* **540, 541** Corbis; **542** Library of Congress; **544** The Granger Collection, New York; **545** Library of Congress; **546, 547** The Granger Collection, New York; **548** Library of Congress; **551** *bottom* The Granger Collection, New York.

Unit 6

552–553 The Granger Collection, New York.

Chapter 19, 554 *Jerked Down* (1907), Charles M. Russell. Oil on canvas, 23" x 36 3/4", accession no. 0137.2246. From the Collection of Gilcrease Museum, Tulsa (Oklahoma); **555** Solomon D. Butcher Collection, Nebraska State Historical Society; **557** Denver Public Library; **559** *top* The Granger Collection, New York; *bottom* Copyright © Dan Suzio/Photo Researchers, Inc.; **560** *California Vaqueros* (1875), James Walker. Oil on canvas, 31" x 46". Courtesy of The Anschutz Collection. Photo by William J. O'Connor; **561** Culver Pictures; **562** *background* The Granger Collection, New York; *foreground* State Historical Society of North Dakota; **564** *Custer's Last Stand* (1899), Edgar S. Paxson. Oil on canvas, 70 1/2" x 106". Buffalo Bill Historical Center, Cody, Wyoming; **565** *top* The Granger Collection, New York; *center* Culver Pictures; *bottom* Library of Congress; **566** Western History Collections, University of Oklahoma Libraries; **567** Library of Congress; **568** The Kansas State Historical Society, Topeka, Kansas; **569** *top, View of San Francisco [Formerly Yerba Buena]* (1847), attributed to Victor Prevost. Oil on canvas, 25" x 30". California Historical Society, gift of the Ohio Historical Society; *bottom* The Bancroft Library, University of California, Berkeley; **570** *top* The Granger Collection, New York; *center* Courtesy of the Arizona Historical Society/Tucson; *bottom* The Granger Collection, New York; **571** Photofest; **572–573** Wyoming Division of Cultural Resources; **573** *top* Library of Congress; **574** *foreground, Northern Pacific Railroad. The Pioneer Route to Fargo Moorhead Town Bismark Dakota and Montana and the Famous Valley of the Yellowstone* (about 1885), Creator–Poole Brothers, Printers. Broadside. Chicago Historical Society; **575** Solomon D. Butcher Collection, Nebraska State Historical Society; **577** The Granger Collection, New York; **578** Culver Pictures; **579** *Rush for the Oklahoma Land* (1894), John Steuart Curry. Department of the Interior.

Chapter 20, 582 Detail of *Forging the Shaft* (1874–1877), John Ferguson Weir. Oil on canvas, 52" x 73 1/4" (132.1 cm x 186.1 cm). Painted at West Point Foundry at Cold Spring, New York. The Metropolitan Museum of Art, Purchase, Lyman G. Bloomingdale Gift, 1901 (01.7.1). Photograph copyright © 1983 The Metropolitan Museum of Art; **583** The Granger Collection, New York; **584** The Newberry Library, Chicago; **585** Corbis-Bettmann; **587** Library of Congress; **588** *background* National Museum of American History, Smithsonian Institution; *left foreground* The Granger Collection, New York; *right foreground* Courtesy of AT&T; **589** *left background* U.S. Dept. of Commerce —Patent & Trademark Office, Washington, D.C.; *left foreground* First Church of Christ, Lynn, Massachusetts; **590** Special Collections Division, University of Washington Libraries. Negative no. 2315; **591** *top* Courtesy Colorado Historical Society; *bottom, The Last Spike* (1869). William T. Garrett Foundry, San Francisco. 17 6/10 carat gold, alloyed with copper. 5 9/16" x 7/16" x 1/2" (shaft including head), 1/2" x 1 3/8" x 1 1/4". Iris & B. Gerald Cantor Center for Visual Arts at Stanford University. Gift of David Hewes, 1998.115; **593** Copyright © United States Postal Service; **594** Corbis-Bettmann; **595** *left, John Davison Rockefeller* (1967), Adrian Lamb, after the 1917 oil by John Singer Sargent. Oil on canvas, 58 3/4" x 45 3/4". National Portrait Gallery, Smithsonian Institution/Art Resource, New York; *right* National Portrait Gallery, Smithsonian Institution/Art Resource, New York; **596** *left* Detail of *Going to the Opera–Family Portrait* (date unknown), Seymour Guy. Used with permission from Biltmore Estate, Asheville, North Carolina; *right, Room in a Tenement Flat* (about 1910), photo by Jessie Tarbox Beals. The Jacob A. Riis Collection, Museum of the City of New York; **598** *top right* Library of Congress; *center left* Chicago Historical Society; *bottom right* Library of Congress; **598** *bottom left,* **599** *center left* Corbis-Bettmann; **599** *bottom* Chicago Historical Society; **600, 601** Corbis-Bettmann; **602** Brown Brothers.

Chapter 21, 606 The Granger Collection New York; **607** Corbis-Bettmann; **609** Copyright © Collection of The New-York Historical Society; **610** Illustration by Patrick Gnan; **611** The Granger Collection, New York; **613** Corbis-Bettmann; **614** The Granger Collection, New York; **616** Brown Brothers; **617** S. Yamashita/Corbis; **618** Illustration by Ronald Himler. Cop, © 1995 HarperCollins Publishers. Bottom background copyrig © Bill Pogue; **619** Corbis; **620** Copyright © Jacques Chenet/Liaison Agency; **621** *top* Joseph Schwartz Collection/Corbis; *bottom* Corbis; **622** Library of Congress; **623** The Granger Collection, New York; **626, 627** Corbis-Bettmann; **628** *top* Lake County Museum/Corbis; *center, bottom* The Granger Collection, New York; **629** Corbis-Bettmann; **631, 632** *top right* The Granger Collection, New York; **632** *bottom left* Culver Pictures; **633** Photos by Sharon Hoogstraten.

Unit 7

634–635 Brown Brothers.

Chapter 22, 636 *top left, Street Arabs in Sleeping Quarters at Night* (date unknown), photo by Jacob A. Riis. The Jacob A. Riis Collection, Museum of the City of New York; *top right* Library of Congress; *bottom left Family Making Artificial Flowers* (about 1910), photo by Jessie Tarbox Beals. The Jacob A. Riis Collection, Museum of the City of New York; *bottom right* Collection of the International Museum of Photography at George Eastman House; **637** Keystone-Mast Collection U.C.R./California Museum of Photography, University of California, Riverside; **638** Brown Brothers; **639, 640, 641** Library of Congress; **642** *bottom* Brown Brothers; *left inset* Doubleday, Page and Company, New York, 1906, second issue; **643** *Theodore Roosevelt* (date unknown), John Singer Sargent. White House Collection. Copyright © 1992 White House Historical Association; **644** *center* Culver Pictures; *bottom left* Copyright © F. Sieb/H. Armstrong Roberts; *bottom right* Copyright © W. Bertsch/H. Armstrong Roberts; **645** *bottom* Copyright © J. Blank/H. Armstrong Roberts; **646** Eugene V. Debs Collection/Tamiment Institute Library, New York University; **648, 649** Culver Pictures; **650** *top, Lillian Wald* (1919), William Valentine Schevill. Oil on cardboard, 71.7 cm x 71.7 cm. (28 1/4" x 28 1/4"). National Portrait Gallery, Smithsonian Institution. Gift of the Visiting Nurse Service of New York; *bottom, A Short Cut over the Roofs of the Tenements; A Henry Street Visiting Nurse* (1908), photo by Jessie Tarbox Beals. Museum of the City of New York. Lent by the Visiting Nurse Service of New York; **651** The Granger Collection, New York; **652** *left* Copyright © 1994 FPG International.

Chapter 23, 656 U.S. Naval Academy Museum; **657** Hawaii State Archives; **658** The Granger Collection, New York; **659** Corbis; **661** *Portrait of Queen Liliuokalani,* date and artist unknown. Bishop Museum; **662** The Granger Collection, New York; **663** Chicago Historical Society; **664, 666** The Granger Collection, New York; **667** From *Puerto Rico: A Political and Cultural History,* Arturo Morales Carrion; **668** U.S. Naval Academy Museum; **671** Illustration by Nick Potondo; *center left* Corbis-Bettmann; **673** Copyright © Caren Firouz/Black Star/PNI; **674, 675** *bottom* The Granger Collection, New York.

Chapter 24, 676, 677, 679 The Granger Collection, New York; **681** *top center* Copyright © Mike Fizer/Check Six; *top right* Copyright © Hulton Getty/Tony Stone Images; *bottom left* Hulton-Deutsch Collection/Corbis; *bottom right* Copyright © Hulton Getty/Tony Stone Images; **682** The Granger Collection, New York; **684–685** Illustration by Patrick Whelan; **684** *top right,* **686, 687** Corbis; **689** *left* From *The Enormous Room* by E. E. Cummings, Penguin Twentieth Century Classics Edition. Cover painting *Prisoner's Round* (1890), Vincent Van Gogh, after Dore. Pushkin Museum of Fine Arts, Moscow. Photo courtesy Scala/Art Resource, New York; *right* From *A Farewell to Arms* by Ernest Hemingway, Scribner Paperback Fiction edition. Cover design by Mary Schuck. Cover illustration by Cathie Black; **691** Wyoming Division of Cultural Resources; **692** Culver Pictures; **693** Panel no. 1: "During the World War There Was a Great Migration North by Southern Negroes" from *The Migration of the Negro* mural series (1940–41), Jacob Lawrence. Tempera on masonite, 12" x 18". Acquired through Downtown Gallery, 1942. The Phillips Collection, Washington, D.C.; **695** Corbis; **696** The Granger Collection, New York; **697** *right,* **701** *bottom* The Granger Collection, New York; **702** Culver Pictures; **703** Photos by Sharon Hoogstraten.

Unit 8

704–705 Morton Beebe/Corbis.

Chapter 25, 706 Copyright © Illustration House, Inc.; **707** *top right* Long Island Auto Museum, New York; *center* Smithsonian Institution, Washington, D.C.; *bottom left* Henry Ford Museum and Greenfield Village, Dearborn, Michigan; *bottom right* Courtesy Montgomery Ward; **708, 709** Brown Brothers; **710** Copyright © 1923 by the New York Times Co. Reprinted by perission; **711** *center right* Copyright © Stock Montage; **713** Underwood & Underwood/Corbis; **714** Culver Pictures; **715** Brown Brothers; **716** UPI/Corbis-Bettmann; **717** Brown Brothers; **718** *from top to bottom* Movie Still Archives; Copyright © Blank Archives/Archive Photos; Copyright © Archive Photos; Photofest; **719** *left* Brown Brothers; *right* Copyright © Albert L. Bresnik; **720** *The Ascent of Ethiopia* (1932), Lois Mailou Jones. Oil on canvas, 23 1/2" x 17 1/4". Accession no. M1993.191. Milwaukee Art Museum, Purchase, African-American Art Acquisition Fund, matching funds from Suzanne and Richard Pieper, with additional support from Arthur and Dorothy Nelle Sanders; **721** *top* Copyright © Stock Montage; *bottom* Book cover from first edition of *The Great Gatsby* by F. Scott Fitzgerald (New York: Charles Scribner's Sons, 1925). Used by permission of Scribner, a division of Simon & Schuster; **722, 723** *bottom* National Baseball Hall of Fame Library, Cooperstown, New York.

Chapter 26, 726 Museum of the City of New York; **727** The Granger Collection, New York; **728** From the Collection of Janice L. and David J. Frent; **729** Nebraska Historical Society; **731** Copyright © 1929 Icon Comm./FPG International; **732** From the collections of the Minnesota Historical Society; **733** Corbis; **734** Library of Congress; **735** Photo by Margaret Suckley/Franklin D. Roosevelt Library; **736** *inset* Courtesy of the Tennessee Valley Authority; **738** F.D.R. Presidential Library; **739** Detail of *Mary McLeod Bethune* (1943–1944), Betsy Graves Reyneau. National Portrait Gallery, Smithsonian Institution/Art Resource, New York; **740** *inset* Corbis; **741** *left* Photofest; *center, right* The Kobal Collection; **742** Oscar White/Corbis; **743** UPI/Corbis-Bettmann; **744** Copyright © McDougal Littell Inc.; **745** Farm Security Administration—Office of War Information Photograph Collection; **746** Library of Congress; **748** *Construction of the Dam* (1937), William Gropper. Mural study, Department of the Interior, National Park Service. National Museum of American Art, Smithsonian Institution/Art Resource, New York; **749** Courtesy Social Security Office; **750** *top* Copyright © 1929 Icon Comm./FPG International; *bottom* From the Collection of Janice L. and David J. Frent; **751** Photo by Dorothea Lange/Library of Congress; **752** Details of *South Texas Panorama* (1939). Warren Hunter. Mural, Alice Texas Post Office. National Museum of American Art, Smithsonian Institution/Art Resource, New York; **753** Photos by Sharon Hoogstraten.

Chapter 27, 754 UPI/Corbis-Bettmann; **755** National Archives; **757** Photo by Hugo Jaeger/Life Magazine, copyright © Time Warner Inc.; **758** Corbis; **759** The Granger Collection, New York; **761** Corbis; **762** Copyright © 1999 Owen H. K./Black Star; **763** The Mariner's Museum/Corbis; **764** *left,* **765** Corbis-Bettmann; **766** *top* National Archives; **767** Hulton-Deutsch Collection/Corbis; **768** Photo by Margaret Bourke White/Life Magazine, copyright © Time Inc.; **770** Corbis; **772** National Archives; **773** The Granger Collection, New York; **774** U.S. Office of War Information, photo no. 208-LV-38BB-3 in the National Archives; **775** Corbis-Bettmann; **776** Wally McNamee/Corbis; **777** Photo by Dorothea Lange/National Archives; **778–779** *background* Copyright © Photolink/PhotoDisc, Inc.; **778** *bottom left* Corbis-Bettmann; *bottom right* The Granger Collection, New York; **779** *top, Captain America* TM and copyright © 1999 Marvel Characters, Inc. Used with permission; **780** Corbis-Bettmann; **782, 783, 785** *bottom* The Granger Collection, New York.

Chapter 28, 786 *background* Photo by Carl Iwasaki/Life Magazine, copyright © Time Inc.; *foreground* Henry Ford Museum and Greenfield Village, Dearborn, Michigan; **787** Photo by J. R. Eyerman/Life Magazine, copyright © Time Warner Inc.; **788** *background* Copyright © R. Walker/H. Armstrong Roberts; *foreground* Cousley Historical Collections. Photo by Stephen Mays, New York; **789** U.S. Army photos; **791** UPI/Corbis-Bettmann; **793** Photo by Walter Sanders/Life Magazine, copyright © 1972 Time Inc.; **794** Photofest; **795** AP/Wide World Photos; **797** Corbis-Bettmann; **798** Photo by Hank Walker/Life Magazine, copyright © Time Inc.; **799** Sovfoto/Eastfoto; **800** Courtesy Brunswick Record Corporation; **801** Copyright © Frank Cezus/Tony Stone Images, Inc.; **802** *from top to bottom* Copyright © Archive Photos; Elvis Presley Enterprises, Inc. Used by permission; Copyright © Harold Lloyd Trust/Archive Photos; The Kobal Collection; **804** *top* Copyright © Tom Bean/Tony Stone Images; *bottom left* Curt Teich Postcard Archives, Lake County Museum, Wauconda, Illinois; **804** *bottom right,* **805** *center left* Copyright © 1996 Terrence Moore; **805** *center right* Copyright © Ferguson & Katzman/Tony Stone Images; *bottom* Copyright © Rob Boudreau/Tony Stone Images; **807** *bottom* Courtesy Ross Lewis and the *Milwaukee Journal*.

Unit 9

808–809 Photo by Francis Miller/Life Magazine, copyright © Time Inc.

Chapter 29, 810 Copyright © James Blair, NGS Image Collection; **811** Photo by John G. Moebes. Reprinted with permission of the *News & Record.* This reprint does not constitute or imply any endorsement or sponsorship of any product, service, company or organization; **812** Copyright © 1950 Elliott Erwitt/Magnum Photos, Inc.; **813** Corbis-Bettmann; **814** AP/Wide World Photos; **815** Black Star/Time Inc.; **816** UPI/Corbis-Bettmann; **818** Corbis-Bettmann; **819** Copyright © 1963 Charles Moore/Black Star; **820** Photo no. KN-C30670 in the John F. Kennedy Library; **822** President Lyndon Baines Johnson Library and Museum; **823** AP/Wide World Photos; **824** Copyright © 1978 George Ballis/Take Stock; **825** Copyright © Robert Fried/Stock Boston/PNI; **826** Corbis-Bettmann; **827** AP/Wide World Photos; **830** *top* UPI/Corbis-Bettmann; *center* Black Star/Time Inc.; *bottom* Photo by Sharon Hoogstraten; **831** *bottom, Freedom* (date unknown), Anthony Gauthier. Property of the Institute of American Indian Arts Museum, no. WIN14.

Chapter 30, 832 Copyright © Seny Norasingh/Light Sensitive; **833** *left* Corbis-Bettmann; *right* Wally McNamee/Corbis; **835** From *The Unquiet American* by Cecil B. Currey. Photo courtesy of the author; **836** AP/Wide World Photos; **837** *right* Sovfoto/Eastfoto; **838** Corbis; **839** Corbis-Bettmann; **840** Corbis; **841** Corbis-Bettmann; **842** *left* Copyright © Popperfoto/Archive Photos; *center* PNI; *right* National Archives; **843** *left* Copyright © 1999 Jim Pickerell/Black Star; *right* Photo by Larry Burrows/Life Magazine, copyright © Time Inc. Courtesy Larry Burrows Collection, New York; **845** Copyright © 1968 Herblock; **846** Copyright © John Paul Filo; **847** *inset* UPI/Corbis-Bettmann; **848** Copyright © 1998 Nik Wheeler/Black Star; **849** Dave G. Houser/Corbis.

Chapter 31, 852 *top left* Copyright © 1999 Robin Moyer/Black Star; *top right* Copyright © Philippe Ledru/Sygma; *center left* Copyright © Jim Moore/Liaison Agency; *center right* Copyright © 1997 Dennis Brack/Black Star; *bottom* Copyright © Archive Photos; **853** *top* Photo by Harry Benson; *center left* Copyright © 1974, The Washington Post. Reprinted with permission; **854** *top left* AP/Wide World Photos; *center left* Copyright © Lary Voigt/Photo Researchers, Inc.; *center right* Copyright © Ted Thai/Time Magazine; **855** Copyright © 1999 Dennis Brack/Black Star; **856** NASA; **858** AP/Wide World Photos; **859** Copyright © Penelope Breese/Liaison Agency; **860** Cartoon by Paul Conrad. Copyright © Los Angeles Times Syndicate; **861** Copyright © 1974 Alex Webb/Magnum Photos, Inc.; **862** AP/Wide World Photos; **863** Copyright © R. Krubner/H. Armstrong Roberts; **864** Copyright © Dirk Halstead/Liaison Agency; **866–867** *background* NASA; **866** *center left* Copyright © 1996 Ariel Skelley/The Stock Market; **867** *top* Copyright © R. Krubner/H. Armstrong Roberts; *bottom right* Photo by Vernon Merritt/Life Magazine, copyright © Time Inc.; **868** From the Collection of David J. and Janice L. Frent.

Chapter 32, 870 Copyright © R. Ian Lloyd/Masterfile; **871** Copyright © 1999 USA Today. Reprinted with permission; **872** Copyright © 1993 Time Inc. Reprinted by permission; **873** From the Collection of Janice L. and David J. Frent; **874** Courtesy Ronald Reagan Library; **875, 876** *top left* AP/Wide World Photos; **876** *top* Copyright © 1998 The Washington Post. Reprinted with permission; **878** Copyright © Paul Soulders/Liaison Agency; **879** Photo by Donna McWilliam; **880** *top right* Corbis-Bettmann; *center* Commodore Business Machines, Inc.; *center right* Copyright © Charles Feil/Stock Boston/PNI; *bottom left* Copyright © Ted Kawalerski/The Image Bank/Chicago Core; **881** Copyright © Peter Charlesworth/ SABA; **882** Photo by Doug Hoke; **883** Photo copyright © Marion Ettlinger; **884** Copyright © Robert Brenner/PhotoEdit/PNI; **887** *bottom* Photograph by Kaku Kurita; **888** *top* Copyright © Peter Simon/Stock Boston/PNI; *bottom left* Copyright © Lambert/Archive Photos; *bottom center* Copyright © Henry/TSI Imaging/Tony Stone Images; **889** Photos by Sharon Hoogstraten.

Presidents of the United States

R40-R42 The Oval Office Collection™ except Clinton AP/Wide World Photos.